S0-AXY-971

Handbook of
PLASTIC AND RUBBER ADDITIVES

Handbook of
PLASTIC AND RUBBER ADDITIVES

An International Guide to More Than 13,000 Products by Trade Name, Chemical, Function, and Manufacturer

Compiled by

Michael and Irene Ash

Gower

Published by
Gower Publishing Limited
Gower House
Croft Road
Aldershot
Hampshire GU11 3HR
England

Gower
Old Post Road
Brookfield
Vermont 05036
U.S.A.

British Library Of Cataloguing in Publication Data
Handbook of Plastic and Rubber Additives: International Guide to More Than 13,000 Products by Trade Name, Chemical, Function, and Manufacturer,
I. Ash, Michael II. Ash, Irene
668.4

ISBN 0-566-07594-6

Library of Congress Cataloging-in-Publication Data
Ash, Michael
Handbook of plastic and rubber additives : an international guide to more than 13,000 products by trade name, chemical, function, and manufacturer / compiled by Michael and Irene Ash.
1340 p. 23.4 cm
ISBN 0-566-07594-6 (hardback) : $195.00 (U.S.)
1. Plastics—Additives. 2. Rubber—Additives. I. Ash, Irene. II. Title
TP1142.A85 1995 96-3822
668.4—dc20 CIP REV

Typeset in Arial Narrow by Synapse Information Resources, Inc.

Printed and bound in Great Britain by
Hartnolls Limited, Bodmin, Cornwall

Contents

Preface

This reference work describes more than 13,000 trade name and chemical additive ingredients that are used for the formulation of plastic and rubber products. The extensive information about these functional materials has been gathered from more than 1200 worldwide manufacturers, their subsidiaries, and distributors.

Plastic additives are a diverse group of specialty chemicals that are either incorporated into the plastic product prior to or during processing, or applied to the surface of the product when processing has been completed. These additives aid in the actual processing of the plastic end product (e.g., blowing agents, mold release agents, lubricants, organic peroxides) or improve the characteristics of the final product (antimicrobials, odorants, colorants, antistatic agents, impact modifiers, and uv stabilizers).

Natural rubbers and synthetic elastomers derive their commercial versatility from the incorporation of additives classified into three categories: antidegradants (including antiozonants and antioxidants), accelerators and vulcanizing agents, specialty additives (e.g., blowing agents, plasticizers, flame retardants, softeners).

Choosing optimal ingredients in the overall process of marketing and developing new and improved plastic and rubber products is a monumental task. This is exacerbated by environmental , health, and safety issues that must be addressed by both additive manufacturers and end product formulators. Chlorinated fluorocarbons have been virtually removed from mold release agents and rigid polyurethane insulation foams, however, the replacements (e.g., pentanes) create performance issues (handling ease), as well as safety problems of their own (flammability and toxicity). Lead and cadmium provide excellent thermal stabilization from a cost-performance perspective, however, there is an increased worldwide environmental pressure to limit their levels in the environment. Certain accelerators used in the tire industry produce harmful levels of nitrosamines, a suspected carcinogen, and are being replaced to reduce the exposure of tire fabricators. Additive manufacturers are faced with the imperative to develop products that not only enhance the overall processability and functionality of the plastic and rubber end products, but to satisfy environmental and health demands as well.

This reference coalesces current and essential information about both trade name and chemical additives into a single source and expedites material selection for the user by cross-referencing trade name products by chemical composition, function, CAS number and EINECS number. Products that have been discontinued by companies are noted in a section of the Appendix.

The book is divided into four sections:

Part I—*Trade Name Reference* contains over 11,000 alphabetical entries of trade name plastic and rubber additives. Each entry references its manufacturer, chemical composition, associated CAS and EINECS identifying numbers, general properties, applications and functions, toxicology, and compliance and regulatory information as provided by the manufacturer and other sources.

Part II—*Chemical Dictionary/Cross-Reference* contains an alphabetical listing of more than 2400 plastic and rubber additive chemicals. Each entry lists the trade name products that are equivalent to the chemical or contains that chemical as one of the trade name product's constituents. Also included in the entry are: CAS (Chemical Abstract Service) numbers, EINECS (European Inventory of Existing Commercial Chemical Substances) numbers, synonyms, molecular and empirical formulas, definition, classification, general properties, uses, regulatory, toxicological, and precaution information wherever possible. Synonyms are thoroughly cross-referenced back to the main entry.

Part III—*FunctionalCross-Reference* contains an alphabetical listing of major plastic and rubber additive functional categories. Over 80 categories are included, e.g., abrasives, accelerators, blowing agents, fillers, flame retardants, impact modifiers, nucleating agents, processing aids, release agents, uv absorbers, waxes, etc. Each functional category entry is followed by an alphabetical listing of the trade name products and chemicals that have that functional attribute.

Part IV—*Manufacturers Directory* contains detailed contact information for the manufacturers of the more than 11,000 trade name products and 2400 generic chemicals that are referenced in this handbook. Wherever possible telephone, telefax, and telex numbers, toll-free 800 numbers, and complete mailing addresses are included for each manufacturer.

The **Appendices** contain the following cross-references:

CAS Number-to-Trade Name Cross-Reference orders many trade names found in Part I by identifying CAS numbers; it should be noted that trade names contain more than one chemical component and the associated CAS numbers in this section refer to each trade name product's primary chemical component.

CAS Number-to-Chemical Cross-Reference orders chemical compounds found in Part II by CAS numbers.

EINECS Number-to-Trade Name Cross-Reference orders many trade names found in Part I by identifying EINECS numbers that refer to each trade name product's primary chemical component.

EINECS Number-to-Chemical Cross-Reference orders chemical compounds found in Part II by EINECS numbers.

This book is the culmination of many months of research, investigation of product sources, and sorting through a variety of technical data sheets and brochures acquired through personal contacts and correspondences with major chemical manufacturers worldwide. We are especially grateful to Roberta Dakan for her skills in chemical information database management. Her tireless efforts have been instrumental in the production of this reference.

M. & I. Ash

NOTE:

The information contained in this reference is accurate to the best of our knowledge; however, no liability will be assumed by the publisher or the authors for the correctness or comprehensiveness of such information. The determination of the suitability of these products for prospective use is the responsibility of the user. It is herewith recommended that those who plan to use any of the products referenced seek the manufacturers instructions for the handling of that chemical.

Abbreviations

ABS	acrylonitrile-butadiene-styrene
absorp.	absorption
ACGIH	American Conference of Governmental Industrial Hygienists
ACM	polyacrylate rubber
ACN	acrylonitrile
act.	active
AEEA	aminoethylethanolamine
AEP	aminoethylpiperazine
agric.	agricultural
alc.	alcohol
amts.	amounts
anhyd.	anhydrous
APHA	American Public Health Association
applic(s).	application(s)
aq.	aqueous
ASA	acrylic-styrene-acrylonitrile
atm	atmosphere
at.wt.	atomic weight
aux.	auxiliary
avail.	available
avg.	average
a.w.	atomic weight
BATF	U.S. Bureau of Alcohol, Tobacco, and Firearms
BDO	1,4-butanediol
BGA	Federal Republic of Germany Health Dept. certification
BHA	butylated hydroxyanisole
BHT	butylated hydroxytoluene
biodeg.	biodegradable
blk.	black
BMC	bulk molding compound
b.p.	boiling point
BR	butadiene rubbers, polybutadienes
B&R	Ball & Ring
br., brn.	brown
brnsh.	brownish
BS	British Standards
B/S	butadiene/styrene
C	degrees Centigrade
CAB	cellulose acetate butyrate
cap.	capillary
CAP	cellulose acetate propionate
CAS	Chemical Abstracts Service
CC	closed cup
cc	cubic centimeter(s)
CCl_4	carbon tetrachloride
CFC	chlorofluorocarbon
CFR	Code of Federal Regulations
char.	characteristic
chem.	chemical
CI	color index
CIIR	chlorobutyl rubber
CIR	Cosmetic Ingredient Review
cks	centistoke(s)

cm	centimeter(s)
cm^3	cubic centimeter(s)
CMC	critical Micelle concentration
CNS	central nervous system
CO	carbon monoxide, epichlorohydrin rubber
COC	Cleveland Open Cup
COF	coefficient of friction
compat.	compatible
compd(s).	compound(s)
compr.	compression
conc.	concentrated, concentration
contg.	containing
cosolv.	cosolvent
cp	centipoise(s)
CPE	chlorinated polyethylene
cps	centipoise(s)
CPVC	chlorinated polyvinyl chloride
CR	chloroprene rubber, polychloroprene
cryst.	crystalline, crystallization
cs or cSt	centistoke(s)
CSBR	carboxylated styrene-butadiene rubber
CSM	chlorosulfonyl polyethylene rubber
ctks	centistoke(s)
DAP	diallyl phthalate
DBP	dibutyl phthalate
DEA	diethanolamide, diethanolamine
dec.	decomposes
decomp.	decomposition
DEDM	diethylol dimethyl
deliq.	deliquescent
dens.	density
deriv.	derivative(s)
DETA	diethylenetriamine
DGE	diglycidyl ether
diam.	diameter
dielec.	dielectric
dil.	dilute
disp.	dispersible, dispersion
dissip.	dissipation
dist.	distilled
distort.	distortion
dk.	dark
DMF	dimethyl formamide
DMSO	dimethyl sulfoxide
DOP	dioctyl phthalate
DOT	U.S. Department of Transportation
DPG	dipropylene glycol
DWV	drainage, waste and vent
EAA	ethylene acrylic acid
eb, EB	electron beam
EBA	ethylene butyl acrylate
EBS	ethylene bis-stearamide
EC	European Community
ECO	polyepichlorohydrin, epichlorohydrin copolymer rubber
ECTFE	ethylene/chlorotrifluoroethylene copolymer
EDTA	ethylenediamine tetraacetic acid

EE	epoxy equivalent
EEC	European Economic Community
EEW	epoxide equivalent weight
EINECS	European Inventory of Existing Commercial Chemical Substances
elec.	electrical
ELINCS	European List of Notified Chemical Substances
elong.	elongation
EMA	ethylene-methyl acrylate
EMI	electromagnetic interference
ENB	5-ethylidene-2-norbornene
EO	ethylene oxide
E/P	ethylene/propylene
EPA	U.S. Environmental Protection Agency
EPC	ethylene-propylene copolymer
EPDM	ethylene-propylene-diene rubber, ethylene-propylene terpolymer
EPM	ethylene-propylene (copolymer) rubber
EPR	ethylene-propylene rubber
EPS	expandable polystyrene
EPT	ethylene-propylene terpolymer (EPDM)
equip.	equipment
esp.	especially
ETFE	ethylene tetrafluoroethylene
ETU	ethylene thiourea
EVA	ethylene vinyl acetate
EVCL	ethylene-vinyl chloride
exc.	excellent
F	degrees Fahrenheit
FDA	Food and Drug Administration (U.S.)
FEMA	Flavor and Extract Manufacturers' Association (U.S.)
FEP	fluorinated ethylene propylene
FKM	fluoroelastomer
flamm.	flammable, flammability
flex.	flexural
f.p.	freezing point
FPM	vinylidene fluoride/hexafluoro propylene copolymer
FR-ABS	flame retardant ABS
FRP	fiberglass-reinforced plastics
ft	foot, feet
f.w.	formula weight
G	giga
g	gram(s)
gal	gallon(s)
g/d	gram/dyne
GFRP	glass fiber-reinforced plastic
G-H	Gardner-Holdt
GI	gastro-intestinal
glac.	glacial
gr.	gravity
gran.	granules, granular
GRAS	generally regarded as safe
grn.	green
GRP	glass-reinforced plastics, glass-reinforced polyester
h	hour(s)
HAF	high abrasion furnace carbon black
HALS	hindered amine light stabilizer
HC	hydrocarbon

HCl	hydrochloride, hydrochloric acid
HDPE	high-density polyethylene
HDT	heat distortion (deflection) temp.
Hg	mercury
HIPS	high-impact polystyrene
HLB	hydrophilic lipophilic balance
HNBR	hydrogenated butadiene acrylonitrile
HTV	high temperature vulcanizing
hyd.	hydroxyl
hydrog.	hydrogenated
Hz	hertz
i.b.p.	initial boiling point
IIR	isobutylene-isoprene rubber, butyl rubber
immisc.	immiscible
in.	inch(es)
incl.	including
incompat.	incompatible
ing.	ingestion
ingred.	ingredient(s)
inh.	inhalation
inj.	injection
inorg.	inorganic
insol.	insoluble
Int'l.	International
IP	intraperitoneal
IPA	isopropyl alcohol
IPM	isopropyl myristate
IPP	isopropyl palmitate
IR	isoprene rubber (synthetic), polyisoprene
i.v.	iodine value
IV	intravenous
JHOSPA	Japan Hygienic Olefine and Styrene Plastics Association
JHPA	Japan Hygienic PVC Association
JSCI	Japanese Standard of Cosmetic Ingredients
k	kilo
kg	kilogram(s)
l	liter(s)
lb	pound(s)
LC50	lethal concentration 50%
LD50	lethal dose 50%
LDPE	low-density polyethylene
liq.	liquid
LLDPE	linear low-density polyethylene
lt.	light
Ltd.	Limited
M	mole
M	mega
m	milli or meter(s)
m-	meta
MA	maleic anhydride, methacrylic acid
max.	maximum
MBTS	2-mercaptobenzothiazole disulfide
MDI	methylene diphenylene isocyanate
MDPE	medium density polyethylene
MEA	monoethanolamine, monoethanolamide
mech.	mechanial

med.	medium
MEK	methyl ethyl ketone
mfg.	manufacture
mg	milligram(s)
MIBK	methyl isobutyl ketone
microcryst.	microcrystalline
MID	Meat Inspection Division
MIL	Military Specifications
min	minute(s), mineral, minimum
misc.	miscible, miscellaneous
mixt.	mixture(s)
ml	milliliter(s)
MLD	minimum lethal dose
mm	millimeter(s)
MMW-HDPE	medium molecular weight high density polyethylene
mo, mos	month(s)
mod.	moderately, modulus
m.p.	melting point
mPa•s	millipascal-second(s)
MT	medium thermal
m.w.	molecular weight
N	normal
nat.	natural
N/B	nitrile-butadiene
NBR	nitrile rubber, nitrile-butadiene rubber
NC	nitrocellulose
NCR	nitrile-chloroprene rubber
need.	needles
NF	National Formulary
nm	nanometer
NMA	nadic methyl anhydride
NMP	N-methyl pyrrolidone
no.	number
nonalc.	nonalcoholic
nonflamm.	nonflammable
nonyel.	nonyellowing
NPG	neopentyl glycol
NR	natural rubber, isoprene rubber (natural)
NSF	National Sanitation Foundation
NTA	nitrilotriacetic acid
NV	nonvolatiles
o-	ortho
OC	open cup
ODC	ozone-depleting compound
OMS	odorless mineral spirits
OPP	oriented polypropylene
org.	organic
OSHA	Occupational Safety and Health Administration
OTC	over-the-counter
o/w	oil-in-water
p-	para
PA	polyamide
Pa	Pascal
PAN	polyacrylonitrile
PB	polybutene-1
PBT	polybutylene terephthalate

PC	polycarbonate
pcf	pounds per cubic foot
PCT	polycyclohexylene terephthalate
PCTFE	polychlorotrifluoroethylene
PE	polyethylene
PEEK	polyetheretherketone
PEG	polyethylene glycol
PEI	polyethylenimine, polyetherimide
PEK	polyetherketone
PEL	permissible exposure level
PES	polyether sulfone
PET	polyethylene terephthalate
petrol.	petroleum
pH	hydrogen-ion concentration
phr	parts per hundred of rubber or resin
PIB	polyisobutylene
PIR	polyisocyanurate
pkg.	packaging
P-M	Pensky-Martens
PMCC	Pensky-Martens closed cup
PMMA	polymethyl methacrylate
PNF	phosphonitrilic fluoroelastomer
PO	propylene oxide
POE	polyoxyethylene, polyoxyethylated
POM	polyoxymethylene
POP	polyoxypropylene, polyoxypropylated
powd.	powder
PP	polypropylene
PPE	polyphenylene ether
PPFA	Plastic Pipe and Fittings Association
PPG	polypropylene glycol
PPI	Plastics Pipe Institute
ppm	parts per million
PPO	polyphenylene oxide
PPS	polyphenylene sulfide
pract.	practically
prep.	preparation(s)
prod.	product(s), production
props.	properties
ps	poise
PS	polystyrene
psi	pounds per square inch
psig	pounds per square inch gauge
pt.	point
PTFE	polytetrafluoroethylene
PTMEG	polytetramethylene ether glycol
PU	polyurethane
PUR	polyurethane
PVA	polyvinyl alcohol
PVAc	polyvinyl acetate
PVAL	polyvinyl alcohol
PVB	polyvinyl butyral
PVC	polyvinyl chloride
PVDC	polyvinylidene chloride
PVDF	polyvinylidene fluoride
PVE	polyvinyl ethyl ether

PVF	polyvinyl fluoride
PVM	polyvinyl methyl ether
PVM/MA	polyvinyl methyl ether/maleic anhydride
PVP	polyvinylpyrrolidone
quat.	quaternary
R&B	Ring & Ball
R.T.	room temperature
rdsh.	reddish
ref.	refractive
resist.	resistance, resistant, resistivity
resp.	respectively
RFI	radio frequency interference
r.h.	relative humidity
RIM	reaction injection molded/molding
RT	room temperature
RTM	resin transfer molding
RTV	room temperature vulcanizing
s	second(s)
SADT	self accelerating decomposition temp.
SAN	styrene-acrylonitrile
sapon.	saponification
sat.	saturated
S/B	styrene/butadiene
SBR	styrene/butadiene rubber
SBS	styrene-butadiene-styrene
SCR	styrene-chloroprene rubber
SE	self-emulsifying
SEBS	styrene-ethylene/butylene-styrene
sec.	secondary
sl.	slightly
sm.	small
SMA	styrene maleic anhydride
SMC	sheet molding compound
soften.	softening
sol.	soluble, solubility
sol'n.	solution
solid.	solidification
solv(s).	solvent(s)
sp.	specific
spec.	specification
SQ	subcutaneous
SR	styrene rubber
SRF	semireiniforced furnace
SS	stainless steel
SSU	Saybolt Universal Seconds
std.	standard
STEL	short term exposure limit
Stod.	Stoddard solvent
STP	standard temperature and pressure
str.	strength
subcut.	subcutaneous
subl.	sublimes
surf.	surface
SUS	Saybolt Universal Seconds
susp.	suspension
syn.	synthetic

t	tertiary
TAPPI	Technical Association of the Pulp & Paper Industry
TBHQ	t-butyl hydroquinone
TCC	Tag closed cup
TDI	toluene diisocyanate
TEA	triethanolamine, triethanolamide
tech.	technical
TEDA	triethylenediamine
TETA	triethylenetetramine
temp.	temperature
tens.	tensile or tension
tert	tertiary
TFE	tetrafluoroethylene
TGE	triglycidyl ether
THF	tetrahydrofuran
thru	through
TIPA	triisopropanolamine
TLV	Threshold Limit Value
TMA	trimellitic anhydride
TMC	thick molding compound
TMTD	tetramethylthiuram disulfide
TOC	Tag open cup
TPE	thermoplastic elastomer
TPO	thermoplastic polyolefin
TPR	thermoplastic rubber
TPU	thermoplastic polyurethane
TSCA	Toxic Substances Control Act
TWA	time weighted average
typ.	typical
UF	urea formaldehyde
UHMW	ultra high molecular weight
UHMWPE	ultra high molecular weight polyethylene
UL	Underwriter's Laboratory
unsat.	unsaturated
UPVC	unplasticized polyvinyl chloride
USDA	U.S. Department of Agriculture
USP	Unites States Pharmacopeia
uv	ultraviolet
V	volt
VA	vinyl acetate
VAE	vinyl acetate ethylene
VC	vinyl chloride
VCA	vinyl chloride acetate
VdC, VDC	vinylidene chloride
veg.	vegetable
visc.	viscous, viscosity
VLDPE	very low density polyethylene
VM&P	Varnish Makers and Painters
VOC	volatile organic compounds
vol.	volume
VSI	Vinyl Siding Institute
VWD	Vinyl Window & Door Institute
wh.	white
w/o	water-in-oil
wt.	weight
XLPE	crosslinked polyethylene

X-PE	crosslinked polyethylene
yel.	yellow
ylsh.	yellowish
yr	year
#	number
%	percent
<	less than
>	greater than
≤	less than or equal to
≥	greater than or equal to
@	at
≈	approximately
α	alpha
β	beta
ε	epsilon
γ	gamma
δ	delta
ω	omega
μ	micron, micrometer
μg	microgram

Part I
Trade Name Reference

A-1. [Harwick] N,N′-Diphenylthiourea (thiocarbanilide); CAS 102-08-9; primary accelerator for latex and repair stocks, CR, CR latex, and EPDM sponge compds.; powd.

A-2. [LaRoche Chem.] Activated alumina; CAS 1344-28-1; EINECS 215-691-6; used in scavenging, dehydration and catalytic applics.; in hydrogen peroxide, polyethylene and drying industries, and for scavenging fluoride, chloride, and trace heavy metals; gran.; bulk dens. 42 lb/ft^3 (packed); surf. area 310 m^2/g; 93.6% Al_2O_3.

A-201. [LaRoche Chem.] Activated alumina; CAS 1344-28-1; EINECS 215-691-6; desiccant with high adsorptive capacity and hardness; catalyst carrier; scavenger for impurities in gas or liq. process streams, in hydrogen peroxide and polyethylene prod.; sphere; bulk dens. 48 lb/ft^3 (packed); surf. area 325 m^2/g; 93.6% Al_2O_3.

AA Standard. [CasChem] Castor oil; emollient for industrial applics. where lt. color, high purity, and low acidity are desirable; plasticizer, wetting agent, lubricant for rubber compding.; FDA approval; Gardner 1+; sol. in alcohols, esters, ethers, ketone, and aromatic solvs.; sp.gr. 0.959; visc. 7.5 stokes; pour pt. -10 F; acid no. 2; iodine no. 86; sapon. no. 180; hyd. no. 164.

Abalyn®. [Hercules] Methyl rosinate; resin with compatibility, surf.-wetting properties, visc., and tack used in lacquers, inks, paper coatings, varnishes, adhesives, sealing compds., plastics, wood preservatives, and perfumes; amber visc. liq.; sol. in esters, ketones, alcohols, ethers, petrol. hydrocarbons, and veg. and min. oils; water-insol.; sp.gr. 1.03; dens. 1.03 kg/l; visc. (G-H) Z1; b.p. 352-356 C; acid no. 6; sapon. no. 160; ref. index 1.5300; flash pt. (COC) 180.

Abcure S-40-25 [Abco Industries] Benzoyl peroxide disp. in an inorg. medium; CAS 94-36-0; EINECS 202-327-6; catalyst; alternative to MEK peroxide, benzoyl peroxide pastes or granules for catalyzation of unsat. polyester resins; well suited for spray applic.; red or wh. pourable liq.; sp.gr. 1.2; dens. 11 lb/gal; visc. 100-300 cps; source of bivalent oxygen which is very reactive with strong acids, oxidizers, reducing agents, promoters; nonflamm. at ambient conditions; 40% active.

Abex® 12S. [Rhone-Poulenc Surf. & Spec.] Proprietary/sodium; anionic; emulsifier for vinyl acetate, acrylates, methacrylates, styrene, butadiene polymerization; for paper coatings, textile coatings, paints, adhesives, industrial coatings; liq.; 30% act.

Abex® 18S. [Rhone-Poulenc Surf. & Spec.] Proprietary/sodium; anionic; detergent, low foaming emulsifier for polymerization of vinyl acetate, acrylates, styrene, butadiene; mineral oil emulsifier; rewetting agent; for paper coatings, textile coatings, paints, adhesives, industrial coatings; FDA 21CFR §175.105, 176.170, 176.180; liq.; surf. tens. 48 dynes/cm (@ CMC); 35% conc.

Abex® 22S. [Rhone-Poulenc Surf. & Spec.] Proprietary/sodium; anionic; detergent, low foaming emulsifier for use in polymerization of vinyl acetate and acrylates; mineral oil emulsifier; rewetting agent; for paper coatings, textile coatings, paints, adhesives, industrial coatings; FDA 21CFR §175.105, 176.170, 176.180; liq.; surf. tens. 55 dynes/cm (@ CMC); 25% conc.

Abex® 23S. [Rhone-Poulenc Surf. & Spec.] Sodium laureth sulfate; CAS 9004-82-4; anionic; emulsifier for emulsion polymerization of vinyl acetate, acrylates, methacrylates, styrene, butadiene; for paper coatings, textile coatings, paints, adhesives, industrial coatings; liq.; surf. tens. 40 dynes/cm (@ CMC); 60% conc.

Abex® 26S. [Rhone-Poulenc Surf. & Spec.] Proprietary/sodium; anionic; emulsifier for emulsion polymerization of vinyl acetate, acrylates, methacrylates, styrene, butadiene; for paper coatings, textile coatings, paints, adhesives, industrial coatings; FDA 21CFR §175.105, 176.170, 176.180; liq.; surf. tens. 55 dynes/cm (@ CMC); 33% conc.

Abex® 33S. [Rhone-Poulenc Surf. & Spec.] Proprietary/sodium; anionic; emulsifier for emulsion polymerization of vinyl acetate, acrylates, methacrylates, styrene, butadiene; for paper coatings, textile coatings, paints, adhesives, industrial coatings; liq.; surf. tens. 37 dynes/cm (@ CMC); 27% conc.

Abex® 1404. [Rhone-Poulenc Surf. & Spec.] Proprietary surfactant; anionic; surfactant for emulsion polymerization; liq.; 42% act.; Unverified

Abex® AAE-301. [Rhone-Poulenc Surf. & Spec.] Octylphenol/sodium; anionic; emulsifier for emulsion polymerization of vinyl acetate, acrylates, methacrylates, styrene, butadiene; for paper coatings, textile coatings, paints, adhesives, industrial coatings; liq.; 20% conc.

Abex® EP-100. [Rhone-Poulenc Surf. & Spec.] Ammonium nonoxynol-4 sulfate; CAS 9051-57-4; anionic; emulsifier for emulsion polymerization of vinyl acetate, acrylates, methacrylates, styrene, butadiene; for adhesives, paints, paper, textile, and industrial coatings; surfactant for mild light-duty liqs.; liq.; surf. tens. 33 dynes/cm (@ CMC); 30% act.

Abex® EP-110. [Rhone-Poulenc Surf. & Spec.] Ammonium nonoxynol-9 sulfate; CAS 9051-57-4; anionic; primary emulsifier and stabilizing agent for the preparation of vinyl acetate, vinyl acetate/acrylic, all acrylic, styrene/acrylic, and S/B emulsion copolymers; wetting agent, dispersant for agric. formulations; for adhesives, paints, paper, textile, and industrial coatings; FDA, EPA compliance; pale yel. liq.; sp.gr. 1.04; visc. 91.0 cks; pour pt. 0 C; surf. tens. 38.3 dynes/cm (1%); 30% act.

Abex® EP-115. [Rhone-Poulenc Surf. & Spec.] Ammonium nonoxynol-20 sulfate; CAS 9051-57-4; anionic; emulsifier for agric. formulations and polymerization of most monomer systems; for use in latexes where extended shear stability and freeze/thaw stability are desired; liq.; surf. tens. 40 dynes/cm (@ CMC); 30% conc.

Abex® EP-120. [Rhone-Poulenc Surf. & Spec.] Ammonium nonoxynol-30 sulfate; CAS 9051-57-4; anionic; emulsifier for emulsion polymerization of vinyl acetate, acrylates, methacrylates, styrene, butadiene; for paints, adhesives, paper, textile, and industrial coatings; provides mech. and freeze/thaw stability; wetting agent, dispersant for agric. formulations; FDA compliance; pale yel. liq.; sp.gr. 1.06; visc. 111 cks; pour pt. 8 C; surf. tens. 40 dynes/cm (1%); 30% solids.

Abex® EP-227. [Rhone-Poulenc Surf. & Spec.] Ammonium nonoxynol-77 sulfate; anionic; emulsifier, stabilizer for prep. of vinyl acetate, vinyl acetate/acrylic, acrylic, styrene/acrylic, and styrene/butadiene emulsion copolymers; emulsifier for agric. formulation; liq.; surf. tens. 44 dynes/cm (@ CMC); 30% act.

Abex® EP-277. [Rhone-Poulenc Surf. & Spec.] Sodium octyl phenol sulfate; emulsifier for emulsion polymerization of vinyl acetate, acrylates, methacrylates, styrene, butadiene; for adhesives, paints, paper, textile, and industrial coatings.

Abex® JKB. [Rhone-Poulenc Surf. & Spec.] Proprietary ammonium surfactant; anionic; emulsifier for high-acid polymerization systems; FDA 21CFR §175.105; liq.; surf. tens. 40 dynes/cm (@ CMC); 30% conc.

Abex® VA 50. [Rhone-Poulenc Surf. & Spec.] Octoxynol-33, sodium laureth sulfate; anionic; emulsifier for high solids vinyl acetate emulsions; FDA 21CFR §175.105, 176.170, 176.180; liq.; surf. tens. 44 dynes/cm (@ CMC); 46% conc.

Abitol®. [Hercules] Dihydroabietyl alcohol; resinous plasticizer and tackifier in plastics, lacquers, inks, and adhesives; chemical intermediate; colorless visc. liq.; low odor; sol. in alcohols, esters, ketones, chlorinated solvs., and in aliphatic, aromatic, or terpene hydrocarbons; insol. in water; sp.gr. 1.008; dens. 8.4 lb/gal; visc. 400 poises (40 C); acid no. 0.3; sapon. no. 15; ref. index 1.5262; flash pt. (COC) 185 C; 100% liq. or 90% solids toluene or xylene sol'ns.

Ablumol BS. [Taiwan Surf.] Butyl stearate; CAS 123-95-5; EINECS 204-666-5; lubricant for transparent PVC prods.

Ablumol PSS. [Taiwan Surf.] Styrenated phenol; CAS 61788-44-1; nonstaining antioxidant for natural and syn. rubber.

Ablumol WTI. [Taiwan Surf.] Surfactant, resin quality improver which improves flow-out of resin finish.

Ablunol NP30. [Taiwan Surf.] Nonoxynol-30; CAS 9016-45-9; nonionic; emulsifier for vinyl acetate and acrylate emulsion polymerization; stabilizer; used in latex paints, floor finishes, paper coatings, textiles; solid; sp.gr. 1.08 (50 C); HLB 17.1; cloud pt. 74 C (1% in 10% NaCl); 100% act.

Ablunol NP30 70%. [Taiwan Surf.] Nonoxynol-30; CAS 9016-45-9; nonionic; emulsifier for vinyl acetate and acrylate emulsion polymerizations; stabilizer; used in latex paints, floor finishes, paper coatings, textiles; liq.; HLB 17.1; cloud pt. 74 C (1% in 10% NaCl); 70% act.

Ablunol NP40. [Taiwan Surf.] Nonoxynol-40; CAS 9016-45-9; nonionic; emulsifier for vinyl acetate and acrylate emulsion polymerizations; stabilizer; used in latex paints, floor finishes, paper coatings, textiles; solid; sp.gr. 1.08 (50 C); HLB 17.8; cloud pt. 76 C (1% in 10% NaCl); 100% act.

Ablunol NP40 70%. [Taiwan Surf.] Nonoxynol-40; CAS 9016-45-9; nonionic; emulsifier for vinyl acetate and acrylate emulsion polymerizations; stabilizer; used in latex paints, floor finishes, paper coatings, textiles; liq.; HLB 17.8; cloud pt. 76 C (1% in 10% NaCl); 70% act.

Ablunol NP50. [Taiwan Surf.] Nonoxynol-50; CAS 9016-45-9; nonionic; emulsifier for vinyl acetate and acrylate emulsion polymerizations; stabilizer; used in latex paints, floor finishes, paper coatings, textiles; solid; sp.gr. 1.08 (60 C); HLB 18.2; cloud pt. 76 C (1% in 10% NaCl); 100% act.

Ablunol NP50 70%. [Taiwan Surf.] Nonoxynol-50; CAS 9016-45-9; nonionic; emulsifier for vinyl acetate and acrylate emulsion polymerizations; stabilizer; used in latex paints, floor finishes, paper coatings, textiles; liq.; HLB 18.2; cloud pt. 76 C (1% in 10% NaCl); 70% act.

Ablusol C-78. [Taiwan Surf.] Sodium dioctyl sulfosuccinate; anionic; wetting agent for textile, PU leather; emulsion polymerization, agric. emulsion, and other industrial applics.; liq.; 70% conc.

Ablusol DA. [Taiwan Surf.] Ethoxylated decyl alcohol sulfosuccinate monoester; anionic; emulsifier for polyacrylate emulsion polymerization; liq.; water-sol.; 30% act.

Ablusol M-75. [Taiwan Surf.] Dioctyl sulfosuccinate; anionic; wetting agent for industrial applics.; suitable for adding to PU resin; liq.; 75% act.

Ablusol OA. [Taiwan Surf.] Oleyl sulfosuccinamate; anionic; foaming agent for noncarboxylated SBR latexes where stable uniform cell structure is necessary; for foam carpet underlay, foam carpet backing; liq.; 35% act.

Ablusol SF Series. [Taiwan Surf.] Anionic; emulsifier for emulsion polymerization; foaming agent, scouring and dyeing assistant for textiles.

Ablusol SN Series. [Taiwan Surf.] Anionic; emulsifier for emulsion polymerization; foaming agent, scouring and dyeing assistant for textiles.

Ablusol TA. [Taiwan Surf.] Octadecyl sulfosuccinamate; anionic; foaming agent for noncarboxylated SBR latexes where stable uniform cell structure is necessary; for foam carpet underlay, foam carpet backing; paste; 35% act.

Abluwax EBS. [Taiwan Surf.] Ethylene bisstearamide; lubricant for ABS, PS, and PVC; defoamer and mold-releasing agent.

ABT-2500®. [Specialty Minerals] Talc; CAS 14807-96-6; EINECS 238-877-9; antiblocking agent for polyolefin blown films; 2.3 μ avg. particle size; < 0.05% 325 mesh residue; dry brightness 89 ± 1.5; sp.gr. 2.7; bulk dens. 13 lb/ft^3 (loose), 37 lb/ft^3 (tapped); 60% SiO_2, 33% MgO.

A-C® 6. [AlliedSignal/Perf. Addit.] Polyethylene homopolymer; CAS 9002-88-4; EINECS 200-815-3; additive wax for use in adhesives, ink, floor finishes, paper coatings, personal care, plastics, rubber, textiles, wax blends; gellant for oils in personal care prods.; increases permanency, emolliency, moisture retention, water resist., and thermal stability; film-former, oil or fragrance encapsulator, nonirritating abrasive; wh. waxy prills, powd., char. waxy odor; negligible sol. in water; dens. 0.92 g/cc; visc. 375 cps (140 C); drop pt. 106 C; acid no. nil; flash pt. > 231 C; hardness 4.0 dmm; *Toxicology:* LD50 (oral, rat) > 2000 mg/kg; mild skin irritant; dust may cause mech. eye and respiratory tract irritation; OSHA TWA 5 mg/m^3 (respirable dust), 15 mg/m^3 (total dust) recommended; *Precaution:* incompat. with strong oxidizing agents; hazardous decomp. prods.: oxides of carbon, various oxidized and nonoxidized hydrocarbons.

A-C® 6A. [AlliedSignal/Perf. Addit.] Polyethylene homopolymer; CAS 9002-88-4; EINECS 200-815-3; additive wax for use in adhesives, inks, floor finishes, paper coatings, personal care, plastics, rubber, textiles, and wax blends; gellant for oils in personal care prods.; increases permanency, emolliency, moisture retention, water resist., and thermal stability; film-former, oil or fragrance encapsulator, nonirritating abrasive; wh. waxy prills, powd., char. waxy odor; negligible sol. in water; dens. 0.92 g/cc; visc. 375 cps (140 C); drop pt. 106 C; acid no. nil; flash pt. > 231 C; hardness 4.0 dmm; *Toxicology:* LD50 (oral, rat) > 2000 mg/kg; mild skin irritant; dust may cause mech. eye and respiratory tract irritation; OSHA TWA 5 mg/m^3 (respirable dust), 15 mg/m^3 (total dust) recommended; *Precaution:* incompat. with strong oxidizing agents; hazardous decomp. prods.: oxides of carbon, various oxidized and nonoxidized hydrocarbons.

A-C® 7, 7A. [AlliedSignal/Perf. Addit.] Polyethylene homopolymer wax; CAS 9002-88-4; EINECS 200-815-3; processing lubricant, melt index modifier, pigment dispersant, mold release aid; external lubricant PVC, color concs., polyolefin flow modifiers; thickener for cosmetic and pharmaceutical gels; prills, powd.; sol. in hot min. oil and fatty esters; dens. 0.92 g/cc; visc. 450 cps (140 C); drop pt. 109 C; acid no. nil; hardness 2.5 dmm.

A-C® 8, 8A. [AlliedSignal/Perf. Addit.] Polyethylene homopolymer; CAS 9002-88-4; EINECS 200-815-3; gellant for oils in personal care prods.; increases permanency, emolliency, moisture retention, water resist., and thermal stability; film-former, oil or fragrance encapsulator, nonirritating abrasive; processing lubricant, melt index modifier, pigment dispersant, mold release aid; external lubricant PVC, color concs., polyolefin flow modifiers; prills, powd.; dens. 0.93 g/cc; visc. 400 cps (140 C); drop. pt. 116 C; acid no. nil; hardness 1.0 dmm.

A-C® 9, 9A, 9F. [AlliedSignal/Perf. Addit.] Polyethylene homopolymer; CAS 9002-88-4; EINECS 200-815-3; gellant for oils in personal care prods.; increases permanency, emolliency, moisture retention, water resist., and thermal stability; film-former, oil or fragrance encapsulator, nonirritating abrasive; processing lubricant, melt index modifier, pigment dispersant, mold release aid; external lubricant PVC, color concs., polyolefin flow modifiers; prills, powd., fine powd.; avg. particle size 220 μ (9A), 110 μ (9F); dens. 0.94 g/cc; visc. 450 cps (140 C); drop. pt. 117 C; acid no. nil; hardness 0.5 dmm.

A-C® 15. [AlliedSignal/Perf. Addit.] Polyethylene homopolymer; CAS 9002-88-4; EINECS 200-815-3; additive wax for use in inks, personal care prods., adhesives, coatings, plastics, paper, rubber, textiles; gellant for oils in personal care prods.; increases permanency, emolliency, moisture retention, water resist., and thermal stability; film-former, oil or fragrance encapsulator, nonirritating abrasive; prills; dens. 0.93 g/cc; visc. 125 cps (140 C); drop pt. 109 C; acid no. nil; hardness 2.5 dmm.

A-C® 16. [AlliedSignal/Perf. Addit.] Polyethylene homopolymer; CAS 9002-88-4; EINECS 200-815-3; gellant for oils in personal care prods.; increases permanency, emolliency, moisture retention, water resist., and thermal stability; film-former, oil or fragrance encapsulator, nonirritating abrasive; additive for plastics, rubber, adhesives, coatings, inks, paper, polishes, textiles; prills; dens. 0.91 g/cc; visc. 525 cps (140 C); drop. pt. 102 C; acid no. nil; hardness 5.5 dmm.

Ac35. [Sovereign] Hindered phenol; economical nonstaining, nondiscoloring antioxidant; ultra high m.w. for high heat stability; works in all polymers; esp. good in sol'n. polymers.

A-C® 307, 307A. [AlliedSignal/Perf. Addit.] Oxidized HDPE homopolymer; CAS 68441-17-8; additive wax for use in inks, personal care prods., adhesives, coatings, plastics, rubber, textiles; gran., powd.; dens. 0.98 g/cc; visc. 85,000 cps (150 C); drop pt. 140 C; acid no. 5-9; hardness < 0.5 dmm.

A-C® 316, 316A. [AlliedSignal/Perf. Addit.] Oxidized

HDPE homopolymer; CAS 68441-17-8; additive wax for ink, floor finishes, personal care, plastics, rubber, textiles, wax blends; gellant for oils in personal care prods.; increases permanency, emolliency, moisture retention, water resist., and thermal stability; film-former, oil or fragrance encapsulator, nonirritating abrasive; gran., powd.; dens. 0.98 g/cc; visc. 8500 cps (150 C); drop pt. 140 C; acid no. 16; hardness < 0.5 dmm.

A-C® 325. [AlliedSignal/Perf. Addit.] Oxidized HDPE; CAS 68441-17-8; additive wax for polishes, finishes, and emulsions; heel-mark resistance and compatibility; also for adhesives, coatings, inks, plastics, rubber, textiles; gran.; sp.gr. 0.99; visc. 4400 cps (150 C); drop pt. 136 C; acid no. 25; hardness < 0.5 dmm.

A-C® 330. [AlliedSignal/Perf. Addit.] Oxidized HDPE; CAS 68441-17-8; additive wax for polishes, finishes, and emulsions; heel-mark resistance and compatibility; also for adhesives, coatings, plastics, rubber, textiles; gran.; sp.gr. 0.99; visc. 3600 cps (150 C); drop pt. 137 C; acid no. 30; hardness < 0.5 dmm.

A-C® 392. [AlliedSignal/Perf. Addit.] Oxidized HDPE; CAS 68441-17-8; additive wax for polishes, finishes, and emulsions; heel-mark resistance and compatibility; also for adhesives, coatings, plastics, rubber, textiles; gran.; sp.gr. 0.99; visc. 4500 cps (150 C); drop pt. 138 C; acid no. 30; hardness < 0.5 dmm.

A-C® 395, 395A. [AlliedSignal/Perf. Addit.] Oxidized HDPE homopolymer; CAS 68441-17-8; additive wax for use in inks, personal care prods., adhesives, coatings, plastics, rubber, textiles; gran.; dens. 1.00 g/cc; visc. 2500 cps (150 C); drop pt. 137 C; acid no. 41; hardness < 0.5 dmm.

A-C® 400. [AlliedSignal/Perf. Addit.] Ethylene/VA copolymer; CAS 24937-78-8; additive wax for use in adhesives, inks, coatings, personal care, wax blends, plastics, rubber, textiles; gellant for oils in personal care prods.; increases permanency, emolliency, moisture retention, water resist., and thermal stability; film-former, oil or fragrance encapsulator, nonirritating abrasive; prills, powd.; dens. 0.92 g/cc; visc. 575 cps (140 C); drop pt. 92 C; hardness 9.5 dmm; 13% VA.

A-C® 400A. [AlliedSignal/Perf. Addit.] Low m.w. EVA copolymer; CAS 24937-78-8; wax additive for use in adhesives, inks, personal care prods., wax blends; pigment dispersant; PS color concs.; gellant for oils in personal care prods.; increases permanency, emolliency, moisture retention, water resist., and thermal stability; film-former, oil or fragrance encapsulator, nonirritating abrasive; prills, powd.; dens. 0.92 g/cc; visc. 575 cps (140 C); drop pt. 92 C; hardness 9.5 dmm; 13% VA.

A-C® 405M. [AlliedSignal/Perf. Addit.] EVA copolymer; CAS 24937-78-8; wax additive for adhesives, inks, coatings, personal care, wax blends, plastics, rubber, textiles; prills; dens. 0.92 g/cc; visc. 600 cps (140 C); drop pt. 101 C; hardness 5.0 dmm; 8% VA.

A-C® 405S. [AlliedSignal/Perf. Addit.] EVA copolymer; CAS 24937-78-8; wax additive for adhesives, inks, coatings, personal care, wax blends, plastics, rubber, textiles; prills; dens. 0.92 g/cc; visc. 600 cps (140 C); drop pt. 95 C; hardness 7.0 dmm; 11% VA.

A-C® 405T. [AlliedSignal/Perf. Addit.] EVA copolymer; CAS 24937-78-8; wax additive for adhesives, inks, coatings, personal care, wax blends, plastics, rubber, textiles; gellant for oils in personal care prods.; increases permanency, emolliency, moisture retention, water resist., and thermal stability; film-former, oil or fragrance encapsulator, nonirritating abrasive; prills; dens. 0.92 g/cc; visc. 600 cps (140 C); drop pt. 103 C; hardness 4.0 dmm; 6% VA.

A-C® 430. [AlliedSignal/Perf. Addit.] EVA copolymer; CAS 24937-78-8; gellant for oils in personal care prods.; increases permanency, emolliency, moisture retention, water resist., and thermal stability; film-former, oil or fragrance encapsulator, nonirritating abrasive; wax additive for adhesives, coatings, personal care, wax blends, plastics, rubber, textiles; suspending agent for active ingreds. in cosmetic formulations; grease-like; dens. 0.93 g/cc; visc. 600 cps (140 C); drop pt. 80 C.

A-C® 540, 540A. [AlliedSignal/Perf. Addit.] Ethylene-acrylic acid copolymer; CAS 9010-77-9; plastics lubricant and processing aid, pigment dispersant; internal lubricant PVC, nylon 6, nylon color concs.; for adhesives, floor finishes, personal care, plastics, wax blends; gellant for oils in personal care prods.; increases permanency, emolliency, moisture retention, water resist., and thermal stability; film-former, oil or fragrance encapsulator, nonirritating abrasive; prills, powd.; dens. 0.93 g/cc; visc. 575 cps (140 C); drop pt. 105 C; acid no. 40; hardness 2.0 dmm.

A-C® 580. [AlliedSignal/Perf. Addit.] Ethylene/acrylic acid copolymer; CAS 9010-77-9; alkali-dispersible additive for recyclable hot-melt and aq. adhesives and coatings; plastics and rubber processing; gellant for oils in personal care prods.; increases permanency, emolliency, moisture retention, water resist., and thermal stability; film-former, oil or fragrance encapsulator, nonirritating abrasive; prills; dens. 0.94 g/cc; visc. 650 cps (140 C); drop pt. 102 C; acid no. 75; hardness 4.0 dmm.

A-C® 617, 617A. [AlliedSignal/Perf. Addit.] Polyethylene homopolymer; CAS 9002-88-4; EINECS 200-815-3; additive wax for use in inks, plastics, rubber, and personal care prods.; thickener for cosmetic and pharmaceutical gels; increases permanency, emolliency, moisture retention, water resist., and thermal stability; film-former, oil or fragrance encapsulator, nonirritating abrasive; wh. waxy prills, powd., char. waxy odor; sol. in hot min. oil and fatty esters; negligible sol. in water; dens. 0.91 g/cc; visc. 200 cps (140 C); drop pt. 101 C; acid no. nil; flash pt. > 231 C; hardness 7.0 dmm; *Toxicology:* LD50 (oral, rat) > 2000 mg/kg; mild skin irritant; dust may cause mech. eye and respiratory tract irritation; OSHA TWA 5 mg/m^3 (respirable dust), 15 mg/m^3 (total dust) recommended; *Precaution:* incompat. with strong oxidizing agents; hazardous decomp. prods.: oxides of carbon, various oxidized and nonoxidized hydrocarbons.

A-C® 629. [AlliedSignal/Perf. Addit.] Oxidized polyethylene homopolymer; CAS 68441-17-8; additive wax for adhesives, ink, floor finishes, paper coatings, personal care, plastics, textiles, wax blends; processing lubricant, mold release aid; PVC lubricant; gellant for oils in personal care prods.; increases permanency, emolliency, moisture retention, water resist., and thermal stability; film-former, oil or fragrance encapsulator, nonirritating abrasive; wh. waxy prills, powd., char. waxy odor; negligible sol. in water; dens. 0.93 g/cc; visc. 200 cps (140 C); drop pt. 101 C; acid no. 15; flash pt. (OC) > 260 C; hardness 5.5 dmm; *Toxicology:* LD50 (oral, rat) > 2500 mg/kg; nonirritating to skin and eyes; dust may cause mech. irritation; OSHA TWA 5 mg/m^3 (respirable dust), 15 mg/m^3 (total dust) recommended; *Precaution:* avoid strong oxidizing agents and amines; hazardous decomp. prods.: oxides of carbon, various oxidized and nonoxidized hydrocarbons; very slippery on floors; *Storage:* store away from heat sources.

A-C® 629A. [AlliedSignal/Perf. Addit.] Oxidized polyethylene homopolymer; CAS 68441-17-8; additive wax for adhesives, ink, floor finishes, paper coatings, personal care, plastics, textiles, wax blends; processing lubricant, mold release aid; PVC lubricant; gellant for oils in personal care prods.; increases permanency, emolliency, moisture retention, water resist., and thermal stability; film-former, oil or fragrance encapsulator, nonirritating abrasive; prills, powd.; dens. 0.93 g/cc; visc. 200 cps (140 C); drop pt. 101 C; acid no. 15; hardness 5.5 dmm.

A-C® 655. [AlliedSignal/Perf. Addit.] Oxidized polyethylene homopolymer; CAS 68441-17-8; additive wax for use in adhesives, inks, floor finishes, paper coatings, personal care, plastics, rubber, textiles, and wax blends; prills; dens. 0.93 g/cc; visc. 210 cps (140 C); drop pt. 107 C; acid no. 16; hardness 2.5 dmm.

A-C® 656. [AlliedSignal/Perf. Addit.] Oxidized polyethylene homopolymer; CAS 68441-17-8; additive wax for use in adhesives, inks, floor finishes, paper coatings, personal care, plastics, and rubber; prills; dens. 0.92 g/cc; visc. 185 cps (140 C); drop pt. 98 C; acid no. 15; hardness 9.0 dmm.

A-C® 712. [AlliedSignal/Perf. Addit.] Polyethylene homopolymer; CAS 9002-88-4; EINECS 200-815-3; additive for plastics, rubber, adhesives, coatings, inks, paper, polishes, textiles; diced form; sp.gr. 0.92; visc. 1600 cps (140 C); drop pt. 108 C; acid no. nil; hardness 3.5 dmm.

A-C® 715. [AlliedSignal/Perf. Addit.] Polyethylene homopolymer; CAS 9002-88-4; EINECS 200-815-3; additive for plastics, rubber, adhesives, coatings, inks, paper, polishes, textiles; diced form; sp.gr. 0.92; visc. 4000 cps (140 C); drop pt. 109 C; acid no. nil; hardness 2.5 dmm.

A-C® 725. [AlliedSignal/Perf. Addit.] Polyethylene homopolymer; CAS 9002-88-4; EINECS 200-815-3; additive for plastics, rubber, adhesives, coatings, inks, paper, polishes, textiles; diced form; sp.gr. 0.92; visc. 1400 cps (140 C); drop pt. 108 C; acid no. nil; hardness 3.5 dmm.

A-C® 735. [AlliedSignal/Perf. Addit.] Polyethylene homopolymer; CAS 9002-88-4; EINECS 200-815-3; additive for plastics, rubber, adhesives, coatings, inks, paper, polishes, textiles; diced form; sp.gr. 0.92; visc. 6000 cps (140 C); drop pt. 110 C; acid no. nil; hardness 2.5 dmm.

A-C® 1702. [AlliedSignal/Perf. Addit.] Polyethylene homopolymer; CAS 9002-88-4; EINECS 200-815-3; gellant for oils in personal care prods.; increases permanency, emolliency, moisture retention, water resist., and thermal stability; film-former, oil or fragrance encapsulator, nonirritating abrasive; additive for plastics, rubber, adhesives, coatings, inks, paper, polishes, textiles; grease-like; negligible sol. in water; dens. 0.88 g/cc; visc. 40 cps (140 C); drop. pt. 92 C; acid no. nil; flash pt. > 231 C; hardness 90.0 dmm (D1321); *Toxicology:* LD50 (oral, rat) > 2000 mg/kg; mild skin irritant; *Precaution:* incompat. with strong oxidizing agents; hazardous decomp. prods.: oxides of carbon, various oxidized and nonoxidized hydrocarbons.

A-C® 5120. [AlliedSignal/Perf. Addit.] Low m.w. ethylene-acrylic acid copolymer; CAS 9010-77-9; alkali-dispersible additive for recyclable hot-melt and aq. adhesives and coatings; gellant for personal care and household prods. (cosmetics, protective moisturizers, medications, dental adhesives, deodorizers); plastics and rubber processing; prills; dens. 0.94 g/cc; visc. 650 cps (140 C); drop pt. 92 C; acid no. 120; hardness 11.5 dmm.

A-C® 5180. [AlliedSignal/Perf. Addit.] Low m.w. ethylene-acrylic acid copolymer; CAS 9010-77-9; alkali-dispersible additive for recyclable hot-melt and aq. adhesives and coatings; plastics and rubber processing; grease-like; dens. 0.96 g/cc; visc. 650 cps (140 C); drop pt. 76 C; acid no. 180; hardness 50 dmm.

A-C® 6702. [AlliedSignal/Perf. Addit.] Oxidized polyethylene; CAS 68441-17-8; gellant for oils in personal care prods.; increases permanency, emolliency, moisture retention, water resist., and thermal stability; film-former, oil or fragrance encapsulator, nonirritating abrasive; additive for plastics, rubber, adhesives, coatings, inks, paper, polishes, textiles; grease-like; dens. 0.85 g/cc; visc. 35 cps (140 C); drop pt. 85 C; acid no. 15; hardness 90.0 dmm (D1321).

Accelerator 399. [Huntsman] Epoxy curing promoter for use with amine hardeners and curing agents; pale yel. clear sl. visc. liq.; sp.gr. 1.089; visc. 880 cps; pour pt. 5 F; flash pt (PMCC) 180 F; *Toxicology:* LD50 (oral, rat) 3.3 g/kg; sl. toxic by ingestion; may cause severe eye irritation; moderately irritating to skin.

Accelerator 55028. [Akzo Nobel] Cobalt octoate, hydrocarbon solv.; accelerator developed for filled systems to decrease cobalt absorption and geltime drift; sol'n.; 3% cobalt.

Accelerator D. [Air Prods./Perf. Chems.] Cyclo amine blended with organo metallic salt; accelerator recommended for use in lt.-colored compds., esp. shoe soling industry; wh. to beige nondusting powd.; Unverified

Accelerator E. [Scott Bader] 0.4% active cobalt sol'n. in styrene; accelerator.

Accelerator NL-6. [Akzo Nobel] Cobalt octoate, hydrocarbon solv.; accelerator mainly used in combination with ketone peroxides for cure of unsat. polyester resins at R.T. and elevated temps.; sol'n.; 6% cobalt.

Accelerator NL-12. [Akzo Nobel] Cobalt octoate, hydrocarbon solv.; accelerator mainly used in combination with ketone peroxides for cure of unsat. polyester resins at R.T. and elevated temps.; sol'n.; 12% cobalt.

Accelerator NL49P. [Akzo Nobel] Cobalt octoate, phthalate solv.; accelerator mainly used in combination with ketone peroxides for cure of unsat. polyester resins at R.T. and elevated temps.; sol'n.; 1% cobalt.

Accelerator NL51P. [Akzo Nobel] Cobalt octoate, phthalate solv.; accelerator mainly used in combination with ketone peroxides for cure of unsat. polyester resins at R.T. and elevated temps.; sol'n.; 6% cobalt.

Accelerator NL53. [Akzo Nobel] Cobalt octoate, hydrocarbon solv.; accelerator mainly used in combination with ketone peroxides for cure of unsat. polyester resins at R.T. and elevated temps.; sol'n.; 10% cobalt.

Accelerator VN-2. [Akzo Nobel] Vanadium salt; accelerator for ketone peroxides, hydroperoxides, and peroxy esters; short gel times and a very high speed of cure can be achieved; sol'n.; 0.2% vanadium.

Acconon 300-MO. [Karlshamns] PEG 300 oleate; CAS 9004-96-0; nonionic; emulsifier, lubricant, chemical intermediate; for cosmetics, food, agriculture, plastics; Gardner 6 liq.; sol. in org. solv.; sp.gr. 0.99; dens. 8.3 lb/gal; m.p. < -5 C; 99% act.; Unverified

Acconon 400-MO. [Karlshamns] PEG-8 oleate; CAS 9004-96-0; nonionic; emulsifier, dispersant, lubricant, chemical intermediate, solubilizer, visc. control agent; for cosmetics, pharmaceuticals, food, agric., plastics; Gardner 4 max. liq.; sol. in org. solv.; water-disp.; sp.gr. 1.01;dens. 8.4 lb/gal; m.p. < 10 C; HLB 12; pH 5.5-6.5; 99% act.

Aceto. [Aceto] 3,5-Di-t-butyl p-hydroxybenzoic acid; uv stabilizer for PE, PP; powd.; m.p. 200-210 C.

Acetorb A. [Aceto] 2-Hydroxy-4-methoxy benzophenone; CAS 131-57-7; EINECS 205-031-5; uv stabilizer for ABS, cellulosics, polyesters, PS, flexible and rigid PVC, VDC; powd.; m.p. 62-63.5 C.

Acintol® 208. [Arizona] Tall oil acid isooctyl ester; EINECS 269-788-3; plasticizer for chloroprene, nitrile and Hypalon® elastomer systems; good plasticizer retention at elevated temps.; Gardner 2+; sp.gr. 0.87; dens. 7.23 lb/gal; visc. 6-7 cps; flash pt. (CC) > 200 F; acid no. 0.5; sapon. no. 143.

Acintol® DFA. [Arizona] Tall oil acid; CAS 61790-12-3; EINECS 263-107-3; for metalworking fluids, lubricant additives, oilfield chems., asphalt emulsifiers, alkyd resins, industrial/household cleaners, plasticizers, textile drawing lubricants, surf. coatings, rubber prods.; Gardner 8; sp.gr. 0.91; visc. 20 cps; acid no. 190; iodine no. 135; sapon. no. 197; flash pt. (OC) 204 C.

Acintol® EPG. [Arizona] Tall oil acid; CAS 61790-12-3; EINECS 263-107-3; surfactant for metalworking fluids, lubricant additives, oilfield chems., asphalt emulsifiers, alkyd resins, industrial/household cleaners, plasticizers, textile drawing lubricants, surf. coatings, rubber prods.; epoxy grade; Gardner 1+; sp.gr. 0.897; dens. 7.45 lb/gal; visc. 20 cps; acid no. 199; iodine no. 130; sapon. no. 200; flash pt. (OC) 204 C; 99.0% fatty acids.

Acintol® FA-1. [Arizona] Tall oil acid; CAS 61790-12-3; EINECS 263-107-3; surfactant for metalworking fluids, lubricant additives, oilfield chems., asphalt emulsifiers, alkyd resins, industrial/household cleaners, plasticizers, textile drawing lubricants, surf. coatings, rubber prods.; Gardner 5; sp.gr. 0.906; dens. 7.53 lb/gal; visc. 20 cps; acid no. 194; iodine no. 131; sapon. no. 197; flash pt. (OC) 204 C; 92.8% fatty acids.

Acintol® FA-1 Special. [Arizona] Tall oil acid; CAS 61790-12-3; EINECS 263-107-3; surfactant for metalworking fluids, lubricant additives, oilfield chems., asphalt emulsifiers, alkyd resins, industrial/household cleaners, plasticizers, textile drawing lubricants, surf. coatings, rubber prods.; Gardner 4+; sp.gr. 0.91; dens. 7.50 lb/gal; visc. 20 cps; acid no. 195; iodine no. 131; sapon. no. 198; flash pt. (OC) 204 C; 95.2% fatty acids.

Acintol® FA-2. [Arizona] Tall oil acid; CAS 61790-12-3; EINECS 263-107-3; surfactant for metalworking fluids, lubricant additives, oilfield chems., asphalt emulsifiers, alkyd resins, industrial/household cleaners, plasticizers, textile drawing lubricants, surf. coatings, rubber prods.; Gardner 3+; sp.gr. 0.898; dens. 7.47 lb/gal; visc. 20 cps; acid no. 197; iodine no. 130; sapon. no. 199; flash pt. (OC) 204 C; 97.8% fatty acids.

Acintol® FA-3. [Arizona] Tall oil acid; CAS 61790-12-3; EINECS 263-107-3; surfactant for metalworking fluids, lubricant additives, oilfield chems., asphalt emulsifiers, alkyd resins, industrial/household cleaners, plasticizers, textile drawing lubricants, surf. coatings, rubber prods.; Gardner 2+; sp.gr. 0.897; dens. 7.45 lb/gal; visc. 20 cps; acid no. 198; iodine no. 130; sapon. no. 200; flash pt. (OC) 204 C; 98.8% fatty acids.

Acintol® Liquaros. [Arizona] Tall oil rosin; used in mfg. of paper size, intermediate in rosin deriv. prod., printing ink binders, tackifier resin in sealants and mastics, starting point rosin for resin esters; FDA 21CFR §175.105, 175.125, 175.300, 176.170, 176.180, 176.210, 177.1200, 177.1210, 177.2600, 178.3120, 178.3800, 178.3850, 178.3870; Gardner 10+ color; sp.gr. 1.02; visc. > 20,000 cps; acid no. 157; flash pt. (OC) > 190 C;

60% rosin acids content.

Acintol® R Type S. [Arizona] Tall oil rosin; printing ink binder as resin or salt, paper sizing agent, emulsifier for SBR polymerization as soap, tackifier resin in adhesives, imidazoline modifier in corrosion inhibitors, elastomer modifier in emulsion polymerization, dust control additive; film former/plasticizer in lacquers and varnishes; FDA 21CFR §175.105, 175.125, 175.300, 176.170, 176.180, 176.210, 177.1200, 177.1210, 177.2600, 178.3120, 178.3800, 178.3850, 178.3870; Gardner 7 color; sp.gr. 1.01; soften. pt. (R&B) 76 C; acid no. 175-178; flash pt. (OC) 226 C.

Acintol® R Type SB. [Arizona] Tall oil rosin; printing ink binder as resin or salt, paper sizing agent, emulsifier for SBR polymerization as soap, tackifier resin in adhesives, imidazoline modifier in corrosion inhibitors, elastomer modifier in emulsion polymerization, dust control additive; film former/plasticizer in lacquers and varnishes; FDA 21CFR §175.105, 175.125, 175.300, 176.170, 176.180, 176.210, 177.1200, 177.1210, 177.2600, 178.3120, 178.3800, 178.3850, 178.3870; Gardner 8 color; sp.gr. 1.01; soften. pt. (R&B) 75 C; acid no. 165-166; flash pt. (OC) 226 C.

Acintol® R Type SFS. [Arizona] Tall oil rosin, formaldehyde-treated; used in mfg. of paper size, intermediate in rosin deriv. prod., printing ink binders, tackifier resin in sealants and mastics, starting point rosin for resin esters; FDA 21CFR §175.105, 175.125, 175.300, 176.170, 176.180, 176.210, 177.1200, 177.1210, 177.2600, 178.3120, 178.3800, 178.3850, 178.3870; Gardner 11 color; sp.gr. 1.06; soften. pt. (R&B) 77 C; acid no. 155; flash pt. (OC) 226 C; 91.5-93.4% rosin acids.

Acintol® R Type SM4. [Arizona] Tall oil rosin, maleated; used in mfg. of paper size, intermediate in rosin deriv. prod., printing ink binders, tackifier resin in sealants and mastics, starting point rosin for resin esters; FDA 21CFR §175.105, 175.125, 175.300, 176.170, 176.180, 176.210, 177.1200, 177.1210, 177.2600, 178.3120, 178.3800, 178.3850, 178.3870; Gardner 9+ color; sp.gr. 1.01; soften. pt. (R&B) 86 C; acid no. 200; flash pt. (OC) 226 C; 93.5% rosin acids.

Aclarat 8678 Granules, Liq. [Sandoz] Fluorescent whitener for commerical/industrial laundry detergents, rug/upholstery cleaners, fabric softeners, laundry bleach, whitening soap, brightening polymers and plastics; esp. for synthetics and wool.

AClyn® 201A. [AlliedSignal/Perf. Addit.] Ethylene/calcium acrylate copolymer; CAS 26445-96-5; low m.w. ionomers used as processing and performance additives; improves dispersion of additives in plastics; adhesion to variety of substrates; encapsulant for personal care prods.; visc. 5500 cps (190 C); m.p. 102 C; acid no. 42.

AClyn® 246A. [AlliedSignal/Perf. Addit.] Ethylene/magnesium acrylate copolymer; low m.w. ionomers used as processing and performance additives; improves dispersion of additives in plastics; adhesion to variety of substrates; encapsulant for personal care prods.; visc. 7000 cps (190 C); m.p. 95 C; acid no. nil.

ACM-MW. [AluChem] Magnesium hydroxide; CAS 1309-42-8; EINECS 215-170-3; flame retardant for ABS, acrylics, cellulosics, epoxy, phenolic, PC, thermoset polyester, PS, PE, PP, PVAc, PVC, PU.

Acofor. [Reichhold] Dist. tall oil fatty acids; CAS 61790-12-3; EINECS 263-107-3; latex stabilizer, dispersant (as soap) for pigments and fillers; Gardner 3 color; sp.gr. 0.903; flash pt. (COC) 400 C; Unverified

Acrawax® C. [Lonza] N,N′-ethylene bisstearamide; CAS 110-30-5; internal and surf. lubricant in PVC, PP, PE, ABS, nylon, acetal, thermoplastic polyester, PS, phenolic, PU; processing aid, plasticizer, compatibilizer, slip/antiblock, flow improver, pigment dispersant; used in hot-melt adhesives and coatings; powd. and atomized grades as lubricant and binder for cold compaction of powd. metal parts; FDA clearance for food contact applics.; cream hard waxy solid; 10% max. on 10 mesh (beads); 2% max. on 40 mesh (prilled); 1% max. on 100 mesh (powd.); 0.1% max. on 325 mesh (atomized); insol. in water; sp.gr. 0.97; m.p. 140-145 C; acid no. 8 max.; flash pt. 285 C.

Acrawax® C CG. [Lonza] Amide wax; compatibilizer for polyethylene; Gardner 3 prilled; m.p. 143 C; acid no. 5.

Acrawax® C SG. [Lonza] Ethylenebisstearamide; lubricant for thermoplastics; vinyl siding grade; Gardner 3 prilled; m.p. 143 C; acid no. 5.

Actafoam® F-2. [Uniroyal] Activator-stabilizer for vinyl foams containing Kempore blowing agents; gas release accelerator; paste, powd.

Actafoam® R-3. [Uniroyal] Activator-stabilizer for vinyl foams containing Kempore blowing agents; gas release accelerator; liq.

Actimer FR-803. [AmeriBrom; Dead Sea Bromine] Tribromostyrene; CAS 61368-34-1; reactive monomer which can be copolymerized with styrene, acrylonitrile and maleic anhydride to impart flame retardancy to HIPS, ABS, SAN, SMA; crosslinking agent providing flame retardancy to thermoset systems such as PU; off-wh. free-flowing powd.; m.p. 65-67 C; 94% assay, 70.4% Br; *Toxicology:* LD50 (rat) > 5 g/kg.

Actimer FR-1025M. [AmeriBrom; Dead Sea Bromine] Pentabromobenzyl acrylate; CAS 594477-55-11; flame retardant for engineering thermoplastics (PET, PBT, PC, nylon 6 and 6/6); processing aid; maintains transparency of PC and HIPS resins; wh. free-flowing powd.; sol. in MEK, MIBK, styrene; insol. in water; m.p. 122 C; 98% purity, 71.8% Br; *Toxicology:* LD50 (rat) > 5 g/kg.

Actimer FR-1033. [AmeriBrom; Dead Sea Bromine] Tribromophenyl maleimide; CAS 59789-51-4; reactive flame retardant for polymeric systems which undergo crosslinking (XLPE, EPDM); can be grafted onto unsat. sites in ABS or copolymerized with styrenics to yield flame retardant systems with improved thermal props.; off-wh. free-flowing cryst.; sol. in aromatic and chlorinated hydrocarbons, actone, DMF, DMSO, ethanol; insol. in water, aliphatic hydrocarbons; m.p. 138 C; 96% min. assay, 57.3% Br; *Toxicology:* LD50 > 5 g/kg.

Actiron NX 3. [Protex] 2,4,6-Tri (dimethylaminomethyl) phenol; CAS 90-72-2; accelerator, hardener, and catalyst for epoxy resins, incl. flexible epoxy resins, anhydride epoxy systems, and epoxy resins containing polyamide hardeners; polyurethane additive; brn. to red-brn. liq., amine type odor; sol. in org. solvs. and cold water; only sl. sol. in hot water; sp.gr. 0.98 g/cc; visc. 375 ± 125 mPa•s; distill. pt. 316 C; flash pt. (CC) 157 C; ref. index 1.517 (20 C); *Toxicology:* LD50 (oral, rat) 2500 mg/kg; noxious by ingestion; irritant to respiratory system, skin, eyes; *Precaution:* incompat. with strong oxidizers; hazardous decomp. prods.: CO, CO_2, NO_2; *Storage:* keep container tightly closed.

Activ-8. [R.T. Vanderbilt] Sol'n. forms containing 1,10-phenanthroline; CAS 66-71-7; drier accelerator and stabilizer used in combination with manganese and/or cobalt in coating systems that cure by oxidative polymerization; dens. 0.95 mg/m^3; flash pt. 36 C; 38% act.

Activator 101. [Sovereign] Zinc soap; activator for all millable urethanes.

Activator 1102. [Air Prods./Perf. Chems.] Dibutyl ammonium oleate; accelerator/activator for natural and syn. rubbers; lubricant; dk. red/brn. low-visc. liq.; sp.gr. 0.87; Unverified

Activator STAG. [Akrochem] Complex sec. amine, surface treated; relatively nonstaining, nondiscoloring activator for thiazole-type accelerators; primary accelerator for natural rubber; strong sec. accelerator in SBR; lt. blue nondusting powd.; sp.gr. 1.26 ± 0.03; m.p. 130 C min.

Activex 233 25%. [J.M. Huber/Chems.] Proprietary; endothermic chem. foaming agent for ABS, acetal, EVA, flexible PVC; processing range 300+ F; pellets; gas yield 45 cc/g.

Activex 235 25%. [J.M. Huber/Chems.] Proprietary; endothermic chem. foaming agent for ABS, acetal, EVA, HDPE, LDPE, PP, flexible PVC, TPE; processing range 300+ F; pellets; gas yield 45 cc/g.

Activex 236 25%. [J.M. Huber/Chems.] Proprietary; endothermic chem. foaming agent for LDPE, PP; processing range 300+ F; pellets; gas yield 45 cc/g.

Activex 237 25%. [J.M. Huber/Chems.] Proprietary; endothermic chem. foaming agent for ABS, acetal, EVA, HDPE, LDPE, PP, TPE; processing range 300+ F; pellets; gas yield 45 cc/g.

Activex 436 25%. [J.M. Huber/Chems.] Proprietary; endothermic chem. foaming agent for PP; processing range 300+ F; pellets; gas yield 36 cc/g.

Activex 437 25%. [J.M. Huber/Chems.] Proprietary; endothermic chem. foaming agent for HDPE, LDPE, PPO, PP, PS, HIPS, TPE; processing range 300+ F; pellets; gas yield 36 cc/g.

Activex 447. [J.M. Huber/Chems.] Proprietary; endothermic chem. foaming agent for HDPE, LDPE, PPO, PP, PS, HIPS, TPE; processing range 300-600 F; pellets; gas yield 53 cc/g.

Activex 533 25%. [J.M. Huber/Chems.] Proprietary; endothermic chem. foaming agent for ABS, acetal, EVA, flexible PVC; processing range 300-600 F; pellets; gas yield 30 cc/g.

Activex 535 25%. [J.M. Huber/Chems.] Proprietary; endothermic chem. foaming agent for ABS, acetal, EVA, HDPE, PP, TPE; processing range 300-600 F; pellets; gas yield 30 cc/g.

Activex 536 25%. [J.M. Huber/Chems.] Proprietary; endothermic chem. foaming agent for HDPE, PP, TPE, PET, PBT; processing range 300-600 F; pellets; gas yield 30 cc/g.

Activex 537 25%. [J.M. Huber/Chems.] Proprietary; endothermic chem. foaming agent for HDPE, PP, HIPS, PS, TPE, polyamide, PBT; processing range 300-600 F; pellets; gas yield 30 cc/g.

Activex 539 25%. [J.M. Huber/Chems.] Proprietary; endothermic chem. foaming agent for PPO, PS, HIPS; processing range 300-600 F; pellets; gas yield 30 cc/g.

Activex 545. [J.M. Huber/Chems.] Proprietary; endothermic chem. foaming agent for ABS, acetal, EVA, HDPE, PP, TPE; processing range 300-600 F; pellets; gas yield 36 cc/g.

Activex 736 25%. [J.M. Huber/Chems.] Proprietary; endothermic chem. foaming agent for HDPE, PC, high temp. resins; processing range 300-600 F; pellets; gas yield 30 cc/g.

ACtol® 60. [AlliedSignal/Perf. Addit.] Resin modifier with hydroxyl functionality; reactive flexibilizer and impact modifier; for polyester powd. coatings, as urethane prepolymers; m.w. 2400; m.p. 71-86 C; hyd. no. 130-170.

ACtol® 65. [AlliedSignal/Perf. Addit.] Resin modifier with hydroxyl ester functionality; resin modifier for increased solv. solubility; for polyester powd. coatings, as urethane prepolymer; m.w. 2400; m.p. 40-65 C; hyd. no. 85-120.

ACtol® 70. [AlliedSignal/Perf. Addit.] Resin modifier with hydroxyl functionality; harder version of ACtol 65; resin modifier for polyester powd. coatings, as urethane prepolymer; m.w. 2500; m.p. 95-105 C; hyd. no. 75-90.

ACtol® 80. [AlliedSignal/Perf. Addit.] Resin modifier with hydroxyl and carboxyl functionality; resin modifier for polyester powd. coatings, as urethane prepolymer; water-disp.; m.w. 2500; m.p. 74-86 C; acid no. 40; hyd. no. 75-115.

ACtone® 1. [AlliedSignal/Perf. Addit.] Low m.w. proprietary ionomer; pigment wetting agent, dispersion aid to increase color strength; used in thermoplastic and thermoset resins; free-flowing powd.; visc. 5500 cps (190 C); bulk dens. 410 kg/m^3; soften. pt. 101 C; acid no. 37; hardness 1.5 dmm.

ACtone® 2000V. [AlliedSignal/Polymer Processing Tech.] color enhancer, dispersion aid for PVC colorants, esp. difficult-to-disperse organics; visc. 9400-14,500 cps (190 C); bulk dens. 26.8 lb/ft^3; hardness 3.8 ± 0.7 dmm.

ACtone® 2010. [AlliedSignal/Perf. Addit.] Low m.w. proprietary ionomer resin; pigment wetting agent, dispersion aid for color concs. and masterbatches for polyester and styrenics; gran.; visc. 500,000 cps (190 C); melt flow 1.0 g/10 min.

ACtone® 2010P. [AlliedSignal/Polymer Processing Tech.] Low m.w. proprietary ionomer resin; pigment wetting agent, dispersion aid for color concs. and masterbatches for polyester and styrenics; powd.; visc. 500,000 cps (190 C); melt flow 1.0 g/10 min.

ACtone® 2316. [AlliedSignal/Perf. Addit.] Color enhancer improving performance of hard-to-disperse org. and inorg. colorants; compat. with engineering thermoplastics and polyolefins; acid no. 3-10; soften. pt. 125-132 C.

ACtone® 2461. [AlliedSignal/Polymer Processing Tech.] Color enhancer, dispersion aid for PE colorants, esp. difficult-to-disperse organics.

ACtone® N. [AlliedSignal/Perf. Addit.] Low m.w. proprietary ionomer; pigment wetting agent, dispersion aid, color enhancer for color concs. for nylon and polyester; gran.; visc. 30,000 cps (190 C); bulk dens. 30.21 lb/ft^3; acid no. 5; soften. pt. 100 C.

ACumist® A-6. [AlliedSignal/Perf. Addit.] Micronized polyethylene; CAS 9002-88-4; EINECS 200-815-3; wax additive for inks, coatings, rubber applics.; micronized powd.; 6 μ avg. particle size; dens. 0.99 g/cc; drop pt. 137 C; acid no. 26-40; hardness < 0.5 dmm.

ACumist® A-12. [AlliedSignal/Perf. Addit.] Micronized polyethylene; CAS 9002-88-4; EINECS 200-815-3; wax additive for adhesives, inks, personal care, rubber applics.; suspension aid, flatting and texturizing agent, and binder for personal care prods.; micronized powd.; 12 μ avg. particle size; dens. 0.99 g/cc; drop pt. 136 C; acid no. 26-40; hardness < 0.5 dmm.

ACumist® A-18. [AlliedSignal/Perf. Addit.] Micronized polyethylene; CAS 9002-88-4; EINECS 200-815-3; wax additive for adhesives, inks, personal care, and rubber applics.; suspension aid, flatting and texturizing agent, and binder for personal care prods.; micronized powd.; 18 μ avg. particle size; dens. 0.99 g/cc; drop pt. 136 C; acid no. 26-40; hardness < 0.5 dmm.

ACumist® A-45. [AlliedSignal/Perf. Addit.] Micronized polyethylene; CAS 9002-88-4; EINECS 200-815-3; wax additive for inks, coatings, rubber applics.; micronized powd.; 45 μ avg. particle size; dens. 0.99 g/cc; drop pt. 137 C; acid no. 26-40; hardness < 0.5 dmm.

ACumist® B-6. [AlliedSignal/Perf. Addit.] Micronized polyethylene; CAS 9002-88-4; EINECS 200-815-3; wax additive for adhesives, inks, personal care, rubber applics.; suspension aid, flatting and texturizing agent, and binder for personal care prods.; provides smoother, silkier afterfeel; micronized powd.; 6 μ avg. particle size; dens. 0.96 g/cc; drop pt. 126 C; acid no. nil; hardness < 0.5 dmm.

ACumist® B-9. [AlliedSignal/Perf. Addit.] Micronized polyethylene; CAS 9002-88-4; EINECS 200-815-3; wax additive for adhesives, inks, personal care, and rubber applics.; micronized powd.; 9 μ avg. particle size; dens. 0.96 g/cc; drop pt. 126 C; acid no. nil; hardness < 0.5 dmm.

ACumist® B-12. [AlliedSignal/Perf. Addit.] Micronized polyethylene; CAS 9002-88-4; EINECS 200-815-3; wax additive for adhesives, inks, personal care, and rubber applics.; suspension aid, flatting and texturizing agent, and binder for personal care prods.; micronized powd.; 12 μ avg. particle size; dens. 0.96 g/cc; drop pt. 126 C; acid no. nil; hardness < 0.5 dmm.

ACumist® B-18. [AlliedSignal/Perf. Addit.] Micronized polyethylene; CAS 9002-88-4; EINECS 200-815-3; wax additive for adhesives, inks, personal care, and rubber applics.; suspension aid, flatting and texturizing agent, and binder for personal care prods.; micronized powd.; 18 μ avg. particle size; dens. 0.96 g/cc; drop pt. 126 C; acid no. nil; hardness < 0.5 dmm.

ACumist® C-5. [AlliedSignal/Perf. Addit.] Micronized polyethylene; CAS 9002-88-4; EINECS 200-815-3; wax additive for adhesives, inks, personal care, and rubber applics.; suspension aid, flatting and texturizing agent, and binder for personal care prods.; micronized powd.; 5 μ avg. particle size; dens. 0.95 g/cc; drop pt. 121 C; acid no. nil; hardness 1.0 dmm.

ACumist® C-9. [AlliedSignal/Perf. Addit.] Micronized polyethylene; CAS 9002-88-4; EINECS 200-815-3; wax additive for adhesives, inks, personal care, and rubber applics.; suspension aid, flatting and texturizing agent, and binder for personal care prods.; micronized powd.; 9 μ avg. particle size; dens. 0.95 g/cc; drop pt. 121 C; acid no. nil; hardness 1.0 dmm.

ACumist® C-12. [AlliedSignal/Perf. Addit.] Micronized polyethylene; CAS 9002-88-4; EINECS 200-815-3; wax additive for adhesives, inks, personal care, and rubber applics.; suspension aid, flatting and texturizing agent, and binder for personal care prods.; micronized powd.; 12 μ avg. particle size; dens. 0.95 g/cc; drop pt. 121 C; acid no. nil; hardness 1.0 dmm.

ACumist® C-18. [AlliedSignal/Perf. Addit.] Micronized polyethylene; CAS 9002-88-4; EINECS 200-815-3; wax additive for adhesives, inks, personal care, and rubber applics.; suspension aid, flatting and texturizing agent, and binder for personal care prods.; micronized powd.; 18 μ avg. particle size; dens. 0.95 g/cc; drop pt. 121 C; acid no. nil; hardness 1.0 dmm.

ACumist® C-30. [AlliedSignal/Perf. Addit.] Micronized polyethylene; CAS 9002-88-4; EINECS 200-815-3; wax additive for inks, coatings, rubber applics.; micronized powd.; 30 μ avg. particle size; dens. 0.95 g/cc; drop pt. 121 C; acid no. nil; hardness 1.0 dmm.

ACumist® D-9. [AlliedSignal/Perf. Addit.] Micronized polyethylene; CAS 9002-88-4; EINECS 200-815-3; wax additive for adhesives, inks, personal care, and rubber applics.; micronized powd.; 10 μ avg. particle size; dens. 0.95 g/cc; drop pt. 118 C; acid no. nil; hardness 1.5 dmm.

Additin 30. [Bayer/Fibers, Org., Rubbers] Phenyl-α-naphthylamine; CAS 90-30-3; staining antioxidant for rubber tech. goods and heavily stressed goods; antiflexcracking agent for NR and IR; storage stabilizer for petrol. prods.; lt. brn. to lt. violet flakes and fused; dens. 1.11 g/cm^3; m.p. ≥ 58 C.

Adeka Carpol MH-50, MH-150, MH-1000. [Asahi Denka Kogyo] Polyoxyalkylene glycol; hydraulic fluid oil, lubricant in high/low temp., heat media, lubricant for textile, compressor; mold lubricant for rubber, plastics, cosmetics, cutting oil, break oil, defoamer; water-sol.; visc. 54.6, 135, and 938 cst resp. (40 C).

Adeka Dipropylene Glycol. [Asahi Denka Kogyo] Dipropylene glycol; EINECS 203-821-4; surfactant for unsat. polyester resin, paint applics.; APHA 15+ color; sp.gr. 1.020-1.030; > 0.2% water.

Adeka Glycilol ED-503. [Asahi Denka Kogyo] DGE of alkylene glycol; epoxy resin diluent; Gardner 2+ color; sp.gr. 1.08; visc. 15-35 cps; EEW 150-180.

Adeka Glycilol ED-505. [Asahi Denka Kogyo] TGE of trimethylol propane; epoxy resin diluent; Gardner 2+ color; sp.gr. 1.16; visc. 125-200 cps; EEW 135-165.

Adeka Glycilol ED-506. [Asahi Denka Kogyo] DGE of polyoxyalkylene glycol; epoxy resin diluent; Gardner 2+ color; sp.gr. 1.07; visc. 40-80 cps; EEW 280-320.

Adeka Lub CB-419. [Adeka Fine Chem.] Chlorinated fatty acid ester; extreme pressure agent for metalworking, in formulation of cutting oil, grinding oils; plasticizer for PVC; liq.; oil-sol.; 36% Cl.

Adeka Lub CB-419R. [Adeka Fine Chem.] Chlorinated fatty acid ester; extreme pressure agent for metalworking, in formulation of cutting oil, grinding oils; plasticizer for PVC; liq.; oil-sol.; 33% Cl.

Adeka Lub CF-40. [Adeka Fine Chem.] Chlorinated fatty acid; extreme pressure agent for metalworking, in formulation of cutting oil, grinding oils; plasticizer for PVC; paste; water-sol.; 38% Cl.

Adeka Lub E-410. [Adeka Fine Chem.] Chlorinated N-paraffin; extreme pressure agent for metalworking, in formulation of cutting oil, grinding oils; plasticizer for PVC; liq.; oil-sol.; 41% Cl.

Adeka Lub E-450. [Adeka Fine Chem.] Chlorinated N-paraffin; extreme pressure agent for metalworking, esp. formulation of cutting oil, grinding oils; plasticizer for PVC; liq.; oil-sol.; 45% Cl.

Adeka Lub E-500. [Adeka Fine Chem.] Chlorinated N-paraffin; extreme pressure agent for metalworking, esp. formulation of cutting oil, grinding oil; plasticizer for PVC; visc. liq.; oil-sol.; 50% Cl.

Adeka Lub S-1. [Adeka Fine Chem.] Chlorinated fatty acid ester; extreme pressure agent for metalworking, in formulation of cutting oil, grinding oils; plasticizer for PVC; visc. liq.; oil-sol.; 35% Cl.

Adeka Lub S-3. [Adeka Fine Chem.] Chlorinated fatty acid ester; extreme pressure agent for metalworking, in formulation of cutting oil, grinding oils; plasticizer for PVC; visc. liq.; oil-sol.; 33% Cl.

Adeka Lub S-7. [Adeka Fine Chem.] Chlorinated fatty acid ester; extreme pressure agent for metalworking, in formulation of cutting oil, grinding oils; plasticizer for PVC; visc. liq.; oil-sol.; 33% Cl.

Adeka Propylene Glycol (T). [Asahi Denka Kogyo] Propylene glycol industrial; CAS 57-55-6; EINECS 200-338-0; surfactant for unsat. polyester resin, paint applics.; APHA 10+ color; sp.gr. 1.037-1.039; > 0.2% water.

ADI 50. [Warenhandels GmbH] Malodor control and freshkeeping masterbatch for LDPE, HDPE, and PP prods. to store food, hygienic prods., disposal bags, and applics. where a gas absorbing effect is needed at the decay of org. materials; FDA, BGA, and EEC compliance for food pkg.; melt flow 2.2 g/10 min; sp.gr. 1.22; *Storage:* store in dry area away from sunlight; 1 yr min. storage life.

Adimoll® BO. [Bayer AG] Benzyloctyl adipate; plasticizer used for PVC articles with resistance to low temps., calendering, coating, extrusion, inj. molding, film, expanded imitation leather; useful in VC copolymers, NC, ethyl cellulose, PS, nat., S/B, N/B, chlorinated, and butyl rubbers; Hazen < 80; dens. 1.002-1.008 g/cc; visc. 16-17 mPa.s; acid no. < 0.1; sapon. no. 305-330; ref. index 1.4805-1.4830; flash pt. 200-220 C (OC); 70:30 PVC:plasticizer suspension: tens. str. 20.5 mPa; tens. elong. 350%; hardness (Shore D) 26.

Adimoll® DN. [Bayer AG] Diisononyl adipate; CAS 33703-08-1; EINECS 251-646-7; plasticizer used in PVC rigid articles which must have stability, calendering, spread coating, extrusion, inj. molding, film, tarpaulins, protective clothing, expanded imitation leather, wire insulation and cables; suitable for VC copolymers, ethyl cellulose, PS, polymethylmethacrylate, nat., S/B, N/B, chlorinated, and butyl rubbers; Hazen < 50; dens. 0.910-0.920 g/cc; visc. 22-25 mPa.s; acid no. < 0.1; sapon. no. 270-290; ref. index 1.4465-1.4490; flash pt. 200-210 (OC); 70:30 PVC:plasticizer suspension: tens. str. 22 mPa; tens. elong. 360%; hardness (Shore D) 32.

Adimoll® DO. [Bayer AG] Dioctyl adipate; CAS 103-23-1; EINECS 203-090-1; plasticizer used for PVC articles with low-temp. resistance, calendering, coating, extrusion, inj. molding, film, tarpaulins, rainwear, tubing, cable sheathing, shoes, shoe soles; used in VC copolymers, PVAc, polyvinyl butyral, NC, ethyl cellulose, PS, nat., S/B, N/B, chlorinated, and butyl rubbers; Hazen < 20; dens. 0.923-0.926 g/cc; visc. 12-14 mPa.s; acid no. < 0.1; sapon. no. 270-290; ref. index 1.4465-1.4480; flash pt. 200-210 C (OC); 70:30 PVC:plasticizer suspension: tens. str. 19.5 mPa; tens. elong. 360%; hardness (Shore D) 26.

Adi-pure®. [DuPont Nylon] High purity adipic acid; CAS 124-04-9; EINECS 204-673-3; chem. intermediate; used in adhesives, coatings (alkyd, PU, gel coat, polyester), nylon 66, PU foams/elastomers/fibers, unsat. polyester resins, plasticizers, lubricants, textile treatments, cosmetic emoolients; pH buffer; wet str. paper resins; food acidulant; FDA 21CRR §184.1009; wh. free-flowing cryst. powd.; m.w. 146.14; dens. 1.36 g/cc; bulk dens. 40-45 lb/ft^3 (loose); melt visc. 4.54 cP (160 C); m.p. 152-153

C; flash pt. (TCC) 196 C; 99.7% assay.

ADK CIZER BF-1000. [Asahi Denka Kogyo] Epoxidized 1,2-polybutadiene; contains epoxy and vinyl groups; for coatings, adhesives, other applics.; JHOSPA compliance for food pkg.; clear liq.; m.w. 1000; visc. 100,000 cps.

ADK CIZER C-8. [Asahi Denka Kogyo] Trioctyl trimellitate; CAS 3319-31-1; heat-resist. PVC plasticizer providing exc. long-term heat stability, good elec. insulating prop., low volatility; easy processing; for elec. cable wire, car upholstery, sealants, gaskets; APHA 60 color; sp.gr. 0.990; visc. 220 cps; flash pt. 242 C.

ADK CIZER D-32. [Asahi Denka Kogyo] Epoxidized octyl stearate; CAS 106-84-3; heat stabilizer and plasticizer having good low temp. props.; for plasticized and paste PVC formulations; provides capture of catalyst residues in polyolefins; JHOSPA compliance for food pkg.; clear liq.; visc. 520 cps.

ADK CIZER E-500. [Asahi Denka Kogyo] 50% Chlorinated n-paraffin; PVC plasticizer with exc. compat., elec. insulating prop.; provides flame retardance; for elec. wire covering, material, cable wire, sheet, leather, gasket; APHA 100 color; sp.gr. 1.250; visc. 1000 cps.

ADK CIZER EP-13. [Asahi Denka Kogyo] Bisphenol A epoxy resin; heat stabilizer for PVC; provides lowest soften. pt. drop in rigid formulation; effective in combination with O-130P in plasticized formulation to prevent discoloration by heat and sticking to roll; exc. halogen capture for flame retardant resins; high corrosion resist.; clear visc. liq.; visc. 1300 cps.

ADK CIZER EP-17. [Asahi Denka Kogyo] Bisphenol A epoxy resin; exc. halogen capture, high corrosion resist. for flame retardant resins; clear visc. liq.; visc. 9600 cps.

ADK CIZER FEP-13. [Asahi Denka Kogyo] Epoxy resin type, blend of EP-13 and O-130P; PVC stabilizer providing lower visc. and improved workability over EP-13 and O-130P.

ADK CIZER O-130P. [Asahi Denka Kogyo] Epoxidized soybean oil; CAS 8013-07-8; EINECS 232-391-0; stabilizer and plasticizer for PVC; provides exc. heat and weathering stability and applicable to all PVC formulations; esp. for improving heat stability in Ba-Zn or Ca-Zn formulation; for film and sheet for agri., transparent food pkg., cable, gasket, hose, tube, paint; stabilizer for org. halogen such as neoprene; approved by FDA and JHPA; APHA 160 color; m.w. 1000; sp.gr. 0.993; visc. 350 cps; flash pt. 293 C.

ADK CIZER O-180A. [Asahi Denka Kogyo] Epoxidized linseed oil; stabilizer providing higher heat weathering stability than O-130P; suitable for rigid calendering sheet and rigid bottle formulation; approved by FDA and JHPA.

ADK CIZER PN-250. [Asahi Denka Kogyo] Polyester adipate; PVC plasticizer providing low volatility, exc. heat aging stability, exc. resist. to extraction and migration; for migration-proof elec. wire, tarpaulin, film, gasket, boot; APHA 90 color; m.w. 2000; sp.gr. 1.091; visc. 4500 cps; flash pt. 278 C.

ADK CIZER PN-650. [Asahi Denka Kogyo] Polyester adipate; PVC plasticizer providing low volatility, exc. heat aging stability, exc. resist. to extraction and migration; for migration-proof elec. wire, tarpaulin, film, gasket, boot; APHA 100 color; m.w. 1800; sp.gr. 1.092; visc. 3000 cps; flash pt. 278 C.

ADK CIZER RS-107. [Asahi Denka Kogyo] Adipic acid ether ester type; plasticizer for NBR, CR, ECO, ACM, SBR, CSM, PU, polysulfide; exc. compat. with PVC, exc. low temp. props., long-term heat stability; for car safety hose resist. to oil and gasoline; FDA approved; APHA 60 color; m.w. 434; sp.gr. 1.020; visc. 20 cps; flash pt. 215 C.

ADK STAB 36. [Asahi Denka Kogyo] Ca-Zn; nontoxic PVC stabilizer providing exc. initial color and transparency; for plasticized sheet, hose, and medical tubing with nontoxic chelator; approved by FDA and JHPA (below 10%).

ADK STAB 37. [Asahi Denka Kogyo] Nontoxic PVC stabilizer providing more heat stability and transparency than ADK STAB 36; approved by FDA and JHPA (below 10%); paste.

ADK STAB 38. [Asahi Denka Kogyo] Nontoxic PVC stabilizer providing more heat stability and transparency than ADK STAB 36; approved by FDA and JHPA (below 10%); paste.

ADK STAB 135A. [Asahi Denka Kogyo] Mono alkyl diaryl phosphite; stabilizer providing exc. heat stability, initial color and transparency with all Ba-Zn stabilizers, esp. with powd. type stabilizers; for ABS, PVC, and engineering plastics; clear liq.; m.w. 375; visc. 15 cps.

ADK STAB 144. [Asahi Denka Kogyo] Zn complex; PVC stabilizer for homogeneous tile flooring; provides exc. heat stability, improves initial color; effective blended with phosphite; powd.

ADK STAB 189E. [Asahi Denka Kogyo] Ba-Cd-Zn; PVC stabilizer providing exc. initial color, transparency and heat stability; for transparent film and sheet; liq.

ADK STAB 260. [Asahi Denka Kogyo] Phosphite; stabilizer for polyolefins, ABS, and engineering plastics; provides high color improvement and process stability; JHOSPA compliance for food pkg.; clear liq.; m.w. 1240; visc. 1100 cps.

ADK STAB 273. [Asahi Denka Kogyo] Epoxy type; stabilizer for flame retardant resins; halogen capture; improves process stability; FDA and JHOSPA compliance for food pkg.; clear liq.; visc. 660 cps.

ADK STAB 328. [Asahi Denka Kogyo] Hindered phenol; effective antioxidant for polyolefins; good compat. with polyolefins; FDA and JHOSPA compliance for food pkg.; wh. powd.; m.p. < 56 C.

ADK STAB 329. [Asahi Denka Kogyo] TNPP; CAS 26523-78-4; EINECS 247-759-6; PVC stabilizer improving initial color and transparency with Ca-Zn stabilizers; approved by FDA and JHPA.

ADK STAB 329K. [Asahi Denka Kogyo] Phosphite; stabilizer for polyolefins, ABS, PS, engineering plastics, etc.; provides good heat and color stability at low cost; FDA and JHOSPA compliance for food pkg.; clear liq.; m.w. ≈ 816; visc. 9000 cps.

ADK STAB 465. [Asahi Denka Kogyo] Di-n-octyltin mercaptide; CAS 58229-88-2; PVC stabilizer providing exc. heat stability, color and transparency; suitable for rigid calendering sheet and blow formulation; approved by FDA and JHPA; liq.

ADK STAB 465E, 465L. [Asahi Denka Kogyo] Di-n-octyltin mercaptide; CAS 58229-88-2; PVC stabilizer providing exc. heat stability, color and transparency; suitable for rigid calendering sheet and blow formulation; approved by FDA and JHPA; liq., low odor.

ADK STAB 466. [Asahi Denka Kogyo] Octyltin mercaptide; PVC stabilizer providing exc. heat stability, color and transparency; suitable for rigid calendering sheet formulation; approved by FDA and JHPA; liq.

ADK STAB 467. [Asahi Denka Kogyo] Octyltin mercaptide; PVC stabilizer providing exc. heat stability, color and transparency; suitable for rigid calendering sheet formulation; approved by FDA and JHPA; liq.

ADK STAB 471. [Asahi Denka Kogyo] Octyltin mercaptide; PVC stabilizer providing exc. heat stability, color and transparency; suitable for rigid calendering sheet formulation; approved by FDA and JHPA; liq.

ADK STAB 517. [Asahi Denka Kogyo] Dialkyl monoaryl phosphite; PVC stabilizer providing exc. heat stability, initial color and transparency with all Ba-Zn stabilizers, esp. with powd. type stabilizers.

ADK STAB 522A. [Asahi Denka Kogyo] Phosphite; heat and color stabilizer for polyolefins, ABS, and engineering plastics; for high temp. processing; JHOSPA compliance for food pkg.; clear liq.; m.w. 1831; visc. 1800 cps.

ADK STAB 593. [Asahi Denka Kogyo] Ca-Zn; nontoxic general purpose PVC stabilizer with chelator; approved by FDA and JHPA (below 6.0%).

ADK STAB 666. [Asahi Denka Kogyo] Ba-Zn; PVC stabilizer for transparent top layer flooring formulation; provides exc. weatherability and prevents whitening in water absorp.; liq.

ADK STAB 1013. [Asahi Denka Kogyo] Mono alkyl diaryl phosphite; PVC stabilizer providing exc. heat stability, initial color and transparency with all Ba-Zn stabilizers, esp. with powd. type stabilizers.

ADK STAB 1178. [Asahi Denka Kogyo] TNPP; CAS 26523-78-4; EINECS 247-759-6; stabilizer improving initial color and transparency with Ca-Zn stabilizers; for polyolefins, ABS, PS, engineering plastics, etc.; low cost; FDA and JHOSPA compliance for food pkg.; clear liq.; m.w. 689; visc. 4000 cps.

ADK STAB 1292. [Asahi Denka Kogyo] Dibutyltin mercaptide; PVC stabilizer for rigid transparent PVC with highest heat stability; applicable to rigid water pipe requiring NSF approval for high resist. to water extraction; liq.

ADK STAB 1413. [Asahi Denka Kogyo] Benzophenone type; light stabilizer, UV absorber for polyolefins, PVC, etc.; good compat. with polymers; FDA and JHOSPA compliance for food pkg.; sl. yel. powd.; m.w. 326; m.p. 48 C.

ADK STAB 1500. [Asahi Denka Kogyo] Special phosphite; stabilizer providing exc. heat, color and weathering stability to PVC and ABS; for rigid calendering sheet and blow bottle; approved by FDA and JHPA @ $\leq$ 1% in rigid prods.; clear liq.; m.w. $\approx$ 1112; visc. 10000 cps.

ADK STAB 1502. [Asahi Denka Kogyo] Special phosphite; low visc. version of ADK STAB 1500; PVC stabilizer providing exc. heat, color and weathering stability; for rigid calendering sheet and blow bottle; approved by FDA and JHPA @ $\leq$ 1% in rigid prods.

ADK STAB 1505-A. [Asahi Denka Kogyo] Special phosphite; PVC stabilizer providing exc. color stability and same as 1500 in heat stability.

ADK STAB 2013. [Asahi Denka Kogyo] Dialkyl monoaryl phosphite; PVC stabilizer providing exc. heat stability, initial color and transparency with all Ba-Zn stabilizers, esp. with powd. type stabilizers.

ADK STAB 2112. [Asahi Denka Kogyo] Phosphite; stabilizer for polyolefins, ABS, PS, engineering plastics, etc.; provides exc. hydrolytic and process stability; FDA and JHOSPA compliance for food pkg.; wh. powd.; m.w. 647; m.p. 183 C; 4.5-5.0% P.

ADK STAB 2335. [Asahi Denka Kogyo] Zinc borate; flame retardant, smoke preventer for flooring and cable formulations.

ADK STAB 3010. [Asahi Denka Kogyo] Phosphite; stabilizer for ABS and engineering plastics; provides color improvement at high temp. processing; clear liq.; m.w. 503; visc. 20 cps.

ADK STAB AC-111. [Asahi Denka Kogyo] Ba-Mg-Zn; general purpose PVC stabilizer providing exc. heat stability, transparency, and thixotropic props.; esp. suitable for formulation of gloves; liq.

ADK STAB AC-113. [Asahi Denka Kogyo] Mg-Zn; general purpose PVC stabilizer providing exc. heat and color stability and transparency; suitable for whole formulation of paste PVC; liq.

ADK STAB AC-116. [Asahi Denka Kogyo] Ca-Zn; general purpose PVC stabilizer providing exc. heat stability, transparency, and thixotropic props.; esp. suitable for formulation of gloves; liq.

ADK STAB AC-118. [Asahi Denka Kogyo] Ba-Zn; PVC stabilizer for transparent top layer flooring formulation; provides exc. weatherability and prevents whitening in water absorp.; also suitable for outdoor applic. such as PVC cameras; liq.

ADK STAB AC-122. [Asahi Denka Kogyo] Ba-Zn; general purpose PVC stabilizer providing good heat stability and color resist.; liq.

ADK STAB AC-133. [Asahi Denka Kogyo] Ba-Cd-Zn; PVC stabilizer providing exc. initial color, transparency and heat stability; for calendering sheet, leather and plasticized PVC compd. by extrusion and inj. molding processes; liq.

ADK STAB AC-133N. [Asahi Denka Kogyo] Ba-Cd-Zn; PVC stabilizer providing exc. initial color, transparency and heat stability; for calendering semi-rigid and plasticized film, sheet, plasticized PVC compd. by extrusion and inj.; liq.

ADK STAB AC-135. [Asahi Denka Kogyo] Ba-Cd-Zn;

one-pkg. PVC stabilizer providing exc. initial color, transparency and heat stability; for calendering sheet, leather and plasticized PVC compds. by extrusion and inj.; liq.

ADK STAB AC-141. [Asahi Denka Kogyo] Ba-Zn; general purpose PVC stabilizer providing good heat stability and color resist.; liq.

ADK STAB AC-143. [Asahi Denka Kogyo] Ba-Zn; general purpose PVC stabilizer providing good heat stability and color resist.; liq.

ADK STAB AC-153. [Asahi Denka Kogyo] Ba-Zn; general purpose PVC stabilizer providing good heat stability and color resist.; liq.

ADK STAB AC-158. [Asahi Denka Kogyo] Ba-Zn; PVC stabilizer for extrusion and inj. molding applics.; suitable for extensive use in conjunction with a small amount of powd. stabilizer; liq.

ADK STAB AC-167. [Asahi Denka Kogyo] Ba-Zn; PVC stabilizer providing weatherability and heat stability; for outdoor-use tarpaulin; liq.

ADK STAB AC-168. [Asahi Denka Kogyo] Ba-Zn; general purpose PVC stabilizer providing good heat stability and color resist.; liq.

ADK STAB AC-169. [Asahi Denka Kogyo] Ba-Zn; one-pkg. PVC stabilizer providing exc. heat stability and transparency; for calender sheets of plasticized PVC; liq.

ADK STAB AC-173. [Asahi Denka Kogyo] Ba-Zn; general purpose PVC stabilizer providing exc. heat and color stability and transparency; suitable for whole formulation of paste PVC; liq.

ADK STAB AC-181. [Asahi Denka Kogyo] Ba-Zn; PVC stabilizer for extrusion and inj. molding applics.; suitable for extensive use in conjunction with a small amount of powd. stabilizer; liq.

ADK STAB AC-182. [Asahi Denka Kogyo] Ba-Zn; one-pkg. PVC stabilizer for high loading filler formulation and TiO_2 formulation; liq.

ADK STAB AC-183. [Asahi Denka Kogyo] Ba-Zn; one-pkg. PVC stabilizer providing exc. heat stability and transparency; for calender sheets of plasticized PVC; liq.

ADK STAB AC-186. [Asahi Denka Kogyo] Ba-Zn; PVC stabilizer providing exc. heat stability and color resist.; can be used from transparency to opaque with filled $CaCO_3$; for general purpose, automobile compds.; liq.

ADK STAB AC-190. [Asahi Denka Kogyo] Ba-Zn; PVC stabilizer providing exc. color retention, heat stability, transparency; for calendered films and sheets; liq.

ADK STAB AC-212. [Asahi Denka Kogyo] Ba-Zn; one-pkg. general purpose PVC stabilizer providing exc. heat and color stability and transparency; for calender sheets of plasticized PVC; suitable for whole formulation of paste PVC; liq.

ADK STAB AC-216. [Asahi Denka Kogyo] Ba-Zn; PVC stabilizer providing exc. color retention, heat stability, transparency; for calendered film and sheet for semirigid applics.; liq.

ADK STAB AC-229. [Asahi Denka Kogyo] Ba-Zn; PVC stabilizer providing weatherability and heat stability; for outdoor-use tarpaulin; liq.

ADK STAB AC-230. [Asahi Denka Kogyo] Ba-Zn; PVC stabilizer providing printability; for calendered film and sheets; liq.

ADK STAB AC-235. [Asahi Denka Kogyo] Ba-Zn; PVC stabilizer providing printability and good heat stability; for calendered film and sheets; liq.

ADK STAB AC-243. [Asahi Denka Kogyo] Ba-Zn; PVC stabilizer providing exc. heat stability, transparency, and weatherability; for extrusion T-die sheets; liq.

ADK STAB AC-244. [Asahi Denka Kogyo] Ba-Zn; PVC stabilizer providing exc. heat stability, transparency, and weatherability; for extrusion T-die sheets; liq.

ADK STAB AC-248. [Asahi Denka Kogyo] Ba-Zn; PVC stabilizer providing exc. color retention, heat stability, transparency; for calendered film and sheet for semirigid applics.; liq.

ADK STAB AC-250. [Asahi Denka Kogyo] Ba-Zn; PVC stabilizer providing exc. heat stability, transparency, and weatherability; for extrusion T-die sheets; liq.

ADK STAB AC-252. [Asahi Denka Kogyo] Ba-Zn; one-pkg. PVC stabilizer providing exc. heat stability and transparency; for calender sheets of plasticized PVC; suitable for semirigid applics.; liq.

ADK STAB AC-303. [Asahi Denka Kogyo] Ba-Zn; PVC stabilizer for transparent top layer flooring formulation; provides exc. weatherability and prevents whitening in water absorp.; also suitable for outdoor applic. such as PVC cameras; liq.

ADK STAB AC-307. [Asahi Denka Kogyo] Ba-Zn; PVC stabilizer for transparent top layer flooring formulation; provides exc. weatherability and prevents whitening in water absorp.; liq.

ADK STAB AC-309. [Asahi Denka Kogyo] Ba-Zn; PVC stabilizer for transparent top layer flooring formulation; provides exc. weatherability and prevents whitening in water absorp.; liq.

ADK STAB AC-310. [Asahi Denka Kogyo] Ba-Zn; stabilizer for foam disp. PVC for wallpaper; provides exc. color stability; used for unfoamed applic.; liq.

ADK STAB AC-311. [Asahi Denka Kogyo] Ca-Zn; general purpose PVC stabilizer providing exc. heat stability, transparency, and thixotropic props.; esp. suitable for formulation of gloves; liq.

ADK STAB AO-15. [Asahi Denka Kogyo] Hindered phenol; nondiscoloring antioxidant for PP fibers, ABS, etc.; FDA and JHOSPA compliance for food pkg.; wh. powd.; m.p. < 210 C.

ADK STAB AO-18. [Asahi Denka Kogyo] Hindered phenol; highly effective antioxidant protecting PP composites and PE cables from thermal degradation; wh. powd.; m.p. < 210 C.

ADK STAB AO-20. [Asahi Denka Kogyo] Hindered phenol; highly effective antioxidant for PP, HDPE, ABS, etc.; high weatherability, low volatility; FDA and JHOSPA compliance for food pkg.; wh. powd.; m.w. 784; m.p. 221 C.

ADK STAB AO-23. [Asahi Denka Kogyo] Thioether;

nondiscoloring, nonstaining antioxidant for elastomers and X-PE; exc. heat stabilizer; yel. liq.; m.w. ≈ 900; visc. 3000 cps.

ADK STAB AO-30. [Asahi Denka Kogyo] Hindered phenol; highly effective antioxidant for polyolefins, ABS, etc.; exc. synergism with thioethers; good extraction resist.; FDA and JHOSPA compliance for food pkg.; wh. powd.; m.w. 545; m.p. 186 C.

ADK STAB AO-37. [Asahi Denka Kogyo] Antioxidant for petrol. resin; wh. powd.; m.p. ≈ 165 C.

ADK STAB AO-40. [Asahi Denka Kogyo] Hindered phenol; effective antioxidant for ABS and rubbers; FDA and JHOSPA compliance for food pkg.; wh. powd.; m.w. 383; m.p. 210 C.

ADK STAB AO-50. [Asahi Denka Kogyo] Hindered phenol; effective antioxidant for all plastics and elastomers; good compat. with common polymers; substitute for BHT; FDA and JHOSPA compliance for food pkg.; wh. powd.; m.w. 531; m.p. 50 C.

ADK STAB AO-60. [Asahi Denka Kogyo] Hindered phenol; versatile and highly effective antioxidant for all plastics, elastomers, and fibers; FDA and JHOSPA compliance for food pkg.; wh. powd.; m.w. 1178; m.p. 115 C.

ADK STAB AO-80. [Asahi Denka Kogyo] Hindered phenol; highly effective antioxidant for all plastics, elastomers, and fibers; synergistic with thioethers; exc. resist. to discoloration; FDA and JHOSPA compliance for food pkg.; wh. powd.; m.w. 741; m.p. 125 C.

ADK STAB AO-412S. [Asahi Denka Kogyo] Thioether; antioxidant for polyolefins, ABS, engineering plastics, etc.; low volatility; exc. synergy with phenolics; wh. powd.; m.w. 1162; m.p. 50 C.

ADK STAB AO-503A. [Asahi Denka Kogyo] Thioether; antioxidant, synergist for elastomers; clear liq.; m.w. 543; visc. 60 cps.

ADK STAB AO-616. [Asahi Denka Kogyo] Hindered phenol; mixt. of AO-60 and PEP-8; antioxidant for polyolefins and others; FDA and JHOSPA compliance for food pkg.; wh. powd.

ADK STAB AP-536. [Asahi Denka Kogyo] Ba-Zn; general purpose PVC stabilizer providing good heat stability, color retention; powd.

ADK STAB AP-539. [Asahi Denka Kogyo] Ba-Zn; general purpose PVC stabilizer providing good heat stability, color retention; powd.

ADK STAB AP-540. [Asahi Denka Kogyo] Ba-Zn; general purpose PVC stabilizer providing good heat stability, color retention; powd.

ADK STAB AP-543. [Asahi Denka Kogyo] Ba-Zn; general purpose PVC stabilizer providing good heat stability, color retention; powd.

ADK STAB AP-548. [Asahi Denka Kogyo] Ba-Zn; general purpose PVC stabilizer providing good heat stability, color retention; powd.

ADK STAB AP-550. [Asahi Denka Kogyo] Ba-Zn; general purpose PVC stabilizer providing good heat stability, color retention; powd.

ADK STAB AP-551. [Asahi Denka Kogyo] Ba-Zn; general purpose PVC stabilizer providing good heat stability, color retention; powd.

ADK STAB ATC-1. [Asahi Denka Kogyo] Light stabilizer and chalking inhibitor for PVC formulation; provides exc. chalking protection and weathering stability.

ADK STAB AX-38. [Asahi Denka Kogyo] Phosphate type; lubricant for EVA and TPE; clear liq.; visc. 700 cps.

ADK STAB BAP-1. [Asahi Denka Kogyo] Auxiliary agent for foam PVC; special cell adjuster for fine cells; provides exc. cell stability blended 0.3-1.0 phr with foam stabilizer; liq.

ADK STAB BAP-2. [Asahi Denka Kogyo] Auxiliary agent for foam PVC; blowing aid acting as kicker; optionally increases expansion coefficient without roughness of cell; powd.

ADK STAB BAP-4. [Asahi Denka Kogyo] Auxiliary agent for foam PVC; special cell adjuster for fine cells; provides exc. cell stability blended 0.3-1.0 phr with foam stabilizer; liq.

ADK STAB BAP-5. [Asahi Denka Kogyo] Auxiliary agent for foam PVC; inhibitor at initial expansion; liq.

ADK STAB BAP-7. [Asahi Denka Kogyo] Auxiliary agent for foam PVC; inhibitor at initial expansion; flake.

ADK STAB BAP-8. [Asahi Denka Kogyo] Auxiliary agent for foam PVC; special cell adjuster for fine cells; provides exc. cell stability blended 0.3-1.0 phr with foam stabilizer; liq.

ADK STAB BT-11. [Asahi Denka Kogyo] Dibutyltin dilaurate; CAS 77-58-7; EINECS 201-039-8; PVC stabilizer providing good heat and weathering resist.; improves processability in rigid transparent formulation; liq.

ADK STAB BT-18. [Asahi Denka Kogyo] Dibutyltin dilaurate; CAS 77-58-7; EINECS 201-039-8; PVC stabilizer providing good heat, weathering, and process stability; for rigid and plasticized PVC formulation; liq.

ADK STAB BT-23. [Asahi Denka Kogyo] Dibutyltin dilaurate; CAS 77-58-7; EINECS 201-039-8; PVC stabilizer for light elec. parts and vehicle parts made by paste PVC; suitable for transparent sleeve due to exc. corrosion resist. and harness clip due to exc. heat stability; liq.

ADK STAB BT-25. [Asahi Denka Kogyo] Dibutyltin dilaurate; CAS 77-58-7; EINECS 201-039-8; PVC stabilizer providing exc. heat and weathering stability; improves transparency and processability in plasticized PVC formulation; liq.

ADK STAB BT-27. [Asahi Denka Kogyo] Dibutyltin dilaurate; CAS 77-58-7; EINECS 201-039-8; PVC stabilizer providing good heat, weathering, and process stability; for rigid and plasticized PVC formulation; liq.

ADK STAB BT-31. [Asahi Denka Kogyo] Dibutyltin maleate; CAS 15535-69-0; PVC stabilizer providing exc. initial color transparency; for whole rigid PVC prods., extrusion, press and blow molding; liq.

ADK STAB BT-52. [Asahi Denka Kogyo] Dibutyltin maleate; CAS 15535-69-0; PVC stabilizer providing heat and weathering stability and transparency to rigid PVC prods.; resists high temp.; improves processability with BT-11, BT-18 and special lubricant; liq.

ADK STAB BT-53A. [Asahi Denka Kogyo] Dibutyltin maleate; CAS 15535-69-0; PVC stabilizer providing heat and weathering stability and transparency to rigid PVC prods.; resists high temp.; improves processability with BT-11, BT-18 and special lubricant; liq.

ADK STAB BT-83. [Asahi Denka Kogyo] Dibutyltin mercaptide; PVC stabilizer with exc. heat resist. and color stability; for inj. molding and rigid PVC extrusion (T-die, contour); powd.

ADK STAB C. [Asahi Denka Kogyo] Mono alkyl diaryl phosphite; stabilizer providing exc. heat stability, initial color and transparency with all Ba-Zn stabilizers, esp. with powd. type stabilizers; for ABS, PVC, and engineering plastics; clear liq.; m.w. 346; visc. 10 cps.

ADK STAB CA-ST. [Asahi Denka Kogyo] Calcium stearate; CAS 1592-23-0; EINECS 216-472-8; nontoxic stabilizer, lubricant, and release aid for PVC, polyolefin, ABS, etc.; reacts with acidic ingreds.; FDA and JHOSPA compliance for food pkg.; wh. powd.

ADK STAB CDA-1. [Asahi Denka Kogyo] Nondiscoloring metal deactivator for polyolefins, ABS, etc.; prevents oxidative degradation caused by metal catalysts; wh. powd.; m.w. 204; m.p. 325 C.

ADK STAB CDA-6. [Asahi Denka Kogyo] Nondiscoloring metal deactivator for polyolefins, ABS, etc.; prevents oxidative degradation caused by metal catalysts; wh. powd.; m.w. 498; m.p. 212 C.

ADK STAB CPL-37. [Asahi Denka Kogyo] Ba complex; PVC stabilizer for calendered sheets and powd. compds. for automobile upholstery; prevents discoloration and loss of physical props. due to amine compd.; liq.

ADK STAB CPL-1551. [Asahi Denka Kogyo] Special phosphite; PVC stabilizer for calendered sheets and powd. compds. for automobile upholstery; provides heat stability, color retention and light stability; liq.

ADK STAB CPS-1. [Asahi Denka Kogyo] Ba-Zn-Mg complex; PVC stabilizer for automotive industry, calendered sheets, semirigid and soft sheets for dashboard; imparts exc. heat aging resist. and color retention against contact with foamed PU; powd.

ADK STAB CPS-2R. [Asahi Denka Kogyo] Ba-Zn-Mg-Al complex; PVC stabilizer for powd. slush compds. for dash board and console boxes in automobile industry; imparts exc. heat stability, heat aging resist., and color resist. in contact with formed PU; powd.

ADK STAB CPS-55R. [Asahi Denka Kogyo] Ba-Zn-Mg-Al complex; PVC stabilizer for powd. slush compds. for dash board and console boxes in automobile industry; imparts exc. heat stability, heat aging resist., and color resist. in contact with formed PU; powd.

ADK STAB CPS-896. [Asahi Denka Kogyo] Ba-Zn-Mg complex; PVC stabilizer for automotive industry, powd. slush compd. for dashboard and console box; provides low antifogging, good heat stability; paste.

ADK STAB CPS-900. [Asahi Denka Kogyo] Ba-Zn-Mg complex; PVC stabilizer for automotive industry, calendered sheets and powd. compds.; imparts exc. heat stability, color retention; prevents loss of mech. props.; paste.

ADK STAB EC-14. [Asahi Denka Kogyo] Zn-phosphite; PVC stabilizer; improves initial color with Ba-Zn stabilizers in formulations containing TiO_2 and $CaCO_3$; suitable for plasticized and rigid calendering formulation and paste formulation.

ADK STAB EXL-5. [Asahi Denka Kogyo] Aliphatic type; PVC lubricant providing better volatization compared to stearic types; suitable for vacuum molding sheet; approved by JHPA; powd.

ADK STAB FC-112. [Asahi Denka Kogyo] Acrylic oligomer type; lubricant for PVC, FR-ABS, etc. with exc. compat. and external lubricity; provides exc. transparency, bleeding, and blooming props.; suitable for formulations contg. 20-30% plasticizers; clear liq.; visc. 3000 cps.

ADK STAB FC-113. [Asahi Denka Kogyo] Acrylic oligomer type; lubricant for rigid PVC formulations, FR-ABS, etc.; wh. powd.; soften. pt. 110 C.

ADK STAB FL-21. [Asahi Denka Kogyo] Na-Zn; stabilizer for foam disp. PVC at low expansion; for wallpaper, leathercloth formulation, and expansion layer formulation for cushion flooring; provides exc. uniformity of expansion cell; liq.

ADK STAB FL-22. [Asahi Denka Kogyo] Ca-Na-Zn; stabilizer for disp. PVC for wallpaper, leathercloth, for expansion layer formulation for cushion flooring; provides exc. uniformity of expansion cell; liq.

ADK STAB FL-23. [Asahi Denka Kogyo] Na-Zn; stabilizer for foam disp. PVC from low to high expansion; for wallpaper; liq.

ADK STAB FL-32. [Asahi Denka Kogyo] K-Zn; stabilizer for disp. PVC for wallpaper; provides exc. plate-out prop. and uniformity of cell; liq.

ADK STAB FL-41. [Asahi Denka Kogyo] K-Zn; stabilizer for disp. PVC for wallpaper; provides exc. plate-out prop. and low sol-visc. for high filler formulation; liq.

ADK STAB FL-43. [Asahi Denka Kogyo] Ba-Zn; stabilizer for foam disp. PVC from low to high expansion; provides exc. plate-out prop.; for wallpaper; liq.

ADK STAB FL-44. [Asahi Denka Kogyo] Ba-Zn; stabilizer for foam disp. PVC for wallpaper; provides quick expansion speed and exc. whiteness; liq.

ADK STAB FL-45. [Asahi Denka Kogyo] Ba-Zn; stabilizer for foam disp. PVC for wallpaper; provides exc. plate-out prop.; liq.

ADK STAB FL-54. [Asahi Denka Kogyo] Ba-Zn; stabilizer for foamed disp. PVC; provides exc. uniformity of expansion cell; suitable for leathercloth formulation; liq.

ADK STAB FL-55. [Asahi Denka Kogyo] Ba-Zn; foam

PVC stabilizer for expanded leathercloth; provides exc. plate-out props.; liq.

ADK STAB FL-64. [Asahi Denka Kogyo] Ba-Zn; foam PVC stabilizer for expanded leathercloth; provides exc. plate-out props.; liq.

ADK STAB FS-12. [Asahi Denka Kogyo] Zn complex; PVC stabilizer for homogeneous tile flooring; provides exc. heat stability, improves initial color; effective blended with phosphite; powd.

ADK STAB GR-16. [Asahi Denka Kogyo] Ca-Zn; PVC stabilizer providing exc. long term heat stability; for extrusion and inj. molding by filler loading formulation; effective with ADK STAB 1500; powd.

ADK STAB GR-18. [Asahi Denka Kogyo] Ca-Zn; PVC stabilizer providing exc. long term heat stability; for extrusion and inj. molding by filler loading formulation; effective with ADK STAB 1500; powd.

ADK STAB GR-22. [Asahi Denka Kogyo] Ca-Zn; PVC stabilizer providing exc. long term heat stability; for extrusion and inj. molding by filler loading formulation; effective with ADK STAB 1500; powd.

ADK STAB HP-10. [Asahi Denka Kogyo] Phosphite; stabilizer for polyolefins, ABS, PS, engineering plastics, etc.; provides exc. hydrolytic stability, process stability, and color improvement at high temp.; JHOSPA compliance for food pkg.; wh. powd.; m.w. 583; m.p. 148 C.

ADK STAB JX-3. [Asahi Denka Kogyo] Phosphite; stabilizer for ABS providing exc. heat and color stability; can be used during polymerization; pale yel. liq.; visc. 7400 cps.

ADK STAB LA-31. [Asahi Denka Kogyo] High m.w. benzotriazole type; light stabilizer, UV absorber for ABS and engineering plastics; low volatility, nonvolatile @ high processing temps.; pale yel. powd.; m.w. 659; mp. 195 C.

ADK STAB LA-32. [Asahi Denka Kogyo] Benzotriazole type; UV absorber for PVC, ABS, PS, PU, NMA, etc.; FDA and JHOSPA compliance for food pkg.; lt. yel. powd.; m.w. 225; m.p. 130 C.

ADK STAB LA-34. [Asahi Denka Kogyo] Benzotriazole type; light stabilizer, UV absorber for polyolefins and other polymers; provides exc. heat and light stability; pale yel. powd.; m.w. 358; m.p. 51 C.

ADK STAB LA-36. [Asahi Denka Kogyo] light stabilizer, UV absorber for polyolefins and others; provides greater absorp. of longer wavelengths than LA-32; FDA and JHOSPA compliance for food pkg.; lt. yel. powd.; m.w. 315; m.p. 139 C.

ADK STAB LA-51. [Asahi Denka Kogyo] High m.w. benzophenone type; light stabilizer for engineering plastics; low volatility, nonvolatile @ high processing temps.; yel. powd.; m.w. 456; m.p. ≈ 225 C.

ADK STAB LA-57. [Asahi Denka Kogyo] Hindered amine light stabilizer; light stabilizer for polyolefins, PVC, ABS, etc.; superior light stability; wh. powd.; m.w. 701; m.p. 132 C.

ADK STAB LA-62. [Asahi Denka Kogyo] Hindered amine light stabilizer; light stabilizer for PP, PU, paint, latex, etc.; good compat.; lt. amber liq.; visc. 7000 cps.

ADK STAB LA-63. [Asahi Denka Kogyo] High m.w. hindered amine light stabilizer; light stabilizer for polyolefins, PVC, ABS, etc.; provides superior light stability; pale yel. powd.; m.p. ≈ 100 C.

ADK STAB LA-67. [Asahi Denka Kogyo] Hindered amine light stabilizer; light stabilizer for PP, PU, paint, latex, etc.; good compat.; lt. amber liq.; visc. 500 cps.

ADK STAB LA-68. [Asahi Denka Kogyo] High m.w. hindered amine light stabilizer; light stabilizer for polyolefins, PVC, ABS, etc.; provides superior light stability; pale yel. powd.; m.p. 70-80 C.

ADK STAB LA-68LD. [Asahi Denka Kogyo] High m.w. hindered amine; light stabilizer for films, fibers, polyolefins, PVC, ABS, etc.; superior light stability; pale yel. flake; m.p. 70-80 C.

ADK STAB LA-77. [Asahi Denka Kogyo] Standard type hindered amine; light stabilizer for polyolefins, ABS, etc.; superior light stability; JHOSPA compliance for food pkg.; wh. powd.; m.w. 481; m.p. 84 C.

ADK STAB LA-82. [Asahi Denka Kogyo] Polymerizable HALS; light stabilizer for acrylate, etc.; clear liq.; m.w. 239; visc. 12 cps.

ADK STAB LA-87. [Asahi Denka Kogyo] Polymerizable HALS; light stabilizer for acrylate, etc.; wh. powd.; m.w. 225; m.p. 58 C.

ADK STAB LS-2. [Asahi Denka Kogyo] Dibutyltin maleate; CAS 15535-69-0; lubricant and stabilizer for PVC; provides exc. transparency and gives good props. with tin maleate stabilizer; for extrusion and blow molding; liq.

ADK STAB LS-3. [Asahi Denka Kogyo] Oligomer type; lubricant for PVC with exc. compat. and external lubricity; provides exc. transparency, bleeding, and blooming props.; suitable for formulations contg. 20-30% plasticizers; liq.

ADK STAB LS-5. [Asahi Denka Kogyo] Oligomer type; lubricant for PVC with exc. compat. and external lubricity; provides exc. transparency, bleeding, and blooming props.; suitable for plasticized formulations, esp. to prevent plate-out in formulations contg. polyester plasticizer; liq.

ADK STAB LS-8. [Asahi Denka Kogyo] Butyl stearate; CAS 123-95-5; EINECS 204-666-5; plastics lubricant; approved by FDA and JHPA; liq.

ADK STAB LS-9. [Asahi Denka Kogyo] Complex; PVC lubricant providing exc. transparency; esp. useful with tin stabilizers; liq.

ADK STAB LS-10. [Asahi Denka Kogyo] Ester type; PVC lubricant providing exc. bleeding and blooming props.; suitable for slush molding; liq.

ADK STAB LX-45. [Asahi Denka Kogyo] Nonstaining antioxidant for rubber latexes, SBR latex, etc.; wh. emulsion; sp.gr. 1.014; visc. 150 cps.

ADK STAB LX-802. [Asahi Denka Kogyo] Nonstaining antioxidant for rubber latexes, SBR latex, etc.; improves NO_x staining and color; wh. suspension; visc. 800 cps.

ADK STAB MG-ST. [Asahi Denka Kogyo] Magnesium stearate; CAS 557-04-0; EINECS 209-150-3; nontoxic stabilizer, lubricant, and release agent for PVC, polyolefin, ABS, etc.; reacts with acidic ingreds.; FDA and JHOSPA compliance for food pkg.; wh. powd.; m.p. ≈ 130 C.

ADK STAB NA-10. [Asahi Denka Kogyo] Nucleating agent for polyolefins, PBT, PET, PA; improves transparency, rigidity, hardness and processability; good compat. and dispersion; FDA and JHOSPA compliance for food pkg.; wh. powd.; m.w. 384; m.p. > 300 C.

ADK STAB NA-11. [Asahi Denka Kogyo] Sodium 2,2′-methylenebis-(4,6-di-t-butylphenyl) phosphate; CAS 85209-91-2; EINECS 286-344-4; nucleating agent upgrading heat deflection temp., flex. mod., impact str. of PP, PET, PBT, polyamide; gives high transparency at low concs.; raises crystallization temp.; FDA 21CFR §178.3295 and JHOSPA compliance for food pkg.; wh. fine powd.; sol. in methanol, ethanol; m.w. 508; m.p. > 400 C; *Toxicology:* LD50 (acute oral, rat) > 7800 mg/kg; nonirritating to skin and eye.

ADK STAB OF-14. [Asahi Denka Kogyo] Ba-Zn; general purpose foam PVC stabilizer for expanded leathercloth and wall paper; gran.

ADK STAB OF-15. [Asahi Denka Kogyo] Ba-Zn; general purpose foam PVC stabilizer for expanded leathercloth and wall paper; expands quickly at first stage; gran.

ADK STAB OF-19. [Asahi Denka Kogyo] Ba-Zn; general purpose foam PVC stabilizer for expanded leathercloth and wall paper; gran.

ADK STAB OF-23. [Asahi Denka Kogyo] Ba-Zn; general purpose foam PVC stabilizer for expanded leathercloth by calendering, shoesole by inj., and wall paper; gran.

ADK STAB OF-30G. [Asahi Denka Kogyo] Mg-Zn; general purpose foam PVC stabilizer for expanded leathercloth and wall paper; provides exc. whiteness and plate-out props.; powd.

ADK STAB OF-31. [Asahi Denka Kogyo] Ba-Zn; general purpose foam PVC stabilizer for expanded leathercloth and wall paper; gran.

ADK STAB OT-1. [Asahi Denka Kogyo] Di-n-octyltin dilaurate; PVC stabilizer providing exc. external lubricity, heat and process stability with Ca-Zn stabilizers; suitable for rigid sheet and blow formulation; approved by JHPA; liq.

ADK STAB OT-9. [Asahi Denka Kogyo] Di-n-octyltin maleate; PVC stabilizer providing exc. heat stability and weatherability; suitable for rigid extrusion and calendering formulation; approved by JHPA; liq.

ADK STAB P. [Asahi Denka Kogyo] Phosphite; stabilizer for polyolefins; provides exc. color improvement and lt. stability; JHOSPA compliance for food pkg.; clear liq.; visc. 2500 cps.

ADK STAB PEP-4C. [Asahi Denka Kogyo] Phosphite; antiscorching agent for PU; clear liq.; visc. 5000 cps.

ADK STAB PEP-8. [Asahi Denka Kogyo] Phosphite; stabilizer for polyolefins, ABS, PS, etc.; provides exc. color improvement, high process stability; FDA and JHOSPA compliance for food pkg.; wh. flake; m.w. 733; soften. pt. 52 C; 7.3-7.9% P.

ADK STAB PEP-8F. [Asahi Denka Kogyo] Phosphite; stabilizer for polyolefins, ABS, PS, etc.; provides exc. color improvement, high process stability; improvement in hydrolysis over PEP-8; FDA and JHOSPA compliance for food pkg.; wh. flake; soften. pt. 50 C.

ADK STAB PEP-8W. [Asahi Denka Kogyo] Mixt. of ADK STAB PEP-8 (phosphite) and CA-ST (calcium stearate); stabilizer for polyolefins, ABS, PS, etc.; easy handling; FDA and JHOSPA compliance for food pkg.; wh. powd.

ADK STAB PEP-11C. [Asahi Denka Kogyo] Phosphite; stabilizer, antiscorching agent for PU; wh. or pale yel. liq.; visc. 500 cps.

ADK STAB PEP-24G. [Asahi Denka Kogyo] Phosphite; stabilizer for polyolefins, ABS, PS, engineering plastics, etc.; provides exc. process stability and color improvement; FDA and JHOSPA compliance for food pkg.; wh. gran.; m.w. 604; m.p. 165 C.

ADK STAB PEP-36. [Asahi Denka Kogyo] Phosphite; stabilizer for polyolefins, ABS, PS, engineering plastics, etc.; provides exc. process stability and color improvement at higher temps.; FDA and JHOSPA compliance for food pkg.; wh. gran.; m.w. 633; m.p. 237 C.

ADK STAB QL. [Asahi Denka Kogyo] Phosphite; stabilizer for polyolefins (PP, LLDPE); provides exc. process stability; JHOSPA compliance for food pkg.; clear liq.; visc. 1000 cps.

ADK STAB RUP-9. [Asahi Denka Kogyo] Ba-Zn; PVC stabilizer for nonmigrating elec. wire using polyester plasticizer; provides exc. heat stability and insulating props. with AC-186; powd.

ADK STAB RUP-10. [Asahi Denka Kogyo] Ba-Zn; PVC stabilizer for nonmigrating elec. wire using polyester plasticizer; provides exc. heat stability and insulating props. with AC-186; powd.

ADK STAB RUP-11. [Asahi Denka Kogyo] Ba-Zn; PVC stabilizer for nonmigrating elec. wire using polyester plasticizer; provides exc. heat stability and insulating props. with AC-186; powd.

ADK STAB RUP-14. [Asahi Denka Kogyo] Ba-Zn; PVC stabilizer providing exc. heat stability with AC-186; for calendered sheet, film, and extrusion prods., nonmigrating wire using polyester plasticizer; powd.

ADK STAB SC-12. [Asahi Denka Kogyo] Ca-Zn; nontoxic PVC stabilizer for rigid calendering sheet, plasticized calendering film, and plasticized extrusion for hose; exc. initial color protection and transparency; approved by FDA and JHPA (below 12%); liq.

ADK STAB SC-24. [Asahi Denka Kogyo] Ca-Zn; nontoxic PVC stabilizer for plasticized prods.; approved by FDA and JHPA (below 13.6%); liq.

ADK STAB SC-26. [Asahi Denka Kogyo] Ca-Zn; nontoxic PVC stabilizer for plasticized prods.; approved by FDA and JHPA (below 7.0%); liq.

ADK STAB SC-32. [Asahi Denka Kogyo] Ca-Zn; general purpose nontoxic PVC stabilizer providing exc. heat, color, and visc. stability; suitable for toy formulation; approved by JHPA; liq.

ADK STAB SC-34. [Asahi Denka Kogyo] Ca-Zn; general purpose PVC stabilizer providing exc. heat, color, and visc. stability; suitable for toy formulation; liq.

ADK STAB SC-35. [Asahi Denka Kogyo] Mg-Zn; general purpose nontoxic PVC stabilizer providing exc. heat, color, and visc. stability; suitable for toy formulation; approved by JHPA; liq.

ADK STAB SC-102. [Asahi Denka Kogyo] TNPP; CAS 26523-78-4; EINECS 247-759-6; PVC stabilizer improving initial color and transparency with Ca-Zn stabilizers; improves chelating effect; used for plasticized and rigid formulations; approved by FDA and JHPA.

ADK STAB SP-55. [Asahi Denka Kogyo] Ca-Zn; nontoxic PVC stabilizer for plasticized sheet and hose; provides exc. long-term heat resist. and lubricity; approved by JHPA (below 11.7%); powd.

ADK STAB SP-69. [Asahi Denka Kogyo] Ca-Zn; nontoxic PVC stabilizer for rigid applics. only; provides exc. continuous heat resist.; approved by FDA and JHPA (below 1.6%); powd.

ADK STAB SP-76. [Asahi Denka Kogyo] Ca-Zn; nontoxic PVC stabilizer providing exc. initial color and blooming props.; approved by JHPA (below 6.0%); powd.

ADK STAB SP-86. [Asahi Denka Kogyo] Ca-Zn; nontoxic PVC stabilizer for rigid and semirigid extrusion and blow molding applics. with SP-83; approved by JHPA (below 6.7%); powd.

ADK STAB TPP. [Asahi Denka Kogyo] Phosphite; stabilizer for PU and rubbers; provides color improvement and heat stability; clear liq.; m.w. 310; visc. 18 cps.

ADK STAB ZN-ST. [Asahi Denka Kogyo] Zinc stearate; CAS 557-05-1; EINECS 209-151-9; nontoxic stabilizer, lubricant, and release agent for PVC, polyolefin, ABS, etc.; reacts with acidic ingreds.; FDA and JHOSPA compliance for food pkg.; wh. powd.; m.p. ≈ 120 C.

ADK STAB ZS-27. [Asahi Denka Kogyo] Effective heat stabilizing metal deactivator for filled PP; JHOSPA compliance for food pkg.; wh. powd.; m.p. > 65 C.

ADK STAB ZS-90. [Asahi Denka Kogyo] Metal deactivator for filled PP; improved compat. with filled PP; JHOSPA compliance for food pkg.; wh. powd.; m.p. > 110 C.

Adma® 8. [Albemarle] Octyldimethylamine; CAS 7378-99-6; cationic; intermediate for quat. ammonium compds., amine oxides, betaines; for household prods., disinfectants, sanitizers, industrial hand cleaners, cosmetics, bubble baths, deodorants, polymer additives, PU foam catalysis, epoxy curing agent; APHA < 10 liq., fatty amine odor; sp.gr. 0.765; f.p. -57 C; amine no. 352; flash pt. (TCC) 64 C; 100% conc.; *Precaution:* corrosive.

Adma® 10. [Albemarle] Decyldimethylamine; CAS 1120-24-7; cationic; intermediate for quat. ammonium compds., amine oxides, betaines; for household prods., disinfectants, sanitizers, industrial hand cleaners, cosmetics, bubble baths, deodorants, polymer additives, PU foam catalysis, epoxy curing agent; clear liq., fatty amine odor; sp.gr. 0.778; f.p. -35 C; amine no. 300; flash pt. (TCC) 91 C; 100% conc.; *Precaution:* corrosive.

Adma® 12. [Albemarle] Dodecyl dimethylamine; CAS 112-18-5; EINECS 203-943-8; cationic; intermediate for quat. ammonium compds., amine oxides, betaines; for household prods., disinfectants, sanitizers, industrial hand cleaners, cosmetics, bubble baths, deodorants, polymer additives, PU foam catalysis, epoxy curing agent; clear liq., fatty amine odor; sp.gr. 0.778; f.p. -22 C; amine no. 259; flash pt. (PM) 114 C; 100% conc.; *Precaution:* corrosive.

Adma® 14. [Albemarle] Tetradecyl dimethylamine; CAS 112-75-4; EINECS 204-002-4; cationic; intermediate for quat. ammonium compds., amine oxides, betaines; for household prods., disinfectants, sanitizers, industrial hand cleaners, cosmetics, bubble baths, deodorants, polymer additives, PU foam catalysis, epoxy curing agent; clear liq.; sp.gr. 0.794; f.p. -6 C; amine no. 229; flash pt. (PM) 132 C; 100% conc.; *Precaution:* corrosive.

Adma® 16. [Albemarle] Hexadecyl dimethylamine; CAS 112-69-6; EINECS 203-997-2; cationic; intermediate for quat. ammonium compds., amine oxides, betaines; for household prods., disinfectants, sanitizers, industrial hand cleaners, cosmetics, bubble baths, deodorants, polymer additives, PU foam catalysis, epoxy curing agent; clear liq., fatty amine odor; sp.gr. 0.800; f.p. 8 C; amine no. 206; flash pt. (PM) 142 C; 100% conc.; *Precaution:* corrosive.

Adma® 18. [Albemarle] Octadecyl dimethylamine; CAS 124-28-7; EINECS 204-694-8; intermediate for mfg. of quaternary ammonium compds. for biocides, textile chems., oilfield chems., amine oxides, betaines, polyurethane foam catalysts, epoxy curing agents; household prods., disinfectants, sanitizers, industrial hand cleaners, cosmetics, bubble baths, deodorants; clear liq., fatty amine odor; sp.gr. 0.807; f.p. 21 C; amine no. 186; flash pt. (PM) 163 C; 98% tert. amine; *Precaution:* corrosive.

Adma® 246-451. [Albemarle] Dodecyl dimethylamine (40%), tetradecyl dimethylamine (50%), hexadecyl dimethylamine (10%); cationic; intermediate for quat. ammonium compds., amine oxides, betaines; for household prods., disinfectants, sanitizers, industrial hand cleaners, cosmetics, bubble baths, deodorants, polymer additives, PU foam catalysis, epoxy curing agent; clear liq., fatty amine odor; sp.gr. 0.792; f.p. -13 C; amine no. 238; flash pt. (PM) 114 C; 100% conc.; *Precaution:* corrosive.

Adma® 246-621. [Albemarle] Dodecyl dimethylamine (65%), tetradecyl dimethylamine (25%), hexadecyl dimethylamine (10%); cationic; intermediate for quat. ammonium compds., amine oxides, betaines; for household prods., disinfectants, sanitizers, industrial hand cleaners, cosmetics, bubble baths, deodorants, polymer additives, PU foam catalysis, epoxy curing agent; clear liq., fatty amine odor; sp.gr. 0.791; f.p. -18 C; amine no. 245; flash pt. (PM) 121 C; 100% conc.; *Precaution:* corrosive.

Adma® 1214. [Albemarle] Dodecyl dimethylamine

(65%), tetradecyl dimethylamine (35%); cationic; intermediate for quat. ammonium compds., amine oxides, betaines; for household prods., disinfectants, sanitizers, industrial hand cleaners, cosmetics, bubble baths, deodorants, polymer additives, PU foam catalysis, epoxy curing agent; clear liq., fatty amine odor; sp.gr. 0.791; f.p. -18 C; amine no. 249.2; flash pt. (PM) 121 C; 100% conc.; *Precaution:* corrosive.

Adma® 1416. [Albemarle] Dodecyl dimethylamine (5%), tetradecyl dimethylamine (60%), hexadecyl dimethylamine (30%), octadecyl dimethylamine (5%); cationic; intermediate for quat. ammonium compds., amine oxides, betaines; for household prods., disinfectants, sanitizers, industrial hand cleaners, cosmetics, bubble baths, deodorants, polymer additives, PU foam catalysis, epoxy curing agent; APHA < 10 liq.; sp.gr. 0.796; amine no. 220; 100% conc.

Adma® WC. [Albemarle] Octyl dimethylamine (7%), decyl dimethylamine (6%), dodecyl dimethylamine (53%), tetradecyl dimethylamine (19%), hexadecyl dimethylamine (9%), octadecyl dimethylamine (6%); cationic; intermediate for quat. ammonium compds., amine oxides, betaines; for household prods., disinfectants, sanitizers, industrial hand cleaners, cosmetics, bubble baths, deodorants, polymer additives, PU foam catalysis, epoxy curing agent; clear liq., fatty amine odor; sp.gr. 0.798; f.p. -22 C; amine no. 249; flash pt. (PM) 102 C; 100% conc.; *Precaution:* corrosive.

Admex® 433. [Hüls Am.] Low m.w. polymeric polyester; plasticizer for wide range of polymers incl. polymethyl methacrylates, PU, PS; provides exc. structure in foam moldings; lt. colored clear bright fluid to visc. liq., mild ester odor; m.w. 800; sp.gr. 1.080; visc. 1600 cps; pour pt. 10 F; acid no. 1.5; hyd. no. 20; flash pt. 460 F; ref. index 1.508.

Admex® 515. [Hüls Am.] Med.-low m.w. polymeric adipate polyester; plasticizer for plastisols due to exc. rheological chars.; low visc. for oil filter plastisols; nonmigratory in paper coatings; lt. colored clear bright fluid to visc. liq., mild ester odor; m.w. 2300; sp.gr. 1.060; visc. 600 cps; pour pt. 35 F; acid no. 2; hyd. no. 17; flash pt. 475 F; ref. index 1.463.

Admex® 522. [Hüls Am.] Very low m.w. polymeric polyester; plasticizer providing exc. fusion chars. to PVC compositions; oil resist. for oil filter construction; flexible in inner shoe soles; toughness for molded sporting goods; low visc. for plastisol toys and dolls; lt. colored clear bright fluid to visc. liq., mild ester odor; m.w. 700; sp.gr. 1.060; visc. 900 cps; pour pt. -5 F; acid no. 1.5; hyd. no. 18; flash pt. 465 F; ref. index 1.503.

Admex® 523. [Hüls Am.] Low m.w. polymeric polyester; plasticizer providing exc. processability to PVC compositions; compat. with wide range of other polymers; good adhesion in metal coatings; high gel str. in plastisols at low fusion temps.; lt. colored clear bright fluid to visc. liq., mild ester odor; m.w. 900; sp.gr. 1.100; visc. 4400 cps; pour pt. 15 F; acid no. 2; hyd. no. 23; flash pt. 460 F; ref. index 1.514.

Admex® 525. [Hüls Am.] Low m.w. polymeric adipate polyester; plasticizer compat. with cellulosic polymers and PVAc; migration-resist. for pressure-sensitive decorative coatings; permanent and tough for quality luggage; lt. colored clear bright fluid to visc. liq., mild ester odor; m.w. 1300; sp.gr. 1.037; visc. 350 cps; pour pt. 15 F; acid no. 1.5; hyd. no. 23; flash pt. 540 F; ref. index 1.461.

Admex® 760. [Hüls Am.] Ultra high m.w. polymeric adipate polyester; plasticizer with max. resist. to extraction, migration, and volatility; high m.w. for solv.-resist. in print rollers; nonmigratory for skin patch drug delivery; permanent; lt. colored clear bright fluid to visc. liq., mild ester odor; m.w. 13,500; sp.gr. 1.130; visc. 79,250 cps; pour pt. 35 F; acid no. 3; hyd. no. 2; flash pt. 575 F; ref. index 1.470.

Admex® 761. [Hüls Am.] Med. m.w. polymeric adipate polyester; plasticizer with exc. resist. to org. extractants; exc. foam structure for surgical tape; nonextractable in furniture upholstery; lt. colored clear bright fluid to visc. liq., mild ester odor; m.w. 3500; sp.gr. 1.112; visc. 3650 cps; pour pt. 30 F; acid no. 2.5; hyd. no. 3; flash pt. 515 F; ref. index 1.480.

Admex® 770. [Hüls Am.] Med.-high m.w. polymeric adipate polyester; plasticizer with exc. extraction resist. and weatherability; good adhesion for decorative decals; nonvolatile in automotive instrument panels; compat. with nitrile rubber; ideal for exterior graphics; lt. colored clear bright fluid to visc. liq., mild ester odor; m.w. 4200; sp.gr. 1.110; visc. 4300 cps; pour pt. 15 F; acid no. 1.5; hyd. no. 4; flash pt. 540 F; ref. index 1.468.

Admex® 775. [Hüls Am.] High m.w. polymeric adipate polyester; plasticizer with max. resist. to aq. and org. extractants; exc. permanence and flexibility in conveyor belts; impervious to aq. environment as dishwasher gasketing; very lt. colored clear bright fluid to visc. liq., mild ester odor; m.w. 5100; sp.gr. 1.110; visc. 5000 cps; pour pt. 0 F; acid no. 0.5; hyd. no. 17; flash pt. 530 F; ref. index 1.467.

Admex® 910. [Hüls Am.] Med. m.w. polymeric adipate polyester; plasticizer with permanence and outstanding organoleptic props. in vinyl compositions; hydrolytic resist. in dishwasher gaskets; low odor and taste in refrigerator gaskets; lt. colored clear bright fluid to visc. liq., mild ester odor; m.w. 4000; sp.gr. 1.058; visc. 4350 cps; pour pt. 0 F; acid no. 1; hyd. no. 3; flash pt. (COC) 535 F; ref. index 1.470.

Admex® 1663. [Hüls Am.] Polymeric adipate ester; plasticizer offering printability and flexibility in calendered film and sheet, resist. to extraction by aliphatic hydrocarbons such as gasoline, uv resist. and weatherability; lt. color, low odor; m.w. 2400; sp.gr. 1.080; visc. 1000 cps; pour pt. -15 F; acid no. 1; hyd. no. 6; flash pt. (COC) 480 F; ref. index 1.464.

Admex® 1665. [Hüls Am.] Med. m.w. polymeric plasticizer; plasticizer imparting optimum combination of permanence, processability, and compatibility; exc. resist. to migration in polymer-based pressure-sensitive adhesives; outstanding lt. stability; allows max. output in calendering and extrusion operations; clear, bright fluid to visc. liq.; sp.gr. 1.096; visc. 6200 cps; pour pt. 15 F; acid no. 1.9; hyd. no. 6; flash pt. (COC) 525 F; ref. index 1.468; 0.1% moisture.

Admex® 1723. [Hüls Am.] Med.-low m.w. polymeric adipate plasticizer; plasticizer with exc. processing for calendered sheet; printable for decorative graphics; permanent in single-ply roofing; nonvolatile and flexible in vheicle and furniture upholstery; lt. colored clear, bright fluid to visc. liq., mild ester odor; m.w. 2500; sp.gr. 1.107; visc. 4800 cps; pour pt. 15 F; acid no. 2.0; hyd. no. 16; flash pt. 525 F; ref. index 1.475; 0.1% moisture.

Admex® 2632. [Hüls Am.] Med.-low m.w. polymeric adipate; plasticizer for polymers in contact with food; nonextractable by lipids in medical applics.; washable in hospital sheeting; permanent in beverage tubing and food equip. coatings; FDA sanctioned; lt. colored clear bright fluid to visc. liq., mild ester odor; m.w. 2200; sp.gr. 1.080; visc. 3380 cps; pour pt. 40 F; acid no. 1; hyd. no. 20; flash pt. 545 F; ref. index 1.465.

Admex® 3752. [Hüls Am.] Med.-low m.w. polymeric adipate; plasticizer featuring softness and low temp. impact resist.; resist. to extraction by aliphatic hydrocarbons such as gasoline; provides printability and flexibility in automotive caulks and sealants, suppleness in handbags, outerwear, sportswear; low-visc. liq.; m.w. 2600; sp.gr. 1.059; visc. 800 cps; pour pt. 10 F; acid no. 2; hyd. no. 20; flash pt. (COC)530 F; ref. index 1.463.

Admex® 6187. [Hüls Am.] High m.w. polymeric adipate polyester; plasticizer with exc. permanence; extraction-resist. in gasoline hose and tubing; oil-resist. in elec. cable insulation; solv.-resist. in industrial boots, gloves, and aprons; permanent and washable in wall and shelf coverings; very lt. colored clear bright fluid to visc. liq., mild ester odor; m.w. 5100; sp.gr. 1.110; visc. 5000 cps; pour pt. 0 F; acid no. 0.5; hyd. no. 17; flash pt. 530 F; ref. index 1.467.

Admex® 6969. [Hüls Am.] Med. m.w. polymeric adipate polyester; plasticizer with good balance between permanence and efficiency; flexible and nonvolatile for elec. tape; permanent and washable in coated fabrics; optimum visc. for pigment grinding; compat. with neoprene rubber; very lt. colored clear bright fluid to visc. liq., mild ester odor; m.w. 4000; sp.gr. 1.090; visc. 3000 cps; pour pt. -10 F; acid no. 0.5; hyd. no. 18; flash pt. 535 F; ref. index 1.465.

Admex® 6985. [Hüls Am.] High m.w. polymeric adipate polyester; plasticizer; chemically stable and extraction-resist. in tank liners; permanent and printable in decorative decals; nonextractable in floor coverings; nonvolatile in automotive headrests; lt. colored clear bright fluid to visc. liq., mild ester odor; m.w. 6400; sp.gr. 1.080; visc. 5250 cps; pour pt. 40 F; acid no. 1; hyd. no. 18; flash pt. 530 F; ref. index 1.466.

Admex® 6994. [Hüls Am.] Med. m.w. polymeric adipate polyester; plasticizer with optimum hydrolytic stability and exc. permanence; nonextractable in industrial conveyor belts; solv.-resist. in printer rolls; oil-resist. in industrial boots, gloves, aprons; chemically stable and extraction-resist. in tank liners; lt. colored clear bright fluid to visc. liq., mild ester odor; m.w. 3700; sp.gr. 1.060; visc. 4400 cps; pour pt. 0 F; acid no. 1; hyd. no. 3; flash pt. (COC) 535 F; ref. index 1.469.

Admex® 6995. [Hüls Am.] Med. m.w. polymeric adipate polyester; plasticizer offering permanence and toughness for highly filled automotive plastisols; migration-resist., low volatility, exc. flexibility, paintable; lt. colored clear liq., mild ester odor; m.w. 3100; sp.gr. 1.090; visc. 2000 cps; pour pt. -15 F; acid no. 1; hyd. no. 5; flash pt. (COC) 480 F; ref. index 1.465.

Admex® 6996. [Hüls Am.] Low m.w. polymeric adipate polyester; plasticizer imparting low temp. flexibility with minimal migration into pressure-sensitive adhesives; nonvolatile, flexible and nonmigratory in elec. tape; soft and efficient in coated fabrics; permanent in high temp. elec. applics.; low visc. for paper coatings; lt. colored clear bright fluid to visc. liq., mild ester odor; m.w. 1800; sp.gr. 1.070; visc. 550 cps; pour pt. -15 F; acid no. 1; hyd. no. 7; flash pt. 480 F; ref. index 1.463.

Admex® 6999. [Hüls Am.] Low m.w. polymeric polyester; plasticizer with high PVC compat. resulting in high toughness; exc. structural foam for sound deadening; high quality molded foam; compat. with cellulosic polymers and PVAc; lt. colored clear bright fluid to visc. liq., mild ester odor; m.w. 950 ; sp.gr. 1.090; visc. 4600 cps; pour pt. 10 F; acid no. 1.5; hyd. no. 21; flash pt. 445 F; ref. index 1.515.

Adogen® 432. [Witco Corp.] Dialkyl (C12-C18) dimethyl ammonium chloride, IPA/water; cationic; fabric softener conc. for home and commercial laundries; textile processing; antistatic coating for cellulose acetate, polyacetal, PE, PP; FDA accepted; Gardner 3 max. liq.; m.w. 521; flash pt. (PM) 70 F; 67-69% quat.

Adogen® 442. [Witco Corp.] Quaternium-18; fabric softener conc. for home and commercial laundries; textile processing; antistatic coating for cellulose acetate, polyacetal, PE, PP; FDA accepted; Gardner 4 max. paste; m.w. 569; flash pt. (PMCC) 65 F; 75% solids.

Adol® 52. [Procter & Gamble] Cetyl alcohol; CAS 36653-82-4; EINECS 253-149-0; external mold release for acrylics; .

Adol® 62 NF. [Procter & Gamble] Stearyl alcohol; CAS 112-92-5; EINECS 204-017-6; nonionic; emollient, glass frit binders, waxes, emulsion stabilizers, esters, tertiary amines, surfactants, polymers, chemical intermediate; emollient, emulsion stabilizer, visc. modifier, conditioner cosmetic formulations, skin care prods.; opacifier; Lovibond 5Y/0.54 max. flakes, odorless; sol. in IPA, acetone, naphtha, lt. min. oil; m.w. 272; sp.gr. 0.817 (60/25 C); visc. 42 SSU (210 F); m.p. 56-60 C; b.p. 337-

360 C (760 mm, 90%); acid no. 1.0 max.; sapon. no. 3.0 max.; 100% conc.

Adol® 63. [Procter & Gamble] Cetearyl alcohol; nonionic; emulsifiers; prime base for detergents; used in plasticizers, tert. amines, lube oil additives, textile auxiliaries, mold lubricants, polymers, org. synthesis, chemical intermediates; emollient, emulsion stabilizer, visc. modifier, conditioner for skin care prods.; opacifier for creams and lotions; Lovibond 5Y/0.5R max. flake; m.w. 268; sp.gr. 0.816 (60/25 C); visc. 44 SSU (210 F); m.p. 48-53 C; b.p. 312-344 C (760 mm, 90%); acid no. 1.0 max.; sapon. no. 3.0 max.; 100% conc.

Advantage™. [Milliken] Colorants for opaque polyolefin applics.; for inj. molded prods. such as lids and closures where dimensional tolerances are tight or where warpage from nucleation is a problem; can be formulated for food contact applics.; pellets; *Toxicology:* nontoxic.

Advapak® LS-203. [Morton Int'l./Plastics Additives] Organotin compd.; one-pack lubricating stabilizer for high-output PVC pipe prod. on multiple screw extruders; furnishes exc. residual color stability for regrind extrusion; NSF approved; off-wh. to lt. yel. EZ-FLO bead; sp.gr. 0.95; bulk dens. 32 lb/ft^3; drop pt. 110 C; 100% solids; *Storage:* store in cool, dry area.

Advapak® LS-203HF. [Morton Int'l./Plastics Additives] Organotin compd.; lubricating stabilizer for high-output multiscrew PVC pipe extrusion; furnishes uniform powd. blends, exc. residual color stability for regrind extrusion; EZ-FLO beads; sp.gr. 1.0; bulk dens. ≈ 31 lb/ft^3; drop pt. 111 C; *Storage:* store in cool, dry area.

Advapak® LS-203HP. [Morton Int'l./Plastics Additives] Organotin compd.; high performance lubricating stabilizer for high output multiscrew rigid PVC extrusion, pipe prod.; furnishes uniform powd. blends, exc. residual color stability for regrind extrusion; EZ-FLO beads; sp.gr. 0.97; bulk dens. 31 lb/ft^3; drop pt. 107 C; 100% solids; *Storage:* store in cool, dry area.

Advapak® ML-1325. [Morton Int'l./Plastics Additives] Ester; lubricant pkg. for inj. molding formulations; suitable in PVC for potable water contact; NSF approved; cream to lt. yel. EZ-FLO beads; sp.gr. 0.98; m.p. 94 C; *Storage:* store in cool, dry area; indoor storage is recommended.

Advapak® ML-2516. [Morton Int'l./Plastics Additives] Lubricant providing all the calcium stearate and oxidized polyethylene required for high-output multiscrew PVC pipe extrusion; synergistic with organotin mercaptides; EZ-FLO beads; sp.gr. 0.97; drop pt. 111 C; 100% solid; *Storage:* store in cool, dry area.

Advapak® SLS-1000. [Morton Int'l./Specialty Chem.] Multifunctional lubricating stabilizer for PVC siding and profile.

Advastab® TM-181. [Morton Int'l./Plastics Additives] Methyltin mercaptide; heat stabilizer for rigid PVC processes and PVC-PVA formulations with prolonged/severe processing temps., for chlorinated PVC compds; tin catalyst in PU polymerizations; ideal for clear bottles, clear rigid sheet, profiles, other critical inj. molding, blow molding, and extrusion applics.; APHA 70 clear liq.; sp.gr. 1.175; dens. 9.8 lb/gal; visc. 50 cs; *Storage:* indoor storage is recommended; avoid iron and nonstainless steel for piping, valves, and pumps used for handling.

Advastab® TM-181-FS. [Morton Int'l./Specialty Chem.] Methyltin mercaptide; heat stabilizer for rigid PVC and rigid PVC copolymers for food pkg., bottles, profile, and inj. moldings; Gardner 1 max. clear liq.; sp.gr. 1.17; dens. 9.7 lb/gal; visc. 50 cs.

Advastab® TM-183-B. [Morton Int'l./Plastics Additives] Methyltin mercaptide; heat stabilizer for rigid and semirigid PVC and copolymer processing incl. extrusion, calendering, inj. and nonfood blow molding; provides good early color hold and exc. dynamic process stability; provides sparkling clarity in clear sheet, film, bottles; exc. for outdoor applics. (siding, trim), profiles; Gardner 2 clear liq.; dens. 1.14 g/ml; visc. 79 cs; ref. index 1.506; *Storage:* indoor storage is recommended; avoid iron and nonstainless steel for piping, valves, and pumps used for handling.

Advastab® TM-183-O. [Morton Int'l./Plastics Additives] Methyltin mercaptide; heat stabilizer for rigid PVC and copolymer blow molding, calendered, extruded formulations for food pkg. applics.; offers lubricity, good early color, dynamic processing stability, good residual stability for efficient use of regrind; FDA, BGA and other approvals; Gardner 2 clear liq.; dens. 1.13 g/ml; visc. 56 cs; ref. index 1.503; *Storage:* store in cool, dry area; indoor storage is recommended; stainless steel recommended for piping, valves, and pumps used for handling.

Advastab® TM-281 IM. [Morton Int'l./Specialty Chem.] Methyltin mercaptide; PVC heat stabilizer for inj. molding applics., e.g., rigid PVC pipe fittings, elec. building prods.

Advastab® TM-281 SP. [Morton Int'l./Specialty Chem.] Methyltin mercaptide; PVC heat stabilizer with exc. color retention for siding and profile extrusion in building trade.

Advastab® TM-692. [Morton Int'l./Plastics Additives] Methyltin mercaptide; heat stabilizer for most rigid PVC applics. for potable water, sewer, irrigation pipe, conduit, and duct, and extrusion applics., profile, foamed profile; NSF approved; liq.; sp.gr. 0.99; dens. 8.3 lb/gal; visc. 60 cs; cloud pt. -7 C; pour pt. -18 C.

Advastab® TM-694. [Morton Int'l./Specialty Chem.] Tin; heat stabilizer for rigid PVC formulations; esp. for extrusion of potable water, sewer and irrigation pipe, elec. conduit and telephone duct, profiles and tubing; exc. initial color stability; NSF-approved; liq.; sp.gr. 0.98; dens. 8.2 lb/gal; visc. 59.2 cs; pour pt. -12 C; cloud pt. -1 C.

Advastab® TM-696. [Morton Int'l./Plastics Additives] Tin; heat stabilizer for rigid PVC formulations, multiscrew extrusion of potable water, sewer, DWV, and irrigation pipe, elec. conduit, telephone duct; provides good initial color stability and ability to maintain color during processing; NSF-approved; liq.; sp.gr. 0.94; dens.7.8 lb/gal; visc. 27 cs; pour pt. -11 C; cloud pt. -3 C.

Advastab® TM-697. [Morton Int'l./Specialty Chem.] Tin mercaptide; stabilizer for rigid PVC applics., esp. for large diameter pipe; also for potable water, sewer, irrigation pipe, telephone duct, elec. conduit, profile and inj. moldings; exc. initial whiteness and ability to maintain good color hold during processin; NSF-approved; liq.; sp.gr. 1.03; dens. 8.6 lb/gal; visc. 56 cs; cloud pt. -2 C; pour pt. -9 C.

Advastab® TM-790 Series. [Morton Int'l./Plastics Additives] Organotin compd.; heat stabilizer/external lubricant for single and multiscrew extrusion of PVC potable water, sewer and irrigation pipe, conduit, telephone duct; exc. early color and regrind stability; NSF-approved; liq.; sp.gr. 0.89; dens. 7.5 lb/gal; visc. 250 cp; pour pt. -9.5 C; *Storage:* store @ R.T.

Advastab® TM-948. [Morton Int'l./Specialty Chem.] Methyltin; heat stabilizer for PVC; exc. early and long-term color retention for processing general-purpose rigid PVC bottles and sheet; virtually no odor in finished prod.; Gardner ≤ 2 liq.; sp.gr. 1.17; dens. 9.7 lb/gal.

Advastab® TM-2080. [Morton Int'l./Specialty Chem.] Methyltin mercaptide; heat stabilizer for rigid PVC and copolymers used for food pkg.; provides good early color, exc. dynamic processing stability, good residual stability for efficient use of regrind; for clear rigid bottles; FDA 21CFR §178.2010; Gardner 1 max. clear liq.; sp.gr. 1.17; dens. 9.7 lb/gal; visc. 50 cs; 0.1% max. trimethyltin compds.

Advawax® 165. [Morton Int'l./Specialty Chem.] Paraffin wax; CAS 8002-74-2; EINECS 232-315-6; external lubricant for opaque rigid PVC molding, single- and twin-screw extrusion; used with internal lubricant such as calcium stearate; FDA accepted; solid.

Advawax® 240. [Morton Int'l./Specialty Chem.] N,N′-ethylene bisoleamide; CAS 110-31-6; EINECS 203-756-1; syn. wax used as plastics processing lubricant and release agent, antistat, m.p. modifier for waxes, industrial asphalts and tar, pigment dispersing agent for resin systems; polyamide-paraffin coupling agent; used in adhesive tapes, coatings, food pkg. materials; FDA 21CFR §176.170, 176.180, 177.1200, 178.3860; Gardner 13 max. sm. bead; 20 mesh; insol. in water and most org. solvs. @ R.T.; sol. hot in CCl_4, DMF, ethanol, heptane, kerosene, MEK, MIBK, naphtha, toluene, xylene; dens. 5 lb/gal; bulk dens. 38 lb/ft^3; m.p. 115-118 C; acid no. 12 max.; flash pt. (COC) 270 C; fire pt. (COC) 310 C; *Precaution:* may form combustible or explosive mixts. of dust and air during handling.

Advawax® 280 [Morton Int'l./Specialty Chem.] N,N′-ethylene bisstearamide; syn. wax used as plastics processing lubricant and release agent, antistat, m.p. modifier for waxes, industrial asphalts and tar, pigment dispersing agent for resin systems; polyamide-paraffin coupling agent; used in adhesive tapes, coatings, food pkg. materials; FDA 21CFR §175.105, 175.300, 176.170, 176.180, 177.1200; NSF approved; Gardner 5 max. sm. bead; 40 mesh; insol. in water and most org. solvs. @ R.T.; sol. hot in DMF, heptane, kerosene, MIBK, naphtha, toluene, xylene; dens. 4.7 lb/gal; bulk dens. 36.8 lb/ft^3; m.p. 144-146 C; acid no. 10 max.; flash pt. (COC) 280 C; fire pt. (COC) 315 C; *Precaution:* may form combustible or explosive mixts. of dust and air during handling.

Advawax® 290. [Morton Int'l./Plastics Additives] N,N′-Ethylene bisstearamide; syn. wax used as plastics processing lubricant and release agent; m.p. modifier for waxes and resin blends and industrial asphalt and tar; pigment dispersing agent for resin systems; paper-making defoamer; used in adhesive tapes, coatings, PP; also for food pkg. materials; FDA 21CFR §175.105, 175.300, 176.170, 176.180, 177.1200; Gardner 9 max. sm. beads; insol. in water and most org. solvs. @ R.T.; sol. hot in Cellosolve, MDF, MIBK, benzene, xylene, kerosene, heptane, naphtha; dens. 4.75 lb/gal; bulk dens. 35.6 lb/ft^3; m.p. 143-146 C; acid no. 10 max.; flash pt. (COC) 290 C; fire pt. (COC) 310 C; *Precaution:* may form combustible or explosive mixts. of dust and air during handling.

AEP. [Air Prods./Perf. Chems.] Aminoethylpiperazine; CAS 140-31-8; EINECS 205-411-0; room or elevated temp. epoxy curing agent providing rapid gel and initial cure @ R.T. to B-stage; post-cure gives rigid castings with very high impact resist.; system requires modification for full cure @ R.T.; for civil engineering adhesives, decorative floorings, decoupage, small elec. pottings/encapsulation; accelerating co-curing agent for other amines; Gardner 1 liq.; sp.gr. 0.984; visc. 10 cps; amine no. 1297; equiv. wt. 43; heat distort. temp. 126 F (264 psi).

Aerosil® 90 [Degussa] Fumed silica; CAS 112945-52-5; reinforcement and rheology aid for acrylic, butyl rubber, polysulfide rubber, PU, RTV silicone rubber sealing compds.; increases filler loading and/or extrudability; wh. fluffy powd.; 20 nm avg. particle size; sp.gr. 2.2; dens. ≈ 100 g/l (densed); surf. area 90 ± 15 m^2/g; pH 3.6-4.5 (4% aq. disp.); > 99.8% assay; *Toxicology:* TLV 10 mg/m^3 total dust; LD50 > 20,000 mg/kg; may cause eye, skin, or respiratory tract irritation on overexposure; *Precaution:* incompat. with strong bases and hydrofluoric acid.

Aerosil® 130. [Degussa; Degussa AG] Fumed silica; CAS 112945-52-5; filler for plastics and silicone rubbers; thickener, structuring and reinforcing agent for R.T. curing sealing compds.; thixotrope and thickener in personal care creams; wh. fluffy powd.; 16 nm avg. particle size; sp.gr. 2.2; dens. ≈ 120 g/l (densed); surf. area 130 ± 25 m^2/g; pH 3.6-4.3 (4% aq. susp.); > 99.8% assay; *Toxicology:* TLV 10 mg/m^3 total dust; LD50 > 20,000 mg/kg; may cause eye, skin, or respiratory tract irritation on overexposure; *Precaution:* incompat. with strong bases and hydrofluoric acid.

Aerosil® 150. [Degussa; Degussa AG] Fumed silica; CAS 112945-52-5; filler for plastics and silicone rubbers; thickener, structuring and reinforcing agent for R.T. curing sealing compds.; wh. fluffy powd.; 14 nm avg. particle size; sp.gr. 2.2; dens. ≈ 120 g/l (densed); surf. area 150 ± 15 m^2/g; pH 3.6-4.3 (4% aq. susp.); > 99.8% assay; *Toxicology:* TLV 10 mg/m^3 total dust; LD50 > 20,000 mg/kg; may cause eye, skin, or respiratory tract irritation on overexposure; *Precaution:* incompat. with strong bases and hydrofluoric acid.

Aerosil® 200. [Degussa; Degussa AG] Fumed silica; CAS 112945-52-5; EINECS 231-545-4; thickener, reinforcement, anticaking and free-flow agent with high absorp. capacity; for adhesive, food, cosmetics, paint, paper, film, pesticides, pharmaceuticals, plastics, silicone rubber, inks, sealants industries; thixotrope for greases and min. oils; FDA 21CFR §133.146(b), 160.105(a)(d), 172.230(a), 172.480, 173.340(a), 175, 176, 177, 573.940; ≤ 3% for cosmetics, internal pharmaceuticals; wh. fluffy powd.; 12 nm avg. particle size; sp.gr. 2.2; dens. ≈ 120 g/l (densed); surf. area 200 ± 25 m^2/g; pH 3.6-4.3 (4% aq. susp.); > 99.8% assay; *Toxicology:* TLV 10 mg/m^3 total dust; LD50 > 20,000 mg/kg; may cause eye, skin, or respiratory tract irritation on overexposure; *Precaution:* incompat. with strong bases and hydrofluoric acid.

Aerosil® 300. [Degussa; Degussa AG] Fumed silica; CAS 112945-52-5; for low thermal syn. powds. used to produce insulation materials; thickener, thixotrope, antisettling agent for paints, plastics, sealants, personal care creams; filler for heat-cured silicone rubber; FDA approved (≤ 3% for cosmetics); wh. fluffy powd.; 7 nm avg. particle size; sp.gr. 2.2; dens. ≈ 120 g/l (densed); surf. area 300 ± 30 m^2/g; pH 3.6-4.3 (4% aq. susp.); > 99.8% assay; *Toxicology:* TLV 10 mg/m^3 total dust; LD50 > 20,000 mg/kg; may cause eye, skin, or respiratory tract irritation on overexposure; *Precaution:* incompat. with strong bases and hydrofluoric acid.

Aerosil® 325. [Degussa] Silica; CAS 7631-86-9; filler for silicone rubber industry; wh. fluffy powd.; sp.gr. 2.2; *Toxicology:* TLV 10 mg/m^3 total dust; LD50 > 20,000 mg/kg; may cause eye, skin, or respiratory tract irritation on overexposure; *Precaution:* incompat. with strong bases and hydrofluoric acid.

Aerosil® 380. [Degussa; Degussa AG] Fumed silica; CAS 112945-52-5; filler for plastics, silicone rubber, cosmetics, pharmaceuticals, direct and indirect food applics., transparent polyester resin coatings, specialty RTV systems; FDA 21CFR §133.146(b), 160.105(a)(d), 172.230(a), 172.480, 173.340(a), 175, 176, 177, 573.940; wh. fluffy powd.; 7 nm avg. particle size; sp.gr. 2.2; dens. ≈ 120 g/l (densed); surf. area 380 ± 30 m^2/g; pH 3.6-4.3 (4% aq. susp.); > 99.8% assay; *Toxicology:* TLV 10 mg/m^3 total dust; LD50 > 20,000 mg/kg; may cause eye, skin, or respiratory tract irritation on overexposure; *Precaution:* incompat. with strong bases and hydrofluoric acid.

Aerosil® COK 84. [Degussa] Fumed silica and highly dispersed aluminum oxide (5:1); filler for plastics; thixotropic and antisedimentation agent in epoxy casting resins, polymethylmethacrylate and acrylic casting, coating and molding materials; thickener for aq. systems and other polar liqs.; wh. fluffy powd.; sp.gr. 2.2; dens. ≈ 50 g/l; surf. area 170 ± 30 m^2/g; pH 3.6-4.3 (4% aq. susp.); 82-86% SiO_2, 14-18% Al_2O_3.

Aerosil® OX50. [Degussa] Fumed silica; CAS 112945-52-5; filler for plastics, silicone rubber; antiblocking agent for PET, PP and PE films and tapes; for coating of elec. light bulbs; wh. fluffy powd.; 40 nm avg. particle size; sp.gr. 2.2; dens. ≈ 130 g/l; surf. area 50 ± 15 m^2/g; pH 3.8-4.5 (4% aq. susp.); > 99.8% assay; *Toxicology:* TLV 10 mg/m^3 total dust; LD50 > 20,000 mg/kg; may cause eye, skin, or respiratory tract irritation on overexposure; *Precaution:* incompat. with strong bases and hydrofluoric acid.

Aerosil® R202. [Degussa] Fumed silica; CAS 67762-90-87; for thickening, thixotropizing, and improved visc. stability of epoxy potting compds., PU systems, greases, paints, plastics; wh. fluffy powd.; 14 nm avg. particle size; sp.gr. 2.2; dens. ≈ 90 g/l (densed); surf. area 100 ± 20 m^2/g; pH 4-6 (in water:ethanol 1:3); > 99.8% assay.

Aerosil® R805. [Degussa] Fumed silica; CAS 92797-60-9; thickener, thixotrope, structuring agent for epoxy-based systems, adhesive and multilayer systems; antisettling, antisag, anticorrosion agent for paints, plastics; wh. fluffy powd.; 12 nm avg. particle size; sp.gr. 2.2; dens. ≈ 90 g/l (densed); surf. area 150 ± 25 m^2/g; pH 3.5-5.5 (in water:ethanol 1:3); > 99.8% assay.

Aerosil® R812. [Degussa] Syn. amorphous fumed silica; CAS 68909-20-6; hydrophobic antisettling, antisag, anticorrosion agent for high-solids coatings; thickening and thixotropy of vinyl ester resins; reinforcement for high strength systems; suspension and redispersability props. in pharmaceutical/cosmetic aerosols; water repellent props. for lipsticks; wh. fluffy powd.; 7 nm avg. particle size; sp.gr. 2.2; dens. ≈ 80 g/l (densed); surf. area 260 ± 30 m^2/g; pH 5.5-7.5 (in water:ethanol 1:3); > 99.8% assay.

Aerosil® R972. [Degussa; Degussa AG] Syn. amorphous fumed silica; CAS 60842-32-2; hydrophobic anticaking and free-flow agent for adhesive, elec., cosmetics, paint, pesticides, pharmaceuticals, plastics, inks industries; improves water resist. of greases; water repellent for lipsticks; reinforcement for RTV silicone rubber, fluoroelastomers; FDA approved (≤ 3% for cosmetics); fluffy wh. powd.; 16 nm avg. particle size; dens. ≈ 90 g/l (densed); surf. area 110 ± 20 m^2/g; pH 3.6-4.3 (in water:acetone or methanol 1:1); > 99.8% assay.

Aerosil® R972V. [Degussa] Fumed silica; reinforces and improves storage stability of RTV compds.; yields softer silicone rubbers.

Aerosil® R974. [Degussa] Silica dimethyl silylate; CAS 60842-32-2; hydrophobic anticaking agent with suspending char.; thickener, thixotrope for water resist. of greases; rheology aid for acrylic, epoxy, PMMA, PU, PVC plastisol coating compds.; reinforcement/rheology for acrylic, PU, butyl, polysulfide, silicone rubber sealants and PU, PVAC, PVC, CR, TPE adhesives; wh. fluffy powd.; 12 nm particle size; dens. ≈ 90 g/l (densed); surf. area 170 ± 20 m^2/g; pH 3.4-4.2 (in water: ethanol 1:3); > 99.8% assay.

Aerosil® TT600. [Degussa] Fumed silica; flattening agent; antiblocking agent for cellophane, linear polyerster, PP, and PVC films; wh. fluffy powd.; 40 nm avg. particle size; dens. ≈ 60 g/l; surf. area 200 ± 50 m^2/g; pH 3.6-4.5 (4% aq. disp.); > 99.8% assay.

Aerosol® 18. [Cytec Industries] Disodium stearyl sulfosuccinamate; CAS 14481-60-8; EINECS 238-479-5; anionic; emulsifier, dispersant, foamer, detergent, solubilizer for soaps and surfactants; alkaline cleaner formulations, brick and tile cleaners, emulsion polymerization of vinyl chloride and SBRs; emulsifying oils and waxes, household detergents, cleaning paper mill felts; foamer for foamed latexes and plastics; biodeg.; FDA 21CFR §176.170, 176.180; Gardner 10 max. creamy paste; m.w. 493; water-disp.; sp.gr. 1.07; dens. 8.9 lb/gal; acid no. 4.0 max.; flash pt. (Seta CC) > 200 F; surf. tens. 39 dynes/cm; 35 ± 1.5% solids; *Toxicology:* LD50 (rat, oral) 2.68 mg/kg; mild skin, mild to moderate eye irritation; *Storage:* store in tightly closed containers.

Aerosol® 22. [Cytec Industries] Tetrasodium dicarboxyethyl stearyl sulfosuccinamate; CAS 38916-42-6; anionic; emulsifier, dispersant, solubilizer, surfactant; emulsion polymerization of vinyl monomers; polishing waxes; surf. tension depressant for writing and drawing inks; demulsifier for w/o emulsions; cleaning of paper mill felts; industrial, household, and metal cleaners; biodeg.; FDA 21CFR §175.105, 176.170, 176.180, 178.3400; EPA exempt 40CFR §180.1001(d); lt. tan clear to cloudy liq.; water-sol.; m.w. 653; sp.gr. 1.12; dens. 9.4 lb/gal; visc. 53 cps; m.p. > 200 C; flash pt. (Seta CC) 143 F; acid no. 2.0; pH 7-8; surf. tens. 41 dynes/cm; 35% act. in water/alcohol; *Toxicology:* LD50 (rat, acute oral) 18.7 mL/kg (sol'n.); mild skin and eye irritant.

Aerosol® 501. [Cytec Industries] Disodium alkyl sulfosuccinate; anionic; dispersant, emulsifier, wetting agent, dispersant, foaming agent; used for acrylic and vinyl acetate emulsions; self-crosslinking latexes; textile wetting and foaming applics.; biodeg.; APHA 60 max. clear liq.; water-sol.; sp.gr. 1.16; dens. 9.66 lb/gal; visc. 260 cps; f.p. -3 C; flash pt. (Seta CC) > 200 F; acid no. 7.0; pH 6.0-7.0; surf. tens. 28 dynes/cm; 50% act. in water; *Toxicology:* nonirritating or sensitizing to skin.

Aerosol® A-102. [Cytec Industries] Disodium deceth-6 sulfosuccinate; CAS 39354-45-5; anionic; emulsifier, solubilizer, foamer, dispersant, surfactant, wetting agent; used in emulsion polymerization of PVAc/acrylics, textiles, cosmetics, shampoos, wallboard, adhesives; biodeg.; stable to acid media; FDA 21CFR §175.105; colorless to lt. yel. clear liq.; sol. in water, dimethyl sulfoxide; m.w. 614; sp.gr. 1.08; dens. 9.01 lb/gal; visc. 40 cps; f.p. -4 C; flash pt. (Seta CC) > 200 F; acid no. 6 max.; pH 4.5-5.5; surf. tens. 33 dynes/cm; 30% act. in water; *Toxicology:* LD50 (rat, oral) 30.8 mL/kg, (rabbit, dermal) > 5.0 mL/kg; moderate skin, minimal eye irritation.

Aerosol® A-103. [Cytec Industries] Disodium nonoxynol-10 sulfosuccinate; CAS 9040-38-4; anionic; emulsifier, solubilizer, wetting agent, surfactant, surf. tens. depressant; used in PVAc acrylic emulsions; textile emulsions, pad-bath additive, textile wetting; cosmetics, shampoos, wallboard, adhesives; FDA 21CFR §175.105; colorless to lt. yel. clear liq.; sol. in water, MEK; partly sol. in other polar solvs.; m.w. 854; sp.gr. 1.09; dens. 9.1 lb/gal; visc. 170-190 cps; f.p. -9 C; flash pt. (Seta CC) > 200 F; acid no. 10 max.; pH 4.5-5.5; surf. tens. 34 dynes/cm; 35% act. in water; *Toxicology:* LD50 (rat, oral) > 10 mL/kg; not appreciably irritating to skin and eyes.

Aerosol® A-196-40. [Cytec Industries] Dicyclohexyl sodium sulfosuccinate; CAS 23386-52-9; EINECS 245-629-3; anionic; dispersant, surfactant; sole emulsifier for modified S/B; post additive to stabilize latex and promote adhesion; biodeg.; FDA 21CFR §178.3400; clear to cloudy liq.; sol. warm in org. solvs.; sol. 25 g/100 ml water; flash pt. (Seta CC) > 200 F; surf. tens. 39 dynes/cm; 40% act. in water.

Aerosol® A-268. [Cytec Industries] Disodium isodecyl sulfosuccinate; CAS 37294-49-8; EINECS 253-452-8; anionic; surfactant, sole emulsifier for PVC latexes-vinyl, vinylidene chloride, acrylics, surf. tens. depressant, solubilizer; EPA exempt 40CFR §180.1001(d); clear liq.; water sol.; dens. 9.96 lb/gal; sp.gr. 1.19; visc. 150 cps; f.p. -5 C; flash pt. (Seta CC) > 200 F; surf. tens. 28 dynes/cm; 50% act. in water; *Toxicology:* LD50 (rat, oral) 5.2 mL/kg (sol'n.), (rabbit, dermal) > 10 mL/kg; mild skin, moderate eye irritation.

Aerosol® AY-65. [Cytec Industries] Diamyl sodium sulfosuccinate; CAS 922-80-5; EINECS 213-085-6; anionic; wetting agent, dispersant, surfactant; used in agriculture, emulsion polymerization, electroplating, ore leaching, cleaning of porcelain, tile, brick, cement; biodeg.; FDA 21CFR §178.3400; clear liq.; sol. in water/ethanol; sp.gr. 1.081; dens. 9.0 lb/gal; f.p. < 18 C; flash pt. (Seta CC) 77 F; surf. tens. 29 dynes/cm; 65% conc. in water/alcohol.

Aerosol® C-61. [Cytec Industries] Ethoxylated alkyl guanidine-amine complex; cationic; antistat, pigment dispersant, flushing agent, wetting agent, settling agent; alkaline, cement, brick, and tile cleaner formulations for crystal growth control, emulsion breaking, alkaline metal and paint brush cleaners, paint removers; textile softener; demulsifying agent; foaming agent; for plastics, paper, textiles, adhesive industries; lt. tan creamy paste; strong ammoniacal odor; sol. in org. solvs. in presence of alcohol; disp. in water; sp.gr. 1.00;

dens. 8 lb/gal; flash pt. (Seta CC) 94 F; surf. tens. 34 dynes/cm; partially biodeg.; 70.4% act.

Aerosol® DPOS-45. [Cytec Industries] Disodium mono- and didodecyl diphenyl oxide disulfonate; CAS 25167-32-2; anionic; emulsifier, dispersant, solubilizer, primary surfactant for emulsion polymerization systems; coupling agent; high electrolyte tolerance; stable in highly acidic and alkaline sol'ns. and at elevated temps.; liq.; water-sol.; insol. in org. solvs.; flash pt. (Seta CC) > 200 F; surf. tens. 34 dyne/cm; 45% act. in water.

Aerosol® IB-45. [Cytec Industries] Sodium diisobutyl sulfosuccinate; CAS 127-39-9; EINECS 204-839-5; anionic; emulsifier, wetting agent; emulsion polymerization of styrene, butadiene and copolymers; dye and pigment dispersant; for leaching, electroplating; biodeg.; FDA 21CFR §178.3400; EPA exempt 40CFR §180.1001(d); clear liq.; water-sol.; extremely hydrophilic; m.w. 332; sol. in water; sp.gr. 1.12; dens. 9.3 lb/gal; f.p. 20-21 C; flash pt. (Seta CC) > 200 F; acid no. 2 max.; pH 5-7 (10%); surf. tens. 49 dynes/cm; biodeg.; 45% act. in water; *Toxicology:* LD50 (rat, oral) 6.16 g/kg (solids); LD50 (rabbit, dermal) > 5 g/kg; nontoxic by single-dose ingestion.

Aerosol® MA-80I. [Cytec Industries] Dihexyl sodium sulfosuccinate; CAS 3006-15-3; EINECS 221-109-1; anionic; dispersant, textile wetting agent, emulsifier, solubilizer, penetrant; used for emulsion polymerization, battery separators, electroplating, ore leaching; germicidal act.; solubilizer for shampoos; not as rapidly biodeg. as Aerosol 18 and 22; FDA 21CFR §175.105, 176.170, 176.180, 177.1210, 178.3400; APHA 50 max. clear slightly visc. liq.; sol. in water, alcohol, and org. solvs.; sp.gr. 1.13; dens. 9.4 lb/gal; f.p. -28 C; m.p. 199-292 C; flash pt. (Seta CC) 115 F; surf. tens. 28 dynes/cm; 80% act. in IPA.

Aerosol® NPES 458. [Cytec Industries] Ammonium salt of sulfated nonylphenoxy POE ethanol; CAS 9051-57-4; anionic; high foaming surfactant for emulsion polymerization of acrylic, styrene and vinyl acetate systems, dishwashing detergents, germicides, pesticides, general purpose cleaners, cosmetics, and textile wet processing applics.; FDA 21CFR §178.3400; EPA exempt 40CFR § 180.1001(d); pale yel. clear liq., alcoholic odor; sol. in water; partly sol. in org. solvs.; m.w. 493; sp.gr. 1.065; dens. 8.9 lb/gal; visc. 100 cps; f.p. < 0 C; flash pt. (PMCC) 83 F; pH 6.5-7.5; surf. tens. 31 dynes/cm; 58% conc. in water/alcohol; *Toxicology:* LD50 (rat, oral) > 5 g/kg; severe eye, moderate skin irritant.

Aerosol® NPES 930. [Cytec Industries] Ammonium salt of sulfated nonylphenoxy POE ethanol; CAS 9051-57-4; anionic; emulsifier for emulsion polymerization of vinyl acetate, acrylic copolymers, styrene acrylic copolymers; imparts superior water resistance in films; FDA 21CFR §175.105, 176.180; EPA exempt 40CFR §180.1001(d); yel. clear liq.; water-sol.; insol. in org. solvs.; m.w. 713; sp.gr. 1.04; dens. 8.7 lb/gal; visc. 90 cps; gel pt. < 55 F; flash pt. (Seta CC) > 200 F; pH 7.0-7.5 (10%); surf. tens. 33 dynes/cm; 30% act. in water; *Toxicology:* LD50 (rat, oral) > 10 g/kg; severe skin and eye irritation.

Aerosol® NPES 2030. [Cytec Industries] Ammonium salt of sulfated nonylphenoxy POE ethanol; CAS 9051-57-4; anionic; surfactant for emulsion polymerization of acrylic monomers where small particle size and water resistant props. are required; yel. clear liq.; water-sol.; insol. in org. solvs.; m.w. 1197; sp.gr. 1.05-1.07; dens. 8.8 lb/gal; visc. 100 cps; gel pt. < 55 F; flash pt. (Seta CC) > 200 F; pH 7.0-7.5 (10%); surf. tens. 43 dynes/cm; 30% act. in water; *Toxicology:* LD50 (rat, oral) > 10 g/kg; moderate skin irritation on prolonged contact; mild eye irritation.

Aerosol® NPES 3030. [Cytec Industries] Ammonium salt of sulfated nonylphenoxy POE ethanol; CAS 9051-57-4; anionic; emulsifier for emulsion polymerization of acrylic, vinyl acetate and styrene-acrylic systems; forms films with superior water resist.; FDA 21CFR §175.105, 176.180; EPA exempt 40CFR §180.1001(d); yel. clear liq.; water-sol.; insol. in org. solvs.; m.w. 1637; sp.gr. 1.05-1.07; dens. 8.8 lb/gal; visc. 110 cps; gel pt. < 55 F; flash pt. (Seta CC) > 200 F; pH 7.0-7.5 (10%); surf. tens. 43 dynes/cm; 30% act. in water; *Toxicology:* LD50 (rat, oral) > 10 g/kg; minimal skin and eye irritation.

Aerosol® OT-75%. [Cytec Industries] Dioctyl sodium sulfosuccinate; CAS 577-11-7; EINECS 209-406-4; anionic; wetting agent and surf. tens. depressant used in textile, rubber, petrol., paper, metal, paint, plastic, and agric. industries; antistat for cosmetics, dry cleaning detergents, emulsion, plastic, pipelines, and suspension polymerization; emulsifier wax for polish, firefighting, germicide, metal cleaner, mold release agent; dispersant in paints and inks, paper, photography, process aid, rust preventative, soldering flux, wallpaper removal; FDA 21CFR §175.105, 175.300, 176.170, 176.210, 177.1200, 177.2800, 178.3400; APHA 100 max. clear visc. liq.; m.w. 444; sol. org. solv.; limited water sol.; sp.gr. 1.09; visc. 200 cps; flash pt. 85 C (OC); acid no. 2.5 max.; surf. tens. 28.7 dynes/cm (0.1% aq.); biodeg.; 75 ± 2% solids in water/alcohol.

Aerosol® OT-100%. [Cytec Industries] Dioctyl sodium sulfosuccinate; anionic; emulsifier, dispersant, lubricant, wetting agent, mold release agent for emulsion and suspension polymerization, drycleaning, emulsifying waxes, industrial cleaners, paints; surfactant for water-free systems; biodeg.; FDA 21CFR §175.105, 175.300, 176.170, 176.210, 177.1200, 177.2800, 178.3400; APHA 100 max. waxy solid; m.w. 444; sol. in polar and nonpolar solv.; sol. in oil, fat, and wax @ 75 C; disp. in water; sp.gr. 1.1; m.p. 153-157 C; acid no. 2.5 max.; surf. tens. 28.7 dynes/cm (0.1% aq.); 100% conc.

Aerosol® OT-B. [Cytec Industries] Dioctyl sodium sulfosuccinate, sodium benzoate; anionic; wetting agent, dispersant, solubilizer, adjuvant for agric.

chem. wettable powds.; pigment dispersant in plastics; used in face powds. and powd. shampoos; FDA 21CFR §175.105, 175.300, 176.170, 176.210, 177.1200, 177.2800, 178.3400; wh. powd.; bulk particle size 15-150 μ; m.w. 444; water-disp.; sp.gr. 1.1; m.p. < 300 C; acid no. 2.5 max.; surf. tens. 28.7 dynes/cm (0.1% aq.); 85% act., 15% sodium benzoate.

Aerosol® OT-S. [Cytec Industries] Dioctyl sodium sulfosuccinate; anionic; wetting agent, surf. tens. depressant, emulsifier for plastics, organosols, lacquers, varnishes, all org. media, dry cleaning, corrosion resistant lubricants, agric. emulsions; biodeg.; FDA 21CFR §175.105, 175.300, 176.170, 176.210, 177.1200, 177.2800, 178.3400; lt. amber, transparent liq.; limited water sol., good org. solv.; sp.gr. 1.0; visc. 200-300 cps; flash pt. (Seta CC) 134 F; surf. tens. 26 dynes/cm; 70% act. in lt. petrol. distillate.

Aerosol® TR-70. [Cytec Industries] Ditridecyl sodium sulfosuccinate; CAS 2673-22-5; EINECS 220-219-7; anionic; emulsifier, surfactant; used in emulsion polymerization of vinyl chloride and vinyl acetate, suspension polymerization of vinyl chloride; dispersant for resins, pigments, polymers, and dyes in org. systems; pigment dispersant in printing inks; rust preventative; biodeg.; FDA 21CFR §175.105, 176.180, 178.3400; clear liq.; sol. in org. media; limited water sol.; sp.gr. 0.995; dens. 8.3 lb/gal; visc. 110 cps; f.p. -40 C; surf. tens. 26 dynes/cm; 70% act. in water/alcohol.

AF 100 IND. [Harcros] Silicone antifoam; antifoamer for aq. and nonaq. systems; used in adhesive, paint, and ink mfg., sizes, industrial cooking processes, vacuum distillations, insecticides, deasphalting, and resin polymerization; FDA compliance; translucent liq.; sol. in aliphatic, aromatic, and chlorinated solvs.; insol. in water; sp.gr. 1.01; dens. 8.4 lb/gal; pour pt. < 0 F; flash pt. > 600 F (PMCC); 100% conc.

AF HL-27. [Harcros] Nonsilicone; antifoam for urea and phenolic resins, latex paints, inks, adhesives; FDA 21CFR §175.105, 176.200, 176.210, 177.1200; creamy yel. liq.; water-insol.; sp.gr. 0.95; dens. 7.9 lb/gal; pour pt. 32 F; flash pt. (PMCC) > 300 F; 100% act.

Afco-Chem B. [Adeka Fine Chem.] Sodium stearate; CAS 822-16-2; EINECS 212-490-5; lubricant for metallic sintering; lubricant and stabilizer for resins; pigment dispersant; mold release; waterproofing agent; lubricant additive.

Afco-Chem CS. [Adeka Fine Chem.] Calcium stearate; CAS 1592-23-0; EINECS 216-472-8; lubricant for metallic sintering; lubricant and stabilizer for resins; pigment dispersant; mold release; waterproofing agent; lubricant additive.

Afco-Chem CS-1. [Adeka Fine Chem.] Calcium stearate; CAS 1592-23-0; EINECS 216-472-8; lubricant for metallic sintering; lubricant and stabilizer for resins; pigment dispersant; mold release; waterproofing agent; lubricant additive.

Afco-Chem CS-S. [Adeka Fine Chem.] Calcium stearate; CAS 1592-23-0; EINECS 216-472-8; lubricant for metallic sintering; lubricant and stabilizer for resins; pigment dispersant; mold release; waterproofing agent; lubricant additive.

Afco-Chem EX-95. [Adeka Fine Chem.] Sodium and calcium stearate; lubricant for metallic sintering; lubricant and stabilizer for resins; pigment dispersant; mold release; waterproofing agent; lubricant additive.

Afco-Chem LIS. [Adeka Fine Chem.] Lithium stearate; CAS 4485-12-5; EINECS 224-772-5; lubricant for metallic sintering; lubricant and stabilizer for resins; pigment dispersant; mold release; waterproofing agent; lubricant additive.

Afco-Chem MGS. [Adeka Fine Chem.] Magnesium stearate; CAS 557-04-0; EINECS 209-150-3; lubricant for metallic sintering; lubricant and stabilizer for resins; pigment dispersant; mold release; waterproofing agent; lubricant additive.

Afco-Chem ZNS. [Adeka Fine Chem.] Zinc stearate; CAS 557-05-1; EINECS 209-151-9; lubricant for metallic sintering; lubricant and stabilizer for resins; pigment dispersant; mold release; waterproofing agent; lubricant additive.

Afflair® Lustre Pigments. [EM Industries] Mica platelets coated with titanium dioxide and/or iron oxide; luster pigments for coatings, inks, and plastics.

Aflux® 12. [Rhein Chemie] Fatty acid esters with inorg. carriers; processing promoter for NR and all SR types; also for non-black and transparent qualities; molded and extruded goods, rubber footwear, soles; off-wh. pellets; sp.gr. 1.2.

Aflux® 32. [Rhein Chemie] Thiodipropionic dilauryl ester; dispersant and internal lubricant for tech. molded and extruded rubber articles based on CM and CSM; improves processability and heat stability; wh. flakes; sp.gr. 1.02.

Aflux® 42. [Rhein Chemie] Blend of fatty acid esters and fatty alcohols; processing promoter for all rubbers incl. EPDM; molded and extruded mech. goods, footwear; suitable for continuous vulcanization, inj. molding; lt. brn. to brn. pellets; sp.gr. 0.9.

Aflux® 54. [Rhein Chemie] Pentaerythrityl tetrastearate; processing promoter for CO, ECO, FPM; molded and extruded mech. goods; decreases visc. of rubber compds., reduces sticking, promotes demolding of press-cured articles; ylsh. flakes or powd.; sp.gr. 1.0.

Aflux® R. [Rhein Chemie] Barium salt of tall oil fatty acid; dispersant and internal lubricant for highly loaded NR and SBR compds. containing carbon blk., for sulfur disp. in hard rubber, tech. molded and extruded articles, V-belts, tires, transparent vulcanizates; sl. activation of vulcanization; lt. brn. fine powd.; sp.gr. 1.2.

Aflux® S. [Rhein Chemie] Fatty acid esters bound to high-activity silica; dispersant and internal lubricant for NR and SR rubbers, molded and extruded goods, footwear, soles, light-colored and transparent goods; ylsh. wh. fine powd.; sp.gr. 1.14.

Ageflex AGE. [CPS] Allyl glycidyl ether; CAS 106-92-

3; EINECS 203-442-4; modifier for elastomer, epoxies, adhesives, fibers; reactive intermediate for coatings, sizing/finishing agent for fiberglass; silane intermediate in elec. coatings; APHA 100 color; m.w. 114.15; sp.gr. 0.970 (20/20 C); f.p. -100 C; flash pt. 57 C; 98% act.

Ageflex AMA. [CPS] Allyl methacrylate; CAS 96-05-9; EINECS 202-473-0; silane monomer intermediate; crosslinker offering two-stage polymerization, abrasion and solv. resist.; polymer modifier for high impact plastics, adhesives, acrylic elastomers, photoresists, optical polymers; AOHA 50 color; m.w. 126.12; sp.gr. 0.929 (20/20 C); flash pt. 37 C; 98% act.

Ageflex 1,3 BGDMA. [CPS] 1,3-Butylene glycol dimethacrylate; CAS 11890-98-8; crosslinker for plastisols, hard rubber rolls, cast acrylic sheet/rods, coagent for rubber compding., impregnant for metal and wood composites, adhesives, glass-reinforced plastics; APHA 50 color; m.w. 226.26; sp.gr. 0.944; 98% act.

Ageflex BGE. [CPS] Butyl glycidyl ether; CAS 2426-08-6; reactive diluent in epoxy resins, laminating, flooring, elec. casting and encapsulants; APHA 250 color; m.w. 130.19; sp.gr. 0.920 (20/20 C); flash pt. 64 C; 93% act.

Ageflex CHMA. [CPS] Cyclohexyl methacrylate; CAS 101-43-9; polymer modifier for optical lens coatings, adhesives, floor polishes, vinyl polymerization, anaerobic adhesives; APHA 150 clear liq.; m.w. 168.24; sp.gr. 0.950 (20/20 C); b.p. 60 C (2 mm); flash pt. 82 C; 98% act.; *Precaution:* combustible.

Ageflex DEGDMA. [CPS] Diethylene glycol dimethacrylate; CAS 2358-84-1; crosslinking for rubber vulcanization, moisture barrier films and coatings, photopolymer printing plates and letterpress inks, conversion coatings and adhesives; APHA 100 color; m.w. 242.27; sp.gr. 1.056 (20/20 C); flash pt. 66 C; 98% act.

Ageflex EGDMA. [CPS] Ethylene glycol dimethacrylate; CAS 97-90-5; EINECS 202-617-2; crosslinker and modifier of ABS, acrylic and PVC, ion exchange resins, encapsulation of smokeless powd., glaze coatings, dental polymers, paper processing aids, rubber modifier, adhesives, optical polymers, leather finishing, moisture barrier films; fiberglass-reinforced polyesters, emulsion polymerization; APHA 50 color; m.w. 198.22; sp.gr. 1.055 (20/20 C); f.p. -40 C; flash pt. 68 C; 98% act.

Ageflex FA-1. [CPS] Dimethylaminoethyl acrylate; CAS 2439-35-2; adhesion promoter in UV and EB cured coatings for metal, plastic, paper, and wood surfs.; catalyst for epoxy molding and extrusion resins; intermediate for water treatment chems., quat. monomers, silane coupling agents, conductive paper coatings; crosslinking agent for polyester-diisocyanate prepolymers in leather finishing; flotation agent for ester purification; APHA 100 clear liq.; m.w. 143.19; sp.gr. 0.940 (20/20 C); f.p. < -60 C; b.p. 94 C (50 mm); flash pt. 66 C; 99% act.; *Toxicology:* cause severe irritation and possible damage on contact with eyes and skin; vapor irritating to eyes, skin, nasal membranes; harmful if swallowed or inhaled; *Precaution:* corrosive.

Ageflex FA-2. [CPS] Diethylaminoethyl acrylate; CAS 2426-54-2; industrial and automotive coatings, electronic photo resists, dye additives, lube oil additives; intermediate for water treatment chems., silane coupling agents, conductive paper coatings; retention aids for paper mfg.; flocculant and coagulant; crosslinking agent for polyester-diisocyanate prepolymers in leather finishing; catalyst for epoxy molding, extrusion resins; in mfg. of copolymers for lipsticks; detergent for jet fuels; clear liq.; m.w. 171.24; sp.gr. 0.939 (20/20 C); f.p. < -60 C; b.p. 81 C (10 mm); flash pt. 90 C; *Toxicology:* cause severe irritation and possible damage on contact with eyes and skin; vapor irritating to eyes, skin, nasal membranes; harmful if swallowed or inhaled; *Precaution:* corrosive.

Ageflex FM-1. [CPS] Dimethylaminoethyl methacrylate; CAS 2867-47-2; EINECS 220-688-8; detergent and sludge dispersant in lubricants; visc. index improver; flocculant for waste water treatment; retention aid for paper mfg.; acid scavenger in PU foams; corrosion inhibitor; resin and rubber modifier; used in acrylic polishes and paints, hair prep. copolymers, sugar clarification, adhesives, water clarification; APHA 50 clear liq.; very sol. in water; sol. in org. solvs.; m.w. 157.21; visc. 1.38 cst; vapor pressure 12 mm Hg (70 C); b.p. 68.5 C (10 mm); f.p. < -60 C; 99% assay; *Toxicology:* poison; harmful if swallowed; severe eye burns and skin irritation; irritating vapor; *Storage:* store in cool area, preferably 75 F or lower away from direct sunlight; exclude moisture.

Ageflex FM-2. [CPS] Diethylaminoethyl methacrylate; CAS 105-16-8; industrial and automotive clear coatings, dye additives, intermediate for water treatment and oilfield chems., stabilizer for fuel oils, sweetening agent for various hydrocarbon oils; acrylic resin modifier for automotive industry; rubber modifier and stabilizer; retention aids for paper mfg.; water clarification of o/w emulsions; clear liq.; m.w. 185.27; sp.gr. 0.922 (20/20 C); f.p. < -60 C; b.p. 114 C (30 mm); flash pt. 77 C; *Toxicology:* severe irritant and possibly damaging to eyes and skin; vapor irritating to eyes, skin, and nasal membranes; harmful if swallowed; *Precaution:* combustible; *Storage:* store in cool, preferably below 75 F away from direct sunlight; exclude moisture.

Ageflex FM-10. [CPS] Isodecyl methacrylate; CAS 29964-84-9; pressure-sensitive adhesives, coatings for leather, textiles, paper, nonwovens, polymer modifier/stabilizer, visc. index improver, dispersion for plastics and rubber, floor waxes, potting compds., sealants, adhesives; APHA 100 clear liq.; m.w. 226.36; sp.gr. 0.878 (20/20 C); b.p. 126 C (10 mm); flash pt. 121 C; 97.5% act.

Ageflex MEA. [CPS] Methoxyethyl acrylate; CAS 3121-61-7; solv.-resist. elastomer, polyacrylate rubber, uv-curable reactive diluent, soft contact lenses, PVC impact modifier, fabric coatings, barrier coatings for polyethylene, textile coatings; APHA 50 color; m.w. 130.15; sp.gr. 1.012 (20/20 C); flash pt. 82 C; 98.5% act.

Ageflex n-HA. [CPS] n-Hexyl acrylate; CAS 2499-95-8; glass coatings, polymer modifiers, adhesives, additives for cements and sealants; APHA 150 clear liq.; m.w. 156.23; sp.gr. 0.880 (20/20 C); f.p. -45 C; b.p. 40 C (1 mm); flash pt. 71 C; 98% assay; Unverified

Ageflex T4EGDA. [CPS] Tetraethyleneglycol diacrylate; CAS 17831-71-9; fast curing crosslinking monomer providing good adhesion and flexibility, low shrinkage, and good impact str. in inks, coatings, adhesives, photo resists, and rubber prods.; APHA 150 color; sl. water-sol.; m.w. 302.33; sp.gr. 1.11 (20/20 C); f.p. < -20 C; flash pt. > 93 C.

Ageflex TBGE. [CPS] t-Butyl glycidyl ether; CAS 7665-72-7; reactive diluent in epoxy resins, corrosion inhibitor in some solvs., modifier for amines, acids and thiols; APHA 30 color; m.w. 130.19; sp.gr. 0.900 (20/20 C); f.p. -70 C; flash pt. 54 C; 99% act.

Ageflex TM 402, 403, 404, 410, 421, 423, 451, 461, 462. [CPS] Trimethylolpropane trimethacrylate blends; CAS 3290-92-4; crosslinking monomers; processing aid for extrusion and molding of plastisols and rubber compds. (improves abrasion resist., adhesion in PVC plastisols, scorch and chem. resist., elevated temp. stability); APHA < 400 color; m.w. 338.39; sp.gr. 1.06 (20/20 C).

Ageflex TMPTA. [CPS] Trimethylolpropane triacrylate; CAS 15625-89-5; crosslinking monomer for uv-cured adhesives, wood fillers, inks, coatings, dry film photo polymer resists, flexographic, offset and screen printing inks, vinyl acrylic latex paint, exterior coatings, highly crosslinked polybutadiene rubber; APHA 200 color; m.w. 296.31; sp.gr. 1.108 (20/20 C); flash pt. > 160 C.

Ageflex TMPTMA. [CPS] Trimethylolpropane trimethacrylate; CAS 3290-92-4; crosslinking monomer for coagents for wire and cable, hard rubber rolls, polybutadiene and polyethylene, moisture barrier films and coatings, plastisols and vinyl acetate latexes, adhesives, molding compds., textile prods.; APHA 100 color; m.w. 338.39; sp.gr. 1.06 (20/20 C); 99% act.

Ageflex TPGDA. [CPS] Tripropylene glycol diacrylate; CAS 68901-05-3; crosslinking monomer for uv-curable inks/coatings, floor tiles, wood coatings and fillers, adhesives, textile finishes and rubber compds.; APHA 150 color; m.w. 300.00; sp.gr. 1.039 (20/20 C); flash pt. > 100 C; 98.5% act.

Agent AT-1190. [Rhone-Poulenc Surf. & Spec.] Octylphenol/sodium; anionic; emulsifier for emulsion polymerization of vinyl acetate, acrylates, methacrylates, styrene, butadiene; for paints, adhesives, textile, paper, and industrial coatings; 30% act.

Agerite® DPPD. [R.T. Vanderbilt] Diphenyl-p-phenylene diamine; CAS 74-31-7; antioxidant for NR, SR, automotive and appliance molded goods, tires, latex; improves environemental flex and stress cracking; sl. discoloration; dk. blue-blksh. powd., 98% thru 100 mesh; sol. in acetone, toluene, chloroform; insol. in water; m.w. 260.34; dens. 1.28 ± 0.03 mg/m^3; m.p. 144-153 C.

Agerite® Geltrol. [R.T. Vanderbilt] Alkylated-arylated bisphenolic phosphite; nonstaining, nonblooming antioxidant and polymer stabilizer for rubbers (NR, SR); gel inhibitor during polymerization; for wire and cable, adhesives and cements, hot melts; amber visc. liq.; sol. in toluene, chloroform, gasoline; insol. in water; dens. 0.94 ± 0.02 mg/m^3 (60 C).

Agerite® Hipar T. [R.T. Vanderbilt] 50% Dioctylated diphenylamine, 25% diphenyl-p-phenylenediamine, 25% diphenylamine-acetone reaction prod.; antioxidant for NR and syn. rubbers; for tires, hose and belting, automotive and appliance molded goods; inhibits oxygen attack, environmental flex and stress cracking, improves heat aging; gray to blksh. powd.; sol. in acetone, toluene, chloroform; insol. in water; dens. 1.15 ± 0.03 mg/m^3; m.p. 70 C min.

Agerite® HP-S. [R.T. Vanderbilt] 65% Dioctylated diphenylamine, 35% diphenyl-p-phenylenediamine; antioxidant for NR, SR, CR for outdoor service; for tires, hose and belting, automotive and appliance molded goods, wire and cable; inhibits oxygen attack, environmental flex and stress cracking, improves heat aging; gray to blksh. rods, powd., 80% min. thru 100 mesh; sol. in acetone, toluene, chloroform; insol. in water; dens. 1.11 ± 0.03 mg/m^3; m.p. 80-100 C.

Agerite® HP-S Rodform. [R.T. Vanderbilt] Octylated diphenylamine, N,N′-diphenyl-p-phenylenediamine; rubber antioxidant; gray to blk. rods, sl. phenolic odor; negligible sol. in water; dens. 1.11 mg/m^3; 1% max. volatiles; *Toxicology:* LD50 (oral, rat) > 7000 mg/kg (octylated diphenylamine), 10,000 mg/kg (diphenyl-p-phenylenediamine); heating may generate vapors which can irritate eyes and respiratory passages; OSHA TWA 5 mg/m^3 (respirable dust), 15 mg/m^3 (total dust) recommended; *Precaution:* incompat. with strong oxidizing agents; hazardous decomp. prods.: oxides of carbon and nitrogen, aromatic and aliphatic hydrocarbons.

Agerite® MA. [R.T. Vanderbilt] Polymerized 1,2-dihydro-2,2,4-trimethylquinoline; CAS 26780-96-1; sl. staining, nonblooming antioxidant providing aging protection to XLPE, NR, SR; for wire and cable, adhesives, cements, latex applics.; inhibits oxygen attack; amber pellets, cream-tan powd., odorless; 99.5% min. thru 200 mesh; sol. in acetone, toluene, chloroform; insol. in water; m.w. 173.26; sp.gr. 1.03-1.09; m.p. 105 C min.; flash pt. (COC) 204 C; *Toxicology:* LD50 (oral, rat) 4900 mg/kg, (dermal, rabbit) > 20,000 mg/kg; nonirritating to skin and eyes; OSHA TWA 5 mg/m^3 (respirable dust), 15 mg/m^3 (total dust) recommended; *Precaution:* dust may present explosion hazard in presence of ignition source; incompat. with strong oxidizers; hazardous decomp. prods.: CO, CO_2, NO_x, aromatic/aliphatic hydrocarbons; *Storage:* keep containers closed.

Agerite® NEPA. [R.T. Vanderbilt] Mixt. of alkylated diphenylamines; nonstaining, nonblooming antioxidant for NR, SR for hot melts and latex applics.; inhibits oxygen attack, improves heat aging; liq.

Agerite® Resin D. [R.T. Vanderbilt] Polymerized 1,2-dihydro-2,2,4-trimethylquinoline; CAS 26780-96-1; sl. staining, nonblooming antioxidant for NR, SR, XLPE for tires, hose, belting, automotive and appliance molded goods, wire and cable, latex applics.; inhibits oxygen attack, improves heat aging; amber small pellets, cream-tan powd., 98% min. thru 100 mesh; sol. in acetone, toluene, chloroform; insol. in water; m.w. 173.26; dens. 1.06 ± 0.03 mg/m³; soften. pt. 74 C min.; flash pt. (COC) 204 C; *Toxicology:* LD50 (oral, rat) 4900 mg/kg, (dermal, rabbit) > 20,000 mg/kg; nonirritating to skin and eyes; OSHA TWA 5 mg/m³ (respirable dust), 15 mg/m³ (total dust) recommended; *Precaution:* dust may present explosion hazard in presence of ignition source; incompat. with strong oxidizers; hazardous decomp. prods.: CO, CO_2, NO_x, aromatic/aliphatic hydrocarbons; *Storage:* keep containers closed.

Agerite® Stalite. [R.T. Vanderbilt] Octylated diphenylamines; nonstaining, nonblooming antioxidant for NR, SR, CR; scorch retarder for CR; gel inhibitor polymer stabilizer; for hose and belting, automotive and appliance molded goods, wire and cable, footwear, adhesives, cements, hot melts, latex; inhibits oxygen attack, improves heat aging; red. brn. visc. liq.; sol. in alcohol, toluene, gasoline; insol. in water; m.w. 281.45; dens. 0.99 ± 0.02 mg/m³.

Agerite® Stalite S. [R.T. Vanderbilt] Octylated diphenylamine; nonstaining, nonblooming antioxidant for NR, SR, CR; scorch retarder for CR; for hose, belting, automotive and appliance molded goods, wire and cable, footwear, adhesives, cements, hot melts, latex; lt. tan powd.; sol. in alcohol, toluene, gasoline; insol. in water; m.w. 393.66; dens. 1.02 ± 0.03 mg/m³; m.p. 89-103 C.

Agerite® Superflex. [R.T. Vanderbilt] Diphenylamine-acetone reaction prod.; CAS 9003-79-6; nonblooming antioxidant for NR, SR, tires, automotive and appliance molded goods, wire and cable; inhibits oxygen attack, environmental flex and stress cracking, improves heat aging; dk. brn. liq.; sol. in acetone, toluene, chloroform; insol. in water; dens. 1.10 ± 0.02 mg/m³.

Agerite® Superflex Solid G. [R.T. Vanderbilt] Diphenylamine-acetone reaction prod. on an inert carrier; CAS 9003-79-6; nonblooming antioxidant for NR, SR, tires, automotive and appliance molded goods, wire and cable; inhibits oxygen attack, environmental flex and stress cracking, improves heat aging; powd.; sol. in acetone, toluene, chloroform; insol. in water; dens. 1.33 ± 0.03 mg/m³; 75% act.

Agerite® Superlite. [R.T. Vanderbilt] Polybutylated bisphenol A; CAS 68784-69-0; nonstaining, nonblooming antioxidant for NR, SR, latex, automotive and appliance molded goods, tires, wire and cable, footwear, adhesives, cements, hot melts; gel inhibitor polymer stabilizer; amber liq.; sol. in toluene, chloroform, gasoline; insol. in water; dens. 0.945-0.965 mg/m³.

Agerite® Superlite® Emulcon®. [R.T. Vanderbilt] Polybutylated bisphenol A, triisobutylene, diisobutylene, potassium oleate; antioxidant; amber liq., phenolic odor; disp. in water; dens. 0.95-0.97 mg/m³; flash pt. (CC) 61.5 C; *Toxicology:* may cause skin and eye irritation; *Precaution:* combustible; incompat. with strong oxidizers; hazardous decomp. prods.: carbon oxides, aromatic/aliphatic hydrocarbons; *Storage:* store in closed containers above 22 C to prevent crystallization; keep away from heat, sparks, open flame.

Agerite® Superlite Solid. [R.T. Vanderbilt] Polybutylated bisphenol A on inert carrier; CAS 68784-69-0; nonstaining, nonblooming antioxidant for NR, SR, automotive and appliance molded goods, tires, wire and cable, footwear, adhesives, cements, hot melts; gray-tan powd.; sol. in toluene, chloroform, gasoline; insol. in water; dens. 1.26 ± 0.03 mg/m³; 73% act.

Agerite® White. [R.T. Vanderbilt] sym-Di-β-naphthyl-p-phenylenediamine; CAS 93-46-9; nonstaining antioxidant for NR, SR, automotive and appliance molded goods, wire and cable, adhesives, cements, hot melts, latex; inhibits oxygen attack; lt. tan to gray powd., sl. amine odor; negligible sol. in water; dens. 1.22-1.28 mg/m³; m.p. 224-230 C; *Toxicology:* LD50 (oral, rat) 4500 mg/kg; skin sensitizer; OSHA TWA 5 mg/m³ (respirable dust), 15 mg/m³ (total dust) recommended; contains trace amts. of a carcinogen; *Precaution:* incompat. with strong oxidizers; hazardous decomp. prods.: oxides of nitrogen and carbon, aromatic/aliphatic hydrocarbons.

Agerite® White White. [R.T. Vanderbilt] sym-Di-β-naphthyl-p-phenylenediamine; CAS 93-46-9; nonstaining antioxidant for NR, SR, automotive and appliance molded goods, wire and cable, adhesives, cements, hot melts, latex; inhibits oxygen attack; lt. tan to gray powd., sl. amine odor; negligible sol. in water; dens. 1.22-1.28 mg/m³; m.p. 224-230 C; *Toxicology:* LD50 (oral, rat) 4500 mg/kg; skin sensitizer; OSHA TWA 5 mg/m³ (respirable dust), 15 mg/m³ (total dust) recommended; contains trace amts. of a carcinogen; *Precaution:* incompat. with strong oxidizers; hazardous decomp. prods.: oxides of nitrogen and carbon, aromatic/aliphatic hydrocarbons.

AgGlad™ Filament 16. [PQ Corp.] "E" glass fibers clad with thin layer of silver; conductive reinforcement providing higher impact str. and dimensional stability, reduced heat distort. to conductive plastic composites, adhesives, coatings, caulks, sealants, putties, gasketing materials, EMI/RFI applics.; off-wh. floccular, odorless; $^1/_{16}$ in. screen size; sp.gr. 2.8-3.0; bulk dens. 40-50 lb/ft³; m.p. 961.9 C (coating); bulk resist. 0.008 ohm-cm; *Toxicology:* absorp. of silver metal causes skin pigmentation (argyria); may cause skin or eye irritation or allergic reactions; dust may cause respiratory irritation;

Precaution: may burn or explode if contacted with oxidizers or dispersed in air.

AgGlad™ Filament 32. [PQ Corp.] "E" glass fibers clad with thin layer of silver; conductive reinforcement providing higher impact str. and dimensional stability, reduced heat distort. to conductive plastic composites, adhesives, coatings, caulks, sealants, putties, gasketing materials, EMI/RFI applics.; off-wh. floccular, odorless; $^1/_{32}$ in. screen size; sp.gr. 2.8-3.0; bulk dens. 65-70 lb/ft^3; m.p. 961.9 C (coating); bulk resist. 0.008 ohm-cm; *Toxicology:* absorp. of silver metal causes skin pigmentation (argyria); may cause skin or eye irritation or allergic reactions; dust may cause respiratory irritation; *Precaution:* may burn or explode if contacted with oxidizers or dispersed in air.

AgGlad™ Platelet 8. [PQ Corp.] Milled flakes of chemical resist. "C" glass with thin layer of silver; conductive reinforcement providing higher impact str. and dimensional stability, reduced heat distort. to conductive plastic composites, adhesives, caulks, sealants, putties, gasketing materials, EMI/RFI applics.; silver flake, odorless; $^1/_8$ in. screen size; sp.gr. 2.9-3.1; bulk dens. 15-20 lb/ft^3; m.p. 961.9 C (coating); bulk resist. 0.008 ohm-cm; *Toxicology:* absorp. of silver metal causes skin pigmentation (argyria); may cause skin or eye irritation or allergic reactions; *Precaution:* may burn or explode if contacted with oxidizers or dispersed in air.

AgGlad™ Platelet 64. [PQ Corp.] Milled flakes of chemical resist. "C" glass with thin layer of silver; conductive reinforcement providing higher impact str. and dimensional stability, reduced heat distort. to conductive plastic composites, adhesives, caulks, sealants, putties, gasketing materials, EMI/RFI applics.; silver flake, odorless; $^1/_{64}$ in. screen size; sp.gr. 2.9-3.1; bulk dens. 25-30 lb/ft^3; m.p. 961.9 C (coating); bulk resist. 0.008 ohm-cm; *Toxicology:* absorp. of silver metal causes skin pigmentation (argyria); may cause skin or eye irritation or allergic reactions; *Precaution:* may burn or explode if contacted with oxidizers or dispersed in air.

AgGlad™ TW Microspheres. [PQ Corp.] Silver coated thick-walled hollow spheres; filler exhibiting high conduct. at one-third the wt. of silver flake or powd.; provides high compr. str. and exc. processing to conductive thermosets, conductive gaskets, sealants, caulks, and adhesives, paints and coatings, inks; applics. where EMI/RFI control is required; gray thick-walled spheres; 3.0 μ avg. particle size; sp.gr. 3.8; bulk dens. 60-70 lb/ft^3; hardness (Moh) 7; dry bulk resist. 0.003 ohm-cm; *Toxicology:* absorp. of silver metal causes skin pigmentation (argyria); may cause skin or eye irritation or allergic reactions; contains < 5% cryst. silica—dust may cause chronic lung disease; *Precaution:* may burn or explode if contacted with oxidizers or dispersed in air.

Agitan 260. [Münzing Chemie GmbH] Silicone-free defoamer for emulsion paints, gloss emulsion paints, syn. renderings, adhesives, silicate paints, aq. epoxy resin systems; BGA compliance.

Agitan 281. [Münzing Chemie GmbH] Silicone-free defoamer for emulsion paints, emulsion polymers, adhesives, aq. systems, silicate paints.

Agitan 295. [Münzing Chemie GmbH] Silicone-free defoamer for emulsion paints, gloss emulsion paints, printing inks, emulsion polymers, adhesives, aq. systems, polymerization processes, wood preservative stains.

Agitan 296. [Münzing Chemie GmbH] Silicone-free defoamer for emulsion paints, emulsion polymers, aq. systems.

Agitan 301. [Münzing Chemie GmbH] Silicone defoamer; biodeg. defoamer for emulsion paints, emulsion polymers, syn. renderings, adhesives; BGA compliance.

Agitan 305. [Münzing Chemie GmbH] Silicone-free defoamer for emulsion paints, emulsion polymers, syn. renderings, adhesives; BGA compliance.

Agitan 650. [Münzing Chemie GmbH] Silicone-free defoamer for emulsion polymers, water-dilutable systems, pigmented and unpigmented systems; easily emulsifiable; BGA compliance.

Agitan 655. [Münzing Chemie GmbH] Silicone-free defoamer for emulsion polymers, water-dilutable systems, pigmented and unpigmented systems; easily emulsifiable; BGA compliance.

Agitan 702. [Münzing Chemie GmbH] Silicone-free defoamer for hydrosols, emulsion paints, emulsion polymers, aq. systems.

Agitan 703 N. [Münzing Chemie GmbH] Silicone-free defoamer for emulsion paints, water paints, emulsion polymers, aq. resins, printing inks; easily emulsifiable; BGA compliance.

Agitan E 255. [Münzing Chemie GmbH] Silicone emulsion; defoamer for emulsion paints, gloss emulsion paints, syn. renderings, adhesives, aq. systems, glazes, aq. printing inks, polymerization processes; BGA compliance.

Agitan E 256. [Münzing Chemie GmbH] Silicone emulsion; defoamer for emulsion paints, gloss emulsion paints, syn. renderings, adhesives, aq. systems, glazes, aq. printing inks, aq. systems, polymerization processes, aq. epoxy resin systems, corrosion prevention.

Airvol® 205. [Air Prods./Polymers] Polyvinyl alcohol, partially hydrolyzed; CAS 25213-24-5; binder, carrier, compounding agent, dispersant, stabilizer, protective colloid in polymerizations; for textiles, paper, adhesives, cement/plaster additive, peelable caulks, ceramics, strippable coatings, mold release, nonwovens; wh. to cream gran. powd.; sp.gr. 1.27-1.31; bulk dens. 40 lb/ft^3; visc. 5.2-6.2 cps (4% aq., 20 C); pH 4.5-6.5 (4% aq.); ref. index 1.55 (20 C); 87-89% hydrolysis.

Airvol® 523. [Air Prods./Polymers] Polyvinyl alcohol, partially hydrolyzed; CAS 25213-24-5; binder, carrier, compounding agent, dispersant, stabilizer, protective colloid in polymerizations; for textiles, paper, adhesives, cement/plaster additive, films, ceraics, molded prods., mold release; visc. 23-27 cps (4% aq., 20 C); pH 4-6 (4% aq.); 87-89% hydrolysis.

Airvol® 540. [Air Prods./Polymers] Polyvinyl alcohol, partially hydrolyzed; CAS 25213-24-5; binder, carrier, compounding agent, dispersant, stabilizer, protective colloid in polymerizations; for textiles, paper, adhesives, cement/plaster additive, films, molded prods., nonwovens; visc. 45-55 cps (4% aq., 20 C); pH 4-6 (4% aq.); 87-89% hydrolysis.

Airvol® 805. [Air Prods./Polymers] Polyvinyl alcohol; improved grade providing processing advantages, superior water sol., lower foaming, and reduced gel content in emulsion polymerization applics.; FDA 21CFR §175.105, 175.300, 176.170, 176.180; BGA compliances; wh. to cream gran. powd.; sp.gr. 1.27-1.31; bulk dens. 40 lb/ft^3; visc. 5.2-6.2 cps (4% aq., 20 C); pH 4.5-6.5 (4% aq.); ref. index 1.55 (20 C); 87-89% hydrolysis.

Airvol® 823. [Air Prods./Polymers] Polyvinyl alcohol; improved grade providing processing advantages, superior water sol., lower foaming, and reduced gel content in emulsion polymerization applics.; FDA 21CFR §175.105, 175.300, 176.170, 176.180; BGA compliances; wh. to cream gran. powd.; sp.gr. 1.27-1.31; bulk dens. 40 lb/ft^3; visc. 23-27 cps (4% aq., 20 C); pH 4-6 (4% aq.); ref. index 1.55 (20 C); 87-89% hydrolysis.

Airvol® 840. [Air Prods./Polymers] Polyvinyl alcohol; improved grade providing processing advantages, superior water sol., lower foaming, and reduced gel content in emulsion polymerization applics.; FDA 21CFR §175.105, 175.300, 176.170, 176.180; BGA compliances; wh. to cream gran. powd.; sp.gr. 1.27-1.31; bulk dens. 40 lb/ft^3; visc. 45-55 cps (4% aq., 20 C); pH 4-6 (4% aq.); ref. index 1.55 (20 C); 87-89% hydrolysis.

Ajicure® MY-24 [Ajinomoto] Proprietary; accelerator for latent epoxy resin systems providing high storage stability, longer pot life; curable at lower temps.; pale yel. powd.; 8 μm avg. particle size; sol. >0.01 g/100 g in water; sp.gr. 1.27; soften. pt. 100-130 C; pH 8.8 (10% suspension); *Toxicology:* LD50 (oral, mice) 20 g/kg; sl. skin irritant; nonsensitizing to skin.

Ajicure® PN-23. [Ajinomoto] Proprietary; accelerator for latent epoxy resin systems providing high storage stability, longer pot life; curable at lower temps.; pale yel. powd.; 8 μm avg. particle size; sol. > 20g/100 g in NMP, DMSO, < 0.01g/100 g in water; sp.gr. 1.28; soften. pt. 100-130 C; pH 9.2 (10% susp.); *Toxicology:* LD50 (oral, mice) 1.23 g/kg; sl. skin irritant; nonsensitizing to skin.

Akrochem® 9930 Zinc Oxide Transparent. [Akrochem] Precipitated basic zinc carbonate; accelerator-activator for transparent nat. and syn. rubber goods, adhesives; wh. powd.; sp.gr. 3.5; m.p. 168 C min.; 70-72% zinc oxide.

Akrochem® Accelerator 40B Liq. [Akrochem] Aldehyde amine; accelerator for rubber; amber to orange liq., aromatic odor; sp.gr. 0.95; visc. 135 cs; flash pt. < 200 F; 40 ± 3% dihydropyridine (act.).

Akrochem® Accelerator BZ Powder. [Akrochem] Zinc dibutyl dithiocarbamate; CAS 136-23-2; nondiscoloring, nonstaining accelerator for EPDM, NR and SR latexes; stabilizer and antioxidant in uncured rubber; off-wh. to cream powd. or pellets; sol. in acetone; sp.gr. 1.25; m.p. 104 C; 95% min. purity.

Akrochem® Accelerator EZ. [Akrochem] Zinc diethyl dithiocarbamate; CAS 14323-55-1; nondiscoloring, nonstaining accelerator for NR and SR latexes; off-wh. powd.; sp.gr. 1.5; m.p. 172 C min.; 98% min. purity.

Akrochem® Accelerator MF. [Akrochem] 2-Benzothiazyl-N-morpholine disulfide; CAS 102-77-2; accelerator in natural and syn. rubbers, e.g., tire and mech. goods; yel-br. powd.; sp.gr. 1.48; m.p. 125 C.

Akrochem® Accelerator MZ. [Akrochem] Zinc dimethyl dithiocarbamate; CAS 137-30-4; nondiscoloring, nonstaining accelerator for NR, IR, BR, SBR, IIR, and EPDM rubbers and NR latex; off-wh. powd. or pellets; sp.gr. 1.70; m.p. 240 C min.; 98% min. purity.

Akrochem® Accelerator R. [Akrochem] 4,4′-Dithiodimorpholine; CAS 103-34-4; accelerator for uses where a nonblooming or nonstaining sulfur donor is required; used for EV or semi-EV compds., in syn. and nat. rubbers; wh. to cream powd., very slight odor; sol. in acetone, benzene, ethyl acetate; sp.gr. 1.36; m.p. 121 ± 5 C.

Akrochem® Accelerator VS. [Akrochem] Zinc salt of dibutyl dithiophosphoric acid on silica carrier; nonblooming and nonstaining accelerator for EPDM compds. and other sulfur-curable elastomers, esp. hose and belt applics.; lt. gray powd.; sol. in naphtha, benzene, min. oils, and ethanol; sp.gr. 1.42; 62% act. on silica.

Akrochem® Accelerator ZIPPAC. [Akrochem] Zinc-amine dithiophosphate complex coated with min. oil; accelerator; used with thiazoles and thiurams for fast cure rates in ENB-type EPDM's; nondiscoloring, nonstaining; lt. gray powd.; sp.gr. 1.05; m.p. 100 C min.

Akrochem® Antioxidant 12. [Akrochem] Butylated reaction prod. of p-cresol and dicyclopentadiene; CAS 68610-51-5; EINECS 271-867-2; nonstaining antioxidant, stabilizer, and antiozonant in polymers incl. nat. and syn. polyisoprene, neoprene, nitrile, SBR rubber, and latex; FDA 21CFR §175.105, 177.2600; amber pellet, flakes or wh. powd.; sp.gr. 1.10; m.p. 115 C.

Akrochem® Antioxidant 16 Liq. [Akrochem] Styrenated phenol; CAS 61788-44-1; low volatility, high efficiency, nonstaining, nondiscoloring antioxidant used in solid rubber, latex and plastic compounding, SBR latex compding. urethanes, ABS, and EPDM; straw-colored liq.; sol. in toluene, ethyl acetate, chloroform, gasoline, hexane; easily emulsifiable in water; sp.gr. 1.08 ± 0.02; visc. 200 poise; *Storage:* exc. storage stability.

Akrochem® Antioxidant 16 Powd. [Akrochem] 65% Antioxidant 16 Liq. (styrenated phenol) on inert binder; low volatility, high efficiency, nonstaining, nondiscoloring antioxidant used in solid rubber, latex and plastic compounding, SBR latex compding. urethanes, ABS, and EPDM; off-wh. powd.; sol. in toluene, ethyl acetate, chloroform, gasoline, hexane; easily emulsifiable in water; sp.gr. 1.23; *Storage:* exc. storage stability.

Akrochem® Antioxidant 32. [Akrochem] Hindered phenolic; low volatility, high efficiency, nonstaining, nondiscoloring antioxidant for wh. or lt. colored rubber compds., latex applics., and plastics; for NR, BR, B/S copolymers, polyisoprene, CR, butyl, nitrile, etc.; pale yel. powd., mild phenolic odor; sol. in aliphatics, aromatics, esters, ketones; sp.gr. 1.05; soften. pt. 125 C; *Storage:* exc. storage stability.

Akrochem® Antioxidant 33. [Akrochem] Hindered phenolic; low volatility, high efficiency, nonstaining, nondiscoloring antioxidant for wh. or lt. colored elastomers, latex, and plastics; for NR, BR, B/S copolymes, IR, CR, butyl, nitrile, etc.; wh. fine powd., pract. odorless; sol. in aliphatics, aromatics, esters, and ketones; sp.gr. 1.04; soften. pt. 100 C min; *Storage:* exc. storage stability.

Akrochem® Antioxidant 36. [Akrochem] Polymeric hindered phenolic; nonstaining antioxidant for rubber compds., SBR, butyl, CR, polyisoprene, nitrile, chlorinated rubber, polyolefins, ethyl cellulose, Elvac, PVC, epoxy, and waxes; wh. fine powd.; sol. in naphtha, toluene, and IPA; sp.gr. 1.05; m.p. (R&B) 148 C.

Akrochem® Antioxidant 43. [Akrochem] Hindered phenolic; low volatility, high efficiency, nonstaining, nondiscoloring antioxidant for wh. or lt. colored elastomers, latex, and plastics; for NR, BR, B/S copolymes, IR, CR, butyl, nitrile, etc.; lt. yel. flakes, pract. odorless; sol. in aliphatics, aromatics, esters, and ketones; sp.gr. 1.04; soften. pt. 100-110 C; *Storage:* exc. storage stability.

Akrochem® Antioxidant 58. [Akrochem] 2-Mercapto-4(5)-methyl benzimidazole, zinc salt; nonstaining antioxidant for use in wh. or dk. colored rubber compds. to improve heat resist.; protects against oxygen, heat crazing, and rubber poisoning, but not effective for antiflexcracking or antiozonant conditions; brightener in wh. goods @ < 0.5 phr; improves dimensional stability of extrudates; synergistic with other phenolic or amine antioxidants; wh. powd.; sp.gr. 1.75; m.p. decomp. > 300 C; *Storage:* 2 yr storage stability.

Akrochem® Antioxidant 60. [Akrochem] 2-Mercapto-4(5)-methyl benzimidazole; nonstaining, nondiscoloring antioxidant for natural and syn. rubbers, except CR; exc. protection against heat and oxygen aging; protects against rubber poisons; imparts steam resist.; brightening effect in transparent goods; synergistic with amine or phenolic antioxidants; sl. retarding effect in sulfur cures; optimum protection in thiuram and dithiocarbamate systems; yel.-wh. powd.; disp. readily in rubber mixes; sp.gr. 1.25; m.p. 290 C min.; *Storage:* > 2 yr storage stability.

Akrochem® Antioxidant 235. [Akrochem] 2,2′-Methylenebis (4-methyl-6-t-butylphenol); CAS 119-47-1; low volatility, nonstaining antioxidant with resist. to migration; very good protection against oxidation; for wh. or lt. colored rubber stocks, such as medical equip., rubber thread, latex compds., NR, BR, B/S copolymers, IR, CR, plastics; off-wh. cryst. powd.; sol. in ethanol, acetone, ethylene chloride, benzene; sl. sol. in n-hexane; insol. in water; sp.gr. 1.04; m.p. 124 C min.; *Storage:* > 2 yr storage stability.

Akrochem® Antioxidant 1010. [Akrochem] Tetrakis [methylene(3,5-di-t-butyl-4-hydroxyhydrocinnamate)] methane; CAS 6683-19-8; EINECS 229-722-6; low volatility, nonstaining antioxidant imparting exc. heat stability and color retention to plastics (PE, PS, acetal, PVC, methacrylic, PC, polyester, polymethylpentene), rubber (SBR, EPR, EPDM, BR, neoprene, nitrile, IR), latex, varnishes; adhesives, TPR, hot melts; wh. powd.; sol. 71% in chloroform, 56% in benzene, 47% in acetone, 46% in ethyl acetate, 0.01% in water; m.w. 1177.7; m.p. 110-125 C.

Akrochem® Antioxidant 1076. [Akrochem] Stearyl 3-(3,5-di-t-butyl-4-hydroxyphenyl) propionate; CAS 2082-79-3; EINECS 218-216-0; low volatility, nonstaining antioxidant imparting exc. heat stability and color retention to plastics (PE, PS, acetal, PVC, methacrylic, PC, polyester, polymethylpentene), rubber (SBR, EPR, EPDM, BR, neoprene, nitrile, IR), latex, varnishes; adhesives, TPR, hot melts; wh. powd.; sol. 57% in chloroform and benzene, 42% in ethyl acetate, 31% in hexane, 26% in acetone, 0.01% in water; m.w. 530.9; m.p. 49-54 C.

Akrochem® Antioxidant BHT. [Akrochem] BHT; CAS 128-37-0; EINECS 204-881-4; nonstaining antioxidant, stabilizer in rubber, plastics (PU, polyvinyls, polyester, ABS, EVA, PP, LDPE, HDPE, HIPS, polyamide, PC), pkg. materials, waxes, insecticides, syn. lubricants, paints; protects and heat and oxygen; resist. to hydrolysis; wh. cryst., sl. odor; sol. in most hydrocarbon solvs.; insol. in water, but easily emulsified; sp.gr. 1.03; bulk dens. 38-42 lb/ft^3; 99% purity; *Storage:* stable under normal storage conditions.

Akrochem® Antioxidant DQ. [Akrochem] 2,2,4-Trimethyl-1,2-dihydroquinoline, polymerized; CAS 26780-96-1; staining, discoloring, nonblooming antioxidant in rubber goods requiring resistance to high temps.; effective for NR, IR, SBR, EPR/EPDM, BR; copper and manganese inhibitor; antiozonant for CR; some antiflexcracking props. in NR; improves scorch resist. but increases cure time at higher concs.; yel. to amber powd. and pellet, very weak, char. odor; sol. in acetone, ethyl acetate, alcohol, benzene, methylene chloride, and CCl_4; insol. in water; sp.gr. 1.095 ± 0.02; m.p. 75 C; *Storage:* exc. storage stability.

Akrochem® Antioxidant DQ-H. [Akrochem] 2,2,4-Trimethyl-1,2-dihydroquinoline, polymerized; CAS 26780-96-1; staining, discoloring, nonblooming antioxidant in rubber goods requiring resistance to high temps.; effective for NR, IR, SBR, EPR/EPDM, NBR, BR; copper and manganese inhibitor; antiozonant for CR; some antiflexcracking props. in NR; improves scorch resist. but increases cure time at higher concs.; yel. to amber powd. and pellet, very weak, char. odor; sol. in acetone, ethyl acetate, alcohol, benzene, methylene chloride, and CCl_4; insol. in water; sp.gr. 1.095 ± 0.02; m.p. 107 C; *Storage:* exc. storage stability.

Akrochem® Antioxidant PANA. [Akrochem] Phenyl-α-naphthylamine; CAS 90-30-3; staining, discoloring, nonblooming antioxidant for rubber prods.; provides antiflexcracking under dynamic stress, protection against oxidation and heat; for NR, IR, SBR, NBR, BR, CR; brn. to dk. violet cryst. lumps or flakes; sol. in benzene, acetone, CCl_4, ethyl acetate, methylene chloride, and ethanol; insol. in water; sp.gr. 1.2; m.p. 50 C min; *Storage:* unlimited storage stability.

Akrochem® Antioxidant S. [Akrochem] Octylated diphenylamine; staining, nondiscoloring antioxidant protecting dynamically stressed rubber prods.; antiozonant; antiflexcracking; for CR, NR, IR, SBR, BR, NBR; increases scorch resist. and extends cure time @ high concs.; tan to brn. powd./gran./pellets; sol. in acetone, benzene, methylene chloride, and naphtha; insol. in water; sp.gr. 1.00; m.p. 88 C min; *Storage:* > 2 yr storage stability if stored in closed container under cool, dry conditions (77 F).

Akrochem® Antiozonant MPD-100. [Akrochem] Mixed diaryl p-phenylene diamine; antiozonant, antidegradant, antioxidant, and antiflexcracking agent for most diene polymers; brn. flakes; sp.gr. 1.20; m.p. 90-105 C.

Akrochem® Antiozonant PD-2. [Akrochem] N-(1,3-dimethyl butyl)-N′-phenyl-p-phenylene diamine; CAS 793-24-8; antiozonant protecting rubber polymers against heat, oxidation, and flex cracking; copper, manganese inhibitor and SBR stabilizer; brn. to violet-brn. mass; aromatic odor; sol. in benzene, CCl_4, methylene chloride, acetone, ethyl acetate, and ethyl alcohol; insol. in water; sp.gr. 1.10; m.p. 45 C min.

Akrochem® BBTS. [Akrochem] N-t-Butyl-2-benzothioazole sulfenamide; CAS 95-31-8; delayed-action accelerator for natural and syn. rubbers; lt. tan powd.; sp.gr. 1.29 ± 3; m.p. 104 C min.

Akrochem® Calcium Stearate. [Akrochem] Calcium stearate; CAS 1592-23-0; EINECS 216-472-8; mold release agent; wh. fine powd.; 98.5% thru 325 mesh; apparent dens. 1.9 lb/gal; m.p. 145-160 C.

Akrochem® CBTS. [Akrochem] N-Cyclohexyl-2-benzothiazole sulfenamide; CAS 95-33-0; delayed-action sulfenamide accelerator for natural, reclaim, and syn. rubbers; nondiscoloring; wh. to lt. gray powd. and pellets; disp. readily at temps. as low as 160 F; sp.gr. 1.28; m.p. 94-102 C.

Akrochem® Cu.D.D. [Akrochem] Copper dimethyl dithiocarbamate; CAS 137-29-1; accelerator for butyl rubber, EPDM rubbers; brn. powd.; partly sol. in acetone and toluene; sp.gr. 1.75; m.p. > 300 C; 21% Cu.

Akrochem® Cu.D.D.-PM. [Akrochem] Copper dimethyl dithiocarbamate disp. in a polymeric binder; CAS 137-29-1; dispersion form of Akrochem Cu.D.D.; brn. pellets; sp.gr. 1.4; 75% disp.

Akrochem® CZ-1. [Akrochem] Dimethyl cyclohexyl ammonium dibutyl dithiocarbamate; accelerator for NR, SBR, or latexes; pale yel. to lt. tan crystals and lumps, strong amine odor; sp.gr. 1.03; m.p. 52.5 C; 98% purity.

Akrochem® DCBS. [Akrochem] Benzothiazyl 1-2-dicyclohexyl sulfenamide; accelerator for rubber industry.

Akrochem® DCP-40C. [Akrochem] Dicumyl peroxide absorbed on calcium carbonate; accelerator for syn. elastomers; wh. powd. or pellets; sp.gr. 1.60; 40 ± 1% peroxide.

Akrochem® DCP-40K. [Akrochem] Dicumyl peroxide absorbed on kaolin clay; accelerator for syn. elastomers; wh. powd. or pellets; sp.gr. 1.56; 40 ± 1% peroxide.

Akrochem® D.E.T.U. Accelerator. [Akrochem] N,N′-diethylthiourea; CAS 105-55-5; accelerator for high-temp.-curing of mercaptan-modified CR (W-type neoprenes) for continuous prod. of solid and cellular extruded goods; nondiscoloring, nonstaining; wh. to off-wh. flake; m.w. 132.2; sol. in water and aromatic hydrocarbons; sp.gr. 1.1; m.p. 74 C min.

Akrochem® DOTG. [Akrochem] Diorthotolyl guanidine; CAS 97-39-2; accelerator; provides activation for MBTS, MBT, and other thiazole accelerators in nat. rubber, SBR, NBR, and CR; wh. nondusting powd., weak odor, bitter taste; sol. in acetone, ethyl alcohol, ethyl acetate, and methylene chloride; sp.gr. 1.18; m.p. 167 C min.

Akrochem® DPG. [Akrochem] Diphenyl guanidine; CAS 102-06-7; accelerator-activator for nat. rubber, SBR, and NBR; wh. nondusting powd. or pellets, weak odor, bitter taste; sol. in acetone, methylene chloride, ethyl acetate, ethyl alcohol, and benzene; sp.gr. 1.19; m.p. 142 C min.

Akrochem® DPTT. [Akrochem] Dipentamethylene thiuram hexasulfide; CAS 120-54-7; very act. sulfur-bearing accelerator imparting heat resistance to sulfurless compds.; primary accelerator for Hypalon, butyl, and EPDM; vulcanizing agent for heat-resistant latex; nondiscoloring, nonstaining; lt. gray powd.; sp.gr. 1.50; m.p. 234 F min.; 25% avail. sulfur.

Akrochem® Hydrated Alumina. [Akrochem] Alumina trihydrate; CAS 1333-84-2; inert filler and flame-retardant smoke suppressant for latex foam, rubber, carpet backing, epoxies, reinforced polyesters, phenolics, and urethane foam; wh. powd.; avail. in grades with median particle sizes of 3.0 μ, 4.0 μ, and 8.0 μ resp.; avail. in extra wh. grades noted with suffix EW; sp.gr. 2.42; GE brightness 92-93 (reg. grade), 95-97 (EW); 65% alumina.

Akrochem® MBT. [Akrochem] 2-Mercaptobenzothiazole; CAS 149-30-4; EINECS 205-736-8; very act., nondiscoloring org. accelerator for rubbers; cream to lt. yel. powd.; disp. readily; sp.gr. 1.51; m.p. 165-180 C.

Akrochem® MBTS. [Akrochem] Benzothiazyl disulfide; CAS 120-78-5; nonstaining accelerator; very act. at temps. above 280 F; cream to lt. yel. powd.; disp. readily; sp.gr. 1.50; m.p. 160-170 C.

Akrochem® MBTS Pellets. [Akrochem] Benzothiazyl disulfide; CAS 120-78-5; nonstaining accelerator; very act. at temps. above 280 F; cream to lt. yel. 2-mm pellets; disp. readily; sp.gr. 1.50; m.p. 158 C min.

Akrochem® OBTS. [Akrochem] N-Oxydiethylene-2-benzothiazole sulfenamide; CAS 102-77-2; primary accelerator for natural, SBR, nitrile, and other general-purpose rubbers; tan dustless flake; sp.gr. 1.34 ± 0.03; m.p. 76-88 C.

Akrochem® OMTS. [Akrochem] N-Oxydiethylene-2-benzothiazole-sulfenamide; CAS 102-77-2; delayed-action accelerator for SBR, NR, and nitrile rubbers; tan/brn. powd.; sol. in chloroform; sp.gr. 1.40 ± 0.03; m.p. 70 C min.

Akrochem® P 37. [Akrochem] Cashew nut oil-modified phenol-formaldehyde two-step thermoplastic resin; tackifier and plasticizer for nitrile rubber; dk. brn. visc. liq.; sol. in ketones and blends of alcohols and aromatics; sp.gr. 1.10; visc. ≈ 3,500,000.

Akrochem® P 40. [Akrochem] Thermosetting two-step phenolic resin; tackifier; hardening and reinforcing agent for nitrile, nitrile-SBR blends, Hypalon, and BR polymers; finely pulverized; 98% min. thru 200 mesh; sol. in alcohols and ketones; sp.gr. 1.18 (cured); m.p. 82-93 C.

Akrochem® P 49. [Akrochem] Thermosetting two-step phenolic resin, oil-modified, and supplied in the therm; tackifier recommended for use in SBR rubber; plasticizing action during processing; amber flake; sol. in ketones and blends of alcohols and aromatics; 98% min. thru 200 mesh; sp.gr. 1.16; R&B m.p. 95 C.

Akrochem® P 55. [Akrochem] Thermosetting two-step phenolic resin, oil-modified; tackifier for use in SBR rubber; plasticizer during processing; reinforcing agent after cure; off-wh. finely pulverized; 98% thru 200 mesh; sol. in ketones and blends of alcohol and aromatics; sp.gr. 1.17; Cap. tube m.p. 73 C.

Akrochem® P 82. [Akrochem] Thermosetting two-step phenolic resin; plasticizer; hardening and reinforcing agent for nitrile, nitrile-SBR blends, Hypalon, and BR polymers; xfinely pulverized; 98% min. thru 200 mesh; sol. in alcohols and ketones; sp.gr. 1.18 (cured); m.p. 82-93 C.

Akrochem® P 86. [Akrochem] Thermosetting two-step phenolic resin; thermoplastic as received; tackifier; offers plasticizer action during processing; brn. flakes; sol. in ketones and blends of alcohols and aromatics; sp.gr. 1.16; R&B m.p. 95 C.

Akrochem® P 87. [Akrochem] Cashew nut oil-modified phenol-formaldehyde thermosetting two-step resin; contains the necessary hexa to make it thermosetting; tackifier; modifier for nitrile rubber compds. and solv. cements; plasticizer during processing; brn. finely pulverized powd.; 2% max. retained on 200 mesh; sol. in ketones and blends of alcohols and aromatics; sp.gr. 1.17; Cap. tube m.p. 65 C.

Akrochem® P 90. [Akrochem] Thermoplastic alkyl phenol formaldehyde novalak resin; tackifier compat. with most elastomers; widely used in plied-up constructions; flake; 95% thru 3/8 in. screen; sp.gr. 1.00; R&B m.p. 90 C; acid no. 33.

Akrochem® P 101. [Akrochem] Heat-reactive phenolic resin; used in the resin curing of butyl rubbers, adhesive systems, and sealants; compded. with chloroprene, nitrile, and nat. rubbers for cement applics.; lump; USDA X color; sol. in aromatic and aliphatic solvs. incl. ketones, toluol, benzol, xylene, and naphthas; sp.gr. 1.05; Cap. tube m.p. 63 C.

Akrochem® P 105. [Akrochem] Nonphasing, heat-reactive phenolic resin; reacts with an act. magnesium oxide in solv. solution forming a salt which greatly increases the heat resistance of the adhesive; recommended for use in neoprene solv.-type contact cements; pale ylsh. lump; sol. in aromatic hydrocarbons and ketone solvs., incl. acetone, benzene, MEK, MIBK, toluene, and xylene; partially sol. in aliphatic solvs.; completely sol. when 5% of aromatic solv. is included; sp.gr. 1.10; m.p. 70-90 C.

Akrochem® P 108. [Akrochem] Heat-reactive alkyl phenolic resin; used in neoprene solv. cements; reacts with an act. magnesium oxide in solv. solution forming a salt which greatly increases the heat resistance of the adhesive; pale yel. lump; sol. in aromatic and aliphatic solvs. incl. toluene, xylene, ketones; sp.gr. 1.1; m.p. 76-88 C.

Akrochem® P 124. [Akrochem] Heat-reactive phenolic resin; curing agent for butyl or halogenated butyl polymers; red lumps; faint halogen odor; sol. in aliphatic and aromatic solvs., incl. benzene, xylene, toluene, ketones, and various naphthas; sp.gr. 1.05; Cap. tube m.p. 57 C; 99.5+% act.

Akrochem® P 125. [Akrochem] Heat-reactive bromomethyl alkylated phenolic resin; used in nat. rubber and butyl cement; red lumps; faint halogen odor; sp.gr. 1.05; Cap. tube m.p. 135 F.

Akrochem® P 126. [Akrochem] Heat-reactive alkyl phenolic resin; used in Neoprene AF-based contact adhesives to impart low initial visc. and exc. visc. stability; also used with the adhesive grades of chloroprene; lumps; color X or lighter; sol. in aromatic solvs. incl. toluene, benzene, and xylene, ketones, and chlorinated hydrocarbons; partially sol. in aliphatic hydrocarbons; sp.gr. 1.10; Cap. tube m.p. 185 F.

Akrochem® P 133. [Akrochem] Thermoplastic alkyl phenolic resin; tackifier resin for rubber compds. used in tire construction and mechanical goods; flakes; sol. in esters, ketones, aromatic and aliphatic chlorinated hydrocarbons; insol. in alcohols; sp.gr. 1.04; G-H visc. O (toluene-64% resin); R&B m.p. 97 C; acid no. 35.

Akrochem® P 478. [Akrochem] Thermosetting one-step phenolic resin; plasticizer for rubber compding., adhesives; powd.; sol. in alcohol, ketones, and esters; sp.gr. 1.25; Cap. tube m.p. 160 F.

Akrochem® P 486. [Akrochem] Thermoplastic cashew-modified phenolic resin without curing agent; plasticizer and reinforcing agent in nitrile rubber compds.; brn. flake; sol. in higher alcohols, ketones, and esters; limited tolerance for aromatic hydrocarbons; B&R soften. pt. 90-100 C.

Akrochem® P 487. [Akrochem] Cashew nut oil-modified phenol-formaldehyde two-step thermosetting resin; contains hexa necessary to make it thermosetting; modifier for nitrile rubber compds. and solv. cements; plasticizes the stock during processing; brn. powd.; 2% max. retained on 200 mesh; sol. in ketones, higher alcohols, esters; limited tolerance for aromatic hydrocarbons; sp.gr. 1.17; B&R soften. pt. 65 C.

Akrochem® P 3100. [Akrochem] Wettable zinc stearate; CAS 557-05-1; EINECS 209-151-9; mold release agent for rubber processing; provides uniform antitack and antiblock coatings for uncured rubber slab and extrudates; wh. free-flowing powd.; 99.8% thru 325 mesh; rapidly disp. in water; m.p. 120 C.

Akrochem® P 4000. [Akrochem] Calcium stearate; CAS 1592-23-0; EINECS 216-472-8; mold release agent; wh. to off-wh. fine powd.; 99.8% thru 325 mesh; m.p. 148 C; 2% moisture.

Akrochem® P 4100. [Akrochem] Wettable calcium stearate; CAS 1592-23-0; EINECS 216-472-8; mold release agent; wh. to off-wh. fine powd.; 99.8% thru 325 mesh; m.p. 173 C; 2% moisture.

Akrochem® PEG 3350. [Akrochem] Polyethylene glycol; activator for compounding with silica fillers; process aid, lubricant for rubber compds. (natural and syn.); mold release for foam and mech. goods; wh. flakes; sol. in water; sp.gr. 1.20; m.p. 130 F.

Akrochem® Peptizer 66. [Akrochem] Oil-coated activated 2,2′-dibenzamido diphenyldisulfide absorbed on clay; peptizer, processing aid, visc. reducer for natural and syn. rubbers; improves mold flow, uniformity of extrusions and calendered sheets; FDA 21CFR §177.2600; grayish powd.; partly sol. in org. solvs.; pract. insol. in water; sp.gr. 1.83; 40% act.

Akrochem® Plasticizer LN. [Akrochem] Naphthenic rubber process oil; low-visc. plasticizer; m.w. 320; sp.gr. 0.916 (60 F); dens. 7.6 lb/gal; visc. 154 SUS (100 F); pour pt. -40 F; ref. index 1.5076; COC flash pt. 350 F.

Akrochem® Powder Colors. [Akrochem] Org. and inorg. pigments; colorants for rubber and certain thermoplastic polymers.

Akrochem® Rubbersil RS-150. [Akrochem] Precipitated amorphous silica; highly reinforcing filler for use in syn. and natural rubber compding., for mech. rubber goods, tires, adhesives, footwear, soling; wh. fine powd.; 0.5% max. on 325 mesh; sp.gr. 2.0; bulk dens. 250 g/l; surf. area 150 m^2/g; oil absorp. 220 ml/100 g linseed oil; pH 6.2-7.3 (5 g/100 ml water); 98% SiO_2.

Akrochem® Rubbersil RS-200. [Akrochem] Precipitated amorphous silica; highly reinforcing filler for use in syn. and natural rubber compding., for mech. rubber goods, tires, adhesives, footwear, soling; wh. fine powd.; 0.5% max. on 325 mesh; sp.gr. 2.0; bulk dens. 250 g/l; surf. area 180 m^2/g; oil absorp. 220 ml/100 g linseed oil; pH 6.2-7.3 (5 g/100 ml water); 98% SiO_2.

Akrochem® Silicone Emulsion 1M. [Akrochem] Dimethylpolysiloxane aq. emulsion with sodium nitrite as corrosion inhibitor; nonionic; mold release agent; also for auto/furniture polish; wh. emulsion; sp.gr. 0.98-1.02; dens. 8.25 lb/gal; visc. 1000 cst; pH 7.0; 35% silicone; *Toxicology:* nontoxic; physiologically inert; *Storage:* 6 mos storage @ 70 F; avoid freezing.

Akrochem® Silicone Emulsion 10M. [Akrochem] Dimethylpolysiloxane aq. emulsion with sodium nitrite as corrosion inhibitor; nonionic; mold release agent; also for auto/furniture polish, aerosol spray starch; wh. emulsion; sp.gr. 0.98-1.02; dens. 8.25 lb/gal; visc. 10,000 cst; pH 7.0; 35% silicone; *Toxicology:* nontoxic; physiologically inert; *Storage:* 6 mos storage @ 70 F; avoid freezing.

Akrochem® Silicone Emulsion 60M. [Akrochem] Dimethylpolysiloxane aq. emulsion with sodium nitrite as corrosion inhibitor; nonionic; mold release agent; also for aerosol spray starch; wh. emulsion; sp.gr. 0.98-1.02; dens. 8.25 lb/gal; visc. 60,000 cst; pH 7.0; 35% silicone; *Toxicology:* nontoxic; physiologically inert; *Storage:* 6 mos storage @ 70 F; avoid freezing.

Akrochem® Silicone Emulsion 100. [Akrochem] Dimethylpolysiloxane aq. emulsion with sodium nitrite as corrosion inhibitor; nonionic; mold release agent; also for auto/furniture polish, aerosol window cleaners, cosmetic hand creams/lotions, textile finishing, and other specialty formulations; wh. emulsion; sp.gr. 0.96-1.00; dens. 8.3 lb/gal; visc. 100 cst; pH 7.0; 35% silicone; *Toxicology:* nontoxic; physiologically inert; *Storage:* 6 mos storage stability in unopened containers; store above 90 F; avoid freezing.

Akrochem® Silicone Emulsion 350. [Akrochem] Dimethylpolysiloxane aq. emulsion with sodium nitrite as corrosion inhibitor; nonionic; mold release agent; also for auto/furniture polish, aerosol window cleaners, cosmetic hand creams/lotions, textile finishing, and other specialty formulations; wh. emulsion; sp.gr. 0.98-1.02; dens. 8.25 lb/gal; visc. 350 cst; pH 7.0; 35% silicone; *Toxicology:* nontoxic; physiologically inert; *Storage:* 6 mos storage @ 70 F; avoid freezing.

Akrochem® Silicone Emulsion 350 Conc. [Akrochem] Dimethylpolysiloxane emulsion; nonionic; mold release agent; wh. emulsion; sp.gr. 0.96-1.00; dens. 8.3 lb/gal; visc. 350 cst; pH 7.0; 60% silicone; *Storage:* 6 mos shelf life.

Akrochem® Silicone Fluid 350. [Akrochem] Dimethylpolysiloxane; release and slip agent, additive in rubber and plastics for special surf. props., water repellent treatments; offers inertness, heat and oxidative stability, good elec. props.; general purpose release agent for most tire and mech. goods, e.g., tire molds, shock mounts, fan belts, o-rings, hose, toys, shoe heels, floor tile, rubber runners, wire and cable; FDA approved (10 ppm max.); water-wh. clear liq.; misc. with nonpolar liqs.; sp.gr. 0.95-0.98; dens. 7.8-8.0 lb/gal; visc. 350 cst; *Toxi-*

cology: nontoxic.

Akrochem® TBUT. [Akrochem] Tetrabutyl thiuram disulfide; CAS 1634-02-2; accelerator for rubber industry.

Akrochem® TDEC. [Akrochem] Tellurium diethyl dithiocarbamate; CAS 20941-65-5; fast-curing primary or sec. accelerator for use in natural rubber, SBR, NBR, EPDM, and butyl; orange to yel. pellet, char. odor; sol. in benzene, chloroform, carbon disulfide; sp.gr. 1.42; m.p. 108 C min.; 80% act.

Akrochem® TETD. [Akrochem] Tetraethyl thiuram disulfide; CAS 97-77-8; accelerator for rubber industry.

Akrochem® Thio No. 1. [Akrochem] N,N′ diphenyl thiourea; CAS 102-08-9; accelerator for CR latex, NR latex, and cements, EPDM sponge compds.; activates thiazole accelerators; essentially nondiscoloring; off-wh. powd.; sol. in acetone; sl. sol. in alcohol, ether, toluene, naphtha, CCl_4; sp.gr. 1.30; m.p. 148 C; 98% purity.

Akrochem® TM/ETD. [Akrochem] 60% Tetramethyl thiuram disulfide/40% tetraethyl thiuram disulfide; very act., sulfur-bearing, nondiscoloring org. accelerator; gives nonblooming cures in EV and semi-EV systems; used in nitrile rubber, SBR and EPDM; off-wh. powd.; sp.gr. 1.35; m.p. 66 C min.; 12.1% avail. sulfur.

Akrochem® TMTD. [Akrochem] Tetramethylthiuram disulfide; CAS 137-26-8; very act., sulfur-bearing, nondiscoloring org. accelerator and activator; for curing systems requiring very low or no sulfur and for butyl and EPDM compds.; wh. to lt. gray powd.; sp.gr. 1.42; m.p. 148 C min.

Akrochem® TMTD Pellet. [Akrochem] Tetramethylthiuram disulfide; CAS 137-26-8; pellet form of Akrochem TMTD Powder; wh. to lt. gray pellet; sp.gr. 1.40; m.p. 148 C.

Akrochem® TMTM. [Akrochem] Tetramethyl thiuram monosulfide; CAS 97-74-5; nonstaining, nondiscoloring, fast-curing accelerator for use alone or in combination in NR, SBR, NBR, butyl rubber, Neoprene, and reclaim rubber; yel. powd.; sp.gr. 1.39; m.p. 105 C min.

Akrochem® VC-40C. [Akrochem] Bis (t-butylperoxyisopropyl) benzene absorbed on calcium carbonate; accelerator for syn. elastomers; wh. powd. or pellets, low odor; sp.gr. 1.53; 40 ± 1% peroxide (act.).

Akrochem® VC-40K. [Akrochem] Bis (t-butylperoxyisopropyl) benzene absorbed on kaolin clay; accelerator for syn. elastomers; wh. powd. or pellets, low odor; sp.gr. 1.51; 40 ± 1% peroxide (act.).

Akrochem® Wettable Zinc Stearate. [Akrochem] Wettable zinc stearate; CAS 557-05-1; EINECS 209-151-9; mold release agent; wh. fine powd.; 95% thru 325 mesh; apparent dens.0.3 g/ml; m.p. 120 C.

Akrochem® Z.B.E.D. [Akrochem] Zinc dibenzyl dithiocarbamate; CAS 14726-36-4; nondiscoloring, nonstaining accelerator for rubber; creamy wh. powd.; sp.gr. 1.41; m.p. 180-188 C; 95% min. purity.

Akrochem® ZDA Powd. [Akrochem] Zinc diacrylate; CAS 14643-87-9; activator for rubber compounding.

Akrochem® Zinc Oxide 35. [Akrochem] Precipitated zinc oxide; CAS 1314-13-2; EINECS 215-222-5; accelerator activator for rubber articles based on natural and syn. elastomers, and latex applics., e.g., transparent and translucent rubber goods, dynamically stressed articles; also for adhesive applics.; crosslinking agent for metal oxide-curable elastomers; reinforcing filler; nonblooming; wh. to ylsh. grn. powd.; sp.gr. ≈5.4 g/cm^3; surf. area 35 m^2/g; 93-96% ZnO.

Akrochem® Zinc Stearate. [Akrochem] Zinc stearate; CAS 557-05-1; EINECS 209-151-9; mold release agent; wh. fine powd.; 99.8% thru 325 mesh; apparent dens. 2.8 lb/gal; m.p. 120 C.

Akrochem® ZMBT. [Akrochem] Zinc salt of 2-mercaptobenzothiazole; CAS 155-04-4; semi-ultra accelerator for rubber; ylsh. powd., pract. odorless, bitter taste; sp.gr. 1.70 ± 0.03; m.p. 315 C (dec.).

Akrochem® Z.P.D. [Akrochem] Zinc pentamethylene dithiocarbamate; ultra-accelerator for dry, natural rubber, esp. footwear; nonpigmenting and nonstaining; off-wh. powd.; sp.gr. 1.6; m.p. 225 C min.; 98% min. purity.

Akrochlor™ L-39. [Akrochem] Chlorinated paraffins; low-cost source of halogen for flame retardancy purposes; compat. with most polymers; liq.; sp.gr. 1.09; visc. 3 poise; 39% Cl.

Akrochlor™ L-40. [Akrochem] Chlorinated paraffin; low-cost source of halogen for flame retardancy purposes; compat. with most polymers; liq.; sp.gr. 1.11; visc. 8 poise; 40% Cl.

Akrochlor™ L-42. [Akrochem] Chlorinated paraffin; low-cost source of halogen for flame retardancy purposes; compat. with most polymers; liq.; sp.gr. 1.15; visc. 29 poise; 42% Cl.

Akrochlor™ L-57. [Akrochem] Chlorinated paraffin; low-cost source of halogen for flame retardancy purposes; compat. with most polymers; liq.; sp.gr. 1.30; visc. 13 poise; 57% Cl.

Akrochlor™ L-60. [Akrochem] Chlorinated paraffin; low-cost source of halogen for flame retardancy purposes; compat. with most polymers; liq.; sp.gr. 1.35; visc. 20 poise; 60% Cl.

Akrochlor™ L-61. [Akrochem] Chlorinated paraffin; low-cost source of halogen for flame retardancy purposes; compat. with most polymers; liq.; sp.gr. 1.38; visc. 60 poise; 61% Cl.

Akrochlor™ L-70LV. [Akrochem] Chlorinated paraffin; low-cost source of halogen for flame retardancy purposes; compat. with most polymers; liq.; sp.gr. 1.52; visc. > 10^7 poise; 67% Cl.

Akrochlor™ L-170. [Akrochem] Chlorinated paraffin; low-cost source of halogen for flame retardancy purposes; compat. with most polymers; liq.; sp.gr. 1.49; visc. 202 SUS (210 F); 68% Cl.

Akrochlor™ L-170HV. [Akrochem] Chlorinated paraffin; low-cost source of halogen for flame retardancy purposes; compat. with most polymers; liq.; sp.gr. 1.54; visc. 520 SUS (210 F); 68% Cl.

Akrochlor™ R-70. [Akrochem] Chlorinated paraffin; low-cost source of halogen for flame retardancy purposes; compat. with most polymers; Gardner < 1 solid; 94% thru 50 mesh; sp.gr. 1.60; soften. pt. (B&R) 102 C; 70% Cl.

Akrochlor™ R-70S. [Akrochem] Chlorinated paraffin; low-cost source of halogen for flame retardancy purposes; compat. with most polymers; Gardner < 1 solid; 94% thru 50 mesh; sp.gr. 1.60; soften. pt. (B&R) 102 C; 70% Cl.

Akrodip™ C-100. [Akrochem] Aq. clay dispersion; release agent, slab dip; provides uniform antistick coating to unvulcanized rubber slabs and continuous rubber sheets for batch-off mixing systems; tan paste; disp. in water; sp.gr. 1.29; dens. 10 lb/gal; 50% solids.

Akrodip™ CS-400. [Akrochem] Calcium stearate disp.; CAS 1592-23-0; EINECS 216-472-8; slab dip providing fast drying, waterproof antistick coating on rubber stock; esp. for water-cooled takeaways; wh. soft paste; disp. in water in all proportions; sp.gr. 1.0; 22% solids.

Akrodip™ Dry. [Akrochem] Aluminum silicate aq. disp.; release agent, slab dip providing antistick coating for unvulcanized rubber slabs and continuous sheets for batch-off systems; off-wh. powd.; disp. in water; sp.gr. 2.5; 30% solids.

Akrodip™ V-300. [Akrochem] Vegetable oil soap emulsion; slab dip providing clear antistick coating for unvulcanized rubber sheets and preforms; esp. useful where pigmented dips cannot be tolerated; amber liq.; misc. with water; sp.gr. 1.0; dens. 8.4 lb/gal; pH 9; 30% solids.

Akrodip™ V-301. [Akrochem] Oil soap emulsion; lubricant, slab dip providing clear antistick coating for unvulcanized rubber sheets and preforms; esp. useful where pigmented dips cannot be tolerated; amber paste; misc. with water; sp.gr. 1.0; dens. 8.0 lb/gal; pH 10; 36% solids.

Akrodip™ Z-50/50. [Akrochem] Zinc stearate aq. disp.; CAS 557-05-1; EINECS 209-151-9; release agent, slab dip providing antistick coating for unvulcanized rubber sheets, esp. for batch-off systems using refrigerated water; also for prepped rubber preforms; wh. smooth paste; misc. with water; sp.gr. 1.02; dens. 8.4 lb/gal; 50% solids.

Akrodip™ Z-200. [Akrochem] Zinc stearate aq. disp.; CAS 557-05-1; EINECS 209-151-9; slab dip providing antistick coating for unvulcanized rubber sheets; esp. useful for batch-off systems using refrigerated water; good permanence; also for prepped rubber preforms; wh. smooth paste; misc. with water; sp.gr. 1.01; dens. 8.4 lb/gal; 26% solids.

Akrodip™ Z-250. [Akrochem] Zinc stearate aq./alcohol disp.; CAS 557-05-1; EINECS 209-151-9; slab dip providing antistick coating for unvulcanized rubber sheets; esp. useful for batch-off systems using refrigerated water; good permanence; also for prepped rubber preforms; wh. smooth paste; misc. with water; sp.gr. 0.8-1.0; dens. 9.2 lb/gal; 30% solids.

Akrofax™ 900C. [Akrochem] Vulcanized vegetable oil; rubber processing aid; absorbent for min. oils and other liq. plasticizers on mill and banbury equip.; speeds incorporation of fillers; flow promoter; provides unique surface finish to vulcanized rubber goods; improves ozone resistance; milled crumb; sp.gr. 1.08.

Akrofax™ A. [Akrochem] Vulcanized vegetable oil; rubber processing aid; absorbent for min. oils and other liq. plasticizers on mill and banbury equip.; speeds incorporation of fillers; flow promoter; provides unique surface finish to vulcanized rubber goods; improves ozone resistance; milled crumb; sp.gr. 1.00.

Akrofax™ B. [Akrochem] Sulfur-type vulcanized vegetable oil; extender, processing aid and softener for colored rubber prods., esp. for dimensional stability in tubing and calendered goods; lt. yel. crumb; sp.gr. 1.03.

Akroform® DBTU PM. [Akrochem] Dibutylthiourea on a polymeric binder; CAS 109-46-6; accelerator for rubber.

Akroform® DCP-40 EPMB. [Akrochem] Dicumyl peroxide in EPR rubber; accelerator in masterbatch form; used with syn. elastomers; 1/4 in. diced pellets; sp.gr. 0.92; 40.5-41.0% peroxide (act.).

Akroform® DETU PM. [Akrochem] N,N′-Diethylthiourea on a polymeric binder; CAS 105-55-5; nondiscoloring, nonstaining accelerator for high temp. curing of mercaptan-modified polychloroprenes (W-type neoprenes); for mfg. of solid and cellular extruded goods, e.g., car door and boot seals; off-wh. pellet, sl. typ. odor; sp.gr. 1.1; m.p. 74 C min.; 75% act.; 23.7-24.7% S, 20.5-21.8% N; *Toxicology:* causes cancer in animals; hot vapors from mixes contg. this prod. may cause severe eye irritation; *Storage:* exc. storage stability.

Akroform® ETU-22 PM. [Akrochem] Ethylene thiourea on compat. polymeric binder carrier; CAS 96-45-7; EINECS 202-506-9; accelerator imparting a high state of cure to neoprene compds.; nonstaining and nondiscoloring; off-wh. pellet, sl. typ. odor; disperses in dry polymers; sp.gr. 1.2; m.p. 200 C min. for act.; 75% act.

Akroform® VC-40 EPMB. [Akrochem] Bis (t-butylperoxy-isopropyl) benzene in EPR rubber; accelerator in masterbatch form; used with syn. elastomers; $^{1}/_{4}$ in. diced pellets, low odor; sp.gr. 0.89; 40.5-41.0% peroxide (act.).

Akro-Gel. [Akrochem] Water gel of a sodium salt of a methyl tauride; release agent for use with all types of elastomers to prevent sticking of uncured compds. and extrusions; as slab dip, mold lubricant, to prevent water spotting in open steam cure; amber clear gel, mild odor; sol. in water at 54 C; pH 6.5-8.0 (10%).

Akrolease® E-9410. [Akrochem] Silicone polymer; release agent for PU, polyester, or epoxy materials from plastic or metal molds; corrosion inhibitor for metal; water-wh. clear liq., sl. odor; sol. in hydrocarbon and chlorinated solvs.; sp.gr. 0.968; dens. 8.1 lb/gal; visc. 225 cps; vapor pressure 0.1 mm Hg (68 F); flash pt. (PM) 280 F; 100% act.; *Toxicology:*

LD50 (oral, rat) > 4640 mg/kg, (dermal, rabbit) > 1000 mg/kg; nontoxic; physiologically inert; mild irritant to rabbit skin and eyes; *Precaution:* burns with difficulty, but supports combustion; *Storage:* exc. storage stability; store in cool, dry, well-ventilated area.

Akrolease® E-9491. [Akrochem] Silicone emulsion; nonionic; release agent for plastics in molding operations, esp. epoxy, polyester, and PU from metal or plastic molds; improves release and lubricity; corrosion inhibitor for metals; wh., cream; visc. 225 cSt (of silicone); 35% solids in water.

Akroplast®. [Akrochem] Thermoplastic pigment dispersions; colorants for vinyls, acrylics, polyethylene, cellulose acetate butyrate, rubbers, urethanes, phthalate pastes; avail. as pelletized, diced, gran., and powd. forms.

Akrosperse® Color Masterbatches. [Akrochem] Dispersions of Akrochem® Pigments in various elastomers (SBR, EPR, EPDM, EVA, EP/EVA, NBR, NR, CM, CR, PIB, IIR, silicone).

Akrosperse® D-177. [Akrochem] Tetramethylthiuram disulfide; CAS 137-26-8; 5% oil-treated powder version of Akrochem TMTD; wh. to lt. gray dedusted powd.; sp.gr. 1.39; m.p. 148 C.

Akrosperse® DCP-40 EPMB. [Akrochem] Dicumyl peroxide in EPR rubber; accelerator in masterbatch form; used with syn. elastomers; off-wh. thin sheets; sp.gr. 0.92; 40.5-41.0% peroxide (act.).

Akrosperse® Plasticizer Paste Colors. [Akrochem] Org. and inorg. pigments; plasticizer pigment dispersions.

Akrosperse® VC-40 EPMB. [Akrochem] Bis (t-butylperoxy-isopropyl) benzene in EPR rubber; accelerator in masterbatch form; used with syn. elastomers; off-wh. thin sheets, low odor; sp.gr. 0.89; 40.5-41.0% peroxide (act.).

Akrosperse® Water Paste Colors. [Akrochem] Aq. pigment dispersions; colorants for latex prods. and adhesives.

Akrotak 100. [Akrochem] Tackifier for syn. and natural rubber, adhesives and hot melts; provides wetting for filler/rubber; speeds up incorporation of min. fillers; pale yel. to amber flake; sp.gr. 1.04; dens. 8.2 lb/gal (150 C); acid no. 5-13; soften. pt. (B&R) 100 C.

Akrowax 130. [Akrochem] Fully refined paraffin wax derived from petrol.; CAS 8002-74-2; EINECS 232-315-6; contains an oxidation inhibitor; wax used in rubber to give ozone resistance; inhibits cracking and frosting in NR, SBR, CR, and NBR; also for food pkg. applics.; wh. hard cryst. material, low odor; sp.gr. 0.90 (100 F); visc. 39 SUS (210 F); m.p. 130 F; flash pt. 430 F.

Akrowax 145. [Akrochem] Fully refined paraffin wax derived from petrol.; CAS 8002-74-2; EINECS 232-315-6; contains an oxidation inhibitor; wax used in rubber to give ozone resistance; inhibits cracking and frosting in NR, SBR, CR, and NBR; also for food pkg. applics.; wh. hard cryst. material, low odor; sp.gr. 0.82 (100 F); visc. 41 SUS (210 F); m.p. 141 F; flash pt. 440-450 F.

Akrowax 5025. [Akrochem] Antiozone wax designed to inhibit cracking of rubber vulcanizates incl. nat. rubber, SBR, CR, butyl, low and med. acrylonitrile types of nirile rubber; nonstaining; wh. flakes, prills, or slabs; sp.gr. 0.92; congeal. pt. 130-150 F.

Akrowax 5026. [Akrochem] Petrol. wax; antisunchecking wax designed to inhibit cracking of rubber vulcanizates incl. nat. rubber, SBR, BR, CR, and low and med. acrylonitrile-type nitrile polymers; end-uses incl. mechanical and automotive parts, heels, soles, footwear, and tire sidewalls; with antiozonants for tire tread, conveyor belts, hose, and elec. wire and cable; nonstaining; wh. flakes, prills; sp.gr. 0.92; congeal. pt. 148-157 F.

Akrowax 5030. [Akrochem] Petrol. wax; antisunchecking wax designed to inhibit cracking of rubber vulcanizates incl. nat. rubber, SBR, BR, CR, and low and med. acrylonitrile-type nitrile polymers; end-uses incl. mechanical and automotive parts, heels, soles, footwear, and tire sidewalls; with antiozonants for tire tread, conveyor belts, hose, and elec. wire and cable; nonstaining; wh. flakes, prills; sp.gr. 0.92; congeal. pt. 148-157 F.

Akrowax 5031. [Akrochem] Petrol. wax; antisunchecking wax designed to inhibit cracking of rubber vulcanizates incl. butyl, chlorobutyl, CR, all types of nitrile rubber, nat. rubber, SBR, and BR polymers; nonstaining; for elec. wire and cable, tire sidewalls, off-the-road and agric. tires, and mechanical or automotive parts; wh. flakes, prills; sp.gr. 0.92; congeal. pt. 146-152 F.

Akrowax 5032. [Akrochem] Petrol. wax; antisunchecking wax designed to inhibit cracking of rubber vulcanizates incl. butyl, chlorobutyl, CR, all types of nitrile rubber, nat. rubber, SBR, and BR polymers; nonstaining; for elec. wire and cable, tire sidewalls, off-the-road and agric. tires, and mechanical or automotive parts; wh. flakes, prills; sp.gr. 0.92; congeal. pt. 146-152 F.

Akrowax 5050. [Akrochem] Wax blend for antichecking and antiozonant protection of rubber; nonstaining, nondiscoloring; Gardner 1-2 prills; visc. 13 cps (210 F); congeal. pt. 157 F.

Akrowax 5073. [Akrochem] Antiozone wax designed to inhibit cracking of rubber vulcanizates incl. nat. rubber, SBR, CR, butyl, low and med. acrylonitrile types of nitrile rubber; nonstaining; wh. flakes, prills, or slabs; sp.gr. 0.92; congeal. pt. 130-150 F.

Akrowax Micro 23. [Akrochem] Microcryst. wax composed of mixt. of hydrocarbons incl. normal paraffins, branc; wax used in rubber to give ozone resistance; with Akrowax 130 to inhibit cracking in NR, SBR, NBR, and CR; ASTM 2 wax; sp.gr. 0.92; visc. 86 SUS (@ 210 F); m.p. 176 F; flash pt. (COC) 540 F.

Akrowax PE. [Akrochem] LDPE; process aid in natural and syn. rubbers; wh. powd.; m.w. 5600-6400; sp.gr. 0.92; m.p. 98-108 C; acid no. 0; sapon. no. 0.

Aktiplast®6 N. [Rhein Chemie] Blend of dixylyldisulfides; reclaiming agent for scrap rubber; dk. brn. liq.; sp.gr. 1.12.

Aktiplast® 6 R. [Rhein Chemie] Blend of diaryldisulfides; reclaiming agent for scrap rubber; brn. liq.; sp.gr. 1.12.

Aktiplast® AS. [Rhein Chemie] N,N-Bis-stearoylethylenediamide; CAS 110-30-5; peptizing and dispersing agents for tech. NR articles; decreases visc., provides better dispersion of ingreds.; ylsh. pellets; sp.gr. 1.00.

Aktiplast® F. [Rhein Chemie] Activated zinc salts of higher molecular unsat. fatty acids; processing promoter; peptizing and dispersing agents for molded and extruded rubber, conveyor belts, tire compds.; vulcanization accelerator; retards scorch; brn. pellets; sp.gr. 1.08.

Aktiplast® PP. [Rhein Chemie] Zinc salts of higher molecular fatty acids; peptizing and dispersing agents for tire compds., molded, extruded, and hard rubber; vulcanization accelerator; peptizing effect in NR and IR from 60 C; retards scorch; lt. brn. pellets; sp.gr. 1.08.

Aktiplast® T. [Rhein Chemie] Zinc salts of higher molecular unsat. fatty acids; processing promoter for molded and extruded rubber goods, expanded rubber articles, hard rubber; peptizing effect in NR and IR from 60 C; retards scorch; vulcanization accelerator; lt. yel.-brn. pellets; sp.gr. 1.05.

Aktisil® AM. [Hoffmann Min.] Sillitin Z 86 (quartz-kaolinite) coated with γ-aminopropyltriethoxysilane; filler for thermosets, thermoplastics, thermoplastic PU.

Aktisil® EM. [Hoffmann Min.] Sillitin Z 86 (quartz-kaolinite) coated with γ-glycidoxypropyltrimethoxysilane; filler, min. additive for thermosets, thermoplastic PU.

Aktisil® MM. [Hoffmann Min.] Quartz-kaolinite coated with γ-Mercaptopropyltrimethoxysilane; filler for sulfur and metal oxide-cured systems.

Aktisil® PF 216. [Hoffmann Min.] Quartz-kaolinite coated with bis-(3-(triethoxysilyl)propyl) tetrasulfane; filler for sulfur-cured systems.

Aktisil® PF 231. [Hoffmann Min.] Sillitin Z86 (quartz-kaolinite) coated with stearic acid; filler, min. additive for thermoplastic PU.

Aktisil® VM. [Hoffmann Min.] Quartz-kaolinite coated with vinyl-tris(β-methoxyethoxy)silane; filler for peroxide-cured systems.

Aktisil® VM 56. [Hoffmann Min.] Quartz-kaolinite coated with vinyltriethoxysilane; filler for peroxide-cured systems, polyolefins.

Akypo OP 80. [Chem-Y GmbH] Octoxynol-9 carboxylic acid; CAS 107628-08-0; 72160-13-5; anionic/nonionic; detergent, emulsifier; used in aq. sol'ns, metalworking fluids, emulsion polymerization, film developing baths; lime soap dispersant; moderate foam; Gardner 2 clear liq.; sol. in water; insol. in oils; sp.gr. 1.08; visc. 2000 mPa•s; acid no. 73-83; pH 2-3 (1%); surf. tens. 33 mN/m; 90% act.

Akyporox NP 15. [Chem-Y GmbH] Nonoxynol-1; CAS 27986-36-3; nonionic; emulsifier for emulsion polymerization; liq.; sp.gr. 1.0; HLB 4.7; 100% act.

Akyporox NP 30. [Chem-Y GmbH] Nonoxynol-3; CAS 9016-45-9; nonionic; for mfg. of hair dye formulations; emulsifier for emulsion polymerization; liq.; 100% act.

Akyporox NP 200. [Chem-Y GmbH] Nonoxynol-20; CAS 9016-45-9; nonionic; emulsifier, wetting agent used in textile prods., emulsion polymerization, degreasing baths, electroplating industry; solid; sp.gr. 1.08; m.p. 33 C; HLB 16.0; cloud pt. > 100 C (1%); surf. tens. 42 mN/m; 100% act.

Akyporox NP 300V. [Chem-Y GmbH] Nonoxynol-30; CAS 9016-45-9; nonionic; emulsifier for calcium stearate, emulsion polymerization; liq.; 70% act.

Akyporox NP 1200V. [Chem-Y GmbH] Nonoxynol-120; CAS 9016-45-9; nonionic; emulsifier for emulsion polymerization; Gardner 2 clear liq.; sp.gr. 1.08; visc. 1000 mPa•s; HLB 19.2; cloud pt. > 100 C (1%); pH 7-8; surf. tens. 52 mN/m; 50% act.

Akyporox OP 250V. [Chem-Y GmbH] Octoxynol-25; CAS 9002-93-1; nonionic; emulsifier for emulsion polymerization; perfume solubilizer; Gardner 2 clear liq.; sp.gr. 1.07; visc. 1000 mPa•s; cloud pt. > 100 C (1%); HLB 17.3; pH 8-9 (5%); surf. tens. 40 mN/m; 70% act.

Akyporox OP 400V. [Chem-Y GmbH] Octoxynol-40; CAS 9002-93-1; nonionic; emulsifier for emulsion polymerization; Gardner 2 clear liq.; sp.gr. 1.1; visc. 1200 mPa•s; cloud pt. > 100 C (1%); HLB 17.9; pH 7-8.5 (5%); surf. tens. 45 mM/m; 70% conc.

Akyporox RLM 160. [Chem-Y GmbH] Laureth-16; CAS 9002-92-0; nonionic; emulsifier for emulsion polymerization; liq.; sp.gr. 1.04; m.p. 35 C; cloud pt. > 100 C (1%); HLB 15.6; surf. tens. 45 mN/m; 100% act.

Akyporox RTO 70. [Chem-Y GmbH] Oleth-7; CAS 9004-98-2; nonionic; emulsifier for emulsion polymerization; liq.; 100% conc.

Akyporox SAL SAS. [Chem-Y GmbH] Sodium lauryl sulfate; CAS 142-87-0; anionic; detergent, shampoo base, emulsifier for emulsion polymerization; liq.; 28% conc.

Akyposal 9278 R. [Chem-Y GmbH] Sodium laureth sulfate; CAS 9004-82-4; emulsifier for emulsion polymerization; liq.; 30% act.

Akyposal ALS 33. [Chem-Y GmbH] Ammonium lauryl sulfate; CAS 2235-54-3; EINECS 218-793-9; anionic; detergent, emulsifier; shampoo base; used in emulsion polymerization; liq.; 33% conc.

Akyposal BD. [Chem-Y GmbH] Sodium octoxynol-6 sulfate; CAS 69011-84-3; anionic; emulsifier for emulsion polymerization; Gardner 5 clear liq.; sp.gr. 1.0; visc. 100 mPa•s; pH 2-4; 32% act.

Akyposal EO 20 MW. [Chem-Y GmbH] Sodium laureth sulfate; CAS 9004-82-4; anionic; detergent, base for shampoos and bubble baths, emulsifier for emulsion polymerization; liq.; 28% conc.

Akyposal NAF. [Chem-Y GmbH] Sodium dodecylbenzene sulfonate; emulsifier for emulsion polymerization; liq.; 24% act.

Akyposal NPS 60. [Chem-Y GmbH] Sodium nonoxynol-6 sulfate; anionic; emulsifier for emulsion polymerization; Gardner 3 clear liq.; sp.gr. 1.05; visc. 15,000 mPa•s; pH 6-7 (10%); 32% act.

Akyposal NPS 100. [Chem-Y GmbH] Sodium nonoxynol-10 sulfate; CAS 9014-90-8; anionic; emulsifier for polymerization; Gardner 3 clear liq.; sp.gr. 1.05; visc. 500 mPa•s; pH 6-7; 34% conc.

Akyposal NPS 250. [Chem-Y GmbH] Sodium nonoxynol-25 sulfate; CAS 9014-90-8; anionic; emulsifier for polymerization; liq.; 34% conc.

Akyposal OP 80. [Chem-Y GmbH] Octoxynol-9 carboxylic acid; emulsifier for emulsion polymerization; liq.; 90% act.

Alatac 100-10. [Sovereign] Rosin oil; low-cost tackifier for tires; provides exc. initial tack; usually used in combination with phenolic resins to improve aged tack; liq.

Albacar® 5970. [Specialty Minerals] Precipitated calcium carbonate; functional pigment, filler for plastics, rubber, paints, adhesives, sealants, caulks, paper; unique rosette particle shape makes it useful in applics. where high oil absorp. and high visc. are desired; powd.; 1.9 μ median particle size; 0.03% on 325 mesh; sp.gr. 2.7; dens. 31 lb/ft^3 (tapped); bulk dens. 14 lb/ft^3; oil absorp. 50 g/100 g; brightness 98; 98% $CaCO_3$, 1% $MgCO_3$.

Albafil®. [Specialty Minerals] Precipitated calcium carbonate; filler for powd. coatings and specialty thin film applics.; offers precisely controlled particle size, narrow distribution, symmetrical particle shape, high brightness; powd.; 0.7 μ median particle size; 0.2% on 325 mesh; sp.gr. 2.7; dens. 33 lb/ft^3 (tapped); bulk dens. 17 lb/ft^3; oil absorp. 30 g/100 g; brightness 98; 98% $CaCO_3$, 1% $MgCO_3$.

AlbaFlex 25. [Aspect Minerals] High purity wet ground muscovite mica; CAS 12001-26-2; reinforcing filler for plastics, paints, coatings, and rubber applics.; high aspect ratio; off-wh. fine flakes; 52-58 μ avg. particle size; 75-85% thru 325 mesh; sp.gr. 2.8; bulk dens. 8-12 lb/ft^3; pH 7-8 (28% solids); 45.26% SiO_2, 38.15% Al_2O_3, 7.53% K_2O.

AlbaFlex 50. [Aspect Minerals] High purity wet ground muscovite mica; CAS 12001-26-2; reinforcing filler for plastics, paints, coatings, and rubber applics.; high aspect ratio; off-wh. fine flakes; 42-47 μ avg. particle size; 90% thru 325 mesh; sp.gr. 2.8; bulk dens. 8-11 lb/ft^3; pH 7-8 (28% solids); 45.26% SiO_2, 38.15% Al_2O_3, 7.53% K_2O.

AlbaFlex 100. [Aspect Minerals] High purity wet ground muscovite mica; CAS 12001-26-2; reinforcing filler for plastics, paints, coatings, and rubber applics.; high aspect ratio; off-wh. fine flakes; 34-39 μ avg. particle size; 95% thru 325 mesh; sp.gr. 2.8; bulk dens. 8-11 lb/ft^3; pH 7-8 (28% solids); 45.26% SiO_2, 38.15% Al_2O_3, 7.53% K_2O.

AlbaFlex 200. [Aspect Minerals] High purity wet ground muscovite mica; CAS 12001-26-2; reinforcing filler for plastics, paints, coatings, and rubber applics.; high aspect ratio; off-wh. fine flakes; 22-27 μ avg. particle size; 90% thru 400 mesh; sp.gr. 2.8; bulk dens. 7-10 lb/ft^3; pH 7-8 (28% solids); 45.26% SiO_2, 38.15% Al_2O_3, 7.53% K_2O.

Albaglos®. [Specialty Minerals] Precipitated calcium carbonate; functional pigment, filler, extender, reinforcer, and bodying agent in plastics (thermoset polyester, PVC), rubber, adhesives, sealants, caulks, paints, coatings, printing inks; chem. reagent, carrier, calcium source; controlled particle size; powd.; 0.8 μ median particle size; 0.015% on 325 mesh; sp.gr. 2.7; dens. 33 lb/ft^3 (tapped); bulk dens. 17 lb/ft^3; oil absorp. 30 g/100 g; brightness 98; 98% $CaCO_3$, 1% $MgCO_3$.

Albrite® BTD HP. [Albright & Wilson Am.] Phosphite; antioxidant for PP.

Albrite® DBHP. [Albright & Wilson Am.] Phosphite; nonstaining antioxidant, heat and lt. stabilizer for PP; clear.

Albrite® DIOP. [Albright & Wilson Am.] Phosphite; nonstaining antioxidant, heat and lt. stabilizer for PP; clear.

Albrite® DLHP. [Albright & Wilson Am.] Phosphite; nonstaining antioxidant for cellulosics, nylon, polyester, PE, PP, PVC; clear.

Albrite® DMHP. [Albright & Wilson Am.] Phosphite; antioxidant, metal chelator, heat and lt. stabilizer for polyester, plastic tape.

Albrite® DOHP. [Albright & Wilson Am.] Phosphite; nonstaining antioxidant, heat and lt. stabilizer for cellulosics, nylon, polyester, PP, PVC; clear.

Albrite® PA-75. [Albright & Wilson Am.] 75% Phenyl acid phosphate in butyl alcohol; acid catalyst in resin curing, chemical intermediate in formulation of rust preventatives, antistats, textile lubricants, oil additives, heavy metal extractants; sol'n.; sl. sol. in water; sp.gr. 1.19 (20/4 C); acid no. 270 min.; flash pt. (PMCC) 86 F.

Albrite® T2CEP. [Albright & Wilson Am.] Tris 2-chloroethyl phosphite; nonstaining antioxidant, sec. stabilizer for plastics, esp. acrylic resins; also for cellulosics, nyon, polyester, PP, PVC; colorless mobile liq.; 11.5% P.

Albrite® TBP. [Albright & Wilson Am.] Tributyl phosphate; CAS 126-73-8; EINECS 204-800-2; plasticizer for CR, CPE, Hypalon elastomers, PVC, PVAc, cellulosics, PS, ABS, ink systems; lubricant additive; antifoam; ink solv.; low odor.

Albrite® TBPO. [Albright & Wilson Am.] Tributyl phosphate; CAS 126-73-8; EINECS 204-800-2; flame retardant for cellulosics, epoxy, PS, PVAc, PVC.

Albrite® TIOP. [Albright & Wilson Am.] Triisooctyl phosphite; CAS 25103-12-2; EINECS 246-614-4; nonstaining antioxidant, sec. stabilizer and mold release agent for cast acrylics, cellulosics, nylon, polyester, PP; colorless mobile liq.; 7.4% P.

Albrite® TIPP. [Albright & Wilson Am.] Phosphite; heat and lt. stabilizer, metal chelator for polyester.

Albrite® TMP. [Albright & Wilson Am.] Phosphite; metal chelator for polyester.

Albrite® TPP. [Albright & Wilson Am.] Triphenyl phosphite; CAS 101-02-0; EINECS 202-908-4; costabilizer for PVC; reactive diluent for a variety of resin systems; colorless mobile liq.; 10% P.

Albrite® Tributyl Phosphate. [Albright & Wilson Am.] Tri (n-butyl) phosphate; CAS 126-73-8; EINECS 204-800-2; polymer solvating and plasticizing agent, antifoam for polymer systems, low temp., low flamm. hydraulic fluids, as metal complexing agent in solv. extraction; plasticizer for cellulosic, nitrocellulose, vinyl resins; defoamer in adhesives, textile sizes, vat dyeing; pigment wetter, rheology modifier, coalescing aid in coatings; colorless clear liq., mild odor; sl. sol. in water; sp.gr. 0.98; visc. 3.4 cps; vapor pressure 13.7 mm Hg (20 C); f.p. -80 C; b.p. 289 C (dec.); acid no. < 0.1; flash pt. 160 C; surf. tens. 28 dynes/cm (20 C); 11.7% P; *Toxicology:* LD50 1.8 ml/kg, mod. toxic; avoid prolonged/repeated eye and skin contact.

Alcamizer 1. [Kyowa Chem. Industry] Magnesium aluminum carbonate; CAS 12539-23-0, 11097-59-9; EINECS 234-319-3; heat stabilizer for PVC that reacts in a unique way with HCl in PVC; offers high heat stability, nontoxicity, high transparency, good weatherability; recommended for flexible applics.; JHPA listed; wh. fine powd., odorless; 92.7% > 1 μm; sp.gr. 2.1; bulk dens. 0.42 g/ml; surf. area 17 m^2/g; oil absorp. 50 ml/100 g (linseed oil); ref. index 1.49-1.51; pH 8.5-9.0 (2% ethanol); hardness (Mohs) 2.0.

Alcamizer 2. [Kyowa Chem. Industry] Magnesium aluminum carbonate; CAS 96492-31-8, 11097-59-9; EINECS 234-319-3; heat stabilizer for PVC that reacts in a unique way with HCl in PVC; offers high heat stability, nontoxicity, high transparency, good weatherability; recommended for rigid and flexible applics.; JHPA listed; wh. fine powd., odorless; 92.7% < 1 μm; sp.gr. 2.1; bulk dens. 0.45 g/ml; surf. area 20 m^2/g; oil absorp. 60 ml/100 g (linseed oil); ref. index 1.49-1.51; pH 8.5-9.0; hardness (Mohs) 2.0.

Alcamizer 4. [Kyowa Chem. Industry] Magnesium aluminum carbonate; CAS 119758-00-8; EINECS 234-319-3; heat stabilizer for PVC that reacts in a unique way with HCl in PVC; offers high heat stability, nontoxicity, high transparency, good weatherability; recommended for rigid and flexible applics.; JHPA listed; wh. fine powd., odorless; 89.8% < 1 μm; sp.gr. 2.3; bulk dens. 0.40 g/ml; surf. area 8 m^2/g; oil absorp. 50 ml/100 g (linseed oil); ref. index 1.53-1.55; pH 8.0-9.5; hardness (Mohs) 2.0.

Alcamizer 4-2. [Kyowa Chem. Industry] Magnesium aluminum carbonate; CAS 136618-52-5; EINECS 234-319-3; heat stabilizer for PVC that reacts in a unique way with HCl in PVC; offers high heat stability, nontoxicity, high transparency, good weatherability; recommended for rigid applics.; JHPA listed; wh. fine powd., odorless; 91.2% < 1 μm; sp.gr. 2.3; bulk dens. 0.44 g/ml; surf. area 11 m^2/g; oil absorp. 60 ml/100 g (linseed oil); ref. index 1.53-1.55; pH 8.0-9.5; hardness (Mohs) 2.0.

Alcamizer 5. [Kyowa Chem. Industry] Magnesium aluminum carbonate; CAS 136618-51-4; heat stabilizer for PVC that reacts in a unique way with HCl in PVC; offers high heat stability, nontoxicity, high transparency, good weatherability; recommended for joining part of urethane; wh. fine powd., odorless; 84.3% < 1 μm, 14.2% 1-2 μm; sp.gr. 2.2; bulk dens. 0.55 g/ml; surf. area 11 m^2/g; oil absorp. 70 ml/100 g (linseed oil); ref. index 1.49-1.51; pH 8.5-9.0; hardness (Mohs) 2.0.

Alcan BW53, BW103, BW153. [Alcan] Alumina, hydrate; CAS 1333-84-2; filler for flame retardants, polyester and acrylic resins in syn. marble; high whiteness; also in paper coating and chemical processes; 50, 15, 3 μm avg. grain diam.; ref. index 1.57; 99.8% conc.

Alcan FRF 5. [Alcan] Alumina hydrate; CAS 1333-84-2; flame retardant and smoke suppressant props. for plastics and rubber industries; wh. cryst. powd.; 85% on 325 mesh; sp.gr. 2.42; bulk dens. 1.2 g/cc (untamped); oil absorp. 15 g/100 g; decomp. temp. 200-250 C; ref. index 1.57.

Alcan FRF 10. [Alcan] Alumina hydrate; CAS 1333-84-2; flame retardant and smoke suppressant props. for plastics and rubber industries; wh. cryst. powd.; 15% on 325 mesh; sp.gr. 2.42; bulk dens. 0.9 g/cc (untamped); oil absorp. 20 g/100 g; decomp. temp. 200-250 C; ref. index 1.57.

Alcan FRF 20. [Alcan] Alumina hydrate; CAS 1333-84-2; flame retardant and smoke suppressant props. for plastics and rubber industries; wh. cryst. powd.; 10% on 325 mesh; sp.gr. 2.42; bulk dens. 0.8 g/cc (untamped); oil absorp. 21 g/100 g; decomp. temp. 200-250 C; ref. index 1.57.

Alcan FRF 30. [Alcan] Alumina hydrate; CAS 1333-84-2; flame retardant and smoke suppressant props. for plastics and rubber industries; wh. cryst. powd.; 0.02% on 325 mesh; sp.gr. 2.42; bulk dens. 0.7 g/cc (untamped); oil absorp. 22 g/100 g; decomp. temp. 200-250 C; ref. index 1.57.

Alcan FRF 40. [Alcan] Alumina hydrate; CAS 1333-84-2; flame retardant and smoke suppressant props. for plastics and rubber industries; wh. cryst. powd.; 0.01% on 325 mesh; sp.gr. 2.42; bulk dens. 0.7 g/cc (untamped); oil absorp. 22 g/100 g; decomp. temp. 200-250 C; ref. index 1.57.

Alcan FRF 60. [Alcan] Alumina hydrate; CAS 1333-84-2; flame retardant and smoke suppressant props. for plastics and rubber industries; wh. cryst. powd.; < 0.01% on 325 mesh; sp.gr. 2.42; bulk dens. 0.6 g/cc (untamped); oil absorp. 23 g/100 g; decomp. temp. 200-250 C; ref. index 1.57.

Alcan FRF 80. [Alcan] Alumina hydrate; CAS 1333-84-2; flame retardant and smoke suppressant props. for plastics and rubber industries; wh. cryst. powd.; < 0.01% on 325 mesh; sp.gr. 2.42; bulk dens. 0.6 g/cc (untamped); oil absorp. 24 g/100 g; decomp. temp. 200-250 C; ref. index 1.57.

Alcan FRF 80 S5. [Alcan] Alumina hydrate, coated with silane; CAS 1333-84-2; surf. coated fire retardant, smoke suppressant filler for polyester resin systems, epoxide sytems, PU polyols; wh. cryst. powd.; 0.1% max. on 325 mesh, 6-7 μm median particle size; sp.gr. 2.41; bulk dens. 0.7 g/cc (untamped); oil absorp. 19 g/100 g; decomp. pt. 200-250 C; ref. index 1.57.

Alcan FRF LV1 S5. [Alcan] Alumina hydrate, coated with silane; CAS 1333-84-2; surf. coated fire retardant, smoke suppressant filler for polyester resin systems, epoxide sytems, PU polyols; wh. cryst. powd.; 10% max. on 325 mesh, 8-11 μm median particle size; sp.gr. 2.41; bulk dens. 0.85 g/cc (untamped); oil absorp. 15 g/100 g; decomp. temp. 200-250 C; ref. index 1.57.

Alcan FRF LV2, LV4, LV5, LV6, LV7, LV8, LV9. [Alcan] Alumina hydrate; CAS 1333-84-2; flame retardant and smoke suppressant.

Alcan FRF LV2 S5. [Alcan] Alumina hydrate, coated with silane; CAS 1333-84-2; surf. coated fire retardant, smoke suppressant filler for polyester resin systems, epoxide sytems, PU polyols; wh. cryst. powd.; 60% max. on 325 mesh, 55 μm median particle size; sp.gr. 2.41; bulk dens. 1.0 g/cc (untamped); oil absorp. 13 g/100 g; decomp. pt. 200-250 C; ref. index 1.57.

Alcan H-10. [Alcan] Alumina hydrate; CAS 1333-84-2; for prod. of low-iron aluminum sulfate, alumina-based catalysts, sodium aluminate and other aluminum salts, coated titanium dioxide pigments, ceramic and glass prods., fire-retardant carpet backing; flame retardant for ABS, acrylic, epoxy, nitrile, phenolic, PC, thermoset polyester, PE, PP, PS, PVAc, PVC, PU.

Alcan Superfine 4. [Alcan] Alumina hydrate; CAS 1333-84-2; filler for fire retardants/smoke suppressants for plastics, rubber, paints, adhesives, adhesive tapes and in toothpaste, cosmetics, polishes and waxes; in paper coatings; wh. cryst. powd.; 0.1% max. on 325 mesh; 1.4 μm median particle size; sp.gr. 2.42; bulk dens. 0.25 g/cc (untamped); surf. area 4 m^2/g; oil absorp. 36; ref. index 1.57.

Alcan Superfine 7. [Alcan] Alumina hydrate; CAS 1333-84-2; filler for fire retardants/smoke suppressants for plastics, rubber, paints, adhesives, adhesive tapes and in toothpaste, cosmetics, polishes and waxes; in paper coatings; wh. cryst. powd.; 0.1% max. on 325 mesh, 1.0 μm median particle size; sp.gr. 2.42; bulk dens. 0.2 g/cc (untamped); surf. area 7 m^2/g; oil absorp. 42; ref. index 1.57.

Alcan Superfine 11. [Alcan] Alumina hydrate; CAS 1333-84-2; filler for fire retardants/smoke suppressants for plastics, rubber, paints, adhesives, adhesive tapes and in toothpaste, cosmetics, polishes and waxes; in paper coatings; wh. cryst. powd.; 0.1% max. on 325 mesh, 0.7 μm median particle size; sp.gr. 2.42; bulk dens. 0.2 g/cc (untamped); surf. area 11 m^2/g; oil absorp. 44; ref. index 1.57.

Alcan Ultrafine UF7. [Alcan] Alumina hydrate; CAS 1333-84-2; flame retardant and smoke suppressant for plastics and rubbers, cable, conveyor belting, thermoplastic moldings, PVC, paper coatings, toothpaste, paint (titania extender), adhesives, cosmetics; wh. ultrafine cryst. powd.; 1.2 μm median particle size; sp.gr. 2.42; bulk dens. 0.55 g/cc (untamped); surf. area 7 m^2/g; oil absorp. 26 g/100 g; ref. index 1.57; pH 9.5 (5% slurry).

Alcan Ultrafine UF11. [Alcan] Alumina hydrate; CAS 1333-84-2; flame retardant and smoke suppressant for plastics and rubbers, cable, conveyor belting, thermoplastic moldings, PVC, paper coatings, toothpaste, paint (titania extender), adhesives, cosmetics; wh. ultrafine cryst. powd.; 0.9 μm median particle size; sp.gr. 2.42; bulk dens. 0.55 g/cc (untamped); surf. area 11 m^2/g; oil absorp. 30 g/100 g; ref. index 1.57; pH 9.5 (5% slurry).

Alcan Ultrafine UF15. [Alcan] Alumina hydrate; CAS 1333-84-2; flame retardant and smoke suppressant for plastics and rubbers, cable, conveyor belting, thermoplastic moldings, PVC, paper coatings, toothpaste, paint (titania extender), adhesives, cosmetics; wh. ultrafine cryst. powd.; 0.8 μm median particle size; sp.gr. 2.42; bulk dens. 0.50 g/cc (untamped); surf. area 15 m^2/g; oil absorp. 37 g/100 g; ref. index 1.57; pH 10 (5% slurry).

Alcan Ultrafine UF25. [Alcan] Alumina hydrate; CAS 1333-84-2; flame retardant and smoke suppressant for plastics and rubbers, cable, conveyor belting, thermoplastic moldings, PVC, paper coatings, toothpaste, paint (titania extender), adhesives, cosmetics; wh. ultrafine cryst. powd.; 0.7 μm median particle size; sp.gr. 2.42; bulk dens. 0.55 g/cc (untamped); surf. area 25 m^2/g; oil absorp. 43 g/100 g; ref. index 1.57; pH 10.5 (5% slurry).

Alcan Ultrafine UF35. [Alcan] Alumina hydrate; CAS 1333-84-2; flame retardant and smoke suppressant for plastics and rubbers, cable, conveyor belting, thermoplastic moldings, PVC, paper coatings, toothpaste, paint (titania extender), adhesives, cosmetics; wh. ultrafine cryst. powd.; 0.5 μm median particle size; sp.gr. 2.42; bulk dens. 0.60 g/cc (untamped); surf. area 35 m^2/g; oil absorp. 50 g/100 g; ref. index 1.57; pH 11 (5% slurry).

Aldosperse® MO-50. [Lonza] Glycol monooleate and polysorbate 80; nonionic; antifog for PVC; liq.; 100% conc.

Alfol® 16. [Vista] Cetyl alcohol; CAS 36653-82-4; EINECS 253-149-0; detergent intermediate; emollient used in cosmetics; plastics additive; wh. waxy solid; typ. fatty alcohol odor; sol. in alcohol, acetone, ether; water-insol.; dens. 6.77; sp.gr. 0.813; visc. 6.77; m.p. 117 C; b.p. 604 C; 98.9% act.

Alfol® 18. [Vista] Stearyl alcohol; CAS 112-92-5; EINECS 204-017-6; surfactant intermediate; emollient used in cosmetics; plastics additive; wh. waxy solid; typ. fatty alcohol odor; sol. in alcohol, acetone, ether; water-insol.; dens. 6.71; sp.gr. 0.8075; visc. 13.5; m.p. 135 C; b.p. 640 C; 98.7% act.

Alfol® 20+. [Vista] Blend of C20 and higher even-carbon-number primary linear alcohols; lubricant for plastics, textiles, and metals; defoamer for paper, aq. slurries; water evaporation control; chem. intermediate for emulsifiers, biodeg. surfactants, plastic lubricants, syn. waxes and binders; FDA approved; off-wh. solid; sol. in alcohol, acetone, ether; water-insol.; m.w. 431; sp.gr. 0.817 (140/140 F); dens. 6.80; m.p. 113-129 F; b.p. > 650 F; iodine no. 8.7; sapon. no. 5.7; hyd. no. 157; flash pt. (PM) 390 F; 88.5% act.

Alfol® 1012 HA. [Vista] C10-C12 linear primary alcohol; EINECS 288-117-5; as lubricants in polymer processing, metal rolling and forming, emollients/lubricants for cosmetics, defoamers; intermediate for surfactants for cosmetic, household, and industrial use, esters for additives for plastics, rubber, metalworking, lubricants, and quaternaries for germicides and fabric softeners; colorless clear liq.; m.w. 164; sp.gr. 0.834; visc. 10.4 cSt (100 F); m.p. 35-40 F; b.p. 425-525 F; iodine no. 0.04; sapon. no. 0.1; hyd. no. 343; flash pt. (PM) 237 F; 99.8% conc.

Alfol® 1014 CDC. [Vista] C10-C14 linear primary alcohol; EINECS 283-066-5; as lubricants in polymer processing, metal rolling and forming, emollients/lubricants for cosmetics, defoamers; intermediate for surfactants for cosmetic, household, and industrial use, esters for additives for plastics, rubber, metalworking, lubricants, and quaternaries for germicides and fabric softeners; colorless clear liq.; m.w. 186; sp.gr. 0.836; visc. 12.5 cSt (100 F); m.p. 41-45 F; b.p. 450-545 F; iodine no. 0.07; sapon. no. 0.1; hyd. no. 302; flash pt. (PM) 250 F; 99% conc.

Alfol® 1214. [Vista] C12-C14 linear primary alcohol; EINECS 279-420-3; biodeg. lubricants in polymer processing, metal rolling and forming, emollients/lubricants for cosmetics, defoamers; intermediate for surfactants for cosmetic, household, and industrial use, esters for additives for plastics, rubber, metalworking, lubricants, and quaternaries for germicides and fabric softeners; colorless clear liq.; sweet, typ. fatty alcohol odor; sol. in alcohol, acetone, ether; water-insol.; m.w. 198; sp.gr. 0.838; dens. 7.0; visc. 14.3 cSt (100 F); m.p. 70-75 F; b.p. 518-575 F; iodine no. 0.05; sapon. no. 0.1; hyd. no. 284; flash pt. (PM) 265 F; 99.5% act.

Alfol® 1214 GC. [Vista] C12-C14 linear primary alcohol; EINECS 279-420-3; biodeg. lubricants in polymer processing, metal rolling and forming, emollients/lubricants for cosmetics, defoamers; intermediate for surfactants for cosmetic, household, and industrial use, esters for additives for plastics, rubber, metalworking, lubricants, and quaternaries for germicides and fabric softeners; colorless clear liq.; m.w. 195; sp.gr. 0.838; visc. 14.3 cSt (100 F); m.p. 70-75 F; b.p. 518-575 F; iodine no. 0.05; sapon. no. 0.18; hyd. no. 287; flash pt. (PM) 265 F; 99% act.

Alfol® 1216. [Vista] C12-C16 linear alcohols; CAS 68855-56-1; EINECS 272-490-6; lubricants in polymer processing, metal rolling and forming, emollients/lubricants for cosmetics, defoamers; intermediate for surfactants for cosmetic, household, and industrial use, esters for additives for plastics, rubber, metalworking, lubricants, and quaternaries for germicides and fabric softeners; colorless clear liq., sweet, typ. fatty alcohol odor; sol. in alcohol, acetone, ether; water-insol.; m.w. 203; sp.gr. 0.84; dens. 7.0; visc. 14.5 cSt (100 F); m.p. 63-70 F; b.p. 514-592 F; iodine no. 0.1; sapon. no. 0.5; hyd. no. 276; flash pt. (PM) 265 F; 99% act.

Alfol® 1216 CO. [Vista] C12-C14 linear primary alcohol; EINECS 279-420-3; lubricants in polymer processing, metal rolling and forming, emollients/lubricants for cosmetics, defoamers; intermediate for surfactants for cosmetic, household, and industrial use, esters for additives for plastics, rubber, metalworking, lubricants, and quaternaries for germicides and fabric softeners; colorless clear liq.; m.w. 198; sp.gr. 0.84; visc. 14.5 cSt (100 F); m.p. 63-70 F; b.p. 529-590 F; iodine no. 0.08; sapon. no. 0.18; hyd. no. 284; flash pt. (PM) 265 F; 99.7% act.

Alfol® 1218 DCBA. [Vista] C12-C18 linear primary alcohol; CAS 67762-25-8; EINECS 267-006-5; lubricants in polymer processing, metal rolling and forming, emollients/lubricants for cosmetics, defoamers; intermediate for surfactants for cosmetic, household, and industrial use, esters for additives for plastics, rubber, metalworking, lubricants, and quaternaries for germicides and fabric softeners; colorless clear liq.; m.w. 214; sp.gr. 0.84; visc. 15.0 cSt (100 F); m.p. 68-73 F; b.p. 525-660 F; iodine no. 0.11; sapon. no. 0.18; hyd. no. 262; flash pt. (PM) 275 F; 99.6% act.

Alfol® 1412. [Vista] C12-C14 linear primary alcohol; EINECS 279-420-3; lubricants in polymer processing, metal rolling and forming, emollients/lubricants for cosmetics, defoamers; intermediate for surfactants for cosmetic, household, and industrial use, esters for additives for plastics, rubber, metalworking, lubricants, and quaternaries for germicides and fabric softeners; wh. solid; m.w. 205; sp.gr. 0.839; visc. 14.4 cSt (100 F); m.p. 72-75 F; b.p. 525-585 F; iodine no. 0.1; sapon. no. < 1; hyd. no. 274; flash pt. (PM) 270 F; 99.7% act.

Alfol® 1416 GC. [Vista] C14-16 linear primary alcohol; EINECS 269-790-4; as lubricants in polymer processing, metal rolling and forming, emollients/lubricants for cosmetics, defoamers; intermediate for surfactants for cosmetic, household, and industrial use, esters for additives for plastics, rubber, metalworking, lubricants, and quaternaries for germicides and fabric softeners; wh. solid; m.w. 222; sp.gr. 0.822 (100/100 F); visc. 11.5 cSt (100 F); m.p. 95-99 F; b.p. 582-638 F; iodine no. < 0.04; sapon. no. < 1; hyd. no. 253; flash pt. (PM) 305 F; 99.8% conc.

Alfol® 1418 DDB. [Vista] C14, C16, C18 alcohol blend; EINECS 267-009-1; as lubricants in polymer processing, metal rolling and forming, emollients/lubricants for cosmetics, defoamers; intermediate for surfactants for cosmetic, household, and industrial use, esters for additives for plastics, rubber, metalworking, lubricants, and quaternaries for germicides and fabric softeners; wh. solid; m.w. 243; sp.gr. 0.819 (110/110 F); visc. 14.6 cSt (110 F); m.p. 97-102 F; b.p. 598-659 F; iodine no. 0.6; sapon. no. 0.5; hyd. no. 231; flash pt. (PM) 290 F; 99.9% act.

Alfol® 1418 GBA. [Vista] C14-18 linear primary alcohol; EINECS 267-009-1; as lubricants in polymer processing, metal rolling and forming, emollients/lubricants for cosmetics, defoamers; intermediate for surfactants for cosmetic, household, and industrial use, esters for additives for plastics, rubber, metalworking, lubricants, and quaternaries for germicides and fabric softeners; wh. solid; m.w.

227; sp.gr. 0.835 (100/100 F); m.p. 97-102 F; b.p. 598-660 F; iodine no. < 0.7; sapon. no. < 1; hyd. no. 247; flash pt. (PM) 305 F; 99.8% conc.

Alfol® 1618. [Vista] C16-18 linear primary alcohol; biodeg. lubricants in polymer processing, metal rolling and forming, emollients/lubricants for cosmetics, defoamers; intermediate for surfactants for cosmetic, household, and industrial use, esters for additives for plastics, rubber, metalworking, lubricants, and quaternaries for germicides and fabric softeners; wh. waxy solid, typ. fatty alcohol odor; sol. in water, alcohol, chloroform, ether, benzene, glacial acetic acid; sp.gr. 0.840 (60/60 F); dens. 6.81; visc. 15 cSt (122 F); m.p. 110-120 F; b.p. 628-662 F; iodine no. 0.15; sapon. no. 0.07; hyd. no. 219; flash pt. (PM) 325 F; 99.6% act.

Alfol® 1618 CG. [Vista] C16-18 linear primary alcohol; biodeg. lubricants in polymer processing, metal rolling and forming, emollients/lubricants for cosmetics, defoamers; intermediate for surfactants for cosmetic, household, and industrial use, esters for additives for plastics, rubber, metalworking, lubricants, and quaternaries for germicides and fabric softeners; wh. solid; m.w. 266; sp.gr. 0.820 (120/120 F); visc. 13.7 cSt (140 F); m.p. 110-120 F; b.p. 630-670 F; iodine no. 0.15; sapon. no. 0.07; hyd. no. 211; flash pt. (PM) 340 F; 99.6% act.

Alfol® 1618 GC. [Vista] C16-18 linear primary alcohol; as lubricants in polymer processing, metal rolling and forming, emollients/lubricants for cosmetics, defoamers; intermediate for surfactants for cosmetic, household, and industrial use, esters for additives for plastics, rubber, metalworking, lubricants, and quaternaries for germicides and fabric softeners; wh. solid; m.w. 263; sp.gr. 0.820 (140/140 F); m.p. 110-120 F; b.p. 630-670 F; iodine no. 0.8; sapon. no. 0.5; hyd. no. 213; flash pt. (PM) 325 F; 99.8% conc.

Alfonic® 1412-A Ether Sulfate. [Vista] Sulfated ethoxylated alcohol deriv., ammonium salt; anionic; detergent, wetting agent, emulsifier, foaming agent for detergent formulations, shampoos, emulsion polymerization, oil well drilling, agriculture; clear liq.; mild odor; dens. 8.4; visc. 76 cps; biodeg.; 58% act.; Unverified

Algoflon® L. [Ausimont] PTFE wax; CAS 9002-84-0; internal/external lubricant for thermoplastics, thermosets, elastomers, coatings; solid.

Alkamide® STEDA. [Rhone-Poulenc Surf. Canada] Ethylene bisstearamide; CAS 110-30-5; nonionic; additive in pulp and paper defoamer formulations; lubricant, mold release, antiblocking agent in calendering, extrusion, and inj. molding of PVC and other plastics; plasticizer, antistat, pigment dispersant for resins and plastics; flakes; 95% amide.

Alkamuls® EL-620L. [Rhone-Poulenc Surf. & Spec.] PEG-30 castor oil; CAS 61791-12-6; nonionic; emulsifier, dispersant, softener, rewetting agent, lubricant, emulsion stabilizer, dyeing assistant, antistat, solubilizer for textiles, wet-str. papers, fat liquoring, emulsion paints, oleoresinous binders, glass-reinforced plastics, PU foams, perfumes, cosmetics; low dioxane; EPA compliance; liq.; HLB 12.0; 10% act.

Alkamuls® EL-719. [Rhone-Poulenc Surf. & Spec.] PEG-40 castor oil; CAS 61791-12-6; nonionic; emulsifier, wetting agent for industrial/household cleaners; dispersant for pigments; for pesticides, paper, leather, plastics, paint, textile and cosmetics industries; emulsifier for vitamins and drugs; FDA, EPA compliance; liq.; sol. in water, acetone, CCl_4, alcohols, veg. oil, ether, toluene, xylene; sp.gr. 1.06-1.07; dens. 8.9-9.0 lb/gal; visc. 500-800 cps; HLB 13.6; cloud pt. 80 C (1% aq.); flash pt. 275-279 C; surf. tens. 38 dynes/cm (0.1%); 96% act.

Alkamuls® GMR-55LG. [Rhone-Poulenc Surf. Canada] Glyceryl mono/dioleate; CAS 25496-72-4; nonionic; coemulsifier, lubricant, softener, emollient for mold releases and spin finishes for syn. fibers; rust preventive additive in compounded oils; lubricant, antistat, antifog for PVC film processing; lt. amber liq. to paste, very mild char. odor; sol. in most aromatic solvs., disp. in min. oil, most aliphatic solvs., insol. in water; dens. 0.96 g/ml; m.p. 20-25 C; HLB 3.0; 42% monoglyceride.

Alkamuls® GMS/C. [Rhone-Poulenc Surf. & Spec.] Glyceryl stearate; CAS 31566-31-1; nonionic; emulsifier, wetting agent for cosmetic, agric., textile industries; coupler used to bind waxes together; emollient and thickener in cosmetic creams; internal antistat for PE, PP, PS, flexible and rigid PVC, BMC, SMC, epoxy, phenolic; FDA approved; flake; water-disp.; m.p. 58-63 C; HLB 3.4; 100% conc.

Alkamuls® PSML-20. [Rhone-Poulenc Surf. & Spec.] Polysorbate 20; CAS 9005-64-5; nonionic; emulsifier, solubilizer, antistat, visc. modifier, lubricant for textiles, cosmetics, pharmaceuticals; antistat for rigid PVC; FDA approved as indirect additive; yel. liq.; sol. in water, aromatic solv.; dens. 1.1 g/ml; HLB 16.7; sapon. no. 40-50; 97% act.

Alkamuls® R81. [Rhone-Poulenc Surf. & Spec.] PEG-18 castor oil; CAS 61791-12-6; nonionic; softener, antistat for textile finishing, plastics processing; liq.; 99% conc.

Alkamuls® S-8. [Rhone-Poulenc Surf. & Spec.] PEG-8 stearate; CAS 9004-99-3; nonionic; emulsifier, self-emulsifying lubricant and softener for syn. fibers; internal antistat for PS; FDA approved; solid; HLB 11.2; 100% conc.

Alkamuls® SML. [Rhone-Poulenc Surf. Canada] Sorbitan laurate; CAS 1338-39-2; nonionic; emulsifier for oils and fats in cosmetic, metalworking and industrial oil prods.; corrosion inhibitor; lubricant and antistat for PVC; amber liq.; typ. odor; disp. in water; moderately sol. most alcohols, veg. and min. oils; sp.gr. 1.05 (60 F); dens. 8.330 lb/gal; HLB 8.6; sapon. no. 160-170; hyd. no. 320-350; flash pt. > 200 C; 100% act.

Alkanol® 189-S. [DuPont] Sodium alkyl sulfonate; anionic; wetting agent, detergent, penetrant, foamer for textiles, elastomers, plastics, film, metal cleaning and pickling, hard surf. cleaning, and chemical mfg.; effective in acid and alkali media; reddish-br. liq., alcoholic odor; sol. in water; sp.gr.

1.06 g/mL; dens. 8.8 lb/gal; visc. 30 cps; cloud pt. < 0 C; flash pt. (PMCC) 21 C; pH 7.5-9.0 (1%); surf. tens. 38 dynes/cm (0.1%); flamm.; 31.5% act.; contains IPA.

Alkapol PEG 300. [Rhone-Poulenc Surf. & Spec.] PEG-6; CAS 25322-68-3; EINECS 220-045-1; intermediate for surfactants; binder/lubricant in pharmaceuticals; plasticizer; paper softener; humectant; solvent; antistat; for cosmetics, textile, plastics processing, dyes and inks; liq.; m.w. 285-315; water-sol.; dens. 1.13 g/ml; pH 5-8 (5% DW); 100% conc.

Alkasurf® NP-4. [Rhone-Poulenc Surf. & Spec.] Nonoxynol-4; CAS 9016-45-9; nonionic; emulsifier, detergent, dispersant, intermediate, stabilizer; plasticizer, antistat for plastics, surfactants, household, industrial, and cosmetic use, fat liquoring, cutting and sol. oils; lt. liq.; low odor; water insol.; sp.gr. 1.02; HLB 9; 100% act.

Alkasurf® NP-6. [Rhone-Poulenc Surf. & Spec.] Nonoxynol-6; CAS 9016-45-9; nonionic; emulsifier, coemulsifier, oil-sol. dispersant; used for household and industrial cleaners; intermediate; plasticizer and antistat for plastics; emulsifier for min. oils; insecticides, fungicides, herbicides; fat liquoring; making of cutting and sol. oils; dispersing waxes, pigments, resins; printing; preparation of emulsified paint; lt. liq.; low odor; oil-sol., water insol.; sp.gr. 1.04; HLB 11; 100% act.

Allco BTDA. [Allco] 3,3′,4,4′-Benzophenonetetracarboxylic dianhydride; CAS 2421-28-5; epoxy curing agent; monomer in unsat. and saturated polyester resins; ester derivs. as intermediates in prod. of polyimides, lubricants, plasticizers; intermediate for alkyd, PU resins, polypyrrones; end-uses incl. electronic transfer molding compds., wire coatings, aerospace adhesives, advanced structure composites, uv-cured inks, insulating coatings and foams, flame retardant prods., release agents, molded engineering parts, refinery catalysts; lt. tan free-flowing cryst. powd., odorless, tasteless; sol. > 16 g/100 g in methyl pyrrolidone, dimethylformamide, dimethyl sulfoxide; 0.26 g/100 g water; m.w. 322.22; sp.gr. 1.57 (30 C); vapor pressure 0.5 mm Hg; m.p. 221-225 C; b.p. 380-400 C; pH 2.4 (2.1 g/L); 97-100% act.; *Toxicology:* fumes from molten BTDA are highly toxic; solid considered nontoxic; LD50 (oral, rats) > 12,800 mg/kg (as tetra-acid); may cause respiratory irritation/allergic response on overexposure.

Allco PMDA. [Allco] Pyromellitic acid dianhydride; CAS 2421-28-5; intermediate for polyimides, polyester resins, alkyd and PU resins, polypyrrones; epoxy curing agent; end-uses incl. electronic transfer molding compds., wire coatings, aerospace adhesives, advanced structural composites, uv-cured inks, insulating coatings and foams, flame retardant prods., release agents, molded high-performance engineering parts; sol. 0.4 g/100 g in water, reacts to give tetra-acid; m.w. 218.12; dens. 1.680 (20 C); vapor pressure 63 mm Hg (290 C); m.p. 284-286 C; b.p. 380-400 C; pH 2.2 (3.9 g/L); 98-100% act.; *Toxicology:* LD50 (oral, rat) 2250-2595 mg/kg; skin and eye irritant, but nonsensitizing; may cause respiratory irritation/allergic response on inhalation of dust or fumes.

AlNel A-100. [Advanced Refractory Tech.] Aluminum nitride; CAS 24304-00-5; EINECS 246-140-8; filler for elastomeric, epoxy, polymeric, and other filled systems; provides high thermal conductivity, good dielec. strength, corrosion resistance; gray to wh. powd.; 3.0-4.0 μ avg. particle diam.; dens. 3.26 g/cc; sublimes at 2450 C; conduct. 160-220 W/mK; dielec. const. 8.2-9.0; Unverified

AlNel A-200. [Advanced Refractory Tech.] Aluminum nitride; CAS 24304-00-5; EINECS 246-140-8; filler for elastomeric, epoxy, polymeric, and other filled systems; provides high thermal conductivity, good dielec. strength, corrosion resistance; gray to wh. powd.; 3.0-4.0 μ avg. particle diam.; dens. 3.26 g/cc; sublimes at 2450 C; conduct. 100-150 W/mK; dielec. const. 8.0-9.0; Unverified

AlNel AG. [Advanced Refractory Tech.] Aluminum nitride; CAS 24304-00-5; EINECS 246-140-8; filler for elastomeric, epoxy, polymeric, and other filled systems; provides high thermal conductivity, good dielec. strength, corrosion resistance; fine to coarse grains; Unverified

Alperox-F. [Elf Atochem N. Am./Org. Peroxides] Lauroyl peroxide; CAS 105-74-8; initiator for bulk, sol'n., and suspension polymerization, high-temp. curing of polyester resins, and cure of acrylic syrup; flakes; 98% act.; 3.93% act. oxygen.

Alscoap LN-40, LN-90. [Toho Chem. Industry] Sodium lauryl sulfate; CAS 151-21-3; EINECS 205-788-1; anionic; foaming agent, detergent, base for shampoos, detergents, toothpaste; polymerization emulsifier for syn. resins and latex; liq. and powd. resp.; 40 and 90% conc.

Altax®. [R.T. Vanderbilt] Benzothiazyl disulfide (CAS 120-78-5), zinc stearate (CAS 557-05-1), petrol. process oil (CAS 64742-55-8 or 64742-56-9); accelerator for nat. and syn. rubbers; primary accelerator and scorch modifying sec. accelerator in NR and SBR copolymers; retarder-plasticizer in neoprene (G types); cure modifier in W types; cream to lt. yel. powd.; also avail. in rods, dustless and thread-grade; negligible sol. in water; m.w. 332.48; dens. 1.51 mg/m^3; m.p. 159-170 C; flash pt. (COC) 271 C; *Toxicology:* LD50 (oral, rat) 7940 mg/kg, (dermal, rabbit) 7940 mg/kg; eye and skin irritant; inh. may irritate respiratory tract; skin sensitizer; OSHA PEL/TWA 5 mg/m^3 (respirable dust), 15 mg/m^3 (total dust); petrol. process oil (as mist): OSHA 5 mg/m^3; *Precaution:* may form explosive dust-air mixts.; incompat. with strong oxidizers, acids; hazardous decomp. prods.: oxides of carbon, sulfur, nitrogen; thiazole fragments; aromatic/aliphatic hydrocarbons; *Storage:* store in cool place away from food, acids, oxidizers; keep containers closed when not in use.

Aluminum Oxide C. [Degussa] Alumina; CAS 1344-28-1; EINECS 215-691-6; free-flow and anticaking agent; aids in reducing electrostatic charges of powder substances for personal care and elec. industry; antistat for PVC powd.; wh. fluffy powd., odorless; 13 nm avg. particle size; insol. in water; dens. ≈ 50 g/l (tapped); surf. area 100 m^2/g; m.p. > 1800 C; pH 4.5-5.5 (4% aq. disp.); > 99.6 assay; *Toxicology:* TLV 10 mg/m^3 total dust; LD50 (oral, rat) > 5000 mg/kg; dust may cause respiratory irritation.

Alvinox 100. [3V] Phenolic; antioxidant for PE, PP, EPDM; FDA 21CFR §178.2010.

Alvinox FB. [3V] Phenolic; antioxidant for PE, PP, PVC; synergistic with uv stabilizers; FDA 21CFR §178.2010.

Amax®. [R.T. Vanderbilt] N-Oxydiethylene benzothiazole-2-sulfenamide; CAS 102-77-2; accelerator for SBR, NR, IR, BR; delayed action primary accelerator and scorch modifying sec. accelerator; tan flake; m.w. 252.35; dens. 1.37 mg/m3; m.p. 70-90 C.

Amgard® CHT. [Albright & Wilson Am.] Stabilized encapsulated red phosphorus with dust suppressant; effective flame retardant at low loadings for thermoplastic and thermoset resins; 95% P.

Amgard® CPC 452. [Albright & Wilson Am.] Red phosphorous-based addtive in nylon 6,6 carrier resin; flame retardant additive for nylon resins; for inj. molding, glass-filled, and extrusion resins and elec. applics.; red pellets; insol.; dens. 1.3 g/cc; bulk dens. 0.6 g/cc; 50% P, < 0.5% moisture; *Toxicology:* nonhazardous material; *Storage:* thoroughly dry prior to use.

Amgard® EDAP. [Albright & Wilson Am.] Phosphorous-based additive; flame retardant additive for polypropylene, olefin polymers for extrusion, molding, film, and fiber applics. and for thermosets (polyester, epoxy, epoxy/thermoplastic blends); exc. thermal stability, high efficiency (UL94 VO in PP @ 30% loadings); wh. free-flowing powd.; < 5 μ particle size; dens. 1.22 g/cc; bulk dens. 0.63 g/cc; m.p. dec. above 250 C; 19.6% P, 62% H_3PO_4; *Toxicology:* nonhazardous material; avoid prolonged/repeated skin contact, inhalation of dust, eye contact.

Amgard® MC. [Albright & Wilson Am.] Ammonium polyphosphate; flame retardant additive for intumescent coatings, plastics, and rubber; low water-sol.; 29.5% P.

Amgard® ND. [Albright & Wilson Am.] Dimelamine phosphate; flame retardant for intumescent coatings, plastics, and paper; very low water-sol.; 8.3% P, 45.8% N.

Amgard® NH. [Albright & Wilson Am.] Melamine phosphate; flame retardant for intumescent coatings, plastics, and paper; low water-sol.; 13.4% P, 38.6% N.

Amgard® NK. [Albright & Wilson Am.] Ethylene diamine phosphate; high-performance flame retardant for olefin polymers providing low dens. prods. with exc. impact resist. and elong.; suitable for black or dark colored applics.; contributes to surf. color formation; recommended for EVA polymers.

Amgard® NP. [Albright & Wilson Am.] Ethylene diamine phosphate; flame retardant for thermoplastics (PP homopolymers and copolymers, TPE and TPU used in extrusion, molding and film applics.) and thermosets (polyester, epoxies); produces low dens. flame-retardant polyolefins with high elong. and impact resist.; exc. thermal stability, high efficiency; produces UL94 VO @ 1/16 in. in PP @ 30% loadings; wh. free-flowing powd.; < 8 μ particle size; dens. 1.20 g/cc; bulk dens. 0.63 g/cc; m.p. dec. > 250 C; 15.7% P; *Toxicology:* not considered hazardous, but avoid inhalation of dust, eye contact, prolonged/repeated skin contact.

Amgard® PI. [Albright & Wilson Am.] Ammonium polyphosphate; flame retardant for natural rubber and particle board; limited water-sol.; 25-26% P.

Amgard® TBEP. [Albright & Wilson Am.] Tributoxyethyl phosphate; CAS 78-51-3; EINECS 201-122-9; plasticizer and defoamer for plastics; leveling agent for acrylic and styrenic floor polishes; clear mobile liq.; 7.8% P.

Amgard® $TBPO_4$. [Albright & Wilson Am.] Tributyl phosphate; CAS 126-73-8; EINECS 204-800-2; flame retardant plasticizer in plastics and coatings; solvent extractant; in hydraulic fluids because of its low flamm. props.; 11.6% P.

Amgard® TOF. [Albright & Wilson Am.] Trioctyl phosphate; CAS 78-42-2; EINECS 201-116-6; outstanding low-temp. plasticizer for vinyl resins, syn. rubbers, nitrocellulose; contributes to low flamm. chars.; clear mobile liq.; 7.1% P.

Amical® 85. [Franklin Industrial Minerals] Calcium carbonate; filler designed for max. loading in resin systems incl. polyester (BMC, SMC, RTM, spray-up), PV C (wire and cable insulation, plastisols, floor tile, extrusions), PP, HDPE, LLDPE, nylon, rubber (automotive goods, flooring, footwear), adhesives, caulks, ceramics, sealants; off-wh. free-flowing powd.; 6.0-8.0 μ median particle size; 14-20% on 325 mesh; sp.gr. 2.71; brightness 74-80; pH 8.8-9.5; hardness (Mohs) 2.5; 97% $CaCO_3$, 0.2% moisture.

Amical® 95. [Franklin Industrial Minerals] Calcium carbonate; filler designed for max. loading in resin systems incl. polyester (BMC, SMC, RTM, spray-up), PV C (wire and cable insulation, plastisols, floor tile, extrusions), PP, HDPE, LLDPE, nylon, rubber (automotive goods, flooring, footwear), adhesives, caulks, ceramics, sealants; off-wh. free-flowing powd.; 7 μ mean particle size; 7% on 325 mesh; sp.gr. 2.71; brightness 74-80; pH 8.8-9.5; hardness (Mohs) 2.5; 97% $CaCO_3$, 0.2% moisture.

Amical® 101. [Franklin Industrial Minerals] Calcium carbonate; filler designed for max. loading in resin systems incl. polyester (BMC, SMC, RTM, spray-up), PV C (wire and cable insulation, plastisols, floor tile, extrusions), PP, HDPE, LLDPE, nylon, rubber (automotive goods, flooring, footwear), adhesives, caulks, ceramics, sealants; off-wh. free-flowing powd.; 4.5 μ mean particle size; 0.5% on 325 mesh; sp.gr. 2.71; brightness 74-80; pH 8.8-9.5; hardness (Mohs) 2.5; 97% $CaCO_3$, 0.2% moisture.

Amical® 202. [Franklin Industrial Minerals] Calcium carbonate; filler designed for max. loading in resin systems incl. polyester (BMC, SMC, RTM, spray-up), PVC (wire and cable insulation, plastisols, floor tile, extrusions), PP, HDPE, LLDPE, nylon, rubber (automotive goods, flooring, footwear) adhesives, caulks, ceramics, sealants; NSF approved for PVC systems; off-wh. free-flowing powd.; 3.2 μ mean particle size; 0% on 325 mesh; sp.gr. 2.71; brightness 74-80; pH 8.8-9.5; hardness (Mohs) 2.5; 97% $CaCO_3$, 0.2% moisture.

Amical® 303. [Franklin Industrial Minerals] Calcium carbonate; filler for polyester (BMC, SMC, RTM), rubber (automotive goods, flooring, footwear), PVC (wire/cable insulation, plastisols, extrusions), PP, HDPE, LLDPE, nylon, adhesives, caulks, ceramics, sealants; NSF approved for PVC systems; off-wh. free-flowing powd.; 2.3 μ mean particle size; 0% on 325 mesh; sp.gr. 2.71; brightness 74-80; pH 8.8-9.5; hardness (Mohs) 2.5; 97% $CaCO_3$, 0.2% moisture.

Amical® SC2. [Franklin Industrial Minerals] Stearate surface-coated calcium carbonate; filler for plastics, paints, rubber industries; off-wh. free-flowing powd.; 2.3 μ mean particle size; 0% on 325 mesh; sp.gr. 2.71; brightness 72-80; pH 8.8-9.5; hardness (Mohs) 2.00; 97% $CaCO_3$.

Amical® SC3. [Franklin Industrial Minerals] Stearate surface-coated calcium carbonate; filler for plastics, paints, rubber industries; hydrophobic with enhanced processability; surface coating is compat. with PVC, polyolefins, silicones, and engineering plastics; provides higher loadings, improved filler-polymer bonding, reduced abrasion, improved dispersion; off-wh. free-flowing finely ground powd.; 2.9-3.3 μ median particle size; 0% on 325 mesh; sp.gr. 2.71; brightness 72-80; pH 8.8-9.5; hardness (Mohs) 2.00; 97% $CaCO_3$.

Amical® SC5. [Franklin Industrial Minerals] Stearate surface-coated calcium carbonate; filler for plastics, paints, rubber industries; hydrophobic with enhanced processability; surface coating is compat. with PVC, polyolefins, silicones, and engineering plastics; provides higher loadings, improved filler-polymer bonding, reduced abrasion, improved dispersion; off-wh. free-flowing finely ground powd.; 3.5-5.5 μ median particle size; 0.5% on 325 mesh; sp.gr. 2.71; brightness 72-80; pH 8.8-9.5; hardness (Mohs) 2.00; 97% $CaCO_3$.

Amical® SC7. [Franklin Industrial Minerals] Stearate surface-coated calcium carbonate; filler for plastics, paints, rubber industries; hydrophobic with enhanced processability; surface coating is compat. with PVC, polyolefins, silicones, and engineering plastics; provides higher loadings, improved filler-polymer bonding, reduced abrasion, improved dispersion; off-wh. free-flowing finely ground powd.; 7 μ mean particle size; 7% on 325 mesh; sp.gr. 2.71; brightness 72-80; pH 8.8-9.5; hardness (Mohs) 2.00; 97% $CaCO_3$.

Amicure® 33-LV. [Air Prods./Perf. Chems.] Triethylene diamine in dipropylene glycol; catalyst for PU coatings, adhesives, sealants; 33% act.

Amicure® 101. [Air Prods./Perf. Chems.] Modified aromatic amine; elevated temp. epoxy curing agent; nonstaining, MDA-free; exhibits lower exotherm and higher heat resist. than MDA; for filament-wound pipe, elec. encapsulation, tooling, large castings, adhesives; Gardner 9 liq.; sp.gr. 1.00; visc. 100-200 cps; amine no. 580; equiv. wt. 48.5; heat distort. temp. 311 F (264 psi).

Amicure® CG-325. [Air Prods./Perf. Chems.] Dicyandiamide, 0.5% inert flow control additive; CAS 461-58-5; EINECS 207-312-8; epoxy curing agent for structural laminates, one-component adhesives; powd., < 45 μ; sp.gr. 0.42; m.p. 403 F; heat distort. temp. 250 F (264 psi).

Amicure® CG-1200. [Air Prods./Perf. Chems.] Dicyandiamide, 0.5% inert flow control additive; CAS 461-58-5; EINECS 207-312-8; epoxy curing agent for powd. coatings, structural laminates incl. solv.-free prepregs, adhesives, film adhesives; powd., < 10 μ; sp.gr. 0.30; m.p. 403 F; heat distort. temp. 250 F (264 psi).

Amicure® CG-1400. [Air Prods./Perf. Chems.] Dicyandiamide, 0.5-1% inert flow control additive; CAS 461-58-5; EINECS 207-312-8; epoxy curing agent for powd. coatings, film adhesives; powd., < 5 μ; sp.gr. 0.28; m.p. 403 F; heat distort. temp. 250 F (264 psi).

Amicure® CG-NA. [Air Prods./Perf. Chems.] Dicyandiamide, unpulverized, with no inert flow control agent; CAS 461-58-5; EINECS 207-312-8; epoxy cure agent for elec. laminates, adhesives, powd. coatings where the total resin system is pulverized prior to extrusion; powd.; < 180 μ; sp.gr. 0.60; m.p. 403 F; heat distort. temp. 250 F (264 psi).

Amicure® CL-485. [Air Prods./Perf. Chems.] Aliphatic amine tetrol; crosslinker and reactivity enhancer for two-part PU coatings, adhesives, and sealants; wh. liq.; sp.gr. 1.03; visc. 50,000 cps; hyd. no. 770; 99% min. tert. amine; *Toxicology:* moderate eye irritant; *Storage:* store @ 25-55 C; hygroscopic—cover with dry air or nitrogen atmosphere; handle in heated equip.

Amicure® DBU-E. [Air Prods./Perf. Chems.] Diazabicycloundecene; CAS 6674-22-2; EINECS 229-713-7; high purity electronic grade; elevated temp. highly efficient accelerator for phenolic novolac and other epoxy cures, incl. those cured with anhydrides; for elec. encapsulation and transfer molding powds.; lt. yel. liq.; m.w. 152; sp.gr. 1.11; visc. 14 cps.

Amicure® PACM. [Air Prods./Perf. Chems.] Bis (p-aminocyclohexyl) methane; CAS 1761-71-3; elevated temp. epoxy curing agent providing improved high-temp. performance, increased shear str. at elevated temps. in two-component ambient or heat-cure adhesive formulations for plastic or metal bonding, filament winding, wet lay-up laminating; casting, RIM, pultrusion for general industrial, tooling, and automotive applics.; Gardner 2 liq.; sp.gr. 0.96; visc. 80 cps; amine no. 526; equiv. wt. 52.5; heat distort. temp. 300 F (264 psi).

Amicure® SA-102. [Air Prods./Perf. Chems.] Diazabicycloundecene, 2-ethylhexanoic acid salt; intermediate reactivity elevated temp. epoxy curing agent for elec. insulation applics. incl. casting, potting, dip coating, and impregnation; lt. yel. liq.; m.w. 310; sp.gr. 1.02; visc. 3700 cps.

Amicure® TEDA. [Air Prods./Perf. Chems.] Triethylene diamine; CAS 280-57-9; catalyst for PU coatings, adhesives, sealants; minimal catalytic activity; solid.

Amicure® UR. [Air Prods./Perf. Chems.] 1-Phenyl-3,3-dimethyl urea; CAS 101-42-8; elevated temp. co-curing accelerator for dicyandiamide-cured epoxy resins; exceptional latency and rapid cure above activation temp.; substitute for chlorophenyl ureas; for one-component adhesives, high-performance composites, prepregs; wh. powd.; m.w. 165; m.p. 268 F.

Amicure® UR2T. [Air Prods./Perf. Chems.] Tolyl bis (dimethyl urea); co-curing accelerator for dicyandiamide-cured epoxy resins; faster green str. adhesion build than Amicure UR, used at lower loadings; substitute for chlorphenyl ureas; for one-component adhesives, high-performance composites, prepregs; wh. powd.; m.w. 272; m.p. 360-374 F.

Amine C. [Ciba-Geigy] Imidazoline deriv.; cationic; emulsifier, dispersant, detergent; used in acid cleaners; antistat for textiles, plastics; paints; corrosion inhibitor; cutting and rust preventive oils; yel. liq.; sol. in polar org. solv., hydrocarbons, relatively insol. water; 100% act.; Unverified

Amine HH. [Air Prods./Perf. Chems.] Mixed polyamines; room or elevated temp. epoxy curing agent for general civil engineering applics., patch repair compds., heavily filled mortars, abrasion-resist. floorings; economical alternative; intermediate in reactivity and performance between AEP and TETA; Gardner 5 liq.; sp.gr. 0.999; visc. 10 cps; amine no. 1250; equiv. wt. 37; heat distort. temp. 126 F (264 psi).

Aminox®. [Uniroyal] Diphenylamine-acetone reaction prod.; CAS 9003-79-6; antioxidant protecting polymers from loss of physical props. on extended exposure to heat; FDA approved for EVA and polyamides; FDA approved; flake or powd.

Aminox® Naugard A. [Uniroyal] Antioxidant protecting polymers from loss of physical props. on extended exposure to heat; for EVA and polyamide hot-melt adhesives, nylon 6.

Amoco® H-15. [Amoco] Polybutene (isobutylene butene copolymer); used as tackifier, strengthener, and extender in adhesives, as plasticizer for rubber, as vehicle and fugitive binder for coatings, as cling additive for LLDPE stretch wrap films, as reactive intermediate for specialty chemicals, as leather impregnant, as vehicle or modifier for caulks, sealants, and glazing compds., and in lubricants, paper treatments, elec. compds.; FDA compliance; visc. liq.; sp.gr. 0.860-0.871 (15.5 C); visc. 2441 SUS (38 C); pour pt. -35 C; PMCC flash pt. 138 C; acid no. 0.29; ref. index 1.4847.

Amoco® H-300E. [Amoco] Hydrog. polybutene; CAS 68937-10-0; used for cosmetics (lip gloss, baby oil), metalworking lubricants, cable oils, thermoplastic modifier, rubber modifier; FDA compliance; bright, clear visc. liq.; misc. with min. oils, org. solvs.; sp.gr. 0.86-0.91; visc. 600-655 cSt (99 C); pour pt. 2 C; flash pt. 235 C; acid no. < 0.1.

Amoco® L-14. [Amoco] Polybutene; used as tackifier, strengthener, and extender in adhesives, as plasticizer for rubber, as vehicle and fugitive binder for coatings, as cling additive for LLDPE stretch wrap films, as reactive intermediate for specialty chemicals, as leather impregnant, as vehicle or modifier for caulks, sealants, and glazing compds., and in lubricants, paper treatments, elec. compds.; FDA compliance; visc. liq.; sp.gr. 0.830-0.845 (15.5 C); visc. 139 SUS (38 C); pour pt. -51 C; PMCC flash pt. 121 C; acid no. 0.18; ref. index 1.4680.

Amoco® L-50. [Amoco] Polybutene (isobutylene butene copolymer); used as tackifier, strengthener, and extender in adhesives, as plasticizer for rubber, as vehicle and fugitive binder for coatings, as cling additive for LLDPE stretch wrap films, as reactive intermediate for specialty chemicals, as leather impregnant, as vehicle or modifier for caulks, sealants, and glazing compds., and in lubricants, paper treatments, elec. compds.; FDA compliance; sp.gr. 0.845-0.860 (15.5 C); visc. 504 SUS (38 C); pour pt. -40 C; PMCC flash pt. 135 C; acid no. 0.23; ref. index 1.4758.

Ampacet 10053. [Ampacet] Antistat additive in LDPE carrier for thin gauge film applics., inj. molding; FDA approved for food pkg. use @ up to 2% in film, 3% in inj. molding; 5% act.

Ampacet 10057. [Ampacet] 10% Benzophenone in LDPE carrier; UV absorber for thin gauge film applics.; FDA approved for food pkg. use.

Ampacet 10061. [Ampacet] Slip additive in LDPE carrier; for thin gauge film applics.; FDA approved for food pkg. use; 5% amides.

Ampacet 10063. [Ampacet] 20% Diatomaceous silica in LDPE carrier; antiblock additive for thin gauge film applics.; FDA approved for food pkg. use.

Ampacet 10117. [Ampacet] Slip additive in LDPE carrier for thin gauge film applics.; FDA approved for food pkg. use; 5% amides.

Ampacet 10123. [Ampacet] 10% Azodicarbonamide in LDPE carrier; foam additive for thin gauge film applics.; FDA approved for food pkg. use.

Ampacet 10286. [Ampacet] Purge additive in LDPE carrier for thin gauge film applics.; heavy metals-free.

Ampacet 10478. [Ampacet] 10% Hindered amine light stabilizer, 2.5% hindered phenol antioxidant in LDPE carrier; UV inhibitor for thin gauge film applics.; FDA approved up to 2.5%.

Ampacet 10562. [Ampacet] Process aid in LLDPE carrier; for elimination of melt fracture; FDA approved for food pkg. use.

Ampacet 10886. [Ampacet] Antioxidant in LDPE carrier; for thin gauge film applics.; FDA approved for food pkg. use; 5% act.

Ampacet 10919. [Ampacet] Process aid in LLDPE

carrier; for elimination of melt fracture; FDA approved for food pkg. use.

Ampacet 11040. [Ampacet] 50% Rutile titanium dioxide in LLDPE carrier resin; wh. conc. for blown film, inj. and blow molding applics.; typical letdown 9:1; FDA approved for food pkg. use; sp.gr. 1.51.

Ampacet 11058. [Ampacet] 70% Rutile titanium dioxide in LDPE carrier resin; wh. conc. for blown and cast film, inj. molding applics.; typical letdown 9:1; FDA approved for food pkg. use; sp.gr. 2.02.

Ampacet 11070. [Ampacet] 50% Rutile titanium dioxide in LLDPE carrier resin with stearate; wh. conc. for blown film, inj. and blow molding applics.; typical letdown 19:1; FDA approved for food pkg. use; sp.gr. 1.51.

Ampacet 11078. [Ampacet] 50% Rutile titanium dioxide in LDPE carrier resin with stearate; wh. color conc. for blown film, inj. and blow molding applics.; typical letdown 9:1; FDA approved for food pkg. use; sp.gr. 1.52.

Ampacet 11171. [Ampacet] 50% Rutile titanium dioxide in LDPE carrier resin with antioxidant; wh. conc. for extrusion coating applics.; typical letdown 9:1; FDA approved for food pkg. use; sp.gr. 1.51.

Ampacet 11171-101. [Ampacet] 50% Rutile titanium dioxide in LDPE carrier resin with stearate; wh. conc. for extrusion coating applics.; typical letdown 4:1; FDA approved for food pkg. use; sp.gr. 1.51.

Ampacet 11187. [Ampacet] 70% Rutile titanium dioxide in ethylene methyl acrylate copolymer carrier resin with stearate; wh. conc. for extrusion coating applics.; typical letdown 5:1; FDA approved for food pkg. use; sp.gr. 2.07.

Ampacet 11200. [Ampacet] 70% Rutile titanium dioxide in ethylene methyl acrylate copolymer carrier resin with stearate; wh. conc. for blown and cast film, inj. and blow molding applics.; typical letdown 9:1; FDA approved for food pkg. use; sp.gr. 2.07.

Ampacet 11215. [Ampacet] 50% Rutile titanium dioxide in LDPE/LLDPE carrier resin with antioxidant; wh. conc. for blown film, inj. and blow molding applics.; typical letdown 19:1; FDA approved for food pkg. use; sp.gr. 1.51.

Ampacet 11246. [Ampacet] 51% Rutile titanium dioxide in LDPE carrier resin; wh. conc. for inj. and blow molding applics.; typical letdown 24:1; FDA approved for food pkg. use; sp.gr. 1.54.

Ampacet 11247. [Ampacet] 50% Rutile titanium dioxide in LLDPE carrier resin with antioxidant; wh. conc. for blown film, inj. and blow molding applics.; typical letdown 9:1; FDA approved for food pkg. use; sp.gr. 1.51.

Ampacet 11299. [Ampacet] 70% Rutile titanium dioxide in LDPE carrier resin with stearate; wh. conc. for blown film, inj. molding applics.; typical letdown 9:1; FDA approved for food pkg. use; sp.gr. 2.04.

Ampacet 11338. [Ampacet] 50% Rutile titanium dioxide in LDPE carrier resin with antioxidant, stearate; wh. conc. for extrusion coating applics.; typical letdown 9:1; FDA approved for food pkg. use; sp.gr. 1.51.

Ampacet 11343. [Ampacet] 50% Rutile titanium dioxide in PP carrier resin with stearate; wh. conc. for inj. and blow molding applics.; typical letdown 9:1; FDA approved for food pkg. use; sp.gr. 1.48.

Ampacet 11371. [Ampacet] Flame retardant additive for blown film applics.; 60% act.

Ampacet 11416. [Ampacet] 50% Rutile titanium dioxide in LDPE carrier resin with stearate; wh. conc. for blown and cast film applics.; typical letdown 9:1; FDA approved for food pkg. use; sp.gr. 1.51.

Ampacet 11418-F. [Ampacet] 50% Rutile titanium dioxide in HDPE carrier resin with stearate; wh. conc. for blown film, inj. and blow molding applics.; typical letdown 9:1; FDA approved for food pkg. use; sp.gr. 1.55.

Ampacet 11495-S. [Ampacet] 50% Rutile titanium dioxide in LDPE carrier resin with antioxidant, stearate; wh. conc. for extrusion coating applics.; typical letdown 9:1; FDA approved for food pkg. use; sp.gr. 1.51.

Ampacet 11560. [Ampacet] 50% Rutile titanium dioxide in LDPE/LLDPE carrier resin with antioxidant, stearate; wh. color conc. for blown film, inj. and blow molding applics.; typical letdown 9:1; FDA approved for food pkg. use; sp.gr. 1.53.

Ampacet 11572. [Ampacet] 50% Rutile titanium dioxide in LDPE carrier resin with stearate; wh. conc. for blown film, inj. molding applics.; typical letdown 19:1; FDA approved for food pkg. use; sp.gr. 1.77.

Ampacet 11578-P. [Ampacet] 51% Rutile titanium dioxide in LDPE carrier resin with stearate; wh. conc. for blown film, inj. molding applics.; typical letdown 4:1; FDA approved for food pkg. use; sp.gr. 1.53.

Ampacet 11737. [Ampacet] 80% Rutile titanium dioxide in ethylene methyl acrylate copolymer carrier resin with stearate; wh. conc. for blown film, inj. and blow molding applics.; typical letdown 19:1; FDA approved for food pkg. use; sp.gr. 2.48.

Ampacet 11739. [Ampacet] 50% Rutile titanium dioxide in LLDPE/LDPE carrier resin with antioxidant, stearate; wh. conc. for blown and cast film, inj. molding applics.; typical letdown 4:1; FDA approved for food pkg. use; sp.gr. 1.51.

Ampacet 11744. [Ampacet] 51% Rutile titanium dioxide in LLDPE carrier resin; wh. conc. for inj. and blow molding applics.; typical letdown 24:1; FDA approved for food pkg. use; sp.gr. 1.54.

Ampacet 11748. [Ampacet] 70% Rutile titanium dioxide in LLDPE carrier resin with antioxidant, stearate; wh. conc. for blown film, inj. and blow molding applics.; typical letdown 33:1; FDA approved for food pkg. use; sp.gr. 2.03.

Ampacet 11851. [Ampacet] 50% Rutile titanium dioxide in LLDPE carrier resin with stearate; wh. conc. for inj. molding applics.; typical letdown 9:1; FDA approved for food pkg. use; sp.gr. 1.52.

Ampacet 11853. [Ampacet] 50% Rutile titanium dioxide in LDPE carrier resin with antioxidant, stearate; wh. conc. for blown film, inj. and blow molding applics.; typical letdown 9:1; FDA approved for food pkg. use; sp.gr. 1.51.

Ampacet 11875. [Ampacet] 50% Rutile titanium dioxide in LDPE carrier resin; wh. conc. for blown film, inj. and blow molding applics.; typical letdown 9:1; FDA approved for food pkg. use; sp.gr. 1.53.

Ampacet 11912. [Ampacet] 50% Rutile titanium dioxide in LLDPE carrier resin; wh. conc. for inj. and blow molding applics.; typical letdown 19:1; FDA approved for food pkg. use; sp.gr. 1.77.

Ampacet 11913. [Ampacet] 51% Rutile titanium dioxide in LLDPE carrier resin; wh. conc. for inj. and blow molding applics.; typical letdown 24:1; FDA approved for food pkg. use; sp.gr. 1.73.

Ampacet 11919-S. [Ampacet] 50% Rutile titanium dioxide in LDPE carrier resin with stearate; wh. conc. for extrusion coating applics.; typical letdown 4:1; FDA approved for food pkg. use; sp.gr. 1.52.

Ampacet 11979. [Ampacet] 50% Rutile titanium dioxide in LLDPE carrier resin with stearate; wh. conc. for blown film, inj. and blow molding applics.; typical letdown 19:1; FDA approved for food pkg. use; sp.gr. 1.77.

Ampacet 12083. [Ampacet] Siliver color conc. in LDPE carrier; color conc. for blown film, inj. molding applics.; FDA approved for food pkg. use; 25% pigment.

Ampacet 13134. [Ampacet] Yel. color conc. in LDPE carrier; color conc. for blown film, inj. molding applics.; 46% pigment.

Ampacet 14052. [Ampacet] Orange color conc. in LDPE carrier; color conc. for blown film, inj. molding applics.; 50% pigment.

Ampacet 15114. [Ampacet] Red color conc. in LDPE carrier; color conc. for blown film, inj. molding applics.; heavy metals-free; 30% pigment.

Ampacet 15250. [Ampacet] Red color conc. in LDPE carrier; color conc. for blown film, inj. molding applics.; blue under tone in transmission; 40% pigment.

Ampacet 15391. [Ampacet] Redwood conc. in LDPE carrier; color conc. for blown film, inj. molding applics.; heavy metals-free; 45% pigment.

Ampacet 16180. [Ampacet] Dark blue color conc. in LDPE carrier; color conc. for blown film applics.; FDA approved for food pkg. use; 25% pigment.

Ampacet 16192. [Ampacet] Med. blue color conc. in LDPE carrier; color conc. for blown film applics.; heavy metals-free; 30% pigment.

Ampacet 16238. [Ampacet] Lt. blue color conc. in LDPE carrier; color conc. for thin gauge film applics.; heavy metals-free; 50% pigment.

Ampacet 16438. [Ampacet] Lt. blue color conc. in LDPE carrier; color conc. for thin gauge film applics.; FDA approved for food pkg. use; 50% pigment.

Ampacet 17106. [Ampacet] Mint green color conc. in LDPE carrier; color conc. for blown film, inj. molding applics.; FDA approved for food pkg. use; 50% pigment.

Ampacet 17161. [Ampacet] Med. green color conc. in LDPE carrier; color conc. for blown film, inj. molding applics.; 50% pigment.

Ampacet 17491. [Ampacet] Dk. green color conc. in LDPE carrier; color conc. for blown film, inj. molding applics.; heavy metals-free; 40% pigment.

Ampacet 18088. [Ampacet] Lt. buff color conc. in LDPE carrier; color conc. for blown film, inj. molding applics.; heavy metals-free; 50% pigment.

Ampacet 18109. [Ampacet] Dk. brn. color conc. in LDPE carrier; color conc. for blown film, inj. molding applics.; heavy metals-free; 32% pigment.

Ampacet 18402. [Ampacet] Utility buff color conc. in LDPE/LLDPE carrier; color conc. for blown film, inj. molding applics.; 40% pigment.

Ampacet 18537. [Ampacet] Utility brn. color conc. in LDPE/LLDPE carrier; color conc. for blown film, inj. molding applics.; heavy metals-free; 30% pigment.

Ampacet 18659. [Ampacet] Lt. buff color conc. in LDPE carrier; color conc. for blown film, inj. molding applics.; FDA approved for food pkg. use; 50% pigment.

Ampacet 19153. [Ampacet] 30% Carbon black in LDPE carrier resin with antioxidant, stearate; black conc. for extrusion coating applics.; can be used at processing temps. up to 620 F; sp.gr. 1.08.

Ampacet 19153-S. [Ampacet] 30% Carbon black in LDPE carrier resin with antioxidant, stearate; photographic grade black conc. for extrusion coating applics.; can be processed up to 620 F; dry before use for optimum performance; sp.gr. 1.08.

Ampacet 19200. [Ampacet] 41% Carbon black (SRF) in LDPE carrier resin with antioxidant, stearate; black conc. for blown thin-gauge and critical cast film applics., mulch film; heavy metals-free; sp.gr. 1.16.

Ampacet 19238. [Ampacet] 45% Carbon black (SRF) in ethylene methyl acrylate copolymer carrier resin with stearate; black conc. for inj. molding, sheeting, engineering resins (polyethylene, PP, polyamide, PC, ionomer, acrylic, acetal); heavy metals-free; sp.gr. 1.20.

Ampacet 19252. [Ampacet] Gray color conc., copolymer type, for blown film applics.; heavy metals-free; 50% pigment.

Ampacet 19258. [Ampacet] 50% Carbon black (SRF) in ethylene vinyl acetate carrier resin with antioxidant, stearate; black conc. for noncritical inj. molding, sheeting, engineering resins, compounding applics. using polyethylene, HIPS, ABS, or PVC; not for use above 400 F; heavy metals-free; sp.gr. 1.23.

Ampacet 19270. [Ampacet] 50% Carbon black (SRF) in LDPE carrier resin with antioxidant, stearate; black conc. for high-quality blown thin gauge film and other critical applics.; heavy metals-free; sp.gr. 1.22.

Ampacet 19278. [Ampacet] 35% Carbon black in LDPE/EVA carrier resin with antioxidant, stearate; black conc. for inj. molding of engineering resins (polyethylene, HIPS, ABS, PVC); not for use above 400 F; sp.gr. 1.12.

Ampacet 19400. [Ampacet] 40% Carbon black in LDPE/EVA carrier resin with stearate; utility black color conc. for sheeting, inj. molding, compounding of polyolefins, noncritical applics.; sp.gr. 1.30.

Ampacet 19458. [Ampacet] 50% Carbon black (SRF) in LDPE carrier resin with antioxidant, stearate; black conc. for thin-gauge blown film, sheeting, molding applics.; heavy metals-free; sp.gr. 1.31.

Ampacet 19475-S. [Ampacet] 41% Carbon black in LDPE carrier resin with stearate; black conc. for blown and cast film, critical mulch film; sp.gr. 1.15.

Ampacet 19492. [Ampacet] 37% Carbon black in LLDPE/ethylene methyl acrylate copolymer carrier resin with stearate; black conc. for sheeting, molding, conductive molding applics.; sp.gr. 1.14.

Ampacet 19500. [Ampacet] 50% Carbon black in PE/EVA carrier resin with stearate; black conc. for sheeting, molding, noncritical applics.; utility grade; sp.gr. 1.27.

Ampacet 19552. [Ampacet] 40% Carbon black in PE/EVA carrier resin with stearate; black conc. for sheeting, molding, drainage tile, pipe, noncritical applics.; utility grade; sp.gr. 1.25.

Ampacet 19584. [Ampacet] 41% Carbon black (SRF) in LLDPE carrier resin with antioxidant, stearate; black conc. for sheeting, molding, thin-gauge HDPE blown film, thin wall profile extrusion; heavy metals-free; sp.gr. 1.15.

Ampacet 19614. [Ampacet] 40% Carbon black in HDPE carrier resin with stearate; black conc. for HDPE drums and sheeting; sp.gr. 1.17.

Ampacet 19673. [Ampacet] 50% Carbon black (SRF) in LDPE carrier resin with antioxidant, stearate; black conc. for thick blown films, sheeting, compounding, noncritical applics.; heavy metals-free; sp.gr. 1.22.

Ampacet 19699. [Ampacet] 10% Carbon black in LLDPE carrier resin; black conc. for blown film, sheeting, molding polyolefin applics.; FDA approved; sp.gr. 0.97.

Ampacet 19709. [Ampacet] 36% Carbon black in LLDPE carrier resin with antioxidant; N-550 black conc. for critical pipe applics.; sp.gr. 1.12.

Ampacet 19716. [Ampacet] 41% Carbon black in LDPE carrier resin with antioxidant, stearate; black conc. for thin gauge blown film, sheeting, molding applics.; sp.gr. 1.35.

Ampacet 19717. [Ampacet] 36% Carbon black in LLDPE carrier resin with antioxidant, stearate; black conc. for thin-gauge blown film; for outdoor applics. requiring good weatherability; sp.gr. 1.12.

Ampacet 19786. [Ampacet] 26% Carbon black in EVA carrier resin with stearate; black conc. for conductive blown film applics.; sp.gr. 1.07.

Ampacet 19832. [Ampacet] 40% Carbon black in LDPE carrier resin with antioxidant, stearate; black color conc. for blown film, inj. molding, sheeting for construction and agric. applics.; heavy metals-free; sp.gr. 1.37.

Ampacet 19873. [Ampacet] 35% Carbon black in PP carrier resin with stearate; black conc. for conductive sheeting, molding applics.; sp.gr. 1.10.

Ampacet 19897. [Ampacet] 41% Carbon black in LLDPE carrier resin with antioxidant, stearate; black conc. for LLDPE and HMWHDPE blown film; sp.gr. 1.35.

Ampacet 19975. [Ampacet] 50% Carbon black in LLDPE carrier resin with antioxidant, stearate; N-550 black conc. for blown film for drip irrigation; sp.gr. 1.22.

Ampacet 19991. [Ampacet] 50% Carbon black in LLDPE carrier resin with stearate; black conc. for thin wall inj. molding applics.; sp.gr. 1.22.

Ampacet 19999. [Ampacet] 40% Carbon black in PP carrier resin with stearate; black conc. for inj. molding and compounding; sp.gr. 1.13.

Ampacet 39270. [Ampacet] 50% Carbon black in LDPE carrier resin; black conc. for thin gauge film where degradability is desired; sp.gr. 1.23.

Ampacet 41312. [Ampacet] 50% Anatase titanium dioxide in PP carrier resin with stearate; wh. conc. for fiber applics.; typical letdown 49:1; FDA approved for food pkg. use; sp.gr. 1.49.

Ampacet 41424. [Ampacet] 50% Anatase titanium dioxide in LDPE carrier resin with stearate; wh. conc. for fiber applics.; typical letdown 49:1; FDA approved for food pkg. use; sp.gr. 1.51.

Ampacet 41438. [Ampacet] 50% Rutile titanium dioxide in PP carrier resin with stearate; wh. conc. for fiber applics.; typical letdown 49:1; FDA approved for food pkg. use; sp.gr. 1.49.

Ampacet 41483. [Ampacet] 50% Rutile titanium dioxide in PP carrier resin with stearate; wh. conc. for fiber applics.; typical letdown 49:1; FDA approved for food pkg. use; sp.gr. 1.49.

Ampacet 41495. [Ampacet] 50% Rutile titanium dioxide in LDPE carrier resin with antioxidant, stearate; wh. conc. for fiber applics.; typical letdown 4:1; FDA approved for food pkg. use; sp.gr. 1.51.

Ampacet 49270. [Ampacet] 50% Carbon black in LDPE carrier resin with antioxidant, stearate; black conc. for monofilament and slit tape applics.; sp.gr. 1.22.

Ampacet 49315. [Ampacet] 25% Carbon black in PP carrier resin with antioxidant, stearate; N-220 black conc. for multifilament and nonwovens; sp.gr. 1.03.

Ampacet 49370. [Ampacet] 40% Carbon black in PP carrier resin with stearate; black conc. for general purpose molding applics.; sp.gr. 1.13.

Ampacet 49419. [Ampacet] 40% Carbon black in LLDPE carrier resin with antioxidant, stearate; N-110 black conc. for multifilament and staple applics.; sp.gr. 1.12.

Ampacet 49842. [Ampacet] 36% Carbon black in PP carrier resin with stearate; N-110 black conc. for multifilament and staple applics.; sp.gr. 1.11.

Ampacet 49882. [Ampacet] 30% Carbon black in PP carrier resin with antioxidant; N-110 black conc. for geotextiles and strapping applics.; sp.gr. 1.06.

Ampacet 49920. [Ampacet] 40% Carbon black in PP carrier resin with antioxidant, stearate; N-550 black conc. for monofilament and staple applics.; sp.gr. 1.13.

Ampacet 49984. [Ampacet] 35% Carbon black in HDPE carrier resin with antioxidant; N-110 black conc. for geotextiles and strapping; sp.gr. 1.15.

Ampacet 110017. [Ampacet] 50% Rutile titanium dioxide in LLDPE carrier resin with stearate; wh. conc. for blown film, inj. and blow molding applics.; typical letdown 19:1; FDA approved for food pkg. use; sp.gr. 2.08.

Ampacet 110025. [Ampacet] 80% Rutile titanium dioxide in LLDPE carrier resin with stearate; wh. conc. for cast film applics.; typical letdown 15:1; FDA approved for food pkg. use; sp.gr. 2.46.

Ampacet 110041. [Ampacet] 50% Rutile titanium dioxide in LDPE carrier resin with stearate; wh. conc. for blown and cast film applics.; typical letdown 9:1; FDA approved for food pkg. use; sp.gr. 1.54.

Ampacet 110052. [Ampacet] 50% Anatase titanium dioxide in LDPE carrier resin with stearate; wh. conc. for extrusion coating applics.; typical letdown 4:1; FDA approved for food pkg. use; sp.gr. 1.49.

Ampacet 110070. [Ampacet] 50% Rutile titanium dioxide in LLDPE carrier resin with stearate; wh. conc. for inj. and blow molding applics.; typical letdown 24:1; FDA approved for food pkg. use; sp.gr. 2.09.

Ampacet 110181-A. [Ampacet] Wh. color conc. in LDPE carrier; wh. color conc. for extrusion coating and cast film applics.; heat stable to 630 F; FDA 21CFR §177.1520(c) 2.1, 178.2010, 178.3297; sp.gr. 2.46; bulk dens. 87-97; melt index 20 g/10 min.

Ampacet 110313. [Ampacet] Wh. color conc. in LDPE carrier; wh. color conc. for extrusion coating and cast film applics.; heat stable to 630 F; FDA 21CFR §177.1520(c) 2.1, 178.2010, 178.3297; sp.gr. 2.03; bulk dens. 74-84; melt index 5.00 g/10 min.

Ampacet 110370. [Ampacet] Wh. color conc. in LDPE carrier; wh. color conc. for cast and blown film applics.; heat stable to 630 F; FDA 21CFR §177.1520(c) 2.1, 178.3297; sp.gr. 1.51; bulk dens. 55-65; melt index 4.80 g/10 min.

Ampacet 190014. [Ampacet] 35% Carbon black in LLDPE carrier resin with antioxidant; N-550 black conc. for critical pipe applics.; sp.gr. 1.11.

Ampacet 190015. [Ampacet] 50% Carbon black in HDPE carrier resin with stearate; black conc. for critical sheet, HDPE drums, and compounding applics.; sp.gr. 1.25.

Amsco Rubber Solvent. [Unocal] Aliphatic solv. consisting primarily of C4-9 hydrocarbons; low-boiling, flamm. liq. solv.; meets air quality requirements of Los Angeles Rule 66/San Francisco Reg. 3; colorless clear liq.; lt. aliphatic odor; b.p. 121 F; misc. with most common org. solvs.; sp.gr. 0.697; dens. 5.80 lb/gal; visc. 0.57 cSt; vapor pressure 180 mm Hg; ref. index 1.3934; flash pt. (TCC) < 0 F; 73.0 vol.% paraffins, 19.4 vol.% cycloparaffins, 7.6 vol.% aromatics.

Amsperse. [Amspec] Antimony oxide disp.; CAS 1309-64-4; EINECS 215-175-0; flame retardant for ABS, acrylics, cellulosics, epoxy, nitrile, phenolic, PC, thermoset polyester, PE, PP, PS, PVAc, PVC, PU, EVA, PBT, SAN, CPE, TPR, nylon.

Amyl Ledate®. [R.T. Vanderbilt] Lead diamyldithiocarbamate; CAS 36501-84-5; ultra accelerator for nat. and polyisoprene rubber; lt. amber liq.; m.w. 672.05; dens. 1.10 mg/m^3; 50% conc. in oil.

Amyl Zimate®. [R.T. Vanderbilt] Zinc diamyldithiocarbamate; CAS 15337-18-5; ultra accelerator for nat. and syn. rubbers; lt. amber liq.; m.w. 530.22; dens. 0.99 mg/m^3; 50% conc.

Ancamide 100-IT-60. [Air Prods./Perf. Chems.] 60% Polyamide sol'n. in isopropanol and toluene (1:1); R.T. epoxy curing agent for solv.-based flexible coatings, adhesion additive for nonreactive polyamide-based inks and hot-melt adhesives; toughening additive for epoxy adhesives; semisolid resin as film-fomer on evaporation of solv.; FDA 21CFR §175.105, 175.300, 176.170; Gardner 9 liq.; sp.gr. 0.910; visc. 1400 cps; amine no. 55; equiv. wt. 733.

Ancamide 220. [Air Prods./Perf. Chems.] Polyamide; R.T. epoxy curing agent providing high flexibility, long pot life; for solv.-based maintenance coatings, primers, sealers, coatings for concrete floorings; cure can be accelerated with up to 5 phr Ancamine K54; FDA 21CFR §175.105, 175.300, 176.170; Gardner 7 liq.; sp.gr. 0.970; visc. 330,000 cps; amine no. 245; equiv. wt. 185.

Ancamide 220-IPA-73. [Air Prods./Perf. Chems.] 73% Polyamide sol'n. in isopropanol; R.T. epoxy curing agent providing high flexibility, long pot life; for solv.-based maintenance coatings, primers, sealers, coatings for concrete floorings; FDA 21CFR §175.105, 175.300, 176.170; Gardner 7 liq.; sp.gr. 0.940; visc. 2100 cps; amine no. 180; equiv. wt. 253.

Ancamide 220-X-70. [Air Prods./Perf. Chems.] 70% Polyamide sol'n. in xylene; R.T. epoxy curing agent providing high flexibility, long pot life; for solv.-based maintenance coatings, primers, sealers, coatings for concrete floorings; FDA 21CFR §175.105, 175.300, 176.170; Gardner 7 liq.; sp.gr. 0.940; visc. 1100 cps; amine no. 170; equiv. wt. 264.

Ancamide 260A. [Air Prods./Perf. Chems.] Amide/imidazoline; R.T. epoxy curing agent for coatings, concrete repairs, sealants, cable-jointing compds., consumer adhesives; 1:1 vol. ratios with std. liq. epoxy resin possible; improved reactivity, chem. resist., or visc. reduction in blends; FDA 21CFR §175.105, 175.300, 176.170; Gardner 7 liq.; sp.gr. 0.960; visc. 40,000 cps; amine no. 350; equiv. wt. 120.

Ancamide 260TN. [Air Prods./Perf. Chems.] Amide/imidazoline; R.T. epoxy curing agent for coatings, concrete repairs, sealants, cable-jointing compds., consumer adhesives, coal tar-extended coatings; higher imidazoline version of Ancamide 260A for improved compatibility; FDA 21CFR §175.105, 175.300, 176.170; Gardner 7 liq.; sp.gr. 0.960; visc. 35,000 cps; amine no. 350; equiv. wt. 120.

Ancamide 350A. [Air Prods./Perf. Chems.] Imidazoline/amide; R.T. epoxy curing agent for high-solids coatings, adhesives, sealants, putties, flexible cable-jointing compds.; improved chem. solv. resist.; FDA 21CFR §175.105, 175.300, 176.170;

Gardner 7 liq.; sp.gr. 0.970; visc. 11,000 cps; amine no. 380; equiv. wt. 100.

Ancamide 375A. [Air Prods./Perf. Chems.] Imidazoline/amide; R.T. epoxy curing agent for low VOC coatings, adhesives, sealants, putties, flexible cable-jointing compds.; FDA 21CFR §175.105, 175.300, 176.170; Gardner 7 liq.; sp.gr. 0.960; visc. 150 cps; amine no. 379; equiv. wt. 100.

Ancamide 400. [Air Prods./Perf. Chems.] Polyamide; R.T. epoxy curing agent providing rapid cure; compat. with bisphenol F and conventional epoxy resins without induction; good color; for coatings, floorings, adhesives, sealants, putties, elec. potting, water-wipeable grouts; Gardner 7 liq.; sp.gr. 0.970; visc. 1600 cps; amine no. 405; equiv. wt. 95.

Ancamide 400-BX-60. [Air Prods./Perf. Chems.] Polyamide adduct, 60% solids in butanol/xylene (1:4); R.T. epoxy curing agent providing good epoxy resin compat. without induction and better cure under adverse conditions; for general-purpose maintenance paints and primers, fast touch dry coatings with solid resins; Gardner 7 liq.; sp.gr. 0.950; visc. 1800 cps; amine no. 122; equiv. wt. 370.

Ancamide 500. [Air Prods./Perf. Chems.] Amide/imidazoline; R.T. epoxy curing agent providing low visc., moderate pot life, good adhesion to concrete; used alone or with other curing agents; noncritical mixing ratio; for concrete coatings, tile grouts, crack injection, floorings, riverstone mortars, elec. encapsulation, tooling, general-purpose adhesives; FDA 21CFR §175.300, 176.170; Gardner 7 liq.; sp.gr. 0.950; visc. 250 cps; amine no. 445; equiv. wt. 90; heat distort. temp. 113 F (264 psi).

Ancamide 501. [Air Prods./Perf. Chems.] Modified amide/imidazoline; accelerated version of Ancamide 500; R.T. epoxy curing agent providing cures under humid conditions, very good adhesion, exc. adhesion to concrete, good chem. resist.; for matt floorings, concrete coatings, patching compds., old-to-new concrete adhesives; FDA 21CFR §175.300, 176.170; Gardner 7 liq.; sp.gr. 0.980; visc. 600 cps; amine no. 550; equiv. wt. 68; heat distort. temp. 116 F (264 psi).

Ancamide 502. [Air Prods./Perf. Chems.] Amide/imidazoline; short pot life version of Ancamide 500; R.T. epoxy curing agent providing short pot life; for matt floorings, concrete coatings, patching compds., old-to-new concrete adhesives; FDA 21CFR §175.300, 176.170; Gardner 7 liq.; sp.gr. 0.950; visc. 300 cps; amine no. 450; equiv. wt. 90; heat distort. temp. 113 F (264 psi).

Ancamide 503. [Air Prods./Perf. Chems.] Amide; R.T. epoxy curing agent providing faster pot life and thin film cure time than Ancamide 500 or 502; for concrete coatings, floorings, old-to-new concrete adhesives, tile grouts; FDA 21CFR §175.300, 176.170; Gardner 6 liq.; sp.gr. 0.950; visc. 350 cps; amine no. 500; equiv. wt. 90; heat distort. temp. 119 F (264 psi).

Ancamide 506. [Air Prods./Perf. Chems.] Imidazoline/amide; R.T. epoxy curing agent providing highest imidazoline content of amidoamine range giving lowest reactivity/visc., good thru-cure with very little exotherm; for civil engineering, building and construction, wet lay-up laminates, elec. encapsulation, high-solids coatings (often with Ancamine 1618 or 1693); pot life extender for cycloaliphatic amine; FDA 21CFR §175.300, 176.170; Gardner 7 liq.; sp.gr. 0.940; visc. 300 cps; amine no. 420; equiv. wt. 105; heat distort. temp. 113 F (264 psi).

Ancamide 507. [Air Prods./Perf. Chems.] Amidoamine adduct; R.T. epoxy curing agent providing fast cure, improved water resist.; for use alone or with polyamide adducts; for high-solids, anticorrosive marine coatings, concrete patching and flooring mortars, adhesives and patch kits; FDA 21CFR §175.300, 176.170; Gardner 6 liq.; sp.gr. 0.990; visc. 950 cps; amine no. 600; equiv. wt. 65; heat distort. temp. 127 F (264 psi).

Ancamide 700-B-75. [Air Prods./Perf. Chems.] Polyamide adduct, 75% solids in butanol; R.T. epoxy curing agent providing exc. adhesion and cure under adverse conditions; no induction period required; for high-solids anticorrosive coatings for marine/industrial use; meets US Navy MIL-P-24441; FDA 21CFR §175.105, 175.300, 176.170; Gardner 7 liq.; sp.gr. 0.960; visc. 5000 cps; amine no. 240; equiv. wt. 170.

Ancamide 2050. [Air Prods./Perf. Chems.] Polyamide adduct; R.T. epoxy curing agent achieving high gloss, flexibility, hardness, and reverse impact resist. with liq. epoxy resins; no induction required; noncritical loading (70-100 phr); for high-solids coatings, primers and coatings for concrete adhesives, sealants, putties, flexible cable-jointing compds.; Gardner 7 liq.; sp.gr. 1.02; visc. 4000 cps; amine no. 225; equiv. wt. 150.

Ancamide 2137. [Air Prods./Perf. Chems.] Amidoamine adduct; R.T. epoxy curing agent providing improved resist. to blush and exudation, exc. flexibility and reverse impact resist., noncritical loading (70-100 phr); for high-solids coatings, adhesives, sealants, putties, flexible cable-jointing compds., coatings and primers for concrete; Gardner 7 liq.; sp.gr. 1.039; visc. 1800 cps; amine no. 300; equiv. wt. 150; heat distort. temp. 112 F (264 psi).

Ancamide 2349. [Air Prods./Perf. Chems.] Modified amide/imidazoline; noncorrosive version of Ancamide 501; R.T. epoxy curing agent providing good adhesion, exc. adhesion to concrete, good chem. resist.; for civil engineering applics., high-solids coatings, concrete adhesive; Gardner 7 liq.; sp.gr. 0.980; visc. 800 cps; amine no. 550; equiv. wt. 68.

Ancamine® 1110. [Air Prods./Perf. Chems.] Dimethylaminomethyl phenol; CAS 25338-55-0; R.T. epoxy curing agent for wet concrete adhesives and coatings in combination with polysulfide and polymercaptan; elec. castings in combination with anhydrides; reduced reactivity variant of Ancamine K54; pale brn. liq.; sp.gr. 1.03; visc. 300 cps; amine no. 370.

Ancamine® 1482. [Air Prods./Perf. Chems.] Modified

aromatic amine; elevated temp. epoxy curing agent; permits mixing with resin and curing at lower temps. thus extending pot life; exc. mech. props., chem. and heat resist.; for wet lay-up laminates, filament-wound pipe and tanks, elec. potting, casting, and elec./electronic encapsulation; Gardner 18 liq.; sp.gr. 1.16; visc. 0-1500 cps; amine no. 770; equiv. wt. 37; heat distort. temp. 320 F (264 psi).

Ancamine® 1483. [Air Prods./Perf. Chems.] Modified aliphatic amine; R.T. epoxy curing agent providing rigidity and high physical str.; for wet lay-up laminating, gel coats, patch repair kits; phenol-free version (Ancamine 1916) avail.; Gardner 3 liq.; sp.gr. 1.09; visc. 6000 cps; amine no. 840; equiv. wt. 44; heat distort. temp. 136 F (264 psi).

Ancamine® 1561. [Air Prods./Perf. Chems.] Modified cycloaliphatic amine; R.T. epoxy curing agent providing good color, relatively fast cure at low temps.; for solv.-free coatings, self-leveling floorings, injection repair compds.; Gardner 1 liq.; sp.gr. 1.0; visc. 50 cps; amine no. 325; equiv. wt. 85; heat distort. temp. 120 F (264 psi).

Ancamine® 1608. [Air Prods./Perf. Chems.] Modified aliphatic amine; R.T. epoxy curing agent providing chem. and solv. resist.; for high-solids, solv.-free maintenance coatings, tank linings, wet lay-up laminating, adhesives; accelerator for amidoamines; FDA 21CFR §175.300; Gardner 3 liq.; sp.gr. 1.08; visc. 3500 cps; amine no. 800; equiv. wt. 44; heat distort. temp. 147 F (264 psi).

Ancamine® 1617. [Air Prods./Perf. Chems.] Modified aliphatic amine; R.T. epoxy curing agent providing exc. gloss, surf. appearance, solv. resist., and low temp. cure; for high-gloss chem./solv.-resist. coatings, tank linings for steel and concrete, floorings; accelerator for amidoamines and polyamides; Gardner 2 liq.; sp.gr. 1.00; visc. 4000 cps; amine no. 405; equiv. wt. 77; heat distort. temp. 120 F (264 psi).

Ancamine® 1618. [Air Prods./Perf. Chems.] Modified cycloaliphatic amine; R.T. epoxy curing agent providing exc. color, color stability, high gloss, nonblushing surf. appearance, good chem. resist.; for solv.-free and high-solids coatings, gel coats, self-leveling floorings, mortars, water-wipeable tile grouts; FDA 21CFR §175.105; Gardner 1 liq.; sp.gr. 1.03; visc. 400 cps; amine no. 275; equiv. wt. 113; heat distort. temp. 115 F (264 psi).

Ancamine® 1636. [Air Prods./Perf. Chems.] Modified aliphatic amine; R.T. epoxy curing agent providing low visc., long pot life to ambient-cured, two-phase toughened systems contg. CTBN; provides high peel str. in adhesives and high impact str. in composites; for filament-wound structures, acid-resist. coatings; Gardner 2 liq.; sp.gr. 0.980; visc. 20 cps; amine no. 950; equiv. wt. 38; heat distort. temp. 149 F (264 psi).

Ancamine® 1637. [Air Prods./Perf. Chems.] Modified aliphatic amine; R.T. epoxy curing agent providing good chem. resist., exc. resist. to gasohol, good compat. with epoxy novolacs; cures down to 35 F under adverse conditions; for laminates, adhesives, solv.-free coatings, floorings, machinery grouts, coal tar-extended coatings and castings; Gardner 6 liq.; sp.gr. 1.09; visc. 4000 cps; amine no. 750; equiv. wt. 50; heat distort. temp. 129 F (264 psi).

Ancamine® 1637-LV. [Air Prods./Perf. Chems.] Modified aliphatic amine; lower visc. version of Ancamine 1637; R.T. epoxy curing agent for solv.-free, high-solids coatings; accelerator for other amines; Gardner 6 liq.; sp.gr. 1.08; visc. 1500 cps; amine no. 710; equiv. wt. 50; heat distort. temp. 115 F (264 psi).

Ancamine® 1638. [Air Prods./Perf. Chems.] Modified aliphatic amine; R.T. epoxy curing agent providing very low visc., fast cure rate, good chem. resist., high heat distort.; for adhesives, grouts, floor coatings, wet lay-up laminating; accelerator and visc. reducer for other amines; Gardner 2 liq.; sp.gr. 1.03; visc. 100 cps; amine no. 1070; equiv. wt. 31; heat distort. temp. 150 F (264 psi).

Ancamine® 1644. [Air Prods./Perf. Chems.] Modified aliphatic amine; R.T. epoxy curing agent providing resist. to blushing and formation of greasy surfs. during thin film cure, fast cure, lt. color, moisture insensitivity, tough cures; for floor coatings, fast-setting adhesives, cold weather patching compds.; accelerator for cycloaliphatic amines, aliphatic amines, polyamides, and amidoamines; Gardner 2 liq.; sp.gr. 0.970; visc. 2500 cps; amine no. 385; equiv. wt. 154; heat distort. temp. 124 F (264 psi).

Ancamine® 1693. [Air Prods./Perf. Chems.] Modified cycloaliphatic amine; R.T. epoxy curing agent providing good color stability, exc. chem. resist. esp. to acids, org. solvs., and sour crude oil; for solv.-free and high-solids coatings, secondary containment coatings, self-leveling floorings; lt. amber liq.; sp.gr. 1.04; visc. 100 cps; amine no. 310; equiv. wt. 96; heat distort. temp. 95 F (264 psi).

Ancamine® 1767. [Air Prods./Perf. Chems.] Modified aliphatic amine; R.T. epoxy curing agent providing resist. to blushing and formation of greasy surfs. during thin film cure, fast cure, lt. color, moisture insensitivity, low modulus, high flexibility; for floor coatings, fast-setting adhesives, cold weather patching compds.; accelerator for cycloaliphatic amines, aliphatic amines, polyamides, and amidoamines; Gardner 2 liq.; sp.gr. 0.970; visc. 6000 cps; amine no. 310; equiv. wt. 180; heat distort. temp. 83 F (264 psi).

Ancamine® 1768. [Air Prods./Perf. Chems.] Modified aliphatic amine; R.T. epoxy curing agent providing resist. to blushing and formation of greasy surfs. during thin film cure, fast cure, lt. color, moisture insensitivity, harder more rigid cures; for floor coatings, fast-setting adhesives, cold weather patching compds.; accelerator for cycloaliphatic amines, aliphatic amines, polyamides, and amidoamines; Gardner 2 liq.; sp.gr. 0.972; visc. 250 cps; amine no. 610; equiv. wt. 95; heat distort. temp. 147 F (264 psi).

Ancamine® 1769. [Air Prods./Perf. Chems.] Modified aliphatic amine; room or elevated temp. epoxy curing agent providing low visc., shrinkage, vapor pressure; yields good mech. and elec. props.; for small elec. potting and castings, wet lay-up laminating, gel coats, repair kits, adhesives; Gardner 1 liq.; sp.gr. 1.01; visc. 600 cps; amine no. 965; equiv. wt. 48; heat distort. temp. 125 F (264 psi).

Ancamine® 1770. [Air Prods./Perf. Chems.] 1,2-Cyclohexanediamine; elevated temp. epoxy curing agent providing low visc., low loading, enhanced modulus; for potting, casting, and wet lay-up laminating for elec. and general industrial applics.; Gardner 2 liq.; sp.gr. 0.947; visc. 10 cps; amine no. 974; equiv. wt. 29; heat distort. temp. 310 F (264 psi).

Ancamine® 1784. [Air Prods./Perf. Chems.] Modified aliphatic amine; R.T. epoxy curing agent providing long pot life, low visc., lt. color, good color stability; cured film props. incl. high gloss, flexibility, freedom from blush; for high-solids and solv.-free coatings, floorings, riverstone mortars, laminates, castings, adhesives; Gardner 1 liq.; sp.gr. 0.950; visc. 50 cps; amine no. 320; equiv. wt. 86; heat distort. temp. 122 F (264 psi).

Ancamine® 1833. [Air Prods./Perf. Chems.] Modified aromatic amine; room or elevated temp. epoxy curing agent providing low visc., high chem. resist.; compat. with aliphatic amines to achieve faster cure and enhance solv. resist.; for high-solids and solv.-free maintenance and marine coatings, tank laminate repairs; floorings, grouts; Gardner 8 liq.; sp.gr. 1.08; visc. 500 cps; amine no. 875; equiv. wt. 36; heat distort. temp. 122 F (264 psi).

Ancamine® 1856. [Air Prods./Perf. Chems.] Modified aliphatic amine; R.T. epoxy curing agent providing good cures in cold, damp conditions with rapid development of hardness; exc. chem. resist.; for grouts, mortars, concrete repair compds., floorings; usually used with an extender; Gardner 5 liq.; sp.gr. 1.12; visc. 3000 cps; amine no. 490; equiv. wt. 73; heat distort. temp. 132 F (264 psi).

Ancamine® 1882. [Air Prods./Perf. Chems.] Modified cycloaliphatic amine; R.T. epoxy curing agent providing relatively rapid thin film cure in adverse conditions (high humidity, low ambient temp.), blush-free, nongreasy, high-gloss coatings with exc. abrasion resist.; bonds to damp concrete; for solv.-free and high-solids coatings, floorings, mortars; lt. amber liq.; sp.gr. 0.990; visc. 60 cps; amine no. 340; equiv. wt. 92; heat distort. temp. 115 F (264 psi).

Ancamine® 1884. [Air Prods./Perf. Chems.] Modified cycloaliphatic amine; R.T. epoxy curing agent providing long pot life, bonding to damp concrete; produces glossy, nongreasy films when cured in high humidity despite slow thin film set time; for solv.-free and high-solids coatings, floorings, grouts, exterior patching mortar and overlay binder; Gardner 2 liq.; sp.gr. 1.04; visc. 300 cps; amine no. 360; equiv. wt. 86; heat distort. temp. 108F (264 psi).

Ancamine® 1895. [Air Prods./Perf. Chems.] Modified cycloaliphatic amine; R.T. epoxy curing agent providing films with high gloss and hardness; cures down to 35-40 F; for high-solids and solv.-free coatings, mortars, grouts, floorings; Gardner 1 liq.; sp.gr. 1.02; visc. 1200 cps; amine no. 410; equiv. wt. 75.

Ancamine® 1916. [Air Prods./Perf. Chems.] Modified aliphatic amine; phenol-free version of Ancamine 1483; R.T. epoxy curing agent for heat-resist., solv.-free and sol'n. coatings; FDA 21CFR §175.300, 176.170; Gardner 4 liq.; sp.gr. 1.09; visc. 6000 cps; amine no. 840; equiv. wt. 43; heat distort. temp. 136 F (264 psi).

Ancamine® 1922. [Air Prods./Perf. Chems.] Diethylene glycol di(aminopropyl) ether; room or elevated temp. epoxy curing agent providing exceptional toughness, resiliency, thermal shock resist., outstanding impact resist., good elec. props.; for structural adhesives, elec. potting/encapsulation for aerospace, automotive, and other high-performance applics.; Gardner 1 liq.; sp.gr. 1.00; visc. 10 cps; amine no. 507; equiv. wt. 55; heat distort. temp. 109 F (264 psi).

Ancamine® 1934. [Air Prods./Perf. Chems.] Modified cycloaliphatic amine; R.T. epoxy curing agent; slow version of Ancamine 1618, having lower solv. resist. but better film appearance when cured at low temps. (40-45 F); for decorative tile grouts, self-leveling and pebble-finish floorings, swimming pool paints, decorative high-solids coatings; Gardner 2 liq.; sp.gr. 1.02; visc. 300 cps; amine no. 272; equiv. wt. 100; heat distort. temp. 104 F (264 psi).

Ancamine® 2014AS. [Air Prods./Perf. Chems.] Modified aliphatic amine; elevated temp. latent epoxy curing agent providing extremely long shelf stability in undiluted resins, exc. low temp. reactivity; cures rapidly above its activation temp. (167 F); dicyandiamide cure accelerator; for one-component adhesives, small-scale potting and casting, prepregs, film adhesives; wh. micronized powd.; avg. particle size 10 μ; easily dispersed in liq. epoxy resin; bulk dens. 0.253 g/cm^3; m.p. 205 F; amine no. 184; equiv. wt. 52.

Ancamine® 2014FG. [Air Prods./Perf. Chems.] Modified aliphatic amine; finer version of Ancamine 2014AS; elevated temp. latent epoxy curing agent providing faster development of props.; finer particles reduce the formulation shelf stability when used as sole curing agent; for one-component adhesives, small-scale potting and casting, prepregs, film adhesives; accelerator for dicyandiamide; wh. micronized powd.; 90% < 6 μ particle size; bulk dens. 0.344 g/cm^3; m.p. 205 F; amine no. 184; equiv. wt. 52.

Ancamine® 2021. [Air Prods./Perf. Chems.] Modified aliphatic amine; accelerated version of Ancamine 1784; R.T. epoxy curing agent providing lt. color, low visc., moderately long pot life, high-gloss finishes, moderate chem. resist.; for high-solids, solv.-free coatings, floorings, laminates; Gardner 3 liq.; sp.gr. 0.990; visc. 400 cps; amine no. 380; equiv. wt. 85; heat distort. temp. 110 F (264 psi).

Ancamine® 2049. [Air Prods./Perf. Chems.] 3,3′-

Dimethylmethylenedi(cyclohexylamine); elevated temp. epoxy curing agent providing low color, longer pot life; for casting, potting encapsulation, wet lay-up laminating and filament winding for tooling, elec., and general industrial applics.; Gardner 2 liq.; sp.gr. 0.947; visc. 120 cps; amine no. 458; equiv. wt. 60; heat distort. temp. 296 F (264 psi).

Ancamine® 2056. [Air Prods./Perf. Chems.] Modified aromatic amine; R.T. epoxy curing agent providing high visc., low odor, exc. chem. resist. esp. with long-term immersion in org. acids, veg. oils, wines; for coatings for corrosive environments (chem. and oil tanks), floorings, coal-tar epoxy coatings; not suitable for use in food plants; Gardner 18 liq.; sp.gr. 1.12; visc. 250 cps; amine no. 285; equiv. wt. 95; heat distort. temp. 119 F (264 psi).

Ancamine® 2071. [Air Prods./Perf. Chems.] Modified aliphatic amine; phenol-free alternative to Ancamine AD; R.T. epoxy curing agent providing fast cure under cold, damp conditions; bonds well to variety of substrates; for metal and concrete bonding, patch repair compds.; accelerator for other amines; Gardner 3 liq.; sp.gr. 1.018; visc. 1000 cps; amine no. 506; equiv. wt. 95; heat distort. temp. 138 F (264 psi).

Ancamine® 2072. [Air Prods./Perf. Chems.] Modified cycloaliphatic amine; R.T. epoxy curing agent; phenol-free version of Ancamine MCA; produces higher HDT and harder films, which cure at low temps. under high humidity conditions; for industrial and semi-decorative coatings, general-purpose concrete primer (with Ancamide 506 and Ancamine K54), industrial floorings, mortars; Gardner 3 liq.; sp.gr. 0.998; visc. 200 cps; amine no. 350; equiv. wt. 102; heat distort. temp. 136 F (264 psi).

Ancamine® 2074. [Air Prods./Perf. Chems.] Modified cycloaliphatic amine; R.T. epoxy curing agent with exc. color and low temp. cure chars.; gives good film flexibility and bonds well to damp concrete; for solv.-free and high-solids decorative coatings, self-leveling floorings, grouts, thermal shock-resist. floorings in combination with Epodil L diluent; Gardner 1 liq.; sp.gr. 0.996; visc. 60 cps; amine no. 345; equiv. wt. 92; heat distort. temp. 120 F (264 psi).

Ancamine® 2089M. [Air Prods./Perf. Chems.] Modified aliphatic amine; phenol-free; R.T. epoxy curing agent providing very rapid cure at low temps.; in combination with extenders gives high-gloss films; accelerator for cycloaliphatic amine, aliphatic amine, and amidoamine cured coatings and floorings; for solv.-free, high-solids coatings, adhesives, concrete patching compds.; Gardner 4 liq.; sp.gr. 1.00; visc. 100 cps; amine no. 380; equiv. wt. 75; heat distort. temp. 133 F (264 psi).

Ancamine® 2136. [Air Prods./Perf. Chems.] Modified cycloaliphatic amine; R.T. epoxy curing agent with improved color and color stability over Ancamine 1884; for solv.-free and high-solids coatings, floorings, grouts; Gardner 2 liq.; sp.gr. 1.024; visc. 300 cps; amine no. 300; equiv. wt. 92.

Ancamine® 2143. [Air Prods./Perf. Chems.] Modified cycloaliphatic amine; R.T. epoxy curing agent providing exc. color and color stability, good resist. to carbamation at ambient and low temps., chem. resist.; for solv.-free and high-solids coatings, floorings, tile grouts; Gardner 2 liq.; sp.gr. 1.03; visc. 600 cps; amine no. 313; equiv. wt. 115; heat distort. temp. 115 F (264 psi).

Ancamine® 2167. [Air Prods./Perf. Chems.] Polycycloaliphatic amine; elevated temp. epoxy curing agent providing low visc., improved tens. str., toughness, and elong.; for filament winding, casting, potting, wet lay-up laminating for tooling and general industrial applics.; Gardner 3 liq.; sp.gr. 0.975; visc. 210 cps; amine no. 520; equiv. wt. 53.

Ancamine® 2168. [Air Prods./Perf. Chems.] Polycycloaliphatic amine; elevated temp. epoxy co-curing agent providing high visc., improved glass transition temp., chem. resist. esp. to solvs.; for filament winding, RTM, and casting; Gardner 12-15 liq.; sp.gr. 1.01; visc. 51,000 cps (104 F); amine no. 505; equiv. wt. 55.

Ancamine® 2205. [Air Prods./Perf. Chems.] Modified aliphatic amine; phenol-free version of Ancamine AD; R.T. epoxy curing agent providing good low temp. cure and adhesion to concrete; accelerator for polyamides and amidoamines; for adhesives, concrete bonding, crack filling; Gardner 4 liq.; sp.gr. 1.05; visc. 3600 cps; amine no. 600; equiv. wt. 95.

Ancamine® 2264. [Air Prods./Perf. Chems.] Polycycloaliphatic amine; elevated temp. epoxy curing agent providing higher glass transition temp., good toughness, and good chem. resist.; for filament winding, RTM, casting, potting for tooling, elec., and general industrial applics.; Gardner 4 liq.; sp.gr. 0.9995; visc. 2600 cps; amine no. 512; equiv. wt. 54.

Ancamine® 2280. [Air Prods./Perf. Chems.] Modified cycloaliphatic amine; R.T. epoxy curing agent providing exc. resist. to carbamation, even when cured under adverse conditions incl. high humidity; good chem. resist.; for industrial floorings, high-solids coatings, chem. resist. mortars and tank linings; Gardner 8 liq.; sp.gr. 1.06; visc. 450 cps; amine no. 250; equiv. wt. 110.

Ancamine® 2286. [Air Prods./Perf. Chems.] Modified cycloaliphatic amine; R.T. epoxy curing agent providing low visc., good color and color stability, good low temp. cure rate, resist. to carbamation; for solv.-free coatings, self-leveling floorings, flooring mortars, grouts; Gardner 1 liq.; sp.gr. 1.01; visc. 100 cps; amine no. 325; equiv. wt. 95.

Ancamine® 2337XS. [Air Prods./Perf. Chems.] Modified aliphatic amine; latent curing agent for epoxy resins; very rapid reactivity above 70 C; provides rapid green str. development; good adhesion; for one-component conduction or induction heat-cure adhesives, potting, coatings, hot-melt prepregs; also as co-cure accelerator with Ancamine 2014AS and 2014FG curing agents; lt. yel. micronized powd.; 90% ≤ 10 μ; m.p. 63-78 C; amine no. 250-270; *Storage:* 6 mos min. storage life in sealed containers @ ambient temp.

Ancamine® AD. [Air Prods./Perf. Chems.] Modified aliphatic amine; R.T. epoxy curing agent providing rapid cure in thin films; cures in cold, damp conditions with exc. adhesion to variety of substrates; for adhesives, patch epoxy coatings, concrete bonding, crack filling; accelerator for other amines; Gardner 6 liq.; sp.gr. 1.08; visc. 1500 cps; amine no. 485; equiv. wt. 107; heat distort. temp. 118 F (264 psi).

Ancamine® DL-50. [Air Prods./Perf. Chems.] Methylene dianiline; CAS 101-77-9; elevated temp. epoxy curing agent providing good mech., thermal, and elec. props., exc. chem. resist.; for filament winding, laminating, tooling, and castings; Gardner 18 liq.; sp.gr. 1.15; visc. 6600 cps (122 F); amine no. 550; equiv. wt. 50; heat distort. temp. 320 F (264 psi).

Ancamine® K54. [Air Prods./Perf. Chems.] 2,4,6-Tris (dimethylaminomethyl) phenol; CAS 90-72-2; R.T. epoxy curing agent; efficient activator for epoxy resins cured with wide variety of hardener types incl. polyamide, amidoamines, polymercaptans, and anhydrides; for coatings, adhesives, castings, potting, encapsulation; amber liq.; sp.gr. 0.980; visc. 200 cps; amine no. 630.

Ancamine® K61B. [Air Prods./Perf. Chems.] 2,4,6-Tris (dimethylaminomethyl) phenol, 2-ethylhexanoic acid salt; elevated temp. epoxy curing agent providing extended pot life; designed to give less exothermic cure than Ancamine K54; for small and med.-sized castings, potting and impregnation varnishes; amber liq.; sp.gr. 0.970; visc. 600 cps; amine no. 230; heat distort. temp. 72 F.

Ancamine® LO. [Air Prods./Perf. Chems.] Modified aromatic amine; R.T. epoxy curing agent providing low visc., low odor, good chem. resist. esp. to org. acids; for floorings, coatings and castings in dark colors; Gardner 16 liq.; sp.gr. 1.13; visc. 1500 cps; amine no. 285; equiv. wt. 95; heat distort. temp. 118 F (264 psi).

Ancamine® LT. [Air Prods./Perf. Chems.] Modified aromatic amine; R.T. epoxy curing agent combining acid resist. more typical of aromatic amines with solv. resist., fast initial cure and low phr of aliphatic amines; for tooling, wet lay-up laminating, tank linings; Gardner 18 liq.; sp.gr. 1.16; visc. 2300 cps; amine no. 280; equiv. wt. 98; heat distort. temp. 109 F (264 psi).

Ancamine® MCA. [Air Prods./Perf. Chems.] Modified cycloaliphatic amine; R.T. epoxy curing agent providing exc. adhesion to cold damp concrete, good chem. resist.; fast curing even in cold (40 F) damp conditions and under water; for industrial flooring mortars with exc. trowelability, coatings and concrete repair mortars; concrete bonding in combination with polyamides and amidoamines; Gardner 3 liq.; sp.gr. 1.03; visc. 150 cps; amine no. 305; equiv. wt. 101; heat distort. temp. 109 F (264 psi).

Ancamine® T. [Air Prods./Perf. Chems.] Modified aliphatic amine; R.T. or elevated temp. epoxy curing agent providing exc. color, good resist. to solvs.; for wet lay-up laminating, tooling, adhesives, boat and auto body patch repair kits; FDA 21CFR §175.105; Gardner 1 liq.; sp.gr. 1.03; visc. 300 cps; amine no. 1145; equiv. wt. 36; heat distort. temp. 129 F (264 psi); *Toxicology:* reduced level of skin irritation relative to most aliphatic polyamines.

Ancamine® T-1. [Air Prods./Perf. Chems.] Modified aliphatic amine; accelerated version of Ancamine T; R.T. epoxy curing agent providing faster thin film cure and reduced sensitivity to moisture during cure; for wet lay-up laminating, tooling, adhesives, boat and auto body patch repair kits; FDA 21CFR §175.105; Gardner 1 liq.; sp.gr. 1.07; visc. 1500 cps; amine no. 885; equiv. wt. 47; heat distort. temp. 140 F (264 psi); *Toxicology:* reduced level of skin irritation relative to most aliphatic polyamines.

Ancamine® TL. [Air Prods./Perf. Chems.] Modified aromatic amine; R.T. epoxy curing agent curing down to 40 F and under water; outstanding resist. to org. acids, min. acids, alkalies; for coatings for corrosive environments (chem. and oil tanks), coal-tar epoxy coatings; not suitable for use in food plants; Gardner 18 liq.; sp.gr. 1.12; visc. 5000 cps; amine no. 245; equiv. wt. 118; heat distort. temp. 118 F (264 psi).

Ancamine® XT. [Air Prods./Perf. Chems.] Modified aliphatic amine; R.T. epoxy curing agent providing fast cure rate and good chem. resist.; co-curing agent for slower amines; patch repair; adhesives; Gardner 4 liq.; sp.gr. 1.04; visc. 60 cps; amine no. 805; equiv. wt. 41; heat distort. temp. 156 F (264 psi).

Anchor® 1040. [Air Prods./Perf. Chems.] Boron trifluoride-amine complex; epoxy curing agent for prepregs with solid or liq. epoxy resins, sometimes modified with Ancamine DL50; molding powds.; orange-red liq.; good sol. in epoxy resins; sp.gr. 1.13; visc. 20,000 cps; heat distort. temp. 266 F (264 psi).

Anchor® 1115. [Air Prods./Perf. Chems.] Boron trifluoride-amine complex; epoxy curing agent for prepregs with solid or liq. epoxy resins, sometimes modified with Ancamine DL50; molding powds.; heat cure insulating varnishes; dk. liq.; good sol. in epoxy resins; sp.gr. 1.15; visc. 1700 cps; heat distort. temp. 275F (264 psi).

Anchor® 1170. [Air Prods./Perf. Chems.] Boron trifluoride-amine complex; epoxy curing agent for laminates, adhesives, body solders; dk. visc. liq.; good sol. in epoxy resins; sp.gr. 1.25; visc. 8000 cps; heat distort. temp. 230 (264 psi).

Anchor® 1171. [Air Prods./Perf. Chems.] Boron trifluoride-amine complex; epoxy curing agent for hot-dip coatings of electronic components, low-temp. cure prepregs, lacquers; dk. brn. liq.; good sol. in epoxy resins; sp.gr. 1.23; visc. 12,000 cps; heat distort. temp. 262 (264 psi).

Anchor® 1222. [Air Prods./Perf. Chems.] Boron trifluoride-amine complex; epoxy curing agent for one-component coatings, laminates, molding compds., vacuum impregnation of elec. motor coils; dk. amber liq.; good sol. in epoxy resins; sp.gr.

1.10; visc. 600 cps; heat distort. temp. 223 (264 psi).

Anchor® DBD. [Air Prods./Perf. Chems.] Activated dithiocarbamate; nonstaining, very fast accelerator used with dry nat. rubber, SBR stocks, and nat. rubber latex; recommended for cements, adhesives, and calendered and extruded compds.; brn. visc. liquid; sol. in hydrocarbons, ketones, alcohols, carbon disulfide, and chlorinated solvs.; sp.gr. 1.10; Unverified

Anchor® DNPD. [Neochem] N,N′-Di-β-naphthyl-p-phenylene diamine; CAS 93-46-9; nonstaining antioxidant for rubber, SBR, NBR, BR, and polyethylene, latex foam, thread, and dipped goods, in adhesives, cable insulation and sheathing; brnsh.-gray fine powd.; sol. in acetone, chloroform and benzene; insol. in petrol. solvents, ethanol, CCl_4, water and aq. alkalis; sp.gr. 1.28; m.p. 230 C.

Anchor® DOTG. [Air Prods./Perf. Chems.] Di-o-tolylguanidine; CAS 97-39-2; med.-speed accelerator; best as sec. accelerator in CR; off-wh. powd.; sol. in chloroform, methanol; sp.gr. 1.10; m.p. 175 C; Unverified

Anchor® DPG. [Air Prods./Perf. Chems.] Diphenyl guanidine; CAS 102-06-7; med.-speed accelerator used to activate thiazoles and sulfenamides; for vulcanization of thick articles, footwear applics.; sec. gelling agent in latex foam; peptizer for sulfur-modified CR; wh. powd.; sol. in benzene hydrocarbons, alcohol, acetone, chloroform; sp.gr. 1.19; m.p. 147 C; Unverified

Anchor® HDPA. [Anhydrides & Chems.] Alkylated diphenylamines, mainly heptylated; antioxidant for use in foamed latex carpet backings; dk. brn. visc. liq.; sol. in acetone, CCl_4, most esters and min. oils; insol. in water; sp.gr. 0.96.

Anchor® HDPA/SE. [Anhydrides & Chems.] Alkylated diphenylamines (principally heptylated diphenylamine) with selected surface-active agents; antioxidant for the compding. of nat. and syn. rubber latices; for latex carpet backings, latex-based adhesives; dk. brn. visc. liq.; mild char. odor; sol. in acetone, CCl_4, most esters and min. oils; insol. in water; readily emulsified in cold water; sp.gr. 0.98; visc. 20 poise.

Anchor® ODPA. [Air Prods./Perf. Chems.] Octylated diphenylamine; low staining antioxidant for use in nat. and syn. rubbers, for footwear, cable insulation, mechanical rubber goods; lt. brn. flakes; sol. in petrol, benzene, ethylene dichloride, and acetone; insol. in water; sp.gr. 0.99; m.p. 82 C min.; Unverified

Anchor® ZDBC. [Air Prods./Perf. Chems.] Zinc diethyl dithiocarbamate; CAS 14323-55-1; ultra fast, nonstaining accelerator used in latex and dry rubber compding; antioxidant in pressure-sensitive and hot-melt adhesives; wh. to cream powd.; sol. in petrol. and benzene hydrocarbons, carbon disulfide and chlorinated solvs.; sp.gr. 1.24; m.p. 106 C; Unverified

Anchor® ZDEC. [Air Prods./Perf. Chems.] Zinc diethyl dithiocarbamate; CAS 14323-55-1; ultra fast, nonstaining accelerator used in nat. and syn. rubbers incl. SBR, NBR, polyisoprene, butyl, chlorobutyl, and EPDM for foam, cable insulation, molded goods, etc.; wh. to ylsh. powd.; sol. in benzene hydrocarbons, carbon disulfide, chlorinated solvs., and dilute alkalis; sp.gr. 1.48; m.p. 178 C; Unverified

Anoxsyn 442. [Elf Atochem N. Am./Plastics Addit.] Bis alkyl sulfide; nonhydrolyzable antioxidant for ABS, EVA, polyester, PE, PP, PS; wh. powd.

Anstac M, 2M. [CDC Int'l.] Proprietary; antistatic coating for ABS, acrylic, cellulosics, nylon, polyacetal, PC, PE, PP, PS, flexible and rigid PVC, etc.

Anstex AK-25. [Toho Chem. Industry] Special phosphate; antistat for syn. fibers, plastics; antifog for PVC film for agric. use; liq.; 55% act.

Anstex SA. [Toho Chem. Industry] Antistatic agent for PP film.

Antarox® BA-PE 70. [Rhone-Poulenc Surf. & Spec.] Aliphatic alcohol alkoxylate; nonionic; moderate foaming surfactant, emulsifier, dispersant for insecticides, herbicides, latex polymerization, leather finishing applics.; biodeg.; soft solid; HLB 16.1; 100% conc.

Antarox® BA-PE 80. [Rhone-Poulenc Surf. & Spec.] Aliphatic alcohol alkoxylate; nonionic; moderate foaming surfactant, emulsifier, dispersant for insecticides, herbicides, latex polymerization, leather finishing applics.; biodeg.; hard solid; HLB 26.1; 100% conc.

Antarox® F88. [Rhone-Poulenc Surf. & Spec.] Poloxamer-238; CAS 9003-11-6; nonionic; defoamer, dispersant, wetting agent, emulsifier, demulsifier, leveling agent, detergent, foam booster, viscosifier for industrial/household cleaners, lt. duty liqs., syndet bars, toilet tank blocks; latex stabilizer; flakes; HLB 28.0; pour pt. 54 C; cloud pt. > 100 C (1% aq.); 100% conc.

Antarox® L-62. [Rhone-Poulenc Surf. & Spec.] Poloxamer 182; CAS 9003-11-6; nonionic; defoamer, dispersant, wetting agent, emulsifier, demulsifier, leveling agent, detergent for household/industrial cleaners, hard surf. cleaners, laundry, skin care prods., emulsion polymerization; liq.; HLB 7.0; pour pt. -4 C; cloud pt. 32 C (1% aq.); 100% conc.

Antarox® L-62 LF. [Rhone-Poulenc Surf. & Spec.] EO/PO block copolymer; CAS 9003-11-6; nonionic; defoamer, dispersant, wetting agent, emulsifier, demulsifier, leveling agent, detergent for industrial/household cleaners, hard surf. cleaning, laundry, skin care prods., emulsion polymerization; liq.; HLB 7.0; pour pt. -10 C; cloud pt. 28 C (1% aq.); 100% conc.

Antarox® L-64. [Rhone-Poulenc Surf. & Spec.] Poloxamer 184; CAS 9003-11-6; nonionic; defoamer, dispersant, wetting agent, emulsifier, oil demulsifier, leveling agent, detergent for industrial/household cleaners, hard surf. cleaning, metalworking fluids, laundry, skin care, emulsion polymerization; plasticizer for resin compositions; defoamer in paper coatings; liq.; HLB 15.0; pour pt. 16 C; cloud pt. 59 C (1% aq.); 100% conc.

Antarox® L-68/LF. [Rhone-Poulenc Surf. & Spec.] EO/PO block copolymer; nonionic; moderate foamer; latex stabilizer, leveling agent/solubilizer for chrome, acid, direct, vat, and caprosyl dyes; dye stripping agent; in hard surf. and laundry detergents, toilet bowl and block detergents, syndet bar soaps; solubilizer/emollient in shampoos; acid pickling of sheet steel, in auto radiator cleaners; stabilizer in mouthwash; liq.; HLB 26.4; cloud pt. < 100 C.

Antarox® PGP 18-2. [Rhone-Poulenc Surf. Canada] Poloxamer-182; nonionic; low-foaming surfactant, detergent, emulsifier for rinse aids, dishwashes, hard surf. cleaners, laundry compds., cosmetics, shampoos, emulsion polymers; liq.; HLB 7.0; cloud pt. 30-34 C; 100% conc.

Antarox® PGP 18-2LF. [Rhone-Poulenc Surf. Canada] Ethoxylated propoxylated glycol; nonionic; surfactant, detergent, emulsifier for sulfate pulp washing, rinse aids, dishwash, hard surf. cleaners, laundry compds., cosmetics, shampoos, emulsion polymers; liq.; HLB 7.0; cloud pt. 26-30 C; 100% conc.

Antarox® PGP 18-8. [Rhone-Poulenc Surf. Canada] Poloxamer 188; nonionic; moderate foamer; latex stabilizer, leveling agent/solubilizer for chrome, acid, direct, vat, and caprosyl dyes; dye stripping agent; in hard surf. and laundry detergents, toilet bowl and block detergents, syndet bar soaps; solubilizer/emollient in shampoos; acid pickling of sheet steel; in auto radiator cleaners; stabilizer in mouthwash; flakes; HLB 29.0; cloud pt. < 100 C; 100% conc.

Antiblaze® 19. [Albright & Wilson Am.] Neutral cyclic diphosphonate ester; flame retardant for textiles and plastics incl. rigid RIM; achieves equivalent flame retardance in many systems with significantly reduced loadings; liq.; 21% P.

Antiblaze® 78. [Albright & Wilson Am.] Chlorinated phosphonate ester; flame retardant for use in rigid and flexible urethanes, bonded foam, semidurable textile applics., and phenolic-based laminates; liq.; 12% P, 34% Cl.

Antiblaze® 80. [Albright & Wilson Am.] Tris (2-chloropropyl) phosphate; flame retardant for use in rigid urethane systems; provides exc. storage life and enhanced fluorocarbon compat. in combination with either A or B components; liq.; 9.3% P, 33% Cl.

Antiblaze® 100. [Albright & Wilson Am.] Chlorinated diphosphate ester; flame retardant for virgin and bonded flexible urethane foam; low volatility allows its use in textiles, paper, adhesives, and epoxy and phenolic-based laminates; cost-effective means to comply with Cal. 117 and other flamm. tests; liq.; 10.6% P, 36% Cl.

Antiblaze® 125. [Albright & Wilson Am.] Chlorinated phosphorus ester; low visc. flame retardant for bonded flexible urethane foams and fabric backcoatings required to met the pill test (CPSC FF1-70) and automotive MVSS 302 stds.; liq.; 11.2% P, 34% Cl.

Antiblaze® 150. [Albright & Wilson Am.] Chlorinated phosphorus ester; flame retardant for flexible PU foams; provides cost-effective means to comply with MVSS-302; liq.; 11.3% P, 35% Cl.

Antiblaze® 175. [Albright & Wilson Am.] Chlorinated phosphorus ester; low visc. flame retardant for molded flexible PU foams required to meet MVSS 302 stds.; provides exc. stability in flexible pkg. components; liq.; 10.1% P, 35% Cl.

Antiblaze® 195. [Albright & Wilson Am.] Tris (dichloropropyl) phosphate; flame retardant for flexible PU foams; exc. hydrolytic stability; liq.; 7.2% P, 49.1% Cl.

Antiblaze® 519. [Albright & Wilson Am.] Triisopropylphenyl phosphate; flame retardant for acrylics, cellulosics, epoxy, nitrile, phenolic, thermoset polyester, PP, PS, PVAc, PVC, flexible PU foam.

Antiblaze® 1045. [Albright & Wilson Am.] Neutral cyclic diphosphonate ester; cost-effective flame retardant for thermoplastics; liq.; 20% P.

Antiblaze® DMMP. [Albright & Wilson Am.] Dimethyl methylphosphonate; CAS 756-79-6; flame retardant for rigid urethane foams, unsat. polyester resins, and water-borne latexes; also as a solv.; liq.; 25% P.

Antiblaze® TCP. [Albright & Wilson Am.] Tricresyl phosphate; flame retardant for acrylics, cellulosics, epoxy, nitrile, thermoset polyester, PS, PVAc, PVC, flexible PU foam.

Antiblaze® TDCP/LV. [Albright & Wilson Am.] Tris(dichloropropyl) phosphate; low visc. version of Antiblaze 195; flame retardant for flexible PU foams; exc. hydrolytic stability; liq.; 7.4% P, 47% Cl.

Antiblaze® TXP. [Albright & Wilson Am.] Trixylenyl phosphate; CAS 25155-23-1; flame retardant for acrylics, cellulosics, epoxy, thermoset polyester, PS, PVAc, PVC, flexible PU foam.

Antiblock-System Hoechst B1980. [Hoechst AG] Antiblocking agent in LDPE carrier; antiblock conc. for LDPE, LLDPE, HDPE, PP, EVA; FDA and BGA approvals.

Antiblock-System Hoechst B1981. [Hoechst AG] Antiblocking agent in LDPE carrier; antiblock conc. for LDPE, LLDPE, HDPE, PP, EVA; BGA approvals.

Antifoam 7800 New. [Bayer/Fibers, Org., Rubbers] Higher hydrocarbons and their sulfonic acid derivs.; antifoam for aq. systems in sugar, fertilizer, phosphoric acid, dyestuffs, paper, leather, plastics, and chemical industries; resistant to acid and weak alkalis; faint yel. liq., sl. turbid @ R.T.; sp.gr. 0.87; b.p. 80 C; flash pt. > 100 C.

Antilux® 110. [Rhein Chemie] Paraffin and microcrystalline waxes; sun-checking, anti-weathering and ozone protective wax for tech. molded and extruded rubber goods, lt. colored rubber goods, cellular rubber, cables; pale blue flakes; sp.gr. 0.92; solid. pt. 60-64 C.

Antilux® 111. [Rhein Chemie] Paraffins and microcrystalline waxes; sun-checking, anti-weathering and ozone protective wax for tires, conveyor belts, tech. rubber goods; pale grn. flakes; sp.gr. 0.92; solid. pt. 64-68 C.

Antilux® 500. [Rhein Chemie] Paraffins and microcrystalline waxes; sun-checking, anti-weathering and ozone protective wax for tech. molded and extruded rubber goods, cellular rubber, cables; wh. to ylsh. flakes; sp.gr. 0.91; solid. pt. 53-57 C.

Antilux® 550. [Rhein Chemie] Paraffins and microcrystalline waxes; sun-checking, anti-weathering and ozone protective wax for tech. molded and extruded rubber goods, cellular rubber, cables; pale yel. flakes; sp.gr. 0.91; solid. pt. 56-60 C.

Antilux® 600. [Rhein Chemie] Paraffins and microcrystalline waxes; sun-checking, anti-weathering and ozone protective wax for tech. molded and extruded articles, cellular rubber, cable coverings; pale yel. flakes; sp.gr. 0.91; solid. pt. 58-62 C.

Antilux® 620. [Rhein Chemie] Paraffins and microcrystalline waxes; sun-checking, anti-weathering and ozone protective wax for tech. molded and extruded rubber goods, cellular rubber, tires, cables; pale yel. flakes; sp.gr. 0.92; solid. pt. 58-62 C.

Antilux® 654. [Rhein Chemie] Paraffins and microcrystalline waxes; sun-checking, anti-weathering and ozone protective wax for tech. molded and extruded rubber goods, tires, conveyor belts; wh. to pale yel. flakes; sp.gr. 0.92; solid. pt. 63-67 C.

Antilux® 660. [Rhein Chemie] Paraffins and microcrystalline waxes; sun-checking, anti-weathering and ozone protective wax for tires, conveyor belts, tech. rubber goods; blue flakes; sp.gr. 0.92; solid. pt. 63-67 C.

Antilux® 750. [Rhein Chemie] Microcrystalline waxes; sun-checking, anti-weathering and ozone protective wax for inj. molded and extruded rubber goods; lubricant for cables; med. brn. flakes; sp.gr. 0.9; solid. pt. 75 C.

Antilux® L. [Rhein Chemie] Paraffins and microcrystalline waxes; sun-checking, anti-weathering and ozone protective wax for rubber articles in contact with foodstuffs, toys, surgical and pharmaceutical rubber articles; wh. flakes; sp.gr. 0.91; solid. pt. 53-57 C.

Antioxidant 235. [Akrochem] 2,2′-Methylene-bis (4 methyl-6 tert. butylphenol); CAS 119-47-1; nonstaining antioxidant in elastomers for medical equipment, rubber thread, and latex compds.; off-wh. cryst. powd.; sol. in ethanol, acetone, methylene chloride, and benzene; insol. in water; sp.gr. 1.04; m.p. > 124 C.

Antioxidant 425. [Cytec Industries] 2,2′-Methylenebis (4-ethyl-6-tert-butylphenol); CAS 88-24-4; antioxidant protecting impact molding resins from oxidation degradation; stabilizer for acrylics and ABS; used in molded prods.; cream to wh. powd.; m.w. 368.5; sol. in acetone, dioxane, ethyl acetate, chloroform, ethanol, benzene, n-heptane, water; sp.gr. 1.10; m.p. 117-123 C.

Antioxydant NV 3. [BASF AG] Alkylphenol; used in tire industry for tech. goods.

Antislip-System Hoechst S1990. [Hoechst AG] Antislip agent in LDPE carrier; antislip conc. for LDPE, LLDPE, HDPE; may also be used in EVA, PP; FDA and BGA approvals.

Antistat A21750. [Polycom Huntsman] Amine; internal antistat for PE, PP; limited FDA acceptance.

Antistat A21800. [Polycom Huntsman] Proprietary; internal antistat for PE, PP; FDA approved.

Antistatic KN. [3V] Quat. ammonium compd.; internal/external antistat for ABS, acrylic, cellulosics, nylon, acetal, PC, PE, PP, PS, PVC.

Antistaticum RC 100. [Rhein Chemie] Fatty alkyl ether of polyethylene glycol; antistatic plasticizer for NR and syn. rubber and PVC plastics; whitish visc. liq.; sp.gr. 0.94.

Antisun®. [Frank B. Ross] Complex hydrocarbon mixt.; antiozonant, anticracking and sunchecking wax for the rubber industry; FDA 21CFR §177.2600; wh. solid, odorless; insol. in water; sp.gr. 0.917-0.939; m.p. 70-74 C; acid no. nil; sapon. no. 1.0 max.; flash pt. (COC) 400 F min.; *Toxicology:* may irritate eyes, skin; molten wax can cause burns; fumes from decomp. may cause respiratory irritation; dust may irritate mucous membranes and respiratory tract; TLV/TWA 5 mg/m^3 (respirable dust); *Precaution:* incompat. with strong oxidizing agents.

Anti-Terra®-204. [Byk-Chemie USA] Sol'n. of higher m.w. carboxylic acid salts of polyamine amides; wetting and dispersing additive to improve antisedimentation and sagging props.; used in alkyds, PVC copolymers, chlorinated rubber, epoxy systems; gellant for organophilic bentonites; sp.gr. 0.92-0.94 g/cc; dens. 7.67-7.84 lb/gal; acid no. 37-44; flash pt. (Seta) 31 C; ref. index 1.466-1.473; 50-54% NV in methoxypropanol/naphtha (3/2).

Anti-Terra®-P. [Byk-Chemie USA] Phosphoric acid salt of long chain carboxylic acid polyamine amides; cationic; wetting and dispersing additive to prevent settling and flooding of pigments in alkyds, alkyd-melamine, chlorinated rubber systems, PVC copolymers, acid catalyzed paints; sp.gr. 0.98-1.00 g/cc; dens. 8.18-8.35 lb/gal; acid no. 160-180; flash pt. (Seta) 26 C; 39-44% NV in isobutanol/xylene/water (3/1/1).

Antozite® 1. [R.T. Vanderbilt] N,N′-di(2-octyl)-p-phenylenediamine; antiozonant for NR, IR, BR, and SBR; dk. redsh. brn. liq.; m.w. 332.57; sol. in chloroform, toluene, petrol. ether, ethyl alcohol; insol. in water; dens. 0.90 ± 0.02 mg/m^3; visc. 110-145 SUS (40 C).

Antozite® 2. [R.T. Vanderbilt] N,N′-di-3(5-methylheptyl)-p-phenylenediamine; antiozonant for NR, IR, BR, and SBR; dk. redsh. brn. liq.; m.w. 332.57; sol. in chloroform, toluene, petrol. ether, ethyl alcohol; insol. in water; dens. 0.90 ± 0.02 mg/m^3; visc. 180-270 SUS (40 C).

Antozite® 67P. [R.T. Vanderbilt] N-(1,3-Dimethylbutyl)-N′-phenyl-p-phenylenediamine; CAS 793-24-8; antiozonant for rubber; m.w. 268.40; sol. in toluene, acetone, ethyl acetate, ethylene dichloride; insol. in water; dens. 0.986-1.00 mg/m^3.

AO. [Climax Performance] Antimony oxide; CAS 1309-64-4; EINECS 215-175-0; flame retardant synergist (with halogen) for use in a broad range of polymers; wh. cryst. powd., odorless; insol. in water; sol. in conc. HCl and sulfuric acids, and strong alkalies; 1.5 μ avg. particle size; 99.9% thru 325 mesh; sp.gr. 5.5; 99.5% Sb_2O_3.

AO47L. [Sovereign] Styrenated phenol; CAS 61788-44-1; antioxidant for rubber articles in contact with food; nonstaining in latex or dry rubber; can be emulsified for latex; liq.

AO47P. [Sovereign] Styrenated phenol; CAS 61788-44-1; antioxidant for rubber articles in contact with food; nonstaining in latex or dry rubber; can be emulsified for latex; powd.

AO872. [Sovereign] Modified phenolic blend; nonstaining, nondiscoloring, moderate cost antioxidant with exc. heat resist. for all emulsion polymers; replaces methylene bisphenol and butylated p-cresol antioxidant types.

AO924. [Sovereign] Diphenylamine blend; sl. staining high heat resist. general purpose antioxidant for all polymers.

AOM. [Climax Performance] Ammonium octamolybdate; smoke suppressant for PVC and PVC alloys; also for use in polyesters and thermoplastic elastomers; for transportation, construction, wire and cable industries that must meet stringent smoke standards; wh. to off-wh. free-flowing powd.; 2 μ avg. particle size; 99.5% < 12 μ; sol. 4 g/100 ml water; sp.gr. 3.18; bulk dens. 30 lb/ft^3; oil absorp. 20; dec. temp. 482 F.

Aqualease™ 2802. [George Mann] High m.w. silicone polymer; water-based heavy-duty release agent for cast urethane elastomer and integral skin urethane foams; milky liq.; sp.gr. 0.96; f.p. 0 C; 12% NV; *Precaution:* do not take internally; *Storage:* keep from freezing; store in warm place; if frozen, thaw and mix well before using.

Aqualease™ 6100. [George Mann] Aq.-based; release agent for urethane elastomers and foams.

Aqualease™ 6101. [George Mann] Aq. emulsion; release agent for thermosetting resins, e.g., urethane elastomers, epoxies, filled polyesters, vinyl ester polymers; suggested for hot molding (> 150 F) by spraying on hot surfaces; milky emulsion, low odor; dens. 8.0-8.2 lb/gal; f.p. 0 C; *Toxicology:* low toxicity; may cause temporary eye irritation; *Precaution:* do not take internally; *Storage:* 6 mos shelf life; store in warm area above 0 C.

Aqualease™ 6102. [George Mann] Silicone aq. emulsion; release agent with better corrosion protection and exc. film formation; releases syn. microcellular and solid elastomers, epoxies and polyester systems; milky emulsion, low odor; dens. 8.0-8.2 lb/gal; f.p. 0 C; *Toxicology:* low toxicity; may cause temporary eye irritation; *Precaution:* do not take internally; *Storage:* 6 mos shelf life; store in warm area above 0 C; mix well before use.

Aqualease™ 6201. [George Mann] Aq. emulsion; release agent for thermosetting resins, e.g, urethane elastomers, epoxies, filled polyesters, vinyl ester polymers; exc. for hot molding over 150 F; milky emulsion, low odor; dens. 8.0-8.2 lb/gal; f.p. 0 C; *Toxicology:* low toxicity; may cause temporary eye irritation; *Precaution:* do not take internally; *Storage:* 6 mos shelf life; store in warm area above 0 C.

Aqualease™ 6202. [George Mann] Silicone aq. emulsion; release agent for syn. foam and solid urethane elastomers, epoxies, and polyester systems; provides corrosion protection for steel molds; exc. film formation; milky emulsion, low odor; dens. 8.0-8.2 lb/gal; f.p. 0 C; *Toxicology:* low toxicity; may cause temporary eye irritation; *Precaution:* do not take internally; *Storage:* 6 mos shelf life; store in warm area above 0 C; mix well before use.

Aqualine GP100. [Dexter/Frekote] Water-based release agent; semipermanent release agent for applic. to mold surfaces @ 20-205 C for molding thermoset resins and rotomoldable thermoplastics; chemically bonds to surf. to form a durable interface providing multiple releases with no contaminating transfer; non-CFC, very low VOC; wh. emulsion, mild odor; sp.gr. 1.00; flash pt. (PMCC) none; *Toxicology:* nonirritating to eyes and skin; *Storage:* 6 mos shelf life; store below 40 C.

Aqualine R-110. [Dexter/Frekote] Semi-permanent water-based release agent for all nat. and syn. organic rubber compds. except silicone elastomers; for applic. to mold surfaces which are preheated to 60-182 C; chemically bonds to mold surface forming thin, chemically resist. coating; nonflamm.; high thermal stability; wh. aq. emulsion, sl. mild odor; sp.gr. 1.00 ± 0.02; flash pt. (PMCC) none; *Toxicology:* very low toxicity; *Storage:* store below 40 C; shelf life 12 mos in unopened can.

Aqualine RC-321. [Dexter/Frekote] Water-dilutable mold release conc. providing durable release film for molding of all natural and syn. polymers except silicone elastomers; effective when applied between 60 and 182 C; fast curing; high thermal stability.

Aquastab PA 42M. [Eastman] Irganox 1010 (tetrakis [methylene (3,5-di-tert-butyl-4-hydroxyhydrocinnamate)] methane) in aq. emulsion of low m.w. oxidized polyolefin wax, with nonionic surfactants, antifoam, biocides; stabilizer/antioxidant for polymers; off-wh. powd.; 90% ≤ 20.5 μ; dens. 1.03-1.04 g/ml; pH 8-9; 37-38.5% water.

Aquastab PA 43. [Eastman] Irganox 1076 (octadecyl 3,5-di-t-butyl-4-hydroxyhydrocinnamate) in aq. emulsion of low m.w. oxidized polyolefin wax, with nonionic surfactants, antifoam, biocides; stabilizer/antioxidant for polymers; off-wh. powd.; 90% ≤ 31.6 μ; dens. 1.00-1.01 g/ml; pH 8.5-9.5; 37-38.5% water.

Aquastab PA 48. [Eastman] Dilauryl thiodipropionate in aq. emulsion of low m.w. oxidized polyolefin wax, with nonionic surfactants, antifoam, biocides; stabilizer/antioxidant for polymers; off-wh. powd.; 90% ≤ 46.4 μ; dens. 1.01-1.02 g/ml; pH 8.5-9.5; 45.8-47.3% water.

Aquastab PA 52. [Eastman] Irgafos 168 (tris (2,4-di-t-butylphenyl) phosphite) in aq. emulsion of low m.w. oxidized polyolefin wax, with nonionic surfactants, antifoam, biocides; stabilizer/antioxidant for polymers; off-wh. powd.; 90% ≤ 25.7 μ; dens. 1.00-1.01 g/ml; pH 8.3-9.3; 37-38.5% water.

Aquastab PA 58. [Eastman] BHT in aq. emulsion of low m.w. oxidized polyolefin wax, with nonionic surfactants, antifoam, biocides; stabilizer/antioxidant for polymers; off-wh. powd.; 90% ≤ 25.8 μ; dens. 1.01-1.02 g/ml; pH 8.4-9.4; 37-38.5% water.

Aquastab PB 59. [Eastman] UV-Chek AM-340 (2,4-Di-t-butylphenyl 3,5-di-t-butyl-4-hydroxybenzoate) in aq. emulsion of low m.w. oxidized polyolefin wax, with nonionic surfactants, antifoam, biocides; stabilizer/uv inhibitor for polymers; off-wh. powd.; 90% ≤ 21.4 μ; dens. 1.025-1.035 g/ml; pH 8.2-9.2; 37-38.5% water.

Aquastab PB 503. [Eastman] Tinuvin P (drometrizole) in aq. emulsion of low m.w. oxidized polyolefin wax, with nonionic surfactants, antifoam, biocides; stabilizer/uv inhibitor for polymers; off-wh. powd.; 90% ≤ 22.8 μ; dens. 1.13-1.14 g/ml; pH 8.2-9.2; 37-38.5% water.

Aquastab PC 44. [Eastman] Chimassorb 944 in aq. emulsion of low m.w. oxidized polyolefin wax, with nonionic surfactants, antifoam, biocides; lt. stabilizer for polymers; off-wh. powd.; 90% ≤ 19.6 μ; dens. 1.00-1.01 g/ml; pH 9.5-10.5; 47.1-48.6% water.

Aquastab PC 45M. [Eastman] Tinuvin 770 (bis (2,2,6,6-tetramethyl-4-piperidinyl) sebacate) in aq. emulsion of low m.w. oxidized polyolefin wax, with nonionic surfactants, antifoam, biocides; lt. stabilizer for polymers; off-wh. powd.; 90% ≤ 42.6 μ; dens. 1.025-1.035 g/ml; pH 9.5-10.5; 37-38.5% water.

Aquastab PD 31M. [Eastman] Calcium stearate in aq. emulsion of low m.w. oxidized polyolefin wax, with nonionic surfactants, antifoam, biocides; stabilizer/antacid for polymers; off-wh. powd.; 90% ≤ 22.7 μ; dens. 1.015-1.025 g/ml; pH 9.4-10.4; 47-48.5% water.

Aquastab PD 412M. [Eastman] Hydrotalcite in aq. emulsion of low m.w. oxidized polyolefin wax, with nonionic surfactants, antifoam, biocides; stabilizer/antacid for polymers; off-wh. powd.; 90% ≤ 42.1 μ; dens. 1.2-1.3 g/ml; pH 8.8-9.8; 45.1-46.6% water.

Aquastab PH 502. [Eastman] Oxalylbis (benzylidene hydrazide) in aq. emulsion of low m.w. oxidized polyolefin wax, with nonionic surfactants, antifoam, biocides; stabilizer/metal deactivator for polymers; off-wh. powd.; 90% ≤ 8.8 μ; dens. 1.08-1.18 g/ml; pH 8-9; 37-38.5% water.

Aquatac® 5527. [Arizona] Water-based resin; freeze/thaw-stable tackifier for removable and permanent label adhesive applics.; features unique surfactant system; provides better specific adhesion to a variety of substrates; produces more stable adhesives; suitable for carboxylated S/B, neoprene, SBR, NR, 2-ethylhexyl acrylate, butyl acrylic, and vinyl acetate ethylene; FDA 21CFR §175.105; wh.; dens. 1.02 g/cc; visc. 1250 cps; soften. pt. (R&B) 27 C; flash pt. (COC) 212 F; pH 8-10; 55% solids.

Aquatac® 5590. [Arizona] Aq. resin dispersion; tackifier for carboxylated S/B, SBR, NR, 2-ethylhexyl acrylate, butyl acrylic, VAE, and neoprene systems; provides high heat resist, improved adhesion to PE, faster drying adhesives; features high soften. pt., unique surfactant system, high solids; FDA 21CFR §175.105; visc. 3000 cps; soften. pt. (R&B) 90 C; pH 8.8; 55% solids.

Aquatac® 6025. [Arizona] Aq. resin dispersion; tackifier for adhesives incl. carboxylated SBR, 2-ethylhexyl acrylate, butyl acrylic, SBR, NR, neoprene, VAE, urethane systems; provides improved aging props., enhanced adhesive props.; FDA 21CFR §175.105, 176.210; visc. 5350 cps; soften. pt. (R&B) 25 C; pH 7.0; 60% solids.

Aquatac® 6085. [Arizona] Glyceryl rosinate aq. disp.; waterborne tackifier for pressure-sensitive adhesives; produces formulations with aggressive tack and peel, exc. shear props. after aging; for waterborne labels, decals, shelf liners, construction adhesives, tapes; suitable for carboxylated SBR, 2-ethylhexyl acrylate, butyl acrylic, SBR, NR, neoprene, VAE, urethane systems; FDA 21CFR §175.105, 176.210; milky beige disp.; 2.5 μ particle size; dens. 8.8 lb/gal; visc. 7500 cps; soften. pt. (R&B) 82 C (of base resin); pH 7.0; 60% solids; *Storage:* prevent freezing, mold and bacteria contamination.

Aquatac® 6085B-1. [Bergvik Kemi] Aq. resin dispersion; tackifier for adhesives incl. carboxylated SBR, 2-ethylhexyl acrylate, butyl acrylic, SBR, NR, neoprene, VAE, urethane systems; provides improved aging props., enhanced adhesive props.; FDA 21CFR §175.105, 176.210; visc. 1000 cps; soften. pt. (R&B) 85 C; pH 7.0; 59% solids.

Aquatac® 6085B-7. [Bergvik Kemi] Aq. resin dispersion; tackifier for adhesives incl. carboxylated SBR, 2-ethylhexyl acrylate, butyl acrylic, SBR, NR, neoprene, VAE, urethane systems; provides improved aging props., enhanced adhesive props.; FDA 21CFR §175.105, 176.210; visc. 7000 cps; soften. pt. (R&B) 85 C; pH 7.0; 60% solids.

Aquatac® 9027. [Arizona] Aq. resin dispersion; tackifier suitable for carboxylated S/B, SBR, and NR systems; for removable and permanent label adhesive applics.; high solids; FDA 21CFR §175.105; Gardner 8-10 color; dens. 1.00 g/cc; soften. pt. (R&B) 27 C; flash pt. (COC) 212 F; pH 9.0; 90% solids.

Aquatac® 9041. [Arizona] Aq. resin dispersion; higher peel version of Aquatac 9027; tackifier for carboxylated SBR, SBR, NR systems; recommended for label applics.; Gardner 9-11 color; dens. 1.00 g/cc; soften. pt. (R&B) 41 C; flash pt. (COC) 212 F; pH 9.0; 90% solids.

Aquathene™ CM 92830. [Quantum/USI] Blk. catalyst masterbatch; catalyst masterbatch designed for use with Aquathene HR 92821 ethylene vinylsilane copolymer compd. to produce low voltage power cable insulation; added at 15% to produce a 16% carbon blk.-filled crosslinkable insulation suitable for UL44 applics.; dens. 0.94; *Storage:* store separately from Aquathene HR 92821 until extrusion; does not require moisture-proof pkg.

Aquathene™ CM 92832. [Quantum/USI] N110 Carbon blk. catalyst masterbatch; CAS 1333-86-4; EINECS 215-609-9; catalyst masterbatch designed for use with Aquathene AQ 120-000 ethylene vinylsilane copolymer compd. to produce low voltage power cable insulation; added at 15% to produce a 2.5% carbon blk.-filled crosslinkable insulation suitable for UL854 applics; dens. 1.01; *Storage:* store separately from Aquathene AQ 120-000 until extrusion; prevent exposure to moisture.

Aquatreat DNM-9. [Alco] Dithiocarbamate salts; short-stop in emulsion polymerization of rubber; biocide, fungicide, and algicide used in water treatment, paper, sugar, and petrol. applics.; 9% act. in water.

Aquatreat DNM-30 [Alco] Sodium dimethyldithiocarbamate (15-16.6%), nabam (15-16.6%), inert; short-stop in emulsion polymerization of rubber; fungicide and bactericide for use in pulp/paper mills, sugar mills, drilling fluids, petrol. recovery; algicide for use in industrial recirculating water cooling towers, etc.; yel.-grn.; sulfurous odor; sp.gr. 1.18; dens. 9.6 lb/gal; pH 10.5-12.4; 30-33.2% act.; *Toxicology:* eye and skin irritant.

Aquatreat KM. [Alco] Potassium dimethyldithiocarbamate; CAS 128-03-0; short-stop in emulsion polymerization of rubber; biocide, fungicide, and algicide used in water treatment, paper, sugar, and petrol. applics.; 50% act. in water.

Aquatreat SDM. [Alco] Sodium dimethyldithiocarbamate; short-stop in emulsion polymerization of rubber; biocide, fungicide, and algicide used in water treatment, paper, sugar, and petrol. applics.; 40% act. in water.

Araldite® PY 306. [Ciba-Geigy/Plastics, Plastics UK] Bisphenol F epoxy liq.; modifier of other resins for lower visc., higher solids coatings; Unverified

Araldite® XD 897. [Ciba Plastics UK] Epoxy resin; stabilizer for chlorinated vinyl resins; Unverified

Aranox®. [Uniroyal] p-(p-Toluenesulfonyl amido) diphenylamine; antioxidant protecting EVA and polyamide hot-melt adhesives, PP, LDPE, LLDPE, HDPE against thermal degradation; FDA approvals.

Arazate®. [Uniroyal] Zinc dibenzyldithiocarbamate; CAS 14726-36-4; activator.

Arconate® 1000. [Arco] Propylene carbonate; CAS 108-32-7; EINECS 203-572-1; solv. with high boiling pt., low toxicity, broad range of applics.; reactive diluent for woodbinders, urethane foams and coatings, foundry sand binders, in textile and syn. fiber industry, natural gas treating; lubricant in cosmetics; mild odor; sp.gr. 1.203-1.210 (20/20 C); dens. 10.1 lb/gal (20 C); visc. 4 cs; vapor pressure 0.02 mm Hg (20 C); f.p. -50 C; b.p. 242 C; flash pt. (TOC) 132 C; ref. index 1.419; pH 4.5-6.5 (10% aq.); 99% min. assay.

Arconate® HP. [Arco] Propylene carbonate; CAS 108-32-7; EINECS 203-572-1; high purity solv. with high boiling pt., low toxicity, broad range of applics.; reactive diluent for woodbinders, urethane foams and coatings, foundry sand binders, in textile and syn. fiber industry, natural gas treating; lubricant in cosmetics; crystal clear, mild odor; sp.gr. 1.203-1.210 (20/20 C); dens. 10.1 lb/gal (20 C); visc. 4 cs; vapor pressure 0.02 mm Hg (20 C); f.p. -50 C; b.p. 242 C; flash pt. (TOC) 132 C; ref. index 1.419; pH 6.5-7.5 (10% aq.); 99.6% min. assay.

Arconate® PC. [Arco] Propylene carbonate; CAS 108-32-7; EINECS 203-572-1; solv. with high boiling pt., low toxicity, broad range of applics.; reactive diluent for woodbinders, urethane foams/coatings/elastomers, foundry sand binders, in textile and syn. fiber industry, natural gas treating, petrol. fraction extraction; in surf. coatings, electrolytes, lubricants for cosmetics; also as accelerator, reactive intermediate, compatibilizer, dispersant, wetting agent, or scavenger in resin systems; exempt from Rule 66; sol. 8% in water (20 C); m.w. 102.07; sp.gr. 1.203 (20/20 C); dens. 10.0 lb/gal (20 C); visc. 6.6 cs (0 F); vapor pressure < 0.1 mm Hg (20 C); f.p. -50 C; b.p. 242 C (760 mm); flash pt. (TOC) 270 F; ref. index 1.421 (20 C0; *Toxicology:* pract. nontoxic based on oral and dermal acute testing; moderate eye irritant, sl. skin irritant; *Precaution:* may decompose in strong acid or base.

Arcosolv® DPM. [Arco] Dipropylene glycol monomethyl ether; CAS 34590-94-8; solv. for coatings, cleaners, inks, agric. prods., cosmetics, chemical intermediate applics., epoxy laminates, adhesives, floor polish, fuel additives, oil field, mining, and electronic chemicals; APHA 15 max. liq.; mild, pleasant char. odor; sol. in water; misc. with a number of org. solvs.; m.w. 148.2; sp.gr. 0.950-0.953; dens. 7.91 lb/gal; visc. 3.6 cstk; f.p. -80 C; b.p. 188.3 (760 mm); flash pt. (TCC) 75 C; ref. index 1.422; surf. tens. 28.2 dynes/cm; sp. heat 0.54 cal/g/°C; *Precaution:* combustible.

Arcosolv® DPMA. [Arco] Dipropylene glycol methyl ether acetate; CAS 88917-22-0; solv. where a slow evaporating nonhydroxylic solv. is required; effective in coatings; as a coalescent in waterborne emulsion systems; also for cleaners, inks, epoxy laminates, agric., adhesives, floor polish, fuel additives, oil field, mining; electronic chemicals; APHA 15 max. color; sol. 12.3% in water; m.w. 190.2; sp.gr. 0.972; dens. 8.14 lb/gal; visc. 2.2 cstk; b.p. 205 C (760 mm Hg); ref. index 1.4142; flash pt. (Seta) 186 F; surf. tens. 28.3 dynes/cm.

Arcosolv® PM. [Arco] Propylene glycol methyl ether; CAS 107-98-2; solv. for coatings, cleaners, inks, agric. prods., cosmetics, chemical intermediate applics., epoxy laminates, adhesives, floor polish, fuel additives, oil field, mining, and electronic chemicals; APHA 10 max. liq.; mild, pleasant char. odor; sol. in water; misc. with a number of org. solvs.; m.w. 90.1; sp.gr. 0.918-0.921; dens. 7.65 lb/gal; visc. 1.8 cstk; f.p. -95 C; b.p. 120.1 C (760 mm); ref. index 1.404; flash pt. (TCC) 32 C; surf. tens. 26.5 dynes/cm; sp. heat 0.57 cal/g/°C; *Precaution:* flamm.

Arcosolv® PMA. [Arco] Propylene glycol methyl ether acetate; CAS 108-65-6; slow-evaporating solv. with good solvency for many commonly used coating resins, e.g., acrylics, NC, and urethanes; used in lacquers, water-based paints; APHA 10 max. liq.; mild, ester-like odor; sol. 18.5% in water; m.w. 132.2; sp.gr. 0.963-0.966; dens. 8.02 lb/gal; visc. 1.1 cstk; vapor pressure 3.7 mm Hg (20 C); b.p. 146.0 C (760 mm); flash pt. (TCC) 47 C; ref. index 1.400; surf. tens. 27.4 dynes/cm; *Precaution:* combustible.

Arcosolv® PTB. [Arco] Propylene glycol mono-t-butyl ether; CAS 57018-52-7; solv. offering a blend of hydrophobicity and hydrophilicity for coatings, cleaners, electronic, and ink applics.; strong coupling ability; used in lt.-duty and hard-surf. cleaners, water-reducible polyester and alkyd resin prod.; chemical intermediate in the synthesis of monomeric and polymeric prods.; cosolv.; APHA 15 max. liq.; partial water sol. at ambient temps.; m.w. 132.2; sp.gr. 0.870-0.874; visc. 3.8 cstk; vapor pressure 1.9 mm Hg (20 C); f.p. -56 C; b.p. 151 C (760 mm Hg); flash pt. (TCC) 45 C; ref. index 1.4116; surf. tens. 24.2 dynes/cm; *Toxicology:* LD50 (oral) 3771 mg/kg, (dermal) > 2.0 g/kg; sl. toxic orally; mildly irritating to skin, severe-but-reversible eye irritant in pure form.

Arcosolv® TPM. [Arco] PPG-3 methyl ether; CAS 25498-49-1; slow evaporating solv. for coatings, cleaners, inks, agric. prods., cosmetics, chemical intermediate applics., epoxy laminates, adhesives, floor polish, fuel additives, oil field, mining, and electronic chemicals; APHA 15 max. liq.; mild, pleasant char. odor; sol. in water; misc. with a number of org. solvs.; m.w. 206.3; sp.gr. 0.962-0.965; dens. 8.03 lb/gal; visc. 5.8 cstk; f.p. -79 C; b.p. 242.4 C (760 mm); ref. index 1.430; flash pt. (PMCC) 114 C; surf. tens. 29.0 dynes/cm; sp. heat 0.51 cal/g/°C.

Arizona 208. [Arizona] Tall oil fatty acid ester; for plasticizers, extenders, surface-act. agents in grinding and cutting oils, specialty lubricant additives, corrosion inhibitors, specialty solvs. for printing inks, metalworking, and oil well servicing; plasticizer for CR, NBR, NR, BR, CSM; good low temp. flex; resist. to heat extraction; Gardner 2+ color; sp.gr. 0.87; visc. 6-7 cps; acid no. 0.5; iodine no. 98; cloud pt. < 4 C; flash pt. (CC) > 94 C.

Arizona DR-22. [Arizona] Disproportionated rosin; emulsifier, detergent, wetting agent; used to prepare emulsifiers for SBR polymerization, as short-stop for solv. polymerizations of rubber, as plasticizer/tackifier; used to make Arizona disproportionated tall oil rosin soaps (DRS-40, 42, 43,44); FDA 21CFR §175.105, 175.125, 175.300, 176.170, 176.180, 176.210, 177.1200, 177.1210, 177.2600, 178.3120, 178.3800, 178.3850, 178.3870; Gardner 6+ color; sp.gr. 1.06; dens. 8.2 lb/gal (150 C); soften. pt. (R&B) 59 C; acid no. 161; sapon. no. 165; flash pt. (OC) > 204 C; 100% solids.

Arizona DR-25. [Arizona] Sodium soap of disproportionated rosin; CAS 61790-51-0; EINECS 263-144-5; for preparing emulsifiers for SBR polymerization; FDA 21CFR §175.105, 175.125, 175.300, 176.170, 176.180, 176.210, 177.1200, 177.1210, 177.2600, 178.3120, 178.3800, 178.3850, 178.3870; Gardner 8 color; sp.gr. 1.06; dens. 8.2 lb/gal; (150 C); soften. pt. (R&B) 73 C; acid no. 138; flash pt. (OC) > 204 C.

Arizona DR Mix-26. [Arizona] Emulsifier-mixed disproportionated rosin and fatty acid; emulsifier for SBR industry; paste; 100% conc.; Unverified

Arizona DRS-40. [Arizona] Potassium soap of disproportionated tall oil rosin; anionic; emulsifier, detergent, wetting agent; for ABS, SBR, other syn. elastomers; FDA 21CFR §175.105, 175.125, 176.170, 176.180, 176.210, 177.1200, 177.2600, 178.3120, 178.3800, 1787.3850, 178.3870; tan paste; bland odor; water disp.; sp.gr. 1.1; dens. 9.08 lb/gal (80 C); visc. 1800 cps (65 C); acid no. 16; flash pt. (OC) > 204 C; 79.5% solids.

Arizona DRS-42. [Arizona] Potassium soap of disproportionated tall oil rosin; anionic; emulsifier, detergent, wetting agent; for ABS, SBR, other syn. elastomers; FDA 21CFR §175.105, 175.125, 176.170, 176.180, 176.210, 177.1200, 177.2600, 178.3120, 178.3800, 1787.3850, 178.3870; tan paste, bland odor; water disp.; dens. 9.08 lb/gal (80 C); visc. 1500 cps (65 C); acid no. 12.5; flash pt. (OC > 204 C; 80% solids.

Arizona DRS-43. [Arizona] Sodium salt of disproportionated tall oil rosin; CAS 61790-51-0; EINECS 263-144-5; anionic; emulsifier, detergent, wetting agent; emulsifier for ABS, SBR, syn. elastomers; FDA 21CFR §175.105, 175.125, 176.170, 176.180, 176.210, 177.1200, 177.2600, 178.3120, 178.3800, 1787.3850, 178.3870; tan paste; bland odor; water disp.; sp.gr. 1.1; dens. 8.92 lb/gal (80 C); visc. 2500 cps (65 C); acid no. 11.5; flash pt. (OC) > 204 C; 71% solids.

Arizona DRS-44. [Arizona] Sodium soap of disproportionated tall oil rosin; CAS 61790-51-0; EINECS 263-144-5; anionic; emulsifier for prod. of ABS, SBR and other syn. elastomers; FDA 21CFR §175.105, 175.125, 176.170, 176.180, 176.210, 177.1200, 177.2600, 178.3120, 178.3800, 1787.3850, 178.3870; paste; 70% conc.

Arizona DRS-50. [Arizona] Potassium soap of disproportioned tall oil rosin; anionic; polymerization emulsifier for the SBR industry; FDA 21CFR §175.105, 175.125, 176.170, 176.180, 176.210, 177.1200, 177.2600, 178.3120, 178.3800, 1787.3850, 178.3870; paste; water-disp.; dens. 9.09 lb/gal (80 C); visc. 1800 cps (65 C); flash pt. > 400 F; acid no. 26; 84.5% solids.

Arizona DRS-51E. [Arizona] Potassium soap of disproportionated tall oil rosin; anionic; polymerization emulsifier for SBR industry; FDA 21CFR §175.105, 175.125, 176.170, 176.180, 176.210, 177.1200, 177.2600, 178.3120, 178.3800, 1787.3850, 178.3870; paste; water-disp.; dens. 9.08 lb/gal (80 C); visc. 800 cps (65 C); flash pt. > 400 F; acid no. 19; 79.5% solids.

Armeen® 18. [Akzo Nobel; Akzo Nobel BV]

Stearamine (primary amine); CAS 124-30-1; EINECS 204-695-3; cationic; emulsifier, flotation agent, corrosion inhibitor, anticaking agent; hard rubber mold release agent; also for cosmetics; Gardner 3 max. solid; sol. in ethanol, IPA, chloroform, toluene, CCl_4, slightly sol. in acetone, kerosene; sp.gr. 0.792 (60/4 C); visc. 45.6 SSU; m.p. 122-133 F; pour pt. 115 F; iodine no. 3; amine no. 202; flash pt. (PMCC) > 149 C; 97% primary amine.

Armeen® 18D. [Akzo Nobel] Stearamine, dist.; CAS 124-30-1; EINECS 204-695-3; cationic; emulsifier, flotation agent, corrosion inhibitor, anticaking agent; rubber processing auxiliary; mold release agent for plastics and rubber; also for cosmetics; Gardner 1 max. solid; sol. in ethanol, IPA, chloroform, toluene, CCl_4, slightly sol. in methanol, kerosene; sp.gr. 0.791-0.792 (60/4 C); visc. 43.7 SSU; m.p. 122-133 F; pour pt. 110 F; iodine no. 3; amine no. 204; flash pt. (PMCC) > 149 C; 98% primary amine.

Armeen® DM8. [Akzo Nobel BV] Tert. amine; cationic; polyurethane catalyst; corrosion inhibitor; chemical intermediate; also for cosmetics; liq.; 98% conc.

Armeen® DM10. [Akzo Nobel BV] Tert. amine; cationic; polyurethane catalyst; corrosion inhibitor; chemical intermediate; also for cosmetics; liq.; 98% conc.

Armeen® DM12. [Akzo Nobel BV] Tert. amine; cationic; polyurethane catalyst; corrosion inhibitor; chemical intermediate; also for cosmetics; liq.; 98% conc.

Armeen® DM14. [Akzo Nobel BV] Tert. amine; cationic; polyurethane catalyst; corrosion inhibitor; chemical intermediate; also for cosmetics; liq.; 98% conc.

Armeen® DM16. [Akzo Nobel BV] Tert. amine; cationic; polyurethane catalyst; corrosion inhibitor; chemical intermediate; also for cosmetics; liq.; 98% conc.

Armeen® DMC. [Akzo Nobel BV] Tert. amine; cationic; polyurethane catalyst; corrosion inhibitor; chemical intermediate; also for cosmetics; liq.; 98% conc.

Armeen® DMHT. [Akzo Nobel BV] Tert. amine; cationic; polyurethane catalyst; corrosion inhibitor; chemical intermediate; also for cosmetics; liq.; 98% conc.

Armeen® DMO. [Akzo Nobel BV] Tert. amine; cationic; polyurethane catalyst; corrosion inhibitor; chemical intermediate; also for cosmetics; liq.; 98% conc.

Armeen® DMT. [Akzo Nobel BV] Tert. amine; cationic; polyurethane catalyst; corrosion inhibitor; chemical intermediate; also for cosmetics; liq.; 98% conc.

Armeen® M2-10D. [Akzo Nobel] Didecyl methylamine; CAS 7396-58-9; EINECS 230-990-1; cationic; chemical intermediate for water-sol. betaines; catalyst for urethane resins; also for cosmetics; Gardner 1 max. liq.; sp.gr. 0.80; m.p. -4 C; iodine no. 0.5; amine no. 175; flash pt. (PMCC) > 132 C; 97% tert. amine.

Armeen® N-CMD. [Akzo Nobel] N-coco morpholine; catalyst in PU foams; also for cosmetics; Gardner 3 liq.; sol. in acetone, IPA; sp.gr. 0.87; visc. 50 SSU (55 C); m.p. -29/-10 C; pour pt. -10 C; flash pt. 155 C (COC); 97% min. act.

Armeen® Z. [Akzo Nobel; Akzo Nobel BV] Cocaminobutyric acid; CAS 68649-05-8; EINECS 272-021-5; amphoteric; pigment softening, dispersing agent; antifogging agent, foam booster, stabilizer, wetting agent in alkaline paint strippers, latex emulsions, latex rubber reclamation, inks, plastic films, cosmetics; cooling tower corrosion inhibitor; Gardner 8 pumpable slurry; sol. in water, IPA, ethyl acetate; sp.gr. 0.98; visc. 247 SSU; flash pt. (TCC) 175 F; pour pt. 65 F; pH 6.5-7.5 (10% aq.); 100% conc.

Armid® 18. [Akzo Nobel] Stearamide, antiblock agent; CAS 124-26-5; EINECS 204-693-2; internal lubricant and slip agent for processed plastics, coatings, and films; builder, foam visc. stabilizer, and foam booster in syn. detergent formulations, cosmetics; water repellent for textiles; improves dye solubility in printing inks, dyes, carbon paper coatings, and fusible coatings for glassware and ceramics; intermediate for syn. waxes; pigment dispersant; thickener for paint; Gardner 7 flake; water-insol.; sp.gr. 0.52 (100 C); m.p. 99-109 C; flash pt. 225 C; 90% act.

Armid® E. [Akzo Nobel] Erucamide; CAS 112-84-5; EINECS 204-009-2; mold release agent for rubber and plastics; auxiliary for processing rubber; Gardner 7 max. flakes, pellets; m.p. 81 C; iodine no. 80; 90% min. amide.

Armid® HT. [Akzo Nobel] Hydrog. tallow amide; CAS 61790-31-6; EINECS 263-123-0; antiblock, lubricant, slip agent for plastics, coatings, films; builder, foam visc. stabilizer, foam booster in syn. detergents; water repellent for textiles; improves dye sol. in inks, dyes; intermediate for syn. waxes; pigment dispersant; also antifoam in steam generator systems, lubricant additive; auxiliary for rubber processing; Gardner 7 flakes, powd. 99% -60 mesh; m.w. 277; insol. in water; sp.gr. 0.851 (100 C); visc. 16 cps; m.p. 98-103 C; flash pt. 225 C; 90% act.

Armid® HTD. [Akzo Nobel] Stearylamine; CAS 124-30-1; EINECS 204-695-3; processing aid for high visc. rubber compds.; facilitating flow behavior and improving mold release; powd.

Armid® O. [Akzo Nobel] Oleamide; CAS 301-02-0; EINECS 206-103-9; internal lubricant and slip agent for processed plastics, coatings, and films; builder, foam visc. stabilizer, and foam booster in syn. detergent formulations; water repellent for textiles; also release agent in cosmetics, penetrant in paper manufacture; Gardner 7 flake, solid; bland odor; m.w. 279; sol (g/100 ml solv. with heating) 59 g in 95% IPA; 30 g in 95% ethanol; 15 g in acetone and trichloroethylene; 11 g in ethyl acetate and MIBK; insol. in water; sp.gr. 0.830 (100 C); visc. 25 cps; m.p. 68 C; flash pt. 207 C; 90% act.

Armocure® 100. [Akzo Nobel] Aliphatic polyamine; curing agent for flexible epoxy resins used in industrial applic.; adhesives and preparation of laminating resins; Gardner 3 max.; sp.gr. 0.831 (40 C); visc. 19 cP; cloud pt. 15 C max.; 89% min. act.

Armoflo® 48, 65. [Akzo Nobel] Anticaking and antidusting agents for hygroscopic materials such as ammonium sulfate, ammonium nitrate, diammonium phosphate, and urea; also for treating thermosetting plastics, detergents, and insecticides; liq.; sp.gr. 0.854 (60 F) and 0.824 (24 C) resp.; m.p. -4 C and 5 F resp.

Armoslip® 18. [Akzo Nobel] Stearamide; CAS 124-26-5; EINECS 204-693-2; internal lubricant, mold release agent, antiblock and slip agent for PE, PP, cellophane; flakes, pellets.

Armoslip® EXP. [Akzo Nobel] Erucamide; CAS 112-84-5; EINECS 204-009-2; internal lubricant, mold release agent, slip agent in polyolefins; flakes, pellets.

Armostat® 310. [Akzo Nobel] N-Tallow alkyl-2,2′-iminobisethanol; CAS 61791-44-4; permanent internal antistat for HDPE, LDPE; yel. paste; typ. odor; m.w. 350; sp.gr. 0.916; i.b.p. > 300 C (760 mm Hg); f.p. 32 C; flash pt. (PMCC) > 204 C; 97% act.

Armostat® 350. [Akzo Nobel] Ethoxylated amine; internal antistat for LDPE; pellets.

Armostat® 375. [Akzo Nobel] Bis (2-hydroxyethyl) tallow amine (73%) and polyethylene (25%); internal antistat for high and low dens. polyethylene; FDA approved; off-wh. pellets; very slight amine odor.

Armostat® 410. [Akzo Nobel] N-Coco alkyl-2,2′-iminobisethanol; CAS 61791-31-9; permanent internal antistat for PP, PS, ABS, SAN resins; FDA approved; clear yel. liq.; typ. odor; m.w. 285; sp.gr. 0.874; f.p. 12 C; i.b.p. 170 C; flash pt. (PMCC) > 204 C; 97% min. act.

Armostat® 450. [Akzo Nobel] Ethoxylated amine; internal antistat for PS; pellets; 50% conc.

Armostat® 475. [Akzo Nobel] Bis (2-hydroxyethyl) coco amine (73%) CAS #61791-31-9 and polypropylene (25%) CAS #9003-07-0; internal antistat for PP, HDPE; off-wh. pellets; slight amine odor; sp.gr. 0.95; m.p. > 180 C; flash pt. (PMCC) > 200 C.

Armostat® 550. [Akzo Nobel] Ethoxylated amine; internal antistat for SAN, ABS; pellets.

Armostat® 710. [Akzo Nobel] Amine; internal antistat for PE, PP; FDA approval.

Armostat® 801. [Akzo Nobel] Glyceryl stearate; internal antistat for polyethylene, PP, and PVC; GRAS for food additive use; flakes.

Armostat® 810. [Akzo Nobel] Glyceryl oleate; internal antistat for PE, PP, and PVC; liq.

Arneel® 18 D. [Akzo Nobel] Stearyl nitrile; detergent, wetting agent, rust inhibitor; metal wetting with min. oils; plasticizer for syn. rubbers and plastics; Gardner 6; f.p. 39 C; b.p. 330-360 C.

AroCy® B-10. [Rhone-Poulenc] Bisphenol A dicyanate monomer; co-reacts with and cures epoxy resins; useful in formulating low visc. impregnants for RTM, pultrusion, and filament winding applics., for alloying with thermoplastic tougheners, and as reactive diluent for difficult-to-process hot-melt prepregs and adhesives; wh. cryst. powd.; sp.gr. 1.259; melt visc. 20 cps (> 80 C); m.p. 79 C; ref. index 1.5395 (90 C).

AroCy® F-10. [Rhone-Poulenc] Hexafluorobisphenol A dicyanate monomer; flame-retardant; offers exc. mech. str., adhesion, toughness, and low shrinkage; used for low-visc. hot-melt impregnations, dissolving thermoplastic tougheners, and as reactive diluent for difficult-to-process hot-melt matrix and adhesive resins; co-reacts with and cures epoxy resins; wh. cryst. powd.; sp.gr. 1.497; melt visc. 20 cfps (> 90 C); m.p. 87 C; ref. index 1.476 (110 C); 29.5% F; > 99% purity.

Arodet BN-100. [Arol Chem. Prods.] Ethoxylated alcohol; detergent, wetting agent, emulsifier, dyeing assistant, dispersant for dyeing, finishing, textiles, pigments, resins; clear liq.; mild odor; water sol.; 100% act.

Aromatic Oil 745. [Neville] Aromatic plasticizer; plasticizer used in adhesives, rubber (cements, mech. and molded goods, tires), caulking compds.; FDA 21CFR §175.105, 177.2600; dk. brn. liq.; sol. in ethers and chlorinated, aromatic, naphthenic, and terpene hydrocarbons; m.w. 200; sp.gr. 1.03; flash pt. (COC) 260 F.

Aromox® C/12. [Akzo Nobel] Dihydroxyethyl cocamine oxide, IPA; CAS 61791-47-7; EINECS 263-180-1; cationic; wetting agent, emulsifier, stabilizer, antistat, foaming agent for detergents, shampoos, cosmetics, textiles, metal plating, petrol. additives, paper, plastics, rubber; Gardner 2 clear liq.; sp.gr. 0.949; visc. 52 cp; cloud pt. 18 F; flash pt. 82 F; pour pt. 0 F; surf. tens. 33 dynes/cm; biodeg.; 50% act. in aq. IPA.

Aromox® C/12-W. [Akzo Nobel; Akzo Nobel BV] Dihydroxyethyl cocamine oxide; CAS 61791-47-7; EINECS 263-180-1; cationic; wetting agent, emulsifier, stabilizer, antistat, foaming agent for detergents, shampoos, cosmetics, textiles, metal plating, petrol. additives, paper, plastics, rubber; gel sensitizer for latex foam; biodeg.; Gardner 2 clear liq.; sp.gr. 0.997; visc. 2097 cp; HLB 18.4; flash pt. 212 F; pour pt. 35 F; surf. tens. 30.8 dynes/cm; 40% act. in water.

Arosurf® TA-100. [Witco/H-I-P] Distearyl dimonium chloride; CAS 107-64-2; EINECS 203-508-2; cationic; fabric softener conc., conditioner for home and commerical laundry and textile processing; personal care prods.; antistatic coating for ABS, acrylic, cellulosics, nylon, polyacetal, PP, PS, PVC; Gardner 3 max. powd.; water-disp.; m.w. 583; flash pt. (PMCC) > 200 F; 93% quat. min.

Arquad® 2C-75. [Akzo Nobel; Akzo Nobel BV] Dicocodimonium chloride, aq. IPA; CAS 61789-77-3; EINECS 263-087-6; cationic; biodeg. emulsifier, foaming, wetting, dispersing agents, corrosion inhibitor, softener, dyeing aid, antistat for textiles, paper, cosmetics; industrial, agriculture, plastics, petrol. industry, acid pickling baths; bactericide, algicide; Gardner 7 semiliq.; sol. in alcohols, benzene, chloroform, CCl_4; disp. in water; m.w. 447; sp.gr.0.89; HLB 11.4; flash pt. < 80 F; pour pt. 10 F; surf. tens. 30 dynes/cm (0.1%); pH 9; 75% act. in aq. IPA; *Precaution:* flamm.

Arquad® 12-50. [Akzo Nobel] Laurtrimonium chloride, IPA; CAS 112-00-5; EINECS 203-927-0; cationic; biodeg. emulsifier, foaming, wetting, dispersing agents, corrosion inhibitor, softener, dyeing aid, antistat for textiles, paper, cosmetics; industrial, agriculture, plastics, petrol. industry, acid pickling baths; bactericide, algicide; gel sensitizer for latex foam; Gardner 1 liq.; sol. in water, alcohols, chloroform, CCl_4; m.w. (act.) 263; sp.gr. 0.89; f.p. 13 F; HLB 17.1; flash pt. < 80 F; surf. tens. 33 dynes/cm; pH 5-8 (10% aq.); 50% act. in aq. IPA.

Arquad® 16-50. [Akzo Nobel; Akzo Nobel BV] Cetrimonium chloride, IPA; CAS 112-02-7; EINECS 203-928-6; cationic; emulsifier, foaming, wetting, dispersing agents, corrosion inhibitor, softener, dyeing aid, antistat for textiles, paper, cosmetics; industrial, agriculture, plastics, petrol. industry, acid pickling baths; bactericide, algicide; rubber to textile bonding agent; biodeg.; Gardner 6 liq.; sol. in water, alcohols, chloroform, CCl_4; m.w. (act.) 319; sp.gr. 0.88; f.p. 61 F; HLB 15.8; flash pt. < 80 F; surf. tens. 34 dynes/cm; pH 5-8 (10% aq.); 49-52% quat. in aq. IPA.

Arquad® 18-50. [Akzo Nobel; Akzo Nobel BV] Steartrimonium chloride, IPA; CAS 112-03-8; EINECS 203-929-1; cationic; emulsifier, foaming, wetting, dispersing agents, corrosion inhibitor, softener, dyeing aid, antistat for textiles, paper, cosmetics; industrial, agriculture, plastics, petrol. industry, acid pickling baths; bactericide, algicide; dye leveling agent, visc. stabilizer, in lubricant compding.; biodeg.; Gardner 7 liq.; sol. in water, alcohols, chloroform, CCl_4; m.w. (act.) 347; sp.gr. 0.88; HLB 15.7; flash pt. < 80 F; surf. tens. 34 dynes/cm; pH 5-8 (10% aq.); 50% act. in aq. IPA.

Arquad® C-50. [Akzo Nobel; Akzo Nobel BV] Cocotrimonium chloride, IPA; CAS 61789-18-2; EINECS 263-038-9; cationic; biodeg. emulsifier, foaming, wetting, dispersing agents, corrosion inhibitor, softener, dyeing aid, antistat for textiles, paper, cosmetics; industrial, agriculture, plastics, petrol. industry, acid pickling baths; bactericide, algicide; gel sensitizer for latex foam; Gardner 7 liq.; sol. in water, alcohols, chloroform, CCl_4; m.w. (active) 278; sp.gr. 0.89; f.p. 5 F; HLB 16.5; flash pt. < 80 F; surf. tens. 31 dynes/cm (0.1%); pH 5-8 (10% aq.); 50% act. in aq. IPA.

Arquad® T-27W. [Akzo Nobel] Tallow trimonium chloride; CAS 8030-78-2; EINECS 232-447-4; biodeg. emulsifier, foaming, wetting, dispersing agents, corrosion inhibitor, softener, dyeing aid, antistat for textiles, paper, cosmetics; industrial, agriculture, plastics, petrol. industry, acid pickling baths; bactericide, algicide; Gardner 3 max. liq.; sol. in water, alcohols, chloroform, CCl_4; m.w. 343; HLB 14.2; pH 5-8 (10% aq.); 26-29% act. in water.

Artic Mist. [Luzenac Am.] Platy talc; CAS 14807-96-6; EINECS 238-877-9; reinforcing filler for superior processability in rubber applics.; flatting agent, visc. control for coating formulations; provides dry hiding, imparts good flow, leveling and superior film smoothness; powd.; 2.2 μ median particle size; 100% thru 325 mesh; fineness (Hegman) 6; bulk dens. 13 lb/ft^3 (loose); bulking value 0.0436; surf. area 12 m^2/g; oil absorp. 53; brightness (GE) 88; 98-99% talc; 61% SiO_2, 31% MgO.

Arubren®. [Bayer AG] Plasticizer for rubber and latex.

Arylan® SBC25. [Akcros] Sodium dodecylbenzene sulfonate; anionic; biodeg. wetting agent, detergent base, emulsifier for emulsion polymerization; clear liq., mild odor; water-sol.; sp.gr. 1.036; visc. 150 cst; pour pt. 0 C; flash pt. (COC) > 95 C; pH 7.5 (1% aq.); surf. tens. 33 dynes/cm (0.1%); 25% act. in water.

Arylan® SC15. [Akcros] Sodium dodecylbenzene sulfonate; anionic; biodeg. wetting agent, detergent base, emulsifier for emulsion polymerization; clear liq., mild odor; water-sol.; sp.gr. 1.008; visc. 300 cst; pour pt. 5 C; flash pt. (COC) > 95 C; pH 7.5 (1% aq.); 15% act. in water.

Arylan® SC30. [Akcros] Sodium dodecylbenzene sulfonate; anionic; biodeg. wetting agent, detergent base, emulsifier for emulsion polymerization; liq./soft paste, mild odor; water-sol.; sp.gr. 1.040; visc. 5100 cst; pour pt. 12 C; flash pt. (COC) > 95 C; pH 7.5 (1% aq.); 30% act.

Arylan® SO60 Acid. [Akcros] Branched chain tridecyl benzene sulfonic acid; CAS 25496-01-9; EINECS 247-036-5; detergent, emulsifier, wetting agent, emulsion polymerization aids; br., visc. liq.; char. odor; water sol.; sp.gr. 1.045; visc. > 25,000 cs; flash pt. > 250 F (COC); pour pt. < 0 C; pH 2.0; > 96% act.; Unverified

Arylan® SX85. [Akcros] Sodium dodecylbenzene sulfonate; anionic; heavy-duty industrial detergent for industrial cleaning, hard surf. cleaning, textile scouring; base surfactant for powd. detergents; wetting agent for powds.; emulsifier for emulsion polymerization; off-wh. powd., mild odor; water-sol.; bulk dens. 0.50 g/cc; pH 9.5 (1% aq.); surf. tens. 35 dynes/cm (0.1%); 85% act.; *Toxicology:* avoid inhalation of dust, exposure to eyes.

Arylan® SY30. [Akcros] Sodium alkylbenzene sulfonate; anionic; biodeg. emulsifier for emulsion polymerization; yel. clear liq.; surf. tens. 37 dynes/cm (0.1%); 30% act.

Arylan® SY Acid. [Akcros] Alkylbenzene sulfonic acid; anionic; biodeg. emulsifier for emulsion polymerization; brn. visc. liq.; surf. tens. 36 dynes/cm (0.1%); 96% act.

Arylan® TE/C. [Akcros] Benzene sulfonate, TEA salt; anionic; detergent, emulsifier, wetting agent, emulsion polymerization aids; clear liq.; mild odor; water sol.; sp.gr. 1.060; visc. 900 cs; flash pt. > 200 F (COC); pour pt. < 0 C; pH 7-8.5; biodeg.; 40% act.; Unverified

ASP® 072. [Engelhard] Hydrous aluminum silicate; pretreated to form a deflocculated suspension of high fluidity in water vehicles; extender pigment in paints (improves suspension of other pigments); used in adhesives for corrugated board, laminated fiber board, paper bag seams, paper tubes, cores; used in plastics as unmatched min. fillers providing opacity; extender in printing inks; useful in rubber industry, in mastics, putties, caulking chemicals, textile and chemical conditioning; wh. particulate, particle size 0.3 μ; sp.gr. 2.58; dens. 42-46 lb/ft^3; oil absorp. 37-41; ref. index 1.56; pH 6.3-7.0; Unverified

ASP® 100. [Engelhard] Hydrous aluminum silicate; unmodified; extender pigment in paints (improves suspension of other pigments); used in adhesives for corrugated board, laminated fiber board, paper bag seams, paper tubes, cores; used in plastics as unmatched min. fillers providing opacity; extender in printing inks; useful in rubber industry, in mastics, putties, caulking chemicals, textile and chemical conditioning; wh. particulate, particle size 0.55 μ; sp.gr. 2.58; dens. 10-15 lb/ft^3; ref. index 1.56; pH 3.8-4.6; 45.4% SiO_2; 38.8% Al_2O_3.

ASP® 101. [Engelhard] Hydrous aluminum silicate with 0.5% stearate; pretreated to form a deflocculated suspension of high fluidity in water vehicles; extender pigment in paints, adhesives, plastics, printing inks, rubber, mastics, putties, caulking chemicals, textile and chemical conditioning; wh. particulate, particle size 0.55 μ; sp.gr. 2.58; dens. 15-20 lb/ft^3; oil absorp. 41-44; ref. index 1.56; pH 3.8-4.6.

ASP® 102. [Engelhard] Hydrous aluminum silicate; pretreated to form a deflocculated suspension of high fluidity in water vehicles; extender pigment in paints, adhesives, plastics, printing inks, rubber, mastics, putties, caulking chemicals, textile and chemical conditioning; wh. particulate, particle size 0.55 μ; sp.gr. 2.58; dens. 45-50 lb/ft^3; oil absorp. 37-41; ref. index 1.56; pH 6.3-7.0.

ASP® 170. [Engelhard] Hydrous aluminum silicate; produced by ultraflotation process; extender pigment in paints, adhesives, plastics, rubber, textiles, etc.; wh. particulate; particle size 0.55 μ; sp.gr. 2.58; dens. 5-10 lb/ft^3; oil absorp. 37-41; ref. index 1.56; pH 6.5-7.5.

ASP® 200. [Engelhard] Hydrous aluminum silicate; unmodified; extender pigment in paints, adhesives, plastics, rubber, textiles, etc.; wh. particulate, particle size 0.55 μ; sp.gr. 2.58; dens. 10-15 lb/ft^3; oil absorp. 37-41; ref. index 1.56; pH 3.8-4.6.

ASP® 400. [Engelhard] Hydrous aluminum silicate; unmodified; extender pigment in paints, adhesives, plastics, rubber, textiles, etc.; also offers antipenetration chars. making it effective as blocking agent in adhesives; wh. particulate, particle size 4.8 μ; sp.gr. 2.58; dens. 30-35 lb/ft^3; oil absorp. 28-32; ref. index 1.56; pH 3.8-4.6; 45.4% SiO_2, 38.8% Al_2O_3.

ASP® 400P. [Engelhard] Hydrous aluminum silicate; unmodified; extender pigment in paints, adhesives, plastics, rubber, textiles, etc.; also offers antipenetration chars. making it effective as blocking agent in adhesives; highly pulverized form for use in org. systems; wh. particulate; particle size 4.8 μ; sp.gr. 2.58; dens. 30-35 lb/ft^3; oil absorp. 28-32; ref. index 1.56; pH 3.8-4.6; 45.4% SiO_2; 38.8% Al_2O_3.

ASP® 600. [Engelhard] Hydrous aluminum silicate; unmodified; extender pigment in paints, adhesives, plastics, rubber, textiles, etc.; wh. particulate, particle size 0.8 μ; sp.gr. 2.58; dens. 15-20 lb/ft^3; oil absorp. 34-38; ref. index 1.56; pH 3.8-4.6.

ASP® 602. [Engelhard] Hydrous aluminum silicate; pretreated to form a deflocculated suspension of high fluidity in water vehicles; highly pulverized form for use in org. systems; extender pigment in paints, adhesives, plastics, printing inks, rubber, mastics, putties, caulking chemicals, textile and chemical conditioning; wh. particulate, particle size 0.8 μ; sp.gr. 2.58; dens. 45-50 lb/ft^3; oil absorp. 34-38; ref. index 1.56; pH 6.3-7.0.

ASP® 900. [Engelhard] Hydrous aluminum silicate; unmodified; extender pigment in paints, adhesives, plastics, rubber, textiles, etc.; wh. particulate, particle size 1.5 μ; sp.gr. 2.58; dens. 30-35 lb/ft^3; oil absorp. 31-35; ref. index 1.56; pH 4.2-5.0; 45.4% SiO_2; 38.8% Al_2O_3.

ASP® NC. [Engelhard] Delaminated hydrous aluminum silicate; reinforcing agent for rubber and polymer systems; wh. highly pulverized powd.; 0.7 μm avg. particle size; 0.01% max. residue +325 mesh; sp.gr. 2.58 g/cc; bulk dens. 18 lb/ft^3 (loose), 23 lb/ft^3 (tamped); bulking value 0.047 gal/lb; oil absorp. 41-46 lb/100 lb; ref. index 1.56; pH 6-8; brightness (GE) 87-89; 45.4% SiO_2; 38.5% Al_2O_3.

AST-1001. [Merix] Light stabilizer, antistat for plastics; nonflamm.; faint pleasant odor; neutral pH.

Astrawax 23. [Astor Wax] Bisstearamide-based wax; high melting wax for inks, paints, rubber, and plastics applics.; powd., lumps; visc. 12-14 cps (148 C); congeal pt. 138-143 C; penetration 4-6; acid no. 6; sapon. no. 6.

Astrowet B-45. [Alco] Sodium diisobutyl sulfosuccinate; CAS 127-39-9; EINECS 204-839-5; anionic; polymerization emulsifier; liq.; 45% conc.; Unverified

Astrowet H-80. [Alco] Sodium dihexyl sulfosuccinate; anionic; polymerization emulsifier; liq.; 80% conc.; Unverified

AT-1. [Sovereign] Anatase titanium dioxide; CAS 13463-67-7; EINECS 236-675-5; economical pigment for rubber, paper, ceramics, and road paints.

Atlas 1500. [ICI Surf. Am.] Glyceryl stearate; food emulsifier; emulsion stabilizer and film-former in caramels; retards starch crystallization in starch jellies; inhibits oil separation in peanut butter; in dehydrated potatoes; antistat for PP, PS useful in food pkg.; FDA 21CFR §182.1324, GRAS, 184.1505; Canada compliance; ivory wh. flakes, bland odor and taste; sol. above its m.p. in IPA, veg., min. oils; m.p. 140 F; HLB 3.5; iodine no. < 5; flash pt. > 300 F; 52% min. alpha monoglyceride; *Toxicology:* nonirritating to skin, noncorrosive.

Atmer® 100. [ICI Surf. Am.] Sorbitan laurate; antifog agent for PE and EVA food-wrapping films; plastics antistat; cling additive; red-brn. liq.; visc. 3900-5000 mPa•s.

Atmer® 102. [ICI Surf. Am.] Sorbitan ester; antifog for thermoplastics; tan solid; m.p. 44-47 C.

Atmer® 103. [ICI Surf. Am.] Sorbitan ester; antifog

agent for long-lasting properties in LDPE, EVA, and PVC agric. film; wetting agent for PP and PE films; pale cream solid; m.p. 50-53 C.

Atmer® 104. [ICI Surf. Am.] Sorbitan ester; antifog, lubricant for thermoplastics; cream solid; m.p. 48-53 C.

Atmer® 105. [ICI Surf. Am.] Sorbitan oleate; antifog, antistat, cling additive for low-temp. LDPE film applics.; amber liq.; visc. 950-1100 mPa•s.

Atmer® 106. [ICI Surf. Am.] Sorbitan trioleate; antifog, lubricant, cling additive for LDPE film; amber liq.; visc. 170-230 mPa•s.

Atmer® 110. [ICI Surf. Am.] POE sorbitan laurate; antifog, external antistat, wetting agent for thermoplastics incl. PE, PP, PVC, styrenic copolymers; FDA 21CFR §178.3400; yel. liq.; water-sol.; visc. 250-450 mPa•s.

Atmer® 111. [ICI Surf. Am.] POE sorbitan ester; antifog for thermoplastics; yel. liq.; visc. 650 mPa•s.

Atmer® 112. [ICI Surf. Am.] POE sorbitan ester; antifog for thermoplastics; pale yel. liq.; visc. 400-650 mPa•s.

Atmer® 113. [ICI Surf. Am.] POE sorbitan stearate; antifog, external antistat, wetting agent for thermoplastics; FDA 21CFR §178.3400; pale yel. liq.; water-sol.; visc. 75-175 mPa•s (50 C).

Atmer® 114. [ICI Surf. Am.] POE sorbitan ester; antifog agent for thermoplastics; ivory solid; m.p. 36-40 C.

Atmer® 115. [ICI Surf. Am.] POE sorbitan ester; antifog for thermoplastics; pale yel. solid; m.p. 30-35 C.

Atmer® 116. [ICI Surf. Am.] POE sorbitan ester; antifog agent for flexible PVC food-wrapping film; kosher; yel.-brn. liq.; visc. 375-480 mPa•s.

Atmer® 117. [ICI Surf. Am.] POE sorbitan ester; antifog for thermoplastics; yel.-brn. liq.; visc. 400-500 mPa•s.

Atmer® 118. [ICI Surf. Am.] POE sorbitan ester; antifog for thermoplastics; yel.-brn. liq.; visc. 250-450 mPa•s.

Atmer® 121. [ICI Surf. Am.] Glyceryl oleate; antifog agent and cling additive for PVC food-wrapping film; pale yel. liq.; visc. 140 mPa•s.

Atmer® 122. [ICI Surf. Am.] Glyceryl stearate; internal antistat for PP, LDPE, flexible PVC; antifog, processing aid, lubricant; expandable polymers; rapid action, good thermal stability; FDA 21CFR §184.1505; wh. powd.; m.p. 63 C.

Atmer® 124. [ICI Surf. Am.] Glyceryl ester; antifog, antistat for expandable polymers; wh. powd.; m.p. 63 C.

Atmer® 125. [ICI Surf. Am.] Glyceryl stearate; antifog, internal antistat, lubricant for polyolefins esp. PP and expandable polymers; rapid action, good thermal stability; FDA 21CFR §184.1505; wh. powd.; m.p. 63 C.

Atmer® 129. [ICI Surf. Am.] Glyceryl stearate; processing aid, antifog, lubricant, internal antistat for PP, LDPE, plasticized PVC; outstanding thermal stability; FDA 21CFR §184.1324; wh. microbeads; m.p. 69 C; > 90% monoester.

Atmer® 130. [ICI Surf. Am.] POE alkylaryl ether; antifog for expandable polymers; wh. solid; m.p. 40 C.

Atmer® 131. [ICI Surf. Am.] POE alkylaryl ether; antifog for expandable polymers; colorless liq.; visc. 360 cps (20 C).

Atmer® 132. [ICI Surf. Am.] Ethoxylated nonyl phenol; antifog agent for flexible PVC food-wrapping film, expandable polymers; colorless liq.; visc. 360 mPa•s (20 C).

Atmer® 133. [ICI Surf. Am.] POE oil; antifog for expandable polymers; colorless liq.; visc. 350-550 cps.

Atmer® 134. [ICI Surf. Am.] POE alkylaryl ether; antifog for expandable polymers; colorless paste; visc. 84 cps (50 C).

Atmer® 135. [ICI Surf. Am.] POE alkyl ether; plastics additive for expandable polymers; colorless liq.; visc. 25-100 cps.

Atmer® 136. [ICI Surf. Am.] POE alkyl ether; plastics additive for expandable polymers; wh. solid; m.p. 35-40 C.

Atmer® 137. [ICI Surf. Am.] POE alkyl ether; plastics additive for expandable polymers; cream solid; m.p. 32 C.

Atmer® 138. [ICI Surf. Am.] POE ester; internal antistat for HDPE, expandable polymers; rapid action, good thermal stability; wh. cream solid; m.p. 38-40 C.

Atmer® 139. [ICI Surf. Am.] POE ester; antistat for expandable polymers; wh. cream solid; m.p. 46 C.

Atmer® 150. [ICI Surf. Am.] Alkoxylated alcohol; antiozonant for thermoplastics; tan liq.; visc. 200 cps.

Atmer® 153. [ICI Surf. Am.] Alkoxylated fatty derivs.; plasticizer; PVC plastisol visc. depressant; pale yel. liq.; visc. 150 mPa•s.

Atmer® 154. [ICI Surf. Am.] POE fatty acid ester; visc. depressant; internal antistat for flexible PVC; rapid action, exc. thermal stability; pale yel. liq.; visc. 90 mPa s.

Atmer® 163. [ICI Surf. Am.] POE alkyl amine; internal and external antistat for wide range of polymers, esp. polyolefins and styrenics (ABS, HIPS); long lasting, exc. heat and color stability; FDA 21CFR §178.3130; clear liq.; visc. 150 mPa•s.

Atmer® 164. [ICI Surf. Am.] Syn. quat. ammonium compd. in 1,4-butanediol; internal antistat for PU; yel. liq.; visc. 150 mPa•s (40 C).

Atmer® 171. [ICI Surf. Am.] POE triglyceride; antifog for thermoplastics; cell regulator; pale yel. liq.; visc. 1000-1300 mPa•s.

Atmer® 184. [ICI Surf. Am.] Glyceryl fatty acid ester; antifog agent for long-lasting properties in EVA and LDPE agric. film; lt. tan powd.; m.p. 40-53 C.

Atmer® 185. [ICI Surf. Am.] Glycerol fatty acid ester; antifog for thermoplastics; lt. tan powd.; m.p. 40-53 C.

Atmer® 190. [ICI Surf. Am.] Alkyl sulfonate; internal and external antistat for PS, rigid PVC, and other polymers with high dielec. const.; exc. thermal stability; opacity development in clear materials;

FDA 21CFR §178.3400; pale yel. flakes; m.p. 140 C.

Atmer® 502. [ICI Surf. Am.] POE alkyl ether; antifog, wetting agent for thermoplastics; wh. semisolid.

Atmer® 505. [ICI Surf. Am.] Fatty alcohol ethoxylate; visc. depressant for thermoplastics; wh. liq.; visc. 60 cps.

Atmer® 508. [ICI Surf. Am.] POE alkylaryl ether; nonionic; surfactant; plasticizer; PVC plastisol visc. depressant; pale yel. liq.; sol. in lower alcohols, cellosolve, many petrol. solv.; sp.gr. 1.0; visc. 220 cps; flash pt. > 300 F.

Atmer® 645. [ICI Surf. Am.] Blend; nonionic; antifog agent, wetting agent for PE, EVA, and PVC food-wrapping films; yel. liq.; visc. 550 cps.

Atmer® 646. [ICI Surf. Am.] Blend; nonionic; antifog agent for thermoplastics; yel. liq.; visc. 220 cps.

Atmer® 647. [ICI Surf. Am.] Blend; nonionic; antifog agent for PE, EVA, and PVC food-wrapping films; yel. paste.

Atmer® 648. [ICI Surf. Am.] Blend; nonionic; antifog agent for PE, EVA, and PVC food-wrapping films; wh.-brn. solid.

Atmer® 649. [ICI Surf. Am.] Blend; nonionic; antifog agent for PE, EVA, and PVC food-wrapping films; yel. liq.; visc. 180 cps.

Atmer® 650. [ICI Surf. Am.] Blend; nonionic; antifog agent for PE, EVA, and PVC food-wrapping films; yel. liq.; visc. 240 cps.

Atmer® 654. [ICI Surf. Am.] Blend; nonionic; antifog agent for PE, EVA, and PVC food-wrapping films; yel. paste.

Atmer® 655. [ICI Surf. Am.] Blend; nonionic; antifog for thermoplastics; yel. liq.; visc. 650 mPa•s.

Atmer® 656. [ICI Surf. Am.] Blend; nonionic; antifog for thermoplastics; yel. solid.

Atmer® 657. [ICI Surf. Am.] Blend; nonionic; antifog for thermoplastics; pale yel. liq.; visc. 230 cps.

Atmer® 658. [ICI Surf. Am.] Blend; nonionic; antifog for thermoplastics; wh.-brn. solid.

Atmer® 685. [ICI Surf. Am.] Blend; nonionic; antifog agent, wetting agent for PE, EVA, and PVC food-wrapping films; pale yel. liq.; visc. 170 mPa•s.

Atmer® 687. [ICI Surf. Am.] Blend; nonionic; antifog for thermoplastics; pale yel. liq.; visc. 500 cps.

Atmer® 688. [ICI Surf. Am.] Blend; nonionic; antifog for thermoplastics; pale yel. liq.; visc. 500 cps.

Atmer® 1002. [ICI Surf. Am.] Quat. ammonium salt; internal antistat for PU; long lasting; yel. liq.; visc. 420 cps.

Atmer® 1004. [ICI Surf. Am.] Quat. ammonium salt; external antistat for all polymers; rapid action, effective at low usage levels; amber liq.; visc. 100 cps (50 C).

Atmer® 1005. [ICI Surf. Am.] Quat. ammonium salt; external antistat for all polymers; rapid action, effective at low usage levels; yel. liq.; visc. 500 cps.

Atmer® 1006. [ICI Surf. Am.] Quat. ammonium salt; antistat for thermoplastics; red-brn. liq.; visc. 1000 cps.

Atmer® 1007. [ICI Surf. Am.] Glyceryl oleate; antifog, cling agent, lubricant for LDPE and PVC films; pale yel. paste; visc. 530 mPa•s.

Atmer® 1010. [ICI Surf. Am.] Glyceryl oleate; processing aid, antifog agent and cling additive for PVC food-wrapping film, LDPE; kosher; pale yel. liq., bland odor; sol. in lower alcohols, min. oils, veg. oils; insol. in water; sp.gr. 0.94; HLB 3.0; pour pt. 11 C; b.p. > 100 C; flash pt. (COC) > 300 F.

Atmer® 1012. [ICI Surf. Am.] Glyceryl ester; antifog, cling agent, lubricant for thermoplastics; pale yel. liq.; visc. 140 mPa•s.

Atmer® 1021. [ICI Surf. Am.] POE glyceride ester; antistat, lubricant for thermoplastics; tan solid.

Atmer® 1024. [ICI Surf. Am.] Alcohol phosphate ester; antistat for thermoplastics; colorless liq.; visc. 25 mPa•s.

Atmer® 1025. [ICI Surf. Am.] POE oleate; lubricant, internal antistat for flexible PVC; rapid action, exc. thermal stability; amber liq.; visc. 90 mPa•s.

Atmer® 1026. [ICI Surf. Am.] POE ester; lubricant for thermoplastics; lt. yel. liq.; visc. 100 cps.

Atmer® 1027. [ICI Surf. Am.] Quat. ammonium compd.; internal antistat for flexible PVC; rapid action, high thermal stability; amber-brn. liq.; visc. 2500 mPa•s.

Atmer® 1040. [ICI Surf. Am.] EO-PO block copolymer; lubricant for thermoplastics; wh. flake; visc. 2800 mPa•s (77 C).

Atmer® 1041. [ICI Surf. Am.] EO-PO block copolymer; lubricant for thermoplastics; colorless liq.; visc. 800 cps.

Atmer® 1042. [ICI Surf. Am.] EO-PO block copolymer; lubricant for thermoplastics; colorless paste; visc. 305 cps (60 C).

Atmer® 7001. [ICI Surf. Am.] 50% Atmer 129/163 (2:1) in PP; antistat for inj. molding PP; wh. pellets.

Atmer® 7002. [ICI Surf. Am.] 50% Atmer 129 in PP; antistat for inj. molding PP; wh. pellets.

Atmer® 7006. [ICI Surf. Am.] 50% Atmer 129/163 (2:1) in PP; antistat for PP film; wh. pellets.

Atmer® 7007. [ICI Surf. Am.] 50% Atmer 129/163 (4:1) in PP; antistat for inj. molding of filled PP; wh. pellets.

Atmer® 7101. [ICI Surf. Am.] 50% Atmer 129 in polyethylene; antifog, antistat for PE film; wh. pellets.

Atmer® 7108. [ICI Surf. Am.] 50% Atmer 685 in LLDPE; antifog for LLDPE film; wh. pellets.

Atmer® 7109. [ICI Surf. Am.] 50% Atmer 129/163 (4:1); antistat for inj. molding PE; pellets/powd.

Atmer® 7202. [ICI Surf. Am.] 50% Atmer 190 in HIPS; antistat for inj. molding HIPS; wh. pellet.

Atmer® 7203. [ICI Surf. Am.] 50% Atmer 190 in ABS; antistat for inj. molding ABS; wh. pellet.

Atmer® 8102. [ICI Surf. Am.] Antistat for plastic films; wh. pellet; 40% act. in polyethylene.

Atmer® 8103. [ICI Surf. Am.] Antistat for plastic films; wh. pellet; 50% act. in polyethylene.

Atmer® 8103-35. [ICI Surf. Am.] Antistat for molding; wh. pellet; 35% act. in HDPE.

Atmer® 8112. [ICI Surf. Am.] 20% in PE; antistat for plastic films; wh. pellet.

Atmer® 8145. [ICI Surf. Am.] Cling agent for films; wh.

pellet; 20% in polyethylene.

Atmer® 8158. [ICI Surf. Am.] Antifog for greenhouse films; wh. pellet; 30% in polyethylene.

Atmer® 8163. [ICI Surf. Am.] Antifog for films; wh. pellet; 20% in polyethylene.

Atmer® 8215. [ICI Surf. Am.] Antifog for film/sheet; wh. pellet; 20% in S/B copolymer.

Atmer® 8216. [ICI Surf. Am.] Antifog for film/sheet; wh. pellet; 20% in S/B copolymer.

Atmos® 150. [ICI Am.; ICI Atkemix; ICI Surf. UK; Witco/Oleo-Surf.] Glyceryl stearate; nonionic; surfactant, emulsifier for cosmetic creams and lotions; food emulsifier for puddings, frozen desserts; emulsion stabilizer for icings; provides lubrication for taco shells; extrusion aid for pasta; also for coffee whiteners; antistat for plastics (PP, PS) useful in food pkg.; FDA 21CFR §182.1324 (GRAS), 184.1505; Canada compliance; ivory wh. powd., bland odor and taste; sol. above its m.p. in veg. oils, min. oil, IPA; insol. in water, cottonseed oil; m.p. 140 F; HLB 3.5; iodine no. $\leq$5; flash pt. > 300 F; 52% min. alpha monoglyceride; *Toxicology:* nonirritating to skin, noncorrosive.

Atmos® 300. [ICI Am.; ICI Atkemix; ICI Surf. UK; Witco/Oleo-Surf.] Glyceryl oleate and propylene glycol; nonionic; surfactant, emulsifier for cosmetic creams, lotions; food emulsifier; antifoaming agent for starch jellies, sugar-protein syrup systems; provides dispersibility in coffee whiteners, flavors; antistat for PE, flexible PVC, plastisols; FDA 21CFR §184.1505; Canada compliance; lt. amber clear liq.; sol. in cottonseed oil, IPA; insol. in water; sp.gr. 0.96; visc. 130 cps; m.p. 70 F; HLB 2.8; iodine no. 68$\pm$3; flash pt. > 300 F; 46% min. alpha monoglyceride.

Atmul® 84. [ICI Surf. Am.; ICI Atkemix; Witco/Oleo-Surf.] Glyceryl stearate; food emulsifier for puddings; improves volume and texture of cakes; emulsion stabilizer and film-former in caramels; inhibits oil separation in peanut butter; in chewing gum, starch jellies, dehydrated potatoes, coffee whiteners, whipped toppings; internal antistat for PE, PP, PS, flexible and rigid PVC, plastisols; FDA 21CFR §184.1505; Canada compliance; ivory wh. beads or flakes, bland odor and taste; sol. above its m.p. in veg., min. oils; m.p. 140 F; HLB 2.8; iodine no. < 5; flash pt. > 300 F; 40% min. alpha monoglyceride.

Atmul® 124. [ICI Surf. Am.; ICI Atkemix; Witco/Oleo-Surf.] Glyceryl stearate; food emulsifier; emulsion stabilizer and film-former in caramels; retards starch crystallization in starch jellies; inhibits oil separation in peanut butter; in dehydrated potatoes; antistat for PP, PS useful in food pkg.; FDA 21CFR §182.1324, GRAS, 184.1505; Canada compliance; ivory wh. flakes, bland odor and taste; sol. above its m.p. in IPA, veg., min. oils; m.p. 140 F; HLB 3.5; iodine no. < 5; flash pt. > 300 F; 52% min. alpha monoglyceride; *Toxicology:* nonirritating to skin, noncorrosive.

Atmul® 695. [Witco/H-I-P] Mono- and diglycerides; food emulsifier for puddings; internal/external antistat for PE, flexible PVC, plastisols; FDA 21CFR §184.1505; amber semiliq.; m.p. 76 F; HLB 3.0; iodine no. 76$\pm$3; 52% min. alpha monoglyceride.

Atomite®. [ECC Int'l.] Calcium carbonate; CAS 1317-65-3; EINECS 207-439-9; pigment, filler, reinforcement; high brightness, controlled particle size, easy dispersing grade for water and solv.-based coatings, paper, food, rubber, plastics, caulks, sealants, adhesives, etc.; NSF compliance; wh. powd., odorless; 3.0 μ mean particle size; fineness (Hegman) 6.0; negligible sol. in water; sp.gr. 2.71; bulk dens. 45 lb/ft^3 (loose); surf. area 2.8 m^2/g; oil absorp. 15; pH 9.5 (5% slurry); nonflamm.; 97.6% $CaCO_3$, 0.2% max. moisture; *Toxicology:* TLV/TWA 10 mg/m^3, considered nuisance dust; *Precaution:* incompat. with acids.

Atpol HD722. [ICI Surf. Am.; ICI Surf. UK] PEG-44 allyl alcohol; emulsifier for polymerization, metalworking; yel. to wh. solid.

Atpol HD745. [ICI Surf. Am.; ICI Surf. UK] PEG-23.4 allyl alcohol; emulsifier for polymerization, metalworking; cream solid; HLB 18.9.

Atpol HD861. [ICI Surf. UK] POE alkyl alcohol; emulsifier for polymerization, metalworking; lt. brn. solid; HLB 16.8.

Atpol HD863. [ICI Surf. Am.; ICI Surf. UK] PEG-7 allyl alcohol; emulsifier for polymerization, metalworking; lt. brn. waxy liq.; HLB 16.8.

Atpol HD975. [ICI Surf. Am.; ICI Surf. UK] PEG-4 allyl alcohol; emulsifier for polymerization, metalworking; yel. liq.; HLB 15.0.

Avirol® A. [Henkel/Functional Prods.] Ammonium lauryl sulfate; anionic; emulsifier for emulsion polymerization; additive for mech. latex foaming; foaming agent for acrylate disps., carpet and upholstery cleaners; air entraining agent for mortars; liq.; 30-32% solids.

Avirol® AE 3003. [Henkel/Functional Prods.] Ammonium laureth sulfate; anionic; emulsifier for vinyl acetate copolymers, S/B latexes, vinyl chloride copolymers, acrylate homo- and copolymers; liq.; 31-33% solids.

Avirol® AOO 1080. [Henkel/Functional Prods.] Oxidized oleic acid, ammonium salt; anionic; emulsifier for vinyl chloride and acrylic polymers; liq.; 40% conc.

Avirol® FES 996. [Henkel/Functional Prods.] Sodium laureth sulfate; CAS 9004-82-4; anionic; emulsifier for vinyl acetate copolymers, S/B latexes, vinyl chloride copolymers, acrylate homo- and copolymers; liq./paste; 62-65% solids.

Avirol® SA 4106. [Henkel/Functional Prods.] Sodium 2-ethylhexyl sulfate; anionic; wetting agent, stabilizer for plastics, rubber, adhesives, food contact paper; coemulsifier for vinyl chloride, acrylics, vinyl acetate copolymers; biodeg.; amber liq.; sp.gr. 1.1; dens. 9.2 lb/gal; visc. 200 cps; cloud pt. 5 C max.; surf. tens. 38 dynes/cm; pH 7-10 (10%); 43-46% solids.

Avirol® SA 4108. [Henkel/Functional Prods.] Sodium n-octyl sulfate; anionic; emulsifier, low foaming surfactant, wetting agent; liq.; 35-37% solids.

Avirol® SA 4110. [Henkel/Functional Prods.] Sodium

n-decyl sulfate; anionic; wetting and emulsifying agent for plastics; amber liq.; sp.gr. 1.065; dens. 8.8 lb/gal; visc. 100 cps. max.; pH 8-10 min. (10%); cloud pt. -5 C; surf. tens. 31 dynes/cm (@ CMC); 31-33% solids.

Avirol® SA 4113. [Henkel/Functional Prods.] Sodium tridecyl sulfate; CAS 3026-63-9; EINECS 221-188-2; anionic; emulsifier for S/B and vinyl chloride copolymers; liq.; 29-31% solids.

Avirol® SE 3002. [Henkel/Functional Prods.] Sodium laureth sulfate; CAS 9004-82-4; 1335-72-4; anionic; emulsifier for vinyl acetate copolymers, S/B latexes, vinyl chloride copolymers, acrylate homo- and copolymers; liq.; 27-30% solids.

Avirol® SE 3003. [Henkel/Functional Prods.] Sodium laureth sulfate; CAS 9004-82-4; 1335-72-4; anionic; emulsifier for vinyl acetate copolymers, S/B latexes, vinyl chloride copolymers, acrylate homo- and copolymers; liq.; 27-30% solids.

Avirol® SL 2010. [Henkel/Functional Prods.] Sodium lauryl sulfate; CAS 151-21-3; anionic; dispersing agent and emulsifier for acrylates, styrene acrylic, vinyl chloride, vinylidene chloride and vinyl acetate copolymers; foaming agent for mech. latex foaming, carpet and upholstery cleaners; air entraining agent for mortars; liq.; 30-32% solids.

Avirol® SL 2020. [Henkel/Functional Prods.] Sodium lauryl sulfate; CAS 151-21-3; anionic; emulsifier for emulsion polymerization; additive for mech. latex foaming; foaming agent for acrylate disps., carpet and upholstery cleaners; air entraining agent for mortars; liq.; 30-32% solids.

Avirol® SO 70P. [Henkel/Functional Prods.] Sodium dioctyl sulfosuccinate; anionic; wetting agent, emulsifier for antifog compositions, cleaners for automobiles, dry cleaning formulations, and glass surfaces; pigment dispersion in paints and inks; latex paint stabilization; used in pesticides; APHA 100 clear liq.; sol. in polar and nonpolar solv.; sp.gr. 1.06; dens. 8.8 lb/gal; flash pt. 280 F (PMCC); cloud pt. -5 C; 70% solids; Unverified

Azo D. [Dong Jin] Azodicarbonamide; CAS 123-77-3; EINECS 204-650-8; chem. foaming agent for ABS, acetal, acrylic, EVA, HDPE, LDPE, PPO, PP, PS, HIPS, flexible PVC, TPE; processing range 300-450 F; gas yield 220 cc/g.

Azo DX72. [Dong Jin] Activated azodicarbonamide; CAS 123-77-3; EINECS 204-650-8; chem. foaming agent for ABS, EVA, LDPE, PS, HIPS, flexible PVC, TPE; processing temps. 300-450 F; gas yield 120-215 cc/g.

Azocel OBSH. [Fairmount] 4,4′-Oxybis (benzenesulfonyl) hydrazine; CAS 80-51-3; chem. foaming agent for EVA, LDPE, PS, flexible PVC; processing temps. 250-375 F; gas yield 120-125 cc/g.

Azofoam. [Biddle Sawyer] Azodicarbonamide; CAS 123-77-3; EINECS 204-650-8; chem. foaming agent for ABS, acetal, acrylic, EVA, HDPE, LDPE, PPO, PP, PS, HIPS, flexible PVC, TPE; processing range 300-450 F; gas yield 220 cc/g.

Azofoam B-95. [Biddle Sawyer] 4,4′-Oxybis (benzenesulfonyl) hydrazine; CAS 80-51-3; chem. foaming agent for EVA, LDPE, PS, flexible PVC; processing temps. 250-375 F; gas yield 120-125 cc/g.

Azofoam B-520. [Biddle Sawyer] Proprietary; endothermic chem. foaming agent for PS and rigid PVC; processing range 300-360 F; gas yield 115-120 cc/g.

Azofoam DS-1, DS-2. [Biddle Sawyer] Modified azodicarbonamide; CAS 123-77-3; EINECS 204-650-8; chem. foaming agent for rigid PVC, PP; processing temps. 300-390 F; gas yield 155-230 cc/g.

Azofoam M-1, M-2, M-3. [Biddle Sawyer] Activated azodicarbonamide; CAS 123-77-3; EINECS 204-650-8; chem. foaming agent for ABS, EVA, LDPE, PS, HIPS, flexible PVC, TPE; processing temps. 300-450 F; gas yield 120-215 cc/g.

Azofoam UC. [Biddle Sawyer] Modified azodicarbonamide; CAS 123-77-3; EINECS 204-650-8; nonplateout chem. foaming agent for ABS, acetal, acrylic, EVA, HDPE, LDPE, PPO, PP, PS, HIPS, flexible PVC, TPE; processing range 300-450 F; gas yield 180-190 cc/g.

Aztec® AAP-LA-M2. [Aztec Peroxides] Acetyl acetone peroxide; CAS 37187-22-7; EINECS 215-661-2; curing agent, low activity; liq. mixt.; SADT 80 C; flash pt. (Seta) 44 C; *Storage:* store @ 5-25 C.

Aztec® AAP-NA-2. [Aztec Peroxides] Acetyl acetone peroxide; CAS 37187-22-7; EINECS 215-661-2; curing agent, normal activity; liq. mixt.; SADT 70 C; flash pt. (Seta) > 110 C; *Storage:* store @ 5-25 C.

Aztec® AAP-SA-3. [Aztec Peroxides] Acetyl acetone peroxide; CAS 37187-22-7; EINECS 215-661-2; curing agent, super activity; liq. mixt.; SADT 70 C; flash pt. (CC) 64 C; *Storage:* store @ 5-25 C.

Aztec® ACSP-28-FT. [Aztec Peroxides] Acetyl cyclohexane sulfonyl peroxide; CAS 3179-56-4; polymerization initiator; liq.; SADT 5 C; flash pt. (Seta) < 0 C; 28% act. in phthalate; *Storage:* store @ ≈ 12 C.

Aztec® ACSP-55-W1. [Aztec Peroxides] Acetyl cyclohexane sulfonyl peroxide; CAS 3179-56-4; polymerization initiator; water-damped gran.; SADT 15 C; *Storage:* store below -10 C.

Aztec® BCHPC. [Aztec Peroxides] Di (4-t-butyl cyclohexyl) peroxydicarbonate; CAS 15520-11-3; polymerization initiator, curing agent; powd.; SADT 45 C; *Storage:* store below 20 C.

Aztec® BCHPC-40-SAQ. [Aztec Peroxides] Di (4-t-butyl cyclohexyl) peroxydicarbonate; CAS 15520-11-3; polymerization initiator; liq.; SADT 40 C; 40% susp. in water; *Storage:* store @ 5-20 C.

Aztec® BCHPC-75-W. [Aztec Peroxides] Di (4-t-butyl cyclohexyl) peroxydicarbonate; CAS 15520-11-3; polymerization initiator; water-damped powd.; SADT 45 C; *Storage:* store @ 5-20 C.

Aztec® BCUP. [Aztec Peroxides] t-Butylcumyl peroxide; CAS 3457-61-2; polymerization initiator, crosslinking agent; liq.; SADT 90 C; flash pt. (Seta) 55 C; *Storage:* store @ 20-30 C.

Aztec® 1,1-BIS-50-AL. [Aztec Peroxides] 1,1-Di (t-butylperoxy) cyclohexane; CAS 3006-86-8; polymerization initiator, curing agent; liq.; SADT 70 C; flash pt. (Seta) 43 C; 50% sol'n. in aliphatics; *Storage:* store below 30 C.

Aztec® 1,1-BIS-50-WO. [Aztec Peroxides] 1,1-Di (t-butylperoxy) cyclohexane; CAS 3006-86-8; polymerization initiator; liq.; SADT 70 C; flash pt. (Seta) 49 C; 50% sol'n. in wh. oil; *Storage:* store @ 10-30 C.

Aztec® 1,1-BIS-80-BBP. [Aztec Peroxides] 1,1-Di (t-butylperoxy) cyclohexane in butyl benzyl phthalate sol'n.; polymerization initiator; liq.; SADT 65 C; flash pt. (Seta) 89 C; 80% act.; *Storage:* store below 35 C.

Aztec® BP-05-FT. [Aztec Peroxides] Dibenzoyl peroxide; CAS 94-36-0; EINECS 202-327-6; polymerization initiator, curing agent; liq.; SADT > 70 C; flash pt. (Seta) 110 C; 5% sol'n. in phthalate; *Storage:* store @ 5-30 C.

Aztec® BP-20-GY. [Aztec Peroxides] Dibenzoyl peroxide; CAS 94-36-0; EINECS 202-327-6; curing agent; powd.; SADT 70 C; 20% act. in gypsum; *Storage:* store below 30 C.

Aztec® BP-40-S. [Aztec Peroxides] Dibenzoyl peroxide; CAS 94-36-0; EINECS 202-327-6; curing agent; liq.; SADT 60 C; 40% susp. in phthalate; *Storage:* store @ 5-30 C.

Aztec® BP-50-FT. [Aztec Peroxides] Dibenzoyl peroxide; CAS 94-36-0; EINECS 202-327-6; polymerization initiator, curing agent; powd.; SADT 60 C; 50% act. in phthalate; *Storage:* store below 30 C.

Aztec® BP-50-P1. [Aztec Peroxides] Dibenzoyl peroxide; CAS 94-36-0; EINECS 202-327-6; curing agent; paste; SADT 50 C; 50% act. in phthalate; *Storage:* store @ 5-30 C.

Aztec® BP-50-PSI. [Aztec Peroxides] Dibenzoyl peroxide; CAS 94-36-0; EINECS 202-327-6; crosslinking agent; paste; SADT 60 C; 50% act. in silicone oil; *Storage:* store @ 5-30 C.

Aztec® BP-50-SAQ. [Aztec Peroxides] Dibenzoyl peroxide; CAS 94-36-0; EINECS 202-327-6; special purpose peroxide; liq.; SADT 60 C; 50% susp. in water; *Storage:* store @ 5-30 C.

Aztec® BP-60-PCL. [Aztec Peroxides] Dibenzoyl peroxide; CAS 94-36-0; EINECS 202-327-6; curing agent; paste; 60% act. in chlorinated hydrocarbons; *Storage:* store @ 5-30 C.

Aztec® BP-70-W. [Aztec Peroxides] Dibenzoyl peroxide; CAS 94-36-0; EINECS 202-327-6; polymerization initiator; water-damped powd.; SADT 70 C; *Storage:* store @ 5-30 C.

Aztec® BP-77-W. [Aztec Peroxides] Dibenzoyl peroxide; CAS 94-36-0; EINECS 202-327-6; polymerization initiator; water-damped powd.; SADT 70 C; *Storage:* store @ 5-30 C.

Aztec® BU-50-AL. [Aztec Peroxides] 2,2-Di (t-butylperoxy) butane; CAS 2167-23-9; polymerization initiator, curing agent; liq.; SADT 70 C; flash pt. (Seta) 32 C; 50% sol'n. in aliphatics; *Storage:* store below 30 C.

Aztec® BU-50-WO. [Aztec Peroxides] 2,2-Di (t-butylperoxy) butane; CAS 2167-23-9; polymerization initiator; liq.; SADT 70 C; flash pt. (Seta) 32 C; 50% sol'n. in wh. oil; *Storage:* store below 30 C.

Aztec® CEPC. [Aztec Peroxides] Dicetyl peroxydicarbonate; CAS 26332-14-5; polymerization initiator; flakes; SADT 40 C; *Storage:* store below 20 C.

Aztec® CEPC-40-SAQ. [Aztec Peroxides] Dicetyl peroxydicarbonate; CAS 26332-14-5; polymerization initiator; liq.; SADT 40 C; 40% susp. in water; *Storage:* store @ 5-20 C.

Aztec® CHP-50-P1. [Aztec Peroxides] Cyclohexanone peroxide; CAS 12262-58-7; EINECS 235-527-7; curing agent; paste; SADT 50 C; 50% act. in phthalate; *Storage:* store @ 5-25 C.

Aztec® CHP-80. [Aztec Peroxides] Cumene hydroperoxide; CAS 80-15-9; polymerization initiator; liq.; SADT > 80 C; flash pt. (Seta) 57 C; 80% sol'n. in cumene; *Storage:* store @ 5-30 C.

Aztec® CHP-90-W1. [Aztec Peroxides] Cyclohexanone peroxide; CAS 12262-58-7; EINECS 235-527-7; polymerization initiator, curing agent; water-damped powd.; SADT 60 C; *Storage:* store @ 5-25 C.

Aztec® CHP-HA-1. [Aztec Peroxides] Cyclohexanone peroxide; CAS 12262-58-7; EINECS 235-527-7; curing agent, high activity; liq. mixt.; SADT 50 C; flash pt. (Seta) > 110 C; *Storage:* store below 30 C.

Aztec® CHPC. [Aztec Peroxides] Dicyclohexyl peroxydicarbonate; CAS 1561-49-5; polymerization initiator; powd.; SADT 20 C; *Storage:* store below 0 C.

Aztec® CHPC-90-W. [Aztec Peroxides] Dicyclohexyl peroxydicarbonate; CAS 1561-49-5; polymerization initiator; water-damped powd.; SADT 25 C; *Storage:* store @ 0-5 C.

Aztec® CUPND-75-AL. [Aztec Peroxides] Cumylperoxy neodecanoate; CAS 26748-47-0; polymerization initiator; liq.; SADT 15 C; flash pt. (Seta) 49 C; 75% sol'n. in aliphatics; *Storage:* store below -15 C.

Aztec® DCLBP-50-PSI [Aztec Peroxides] Di (2,4-dichloro benzoyl) peroxide; crosslinking applics.; paste; SADT 60 C; 50% act. in silicone oil; *Storage:* store @ 5-30 C.

Aztec® DCP-40-G. [Aztec Peroxides] Dicumyl peroxide; CAS 80-43-3; crosslinking agent; gran.; SADT > 70 C; 40% act. in kaolin; *Storage:* store below 30 C.

Aztec® DCP-40-IC. [Aztec Peroxides] Dicumyl peroxide; CAS 80-43-3; crosslinking agent; powd.; SADT > 70 C; 40% act. in chalk; *Storage:* store below 30 C.

Aztec® DCP-40-IC1. [Aztec Peroxides] Dicumyl peroxide; CAS 80-43-3; crosslinking agent; powd.; SADT > 70 C; 40% act. in kaolin; *Storage:* store below 30 C.

Aztec® DCP-R. [Aztec Peroxides] Dicumyl peroxide; CAS 80-43-3; crosslinking agent; powd.; SADT 80 C; 98% purity; 5.92% act. oxygen; *Storage:* store below 30 C.

Aztec® DHPBZ-75-W. [Aztec Peroxides] 2,5-Dimethyl 2,5-di (benzoylperoxy) hexane; polymerization initiator; water-damped powd.; SADT 80 C; *Storage:* store @ 5-30 C.

Aztec® DHPEH. [Aztec Peroxides] 2,5-Dimethyl-2,5-di (2-ethylhexanoyl peroxy) hexane; CAS 13052-09-0; polymerization initiator, curing agent; liq.; SADT 40 C; flash pt. (Seta) 74 C; *Storage:* store below 10 C.

Aztec® 2,5-DI. [Aztec Peroxides] 2,5-Dimethyl-2,5-di(t-butylperoxy)hexane; CAS 78-63-7; crosslinking agent; liq.; SADT 90 C; flash pt. (Seta) 50 C; 90% min. act.; 9.92% act. oxygen; *Storage:* store @ 10-30 C.

Aztec® 2,5-DI-45-G. [Aztec Peroxides] 2,5-Dimethyl-2,5-di(t-butylperoxy)hexane; CAS 78-63-7; crosslinking agent; gran.; SADT 80 C; 45% act. in chalk; *Storage:* store below 30 C.

Aztec® 2,5-DI-45-IC. [Aztec Peroxides] 2,5-Dimethyl-2,5-di(t-butylperoxy)hexane; CAS 78-63-7; crosslinking agent; powd.; SADT 90 C; 45% act. in chalk; *Storage:* store below 30 C.

Aztec® 2,5-DI-45-IC1. [Aztec Peroxides] 2,5-Dimethyl-2,5-di(t-butylperoxy)hexane; CAS 78-63-7; crosslinking agent; powd.; SADT 90 C; 45% act. in kaolin; *Storage:* store below 30 C.

Aztec® 2,5-DI-45-PSI. [Aztec Peroxides] 2,5-Dimethyl-2,5-di(t-butylperoxy)hexane; CAS 78-63-7; crosslinking agent; paste; SADT 80 C; 45% act. in silicone oil; *Storage:* store @ 5-30 C.

Aztec® 2,5-DI-50-C. [Aztec Peroxides] 2,5-Dimethyl-2,5-di(t-butylperoxy)hexane with calcium carbonate; crosslinking agent; powd.; SADT 90 C; 50% act.; *Storage:* store below 30 C.

Aztec® 2,5-DI-70-S. [Aztec Peroxides] 2,5-Dimethyl-2,5-di(t-butylperoxy)hexane with silica; crosslinking agent; powd.; SADT 90 C; 70% act.; *Storage:* store below 30 C.

Aztec® DIHP-55-W. [Aztec Peroxides] 1,4-Di (2-hydroperoxy isopropyl) benzene; specialty peroxide; water-damped powd.; SADT 60 C; *Storage:* store @ 5-30 C.

Aztec® DIPND-50-AL. [Aztec Peroxides] 1,4-Di(2-neodecanoyl peroxy isopropyl) benzene; polymerization initiator; liq.; SADT 10 C; flash pt. (Seta) 47 C; 50% sol'n. in aliphatics; *Storage:* store below -15 C.

Aztec® DIPP-2. [Aztec Peroxides] 1,3-Di (2-t-butylperoxy isopropyl) benzene; CAS 25155-25-3; polymerization initiator, crosslinking agent; flakes; SADT 90 C; *Storage:* store below 30 C.

Aztec® DIPP-40-G. [Aztec Peroxides] Di (2-t-butylperoxy isopropyl) benzene; CAS 25155-25-3; crosslinking agent; gran.; SADT 90 C; 40% act. in chalk; *Storage:* store below 30 C.

Aztec® DIPP-40-IC. [Aztec Peroxides] Di (2-t-butylperoxy isopropyl) benzene; CAS 25155-25-3; crosslinking agent; powd.; SADT 90 C; 40% act. in chalk; *Storage:* store below 30 C.

Aztec® DIPP-40-IC1. [Aztec Peroxides] Di (2-t-butylperoxy isopropyl) benzene, kaolin; crosslinking agent; powd.; SADT 90 C; 40% act. in kaolin; *Storage:* store below 30 C.

Aztec® DIPP-40-IC5. [Aztec Peroxides] Di (2-t-butylperoxy isopropyl) benzene with PP; crosslinking agent; powd.; SADT 90 C; 40% act. in PP; *Storage:* store below 30 C.

Aztec® DP. [Aztec Peroxides] Didecanoyl peroxide; CAS 762-12-9; polymerization initiator; flakes; SADT 40 C; *Storage:* store below 15 C.

Aztec® DTAP. [Aztec Peroxides] Di (t-amyl) peroxide; CAS 10508-09-5; polymerization initiator; liq.; SADT 90 C; flash pt. (Seta) < 0 C; *Storage:* store below 30 C.

Aztec® DTBP. [Aztec Peroxides] Di (t-butyl) peroxide; CAS 110-05-4; EINECS 203-733-6; polymerization initiator, crosslinking agent; liq.; SADT > 80 C; flash pt. (Seta) 4 C; *Storage:* store below 30 C.

Aztec® EBU-40-G. [Aztec Peroxides] Ethyl 3,3-di (t-butylperoxy) butyrate; crosslinking agent; gran.; SADT 80 C; 40% act. in chalk; *Storage:* store below 30 C.

Aztec® EBU-40-IC. [Aztec Peroxides] Ethyl 3,3-di (t-butylperoxy) butyrate; crosslinking agent; powd.; SADT 80 C; 40% act. in chalk; *Storage:* store below 30 C.

Aztec® EHPC-40-EAQ. [Aztec Peroxides] Di (2-ethylhexyl) peroxydicarbonate; CAS 16111-62-4; polymerization initiator; emulsion, frozen flakes; SADT 5 C; 40% act.; *Storage:* store below -15 C.

Aztec® EHPC-40-ENF. [Aztec Peroxides] Di (2-ethylhexyl) peroxydicarbonate; CAS 16111-62-4; polymerization initiator; nonfreezing emulsion; SADT 5 C; 40% act.; *Storage:* store @ -25 to -15 C.

Aztec® EHPC-65-AL. [Aztec Peroxides] Di (2-ethylhexyl) peroxydicarbonate; CAS 16111-62-4; polymerization initiator; liq.; SADT 5 C; flash pt. (Seta) 53 C; 65% sol'n. in aliphatics; *Storage:* store below -20 C.

Aztec® EHPC-75-AL. [Aztec Peroxides] Di (2-ethylhexyl) peroxydicarbonate; CAS 16111-62-4; polymerization initiator; liq.; SADT 5 C; flash pt. (Seta) 55 C; 75% sol'n. in aliphatics; *Storage:* store below -20 C.

Aztec® HMCN-30-AL. [Aztec Peroxides] 3,3,6,6,9,9-Hexamethyl 1,2,4,5-tetraoxa cyclononane; polymerization initiator, crosslinking agent; liq.; SADT > 80 C; flash pt. (Seta) 43 C; 30% sol'n. in aliphatics; *Storage:* store @ 5-30 C.

Aztec® HMCN-30-WO-2. [Aztec Peroxides] 3,3,6,6,9,9-Hexamethyl 1,2,4,5-tetraoxa cyclononane; polymerization initiator, crosslinking agent; liq.; SADT > 80 C; flash pt. (Seta) 22 C; 30% sol'n. in wh. oil; *Storage:* store @ 15-30 C.

Aztec® HMCN-40-IC3. [Aztec Peroxides] 3,3,6,6,9,9-Hexamethyl 1,2,4,5-tetraoxa cyclononane with polyethylene; crosslinking agent; powd.; SADT > 80 C; 40% act. in PE; *Storage:* store below 30 C.

Aztec® INP-37-AL. [Aztec Peroxides] Di (3,5,5-trimethyl hexanoyl) peroxide; polymerization initiator; liq.; SADT > 40 C; flash pt. (CC) 43 C; 37% act. in aliphatics; *Storage:* store @ -10 to 0 C.

Aztec® INP-75-AL. [Aztec Peroxides] Di (3,5,5-trimethyl hexanoyl) peroxide; polymerization initiator; liq.; SADT 30 C; flash pt. (Seta) 58 C; 75% act. in aliphatics; *Storage:* store @ -10 to 0 C.

Aztec® LP. [Aztec Peroxides] Dilauroyl peroxide; CAS 105-74-8; polymerization initiator; flakes;

SADT 50 C; *Storage:* store below 30 C.

Aztec® LP-25-SAQ. [Aztec Peroxides] Dilauroyl peroxide; CAS 105-74-8; polymerization initiator; liq.; SADT 50 C; 25% act. susp. in water; *Storage:* store @ 5-30 C.

Aztec® LP-40-SAQ. [Aztec Peroxides] Dilauroyl peroxide; CAS 105-74-8; polymerization initiator; liq.; SADT 50 C; 40% act. susp. in water; *Storage:* store @ 5-30 C.

Aztec® MEKP-HA-1. [Aztec Peroxides] Methyl ethyl ketone peroxide; CAS 1338-23-4; EINECS 215-661-2; polymerization initiator, curing agent, high activity; liq. mixt.; SADT 60 C; flash pt. (CC) 81 C; *Storage:* store below 30 C.

Aztec® MEKP-HA-2. [Aztec Peroxides] Methyl ethyl ketone peroxide; CAS 1338-23-4; EINECS 215-661-2; polymerization initiator, curing agent, high activity; liq. mixt.; SADT 60 C; flash pt. (Seta) 52 C; *Storage:* store below 30 C.

Aztec® MEKP-LA-2. [Aztec Peroxides] Methyl ethyl ketone peroxide; CAS 1338-23-4; EINECS 215-661-2; curing agent, low activity; liq. mixt.; SADT 60 C; flash pt. (CC) 76 C; *Storage:* store below 30 C.

Aztec® MEKP-SA-2. [Aztec Peroxides] Methyl ethyl ketone peroxide; CAS 1338-23-4; EINECS 215-661-2; curing agent, super activity; liq. mixt.; SADT 60 C; flash pt. (Seta) 74 C; *Storage:* store below 30 C.

Aztec® MIKP-LA-M1. [Aztec Peroxides] Methyl isobutyl ketone peroxide; CAS 37206-20-5; curing agent; low activity; liq. mixt.; SADT 60 C; flash pt. (Seta) 37 C; *Storage:* store below 30 C.

Aztec® MYPC. [Aztec Peroxides] Dimyristyl peroxydicarbonate; CAS 53220-22-7; polymerization initiator, curing agent; flakes; SADT 40 C; *Storage:* store below 20 C.

Aztec® MYPC-40-SAQ. [Aztec Peroxides] Dimyristyl peroxydicarbonate; CAS 53220-22-7; polymerization initiator; liq.; SADT 35 C; 40% susp. in water; *Storage:* store @ 5-20 C.

Aztec® NBV-40-G. [Aztec Peroxides] n-Butyl 4,4-di (t-butylperoxy) valerate; CAS 995-33-5; crosslinking agent; gran.; SADT 70 C; 40% act. in chalk; *Storage:* store below 30 C.

Aztec® NBV-40-IC. [Aztec Peroxides] n-Butyl 4,4-di (t-butylperoxy) valerate; CAS 995-33-5; crosslinking agent; powd.; SADT 70 C; 40% act. in chalk; *Storage:* store below 30 C.

Aztec® PMBP-50-PSI. [Aztec Peroxides] Di (4-methylbenzoyl) peroxide; crosslinking agent; paste; SADT > 90 C; 50% act. in silicone oil; *Storage:* store @ 5-30 C.

Aztec® SUCP-70-W. [Aztec Peroxides] Disuccinic acid peroxide; polymerization initiator; water-damped powd.; SADT 30 C; *Storage:* store below -10 C.

Aztec® TAHP-80-AL. [Aztec Peroxides] t-Amyl hydroperoxide; CAS 3425-61-4; polymerization initiator; liq.; SADT > 90 C; flash pt. (Seta) 42 C; 80% sol'n. in aliphatics; *Storage:* store @ 5-30 C.

Aztec® TAPB. [Aztec Peroxides] t-Amylperoxy benzoate; CAS 4511-39-1; curing agent; liq.; SADT > 60 C; flash pt. (Seta) 43 C; 94% act.; *Storage:* store below 35 C.

Aztec® TAPB-90-OMS. [Aztec Peroxides] t-Amylperoxy benzoate, odorless min. spirits; polymerization initiator; liq.; SADT 60 C; flash pt. (Seta) 58 C; 90% act.; *Storage:* store below 30 C.

Aztec® TAPEH. [Aztec Peroxides] t-Amylperoxy 2-ethylhexanoate; CAS 686-31-7; polymerization initiator, curing agent; liq.; SADT 40 C; flash pt. (Seta) 58 C; *Storage:* store below 10 C.

Aztec® TAPEH-75-OMS. [Aztec Peroxides] t-Amylperoxy 2-ethylhexanoate, odorless min. spirits; polymerization initiator, curing agent; liq.; SADT 40 C; flash pt. (Seta) 36 C; 75% act.; *Storage:* store below 10 C.

Aztec® TAPEH-90-OMS. [Aztec Peroxides] t-Amylperoxy 2-ethylhexanoate, odorless min. spirits; polymerization initiator, curing agent; liq.; SADT 40 C; flash pt. (Seta) 36 C; 90% act.; *Storage:* store below 10 C.

Aztec® TAPND-75-AL. [Aztec Peroxides] t-Amylperoxyneodecanoate; CAS 68299-16-1; polymerization initiator; liq.; SADT 20 C; flash pt. (Seta) 48 C; 75% sol'n. in aliphatics; *Storage:* store below -10 C.

Aztec® TAPPI-75-OMS. [Aztec Peroxides] t-Amylperoxy pivalate, odorless min. spirits; polymerization initiator; liq.; SADT 25 C; flash pt. (Seta) 47 C; 75% act.; *Storage:* store below -5 C.

Aztec® TBHP-22-OL1. [Aztec Peroxides] t-Butyl hydroperoxide; CAS 75-91-2; EINECS 200-915-7; specialty peroxide; liq.; SADT 70 C; flash pt. (Seta) 61 C; 22% sol'n. in alcohols; *Storage:* store @ 5-30 C.

Aztec® TBHP-70. [Aztec Peroxides] t-Butyl hydroperoxide; CAS 75-91-2; EINECS 200-915-7; polymerization initiator; liq.; SADT > 90 C; flash pt. (CC) 42 C; 70% aq. sol'n.; *Storage:* store @ 5-30 C.

Aztec® TBPA-50-OMS. [Aztec Peroxides] t-Butylperoxy acetate in odorless min. spirits; polymerization initiator; liq.; SADT > 70 C; flash pt. (Seta) 37 C; 50% act.; *Storage:* store below 35 C.

Aztec® TBPA-75-OMS. [Aztec Peroxides] t-Butylperoxy acetate in odorless min. spirits; polymerization initiator; liq.; SADT > 70 C; flash pt. (Seta) 38 C; 75% act.; *Storage:* store below 35 C.

Aztec® TBPB. [Aztec Peroxides] t-Butylperoxy benzoate; CAS 614-45-9; polymerization initiator, curing agent; liq.; SADT 60 C; flash pt. (Seta) 100 C; *Storage:* store @ 10-30 C.

Aztec® TBPB-50-FT. [Aztec Peroxides] t-Butylperoxy benzoate; CAS 614-45-9; curing agent; liq.; SADT 80 C; flash pt. (Seta) > 100 C; 50% sol'n. in phthalate; *Storage:* store below 30 C.

Aztec® TBPB-50-IC. [Aztec Peroxides] t-Butylperoxy benzoate; CAS 614-45-9; curing agent, crosslinking agent; powd.; SADT 70 C; 50% act. in chalk; *Storage:* store below 30 C.

Aztec® TBPB-SA-M2. [Aztec Peroxides] t-Butylperoxy benzoate; CAS 614-45-9; curing agent, super activity; liq. mixt.; SADT 70 C; flash pt. (Seta) > 110 C; *Storage:* store below 30 C.

Aztec® TBPEH. [Aztec Peroxides] t-Butylperoxy 2-ethylhexanoate; polymerization initiator, curing agent; liq.; SADT 40 C; flash pt. (Seta) 104 C; *Storage:* store below 15 C.

Aztec® TBPEH-30-AL. [Aztec Peroxides] t-Butylperoxy 2-ethylhexanoate; polymerization initiator; liq.; SADT 40 C; flash pt. (Seta) 55 C; 30% sol'n. in aliphatics; *Storage:* store below 15 C.

Aztec® TBPEH-50-DOP. [Aztec Peroxides] t-Butylperoxy 2-ethylhexanoate in DOP; polymerization initiator, curing agent; liq.; SADT 40 C; flash pt. (Seta) 46 C; 50% act.; *Storage:* store below 15 C.

Aztec® TBPEH-50-OMS. [Aztec Peroxides] t-Butylperoxy 2-ethylhexanoate, odorless min. spirits; polymerization initiator, curing agent; liq.; SADT 40 C; flash pt. (Seta) 46 C; 50% act.; *Storage:* store below 15 C.

Aztec® TBPEHC. [Aztec Peroxides] t-Butylperoxy 2-ethylhexyl carbonate; CAS 34443-12-4; crosslinking agent; liq.; SADT 70 C; flash pt. (Seta) 66 C; *Storage:* store below 30 C.

Aztec® TBPIB-75-AL. [Aztec Peroxides] t-Butylperoxy isobutyrate; CAS 109-13-7; polymerization initiator; liq.; SADT 40 C; flash pt. (Seta) 50 C; 75% sol'n. in aliphatics; *Storage:* store below 15 C.

Aztec® TBPIN. [Aztec Peroxides] t-Butylperoxy 3,5,5-trimethyl hexanoate; CAS 13122-18-4; polymerization initiator, curing agent; liq.; SADT 70 C; flash pt. (Seta) 88 C; *Storage:* store below 30 C.

Aztec® TBPIN-30-AL. [Aztec Peroxides] t-Butylperoxy 3,5,5-trimethyl hexanoate; CAS 13122-18-4; polymerization initiator; liq.; SADT > 70 C; flash pt. (Seta) 57 C; 30% sol'n. in aliphatics; *Storage:* store below 30 C.

Aztec® TBPM-50-FT. [Aztec Peroxides] t-Butyl monoperoxy maleate; CAS 1931-62-0; polymerization initiator, curing agent; powd.; SADT 50 C; 50% act. in phthalate; *Storage:* store below 30 C.

Aztec® TBPM-50-P. [Aztec Peroxides] t-Butyl monoperoxy maleate; CAS 1931-62-0; polymerization initiator, curing agent; paste; SADT 50 C; 50% act. in phthalate; *Storage:* store @ 5-25 C.

Aztec® TBPND. [Aztec Peroxides] t-Butylperoxy neodecanoate; CAS 26748-41-4; polymerization initiator; liq.; SADT 15 C; flash pt. (Seta) 52 C; *Storage:* store below -10 C.

Aztec® TBPND-40-EAQ. [Aztec Peroxides] t-Butylperoxy neodecanoate; CAS 26748-41-4; polymerization initiator; emulsion, frozen flakes; SADT 20 C; 40% act.; *Storage:* store below -10 C.

Aztec® TBPND-75-OMS. [Aztec Peroxides] t-Butylperoxy neodecanoate; CAS 26748-41-4; polymerization initiator; liq.; SADT 15 C; flash pt. (Seta) 50 C; 75% sol'n. in aliphatics; *Storage:* store below -10 C.

Aztec® TBPPI-25-AL. [Aztec Peroxides] t-Butylperoxy pivalate; CAS 927-07-1; polymerization initiator; liq.; SADT 20 C; flash pt. (Seta) 49 C; 25% sol'n. in aliphatics; *Storage:* store below 0 C.

Aztec® TBPPI-50-OMS. [Aztec Peroxides] t-Butylperoxy pivalate, odorless min. spirits; polymerization initiator; liq.; SADT 25 C; flash pt. (Seta) 49 C; 50% act.; *Storage:* store @ -15 to -5 C.

Aztec® TBPPI-75-OMS. [Aztec Peroxides] t-Butylperoxy pivalate, odorless min. spirits; polymerization initiator; liq.; SADT 25 C; flash pt. (Seta) 44 C; 75% act.; *Storage:* store @ -15 to -5 C.

Aztec® 2,5-TRI. [Aztec Peroxides] 2,5-Dimethyl-2,5-di(t-butylperoxy)hexyne-3; CAS 1068-27-5; crosslinking agent; liq.; SADT 90 C; flash pt. (Seta) 55 C; 90% min. act.; 9.92% act. oxygen; *Storage:* store @ 10-30 C.

Aztec® 3,3,5-TRI. [Aztec Peroxides] 1,1-Di (t-butylperoxy) 3,3,5-trimethyl cyclohexane; CAS 6731-36-8; polymerization initiator; liq.; SADT 70 C; flash pt. (Seta) 42 C; 92% act.; *Storage:* store below 30 C.

Aztec® 3,3,5-TRI-28-IC3. [Aztec Peroxides] 1,1-Di (t-butylperoxy) 3,3,5-trimethyl cyclohexane in polyethylene; crosslinking agent; powd.; SADT 70 C; 28% act. in polyethylene; *Storage:* store below 30 C.

Aztec® 3,3,5-TRI-40-G. [Aztec Peroxides] 1,1-Di (t-butylperoxy) 3,3,5-trimethyl cyclohexane; CAS 6731-36-8; crosslinking agent; gran.; SADT 60 C; 40% act. in chalk; *Storage:* store below 30 C.

Aztec® 3,3,5-TRI-40-IC. [Aztec Peroxides] 1,1-Di (t-butylperoxy) 3,3,5-trimethyl cyclohexane; CAS 6731-36-8; crosslinking agent; powd.; SADT 60 C; 40% act. in chalk; *Storage:* store below 30 C.

Aztec® 3,3,5-TRI-50-AL. [Aztec Peroxides] 1,1-Di (t-butylperoxy) 3,3,5-trimethyl cyclohexane; CAS 6731-36-8; polymerization initiator, curing agent; liq.; SADT 70 C; flash pt. (Seta) 40 C; 50% sol'n. in aliphatics; *Storage:* store below 30 C.

Aztec® 3,3,5-TRI-50-FT. [Aztec Peroxides] 1,1-Di (t-butylperoxy) 3,3,5-trimethyl cyclohexane; CAS 6731-36-8; polymerization initiator, curing agent; liq.; SADT 60 C; flash pt. (Seta) 51 C; 50% sol'n. in phthalate; *Storage:* store below 30 C.

Aztec® 3,3,5-TRI-55-AL. [Aztec Peroxides] 1,1-Di (t-butylperoxy) 3,3,5-trimethyl cyclohexane; CAS 6731-36-8; crosslinking agent; liq.; SADT 60 C; flash pt. (Seta) 42 C; 55% sol'n. in aliphatics; *Storage:* store below 30 C.

Aztec® 3,3,5-TRI-75-DBP. [Aztec Peroxides] 1,1-Di (t-butylperoxy) 3,3,5-trimethyl cyclohexane in dibutyl phthalate; polymerization initiator; liq.; SADT 70 C; flash pt. (Seta) 42 C; 75% sol'n. in dibutyl phthalate; *Storage:* store below 30 C.

Azthane® I-100. [Shell] Crosslinking agent; sp.gr. 1.15; soften. pt. 105 C; EEW 350; 12% total NCO.

B

Baco DH101. [Alcan; BA Chem. Ltd.] Alumina trihydate; CAS 21645-51-2; for mfg. of aluminum sulfate and other aluminum compds., coating of TiO_2, catalysts, additives to glass, vitreous enamel, and glazes; flame retardant and smoke suppressant in plastics and rubber; very low insol. content; wh. cryst. powd.; 33% on 75 μm screen, 60% on 53 μm screen; sp.gr. 2.42; bulk dens. 1.1 g/cc (untamped); *Toxicology:* nontoxic; ACGIH 10 mg/m^3 total airborne particulate.

Bantox®. [Monsanto] Zinc mercaptobenzylthiazol; CAS 155-04-4; fast cure accelerator for latex; powd.

Bara-200C. [Cimbar Perf. Minerals] Industrial grade barytes; CAS 7727-43-7; EINECS 231-784-4; filler for rubber and plastics, acoustical compds., urethane foams, mold release agents, and friction materials; powd.; 97% thru 200 mesh, 85% thru 325 mesh; sp.gr. 4.25; oil absorp. 8-9; pH 7.0; 94% $BaSO_4$, < 0.1% moisture.

Bara-200N. [Cimbar Perf. Minerals] Industrial grade barytes; CAS 7727-43-7; EINECS 231-784-4; filler for rubber and plastics, acoustical compds., urethane foams, mold release agents, and friction materials; powd.; 97% thru 200 mesh, 85% thru 325 mesh; sp.gr. 4.25; oil absorp. 8-9; pH 7.0; 92% $BaSO_4$, < 0.1% moisture.

Bara-325C. [Cimbar Perf. Minerals] Industrial grade barytes; CAS 7727-43-7; EINECS 231-784-4; filler for rubber and plastics, acoustical compds., urethane foams, mold release agents, and friction materials; powd.; 99.8% thru 200 mesh, 98% thru 325 mesh; sp.gr. 4.25; oil absorp. 9-10; pH 7.0; 94% $BaSO_4$, < 0.1% moisture.

Bara-325N. [Cimbar Perf. Minerals] Industrial grade barytes; CAS 7727-43-7; EINECS 231-784-4; filler for rubber and plastics, acoustical compds., urethane foams, mold release agents, and friction materials; powd.; 99.8% thru 200 mesh, 98% thru 325 mesh; sp.gr. 4.25; oil absorp. 9-10; pH 7.0; 92% $BaSO_4$, < 0.1% moisture.

Barafoam 200. [Cimbar Perf. Minerals] Natural unbleached barytes; CAS 7727-43-7; EINECS 231-784-4; filler for reinforced urethane, filled foam, rubber, and plastics, acoustical compds., mold release agents, adhesives, coatings, and sealants; powd.; 98% thru 200 mesh, 90% thru 325 mesh; sp.gr. 4.0-4.2; oil absorp. 8-9; pH 7.0; 85-88% $BaSO_4$, < 0.1% moisture.

Barafoam 325. [Cimbar Perf. Minerals] Natural unbleached barytes; CAS 7727-43-7; EINECS 231-784-4; filler for reinforced urethane, filled foam, rubber, and plastics, acoustical compds., mold release agents, adhesives, coatings, and sealants; powd.; 99.8% thru 200 mesh, 98% thru 325 mesh; sp.gr. 4.0-4.2; oil absorp. 9-10; pH 7.0; 85-88% $BaSO_4$, < 0.1% moisture.

Barimite™. [Cimbar Perf. Minerals] Natural unbleached barytes; CAS 7727-43-7; EINECS 231-784-4; filler in primers (automotive and appliance), brake linings, liners for jar lids, plastisols, acoustical compds., urethane foams, athletic goods (bowling/golf/tennis balls); powd.; 4 μ mean particle size; fineness (Hegman) 4; sp.gr. 4.35; bulk dens. 80 lb/ft^3 (loose); oil absorp. 11; 95% $BaSO_4$, 0.15% moisture.

Barimite™ 200. [Cimbar Perf. Minerals] Natural unbleached barytes; CAS 7727-43-7; EINECS 231-784-4; filler in acoustical compds., friction prods., urethane foams, molded compds.; powd.; 12 μ mean particle size; 8% on 325 mesh; sp.gr. 4.35; bulk dens. 120 lb/ft^3 (loose); oil absorp. 7; pH 7.0; 95% $BaSO_4$, 0.15% moisture.

Barimite™ 1525P. [Cimbar Perf. Minerals] Barytes; CAS 7727-43-7; EINECS 231-784-4; plastics grade processing aid for compounding, extrusion, molding; increased prod.; improved tens. and elong. props.; blocks radiation; low abrasion, low resin demand; buff powd.; 2-3 μ mean particle size; 0% on 325 mesh; sp.gr. 4.35-4.40; surf. area 1.9 m^2/g; oil absorp. 11-12; ref. index 1.65; pH 7.5; hardness (Mohs) 2.5; conduct. 6 x 10^{-3} Cal/cm; sp. heat 0.11 Cal/g °C; dielec. const. 7.3; 97-98% $BaSO_4$, 0.1% moisture.

Barimite™ 4009P. [Cimbar Perf. Minerals] Barytes; CAS 7727-43-7; EINECS 231-784-4; plastics grade processing aid for compounding, extrusion, molding; increased prod.; improved tens. and elong. props.; blocks radiation; low abrasion, low resin demand; buff powd.; 8.5-9.5 μ mean particle size; 0.5% on 325 mesh; sp.gr. 4.35-4.40; surf. area 0.8 m^2/g; oil absorp. 7-9; ref. index 1.65; pH 7.5; hardness (Mohs) 2.5; conduct. 6 x 10^{-3} Cal/cm; sp. heat 0.11 Cal/g °C; dielec. const. 7.3; 97-98% $BaSO_4$, 0.06% moisture.

Barimite™ UF. [Cimbar Perf. Minerals] Natural unbleached barytes; CAS 7727-43-7; EINECS 231-784-4; additive in coatings (automotive primers/topcoats, gloss enamels, powd. coatings, semi-gloss/gloss latexes, architectural coatings); fillers in plastics, ceramics, friction materials, rubber goods, plastisols, adhesives, latex prods., urethane foams, acoustical compds., insulating materials, sound attenuation prods.; powd.; 1.8 μ mean particle size; fineness (Hegman) 7; sp.gr. 4.35; bulk dens. 60 lb/ft^3 (loose); oil absorp. 13; pH 7; 95% $BaSO_4$, 0.15% moisture.

Barimite™ XF. [Cimbar Perf. Minerals] Natural unbleached barytes; CAS 7727-43-7; EINECS 231-784-4; filler in primers (automotive and appliance), plastisols, acoustical compds., urethane foams, athletic goods (bowling/golf/tennis balls); powd.; 2.5 μ mean particle size; fineness (Hegman) 5.5; sp.gr. 4.35; bulk dens. 75 lb/ft^3 (loose); oil absorp. 12; pH 7; 95% $BaSO_4$, 0.15% moisture.

Bäropan MC 8046 SP. [Bärlocher GmbH] Ca/Zn; stabilizer for shoe inj. molding of plasticized PVC.

Bäropan TX 296 KA. [Bärlocher GmbH] Stabilizer for cable compds.

Bäropol. [Bärlocher GmbH] Additives for polyolefins and polystyrenes (antioxidants, lubricants, release agents, uv stabilizers, antiblocking agents, antistats, flame retardants, fillers and pigments).

Bärostab® CT 901. [Bärlocher GmbH] Ca/Zn prod.; stabilizer offering very good color props. and exc. transparency for inj. molding of plasticized PVC; for extrusion and shoe inj. molding.

Bärostab® CT 926 HV. [Bärlocher GmbH] Ca/Zn; stabilizer for flooring base coats, artificial leather, wallpaper, conveyor belts; low volatility; low toxicity; compat. with isocyanate adhesion promoter; liq., low odor.

Bärostab® CT 1500. [Bärlocher GmbH] Mg/Al/Zn; stabilizer for artificial leather, wallpaper, conveyor belts, dashboard panels, esp. low fogging applics. in automobile mfg.; phenol-free; good thermal aging props.; stable to 120 C; paste, low odor.

Bärostab® CT 9140. [Bärlocher GmbH] Ca/Zn prod.; stabilizer for flooring top coats; extreme low volatility, good stability and thermal aging props., superior transparency, free of dk. yellowing, resist. to water staining; liq.

Bärostab® L 230. [Bärlocher GmbH] Zinc octoate; fast-kicking stabilizer for foam processing in prod. of sealings for caps; good organoleptic props., exc. color; worldwide approval for contact with foodstuffs; FDA approved for capliners; liq.

Bärostab® NT 210. [Bärlocher GmbH] Ca/Zn; stabilizer for toys, PVC coating through dip molding (tool handles, bicicyle handles, PVC boots and gloves, wire baskets); for transparent and brightly pigmented goods; BGA approved; liq.; *Toxicology:* nontoxic.

Bärostab® NT 211. [Bärlocher GmbH] Ca/Zn; stabilizer for puppets, balls, toy animals mfg. by rotational molding; FDA approved for foodstuffs; liq.; *Toxicology:* nontoxic.

Bärostab® NT 212. [Bärlocher GmbH] Ca/Zn; stabilizer for puppets, balls, toy animals mfg. by rotational molding; approved for foodstuffs in France; liq.; *Toxicology:* nontoxic.

Bärostab® NT 213. [Bärlocher GmbH] Mg/Zn; stabilizer for puppets, balls, toy animals mfg. by rotational molding; approved for foodstuffs in Italy; liq.; *Toxicology:* nontoxic.

Bärostab® NT 224. [Bärlocher GmbH] Ca/Zn; stabilizer for toys, PVC coating through dip molding (tool handles, bicicyle handles, PVC boots and gloves, wire baskets); for pigmented goods; BGA approval; liq., low odor; *Toxicology:* nontoxic.

Bärostab® NT 1005. [Bärlocher GmbH] Mg/Al/Zn complex; stabilizer for plastisol processing; esp. for artificial leather, wallpaper, conveyor belts, toys, nontoxic applics.; approved for food contact (Germany, France, USA); paste, odor-free; *Toxicology:* nontoxic.

Bärostab® UBZ 76 BX. [Bärlocher GmbH] Ba/Zn complex with self-lubrication; stabilizer for transparent cable sheathing compds.; hydrolysis-resist.; imparts high transparency and exc. light stability.

Bärostab® UBZ 614. [Bärlocher GmbH] Ba/Zn; stabilizer for flooring base coats, artificial leather, wallpaper, conveyor belts; liq.

Bärostab® UBZ 630. [Bärlocher GmbH; R.T. Vanderbilt] Barium zinc complex; stabilizer for PVC plastisols, transparent topcoats; Gardner 2 max. liq.; dens. 1.064 ± 0.01 mg/m^3; Unverified

Bärostab® UBZ 632. [Bärlocher GmbH; R.T. Vanderbilt] Barium zinc complex; stabilizer for PVC plastisols, esp. topcoats for lt. pigmented cushion vinyl flooring; features good color; esp. suitable in presence of inhibitors such as TMA; Gardner 3 max. liq.; dens. 1.01-1.04 mg/m^3.

Bärostab® UBZ 637. [Bärlocher GmbH] Ba/Zn; stabilizer for artificial leather, wallpaper, conveyor belts; superior stabilizing and color for transparent and white pigmented applics.; liq.

Bärostab® UBZ 655 X. [Bärlocher GmbH] Ba/Zn; stabilizer for artificial leather, wallpaper, conveyor belts; high quality plastisol stabilizer, extremely low volatility; liq., low odor.

Bärostab® UBZ 667. [Bärlocher GmbH] Ba/Zn; stabilizer for artificial leather, wallpaper, conveyor belts, low fogging automotive applics.; liq.

Bärostab® UBZ 671. [Bärlocher GmbH] Ba/Zn; stabilizer for external coatings, roofing sheets, canvas material, tents, portable inflatable structures, coated metal surfaces; good thermal, photostability, and aging props.; ideal for coating of canvas material; liq., low odor.

Bärostab® UBZ 776 X. [Bärlocher GmbH] Ba/Zn; stabilizer for external coatings, roofing sheets, canvas material, tents, portable inflatable structures, coated metal surfaces; good weathering props. and photostability, good thermal aging props.; high resist. to environmental influences; liq., low odor.

Bärostab® UBZ 791. [Bärlocher GmbH; R.T. Vanderbilt] Barium zinc; stabilizer for flexible and semirigid PVC, esp. sheet and film, and where color hold under dynamic stress is important; Gardner 4-5 liq.;

dens. 1.05 ± 0.01 mg/m^3; ref. index 1.490 ± 0.002; Unverified

Bärostab® UBZ 793. [Bärlocher GmbH; R.T. Vanderbilt] Barium zinc; general purpose stabilizer for transparent and pigmented, flexible and semirigid PVC and esp. suspension PVC; Gardner 10 max. liq.; dens. 1.08 ± 0.02 mg/m^3; Unverified

Bärostab® UBZ 820 KA. [Bärlocher GmbH] Ba/Zn complex with self-lubrication; stabilizer for transparent cable compds.; imparts exc. heat stability.

Barquat® CME-35. [Lonza] Cetethyl morpholinium ethosulfate; CAS 78-21-7; EINECS 201-094-8; antistat, combing aid and detangling agent, textile lubricant, odor counteractant; antistatic coating for cellulose acetate; liq.; 35% act.

Bartex® 10. [Hitox] Barium sulfate; CAS 7727-43-7; EINECS 231-784-4; extender pigment for gloss enamels, powd. coatings, semigloss latexes, house paints, primers; filler in many plastics applics., in rubber goods, and in ceramics; wh. fine powd.; 1 μ avg. particle size; trace on 325 mesh; fineness (Hegman) 7+; insol. in water; sp.gr. 4.36; dens. 36.64 lb/gal; bulk dens. 61.3 lb/ft^3 (loose); oil absorp. 11-13; dry brightness 92+; m.p. ≈ 1580 C; pH 7; 98% $BaSO_4$; *Toxicology:* ACGIH TLV 10 mg/m^3 (total); excessive exposure to dust may produce baritosis, mild bronchial irritation; skin and eye contact may cause mech. abrasion irritation; avoid breathing dust; *Precaution:* nonflamm.

Bartex® 65. [Hitox] Barium sulfate; CAS 7727-43-7; EINECS 231-784-4; extender pigment for gloss enamels, powd. coatings, semigloss latexes, house paints, primers; filler in many plastics applics., in rubber goods, and in ceramics; wh. fine powd.; 1.6 μ avg. particle size; trace on 325 mesh; fineness (Hegman) 6.5-8.0; insol. in water; sp.gr. 4.36; dens. 36.6 lb/gal; bulk dens. 67.8 lb/ft^3 (loose); oil absorp. 11-13; dry brightness 89-90; m.p. ≈ 1580 C; pH 7; 98% $BaSO_4$; *Toxicology:* ACGIH TLV 10 mg/m^3 (total); excessive exposure to dust may produce baritosis, mild bronchial irritation; skin and eye contact may cause mech. abrasion irritation; avoid breathing dust; *Precaution:* nonflamm.

Bartex® 80. [Hitox] Barium sulfate; CAS 7727-43-7; EINECS 231-784-4; extender pigment for gloss enamels, powd. coatings, semigloss latexes, house paints, primers; filler in many plastics applics., in rubber goods, and in ceramics; wh. fine powd.; 7-8 μ avg. particle size; 0.1% on 325 mesh; fineness (Hegman) 4.5-5.0; insol. in water; sp.gr. 4.36; dens. 36.6 lb/gal; bulk dens. 93.9 lb/ft^3 (loose); oil absorp. 11; dry brightness 89-90; m.p. ≈ 1580 C; pH 7; 98% $BaSO_4$; *Toxicology:* ACGIH TLV 10 mg/m^3 (total); excessive exposure to dust may produce baritosis, mild bronchial irritation; skin and eye contact may cause mech. abrasion irritation; avoid breathing dust; *Precaution:* nonflamm.

Bartex® FG-2. [Hitox] Barium sulfate; CAS 7727-43-7; EINECS 231-784-4; filler grade; wh. fine powd., odorless; < 2 μ avg. particle size; trace on 325 mesh; fineness (Hegman) 6.5; insol. in water; sp.gr. 4.36; dens. 36.6 lb/gal; bulk dens. 72 lb/ft^3 (loose); oil absorp. 11.5; dry brightness 85.3; m.p. ≈ 1580 C; pH 8; 94% $BaSO_4$; *Toxicology:* ACGIH TLV 10 mg/m^3 (total); excessive exposure to dust may produce baritosis, mild bronchial irritation; skin and eye contact may cause mech. abrasion irritation; avoid breathing dust; *Precaution:* nonflamm.

Bartex® FG-10. [Hitox] Barium sulfate; CAS 7727-43-7; EINECS 231-784-4; filler grade; wh. fine powd., odorless; 10 μ avg. particle size; 2.7% on 325 mesh; fineness (Hegman) 1.5; insol. in water; sp.gr. 4.36; dens. 36.6 lb/gal; bulk dens. 99.6 lb/ft^3 (loose); oil absorp. 10.7; dry brightness 85.3; m.p. ≈ 1580 C; pH 8; 94% $BaSO_4$; *Toxicology:* ACGIH TLV 10 mg/m^3 (total); excessive exposure to dust may produce baritosis, mild bronchial irritation; skin and eye contact may cause mech. abrasion irritation; avoid breathing dust; *Precaution:* nonflamm.

Bartex® OWT. [Hitox] Barium sulfate; CAS 7727-43-7; EINECS 231-784-4; extender pigment for gloss enamels, powd. coatings, semigloss latexes, house paints, primers; filler in many plastics applics., in rubber goods, and in ceramics; off-wh. fine powd.; 1.6 μ avg. particle size; trace on 325 mesh; fineness (Hegman) 6.5-7; insol. in water; sp.gr. 4.36; dens. 36.6 lb/gal; bulk dens. 67 lb/ft^3 (loose); oil absorp. 11-13; dry brightness 80-89; m.p. ≈ 1580 C; pH 7; 98% $BaSO_4$; *Toxicology:* ACGIH TLV 10 mg/m^3 (total); excessive exposure to dust may produce baritosis, mild bronchial irritation; skin and eye contact may cause mech. abrasion irritation; avoid breathing dust; *Precaution:* nonflamm.

Barytes Microsupreme 2410/M. [Colin Stewart Minchem Ltd] Barytes; CAS 7727-43-7; EINECS 231-784-4; filler; wh. powd., odorless; 100% < 20 μ particle size; fineness (Hegman) 6.5; sp.gr. 4.4; bulk dens. 2.0 g/ml (tapped); oil absorp. 13.5; brightness 95; hardness (Mohs) 3.0; 96.89-99.9% $BaSO_4$; *Precaution:* avoid unnecessary dust.

Barytes Supreme. [Colin Stewart Minchem Ltd] Barytes; CAS 7727-43-7; EINECS 231-784-4; filler; wh. powd., odorless; 98% < 45 μ particle size; fineness (Hegman) 2.5; sp.gr. 4.4; bulk dens. 2.5 g/ml (tapped); oil absorp. 11.5; brightness 94; hardness (Mohs) 3.0; 96.89-99.9% $BaSO_4$; *Precaution:* avoid unnecessary dust.

BAX 1091FM. [Alcan; BA Chem. Ltd.] Magnesium hydroxide; CAS 1309-42-8; EINECS 215-170-3; flame retardant, filler, smoke suppressant for plastics and rubber incl. PP, polyamide, PE, PVC, elastomers; for loading up to 70% and processing to 300 C; recommended for use with surf. coating; high thermal stability, low level of ionic impurities; for cable sheathing, pipes, ducting, elec. conduit, inj. molded components; wh. free-flowing powd.; 0.7 μm median particle size; dens. 2.38 g/cc; bulk dens. 0.53 g/cc (untamped); surf. area 12 m^2/g; pH 10 (5% aq.); 97% min. act.; *Toxicology:* low toxic hazard on handling; low toxicity degradation prods.; ACGIH 10 mg/m^3 total airborne particulate; *Storage:* store in dry place in air-tight containers to avoid moisture pick-up.

Bayer UV Absorber 325, 340. [Bayer AG] UV absorbers for the protection of rigid and plasticized PVC, cellulose derivs., PS, polyacrylate; used in clear finishes to protect the substrate against light.

Baymod® A 50. [Bayer AG] ABS; CAS 9003-56-9; impact modifier for improving impact str. of PVC.

Baymod® A 63 A. [Bayer AG] ABS/nitrile rubber combination; impact modifier for use in semirigid PVC sheet for thermoforming (crash pads).

Baymod® A 72 A/74 A. [Bayer AG] ABS/nitrile rubber combination; impact modifier for use in semirigid PVC sheet for thermoforming (crash pads).

Baymod® A 90/92. [Bayer AG] ABS; CAS 9003-56-9; impact modifier for use in rigid and plasticized PVC where high heat stability is required.

Baymod® A 95. [Bayer AG; Bayer/Fibers, Org., Rubbers] SAN copolymer; CAS 9003-54-7; modifier improving the thermoforming props. and heat resist. of semirigid and rigid PVC; suitable for weatherable applics.; high heat stability; powd.; sp.gr. 1.08; bulk dens. 0.37 g/cm^3.

Baymod® A KA 8572. [Bayer/Fibers, Org., Rubbers] ASA; modifier improving the thermal stability and impact resist. of weatherable rigid PVC; powd.; sp.gr. 1.08; bulk dens. 0.35 g/cm^3.

Baymod® A KU3-2086. [Bayer/Fibers, Org., Rubbers] ABS; CAS 9003-56-9; high efficiency modifier improving the thermal stability and impact resist. of nonweatherable rigid PVC; powd.; sp.gr. 1.05; bulk dens. 0.35 g/cm^3.

Baymod® L450 N. [Bayer AG; Bayer/Fibers, Org., Rubbers] EVA copolymer; CAS 24937-78-8; impact modifier for rigid weatherable PVC, molded and extruded articles such as siding and window profiles, heat and light aging resist. thermoformed PVC instrument panel skins, aging resist. plasticized vinyls; pellet; sp.gr. 0.98; melt flow 1-4 g/10 min; 45% VA.

Baymod® L450 P. [Bayer AG; Bayer/Fibers, Org., Rubbers] EVA copolymer; CAS 24937-78-8; impact modifier for rigid weatherable PVC.

Baymod® L2450. [Bayer/Fibers, Org., Rubbers] EVA copolymer; CAS 24937-78-8; impact modifier for rigid weatherable PVC, plasticizer in flexible PVC; for molded and extruded articles such as siding and window profiles, heat and lt. aging resist. thermoformed PVC instrument panel skins, aging resist. plasticized vinyls; coarse powd.; sp.gr. 0.98; melt flow 2-5 g/10 min; 45% VA.

Baymod® PU. [Bayer AG] Aliphatic polyurethane; plasticizing polymer esp. for use in PVC; good low-temp. stability, solv. resist., grease resist.; no plasticizer migration; high abrasion resist.; used in PVC shoe soles, cables, profiles, inj. molding, film; impact modifier for polyoxymethylene; BGA approved.

Baymoflex A KU3-2069.A. [Bayer/Fibers, Org., Rubbers] ASA; modifier for semirigid halogen-free sheeting, esp. for automotive instrument panels; powd.; sp.gr. 1.07; bulk dens. 0.30 g/cm^3.

BDO. [Arco] 1,4-Butanediol; CAS 110-63-4; EINECS 203-786-5; used to mfg. PBT, thermoplastic copolyester elastomers, as chain extender in polyester polyurethanes, for polyester plasticizers; intermediate for THF, γ-butyrolactone, n-methyl pyrrolidone; used in polymer systems such as PU for automotive parts, sealants, adhesives, appliance housings; colorless liq.; sol. in water, most alcohols, esters, ketones, glycol ethers, acetates; m.w. 90.12; sp.gr. 1.015 (20/20 C); dens. 8.4 lb/gal (20 C); vapor pressure < 1 mm Hg (20 C); f.p. 19-20 C; b.p. 228 C; hyd. no. 1280; flash pt. (COC) 155 C; 99.9+% assay; *Toxicology:* low toxicity; LD50 (oral, rat) 1780 mg/kg; *Precaution:* in presence of strong acids, dehydrates to tetrahydrofuran; *Storage:* keep away from heat, flame; solidifies below 19 C; store under dry nitrogen in tightly closed containers.

Bearflex® LAO. [Witco/Golden Bear] Extender oil offering lt. color and low aniline pts.; for elastomer compounding, esp. nitrile, neoprene, SBR, and natural rubber; m.w. 238; sp.gr. 0.9725 (15.6 C); dens. 8.1 lb/gal; visc. 46 cSt (40 C); pour pt. -21 C; flash pt. (COC) 174 C; ref. index 1.5377.

Beaverwhite 200. [Luzenac Am.] Talc; CAS 14807-96-6; EINECS 238-877-9; high purity platy talc pigment for paints, rubber dusting, building prods., adhesives, caulks, and sealants; powd.; 10 μ median diam.; 99% thru 200 mesh, 97% thru 325 mesh; fineness (Hegman) 2; sp.gr. 11 m^2/g; dens. 25 lb/ft^3 (loose); bulking vlaue 23.3 lb/gal; surf. area 11 m^2/g; oil absorp. 28; brightness (TAPPI) 84; pH 9.5 (10% slurry); 98% talc; 61% SiO_2, 31% MgO; 0.5% moisture.

Beckacite® 4900. [Arizona] Modified tall oil rosin; for mfg. of paper size, intermediate in rosin deriv. prod., printing ink binders as resins, tackifier resin in sealants and mastics, as starting point rosin for resin esters; Gardner 7 color (60% ethanol); soften. pt. (R&B) 100 C; acid no. 240; flash pt. (OC) > 204 C.

Benetex™ OB. [Mayzo] Optical brightener for polymers.

Benox® L-40LV. [Norac] Benzoyl peroxide disp.; CAS 94-36-0; EINECS 202-327-6; catalyst for unsat. polyester and vinyl ester resins; faster gel to peak and lower exotherm than MEK peroxide systems; polymerization initiator for vinyl monomers, e.g., styrene, acrylate; wh. liq.; sp.gr. 1.1; visc. 750-1000 cps; 40% benzoyl peroxide, 12% water, 2.64% act. oxygen; *Storage:* store below 27 C away from heat sources or direct sunlight; avoid contact with promoters, accelerators, strong reducing agents.

Benzoflex® 2-45. [Velsicol] Diethylene glycol dibenzoate; CAS 120-55-8; EINECS 204-407-6; plasticizer compatible with PVAc homopolymer and copolymer emulsions; produces adhesives with quick grab and set times; solvator and compatible with PVC; APHA 100 max., mild ester odor; sol. in aliphatic and aromatic hydrocarbons; sol. < 0.01% in water; m.w. 314.3; f.p. 16 and 28 C; b.p. 240 C; sp.gr. 1.178; dens. 9.8 lb/gal; flash pt. (COC) 199 C; ref. index 1.5424; 98.0% min. assay.

Benzoflex® 9-88. [Velsicol] Dipropylene glycol dibenzoate; high solvating monomeric plasticizer for elastomers, PVAc emulsion adhesives, plastisols, caulks, vinyl floor coverings, coating applics.; enhances film formation and surf. wetting in PVAc homopolymer emulsion adhesives; APHA 100 max. clear oily liq., mild ester odor; sol. in aliphatic and aromatic hydrocarbons; sol. < 0.01% in water; m.w. 342.3; sp.gr. 1.120; dens. 9.346 lb/gal; f.p. -40 C; b.p. 232 C; pour pt. -19 C; flash pt. (COC) 199 C; ref. index 1.5282; 98% min. assay.

Benzoflex® 9-88 SG. [Velsicol] Dipropylene glycol dibenzoate; plasticizer for cast urethane applics., graphic art printing rolls; APHA 150 max. clear oily liq.; mild ester odor; sol. in aliphatic and aromatic hydrocarbons; sol. < 0.01% in water; m.w. 342.3; f.p. -40 C; b.p. 232 C; sp.gr. 1.120; dens. 9.346 lb/gal; pour pt. -19 C; flash pt. (COC) 199 C; ref. index 1.5282; 99.0% min. assay.

Benzoflex® 50. [Velsicol] Diethylene glycol dibenzoate, dipropylene glycol dibenzoate (1:1); monomeric plasticizer used in PVAc adhesives, acrylic latex caulk formulations, plastisols, and dry-blended vinyl formulations; APHA 100 max., mild ester odor; sol. in aliphatic and aromatic hydrocarbons; sol. < 0.01% in water; m.w. 328.3; sp.gr. 1.154; dens. 9.6 lb/gal; f.p. < 0 C; b.p. 240 C; pour pt. -21 C; flash pt. (COC) 199 C min.; ref. index 1.535; 98.0% min. assay.

Benzoflex® 131. [Velsicol] Isodecyl benzoate; plasticizer for plastisols, adhesives, sealants, caulks.

Benzoflex® 284. [Velsicol] Propylene glycol dibenzoate; solvating plasticizer for PVC applic., stain-resist. flooring, latex caulk formulations, PVAc emulsion adhesives, and castable PU; used in latex paints as coalescing agent; APHA 100 max., mild ester odor; sol. in aliphatic and aromatic hydrocarbons; sol. 0.01% in water; m.w. 284.3; sp.gr. 1.146; dens. 9.55 lb/gal; f.p. -3 C; b.p. 233 C; pour pt. -16 C; flash pt. (COC) 199 C min.; ref. index 1.544; 98.0% min. assay.

Benzoflex® 400. [Velsicol] Polypropylene glycol dibenzoate; plasticizer for PU and polysulfide sealants, acrylic coatings, caulks, PVC, adhesives.

Benzoflex® P-200. [Velsicol] PEG 200 dibenzoate; plasticizer for PVAc adhesive formulations, phenol-formaldehyde resins, alkyd-modified phenol-formaldehyde varnishes; limited compat. with PVC; APHA 100 max.; m.w. 408.2; dens. 9.7 lb/gal; f.p. -30 C; b.p. 217 C; flash pt. (TOC) 478 F; ref. index 1.5252; 98.0% min. assay.

Benzoflex® S-312. [Velsicol] Neopentyl glycol dibenzoate; CAS 4196-89-8; process aid, modifier, plasticizer for thermoplastics, hot-melt adhesives, coatings.

Benzoflex® S-404. [Velsicol] Glyceryl tribenzoate; CAS 614-33-5; process aid, modifier for thermoplastics, hot-melt adhesives.

Benzoflex® S-552. [Velsicol] Pentaerythritol tetrabenzoate; CAS 4196-86-5; plasticizer/extender for coatings, modifier for hot-melt adhesives, aq. adhesives, delayed tack adhesives, process aid for thermoplastics.

Benzoic Acid K. [Bayer AG] Gloss and flow promoter for use in oil and synthetic resin-based topcoats; hardener for rubber soles, heels, floor coverings.

Benzyl Tuads®. [R.T. Vanderbilt] Tetrabenzylthiuram disulfide; CAS 10591-85-2; accelerator for rubber; lt. yel. powd.; negligible sol. in water; dens. 1.31 mg/m^3; vapor pressure 0.82-0.90 Pa; m.p. 127-135 C; *Toxicology:* LD50 (oral, rat) > 5000 mg/kg; nonirritating to skin, eyes; nonsensitizing to skin; avoid inh. of dust; *Precaution:* incompat. with oxidizing agents; hazardous decomp. prods.: oxides of sulfur, nitrogen, and carbon; *Storage:* good storage stability when stored in dry area, in closed containers, away from heat sources.

Benzyl Tuads® Solid. [R.T. Vanderbilt] Tetrabenzylthiuram disulfide; CAS 10591-85-2; accelerator for rubber; 68% act. on inert filler.

Benzyl Tuex®. [Uniroyal] N,N,N′,N′-Tetrabenzylthiuram disulfide; CAS 10591-85-2; nondiscoloring, nonstaining primary or sec. accelerator for natural and syn. rubbers; for tire compds., wire insulation, mech. goods, sponge, footwear, sheeting, hose; activator for sulfenamide and thiazole accelerators; sulfur donor minimizing reversion in natural rubber; off-wh. to pale yel. powd., char. odor; sol. in toluene, chloroform, methylene chloride; sl. sol. in acetone, benzene; insol. in water; sp.gr. 1.31; m.p. 127 C; *Toxicology:* LD50 (oral, rat) > 5 g/kg; pract. nontoxic; minimally irritating to eyes, nonirritating to skin; *Precaution:* incompat. with strong oxidizers; thermal decomp. prods.: sulfur, alkylamines, organo-sulfur compds.; combustion decomp. prods.: sulfuric acid, oxides of carbon, nitrogen, sulfur; *Storage:* good storage stability when stored in dry area, in closed containers, away from heat sources.

Berol 02. [Berol Nobel AB] Nonylphenol ethoxylate; CAS 68412-54-4; nonionic; emulsifier; solv. cleaner; emulsion polymerization; biodeg.; Hazen < 200 clear liq.; sol. in ethanol, xylene, trichloroethylene, lt. fuel oil; disp. in water, paraffin oil, wh. spirit; sp.gr. 1.04; visc. 350 cps; HLB 10.9; cloud pt. 60-70 C; flash pt. > 100 C; pour pt. < 0 C; surf. tens. 31 dynes/cm; pH 6-7 (1% aq.); surf. tens. 31 mN/m (0.1%); Draves wetting 1.6 g/l (25 s); Ross-Miles foam 20 mm (initial, 0.05%, 50 C); 100% conc.

Berol 28. [Berol Nobel AB] PEG-7 oleamine; CAS 26635-93-8; emulsifier, dispersant, wetting agent, antistat, anticorrosive for agric., leather, textiles, metalworking and plastic industries; Gardner ≤ 12 liq.; sol. @ 5% in alcohols, low aromatic solvs., propylene glycol, wh. spirit, xylene, water; sp.gr. 0.980 g/cc; visc. 160 mNs/m; HLB 10.6; pour pt. -15 C; cloud pt. 42 C (1% in 10% NaCl); pH 10 (1% aq.); surf. tens. 36.5 mN/m (0.1%); Draves wetting 10 g/l (25 s); Ross-Miles foam 50 mm (initial, 0.05%, 50 C); 100% conc.; *Precaution:* corrosive.

Berol 108. [Berol Nobel AB] PEG-40 castor oil; CAS 61791-12-6; nonionic; surfactant, emulsifier for chemical industry; as softener, rewetting agent, pigment dispersant, dye assistant, leveling agent for paints, textiles, leather; lubricant additive and emulsifier in lubricants for plastics, metals, textiles;

Gardner ≤ 5 cloudy visc. liq.; sol. @ 5% in ethanol, IPA, water, xylene; disp. in propylene glycol; sp.gr. 1.06; visc. 700 mPa•s; HLB 13.3; sapon. no. 60-64; pour pt. 20 C; cloud pt. 73-77 C (5 g in 25 ml 25% butyl diglycol); clear pt. 34 C; flash pt. > 100 C; pH 5-7 (1% aq.); surf. tens. 45 mN/m (0.1%); Draves wetting > 10 g/l (25 s); Ross-Miles foam 70 mm (initial, 0.05%, 50 C); 100% act.

Berol 190. [Berol Nobel AB] PEG-75 castor oil; CAS 61791-12-6; nonionic; surfactant, emulsifier for chemical industry; as softener, rewetting agent, pigment dipsersant, dye assistant, leveling agent for paint, textile, leather; lubricant additive/emulsifier in lubricants for plastics, metals, textiles; Gardner ≤ 5 wax; sol. @ 5% in ethanol, water, xylene; partly sol. in IPA; sp.gr. 1.06; visc. 260 mPa•s; HLB 15.7; sapon. no. 37.5-42.5; pour pt. 37 C; cloud pt. 65 C (1% in 10% NaCl); flash pt. > 150 C; pH 507 (1% aq.); surf. tens. 45 mN/m (0.1%); Draves wetting > 10 g/l (25 s); Ross-Miles foam 40 mm (initial, 0.05%, 50 C); 100% act.

Berol 191. [Berol Nobel AB] PEG-200 castor oil; CAS 61791-12-6; nonionic; surfactant, emulsifier for chemical industry; as softener, rewetting agent, pigment dipsersant, dye assistant, leveling agent for paint, textile, leather; lubricant additive/emulsifier in lubricants for plastics, metals, textiles; Gardner ≤ 5 wax; sol. @ 5% in ethanol, water; sp.gr. 1.08; visc. 588 mPa•s; HLB 18.1; sapon. no. 16.2-18.4; pour pt. 45 C; cloud pt. 65 C (1% in 10% NaCl); flash pt. > 150 C; pH 5-7 (1% aq.); surf. tens. 46 mN/m (0.1%); Draves wetting > 10 g/l (25 s); Ross-Miles foam 45 mm (initial, 0.05%, 50 C); 100% act.

Berol 198. [Berol Nobel AB] PEG-160 castor oil; CAS 61791-12-6; nonionic; surfactant, emulsifier for chemical industry; as softener, rewetting agent, pigment dispersant, dye assistant, leveling agent for paint, textile, leather; lubricant additive/emulsifier in lubricants for plastics, metals, and textiles; Gardner ≤ 5 wax; sol. @ 5% in water, ethanol; disp. in propylene glycol; sp.gr. 1.08; visc. 510 mPa•s; HLB 17.7;sapon. no. 19-23; pour pt. 43 C; cloud pt. 61 C (1% in 10% NaCl); flash pt. > 150 C; pH 6-8 (1% aq.); surf. tens. 46 mN/m (0.1%); Draves wetting > 10 g/l (25 s); Ross-Miles foam 45 mm (initial, 0.05%, 50 C); 100% act.

Berol 199. [Berol Nobel AB] PEG-32 castor oil; CAS 61791-12-6; nonionic; surfactant, emulsifier for chemical industry; as softener, rewetting agent, pigment dipsersant, dye assistant, leveling agent for paint, textile, leather; lubricant additive/emulsifier in lubricants for plastics, metals, textiles; Gardner ≤ 5 liq.; sol. @ 5% in ethanol, IPA, water, xylene; disp. in propylene glycol; sp.gr. 1.05; visc. 700 mPa•s; HLB 12.3; sapon. no. 68-76; pour pt. 17 C; cloud pt. 70-74 C (5 g in 25 ml 25% butyl diglycol); flash pt. > 100 C; pH 7-8 (5% aq.); surf. tens. 45 mN/m (0.1%); Draves wetting > 10 g/l (25 s); Ross-Miles foam 40 mm (initial, 0.05%, 50 C); 100% act.

Berol 269. [Berol Nobel AB] Dinonylphenol ethoxylate; CAS 68891-21-4; nonionic; emulsifier for polymerization; detergent; biodeg.; Hazen < 500 opalescent liq.; sol. in ethanol, xylene, trichloroethylene, lt. fuel oil, disp. in water, wh. spirit; sp.gr. 1.02; visc. 550 cps; HLB 10.5; cloud pt. 60 C; flash pt. > 100 C; pour pt. < 0 C; pH 6-7 (1% aq.); 100% conc.; *Toxicology:* degreasing to skin.

Berol 277. [Berol Nobel AB] Nonylphenol ethoxylate; CAS 68412-54-4; nonionic; emulsifier for emulsion polymerization; biodeg.; Hazen < 200 clear liq.; sol. in water, ethanol; sp.gr. 1.09; visc. 1000 cps; HLB 17; cloud pt. 77 C (1% in 10/NaCl); flash pt. >100 C; pour pt. 12 C; pH 6-7 (1% aq.); surf. tens. 41 dynes/cm (0.1%); Draves wetting > 10 g/l (25 s); Ross-Miles foam 100 mm (initial, 0.05%, 50 C); 70% act.; *Toxicology:* degreasing to skin.

Berol 278. [Berol Nobel AB] Nonylphenol ethoxylate; CAS 68412-54-4; nonionic; emulsifier for emulsion polymerization; biodeg.; Hazen < 200 clear liq.; sol. in water, ethanol; sp.gr. 1.08; visc. 625 cps; HLB 15.2; cloud pt. 97-99 C (1% aq.); flash pt. > 100 C; pour pt. 0 C; pH 6-7 (1% aq.); surf. tens. 37 dynes/cm (0.1%); Draves wetting 10 g/l (25 s); Ross-Miles foam 130 mm (initial, 0.5%, 50 C); 80% act.

Berol 281. [Berol Nobel AB] Nonylphenol ethoxylate; CAS 68412-54-4; nonionic; emulsifier for emulsion polymerization; biodeg.; Hazen < 200 clear liq.; sol. in water, ethanol; sp.gr. 1.09; visc. 625 cps; HLB 16; cloud pt. 73 C (1% in 10/NaCl); flash pt. >100 C; pour pt. 7 C; pH 6-7 (1% aq.); surf. tens. 39 mN/m; Draves wetting > 10 g/l (25 s); Ross-Miles foam 130 mm (initial, 0.05%, 50 C); 80% conc.; *Toxicology:* LD50 (oral, rat) > 4 g/kg; degreasing to skin.

Berol 291. [Berol Nobel AB] Nonylphenol ethoxylate; CAS 68412-54-4; nonionic; emulsifier for emulsion polymerization; Hazen < 200 hard wax; sol. in water, ethanol, xylene, trichloroethylene; sp.gr. 1.07 (60 C); visc. 300 cps (60 C); HLB 18.2; cloud pt. 77 C (1% in 10/NaCl); flash pt. >100 C; pour pt. 51 C; pH 6-7 (1% aq.); surf. tens. 45 mN/m (0.1%); Draves wetting > 10 g/l (25 s); Ross-Miles foam 100 mm (initial, 0.05%, 50 C); 100% act.; *Toxicology:* LD50 (oral, rat) > 4 g/kg; non skin sensitizing.

Berol 292. [Berol Nobel AB] Nonylphenol ethoxylate; CAS 68412-54-4; nonionic; emulsifier for emulsion polymerization; biodeg.; Hazen < 200 soft wax; sol. in water, ethanol, xylene, trichloroethylene, disp. in wh. spirit; sp.gr. 1.07 (40 C); visc. 200 cps (40 C); HLB 16.0; cloud pt. 73 C (1% in 10/NaCl); flash pt. > 100 C; pour pt. 33 C; pH 6-7 (1% aq.); surf. tens. 39 dynes/cm; Draves wetting > 10 g/l (25 s); Ross-Miles foam 130 mm (initial, 0.05%, 50 C); 100% act.; *Toxicology:* degreasing to skin.

Berol 295. [Berol Nobel AB] Nonylphenol ethoxylate; CAS 68412-54-4; nonionic; emulsifier for emulsion polymerization; Hazen < 300 clear liq.; sol. in water, ethanol; sp.gr. 1.10; visc. 1500 cps; HLB 17.8; cloud pt. 77 C (1% in 10/NaCl); flash pt. > 100 C; pour pt. 14 C; pH 5-7 (1% aq.); surf. tens. 43 mN/m (0.1%); Draves wetting > 10 g/l (25 s); Ross-Miles foam 90 mm (initial, 0.05%, 50 C); 70% act.; *Toxicology:* LD50 (oral, rat) > 4 g/kg; non skin sensitizing.

Berol 302. [Berol Nobel AB] PEG-2 oleamine; CAS

13127-82-7; cationic; emulsifier, dispersant, wetting agent, antistat, anticorrosive for agric., leather, textiles, metalworking and plastic industries; Gardner ≤ 10 clear to cloudy liq.; sol. @ 5% in ethanol, IPA, low aromatic solvs., wh. spirit, xylene; disp. in propylene glycol; sp.gr. 0.904; visc. 140 mNs/m; HLB 4.9; pour pt. 14 C; clear pt. 24 C; pH 9 (1% aq.); 100% act.

Berol 303. [Berol Nobel AB] PEG-12 oleamine; CAS 26635-93-8; emulsifier, wetting agent, antistat, anticorrosive for agric., leather, textiles, metalworking and plastic industries; Gardner ≤ 15 clear liq.; sol. @ 5% in water, ethanol, IPA, propylene glycol, xylene; disp. in wh. spirit; sp.gr. 1.02; visc. 220 mPa•s; HLB 13.2; pour pt. -5 C; cloud pt. 75 C (1% in 10% NaCl); flash pt. > 150 C; pH 10 (1% aq.); surf. tens. 41 mN/m (0.1%); Draves wetting > 10 g/l (25 s); Ross-Miles foam 100 mm (initial, 0.05%, 50 C); 100% act.

Berol 307. [Berol Nobel AB] PEG-2 cocamine; CAS 61791-14-8; 61791-31-9; emulsifier, wetting agent, antistat, anticorrosive for agric., leather, textiles, metalworking and plastic industries; Gardner ≤ 6 clear liq.; sol. @ 5% in ethanol, IPA, low aromatic solvs., propylene glycol, wh. spirit, xylene; disp. in water; sp.gr. 0.910; visc. 150 mPa•s; HLB 5.9; cloud pt. 47 C (5 g in 25 ml 25% butyl diglycol); clear pt. 11 C; surf. tens. 30 mN/m (0.1%); Draves wetting 3 g/l (25 s); Ross-Miles foam 20 mm (initial, 0.05%, 50 C); 100% act.; *Precaution:* corrosive.

Berol 374. [Berol Nobel AB] EO/PO block polymer; CAS 9003-11-6; nonionic; low-foaming surfactant, emulsifier for emulsion polymerization esp. for latex paints; foam depressant/detergent for foodstuffs industry; emollient; wh. flakes; m.w. 2200; sol. in ethanol, xylene, trichloroethylene, wh. spirit; disp. water; sp.gr. 1.05; visc. 450 cps; cloud pt. 24-26 C (1% aq.); flash pt. > 100 C; pour pt. < -10 C; pH 5-7 (1% aq.); surf. tens. 40 dynes/cm (0.1%); Draves wetting 2 g/l (25 s); Ross-Miles foam 5 mm (initial, 0.05%, 50 C); 100% act.

Berol 381. [Berol Nobel AB] PEG-15 tallowamine; CAS 61791-26-2; emulsifier, wetting agent, antistat, anticorrosive for agric., leather, textiles, metalworking and plastic industries; Gardner ≤ 10 clear to cloudy liq.; sol. @ 5% in water, ethanol, IPA, propylene glycol, xylene; sp.gr. 1.03; visc. 600 mPa•s; HLB 14.2; pour pt. 5 C; cloud pt. 77-87 C (1% in 10% NaCl); clear pt. 11 C; flash pt. > 150 C; pH 9 (1% aq.); surf. tens. 40 mN/m (0.1%); Draves wetting > 10 g/l (25 s); Ross-Miles foam 55 mm (initial, 0.05%, 50 C); 100% act.

Berol 386. [Berol Nobel AB] PEG-20 tallowamine; CAS 61791-26-2; emulsifier, wetting agent, antistat, anticorrosive for agric., leather, textiles, metalworking and plastics industries; red/brn. liq.; sol. @ 5% in water, ethanol, IPA, propylene glycol, xylene; sp.gr. 1.04; visc. 470 mPa•s; HLB 15.3; pour pt. 18 C; cloud pt. 85 C (1% in 10% NaCl); clear pt. 36 C; pH 9 (1% aq.); surf. tens. 42 mN/m (0.1%); Draves wetting > 10 g/l (25 s); Ross-Miles foam 60 mm (initial, 0.05%, 50 C); 100% act.; *Precaution:* corrosive.

Berol 387. [Berol Nobel AB] PEG-40 tallowamine; CAS 61791-26-2; emulsifier, wetting agent, antistat, anticorrosive for agric., leather, textiles, metalworking and plastics industries; Gardner ≤ 15 wax; sol. @ 5% in water, ethanol, IPA, propylene glycol, xylene; sp.gr. 1.06; visc. 144 mPa•s; HLB 17.3; pour pt. 36 C; cloud pt. 81-89 C (1% in 10% NaCl); flash pt. > 150 C; pH 9 (1% aq.); surf. tens. 49 mN/m (0.1%); Draves wetting > 10 g/l (25 s); Ross-Miles foam 70 mm (initial, 0.05%, 50 C); 100% act.; *Precaution:* corrosive.

Berol 389. [Berol Nobel AB] PEG-10 tallowamine; CAS 61791-26-2; emulsifier, wetting agent, antistat, anticorrosive for agric., leather, textiles, metalworking and plastics industries; Gardner ≤ 12 clear to cloudy liq.; sol. @ 5% in water, ethanol, IPA, propylene glycol, xylene; sp.gr. 1.00; visc. 220 mPa•s; HLB 12.4; pour pt. < 2 C; cloud pt. 65-70 C (1% in 10% NaCl); flash pt. > 150 C; pH 9-11 (1% aq.); surf. tens. 40 mN/m (0.1%); Draves wetting > 10 g/l (25 s); Ross-Miles foam 60 mm (initial, 0.05%, 50 C); 100% act.; *Precaution:* corrosive.

Berol 391. [Berol Nobel AB] PEG-5 tallowamine; CAS 61791-26-2; emulsifier, wetting agent, antistat, anticorrosive for agric., leather, textiles, metalworking and plastics industries; yel./brn. liq.; sol. @ 5% in water, ethanol, IPA, low aromatic solvs., propylene glycol, wh. spirit, xylene; sp.gr. 0.96; visc. 160 mPa•s; HLB 9.0; pour pt. 8 C; cloud pt. 24-31 C (1% aq.); flash pt. >> 150 C; pH 9 (1% aq.); surf. tens. 34 mN/m (0.1%); Draves wetting 10 g/l (25 s); Ross-Miles foam 15 mm (initial, 0.05%, 50 C); 100% act.; *Precaution:* corrosive.

Berol 392. [Berol Nobel AB] PEG-15 tallowamine; CAS 61791-26-2; emulsifier, wetting agent, antistat, anticorrosive for agric., leather, textiles, metalworking and plastics industries; red/brn. liq.; sol. @ 5% in water, ethanol, IPA, propylene glycol, xylene; sp.gr. 1.03; visc. 270 mPa•s; HLB 14.2; pour pt. 8 C; cloud pt. 77-87 C (1% aq.); flash pt. >> 150 C; pH 9 (1% aq.); surf. tens. 41 mN/m (0.1%); Draves wetting > 10 g/l (25 s); Ross-Miles foam 45 mm (initial, 0.05%, 50 C); 100% act.; *Precaution:* corrosive.

Berol 397. [Berol Nobel AB] PEG-15 cocamine; CAS 61791-14-8; emulsifier, wetting agent, antistat, anticorrosive for agric., leather, textiles, metalworking and plastic industries; Gardner ≤ 12 clear to cloudy liq.; sol. @ 5% in water, ethanol, IPA, propylene glycol, xylene; sp.gr. 1.043; visc. 193 mPa•s; HLB 15.3; cloud pt. 85 C (1% in 10% NaCl); clear pt. 10 C; surf. tens. 40 mN/m (0.1%); Draves wetting > 10 g/l (25 s); Ross-Miles foam 110 mm (0.05%, 50 C); 100% act.; *Precaution:* corrosive.

Berol 452. [Berol Nobel AB] Sodium lauryl polyglycol ether sulfate; CAS 68891-38-3; foaming agent, degreaser for hair shampoos, foam baths, manual dishwash, general detergents; emulsifier for polymerization of vinyl chloride, vinyl acetate, acrylic acids; Hazen ≤ 200 clear liq.; sol. @ 5% in water, propylene glycol; disp. in ethanol, IPA; sp.gr. 1.040;

visc. 200 mPa•s; clear pt. < 12 C; flash pt. >> 100 C; pH 6.5-8.0 (3% aq.); surf. tens. 37 mN/m (0.1%); Draves wetting > 10 g/l (25 s); Ross-Miles foam 185 mm (initial, 0.05%, 50 C); 27-29% act.; *Precaution:* corrosive.

Berol 455. [Berol Nobel AB] PEG-3 tallow diamine; emulsifier, wetting agent, antistat, anticorrosive for agric., leather, textiles, metalworking and plastic industries; liq.; HLB 20.1; 100% conc.

Berol 456. [Berol Nobel AB] PEG-2 tallowamine; CAS 61791-44-4; emulsifier, wetting agent, antistat, anticorrosive for agric., leather, textiles, metalworking and plastic industries; Gardner ≤ 8 paste; sol. @ 5% in ethanol, IPA, low aromatic solvs., propylene glycol, wh. spirit, xylene; disp. in water; sp.gr. 0.90; visc. 90 mPa•s; HLB 4.9; pour pt. 24 C; clear pt. 27 C; pH 9 (1% aq.); surf. tens. 33 mN/m (0.1%); Draves wetting > 10 g/l (25 s); Ross-Miles foam 10 mm (initial, 0.5%, 50 C); 100% act.

Berol 457. [Berol Nobel AB] PEG-5 tallowamine; CAS 61791-26-2; emulsifier, wetting agent, antistat, anticorrosive for agric., leather, textiles, metalworking and plastic industries; yel./brn. liq.; sol. @ 5% in water, ethanol, IPA, low aromatic solvs., propylene glycol, wh. spirit, xylene; sp.gr. 0.95; visc. 100 mPa•s; HLB 9.0; pour pt. 6 C; cloud pt. 22 C (1% aq.); clear pt. 18 C; flash pt. >> 150 C; pH 10 (1% aq.); surf. tens. 34 mN/m (0.1%); Ross-Miles foam 20 mm (initial, 0.05%, 50 C); 100% act.; *Precaution:* corrosive.

Berol 458. [Berol Nobel AB] PEG-10 tallowamine; CAS 61791-26-2; emulsifier, wetting agent, antistat, anticorrosive for agric., leather, textiles, metalworking and plastic industries; Gardner ≤ 9 clear liq.; sol. @ 5% in water, ethanol, IPA, propylene glycol, wh. spirit, xylene; sp.gr. 1.03; visc. 190 mPa•s; HLB 12.4; pour pt. 3 C; cloud pt. 71 C (1% in 10% NaCl); clear pt. 7 C; flash pt. > 150 C; pH 10 (1% aq.); surf. tens. 39 mN/m (0.1%); Ross-Miles foam 80 mm (initial, 0.05%, 50 C); 100% act.

Be Square® 175. [Petrolite] Microcryst. wax; CAS 63231-60-7; EINECS 264-038-1; plastic wax offering high ductility, flexibility at very low temps.; provides protective barrier properties against moisture vapor and gases; uses incl. hot-melt laminating adhesives and coatings, in antisunchecking agents in rubber goods, elec. insulating agents, leather treating agents, water repellents for textiles, rustproof coatings, cosmetic ingreds., as plasticizer in crayons, dental compds., chewing gum base, and candles; incl. FDA §172.230, 172.615, 175.105, 175.300, 176.170, 176.180, 176.200, 177.1200, 178.3710, 179.45; amber wax; dens. 0.93 g/cc; visc. 13 cps (99 C); m.p. 83 C; flash pt. 293 C.

Be Square® 185. [Petrolite] Hard microcryst. wax consisting of n-paraffinic, branched paraffinic, and naphthenic hydrocarbons; CAS 63231-60-7; EINECS 264-038-1; wax used in hot-melt coatings and adhesives, cup and paper coatings, printing inks, plastic modification (as lubricant and processing aid), lacquers, paints, and varnishes, as binder in ceramics, for potting in elec./electronic; components, in investment casting, rubber and elastomers (plasticizer, antisunchecking, antiozonant), as emulsion wax size in papermaking, as fabric softener ingred., in cosmetic hand creams and lipsticks, chewing gum base; incl. FDA §172.230, 172.615, 175.105, 175.300, 176.170, 176.180, 176.200, 177.1200, 178.3710, 179.45; amber wax; very low sol. in org. solvs.; sp.gr. 0.93; visc. 15 cps (99 C); m.p. 90.5 C.

Be Square® 195. [Petrolite] Hard microcryst. wax consisting of n-paraffinic, branched paraffinic, and naphthenic hydrocarbons; CAS 63231-60-7; EINECS 264-038-1; wax used in adhesives, ceramics, chewing gum, cosmetics, elec., explosives, pkg., paints, polish, printing inks, plastics processing, rustproofing; FDA §172.230, 172.615, 175.105, 175.300, 176.170, 176.180, 176.200, 177.1200, 178.3710, 179.45, etc.; wh., amber wax; very low sol. in org. solvs.; sp.gr. 0.93; visc. 15.5 cps (99 C); m.p. 93 C.

Beta W 7. [Wacker-Chemie GmbH] β-Cyclodextrin; CAS 7585-39-9; EINECS 231-493-2; complex hosting guest molecules; increases the sol. and bioavailability of other substances; masks flavor, odor, or coloration; stabilizes against light, oxidation, heat, and hydrolysis; turns liqs. or volatiles into stable solid powds.; for use in pharmaceuticals, cosmetics, toiletries, foods, tobacco, pesticides, textiles, paints, plastics, synthesis, polymers; wh. cryst. powd.; sol. 1.85 g/100 ml in water; m.w. 1135.

Beta W7 HP 0.9. [Wacker-Chemie GmbH] Hydroxypropyl-β-cyclodextrin; CAS 94035-02-6; complex hosting guest molecules; increases the sol. and bioavailability of other substances; masks flavor, odor, or coloration; stabilizes against light, oxidation, heat, and hydrolysis; turns liqs. or volatiles into stable solid powds.; for use in pharmaceuticals, cosmetics, toiletries, foods, tobacco, pesticides, textiles, paints, plastics, synthesis, polymers; m.w. 1507.

Beta W7 M1.8. [Wacker-Chemie GmbH] Methyl-β-cyclodextrin; complex hosting guest molecules; increases the sol. and bioavailability of other substances; masks flavor, odor, or coloration; stabilizes against light, oxidation, heat, and hydrolysis; turns liqs. or volatiles into stable solid powds.; for use in pharmaceuticals, cosmetics, toiletries, foods, tobacco, pesticides, textiles, paints, plastics, synthesis, polymers; m.w. 1311.

Beta W7 P. [Wacker-Chemie GmbH] β-Cyclodextrin polymer; complex hosting guest molecules; increases the sol. and bioavailability of other substances; masks flavor, odor, or coloration; stabilizes against light, oxidation, heat, and hydrolysis; turns liqs. or volatiles into stable solid powds.; for use in pharmaceuticals, cosmetics, toiletries, foods, tobacco, pesticides, textiles, paints, plastics, synthesis, polymers.

Betaprene® 253. [Arizona] Hydrocarbon resin; resin for paste ink applics.; tackifier for adhesives; Gardner 11 color; soften. pt. (R&B) 100 C.

Betaprene® 255. [Arizona] Hydrocarbon-based resin;

tackifier for adhesives; Gardner 12+ color; soften. pt. (R&B) 132 C.

Betaprene® BC115. [Arizona] Hydrocarbon-based resin; heat-reactive tackifier for mastics, construction adhesives (in noncritical areas); FDA 21CFR §175.105, 175.300, 176.170, 176.180, 177.1210; Gardner 10+ color; soften. pt. (R&B) 117 C.

Betaprene® BR100. [Arizona] Polyolefinic hydrocarbon resin; heat-reactive tackifying resin for mastics, construction adhesives (in noncritical areas); FDA 21CFR §175.105, 175.300, 176.170, 176.180, 177.1210; Gardner 11 color; soften. pt. (R&B) 100 C.

Betaprene® BR100/60MS. [Arizona] Hydrocarbon-based resin in min. spirits; tackifier for adhesives; FDA 21CFR §175.105, 175.300, 176.170, 176.180, 177.1210; Gardner 10+ color; 60% solids.

Betaprene® BR100/70MS. [Arizona] Hydrocarbon-based resin in min. spirits; tackifier for adhesives; FDA 21CFR §175.105, 175.300, 176.170, 176.180, 177.1210; Gardner 12 color; 70% solids.

Betaprene® BR105. [Arizona] Hydrocarbon-based resin; tackifier for adhesives; FDA 21CFR §175.105, 175.300, 176.170, 176.180, 177.1210; Gardner 11+ color; soften. pt. (R&B) 103 C.

Betaprene® BR110. [Arizona] Hydrocarbon-based resin; tackifier for adhesives; FDA 21CFR §175.105, 175.300, 176.170, 176.180, 177.1210; Gardner 11+ color; soften. pt. (R&B) 111 C.

Beta-Tac® 160. [Arizona] Polyolefinic hydrocarbon resin; tackifier used for inks, mastics, construction adhesives; FDA 21CFR §175.105, 175.300, 176.170, 176.180, 177.1210; Gardner 14 color; visc. Z5-Z6 (50% NV in Magie 535); soften. pt. (R&B) 160 C; acid no. 1.

Beycostat 656 A. [Ceca SA] PEG-8 alkylphenol phosphate ester; anionic/nonionic; emulsifier for emulsion polymerization; liq.; 100% conc.

Beycostat LP 9 A. [Ceca SA] PEG-9 C12 fatty alcohol phosphate ester; anionic/nonionic; emulsifier for emulsion polymerization; paste; sol. hot in water; acid no. 95; pH 2 (5%); surf. tens. 34 dynes/cm; 100% conc.

Beycostat LP 12 A. [Ceca SA] Phosphate ester; anionic; surfactant for acid or basic detergent formulation, emulsion polymerization, min. flotation; solid; sol. in water; acid no. 93; pH 2 (5%); 100% conc.

Beycostat NA. [Ceca SA] PEG-9 alkylphenol phosphate ester; anionic/nonionic; emulsifier for emulsion polymerization; liq.; sol. in water; acid no. 86; pH 2 (5%); surf. tens. 32 dynes/cm; 100% conc.

Beycostat NE. [Ceca SA] Alkyl ether sulfate and solv.; anionic; antistatic for PVC, wetting agent; liq.; pH 7.5.

BHMT Amine. [DuPont Nylon] 25-75% Bis-hexamethylene triamine with 5-70% oligomeric amines; CAS 68411-90-5; asphalt antistripping agents, cationic emulsifiers, ore flotation collectors, chelating agent, corrosion/scale inhibitor, curing agent for epoxy resins and urethanes, flocculating agents, wet str. paper resins; liq.; sp.gr. 0.93-0.97; f.p. 33 C; b.p. 249 C (100 mm); amine no. 525-700; flash pt. (OC) 121 C; *Precaution:* corrosive liq.; partial decomp. @ 200 C evolves ammonia.

BIBBS. [Akrochem] N,N-diisopropyl benzothiazole-2-sulfenamide; delayed-action sulfenamide accelerator; off-wh. to lt. tan powd.; sol. in benzene, acetone, ethanol, methanol, and CCl_4; sp.gr. 1.21; m.p. 57 C.

BiCAT™ 8. [Shepherd] Metal carboxylate; amine-free polyurethane catalyst for water-blown flexible foam; synergistic with stannous octoate; provides improved props., better long-term color stability of foam, elimination of amine odor; Gardner 6 max. color; sp.gr. 1.10-1.16; 80% min. solids; 7.8-8.2% Bi, 7.8-8.2% Zn.

BiCAT™ H. [Shepherd] Metal carboxylate; amine-free polyurethane catalyst for water-blown flexible foam; synergistic with stannous octoate; provides improved props., better long-term color stability of foam, elimination of amine odor; Gardner 10 max. color; sp.gr. 1.30-1.36; 85% min. solids; 27.7-28.3% Bi.

BiCAT™ V. [Shepherd] Metal carboxylate; amine-free polyurethane catalyst for water-blown flexible foam; synergistic with stannous octoate; provides improved props., better long-term color stability of foam, elimination of amine odor; Gardner 6 max. color; sp.gr. 1.15-1.21; visc. (G-H) Z max.; 80% min. solids; 19.7-20.3% Bi.

BiCAT™ Z. [Shepherd] Metal carboxylate; catalyst for water-blown flexible foam in absence of amines; Gardner 5 max. color; sp.gr. 1.10-1.16; visc. (G-H) Z10 max.; 95% min. solids; 18.7-19.3% Zn.

Bilt-Plates® 156. [R.T. Vanderbilt] Kaolin; CAS 1332-58-7; EINECS 296-473-8; grit-free reinforcing min. filler and extender for elastomers; provides max. reinforcement; wh. to cream fine powd.; 0.2 μm median particle size; 99.95% thru 325 mesh; dens. 2.62 mg/m^3; bulk dens. 43 lb/ft^3 (compacted); oil absorp. 41; brightness 76.

bioMeT 14. [Elf Atochem N. Am.] 10% Diphenylstibine 2-ethyl hexoate sol'n. in dioctylphthalate; antimicrobial compd. for protection of PVC systems; pale yel. clear liq., odorless; sp.gr. 1.0346; dens. 8.634 lb/gal; visc. 66.7 cs; f.p. -37 C; acid no. 13.4; flash pt. 200 C; ref. index 1.4926; 10% act.

bioMeT TBTO. [Elf Atochem N. Am.] Bis (tributyltin) oxide; CAS 56-35-9; antimicrobial for paper mill slime control, industrial cooling water, sec. oil recovery, hospital use, textiles, plastics, urethane foam, paper preservation; antifoulant for ship-bottom paints; wood preservative; straw-colored clear liq.; sol. in org. solvs.; relatively insol. in water; m.w. 596; sp.gr. 1.17; dens. 9.7 lb/gal; f.p. < -45 C; b.p. 180 C (2 mm); flash pt. (TCC) > 100 C; 96% assay, 38.8% tin; *Toxicology:* extremely irritating to eyes and skin; can cause eye damage; prolonged/repeated skin contact can cause chemical burns.

Bio-Pruf®. [Morton Int'l./Plastics Additives] Antimicrobials for plastic prods. for healthcare, sanitary maintenance, construction, transportation and agric. markets; protects against cosmetic/hygienic and structural degradation.

Bismate® Powd. [R.T. Vanderbilt] Bismuth dimethyldithiocarbamate; CAS 21260-46-8; ultra accelerator for NR, IR, BR, and SBR; for high temp., high speed vulcanization; lemon yel. powd., 99.9% thru 100 mesh; sol. in chloroform; practically insol. in water; m.w. 569.66; dens. 2.04 ± 0.03 mg/m^3; m.p. > 230 C with dec.; 35-38% Bi; *Toxicology:* toxicity (oral, rat) > 3000 mg/kg; dust may cause mech. irritation to eyes, skin, upper respiratory tract; OSHA TWA 5 mg/m^3 (respirable dust), 15 mg/m^3 (total dust) recommended; *Precaution:* incompat. with acids, oxidizing agents; hazardous decomp. prods.: bismuth oxide, carbon and sulfur dioxide, carbon disulfide.

Bismate® Rodform®. [R.T. Vanderbilt] Bismuth dimethyldithiocarbamate; CAS 21260-46-8; ultra accelerator for NR, IR, BR, and SBR; for high temp., high speed vulcanization; rods; sol. in chloroform; practically insol. in water; m.w. 569.66; dens. 2.04 ± 0.03 mg/m^3; m.p. > 228 C with dec.; 32-34% Bi (rods); *Toxicology:* toxicity (oral, rat) > 3000 mg/kg; dust may cause mech. irritation to eyes, skin, upper respiratory tract; OSHA TWA 5 mg/m^3 (respirable dust), 15 mg/m^3 (total dust) recommended; *Precaution:* incompat. with acids, oxidizing agents; hazardous decomp. prods.: bismuth oxide, carbon and sulfur dioxide, carbon disulfide.

Bismet. [Akrochem] Bismuth dimethyldithiocarbamate; CAS 21260-46-8; accelerator for SBR, NR, IR, and BR compds. that are high-temp. cured; nonstaining; yel. powd.; sp.gr. 1.9; m.p. > 230 F.

BL 3. [Releasomers] Semipermanent release agent for applic. to flexible molds and metal molds; effective for halogenated or peroxide-cured elastomers; can be applied to hot or cold mold surfs.; solv. sol'n.; *Toxicology:* avoid prolonged/repeated skin contact, breathing of vapors; *Precaution:* keep away from heat, open flame, sparks; *Storage:* keep containers closed when not in use.

Black Out® Black. [R.T. Vanderbilt] Toluene (77.5%), 2-butanone (10.9%), carbon black (2.2%); decorative and protective finishing agent for rubber; deposited films are resistant to ozone, acids, alkalies, and paraffinic hydrocarbons; avail. clear, blk. and wh. liq., toluene/ketone-type odor; mod. sol. in water; dens. 0.88, 0.88, and 0.91 ± 0.03 mg/m^3 resp.; flash pt. (TCC) 5 C (all); 89% volatiles; 11-12, 11-12, and 15-16% total solids resp.; *Toxicology:* harmful if inhaled; skin and eye irritant; toluene: LD50 (oral, rat) 5000 mg/kg, (dermal, rabbit) 14,000 g/kg); toxic by inh., severe eye irritant, reproductive effects; OSHA TWA 100 ppm; 2-butanone: OSHA TWA 200 ppm; carbon blk.: OSHA TWA 3.5 mg/m^3; *Precaution:* flamm.; incompat. with strong oxidizers; reacts violently with chlorine, permanganate, dicrhomate; hazardous decomp. prods.: carbon oxides, chlorhydrocarbons; *Storage:* keep container tightly closed to prevent evaporation; keep away from heat, flame.

Black Pearls® 120. [Cabot/Special Blacks Div] Carbon black; CAS 1333-86-4; EINECS 215-609-9; blue toned colorant, uv stabilizer for plastics, inks, coatings, cement; low oil absorp., easy dispersion; exc. for coloring or tinting vs. lampblack; pellets; 75 nm particle size; dens. 33 lb/ft^3; surf. area 25 m^2/g; pH 6-8; 0.5% volatile.

Black Pearls® 130. [Cabot/Special Blacks Div] Carbon black; CAS 1333-86-4; EINECS 215-609-9; colorant for masterbatches for many end uses in plastics; bluest tone lamp plack replacement for coatings, gravure inks; pellets; 75 nm particle size; dens. 31 lb/ft^3; surf. area 25 m^2/g; pH 6-8; 0.5% volatile.

Black Pearls® 160. [Cabot/Special Blacks Div] Carbon black; CAS 1333-86-4; EINECS 215-609-9; colorant for masterbatches for many end uses in plastics; bluest tone lamp plack replacement for coatings, gravure inks; pellets; 50 nm particle size; dens. 27 lb/ft^3; surf. area 35 m^2/g; pH 6-8; 1.0% volatile.

Black Pearls® 170. [Cabot/Special Blacks Div] Carbon black; CAS 1333-86-4; EINECS 215-609-9; blue toned black for lampblack replacement; colorant for plastics, coatings, matte finish inks; pellets; 50 nm particle size; dens. 23 lb/ft^3; surf. area 35 m^2/g; pH 6-8; 1.0% volatile.

Black Pearls® 280. [Cabot/Special Blacks Div] Carbon black; CAS 1333-86-4; EINECS 215-609-9; blue toned black for lampblack replacement; colorant, uv stabilizer for plastics, coatings, matte finish inks; pellets; 41 nm particle size; dens. 22 lb/ft^3; surf. area 42 m^2/g; pH 6-8; 1.0% volatile.

Black Pearls® 450. [Cabot/Special Blacks Div] Carbon black; CAS 1333-86-4; EINECS 215-609-9; colorant providing sl. less jetness and str. but easier dispersion than Regal 330; for inks, coatings, plastics, paper; pellets; 27 nm particle size; dens. 29 lb/ft^3; surf. area 80 m^2/g; pH 6-8; 1.0% volatile.

Black Pearls® 460. [Cabot/Special Blacks Div] Carbon black; CAS 1333-86-4; EINECS 215-609-9; blue toned high structure colorant offering good tint str.; for inks, plastics, coatings, fibers; pellets; 25 nm particle size; dens. 23 lb/ft^3; surf. area 84 m^2/g; pH 6-8; 1.0% volatile.

Black Pearls® 470. [Cabot/Special Blacks Div] Carbon black; CAS 1333-86-4; EINECS 215-609-9; blue toned high structure colorant offering good tint str.; for inks, plastics, coatings, fibers; pellets; 25 nm particle size; dens. 23 lb/ft^3; surf. area 85 m^2/g; pH 6-8; 1.0% volatile.

Black Pearls® 480. [Cabot/Special Blacks Div] Carbon black; CAS 1333-86-4; EINECS 215-609-9; blue toned high structure colorant offering good tint str.; for inks, plastics, coatings, fibers; pellets; 25 nm particle size; dens. 21 lb/ft^3; surf. area 85 m^2/g; pH 6-8; 1.0% volatile.

Black Pearls® 490. [Cabot/Special Blacks Div] Carbon black; CAS 1333-86-4; EINECS 215-609-9; blue toned high structure colorant offering good tint str.; for inks, plastics, coatings, fibers; pellets; 25 nm particle size; dens. 21 lb/ft^3; surf. area 87 m^2/g; pH 6-8; 1.0% volatile.

Black Pearls® 520. [Cabot/Special Blacks Div] Carbon black; CAS 1333-86-4; EINECS 215-609-9; colorant providing high tint str. and good jetness for inks and plastics; pellets; 24 nm particle size; dens. 25 lb/ft^3; surf. area 110 m^2/g; pH 6-8; 1.0% volatile.

Black Pearls® 570. [Cabot/Special Blacks Div] Carbon black; CAS 1333-86-4; EINECS 215-609-9; colorant providing high tint str. and good dispersibility; for inks, plastics, fibers; pellets; 24 nm particle size; dens. 22 lb/ft^3; surf. area 110 m^2/g; pH 6-8; 1.0% volatile.

Black Pearls® 700. [Cabot/Special Blacks Div] Carbon black; CAS 1333-86-4; EINECS 215-609-9; med. jetness colorant, uv stabilizer for plastics and coatings; exc. elec. conductivity; pellets; 18 nm particle size; dens. 19 lb/ft^3; surf. area 200 m^2/g; pH 6-8; 1.5% volatiles.

Black Pearls® 800. [Cabot/Special Blacks Div] Carbon black; CAS 1333-86-4; EINECS 215-609-9; med. jetness colorant, uv stabilizer for plastics and coatings; pellets; 17 nm particle size; dens. 28 lb/ft^3; surf. area 210 m^2/g; pH 6-8; 1.5% volatiles.

Black Pearls® 880. [Cabot/Special Blacks Div] Carbon black; CAS 1333-86-4; EINECS 215-609-9; med. jetness colorant, uv stabilizer for with exc. dispersion chars. for plastics and coatings; pellets; 16 nm particle size; dens. 21 lb/ft^3; surf. area 220 m^2/g; pH 6-8; 1.5% volatiles.

Black Pearls® 900. [Cabot/Special Blacks Div] Carbon black; CAS 1333-86-4; EINECS 215-609-9; med. high jetness colorant for plastic compds., lacquers, and enamels; pellets; 15 nm particle size; dens. 29 lb/ft^3; surf. area 230 m^2/g; pH 6-8; 2.0% volatiles.

Black Pearls® 1000. [Cabot/Special Blacks Div] Carbon black; CAS 1333-86-4; EINECS 215-609-9; med. high jetness colorant with good stability for coatings and resistive plastics; pellets; 16 nm particle size; dens. 24 lb/ft^3; surf. area 343 m^2/g; pH 6-8; 9.5% volatiles.

Black Pearls® 1100. [Cabot/Special Blacks Div] Carbon black; CAS 1333-86-4; EINECS 215-609-9; colorant; high jetness furnace black for plastics and coatings; pellets; 14 nm particle size; dens. 29 lb/ft^3; surf. area 240 m^2/g; pH 6-8; 2.0% volatiles.

Black Pearls® 3700. [Cabot/Special Blacks Div] Carbon black; CAS 1333-86-4; EINECS 215-609-9; extra clean, extra smooth furnace carbon blk. for conductor and insultion shielding of high and medium voltage cables; provides very low compd. moisture absorp. props.; minimizes risk of premature cure in silane-treated systems; improves rheology; superior elec. performance in highly loaded polymer systems; 0 ppm +325 mesh residue; 2.0% volatiles.

Black Pearls® L. [Cabot/Special Blacks Div] Carbon black; CAS 1333-86-4; EINECS 215-609-9; colorant, uv stabilizer for elec. resistivity in plastics, optimum flow in offset inks, carbon paper, and ribbons; good for dry toners; pellets; 24 nm particle size; dens. 32 lb/ft^3; surf. area 138 m^2/g; pH 6-8; 5.0% volatile.

Blanc fixe F. [Sachtleben Chemie GmbH] Syn. precipitated barium sulfate; CAS 7727-43-7; EINECS 231-784-4; inert filler resistant to weathering; improves hardness and stiffness of plastics and can render plastics opaque to x-rays; wh. powd.; 1 μm particle size; dens. 4.4 g/ml; pH 9.

Blanc fixe micro®. [Sachtleben Chemie GmbH] Syn. precipitated barium sulfate; CAS 7727-43-7; EINECS 231-784-4; micronized inert filler resistant to weathering; improves hardness and stiffness of plastics and can render plastics opaque to x-rays; wh. powd.; 0.7 μm particle size; dens. 4.4 g/ml; pH 9.

Blanc fixe N. [Sachtleben Chemie GmbH] Syn. precipitated barium sulfate; CAS 7727-43-7; EINECS 231-784-4; inert filler resistant to weathering; improves hardness and stiffness of plastics and can render plastics opaque to x-rays; wh. powd.; 3 μm particle size; dens. 4.4 g/ml; pH 9.

Blendex® 131. [GE Specialty] ABS; CAS 9003-56-9; modifier resin for calendered films requiring good thermoforming properties, semiflexible PVC film applics.; base resin in calendered ABS sheet; wh. powd.; sp.gr. 1.01; bulk dens. 17.5 lb/ft^3; tens. str. 4200 psi; Izod impact str. 6.9 ft lb/in. (1/4 in.); hardness Rockwell R81; distort. temp. 215 F (264 psi).

Blendex® 336. [GE Specialty] ABS; CAS 9003-56-9; impact modifier for clear and opaque PVC, incl. pipe and conduit, calendered or extruded film and sheet, inj. molded prods.; wh. powd.; sp.gr. 0.98; bulk dens. 14.0 lb/ft^3; tens. str. 600 psi; Izod impact str. 5.5 ft lb/in. (1/4 in.); hardness Shore D42; distort. temp. 114 F (264 psi).

Blendex® 338. [GE Specialty] ABS; CAS 9003-56-9; impact modifier providing toughness to opaque rigid sheet, profile, and inj. molding PVC applics.; also for PC and polyesters; sp.gr. 0.97; bulk dens. 17.0 lb/ft^3; tens. str. 530 psi; impact str. (Izod) 5.2 ft lb/in. (1/4 in.); hardness Shore D36; distort. temp. 127 F (264 psi).

Blendex® 340. [GE Specialty] CAS 26741-53-7; EINECS 247-952-5; high efficiency impact modifier for PVC siding and profile substrate applics.; superior high impact performance, exc. low temp. brittleness resist.; powd.; sp.gr. 0.97; bulk dens. 17 lb/ft^3; tens. str. 530 psi; tens. mod. 0.22 x 10^5 psi; impact str. (Izod) 5.2 ft/in. (1/4 in.); hardness Shore D36.

Blendex® 424. [GE Specialty] Impact modifier for high transparency, rigid PVC applics, incl. extruded and calendered sheet requiring clarity and toughness with good vacuum formability; particulate 99% < 8 mesh; sp.gr. 0.97; bulk dens. 15.8 lb/ft^3; ref. index 1.539; tens. str. 940 psi; hardness Shore D57.

Blendex® 467. [GE Specialty] ABS; CAS 9003-56-9; modifier resin for transparent vinyl sheet and bottle applics.; sp.gr. 0.96; bulk dens. 15.6 lb/ft^3; tens. str. 1425 psi; Izod impact str. 6.5 ft lb/in. (1/4 in.); hardness Shore D50; distort. temp. 145 F (264 psi).

Blendex® 586. [GE Specialty] Poly (α-methylstyrene-styrene acrylonitrile); high heat modifier resin used to upgrade rigid PVC compds. providing low cost, toughness, rigidity, flame resist., chem. inertness, and heat resist.; for pipe, appliance and business machine housings, elec. devices, industrial sheeting, dk. colored siding and exterior profiles, thermoformed parts; sp.gr. 1.09; bulk dens. 20.0 lb/ft^3; ref. index 1.57; tens. str. 9000 psi; Izod impact str. 0.5 ft lb/in. (1/8 in.); hardness Rockwell R120; distort. temp. 124 C (264 psi annealed).

Blendex® 587. [GE Specialty] Poly(α-methylstyrene-acrylonitrile; modifier resin that increases the service temp. of PVC homopolymer and copolymer prods., esp. rigid PVC, outdoor applics., inj. moldings; for pipe and fittings for conveying hot fluids, appliance and business machine housings, elec./telecommunications devices, industrial sheeting, thermoformed parts, inj. molded tech. parts; wh. free-flowing powd.; ≈ 2.38 mm max. particle size; melt flow 6 g/10 min (220 C, 10 kg); sp.gr. 1.08; bulk dens. 21.8 lb/ft^3; soften. pt. (Vicat) 126 C.

Blendex® 590. [GE Specialty] Methyl methacrylate/styrene acrylonitrile copolymer; process aid for polymer applics. requiring higher molten elasticity; features good melt stability, superior hot elong., exc. color stability, high clarity and gloss, low odor; for rigid and semirigid vinyl, foamed PVC and PVC/NBR, clear vinyl for blow molding and thermoforming; wh. free-flowing powd.; sp.gr. 1.14; ref. index 1.538.

Blendex® 703. [GE Specialty] Modifier to raise the heat distort. temp. and impact resistance of PVC compds.; sp.gr. 1.05; bulk dens. 17.0 lb/ft^3; tens. str. 6800 psi; Izod impact str. 3.3 ft lb/in. (1/8 in.); hardness Rockwell R114; distort. temp. 242 F (264 psi).

Blendex® 869. [GE Specialty] SAN copolymer; CAS 9003-54-7; process aid for rigid and semirigid PVC compds.; provides increased PVC melt str. at low concs., good dimensional stability of extruded profiles, good processing/thermoforming props.; for low-dens. PVC foamed profiles and sheeting, blow molded containers, calendered rigid/semirigid PVC film; wh. free-flowing powd.; ≈ 0.1 mm particle size; m.w. > 6 x 10^6; sp.gr. 1.07; bulk dens. ≈ 350 kg/m^3; ref. index 1.57.

Blendex® 975. [GE Specialty] ASA copolymer; weatherable modifier for use in rigid polymer applics. requiring good outdoor aging; can be alloyed with PVC and other miscible thermoplastics in molded, extruded or calendered applics.; provides impact and higher heat distort. props.; sp.gr. 1.06; tens. str. 5800 psi; tens. elong. 30% (break); flex. str. 7000 psi; flex. mod. 280,000 psi; impact str. (Izod) 9.0 ft lb/in. (1/8 in.); hardness Rockwell R87.

Blendex® 977. [GE Specialty] ASA copolymer with uv inhibitor; weatherable modifier for use in rigid polymer applics. requiring good outdoor aging; can be alloyed with PVC and other miscible thermoplastics in molded, extruded or calendered applics.; provides impact and higher heat distort. props.; sp.gr. 1.06; tens. str. 5800 psi; tens. elong. 30% (break); flex. str. 7000 psi; flex. mod. 280,000 psi; impact str. (Izod) 9.0 ft lb/in. (1/8 in.); hardness Rockwell R87.

Blendex® HPP 801. [GE Specialty] Polycarbonate resin; modifier for other polymers; base for color concs.; reinforcing modifier for recycled polymers and alloys; provides improved pigment absorption, greater compounding flexibility; wh. gran. powd.; melt flow 6.5 g/10 min.; sp.gr. 1.2; tens. str. 9000 psi; flex. str. 14,200 psi; impact str. (Izod) 17 ft lb/in. (1/8 in.); distort. temp. 270 F (264 psi); *Precaution:* sensitive to acids and bases.

Blendex® HPP 802. [GE Specialty] Polycarbonate resin; modifier for other polymers; base for color concs.; reinforcing modifier for recycled polymers and alloys; provides improved pigment absorption, greater compounding flexibility; wh. gran. powd.; melt flow 9.5 g/10 min.; sp.gr. 1.2; tens. str. 9000 psi; flex. str. 14,000 psi; impact str. (Izod) 15 ft lb/in. (1/8 in.); distort. temp. 270 F (264 psi).

Blendex® HPP 803. [GE Specialty] Polycarbonate modifier resin; modifier for other polymers; base for color concs.; reinforcing modifier for recycled polymers and alloys; provides improved pigment absorption, greater compounding flexibility; wh. gran. powd.; melt flow 16.5 g/10 min.; sp.gr. 1.2; tens. str. 9000 psi; flex. str. 14,000 psi; impact str. (Izod) 13 ft lb/in. (1/8 in.); distort. temp. 265 F (264 psi).

Blendex® HPP 804. [GE Specialty] Polycarbonate modifier resin; modifier for other polymers; base for color concs.; reinforcing modifier for recycled polymers and alloys; provides improved pigment absorption, greater compounding flexibility; wh. gran. powd.; melt flow 22 g/10 min.; sp.gr. 1.2; tens. str. 9000 psi; flex. str. 13,500 psi; impact str. (Izod) 12 ft lb/in. (1/8 in.); distort. temp. 260 F (264 psi).

Blendex® HPP 820. [GE Specialty] Polyphenylene ether resin; modifier resin used to strengthen and increase the heat resist. of compat. polymers; upgrades heat resist. of polymers for critical elec. applics. (resistors, circuit boards, etc.) with no deterioration in elec. props.; features good mech. str., elec. props., hydrolytic stability, flammability; suitable for styrenics, epoxies; amber powd.; readily dissolved in chloroform, chlorobenzene, trichloroethylene, benzene, styrene monomer, toluene, tetrahydrofuran, 1,1,2-trichloroethane; sp.gr. 1.06 g/cc; water absorp. 0.1%; tens. str. 11,000 psi (yield); tens. elong. 35%; flex. strt. 15,000 psi; flex. mod. 3.35 x 10^5 psi; impact str. (Izod) 1.0 ft lb/in. notched; distort. temp. 215 C (264 psi).

BLE® 25. [Uniroyal] Acetone/diphenylamine reaction prod.; superaging and flex resist. antioxidant for natural and syn. rubber; Unverified

BLO®. [ISP] γ-Butyrolactone; solv. for PAN, PS, fluorinated hydrocarbons, cellulose triacetate, shellac; used in paint removers, petrol. processing, hectograph process, speciality inks; intermediate for aliphatic and cyclic compds.; reaction and diluent solv. for pesticides; used in dyeing of acetate; wetting agent for cellulose acetate films, fibers, solv. welding of plastic films in adhesive applics.; liq.; f.p. -44 C; b.p. 204 C; flash pt. 98 C (OC).

Blo-Foam® 5PT. [Rit-Chem] 5-Phenyltetrazole; CAS 3999-10-8; nitrogen-producing blowing agent for use in high temp. foams of engineering plastics such as PC, nylon 6/6, ABS, PET, PPE, PA, PBT; suitable for expansion of ammonia-sensitive polymers; for extrusion, inj., compr. and rotational molding; wh. cryst. needles, odorless; sol. in common org. solvs.; m.p. 210-220 C; decomp. temp. 234-254 C; gas vol. ≥ 190 ml/g @ STP.

Blo-Foam® 715. [Rit-Chem] Modified azodicarbonamide; CAS 123-77-3; EINECS 204-650-8; nitrogen-releasing blowing agent for expansion of EVA, LDPE, EPDM, SBR, NBR, neoprene, neoprene blends, and natural rubber; higher gas yield than other low temp. blowing agents; nondiscoloring, nonstaining; off-wh., odorless; decomp. temp. 158 C; gas vol. 190-190 @ STP.

Blo-Foam® 754. [Rit-Chem] Modified azodicarbonamide; CAS 123-77-3; EINECS 204-650-8; nitrogen-releasing blowing agent for expansion of EVA, LDPE, EPDM, SBR, NBR, neoprene, neoprene blends, and natural rubber; higher gas yield than other low temp. blowing agents; nondiscoloring, nonstaining; off-wh., odorless; decomp. temp. 155-165 C; gas vol. ≥ 155 @ STP.

Blo-Foam® 765. [Rit-Chem] Modified azodicarbonamide; CAS 123-77-3; EINECS 204-650-8; nitrogen-releasing blowing agent for expansion of EVA, LDPE, EPDM, SBR, NBR, neoprene, neoprene blends, and natural rubber; higher gas yield than other low temp. blowing agents; nondiscoloring, nonstaining; off-wh., odorless; decomp. temp. 154-160 C; gas vol. 180-200 @ STP.

Blo-Foam® ADC 150. [Rit-Chem] Azodicarbonamide; CAS 123-77-3; EINECS 204-650-8; chemical blowing agent for foamed plastics and rubbers; esp. for unicellular prods. where low gas loss by diffusion is required since it blows by releasing nitrogen; high decomp. exotherm allows self-nucleation; pale yel. fine powd.; 1.5 μ particle size; sol. in DMSO, with decomp. in alkaline sol'ns.; insol. in benzene, ethylene dichloride; relatively insol. in water; m.w. 116.08; sp.gr. 1.65; decomp. temp. 200 C; gas vol. 220 ml/g; pH 6.5-7; *Toxicology:* avoid breathing dust; *Storage:* store in cool, dry place away from sources of heat close to decomp. temp. (190 C); protect from strong alkalis.

Blo-Foam® ADC 150FF. [Rit-Chem] Modified azodicarbonamide; CAS 123-77-3; EINECS 204-650-8; chemical blowing agent for foamed plastics and rubbers; free-flowing grade for automatic hopper metering and blending equip.; for inj. molding structural foam, extrusion, vinyl plastisols; powd.; 1.5 μ particle size; decomp. temp. 190-210 C; gas vol. 220 @ STP.

Blo-Foam® ADC 300. [Rit-Chem] Azodicarbonamide; CAS 123-77-3; EINECS 204-650-8; chemical blowing agent for foamed plastics and rubbers; esp. for unicellular prods. where low gas loss by diffusion is required since it blows by releasing nitrogen; high decomp. exotherm allows self-nucleation; pale yel. fine powd.; 3 μ particle size; sol. in DMSO, with decomp. in alkaline sol'ns.; insol. in benzene, ethylene dichloride; relatively insol. in water; m.w. 116.08; sp.gr. 1.65; decomp. temp. 200 C; gas vol. 220 ml/g; pH 6.5-7; *Toxicology:* avoid breathing dust; *Storage:* store in cool, dry place away from sources of heat close to decomp. temp. (190 C); protect from strong alkalis.

Blo-Foam® ADC 300FF. [Rit-Chem] Modified azodicarbonamide; CAS 123-77-3; EINECS 204-650-8; chemical blowing agent for foamed plastics and rubbers; free-flowing grade for automatic hopper metering and blending equip.; for inj. molding structural foam, extrusion, vinyl plastisols; powd.; 3.0 μ particle size; decomp. temp. 190-210 C; gas vol. 220 @ STP.

Blo-Foam® ADC 300WS. [Rit-Chem] Modified azodicarbonamide; CAS 123-77-3; EINECS 204-650-8; dispersible chemical blowing agent for thermoplastics, esp. PVC plastisol foams for prod. of cellular leathercloth, floor coverings, wall coverings, carpet underlay, door-sealing strips, sealing compds.; lt. yel. fine powd.; 6.5-7.5 μ particle size; disp. easily in plastic compds.; insol. in common solvs. and plasticizers; decomp. temp. 200-210 C; gas vol. 215-220 ml/g @ STP; *Toxicology:* avoid breathing dust; *Precaution:* will burn on exposure to flame; self-extinguishing when flame is removed; *Storage:* store in cool, dry place away from heat sources close to decomp. temp.; protect from strong alkalis.

Blo-Foam® ADC 450. [Rit-Chem] Azodicarbonamide; CAS 123-77-3; EINECS 204-650-8; chemical blowing agent for foamed plastics and rubbers; esp. for unicellular prods. where low gas loss by diffusion is required since it blows by releasing nitrogen; high decomp. exotherm allows self-nucleation; lt. yel. fine powd.; 4.5 μ particle size; sol. in DMSO, with decomp. in alkaline sol'ns.; insol. in benzene, ethylene dichloride; relatively insol. in water; m.w. 116.08; sp.gr. 1.65; decomp. temp. 200 C; gas vol. 220 ml/g; pH 6.5-7; *Toxicology:* avoid breathing dust; *Storage:* store in cool, dry place away from sources of heat close to decomp. temp. (190 C); protect from strong alkalis.

Blo-Foam® ADC 450FFNP. [Rit-Chem] Modified azodicarbonamide; CAS 123-77-3; EINECS 204-650-8; chemical blowing agent for foamed plastics and rubbers; non-plateout grade for use where problems of steel mold corrosion, die blockage, or screw buildup are experienced; free-flowing powd.; 4.5 μ particle size; decomp. temp. 191-210 C; gas vol. 150-180 @ STP.

Blo-Foam® ADC 450LT. [Rit-Chem] Low temp. chemical blowing agent for foamed plastics and rubbers; allows use of normal processing temps. (182-193 C), does not permit plate-out; decomp. at increased rate in calendering or plastisol process; powd.; 4.5 μ particle size; decomp. temp. 180-190 C.

Blo-Foam® ADC 450WS. [Rit-Chem] Modified azodicarbonamide; CAS 123-77-3; EINECS 204-650-8; dispersible chemical blowing agent for thermoplastics, esp. PVC plastisol foams for prod. of cellular leathercloth, floor coverings, wall coverings, carpet underlay, door-sealing strips, sealing compds.; lt.

yel. fine powd.; 8-10 μ particle size; disp. easily in plastic compds.; insol. in common solvs. and plasticizers; decomp. temp. 200-210 C; gas vol. 215-220 ml/g @ STP; *Toxicology:* avoid breathing dust; *Precaution:* will burn on exposure to flame; self-extinguishing when flame is removed; *Storage:* store in cool, dry place away from heat sources close to decomp. temp.; protect from strong alkalis.

Blo-Foam® ADC 550. [Rit-Chem] Azodicarbonamide; CAS 123-77-3; EINECS 204-650-8; chemical blowing agent for foamed plastics and rubbers; esp. for unicellular prods. where low gas loss by diffusion is required since it blows by releasing nitrogen; high decomp. exotherm allows self-nucleation; orange-yel. fine powd.; 5.5 μ particle size; sol. in DMSO, with decomp. in alkaline sol'ns.; insol. in benzene, ethylene dichloride; relatively insol. in water; m.w. 116.08; sp.gr. 1.65; decomp. temp. 200 C; gas vol. 220 ml/g; pH 6.5-7; *Toxicology:* avoid breathing dust; *Storage:* store in cool, dry place away from sources of heat close to decomp. temp. (190 C); protect from strong alkalis.

Blo-Foam® ADC 550FFNP. [Rit-Chem] Modified azodicarbonamide; CAS 123-77-3; EINECS 204-650-8; chemical blowing agent for foamed plastics and rubbers; non-plateout grade for use where problems of steel mold corrosion, die blockage, or screw buildup are experienced; free-flowing powd.; 5.5 μ particle size; decomp. temp. 191-210 C; gas vol. 150-180 @ STP.

Blo-Foam® ADC 800. [Rit-Chem] Azodicarbonamide; CAS 123-77-3; EINECS 204-650-8; chemical blowing agent for foamed plastics and rubbers; esp. for unicellular prods. where low gas loss by diffusion is required since it blows by releasing nitrogen; high decomp. exotherm allows self-nucleation; orange-yel. fine powd.; 8 μ particle size; sol. in DMSO, with decomp. in alkaline sol'ns.; insol. in benzene, ethylene dichloride; relatively insol. in water; m.w. 116.08; sp.gr. 1.65; decomp. temp. 200 C; gas vol. 220 ml/g; pH 6.5-7; *Toxicology:* avoid breathing dust; *Storage:* store in cool, dry place away from sources of heat close to decomp. temp. (190 C); protect from strong alkalis.

Blo-Foam® ADC 1200. [Rit-Chem] Azodicarbonamide; CAS 123-77-3; EINECS 204-650-8; chemical blowing agent for foamed plastics and rubbers; esp. for unicellular prods. where low gas loss by diffusion is required since it blows by releasing nitrogen; high decomp. exotherm allows self-nucleation; orange-yel. fine powd.; 12 μ particle size; sol. in DMSO, with decomp. in alkaline sol'ns.; insol. in benzene, ethylene dichloride; relatively insol. in water; m.w. 116.08; sp.gr. 1.65; decomp. temp. 200 C; gas vol. 220 ml/g; pH 6.5-7; *Toxicology:* avoid breathing dust; *Storage:* store in cool, dry place away from sources of heat close to decomp. temp. (190 C); protect from strong alkalis.

Blo-Foam® ADC-MP. [Rit-Chem] Modified azodicarbonamide; CAS 123-77-3; EINECS 204-650-8; chemical blowing agent for extrusion foam of chemically crosslinked polyolefins, esp. polyethylene, which can be made less dense than other foams; for trimming, upholstery, insulation, sealing, sound deadening in construction, linings, protective padding, flotation devices, sport shoes, protective pkg.; orange-yel. fine powd.; 10-12 μ particle size; decomp. temp. 193-198 C; gas vol. 205-210 ml/g @ STP; pH 9-10.

Blo-Foam® ADC-NP. [Rit-Chem] Modified azodicarbonamide; CAS 123-77-3; EINECS 204-650-8; chemical blowing agent for extrusion foam of chemically crosslinked polyolefins, esp. polyethylene, which can be made less dense than other foams; for trimming, upholstery, insulation, sealing, sound deadening in construction, linings, protective padding, flotation devices, sport shoes, protective pkg.; orange-yel. fine powd.; 10-12 μ particle size; decomp. temp. 180-185 C; gas vol. 195-200 ml/g @ STP; pH 6-7.

Blo-Foam® BBSH. [Rit-Chem] p,p′-Oxybis (benzene sulfonyl hydrazide); CAS 80-51-3; chemical blowing agent for processing of natural and syn. sponge rubber and expanded plastics, esp. EVA copolymers, LDPE, and PVC; for cellular cable insulation, syn. rubber extradates modling, rug underlay, gaskets, upholstery material, etc.; wh. fine cryst. powd.; 5 μ avg. particle diam.; very sol. in DMSO, DMF; mod. sol. in alcohol, polyalkylene glycols; insol. in water, benzene; reacts with acetone, MEK; m.w. 358.39; sp.gr. 1.55; decomp. temp. 156-162 C; gas vol. 120-125 ml/g @ STP; pH 7-7.5; *Toxicology:* avoid ingestion, inhalation of dust, excessive skin contact; *Precaution:* flamm. solid; will continue to burn once ignited, giving off intense heat and large volumes of dense smoke; *Storage:* store in cool, dry place away from heat sources, flames, direct sunlight.

Blo-Foam® DNPT 80. [Rit-Chem] Dinitroso pentamethylene tetramine; CAS 101-25-7; nondiscoloring, nonstaining blowing agent for rubber industry; esp. useful for compr. molding and blow molding of plasticized PVC; lt. yel. fine powd.; sol. in dimethylformamide, dimethylsulfoxide; sl. sol. in pyridine, ethyl acetoacetate; virtually insol. in water, alcohols; m.w. 186.17; sp.gr. 1.45; decomp. temp. 185-195 C; gas vol. 240 ml/g @ STP; pH 6.5-7; *Precaution:* flamm. solid which will continue to burn once ignited, giving off intense heat; *Storage:* store in cool, dry place away from heat sources, flames, direct sunlight, org. and inorg. acids; reclose partially full containers.

Blo-Foam® DNPT 93. [Rit-Chem] Dinitroso pentamethylene tetramine; CAS 101-25-7; nondiscoloring, nonstaining blowing agent for rubber industry; esp. useful for compr. molding and blow molding of plasticized PVC; lt. yel. fine powd.; 25-30 μ particle size; sol. in dimethylformamide, dimethylsulfoxide; sl. sol. in pyridine, ethyl acetoacetate; virtually insol. in water, alcohols; m.w. 186.17; sp.gr. 1.45; decomp. temp. 195-203 C; gas vol. 240 ml/g @ STP; pH 6.5-7; *Precaution:* flamm. solid which will continue to burn once ignited, giving off intense heat; *Storage:* store in cool, dry place away from heat sources, flames, direct sunlight, org. and inorg. acids; reclose partially full containers.

Blo-Foam® DNPT 100. [Rit-Chem] Dinitroso pentamethylene tetramine; CAS 101-25-7; nondiscoloring, nonstaining blowing agent for rubber industry; esp. useful for compr. molding and blow molding of plasticized PVC; lt. yel. fine powd.; 30-40 μ particle size; sol. in dimethylformamide, dimethylsulfoxide; sl. sol. in pyridine, ethyl acetoacetate; virtually insol. in water, alcohols; m.w. 186.17; sp.gr. 1.45; decomp. temp. 195-203 C; gas vol. 250 ml/g @ STP; pH 7-7.5; *Precaution:* flamm. solid which will continue to burn once ignited, giving off intense heat; *Storage:* store in cool, dry place away from heat sources, flames, direct sunlight, org. and inorg. acids; reclose partially full containers.

Blo-Foam® KL9. [Rit-Chem] Modified azodicarbonamide; CAS 123-77-3; EINECS 204-650-8; one-pkg. blowing agent system for low temp. thermoplastic polymer foam applics., esp. low temp. PVC plastisol and rubber foam; yel. free-flowing fine powd.; 6-7 μ particle size; decomp. temp. 147-155 C; gas vol. 135-145 ml/g @ STP; pH 6.5-7.5; *Toxicology:* avoid breathing dust; *Precaution:* decomp. can be initiated by hot objects, such as welding sparks or a lighted match; *Storage:* store in dry, cool area away from heat sources in closed containers; protect from strong alkalis.

Blo-Foam® KL10. [Rit-Chem] Modified azodicarbonamide; CAS 123-77-3; EINECS 204-650-8; preactivated one-pkg. blowing agent system for low temp. or high speed thermoplastic polymer foam applics., esp. PVC or sponge rubber; pale yel. free-flowing fine powd.; decomp. temp. 127-133 C; gas vol. 135-145 ml/g @ STP; pH 6.5-7.5; *Toxicology:* avoid breathing dust; *Precaution:* decomp. can be initiated by hot objects, such as welding sparks or a lighted match; *Storage:* store in dry, cool area away from heat sources in closed containers; protect from strong alkalis.

Blo-Foam® RA. [Rit-Chem] p-Toluene sulfonyl semicarbazide; blowing agent for plastics expanding at high temps., in extrusion and banbury compounding; also for vinyl plastisols, calendering and rigid PVC extrusion powd. blends, rubber expansion; for inj. molding structural foam for improved surf. appearance in HIPS, ABS, PP, nylon; also for PPO, HDPE, EVA; wh. fine powd.; 5 μ avg. particle diam.; sol. in DMSO; insol. in water, toluene, benzene, acetone; reacts with DMF; m.w. 229.25; sp.gr. 1.44; decomp. temp. 230-234 C, 213-225 C in plastics; gas vol. 140 ml/g @ STP; pH 6.5-7.

Blo-Foam® SH. [Rit-Chem] p-Toluene sulfonyl hydrazide; CAS 877-66-7; highly active nitrogen-type blowing agent for expansion of odorless cellular rubber goods, e.g., open cell natural rubber, molded closed cell SBR soling, neoprene rubber, epoxy foam, liq. polysulfide rubber; cream-colored powd.; 12 μ avg. particle diam.; very sol. in DMSO; sol. in alcohol; insol. in water, toluene, benzene; reacts with DMF, MEK, acetone; m.w. 186.23; sp.gr. 1.42; decomp. temp. 105-109 C; gas vol. 115 ml/g @ STP; pH 7-7.5; *Toxicology:* avoid ingestion, inhalation of dust, excessive skin contact; *Precaution:* will burn rapidly when ignited; *Storage:* store in cool, dry area away from heat sources, flames, direct sunlight; minimize dusting.

BLS™ 531. [Mayzo] 2-Hydroxy-4-n-octoxybenzophenone; CAS 1843-05-6; EINECS 217-421-2; light stabilizer, uv absorber for plastics and other org. polymers incl. PP, PE, PVC, acrylics, PC, EVA, polyester, and in adhesives and sealants; retards yellowing, contributes low color; FDA 21CFR §178.2010, 178.3710; lt. yel. powd.; sol. (g/100 ml): 72 g benzene, 65 g MEK, 35 g acetone, 25 g DOP; insol. in water; m.w. 326; sp.gr. 1.16; m.p. 48-49 C; dec. pt. > 300 C; *Toxicology:* LD50 (oral, rat) > 10 g/kg, (dermal, rabbit) > 10 g/kg; nonirritating to skin and eyes; *Precaution:* incompat. with strong oxidizing agents; *Storage:* 1 yr storage life in sealed containers stored in cool, dry areas; keep sealed when not in use.

BLS™ 1770. [Mayzo] Bis (2,2,6,6-tetramethyl-4-piperidinyl) decanedioate; CAS 52829-07-0; EINECS 258-207-9; HALS stabilizer for polyolefins incl. natural and pigmented PP multifilament slit film, PE, ethylene-propylene co- and terpolymers, PU, EPDM, ABS, SAN, ASA, other styrenics; synergistic with phenolic antioxidants; wh. to sl. ylsh. cryst. powd.; sol. 56% in methylene chloride, 50% in ethanol, 46% in benzene, 45% in chloroform, 38% in methanol, 19% in acetone, < 0.01% in water; m.w. 481; sp.gr. 1.05 (20 C); vapor pressure 1×10^{-10} mm Hg (20 C); m.p. 81-86 C; dec. pt. > 220 C; flash pt. (Marcusson) > 302 F; *Toxicology:* LD50 (oral, rat) 3700 mg/kg; hazardous; severe eye irritant; may cause irritation to skin, nasal membranes, respiratory tract; may leave bitter taste in mouth; *Storage:* 1 yr storage life in sealed containers in cool, dry areas away from light; keep sealed when not in use.

Blue Label Silicone Spray No. SB412-A. [IMS] < 5% Silicone release agent (CAS 63148-62-9) with < 5% aminofunctional silicone and 50-60% HFC-152A, 30-40% dimethyl ether, 5-15% aliphatic petrol. distillate (CAS 64742-89-8); non-CFC, non-ozone depleting mold release forming a thin, tough film on the most highly polished mold surfs.; effective for ABS, acetal, acrylic, nylon, PC, polyester, polyethylene, PS, rubber; gives multiple releases; may cause film buildup on warm molds if used too heavily; clear mist with sl. ethereal odor as dispensed from aerosol pkg.; sl. sol. in water; sp.gr. > 1; vapor pressure 60 ± 10 psig; flash pt. < 0 F; > 90% volatile; *Toxicology:* inh. may cause CNS depression with dizziness, headache; higher exposure may cause heart problems; gross overexposure may cause fatality; eye and skin irritation; direct contact with spray can cause frostbite; *Precaution:* flamm. limits 1-18%; at elevated temps. (> 130 F) aerosol containers may burst, vent, or rupture; incompat. with strong oxidizers, strong caustics, reactive metals (Na, K, Zn, Mg, Al); contains methylpolysiloxanes which can generate formaldehyde; *Storage:* store in cool, dry area out of direct sunlight; do not puncture, incinerate, or store above 120 F.

BM-723. [Rohm Tech] Methacrylic anhydride; hydrolyzable crosslinker, photoresists, prod. of methacrylates and methacrylamides.

BM-729. [Rohm Tech] Tetrahydrofurfuryl-2-methacrylate; CAS 2455-24-5; functional monomer; reactive thinner for radiation curing; adhesion promoters; adhesives; rubber modifiers.

BM-801. [Rohm Tech] Methacrylamide; CAS 79-39-0; EINECS 201-202-3; used in self-crosslinking emulsions, heat-curing coatings, silk grafting, crosslinked acrylic sheets.

BM-818. [Rohm Tech] N-Methylol methacrylamide; used in self-crosslinking emulsions, heat-curing coatings; 60% solids in water.

BM-903. [Rohm Tech] 2-Hydroxyethyl methacrylate; CAS 868-77-9; EINECS 212-782-2; thermally crosslinkable paint resins and emulsions, binders for textiles and paper, adhesives, urethane methacrylates, reactive thinners for radiation curing, grafting of textile fibers, scale inhibitors, rubber modifiers, contact lenses; photopolymer plastes, photoresists; 97% purity.

BM-951. [Rohm Tech] Hydroxypropyl methacrylate; CAS 27813-02-1; EINECS 248-666-3; thermally crosslinkable paint resins and emulsions, binders for textiles and paper, adhesives, urethane methacrylates, reactive thinners for radiation curing, grafting of textile fibers, scale inhibitors, rubber modifiers, contact lenses; photopolymer plastes, photoresists; 98% purity.

BNX® 1000. [Mayzo] Proprietary; nondiscoloring, nonstaining antioxidant system for protection of adhesives (esp. hot-melt and water-based), rubber, resins, other polymer systems; synergistic with benzophenone, benzotriazole, HALS; FDA 21CFR §174.178, 175.105, 175.300, 177.1010, 177.1210, 177.1350, 177.1420(a)(3), 177.1480, 177.1520(c), 177.1580, 177.1640, 177.1810, 177.2600, 178.2010, 178.3790, 179.45; milky wh. liq., essentially odorless; dens. 8.2 lb/gal; flash pt. (OC) > 350 F; *Toxicology:* LD50 (oral, rat) > 5000 mg/kg; nonirritating to skin, eyes, on inhalation; *Storage:* 1 yr storage life in sealed containers in cool, dry areas; keep sealed when not in use.

BNX® 1010, 1010G. [Mayzo] Tetrakis [methylene-3(3′,5′-di-t-butyl-4-hydroxyphenyl) propionate] methane; CAS 6683-19-8; EINECS 229-722-6; nonstaining, nondiscoloring antioxidant for PE, PS, PP, polyacetal, AS resin, methacrylic resin, PC, polyester, polymethylpentene, butadiene resin, polybutene-1, EPM, EPD, PVC; stabilizer for org. and polymeric materials; also for petrol. prods., syn. rubbers, latex, varnish, adhesives, wire and cable insulation, syn. diester fluids, oils, fats, waxes; good extract resist.; FDA 21CFR §175.105, 175.125, 175.300, 175.320, 176.170, 176.180, 176.210, 177.1210, 177.1310, 177.1330, 177.1350, 177.1420, 177.1520, 177.1570, 177.1630, 177.1640, 177.2470, 177.2480, 177.2600, 178.2010, 178.3800, 178.3850, 179.45; wh. cryst. powd. (1010) or gran. (1010G), odorless, tasteless; sol. 71% in chloroform, 56% in benzene, 47% acetone, 46% ethyl acetate, 1% methanol, 0.01% water; m.w. 1177.7; sp.gr. 1.15; vapor pressure 10^{-12} mm Hg (20 C); m.p. 110-125 C; dec. pt. > 220 C; flash pt. (Marcusson) 567 F; *Toxicology:* LD50 (oral, rat) > 5000 mg/kg, (dermal, rabbit) > 3160mg/kg; low oral and dermal toxicity; nonsensitizing; *Precaution:* incompat. with strong oxidizing agents; thermal decomp. may produce CO, CO_2; *Storage:* 1 yr storage in sealed containers kept in cool, dry areas.

BNX® 1076, 1076G. [Mayzo] Stearyl-3-(3′,5′-di-t-butyl-4-hydroxyphenyl)propionate; CAS 2082-79-3; EINECS 218-216-0; nonstaining, nondiscoloring antioxidant for org. and polymeric materials incl. PE, PP, polymethylpentene, PS, ABS, methacrylic resin, PVC, EPDM, polyester, rubbers, latex, varnish, adhesives, urethane, petrol. prods.; good extraction resist.; FDA 21CFR §174.178, 175.105, 177.1010, 177.1350, 177.1420(a)(3), 177.1480, 177.1520(c), 177.1580, 177.1640, 177.1810, 178.2010, 178.3790, 179.45; wh. cryst. powd. (1076) or gran. (1076G), odorless, tasteless; sol. 57% in chloroform, benzene, 42% in ethyl acetate, 31% in hexane, 26% in acetone, < 0.01% in water; m.w. 530.9; sp.gr. 1.02; vapor pressure 2×10^{-9} mm Hg (20 C); m.p. 49-54 C; dec. pt. > 220 C; flash pt. (Marcusson) 523 F; partly biodeg.; *Toxicology:* LD50 (oral, rat) > 10,000 mg/kg, (dermal, rabbit) > > 2000 mg/kg, (IP, rat) > 1000 mg/kg; low oral and dermal toxicity; nonsensitizing; *Precaution:* incompat. with strong oxidizing agents; thermal decomp. may produce CO, CO_2; *Storage:* 1 yr storage in sealed containers kept in cool, dry areas.

Bonding Agent 2001. [Bayer AG; Bayer/Fibers, Org., Rubbers] Polymethylene polyphenyl isocyanate sol'n. in dibutyl phthalate; one-component bonding agent for PVC coatings; used for tarpaulins, awnings, flexible containers, protective clothing, conveyor belts, air domes, tents; sol'n.; visc. 10,000 cps (23 C); 30% solids, 4.7% NCO.

Bonding Agent 2005. [Bayer AG; Bayer/Fibers, Org., Rubbers] Polymethylene polyphenyl isocyanate sol'n. in butyl acetate; one-component bonding agent improving adhesion of PVC to syn. fiber substrates; used for tarpaulins, awnings, flexible containers, protective clothing, conveyor belts, air domes, tents; sol'n.; visc. 85 cps (23 C); 40% solids, 6.5% NCO.

Bonding Agent TN/S 50. [Bayer AG; Bayer/Fibers, Org., Rubbers] Hydroxyl polyester sol'n. in benzyl butyl phthalate; two-component bonding agent improving adhesion of PVC to syn. fiber substrates; used for tarpaulins, awnings, flexible containers, protective clothing, conveyor belts, air domes, tents; produces optimum adhesion and lightfastness; sol'n.; visc. 1100 cps (23 C); 50% solids.

Bondogen E®. [King Industries; Struktol] High m.w.

sulfonic acid and low visc. min. oil; high activity plasticizer, processing aid, and peptizer for elastomers; effective scorch retarder; esp. useful in solv. cements; dk. mahogany liq.; sp.gr. 0.93 (15.6 C); dens. 7.8 lb/gal; visc. 200 SUS (37.8 C); acid no. 41; flash pt. (COC) 120 C.

Borax. [U.S. Borax] Sodium tetraborate decahydrate; CAS 1303-96-4; dispersant, wetting agent for NR, SR latexes; mold lubricant for general dry rubber molding; wh. powd., odorless; sol. in water, glycerin; sp.gr. 1.73.

Bowax 2015. [Sovereign] Polyethylene wax; CAS 9002-88-4; EINECS 200-815-3; process aid for rubber; low melt visc. provides improved pigment wetting and dispersion; enhances flow and release; lower cost than other PE waxes; m.p. 114 C.

Britol® 6NF. [Witco/Petroleum Spec.] White min. oil NF; white oil functioning as binder, carrier, conditioner, defoamer, dispersant, extender, heat transfer agent, lubricant, moisture barrier, plasticizer, protective agent, release agent, and/or softener; used in cosmetics, pharmaceuticals, food, plastics, agric., and paper making applics.; sp.gr. 0.830-0.858; visc. 8.5-10.8 cst (40 C); pour pt. -24 C max.; flash pt. 166 C min.

Bromoklor™ 50. [Ferro/Keil] Halogenated aliphatic liqs. containing bromine and chlorine; flame retardant used in plasticized PVC, PU, latex, fabric; uses incl. disp. molded or coated automotive parts, interior trim, pkg. closures, boots, hand grips, flooring, upholstery, carpet backing, furniture and wall covering, and pkg. film; plasticizing properties in PVC (not primary); lt. straw liq.; sp.gr. 1.36-1.38; visc. 270 cps; 33% Br, 19% Cl.

Bromoklor™ 70. [Ferro/Keil] Halogenated aliphatic liq. containing bromine and chlorine; flame retardant for PVC, PU, latex, fabric; sec. plasticizer; low volatility; straw liq.; sp.gr. 1.65-1.67; visc. 5000 cps; 35% Br, 35% Cl.

BSWL 201. [Eagle-Picher] Basic silicate wh. lead with lt. min. oil added for dust control; wh. pigment acting as heat stabilizer for chlorinated polyethylene, chlorosulfonated polyethylene, PVC, and polyepichlorohydrin; rust-inhibitive pigment in the automobile industry; used in industrial or maintenance paints; not for aq. systems; wh. 99.5% -325 mesh; sp.gr. 5.0; 75% Pb.

BSWL 202. [Eagle-Picher] Basic silicate wh. lead; CAS 10099-76-0; wh. pigment acting as heat stabilizer for chlorinated polyethylene, chlorosulfonated polyethylene, PVC, and polyepichlorohydrin; rust-inhibitive pigment in the automobile industry; used in industrial or maintenance paints; wh.; 99.5% -325 mesh; sp.gr. 6.0; dens. 53.31 lb/solid gal; apparent dens. 13 g/in.^{3}; bulking value 0.0188 gal/lb; oil absorp. 16.0; brightness in oil 86.0% (Beckman); 78.6% Pb; 15.4% SiO_2.

Buca®. [Engelhard] Hydrous aluminum silicate; reinforcing extender for rubber and polymer systems; wh. highly pulverized powd.; 0.4 μm avg. particle size; 0.01% max. residue +325 mesh; sp.gr. 2.58 g/cc; bulk dens. 13 lb/ft^3 (loose), 18 lb/ft^3 (tamped); bulking value 0.047 gal/lb; oil absorp. 45-50 lb/100 lb; ref. index 1.56; pH 6-8; G.E. brightness (GE) 86-88; 45.4% SiO_2; 38.5% Al_2O_3.

Bulab® Flamebloc. [Buckman Labs] Hydrated borate compd.; synergist in formulations contg. phosphorous compds. and/or halogen donors; substitute for antimony trioxide; adds fire resist. to plastics (flexible PVC, plastisols, polyolefins), paints, textile finishes, rubber compds. (neoprene, S/B); provides reduced toxicity of fumes emitted during combustion, reduced smoke emissions, greater transparency and brighter colors; wh. powd.; sol. 0.3-0.4% in water @ 21 C: sp.gr. 3.25-3.35; dens. 27-29 lb/gal; ref. index 1.55-1.60; pH 9.8-10.3 (sat. sol'n. @ 21 C).

Buna AP 147. [Hüls Am.] Sequence EPDM; blend component for improving flowability and grn. str.

Buna AP 331. [Hüls Am.] Random EPDM; modifier for polyolefins; peroxide-cured articles.

Buna AP 437. [Hüls Am.] Sequence EPDM; modifier for polyolefins; also for uncured articles such as cable fillers, sound dampening sheets, roofing sheets.; Unverified

Buna BL 6533. [Bayer AG; Bayer/Fibers, Org., Rubbers] SBR block copolymer, lithium catalyst, nonstaining stabilizer; styrene impact modifier (HIPS) alone or in combination with polybutadiene; used in polymerization process and mech. blends; imparts higher gloss and transparency than Buna CB; FDA 21CFR §177.2600; lt. colored bales; visc. 38 cps (5% in styrene); 40% bound styrene.

Buna CB 14. [Bayer AG; Bayer/Fibers, Org., Rubbers] Butadiene rubber, organometalic cobalt; nonstaining; impact modifier for HIPS; lt. brn. bales; sp.gr. 0.92; Mooney visc. 45-50; 97% cis 1,4 content; Unverified

Buna CB HX 529. [Bayer/Fibers, Org., Rubbers] Butadiene rubber, lithium catalyst, nonstaining stabilizer; high impact modifier for PS; imparts resist. to aging, reversion, abrasion, flexcracking; for tires, retreads, sporting goods, conveyor belting, footwear soles, blends with NR; FDA 21CFR §177.2600; lt. colored bales; sp.gr. 0.91; visc. 160 cps (5.43% in toluene); 38% cis-1,4 content.

Buna CB HX 530. [Bayer/Fibers, Org., Rubbers] Butadiene rubber, lithium catalyst, nonstaining stabilizer; high impact modifier for PS; imparts resist. to aging, reversion, abrasion, flexcracking; for tires, retreads, sporting goods, conveyor belting, footwear soles, blends with NR; FDA 21CFR §177.2600; lt. brn. bales; sp.gr. 0.91; visc. 250 cps (5.43% in toluene); 68% cis-1,4 content.

Buna CB HX 565. [Bayer/Fibers, Org., Rubbers] Butadiene rubber, lithium catalyst, nonstaining stabilizer; high impact modifier for PS; imparts resist. to aging, reversion, abrasion, flexcracking; for tires, retreads, sporting goods, conveyor belting, footwear soles, blends with NR; FDA 21CFR §177.2600; lt. colored bales; sp.gr. 0.91; visc. 42 cps (5.43% in toluene); 38% cis-1,4 content.

Bunatak™ N. [C.P. Hall] Tackifier and nonvolatile plasticizer for nitrile rubbers, Hypalon® for gasket, hose, roll, specialty footwear stocks; amber visc. liq.; sp.gr. 1.02 ± 0.04.

Bunatak™ U. [C.P. Hall] Tackifier and nonvolatile plasticizer for nitrile rubbers, SBR, neoprene for mech. goods, pharmaceutical prods., footwear, belting, adhesives, and coatings; improves processing and handling; amber visc. liq.; sp.gr. 0.96 ± 0.04.

Bunaweld™ 780. [C.P. Hall] Polymeric resin; tackifier and nonvolatile plasticizer for Hypalon®, natural and syn. polymers incl. SBR, neoprene for mfg. of hose, belts, mech. goods; binder; helps stock dissolve in solvs. more rapidly, contributing to smoother knife or spreader applic.; also for friction and coating stocks; amber very visc. liq.; sp.gr. 0.97 ± 0.04.

Burez K20-505A. [Akzo Nobel/Eka Nobel] Potassium soap of disproportionated rosin; CAS 61790-50-9; process/polymerization aid in the prod. of syn. nitrile rubbers; air entrainment agent/mortar plasticizer in prod. of concrete; FDA 21CFR §175.105 and BGA 14 and 21 for indirect food contact; amber clear liq.; acid no. < 1; pH 10.3 ± 0.3 (20 C); 20 ± 1% solids.

Burez K25-500D. [Akzo Nobel/Eka Nobel] Potassium soap of disproportionated rosin; CAS 61790-51-0; EINECS 263-144-5; emulsifier for emulsion polymerization, mainly for SBR; primary emulsifier for prod. of other syn. rubbers (ABS, CR); yields a latex of good color stability; FDA 21CFR §175.105, 178.3870; lt. yel. sol'n.; sp.gr. 1.13 (20 C); pH 10.5; 25 ± 1% solids; *Storage:* store in SS tanks, preferably under nitrogen and slow interval stirring; protect from freezing.

Burez K50-505A. [Akzo Nobel/Eka Nobel] Potassium soap of disproportionated rosin; process/polymerization aid in prod. of syn. nitrile rubbers; air entrainment agent/mortar plasticizer for prod. of concrete; FDA 21CFR §175.105 and BGA 14 and 21 for indirect food contact; amber clear liq.; acid no. < 1; pH 11 ± 1 (20 C); 50 ± 1% solids.

Burez K80-500. [Akzo Nobel/Eka Nobel] Potassium soap of disproportionated rosin; CAS 61790-51-0; EINECS 263-144-5; emulsifier for emulsion polymerization, mainly for SBR; primary emulsifier for prod. of other syn. rubbers (ABS, CR); yields latex of good stability; FDA 21CFR §175.105, 178.3870; lt. yel. to amber paste; sp.gr. 1.13 (20 C); acid no. 11 ± 1; 80 ± 1% solids; *Storage:* store in SS tanks, preferably under nitrogen and slow interval stirring; protect from freezing.

Burez K80-500D. [Akzo Nobel/Eka Nobel] Potassium soap of disproportionated rosin; CAS 61790-51-0; EINECS 263-144-5; emulsifier for emulsion polymerization, mainly for SBR; primary emulsifier for prod. of other syn. rubbers (ABS, CR); yields latex of good stability and light color; FDA 21CFR §175.105, 178.3870; lt. yel. paste; sp.gr. 1.13 (20 C); acid no. 12 ± 1; 80 ± 1% solids; *Storage:* store in SS tanks, preferably under nitrogen and slow interval stirring; protect from freezing.

Burez K80-2500. [Akzo Nobel/Eka Nobel] Potassium soap of disproportionated tall oil rosin; CAS 61790-51-0; EINECS 263-144-5; emulsifier for emulsion polymerization, mainly for SBR; primary emulsifier for prod. of other syn. rubbers (ABS, CR); yields latex of good stability; FDA 21CFR §175.105, 178.3870; lt. yel. to amber paste; sp.gr. 1.13 (20 C); acid no. 12 ± 1; 80 ± 1% solids; *Storage:* store in SS tanks, preferably under nitrogen and slow interval stirring; protect from freezing.

Burez NA 70-500. [Akzo Nobel/Eka Nobel] Sodium soap of disproportionated gum rosin; CAS 61790-51-0; EINECS 263-144-5; emulsifier for emulsion polymerization, mainly for SBR; primary emulsifier for prod. of other syn. rubbers (ABS, CR); yields latex of good stability; FDA 21CFR §175.105, 178.3870; lt. yel. to amber paste; sp.gr. 1.13 (20 C); acid no. 11 ± 1; 70 ± 1% solids; *Storage:* store in SS tanks, preferably under nitrogen and slow interval stirring; protect from freezing.

Burez NA 70-500D. [Akzo Nobel/Eka Nobel] Sodium soap of disproportionated gum rosin; CAS 61790-51-0; EINECS 263-144-5; emulsifier for emulsion polymerization, mainly for SBR; primary emulsifier for prod. of other syn. rubbers (ABS, CR); yields latex of good stability and light color; FDA 21CFR §175.105, 178.3870; yel. paste; sp.gr. 1.13 (20 C); acid no. 10 ± 1; 70 ± 1% solids; *Storage:* store in SS tanks, preferably under nitrogen and slow interval stirring; protect from freezing.

Burez NA 70-2500. [Akzo Nobel/Eka Nobel] Sodium soap of disproportionated tall oil rosin; CAS 61790-51-0; EINECS 263-144-5; emulsifier for emulsion polymerization, mainly for SBR; primary emulsifier for prod. of other syn. rubbers (ABS, CR); yields latex of good stability; FDA 21CFR §175.105, 178.3870; lt. yel. to amber paste; sp.gr. 1.13 (20 C); acid no. 11 ± 1; 70 ± 1% solids; *Storage:* store in SS tanks, preferably under nitrogen and slow interval stirring; protect from freezing.

Burgess 30. [Burgess Pigment] Aluminum silicate, anhyd.; pigment providing elec. insulation props. to PVC compds.; also provides exc. dispersion and uniform neutral color; provides low compr. set, low permanent set, and acts as extrusion aid in rubber compds.; useful in glaze and body applics. in ceramics; cream wh. thin flat plate; 0.2% on 325 mesh; 1.5 μ avg. particle size; sp.gr. 2.63; oil absorp. 55-60; pH 5.0-6.0; 51.0-52.4% silica, 42.1-44.3% alumina, 0.5% max. moisture.

Burgess 30-P. [Burgess Pigment] Aluminum silicate, thermo-optic; used in PVC compds.; combines ease of incorporation, uniformity of compd. color, exc. elec. props., low sp.gr.; U.S. patent 3,021,195; wh. amorphous particles; 0.3% max. on 325 mesh; sp.gr. 2.2; oil absorp. 65; ref. index 1.62; pH 4.0-4.5; 51.0-52.4% silica, 42.1-44.3% alumina, 0.5% max. moisture.

Burgess 2211. [Burgess Pigment] Aluminum silicate, anhyd., silane surf. treated; filler used in min.-filled nylon applics. and polyterephthalate, urethane, PVC, thermoplastic polyester; features low warpage and high impact str.; wh. powd.; 1.4 μ avg.

particle size; 0.02% max. on 325 mesh; sp.gr. 2.63; ref. index 1.625; pH 8.0-9.6; 51.0-52.4% silica, 42.1-44.3% alumina, 0.5% max. moisture.

Burgess 5178. [Burgess Pigment] Aluminum silicate, anhyd., surf. treated; pigment providing exc. wet and dry elec. props. in a range of polymers incl. crosslinked PE, EPR, EPT, EPDM, etc.; wh. powd.; 1.0 μ avg. particle size; 0.03-0.009% max. on 325 mesh; sp.gr. 2.63; oil absorp. 55-60; ref. index 1.625; pH 6.5-7.5; 51.0-52.4% silica, 42.1-44.3% alumina, 0.5% max. moisture; *Toxicology:* nontoxic.

Burgess CB. [Burgess Pigment] Aluminum silicate, anhyd., surf. treated; additive providing exc. wet and dry electricals, exc. dispersion in EPR, EPT, crosslinked PE, and polyester; allows for higher filler loadings in polyester BMC, SMC, and clay-filled epoxies; wh. powd.; 1.0 μ avg. particle size; 0.03-0.009% max. on 325 mesh; sp.gr. 2.63; oil absorp. 55-60; ref. index 1.625; pH 6.5-7.5; 51.0-52.4% silica, 42.1-44.3% alumina, 0.5% max. moisture; *Toxicology:* nontoxic.

Burgess KE. [Burgess Pigment] Aluminum silicate, anhyd., surf. treated; very pure, high brightness clay for use in EPR, EPT, crosslinked polyethylene and polyester systems; exc. wet and dry, initial and long term elec. chars.; increases tens. str. and compr. set; powd.; 1.0 μ avg. particle size; 0.009-0.03% on 325 mesh; sp.gr. 2.63; oil absorp. 55-60; ref. index 1.625; pH 6.5-7.5; 51.0-52.4% silica, 42.1-44.3% alumina, 0.5% max. moisture.

Busan® 11-M1. [Buckman Labs] Barium metaborate; CAS 13701-59-2; flame retardant for ABS, acrylics, epoxy, nitrile, phenolic, PC, thermoset polyester, PS, PE, PP, PVAc, PVC, PU.

Busan® IIMI. [Buckman Labs] Proprietary; uv stabilizer for ABS, epoxy, polyesters, PE, PP, flexible and rigid PVC, VDC; powd.; m.p. 1000 C.

Buspense 047. [Buckman Labs] Fatty acid amide; internal/external lubricant for PE, PP, PVC, mold release; liq.

Butasan®. [Monsanto] Zinc dibutyldithiocarbamate; CAS 136-23-2; accelerator for EPDM; rapid accelerator for latexes, NR, and SBR; antioxidant for noncuring or adhesive applics.; powd. or pellets.

Butyl Dioxitol. [Shell] Butoxydiglycol; CAS 112-34-5; EINECS 203-961-6; solv. for nitrocellulose, phenolics, Epon resins, alkyds, and acrylics, oils, and dyes; used in lacquers and inks which require slow evaporating solv.; lacquer and enamel formulations requiring improved flow-out and gloss; coupling solv. in preparation of specialized cleaning sol'ns. and cutting oils; coalescing agent in latex paints; component of brake fluids in automotive industry; colorless liq.; mild odor; water-misc.; sp.gr. 0.949-0.952; 0.10% max. water.

Butyl Eight®. [R.T. Vanderbilt] Activated dithiocarbamate; ultra accelerator; for accelerating vulcanization of spread solv. cements, calendered and extruded stocks at room or slightly elevated temps.; accelerated stocks should be used within 8-12 h to prevent precure; redsh. brn. liq.; sol. in acetone, toluene, chloroform, alcohol, carbon disulfide; dens. 1.01 ± 0.02 mg/m^3; flash pt. 40 C.

Butyl Namate®. [R.T. Vanderbilt] Sodium di-n-butyldithiocarbamate aq. sol'n.; CAS 136-30-1; accelerator for latexes of all elastomers; pale amber liq.; misc. with water; m.w. 227.36; dens. 1.09 ± 0.02 mg/m^3; 47% min. assay; *Toxicology:* discolors skin in presence of copper and iron; *Precaution:* incompat. with acids; hazardous decomp. prods.: oxides of sulfur, nitrogen, sodium and carbon; *Storage:* prod. may crystallize below f.p.; stir and warm to 60 F to dissolve.

Butyl Oleate C-914. [C.P. Hall] Butyl oleate; CAS 142-77-8; EINECS 205-559-6; plasticizer.

Butyl Oxitol. [Shell] Butoxyethanol; CAS 111-76-2; EINECS 203-905-0; solv. for nitrocellulose, alkyds, phenolics, acrylics, Epon resins, nat. resins, dyes, waxes, oils, and org. materials; used in surf. coating formulations such as lacquers and enamels where it imparts gloss and improved leveling chars; coupling agent in cleaners and cutting oils; colorless liq.; mild odor; water-misc.; sp.gr. 0.898-0.901; 0.20% max. water.

Butyl Stearate C-895. [C.P. Hall] Butyl stearate; CAS 123-95-5; EINECS 204-666-5; plasticizer.

Butyl Tuads®. [R.T. Vanderbilt] Tetrabutylthiuram disulfide; CAS 1634-02-2; ultra accelerator; sulfur donor accelerator for sol. cure systems in natural and polyisoprene rubbers; used in conjunction with Amyl Ledate in EPDM; dk. amber liq.; m.w. 408.76; sol. in toluene, chloroform, carbon disulfide; practically insol. in water; dens. 1.06 ± 0.02 mg/m^3; 7.8% avail. sulfur.

Butyl Zimate®. [R.T. Vanderbilt] Zinc di-n-butyldithiocarbamate; CAS 136-23-2; ultra accelerator for EPDM and natural and syn. latexes; provides fast, flat cures in SBR, nitrile, and neoprene latexes; nonstaining, nonblooming antioxidant in noncuring applics. and stabilizer in IIR; antioxidant in thermoplastic rubbers, NR, SR, IIR, adhesives, cements, hot melts, latex; improves heat aging; wh.-cream powd.; 99.9% thru 100 mesh; also avail. as 50% act. slurry; sol. in toluene, carbon disulfide, chloroform, gasoline; m.w. 474.13; dens. 1.21 ± 0.03 mg/m^3; m.p. 104-112 C; 13-15% Zn.

Butyl Zimate® Dustless. [R.T. Vanderbilt] Zinc di-n-butyldithiocarbamate; CAS 136-23-2; rubber accelerator.

Butyl Zimate® Slurry. [R.T. Vanderbilt] Zinc di-n-butyldithiocarbamate slurry; CAS 136-23-2; ultra accelerator for EPDM and natural and syn. latexes; provides fast, flat cures in SBR, nitrile, and neoprene latexes; slurry; 50% act.

BVA ND-205 Solv. [Ridge Tech.] Highly refined naphthalene-depleted aromatic hydrocarbon solv.; CAS 64742-94-5; solv. for textile, agric., resin, industrial degreasing, and fuel oil additive applics.; exc. solvency; EPA clearance; pale yel. clear liq., typ. aromatic odor; sp.gr. 0.99; dens. 8.24 lb/gal; visc. 6.1 cs (32 F); i.b.p. 440 F; pour pt. -10 F; flash pt. (TCC) 215 F; > 98% total aromatics.

BXA Flake. [Uniroyal] Diphenylamine-acetone reaction prod.; CAS 9003-79-6; antioxidant.

BYK®-034. [Byk-Chemie USA] Mixt. of hydrophobic components in paraffin-based min. oil; contains silicone; defoamer for emulsion paints, gloss and semigloss latex systems, emulsion plasters, emulsion adhesives, industrial emulsions (water-thinnable polyurethane resins, alkyd resin emulsions); sp.gr. 0.83-0.89 g/cc; dens. 6.92-7.42 lb/gal; flash pt. (Seta) > 110 C; ref. index 1.472-1.482; 100% NV.

BYK®-065. [Byk-Chemie USA] Polysiloxane sol'n.; defoamer for org. systems incl. chlorinated rubber, vinyl resin, acrylic resin, alkyds; sp.gr. 0.94-0.96 g/cc; dens. 7.76-8.09 lb/gal; flash pt. (Seta) 43 C; ref. index 1.445-1.455; <1% NV in cyclohexanone.

BYK®-066. [Byk-Chemie USA] Polysiloxane sol'n.; defoamer for org. systems incl. chlorinated rubber, epoxy, 2-part polyurethane, alkyd, alkyd/melamine, self-crosslinking acrylates, polyesters; sp.gr. 0.80-0.82 g/cc; dens. 6.76-6.93 lb/gal; flash pt. (Seta) 47 C; ref. index 1.411-1.417; < 1% NV in diisobutyl ketone.

BYK®-080. [Byk-Chemie USA] Polysiloxane copolymer nonaq. emulsion; defoamer for polar org. coating systems, esp. 2-part polyurethane paints and furniture finishes, acrylic paints, chlorinated rubber paints; sp.gr. 1.02-1.05 g/cc; dens. 8.50-8.76 lb/gal; flash pt. (Seta) > 110 C; 85% NV in propylene glycol.

BYK®-141. [Byk-Chemie USA] Polysiloxanes/polymer mixt.; defoamer for solv.-based coatings, esp. wood finishes based on nitrocellulose, acid curing, polyester, or polyurethane, epoxy coatings, chlorinated rubber, or vinyl resin systems; sp.gr. 0.85-0.89 g/cc; dens. 7.09-7.43 lb/gal; flash pt. (Seta) 28 C; ref. index 1.478-1.488; 2-4% NV in naphtha/isobutanol (11/2).

BYK®-2615. [Byk-Chemie USA] 76% Calcium oxide, 20% sat. hydrocarbons, 1% amorphous silica, 1% bis(2-ethylhexyl) phthalate; additive eliminating moisture problems in PVC plastisols; for artificial leather, carpetbacking, casting, coil coating, conveyor belts, dipping, flooring, rotational molding, inks, tarps, wallpapers; wetting and dispersing carriers activate and stabilize the moisture absorber CaO; prevents blistering, bubbles, craters; lt. gray paste, mild hydrocarbon odor; insol. in water; sp.gr. 1.90; dens. 16.67 lb/gal; flash pt. (Seta CC) 96 C; *Toxicology:* LD50 (oral, rat) 7150 mg/kg; sl. skin irritant; severe eye irritant; inh. of vapors may cause respiratory tract irritation, headaches, nausea, vomiting, CNS depression; chronic absorp. may injure kidney; *Precaution:* flamm. limits 0.6-6.5% vol. in air; incompat. with water, strong oxidizers, fluorine; hazardous combustion prods.: CO, CO_2, NO_x, sulfur oxides; *Storage:* moisture sensitive; keep container tightly closed when not in use.

BYK®-3105. [Byk-Chemie USA] Methylalkylpolysiloxane; air release additive for PVC plastisols; for artificial leather, carpetbacking, casting, coil coating, conveyor belts, dipping, flooring, rotational molding, inks, tarps, wallpapers; eliminates or reduces degassing; lt. yel. liq., ester-like odor; insol. in water; sp.gr. 0.91; dens. 7.58 lb/gal; flash pt. (Seta CC) > 65 C; 99% NV; *Toxicology:* LD50 (oral, rat) > 20,000 mg/kg; nonirritating to skin and eyes; inh. of high concs. of vapor may cause respiratory tract irritation; *Precaution:* flamm. limits 0.8-6.8% vol. in air; incompat. with strong acids/oxidizing agents; hazardous combustion prods.: CO, CO_2, silicon compds., formaldehyde; *Storage:* keep container tightly closed when not in use.

BYK®-3155. [Byk-Chemie USA] Polyoxyalkylene derivs.; air release additive for PVC plastisols; for artificial leather, carpetbacking, casting, coil coating, conveyor belts, dipping, flooring, rotational molding, inks, tarps, wallpapers; eliminates or reduces degassing; colorless liq., odorless; sol. in water; sp.gr. 1.26; dens. 9.39 lb/gal; flash pt. (Seta CC) > 150 C; > 80% NV; *Toxicology:* repeated/prolonged contact may cause eye/skin irritation; repeated ingestion may irritate digestive tract; *Precaution:* incompat. with strong oxidizing agents; hazardous combustion prods.: CO, CO_2; *Storage:* keep container tightly closed when not in use.

BYK®-A 500. [Byk-Chemie USA] Silicone-free polymeric sol'n. with 84% lt. aromatic naphtha (CAS 64742-95-6), 6% 1-methoxy-2-propanol acetate; defoamer, air release agent for unsat. polyester laminating, spray-up, hand lay-up molding, gel coats, and solv.-free epoxy flooring systems; prevents air entrapment and porosity; ylsh. clear liq., aromatic odor; insol. in water; sp.gr. 0.88; dens. 7.43 lb/gal; vapor pressure 4 mm Hg (20 C0; b.p. 135-185 C; flash pt. (Seta CC) 45 C; ref. index 1.490-1.500; 6.5% NV; *Toxicology:* LD50 (oral, rat) 4200 mg/kg; nonirritant to skin; sl. eye irritant; inh. of vapors may cause respiratory tract irritation, headache, nausea, vomiting, CNS depression; *Precaution:* flamm.; flamm. limits 1-12% vol. in air; incompat. with strong oxidizing agents; hazardous combustion prods.: CO, CO_2; 1-methoxy-2-propanol acetate may form explosive peroxides on exposure to air; *Storage:* keep container tightly closed when not in use; avoid heat, flame, ignition sources.

BYK®-A 501. [Byk-Chemie USA] Silicone-free polymeric sol'n. with 50% lt. aromatic naphtha (CAS 64742-95-6), 7% 1-methoxy-2-propanol acetate; defoamer, air release agent for unsat. polyester laminating, spray-up, hand lay-up molding, gel coats, and solv.-free epoxy flooring systems; prevents air entrapment and porosity; ylsh. clear liq., aromatic odor; insol. in water; sp.gr. 0.89; dens. 7.43 lb/gal; vapor pressure 4 mm Hg (20 C); b.p. 107-185 C; flash pt. (Seta CC) 46 C; ref. index 1.497-1.503; 44% NV; *Toxicology:* LD50 (oral, rat) 4400 mg/kg; nonirritating to skin and eyes; inh. of vapors may cause respiratory tract irritation, headaches, nausea, vomiting, CNS depression; *Precaution:* flamm.; flamm. limits 1-12% vol. in air; incompat. with strong oxidizing agents; hazardous combustion prods.: CO, CO_2; 1-methoxy-2-

propanol acetate may form explosive peroxides on exposure to sir; *Storage:* keep container tightly closed when not in use; avoid heat, flame, ignition sources.

BYK®-A 510. [Byk-Chemie USA] Silicone-free polymeric sol'n. with 50% lt. aromatic naphtha (CAS 64742-95-6), 29% 1-methoxy-2-propanol acetate, 5% Stod., 4% isobutanol; defoamer, air release agent for unsat. polyester resins; ylsh. clear liq., min. spirit-like odor; insol. in water; sp.gr. 0.90; dens. 7.51 lb/gal; vapor pressure 4 mm Hg (20 C); b.p. 107-196 C; flash pt. (Seta CC) 37 C; ref. index 1.465; 12% NV; *Toxicology:* LD50 (oral, rat) 4300 mg/kg; nonirritating to skin and eyes; inh. of vapors may cause respiratory tract irritation, headaches, nausea, vomiting, CNS depression, blood/bone marrow/liver/kidney damage; *Precaution:* flamm.; flamm. limits 0.6-12% vol. in air; incompat. with strong oxidizing agents; hazardous combustion prods.: CO, CO_2, phosphorous oxides; 1-methoxy-2-propanol acetate may form explosive peroxides on exposure to sir; *Storage:* keep container tightly closed when not in use; avoid heat, flame, ignition sources.

BYK®-A 515. [Byk-Chemie USA] Silicone-free polymeric sol'n. with 71% Stod., 8% hydroxyacetic acid butyl ester, 1% 2-butoxyethanol; defoamer, air release agent for unsat. polyester resins (gel coats, fiber wetting in reinforced applics.); colorless liq., hydrocarbon-like odor; insol. in water; sp.gr. 0.82; dens. 6.84 lb/gal; vapor pressure 4 mm Hg (20 C); b.p. 143-199 C; flash pt. (Seta CC) 34 C; ref. index 1.440; 20% NV; *Toxicology:* LD50 (oral, rat) 14,400 mg/kg; nonirritating to eyes; sl. skin irritant; inh. of 2-butoxyethanol vapors may cause respiratory tract irritation, headaches, nausea, vomiting, CNS depression; *Precaution:* flamm.; flamm. limits 0.6-10.6% vol. in air; incompat. with strong oxidizing agents; hazardous combustion prods.: CO, CO_2, phosphorous oxides; *Storage:* keep container tightly closed when not in use; avoid heat, flame, ignition sources.

BYK®-A 525. [Byk-Chemie USA] 52% Polyether-modified methylalkylpolysiloxane copolymer with 43% Stod., 5% 1-emthoxy-2-propanol acetate; air release additive for thermoset resin systems sucyh as unfilled/filled epoxy, PU, or phenolic resins; prevents air inclusions and porosity; colorless clear liq., ether-like odor; insol. in water; sp.gr. 0.85; dens. 7.17 lb/gal; vapor pressure 2 mm Hg (20 C); b.p. 135-196 C; flash pt. (Seta CC) 38 C; ref. index 1.441; 52% NV; *Toxicology:* LD50 (oral, rat) > 10,000 mg/kg; nonirritating to skin and eyes; inh. of vapors may cause respiratory tract irritation, headaches, nausea, vomiting, CNS depression; *Precaution:* flamm. limits 0.6-12% vol. in air; incompat. with strong oxidizing agents; hazardous combustion prods.: CO, CO_2, silicon compds., formaldehyde; 1-methoxy-2-propanol acetate may form explosive peroxides on exposure to air; *Storage:* keep container tightly closed when not in use; avoid heat, flame, ignition sources.

BYK®-A 530. [Byk-Chemie USA] Polyether-modified methylalkylpolysiloxane copolymer with 95% sat. hydrocarbons; defoamer, air release additive for ambient cure epoxy resin systems, esp. floorings; colorless-ylsh. clear-sl. muddy liq., hydrocarbon-like odor; insol. in water; sp.gr. 0.81 (20 C); dens. 6.76 lb/gal; b.p. 235-257 C; flash pt. (Seta CC) 95 C; ref. index 1.448; 5% NV; *Toxicology:* inh. of vapors may cause respiratory tract irritation, headaches, nausea, vomiting, CNS depression; repeated/prolonged contact may cause eye and skin irritation; repeated ingestion may irritate digestive tract; *Precaution:* flamm. limits 0.6-6.5% vol. in air; incompat. with strong oxidizing agents; hazardous combustion prods.: CO, CO_2, silicon compds., formaldehyde; *Storage:* keep container tightly closed when not in use.

BYK®-A 555. [Byk-Chemie USA] Silicone-free polymeric sol'n. with 62% lt. aromatic naphtha (CAS 64742-95-6); defoamer, air release agent for unsat. polyester resins (gel coats, putty, nonreinforced and reinforced applics., continuous panels); ylsh. clear liq., aromatic odor; insol. in water; sp.gr. 0.88; dens. 7.34 lb/gal; vapor pressure 4 mm Hg (20 C); b.p. 165-185 C; flash pt. (Seta CC) 43 C; ref. index 1.507; 38% NV; *Toxicology:* LD50 (oral, rat) 2275 mg/kg (act. ingred.); inh. of vapors may cause respiratory tract irritation, headaches, nausea, vomiting, CNS depression; *Precaution:* flamm. limits 1.0-7.5% vol. in air; incompat. with strong oxidizing agents; hazardous combustion prods.: CO, CO_2; *Storage:* keep container tightly closed when not in use; avoid heat, flame, ignition sources.

BYK®-LP R 6237. [Byk-Chemie USA] Urea urethane with 45% 1-methyl-2-pyrrolidone and 2% lithium chloride; patented thixotrope, antisag agent for PVC plastisols and bituminous prods.; for artificial leather, carpetbacking, casting, coil coating, conveyor belts, dipping, flooring, rotational molding, inks, tarps, wallpapers; yel. liq., amine-like odor; insol. in water; sp.gr. 1.13 (20 C); dens. 9.42 lb/gal; b.p. 202-205 C; flash pt. (Seta CC) 95 C; ref. index 1.523; *Toxicology:* LD50 (oral, rat) 8000 mg/kg; nonirritating to skin/eyes; repeated ingestion may cause digestive tract irritation; prolonged absorp. of lithium chloride may injure blood and/or kidney; *Precaution:* flamm. limits 1.3-9.5% vol. in air; incompat. with strong oxidizing agents; hazardous combustion prods.: CO, CO_2, NO_x, chlorinated compds.; *Storage:* keep container tightly closed when not in use; avoid heat, flame, ignition sources.

BYK®-LP W 6246. [Byk-Chemie USA] Polyether-modified dimethylpolysiloxane with 48% 2-methoxymethylethoxypropanol; additive improving substrate wetting of plastic surfs. for water-based systems used for printing and finishing of flexible PVC; reduces surf. tens.; ylsh. clear liq., ether-like odor; insol. in water; sp.gr. 1.00 (20 C); b.p. 184 C (solv.); flash pt. (Seta CC) 81 C; ref. index 1.435; 46% NV; *Toxicology:* high concs. of vapors may cause respiratory tract irritation, CNS depression; repeated/prolonged contact may irritate skin/eyes;

repeated ingestion may irritate digestive tract; *Precaution:* flamm. limits 1.3-8.7% vol. in air; incompat. with strong oxidizing agents; hazardous combustion prods.: CO, CO_2, silicon compds., formaldehyde; 2-methoxymethylethoxypropanol may form peroxides on exposure to air; *Storage:* keep container tightly closed when not in use; avoid heat, flame, ignition sources.

BYK®-R 605. [Byk-Chemie USA] Polyhydroxy carboxylic acid amides with 20% lt. aromatic naphtha (CAS 64742-95-6), 19% xylene, 5% isobutanol, 5% ethylbenzene; rheology additive for fumed silica-modified thermoset resin systems, e.g., unsat. polyesters, vinyl esters, epoxy resins; wetting and dispersing agent for fumed silica; stabilizes thixotropic behavior; reduces or prevents separation; brnsh. liq., aromatic odor; insol. in water; sp.gr. 0.93; dens. 7.67 lb/gal; vapor pressure 5 mm Hg (20 C); b.p. 107-185 C; flash pt. (Seta CC) 29 C; ref. index 1.496; 52% NV; *Toxicology:* LD50 (oral, rat) > 10,000 mg/kg; nonirritant to skin and eyes; inh. of vapors may cause respiratory tract irritation, headache, nausea, vomiting, CNS depression, cardiac injury, blood/blome marrow/liver/kidney damage, etc.; *Precaution:* flamm.; flamm. limits 1.0-10.7% vol. in air; incompat. with strong oxidizing agents; hazardous combustion prods.: CO, CO_2, NO_x, sulfur oxides; *Storage:* keep container tightly closed when not in use; avoid heat, flame, ignition sources.

BYK®-S 706. [Byk-Chemie USA] Polyacrylate with 45% lt. aromatic naphtha (CAS 64742-95-6) and 2,6-dimethyl-4-heptanone (CAS 108-83-8); leveling agent for gel coats based on unsat. polyester resins and in solv.-free epoxy resin systems; air release additive in PU resins; colorless clear liq., nonspecific odor; insol. in water; sp.gr. 0.95; dens. 7.93 lb/gal; vapor pressure 3 mm Hg (20 C); b.p. 160-185 C (solv.); flash pt. (Seta) 45 C; ref. index 1.480; 51% NV; *Toxicology:* inh. of vapors may cause respiratory irritation, headaches, nausea, vomiting, CNS depression; repeated/prolonged contact may cause skin/eye irritation; repeated ing. may irritate digestive tract; chronic absorp. may injure kidneys, liver; *Precaution:* flamm.; flamm. limits 1.0-7.5% vol. in air; hazardous combustion prods.: CO, CO_2; *Storage:* keep container tightly closed when not in use; avoid heat, flame, ignition sources.

BYK®-S 715. [Byk-Chemie USA] Blend of high boiling aromatics, ketones, esters; contains 65% lt. aromatic naphtha (CAS 64742-95-6), 25% 2,6-dimethyl-4-heptanone, 5% dipentene, 3% ethyl lactate, 3% alkyl glycolate; flow additive for solv.-free epoxy resin systems; usually used in combination with air release additives; colorless clear liq., aromatic odor; insol. in water; sp.gr. 0.86; dens. 7.18 lb/gal; vapor pressure 3 mm Hg (20 C); b.p. 166-185 C (solv.); flash pt. (Seta) 43 C; ref. index 1.472; *Toxicology:* LD50 (oral, rat) 5000 mg/kg; sl. skin irritant; nonirritating to eyes; inh. of vapors may cause respiratory irritation, headaches, nausea, vomiting, CNS depression; chronic absorp. may injure liver, kidneys; *Precaution:* flamm.; flamm. limits 1.0-7.5% vol. in air; hazardous combustion prods.: CO, CO_2; *Storage:* keep container tightly closed when not in use; avoid heat, flame, ignition sources.

BYK®-S 740. [Byk-Chemie USA] Hydroxypolyesters with 10% paraffin wax (CAS 64742-43-4); patented styrene emission suppressant for unsat. polyester and vinyl ester resins; used in open mold applics., e.g., hand lay-up, spray-up, filament winding; sl. brnsh. paste, paraffinic odor; sp.gr. 0.85; dens. 7.09 lb/gal; acid no. < 7; flash pt. (Seta CC) 46 C; *Toxicology:* LD50 (oral, rat) 15,900 mg/kg; nonirritating to skin; sl. eye irritant; inh. of high concs. of vapor may cause respiratory tract irritation; repeated ingestion may irritate digestive tract; *Precaution:* flamm.; incompat. with strong oxidizing agents; hazardous combustion prods.: CO, CO_2, NO_x, phosphorous oxides; *Storage:* keep container tightly closed when not in use; avoid heat, flame, ignition sources.

BYK®-W 900. [Byk-Chemie USA] Bicyclic boric acid esters (80%) with 16% xylene, 4% ethylbenzene; wetting and dispersing additive for fillers in filled FRP composites; insures better filler/resin compat.; colorless clear liq., aromatic odor; insol. in water; sp.gr. 0.96-1.00; dens. 8.18 lb/gal; vapor pressure 7 mm Hg (20 C0; b.p. 135-143 C; flash pt. (Seta CC) 43 C; ref. index 1.460-1.470; 65-85% NV; *Toxicology:* LD50 (oral, rat) 2465 mg/kg; sl. eye irritant, nonirritating to skin; inh. of vapor may cause respiratory tract irritation, headaches, nausea, vomiting, CNS depression, cardiac injury, blood/liver/kidney damage; *Precaution:* flamm.; incompat. with strong oxidizing materials; hazardous combustion prods.: CO, CO_2; *Storage:* store in tightly closed container; avoid heat, flame, ignition sources.

BYK®-W 920. [Byk-Chemie USA] Salt of unsat. carboxylic acid polyamine amides and higher m.w. acidic polyesters (49%) with 33% xylene, 8% ethylbenzene, 5% 1,2-propanediol and isobutanol; wetting and dispersing additive for plastic composites; recommended for fillers such as calcium carbonate, alumina trihydrate, most common pigments; improves suspension, reduces visc., improves physicals; ylsh. clear liq., aromatic odor; insol. in water; sp.gr. 0.93-0.95; dens. 7.76-7.93 lb/gal; vapor pressure 8 mm Hg (20 C); b.p. 99-143 C; acid no. 20-28; amine no. 16-22; flash pt. (Seta) 25 C; ref. index 1.484-1.494; 48-52% NV; *Toxicology:* LD50 (oral, rat) 12,000 mg/kg; sl. skin irritant; eye irritant; inh. of vapors may cause respiratory tract irritation, headaches, nausea, vomiting, CNS depression, cardiac injury, blood/liver/kidney damage, etc.; *Precaution:* flamm.; incompat. with strong oxidizing materials; hazardous combustion prods.: CO, CO_2, NO_x; *Storage:* store in tightly closed container; avoid heat, flame, ignition sources.

BYK®-W 935. [Byk-Chemie USA] Unsat. polycarboxylic acid polymer with 36% xylene, 9% ethylbenzene, 3% 2,6-Dimethyl-4-heptanone, 2% 4,6-Dimethyl-2-heptanone; wetting, dispersing, and antiflooding additive for unsat. polyester resin systems, esp. gel coats, polyester pastes, other filled or pigmented paste systems; prevents flooding of TiO_2 with colored pigments; prevents sedimentation of inorg. and basic pigments and carbonate extenders; yel. clear liq., aromatic odor; insol. in water; sp.gr. 0.95; dens. 7.92 lb/gal (20 C); vapor pressure 7 mm Hg (20 C); b.p. 138-166 C; acid no. 180; flash pt. (Seta CC) 28 C; ref. index 1.490; *Toxicology:* LD50 (oral, rat) > 5000 mg/kg; sl. skin and eye irritant; inh. of vapors may cause respiratory tract irritation, headaches, nausea, vomiting, CNS depression, cardiac injury, blood/liver/kidney damage, etc.; *Precaution:* flamm.; incompat. with strong oxidizing materials; hazardous combustion prods.: CO, CO_2, phosphorous oxides; *Storage:* store in tightly closed container; avoid heat, flame, ignition sources.

BYK®-W 940. [Byk-Chemie USA] Unsat. polycarboxylic acid polymer with a polysiloxane copolymer, with 36% xylene, 9% ethylbenzene, 3% 2,6-Dimethyl-4-heptanone, 2% 4,6-Dimethyl-2-heptanone; wetting, dispersing, and antiflooding additive for unsat. polyester resin systems; prevents settling of colored pigments, TiO_2, and carbonate extenders; improves color development; sl. brn. clear liq., aromatic odor; insol. in water; sp.gr. 0.95; dens. 7.92 lb/gal (20 C); vapor pressure 7 mm Hg (20 C); b.p. 138-166 C; acid no. 150; flash pt. (Seta CC) 28 C; ref. index 1.489; *Toxicology:* LD50 (oral, rat) > 5000 mg/kg; sl. skin and eye irritant; inh. of vapors may cause respiratory tract irritation, headaches, nausea, vomiting, CNS depression, cardiac injury, blood/liver/kidney damage, etc.; *Precaution:* flamm.; incompat. with strong oxidizing materials; hazardous combustion prods.: CO, CO_2, phosphorous oxides; *Storage:* store in tightly closed container; avoid heat, flame, ignition sources.

BYK®-W 965. [Byk-Chemie USA] Salt of long chain polyamine amides and a polar acidic ester with 41% Stod. solv., 5% 2-butoxyethanol, 2% xylene, 1% 1,2-propanediol, 0.4% ethylbenzene; wetting and dispersing additive for filled, unsat. polyester resin systems; esp. for use with alumina trihydrate, calcium carbonate, calcium sulfate, and talc; widely used in putty compds.; reduces visc., improves flowability yielding higher filler loading; brn. liq., aromatic odor; insol. in water; sp.gr. 0.90; dens. 7.51 lb/gal (20 C); vapor pressure 5 mm Hg (20 C); b.p. 138-199 C; acid no. 30; amine no. 14; flash pt. (Seta CC) 35 C; ref. index 1.467; 52% NV; *Toxicology:* LD50 (oral, rat) > 5000 mg/kg; sl. skin and eye irritant; inh. of vapors may cause respiratory tract irritation, headaches, nausea, vomiting, CNS depression, cardiac injury, blood/liver/kidney damage, etc.; *Precaution:* flamm.; flamm. limits 0.6-10.6% vol. in air; incompat. with strong oxidizing materials; hazardous combustion prods.: CO, CO_2, NO_x; *Storage:* store in tightly closed container; avoid heat, flame, ignition sources.

BYK®-W 968. [Byk-Chemie USA] > 35% Alkylol ammonium salt of an acidic polyester with 60% butyl benzyl phthalate, < 5% hydroxyethyl alkenyl imidazoline; wetting and dispersing additive for min. fillers in amine-accelerated unsat. polyester putty compds.; reduces visc. yielding higher filler loading; lt. brn. clear liq., amine-like odor; sol. in water; sp.gr. 1.10; dens. 9.18 lb/gal (20 C); acid no. 30; amine no. 28; flash pt. (Seta CC) > 100 C; ref. index 1.520; > 99% NV; *Toxicology:* LD50 (oral, rat) > 8000 mg/kg; eye irritant; corrosive to skin; inh. of vapors may cause respiratory tract irritation; butyl benzyl phthalate may have carcinogenic potential; *Precaution:* corrosive liq.; incompat. with strong oxidizers; hazardous combustion prods.: CO, CO_2, NO_x, sulfur oxides, phosphorous oxides; *Storage:* may become turbid below 0 C; warm and homogenize to reverse effect; protect against moisture; keep container tightly closed when not in use.

BYK®-W 972. [Byk-Chemie USA] Higher m.w. block copolymer with 58% 1-methoxy-2-propanol acetate, 10% butyl acetate, 1% 2-methoxy-1-propanol acetate, 0.1% ethylbenzene; additive to reduce separation in filled, low shrink SMC/BMC, wet molding, and other thermoset molding compds. and liq. colorants for pigmented molding compds.; improves pigment stability and color uniformity; ylsh. clear liq., ester-like odor; insol. in water; sp.gr. 1.00-1.03; vapor pressure 5 mm Hg (20 C); b.p. 124-146 C; amine no. 9-12; flash pt. (Seta CC) 39 C; ref. index 1.434-1.442; 29-31% NV; *Toxicology:* nonirritating to skin and eyes; inh. of vapors may cause respiratory tract irritation, CNS depression, narcosis; chronic absorp. may injur liver, kidney, respiratory system; *Precaution:* flamm.; flamm. limits 1.2-12% vol. in air; incompat. with strong oxidizing agents; hazardous combustion prods.: CO, CO_2, NO_x; 1-methoxy-2-propanol may form explosive peroxides on exposure to air; *Storage:* keep container tightly closed when not in use; avoid heat, flame, sources of ignition.

BYK®-W 980. [Byk-Chemie USA] 77% Salt of unsat. polyamine amides and acidic polyesters with 20% 2-butoxyethanol, 2% xylene, 0.5% ethylbenzene; wetting and dispersing additive for filled, unsat. polyester and epoxy resin systems, esp. those with alumina trihydrate or calcium carbonate fillers and silica-filled epoxy flooring compds.; reduces visc. yielding higher filler loading; brn. liq., glycol-like odor; insol. in water; sp.gr. 0.99; dens. 8.26 lb/gal (20 C); vapor pressure 1 mm Hg (20 C); b.p. 132-171 C; acid no. 40; amine no. 30; flash pt. (Seta CC) 66 C; ref. index 1.478; 80% NV; *Toxicology:* LD50 (oral, rat) 8000 mg/kg; sl. skin irritant; eye irritant; inh. of vapors may cause respiratory tract irritation, headaches, nausea, vomiting, CNS depression, cardiac injury, liver/kidney damage, etc.; *Precaution:* combustible; flamm. limits 1.2-10.7% vol. in air; incompat. with strong oxidizers; hazardous combustion prods.: CO, CO_2, NO_x; *Storage:* keep container tightly closed when not in use; avoid heat, flame, sources of ignition.

BYK®-W 990. [Byk-Chemie USA] Partial salt of acidic

polyesters with 18% lt. aromatic naphtha, 13% xylene, 3% ethylbenzene, 2% 1,2-propanediol, 2% isobutanol; wetting and dispersing additive for min. fillers in heat curing, glass fiber-reinforced unsat. polyester formulations, SMC, BMC, DMC, ZNC formulations with fillers such as calcium carbonate and alumina trihydrate; reduces visc. yielding higher filler loading; does not interfere with paintability; ylsh. clear liq., strong hydrocarbon odor; insol. in water; sp.gr. 0.98-1.02; dens. 8.1-8.3 lb/gal; vapor pressure 5 mm Hg (20 C); b.p. 135-185 C; acid no. 50-70; amine no. 5-10; flash pt. (Seta CC) 34 C; 60-64% NV; *Toxicology:* LD50 (oral, rat) 9100 mg/kg; sl. skin irritant; eye irritant; inh. of vapors may cause respiratory tract irritation, headaches, nausea, vomiting, CNS depression, cardiac injury, blood/liver/kidney damage, etc.; *Precaution:* flamm.; flamm. limits 1.0-7.5% vol. in air; incompat. with strong oxidizers; hazardous combustion prods.: CO, CO_2, NO_x, phosphorous oxides; *Storage:* keep container tightly closed when not in use; avoid heat, flame, sources of ignition.

BYK®-W 995. [Byk-Chemie USA] Sat. polyester with acid groups, with 23.5% 1-methoxy-2-propanol acetate, 23% lt. aromatic naphtha, 1.5% phosphoric acid (residual), 0.5% 2-methoxy-1-propanol acetate; wetting and dispersing additive for hot curing, fiber glass-reinforced UP resin systems, low-shrink SMC/BMC/DMC formulations based on unsat. polyester and PS or unsat. pllyester and S/B copolymers filled with calcium carbonate, alumina trihydrate; increase filler loading, improves flow behavior; yel. clear to sl. cloudy liq., aromatic odor; insol. in water; sp.gr. 0.98-1.06; vapor pressure 4 mm Hg (20 C); b.p. 146-185 C; acid no. 48-54; flash pt. (Seta CC) 44 C; 49-51% NV; *Toxicology:* eye irritant; nonirritating to skin; inh. of vapors may cause respiratory tract irritation, headaches, nausea, vomiting, CNS depression; chronic absorp. may cause liver/kidney/respiratory system injury; *Precaution:* flamm.; flamm. limits 1.0-10.8% vol. in air; incompat. with strong oxidizers; hazardous combustion prods.: CO, CO_2, sulfur oxides, phosphorus oxides; 1-methoxy-2-propanol acetate may form explosive peroxides on exposure to air; *Storage:* keep container tightly closed when not in use; avoid heat, flame, sources of ignition.

BYK®-Catalyst 450. [Byk-Chemie USA] Sol'n. of an amine salt of p-toluene sulfonic acid in 1-methoxy-2-propanol/1,2-propylene glycol/water (29:2:1); additive to improve curing in acid catalyzable org. systems incl. polyesters, acrylics; sol'n., char. odor of pyridine; sp.gr. 1.01-1.03; dens. 8.43-8.60 lb/gal; flash pt. (Seta) 35 C; 25-28% NV.

BYK®-Catalyst 451. [Byk-Chemie USA] Sol'n. of an amine salt of p-toluene sulfonic acid in n-propanol/methanol (6:1); additive to improve curing in acid catalyzable org. systems incl. polyesters, acrylics; sol'n., char. odor of pyridine; sp.gr. 0.90-0.92; dens. 7.51-7.68 lb/gal; flash pt. (Seta) 12 C; 25-27% NV.

BYK®-Catalyst 460. [Byk-Chemie USA] Sol'n. of an amine salt of sulfonic acid in methanol; acid catalyst for aq. and solv.-based baking enamels; used in polyester, acrylic, and alkyd resins which are crosslinked with melamine resins; sol'n.; sp.gr. 1.03-1.09; dens. 8.59-9.09 lb/gal; acid no. 135; amine no. 20; flash pt. (Seta) 20 C; 60-65% NV.

Byketol®-OK. [Byk-Chemie USA] Mixt. of high boiling aromatics, ketones and esters; additive to counteract surface defects and improve leveling for solv.-based coatings, chlorinated rubber systems, silk screen inks.

C

Cab-O-Sil® EH-5. [Cabot/Cab-O-Sil] Fumed silica, undensed; CAS 112945-52-5; rheology control, reinforcing, and free-flow agent for adhesives, coatings, powd. coatings, cosmetics, defoamers, inks, insecticides, lubricants, plastisols, liq. resins, and sealants; 0.02% 325 mesh residue; sp.gr. 2.2; bulk dens. 2.5 lb/ft^3; surf. area 380 ± 30 m^2/g; ref. index 1.46; pH 3.7-4.3 (4% aq. slurry); > 99.8% assay; *Toxicology:* LD50 (oral, rat) > 5 g/kg; inert to mildly irritating to skin; inert to very mildly irritating to eyes; *Storage:* store in dry environment away from chemical vapors.

Cab-O-Sil® H-5. [Cabot/Cab-O-Sil] Fumed silica, undensed; CAS 112945-52-5; reinforcing agent for adhesives, coatings, defoamers, elastomers, lubricants, plastisols, liq. resins, sealants, silicones; 0.02% 325 mesh residue; sp.gr. 2.2; bulk dens. 2.5 lb/ft^3; surf. area 300 ± 25 m^2/g; ref. index 1.46; pH 3.7-4.3 (4% aq. slurry); > 99.8% assay; *Toxicology:* LD50 (oral, rat) > 5 g/kg; inert to mildly irritating to skin; inert to very mildly irritating to eyes; *Storage:* store in dry environment away from chemical vapors.

Cab-O-Sil® HS-5. [Cabot/Cab-O-Sil] Fumed silica, undensed; CAS 112945-52-5; rheology control and reinforcing agent for adhesives, coatings, cosmetics, defoamers, elastomers, foods, inks, insecticides, lubricants, pharmaceuticals, plastisols, liq. resins, sealants, silicones, thermal insulation; 0.02% 325 mesh residue; bulk dens. 2.5 lb/ft^3; surf. area 325 ± 25 m^2/g; pH 3.7-4.3 (4% aq. slurry); > 99.8% assay; *Toxicology:* LD50 (oral, rat) > 5 g/kg; inert to mildly irritating to skin; inert to very mildly irritating to eyes; *Storage:* store in dry environment away from chemical vapors.

Cab-O-Sil® L-90. [Cabot/Cab-O-Sil] Fumed silica, undensed; CAS 112945-52-5; dispersant, anticaking agent, rheology control, reinforcement for adhesives, elastomers, sealants, silicones; 0.02% 325 mesh residue; sp.gr. 2.2; bulk dens. 3 lb/ft^3; surf. area 100 ± 15 m^2/g; ref. index 1.46; pH 3.7-4.3 (4% aq. slurry); > 99.8% assay; *Toxicology:* LD50 (oral, rat) > 5 g/kg; inert to mildly irritating to skin; inert to very mildly irritating to eyes; *Storage:* store in dry environment away from chemical vapors.

Cab-O-Sil® LM-130. [Cabot/Cab-O-Sil] Fumed silica, undensed; CAS 112945-52-5; dispersant, anticaking agent, rheology control, reinforcement for adhesives, elastomers, sealants, silicones; 0.02% 325 mesh residue; sp.gr. 2.2; bulk dens. 3 lb/ft^3; surf. area 130 ± 15 m^2/g; ref. index 1.46; pH 3.7-4.3 (4% aq. slurry); > 99.8% assay; *Toxicology:* LD50 (oral, rat) > 5 g/kg; inert to mildly irritating to skin; inert to very mildly irritating to eyes; *Storage:* store in dry environment away from chemical vapors.

Cab-O-Sil® LM-150. [Cabot/Cab-O-Sil] Fumed silica, undensed; CAS 112945-52-5; dispersant, anticaking agent, rheology control, reinforcement for adhesives, elastomers, sealants, silicones; 0.02% 325 mesh residue; sp.gr. 2.2; bulk dens. 2.5 lb/ft^3; surf. area 160 ± 15 m^2/g; ref. index 1.46; pH 3.7-4.3 (4% aq. slurry); > 99.8% assay; *Toxicology:* LD50 (oral, rat) > 5 g/kg; inert to mildly irritating to skin; inert to very mildly irritating to eyes; *Storage:* store in dry environment away from chemical vapors.

Cab-O-Sil® LM-150D. [Cabot/Cab-O-Sil] Fumed silica, densed; CAS 112945-52-5; rheology control and reinforcing agent for adhesives, elastomers, sealants, silicones; 0.02% 325 mesh residue; sp.gr. 2.2; bulk dens. 5 lb/ft^3; surf. area 160 ± 15 m^2/g; ref. index 1.46; pH 3.7-4.3 (4% aq. slurry); > 99.8% assay; *Toxicology:* LD50 (oral, rat) > 5 g/kg; inert to mildly irritating to skin; inert to very mildly irritating to eyes; *Storage:* store in dry environment away from chemical vapors.

Cab-O-Sil® M-5. [Cabot/Cab-O-Sil] Fumed silica; CAS 112945-52-5; rheology control, reinforcement, free flow agent for adhesives, coatings, powd. coatings, cosmetics, defoamers, elastomers, foods, inks, insecticides, lubricants, pharmaceuticals, plastisols, liq. resins, sealants, thermal insulation; wh. fine powd., odorless; 0.2-0.3 μ avg. particle size; 0.02% 325 mesh residue; insol. in water; sp.gr. 2.2; dens. 18.3 lb/gal; bulk dens. 2.5 lb/ft^3; surf. area 200 ± 25 m^2/g; ref. index 1.46; pH 3.7-4.3 (40% aq. slurry); > 99.8% assay; *Toxicology:* LD50 (oral, rat) > 5 g/kg; nontoxic by ingestion; inert to mildly irritating to skin; inert to very mildly irritating to eyes; *Storage:* store in dry environment away from chemical vapors.

Cab-O-Sil® M-7D. [Cabot/Cab-O-Sil] Fumed silica, densed; CAS 112945-52-5; reinforcing agent for adhesives, elastomers, sealants, silicones; 0.02% 325 mesh residue; sp.gr. 2.2; bulk dens. 5 lb/ft^3; surf. area 200 ± 25 m^2/g; ref. index 1.46; pH 3.7-4.3 (4% aq. slurry); > 99.8% assay; *Toxicology:* LD50 (oral, rat) > 5 g/kg; inert to mildly irritating to skin; inert to very mildly irritating to eyes; *Storage:* store in dry environment away from chemical vapors.

Cab-O-Sil® MS-55. [Cabot/Cab-O-Sil] Fumed silica, undensed; CAS 112945-52-5; rheology control and free-flow agent for elastomers, silicones, thermal insulation; 0.02% 325 mesh residue; sp.gr. 2.2; bulk dens. 2.5 lb/ft^3; surf. area 255 ± 25 m^2/g; ref. index 1.46; pH 3.7-4.3 (4% aq. slurry); > 99.8% assay; *Toxicology:* LD50 (oral, rat) > 5 g/kg; inert to mildly irritating to skin; inert to very mildly irritating to eyes; *Storage:* store in dry environment away from chemical vapors.

Cab-O-Sil® MS-75D. [Cabot/Cab-O-Sil] Fumed silica, densed; CAS 112945-52-5; reinforcing agent for elastomers, sealants, silicones; 0.02% 325 mesh residue; sp.gr. 2.2 bulk dens. 5 lb/ft^3; surf. area 255 ± 25 m^2/g; ref. index 1.46; pH 3.7-4.3 (4% aq. slurry); > 99.8% assay; *Toxicology:* LD50 (oral, rat) > 5 g/kg; inert to mildly irritating to skin; inert to very mildly irritating to eyes; *Storage:* store in dry environment away from chemical vapors.

Cab-O-Sil® PTG. [Cabot/Cab-O-Sil] Fumed silica, undensed; CAS 112945-52-5; rheology control and free-flow agent for adhesives, coatings, inks, lubricants, pharmaceuticals, plastisols, liq. resins, sealants; 0.02% 325 mesh residue; sp.gr. 2.2; bulk dens. 2.5 lb/ft^3; surf. area 200 ± 25 m^2/g; ref. index 1.46; pH 3.7-4.3 (4% aq. slurry); > 99.8% assay; *Toxicology:* LD50 (oral, rat) > 5 g/kg; inert to mildly irritating to skin; inert to very mildly irritating to eyes; *Storage:* store in dry environment away from chemical vapors.

Cab-O-Sil® TS-530. [Cabot/Cab-O-Sil] Fumed silica, hexamethyldisilazane-surface treated (CAS 68909-20-6); rheology control agent; reinforcing filler for elastomers, dental compds.; free-flow agent for toners, powd. coatings; antisettling agent in coatings; adsorptive carrier base for perfumes, pesticides, veterinary prods., etc.; also in defoamers, liq. resins, silicones, thermal insulation; wh. fluffy powd., odorless; 0.2-0.3 μ avg. particle size; insol. in water; sp.gr. 2.2; bulk dens. 3 lb/ft^3; surf. area 215 ± 30 m^2/g; pH 4.8-7.5 (4% aq. slurry); 4.2 ± 0.5% carbon content; *Toxicology:* LD50 (oral) > 5000 mg/kg; inh. may cause pulmonary inflammation; not considered a carcinogen; *Precaution:* dry powds. can build static elec. charges subjected to friction; keep away from flamm. or explosive liqs.; *Storage:* store in dry environment away from chemical vapors.

Cab-O-Sil® TS-610. [Cabot/Cab-O-Sil] Fumed silica, dimethyldichlorosilane-surface treated (CAS 68611-44-9); rheology control, reinforcing, free-flow, and hydrophobicity agent for adhesives, coatings, powd. coatings, defoamers, elastomers, inks, insecticides, lubricants, sealants, silicones, thermal insulation, toners; wh. fluffy powd., odorless; 0.2-0.3 μ avg. particle size; insol. in water; sp.gr. 2.2; dens. 18.3 lb/gal; bulk dens. 3 lb/ft^3; surf. area 120 ± 20 m^2/g; ref. index 1.461; pH 4-5 (4% aq. slurry); 0.85 ± 0.15% carbon content; *Toxicology:* LD50 (oral) > 5000 mg/kg; inh. of dust may cause pulmonary inflammation; not considered a carcinogen; *Precaution:* dry powds. can build static elec. charges subjected to friction; keep away from flamm. or explosive liqs.; *Storage:* store in dry environment away from chemical vapors.

Cab-O-Sil® TS-720. [Cabot/Cab-O-Sil] Fumed silica, polydimethylsiloxane fluid-surface treated (CAS 67762-90-7); rheology control, reinforcing, free-flow, and hydrophobicity agent for adhesives, coatings, powd. coatings, defoamers, elastomers, inks, lubricants, liq. resins, sealants, silicones, toners; thixotrope for vinyl ester resins; wh. fluffy powd., odorless; 0.2-0.3 μ avg. particle size; insol. in water; sp.gr. 1.8; dens. 15 lb/gal; bulk dens. 3 lb/ft^3; surf. area 100 ± 20 m^2/g; flash pt. (COC) > 535 F; ref. index 1.452; 5.3 ± 0.8% carbon content; *Toxicology:* LD50 (oral) > 5000 mg/kg; mild eye irritant; inh. of dust may cause pulmonary inflammation; not considered a carcinogen; *Precaution:* dry powds. can build static elec. charges subjected to friction; keep away from flamm. or explosive liqs.; *Storage:* store in dry environment away from chemical vapors.

Cabot® CP 9396. [Cabot Plastics Int'l.] LDPE-based processing aid masterbatch; CAS 9002-88-4; EINECS 200-815-3; improves productivity and processability in LLDPE, HDPE, HMW-HDPE blown and cast films, pipe extrusion, and HDPE blow molding; reduces or eliminates melt fracture, reduces melt temp., improves clarity and gloss, etc.; pellets; dens. 925 kg/m^3; melt flow 3 g/10 min.; Unverified

Cabot® LL 9219. [Cabot Plastics Int'l.] Antifibrillant-deluster in LLDPE masterbatch; antifibrillant-deluster masterbatch for tape extrusion; compat. with LDPE, LLDPE, PP, HDPE; good color.

Cabot® PE 6008. [Cabot Plastics Int'l.] 30% Trimethyl quinoline antioxidant in LDPE carrier; antioxidant masterbatch for polymeric cable applics. incl. cable sheath insulation layers and semiconductive screens; compatible with LDPE, EVA, EBA, LLDPE, HDPE, EPDM, ACN; dk. brn./beige pellets; dens. 970 kg/m^3; melt flow 14 g/10 min.

Cabot® PE 9006. [Cabot Plastics Int'l.] 75% Calcium carbonate plus antioxidant in LDPE carrier; antifibrilation masterbatch for use in PP weaving tapes, PP strapping bands, PP string and twine, LDPE shrink film; effective antiblock; compatible with LPDE, LLDPE, HDPE, PP; pellets; dens. 1820 kg/m^3; melt flow 12 g/10 min.

Cabot® PE 9007. [Cabot Plastics Int'l.] LDPE polymer with mildly abrasive fine natural silica; antiblock masterbatch for blown LDPE and LLDPE film extrusion; also as cleaning compd. for extruders; compatible with LDPE, LLDPE, HDPE, PP, EVA; pellets; dens. 1300 kg/m^3; melt flow 0.5 g/10 min.

Cabot® PE 9017. [Cabot Plastics Int'l.] LDPE polymer with ethoxylated amine, 10% internal antistat; antistatic masterbatch for film extrusion, inj. molding, and blow molding; compatible with LDPE; limited compatibility with LLDPE, HDPE, PP, EVA; pellets;

dens. 920 kg/m^3; melt flow 12 g/10 min.

Cabot® PE 9020. [Cabot Plastics Int'l.] LDPE, 7.5% erucamide slip agent; slip masterbatch offering slow migration; designed for blow molding and extrusion; slow migration reduces problems in sealing and printing; compatible with LDPE LLDPE, HDPE, EVA, PP; pellets; dens. 910 kg/m^3; melt flow 15 g/10 min.; Unverified

Cabot® PE 9026. [Cabot Plastics Int'l.] 75% calcium carbonate in LDPE carrier; antifibrilation masterbatch for PP weaving tapes, PP strapping bands, PP string and twine, LDPE shrink film; effective antiblock; compatible with LDPE, LLDPE, HDPE, PP; pellets; dens. 1820 kg/m^3; melt flow 10 g/10 min.; Unverified

Cabot® PE 9041. [Cabot Plastics Int'l.] LDPE, 7.5% oleamide slip agent; slip masterbatch offering fast migration; designed for blow molding and extrusion; compatible with LDPE, LLDPE, HDPE, EVA, PP; pellets; dens. 910 kg/m^3; melt flow 15 g/10 min.; Unverified

Cabot® PE 9131. [Cabot Plastics Int'l.] HALS and absorber in LDPE masterbatch; uv stabilizer masterbatch for film extrusion, blow molding, inj. molding; compat. with LDPE, LLDPE, HDPE, EVA; good uv stability; recommended for critical sealing and printing applics.

Cabot® PE 9135. [Cabot Plastics Int'l.] HALS and absorber in LDPE masterbatch; uv stabilizer masterbatch for film extrusion, blow molding, inj. molding; compat. with LDPE, LLDPE, HDPE, EVA; exc. uv stability in thin section prods.

Cabot® PE 9136. [Cabot Plastics Int'l.] Nickel quencher and absorber in LDPE masterbatch; uv stabilizer masterbatch for film extrusion; compat. with LDPE, LLDPE, EVA; good uv stability, esp. in greenhouse film subject to high levels of pesticides.

Cabot® PE 9138. [Cabot Plastics Int'l.] Polymer masterbatch with min. additive; infra-red absorber masterbatch; used to improve thermal barrier properties of polyethylene film in greenhouse and other agric. applics.; optimum addition level is 11-12% in polyethylene; pellets; dens. 1710 kg/m^3; melt flow 0.37 g/10 min.

Cabot® PE 9166. [Cabot Plastics Int'l.] 5% Erucamide slip agent in LPDE carrier; slip additive masterbatch providing exc. slip and good antiblock effect to LDPE and LLDPE films; slower migration of slip agent is advantageous when printing, sealing, or laminating processes are carried out immediately after film prod.; pellets; dens. 915 kg/m^3; melt flow 10 g/10 min.

Cabot® PE 9171. [Cabot Plastics Int'l.] 5% Oleamide slip agent in LDPE carrier; slip additive masterbatch providing exc. slip but no antiblocking effect to LDPE and LLDPE films; relatively rapid migration; pellets; dens. 915 kg/m^3; melt flow 10 g/10 min.

Cabot® PE 9172. [Cabot Plastics Int'l.] 5% Erucamide slip agent in LDPE carrier; slip and antiblock additive masterbatch for LDPE and LLDPE films; esp. suitable for high temp. processing or where printing, sealing, or lamination occurs immediately after film prod.; pellets; dens. 945 kg/m^3; melt flow 7 g/10 min.

Cabot® PE 9174. [Cabot Plastics Int'l.] 5% Oleamide slip agent in LDPE carrier; slip and antiblock additive masterbatch for LDPE and LLDPE films; rapid migration; pellets; dens. 945 kg/m^3; melt flow 7 g/10 min.

Cabot® PE 9220. [Cabot Plastics Int'l.] Slip and antiblock agents in LDPE carrier; antiblock and slip masterbatch for LDPE and LLDPE blown film extrusion; compatible with LDPE, LLDPE, HDPE, PP, EVA; limited compatibility with PS; pellets; dens. 1470 kg/m^3; melt flow 23.6 g/10 min.; Unverified

Cabot® PE 9229. [Cabot Plastics Int'l.] 70% Calcium carbonate in LDPE carrier; antiblock masterbatch for film extrusion; enhances printing chars.; compatible with LDPE, LLDPE, HDPE; pellets; dens. 1710 kg/m^3; melt flow 8 g/10 min.

Cabot® PE 9247. [Cabot Plastics Int'l.] Antioxidant in LDPE masterbatch; antioxidant masterbatch for film extrusion; compat. with LDPE, LLDPE; confers exc. long-term thermal stability.

Cabot® PE 9269. [Cabot Plastics Int'l.] Antiblock/deluster/extender in LDPE masterbatch; antiblock and extender with good color for film extrusion; compat. with LDPE, LLDPE, HDPE, EVA.

Cabot® PE 9287. [Cabot Plastics Int'l.] UV screener/stabilizer in LDPE masterbatch; uv stabilizer masterbatch for film extrusion; compat. with LDPE, LLDPE, EVA; reduces transmission of uv through films to aid pkg. of uv-sensitive prods.

Cabot® PE 9321. [Cabot Plastics Int'l.] Starch/LDPE masterbatch; increases degradability of polyolefin films; for film extrusion; compat. with LDPE, LLDPE, EVA.

Cabot® PE 9323. [Cabot Plastics Int'l.] Photodegradant/LDPE masterbatch; confers rapid, sunlight activated, degradability at only 2% addition level; for film extrusion; compat. with LDPE, LLDPE.

Cabot® PE 9324. [Cabot Plastics Int'l.] LDPE-based masterbatch based on nonfluoropolymer system, containing a multifunctional surfactant; CAS 9002-88-4; EINECS 200-815-3; processing aid masterbatch improving productivity and processability by eliminating melt fracture, reducing melt temp., improving bubble stability, reducing screw and die deposits, etc.; used for LLDPE, HDPE, HMW-HDPE blown film, PP film and extrusion, HDPE blow molding; pellets; dens. 950 kg/m^3; melt flow 5 g/10 min.

Cabot® PE 9396. [Cabot Plastics Int'l.] Fluoropolymer/LLDPE masterbatch; processing aid masterbatch for film extrusion; compat. with LDPE, LLDPE; reduces melt fracture, increases output, lowers power consumption.

Cabot® PP 9129. [Cabot Plastics Int'l.] HALS additive/PP masterbatch; uv stabilizer masterbatch for tape and fiber extrusion, inj. molding of PP; good extraction resist. and thermal stability; exc. uv stability in thin section prods.

Cabot® PP 9130. [Cabot Plastics Int'l.] HALS additive/

PP masterbatch; uv stabilizer masterbatch for tape extrusion, inj. molding of PP; exc. uv stability at low addition levels.

Cabot® PP 9131. [Cabot Plastics Int'l.] HALS additive/ PP masterbatch; uv stabilizer masterbatch for tape and fiber extrusion, inj. molding of PP; features reduced water carry-over, wide food contact approvals.

Cachalot® S-53. [M. Michel] Stearyl alcohol; CAS 112-92-5; EINECS 204-017-6; mold release and processing aid for vinyl resins and rubber; Hazen 80 color; sol. in aromatic hydrocarbons and alcohol; m.w. 255-275; sp.gr. 0.815 (60 C); visc. 8 cps (70 C); m.p. 51-54 C; b.p. 315-350 C; acid no. 0.3 max.; iodine no. 2 max.; sapon. no. 1 max.; hyd. no. 205-220; flash pt. 170 C; ref. index 1.4339-1.4341 (70 C).

Cadet® BPO-70. [Akzo Nobel] Benzoyl peroxide; CAS 94-36-0; EINECS 202-327-6; initiator for elevated temp. curing of unsat. polyester resins in applics. for matched metal die molding; pultrusion, vacuum bag molding, continuous laminating, hot-cure casting, inj. molding; gran.; 70% conc.

Cadet® BPO-70W. [Akzo Nobel] Benzoyl peroxide; CAS 94-36-0; EINECS 202-327-6; initiator for elevated-temp. curing of unsat. polyester resins in applics. such as matched metal die molding, pultrusion, vacuum bag molding, continuous laminating, hot-cure casting, inj. molding; also for PS, specialized PVC resin, acrylic polymers; gran.; 70% act. with 30% water; 4.62% act. oxygen; *Storage:* store @ 40 C max.

Cadet® BPO-75W. [Akzo Nobel] Benzoyl peroxide; CAS 94-36-0; EINECS 202-327-6; initiator for acrylates, PS; gran.; 75% assay; 4.95% act. oxygen; *Storage:* store @ 40 C max.

Cadet® BPO-78. [Akzo Nobel] Benzoyl peroxide; CAS 94-36-0; EINECS 202-327-6; initiator for elevated temp. curing of unsat. polyester resins in applics. for matched metal die molding; pultrusion, vacuum bag molding, continuous laminating, hot-cure casting, inj. molding; powd.; 78% assay; 5.15% act. oxygen; *Storage:* store @ 40 C max.

Cadet® BPO-78W. [Akzo Nobel] Benzoyl peroxide; CAS 94-36-0; EINECS 202-327-6; initiator for elevated-temp. curing of unsat. polyester resins in applics. such as matched metal die molding, pultrusion, vacuum bag molding, continuous laminating, hot-cure casting, inj. molding; gran.; 78% act. with 22% water; 5.15% act. oxygen; *Storage:* store @ 40 C max.

Cadox® 40E. [Akzo Nobel] Benzoyl peroxide with plasticizer; CAS 94-36-0; EINECS 202-327-6; initiator for ambient-temp. polyester cures; susp.; 40% assay; 2.64% act. oxygen; *Storage:* store @ 40 C max.

Cadox® BCP. [Akzo Nobel] Benzoyl peroxide with di- and tri-calcium phosphate; initiator for ambient-temp. polyester cures; used in adhesives and repkg. applic.; powd.; 35% conc.

Cadox® BEP-50. [Akzo Nobel] Benzoyl peroxide with phlegmatizer; initiator for ambient-temp. polyester cures; used in mine roof bolts and autobody repair putties; red, wh., and blue paste; 50% conc.

Cadox® BFF-50. [Akzo Nobel] Benzoyl peroxide with dicyclohexyl phthalate; EINECS 202-327-6; initiator for elevated temp. curing of unsat. polyester resins in applics. for matched metal die molding; pultrusion, vacuum bag molding, continuous laminating, hot-cure casting, inj. molding; exc. sol. chars. for sol'n. polymerization applics.; powd.; m.p. 56 C; 50% assay; 3.30% act. oxygen; *Storage:* store @ 25 C max.

Cadox® BFF-50L. [Akzo Nobel] Benzoyl peroxide; CAS 94-36-0; EINECS 202-327-6; initiator for cure of unsaturated polyester resins; dry gran.; 50% peroxide; 3.3% act. oxygen.

Cadox® BFF-60W. [Akzo Nobel] Benzoyl peroxide with dicyclohexyl phthalate; initiator; gran.; 60% assay; 3.96% act. oxygen; *Storage:* store @ 40 C max.

Cadox® BP-55. [Akzo Nobel] Benzoyl peroxide with phlegmatizer; initiator for elevated temp. curing of unsat. polyester resins in applic. for matched metal die molding, pultrusion, vacuum bag molding, continuous laminating, hot-cure casting, inj. molding; ambient-temp. polyester cures; disperses rapidly in polyester formulations; paste; 55% conc.

Cadox® BPO-W40. [Akzo Nobel] Benzoyl peroxide; CAS 94-36-0; EINECS 202-327-6; initiator; suspension; 40% assay; 2.64% act. oxygen.

Cadox® BS. [Akzo Nobel] Benzoyl peroxide on silicone oil/wax carrier; CAS 94-36-0; EINECS 202-327-6; cross-linking agent used for curing silicone rubbers; paste; 50% act.

Cadox® BT-50. [Akzo Nobel] Benzoyl peroxide; CAS 94-36-0; EINECS 202-327-6; initiator for ambient-temp. polyester cures; esp. for automotive body putty pastes; paste; 50% act. in plasticizer.

Cadox® BTA. [Akzo Nobel] Benzoyl peroxide, calcium sulfate; initiator for ambient-temp. polyester cures; powd.; 35% act. on calcium sulfate; 2.31% act. oxygen.

Cadox® BTW-50. [Akzo Nobel] Benzoyl peroxide; CAS 94-36-0; EINECS 202-327-6; initiator for cure of unsaturated polyester resins; paste; 50% peroxide; 3.3% act. oxygen.

Cadox® BTW-55. [Akzo Nobel] Benzoyl peroxide; CAS 94-36-0; EINECS 202-327-6; initiator for cure of unsaturated polyester resins; paste; 55% peroxide; 3.6% act. oxygen.

Cadox® F-85. [Akzo Nobel] Ketone peroxide; initiator for ambient-temp. polyester cures; liq.; 8.5% act. oxygen.

Cadox® HBO-50. [Akzo Nobel] MEK peroxide; CAS 1338-23-4; EINECS 215-661-2; initiator for cure of unsaturated polyester resins; sol'n.; 9.0% act. oxygen.

Cadox® L-30. [Akzo Nobel] MEK peroxide; CAS 1338-23-4; EINECS 215-661-2; initiator for cure of unsaturated polyester resins; sol'n.; 5.3% act. oxygen.

Cadox® L-50. [Akzo Nobel] MEK peroxide; CAS 1338-23-4; EINECS 215-661-2; initiator for ambient-temp. polyester cures for hand lay-up, spray-up, resin transfer molding, polymer concrete, gelcoats; liq.; 9% act. oxygen.

Cadox® M-30. [Akzo Nobel] MEK peroxide in DMP/DAP; initiator for ambient-temp. polyester cures; used for spray-up applics.; liq.; 5.3% act. oxygen.

Cadox® M-50. [Akzo Nobel] MEK peroxide in dimethyl phthalate; initiator for ambient-temp. polyester cures; excellent batch-to-batch consistency; versatile ambient-temp. curing catalyst; liq.; 9.0% act. oxygen.

Cadox® M-105. [Akzo Nobel] MEK peroxide; CAS 1338-23-4; EINECS 215-661-2; initiator for ambient-temp. polyester cures; excellent batch-to-batch consistency; versatile ambient-temp. curing catalyst; liq.

Cadox® MDA-30. [Akzo Nobel] MEK peroxide in DAP; initiator for ambient-temp. polyester cures; low visc.; used for airless spray-gun applic.; liq.; 5.3% act. oxygen.

Cadox® OS. [Akzo Nobel] Bis(o-chlorobenzoyl)-peroxide; cross-linking agent used for curing silicone rubbers; paste; 50% act. BPO with silicone fluid.

Cadox® PS. [Akzo Nobel] Bis(p-chlorobenzoyl)-peroxide; cross-linking agent used for curing silicone rubbers; paste; 50% act. with silicone fluid.

Cadox® TDP. [Akzo Nobel] 2,4-Dichlorobenzoyl peroxide in dibutyl phthalate plasticizer; CAS 94-17-7; highly reactive initiator used with amine-accelerated polyester resins; crosslinking agent for rubber; paste; 50% act.

Cadox® TS-50. [Akzo Nobel] Bis(2,4-dichlorobenzoyl)peroxide, dibutyl phthalate; cross-linking agent used for curing silicone rubbers; paste; 50% act.

Cadox® TS-50S. [Akzo Nobel] 2,4-Dichlorobenzoyl peroxide on silicone oil carrier, phthalate-free; CAS 94-17-7; crosslinking agent for polymers; co-vulcanizing agent; paste; 50% act.

Calfoam ES-30. [Pilot] Sodium laureth sulfate; CAS 9004-82-4; anionic; detergent, foam stabilizer, flash foamer, wetter for detergent systems, personal care prods., wool washing; emulsion polymerization; yel. clear liq.; mild odor; dens. 8.8 lb/gal; pH 8.0; 30% solids.

Calfoam NLS-30. [Pilot] Ammonium lauryl sulfate; CAS 2235-54-3; EINECS 218-793-9; anionic; mild surfactant base for neutral pH shampoos, bubble baths, rug cleaner formulations, cosmetic emulsifiers, emulsion polymerization; liq.; 30% act.

Calfoam SLS-30. [Pilot] Sodium lauryl sulfate; CAS 151-21-3; EINECS 205-788-1; anionic; mild detergent, foamer for personal care prods.; rug/upholstery shampoos; emulsifier for cosmetics, emulsion polymerization of latex, SBR rubber, polyacrylates, elastomers; foaming agent for foamed rubber; wh. paste; mild odor; pH 8.0; 30% act.

Califlux® 90. [Witco/Golden Bear] Aromatic oil; extender oil offering low aniline pts.; for elastomer compounding, esp. nitrile, neoprene, SBR, and natural rubber; m.w. 325; sp.gr. 0.9840 (15.6 C); dens. 8.2 lb/gal; visc. 519 cSt (40 C); flash pt. (COC) 208 C; pour pt. -6 C; ref. index 1.5534.

Califlux® 510. [Witco/Golden Bear] Aromatic oil; extender oil offering low aniline pts.; for elastomer compounding, esp. nitrile, neoprene, SBR, and natural rubber; m.w. 275; sp.gr. 0.9895 (15.6 C); dens. 8.2 lb/gal; visc. 138 cSt (40 C); flash pt. (COC) 168 C; pour pt. -3 C; ref. index 1.5516.

Califlux® GP. [Witco/Golden Bear] Aromatic oil; extender oil offering low aniline pts.; for elastomer compounding, esp. nitrile, neoprene, SBR, and natural rubber; m.w. 311; sp.gr. 1.0035 (15.6 C); dens. 8.4 lb/gal; visc. 1050 cSt (40 C); flash pt. (COC) 212 C; pour pt. 6 C; ref. index 1.5707.

Califlux® LP. [Witco/Golden Bear] Aromatic oil; extender oil offering low aniline pts.; for elastomer compounding, esp. nitrile, neoprene, SBR, and natural rubber; m.w. 246; sp.gr. 0.9826 (15.6 C); dens. 8.2 lb/gal; visc. 36 cSt (40 C); flash pt. (COC) 160 C; pour pt. -24 C; ref. index 1.5503.

Califlux® SP. [Witco/Golden Bear] Aromatic oil; extender oil offering low aniline pts.; for elastomer compounding, esp. nitrile, neoprene, SBR, and natural rubber; m.w. 318; sp.gr. 1.0107 (15.6 C); dens. 8.4 lb/gal; visc. 2750 cSt (40 C); flash pt. (COC) 216 C; pour pt. 12 C; ref. index 1.5750.

Califlux® TT. [Witco/Golden Bear] Aromatic oil; extender oil offering low aniline pts.; for elastomer compounding, esp. nitrile, neoprene, SBR, and natural rubber; m.w. 325; sp.gr. 1.0180 (15.6 C); dens. 8.5 lb/gal; visc. 6000 cSt (40 C); flash pt. (COC) 216 C; pour pt. 18 C; ref. index 1.5829.

Calight RPO. [Calumet Lubricants] Naphthenic process oil; for rubber industry, resin extending, PVC, textiles, caulking compds., etc.; FDA 21CFR §178.3620(c); m.w. 315; sp.gr. 0.920 (60 F); dens. 7.66 lb/gal; visc. 152 SUS (100 F); pour pt. -30 F; flash pt. (COC) 360 F; ref. index 1.5034.

Calimulse EM-30. [Pilot] Sodium alkylbenzene sulfonate (hard); emulsifier for emulsion polymers and other prods. which do not enter sewage streams; paste; 30% conc.

Calimulse EM-99. [Pilot] Alkylbenzene sulfonic acid (hard); neutralized as emulsifiers for agric., emulsion polymers, and other prods. which do not enter sewage streams; thick liq.; 97% conc.

Calimulse PRS. [Pilot] Isopropylamine dodecylbenzene sulfonate; anionic; biodeg. emulsifier, solubilizer; dry cleaning; degreasers; latex emulsifier; pigment dispersant; agric. sprays, oil slick emulsifiers; clear amber liq.; dens. 8.5 lb/gal; pH 4.8; 90% conc.

Calsoft F-90. [Pilot] Sodium dodecylbenzene sulfonate; CAS 25155-30-0; EINECS 246-680-4; anionic; detergent, emulsifier, wetter for all-purpose and hard surface cleaners, bubble baths, degreasers, laundry powds., textile scouring aids, emulsion polymers, sanitation, emulsion paints, wettable powds., ore flotation, metal pickling; biodeg.; wh. free-flowing flake; water-sol.; dens. 0.45 g/cc; pH 8.0 (1%); 90% act.

Calsoft L-60. [Pilot] Sodium dodecylbenzene sulfonate; CAS 25155-30-0; EINECS 246-680-4; anionic; biodeg. emulsion stabilizer; wetting and foaming agent, detergent, emulsifier for household and industrial detergents, agric. emulsions, dye bath leveling, rug cleaners, bubble baths, ore flotation, and air entrainment in concrete and gypsum board; washing fruits and vegetables; emulsion polymerization; water-wh. pasty liq.; odorless; dens. 8.7 lb/gal; pH 7.4; 60% solids.

Calsol 510. [Calumet Lubricants] Naphthenic process oil; nonstaining oil for rubber industry, resin extending, PVC, textiles; provides exc. initial color and color stability for thermoplastics, radial and styrene block elastomers; FDA 21CFR §178.3620(c); m.w. 300; sp.gr. 0.8984 (60 F); dens. 7.48 lb/gal; visc. 110 SUS (100 F); pour pt. -50 F; flash pt. (COC) 330 F; ref. index 1.4894.

Calsol 804. [Calumet Lubricants] Naphthenic process oil; provides exc. color stability in resin extending, PVC, textiles, caulking compds. and other applics.; m.w. 208; sp.gr. 0.8911 (60 F); dens. 7.37 lb/gal; visc. 42.7 SUS (100 F); pour pt. -60 F; flash pt. (COC) 260 F; ref. index 1.4850.

Calsol 806. [Calumet Lubricants] Naphthenic process oil; provides exc. color stability in resin extending, PVC, textiles, caulking compds. and other applics.; FDA 21CFR §178.3620(c); m.w. 249; sp.gr. 0.8996 (60 F); dens. 7.49 lb/gal; visc. 61.0 SUS (100 F); pour pt. -60 F; flash pt. (COC) 300 F; ref. index 1.4895.

Calsol 810. [Calumet Lubricants] Naphthenic process oil; provides exc. color stability in resin extending, PVC, textiles, caulking compds. and other applics.; FDA 21CFR §178.3620(c); m.w. 311; sp.gr. 0.9111 (60 F); dens. 7.59lb/gal; visc. 110 SUS (100 F); pour pt. -40 F; flash pt. (COC) 340 F; ref. index 1.4970.

Calsol 815. [Calumet Lubricants] Naphthenic process oil; provides exc. color stability in resin extending, PVC, textiles, caulking compds. and other applics.; FDA 21CFR §178.3620(c); m.w. 315; sp.gr. 0.9080 (60 F); dens. 7.56 lb/gal; visc. 155 SUS (100 F); pour pt. -35 F; flash pt. (COC) 345 F; ref. index 1.4964.

Calsol 830. [Calumet Lubricants] Naphthenic process oil; CAS 67254-74-4; process oil for rubber applics.; FDA 21CFR §178.3620(c); m.w. 355; sp.gr. 0.9242 (60 F); dens. 7.697 lb/gal; visc. 329 SUS (100 F); pour pt. -20 F; flash pt. (COC) 340 F; ref. index 1.5055.

Calsol 850. [Calumet Lubricants] Naphthenic process oil; provides exc. color stability in resin extending, PVC, textiles, caulking compds. and other applics.; m.w. 373; sp.gr. 0.9291 (60 F); dens. 7.74 lb/gal; visc. 515 SUS (100 F); pour pt. 0 F; flash pt. (COC) 400 F; ref. index 1.5082.

Calsol 875. [Calumet Lubricants] Naphthenic process oil; provides exc. color stability in resin extending, PVC, textiles, caulking compds. and other applics.; m.w. 385; sp.gr. 0.9285 (60 F); dens. 7.73 lb/gal; visc. 800 SUS (100 F); pour pt. 5 F; flash pt. (COC) 410 F; ref. index 1.5090.

Calsol 5120. [Calumet Lubricants] Naphthenic process oil; nonstaining oil providing exc. initial color and color stability for thermoplastics, radial and styrene block elastomers; m.w. 510; sp.gr. 0.9100 (60 F); dens. 7.58 lb/gal; visc. 1225 SUS (100 F); pour pt. -10 F; flash pt. (COC) 450 F; ref. index 1.4970.

Calsol 5550. [Calumet Lubricants] Naphthenic process oil; nonstaining oil providing exc. initial color and color stability for thermoplastics, radial and styrene block elastomers; FDA 21CFR § 178.3620(c); m.w. 425; sp.gr. 0.9047 (60 F); dens. 7.53 lb/gal; visc. 510 SUS (100 F); pour pt. -30 F; flash pt. (COC) 430 F; ref. index 1.4950.

Calsol 8120. [Calumet Lubricants] Naphthenic process oil; provides exc. color stability in resin extending, PVC, textiles, caulking compds. and other applics.; m.w. 398; sp.gr. 0.9371 (60 F); dens. 7.80 lb/gal; visc. 1300 SUS (100 F); pour pt. 10 F; flash pt. (COC) 450 F; ref. index 1.5120.

Calsol 8200. [Calumet Lubricants] Naphthenic process oil; CAS 67254-74-4; process oil for rubber applics.; FDA 21CFR §178.3620(c); m.w. 393; sp.gr. 0.9415 (60 F); dens. 7.84 lb/gal; visc. 2022 SUS (100 F); pour pt. 12 F; flash pt. (COC) 420 F; ref. index 1.5135.

Calsol 8240. [Calumet Lubricants] Naphthenic process oil; provides exc. color stability in resin extending, PVC, textiles, caulking compds. and other applics.; m.w. 400; sp.gr. 0.9433 (60 F); dens. 7.86 lb/gal; visc. 2400 SUS (100 F); pour pt. 20 F; flash pt. (COC) 460 F; ref. index 1.5130.

Calsolene Oil HSA. [ICI Am.; ICI Surf. UK] Sulfated fatty acid ester; anionic; emulsifier for polymerization; low foaming wetting and penetrating agent for textiles; stable in acid and alkalies; gold-yel. clear liq. (may be turbid at room temp.); dissolves in hard and soft water and in salt sol'ns.; flash pt. (OC) 200 F; 45% conc.

Calsolene Oil HSAD. [ICI Am.] Sulfated fatty ester; anionic; wetting agent for textiles; emulsifier for polymerization, lubricant, leveling agent; yel. liq.; 45% conc.

Calwhite®. [Georgia Marble] Ground calcium carbonate; filler for exterior/interior paints and coatings where flat and med. gloss surfs. are required; adhesives, caulks and sealants based on acrylic, polyolefin, silicone, and PU; plastics (polyolefin, PU, compr. and inj. molded, plastisols), rubber, thermosets (BMC/SMC, pultrusion); powd.; 6.0 μ median particle size; 0.008% on 325 mesh; fineness (Hegman) 5; oil absorp. 10; brightness 94; pH 9.0-9.5; 95% min. $CaCO_3$.

Calwhite® II. [Georgia Marble] Calcium carbonate; economical extender filler for BMC, SMC, TMC, pultrusions, transfer moldings, wet mat molding, and other glass fiber-reinforced thermoset polyester compds.; also for polyolefins, urethanes, epoxies, moldings, adhesives, caulks, sealants, mastics, foamed compositions, paints and coatings (flat and gloss, enamels, varnishes); very wh. free-flowing powd.; 7 μ mean particle size; 0.05% on 325 mesh; oil absorp. 12-14%; brightness 94; 95%

min. $CaCO_3$.

Camel-CAL®. [Genstar Stone Prods.] Calcium carbonate; filler for water-based coatings and inks, paper, paper coatings, thermosets (glass-reinforced polyester), thermoplastics (PP, ABS), PVC pipe, siding, profiles, extrusion, film, sheeting, color concs., paint (gloss, semigloss, flat latex); 0.7 μ avg. particle diam.; 100% finer than 7 μ; 65% finer than 1 μ; 0.14% sol. in water; sp.gr. 2.70-2.71; dens. 22.57 lb/gal solid; bulk dens. 35 lb/ft^3 (loose); oil absorp. 28 cc/100 g; brightness (Hunter) 96; ref. index 1.6; pH 9.5 (sat. sol'n.).

Camel-CAL® ST. [Genstar Stone Prods.] Stearate-coated calcium carbonate; filler for water-based coatings and inks, paper, paper coatings, thermosets (glass-reinforced polyester), thermoplastics (PP, ABS), PVC pipe, siding, profiles, extrusion, film, sheeting, color concs., paint (gloss, semigloss, flat latex); 0.7 μ avg. particle diam.; 100% finer than 7 μ; 65% finer than 1 μ; 0.14% sol. in water; sp.gr. 2.70-2.71; dens. 22.57 lb/gal solid; bulk dens. 25 lb/ft^3 (loose); oil absorp. 28 cc/100 g; brightness (Hunter) 96; ref. index 1.6; pH 9.5 (sat. sol'n.); hardness (Moh) 3.0; 1% surface treatment.

Camel-CAL® Slurry. [Genstar Stone Prods.] Calcium carbonate slurry; filler for water-based coatings and inks, paper, paper coatings, thermosets (glass-reinforced polyester), thermoplastics (PP, ABS), PVC pipe, siding, profiles, extrusion, film, sheeting, color concs., paint (gloss, semigloss, flat latex); slurry; 0.7 μ avg. particle diam.; 100% finer than 7 μ; 90% finer than 2 μ; 0.14% sol. in water; sp.gr. 2.70-2.71; dens. 15.87 lb/gal; visc. 90-150 cps; brightness (Hunter) 96; ref. index 1.6; pH 9.5 (sat. sol'n.); hardness (Moh) 3.0; 75 ± 0.5% solids.

Camel-CARB®. [Genstar Stone Prods.] Calcium carbonate; filler/extender used in interior flat paint and exterior house paints, rubber compds., putty and caulk, ceramics, adhesives, linoleum, floor tile, and textile coatings; 7.0 μ avg. particle diam.; 99.5% finer than 44 μ; 50.0% finer than 7 μ; 0.03% sol. in water; sp.gr. 2.70-2.71; dens. 22.57 lb/gal solid; bulk dens. 58 lb/ft^3 (loose); oil absorp. 13 cc/100 g; brightness (Hunter) 93; ref. index 1.6; pH 9.5 (sat.); hardness (Moh) 3.0.

Camel-FIL™. [Genstar Stone Prods.] Calcium carbonate; filler designed for high filler loading in glass-reinforced polyester (SMC, TMC, BMC); also for epoxies, PVC (rigid and flexible), PP, rubber automotive goods and floor tiles, caulks, sealants, and adhesives; 5.5-6.0 μ avg. particle diam.; 97% finer than 25 μ; 46% finer than 6 μ; 0.06% sol. in water; sp.gr. 2.70-2.71; dens. 22.57 lb/gal solid; bulk dens. 55 lb/ft^3 (loose); oil absorp. 17 cc/100 g; brightness (Hunter) 93; ref. index 1.6; pH 9.5 (sat.); hardness (Moh) 3.0.

Camel-FINE. [Genstar Stone Prods.] Calcium carbonate; powd.; 2 μ avg. particle diam.; 99.9% finer than 8 μ; 0.15% sol. in water; sp.gr. 2.70-2.71; dens. 22.57 lb/gal; bulk dens. 38 lb/ft^3 (loose); oil absorp. 23 ± 1 cc/100 g; brightness (Hunter) 95; ref. index 1.6; pH 9.5 (sat.); hardness (Moh) 3.0.

Camel-FINE ST. [Genstar Stone Prods.] Calcium carbonate; powd.; 2 μ avg. particle diam.; 99.9% finer than 8 μ; 0.15% sol. in water; sp.gr. 2.70-2.71; dens. 22.57 lb/gal; bulk dens. 38 lb/ft^3 (loose); oil absorp. 23 ± 1 cc/100 g; brightness (Hunter) 95; ref. index 1.6; pH 9.5 (sat.); hardness (Moh) 3.0; 1% surface treatment.

Camel-TEX®. [Genstar Stone Prods.] Calcium carbonate; fine-ground, general-purpose filler used in interior flat paints, primers, and sealers, polyester fiberglass premixes, preforms, and hand lay-up gel coats, rubber automotive prods., household prods., tubing, medical prods., and closures, putty, caulk, bath tub sealers, body deadeners, and adhesives; 5.0 μ avg. particle diam.; 99.7% finer than 25 μ; fineness (Hegman) 4-5; 0.04% sol. in water; sp.gr. 2.70-2.71; dens. 22.57 lb/gal solid; bulk dens. 50 lb/ft^3 (loose); oil absorp. 14 cc/100 g; brightness (Hunter) 93; ref. index 1.6; pH 9.5 (sat. sol'n.); hardness (Moh) 3.0.

Camel-WITE®. [Genstar Stone Prods.] Calcium carbonate; CAS 471-34-1; EINECS 207-439-9; filler for paint, paper, paper coating, PVC, rubber (automotive goods, footwear, medical supplies), thermoplastics (PP, nylon, urethanes, HDPE, LDPE, ABS, PS), thermosets (SMC, BMC, TMC glass reinforced polyesters, epoxy, alkaline phenolics), caulks, glazing compds., ceramics, adhesives, food processing; very wh. dry powd.; 3.0 μ avg. particle diam.; 99.9% finer than 12 μ; 50% finer than 3 μ; 0.08% sol. in water; sp.gr. 2.70-2.71; dens. 22.57 lb/gal solid; bulk dens. 40 lb/ft^3 (loose); oil absorp. 15 cc/100 g; brightness (Hunter) 95; ref. index 1.6; pH 9.5 (sat. sol'n.); hardness (Moh) 3.0.

Camel-WITE® Slurry. [Genstar Stone Prods.] Calcium carbonate aq. slurry; filler for paint, paper, paper coating, PVC, rubber (automotive goods, footwear, medical supplies), thermoplastics (PP, nylon, urethanes, HDPE, LDPE, ABS, PS), thermosets (SMC, BMC, TMC glass reinforced polyesters, epoxy, alkaline phenolics), caulks, glazing compds., ceramics, adhesives, food processing; very wh. slurry; 3.0 μ avg. particle diam.; 99.9% finer than 12 μ; 50% finer than 3 μ; 0.08% sol. in water; sp.gr. 2.70-2.71; dens. 15.3 lb/gal; bulk dens. 40 lb/ft^3 (loose); visc. 100-200; brightness (Hunter) 95; ref. index 1.6; pH 9.5 (sat. sol'n.; hardness (Moh) 3.0; 72.0 ± 0.5% solids.

Camel-WITE® ST. [Genstar Stone Prods.] Surface-treated calcium carbonate; filler for paint, paper, paper coating, PVC, rubber (automotive goods, footwear, medical supplies), thermoplastics (PP, nylon, urethanes, HDPE, LDPE, ABS, PS), thermosets (SMC, BMC, TMC glass reinforced polyesters, epoxy, alkaline phenolics), caulks, glazing compds., ceramics, adhesives, food processing; very wh. dry powd.; 3.0 μ avg. particle diam.; 99% finer than 12 μ; 50% finer than 3 μ; 0.08% sol. in water; sp.gr. 2.70-2.71; dens. 22.57 lb/gal solid; bulk dens. 35 lb/ft^3 (loose); oil absorp. 15 cc/100 g; brightness (Hunter) 95; ref. index 1.6; pH 9.5 (sat. sol'n.); hardness (Moh) 3.0; 1% surface treatment.

CAO®-3. [PMC Specialties] 2,6-Ditert.-butyl-p-cresol (BHT); CAS 128-37-0; EINECS 204-881-4; food-grade antioxidant, stabilizer, processing aid for rubber, elastomers, polymeric materials (polyolefins, polyurethane, polyester, polyvinyls), pkg. materials, waxes, insecticides, syn. lubricants, paints, foods/feeds, pharmaceuticals, cosmetics; approved for food and feed prods., pkg., food contact uses; retards oxidative deterioration in animal, veg., min. oils, fats, greases; FDA 21CFR §137.350, 166.110, 172.115, 172.615, 172.878, 172.880, 172.882, 172.886, 173.340, 174, 175, 176, 177, 178, 181.24, 182.3173, 582.3173, GRAS, USDA; kosher; wh. cryst., sl. odor; sol. (g/100 ml): 45-55 g aliphatic/aromatic solvs., 40-45 g animal fats, ketone, acetone, 25-30 g veg. oils; m.w. 220.36; sp.gr. 0.899 (80/4 C); bulk dens. 1.048; visc. 3.47 cst (80 C); f.p. 69.2 C min.; m.p. 157 F; b.p. 260-262 C (760 mm); flash pt. (COC) 275 C; ref. index 1.4859; 99% min. purity; *Toxicology:* ACGIH TLV-TWA 10 mg/m^3; may cause skin irritation, respiratory passage irritation, severe eye irritation/burns; *Storage:* stable when stored below 77 F under dry conditions in unopened, original containers; may yel. sl. on aging.

CAO®-5. [PMC Specialties] 2,2′-Methylene-bis (4-methyl-6-t-butylphenol); CAS 119-47-1; nonstaining, nondiscoloring antioxidant used in adhesives (hot-melt and rubber-based), rubbers (polybutadiene, B/S, neoprene, chlorinated butyl, etc.), high polymers (polyolefins, POM, PU, ABS), latexes; FDA 21CFR §175.105, 177.2600, 178.2010; wh. to off-wh. cryst. powd., sl. odor; 95% thru 100 mesh; sol. (g/100 ml): 172 g acetone, 155 g dioxane, 114 g ethyl acetate, 80 g chloroform, 46 g benzene, 40 g 95% ethanol; m.w. 340.5; m.p. 125-132 C; *Storage:* stable over long periods when stored under normal (below 77 F) dry conditions in unopened original containers.

CAO®-14. [PMC Specialties] 2,2′-Methylene-bis (4-methyl-6-tert.-butyl-phenol); CAS 119-47-1; nonstaining, nondiscoloring antioxidant used in adhesives (hot-melt and rubber-based), rubbers (polybutadiene, B/S, neoprene, chlorinated butyl, etc.), high polymers (polyolefins, POM, PU, ABS), latexes; Unverified

CAO®-92. [PMC Specialties] Hindered phenolic; low-freeze antioxidant for petrol., rubber, fuel, hydrocarbon printing inks, cutting oils, lubricants, waxes, polymers, etc.; dk. brn. visc. liq.; sol. in aliphatic/aromatic solvs., ketone, acetone, monohydric alcohols (C1-3); sp.gr. 0.92; visc. 5000 cps max.; f.p. 14 F; flash pt. 250 F; *Toxicology:* can cause skin and severe eye irritation.

Capcure® 3-800. [Henkel/Coatings & Inks] Uncatalyzed mercaptan polymer; accelerator for amides, amines, and amido-amines in conventional and high-solids applics., fast-set adhesives, castings, potting, laminates; requires amine catalyst for activation; Gardner 2 max. color; dens. 9.6 lb/gal; visc. 100-150 poise; gel time 4.5-6.5 min (30-50 g); After 7 day 25 C cure: tens. str. 5900 psi; elong. 3.75%; flex. str. 10,000 psi; 100% solids.

Capcure® 3830-81. [Henkel/Coatings & Inks] Precatalyzed mercaptan polymer (combination of Capcure 3-800 and EH-30 at 8:1); accelerator for conventional and high-solids applics., fast-set adhesives, castings, potting, laminates; Gardner 3 max. color; dens. 9.5 lb/gal; visc. 100-160 poise; gel time 2.75-3.75 min (30-50 g); 100% solids.

Capcure® AF. [Henkel/Coatings & Inks] Aromatic amine; co-curing agent with other epoxy hardeners to improve heat and chem. resist.; for high-performance structural engineering applics., filament winding, high-temp. adhesives, corrosion-resist. coatings, elec. encapsulations, tooling compds.; nonstaining; dk. clear liq.; sp.gr. 1.04; visc. 150 cp.

Capcure® EH-30. [Henkel/Coatings & Inks] 2,4,6-Tri(dimethylaminomethyl) phenol; CAS 90-72-2; accelerator for mercaptan-cured epoxies, low-temp. acceleration of polyamides, amido-amines; seldom used as sole curing agent; Gardner 6 max. color; sp.gr. 0.97; dens. 8.1 lb/gal; visc. 1.8-3.2 poise; amine no. 600; gel time 40 min (200 g); 100% solids.

Capcure® Emulsifier 37S. [Henkel/Functional Prods.] Ethoxylate; nonionic; epoxy resin emulsifier; waxy solid; m.p. 50 C; HLB 18.0; 100% act.

Capcure® Emulsifier 65. [Henkel/Functional Prods.] Ethoxylated aq. sol'n.; nonionic; epoxy resin emulsifier; lt. colored clear liq.; sp.gr. 1.08; visc. 12,000 cps; HLB 18.0; 65% act.

Captax®. [R.T. Vanderbilt] 2-Mercaptobenzothiazole; CAS 149-30-4; EINECS 205-736-8; thiazole accelerator; primary accelerator for natural and syn. rubbers; nonstaining and nondiscoloring; lt. yel. to buff powd., 99.8% thru 100 mesh; sol. in dil. caustic, toluene, carbon disulfide, chloroform, alcohol; m.w. 167.25; dens. 1.50 ± 0.03 mg/m^3; m.p. 164-175 C.

Carbowax® PEG 200. [Union Carbide] PEG-4; CAS 25322-68-3; intermediate for surfactants, lubricants, urethanes; antistat, humectant, mold release agent, plasticizer for adhesives, inks, lubricants; water-wh. visc. liq.; sol. in water; m.w. 190-210; sp.gr. 1.1239; dens. 9.38 lb/gal; visc. 4.3 cSt (210 F); flash pt. (CCC) > 300 F; ref. index 1.459; pH 4.5-7.5 (5% aq.); surf. tens. 44.5 dynes/cm.

Carbowax® PEG 300. [Union Carbide] PEG-6; CAS 25322-68-3; EINECS 220-045-1; coupling agent, solv., vehicle, humectant, lubricant, binder, base; used in adhesives, agric., ceramics, chem. intermediates, cosmetics/toiletries, electroplating/electropolishing, food processing, household prods., lubricants, metal fabrication, paints, paper, pharmaceuticals, printing, rubber and elastomers, textiles, wood processing; water-wh. visc. liq.; sol. in water, alcohols, glycerin, glycols; m.w. 285-315; sp.gr. 1.1250; dens. 9.38 lb/gal; visc. 5.8 cSt (99 C); f.p. -15 to -8 C; hyd. no. 356-394; flash pt. (PMCC) > 180 C; ref. index 1.463; pH 4.5-7.5 (5% aq.); surf. tens. 44.5 dynes/cm.

Carbowax® PEG 400. [Union Carbide] PEG-8; CAS 25322-68-3; EINECS 225-856-4; antistat, surfactant intermediate, dye carrier, humectant, lubricant, release agent, plasticizer for adhesives, capsules, ceramic glazes, creams and lotions, deodorant sticks, inks, lipsticks; liq.; sol. in water, methanol, ethanol, acetone, trichloroethylene, Cellosolve®, Carbitol®, dibutyl phthalate, toluene; m.w. 380-420; sp.gr. 1.1254; dens. 9.39 lb/gal; visc. 7.3 cSt (210 F); f.p. 4-8 C; flash pt. (CCC) > 350 F; ref. index 1.465; pH 4.5-7.5 (5% aq.); surf. tens. 44.5 dynes/cm.

Carbowax® PEG 540 Blend. [Union Carbide] PEG-6 and PEG-32 (41:59); base for ointments and suppositories; also for adhesives, agric., ceramics, chem. intermediates, electroplating, household prods., lubricants, metal fabrication, paints, paper, printing, rubber, textiles, wood processing; soft solid; sol. in methylene chloride, 73% in water, 50% in trichloroethylene, 48% in methanol; m.w. 500-600; sp.gr. 1.0930 (60 C); dens. 9.17 lb/gal (55 C); visc. 15.1 cSt (210 F); f.p. 38-41; flash pt. (CCC) > 350 F; pH 4.5-7.5 (5% aq.).

Carbowax® PEG 600. [Union Carbide] PEG-12; CAS 25322-68-3; EINECS 229-859-1; antistat, surfactant intermediate, humectant, lubricant, release agent, plasticizer for adhesives, capsules, ceramic glaze, creams and lotions, dentifrices, deodorant sticks, inks, lipsticks, wood treatment; liq.; sol. in water; m.w. 570-630; sp.gr. 1.1257; dens. 9.40 lb/gal; visc. 10.8 cSt (210 F); f.p. 20-25 C; flash pt. (CCC) > 350 F; ref. index 1.46; pH 4.5-7.5 (5% aq.); surf. tens. 44.5 dynes/cm.

Carbowax® PEG 900. [Union Carbide] PEG-20; CAS 25322-68-3; antistat, surfactant intermediate, lubricant, release agent, ointment and suppository base, plasticizer for adhesives, ceramic glaze, creams and lotions, dentifrices, deodorant sticks, wood treatment; soft solid; sol. 86% in water; m.w. 855-900; sp.gr. 1.0927 (60 C); dens. 9.16 lb/gal (55 C); visc. 15.3 cSt (210 F); f.p. 32-36; flash pt. (CCC) > 350 F; pH 4.5-7.5 (5% aq.).

Carbowax® PEG 1000. [Union Carbide] PEG-20; CAS 25322-68-3; antistat, surfactant intermediate, lubricant, release agent, ointment and suppository base, plasticizer for adhesives, ceramic glaze, creams and lotions, dentifrices, deodorant sticks, wood treatment; soft solid; sol. 80% in water; m.w. 950-1050; sp.gr. 1.0926 (60 C); dens. 9.16 lb/gal (55 C); visc. 17.2 cSt (210 F); f.p. 37-40; flash pt. (CCC) > 350 F; pH 4.5-7.5 (5% aq.).

Cardis® 10. [Petrolite] Oxidized microcryst. wax; external lubricant, process aid, slip, antiblock, scuff, mar resist. aid in styrenics, polyolefins, TPOS; used in the formulation of emulsions, textile finish softeners, floor, auto, and liq. solv.-based polishes; solid; m.p. 97 C; acid no. 16; sapon. no. 29.

Cardis® 36. [Petrolite] Oxidized microcryst. wax; external lubricant, process aid, slip, antiblock, scuff, mar resist. aid in styrenics, polyolefins, TPOs; used in the formulation of emulsions, textile finish softeners, floor, auto, and liq. solv.-based polishes; color 1.50 (D1500) wax; m.p. 92 C; acid no. 33; sapon. no. 70.

Cardis® 314. [Petrolite] Oxidized microcryst. wax; external lubricant, process aid, slip, antiblock, scuff, mar resist. aid in styrenics, polyolefins, TPOs; used in the formulation of emulsions, polishes, and coatings; modifier in solv. polish systems; color 2.0 (D1500) wax; m.p. 87 C; acid no. 18; sapon. no. 40.

Cardis® 319. [Petrolite] Oxidized microcryst. wax; external lubricant, process aid, slip, antiblock, scuff, mar resist. aid in styrenics, polyolefins, TPOs; used in the formulation of emulsions, coatings, textile finish softeners, floor, auto, and liq. solv.-based polishes; color 1.5 (D1500) wax; m.p. 92 C; acid no. 28; sapon. no. 58.

Cardis® 320. [Petrolite] Oxidized microcryst. wax; external lubricant, process aid, slip, antiblock, scuff, mar resist. aid in styrenics, polyolefins, TPOs; used in the formulation of emulsions, coatings, textile finish softeners, floor, auto, and liq. solv.-based polishes; color 2.0 (D1500) wax; m.p. 91 C; acid no. 36; sapon. no. 70.

Cardis® 370. [Petrolite] Oxidized microcryst. wax; external lubricant, process aid, slip, antiblock, scuff, mar resist. aid in styrenics, polyolefins, TPOs; used in the formulation of emulsions, textile finish softeners, floor, auto, and liq. solv.-based polishes; m.p. 85 C; acid no. 36; sapon. no. 74.

Cardolite® NC-507. [Cardolite] 3-(n-Pentadecyl) phenol; CAS 501-24-6, 3158-56-3; starting raw material for surfactants, antioxidants, anticorrosives; lubricant additive; cosolv. for insecticides, germicides; resin modifier; tan waxy solid; sol. in org. solvs. incl. aliphatic hydrocarbons; m.w. 304; sp.gr. > 1.0; m.p. 44-48 C (760 mm); b.p. 243-249 C (10 mm); ref. index 1.494.

Cardolite® NC-510. [Cardolite] 3-(n-Pentadecyl) phenol; CAS 501-24-6; starting raw material for surfactants; coupling agent for pigments and dyes; cosolv. for insecticides, germicides; resin modifier for phenolic-aldehyde, polyester, PC polymers; also in photographic industry; sl. pink to wh. waxy solid; sol. in org. solvs. incl. aliphatic hydrocarbons; m.w. 304; sp.gr. > 1.0; m.p. 49-51 C (760 mm); b.p. 190-195 C (1 mm); ref. index 1.4750.

Cardolite® NC-511. [Cardolite] 3-(n-Penta-8′-decenyl) phenol; CAS 8007-24-7; starting raw material for surfactants, antioxidants, anticorrosives; lubricant additive; cosolv. for insecticides; in drilling muds; resin modifier; brn. to amber oily liq.; sol. in org. solvs. incl. aliphatic hydrocarbons; m.w. 302; sp.gr. 0.930 (20 C); b.p. 223-227 C (10 mm); ref. index 1.5112.

Cardolite® NC-513. [Cardolite] Reactive flexibilizer and diluent for epoxy; reactive intermediate in polyester, alkyl, and styrene maleic resins; improves acid resist., reduces visc.; for floor and surfacing compds., VOC-compliant surf. coatings, tooling resins, adhesives, electrical applics.; Gardner 13 low-visc. liq.; visc. 50 cps; flash pt. 400 F; EEW 490; *Toxicology:* may cause skin irritation to individuals allergic to poison ivy family of plants; avoid prolonged breathing of vapors; *Storage:* unlimited storage stability; store in cool, dry area in closed containers.

Cardolite® NC-540. [Cardolite] Phenalkamine; solv.-free curing agent for epoxy coatings, castings, adhesives, potting and encapsulation, laminates, primers, marine coatings, surf. tolerant coatings, pipe and tank linings over concrete, concrete repair, automotive adhesives; Gardner 17 color; visc. 2500 cps; dens. 8.27 lb/gal; amine no. 520; flash pt. (PM) 107 C; > 96% solids; *Storage:* store in tightly sealed containers in dry location at ambient temps.; keep exposure to air and moisture to a minimum.

Cardolite® NC-541. [Cardolite] Phenalkamine with ethylene diamine; curing agent for epoxy coatings, castings, adhesives, potting and encapsulation, laminates; reddish-brn. visc. liq., sl. amine odor; insol. in water; sp.gr. 1.0; flash pt. (PMCC) 145 F; pH 10.5; *Toxicology:* LD50 (oral, rat) > 5 g/kg; pract. nontoxic orally; mod. irritating to eyes; may cause severe skin irritation, burns, blistering, scar formation; irritating to respiratory tract; *Precaution:* incompat. with strong acids, strong oxidizers and reactive organometallic compds.; hazardous decomp. prods.: CO, CO_2, NO_x; may evolve ammonia on exposure to atmospheric moisture in uncured state; *Storage:* open containers cautiously in a well-ventilated place; reclose tightly.

Cardolite® NC-700. [Cardolite] Cardanol; accelerator for epoxy systems; diluent; flexibilizer for phenolic systems.

Cardolite® NC-1307. [Cardolite] Terpene-based extender/flexibilizer/diluent for epoxy mfg.; also for concrete coatings.

Cardura® E-10. [Shell] Neodecanoic acid glycidyl ester; reactive epoxy modifier/diluent providing good chem. and crystallization resist., while lowering visc. and V.O.C.; Gardner 1 max. color; dens. 8.02 lb/gal; visc. 5-10 cps; EEW 244-256.

Cariflex BR 1220. [H. Muehlstein] Butadiene rubber; general purpose stabilizer, nonstaining; dens. 0.90 mg/m^3; Mooney visc. 45 (ML1+4, 100 C); 96.5% cis content; Unverified

Carsonol® ALS. [Lonza] Ammonium lauryl sulfate; CAS 2235-54-3; EINECS 218-793-9; anionic; biodeg. high foaming detergent, wetting and emulsifying agent for cosmetics (shampoos, bubble baths, shaving creams, cleansing creams, skin lotions), industrial cleaners, rug/upholstery cleaners, household detergents, veg. scrubbing, auto shampoo, pet shampoo, emulsion polymerization; low color visc. liq., low odor; water-sol.; dens. 8.4 lb/gal; pH 6.5-7.0 (10%); 30% act.

Carsonol® ALS-S. [Lonza] Ammonium lauryl sulfate; CAS 2235-54-3; EINECS 218-793-9; anionic; biodeg. high foaming surfactant, wetting agent, emulsifier for shampoos, bubble baths, shaving creams, industrial cleaners, foam/dust control, liq. household detergents, automobile shampoos, emulsion polymerization; low salt grade; liq.; 29% act.

Carsonol® ALS Special. [Lonza] Ammonium lauryl sulfate; CAS 2235-54-3; EINECS 218-793-9; anionic; biodeg. high foaming detergent, wetting and emulsifying agent for cosmetics (shampoos, bubble baths, shaving creams, cleansing creams, skin lotions), industrial cleaners, rug/upholstery cleaners, household detergents, veg. scrubbing, auto shampoo, pet shampoo, emulsion polymerization; low color clear liq., low odor; water-sol.; dens. 8.4 lb/gal; pH 6.5-7.0 (10%); 30% act.

Carsonol® DLS. [Lonza] DEA lauryl sulfate; CAS 143-00-0; EINECS 205-577-4; anionic; biodeg. detergent, foaming agent, wetting agent, emulsifier for personal care prods. (shampoos, bubble baths, shaving preps., cleansing creams), industrial cleaners, household detergents, pet shampoo, veg. scrubbing, emulsion polymerization; clear liq.; low odor; water sol.; dens. 8.49 lb/gal; pH 7.5-8.5 (10%); 35% act.

Carsonol® SLS. [Lonza] Sodium lauryl sulfate; CAS 151-21-3; EINECS 205-788-1; anionic; biodeg. detergent with high foam, good wetting and emulsifying properties used for cosmetics (shampoos, bubble baths, shaving preps.), industrial cleaners, rug/upholstery shampoos, household detergents, veg. scrubbing, emulsion polymerization; low color liq., low odor; water0sol.; dens. 8.7 lb/gal; pH 7.5-8.5 (10%); 30% act.

Carsonol® SLS-R. [Lonza] Sodium lauryl sulfate; CAS 151-21-3; EINECS 205-788-1; anionic; biodeg. detergent, foaming agent, wetting agent, emulsifier for personal care prods., household and industrial cleaners, emulsion polymerization; liq.; 29% act.

Carsonol® SLS-S. [Lonza] Sodium lauryl sulfate; CAS 151-21-3; EINECS 205-788-1; anionic; biodeg. detergent, foaming agent, wetting agent, emulsifier for personal care prods., household and industrial cleaners, emulsion polymerization; liq.; 29% act.

Carsonol® SLS Special. [Lonza] Sodium lauryl sulfate; CAS 151-21-3; EINECS 205-788-1; anionic; biodeg. detergent with high foam, good wetting and emulsifying properties used for cosmetics, industrial cleaners, emulsion polymerization, veg. scrubbing; liq., low color; low odor; water-sol.; dens. 8.7 lb/gal; pH 7.5-8.5 (10%); 30% act.

Carsonol® TLS. [Lonza] TEA-lauryl sulfate; CAS 139-96-8; EINECS 205-388-7; anionic; biodeg. detergent with high foam, good wetting and emulsifying properties used for cosmetics, mild shampoos, bubble baths, chemical specialties, industrial cleaners, emulsion polymerization; clear liq.; low color, odor; water-sol.; dens. 8.7 lb/gal; pH 7.0-7.5 (10%); 40% act.

Carsonon® N-30. [Lonza] Nonoxynol-30; CAS 9016-45-9; nonionic; emulsifier, stabilizer for emulsion polymerization, oils, fats, waxes, essential oils; pale yel. liq. to semisolid; sp.gr. 1.09; cloud pt. 167-171 F (1%); pH 6.0-7.0 (3%); HLB 17.2; 70% act., 30% water.

Carstab® DLTDP. [Morton Int'l./Plastics Additives] Dilauryl thiodipropionate; CAS 123-28-4; EINECS 204-614-1; long-term heat aging stabilizer in conjunction with primary antioxidants; synergist; for

PP, HDPE, ABS; FDA 21CFR §175.300, 181.24, 182.3280, GRAS; wh. cryst. flake; sol. (g/100 g): 65 g toluene, 60 g ethyl acetate, 55 g acetone, 52 g heptane; insol. in water; m.w. 514; sp.gr. 0.896 (80 C); f.p. 40 C; acid no. 0.5; 99% purity; *Toxicology:* LD50 (oral, rat) 10,3 g/kg; low oral and dermal toxicity.

Carstab® DSTDP. [Morton Int'l./Plastics Additives] Distearyl thiodipropionate; CAS 693-36-7; EINECS 211-750-5; long-term heat aging stabilizer in conjunction with primary antioxidants; synergist; for PP, HDPE, ABS; FDA 21CFR §175.105, 175.300, 181.24; wh. cryst. flake or powd.; sol. (g/100 g): 13 g toluene, 2 g heptane, acetone; insol. in water; m.w. 682; sp.gr. 0.858 (80 C); f.p. 65 C; acid no. 1.0; 98% purity; *Toxicology:* low oral and dermal toxicity.

Casamid® 350PM. [Air Prods./Perf. Chems.] Modified polyamide in 25% propylene glycol monomethyl ether; R.T. epoxy curing agent designed to emulsify and cure epoxy resin systems contg. water in the continuous phase; yields fast dry times when cured with solid epoxy resin emulsions; for decorative and protective coatings for applic. to damp surfs., water-based adhesives for civil engineering use (mortars, stopping compds., cement floorings), waterproof structures subject to hydrosatic pressures; Gardner 10 max. liq.; water-sol.; sp.gr. 1.0; visc. 8500 cps; amine no. 250; equiv. wt. 110.

Casamid® 360. [Air Prods./Perf. Chems.] Aq. sol'n. of modified polyamide; curing agent capable of emulsifying and curing epoxy resins at R.T. without added surfactants; designed for use where absence of org. solvs. is desired; for water-based coatings for concrete walls and floors of food storage areas, hospitals, laboratories, nuclear power plants; antigraffiti coatings; Gardner 14 max. liq.; water-sol.; sp.gr. 1.05; visc. 45,000 cps; amine no. 145; equiv. wt. 280.

Casamid® 362W. [Air Prods./Perf. Chems.] Aq. sol'n. of modified polyamide; R.T. epoxy curing agent; long pot life version of Casamid 360; allows prod. of wh. and pastel coatings with good color retention; for water-based coatings for concrete walls and floors of food storage areas, hospitals, laboratories, nuclear power plants; antigraffiti coatings; Gardner 11 max. liq.; water-sol.; sp.gr. 1.05; visc. 55,000 cps; amine no. 145; equiv. wt. 240.

Castorwax® MP-70. [CasChem] Hydrog. castor oil; CAS 8001-78-3; EINECS 232-292-2; wax for anhyd. prods. requiring a soft creamy texture; for cosmetics and pharmaceuticals; processing aid, coupling agent, lubricant, mold release for rubber and plastics; also in metal drawing lubricants, sealants, strippable coatings, dielec. compds., wax polishes, crayons, potting compds.; FDA approval; m.p. 70 C; acid no. 2; iodine no. (Wijs) 38; sapon. no. 180; hyd. no. 158; penetration hardness 42.

Castorwax® MP-80. [CasChem] Hydrog. castor oil; CAS 8001-78-3; EINECS 232-292-2; release agent; wax used for formulating antiperspirant sticks, suspending aid for aluminum chlorhydrate; processing aid, coupling agent, lubricant, mold release for rubber and plastics; also in metal drawing lubricants, sealants, strippable coatings, dielec. compds., wax polishes, crayons, potting compds.; FDA approval; wh. flakes; m.p. 80 C; acid no. 2; iodine no. (Wijs) 29; sapon. no. 180; hyd. no. 158; penetration hardness 9; 100% act.

Cata-Chek® 820. [Ferro/Bedford] Dibutyltin dilaurate; CAS 77-58-7; EINECS 201-039-8; heat and lt. stabilizer in flexible vinyl formulations requiring a high degree of clarity; nonsulfur staining; furnishes good lubrication and provides min. water blushing; Gardner 3 clear liq.; sp.gr. 1.045; visc. 34 cps.

Catalpo®. [Engelhard] Hydrous aluminum silicate; reinforcing extender for rubber and polymer systems; wh. highly pulverized powd.; 0.6 μm avg. particle size; 0.01% max. residue +325 mesh; sp.gr. 2.58 g/cc; bulk dens. 18 lb/ft^3 (loose), 33 lb/ft^3 (tamped); bulking value 0.047 gal/lb; oil absorp. 40-45 lb/100 lb; ref. index 1.56; pH 6-8; brightness (GE) 85-87; 45.4% SiO_2; 38.5% Al_2O_3.

Catalpo® X-1. [Engelhard] Hydrous aluminum silicate; reinforcing extender for rubber and polymer systems; wh. spray-dried beads; 0.6 μm avg. particle size; 0.01% max. residue +325 mesh; sp.gr. 2.58 g/cc; bulk dens. 48 lb/ft^3 (loose), 55 lb/ft^3 (tamped); bulking value 0.047 gal/lb; oil absorp. 40-45 lb/100 lb; ref. index 1.56; pH 6-8; brightness (GE) 85-87; 45.4% SiO_2; 38.5% Al_2O_3.

Catalyst CC. [CNC Int'l. L.P.] Amine hydrochloride; catalyst for urea-formaldehyde resins.

Catalyst RD Liq. [Eastern Color & Chem.] Buffered inorg. salt catalyst; fast-acting resin catalyst; liq.

Catigene® CA 56. [Stepan Europe] Alkyl amidopropyl trimethyl ammonium methoxysulfate; cationic; antistat additive for carpet latex and bitumen; pale yel. liq.; 56% act.

Catinex KB-50. [Pulcra SA] Octylphenol, ethoxylated; nonionic; emulsifier, detergent, wetting agent; co-emulsifier and stabilizer for vinyl-acrylic latex; emulsifier for fats and waxes; solid; HLB 16.2; 100% conc.; Unverified

CB-4-34. [Neville] Aromatic plasticizer; plasticizer used in adhesives, rubber (cements, mech. and molded goods, tires), caulking compds.; FDA 21CFR §175.105, 177.2600; Gardner 14 (50% in toluene) semisolid; sol. in ethers and chlorinated, aromatic, naphthenic, and terpene hydrocarbons; m.w. 515; sp.gr. 1.003; soften. pt. (R&B) 28 C; iodine no. (Wijs) 50; flash pt. (COC) 450 F.

CB2408. [Hüls Am.] Bis (hydroxyethyl) aminopropyltriethoxy silane, ethanol; CAS 7538-44-5; coupling agent, release agent, lubricant, blocking agent, chemical intermediate; chemically bonds, strengthens thermoset melamine and urea-formaldehyde composites; liq.; m.w. 309.5; sp.gr. 0.92 (20 C); flash pt. 24 C; ref. index 1.409; 62% in ethanol.

CC-01. [Cardinal Stabilizers] Tin; high performance heat and lt. stabilizer for rigid and semirigid PVC, plastisols, organosols, foams, records, bottles, PU foam.

CC-68. [Cardinal Stabilizers] Tin; heat stabilizer for

rigid PVC, extruded and blown film.

CC™-101. [ECC Int'l.] Calcium carbonate; CAS 471-34-1, 1317-65-3; EINECS 207-439-9; pigment, filler, reinforcer in paper, paint, polymers, food, caulks, etc., esp. traffic paint, joint cement, rubber compounding, carpet underlay and backing, putties and mastics, flooring prods.; wh. powd., odorless; 93% thru 325 mesh; negligible sol. in water; sp.gr. 2.71; dens. 22.57 lb/gal; bulk dens. 70 lb/ft^3 (loose); bulking value 0.044; ref. index 1.59; pH 9.5; nonflamm.; *Toxicology:* TLV/TWA 10 mg/m^3, considered nuisance dust; *Precaution:* incompat. with acids.

CC-101-FC. [Cardinal Stabilizers] Tin; heat stabilizer for semirigid and rigid PVC for inj. molding.

CC™-103. [ECC Int'l.] Calcium carbonate; CAS 471-34-1, 1317-65-3; EINECS 207-439-9; coarse ground pigment, filler, reinforcement for paper, paint, polymers, food, esp. polyolefins, carpet backing, caulks, sealants, putties, as mild abrasives in cleaners; wh. powd., odorless; 80% thru 325 mesh; negligible sol. in water; sp.gr. 2.71; dens. 22.57 lb/gal; bulk dens. 72 lb/ft^3 (loose); bulking value 0.044; ref. index 1.59; pH 9.5; nonflamm.; *Toxicology:* TLV/TWA 10 mg/m^3, considered nuisance dust; *Precaution:* incompat. with acids.

CC-104-S. [Cardinal Stabilizers] Tin; heat and lt. stabilizer for semirigid and rigid PVC, EVA, bottles, slush molding.

CC™-105. [ECC Int'l.] Calcium carbonate; CAS 471-34-1, 1317-65-3; EINECS 207-439-9; coarse ground pigment, filler, reinforcement for paper, paint, polymers, food, esp. polyolefins, carpet backing, caulks, sealants, putties, as mild abrasives in cleaners; wh. powd., odorless; 65% thru 325 mesh; negligible sol. in water; sp.gr. 2.71; dens. 22.57 lb/gal; bulk dens. 80 lb/ft^3 (loose); bulking value 0.044; ref. index 1.59; pH 9.5; nonflamm.; *Toxicology:* TLV/TWA 10 mg/m^3, considered nuisance dust; *Precaution:* incompat. with acids.

CC-401. [Cardinal Stabilizers] Tin; heat stabilizer for semirigid and rigid PVC bottles, film, and sheet.

CC-811. [Cardinal Stabilizers] Tin; heat and lt. stabilizer for semirigid and rigid PVC, plastisols, organosols, foams, bottles; food grade.

CC-1015-M. [Cardinal Stabilizers] Tin; heat and lt. stabilizer for semirigid and rigid PVC siding and windows, organosols.

CC-7710. [Cardinal Stabilizers] Tin; heat stabilizer for semirigid and rigid PVC pipe and foam core, organosols.

Cecavon CA 30. [Ceca SA] Calcium stearate; CAS 1592-23-0; EINECS 216-472-8; anticaking agent for rubber, concrete, composites, foundry, stamping, molding, and paper industries; powd.; dens. 0.20 max.; m.p. 155-165 C; 6.6-7.4% Ca.

Cecavon ZN 70. [Ceca SA] Zinc stearate; CAS 557-05-1; EINECS 209-151-9; waterproofing agent, lubricant, gellant, opacifier for rubber, concrete, composites, paints, varnish, cosmetics, pharmaceuticals, molding, and paper applics.; powd.; dens. 0.20 max.; m.p. 125-130 C; 10.2-11% Zn.

Cecavon ZN 71. [Ceca SA] Zinc stearate; CAS 557-05-1; EINECS 209-151-9; waterproofing agent, lubricant, gellant, opacifier for rubber, concrete, composites, paints, varnish, cosmetics, pharmaceuticals, molding, and paper applics.; powd.; dens. 0.20 max.; m.p. 125-130 C; 10.2-11% Zn.

Cecavon ZN 72. [Ceca SA] Zinc stearate; CAS 557-05-1; EINECS 209-151-9; waterproofing agent, lubricant, gellant, opacifier for rubber, concrete, composites, paints, varnish, cosmetics, pharmaceuticals, molding, and paper applics.; powd.; dens. 0.20 max.; m.p. 125-130 C; 9.8-10.6% Zn.

Cecavon ZN 73. [Ceca SA] Zinc stearate; CAS 557-05-1; EINECS 209-151-9; waterproofing agent, lubricant, gellant, opacifier for rubber, concrete, composites, paints, varnish, cosmetics, pharmaceuticals, molding, and paper applics.; powd.; dens. $<$ 0.20; m.p. 125-130 C; 10.2-10.8% Zn.

Cecavon ZN 735. [Ceca SA] Zinc stearate; CAS 557-05-1; EINECS 209-151-9; waterproofing agent, lubricant, gellant, opacifier for rubber, concrete, composites, paints, varnish, cosmetics, pharmaceuticals, molding, and paper applics.; liq.; sp.gr. 1 $\pm$ 0.10; visc. 50-200 mPa•s; 35 $\pm$ 2% dry content.

Cedepal CA-890. [Stepan Canada] Octoxynol-40; CAS 9002-93-1; nonionic; primary emulsifier for vinyl acetate and acrylate polymerization; wax; HLB 18.0; 100% conc.

Cedephos® FA600. [Stepan; Stepan Canada] Deceth-4 phosphate; anionic; detergent, emulsifier, wetting agent, hard surf. detergent for industrial cleaners, metal cleaners, janitorial prods., agric., textile wetting, emulsion polymerization, oil emulsification, lubricants, corrosion inhibitor, dedusting agent; coupling agent in highly alkaline industrial detergent systems; straw colored clear liq.; 100% act.

Cedephos® RA600. [Stepan; Stepan Canada] Alkyl ether phosphate, acid form; hydrotrope for nonionics in alkali cleaners; emulsifier for agric., emulsion polymerization, oils; lubricants, corrosion inhibitors, pigment dispersants; esp. for heavy-duty industrial/household alkaline cleaners, soak tank formulations; amber clear liq.; 100% act.

Celite® 270. [Celite] Diatomaceous silica, calcined; functional filler for rubber industry; processing aid in rubber compds. esp. silicone rubber; semireinforcing filler in mechanical rubber goods; pinkish particulate; 5.6 μ median particle size, 1.5% 325-mesh residue; sp.gr. 2.2; dens. 8.0 lb/ft^3; surf. area 4-6 m^2/g; oil absorp. 175%; ref. index 1.45; pH 7.0; 92% SiO_2, 3.3% Al_2O_3.

Celite® 292. [Celite] Diatomaceous silica; functional filler for rubber industry; processing aid in rubber compds.; semireinforcing filler in mechanical rubber goods; gray particulate; 7.7 μ median particle size, 4.0% 325-mesh residue; sp.gr. 2.10; dens. 6.4 lb/ft^3; surf. area 20-30 m^2/g; oil absorp. 155%; ref. index 1.43; pH 7.0 max.; 86.8% SiO_2, 3% Al_2O_3.

Celite® 350. [Celite] Diatomaceous silica, calcined;

functional filler for rubber industry; processing aid in rubber compds. esp. silicone rubber; semireinforcing filler in mechanical rubber goods; pink particulate; 8.0 μ median particle size, 0.3% 325-mesh residue; sp.gr. 2.2; dens. 8.0 lb/ft^3; surf. area 4.6 m^2/g; oil absorp. 170%; ref. index 1.45; pH 6.5; 92.8% SiO_2, 3.3% Al_2O_3.

Celite® Super Fine Super Floss. [Celite] Diatomaceous silica, flux calcined; extender pigment and flatting agent for solv.- and water-thinned paints; used in polishes and cleaners; antiblocking agents in LDPE film; wh. particulate; median particle size 3.5 μ, trace 325-mesh residue; sp.gr. 2.30; dens. 7.5 lb/ft^3; dens. 19.2 lb/solid gal; oil absorp. (Spatula) 100; ref. index 1.46; pH 9.4 max.; 93% SiO_2, 2% Al_2O_3.

Celite® Super Floss. [Celite] Diatomaceous silica, flux calcined; extender pigment and flatting agent for solv.- and water-thinned paints; also used in polishes and cleaners; processing aid for rubber goods esp. silicone rubber; antiblocking agent in LDPE film; particulate; median particle size 5.5 μ, 0.1% retained on 325-mesh screen; dens. 19.2 lb/solid gal; oil absorp. (Spatula) 105; ref. index 1.46; 93% SiO_2, 2% Al_2O_3.

Celite® White Mist. [Celite] Diatomaceous silica, flux calcined; extender pigment and flatting agent for solv.- and water-thinned paints; antiblocking agent for PE films; wh. particulate; median particle size 5.5 μ, 0.05% 325-mesh residue; sp.gr. 2.30; dens. 7.5 lb/ft^3; dens. 19.2 lb/solid gal; oil absorp. (Spatula) 105; ref. index 1.46; pH 9.3 max.; 93% SiO_2, 2% Al_2O_3.

Cellcom. [Plastics & Chem.] Azodicarbonamide; CAS 123-77-3; EINECS 204-650-8; chem. foaming agent for ABS, acetal, acrylic, EVA, HDPE, LDPE, PPO, PP, PS, HIPS, flexible PVC, TPE; processing range 300-450 F; gas yield 220 cc/g.

Cellcom OBSH-ASA2. [Plastics & Chem.] 4,4′-Oxybis (benzenesulfonyl) hydrazine; CAS 80-51-3; chem. foaming agent for EVA, LDPE, PS, flexible PVC; processing temps. 250-375 F; gas yield 120-125 cc/g.

Cellex-TS. [Rit-Chem] Zinc ditolyl sulfinate dihydrate; activator improving decomp. chars. of Blo-Foam ADC; imparts no odor; remains nonstaining in finished foam prod.; wh. powd.; m.p. 265-275 C; 95% min. assay; *Storage:* exc. storage stability under normal temp. conditions.

Celogen® AZ 120. [Uniroyal] Azodicarbonamide; CAS 123-77-3; EINECS 204-650-8; chemical blowing agent for thermoset and thermoplastic polymers and rubber; for inj.-molding structural foam, extrusion of profiles, sheet, pipe, and wire coatings, and vinyl plastisol, coating, and calendering; offers fastest decomp. rate; useful for relatively low temp. precures and expansions; for SBR closed cell sheets; maturing agent for flour; FDA 21CFR §172.806 (45 ppm), 175.300, 177.1210 (2% max.), 177.1520(c), 177.2600 (5% max.); yel.-orange fine powd.; 2 μ avg. particle size; sol. with decomp. in alkaline sol'ns.; sl. sol. in water; insol. in benzene, ethylene dichloride; sp.gr. 1.66; decomp. pt. 190-220 C; gas yield 220 cc/g STP; gas: 65% N_2, 24% CO, 5% CO_2, 5% NH_3.

Celogen® AZ 130. [Uniroyal] Azodicarbonamide; CAS 123-77-3; EINECS 204-650-8; chemical blowing agent for thermoset and thermoplastic polymers and rubber; for inj.-molding structural foam, extrusion of profiles, sheet, pipe, and wire coatings, coating, and calendering; suitable for HIPS, ABS, HDPE, PP, PVC, TPE, acetal, PPO, vinyl plastisol; choice for most closed cell sheet stocks because it provides rapid controlled foam development; maturing agent in flour; FDA 21CFR §172.806 (45 ppm), 175.300, 177.1210 (2% max.), 177.1520(c), 177.2600 (5% max.); yel.-orange fine powd.; 3 μ avg. particle size; sol. with decomp. in alkaline sol'ns.; sl. sol. in water; insol. in benzene, ethylene dichloride; sp.gr. 1.66; decomp. pt. 190-220 C; gas yield 220 cc/g STP; gas: 65% N_2, 24% CO, 5% CO_2, 5% NH_3; *Toxicology:* LD50 (oral, rat) 6.8 g/kg, (dermal, rabbit) > 2 g/kg; sl. eye irritant; repeated min. inh. may cause respiratory sensitization and asthma; decomp. gases may cause irritation to eyes, lungs, mucous membranes; *Precaution:* evolves large vols. of gas during decomp.; may form explosive dust-air mixts.; *Storage:* store in cool, dry area in closed containers; avoid any heat source close to 180 C.

Celogen® AZ 150. [Uniroyal] Azodicarbonamide; CAS 123-77-3; EINECS 204-650-8; chemical blowing agent for thermoset and thermoplastic polymers and rubber; for inj.-molding structural foam, extrusion of profiles, sheet, pipe, and wire coatings, and vinyl plastisol, coating, and calendering; useful in extruded closed cell profiles; maturing agent in flour; FDA 21CFR §172.806 (45 ppm), 175.300, 177.1210 (2% max.), 177.1520(c), 177.2600 (5% max.); yel.-orange powd.; 5 μ avg. particle size; sol. with decomp. in alkaline sol'ns.; sl. sol. in water; insol. in benzene, ethylene dichloride; sp.gr. 1.66; decomp. pt. 190-220 C; gas yield 220 cc/g STP; gas: 65% N_2, 24% CO, 5% CO_2, 5% NH_3; *Toxicology:* LD50 (oral, rat) 6.8 g/kg, (dermal, rabbit) > 2 g/kg; sl. eye irritant; repeated min. inh. may cause respiratory sensitization and asthma; decomp. gases may cause irritation to eyes, lungs, mucous membranes; *Precaution:* evolves large vols. of gas during decomp.; may form explosive dust-air mixts.; *Storage:* store in cool, dry area in closed containers; avoid any heat source close to 180 C.

Celogen® AZ 180. [Uniroyal] Azodicarbonamide; CAS 123-77-3; EINECS 204-650-8; chemical blowing agent for thermoset and thermoplastic polymers and rubber; for inj.-molding structural foam, extrusion of profiles, sheet, pipe, and wire coatings, and vinyl plastisol, coating, and calendering; suggested for press molded and continuous sheet applics. where slow cure rates are desirable; maturing agent in flour; FDA 21CFR §172.806 (45 ppm), 175.300, 177.1210 (2% max.), 177.1520(c), 177.2600 (5% max.); yel.-orange powd.; 8 μ avg. particle size; sol. with decomp. in alkaline sol'ns.; sl.

sol. in water; insol. in benzene, ethylene dichloride; sp.gr. 1.66; decomp. pt. 190-220 C; gas yield 220 cc/g STP; gas: 65% N_2, 24% CO, 5% CO_2, 5% NH_3; *Toxicology:* LD50 (oral, rat) 6.8 g/kg, (dermal, rabbit) > 2 g/kg; sl. eye irritant; repeated min. inh. may cause respiratory sensitization and asthma; decomp. gases may cause irritation to eyes, lungs, mucous membranes; *Precaution:* evolves large vols. of gas during decomp.; may form explosive dust-air mixts.; *Storage:* store in cool, dry area in closed containers; avoid any heat source close to 180 C.

Celogen® AZ 199. [Uniroyal] Azodicarbonamide; CAS 123-77-3; EINECS 204-650-8; chemical blowing agent for thermoset and thermoplastic polymers and rubber; for inj.-molding structural foam, extrusion of profiles, sheet, pipe, and wire coatings, and vinyl plastisol, coating, and calendering; suggested for automotive extruded closed cell profiles based on EPDM, EPDM/neoprene or SBR blends; usaed for very fast prod. rates; maturing agent in flour; FDA 21CFR §172.806 (45 ppm), 175.300, 177.1210 (2% max.), 177.1520(c), 177.2600 (5% max.); yel.-orange powd.; 10 μ avg. particle size; sol. with decomp. in alkaline sol'ns.; sl. sol. in water; insol. in benzene, ethylene dichloride; sp.gr. 1.66; decomp. pt. 190-220 C; gas yield 220 cc/g STP; gas: 65% N_2, 24% CO, 5% CO_2, 5% NH_3; *Toxicology:* LD50 (oral, rat) 6.8 g/kg, (dermal, rabbit) > 2 g/kg; sl. eye irritant; repeated min. inh. may cause respiratory sensitization and asthma; decomp. gases may cause irritation to eyes, lungs, mucous membranes; *Precaution:* evolves large vols. of gas during decomp.; may form explosive dust-air mixts.; *Storage:* store in cool, dry area in closed containers; avoid any heat source close to 180 C.

Celogen® AZ 754. [Uniroyal] Activated azodicarbonamide; CAS 123-77-3; EINECS 204-650-8; Chemical blowing agent for plastics processed at 330-450 F; for EVA and LDPE inj. molding structural foam applics.

Celogen® AZ-760-A. [Uniroyal] Modified azodicarbonamide; CAS 123-77-3; EINECS 204-650-8; blowing agent for plastics; yel. fine free-flowing powd.; 4 μ avg. particle size; decomp. pt. 200-215 C; gas yield 140-190 cc/g @ STP; gas: 42% N_2, 21% CO_2, 21% NH_3, 16% CO; *Toxicology:* dust may be irritating; avoid inh. of dust, prolonged skin contact, ingestion; *Storage:* exc. storage stability under normal conditions; store in cool, dry area wasy from heat sources, direct sunlight.

Celogen® AZ-2100. [Uniroyal] Modified azodicarbonamide; CAS 123-77-3; EINECS 204-650-8; flow-treated dispersible chemical blowing agent for cellular vinyl; free-flowing for Hopper blender units; produces blown vinyl prod. with fine uniform cell structure; eliminates settling problems; for plasticized vinyl, inj. molding structural foam, extrusion applics.; FDA 21CFR §175.300 (2% max.), 177.1210 (2% max.), 177.2600 (5% max.); yel. fine powd.; 2 μ avg. particle size; somewhat sol. in polyalkylene glycols, dimethylformamide; insol. in benzene, acetone, water; sp.gr. 1.65; decomp. pt. 205-215 C; gas yield 220 cc/g STP; gas: 65% N_2, 24% CO, 5% CO_2, 5% NH_3; *Toxicology:* LD50 (oral) > 6 g/kg; dust may be irritating; *Storage:* exc. storage stability under normal conditions; store in cool, dry area away from heat sources, direct sunlight.

Celogen® AZ-2990. [Uniroyal] Modified azodicarbonamide; CAS 123-77-3; EINECS 204-650-8; flow-treated dispersible chemical blowing agent for cellular vinyl; free-flowing for Hopper blender units; produces blown vinyl prod. with fine uniform cell structure; eliminates settling problems; for plasticized vinyl, inj. molding structural foam, extrusion applics.; FDA 21CFR §175.300 (2% max.), 177.1210 (2% max.), 177.2600 (5% max.); yel. fine powd.; 2 μ avg. particle size; somewhat sol. in polyalkylene glycols, dimethylformamide; insol. in benzene, acetone, water; sp.gr. 1.65; decomp. pt. 205-215 C; gas yield 220 cc/g STP; gas: 65% N_2, 24% CO, 5% CO_2, 5% NH_3; *Toxicology:* LD50 (oral) > 6 g/kg; dust may be irritating; *Storage:* exc. storage stability under normal conditions; store in cool, dry area away from heat sources, direct sunlight.

Celogen® AZ 3990. [Uniroyal] Modified azodicarbonamide; CAS 123-77-3; EINECS 204-650-8; chemical blowing agent for PVC profiles, wire and cable, inj.-molding structural foam and extrusion, plastisols, calendering, HDPE, PP, LDPE rotational and expansion casting; for processing at 330-450 F; FDA 21CFR §175.300 (2% max.), 177.1210 (2% max.), 177.2600 (5% max.); yel. fine powd.; 3 μ avg. particle size; somewhat sol. in polyalkylene glycols, dimethylformamide; insol. in benzene, acetone, water; sp.gr. 1.65; decomp. pt. 205-215 C; gas yield 220 cc/g STP; gas: 65% N_2, 24% CO, 5% CO_2, 5% NH_3; *Toxicology:* LD50 (oral) > 6 g/kg; dust may be irritating; *Storage:* exc. storage stability under normal conditions; store in cool, dry area away from heat sources, direct sunlight.

Celogen® AZ-5100. [Uniroyal] Modified azodicarbonamide; CAS 123-77-3; EINECS 204-650-8; flow-treated dispersible chemical blowing agent for cellular vinyl; free-flowing for Hopper blender units; produces blown vinyl prod. with fine uniform cell structure; eliminates settling problems; for plasticized vinyl, inj. molding structural foam, extrusion applics.; FDA 21CFR §175.300 (2% max.), 177.1210 (2% max.), 177.2600 (5% max.); yel. fine powd.; 5 μ avg. particle size; somewhat sol. in polyalkylene glycols, dimethylformamide; insol. in benzene, acetone, water; sp.gr. 1.65; decomp. pt. 205-215 C; gas yield 220 cc/g STP; gas: 65% N_2, 24% CO, 5% CO_2, 5% NH_3; *Toxicology:* LD50 (oral) > 6 g/kg; dust may be irritating; *Storage:* exc. storage stability under normal conditions; store in cool, dry area away from heat sources, direct sunlight.

Celogen® AZNP 130. [Uniroyal] Modified azodicarbonamide; CAS 123-77-3; EINECS 204-650-8; non-plateout chemical blowing agent for plastics, inj. molding structural foam, extrusion, cellular cable insulation, applics. where mold plateout is a problem; operating temps. 199-232 C, from 166 C with activation; yel. fine free-flowing powd.; 3 μ avg. particle size; decomp. pt. 200-215 C; gas yield 140-190 cc/g @ STP; gas: 42% N_2, 21% CO_2, 21% NH_3, 16% CO; *Toxicology:* dust may be irritating; avoid inh. of dust, prolonged skin contact, ingestion; *Storage:* exc. storage stability under normal conditions; store in cool, dry area away from heat sources, direct sunlight.

Celogen® AZNP 199. [Uniroyal] Modified azodicarbonamide; CAS 123-77-3; EINECS 204-650-8; non-plateout chemical blowing agent for plastics, inj. molding structural foam, extrusion, cellular cable insulation, applics. where mold plateout is a problem; operating temps. 199-232 C, from 166 C with activation; yel. fine free-flowing powd.; 12 μ avg. particle size; decomp. pt. 200-215 C; gas yield 140-190 cc/g @ STP; gas: 42% N_2, 21% CO_2, 21% NH_3, 16% CO; *Toxicology:* dust may be irritating; avoid inh. of dust, prolonged skin contact, ingestion; *Storage:* exc. storage stability under normal conditions; store in cool, dry area away from heat sources, direct sunlight.

Celogen® AZRV. [Uniroyal] Modified azodicarbonamide; CAS 123-77-3; EINECS 204-650-8; Chemical blowing agent for plastics processed at 320-390 F; for rigid PVC profiles, pipe, sheet, inj. molding structural foam; also for PP wire and cable.

Celogen® OT. [Uniroyal] p,p′-Oxybis benzene sulfonyl hydrazide; CAS 80-51-3; nondiscoloring, nonstaining chemical blowing agent for sponge rubber and expanded plastics (LDPE wire/cable, structural foam inj. moldings, and rotational casting, flexible PVC structural foam inj. molding) processed at 149-177 C; FDA 21CFR §175.300 (0.5% max.), 177.1200 (0.5% max.), 177.2600 (5% max.); wh. fine cryst. powd., odorless; very sol. in dimethyl sulfoxide, dimethyl formamide; sol. in acetone (reacts); mod. sol. in ethanol, polyalkylene glycols; insol. in benzene, ethylene dichloride, gasoline, cold water; sp.gr. 1.55; decomp. pt. 158-160 C; gas yield 125 cc/g @ STP; gas: 91% N_2, 9% H_2O; *Toxicology:* LD50 (oral, rat) > 5.2 g/kg; nontoxic; avoid ingestion, inh. of dust, excessive skin contact; *Precaution:* flamm. solid which will continue to burn once ignited producing dense smoke and intense heat; oxidizing agents will reduce decomp. temp.; gaseous decomp. prods.: N_2, CO_2; *Storage:* exc. storage stability under normal conditions; store in cool, dry place away from heat sources, direct sunlight; reclose containers.

Celogen® RA. [Uniroyal] p-Toluene sulfonyl semicarbazide; chemical blowing agent for polymers processing at 216-260 C range (191 C with activation); for inj. molding structural foam (HIPS, ABS, PP), extrusion (rigid PVC), plasticized vinyl; wh. fine powd.; sol. in dimethylsulfoxide, dimethylformamide (reacts), conc. ammonium hydroixde; insol. in water, acetic acid, acetone, CCl_4, IPA, toluene; sp.gr. 1.44; decomp. pt. 228-235 C; gas yield 140 cc/g @ STP; gas: 55% N_2, 37% CO_2, 2% CO, 3% NH_3; *Toxicology:* LD50 (oral, rat) > 10 g/kg; nonirritating to skin and eyes; avoid ingestion, inh. of dust, excessive skin contact; *Precaution:* will continue to burn once ignited giving off dense smoke; *Storage:* good storage stability under normal conditions; store in cool, dry area away from heat sources and direct sunlight.

Celogen® TSH. [Uniroyal] p-Toluene sulfonyl hydrazide; CAS 877-66-7; chemical blowing agent for low temp. processing (105-132 C), thermoset polyester, vinyl, epoxy foam; cream-colored powd.; very sol. in dimethyl sulfoxide; reacts with dimethyl formamide, MEK, acetone; insol. in water; sp.gr. 1.42; gas yield 115 cc/g @ STP; gas: nitrogen and steam; *Toxicology:* avoid ingestion, inh. of dust, excessive skin contact; *Precaution:* burns rapidly when ignited producing smoke; *Storage:* good storage stability under normal conditions; store in cool, dry place away from heat sources, direct sunlight.

Celogen® XP-100. [Uniroyal] Sulfonyl hydrazide; chemical blowing agent.

Ceramtex. [Colloids Ltd] Specialty masterbatches which provides ceramic-type speckle effect in polyolefin and styrenic moldings; suitable for food contact and cosmetics; applics. incl. housewares, picnic ware, cosmetic containers, decorative horticultural ware.

Cereclor 42. [StanChem] Halogenated organic; flame retardant for most polymers.

CeriDust 9610F. [Hoechst Celanese] PTFE wax; CAS 9002-84-0; internal/external lubricant for thermoplastics, coatings; solid.

Cetamoll®. [BASF AG] Plasticizers for polyamide-based surface coating resins and for polyamides.

Cetodan® 95 CO. [Grindsted Prods.] Acetylated palm kernel glycerides; food-grade plasticizer for PVC; FDA 21CFR §172.828.

Cetodan® 95-ML. [Grindsted Prods.] Acetylated monoglycerides; food-grade plasticizer for PVC; FDA 21CFR §172.828.

CF 1500. [Custom Fibers] Cellulose fibers; CAS 65996-61-4; reinforcing filler; contributes exc. mech. props. in V-belt formulations, etc.; high oil absorp.; recommended where high tens. str. reinforcement, max. visc. increase and drag resist. are required; gray, wh. or off-wh. super extra coarse fibers, no discernible odor; sp.gr. 1.58; dens. 13.2 l b/gal; oil absorp. 1000%; pH 7.0; *Toxicology:* considered nuisance dust; avoid inhalation; may cause irritation to eyes and respiratory system; ingestion may cause vomiting or diarrhea; *Precaution:* combustible; avoid extreme heat and open flame.

CF 32,500T Coarse. [Custom Fibers] Cellulose fibers; CAS 65669-61-4; visc. control, sag resistance, fiber reinforcement, improved str. for solv. or water-based systems incl. asphalt plastic roof cement, block filler, gasket material, epoxy adhesives, tile cement, phenolic molding compds.; gray; sp.gr. 1.58; dens. 13.2 lb/gal; oil absorp. 720-800%; pH 7.0; Unverified

CFE-50. [Ferro/Keil] Chlorinated fatty material; additive for cutting oils and other metalworking applics.; outstanding EP and antiwipe/antiweld props. at relative low Cl content; provides corrosion protection; plasticizer component for polymeric systems; sol. in naphthenic oils, paraffinic oils; sp.gr. 1.14; dens. 9.5 lb/gal; visc. 2.3 poise; acid no. 7; pour pt. -29 C.

CGA 12. [Ciba-Geigy/Additives] 2,2′,2′′-Nitrilo [triethyl-tris[3,3′,5,5′-tetra-t-butyl-1,1′-biphenyl-2,2′-diyl]phosphite] CAS 80410-33-9; process stabilizer, sec. antioxidant protecting polymers against thermal degradation and discoloration during processing; used with Irganox® primary antioxidants provides long-term protection against thermo-oxidative degradation after processing; exc. hydrolytic and thermal stability; for LLDPE, HDPE, PP, PC, polyamides, linear polyesters, HIPS, ABS, SAN, adhesives, elastomers (BR, IR); wh. free-flowing powd., odorless, tasteless; sol. (mg/ml @ 20 C): 780 mg in toluene, 370 mg in cyclohexane; m.w. 1465; m.p. 120-220 C; *Precaution:* fire and explosive hazard.

CHDA. [Eastman] 1,4-Cyclohexanedicarboxylic acid; specialty monomer for polymer applics.

CHDM. [Eastman] 1,4-Cyclohexanedimethanol; CAS 105-08-8; specialty monomer for polymer applics.

Chel DTPA-41. [Ciba-Geigy] Pentasodium diethylenetriaminepentaacetic acid, aq. sol'n.; CAS 140-01-2; EINECS 205-391-3; chelating agent for detergents, textiles (reduces catalytic degradation and fiber tenderization), polymerization (controls metal ions in phenolic resin emulsion systems); straw clear liq.; sp.gr. 1.30-1.32; chel. value 80; pH 11.0-11.8 (1% aq.); 41% act.

Chemax DOSS/70E. [Chemax] Sodium dioctyl sulfosuccinate; wetting agent for textile, agric., detergent formulations, emulsion polymerization; pigment dispersant in paints and inks; solubilizer for drycleaning solvs.; liq.; flash pt. > 100 F; 70% act.

Chemax DOSS/70HFP. [Chemax] Sodium dioctyl sulfosuccinate; wetting agent for textile, agric., detergent formulations, emulsion polymerization; pigment dispersant in paints and inks; solubilizer for drycleaning solvs.; liq.; flash pt. > 200 F; 70% act.

Chemax DOSS-75E. [Chemax] Dioctyl sodium sulfosuccinate; CAS 577-11-7; EINECS 209-406-4; anionic; wetting and rewetting agent, detergent, emulsifier for emulsion polymerization, cosmetics, textile, agric., detergent formulations; pigment dispersant in paints and inks; solubilizer for drycleaning solvs.; liq.; 75% conc.

Chemax HCO-5. [Chemax] PEG-5 hydrog. castor oil; CAS 61788-85-0; nonionic; emulsifier, lubricant, softener, dispersant; coemulsifier for syn. esters; for cosmetics, plastics, metals, textiles, leather, paint, and paper indutries; liq.; oil-sol.; sapon. no. 142; HLB 3.8; 100% conc.

Chemax HCO-16. [Chemax] PEG-16 hydrog. castor oil; CAS 61788-85-0; nonionic; emulsifier, lubricant, dispersant, softener; for cosmetics, textiles, plastics, metalworking, paint, paper, leather; liq.; sapon. no. 100; HLB 8.6; 100% conc.

Chemax HCO-25. [Chemax] PEG-25 hydrog. castor oil; CAS 61788-85-0; nonionic; emulsifier, lubricant, and softener for cosmetics, plastics, metals, textiles, leather, paint, and paper industries; liq.; sapon. no. 80; HLB 10.8; 100% conc.

Chemax HCO-200/50. [Chemax] PEG-200 hydrog. castor oil; CAS 61788-85-0; nonionic; emulsifier and lubricant for cosmetics, plastics, metals, textiles, paint, paper, leather industries; liq.; water-sol.; sapon. no. 17; HLB 18.1; 50% act.

Chemax NP-40. [Chemax] Nonoxynol-40; CAS 9016-45-9; nonionic; polymerization emulsifier for vinyl acetate and acrylic emulsions; stabilizer for syn. latices; wetting agent in electrolyte sol'ns.; cosmetics surfactant; solid; pour pt. 112 F; cloud pt. > 100 C (1% aq.); HLB 17.8; 100% conc.

Chemax OP-40/70. [Chemax] Octoxynol-40; CAS 9002-93-1; nonionic; emulsifier for vinyl acetate and acrylate polymerization; cosmetics; liq.; pour pt. 25 F; cloud pt. > 100 C (1% aq.); HLB 17.9; 70% conc.

Chemeen 18-2. [Chemax] PEG-2 stearamine; CAS 10213-78-2; EINECS 233-520-3; mild cationic; emulsifier and antistat in cosmetics, textiles, metal buffing, and rubber compds.; lubricant for fiber glass; solid; m.w. 365; HLB 4.8.

Chemeen 18-5. [Chemax] PEG-5 stearamine; CAS 26635-92-7; emulsifier and antistat in cosmetics, textiles, metal buffing, and rubber compds.; lubricant for fiber glass; solid; m.w. 540.

Chemeen 18-50. [Chemax] PEG-50 stearamine; CAS 26635-92-7; mild cationic; emulsifier and antistat in cosmetics, textiles, metal buffing, and rubber compds.; lubricant for fiber glass; solid; m.w. 2400; HLB 17.8.

Chemeen HT-5. [Chemax] PEG-5 hydrog. tallow amine; mild cationic; emulsifier and antistat in textiles, metal buffing and rubber compds., lubricant for fiber glass; paste; m.w. 495; HLB 9.0.

Chemeen HT-50. [Chemax] PEG-50 hydrog. tallow amine; emulsifier and antistat in textiles, metal buffing and rubber compds., lubricant for fiber glass; solid; m.w. 2470; 100% conc.; Unverified

Chemfac 100. [Chemron] Proprietary; nonionic; defoamer for vacuum filtration fluids; esp. for processing yel. cake and waste concs.; defoamer for guar, HPG, CMHPG; surfactant for polymerization; lt. yel. liq.; 100% act.

Chemfac NC-0910. [Chemax] POE alkyl phenol phosphate; wetting agent, detergent, hydrotrope, emulsifier, rust inhibitor, EP additive for alkaline detergents, metal cleaners, hard surf. cleaners, textile scours, metal and textile lubricants, drycleaning soaps, emulsion polymerization, agric. formulations; liq.; acid no. 145.

Chemfac PA-080. [Chemax] Alcohol phosphate ester; anionic; detergent, emulsifier, wetting agent, lubricant, antistat for alkaline detergents, metal cleaners, hard surf. cleaners, textile scours, metal and textile lubricants, drycleaning soaps, emulsion polymerization, agric. formulations; liq.; acid no.

345; 100% conc.

Chemfac PB-082. [Chemax] Phosphate esters; anionic; detergent, wetting, and coupling agent, antistat, emulsifier for alkaline detergents, metal cleaners, hard surf. cleaners, textile scours, metal and textile lubricants, drycleaning soaps, emulsion polymerization, agric. formulations; liq.; acid no. 195; 99% min. act.

Chemfac PB-106. [Chemax] POE alkyl ether phosphate; anionic; wetting agent, detergent, hydrotrope, emulsifier, rust inhibitor, EP additive for alkaline detergents, metal cleaners, hard surf. cleaners, textile scours, metal and textile lubricants, drycleaning soaps, emulsion polymerization, agric. formulations; liq.; acid no. 120.

Chemfac PB-135. [Chemax] POE alkyl ether phosphate; anionic; wetting agent, detergent, hydrotrope, emulsifier, rust inhibitor, EP additive for alkaline detergents, metal cleaners, hard surf. cleaners, textile scours, metal and textile lubricants, drycleaning soaps, emulsion polymerization, agric. formulations; liq.; acid no. 115.

Chemfac PB-184. [Chemax] Oleth-4 phosphate; anionic; wetting agent, detergent, hydrotrope, emulsifier, rust inhibitor, EP additive for alkaline detergents, metal cleaners, hard surf. cleaners, textile scours, metal and textile lubricants, drycleaning soaps, emulsion polymerization, agric. formulations; liq.; acid no. 140; 100% conc.

Chemfac PB-264. [Chemax] Phosphate ester; anionic; wetting agent, detergent, hydrotrope, emulsifier, rust inhibitor, EP additive for alkaline detergents, metal cleaners, hard surf. cleaners, textile scours, metal and textile lubricants, drycleaning soaps, emulsion polymerization, agric. formulations; liq.; acid no. 165; 100% conc.

Chemfac PC-099E. [Chemax] POE alkyl phenol phosphate; anionic; detergent, wetting agent, primary emulsifier in emulsion polymerization; liq.; acid no. 115; 99% min. act.

Chemfac PC-188. [Chemax] Phosphate ester; anionic; wetting agent, detergent, hydrotrope, emulsifier, rust inhibitor, EP additive for alkaline detergents, metal cleaners, hard surf. cleaners, textile scours, metal and textile lubricants, drycleaning soaps, emulsion polymerization, agric. formulations; liq.; acid no. 85; 100% conc.

Chemfac PD-600. [Chemax] POE alkyl ether phosphate; CAS 52019-36-0; anionic; wetting agent, detergent, hydrotrope, emulsifier, rust inhibitor, EP additive for alkaline detergents, metal cleaners, hard surf. cleaners, textile scours, metal and textile lubricants, drycleaning soaps, emulsion polymerization, agric. formulations; liq.; acid no. 210.

Chemfac PF-623. [Chemax] POE alkyl ether phosphate; anionic; wetting agent, detergent, hydrotrope, emulsifier, rust inhibitor, EP additive for alkaline detergents, metal cleaners, hard surf. cleaners, textile scours, metal and textile lubricants, drycleaning soaps, emulsion polymerization, agric. formulations; liq.; acid no. 170.

Chemfac PF-636. [Chemax] POE phosphate; anionic; wetting agent, detergent, hydrotrope, emulsifier, rust inhibitor, EP additive for alkaline detergents, metal cleaners, hard surf. cleaners, textile scours, metal and textile lubricants, drycleaning soaps, emulsion polymerization, agric. formulations; liq.; acid no. 400; 90% conc.

Chemfac PN-322. [Chemax] Phosphate ester, neutralized; anionic; hydrotrope for solubilizing nonionic surfactants in high concs. of alkali or other electrolytes; wetting agent, detergent, emulsifier, rust inhibitor, EP additive for alkaline detergents, textile scours, emulsion polymerization, lubricants, agric.; liq.; 50% conc.

Chemiflex 315XA(80). [Sanyo Chem. Industries] Blocked isocyanate; adhesion promoter for PVC plastisols to ED paint; liq.; 80% conc; Unverified

Chemigum® N8. [Goodyear] NBR, hot polymer, nonstaining antioxidant; low shrinkage polymer for blending with other Chemigums; modifies PVC and ABS resins; imparts flexibility for vacuum forming; cream-tan solid gum, mild pleasant odor; sol. 20% in MEK; sp.gr. 0.98; Mooney visc. 72-86 (ML4, 212 F); 50 part HAF blk. compd. properties: tens. str. 3000 psi; tens. elong. 240%; Shore hardness A 79; 33% ACN.

Chemigum® P83. [Goodyear] NBR powd. partitioned with 9% PVC, lightly precrosslinked, stabilized; designed for PVC modification; improves hot-melt stability, plasticizer permanence, and processability for calendered and extruded film, sheet, gaskets and seals, hose and tubing, tech. extrusion profiles; 0.25-0.50 mm median particle size; sp.gr. 1.00; Mooney visc. 45 (ML4, 100 C); 33% bound ACN.

Chemlease 55. [Chemlease] Silicone blend; mold release for urethane elastomers.

Chemlease 80W. [Chemlease] Water-based; semi-permanent release system for inj., transfer, and compr. molding of rubbers, tires, thermoset prepregs and laminates, inj. molding of thermoplastic urethanes and TPEs, polyurea hybrids, thermoplastics.

Chemlease 88. [Chemlease] Fluoropolymer blend; dry lubricant mold release for elec. potting, filament winding, phenolic board, adhesive laminate.

Chemlease 158R. [Chemlease] Nonsilicone; water-based release system for epoxy composites, compr. molding, RIM urethane molding, rubbers.

Chemlease 906E. [Chemlease] Silicone water-based emulsion; mold release for rubber and plastics industries; Unverified

Chemlease PMR. [Chemlease] Solv.-based; semi-permanent release agent for polyester gel coat laminates; also for resin transfer, thermoforming.

Chemlease SP 40. [Chemlease] General-purpose semi-permanent release system for plastic and rubber, elec. potting, filament winding, phenolic board, roto molding; Unverified

Chem-Master R-11. [Syn. Prods.] 90% Litharge, 10% EPM; elastomeric dispersion.

Chem-Master R-13. [Syn. Prods.] 85% French zinc oxide, 15% EPM; elastomeric dispersion.

Chem-Master R-15. [Syn. Prods.] 75% Ethylene

thiourea, 25% EPM; elastomeric dispersion.

Chem-Master R-51. [Syn. Prods.] 90% Litharge, 10% EPDM; elastomeric dispersion.

Chem-Master R-74. [Syn. Prods.] 75% Ethylene thiourea, 25% ECO; elastomeric dispersion.

Chem-Master R-102. [Syn. Prods.] 85% French zinc oxide, 15% NBR; elastomeric dispersion.

Chem-Master R-103. [Syn. Prods.] 85% French zinc oxide, 15% polyisobutylene; elastomeric dispersion.

Chem-Master R-135. [Syn. Prods.] 85% Antimony oxide/15% SBR; elastomeric dispersion.

Chem-Master R-215. [Syn. Prods.] 85% Cadmium oxide, 15% NBR; elastomeric dispersion.

Chem-Master R-231. [Syn. Prods.] 90% Litharge, 10% polyisobutylene; elastomeric dispersion.

Chem-Master R-301. [Syn. Prods.] 75% Ethylene thiourea, 25% EPMD; elastomeric dispersion.

Chem-Master R-303. [Syn. Prods.] 85% Cadmium oxide, 15% EPDM; elastomeric dispersion.

Chem-Master R-321. [Syn. Prods.] 85% Antimony oxide/15% EPDM; elastomeric dispersion.

Chem-Master R-354. [Syn. Prods.] 80% Barium carbonate, 20% ECO; elastomeric dispersion.

Chem-Master R-482A. [Syn. Prods.] 65% Cadmium stearate, 35% EPM; elastomeric dispersion.

Chem-Master R-486. [Syn. Prods.] 75% Urea, 25% EPM; elastomeric dispersion.

Chem-Master R-524. [Syn. Prods.] Ethyl cadmate, EPM; elastomeric dispersion.

Chem-Master R-936. [Syn. Prods.] 50% Dipropylene glycol, 50% NBR; elastomeric dispersion.

Chem-Master R-8414. [Syn. Prods.] 85% Antimony oxide/15% EPM; elastomeric dispersion.

Chem-Master RD-50. [Syn. Prods.] 50% Zinc peroxide, 50% NBR; elastomeric dispersion.

Chemphos TC-227. [Chemron] Aromatic phosphate ester; detergent, wetting agent, emulsifier, coupling agent, surf. tension reducer; for alkaline cleaners, heavy-duty all-purpose metalworking detergents, steam cleaning, dairy cleaners, bottle washing compds., floor strippers; lubricant and detergent for drilling fluids; oil treating chemicals; emulsion polymerization of vinyl acetate, acrylates, SBR; clear visc. liq.; sol. in water, most oxygenated solvs., aromatic solvs., chlorinated solvs.; 100% act.

Chemphos TC-310. [Chemron] Aromatic phosphate ester; anionic; detergent, wetting agent, emulsifier, dispersant, surf. tension reducer, rust inhibitor; emulsion polymerization surfactant for vinyl acetate, acrylates, SBR; clear visc. liq.; sol. in water, alcohols, hydrocarbon solvs., chlorinated solvs.; 100% act.

Chemphos TC-337. [Chemron] Nonoxynol-20 phosphate; anionic; emulsion polymerization surfactant for vinyl acetate, acrylates, SBR; emulsifier, solubilizer, antistat, substantivity agent for hair care prods., perms, straighteners, depilatories; resistant to hydrolysis; clear visc. liq.; 100% act.

Chemprene R-10. [Chemfax] Thermoplastic isoprenoidal polymer; softener, tackifier, and reinforcing agent used in compounding of syn., natural, and reclaim rubber; used in calendering and extruding, hard rubber, molded goods, in camelback, tubes, and tire stocks, in rubber cements, wire and cable insulation, floor tile, shoe compds., caulks; Barrett 2 visc. liq. to semisolid; sol. in min. spirits, ketones, cyclic alchols, etc.; sp.gr. 1.02; soften. pt. 10 C.

Chemprene R-100. [Chemfax] Thermoplastic isoprenoidal polymer; softener, tackifier, and reinforcing agent used in compounding of syn., natural, and reclaim rubber; used in calendering and extruding, hard rubber, molded goods, in camelback, tubes, and tire stocks, in rubber cements, wire and cable insulation, floor tile, shoe compds., caulks; Barrett 2 solid, flakes; sol. in min. spirits, ketones, cyclic alchols, etc.; sp.gr. 1.10; soften. pt. 100 C.

Chemprene R-115. [Chemfax] Thermoplastic isoprenoidal polymer; softener, tackifier, and reinforcing agent used in compounding of syn., natural, and reclaim rubber; used in calendering and extruding, hard rubber, molded goods, in camelback, tubes, and tire stocks, in rubber cements, wire and cable insulation, floor tile, shoe compds., caulks; solid and flakes; sol. in min. spirits, ketones, cyclic alchols, etc.; sp.gr. 1.10; G-H visc. Z1 (70% in min. spirits); soften. pt. 115 C; acid no. 0; sapon. no. 0.

Chemquat 12-33. [Chemax] Laurtrimonium chloride; CAS 112-00-5; EINECS 203-927-0; cationic; surfactant, corrosion inhibitor, antistat for plastics, textile dyeing aid, gel sensitizer in latex foam prod.; visc. depressant in paper and textile softener formulations; 33% act.

Chemquat 12-50. [Chemax] Laurtrimonium chloride; CAS 112-00-5; EINECS 203-927-0; cationic; surfactant, corrosion inhibitor, antistat for plastics, textile dyeing aid, gel sensitizer in latex foam prod.; visc. depressant in paper and textile softener formulations; liq.; 50% act.

Chemquat 16-50. [Chemax] Cetrimonium chloride; CAS 112-02-7; EINECS 203-928-6; cationic; surfactant, corrosion inhibitor, antistat for plastics, textile dyeing aid, gel sensitizer in latex foam prod.; visc. depressant in paper and textile softener formulations; liq.; 50% act.

Chemquat C/33W. [Chemax] Coco trimethyl ammonium chloride; CAS 61789-18-2; EINECS 263-038-9; cationic; surfactant, corrosion inhibitor, antistat for plastics, textile dyeing aid, gel sensitizer in latex foam prod.; visc. depressant in paper and textile softener formulations; liq.; 33% act.

Chemstat® 106G/60DC. [Chemax] Bis (2-hydroxyethyl) octyl methylammonium p-toluene sulfonate; CAS 58767-50-3; permanent antistatic agent for use in expandable PS and other styrenic polymers where high processing temps. are required; lubricating aid in EPS bead screening process; stable to 270-300 C; wh. powd.; dens. 0.44 g/cc (loose); 60% act.

Chemstat® 106G/90. [Chemax] Bis (2-hydroxyethyl) octyl methylammonium p-toluene sulfonate; CAS 58767-50-3; permanent internal antistat for polystyrene and other thermoplastics where high processing temps. required; also as external antistat; for extrusion, inj. and blow molding of thermoplastics esp. PS, ABS; stable to 300 C; APHA 300 max. (25% aq.) visc. liq., char odor; sol. in water; sp.gr. 1.1; b.p. 212 F; flash pt. (COC) > 200 F; pH 5-8 (1% aq.); 8-12% water; *Toxicology:* LD50 (oral, rat) 1.41 g/kg; corrosive to eyes, skin, mucous membranes; can cause nausea, diarrhea, lethargy; *Precaution:* incompat. with strong oxidizing agents; *Storage:* store in tightly closed containers away from incompat. materials; avoid extreme heat or cold.

Chemstat® 122. [Chemax] N,N-Bis (2-hydroxyethyl) coco amine; cationic; surfactant; permanent internal antistat for polyolefin film and molded prods.; effective in extrusion, inj. molding, and blow molding of thermoplastics; for HDPE, LDPE, PP, ABS; FDA 21CFR §178.3130; clear liq.; sp.gr. 0.88; flash pt. (COC) > 350 F; fire pt. (COC) > 400 F; pH 9-11 (10% in 1:1 IPA/water); 97% min. tert. amine, 1% max. water.

Chemstat® 122/60DC. [Chemax] N,N-Bis (2-hydroxyethyl) coco amine; permanent internal antistat for thermoplastic films, sheets, and molded prods.; for HDPE, LDPE, PP, ABS; FDA 21CFR §178.3130; wh. ivory powd.; dens. 0.31 g/cc (loose); 60% act.; *Toxicology:* irritating to skin, extremely irritating to eyes; airborne particles are extremely irritating to mucous membranes if inhaled; *Storage:* store in cool, dry area away from strong oxidizing and reducing agents.

Chemstat® 172T. [Chemax] Bis (2-hydroxyethyl) oleamine; CAS 13127-82-7; internal antistat for polyolefins (HDPE, LDPE); FDA 21CFR § 178.3130; clear liq.; sp.gr. 0.90; pour pt. -4 C; pH 9-11 (10% in 1:1 IPA/water); 97% min. tert. amine, 1% max. water.

Chemstat® 182. [Chemax] N,N Bis (2-hydroxyethyl) tallow amine; CAS 61791-44-4; cationic; surfactant, permanent internal antistat for polyolefin film and molded prods.; effective in extrusion, inj. and blow molding; for HDPE, LDPE; FDA 21CFR §178.3130; clear liq./soft paste; sp.gr. 0.92; flash pt. (COC) > 390 F; fire pt. (COC) > 450 F; pH 9-11 (10% in 1:1 IPA/water); 97% min. tert. amine, 1% max. water.

Chemstat® 182/67DC. [Chemax] N,N-Bis (2-hydroxyethyl) tallow amine; CAS 61791-44-4; permanent internal antistat for thermoplastic film, sheets, and molded prods. (HDPE, LDPE); FDA 21CFR §178.3130; wh.-ivory powd.; dens. 0.39 g/cc (loose); 67% act.; *Toxicology:* irritating to skin, extremely irritating to eyes; airborne particles are extremely irritating to mucous membranes if inhaled; *Storage:* store in cool, dry area away from strong oxidizing and reducing agents.

Chemstat® 192. [Chemax] N,N-Bis (2-hydroxyethyl) stearyl amine; CAS 10213-78-2; EINECS 233-520-3; cationic; surfactant, permanent internal antistat for polyolefin film and molded prods.; effective in extrusion, inj. and blow molding for PP, HDPE, ABS, PS, nylon; FDA 21CFR §178.3130; wh. solid; flash pt. (COC) > 390 F; fire pt. (COC) > 450 F; 97% min. tert. amine, 0.5% max. moisture.

Chemstat® 192/NCP. [Chemax] N,N-Bis (2-hydroxyethyl) stearyl amine; CAS 10213-78-2; EINECS 233-520-3; cationic; permanent internal antistat effective in eliminating electrostatic problems in extrusion, inj. and blow molding of thermoplastics (PP, HDPE, ABS, PS, nylon); FDA 21CFR §178.3130; off-wh. free-flowing powd., mild odor; insol. in water; sp.gr. 0.97-1.00; flash pt. (COC) > 390 F; fire pt. (COC) > 450 F; 93% act., 97% min. tert. amine, 1.5% max. water; *Toxicology:* corrosive to eyes, skin, mucous membranes; may cause gastrointestinal irritation, nausea, diarrhea; *Precaution:* incompat. with strong oxidizing agents; *Storage:* store in tightly closed containers away from incompat. materials; avoid extreme heat or cold.

Chemstat® 273-C. [Chemax] N,N-Bis (2-hydroxyethyl) coco amine; CAS 61791-31-9; permanent internal antistat for plastic film and molded prods.; eliminates electrostatic problems in extrusion, inj. and blow molding; FDA 21CFR §178.3130; clear liq.; sp.gr. 0.88; flash pt. (COC) > 350 F; fire pt. (COC) > 400 F; 0.5% max. water.

Chemstat® 273-E. [Chemax] N,N-Bis (2-hydroxyethyl) stearyl amine; CAS 10213-78-2; EINECS 233-520-3; cationic; surfactant, permanent internal antistat for plastic film and molded prods.; eliminates electrostatic problems in extrusion, inj. and blow molding; FDA 21CFR §178.3130; wh. solid; flash pt. (COC) > 390 F; fire pt. (COC) > 450 F; 97% min. tert. amine, 0.8% max. water.

Chemstat® 327. [Chemax] Ethylene bisstearamide; lubricant and processing aid for PVC, PS, ABS, nylon; FDA 21CFR §175.300; powd. or beads.

Chemstat® AC-100. [Chemax] Antistat in PE carrier resin; antistat for HDPE, LDPE; FDA 21CFR §184.4324; pellets; 50% act.

Chemstat® AC-101. [Chemax] Antistat in HIPS carrier resin; antistat for PS, ABS, HIPS; FDA 21CFR §178.3400; pellets; 50% act.

Chemstat® AC-200. [Chemax] Antistat in PE carrier resin; antistat for HDPE, LDPE; FDA 21CFR §178.3130; pellets; 40% act.

Chemstat® AC-201. [Chemax] Antistat in PE carrier resin; antistat for HDPE, LDPE; FDA 21CFR §184.4324; pellets; 50% act.

Chemstat® AC-202. [Chemax] Antistat in PE carrier resin; antistat for HDPE, LDPE; FDA 21CFR §178.3130; pellets; 20% act.

Chemstat® AC-1000. [Chemax] Lauramide DEA in LLDPE carrier resin; antistat for LDPE, HDPE, LLDPE; FDA 21CFR §178.3130; pellets; 20% act.

Chemstat® AC-2000. [Chemax] Lauramide DEA in PP carrier resin; antistat for PP; FDA 21CFR §178.3130; pellets; 20% act.

Chemstat® AF-476. [Chemax] Blend of fatty ester derivs.; permanent internal antifog agent for use in PVC film food pkg.; FDA 21CFR §178.3400; clear liq. to soft paste; sp.gr. 1.00.

Chemstat® AF-700. [Chemax] Surfactant blend; nonionic; antifog for EVA film, esp. one-season agric. film; solid.

Chemstat® AF-710. [Chemax] Proprietary blend; antifog for PE, EVA, and PVC food pkg. film; FDA 21CFR §178.3400, 182.4505; liq.

Chemstat® AF-806. [Chemax] Ester blend; antifog for HDPE, LDPE greenhouse film; FDA 21CFR §178.3400, 182.4505; pellet; 30% act.

Chemstat® AF-815. [Chemax] Ester blend; long-lasting antifog for PVC food pkg. and agric. films; FDA 21CFR §178.3400, 182.4505; liq.

Chemstat® AF-906. [Chemax] Proprietary blend; antifog agent for polyethylene film for multiple season agric. use; FDA 21CFR §177.1210, 178.3400; wh. pellets; 30% act.

Chemstat® AF-1006. [Chemax] Proprietary blend; antifog agent for PP food pkg. film; FDA 21CFR §178.3400; pellets.

Chemstat® G-118/52. [Chemax] Glyceryl monostearate; lubricant and flow modifier for PVC, PS, ABS, PE, PP; FDA 21CFR §184.1324, 1505, 174.5; flakes; 52% min. act.

Chemstat® G-118/95. [Chemax] Glyceryl monostearate; lubricant and flow modifier for PVC, PS, ABS, PE, PP; FDA 21CFR §184.1324, 174.5; beads; 90% min. act.

Chemstat® HTSA #1A. [Chemax] Oleyl palmitamide; CAS 16260-09-6; slip/antiblock agent, antistat, mold release with high thermal stability; for PP film, LDPE, HDPE; FDA 21CFR §178.3860; wh. to off-wh. flake, fatty odor; insol. in water; m.p. 63-80 C; flash pt. (COC) > 200 F; 100% act.; *Toxicology:* LD50 (oral, rat) > 5 g/kg; irritant to eyes, skin, respiratory passages; may cause gastrointestinal irritation, nausea, diarrhea; *Precaution:* incompat. with strong oxidizing agents; *Storage:* store in tightly closed containers away from incompat. materials; avoid extreme heat or cold.

Chemstat® HTSA #3B. [Chemax] Stearyl erucamide; CAS 10094-45-8; slip/antiblock agent, mold release with high thermal stability; effective in thermoplastics incl. cellulosics, polyolefins, polyvinyl polymers and copolymers, polyamides; FDA 21CFR §178.3860; wh. to off-wh. free-flowing powd., fatty odor; insol. in water; m.p. 69-77 C; flash pt. (TCC) > 200 F; 100% act.; *Toxicology:* irritant to eyes, skin, mucous membranes; may cause gastrointestinal irritation, nausea, diarrhea; *Precaution:* incompat. with strong oxidizing agents; *Storage:* store in tightly closed containers away from incompat. materials; avoid extreme heat or cold.

Chemstat® HTSA #18. [Chemax] Oleamide; CAS 301-02-0; EINECS 206-103-9; slip/antiblock agent for LDPE, HDPE, PP; reduces coeff. of friction; FDA 21CFR §181.28; beads.

Chemstat® HTSA #18S. [Chemax] Stearamide; CAS 124-26-5; EINECS 204-693-2; slip/antiblock agent for LDPE, HDPE, PP; reduces coeff. of friction; FDA 21CFR §181.28; beads.

Chemstat® HTSA #22. [Chemax] Erucamide; CAS 112-84-5; EINECS 204-009-2; slip/antiblock agent for LDPE, HDPE, PP; reduces coeff. of friction; FDA 21CFR §178.3860; beads.

Chemstat® LD-100. [Chemax] Lauramide DEA; internal antistat for most thermoplastics, esp. LDPE, HDPE, PP; FDA 21CFR §178.3750; wh. waxy solid; sp.gr. 0.96; m.p. 39 C.

Chemstat® LD-100/60. [Chemax] Lauramide DEA; antistat for LDPE, HDPE, PP; FDA 21CFR §178.3130; powd.; 60% act.

Chemstat® LH-305. [Chemax] Antistat in PS carrier; low humidity antistat for molding applics. for PS, ABS; lt. tan pellets; 20% act.

Chemstat® LH-306. [Chemax] Antistat in PE carrier; antistat for HDPE, LDPE, designed for as low as 15% r.h.; FDA 21CFR §178.3130; wh. pellets; 20% act.

Chemstat® P-400. [Chemax] Polyethylene glycol; CAS 25322-68-3; EINECS 203-473-3; permanent internal antistat for HDPE, LDPE; effective for extrusion, inj. and blow molding; stable to 300 C; FDA 21CFR §178.3130, 178.3750; clear liq., mild odor; sol. in water; m.w. 380-420; sp.gr. 1.13; visc. 6.8-8.0 cst (210 F); flash pt. (COC) 430 F; pH 4.5-7.5 (5% aq.); *Toxicology:* LD50 (oral, rat) > 43 mg/kg; mild transient eye irritant; nonirritating to skin; inh. of mist may cause breathing difficulty; ingestion may cause gastrointestinal irritation, nausea, diarrhea; *Precaution:* incompat. with strong oxidizing and reducing agents; *Storage:* store in tightly closed containers away from incompat. materials.

Chemstat® PS-101. [Chemax] Sodium C10-18 alkyl sulfonate; CAS 68037-49-0; permanent internal antistat for thermoplastic film and molded prods.; temporary external antistat; improves pigment dispersibility and gloss; recommended for PS, rubber-modified PS, ABS, flexible and rigid PVC, PE, engineering resins; biodeg.; FDA 21CFR § 178.3130, 178.3400; wh. powd., mild odor; sol. in water, chlorinated hydrocarbons, IPA; sp.gr. 0.6 (loose); flash pt. (COC) > 200 F; 100% act.; *Toxicology:* corrosive to eyes, skin, mucous membranes; will cause gastrointestinal irritation, nausea, diarrhea; *Precaution:* incompat. with strong oxidizing agents; *Storage:* unlimited shelf life; store in tightly closed containers away from incompat. materials; avoid extreme temps.

Chemstat® PS-101/PLT. [Chemax] Sodium alkyl sulfonate in PS carrier; antistat for PS, ABS; high thermal stability; FDA approved for PS; FDA 21CFR §178.3130, 178.3400; pellets; 50% act.

Chemstat® SE-5. [Chemax] Ethoxylated sorbitan ester; antistat for high temp. thermoplastics; liq.

Chimassorb® 119. [Ciba-Geigy/Additives] 1,3,5-Triazine-2,4,6-triamine, N,N'''-[1,2-ethanediylbis[[[4,6-bis[butyl(1,2,2,6,6-pentamethyl-4-piperidinyl)amino]-1,3,5-triazine-2-yl]imino]-3,1-propanediyl]]-bis[N',N''-dibutyl-N',N''-bis(1,2,2,6,6-pentamethyl-4-piperidinyl); CAS 106990-43-6; HALS; lt. stabilizer for PP fiber/carpet applics.; lt. and heat stabilizer for greenhouse film; preferred for color critical outdoor applics.; low volatility; lt. yel. gran.; m.w. 2286; sp.gr. 1.03-1.05 (20 C); m.p. 115-150 C.

Chimassorb® 119FL. [Ciba-Geigy/Additives] 1,3,5-

Triazine-2,4,6-triamine, N,N'''-[1,2-ethanediylbis [[[4,6-bis[butyl(1,2,2,6,6-pentamethyl-4-piperidin-yl)amino]-1,3,5-triazine-2-yl]imino]-3,1-propane-diyl]]-bis[N',N''-dibutyl-N',N''-bis(1,2,2,6,6-pentamethyl-4piperidinyl); CAS 106990-43-6; light and thermal stabilizer for PP fiber applics. in automotive, marine and residential carpets, agric. films, fertilizer bags, thick section pigmented applics., rotational molding applics.; lt. yel. gran.; m.w. 2286; sp.gr. 1.03-1.05 (20 C); m.p. 115-150 C; *Toxicology:* sensitizer; may case allergic skin reactions; may cause liver, lymphatic system, spleen, and kidney damage.

Chimassorb® 944. [Ciba-Geigy/Additives] Poly[[6-[(1,1,3,3-tetramethylbutyl)amino]-s-triazine-2,4-diyl][2,2,6.6-tetramethyl-4-piperidyl) imino] hexamethylene[(2,2,6,6-tetramethyl-4-ipperidyl) imino]]; CAS 70624-18-9; light stabilizer for polymers; broad performance effectiveness in thick and thin sections; suitable for food contact applics.; 21CFR §178.2010; U.K., Japan, Germany compliance; off-wh. powd.; sp.gr. 1.01 (20 C); m.p. 120-150 C.

Chimassorb® 944FD. [Ciba-Geigy/Additives] Poly[[6-[(1,1,3,3-tetramethylbutyl)amino]-s-triazine-2,4-diyl][2,2,6.6-tetramethyl-4-piperidyl) imino]hexamethylene[(2,2,6,6-tetramethyl-4-ipperidyl) imino]]; CAS 70624-18-9; light stabilizer for polymers; suitable for food contact applics.; 21CFR §177.1520, 178.2010; off-wh. powd.; sp.gr. 1.01; m.p. 120-150 C.

Chimassorb® 944FL. [Ciba-Geigy/Additives] Poly[[6-[(1,1,3,3-tetramethylbutyl)amino]-s-triazine-2,4-diyl][2,2,6.6-tetramethyl-4-piperidyl) imino]hexamethylene[(2,2,6,6-tetramethyl-4-ipperidyl) imino]]; CAS 70624-18-9; polymeric hindered amine lt. stabilizer; heat stabilizer for polyolefins; suitable for food contact applics.; 21CFR §177.1520, 178.2010; wh. to off-wh. powd.; sol. > 50% in acetone, benzene, chloroform, methylene chloride; 40% in hexane; m.w. > 2500; sp.gr. 1.01 (20 C); m.p. 100-135 C.

Chimassorb® 944LD. [Ciba-Geigy/Additives] Poly[[6-[(1,1,3,3-tetramethylbutyl)amino]-s-triazine-2,4-diyl][2,2,6.6-tetramethyl-4-piperidyl) imino]hexamethylene[(2,2,6,6-tetramethyl-4-ipperidyl) imino]]; CAS 70624-18-9; polymeric hindered amine lt. stabilizer for use in HDPE and PP; antioxidant; suitable for food contact applics.; 21CFR §177.1520, 178.2010; wh. powd.; very sol. in chloroform and hydrocarbons; slightly sol. in low alcohols; insol. in water; m.w. > 2500; soften. pt. 115-125 C.

Chimin P1A. [Auschem SpA] Complex org. phosphate ester, free acid; anionic; surfactant for drycleaning, hand cleaners, corrosion inhibitor, heavy-duty liq. formulations, polymerization; liq.; 100% conc.

Chimipal APG 400. [Auschem SpA] Polyglycol laurate; CAS 9004-81-3; nonionic; pigment dispersant and wetter, emulsion polymerization; liq.; HLB 13.0; 100% conc.

Chimipon GT. [Auschem SpA] Sodium alkylaryl polyether sulfate; anionic; detergent, emulsifier in emulsion polymerization; liq.; 40% conc.

Chimipon NA. [Auschem SpA] Alkylphenol ether sulfate, ammonium salt; anionic; liq. detergent base, frothing agent, emulsifier in emulsion polymerization; paste; 60% conc.

Chlorez® 700. [Dover] Resinous chlorinated paraffin; flame retardant for LDPE, NR, SR, in coatings, inks, plastics, foams, adhesives, paper, and fabrics; wh. powd., odorless, tasteless; 90% thru 50 mesh; sp.gr. 1.6; bulk dens. 13.5 lb/gal; soften. pt. (B&R) 95-110 C; 70% Cl.

Chlorez® 700-DF. [Dover] Resinous chlorinated paraffin; flame retardant for paints, printing inks, plastics, foams, adhesives, paper and fabric coatings; wh. flakes; sp.gr. 1.6; soften. pt. (B&R) 95-110 C; ref. index 1.54-1.55 (105 C); 70% Cl.

Chlorez® 700-S. [Dover] Resinous chlorinated paraffin; flame retardant in coatings, inks, plastics, foams, adhesives, paper, and fabrics; wh. powd.; 90% thru 50 mesh; sp.gr. 1.6; bulk dens. 13.5 lb/gal; soften. pt. (B&R) 95-110 C; 70% Cl.

Chlorez® 700-SS. [Dover] Resinous chlorinated paraffin; flame retardant with improved thermal stability for use in PE, PP, PS; wh. powd.; 90% thru 50 mesh; sp.gr. 1.6; bulk dens. 14. lb/gal; soften. pt. (B&R) 105-120 C; 71% Cl.

Chlorez® 725-S. [Dover] Resinous chlorinated paraffin; flame retardant with improved stability and higher soften. pt.; for coatings, inks, plastics, foams, adhesives, paper, and fabrics; wh. powd.; 90% thru 50 mesh; sp.gr. 1.6; bulk dens. 14 lb/gal; soften. pt. (B&R) 120-140 C; 71% Cl.

Chlorez® 760. [Dover] Resinous chlorinated paraffin; flame retardant for LDPE, PP, olefins, styrenes, adhesives, wire and cable and other applics.; wh. powd.; 90% thru 50 mesh; sp.gr. 1.7; bulk dens. 14 lb/gal; soften. pt. (B&R) 160 C; 74% Cl.

Chloroflo® 40. [Dover] Chlorinated paraffin; used in adhesives (mastic, pressure sensitive), coatings (aluminum, can & drum, emulsion, fabric, industrial, marine, traffic, wire & cable), rubber (cements, mechanical and molded goods), flame-retardant compds., concrete-curing compds., caulking compds., and cutting oils; Gardner 3 liq.; sol. in esters, ethers, ketones (except acetone), and chlorinated and aromatic hydrocarbons; sp.gr. 1.110; dens. 9.20 lb/gal; visc. 64 SUS (210 F); 38.4% chlorine content, by wt.

Chloroflo® 42. [Dover] Chlorinated paraffin; lubricant additive; plasticizer for polychloroprene.

Chloropren-Faktis A. [Rhein Chemie] Sulfur factice; CAS 7704-34-9; EINECS 231-722-6; processing aid for tech. molded, extruded, and calendered rubber articles, CR; dk. brn. crushed lumps; sp.gr. 1.03; Unverified

Chloropren-Faktis NW. [Rhein Chemie] Sulfur chloride factice; CAS 10025-67-9; processing aid for oil and petrol-resistant molded and extruded CR, NBR rubber goods; wh. fine powd.; sp.gr. 1.10; Unverified

CHP-5. [Witco/PAG] Cumene hydroperoxide sol'n.; CAS 80-15-9; initiator for R.T. curing of vinyl ester of other unsat. polyester resins; esp. useful used with vinyl ester resins that are cobalt promoted; reacts faster than CHP-158; suggested for gel coats, water-sensitive resins, heat-assisted cures, two-pot systems; clear to sl. yel. liq.; sol. in most org. solvs.; sl. sol. in water; sp.gr. 1.039 (24 C); visc. 6 cps; flash pt. (Seta CC) 165 F; 50% conc., 5.1% act. oxygen; *Toxicology:* moderate skin irritant; severe eye irritant; *Precaution:* never mix directly with promoters or accelerators; may cause violent decomp. or explosion; once ignited, will burn intensely; *Storage:* store in closed original containers below 25 C, away from heat, flame, direct sunlight, contamination, strong acids, reducing agents, promoters, accelerators, heavy metals, easily oxidized compds.

CHP-158. [Witco/PAG] 88% sol'n. of cumene hydroperoxide; CAS 80-15-9; polymerization initiator for vinyl monomers and copolymers and the crosslinking of unsat. polyester resins; initiator for the styrenation or modification of alkyds and oils; colorless to pale yel. clear liq., aromatic odor; sl. sol. in water; m.w. 152.18; sp.gr. 1.03; flash pt. (CC) 175 F; pH 4-6; 88% conc.; 9.3% act. oxygen; *Toxicology:* LD50 (oral, rat) 382 mg/kg (73% material); moderate skin irritant; eye irritant; may cause blindness; ingestion may be fatal; *Precaution:* never mix directly with promoters; may cause violent explosion; once ignited, burns sluggishly until thermal decomp. pt. reached; *Storage:* store in original, closed container below 38 C, away from heat, fire, direct sunlight, min. acids, reducing agents, accelerators, easily oxidized compds.

CHPTA 65%. [Chem-Y GmbH] 3-Chloro-2-hydroxypropyltrimethyl ammonium chloride; CAS 3327-22-8; cationic; cationizing reagent for natural and syn. polymers; starch modifier for textiles; liq.; 65% act.

Chroma-Chem® 844. [Hüls Am.] Industrial colorants based on thermoplastic acrylic resin; colorant for in-plant and machine tinting of high performance, nonaq., industrial, and maintenance coatings; recommended for acrylics, alkyds, cellulosic lacquers, chlorinated rubber, epoxies, polyesters, PU, vinyl lacquers.

Chromaflo®. [Plasticolors] Pourable dispersions for high pigment loading, controlled visc. in polymer systems; for pumping and metering.

CI7840. [Hüls Am.] Isocyanatopropyltriethoxy silane; CAS 24801-88-5; coupling agent, chem. intermediate, blocking agent, release agent, lubricant, primer, reducing agent; chemically bonds, strengthens thermoset and thermoplastic composites; for urethane, cellulosics, PBT, polyethylene oxide, polysaccharide, PVAL hydrophilic polymers; liq.; m.w. 247.4; sp.gr. 0.99 (20 C); b.p. 130 C (20 mm); flash pt. 98 C; ref. index 1.419; 95% purity.

Cimbar™ 325. [Cimbar Perf. Minerals] Natural barytes; CAS 7727-43-7; EINECS 231-784-4; additive in coatings (automotive primers/topcoats, gloss enamels, powd. coatings, semigloss/gloss latexes, architectural coatings); fillers in plastics, ceramics, friction materials, rubber goods, plastisols, adhesives, latex prods., urethane foams, acoustical compds., insulating materials, sound attenuation prods.; bright wh. powd.; 7-9 μ mean particle size; 0.025% on 325 mesh; fineness (Hegman) 4; sp.gr. 4.45; bulk dens. 100 lb/ft^3 (loose); oil absorp. 9; pH 7; 98% $BaSO_4$, 0.1% moisture.

Cimbar™ 1025P. [Cimbar Perf. Minerals] Barytes; CAS 7727-43-7; EINECS 231-784-4; plastics grade processing aid for compounding, extrusion, molding; increased prod.; improved tens. and elong. props.; blocks radiation; low abrasion, low resin demand; wh. powd.; 2-3 μ mean particle size; 0% on 325 mesh; sp.gr. 4.40-4.45; surf. area 2 m^2/g; oil absorp. 11-12; ref. index 1.65; pH 7.0; hardness (Mohs) 3; conduct. 6 x 10^{-3} Cal/cm; sp. heat 0.11 Cal/g °C; dielec. const. 7.3; 97-98% $BaSO_4$, 0.15% moisture.

Cimbar™ 1536P. [Cimbar Perf. Minerals] Barytes; CAS 7727-43-7; EINECS 231-784-4; plastics grade processing aid for compounding, extrusion, molding; increased prod.; improved tens. and elong. props.; blocks radiation; low abrasion, low resin demand; wh. powd.; 3.3-4.0 μ mean particle size; 0% on 325 mesh; sp.gr. 4.40-4.45; surf. area 1.4 m^2/g; oil absorp. 10-11; ref. index 1.65; pH 7.0; hardness (Mohs) 3; conduct. 6 x 10^{-3} Cal/cm; sp. heat 0.11 Cal/g °C; dielec. const. 7.3; 97-98% $BaSO_4$, 0.12% moisture.

Cimbar™ 3508P. [Cimbar Perf. Minerals] Barytes; CAS 7727-43-7; EINECS 231-784-4; plastics grade processing aid for compounding, extrusion, molding; increased prod.; improved tens. and elong. props.; blocks radiation; low abrasion, low resin demand; wh. powd.; 6.5-7.5 μ mean particle size; 0.5% on 325 mesh; sp.gr. 4.40-4.45; surf. area 0.33 m^2/g; oil absorp. 7-9; ref. index 1.65; pH 7.0; hardness (Mohs) 3; conduct. 6 x 10^{-3} Cal/cm; sp. heat 0.11 Cal/g °C; dielec. const. 7.3; 97-98% $BaSO_4$, 0.07% moisture.

Cimbar™ CF. [Cimbar Perf. Minerals] Natural barytes; CAS 7727-43-7; EINECS 231-784-4; additive in coatings (automotive primers/topcoats, gloss enamels, powd. coatings, semigloss/gloss latexes, architectural coatings); fillers in plastics, ceramics, friction materials, rubber goods, plastisols, adhesives, latex prods., urethane foams, acoustical compds., insulating materials, sound attenuation prods.; bright wh. powd.; 4.8 μ mean particle size; fineness (Hegman) 5.5; sp.gr. 4.45; bulk dens. 70 lb/ft^3 (loose); oil absorp. 10; pH 7; 98% $BaSO_4$, 0.15% moisture.

Cimbar™ UF. [Cimbar Perf. Minerals] Natural barytes; CAS 7727-43-7; EINECS 231-784-4; additive in coatings (automotive primers/topcoats, gloss enamels, powd. coatings, semigloss/gloss latexes, architectural coatings); fillers in plastics, ceramics, friction materials, rubber goods, plastisols, adhesives, latex prods., urethane foams, acoustical compds., insulating materials, sound attenuation prods.; bright wh. powd.; 2.5 μ mean particle size; fineness (Hegman) 7; sp.gr. 4.45; bulk dens. 60 lb/ft^3 (loose); oil absorp. 11; pH 7; 98% $BaSO_4$, 0.15% moisture.

Cimbar™ XF. [Cimbar Perf. Minerals] Natural barytes; CAS 7727-43-7; EINECS 231-784-4; additive in coatings (automotive primers/topcoats, gloss enamels, powd. coatings, semigloss/gloss latexes, architectural coatings); fillers in plastics, ceramics, friction materials, rubber goods, plastisols, adhesives, latex prods., urethane foams, acoustical compds., insulating materials, sound attenuation prods.; bright wh. powd.; 3.6 μ mean particle size; fineness (Hegman) 6; sp.gr. 4.45; bulk dens. 75 lb/ft^3 (loose); oil absorp. 11; pH 7; 98% $BaSO_4$, 0.15% moisture.

Cimflx 606. [Luzenac Am.] Talc; CAS 14087-96-6; EINECS 238-877-9; reinforcement in PP homopolymer, engineering resins; offers unique color and brightness in thermoplastics, good long-term heat resist. in polyolefins; 15.8 μ median particle size; 97.5% thru 200 mesh; sp.gr. 2.8; dens. 24 lb/ft^3 (loose); surf. area 8.5 m^2/g; bulking value 23.3 lb/gal; oil absorp. 40; brightness 82; pH 8.5 (10% slurry); 0.5% moisture.

Cimpact 600. [Luzenac Am.] Talc; CAS 14087-96-6; EINECS 238-877-9; reinforcing agent for plastics applics. where good color and brightness are required; also provides superior flex. mod. and impact props.; powd.; 2.3 μ median particle size; fineness (Hegman) 6.0; sp.gr. 2.8; dens. 9 lb/ft^3 (loose); bulking value 23.3 lb/gal; surf. area 10.2 m^2/g; oil absorp. 50; pH 9.0 (10% slurry).

Cimpact 610. [Luzenac Am.] Talc; CAS 14087-96-6; EINECS 238-877-9; reinforcing agent for plastics applics. where good color and brightness are required; provides superior flex. mod. and impact props.; powd.; 2.3 μ median particle size; fineness (Hegman) 6.0; sp.gr. 2.8; dens. 9 lb/ft^3 (loose); bulking value 23.3 lb/gal; surf. area 10.2 m^2/g; oil absorp. 50; pH 9.0 (10% slurry).

Cimpact 699. [Luzenac Am.] Platy talc; CAS 14807-96-6; EINECS 238-877-9; reinforcer giving superior impact str. in polyolefins, homopolymer/rubber blends; exc. stability in gel coats; can be used in automotive and nucleation applics.; ultrafine powd.; 1.2 μ median diam.; fineness (Hegman) 6.5; sp.gr. 2.8; dens. 5-9 lb/ft^3 (loose), 18.9 lb/ft^3 (tapped); bulking value 23.3 lb/gal; surf. area 17 m^2/g; oil absorp. 54; brightness 87; pH 9.5 (10% slurry); 1% moisture.

Cimpact 705. [Luzenac Am.] Talc; CAS 14807-96-6; EINECS 238-877-9; plastics additive for high impact, TPO, auto, and appliance applics.; powd.; 1.5 μ median particle size; fineness (Hegman) 7.0; dens. 6-9 lb/ft^3 (loose); brightness 91.

Cimpact 710. [Luzenac Am.] Platy talc; CAS 14807-96-6; EINECS 238-877-9; provides superior impact str. in polyolefins, homopolymer/rubber blends; exc. stability to gel coats; for high impact, TPO, auto, appliance, and nucleation applics.; ultrafine powd.; 1.4 μ median diam.; fineness (Hegman) 7.0; sp.gr. 2.8; dens. 6.0 lb/ft^3 (loose); bulking value 23.3 lb/gal; surf. area 14 m^2/g; oil absorp. 54; pH 9.5 (10% slurry); 0.5% moisture.

Ciptane® 255LD. [PPG Industries] Amorphous precipitated silica with ≈ 3% mercaptosilane coupling agent; base silica is a higher surf. area, more bead-like, Na_2SO_4 residual salt than Ciptane I.

Ciptane® I. [PPG Industries] Amorphous precipitated silica with ≈ 3% mercaptosilane coupling agent; similar to Hi-Sil® 210.

Cirrasol® AEN-XB. [ICI Surf. Am.; ICI Surf. Belgium] Fatty alcohol ethoxylate; nonionic; fiber processing aid; antistat for PVC belting; emulsifier for low m.p. paraffin waxes; colorless liq.; HLB 10.2; 100% conc.

Cirrasol® ALN-GM. [ICI Surf. Am.; ICI Surf. Belgium] POE polyol ester; nonionic; fiber processing aid; antistat for plastics and rubber; wetting agent; yel. liq.; HLB 15.7; 100% conc.

Citroflex® 2. [Morflex] Triethyl citrate; CAS 77-93-0; EINECS 201-070-7; plasticizer for cellulose acetate, cellulose acetate butyrate, cellulose nitrate, chlorinated rubber, ethyl cellulose, PVAc, PVB, PVC, PVP, PVDC, PET, poly(acrylate/methacrylate); also recommended for natural resins such as dammar and ester gums; enhances grease resist.; lightfastness makes it desirable for lacquer formulations; fragrance carrier in perfumes, deodorants, shampoos, detergents; FDA 21CFR §175.300, 175.320, 175.380, 175.390, 176.170, 177.1210, 181.27, 182.1911, 184.1911; kosher; clear liq., essentially odorless; sol. in toluene, 6.5 g/100 ml in water; insol. in heptane; m.w. 276.3; sp.gr. 1.135-1.139; visc. 35 cps; pour pt. -45 C; flash pt. (COC) 155 C; ref. index 1.440-1.442; *Toxicology:* LD50 (oral, rat) 7 cc/kg, (IP, mice) 1750 mg/kg.

Citroflex® 4. [Morflex] Tri-n-butyl citrate; CAS 77-94-1; EINECS 201-071-2; plasticizer for cellulose nitrate, ethyl cellulose, chlorinated rubber, PVAc, PVB, PVC, PVP, PVDC, poly(acrylate/methacrylate); defoamer in proteinaceous sol'ns.; improved lt. stability in cellulose acetate; FDA 21CFR §175.105; clear liq., essentially odorless; sol. in toluene, heptane, < 0.1 g/100 ml water; m.w. 360; sp.gr. 1.037-1.045; visc. 32 cps; pour pt. -62 C; flash pt. (COC) 185 C; ref. index 1.443-1.445; *Toxicology:* LD50 (oral, rat) 7 cc/kg, (IP, mice) 2900 mg/kg.

Citroflex® A-2. [Morflex] Acetyltriethyl citrate; CAS 77-89-4; EINECS 201-066-5; plasticizer for more polar resins such as cellulosics, polyacrylates, PVAc, polyvinyl butyral, PVC, PVP, PVDC, chlorinated rubber; lightfastness for lacquer formulations; plasticizer in food wraps; FDA 21CFR §175.105, 175.300, 175.320, 175.380, 176.170, 177.1210, 178.3910; kosher; clear liq., essentially odorless; sol. in toluene, 0.72 g/100 ml water; m.w. 318.3; sp.gr. 1.135-1.139; visc. 54 cps; pour pt. -43 C; flash pt. (COC) 188 C; ref. index 1.432-1.441; *Toxicology:* LD50 (oral, rat) 7 cc/kg, (IP, mice) 1150 mg/kg; sl. but transient eye irritant; nonirritating to skin.

Citroflex® A-4. [Morflex] Acetyl tri-n-butyl citrate; CAS 77-90-7; EINECS 201-067-0; plasticizer for PVC, PVDC, esp. food films, medical articles; also suitable for cellulose nitrate, chlorinated rubber, ethyl cellulose, PVAc, PVB, PVP, PU, poly(acrylate/methacrylate); sol'n. coatings on paperboard and foil; vinyl toys; adhesives; FDA 21CFR §172.515, 175.105, 175.300, 175.320, 175.380, 175.390, 176.170, 177.1210, 178.3910, 181.27; kosher; clear iq., essentially odorless; sol. in toluene, heptane, < 0.1 g/100 ml water; m.w. 402.5; sp.gr. 1.045-1.055; visc. 33 cps; pour pt. -59 C; flash pt. (COC) 104 C; ref. index 1.441-1.4425; 99% min. ester content; *Toxicology:* LD50 (oral, rat) 30 cc/kg, (IP, mice) 4000 mg/kg; sl. but transient eye irritant; nonirritating to skin.

Citroflex® A-4 Special. [Morflex] Acetyl tri-n-butyl citrate; CAS 77-90-7; EINECS 201-067-0; plasticizer for vinyls, adhesives, coatings; improved long-term stability; recommended for medical articles and other sensitive applics.; *Toxicology:* LD50 (oral, rat) 30 cc/kg, (IP, mice) 4000 mg/kg.

Citroflex® A-6. [Morflex] Acetyl tri-n-hexyl citrate; CAS 24817-92-3; vinyl plasticizer for medical applics. and other toxicologically sensitive areas; also suitable for chlorinated rubber, ethyl cellulose, PU, poly(acrylate/methacrylate); APHA 100 max. clear liq., essentially odorless; sol. in toluene, heptane, < 0.1 g/100 ml water; m.w. 486; sp.gr. 1.003-1.007; visc. 36 cps; pour pt. -57 C; flash pt. (COC) 240 C; ref. index 1.445-1.449; 99% min. ester content.

Citroflex® B-6. [Morflex] n-Butyryl tri-n-hexyl citrate; CAS 82469-79-2; vinyl plasticizer for medical applics. (film to produce blood bags, intravenous tubing); also suitable for chlorinated rubber, ethyl cellulose, PU, poly(acrylate/methacrylate); APHA 100 max. clear liq., essentially odorless; sol. in toluene, heptane, < 0.1 g/100 ml water; m.w. 514; sp.gr. 0.991-0.995; visc. 28 cps; pour pt. -55 C; flash pt. (COC) 204 C; ref. index 1.444-1.448; 99% min. ester content.

ClearTint®. [Milliken] Oligomeric colorants; transparent colorants for polyolefins, esp. grades containing Millad® clarifying agents; nonnucleating; do not cause warpage or shrinkage; combines aesthetics of dyes with migration resist. of pigments; can be formulated for food contact applics.; pellets; *Toxicology:* nontoxic.

ClearTint® PC Green. [Milliken] Liq. color blend; colorant for soft drink bottle mfg. (PET); added during inj. molding of the preforms; homogeneous liq.; sp.gr. 1.14; visc. 5000 cps; < 0.5% water; *Toxicology:* nontoxic.

Clorafin® 40. [Hercules] Chlorinated paraffin; used to flameproof and waterproof textiles; plasticizer for vinyl, resins or filmformers; additive in cutting oils and drawing compds.; Gardner 6 max. visc. liq.; misc. with common solvs. except lower aliphatic alcohols and water; sp.gr. 1.16; dens. 9.6 lb/gal; visc. 26-29 poise; 42% chloride content.

CMD 5185. [Shell] Tetrabromobisphenol A epoxy resin; effective modifier in the formulation of flame-retardant epoxy compds., e.g., semiconductor molding compds., potting and casting compds., powd. coatings; Gardner 3 max. flaked resin; dens. 15.1 lb/gal; melt visc. 375 cps (300 F); m.p. 82 C; 49.5% Br.; Unverified

Coad 10. [Norac] Calcium stearate; CAS 1592-23-0; EINECS 216-472-8; internal and external lubricant for PVC processing; mold release in thermoplastic and thermoset molding; internal lubricant or partitioning in rubber where processing temps. exceed the m.p. of 155 C.

Coad 20. [Norac] Zinc stearate; CAS 557-05-1; EINECS 209-151-9; partitioning agent in dry form where thick uniform coatings are required; promotes mold release; internal lubricant; superior heat stability; controlled particle size and high bulk dens. improves dispersion, provides faster, more uniform mixing; FDA accepted; solid.

Coad 21. [Norac] Zinc stearate; CAS 557-05-1; EINECS 209-151-9; internal/external lubricant for polyolefins, ABS, PS, IPS, rigid/flexible PVC, SMC/TMC/BMC polyesters, phenolics, melamine, alkyd, U-F, polyesters, color concs.; FDA accepted; solid.

Coad 27B. [Norac] Zinc stearate; CAS 557-05-1; EINECS 209-151-9; metal die molding grade with particle size engineered for optimum performance in SMC/BMC applics.; FDA accepted; solid.

Coagulant AW. [Bayer AG] Coagulant for latex compds.

Coagulant CHA. [Bayer AG] Cyclohexyl amine acetate; coagulant used in dipped goods from natural latex; wh. cryst.; dens. 1.12 g/cc; pH 7.5 (10% aq.).

Coagulant WS. [Bayer AG; Bayer/Fibers, Org., Rubbers] Functional polyorganosiloxane; heat sensitizer permitting coagulation pt. of 32-60 C in mixes based on natural or syn. latex, preferably nitrile and polychloroprene latexes, but also S/B and acrylic; ylsh. liq.; dens. 1.03 g/cc; pH 8 (15% aq.).

Coathylene HA 1591. [Hoechst Celanese] Polyethylene wax; CAS 9002-88-4; EINECS 200-815-3; mold release; powd.

Coathylene HA 2454. [Hoechst Celanese] Polyethylene wax; CAS 9002-88-4; EINECS 200-815-3; internal/external lubricant for PVC pipe extrusion, pultrusion, SMC/BMC, antishrink agent; FDA accepted; powd.

Cohedur® A. [Bayer AG] Methylene donor; bonding agent for fabrics and cord bonded to rubber; colorless visc. liq.; dens. 1.2 g/cc.; Unverified

Cohedur® A Solid. [Bayer AG] Methylene donor with filler; bonding agent for fabrics and cord bonded to rubber; wh. powd.; dens. 1.51 g/cc.

Cohedur® RK. [Bayer AG; Bayer/Fibers, Org., Rubbers] Resorcinol diacetate absorbed on silica (1:1); adhesion promoter for direct bonding of fabric or steel cord to rubber; esp. developed to reduce scorching; suitable for all types of rubber except silicone; wh. to ylsh. powd.; sp.gr. 1.55.

Cohedur® RL. [Bayer/Fibers, Org., Rubbers] Sol'n. of resorcinol and methylene donor (1:1); single-pkg.

bonding agent for fabrics and cord bonded to rubber; orange-brn. high visc. liq.; dens. 1.2 g/cc.

Cohedur® RS. [Bayer AG; Bayer/Fibers, Org., Rubbers] Homogeneous solidified melt of resorcinol and stearic acid (2:1); adhesion promoter for fabrics and cord bonded to rubber; to be used with methylene donors; gray to reddsh. brn. flakes; sp.gr. 1.19; m.p. ≈ 60 C.

Colloid™ 581B. [Rhone-Poulenc Surf. & Spec.] Silicone-free; defoamer providing quick foam knock-down in high-surfactant polymerizations; low carryover into compounding; exc. compat.; FDA 21CFR §175.105, 176.170, 176.180, 176.200, 176.210; 100% act.

Colloid™ 635. [Rhone-Poulenc Surf. & Spec.] Silicone-free; antifoam for general stripping and polymer compounding; exc. high shear, pH and temp. stability; FDA 21CFR §175.105, 176.170, 176.180, 176.200, 176.210; 100% act.

Colloid™ 675. [Rhone-Poulenc Surf. & Spec.] Easily dispersible multi-component defoamer for stripping, foam knock-down, and latex handling; very effective with ethylene copolymers and compds.; FDA 21CFR §175.105, 176.170, 176.180, 176.200, 176.210; 100% act.

Colloid™ 681F. [Rhone-Poulenc Surf. & Spec.] Easily dispersible antifoam for vinyl-acrylic polymerization; remains homogeneous; FDA 21CFR §175.105, 176.170, 176.180, 176.200, 176.210; kosher; 96% act.

Colloid™ 685. [Rhone-Poulenc Surf. & Spec.] Readily dispersible defoamer for stripping and processing carboxylated SBR and PVC emulsions; more efficient than high-silica defoamers; FDA 21CFR §175.105, 176.170, 176.180, 176.200, 176.210; 100% act.

Colloid™ 961. [Rhone-Poulenc Surf. & Spec.] Extremely dispersible defoamer with exc. foam knock-down, long-term persistence; compat. with most emulsions, PVA, starches, cellulosics; FDA 21CFR §175.105, 176.170, 176.180, 176.200, 176.210; 100% act.

Colloid™ 985. [Rhone-Poulenc Surf. & Spec.] Silicone-free; easily dispersible defoamer for vinyl-acrylic latexes, sealants; effective with starches, gums, and cellulosics; nonsettling in drum; 100% act.

Colloid™ 994. [Rhone-Poulenc Surf. & Spec.] High silicone-silica levels with other active defoamers; dispersible defoamer for stripping highly visc. emulsions and dispersions; 100% act.

Colloid™ 999. [Rhone-Poulenc Surf. & Spec.] Silicone-free, mineral oil-free; easily dispersible defoamer for compounded systems contg. PVA, gelatin, protein, starch, and cellulosics; FDA 21CFR §175.105, 176.170, 176.180, 176.200, 176.210 with use level limitations; 100% act.

Colloidal Sulphur 95. [Bayer AG] Rubber chemicals for latex processing.

Colloids FR 47 Series. [Colloids Ltd] Flame retardant/PS masterbatch; flame retardant masterbatch for PS extrusion and molding applics.

Colloids FR 48 Series. [Colloids Ltd] Flame retardant/LDPE masterbatch; flame retardant masterbatch for extrusion and molding applics.; compat. with LDPE, HDPE, PP.

Colloids FR 50 Series. [Colloids Ltd] Flame retardant/PP masterbatch; flame retardant masterbatch for PP extrusion and molding applics.

Colloids FR 52 Series. [Colloids Ltd] Flame retardant/ABS masterbatch; flame retardant masterbatch for ABS extrusion and molding applics.

Colloids FR 54 Series. [Colloids Ltd] Flame retardant/nylon masterbatch; flame retardant masterbatch for nylon extrusion and molding applics.

Colloids FR 55 Series. [Colloids Ltd] Flame retardant/EVA masterbatch; flame retardant masterbatch for extrusion and molding applics.; compat. with LDPE, HDPE, PS, PP, ABS, and nylon.

Colloids N 54/1/01. [Colloids Ltd] 50% Additive nylon masterbatch; wh. masterbatch for general compounding and extrusion.

Colloids N 54/1/05. [Colloids Ltd] 40% Additive nylon masterbatch; blue-wh. masterbatch for general compounding and extrusion.

Colloids N 54/1033. [Colloids Ltd] 40% additive nylon masterbatch; blk. masterbatch with very high intensity black dyestuff for use in compounding, esp. min. and glass-filled nylon 6 and 66.

Colloids N 54/1044. [Colloids Ltd] 25% Additive nylon masterbatch; high jet blk. masterbatch for compounding and extrusion.

Colloids N 54/1088. [Colloids Ltd] 25% Additive nylon masterbatch; blk. masterbatch for extrusion and compounding; suitable where uv resist. is required.

Colloids N 54/1099. [Colloids Ltd] 30% Additive nylon masterbatch; general purpose furnace black masterbatch for compounding and extrusion.

Colloids N 54/14/01. [Colloids Ltd] 50% Calcium carbonate/nylon masterbatch; masterbatch for compounding of min.-filled nylon.

Colloids N 54/14/02. [Colloids Ltd] 50% Kaolin/nylon masterbatch; masterbatch for compounding of min.-filled nylon.

Colloids PE 48/10/02 UV. [Colloids Ltd] Additive masterbatch in PE carrier; uv protectant for polyolefin and PP; let-down 2-5%.

Colloids PE 48/10/03. [Colloids Ltd] Selected min. fillers masterbatch in PE carrier; purging agent for process machines.

Colloids PE 48/10/04 AS. [Colloids Ltd] Additive masterbatch in PE carrier; antistat for elimination of electrostatic charge of polyolefin and PP prods. with minimal effect on physical props.; let-down 1-3%.

Colloids PE 48/10/06. [Colloids Ltd] Oleamide-based masterbatch in PE carrier; antiblock/slip masterbatch for film applics.

Colloids PE 48/10/09. [Colloids Ltd] Additive masterbatch in PE carrier; antiblock; antifibrillation agent for PP tape extrusion; improves tape stiffness, opacity, and cloth stability.

Colloids PE 48/10/10. [Colloids Ltd] Silica-based additive masterbatch in PE carrier; antiblock giving improved film clarity.

Colloids PE 48/10/11. [Colloids Ltd] Silica-based additive masterbatch in PE carrier; antiblock for high clarity polyolefin film.

Colloids PE 48/10/12. [Colloids Ltd] TMQ-based additive masterbatch in PE carrier; antioxidant masterbatch for heat protection of silane cross-linked compds. for elec. and other applics.

Colloids PE 48/10/13. [Colloids Ltd] Erucamide-based additive masterbatch in PE carrier; antiblock and slip masterbatch for polyolefins, esp. where higher process temps. are required.

Colloids PE 48/32 C. [Colloids Ltd] 40% Med. color carbon black/PE masterbatch; blk. masterbatch for compounding LDPE, HDPE, PP; let-down 2-4%.

Colloids PE 48/32 F. [Colloids Ltd] 40% Med. color carbon black/PE masterbatch; blk. masterbatch for LDPE, HDPE, PP film down to 20 μ; let-down 2-4%.

Colloids PE 48/58. [Colloids Ltd] 50% Furnace carbon black/PE masterbatch; blk. masterbatch for LDPE, HDPE, PP extrusion, molding, and compounding applics.; let-down 2-4%.

Colloids PE 48/61 A. [Colloids Ltd] 40% Carbon black/PE masterbatch; blk. masterbatch for LDPE, HDPE, PP film for food applics., pipe molding; high uv resist.; let-down 2-5%.

Colloids PE 48/61 F. [Colloids Ltd] 40% Carbon black/PE masterbatch; blk. masterbatch for LDPE, HDPE, PP film down to 15 μ; suitable for food applics.; high uv resist.; let-down 2-6%.

Colloids PE 48/61/30. [Colloids Ltd] 30% Carbon black/PE masterbatch; blk. masterbatch for LDPE, HDPE, PP film down to 15 μ, coating applics.; suitable for food applics.; high uv resist.; let-down 3-6%.

Colloids PE 48/62 A. [Colloids Ltd] 40% Carbon black/PE masterbatch; blk. masterbatch for LDPE, HDPE, PP pipe and water cisterns; suitable for food applics.; high uv resist.; let-down 5-6%.

Colloids PE 48/67. [Colloids Ltd] 30% Furnace carbon black/PE masterbatch; blk. masterbatch for LDPE, HDPE, PP film down to 30 μ; let-down 3-5%.

Colloids PE 48/72. [Colloids Ltd] 50% Furnace carbon black/PE masterbatch; blk. masterbatch for LDPE, HDPE, PP film down to 20 μ; let-down 2-4%.

Colloids PE 48/73. [Colloids Ltd] 28% High color carbon black/PE masterbatch; blk. masterbatch for LDPE, HDPE, PP film down to 30 μ; let-down 3-5%.

Colloids PE 48/76. [Colloids Ltd] 40% Furnace carbon black/PE masterbatch; blk. masterbatch for LDPE, HDPE, PP film down to 30 μ; let-down 3-5%.

Colloids PE 48/80. [Colloids Ltd] 40% Furnace carbon black/PE masterbatch; blk. masterbatch for LDPE, HDPE, PP film down to 20 μ; let-down 3-5%.

Colloids PE 48/85. [Colloids Ltd] 50% Furnace carbon black/PE masterbatch; blk. masterbatch for LDPE, HDPE, PP extrusion, molding, non-spec. pipe; let-down 2-3%.

Colloids PP 50/5. [Colloids Ltd] 25% Carbon black/PP masterbatch; blk. masterbatch for PP extrusion and molding applics.; suitable for food applics.; high uv resist.; let-down 2-5%.

Colloids PP 50/7. [Colloids Ltd] 40% Furnace black/PP masterbatch; blk. masterbatch for PP extrusion and molding applics.; let-down 2-4%.

Colloids PP 50/9. [Colloids Ltd] 30% Carbon black/PP masterbatch; blk. masterbatch for PP extrusion and molding applics.; meets BS specs. for potable water applics.; suitable for food applics.; high uv resist.; let-down 2-7%.

Colloids PS 47/31. [Colloids Ltd] 30% High jet carbon black/PS masterbatch; blk. masterbatch for PS, ABS, SAN extrusion and molding applics.; suitable for food applics.; let-down 3-5%.

Colloids PS 47/32. [Colloids Ltd] 40% Med. color carbon black/PS masterbatch; blk. masterbatch for PS, ABS, SAN extrusion and molding applics.; let-down 2-5%.

Colloids PS 47/38. [Colloids Ltd] 40% Furnace carbon black/PS masterbatch; blk. masterbatch for PS, ABS, SAN extrusion and molding applics.; let-down 2-5%.

Colloids SAN 52/8. [Colloids Ltd] 25% High jet carbon black/SAN masterbatch; blk. masterbatch for ABS and SAN extrusion, molding, and compounding applics.; suitable for food applics.; let-down 2-5%.

Colloids UN 53/9D. [Colloids Ltd] 50% High jet carbon black/universal carrier masterbatch; blk. masterbatch for LDPE, HDPE, PP, PS, ABS, SAN, PA extrusion, molding, and compounding applics.; suitable for food applics.; let-down 1-2%.

Colloids UN 53/10. [Colloids Ltd] 50% Furnace carbon black/universal carrier masterbatch; general purpose blk. masterbatch for LDPE, HDPE, PP, PS, ABS, SAN, PA extrusion, molding, and compounding applics.; let-down 1-2%.

Colloids UN 53/12. [Colloids Ltd] 40% High jet carbon black/universal carrier masterbatch; blk. masterbatch for LDPE, HDPE, PP, PS, ABS, SAN, PA extrusion, molding, and compounding applics.; suitable for food applics.; let-down 2-3%.

Colloids UN 53/15. [Colloids Ltd] 50% Med. color carbon black/universal carrier masterbatch; blk. masterbatch for LDPE, HDPE, PP, PS, ABS, SAN, PA extrusion, molding, and compounding applics.; let-down 1-2%.

Colloids UN 53/17. [Colloids Ltd] 40% Med. color carbon black/universal carrier masterbatch; blk. masterbatch for LDPE, HDPE, PP, PS, ABS, SAN, PA molding and compounding applics.; economical grade; let-down 1-3%.

Colloids UN 53/18. [Colloids Ltd] 50% High jet carbon black/universal carrier masterbatch; blk. masterbatch for LDPE, HDPE, PP, PS, ABS, SAN, PA extrusion, molding; suitable for food applics.; let-down 1-2%.

Colok® 265. [Henkel/Coatings & Inks] High m.w. diether of propoxylated bisphenol A; polyester, urethane, and alkyd modifier providing unique combination of chem. resist., adhesive props., and flexibility; wh. solid; sp.gr. 1.09; visc. 14000 cps (60 C); hyd. no. 318; 100% act.

Colormatch®. [Plasticolors] High pigment loading dispersions for exc. color control and economy of use; for polyester, epoxy, vinyl, urethane, vinyl ester, gel coats.

Colortech 10007-11. [Colortech] UV absorber conc. for PE, PP.

Colortech 10043-12. [Colortech] Phosphite; heat stabilizer for PE, PP; hydrolytically stable.

Colortech 10044-12. [Colortech] Phosphite blend in PE conc.; high-performance antioxidant for PE, PP; FDA 21CFR §178.2010.

Colortech 10045-11. [Colortech] Hindered phenolic conc. in PE; nondiscoloring antioxidant, heat stabilizer for PE, PP; FDA 21CFR §178.2010.

Colortech 10310-12. [Colortech] Amine; internal antistat for PE, PP; FDA accepted.

Colortech 10942-12. [Colortech] Proprietary HALS/uv absorber; uv stabilizer for PE, PP; pellets; m.p. 125 C.

Colortech 10944-11. [Colortech] HALS/antioxidant conc.; high performance heat and lt. stabilizer/antioxidant for PE, PP.

Comboloob 0609. [Astor Wax] Syn. and hydrocarbon wax system; lubricant for processing under moderate to high temp. and shear conditions; for rigid PVC pipe, window profiles, inj. molding.

Comboloob 0827. [Astor Wax] Combination lubricant promoting impact resist. in highly filled compds. by aiding and controlling fusion; for sewer pipe, siding subtrate.

Comboloob 0827/1. [Astor Wax] Comboloob 827 with polyethylene wax; combination lubricant promoting impact resist. in highly filled compds. by aiding and controlling fusion; for sewer pipe, siding subtrate.

Cometals Aluminum Stearate. [Cometals] Aluminum stearate; CAS 7047-84-9; EINECS 230-325-5; internal/external lubricant for nylon, PVC, polyesters; solid.

Cometals Barium Stearate. [Cometals] Barium stearate; CAS 6865-35-6; internal/external lubricant for vinyls; solid.

Cometals Cadmium Stearate. [Cometals] Cadmium stearate; CAS 2223-93-0; internal/external lubricant for PVC; solid.

Cometals Calcium Stearate. [Cometals] Calcium stearate; CAS 1592-23-0; EINECS 216-472-8; internal/external lubricant for PVC; solid.

Cometals Magnesium Stearate. [Cometals] Magnesium stearate; CAS 557-04-0; EINECS 209-150-3; internal/external lubricant for ABS, vinyls; solid.

Cometals Sodium Stearate. [Cometals] Sodium stearate; CAS 822-16-2; EINECS 212-490-5; internal/external lubricant for nylon, PP, vinyls; solid.

Cometals Zinc Stearate. [Cometals] Zinc stearate; CAS 557-05-1; EINECS 209-151-9; internal/external lubricant for BMC/SMC polyester, phenolics, PVC; FDA accepted; solid.

Compimide® TM-121. [Shell] Bisallyl phenyl compd.; toughening modifier for use with Compimide bismaleimide resins for structural applics.; visc. 1.2-2.5 poise (160 F).

Compimide® TM-123. [Shell] Substituted benzophenone; toughening modifier for use with Compimide bismaleimide resins for structural applics.; visc. 10-16 poise (160 F).

Conap® MR-5014. [Conap] Water-based mold release; external mold release for all PU materials, epoxy, and other resin systems; recommended for cast elastomers and epoxies @ 85-150 C mold temps.; off-wh.; sp.gr. 0.99; visc. < 5 cps; 10% solids; *Storage:* mix prior to use; store @ 13-27 C; keep from freezing.

Conap® MR-5015. [Conap] Water-based mold release; external mold release for all PU materials, epoxy, and other resin systems; does not leave silicone residue on finished part; recommended for cast elastomers and epoxies @ 85-150 C mold temps.; off-wh.; sp.gr. 0.99; visc. < 5 cps; 1% solids; *Storage:* mix prior to use; store @ 13-27 C; keep from freezing.

Continex® LH-10. [Witco/Concarb] Carbon black; CAS 1333-86-4; EINECS 215-609-9; rubber reinforcement providing improved abrasion resist. props. with lower hysteresis; for truck and high performance tread black, plastics, paper, and printing ink industries; blk. amorphous pellets, odorless; insol. in water; sp.gr. 1.7-1.9 (16 C); bulk dens. 20 lb/ft^3; vapor pressure negligible; nitrogen surf. area 127 m^2/g; fire pt. 500-700 F; 99% C; *Toxicology:* OSHA and ACGIH PEL 3.5 mg/m^3; nuisance dust; inh. of dust may cause temporary discomfort to nose and respiratory tract; *Precaution:* incompat. with strong oxidizers; hazardous decomp. prods.: CO, CO_2 from burning.

Continex® LH-20. [Witco/Concarb] Carbon black; CAS 1333-86-4; EINECS 215-609-9; rubber reinforcement providing similiar abrasion props. and lower hysteresis than N234; for truck treads; blk. amorphous pellets, odorless; insol. in water; sp.gr. 1.7-1.9 (16 C); bulk dens. 19.5 lb/ft^3; vapor pressure negligible; nitrogen surf. area 108 m^2/g; fire pt. 500-700 F; 99% C; *Toxicology:* OSHA and ACGIH PEL 3.5 mg/m^3; nuisance dust; inh. of dust may cause temporary discomfort to nose and respiratory tract; *Precaution:* incompat. with strong oxidizers; hazardous decomp. prods.: CO, CO_2 from burning.

Continex® LH-30. [Witco/Concarb] Carbon black; CAS 1333-86-4; EINECS 215-609-9; rubber reinforcement providing similar abrasion resist. and lower hysteresis than N339; for passenger treads; blk. amorphous pellets, odorless; insol. in water; sp.gr. 1.7-1.9 (16 C); bulk dens. 19 lb/ft^3; vapor pressure negligible; nitrogen surf. area 92 m^2/g; fire pt. 500-700 F; 99% C; *Toxicology:* OSHA and ACGIH PEL 3.5 mg/m^3; nuisance dust; inh. of dust may cause temporary discomfort to nose and respiratory tract; *Precaution:* incompat. with strong oxidizers; hazardous decomp. prods.: CO, CO_2 from burning.

Continex® LH-35. [Witco/Concarb] Carbon black; CAS 1333-86-4; EINECS 215-609-9; rubber reinforcement providing sl. better abrasion resist. with same level of hysteresis than N351; for tire treads; blk. amorphous pellets, odorless; insol. in water; sp.gr. 1.7-1.9 (16 C); bulk dens. 23 lb/ft^3; vapor pressure negligible; nitrogen surf. area 81 m^2/g; fire pt. 500-700 F; 99% C; *Toxicology:* OSHA and ACGIH PEL 3.5 mg/m^3; nuisance dust; inh. of dust may cause temporary discomfort to nose and respiratory tract; *Precaution:* incompat. with strong oxidizers; hazardous decomp. prods.: CO, CO_2 from burning.

Continex® N110. [Witco/Concarb] Carbon black; CAS 1333-86-4; EINECS 215-609-9; rubber reinforcement giving high level of reinforcement, superior abrasion resist.; for truck tire treads, compr. pads, tank treads, heavy belting; bulk dens. 21 lb/ft^3; nitrogen surf. area 137 m^2/g; iodine no. 145.

Continex® N220. [Witco/Concarb] Carbon black; CAS 1333-86-4; EINECS 215-609-9; rubber reinforcement providing high level of reinforcement, exc. abrasion resist.; for truck tire and passenger treads, belting, mech. rubber goods; blk. amorphous pellets, odorless; insol. in water; sp.gr. 1.7-1.9 (16 C); bulk dens. 21.5 lb/ft^3; vapor pressure negligible; nitrogen surf. area 119 m^2/g; iodine no. 121; fire pt. 500-700 F; 99% C; *Toxicology:* OSHA and ACGIH PEL 3.5 mg/m^3; nuisance dust; inh. of dust may cause temporary discomfort to nose and respiratory tract; *Precaution:* incompat. with strong oxidizers; hazardous decomp. prods.: CO, CO_2 from burning.

Continex® N234. [Witco/Concarb] Carbon black; CAS 1333-86-4; EINECS 215-609-9; rubber reinforcement for truck and high performance tire treads; premium precure retreading for truck, passenger, aircraft; blk. amorphous pellets, odorless; insol. in water; sp.gr. 1.7-1.9 (16 C); bulk dens. 21.5 lb/ft^3; vapor pressure negligible; nitrogen surf. area 121 m^2/g; iodine no. 120; fire pt. 500-700 F; 99% C; *Toxicology:* OSHA and ACGIH PEL 3.5 mg/m^3; nuisance dust; inh. of dust may cause temporary discomfort to nose and respiratory tract; *Precaution:* incompat. with strong oxidizers; hazardous decomp. prods.: CO, CO_2 from burning.

Continex® N299. [Witco/Concarb] Carbon black; CAS 1333-86-4; EINECS 215-609-9; rubber reinforcement; blk. amorphous pellets, odorless; insol. in water; sp.gr. 1.7-1.9 (16 C); bulk dens. 21 lb/ft^3; vapor pressure negligible; iodine no. 108; fire pt. 500-700 F; 99% C; *Toxicology:* OSHA and ACGIH PEL 3.5 mg/m^3; nuisance dust; inh. of dust may cause temporary discomfort to nose and respiratory tract; *Precaution:* incompat. with strong oxidizers; hazardous decomp. prods.: CO, CO_2 from burning.

Continex® N326. [Witco/Concarb] Carbon black; CAS 1333-86-4; EINECS 215-609-9; rubber reinforcement providing good resist. to cut and tear; used in OTR tires, wire belt coating, motor mounts, shoe soles and heels; blk. amorphous pellets, odorless; insol. in water; sp.gr. 1.7-1.9 (16 C); bulk dens. 28.5 lb/ft^3; vapor pressure negligible; nitrogen surf. area 82 m^2/g; iodine no. 82; fire pt. 500-700 F; 99% C; *Toxicology:* OSHA and ACGIH PEL 3.5 mg/m^3; nuisance dust; inh. of dust may cause temporary discomfort to nose and respiratory tract; *Precaution:* incompat. with strong oxidizers; hazardous decomp. prods.: CO, CO_2 from burning.

Continex® N330. [Witco/Concarb] Carbon black; CAS 1333-86-4; EINECS 215-609-9; rubber reinforcement with economical wear props. and good processing; used in truck tire carcass stocks, retread, curing bladders, industrial goods, bicycle tires; bulk dens. 23.5 lb/ft^3; nitrogen surf. area 83 m^2/g; iodine no. 82.

Continex® N339. [Witco/Concarb] Carbon black; CAS 1333-86-4; EINECS 215-609-9; rubber reinforcement providing good abrasion and extrusion chars.; for passenger tire treads, motor mounts, solid tires, conveyor belts; blk. amorphous pellets, odorless; insol. in water; sp.gr. 1.7-1.9 (16 C); bulk dens. 21.5 lb/ft^3; vapor pressure negligible; nitrogen surf. area 96 m^2/g; iodine no. 90; fire pt. 500-700 F; 99% C; *Toxicology:* OSHA and ACGIH PEL 3.5 mg/m^3; nuisance dust; inh. of dust may cause temporary discomfort to nose and respiratory tract; *Precaution:* incompat. with strong oxidizers; hazardous decomp. prods.: CO, CO_2 from burning.

Continex® N343. [Witco/Concarb] Carbon black; CAS 1333-86-4; EINECS 215-609-9; rubber reinforcement providing good reinforcement and extrusion chars.; for passenger tire treads, solid tires, motor mounts; blk. amorphous pellets, odorless; insol. in water; sp.gr. 1.7-1.9 (16 C); bulk dens. 20 lb/ft^3; vapor pressure negligible; nitrogen surf. area 100 m^2/g; iodine no. 92; fire pt. 500-700 F; 99% C; *Toxicology:* OSHA and ACGIH PEL 3.5 mg/m^3; nuisance dust; inh. of dust may cause temporary discomfort to nose and respiratory tract; *Precaution:* incompat. with strong oxidizers; hazardous decomp. prods.: CO, CO_2 from burning.

Continex® N351. [Witco/Concarb] Carbon black; CAS 1333-86-4; EINECS 215-609-9; rubber reinforcement providing high structure with low hysteresis; fast mixing compds. for passenger tread, cable jackets, plastics, paper, and printing ink industries; blk. amorphous pellets, odorless; insol. in water; sp.gr. 1.7-1.9 (16 C); bulk dens. 21.5 lb/ft^3; vapor pressure negligible; nitrogen surf. area 72 m^2/g; iodine no. 68; fire pt. 500-700 F; 99% C; *Toxicology:* OSHA and ACGIH PEL 3.5 mg/m^3; nuisance dust; inh. of dust may cause temporary discomfort to nose and respiratory tract; *Precaution:* incompat. with strong oxidizers; hazardous decomp. prods.: CO, CO_2 from burning.

Continex® N550. [Witco/Concarb] Carbon black; CAS 1333-86-4; EINECS 215-609-9; rubber reinforcement providing med. abrasion resist.; imparts smooth surf. and dimensional stability to extruded profiles, cable jackets, hose and belting, brake cups, piping; blk. amorphous pellets, odorless; insol. in water; sp.gr. 1.7-1.9 (16 C); bulk dens. 22 lb/ft^3; vapor pressure negligible; nitrogen surf. area 41 m^2/g; iodine no. 43; fire pt. 500-700 F; 99% C; *Toxicology:* OSHA and ACGIH PEL 3.5 mg/m^3; nuisance dust; inh. of dust may cause temporary discomfort to nose and respiratory tract; *Precaution:* incompat. with strong oxidizers; hazardous decomp. prods.: CO, CO_2 from burning.

Continex® N650. [Witco/Concarb] Carbon black; CAS 1333-86-4; EINECS 215-609-9; rubber reinforcement providing med. reinforcement; good extrusions for weatherstripping, single-ply roofing, vacuum and heater hose, o-rings, inner tubes, body mounts, cable; blk. amorphous pellets, odorless; insol. in water; sp.gr. 1.7-1.9 (16 C); bulk dens. 22 lb/ft^3; vapor pressure negligible; nitrogen surf. area 38 m^2/g; iodine no. 36; fire pt. 500-700 F; 99% C; *Toxicology:* OSHA and ACGIH PEL 3.5 mg/m^3; nuisance dust; inh. of dust may cause temporary discomfort to nose and respiratory tract; *Precaution:* incompat. with strong oxidizers; hazardous decomp. prods.: CO, CO_2 from burning.

Continex® N660. [Witco/Concarb] Carbon black; CAS 1333-86-4; EINECS 215-609-9; rubber reinforcement for tire carcass compds., inner tubes, bicycle tires, cable insulation, body mounts; blk. amorphous pellets, odorless; insol. in water; sp.gr. 1.7-1.9 (16 C); bulk dens. 26.5 lb/ft^3; vapor pressure negligible; nitrogen surf. area 35 m^2/g; iodine no. 36; fire pt. 500-700 F; 99% C; *Toxicology:* OSHA and ACGIH PEL 3.5 mg/m^3; nuisance dust; inh. of dust may cause temporary discomfort to nose and respiratory tract; *Precaution:* incompat. with strong oxidizers; hazardous decomp. prods.: CO, CO_2 from burning.

Continex® N683. [Witco/Concarb] Carbon black; CAS 1333-86-4; EINECS 215-609-9; rubber reinforcement providing high structure for good extrusion profiles; used in weatherstripping and mech. rubber goods; blk. amorphous pellets, odorless; insol. in water; sp.gr. 1.7-1.9 (16 C); bulk dens. 21 lb/ft^3; vapor pressure negligible; nitrogen surf. area 36 m^2/g; iodine no. 35; fire pt. 500-700 F; 99% C; *Toxicology:* OSHA and ACGIH PEL 3.5 mg/m^3; nuisance dust; inh. of dust may cause temporary discomfort to nose and respiratory tract; *Precaution:* incompat. with strong oxidizers; hazardous decomp. prods.: CO, CO_2 from burning.

Continex® N762. [Witco/Concarb] Carbon black; CAS 1333-86-4; EINECS 215-609-9; rubber reinforcement providing low reinforcement, high loading capacity, low hysteresis; for tire beads, hydraulic hose, molded goods, sol'ns.; blk. amorphous pellets, odorless; insol. in water; sp.gr. 1.7-1.9 (16 C); bulk dens. 28.5 lb/ft^3; vapor pressure negligible; nitrogen surf. area 28 m^2/g; iodine no. 27; fire pt. 500-700 F; 99% C; *Toxicology:* OSHA and ACGIH PEL 3.5 mg/m^3; nuisance dust; inh. of dust may cause temporary discomfort to nose and respiratory tract; *Precaution:* incompat. with strong oxidizers; hazardous decomp. prods.: CO, CO_2 from burning.

Continex® N774. [Witco/Concarb] Carbon black; CAS 1333-86-4; EINECS 215-609-9; rubber reinforcement providing easy mixing, high loading, smooth extrusions; used in belts, hoses, automotive molded goods, footwear, innerliners; blk. amorphous pellets, odorless; insol. in water; sp.gr. 1.7-1.9 (16 C); bulk dens. 27.5 lb/ft^3; vapor pressure negligible; nitrogen surf. area 29 m^2/g; iodine no. 29; fire pt. 500-700 F; 99% C; *Toxicology:* OSHA and ACGIH PEL 3.5 mg/m^3; nuisance dust; inh. of dust may cause temporary discomfort to nose and respiratory tract; *Precaution:* incompat. with strong oxidizers; hazardous decomp. prods.: CO, CO_2 from burning.

Cordex AT-172. [Finetex] cationic; antistat for acrylics, polyamides, polyesters, and acetate fibers.

Cordon NU 890/75. [Finetex] Sulfated castor oil; CAS 8002-33-3; EINECS 232-306-7; anionic; emulsifier for pine oil; plasticizer, lubricant, textile dyeing and kier bleaching assistant; dispersant for vinyl pigments; liq.; 70% conc.

Cosmopon BN. [Auschem SpA] Sodium oleyl sulfosuccinamate; anionic; emulsifier for emulsion polymerization; foaming agent for latex emulsions; antigelling and cleaning agents for paper mill felts; paste; 35% conc.

Coupsil VP 6109. [Degussa AG; Struktol] Ultrasil VN 2 (silica) reacted with Si 69 (bis (3-triethoxysilylpropyl) tetrasulfane); reinforcing filler for rubber compds. incl. NR, IR, SBR, BR, NBR, EPDM, and EPM; for tires and dynamically stressed mech. goods; wh. to lt. yel. powd.; insol. in all common solvs.; dens. 310 g/l (tapped); pH 7.2; 8.25% silane; *Storage:* 12 mos storage stability when stored in dry conditions below 50 C.

Coupsil VP 6411. [Degussa AG; Struktol] Ultrasil VN 2 (silica) reacted with Si 264 (thiocyanato propyl triethoxy silane); reinforcing filler for rubber compds. incl. NR, IR, SBR, BR, NBR, EPDM, and EPM; for mech. rubber goods; wh. to lt. yel. powd.; insol. in all common solvs.; dens. 310 g/l (tapped); pH 7.2; 9.7% silane; *Storage:* 12 mos storage stability when stored in dry conditions below 50 C.

Coupsil VP 6508. [Degussa AG; Struktol] Ultrasil VN 2 (silica) reacted with Si 225 (vinyl triethoxy silane); reinforcing filler for rubber compds. incl. NR, IR, SBR, BR, NBR, EPDM, and EPM; wh. to lt. yel. powd.; insol. in all common solvs.; dens. 310 g/l (tapped); pH 7.2; 7.2% silane; *Storage:* 12 mos storage stability when stored in dry conditions below 50 C.

Coupsil VP 8113. [Degussa AG; Struktol] Ultrasil VN 3 (silica) reacted with Si 69 (bis (3-triethoxysilylpropyl) tetrasulfane); reinforcing filler for rubber compds. incl. NR, IR, SBR, BR, NBR, EPDM, and EPM; for tires and dynamically stressed mech. goods; wh. to lt. yel. powd.; insol. in all common solvs.; dens. 300 g/l (tapped); pH 6.5; 11.3% silane; *Storage:* 12 mos storage stability when stored in dry conditions below 50 C.

Coupsil VP 8415. [Degussa AG; Struktol] Ultrasil VN 3 (silica) reacted with Si 264 (thiocyanato propyl triethoxy silane); reinforcing filler for rubber compds. incl. NR, IR, SBR, BR, NBR, EPDM, and EPM; for mech. rubber goods; wh. to lt. yel. powd.; insol. in all common solvs.; dens. 300 g/l (tapped); pH 6.5; 13% silane; *Storage:* 12 mos storage stability when stored in dry conditions below 50 C.

CP0320. [Hüls Am.] Phenyltriethoxy silane; CAS 780-69-8; coupling agent, chem. intermediate, blocking agent, release agent, lubricant, primer, reducing agent; chemically bonds, strengthens thermoplastic composites, e.g., PP, PS; liq.; m.w. 240.4; sp.gr. 0.99 (20 C); b.p. 233-234 C; flash pt. 121 C; ref. index 1.461; 99% purity.

CP-343-3 (25% in xylene). [Eastman] Chlorinated polyolefin; used as primer and stir-in additive for PP, selected plastics, and metals; emulsifiable; Gardner 6-7 liq.; reducible to 5% CPO with xylene, toluene, trichloroethylene, 1,1,1-trichloroethane, methylene chloride, ethyl acetate, MEK, methyl amyl ketone, tetrahydrofuran, n-butyl acetate, isopropyl acetate, n-propyl acetate; sp.gr. 0.906; visc. 12 cP; flash pt. (TOC) 34 C; Solid CPO properties: sp.gr. 0.90; hardness (Tukon Knoops) 2; 25% in xylene.

CP Filler™. [ECC Int'l.] Calcium carbonate; CAS 471-34-1, 1317-65-3; competitively priced filler/extender/reinforcement/pigment for construction prods. where color is of secondary importance; for joint compds., patching compds., antiblocking powd., rubber, putty, scrubbing compds.; off-wh. powd., odorless; 13.5 μ mean particle size; 7% on 325 mesh; negligible sol. in water; sp.gr. 2.71; bulk dens. 60 lb/ft^3 (loose); pH 9.5 (5% slurry); nonflamm.; 96.5% $CaCO_3$, 0.2% max. moisture.

CPH-31-N. [C.P. Hall] Glyceryl oleate; nonionic; antiblocking agent; amber liq.; sol. in toluene, kerosene, ethanol, acetone, propylene glycol, min. oil; partly sol. in hexane; insol. in water; m.w. 340; sp.gr. 0.943; f.p. 9 C; HLB 3.7; acid no. 3.0; iodine no. 88; sapon. no. 168; hyd. no. 230; ref. index 1.467; 100% conc.

CPH-37-NA. [C.P. Hall] Glycol stearate; CAS 111-60-4; EINECS 203-886-9; internal/external lubricant for plasticized PVC compds.; emulsifier, opacifier, pearlescent for cosmetics; wh. flake; sol. in toluene, ethanol, acetone, min. oil; m.w. 300; m.p. 57 C; acid no. 3.0; sapon. no. 187.

CPH-376N. [C.P. Hall] Internal antistat for rigid and flexible PVC.

CPS 076. [Hüls Am.] (N-Trimethoxysilylpropyl)-polyethylenimine; coupling agent esp. for min.-filled and adhesive bonding applics. of high m.w. thermoplastic polyamides and polyesters; sol'n.; m.w. 1000; sp.gr. 0.91; visc. 175-250 cSt; ref. index 1.442; flamm.; 50% act. in IPA.

CPS 076.5. [Hüls Am.] Polyamine dialkoxysilane; polymeric coupling agent for thermosets (urea-formaldehyde), thermoplastics (polyacrylate, nylon 6, PC, polyether ketone, PEEK, PPO, PPS, polysulfone, PES, PVC), nitrile and SAN sealants, copper, iron substrates; dens. 0.92; visc. 125-175 ctsk; 50% act. in IPA.

CPS 078.5. [Hüls Am.] Triethoxysilyl modified poly(1,2-butadiene); CAS 72905-90-9; coupling agent esp. for polyolefins and polyolefin elastomers, min.- and glass-filled formulations, sealants (acrylic, SBR), rubbers (butyl, neoprene, isoprene, fluorocarbon), as primer coat for metals in insert molding; sol'n.; m.w. 3500-4000; sp.gr. 0.90; visc. 35-50 cSt; 50% act. in toluene; *Precaution:* flamm.

CPS 078.9. [Hüls Am.] Vinyl methoxy-siloxane oligomer; polymeric coupling agent for thermosets (parylene), thermoplastics (polyether ketone, PEEK, PP), rubbers (butyl, neoprene, isoprene); dens. 1.1; visc. 8-11 ctsk.

CPS 120. [Hüls Am.] Polymethylhydrosiloxane; CAS 63148-57-2; crosslinker, waterproofing agent; sp.gr. 0.99; visc. 30 cSt.

CPS 123. [Hüls Am.] Polymethylhydrosiloxane; CAS 63148-57-2; crosslinker, waterproofing agent; sp.gr. 0.99; visc. 25-30 Cst.

CPS 925. [Hüls Am.] Polyvinylmethylsiloxane; CAS 68037-87-6; coupling agent.

Crestomer® 1080. [Scott Bader] Tough, flexible base resin in styrene sol'n.; for formulation of high performance, high impact adhesives and filled casting resins; impact modifier for polyester resins.

Cri-Spersion CRI-ACT-45. [Cri-Tech] 45% disp. of calcium hydroxide/magnesium oxide (2:1) on a fluorocarbon elastomer base; activator for fluoroelastomer compounding; opaque brownish chopped or slab; sol. in esters and ketones; sp.gr. 1.85-1.95.

Cri-Spersion CRI-ACT-45-1/1. [Cri-Tech] 45% disp. of calcium hydroxide/magnesium oxide (1:1) on a fluorocarbon elastomer base; activator for fluoroelastomer compounding; opaque, brownish yel. chopped or slab; sol. in esters and ketones; sp.gr. 1.89-1.99.

Cri-Spersion CRI-ACT-45-LV. [Cri-Tech] 45% disp. of calcium hydroxide/magnesium oxide (2:1) on a fluorocarbon elastomer base; activator for fluoroelastomer compounding in low-visc. compounds; opaque brownish yel. chopped or slab; sol. in esters and ketones; sp.gr. 1.89-1.99.

Crodamide 203. [Croda Universal] Amide wax; heat-stable internal/external lubricant, slip and antiblock for polyolefins and engineering thermoplastics; decreases coeff. of friction, improves melt flow, torque release for sealing gaskets, improves mold release; FDA accepted; solid.

Crodamide 212. [Croda Universal] Mold release, lubricant, antiblocking agent.

Crodamide ER. [Croda Universal] Erucamide; CAS 112-84-5; EINECS 204-009-2; internal slip and antiblock agent for polyolefins; internal release agent for molded thermoplastic polymers; decreases coeff. of friction, improves melt flow, torque release for sealing gaskets, improves mold release; wax.

Crodamide O. [Croda Inc.; Croda Chem. Ltd.] Oleamide; CAS 301-02-0; EINECS 206-103-9; internal slip and antiblock agent for polyolefins; internal release agent for moulded thermoplastic polymers.

Crodamide OR. [Croda Universal] Oleamide; CAS 301-02-0; EINECS 206-103-9; internal slip and antiblock agent for polyolefins; internal release agent for molded thermoplastic polymers; decreases coeff. of friction, improves melt flow, torque release for sealing gaskets, improves mold release; wax.

Crodamide S. [Croda Chem. Ltd.] Stearamide; CAS 124-26-5; EINECS 204-693-2; internal slip and antiblock agent for polyolefins; internal release agent for molded thermoplastic polymers; wax.

Crodamide SR. [Croda Universal] Stearamide; CAS 124-26-5; EINECS 204-693-2; internal slip and antiblock agent for polyolefins; internal release agent for molded thermoplastic polymers; improves mold release and lubricity; wax.

Crodamine 1.16D. [Croda Universal Ltd.] Primary cetyl amine; CAS 143-27-1; EINECS 205-596-8; cationic; emulsifier for herbicides, ore flotation, pigment dispersion; aux. for textiles, leather, rubber, plastics, and metal industries; liq.; 100% conc.

Crodamine 1.18D. [Croda Universal Ltd.] Stearylamine; CAS 124-30-1; EINECS 204-695-3; cationic; emulsifier for herbicides, ore flotation, pigment dispersion; aux. for textiles, leather, rubber, plastics, and metal industries; solid; 100% conc.

Crodamine 1.HT. [Croda Universal Ltd.] Hydrog. tallow amine; CAS 61788-45-2; EINECS 262-976-6; cationic; anticaking agent for fertilizers; emulsifier for herbicides, ore flotation, pigment dispersion; aux. for textiles, leather, rubber, plastics, and metal industries; waxy solid; 100% conc.

Crodamine 1.O, 1.OD. [Croda Universal Ltd.] Oleyl amine; CAS 112-90-3; EINECS 204-015-5; cationic; emulsifier for herbicides, ore flotation, pigment dispersion; aux. for textiles, leather, rubber, plastics, and metal industries; solid and liq. resp.; 100% conc.

Crodamine 1.T. [Croda Universal Ltd.] Tallow amine; CAS 61790-33-8; EINECS 263-125-1; cationic; emulsifier for herbicides, ore flotation, pigment dispersion; aux. for textiles, leather, rubber, plastics, and metal industries; liq.; 100% conc.

Crodamine 3.A16D. [Croda Universal Ltd.] Palmityl dimethylamine; CAS 112-69-6; cationic; emulsifier for herbicides, ore flotation, pigment dispersion; aux. for textiles, leather, rubber, plastics, and metal industries; liq.; 100% conc.

Crodamine 3.A18D. [Croda Universal Ltd.] Stearyl dimethylamine, dist.; CAS 124-28-7; EINECS 204-694-8; cationic; emulsifier for herbicides, ore flotation, pigment dispersion; aux. for textiles, leather, rubber, plastics, and metal industries; paste; 100% conc.

Crodamine 3.ABD. [Croda Universal Ltd.] Dimethyl behenylamine; CAS 21542-96-1; cationic; emulsifier for herbicides, ore flotation, pigment dispersion; aux. for textiles, leather, rubber, plastics, and metal industries; solid; 100% conc.

Crodamine 3.AED. [Croda Universal Ltd.] Dimethyl erucylamine; cationic; emulsifier for herbicides, ore flotation, pigment dispersion; aux. for textiles, leather, rubber, plastics, and metal industries; intermediate; liq.; 100% conc.

Crodamine 3.AOD. [Croda Universal Ltd.] Dimethyl oleylamine; CAS 28061-69-0; cationic; emulsifier for herbicides, ore flotation, pigment dispersion; aux. for textiles, leather, rubber, plastics, and metal industries; liq.; 100% conc.

Cryoflex® 660. [Sartomer] Dibutoxyethoxyethyl formal; plasticizer for elastomers, TPE; enhances low temp. cure props., reduces processing visc.; clear liq.; visc. 8 cps; 98% reactive esters.

Crystal 1000. [TSE Industries] < 2% Release blend (CAS 63148-62-9) in 97% water/< 1% ethanol; semi-permanent mold release providing high slip for inj., compr. and transfer molding of urethane parts, EPDM, NR, and syn. elastomers except silicone; multiple releases; environmentally friendly; wh. liq., mild pleasant odor; completely sol. in water; sp.gr. 0.99; vapor pressure 17 mm Hg (68 F); b.p. 99 C; flash pt. none; *Storage:* store in sheltered area; protect from freezing, direct heat.

Crystal 1053. [TSE Industries] < 5% Release blend (CAS 63148-55-0) in 94% water/< 1% ethanol; water-based mold release for fluoroelastomers, specialty elastomers, difficult-to-release applics., SBR, urethane, natural, peroxide-cured EPDM, and polyacrylic rubber; multiple releases; environmentally friendly; wh. liq., mild pleasant odor; completely sol. in water; sp.gr. 0.99; vapor pressure 17 mm Hg (68 F); b.p. 99 C; flash pt. none; *Storage:* store in sheltered area; protect from freezing, direct heat.

Crystal 3000. [TSE Industries] Mold release agent for silicone elastomers; will not work in combination with any water-based prods.

Crystal 4000. [TSE Industries] Water-based semipermanent mold release agent for cast urethanes where max. slip is required to demold from deep groove molds; multiple releases; environmentally friendly.

Crystal 7000. [TSE Industries] < 7% Release blend (CAS 63148-62-9) in 92% water with < 1% ethanol; semipermanent mold release for urethane integral skin foam and flexible foam; provides multiple release, high slip; environmentally friendly; wh. liq., mild pleasant odor; completely sol. in water; sp.gr. 0.99; vapor pressure 17 mm Hg (68 F); b.p. 99 C; flash pt. none; *Storage:* store in sheltered area; protect from freezing, direct heat.

Crystal HMT. [TSE Industries] Proprietary blend; water-based mold release agent for hot mold touch-ups in rubber industry; wh. liq., mild pleasant odor; completely sol. in water; sp.gr. 0.99; vapor pressure 17 mm Hg (68 F); b.p. 99 C; flash pt. none; *Storage:* store in sheltered area; protect from freezing.

Crystex® 60. [Akzo Nobel] Insoluble sulfur (polymeric sulfur and sulfur); CAS 9035-99-8, 7704-34-9; rubber vulcanizing agent for unsat. elastomers to produce polysulfidic cross-links; bright yel. powd.; 99% min. thru 80 mesh, 98.5% min. thru 100 mesh; dens. 2000 kg/m^3; bulk dens. 670-750 kg/m^3; 99.7% S, 60% insol. sulfur.

Crystex® 60 OT 10. [Akzo Nobel] Insol. and rhombic sulfur, oil-treated; prevents cryst. of sulfur on uncured rubber surfs. (sulfur bloom); retards bin scorch, minimizes sulfur migration; 90% S.

Crystex® 90 OT 20. [Akzo Nobel] Insoluble sulfur (polymeric sulfur and sulfur), oil treated; prevents crystallization of sulfur on uncured rubber surfaces (sulfur bloom); detackifier; retards bin scorch; minimizes sulfur migration; 20% oil, 80% sulfur.

Crystex® HS. [Akzo Nobel] Insoluble sulfur (polymeric sulfur and sulfur), oil treated; prevents crystallization of sulfur on uncured rubber surfaces (sulfur bloom); detackifier; retards bin scorch; minimizes sulfur migration; high temp. stability grade; 20% oil, 80% sulfur.

Crystex® HS OT 10. [Akzo Nobel] Insol. sulfur, oil-treated; prevents cryst. of sulfur on uncured rubber surfs. (sulfur bloom); retards bin scorch, minimizes sulfur migration; high-temp. stable; 90% S.

Crystex® HS OT 20. [Akzo Nobel] Insol. sulfur, oil-treated; prevents cryst. of sulfur on uncured rubber surfs. (sulfur bloom); retards bin scorch, minimizes sulfur migration; high-temp. stable; 80% S.

Crystex® OT 10. [Akzo Nobel] Insol. sulfur, oil-treated; prevents cryst. of sulfur on uncured rubber surfs. (sulfur bloom); retards bin scorch, minimizes sulfur migration; 90% S.

Crystex® OT 20. [Akzo Nobel] Insol. sulfur, oil-treated; prevents cryst. of sulfur on uncured rubber surfs. (sulfur bloom); retards bin scorch, minimizes sulfur migration; 80% S.

Crystex® Regular. [Akzo Nobel] Insoluble sulfur (polymeric sulfur and sulfur); CAS 9035-99-8, 7704-34-9; rubber vulcanizing agent; prevents crystallization of sulfur on uncured rubber surfaces (sulfur bloom); detackifier; retards bin scorch; minimizes sulfur migration; bright yel. powd.; 99% min. thru 80 mesh, 98.5% min. thru 100 mesh; > 10 μ avg. particle size; dens. 2000 kg/m^3; bulk dens. 640-720 kg/m^3; 99.8% S, 90% min. insol. sulfur.

Crystic Pregel 17. [Scott Bader] Silica in a general purpose polyester resin disp.; resin additive to confer thixotropic properties to general purpose polyester resins for laminating or gelcoat applics.; pale yel. paste; flash pt. < 32 C.

Crystic Pregel 27. [Scott Bader] Silica in an isophthalic polyester resin disp.; resin additive to confer thixotropic properties to Crystic 199 and 625MV; pale yel. paste; flash pt. < 32 C.

CS1590. [Hüls Am.] 3-(N-Styrylmethyl-2-aminoethylamino) propyltrimethoxy silane hydrochloride, methanol; CAS 52783-38-7; coupling agent, chem. intermediate, blocking agent, release agent, lubricant, primer, reducing agent; chemically bonds, strengthens thermoset and thermoplastic composites; for DAP, polyacetals, fluorocarbon rubber; liq.; m.w. 375.0; sp.gr. 0.92; flash pt. 13 C; ref. index 1.395; 40% in methanol.

CT-88 Aerosol. [Chem-Trend] Fluoropolymer/organic binder disp.; dry-film release agent and lubricant for molding operations, esp. where abrasion is a problem; effective on vinyls, polyesters, epoxies, phenolics, and PP; and for releasing silicone rubber from metal or plastic molds; applics. incl. casting, potting, hand layup, encapsulating, filament winding, inj. molding, compression molding; off-wh. aerosol; cloudy disp.; ethereal odor; b.p. ≈ 165 F; nil sol. in water; 97.3% volatile.

CT2902. [Hüls Am.] 1-Trimethoxysilyl-2-(chloromethyl) phenylethane; CAS 68128-25-6; coupling agent, chem. intermediate, blocking agent, release agent, lubricant, primer, reducing agent; chemically bonds, strengthens thermoset and thermoplastic composites; for polyimide, polyamide-imide; amber liq.; m.w. 274.8; b.p. 161 C (1.5 mm); flash pt. 130 C; 95% purity.

CTA. [Wako Pure Chem. Ind.; Wako Chem. USA] 2-Ethylhexyl thioglycolate; CAS 7659-86-1; EINECS 231-626-4; chain transfer agent; colorless clear liq., mercaptan odor; sl. sol. in water; sp.gr. 0.97-0.98 (20/4 C); b.p. 130-140 C (2000 Pa); flash pt. 136 C; *Toxicology:* LD50 (oral, rat) 303 mg/kg; eye and skin irritant; harmful if inhaled and ingested; ing. may cause anorexia, headache, vomiting, fever, etc.; prolonged exposure can cause shock, narcosis, death; *Precaution:* may emit toxic, irritating fumes on combustion; incompat. with oxidizers; hazardous decomp. prods.: CO, sulfur oxides; *Storage:* store in cool (< 25 C), well-ventilated area in tightly closed container away from sunlight, heat sources.

Cure-Rite® 18. [Akrochem; R.T. Vanderbilt] Thiocarbamyl sulfenamide; nonstaining primary accelerator for EPDM, SBR, nitrile, natural and butyl rubbers; wh. powd. and pellets; sol. in benzene, chloroform, CCl_4, ether, and alcohol; m.w. 248; sp.gr. 1.35 ± 0.05; m.p. 133 C min.

Curezol® 1B2MZ. [Air Prods./Perf. Chems.] Imidazole; effective elevated temp. accelerator for anhydrides; high reactivity, noncrystallizing; for casting, potting, and encapsulation in elec. and electronic applics.; pale yel. liq.; m.w. 172; sp.gr. 1.07; visc. 700 cps; heat distort. temp. 310 F.

Curezol® 2E4MZ. [Air Prods./Perf. Chems.] 2-Ethyl 4-methyl imidazole; CAS 931-36-2; EINECS 213-234-5; elevated temp. epoxy curing agent for structural adhesives, elec. laminates, molding powds., composites, filament winding, solder-resistant inks, potting compds.; accelerator for dicyandiamide or anhydride curing agents; pale yel. liq./solid; m.w. 110; sp.gr. 0.990; dens. 8.2 lb/gal; visc. 95 poise; m.p. 68 F; heat distort. temp. 313 F.

Curezol® 2MA-OK. [Air Prods./Perf. Chems.] Imidazole; elevated temp. accelerator for dicyandiamide, anhydride, and phenolic curing agents; best combination of latency and low-temp. cure; for elec. and electronic insulation, solder resist. inks and structural adhesives; wh. powd.; m.w. 384; m.p. 250 F.

Curezol® 2MZ-Azine. [Air Prods./Perf. Chems.] Imidazole; CAS 38668-46-1; elevated temp. epoxy curing agent for electronic applics., solder-resist. inks, insulating powds., structural adhesives; accelerator for dicyandiamide and anhydride curing agents; wh. fine powd.; m.w. 219; m.p. 480 F; heat distort. temp. 313 F.

Curezol® 2PHZ. [Air Prods./Perf. Chems.] Imidazole; CAS 61698-32-6; epoxy curing agent for adhesives, casting, potting, encapsulation; accelerator for dicyandiamide and anhydrides; adhesive for surf. mounting of devices onto circuit boards; ylsh.-

pink powd.; m.p. 415-437 F (decomp.).

Curezol® 2PHZ-S. [Air Prods./Perf. Chems.] Imidazole; elevated temp. accelerator for dicyandiamide and anhydride curing agents; for surf.-mounted electronic adhesives, elec. encapsulation and transfer molding powds.; intermediate in mfg. of curing agents by reaction through methylol groups; yel-pink powd.; m.w. 204; m.p. 420 F; heat distort. temp. 313 F.

Curezol® 2PZ. [Air Prods./Perf. Chems.] 2-Phenyl imidazole; CAS 670-96-2; epoxy curing agent for elec. laminates, printed circuit boards, molding compds., potting; accelerator for dicyandiamide, anhydride, and phenolic curing agents; pale pink powd.; m.w. 144; m.p. 279-280 F; heat distort. temp. 316 F.

Curezol® 2PZ-CNS. [Air Prods./Perf. Chems.] Imidazole; CAS 68083-35-2; epoxy curing agent for transfer molding powds., elec. powd. coatings; accelerator for anhydrides, solid epoxy resins; wh. powd.; m.p. 356-360 F; Unverified

Curezol® 2PZ-OK. [Air Prods./Perf. Chems.] Imidazole; epoxy curing agent for molding compds., powd. coatings; accelerator for dicyandiamide; longer pot life than Curezol 2PZ; wh. powd.; m.p. 284 F (decomp.); Unverified

Curezol® AMI-2. [Air Prods./Perf. Chems.] Imidazole; elevated temp. accelerator for dicyandiamide, anhydride, and phenolic curing agents; for elec. laminates, powd. coatings, molding powds., structural adhesives; pale yel. powd.; m.w. 82; m.p. 140 F; heat distort. temp. 300 F.

Curezol® C17Z. [Air Prods./Perf. Chems.] Imidazole; elevated temp. epoxy curing agent with long latency; for structural adhesives, powd. coatings, molding powds., structural laminates; accelerator for anhydride, phenolic, and dicyandiamide curing agents; wh. powd.; m.w. 306; m.p. 187-190 F; heat distort. temp. 312 F.

Curithane® 52. [Air Prods./Perf. Chems.] Trimerization co-catalyst for isocyanate rigid PU systems; liq.; sl. sol. in water; sp.gr. 1.1; visc. 13,500 cps; hyd. no. 501; flash pt. (CC) 152 C.

Curithane® 97. [Air Prods./Perf. Chems.] Trimerization co-catalyst for isocyanate rigid PU systems; liq.; sl. sol. in water; sp.gr. 1.1; visc. 850 cps; flash pt. (CC) 176.7 C.

C-Wax-140. [Cardinal Carolina] Glycerol ester; internal lubricant for rigid PVC bottles, sheet, film, inj. molding, profiles, highly filled PVC pipe; FDA accepted; solid.

Cyanamer N300LMW. [Cytec Industries] Polyacrylamide; CAS 9003-05-8; nonionic; thickener, binder, sizing, flocculating, suspending, crosslinking agent, filtering aid, lubricant; used in adhesives, agric., cement, coatings, cosmetics, detergents, fire fighting, graphic arts, insulation, latex mfg., plaster, printing ink, processing, textile and leather, ceramics, dyes and pigments, used for polymer recovery in adhesive tapes, paints, latex mfg. and processing; wh. powd.; m.w. 5-6 x 10^6; water-sol.; visc. 1.8-2.2 cps (0.1%); pH 6.0-6.5 (1%).

Cyanamer P-21. [Cytec Industries] Modified polyacrylamide; dispersant, solubilizer, antiprecipitant, thickener, binder, flocculating, suspending, and crosslinking agent, lubricant; used in adhesives, agriculture, cosmetics, detergents, foundry molds, metal processing, plaster, printing inks, processing, textile and leather, latex mfg., cements; wh. powd.; m.w. 200,000; water-sol.; visc. 300-800 cps (10%); pH 10-12.

Cyanox® 425. [Cytec Industries] 2,2′-Methylenebis (4-ethyl-6-tert-butylphenol); CAS 88-24-4; antioxidant for rubber-modified plastics, impact molding resins; stabilizes acrylics and ABS; end-uses incl. extruded and molded prods.; FDA 21CFR §175.105, 177.1010, 178.2010; cream to wh. powd., phenolic odor; sol. (g/100 g): > 60.0 g in acetone and dioxane, 56.5 g in ethyl acetate, 50.0 g in chloroform, 42.8 g in 95% ethanol, 33.3 g in benzene, 9.1 g in n-heptane; < 0.1 g in water; m.w. 368.5; sp.gr. 1.10; bulk dens. 19.7 lb/ft^3; m.p. 123 C; *Toxicology:* LD50 (oral, rat) > 15 g/kg, (dermal, rabbit) > 8 g/kg; minimal eye and skin irritant; *Precaution:* may form explosive dust-air mixts.

Cyanox® 711. [Cytec Industries] Ditridecyl thiodipropionate; CAS 10595-72-9; sec. antioxidant for stabilizing polyolefins, ABS, petrol. lubricants, SBR latex compositions; synergistic with phenolic antioxidants; colorless to sl. yel. clear liq., mild char. odor; sol. (g/100 g): 13.8 g in 95% ethanol; misc. with ethyl acetate, n-heptane, MEK, and toluene; m.w. 543; sp.gr. 0.936; dens. 7.8 lb/gal; acid no. 3.0; sapon. no. 200-210; flash pt. (CC) > 230 F; *Toxicology:* LD50 (oral, rat) > 10 ml/kg, (dermal, rabbit) > 5 ml/kg; minimal skin and eye irritant; *Precaution:* incompat. with strong oxidizing agents; hazardous decomp. prods.: CO, CO_2, oxides of sulfur.

Cyanox® 1212. [Cytec Industries] Mixed lauryl-stearyl thiodipropionate; sec. antioxidant for stabilizing polyolefins; applications incl. pipe, hot-melt adhesives, and molded olefin prods.; synergistic with phenolic antioxidants; FDA GRAS (0.02% max. total antioxidants); wh. waxy cryst. flakes, char. sweet ester odor; sol. (g/100 g): 30.0 g in toluene, 4.2 g in n-heptane; 0.05 g in water and 95% ethanol; sp.gr. 1.018; bulk dens. 27.6 lb/ft^3; vapor pressure 2.7 mm Hg (163 C); m.p. 50 C; acid no. 1.0; *Toxicology:* LD50 (oral, rat) > 5 g/kg (estimate), (dermal, rabbit) > 2 g/kg (estimate); minimal eye and skin irritant; *Precaution:* hazardous decomp. prods.: CO, CO_2, oxides of sulfur.

Cyanox® 1790. [Cytec Industries] 1,3,5-Tris (4-tert-butyl-3-hydroxy-2,6-dimethylbenzyl)-1,3,5-triazine-2,4,6-(1H,3H,5H)-trione; CAS 40601-76-1; high performance, gas fade-resist. antioxidant for use in polyolefin pipe, film, household appliances, olefin and urethane fibers, styrenics, polyesters, PU elastomers; FDA 21CFR §177.1640, 178.2010(b); off-wh. powd., odorless; sol. (g/100 g sol'n.) > 10 g in cyclohexane, styrene, toluene, MEK; 4.6 g in ethanol; insol. in water; m.w. 699; m.p. 145-155 C; flash pt. 276 C; *Toxicology:* LD50 (oral, rat) > 10 g/kg, (dermal, rabbit) > 5 g/kg; minimal skin and eye irritant; *Precaution:* airborne dust may present explosion hazard; incompat. with strong oxidizing agents and min. acids; hazardous decomp. prods.: CO, CO_2, ammonia, oxides of nitrogen and/or hydrogen cyanide.

Cyanox® 2246. [Cytec Industries] 2,2′-Methylenebis (4-methyl-6-tert-butylphenol); CAS 119-47-1; antioxidant preventing thermal oxidation of ABS, polyethylene, PP, and EVA; oxidation inhibitor for fats, oils, and paraffin wax; polymerization inhibitor in chemical processes; for ABS and SBR latex, hotmelt adhesives, latex carpet backing, and specialty olefin applics.; FDA 21CFR §175.105, 177.2600, 178.2010; cream to wh. powd., phenolic odor; sol. (g/100 g): > 60.0 g in acetone, 59.9 g in dioxane, 54.4 g in ethyl acetate, 47.6 g in chloroform, 36.7 g in 95% ethanol, 35.8 g in benzene, 5.8 g in heptane, < 0.1 g in water; m.w. 340.5; sp.gr. 1.08; bulk dens. 19 lb/ft^3; m.p. 125-128 C; *Toxicology:* LD50 (oral, rat) > 10 g/kg, (dermal, rabbit) > 10 g/kg; nonirritating to skin and eyes; *Precaution:* may form explosive dust-air mixts.; incompat. with strong oxidizing agents.

Cyanox® 2777. [Cytec Industries] 1:2 blend of 1,3,5-tris (4-t-butyl-3-hydroxy-2,6-dimethylbenzyl)-1,3,5-triazine-2,4,6-(1H,3H,5H) trione and tris (2,4-di-t-butylphenyl) phosphite; antioxidant, stabilizer for polymers, esp. polyolefins in high-temp. processing; low volatility, low color contribution, low gas yellowing props.; provides outstanding melt processing stabilization without pinking problems; FDA 21CFR §177.1640, 178.2010; off-wh. powd.; sol. (g/100 g): 13.7 g ethyl acetate, 3.2 g acetone, 2.5 g hexane; sp.gr. 1.07; bulk dens. 39.6 lb/ft^3; m.p. 166-171 C; *Toxicology:* LD50 (oral, rat) > 10 g/kg, (dermal, rabbit) > 5 g/kg; nonirritating to skin and eyes; *Precaution:* incompat. with strong oxidizing agents; hazardous decomp. prods.: CO, CO_2.

Cyanox® LTDP. [Cytec Industries] Dilauryl thiodipropionate; CAS 123-28-4; EINECS 204-614-1; sec. antioxidant in ABS, PP, and polyethylene; used in food pkg. materials, automotive, appliance, battery casing, pipe; stabilization of oils, lubricants, sealants, and adhesives; synergistic with phenolic antioxidants; FDA 21CFR §181.24, 182.3280, GRAS; wh. waxy cryst. flakes, char. sweet ester odor; sol. (g/100 g): 51.2 g in MEK and acetone, 40.5 g in n-heptane, 39.4 g in ethyl acetate; < 0.5 g in 95% ethanol; m.w. 514; sp.gr. 0.915; bulk dens. 30 lb/ft^3; vapor pressure 0.2 mm Hg (163 C); m.p. 40 C; acid no. 1.0; *Toxicology:* LD50 (oral, rat) > 25 g/kg (estimate), (dermal, rabbit) > 10 g/kg (estimate); direct contact may cause mild eye and skin irritation; *Precaution:* hazardous decomp. prods.: CO, CO_2, oxides of sulfur.

Cyanox® MTDP. [Cytec Industries] Dimyristyl thiodipropionate; CAS 16545-54-3; sec. antioxidant for long-term abient temp. protection of polyolefins incl. PP, LDPE, HDPE; synergistic with phenolic antioxidants; FDA 21CFR §178.2010; wh. flakes or powd., char. sweet odor; negligible sol. in water; m.w. 570; sp.gr. 0.967; bulk dens. 31 lb/ft^3; m.p. 47 C; acid no. 0.5; 97% purity; *Toxicology:* LD50 (oral, rat) > 5000 mg/kg (estimate), (dermal, rabbit) > 2000 mg/kg (estimate); direct contact may cause minimal eye and skin irritation; *Precaution:* hazardous decomp. prods.: CO, CO_2, oxides of sulfur.

Cyanox® STDP. [Cytec Industries] Distearyl thiodipropionate; CAS 693-36-7; EINECS 211-750-5; sec. antioxidant used in polyolefins and other polymers; used in automotive, appliance, container film, sealant, and adhesive applics.; synergistic with phenolic antioxidants; FDA 21CFR §181.24; GRAS (200 ppm max.); wh. waxy cryst. flakes, char. sweet ester odor; sol. (g/100 g): 10.7 g in toluene, 1.5 g in n-heptane; < 0.5 g in ethyl acetate, 95% ethanol, and MEK; m.w. 683; sp.gr. 1.027; dens. 4.18 lb/gal; bulk dens. 26.9 lb/ft^3; vapor pressure 0.15 mm (162 C); m.p. 64 C; acid no. 1.5 max.; *Toxicology:* LD50 (oral, rat) > 5 g/kg (estimate), (dermal, rabbit) > 2 g/kg (estimate); direct contact may cause minimal eye and skin irritation; *Precaution:* hazardous decomp. prods.: CO, CO_2, oxides of sulfur.

Cyasorb® UV 9. [Cytec Industries] 2-Hydroxy-4-methoxybenzophenone; CAS 131-57-7; EINECS 205-031-5; lt. stabilizer and uv absorber for plastics and coatings; esp. for flexible and rigid PVC, unsat. polyesters, and acrylics; used in outdoor sheeting and glazing applics., molded products, adhesives; FDA 21CFR §177.1010; pale cream to wh. powd., odorless; sol. (g/100 g): 59.5 g methylene chloride, 56.2 g benzene, 51.2 g styrene, 34.5 g CCl_4, 18.7 g di-2-ethylhexyl phthalate; m.w. 228; sp.gr. 1.324; bulk dens. 3.2 lb/gal; set pt. 62 C; b.p. 150-160 C (5 mm); *Toxicology:* estimated LD50 (oral, rat) > 10 g/kg, (dermal, rabbit) > 16 g/kg; direct contact may cause minimal eye and skin irritation; *Precaution:* incompat. with strong oxidizing agents; hazardous decomp. prods.: CO, CO_2.

Cyasorb® UV 24. [Cytec Industries] 2,2′-Dihydroxy-4-methoxybenzophenone; CAS 131-53-3; EINECS 205-026-8; lt. stabilizer and uv absorber for coatings and plastics, e.g., alkyds, phenolics, PU coatings; stabilizer for polyester film and PVC formulations; yel. powd., pract. odorless; sol. (g/100 g): 55.3 g MEK, 46.6 g benzene, 31.1 g di-2-ethylhexyl phthalate, 22.2 g CCl_4, 21.4 g 95% ethanol; m.w. 244.2; sp.gr. 1.382; bulk dens. 3.5 lb/gal; m.p. 68 C; b.p. 160-170 C (0.5-1.0 mm); > 99% act.; *Toxicology:* LD50 (oral, rat) > 10 g/kg, (dermal, rabbit) > 10 g/kg; pract. nontoxic orally; repeated/prolonged skin contact may cause allergic reactions; *Precaution:* incompat. with strong oxidizing agents; hazardous decomp. prods.: CO, CO_2.

Cyasorb® UV 416. [Cytec Industries] 2-Hydroxy-4-acryloxyethoxybenzophenone; uv absorber recommended for bonding into acrylic polymers; lt. yel. powd.; sol. in common org. solvs.; insol. in water; m.w. 312; m.p. 77-80 C.

Cyasorb® UV 500. [Cytec Industries] 1,5-Dioxaspiro (5,5) undecane 3,3-dicarboxylic acid, bis (2,2,6,6-tetramethyl-4-piperidinyl) ester; lt. stabilizer for high performance PP fiber applics.; m.w. 522; bulk dens. 29.5 lb/ft^3; m.p. 164-168 C; 95-100% purity.

Cyasorb® UV 531. [Cytec Industries] 2-Hydroxy-4-n-octoxybenzophenone; CAS 1843-05-6; EINECS 217-421-2; lt. stabilizer and uv absorber for plastics and coatings, e.g., polyethylene, PP, PVC, and EVA; uses incl. pipe, storage tanks, and auto, marine, garden prods., auto refinish and industrial coatings, adhesives and sealants; FDA 21CFR §178.2010 (0.5% of polymer), 178.3710 (0.01% max.); lt. yel. powd., odorless; sol. (g/100 g sol'n.) 74.3 g acetone, 72.7 g benzene, 69.8 g methylene chloride, 40.1 g n-hexane, 20.5 g di-2-ethylhexyl phthalate; m.w. 326.1; sp.gr. 1.160; bulk dens. 3 lb/gal; m.p. 48-49 C; *Toxicology:* very low acute toxicity; LD50 (oral, rat) > 10 g/kg, (dermal, rabbit) > 10 g/kg; nonirritating to skin and eyes; *Precaution:* incompat. with strong oxidizing agents; hazardous decomp. prods.: CO, CO_2.

Cyasorb® UV 1084. [Cytec Industries] 2,2′-Thiobis (4-t-octylphenolato)-n-butylamine nickel II; CAS 14516-71-3; lt. and heat stabilizer for polyolefins, e.g., PP fiber, LDPE agric. films and pool liners, and molded prods.; pale grn. powd., odorless; sol. (g/100 g): 51.2 g n-heptane, 48.8 g tetrahydrofuran, 42.8 g toluene; m.w. 571; sp.gr. 1.13; bulk dens. 30.2 lb/ft^3; m.p. 258-261 C; *Toxicology:* estimated LD50 (oral, rat) > 5 g/kg, (dermal, rabbit) > 2 g/kg; direct contact may cause minimal eye and skin irritation; *Precaution:* incompat. with strong oxidizing agents; hazardous decomp. prods.: CO, CO_2, oxides of sulfur, n-butyl amine, ammonia, oxides of nitrogen sulfur, nickel fumes.

Cyasorb® UV 1164. [Cytec Industries] 2-[4,6-Bis(2,4-dimethylphenyl)-1,3,5-triazin-2-yl]-5-(octyloxy) phenol; uv absorber recommended for nylon and other engineering resins; low volatility; lt. yel. powd.; m.w. 510; bulk dens. 20 lb/ft^3; m.p. 87-89 C.

Cyasorb® UV 2098. [Cytec Industries] 2-Hydroxy-4-acryloyloxyethoxy benzophenone; lt. stabilizer, uv absorber which may be chemically bonded with monomers or polymers; lt. yel. powd. or flake, odorless; sol. in common org. solvs.; insol. water; m.w. 312; sp.gr. 1.36; m.p. 77-80 C; *Toxicology:* LD50 (oral, rat) > 5 g/kg, (dermal, rabbit) > 2 g/kg; nonirritating to skin; minimal eye irritant; *Precaution:* incompat. with strong oxidizing agents; hazardous decomp. prods.: CO, CO_2.

Cyasorb® UV 2126. [Cytec Industries] Polymer of 4-(2-acryloyloxyethoxy)-2-hydroxybenzophenone; lt. stabilizer and uv absorber for films and plastics in automotive, greenhouse, home siding, and solar applics.; pale cream to lt. yel. powd., odorless; sol. (g/100 g): > 50 g in MEK and dimethylacethamide, 30-40 g in benzene; m.w. ≈ 50,000; bulk dens. 35 lb/ft^3; m.p. 85-95 C; *Toxicology:* LD50 (oral, rat) > 10 g/kg, (dermal, rabbit) > 5 g/kg; direct contact may cause minimal eye and skin irritation; *Precaution:* may form explosive dust-air mixts.; incompat. with strong oxidizing agents; hazardous decomp. prods.: CO, CO_2.

Cyasorb® UV 2908. [Cytec Industries] 3,5-Di-t-butyl-4-hydroxybenzoic acid, n-hexadecyl ester; lt. stabilizer, free radical scavenger for polyolefins, esp. pigmented opaque formulations; antioxidant; for pipe, crates, drums, auto, marine, garden, recreational prods.; FDA 21CFR §178.2010 (0.5% max.); wh. to off-wh. free-flowing cryst. powd., odorless; sol. (g/100 g): 61.1 g chloroform, 53.5 g toluene, 35.9 g hexane; m.w. 475; sp.gr. 1.07; bulk dens. 16 lb/ft^3; m.p. 60 C; *Toxicology:* LD50 (oral, rat) > 5 g/kg, (dermal, rabbit) > 5 g/kg; direct contact may cause minimal eye and skin irritation; *Precaution:* incompat. with strong oxidizing agents; hazardous decomp. prods.: CO, CO_2.

Cyasorb® UV 3346. [Cytec Industries] Oligomeric hindered amine; CAS 82451-48-7; stabilizer for polymers alone or in combination with UV absorbers; provides thermal antioxidant protection at elevated service temps.; wh. to pale yel. powd., odorless; sol. in common org. solvs.; insol. in water; sp.gr. 1.14; soften. pt. 110-130 C; 99% act.; *Toxicology:* LD50 (oral, rat) 2.1 g/kg, (dermal, rabbit) > 2 g/kg; causes eye burns; may cause skin irritation; dust is irritating; STEL 5 mg/m^3; *Precaution:* incompat. with strong oxidizing agents, acids, acid halides, certain halogens; hazardous decomp. prods.: CO, CO_2, ammonia, oxides of nitrogen and/or hydrogen cyanide; *Storage:* keep container closed.

Cyasorb® UV 3346 LD. [Cytec Industries] Poly [(6-morpholino-s-triazine-2,4-diyl) [2,2,6,6-tetramethyl-4-piperidyl) imino]-hexamethylene [(2,2,6,6-tetramethyl-4-piperidyl) imino]]; lt. stabilizer, free radical scavenger, thermal antioxidant for polymers, esp. polyolefins requiring weatherability; FDA 21CFR §178.2010; off-wh. powd.; sol. (g/100 g): 76 g tetrahydrofuran, 67 g acetone, 62 g ethyl acetate, 25 g toluene, 23 g ethanol; m.w. 1600 ± 10%; bulk dens. 32.3 lb/ft^3; m.p. 110-130 C; *Toxicology:* LD50 (oral, rat) 2100 mg/kg, (dermal, rabbit) > 2000 mg/kg, (inh., rat) 2.8 mg/l; may be irritating to skin and respiratory tract; eye irritant.

Cyasorb® UV 3581. [Cytec Industries] Nonpolymeric HALS; lt. stabilizer for plastics; pale yel. liq.; sol. in common org. solvs.; insol. in water; m.w. 407; 100% act.; *Toxicology:* LD50 (oral, rat) 2 g/kg, (dermal, rabbit) 2 g/kg; irritating to skin and eyes; *Precaution:* moderately strong base.

Cyasorb® UV 3604. [Cytec Industries] Nonpolymeric methylated HALS; lt. stabilizer for plastics where lower basicity is desired; pale yel. liq.; sol. in common org. solvs.; insol. in water; m.w. 420; 100% act.; *Toxicology:* direct contact may cause severe eye and skin irritation.

Cyasorb® UV 3638. [Cytec Industries] 2.2′-(1,4-Phenylene)bis[4H-3,1-benzoxazin-4-one] uv absorber recommended for thermoplastic polyester and polycarbonate; low volatility; off-wh. powd.; m.w. 368; bulk dens. 10.5 lb/ft^3; m.p. 310 C.

Cyasorb® UV 3668. [Cytec Industries] Nonpolymeric acetylated HALS; lt. stabilizer for plastics which are to be painted with acid-catalyzed acrylic coatings; pale yel. liq.; sol. in common org. solvs.; insol. in water; m.w. 449; 100% act.; *Toxicology:* LD50 (oral, rat) and (dermal, rabbit) > 5 g/kg for structurally similar material.

Cyasorb® UV 3853. [Cytec Industries] Hindered amine; lt. stabilizer, thermal antioxidant for polyolefins; also suitable for ABS copolymers, PS, polyamides, and PU; exc. resist. to extraction by hot water and detergents; wh.-lt. yel. wax-like solid; very sol. in acetone, IPA; sol. in hydrocarbons, aromatic solvs., chlorinated solvs.; insol. in water; m.w. 423.7; dens. 55.7 lb/ft^3 (20 C); m.p. 28-32 C; *Toxicology:* LD50 (oral, rat) > 15 g/kg, (dermal, rat) > 5 g/kg; mild eye irritant on direct contact; nonirritating to skin; nonmutagenic.

Cyasorb® UV 5411. [Cytec Industries] 2-(2-Hydroxy-5-t-octylphenyl)-benzotriazole; CAS 3147-75-9; lt. stabilizer and uv absorber for polymeric systems incl. polyester, PVC, styrenics, acrylics, PC, and polyvinyl butyral; end-uses incl. molding, sheet, and glazing materials for window, marine, and auto applics.; also in coatings, photo prods., sealants, and elastomeric materials; FDA 21CFR § 178.2010; wh. powd., very mild char. odor; sol. (g/100 g): 41.5 g in methylene chloride, 39.4 g in benzene, 33.2 g in styrene, 21.4 g in ethyl acetate; m.w. 325; sp.gr. 1.18; bulk dens. 25 lb/ft^3; m.p. 101-105 C; *Toxicology:* LD50 (oral, rat) > 10 g/kg, (dermal, rabbit) > 5 g/kg; direct contact may cause eye or skin irritation; *Precaution:* incompat. with strong oxidizing agents; hazardous decomp. prods.: CO, CO_2, NO_x.

Cyastat® 609. [Cytec Industries] N,N-Bis (2-hydroxyethyl)-N-(3-dodecyloxy-2-hydroxypropyl) methyl ammonium methosulfate in IPA/water (1:1); antistatic agent with good heat stability; for PVC phonograph records, specialty pkg.; lt. yel. to amber liq., sl. IPA odor; sol. in water, acetone, alcohol, and other polar solvents of low m.w.; m.w. 473.6; sp.gr. 0.964; dens. 8.2 lb/gal; b.p. 79 C; flash pt. (CC) 14.4 C; pH 4-6 (10%); 50% solids; *Toxicology:* LD50 (oral, rat) 4.1 ml/kg, (dermal, rabbit) 2.8 ml/kg; sl. toxic by ingestion; causes skin and eye burns; chronic overexposure to IPA vapors may cause CNS depression, headaches, nausea, staggered gait; *Precaution:* flamm. liq. and vapor; corrosive; incompat. with strong oxidizing agents and min. acids; hazardous decomp. prods.: CO, CO_2, ammonia, oxides of nitrogen and sulfur; *Storage:* store @ R.T.; keep container tightly closed; keep away from heat, sparks, flame; solids may separate if cooled below 60 F.

Cyastat® LS. [Cytec Industries] (3-Lauramidopropyl) trimethylammonium methyl sulfate; CAS 10595-49-0; antistatic agent for polymeric materials incl. PVC, PS, polyolefins, ABS; for internal or external use in thermoplastics; for pkg., electronic and polymer deflashing applics.; off-wh. to lt. tan cryst. powd., char. odor; sol. (g/100 g): 133 g ethanol, 75 g water and ethyl Cellosolve, 11.4 g acetone; m.w. 410; sp.gr. 1.121; bulk dens. 28.5 lb/ft^3; m.p. 99-103 C; > 99% act.; *Toxicology:* LD50 (oral, rat) 1.8 g/kg, (dermal, rabbit) 2.8 g/kg; relatively low toxicity, but moderate to severe skin and eye irritant; may cause allergic skin reaction; *Precaution:* incompat. with strong oxidizing agents; hazardous decomp. prods.: CO, CO_2, ammonia, oxides of nitrogen and sulfur.

Cyastat® SN. [Cytec Industries] Stearamidopropyl dimethyl-2-hydroxyethyl ammonium nitrate, IPA/water (1:1); antistatic agent for polymers; used for plastics, surface coatings, paper, glass, and other materials; dispersant in coatings; lt. yel. to amber liq., sl. IPA odor; sol. in water, acetone, alcohol, and other polar solvents of low m.w.; m.w. 476; sp.gr. 0.95; dens. 7.9 lb/gal; b.p. 83 C; flash pt. (CC) 12.2 C; pH 4-6; 50% solids; *Toxicology:* LD50 (oral, rat) 3300 mg/kg, (dermal, rabbit) > 5 ml/kg; direct contact may cause mod. eye and skin irritation; overexposure to vapor may cause respiratory tract irritation and CNS depression; *Precaution:* flamm. liq. and vapor; incompat. with strong oxidizing agents and min. acids; hazardous decomp. prods.: CO, CO_2, ammonia, NO_x; *Storage:* keep away from heat, sparks, and flame; keep container closed.

Cyastat® SP. [Cytec Industries] Stearamidopropyl dimethyl-2-hydroxyethyl ammonium dihydrogen phosphate, IPA/water; cationic; antistatic agent with surface-active properties; for plastics, waxes, textiles, and glass; emulsifier, settling, dispersing, and rewetting agent; pale yel. clear liq., sl. IPA odor; misc. with water, acetone, alcohol, and other polar solvents of low m.w.; m.w. 509; sp.gr. 0.94; dens. 7.8 lb/gal; b.p. 80 C; flash pt. (CC):14.4 C; pH 6-8; 35% solids; *Toxicology:* estimated LD50 (oral, rat) 3.3 g/kg, (dermal, rabbit) > 5 g/kg; mod. eye irritant; overexposure to vapor can cause respiratory tract irritation, CNS depression; *Precaution:* flamm. liq. and vapor; incompat. with strong oxidizing agents; hazardous decomp. prods.: CO, CO_2, ammonia, oxides of nitrogen and phosphorus; *Storage:* keep container closed; keep away from heat, sparks, flame.

Cyclolube® 85. [Witco/Golden Bear] Low aromatic naphthenic oil; process oil offering exc. color and heat stability for rubber compounding, esp. EPDM and butyl; also for SBR, isoprene, and natural rubber; m.w. 450; sp.gr. 0.9129 (15.6 C); dens. 7.6lb/gal; visc. 290 cSt (40 C); flash pt. (COC) 244 C; pour pt. -18 C; ref. index 1.4954.

Cyclolube® 120. [Witco/Golden Bear] Low aromatic naphthenic oil; process oil offering exc. color and heat stability for rubber compounding, esp. EPDM and butyl; also for SBR, isoprene, and natural rubber; m.w. 468; sp.gr. 0.9159 (15.6 C); dens. 7.6 lb/gal; visc. 386 cSt (40 C); flash pt. (COC) 254 C; pour pt. -12 C; ref. index 1.5006.

Cyclolube® 132. [Witco/Golden Bear] Naphthenic oil distillate; process oils offering lt. color, low pour pts. for general-purpose compounding of elastomers, e.g. neoprene, SBR, isoprene, EDPM, butyl, and natural rubber; m.w. 359; sp.gr. 0.9593 (15.6 C); dens. 8.0 lb/gal; visc. 425 cSt (40 C); flash pt. (COC) 214 C; pour pt. -3 C; ref. index 1.5337.

Cyclolube® 210. [Witco/Golden Bear] Low aromatic

naphthenic oil; process oil offering exc. color and heat stability for rubber compounding, esp. EPDM and butyl; also for SBR, isoprene, and natural rubber; m.w. 305; sp.gr. 0.9042 (15.6 C); dens. 7.5 lb/gal; visc. 20 cSt (40 C); flash pt. (COC) 158 C; pour pt. -42 C; ref. index 1.4927.

Cyclolube® 213. [Witco/Golden Bear] Low aromatic naphthenic oil; process oil offering exc. color and heat stability for rubber compounding, esp. EPDM and butyl; also for SBR, isoprene, and natural rubber; m.w. 320; sp.gr. 0.9100 (15.6 C); dens. 7.6 lb/gal; visc. 42 cSt (40 C); flash pt. (COC) 180 C; pour pt. -36 C; ref. index 1.4997.

Cyclolube® 270. [Witco/Golden Bear] Low aromatic naphthenic oil; process oil offering exc. color and heat stability for rubber compounding, esp. EPDM and butyl; also for SBR, isoprene, and natural rubber; m.w. 368; sp.gr. 0.9218 (15.6 C); dens. 7.7 lb/gal; visc. 133 cSt (40 C); flash pt. (COC) 206 C; pour pt. -21 C; ref. index 1.4995.

Cyclolube® 413. [Witco/Golden Bear] Low aromatic naphthenic oil; process oil offering exc. color and heat stability for rubber compounding, esp. EPDM and butyl; also for SBR, isoprene, and natural rubber; m.w. 327; sp.gr. 0.8956 (15.6 C); dens. 7.5 lb/gal; visc. 29 cSt (40 C); flash pt. (COC) 168 C; pour pt. -45 C; ref. index 1.4875.

Cyclolube® 2290. [Witco/Golden Bear] Low aromatic naphthenic oil; process oil offering exc. color and heat stability for rubber compounding, esp. EPDM and butyl; also for SBR, isoprene, and natural rubber; m.w. 453; sp.gr. 0.9340 (15.6 C); dens. 7.8 lb/gal; visc. 515 cSt (40 C); flash pt. (COC) 238 C; pour pt. -6 C; ref. index 1.5077.

Cyclolube® 2310. [Witco/Golden Bear] Low aromatic naphthenic oil; process oil offering exc. color and heat stability for rubber compounding, esp. EPDM and butyl; also for SBR, isoprene, and natural rubber; m.w. 414; sp.gr. 0.9248 (15.6 C); dens. 7.7 lb/gal; visc. 243 cSt (40 C); flash pt. (COC) 226 C; pour pt. -9 C; ref. index 1.5059.

Cyclolube® 4053. [Witco/Golden Bear] Low aromatic naphthenic oil; process oil offering exc. color and heat stability for rubber compounding, esp. EPDM and butyl; also for SBR, isoprene, and natural rubber; m.w. 400; sp.gr. 0.9042 (15.6 C); dens. 7.5 lb/gal; visc. 97 cSt (40 C); flash pt. (COC) 200 C; pour pt. -30 C; ref. index 1.4902.

Cyclolube® NN-1. [Witco/Golden Bear] Naphthenic oil distillate; process oils offering lt. color, low pour pts. for general-purpose compounding of elastomers, e.g. neoprene, SBR, isoprene, EDPM, butyl, and natural rubber; m.w. 266; sp.gr. 0.9218 (15.6 C); dens. 7.7 lb/gal; visc. 19 cSt (40 C); flash pt. (COC) 158 C; pour pt. -42 C; ref. index 1.5063.

Cyclolube® NN-2. [Witco/Golden Bear] Naphthenic oil distillate; process oils offering lt. color, low pour pts. for general-purpose compounding of elastomers, e.g. neoprene, SBR, isoprene, EDPM, butyl, and natural rubber; m.w. 295; sp.gr. 0.9279 (15.6 C); dens. 7.7 lb/gal; visc. 28 cSt (40 C); flash pt. (COC) 166 C; pour pt. -36 C; ref. index 1.5114.

Cyclonox® BT-50. [Akzo Nobel] Cyclohexanone peroxide; CAS 12262-58-7; EINECS 235-527-7; initiator for ambient-temp. polyester cures; for automotive body putty, hobby and automotive kits; paste; 50% ketone peroxide in plasticizer.

Cycom® NCG 1200 Unsized. [Cytec Industries] Continuous nickel-coated graphite fiber; fiber used in filament winding to mfg. structural parts for aerospace, automotive, and other applics.; provides lightening strike protection of composite parts; can be sized with binders for compat. with almost all plastic resins then chopped to provide ESD, EMI/RFI protection.

Cycom® NCG 1204 p8. [Cytec Industries] 80-85% Nickel-coated graphite fiber with proprietary binder; fiber for compounding operations; chopped to specified length (typically 1/4 in.).

Cylacell. [Cylatec] Proprietary mixt. of sodium and magnesium silicates with borates and water; single-component environmentally safe endothermic blowing agent for thermoset elastomers and polymers (EPDM, HNBR, SBR, silicone, neoprene, BMC); reacts completely at operating temps. above 300 F without need for activators, generating ≈ 425 cc blowing vapor per gram of material at 100 C and std. pressure; wh. free-flowing ultrafine powd., odorless; partly sol. in water; sp.gr. 0.8-1.0; m.p. 649 C; pH 10.4-11.3 (1% aq.); *Toxicology:* dust may irritate respiratory tract and eyes; may cause skin irritation; may cause irritation on ingestion; *Precaution:* nonflamm.; will not support combustion; incompat. with conc. acids; *Storage:* unlimited shelf life.

Cylink HPC-75. [Cytec Industries] Melamine-based polyol; crosslinking agent for isocyanate, phenolic, amino, and carboxylic resins; provides high hydroxyl functionality, flame retardancy, improved flexibility/moisture resist.; for water-based systems, phenolic resins/fiber glass binders, melamine-formaldehyde resins, binders for nonwovens, laminates, textiles; water-wh. soln.; sol. in water, methanol, ethanol, polar solvs.; m.w. 912 (of monomer); visc. 1900 cps; pH 4.3; 75% aq.; *Toxicology:* LD50 (oral, rat) > 5000 mg/kg, (dermal, rabbit) > 2000 mg/kg; minimal eye irritant.

Cylink HPC-90. [Cytec Industries] Melamine-based polyol; crosslinking agent for isocyanate, phenolic, amino, and carboxylic resins; provides high hydroxyl functionality, flame retardancy, improved flexibility/moisture resist.; visc. liq.; sol. in water, methanol, ethanol, polar solvs.; m.w. 912 (of monomer); visc. 187,000 cps; pH 4.0.

Cylink HPC-100. [Cytec Industries] Melamine-based polyol; crosslinking agent for isocyanate, phenolic, amino, and carboxylic resins; provides high hydroxyl functionality, flame retardancy, improved flexibility/moisture resist.; for PU rigid foams, urethane coating applics.; pale yel. dry powd.; sol. in methanol, ethanol, ethylene glycol, diethylene glycol, propylene glycol, 1,4-butanediol, acetate; m.w. 912 (of monomer); soften. pt. 85-100 C.

Cymel 303. [Cytec Industries] Hexamethoxymethylmelamine; CAS 3089-11-0; crosslinking agent used in melamine resin coating systems; used in general industrial finishes, coil-coating enamels, appliance finishes; also useful with alkyd, polyester, thermosetting acrylic, epoxy, and cellulose resins; Gardner 2 max. clear, visc. liq.; limited water-sol.; dens. 10.0 lb/gal; visc. (G-H) X-Z2; flash pt. > 180 F; 98% NV.

Cymel 370. [Cytec Industries] Methylated melamine resin; crosslinking agent for solv. and water-based coatings, e.g., general purpose spray coatings, coil coatings, metal decorating enamels, varnishes; Gardner 1 max. clear liq.; sol. in methanol, ethanol, isobutanol, ethylene and propylene glycols, MEK, MIBK, toluene, xylene; partly sol. in water; sp.gr. 1.18; dens. 9.8 lb/gal; visc. (G-H) Z2-Z4; flash pt. (CC) 72 F; 88 ± 2% solids in IPA/butanol.

Cymel 373. [Cytec Industries] Partially methylated melamine formaldehyde resin; CAS 9003-08-1; crosslinking agent for water-based coatings, e.g., emulsions or sol'ns.; Gardner 1 max. clear liq.; sol. in methanol, ethanol, IPA, ethylene and propylene glycols, MEK; sp.gr. 1.26; dens. 10.5 lb/gal; visc. (G-H) Z-Z4; 85 ± 2% solids in water.

Cymel 380. [Cytec Industries] Partially methylated melamine formaldehyde resin; CAS 9003-08-1; crosslinking agent for solv. and water-based coatings; Gardner 1 max. clear liq.; sol. in methanol, ethanol, isobutanol, ethylene and propylene glycols, MEK, MIBK, toluene, xylene; partly sol. in water; sp.gr. 1.15; dens. 9.6 lb/gal; visc. (G-H) V-Z; flash pt. (CC) 62 F; 80 ± 2% solids in IPA/butanol.

Cymel 1141. [Cytec Industries] Alkylated carboxyl-modified melamine formaldehyde resin; crosslinking agent for cationic resins; used for emulsions, water-sol. coatings, and cationic electroplating applics.; Gardner 3 max. clear, visc. liq.; water-disp.; sol. in common org. solvs.; sp.gr. 1.07; dens. 8.95 lb/gal; visc. (G-H) W-Y; acid no. 22 ± 3; flash pt. (Seta CC) 93 F; 85 ± 2% solids in isobutanol.

D

D-400. [Olin] PPG diol; lubricant in two- and four-cycle engines; used in the processing of rubber, to inhibit foam; used in brake fluids, cosmetics, oil and grease compds., pesticides, urethane foams, urethane coatings, adhesives, elastomers, and sealants/caulks; liq.; m.w. 400; f.p. -26 C; sp.gr. 1.009; dens. 8.40 lb/gal; visc. 70 cs; ref. index 1.445; flash pt. 204 C; Unverified

D-1000. [Olin] PPG diol; lubricant in two- and four-cycle engines; used in the processing of rubber, to inhibit foam; used in brake fluids, cosmetics, oil and grease compds., pesticides, urethane foams, urethane coatings, adhesives, elastomers, and sealants/caulks; liq.; m.w. 1000; f.p. -36 C; sp.gr. 1.005; dens. 8.37 lb/gal; visc. 150 cs; ref. index 1.448; flash pt. 230 C; Unverified

D-1200. [Olin] PPG diol; lubricant in two- and four-cycle engines; used in the processing of rubber, to inhibit foam; used in brake fluids, cosmetics, oil and grease compds., pesticides, urethane foams, urethane coatings, adhesives, elastomers, and sealants/caulks; liq.; m.w. 1200; f.p. -36 C; sp.gr. 1.006; dens. 8.37 lb/gal; visc. 190 cs; ref. index 1.448; flash pt. 232 C; Unverified

D-1300. [Olin] PPG diol; lubricant in two- and four-cycle engines; used in the processing of rubber, to inhibit foam; used in brake fluids, cosmetics, oil and grease compds., pesticides, urethane foams, urethane coatings, adhesives, elastomers, and sealants/caulks; liq.; m.w. 1300; f.p. -34 C; sp.gr. 1.005; dens. 8.37 lb/gal; visc. 225 cs; ref. index 1.449; flash pt. 235 C; Unverified

D-2000. [Olin] PPG diol; lubricant in two- and four-cycle engines; used in the processing of rubber, to inhibit foam; used in brake fluids, cosmetics, oil and grease compds., pesticides, urethane foams, urethane coatings, adhesives, elastomers, and sealants/caulks; liq.; m.w. 2000; f.p. -29 C; sp.gr. 1.005; dens. 8.37 lb/gal; visc. 325 cs; ref. index 1.449; flash pt. 238 C; Unverified

D-3000. [Olin] PPG diol; lubricant in two- and four-cycle engines; used in the processing of rubber, to inhibit foam; used in brake fluids, cosmetics, oil and grease compds., pesticides, urethane foams, urethane coatings, adhesives, elastomers, and sealants/caulks; liq.; m.w. 3000; f.p. -32 C; sp.gr. 1.004; dens. 8.36 lb/gal; visc. 580 cs; ref. index 1.449; flash pt. 238 C; Unverified

D-4000. [Olin] PPG diol; lubricant in two- and four-cycle engines; used in the processing of rubber, to inhibit foam; used in brake fluids, cosmetics, oil and grease compds., pesticides, urethane foams, urethane coatings, adhesives, elastomers, and sealants/caulks; liq.; m.w. 4000; f.p. -30 C; sp.gr. 1.003; dens. 8.36 lb/gal; visc. 895 cs; ref. index 1.450; flash pt. 343 C; Unverified

Dabco® 33-LV®. [Air Prods./Perf. Chems.] 33% Triethylenediamine in 67% dipropylene glycol; amine-based catalyst for PU high resiliency flexible and rigid slabstock and molded foam, elastomers, shoe soles, coating; liq.; sol. in water; sp.gr. 1.13 (24 C); visc. 100 cps (24 C); vapor pressure 2 mm Hg (38 C); hyd. no. 560; flash pt. (PMCC) > 110 C.

Dabco® 120. [Air Prods./Perf. Chems.] Tetravalent tin catalyst; metal-based catalyst for PU flexible molded foam, semiflexible, rigid, elastomer, shoe sole applics.; liq.; insol. in water; sp.gr. 0.995; visc. 19 cps; f.p. -20 C; flash pt. (PMCC) 121 C.

Dabco® 125. [Air Prods./Perf. Chems.] Tetravalent tin catalyst; metal-based catalyst for PU semiflexible, rigid, elastomer, shoe sole applics.; liq.; insol. in water; sp.gr. 1.15; visc. 282 cps; f.p. -25 C; flash pt. (PMCC) 123 C.

Dabco® 131. [Air Prods./Perf. Chems.] Tetravalent tin catalyst; metal-based catalyst for PU rigid applics.; liq.; insol. in water; sp.gr. 1.11; visc. 33 cps; vapor pressure 0.04 mm Hg (149 C); f.p. -23 C; flash pt. (PMCC) 130 C.

Dabco® 1027. [Air Prods./Perf. Chems.] Amine-based catalyst providing delayed cream and/or faster cure, improved processing in microcellular polyurethane foams, elastomers, shoe soles; blend with Dabco EG; liq.; sol. in water; sp.gr. 1.09; visc. 72 cps; vapor pressure 0 mm Hg (21 C); f.p. < -78 C; hyd. no. 1140; flash pt. (Seta) 95 C; 37% in ethylene glycol.

Dabco® 1028. [Air Prods./Perf. Chems.] Amine-based catalyst providing delayed cream and/or faster cure, improved processing in microcellular polyurethane foams, elastomers, shoe soles; blend with Dabco S-25; liq.; sol. in water; sp.gr. 1.03; visc. 134 cps; vapor pressure 0 mm Hg (25 C); f.p. < -78 C; hyd. no. 895; flash pt. (Seta) 103.9 C; 30% in 1,4-butanediol.

Dabco® 2039. [Air Prods./Perf. Chems.] Amine; low odor catalyst for polyester PU flexible slabstock foam; suitable for textile and industrial grades; amber liq., ammonia-like odor; sol. in water; sp.gr. 1.006 (21 C); b.p. 198 C; flash pt. (PMCC) 95 C.

Dabco® 7928. [Air Prods./Perf. Chems.] Amine-based catalyst for flexible slabstock PU foam; for early cream, initial visc. increase in continuous blockfoam; optimized for conventional equip.; liq.; sol. in water; sp.gr. 0.910; vapor pressure 17 mm Hg (25 C); f.p. -60 C; hyd. no. 586; flash pt. 43.3 C.

Dabco® 8136. [Air Prods./Perf. Chems.] Amine-based delayed-action catalyst for flexible molded and semiflexible PU applics.; liq.; sol. in water; sp.gr. 1.06; visc. 185 cps (23 C); vapor pressure 10 mm Hg (38 C); f.p. -52 C; hyd. no. 395; flash pt. (Seta) 46.1 C.

Dabco® 8154. [Air Prods./Perf. Chems.] Amine-based, delayed-action, acid-blocked catalyst for improved processing of flexible slabstock PU foam, flexible molded and rigid applics.; liq.; sol. in water; sp.gr. 1.058; visc. 186 cps (23 C); vapor pressure 5 mm Hg (38 C); f.p. -53 C; hyd. no. 444; flash pt. (Seta) > 110 C.

Dabco® 8264. [Air Prods./Perf. Chems.] Amine-based catalyst for improved processing of flexible slabstock polyether PU foam; provides wide formulating latitude; liq.; sol. in water; sp.gr. 0.9634; visc. 20.5 cps (23 C); vapor pressure 2 mm Hg (0 C); f.p. -57 C; hyd. no. 592; flash pt. (Seta) 54.4 C.

Dabco® B-16. [Air Prods./Perf. Chems.] N-Cetyl, N,N-dimethylamine; amine-based catalyst for flexible slabstock PU foam; controls crosslinking; liq.; insol. in water; sp.gr. 0.804; vapor pressure 17 mm Hg (21 C); flash pt. > 93 C.

Dabco® BDMA. [Air Prods./Perf. Chems.] Benzyl dimethylamine; CAS 103-83-3; EINECS 203-149-1; amine-based catalyst for rigid PU foam; liq.; sl. sol. in water; sp.gr. 0.900; visc. < 0.1 cps; vapor pressure 1.8 mm Hg (20 C); f.p. -75 C; flash pt. (CC) 53.6 C.

Dabco® BDO. [Air Prods./Perf. Chems.] 1,4-Butanediol; CAS 110-63-4; EINECS 203-786-5; crosslinker for PU microcellular and elastomeric applics.; curative, chain extender; provides reactive H-source in prepolymer prod.; used to provide hard segments in PU; liq.; sol. in water; sp.gr. 1.015; visc. 70-73 cps; vapor pressure < 1 mm Hg (37.8 C); hyd. no. 1245; flash pt. 121 C.

Dabco® BL-11. [Air Prods./Perf. Chems.] 70% Bis (dimethylaminoethyl) ether, 30% dipropylene glycol; amine-based catalyst for flexible slabstock PU foam; exc. blowing catalyst; liq.; sol. in water; sp.gr. 0.902 (20 C); visc. 4.1 cps (20 C); vapor pressure 0.58 mm Hg (20 C); hyd. no. 251; flash pt. (TCC) 70.6 C.

Dabco® BL-16. [Air Prods./Perf. Chems.] Bis (dimethylaminoethyl) ether; amine-based active blowing catalyst for flexible PU slabstock; liq.; sol. in water; sp.gr. 1.003 (23 C); visc. 16 cps; vapor pressure 6 mm Hg (38 C); flash pt. (Seta) 102 C.

Dabco® BL-17. [Air Prods./Perf. Chems.] Delayed action, acid-blocked version of Dabco BL-11; amine-based catalyst for flexible PU slabstock; liq.; sol. in water; sp.gr. 1.04 (20 C); vapor pressure 0.61 mm Hg (20 C); hyd. no. 476; flash pt. (TCC) 65 C.

Dabco® BL-22. [Air Prods./Perf. Chems.] Blowing co-catalyst for TDI-based molded PU, rigid applics.; designed for use as co-catalyst with Dabco 33-LV, etc.; liq.; misc. with water; dens. 0.8304 (21 C); visc. 255 cps; vapor pressure 5 mm Hg (21 C); f.p. < -78 C; b.p. 204 C; hyd. no. 0; flash pt. (CC) 70 C; pH 11.3 (3% aq.).

Dabco® BLV. [Air Prods./Perf. Chems.] Optimized 3:1 blend of Dabco 33-LV and Dabco BL-11; amine-based catalyst for flexible slabstock PU foam; exc. blowing catalyst; liq.; sol. in water; sp.gr. 1.0; visc. 60 cps (23 C); vapor pressure < 2 mm Hg (20 C); hyd. no. 260; flash pt. (Seta) 79 C.

Dabco® CL-485. [Air Prods./Perf. Chems.] Tetrafunctional hydroxyl crosslinker with amine functionality; crosslinker for PU elastomers, shoe soles; liq.; sol. in water; sp.gr. 1.03 (20 C); visc. 52,000 cps; vapor pressure > 1 mm Hg; hyd. no. 770; flash pt. (CC) > 200 C.

Dabco® Crystalline. [Air Prods./Perf. Chems.] Triethylenediamine; CAS 280-57-9; amine-based catalyst for PU coatings, rigid applics., elastomeric applics., shoe soles; solid; sol. 127 g/100 g water (45 C); sp.gr. 1.14 (28 C); vapor pressure 2.9 mm Hg (50 C); f.p. 159.8 C; flash pt. (PMCC) > 38 C.

Dabco® CS90. [Air Prods./Perf. Chems.] Trichloroethane; CAS 71-55-6; catalyst for chlorinated solv.-blown PU slabstock foams; liq.; sol. in water; sp.gr. 0.958; f.p. -75 C; hyd. no. 358; flash pt. 77.2 C.

Dabco® DC-2®. [Air Prods./Perf. Chems.] Proprietary blend of delayed-activity tin and a tert. amine; catalyst for PU coatings, elastomers, shoe soles; liq.; partly sol. in water; sp.gr. 1.23; visc. 270 cps (38 C); hyd. no. 541; flash pt. (PMCC) 65.7 C; 45% act.

Dabco® DC193. [Air Prods./Perf. Chems.] Silicone glycol copolymer; surfactant used for prod. of rigid polyurethane foam, elastomers; suitable for board, laminates, foam-in-place, spray foam, and shoe soles; nonhydrolyzable; Gardner 2 clear visc. liq.; sol. in water; sp.gr. 1.07; visc. 425 cps; f.p. 15 C; hyd. no. 75; pour pt 11 C; flash pt. (CC) 92 C; ref. index 1.45.

Dabco® DC197. [Air Prods./Perf. Chems.] Silicone glycol copolymer; surfactant for prod. of high-density rigid molded polyurethane foams; exc. emulsifying ability, good dimensional stability; premix stable; low f.p.; nonhydrolyzable; Gardner 4 clear to sl. hazy liq.; sol. in water; sp.gr. 1.051; visc. 340 cps; f.p. -14 C; hyd. no. 17; flash pt. 88 C.

Dabco® DC198. [Air Prods./Perf. Chems.] Silicone glycol copolymer; high-potency surfactant used in prod. of flexible slabstock polyurethane foam, rigid, elastomeric, shoe sole applics.; Gardner 3 clear to hazy liq.; sol. in water, polyol, fluorocarbon, water-amine streams; sp.gr. 1.039; visc. 1600 cps; vapor pressure < 5 mm Hg; flash pt. (CC) 84 C; pH 5.3 (30% aq.).

Dabco® DC1315. [Air Prods./Perf. Chems.] Silicone glycol copolymer; wide processing-latitude surfactant for use in flame-retarded foam systems such as

PU slabstock and hot molded foam; Gardner 2-4 clear to sl. hazy liq.; sol. in polyol, fluorocarbon, water-amine streams; sp.gr. 1.03; visc. 1000 cs; flash pt. (CC) 60.6 C.

Dabco® DC1536. [Air Prods./Perf. Chems.] Silicone surfactant; surfactant for reduced shrinkage in PU elastomers, shoe soles; liq.; insol. in water; sp.gr. 0.965; visc. 29.9 cps; vapor pressure 4 mm Hg (21 C); f.p. -68 C; flash pt. (CC) 177 C.

Dabco® DC1537. [Air Prods./Perf. Chems.] Silicone surfactant; surfactant for reduced shrinkage in PU elastomers, shoe soles; liq.; insol. in water; sp.gr. 0.915; visc. 5 cps; vapor pressure 6 mm Hg (21 C); f.p. < 77 C; flash pt. (CC) 154 C.

Dabco® DC1538. [Air Prods./Perf. Chems.] Silicone surfactant; high-efficiency all-purpose surfactant for PU flexible slabstock; liq.; sol. in water; sp.gr. 1.036; visc. 1300 cps; vapor pressure < 5 mm Hg; hyd. no. 167; flash pt. > 110 C.

Dabco® DC1630. [Air Prods./Perf. Chems.] Specially formulated dimethyl polysiloxane; surfactant for use in high-resiliency, flexible polyurethane foam based on high m.w. polyols and polymeric isocyanates; clear liq.; sp.gr. 0.919; visc. 5.0 cs; flash pt. (CC) 138 C; ref. index 1.3966.

Dabco® DC5043. [Air Prods./Perf. Chems.] Silicone glycol copolymer; med.-efficiency stabilizer/surfactant for prod. of flexible and semiflexible molded and slabstock high-resiliency polyurethane foam; nonhydrolyzable; broad processing latitude; clear to lt. straw liq.; insol. in water; sp.gr. 1.002 (21 C); visc. 300 cps; vapor pressure 17.7 mm Hg (21 C); f.p. -53 C; flash pt. (CC) 60 C.

Dabco® DC5098. [Air Prods./Perf. Chems.] Silicone glycol copolymer; nonhydrolyzable surfactant for prod. of polyisocyanurate and conventional rigid PU foam; lt. straw clear to sl. hazy liq.; sol. in water; sp.gr. 1.08; visc. 250 cps; f.p. 12 C; hyd. no. < 10; flash pt. (CC) 61 C.

Dabco® DC5103. [Air Prods./Perf. Chems.] Silicone glycol copolymer; nonhydrolyzable surfactant for prod. of all types of rigid PU foam; low freeze pt.; premix-stable; clear to lt. straw liq.; sol. in water; sp.gr. 1.05; visc. 250 cps; f.p. -7 C; hyd. no. 104; pour pt. -5 C; flash pt. (CC) 61 C.

Dabco® DC5125. [Air Prods./Perf. Chems.] Silicone glycol copolymer; wide-processing-latitude surfactant for prod. of conventional and flame-retarded flexible polyurethane slabstock foam and hot molded foam systems; Gardner 2-4 clear to sl. hazy liq.; sol. in water, polyol, fluorocarbon, and water-amine streams; sp.gr. 1.03; visc. 1000 cps; vapor pressure < 5 mm Hg; flash pt. (CC) 60.6 C.

Dabco® DC5160. [Air Prods./Perf. Chems.] Silicone glycol copolymer; med.-potency, wide processing-latitude mutli-purpose surfactant used in prod. of conventional and flame-retarded hot molded and flexible slabstock polyurethane foam; liq.; sol. in water, polyol, fluorocarbon, water-amine streams; sp.gr. 1.03; visc. 950 cps; vapor pressure < 5 mm Hg; flash pt. (CC) 61 C.

Dabco® DC5164. [Air Prods./Perf. Chems.] Silicone surfactant; high-efficiency stabilizer for prod. of difficult-to-stabilize high-resiliency, low-density flexible molded PU foam formulations; clear to sl. hazy liq.; insol. in water; sp.gr. 1.005; visc. 305 cps; vapor pressure 20.8 mm Hg (21 C); f.p. -53 C; hyd. no. 20; flash pt. 92 C.

Dabco® DC5169. [Air Prods./Perf. Chems.] Silicone surfactant; surfactant for prod. of high-resiliency MDI-based PU flexible molded foams; for prevention of basal cell formation; clear to sl. hazy liq.; insol. in water; sp.gr. 0.968; visc. 29.2 cps; vapor pressure 8.6 mm Hg (21 C); f.p. -54 C; hyd. no. 0; flash pt. > 113 C.

Dabco® DC5180. [Air Prods./Perf. Chems.] Surfactant for flexible slabstock PU foam.

Dabco® DC5244. [Air Prods./Perf. Chems.] Silicone prod.; cell-opening surfactant for prod. of all types of high-resiliency foams; clear liq.; sp.gr. 1.05; visc. 800-1200 cs; flash pt. (CC) 87 C.

Dabco® DC5258. [Air Prods./Perf. Chems.] Silicone surfactant; surfactant which improves cell opening in high-density MDI or MDI/TDI foam; provides for max. airflow without sacrificing surfactant stability; liq.; insol. in water; sp.gr. 1.020; visc. 289 cps; vapor pressure 8.9 mm Hg (21 C); f.p. -55 C; hyd. no. 46; flash pt. 92 C.

Dabco® DC5270. [Air Prods./Perf. Chems.] Silicone prod.; surfactant for low-density, flexible slabstock PU foams.

Dabco® DC5357. [Air Prods./Perf. Chems.] Silicone surfactant; surfactant designed for water co-blown rigid PU systems; liq.; sol. hazy in water; sp.gr. 1.079; visc. 300 cps; f.p. 3 C; hyd. no. 54; flash pt. > 61 C.

Dabco® DC5365. [Air Prods./Perf. Chems.] Silicone surfactant; low-to-med. efficiency surface stabilizer for TDI/MDI flexible mold foams; liq.; insol. in water; sp.gr. 1.000; visc. 290 cps; vapor pressure 7.7 mm Hg (21 C); f.p. -60 C; hyd. no. 170; flash pt. > 113 C.

Dabco® DC5367. [Air Prods./Perf. Chems.] Silicone surfactant; surfactant designed for water co-blown rigid PU systems; liq.; sol. hazy in water; sp.gr. 1.033; visc. 300 cps; f.p. 0 C; hyd. no. 31; flash pt. > 61 C.

Dabco® DC5374. [Air Prods./Perf. Chems.] Silicone surfactant; surfactant designed for HCFC-123 or 141b-blown rigid PU systems; liq.; insol. in water; sp.gr. 1.050; visc. 300 cps; f.p. -4 C; hyd. no. 73; flash pt. > 61 C.

Dabco® DC5418. [Air Prods./Perf. Chems.] Silicone glycol copolymer; surfactant for MDI-based molded polyurethane foam systems, providing efficient cell opening and bulk stabilization in all-water-blown foams; nonhydrolyzable; clear to straw low-visc. liq.; insol. in water; sp.gr. 0.9658 (21 C); visc. 28.9 cps; vapor pressure 7.6 mm Hg (21 C); f.p. -65 C; b.p. > 149 C; hyd. no. 226; flash pt. (CC) 113 C; *Toxicology:* eye irritant; may cause skin irritation on repeated/prolonged contact, respiratory irritation on inhalation; *Precaution:* noncombustible; *Storage:* 12 mos shelf life stored below 32 C.

Dabco® DC5425. [Air Prods./Perf. Chems.] Silicone

surfactant; med.-efficiency stabilizer for water-blown TDI-based flexible molded PU foams; liq.; insol. in water; sp.gr. 0.980; visc. 57.1 cps; vapor pressure 6.0 mm Hg (21 C); f.p. -65 C; hyd. no. 14; flash pt. 78 C.

Dabco® DC5450. [Air Prods./Perf. Chems.] Silicone surfactant; strong cell opening surfactant for water-blown MDI-based flexible molded PU foams; provides enhanced bulk stability; liq.; insol. in water; sp.gr. 1.019; visc. 302 cps; vapor pressure 9.0 mm Hg (21 C); f.p. -35 C; hyd. no. 85; flash pt. > 113 C.

Dabco® DC5454. [Air Prods./Perf. Chems.] Silicone surfactant; surfactant designed for HCFC-123 or 141b-blown rigid PU systems; liq.; insol. in water; sp.gr. 1.060; visc. 250 cps; f.p. -7 C; hyd. no. 81; flash pt. > 61 C.

Dabco® DC5526. [Air Prods./Perf. Chems.] Silicone surfactant; surfactant designed for polyester slabstock PU foams; provides exc. foam stability, fine regular cell structured foam; liq.; sp.gr. 0.998 (20 C); visc. 175 cps; flash pt. (Seta CC) 104 C; 0.5% water.

Dabco® DC5885. [Air Prods./Perf. Chems.] Silicone surfactant; med.-efficiency stabilizer for water-blown TDI-based PU flexible molded foams; provides balanced cell opening capability; liq.; insol. in water; sp.gr. 0.984 (21 C); visc. 68.5 cps; vapor pressure 5.0 mm Hg (21 C); f.p. -63 C; hyd. no. 123; flash pt. 74 C.

Dabco® DC5890. [Air Prods./Perf. Chems.] Silicone surfactant; melamine-filled surfactant for flexible slabstock PU foam processing; liq.; insol. in water; sp.gr. 1.0; visc. 534 cps; vapor pressure < 5 mm Hg; f.p. -32 C; hyd. no. 17; flash pt. 78 C.

Dabco® DC5895. [Air Prods./Perf. Chems.] Silicone surfactant; melamine-filled surfactant for flexible slabstock PU foam processing; liq.; insol. in water; sp.gr. 1.0; visc. 46.8 cps; vapor pressure < 5 mm Hg; f.p. -69 C; flash pt. 156 C.

Dabco® DEOA-LF. [Air Prods./Perf. Chems.] Diethanolamine plus 15% water to reduce the f.p.; CAS 111-42-2; EINECS 203-868-0; crosslinker for flexible slabstock PU foam processing and flexible molds; liq.; sol. in water; sp.gr. 1.094; vapor pressure 3.5 mm Hg (20 C); f.p. 28 C; hyd. no. 1374.

Dabco® DM9534. [Air Prods./Perf. Chems.] Metal-based catalyst for flexible slabstock PU foam.

Dabco® DM9793. [Air Prods./Perf. Chems.] Metal-based catalyst for flexible slabstock PU foam.

Dabco® DMEA. [Air Prods./Perf. Chems.] Dimethylethanolamine; CAS 108-01-0; EINECS 203-542-8; amine-based catalyst for flexible slabstock PU foam and rigid applics.; liq.; sol. in water; sp.gr. 0.89 (20 C); vapor pressure 4 mm Hg (20 C); f.p. -58.6 C; hyd. no. 629.5; flash pt. (OC) 40 C.

Dabco® EG. [Air Prods./Perf. Chems.] 33% Triethylenediamine, 67% ethylene glycol; amine-based catalyst for PU elastomers, shoe soles; liq.; sol. in water; sp.gr. 1.09; visc. 65 cps (23 C); hyd. no. 1207; flash pt. (Seta) 105 C.

Dabco® FF-2002. [Air Prods./Perf. Chems.] Amine-based delayed-action catalyst; for MDI-based flexible molded foam; liq.; sol. in water; sp.gr. 1.084; visc. 150 cps; vapor pressure 2.0 mm Hg (21 C); f.p. -66 C; hyd. no. 597; flash pt. 100 C.

Dabco® FF-2003. [Air Prods./Perf. Chems.] Amine-based delayed-action catalyst; catalyst for high water MDI-based flexible molded foam; liq.; sol. in water; sp.gr. 1.089; visc. 167 cps; vapor pressure 4.0 mm Hg (21 C); f.p. -40 C; hyd. no. 582; flash pt. 103 C.

Dabco® H-1010. [Air Prods./Perf. Chems.] Amine-based delayed-action catalyst; low-corrosivity, low-odor catalyst for flexible molded PU; liq.; sol. in water; sp.gr. 1.158; visc. 9.7 cps; vapor pressure 16.5 mm Hg (21 C); f.p. -9 C; hyd. no. 3117; flash pt. > 105 C.

Dabco® K-15. [Air Prods./Perf. Chems.] Potassium 2-ethylhexanoate and diethylene glycol; metal-based catalyst for rigid PU applics.; liq.; sol. in water; sp.gr. 1.1 (15.6 C); visc. 7200 cps (27 C); vapor pressure < 0.1 mm Hg (23 C); hyd. no. 286; flash pt. (CC) 137.8 C.

Dabco® MC. [Air Prods./Perf. Chems.] Amine-based; acid-blocked blowing catalyst for methylene-chloride blown flexible slabstock PU foam; improves processing; liq.; sol. in water; sp.gr. 1.023; visc. 63.5 cps; vapor pressure 1.0 mm Hg (0 C); f.p. -66.7 C; hyd. no. 757; flash pt. 84 C.

Dabco® NCM. [Air Prods./Perf. Chems.] N-Cocomorpholine; amine-based catalyst for flexible polyester PU slabstock foam; improves die-cuttability; liq.; insol. in water; sp.gr. 0.872; vapor pressure 7 mm Hg (21 C); flash pt. > 93 C.

Dabco® NEM. [Air Prods./Perf. Chems.] N-Ethylmorpholine; CAS 100-74-3; EINECS 202-885-0; amine-based catalyst for flexible slabstock PU foam; liq.; sol. in water; sp.gr. 0.914; visc. 2.5 cps (23 C); vapor pressure 5 mm Hg (20 C); f.p. -60 C; flash pt. (TCC) 32 C.

Dabco® NMM. [Air Prods./Perf. Chems.] N-Methylmorpholine; CAS 109-02-4; EINECS 203-640-0; amine-based catalyst for flexible polyester PU slabstock foam; liq.; sol. in water; sp.gr. 0.92 (23 C); visc. 2.5 cps (23 C); vapor pressure 16.6 mm Hg (20 C); f.p. -57 C; flash pt. (CC) 23 C.

Dabco® R-8020®. [Air Prods./Perf. Chems.] Blend of Dabco DMEA and Dabco Crystalline; amine-based catalyst for rigid and flexible PU applics.; liq.; sol. in water; sp.gr. 0.916 (20 C); visc. 2.7 cps; vapor pressure 4.65 mm Hg (21 C); hyd. no. 510; flash pt. (TOC) 51 C.

Dabco® RC. [Air Prods./Perf. Chems.] 2:1 Blend of Dabco 33-LV and Dabco BL-11; amine-based catalyst for flexible slabstock PU applics.; liq.; sol. in water; sp.gr. 0.987; vapor pressure 5.7 mm Hg (21 C); f.p. -70 C; hyd. no. 437; flash pt. 77 C.

Dabco® S-25. [Air Prods./Perf. Chems.] 25% Triethylenediamine, 75% 1,4-butanediol; amine-based catalyst for PU elastomers, shoe soles; liq.; sol. in water; sp.gr. 1.024; visc. 132 cps (23 C); hyd. no. 934; flash pt. (Seta) 108 C.

Dabco® T. [Air Prods./Perf. Chems.] Reactive tert. amine blowing catalyst for semiflexible, flexible slabstock, and rigid PU applics.; liq.; sol. in water; sp.gr. 0.91 (20 C); visc. 5 cps (21 C); vapor pressure 81.4 mm Hg (38 C); hyd. no. 384; flash pt. (TCC) 99 C.

Dabco® T-1. [Air Prods./Perf. Chems.] Dibutyltin diacetate; CAS 1067-33-0; catalyst for PU coatings.

Dabco® T-5. [Air Prods./Perf. Chems.] Dibutyltin disulfide; catalyst for PU coatings; temp.-activated.

Dabco® T-9. [Air Prods./Perf. Chems.] Stannous octoate; CAS 301-10-0; metal-based catalyst for PU flexible slabstock and molded foam; liq.; insol. in water; sp.gr. 1.25; visc. 312 cps; flash pt. (PM) 142 C.

Dabco® T-10. [Air Prods./Perf. Chems.] 50% Stannous octoate in dioctylphthalate; metal-based catalyst for flexible slabstock PU foam; liq.; insol. in water; sp.gr. 1.25; visc. 82 cps; flash pt. (TCC) > 126.7 C.

Dabco® T-11. [Air Prods./Perf. Chems.] 50% Stannous octoate in a special grade of mineral oil; metal-based catalyst for flexible slabstock PU foam; liq.; insol. in water; sp.gr. 1.02; visc. 102 cps; flash pt. (CC) 93.3 C.

Dabco® T-12. [Air Prods./Perf. Chems.] Dibutyltin dilaurate; CAS 77-58-7; EINECS 201-039-8; metal-based catalyst for PU flexible molded foam, semiflexible, flexible slabstock, rigid, elastomer, shoe sole, and coatings; liq. above 65 F; insol. in water; sp.gr. 1.05; visc. 43 cps; vapor pressure 0.15 mm Hg (160 C); f.p. 18 C; flash pt. (COC) 235 C.

Dabco® T-16. [Air Prods./Perf. Chems.] 50% Stannous octoate in diisononyl phthalate; metal-based catalyst for PU flexible slabstock; liq.; insol. in water; sp.gr. 1.023; vapor pressure 20 mm Hg; flash pt. > 127 C.

Dabco® T-45. [Air Prods./Perf. Chems.] Potassium octoate in polyol; metal-based catalyst for PU rigid applics.; liq.; sol. in water; sp.gr. 1.09; visc. 20,000 cps; f.p. -5 C; hyd. no. 86; flash pt. (TCC) 102 C.

Dabco® T-95. [Air Prods./Perf. Chems.] 33% Stannous octoate in dioctyl phthalate; metal-based catalyst for flexible slabstock PU foam; liq.; sl. sol. in water; sp.gr. 1.10; visc. 101 cps; flash pt. (CC) > 130 C.

Dabco® T-96. [Air Prods./Perf. Chems.] 33% Stannous octoate in diisononylphthalate; metal-based catalyst for flexible slabstock PU foam; liq.; insol. in water; sp.gr. 1.023; vapor pressure 20 mm Hg; flash pt. > 93 C.

Dabco® T-120. [Air Prods./Perf. Chems.] Butyltin mercaptide; catalyst for PU coatings.

Dabco® T-125. [Air Prods./Perf. Chems.] Dibutyltin diester; catalyst for PU coatings.

Dabco® T-131. [Air Prods./Perf. Chems.] Dibutyltin (bis) mercaptide; catalyst for PU coatings; ambient temp. cure; provides strong exotherm.

Dabco® TETN. [Air Prods./Perf. Chems.] Highly volatile, cost-effective amine catalyst providing balanced blow/gel for flexible molded and rigid PU applics.; liq.; sol. 1.5 g/100 cc water (20 C); sp.gr. 0.7275; vapor pressure 53.5 mm Hg (20 C); flash pt. (TOC) -4 C.

Dabco® TL. [Air Prods./Perf. Chems.] Amine-based catalyst for improved processing of flexible slabstock polyether PU foam; for methylene chloride systems; liq.; sol. in water; sp.gr. 0.988; visc. 40 cps (24 C); vapor pressure 20.1 mm Hg (38 C); f.p. < -20 C; hyd. no. 490; flash pt. (Seta) 89 C.

Dabco® TMR®. [Air Prods./Perf. Chems.] Amine-based trimerization catalyst for rigid PU applics.; liq.; sol. in water; sp.gr. 1.050; hyd. no. 463; flash pt. (TCC) 121 C.

Dabco® TMR-2®. [Air Prods./Perf. Chems.] Amine-based trimerization catalyst for rigid PU applics.; back-end cure; liq.; sol. in water; sp.gr. 1.055; hyd. no. 463; flash pt. (TCC) 121 C.

Dabco® TMR-3. [Air Prods./Perf. Chems.] Amine-based trimerization catalyst for rigid bunstock applics.; liq.; sol. in water; sp.gr. 1.066; visc. 60 cps (23 C); vapor pressure 8 mm Hg; f.p. -56.7 C; hyd. no. 563; flash pt. (Seta) > 110 C.

Dabco® TMR-4. [Air Prods./Perf. Chems.] Amine-based trimerization catalyst for pour-in-place rigid isocyanurate systems; liq.; sol. in water; sp.gr. 1.06 (20 C); vapor pressure 5 mm Hg (38 C); hyd. no. 476; flash pt. 100.6 C.

Dabco® TMR-30. [Air Prods./Perf. Chems.] Amine-based polyisocyanurate lamination co-catalyst (rigid applics.); liq.; sol. in water; sp.gr. 0.97; hyd. no. 213; flash pt. (Cleveland) 157 C.

Dabco® X-542. [Air Prods./Perf. Chems.] Amine-based catalyst for PU flexible molded foam; provides fast cure.

Dabco® X-543. [Air Prods./Perf. Chems.] Amine-based catalyst for PU flexible molded foam; provides delayed action, fast cure, reduces demold time; liq.; sol. in water; sp.gr. 1.048; vapor pressure 8 mm Hg (38 C); hyd. no. 321; flash pt. (Seta) 88.3 C.

Dabco® X-8161. [Air Prods./Perf. Chems.] Blend of aliphatic and hydroxy-containing amines; catalyst for semiflexible PU applics.; liq.; sol. in water; sp.gr. < 1.0; vapor pressure < 1.0 mm Hg; hyd. no. 307; flash pt. > 93 C.

Dabco® X2-5357. [Air Prods./Perf. Chems.] Surfactant for reduced CFC and non-CFC blown rigid polyurethane foam systems; Gardner 3 color; sp.gr. 1.04; visc. 400 cps; flash pt. (CC) > 61 C; hyd. no. 57.

Dabco® X2-5367. [Air Prods./Perf. Chems.] Surfactant for reduced CFC and non-CFC blown rigid polyurethane foam systems; Gardner 4 color; sp.gr. 1.04; visc. 280 cps; flash pt. (CC) > 61 C; hyd. no. 31.

Dabco® XDM™. [Air Prods./Perf. Chems.] Amine-based co-catalyst for improved demolding props., surface cure; for flexible molded and slabstock PU applics.; liq.; sol. in water; sp.gr. 0.942; visc. 2.7 cps (24 C); vapor pressure 3 mm Hg (20 C); flash pt. (TOC) 76 C.

Dabco® XF-C10-40. [Air Prods./Perf. Chems.] Amine-based catalyst for PU flexible molded foam; provides fast cure.

Dabco® XF-F2002, XF-F2003. [Air Prods./Perf. Chems.] Amine catalyst; for all-water-blown MDI cold cure seating process in polyurethane prod.

Dama® 810. [Albemarle] Dioctyl/octyldecyl/didecyl methyl amines; cationic; intermediate for mfg. of quaternary ammonium compds. for biocides, textile chemicals, oil field chemicals, amine oxides, betaines, polyurethane foam catalysis, epoxy curing agent; in fabric softeners, disinfectants, laundry detergents; clear liq., fatty amine odor; sp.gr. 0.801; f.p. -38.6 C; amine no. 197; flash pt. (PM) 166 C; *Precaution:* corrosive.

Dama® 1010. [Albemarle] Didecyl methylamine; CAS 7396-58-9; EINECS 230-990-1; cationic; intermediate for mfg. of quaternary ammonium compds. for biocides, textile chemicals, oil field chemicals, amine oxides, betaines, polyurethane foam catalysis, epoxy curing agent; in fabric softeners, disinfectants, laundry detergents; clear liq., fatty amine odor; sp.gr. 0.807; f.p. -6.3 C; amine no. 177; flash pt. (PM) > 93 C; 100% conc.; *Precaution:* corrosive.

Dantocol® DHE. [Lonza] DEDM hydantoin; CAS 26850-24-8; crosslinker in acrylic, epoxy, PU resins, hot-melt adhesives; intermediate for esterification/ethoxylation prods. in emulsifiers, textile finishes, cosmetics; for high-temp. resist. polyester or imide-amide wire enamels, acrylic, PU spec. coatings; film-former and lubricant in permanent waves, shaving preps., spec. bar soap formulations in cosmetics; Gardner 1 clear liq. to solid; m.p. 63 C; hyd. no. 485-515; pH 6.5 (5%); 0.5% max. moisture.

Dantoest® DHE DL. [Lonza] DEDM hydantoin dilaurate; fiber lubricant; intermediate for epoxies, urethane resins, and antistatic lubricants for the textiles industry; Gardner 3 liq.; acid no. 2; hyd. no. 10.

Dantosperse® DHE (15) MO. [Lonza] PEG-15 DEDM hydantoin oleate; fiber lubricant; intermediate for epoxies, urethane resins, and antistatic lubricants for the textiles industry; Gardner 1-2 liq.; acid no. 2, 2, 2, and 1 resp.; hyd. no. 60, 45, 40, 31 resp.

Dapro 5005. [Daniel Prods.] Special drier, chelating catalyst replacing cobalt drier in modified polyurethanes and alkyds.

Darathane® WB-4000. [W.R. Grace/Organics] Waterborne urethane prepolymer; cationic; dispersant; emulsifier; latex polymerization; yields cationic, cross linkable, emulsifier-free polymers on curing; waste water treatment; liq.; 35% act.

Darex® 632L. [W.R. Grace/Organics] S/B latex; CAS 9003-55-8; high styrene latex for blending with natural and syn. latices; increases stiffness, hardness, and tear resistance; used in rug backing cmpds. and other textile goods; reinforces dipped goods, foam, rubber prod.; saturates felt for brittle-free box toes; particle size 0.15 μ; dens. 8.6 lb/gal; visc. 60 cps; surf. tens. 50 dynes/cm; pH 11.0; 52% solids; 90% styrene, 10% butadiene.

Darex® 636L. [W.R. Grace/Organics] High styrene S/B latex; CAS 9003-55-8; nonfilm forming latex used as box toe fabric saturant; for blending with natural or syn. latexes to give increased stiffness, hardness, tear resist.; particle size 0.15 μ; dens. 8.6 lb/gal; visc. 30-110 cps; surf. tens. 35-45 dynes/cm; pH 10.5-12.0; 50-52% solids; 97% styrene, 3% butadiene.

Darex® 670L. [W.R. Grace/Organics] High styrene latex; CAS 9003-53-6; nonfilm forming latex used as a saturant; antioxidants not required; can be blended with natural or syn. latexes to give increased stiffness, hardness, tear resist.; particle size 0.5 μ; dens. 8.6 lb/gal; visc. 80 cps; surf. tens. 40 dynes/cm; pH 11.5; 50.5% total solids.

Dark Green No. 2™. [Witco/Petroleum Spec.] Petrolatum tech.; binder, carrier, lubricant, moisture barrier, plasticizer, protective agent, softener for tech. applics. (elec., paper and pkg., plastics, rubber, textiles, industrial use); dk. brn.; m.p. 60-68 C.

Darvan® L. [R.T. Vanderbilt] Ammonium salts of alkyl phosphate; dispersant; mold lubricant for elastomers; corrosion inhibitor; cream to amber paste; dens. 1.04 ± 0.03 mg/m^3; pH 7.0-8.0 (5%).

Darvan® ME. [R.T. Vanderbilt] Sodium alkyl sulfates; anionic; latex stabilizer; NR, SR mold and stock lubricant; foam modifier, wetting agent, and emulsifier for latexes; wh. to cream powd.; 99.9% min. thru 10 mesh; dens. 1.19 ± 0.03 mg/m^3; pH 9.0-11.0 (10%).

Darvan® No. 1. [R.T. Vanderbilt] Sodium polynaphthalene sulfonate; latex dispersant; water-sol.; pH 8.0-10.5 (1%); 87% min. act.

Darvan® No. 2. [R.T. Vanderbilt] Sodium lignosulfonate; CAS 8061-51-6; dispersing and emulsifying agent for rubber industry, esp. for zinc oxide, clays, and sulfur; dk. brn. powd.; water-sol.; dens. 1.25 ± 0.03 mg/m^3; pH 7.0-8.5 (1%); 84.5% min. act.

Darvan® No. 6. [R.T. Vanderbilt] Polymerized alkyl naphthalene sulfonic acid, sodium salt; dispersant for NR and SR latexes; water-sol.; pH 7.8-10.4 (1%); 85% min. act.

Darvan® No. 7. [R.T. Vanderbilt] Sodium polymethacrylate; dispersant for NR, SR latexes; surf. act. agent for preparing high solids disps.; water-wh. clear to slightly opalescent liq.; dens. 1.16 ± 0.02 mg/m^3; visc. 75 cps max.; pH 9.5-10.5; 25.0 ± 1% total solids.

Darvan® No. 31. [R.T. Vanderbilt] Carboxylated polyelectrolyte, sodium salt; pigment dispersant for aq. systems; latex coatings; yel. liq.; sol. in water systems; dens. 1.11 ± 0.02 mg/m^3; visc. 75 cps max.; pH 9-11; 25% min. total solids.

Darvan® NS. [R.T. Vanderbilt] c-Cetyl betaine and c-decyl betaine aq. sol'n.; nonionic; stabilizer for low alkaline latex systems; latex foam modifier and wetting agent; clear to lt. amber liq.; dens. 1.05 ± 0.02 mg/m^3; pH 8.5-11.5 (10%); 32.0-36.0% total solids.

Darvan® SMO. [R.T. Vanderbilt] Sodium salt of sulfated methyl oleate; dispersant; with Darvan WAQ improves smoothness and gloss of dipped CR latex films; clear to lt. amber liq.; dens. 1.08 ± 0.02 mg/m^3; pH 6.0-7.5 (10%); 30-35% total solids.

Darvan® WAQ. [R.T. Vanderbilt] Sodium alkyl sulfates; anionic; latex stabilizer for NR and syn. latexes; wetting agent and emulsifier for latexes; mold and internal lubricant; clear water-wh. paste (> 30 C); dens. 1.04 ± 0.03 mg/m³; pH 7.0-9.0 (10%).

Daxad® 11G. [W.R. Grace/Organics] Low m.w. naphthalene sulfonate formaldehyde condensate, sodium salt; dispersant for pigments in aq. media; used in agric. chemicals, mastics, caulks, sealants, pigment slurries and disps.; also for inks, syn. polymers, paper coating disps., paper mill slime control, pitch control, pulp digestion, and tall oil separation; buff fine gran.; dens. 42 lb/ft³; pH 8.0-10.5 (1%); surf. tens. 70-71 dynes/cm (1%); 87% min. act.

Daxad® 14B. [W.R. Grace/Organics] Low m.w. naphthalene sulfonate formaldehyde condensate, sodium salt sol'n.; dispersant for emulsion polymerization, dyestuffs, tanning, herbicides, pesticides, and pitch; dk. brn. liq.; 45% total solids.

Daxad® 14C. [W.R. Grace/Organics] Low m.w. naphthalene sulfonate formaldehyde condensate, sodium and potassium salts in sol'n.; dispersant for emulsion polymerization, dyestuffs, tanning, herbicides, pesticides, and pitch; designed for cold weather stability; dk. brn. liq.; 45% total solids.

Daxad® 15. [W.R. Grace/Organics] Low m.w. naphthalene sulfonate formaldehyde condensate, sodium salt; industrial grade, general purpose dispersant for emulsion polymerization, dyestuffs, tanning, herbicides, pesticides, and pitch; amber powd.; dens. 35 lb/ft³; pH 9.5 (1%); surf. tens. 70 dynes/cm (1%); 85% total solids.

Daxad® 16. [W.R. Grace/Organics] Low m.w. naphthalene sulfonate formaldehyde condensate, sodium salt; dispersant for emulsion polymerization, concrete, dyestuffs, tanning, herbicides, pesticides, and pitch; dk. brn. liq.; dens. 10.4 lb/gal; pH 9.5 (1%); 47.5% total solids.

Daxad® 17. [W.R. Grace/Organics] Low m.w. naphthalene sulfonate formaldehyde condensate, sodium salt; industrial grade dispersant for emulsion polymerization, dyestuffs, tanning, herbicides, pesticides, and pitch; amber gran.; ≤ 3% passes thru 100 mesh; dens. 42 lb/ft³; pH 7.8-10.4 (1%); surf. tens. 70 dynes/cm (1%); 85% act.

Daxad® 30. [Hampshire] Sodium polymethacrylate sol'n.; CAS 25086-62-8; dispersant esp. for pigments in aq. sol'ns.; stabilizer in latex emulsions; used in paint formulations, emulsion polymerization, water treatment, agriculture, cosmetic base makeup, industrial cleaners, in large particle suspensions; water-wh. clear liq.; sol. in water systems; sp.gr. 1.15; dens. 9.6 lb/gal; visc. 75 cps max.; pH 10.0; surf. tens. 70 dynes/cm (1%); 25% solids.

Daxad® 30-30. [W.R. Grace/Organics] Sodium polymethacrylate sol'n.; dispersant esp. for pigments in aq. sol'ns.; used in paint formulations, emulsion polymerization, water treatment (as a scale control agent for boiler systems), in large particle suspensions; water-wh. clear liq.; sol. in water systems; sp.gr. 1.21; dens. 10.1 lb/gal; visc. 150 cps max.; pH 10.0; surf. tens. 70 dynes/cm (1%); 30% solids.

Daxad® 31. [Hampshire] Isobutylene maleic anhydride copolymer sol'n.; CAS 26426-80-2; dispersant for aq. systems; used in cosmetic base makeup, latex paints and coatings, enamels, polymerization, leather tanning, and water treatment; crumb control in polybutadiene rubber mfg.; pale amber clear liq.; sol. in water systems; sp.gr. 1.11; dens. 9.2 lb/gal; visc. 30 cps; pH 10.0; surf. tens. 50 dynes/cm (1%); 25% total solids.

Daxad® 31S. [W.R. Grace/Organics] Sodium polyisobutylene maleic anhydride copolymer; dispersant for aq. systems; used in cosmetic base makeup, latex paints and coatings, enamels, polymerization, leather tanning, and water treatment; crumb control in polybutadiene rubber mfg.; wh. fine powd.; sol. in water systems; dens. 32-35 lb/ft³ (tamped); pH 10.0 (1%); 90% act.

Daxad® 34. [W.R. Grace/Organics] Polymethacrylic acid sol'n.; CAS 25087-26-7; dispersant for pigments and fillers in ceramics, polymerization; pale amber clear liq.; sol. in water systems; sp.gr. 1.09; dens. 9.1 lb/gal; visc. 400 cps max.; pH 3; surf. tens. 70 dynes/cm (1%); 25% total solids.

Daxad® 34A9. [W.R. Grace/Organics] Ammonium polymethacrylate sol'n.; dispersant for pigments and fillers commonly used in latex paint systems; also for polymerization and clay coating; clear liq.; sol. in water systems; sp.gr. 1.11; dens. 9.2 lb/gal; visc. 25 cps max.; pH 9; surf. tens. 76 dynes/cm (1%); 24% total solids.

DB-9 Paintable Silicone Emulsion. [Genesee Polymers] Mercaptoalkyl functional methylalkylaryl polysiloxane emulsion; nonionic; release agent in plastic and rubber molded parts which are to be painted or bonded in subsequent operations; mandrel release agent (hose mfg.); textile lubricants; internal rubber lubricant/release; metalforming lubricant; wh. opaque fluid; sp.gr. 0.97; dens. 8.0 lb/gal; pH 8.5; 50% silicone; *Storage:* good freeze-thaw stability, but guard against freezing; store below 100 F.

DBE. [DuPont] Dibasic ester mixt. (60% dimethyl glutarate, 15% dimethyl adipate, 24% dimethyl succinate); solv. for coatings, cleaners, inks, textile lubricants, urethane prod.; plasticizer for flexible thermoset polyester; polymer intermediate for polyester polyols for urethanes, wet-str. paper resins, polyester resins; specialty chemical intermediate; clear colorless liq., mild sweet odor; m.w. 159; sol. in alcohols, ketones, ethers, most hydrocarbons; sl. sol. in water and higher paraffinic hydrocarbons; sp.gr. 1.092; dens. 9.09 lb/gal; visc. 2.4 cSt; f.p. -20 C; b.p. 196-225 C; acid no. 0.3 max.; flash pt. (TCC) 100 C; surf. tens. 35.6 dynes/cm.; 99% min. esters; *Toxicology:* LD50 (oral, rat) 8191 mg/kg, (dermal, rabbit) > 2250 mg/kg; mild to severe skin irritant, moderate eye irritant; may cause nose and throat irritation and blurry vision; *Precaution:* do not mix with strong oxidants, acids, or alkalies; *Storage:* store in well ventilated place in tightly closed container.

DBE-2, -2SPG. [DuPont] Dibasic ester mixt. (76% dimethyl glutarate, 24% dimethyl adipate, 0.3% dimethyl succinate); solv. for coatings, cleaners, inks, textile lubricants, urethane prod.; plasticizer for flexible thermoset polyester; polymer intermediate for polyester polyols for urethanes, wet-str. paper resins, polyester resins; specialty chemical intermediate; clear colorless liq., mild odor; m.w. 163; sol. in alcohols, ketones, ethers, most hydrocarbons; sl. sol. in water and higher paraffinic hydrocarbons; sp.gr. 1.081; dens. 9.00 lb/gal; visc. 2.5 cSt; f.p. -13 C; b.p. 210-225 C; acid no. 1 max.; flash pt. (TCC) 104 C; 99% min. esters; *Toxicology:* may cause eye irritation.

DBE-3. [DuPont] Dibasic ester mixt. (89% dimethyl adipate, 10% dimethyl glutarate, 0.2% dimethyl succinate); solv. for coatings, cleaners, inks, textile lubricants, urethane prod.; plasticizer for flexible thermoset polyester; polymer intermediate for polyester polyols for urethanes, wet-str. paper resins, polyester resins; specialty chemical intermediate; clear colorless liq., mild odor; m.w. 173; sol. in alcohols, ketones, ethers, most hydrocarbons; sl. sol. in water and higher paraffinic hydrocarbons; sp.gr. 1.068; dens. 8.89 lb/gal; visc. 2.5 cSt; f.p. 8 C; b.p. 215-225 C; acid no. 1 max.; flash pt. (TCC) 102 C; 99% min. esters; *Toxicology:* may cause eye irritation.

DBE-4. [DuPont] Dimethyl succinate; CAS 106-65-0; EINECS 203-419-9; solv. for coatings, cleaners, inks, textile lubricants, urethane prod.; plasticizer for flexible thermoset polyester; polymer intermediate for polyester polyols for urethanes, wet-str. paper resins, polyester resins; specialty chemical intermediate; clear colorless liq., mild odor; m.w. 146; sol. in alcohols, ketones, ethers, most hydrocarbons; sl. sol. in water and higher paraffinic hydrocarbons; sp.gr. 1.121; dens. 9.33 lb/gal; visc. 2.5 cSt; f.p. 19 C; b.p. 196 C; flash pt. (TCC) 94 C; 98.5% min. esters; *Toxicology:* may cause eye irritation.

DBE-5. [DuPont] Dimethyl glutarate; CAS 1119-40-0; EINECS 214-277-2; solv. for coatings, cleaners, inks, textile lubricants, urethane prod.; plasticizer for flexible thermoset polyester; polymer intermediate for polyester polyols for urethanes, wet-str. paper resins, polyester resins; specialty chemical intermediate; clear colorless liq., mild odor; m.w. 160; sol. in alcohols, ketones, ethers, most hydrocarbons; sl. sol. in water and higher paraffinic hydrocarbons; sp.gr. 1.091; dens. 9.08 lb/gal; visc. 2.5 cSt; f.p. -37 C; b.p. 210-215 C; acid no. 1 max.; flash pt. (TCC) 107 C; 99% min. esters; *Toxicology:* may cause eye irritation.

DBE-6. [DuPont] Dimethyl adipate; CAS 627-93-0; EINECS 211-020-6; solv. for coatings, cleaners, inks, textile lubricants, urethane prod.; plasticizer for flexible thermoset polyester; polymer intermediate for polyester polyols for urethanes, wet-str. paper resins, polyester resins; specialty chemical intermediate; clear colorless liq., mild odor; m.w. 174; sol. in alcohols, ketones, ethers, most hydrocarbons; sl. sol. in water and higher paraffinic hydrocarbons; sp.gr. 1.064; dens. 8.86 lb/gal; visc. 2.5 cSt; f.p. 10 C; b.p. 227-230 C; acid no. 1 max.; flash pt. (TCC) 113 C; 99% min. esters; *Toxicology:* may cause eye irritation.

DBE-9. [DuPont] Dibasic ester mixt. (66% dimethyl glutarate, 33% dimethyl succinate, 0.2% dimethyl adipate); solv. for coatings, cleaners, inks, textile lubricants, urethane prod.; plasticizer; polymer intermediate for polyester polyols for urethanes, wet-str. paper resins, polyester resins; specialty chemical intermediate; clear colorless liq., mild odor; m.w. 156; sol. in alcohols, ketones, ethers, most hydrocarbons; sl. sol. in water and higher paraffinic hydrocarbons; sp.gr. 1.099; dens. 9.15 lb/gal; visc. 2.4 cSt; f.p. -10 C; b.p. 196-215 C; flash pt. (TCC) 94 C; 99% min. esters; *Toxicology:* may cause eye irritation.

DBQDO®. [Lord] Dibenzoyl-p-quinone dioxime; nonsulfur vulcanizing agent for natural and syn. elastomers; used for curing butyl comps. which require heat resist. and elec. resistivity.

DCH-99. [DuPont Nylon] 1,2-Diaminocyclohexane; CAS 694-83-7; EINECS 211-776-7; high-quality polyamine; epoxy curing agent; chelating agent for oilfield, textile, water treatment, detergent fields; herbicide intermediate; polyamide resins for adhesives, films, plastics, inks; PU for extenders, catalysts; scale/corrosion inhibitors; colorless clear liq., amine-like odor; misc. with water; m.w. 114; sp.gr. 0.94 (20 C); vapor pressure 0.387 mm Hg (20 C); m.p. 2 C (cis), 15 C (trans); b.p. 183 C (760 mm); amine no. 967; flash pt. (CC) 75 C; ref. index 1.487; pH 12 (5%); *Toxicology:* LD50 (oral, rat) 2300 mg/kg; may cause burns and ulceration of skin and eyes, allergic skin reactions; *Precaution:* corrosive liq.; combustible; incompat. with acids, oxidizing agents; thermal dec. may produce CO, CO_2, NO_2; *Storage:* store in cool place in tightly closed container; keep away from heat, sparks and flames.

DD-8126. [Dover] Brominated paraffin; flame retardant for PUF, foam-in-place for pkg., rubber, textiles; visc. reducer; Gardner 2 liq.; sp.gr. 1.19; visc. 10 cps; 44% bromine.

DD-8133. [Dover] Aromatic bromine and aliphatic chlorine; flame retardant for PP, PE, SBR, other rubbers, unsat. polyesters, and coatings; wh. solid; sp.gr. 2.3; soften. pt. 130-300 C; 41% bromine; 35% chlorine.

DD-8307. [Dover] Bromochlorinated paraffin with phosphorus; flame retardant for rigid and flexible PUF, RIM, textiles, rubber, and PVC for general purpose applics. in construction, agric., consumer, and industrial markets where UL rated material is required; Gardner 2 clear liq.; pleasant odor; sol. in aliphatic, aromatic, and chlorianted hydrocarbons, esters, ketones, higher alcohols (except methyl and ethyl); insol. in water; sp.gr. 1.4; visc. 1000 cps; 26% bromine; 22% chlorine; 5% phosphorus.

DEA Commercial Grade. [Dow] Diethanolamine; CAS 111-42-2; EINECS 203-868-0; used in surfactants, cosmetics/toiletries, metalworking fluids, textile chemicals, gas conditioning chemicals, agric. intermediates, adhesives, antistats, coatings, petroleum, polymers, rubber processing, and cement grinding aids; sp.gr. 1.0881 (30/4 C); dens. 9.09 lb/gal (30 C); visc. 351.9 cps (30 C); f.p. 28 C; b.p. 268 C (760 mm Hg); flash pt. (Seta CC) 325 F; fire pt. 300 F; ref. index 1.4750 (30 C).

DEA Low Freeze Grade. [Dow] Diethanolamine; CAS 111-42-2; EINECS 203-868-0; used in surfactants, cosmetics/toiletries, metalworking fluids, textile chemicals, gas conditioning chemicals, agric. intermediates, adhesives, antistats, coatings, petroleum, polymers, rubber processing, and cement grinding aids; sp.gr. 1.0881 (30/4 C); dens. 9.09 lb/gal (30 C); visc. 351.9 cps (30 C); f.p. 28 C; b.p. 268 C (760 mm Hg); flash pt. (Seta CC) 325 F; fire pt. 300 F; ref. index 1.4750 (30 C).

Decalin. [DuPont] Decahydronaphthalene; CAS 91-17-8; solv. and stabilizer for shoe creams and floor waxes; solv. in paint and lacquers, oils, resins, rubber, and asphalt; colorless to pale yel. liq.; m.w. 138.2; f.p. -45 C; b.p. 186 C; negligible sol. in water; sp.gr. 0.876; dens. 7.39 lb/gal; flash pt. 57 C (TCC); 97.0% min. conc.

Decanox-F. [Elf Atochem N. Am./Org. Peroxides] Decanoyl peroxide; initator for bulk, sol'n., and suspension polymerization, curing elastomers, and high-temp. cure of polyester resins; flaked solid; 98.5% act.; 3.93% min. act. oxygen.

Decap. [Dynaloy] DMSO-based; CAS 67-68-5; EINECS 200-664-3; solv. for depotting, deflashing, decapsulation of epoxy castings, transfer moldings; used hot (150 C); nonselective.

Dechlorane® Plus 25. [Occidental] Chlorine-containing cycloaliphatic compd.; 1,2,3,4,7,8,9,10,13,13,14,14-dodecachloro-1,4,4a,5,6,6a,7,10,10a,11,12,12a-dodecahydro-1, 4:7,10-dimethanodibenzo (a,e) cyclooctene; CAS 13560-89-9; flame retardant in thermoplastic, thermoset, elastomeric polymers (ABS, CR, DAP, EEA, EPDM, epoxy, EPR, EVA, NR, neoprene, nylon, PBT, phenolics, polyester, PE, PP, PU, SBR, silicone rubber, TPE, TPU); synergistic with antimony oxide; wh. cryst. powd., odorless; 5 μ avg. particle diam.; low sol. in most solvs. and polymer systems; insol. in water; m.w. 653.77; dens. 1.8 g/cc; bulk dens. 38-42 lb/ft^3; m.p. 350 C (dec.); pH 6.0-8.0 (methanol-water extract); flash pt. none; 65% chlorine.

Dechlorane® Plus 35. [Occidental] Chlorine-containing cycloaliphatic compd.; 1,2,3,4,7,8,9, 10,13,13,14,14-dodecachloro-1,4,4a,5,6, 6a,7,10,10a,11,12,12a-dodecahydro-1, 4:7,10-dimethanodibenzo (a,e) cyclooctene; CAS 13560-89-9; flame retardant in thermoplastic, thermoset, elastomeric polymers (ABS, CR, DAP, EEA, EPDM, epoxy, EPR, EVA, NR, neoprene, nylon, PBT, phenolics, polyester, PE, PP, PU, SBR, silicone rubber, TPE, TPU); synergistic with antimony oxide; wh. cryst. powd., odorless; 2 μ avg. particle diam.; low sol. in most solvs. and polymer systems; insol. in water; m.w. 653.77; dens. 25-30 lb/ft^3; pH 6-8 (methanol-water extract); 65% chlorine.

Dechlorane® Plus 515. [Occidental] Chlorine-containing cycloaliphatic compd.; 1,2,3,4,7,8,9,10,13,13,14,14-dodecachloro-1,4,4a,5,6, 6a,7,10,10a,11,12,12a-dodecahydro-1,4:7,10-dimethanodibenzo (a,e) cyclooctene; CAS 13560-89-9; flame retardant in thermoplastic, thermoset, elastomeric polymers (ABS, CR, DAP, EEA, EPDM, epoxy, EPR, EVA, NR, neoprene, nylon, PBT, phenolics, polyester, PE, PP, PU, SBR, silicone rubber, TPE, TPU); synergistic with antimony oxide; wh. cryst. powd., odorless; 9 μ avg. particle diam.; low sol. in most solvs. and polymer systems; insol. in water; m.w. 653.77; dens. 1.8 g/cc; bulk dens. 38-42 lb/ft^3; m.p. 350 C (dec.); pH 6.0-8.0 (methanol-water extract); flash pt. none; 65% chlorine.

Dee-Tac. [Olin] Detackifier for uncured rubber; paste.; Unverified

Defomax. [Toho Chem. Industry] Nonionic surfactant/wax emulsions; nonionic; antifoaming agents for rubbers and plastics; liq./paste.

Degressal® SD 40. [BASF AG] Phosphoric acid ester; defoamer for cleaners; defoamer and plasticizer for dry-bright emulsions; liq.; 100% act.

D.E.H. 20. [Dow] Diethylenetriamine; CAS 111-40-0; EINECS 203-865-4; aliphatic polyamine curing agent for epoxy resins; used for civil engineering, adhesives, grouts, casting and elec. encapsulation; D.E.H. 20 is a general purpose curing agent; liq.; misc. with polar solvs. (water, alcohols, acetone, benzene, ethyl ethers); sp.gr. 0.949; dens. 7.89 lb/gal; visc. 6 cps; flash pt. 208 F.

D.E.H. 24. [Dow] Triethylenetetramine; CAS 112-24-3; EINECS 203-950-6; aliphatic polyamine curing agent for epoxy resins; used for civil engineering, adhesives, grouts, casting and elec. encapsulation; liq.; misc. with polar solvs. (water, alcohols, acetone, benzene, ethyl ethers); sp.gr. 0.978; dens. 8.13 lb/gal; visc. 22 cps; flash pt. 245 F.

D.E.H. 26. [Dow] Tetraethylenepentamine; CAS 112-57-2; EINECS 203-986-2; aliphatic polyamine curing agent for epoxy resins; used for civil engineering, adhesives, grouts, casting and elec. encapsulation; R.T. curing agent often used in 2-pkg. protective coating systems; liq.; misc. with polar solvs. (water, alcohols, acetone, benzene, ethyl ethers); sp.gr. 0.993; dens. 8.26 lb/gal; visc. 55 cps; flash pt. 310 F.

D.E.H. 29. [Dow] Aliphatic polyamine blend; aliphatic polyamine curing agent for epoxy resins; used for civil engineering, adhesives, grouts, casting and elec. encapsulation; liq.; sp.gr. 1.01; dens. 8.46 lb/gal; visc. 300 cps; flash pt. 330 F.

D.E.H. 39. [Dow] Aminoethylpiperazine; CAS 140-31-8; EINECS 205-411-0; aliphatic polyamine curing agent for epoxy resins; used for civil engineering, adhesives, grouts, casting and elec. encapsulation; liq.; sp.gr. 0.982; dens. 8.17 lb/gal; visc. 10 cps; flash pt. 175 F.

D.E.H. 40. [Dow] Accelerated dicyandiamide; epoxy curing agent for powd. coating formulations; wh. powd.; sp.gr. 1.33; dens. 0.31 g/cc.

D.E.H. 52. [Dow] Diethylenetriamine adduct; CAS 111-40-0; EINECS 203-865-4; adducted aliphatic polyamine curing agent for epoxy resins; liq.

D.E.H. 58. [Dow] Diethylenetriamine modified with Bisphenol A; CAS 111-40-0; EINECS 203-865-4; adducted aliphatic polyamine curing agent for epoxy resins; liq.

Dehscofix 920. [Albright & Wilson Am.] Sodium naphthalene-formaldehyde sulfonate; anionic; wetting agents for agric. chemicals; pigment dispersant for leather processing; dispersant/emulsifier for emulsion polymerization, rubber; powd.; 92% conc.

Dehscofix 930. [Albright & Wilson Am.] Ammonium naphthalene-formaldehyde sulfonate; anionic; pigment dispersant for leather processing; dispersant/emulsifier for emulsion polymerization, rubber; powd.; 92% conc.

Dehydat 20. [Henkel] Internal antistat for ABS, PE, PP, flexible PVC, SAN, SBR; FDA approved.

Dehydat 22. [Henkel] Internal antistat for ABS, PE, PP, flexible PVC, SAN, SBR; FDA approved.

Dehydat 80X. [Henkel] Anionic; internal/external antistat for ABS, PE, PP, PS, flexible and rigid PVC, PET, PBT, PMMA, SAN, SBR; FDA approved.

Dehydat 93P. [Henkel] Anionic; internal/external antistat for ABS, PE, PP, PS, flexible and rigid PVC, PET, PBT, PMMA, SAN, SBR; FDA approved.

Dehydat 3204. [Henkel] Internal antistat for ABS, PS, flexible PVC, SAN, SBR; FDA approved.

Dehydat 7882. [Henkel] Anionic; internal/external antistat for ABS, PS, rigid PVC, PET, PBT, PMMA, SAN, SBR; FDA approved.

Dehydran 520. [Henkel/Functional Prods.] Fatty acid ester; defoamer for polymerization, esp. during monomer stripping in mfg. of suspension PVC; biodeg.; liq.; 100% act.

Dehydran 1019. [Henkel/Functional Prods.] Nonsilicone defoamer; defoamer for prod. and processing of polymer dispersions; liq.; 100% act.

Dehydran P 4. [Henkel/Functional Prods.] Nonsilicone defoamer; defoamer for polymerization esp. during removal of remaining monomers in prod. of suspension, microsuspension, and emulsion PVC; liq.; 100% act.

Dehydran P 11. [Henkel/Functional Prods.] Trace silicone defoamer; defoamer for polymerization esp. during removal of remaining monomers in prod. of suspension, microsuspension, and emulsion PVC; liq.; 100% act.

Dehydran P 12. [Henkel/Functional Prods.] Nonsilicone defoamer; for monomer recovery in prod. of syn. rubber; antifoaming agent for syn. polymer disps., water-based paints, adhesives, glues; liq.; 100% act.

Dehydrophen PNP 30. [Pulcra SA] Nonoxynol-30; CAS 9016-45-9; nonionic; all-purpose high-temp. wetting agent, dispersant, emulsifier; emulsion polymerization; solid; HLB 17.1; pH 5.5-7.5 (5%); 100% conc.

Dehydrophen PNP 40. [Pulcra SA] Nonoxynol-40; CAS 9016-45-9; nonionic; emulsifier for fats, oils, waxes, org. solvs., emulsion polymerization; latex stabilizer; flakes; HLB 17.8; 100% conc.

Dehyquart® A. [Henkel/Cospha; Henkel/Functional Prods.; Henkel KGaA/Cospha] Cetrimonium chloride; CAS 112-02-7; EINECS 203-928-6; cationic; emulsifier for emulsion polymerization, cosmetic creams and lotions; softener, conditioner, bactericide, fungicide, and odor inhibitor in personal care prods.; antistat for hair and fibers; substantive to hair and skin; pale yel. clear liq., typ. odor; pH 5-8; 24% act.

Dehyquart® DAM. [Henkel/Functional Prods.] Distearyl dimethyl ammonium chloride; CAS 107-64-2; EINECS 203-508-2; cationic; emulsifier for plastics industry; conditioning component for hair care preparations; antistat; paste; 70-80% act.

Dehyquart® LDB. [Henkel/Functional Prods.] Lauralkonium chloride; CAS 139-07-1; EINECS 205-351-5; cationic; bactericide and fungicide for disinfectants; emulsifier; external antistat for plastics; liq.; 34-36% conc.

Dehyquart® LT. [Henkel/Functional Prods.] Laurtrimonium chloride; CAS 112-00-5; EINECS 203-927-0; cationic; emulsifier for plastics industry, wetting agent, antistat, bactericide, demulsifier, deodorant, conditioning component for hair care prods.; liq.; 34-36% conc.

Delac® MOR. [Uniroyal] N-Oxydiethylene benzothiazole-2-sulfenamide; CAS 102-77-2; delayed action accelerator for SBR, natural rubber, polyisoprene, polybutadiene, and nitrile rubber; tan flakes or pellets; sol. in acetone, benzene, CCl_4, methanol; insol. in water; sp.gr. 1.37; m.p. 75-90 C; Unverified

Delac® NS. [Uniroyal] N-t-butyl-2-benzothiazole sulfenamide; delayed action accelerator for natural and syn. rubbers; used in tire treads, carcass, mechanicals, and wire jackets; lt. tan powd. or flake; sol. in acetone, benzol, alcohol, chloroform, ether, naphtha; insol. in water; sp.gr. 1.29; m.p. $\geq$ 104 C; Unverified

Dellatol® BBS. [Bayer AG] n-Butylbenzene sulfonamide; CAS 3622-84-2; EINECS 222-823-6; plasticizer for polyamide 6, 66, 11, and 12; also suitable for copolyamides; flexibilizes cellulose derivs., esp. flame-retardant cable coatings based on cellulose acetate and cellulose acetobutyrate.

Delvet 65. [Occidental] Halogenated org. disp.; flame retardant for acrylic, cellulose acetate, epoxy, ethyl cellulose, nitrile, PE, PVAc, PVC.

Deodorant #4761-F, OS. [Andrea Aromatics] Mixture of fragrance materials; masks sharper, more volatile odors during polymer processing and in finished prod.; Unverified

Depasol CM-41. [Pulcra SA] Dialkyl sodium sulfosuccinate; anionic; dispersant for pigments; wetting and penetrating agent; used in drycleaning detergents and emulsion polymerization; rewetting agent for textile and paper industries; liq.; 50 $\pm$ 2% conc.; Unverified

D.E.R. 337. [Dow] Bisphenol-A epoxy resin; ideal for adhesive, casting, potting, encapsulation, and wet lay-up applics.; after cure, yields highly crosslinked thermoset polymers; used in adhesives and coatings, or as a modifier for other epoxy resins; Gardner 3 max. semisolid; sp.gr. 1.16; dens. 9.7 lb/gal; visc. 400-800 cps; flash pt. 252 C; EEW 230-250; Cured with 21 phr MDA:; tens. str. 69 MPa; tens. elong. 5.5% (break); flex. str. 107.1 MPa; Izod impact 28.8 J/m notch; Rockwell hardness M100; distort. temp. 133 C (1.8 MPa, unannealed); dielec. const. 4.02 (1 kHz); vol. resist. 8.11 x 10^{15} ohm-cm.; 100% NV.

D.E.R. 732. [Dow] Epichlorohydrin-polyglycol reaction prod.; flexible epoxy resin; used as additives to base epoxy systems; increases flexibility in adhesives, construction and civil engineering applics. (aggregates, seamless floors, machine grouts), industrial maintenance coatings, and structural prepegs; Gardner 1 max. liq.; sp.gr. 1.06; dens. 8.9 lb/gal; visc. 55-100 cps; COC flash pt. 205 C; EEW 305-335; After cure, D.E.R. 732/D.E.R. 331 (3:7)-curing agent D.E.H. 24:; tens. str. 5985 psi; tens. elong. (ultimate) 5.53%; flex. str. 10,825 psi; Izod impact 0.54 ft lb/in. notched; Rockwell hardness M87; distort. temp. 58 C; vol. resist. 1.97 x 10^{15}; 100% NV.

Deriphat® 154. [Henkel/Cospha; Henkel Canada] Disodium N-tallow-β iminodipropionate; CAS 61791-56-8; EINECS 263-190-6; amphoteric; detergent, solubilizer for personal care prods., hard surface cleaning, textiles, emulsion polymerization; good substantivity; wh. powd.; sol. in strong acids, alkalies, and ionic systems; dens. 2 lb/gal; pH 11; 98% solids.

Deriphat® 160. [Henkel/Cospha; Henkel/Functional Prods.; Henkel Canada] Disodium N-lauryl β-iminodipropionate; CAS 3655-00-3; EINECS 222-899-0; amphoteric; detergent, solubilizer, primary emulsifier used in org. and inorg. compds.; emulsion polymerization and stabilization; wetting agent; mild surfactant for hair and skin prods.; wh. powd.; dens. 2.0 lb/gal; 98% solids.

Deriphat® 160C. [Henkel/Cospha; Henkel/Functional Prods.; Henkel Canada] Sodium-N-lauryl β-iminodipropionate; CAS 26256-79-1; amphoteric; detergent, solubilizer, stabilizer; used in petrol. processing, emulsion polymerization, foaming cleaners, personal care prods.; amber clear liq.; sol. in strong acid, alkali, and ionic systems; sp.gr. 1.04; dens. 8.6 lb/gal; pH 7.5; 30% solids.

Desmodur® DA. [Bayer/Fibers, Org., Rubbers] Aliphatic polyisocyanate; crosslinking agent improving the resist. to heat, greases, oils, plasticizers, org. solvs., and the adhesion to many substrates; for aq. adhesive systems; ylsh. clear liq.; disp. in water; dens. 1.2 g/cm^3; visc. 3000 cps; 100% solids, 19.5% NCO.

Desmodur® L75. [Bayer/Fibers, Org., Rubbers] Toluene diisocyanate/polyol adduct sol'n. in ethyl acetate; bonding agent for PVC coatings in prod. of tarpaulins, roofing membranes, tenting, protective clothing, conveyor belts; pale yel. clear liq.; dens. 1.2 g/cm^3; visc. 1600 cps (23 C); 75% solids, 13% NCO.

Desmodur® L75A. [Bayer/Fibers, Org., Rubbers] Toluene diisocyanate/polyol adduct sol'n. in ethyl acetate; cross-linking agent that improves resistance to heat, greases, oils, plasticizers, org. solvs., and the adhesion to many substrates; used in adhesives; pale yel. clear liq.; dens. 1.2 g/cm^3; 75% solids, 13% NCO (of sol'n.).

Desmodur® MP-225. [Bayer/Fibers, Org., Rubbers] Modified polyisocyanate based on 4,4′-diphenylmethane diisocyanate; CAS 101-68-8; crosslinking agent improving resist. to heat, greases, oils, plasticizers, org. solvs, and the adhesion to many substrates; for adhesives based on PU, polyols; also as building blocks for PU; pale yel. liq.; sol. in esters, ketones, aromatic hydrocarbons, chlorinated solvs.; sp.gr. 1.2; visc. 700 cps; 100% solids, 22.8% NCO.

Desmodur® N100. [Bayer/Fibers, Org., Rubbers] Aliphatic polyisocyanate based on hexamethylene diisocyanate; two-component bonding agent for PVC coatings in prod. of tarpaulins, roofing membranes, tenting, protective clothing, conveyor belts; pale yel. liq.; visc. 10,000 cps (23 C); 100% solids, 22% NCO.

Desmodur® RC. [Bayer/Fibers, Org., Rubbers] Toluene diisocyanate-based polyisocyanate in ethyl acetate; crosslinking agent improving the resist. to heat, greases, oils, plasticizers, org. solvs., and the adhesion to many substrates; for adhesives based on PU, polyols, or rubber; pale yel. clear liq.; dens. 1.0 g/cm^3; visc. 3 cps; 30% solids, 7% NCO (of sol'n.).

Desmodur® RE. [Bayer/Fibers, Org., Rubbers] Triphenyl methane 4,4′,4′′-triisocyanate in ethyl acetate; adhesion promoter for fabric and metal treatment for rubber-to-fabric and rubber-to-metal bonding; suitable for nat. and syn. rubbers except silicone rubber; crosslinking agent improving resist. to heat, greases, oils, plasticizers, org. solvs, and the adhesion to many substrates; for adhesives based on PU, polyols, or rubber; ylsh.-grn. or red-brn. to dk. violet mobile liq.; dilutable with methylene chloride, trichloroethylene, acetone, MEK, ethyl acetate, toluene; limited compat. with aliphatic hydrocarbons; sp.gr. 1.0; visc. 3 cps; flash pt. (CC) -3 C; 27% solids, 9.3% NCO (of sol'n.).

Desmodur® RFE. [Bayer/Fibers, Org., Rubbers] Tris (p-isocyanato-phenyl)-thiophosphate in ethyl acetate; adhesion promoter for fabric and metal treatment for rubber-to-fabric and rubber-to-metal bonding; suitable for nat. and syn. rubbers except silicone rubber; crosslinking agent improving resist. to heat, greases, oils, plasticizers, org. solvs, and the adhesion to many substrates; for adhesives based on PU, polyols, or rubber; ylsh.-brn. mobile liq.; dilutable with methylene chloride, trichloroethylene, acetone, MEK, ethyl acetate, toluene; limited compat. with aliphatic hydrocarbons; sp.gr. 1.0; visc. 3 cps; flash pt. (CC) -3 C; 27% solids, 7.2% NCO (of sol'n.).

Desmodur® TT. [Bayer/Fibers, Org., Rubbers] Dimer of 2,4-toluenediisocyanate; two-component bonding agent for PVC coatings (in tarpaulins, roofing membranes, tenting, protective clothing, conveyor belts); adhesion promoter for fabric treatment for rubber-to-fabric bonding; esp. for polyester fiber; wh. to pale yel. powd.; sp.gr. 1.48; 24.3% NCO content; 100% solids.

Desmodur® VK-5. [Bayer/Fibers, Org., Rubbers] Polymethylene polyphenyl isocyanate; CAS 101-68-8; crosslinking agent improving the resist. to heat, greases, oils, plasticizers, org. solvs., and the adhesion to many substrates; for adhesives based on PU, polyols, or rubber; also as building blocks for PU; dk. brn. liq.; dens. 1.2 g/cm^3; visc. 55 cps; 100% solids, 32% NCO.

Desmodur® VK-18 [Bayer/Fibers, Org., Rubbers] Polymethylene polyphenyl isocyanate; CAS 101-68-8; adhesion promoter for fabric treatment for rubber-to-fabric bonding; for polyester and other textiles; crosslinking agent improving the resist. to heat, greases, oils, plasticizers, org. solvs., and the adhesion to many substrates; for adhesives based on PU, polyols, or rubber; as building blocks for PU; dk. brn. liq.; dens. 1.24 g/cm^3; visc. 200 cps; 100% solids; 31.5% NCO.

Desmodur® VK-70. [Bayer/Fibers, Org., Rubbers] Polymethylene polyphenyl isocyanate; CAS 101-68-8; crosslinking agent improving the resist. to heat, greases, oils, plasticizers, org. solvs., and the adhesion to many substrates; for adhesives based on PU, polyols, or rubber; also as building blocks for PU; dk. brn. liq.; dens. 1.2 g/cm^3; visc. 700 cps; 100% solids, 31% NCO.

Desmodur® VK-200. [Bayer/Fibers, Org., Rubbers] Polymethylene polyphenyl isocyanate; CAS 101-68-8; crosslinking agent improving the resist. to heat, greases, oils, plasticizers, org. solvs., and the adhesion to many substrates; for adhesives based on PU, polyols, or rubber; also as building blocks for PU; dk. brn. liq.; dens. 1.2 g/cm^3; visc. 2000 cps; 100% solids, 30% NCO.

Desmodur® VKS-2. [Bayer/Fibers, Org., Rubbers] Polymethylene polyphenyl isocyanate; CAS 101-68-8; crosslinking agent improving the resist. to heat, greases, oils, plasticizers, org. solvs., and the adhesion to many substrates; for adhesives based on PU, polyols, or rubber; as building blocks for PU; dk. brn. liq.; sol. in methylene chloride, trichloroethylene, acetone, MEK, ethyl acetate, toluene; insol. in aliphatic hydrocarbons; sp.gr. 1.2; visc. 20 cps; 100% solids, 33% NCO.

Desmodur® VKS-4. [Bayer/Fibers, Org., Rubbers] Polymethylene polyphenyl isocyanate; CAS 101-68-8; crosslinking agent improving the resist. to heat, greases, oils, plasticizers, org. solvs., and the adhesion to many substrates; for adhesives based on PU, polyols, or rubber; as building blocks for PU; dk. brn. liq.; sol. in methylene chloride, trichloroethylene, acetone, MEK, ethyl acetate, toluene; insol. in aliphatic hydrocarbons; sp.gr. 1.2; visc. 40 cps; 100% solids, 32% NCO.

Desmodur® VKS-5. [Bayer/Fibers, Org., Rubbers] Polymethylene polyphenyl isocyanate; CAS 101-68-8; crosslinking agent improving the resist. to heat, greases, oils, plasticizers, org. solvs., and the adhesion to many substrates; for adhesives based on PU, polyols, or rubber; as building blocks for PU; dk. brn. liq.; sol. in methylene chloride, trichloroethylene, acetone, MEK, ethyl acetate, toluene; insol. in aliphatic hydrocarbons; sp.gr. 1.2; visc. 50 cps; 100% solids, 32% NCO.

Desmodur® VKS-18. [Bayer/Fibers, Org., Rubbers] Polymethylene polyphenyl isocyanate; CAS 101-68-8; crosslinking agent improving the resist. to heat, greases, oils, plasticizers, org. solvs., and the adhesion to many substrates; for adhesives based on PU, polyols, or rubber; as building blocks for PU; dk. brn. liq.; sol. in methylene chloride, trichloroethylene, acetone, MEK, ethyl acetate, toluene; insol. in aliphatic hydrocarbons; sp.gr. 1.2; visc. 200 cps; 100% solids, 30.5% NCO.

Desmorapid. [Bayer AG] Catalyst for polyurethanes.

Desonic® S-100. [Witco/New Markets] Octoxynol-9 (9-10 EO); CAS 9002-93-1; nonionic; metal and textile processing surfactant; household and industrial cleaners; emulsifier for vinyl and acrylic polymerization; liq.; cloud pt. 140-158 F; 100% act.

Desonic® S-405. [Witco/New Markets] Octoxynol-40; CAS 9002-93-1; nonionic; coemulsifier for vinyl and acrylic polymerization; dye assistant; liq.; cloud pt. 165-176 F (10% NaCl); 70% act.

Desophos® 4 CP. [Witco/New Markets] Phosphate ester; softener, antistat for textile finishing; in lubricants for filament yarns, syn. fibers, wool; emulsifier for cosmetic oils and creams, polymerization of latexes; liq.; 95% act.

Desophos® 4 NP. [Witco/New Markets] Phosphate ester; nonionic; emulsifier for emulsion polymerization; liq.; most oil-sol. of series; water-sol. when neutralized; 100% act.

Desophos® 6 NP. [Witco/New Markets] Phosphate ester, free acid; anionic; emulsifier, detergent; for emulsion polymerization; liq.; 100% act.

Desophos® 6 NP4. [Witco/New Markets] Phosphate ester; polymerization emulsifier for PVAc and acrylic films; also in waterless hand cleaners and laundry detergents; liq.; 100% act.

Desophos® 9 NP. [Witco/New Markets] Phosphate ester; nonionic; dedusting agent for dry cleaning detergent, alkaline powders, water-repellent fabric finishes; emulsion polymerization of PVAc and acrylic films; liq.; 100% act.

DETA. [Air Prods./Perf. Chems.] Diethylenetriamine; CAS 111-40-0; EINECS 203-865-4; R.T. epoxy curing agent providing low equiv. wt., high nitrogen content; for intermediates for amidoamines, imidazolines, epoxy adducts, and other addition compds.; occasionally used for floorings, sol'n. coatings, and encapsulation; FDA 21CFR §175.300, 176.170; Gardner 1 liq.; sp.gr. 0.948; visc. < 10 cps; amine no. 1640; equiv. wt. 21; heat distort. temp. 127 F (264 psi).

DETDA 80. [Lonza Ltd] 80:20 Mixt. of 3,5-diethyl-2,4-diaminotoluene and 3,5-diethyl-2,6-diaminotoluene; difunctional chain extender in RIM and

RRIM processes for PU used in high-load components in industrial applics., esp. automotive industry; hardener for epoxy resins; m.w. 178.28; b.p. 132 C; *Toxicology:* LD50 (oral, rat) 738 mg/kg; pract. nonirritating to skin.

Dextrol OC-15. [Dexter] Complex org. phosphate ester free acid; anionic; detergent, wetting agent, emulsifier; for pesticides, PVA and acrylic polymerization; corrosion inhibitor; dispersant for magnetic oxide in aromatic solvs.; lt.-colored liq.; 100% act.

Dextrol OC-20. [Dexter] Complex org. phosphate ester free acid; CAS 51811-79-1; anionic; detergent, wetting agent, emulsifier; for pesticides, emulsion polymerization; corrosion inhibitor; dedusting agent for alkaline powd.; liq.; water-sol.; 100% act.

Dextrol OC-40. [Dexter] Phosphate ester free acid; CAS 9046-01-9; anionic; wetting agent, penetrant for nonwoven fabrics; stabilizer for resins; liq.; 100% conc.

Dextrol OC-50. [Dexter] Alkyl phenoxy ethoxylate, partially sodium neutralized; anionic; wetting and dispersing agent for latex coatings, colorant systems, detergent formulations; Gardner 2 max. clear visc. liq.; sol. in water and solvs.; sp.gr. 1.11; dens. 9.25 lb/gal; visc. 6000 $\pm$ 1500 cps; flash pt. (TCC) > 200 F; pH 5.55 $\pm$ 0.5; surf. tens. 35.2 dynes/cm (0.1%); 92 $\pm$ 1% act.

Dextrol OC-70. [Dexter] Phosphated aliphatic ethoxylate; anionic; detergent, wetting agent, dispersant, emulsifier; for alkyd paints, solv.-based coatings, colorant systems, pesticides, PVA and acrylic polymerization; corrosion inhibitor; dispersant for magnetic oxide in aromatic solvs.; Gardner 6 max. clear to hazy liq.; sp.gr. 1.06; dens. 8.83 lb/gal; visc. 800 $\pm$ 200 cps; flash pt. (TCC) > 300 F; pH 1.5 $\pm$ 0.3; surf. tens. 32.4 dynes/cm (0.1%); 100% act.

DHT-4A. [Kyowa Chem. Industry] Halogen scavenger for polyolefins; superfine particles; sp.gr. 0.56 m/100 g; dens. 2.36; ref. index 1.56; hardness (Moh) 2.5; 98.8% $Mg(OH)_2$.

Diablack® A. [Mitsubishi Kasei] SAF (N110) carbon black; CAS 1333-86-4; EINECS 215-609-9; blk. for the rubber industry; 18 μ particle diam.; surf. area 163 m^2/g; oil absorp. 116.

Diablack® E. [Mitsubishi Kasei] FEF (N550) Carbon black; CAS 1333-86-4; EINECS 215-609-9; blk. for the rubber industry; 48 μ particle diam.; surf. area 41 m^2/g; oil absorp. 115.

Diablack® G. [Mitsubishi Kasei] GPF (N660) Carbon black; CAS 1333-86-4; EINECS 215-609-9; blk. for the rubber industry; 80 μ particle diam.; surf. area 28 m^2/g; oil absorp. 84.

Diablack® H. [Mitsubishi Kasei] HAF (N330) Carbon black; CAS 1333-86-4; EINECS 215-609-9; blk. for the rubber industry; 31 μ particle diam.; surf. area 79 m^2/g; oil absorp. 105.

Diablack® HA. [Mitsubishi Kasei] Carbon black; CAS 1333-86-4; EINECS 215-609-9; blk. for the rubber industry; 32 μ particle diam.; surf. area 74 m^2/g; oil absorp. 103.

Diablack® I. [Mitsubishi Kasei] ISAF (N220) carbon black; CAS 1333-86-4; EINECS 215-609-9; blk. for the rubber industry; 23 μ particle diam.; surf. area 114 m^2/g; oil absorp. 114.

Diablack® II. [Mitsubishi Kasei] IISAF (N285) Carbon black; CAS 1333-86-4; EINECS 215-609-9; blk. for the rubber industry; 24 μ particle diam.; surf. area 98 m^2/g; oil absorp. 124.

Diablack® LH. [Mitsubishi Kasei] HAF-LS (N326) Carbon black; CAS 1333-86-4; EINECS 215-609-9; blk. for the rubber industry; 31 μ particle diam.; surf. area 84 m^2/g; oil absorp. 74.

Diablack® LI. [Mitsubishi Kasei] ISAF-LS (N219) Carbon black; CAS 1333-86-4; EINECS 215-609-9; blk. for the rubber industry; 23 μ particle diam.; surf. area 107 m^2/g; oil absorp. 78.

Diablack® LR. [Mitsubishi Kasei] Carbon black; CAS 1333-86-4; EINECS 215-609-9; blk. for the rubber industry; 85 μ particle diam.; surf. area 24 m^2/g; oil absorp. 48.

Diablack® M. [Mitsubishi Kasei] MAF Carbon black; CAS 1333-86-4; EINECS 215-609-9; blk. for the rubber industry; 43 μ particle diam.; surf. area 48 m^2/g; oil absorp. 120.

Diablack® N220M. [Mitsubishi Kasei] Carbon black; CAS 1333-86-4; EINECS 215-609-9; blk. for the rubber industry; 22 μ particle diam.; surf. area 123 m^2/g; oil absorp. 115.

Diablack® N234. [Mitsubishi Kasei] N234 Carbon black; CAS 1333-86-4; EINECS 215-609-9; blk. for the rubber industry; 22 μ particle diam.; surf. area 123 m^2/g; oil absorp. 124.

Diablack® N339. [Mitsubishi Kasei] N 339 Carbon black; CAS 1333-86-4; EINECS 215-609-9; blk. for the rubber industry; 26 μ particle diam.; surf. area 96 m^2/g; oil absorp. 124.

Diablack® N550M. [Mitsubishi Kasei] MAF Carbon black; CAS 1333-86-4; EINECS 215-609-9; blk. for the rubber industry; 43 μ particle diam.; surf. area 47 m^2/g; oil absorp. 131.

Diablack® N760M. [Mitsubishi Kasei] SRF-LM (N762) Carbon black; CAS 1333-86-4; EINECS 215-609-9; blk. for the rubber industry; 85 μ particle diam.; surf. area 27 m^2/g; oil absorp. 62.

Diablack® R. [Mitsubishi Kasei] SRF (N774) Carbon black; CAS 1333-86-4; EINECS 215-609-9; blk. for the rubber industry; 85 μ particle diam.; surf. area 28 m^2/g; oil absorp. 69.

Diablack® SF. [Mitsubishi Kasei] FEF-HS Carbon black; CAS 1333-86-4; EINECS 215-609-9; blk. for the rubber industry; 40 μ particle diam.; surf. area 50 m^2/g; oil absorp. 127.

Diablack® SH. [Mitsubishi Kasei] HAF-HS (N347) Carbon black; CAS 1333-86-4; EINECS 215-609-9; blk. for the rubber industry; 31 μ particle diam.; surf. area 78 m^2/g; oil absorp. 128.

Diablack® SHA. [Mitsubishi Kasei] N351 Carbon black; CAS 1333-86-4; EINECS 215-609-9; blk. for the rubber industry; 32 μ particle diam.; surf. area 74 m^2/g; oil absorp. 128.

Diak No. 4. [R.T. Vanderbilt] 4,4´-Methylenebis (cyclohexylamine)carbamate; CAS 13253-82-2;

EINECS 236-239-4; processing aid for Viton B and B-50; wh. fine powd., sl. amine odor; 0% on 100 mesh; sol. in nonpolar solvs.; sl. sol. in water; dens. 1.23 mg/m^3; m.p. 155 C; flash pt. (OC) > 149 C; 94% min. assay; *Toxicology:* LD50 (oral, rat) 1000 mg/kg; may cause irritation of eyes, skin, nose, and throat; *Precaution:* may form flamm. dust/air mixts.; keep dust away from heat, open flame; incompat. with acidic materials; hazardous combustion prods.: carbon oxides, nitrogen oxides, ammonia; *Storage:* exc. storage stability; store in cool, dry area.

Dialead®. [Mitsubishi Kasei] Coal tar pitch-based carbon fiber; reinforcement for thermoset plastics, cements, metal, rubber applics.

Dianol® 220. [Akzo Nobel] Ethoxylated bisphenol A diol; CAS 32492-61-8, 901-44-0; reactive modifier for sat. and unsat. polyesters, vinyl esters, and PU resin formulations; Gardner < 1 color; m.w. 310-330; visc. 21,500 cps (125 C); m.p. 105-112 C; hyd. no. 350; pH ≤ 12.

Dianol® 240. [Akzo Nobel] Ethoxylated bisphenol A diol; CAS 32492-61-8; reactive modifier in sat. and unsat. polyesters, vinyl esters, and PU resin formulations; Gardner ≤ 4 color; m.w. 395-415; visc. 10,000 cps; hyd. no. 270-285; pH ≤ 12.

Dianol® 240/1. [Akzo Nobel] Ethoxylated bisphenol A diol; CAS 32492-61-8; reactive modifier in sat. and unsat. polyesters, vinyl esters, and PU resin formulations; Gardner ≤ 3 color; m.w. 395-415; visc. 8000 cps; hyd. no. 270-285; pH 9.

Dianol® 265. [Akzo Nobel] Ethoxylated bisphenol A diol; CAS 32492-61-8; reactive modifier in sat. and unsat. polyesters, vinyl esters, and PU resin formulations; Gardner ≤ 2 color; m.w. 510; visc. 3000 cps; hyd. no. 220; pH 9.

Dianol® 285. [Akzo Nobel] Ethoxylated bisphenol A diol; CAS 32492-61-8; reactive modifier in sat. and unsat. polyesters, vinyl esters, and PU resin formulations; Gardner ≤ 2 color; m.w. 590; visc. 2000 cps; hyd. no. 190; pH 9.

Diapol® WMB® 1808. [Mitsubishi Kasei] SBR carbon black masterbatch with 75 phr N330 carbon blk, 50 phr high aromatic extender oil; blk. masterbatch for rubber industry.

Diapol® WMB® 1833. [Mitsubishi Kasei] SBR carbon black masterbatch with 82.5 phr N347 blk. and 62.5 phr high aromatic extender oil; for rubber industry.

Diapol® WMB® 1847. [Mitsubishi Kasei] SBR carbon black masterbatch with 75 phr N339 blk. and 50 phr high aromatic extender oil; for rubber industry.

Diapol® WMB® 1848. [Mitsubishi Kasei] SBR carbon black masterbatch with 82.5 phr N339 blk. and 62.5 phr high aromatic extender oil; for rubber industry.

Diapol® WMB® 1849. [Mitsubishi Kasei] SBR carbon black masterbatch with 82.5 phr N351 blk. and 62.5 phr high aromatic extender oil; for rubber industry.

Diapol® WMB® 1939. [Mitsubishi Kasei] SBR carbon black masterbatch with 75 phr N285 blk. and 50 phr high aromatic extender oil; for rubber industry.

Diapol® WMB® S900. [Mitsubishi Kasei] SBR carbon black masterbatch with 52 phr N330 blk. and 10 phr naphthene extender oil; for rubber industry.

Diapol® WMB® S903. [Mitsubishi Kasei] SBR carbon black masterbatch with 90 phr N330 blk. and 20 phr naphthene extender oil; for rubber industry.

Diapol® WMB® S910. [Mitsubishi Kasei] SBR carbon black masterbatch with 52 phr N330 blk.; for rubber industry.

Diapol® WMB® S920. [Mitsubishi Kasei] SBR carbon black masterbatch with 90 phr N660 blk. and 20 phr naphthene extender oil; for rubber industry.

Diapol® WMB® S960. [Mitsubishi Kasei] SBR carbon black masterbatch with 60 phr N220 blk. and 20 phr high aromatic extender oil; for rubber industry.

Diaresin Dye®. [Mitsubishi Kasei] Solvent dyes for coloring thermoplastics and thermosets; provides high brilliancy and transparency with good fastness props.

Di-Cup 40C. [Hercules] Dicumyl peroxide supported on precipitated calcium carbonate; vulcanizing agent, and high-temp. polymerization catalyst for rubber and plastics; can be used to cross-link a wide variety of polymers; can be emulsified for applics. involving aq. systems; off-wh. free-flowing powd.; m.w. 270; sp.gr. 1.57; 39.5-41.5% act.

Di-Cup 40KE. [Hercules] Dicumyl peroxide supported on Burgess KE clay; vulcanizing agent, and high-temp. polymerization catalyst for rubber and plastics; can be used to cross-link a wide variety of polymers; can be emulsified for applics. involving aq. systems; off-wh. free-flowing powds.; m.w. 270; sp.gr. 1.55; 39.5-41.5% act.

Di-Cup R, T. [Hercules] Dicumyl peroxide; CAS 80-43-3; vulcanizing agent, and high-temp. polymerization catalyst for rubber and plastics; can be used to cross-link a wide variety of polymers; can be emulsified for applics. involving aq. systems; pale yel. to wh. granular solid and pale yel., fused, semicryst. solid resp.; m.w. 270; sol. or disp. in natural and syn. rubber compds., silicone gums, and polyester resins, veg. oils; water-insol.; sp.gr. 1.018 and 1.023 resp.; m.p. slowly melt at 38 C and 30 C resp.; 98-100% and 92-95% act. resp.

Dicyanex® 200X. [Air Prods./Perf. Chems.] Dicyandiamide; CAS 461-58-5; EINECS 207-312-8; epoxy curing agent for printed circuit board laminates; 200 mesh powd.; dens. 4.0 lb/gal; m.p. 410 F; Unverified

Dicyanex® 325. [Air Prods./Perf. Chems.] Dicyanamide with 1-3% inert flow control additive; CAS 461-58-5; EINECS 207-312-8; epoxy curing agent for prepregs, composites; 325 mesh powd.; dens. 3.5 lb/gal; m.p. 410 F; Unverified

Dicyanex® 1200. [Air Prods./Perf. Chems.] Dicyandiamide with 1-3% inert flow control additive; CAS 461-58-5; EINECS 207-312-8; epoxy curing agent for prepregs, composites; 10 μ powd.; dens. 2.5 lb/gal; m.p. 410 F; Unverified

Diene® 35AC3. [Firestone Syn. Rubber] Alkyl lithium polymerized polybutadiene; CAS 9003-17-2; impact modifier for thermoplastic resins; features exc. color and low gel; APHA 2.5 color; sp.gr. 0.90; Mooney visc. 37 (ML4, 100 C); ref. index 1.5167.

Diene® 35AC10. [Firestone Syn. Rubber] Low visc., low gel, med. cis polybutadiene rubber, alkyl lithium polymerized; CAS 9003-17-2; graftable impact modifier for PS, ABS, and other engineering resins; features exc. color and low gel; APHA 10 max. amorphous solid, low or no odor; nil sol. in water; sp.gr. 0.90; visc. 97 cP (5.43% in toluene); Mooney visc. 37 (ML4, 100 C); ref. index 1.5167; *Precaution:* combustion byprods.: CO, CO_2; *Storage:* store below 150 F.

Diene® 55AC10. [Firestone Syn. Rubber] Med. cis alkyl lithium polymerized polybutadiene; CAS 9003-17-2; graftable impact modifier for PS, ABS, and other engineering resins; features exc. color and low gel; APHA 10 max. amorphous solid, low or no odor; nil sol. in water; sp.gr. 0.90; visc. 162 cP (5.43% in toluene); Mooney visc. 52 (ML4, 100 C); ref. index 1.5167; *Precaution:* combustion byprods.: oxides of carbon; *Storage:* store below 150 F.

Diene® 55NF. [Firestone Syn. Rubber] Polybutadiene rubber, sol'n. polymerized, nonstaining stabilizer; CAS 9003-17-2; used in tire treads and mech. goods; exc. resist. to abrasion, high resilience and rebound, exc. low temp. flexibility, good resist. to fatigue cracking; processing aid to reduce tack in overly tacky compds.; colorless to very lt. appearance; sp.gr. 0.90; Mooney visc. 47-57 (ML4, 100 C).

Diene® 70AC. [Firestone Syn. Rubber] High visc. alkyl lithium polymerized med. cis polybutadiene rubber; CAS 9003-17-2; graftable impact modifier for PS, ABS, and other engineering resins; features exc. color and low gel; imparts superior toughness and lower gloss; APHA 10 max. amorphous solid, low or no odor; sp.gr. 0.90; visc. 250 cP (5.43% in toluene); Mooney visc. 71 (ML4, 100 C); ref. index 1.5167; *Precaution:* combustion byprods.: oxides of carbon; *Storage:* store below 150 F.

Diglycolamine® Agent (DGA®). [Huntsman] 2-(2-Aminoethoxy) ethanol; CAS 929-06-6; EINECS 213-195-4; solv. for removal of CO_2 or H_2S from gases, for recovery of aromatics from refinery streams, for prep. of foam stabilizers, wetting agents, emulsifiers, and condensation polymers; clear colorless slightly visc. liq.; mild amine odor; misc. with water, alcohols, aromatic hydrocarbons, ethyl ether; m.w. 105.14; sp.gr. 1.0572; dens. 8.7 lb/gal; b.p. 221 C; f.p. -12.5 C; flash pt. (PMCC) 255 F; ref. index 1.4598; 98% min. act.

Diglyme. [Ferro/Grant] Diethylene glycol dimethyl ether; CAS 111-96-6; EINECS 203-924-4; solv. which tends to solvate cations; used in electrochemistry, polymer, and boron chemistry; physical processes such as gas absorption, extraction, stabilization; used in industrial prods. such as fuels, lubricants, textiles, pharmaceuticals, pesticides; colorless clear; ethereal, nonresidual odor; m.w. 134.17; f.p. -64 C; b.p. 162 C (760 mm Hg); water-sol.; sp.gr. 0.9451; dens. 7.88 lb/gal; visc. 2.0 cP; ref. index 1.4078; pH neutral; flash pt. 57 (CC); 99.6% min. purity.

Dimodan LS. [Grindsted Prods.] Distilled monoglycerides; food-grade additive for cosmetics/toiletries, cutting oils, lubricants, printing inks, resins and plastics; antistat, lubricant, cling aid for PP, PE; antistat, lubricant for EPS; processing aid, lubricant, antistat, antifog for PVC; also for nonwoven fibers, thermosets; dispersant for color concs.; FDA 21CFR §182.4505 (GRAS).

Dimodan LS Kosher. [Grindsted Prods.; Grindsted Prods. Denmark] Glyceryl linoleate with 200 ppm max. BHT, 200 ppm max. citric acid as antioxidants; CAS 2277-28-3; EINECS 218-901-4; nonionic; emulsifier for emulsions, face creams and masks; w/o food emulsifier for low-calorie spreads, icing shortenings, and cake shortenings; internal lubricant for PE, PP, PVC; EEC, FDA §184.1505 (GRAS); soft plastic; m.p. 50 C; iodine no. 110; 90% min. monoester.; *Storage:* store under cool, dry conditions.

Dimodan PM. [Grindsted Prods.; Grindsted Prods. Denmark] Dist. monoglyceride from edible refined hydrog. lard or tallow; nonionic; food emulsifier; starch complexing agent; for margarine, cake shortenings, confectionery coatings; softener for bread; peanut butter stabilizer; amylose complexing agent for dehydrated potatoes; also for cosmetics/toiletries, adhesives, cutting oils, fabric softeners, pharmaceuticals, printing inks, fertilizers; as antistat, lubricant, processing aid in PP, PE, EPS, PVC, nonwoven fibers, thermoset resins; dispersant for color concs.; EEC E471, FDA 21CFR §184.1505, 184.4505, GRAS; beads; m.p. 70 C; iodine no. 2 max.; 90% min. monoester.; *Storage:* store in cool, dry area.

Dimodan PV. [Grindsted Prods.; Grindsted Prods. Denmark] Hydrog. soybean oil dist. monoglyceride, unsat.; nonionic; food emulsifier, starch complexing agent, antisticking agent; crumb softener for bread; aerating agent in cake mixes and frozen desserts; also for cosmetics/toiletries, pharmaceuticals, printing inks, fertilizers, adhesives; as antistat, lubricant, processing aid in PP, PE, EPS, PVC, nonwoven fibers, thermoset resins; dispersant for color concs.; EEC, FDA §184.1505 (GRAS), 184.4505 (GRAS); beads, powd.; m.p. 72 C; iodine no. 2 max.; 90% min. monoester.

Dimodan PV 300. [Grindsted Prods.] Distilled monoglycerides; food-grade additive for resins and plastics; antistat, lubricant, cling aid for PP, PE; antistat, lubricant for EPS; processing aid, lubricant, antistat, antifog for PVC; also for nonwoven fibers, thermosets; dispersant for color concs.; FDA 21CFR §182.4505 (GRAS).

Dimul S. [Witco/H-I-P] Monoglycerides; food emulsifier for puddings; inhibits oil separation in peanut butter; improves palatability in dehydrated potatoes; extrusion aid in pasta; internal lubricant, slip agent, mold release, antiblock agent for most plastics, processing; FDA 21CFR §184.1505; wh. flowable powd.; m.p. 151 F; HLB 4.3; iodine no. < 4; 90% min. alpha monoglyceride.

Dinoram C. [Ceca SA] N-Coco propylene diamine;

CAS 61791-63-7; EINECS 263-195-3; cationic; chemical intermediate; reaction catalyst for epoxy resins; paste.

Dioxitol-High Gravity. [Shell] Diethylene glycol monoethyl ether, ethylene glycol; solv. for natural and syn. resins, dyes, fats used as component of heavy duty automotive brake fluids; in paint industry in nonaq. stains, and nongrain raising stains for furniture and wood; yarn and cloth conditioner in textile industry; dye solv. where deep penetration and bright shades are obtained; component of industrial cleaners; coupling agent; mfg. of esters; intermediate in mfg. of plasticizers, solv.; colorless liq.; water-misc.; sp.gr. 1.021-1.027; 72% Dioxitol-Low Gravity, 28% ethylene glycol.; Unverified

Dioxitol-Low Gravity. [Shell] Diethylene glycol monoethyl ether; CAS 111-90-0; EINECS 203-919-7; solv. for resins, dyes, fats, in auto brake fluids, paint industry, industrial cleaners, mfg. of esters; intermediate in mfg. of plasticizers; yarn and cloth conditioner in textiles; dye solv. where deep penetration and bright shades are obtained; colorless liq.; water-misc.; sp.gr. 0.986-0.990; Unverified

DIPA Commercial Grade. [Dow] Diisopropanolamine; CAS 110-97-4; EINECS 203-820-9; used to produce soaps with good hard surf. detergency, shampoos, pharmaceuticals, emulsifiers, textile specialties, agric. and polymer curing chemicals, adhesives, antistats, coatings, metalworking, petroleum, rubber, gas conditioning chemicals; sp.gr. 0.992 (40/4 C); dens. 8.27 lb/gal (40 C); visc. 870 cps (30 C); f.p. 44 C; b.p. 249 C (760 mm Hg); flash pt. (Seta CC) 276 F; fire pt. 275 C; ref. index 1.4595 (30 C).

DIPA Low Freeze Grade 85. [Dow] Diisopropanolamine; CAS 110-97-4; EINECS 203-820-9; used to produce soaps with good hard surf. detergency, shampoos, pharmaceuticals, emulsifiers, textile specialties, agric. and polymer curing chemicals, adhesives, antistats, coatings, metalworking, petroleum, rubber, gas conditioning chemicals; sp.gr. 0.992 (40/4 C); dens. 8.27 lb/gal (40 C); visc. 870 cps (30 C); f.p. 44 C; b.p. 249 C (760 mm Hg); flash pt. (Seta CC) 276 F; fire pt. 275 C; ref. index 1.4595 (30 C); 15% water.

DIPA Low Freeze Grade 90. [Dow] Diisopropanolamine; CAS 110-97-4; EINECS 203-820-9; used to produce soaps with good hard surf. detergency, shampoos, pharmaceuticals, emulsifiers, textile specialties, agric. and polymer curing chemicals, adhesives, antistats, coatings, metalworking, petroleum, rubber, gas conditioning chemicals; sp.gr. 0.992 (40/4 C); dens. 8.27 lb/gal (40 C); visc. 870 cps (30 C); f.p. 44 C; b.p. 249 C (760 mm Hg); flash pt. (Seta CC) 276 F; fire pt. 275 C; ref. index 1.4595 (30 C); 10% water.

DIPA NF Grade. [Dow] Diisopropanolamine; CAS 110-97-4; EINECS 203-820-9; used to produce soaps with good hard surf. detergency, shampoos, pharmaceuticals, emulsifiers, textile specialties, agric. and polymer curing chemicals, adhesives, antistats, coatings, metalworking, petroleum, rubber, gas conditioning chemicals; sp.gr. 0.992 (40/4 C); dens. 8.27 lb/gal (40 C); visc. 870 cps (30 C); f.p. 44 C; b.p. 249 C (760 mm Hg); flash pt. (Seta CC) 276 F; fire pt. 275 C; ref. index 1.4595 (30 C).

Diplast® A. [Lonza SpA] Di-n-butyl phthalate; CAS 84-74-2; EINECS 201-557-4; monomeric plasticizer for PVC, cellulosics, paints, as vehicles for pigment dispersions; oily limpid liq.; m.w. 278.34; sp.gr. 1.047-1.049; b.p. 340 C (760 mm); sapon. no. 403 ± 3.

Diplast® B. [Lonza SpA] Diisobutyl phthalate; CAS 84-69-5; monomeric plasticizer for PVC, cellulosics, paints, as vehicles for pigment dispersions; oily limpid liq.; m.w. 278.34; sp.gr. 1.040-1.042; b.p. 327 C (760 mm); sapon. no. 403 ± 3.

Diplast® DOA. [Lonza SpA] Di-2-ethylhexyl adipate; CAS 103-23-1; EINECS 203-090-1; monomeric plasticizer for PVC, cellulosics, paints, as vehicles for pigment dispersions; oily limpid liq.; m.w. 370.6; sp.gr. 0.925-0.927; b.p. 208-218 C (4 mm); sapon. no. 304 ± 2.

Diplast® E. [Lonza SpA] Di-n-heptyl phthalate; CAS 3648-21-3; monomeric plasticizer for PVC, cellulosics, paints, as vehicles for pigment dispersions; oily limpid liq.; m.w. 362; sp.gr. 0.991-0.993; b.p. 370 C (760 mm); sapon. no. 309 ± 2.

Diplast® L8. [Lonza SpA] Di-n-octyl phthalate; CAS 117-84-0; monomeric plasticizer for PVC, cellulosics, paints, as vehicles for pigment dispersions; oily limpid liq.; m.w. 390; sp.gr. 0.979-0.981; b.p. 220 C (5 mm); sapon. no. 287 ± 3.

Diplast® N. [Lonza SpA] Diisononyl phthalate; CAS 14103-61-8; monomeric plasticizer for PVC, cellulosics, paints, as vehicles for pigment dispersions; oily limpid liq.; m.w. 418.6; sp.gr. 0.985-0.897; b.p. 413 C (760 mm); sapon. no. 269 ± 2.

Diplast® O. [Lonza SpA] Di-2-ethylhexyl phthalate; CAS 117-81-7; EINECS 204-211-0; monomeric plasticizer for PVC, cellulosics, paints, as vehicles for pigment dispersions; oily limpid liq.; m.w. 390.56; sp.gr. 1.486-1.487; b.p. 327 C (760 mm); sapon. no. 287 ± 2.

Diplast® R. [Lonza SpA] Diisodecyl phthalate; CAS 26761-40-0; monomeric plasticizer for PVC, cellulosics, paints, as vehicles for pigment dispersions; oily limpid liq.; m.w. 446.7; sp.gr. 0.968-0.971; b.p. 248 C (760 mm); sapon. no. 251 ± 2.

Diplast® TM. [Lonza SpA] Tri-2-ethylhexyl trimellitate; CAS 3319-31-1; monomeric plasticizer for PVC, cellulosics, paints, as vehicles for pigment dispersions; oily limpid liq.; m.w. 546; sp.gr. 0.989-0.992; b.p. 282 C (3 mm); sapon. no. 305 ± 3.

Diplast® TM8. [Lonza SpA] Tri-n-octyl trimellitate; CAS 89-04-3; monomeric plasticizer for PVC, cellulosics, paints, as vehicles for pigment dispersions; oily limpid liq.; m.w. 546; sp.gr. 0.985-0.987; m.p. -75 C; sapon. no. 305 ± 3.

Diprosin A-100. [Toho Chem. Industry] Disproportionated rosin; CAS 8050-09-7; polymerization emulsifier for syn. rubbers and plastics (SBR, NBR, CR, ABS, etc.); pale yel. transparent solid; sol. in alcohol, benzene, ether, acetic acid, acetone;

partly sol. in petrol. spirit; insol. in water; sp.gr. 1.08; soften. pt. 78-82 C; acid no. 155-160; flash pt. 190 C; pH acidic; *Toxicology:* may affect respiration; may cause allergic reaction; repeated/prolonged contact may cause skin sensitization; *Precaution:* may form flamm. or explosive dust-air mixts.; sl. fire hazard when exposed to heat or flame; thermal decomp. prods. incl. toxic/hazardous fumes of formaldehyde and oxides of carbon; *Storage:* store in cool, dry, well-ventilated area; keep container tightly closed; avoid creation of dust.

Diprosin K-80. [Toho Chem. Industry] Disproportionated rosin potassium soap; polymerization emulsifier for syn. rubbers and plastics (SBR, NBR, CR, ABS, etc.); paste.

Diprosin N-70. [Toho Chem. Industry] Disproportionated rosin sodium soap; CAS 61790-51-0; EINECS 263-144-5; polymerization emulsifier for syn. rubbers and plastics (SBR, NBR, CR, ABS, etc.); paste.

Disflamoll® DPK. [Bayer AG; Bayer/Fibers, Org., Rubbers] Diphenylcresyl phosphate; CAS 26444-49-5; flame retardant plasticizer for plasticized PVC prods.; used in air ducts, tarpaulins, driving and conveyor belts, imitation leather, coatings, hoses and extruded goods, cable sheathing and insulation, soles and inj. molded items; improves bitumen resist.; colorless liq.; sol. in solvs.; insol. in water; dens. 1.20-1.21 g/cm^3; visc. 45-50 mPa•s; b.p. 225-235 C; acid no. < 0.2; flash pt.(OC) > 230 C; ref. index 1.562-1.564.

Disflamoll® DPO. [Bayer AG; Bayer/Fibers, Org., Rubbers] Diphenyloctyl phosphate; CAS 1241-94-7; flame retardant plasticizer for PVC applics., dip, rotationally, extruded and inj. molded parts, mechanical foam; low temp. flexibility and weatherability; colorless liq.; sol. in solvs.; insol. in water; dens. 1.08-1.09 g/cc; visc. 21-23 mPa•s; b.p. 225 C; acid no. < 0.05; flash pt. > 190 C; ref. index 1.508-1.511.

Disflamoll® TCA. [Bayer AG] Trichloroethyl phosphate; CAS 306-52-5; flame retardant for PVC; good light stability; for coatings (PVA, acrylics, chlorinated rubber, NC lacquers); flame retardant plasticizer in celluloid, cellulose acetate, cellulose acetate butyrate compds.

Disflamoll® TKP. [Bayer AG; Bayer/Fibers, Org., Rubbers] Tricresyl phosphate; flame retardant plasticizer for plasticized PVC prods.; used in air ducts, tarpaulins, driving and conveyor belts, imitation leather, coatings, hoses and extruded goods, cable sheathing and insulation, soles and inj. molded items; colorless to sl. ylsh. liq.; sol. in org. solvs.; insol. in water; dens. 1.175-1.185 g/cc; visc. 65-75 mPa•s; b.p. 240-250 C; acid no. < 0.2; flash pt. (OC) > 230 C; ref. index 1.558-1.560.

Disflamoll® TOF. [Bayer AG; Bayer/Fibers, Org., Rubbers] Trioctyl phosphate; CAS 78-42-2; EINECS 201-116-6; flame retardant plasticizer for PVC with extreme low temp. flexibility (to -65 C); good resist. to saponification; also for solv.-free PU systems for joint sealants; exc. solv. power for PU; dens. 0.92 g/cm^3; visc. 14 cps (23 C).

Disflamoll® TP. [Bayer AG; Bayer/Fibers, Org., Rubbers] Triphenyl phosphate; CAS 115-86-6; EINECS 204-112-2; flame retardant; gelatinizing and plasticizing agent for collodion cotton; plasticizer w/o gelatinizing properties for acetyl cellulose; reduces flamm. of NC and acetyl cellulose-based plastic compds. and lacquer films; mfg. of photographic film materials, surf. coatings; for phenolic laminates, cellullose acetate film/compds., rubber articles made from acrylonitrile-butadiene, polychloroprene rubber; not compat. with PVC; wh. crystals, odorless; sol. in org. solvs.; insol. in water; dens. 1.205 g/cm^3 (50 C); visc. 11 cps (50 C); b.p. 250 C; acid no. 0.05 max.; flash pt. 235 C (OC); ref. index 1.555 (50 C).

Disperplast®1142. [Byk-Chemie USA] Polar acidic ester of long chain alcohols; anionic; wetting and dispersing additive for fillers and pigments in PVC plastisols, filler surf. treatments, prod. of masterbatches; DOP-free; provides lower visc., higher loadings, better color development; lt. brn. visc. liq., alcoholic odor; insol. in water; sp.gr. 1.07; dens. 8.83 lb/gal; acid no. 85; flash pt. (PMCC) > 110 C; ref. index 1.468; 100% act.; *Toxicology:* LD50 (oral, rat) 2600 mg/kg; eye irritant; nonirritating to skin; repeated ingestion may irritate digestive tract; *Precaution:* incompat. with strong oxidizers; hazardous combustion prods.: CO, CO_2; *Storage:* may become turbid below 10 C; warm to R.T. to reverse; keep container tightly closed when not in use.

Disperplast®-1150. [Byk-Chemie USA] Polar acidic ester of long chain alcohols; anionic; wetting and dispersing additive for TiO_2, ZnO, $CaCO_3$, and azodicarbonamide pastes in plasticizers for prod. of chemically foamed PVC plastisols; promotes more uniform and smaller cell structure; improves long-term color stability; colorless liq., alcoholic odor; sol. in water; sp.gr. 1.02; dens. 8.43 lb/gal; acid no. 95; flash pt. (PMCC) > 200 C; ref. index 1.459; 100% act.; *Toxicology:* LD50 (oral, rat) > 4000 mg/kg; sl. eye irritant; nonirritating to skin; repeated ingestion may irritate digestive tract; *Precaution:* incompat. with strong oxidizers; hazardous combustion prods.: CO, CO_2; *Storage:* may crystallize below 18 C; warm to 25-30 C with stirring to reverse; keep container tightly closed when not in use.

Disperplast®-I. [Byk-Chemie USA] Partial amide of higher m.w. unsat. polycarboxylic acid in 50% diisodecyl phthalate plasticizer; anionic; wetting, dispersing, and antiflooding additive for inorg. pigments in PVC plastisols and azodicarbonamide in nonfogging plasticizers; DOP-free; lt. brn. visc. liq., odorless; insol. in water; sp.gr. 0.97; dens. 8.08 lb/gal; acid no. 58; amine no. 11; flash pt. (PMCC) > 200 C; ref. index 1.485; *Toxicology:* LD50 (oral, rat) > 5000 mg/kg; sl. skin and eye irritant; inh. of vapors may cause irritation to respiratory tract and mucous membranes; *Precaution:* incompat. with strong oxidizers; hazardous combustion prods.: CO, CO_2, NO_x, phosphorous oxides, silicon compds., formaldehyde; *Storage:* may become turbid below 10 C; warm to 30 C to reverse; keep container tightly closed when not in use.

Disperplast®-P. [Byk-Chemie USA] Unsat. polycarboxylic acid in 50% diisodecyl phthalate plasticizer; anionic; wetting, dispersing, and antiflooding additive for org. pigments and carbon blk. in PVC plastisols, nonfogging plasticizers; DOP-free; lt. brn. visc. liq., odorless; insol. in water; sp.gr. 1.01; dens. 8.08 lb/gal; acid no. 180; flash pt. (PMCC) 120 C; ref. index 1.487; *Toxicology:* LD50 (oral, rat) > 5000 mg/kg; sl. skin irritant; eye irritant; inh. of vapors may cause irritation to respiratory tract and mucous membranes; *Precaution:* incompat. with strong oxidizers; hazardous combustion prods.: CO, CO_2, phosphorous oxides; *Storage:* may become turbid below 10 C; warm to 30 C to reverse; keep container tightly closed when not in use.

Disperso II. [Sovereign] Zinc stearate; CAS 557-05-1; EINECS 209-151-9; Coad 20 with emulsifying agent for water-based slab dips.

Disponil AAP 307. [Henkel/Functional Prods.] Alkylaryl polyglycol ether; nonionic; coemulsifier for polyacrylates, acrylate-vinyl acetate copolymers, other applics.; dispersant for emulsion paints; liq.; HLB 17.3; 70% act.

Disponil AAP 436. [Henkel/Functional Prods.] Alkylaryl polyglycol ether; nonionic; coemulsifier for polyacrylates, acrylate vinyl acetate copolymers; dispersant for emulsion paints; liq.; HLB 17.9; 60% act.

Disponil AAP 437. [Pulcra SA] Octylphenol, ethoxylated; nonionic; primary emulsifier for vinyl acetate and acrylate polymerization; post stabilizer for syn. latex; dyeing assistant and emulsifier for fats and waxes; solid; HLB 17.9; 100% conc.

Disponil AEP 5300. [Henkel/Functional Prods.] Ether phosphate, acid ester; anionic; emulsifier for rosin, vinyl acetate and acrylate systems; liq.; 100% conc.

Disponil AEP 8100. [Henkel/Functional Prods.] Ether phosphate, acid ester; anionic; emulsifier for rosin, vinyl acetate and acrylate systems; liq.; 100% solids.

Disponil AEP 9525. [Henkel/Functional Prods.] Ether phosphate, sodium salt; anionic; emulsifier for rosin, vinyl acetate and acrylate systems; liq.; 25% solids.

Disponil AES 13. [Henkel/Functional Prods.] Sodium alkylaryl ether sulfate; anionic; emulsifier for vinyl acetate homopolymers, acrylate homo and copolymers, styrene acrylate copolymers, vinyl acetate-acrylate copolymers, VAE copolymers, PVDC latexes, vinyl chloride homo and copolymer latexes; liq.; 31-34% conc.

Disponil AES 21. [Henkel/Functional Prods.] Sodium alkylaryl ether sulfate; anionic; emulsifier for vinyl acetate homo and copolymers, acrylate homo and copolymers, styrene-acrylate, S/B latexes, vinyl propionate copolymers, PVDC, vinyl chloride homo and copolymer latexes; liq.; 30-35% solids.

Disponil AES 42. [Henkel/Functional Prods.] Sodium alkylaryl ether sulfate; anionic; emulsifier for vinyl acetate homo and copolymers, acrylate homo and copolymers, styrene-acrylate, S/B latexes, vinyl propionate copolymers, PVDC, vinyl chloride homo and copolymer latexes; liq.; 40% solids.

Disponil AES 48. [Henkel/Functional Prods.] Ammonium alkylaryl ether sulfate; anionic; emulsifier for emulsion polymerization, natural fats; dispersant for chrome and lime soaps; liq./paste; 68-72% solids.

Disponil AES 60. [Henkel/Functional Prods.] Sodium alkylaryl ether sulfate; anionic; emulsifier for vinyl acetate homo and copolymers, acrylate homo and copolymers, styrene-acrylate, S/B latexes, vinyl propionate copolymers, PVDC, vinyl chloride homo and copolymer latexes; liq.; 34-36% solids.

Disponil AES 60 E. [Pulcra SA] Sodium nonoxynol-8 sulfate; anionic; emulsifier for emulsion polymerization; liq.; 30% conc.

Disponil AES 72. [Henkel/Functional Prods.] Sodium alkylaryl ether sulfate; anionic; low foaming emulsifier for vinyl acetate homo and copolymers, acrylate homo and copolymers, styrene-acrylate, S/B latexes, vinyl propionate copolymers, PVDC, vinyl chloride homo and copolymer latexes; liq.; 34-36% solids.

Disponil APG 110. [Henkel/Functional Prods.] Linear alkyl diol polyglycol ether; nonionic; coemulsifier; latex stabilizer; liq.; HLB 14.1; 100% act.

Disponil CSL 100 K. [Henkel/Functional Prods.] Complex ester of natural fatty acid, calcium salt; sec. dispersant for S-PVC; beads/powd.; 100% act.

Disponil FES 32. [Henkel/Functional Prods.] Sodium laureth sulfate; CAS 9004-82-4; anionic; emulsifier for vinyl acetate copolymers, S/B latexes, vinyl chloride copolymers, acrylate homo- and copolymers; liq.; 29-32% solids.

Disponil FES 61. [Henkel/Functional Prods.] Sodium laureth sulfate; CAS 9004-82-4; anionic; emulsifier for vinyl acetate copolymers, S/B latexes, vinyl chloride copolymers, acrylate homo- and copolymers; liq.; 29-32% solids.

Disponil FES 77. [Henkel/Functional Prods.] Sodium laureth sulfate; CAS 9004-82-4; anionic; emulsifier for polymerization of vinyl acetate copolymers, S/B latexes, vinyl chloride copolymers, acrylate homo- and copolymers; liq.; 30-34% conc.

Disponil FES 92E. [Pulcra SA] Sodium laureth-12 sulfate; CAS 9004-82-4; anionic; surfactant for low-irritation shampoos, emulsion polymerization; liq.; m.w. 830; pH 8.0-9.0 (10%); 29-31% act.

Disponil MGS 777. [Henkel/Functional Prods.] Surfactant blend; rheology modifier for PVC-plastisol resins; liq.; 100% act.

Disponil MGS 935. [Henkel/Functional Prods.] Surfactant blend; anionic; emulsifier for vinyl acetate homo and copolymers; foaming and wetting agent for highly water absorbent foam resins; liq.; 50% solids.

Disponil O 5. [Henkel/Functional Prods.] Cetoleth-5; nonionic; emulsifier for emulsion polymerization; wetting agent; liq.; HLB 9.5; 100% act.

Disponil O 10. [Henkel/Functional Prods.] Cetoleth-10; nonionic; emulsifier for emulsion polymerization; wetting agent; paste; HLB 12.5; 100% act.

Disponil O 20. [Henkel/Functional Prods.] Ceteth-20; nonionic; emulsifier for emulsion polymerization; solubilizer, dispersant, latex stabilizer; solid; HLB 16.5; 100% act.

Disponil O 250. [Henkel/Functional Prods.] Ceteth-25; nonionic; emulsifier for emulsion polymerization; solubilizer, dispersant, latex stabilizer; flakes; HLB 18.4; 100% act.

Disponil PNP 208. [Pulcra SA] Ethoxylated nonylphenol; nonionic; cosurfactant for emulsion and dispersion mfg.; transparent liq., char. odor; sol. in water; forms gels @ 40-70%; dens. 1.08 g/cc; visc. < 1000 mPa•s; solid. pt. < -5 C; flash pt. > 150 C; pH 5.5-75. (1%); 80% conc.; *Storage:* store at ambient temps. in original pkg.; protect against heat and cold.

Disponil SML 100 F1. [Henkel/Functional Prods.] Sorbitan laurate; CAS 1338-39-2; nonionic; surfactant for polymerization; liq.; HLB 8.6; 100% act.

Disponil SML 104 F1. [Henkel/Functional Prods.] Polysorbate 21; CAS 9005-64-5; nonionic; surfactant for polymerization; liq.; HLB 13.3; 100% act.

Disponil SML 120 F1. [Henkel/Functional Prods.] Polysorbate 20; CAS 9005-64-5; nonionic; surfactant for polymerization; liq.; HLB 16.7; 100% act.

Disponil SMO 100 F1. [Henkel/Functional Prods.] Sorbitan oleate; CAS 1338-43-8; nonionic; surfactant for polymerization; liq.; HLB 4.3; 100% act.

Disponil SMO 120 F1. [Henkel/Functional Prods.] Polysorbate 80; CAS 9005-65-6; nonionic; surfactant for polymerization; liq.; HLB 15.0; 100% act.

Disponil SMP 100 F1. [Henkel/Functional Prods.] Sorbitan palmitate; CAS 26266-57-9; EINECS 247-568-8; nonionic; surfactant for polymerization; flakes; HLB 6.7; 100% act.

Disponil SMP 120 F1. [Henkel/Functional Prods.] Polysorbate 40; CAS 9005-66-7; nonionic; surfactant for polymerization; liq.; HLB 15.6; 100% act.

Disponil SMS 100 F1. [Henkel/Functional Prods.] Sorbitan stearate; CAS 1338-41-6; EINECS 215-664-9; nonionic; surfactant for polymerization; flakes; HLB 4.7; 100% act.

Disponil SMS 120 F1. [Henkel/Functional Prods.] Polysorbate 60; CAS 9005-67-8; nonionic; surfactant for polymerization; liq.; HLB 14.9; 100% act.

Disponil SSO 100 F1. [Henkel/Functional Prods.] Sorbitan sesquioleate; CAS 8007-43-0; EINECS 232-360-1; nonionic; surfactant for polymerization; liq.; HLB 3.7; 100% act.

Disponil STO 100 F1. [Henkel/Functional Prods.] Sorbitan trioleate; CAS 26266-58-0; nonionic; surfactant for polymerization; liq.; HLB 1.8; 100% act.

Disponil STO 120 F1. [Henkel/Functional Prods.] Polysorbate 85; CAS 9005-70-3; nonionic; surfactant for polymerization; liq.; HLB 11.0; 100% act.

Disponil STS 100 F1. [Henkel/Functional Prods.] Sorbitan tristearate; CAS 26658-19-5; nonionic; surfactant for polymerization; flakes; HLB 2.1; 100% act.

Disponil STS 120 F1. [Henkel/Functional Prods.] Polysorbate 65; CAS 9005-71-4; nonionic; surfactant for polymerization; solid; HLB 10.5; 100% act.

Disponil SUS 29 L. [Henkel/Functional Prods.] Sodium sulfosuccinamate; anionic; emulsifier for vinyl acetate copolymer, acrylate homo and copolymer, S/B latexes, vinyl chloride copolymers; liq.; 35% solids.

Disponil SUS 65. [Henkel/Functional Prods.] Fatty alcohol ethoxylate sulfosuccinate; anionic; emulsifier for vinyl acetate copolymer, acrylate homo and copolymer, S/B latexes, vinyl chloride copolymers; liq.; 40% solids.

Disponil SUS 87 Special. [Henkel/Functional Prods.] Disodium sulfosuccinate; anionic; emulsifier in polymerization of vinyl acetate copolymers, acrylate homo and copolymers, S/B latexes, and vinyl chloride copolymers; liq.; 30-32% conc.

Disponil SUS 90. [Henkel/Functional Prods.] Sodium alkylaryl EO sulfosuccinate; anionic; emulsifier for vinyl acetate copolymer, acrylate homo and copolymer, S/B latexes, vinyl chloride copolymers; liq.; 32-35% solids.

Disponil SUS IC 8. [Henkel/Functional Prods.] Dioctyl sodium sulfosuccinate; wetting agent, coemulsifier for plastics industry; liq.; 70-75% conc.

Disponil TA 5. [Henkel/Functional Prods.] Linear sat. fatty alcohol polyglycol ether; nonionic; wetting agent, polymerization emulsifier, dispersant; liq.; HLB 9.2; 100% act.

Disponil TA 25. [Henkel/Functional Prods.] Linear sat. fatty alcohol polyglycol ether; nonionic; wetting agent, polymerization emulsifier, dispersant; solid; HLB 16.2; 100% act.

Disponil TA 430. [Henkel/Functional Prods.] Linear sat. fatty alcohol polyglycol ether; nonionic; wetting agent, polymerization emulsifier, dispersant; flake; HLB 17.4; 100% act.

Dixie® Clay. [R.T. Vanderbilt] Kaolin; CAS 1332-58-7; EINECS 296-473-8; hard reinforcing min. filler and extender for elastomers, polyester, styrene, epoxies, phenol formaldehyde resins; better flexural modulus and improved flow chars.; wh. to cream powd., 0.2 μm median particle size; 99.8% min. thru 325 mesh; dens. 2.62 mg/m^3; bulk dens. 42 lb/ft^3 (compacted); oil absorp. 42; brightness 70; pH 4.7 (10% aq. slurry); 44.87% SiO_2, 33.77% Al_2O_3.

DMPA®. [IMC/Americhem] Dimethylolpropionic acid; CAS 4767-03-7; in prep. of water-sol. alkyd resins, conventional alkyds, polyester resins, surfactants, chemical intermediates, syn. lubricants, plasticizers, pharmaceuticals, cosmetics; produces coatings with outstanding thermal, hydrolytic, and color stability; off-wh. free-flowing gran.; sol. in water, methanol; sl. sol. in acetone; insol. in benzene; m.w. 134; sp.gr. 1.355; m.p. 170-180 C; *Toxicology:* essentially nontoxic; LD50 (mouse, oral) > 5000 mg/kg; sl. irritating to abraded skin; mod. eye irritant.

Doittol K21. [Henkel] Fatty acids and alkylpolyglycol ethers; antiblocking and mold release agent; liq.; Unverified

Dolocron® 15-16. [Specialty Minerals] Pulverized dolomitic limestone; CAS 1317-65-3; filler for building prods., cultured marble, putties, caulking compds., floor tile, rubber, cement, plastics, ceramics, coatings; powd.; 15 μ max. particle size; sp.gr. 2.8; bulk dens. 55 lb/ft^3; oil absorp. 16 g/100 g; brightness 90; 55% $CaCO_3$, 43% $MgCO_3$.

Dolocron® 32-15. [Specialty Minerals] Pulverized dolomitic limestone; CAS 1317-65-3; filler for building prods., cultured marble, putties, caulking compds., floor tile, rubber, cement, plastics, ceramics, coatings; powd.; 32 μ max. particle size; sp.gr. 2.8; bulk dens. 65 lb/ft^3; oil absorp. 15 g/100 g; brightness 89; 55% $CaCO_3$, 43% $MgCO_3$.

Dolocron® 40-13. [Specialty Minerals] Pulverized dolomitic limestone; CAS 1317-65-3; filler for building prods., cultured marble, putties, caulking compds., floor tile, rubber, cement, plastics, ceramics, coatings; powd.; 40 μ max. particle size; sp.gr. 2.8; bulk dens. 65 lb/ft^3; oil absorp. 13 g/100 g; brightness 89; 55% $CaCO_3$, 43% $MgCO_3$.

Dolocron® 45-12. [Specialty Minerals] Pulverized dolomitic limestone; CAS 1317-65-3; filler for building prods., cultured marble, putties, caulking compds., floor tile, rubber, cement, plastics, ceramics, coatings; powd.; 45 μ max. particle size; sp.gr. 2.8; bulk dens. 70 lb/ft^3; oil absorp. 12 g/100 g; brightness 88; 55% $CaCO_3$, 43% $MgCO_3$.

Doverguard® 152. [Dover] Chlorinated paraffin; flame retardant for plastics, rubbers, adhesives, air filters, paints, fabric coatings; Gardner 1 liq.; sp.gr. 1.270; visc. 15 poise; 51% Cl.

Doverguard® 170. [Dover] Chlorinated paraffin; flame retardant for plastics, rubbers, adhesives, paints, fabric coatings; Gardner 1 liq.; sp.gr. 1.520; visc. 1500 poise; 70% Cl.

Doverguard® 700. [Dover] Resinous chlorinated paraffin; flame retardant for plastics esp. LDPE, reinforced polyester, rubbers, adhesives, paints, fabric coatings; Gardner < 1 solid; sp.gr. 1.6; soften. pt. 95-110 C; 70% Cl.

Doverguard® 700-S. [Dover] Resinous chlorinated paraffin; flame retardant for plastics esp. linear HDPE, PP, rubbers, adhesives, paints, fabric coatings; Gardner < 1 solid; sp.gr. 1.6; soften. pt. 95-110 C; 70% Cl.

Doverguard® 700-SS. [Dover] Resinous chlorinated paraffin; flame retardant for plastics esp. linear HDPE, PP, PS, rubbers, adhesives, paints, fabric coatings; Gardner < 1 solid; sp.gr. 1.6; soften. pt. 105-120 C; 71% Cl.

Doverguard® 760. [Dover] Resinous chlorinated paraffin; flame retardant for plastics esp. linear HDPE, PP, PS, rubbers, adhesives, paints, fabric coatings; Gardner < 1 solid; sp.gr. 1.7; soften. pt. 160 C; 74% Cl.

Doverguard® 5761. [Dover] Chlorinated paraffin; flame retardant for plastics, rubbers, adhesives, paints, fabric coatings; Gardner 1 liq.; sp.gr. 1.355; visc. 20 poise; 59% Cl.

Doverguard® 8133. [Dover] Resinous bromochlorinated paraffin; flame retardant for polyolefins, PS, reinforced polyesters, and fabrics; Gardner 1 solid; sp.gr. 2.3 (50 C); soften. pt. 130-300 C; 41% Br, 35% Cl.

Doverguard® 8207-A. [Dover] Bromochlorinated paraffin; flame retardant, plasticizer for PVC, PU foams, textiles, carpet backing, rubbers, adhesives, paints; Gardner 1 liq.; sp.gr. 1.42 (50 C); visc. 22 poise; 30% Br, 29% Cl.

Doverguard® 8208-A. [Dover] Bromochlorinated paraffin; flame retardant, plasticizer for PVC, PU foams, textiles, carpet backing, rubbers, adhesives, paints; Gardner 2 liq.; sp.gr. 1.38 (50 C); visc. 15 poise; 32% Cl, 26% Br, 1.2% P.

Doverguard® 8307-A. [Dover] Bromochlorinated paraffin; flame retardant, plasticizer for PVC, PU foams, textiles, carpet backing, rubbers, adhesives, paints; Gardner 1 liq.; sp.gr. 1.37 (50 C); visc. 1 poise; 24% Br, 23% Cl, 5% P.

Doverguard® 8410. [Dover] Brominated paraffin; flame retardant, plasticizer for PVC, PU foams, textiles, carpet backing, rubbers, adhesives, paints; Gardner 3 liq.; sp.gr. 1.52 (50 C); visc. 0.5 poise; 57% Br.

Doverguard® 8426. [Dover] Brominated paraffin; flame retardant, plasticizer for PVC, PU foams, textiles, carpet backing, rubbers, adhesives, paints; Gardner 1 liq.; sp.gr. 1.16 (50 C); visc. 0.15 poise; 42% Br.

Doverguard® 9021. [Dover] Bromochlorinated paraffin/phosphorus blend; flame retardant for flexible PU foam formulations.

Doverguard® 9119. [Dover] Bromochlorinated paraffin; flame retardant, plasticizer for PVC, PU foams, textiles, carpet backing, rubbers, adhesives, paints; Gardner 1 liq.; sp.gr. 1.58 (50 C); visc. 65 poise; 33% Br, 33% Cl.

Doverguard® 9122. [Dover] Bromochlorinated paraffin/phosphorus blend; flame retardant, plasticizer for PVC, PU foams, textiles, carpet backing, rubbers, adhesives, paints; Gardner 1 liq.; sp.gr. 1.35 (50 C); visc. 3 poise; 16% Br, 16% Cl, 4% P.

Dovernox 10. [Dover] Tetrakis methylene (3,5-di-t-butyl-4-hydroxyhydrocinnamate) methane; CAS 6683-19-8; EINECS 229-722-6; nonstaining, nondiscoloring antioxidant for polymers providing good long term heat aging and high temp. stability to polyolefins, styrenics, block copolymers, elastomers, adhesives, PVC, PU; low volatility; broad FDA acceptance; wh. free-flowing powd.; sol. (g/100 ml, 20 C): 47 g acetone, 46 g ethyl acetate, 1 g methanol, < 0.01 g water; m.w. 1178; m.p. 110-125 C; *Toxicology:* low oral and dermal toxicity.

Dovernox 76. [Dover] Octadecyl 3,5-di-t-butyl-4-hydroxyhydrocinnamate; CAS 2082-79-3; EINECS 218-216-0; nonstaining, nondiscoloring antioxidant for polymers providing good long term heat aging and high temp. processing stability to polyolefins, styrenics, block copolymers, elastomers, adhesives, PVC, PU; low volatility; broad FDA approval; wh. cryst. powd.; sol. (g/100 ml, 20 C): 38 g ethyl acetate, 32 g hexane, 19 g acetone, < 0.01 g water; m.w. 531; m.p. 49-54 C; *Toxicology:* low oral and dermal toxicity.

Dovernox 3114. [Dover] Tris (3,5-di-t-butyl-4-

hydroxybenzyl) isocyanurate; CAS 27676-62-6; nonstaining, nondiscoloring antioxidant protecting polyolefins and polymers against oxidative degradation by heat, lt., and mech. shear; esp. for PP film, fiber applics., and in talc- and other min.-filled PP compds.; nonvolatile, nonextractable; wh. flowable powd.; sol. (g/100 ml, 20 C): 32 g ethyl acetate, 27 g acetone, < 0.01 g water; m.w. 784; sp.gr. 1.03; m.p. 218 C min.

Doverphos® 4. [Dover] Trisnonylphenyl phosphite; CAS 26523-78-4; EINECS 247-759-6; heat stabilizer for PVC, ABS, polyolefins, some rubber prods.; FDA 21CFR §175.105, 175.300, 177.1210, 177.2600, 178.2010; Gardner 2 max. clear liq.; sp.gr. 0.980-0.992 (25/15.5 C); dens. 8.2 lb/gal; visc. 4600 cps; acid no. 0.1 max.; flash pt. (COC) 405 F; ref. index 1.5255-1.5280; 4.3% P; *Toxicology:* skin irritant; *Storage:* keep containers tightly closed; protect from humidity.

Doverphos® 4-HR. [Dover] Trisnonylphenyl phosphite, 0.75% triisopropanolamine; hydrolysis-resistant antioxidant for elastomer mfg.; FDA 21CFR §178.2010; Gardner 3 max. clear liq.; sp.gr. 0.980-0.992 (25/15.5 C); dens. 8.2 lb/gal; visc. 4900 cps; acid no. 0.5 max.; flash pt. (COC) 383 F; ref. index 1.5250-1.5280; 4.3% P; *Toxicology:* skin irritant; *Storage:* keep containers tightly closed; protect from humidity.

Doverphos® 4-HR Plus. [Dover] Trisnonylphenyl phosphite, 1.0% triisopropanolamine; enhanced hydrolysis resist. antioxidant for elastomer mfg.; FDA 21CFR §178.2010; Gardner 3 max. clear liq.; sp.gr. 0.980-0.992 (25/15.5 C); dens. 8.2 lb/gal; visc. 5900 cps; acid no. 0.5 max.; flash pt. (COC) 383 F; ref. index 1.5250-1.5280; 4.3% P; *Toxicology:* skin irritant; *Storage:* keep containers tightly closed; protect from humidity.

Doverphos® 4 Powd. [Dover] Trisnonylphenyl phosphite; CAS 26523-78-4; EINECS 247-759-6; heat stabilizer for PVC, ABS, polyolefins, some rubber prods.; FDA compliance; wh. powd.; sp.gr. 1.15; 73% TNPP, 27% SiO_2.

Doverphos® 6. [Dover] Triisodecyl phosphite; CAS 25448-25-3; EINECS 246-998-3; heat stabilizer for polyolefins, PU, PVC, PET, coatings, lubricants; improves color in rigid and plasticized PVC; chelating agent with metal carboxylates as polymer additives, esp. for chlorinated polymers such as PVC and chlorinated PE; antioxidant and EP additive for lubricants; APHA 50 max. clear liq.; sp.gr. 0.884-0.904 (25/15.5 C); dens. 7.4 lb/gal; visc. 15 cps; acid no. 0.05 max.; flash pt. (COC) 420 F; ref. index 1.4530-1.4610; 6.2% P; *Toxicology:* skin irritant; *Storage:* keep containers tightly closed; protect from humidity.

Doverphos® 7. [Dover] Phenyl diisodecyl phosphite; CAS 25550-98-5; EINECS 247-098-3; stabilizer for rigid and flexible PVC; chelating agent with metal carboxylates as polymer additives, esp. for chlorinated polymers; APHA 50 max. clear liq.; sp.gr. 0.938-0.947 (25/15.5 C); dens. 7.8 lb/gal; visc. 17 cps; acid no. 0.05 max.; flash pt. (COC) 420 F; ref. index 1.4780-1.4810; 7.1% P; *Toxicology:* skin irritant; *Storage:* keep containers tightly closed; protect from humidity.

Doverphos® 8. [Dover] Diphenyl isodecyl phosphite; CAS 26544-23-0; stabilizer for rigid and flexible PVC; chelating agent with metal carboxylates as polymer additives, esp. for chlorinated polymers such as PVC and chlorinated PE; improves color, heat and light stability; APHA 50 max. clear liq.; sp.gr. 1.022-1.032 (25/15.5 C); dens. 8.6 lb/gal; visc. 14 cps; acid no. 0.05 max.; flash pt. (PMCC) 320 F; ref. index 1.5160-1.5190; 8.3% P; *Toxicology:* skin irritant; *Storage:* keep containers tightly closed; protect from humidity.

Doverphos® 9. [Dover] Diphenyl isooctyl phosphite; CAS 26401-27-4; EINECS 247-658-7; stabilizer for rigid and flexible PVC; chelating agent with metal carboxylates as polymer additives, esp. for chlorinated polymers such as PVC and chlorinated PE; improves color, heat and light stability; APHA 50 max. clear liq.; sp.gr. 1.040-1.047 (25/15.5 C); dens. 8.7 lb/gal; visc. 9 cps; acid no. 0.05 max.; flash pt. (PMCC) 358 F; ref. index 1.5210-1.5230.

Doverphos® 10. [Dover] Triphenyl phosphite; CAS 101-02-0; EINECS 202-908-4; heat stabilizer for engineering thermoplastics, epoxies, PVC, ABS, coatings, adhesives; improves color in polyesters; aids curing and hardening in epoxies; APHA 50 max. clear liq.; sp.gr. 1.180-1.186 (25/15.5 C); dens. 9.8 lb/gal; acid no. 0.50 max.; ref. index 1.5880-1.5900; 10% P; *Toxicology:* skin irritant; *Storage:* keep containers tightly closed; protect from humidity.

Doverphos® 10-HR. [Dover] Triphenyl phosphite, 0.5% triisopropanolamine; improved hydrolysis-resist. heat stabilizer for engineering thermoplastics, epoxies, PVC, ABS, coatings, adhesives; improves color in polyesters; aids curing and hardening in epoxies; APHA 50 max. clear liq.; sp.gr. 1.180-1.186 (25/15.5 C); dens. 9.8 lg/gal; acid no. 0.50 max.; ref. index 1.5860-1.5890; 10% P; *Toxicology:* skin irritant; *Storage:* keep containers tightly closed; protect from humidity.

Doverphos® 10-HR Plus. [Dover] Triphenyl phosphite, 1.0% triisopropanolamine; improved hydrolysis-resist. heat stabilizer for engineering thermoplastics, epoxies, PVC, ABS, coatings, adhesives; improves color in polyesters; aids curing and hardening in epoxies; APHA 50 max. clear liq.; sp.gr. 1.180-1.186 (25/15.5 C); dens. 9.8 lb/gal; acid no. 0.50 max.; ref. index 1.5860-1.5890; 10% P; *Toxicology:* skin irritant; *Storage:* keep containers tightly closed; protect from humidity.

Doverphos® 11. [Dover] Tetraphenyl dipropylene glycol diphosphite; CAS 80584-85-6; EINECS 279-498-9; stabilizer in rigid and plasticized PVC applics. where high gloss is important; results in low color in film applics.; APHA 50 liq.; sp.gr. 1.164-1.188 (25/15.5 C); visc. 80 cps; acid no. 0.20; flash pt. (PM) 350 F; ref. index 1.5570-1.5620; 10.9% P.

Doverphos® 12. [Dover] Poly (dipropylene glycol) phenyl phosphite; CAS 80584-86-7; EINECS 279-

499-4; stabilizer for rigid and plasticized PVC where early color hold is important; chelating agent with metal carboxylates as polymer additives, esp. for chlorinated polymers such as PVC and chlorinated PE; improves color, heat and light stability; APHA 50 max. clear liq.; sp.gr. 1.168-1.180 (25/15.5 C); dens. 9 lb/gal; acid no. 0.10 max.; flash pt. (COC) 395 F; ref. index 1.5340-1.5380; 11.8% P; *Toxicology:* skin irritant; *Storage:* keep containers tightly closed; protect from humidity.

Doverphos® 49. [Dover] Trisisotridecyl phosphite; stabilizer in conjunction with mixed metal heat stabilizers used in PVC compounding; good hydrolytic stability, low volatility; liq.

Doverphos® 53. [Dover] Trilauryl phosphite; CAS 3076-63-9; EINECS 221-356-5; stabilizer in conjunction with mixed metal heat stabilizers used in PVC compounding; liq.; 5.3% P.

Doverphos® 213. [Dover] Diphenyl phosphite; CAS 4712-55-4; EINECS 225-202-8; polymer stabilizer; APHA 50 liq.; sp.gr. 1.200-1.220 (25/15.5 C); visc. 12 cps; acid no. 15; flash pt. (COC) 345 F; ref. index 1.550-1.5575; 13.3% P.

Doverphos® 613. [Dover] Alkyl (C12-C15) bisphenol A phosphite; CAS 93356-94-6; stabilizer improving color, lt. stability, and clarity of rigid PVC and vinyl copolymers used in food pkg. applics. when used in conjunction with primary metal and epoxy stabilizers; liq.; 6.1% P.

Doverphos® 675. [Dover] Tetrakis isodecyl 4,4′-isopropylidene diphosphite; CAS 61670-79-9; stabilizer for plastics; liq.; 6.7% P.

Doverphos® S-480. [Dover] Tris (2,4-di-t-butylphenyl) phosphite; CAS 31570-04-4; processing stabilizer for wide range of polymers; hydrolytically stable; solid; 4.8% P.

Doverphos® S-680. [Dover] Distearylpentaerythritol diphosphite; CAS 3806-34-6; EINECS 223-276-6; color stabilizer and melt flow aid for polymer processing (polyolefins, PS, rubber, ABS, PVC, polyester, PC); synergistic with hindered phenol antioxidants and HALS; FDA compliance; wh. gran.; insol. in water; partly sol. in toluene, chloroform; m.w. 732; sp.gr. 0.920-0.935; m.p. 40-70 C; acid no. 0.50; flash pt. (COC) 450 F; 50% act., 7.9% P.

Doverphos® S-682. [Dover] Distearylpentaerythritol diphosphite, 1% triisopropanolamine; enhanced hydrolytic resist. color stabilizer and melt flow aid for polymer processing (polyolefins, PS, rubber, ABS, PVC, polyester, PC); FDA compliance; wh. granules; insol. in water; partly sol. in toluene, chloroform; m.w. 732; sp.gr. 0.920-0.935; m.p. 40-70 C; acid no. 0.50; flash pt. (COC) 450 F; 50% act., 7.1% P.

Doverphos® S-686, S-687. [Dover] Distearylpentaerythritol diphosphite; CAS 3806-34-6; EINECS 223-276-6; color stabilizer and melt flow aid for polyolefins, PS, ABS, rubber, PVC, polyester, PC, and cellulosics; improves uv resist. used with light stabilizers; improved hydrolytic stability; FDA 21CFR §178.2010; wh. granules; insol. in water; partly sol. in toluene, chloroform; sp.gr. 0.920-0.935; m.p. 50-80 C; 50% act. in inert ingreds.; *Storage:* store in cool, dry area; minimize contact with humid air.

Doversperse 3. [Dover] Aq. emulsion of Paroil 170HV (chlorinated flame retardant); flame retardant, plasticizer, tackifier improving adhesion, imparting chem. and water resist.; for adhesives, rubber coatings, inks, carpet backings, paper and fabric coatings; cream wh. liq.; sp.gr. 1.54; dens. 11.3 lb/gal; visc. 150-300 poise; pH 7; 66.5% solids, 45% Cl.

Doversperse 8843. [Dover] Aq. flame retardant; cream wh. liq.; sp.gr. 1.70; visc. 40-70 poise; 65% solids, 20% Br, 17% Cl, 16% Sb.

Doversperse 8929. [Dover] Aq. flame retardant; cream wh. liq.; sp.gr. 1.50; visc. 40-70 poise; 65% solids, 34% Cl, 16% Sb.

Doversperse A-1. [Dover] Aq. disp. of Chlorez 700; nonionic; flame retardant improving adhesion, imparting chem. and water resist.; for adhesives, rubber coatings, inks, carpet backings, paper and fabric coatings; recommended where increased hardness is required; cream wh. liq.; sp.gr. 1.60; dens. 11 lb/gal; visc. 48 poise; pH 7.5; 65% solids, 45% Cl.

Dow E200. [Dow] PEG-4; CAS 25322-68-3; EINECS 203-989-9; carrier, modifier with exc. solvency, lubricity for personal care prods. (makeup, bath preps., toothpaste), foam control agents, syn. lubricants, rubber and plastics, metalworking fluids, chem. intermediates; controls cure rates in clay-contg. rubber formulations; improves flexibility, imparts antistatic props. to thermoplastics; as heat transfer fluids in plastic thermoforming; mold release; liq.; m.w. 200; sp.gr. 1.124; dens. 9.35 lb/gal; visc. 40 cSt; f.p. supercools; flash pt. (PMCC) 340 F; ref. index 1.459; sp. heat 0.524 cal/g/°C.

Dow E300. [Dow] PEG-6; CAS 25322-68-3; EINECS 203-989-9; carrier, modifier with exc. solvency, lubricity for personal care prods. (makeup, bath preps., toothpaste), foam control agents, syn. lubricants, rubber and plastics, metalworking fluids, chem. intermediates; controls cure rates in clay-contg. rubber formulations; improves flexibility, imparts antistatic props. to thermoplastics; as heat transfer fluids in plastic thermoforming; mold release; liq.; m.w. 300; sp.gr. 1.125; dens. 9.36 lb/gal; visc. 69 cSt; f.p. -10 C; flash pt. (PMCC) > 400 F; ref. index 1.463; sp. heat 0.508 cal/g/°C.

Dow E400. [Dow] PEG-8; CAS 25322-68-3; EINECS 203-989-9; carrier, modifier with exc. solvency, lubricity for personal care prods. (makeup, bath preps., toothpaste), foam control agents, syn. lubricants, rubber and plastics, metalworking fluids, chem. intermediates; controls cure rates in clay-contg. rubber formulations; improves flexibility, imparts antistatic props. to thermoplastics; as heat transfer fluids in plastic thermoforming; mold release; liq.; m.w. 400; sp.gr. 1.125; dens. 9.36 lb/gal; f.p. 6 C; visc. 90 cSt; flash pt. (PMCC) > 450 F; ref. index 1.465; sp. heat 0.498 cal/g/°C.

Dow E600. [Dow] PEG-12; CAS 25322-68-3; EINECS

203-989-9; carrier, modifier with exc. solvency, lubricity for personal care prods. (makeup, bath preps., toothpaste), foam control agents, syn. lubricants, rubber and plastics, metalworking fluids, chem. intermediates; controls cure rates in clay-contg. rubber formulations; improves flexibility, imparts antistatic props. to thermoplastics; as heat transfer fluids in plastic thermoforming; mold release; liq.; sol. in water, ethanol, cyclomethicone, sunscreens, lactic acid; m.w. 600; sp.gr. 1.126; dens. 9.37 lb/gal; f.p. 22 C; visc. 131 cSt; flash pt. (PMCC) > 450 F; ref. index 1.466; sp. heat 0.490 cal/g/°C.

Dow E900. [Dow] CAS 25322-68-3; EINECS 203-989-9; carrier, modifier with exc. solvency, lubricity for personal care prods. (makeup, bath preps., toothpaste), foam control agents, syn. lubricants, rubber and plastics, metalworking fluids, chem. intermediates; controls cure rates in clay-contg. rubber formulations; improves flexibility, imparts antistatic props. to thermoplastics; as heat transfer fluids in plastic thermoforming; mold release; wh. waxy solid; m.w. 900; sp.gr. 1.204; f.p. 34 C; visc. 100 cSt (100 F); flash pt. (PMCC) > 450 F.

Dow E1000. [Dow] PEG-20; CAS 25322-68-3; EINECS 203-989-9; carrier, modifier with exc. solvency, lubricity for personal care prods. (makeup, bath preps., toothpaste), foam control agents, syn. lubricants, rubber and plastics, metalworking fluids, chem. intermediates; controls cure rates in clay-contg. rubber formulations; improves flexibility, imparts antistatic props. to thermoplastics; as heat transfer fluids in plastic thermoforming; mold release; wh. waxy solid; m.w. 1000; sp.gr. 1.214; f.p. 37 C; visc. 18 cSt (210 F); flash pt. (PMCC) > 450 F.

Dow E1450. [Dow] PEG-32; CAS 25322-68-3; EINECS 203-989-9; carrier, modifier with exc. solvency, lubricity for personal care prods. (makeup, bath preps., toothpaste), foam control agents, syn. lubricants, rubber and plastics, metalworking fluids, chem. intermediates; controls cure rates in clay-contg. rubber formulations; improves flexibility, imparts antistatic props. to thermoplastics; as heat transfer fluids in plastic thermoforming; mold release; wh. waxy solid; m.w. 1450; sp.gr. 1.214; f.p. 44 C; visc. 29 cSt (210 F); flash pt. (PMCC) > 450 F.

Dow E3350. [Dow] PEG-75; CAS 25322-68-3; EINECS 203-989-9; carrier, modifier with exc. solvency, lubricity for personal care prods. (makeup, bath preps., toothpaste), foam control agents, syn. lubricants, rubber and plastics, metalworking fluids, chem. intermediates; controls cure rates in clay-contg. rubber formulations; improves flexibility, imparts antistatic props. to thermoplastics; as heat transfer fluids in plastic thermoforming; mold release; wh. waxy solid, pract. odorless; sol. > 100 g/100 g in water; m.w. 3350; sp.gr. 1.224; visc. 93 cSt (210 F); f.p. 54 C; b.p. dec.; flash pt. (PMCC) > 232 C; pH 4.5-7.5 (5% aq.).

Dow E4500. [Dow] PEG-100; CAS 25322-68-3; EINECS 203-989-9; carrier, modifier with exc. solvency, lubricity for personal care prods. (makeup, bath preps., toothpaste), foam control agents, syn. lubricants, rubber and plastics, metalworking fluids, chem. intermediates; controls cure rates in clay-contg. rubber formulations; improves flexibility, imparts antistatic props. to thermoplastics; as heat transfer fluids in plastic thermoforming; mold release; wh. waxy solid, sl. polyether odor; colorless visc. liq. above 136 F; sol. > 100 g/100 g in water; m.w. 4500; sp.gr. 1.224; visc. 180 cSt (210 F); f.p. 58 C; b.p. dec.; flash pt. (PMCC) > 232 C; pH 4.5-7.5 (5% aq.).

Dow E8000. [Dow] PEG-150; CAS 25322-68-3; EINECS 203-989-9; carrier, modifier with exc. solvency, lubricity for personal care prods. (makeup, bath preps., toothpaste), foam control agents, syn. lubricants, rubber and plastics, metalworking fluids, chem. intermediates; controls cure rates in clay-contg. rubber formulations; improves flexibility, imparts antistatic props. to thermoplastics; as heat transfer fluids in plastic thermoforming; mold release; wh. waxy solid, clear liq. above 65 C, pract. odorless; sol. > 100 g/100 g in water; m.w. 8000; sp.gr. 1.224; f.p. 60 C; visc. 800 cSt (210 F); flash pt. (PMCC) > 260 C; pH 4.5-7.5 (5% aq.).

Dow Corning® 7 Compound. [Dow Corning] Silicone compd.; lubricant, release agent for mold break-in; preservative and lubricant for rubber.

Dow Corning® 20 Release Coating. [Dow Corning] Silicone blend of resin and fluid in solv.; multiple release agent for most plastics, urethane mold castings, metal, rubber; heat cured; 50% act.

Dow Corning® 24 Emulsion. [Dow Corning] Silicone emulsion; nonionic; release agent, lubricant for indirect food contact, pkg., melamines, acrylics, phenolics, polyethyl, coated paper, rubber; FDA approved; water-dilutable; 35% conc.

Dow Corning® 190 Surfactant. [Dow Corning] Dimethicone copolyol; nonionic; silicone surfactant, surf. tens. depressant, wetting agent, emulsifier, foam builder, humectant, softener; used for producing flexible slabstock urethane foam; plasticizer for hair resins; imparts spreadability, lt. nongreasy feel, and detackification to hair sprays, shampoos, skin care lotions, perfumes, shaving soaps; Gardner 2 hazy, low-visc. liq.; sol. in water, ethanol, IPM; disp. in propylene glycol; sp.gr. 1.037; visc. 1500 cst; b.p. > 200 C; HLB 14.4; flash pt. (COC) 121 C; ref. index 1.448; 100% conc.; *Toxicology:* nontoxic.

Dow Corning® 193 Surfactant. [Dow Corning] Dimethicone copolyol; nonionic; silicone surfactant, surf. tens. depressant, wetting agent, emulsifier, foam builder, humectant, softener; used for producing flexible slab stock urethane foam; plasticizer for hair resins; imparts spreadability, lt. nongreasy feel, and detackification to hair sprays, shampoos, skin care lotions, perfumes, shaving soaps; Gardner 2 hazy, visc. liq.; sol. in water, alcohol, hydroalcoholic systems, propylene glycol, IPM; sp.gr. 1.07; visc. 465 cs; HLB 13.6; f.p. 50 F; b.p. >

200 C; flash pt. (COC) 149 C; pour pt. 52 F; ref. index 1.454; 100% conc.; *Toxicology:* nontoxic.

Dow Corning® 197 Surfactant. [Dow Corning] Silicone surfactant; for polyurethane foam industry; sp.gr. 1.05; visc. 230 cSt; flash pt. (CC) > 60.6 C.

Dow Corning® 198 Surfactant. [Dow Corning] Silicone surfactant; for polyurethane foam industry; sp.gr. 1.04; visc. 1600 cSt; flash pt. (CC) > 60.6 C.

Dow Corning® 200 Fluid. [Dow Corning] Dimethicone; foam control agent for nonaq. systems, distillation, resin mfg., asphalt, oil refining, gas-oil separation; water barrier; internal/external lubricant/release agent for plastics and elastomers; lubricant, gloss, water repellency, softness agent for hair care prods.; emollient, lubricant in antiperspirants/deodorants; reduces valve clogging in aerosol prods.; emollient for skin lotions and creams; skin protectant; aids spreading and rub-in; clear fluid; sol. in IPM; dilutable in aliphatic, aromatic or chlorinated solvs.; 10 visc. grades: sp.gr. 0.761-0.975; visc. 0.65-12,5000 cst; m.p. < -40 C; b.p. 100- > 200 C; flash pt. (CC) -2 to > 321 C; ref. index 1.375-1.403; 100% act.; *Toxicology:* LD50 (oral, rat) > 35 g/kg; nonirritating to skin on repeated/prolonged contact; transitory eye irritant.

Dow Corning® 203 Fluid. [Dow Corning] Silicone fluid; release film providing internal release and lubrication for epoxies, melamines, urethanes, metals, rubber; 100% act.

Dow Corning® 236 Dispersion. [Dow Corning] Silicone disp.; release agent; applied to surfaces to provide a durable, rubbery coating; prevents adhesion of most sticky materials; protects plastics from weather and shields metals from corrosion.

Dow Corning® 290 Emulsion. [Dow Corning] Silicone emulsion; high-solids, paintable multiple release agent for plastics, epoxies, melamines, polyethyl, elastomers, and die castings; water-dilutable; 50% act.

Dow Corning® 346 Emulsion. [Dow Corning] Silicone emulsion; release agent and general lubricant for plastics and rubber, e.g., polyethylene bags; water-dilutable; 60% solids.

Dow Corning® 347 Emulsion. [Dow Corning] Silicone emulsion; nonionic; release agent for plastics and elastomers, esp. effective on hot molds; for acrylics, foundry castings, rubber, fibers; low visc. fluid base; water-dilutable; high solids; 60% act.

Dow Corning® 1248 Fluid. [Dow Corning] Silicone polycarbinols (graft copolymer type); nonionic; surfactant for org. systems and urethanes; clear liq.; sol. in hydrocarbons; visc. 160 cs; HLB 1.5; flash pt. 250 F; pour pt. -130 F; 100% conc.

Dow Corning® 1250. [Dow Corning] Silicone in solv. (xylene); for polyurethane foam industry; profoamer for mechanical frothing; allows foam fusing at low temp.; very lt. straw-colored thin liq.; sp.gr. 1.00; visc. 6 cst; flash pt. 85 F (OC); 50% silicone.

Dow Corning® 1252. [Dow Corning] Silicone in solv. (texanol isobutyrate); for polyurethane foam industry; profoamer for mechanical frothing; allows foam fusing at low temp.; very lt. straw-colored thin liq.; sp.gr. 1.06; visc. 250 cs; flash pt. 135 F (OC); 50% silicone.

Dow Corning® 1315 Surfactant. [Dow Corning] Silicone glycol copolymer; nonionic; surfactant for PU foams; wetting agent; detackifier and plasticizer for hair fixative resins; liq.; sol. in water, ethanol; sp.gr. 1.03; visc. 1000 cSt; HLB 15.3; flash pt. (CC) > 60.6 C; 100% conc.

Dow Corning® 5043 Surfactant. [Dow Corning] Silicone surfactant; for polyurethane foam industry; sp.gr. 1.00; visc. 265 cSt; flash pt. (CC) > 60.6 C.

Dow Corning® 5098. [Dow Corning] Silicone surfactant; for polyurethane foam industry; sp.gr. 1.07; visc. 195 cSt; flash pt. (CC) > 60.6 C.

Dow Corning® 5103 Surfactant. [Dow Corning] Silicone glycol copolymer; nonionic; surfactant for PU foams; wetting agent; detackifier and plasticizer for hair fixative resins; liq.; sol. in water, ethanol; sp.gr. 1.05; visc. 250 cSt; HLB 9.7; flash pt. (CC) > 60.6 C; 100% conc.

Dow Corning® 7119 Release/Parting Agent. [Dow Corning] Silicone; internal release agent for paintable parts in RIM urethanes; sol. in nonpolar org. solvs.; sol. in nonpolar org. solvs.

Dow Corning® HV-490 Emulsion. [Dow Corning] High m.w. silicone emulsion; anionic; release agent for rubber and plastics fabrication, yarn thread; very stable; milky wh. emulsion; fine particle size (< 5 μ avg.); water sol.; sp.gr. 0.995; flash pt. (CC) 100 C; pH 7.5; 36% act.

Dow Corning® Q1-6106. [Dow Corning] Coupling agent designed to promote adhesion of two dissimilar materials; in reinforced plastic systems, provides coupling of most reinforcing agents, e.g., glass, Kevlar to most polar org. polymers and engineering plastics, e.g., epoxies, urethanes, acrylics, polysulfones, PPS, melamines, polyimides, PC, and thermoplastic polyesters; adhesion promoter for plastics and metals; wh. to slightly yel. visc. liq.; sol. in polar solvs. such as methanol and Dowanol PM; sol. in water; sp.gr. 1.19; visc. 850 cSt; ref. index 1.5104; flash pt. (CC) 155 F; 100% act.

Dow Corning® Z-6020. [Dow Corning] N-(2-aminoethyl-3-aminopropyl trimethoxy silane; CAS 1760-24-3; EINECS 212-164-2; coupling agent used for epoxies, phenolics, melamines, nylons, PVC, acrylics, polyolefins, polyurethanes, nitrile rubbers.

Dow Corning® Z-6030. [Dow Corning] 3-Methacryloxypropyl trimethoxysilane; CAS 2530-85-0; EINECS 219-785-8; coupling agent used for free-radical cross-linked polyester, rubber, polyolefins, styrenics, acrylics.

Dow Corning® Z-6032. [Dow Corning] N- [2-(Vinylbenzylamino) -ethyl] -3-aminopropyl trimethoxysilane; coupling agent used for most thermoset and thermoplastic resins.

Dow Corning® Z-6040. [Dow Corning] 3-Glycidoxypropyltrimethoxysilane; CAS 2530-83-8; EINECS 219-784-2; coupling agent used for epoxies, urethane, acrylic and polysulfide sealants.

Dow Corning® Z-6075. [Dow Corning] Vinyltriacetoxy silane; CAS 4130-08-9; coupling agent used for polyesters, polyolefins, EPDM, EPM (peroxide cured).

Dow Corning® Z-6076. [Dow Corning] 3-Chloropropyl trimethoxy silane; CAS 2530-87-2; EINECS 219-787-9; coupling agent used for epoxy, styrenics, nylon.

Dow Corning® Antifoam 1400. [Dow Corning] Silicone compd.; foam control agent for glycol scrubbing, resin mfg., oil refining, asphalts, adhesives/coatings, metalworking, pesticide/fertilizer, and detergents industries; for extreme pH conditions; med. off-wh. to wh. liq.; 100% act.

Dow Corning® Antifoam 1410. [Dow Corning] Silicone emulsion; nonionic; foam control agent for inks, textile starching/sizing, cutting oils, resin mfg., gas processing, adhesives/coatings, waste water treatment, pesticide/fertilizer industries; for extreme pH conditions; thin wh. cream; 10% act.

Dow Corning® Antifoam 1430. [Dow Corning] Silicone emulsion; nonionic; foam control agent for detergents, distillation, glycol scrubbing, latex mfg., metalworking, waste water treatment, adhesives/coatings, and textile industries; for extreme pH conditions; med. wh. cream; 30% act.

Dowfax 2A1. [Dow; Dow Europe] Sodium dodecyl diphenyloxide disulfonate; anionic; detergent, emulsifier, wetting agent, solubilizer, dispersant, spreading agent, penetrant for detergent formulation, emulsion polymerization, agric., electroplating, ore flotation, drilling muds; leveling agent for acid dyeing of nylon, dyeing assistant, emulsifier for dye carriers; indirect food additive; FDA 21CFR §178.3400 and EPA compliance; amber clear liq.; sol. in water, aq. sol'ns. of acids, alkalis, electrolytes; m.w. 575; sp.gr. 1.16; visc. 145 cps (0.10%); HLB 16.7; surf. tens. 31 dynes/cm (0.1%); 45% act.; *Toxicology:* toxic to fish.

Dowfax 2EP. [Dow; Dow Europe] Sodium dodecyl diphenyloxide disulfonate; anionic; detergent, emulsifier, wetting agent, solubilizer, dispersant, spreading agent, penetrant for detergent formulation, emulsion polymerization, agric., electroplating, ore flotation, drilling muds; leveling agent for acid dyeing of nylon, dyeing assistant, emulsifier for dye carriers; indirect food additive; FDA 21CFR §178.3400 and EPA compliance; pale brn. liq.; sol. in water, 20% caustic and HCl, 35% TKPP; sl. sol. in ethanol; m.w. 575; sp.gr. 1.16; visc. 145 cps; HLB 16.7; surf. tens. 31 dynes/cm (0.1%); 45% conc.; *Toxicology:* toxic to fish.

Dowfax 3B2. [Dow; Dow Europe] Sodium n-decyl diphenyloxide disulfonate; CAS 36445-71-3; anionic; detergent, emulsifier, wetting agent, solubilizer, dispersant, spreading agent, penetrant for detergent formulation, emulsion polymerization, agric., electroplating, ore flotation, drilling muds; leveling agent for acid dyeing of nylon, dyeing assistant, emulsifier for dye carriers; biodeg.; FDA 21CFR §178.3400 and EPA compliance; red-brn. liq.; sol. in water, 25% caustic, 15% HCl, 35% TKPP; sl. sol. in ethanol; sp.gr. 1.16; visc. 120 cps; HLB 17.8; surf. tens. 38 dynes/cm (0.1%); 45% act.; *Toxicology:* toxic to fish.

Dowfax 8390. [Dow; Dow Europe] Sodium n-hexadecyl diphenyloxide disulfonate; anionic; detergent, emulsifier, wetting agent, solubilizer, dispersant, spreading agent, penetrant for detergent formulation, emulsion polymerization, agric., electroplating, ore flotation, drilling muds; leveling agent for acid dyeing of nylon, dyeing assistant, emulsifier for dye carriers; indirect food additive; biodeg.; FDA 21CFR §178.3400 and EPA compliance; amber clear liq.; sol. in water, 25% caustic, 20% HCl, 35% TKPP; sl. sol. in ethanol; m.w. 642; sp.gr. 1.11; visc. 10 cps; surf. tens. 42 dynes/cm (0.1%); 35% conc.; *Toxicology:* toxic to fish.

Dowfax C6L. [Dow] Sodium hexyl diphenyloxide disulfonate; anionic; lowest foaming and highest charge density prod. in Dowfax series; high solubilizing capabilities in acids, alkalies, and electrolytes; for cleaning, emulsion polymerization, latex mfg., paints, adhesives, min. and metal processing, textile applics.; amber clear liq.; sol. in water, 15% HCl, 45% TKPP, 32% NaOH; sl. sol. in ethanol; m.w. 474; sp.gr. 1.1954; visc. 50.4 cps; HLB 19.0; surf. tens. 41 dynes/cm (0.1%); 45% act.

Dowfax C12L. [Dow] Alkylated disulfonated diphenyl oxide; detergent, emulsifier, wetting agent, solubilizer, dispersant, spreading agent, penetrant for detergent formulation, emulsion polymerization, agric., electroplating, ore flotation, drilling muds; leveling agent for acid dyeing of nylon, dyeing assistant, emulsifier for dye carriers; indirect food additive; biodeg.; FDA 21CFR §178.3400 and EPA compliance; amber clear liq.; sol. in 20% caustic; m.w. 575; visc. 131 cps; surf. tens. 41.1 dynes/cm (0.1%); 45% act.; *Toxicology:* toxic to fish.

Dowicil® 75. [Dow] 1-(3-chloroallyl)-3,5,7-triaza-1-azoniaadamantane chloride (act. ingred.) and sodium bicarbonate (stabilizer); preservative for aq. end prods. such as adhesives, latex emulsions, paints, metal-cutting fluids, drilling muds, biodeg. detergents, paper coatings, PU resins, agric., textile and construction materials, inks, starch, polishes and waxes; antimicrobial activity; indirect food additive; FDA 21 CFR §175.105, 176.170, 176.180, 177.1680; off-wh. powd., 100% thru #5 US sieve; sol. (act. ingred. g/100 g) in water (222.0 g), anhyd. IPA (39.5 g), propylene glycol USP (20.6 g); m.w. 251.2; sp.gr. 1.54 g/cc; dens. 43 lb/ft^3; b.p. dec. > 60 C; 67.5% act.; *Toxicology:* LD50 (oral, rat) 1190 mg/kg, (dermal, rabbit) 2000 mg/kg; sl. eye and skin irritant.

DPG. [Monsanto] Diphenylguanidine; CAS 102-06-7; sec. accelerator for rubber industry; also used in differential ore flotation of copper/nickel deposits, security paper, and thermal print paper developing; 2 mm pellets or dust-suppressed powd.

DPTT-S. [Akrochem] Dipentamethylene thiuram hexasulfide; CAS 120-54-7; very act. sulfur-bearing accelerator imparting heat resistance to sulfurless compds.; primary accelerator for

Hypalon, butyl, and EPDM; vulcanizing agent for heat-resistant latex; nondiscoloring, nonstaining; buff powd.; sp.gr. 1.52; m.p. 117 C min.; 25% avail. sulfur.

Drakeol 9. [Penreco] Lt. min. oil NF; CAS 8042-47-5; base material, carrier for cosmetics; emulsified lubricant for laxatives; pigment dispersant, lubricant for plastics; plasticizer for PS; textile and paper lubricant; divider oil, mold release lubricant for food industry, coating for fruits and vegetables, food pkg. materials; FDA 21CFR §172.878, 178.3620, 573.680; sp.gr. 0.838-0.854; dens. 7.03-7.16 lb/gal; visc. 14.2-17.0 cSt (40 C); pour pt. 09 C; flash pt. 179 C; ref. index 1.4665.

Drakeol 10. [Penreco] Lt. min. oil NF; CAS 8042-47-5; plasticizer, lubricant for plastics; food additive; emollient, cosmetic raw material; FDA 21CFR §172.878, 178.3620, 573.680; sp.gr. 0.838-0.864; dens. 7.08-7.25 lb/gal; visc. 17.7-20.2 cSt (40 C); pour pt. -9 C; flash pt. 182 C; ref. index 1.4692.

Drakeol 19. [Penreco] Min. oil USP; CAS 8042-47-5; primary plasticizer for ethyl cellulose; lubricant for textile/paper; ingred. in cosmetics (creams, lotions, sunscreens), pharmaceuticals (laxatives, topical ointments), foods (lubes/greases, food pkg., divider oils), plastics (catalyst carriers); FDA 21CFR §172.878, 178.3620, 573.680; sp.gr. 0.852-0.876; dens. 7.14-7.31 lb/gal; visc. 34.9-37.3 cSt (40 C); pour pt. -12 C; flash pt. 188 C; ref. index 1.4725.

Drakeol 21. [Penreco] Min. oil USP; CAS 8042-47-5; primary plasticizer for ethyl cellulose; ingred. in cosmetics (creams, lotions, sunscreens), pharmaceuticals (laxatives), foods (lubes/greases, food pkg.), plastics (lubes for PS, PVC, annealing, catalyst carriers), adhesives; FDA 21CFR §172.878, 178.3620, 573.680; sp.gr. 0.853-0.876; dens. 7.15-7.32 lb/gal; visc. 38.4-41.5 cSt (40 C); pour pt. -12 C; flash pt. 193 C; ref. index 1.4733.

Drakeol 32. [Penreco] Min. oil USP; CAS 8042-47-5; plasticizer; food additive; FDA 21CFR §172.878, 178.3620, 573.680; sp.gr. 0.856-0.876; dens. 7.18-7.35 lb/gal; visc. 60.0-63.3 cSt (40 C); pour pt. -12 C; flash pt. 213 C; ref. index 1.4770.

Drakeol 34. [Penreco] Min. oil USP; CAS 8042-47-5; plasticizer/lube for PS, PVC, annealing, catalyst carriers, thermoplastic rubber extender oils; for bakery pan oils, food pkg., lubes/greases in food industry; adhesives; household cleaners and polishes; FDA 21CFR §172.878, 178.3620, 573.680; sp.gr. 0.858-0.872; dens. 7.19-7.31 lb/gal; visc. 72.0-79.5 cSt (40 C); pour pt. -9 C; flash pt. 238 C; ref. index 1.4760.

Drakeol 35. [Penreco] Min. oil USP; CAS 8042-47-5; ingred. in laxatives, cosmetic creams, lotions, sunscreens, foods (bakery pan oils, lubes/greases, food pkg.); plasticizer/lubricant for PS, ethyl cellulose, PVC, annealing, catalyst carriers, thermoplastic rubber; water repellant for paper; adhesives, household cleaners and polishes; FDA 21CFR §172.878, 178.3620, 573.680; sp.gr. 0.864-0.881; dens. 7.25-7.35 lb/gal; visc. 65.8-71.0 cSt (40 C); pour pt. -15 C; flash pt. 216 C; ref. index 1.4785.

Drapex® 4.4. [Witco/PAG] Octyl epoxy tallate; CAS 61788-72-5; plasticizer for vinyl compds. providing exc. low-temp. flexibility, low volatility, heat and light stabilizing props.; low solvating action; esp. suited for plastisols and organosols; Gardner 1 color, faintly fatty odor; sol. < 0.01% in water; m.w. 420; sp.gr. 0.920; dens. 7.67 lb/gal; visc. 25 cps; pour pt. -22 C; acid no. 0.5; iodine no. 2.5; flash pt. (COC) 220 C; ref. index 1.4570.

Drapex® 6.8. [Witco/PAG] Epoxidized soybean oil; CAS 8013-07-8; EINECS 232-391-0; plasticizer for vinyl compds.; offers resist. to extraction by oils and solvs., resist. to migration, low volatility; imparts exc. heat and lt. stability; food contact use; FDA 21CFR §181.27; Gardner 1 color, faintly fatty odor; sol. < 0.01% in water; m.w. 1000; sp.gr. 0.992; dens. 8.27 lb/gal; visc. 320 cps; pour pt. 0 C; acid no. 0.5; iodine no. 1.6; flash pt. (COC) 290 C; ref. index 1.45710.

Drapex® 10.4. [Witco/PAG] Epoxidized linseed oil; plasticizer for vinyl compds.; compat. with primary plasticizers; provides optimum heat stability, exc. extraction resist.; suitable for food contact; FDA 21CFR §178.3740; Gardner 1 color, faintly fatty odor; sol. < 0.01% in water; m.w. 1000; sp.gr. 1.030; dens. 8.6 lb/gal; visc. 900 cps; pour pt. -5 C; acid no. 0.65; iodine no. 2.5; flash pt. (COC) 290 C; ref. index 1.4765.

Drapex® 334F. [Witco/PAG] Med. m.w. polyester plasticizer; permanent plasticizer for use in PVC compds. for food pkg. and medical devices (tubing, blood bags, prosthetic devices); FDA 21CFR §178.3740; very lt. straw colored liq., extremely low odor; sp.gr. 1.082; dens. 9.0 lb/gal; visc. 3465 cps; pour pt. 3 C; acid no. 1.0; flash pt. (CC) 230 C; ref. index 1.465; 0.15% moisture.

Drapex® 409. [Witco/PAG] Med. m.w. polymeric plasticizer; plasticizer for PVC and rubber formulations with exc. permanence in finished goods, good elec. props., migration resist. to PS and ABS, oil, and hexane; for high-temp. wire and cable coatings, refrigerator gasketing, film, tape, coated fabrics; lt. straw colored liq., extremely low odor; sp.gr. 1.082; dens. 9.0 lb/gal; visc. 3465 cps; pour pt. 4 C; acid no. 1.0; flash pt. (CC) 230 C; ref. index 1.465; 0.15% moisture.

Drapex® 411. [Witco/PAG] High m.w. polyester plasticizer; plasticizer for PVC providing superior permanence, resist. to extraction and migration; for decorative appliques, wire insulation, molded parts for under-hood use, dry-cleanable applics.; lt. straw colored liq., extremely low odor; sp.gr. 1.107; dens. 9.2 lb/gal; visc. 8850 cps; pour pt. 21.1 C; acid no. 1.0; flash pt. (CC) 239 C; ref. index 1.477; 0.15% moisture.

Drapex® 412. [Witco/PAG] Low m.w. polyester plasticizer; plasticizer for PVC providing balance of permanence, low-temp. flexibility, and fast fusion props.; esp. suited for plastisol compds. ranging from coated fabric to slush molded and dipped goods; lt. straw colored liq., extremely low odor; sp.gr. 1.045; dens. 8.7 lb/gal; visc. 380 cps; pour pt.

-20 C; acid no. 2.0; flash pt. (CC) 188 C; ref. index 1.458; 0.2% moisture.

Drapex® 420. [Witco/PAG] Med.-high m.w. polyester plasticizer; plasticizer for PVC compds. requiring exc. durability, exc. resist. to migration, good elec. props.; for elec. tape and insulation, refrigerator gasketing, oil-resist. wire insulation, PVC film and sheeting; also plasticizes ethyl cellulose, nitrocellulose, acrylic caulking compds., adhesives (PVAc, S/B, acrylic latex); lt. straw colored liq., extremely low odor; sp.gr. 1.095; dens. 9.1 lb/gal; visc. 5500 cps; pour pt. -18 C; acid no. 2.0; flash pt. (CC) 200 C; ref. index 1.465; 0.2% moisture.

Drapex® P-1. [Witco/PAG] Med.-high m.w. polyester plasticizer; plasticizer for use in PVC compds. requiring durability, low-temp. flexibility; resist. to migration, high dielec. props., e.g., for elec. tape compds., refrigerator gasketing, wall covering, automotive interiors, oil-resist. wire insulation; lt. straw colored liq., extremely low odor; sp.gr. 1.085; dens. 9.06 lb/gal; visc. 5600 cps; pour pt. 10 C; acid no. 1.0; flash pt. (CC) 174 C; ref. index 1.4670; 0.05% moisture.

Drapex® P-7. [Witco/PAG] Med. m.w. polyester plasticizer; plasticizer for use in PVC compds. requiring permanence and exc. resist. to migration; very lt. straw colored liq., extremely low odor; sp.gr. 1.110; dens. 9.26 lb/gal; visc. 4400 cps; pour pt. -27 C; acid no. 1.95; flash pt. (CC) 135 C; ref. index 1.4807; 0.12% moisture.

Dresinate® 81. [Hercules] Rosin sodium soap; CAS 61790-51-0; EINECS 263-144-5; anionic; emulsifier ingred. and/or stabilizer in solv. cleaners and sol. oils for metalworking, disinfectants, oil-well drilling muds, drawing and grinding compds.; plasticizer; pale-colored liq.; dilutable with water and aq. sol'ns. of alcohols and glycols; dens. 8.7 lb/gal; visc. 5.7 poises (140 F); acid no. 15; 87% solids.

Dresinate® 91. [Hercules; Hercules BV] Rosin potassium soap; anionic; emulsifier ingred. and/or stabilizer in solv. cleaners and sol. oils for metalworking, disinfectants, oil-well drilling muds, drawing and grinding compds.; plasticizer; pale liq.; dilutable with water and aq. sol'ns. of alcohols and glycols; dens. 8.8 lb/gal; visc. 5.3 poises (140 F); acid no. 15; 88% solids.

Dresinate® 214. [Hercules; Hercules BV] Potassium soap of a pale modified rosin; anionic; emulsifier, pigment wetting agent and dispersant, foaming agent, used in mfg. of adhesives, polymer emulsion syn. latices; paste; dilutable with water and aq. sol'ns. of alcohols and glycols; dens. 9.22 lb/gal (60 C); visc. 23 poises (140 F); 80% solids.

Dresinate® 731. [Hercules; Hercules BV] Sodium soap of a pale modified rosin; CAS 61790-51-0; EINECS 263-144-5; nonionic; emulsifier, pigment wetting agent and dispersant, foaming agent, used in mfg. of adhesives, polymer emulsion syn. latices; pale paste; dilutable with water and aq. sol'ns. of alcohols and glycols; dens. 9.1 lb/gal (60 C); visc. 11 poises (140 F); acid no. 11; 70% solids.

Dresinate® TX. [Hercules] Tall oil sodium soap; anionic; emulsifier for oils and asphalts; detergent aid; dispersant for pigments; stabilizer for syn. rubber latices, used for industrial and household cleaners; conditioner for sulfur dusts; dk. brn. powd.; sol. in water and aq. sol'ns. of alcohols and glycols; dens. 26 lb/ft^3; 96% min. solids.

Dresinate® X. [Hercules] Sodium soap of a pale rosin; CAS 61790-51-0; EINECS 263-144-5; anionic; emulsifier for oils and asphalts; detergent aid; dispersant for pigments; stabilizer for syn. rubber latices, used for industrial and household cleaners; conditioner for sulfur dusts; pale cream, powd.; sol. in water and aq. sol'ns. of alcohols and glycols; dens. 26 lb/ft^3; 95% min. solids.

Dresinate® XX. [Hercules] Sodium soap of a dk. rosin; CAS 61790-51-0; EINECS 263-144-5; anionic; emulsifier for oils and asphalts; detergent aid; dispersant for pigments; stabilizer for syn. rubber latices, used for industrial and household cleaners; conditioner for sulfur dusts; lt. br. powd.; sol. in water and aq. sol'ns. of alcohols and glycols; dens. 28 lb/ft^3; 96% min. solids.

Dresinol® 42. [Hercules] Partially decarboxylated rosin aq.-based disp.; modifier for polymer film-formers in adhesives, sizings; improves wetting of pigments and fillers, penetration, adhesion to substrates, surface tack and bond strength; cream liq.; visc. 600 cps; soften. pt. 46 C; pH 9.4; 40% solids.

Drewfax® 0007. [Drew Ind. Div.] Sodium dioctyl sulfosuccinate; CAS 577-11-7; EINECS 209-406-4; wetting, penetrating, surf. tens. reducing agent, dispersant for industrial coatings, adhesives, inks, pigments, textile, cosmetic, paper, metal, paint, rubber, plastics, petrol. and agric. industries; FDA compliance; colorless clear visc. liq.; sol. in water and various solvs.; sp.gr. 1.08; dens. 9.0 lb/gal; visc. 400 cps; pH 5-6 (1%); 75% solids.

Drewplast® 017. [Stepan/PVO; Harwick] Glycerol mono and diesters of fatty acids; food-grade plastics additive; antistat for LDPE, lubricant for rigid and flexible PVC; good thermal props., stable to oxidative degradation; FDA-sanctioned.

Drewplast® 030. [Stepan/PVO; Harwick] Glycerol mono and diesters of fatty acids; food-grade plastics additive; antifog for PVC sheet and film; good thermal props., stable to oxidative degradation; FDA-sanctioned.

Drewplast® 051. [Stepan/PVO; Harwick] Glycerol mono and diesters of fatty acids; food-grade plastics additive; antifog for PVC sheet and film; cling agent for LDPE; lubricant for rigid and flexible PVC; good thermal props., stable to oxidative degradation; FDA-sanctioned.

Drewplus® L-108. [Drew Ind. Div.] Blend of min. oils, emulsifiers, silica derivs.; defoamer for latex/rubber applics. esp. monomer stripping, acrylic, PVAc, NBR, SBR, PVC; FDA 21CFR §175.105, 176.210, 176.200, 178.3120, 177.120; straw-colored opaque liq.; emulsifiable in water; sp.gr. 0.89; dens. 7.58 lb/gal; visc. 800 cps; pour pt. 9 C; flash pt. (PMCC) 93 C; 100% act.

Drewplus® L-123. [Drew Ind. Div.] Nonsilicone defoamer; defoamer for SBR monomer stripping, ABS, PVAC latexes; FDA 21CFR §175.105, 176.210, 178.3120; off-wh. opaque liq.; disp. in water; sp.gr. 0.92; dens. 7.68 lb/gal; flash pt. > 300 F; *Storage:* 1-2 mos storage stability; if settling occurs, reverse with mild agitation prior to use.

Drewplus® L-131. [Drew Ind. Div.] Proprietary blend of min. oil, silica derivs. and surfactants; defoamer for latex/rubber applics. (SBR, PVAC, PVA, NBR, acrylic), esp. for paper coating and carpet backing; FDA 21CFR §175.105, 176.200, 176.210; yel. opaque liq.; disp. in water; sp.gr. 0.92; visc. 600 cps; pour pt. -4 C; flash pt. (PMCC) > 93 C.

Drewplus® L-139. [Drew Ind. Div.] Blend of min. oil, silica deriv., surfactants; defoamer for latexes (SBR, NBR, PVC, PVAc, PVA, SAN, polybutadiene, vinyl acrylic), monomer stripping, degassing, adhesives; FDA 21CFR §175.105, 176.210, 176.200, 178.3120, 177.120; off-wh. translucent liq.; emulsifiable in water; sp.gr. 0.89; dens. 7.43 lb/gal; visc. 400 cps; pour pt. 8 C; flash pt. (PMCC) 94 C; *Storage:* 1 yr storage stability @ R.T.; stir prior to use; freezing does not affect prod. performance.

Drewplus® L-140. [Drew Ind. Div.] Blend of minerals, silica derivs., and surfactants; defoamer for latex emulsions (SBR, NBR, PVC, PVA, vinyl-acrylic), adhesives, inks, paint and industrial coatings; FDA 21CFR §175.105, 175.300, 176.170, 176.180, 176.200, 176.210, 176.200, 173.340, 175.300, 178.3120, 177.1200; EPA 40CFR §180.1001; gray-wh. liq.; disp. in water; sp.gr. 0.88; dens. 7.34 lb/gal; visc. 1150 cps; flash pt. 150 F; *Storage:* 1 yr storage stability @ R.T.; stir prior to use; freezing does not affect prod. performance.

Drewplus® L-191. [Drew Ind. Div.] Blend of min. oils, surfactants, silica derivs.; defoamer for latexes (SBR, NBR, acrylic, vinyl acrylic); FDA 21CFR §175.105, 176.210, 176.200, 178.3120, 177.120; straw-colored opaque liq.; disp. in water; sp.gr. 0.92; dens. 7.68 lb/gal; visc. 800 cps; pour pt. 0 C; flash pt. > 300 F; *Storage:* mix before using.

Drewplus® L-198. [Drew Ind. Div.] Blend of min. oils, silica derivs., and esters; defoamer for polymerization, latex stripping (SBR, PVC, EPDM, PVDC), paper coatings; FDA 21CFR §175.105, 176.210, 176.200, 178.3120, 177.120; straw-colored opaque liq.; emulsifiable in water; sp.gr. 0.95; dens. 7.93 lb/gal; visc. 800 cps; pour pt. 10 C; flash pt. (PMCC) 148 C; *Storage:* store in warm place; stir prior to use; freezing does not affect prod. performance.

Drewplus® L-468. [Drew Ind. Div.] Blend of silica and organic hydrophobic solids in a min. oil carrier; foam control agent for latex paints based on vinyl acrylic, acrylic and terpolymer emulsions, aq. industrial coatings (base coats for wood, metal primers), aq. adhesives; FDA compliance; off-wh. opaque liq.; disp. in surfactant systems; sp.gr. 0.88; dens. 7.2-7.4 lb/gal.

Drewplus® L-474. [Drew Ind. Div.] Blend of alcohols, silica derivs., surfactants; defoamer for latexes (PVC, PVAc, styrene acrylic, NR), adhesives, dyestuffs, paper coatings, paints; FDA 21CFR §175.105, 176.210, 178.3120; yel. opaque liq.; disp. in water; sp.gr. 0.94; dens. 7.84 lb/gal; visc. 320 cps; pour pt. 0 C; flash pt. 200 F.

Drewplus® L-475. [Drew Ind. Div.] Blend of min. oils and silica derivs.; nonionic; defoamer for paints/coatings, latexes (acrylic, PVAc, styrene acrylic, PVDC), adhesives, paper coatings; off-wh. opaque liq.; disp. in surfactant systems; insol. in water; sp.gr. 0.91; dens. 7.59 lb/gal; visc. 950 cps; *Storage:* 1-2 mos storage stability under normal conditions; if settling occurs, use mild agitation before use; stored at or below freezing, visc. will increase and some phase separation may occur.

Drewplus® L-523. [Drew Ind. Div.] Blend of fatty oils, surfactants, and silica derivs.; defoamer for latexes (SBR, EVA), food/fermentation applics. incl. chewing gum and potato processing; FDA 21CFR §'173.340, 175.105, 176.200, 176.210; off-wh. opaque liq.; disp. in water; sp.gr. 0.99; dens. 8.26 lb/gal; visc. 800 cps; pour pt. 10 C.

Drewplus® Y-125. [Drew Ind. Div.] Blend of min. oils and fatty acid derivs.; defoamer for controlling entrained air during latex stripping, paints/coatings mfg.; suitable for SBR, PVAc, PCA emulsions, ACN resin systems; FDA 21CFR §175.105, 176.210, 176.200, 178.3120, 177.120; tan opaque liq.; emulsifiable in water; sp.gr. 0.93; dens. 7.76 lb/gal; visc. 2200 cps; pour pt. 0 C; flash pt. (PMCC) 93 C.

Drewplus® Y-250. [Drew Ind. Div.] Blend of min. oils, silica derivs., and esters; defoamer for aq. industrial systems requiring quick foam knockdown, long lasting foam prevention, paints/coatings, latex (SBR, acrylic, PVAc, PVA, styrene acrylic, PVDC, vinyl acrylic), inks, adhesives, cutting oils, food/fermentation applics.; off-wh. opaque liq.; disp. in water; sp.gr. 0.91; dens. 7.59 lb/gal; visc. 1800 cps; flash pt. 240 F.

Drewplus® Y-281. [Drew Ind. Div.] Blend of min. oils, silica derivs., and surfactants; defoamer for paints, latex/rubber, industrial coatings, adhesives; FDA compliance; off-wh. liq.; disp. in water; sp.gr. 0.93; dens. 7.76 lb/gal; visc. 1000 cps; pour pt. 0 C.

Drewplus® Y-601. [Drew Ind. Div.] Blend of min. oils, organic solids, surfactants; defoamer for latex (SBR, PVC, B/S, PVAc), paints/coatings, adhesives, water-borne flexographic or gravure inks; off-wh. opaque liq.; disp. in water; sp.gr. 0.83; dens. 6.93 lb/gal; visc. 400 cps; pour pt. 0 C; *Storage:* mild agitation may be necessary if sl. separation occurs.

Drikalite®. [ECC Int'l.] Calcium carbonate; CAS 471-34-1, 1317-65-3; extender pigment/filler/reinforcement in mech. rubber goods, flooring compds., mastics, paints, adhesives, patching compds., putties, caulks, glazes, tile grouts, joint systems; wh. powd., odorless; 7 μ mean particle size; 0.35% on 325 mesh; negligible sol. in water; sp.gr. 2.71; dens. 50 lb/ft^3 (loose); oil absorp. 9; pH 9.5 (5% slurry); nonflamm.; 97.4% $CaCO_3$, 0.2% max. moisture; *Toxicology:* TLV/TWA 10 mg/m^3, considered nuisance dust; *Precaution:* incompat. with acids.

Drimix®. [Kenrich Petrochemicals] Free-flowing

powd. forms of liq. additives used in polymer compounding; provides faster incorporation; powd.

Dry-Blend® NCG dB910. [Cytec Industries] 65% Nickel-coated graphite fibers sized with a proprietary binder; fibers designed for dry-blending with ABS, PC, ABS/PC blends, and nylon; yields conductive plastics for demanding elec. and thermal applics., e.g., lightning strike protection, aerospace composites, architectural shielding, conductive adhesives; fibers, 12,000 filaments per tow, chopped to specified length (typically 1/4 or 3/16 in.).

Dry-Blend® NCG dB920. [Cytec Industries] 65% Nickel-coated graphite fiber sized with proprietary binder (35% binder); fiber specifically for dry-blending with PPO resins; for conductive plastics; fibers, 12,000 filaments per tow, chopped to specified length (typically 1/4 in.).

Dryspersion®. [Kenrich Petrochemicals] Oil-coated powd. dispersions for polymer processing; provides non-dusty, easy dispersing form, ideal for automated systems.

DS-207. [Syn. Prods.] Lead stearate; CAS 1072-35-1; external lubricant and costabilizer for rigid and flexible PVC; solid.

Dualite® M6001AE. [Pierce & Stevens] Hollow composite microspheres (shell: PVDC copolymer; coating: calcium carbonate); low-dens. filler for use in plastics, coatings, adhesives, BMC, SMC, rubber compding., paper mfg., building materials, explosives, etc.; for moderate exposure to nonaggressive solvs.; heat resist. to 250 F; good chem. and pressure resist.; 40 μ mean particle size; dens. 2.71 ($CaCO_3$), 0.02 (microspheres), 0.13 ± 0.02 (composite); 85% $CaCO_3$; 15% microspheres.

Dualite® M6017AE. [Pierce & Stevens] Hollow composite microspheres (shell: acrylonitrile copolymer; coating: calcium carbonate); low-dens. filler for use in plastics, coatings, adhesives, BMC, SMC, rubber compding., paper mfg., building materials, explosives, etc.; for prolonged exposure to aggressive solvs.; heat resist. to 300 F; better chem. and pressure resist.; 70 μ mean particle size; dens. 2.71 ($CaCO_3$), 0.02 (microspheres), 0.13 ± 0.02 (composite); 85% $CaCO_3$; 15% microspheres.

Dualite® M6032AE. [Pierce & Stevens] Hollow composite microspheres with calcium carbonate coating; low-dens. filler for use in plastics, coatings, adhesives, BMC, SMC, rubber compding., paper mfg., building materials, explosives, etc.; for moderate exposure to nonaggressive solvs.; heat resist. to > 350 F; best chem. and pressure resist.; 70 μ mean particle size; dens. 2.71 ($CaCO_3$), 0.02 (microspheres), 0.13 ± 0.02 (composite); 85% $CaCO_3$; 15% microspheres.

Dualite® M6033AE. [Pierce & Stevens] Hollow composite microspheres with calcium carbonate coating; low-dens. filler for use in plastics, coatings, adhesives, BMC, SMC, rubber compding., paper mfg., building materials, explosives, etc.; for moderate exposure to nonaggressive solvs.; heat resist. to 250 F; good chem. and pressure resist.; 25 μ mean particle size; dens. 2.71 ($CaCO_3$), 0.02 (microspheres), 0.13 ± 0.02 (composite); 85% $CaCO_3$; 15% microspheres.

Duomeen® C. [Akzo Nobel; Akzo Nobel BV] N-Coco-1,3-diaminopropane; CAS 61791-63-7; EINECS 263-195-3; cationic; chemical intermediate; corrosion inhibitor, fuel oil additive, flotation agent; used in metals, textiles, plastics, herbicides; epoxy curing agent; dk. amber. liq.; ammonia odor; sol. in naphtha, min. oil, IPA; m.w. 276; sp.gr. 0.836; visc. 10 cP; m.p. 71 F; b.p. > 300 C; iodine no. 8; amine no. 410; flash pt. (PMCC) > 149 C; 100% conc.

Duradene® 710. [Firestone Syn. Rubber] Sol'n.-polymerized B/S copolymer, nonstaining; CAS 9003-55-8; processing and extrusion aid to be used with other elastomers for such applics. as shoe soling; also for asphalt modification and adhesives; Mooney visc. 45; 27% bound styrene.

Duralink® HTS. [Monsanto] Hexamethylene bisthiosulfate disodium salt dihydrate; CAS 5719-73-3; post vulcanization stabilizer for sulfur cures of NR, IR, SBR, and NBR; used in tire treads, sidewalls, and general industrial prods. incl. belting and inj. molded goods; bonding promoter for rubber-based steel adhesion; nondiscoloring, nonstaining; wh. dust-suppressed powd.; very sol. in water; sp.gr. 1.4.

Duramite®. [ECC Int'l.] Calcium carbonate; CAS 1317-65-3; pigment, filler, reinforcement with unique particle size distribution, easy dispersion; for mix-in applic. in carpet backing, roofing compds., spackles, joint compds., paint, coatings, low-visc. plastisols, foam rubber, plastics; wh. powd., odorless; 11 μ mean particle size; 0.2% on 325 mesh; negligible sol. in water; sp.gr. 2.71; dens. 75 lb/ft^3 (loose); oil absorp. 8; pH 9.5 (5% slurry); nonflamm.; 97.4% $CaCO_3$, 0.2% max. moisture; *Toxicology:* TLV/TWA 10 mg/m^3, considered nuisance dust; *Precaution:* incompat. with acids.

Durastrength 200. [Elf Atochem N. Am.] 2-Propenoic acid, 2-methylmethyl ester, polymer with 1,3-butadiene and butyl 2-propenoate; CAS 33031-74-2; impact modifier for PVC, durable exterior building prods.; exc. weatherability, color retention; improves processing, provides superior surf. finish; wh. powd., char. odor; insol. in water; sp.gr. 1.09; > 99% assay; *Toxicology:* nuisance dust: ACGIH TLV 10 mg/m^3 (total dust); may cause skin irritation or dermatitis, mild to mod. eye irritation, irritation to respiratory tract by inh.; *Precaution:* dust can form explosive mixts. with air; may produce irritating fumes on heating; incompat. with oxidizers.

Durax®. [R.T. Vanderbilt] N-Cyclohexyl-2-benzothiazolesulfenamide; CAS 95-33-0; nondiscoloring primary delayed-action accelerator for NR and syn. rubbers; safe at processing temps., fast at curing temps.; cream to buff powd., sl. amine odor; sol. in chloroform, ether, toluene; insol. in water; dens. 1.30 mg/m^3; m.p. 100-105 C; flash pt. (COC) 177 C (liquefied); *Toxicology:* LD50 (oral, rat) 5300 mg/kg, (dermal, rabbit) > 7940 mg/kg, pract. nontoxic; causes irritation to eyes, skin, respiratory tract; allergic sensitizer; recommended OSHA PEL/TWA 5 mg/m^3 (respirable dust), 15 mg/m^3 (total dust); *Precaution:* combustible dust-explosion potential; stable to 218 C; decomp. yields mercaptobenzothiazole and cyclohexylamine; incompat. with reducing agents, acids; hazardous decomp. prods. incl.: CO, nitrogen oxides, sulfur dioxide fumes; *Storage:* degrades on storage esp. under warm, humid conditions; store below 40 C, dry, away from sunlight in closed containers; 6 mos storage life.

Durax® Rodform. [R.T. Vanderbilt] N-Cyclohexyl-2-benzothiazolesulfenamide; CAS 95-33-0; vulcanization accelerator for rubber; cream to lt. tan pellets, sl. amine odor; negligible sol. in water; dens. 1.28 mg/m^3; flash pt. (COC) 177 C (liquefied); *Toxicology:* LD50 (oral, rat) 5300 mg/kg, (dermal, rabbit) > 7940 mg/kg; pract. nontoxic by ing.; eye, skin, respiratory tract irritant; may cause allergic skin reaction; OSHA PEL/TWA 5 mg/m^3 (respirable dust), 15 mg/m^3 (total dust) recommended; *Precaution:* dust may form explosive mixts. in air; incompat. with reducing agents, acids; hazardous decomp. prods.: CO, NO_x, cyclohexylamine, SO_2 fumes, mercaptobenzothiazole disulfide, 2-mercaptobenzothiazole, benzothiazole; *Storage:* store in cool (under 40 C), well-ventilated place in closed containers, away from sunlight, foodstuffs, reducing agents, acids; slowly degrades on storage; use within 6 mos.

Durazone® 37. [Uniroyal] 2,4,6-Tris-(N-1,4-dimethylpentyl-p-phenylenediamino)-1,3,5-triazine; CAS 121246-28-4; nonstaining antiozonant/antioxidant for natural and syn. rubbers, e.g., tires, hose, footwear, mech. goods, roofing, wire and cable; buff to lt. pink cryst. powd., char. odor; m.w. 694; sol. in org. solvs.; sp.gr. 1.13; m.p. 130 C min.; Unverified

Durelease HS1. [Struktol] Solv.-based; high performance semipermanent release system providing a rapid curing film on mold surfs. @ 200-475 F; releases wide range of molding resins; multiple releases; ideal for metal-to-rubber bonding.

Durelease HS2. [Struktol] Solv.-based; semipermanent release system providing a rapid curing film on mold surfs. @ 200-475 F; releases wide range of molding resins; multiple releases; ideal for metal-to-rubber bonding; extra fine film forming formulation; ideal for fine detail molding such as golf balls and seals where exact details must be maintained.

Dur-Em® 117. [Van Den Bergh Foods] Mono- and diglycerides with citric acid (preservative); nonionic; food emulsifier for margarine, bread, frozen desserts, veg. dairy prods., danish, candies, chewing gum, reduced fat foods; textile lubricant and finishing agent; emulsifier for cosmetic and pharmaceutical creams and lotions; lubricant for thermoplastics; dispersant for inorg. pigments; FDA 21CFR §182.4505, GRAS; kosher; wh. beads, flakes, typ. odor/flavor; m.p. 62-65 C; HLB 2.8; iodine no. 5 max.; flash pt. 300 F min.; 100% conc.; 40% min. alpha monoglyceride; *Storage:* store in cool, dry place away from odor-producing substances.

Dur-Em® 300K. [Van Den Bergh Foods] Mono- and diglycerides, propylene glycol; emulsifier; dispersant, solubilizer in flavor and color systems; wetting agent in spray-dried foods; antifoam for high-sugar prods. (juices, jellies) and high protein systems; processing aid for yeast mfg.; antifog in polyester films; FDA 21CFR §182.4505, 182.1666; kosher; yel. liq., may cloud at low temp., bland flavor; HLB 2.8; iodine no. 63 min.; flash pt. > 300 F; fire pt. > 300 F; 46% min. monoglyceride; *Storage:* store sealed in dry place away from odor-producing substances.

Durfax® 80. [Van Den Bergh Foods] Polysorbate 80; CAS 9005-65-6; nonionic; food emulsifier for frozen desserts, whipped toppings, shortenings; solubilizer, dispersant in pickles, vitamins; yeast defoamer; personal care prods.; antifog agent in plastics and aerosol furniture polish; FDA 21CFR §172.840; kosher; yel. liq.; water-sol.; HLB 15.0; acid no. 3 max.; sapon. no. 45-55; hyd. no. 65-80; 100% conc.; *Storage:* store sealed in a cool, dry place away from odor-producing substances; 180 day storage life @ 40-95 F.

Durosil. [Degussa] Precipitated silica; hydrophilic low surf. area reinforcing silica for thermoplastic rubber compds.; powd.; 40 nm avg. particle size; dens. 210 g/l (tapped); surf. area 60 m^2/g; pH 9 (5% aq. susp.); 98% assay.

Durtan® 80. [Van Den Bergh Foods] Sorbitan oleate; CAS 1338-43-8; nonionic; used in water softening compds., auto polishes, pesticide formulations; antifog in plastics; rust inhibitor additive; ink pigment dispersant; HLB 4.3; sapon. no. 148-161.

Dutral® CO-059. [Ausimont] EPM/EPDM copolymer; elastomer with low grn. str., high Mooney visc. for modifying polyolefin plastics and a variety of vulcanized goods; nonstaining.; Unverified

Dutral® PM-06PLE. [Ausimont] EPM/EPDM copolymer/HDPE blend; for polyolefin modification, esp. for impact enhancement of PP; can be used in a continuous mixer or an extruder; FDA approved for food contact applics.; pellets.; Unverified

DX-5114. [Shell] Brominated epoxy resin with epoxy end groups; flame retardant for ABS, HIPS; m.p. 103 ± 5 C; 50% Br.

DX-5119. [Shell] Brominated epoxy resin with epoxy end groups; general purpose flame retardant for thermoplastics; m.p. 82 ± 5 C; 50% Br.

DY 023. [Ciba-Geigy/Plastics] Cresyl glycidyl ether; CAS 2210-79-9; diluent for epoxy; Unverified

DY 025. [Ciba-Geigy/Plastics] Aliphatic glycidyl ether consisting primarily of C12 and C14 alkyl groups; diluent for epoxy; Unverified

DY 027. [Ciba-Geigy/Plastics] Alkyl glycidyl ether consisting primarily of C8 and C10 alkyl groups; diluent for epoxy; Unverified

Dymsol® 38-C. [Henkel/Functional Prods.] Fatty acid sulfate, sodium salt; anionic; emulsifier for polymerization; dk. amber liq.; 70% act.

Dymsol® 2031. [Henkel/Functional Prods.; Henkel-Nopco] Sulfonated fatty acid, sodium salt; anionic;

primary or sec. emulsifier for emulsion polymerization; dk. amber liq.; biodeg.; 60% act.

Dymsol® LP. [Henkel/Functional Prods.] Alkylaryl sulfonate; anionic; emulsifier for polymerization, large particle size latices; clear yel. liq.; biodeg.; 33% act.

Dymsol® PA. [Henkel/Functional Prods.] Sulfated ester; anionic; primary emulsifier for emulsion polymerization; liq.; 75% solids.

Dynaflush. [Dynaloy] Nonflamm., noncorrosive, noncarcinogenic, non-ozone depleting solv. for flushing and cleaning residues from urethane processing equip. at R.T.; can be used hot (95 C) for faster results; clear liq.; sp.gr. 1.06; b.p. 202 C; flash pt. (CC) 191 F; pH 4-6.

Dynamar® FC-5157. [3M/Spec. Fluoropolymers] Vulcanizing agent for polyepichlorohydrin elastomers in conj. with Dynamar FX 5166; pink 1 in. sq. chips; sp.gr. 1.2; 75% act.

Dynamar® FC-5158. [3M/Spec. Fluoropolymers] Water-based fluorochemical aq. emulsion; mold release agent designed for hot molding operations, provides multiple releases of syn. elastomers @ normal molding temps. (200-400 F); effective in low concs.; suitable for silicone, ethylene acrylic, fluorosilicone, chlorosulfonated PE, EPDM, nitrile, epichlorohydrin, fluoroelastomers, PU elastomers; translucent off-wh. liq.; fused film is sol. in Freon 113; sp.gr. 1.0; 4.5% solids; *Toxicology:* avoid prolonged/repeated skin contact, breathing of fumes.

Dynamar® FC-5163. [3M/Spec. Fluoropolymers] Water-based fluorochemical aq. emulsion; mold release agent designed for hot molding operations, provides multiple releases of syn. elastomers @ normal molding temps. (200-400 F); effective in low concs.; suitable for silicone, ethylene acrylic, fluorosilicone, chlorosulfonated PE, EPDM, nitrile, epichlorohydrin, fluoroelastomers, PU elastomers; milky wh. liq.; fused film is sol. in Freon 113; sp.gr. 1.1; *Toxicology:* avoid prolonged/repeated skin contact, breathing of fumes.

Dynamar® FX 5166. [3M/Spec. Fluoropolymers] Cure accelerator for polyepichlorohydrin elastomers in conj. with Dynamar FC 5157; wh. free-flowing powd.; sp.gr. 1.3; 70% act.

Dynamar® FX-5170. [3M/Spec. Fluoropolymers] Water-based fluorochemical aq. susp.; mold release agent designed for hot molding operations, provides multiple releases of syn. elastomers @ normal molding temps. (200-400 F); suitable for peroxide-cured elastomers (silicone, fluorosilicone, nitrile, ethylene acrylic, fluoroelastomer, tetrafluoroethylene/propylene, EPDM, chlorinated polyethylene, chlorosulfonated polyethylene); translucent off-wh. liq.; fused film is sol. in 50/50 mixt. of butyl Cellosolve and water; sp.gr. 1.0; 1% solids.

Dynamar® FX-5910. [3M/Spec. Fluoropolymers] Multipurpose extrusion aid for polyolefins (LDPE, VLDPE, LLDPE, HDPE, HMWHDPE, EVA, PP) and other thermoplastics; esp. suitable for preblending with pelletized resin for prod. of masterbatches; pellet; 98% act.

Dynamar® FX-5920. [3M/Spec. Fluoropolymers] Multipurpose extrusion aid for blown film applics. contg. titanium dioxide-based pigments and silica-based antiblocking agents; also for extrusion applics. contg. other nonreactive additives; offers faster equilibration time, lower dosage levels; powd.; 97% act.

Dynamar® FX-9613. [3M/Spec. Fluoropolymers] Multipurpose extrusion aid for polyolefins (LDPE, VLDPE, LLDPE, HDPE, HMWHDPE, EVA, PP) and other thermoplastics; esp. suitable for direct addtiion via weight loss feeders or for preblending with resins; provides uniform melt disp.; free-flowing 25 mesh powd.; 90% act.

Dynamar® PPA-790. [3M/Spec. Fluoropolymers] Polymer processing additive providing enhanced release and flow in molding and extrusion of elastomer-based compds.; improves mold release; effective over broad temp. range (up to 200 C); lt. brn. flakes; sp.gr. 2.0; m.p. 70-100 C; *Toxicology:* LD50 (oral, rat) > 5 g/kg (pract. nontoxic); nonirritating dermally; mild transient eye irritant.

Dynamar® PPA-791. [3M/Spec. Fluoropolymers] Polymer processing additive providing enhanced release and flow in molding and extrusion of elastomer-based compds.; improves mold release; effective over broad temp. range (up to 200 C); lt. brn. flakes; sp.gr. 1.75; m.p. 75-95 C; *Toxicology:* LD50 (oral, rat) > 5 g/kg (pract. nontoxic); nonirritating dermally; mild transient eye irritant.

Dynamar® PPA-2231. [3M/Spec. Fluoropolymers] Multipurpose extrusion aid for polyolefins (LDPE, VLDPE, LLDPE, HDPE, HMWHDPE, EVA, PP) and other thermoplastics; esp. suitable for preblending with pelletized resin for prod. of masterbatches; porous pellet; 98% act.

Dynapol® Catalyst 1203. [Hüls Am.] Blocked acid catalyst; nonionic; catalyst for thermosetting polyesters; aids flow and leveling; 50% solids.

Dynasolve XD 16-3. [Dynaloy] Aq. sol'n.; solv. for removal of silicone oils and uncured silicone polymers from molds, molded parts, and processing equip.; esp. effective for cleaning mold release residue from plastic molded parts; lt. amber liq.; sp.gr. 1.010; b.p. 212 F; flash pt. none; pH 6-8; *Toxicology:* avoid contact with eyes and skin, breathing of vapors.

Dynasylan® 0116. [Hüls Am.] Bis-(N-methylbenzamide) ethoxymethyl silane; CAS 16230-35-6; coupling agent, release agent, lubricant, blocking agent, chemical intermediate; amber visc. liq.; m.w. 356.5; sp.gr. 1.13 (20 C); b.p. 210 C (1 mm); flash pt. > 110 C; 90% purity.

Dynasylan® 1110. [Hüls Am.] N-Methylaminopropyltrimethoxysilane; CAS 3069-25-8; coupling agent, chem. intermediate, blocking agent, release agent, lubricant, primer, reducing agent; liq.; m.w. 193.3; sp.gr. 0.98 (20 C); b.p. 106 C (30 mm); flash pt. 70 C; ref. index 1.419 (20 C); 96% purity.

Dynasylan® 1411. [Hüls Am.] N-(2-Aminoethyl)-3-

aminopropylmethyldimethoxy silane; CAS 3069-29-2; coupling agent, chem. intermediate, blocking agent, release agent, lubricant, primer, reducing agent; liq.; m.w. 206.4; sp.gr. 0.975; b.p. 129-130 C (10 mm); ref. index 1.4447; 95% purity.

Dynasylan® 1505. [Hüls Am.] 3-Aminopropylmethyldiethoxysilane; CAS 3179-76-8; coupling agent, chem. intermediate, blocking agent, release agent, lubricant, primer, reducing agent; chemically bonds, strengthens thermoset furan and melamine composites; liq.; m.w. 191.4; sp.gr. 0.92 (20 C); b.p. 85-88 C (8 mm); flash pt. 125 C; ref. index 1.427; 97% purity.

Dynasylan® 3403. [Hüls Am.] 3-Mercaptopropylmethyldimethoxysilane; CAS 31001-77-1; coupling agent, chem. intermediate, blocking agent, release agent, lubricant, primer, reducing agent; liq.; m.w. 180.4; sp.gr. 0.99 (20 C); b.p. 96 C (30 mm); flash pt. 97 C; ref. index 1.4502; 95% purity.

Dynasylan® AMEO. [Hüls Am.] Aminopropyltriethoxysilane; CAS 919-30-2; EINECS 213-048-4; coupling agent, chem. intermediate, blocking agent, release agent, lubricant, primer, reducing agent; chemically bonds, strengthens thermosets and thermoplastic composites; for DAP, epoxy, furan, phenol-formaldehyde, U-F, urethane, cellulosics, nylon 6, polyamide-imide, PBT, PC, PVAc, PPS, polysulfone, PES, PVB, PVC, sealants (polysulfide, nitrile, SAN), rubber (fluorocarbon, epichlorohydrin, silicone); liq.; m.w. 221.4; sp.gr. 0.95 (20 C); b.p. 122-123 C (30 mm); flash pt. 96 C; ref. index 1.423 (20 C); 98.5% purity.

Dynasylan® AMEO-40. [Hüls Am.] Aminoalkyl silicone sol'n.; organo-silicon compds. for bonding between org. and inorg. components of a system; liq.; water-sol.; Unverified

Dynasylan® AMEO-P. [Hüls Am.] Aminopropyltriethoxy silane; CAS 919-30-2; EINECS 213-048-4; organo-silicon compds. for bonding between org. and inorg. components of a system; liq.; water-sol.

Dynasylan® AMEO-T. [Hüls Am.] Aminopropyltriethoxy silane, tech. grade; CAS 919-30-2; EINECS 213-048-4; organo-silicon compds. for bonding between org. and inorg. components of a system; liq.; water-sol.

Dynasylan® AMMO. [Hüls Am.] Aminopropyltrimethoxysilane; CAS 13822-56-3; EINECS 237-511-5; coupling agent, chem. intermediate, blocking agent, release agent, lubricant, primer, reducing agent; liq.; water-sol.; m.w. 179.2; sp.gr. 1.01 (20 C); b.p. 80 C (8 mm); flash pt. 104 C; ref. index 1.420; 98% purity.

Dynasylan® AMTC. [Hüls Am.] Amyltrichlorosilane; CAS 107-72-2; coupling agent, chem. intermediate, blocking agent, release agent, lubricant, primer, reducing agent; liq.; m.w. 205.6; sp.gr. 1.13 (20 C); b.p. 171-172 C; flash pt. 30 C; ref. index 1.438; 97% purity.

Dynasylan® BDAC. [Hüls Am.] Di-t-butoxydiacetoxysilane; CAS 13170-23-5; coupling agent, chem. intermediate, blocking agent, release agent, lubricant, primer, reducing agent; liq.; m.w. 292.4; sp.gr. 1.0196 (20 C); b.p. 102 C (5 mm); flash pt. 95 C; ref. index 1.4040 (20 C); 97% purity.

Dynasylan® BSA. [Hüls Am.] Bis(trimethylsilyl) acetamide; CAS 10416-59-8; coupling agent, chem. intermediate, blocking agent, release agent, lubricant, primer, reducing agent; liq.; m.w. 203.4; sp.gr. 0.83 (20 C); b.p. 71-73 C (35 mm); flash pt. 17 C; ref. index 1.418 (20 C); 95% purity.

Dynasylan® CPTEO. [Hüls Am.] 3-Chloropropyltriethoxysilane; CAS 5089-70-3; EINECS 225-805-6; coupling agent, chem. intermediate, blocking agent, release agent, lubricant, primer, reducing agent; chemically bonds, strengthens PVC composites; liq.; m.w. 240.8; sp.gr. 1.009 (20 C); b.p. 98-102 C (10 mm); flash pt. 61 C; ref. index 1.420 (20 C); 98% purity.

Dynasylan® CPTMO. [Hüls Am.] 3-Chloropropyltrimethoxysilane; CAS 2530-87-2; EINECS 219-787-9; coupling agent, chem. intermediate, blocking agent, release agent, lubricant, primer, reducing agent; liq.; m.w. 198.7; sp.gr. 1.081; b.p. 183 C; flash pt. 66 C; ref. index 1.464; 98% purity.

Dynasylan® DAMO. [Hüls Am.] N-2-aminoethyl-3-aminopropyltrimethoxysilane; CAS #1760-24-3; CAS 1760-24-3; EINECS 212-164-2; coupling agent, chem. intermediate, blocking agent, release agent, lubricant, primer, reducing agent; liq.; m.w. 222.4; sp.gr. 1.01; b.p. 140 C (15 mm); flash pt. 121 C; ref. index 1.442; 95% purity.

Dynasylan® DAMO-P. [Hüls Am.] Aminoethylamino propyltrimethoxy silane; CAS 1760-24-3; EINECS 212-164-2; coupler for epoxy, phenolic, melamine, nylons, PVC, acrylics, polyolefins, PU and nitrile rubber; liq.; water-sol.; m.w. 222.4; sp.gr. 1.01; b.p. 140 C (15 mm); flash pt. 121 C; ref. index 1.442; 95% purity.

Dynasylan® DAMO-T. [Hüls Am.] Aminoethylamino propyltrimethoxy silane, tech. grade; CAS 1760-24-3; EINECS 212-164-2; coupler for epoxy, phenolic, melamine, nylons, PVC, acrylics, polyolefins, PU and nitrile rubber; liq.; water-sol.

Dynasylan® ETAC. [Hüls Am.] Ethyltriacetoxysilane; CAS 17689-77-9; coupling agent, chem. intermediate, blocking agent, release agent, lubricant, primer, reducing agent; liq.; m.w. 234.3; sp.gr. 1.14 (20 C); b.p. 107 C (8 mm); flash pt. 8 C; ref. index 1.412 (20 C); 95% purity.

Dynasylan® GLYMO. [Hüls Am.] 3-Glycidoxypropyltrimethoxysilane; CAS 2530-83-8; EINECS 219-784-2; coupling agent, chem. intermediate, blocking agent, release agent, lubricant, primer, reducing agent; chemically bonds, strengthens thermoset and thermoplastic composites; liq.; m.w. 236.3; sp.gr. 1.070 (20 C); b.p. 120 C (2 mm); flash pt. 79 C; ref. index 1.428; 98% purity.

Dynasylan® HMDS. [Hüls Am.] Hexamethyldisilazane; CAS 999-97-3; EINECS 213-668-5; coupling agent, chem. intermediate, blocking agent, release agent, lubricant, primer, reducing agent; liq.; m.w. 161.4; sp.gr. 0.77 (20 C); b.p. 126-127 C; flash pt. 27 C; ref. index 1.408 (20 C); 98% purity.

Dynasylan® IBTMO. [Hüls Am.] Isobutyltrimethoxysilane; CAS 18395-30-7; coupling agent, chem. intermediate, blocking agent, release agent, lubricant, primer, reducing agent; liq.; m.w. 178.3; sp.gr. 0.93 (20 C); b.p. 154-157 C; flash pt. 14 C; ref. index 1.396 (20 C); 98% purity.

Dynasylan® IMEO. [Hüls Am.] N-[3-(Triethoxysilyl)-propyl] 4,5-dihydroimidazole; CAS 58068-97-6; coupling agent, chem. intermediate, blocking agent, release agent, lubricant, primer, reducing agent; chemically bonds, strengthens thermoset phenol-formaldehyde composites; liq.; m.w. 274.1; sp.gr. 1.00 (20 C); b.p. 134 C (2 mm); flash pt. 78 C; ref. index 1.45; 90% purity.

Dynasylan® MEMO. [Hüls Am.] 3-Methacryloxypropyltrimethoxysilane; CAS 2530-85-0; EINECS 219-785-8; coupling agent, chem. intermediate, blocking agent, release agent, lubricant, primer, reducing agent; chemically bonds, strengthens thermoplastic and thermoset composites; for methyl methacrylate, unsat. polyester, polyacrylate, polyether ketone, PEEK, PP, PS, sealants (acrylic, SBR), rubbers (vinyl-terminated silicone); liq.; m.w. 248.4; sp.gr. 1.045; b.p. 190 C; flash pt. 92 C; ref. index 1.429; 97% purity.

Dynasylan® MTAC. [Hüls Am.] Methyltriacetoxysilane; CAS 4253-34-3; coupling agent, chem. intermediate, blocking agent, release agent, lubricant, primer, reducing agent; solid; m.w. 220.3; sp.gr. 1.17 (20 C); b.p. 87-88 C (3 mm); flash pt. 5 C; ref. index 1.408 (20 C); 96% purity.

Dynasylan® MTES. [Hüls Am.] Methyltriethoxysilane; CAS 2031-67-6; EINECS 217-983-9; coupling agent, chem. intermediate, blocking agent, release agent, lubricant, primer, reducing agent; liq.; m.w. 178.3; sp.gr. 0.90 (20 C); b.p. 141-143 C; flash pt. 38 C; ref. index 1.383 (20 C); 98% purity.

Dynasylan® MTMO. [Hüls Am.] 3-Mercaptopropyltrimethoxysilane; CAS 4420-74-0; EINECS 224-588-5; coupling agent, chem. intermediate, blocking agent, release agent, lubricant, primer, reducing agent; chemically bonds, strengthens thermoset and thermoplastic composites; for epoxy, polysulfide sealants, neoprene, isoprene, and epichlorohydrin rubbers, gold and precious metals; liq.; m.w. 196.3; sp.gr. 1.05; b.p. 219-220 C; flash pt. 93 C; ref. index 1.440; 97% purity.

Dynasylan® MTMS. [Hüls Am.] Methyltrimethoxysilane; CAS 1185-55-3; EINECS 214-685-0; coupling agent, chem. intermediate, blocking agent, release agent, lubricant, primer, reducing agent; liq.; m.w. 136.2; sp.gr. 0.955 (20 C); b.p. 102-103 C; flash pt. 8 C; ref. index 1.3696 (20 C); 98% purity.

Dynasylan® OCTEO. [Hüls Am.] Octyltriethoxysilane; CAS 2943-75-1; EINECS 220-941-2; coupling agent, chem. intermediate, blocking agent, release agent, lubricant, primer, reducing agent; liq.; m.w. 276.5; sp.gr. 0.88 (20 C); b.p. 98-99 C (2 mm); flash pt. 100 C; ref. index 1.415 (20 C); 98% purity.

Dynasylan® PETCS. [Hüls Am.] 2-Phenethyltrichlorosilane; CAS 940-41-0; coupling agent, chem. intermediate, blocking agent, release agent, lubricant, primer, reducing agent; liq.; m.w. 239.6; sp.gr. 1.24 (20 C); b.p. 92-96 C (3 mm); flash pt. 78 C; ref. index 1.518 (20 C); 96% purity.

Dynasylan® PTMO. [Hüls Am.] n-Propyltrimethoxysilane; CAS 1067-25-0; coupling agent, chem. intermediate, blocking agent, release agent, lubricant, primer, reducing agent; liq.; m.w. 164.3; sp.gr. 0.932; b.p. 142 C; flash pt. 35 C; ref. index 1.388; 98% purity.

Dynasylan® TCS. [Hüls Am.] Trichlorosilane; CAS 10025-78-2; EINECS 233-042-5; coupling agent, chem. intermediate, blocking agent, release agent, lubricant, primer, reducing agent; liq.; m.w. 135.5; sp.gr. 1.342 (20 C); b.p. 31-32 C; flash pt. -8 C; ref. index 1.402 (20 C); 99.9% purity.

Dynasylan® TRIAMO. [Hüls Am.] Trimethoxysilylpropyldiethylene triamine; CAS 35141-30-1; coupling agent, chem. intermediate, blocking agent, release agent, lubricant, primer, reducing agent; chemically bonds, strengthens thermoset and thermoplastic composites; liq.; m.w. 265.4; sp.gr. 1.03 (20 C); b.p. 114-168 C (2 mm); flash pt. 137 C; ref. index 1.463; 91% purity.

Dynasylan® VTC. [Hüls Am.] Vinyltrichlorosilane; CAS 75-94-5; EINECS 200-917-8; coupling agent, chem. intermediate, blocking agent, release agent, lubricant, primer, reducing agent; liq.; m.w. 161.5; sp.gr. 1.24 (20 C); b.p. 92-93 C; flash pt. 21 C; ref. index 1.43 (20 C); 97% purity.

Dynasylan® VTEO. [Hüls Am.] Vinyl triethoxysilane; CAS 78-08-0; EINECS 201-081-7; coupling agent, chem. intermediate, blocking agent, release agent, lubricant, primer, reducing agent; liq.; m.w. 190.3; sp.gr. 0.90 (20 C); b.p. 160-161 C; flash pt. 34 C; ref. index 1.396 (20 C); 98% purity.

Dynasylan® VTMO. [Hüls Am.] Vinyltrimethoxysilane; CAS 2768-02-7; EINECS 220-449-8; coupling agent, chem. intermediate, blocking agent, release agent, lubricant, primer, reducing agent; chemically bonds, strengthens thermoset and thermoplastic composites; for methyl methacrylate, unsat. polyester, polyacrylate, PVAc, acrylic and SBR sealants, silicone (hydroxyl- or vinyl-terminated) rubbers; liq.; m.w. 148.2; sp.gr. 1.13 (20 C); b.p. 123 C; flash pt. 23 C; ref. index 1.393; 98% purity.

Dynasylan® VTMOEO. [Hüls Am.] Vinyl tris (methoxyethoxy) silane; CAS 1067-53-4; EINECS 213-934-0; coupling agent, chem. intermediate, blocking agent, release agent, lubricant, primer, reducing agent; chemically bonds, strengthens thermoset methyl methacrylate composites; liq.; m.w. 280.4; sp.gr. 1.04 (20 C); b.p. 284-286 C; flash pt. 71 C; ref. index 1.427; 97% purity.

Dyphene® 595. [PMC Specialties] Octylphenol formaldehyde (alkyl phenolic); heat-reactive curing resin recommended for vulcanization of butyl bladders used in tire curing applics.; provides superior heat resist., high mech. str., chem. and solv. resist., low temp. flexibility, good elec. and heat-aging props.; nonstaining; also for SBR and NR, rubber-based solv. adhesives, cements, and pressure-sensitive adhesives; yel.-amber clear to turbid crushed lumps; sol. in benzene, toluene, xylene, hexane, VM&P naphtha, acetone, MEK, ethyl acetate, chlorinated HC, drying oils; sp.gr. 1.030-1.050; visc. 1.4-3.2 stokes (64% in toluene); soften. pt. (R&B) 80-90 C; acid no. 10-30; *Storage:* store in cool, dry place; 1 yr min. shelf life stored @ 100 F or lower.

Dyphene® 877PLF. [PMC Specialties] Thermosetting two-stage phenolic resin with hexa; reinforcing and stiffening for adhesives, cements, bonded abrasives, electronic encapsulating/potting, molding applics.; powd.

Dyphene® 6745-P. [PMC Specialties] Thermosetting two-step phenolic resin; stiffener/hardener for rubber reinforcing applics. (NR, SBR, NBR); increases modulus, tear resist.; suggested for tire bead and tread compds., shoe soles and heels, skate wheels, etc.; powd.; 99% min. thru 200 mesh; apparent dens. 0.29 g/cc; m.p. 72-87 C; 7.5-8.5% hexa; *Storage:* 1 yr min. shelf life stored in closed containers at < 90 F; exposure to high temp. or humidity can cause sintering or caking problems.

Dyphene® 6746-L. [PMC Specialties] Thermoplastic novolac phenolic resin; stiffener/hardener for rubber reinforcing applics. (NR, SBR, NBR); increases modulus, tear resist.; for tire bead and tread compds., shoe soles and heels, skate wheels, etc.; base resin for Dyphene 6745-P; recommended where high processing temps. are encountered or to avoid scorch during mixing; crushed; sp.gr. 1.150-1.175; soften. pt. (R&B) 100-110; *Storage:* 1 yr min. shelf life stored in closed containers at < 90 F; exposure to high temp. or humidity can cause sintering or caking problems.

Dyphene® 8318. [PMC Specialties] Octylphenol formaldehyde (alkyl phenolic novolak); nonheat-reactive tackifying resin for rubber compounding and tire cements (NR, SBR, BR); imparts high initial tack in grn. rubber stock; processing and dispersion aid; reduces mixing visc.; yel. to amber clear flakes; sol. in oils, acetone, MEK, toluene, xylene, VM&P naphtha, hexane, CCl_4; sp.gr. 1.00-1.05; soften. pt. (R&B) 85-105 C; acid no. 40-60; *Storage:* store in dry area; 1 yr min. shelf life @ temps. of 100 F or lower.

Dyphene® 8320. [PMC Specialties] Alkyl phenolic novolac resin; nonheat-reactive tackifying resin for rubber compounding (carcass, innerliners) and splicing cements; imparts exceptional initial tack and tack retention; processing and dispersion aid; reduces mixing visc.; for SBR or blends of NR, BR, SBR; yel. to amber clear flakes; sol. in acetone, MEK, benzene, toluene, xylene, VM&P naphtha, gasoline, CCl_4; sp.gr. 1.020-1.050; soften. pt. (R&B) 108-124 C; acid no. 25-42; *Storage:* store in dry area at 100 F or lower to obtain 1 yr min. shelf life.

Dyphene® 8330. [PMC Specialties] Octylphenol formaldehyde (alkyl phenolic novolak); nonheat-reactive tackifying resin for rubber compounding and tire cements (NR, SBR, BR); imparts high initial tack in grn. rubber stock; processing and dispersion aid; reduces mixing visc.; yel. to amber clear flakes; sol. in oils, acetone, MEK, toluene, xylene, VM&P naphtha, hexane, CCl_4; sp.gr. 1.00-1.06; soften. pt. (R&B) 85-110 C; acid no. 30-70; *Storage:* store in dry area; 1 yr min. shelf life @ temps. of 100 F or lower.

Dyphene® 8340. [PMC Specialties] Modified octylphenol formaldehyde (alkyl phenolic novolak); nonheat-reactive tackifying resin with improved wire adhesion performance; imparts high initial tack in unvulcanized rubber stocks; processing aid; reduces mixing visc.; for tire carcass (NR/SBR, NR/SBR/BR, esp. with wire reinforcement); rubber compounding; amber clear flakes; sol. in ketones, esters, aromatic and aliphatic hydrocarbons; sp.gr. 1.023; soften. pt. (R&B) 110 C; acid no. 48; *Storage:* store in dry area; 1 yr min. shelf life @ temps. of 100 F or lower.

Dyphene® 8400. [PMC Specialties] Alkyl phenolic novolac resin; nonheat-reactive tackifying resin for rubber compounding (tire, carcass, innerliners) and splicing cements; imparts high initial tack in grn. rubber stocks; processing and dispersion aid; reduces mixing visc.; for blends of NR, BR, SBR, chlorobutyl; yel. to amber clear flakes; sol. in oil, acetone, MEK, toluene, xylene, VM&P naphtha, hexane, CCl_4; sp.gr. 1.00-1.06; soften. pt. (R&B) 85-115 C; acid no. 30-70; *Storage:* store in dry area at 100 F or lower to obtain 1 yr min. shelf life.

Dyphene® 8787. [PMC Specialties] Modified octylphenol formaldehyde (alkyl phenolic novolak); nonheat-reactive tackifying resin for rubber compounding (NR, SBR, SR); imparts high initial tack and tack retention; processing and dispersion aid; reduces mixing visc.; for tire carcass, bead and innerliners; effective for IIR; amber clear flakes; sol. in oil, benzene, toluene, xylene; sp.gr. 1.020-1.035; visc. 0.9-2.25 stokes (60% in toluene); soften. pt. (R&B) 95-105 C; acid no. 32-48; *Storage:* store in dry area; 1 yr min. shelf life @ temps. of 100 F or lower.

Dyphene® 8845P. [PMC Specialties] Modified two-stage phenolic resin with hexa; reinforcing and stiffening resin for adhesives, cements, bonded abrasives, electronic encapsulating/potting, molding applics.; powd.

Dyphene® 9273. [PMC Specialties] Octylphenol formaldehyde (alkyl phenolic); heat-reactive curing resin recommended for vulcanization of butyl bladders used in tire curing applics.; provides superior heat resist., high mech. str., chem. and solv. resist., low temp. flexibility, good elec. and heat-aging props.; nonstaining; also for SBR and NR, rubber-based solv. adhesives, cements, and pressure-sensitive adhesives; yel.-amber clear to turbid crushed lumps; sol. in benzene, toluene, xylene, hexane, VM&P naphtha, acetone, MEK, ethyl acetate, chlorinated HC, drying oils; sp.gr. 1.035-1.055; soften. pt. (R&B) 80-95 C; acid no. 30-40; *Storage:* store in cool, dry place; 1 yr min. shelf life stored @ 100 F or lower.

Dyphene® 9474. [PMC Specialties] Butylphenol formaldehyde; nonheat-reactive tackifying resin compat. with acrylic resins, neoprene, cis-polyisoprene, SBR, and butyl rubbers; for neoprene cements requiring heat reactivation, acrylic-based pressure-sensitive adhesives; lumps; sol. in oil, aliphatic and aromatic hydrocarbons, ketones, ketone/alcohol mixts.; sp.gr. 1.040-1.050; soften. pt. (R&B) 115-130 C; acid no. 55-105; *Storage:* 1 yr min. shelf life stored in closed shipping containers @ 75-85 F.

Dyphos. [Syn. Prods.] Dibasic lead phosphite; CAS 1344-40-7; high level heat, lt., and outdoor weathering stabilizer for PVC, plastisols, extrusion, calendering, foams, chlorinated hydrocarbons of rubbers, elec. insulation; wh. fine acicular cryst.; sp.gr. 6.1; 90% PbO.

Dyphos Envirostab. [Syn. Prods.] Dibasic lead phosphite; CAS 1344-40-7; heat and lt. stabilizer for flexible and rigid PVC; good outdoor weathering; dust-free.

Dyphos XL. [Syn. Prods.] Dibasic lead phosphite; CAS 1344-40-7; high level heat, lt., and outdoor weathering stabilizer for PVC, plastisols, extrusion, calendering, foams, chlorinated hydrocarbons of rubbers, elec. insulation; wh. coated fine acicular cryst.; sp.gr. 4.5; 81.2% PbO.

Dytek® A. [DuPont Nylon] 2-Methylpentamethylenediamine; CAS 15520-10-2; epoxy curing agent; also used in PUs, wet strength resins, scale and corrosion inhibitors, motor oil and gasoline additives, polyamide plastics, films, adhesives, and inks; colorless liq., weak ammonia, fish odor; misc. with water, acetone, ethanol, cyclohexane, n-heptane, CCl_4, xylene; m.w. 116.2; sp.gr. 0.86; visc. 2.9 cp; vapor pressure 130 mm Hg (135 C); f.p. < -70 C; m.p. -56 to -60 C; b.p. 193 C (760 mm); flash pt. (OC) 82 C; surf. tens. 34.9 dynes/g; > 98.5% act.; *Toxicology:* LD50 (oral, rat) 1690 mg/kg; corrosive to eyes and skin; inh. can irritate nose and throat; ingestion may cause gastrointestinal irritation; *Precaution:* combustible; incompat. with strong oxidants; dec. with heat; emits toxic fumes of nitrogen oxides on dec.; *Storage:* store in tightly closed container, under nitrogen padding to prevent air intrusion; keep away from heat, flame.

Dytek® EP. [DuPont Nylon] 1,3-Pentanediamine; CAS 589-37-7; epoxy curative; solv. for gas treatment; surfactant (asphalt emulsifiers, textiles); scale/corrosion inhibitor; additive for fuel, oil, plastic, petrol. and chem. processing; extender, catalyst for PU; in coatings, adhesives, sealants, elastomers; liq., sl. odor; misc. with water; m.w. 102; dens. 0.855 g/ml; visc. 1.89 cp; vapor pressure 1 mm Hg (20 C); b.p. 164 C; flash pt. (CC) 59 C; surf. tens. 32.2 dyn/cm; *Precaution:* corrosive, flamm.; incompat. with strong oxidants; thermal decomp. may emit NO_x.

Dythal. [Syn. Prods.] Dibasic lead phthalate; CAS 69011-06-9; heat stabilizer for flexible and semi-rigid PVC, plastisols, organosols; processability allows high extrusion rates; wh. fluffy powd.; sp.gr. 4.5; 79.8% PbO.

Dythal XL. [Syn. Prods.] Dibasic lead phthalate; CAS 69011-06-9; heat stabilizer for flexible and semi-rigid PVC, plastisols, organosols; processability allows high extrusion rates; wh. coated fluffy powd.; sp.gr. 3.5; 71.5% PbO.

E

Ease Release™ 200 Series. [George Mann] General purpose release agent for release of most types of casting and molding systems, e.g., PU elastomers, PU foam, epoxy resin, polyester, RTV silicones, rubber, and thermoplastic polymers; effective on aluminum, chrome, RTV silicone, epoxy, polyester, rubber, and steel molds; grades avail.: 200—solv. trichloromonofluoromethane, dichlorodifluoromethane; 201—solv. trichloromonofluoromethane, dens. 12 lb/gal; 202—solv. methylene chloride, dens. 10.8 lb/gal; 203—solv. Chlorothene® SM, dens. 10.8 lb/gal; 204—solv. trichlorotrifluoroethane, dens. 12.6 lb/gal; 205—solv. petrol. hydrocarbons, dens. 6 lb/gal; 206—solv. hexane, dens. 5.7 lb/gal.

Ease Release™ 300 Series. [George Mann] Release agent for urethane elastomers and foams, and epoxy resins; effective on steel, aluminum, chrome, epoxy, RTV urethane, and silicone rubber mold surfaces; grades avail.: 300—solv. trichloromonofluoromethane, dichlorodifluoromethane; 301—solv. trichloromonofluoromethane, dens. 12 lb/gal; 302—solv. methylene chloride, dens. 10.8 lb/gal; 303—solv. Chlorothene® SM, dens. 10.8 lb/gal; 304—solv. trichlorotrifluoroethane, dens. 12.6 lb/gal; 305—solv. petrol. hydrocarbons, dens. 6 lb/gal; 306—solv. hexane, dens. 5.7 lb/gal.

Ease Release™ 400 Series. [George Mann] Release agents for PU elastomers, thiokol rubbers, EPDM, and other syn. elastomers from aluminum, chrome, epoxy, polyester, rubber, and steel molds; grades avail.: 400—solv. trichloromonofluoromethane, dichlorodifluoromethane; 401—solv. trichloromonofluoromethane, dens. 12.1 lb/gal; 402—solv. methylene chloride, dens. 10.9 lb/gal; 403—solv. Chlorothene® SM, dens. 10.9 lb/gal; 404—solv. trichlorotrifluoroethane, dens. 12.8 lb/gal; 405—solv. petrol. hydrocarbons, dens. 6 lb/gal; 406—solv. hexane, dens. 5.7 lb/gal.

Ease Release™ 500 Series. [George Mann] General purpose release and dry lubricant containing no silicone or oils; releases most casting and molding systems, e.g., PU elastomers, PU foams, epoxy resins, polyester, rubbers, thermoplastic polymers, phenolic resins from aluminum, chrome, epoxy, polyester, rubber, and steel molds; grades avail.: 500—solv. trichloromonofluoromethane, dichlorodifluoromethane; 504—solv. trichlortrifluoroethane, dens. 13 lb/gal; 507—solv. water, dens. 8.6 lb/gal.

Ease Release™ 700. [George Mann] Silicone; general purpose release agent and lubricant for inj. molded plastics; also as assembly lubricant for plastic parts; FDA acceptable active ingreds.

Ease Release™ 900. [George Mann] Release agent for unitsole and shoe components using rigid and flexible microcellular foam urethanes; reduces rejects due to poor adhesion of stai coats and adhesives; reduces buildup caused by silicone release agents.

Ease Release™ 1700. [George Mann] Nonsilicone; release agent for thermoplastics, composites, epoxies, phenolic, and melamines; paintable.

Ease Release™ 2040 Series. [George Mann] Semipermanent release coating providing long lasting release for urethane, epoxy, rubber, aluminum, and steel molds, molding presses, coating machinery; effective with urethanes, rubbers, epoxies, hot-melt adhesives, polyesters, and other polymers and gums; grades avail.: 2040; 2044—dens. 12.3 lb/gal; 2045—solv. petrol. hydrocarbons, dens. 6.2 lb/gal.

Ease Release™ 2110. [George Mann] General-purpose water-washable release for rigid or flexible foams; water-alcohol solv. carrier for fast drying.

Ease Release™ 2116. [George Mann] Release agent providing harder film and higher m.p. for higher temp. resins; used for rubber and composite materials.

Ease Release™ 2148. [George Mann] Silicone; general-purpose water-washable release agent for inj. molding, vacuum forming, and calendaring of thermoplastics.

Ease Release™ 2181. [George Mann] Nonsilicone; water-based release agent for phenolic, melamine, and other resin fiber composites such as brake pads and tubing; water-misc.

Ease Release™ 2191. [George Mann] Silicone compd. blend; release agent for casting wax models and molds, inj. molding, assembly lube; provides multiple releases.

Ease Release™ 2197. [George Mann] Silicone; general-purpose paintable release agent for inj. molding thermoplastics; parts can be painted or stamped with minimal cleaning.

Ease Release™ 2249. [George Mann] Nonsilicone;

general-purpose release agent for urethane, epoxy, and thermoplastics up to 400 F; can be used for casting, inj. molding, and RIM processing.

Ease Release™ 2251. [George Mann] Release agent for urethane elastomers; leaves no buildup; can be used on hot or cold molds and MDI or TDI-based elastomers.

Ease Release™ 2300. [George Mann] General-purpose release agent for urethanes; improved slip; exc. for deep molds and low green str. elastomers; parts are easily cleaned for finishing or bonding.

Ease Release™ 2400. [George Mann] Release agent for unitsole and mold-on-last shoe applics.; provides better release, drier surface, and fewer adhesion rejects than Ease Release 900 series.

Eastman® DTBHQ. [Eastman] 2,5-Di-t-butylhydroquinone; CAS 88-58-4; polymerization inhibitor, antioxidant, and stabilizer for syn. rubbers, plastics, unsat. polyesters; in rubber industry as stopping agent, antioxidant, color stabilizer; wh. to tan cryst.; sol. (g/100 g): 44 g ethyl acetate, 34 g acetone, 32 g acetone; insol. in water and caustic; m.w. 222.31; m.p. 210 C; flash pt. (COC) 216 C; fire pt. (COC) 216 C.

Eastman® HQEE. [Eastman] Hydroquinone bis(2-hydroxyethyl) ether; CAS 104-38-1; EINECS 203-197-3; chain extender with exc. hardness, tear str. for polyurethane elastomers; used in elastomers for oil well seals, forklift tires, seals for hydraulic cylinders, conveyor liners, skateboard wheels; chemical intermediate; wh.-tan flake, odorless; sol. 4% in acetone, ethanol, 1% ethyl acetate, < 1% water; m.w. 198.2; sp.gr. 1.15; bulk dens. 0.51 g/cc (loose); visc. 15 cP (110 C); m.p. 98 C; b.p. 190 C (0.3 mm); hyd. no. 555; flash pt. (COC) 224 C; sp. heat 0.4 cal/g/°C; 89-99% assay; *Toxicology:* LD50 (oral, rat) > 3200 mg/kg, (dermal, guinea pig) > 1000 mg/kg; sl. skin irritant; *Precaution:* powd. material may form explosive dust-air mixts.; molten material will produce thermal burns; incompat. with oxidizers.

Eastman® HQMME. [Eastman] Hydroquinone monomethyl ether; CAS 150-76-5; antioxidant for chlorinated hydrocarbons, aldehydes, oils of turpentine; polymerization inhibitor in vinylidene chloride, acrylonitrile, acrylic and methacrylic esters, other vinyl monomers; stabilizer for urethane-type polyether polyols, ethylcellulose; chemical intermediate for mfg. stabilizers, pharmaceuticals, dyes, BHA; wh. flakes; sol. (g/100 g): > 350 g actone and ethanol, > 200 g ethyl acetate, 35 g cottonseed oil, 32 g toluene, 4 g water; m.w. 124.14; bulk dens. 36.5 lb/ft^3; b.p. 246 C; solid. pt. 54 C; flash pt. (COC) 132 C; fire pt. 135 C; 99% act.

Eastman® MTBHQ. [Eastman] Mono-t-butylhydroquinone; CAS 1948-33-0; EINECS 217-752-2; polymerization inhibitor/retarder for unsat. polyesters and other polymers during storage and shipping; antioxidant for rubber, monomers; chemical intermediate; wh. to lt. tan cryst.; sol. (g/100 g): 190 g acetone, 120 g ethanol, 100 g ethyl acetate; sl. sol. in water; m.w. 166.22; m.p. 125 C; flash pt. (COC) 171 C; fire pt. (COC) 174 C.

Eastman® PA-208. [Eastman] Glyceryl stearate; antistat for PC, PP; antifog for PE; FDA 21CFR §182.1324, 182.4505; beads; sol. @ 3% in N-methyl-2-pyrrolidone; insol. in water, acetone, ethanol, metanol, MEK; m.p. 73 C; acid no. 3 max.; iodine no. 5 max.; 90% min. monoester.

Eastman® TEP. [Eastman] Triethyl phosphate; CAS 78-40-0; EINECS 201-114-5; intermediate for insecticides/pesticides, in floor polishes, as plasticizer for cellulose acetate, as solv. for textiles, dyeing assistant, in sizes, flame retardant synergist in unsat. polyesters, in lubricants, hydraulic fluids, as catalyst; visc. suppressant for general polymer applics.; colorless clear liq., mild pleasant cider-like odor; misc. with water, ethanol, ethyl acetate, ethyl ether, acetone, chloroform, benzene, xylene, castor oil; partly sol. in min. oil, kerosene, CS_2; sp.gr. 1.067-1.072 (20/20 C); dens. 8.91 lb/gal; visc. 1.76 cP (20 C); vapor pressure 1 mm Hg (40 C); m.p. -56.4 C; b.p. 209 C; flash pt. (PMCC) 99 C; fire pt. (COC) 141 C; ref. index 1.4019; *Toxicology:* LD50 (oral, rat) 1311 mg/kg (sl. toxic), (dermal, guinea pig) > 20 ml/kg; causes eye irritation; sl. skin irritant; emits highly toxic fumes of phosphorus oxides when heated to decomp.; *Precaution:* incompat. with oxidizing materials.

Eastman® Inhibitor OABH. [Eastman] Oxalyl bis(benzylidenehydrazide); copper deactivator, stabilizer used in polyolefins in contact w/ copper or copper-containing alloys; wh. crys. powd., 99% min. thru 200 mesh; sol. 5540 mg in n-methyl-2-pyrrolidinone (105 C); 1990 mg in dimethylformamide (88 C); 70.4 mg in 1,4-butanediol (102 C); 59.9 g in ethylene glycol (98 C); 67.4 mg in 1,3-propanediol (100 C); 191 mg in acetic acid (81 C); m.w. 294.16 (calcd.); dens. 0.216 kg/l; sapon. no. 144-150; 97.0% min. assay.

Eastman® Inhibitor Poly TDP 2000. [Eastman] Thiodipropionate polyester; CAS 63123-11-5; sec. antioxidant used with phenolic primary antioxidants in polypropylene and other polyolefins; paste-like solid; m.w. 2000; sp.gr. 1.09 (90 C); visc. 700-1100 cP (100 C); m.p. 30 C; acid no. 2.5; hyd. no. 30; flash pt. 249 C.

Eastman® Inhibitor RMB. [Eastman] Resorcinol monobenzoate; CAS 136-36-7; industrial grade UV absorber/stabilizer for cellulosic plastics and PVC formulations; wh. cryst. solid, odorless; sol. (g/100 g): 75.2 g in acetone; 27.7 g in ethanol; > 0.1 g in benzene, n-hexane, DOP, and water; m.w. 214.2; sp.gr. > 1 (20/20 C); dens. 5.7 lb/gal; vapor pressure 0.15 mm Hg (140 C); m.p. 132-135 C; b.p. 140 C (0.15 mm); flash pt. (COC) 214 C; *Toxicology:* LD50 (oral, rat) 1600-3200 mg/kg (sl. toxic), (dermal, guinea pig) > 20 mL/kg; causes skin irritation; may cause allergic skin reaction; *Precaution:* incompat. with oxidizing materials.

Eastobrite® OB-1. [Eastman] 2,2′-(1,2-Ethenediyldi-4,1-phenylene)bisbenzoxazole; CAS 1533-45-5; optical brightener; fluorescent whitening agent for use in linear polyester, PET, nylon and polyester fibers; yel. cryst. solid, odorless; water-insol.; m.w. 414.4; sp.gr. 1.39; m.p. 359 C; 97% min. assay.

Eastoflex E1060. [Eastman] Amorphous polyolefin; processing aid, dispersant for polyolefins and rubbers.

Eastoflex P1023. [Eastman] Amorphous polyolefin; processing aid, dispersant for polyolefins and rubbers.

Eastotac™ H-100. [Eastman] Hydrog. C_5 aliphatic hydrocarbon resin; tackifier, processing aid for adhesives, caulks, sealants, elastomers, polymers, and other resins incl. EVA, amorphous polyolefins, PE waxes, butyl rubbers, inks, paints, varnishes, textile sizes in dry-cleaning; FDA 21CFR §175.105; Gardner 4 flake; bulk dens. 1.04 g/cc (23 C); visc. 150 cP (190 C); soften. pt. (R&B) 100 C; acid no. < 0.1; flash pt. (COC) 242 C.

Eastotac™ H-100E. [Eastman] Hydrog. C_5 aliphatic hydrocarbon resin; tackifier, processing aid for adhesives, caulks, sealants, elastomers, polymers, and other resins incl. EVA, amorphous polyolefins, PE waxes, butyl rubbers, inks, paints, varnishes, textile sizes in dry-cleaning; FDA 21CFR §175.105; Gardner 8 flake; sol. in chloroform, heptane, hexane, methylene chloride, toluene, perchloroethylene, xyelne; sl. sol. in acetone, IPA, MEK; sp.gr. 1.04; visc. 200 cP (190 C); soften. pt. (R&B) 100 C; acid no. < 0.1; flash pt. (COC) 242 C; fire pt. (COC) 266 C.

Eastotac™ H-100L. [Eastman] Hydrog. C_5 aliphatic hydrocarbon resin; tackifier, processing aid for adhesives, caulks, sealants, elastomers, polymers, and other resins incl. EVA, amorphous polyolefins, PE waxes, butyl rubbers, inks, paints, varnishes, textile sizes in dry-cleaning; FDA 21CFR §175.105; Gardner 2 flake; sol. in chloroform, heptane, hexane, methylene chloride, toluene, perchloroethylene, xyelne; sl. sol. in acetone, IPA, MEK; sp.gr. 1.04; visc. 200 cP (190 C); soften. pt. (R&B) 100 C; acid no. < 0.1; flash pt. (COC) 242 C; fire pt. (COC) 266 C.

Eastotac™ H-100R. [Eastman] Hydrog. C_5 aliphatic hydrocarbon resin; tackifier, processing aid for adhesives, caulks, sealants, elastomers, polymers, and other resins incl. EVA, amorphous polyolefins, PE waxes, butyl rubbers, inks, paints, varnishes, textile sizes in dry-cleaning; FDA 21CFR §175.105; Gardner 4 flake; sol. in chloroform, heptane, hexane, methylene chloride, toluene, perchloroethylene, xyelne; sl. sol. in acetone, IPA, MEK; sp.gr. 1.04; visc. 200 cP (190 C); soften. pt. (R&B) 100 C; acid no. < 0.1; flash pt. (COC) 242 C; fire pt. (COC) 266 C.

Eastotac™ H-100W. [Eastman] Hydrog. C_5 aliphatic hydrocarbon resin; tackifier, processing aid for adhesives, caulks, sealants, elastomers, polymers, and other resins incl. EVA, amorphous polyolefins, PE waxes, butyl rubbers, inks, paints, varnishes, textile sizes in dry-cleaning; FDA 21CFR §175.105; Gardner < 1 flake; sol. in chloroform, heptane, hexane, methylene chloride, toluene, perchloroethylene, xyelne; sl. sol. in acetone, IPA, MEK; sp.gr. 1.04; visc. 200 cP (190 C); soften. pt. (R&B) 100 C; acid no. < 0.1; flash pt. (COC) 242 C; fire pt. (COC) 266 C.

Eastotac™ H-115E. [Eastman] Hydrog. C_5 aliphatic hydrocarbon resin; tackifier, processing aid for adhesives, caulks, sealants, elastomers, polymers, and other resins incl. EVA, amorphous polyolefins, PE waxes, butyl rubbers, inks, paints, varnishes, textile sizes in dry-cleaning; FDA 21CFR §175.105; Gardner 8 flake; sol. in chloroform, heptane, hexane, methylene chloride, toluene, perchloroethylene, xyelne; sl. sol. in acetone, IPA, MEK; sp.gr. 1.04; visc. 400 cP (190 C); soften. pt. (R&B) 115 C; acid no. < 0.1; flash pt. (COC) 257 C; fire pt. (COC) 283 C.

Eastotac™ H-115L. [Eastman] Hydrog. C_5 aliphatic hydrocarbon resin; tackifier, processing aid for adhesives, caulks, sealants, elastomers, polymers, and other resins incl. EVA, amorphous polyolefins, PE waxes, butyl rubbers, inks, paints, varnishes, textile sizes in dry-cleaning; FDA 21CFR §175.105; Gardner 2 flake; sol. in chloroform, heptane, hexane, methylene chloride, toluene, perchloroethylene, xyelne; sl. sol. in acetone, IPA, MEK; sp.gr. 1.04; visc. 400 cP (190 C); soften. pt. (R&B) 115 C; acid no. < 0.1; flash pt. (COC) 257 C; fire pt. (COC) 283 C.

Eastotac™ H-115R. [Eastman] Hydrog. C_5 aliphatic hydrocarbon resin; tackifier, processing aid for adhesives, caulks, sealants, elastomers, polymers, and other resins incl. EVA, amorphous polyolefins, PE waxes, butyl rubbers, inks, paints, varnishes, textile sizes in dry-cleaning; FDA 21CFR §175.105; Gardner 4 flake; sol. in chloroform, heptane, hexane, methylene chloride, toluene, perchloroethylene, xyelne; sl. sol. in acetone, IPA, MEK; sp.gr. 1.04; visc. 400 cP (190 C); soften. pt. (R&B) 115 C; acid no. < 0.1; flash pt. (COC) 257 C; fire pt. (COC) 283 C.

Eastotac™ H-115W. [Eastman] Hydrog. C_5 aliphatic hydrocarbon resin; tackifier, processing aid for adhesives, caulks, sealants, elastomers, polymers, and other resins incl. EVA, amorphous polyolefins, PE waxes, butyl rubbers, inks, paints, varnishes, textile sizes in dry-cleaning; FDA 21CFR §175.105; Gardner < 1 flake; sol. in chloroform, heptane, hexane, methylene chloride, toluene, perchloroethylene, xyelne; sl. sol. in acetone, IPA, MEK; sp.gr. 1.04; visc. 400 cP (190 C); soften. pt. (R&B) 115 C; acid no. < 0.1; flash pt. (COC) 257 C; fire pt. (COC) 283 C.

Eastotac™ H-130E. [Eastman] Hydrog. C_5 aliphatic hydrocarbon resin; tackifier, processing aid for adhesives, caulks, sealants, elastomers, polymers, and other resins incl. EVA, amorphous polyolefins, PE waxes, butyl rubbers, inks, paints, varnishes, textile sizes in dry-cleaning; FDA 21CFR §175.105; Gardner 8 flake; sol. in chloroform, heptane, hexane, methylene chloride, toluene, perchloroethylene, xyelne; sl. sol. in acetone, IPA, MEK; sp.gr. 1.04; visc. 1200 cP (190 C); soften. pt. (R&B) 130

C; acid no. < 0.1; flash pt. (COC) 299 C; fire pt. (COC) 325 C.

Eastotac™ H-130L. [Eastman] Hydrog. C_5 aliphatic hydrocarbon resin; tackifier, processing aid for adhesives, caulks, sealants, elastomers, polymers, and other resins incl. EVA, amorphous polyolefins, PE waxes, butyl rubbers, inks, paints, varnishes, textile sizes in dry-cleaning; FDA 21CFR §175.105; Gardner 2 flake; sol. in chloroform, heptane, hexane, methylene chloride, toluene, perchloroethylene, xyelne; sl. sol. in acetone, IPA, MEK; sp.gr. 1.04; visc. 1200 cP (190 C); soften. pt. (R&B) 130 C; acid no. < 0.1; flash pt. (COC) 299 C; fire pt. (COC) 325 C.

Eastotac™ H-130R. [Eastman] Hydrog. C_5 aliphatic hydrocarbon resin; tackifier, processing aid for adhesives, caulks, sealants, elastomers, polymers, and other resins incl. EVA, amorphous polyolefins, PE waxes, butyl rubbers, inks, paints, varnishes, textile sizes in dry-cleaning; FDA 21CFR §175.105; Gardner 4 flake; sol. in chloroform, heptane, hexane, methylene chloride, toluene, perchloroethylene, xyelne; sl. sol. in acetone, IPA, MEK; sp.gr. 1.04; visc. 1200 cP (190 C); soften. pt. (R&B) 130 C; acid no. < 0.1; flash pt. (COC) 299 C; fire pt. (COC) 325 C.

Eastotac™ H-130W. [Eastman] Hydrog. C_5 aliphatic hydrocarbon resin; tackifier, processing aid for adhesives, caulks, sealants, elastomers, polymers, and other resins incl. EVA, amorphous polyolefins, PE waxes, butyl rubbers, inks, paints, varnishes, textile sizes in dry-cleaning; FDA 21CFR §175.105; Gardner < 1 flake; sol. in chloroform, heptane, hexane, methylene chloride, toluene, perchloroethylene, xyelne; sl. sol. in acetone, IPA, MEK; sp.gr. 1.04; visc. 1200 cP (190 C); soften. pt. (R&B) 130 C; acid no. < 0.1; flash pt. (COC) 299 C; fire pt. (COC) 325 C.

Eastotac™ H-142R. [Eastman] Hydrog. C_5 aliphatic hydrocarbon resin; tackifier, processing aid for adhesives, caulks, sealants, elastomers, polymers, and other resins incl. EVA, amorphous polyolefins, PE waxes, butyl rubbers, inks, paints, varnishes, textile sizes in dry-cleaning; FDA 21CFR §175.105; Gardner 4 flake; sol. in chloroform, heptane, hexane, methylene chloride, toluene, perchloroethylene, xyelne; sl. sol. in acetone, IPA, MEK; sp.gr. 1.04; visc. 3000 cP (190 C); soften. pt. (R&B) 142 C; acid no. < 0.1; flash pt. (COC) 321 C; fire pt. (COC) 335 C.

Ebal. [Mitsubishi Gas] p-Ethylbenzaldehyde; CAS 4748-78-1; EINECS 225-268-8; additive for resins; intermediate for pharmaceuticals, fragrances; colorless liq., aromatic odor; sol. in ethanol, ether, toluene; insol. in water; m.w. 134.2; sp.gr. 1.000; b.p. 221 C; m.p. -33 C; acid no. 0.3; flash pt. (COC) 98 C; 97.5% purity; *Toxicology:* LD50 (oral, rat) 1700 mg/kg; eye and skin irritant.

Ebecryl® 220. [UCB Radcure] Mutlifunctional aromatic urethane acrylate contg. an acrylated polyol diluent; oligomer providing very fast cure with exc. hardness and solv. resist.; for wood coatings, fillers, lithographic inks, scratch resist. coatings on plastics; additive to increase cure speed and gloss; Gardner 2 color; m.w. 1000; visc. 28,000 cps; tens. str. 4400 psi; tens. elong. 1%.

Ebecryl® 230. [UCB Radcure] Aliphatic urethane diacrylate; oligomer which cures to a soft flexible film; for inks and coatings for plastic and metal, laminating adhesives; additive improving flexibility of epoxy and urethane formulations; Gardner 1 color; m.w. 5000; visc. 40,000 cps; tens. str. 150 psi; tens. elong. 83%.

Ebecryl® 1360. [UCB Radcure] Silicone hexaacrylate; additive contributing slip, substrate wetting and flow props. to overprint varnishes, clear coatings on paper, plastics and metals.

Eccosphere® FTD-200. [Emerson & Cuming] Hollow glass spheres; filler for polymers providing reduced weight, improved thermal insulation and mech. str., lower dielec. constant; for aerospace (ablative barrier coatings, syntactic foam core materials, airframe composites), electronics (potting compds.), medicine, recreation (tennis racquets, bowling and golf balls, high tech bicycles), military, and manufacturing applics.; 70 μm median particle size; true particle dens. 0.20 g/cc; bulk dens. 0.13 g/cc; soften. pt. 750 C; 93% SiO_2, 3% B_2O_3, 2% Na_2O, 1% BaO.

Eccosphere® FTD-202. [Emerson & Cuming] Hollow glass spheres; filler for polymers providing reduced weight, improved thermal insulation and mech. str., lower dielec. constant; for aerospace (ablative barrier coatings, syntactic foam core materials, airframe composites), electronics (potting compds.), medicine, recreation (tennis racquets, bowling and golf balls, high tech bicycles), military, and manufacturing applics.; 55 μm median particle size; true particle dens. 0.25 g/cc; bulk dens. 0.16 g/cc; soften. pt. 750 C; 93% SiO_2, 3% B_2O_3, 2% Na_2O, 1% BaO.

Eccosphere® FTD-235. [Emerson & Cuming] Hollow glass spheres; filler for polymers providing reduced weight, improved thermal insulation and mech. str., lower dielec. constant; for aerospace (ablative barrier coatings, syntactic foam core materials, airframe composites), electronics (potting compds.), medicine, recreation (tennis racquets, bowling and golf balls, high tech bicycles), military, and manufacturing applics.; 40 μm median particle size; true particle dens. 0.35 g/cc; bulk dens. 0.22 g/cc; soften. pt. 750 C; 93% SiO_2, 3% B_2O_3, 2% Na_2O, 1% BaO.

Eccosphere® IG-25. [Emerson & Cuming] Sodium borosilicate hollow glass spheres, industrial grade; filler for polymers providing reduced weight, improved thermal insulation and mech. str., lower dielec. constant; for aerospace (ablative barrier coatings, syntactic foam core materials, airframe composites), electronics (potting compds.), medicine, recreation (tennis racquets, bowling and golf balls, high tech bicycles), military, and manufacturing applics.; 60 μm median particle size; true particle dens. 0.27 g/cc; bulk dens. 0.17 g/cc; soften.

pt. 480 C.

Eccosphere® IG-101. [Emerson & Cuming] Sodium borosilicate hollow glass spheres, industrial grade; filler for polymers providing reduced weight, improved thermal insulation and mech. str., lower dielec. constant; for aerospace (ablative barrier coatings, syntactic foam core materials, airframe composites), electronics (potting compds.), medicine, recreation (tennis racquets, bowling and golf balls, high tech bicycles), military, and manufacturing applics.; 55 μm median particle size; true particle dens. 0.35 g/cc; bulk dens. 0.22 g/cc; soften. pt. 480 C.

Eccosphere® IGD-101. [Emerson & Cuming] Sodium borosilicate hollow glass spheres, industrial grade; filler for polymers providing reduced weight, improved thermal insulation and mech. str., lower dielec. constant; for aerospace (ablative barrier coatings, syntactic foam core materials, airframe composites), electronics (potting compds.), medicine, recreation (tennis racquets, bowling and golf balls, high tech bicycles), military, and manufacturing applics.; 55 μm median particle size; true particle dens. 0.30 g/cc; bulk dens. 0.19 g/cc; soften. pt. 480 C.

Eccosphere® SDT-28. [Emerson & Cuming] Hollow glass spheres; filler for polymers providing reduced weight, improved thermal insulation and mech. str., lower dielec. constant; for aerospace (ablative barrier coatings, syntactic foam core materials, airframe composites), electronics (potting compds.), medicine, recreation (tennis racquets, bowling and golf balls, high tech bicycles), military, and manufacturing applics.; 16 μm median particle size; true particle dens. 0.28 g/cc; bulk dens. 0.18 g/cc; soften. pt. 850 C; dissip. factor 0.001 (1-8.6 Ghz); dielec. const. 1.17 (1-8.6 GHz); 93% SiO_2, 3% B_2O_3, 2% Na_2O, 1% BaO.

Eccosphere® SDT-40. [Emerson & Cuming] Hollow glass spheres; filler for polymers providing reduced weight, improved thermal insulation and mech. str., lower dielec. constant; for aerospace (ablative barrier coatings, syntactic foam core materials, airframe composites), electronics (potting compds.), medicine, recreation (tennis racquets, bowling and golf balls, high tech bicycles), military, and manufacturing applics.; 12 μm median particle size; true particle dens. 0.40 g/cc; bulk dens. 0.25 g/cc; soften. pt. 850 C; dissip. factor 0.002 (1-8.6 Ghz); dielec. const. 1.28 (1-8.6 GHz); 93% SiO_2, 3% B_2O_3, 2% Na_2O, 1% BaO.

Eccosphere® SDT-60. [Emerson & Cuming] Hollow glass spheres; filler for polymers providing reduced weight, improved thermal insulation and mech. str., lower dielec. constant; for aerospace (ablative barrier coatings, syntactic foam core materials, airframe composites), electronics (potting compds.), medicine, recreation (tennis racquets, bowling and golf balls, high tech bicycles), military, and manufacturing applics.; 6 μm median particle size; true particle dens. 0.60 g/cc; bulk dens. 0.38 g/cc; soften. pt. 850 C; dissip. factor 0.003 (1-8.6 Ghz); dielec. const. 1.43 (1-8.6 GHz); 93% SiO_2, 3% B_2O_3, 2% Na_2O, 1% BaO.

Eccowet® W-50. [Eastern Color & Chem.] Sodium aliphatic ester sulfonate; anionic; wetting agent, penetrant, dispersant, solubilizer, emulsifier, detergent for textiles, metal processing, disinfectants, paints, pigments, wallpaper, rubber cements, adhesives, drycleaning detergents, topical pharmaceuticals, cosmetics; colorless liq.; misc. with water; sol. in alcohol, glycols, acetones, dilute electrolyte sol'ns. (5%); visc. 55 cps; pH 8.0 ± 0.3 (1%); 50% conc.

Eccowet® W-88. [Eastern Color & Chem.] Sulfonated organic ester; wetting agent, penetrant, dispersant, solubilizer, emulsifier, detergent for textiles, metal processing, disinfectants, paints, pigments, wallpaper, rubber cements, adhesives, drycleaning detergents, topical pharmaceuticals, cosmetics; amber visc. liq.; pH 8.0 ± 0.3 (1%); 21% act.

Ecco White® FW-5. [Eastern Color & Chem.] Distyrylphenyl; optical brightener for nylon and blends; water-sol.

Ecco White® OP. [Eastern Color & Chem.] Benzoxazole; optical brightener for polyesters and blends; disp. in water; Unverified

Ekaland CBS. [Sovereign] N-Cyclohexyl-2-benzothiazylsulfenamide; CAS 95-33-0; delayed-action accelerator for rubber; provides fast mixing of difficult, black reinforced compds. without sacrificing scorch safety or physical props.; oiled powd.

Ekaland CMBT. [Sovereign] Copper 2-mercaptobenzothiazolate; fast curing accelerator for EPDM; exc. for compr. set in hose compds.; powd.

Ekaland DBTU. [Sovereign] Dibutyl thiourea; CAS 109-46-6; ultra-accelerator for use in CR; sec. accelerator for NR and SBR; cryst.

Ekaland DETU. [Sovereign] Diethyl thiourea; CAS 105-55-5; ultra-accelerator for use in CR; sec. accelerator for NR and SBR; cryst.

Ekaland DOTG. [Sovereign] Diorthotolyl guanidine; CAS 97-39-2; nonstaining thiazole booster; curing agent for Vamac; more active than Ekaland DPG; oiled powd.

Ekaland DPTT. [Sovereign] Dipentamethylenethiuram tetradisulfide; sec. accelerator for NR, SBR, EPDM, NBR; oiled powd.

Ekaland ETU. [Sovereign] Ethylene thiourea; CAS 96-45-7; EINECS 202-506-9; ultra-accelerator for use in CR; oiled powd.

Ekaland MBT. [Sovereign] 2-Mercaptobenzothiazole; CAS 149-30-4; EINECS 205-736-8; fast curing accelerator suitable for most polymers; optimal results obtained with thiuram, dithiocarbamate, or guanidine boosters; oiled powd.

Ekaland MBTS. [Sovereign] 2-Mercaptobenzothiazole disulfide; CAS 120-78-5; fast curing accelerator suitable for most polymers; provides better scorch resist. than Ekaland MBT; optimal results obtained with thiuram, dithiocarbamate, or guanidine boosters; oiled powd.

Ekaland NDBC. [Sovereign] Nickel dibutyldithiocarbamate; CAS 13927-77-0; antiozonant for NBR and SBR; antioxidant for Hypalon; usually non-staining; oiled powd.

Ekaland TETD. [Sovereign] Tetraethylthiuram disulfide; CAS 97-77-8; fast curing primary or sec. accelerator for NR, SBR, nitrile, and EPDM; gran.

Ekaland TMTM. [Sovereign] Tetramethylthiuram monosulfide; CAS 97-74-5; fast curing ultra accelerator for nitrile, EPDM, and most other polymers; for use alone or with thiazoles; gives tight cure with good compr. set, high modulus, good color and age resist.; oiled owd. or gran.

Ekaland ZMBT. [Sovereign] Zinc 2-mercaptobenzothiazole; CAS 155-04-4; accelerator for latex; gives very high modulus in latex films; good compr. set in latex foams; less scorchy than MBTS; powd.

Ektapro® EEP Solvent. [Eastman] Ethyl 3-ethoxypropionate; CAS 763-69-9; high-performance retarder solv. for formulating enamels, lacquers, topcoats, and primers; urethane-grade quality; polymerization solv.; Pt-Co 15 max. color; moderate odor; m.w. 146.19; f.p. < -50 C; b.p. 165 C (760 mm); 2.9% sol. in water; sp.gr. 0.95; dens. 7.91 lb/gal; visc. 1.0 cP; flash pt. (Seta) 58 C; surf. tens. 27.0 dynes/cm (23 C); elec. resist. 20 megohms.

Ektasolve® DP. [Eastman] Diethylene glycol monopropyl ether; CAS 6881-94-3; evaporating, water-misc. solv. used in sol'n. and water-dilutable coatings, cosmetics; act. for many coating materials incl. NC, acrylic copolymers, epoxy resins, chlorinated rubber, and alkyd resins; strong coupling agent with some resin systems in water-dilutable coatings; colorless clear liq.; mild odor; m.w. 148.2; f.p. < -90 C; b.p. 202 C min.; water-sol.; sp.gr. 0.963; dens. 0.96 kg/l; ref. index 1.429; flash pt. 93 C (TCC); fire pt. 103 C; > 99% volatiles by vol.

Ektasolve® EB. [Eastman] Butoxyethanol; CAS 111-76-2; EINECS 203-905-0; solv. for alkyd, phenolic, maleic, and cellulose nitrate resins; excellent retarder for lacquers, improving gloss and flow-out, blush resistance, and reducing the formation of orange peel; useful in formulating hot-spray, brushing, flow-coat, and aerosol lacquers; colorless liq.; faint odor; m.w. 118.17; f.p. -75 C; b.p. 169.0 C min.; water-sol.; sp.gr. 0.902; dens. 0.90 kg/l; ref. index 1.4193; flash pt. 62 C (TCC); fire pt. 70 C; > 99% volatiles by vol.

Ektasolve® EP. [Eastman] Ethylene glycol monopropyl ether; CAS 2807-30-9; EINECS 220-548-6; slow evaporating solv. used in coatings and cosmetics; useful in waterborne coating systems; coupling solv. for resin/water systems; controls visc. of waterborne resins; effective for NC, acrylic, epoxy, polyamide, and alkyd resins; retarder in coating systems; colorless clear liq.; mild odor; m.w. 104.15; f.p. < -90 C; b.p. 149.5 min.; water-sol.; sp.gr. 0.9125; dens. 0.91 kg/l; ref. index 1.4136; flash pt. 49 C (TCC); fire pt. 56 C; 100% volatiles by vol.

Elastocarb® Tech Heavy. [Morton Int'l./Plastics Additives] Basic magnesium carbonate; inorg. filler providing flame retardancy and smoke suppression to elastomers, plastics, and thermosets incl. EPDM, PP, PE, PVC; used in wire and cable compds., conduit/tubing, film and sheet; wh. powd. (flat platelets); 3 μ mean particle size; 0.5% retained 325 mesh; sp.gr. 2.16; bulk dens. 10 lb/ft^3 (aerated); surf. area 24 m^2/g; ref. index 1.58; 40% min. magnesium oxide, 34% max. carbon dioxide.

Elastocarb® Tech Light. [Morton Int'l./Plastics Additives] Basic magnesium carbonate; inorg. filler providing flame retardancy and smoke suppression to elastomers, plastics, and thermosets incl. EPDM, PP, PE, PVC; used in wire and cable compds., conduit/tubing, film and sheet; wh. powd. (flat platelets); 2 μ mean particle size; 0.5% retained 325 mesh; sp.gr. 2.16; bulk dens. 8 lb/ft^3 (aerated); surf. area 20 m^2/g; ref. index 1.58; 40% min. magnesium oxide, 34% max. carbon dioxide.

Elastocarb® UF. [Morton Int'l./Plastics Additives] Basic magnesium carbonate; precipitated ultrafine grade filler for use in rigid PVC compds., EPDM, CSM, conduit, wire and cable compds., aircraft interior components, other transportation uses; reduces smoke dens. and increases physical properties; wh. ultra fine powd.; 90% < 5 μ, 40% < 1 μ; 1 μ mean particle size; sp.gr. 2.19 bulk dens. 9 lb/ft^3 (aerated); ref. index 1.58; 40% min. magnesium oxide, 34% max. carbon dioxide.

Elastomag® 100. [Akrochem] Magnesium oxide; CAS 1309-48-4; EINECS 215-171-9; chemical thickener for polyester resins; anticaking agent; used in syn. rubber compding., adhesives, fuel oil additives, and as acid acceptor for specialty plastics; powd.; 99.9% thru 325 mesh; sp.gr. 3.2; bulk dens. 18 lb/ft^3 (aerated); surf. area 104-141 m^2/g; 98% MgO.

Elastomag® 100R. [Akrochem] Magnesium oxide; CAS 1309-48-4; EINECS 215-171-9; chemical thickener for polyester resins; anticaking agent; used in syn. rubber compding., adhesives, fuel oil additives, and as acid acceptor for specialty plastics; powd.; 99.9% thru 325 mesh; sp.gr. 3.2; bulk dens. 18 lb/ft^3 (aerated); surf. area 86-113 m^2/g; 98% MgO.

Elastomag® 170. [Akrochem] Magnesium oxide; CAS 1309-48-4; EINECS 215-171-9; chemical thickener for polyester resins; anticaking agent; used in syn. rubber compding., adhesives, fuel oil additives, and as acid acceptor for specialty plastics; highest activity of the Elastomag MgOs; powd.; 99.9% thru 325 mesh; sp.gr. 3.2; bulk dens. 18 lb/ft3 (aerated); surf. area 141-188 m^2/g; 98% MgO.

Elastomag® 170 Micropellet. [Akrochem] Magnesium oxide; CAS 1309-48-4; EINECS 215-171-9; chemical thickener for polyester resins; anticaking agent; used in syn. rubber compding., adhesives, fuel oil additives, and as acid acceptor for specialty plastics; dust-free micropellet; 40.0% max. < 70 mesh; 0.1% max. > 20 mesh; sp.gr. 3.2; bulk dens. 24 lb/ft^3 (aerated); surf. area 141-188 m^2/g; 98% MgO.

Elastomag® 170 Special. [Akrochem] Magnesium oxide; CAS 1309-48-4; EINECS 215-171-9; chemical thickener for polyester resins; used in adhesives, fuel oil additives, and as acid acceptor for specialty plastics and rubber; powd.; 99.9% thru 325 mesh; 50.0% min. < 1 μ; sp.gr. 3.2; bulk dens. 18 lb/ft^3 (aerated); surf. area 132 m^2/g; 98% MgO.

Electrofine S-70. [Elf Atochem N. Am.] Chlorinated paraffin; flame retardant for plastics, textiles, paper and cardboard in conjunction with antimony trioxide; improves hardness, gloss and resist. to acids and bases in paints; wh. powd., odorless; 90% < 200 μm; sol. in hydrocarbons, ketones, esters; insol. in water, light alcohols; sp.gr. 1.615; apparent bulk dens. 600 kg/m^3; soften. pt. 95-105 C; flash pt. (CC) none; 69-71% Cl; *Toxicology:* LD50 (oral) > 109 g/kg; nonirritating to skin and eyes; *Storage:* store in dry place away from effects of weather and heat; avoid contact with ferrous metals.

Electrosol 325. [Alframine] External antistat for acrylic, nylon, flexible and rigid PVC; FDA approved as indirect additive.

Electrosol D. [Alframine] Internal antistat for PE, PP, polyester; FDA approved as indirect additive.

Electrosol M. [Alframine] Internal antistat for PE, PP, polyester, flexible and rigid PVC; FDA approved as indirect additive.

Electrosol S-1-X. [Alframine] Internal antistat for nylon, acetal, PC, PE, PP, PS, flexible and rigid PVC; FDA approved as indirect additive.

Elecut S-507. [Takemoto Oil & Fat] Alkylaryl sulfonate salt; anionic; detergent; antistat for plastics; liq.; 75% conc.

Eleminol ES-70. [Sanyo Chem. Industries] POE nonylphenyl ether sulfate; anionic; primary emulsifier for emulsion polymerization; used in textile prods.; liq.; 80% conc.

Eleminol HA-100. [Sanyo Chem. Industries] POE nonylphenyl ether; nonionic; emulsifier for emulsion polymerization; protective colloidal props.; flake; HLB 18.2; 100% conc.

Eleminol HA-161. [Sanyo Chem. Industries] POE nonylphenyl ether; nonionic; primary emulsifier for emulsion polymerization; protective colloidal props.; flake; HLB 18.8; 100% conc.

Eleminol JS-2. [Sanyo Chem. Industries] Sodium alkylaryl sulfosuccinate; coemulsifier for emulsion polymerization of vinyl acetate, acrylate, vinyl chloride, and styrene/butadiene resins; liq.; 40% conc.

Eleminol MON-2. [Sanyo Chem. Industries] Sodium alkyldiphenyl ether disulfonate; emulsion polymerization of S/B and acrylate resins; liq.; 48% conc.

Eleminol MON-7. [Sanyo Chem. Industries] Sodium dodecyl diphenyl ether disulfonate; CAS 28519-02-0; anionic; emulsifier for emulsion polymerization; liq.; sol. in acidic, alkaline and electrolyte sol'ns.; 49% conc.

Elfan® 260 S. [Akzo Nobel] Sodium lauryl sulfate; CAS 151-21-3; EINECS 205-788-1; detergent for textiles, skins, cosmetics, emulsifier for emulsion polymerization; yel. liq., paste; water sol. (30 C); sp.gr. 1.04; visc. 200 cps (30 C); f.p. 15 C; 30% act.

Elfan® WA. [Akzo Nobel BV] Sodium dodecylbenzene sulfonate; anionic; heavy-duty detergent for textiles, household; emulsifier for emulsion polymerization; yel. paste; sol. 250 g/l water; sp.gr. 1.08; f.p. < 0 C; m.p. 5 C; pH 8-10; 50% act.

Elfan® WA Powder. [Akzo Nobel BV] Sodium dodecylbenzene sulfonate; anionic; heavy-duty detergent for textiles, household; emulsifier for emulsion polymerization; lt. yel. beads, powd.; sol. 100 g/l water; dens. 300 g/l; pH 8-9; 80% act.

Elftex® 120. [Cabot Europe/Special Blacks] Carbon black; CAS 1333-86-4; EINECS 215-609-9; utility grade colorant, uv stabilizer for highly loaded conc. plastics; low cost; pellets; 60 nm particle size; dens. 32 lb/ft^3; surf. area 30 m^2/g; 1.0% volatile.

Elftex® 125. [Cabot Europe/Special Blacks] Carbon black; CAS 1333-86-4; EINECS 215-609-9; utility grade colorant, uv stabilizer for highly loaded conc. plastics; low cost; fluffy; 60 nm particle size; dens. 22 lb/ft^3; surf. area 30 m^2/g; 1.0% volatile.

Elftex® 160. [Cabot Europe/Special Blacks] Carbon black; CAS 1333-86-4; EINECS 215-609-9; utility grade colorant for plastics providing very blue tone, easy dispersion; pellets; 50 nm particle size; dens. 27 lb/ft^3; surf. area 35 m^2/g; 1.0% volatile.

Elftex® 280. [Cabot Europe/Special Blacks] Carbon black; CAS 1333-86-4; EINECS 215-609-9; utility grade colorant, uv stabilizer for plastics providing blue tone, exc. dispersion, smooth surface; fluffy; 41 nm particle size; dens. 23 lb/ft^3; surf. area 42 m^2/g; 1.0% volatile.

Elftex® 285. [Cabot Europe/Special Blacks] Carbon black; CAS 1333-86-4; EINECS 215-609-9; utility grade colorant for plastics providing blue tone, exc. dispersion, smooth surface; fluffy; 41 nm particle size; surf. area 42 m^2/g; 1.0% volatile.

Elftex® 415. [Cabot Europe/Special Blacks] Carbon black; CAS 1333-86-4; EINECS 215-609-9; colorant for plastics providing high tint with blue tone; fluffy; 25 nm particle size; dens. 22 lb/ft^3; surf. area 90 m^2/g; 1.0% volatile.

Elftex® 430. [Cabot Europe/Special Blacks] Carbon black; CAS 1333-86-4; EINECS 215-609-9; low structure colorant, uv stabilizer for plastics; pellets; 27 nm particle size; dens. 27 lb/ft^3; surf. area 80 m^2/g; 1.0% volatile.

Elftex® 435. [Cabot Europe/Special Blacks] Carbon black; CAS 1333-86-4; EINECS 215-609-9; low structure colorant for plastics; fluffy; 27 nm particle size; dens. 19 lb/ft^3; surf. area 80 m^2/g; 1.0% volatile.

Elftex® 460. [Cabot Europe/Special Blacks] Carbon black; CAS 1333-86-4; EINECS 215-609-9; colorant for plastics; high tint str., good dispersion; pellets; 27 nm particle size; dens. 24 lb/ft^3; surf. area 74 m^2/g; 1.0% volatile.

Elftex® 465. [Cabot Europe/Special Blacks] Carbon black; CAS 1333-86-4; EINECS 215-609-9; colorant for plastics; high tint str., good dispersion; fluffy; 27 nm particle size; surf. area 74 m^2/g; 1.0% volatile.

Elftex® 570. [Cabot Europe/Special Blacks] Carbon black; CAS 1333-86-4; EINECS 215-609-9; colorant, uv stabilizer for plastics; good dispersion; pellets; 23 nm particle size; dens. 23 lb/ft^3; surf. area 115 m^2/g; 1.5% volatile.

Elftex® 670. [Cabot Europe/Special Blacks] Carbon black; CAS 1333-86-4; EINECS 215-609-9; for coloring plastics; very good uv protection; pellets; 20 nm particle size; dens. 22 lb/ft^3; surf. area 140 m^2/g; 1.5% volatile.

Elftex® 675. [Cabot Europe/Special Blacks] Carbon black; CAS 1333-86-4; EINECS 215-609-9; for coloring plastics; very good uv protection; fluffy; 20 nm particle size; dens. 14 lb/ft^3; surf. area 140 m^2/g; 1.5% volatile.

ELP-3. [Morton Int'l./Polymer Systems] Epoxy terminated polysulfide polymer; modifier for epoxy resins; used in concrete adhesives, chemically resist. linings or coatings, bonding to metallic substrates with oily finishes.

Eltesol® 4009, 4018. [Albright & Wilson UK] Xylene sulfonic acid modified with methanol and sulfuric acid; catalysts for curing cold-setting phenol-formaldehyde and phenol-furane resins used in the foundry industry as binders for sand in the prod. of molds and cores; clear brn. liq.; sp.gr. 1.25 and 1.30 resp.; visc. 165-195 cs and 140-170 cs; flash pt. (CC) 72 and 93 C.

Eltesol® 4200. [Albright & Wilson UK] Xylene sulfonic acid blend; CAS 25321-41-9; EINECS 246-839-8; catalyst for foundry resins.

Eltesol® 4402. [Albright & Wilson UK] Aromatic benzenesulfonic acid blend; anionic; highly reactive catalyst for foundry resins; liq.; 88% conc.

Eltesol® CA 65. [Albright & Wilson UK] Cumene sulfonic acid; CAS 28631-63-2; anionic; catalyst for foundry resins; descaling agent for metal cleaning; antistress additive and plating aid in electroplating bath; curing aid in the plastics industry; raw material in the mfg. of dyes and pigments; detergents industry; dens. 1.15 g/cm^3; visc. 30 cs; 65 $\pm$ 1.0% act. in water.

Eltesol® CA 96. [Albright & Wilson UK] Cumene sulfonic acid conc.; CAS 28631-63-2; anionic; catalyst for foundry resins; descaling agent for metal cleaning; antistress additive and plating aid in electroplating bath; curing aid in the plastics industry; raw material in the mfg. of dyes and pigments; detergents industry; brn. visc. liq.; dens. 1.25 g/cm^3; visc. 1000 cs; 96.0 $\pm$ 1.0% act.

Eltesol® PSA 65. [Albright & Wilson UK] Phenol sulfonic acid; CAS 1333-39-7; anionic; catalyst for foundry resins; descaling agent for metal cleaning; antistress additive and plating aid in electroplating bath; curing aid in the plastics industry; raw material in the mfg. of dyes and pigments; detergents industry; pharmaceutical chemicals and disinfectants; liq.; dens. 1.3 g/cm^3; visc. 50 cs; 65.0 $\pm$ 1.0% act. in water.

Eltesol® TA Series. [Albright & Wilson UK] Toluene sulfonic acid; anionic; catalyst for foundry resins; descaling agent for metal cleaning; antistress additive and plating aid in electroplating bath; curing aid in the plastics industry; raw material in the mfg. of dyes and pigments; detergents industry; hydrotrope, intermediate; used in mfg. of acrylonitrile; agric. formulations; liq.; dens. 1.2 g/cm^3; visc. 15 cs.

Eltesol® TA 65. [Albright & Wilson UK] 65% Toluene sulfonic acid and 1.4% sulfonic acid aq. sol'n.; anionic; curing agent for resins in foundry cores, plastics, coatings; intermediate; catalyst in foundry and chemical industries; hardening agent in plastics; activator for nicotine insecticides; descaling agent in metal cleaning; in electroplating baths; lt. amber clear liq.; dens. 1.2 g/cc; visc. 9-12 cs; 65% conc.

Eltesol® TA 96. [Albright & Wilson UK] Toluene sulfonic acid; anionic; curing agent for resins in foundry cores, plastics, coatings; intermediate; catalyst in foundry and chemical industries; hardening agent in plastics; activator for nicotine insecticides; descaling agent in metal cleaning; in electroplating baths; brn. thick cryst. liq.; dens. 1.3 g/cm^3; 95% act.

Eltesol® TA/E. [Albright & Wilson UK] 52% Toluene sulfonic acid and 11.5% sulfuric acid aq. sol'n.; catalyst for curing cold-setting foundry resins; clear amber/brn. liq.; sp.gr. 1.26; visc. 10 cs.

Eltesol® TA/F. [Albright & Wilson UK] 63% Toluene sulfonic acid and 1.2% sulfuric acid aq. sol'n.; catalyst for curing cold-setting foundry resins; descaling agent for metal cleaning; antistress additive and plating aid in electroplating bath; curing aid in the plastics industry; raw material in the mfg. of dyes and pigments; detergents industry; hydrotrope, intermediate; clear amber/brn. liq.; sp.gr. 1.22; visc. 15 cs.

Eltesol® TA/H. [Albright & Wilson UK] 54% Toluene sulfonic acid and 9% sulfuric acid aq. sol'n.; catalyst for curing cold-setting foundry resins; clear amber/brn. liq.; sp.gr. 1.26; visc. 10 cs.

Eltesol® TSX. [Albright & Wilson UK] p-Toluene sulfonic acid monohydrate BP; CAS 70788-37-3; anionic; catalyst for org. synthesis, syn. resins, mfg. of p-cresol, toluene derivs., pharmaceutical prods., dyestuffs; chemical intermediate; wh. to off-wh. cryst.; m.p. 103 C; 97% conc.

Eltesol® TSX/A. [Albright & Wilson Am.; Albright & Wilson UK] p-Toluene sulfonic acid monohydrate; CAS 104-15-4; EINECS 203-180-0; anionic; catalyst for org. synthesis, esp. esterification, syn. resins, mfg. of p-cresol, toluene derivs., pharmaceutical prods., dyestuffs; chemical intermediate; curing agent for resins and coatings; wh. crystals; m.p. 103.5 C; 96% act.

Eltesol® TSX/SF. [Albright & Wilson UK] p-Toluene sulfonic acid monohydrate; CAS 104-15-4; EINECS 203-180-0; anionic; catalyst for org. synthesis, syn. resins, mfg. of p-cresol, toluene derivs., pharmaceutical prods., dyestuffs; chemical intermediate; wh. cryst.; m.p. 103.5 C; 98% act.

Eltesol® XA65. [Albright & Wilson UK] Xylene sulfonic acid; CAS 25321-41-9; EINECS 246-839-8; anionic; intermediate, catalyst in preparation of esters, hardening agent in plastics, activator for nicotine insecticides; lt. amber, clear to slightly hazy liq.; dens. 1.2 g/cc; visc. 9-12 cs; 65% act.

Eltesol® XA90. [Albright & Wilson UK] Xylene sulfonic acid conc.; CAS 25321-41-9; EINECS 246-839-8; anionic; catalyst for foundry resins; descaling agent for metal cleaning; antistress additive and plating aid in electroplating bath; curing aid in the plastics industry; raw material in the mfg. of dyes and pigments; detergents industry; brn. visc. liq.; dens. 1.35 g/cm^3; visc. 3000 cs; 95.0 ± 2.0% act.

Eltesol® XA/M65. [Albright & Wilson UK] Xylene sulfonic acid; CAS 25321-41-9; EINECS 246-839-8; anionic; intermediate, catalyst in preparation of esters, hardening agent in plastics, activator for nicotine insecticides.

Elvace® 1870. [Reichhold] VAE copolymer emulsion; adhesive base with improved plasticizer response, quick tack, and adhesion compared with homopolymer; dens. 1078 g/l; visc. 1700 cps; pH 4.5; 55% solids; Unverified

Emcol® 4300. [Witco/New Markets] Disodium C12-15 pareth sulfosuccinate; CAS 39354-47-5; anionic; dispersant, wetting, foaming, detergent, emulsifying agent for bubble bath, shampoo, cosmetics and toiletries; emulsifier for acrylic, vinyl acetate, vinyl acrylic polymerization; lt. clear liq.; sol. in water; sp.gr. 1.08; visc. 50 cps; acid no. 5.0; pH 6.2 (3% aq.); surf. tens. 29.3 dynes/cm (1%); Ross-Miles foam 162 mm (initial, 1%, 49 C); 33% solids.

Emcol® 4910. [Witco/New Markets] Sodium lauryl/propoxy sulfosuccinate; emulsifier for acrylic, vinyl acrylic polymerization; visc. 16,000 cps; surf. tens. 31.0 dynes/cm (1%); Ross-Miles foam 163 mm (initial, 1%, 49 C); 60% solids.

Emcol® 4930, 4940. [Witco/New Markets] Asymmetrical sulfosuccinate; anionic; surfactants for emulsion polymerization, latex stabilization, wetting, emulsification; liq.

Emcol® CC-42. [Witco/New Markets] PPG-40 diethylmonium chloride; CAS 9076-43-1; cationic; pigment dispersant, particle suspension aid, emulsifier, solv., conditioner, antistat, lubricant, corrosion inhibitor for toiletries, cosmetics, germicides, syn. fibers and plastics, textiles, industrial processes; ore flotation additive; lt. amber oily liq.; sol. in IPA, acetone, MEK, min. spirits, ethanol; partly sol. in water; sp.gr. 1.01; flash pt. > 200 C (PMCC); pH 6.5; 100% conc.

Emcol® CC-55. [Witco/New Markets] Polypropoxy quat. ammonium acetate; cationic; antistat; conditioner for hair rinse preparations; emulsifier for cosmetics and textile flame retardants; solv. for phenolic-type germicides for cosmetics and toiletries; antistat for syn. fibers and plastics; fabric conditioner; lubricant for textile and industrial formulations; solv. cleaning and scouring agent; corrosion inhibitor in protective coatings; pigment dispersant in nonaq. media; o/w emulsifier; lt. amber oily liq.; sol. @ 25% in ethanol, IPA, acetone, MEK; water-disp.; sp.gr. 1.02; flash pt. > 93 C (PMCC); pH 6.5; 99% solids.

Emcol® CC-57. [Witco/New Markets] Polypropoxy quat. ammonium phosphate; cationic; antistat, conditioner, emulsifier, solv., lubricant, solv. cleaning and scouring agent, corrosion inhibitor, and dispersant; used in syn. fibers and plastics, personal care prods., germicides, flame retardants, textile and industrial applic., protective coatings, and pigments; lt. amber oily liq.; sol. in water, ethanol, IPA, acetone, MEK; sp.gr. 1.12; flash pt. > 93 C (PMCC).

Emcol® DOSS. [Witco/New Markets] Sodium dioctyl sulfosuccinate; emulsifier for acrylonitrile polymerization; improves surf. wetting; FDA compliance; visc. 250 cps; surf. tens. 25.7 dynes/cm (1%); Ross-Miles foam 243 mm (initial, 1%, 49 C).

Emcol® K8300. [Witco/New Markets] Disodium oleamido-MIPA sulfosuccinate; CAS 43154-85-4; EINECS 256-120-0; anionic; dispersant, particle suspension aid, wetting agent, foam booster/stabilizer, detergent; emulsifier in emulsion polymerization of acrylic, styrene acrylic, vinyl acrylic; FDA compliance; amber liq.; sol. in water; sp.gr. 1.10; visc. 1200 cps; pH 6.5 (3% aq.); surf. tens. 30.5 dynes/cm (1%); Ross-Miles foam 158 mm (initial, 1%, 49 C); 38% solids.

Emerest® 2301. [Henkel/Emery] Methyl oleate; nonionic; base for industrial lubricants; mold release agent, defoamer, flotation agent, plasticizer for cellulosic plastics, needle lubricants; when sulfated is useful as wetting, rewetting, and dye leveling agent in textile and leather industries; Gardner < 6 liq.; sol. 5% in min. oil, toluene, IPA, xylene; insol. in water; dens. 7.3 lb/gal; visc. 5 cSt (100 F); pour pt. -16 C; flash pt. 350 F; 100% act.

Emerest® 2302. [Henkel/Emery; Henkel/Textile] Propyl oleate; CAS 1330-80-9; nonionic; base for industrial lubricants; mold release agent, defoamer, flotation agent, plasticizer for cellulosic plastics, needle lubricants; when sulfated is useful as wetting, rewetting, and dye leveling agent in textile and leather industries; Gardner < 6 liq.; sol. @ 5% in min. oil, toluene, IPA, xylene; dens. 7.3 lb/gal; visc. 5 cSt (100 F); pour pt. -16 C; flash pt. 350 F; 100% act.

Emerest® 2650. [Henkel/Emery; Henkel/Cospha; Henkel/Textile; Henkel Canada] PEG-8 laurate; CAS 9004-81-3; nonionic; hydrophilic surfactant functioning as leveling and wetting agent; defoamer in latex paints; dispersant in pigment grinding; solubilizer for oils and solvs.; antiblock agent in vinyls; textile processing; agric.; EPA-exempt; Gardner 1 liq.; sol. 5% in water, xylene; dens. 8.6 lb/gal; visc. 41 cSt (100 F); HLB 13.2; pour pt. 12 C; flash pt. 450 F; cloud pt. 33 C; 100% act.

Emerest® 2675. [Henkel/Emery] PEG-50 stearate; CAS 9004-99-3; nonionic; hydrophilic emulsifier used for preparing solubilized oils; visc. modifier, softener or plasticizer in acrylic or vinyl resin emulsions; Gardner 1 liq.; sol. @ 5% in water; dens. 8.5 lb/gal; visc. 671 cSt (100 F); HLB 17.8; pour pt. 0 C; cloud pt. 81 C (5% saline); flash pt. 540 F; 30% act. in water; 30% aq.

Emerwax® 9380. [Henkel/Emery] Fatty acid amide; slip and antiblock agent for PE and PP; liq.

Emery® 5700. [Henkel/Emery] Phenyl ethanolamine; CAS 122-98-5; intermediate for basic dyes for syn.

fibers; stabilizer for colloidal disps. of metal oxides in lubricants; in photographic color developer, emulsion polymerization of diene monomers; lt. yel. to amber liq.; m.w. 137.2; sol. in water (4%), acetone, benzene, CCl_4, ethyl acetate, ethyl ether, methanol; flash pt. (COC) 305 F; 98% min. act.; Unverified

Emery® 5702. [Henkel/Emery] PEG-3 aniline; CAS 36356-83-9; cure promoter for polyester resins; amber to brn. liq.; m.w. 225; flash pt. (COC) 410 F; 50% min. act.; Unverified

Emery® 5703. [Henkel/Emery] Phenyl diethanolamine; CAS 120-07-0; coupling agent for disperse dyes for syn. fibers; intermediate for dyes; cure promoter for polyester resins; curing agent for urethane elastomers; intermediate; also in hair dyes, heavy-duty detergents, paint strippers; wh. to off-wh. cryst. solid; m.w. 181.2; sol. in water (2.5%), acetone, methanol; m.p. 55.5 C; flash pt. (COC) 380 F; 98% min. act.; Unverified

Emery® 5707. [Henkel/Emery] Ethyl phenyl ethanolamine; CAS 92-50-2; coupling agent for dyes for syn. fibers; cure promoter for polyester resins; photosensitive chemical for paper coatings; intermediate for oil-sol. dyestuffs; lt. tan cryst. solid; m.w. 165.2; sol. in acetone, benzene, CCl_4, ethyl ether, methanol; flash pt. (COC) 280 F; 97% min. act.; Unverified

Emery® 5709. [Henkel/Emery] m-Tolyl diethanolamine; CAS 91-99-6; coupling agent for dyes for syn. fibers; cure promoter for polyester resins; also in mfg. of nonporous PU elastomers; tan cryst. solid; m.w. 195.2; sol. in acetone, ethanol; m.p. 67.5 C; flash pt. (COC) 385 F; 98% min. act.; Unverified

Emery® 5710. [Henkel/Emery] p-Tolyl diethanolamine; CAS 3077-12-1; cure promoter for polyester resins; intermediate for plastics; brn. cryst. solid; m.w. 195.2; m.p. 53 C; flash pt. (COC) 380 F; 98% min. act.; Unverified

Emery® 5714. [Henkel/Emery] N-Ethyl-N-hydroxyethyl-m-toluidine; CAS 91-88-3; coupling agent for disperse dyes for syn. fibers; photosensitive chemical for paper coatings; accelerator for polyester resins; yel.-brn. liq.; m.w. 179.2; flash pt. (COC) 280 F; 96.5% min. act.; Unverified

Emery® 6686. [Henkel/Emery] PEG 600; CAS 25322-68-3; chemical intermediate for coatings, adhesives, lubricants, metalworking, paper mfg., petrol. prod., ceramics, printing, electronics, solvs., cleaners, latex paints, mold release agent, rubber; Gardner < 1 liq.; sol. in water, xylene; disp. in Stod.; dens. 9.4 lb/gal; visc. 63 cSt (100 F); pour pt. 22 C; flash pt. 475 F.

Emery® 6687. [Henkel/Emery] PEG-300; CAS 25322-68-3; chemical intermediate for coatings, adhesives, lubricants, metalworking, paper mfg., petrol. prod., ceramics, printing, electronics, solvs., cleaners, latex paints, mold release agent, rubber, textiles; Gardner < 1 liq.; sol. in water, xylene; disp. in Stod.; dens. 9.4 lb/gal; visc. 33 cSt (100 F); pour pt. -10 C; flash pt. > 400 F.

Emery® 6705. [Henkel/Emery] Phenoxyethanol; CAS 122-99-6; EINECS 204-589-7; solv. for NC, cellulose acetate, ethylcellulose, and vinyl, alkyl, and ester type resins; used in printing inks and special formulations for removing industrial finishes and coatings; Gardner 2 liq.; sol. 5% in Stod., xylene; water-insol.; dens. 9.1 lb/gal; visc. 10 cSt (100 F); pour pt. 13 C; flash pt. 250 F; Unverified

Emery® 6724. [Henkel/Emery; Henkel/Textile] Alkyl capped pelargonate; lubricant for nylon, polyester, and PP fibers; Gardner 1 liq.; sol. @ 5% in water, butyl stearate, glyceryl trioleate, xylene; disp. in min. oil, Stod.; dens. 8.6 lb/gal; visc. 20 cSt (100 F); pour pt. 6 C; flash pt. 470 F.

Emery® 6773. [Henkel/Emery] PEG-200; CAS 25322-68-3; chemical intermediate for coatings, adhesives, lubricants, metalworking, paper mfg., petrol. prod., ceramics, printing, electronics, solvs., cleaners, latex paints, mold release agent, rubber; Gardner < 1 clear visc. liq.; sol. @ 5% in water, xylene; m.w. 200; dens. 9.4 lb/gal; visc. 25 cSt (100 F); pour pt. < -15 C; flash pt. > 330 F.

Emka Catalyst P-35. [Emkay] Org. amine; catalyst giving exc. bath stability for thermosetting resins in textile industry.

Emkapyl 400, 600, 1200, 2000, 4000. [ICI Surf. UK] Polypropylene glycol; modifier for resins; colorless liq.

Empee® PE Conc. 1. [Monmouth Plastics] Imparts flame retardancy to low and high dens. polyethylene.

Empee® PO Conc. 61. [Monmouth Plastics] Imparts flame retardancy to low and high dens. polyethylene.

Empee® PP Conc. 4. [Monmouth Plastics] PP, flame retardant; flame retardant conc. for let-down with PP homopolymer.

Empee® PP Conc. 33. [Monmouth Plastics] PP, flame retardant; imparts flame retardancy to most PP homopolymers; for use in fiber and film applics.

Empee® PP Conc. 43. [Monmouth Plastics] PP, flame retardant; imparts flame retardancy to PP homo and copolymers; suited for fiber and film and extrusion applics.

Emphos™ CS-136. [Witco/New Markets] Nonoxynol-6 phosphate; anionic; lubricant; antistat; emulsifier for cutting fluids, PVAc, and acrylic film formation; detergent for hard surfaces, metal cleaners, and dry cleaning systems; waterless hand cleaner component; personal care formulations; FDA compliance; clear visc. liq.; sol. in water, ethanol, CCl_4, perchloroethylene, heavy aromatic naphtha, kerosene; sp.gr. 1.09; visc. 4500 cps; pour pt. 18 C; acid no. 28; pH 5.0 (3% aq.); surf. tens. 30.0 dynes/cm (1%); Ross-Miles foam 158 mm (initial, 1%, 49 C); 99% solids.

Emphos™ PS-222. [Witco/New Markets] Aliphatic phosphate ester; anionic; emulsifier for pesticide formulations, detergents, hydrotropes and emulsion polymerization; corrosion inhibitor; detergent; dispersant; lubricant; personal care formulations; liq.; sol. in oil, disp. in water; sp.gr. 1.02; acid no. 105

(to pH 5.5), 155 (to pH 9.5); pH 2.4 (3% aq.).

Emphos™ PS-400. [Witco/New Markets] Linear alcohol ethoxylate phosphate ester, acid form; anionic; polymer particle dispersant; monomer/water emulsifier; textile antistat; also for pesticides, detergents, emulsion polymerization, personal care formulations; lubricant; wetting agent; clear liq.; sol. @ 25% in ethanol, CCl_4, perchloroethylene, xylene, heavy aromatic naphtha, kerosene, min. oil; water-disp.; sp.gr. 1.00; pour pt. > 10 C; acid no. 220 (to pH 5.5), 300 (to pH 9.5); surf. tens. 30.6 dynes/cm (0.05% aq.); pH 2.0 (3% aq.).

Emphos™ PS-410. [Witco/New Markets] Aliphatic phosphate ester; anionic; emulsifier for pesticide formulations, detergents, hydrotropes and emulsion polymerization; liq.

Empicol® 0303. [Albright & Wilson UK] Sodium lauryl sulfate BP; CAS 151-21-3; EINECS 205-788-1; anionic; surfactant in toothpaste and pharmaceutical preps.; emulsion polymerization; wh. fine powd.; dens. 0.35 g/cc; pH 8-10 (1% aq.); 95% act.

Empicol® 0303V. [Albright & Wilson UK] Sodium lauryl sulfate BP; CAS 151-21-3; EINECS 205-788-1; anionic; surfactant in toothpaste and pharmaceutical preps.; emulsion polymerization; BP and USP compliance; wh. low-dusting needles; pH 8-10 (1% aq.); 95% conc.

Empicol® ESB. [Albright & Wilson Australia] Sodium laureth sulfate; CAS 9004-82-4; 68585-34-2; anionic; shampoo ingred.; mild detergent for textile and leather processing; dispersant/emulsifier for emulsion and suspension polymerization; liq.; 28% conc.

Empicol® L Series. [Albright & Wilson UK] Alkyl sulfate; dispersant/emulsifier for emulsion and suspension polymerization, agric. formulations; foaming agent for carpet backing; air entraining agent for construction; flotation aid.

Empicol® LQ33/T. [Albright & Wilson UK] MEA-lauryl sulfate; CAS 4722-98-9; EINECS 225-214-3; anionic; surfactant in the mfg. of personal care prods.; emulsifier in the mfg. of rubber latices and for resins; bactericidal detergents; pale amber visc. liq.; dens. 1.03 g/cc visc. 6000 cs; pH 7.0 ± 0.5 (5% aq.); 33.5 ± 1.0% act.

Empicol® LX. [Albright & Wilson Am.; Albright & Wilson UK] Sodium lauryl sulfate; CAS 68585-47-7; anionic; emulsifier in the mfg. of plastics, resins, and syn. rubbers; foaming agent for rubber foams, personal care prods., pharmaceuticals, and carpet and upholstery shampoos; lubricant in mfg. of molded rubber goods; wh. powd.; dens. 0.35 g/cc; pH 9.5-10.5 (1% aq.); 90% act.

Empicol® LX28. [Albright & Wilson Am.; Albright & Wilson UK] Sodium lauryl sulfate; CAS 151-21-3; EINECS 205-788-1; anionic; emulsifier in the mfg. of plastics, resins, and syn. rubbers; foaming agent for rubber foams, personal care prods., toothpaste, and carpet and upholstery shampoos; lubricant in mfg. of molded rubber goods; pale yel. liq., paste; dens. 1.05 g/cc; pH 8.0-9.5 (5% aq.); 28% act. aq. sol'n.

Empicol® LXS95. [Albright & Wilson UK] Sodium lauryl sulfate; CAS 151-21-3; EINECS 205-788-1; anionic; detergent, foamer, emulsifier used in toothpastes, emulsion polymerization, rubber, pharmaceuticals; wh. powd.; dens 0.35 g/cc; pH 9.6-10.5 (1% aq.); 94.0% min. act.

Empicol® LXV. [Albright & Wilson Am.; Albright & Wilson UK] Sodium lauryl sulfate; CAS 151-21-3; EINECS 205-788-1; anionic; emulsifier in the mfg. of plastics, resins, and syn. rubbers; foaming agent for rubber foams, personal care prods., toothpaste, and carpet and upholstery shampoos; lubricant in mfg. of molded rubber goods; wh. needles; dens. 0.5 g/cc; pH 9.5-10.5 (1% aq.); 85% act.

Empicol® LXV/D. [Albright & Wilson UK] Sodium lauryl sulfate USP, BP; CAS 151-21-3; EINECS 205-788-1; anionic; emulsifier in the mfg. of plastics and rubbers; foaming agent for rubber foams; lubricant for mfg. of molded rubber goods; also in mfg. of carpet and upholstery shampoos, toiletries, toothpastes; wh. needles; sp.gr. 0.5 g/cc; pH 9.5-10.5 (5% aq.); 90% act.

Empicol® LY28/S. [Albright & Wilson UK] Sodium lauryl sulfate; CAS 151-21-3; coprecipitant in mfg. of photographic film; emulsion polymerization in the plastic and rubber industries; foaming agent in carpet processes; pale yel. liq. to paste; dens. 1.05 g/cc; visc. 28 cs; cloud pt. -1 C; pH 8.5-9.5 (5% aq.); 29.0 ± 1.0% act.; Unverified

Empicol® LZ. [Albright & Wilson UK] Sodium lauryl sulfate; CAS 68955-19-1; anionic; detergent, wetting and foaming agent in personal care prods., pharmaceuticals; emulsifier in mfg. of rubbers, plastics, and resins by emulsion polymerization; foaming agent in mfg. of foam rubber goods; lubricant used in plastic goods; wh. powd.; dens. 0.35 g/cc; pH 9.5-10.5 (1% aq.); 89.0% min. act.

Empicol® LZ/D. [Albright & Wilson UK] Sodium lauryl sulfate BP/USP; CAS 68955-19-1; anionic; surfactant, foaming agent, emulsifier, wetting agent for dental preps., toiletries, rubber, plastics, foam rubber; lubricant in extrusion of plastic goods, e.g., PVC; wh. powd.; sp.gr. 0.35; pH 9.5-10.5 (1% aq.); 90% min. act.

Empicol® LZG 30. [Albright & Wilson UK] Sodium lauryl sulfate; CAS 151-21-3; anionic; surfactant, foaming and wetting agent used in industrial processes; emulsifier in mfg. of syn. rubbers, plastics, and resins by emulsion polymerization; cream paste; dens. 1.05 g/cc; pH 8.0-9.5 (2% aq.); 30% act.; Unverified

Empicol® LZGV. [Albright & Wilson UK] Sodium lauryl sulfate; CAS 151-21-3; anionic; surfactant, foaming and wetting agent used in industrial processes; emulsifier in mfg. of syn. rubbers, plastics, and resins by emulsion polymerization; cream needles; dens. 0.5 g/cc; pH 9.5-10.5 (1% aq.); 85% act.; Unverified

Empicol® LZGV/C. [Albright & Wilson UK] Sodium lauryl sulfate; CAS 151-21-3; surfactant, foaming and wetting agent used in industrial processes; emulsifier in mfg. of syn. rubbers, plastics, and resins by emulsion polymerization; cream needles; dens. 0.5 g/cc; pH 9.5-10.5 (1% aq.); 89% act.; Unverified

Empicol® LZP. [Albright & Wilson UK] Sodium lauryl sulfate; CAS 151-21-3; EINECS 205-788-1; detergent, wetting and foaming agent in personal care prods., pharmaceuticals; emulsifier in mfg. of rubbers, plastics, and resins by emulsion polymerization; foaming agent in mfg. of foam rubber goods; lubricant used in plastic goods; wh. powd.; dens. 0.35 g/cc; pH 9.5-10.5 (1% aq.); 89.0% min. act.; Unverified

Empicol® LZV/D. [Albright & Wilson UK] Sodium lauryl sulfate; CAS 151-21-3; EINECS 205-788-1; anionic; surfactant, foaming agent, emulsifier, wetting agent for dental preps., toiletries, rubber, plastics, foam rubber; lubricant in extrusion of plastic goods, e.g., PVC; wh. fine needles; sp.gr. 0.5; pH 9.5-10.5 (1% aq.); 90% min. act.

Empicol® TL40/T. [Albright & Wilson UK] TEA-lauryl sulfate; CAS 139-96-8; EINECS 205-388-7; anionic; surfactant for personal care prods.; emulsifier in the mfg. of rubber latices; pale yel. clear liq.; dens. 1.08 g/cc; visc. 45 cs; cloud pt. < 0 C; pH 7.0 ± 0.5 (5% aq.); 41% act. in water.

Empicol® WA. [Albright & Wilson UK] Sodium lauryl sulfate; CAS 151-21-3; industrial surfactant used in rubber foams for carpet backing; pale yel. clear liq.; m.w. 292 (of act.); sp.gr. 1.04; pH 8.5-9.5 (5%); 27% act.; Unverified

Empicryl® 6030. [Albright & Wilson Am.] Cetyl stearyl methacrylate, MEHQ stabilizer; monomer for mfg. of polymers used as oil additives, visc. index improvers, and pour pt. depressants; internal plasticizer for adhesives, uv-curable resins, and coatings.

Empicryl® 6045. [Albright & Wilson UK] Alkyl methacrylate polymer in hydrocarbon oil; visc. index improver for formulation of high visc. index hydraulic fluids; emulsifier, dispersant for emulsion polymerization; reactive diluent for adhesives and coatings; yel. visc. fluid; sp.gr. 0.90; visc. 1100 mm^2/s; pour pt. -3 C; flash pt. > 100 C; 60% min. oil content.

Empicryl® 6047. [Albright & Wilson Am.; Albright & Wilson UK] Lauryl-myristyl methacrylate, MEHQ stabilizer; emulsifier, dispersant for emulsion polymerization; monomer for mfg. of methacrylate polymers/copolymers, polymers used as oil additives, visc. index improvers, pour pt. depressants; internal plasticizer for adhesives, uv-curable resins, and waterproofing coatings; pale yel. liq.; sp.gr. 0.9; flash pt. 80 C; acid no. 0.15 max.; hyd. no. 12 max.

Empicryl® 6052. [Albright & Wilson UK] Alkyl methacrylate polymer in hydrocarbon oil; visc. index improver for formulation of high visc. index hydraulic fluids; emulsifier, dispersant for emulsion polymerization; reactive diluent for adhesives and coatings; yel. visc. fluid; sp.gr. 0.91; visc. 1000 mm^2/s (100 C); pour pt. -3 C; flash pt. > 100 C; 33% min. oil content.

Empicryl® 6054. [Albright & Wilson UK] Alkyl methacrylate polymer in hydrocarbon oil; visc. index improver for formulation of high visc. index hydraulic fluids; emulsifier, dispersant for emulsion polymerization; reactive diluent for adhesives and coatings; yel. visc. fluid; sp.gr. 0.93; visc. 1600 mm^2/s (100 C); pour pt. -12 C; flash pt. > 100 C; 20% min. oil content.

Empicryl® 6058. [Albright & Wilson UK] Alkyl methacrylate polymer in hydrocarbon oil; visc. index improver for formulation of high visc. index hydraulic fluids; emulsifier, dispersant for emulsion polymerization; reactive diluent for adhesives and coatings; pale yel. visc. fluid; sp.gr. 0.92; visc. 1100 mm^2/s (100 C); pour pt -3 C; flash pt. > 100 C; 37% min. oil content.

Empicryl® 6059. [Albright & Wilson UK] Polyalkyl methacrylate copolymer in hydrocarbon oil; visc. index improver for formulation of high visc. index hydraulic fluids; emulsifier, dispersant for emulsion polymerization; reactive diluent for adhesives and coatings; yel. visc. fluid; sp.gr. 0.91; visc. 900 mm^2/s (100 C); pour pt -3 C; flash pt. > 100 C.

Empicryl® 6070. [Albright & Wilson UK] Polyalkyl methacrylate copolymer in hydrocarbon oil; visc. index improver for formulation of hydraulic oils; emulsifier, dispersant for emulsion polymerization; reactive diluent for adhesives and coatings; yel. visc. fluid; sp.gr. 0.94; visc. 800 mm^2/s (100 C); pour pt 15 C; flash pt. > 100 C.

Empigen® AM. [Albright & Wilson UK] Alkyl dimethyl amine, dist.; cationic; intermediate for mfg. of high quality derivs. such as quat. ammonium compds., betaines, amine oxides; catalyst for PU foam, resin curing agent, corrosion inhibitor, and flotation aid; pale straw waxy solid; char. odor; dens. 0.84 g/cc; 96.0% min. tert. amine; Unverified

Empigen® AY. [Albright & Wilson UK] Tert. alkyl ethoxy dimethyl amine; cationic; intermediate for mfg. of high quality derivs. such as quat. ammonium compds., betaines, amine oxides; catalyst for PU foam, resin curing agent, corrosion inhibitor, and flotation aid; amber liq.; dens. 0.89 g/cc; 94.0% min. tert. amine; Unverified

Empigen® OB. [Albright & Wilson Am.; Albright & Wilson UK] Lauramine oxide; CAS 1643-20-5; EINECS 216-700-6; nonionic; coactive, detergent, antistat, foam booster/stabilizer and visc. modifier for personal care prods., surgical scrubs, fire fighting foam concs., foamed rubbers, bleach additive; solubilizer; pale straw liq.; dens. 0.98 g/cc; visc. 25 cs (20 C); pH 7.5 ± 0.5 (5% aq.); 30.0 ± 1.5% act.

Empigen® OY. [Albright & Wilson Am.; Albright & Wilson UK] PEG-3 lauramine oxide; CAS 59355-61-2; nonionic; detergent, antistat, foam booster/stabilizer for foamed rubbers, fire fighting, bleach additive, shampoos, hair and bath prods.; pale straw liq.; dens. 1.0 g/cc; visc. 23 cs; pH 6.5 ± 0.5 (5% aq.); 25.0 ± 1.0% act.

Empilan® 2020. [Albright & Wilson UK] Lauryl alcohol ethoxylate; stabilizer for latices produced by emulsion polymerization; wh./cream solid; dens. 1.05 g/cc; m.p. 35 C; HLB 15.7; hyd. no. 64 ± 4; Unverified

Empilan® CDE. [Albright & Wilson UK] Cocamide

DEA (1:1); CAS 61791-31-9; EINECS 263-163-9; nonionic; foam boosting/stabilizing agent, solubilizer, detergent for use in personal care and detergent prods.; antistat in plastics; softener for leather processing; pale yel. visc. liq.; dens. 1.0 g/cm^3; 100% act.

Empilan® CDX. [Albright & Wilson UK] Cocamide DEA; CAS 68603-42-9; EINECS 263-163-9; nonionic; foam booster/stabilizer, solubilizer, and visc. modifier for toiletry and detergent formulations; antistatic agent in plastics; pale yel. visc. liq.; dens. 1.0 g/cm^3.

Empilan® DL 40. [Albright & Wilson UK] Fatty alcohol ethoxylate; nonionic; emulsifier for fatty acids and min. oils for fiber lubricants, wetting and stabilizing agent for syn. rubber latices, textile dye leveling and dispersing agent in textile industry; antistat; pale yel. clear liq.; misc. with water; dens. 1.046 g/cc; cloud pt. > 100 C in dist. water; 40% act.; Unverified

Empilan® DL 100. [Albright & Wilson UK] Fatty alcohol ethoxylate; emulsifier for fatty acids and min. oils for fiber lubricants, wetting and stabilizing agent for syn. rubber latices, textile dye leveling and dispersing agent in textile industry; antistat; stiff waxy solid; warm water sol.; m.p. 36 C; cloud pt. > 100 C (1% aq.); 100% act.; Unverified

Empilan® GMS LSE40. [Albright & Wilson UK] Glyceryl stearate SE; nonionic; emulsifier in the baking and food industry; lubricant in prod. of PVC sheets and expanded PS; wh. microbead powd.; dens. 0.5 g/cm^3; m.p. 55-65 C; 40.0% min. monoglyceride.

Empilan® GMS LSE80. [Albright & Wilson UK] Glyceryl stearate SE; nonionic; emulsifier in the baking and food industry; lubricant in prod. of PVC sheets and expanded PS; wh. microbead powd.; dens. 0.5 g/cm^3; m.p. 64 C; 80.0% min. monoglyceride.

Empilan® GMS MSE40. [Albright & Wilson UK] Glyceryl stearate SE; nonionic; emulsifier in the baking and food industry; lubricant in prod. of PVC sheets and expanded PS; wh. microbead powd.; dens. 0.5 g/cm^3; m.p. 55-60 C; 36.0% min. monoglyceride.

Empilan® GMS NSE40. [Albright & Wilson UK] Glyceryl stearate SE; CAS 31566-31-1; nonionic; emulsifier in the baking and food industry; lubricant and antistat in prod. of PVC sheets and expanded PS; stabilizer; wh. microbead powd.; dens. 0.5 g/cm^3; m.p. 55-60 C; pH 6-8 (10% aq.); 36-40% act.

Empilan® GMS NSE90. [Albright & Wilson UK] Glyceryl stearate SE; nonionic; emulsifier in the baking and food industry; lubricant and antistat in prod. of PVC sheets and expanded PS; stabilizer; wh. microbead powd.; dens. 0.5 g/cm^3; m.p. 65 C; 90.0% min. monoglyceride.

Empilan® GMS SE70. [Albright & Wilson UK] Glyceryl stearate SE; nonionic; emulsifier in the baking and food industry; lubricant in prod. of PVC sheets and expanded PS; wh. microbead powd.; dens. 0.5 g/cm^3; m.p. 61 C; 70.0% min. monoglyceride.

Empilan® K Series. [Albright & Wilson UK] Alkyl ethoxylates; emulsifier, wetting agent for agric. emulsifiable and suspension concs., spinning oils in textile processing; emulsifier, dispersant for emulsion polymerization; rubber compding. aid; antistat for plastics.

Empilan® LDE. [Albright & Wilson UK] Lauramide DEA (1:1); CAS 120-40-1; EINECS 204-393-1; nonionic; foam booster/stabilizer, solubilizer, thickener, detergent used in shampoos and liq. detergent formulations; antistat for plastics; cream solid; dens. 1.0 g/cm^3; 100% act.

Empilan® LDX. [Albright & Wilson UK] Lauramide DEA and diethanolamine; nonionic; foam booster/stabilizer, solubilizer, and visc. modifier for toiletry and detergent formulations; antistatic agent in plastics; cream solid; dens. 1.0 g/cm^3; 100% conc.

Empilan® NP9. [Albright & Wilson UK] Nonoxynol-9; CAS 9016-45-9; nonionic; wetting agent, detergent, emulsifier, solubilizer; for agric. emulsifiable and suspension concs., leather processing, emulsion and suspension polymerization; mortar plasticizer; plastics antistat; emulsifier for cosmetic creams and lotions; pale straw soft paste/liq.; dens. 1.0 g/cc; visc. 300 cs; cloud pt. 55 C (1% aq.); 100% act.

Empimin® KSN27. [Albright & Wilson UK] Sodium laureth sulfate (3 EO); CAS 9004-82-4; anionic; detergent raw material for high-quality liq. detergents, textile and leather processing; emulsifier, dispersant for emulsion polymerization; biodeg.; pale straw clear or slightly turbid liq.; dens. 1.05 g/cc; visc. 100 cs; cloud pt. 0 C; pH 6.5-7.5 (5% aq.); 27% aq. sol'n.

Empimin® KSN70. [Albright & Wilson UK] Sodium laureth sulfate (3 EO); CAS 9004-82-4; anionic; detergent raw material for high-quality liq. detergents, textile and leather processing; wetting agent for agric. wettable powds.; emulsifier, dispersant for emulsion polymerization; straw opaque pourable visc. liq.; dens. 1.1 g/cc; set pt. 8 C; pH 7.0-9.0 (2% aq.); 70% aq. sol'n.

Empimin® LR28. [Albright & Wilson UK] Sodium lauryl sulfate; CAS 151-21-3; anionic; foaming agent in rubber latex systems; emulsifier for emulsion polymerization; yel./amber liq.; dens. 1.04 g/cm^3; cloud pt. < 0 C; pH 7.0-9.0 (5% aq.); 28.0% min. act.

Empimin® MA. [Albright & Wilson UK] Sodium dihexyl sulfosuccinate; CAS 3006-15-3; anionic; emulsifier, dispersant, wetting agent for emulsion polymerization, agric. emulsifiable concs.; straw clear liq.; sol. in water; dens. 1.1 g/cm^3; visc. 80 cs; flash pt. 30 C (CC); pH 6.0 ± 1.0 (5%); 63.0% min. act.

Empimin® MHH. [Albright & Wilson UK] Disodium N-cocoyl sulfosuccinate; foaming agent for rubber latex compds.; emulsifier in the mfg. of polymers by emulsion polymerization; dens. 1.1 g/cm^3; pH 8.0 ± 0.5 (5% aq.); 32.5% min. act.; Unverified

Empimin® MKK98. [Albright & Wilson UK] Disodium N-cetyl stearyl sulfosuccinate; anionic; foaming agent for rubber latex compds.; emulsifier in the mfg. of polymers by emulsion polymerization; pale cream powd.; dens. 0.3 g/cm^3; pH 8.5 ± 0.5 (5% aq.); 98% act.; Unverified

Empimin® MKK/AU. [Albright & Wilson Australia]

Sodium N-tallow sulfosuccinamate; CAS 68988-69-2; anionic; foaming agent for rubber latices used in carpet backing; liq.; 30% conc.

Empimin® MKK/L. [Albright & Wilson UK] Disodium N-cetyl stearyl sulfosuccinamate; anionic; foaming agent for rubber latex compds.; emulsifier in the mfg. of polymers by emulsion polymerization; pale cream opaque liq.; dens. 1.1 g/cm³; pH 8.25 ± 0.25 (5% aq.); 28.0% min. act.; Unverified

Empimin® MSS. [Albright & Wilson UK] Diammonium N-cocoyl sulfosuccinate; foaming agent for rubber latex compds.; emulsifier in the mfg. of polymers by emulsion polymerization; yel./pale cream clear/opaque liq.; dens. 1.1 g/cm³; pH 6.5 ± 0.5 (5% aq.); 31.0% min. act.; Unverified

Empimin® MTT. [Albright & Wilson UK] Disodium N-oleyl sulfosuccinamate; foaming agent for rubber latices; pale cream liq.; pH 8 (5% aq.); 40% act.; Unverified

Empimin® MTT/A. [Albright & Wilson UK] Disodium N-oleyl sulfosuccinate; anionic; foaming agent for rubber latex compds.; emulsifier in the mfg. of polymers by emulsion polymerization; yel./pale cream clear/opaque liq.; dens. 1.1 g/cm³; pH 8.0 ± 0.5 (5% aq.); 28.0% min. act.; Unverified

Empimin® OP70. [Albright & Wilson UK] Sodium dioctyl sulfsuccinate; CAS 1369-66-3; anionic; emulsifier, dispersant, wetting agent used for emulsion polymerization, oil slicks, textiles, agrochem.; straw clear liq.; sol. in water; dens. 1.10 g/cm³; visc. 450 cs; flash pt. > 100 C (CC); pH 6.0 ± 1.0 (5%); 72.0 ± 2.0% act.

Empimin® OT. [Albright & Wilson UK] Sodium dioctyl sulfosuccinate; CAS 577-11-7; anionic; dispersant, emulsifier, detergent, wetting agent for o/w emulsions, agrochem., emulsion polymerization, filler and extender dispersions, leather processing; straw clear liq.; dens. 1.0 g/cm³; visc. 55 cs; flash pt. 25 C (CC); pH 6.0 ± 1.0 (5%); 60.0% min. act.

Empimin® OT75. [Albright & Wilson UK] Dioctyl sodium sulfosuccinate; CAS 577-11-7; anionic; wetting agent and emulsifier for emulsion polymerization; liq.; 75% conc.

Empiphos 4KP. [Albright & Wilson UK] Tetrapotassium pyrophosphate; CAS 7320-34-5; EINECS 230-785-7; detergent builder for liq. detergents; pigment dispersant and stabilizer in emulsion paints; clarifying agent in liq. soaps; mfg. of syn. rubber; boiler water treatment; sol. in water; dens. 75 lb/ft³; pH 10.3.

Empol® 1004. [Henkel/Emery] Dimer acid, hydrog.; CAS 68783-41-5; surfactant for industrial applics., thermographic resins, hot-melt adhesives, lt.-colored epoxy curing agents and ester derivs. for lubricant applics.; FDA 21CFR §175.105, 175.300, 175.320, 176.170, 176.200, 176.210, 177.1200, 177.1210, 177.2420, 178.3910; Gardner 2 max. color; sp.gr. 0.940 (25/20 C); visc. 50 poise; acid no. 190 min.; iodine no. 10 max.; sapon. no. 195; pour pt. 5 F; flash pt. 555 F; fire pt. 650 F; 79% dibasic, 5% monobasic, 16% polybasic.

Empol® 1008. [Henkel/Emery] Hydrog. and dist. dimer acid; CAS 68783-41-5; polymer grade acid suitable for use as a reactant in high polymers of the polyamide, polyester, and urethane types; exc. oxidation stability, color stability; FDA 21CFR §175.105, 175.300, 175.320, 176.170, 176.200, 176.210, 177.1200, 177.1210, 177.2420, 178.3910; Gardner 1 max. color; sp.gr. 0.940 (25/20 C); visc. 47 poise; acid no. 194; iodine no. 7; sapon. no. 197; pour pt. 5 F; flash pt. 575 F; fire pt. 630 F; 94% dibasic, 3% polybasic, 3% monobasic.

Empol® 1016. [Henkel/Emery] Dilinoleic acid; CAS 61788-89-4; surfactant for industrial applics., lubricants; offers good control of reactions and little chain-stopping effect in polymer systems; FDA 21CFR §175.105, 175.300, 175.320, 176.170, 176.200, 176.210, 177.1200, 177.1210, 177.2420, 178.3910; Gardner 6 max. color; sp.gr. 0.945 (25/20 C); visc. 50 poise; acid no. 194; sapon. no. 200; pour pt. 5 F; flash pt. 530 F; fire pt. 600 F; 80% dibasic, 16% polybasic, 4% monobasic.

Empol® 1018. [Henkel/Emery] Dilinoleic acid; CAS 61788-89-4; surfactant for industrial applics., lubricants; used in prod. of solid and liq. reactive polyamide resin compositions and for cross-linking polyesters to enhance toughness and flexibility; FDA 21CFR §175.105, 175.300, 175.320, 176.170, 176.200, 176.210, 177.1200, 177.1210, 177.2420, 178.3910; Gardner 8 max. color; sp.gr. 0.956 (25/20 C); visc. 92 poise; acid no. 192; sapon. no. 200; pour pt. 15 F; flash pt. 530 F; fire pt. 600 F; 79% dibasic, 17% polybasic, 4% monobasic.

Empol® 1020. [Henkel/Emery] Dimer acid; CAS 61788-89-4; surfactant for industrial applics., lubricants; used in liq. and solid polyamides and polymers for chain termination; FDA 21CFR §175.105, 175.300, 175.320, 176.170, 176.200, 176.210, 177.1200, 177.1210, 177.2420, 178.3910; Gardner 7 max. color; sp.gr. 0.949 (25/20 C); visc. 45 poise; acid no. 190; sapon. no. 194; pour pt. 15 F; flash pt. 500 F; fire pt. 545 F; 77% dibasic, 12% monobasic, 11% polybasic.

Empol® 1022. [Henkel/Emery] Dilinoleic acid; CAS 61788-89-4; surfactant for industrial applics., lubricants; used for liq. and solid polyamide applics.; FDA 21CFR §175.105, 175.300, 175.320, 176.170, 176.200, 176.210, 177.1200, 177.1210, 177.2420, 178.3910; Gardner 8 max. color; sp.gr. 0.953 (25/20 C); visc. 75 poise; acid no. 192; sapon. no. 195; pour pt. 15 F; flash pt. 530 F; fire pt. 600 F; 78% dibasic, 15% polybasic, 7% monobasic.

Empol® 1026. [Henkel/Emery] Dimer acid; CAS 61788-89-4; surfactant for industrial applics., lubricants; used in resin applics. where lt. color is not necessary; FDA 21CFR §175.105, 175.300, 175.320, 176.170, 176.200, 176.210, 177.1200, 177.1210, 177.2420, 178.3910; Gardner 8 max. color; sp.gr. 0.951 (25/20 C); visc. 65 poise; acid no. 191; sapon. no. 195; pour pt. 25 F; flash pt. 555 F; fire pt. 605 F; 82% dibasic, 11% polybasic, 7% monobasic.

Empol® 1040. [Henkel/Emery] Trilinoleic acid; CAS 68937-90-6; surfactant for industrial applics., lubricants; used primarily as a low-cost flexibilizing curing agent for epoxy resins; sp.gr. 0.968 (25/20 C); visc. 530 poise; acid no. 185; sapon. no. 195; pour pt. 25 F; flash pt. 595 F; fire pt. 680 F; 67% polybasic, 31% dibasic, 2% monobasic.

Emsorb® 2515. [Henkel/Emery; Henkel/Textiles] Sorbitan laurate; CAS 1338-39-2; nonionic; coupler, emulsifier; used in household specialities, industrial oils, agric., cosmetics, and emulsion polymerization; antifoam properties; EPA-exempt; Gardner 7 liq.; sol. in min. oil, butyl stearate, glycerol trioleate, Stod. solv., xylene; water-disp.; dens. 8.8 lb/gal; visc. 1000 cSt; HLB 8.0; pour pt. 15 C; flash pt. 430 F; cloud pt. < 25 C; 100% conc.

Emsorb® 2516. [Henkel/Emery] Sorbitan isostearate; nonionic; aux. emulsifier, solubilizer, corrosion inhibitor in lubricants, metal protectants and cleaners, emulsion polymerization; Gardner 2 solid; sol. @ 5% in min. oil, Stod., xylene; disp. in water, butyl stearate, glyceryl trioleate; dens. 8.2 lb/gal; visc. 1200 cSt (100 F); HLB 4.6; m.p. 50 C; flash pt. 460 F.

Emsorb® 2518. [Henkel/Organic Prods.] Sorbitan diisostearate; CAS 68238-87-9; nonionic; aux. emulsifier, solubilizer, corrosion inhibitor in lubricants, metal protectants and cleaners, emulsion polymerization; Gardner 5 liq.; sol. @ 5% in min. oil, butyl stearate, glyceryl trioleate, Stod., xylene; dens. 8.2 lb/gal; visc. 730 cSt (100 F); HLB 3.0; pour pt. -4 C; flash pt. 480 F; 100% conc.

Emsorb® 6901. [Henkel/Emery; Henkel/Textiles] PEG-5 sorbitan oleate; CAS 9005-65-6; nonionic; o/w emulsifier and lubricant in industrial lubricants, textile lubricants and softeners, metal treatment, paints, emulsion polymerization, agric.; color dispersant in plastics; EPA-exempt; Gardner 6 liq.; sol. in glycerol trioleate, Stod. solv.; water-disp.; dens. 8.5 lb/gal; visc. 450 cs; HLB 10.0; cloud pt. < 25 C; 100% conc.

Emsorb® 6909. [Henkel/Emery] PEG-4 sorbitan stearate; CAS 9005-67-8; nonionic; emulsifier for hydraulic fluids, metal treatment, emulsion polymerization, paints; color dispersants for plastics; Gardner 6 solid; sol. in min. oil, glyceryl trioleate, Stod.; disp. in water, butyl stearate, xylene; m.p. 35 F; HLB 9.6; flash pt. 520 F.

Emsorb® 6916. [Henkel/Emery] PEG-4 sorbitan laurate; CAS 9005-64-5; nonionic; o/w emulsifier in emulsion polymerization of PVC; Gardner 7 liq.; disp. in water, butyl stearate, perchloroethylene, Stod.; dens. 9.0 lb/gal; visc. 700 cs; HLB 12.1; cloud pt. < 25 C; Unverified

Emulan® NP 2080. [BASF AG] Alkylphenol + EO; nonionic; emulsifier for o/w or w/o emulsions, emulsion polymerization; liq.; HLB 16.0; 80% conc.

Emulan® NP 3070. [BASF AG] Alkylphenol + EO; nonionic; emulsifier for emulsion polymerization; liq.; HLB 17.0; 70% conc.

Emulan® OP 25. [BASF AG] Alkylphenol + EO; nonionic; emulsifier in emulsion polymerization; wax; HLB 16.5; 100% conc.

Emulbon LB-78. [Toho Chem. Industry] Six-membered borate with monoalkyl POE group; lubricant, plasticizer, solv.; liq.

Emuldan DG 60. [Grindsted Prods.] Diglycerides; food-grade additive for resins and plastics; processing aid, lubricant for PVC; dispersant for color concs.; FDA 21CFR §182.4505 (GRAS).

Emuldan DO 60. [Grindsted Prods.] Mono-dioleate; food-grade additive for lubricants, resins and plastics; processing aid, lubricant for PVC; dispersant for color concs.; FDA 21CFR §182.4505 (GRAS).

Emuldan HV 40. [Grindsted Prods.] Diglycerides; food-grade additive for resins and plastics; processing aid, lubricant, antistat for PVC; dispersant for color concs.; FDA 21CFR §182.4505 (GRAS).

Emuldan HV 52. [Grindsted Prods.] Diglycerides; food-grade additive for resins and plastics; processing aid, lubricant, antistat for PVC; dispersant for color concs.; FDA 21CFR §182.4505 (GRAS).

Emulsifiant 33 AD. [Seppic] Ethoxylated/sulfated alkylphenol; CAS 68649-55-8; anionic; emulsifier for emulsion polymerization; liq.; 40% conc.

Emulsifier K 30 40%. [Bayer AG; Bayer/Fibers, Org., Rubbers] Sodium alkane sulfonates based on n-paraffin; anionic; emulsifier for emulsion polymerization, esp. acrylic acid esters, VC, VA, vinylidiene chloride, butadiene, styrene, processing natural and syn. latexes, plastic dispersions; effective over wide pH range; antistat; up to 99% biodeg.; FDA 21CFR §176.170, 176.180, 178.3400; clear sol'n., virtually odorless; sol. in water; low sol. in alcohols; m.w. 330; visc. 5 mPa•s (25%); HLB 11-12; cloud pt. < 20 C (40%); pH alkaline; surf. tens. 34.9 dynes/cm (0.1%); 40% act.; *Toxicology:* LD50 (oral, rats) 2 g/kg; *Storage:* to prevent cloudiness and separation of sol'n. at temps. below 20 C, it should be warmed and homogenized.

Emulsifier K 30 68%. [Bayer/Fibers, Org., Rubbers] Sodium alkane sulfonates based on n-paraffin; anionic; emulsifier for emulsion polymerization, esp. acrylic acid esters, VC, VA, vinylidiene chloride, butadiene, styrene, processing natural and syn. latexes, plastic dispersions; effective over wide pH range; antistat; up to 99% biodeg.; FDA 21CFR §176.170, 176.180, 178.3400; colorless to ylsh. pumpable paste, virtually odorless; sol. in water; low sol. in alcohols; m.w. 330; visc. 5 mPa•s (25%); HLB 11-12; cloud pt. < 20 C (40%); pH alkaline; surf. tens. 34.9 dynes/cm (0.1%); 68% act.; *Toxicology:* LD50 (oral, rats) 2 g/kg; *Storage:* separates on storage; stir before use to homogenize.

Emulsifier K 30 76%. [Bayer/Fibers, Org., Rubbers] Sodium alkane sulfonates based on n-paraffin; anionic; emulsifier for emulsion polymerization, esp. acrylic acid esters, VC, VA, vinylidiene chloride, butadiene, styrene, processing natural and syn. latexes, plastic dispersions; effective over wide pH range; antistat; up to 99% biodeg.; FDA 21CFR §176.170, 176.180, 178.3400; wh.-ylsh. stiff paste, virtually odorless; sol. in water; low sol. in alcohols; m.w. 330; visc. 5 mPa•s (25%); HLB 11-12; cloud pt. < 20 C (40%); pH alkaline; surf. tens.

34.9 dynes/cm (0.1%); 76% act.; *Toxicology:* LD50 (oral, rats) 2 g/kg.

Emulsifier K 30 95%. [Bayer/Fibers, Org., Rubbers] Sodium alkane sulfonates based on n-paraffin; anionic; emulsifier for emulsion polymerization, esp. acrylic acid esters, VC, VA, vinylidiene chloride, butadiene, styrene, processing natural and syn. latexes, plastic dispersions; effective over wide pH range; antistat; up to 99% biodeg.; FDA 21CFR §176.170, 176.180, 178.3400; wh.-ylsh. flakes, virtually odorless; sol. in water; low sol. in alcohols; m.w. 330; visc. 5 mPa•s (25%); HLB 11-12; cloud pt. < 20 C (40%); pH alkaline; surf. tens. 34.9 dynes/cm (0.1%); 95% act.; *Toxicology:* LD50 (oral, rats) 2 g/kg; *Storage:* hygroscopic; use up completely when opened.

Emulvin® W. [Bayer/Fibers, Org., Rubbers] Aromatic polyglycol ether; nonionic; emulsifier, latex and foam stabilizer, wetting agent; protects latex mixes from the sensitizing effects of fillers and electrolytes; ylsh. brn. oily liq.; sp.gr. 1.13; pH 7 (10% aq.).

Epal® 6. [Albemarle] Hexanol; CAS 111-27-3; EINECS 203-852-3; detergent and emulsifier intermediate for household and industrial prods.; emulsion polymerization; FDA compliance; water-wh. mobile liq., lt. fatty alcohol odor; m.w. 102; sp.gr. 0.815 g/ml; dens. 6.8 lb/gal; f.p. -49 C; b.p. 151-160 C (1 atm); flash pt. (CC) 61 C; hyd. no. 548; combustible; 100% conc.; *Toxicology:* temporary irritation on direct skin or eye contact; *Storage:* protect against water contamination by blanketing with inert as or dry air.

Epal® 8. [Albemarle] Octanol; CAS 111-87-5; EINECS 203-917-6; detergent and emulsifier intermediate for household and industrial prods.; emulsion polymerization; water-wh. mobile liq., lt. fatty alcohol odor; m.w. 130.2; sp.gr. 0.821 g/ml; dens. 6.85 lb/gal; f.p. -14 C; b.p. 184-195 C (1 atm); flash pt. (CC) 88 C; hyd. no. 431; combustible; 100% conc.; *Toxicology:* temporary irritation on direct skin or eye contact; *Storage:* protect against water contamination by blanketing with inert as or dry air.

Epal® 10. [Albemarle] Decanol; CAS 112-30-1; EINECS 203-956-9; detergent and emulsifier intermediate for household and industrial prods.; emulsion polymerization; water-wh. mobile liq., mild char. odor; m.w. 158; sp.gr. 0.827 g/ml; dens. 6.90 lb/gal; f.p. 8 C; b.p. 226-230 C (1 atm); flash pt. (PMCC) 113 C; hyd. no. 354; 100% conc.; *Toxicology:* temporary irritation on direct skin or eye contact; *Storage:* protect against water contamination by blanketing with inert as or dry air.

Epal® 12. [Albemarle] Dodecanol; CAS 112-53-8; EINECS 203-982-0; detergent and emulsifier intermediate for household prods., industrial and institutional cleaning, emulsion polymerization, and personal care prods.; water-wh. mobile liq., mild char. odor; m.w. 186; sp.gr. 0.830 g/ml; dens. 6.93 lb/gal; f.p. 24 C; b.p. 258-264 C (1 atm); flash pt. (PMCC) 132 C; hyd. no. 301; 100% conc.; *Toxicology:* temporary irritation on direct skin or eye contact; *Storage:* protect against water contamination by blanketing with inert as or dry air.

Epal® 12/70. [Albemarle] Dodecanol (70%), tetradecanol (29%); detergent and emulsifier intermediate for household prods., industrial and institutional cleaning, emulsion polymerization, and personal care prods.; biodeg.; clear liq., mild char. odor; m.w. 195; sp.gr. 0.832 g/ml; dens. 6.94 lb/gal; f.p. 18 C; b.p. 126-129 C (760 mm); flash pt. (PMCC) 135 C; hyd. no. 288; 100% conc.; *Toxicology:* temporary irritation on direct skin or eye contact; *Storage:* protect against water contamination by blanketing with inert as or dry air.

Epal® 12/85. [Albemarle] Dodecanol (86%), tetradecanol (14%); CAS 67762-41-8; EINECS 267-019-6; detergent and emulsifier intermediate for household prods., industrial and institutional cleaning, emulsion polymerization, and personal care prods.; biodeg.; clear liq., mild char. odor; m.w. 190; sp.gr. 0.830 g/ml; dens. 6.93 lb/gal; f.p. 20 C; b.p. 126-129 C (760 mm);flash pt. (PMCC) 138 C; hyd. no. 297; 100% conc.; *Toxicology:* temporary irritation on direct skin or eye contact; *Storage:* protect against water contamination by blanketing with inert as or dry air.

Epal® 14. [Albemarle] Tetradecanol; CAS 112-72-1; EINECS 204-000-3; detergent and emulsifier intermediate for household prods., industrial and institutional cleaning, emulsion polymerization, and personal care prods.; biodeg.; wh. waxy solid, mild char. odor; m.w. 215; sp.gr. 0.822 g/ml; dens. 6.86 lb/gal; f.p. 35 C; flash pt. (PMCC) 149 C; hyd. no. 261; 100% conc.; *Toxicology:* temporary irritation on direct skin or eye contact; *Storage:* protect against water contamination by blanketing with inert as or dry air.

Epal® 16NF. [Albemarle] Hexadecanol NF; CAS 36653-82-4; EINECS 253-149-0; USP grade biodeg. detergent and emulsifier intermediate for household prods., industrial and institutional cleaning, emulsion polymerization, and personal care prods.; wh. waxy solid, mild char. odor; m.w. 242; sp.gr. 0.818 g/ml; dens. 6.83 lb/gal; f.p. 44 C; flash pt. (PMCC) 175 C; hyd. no. 232; 100% conc.; *Toxicology:* temporary irritation on direct skin or eye contact; *Storage:* protect against water contamination by blanketing with inert as or dry air.

Epal® 1214. [Albemarle] Dodecanol (66%), tetradecanol (27%), hexadecanol (6%); CAS 67762-41-8; EINECS 267-019-6; biodeg. detergent and emulsifier intermediate for household prods., industrial and institutional cleaners, emulsion polymerization, personal care prods.; clear sl. visc. liq., mild char. odor; m.w. 197; sp.gr. 0.830 g/ml; dens. 6.93 lb/gal; f.p. 22 C; b.p. 233-299 C (1 atm); flash pt. (PMCC) 138 C; hyd. no. 284; 100% conc.; *Toxicology:* temporary irritation on direct skin or eye contact; *Storage:* protect against water contamination by blanketing with inert gas or dry air.

Epan 710. [Dai-ichi Kogyo Seiyaku] PPG PEG ether; CAS 9003-11-6; nonionic; low foaming detergent, emulsifier, dispersant for emulsion polymerization of syn. resins, latex; liq.; 100% conc.

Epan 720. [Dai-ichi Kogyo Seiyaku] PPG PEG ether; CAS 9003-11-6; nonionic; low foaming detergent, emulsifier, dispersant for emulsion polymerization of syn. resins, latex; liq.; 100% conc.

Epan 740. [Dai-ichi Kogyo Seiyaku] PPG PEG ether; CAS 9003-11-6; nonionic; low foaming detergent, emulsifier, dispersant for emulsion polymerization of syn. resins, latex; paste; 100% conc.

Epan 750. [Dai-ichi Kogyo Seiyaku] PPG PEG ether; CAS 9003-11-6; nonionic; low foaming detergent, emulsifier, dispersant for emulsion polymerization of syn. resins, latex; paste; 100% conc.

Epan 785. [Dai-ichi Kogyo Seiyaku] PPG PEG ether; CAS 106392-12-5; nonionic; low foaming detergent, emulsifier, dispersant for emulsion polymerization of syn. resins, latex; solid; 100% conc.

Epi-Cure® 470. [Shell] Epoxy curing agent for composites; visc. 120-170 cP; 11.5-13.5% amine nitrogen content.

Epi-Cure® 537. [Shell] Accelerator for cure of composites; visc. 6-70 poise.

Epi-Cure® 3010. [Shell] Amido amine; epoxy curing agent for adhesives, high solids coatings, castings; Gardner > 12 color; dens. 8.0 lb/gal; visc. 400-900 cps; amine no. 375-475; flash pt. > 200 F; pot life 70 min.

Epi-Cure® 3015. [Shell] Amido amine; epoxy curing agent for high solids coatings, elec., castings, grouts, laminates; Gardner 9 max. color; dens. 8.0 lb/gal; visc. 500-900 cps; amine no. 420-450; flash pt. > 220 F; pot life 45 min.

Epi-Cure® 3025. [Shell] Amidoamine; epoxy curing agent for adhesives, high solids coatings, potting, laminates, castings; Gardner 9 max. color; dens. 7.9 lb/gal; visc. 200-400 cps; amine no. 450-500; flash pt. > 230 F; pot life 150 min.

Epi-Cure® 3030. [Shell] Amidoamine; curing agent; FDA acceptable; Gardner 9 max. color; dens. 7.9 lb/gal; visc. 300-600 cP; amine no. 400-425; EEW 95; gel time 100 min (100 g mass with Epon Resin 828).

Epi-Cure® 3035. [Shell] Amidoamine; curing agent; Gardner 9 max. color; dens. 7.9 lb/gal; visc. 300-600 cP; amine no. 400-425; EEW 95; gel time 100 min (100 g mass with Epon Resin 828).

Epi-Cure® 3046. [Shell] Amidoamine; epoxy curing agent for toughened systems for structural applics.; long pot life; Gardner 13 max. color; dens. 7.8 lb/gal; visc. 120-280 cps; EEW 90.

Epi-Cure® 3055. [Shell] Amidoamine; curing agent; Gardner 13 max. color; dens. 7.8 lb/gal; visc. 150-300 cP; amine no. 449-473; EEW 90; gel time 255 min (100 g mass with Epon Resin 828).

Epi-Cure® 3060. [Shell] Amidoamine; curing agent; Gardner 10 max. color; dens. 8.1 lb/gal; visc. 900-2000 cP; amine no. 580-620; EEW 66; gel time 45 min (100 g mass with Epon Resin 828).

Epi-Cure® 3061. [Shell] Amidoamine; curing agent; Gardner 13 max. color; dens. 7.8 lb/gal; visc. 220-430 cP; amine no. 313-345; EEW 115; gel time 465 min (100 g mass with Epon Resin 828).

Epi-Cure® 3070. [Shell] Amidoamine; curing agent; Gardner 13 max. color; dens. 8.1 lb/gal; visc. 300-600 cP; amine no. 517-553; EEW 65; gel time 41 min (100 g mass with Epon Resin 828).

Epi-Cure® 3072. [Shell] Amidoamine; curing agent; Gardner 12 max. color; dens. 8.1 lb/gal; visc. 500-900 cP; amine no. 560-575; EEW 65; gel time 40 min (100 g mass with Epon Resin 828).

Epi-Cure® 3090. [Shell] Amidoamine; curing agent; Gardner 11 max. color; dens. 8.2 lb/gal; visc. 3000-6000 cP; amine no. 230-260; EEW 190; gel time 120 min (100 g mass with Epon Resin 828).

Epi-Cure® 3100-ET-60. [Shell] Polyamide; curing agent; Gardner 9 max. sol'n.; dens. 7.6 lb/gal; visc. (G-H) W-Y; amine no. 51-57; pot life 18 h (250 g); 59-61% solids in IPA/toluene.

Epi-Cure® 3100-HX-60. [Shell] Polyamide; curing agent; Gardner 9 max. sol'n.; dens. 7.7 lb/gal; visc. (G-H) Z-Z3; amine no. 51-57; pot life > 12 h (250 g); 59-61% solids in propylene glycol methyl ether/ xylene.

Epi-Cure® 3100-XY-60. [Shell] Polyamide; curing agent; Gardner 12 max. sol'n.; dens. 7.7 lb/gal; visc. (G-H) X-Z2; amine no. 350; 60% solids in xylene/ethylene glycol propyl ether.

Epi-Cure® 3115. [Shell] Polyamide; epoxy curing agent for toughened systems for structural applics.; provides toughness, flexibility, and water resist.; long pot life; Gardner 9 max. color; dens. 8.1 lb/gal; visc. 50,000-75,000 cps (40 C); EEW 156.

Epi-Cure® 3115-E-73. [Shell] Polyamide; curing agent; Gardner 9 max. sol'n.; dens. 7.6 lb/gal; visc. (G-H) Z-Z2; amine no. 170-185; pot life > 12 h (250 g); 72-74% solids in IPA.

Epi-Cure® 3115-X-70. [Shell] Polyamide; curing agent; Gardner 9 max. sol'n.; dens. 7.8 lb/gal; visc. (G-H) V-Z; amine no. 161-173; 69-71% solids in xylene.

Epi-Cure® 3123. [Shell] Polyamide; curing agent; Gardner 9 max. color; dens. 8.3 lb/gal; visc. 50-200 cP; amine no. 320-360; EEW 124; pot life 3-4 h (250 g).

Epi-Cure® 3125. [Shell] Polyamide; epoxy curing agent for adhesives, coatings, grouts, elec., laminates, casting, flooring; Gardner 9 max. color; dens. 8.1 lb/gal; visc. 8000-12,000 cps (40 C); amine no. 330-360; flash pt. > 230 F; pot life 120 min.

Epi-Cure® 3140. [Shell] Polyamide; epoxy curing agent for adhesives, coatings, grouts, elec., laminates, casting, flooring; Gardner 9 max. color; dens. 8.1 lb/gal; visc. 3000-6000 cps (40 C); amine no. 350-400; flash pt. > 230 F; pot life 120 min.

Epi-Cure® 3141. [Shell] Polyamide; curing agent; Gardner 9 max. color; dens. 8.1 lb/gal; visc. 2000-3000 cP; amine no. 370-400; EEW 103; pot life 2.8 h (250 g).

Epi-Cure® 3150. [Shell] Polyamide; curing agent; Gardner 10 max. color; dens. 7.9 lb/gal; visc. 5-10 poise; amine no. 425-450; EEW 130.

Epi-Cure® 3155. [Shell] Polyamide; curing agent; Gardner 9 max. color; dens. 8.2 lb/gal; visc. 3000-

6000 cP; amine no. 200-230; EEW 133; pot life 2 h (250 g).

Epi-Cure® 3175. [Shell] Polyamide; curing agent; Gardner 9 max. color; dens. 8.1 lb/gal; visc. (G-H) Z; amine no. 300-380; EEW 103.

Epi-Cure® 3180-F-75. [Shell] Polyamide; curing agent; Gardner 10 max. sol'n.; dens. 8.0 lb/gal; visc. 4500-9000 cP; amine no. 237-258; pot life > 12 h (250 g); 75% solids in n-butanol.

Epi-Cure® 3185-FX-60. [Shell] Polyamide; curing agent; Gardner 9 max. sol'n.; dens. 7.8 lb/gal; visc. 9-14 poise; amine no. 115-130; 56-61% solids in butanol/xylene.

Epi-Cure® 3192. [Shell] Polyamide; curing agent; Gardner 8 max. color; thixotropic; dens. 7.8 lb/gal; EEW 133; pot life 1 h (250 g).

Epi-Cure® 3200. [Shell] Aliphatic; epoxy co-curing agent for other amines for flooring, coatings, and encapsulation; Gardner 2 color; dens. 8.2 lb/gal; visc. 20 cps (20 C); flash pt. 419 F; pot life 19 min.

Epi-Cure® 3213. [Shell] Aliphatic amine; curing agent; Gardner 4 max. color; dens. 8.2 lb/gal; visc. (G-H) Q-U; amine no. 176-216; EEW 200; pot life 5-18 h (R.T.).

Epi-Cure® 3214. [Shell] Aliphatic amine; curing agent; Gardner 4 max. color; dens. 8.2 lb/gal; visc. (G-H) K-R; amine no. 176-216; EEW 200; pot life 5-18 h (R.T.).

Epi-Cure® 3218. [Shell] Aliphatic amine; curing agent; Gardner 12 max. color; dens. 8.2 lb/gal; visc. 3000 cP; amine no. 960; EEW 44; gel time 18 min (100 g).

Epi-Cure® 3223. [Shell] Aliphatic; epoxy co-curing agent for other amines for flooring, coatings, and encapsulation; Gardner 1 color; dens. 7.9 lb/gal; visc. 10 cps (20 C); flash pt. 410 F; pot life 25 min.

Epi-Cure® 3234. [Shell] Aliphatic; epoxy co-curing agent for other amines for flooring, coatings, and encapsulation; Gardner 2 color; dens. 8.2 lb/gal; visc. 25 cps (20 C); flash pt. 527 F; pot life 30 min.

Epi-Cure® 3245. [Shell] Aliphatic; epoxy co-curing agent for other amines for flooring, coatings, and encapsulation; Gardner 3 color; dens. 8.3 lb/gal; visc. 100 cps (20 C); flash pt. 536 F; pot life 35 min.

Epi-Cure® 3251. [Shell] Aliphatic amine; curing agent; Gardner 5 max. color; dens. 8.3 lb/gal; visc. 400-700 cP; amine no. 350-390; EEW 76; gel time 14 min (100 g).

Epi-Cure® 3253. [Shell] Aliphatic tert. amine; epoxy accelerator; Gardner 8 max. color; dens. 8.1 lb/gal; visc. 200-260 cps; amine no. 575-625; flash pt. > 230 F; pot life 38 min.

Epi-Cure® 3254. [Shell] Aliphatic amine; curing agent; Gardner 9 max. color; dens. 7.8 lb/gal; visc. 700-900 cP; amine no. 230-250; EEW 152; gel time 48 min (100 g).

Epi-Cure® 3255. [Shell] Aliphatic; epoxy co-curing agent for low temp. applics.; Gardner 10 max. color; dens. 8.9 lb/gal; visc. 1000-1600 cps; amine no. 870-930; flash pt. 284 F; pot life 10 min.

Epi-Cure® 3260. [Shell] Polyether urethane amine; epoxy flexibilizer; Gardner 5 max. color; dens. 8.5 lb/gal; visc. 13,000-20,000 cps; amine no. 65-75; flash pt. 212 F; pot life 8-9 hr.

Epi-Cure® 3262. [Shell] Modified aliphatic amine; epoxy curing agent for toughened systems for structural applics.; internally flexibilized; provides thermal shock resist.; Gardner 3 max. color; dens. 8.6 lb/gal; visc. 1800-3400 cps; EEW 157.

Epi-Cure® 3265. [Shell] Accelerated polyether urethane amine; epoxy flexibilizer; Gardner 4 max. color; dens. 8.1 lb/gal; visc. 3000-5000 cps; amine no. 120-135; flash pt. 248 F; pot life 45 min.

Epi-Cure® 3266. [Shell] Accelerated aliphatic amine; moderately fast setting epoxy curing agent, flexibilizer imparting rubber-like elastomeric and tear resist. props. with epoxy toughness and adhesive qualities at and below freezing; for traffic wear and bridge deck coatings, expansion-joint adhesives; swimming pool sealants, stress-relaxing potting compds.; Gardner 9 max. clear low-visc. liq.; dens. 8.1 lb/gal; visc. 1200-1700 cps; amine no. 148; flash pt. (CC) 251 F; pot life 120 min; 100% solids; *Toxicology:* causes severe skin and eye irritation; can cause burns on prolonged contact; do not breathe vapors; *Storage:* store in cool, dry place in tightly closed containers; prod. will absorb moisture and CO_2 which may affect visc. or create foaming with epoxy resins; 1 yr min. shelf life.

Epi-Cure® 3270. [Shell] Aliphatic; epoxy curing agent for potting and encapsulation, flooring, laminates, patching compds.; Gardner 3 max. color; dens. 8.1 lb/gal; visc. 4000-5000 cps; amine no. 300-350; flash pt. > 230 F; pot life 8 min.

Epi-Cure® 3271. [Shell] Aliphatic; epoxy curing agent for castings, adhesives, tooling, gel coats, laminates, flooring, coatings; Gardner 6 max. color; dens. 8.5 lb/gal; visc. 100-200 cps; amine no. 950-1050; flash pt. 220 F; pot life 13 min.

Epi-Cure® 3273. [Shell] Aliphatic amine; curing agent; Gardner 6 max. color; dens. 8.1 lb/gal; visc. 1200-2000 cP; amine no. 420-480; EEW 90; gel time 11 min (100 g).

Epi-Cure® 3274. [Shell] Aliphatic amine; curing agent; Gardner 1 max. color; dens. 7.9 lb/gal; visc. 40-60 cP; amine no. 275-325; EEW 76; gel time 135 min (100 g).

Epi-Cure® 3275. [Shell] Aliphatic amine; curing agent; Gardner 3 max. color; dens. 8.1 lb/gal; visc. 200-400 cP; amine no. 600-650; EEW 95; gel time 8 min (100 g).

Epi-Cure® 3277. [Shell] Aliphatic amine; curing agent; Gardner 5 max. color; dens. 8.0 lb/gal; visc. 275 cP; amine no. 280-312; EEW 92; gel time 47 min (100 g).

Epi-Cure® 3278. [Shell] Aliphatic amine; curing agent; Gardner 7 max. color; visc. 10-25 cP; amine no. 329-360.

Epi-Cure® 3281-H-60. [Shell] Aliphatic amine; curing agent; Gardner 6 max. color; dens. 9.1 lb/gal; visc. 1200 cP; amine no. 254-293; EEW 142.

Epi-Cure® 3282. [Shell] Aliphatic; epoxy curing agent for castings, adhesives, tooling, gel coats, laminates, flooring, coatings; Gardner 6 max. color;

dens. 9.0 lb/gal; visc. 3000-5000 cps; amine no. 800-900; flash pt. > 230 F; pot life 15 min.

Epi-Cure® 3290. [Shell] Aliphatic; epoxy curing agent for elec., laminates, molding, flooring applics.; Gardner 4 max. color; dens. 8.5 lb/gal; visc. 300-500 cps; amine no. 920-1020; flash pt. > 230 F; pot life 25 min.

Epi-Cure® 3292-FX-60. [Shell] Aliphatic amine; curing agent; Gardner 9 max. color; dens. 8.5 lb/gal; visc. (G-H) Z-Z2; amine no. 225-227; EEW 140.

Epi-Cure® 3293. [Shell] Aliphatic amine; curing agent; Gardner 9 max. color; visc. 36-38 poise.

Epi-Cure® 3295. [Shell] Aliphatic amine; curing agent; Gardner 3 max. color; dens. 8.3 lb/gal; visc. 125-205 cP; amine no. 881-961; EEW 45; gel time 26 min (100 g).

Epi-Cure® 3370. [Shell] Cycloaliphatic amine; curing agent; Gardner 1 max. color; dens. 8.3 lb/gal; visc. 80-145 cP; amine no. 384-407; EEW 72; gel time 25 min (100 g).

Epi-Cure® 3371. [Shell] Cycloaliphatic amine; curing agent; Gardner 8 max. color; dens. 7.9 lb/gal; visc. 10-30 cP; amine no. 1074-1154; EEW 27; gel time 90 min (100 g).

Epi-Cure® 3373. [Shell] Cycloaliphatic amine; curing agent; Gardner 1 max. color; dens. 8.3 lb/gal; visc. 25-30 cP; amine no. 310-325; EEW 87; gel time 40 min (100 g).

Epi-Cure® 3374. [Shell] Cycloaliphatic amine; curing agent; Gardner 1 max. color; dens. 8.4 lb/gal; visc. 45-60 cP; amine no. 320-335; EEW 87; gel time 20 min (1 lb).

Epi-Cure® 3378. [Shell] Cycloaliphatic amine; curing agent; Gardner 4 max. color; dens. 8.6 lb/gal; visc. 2200-2800 cP; amine no. 240-265; EEW 114; gel time 23 min (1 lb).

Epi-Cure® 3379. [Shell] Cycloaliphatic amine; curing agent; Gardner 12 max. color; dens. 8.2 lb/gal; visc. 40-60 cP; EEW 94; gel time 30 min (100 g).

Epi-Cure® 3380. [Shell] Cycloaliphatic amine; curing agent; Gardner 2 max. color; dens. 8.5 lb/gal; visc. 200-400 cP; amine no. 235-295; EEW 114; gel time 40 min (25 g).

Epi-Cure® 3381. [Shell] Cycloaliphatic amine; curing agent; Gardner 1 max. color; dens. 8.3 lb/gal; visc. 50-100 cP; amine no. 290-360; EEW 95; gel time 35 min (100 g).

Epi-Cure® 3382. [Shell] Cycloaliphatic amine; curing agent; Gardner 5 max. color; dens. 8.4 lb/gal; visc. 1000-1500 cP; amine no. 239-250; EEW 118; gel time 23 min (100 g).

Epi-Cure® 3383. [Shell] Cycloaliphatic amine; curing agent; Gardner 4 max. color; dens. 8.5 lb/gal; visc. 250-350 cP; amine no. 255-285; EEW 114; gel time 34 min (100 g).

Epi-Cure® 3384. [Shell] Cycloaliphatic amine; curing agent; Gardner 4 max. color; dens. 8.7 lb/gal; visc. 400-800 cP; amine no. 255-285; EEW 114; gel time 15 min (100 g).

Epi-Cure® 3484. [Shell] Aromatic amine; curing agent; dark color; dens. 8.6-8.7 lb/gal; visc. 6 poise; EEW 38; gel time 11.9 min (100 g, 150 C).

Epi-Cure® 3501. [Shell] Ketimine; curing agent; Gardner 5 max. color; dens. 7.3 lb/gal; visc. 10 cP max.; EEW 52; 15.5-17.5% amine nitrogen content.

Epi-Cure® 3502. [Shell] Ketimine; curing agent; Gardner 8 max. color; dens. 7.1 lb/gal; visc. 2-5 cP; EEW 55; 11-14% amine nitrogen content.

Epi-Cure® 3503. [Shell] Ketimine; curing agent; Gardner 8 max. color; dens. 8.1 lb/gal; visc. 2-5 poise; EEW 101; 9.5-11.5% amine nitrogen content.

Epi-Cure® 8290-Y-60. [Shell] Waterborne curing agent; Gardner 9 max. color; dens. 8.8 lb/gal; visc. (G-H) Z3-Z5; amine no. 361-421; EEW 163; 60% NV in ethylene glycol propyl ether.

Epi-Cure® 8292-Y-60. [Shell] Waterborne curing agent; Gardner 9 max. color; dens. 8.8 lb/gal; visc. (G-H) Z2; amine no. 490-520; EEW 95; 60% NV in ethylene glycol propyl ether.

Epi-Cure® 8535-W-50. [Shell] Waterborne curing agent; Gardner 12 max. color; dens. 8.8 lb/gal; visc. (G-H) Z4-Z6; amine no. 373-453; EEW 102; 50% NV in water.

Epi-Cure® 8536-MY-60. [Shell] Waterborne curing agent; Gardner 13 max. color; dens. 8.3 lb/gal; visc. (G-H) X-Z; amine no. 110-130; EEW 324; 60% NV in ethylene glycol butyl ether/ethylene glycol propyl ether.

Epi-Cure® 8537-WY-60. [Shell] Waterborne curing agent; Gardner 9 max. color; dens. 9.0 lb/gal; visc. (G-H) Z-Z4; amine no. 310-345; EEW 174; 60% NV in water/ethylene glycol propyl ether.

Epi-Cure® 9150. [Shell] Epoxy curing agent for composites; 93.5% min. anhydride.

Epi-Cure® 9350. [Shell] Epoxy curing agent for composites; Gardner 6 max. color; visc. 30-45 poise; 12.4-14.4% amine nitrogen content.

Epi-Cure® 9360. [Shell] Epoxy curing agent for composites; visc. 100-400 poise; 9.3-10.5% amine nitrogen content.

Epi-Cure® 9850. [Shell] Epoxy curing agent for elec. laminating applics.; 93.5% anhydride.

Epi-Cure® HPT Curing Agent 1061-M. [Shell] Aromatic amine; curing agent for high performance composites; rapid gel time; good hot/wet performance; fine powd.; 100% < 30 μ; m.p. 322-329 F.

Epi-Cure® HPT Curing Agent 1062-M. [Shell] Aromatic amine; curing agent for high performance composites; moderate gel time; improved hot/wet performance; fine powd.; 100% < 30 μ; m.p. 300-310 F.

Epi-Cure® P-101. [Shell] Curing agent; powd.; dens. 1.16 g/ml; melt visc. 10-30 poise (150 C); m.p. 85-105 C.

Epi-Cure® P-104. [Shell] Curing agent; powd.; dens. 1.26 g/ml.

Epi-Cure® P-108. [Shell] Curing agent; powd.; dens. 1.38 g/ml.

Epi-Cure® P-187. [Shell] Curing agent; powd.; dens. 1.4 g/ml; m.p. > 250 C.

Epi-Cure® P-201. [Shell] Curing agent; Pt-Co 100 max. powd.; dens. 1.18 g/ml; melt visc. 2-4 poise

(150 C); m.p. 75-80 C; EEW 235-265.

Epi-Cure® P-202. [Shell] Curing agent; Pt-Co 100 max. powd.; dens. 1.18 g/ml; melt visc. 2-4 poise (150 C); m.p. 75-80 C; EEW 235-265.

Epi-Cure® W. [Shell] Aromatic amine; curing agent; Gardner 7 max. color; dens. 8.5 lb/gal; visc. 3-3.5 poise (20 C); EEW 44; gel time 4.1 min (177 C, with epoxy resin of EEW = 170); 15.7-15.9% amine nitrogen content.

Epi-Cure® Y. [Shell] Aromatic amine; curing agent; dens. 9.4 lb/gal; visc. 5-20 poise; EEW 48; gel time 3.3 min (177 C, with epoxy resin of EEW = 170); 13-16% amine nitrogen content.

Epi-Cure® Z. [Shell] Mixt. of aromatic diamines; curing agent for liq. epoxy resins used in structural applics. which require exc. chem. and solv. resist. and high deflection temps.; for castings for plastic tooling, elec./electronic encapsulation, adhesives, laminates, composites for aircraft; filament wound pipe and tanks; sp.gr. 1.14; dens. 9.5 lb/gal; visc. 15-40 poise; flash pt. (Seta) 92.8 C; EEW 38; gel time 8 h (150 g); 17-19% amine nitrogen content; *Toxicology:* LD50 (oral, rat) 0.7 g/kg; highly irritating to eyes; moderate skin irritant; skin sensitizer; may cause cancer and induce toxic hepatitis; contains p,p′-methylenedianiline (TLV 0.1 ppm TWA, 0.5 ppm STEL); *Precaution:* reacts violently with stong oxidizing agents; thermal decomp. produces CO, NO_x; *Storage:* store away from direct sunlight; prolonged storage may cause crystallization or solidfication; warm to reliquefy.

Epikote® 5050. [Shell] Brominated epoxy resin with epoxy end groups; general purpose flame retardant for thermoplastics; m.p. 64 ± 5 C; 50% Br.

Epikote® 5051 H. [Shell] Brominated epoxy resin with epoxy end groups; flame retardant for ABS; m.p. 128 ± 5 C; 50% Br.

Epikote® 5057. [Shell] Brominated epoxy resin with epoxy end groups; flame retardant for PBT and PET; m.p. 160 ± 5 C; 50% Br.

Epikote® 5201. [Shell] Brominated epoxy resin with tribromophenol end groups; flame retardant for HIPS; m.w. 1450; m.p. 100 C; 60% Br.

Epikote® 5201S. [Shell] Brominated epoxy resin with tribromophenol end groups; flame retardant for HIPS; m.w. 1350; m.p. 97 C; 60% Br.

Epikote® 5203. [Shell] Brominated epoxy resin with tribromophenol end groups; flame retardant for ABS; m.w. 1700; m.p. 113 C; 60% Br.

Epikote® 5205. [Shell] Brominated epoxy resin with tribromophenol end groups; flame retardant for ABS; m.w. 2400; m.p. 136 C; 60% Br.

EPlstatic® 100. [Eagle-Picher] Tribasic lead sulfate; heat stabilizer for flexible and rigid PVC compds.; activator/stabilizer in PVC foams; exc. for phonograph record compds.; high surf. area, exc. dispersibility; wh.; 99.9% 325 mesh; sp.gr. 6.7; surf. area 2.6 m^2/g; 82% Pb.

EPlstatic® 101. [Eagle-Picher] Tribasic lead sulfate with lt. min. oil added for dust control; heat stabilizer for flexible and rigid PVC compds.; wh.; 99.9% 325 mesh; sp.gr. 6.0; surf. area 2.6 m^2/g; 80% Pb.

EPlstatic® 102. [Eagle-Picher] Org. modified to provide greater lubricity for reduced power requirements; heat stabilizer for flexible and rigid PVC compds.; wh.; 99.9% 325 mesh; sp.gr. 5.3; surf. area 2.7 m^2/g; 76% Pb.

EPlstatic® 103. [Eagle-Picher] EPlstatic 102 with lt. min. oil added for dust control; heat stabilizer for flexible PVC; wh.; 99.9% 325 mesh; sp.gr. 4.8; surf. area 2.7 m^2/g; 74% Pb.

EPlstatic® 110. [Eagle-Picher] Basic lead silicosulfate; heat stabilizer for use in low-temp. flexible PVC; wh.; 99.9% 325 mesh; sp.gr. 5.1; surf. area 2.0 m^2/g; 66% Pb.

EPlstatic® 111. [Eagle-Picher] Basic lead silicosulfate with lt. min. oil for dust control; heat stabilizer for use in low-temp. flexible PVC; wh. 99.9% 325 mesh; sp.gr. 6.0; surf. area 2.6 m^2/g; 80% Pb.

EPlstatic® 112. [Eagle-Picher] Org. modified to provide greater lubricity for lower torque requirements; enhances heat stabilization in low-temp. flexible PVC; wh.; 99.9% -325 mesh; sp.gr. 4.2; surf. area 2.1 m^2/g; 62% Pb.

EPlstatic® 113. [Eagle-Picher] Basic lead silicosulfate, org. suface treated with lt. min. oil for dust control; heat stabilizer for low-temp. flexible PVC; wh.; 99.9% 325 mesh; sp.gr. 3.7; surf. area 2.1 m^2/g; 60% Pb.

Epi-Tex® 611-Q. [Shell] Epoxy ester emulsion; modifier for PVAc emulsions and thermoplastic latex resins, providing films with increased water and wear resist.; used in coatings for floors, structural masonry, roofs, patios, driveways, and over concrete, wood, and asphalt; milky-wh. appearance; dens. 8.47 lb/gal (sol'n.); visc. 100 KU; Seta flash pt. > 200 F; pH 8.3; 60% NV in 77.5% water, 16.9% aromatic naphtha, 5.6% xylene.

EPlthal 120. [Eagle-Picher] Dibasic lead phthalate; CAS 69011-06-9; heat stabilizer in 90 and 105 C PVC wire and cable compds.; wh.; 99.9% 325 mesh; sp.gr. 4.5; surf. area 2.9 m^2/g; 76% Pb.

EPlthal 121. [Eagle-Picher] Basic lead phthalate with lt. min. oil added for dust control; heat stabilizer in 90 and 105 C PVC wire and cable compds.; wh.; 99.9% 325 mesh; sp.gr. 4.1; surf. area 2.9 m^2/g; 74% Pb.

EPlthal 122. [Eagle-Picher] Org.; heat stabilizer in 90 and 105 C PVC wire and cable compds.; modified to provide additional heat stability and processing ease; wh.; 99.9% 325 mesh; sp.gr. 3.3; surf. area 3.0 m^2/g; 66% Pb.

EPlthal 123. [Eagle-Picher] Epithal 122 with lt. min. oil added for dust control; heat stabilizer in 90 and 105 C PVC wire and cable compds.; wh.; 99.9% 325 mesh; sp.gr. 3.0; surf. area 3.0 m^2/g; 64% Pb.

Epodil® 741. [Air Prods./Perf. Chems.] Butyl glycidyl ether; CAS 2426-08-6; reactive epoxy resin diluent for tooling, elec. applics., floorings, highly filled coatings; high vapor pressure; Gardner 1 liq.; sp.gr. 0.910; dens. 7.7 lb/gal; visc. 2 cps; flash pt. (Seta) 137 F; 0.2% max. moisture.

Epodil® 742. [Air Prods./Perf. Chems.] Cresyl glycidyl

ether; CAS 2210-79-9; reactive epoxy diluent for tooling, elec. applics., coatings, flooring; good chem. resist., esp. against acids and solvs., low volatility, moisture tolerant; Gardner 2 liq.; sp.gr. 1.08; dens. 9.0 lb/gal; visc. 12 cps; flash pt. (Seta) > 200 F; 0.2% max. moisture.

Epodil® 743. [Air Prods./Perf. Chems.] Phenyl glycidyl ether; CAS 122-60-1; EINECS 204-557-2; reactive epoxy diluent for elec. applics.; Gardner 2 color; dens. 9.2 lb/gal; visc. 4-7 cps; flash pt. (Seta) > 200 F.

Epodil® 745. [Air Prods./Perf. Chems.] p-t-Butyl phenyl glycidyl ether; CAS 3101-60-8; reactive epoxy diluent for tooling, elec. applics., flooring; Gardner 3 color; dens. 8.5 lb/gal; visc. 20-40 cps; flash pt. (Seta) > 200 F.

Epodil® 746. [Air Prods./Perf. Chems.] 2-Ethylhexyl glycidyl ether; CAS 2461-15-6; EINECS 219-553-6; reactive epoxy diluent for exposed aggregates, potting, floorings; Gardner 2 liq.; sp.gr. 0.910; dens. 7.6 lb/gal; visc. 9 cps; flash pt. (Seta) > 200 F; 0.1% max. moisture.

Epodil® 747. [Air Prods./Perf. Chems.] Alkyl (C8-C10) glycidyl ether; CAS 68608-96-1; reactive epoxy diluent for exposed aggregates, potting, coatings, floorings; Gardner 1 liq.; sp.gr. 0.900; dens. 7.5 lb/gal; visc. 9 cps; flash pt. (Seta) > 200 F; 0.1% max. moisture.

Epodil® 748. [Air Prods./Perf. Chems.] Alkyl (C12-C14) glycidyl ether; CAS 68609-97-2; reactive diluent for exposed aggregates, coatings, flooring; Gardner 1 liq.; sp.gr. 0.890; dens. 7.4 lb/gal; visc. 12 cps; flash pt. (Seta) > 200 F; 0.1% max. moisture.

Epodil® 749. [Air Prods./Perf. Chems.] Neopentyl glycol diglycidyl ether; CAS 17557-23-2; reactive epoxy diluent for civil engineering applics., elec. potting and encapsulation; Gardner 1 liq.; sp.gr. 1.04; dens. 8.7 lb/gal; visc. 18 cps; flash pt. (Seta) > 200 F; 0.2% max. moisture.

Epodil® 750. [Air Prods./Perf. Chems.] 1,4-Butanediol diglycidyl ether; CAS 2425-79-8; reactive epoxy diluent for elec. potting, casting, and encapsulation; low volatility; Gardner 1 liq.; sp.gr. 1.10; dens. 9.2 lb/gal; visc. 18 cps; flash pt. (Seta) > 200 F; 0.2% max. moisture.

Epodil® 757. [Air Prods./Perf. Chems.] Cyclohexanedimethanol diglycidyl ether; CAS 14228-73-0; reactive epoxy diluent for laminates, civil engineering applics.; Gardner 2 liq.; sp.gr. 1.10; dens. 8.7 lb/gal; visc. 65 cps; flash pt. (Seta) > 200 F; 0.2% max. moisture.

Epodil® 759. [Air Prods./Perf. Chems.] Alkyl (C12-C13) glycidyl ether; CAS 68609-97-2; reactive epoxy diluent for coatings; less likely to crystallize than Epodil 748; Gardner 1 liq.; sp.gr. 0.890; dens. 7.4 lb/gal; visc. 8 cps; flash pt. (Seta) > 200 F; 0.2% max. moisture.

Epodil® 769. [Air Prods./Perf. Chems.] DGE of resorcinol; CAS 101-90-6; reactive epoxy diluent for elec. applics.; Gardner 6 color; dens. 10.1 lb/gal; visc. 300-500 cps; flash pt. (Seta) > 200 F.

Epodil® L. [Air Prods./Perf. Chems.] Low m.w. aromatic hydrocarbon; general-purpose low-toxicity diluent for epoxy systems; used for solv.-free coatings and floorings; improves adhesion to most substrates incl. oily metal (@ 5-10 phr); improves trowelability of epoxy mortars; add to resin or curing agent; Gardner 7 max. liq.; sp.gr. 1.04; dens. 8.6 lb/gal; visc. 100 cps; 0.2% max. moisture.

Epodil® ML. [Air Prods./Perf. Chems.] Fatty ester; low-toxicity diluent/plasticizer/wetting agent for epoxy resins; assists dispersion or displacement of moisture during cure; not compat. with all curing agents; for use in epoxy component for coatings, decorative floorings, and exterior patching; improves gloss and substrate wetting (incl. damp surfaces); Gardner 8 liq.; sp.gr. 0.89; visc. < 10 cps.

Epolene® C-10. [Eastman] Highly branched, med. m.w. polyethylene wax; CAS 9002-88-4; EINECS 200-815-3; nonemulsifiable low-density wax used in hot-melt adhesives and coatings for papers and packaging materials, as paraffin modifiers in slush and cast molding, candles, inks; processing aid for rubber compounding; dispersant for color concs.; FDA 21CFR §175.105; Gardner 1 pellet, sl. odor; negligible sol. in water; m.w. 35,000; dens. 0.906 g/cc; visc. 7800 cps (150 C); soften. pt. (R&B) 104 C; acid no. < 0.05; cloud pt. 77 C (2% in 130 F paraffin); flash pt. (COC) 316 C; *Toxicology:* molten material may produce thermal burns; *Precaution:* incompat. with oxidizers.

Epolene® C-10P. [Eastman] Highly branched, med. m.w. polyethylene wax; CAS 9002-88-4; EINECS 200-815-3; lubricant for flexible and rigid PVC; processing aid for rubber; dispersing aid, mold release agent; powd.; *Precaution:* incompat. with oxidizers; powd. material may form explosive dust-air mixts.

Epolene® C-13. [Eastman] Highly branched, med. m.w. polyethylene wax; CAS 9002-88-4; EINECS 200-815-3; wax designed for use with Epolene waxes or blends containing lower m.w. materials; as paraffin wax modifier, e.g., for paper coatings; additive for inks, hot-melt adhesives; base polymer for color concs.; processing aid, mold release, dispersion aid, coupling agent for plastics; FDA 21CFR §175.105; Gardner 1 pellet; m.w. 76,000; melt index 200 (190 C); dens. 0.913 g/cc; soften. pt. (R&B) 110 C; acid no. < 0.05; cloud pt. 81 C (2%); *Toxicology:* molten material may produce thermal burns.

Epolene® C-13P. [Eastman] Highly branched, med. m.w. polyethylene wax; CAS 9002-88-4; EINECS 200-815-3; extrusion lubricant for PVC; processing aid for rubber; dispersing aid, mold release agent; powd.; *Precaution:* incompat. with oxidizers; powd. material may form explosive dust-air mixts.

Epolene® C-14. [Eastman] Highly branched, med. m.w. polyethylene wax; CAS 9002-88-4; EINECS 200-815-3; wax designed for use with Epolene waxes or blends containing lower m.w. materials; as paraffin wax modifier, e.g., for paper coatings; additive for inks, hot-melt adhesives; base polymer for color concs.; processing aid, mold release, dispersion aid, coupling agent for plastics; FDA 21CFR §175.105; Gardner 1 pellet; melt index 1.6 (190 C); m.w. 143,000; dens. 0.918 g/cc; acid no. < 0.05; cloud pt. 84 C; *Toxicology:* molten material may produce thermal burns.

Epolene® C-15. [Eastman] Highly branched, med. m.w. polyethylene wax; CAS 9002-88-4; EINECS 200-815-3; nonemulsifiable low-density wax used in hot-melt adhesives and coatings for papers and packaging materials, as paraffin modifiers in slush and cast molding, candles, inks; processing aid for rubber compounding; dispersant for color concs.; FDA 21CFR §175.105; Gardner 1 pellet; m.w. 17,000; dens. 0.906 g/cc; visc. 3900 cps (150 C); soften. pt. (R&B) 102 C; acid no. < 0.05; cloud pt. 75 C; *Toxicology:* molten material may produce thermal burns.

Epolene® C-15P. [Eastman] Highly branched, med. m.w. polyethylene wax; CAS 9002-88-4; EINECS 200-815-3; nonemulsifiable low-density wax used in hot-melt adhesives and coatings for papers and packaging materials, as paraffin modifiers in slush and cast molding, candles, inks; processing aid for rubber compounding; dispersant for color concs.; powd.; *Precaution:* incompat. with oxidizers; powd. material may form explosive dust-air mixts.

Epolene® C-16. [Eastman] Chemically modified polyethylene wax; CAS 9002-88-4; EINECS 200-815-3; nonemulsifiable wax; as hot melt coatings for paper (glossy barrier coatings, readily heat sealable to paper prods., metal foils, and plastic films); in parafin wax coatings and in paraffin-copolymer coatings; good wetting and dispersing props. for highly filled compositions; tolerates high levels of inorganic fillers without drastic increases in melt visc.; dispersing and processing aid for plastics; FDA 21CFR §175.105; Gardner 1 pellet; m.w. 26,000; dens. 0.908 g/cc; visc. 8500 cps (150 C); soften. pt. (R&B) 106 C; sapon. no. 5; cloud pt. 78 C (2% in 130 F paraffin).

Epolene® C-17. [Eastman] Highly branched, med. m.w. polyethylene wax; CAS 9002-88-4; EINECS 200-815-3; wax designed for use with Epolene waxes or blends containing lower m.w. materials; as paraffin wax modifier, e.g., for paper coatings; additive for inks, hot-melt adhesives; base polymer for color concs.; processing aid, mold release, dispersion aid, coupling agent for plastics; FDA 21CFR §175.105; Gardner 1 pellet; m.w. 100,000; melt index 20 (190 C); dens. 0.917 g/cc; acid no. < 0.05; cloud pt. 81 C (2% in 130 F paraffin); *Toxicology:* molten material may produce thermal burns.

Epolene® C-17P. [Eastman] Highly branched, med. m.w. polyethylene wax; CAS 9002-88-4; EINECS 200-815-3; extrusion lubricant for PVC; base polymer for color concs.; processing aid for rubber; dispersing aid, mold release agent; powd.; *Precaution:* incompat. with oxidizers; powd. material may form explosive dust-air mixts.

Epolene® C-18. [Eastman] Chemically modified polyethylene; CAS 9002-88-4; EINECS 200-815-3; nonemulsifiable wax; as hot melt coatings for paper (glossy barrier coatings, readily heat sealable to paper prods., metal foils, and plastic films); in parafin wax coatings and in paraffin-copolymer coatings; good wetting and dispersing props. for highly filled compositions; tolerates high levels of inorganic fillers without drastic increases in melt visc.; dispersing and processing aid for plastics; FDA 21CFR §175.105; Gardner 1 pellet; m.w. 15,000; dens. 0.905 g/cc; visc.: 4000 cps (150 C); sapon. no. 5; soften. pt. (R&B) 102 C.

Epolene® E-10. [Eastman] Oxidized low m.w. polyethylene homopolymer resin; CAS 68441-17-8; emulsifiable wax for water-emulsion for floor polishes, imparting excellent slip resistance, toughness, and durability to polish films; finishing agent for cotton and rayon fabrics; textile softener; processing aid, mold release agent, dispersion aid, coupling agent for plastics; lubricant in clay coatings for paper to reduce dusting during calendering; Gardner 2 pellet, odorless; negligible sol. in water; m.w. 3000; sp.gr. 0.942; visc. 800 cps (125 C); soften. pt. (R&B) 106 C; acid no. 15; flash pt. (COC) > 204 C; *Toxicology:* LD50 (oral, rat) > 6400 mg/kg, (dermal, rabbit) > 2000 mg/kg; sl. skin irritant; molten material may produce thermal burns; *Precaution:* incompat. with oxidizers.

Epolene® E-14. [Eastman] Low m.w. oxidized polyethylene homopolymer resin; CAS 68441-17-8; low-density and low soften. pt. emulsifiable wax, imparts slip resistance to floor polish films; lubricant for rigid and flexible PVC; processing aid for color concs.; Gardner 2 powd.; m.w. 3600; dens. 0.939 g/cc; visc. 250 cps (125 C); soften. pt. (R&B) 104 C; acid no. 16; *Toxicology:* LD50 (oral, rat) > 6400 mg/kg, (dermal, rabbit) > 2000 mg/kg; sl. skin irritant; molten material may produce thermal burns; *Precaution:* incompat. with oxidizers.

Epolene® E-14P. [Eastman] Low m.w. oxidized polyethylene wax; CAS 68441-17-8; extrusion lubricant for rigid PVC; powd.; *Toxicology:* LD50 (oral, rat) > 6400 mg/kg, (dermal, rabbit) > 2000 mg/kg; sl. skin irritant; molten material may produce thermal burns; *Precaution:* incompat. with oxidizers; powd. material may form explosive dust-air mixts.

Epolene® E-15. [Eastman] Low m.w. oxidized polyethylene homopolymer resin; CAS 68441-17-8; low-density and low soften. pt. emulsifiable wax, imparts slip resistance to floor polish films; processing aid, mold release agent, dispersion aid, coupling agent for plastics; lubricant in clay coatings for paper to reduce dusting during calendering; Gardner 2 pellet; m.w. 4200; dens. 0.925 g/cc; visc. 350 cps (125 C); soften. pt. (R&B) 100 C; acid no. 16; *Toxicology:* LD50 (oral, rat) > 6400 mg/kg, (dermal, rabbit) > 2000 mg/kg; sl. skin irritant; molten material may produce thermal burns; *Precaution:* incompat. with oxidizers.

Epolene® E-20. [Eastman] Low m.w. oxidized polyethylene wax; CAS 68441-17-8; offers emulsification props. of a low-dens. wax; used in high-speed, buffable floor polish, textile lubricant/softener, fruit coating applics.; processing aid, mold release agent, dispersion aid, coupling agent for plastics; lubricant in clay coatings for paper to reduce dusting during calendering; Gardner 2 pellet; m.w. 7500; dens. 0.960 g/cc; visc. 1500 cP (125 C); soften. pt. (R&B) 111 C; acid no. 17; *Toxicology:* LD50 (oral, rat) > 6400 mg/kg, (dermal, rabbit) > 2000 mg/kg; sl. skin irritant; molten material may produce thermal burns; *Precaution:* incompat. with oxidizers.

Epolene® E-20P. [Eastman] Low m.w. oxidized polyethylene wax; CAS 68441-17-8; extrusion lubricant for clear, rigid PVC; powd.; *Toxicology:* LD50 (oral, rat) > 6400 mg/kg, (dermal, rabbit) > 2000 mg/kg; sl. skin irritant; molten material may produce thermal burns; *Precaution:* incompat. with oxidizers; powd. material may form explosive dust-air mixts.

Epolene® E-43. [Eastman] Chemically modified PP resin; emulsifiable wax, imparts slip resistance to floor polishes; coupling agent for filled polyolefins; compatibilizer for plastic alloy systems; pigment dispersant and processing aid for single plastic systems, e.g., ABS; Gardner 11 solid; m.w. 9100; dens. 0.934 g/cc; visc. 400 cps (190 C); soften. pt. (R&B) 157 C; acid no. 47.

Epolene® G-3002. [Eastman] Chemically modified polymer of propylene; coupling agent for glass-, talc-, or calcium carbonate-filled PP, HDPE; visc. 15,000 cP (225 C); soften. pt. (Vicat) 133 C; acid no. 60.

Epolene® G-3003. [Eastman] Chemically modified polymer of propylene; coupling agent for glass-, talc-, or calcium carbonate-filled PP, HDPE; melt index 12.7 (230 C); soften. pt. (Vicat) 150 C; acid no. 7.

Epolene® N-10. [Eastman] Low m.w. polyethylene homopolymer resin; CAS 9002-88-4; EINECS 200-815-3; nonemulsifiable wax, easily blended with waxes to improve tens. str., abrasion resistance, and adhesion to fibrous substrates; for paper-coating applics., printing inks; processing aid and pigment dispersant for polyolefin color concs.; FDA 21CFR §175.105; Gardner 1 pellet, odorless; negligible sol. in water; m.w. 10,000; dens. 0.925 g/cc; visc. 1500 cps (125 C); soften. pt. (R&B) 111 C; acid no. < 0.05; cloud pt. 85 C (2% in 130 F paraffin); *Toxicology:* LD50 (oral, rat) > 6400 mg/kg, (dermal, rabbit) > 2000 mg/kg; molten material will produce thermal burns; *Precaution:* incompat. with oxidizers.

Epolene® N-11. [Eastman] Low m.w. polyethylene wax; CAS 9002-88-4; EINECS 200-815-3; similar to Epolene N-10; also used as solid lubricant in corrugated board manufacture; mold release additive and lubricant in rubber processing; extrusion and calendering aid for vinyl; FDA 21CFR §175.105; Gardner 1 pellet, odorless; negliglbe sol. in water; m.w. 6000; dens. 0.921 g/cc; visc. 350 cps (125 C); soften. pt. (R&B) 108 C; acid no. < 0.05; cloud pt. 79 C (2% in 130 F paraffin); *Toxicology:* LD50 (oral, rat) > 6400 mg/kg, (dermal, rabbit) > 2000 mg/kg; molten material will produce thermal burns; *Precaution:* incompat. with oxidizers.

Epolene® N-11P. [Eastman] Low m.w. polyethylene wax; CAS 9002-88-4; EINECS 200-815-3; external lubricant for flexible PVC; dispersing aid for color concs. for plastics; powd.; *Toxicology:* LD50 (oral, rat) > 6400 mg/kg, (dermal, rabbit) > 2000 mg/kg; *Precaution:* incompat. with oxidizers; powd. material may form explosive dust-air mixts.

Epolene® N-14. [Eastman] Low m.w. polyethylene wax; CAS 9002-88-4; EINECS 200-815-3; similar to Epolene N-10 with lower m.w. and density; mold release additive and lubricant in rubber processing; extrusion and calendering aid for vinyl; FDA 21CFR §175.105; Gardner 1 pellets, odorless; negligible sol. in water; m.w. 4000; dens. 0.920 g/cc; visc. 150 cps (125 C); soften. pt. (R&B) 106 C; acid no. < 0.05; cloud pt. 77 C (2% in 130 F paraffin); *Toxicology:* LD50 (oral, rat) > 6400 mg/kg, (dermal, rabbit) > 2000 mg/kg; molten material will produce thermal burns; *Precaution:* incompat. with oxidizers.

Epolene® N-14P. [Eastman] Low m.w. polyethylene wax; CAS 9002-88-4; EINECS 200-815-3; lubricant for flexible and rigid PVC; dispersing aid for color concs. for plastics; powd.; *Toxicology:* LD50 (oral, rat) > 6400 mg/kg, (dermal, rabbit) > 2000 mg/kg; *Precaution:* incompat. with oxidizers; powd. material may form explosive dust-air mixts.

Epolene® N-15. [Eastman] Low m.w. PP homopolymer resin; high melting pt. and hardness wax; modifier for petrol. waxes to increase resistance to blocking, scuffing, and abrasion, component for hot melt adhesives; pigment dispersant for color concs., reprographic toner compds., esp. compat. with PP fiber; FDA 21CFR §175.105; Gardner 1 pellets; m.w. 12,000; dens. 0.902 g/cc; visc. 600 cps (190 C); soften. pt. (R&B) 163 C; acid no. < 0.05; cloud pt. 104 C.

Epolene® N-20. [Eastman] Low m.w. polyethylene homopolymer wax; CAS 9002-88-4; EINECS 200-815-3; nonemulsifiable wax with improved resist. to solvs. and oils, and good hardness; used in cosmetics, hot-melt adhesives, dispersing aids for color concs., cable filling composition, slip additives for printing inks, modifiers for hot-melt highway marking; processing aid, mold release, dispersant for plastics; FDA 21CFR §175.105; Gardner 1 pellets, odorless; negligible sol. in water; m.w. 15,000; dens. 0.930 g/cc; visc. 3725 cps (150 C); soften. pt. (R&B) 119 C; acid no. < 0.05; cloud pt. 86 C; *Toxicology:* LD50 (oral, rat) > 6400 mg/kg, (dermal, rabbit) > 2000 mg/kg; molten material will produce thermal burns; *Precaution:* incompat. with oxidizers.

Epolene® N-21. [Eastman] Low m.w. polyethylene homopolymer wax; CAS 9002-88-4; EINECS 200-815-3; nonemulsifiable wax with improved resist. to solvs. and oils, and good hardness; used in cosmetics, hot-melt adhesives, dispersing aids for color concs., cable filling composition, slip additives for printing inks, modifiers for hot-melt highway marking; processing aid, mold release, dispersant for plastics; FDA 21CFR §175.105; Gardner 1 pellets, odorless; negligible sol. in water; m.w. 6500; dens. 0.950 g/cc; visc. 350 cps (150 C); soften. pt. (R&B) 120 C; acid no. < 0.05; cloud pt. 87 C; *Toxicology:* LD50 (oral, rat) > 6400 mg/kg, (dermal, rabbit) > 2000 mg/kg; molten material will produce thermal burns; *Precaution:* incompat. with oxidizers.

Epolene® N-34. [Eastman] Low m.w. polyethylene wax; CAS 9002-88-4; EINECS 200-815-3; extrusion lubricant for polyolefins; mold release additive and lubricant in rubber processing; extrusion and calendering aid for vinyl; FDA 21CFR §175.105; Gardner 1 pellets, odorless; negligible sol. in water; m.w. 6200; dens. 0.910 g/cc; visc. 450 cps (125 C); soften. pt. (R&B) 103 C; acid no. < 0.05; cloud pt. 69 C (2% in 130 F paraffin); *Toxicology:* LD50 (oral, rat) > 6400 mg/kg, (dermal, rabbit) > 2000 mg/kg; molten material will produce thermal burns; *Precaution:* incompat. with oxidizers.

Epon® Resin 1183. [Shell] Brominated epoxy resin; imparts flame retardance to molding compds., powd. coatings, and laminating and casting systems; Gardner 3 max. color; sol. in ketone, aromatic HC, ether alcohol solvs.; dens. 14.4 lb/gal; visc. A (40% in 2-(2-butoxyethoxy) ethanol); m.p. 97 C; wt./epoxide 675; 42% Br.

Epotuf Hardener 37-605. [Reichhold] Modified polyamine; epoxy hardener; good pot life for elec. potting, encapsulating, and casting, hand lay-up (laminating, tooling), flooring, surfacing, and adhesives; liq.; visc. 700-900 cps.; Unverified

Epotuf Hardener 37-610. [Reichhold] Modified polyamine; epoxy hardener; imparts resiliency; used in elec. potting, encapsulating, and casting, hand lay-up (laminating, tooling), flooring, surfacing, and adhesives; liq.; visc. 200-300 cps.; Unverified

Epotuf Hardener 37-611. [Reichhold] Modified polyamine; epoxy hardener; high flexibility; used in elec. potting, encapsulating, casting, hand lay-up (laminating, tooling), flooring, surfacing, and adhesives; liq.; visc. 3500-4500 cps.; Unverified

Epotuf Hardener 37-612. [Reichhold] Polyamide; epoxy hardener; general purpose hardener, good resiliency and adhesion; used in elec. potting, encapsulating, and casting, flooring, surfacing, coatings, and adhesives; liq.; visc. 10,000-19,000 cps.

Epotuf Hardener 37-614. [Reichhold] Modified polyamine; epoxy hardener; wide use range; used in elec. potting, encapsulating, and casting, hand lay-up (laminating, tooling), flooring, surfacing, coatings, and adhesives; liq.; visc. 3500-5500 cps.

Epotuf Hardener 37-618. [Reichhold] Polyamide; sol'n. of Epotuf Hardener 37-617; liq.; visc. (G-H) V-Z.

Epotuf Hardener 37-620. [Reichhold] Amidoamine; epoxy hardener; good adhesion and resiliency; used in elec. potting, encapsulating, and casting, hand lay-up (laminating, tooling), filament winding, flooring, surfacing, coatings, and adhesives; liq.; visc. 400-700 cps.

Epotuf Hardener 37-621. [Reichhold] Polyamide; epoxy hardener; sol'n. form; used in coatings; liq.; visc. (G-H) W-Z1.

Epotuf Hardener 37-622. [Reichhold] Modified polyamine; epoxy hardener; versatile, general purpose hardener; used in elec. potting, encapsulating, and casting, hand lay-up (laminating, tooling), flooring, surfacing, coatings, and adhesives; liq.; visc. 80-150 cps.

Epotuf Hardener 37-624. [Reichhold] Anhydride; epoxy hardener; used in elec. potting, encapsulating, and casting, and filament winding; liq.; visc. 100-200 cps.; Unverified

Epotuf Hardener 37-625. [Reichhold] Polyamide; epoxy hardener; faster-reacting hardener used in coatings and adhesives; liq.; visc. 25,000-38,000 cps.

Epotuf Hardener 37-640. [Reichhold] Polyamide; epoxy hardener; versatile, with instant resin compatibility; used in elec. potting, encapsulating, and casting, flooring, surfacing, coatings, and adhesives; liq.; visc. 9000-15,000 cps.

Epoxol 7-4. [Am. Chem. Services] Epoxidized soybean oil; CAS 8013-07-8; EINECS 232-391-0; aux. plasticizer, acid scavenger, stabilizer for PVC compds.; food pkg. materials; Co-Pt 70; low odor; misc. with esters, ketones, heptane, ethyl ether, vinyl plasticizers, aromatic and chlorinated hydrocarbons; sp.gr. 0.994; dens. 8.28 lb/gal; visc. 314 cps; pour pt. 25 F; acid no. 0.10; sapon. no. 178.1; ref. index 1.4705; flash pt. 590 F (COC).

Epoxol 9-5. [Am. Chem. Services] Epoxidized linseed oils; stabilizing plasticizer; food pkg. materials; Co-Pt 70; low odor; misc. with esters, ketones, heptane, ethyl ether, vinyl plasticizers, aromatic and chlorinated hydrocarbons; sp.gr. 1.03; dens. 8.58 lb/gal; visc. 619 cps; pour pt. 30 F; acid no. 0.12; sapon. no. 172; ref. index 1.4715; flash pt. (COC) 595 F.

Equex SP. [Procter & Gamble] Sodium lauryl sulfate; CAS 151-21-3; anionic; polymerization grade emulsifier; pale yel. clear liq.; clean, pleasant odor; dens. 8.7 lb/gal; sp.gr. 1.04; visc. 90 cps (80 F); cloud pt. 55 F; pH 8 (10%); 29.8% act.; Unverified

Escacure® KB1. [Sartomer] Benzyldimethyl ketal; photoinitiator for uv-cured adhesives (pressure sensitive, structural), coatings (metal, optical, paper, plastic, PVC floor, textile, wood), photopolymers, electronics (photoresists, conformal, encapsulants, solder masks); inks (flexo, gravure, litho, offset, screen); optimum absorp. 250-350 nm; fast cure response; good through-cure; wh. cryst. powd.; m.w. 256; sp.gr. 1.176 (20 C); m.p. 63-66 C.

Escacure® KB60. [Sartomer] 60% sol'n. of benzyldimethyl ketal; photoinitiator for uv-cured adhesives (pressure sensitive, structural), coatings (metal, optical, paper, plastic, PVC floor, textile, wood), photopolymers, electronics (photoresists, conformal, encapsulants, solder masks); inks (flexo, gravure, litho, offset, screen); optimum absorp. 250-350 nm; fast cure response; good through-cure; yel. clear liq.; sp.gr. 1.200 (20 C); visc. 100 cps (20 C); nonflamm.

Escoflex A-122. [Plastics & Chem.] Phenol; antioxidant for ABS, cellulosics, nylon, polyester, PE, PP, PS, PVC, EPDM, olefin copolymers.

Escor® ATX-310. [Exxon] Ethylene-methyl acrylate-acrylic acid terpolymer; impact modifier for engineering thermoplastics; compatibilizer for polar and nonpolar polymers; adhesive film for industrial laminations or bonding elastomers to fabrics and metals; for extrusion coating/laminating; in heat-activated adhesive sealant; FDA 21CFR §175.105, 176.180; melt index 6.0 g/10 min; dens. 0.941 g/cc; m.p. 89 C; acid no. 45; tens. str. 3.3 MPa (yield, 14 MPa (break); tens. elong. > 800% (break); flex. mod. 66 MPa; tens. impact 530 kJ/m^2; hardness (Shore A) 90; soften. pt. (Vicat) 86 C.

Escor® ATX-320. [Exxon] Ethylene-methyl acrylate-acrylic acid terpolymer; adhesion modifier for rubber compds., TPOs, adhesive sealants; improves adhesion to metals, syn. yarns, paints; impact modifier for engineering thermoplastics; compatibilizer for dissimilar polymers; extrusion laminates; FDA 21CFR §175.105, 176.180; melt index 5.0 g/10 min; dens. 0.950 g/cc; m.p. 69 C; acid no. 45; tens. str. 12 MPa (yield, 12 MPa (break); tens. elong. > 800% (break); flex. mod. 25 MPa; tens. impact 760 kJ/m^2; hardness (Shore A) 83; soften. pt. (Vicat) 66 C.

Escor® ATX-325. [Exxon] Ethylene-methyl acrylate-acrylic acid terpolymer; impact modifier for engineering thermoplastics; compatibilizer for dissimilar polymers; adhesion promoter for molded polyolefin parts; improrves toughness, flexibility, adhesion chars.; FDA 21CFR §175.105, 176.180; melt index 20 g/10 min; dens. 0.950 g/cc; m.p. 67 C; acid no. 45; tens. str. 7.8 MPa (yield, 7.8 MPa (break); tens. elong. > 800% (break); flex. mod. 21 MPa; tens. impact 590 kJ/m^2; hardness (Shore A) 80; soften. pt. (Vicat) 60 C.

Esperal® 115RG. [Witco/PAG] Dicumyl peroxide; CAS 80-43-3; for med.-temp. applics. as a polymerization and crosslinking agent; gran.; 99% min. peroxide; 5.9% act. oxygen.

Esperal® 120. [Witco/PAG] 2,5-Dimethyl-2,5-di(t-butylperoxy)hexane; CAS 78-63-7; initiator in crosslinking of polymers and elastomers; melt flow modifier for PP; vulcanization agent for silicone rubber, fluoro elastomers, EPDM, EVA; crosslinking agent for LDPE; colorless liq.; m.w. 290.45; sp.gr. 0.88 (20 C); f.p. < 8 C; flash pt. (Seta CC) 43 C; 90% min. purity, 9.92% min. act. oxygen; *Precaution:* incompat. with strong acids or bases, oxidizing or reducing agents, rust, promoters, accelerators; once ignited, will burn intensely and rapidly; *Storage:* store in original container in cool (< 38 C), place away from heat sources, flame, direct sunlight; Custom

Esperal® 230. [Witco/PAG] 2,5-Dimethyl-2,5-di(t-butylperoxy)hexyne-3; CAS 1068-27-5; initiator in crosslinking of polyolefins and other polymers; crosslinking agent for HDPE and LLDPE at temps. above 180 C; pale yel. liq.; m.w. 286.4; sp.gr. 0.89 (20 C); f.p. < 8 C; flash pt. (Seta CC) 65 C; 90% min. act., 10.05% min. act. oxygen; *Precaution:* incompat. with reducing or oxidizing agents, strong acids or bases, promoters, accelerators; *Storage:* store in original containers in cool (< 38 C) area away from heat sources, flame, direct sunlight; Custom

Espercarb® 438M-60. [Witco/PAG] 60% sol'n. di-s-butyl peroxydicarbonate in odorless min. spirits; CAS 19910-65-7; initiator, crosslinking agent for polymers; sol'n.; 60% conc.; 4.10% act. oxygen; *Storage:* requires refrigerated shipment and storage.

Espercarb® 840. [Witco/PAG] Di-2-ethylhexyl peroxydicarbonate; CAS 16111-62-9; initiator; liq.; 98% min. peroxide; 4.5% act. oxygen; *Storage:* requires refrigerated shipment and storage.

Espercarb® 840M. [Witco/PAG] 75% sol'n. of di-2-ethylhexyl peroxydicarbonate in odorless min. spirits; CAS 16111-62-9; initiator; sol'n.; 75% conc.; 3.5% act. oxygen; *Storage:* requires refrigerated shipment and storage.

Espercarb® 840M-40. [Witco/PAG] 40% sol'n. of di-2-ethylhexyl peroxydicarbonate in odorless min. spirits; CAS 16111-62-9; initiator; sol'n.; 40% conc.; 1.85% act. oxygen; *Storage:* requires refrigerated shipment and storage.

Espercarb® 840M-70. [Witco/PAG] 70% sol'n. of di-2-ethylhexyl peroxydicarbonate in odorless min. spirits; CAS 16111-62-9; initiator; sol'n.; 70% conc.; 3.2% act. oxygen; *Storage:* requires refrigerated shipment and storage.

Esperfoam® FR. [Witco/PAG] MEK peroxide and acetylacetone peroxide; initiator for rapid cures of polyester resins; DOT org. peroxide label is not required; sol'n.; 9.0% act. oxygen.

Esperox® 10. [Witco/PAG] t-Butyl peroxybenzoate; CAS 614-45-9; initiator for polymerization of ethylene and styrene, and for high temp. molding of polyesters; liq.; 98% min. peroxide; 8.1% act. oxygen.

Esperox® 12MD. [Witco/PAG] 50% sol'n. of t-butyl peroxy acetate in odorless min. spirits; CAS 107-71-1; initiator for polymerization of styrene, ethylene, acrylates, etc.; sol'n.; 50% conc.

Esperox® 13M. [Witco/PAG] 75% sol'n. of t-butyl peroxycrotonate in odorless min. spirits; CAS 23474-91-1; initiator for polymerization applics.; sol'n.; 7.6% act. oxygen.

Esperox® 28. [Witco/PAG] t-Butyl peroxy 2-ethyl hexanoate; CAS 3006-82-4; initiator recommended for polymerization of ethylene plus use in med. temp. molding of polyester resin systems; liq.; 98% min. peroxide; 7.2% act. oxygen.

Esperox® 28MD. [Witco/PAG] t-Butyl peroxy 2-ethyl hexanoate in min. spirit diluent; CAS 3006-82-4; initiator used in high pressure polymerization of ethylene; liq.; 50% conc.; 3.7% act. oxygen.

Esperox® 28PD. [Witco/PAG] t-Butyl peroxy 2-ethyl hexanoate in butyl benzyl phthalate; CAS 3006-82-4; initiator used in high speed, heated cures of polyester resin systems; liq.; 50% conc.; 3.7% act. oxygen.

Esperox® 31M. [Witco/PAG] t-Butyl peroxypivalate in min. spirit diluent; CAS 927-07-1; initiator used in polymerization of ethylenically unsat. monomers; liq.; 75% conc.; 6.8% act. oxygen.

Esperox® 33M. [Witco/PAG] t-Butyl peroxyneodecanoate in min. spirit diluent; CAS 26748-41-4; efficient and reactive initiator for polymerization of ethylenically unsat. monomers; liq.; 75% conc.; 4.9% act. oxygen.

Esperox® 41-25. [Witco/PAG] t-Butyl peroxymaleic acid; CAS 1931-62-0; initiator for polymerization of ethylenically unsat. resins; paste; 25% conc.; 2.2% act. oxygen.

Esperox® 41-25A. [Witco/PAG] 25% t-Butyl peroxy maleate disp.; CAS 1931-62-0; initiator for polymerization of various ethylenically unsat. resins and monomers; paste; 25% conc.

Esperox® 497M. [Witco/PAG] t-Butyl peroxy 2-methyl benzoate sol'n. in min. spirit diluent; CAS 22313-62-8; initiator for polymerization of ethylene and styrene, high temp. molding of polyester resin systems, and vulcanization of silicon rubber; liq.; 75% conc.; 5.8% act. oxygen.

Esperox® 545M. [Witco/PAG] t-Amyl peroxyneodecanoate in min. spirit diluent; CAS 68299-16-1; efficient and reactive initiator for polymerization of ethylenically unsat. monomer; liq.; 75% conc.; 4.6% act. oxygen.

Esperox® 551M. [Witco/PAG] t-Amyl peroxypivalate in min. spirit diluent; CAS 29240-17-3; initiator used in polymerization of ethylenically unsat. monomers; liq.; 75% conc.; 6.4% act. oxygen.

Esperox® 570. [Witco/PAG] 95% t-Amyl peroxy 2-ethylhexanoate; CAS 686-31-7; initiator for polymerization of ethylene, acrylates, and unsat. polyester resins; liq.; 95% min. peroxide.

Esperox® 570P. [Witco/PAG] 75% sol'n. of t-amyl peroxy 2-ethyl hexanoate in butyl benzyl phthalate; CAS 686-31-7; initiator; replacement for benzoyl peroxide, t-butyl peroctoate, and diperoctoates; liq.; 75% conc.; 5.2% act. oxygen.

Esperox® 740M. [Witco/PAG] 75% sol'n. of cumyl peroxyneoheptanoate in odorless min. spirits; CAS 130097-36-8; initiator for polymerization of vinyl chloride.; sol'n.; 75% conc.

Esperox® 747M. [Witco/PAG] 75% sol'n. of t-amyl peroxyneoheptanoate in odorless min. spirits; initiator for polymerization of ethylenically unsat. monomers.; sol'n.; 75% conc.

Esperox® 750M. [Witco/PAG] 75% sol'n. of t-butyl peroxyneoheptanoate in odorless min. spirits; CAS 26748-38-9; initiator for polymerization of ethylenically unsat. monomers; sol'n.; 75% conc.

Esperox® 939M. [Witco/PAG] Cumyl peroxyneodecanoate in min. spirit diluent; CAS 26748-47-0; efficient and reactive initiator for polymerization of vinyl chloride; liq.; 75% conc.; 3.9% act. oxygen.

Esperox® 5100. [Witco/PAG] t-Amyl peroxybenzoate; CAS 4511-39-1; initiator for polymerization of ethylene, styrene, acrylates, and curing of unsat. polyester resins; clear to sl. yel. liq.; m.w. 208.25; sp.gr. 1.01; f.p. $\approx$-4 C; flash pt. (Seta CC) 49 C; 95% purity, 7.3% act. oxygen; *Precaution:* incompat. with promoters, accelerators, strong acids or bases; *Storage:* store in original containers in cool (< 38 C) area away from heat sources, flame, direct sunlight.

Esperox® C-496. [Witco/PAG] t-Butyl peroxy-2-ethylhexyl carbonate; CAS 34443-12-4; initiator for polymerization of vinyl monomers, styrene, acrylates, and unsat. polyester resins; crosslinking and curing agent for thermoplastics, silicones, elastomers; clear liq.; m.w. 246.34; sp.gr. 0.93 (20 C); f.p. < -75 C; flash pt. > 93 C; 97% purity, 6.3% act. oxygen; *Precaution:* incompat. with promoters, accelerators, strong acids or bases; *Storage:* store in original containers in cool (< 38 C) area away from heat sources, flame, direct sunlight.

Estabex ABF. [Akzo Nobel] Proprietary; biologically active heat stabilizer for flexbile and semirigid PVC and PVC copolymers, plastisols, organosols, foams, tile, PU.

Estabex® 138-A. [Akzo Nobel] Epoxidized soybean oil compd.; CAS 8013-07-8; EINECS 232-391-0; plasticizer for PVC homopolymer and copolymer resin formulations; FDA food grade applic; Gardner 2 clear liq.; sol. in esters, ethers, ketones, higher alcohols, aromatic and aliphatic hydrocarbons; sp.gr. 0.982; dens. 8.18 lb/gal; Gardner-Holdt visc. G.

Estabex® 2307. [Akzo Nobel] Epoxidized soya bean oil; CAS #8013-07-8; CAS 8013-07-8; EINECS 232-391-0; plasticizer/stabilizer for plastisol applics., organosols, surf. coatings and inks, extruded prods., rigid PVC, chlorinated paraffins, halogenated rubbers, ethyl cellulose, PVC/PVA emulsions; processing aid for rigid PVC; pigment wetter; corrosion inhibitor for agrochemicals; approved for food contact and medical applics.; slightly yel. liq.; dens. 0.99 g/ml; visc. 550 mPa•s; flash pt. 295 C; acid no. 0.8; iodine no. 3; ref. index 1.473; 6.4% oxygen content.

Estabex® 2307 DEOD. [Akzo Nobel] Epoxidized soya bean oil; CAS 8013-07-8; EINECS 232-391-0; deodorized version of Estabex 2307 for sensitive food pkg. applics.; slightly yel. liq.; dens. 0.99 g/ml; visc. 550 mPa•s; acid no. 0.7; iodine no. 2; ref. index 1.473; 6.4% oxygen content.

Estabex® 2381. [Akzo Nobel] Monooctyl ester; CAS 141-38-8; plasticizer/stabilizer for plastisol applics., organosols, surf. coatings and inks, extruded prods.; sec. stabilizer for rigid and flexible PVC compds.; also for pigment wetting; yel. clear liq.; dens. 900 kg/m3; visc. 50 mPa•s; acid no. 0.8; iodine no. 3; ref. index 1.457; 3.8% oxygen content.

Estaflex® ATC. [Akzo Nobel] Acetyl tributyl citrate; CAS 77-90-7; EINECS 201-067-0; primary plasticizer for plastics incl. PP, PS, PVAc, PVB, PVC, vinyl acetate copolymers, chlorinated rubber, ethyl cellulose, nitrocellulose; suitable for coatings in contact with foodstuffs; clear liq., odorless; dens. 1050 kg/m^3; visc. 40 mPa•s; acid no. 0.1; sapon. no. 555; ref. index 1.441.

Estol 1406. [Unichema] Isopropyl oleate; CAS 112-

11-8; EINECS 203-935-4; for plastics, lubricants, cosmetic emollients.

Estol 1407. [Unichema Int'l.] Glyceryl oleate; antistat for polyethylene and polypropylene; antifog agent for LDPE and PP; for lubricants, cosmetic emollients.

Estol 1427. [Unichema] Trimethylolpropane trioleate; for plastics, lubricants, cosmetic emollients.

Estol 1445. [Unichema] Pentaerythrityl tetraoleate; CAS 19321-40-5; EINECS 242-960-5; for plastics, lubricants, cosmetic emollients.

Estol 1461. [Unichema] Glyceryl monostearate SE; emulsifier for polymers; Lovibond 7.0R/1.5R (1 in.) color; sp.gr. 0.92 (90/20 C); visc. 36.1 cSt (80 C); m.p. 60 C; acid no. 3 max.; iodine no. 3 max.; sapon. no. 156-170; flash pt. (COC) 225 C; fire pt. 270 C; 36% min. monoglycerides.

Estol 1467. [Unichema] Glyceryl stearate; emulsifier for polymers, cosmetics; Japan approval for cosmetics; Lovibond 3.0Y/1.0R (1 in.) color; m.p. 70 C; acid no. 3 max.; iodine no. 3 max.; sapon. no. 162; 90% min. monoglycerides.

Estol 1474. [Unichema] Glyceryl stearate; ester for polymers, personal care prods.; Japan approval for cosmetics; Lovibond 3.0Y/1.0R (1 in.) color; sp.gr. 0.92 (80/20 C); visc. 22.2 cSt (80 C); m.p. 60 C; acid no. 3 max.; iodine no. 3 max.; sapon. no. 168-184; flash pt. (COC) 225 C; fire pt. 270 C.

Estol 1476. [Unichema Int'l.] Isobutyl stearate; CAS 646-13-9; EINECS 211-466-1; lubricant for PVC processing; cosmetic emollients.

Estol 1481. [Unichema] Cetostearyl stearate; CAS 93820-97-4; lubricant for PVC processing; cosmetic emollients; Lovibond 20Y/3.0R ($5^1/_4$ in.) color; sp.gr. 0.82 (80/20 C); visc. 4.9 mPa•s (80 C); solid. pt. 50-56 C; acid no. 2 max.; iodine no. 1 max.; sapon. no. 105-115; flash pt. (COC) 250 C.

Estol 1502. [Unichema] Methyl laurate; CAS 111-82-0; EINECS 203-911-3; for plastics, lubricants, cosmetic emollients.

Estol 1574. [Unichema Int'l.] Ethylene glycol diacetate; CAS 111-55-7; EINECS 203-881-1; for plastics, lubricants, cosmetic emollients.

Estol 1579. [Unichema] Triacetin; CAS 102-76-1; EINECS 203-051-9; for plastics, lubricants, cosmetic emollients.

Estol 1593. [Unichema Int'l.] Triethylene glycol diacetate; for plastics, lubricants, cosmetic emollients.

Estol GTCC 1527. [Unichema] Glyceryl tricaprylate/caprate; for plastics, lubricants, cosmetic emollients; APHA 50 max. color; visc. 24 mPa•s; acid no. 0.1 max.; iodine no. 1 max.; sapon. no. 335-360; hyd. no. 5 max.; cloud pt. -12 C.

Estol IPM 1514. [Unichema] Isopropyl myristate; CAS 110-27-0; EINECS 203-751-4; for plastics, lubricants, cosmetic emollients; APHA 30 max. color; acid no. 0.5 max.; iodine no. 1 max.; sapon. no. 206-211; cloud pt. 0 C; 92% min. purity.

Estol IPP 1517. [Unichema] Isopropyl palmitate; CAS 142-91-6; EINECS 205-571-1; for plastics, lubricants, cosmetic emollients; APHA 20 max. color; acid no. 0.5 max.; iodine no. 1 max.; sapon. no. 185-191; cloud pt. 12 C; 90% min. purity.

Ethacure® 100. [Albemarle] Diethyl toluene diamine; high-performance curing agent for epoxy resins; used in filament winding, elec. encapsulation, prepregs, tooling, potting and casting, laminating, coating, molding, and adhesive applics.; lt. red, clear, low-visc. liq.; b.p. 308 C; dens. 1.022 g/ml (20 C); visc. 326 cs (20 C); pour pt. -9 C; equiv. wt. 44.6 g/eq; flash pt.(TCC) > 135 C.

Ethacure® 300. [Albemarle] Curing agent for polyurethane and epoxy resins; sp.gr. 1.208 (20 C); visc. 691 cps (20 C); pour pt. 4 C; flash pt. (TCC) 176 C.

Ethanox® 330. [Albemarle] 1,3,5-Trimethyl-2,4,6-tris (3,5-di-tert-butyl-4-hydroxybenzyl) benzene; CAS 1709-70-2; noncoloring, nonstaining antioxidant and stabilizer for plastic, resin, rubber, wax, indirect food contact, plastic food bottles and wrap, bottle caps, slit film for carpet fibers; nonextractable, low volatility; FDA 21CFR §178.2010; EEC and BGA acceptance; wh. cryst. powd., odorless, tasteless; sol. in methylene chloride, toluene, benzene, acetone, styrene monomer; insol. in water; m.w. 775.2; bulk dens. 33 lb/ft^3; m.p. 244 C; *Toxicology:* LD50 (acute oral) > 15 g/kg; relatively harmless orally; pract. nontoxic by skin absorp.

Ethanox® 398. [Albemarle] 2,2′-Ethylidenebis (4,6-di-t-butylphenyl) fluorophosphonite; CAS 118337-09-0; antioxidant for PP, LLDPE; provides exc. processing stabilization and color control, hydrolytic and thermal stability; nonhygroscopic; wh. cryst. powd.; m.w. 486.7; sol. (g/100 g): 18 g in xylene, 14.7 g in cyclohexane, 6.1 g in hexane; m.p. 200 C; 6.4% phosphorus.

Ethanox® 702. [Albemarle] 4,4′-Methylenebis(2,6-di-t-butylphenol); CAS 118-82-1; antioxidant for natural and syn. elastomers, polyolefin plastics, resins, adhesives, petrol. oil, and waxes; FDA 21CFR §175.105, 178.2010; wh. to lt. straw cryst. powd.; sol. in toluene, petrol. ether, ethyl alcohol; insol. in water; m.w. 424.7; bulk dens. 33 lb/ft^3; m.p. 154 C; b.p. 250 C (10 mm); flash pt. (COC) 400 F; *Toxicology:* LD50 (oral, rat) > 24 g/kg, (dermal, rabbit) > 1.5 g/kg; pract. nontoxic orally, very low dermal toxicity; nonirritating to eyes and skin; *Precaution:* severe dust explosion hazard.

Ethanox® 703. [Albemarle] 2,6-Di-t-butyl-N,N-dimethylamino-p-cresol; CAS 88-27-7; antioxidant for natural and syn. elastomers, polyolefin plastics, resins, adhesives, petrol. oil, and waxes; lt. yel. cryst. powd.; sol. in ethyl alcohol, toluene; insol. in water; m.w. 263.4; m.p. 94 C; b.p. 179 C (40 mm); flash pt. (COC) < 93 C; *Toxicology:* LD50 (oral, rat) 1.03 g/kg (sl. toxic), (dermal, rabbit) > 4 g/kg; nonirritating to skin; severe, but reversible eye irritant.

Ethoduoquad® T/15-50. [Akzo Nobel] N,N,N′,N′,N′-Penta(2-hydroxyethyl)-N-tallowalkyl-1,3-propane diammonium diacetate, aq. IPA; industrial surfactant for agric., textiles, protective coatings, inks, pigment dispersions, acid pickling baths, metalworking, electroplating, plastics mfg.; EPA listed;

Gardner 16 max. liq.; flash pt. (PMCC) 18 C; pH 6-9; 48-52% solids in aq. IPA.

Ethomeen® 18/60. [Akzo Nobel] PEG-50 stearamine; CAS 26635-92-7; cationic; emulsifier, dispersant used in textile processing, cosmetics; prevents premature coagulation of latex rubber; wh. solid; sol. in acetone, IPA, CCl_4, benzene, water; sp.gr. 1.12; flash pt. (COC) 579 F; surf. tens. 49 dynes/cm (0.1%).

Ethoquad® C/12 Nitrate. [Akzo Nobel] PEG-2 cocomethyl ammonium nitrate, IPA; CAS 71487-00-8; cationic; industrial surfactant for cosmetics, agric., textiles, protective coatings, inks, pigment dispersions, acid pickling baths, metalworking, electroplating, plastics mfg.; Gardner 8 max. liq.; sol. in water, alcohols, acetone, benzene, CCl_4, hexylene glycol; sp.gr. 0.975 (20 C); pH 6-7.8; flash pt. (PMCC) 20 C; 59-62% quat. in IPA.

Ethoquad® CB/12. [Akzo Nobel] PEG-2 cocobenzonium chloride, IPA; CAS 61789-68-2; cationic; industrial surfactant for cosmetics, agric., textiles, protective coatings, inks, pigment dispersions, acid pickling baths, metalworking, electroplating, plastics mfg.; Gardner 12 max. liq.; sol. in water, alcohols, acetone, benzene, CCl_4, hexylene glycol; sp.gr. 0.970; flash pt. (PMCC) 26 C; pH 6-9; 73-77% solids in IPA.

Ethoquad® T/12. [Akzo Nobel] PEG-2 tallowalkyl methyl ammonium chloride, ethanol; CAS 67784-77-4; cationic; industrial surfactant for cosmetics, agric., textiles, protective coatings, inks, pigment dispersions, acid pickling baths, metalworking, electroplating, plastics mfg.; Gardner 7 max. liq. to paste; sol. in water, alcohols, acetone, benzene, CCl_4, hexylene glycol; pH 7.0-9.0; flash (Seta) 23 C; 74% quat. in ethanol.

Ethoquad® T/13-50. [Akzo Nobel] PEG-3 tallow alkyl ammonium acetate, aq. IPA; industrial surfactant for cosmetics, agric., textiles, protective coatings, inks, pigment dispersions, acid pickling baths, metalworking, electroplating, plastics mfg.; EPA listed; Gardner 8 max. liq.; sol. in water, alcohols, acetone, benzene, CCl_4, hexylene glycol; sp.gr. 0.952 (20 C); flash pt. (PMCC) 18 C; 48-52% solids in aq. IPA.

Ethosperse® LA-4. [Lonza] Laureth-4; CAS 9002-92-0; EINECS 226-097-1; nonionic; emulsifier for cosmetic, pharmaceutical and industrial uses (hair and skin care prods., antiperspirants, household waxes, polishes, and cleaners, silicone-based lubricants); internal antistat for PE, PS; water-wh. liq.; sol. in methanol, ethanol, acetone, toluol, min. oil, misc. with water, veg. oils; sp.gr. 0.95; visc. 30 cps; HLB 10.0; acid no. 0.3 max.; hyd. no. 145-160; 100% act.

Ethylan® A4. [Harcros UK] PEG-8 oleate; CAS 9004-96-0; nonionic; emulsifier, antifoam agent, dispersing agent for industrial uses; plastics antistat; lt. amber liq., mild fatty odor; disp. in water; sp.gr. 0.963; visc. 120 cs; HLB 10.3; flash pt. (COC) > 150 C; pH 7.0 (1% aq.); 100% act.

Ethylan® A6. [Harcros UK] PEG-12 oleate; CAS 9004-96-0; nonionic; dispersing agent for industrial uses; emulsifier; antistat for plastics; lt. amber liq., mild fatty odor; sol. in water; sp.gr. 1.037; visc. 340 cs; HLB 12.3; cloud pt. 42 C (1% aq.); flash pt. (COC) > 175 C; pH 7.0 (1% aq.); 100% act.

Ethylan® BCD42. [Harcros UK] Modified phenol ethoxylate; nonionic; emulsifier for emulsion polymerization; yel. liq.; HLB 12.4; cloud pt. 55 C (1% aq.); surf. tens. 45 dynes/cm (0.1%); 100% act.

Ethylan® BV. [Harcros UK] Nonoxynol-14; CAS 9016-45-9; nonionic; foam stabilizer and booster, solubilizer, emulsifier used in pesticides, perfumes, emulsion polymerization; EPA approved; wh. soft paste, negligible odor; water sol.; sp.gr. 1.083; visc. 380 cs; HLB 14.5; cloud pt. 90 C (1% aq.); flash pt. > (COC) > 200 C; pour pt. 20 C; pH 7.0 (1% aq.); surf. tens. 37 dynes/cm (0.1%); 100% act.

Ethylan® CD1210. [Harcros UK] C12 fatty alcohol ethoxylate; nonionic; emulsifier, wetting agent, detergent, foaming agent; textile scouring agent; coemulsifier for aromatic solvs., waxes, toxicants; detergent sanitizers; essential oil solubilizer; emulsion polymerization; biodeg.; colorless clear liq., char. odor; sol. in water; sp.gr. 1.030; visc. 165 cs; HLB 14.1; pour pt. < 0 C; cloud pt. 82 C (1% aq.); flash pt. (PMCC) > 95 C; pH 7.0 (1% aq.); surf. tens. 32 dynes/cm (0.1%); 90% act.

Ethylan® CD1260. [Harcros UK] C12 fatty alcohol ethoxylate; nonionic; emulsifier, wetting agent, detergent; emulsion polymerization; wh. flake; HLB 18.6; cloud pt. > 100 C (1% aq.); surf. tens. 50 dynes/cm (0.1%); 100% conc.

Ethylan® CD9112. [Harcros UK] Syn. lower fraction primary alcohol EO condensate (12 EO); nonionic; wetting agent and detergent; scouring and dye leveling agent; coemulsifier for aromatic solvs., waxes, toxicants; detergent sanitizers; emulsion polymerization; essential oil solubilizer; wh. waxy solid; char. fatty alcohol odor; water sol.; sp.gr. 1.029 (40 C); visc. 53 cs (40 C); HLB 15.5; pour pt. 23 C; cloud pt. 100 C (1% aq.); flash pt. (COC) > 200 C; pH 7.0 (1% aq.); surf. tens. 35 dynes/cm (0.1%); 100% act.

Ethylan® D253. [Harcros UK] Syn. primary alcohol ethoxylate (3 EO); nonionic; o/w and w/o emulsifier for min. oils and waxes, visc. depressant for PVC plastisols; intermediate for sulfation; colorless clear liq.; faint odor; oil sol.; sp.gr. 0.920; visc. 35 cs; HLB 7.8; pour pt. 3 C; flash pt. (COC) 168 C; pH 6-8 (1% aq.); 100% act.

Ethylan® DP. [Harcros UK] Nonoxynol-12; CAS 9016-45-9; nonionic; foam stabilizer and booster, emulsifier, solubilizer for essential oils, perfumes; in liq. detergents, pesticides, emulsion polymerization; slight hazy visc. liq.; negligible odor; water sol.; sp.gr. 1.068; visc. 400 cs; HLB 14.0; pour pt. 13 C; cloud pt. 80 C (1% aq.); flash pt. (COC) > 200 C; pH 7.0 (1% aq.); surf. tens. 34 dynes/cm (0.1%); 100% act.

Ethylan® GEL2. [Harcros UK] Polysorbate 20; CAS 9005-64-5; nonionic; w/o emulsifier, solubilizer esp. with sorbitan esters; used in cosmetics, agric., perfumes, fiber and textile lubricants, textile antistats, polymer additives, suspension and emulsion polymerization; amber clear liq., mild odor; sol. in water, alcohols, hydrocarbons; sp.gr. 1.10; visc. 350 cs; ; HLB 16.5; pour pt. -10 C; flash pt. (PMCC) > 150 C; 97% act.

Ethylan® GEO8. [Harcros UK] Polysorbate 80; CAS 9005-65-6; nonionic; emulsifier for cosmetics, pharmaceuticals, agrochem. formulations, textile lubricants, plastic additives, emulsion and suspension polymerization; solubilizer for perfume, flavors, essential oils; amber clear liq., mild fatty odor; sol. in water, alcohols, hydrocarbons; sp.gr. 1.10; visc. 720 cs; HLB 15.0; pour pt. -20 C; flash pt. (PMCC) > 150 C; 97% act.

Ethylan® GEO81. [Harcros UK] Polysorbate 81; CAS 9005-65-6; nonionic; emulsifier for cosmetics, agric., plastic additives, textile fiber lubricants and softeners, suspension and emulsion polymerization; solubilizer for perfume, flavors, essential oils; amber clear liq., mild fatty odor; sol. in alcohols, hydrocarbons; disp. in water; sp.gr. 1.15; visc. 465 cs; HLB 10.0; pour pt. -10 C; flash pt. (PMCC) > 150 C; 100% act.

Ethylan® GEP4. [Harcros UK] Polysorbate 40; CAS 9005-66-7; nonionic; w/o emulsifier, solubilizer esp. with sorbitan esters; used in cosmetics, agric., perfumes, fiber and textile lubricants, textile antistats, polymer additives; liq.; HLB 15.5; 100% conc.; Unverified

Ethylan® GES6. [Harcros UK] Polysorbate 60; CAS 9005-67-8; nonionic; general purpose, low toxicity emulsifier for cosmetics and agrochem.; textile lubricant; plastics additive; emulsion and suspension polymerization; solubilizer for perfume, flavors, essential oils; pale brn. liq./paste; sol. in water, alcohols, hydrocarbons; sp.gr. 1.07; visc. 190 cs (40 C); HLB 15.0; pour pt. 22 C; flash pt. (PMCC) > 150 C; 97% act.

Ethylan® GL20. [Harcros UK] Sorbitan laurate; CAS 1338-39-2; nonionic; emulsifier for cosmetics, pharmaceuticals, agric., plastic antifog, textile fiber lubricant/softener, suspension and emulsion polymerization; amber visc. liq., mild odor; sol. in alcohols, hydrocarbons, natural and paraffinic oils; sp.gr. 1.04; visc. 5250 cs; HLB 8.0; pour pt. 15 C; flash pt. (PMCC) > 150 C; 100% act.

Ethylan® GLE-21. [Harcros UK] Polysorbate 21; CAS 9005-64-5; nonionic; w/o emulsifier, solubilizer esp. with sorbitan esters; used in cosmetics, agric., perfumes, fiber and textile lubricants, textile antistats, polymer additives; liq.; HLB 13.3; 100% conc.; Unverified

Ethylan® GMF. [Harcros UK] Alkylphenol ethoxylate; nonionic; wetting agent, emulsifier for hydrocarbon solv., emulsion polymerization; straw clear liq.; faint odor; water sol.; sp.gr. 1.050; visc. 330 cs; HLB 12.5; cloud pt. 47 C (1% aq.); flash pt. > 350 C (COC); pour pt. 7 C; pH 6-7.5 (1% aq.); 100% act.; Unverified

Ethylan® GO80. [Harcros UK] Sorbitan oleate; CAS 1338-43-8; EINECS 215-665-4; nonionic; emulsifier for cosmetics, pharmaceuticals, agric., plastic antifog, textile fiber lubricant/softener, suspension and emulsion polymerization; amber visc. liq., mild fatty odor; sol. in alcohols, hydrocarbons, natural and paraffinic oils; sp.gr. 1.00; visc. 1100 cs; HLB 4.3; pour pt. -20 C; flash pt. (PMCC) -5 C; 100% act.

Ethylan® GOE-21. [Harcros UK] Polysorbate 81; CAS 9005-65-6; nonionic; w/o emulsifier, solubilizer esp. with sorbitan esters; used in cosmetics, agric., perfumes, fiber and textile lubricants, textile antistats, polymer additives; liq.; HLB 10.0; 100% conc.; Unverified

Ethylan® GPS85. [Harcros UK] Polysorbate 85; CAS 9005-70-3; nonionic; emulsifier for cosmetics, agric., plastic additives, textile fiber lubricants and softeners, suspension and emulsion polymerization; solubilizer for perfume, flavors, essential oils; amber clear liq., mild fatty odor; sol. in alcohols, hydrocarbons; disp. in water; sp.gr. 1.15; visc. 270 cs; HLB 11.0; pour pt. -20 C; flash pt. (PMCC) > 150 C; 100% act.

Ethylan® GS60. [Harcros UK] Sorbitan stearate; CAS 1338-41-6; EINECS 215-664-9; nonionic; emulsifier for cosmetics, pharmaceuticals, agric., plastic antifog, textile fiber lubricant/softener, suspension and emulsion polymerization; tan waxy solid, mild odor; sol. in alcohols, hydrocarbons, natural and paraffinic oils; sp.gr. 0.98; HLB 4.7; pour pt. 50 C; flash pt. (PMCC) > 150 C; 100% act.

Ethylan® GT85. [Harcros UK] Sorbitan trioleate; CAS 26266-58-0; EINECS 247-569-3; nonionic; emulsifier for cosmetics, pharmaceuticals, agric., plastic antifog, textile fiber lubricant/softener, suspension and emulsion polymerization; amber visc. liq., mild fatty odor; sol. in alcohols, hydrocarbons, natural and paraffinic oils; sp.gr. 1.00; visc. 230 cs; HLB 1.5; pour pt. -10 C; flash pt. (PMCC) -10 C; 100% act.

Ethylan® HA Flake. [Harcros UK] Nonoxynol-35; CAS 9016-45-9; nonionic; detergent, wetting agent, latex stabilizer, emulsifier in waxes, resins, hand cleaning gels, emulsion polymerization; wh. waxy solid; negligible odor; water sol.; sp.gr. 1.064 (60 C); visc. 120 cs (60 C); HLB 17.4; cloud pt. > 100 C (1% aq.); flash pt. (COC) > 100 C; pour pt. 43 C; pH 7.0 (1% aq.); surf. tens. 42 dynes/cm (0.1%); 100% act.

Ethylan® HB1. [Harcros UK] Short chain ethoxylate; nonionic; coalescing agent for surface coatings; solv. for acrylic and vinyl acetate copolymers; plasticizer; preservative effective against a variety of microorganisms; lt. amber clear liq., mild odor; sol. hazy in water; sp.gr. 1.098; visc. 27 cs; pour pt. -15 C; flash pt. (COC) > 100 C; pH 8.8 (1% aq.); 100% act.; *Toxicology:* sl. toxic; LD50 (rat, oral) 2-4 ml/kg; sl. skin irritant, severe eye irritant.

Ethylan® HP. [Harcros UK] Nonyl phenol ethoxylate; nonionic; detergent, wetting agent, latex stabilizer, emulsifier in waxes, resins, hand cleaning gels, emulsion polymerization; wh. waxy solid; negligible odor; water sol.; sp.gr. 1.072 (40 C); visc. 78 cs (60 C); HLB 16.6; cloud pt. > 100 C (1% aq.); flash pt. > 400 F (COC); pour pt. 36 C; pH 6-8 (1% aq.);

100% act.; Unverified

Ethylan® KEO. [Harcros UK] Nonoxynol-9 (9.5 EO); CAS 9016-45-9; nonionic; detergent, wetting agent, emulsifier, foam stabilizer, solubilizer, used in pesticides, perfumes, emulsion polymerization; EPA approved; clear colorless liq.; negligible odor; water sol.; sp.gr. 1.060; visc. 331 cs; HLB 13; pour pt. 3 C; cloud pt. 54 C (1% aq.); flash pt. (COC) > 200 C; pH 6-8 (1% aq.); surf. tens. 34 dynes/cm (0.1%); 100% act.

Ethylan® LD. [Harcros UK] Cocamide DEA; CAS 61791-31-9; EINECS 263-163-9; nonionic; foam stabilizer, emulsifier for hand cleaning gels, hard surf. cleaners, shampoos; plastics antistat; clear straw liq.; mild odor; disp. in water; sp.gr. 0.981; visc. 1408 cs; flash pt. (COC) > 350 F; pour pt. 15 C; flash pt. (COC) > 175 C; pH 9.5 (1% aq.); 90% act.

Ethylan® LDA-37. [Harcros UK] Cocamide DEA; nonionic; industrial degreasing, metal cleaning; emulsifier for solvs. and polyethylene waxes; amber clear liq., mild odor; water-sol.; sp.gr. 1.005; visc. 1469 cs; pour pt. -14 C; flash pt. (COC) > 110 C; pH 9.5 (1% aq.); 100% conc.

Ethylan® LDS. [Harcros UK] Cocamide DEA; CAS 61791-31-9; EINECS 263-163-9; nonionic; foam stabilizer, emulsifier for hand cleaning gels, hard surf. cleaners, shampoos; plastics antistat; straw clear liq.; mild odor; water-disp.; sp.gr. 0.990; visc. 810 cs; flash pt. > 350 F (COC); pour pt. 10 C; pH 8.5-10.5 (1% aq.); 87% act.; Unverified

Ethylan® ME. [Harcros UK] Cetoleth-5.5; nonionic; emulsifier and solubilizer for org. solv., veg. oils, paraffin waxes, essential oils, perfumes for toiletries and cosmetics; plastics antistat; straw liq./soft paste, mild fatty odor; disp. in water; sp.gr. 0.970; visc. 81 cs; HLB 10.0; pour pt. 19 C; flash pt. (COC) > 175 C; pH 7.0 (1% aq.); 100% act.

Ethylan® MLD. [Harcros UK] Lauramide DEA; CAS 120-40-1; EINECS 204-393-1; nonionic; foam booster and stabilizer in toiletries and detergent formulations, antistat for plastics; pale cream waxy flake; negligible odor; disp. in water; visc. 107 cs (60 C); pour pt. 40 C; flash pt. (COC) > 150 C; pH 8.5 (1% aq.); 95% act.

Ethylan® N30. [Harcros UK] Nonoxynol-30; CAS 9016-45-9; nonionic; detergent, wetting agent, latex stabilizer, emulsifier in waxes, resins, hand cleaning gels; wh. waxy solid; negligible odor; water sol.; sp.gr. 1.064 (60 C); visc. 90 cs (60 C); HLB 17.0; cloud pt. > 100 C (1% aq.); flash pt. (COC) > 200 C; pour pt. 39 C; pH 7.0 (1% aq.); 100% act.

Ethylan® N50. [Harcros UK] Nonyl phenol ethoxylate; nonionic; detergent, wetting agent, latex stabilizer, emulsifier in waxes, resins, hand cleaning gels; wh. waxy solid; negligible odor; water sol.; sp.gr. 1.073 (60 C); visc. 135 cs (60 C); HLB 18.2; cloud pt. > 100 C (1% aq.); flash pt. > 400 F (COC); pour pt. 43 C; pH 6-8 (1% aq.); 100% act.

Ethylan® N92. [Harcros UK] Nonyl phenol ethoxylate; nonionic; detergent, wetting agent, latex stabilizer, emulsifier in waxes, resins, hand cleaning gels; wh. waxy solid; negligible odor; water sol.; sp.gr. 1.078 (60 C); visc. 340 cs (60 C); HLB 19.0; cloud pt. > 100 C (1% aq.); flash pt. > 400 F (COC); pour pt. 50 C; pH 6-8 (1% aq.); 100% act.; Unverified

Ethylan® OE. [Harcros UK] Cetoleth-13; nonionic; emulsifier for fatty acids, alcohols, and waxes, oil and latex stabilizer, dye leveling agent; emulsifier/solubilizer for essential oils, perfumes, and waxes for toiletries; mfg. of polishes; emulsion polymerization; cream waxy solid; negligible odor; water sol.; sp.gr. 1.009 (40 C); visc. 55 cs (40 C); HLB 14; pour pt. 31 C; cloud pt. 90 C (1% aq.); flash pt. (COC) > 175 C; pH 7.0 (1% aq.); 100% act.

Ethylan® R. [Harcros UK] Cetoleth-19; nonionic; biodeg. emulsifier for fatty acids, waxes; solubilizer for essential oils, perfumes, and waxes for toiletries; latex stabilizer; polish mfg.; emulsion polymerization; dye leveling agent; cream waxy solid; negligible odor; water sol.; sp.gr. 1.023 (40 C); visc. 141 cs (40 C); HLB 17.5; pour pt. 36 C; cloud pt. > 100 C (1% aq.); flash pt. (COC) > 175 C; pH 7.0 (1% aq.); surf. tens. 40 dynes/cm (0.1%); 100% act.

Ethyl Cadmate®. [R.T. Vanderbilt] Cadmium diethyldithiocarbamate; CAS 14239-68-0; ultra accelerator for IIR, EPDM, and SBR; primary accelerator with a thiazole; gives heat resistant, low compr. set properties to NBR and IIR; wh. to lt. gray powd., rods, or gran.; m.w. 408.94; practically insol. in water; dens. 1.48 ± 0.03 mg/m^3 (powd.); 1.39 ± 0.03 mg/m^3 (rods or gran.); m.p. 68 C min.; 11.5-12.9% cadmium content.

Ethyl Diglyme. [Ferro/Grant] Diethylene glycol diethyl ether; CAS 112-73-2; solv. which tends to solvate cations; used in electrochemistry, polymer and boron chemistry; physical processes such as gas absorption, extraction, stabilization; used in industrial prods. such as fuels, lubricants, textiles, pharmaceuticals, pesticides; colorless; mild, nonresidual odor; m.w. 162.23; f.p. -44.3 C; b.p. 189 C; water-sol.; sp.gr. 0.9082; dens. 7.56 lb/gal; visc. 1.4 cP; ref. index 1.4115; pH netural; flash pt. 90 C (CC); surf. tens. 27.2 dynes/cm; 98.0% min. purity.

Ethyl Glyme. [Ferro/Grant] Ethylene glycol diethyl ether; CAS 629-14-1; solv. which tends to solvate cations; used in electrochemistry, polymer and boron chemistry; physical processes such as gas absorption, extraction, stabilization; used in industrial prods. such as fuels, lubricants; colorless; mild ethereal, nonresidual odor; m.w. 118.18; f.p. -74 C; b.p. 121 C; water-sol.; sp.gr. 0.8417; dens. 7.0 lb/gal; visc. 0.7 cP; ref. index 1.3922; pH neutral; flash pt. 27 C (CC); 97.0% min. purity.

Ethyl Namate®. [R.T. Vanderbilt] Sodium diethyldithiocarbamate aq. sol'n.; CAS 148-18-5; rubber preservative; yel. liq.; misc. with water; dens. 1.08 mg/m^3; *Toxicology:* LD50 (oral, rat) 1500 mg/kg, (subcut., rabbit) 500 mg/kg; skin irritant; *Precaution:* acids produce turbidity due to liberation of CS_2; hazardous decomp. prods.: oxides of sodium, sulfur, nitrogen, and carbon.

Ethyl Tellurac®. [R.T. Vanderbilt] Tellurium diethyldithiocarbamate; CAS 20941-65-5; ultra accelerator for NR, SBR, NBR, EPDM; used with thiazole modifiers; produces high modulus vulcanization; particulary act. in IIR compds.; orange-yel. powd., 100% thru 30 mesh; sol. in toluene, carbon disulfide, chloroform; practically insol. in water; m.w. 720.69; dens. 1.44 ± 0.03 mg/m^3; m.p. 108-118 C; 17.5-19.5% tellurium content (powd.); 14.0-16.0% tellurium content (rods).

Ethyl Tellurac® Rodform. [R.T. Vanderbilt] Tellurium diethyldithiocarbamate; CAS 20941-65-5; rubber accelerator; orange-yel. rodform; negligible sol. in water; dens. 1.40 mg/m^3; m.p. 106 C min.; *Toxicology:* LD50 (oral, rat) > 5000 mg/kg, (dermal, rabbit) > 2 g/kg; may be harmful if inhaled; possible skin and eye irritant; may cause garlic breath and sweat, anorexia; nuisance dust TLV-TWA 5 mg/m^3 (8 h, respirable), 10 mg/m^3 (8 hr, total) recommended; *Precaution:* incompat. with acids, strong oxidizers; hazardous decomp. prods.: oxides of tellurium, nitrogen, sulfur, and carbon.

Ethyl Tuads® Rodform. [R.T. Vanderbilt] Tetraethylthiuram disulfide; CAS 97-77-8; ultra accelerator; for NR and syn. rubbers; accelerator and vulcanizing agent; cure modifier for Neoprene (retards G types, accelerates W types); nondiscoloring in lt. stocks; general applics. are same as Methyl Tuads; buff to lt. gray cubes; sol. in toluene, carbon disulfide, chloroform; practically insol. in water; m.w. 296.54; dens. 1.27 ± 0.3 mg/m^3; m.p. 63-75 C; 10.8% avail. sulfur.

Ethyl Zimate®. [R.T. Vanderbilt] Zinc diethyldithiocarbamate; CAS 14324-55-1; ultra accelerator; primary accelerator in NR and SBR; requires a thiazole modifier for safe processing and wide cure range; nondiscoloring in lt. colored stocks; stabilizer in thermoplastic rubbers and hot melts; 1:1 combination (dry wt.) with Zetax suggested for latex foam acceleration; accelerator for butyl latex vulcanization; wh. powd., 99.9% thru 100 mesh; m.w. 361.92; practically insol. in water; dens. 1.48 ± 0.03 mg/m^3; m.p. 171-182.5 C; 15.0-20.5% Zn.

Ethyl Zimate® Slurry. [R.T. Vanderbilt] Zinc diethyldithiocarbamate slurry; CAS 14324-55-1; ultra accelerator; primary accelerator in NR and SBR; slurry; 50% act.

Etocas 10. [Croda Chem. Ltd.] PEG-10 castor oil; CAS 61791-12-6; nonionic; cosmetic and essential oil solubilizer, emulsifier, lubricant, softener, leveling agent, emollient, superfatting agent, antistat, softener, detergent; used in personal care prods., fiber processing, metalworking fluids, emulsion polymerization, insecticides; pale yel. liq.; sol. in ethanol, naphtha, MEK, trichlorethylene, disp. in water; HLB 6.3; cloud pt. < 20 C; acid no. 1.0 max.; sapon. no. 120-130; pH 6-7.5; 97% act.

Eureslip 58. [Witco GmbH; Shell] Amide wax; internal lubricant, slip, antiblock, and mold release agent for PVC, PE, PP; FDA accepted; solid.

Eurestat 10. [Witco GmbH; Shell] Proprietary; internal antistat for flexible and rigid PVC.

Eurestat 66. [Witco GmbH; Shell] Quat. ammonium compd.; external antistat for ABS, acrylic, cellulosics, nylon, PC, PE, PP, PS, flexible and rigid PVC.

Eurestat K22. [Witco GmbH; Shell] Amine; external antistat for PE, PP, PS, flexible PVC; FDA accepted.

Eurestat T22. [Witco GmbH; Shell] Amine; internal antistat for PE, PP, PS, flexible PVC; FDA accepted.

Evangard® 18MP. [Evans Chemetics] Octadecyl 3-mercaptopropionate; CAS 31778-15-1; antioxidant and stabilizer to prevent and retard degradation.

Evangard® DTB. [Evans Chemetics] Distearyl dithiodipropionate; CAS 6729-96-0; antioxidant and stabilizer to prevent and retard degradation.

Evanstab® 12. [Evans Chemetics] Dilauryl thiodipropionate; CAS 123-28-4; EINECS 204-614-1; antioxidant for polyethylene, PP, and polyolefins, ABS; stabilizer for polyolefins, oils and fats, food applic.; plasticizer for rubber prods.; lubricating oil additive; syn. lubricant; chemical preservative in fats and oils; wh. cryst. flakes or powd.; m.w. 514; f.p. 40.0 C min.; sol. in acetone, MEK, n-heptane, toluene, ethyl acetate, ethanol; acid no. 1.0 max.; 99.0% min. assay.

Evanstab® 13. [Evans Chemetics] Ditridecyl thiodipropionate; CAS 10595-72-9; sec. antioxidant in ABS, polyolefins, and other polymer systems; colorless to sl. yel. clear liq.; m.w. 543; b.p. 265 C (0.25 mm); sol. (g/100 g sol'n.) 14 g 95% ethanol; miscible with toluene, ethyl acetate, n-heptane, MEK; acid no. 3.0 max.; sapon. no. 200-210; 99.0% min. assay.

Evanstab® 14. [Evans Chemetics] Dimyristyl thiodipropionate; CAS 16545-54-3; sec. antioxidant for polyolefins; wh. cryst. flakes or powd.; m.w. 570; m.p. 48-50 C; acid no. 1.0 max.; sapon. no. 280-290; 98.5% min. assay.

Evanstab® 18. [Evans Chemetics] Distearyl thiodipropionate; CAS 693-36-7; EINECS 211-750-5; sec. antioxidant for use in polyolefins; used in food-pkg. materials and edible fats and oils; wh. cryst. flakes or powd.; m.w. 683; f.p. 64.0 C min.; sol. (g/100 g sol'n.) 11 g toluene, 2 g n-heptane; < 1 g ethyl acetate, 95% ethanol, MEK, water; 98.0% min. assay.

Evanstab® 1218. [Evans Chemetics] Stearyl/lauryl thiodipropionate; CAS 13103-52-1; antioxidant and stabilizer to prevent and retard degradation.

EW-POL 8021. [Henkel/Functional Prods.] Aryl polyglycol ether; surface-active plasticizer, thickener for PVAc adhesive dispersions; liq.; 100% act.

EX LBA-30. [Boehringer Ingelheim; Henley] Exothermic chemical blowing agent for thermoplastics (PP, PE, PPO, styrenics, etc.) in structural foam molding and extrusion applics.; recommended for removal of sink holes and foaming for wt. reduction.

Exact™ Plastomers. [Exxon] Ethylene/α-olefin copolymers; modifiers for polyolefins combining chars. of plastics and elastomers and providing low temp. impact modification, toughness, str., better ESCR, clarity, flexibility; FDA compliant grades avail.; free-flowing pellets.

EXC-33. [Releasomers] Water-based fluorochemical

copolymer; release agent for rubber, plastic, and metal industries; milky wh. emulsion; sp.gr. < 1; pH > 5; *Storage:* store below 32 C; protect from freezing.

Exocerol® LBA-39. [Boehringer Ingelheim; Henley] Blend of endothermic multicomponent system and an azodicarbonamide-type blowing agent; endothermic/exothermic chemical blowing agent for thermoplastics (HDPE, PP, ABS, PPO, polyolefins, styrenics, etc.) in structural foam molding and extrusion; for removal of sink marks, wt. reduction; food contact approvals; bright yel. tacky liq.; sp.gr. 1.04; dec. 150 C.

Exocerol® OM. [Boehringer Ingelheim; Henley] Blowing system based on citric and azodicarbonamide; chemical blowing agent for foaming of thermoplastics in inj. molding; FDA 21CFR §177.1210, 178.3010; German Plastics Commission compliance; ylsh. free-flowing powd.; dec. 180 C; *Storage:* exc. storage stability when kept in cool, dry place.

Exocerol® OM 70 A. [Henley] Hydrocerol® and azodicarbonamide blowing system in styrenic carrier; enexothermal chemical blowing agent for structural foam molding of PS and engineering resins (e.g., PPO); predominantly used for housings which are subsequently painted; German approval; FDA 21CFR §177.1210, 178.3010; ylsh. cylindrical gran.; dec. 180 C; 70% act.

Exocerol® Spezial 70. [Henley] Hydrocerol® and azodicarbonamide blowing system in PE wax carrier; enexothermal chemical blowing agent for foaming of thermoplastics in inj. molding, preferably at processing temps. from 200-240 C; German approval; FDA 21CFR §177.1210, 178.3010; ylsh. cylindrical gran.; dec. 160 C; 70% act.

Exolit® 422. [Hoechst Celanese/Spec. Chem.] Ammonium polyphosphate; halogen-free flame retardant for PU foam, hot-melt adhesives, coatings, roofing; powd.

Exolit® 462. [Hoechst Celanese/Spec. Chem.] Ammonium polyphosphate; halogen-free flame retardant for PU foam, modified polyester.; powd.

Exolit® IFR-10. [Hoechst Celanese/Spec. Chem.] Ammonium polyphosphate/synergist blend; halogen-free flame retardant for hot-melt adhesives, epoxy, elastomers; for processing temps. to 230 C; powd.

Exolit® IFR-15. [Hoechst Celanese/Spec. Chem.] Ammonium polyphosphate/synergist blend; halogen-free flame retardant for thermoplastic PU (polyether), hot-melt adhesives, epoxy, elastomers; for processing temps. to 230 C; powd.

Exolit® IFR-23. [Hoechst Celanese/Spec. Chem.] Ammonium polyphosphate/synergist blend; intumescent halogen-free flame retardant for PE, PP, TPE, EPDM, EVA, elastomers for molding and extrusion; provides char forming protection, low smoke generation, heat stability > 230 C; off-wh. free-flowing powd.; 12 μ avg. particle size; sp.gr. 1.8; dec. > 230 C; > 23% P.

EXP-24-LS Silicone Wax Emulsion. [Genesee Polymers] Silicone wax copolymer aq. emulsion; nonionic; release, lubricant, detackifier for rubber industry; release, internal lubricant for plastics; antiblocking for paper, carbon paper, typewriter ribbons; release, water repellent, thread/fiber lubricant for textiles; release, water repellent, antiblocking for printing inks; also for candles, glass bottle molding, aluminum prod., machine shop use; dries to continuous wax film, acts as rust protective coating; wh. to tan opaque liq.; sp.gr. 0.98; dens. 8.0 lb/gal; pH 6.0; 13.5% silicone wax, 15% total solids; *Storage:* protect from freezing; store below 100 F.

EXP-36-X20 Epoxy Functional Silicone Sol'n. [Genesee Polymers] Epoxy functional silicone sol'n. in xylene; cures to highly adhesive, flexible coatings (release coatings, marine and industrial maintenance coatings, rust inhibiting coatings, coil coatings, conductive coatings, adhesives); internal release agent, impact modifier for plastics; concrete and foundry mold release; automotive rustproofing; fiberglass sizing agent; filler treatment; electronics; clear thin liq.; sp.gr. 0.90; dens. 7.5 lb/gal; visc. 5 cst; flash pt. 79 F; EEW 2847 g; 20% act.; *Precaution:* flamm.; keep away from fire and flame; *Storage:* keep container tightly sealed.

EXP-38-X20 Epoxy Functional Silicone Sol'n. [Genesee Polymers] Epoxy functional silicone sol'n. in xylene; forms clear, flexible, durable coating with or without a curing agent providing good rust protection; for temporary metal coatings, conductive coatings, tarnish/rust inhibiting coatings, marine and industrial maintenance coatings, flexible coatings; adhesives, electronics; plastic additives; noncured film removable with solv.; lt. straw hazy liq.; sp.gr. 0.91; dens. 7.5 lb/gal; visc. 162 cst; flash pt. (TCC) 79 F; EEW 5180 g; 20% act.; *Precaution:* flamm.; keep away from fire and flame; *Storage:* keep container tightly sealed; do not weld or cut drum, even when empty.

EXP-58 Silicone Wax. [Genesee Polymers] Silicone wax copolymer; paintable lubricant, mold release agent and internal lubricant for plastics; release, lubricant, detackifier for rubber; forms dry, colorless film on substrates, providing rust protection; antiblocking agent for paper, printing inks; release, water repellent, thread/fiber lubricant for textiles; also for aluminum prod., glass bottle molding, candles, machine shop use; wh. hard wax, odorless; sol. in aliphatic hydrocarbons; m.w. 12,000; sp.gr. 0.85; dens. 7.1 lb/gal; m.p. 43 C; flash pt. > 300 F; 100% act.; *Storage:* indefinite storage stability.

EXP-61 Amine Functional Silicone Wax. [Genesee Polymers] Stearyl/aminopropyl methicone copolymer; CAS 110720-64-4; lubricant, water repellent, antiblocking agent, internal processing aid; forms dry, colorless film on substrates providing rust protection, good bonding; lubricant, detackifier for rubber; plastics release; antiblocking for paper; fiber/thread lubricant for textiles; also for printing inks, candles, glass bottle molding, aluminum prod., machine shop use; wh. waxy solid; m.w.

12,000; sp.gr. 0.88; dens. 7.4 lb/gal; m.p. 44 C; flash pt. > 300 F; 100% act.

EXP-77 Mercapto Functional Silicone Wax. [Genesee Polymers] Mercapto functional silicone wax; lubricant and release agent for molded metal, plastic, and rubber parts; release, lubricant, detackifier, internal processing aid for rubber; water repellent, release, thread/fiber lubricant for textiles; antiblocking for printing inks, paper; also for candles, glass bottle molding, machine shop use, aluminum prod.; good high temp. stability; wh. waxy soli, sulfur odor; sol. in aliphatic and aromatic hydrocarbons; m.w. 12,000; sp.gr. 0.88; dens. 7.4 lb/gal; m.p. 44 C; flash pt. > 300 F; 100% act.

Expancel® 091 WU. [Expancel] Spherically formed particles with a thermoplastic shell encapsulating a gas; wet, unexpanded microspheres used as blowing agents in PU (footwear, sporting goods), PVC plastisols (automotive undercoatings), printing ink, fabrics, paper and board, nonwovens; exc. solv. resist.; wh.; 8-11 μm avg. particle size; dens. < 17 kg/m^3; 65% solids.

Expancel® 551 WU. [Expancel] Spherically formed particles with a thermoplastic shell encapsulating a gas; wet, unexpanded microspheres used as blowing agents in PU (footwear, sporting goods), PVC plastisols (automotive undercoatings), printing ink, fabrics, paper and board, nonwovens; good solv. resist.; wh.; 5-8 μm avg. particle size; dens. < 17 kg/m^3; 65% solids.

Expancel® 642 WU. [Expancel] Spherically formed particles with a thermoplastic shell encapsulating a gas; wet, unexpanded microspheres used as blowing agents in PU (footwear, sporting goods), PVC plastisols (automotive undercoatings), printing ink, fabrics, paper and board, nonwovens; good solv. resist.; wh.; 5-8 μm avg. particle size; dens. < 25 kg/m^3; 65% solids.

Expancel® 820 DU. [Expancel] Spherically formed particles with a thermoplastic shell encapsulating a gas; dry, unexpanded microspheres used as blowing agents in PU (footwear, sporting goods), PVC plastisols (automotive undercoatings), printing ink, fabrics, paper and board, nonwovens; wh.; 5-8 μm avg. particle size; dens. < 17 kg/m^3; > 99% solids; Custom

Expandex® 5PT. [Uniroyal] 5-Phenyltetrazole; CAS 3999-10-8; chemical blowing agent for foaming plastics and elastomers at elevated temps. (450-550 F); for nylon, polyester, PC, PPO, polysulfone inj. molding structural foams; wh. needle-like material; 18 μ avg. particle size; sol. in ethanol and other common org. solvs.; sp.gr. 1.42; m.p. 212 C; gas yield 200 cc/g @ STP; gas: N_2; *Storage:* exc. storage stability under normal conditions; store away from heat sources and flames.

Expandex® 175. [Uniroyal] 5-Phenyltetrazole, barium salt; high temp. chemical blowing agent for engineering resins, esp. polysulfone; operating temps. 650-750 F; wh. cryst. powd.; sol. (g/l): 150 g in water, 80 g in DMSO, 38 g in methanol, 2 g in acetone and toluene; sp.gr. 1.671; decomp. pt. 375 C; gas yield 177 ml/g.

Extend. [Hughes Ind.] Blowing agent for cellular rigid vinyl extrusion process; produces prods. with more uniform cell structure and optimum dens. reduction, increased dimensional stability.

Extendospheres® CG. [PQ Corp.] Hollow microspheres; lightweight extender for resin systems; suitable for high-build industrial coatings, spray-up applics., applics. requiring med. particle range; off-wh. free-flowing spherical powd.; 105-115 μ mean diam.; sp.gr. 0.70-0.75; dens. 0.7-0.8 g/cc; bulk dens. 25 lb/ft^3; soften. pt. 1200 C; pH 7.0-7.2 (aq.); hardness (Moh) 5; shell material: 58-65% SiO_2, 28-33% Al_2O_3; *Toxicology:* may contain < 5% cryst. silica; dust may create a respiratory hazard.

Extendospheres® SF-10. [PQ Corp.] Hollow microspheres; lightweight extender for plastisol and asphaltic underbody coatings, high mil-build industrial maintenance coatings, adhesives, spackles, caulks, sealants, high temp.-resist. cementitious and plastics composites; gray free-flowing powd.; 65 μ mean particle diam.; dens. 0.7-0.8 g/cc; bulk dens. 25 lb/ft^3; soften. pt. 1200 C; hardness (Moh) 5; shell material: 58-65% SiO_2, 28-33% Al_2O_3; *Toxicology:* may contain < 5% cryst. silica; dust may create a respiratory hazard.

Extendospheres® SF-12. [PQ Corp.] Hollow microspheres; lightweight extender for plastisol and asphaltic underbody coatings, high mil-build industrial maintenance coatings, adhesives, spackles, caulks, sealants, high temp.-resist. cementitious and plastics composites; gray free-flowing powd.; 60 μ mean particle diam.; dens. 0.7-0.8 g/cc; bulk dens. 25 lb/ft^3; soften. pt. 1200 C; pH 7.0-7.2 (aq. susp.); hardness (Moh) 5; shell material: 58-65% SiO_2, 28-33% Al_2O_3; *Toxicology:* may contain < 5% cryst. silica; dust may create a respiratory hazard.

Extendospheres® SF-14. [PQ Corp.] Hollow microspheres; lightweight extender for plastisol and asphaltic underbody coatings, high mil-build industrial maintenance coatings, adhesives, spackles, caulks, sealants, high temp.-resist. cementitious and plastics composites; gray free-flowing powd.; 55 μ mean particle diam.; dens. 0.7-0.8 g/cc; bulk dens. 25 lb/ft^3; soften. pt. 1200 C; hardness (Moh) 5; shell material: 58-65% SiO_2, 28-33% Al_2O_3; *Toxicology:* may contain < 5% cryst. silica; dust may create a respiratory hazard.

Extendospheres® SG. [PQ Corp.] Hollow microspheres; lightweight, low sp.gr. extender in resin matrices and concrete applics. to reduce weight, add thermal insulation and increase impact str.; for cement, grout, sealants, roofing compds., PVC and latex flooring; off-wh. free-flowing spherical powd.; 120-130 μ mean diam.; sp.gr. 0.70; dens. 0.7-0.8 g/cc; bulk dens. 25 lb/ft^3; soften. pt. 1200 C; pH 7.0-7.2 (aq.); hardness (Moh) 5; shell material: 58-65% SiO_2, 28-33% Al_2O_3; *Toxicology:* may contain < 5% cryst. silica; dust may create a respiratory hazard.

Extendospheres® TG. [PQ Corp.] Hollow

microspheres; lightweight extender for resin systems; recommended for applics. requiring a finer particle size; gray free-flowing spherical powd.; 90-100 μ mean diam.; dens. 0.7-0.8 g/cc; bulk dens. 25 lb/ft^3; soften. pt. 1200 C; hardness (Moh) 5; shell material: 58-65% SiO_2, 28-33% Al_2O_3; *Toxicology:* may contain < 5% cryst. silica; dust may create a respiratory hazard.

Extendospheres® XOL-50. [PQ Corp.] Hollow aluminosilicate microspheres; lightweight extender for plastic and coating systems, reducing composite density; recommended for cultured marble, syntactic foams, other low-shear applics.; wh. free-flowing powd.; 50 μ mean particle diam.; dens. 0.33 g/cc; bulk dens. 7 lb/ft^3; soften. pt. 1800 F; pH 7.0-7.2 (aq. susp.); *Toxicology:* prolonged inh. may irritate respiratory tract.

Extendospheres® XOL-200. [PQ Corp.] Hollow aluminosilicate microspheres; lightweight extender for plastic and coating systems, reducing composite density; recommended for cultured marble, syntactic foams, other low-shear applics.; wh. free-flowing powd.; 100 μ mean particle diam.; dens. 0.24 g/cc; bulk dens. 7.5-9.0 lb/ft^3; soften. pt. 1800 F; pH 7.0-7.2 (aq. susp.); *Toxicology:* prolonged inh. may irritate respiratory tract.

Extrusil. [Degussa] Calcium silicate; hydrophilic extender for rubber compds.; provides thixotropy, antisedimentation in CR adhesives; powd.; ≈ 25 nm avg. particle size; dens. 300 g/l (tapped); surf. area 35 m^2/g; pH 10 (5% aq. susp.); 91% SiO_2, 6% CaO, 2% Na_2O.

Exxelor PA23. [Exxon] Elastomer conc.; polymer modifier offering soft, rubber films, exc. toughness, broad heat seal range, controlled flexibility; used with all polyolefins for specialty films/laminates, fabric coatings, medical films, disposable gloves; dens. 0.90 g/cc; melt index 2.7 g/10 min (190 C); flex. mod. 1130 psi; Shore hardness A65 (5 s); Vicat soften. pt. 140 F.

Exxelor PA30. [Exxon] Elastomer conc.; polymer modifier offering high temp. resist., matte surf. films, outstanding toughness, braod FDA coverage, low moisture permeability; used with PP, HDPE for specialty films/laminates, retort pouches, medical bags, soft pkg. films, tapes; dens. 0.91 g/cc; melt index 0.15 g/10 min (190 C); flex. mod. 6150 psi; Shore hardness A85 (5 s); Vicat soften. pt. 310 F.

Exxelor PA50. [Exxon] Elastomer conc.; polymer modifier offering outstanding toughness, matte surface films, esc. ESCR, broad FDA coverage, low moisture permeability; used with HDPE, LLDPE, LDPE for medical pouches/overwraps, food pkg., heavy-duty films, pond liners/tarpaulins, dunnage bags; pellets; dens. 0.92 g/cc; melt index 0.22 g/10 min (190 C); flex. mod. 6000 psi; Shore hardness A78 (5 s); Vicat soften. pt. 253 F; 15 phr in 4-mil HDPE film: tens. str. 3200 psi (yield, MD); Dart Drop impact 230 g; Elmendorf tear str. 132 g (MD).

Exxelor PE 805. [Exxon] Ethylene-propylene elastomer; contains heat processing stabilizer; CAS 9010-79-1; impact modifier for PP and other polyolefins; may be processed at temps. to 300 C; wh. free-flowing pellets; melt flow 5 g/10 min (230 C, 10 kg); dens. 0.87 g/cc; Mooney visc. 33 (ML1+4, 125 C).

Exxelor PE 808. [Exxon] Ethylene-propylene elastomer; contains heat processing stabilizer and anti-agglomeration coating; CAS 9010-79-1; impact modifier for PP and other polyolefins; may be processed at temps. to 300 C; wh. free-flowing pellets; melt flow 3 g/10 min (230 C, 10 kg); dens. 0.87 g/cc; Mooney visc. 46 (ML1+4, 125 C).

Exxelor PO 1015. [Exxon] Maleic anhydride-functionalized polypropylene; chemical coupling agent for reinforcing agents (glass fibers, inorg. fillers) to PP; compatibilizer for polyolefins with polymers capable of interacting with the maleic anhydride functionality for alloying, recycling, and coextrusion; improves PP adhesion to metal; wh. free-flowing pellets; melt flow 22 g/10 min (190 C, 1.2 kg); dens. 0.9 g/cc; m.p. 138 C; soften. pt. (Vicat) 121C; heat distort. temp. 63 C; 0.4% maleic content; *Toxicology:* may contain small amts. of nonbonded maleic anhydride which is an irritant.

Exxelor VA 1801. [Exxon] Maleic anhydride-functionalized ethylene-propylene elastomer contg. heat processing stabilizer; CAS 9010-79-1; impact modifier for engineering thermoplastics; compatiblizer for polyolefins with polymers capable of interacting with maleic anhydride functionality; high thermostability and weatherability; processable to 300 C; wh. free-flowing semicryst. pellets; melt flow 9 g/10 min (230 C, 10 kg); dens. 0.87 g/cc; 0.6% maleic content; *Toxicology:* contains small amts. of nonbonded maleic anhydride (< 500 ppm) which is an irritant.

Exxelor VA 1803. [Exxon] Maleic anhydride-functionalized ethylene-propylene elastomer contg. anti-agglomeration (HDPE) coating; CAS 9010-79-1; impact modifier for engineering thermoplastics (very low temp. toughness requirements); compatiblizer for polyolefins with polymers capable of interacting with maleic anhydride functionality; high thermostability and weatherability; processable to 300 C; amorphous wh. free-flowing pellets; melt flow 22 g/10 min (230 C, 10 kg); dens. 0.86 g/cc; 0.7% maleic content; *Toxicology:* contains small amts. of nonbonded maleic anhydride (< 500 ppm) which is an irritant.

Exxelor VA 1810. [Exxon] Maleic anhydride-functionalized ethylene-propylene elastomer contg. heat processing stabilizer and anti-agglomeration (HDPE) coating; CAS 9010-79-1; impact modifier for engineering thermoplastics; compatibilizer for polyolefins with polymers capable of interacting with maleic anhydride functionality; high thermostability and weatherability; processable to 300 C; wh. free-flowing semicryst. pellets; melt flow 7.5 g/10 min (230 C, 10 kg); dens. 0.87 g/cc; 0.5% maleic content; *Toxicology:* contains small amts. of nonbonded maleic anhydride (< 500 ppm) which is an irritant.

Exxelor VA 1820. [Exxon] Maleic anhydride-functionalized ethylene-propylene elastomer contg. heat processing stabilizer; CAS 9010-79-1; impact modifier for engineering thermoplastics; compatiblizer for polyolefins with polymers capable of interacting with maleic anhydride functionality; high thermostability and weatherability; processable to 300 C; wh. free-flowing semicryst. pellets; melt flow 5 g/10 min (230 C, 10 kg); dens. 0.87 g/cc; 0.3% maleic content; *Toxicology:* contains small amts. of nonbonded maleic anhydride (< 300 ppm) which is an irritant.

EZn-Chek® 601. [Ferro/Bedford] Vinyl heat stabilizer for 105 C dry insulation applics.; provides exc. early color.

EZn-Chek® 788I. [Ferro/Bedford] Vinyl heat stabilizer for 105 C dry insulation applics.; exc. dyamic thermal stability.

F

F-2000 FR. [AmeriBrom; Dead Sea Bromine] Brominated epoxy; flame retardant with exc. performance in engineering plastics (PBT, PET, nylon), styrenics (ABS, PC), and thermosets (epoxy, phenolic, unsat. polyesters); high thermal stability and thermal aging.

F-2001. [AmeriBrom; Dead Sea Bromine] Brominated epoxy oligomer; reactive flame retardant additive for thermosets (epoxy, phenolic, unsat. polyesters); good uv stability; wh. powd.; sol. in DMF, dioxane, MEK, MIBK; insol. in water, alcohol; m.p. 60-70 C; EE 485-545 g/eq; 49-51% Br; *Toxicology:* LD50 (oral, rat) > 1000 mg/kg; *Storage:* stable under normal conditions; store in cool, dry area.

F-2016. [AmeriBrom; Dead Sea Bromine] Brominated epoxy oligomer; flame retardant for ABS and HIPS; nonblooming; high thermal and uv stability; good processability; mixes and disperses readily into polymer systems; wh. powd.; sol. in DMF, dioxane; partly sol. in MEK, MIBK; insol. in water, alcohol; soften. pt. 105-115 C; EE 750-850; 49-51% Br; *Toxicology:* LD50 (oral, rat) > 5000 mg/kg; *Storage:* stable under normal conditions; store in cool, dry area.

F-2200. [AmeriBrom; Dead Sea Bromine] Brominated epoxy resin; flame retardant resin for potting, wet lay-up and pre-preg laminates, adhesives, molding compds., coatings; for epoxy, phenolic, and unsat. polyester resins; lt. yel. semisolid; sol. in DMF, dioxane, MEK, MIBK, xylene; insol. in water, alcohol; m.p. 51-62 C; EE 340-380 g/eq; 48% Br; *Toxicology:* LD50 (oral, rat) > 1000 mg/kg; *Storage:* stable under normal conditions; store in cool, dry area.

F-2300. [AmeriBrom; Dead Sea Bromine] Med. m.w. brominated epoxy oligomer; flame retardant additive for use in thermoplastics (PBT, PC/ABS, ABS, HIPS); high thermal and uv stability; wh. powd.; sol. in DMF, dioxane; insol. in water, alcohol; m.w. 3500; soften. pt. 120-140 C; 50-52% Br; *Toxicology:* LD50 (oral, rat) > 1000 mg/kg; *Storage:* stable under normal conditions; store in cool, dry area.

F-2300H. [AmeriBrom; Dead Sea Bromine] High m.w. brominated epoxy polymer; flame retardant additive for use in thermoplastics (PBT, PET), thermoplastic elastomers, alloys (PC/ABS), etc.; high thermal and uv stability, good melt flow; lt. straw gran.; sol. in DMF, dioxane; insol. in water, alcohol; m.w. 20,000; soften. pt. 130-150 C; 51-53% Br; *Toxicology:* LD50 (oral, rat) > 1000 mg/kg; *Storage:* stable under normal conditions; store in cool, dry area.

F-2310. [AmeriBrom; Dead Sea Bromine] Brominated epoxy oligomer; flame retardant for use in PS, ABS, alloys (PC/ABS); nonblooming; high thermal stability, good uv stability, good processability; wh. powd.; sol. in DMF, dioxane; partly sol. in MEK, MIBK; insol. in water, alcohol; m.w. 2200; soften. pt. 100-110 C; 50-52% Br; *Toxicology:* LD50 (oral, rat) > 1000 mg/kg; *Storage:* stable under normal conditions; store in cool, dry area.

F-2400. [AmeriBrom; Dead Sea Bromine] High m.w. brominated epoxy; flame retardant for use in thermoplastics (PBT, PET, polyamides, PU), alloys (PC/ABS), styrenics (ABS, HIPS); high thermal stability and thermal aging, high uv stability; lt. straw gran.; sol. in DMF, dioxane; insol. in water, alcohol; m.w. 50,000; soften. pt. 145-155 C; 52-54% Br; *Toxicology:* LD50 (oral, rat) > 1000 mg/kg; *Storage:* stable under normal conditions; store in cool, dry area.

F-2400E. [AmeriBrom; Dead Sea Bromine] High m.w. brominated epoxy; flame retardant for use in thermoplastics (PBT, PET, PU), alloys (PC/ABS), styrenics (ABS, HIPS), etc.; high thermal and uv stability, thermal aging; lt. straw gran.; sol. in DMF, dioxane; insol. in water, alcohol; m.w. 30,000; soften. pt. 145-155 C; 52-54% Br; *Toxicology:* LD50 (oral, rat) > 1000 mg/kg; *Storage:* stable under normal conditions; store in cool, dry area.

F-2430. [AmeriBrom; Dead Sea Bromine] Brominated epoxy masterbatch with antimony oxide; flame retardant for use in PS, ABS, PE, PBT, PET, nylon, PC, alloys, rubbers, and elastomers.

F-2430 SA. [AmeriBrom; Dead Sea Bromine] Brominated epoxy masterbatch with sodium antimonate; flame retardant for use in nylon, rubber, and elastomers.

Faktis 10. [Rhein Chemie] Sulfur factice; CAS 7704-34-9; EINECS 231-722-6; processing aid for rubber hoses and profiles, surgical and foodstuff goods; brn. crushed lumps; sp.gr. 1.00.

Faktis 14. [Rhein Chemie] Sulfur factice; CAS 7704-34-9; EINECS 231-722-6; processing aid for rubber hoses and profiles, foodstuff goods, rubberized fabrics, etc.; brn. crushed lumps; sp.gr. 1.02.

Faktis Asolvan, Asolvan T. [Rhein Chemie] Sulfur

factice; CAS 7704-34-9; EINECS 231-722-6; processing aid for NBR, CR, and other special rubbers, petrol-resistant rubbers; brn. ground; sp.gr. 1.06.

Faktis Badenia C. [Rhein Chemie] Sulfur factice; CAS 7704-34-9; EINECS 231-722-6; processing aid for colored molded and extruded rubber articles; lt. yel. fine powd.; sp.gr. 1.03.

Faktis Badenia T. [Rhein Chemie] Sulfur factice, min. oil extended; processing aid for lt. colored tech. molded and extruded rubber goods; yel. fine powd.; sp.gr. 1.01; Unverified

Faktis HF Braun. [Rhein Chemie] Sulfur factice; CAS 7704-34-9; EINECS 231-722-6; processing aid for blended rubber compds., esp. with EPDM; brn. elastic, lumps, semisolid, powd.; sp.gr. 1.00.

Faktis Para extra weich. [Rhein Chemie] Sulfur factice, min. oil extended; processing aid for soft molded and extruded rubber goods of SBR and SBR/NR blends; brn. lumps; sp.gr. 0.98.

Faktis R, Weib MB. [Rhein Chemie] Sulfur chloride factice; CAS 10025-67-9; processing aid for cold curing of rubber compds.; wh. fine powd.; sp.gr. 1.05 and 1.14 resp.

Faktis RC 110, RC 111, RC 140, RC 141, RC 144. [Rhein Chemie] Sulfur factice; CAS 7704-34-9; EINECS 231-722-6; processing aid for hoses and profiles, rubberized fabrics, tech. goods; dk. crushed lumps; sp.gr. 1.05, 1.02, 1.05, 1.05, and 1.05 resp.; Unverified

Faktis Rheinau H, W. [Rhein Chemie] Sulfur chloride factice; CAS 10025-67-9; processing aid for hot curing of wh. and bright colored rubber goods; wh. fine powd.; sp.gr. 1.06 and 1.03 resp.; Unverified

Faktis T-hart. [Rhein Chemie] Sulfur factice, min. oil extended; processing aid for tech. molded and extruded SBR rubbers and blends; brn. crushed lumps; sp.gr. 1.02; Unverified

Faktis ZD. [Rhein Chemie] Sulfur factice, min. oil extended; processing aid for tech. molded and extruded rubber goods; dk. brn. lumps; sp.gr. 1.02.

Faktogel® 10. [Rhein Chemie] Sulfur vulcanized vegetable oil (pure rapeseed oil); extender, processing aid for hoses and profiles for open cure, surgical goods, rubber threads, rolls, hand-made goods; brn. crushed lumps; dens. 1.00 g/cc.

Faktogel® 14. [Rhein Chemie] Sulfur vulcanized vegetable oil (pure rapeseed oil); extender, processing aid for hoses and profiles for open cure, bicycle valve tubings, roll coverings, rubber threads, rolls, hand-made goods, uppers for rubber shoes, rubberized fabrics; brn. crushed lumps; dens. 1.02 g/cc.

Faktogel® 17. [Rhein Chemie] Sulfur vulcanized vegetable oil (pure rapeseed oil); extender, processing aid for hoses and profiles for open cure, bicycle valve tubings, roll coverings, rubber threads, rolls, hand-made goods, uppers for rubber shoes, rubberized fabrics; brn. crushed lumps; dens. 1.04 g/cc.

Faktogel® 110. [Rhein Chemie] Sulfur vulcanized vegetable oil (pretreated natural oils); extender, processing aid for hoses and profiles for open cure, hand-made goods, rubberized fabrics, mech. rubber goods; brn. crushed lumps; dens. 1.05 g/cc.

Faktogel® 111. [Rhein Chemie] Sulfur vulcanized vegetable oil (pretreated natural oils); extender, processing aid for hoses and profiles for open cure, hand-made goods, rubberized fabrics, mech. rubber goods; brn. crushed lumps; dens. 1.02 g/cc.

Faktogel® 140. [Rhein Chemie] Sulfur vulcanized vegetable oil (pretreated natural oils); extender, processing aid for hoses and profiles for open cure, hand-made goods, rubberized fabrics, mech. rubber goods; brn. crushed lumps; dens. 1.05 g/cc.

Faktogel® 141. [Rhein Chemie] Sulfur vulcanized vegetable oil (pretreated natural oils); extender, processing aid for hoses and profiles for open cure, hand-made goods, rubberized fabrics, mech. rubber goods; brn. crushed lumps; dens. 1.05 g/cc.

Faktogel® 144. [Rhein Chemie] Sulfur vulcanized vegetable oil (pretreated natural oils); extender, processing aid for hoses and profiles for open cure, hand-made goods, rubberized fabrics, mech. rubber goods; brn. crushed lumps; dens. 1.05 g/cc.

Faktogel® A. [Rhein Chemie] Sulfur vulcanized vegetable oil, min. oil-extended; extender, processing aid for molded, extruded, and calendered mech. goods, esp. those based on CR; dk. brn. crushed lumps; dens. 1.03 g/cc.

Faktogel® Asolvan. [Rhein Chemie] Sulfur vulcanized vegetable oil (pure castor oil); extender, processing aid for NBR, CR, CSM, and other special rubbers, goods resist. to gasoline and min. oil, e.g., rolls, gasoline hoses; brn. ground; dens. 1.06 g/cc.

Faktogel® Asolvan T. [Rhein Chemie] Sulfur vulcanized vegetable oil (castor oil and vegetable oils); extender, processing aid for NBR, CR, CSM, and other special rubbers, goods resist. to gasoline and min. oil, e.g., rolls, gasoline hoses; brn. ground; dens. 1.06 g/cc.

Faktogel® Badenia C. [Rhein Chemie] Sulfur vulcanized vegetable oil (pretreated pure rapeseed oil); extender, processing aid for colored molded and extruded goods (rubber goods for surgical, household, camping, and sports items); lt. yel. finely ground; dens. 1.03 g/cc.

Faktogel® Badenia T. [Rhein Chemie] Sulfur vulcanized vegetable oil (pretreated pure rapeseed oil), mineral oil-extended; extender, processing aid for lt. colored molded and extruded goods, mech. goods, bicycle tubes, coating compds.; yel. finely ground; dens. 1.01 g/cc.

Faktogel® HF. [Rhein Chemie] Sulfur vulcanized vegetable oil (pure rapeseed oil); extender, processing aid for blended rubber goods, esp. those with EPDM; cellular and microcellular rubber goods based on all elastomers, esp. expanded profiles; brn. elastic small lumps, semisolid, powd.; dens. 1.00 g/cc.

Faktogel® KE 8384. [Rhein Chemie] Sulfur chloride vulcanized vegetable oil; extender, processing aid for hot curing, wh. and bright colored rubber goods, rubberized fabrics, roll coverings, erasers; wh. finely ground; dens. 1.05 g/cc.

Faktogel® KE 8419. [Rhein Chemie] Sulfur vulcanized vegetable oil (natural oils); extender, processing aid for hoses and profiles for open cure, hand-made goods, rubberized fabrics, mech. rubber goods; brn. crushed lumps; dens. 1.04 g/cc.

Faktogel® MB. [Rhein Chemie] Sulfur chloride vulcanized vegetable oil (pure rapeseed oil), inorg. stabilizers, mineral oil-extended; extender, processing aid for cold cured compds., lt.-colored calendered sheets, rubberized fabrics, elastic threads, eraser compds.; suitable for hot cure when using accelerators; wh. finely ground; dens. 1.14 g/cc.

Faktogel® Para. [Rhein Chemie] Sulfur vulcanized vegetable oil (pure rapeseed oil), mineral oil extended; extender, processing aid for soft molded and extruded mech. goods of SBR and SBR/NR, e.g., soft roll coverings, pump diaphragms, seals; brn. very soft lumps; dens. 0.98 g/cc.

Faktogel® R. [Rhein Chemie] Sulfur chloride vulcanized vegetable oil (pure rapeseed oil), inorg. stabilizers; extender, processing aid for cold cured compds., lt.-colored calendered sheets, rubberized fabrics, elastic threads, eraser compds.; suitable for hot cure when using accelerators; wh. finely ground; dens. 1.05 g/cc.

Faktogel® Rheinau H. [Rhein Chemie] Sulfur chloride vulcanized vegetable oil (pure rapeseed oil), inorg. stabilizers; extender, processing aid for hot curing, wh. and bright colored rubber goods, complicated extruded goods with stringent requirements of dimensional stability, e.g., hoses, seals, roll coverings, erasers; wh. finely ground; dens. 1.06 g/cc.

Faktogel® Rheinau W. [Rhein Chemie] Sulfur chloride vulcanized vegetable oil (pure rapeseed oil, paraffin oil), inorg. stabilizers; extender, processing aid for hot curing, wh. and bright colored rubber goods, rubberized fabrics, roll coverings, erasers; wh. finely ground; dens. 1.03 g/cc.

Faktogel® T. [Rhein Chemie] Sulfur vulcanized vegetable oil (pure rapeseed oil), mineral oil extended; extender, processing aid for molded and extruded mech. goods based on SBR or blends, e.g., seals, sleeves, uppers for shoes, cable compds.; brn. crushed lumps; dens. 1.02 g/cc.

Faktogel® ZD. [Rhein Chemie] Sulfur vulcanized vegetable oil (pure rapeseed oil with pretreated oils), mineral oil extended; extender, processing aid for molded and extruded mech. goods, uppers for shoes, cable compds.; dk. brn. crushed very soft lumps; dens. 1.02 g/cc.

Fancol OA-95. [Fanning] Oleyl alcohol; CAS 143-28-2; EINECS 205-597-3; nonionic; plasticizer, emulsion stabilizer, antifoam and coupling agent, aerosol lubricant, petrol. additive, pigment dispersant; rust preventive; detergent, release agent, cosolvent, softener, tackifier, spreading agent used for metalworking, petrochemicals, pulp and paper, paints and coatings, plastics and polymers, inks, food applics., pharmaceuticals, cosmetics; chemical intermediate; liq.; sol. in IPA, acetone, lt. min. oil, trichloroethylene, kerosene, VMP naphtha, benzene, turpentine; acid no. 0.05 max.; iodine no. 90-96; sapon no. 1 max.; hyd. no. 200-212; cloud pt. 5 C max.

Fancor Lanolin. [Fanning] Lanolin; plasticizer for rubber; lubricant for textile, metalworking compds.; sol. in hydrocarbon oils, most solvs.

Fascat® 2000. [Elf Atochem N. Am.] Stannous compd.; esterification catalyst for mfg. of plasticizers, fatty acid esters; effective where catalyst removal is required; source of stannous tin; blk. cryst.; sol. in alcohols and acids with reaction, in caustic sol'ns.; insol. in water and most org. solvs.; sp.gr. 6.5; decomp. temp. > 1000 C; 87.2% Sn; *Toxicology:* minimally irritating to eyes and skin on prolonged/repeated contact; inh. of dust may cause respiratory tract irritation.

Fascat® 2001. [Elf Atochem N. Am.] Stannous compd.; esterification catalyst for prep. of esters and polyesters, ester plasticizers; wh. to off-wh. cryst. powd.; sl. sol. in sol'ns. of oxalic acid, ammonium chloride, ammonium oxalate; insol. in esters, water; sp.gr. 3.56 (18/4 C); dens. 14.7 lb/gal; m.p. dec. 365 C; 57.3% tin.

Fascat® 2003. [Elf Atochem N. Am.] Stannous compd.; catalyst for esterification and transesterification reactions; catalyst for cure of epoxy resins in polymeric coatings for food pkg.; FDA 21CFR §175.300; colorless to pale yel. clear liq.; sol. in most org. solvs.; dens. 10.5 lb/gal; visc. 250 cs; pour pt. < -25 C; flash pt. 142 C; 28% total tin content; *Toxicology:* prolonged skin exposure may cause irritation; may cause eye irritation; sl. toxic on ingestion; vapors produced above 200 C should not be inhaled; *Storage:* handle in an inert atmosphere.

Fascat® 2004. [Elf Atochem N. Am.] Anhydrous stannous chloride; CAS 7772-99-8; EINECS 231-868-0; esterification catalyst for mfg. of plasticizers and polyesters; wh. flake; sol. in water, acetone, ethanol, methyl isobutyl carbinol, IPA, MEK; sl. sol. in diethyl ether, hi-flash naphtha, min. spirits; dens. 3.95 g/ml; m.p. 246 C; b.p. 623 C; 62% stannous tin; *Toxicology:* LD50 (oral, rat) 1745 mg/kg; mod. toxic by ingestion; extremely irritating to eyes; mildly irritating to skin; Unverified

Fascat® 2005. [Elf Atochem N. Am.] Anhydrous stannous chloride; CAS 7772-99-8; EINECS 231-868-0; food grade esterification catalyst for mfg. of plasticizers and polyesters; approved as component in polymeric coatings in food pkg.; FDA 21CFR §175.300, 184.1845, GRAS; wh. flake; sol. in water, acetone, ethanol, methyl isobutyl carbinol, IPA, MEK; sl. sol. in diethyl ether, hi-flash naphtha, min. spirits; dens. 3.95 g/ml; m.p. 246 C; b.p. 623 C; 62% stannous tin; *Toxicology:* LD50 (oral, rat) 1745 mg/kg; mod. toxic by ingestion; extremely irritating to eyes; mildly irritating to skin.

Fascat® 4100. [Elf Atochem N. Am.] Monobutyltin compd.; esterification catalyst for mfg. of unsat. polyesters, esp. isophthalates, polyester polyols, phthalates, adipates, powd. coatings, coil coatings, specialty enamels at temps. of 200-230 C; wh. solid; sol. in most ester prods.; insol. in most org.

solvs.; dissolves with reaction in carboxylic acids; sp.gr. 1.46; 55% tin; *Toxicology:* LD50 (oral, rat) > 10,000 mg/kg; pract. nontoxic by ingestion; minimally irritating to eyes and skin; avoid prolonged skin exposure.

Fascat® 4101. [Elf Atochem N. Am.] Organotin compd.; esterification catalyst for mfg. of unsat. polyesters, trioctyl trimellitate, etc.; wh. powd.; sol. in polar solvs. with reaction or complex formation, aromatic solvs. @ 80 C with dehydration, dil. NaOH, warm HNO_3; insol. in nonpolar solvs.; sp.gr. 1.26; m.p. dehydrates and polymerizes above 100 C; 48.3% tin, 14.5% Cl; *Toxicology:* prolonged skin contact may cause moderate irritation; inh. of dust may cause irritation.

Fascat® 4102. [Elf Atochem N. Am.] Organotin compd.; esterification/transesterification catalyst for mfg. of plasticizers and sat. or unsat. polyesters, powd. coatings, coil coatings, air-drying water-reducible alkyd enamels; pale yel. liq.; sol. in most org. solvs.; insol. in water; dens. 1.13 g/ml; pour pt. -18 C; 19% Sn, < 1% Cl; *Toxicology:* may cause skin and eye irritation.

Fascat® 4200. [Elf Atochem N. Am.] Organotin compd.; esterification, transesterification, or polyesterification catalyst for mfg. of plasticizers, etc.; colorless to pale yel. clear liq.; sol. in most org. solvs.; insol. in water; sp.gr. 1.32; visc. 18.2 cps; f.p. 10 C; b.p. 130 C (2 mm), dec. below atm. b.p.; flash pt. (COC) 290 F; ref. index 1.475; 33% Sn; *Toxicology:* causes burns to eyes and skin; heated vapors may cause respiratory tract irritation.

Fascat® 4201. [Elf Atochem N. Am.] Organotin compd.; esterification, transesterification, or polyesterification catalyst for prep. of esters, polyesters, alkyds; effective for methacrylate esters @ 100 C; used in coatings industry for prep. of alkyd resins and in reactions for curing blocked isocyanate; wh. amorphous powd.; insol. in water and most org. solvs.; may be brought into sol'n. in alcohols and esters by reaction with the solv.; sp.gr. 1.61; bulk dens. 43.5 lb/ft^3; m.p. dec. 350 C; 47% Sn; *Toxicology:* moderately toxic orally or by skin contact; may cause eye and skin burns on contact; avoid inh. of dust or vapors.

Fascat® 9100. [Elf Atochem N. Am.] Esterification, transesterification, and polycondensation catalyst for prep. of polyesters, unsat. polyesters, alkyds; suitable for food and pharmaceutical contact applics., such as coatings, epoxies, hybrid resins; FDA 21CFR §175.105, 175.300, 176.170, 177.1210, 177.2420; amorphous wh. powd.; insol. in water and org. solvs.; dissolves in acids with reaction; sp.gr. 1.46; bulk dens. 0.84 g/cc; m.p. 300 C (dec.); 57% tin; *Toxicology:* minimally irritating to eyes and skin.

Fascat® 9102. [Elf Atochem N. Am.] Esterification, transesterification, and polycondensation catalyst for prep. of polyesters, unsat. polyesters, alkyds; suitable for food and pharmaceutical contact applics., such as coatings, epoxies, hybrid resins; FDA 21CFR §175.105, 175.300, 176.170, 177.1210, 177.2420; pale yel. liq.; sol. in toluene, hydrocarbons, halocarbons, esters, etc.; insol. in water; sp.gr. 1.13; m.p. -18 C; 20% tin; *Toxicology:* may cause eye and skin irritation.

Fascat® 9201. [Elf Atochem N. Am.] Organotin compd.; esterification, transesterification, and polycondensation catalyst for prep. of polyesters, unsat. polyesters, alkyds; suitable for food and pharmaceutical contact applics., such as coatings, epoxies, hybrid resins; FDA 21CFR §175.105, 175.300, 176.170, 177.1210, 177.2420; wh. powd.; insol. in water, org. solvs.; will dissolve in acids, alochols, esters with reaction; sp.gr. 1.62; bulk dens. 0.70 g/cc; m.p. > 290 C (dec.); 48% tin; *Toxicology:* moderately toxic; may cause eye and skin burns on contact.

FC-24. [3M/Industrial Chem. Prods.] Trifluoromethanesulfonic acid; CAS 1493-13-6; EINECS 216-087-5; catalyst or reactant; polymerization of epoxies, styrenes, and THF, alkylation and some acylation reactions; pharmaceuticals, explosives, dyes, and intermediates; electrolytes; formation of biaryls; dehydrating agent; inhibits formation of metal oxides in welding and soldering fluxes; colorless liq.; sol. in polar org. solvs; misc. in water; m.w. 150.02; f.p. -40 C; b.p. 54 C (8 mm Hg); dens. 1.696 g/cc; visc. 2.87 cP; ref. index 1.325.

FC-520. [3M/Industrial Chem. Prods.] Trifluoromethanesulfonic acid amine salt and diethylene glycol monoethyl ether; catalyst for condensation and ring-opening polymerizations; coatings utilizing such resins as epoxides, aminoplasts, phenolics, and acrylics; coatings for wood, paper, and plastic; lemon-yel. clear liq.; f.p. cryst. at 40 F; misc. in water; sp.gr. 1.2; dens. 10.0 lb/gal; visc. 13 cps; pH 4.5-6.0; flash pt. 230 F (PMCC); 60% sol'n.

Ferro 1288. [Ferro/Bedford] Barium, cadmium, zinc stabilizer; heat stabilizer used in flexible vinyls processed by calendering, extrusion, inj. molding, or plastisol applics.; long-term stability; Gardner 8 liq.; dens 8.7 lb/gal; visc. (Gardner) A.

Ferro BP-10. [Ferro/Color] Ultramarine blue; CAS 57455-37-5; pigment for coloring thermoplastic resins, rubber compds., paints, printing inks, artists' colors, roofing granules; FDA 21CFR §178.3970; red shade blue powd.; 3.5 μ avg. particle size; < 0.10% +325 mesh residue; sp.gr. 2.35; dens. 19.58 lb/gal; oil absorp. 36; < 0.05% free S, < 1% moisture; *Toxicology:* nontoxic.

Ferro CP-18. [Ferro/Color] Ultramarine blue; CAS 57455-37-5; pigment for coloring thermoplastic resins, rubber compds., paints, printing inks, artists' colors, roofing granules; FDA 21CFR §178.3970; red shade blue powd.; 3.3 μ avg. particle size; < 0.10% +325 mesh residue; sp.gr. 2.35; dens. 19.58 lb/gal; oil absorp. 39; < 0.05% free S, < 1% moisture; *Toxicology:* nontoxic.

Ferro CP-50. [Ferro/Color] Ultramarine blue; CAS 57455-37-5; pigment for coloring thermoplastic resins, rubber compds., paints, printing inks, artists' colors, roofing granules; FDA 21CFR §178.3970; med. shade blue powd.; 2.3 μ avg. particle size; <

0.10% +325 mesh residue; sp.gr. 2.35; dens. 19.58 lb/gal; oil absorp. 40; < 0.05% free S, < 1.3% moisture; *Toxicology:* nontoxic.

Ferro CP-78. [Ferro/Color] Ultramarine blue; CAS 57455-37-5; pigment for coloring thermoplastic resins, rubber compds., paints, printing inks, artists' colors, roofing granules; FDA 21CFR §178.3970; powd.; < 0.10% +325 mesh residue; sp.gr. 2.35; dens. 19.58 lb/gal; oil absorp. 39; < 0.05% free S, < 1.3% moisture; *Toxicology:* nontoxic.

Ferro DP-25. [Ferro/Color] Ultramarine blue; CAS 57455-37-5; pigment for coloring thermoplastic resins, rubber compds., paints, printing inks, artists' colors, roofing granules; FDA 21CFR §178.3970; red shade blue powd.; 3.0 μ avg. particle size; < 0.10% +325 mesh residue; sp.gr. 2.35; dens. 19.58 lb/gal; oil absorp. 37; < 0.05% free S, < 1.0% moisture; *Toxicology:* nontoxic.

Ferro EP-37. [Ferro/Color] Ultramarine blue; CAS 57455-37-5; pigment for coloring thermoplastic resins, rubber compds., paints, printing inks, artists' colors, roofing granules; FDA 21CFR §178.3970; red shade blue powd.; 2.6 μ avg. particle size; < 0.10% +325 mesh residue; sp.gr. 2.35; dens. 19.58 lb/gal; oil absorp. 38; < 0.05% free S, < 1.0% moisture; *Toxicology:* nontoxic.

Ferro EP-62. [Ferro/Color] Ultramarine blue; CAS 57455-37-5; pigment for coloring thermoplastic resins, rubber compds., paints, printing inks, artists' colors, roofing granules; FDA 21CFR §178.3970; med. shade blue powd.; 1.1 μ avg. particle size; < 0.10% +325 mesh residue; sp.gr. 2.35; dens. 19.58 lb/gal; oil absorp. 39; < 0.05% free S, < 1.3% moisture; *Toxicology:* nontoxic.

Ferro FP-40. [Ferro/Color] Ultramarine blue; CAS 57455-37-5; pigment for coloring thermoplastic resins, rubber compds., paints, printing inks, artists' colors, roofing granules; FDA 21CFR §178.3970; med. shade blue powd.; 2.0 μ avg. particle size; < 0.10% +325 mesh residue; sp.gr. 2.35; dens. 19.58 lb/gal; oil absorp. 38; < 0.05% free S, < 1.3% moisture; *Toxicology:* nontoxic.

Ferro FP-64. [Ferro/Color] Ultramarine blue; CAS 57455-37-5; pigment for coloring thermoplastic resins, rubber compds., paints, printing inks, artists' colors, roofing granules; FDA 21CFR §178.3970; green shade blue powd.; 0.8 μ avg. particle size; < 0.10% +325 mesh residue; sp.gr. 2.35; dens. 19.58 lb/gal; oil absorp. 39; < 0.05% free S, < 1.3% moisture; *Toxicology:* nontoxic.

Ferro RB-30. [Ferro/Color] Ultramarine blue; CAS 57455-37-5; acid-resist. pigment for coloring thermoplastic resins, rubber compds., paints, printing inks, artists' colors, roofing granules; FDA 21CFR §178.3970; red shade blue powd.; 1.7 μ avg. particle size; < 0.25% +325 mesh residue; sp.gr. 2.30; dens. 19.16 lb/gal; oil absorp. 46; < 0.05% free S, < 1.3% moisture; *Toxicology:* nontoxic.

Ferro V-5. [Ferro/Color] Ultramarine violet; CAS 12769-96-9; pigments for coloring thermoplastic resins, rubber compds., paints, printing inks, artists' colors, roofing granules; FDA 21CFR §178.3970; powd.; 1.35 μ avg. particle size; < 0.10% +325 mesh residue; sp.gr. 2.35; dens. 19.58 lb/gal; oil absorp. 34; < 0.1% free S, < 2% moisture; *Toxicology:* nontoxic.

Ferro V-8. [Ferro/Color] Ultramarine violet; CAS 12769-96-9; pigments for coloring thermoplastic resins, rubber compds., paints, printing inks, artists' colors, roofing granules; FDA 21CFR §178.3970; powd.; 1.05 μ avg. particle size; < 0.10% +325 mesh residue; sp.gr. 2.35; dens. 19.58 lb/gal; oil absorp. 34; < 0.1% free S, < 2% moisture; *Toxicology:* nontoxic.

Ferro V-9415. [Ferro/Color] Pigment yellow 53; CAS 8007-18-9; pigment for thermoplastic and thermoset resins, esp. high temp. engineering resins, PVC siding and profile, industrial finishes; yel. powd.; 1.2 μ avg. particle size; trace +325 mesh residue; sp.gr. 4.4; dens. 36.7 lb/gal; bulking value 0.0272 gal/lb; oil absorp. 12; Unverified

Ferro-Char B-44M. [D.J. Enterprises] Synergist for flame retardants in thermoplastics, PVC, urethane, adhesives, epoxies, rubber compds.; effective at low load levels (3-12%); can be used with glass-reinforced, hollow or solid sphere, MgO, mica, and talc compds.; 44-75 μ particle size; sp.gr. 4.85; m.p. 2400 F; pH 6.8-7.2; 64% Fe, 6% TiO_2, 4% Al_2O_3, 2% SiO_2, 1% MgO, 1% V_2O_5.

Ferrosil™ 14. [Kaopolite] Ferroaluminum silicate; CAS 12178-41-5; filler for abrasion-resistant molding plastics, abrasive compds., auto, metal, and plastic polishes, caulks, sealants, cleaning compds., industrial coatings; extender pigment for primers and other coatings; lt. buff powd., odorless; 98% finer than 14 μ; sp.gr. 4.0; dens. 33.32 lb/gal solid; bulking value 0.030 gal/lb; oil absorp. (Gardner-Coleman) 18.0; ref. index 1.83; pH 8.5-9.0 (20%solids); hardness (Moh) 8-9; > 99% act.; 41.24% SiO_2; 22.23% iron oxide; 20.36% Al_2O_3; 12.35% MgO; 2.97% CaO; *Toxicology:* TLV 10 mg/m^3 (total dust); nuisance dust; *Precaution:* nonflamm., nonexplosive.

Fiberfrax® 6000 RPS. [Carborundum] Kaolin ceramic fiber, surf. treated; CAS 1332-58-7; EINECS 296-473-8; reinforcement and filler for phenolic, epoxy, nylon, melamine and polyurethane systems; high temp. stability, low thermal conductivity, superior corrosion resist., exc. sound dampening, high dielec. constant; wh. to lt. gray fibers; 2-4 μ fiber diam.; 400 μ fiber length; sp.gr. 2.7 g/cc; m.p. 1790 C; tens. str. 250 kpsi; hardness (Mohs) 6.

Fiberfrax® 6900-70. [Carborundum] Kaolin ceramic fiber; CAS 1332-58-7; EINECS 296-473-8; reinforcement material for plastic composites (molded thermosets, thermoplastic), friction materials, high-performance coatings, adhesives, and sealants; exc. high temp. stability, superior chem. resist., very low moisture absorp.; wh. to lt. gray fibers; 2-3 μ mean fiber diam.; 70-150 μ avg. fiber length; sp.gr. 2.73 g/m^3; m.p. 1738 C; tens. str. 2.5×10^5 psi; 42-48% Al_2O_3, 49-55% SiO_2.

Fiberfrax® 6900-70S. [Carborundum] Kaolin ceramic fiber, surf. treated; CAS 1332-58-7; EINECS 296-

473-8; reinforcement material for thermoset and thermoplastic plastic composites, intumescent and friction materials, high-performance coatings, adhesives, and sealants; exc. high-temp. stability, superior chem. resist., very low moisture absorp.; surf. treatment improves resin-to-fiber bonding in phenolic, epoxy, nylon, melamine, and PU systems; wh. to lt. gray fiber; 2-3 μ mean fiber diam.; 70-150 μ avg. fiber length; sp.gr. 2.73 g/m^3; m.p. 1738 C; tens. str. 2.5 x 10^5 psi; 42-48% Al_2O_3, 49-55% SiO_2.

Fiberfrax® EF-119. [Carborundum] Kaolin ceramic fiber; CAS 1332-58-7; EINECS 296-473-8; functional reinforcement for coatings, adhesives, caulk, sealants, thermoset and thermoplastic resin systems, friction materials, and elastomers; temp. stability to 2300 F, superior chem. resist., very low moisture absorp.; improves wear and corrosion resist., dimensional stability; off-wh. to gray fiber; 2-3 μ mean fiber diam., 25 μ mean fiber length; sp.gr. 2.73 g/cc; m.p. 1790 C; 45-51% Al_2O_3, 46-52% SiO_2.

Fiberfrax® EF 122S. [Carborundum] Kaolin ceramic fiber, surf. treated; CAS 1332-58-7; EINECS 296-473-8; friction modifier/additive to friction materials; reinforcement for polymers and elastomers; additive to ablative materials; high temp. stability, superior corrosion resist., very low moisture absorp., exc. frictional stability; wh. to lt. gray fiber; 5-10 μ mean fiber diam., 70-150 μ avg. fiber length; sp.gr. 2.73 g/m^3; m.p. 1738 C; tens. str. 2.5 x 10^5 psi; hardness (Mohs) 6.

Fiberfrax® EF 129. [Carborundum] Kaolin ceramic fiber; CAS 1332-58-7; EINECS 296-473-8; reinforcement for thermoset composites; wh. to lt. gray fiber; 2-3 μ mean fiber diam., 70-150 μ avg. fiber length; sp.gr. 2.73 g/m^3; tens. str. 250 kpsi; hardness (Mohs) 6.

Fiberfrax® Milled Fiber. [Carborundum] Kaolin ceramic fiber; CAS 1332-58-7; EINECS 296-473-8; additive for plastics, resins, refractory cement compositions; compact filler insulation; exc. chem. and corrosion resist.; resists oxidation and reduction; high temp. stability; wh. to lt. gray powd.; 2-3 μ mean fiber diam., 14 μ mean fiber length; sp.gr. 2.73 g/cc; dens. 51 lb/ft^3; m.p. 1790 C; conduct. 0.1887 W/m°C (538 C); sp. heat 1130 J/kg°C (1093 C); 43.4% Al_2O_3, 53.9% SiO_2.

Fiberglas® 101C. [Owens-Corning Fiberglas] Chopped glass strand; reinforcement for BMC applics. such as automotive head lamps, appliances, power tools, elec. components, microwave cookware, business equip.; compat. with polyester.

Fiberkal™. [Kaopolite] Fibrous calcined kaolin; CAS 1332-58-7; EINECS 296-473-8; reinforcement for plastics, rubbers, and various coatings when exposed to extreme temps., solv. and mech. shock; filtration aid; texture paints, caulks, ceramics; asbestos replacement in some applics.; long fine fibers.

Fiberkal™ FG. [Kaopolite] Fibrous calcined kaolin; CAS 1332-58-7; EINECS 296-473-8; reinforcing filler for plastics, refractory coatings, specialty coatings, auto polish, caulks, mastics, sealants, ceramics; filtration aid; replacement for asbestos; easy dispersion in many systems such as thru a spray nozzle; improved elec. props.; very short fiber length (< 100 μ), 2-4 μ diam.

Fibra-Cel® BH-70. [Celite] Cellulose fibers; CAS 9004-34-6; EINECS 232-674-9; functional fillers, reinforcement, thickening aid, processing aid, conditioning agent for the rubber industry (shoe soles, v-belts, tires, gaskets), thermoset and thermoplastic resins, building prods., pet food, asphalt, latex paints; recommended for thermosets, rubber; wh. fibers, 20 x 120 μ fiber (width x length); bulk dens. 9.7 lb/ft^3; oil absorp. 155%; pH 4-7; 99.5% cellulose, 4% max. moisture; *Toxicology:* nonirritating, nontoxic.

Fibra-Cel® BH-200. [Celite] Cellulose fibers; CAS 9004-34-6; EINECS 232-674-9; functional fillers, reinforcement, thickening aid, processing aid, conditioning agent for the rubber industry (shoe soles, v-belts, tires, gaskets), thermoset and thermoplastic resins, building prods., pet food, asphalt, latex paints; recommended for thermoplastics; wh. fibers, 20 x 30 μ fiber (width x length); bulk dens. 13.1 lb/ft^3; oil absorp. 120%; pH 4-7; 99.5% cellulose, 8% max. moisture; *Toxicology:* nonirritating, nontoxic.

Fibra-Cel® CBR-18. [Celite] Cellulose fibers; CAS 9004-34-6; EINECS 232-674-9; functional fillers, reinforcement, thickening aid, processing aid, conditioning agent for the rubber industry (shoe soles, v-belts, tires, gaskets), thermoset and thermoplastic resins, building prods., pet food, asphalt, latex paints; recommended for thermoplastics; brn. fibers, 70 x 350 μ fiber (width x length); bulk dens. 14.4 lb/ft^3; oil absorp. 170%; pH 4-7; 75% cellulose, 3% max. moisture; *Toxicology:* nonirritating, nontoxic.

Fibra-Cel® CBR-40. [Celite] Cellulose fibers; CAS 9004-34-6; EINECS 232-674-9; functional fillers, reinforcement, thickening aid, processing aid, conditioning agent for the rubber industry (shoe soles, v-belts, tires, gaskets), thermoset and thermoplastic resins, building prods., pet food, asphalt, latex paints; recommended for thermosets, rubber; yel.-brn. fibers, 45 x 160 μ fiber (width x length); bulk dens. 7.7 lb/ft^3; oil absorp. 210%; pH 4-7; 70% cellulose, 6% max. moisture; *Toxicology:* nonirritating, nontoxic.

Fibra-Cel® CBR-41. [Celite] Cellulose fibers; CAS 9004-34-6; EINECS 232-674-9; functional fillers, reinforcement, thickening aid, processing aid, conditioning agent for the rubber industry (shoe soles, v-belts, tires, gaskets), thermoset and thermoplastic resins, building prods., pet food, asphalt, latex paints; recommended for thermoplastics; yel.-brn. fibers, 45 x 160 μ fiber (width x length); bulk dens. 7.9 lb/ft^3; oil absorp. 230%; pH 4-7; 70% cellulose, 2% max. moisture; *Toxicology:* nonirritating, nontoxic.

Fibra-Cel® CBR-50. [Celite] Cellulose fibers; CAS 9004-34-6; EINECS 232-674-9; functional fillers, reinforcement, thickening aid, processing aid, conditioning agent for the rubber industry (shoe soles, v-belts, tires, gaskets), thermoset and thermoplastic resins, building prods., pet food, asphalt, latex paints; recommended for thermosets; yel.-brn. fibers, 45 x 150 μ fiber (width x length); bulk dens. 10.6 lb/ft^3; oil absorp. 190%; pH 4-7; 75% cellulose, 9% max. moisture; *Toxicology:* nonirritating, nontoxic.

Fibra-Cel® CC-20. [Celite] Cellulose fibers; CAS 9004-34-6; EINECS 232-674-9; functional fillers, reinforcement, thickening aid, processing aid, conditioning agent for the rubber industry (shoe soles, v-belts, tires, gaskets), thermoset and thermoplastic resins, building prods., pet food, asphalt, latex paints; recommended for thermosets; gray fibers, 25 x 200 μ fiber (width x length); bulk dens. 6.9 lb/ft^3; oil absorp. 250%; pH 4-7; 90% cellulose, 6% max. moisture; *Toxicology:* nonirritating, nontoxic.

Fibra-Cel® SW-8. [Celite] Cellulose fibers; CAS 9004-34-6; EINECS 232-674-9; functional fillers, reinforcement, thickening aid, processing aid, conditioning agent for the rubber industry (shoe soles, v-belts, tires, gaskets), thermoset and thermoplastic resins, building prods., pet food, asphalt, latex paints; recommended for thermoplastics, min.-bonded bldg. prods., asphalt mixtures; off-wh. fibers, 35 x 500 μ fiber (width x length); bulk dens. 4.4 lb/ft^3; oil absorp. 320%; pH 6-8; 90% cellulose, 8% max. moisture; *Toxicology:* nonirritating, nontoxic.

Fibrox 030-E. [Industrial Fibers] Mineral wool fiber; filler/reinforcer for bitumen, paper, gaskets, paints, friction prods., plastics, rubber, coatings, caulks and other applics.; used in nylon 6.6, PP, phenolics, PBT, BMC, SMC; off-wh. fibers, low odor; 5-6 μ avg. diam., 6 mm avg. length; negligible sol. in water; dens. 2.7-2.9 g/cc; m.p. ≈ 2200 F; ref. index 1.62-16.4; hardness (Moh) 6.0; 38-46% SiO_2, 20-40% CaO, 6-14% Al_2O_3, 5-10% MgO; *Toxicology:* ACGIH TLV/TWA 10 mg/m^3; nuisance particulate, may cause transitory skin irritation, possible irritation of eyes, upper respiratory tract.

Fibrox 030-ES. [Industrial Fibers] Mineral wool fiber, silane coated; filler/reinforcer for bitumen, paper, gaskets, paints, friction prods., plastics, rubber, coatings, caulks and other applics.; off-wh. fibers, low odor; 5-6 μ avg. diam., 6 mm avg. length; negligible sol. in water; dens. 2.7-2.9 g/cc; m.p. ≈ 2200 F; ref. index 1.62-1.64; hardness (Moh) 6.0; 38-46% SiO_2, 20-40% CaO, 6-14% Al_2O_3, 5-10% MgO; *Toxicology:* ACGIH TLV/TWA 10 mg/m^3; nuisance particulate, may cause transitory skin irritation, possible irritation of eyes, upper respiratory tract.

Fibrox 030 SC. [Industrial Fibers] Mineral wool fiber; min. fiber for bitumen, paper, gaskets, paints, friction prods., plastics, rubber, coatings, caulks and other applics.; off-wh. fibers, low odor; 5-6 μ avg. diam., 6 mm avg. length; negligible sol. in water; dens. 2.7-2.9 g/cc; m.p. ≈ 2200 F; ref. index 1.62-1.64; hardness (Moh) 6.0; 38-46% SiO_2, 20-40% CaO, 6-14% Al_2O_3, 5-10% MgO; *Toxicology:* ACGIH TLV/TWA 10 mg/m^3; nuisance particulate, may cause transitory skin irritation, possible irritation of eyes, upper respiratory tract.

Fibrox 300. [Industrial Fibers] Mineral wool fiber; filler/reinforcer for bitumen, paper, gaskets, paints, friction prods., plastics, rubber, coatings, caulks and other applics.; off-wh. fibers, low odor; 5-6 μ avg. diam., 16-19 mm avg. length; negligible sol. in water; dens. 2.7-2.9 g/cc; m.p. ≈ 2200 F; ref. index 1.62-1.64; hardness (Moh) 6.0; 38-46% SiO_2, 20-40% CaO, 6-14% Al_2O_3, 5-10% MgO; *Toxicology:* ACGIH TLV/TWA 10 mg/m^3; nuisance particulate, may cause transitory skin irritation, possible irritation of eyes, upper respiratory tract.

Ficel® 18 Series. [Witco/PAG] Combination blowing agent/additive systems; for continuous and batch prod. of crosslinked polyolefin foams; processing temps. 180-200 C.

Ficel® 35 Series. [Witco/PAG] Endothermic blowing agent systems for polyolefins, PVC, styrenics, sponge rubber, for extrusion of cellular profiles; processing temps. 160-200 C.

Ficel® 44 Series. [Witco/PAG] Nondusty powd. forms of selected Ficel blowing agents; for PVC, sponge rubber; processing temps. 200-220 C.

Ficel® 46 Series. [Witco/PAG] Paste disps. of selected Ficel blowing agents; for PVC; processing temps. 200-220 C.

Ficel® 49 Series. [Witco/PAG] Rubber masterbatches of selected Ficel blowing agents in an EP-based carrier; for polyolefins, sponge rubber; processing temps. 140-220 C.

Ficel® 50 Series. [Witco/PAG] Thermoplastic masterbatches of selected Ficel blowing agents in an LDPE or universally compat. carrier; for polyolefins, styrenics; processing temps. 150-260 C.

Ficel® AC1. [Witco/PAG] Azodicarbonamide; CAS 123-77-3; EINECS 204-650-8; chem. blowing agent for polyolefins, PVC, sponge rubber processing at 200-220 C; powd.; 20 μ avg. particle size.

Ficel® AC2. [Witco/PAG] Azodicarbonamide; CAS 123-77-3; EINECS 204-650-8; chem. blowing agent for polyolefins, PVC, styrenics, sponge rubber, cushion vinyl floor covering, expanded vinyl coated fabrics, profiles/pipe, sealants; processing temp. 200-220 C; yel. powd.; 10.5 μ avg. particle size; negligible sol. in water; 100% act.

Ficel® AC2M. [Witco/PAG] Azodicarbonamide; CAS 123-77-3; EINECS 204-650-8; chem. blowing agent for sponge rubber processing at 200-220 C; nondusty powd.; 10.5 μ avg. particle size.

Ficel® AC3. [Witco/PAG] Azodicarbonamide; CAS 123-77-3; EINECS 204-650-8; chem. blowing agent for polyolefins, PVC, styrenics, sponge rubber, cushion vinyl floor covering, expanded vinyl coated fabrics, profiles/pipe, sealants processing at 200-220 C; powd.; 6.2 μ avg. particle size.

Ficel® AC3F. [Witco/PAG] Azodicarbonamide; CAS 123-77-3; EINECS 204-650-8; chem. blowing agent for polyolefins, PVC, styrenics, sponge rubber, cushion vinyl floor covering, expanded vinyl coated fabrics, profiles/pipe, sealants processing at 200-220 C; powd. with tightly controlled particle size of 4.65 μ.

Ficel® AC4. [Witco/PAG] Azodicarbonamide; CAS 123-77-3; EINECS 204-650-8; chem. blowing agent for polyolefins, PVC, styrenics, sponge rubber, cushion vinyl floor covering, expanded vinyl coated fabrics, profiles/pipe, sealants processing at 200-220 C; powd.; 4.65 μ avg. particle size.

Ficel® ACSP4. [Witco/PAG] Coated azodicarbonamide; CAS 123-77-3; EINECS 204-650-8; blowing agent for PVC, sponge rubber; disp. in vinyl plastisols; processing temps. 200-220 C; powd.; 5 μ partiocle size; sol. in dimethyl sulfoxide.

Ficel® ACSP5. [Witco/PAG] Coated azodicarbonamide; CAS 123-77-3; EINECS 204-650-8; blowing agent for PVC, sponge rubber; dispersible grade; powd.; 4.2 μ particle size.

Ficel® AFA. [Witco/PAG] Specially formulated blowing agent with an ammonia-free decomp.; blowing agent for polyolefins, styrenics processing at 230-260 C.

Ficel® AZDN-LF. [Witco/PAG] 2,2-Azodiisobutyronitrile; CAS 78-67-1; EINECS 201-132-3; polymerization initiator for a wide range of monomers, acrylics, PVC, styrenics; crosslinking agent for unsat. polyesters; catalyst for free-radical halogenation or sulfochlorination of hydrocarbons; depolymerizing agent in mfg. of chlorinated rubber; off-wh.free-flowing powd., char. odor; insol. in water; sp.gr. 0.4; m.p. 103 C (dec.).

Ficel® AZDN-LMC. [Witco/PAG] 2,2′-Azodiisobutyronitrile; CAS 78-67-1; EINECS 201-132-3; polymerization initiator for a wide range of monomers, acrylics, PVC, styrenics; crosslinking agent for unsat. polyesters; catalyst for free-radical halogenation or sulfochlorination of hydrocarbons; depolymerizing agent in mfg. of chlorinated rubber; low metal and chloride content for applics. demanding high level of purity.

Ficel® EPA. [Witco/PAG] Nonplating azodicarbonamide; CAS 123-77-3; EINECS 204-650-8; chem. blowing agent for polyolefins, styrenics, structural foam molding, wire and cable insulation, profiles, pipe, and film; self-nucleating; processing temps. 190-220 C.

Ficel® EPB. [Witco/PAG] Activated nonplating azodicarbonamide; CAS 123-77-3; EINECS 204-650-8; chem. blowing agent for polyolefins, styrenics, structural foam molding, wire and cable insulation, profiles, pipe, and film; self-nucleating; processing temps. 170-200 C.

Ficel® EPC. [Witco/PAG] Highly activated nonplating azodicarbonamide; CAS 123-77-3; EINECS 204-650-8; chem. blowing agent for polyolefins, styrenics, sponge rubber, structural foam molding, wire and cable insulation, profiles, pipe, and film; self-nucleating; processing temps. 180-200 C.

Ficel® EPD. [Witco/PAG] Nonplating azodicarbonamide; CAS 123-77-3; EINECS 204-650-8; chem. blowing agent for polyolefins, structural foam molding, wire and cable insulation, profiles, pipe, and film; self-nucleating; processing temps. 200-220 C.

Ficel® EPE. [Witco/PAG] Very highly activated nonplating azodicarbonamide; CAS 123-77-3; EINECS 204-650-8; chem. blowing agent for polyolefins, PVC, styrenics, sponge rubber, structural foam molding, wire and cable insulation, profiles, pipe, and film; self-nucleating; processing temps. 150-170 C.

Ficel® LE. [Witco/PAG] Pre-activated azodicarbonamide; CAS 123-77-3; EINECS 204-650-8; blowing agent system for low temp. or high speed expansions; for PVC, sponge rubber; processing temps. 150-200 C; yel. powd.

Ficel® SCE. [Witco/PAG] Pre-activated azodicarbonamide; CAS 123-77-3; EINECS 204-650-8; chemical blowing agent one-pkg. system used chemical embossing of vinyl floor and wall coverings, plastisol and sponge rubber applics.; food contact applics.; processing temps. 160-180 C; yel. powd.

Firebrake® 500. [U.S. Borax] Anhyd. zinc borate; CAS 12513-27-8; flame retardant for polymers which require processing temps. of 290 C or higher; powd.; 5-15 μ mean particle size; sp.gr. 2.6 g/cc; *Toxicology:* respirable dust.

Firebrake® ZB. [U.S. Borax] Zinc borate; CAS 138265-88-0; flame retardant, smoke suppressant, afterglow suppressant, and anti-arcing agent in PVC, nylon, epoxy, polyolefin, polyester, PU, thermoplastic elastomers, and rubbers; suitable for systems processing at high temps.; wh. fine powd., odorless; 7 μ median particle size; sol. 0.3% in water; m.w. 434.66; sp.gr. 2.7; bulk dens. 50 lb/ft^3; vapor pressure negligible @ 20 C; oil absorp. 35; m.p. phase change @ 650 C; ref. index 1.58; 38.2% ZnO, 48.2% B_2O_3, 13.6% water; *Toxicology:* LD50 (oral, rat) > 10,000 mg/kg, (dermal, rabbit) > 10,000 mg/kg; low acute oral and dermal toxicities; inh. of dust may cause irritation; nuisance dust OSHA PEL 15 mg/m^3 total dust, 5 mg/m^3 respirable dust; nonirritating to eyes and skin; *Precaution:* not combustible or explosive; reaction with strong reducing agents will generate hydrogen gas which could create an explsoive hazard; *Storage:* store at ambient temps. in dry area indoors.

Firebrake® ZB 415. [U.S. Borax] Zinc borate; CAS 1332-07-6; high dehydration temp. flame retardant, smoke suppressant for engineering plastics processing at high temps., esp. flexible PVC in combination with antimony oxide and other common fire retardants; wh. powd.; 10 μ typical particle size; sol. < 0.1% in water; sp.gr. 3.7; oil absorp. 28; dehydration temp. 415 C; ref. index 1.65; pH 6.75-7.5 (10% aq. slurry).

Firebrake® ZB-Extra Fine. [U.S. Borax] Zinc borate; CAS 1332-07-6; flame retardant, smoke suppressant for applics. where extra fine particle size is desirable for retention of polymer props.; wh. powd.; 2 μ typical particle size, all < 6 μ; sp.gr. 2.8; dehydration temp. 290 C; *Toxicology:* respirable dust.

Firebrake® ZB-Fine. [U.S. Borax] Zinc borate; CAS 12513-27-8; flame retardant, smoke suppressant for use where fine grade is preferred or physical props. are critical, e.g., in films or adhesives; wh. fine powd., odorless; 2-4 μ mean particle size; sol. 0.3% in water; m.w. 434.66; sp.gr. 2.7; bulk dens. 25-40 lb/ft^3 (loose); vapor pressure negligible @ 20 C; oil absorp. 30; ref. index 1.58; 37.45% ZnO, 48.05% B_2O_3, 14.5% water; *Toxicology:* LD50 (oral, rat) > 10,000 mg/kg, (dermal, rabbit) > 10,000 mg/kg; low acute oral and dermal toxicities; inh. of dust may cause irritation; nuisance dust OSHA PEL 15 mg/m^3 total dust, 5 mg/m^3 respirable dust; nonirritating to eyes and skin; *Precaution:* not combustible or explosive; reaction with strong reducing agents will generate hydrogen gas which could create an explsoive hazard; *Storage:* store at ambient temps. in dry area indoors.

Firemaster® 642. [Great Lakes] Halogenated phosphate ester; flame retardant for rigid and flexible urethane foams; liq.; sol. in methylene chloride, toluene, MEK, > 100 g/100 g in methanol, < 0.1 g/100 g water; sp.gr. 1.7; 49.2% halogen, 6.5% P.

Firemaster® 836. [Great Lakes] Halogenated phosphate ester; flame retardant for EPS foam, rigid and flexible urethane foams; liq.; sol. in methanol, methylene chloride, toluene, MEK, < 0.1 g/100 g in water; m.w. 416.5; sp.gr. 1.6; 44.5% halogen, 7.5% P.

Firemaster® HP-36. [Great Lakes] Halogenated phosphate ester; flame retardant for EPS foam, rigid urethane foam, unsat. polyesters, TPE wire insulation, adhesives, coatings, and textiles; liq.; sol. in methanol, methylene chloride, toluene, MEK, < 0.1 g/100 g in water; m.w. 416.5; sp.gr. 1.6; 44.5% halogen, 7.5% P.

FireShield® H. [Laurel Industries] Antimony trioxide; CAS 1309-64-4; EINECS 215-175-0; general purpose flame-retardant synergist for plastics, rubber, paper and paint; polymerization catalyst for PET resins and fibers; also for electronics, glass, ceramics, petrol. refining, and chem. mfg.; wh. fine cryst. powd., odorless; 1.0-1.8 μ avg. particle size; very sl. sol. in water; dissolves in conc. HCl, sulfuric acid, strong alkalies; m.w. 291.52; sp.gr. 5.3; bulk dens. 60 lb/ft^3; m.p. 656 C; b.p. 1425 C; ref. index 2.1; 99.3-99.9% Sb_2O_3; *Toxicology:* avoid skin or eye contact, inhalation; use only with adequate ventilation; *Storage:* store in dry, well-ventilated area.

FireShield® HPM. [Laurel Industries] Antimony trioxide; CAS 1309-64-4; EINECS 215-175-0; polymerization catalyst for PET resins and fibers, chem. intermediate, flame retardant synergist for plastics, rubber, paint, paper; suitable for sensitive electronic applics.; wh. fine cryst. powd., odorless; very sl. sol. in water; dissolves in conc. HCl, sulfuric acid, strong alkalies; m.w. 291.52; sp.gr. 5.3; bulk dens. 60 lb/ft^3; surf. area 2.5 m^2/g; m.p. 656 C; b.p. 1425 C; ref. index 2.1; 99.3-99.9% Sb_2O_3; *Toxicology:* avoid skin or eye contact, inhalation; use only with adequate ventilation; *Storage:* store in dry, well-ventilated area.

FireShield® HPM-UF. [Laurel Industries] Antimony trioxide; CAS 1309-64-4; EINECS 215-175-0; flame retardant, catalyst for liq. systems such as epoxies, urethanes, adhesives, paints, solvs., colloids, PET polyesters; wh. fine cryst. powd., odorless; 0.2-0.4 μ avg. particle size; very sl. sol. in water; dissolves in conc. HCl, sulfuric acid, strong alkalies; m.w. 291.52; sp.gr. 5.3; bulk dens. 60 lb/ft^3; surf. area 12.5 m^2/g; m.p. 656 C; b.p. 1425 C; ref. index 2.1; 99.3-99.9% Sb_2O_3; *Toxicology:* avoid skin or eye contact, inhalation; use only with adequate ventilation; *Storage:* store in dry, well-ventilated area.

FireShield® L. [Laurel Industries] Antimony trioxide; CAS 1309-64-4; EINECS 215-175-0; general purpose flame-retardant synergist for plastics, rubber, paper and paint; polymerization catalyst for PET resins and fibers; also for electronics, glass, ceramics, petrol. refining, and chem. mfg.; wh. fine cryst. powd., odorless; 2.5-3.5 μ avg. particle size; very sl. sol. in water; dissolves in conc. HCl, sulfuric acid, strong alkalies; m.w. 291.52; sp.gr. 5.3; bulk dens. 90 lb/ft^3; m.p. 656 C; b.p. 1425 C; ref. index 2.1; 99.3-99.9% Sb_2O_3; *Toxicology:* avoid skin or eye contact, inhalation; use only with adequate ventilation; *Storage:* store in dry, well-ventilated area.

Firstcure™ DMMT. [First Chem.] N,N-Dimethyl-m-toluidine; CAS 121-72-2; EINECS 204-495-6; cure accelerator used to promote free radical generation from organic peroxide initiators in addition polymerizations of unsat. polyester and acrylate systems; colorless to lt. amber oil; m.w. 135.21; sp.gr. 0.941 (10/4 C); vapor pressure < 0.1 mm Hg (72 C); b.p. 212 C (760 mm); flash pt. (CC) 85 C; ref. index 1.5492 (20 C); > 99% act.; *Toxicology:* poisonous; may cause eye and skin irritation; may cause cyanosis, headaches, dizziness, nausea, vomiting, unconsciousness, death; *Precaution:* combustible; vapors are flamm. at elevated temps.; *Storage:* darkens on storage; sensitive to air and/or light; commonly stored in steel or PE tanks.

Firstcure™ DMPT. [First Chem.] N,N-Dimethyl-p-toluidine; CAS 99-97-8; EINECS 202-805-4; cure accelerator used to promote free radical generation from organic peroxide initiators in addition polymerizations of unsat. polyester and acrylate systems; colorless to lt. amber oil; m.w. 135.23; sp.gr. 0.992; vapor pressure 5 mm Hg (72 C); b.p. 210-211 C (760 mm); flash pt. (CC) 83 C; ref. index 1.5458; > 99% act.; *Toxicology:* poisonous; may cause eye and skin irritation; vapor or mist is irritating to mucous membranes and respiratory tract; may cause headaches, cyanosis, dizziness; *Precaution:* combustible; vapors are flamm. at elevated temps.; *Storage:* darkens on storage; sensitive to air and/or light; commonly stored in steel tanks.

Fizul 201-11. [Finetex] Ethoxylated alcohol half-ester sulfosuccinate; detergent, wetting agent, solubilizer, dispersant, emulsifier, hydrotrope; surfactant for foaming wallboards, emulsion polymerization; high electrolyte tolerance; liq.; 31% conc.

Fizul M-440. [Finetex] Tetrasodium N-(1,2-dicarboxyethyl)-N-octadecylsulfosuccinamate; anionic; emulsifier, dispersant, solubilizer, stabilizer, suspending agent; for emulsion polymerization of vinyl chloride, styrene/acrylic, styrene/butadiene, acrylic, and vinyl acetate systems; compat. over wide pH range and in presence of high levels of electrolytes; liq.; 35% conc.

Fizul MD-318. [Finetex] Half-ester sulfosuccinamate; anionic; emulsifier for emulsion polymerization of vinyl acetate alone or in combination with other vinyl functional monomers; high tolerance to calcium carbonate and electrolytes; liq.; 35% conc.

FK 140. [Degussa] Precipitated silica; low surf. area filler for silicone rubber industry; provides improved processability and low visc. buildup.

FK 160. [Degussa] Precipitated silica; active filler for hot-vulcanizing silicone rubber; provides increased surf. area, sl. higher tens. str. and durometer than FK 140, exc. heat resist. in finished materials; wh. fluffy powd.; 18 nm avg. particle size; dens. 70 g/l (tapped); surf. area 160 m^2/g; pH 4.5 (5% aq. susp.); 99.4% assay.

FK 300DS. [Degussa] Precipitated silica; hydrophilic filler for plastics, paper and film industry; powd.; 10 nm avg. particle size; dens. 90 g/l (tapped); surf. area 300 m^2/g; pH 6.5 (5% aq. susp.); 99% assay.

FK 310. [Degussa] Precipitated silica; filler for antiblocking of LDPE, PP, and flatting of PP tapes; powd.; 8 nm avg. particle size; dens. 130 g/l (tapped); surf. area 650 m^2/g; pH 7 (5% aq. susp.); 99% assay.

FK 320. [Degussa] Precipitated silica; anti-plate out additive for calendered PVC; hydrophilic; powd.; 18 nm avg. particle size; dens. 240 g/l (tapped); surf. area 170 m^2/g; pH 6.3 (5% aq. susp.); 98% assay.

FK 320DS. [Degussa] Precipitated silica; filler for plastics, silicone rubber, paper, and film industry applics.; wh. fluffy powd.; 18 nm avg. particle size; dens. ≈ 80 g/l; surf. area 170 m^2/g; pH 6.3 (5% aq. susp.); 98% assay.

FK 500LS. [Degussa] Precipitated silica; thickener for cosmetic creams and lotions; also in thermal insulation, paper, films, pesticides, pharmaceuticals; prevents plate out in PVC calendered films; FDA approved; powd.; 7 nm avg. particle size; dens 80 g/l (tapped); surf. area 450 m^2/g; pH 6.3 (5% aq. susp.); 98.5% assay.

Flame-Amine. [Miljac] Pentaerythritol; CAS 115-77-5; EINECS 204-104-9; flame retardant for PVC, PU flexible and rigid foam.

Flamebloc. [Buckman Labs] Proprietary; uv stabilizer for ABS, epoxy, polyesters, PE, PP, flexible and rigid PVC, VDC; powd.; m.p. 1000 C.

Flamegard® 908. [Sybron] Methylolthiourea-based compd.; flame retardant for nylon.

Flamtard H. [Alcan] Zinc hydroxystannate; CAS 12027-96-2; flame retardant for PVC, polychloroprene, chlorosulfonated polyethylene, other halopolymers; wh. free-flowing powd., odolress; 2.5 μm particle size; sol. 0.10 g/l in water; sp.gr. 3.3; dens. 3.6 g/cc; bulk dens. 1.4 g/cc (untamped); surf. area 4.5 m^2/g; oil absorp. 18 g/100 g; m.p. dec. > 180 C; pH 8.6 (5%); *Toxicology:* low toxicity by ingestion; eye irritation by mechanical abrasion; LD50 (rat, oral) > 5000 mg/kg; *Storage:* store in dry, well-ventilated area away from all sources of heat.

Flamtard M7. [Alcan; BA Chem. Ltd.] Magnesium hydroxide; CAS 1309-42-8; EINECS 215-170-3; flame retardant, smoke suppressant, filler for plastic and rubber prods. incl. PP, nylon, ABS, HIPS, PE, PVC, elastomers; high thermal stability allows inclusion in polymers processed at temps. to 300 C; for loadings to 70%; recommended for use with surf. coatings; applics. incl. cable sheathing, conveyor belting, pipes, and ducting; 5 μm median particle size; dens. 2.38 g/cc; bulk dens. 0.6 g/cc (untamped); surf. area 7 m^2/g; oil absorp. 35 g/100 g; pH 9.8 (5%); > 97% act.; *Toxicology:* low toxic hazard on handling; low toxicity degradation prods.; ACGIH 10 mg/m^3 total airborne particulate; *Storage:* store in dry place in air-tight containers to avoid moisture pick-up.

Flamtard S. [Alcan] Zinc stannate; CAS 12036-37-2; flame retardant for PVC, polychloroprene, chlorosulfonated polyethylene, polypropylene, other halopolymers; wh. free-flowing powd., odorless; 1.7 μm median particle size; sol. 0.10 g/l in water; sp.gr. 3.9; dens. 3.9 g/cc; bulk dens. 0.7 g/cc (untamped); surf. area 33 m^2/g; oil absorp. 25 g/100 g; m.p. dec. > 570 C; pH 8.4 (5%); *Toxicology:* low toxicity by ingestion; eye irritation by mechanical abrasion; LD50 (rat, oral) > 5000 mg/kg; *Precaution:* nonflamm.; *Storage:* store in dry, well-ventilated area away from all sources of heat.

Flamtard Z10. [Alcan; BA Chem. Ltd.] Zinc borate, ultrafine; CAS 12767-90-7; EINECS 233-471-8; flame retardant/smoke suppressant for thermoset and thermoplastic resins incl. PVC, thermosetting polyesters, polyamides, EVA; promotes char formation, suppresses afterglow; improved elec. antitracking in polyamides; synergistic with aluminum trihydroxide, antimony trioxide, halogens; esp. effective as afterglow suppressant in EVA; wh. fine powd.; 1 μm median particle size; dens. 2.8 g/cc; bulk dens. 0.6 g/cc (untamped); surf. area 10 m^2/g; oil absorp. 37 g/100 g; ref. index 1.59 (20 C); pH 8.3 (5%); 37.45% ZnO, 48.05% B_2O_3; *Toxicology:* low toxic hazard on handling; ACGIH 10 mg/m^3 total airborne particulate.

Flamtard Z15. [Alcan; BA Chem. Ltd.] Zinc borate, ultrafine; EINECS 233-471-8; flame retardant/smoke suppressant for thermoset and thermoplastic resins incl. PVC, thermosetting polyesters, polyamides, EVA; promotes char formation, suppresses afterglow; improved elec. antitracking in polyamides; synergistic with aluminum trihydroxide, antimony trioxide, halogens; esp. effective in brominated polyester resins and for high transparency prods.; wh. fine powd.; 1 μm median particle size; dens. 2.8 g/cc; bulk dens. 0.65 g/cc (untamped); surf. area 15 m^2/g; oil absorp. 44 g/100 g; ref. index 1.59 (20 C); pH 8.1 (5%); 37.45% ZnO, 48.05% B_2O_3; *Toxicology:* low toxic hazard on handling; ACGIH 10 mg/m^3 total airborne particulate.

Flectol® ODP. [Harwick] Octylated diphenylamine; staining antioxidant for CR; powd.

Flectol® Pastilles. [Monsanto] Polymerized 1,2-

dihydro-2,2,4-trimethylquinoline; CAS 26780-96-1; general purpose high activity antioxidant; metal deactivator, nonblooming; fuel and lube additives, crosslinked polyethylene stabilization; 12 mm pastilles.

Flexol® Plasticizer 4GO. [Union Carbide] PEG di-2-ethylhexoate; CAS 9004-93-7; plasticizer for rubber.

Flexol® Plasticizer EP-8. [Union Carbide] Octyl-epoxy tallate; CAS 61788-72-5; general purpose epoxy plasticizer.; Unverified

Flexol® Plasticizer EPO. [Union Carbide] Epoxidized soybean oil; CAS 8013-07-8; EINECS 232-391-0; general purpose epoxy plasticizer.; Unverified

Flexricin® 13. [CasChem] Glyceryl ricinoleate; CAS 141-08-2; EINECS 205-455-0; nonionic; wetting agent, wax plasticizer, and mold release agent for rubber polymers, antifoam agent, household and cosmetic applics., rewetting dried skins; FDA approval; Gardner 2 min. liq.; sol. in toluene, butyl acetate, MEK, ethanol; sp.gr. 0.985; visc. 8.8 stokes; pour pt. 20 F; acid no. 5; iodine no. 77; sapon. no. 160; hyd. no. 345; 100% act.

Flexricin® 17. [CasChem] Pentaerythritol ricinoleate; plasticizer; chemical intermediate; Gardner 8 liq.; sp.gr. 0.999; visc. 31 stokes; acid no. 3; iodine no. (Wijs) 74; sapon. no. 157; hyd. no. 328; pour pt. 0 F; 100% act.

Flexricin® 115. [CasChem] N(β-Hydroxyethyl) ricinoleamide; lubricant/antistat for plastics, metals; mold release, antiblocking agent for textile coatings; slip agent for varnishes and lacquers; also for elec. potting compds., crayons, wax blends, high-temp. greases; sp.gr. 1.00; m.p. 46 C; acid no. 5; iodine no. (Wijs) 80; sapon. no. 15; hyd. no. 295; penetration hardness 90.

Flexricin® 185. [CasChem] N,N′-Ethylene bis-ricinoleamide; lubricant/antistat for plastics, metals; mold release, antiblocking agent for textile coatings; slip agent for varnishes and lacquers; also for elec. potting compds., crayons, wax blends, high-temp. greases; sp.gr. 1.02; m.p. 89 C; acid no. 5; iodine no. (Wijs) 93; sapon. no. 11; hyd. no. 159; penetration hardness 19.

Flexricin® P-1. [CasChem] Methyl ricinoleate; CAS 141-24-2; EINECS 205-472-3; low temp. lubricant plasticizer for rubber, phenolic and epoxy resins; produces low durometer stock; FDA approval; Gardner 3 liq.; sp.gr. 0.925; visc. 0.3 stokes; acid no. 3; iodine no. (Wijs) 85; sapon. no. 178; hyd. no. 170; pour pt. -20 F; 100% act.

Flexricin® P-3. [CasChem] Butyl ricinoleate; general purpose plasticizer, lubricant for polyvinyl butyral, rosin Vinsol resin, cellulose acetate butyrate, NC, ethylcellulose; FDA approval; Gardner 3 liq.; sp.gr. 0.919; visc. 0.3 stokes; acid no. 3; iodine no. (Wijs) 77; sapon. no. 158; hyd. no. 163; pour pt. -15 F; 100% act.

Flexricin® P-4. [CasChem] Methyl acetyl ricinoleate; CAS 140-03-4; EINECS 205-392-9; all purpose plasticizer, lubricant for vinyls and lacquers; plasticizer and processing aid for low-temp. rubber compds.; confers cold crack resist. to flexible NC lacquers; leather finishes; low volatility; FDA approval; Gardner 3 liq.; sp.gr. 0.936; visc. 0.2 stokes; acid no. 2; iodine no. (Wijs) 75; sapon. no. 301; hyd. no. 5; pour pt. -15 F; 100% act.

Flexricin® P-6. [CasChem] Butyl acetyl ricinoleate; CAS 140-04-5; lubricity additive for textile finishes; low-temp. plasticizer, processing aid for rubber (SBR, NBR, neoprene, most other elastomers); FDA approval; liq.; 100% act.

Flexricin® P-8. [CasChem] Glyceryl (triacetyl ricinoleate); CAS 101-34-8; EINECS 202-935-1; plasticizer for vinyl wire jacketing and semirigid vinyls, flexible NC lacquers; emollient; stabilizer for anhyd. pigmented systems; exc. elec. props., extrusion lubricity, and heat stability; very low volatility; Gardner 2 liq.; sp.gr. 0.967; visc. 2.4 stokes; acid no. 1; iodine no. (Wijs) 76; sapon. no. 300; hyd. no. 5; pour pt. -40 F; 100% act.

Flo-Gard® CC 120, CC 140, CC 160. [PPG Industries] Amorphous silica; absorbent used as chemical carriers for liqs., grinding and suspension aids, or anticaking and flow-control agents in chemical formulations and processes; serves as highly reinforcing fillers in finished compds. such as rubber; permits extended mixing cycles; provides thixotropic action in water disps. permitting formulation of prods. that resist settling in storage and enhances stability of aq. disps. after dilution in the field; wh. ultrafine powd., 70, 15, and 3 μm median agglomerate size resp.; sp.gr. 2.1; ref. index 1.46; pH 7.0; MOH Scale hardness 0; 97.5% silica content, anhyd. basis.; Unverified

Flo-Gard® SP. [PPG Industries] Amorphous precipitated silica; anticaking and flow-control agent in chemical formulations, as grinding and suspension aid, or as absorbent/carrier for liqs.; Na_2SO_4 residual salt; powd.; pH 7 (5% slurry).

Fluon® L169. [ICI Surf. Am.] PTFE wax; CAS 9002-84-0; internal/external lubricant for thermoplastics, thermosets, elastomers, coatings; FDA accepted; solid.

Fluorad® FC-24. [3M/Industrial Chem. Prods.] Trifluoromethanesulfonic acid; CAS 1493-13-6; EINECS 216-087-5; catalyst and reactant increasing yields in polymerization of epoxies, styrenes, THF, in alkylation and acylation reactions; improves octane rating; used with nitric acid for higher yields of pharmaceuticals, explosives, dyes, and intermediates.

Fluorad® FC-118. [3M/Industrial Chem. Prods.] Ammonium perfluorooctanoate; CAS 3825-26-1; anionic; surfactant for emulsion polymerization of fluorinated monomers; lt. colored liq.; sol. in ethanol, acetone; sp.gr. 1.18; pH 5; surf. tens. 44 dynes/cm (0.5% aq.); 20% aq. sol'n.

Fluorad® FC-126. [3M/Industrial Chem. Prods.] Ammonium perfluoroalkane carboxylate; anionic; surfactant for emulsion polymerization of fluorinated monomers; exc. surface activity in alkaline media; lt. colored free-flowing powd.; sol. > 1000 g/1000 g solv. in acetone, methanol, water; m.p. dec.

at 120 C; pH 6.0; surf. tens. 44 dynes/cm (1000 ppm aq.); 100% act.; *Toxicology:* moderately irritating ocularly; nonirritating dermally; LD50 (rat, ingestion) 540 mg/kg, moderatel.

Fluoroglide® FL 1690. [ICI Fluoropolymers UK] PTFE; CAS 9002-84-0; general purpose additive esp. used in thermoplastics; 17 μ mean particle size; bulk dens. 480 g/l.

Fluoroglide® FL 1700. [ICI Fluoropolymers UK] PTFE; CAS 9002-84-0; oils, greases, and in additives for metal (can) coatings; high shear mixing required for sub-micron particle size; 2 μ mean particle size; bulk dens. 530 g/l.

Foamgard 1332. [Rhone-Poulenc Surf. & Spec.] Nonionic surfactants-based; foam suppressant and defoamer for aq. and semiaq. systems; for laundry detergents, industrial cleaners, paper/pulp systems, metalworking fluids, resin emulsification, waste water treatment, circuit board cleaning; solv. and silicone-free for lower VOC; Gardner < 1 sl. hazy liq.; neutral. no. 0.06 mg KOH/g; 100% act.

Foamphos NP-9. [Alzo] Nonoxynol-9 phosphate; CAS 51811-79-1; EINECS 266-231-6; anionic; low to moderate foaming detergent and emulsifier for heavy-duty all-purpose liq. detergents, sanitary prods.; chelating agent; thickener and coupler for solv.-aq. systems; rust and corrosion inhibitor; emulsifier for emulsion polymerization; dry-cleaning detergent; APHA 200 max. clear to sl. hazy liq., mild organoleptic odor; sol. in water and most org. solvs. (alcohols, glycols, triols, glycol ethers, aliphatic, aromatic, and chlorinated hydrocarbons); partly sol. in min. oils, veg. oils; sp.gr. 1.120; dens. 9.3 lb/gal; acid no. 60-75; cloud pt. -5 C min.; pH 2 max. (10% aq.); Ross-Miles foam 155 mm (initial, 0.1%).

Fomrez® B-306. [Witco/PAG] Organo-silicone; coupler used in mfg. of flexible polyester urethane foams; clear liq.; sp.gr. 1.04; acid no. 1.0; pH 5.0 (3.5% aq.).

Fomrez® B-308. [Witco/PAG] Organo-silicone; coupler used in mfg. of flexible polyester urethane foams; clear liq.; sp.gr. 1.04; acid no. 1.0; pH 5.0 (3.5% aq.).

Fomrez® B-320. [Witco/PAG] Organo-silicone; coupler used in mfg. of flexible polyester urethane foams; also used in mfg. of polyester microcellular urethane foams; clear liq.; sp.gr. 1.04; acid no. 1.0; pH 5.0 (3.5% aq.).

Fomrez® C-2. [Witco/PAG] Stannous octoate, stabilized; CAS 301-10-0; catalyst for mfg. of one-shot polyether urethane foams; yel. clear liq.; sp.gr. 1.27; dens. 10.6 lb/gal; flash pt. > 121 C (COC); 28.1% total tin.

Fomrez® C-4. [Witco/PAG] Stannous octoate, stabilized and anhyd. dioctyl phthalate; catalyst for mfg. of one-shot polyether urethane foams; also used for foam producers; yel. clear liq.; sp.gr. 1.10; dens. 9.2 lb/gal; flash pt. > 121 C (COC); 14.1% total tin.

Fomrez® SUL-3. [Witco/PAG] Dibutyltin diacetate; CAS 1067-33-0; catalyst used in rigid urethane spray foam formulations and in R.T.-vulcanized silicone polymers; colorless to pale yel. liq.; acetic acid odor; m.w. 351; f.p. 5 C; sp.gr. 1.31; ref. index 1.4780; flash pt. 143 C (COC); 33.8% tin.

Fomrez® SUL-4. [Witco/PAG] Dibutyltin dilaurate; CAS 77-58-7; EINECS 201-039-8; catalyst used in mfg. of polyether-based, rigid urethane foams and R.T.-vulcanized silicone polymers; Gardner 1; m.w. 631; f.p. 8 C; sp.gr. 1.05; ref. index 1.4700; flash pt. 232 C (COC); 18.9% tin.

Fomrez® UL-1. [Witco/PAG] Tin catalyst; catalyst for prep. of one-shot or prepolymer-type polyether-based, rigid urethane foams for spray or pour-in-place applic.; liq.; water wh. to pale yel.; f.p. -15 C; sp.gr. 1.01.

Fomrez® UL-2. [Witco/PAG] Proprietary organotin carboxylate; catalyst used as replacement for dibutyltin dilaurate; pale yel. liq.; sp.gr. 1.14; pour pt. -25 C.

Fomrez® UL-6. [Witco/PAG] Organotin; catalyst used in elastomer formulations and polyether-based urethane foam systems; water-wh. to pale yel. liq.; sp.gr. 1.13; pour pt. < -25 C.

Fomrez® UL-8. [Witco/PAG] Organotin; catalyst used in fluorocarbon-blown urethane spray foams; co-catalyst in isocyanurate foam formulations; pale yel. liq.; sp.gr. 1.34; pour pt. -14 C; ref. index 1.50; flash pt. 135 C (COC).

Fomrez® UL-22. [Witco/PAG] Organotin; catalyst for water/fluorocarbon-blown urethane foams; pale yel. liq.; f.p. -10 C; sp.gr. 1.03; ref. index 1.50; flash pt. 185 C (COC).

Fomrez® UL-28. [Witco/PAG] Organotin; catalyst used in mfg. of polyether-based, rigid urethane foams and R.T.-vulcanized silicone polymers; yel. liq.; sp.gr. 1.14; pour pt. -6 C; ref. index 1.47; flash pt. 153 C (COC).

Fomrez® UL-29. [Witco/PAG] Organotin; catalyst used in elastomer formulations and polyether-based urethane foam systems; pale yel. liq.; sp.gr. 1.28; pour pt. -40 C; ref. index 1.50; flash pt. 185 C (COC).

Fomrez® UL-32. [Witco/PAG] Organotin; catalyst for prep. of one-shot or prepolymer-type polyether-based, rigid urethane foams for spray or pour-in-place applic.; pale yel. liq.; f.p. -8 C; sp.gr. 0.98; ref. index 1.49; flash pt. 210 C (COC).

Fonoline® White. [Witco/Petroleum Spec.] Petrolatum USP; CAS 8027-32-5; EINECS 232-373-2; soft, low m.p. for consumer use as petrol. jelly, ointments, industrial applics.; as emollient, protective coating, binder, carrier, lubricant, moisture barrier, plasticizer, protective agent, softener; in confectionery lubricants, meat packing, mold release agents and lubricants for food industry; catalyst carrier, extrusion aid, plasticizer for polymers; FDA 21CFR §172.880; wh., odorless; visc. 9-14 cSt (100 C); m.p. 53-58 C; pour pt. 20 F.

Fonoline® Yellow. [Witco/Petroleum Spec.] Petrolatum USP; CAS 8027-32-5; EINECS 232-373-2; soft, low m.p. for consumer use as petrol. jelly, ointments, industrial applics.; as emollient, protective coating, binder, carrier, lubricant, moisture barrier, plasticizer, protective agent, softener; for confectionery lubricants, meat packing, release agents and lubricants, food handling machinery lubricants for food industry; catalyst carrier, extrusion aid, plasticizer for polymers; FDA 21CFR §172.880; yel., odorless; visc. 9-14 cSt (100 C); m.p. 53-58 C; pour pt. 20 F; 99% solids.

Fortex®. [Petrolite] Hard microcryst. wax consisting of n-paraffinic, branched paraffinic, and naphthenic hydrocarbons; CAS 63231-60-7; EINECS 264-038-1; used in hot-melt coatings/adhesives, paper coatings, printing inks, plastic modification (as lubricant and processing aid), lacquers, paints, and varnishes, as binder in ceramics, for potting, filling, and impregnating elec./electronic components; in investment casting, rubber and elastomers (plasticizer, antisunchecking, antiozonant), as emulsion wax size, as fabric softener ingred., in emulsion and latex coatings, and in cosmetic hand creams and lipsticks; wax; very low sol. in org. solvs.; sp.gr. 0.93; visc. 40 cps (99 C); m.p. 95 C.

FR-20. [AmeriBrom; Dead Sea Bromine] Magnesium hydroxide; CAS 1309-42-8; EINECS 215-170-3; nonhalogen flame retardant and smoke suppressant for ABS, PP, PS, nylon, rubbers, cables; high thermal stability to 340 C; wh. free-flowing powd.; sp.gr. 2.36; bulk dens. 0.35 g/ml (untapped); surf. area 3.5 m^2/g; m.p. dec. > 340 C; 99% min. assay; *Toxicology:* LD50 (oral, rat) > 5000 mg/kg; mildly irritating to eyes; nonirritating to intact skin; *Precaution:* nonflamm., nonexplosive; *Storage:* store in cool, dry, well-ventilated area.

FR-513. [AmeriBrom; Dead Sea Bromine] Tribromoneopentyl alcohol; CAS 36483-57-5; flame retardant for flexible and rigid PU; flame retardant intermediate; wh. to off-wh. flakes; sol. (g/100 g): 270 g in methanol (18 C), 230 g in toluene (40 C), 199 g in ethanol (18 C), 0.2 g in water; sp.gr. 2.28; m.p. 62-67 C; hyd. no. 173; 95% min. assay; *Toxicology:* LD50 (oral, rat) 2823 mg/kg; irritant to eyes; mild irritant to skin; *Precaution:* store in dry, cool, ventilated area.

FR-521. [Dead Sea Bromine] Dibromoneopentyl glycol; CAS 3296-90-0; EINECS 221-967-7; flame retardant for rigid PU; flame retardant intermediate; liq.; Unverified

FR-522. [AmeriBrom; Dead Sea Bromine] Dibromoneopentyl glycol; CAS 3296-90-0; EINECS 221-967-7; flame retardant for unsat. polyesers, rigid PU and foams; flame retardant intermediate; wh. cryst. powd.; sol. (g/100 g): 83 g in acetone; 52 g in IPA; 2 g in water; 0.5 g in CCl_4 and xylene; m.w. 261.94; sp.gr. 2.23; m.p. 109.0 C min.; 98.5% min. assay, 60.0% min. bromine.

FR-612. [AmeriBrom; Dead Sea Bromine] Dibromophenol; CAS 615-58-7; flame retardant for epoxy resins, phenolic resins, polyester resins; flame retardant intermediate; pink solid; highly sol. in acetone, benzene, CCl_4, ether, n-heptane, methanol, etc.; < 0.1 g/100 g in water; m.w. 251.92; sp.gr. 2.04 (40/25 C); m.p. 33-36 C; 98% min. assay, 63-64% bromine.; *Toxicology:* LD50 (oral, mouse) 2780 mg/kg; eye irritant; nonirritant to skin; *Storage:* store in dry, cook, dark area; prod. darkens on exposure to light.

FR-613. [AmeriBrom; Dead Sea Bromine] 2,4,6-Tribromophenol; CAS 118-79-6; reactive flame retardant used mainly as an intermediate for polymeric flame retardants; wh. to off-wh. flakes; sol. in acetone, diethyl ether, benzene, toluene, alcohols, chloroform, CCl_4, petrol. ether, pyridine; sol. 1 part in 14,000 parts water (15 C); m.w. 330.8; sp.gr. 2.55; m.p. 91.5 C min.; 99.1% min. assay, 72.47% bromine.; *Toxicology:* LD50 (oral, rat) > 5000 mg/kg; irritant to skin, mucous membranes, upper respiratory tract; corrosive to eyes; *Precaution:* nonflamm., nonexplosive; decomp. on excessive heating yielding toxic fumes of phenol and bromine; *Storage:* store in dry, cool, ventilated area away from light; stable, but tends to cake.

FR-615. [AmeriBrom; Dead Sea Bromine] Pentabromophenol; reactive flame retardant for epoxy, phenolic, rigid PU foam.

FR-705. [AmeriBrom; Dead Sea Bromine] Pentabromotoluene; CAS 87-83-2; flame retardant for unsat. polyesters, polyethylene, PP, PS, SBR latex, textiles, rubbers.

FR-910. [AmeriBrom; Dead Sea Bromine] Nonhalogenic flame retardant for PP.

FR-913. [AmeriBrom; Dead Sea Bromine] Tribromophenyl allyl ether; CAS 3278-89-5; aromatic flame retardant for expandable PS; synergist with hexabromocyclododecane; wh. cryst. free-flowing powd.; sol. (g/100 g): 130 g in methylene chloride, 125 g in benzene, 77 g in dichloroethane, 67 g in CCl_4, 32 g in acetone; m.w. 371; sp.gr. 2.19; m.p. 75-76.5 C; 98% min. assay, 64.7% Br.; *Toxicology:* LD50 (oral, rat) > 5000 mg/kg, (dermal, rabbit) 2000 mg/kg; *Storage:* store in dry, cool, ventilated area.

FR-1025. [AmeriBrom; Dead Sea Bromine] Poly (pentabromobenzyl) acrylate; CAS 594477-57-3; polymeric flame retardant for engineering thermoplastics, PET, PBT, nylon, PP, and PS; wh. to off-wh. free-flowing powd.; insol. in the common org. solvs. (aliphatic, aromatic, and chlorinated hydrocarbons, noncyclic and cyclic ethers, and ketones, DMF, and DMSO); m.w. 80,000; dens. 2.05 g/cc; m.p. 190-220 C; 70-71% bromine.; *Toxicology:* LD50 (oral, rat) > 5000 mg/kg; mildly irritating to skin and eyes; *Storage:* store in cool, dry, ventilated area.

FR-1033. [AmeriBrom; Dead Sea Bromine] Tribromophenyl maleimide; CAS 59789-51-4; reactive flame retardant for ABS, PE, PP, PS, SAN, polyimides.

FR-1034. [AmeriBrom; Dead Sea Bromine] Tetrabromodipentaerythritol; CAS 109678-33-3; flame retardant for PP extruded fibers; processing aid for ABS and HIPS.

FR-1138. [Albemarle] Dibromoneopentyl glycol; CAS 3296-90-0; EINECS 221-967-7; reactive flame retardant for acrylics, cellulosics, thermoset polyester, rigid and flexible PU foam, PU elastomers.

FR-1205. [AmeriBrom; Dead Sea Bromine] Pentabromodiphenyl oxide; CAS 32534-81-9; flame retardant for use in laminates (both epoxy and phenolic), unsat. polyesters, syn. fibers, and flexible PU foams; suitable for textiles; amber visc. liq.; sol. in CCl_4, methylene chloride, benzene, acetone, freon, and most polyols; insol. in water; sl. sol. in methanol; m.w. 565; sp.gr. 2.25; visc. 2650 cps (40 C); 69-72% Br.; *Toxicology:* LD50 (oral, rat) 5000 mg/kg, (dermal, rabbit) > 2000 mg/kg; may cause skin and eye irritation; *Storage:* store in cool, dry, ventilated area.

FR-1206. [AmeriBrom; Dead Sea Bromine] Hexabromocyclododecane; CAS 25637-99-4; fire retardant for wide range of plastics, textiles, adhesives, and coatings; esp. for styrene-based systems; wh. to off-wh. cryst. free-flowing powd.; sol. (g/100 ml): 10 g in styrene, 7 g in acetone, 6.5 g in toluene, < 0.1 g in water; m.w. 641.7; sp.gr. 2.18; m.p. 180-190 C; 73.8% Br.; *Toxicology:* LD50 (oral, rat) > 5000 mg/kg, (dermal, rabbit) > 8000 mg/kg; LD50 (inh., rat) > 200 mg/l/ih; *Storage:* store in cool, dry, ventilated area away from light.

FR-1208. [AmeriBrom; Dead Sea Bromine] Octabromodiphenyl oxide; CAS 32536-52-0; EINECS 251-087-9; flame retardant for thermoplastics, e.g., ABS, HIPS, LDPE, PP random copolymer; recommended for inj. moldings; wh. to off-wh. cryst. powd.; m.w. 802; sp.gr. 2.9; m.p. 125-165 C; 77-79% bromine.; *Toxicology:* LD50 (oral, rat) > 5000 mg/kg, (dermal, rabbit) > 2000 mg/kg; mildly irritating to skin and eyes; considered nontoxic by ingestion, but daily inh. of dust and eye exposure may be hazardous; *Storage:* store in cool, dry, ventilated area.

FR-1210. [AmeriBrom; Dead Sea Bromine] Decabromodiphenyl oxide; CAS 1163-19-5; EINECS 214-604-9; flame retardant used in thermoplastics and fibers, incl. HIPS, glass-reinforced thermoplastic polyester molding resins, LDPE extrusion coatings, PP (homo and copolymers), ABS, nylon, PBT, PET, PU, SBR latex, textiles, rubber; wh. to off-wh. fine cryst. powd.; sol. (g/100 g): < 1 g in acetone, benzene, chlorobenzene, methylene bromide, methylene chloride, o-xylene; m.w. 960; sp.gr. 3.0 (20 C); m.p. 300-305 C; 82% min. bromine.; *Toxicology:* LD50 (oral, rat) > 5000 mg/kg; mild eye irritant, nonirritating to skin; considered nontoxic orally, but daily inh. of dust and eye exposure may be hazardous; *Storage:* store in cool, dry, ventilated area.

FR-1215. [AmeriBrom; Dead Sea Bromine] Flame retardant for flexible PU.

FR-1524. [AmeriBrom; Dead Sea Bromine] Tetrabromobisphenol A; CAS 79-94-7; EINECS 201-236-9; reactive flame retardant used in the mfg. of epoxy, PC, ABS, phenolic, PS, and polyester resins, rubber; flame retardant intermediate; wh. cryst. powd.; sol. (g/100 g): 240 g in acetone; 92 g in methanol; 0.3 g in benzene, 0.01 g in water; m.w. 543.88; m.p. 181-182 C; 99% min. assay, 58.5% bromine.; *Toxicology:* LD50 (oral, rat) > 5000 mg/kg, (dermal, rabbit) > 2000 mg/kg; nonirritant to eyes and skin; *Storage:* store in cool, dry, ventilated area.

FR-1525. [AmeriBrom; Dead Sea Bromine] Tetrabromobisphenol-A ethoxylate; flame retardant for nylon, PBT, PET, unsat. polyesters; flame retardant intermediate.

FR-2124. [AmeriBrom; Dead Sea Bromine] Tetrabromobisphenol-A allyl ether; flame retardant for expandable PS.

Fragrance Compound. [Surco Prods.] Proprietary; perfume for plastics; yel. liq., char. odor; sol. in oil; sp.gr. 0.7-1.2; flash pt. 100-200 F; *Toxicology:* may be irritating to eyes and skin; overexposure may cause nausea, vomiting; *Precaution:* combustible; incompat. with strong oxidizers; *Storage:* store in well-ventilated area avoiding open flames and ignition sources.

Franklin T-11. [Franklin Industrial Minerals] Calcium limestone; CAS 1317-65-3; filler for the plastics, paint, and rubber industries; wh. powd.; 3 μ mean particle size; 0.005% max. retained on US # 325; fineness (Hegman) 5-6; sp.gr. 2.74; pH 9.0-9.5; hardness 3.

Franklin T-12. [Franklin Industrial Minerals] Calcium limestone; CAS 1317-65-3; filler for the plastics, paint, and rubber industries; wh. powd.; 5-6 μ mean particle size; 0.15% max. retained on US # 325; fineness (Hegman) 3-3.5; sp.gr. 2.74; pH 9.0-9.5; hardness 3.

Franklin T-13. [Franklin Industrial Minerals] Calcium limestone; CAS 1317-65-3; filler for the plastics, paint, and rubber industries; neutral beige; 6 μ mean particle size; 3-31/2 Hegman grind; 0.15% max. retained on US # 325; sp.gr. 2.74; pH 9.0-9.5; hardness 3; Unverified

Franklin T-14. [Franklin Industrial Minerals] Calcium limestone; CAS 1317-65-3; filler for the plastics, paint, and rubber industries; wh. powd.; 7 μ mean particle size; fineness (Hegman) 2.5-3; 0.5% max. retained on US # 325; sp.gr. 2.74; pH 9.0-9.5; hardness 3.

Franklin T-325. [Franklin Industrial Minerals] Calcium limestone; CAS 1317-65-3; filler for the plastics, paint, and rubber industries; neutral beige powd.; 10 μ mean particle size; 5.0% max. retained on US # 325; 0.08% max. retained on US # 200; fineness (Hegman) 1.5-2.5; sp.gr. 2.74; pH 9.0-9.5; hardness 3.

FRCROS 480. [Anzon] Ammonium polyphosphate; flame retardant for ABS, acrylics, cellulosics, epoxy, nitrile, phenolic, PC, thermoset polyester, PE, PP, PS, PVAc, PVC, PU, EVA, TPE, TPR.

FR-D. [FMC] Phosphorus-based diol; reactive flame retardant; effective for rigid PU foam; suitable for pour-in-place, slabstock, and spray formulations; flame retardant/friability modifier in PUR-modified isocyanurate foams; does not plasticize or affect catalyst activity; clear liq.; b.p. 235 C; dens. 1.10 g/ml; visc. 12,000 cps; 14% phosphorus.; Unverified

FRE. [J.M. Huber/Engineered Mins.] Alumina trihydrate; filler providing low cost flame and smoke suppression in fiberglass-reinforced polyester applic.; particulate; 12-15 μ particle size; sp.gr. 2.5; dens. 0.8g/cc; oil absorp. 25 ml/100 g.

Frekote® 31. [Dexter/Frekote] Mold release in areas where warpage is a problem, e.g., rotomolding; aerosols and bulk.; Unverified

Frekote® 44-NC. [Dexter/Frekote] Nonchlorofluorocarbon in dibutyl ether solv.; releasing interface for molding applics. incl. polyester, boron and graphite composites for aerospace and sporting goods, fiberglass laminates, adhesive bonding, rubber molding, rotational molding of polyethylenes and polycarbonates, molding of electrical and electronic components; aerosol and bulk; for matte finish; clear liq., sweet ether odor; sp.gr. 0.770 ± 0.01; flash pt. (PMCC) 25 C; *Toxicology:* very low toxicity; *Storage:* moisture-sensitive; keep container tightly closed; 1 yr shelf life.

Frekote® 700-NC. [Dexter/Frekote] Nonchlorofluorocarbon; aliphatic hydrocarbon, dibutyl ether solvents; all-purpose releasing interface for epoxies, polyester, nat. and syn. org. rubber, thermosetting resins, rotomoldable plastics, filament winding, reinforced polyester; R.T. cure; high gloss and high slip; clear liq., hydrocarbon odor; sp.gr. 0.760 ± 0.01; flash pt. (PMCC) 30 C; *Toxicology:* very low toxicity; *Precaution:* contains flamm. solvs.; *Storage:* store in cool, dry place; moisture-sensitive; keep container tightly closed; 1 yr shelf life.

Frekote® 800-NC. [Dexter/Frekote] Release agent for nat. and syn. org. rubber compds.; forms semipermanent release film on mold surfaces at temps. above 150 C; instant curing; high gloss finish.

Frekote® 815-NC. [Dexter/Frekote] Release agent for nat. and syn. org. rubber compds.; forms semipermanent release interface on mold surfaces at temps. from R.T. to 135 C; chemically bonds to mold surf. to form micro thin, chemically resist. coating; high thermal stability.

Frekote® 1711. [Dexter/Frekote] Silicone-based; release agent for urethane elastomers and skinned foams, rubber (incl. RTV silicone rubber), ABS resins, epoxies, phenolic resins; aerosols and bulk.

Frekote® 1711-EM. [Dexter/Frekote] Water-based silicone mold release; release agent for virtually any molded polymer such as epoxy resins, phenolics, acrylics, ABS, org. rubber compds., RTV silicone, and specialty molding compds.; stable to 500 F; nonflamm.; semigloss finish.

Frekote® EXITT®. [Dexter/Frekote] Silicone-based; release agent for urethane elastomers, compression-molded resins, org. rubbers, EPDM rubber, and specialty molded resins; aerosol and bulk.

Frekote® EXITT-EM. [Dexter/Frekote] Water-based silicone mold release; release agent for urethane elastomer systems; also releases urethane foams, cast elastomers, and most elastomeric resin systems; high gloss finish; precise retention of mold detail; nonflamm.

Frekote® FRP-NC. [Dexter/Frekote] Release agent for reinforced plastics, e.g., for molding polyester boats, tubs, shower stalls and sinks, cultured marble prods., table tops, tanks, and housings; aerosols and bulk; for gloss finish.

Frekote® HMT. [Dexter/Frekote] Silicone-based; release agent for hot mold touchup at temps. above 60 C; for plastics (compression/transfer, laminating), rubber (inj./compression, transfer, tire mfg.), electronic and electrical, and foundry applics.; aerosol and bulk.

Frekote® LIFFT®. [Dexter/Frekote] Silicone-based; release agent for direct injection-molded urethane shoe soles, unitsoles, heel wedges, ski boots, furniture and recreational prods. molded from high-dens. skinned urethane foams and cast elastomers; aerosol and bulk.

Frekote® No. 1-EM. [Dexter/Frekote] Fluoropolymer emulsion; water-based release agent for acrylics, epoxies, nylons, phenolics, PC, PS, many rubber and elastomeric compds.; dries/cures instantly @ 200 F +; nonflamm.; translucent wh. liq., mild odor; sp.gr. 1.00 ± 0.02; *Toxicology:* very low toxicity; *Storage:* 1 yr shelf life in unopened containers; do not freeze.

Frekote® S-10. [Dexter/Frekote] Water-based release agent for molding of silicone rubber compds.; for use on molds with temps. of 150 F or higher; nonflamm.; lt. amber translucent liq., mild odor; sp.gr. 1.0 ± 0.02; flash pt. (TCC) none; *Toxicology:* very low toxicity; *Storage:* 1 yr shelf life; keep from freezing.

Ftalidap®. [Lonza SpA] Diallyl phthalate; CAS 131-17-9; EINECS 205-016-3; diester monomer which easily undergoes polymerizations; esp. suitable for unsat. polyester; provides improved hardness, elec. props., and stability against chemicals and hydrolysis; plasticizer for PVC; reactive plasticizer for vinyl resins; oily limpid liq.; m.w. 278.34; sp.gr. 1.047-1.049; b.p. 340 C (760 mm); sapon. no. 403 ± 3.

Fusabond® MB-110D. [DuPont] Chemically modified LLDPE; anhydride-modified resin for use as compatibilizers in blends and alloys, polymeric coupling agents in reinforced or recycled PE or PP, adhesives and sealants; melt index 40 dg/min (190 C); dens. 0.925 g/cc; soften. pt. (Vicat) 90 C; tens. str. 14.1 MPa; tens. elong. 80%; flex. mod. 403 MPa; 0.85% maleic anhydride.

Fusabond® MB-226D. [DuPont] Chemically modified LLDPE; anhydride-modified resin for use as compatibilizers in blends and alloys, polymeric coupling agents in reinforced or recycled PE or PP, adhesives and sealants; melt index 1.5 dg/min (190 C); dens. 0.93 g/cc; soften. pt. (Vicat) 103 C; tens. str. 11 MPa; tens. elong. 950%; flex. mod. 280 MPa; 0.85% maleic anhydride.

Fusabond® MC-189D. [DuPont] Chemically modified ethylene/vinyl acetate; CAS 24937-78-8; anhydride-modified resin for use as compatibilizers in blends and alloys, polymeric coupling agents in reinforced or recycled PE or PP, adhesives and sealants; melt index 20 dg/min (190 C); dens. 0.958 g/cc; soften. pt. (R&B) 122 C; tens. str. 8.9 MPa; tens. elong. 904%; tens. mod. 16 MPa; 33% VA.

Fusabond® MC-190D. [DuPont] Chemically modified ethylene/vinyl acetate; CAS 24937-78-8; anhydride-modified resin for use as compatibilizers in blends and alloys, polymeric coupling agents in reinforced or recycled PE or PP, adhesives and sealants; melt index 20 dg/min (190 C); dens. 0.953 g/cc; soften. pt. (R&B) 146 C; tens. str. 12.6 MPa; tens. elong. 838%; tens. mod. 30 MPa; 28% VA.

Fusabond® MC-197D. [DuPont] Chemically modified ethylene/vinyl acetate; CAS 24937-78-8; anhydride-modified resin for use as compatibilizers in blends and alloys, polymeric coupling agents in reinforced or recycled PE or PP, adhesives and sealants; melt index 2.5 dg/min (190 C); dens. 0.941 g/cc; soften. pt. (R&B) > 190 C; tens. str. 19.5 MPa; tens. elong. 708%; tens. mod. 94 MPa; 18% VA.

Fusabond® MZ-109D. [DuPont] Chemically modified polypropylene homopolymer; anhydride-modified resin for use as compatibilizers in blends and alloys, polymeric coupling agents in reinforced or recycled PE or PP, adhesives and sealants; melt index 300 dg/min (230 C); dens. 0.95 g/cc; soften. pt. (Vicat) 147 C; tens. str. 30 MPa; tens. elong. 10%; flex. mod. 1379 MPa; 0.6% maleic anhydride.

Fusabond® MZ-203D. [DuPont] Chemically modified polypropylene impact copolymer; anhydride-modified resin for use as compatibilizers in blends and alloys, polymeric coupling agents in reinforced or recycled PE or PP, adhesives and sealants; melt index 250 dg/min (230 C); dens. 0.94 g/cc; soften. pt. (Vicat) 132 C; tens. str. 19 MPa; tens. elong. 11%; flex. mod. 1030 MPa; 0.75% maleic anhydride.

Futurethane UR-2175. [H.B. Fuller] Two-component urethane adhesive; flame retardant; mix ratio 100/29; gel time 50-70 s; full cure 2-4 h @ R.T.; Shore hardness D62; flamm. V-0; Unverified

FW 18. [Degussa] Channel-type carbon black; CAS 1333-86-4; EINECS 215-609-9; for high-quality film calendered systems, high quality hi-fi records with low noise level.

FW 200 Beads and Powd. [Degussa] Channel-type carbon black; CAS 1333-86-4; EINECS 215-609-9; for top quality blk. PP, ABS, PA, PVC, paints and coatings.

Fyarestor® 100. [Witco/PAG] Bromochlorinated paraffin; flame retardant which has replaced chlorinated paraffins used in treating outdoor fabrics and in plastics; water resistance, and some plasticizing; excellent weatherability, compatible in water- and solv.-based systems.

Fyraway. [Harwick] Antimony oxide disp.; CAS 1309-64-4; EINECS 215-175-0; flame retardant for epoxy, nitrile, thermoset polyester, PVAc, PVC.

Fyrebloc. [Anzon] Antimony oxide disp.; CAS 1309-64-4; EINECS 215-175-0; flame retardant for ABS, acrylics, cellulosics, epoxy, nitrile, phenolic, PC, thermoset polyester, PS, PE, PP, PVAc, PVC, PU, EVA, PBT, SAN, CPE, TPR, nylon.

Fyrol® 6. [Akzo Nobel] Diethyl N,N-bis(2-hydroxyethyl)aminomethylphosphonate; reactive flame retardant for rigid urethane foams; incorporated into foam structure by reacting as a polyol; foams exhibit good dimensional stability; for spray, froth, or pour-in-place applics.; amber liq.; misc. with water; m.w. 255; sp.gr. 1.15; dens. 9.6 lb/gal; visc. 160 cps; acid no. 4.5; hyd. no. 460; 12% P, 0% Cl.

Fyrol® 25. [Akzo Nobel] Phosphate ester blend; flame retardant for flexible polyether urethane foam; such foams exhibit little or no scorching; dk. gray liq.; sol. < 1% in water; sp.gr. 1.46; dens. 12.4 lb/gal; visc. 3500 cps; acid no. 0.2; hyd. no. 30; 37% Cl, 10% P.

Fyrol® 38. [Akzo Nobel] Tri(β,β′-dichloroisopropyl) phosphate; CAS 13674-87-8; scorch stabilized flame retardant for flexible and rigid polyurethane foams; reduces discoloration caused by high exotherm following processing of flexible polyether urethane foam; pale amber liq.; sol. < 1% in water; m.w. 431; sp.gr. 1.51; dens. 12.7 lb/gal; visc. 1800 cps; acid no. 0.05; 49% Cl, 7% P.

Fyrol® 42. [Akzo Nobel] Proprietary; flame lamination promoter aiding the flame bonding of flexible polyether urethane foam; also contributes flame retardancy due to phosphorous content; colorless liq.; sol. < 1% in water; sp.gr. 1.028; dens. 8.55 lb/gal; visc. 1100 cps; acid no. 0.50; 5.3% P, 0% Cl.

Fyrol® 51. [Akzo Nobel] Oligomeric phosphate; flame retardant for aq. systems that are used as binders for cellulosic prods., e.g., automotive and industrial air filters, upholstery fabrics, thermoset papers, particle and chip boards; lt. straw liq.; misc. with water; sp.gr. 1.37; dens. 11.4 lb/gal; visc. 5600 cps; acid no. 1.8; 20.5% P, 0% Cl.

Fyrol® 99. [Akzo Nobel] Chlorinated oligomeric phosphate ester; flame retardant for thermosetting and thermoplastic resins; colorless liq.; insol. in water; m.w. 750; sp.gr. 1.45; dens. 12.1 lb/gal; visc. 2200 cps; acid no. 0.4; 26% Cl, 14% P.

Fyrol® Bis Beta. [Akzo Nobel] Bis (β-chloroethyl) vinyl phosphonate; reactive flame retardant for ABS, acrylics, thermoset polyester, PVAc, PVC.

Fyrol® CEF. [Akzo Nobel] Tri(β-chloroethyl) phosphate; CAS 306-52-5; flame retardant well suited for transparent or pastel-shaded plastics and coatings; colorless liq.; sol. < 1% in water; m.w. 286; sp.gr. 1.42; dens. 11.8 lb/gal; visc. 38 cps; acid no. 0.05; 36.7% Cl, 10.8% P.

Fyrol® DMMP. [Akzo Nobel] Dimethyl methylphosphonate; CAS 756-79-6; flame retardant for applics. where high phosphorus content, good solvency, and low visc. are desired; lowers visc. of epoxy resins and unsat. polyesters filled with hydrated alumina oxide; colorless liq.; misc. with water; m.w. 124; sp.gr. 1.17; dens. 9.8 lb/gal; visc. 4 cps; acid no. 1.3; 25% P, 0% Cl.

Fyrol® FR-2. [Akzo Nobel] Tri(β,β′-dichloroisopropyl) phosphate; CAS 13674-87-8; flame retardant for flexible urethane foams and other polymers; colorless liq.; sol. < 1% in water; m.w. 431; sp.gr. 1.52; dens. 12.7 lb/gal; visc. 1800 cps; acid no. 0.05; 7.1% P, 49% Cl.

Fyrol® MC. [Akzo Nobel] Melamine cyanurate; flame retardant for thermoset polyester, PP, PBT, nylon.

Fyrol® MP. [Akzo Nobel] Melamine phosphate; flame retardant for ABS, PE, PP, PU, EVA, TPE, TPR.

Fyrol® MPP. [Akzo Nobel] Melamine pyrophosphate; flame retardant for ABS, PE, PP, PU, EVA, TPE, TPR.

Fyrol® PBR. [Akzo Nobel] Pentabromodiphenyl oxide, < 12% triphenyl phosphate, ≈ 13% organo phosphate ester; flame retardant additive for flexible polyurethane foams; low propensity for discoloration caused by high exotherm in the processing of flexible polyether urethane foam; amber clear liq., aromatic odor; pract. insol. in water; sp.gr. 1.83; dens. 15.3 lb/gal; visc. 600 cps; flash pt. (COC) 480 F; fire pt. (COC) 550 F; 51.6% Br, 2.2% P; *Toxicology:* LD50 (rat, oral) > 3000 mg/kg, (rat, dermal) > 2000 mg/kg; mild skin and eye irritant; triphenyl phosphate may cause cholinesterase inhibition, OSHA PEL 3 mg/m^3 (8 h); *Precaution:* under fire conditions, may decomp. to generate toxic materials (phosphorus oxides); *Storage:* store in cool, dry, well-ventilated area away from flamm. materials and heat sources.

Fyrol® PCF. [Akzo Nobel] Tri(β-chloroisopropyl) phosphate; flame retardant for polymers; colorless liq.; sol. in most common org. solvs., < 1% in water; m.w. 325; sp.gr. 1.29; dens. 10.8 lb/gal; visc. 57 cps; acid no. 0.05; 32.5% Cl, 9.5% P.

Fyrolflex RDP. [Akzo Nobel] Resorcinol bis (diphenylphosphate); flame retardant for PC, PE, TPR, nylon.

G

136-G. [J.M. Huber/Engineered Mins.] Alumina trihydrate; flame retardant, smoke suppressing filler for phenolic compr. molding.

331-G. [J.M. Huber/Engineered Mins.] Alumina trihydrate; flame retardant, smoke suppressing filler for FRP continuous panel elec. applics.

431-G. [J.M. Huber/Engineered Mins.] Alumina trihydrate; flame retardant, smoke suppressor, high loading filler for thermoset, thermoplastic, and rubber applics.; high surf. area, exc. elec. props.; for epoxy encapsulating/potting and elec. laminates, FRP elec. laminates, SMC/BMC, phenolic compr. molding; resin extender in polyester, vinyls, PU, latex, neoprene foam systems, wire/cable insulation, vinyl wall and floor coverings; 64.9% Al_2O_3.

G-50 Barytes™. [Cimbar Perf. Minerals] Natural unbleached coarse barytes; CAS 7727-43-7; EINECS 231-784-4; filler for rubber and plastics, acoustical compds., and friction prods.; coarse powd.; 63% thru 325 mesh; sp.gr. 4.3; dens. 268 lb/ft^3; pH 7.0; 93% $BaSO_4$, < 0.1% moisture.

G623. [GE Silicones] Silicone compd.; general-purpose water-repellent coatings or sealants to prevent galvanic corrosion; lubricant for plastics and rubber.

G661 [GE Silicones] Silicone; external lubricant for rubber or plastic on some metals; FDA; solid.

G-1086. [ICI Surf. Am.; ICI Surf. Belgium] PEG-40 sorbitol hexaoleate; CAS 57171-56-9; nonionic; emulsifier and coupling agent for paraffinic, naphthenic and organic ester lubricants, textiles, metalworking and hydraulic fluids, polymerization; pale yel. oily liq.; sol. in toluene, min. spirits, many veg. oils, acetone, CCl_4, cellosolve, methanol, lower alcohols, aniline; sp.gr. 1; visc. 200 cps; HLB 10.2; flash pt. > 300 F; 100% act.

G-1087. [ICI Am.; ICI Surf. UK] POE sorbitol polyoleate; nonionic; emulsifier for textiles, metalworking and hydraulic fluids, polymerization; amber liq.; HLB 9.2; 100% conc.

G-1096. [ICI Surf. Am.; ICI Surf. Belgium] PEG-50 sorbitol hexaoleate; CAS 57171-56-9; nonionic; emulsifier and coupling agent for personal care prods., textiles, hydraulic and metalworking fluids, polymerization; pale yel. liq.; sol. in veg. oils, acetone, cellosolve, lower alcohols, some aromatic solv., tetrachloride; sp.gr. 1.0; visc. 220 cps; HLB 11.4; flash pt. > 300 F; 100% act.

Gamaco®. [Georgia Marble] Calcium carbonate; filler/extender pigment for paints, enamels, epoxy-based coatings, adhesives, caulks, sealants, plastics (polyolefin, PU, epoxy, plastisols, compr. and inj. molded compds); foam and rubber compds., glass fiber-reinforced polyester (BMC/SMC/TMC where paintable parts are required), pultrusion; powd.; 3.8 μ median particle size; 0.006% on 325 mesh; fineness (Hegman) 6; oil absorp. 16; brightness (Hunter) 94; 95% min. $CaCO_3$.

Gamaco® II. [Georgia Marble] Calcium carbonate; extender filler for BMC, SMC, TMC, pultrusion, transfer molding, wet mat and other glass fiber-reinforced thermoset polyesters, polyolefins, urethanes, epoxies, moldings and extrusions, fluid polymer systems; flat and gloss paints, enamels, varnishes, caulks, sealants, adhesives, mastics, fine-celled foams; powd.; 3 μ mean particle size; 0.006% on 325 mesh; sp.gr. 2.7; bulking value 0.443 gal/lb; oil absorp. 14-16%; brightness 94; hardness (Moh) 3.0; 95% min. $CaCO_3$.

Gamaco® Slurry. [Georgia Marble] Calcium carbonate; filler; 2.8 μ median particle size; brightness (GE) 95; 74 ± 1% solids.

Gama-Fil® 40. [Georgia Marble] Calcium carbonate; filler; slurry; 2.4 μ median particle size; brightness (GE) 95; 74 ± 1% solids.

Gama-Fil® 55. [Georgia Marble] Calcium carbonate; filler; very wh. slurry; 1.8 μ median particle size; 55% < 2 μ, 28% < 1 μ; visc. 300-500 cps; brightness (GE) 95; 74 ± 1% solids.

Gama-Fil® 90. [Georgia Marble] Calcium carbonate; filler; very wh. slurry; 0.8 μ median particle size; 92% < 2 μ, 58% < 1 μ; visc. 300-500 cps; brightness (GE) 95; 74 ± 1% solids.

Gama-Fil® D-1. [Georgia Marble] Calcium carbonate; ultrafine filler with high brightness for plastics, BMC, SMC, paint, caulks, sealants, adhesives, paper, foam urethane, acrylics, filled thermosets/thermoplastics, and rubber; NSF approved for potable water compds.; powd.; 0.7 μ mean particle size; 6 μ top cut; 65% finer than 1 μ; nil retained on 325 mesh; oil absorp. 17 ± 1; brightness (Hunter) 95; 96% min. $CaCO_3$.

Gama-Fil® D-1T. [Georgia Marble] Calcium carbonate, stearate surf.-modified; surf.-modified high brightness filler for polyolefins, BMC/SMC, caulks, sealants, adhesives, foam urethane, acrylics, filled thermoplastics, silicones; hydrophobic; NSF approved for potable water compds.; powd.; 0.7 μ mean particle size; 6 μ top cut; 65% finer than 1 μ; nil retained on 325 mesh; brightness (Hunter) 95; 96% min. $CaCO_3$, 1% stearate.

Gama-Fil® D-2. [Georgia Marble] Calcium carbonate, dry, wet ground, nondispersed; ultrafine high brightness filler for use in plastics, BMC/SMC, paint, caulks, sealants, adhesives, paper, foam urethane, modified acrylics, filled thermosets/thermoplastics, and rubber; NSF approved for potable water materials; powd.; 2.0 μ mean particle size; 8 μ top cut; 0.006% retained on 325 mesh; fineness (Hegman) 6; oil absorp. 16±1; brightness (Hunter) 95; 96% min. $CaCO_3$.

Gama-Fil® D-2T. [Georgia Marble] Stearate surf.-modified calcium carbonate; ultrafine high brightness filler for use in polyolefins, BMC/SMC, caulks, sealants, adhesives, foam urethane, acrylics, filled thermoplastics; hydrophobic; enhanced compat.; NSF approved for potable water materials; powd.; 2.0 μ mean particle size; 8 μ top cut; nil retained on 325 mesh; 25% finer than 1 μ; brightness (Hunter) 95; 96% min. $CaCO_3$, 1% stearate.

Gama-Plas®. [Georgia Marble] Calcium carbonate; filler for use in glass fiber-reinforced polyester compds. (BMC, SMC, TMC, wet mat molding), plastisols esp. for dip coating of wire fencing and metal parts, paints, coatings, caulks, mastics, sealants; wh. powd.; 7.0 μ median particle size; 0.20% on 325 mesh; sp.gr. 2.71; visc. 36,000 cps; brightness 94; hardness (Mohs) 3; 95% min. $CaCO_3$.

Gama-Sperse® 2. [Georgia Marble] Calcium carbonate; ultrafine filler with very high whiteness; 2.0 μ median particle size; 0.006% on 325 mesh; fineness (Hegman) 7+; oil absorp. 21; brightness 94.

Gama-Sperse® 5. [Georgia Marble] Calcium carbonate; filler for paint, plastics, caulks, sealants, adhesives, foam urethane, filled thermosets/thermoplastics, BMC, and rubber applics.; 5.0 μ median particle size; 0.005% retained on #325 wet screen; Hegman grind 5.5; oil absorp. 14±1; dry brightness 94; 96% conc.; Unverified

Gama-Sperse® 80. [Georgia Marble] Calcium carbonate, natural ground; general-use filler for paints, enamels, plastisols esp. slush molded parts requiring high gloss, polyolefin compds. for sheet stock, thin films, paper coatings, and rubber prods.; also in upholstery fabrics, credit card stock, pigment extenders, caulks, sealants, and adhesives; powd.; 2.5 μ median particle size; 0.006% on 325 mesh; fineness (Hegman) 7; sp.gr. 2.71; bulking value 0.443 gal/lb; oil absorp. 19 lb oil/100 lb pigment; brightness 94; pH 9.0-9.5; hardness (Moh) 3; 95% min. $CaCO_3$.

Gama-Sperse® 140. [Georgia Marble] Calcium carbonate, natural ground; general-use filler for paints, plastics, paper coatings, and rubber prods.; 0.005% retained on #325 wet screen; 61/4-61/2 Hegman grind; sp.gr. 2.71; bulking value 0.443 gal/lb; oil absorp. 15-17 lb oil/100 lb pigment; dry brightness 96%; pH 9.0-9.5; hardness (Moh) 3; 95% min. total carbonates.; Unverified

Gama-Sperse® 255. [Georgia Marble] Calcium carbonate, natural ground; general-use filler for paints, plastics, paper coatings, and rubber prods.; 12.0 μ median particle size; 0.02% on 325 mesh; fineness (Hegman) 4; sp.gr. 2.71; bulking value 0.443 gal/lb; oil absorp. 9 lb oil/100 lb pigment; brightness 94; pH 9.0-9.5; hardness (Moh) 3; 95.0% min. total carbonates.

Gama-Sperse® 6451. [Georgia Marble] Calcium carbonate, natural ground; general-use filler for exterior/interior paints, paper coatings, rubber prods.; adhesives, caulks, sealants with acrylic, polyolefin, silicone, and urethane bases; plastics (polyolefin, PU, compr. and inj. molded compds., plastisols), rubber compds., thermosets (BMC, SMC, pultrusion); powd.; 6.0 μ median particle size; 0.008% on 325 mesh; fineness (Hegman) 5; sp.gr. 2.71; bulking value 0.443 gal/lb; oil absorp. 10 lb oil/100 lb pigment; brightness 94; pH 9.0-9.5; hardness (Moh) 3; 95% min. $CaCO_3$.

Gama-Sperse® 6532. [Georgia Marble] Calcium carbonate, natural ground; general-use filler for paints, enamels, epoxy-based coatings, paper coatings, and in adhesives, caulks, sealants with acrylic, polyolefin, silicone, or urethane base; for plastics (polyolefin, PU, epoxy, plastisols, compr. and inj. molded compds.), foam and rubber compds., glass fiber-reinforced polyester (BMC/TMC/SMC where paintable parts are required), pultrusion; powd.; 3.8 μ median particle size; 0.006% on 325 mesh; fineness (Hegman) 6; sp.gr. 2.71; bulking value 0.443 gal/lb; oil absorp. 16 lb oil/100 lb pigment; brightness 94; pH 9.0-9.5; hardness (Moh) 3; 95% min. $CaCO_3$.

Gama-Sperse® 6532 NSF. [Georgia Marble] Calcium carbonate; filler for glass-reinforced polyester compds.; NSF approved for potable water materials; powd.; 3.0 μ median particle size; 0.006% on 325 mesh; brightness 94.

Gama-Sperse® CS-11. [Georgia Marble] Calcium carbonte, calcium stearate surface-modified; filler for plastics, nonaq. formulations, PVC, PP, engineering plastics, silicone formulations, for use in food contact applics.; NSF approved for potable water materials; very wh. free-flowing powd.; 3.0 μ median avg. particle size; 0.005% max. on 325 mesh; sp.gr. 2.7; bulking value 0.443 gal/lb; hardness (Moh) 3.0; 96% min. $CaCO_3$, 1% stearate.

Gamma W8. [Wacker-Chemie GmbH] γ-Cyclodextrin; CAS 7585-39-9; EINECS 231-493-2; complex hosting guest molecules; increases the sol. and bioavailability of other substances; masks flavor, odor, or coloration; stabilizes against light, oxidation, heat, and hydrolysis; turns liqs. or volatiles into stable solid powds.; for use in pharmaceuticals, cosmetics, toiletries, foods, tobacco, pesticides, textiles, paints, plastics, synthesis, polymers; wh. cryst. powd.; sol. 23.2 g/100 ml in water; m.w. 1297; *Toxicology:* LD50 (acute IV, rat) > 3750 mg/kg; nonirritating to eye.

Gamma W8 HP0.6. [Wacker-Chemie GmbH] Hydroxypropyl-γ-cyclodextrin; CAS 99241-25-5; complex hosting guest molecules; increases the sol. and bioavailability of other substances; masks flavor, odor, or coloration; stabilizes against light, oxidation, heat, and hydrolysis; turns liqs. or volatiles into stable solid powds.; for use in pharmaceuticals, cosmetics, toiletries, foods, tobacco, pesticides, textiles, paints, plastics, synthesis, polymers; m.w. 1580.

Gamma W8 M1.8. [Wacker-Chemie GmbH] Methyl-γ-cyclodextrin; complex hosting guest molecules; increases the sol. and bioavailability of other substances; masks flavor, odor, or coloration; stabilizes against light, oxidation, heat, and hydrolysis; turns liqs. or volatiles into stable solid powds.; for use in pharmaceuticals, cosmetics, toiletries, foods, tobacco, pesticides, textiles, paints, plastics, synthesis, polymers; m.w. 1499.

Ganex® P-904. [ISP] Butylated PVP; used in cosmetics and toiletries as moisture barrier, adhesive, protective colloid, and microencapsulating resin; as dispersant for pigments; as solubilizer for dyes; in petroleum industry as sludge and detergent dispersant; protective colloid in coatings; suspending aid in polymerization; dyeing assistant; antiredeposition agent in drycleaning; esp. as dispersant in aq. agric. chemicals or pigmented skin care prods.; sol'n.; sol. in water; m.w. 16,000; 45% solids in IPA, M-Pyrol, or water.

Gantrez® AN-119. [ISP] PVM/MA copolymer; CAS 52229-50-2; anionic; thickener, dispersant, stabilizer, gelling agent, coupler, protective colloid, suspending aid used in emulsion polymerization; adhesives, household detergents, liq. hand cleaners, acid bowl cleaners; produces clear films of high tens. and cohesive str.; powd.; sol. in water, acid, caustic, and org. solv.; 100% conc.

Gantrez® AN-139. [ISP] PVM/MA copolymer; CAS 52229-50-2; anionic; thickener, dispersant, foam stabilizer, coupling agent; for emulsion polymerization, pesticides, petrol. prod., heavy-duty liq. detergents, adhesives; powd.; sol. in water, acid, caustic, and org. solv.; 100% conc.

Gantrez® AN-149. [ISP] PVM/MA copolymer; CAS 52229-50-2; anionic; thickener, dispersant, foam stabilizer, coupling agent; for emulsion polymerization, pesticides, petrol. prod., heavy-duty liq. detergents, adhesives; powd.; sol. in water, acid, caustic, and org. solv.; 100% conc.

Gantrez® AN-169. [ISP] PVM/MA copolymer; CAS 52229-50-2; anionic; thickener, dispersant, foam stabilizer, coupling agent; for emulsion polymerization, pesticides, petrol. prod., heavy-duty liq. detergents, adhesives; powd.; sol. in water, acid, caustic, and org. solv.; 100% conc.

Gantrez® AN-179. [ISP] PVM/MA copolymer; CAS 9011-16-9; anionic; thickener, dispersant, foam stabilizer, coupling agent; for emulsion polymerization, pesticides, petrol. prod., heavy-duty liq. detergents, adhesives; powd.; sol. in water, acid, caustic, and org. solv.; 100% conc.

Gantrez® M-154. [ISP] Polyvinyl methyl ether; CAS 9003-09-2; polymer functioning as tackifier, binder, and plasticizer; used in printing inks, textile sizes and finishes, latex modification; liq.; sol. in water and diverse organic solvs.; thermally reversible solubility in aq. systems; 50% aq. sol'n..

Garbefix 05240. [Great Lakes] Phosphite; nondiscoloring antioxidant for PC, ABS, nylon, polyester, PE, PP; FDA 21CFR §178.2010.

Gardilene S25L. [Albright & Wilson Australia] Sodium alkylbenzene sulfonate; CAS 68081-81-2; anionic; wetting, dispersing, and foaming agent, emulsifier in emulsion polymerization; detergent ingred.; pale yel. clear visc. liq.; visc. 3300 cps; clear pt. < 6 C; pH 7.2 (10%); 25% act. in water.

GE 100. [Raschig] Mixt. of di- and trifunctional epoxides on basis of glycidyl glycerin ether; CAS 90529-77-4; EINECS 292-011-4; reactive diluent with other polymers, in solv.-free coating systems, laminating resins, fiber-reinforced composites, coating of tech. polyester and Aramide fibers (as textile reinforcement in tires, conveyor belts); adhesion promoter for rubber and fibers; sl. ylsh. visc. liq., odorless; sol. in most org. solvs.; difficult to dissolve in water; dens. 1.22; flash pt. 118-135 C; EEW 145 g/mol; 12% Cl; *Toxicology:* LD50 (oral, rat) 5000 mg/kg; strongly irritating to eyes and skin; *Precaution:* highly reactive; heating may cause an explosion; incompat. with strong acids/bases, nonferrous heavy metals and their salts; hazardous decomp. prods.: CO, hydrocarbons; *Storage:* store in original, unopened container in cool, dry place.

Gemtex 445. [Finetex] Sodium diisobutyl sulfosuccinate; CAS 127-39-9; EINECS 204-839-5; anionic; rapid wetting agent, penetrant, emulsifier, dispersant, hydrophilic wetter, coemulsifier, surfactant for emulsion polymerization of styrene/butadiene systems; liq.; 45% conc.

Gemtex 680. [Finetex] Sodium dihexyl sulfosuccinate; anionic; emulsifier, rapid wetter, penetrant, surfactant for emulsion polymerization in vinyl acedtate/acrylate, styrene/butadiene, and styrene/acrylamide systems; liq.; 80% conc.

Gemtex 691-40. [Finetex] Sodium dicyclohexyl sulfosuccinate; CAS 23386-52-9; EINECS 245-629-3; anionic; rapid wetting agent, penetrant, emulsifier, dispersant for emulsion polymerization of styrene/butadiene systems; post-additive for latex systems; liq.; 40% conc.

Gemtex PA-75. [Finetex] Dioctyl sodium sulfosuccinate, isopropyl alcohol; CAS 577-11-7; EINECS 209-406-4; anionic; wetting agent, penetrant for textile uses and emulsion polymerization; dye leveler; bubble baths, bath oils; liq.; 75% conc.

Gemtex PA-85P. [Finetex] Dioctyl sodium sulfosuccinate, propylene glycol; CAS 577-11-7; EINECS 209-406-4; anionic; wetting agent, penetrant for textiles, emulsion polymerization; dye leveler; bubble baths, bath oils; liq.; 85% conc.

Gemtex PAX-60. [Finetex] Dioctyl sodium sulfosuccinate, anhyd.; CAS 577-11-7; EINECS 209-406-4; anionic; wetting agent for textile uses and emulsion polymerization; dye leveler; bubble baths, bath oils;

liq.; 60% conc.

Gemtex SC-75E. [Finetex] Dioctyl sodium sulfosuccinate; CAS 577-11-7; EINECS 209-406-4; anionic; wetting agent for textile uses and emulsion polymerization; dye leveler; bubble baths, bath oils; liq.; 75% conc.

Genamid® 151. [Henkel/Coatings & Inks] Amidoamine resin; epoxy curing agent offering superior wetting, exc. chem. resist., better internal plasticizer, fast cure time; used for high solids coatings, castings, laminates, and adhesives; Gardner 12 max. color; sp.gr. 0.95; dens. 7.9 lb/gal; visc. 2.5-5 poise; amine no. 425-450; tens. str. 6300 psi; tens. elong. 9.6%; distort. temp. 75-85 C; gel time 2-3 h; thru cure 13 h @ R.T.; 100% solids.

Genamid® 250. [Henkel/Coatings & Inks] Amidoamine resin; epoxy curing agent offering superior wetting, exc. chem. resist., better internal plasticizer, fast cure time; used for high solids coatings, castings, potting, laminating, and adhesives; Gardner 10 max. color; sp.gr. 0.95; dens. 7.9 lb/gal; visc. 5-10 poise; amine no. 425-450; tens. str. 8500 psi; tens. elong. 12%; distort. temp. 60 C; gel time 40 min; thru cure 10 h @ R.T.; 100% solids.

Genamid® 490. [Henkel/Coatings & Inks] Amidoamine resin; epoxy curing agent offering superior wetting, exc. chem. resist., better internal plasticizer, fast cure time; used for high solids coatings, castings, and flooring; Gardner 7 max. color; sp.gr. 0.95; dens. 7.9 lb/gal; visc. 1-4 poise; amine no. 350-400; tens. str. 8000 psi; tens. elong. 23%; distort. temp. 60 C; gel time 5 h; thru cure 36 h @ R.T.; 100% solids.

Genamid® 491. [Henkel/Coatings & Inks] Amidoamine resin; epoxy curing agent offering superior wetting, exc. chem. resist., better internal plasticizer, fast cure time; for high solids coatings, castings, flooring, adhesives; Gardner 9 max. color; sp.gr. 0.95; dens. 7.9 lb/gal; visc. 6-12 poise; amine no. 500-580; tens. str. 10,500 psi; tens. elong. 9.5%; distort. temp. 87 C; gel time 25-35 min; thru cure 9 h @ R.T.; 100% solids.

Genamid® 747. [Henkel/Coatings & Inks] Amidoamine resin; epoxy curing agent offering superior wetting, exc. chem. resist., better internal plasticizer, fast cure time; used for high solids coatings, castings, potting, laminating, adhesives; Gardner 11 max. color; sp.gr. 0.94; dens. 7.8 lb/gal; visc. 2-5 poise; amine no. 450-475; tens. str. 9500 psi; tens. elong. 12%; distort. temp. 85 C; gel time 2 h; thru cure 12 h @ R.T.; 100% solids.

Genamid® 2000. [Henkel/Coatings & Inks] Amidoamine resin; epoxy curing agent offering superior wetting, exc. chem. resist., better internal plasticizer, fast cure time; used for coatings, flooring, castings, potting, laminating, adhesives; Gardner 10 max. color; sp.gr. 0.98; dens. 8.2 lb/gal; visc. 10-25 poise; amine no. 580-620; tens. str. 11,000 psi; tens. elong. 9%; distort. temp. 80 C; gel time 30 min; thru cure 8 h @ R.T.; 100% solids.

Genetron® 141b. [AlliedSignal] 1,1-Dichloro-1-fluoroethane; CAS 1717-00-6; blowing agent (replacement for CFCs) in foam applics., rigid board, flexible foam.

Genomoll® P. [Hoechst AG] Tris-(2-chloroethyl) orthophosphoric acid ester; flame retardant for plastics.

Geode™. [Ferro/Color] Complex inorg. color pigments; pigments for plastics and coatings, esp. for use in coil coatings for exterior bldg. prods., PVC siding and window profiles, engineering plastics; heat and light-stable, weatherable, chemically resist.

Geolite® Modifier 91. [OSi Specialties] Softening/stabilizing additive for flexible PU slabstock foam; reduces use of CFC's; liq.; completely sol. in water; sp.gr. 1.1285 (20/20 C); dens. 9.39 lb/gal (20 C); visc. 170 cSt; vapor pressure < 10 mm Hg (20 C); f.p. -40 C; b.p. 130 C; flash pt. (PMCC) none; 10% water.

Geronol ACR/4. [Rhone-Poulenc Surf. & Spec.] Disodium laurethsulfosuccinate; anionic; emulsifier for emulsion polymerization of acrylate, polyvinyl acetate; detergent base, foamer, foam stabilizer, dispersant; liq.; insol. in org. solv.; 30% conc.

Geronol ACR/9. [Rhone-Poulenc Surf. & Spec.] Disodium nonoxynol-10 sulfosuccinate; CAS 67999-57-9; anionic; emulsifier for emulsion polymerization of acrylate, polyvinyl acetate; detergent base, foamer, foam stabilizer, dispersant; liq.; insol. in org. solv.; 30% conc.

Geropon® 99. [Rhone-Poulenc Surf. & Spec.] Dioctyl sodium sulfosuccinate and propylene glycol; CAS 577-11-7; EINECS 209-406-4; anionic; detergent; textile scouring and dispersant for dyes; paper rewetting and felt washing surfactant; wetting agent in cosmetics; detergent additive in dry cleaning fluids; dishwashing compds.; wallpaper removers; agric. sprays; emulsion polymerization; water-based paint formulations; antifog; EPA compliance; clear liq.; water-sol.; sp.gr. 1.08-1.13 (70 F); visc. 500-1000 cps (70 F); pour pt. 40 F; 70% act.

Geropon® AB/20. [Rhone-Poulenc Surf. & Spec.] Ammonium laureth-9 sulfate; anionic; emulsifier for emulsion polymerization and min. oils; detergent, dispersant, foam stabilizer; stable to pH and temp.; liq.; 30% conc.

Geropon® ACR/4. [Rhone-Poulenc Surf. & Spec.] Disodium laureth-4 sulfosuccinate; amphoteric; emulsifier for emulsion polymerization, detergent base, foamer, foam stabilizer, dispersant; mild detergent, emulsifier, foaming agent for shampoos, liq. soaps, facial cleansers, bath gels, and bubble baths; liq.; 30% conc.

Geropon® ACR/9. [Rhone-Poulenc Surf. & Spec.] Disodium nonoxynol-10 sulfosuccinate; CAS 67999-57-9; anionic; emulsifier for emulsion polymerization, detergent base, foamer, foam stabilizer, dispersant; liq.; 30% conc.

Geropon® AS-200. [Rhone-Poulenc Surf. & Spec.] Sodium cocoyl isethionate, coconut acid, stearic acid; anionic; detergent, emulsifier, foam booster/stabilizer, visc. builder, wetting agent, dispersant, suspending agent for textile wet processing, industrial/household detergents, cosmetics (shampoos, liq. soaps, bar soaps, facial cleansers, bath gels, bubble baths), agric. pesticides, leather, rubber, etc.; readily biodeg.; APHA 100 max. color.

Geropon® BIS/SODICO-2. [Rhone-Poulenc Surf. & Spec.] Sodium bistridecyl sulfosuccinate; CAS 2673-22-5; EINECS 220-219-7; anionic; emulsifier and visc. depressant for emulsion polymerization of PVC; latex surf. tension stabilizer; dispersant for resins, pigments into plastics and organic media; base for rust inhibitors; liq.; oil-sol., water-disp.; 60% conc.

Geropon® CYA/45. [Rhone-Poulenc Surf. & Spec.] Sodium diisobutyl sulfosuccinate; CAS 127-39-9; EINECS 204-839-5; anionic; surfactant for emulsion polymerization; liq.; 45% conc.

Geropon® CYA/60. [Rhone-Poulenc Surf. & Spec.] Sodium dioctyl sulfosuccinate; anionic; wetting agent, surf. tension reducer, visc. depressant for emulsion polymerization of PVC; liq.; 60% conc.

Geropon® CYA/DEP. [Rhone-Poulenc Surf. & Spec.] Sodium diisooctyl sulfosuccinate; CAS 127-39-9; EINECS 204-839-5; anionic; wetting agent for textile industry, pesticides, emulsion polymerization, printing inks, water paints; stable in acid media; liq.; sol. in org. solvs.; 75% conc.

Geropon® MLS/A. [Rhone-Poulenc Surf. & Spec.] Sodium methallyl sulfonate; CAS 1561-92-8; anionic; dye improver reactive comonomer for acrylic fibers polymerization; reactive emulsifier or co-emulsifier in latex emulsion polymerization; powd.; 100% conc.

Geropon® SDS. [Rhone-Poulenc Surf. & Spec.] Sodium dioctyl sulfosuccinate; anionic; dispersant and wetting agent for pigments and dyes in plastics, pesticide wettable powds.; EPA compliance; powd.; dissolves in water, partly sol. in org. solvs.; 100% conc.

Geropon® SS-O-75. [Rhone-Poulenc Surf. & Spec.] Sodium dioctyl sulfosuccinate; CAS 577-11-7; anionic; wetting agent, emulsifier for industrial, mining, and textile industries, resin treatments, dry cleaning systems; paint pigment dispersant; clear colorless liq.; typ. odor; 1% water sol., sol. in most org. liq.; sp.gr. 1.09; f.p. < -20 C; surf. tens. 29 dynes/cm (0.1% aq.); pH 5-6; 75% act.

Geropon® T/36-DF. [Rhone-Poulenc Surf. & Spec.] Sodium salt of polycarboxylic acid; anionic; dispersant for final solv. stripping phase in polymerization of BR-SBR; liq.; 25% conc.

Geropon® WT-27. [Rhone-Poulenc Surf. & Spec.] Sodium dioctyl sulfosuccinate; CAS 1639-66-3; anionic; high foaming wetting and rewetting agent, emulsifier, dispersant, penetrant; used in drycleaning detergents, emulsion polymerization, glass cleaners, wallpaper removers, battery separators, textiles, paper, metalworking, dyeing, fire fighting; FDA compliance; water-wh. clear liq.; water sol. up to 1.2% solids and 5.5% @ 70 C; sol. in some polar and nonpolar solvs.; flash pt. > 200 C; pH 5-6 (1%); surf. tens. 28 dynes/cm (@ CMC); 70% act.

Getren® 4/200. [Goldschmidt] Silicone-free organic; release agent for foundry and steel industry, plastics incl. unsat. polyester, epoxy, and PU resins; release agent for rubber molding; lubricant for tire prod., radiator hoses; misc. with water; sp.gr. 1.045 ± 0.005 g/ml; visc. 1800 ± 80 mPa•s; 100% act.; *Storage:* 1 yr storage stability in closed containers; protect from frost; Unverified

Getren® BPL 83. [Goldschmidt AG] Aq. silicone; band ply lubricant for tire prod.; 35% act.

Getren® FD 150. [Goldschmidt AG] Aq. organic; antiblocking agent for the coating of crude rubber; 20% act.

Getren® FD 372. [Goldschmidt AG] Silicone-free wax-based; sprayable release suspension for mfg. of thermosetting and thermoplastic resins; Unverified

Getren® FD 411. [Goldschmidt] Organic; release agent for rubber molding; offers exc. wetting props. even @ 1% sol'n.; water-sol.; 100% act.; Unverified

Getren® FD 575. [Goldschmidt] Organic; release agent for the rubber industry; mainly for prod. of radiator hoses; biodeg.; misc. with water; sp.gr. 1.05 ± 0.01 g/ml; visc. 1500 ± 200 mPa•s; 100% act.; *Storage:* 6 mos storage stability in closed containers; protect from frost; stir before use; Unverified

Getren® S 3 Conc. 100. [Goldschmidt] Silicone-free organic; release agent for plastic and rubber release; 100% act.; Unverified

G-Flex™-11. [Goldsmith & Eggleton] Thermoplastic additive; processing aid improving thermal stability, low temp. flexibility, weatherability of thermoset, thermoplastic, adhesive, and asphalt-based prods.; as nonvolatile plasticizer to improve processing and dimensional stability of extruded and molded goods; for PE, PP, XLPE, polybutene resins, peroxide-cured rubber systems incl. EPDM, EP, hot-melt and pressure-sensitive adhesives; lt. amber; melt index 1.13 g/min (190 C, 9kg); sp.gr. 0.86; visc. (Mooney) 27 (1+8, 125 C); hardness (Shore) A44.

G-Flex™-12. [Goldsmith & Eggleton] Thermoplastic additive; processing aid improving thermal stability, low temp. flexibility, weatherability of thermoset, thermoplastic, adhesive, and asphalt-based prods.; as nonvolatile plasticizer to improve processing and dimensional stability of extruded and molded goods; for PE, PP, XLPE, polybutene resins, peroxide-cured rubber systems incl. EPDM, EP, hot-melt and pressure-sensitive adhesives; lt. amber; melt index 1.85 g/min (190 C, 9kg); sp.gr. 0.86; visc. (Mooney) 25 (1+8, 125 C); hardness (Shore) A42.

G-Flex™ HMD. [Goldsmith & Eggleton] Thermoplastic additive; processing aid improving thermal stability, low temp. flexibility, weatherability of thermoset, thermoplastic, adhesive, and asphalt-based prods.; as nonvolatile plasticizer to improve processing and dimensional stability of extruded and molded goods; for PE, PP, XLPE, polybutene resins, peroxide-cured rubber systems incl. EPDM, EP, hot-melt and pressure-sensitive adhesives; lt.

amber; melt index 1.71 g/min (190 C, 9kg); sp.gr. 0.86; visc. 80,000 cps (25% toluene sol'n.); hardness (Shore) A33.

G-Flex™ HMG. [Goldsmith & Eggleton] Thermoplastic additive; processing aid improving thermal stability, low temp. flexibility, weatherability of thermoset, thermoplastic, adhesive, and asphalt-based prods.; as nonvolatile plasticizer to improve processing and dimensional stability of extruded and molded goods; for PE, PP, XLPE, polybutene resins, peroxide-cured rubber systems incl. EPDM, EP, hot-melt and pressure-sensitive adhesives b; lt. amber; melt index 3.73 g/min (190 C, 9kg); sp.gr. 0.86; visc. 130,000 cps (25% toluene sol'n.); hardness (Shore) A38.

Givsorb® UV-1. [Givaudan-Roure] N^2-(4-Ethoxycarbonylphenyl)-N′-methyl-N′-phenylformamidine; CAS 57834-33-0; industrial uv absorber for cellulosics; liq.; m.p. 27-28 C.

Givsorb® UV-2. [Givaudan-Roure] N-(p-Ethoxycarbonylphenyl)-N′-ethyl-N′-phenylformamidine; CAS 65816-20-8; industrial UV absorber, lt. stabilizer, and antioxidant for PU, PVC, polyolefins, ABS, nylon, and acetal resins; photostable, broad spectrum screening agent for protection against the adverse effects of both UV-B and UV-A radiation; used in surf. coatings, polishes, dyestuffs, carpet treatments; wh. to pale yel. fine powd.; m.w. 296.4; sol. (g/100 g solv.): > 200 g in diethyl phthalate, Carbitol acetate, 2-methoxyethyl acetate, 2-ethoxyethyl acetate; > 100 g in toluene, acetone; > 20 g in IPP; > 10 g in IPA; dens. 1.077 g/cc; m.p. 49-52 C and 62-65 C (two allotropic cryst. forms); b.p. 225 C (1 mm); m.p. 49-65 C; flash pt. (TCC) 200 F.

Glacier 200. [Luzenac Am.] Talc; CAS 14807-96-6; EINECS 238-877-9; economical talc for dusting rubber; powd.; 10 μ mean particle size; brightness (GE) 78.

Glycidol Surfactant 10G. [Olin] p-Nonylphenoxypolyglycidol; nonionic; surfactant used in agric. chemical emulsions, leather processing, paint, emulsion polymerization, waste paper deinking, alkaline cleaners, photographic film emulsion and coating formulations; amber liq.; mild odor; sol. in alcohols, polyethers, acids; sp.gr. 1.032-1.114; dens. 8.6-9.3 lb/gal; visc. 110-195 cps; pour pt. -9 to -14 C; b.p. 93-103 C; cloud pt. > 100 C (0.5% aq.); pH 7.1-7.9 (1% aq.); surf. tens. 27-30 dynes/cm (0.1% act.); 50% act.; Unverified

Glycolube® 100. [Lonza] Polyol ester; internal lubricant for PVC for clear rigid applics., extruded film and sheet, calendered sheet, blow molded bottle compds.; antifog for PVC; FDA approved for indirect food contact; yel. liq. to soft solid; acid no. 2 max.; sapon. no. 166-176; 2% max. free glycerin.

Glycolube® 110. [Lonza] Polyol ester; internal lubricant for PVC and PP; antistat for PP, PE, PS; FDA approved as indirect food additive; wh. beads; m.p. 57-61 C; acid no. 3 max.; 7% max. free glycerin.

Glycolube® 110D. [Lonza] Internal lubricant for rigid and semirigid filled PVC compds.; not recommended for clear applics.; FDA approved as indirect food additive; lt. tan beads; m.p. 58 C; acid no. 3.5 max.; 1% max. moisture.

Glycolube® 140. [Lonza] Polyol ester; lubricant, antistat for PP, PE, extruded films and sheet, inj. molded compds.; internal lubricant for PVC esp. rigid systems, clear and opaque applics.; antifog for PC, PS; slip/antiblock for thermoplastic polyester; FDA approved for indirect food contact use in polyolefins; wh. beads; m.p. 64 C; acid no. 2 max.; 1% max. moisture.

Glycolube® 140 Kosher. [Lonza] Proprietary; lubricant, antistat for PP, PE, PS, extruded films and sheet, inj. molded compds.; internal lubricant for PVC esp. rigid systems, clear and opaque applics.; FDA approved for indirect food contact use in polyolefins; kosher certification; wh. beads; m.p. 64 C; acid no. 2 max.; 1% max. moisture.

Glycolube® 150. [Lonza] Ester; antistat for PP, PE, EPS; beads; m.p. 70 C; iodine no. 5 max.; 90% monoester.

Glycolube® 180. [Lonza] Polyol ester; internal lubricant for rigid PVC requiring crystal clarity, exc. heat stability, and freedom from plate-out; ideal for sheet and film which will be subsequently printed, metallized or laminated; for bottles, clear rigid sheet, semirigid wire and cable, flexible blown films; FDA approved as indirect food additive; yel. clear liq.; acid no. 3 max.; iodine no. 78-88.

Glycolube® 315. [Lonza] Polyol ester; general purpose internal/external lubricant for rigid PVC; enhances long-term stability in all tin-stabilized systems; enhances processability; reduces yellowness in weathering tests; improves melt flow, thermal stability; antifog for PP, PE; FDA approved for food contact; ivory solid; m.p. 58 C; acid no. 6 max.; sapon. no. 150-180.

Glycolube® 345. [Lonza] Ester; external lubricant for rigid PVC, ABS, PET, other engineering thermoplastics; antifog for PC, PS; slip/antiblock for thermoplastic polyester; FDA approved; cream-colored beads; m.p. 50 C; acid no. 2 max.; sapon. no 185-210.

Glycolube® 674. [Lonza] Polyol ester; internal lubricant for rigid PVC, ABS and styrenics; provides exc. heat stability and surf., low fusion in extrusion and molding applics.; antifog for PC, PS; slip/antiblock for thermoplastic polyester; Lovibond 3.0R max. beads; m.p. 59 C; HLB 1; acid no. 6 max.; iodine no. 1; sapon. no. 190-200; pH 5 (3%); 0.2% max. moisture.

Glycolube® 740. [Lonza] Ester; internal lubricant/costabilizer for clear PVC compds., film and sheet extrusion, calendered sheet, blow molded bottles; antifog for plasticized PVC film, PC, PS; slip/antiblock for thermoplastic polyesters; suitable for tin and metal soap-stabilized systems; improves long-term thermal stability, retains good clarity; FDA approved for food contact; amber liq.; acid no. 2 max.; sapon. no. 125-140.

Glycolube® 742. [Lonza] Ester; general purpose lubricant for clear and opaque rigid PVC compds.; improves long-term thermal stability and plate-out resist. in clear applics., improves mold filling in inj. molding; antifog for PC, PS; slip/antiblock for thermoplastic polyesters; FDA permitted as indirect food additive; straw beads; m.p. 55 C; acid no. 8 max.; sapon. no. 130-150; 2% max. moisture.

Glycolube® 825. [Lonza] Polyol ester; internal lubricant for rigid PVC esp. clear blow molded bottles, inj. molding formulations; antistat for PP, PE, PS; antifog for PC, PS; slip/antiblock for thermoplastic polyesters; wh. beads; m.p. 60 C; acid no. 3 max.; 1% max. moisture.

Glycolube® 825 Kosher. [Lonza] Ester; internal lubricant, antistat for rigid PVC esp. clear blow molded bottles, inj. molding formulations; wh. beads; m.p. 60 C; acid no. 3 max.; 1% max. moisture, 1.2% max. free glycerin.

Glycolube® 853. [Lonza] Proprietary; plastics lubricant; FDA approved for food contact use; amber liq.; acid no. 6 max.; sapon. no. 130-140.

Glycolube® AFA-1. [Lonza] Ester; antifog and antistat for PVC food wrap film and other films; imparts thermal stability without adverse effect on optical props.; antifog for PP, PE; FDA 21CFR §172.854, 178.3400; amber liq.; acid no. 2 max.; sapon. no. 85-95; hyd. no. 165-195.

Glycolube® CW-1. [Lonza] Hydrog. castor oil; CAS 8001-78-3; EINECS 232-292-2; internal lubricant for PVC compds.; improves heat stability; also as processing aid/flow promoter for PE and PP; flake/wax; m.p. 85 C min.; acid no. 3 max.; sapon. no. 176-182; hyd. no. 154-162; flash pt. 204 C.

Glycolube® P. [Lonza] Ester wax; lubricant for many polymers in extrusion and molding; mold release agent for PC, thermoplastic polyesters, other engineering thermoplastics; external lubricant in rigid PVC; outstanding thermal stability; wh.-tan beads; 100% thru 10 mesh, 10% thru 100 mesh; m.p. 63 C max.; acid no. 2 max.; sapon. no. 195 max.; hyd. no. 12 max.

Glycolube® PG. [Lonza] Food grade plastics lubricant; external lubricant for PVC, rigid calendered or extruded sheet, bottles, film; contribues release props.; provides finished prods. with superior clarity and freedom from haze; FDA permitted as indirect food additive; Gardner 7 max. clear liq.; acid no. 6 max.; sapon. no. 140-155.

Glycolube® PS. [Lonza] Ester; external lubricant and mold release for rigid PVC and engineering thermoplastics; FDA approved; wh. to tan beads; m.p. 52 C; sapon. no. 150-170; hyd. no. 115-135.

Glycolube® SG-1. [Lonza] Proprietary; lubricant for rigid PVC compds. produced on twin screw extrusion equip.; provides reduced plate-out, improved output rates, improved stability and weatherability, reduced surf. gloss; for PVC siding and profiles; Gardner 5 max. beads; sapon. no. 144; hyd. no. 8 max.

Glycolube® TS. [Lonza] Ester wax; external lubricant for engineering thermoplastics incl. styrenics, ABS, thermoplastic polyester; imparts good processability and exc. mold release; antifog for PC, PS; slip/antiblock for thermoplastic polyesters; FDA permitted as indirect food additive; wh. beads; m.p. 53-59 C; acid no. 4 max.

Glycolube® VL. [Lonza] N,N′-Dioleoylethylenediamine; food grade internal and external lubricant for PVC applics. requiring exc. clarity and printability in finished prod.; slip/antiblock agent in printing ink vehicles; mold release agent for thermoplastic PU; also used for textiles; FDA 21CFR §178.3860; tan bead; 100% thru 14 mesh, 5% thru 200 mesh; m.p. 110-117 C; acid no. 10 max.; flash pt. 270 C.

Glycomul® L. [Lonza] Sorbitan laurate; CAS 1338-39-2; nonionic; emulsifier for edible, cosmetic, coatings, industrial, household (polishes, cleaning prods.), pharmaceutical uses; polymerization emulsifier for thermoplastics; antistat, antifog for PVC; emollient for skin care prods.; lubricant, emulsifier in textile spin finishes; amber clear liq.; sol. in methanol, ethanol, naphtha; disp. in water; sp.gr. 1.0; visc. 3100 cps (30 C); HLB 9; acid no. 7 max.; sapon. no. 158-170; hyd. no. 330-358; flash pt. 204 C; pH 6.5 (5%); 99% min. conc.

Glycomul® O. [Lonza] Sorbitan oleate; CAS 1338-43-8; EINECS 215-665-4; nonionic; emulsifier for cosmetic, pharmaceutical, and industrial applics.; polymerization emulsifier for polyacrylamide; antistat and antifog for bottles and films; amber clear liq.; sol. in ethyl acetate, min. and veg. oils, disp. in water; sp.gr. 1.0; visc. 1000 cps.; HLB 4.3; acid no. 7.5 max.; sapon. no. 143-160; hyd. no. 193-209; flash pt. 220 C; 100% conc.

Glycomul® S. [Lonza] Sorbitan stearate (also avail. in veg. and kosher grade); CAS 1338-41-6; EINECS 215-664-9; nonionic; emulsifier for cosmetic, pharmaceutical, and industrial applics.; pigment dispersant and slip agent in thermoplastic color concs.; antistat for PVC; lt. tan solid or beads; sol. in veg. oil; HLB 5.0; acid no. 5-10; sapon. no. 147-157; hyd. no. 235-260; 100% conc.

Glycon® S-70. [Lonza] 70% Stearic acid; CAS 57-11-4; EINECS 200-313-4; external lubricant for extruded PVC pipe, sheeting; lubricant, defoamer, and component of other food additives; FDA 21CFR §172.860; solid; acid no. 203; iodine no. 0.8; sapon. no. 204.

Glycon® S-90. [Lonza] 90% Stearic acid; CAS 57-11-4; EINECS 200-313-4; external lubricant for extruded PVC pipe, sheeting; lubricant, defoamer, and component of other food additives; FDA 21CFR §172.860; solid; acid no. 199; iodine no. 0.7; sapon. no. 201.

Glycon® TP. [Lonza] Stearic acid, triple pressed.; CAS 57-11-4; EINECS 200-313-4; external lubricant for extruded PVC pipe, sheeting; lubricant, defoamer, and component of other food additives; FDA 21CFR §172.860; solid; acid no. 210; iodine no. 0.8; sapon. no. 211.

Glycosperse® L-20. [Lonza] Polysorbate 20; CAS 9005-64-5; nonionic; emulsifier for food, cosmetic (shampoos, conditioners, bath prods.), pharmaceutical, household, and industrial uses; flavor solubilizer and dispersant; o/w emulsifier, visc. modifier, wetting agent for household prods.; lubricant and emulsifier in textile syn. fiber finish oils; antifog in PVC, antistat in polyolefins; yel. clear liq.; sol. in water, alcohol, acetone; sp.gr. 1.1; visc. 400 cps; HLB 16.7; acid no. 2 max.; sapon. no. 40-50; hyd. no. 96-108; pH 7 (5%); 3% max. moisture.

Glycosperse® O-20. [Lonza] Polysorbate 80; CAS 9005-65-6; nonionic; emulsifier for food, cosmetic, pharmaceutical, and industrial uses; flavor solubilizer and dispersant; also antifog for PVC; yel. liq.; sol. in water, alcohol, ethyl acetate, toluol, veg. oil; sp.gr. 1.0; visc. 400 cps; HLB 15; sapon. no. 44-56; 100% conc.

Glycostat®. [Lonza] Ester; nonionic; antistat for thermoplastics incl. PP, HDPE, LDPE, PVC, PS (crystal and impact); FDA approved; wh. beads; m.p. 61-68 C; acid no. 2 max.; sapon. no. 160-175; flash pt. 200 C.

Glycowax® 765. [Lonza] N,N′-Ethylene bisstearamide; syn. wax for kraft brownstock pulp and paper defoaming, plastics processing, powd. metal lubrication, water treatment defoaming, other industrial applics.; Gardner 4 powd.; 87% min. on 325 mesh; m.p. 143 C; acid no. 2; flash pt. 285 C.

Glycowax® S 932. [Lonza] Triester wax; syn. wax for cosmetics; external lubricant for pipe, bottles, siding; FDA accepted; wh. flake; m.p. 62 C; flash pt. 310 C.

Glyso-Lube™. [C.P. Hall] Soap-type lubricant; lubricant providing max. antistick props. to rubber and other polymers, for slab or continuous sheet stocks, soft and difficult-to-handle stocks; straw-colored liq.; misc. with water in all proportions; sp.gr. 1.01 ± 0.02; pH 9 min. (10%); 21 ± 3% solids.

Glyso-Lube™ 3. [C.P. Hall] Soap-type lubricant; lubricant providing max. antistick props. to rubber and other polymers, for slab or continuous sheet stocks; straw-colored liq.; misc. with water in all proportions; sp.gr. 1.01 ± 0.02; pH 9 min. (10%); 21 ± 3% solids.

Good-rite® 3034. [BFGoodrich/Spec. Polymers] Piperazinone-based hindered amine light stabilizer; light stabilizer and antioxidant for polymers; Unverified

Good-rite® 3140. [BFGoodrich/Spec. Polymers] Amine; sl. staining and discoloring antioxidant for PE and PP.

Good-rite® 3150. [BFGoodrich/Spec. Polymers] 1,1′,1′′-(1,3,5-Triazine-2,4,6-triyltris ((cyclohexylimino)-2,1-ethanediyl)tris(3,3,5,5-tetramethylpiperazinone); antioxidant and hindered amine light stabilizer for processing of polyolefin and other polymeric fibers; Unverified

Good-rite® 3159. [BFGoodrich/Spec. Polymers] 1,1′,1′′-(1,3,5-Triazine-2,4,6-triyltris((cyclohexylimino)-2,1-ethanediyl)tris(3,3,4,5,5-tetramethylpiperazinone); hindered amine light stabilizer and antioxidant for polymers; Unverified

GP-4 Silicone Fluid. [Genesee Polymers] Amodimethicone; intermediate for synthesis of silicone/org. copolymers used in textiles, coatings car polishes; also in lubricant, coating formulations; cosmetic and personal care applics. incl. hair care; mold release for rubber, plastics, foundry applics.; internal mold release; lt. straw clear liq.; sp.gr. 0.97; dens. 8.0 lb/gal; visc. 65 cst; amine no. 75; flash pt. (CC) > 200 F; 100% act.; *Toxicology:* alkaline material; prevent contact with skin or eyes; *Storage:* stable indefinitely in closed containers.

GP-5 Emulsifiable Paintable Silicone. [Genesee Polymers] Methylalkylarylpolysiloxane fluid with nonionic emulsifiers in waterless conc.; nonionic; mold release/lubricant for rubber, plastics, die lubricants, metalworking lubricity additive, foundry, textile lubricants; for painting, printing, or bonding of cast or molded parts; readily forms o/w emulsion by mixing with hard or soft water; clear to hazy liq., mild odor; sp.gr. 0.96; dens. 8.0 lb/gal; 80% silicone, 14% nonionic emulsifiers, 6% aliphatic hydrocarbon; *Toxicology:* alkaline material; prevent contact with skin or eyes; *Storage:* stable indefinitely in closed containers.

GP-6 Silicone Fluid. [Genesee Polymers] Amine functional silicone fluid; internal plastic mold release and lubricant; in textile lubricants, detergent-resist. car polishes, rust inhibitor coatings, silicone/organic copolymer synthesis; lt. straw clear liq.; sp.gr. 0.97; dens. 8.0 lb/gal; visc. 178 cst; amine no. 49; flash pt. (CC) > 300 F; 100% act.; *Toxicology:* alkaline material; prevent contact with skin or eyes; *Storage:* stable indefinitely in closed containers.

GP-50-A Modified Silicone Emulsion. [Genesee Polymers] Modified dimethyl silicone emulsion; nonionic; release agent for rubber, plastics, foundry castings, metal castings; wh. opaque liq.; fine particle size emulsion; dilutable in hard or soft water; sp.gr. 0.98; dens. 8.0 lb/gal; 50% silicone; *Storage:* stable indefinitely; do not freeze; store below 100 F.

GP-51-E Dimethyl Silicone Emulsion. [Genesee Polymers] Dimethyl silicone aq. emulsion; nonionic; release agent for rubber, plastics, foundry castings, metal castings; also for car polish formulations; wh. opaque liq.; fine particle size emulsion; dilutable in hard or soft water; sp.gr. 0.99; dens. 8.0 lb/gal; visc. 100 cst (silicone); flash pt. none; 50% silicone; *Storage:* stable indefinitely; do not freeze; store below 100 F.

GP-52-E Dimethyl Silicone Emulsion. [Genesee Polymers] Dimethyl silicone aq. emulsion; nonionic; release agent for rubber, plastics, foundry castings, metal castings; also for car polish formulations; wh. opaque liq.; fine particle size emulsion; dilutable in hard or soft water; sp.gr. 0.99; dens. 8.0 lb/gal; visc. 350 cst (silicone); flash pt. none; pH 9.0; 50% silicone; *Storage:* stable indefinitely; do not freeze; store below 100 F.

GP-53-E Dimethyl Silicone Emulsion. [Genesee Polymers] Dimethyl silicone aq. emulsion; nonionic; release agent for rubber, plastics, foundry castings, metal castings; also for car polish formulations; wh. opaque liq.; fine particle size emulsion; dilutable in hard or soft water; sp.gr. 1.0; dens. 8.3 lb/gal; visc. 1000 cst (silicone); flash pt. none; pH 8.5; 50% silicone; *Storage:* stable indefinitely; do not freeze; store below 100 F.

GP-54-E Dimethyl Silicone Emulsion. [Genesee Polymers] Dimethyl silicone aq. emulsion; nonionic; release agent for rubber, plastics, foundry castings, metal castings; also for car polish formulations; wh. opaque liq.; fine particle size emulsion; dilutable in hard or soft water; sp.gr. 0.99; dens. 8.0 lb/gal; visc. 10,000 cst (silicone); flash pt. none; pH 9.0; 50% silicone; *Storage:* stable indefinitely; do not freeze; store below 100 F.

GP-60-E Silicone Emulsion. [Genesee Polymers] Organo-modified dimethyl silicone aq. emulsion; nonionic; release agent and lubricant for rubber and plastic molding, foundry applics.; band-ply lube formulations, chain lubricants, fiberglass lubricants, cable lubricants; textile treatments; wh. opaque liq.; dilutable in hard or soft water; sp.gr. 0.96; dens. 8.0 lb/gal; visc. 60,000 cst (silicone); flash pt. none; 50% silicone; *Storage:* do not freeze; store below 100 F.

GP-70-E Paintable Silicone Emulsion. [Genesee Polymers] Organo-modified silicone emulsion; nonionic; release agent for molding or casting of articles to be subsequently painted or bonded; for rubber and plastic release, metal-forming and casting, foundry applics., textile lubricants, automotive gasketing lubricant/release; good high temp. stability; wh. opaque liq.; dilutable in hard or soft water; sp.gr. 0.96; dens. 8.0 lb/gal; pH 5; 50% silicone; *Storage:* 4 cycle freeze-thaw stability; if frozen, thaw, then remix; store @ 40-100 F.

GP-70-S Paintable Silicone Fluid. [Genesee Polymers] Methyl alkylaryl silicone fluid; release agent for molded metal, plastic, or rubber parts; internal lubricant for rubber, vinyl plastisols; also for foundry and die lubricant applics., as paint additive; lt. straw clear liq.; sol. in aliphatic hydrocarbons, chlorinated solvs.; sp.gr. 0.87; dens. 7.3 lb/gal; visc. 1200 cst; flash pt. (CC) > 300 F; 100% act.; *Precaution:* combustible; keep away from open flames and welding equip.

GP-71-SS Mercapto Modified Silicone Fluid. [Genesee Polymers] Dimethicone/mercaptopropyl methicone copolymer; plastic and rubber release agent; internal lubricant and release agent for sulfur and peroxide cure rubber; coreactant in vinyl polymerization; synthesis of org./silicone copolymers; heat stabilizer for dimethyl silicone fluids; in corrosion inhibitor coatings, uv and eb-cured inks and coatings; cosmetic and personal care applics. incl. hair care; water-wh. to lt. straw liq., sulfur odor; sol. in aliphatic and aromatic solvs., chlorinated solvs.; sp.gr. 0.97; dens. 8.0 lb/gal; visc. 150 cst; flash pt. (PMCC) > 450 F; 100% act.

GP-74 Paintable Silicone Fluid. [Genesee Polymers] Methyl/alkylaryl silicone fluid; release agent for molded metal, plastic, or rubber parts; internal lubricant in rubber, vinyl plastisols; also for foundry, adhesive, org. coatings applics.; good wetting props., heat stability; straw clear to sl. hazy liq.; sol. in aliphatic hydrocarbons, chlorinated solvs.; sp.gr. 0.97; dens. 8.0 lb/gal; visc. 800 cst; flash pt. (CC) 180 F; 90% act.; *Precaution:* combustible; keep away from open flame or welding torch, even empty drums.

GP-74-E Paintable Silicone Emulsion. [Genesee Polymers] Organo-modified silicone aq. emulsion; nonionic; release agent for molding or casting of articles to be subsequently painted or bonded; for rubber and plastic release, metal-forming and casting, textile lubricants, automotive gasketing lubricant/release, molding on mandrels; wh. opaque liq.; dilutable in hard or soft water; sp.gr. 0.96; dens. 8.0 lb/gal; pH 8.5; 50% silicone; *Storage:* stable indefinitely; store below 100 F, guard against freezing; if frozen, thaw, then remix.

GP-80-AE Silicone Emulsion. [Genesee Polymers] Dimethyl silicone aq. emulsion; nonionic; exhibits high gloss, film durability, weather and detergent resist. for vinyl/leather treatment, filler treatments, water repellents, textile lubricants, high-performance release agents; cosmetic dressing for rubber (tires, etc.); wh. opaque liq.; dilutable in hard or soft water; sp.gr. 0.96; dens. 8.0 lb/gal; pH 9.0; 50% silicone act.; *Storage:* do not freeze; store below 100 F.

GP-83-AE Silicone Emulsion. [Genesee Polymers] Amino functional silicone aq. emulsion; nonionic; cures to glossy, durable, detergent-resist. film for vinyl/leather surf. treatment, filler treatments, water repellents, car polishes, textile lubricants; cosmetic dressing for rubber (tires, etc.); wh. opaque liq.; dilutable in hard or soft water; sp.gr. 0.96; dens. 8.0 lb/gal; pH 9.0; 40% silicone; *Storage:* do not freeze; store below 100 F.

GP-85-AE Silicone Emulsion. [Genesee Polymers] Amino functional and high visc. silicones in aq. emulsion; nonionic; provides high gloss, durability, and detegent resist. to vinyl/leather/rubber surfaces (vinyl tops, tires, upholstery, seats), water repellents, car polishes, high-performance release agent; off-wh. opaque liq., mild odor; dilutable in hard or soft water; sp.gr. 0.96; dens. 8.0 lb/gal; pH 9.0; 40% silicone; *Storage:* do not freeze; store below 100 F.

GP-86-AE Silicone Emulsion. [Genesee Polymers] Amino functional silicone aq. emulsion; nonionic; provides gloss, durability, and detegent resist. to vinyl conditioners, leather treatment, rust preventative coatings, textile lubricants, tire and rubber dressing, car polishes; improved resist. to weathering, fading, and cracking in vinyl tops; wh. opaque liq., mild odor; dilutable in hard or soft water; sp.gr. 0.96; dens. 8.0 lb/gal; pH 9.5; 50% silicone; *Storage:* do not freeze; store below 100 F.

GP-98 Silicone Wax. [Genesee Polymers] Modified methylalkyl polysiloxane wax; high temp. mold release agent for casting of metals; good for parts for subsequent painting or finishing; high temp. lubricant; also for plastic molding, foundry release; high temp. stablity; amber thick paste; sol. in aliphatic hydrocarbons; sp.gr. 0.94; dens. 8.0 lb/gal; flash pt. > 300 F; 100% act.

GP-180 Resin Sol'n. [Genesee Polymers] Proprietary organosilicone resin conc. in 1,1,1-trichlorethane/

IPA solv.; used as a dry film release coating for molding of plastics or rubber; adherent coating for conveyors, chutes, and handling equip. used in rubber, plastics, and adhesive mfg.; release agent for electronic circuit board mfg.; water-wh. clear thin liq., mild solv. odor; sp.gr. 1.2; dens. 10 lb/gal; flash pt. (PMCC) 148 F; *Storage:* store below 90 F in closed containers; vent slowly when opening; mix well before use.

GP-197 Resin Sol'n. [Genesee Polymers] Silicone resin sol'n. in 1,1,1-trichlorethane/IPA/butanol; used as a dry film release coating for molding of plastics or rubber; adherent coating for conveyors, chutes, and handling equip. used in rubber, plastics, and adhesive mfg.; release agent for electronic circuit board mfg.; yel. clear liq., mild solv. odor; sp.gr. 1.1; dens. 9 lb/gal; flash pt. (PMCC) 130 F; *Storage:* store below 90 F in closed containers.

GP-209 Silicone Polyol Copolymer. [Genesee Polymers] Dimethicone copolyol; nonionic; emulsifier, wetting agent, pigment dispersant, leveling agent, profoaming additive for PU foams, hard surf. cleaners, polishes, cosmetic formulations; inverse sol. suggests use as defoamer for hot aq. surfactant sol'ns.; cell control agent in PU foam; lt. straw clear liq.; sol. in water; m.w. 7800; sp.gr. 1.03; dens. 8.5 lb/gal; visc. 2600 cst; f.p. -50 F; flash pt. (PMCC) > 300 F; 100% act., 15% silicone.; *Toxicology:* very low toxicity.

GP-214 Silicone Polyol Copolymer. [Genesee Polymers] Dimethyl silicone EO/PO block copolymer; nonionic; profoaming additive in plastisol and organosol formulations, cell control agent in polyether PU foams; pigment dispersant, wetting agent, internal lubricant for plastics, leveling and flow control agent for solv. coatings; lt. straw clear liq.; insol. in water; m.w. 11,500; sp.gr. 1.03; dens. 8.5 lb/gal; visc. 6500 cst; f.p. -50 F; flash pt. (PMCC) > 300 F; 100% act., 31% silicone.; *Toxicology:* essentially nontoxic.

GP-215 Silicone Polyol Copolymer. [Genesee Polymers] Dimethicone copolyol; emulsifier, wetting agent, pigment dispersant, leveling agent, profoaming additive for PU foams, hard surf. cleaners, polishes, cosmetics; inverse sol. suggests use as defoamer for hot aq. surfactant sol'ns.; cosmetic and personal care applics. incl. hair care; lt. straw clear liq.; sol. in water; m.w. 9800; sp.gr. 1.03; dens. 8.5 lb/gal; visc. 2000 cst; f.p. -50 F; flash pt. (PMCC) > 300 F; 100% act., 18% silicone.; *Toxicology:* essentially nontoxic.

GP-218 Silicone Polyol Copolymer. [Genesee Polymers] Dimethylpolysiloxane PO block copolymer; wetting, leveling, flow control agent, lubricant for solv.-based coatings, industrial finishes; profoamer additive in PU foams; textile and thread lubricants; internal lubricant for plastics; base for aq. defoamers; pigment dispersant; release agent; cosmetic and personal care applics. incl. hair care; colorless clear liq.; sol. in aliphatic, aromatic, and chlorinated hydrocarbons, alcohols; insol. in water; m.w. 11,000; sp.gr. 0.98; dens. 8.0 lb/gal; visc. 1500 cst; f.p. 65 F; flash pt. (CC) > 300 F; 100% act., 32% silicone.

GP-219 Silicone Polyol Copolymer. [Genesee Polymers] Dimethylpolysiloxane EO block copolymer; emulsifier for w/o emulsions; pigment dispersant for thermoplastics or solv.-based coatings; wetting agent; thread lubricant; leveling and flow control agent; lt. straw clear liq.; insol. in water; m.w. 6500; sp.gr. 1.1; dens. 8.4 lb/gal; visc. 440 cst; HLB 8.4; f.p. 65 F; flash pt. (PMCC) > 300 F; 100% act., 58% silicone.; *Toxicology:* very low toxicity.

GP-530 Silicone Fluid. [Genesee Polymers] Modified silicone copolymer; lubricant, release agent for plastic/metal or rubber/metal interfaces, rubber molding, urethane molding, foundry and eposy release, textile lubricant; exc. heat stability; water-wh./lt. straw clear liq.; sp.gr. 0.97; dens. 8.0 lb/gal; visc. 900 cst; flash pt. > 300 F; 100% act.; *Storage:* stable indefinitely.

GP-7100 Silicone Fluid. [Genesee Polymers] Amine-alkyl modified methyl alkylaryl silicone polymer; release agent for plastic, metal, and rubber parts; internal release in rubber, plastics, plastisols; internal lubricant in rubber; lubricant for metal cutting and drilling; foundry release agent; intermediate in silicone/org. copolymer synthesis; lt. straw liq.; sol. in aliphatic hydrocarbons, chlorinated solvs.; sp.gr. 0.97; dens. 8.0 lb/gal; visc. 750 cst; flash pt. (CC) > 200 F; 100% act.

GP-7100-E Paintable Silicone Emulsion. [Genesee Polymers] Amine-alkyl modified methylakylaryl silicone emulsion; nonionic; release agent for rubber and plastic; automotive gasketing lubricant/release; metal-forming and casting; foundry applics.; textile lubricants; good high temp. stability; wh. opaque liq.; dilutable in hard or soft water; sp.gr. 0.96; dens. 8.0 lb/gal; pH 9.5; 50% silicone; *Storage:* 4 cycle freeze-thaw stability; if frozen, thaw, then remix; store @ 40-100 F.

GP-7101 Silicone Wax. [Genesee Polymers] Dimethyl silicone wax; mold release agent, rubber and plastic lubricant, textile lubricant; metal-forming lubricant; protective coating for steel; in waxes and polishes for automotive and furniture applics.; straw soft wax, mild odor; sol. in aliphatic hydrocarbons, chlorinated solvs.; sp.gr. 0.94; dens. 7.8 lb/gal; m.p. 110-120 F; flash pt. > 300 F; 100% act.; *Storage:* indefinite storage stability.

GP-7102 Silicone Wax. [Genesee Polymers] Dimethyl silicone wax; mold release agent, rubber and plastic lubricant and internal release agent, textile lubricant; metal-forming lubricant; protective coating for steel; in waxes and polishes for automotive and furniture applics.; straw soft wax, mild odor; sol. in aliphatic hydrocarbons, chlorinated solvs.; sp.gr. 0.94; dens. 7.8 lb/gal; m.p. 100-110 F; flash pt. > 300 F; 100% act.; *Storage:* indefinite storage stability.

GP-7105 Silicone Fluid. [Genesee Polymers] Silicone wax copolymer with amine functionality; imparts gloss, detergent resist. to car and shoe polish formulations, leather treatment, fiber treatment, textile lubricant; mold release, lubricant for plastics and rubber; rust inhibition props.; lt. straw thick waxy liq., mild ammonia-like odor; sol. in aliphatic hydrocarbons, chlorinated solvs.; sp.gr. 0.96; dens. 8.0 lb/gal; m.p. 100-110 F; flash pt. > 300 F; *Toxicology:* alkaline material; avoid skin and eye contact; *Storage:* indefinite storage stability.

GP-7105-E Silicone Emulsion. [Genesee Polymers] Amine functional silicone wax copolymer emulsion in water and aliphatic hydrocarbon; nonionic; provides high gloss, detegent and weather resist. to vinyl and rubber conditioners, leather treatment, textile lubricants, car polishes, filler treatments, water repellents; cosmetic dressing for rubber (tires, etc.); wh. opaque liq., mild odor; sp.gr. 0.96; dens. 8.0 lb/gal; pH 9.0; 40% silicone; *Storage:* do not freeze; store below 100 F.

GP-7200 Silicone Fluid. [Genesee Polymers] Mercapto-functional methyl alkyl silicone polymer; paintable release for molded metal, plastic, rubber parts; internal release for rubber and plastics; metal forming lubricant; corrosion inhibiting coatings; greater heat stability in release film; reactive with sulfur and peroxide cured elastomers; yel. clear liq., sulfur odor; sol. in aliphatic and aromatic hydrocarbons; sp.gr. 0.97; dens. 8.0 lb/gal; visc. moderate; flash pt. (PMCC) > 300 F; 100% act.

GP-RA-156 Release Agent. [Genesee Polymers] Silicone; release agent for urethane and epoxy molded parts, many elastomeric prods.; also for foundry applics., textile lubricants; exc. heat stability; clear to straw liq.; sp.gr. 0.97; dens. 8.0 lb/gal; visc. 350 cst; flash pt. (COC) 400 F; 100% act.; *Storage:* store in tightly sealed containers; avoid contact with moisture.

GP-RA-158 Silicone Polish Additive. [Genesee Polymers] Modified silicone polymer in min. spirits; high gloss, detergent-resist. polish additive for paste and emulsion auto care prods., vinyl conditioners, rust preventative coatings, fiberglass sizing agents, textile lubricants, leather treating, water/soil repellents; release agents, rubber lubricants; clear mod. visc. liq.; dilutable in aliphatic hydrocarbons; sp.gr. 0.86; dens. 7 lb/gal; visc. 2000 cst; flash pt. (CC) 99 F; 50% act.; *Precaution:* flamm.; keep away from fire and flames; do not use torch on drums, even if empty; *Storage:* store in closed containers; prevent contact with atmospheric moisture.

GP-RA-159 Silicone Polish Additive. [Genesee Polymers] Aminoalkyl silicone polymer in IPA/min. spirits solv.; high gloss, detergent-resist. polish additive for car polish, vinyl conditioners, rust preventative coatings, fiberglass sizing agents, textile lubricants, leather treating, water/soil repellents; release agents, rubber lubricants; fast drying solv. blend; colorless to straw-colored clear liq.; dilutable with aliphatic solvs. or alcohols; sp.gr. 0.86; dens. 7 lb/gal; visc. 200 cst; flash pt. (PMCC) 52 F; 50% silicone; *Toxicology:* moderate to severe eye irritant; *Precaution:* flamm.; keep away from flames, sparks, heat sources; do not weld near drums, even when empty; *Storage:* store in closed containers; prevent contact with atmospheric moisture.

GP-RA-201 Release Agent. [Genesee Polymers] Silicone; release agent for urethane and epoxy molded parts, many elastomeric prods.; also for foundry applics.; colorless clear liq.; sp.gr. 0.98; dens. 8.0 lb/gal; 100% act.; *Storage:* stable indefinitely.

Granuplast™. [Safas] Patented compounding material that gives a granite look to most plastic parts, e.g., appliances, consumer electronics, furniture and accessories, computer housings, extruded panels, automotive components and trim, toys, sports equip., housewares; suitable for use with ABS, PC, PE, PP, PS, vinyl, nylon, acrylic, PCT, PET; can be formulated to meet FDA, NSF, and UL requirements; avail. conc. and precolored form, in 64 std. colors.

Great Lakes BA-43. [Great Lakes] Bis acrylate of ethoxylated tetrabromobisphenol A; functionally reactive flame retardant used in chemical, irradiation, or uv curing systems; enhances FR and crosslinking performance of elastomers; wh. powd; sol. 0.2 g/100 g solv. in toluene; sol. < 0.1 g/100 g in water, methanol, methylene chloride, MEK; m.w. 740.1; m.p. 115-120 C; 43% bromine.; Unverified

Great Lakes BA-50™. [Great Lakes] Bis (2-hydroxyethyl ether) of tetrabromobisphenol A; difunctional alcohol providing flame retardance; for unsat. polyester thermosets, PBT thermoplastics, PU elastomer wire insulation, adhesives, coatings; useful for laminates for electronic circuit boards; corrosion resistance of systems useful in materials for tanks, ducts, and hoods; off-wh. powd.; sol. (g/100 g): 80 g MEK, 62 g methanol, 48 g methylene chloride, 22 g triethylene glycol, 21 g acetone, 17 g acetate, < 0.1 g in water; m.w. 632.0; sp.gr. 1.8; m.p. 113-119 C; acid no. < 3.5; 50.6% Br.

Great Lakes BA-50P™. [Great Lakes] Bis (2-hydroxyethyl ether) of tetrabromobisphenol A; reactive flame retardant for thermoplastics (PBT), thermosets (unsat. polyester), PU elastomer wire insulation, adhesives, and coatings; wh. powd.; sol. (g/100 g): 80 g MEK, 62 g methanol, 48 g methylene chloride, 22 g triethylene glycol, 21 g acetone, 17 g acetate, < 0.1 g in water; m.w. 632.0; sp.gr. 1.8; m.p. 113-119 C; acid no. < 3.5; 50.6% Br.

Great Lakes BA-59. [Great Lakes] Tetrabromobisphenol A; CAS 79-94-7; EINECS 201-236-9; aromatic-based flame retardant for thermoplastic and thermoset resin systems, epoxy systems; wh. powd.; sol. (g/100 g solv.) 225 g acetone, 168 g MEK, 80 g methanol, 0.1 g water; m.w. 543.7; f.p. 180 C; 58.8% Br.

Great Lakes BA-59P™. [Great Lakes] Tetrabromobisphenol A; CAS 79-94-7; EINECS 201-236-9; flame retardant for thermoplastics (ABS, HIPS, PC), thermosets (epoxy, unsat. polyester), adhesives, coatings, textiles; wh. powd.; sol. (g/100 g): 168 g MEK, 80 g methanol, 27 g methylene chloride, 6 g toluene, < 0.1 g water; m.w. 543.7; sp.gr. 2.2; m.p. 179-181 C; 58.8% Br.

Great Lakes BC-52™. [Great Lakes] Phenoxy terminated tetrabromobisphenol A carbonate oligomer; flame retardant for ABS, HIPS, PBT, PC thermoplastics, and thermoplastic elastomer wire insulation; sol. in toluene; sol. (g/100 g): > 170 g methylene chloride and MEK, < 0.1 g in water, methanol; m.w. ≈ 2500; sp.gr. 2.2; m.p. 210-230 C; 51.3% Br.

Great Lakes BC-58™. [Great Lakes] Phenoxy terminated tetrabromobisphenol A carbonate oligomer; flame retardant for ABS, HIPS, PBT, PC thermoplastics, and thermoplastic elastomer wire insulation; sol. in toluene; sol. (g/100 g): > 170 g methylene chloride, MEK, < 0.1 g in water, methanol; m.w. ≈ 3500; sp.gr. 2.2; m.p. 230-260 C; 58.7% Br.

Great Lakes BE-51™. [Great Lakes] Bis (allyl ether) of tetrabromobisphenol A; CAS 25327-89-3; flame retardant used in EPS; wh. solid; sol. (g/100 g): 47 g methylene chloride, 42 g toluene, 33 g styrene, 12 g MEK, 0.1 g in water and methanol; m.w. 624.0; sp.gr. 1.8; m.p. 118-120 C; 51.2% Br.

Great Lakes CD-75P™. [Great Lakes] Hexabromocyclododecane; EINECS 221-695-9; flame retardant used in thermoplastic and thermosetting polymers, HIPS, EPS foam, textile treatments, latex binders, adhesives, unsat. polyesters, and coatings; wh. powd.; sol. (g/100 g): 12 g toluene, 10 g styrene and MEK, 4 g methylene chloride, 0.1 g water; m.w. 641.7; sp.gr. 2.1; m.p. 185-195 C; 90% assay; 74.7% Br.

Great Lakes DBS™. [Great Lakes] Dibromostyrene; flame retardant for thermoplastics (ABS, HIPS), thermosets (unsat. polyester, flexible urethane foam); liq.; sol. in methylene chloride, toluene, MEK; sol. (g/100 g): 50 g methanol, < 0.1 g water; m.w. 2261.9; sp.gr. 1.8; 59% Br.

Great Lakes DE-60F™ Special. [Great Lakes] 85% DE-71 (pentabromodiphenyl oxide), 15% aromatic phosphate; flame retardant additive for rigid and flexible urethane foams, epoxies, laminates, unsat. polyesters, plasticized PVC compds., PU elastomer wire insulation, adhesives, and coatings; amber liq.; sol. in Freon 11, polyol, styrene, MEK, triethyl phosphate, methylene chloride, toluene, dioctylphthalate; sol. 0.1 g/100 g water; sp.gr. 1.95; dens. 16.4 lb/gal; acid no. 0.25 max.; 52% Br.

Great Lakes DE-61™. [Great Lakes] Pentabromodiphenyl oxide blend; flame retardant for rigid and flexible urethane foams, PU elastomer wire insulation, adhesives, and coatings; liq.; sol. in methylene chloride, toluene, MEK, sol. < 0.1 g/100 g in water, methanol; sp.gr. 1.9; 51% Br.

Great Lakes DE-62™. [Great Lakes] Pentabromodiphenyl oxide blend; flame retardant for rigid and flexible urethane foams, PU elastomer wire insulation, adhesives, and coatings; liq.; sol. in methylene chloride, toluene, MEK, sol. < 0.1 g/100 g in water, methanol; sp.gr. 1.9; 51% Br.

Great Lakes DE-71™. [Great Lakes] Pentabromodiphenyl oxide; CAS 32534-81-9; high visc. flame retardant for thermosetting and thermoplastic resin systems; used for unsat. polyester, rigid and flexible urethane foams, epoxies, PU elastomer wire insulation, laminates, adhesives, coatings, and textiles; amber visc. liq.; sol. in Freon 11, polyol, styrene, MEK, triethyl phosphate, methylene chloride, toluene, dioctylphthalate, sol. < 0.1 g/100 g water; m.w. 564.7; sp.gr. 2.27; dens. 19.0 lb/gal; acid no. 0.25 max.; 70.8% Br.

Great Lakes DE-79™. [Great Lakes] Octabromodiphenyl oxide; CAS 32536-52-0; EINECS 251-087-9; flame retardant for ABS, HIPS, and nylon thermoplastic polymers, polyamide/polyimide wire insulation, adhesives, and coatings; off-wh. powd.; sol. (g/100 g): 25 g styrene, 20 g benzene, 19 g toluene, 11 g methylene chloride, 4 g MEK, < 0.1 g water; m.w. 801.4; sp.gr. 2.8; m.p. 70-150 C; 79.8% Br.

Great Lakes DE-83™. [Great Lakes] Decabromobiphenyl oxide; CAS 1163-19-5; EINECS 214-604-9; flame retardant for thermoplastic, elastomeric, and thermoset polymer systems incl. HIPS, PBT, ABS, nylons, PP, PC, PET, polybutylene, PVC, LDPE, EPDM, unsat. polyesters, epoxies, flexible PU foam, wire insulation (polyolefin, TPE, PU elastomer, polyamide/polyimide), adhesives, coatings, textiles; wh. powd.; 3.2 μ particle size; sol. (g/100 g): 0.2 g toluene, 0.1 g acetone and benzene, < 0.1 g. in water, methanol; m.w. 959.2; sp.gr. 3.3; m.p. 300-315 C; 97% assay; 83.3% Br.

Great Lakes DE-83R™. [Great Lakes] Decabromobiphenyl oxide; CAS 1163-19-5; EINECS 214-604-9; flame retardant for thermoplastic, elastomeric, and thermoset polymer systems incl. HIPS, PBT, ABS, nylons, PP, PC, PET, polybutylene, PVC, LDPE, EPDM, unsat. polyesters, epoxies, flexible PU foam, wire insulation (polyolefin, TPE, PU elastomer, polyamide/polyimide), adhesives, coatings, textiles; wh. powd.; 3.2 μ particle size; sol. (g/100 g): 0.2 g toluene, 0.1 g acetone and benzene, < 0.1 g. in water, methanol; m.w. 959.2; sp.gr. 3.3; m.p. 300-315 C; 97% assay; 83.3% Br.

Great Lakes DP-45™. [Great Lakes] Tetrabromophthalate ester; flame retardant for PVC, TPE and PU elastomer wire insulation, adhesives, coatings, and textiles; liq.; sol. in methylene chloride, toluene, MEK; sol. (g/100 g): 10 g in methanol, < 0.1 g in water; m.w. 706.1; sp.gr. 1.5; 45.3% Br.

Great Lakes FB-72™. [Great Lakes] Proprietary brominated flame retardant; flame retardant for thermoplastics (HIPS, PP, polybutylene), adhesives, coatings, and textiles; solid; partly sol. in methylene chloride, toluene, MEK; sol. < 0.1 g/100 g in water, methanol; sp.gr. 2.5; m.p. 190-200 C; 72% Br.

Great Lakes FF-680™. [Great Lakes] Bis (tribromophenoxy) ethane; flame retardant for applic. where thermal stability at high processing temps. is important; for thermoplastics (ABS, HIPS, PC), thermosets (unsat. polyester), thermoplastic elastomer wire insulation, adhesives, coatings, textiles, lt.-stable applics.; wh. cryst. powd.; sol. in boiling dichlorobenzene, p-xylene perchloroethylene; insol. in water, methanol, MEK; m.w. 687.6; sp.gr. 2.6; dens. 46.7 lb/ft^3; m.p. 223-228 C; 70% Br.

Great Lakes GPP-36™. [Great Lakes] Polypropylene-dibromostyrene graft copolymer; flame retardant for PP and textiles; solid; insol. in water, methanol, methylene chloride, toluene, MEK; sp.gr. 1.3; m.p. 160-175 C; 36% Br.

Great Lakes GPP-39™. [Great Lakes] Polypropylene-dibromostyrene graft copolymer; deriv. of GPP-36; flame retardant for PP and textiles; solid; insol. in water, methanol, methylene chloride, toluene, MEK; sp.gr. 1.3; m.p. 150-170 C; 39% Br.

Great Lakes NH-1197™. [Great Lakes] Intumescent flame retardant; flame retardant for epoxy resins, adhesives, coatings, and textiles; solid; sol. (g/100 g): 6 g in methanol, 0.5 g in water; insol. in methylene chloride, toluene, MEK; sp.gr. 1.74; m.p. 205-212 C; 17% P.

Great Lakes NH-1511™. [Great Lakes] Nonhalogenated intumescent flame retardant; flame retardant for PP, PBT, wire insulation (polyolefin, TPE, PU elastomer), adhesives, coatings, and textiles; solid; partly sol. in water, methanol; insol. in methylene chloride, toluene, MEK; sp.gr. 2.6; m.p. 211-215 C; 15% P.

Great Lakes PDBS-10™. [Great Lakes] Poly(dibromostyrene); flame retardant for thermoplastics (ABS, HIPS, PBT, PC, nylon, PET) and polyamide/polyimide wire insulation; solid; sol. in methylene chloride, toluene, partly sol. in MEK, methanol; insol. in water; m.w. ≈ 10,000; sp.gr. 1.9; m.p. 155-165 C; 59% Br.

Great Lakes PDBS-80™. [Great Lakes] Poly(dibromostyrene); flame retardant for thermoplastics incl. HIPS, PBT, nylon 6 and 6/6, PET, S/B latex coatings; exc. cost/performance for UL94V-0 rating; off-wh. powd. or amber pellets; sol. in methylene chloride, toluene, styrene, THF; partly sol. in acetone, DMF, MEK, methanol; insol. in water; m.w. ≈ 80,000; sp.gr. 1.9; m.p. 220-240 C; 59% Br; *Toxicology:* LD50 (oral, rat) > 5000 mg/kg; avoid contact.

Great Lakes PE-68™. [Great Lakes] Bis (2,3-dibromopropyl ether) of tetrabromobisphenol A; CAS 21850-44-2; flame retardant used in PP, polyethylene, polybutylene, and polyolefin copolymers, wire insulation, textiles; effective at low loading levels; off-wh. powd.; sol. (g/100 g): 50 g methylene chloride, 24 g toluene, 10 g acetone, 7 g MEK, 0.1 g in water and methanol; m.w. 943.6; sp.gr. 2.2; m.p. 90-100 C; 67.7% Br.

Great Lakes PH-73™. [Great Lakes] 2,4,6-Tribromophenol; CAS 118-79-6; reactive intermediate for phenol-based reactions; flame retardant, antifungal agent, or chemical intermediate; lt. cream to tan flakes; sol. (g/100 g): 225 g MEK, 84 g methanol, 60 g methylene chloride, 50 g toluene, 0.1 g water; m.w. 330.8; sp.gr. 2.2; m.p. 95-96 C; 99.5% assay; 72.5% Br.

Great Lakes PHE-65™. [Great Lakes] Tribromophenol allyl ether; CAS 3278-89-5; flame retardant for EPS foam, polyolefin and polyamide/polyimide wire insulation; sol. (g/100 g): > 200 g MEK, > 50 g methanol, methylene chloride, > 30 g toluene, < 0.1 g in water; m.w. 371; sp.gr. 2.1; m.p. 74-76 C; 64.2% Br.

Great Lakes PHT4™. [Great Lakes] Tetrabromophthalic anhydride; CAS 632-79-1; EINECS 211-185-4; flame retardant in prod. of unsat. polyester resins and rigid PU polyols; cohardener for epoxy resins; cost efficient additive for latex emulsions; derivs. used as flame retardants in diverse applic. (adhesives, wire coating, and wool, etc.); off-wh. cryst.; sol. in dimethylformamide, nitrobenzene; sol. (g/100 g): 8 g toluene, 7 g MEK, 3 g methylene chloride; insol. in water; m.w. 463.7; dens. 86 lb/ft^3; m.p. 270-276 C; 68.2% Br.

Great Lakes PHT4-Diol™. [Great Lakes] Tetrabromophthalatediol; CAS 20566-35-2; EINECS 243-885-0; reactive intermediate used to produce flame retardant rigid urethane foam, PU elastomer wire insulation, adhesives, coatings, and fibers; can replace chlorinated polyols; lt. tan visc. liq.; sol. in methylene chloride, toluene, MEK; sol. < 0.1 g/100 g in water, methanol; m.w. 627.9; sp.gr. 1.8; visc. 135,000 cps; acid no. 0.01; 46% Br.

Great Lakes PO-64P™. [Great Lakes] Polydibromophenylene oxide; flame retardant which melts into most polymers to optimize physical properites; enhances flow into thin wall sections; permits higher regrind loading levels; for PBT, PC, nylon, PET, polybutylene; off-wh. powd.; sol. (g/100 g): ≤ 0.1 g in water, methanol, MEK; partly sol. in toluene; m.w. ≈ 6000; sp.gr. 2.3; m.p. 210-240 C; 62% Br.

Great Lakes SP-75™. [Great Lakes] Stabilized hexabromocyclododecane; flame retardant for polybutylene, EPS foam, adhesives, coatings, and textiles; solid; sol. (g/100 g): 12 g in toluene, 10 g MEK, 4 g methylene chloride, 1 g mthanol, < 0.1 g water; sp.gr. 2.1; m.p. 185-195 C; 71% Br.

Grindtek MOP 90. [Grindsted Prods.] Lard glyceride; CAS 61789-10-4; EINECS 263-032-6; component in w/o and o/w creams, lubricant, antistat, antifogging agent in plastics; antimicrobial effects reported; wh. block; sol. warm in ethanol, propylene glycol, tolene, wh. spirit; HLB 4.3.

Grindtek MSP 40. [Grindsted Prods.] Glyceryl stearate; component in w/o and o/w creams, lubricant, antistat, antifogging agent in plastics; antimicrobial effects reported; wh. powd.; sol. in toluene; HLB 2.8.

Grindtek MSP 40F. [Grindsted Prods.] Glyceryl stearate; component in w/o and o/w creams, lubricant, antistat, antifogging agent in plastics; antimicrobial effects reported; wh. fine powd.; sol. warm in toluene; HLB 2.8.

Grindtek MSP 90. [Grindsted Prods.] Glyceryl stearate; component in w/o and o/w creams, lubricant, antistat, antifogging agent in plastics; antimicrobial effects reported; wh. powd.; sol. in ethanol, toluene; HLB 4.3.

Grindtek PGE-DSO. [Grindsted Prods.] Oleic/linoleic fatty acid ester of polyglycerol; o/w emulsifier, antifogging agent for plastics, effective for w/o emulsifications with veg. oils; yel.-brn. visc. liq.; sol. in toluene, warm in peanut oil, wh. spirit, paraffin oil; HLB 2.8.

Grindtek PK 60. [Grindsted Prods.] Palm kernel glycerides; component in w/o and o/w creams, lubricant, antistat, antifogging agent for plastics; wh. block; sol. in ethanol, warm in peanut oil, toluene, paraffin oil, propylene glycol, tolene; HLB 4.1.

Grit-O'Cobs®. [Andersons] Corn cob meal; chemically inert plastic extender and filler; replaces wood flour in wood particle molding with phenolic resins, in profile and sheet stock prod.; filler in glue, asphalt, caulking compds., and rubber; also used in industrial abrasives, as industrial absorbent, as agric. chemical carriers, livestock feed roughage; fine grades used in soaps and cosmetics; tan gran.; essentially odorless; 20.9% sol. in 1% sodium hydroxide, 9.5% in hot water, 5.6% in alcohol, 2.5% in acetone and in 10% sulfuric acid; sp.gr. 1.2; bulk dens. 20-30 lb/ft^3; oil absorp. 100%; water absorp. 133.0%; pH 7.4 (surface); flash pt. (OC) 350 F; hardness (Mohs) 4.5; 47.1% cellulose.

GS-3. [Lilly] Blend; external dry lubricant for epoxy and polyester; also exc. metal-to-metal lubricant; liq.

GTO 80. [SVO Enterprises] Crude high oleic sunflower oil; CAS 8001-21-6; EINECS 232-273-9; food grade environmentally friendly veg. oil for use as feedstocks/raw materials, in plastics, polymers, rubber, and pharmaceuticals; oxidative and thermal stability; readily biodeg.; 77-82% oleic acid.

GTO 90. [SVO Enterprises] Crude very high oleic sunflower oil; CAS 8001-21-6; EINECS 232-273-9; food grade environmentally friendly veg. oil for use as feedstocks/raw materials, in plastics, polymers, rubber, and pharmaceuticals; oxidative and thermal stability; readily biodeg.; 85% min. oleic acid.

GTO 90E. [SVO Enterprises] Crude very high oleic sunflower oil; CAS 8001-21-6; EINECS 232-273-9; food grade environmentally friendly veg. oil for use as feedstocks/raw materials, in plastics, polymers, rubber, and pharmaceuticals; oxidative and thermal stability; readily biodeg.; expeller grade; 85% min. oleic acid.

Gulftene® 4. [Chevron] 1-Butene (C4 alpha olefins); CAS 106-98-9; EINECS 203-449-2; intermediate for biodeg. surfactants for personal care and laundry, and specialty industrial chemicals (polyethylene and other polymers; plasticizers; syn. lubricants; gasoline additives; paper sizing; PVC lubricants); gas; sp.gr. 0.602 (60/60 F); dens. 5.01 lb/gal (60 F); 100% conc.; *Precaution:* flamm. gas.

Gulftene® 6. [Chevron] 1-Hexene (C6 alpha olefins); CAS 592-41-6; EINECS 209-753-1; intermediate for biodeg. surfactants for personal care and laundry, and specialty industrial chemicals (polyethylene and other polymers; plasticizers; syn. lubricants; gasoline additives; paper sizing; PVC lubricants); water-wh. bright, clear liq., char. olefinic odor; ; m.w. 84; sp.gr. 0.677 (60/60 F); dens. 5.64 lb/gal (60 F); b.p. 147 F; flash pt. (TOC) < 20 F; *Toxicology:* LD50 (oral, rat) > 10 g/kg (nontoxic); minimal skin and eye irritation; *Precaution:* flamm. liq.

Gulftene® 8. [Chevron] 1-Octene (C8 alpha olefins); CAS 111-66-0; intermediate for biodeg. surfactants for personal care and laundry, and specialty industrial chemicals (polyethylene and other polymers; plasticizers; syn. lubricants; gasoline additives; paper sizing; PVC lubricants); water-wh. bright, clear liq., char. olefinic odor; ; sp.gr. 0.719 (60/60 F); dens. 6.00 lb/gal (60 F); b.p. 240 F; flash pt. (TCC) 55 F; *Precaution:* flamm. liq.

Gulftene® 10. [Chevron] 1-Decene (C10 alpha olefins); CAS 872-05-9; EINECS 212-819-2; intermediate for biodeg. surfactants for personal care and laundry, and specialty industrial chemicals (polyethylene and other polymers; plasticizers; syn. lubricants; gasoline additives; paper sizing; PVC lubricants); water-wh. bright, clear liq., char. olefinic odor; ; m.w. 140; sp.gr. 0.745 (60/60 F); dens. 6.21 lb/gal (60 F); b.p. 338 F; flash pt. (TOC) 128 F; *Toxicology:* LD50 (oral, rat) > 10 g/kg (nontoxic); minimal skin and eye irritation; *Precaution:* combustible liq.

Gulftene® 12. [Chevron] 1-Dodecene (C12 alpha olefins); CAS 112-41-4; intermediate for biodeg. surfactants for personal care and laundry, and specialty industrial chemicals (polyethylene and other polymers; plasticizers; syn. lubricants; gasoline additives; paper sizing; PVC lubricants); water-wh. bright, clear liq., char. olefinic odor; ; sp.gr. 0.762 (60/60 F); dens. 6.36 lb/gal (60 F); f.p. -31 F; b.p. 400 F; pour pt. -33 F; flash pt. (TCC) 171 F; *Precaution:* combustible liq.

Gulftene® 14. [Chevron] 1-Tetradecene (C14 alpha olefins); CAS 1120-36-1; EINECS 272-493-2; intermediate for biodeg. surfactants for personal care and laundry, and specialty industrial chemicals (polyethylene and other polymers; plasticizers; syn. lubricants; gasoline additives; paper sizing; PVC lubricants); water-wh. bright, clear liq., char. olefinic odor; ; sp.gr. 0.775 (60/60 F); dens. 6.46 lb/gal (60 F); f.p. 9 F; b.p. 440 F; pour pt. 10 F; flash pt. (PM) 225 F; *Precaution:* combustible liq.

Gulftene® 16. [Chevron] 1-Hexadecene (C16 alpha olefins); CAS 629-73-2; intermediate for biodeg. surfactants for personal care and laundry, and specialty industrial chemicals (polyethylene and other polymers; plasticizers; syn. lubricants; gasoline additives; paper sizing; PVC lubricants); water-wh. bright, clear liq., char. olefinic odor; m.w. 224; sp.gr. .785 (60/60 F); dens. 6.54 lb/gal (60 F); f.p. 39 F; b.p. 539 F; pour pt. 45 F; flash pt. (TOC) > 200 F; *Toxicology:* LD50 (oral, rat) > 10 g/kg (nontoxic); minimal skin and eye irritation; *Precaution:* combustible liq.

Gulftene® 18. [Chevron] 1-Octadecene (C18 alpha olefins); CAS 112-88-9; intermediate for biodeg. surfactants for personal care and laundry, and specialty industrial chemicals (polyethylene and other polymers; plasticizers; syn. lubricants; gasoline additives; paper sizing; PVC lubricants); water-wh. bright, clear liq., char. olefinic odor; sp.gr. 0.793 (60/60 F); dens. 6.60 lb/gal (60 F); f.p. 64 F; b.p. 165

F; pour pt. 65 F; flash pt. (PM) 310 F.

Gulftene® 20-24. [Chevron] C20-24 alpha olefins; CAS 64743-02-8; intermediate for biodeg. surfactants for personal care and laundry, and specialty industrial chemicals (polyethylene and other polymers; plasticizers; syn. lubricants; gasoline additives; paper sizing; PVC lubricants); wh. bright, clear waxy solid; sp.gr. 0.856 (60/60 F); dens. 6.67 lb/gal (60 F); visc. 2.1 cSt (99 C); m.p. 96 F; b.p. 146 F; flash pt. (PM) 362 F; *Toxicology:* LD50 (oral, rat) > 5 g/kg.

Gulftene® 24-28. [Chevron] C24-28 alpha olefins; intermediate for biodeg. surfactants for personal care and laundry, and specialty industrial chemicals (polyethylene and other polymers; plasticizers; syn. lubricants; gasoline additives; paper sizing; PVC lubricants); wh. bright, clear waxy solid; sp.gr. 0.891 (60/60 F); dens. 6.83 lb/gal (60 F); visc. 2.5 cSt (99 C); m.p. 143 F; congeal pt. 126 F; b.p. 190 F; flash pt. (PM) 425 F.

Gulftene® 30+. [Chevron] C30 alpha olefin; intermediate for biodeg. surfactants for personal care and laundry, and specialty industrial chemicals (polyethylene and other polymers; plasticizers; syn. lubricants; gasoline additives; paper sizing; PVC lubricants); wh. bright, clear waxy solid; sp.gr. 0.919 (60/60 F); dens. 6.95 lb/gal (60 F); visc. 8.0 cSt (99 C); drop m.p. 163 F; congeal pt. 155 F; b.p. 204 F; flash pt. (PM) 485 F; *Toxicology:* LD50 (oral, rat) > 2 g/kg.

G-White. [J.M. Huber/Engineered Mins.] Calcium carbonate; functional filler extender with easy disp. and low binder demand; used in coatings, plastic and rubber fillers, building prods., ceramics flux, paper fillers, adhesives, cleaning compds., and polishing agents; wh. irreg., uniaxial particles; particle size 5.5 μ, 99% thru 500 mesh; water-sol.; sp.gr. 2.71; dens. 22.6 lb/solid gal, 55 lb/ft^3; oil absorp. 13.5 lb oil/100 lb; ref. index 1.6; hardness (Mohs) 3; 97.7% $CaCO_3$.

H-36. [J.M. Huber/Engineered Mins.] Alumina trihydrate; filler used to suppress flame and smoke in latex flexible foam, vinyl, and urethane carpet-backing compds.; particulate; 25 μ median particle size; sp.gr. 2.42; dens. 0.8 g/cc; 64.9% Al_2O_3.

H-46. [J.M. Huber/Engineered Mins.] Alumina trihydrate; filler used to suppress flame and smoke in latex flexible foam.

H-100. [Franklin Industrial Minerals] Aluminum hydroxide; CAS 21645-51-2; EINECS 244-492-7; flame retardant for use in polyester resins, vinyls, latex, epoxies, adhesives, coatings, elec. components, automotive parts, wire and cable insulation, SMC, BMC, RTM, potting compds., appliance parts, truck caps, rubber prods., carpet backing; finely divided cryst. powd.; 3.0-4.0 μ median particle size; 0% on 325 mesh; sp.gr. 2.42; brightness 89-95; hardness (Mohs) 2.5-3.5; 64.9% Al_2O_3, 0.5% moisture.

H-105. [Franklin Industrial Minerals] Aluminum hydroxide; CAS 21645-51-2; EINECS 244-492-7; flame retardant for use in polyester resins, vinyls, latex, epoxies, adhesives, coatings, elec. components, automotive parts, wire and cable insulation, SMC, BMC, RTM, potting compds., appliance parts, truck caps, rubber prods., carpet backing; finely divided cryst. powd.; 4.5-6.0 μ median particle size; 0% on 325 mesh; sp.gr. 2.42; brightness 89-95; hardness (Mohs) 2.5-3.5; 64.9% Al_2O_3, 0.5% moisture.

H-109. [Franklin Industrial Minerals] Aluminum hydroxide; CAS 21645-51-2; EINECS 244-492-7; flame retardant for use in polyester resins, vinyls, latex, epoxies, adhesives, coatings, elec. components, automotive parts, wire and cable insulation, SMC, BMC, RTM, potting compds., appliance parts, truck caps, rubber prods., carpet backing; finely divided cryst. powd.; 6.5-8.5 μ median particle size; trace on 325 mesh; sp.gr. 2.42; brightness 89-95; hardness (Mohs) 2.5-3.5; 64.9% Al_2O_3, 0.5% moisture.

H-600. [Franklin Industrial Minerals] Aluminum hydroxide; CAS 21645-51-2; EINECS 244-492-7; flame retardant for use in polyester resins, vinyls, latex, epoxies, adhesives, coatings, elec. components, automotive parts, wire and cable insulation, SMC, BMC, RTM, potting compds., appliance parts, truck caps, rubber prods., carpet backing; finely divided cryst. powd.; 25-40 μ median particle size; 36-49% on 325 mesh; sp.gr. 2.42; brightness 89-95; hardness (Mohs) 2.5-3.5; 64.9% Al_2O_3, 0.5% moisture.

H-800. [Franklin Industrial Minerals] Aluminum hydroxide; CAS 21645-51-2; EINECS 244-492-7; flame retardant for use in polyester resins, vinyls, latex, epoxies, adhesives, coatings, elec. components, automotive parts, wire and cable insulation, SMC, BMC, RTM, potting compds., appliance parts, truck caps, rubber prods., carpet backing; finely divided cryst. powd.; 18-25 μ median particle size; 17-23% on 325 mesh; sp.gr. 2.42; brightness 89-95; hardness (Mohs) 2.5-3.5; 64.9% Al_2O_3, 0.5% moisture.

H-900. [Franklin Industrial Minerals] Aluminum hydroxide; CAS 21645-51-2; EINECS 244-492-7; flame retardant for use in polyester resins, vinyls, latex, epoxies, adhesives, coatings, elec. components, automotive parts, wire and cable insulation, SMC, BMC, RTM, potting compds., appliance parts, truck caps, rubber prods., carpet backing; finely divided cryst. powd.; 12-18 μ median particle size; 2-12% on 325 mesh; sp.gr. 2.42; brightness 89-95; hardness (Mohs) 2.5-3.5; 64.9% Al_2O_3, 0.5% moisture.

H-990. [Franklin Industrial Minerals] Aluminum hydroxide; CAS 21645-51-2; EINECS 244-492-7; flame retardant for use in polyester resins, vinyls, latex, epoxies, adhesives, coatings, elec. components, automotive parts, wire and cable insulation, SMC, BMC, RTM, potting compds., appliance parts, truck caps, rubber prods., carpet backing; finely divided cryst. powd.; 9.0-11.5 μ median particle size; 0.5-2.0% on 325 mesh; sp.gr. 2.42; brightness 89-95; hardness (Mohs) 2.5-3.5; 64.9% Al_2O_3, 0.5% moisture.

Halbase 10. [Halstab] Lead sulfate, tribasic; heat stabilizer for PVC wire and cable (60-90°), flexible and semirigid profiles, rigid conduit, records, plastisols, organosols, foams, calendering; wh. fine powd.; 99.5-100% thru 325 mesh; nil sol. in water; sp.gr. 6.9; 88.7% PbO.

Halbase 10 EP. [Halstab] Lead sulfate, tribasic, modified; heat stabilizer for PVC wire and cable (60-90°), flexible and semirigid profiles, rigid conduit, records, plastisols, organosols, foams, calendering; wh. fine powd.; 99.5-100% thru 325 mesh;

nil sol. in water; sp.gr. 5.0; 83.2% PbO.

Halbase 10S. [Halstab] Tribasic lead sulfate/binder; dust-free stabilizer for PVC; strands; sp.gr. 3.6 ± 0.1; 75 ± 1.5% PbO.

Halbase 11. [Halstab] Lead sulfate, basic; heat stabilizer for PVC flexible and semirigid profile extrusions, plastisols, organosols, asbestos tile; esp. suited to compds. using polymeric/polyester type plasticizers; wh. fine powd.; 99.5-100% thru 325 mesh; nil sol. in water; sp.gr. 6.7; 84.5% PbO.

Halbase 100. [Halstab] Lead sulfate, basic, modified; heat stabilizer for PVC foams, plastisols, wire and cable (60-80 C), film, sheeting, flexible profiles; wh. fine powd.; 99.5-100% thru 325 mesh; nil sol. in water; sp.gr. 5.0; 65.5% PbO.

Halbase 100 EP. [Halstab] Lead sulfate, basic, modified; heat stabilizer for PVC foams, plastisols, wire and cable (60-80 C), film, sheeting, flexible profiles; wh. fine powd.; 99.5-100% thru 325 mesh; nil sol. in water; sp.gr. 4.3; 66.6% PbO.

Halcarb 20. [Halstab] Basic lead carbonate (white lead); CAS 598-63-0; EINECS 215-290-6; heat and light stabilizer for PVC wire and cable (60-80 C), flexible and rigid calendering, tape, flexible and semirigid profile extrusions, plastisols; provides high elec. props.; wh. fine powd.; 99.5-100% thru 325 mesh; nil sol. in water; sp.gr. 6.9; 87.7% PbO, 9.9-10.2% CO_q2.

Halcarb 20 EP. [Halstab] Leadcarbonate, basic, modified; CAS 598-63-0; EINECS 215-290-6; heat and light stabilizer for PVC wire and cable (60-80 C), flexible and rigid calendering, tape, flexible and semirigid profile extrusions, plastisols; provides high elec. props.; wh. fine powd.; 99.5-100% thru 325 mesh; nil sol. in water; sp.gr. 5.1; 81.5% PbO.

Halcarb 20S. [Halstab] Basic lead carbonate/binder; dust-free stabilizer for PVC; strands; sp.gr. 3.2 ± 0.1; 71 ± 1.5% PbO.

Halcarb 200. [Halstab] Modified basic lead carbonate; CAS 598-63-0; EINECS 215-290-6; economical heat stabilizer for PVC wire and cable (60-80 C), flexible profiles; wh. fine powd.; 99.5-100% thru 325 mesh; nil sol. in water; sp.gr. 5.0; 66.1% PbO.

Halcarb 200 EP. [Halstab] Coated modified basic lead carbonate; CAS 598-63-0; EINECS 215-290-6; economical heat stabilizer for PVC wire and cable (60-80 C), flexible profiles; wh. fine powd.; 99.5-100% thru 325 mesh; nil sol. in water; sp.gr. 3.6; 48.2% PbO.

Hallco® C-7065. [C.P. Hall] POE laurate SE; internal antistat for flexible and rigid PVC ABS; Gardner 2 clear oily liq.; sol. in kerosene, ethanol, acetone, propylene glycol, min. oil; partly sol. in hexane; disp. in water; sp.gr. 0.998; f.p. -1 C; acid no. 2.5; iodine no. 7; sapon. no. 141; hyd. no. 115; flash pt. 202 C; ref. index 1.458.

Hallco® Lube. [C.P. Hall] Rubber slab dip incorporating release and partitioning agents but contg. no fillers; nondusting soft soap when dried; wets and provides release for highly plasticized rubber compds.; grn. liq.; sp.gr. 1.02; dens. 8.5 lb/gal; visc. < 500 cps; pH 11 ± 0.6; 30 ± 1% solids.

Hallcote® 525. [C.P. Hall] Powd. rubber slab dip easily dispersed in water; good nonfoaming chars.; after air drying, the coating offers low visibility, low dusting, and may be suitable as exterior coating of items to be autoclave or mold vulcanized; gray powd.; water-disp.; sp.gr. 1.38; apparent dens. 46 lb/ft^3; visc. < 100 cps (3% disp.); pH 9.6 ± 1.0 (10:1 with water); 95% solids.

Hallcote® 573. [C.P. Hall] Calcium carbonate; powd. pigmented dip providing nondusting release coating on rubber pellets and for slab operations; gray powd.; sp.gr. 1.43; apparent dens. 80 lb/ft^3; visc. < 100 cps (3% disp.); pH 7.9 ± 1.0 (10:1 with water); 95% solids.

Hallcote® 780. [C.P. Hall] Pigmented slab dip which wets highly plasticized or very hydrophobic rubber compds. and provides exc. release for tacky compds.; cream-colored slurry; sp.gr. 1.36; dens. 11.3 lb/gal; visc. < 500 cps (1:1 aq.); pH 7.5 ± 0.5; 50 ± 2% solids.

Hallcote® 910LF. [C.P. Hall] Rubber slab dip featuring extremely low foaming for highly agitated applics.; recommended for less tacky compds. that wet out more easily; economical; cream-colored slurry; sp.gr. 1.29; dens. 10.8 lb/gal; visc. < 500 cps (1:1 aq.); pH 7.0 ± 1.0; 50 ± 2% solids.

Hallcote® CaSt 50. [C.P. Hall] Calcium stearate-based; CAS 1592-23-0; EINECS 216-472-8; rubber slab dip providing effective release; wh. slurry; water-disp.; sp.gr. 1.02; dens. 8.5 lb/gal; visc. < 500 cps (1:1 aq.); pH 8.8 ± 1.0; 50% solids.

Hallcote® CSD. [C.P. Hall] Calcium stearate-based; CAS 1592-23-0; EINECS 216-472-8; slab dip providing exc. coating and release; used in applics. where zinc stearate is floated on the surface; non-water-disp.; wh. paste; sp.gr. 0.95; dens. 7.9 lb/gal; visc. < 500 cps (1:1 aq.); pH 8.8 ± 1.0; 25% solids.

Hallcote® DPD-524, DPD-547. [C.P. Hall] Calcium carbonate-based; dry powd. dips for broad range of elastomers such as EPDM, NR, SBR, NBR, CPE, CSM, CR, butyl, halobutyl, etc.; does not contain talc or clay; zinc and alpha quartz silica levels well below EPA reportable levels; free-flowing powd.; exc. disp. in cold water; 100% solids.

Hallcote® ES-10. [C.P. Hall] Rubber slab dip contg. same fillers and surfactants as Hallcote 780 but free from heavy metals; cream-colored slurry; sp.gr. 1.37; dens. 11.4 lb/gal; visc. < 500 cps (1:1 aq.); pH 8.0 ± 1.0; 50 ± 2% solids.

Hallcote® ZS. [C.P. Hall] Zinc stearate-based; CAS 557-05-1; EINECS 209-151-9; water-resist. slab dip for rubber providing release in water-cooled applics.; wh. paste; sp.gr. 0.78; dens. 6.5 lb/gal; pH 8.8 ± 1.0 (10:1 with water); 25% solids.

Hallcote® ZS 5050. [C.P. Hall] Zinc stearate-based; CAS 557-05-1; EINECS 209-151-9; rubber slab dip providng exc. release with easy dispersion in the rubber compd. during further processing; wh. disp.; water-disp.; sp.gr. 1.03; dens. 8.6 lb/gal; visc. < 500 cps (1:1 aq.); pH 8.5 ± 1.0; 45% solids.

Hal-Lub-D. [Halstab] Lead stearate, dibasic; costabilizer and lubricant for PVC wire and cable, records, extruded profiles; usually used with a primary lead stabilizer; lt. cream/wh. fine powd.; 99.9% thru 200 mesh; nil sol. in water; sp.gr. 1.9; m.p. 280 C (dec.); 55.4% PbO.

Hal-Lub-D S. [Halstab] Dibasic lead stearate/binder; dust-free stabilizer for PVC; strands; sp.gr. 1.7 ± 0.1; 49.5 ± 1.5% PbO.

Hal-Lub-N. [Halstab] Lead stearate, normal; CAS 1072-35-1; lubricant for PVC wire and cable, extruded profiles, molding compds.; lt. cream/wh. powd.; 99.5% thru 70 mesh; sol. 0.006 g/100 ml in water (50 C); sp.gr. 1.4; m.p. 100-110 C; 30% PbO.

Hal-Lub-N S. [Halstab] Normal lead stearate/binder; dust-free stabilizer for PVC; strands; sp.gr. 1.3 ± 0.1; m.p. 105 C; 25.5 ± 1.0% PbO.

Hal-Lub-N TOTM. [Halstab] Lead stearate, normal; CAS 1072-35-1; low dusting PVC stabilizer, lubricant; wh. fine powd.; sp.gr. 1.4; 30.2% PbO.

Halocarbon Grease 19. [Halocarbon Prods.] Polychlorotrifluoroethylene grease thickened with silica; inert lubricant for chemical industry, aerospace, cryogenic gases, elec., hydraulic fluids, life support systems, low-temp. bath fluids, mech. seals, metalworking, nuclear, pulp and paper, spill control, steel, vacuum pump fluids; mold release agent for plastics and rubber; plasticizer for fluorinated plastics; component of specialty greases and engine oil additives; hardest grease; sol. in most org. liqs.; service temp. -25 to 175 C.

Halocarbon Grease 25-5S. [Halocarbon Prods.] Polychlorotrifluoroethylene grease thickened with silica; inert lubricant for chemical industry, aerospace, cryogenic gases, elec., hydraulic fluids, life support systems, low-temp. bath fluids, mech. seals, metalworking, nuclear, pulp and paper, spill control, steel, vacuum pump fluids; mold release agent for plastics and rubber; plasticizer for fluorinated plastics; component of specialty greases and engine oil additives; lowest vapor pressure; sol. in most org. liqs.; service temp. -20 to 175 C.

Halocarbon Grease 25-10M. [Halocarbon Prods.] Polychlorotrifluoroethylene grease thickened with PCTFE polymer; CAS 9002-83-9; inert lubricant for chemical industry, aerospace, cryogenic gases, elec., hydraulic fluids, life support systems, low-temp. bath fluids, mech. seals, metalworking, nuclear, pulp and paper, spill control, steel, vacuum pump fluids; mold release agent for plastics and rubber; plasticizer for fluorinated plastics; component of specialty greases and engine oil additives; softest grease; sol. in most org. liqs.; service temp. 0-135 C; m.p. 150 C.

Halocarbon Grease 25-20M. [Halocarbon Prods.] Polychlorotrifluoroethylene grease thickened with PCTFE polymer; CAS 9002-83-9; inert lubricant for chemical industry, aerospace, cryogenic gases, elec., hydraulic fluids, life support systems, low-temp. bath fluids, mech. seals, metalworking, nuclear, pulp and paper, spill control, steel, vacuum pump fluids; mold release agent for plastics and rubber; plasticizer for fluorinated plastics; component of specialty greases and engine oil additives; hardest grease; sol. in most org. liqs.; service temp. -5 to 150 C; m.p. 160 C.

Halocarbon Grease 28. [Halocarbon Prods.] Polychlorotrifluoroethylene grease thickened with silica; inert lubricant for chemical industry, aerospace, cryogenic gases, elec., hydraulic fluids, life support systems, low-temp. bath fluids, mech. seals, metalworking, nuclear, pulp and paper, spill control, steel, vacuum pump fluids; mold release agent for plastics and rubber; plasticizer for fluorinated plastics; component of specialty greases and engine oil additives; sol. in most org. liqs.; service temp. -25 to 175 C.

Halocarbon Grease 28LT. [Halocarbon Prods.] Polychlorotrifluoroethylene grease thickened with silica; inert lubricant for chemical industry, aerospace, cryogenic gases, elec., hydraulic fluids, life support systems, low-temp. bath fluids, mech. seals, metalworking, nuclear, pulp and paper, spill control, steel, vacuum pump fluids; mold release agent for plastics and rubber; plasticizer for fluorinated plastics; component of specialty greases and engine oil additives; for low-temp. use; sol. in most org. liqs.; service temp. -45 to 95 C.

Halocarbon Grease 32. [Halocarbon Prods.] Polychlorotrifluoroethylene grease thickened with silica; inert lubricant for chemical industry, aerospace, cryogenic gases, elec., hydraulic fluids, life support systems, low-temp. bath fluids, mech. seals, metalworking, nuclear, pulp and paper, spill control, steel, vacuum pump fluids; mold release agent for plastics and rubber; plasticizer for fluorinated plastics; component of specialty greases and engine oil additives; softest grease; sol. in most org. liqs.; service temp. -25 to 175 C.

Halocarbon Grease X90-10M. [Halocarbon Prods.] Polychlorotrifluoroethylene grease thickened with PCTFE polymer; CAS 9002-83-9; inert lubricant for chemical industry, aerospace, cryogenic gases, elec., hydraulic fluids, life support systems, low-temp. bath fluids, mech. seals, metalworking, nuclear, pulp and paper, spill control, steel, vacuum pump fluids; mold release agent for plastics and rubber; plasticizer for fluorinated plastics; component of specialty greases and engine oil additives; for low temp. use; sol. in most org. liqs.; service temp. -40 to 95 C; m.p. 150 C.

Halocarbon Oil 0.8. [Halocarbon Prods.] Polychlorotrifluoroethylene oil; CAS 9002-83-9; inert lubricant for chemical industry, aerospace, cryogenic gases, elec., hydraulic fluids, life support systems, low-temp. bath fluids, mech. seals, metalworking, nuclear, pulp and paper, spill control, steel, vacuum pump fluids; mold release agent for plastics and rubber; plasticizer for fluorinated plastics; component of specialty greases and engine oil additives; sol. in most org. liqs.; dens. 1.71 g/ml (37.8 C); visc. 1.3 cps (37.8 C); pour pt. -129 C; cloud pt. < -129 C; flash pt. none; ref. index 1.383.

Halocarbon Oil 1.8. [Halocarbon Prods.] Polychlorotrifluoroethylene oil; CAS 9002-83-9; inert lubricant for chemical industry, aerospace, cryogenic gases, elec., hydraulic fluids, life support systems, low-temp. bath fluids, mech. seals, metalworking, nuclear, pulp and paper, spill control, steel, vacuum pump fluids; mold release agent for plastics and rubber; plasticizer for fluorinated plastics; component of specialty greases and engine oil additives; sol. in most org. liqs.; dens. 1.82 g/ml (37.8 C); visc. 3.5 cps (37.8 C); pour pt. -93 C; cloud pt. < -93 C; flash pt. none; ref. index 1.395.

Halocarbon Oil 4.2. [Halocarbon Prods.] Polychlorotrifluoroethylene oil; CAS 9002-83-9; inert lubricant for chemical industry, aerospace, cryogenic gases, elec., hydraulic fluids, life support systems, low-temp. bath fluids, mech. seals, metalworking, nuclear, pulp and paper, spill control, steel, vacuum pump fluids; mold release agent for plastics and rubber; plasticizer for fluorinated plastics; component of specialty greases and engine oil additives; sol. in most org. liqs.; dens. 1.85 g/ml (37.8 C); visc. 7.8 cps (37.8 C); pour pt. -73 C; cloud pt. < -87 C; flash pt. none; ref. index 1.401.

Halocarbon Oil 6.3. [Halocarbon Prods.] Polychlorotrifluoroethylene oil; CAS 9002-83-9; inert lubricant for chemical industry, aerospace, cryogenic gases, elec., hydraulic fluids, life support systems, low-temp. bath fluids, mech. seals, metalworking, nuclear, pulp and paper, spill control, steel, vacuum pump fluids; mold release agent for plastics and rubber; plasticizer for fluorinated plastics; component of specialty greases and engine oil additives; sol. in most org. liqs.; dens. 1.87 g/ml (37.8 C); visc. 12 cps (37.8 C); pour pt. -71 C; cloud pt. < -87 C; flash pt. none; ref. index 1.403.

Halocarbon Oil 27. [Halocarbon Prods.] Polychlorotrifluoroethylene oil; CAS 9002-83-9; inert lubricant for chemical industry, aerospace, cryogenic gases, elec., hydraulic fluids, life support systems, low-temp. bath fluids, mech. seals, metalworking, nuclear, pulp and paper, spill control, steel, vacuum pump fluids; mold release agent for plastics and rubber; plasticizer for fluorinated plastics; component of specialty greases and engine oil additives; sol. in most org. liqs.; dens. 1.90g/ml (37.8 C); visc. 51 cps (37.8 C); pour pt. -40 C; cloud pt. < -71 C; flash pt. none; ref. index 1.407.

Halocarbon Oil 56. [Halocarbon Prods.] Polychlorotrifluoroethylene oil; CAS 9002-83-9; inert lubricant for chemical industry, aerospace, cryogenic gases, elec., hydraulic fluids, life support systems, low-temp. bath fluids, mech. seals, metalworking, nuclear, pulp and paper, spill control, steel, vacuum pump fluids; mold release agent for plastics and rubber; plasticizer for fluorinated plastics; component of specialty greases and engine oil additives; sol. in most org. liqs.; dens. 1.92 g/ml (37.8 C); visc. 108 cps (37.8 C); pour pt. -34 C; cloud pt. -34 C; flash pt. none; ref. index 1.409.

Halocarbon Oil 95. [Halocarbon Prods.] Polychlorotrifluoroethylene oil; CAS 9002-83-9; inert lubricant for chemical industry, aerospace, cryogenic gases, elec., hydraulic fluids, life support systems, low-temp. bath fluids, mech. seals, metalworking, nuclear, pulp and paper, spill control, steel, vacuum pump fluids; mold release agent for plastics and rubber; plasticizer for fluorinated plastics; component of specialty greases and engine oil additives; sol. in most org. liqs.; dens. 1.92 g/ml (37.8 C); visc. 182 cps (37.8 C); pour pt. -26 C; cloud pt. -21 C; flash pt. none; ref. index 1.411.

Halocarbon Oil 200. [Halocarbon Prods.] Polychlorotrifluoroethylene oil; CAS 9002-83-9; inert lubricant for chemical industry, aerospace, cryogenic gases, elec., hydraulic fluids, life support systems, low-temp. bath fluids, mech. seals, metalworking, nuclear, pulp and paper, spill control, steel, vacuum pump fluids; mold release agent for plastics and rubber; plasticizer for fluorinated plastics; component of specialty greases and engine oil additives; sol. in most org. liqs.; dens. 1.95 g/ml (37.8 C); visc. 390 cps (37.8 C); pour pt. -12 C; cloud pt. 2 C; flash pt. none; ref. index 1.412.

Halocarbon Oil 400. [Halocarbon Prods.] Polychlorotrifluoroethylene oil; CAS 9002-83-9; inert lubricant for chemical industry, aerospace, cryogenic gases, elec., hydraulic fluids, life support systems, low-temp. bath fluids, mech. seals, metalworking, nuclear, pulp and paper, spill control, steel, vacuum pump fluids; mold release agent for plastics and rubber; plasticizer for fluorinated plastics; component of specialty greases and engine oil additives; sol. in most org. liqs.; dens. 1.95 g/ml (37.8 C); visc. 780 cps (37.8 C); pour pt. -9 C; cloud pt. 10 C; flash pt. none; ref. index 1.412.

Halocarbon Oil 700. [Halocarbon Prods.] Polychlorotrifluoroethylene oil; CAS 9002-83-9; inert lubricant for chemical industry, aerospace, cryogenic gases, elec., hydraulic fluids, life support systems, low-temp. bath fluids, mech. seals, metalworking, nuclear, pulp and paper, spill control, steel, vacuum pump fluids; mold release agent for plastics and rubber; plasticizer for fluorinated plastics; component of specialty greases and engine oil additives; sol. in most org. liqs.; dens. 1.95 g/ml (37.8 C); visc. 1365 cps (37.8 C); pour pt. 5 C; cloud pt. 13 C; flash pt. none; ref. index 1.414.

Halocarbon Oil 1000N. [Halocarbon Prods.] Polychlorotrifluoroethylene oil; CAS 9002-83-9; inert lubricant for chemical industry, aerospace, cryogenic gases, elec., hydraulic fluids, life support systems, low-temp. bath fluids, mech. seals, metalworking, nuclear, pulp and paper, spill control, steel, vacuum pump fluids; mold release agent for plastics and rubber; plasticizer for fluorinated plastics; component of specialty greases and engine oil additives; sol. in most org. liqs.; dens. 1.95 g/ml (37.8 C); visc. 1950 cps (37.8 C); pour pt. 10 C; cloud pt. 18 C; flash pt. none; ref. index 1.415.

Halocarbon Wax 40. [Halocarbon Prods.] Polychlorotrifluoroethylene wax; CAS 9002-83-9; inert lubricant for chemical industry, aerospace, cryogenic gases, elec., hydraulic fluids, life support systems, low-temp. bath fluids, mech. seals, metalworking, nuclear, pulp and paper, spill control, steel, vacuum pump fluids; mold release agent for plastics and rubber; plasticizer for fluorinated plastics; component of specialty greases and engine oil additives; sol. in most org. liqs.; visc. 190 cst (71.1 C).

Halocarbon Wax 600. [Halocarbon Prods.] Polychlorotrifluoroethylene wax; CAS 9002-83-9; inert lubricant for chemical industry, aerospace, cryogenic gases, elec., hydraulic fluids, life support systems, low-temp. bath fluids, mech. seals, metalworking, nuclear, pulp and paper, spill control, steel, vacuum pump fluids; mold release agent for plastics and rubber; plasticizer for fluorinated plastics; component of specialty greases and engine oil additives; sol. in most org. liqs.; visc. 1000 cst (71.1 C); m.p. 57 C.

Halocarbon Wax 1200. [Halocarbon Prods.] Polychlorotrifluoroethylene wax; CAS 9002-83-9; inert lubricant for chemical industry, aerospace, cryogenic gases, elec., hydraulic fluids, life support systems, low-temp. bath fluids, mech. seals, metalworking, nuclear, pulp and paper, spill control, steel, vacuum pump fluids; mold release agent for plastics and rubber; plasticizer for fluorinated plastics; component of specialty greases and engine oil additives; sol. in most org. liqs.; m.p. 110 C.

Halocarbon Wax 1500. [Halocarbon Prods.] Polychlorotrifluoroethylene wax; CAS 9002-83-9; inert lubricant for chemical industry, aerospace, cryogenic gases, elec., hydraulic fluids, life support systems, low-temp. bath fluids, mech. seals, metalworking, nuclear, pulp and paper, spill control, steel, vacuum pump fluids; mold release agent for plastics and rubber; plasticizer for fluorinated plastics; component of specialty greases and engine oil additives; sol. in most org. liqs.; m.p. 132 C.

Halofree 22. [J.M. Huber/Engineered Mins.] Halogen-free flame-retardant additive with improved smoke-suppressing properties over ATH; used in wire and cable, inj.-molded parts, extrusions, coatings, and adhesives; powd.; 1.1 μ avg. particle size; 0.01% retained on 325 mesh; sp.gr. 2.4 g/cc; bulk dens. 0.33 g/cc (loose), 0.73 g/cc (packed); surf. area 13 m^2/g; oil absorp. 42 cc/100 g.; Unverified

Halphos. [Halstab] Lead phosphite, dibasic; CAS 1344-40-7; PVC heat/light stabilizer providing antioxidant props. for exc. outdoor exposure results; for cable jackets, garden house, profiles for outdoor use; wh. fine powd.; 99.5-100% thru 325 mesh; nil sol. in water; sp.gr. 6.9; 90.2% PbO.

Halphos S. [Halstab] Dibasic lead phosphite/binder; dust-free stabilizer for PVC; strands; sp.gr. 3.3 ± 0.1; 73 ± 1.5% PbO.

Halstab 6S. [Halstab] Lead sulfophthalate/binder; dust-free stabilizer for PVC; strands; sp.gr. 3.3 ± 0.1; 71.5 ± 1.5% PbO.

Halstab 30. [Halstab] Complex lead salt; heat stabilizer for PVC elec. tape, rigid inj. molding, high temp. wire and cable, flexible and rigid profiles, rigid DWV pipe, plastisols, organosols; wh. fine powd.; 99.5-100% thru 325 mesh; nil sol. in water; sp.gr. 5.9; 85.0% PbO.

Halstab 30 EP. [Halstab] Coated complex lead salt; heat stabilizer for PVC elec. tape, rigid inj. molding, high temp. wire and cable, flexible and rigid profiles, rigid DWV pipe, plastisols, organosols; wh. fine powd.; 99.5-100% thru 325 mesh; nil sol. in water; sp.gr. 5.0; 79.0% PbO.

Halstab 31. [Halstab] Modified complex lead salt; economical heat stabilizer for PVC wire and cable (105°), flexible and rigid profiles, CPE, CPVC, plastisols; wh. fine powd.; 99.5-100% thru 325 mesh; nil sol. in water; sp.gr. 3.8; 65.2% PbO.

Halstab 32. [Halstab] Complex lead salt; heat stabilizer for PVC; wh. fine powd.; sp.gr. 4.8; 68% PbO; Unverified

Halstab 50. [Halstab] Blend; stabilizer/lubricant for highly filled PVC compds.; generally used with monomeric plasticizer systems; wh. fine powd.; 100% thru 200 mesh; nil sol. in water; sp.gr. 2.4; 60.4% PbO.

Halstab 55. [Halstab] Blend; stabilizer/lubricant for highly filled PVC compds.; generally used with polymeric plasticizer systems; wh. fine powd.; 100% thru 200 mesh; nil sol. in water; sp.gr. 2.2; 56% PbO.

Halstab 60. [Halstab] Complex lead salt; heat stabilizer for PVC wire and cable (60-105 C), flexible and rigid extruded profiles, rigid molding, plastisols, calendering; wh. fine powd.; 99.5-100% thru 325 mesh; nil sol. in water; sp.gr. 5.4; 83.5% PbO.

Halstab 60 EP. [Halstab] Coated complex lead salt; heat stabilizer for PVC wire and cable (60-105 C), flexible and rigid extruded profiles, rigid molding, plastisols, calendering; wh. fine powd.; 99.5-100% thru 325 mesh; nil sol. in water; sp.gr. 4.4; 77.6% PbO.

Halstab 70. [Halstab] Basic lead stabilizer; heat stabilizer for PVC translucent and colored sheet, vinyl film, vinyl upholstery stocks; wh. fine powd.; 100% thru 200 mesh; nil sol. in water; sp.gr. 5.4; 52.3% PbO.

Halstab 600 EP. [Halstab] Coated modified complex lead salt; economical heat stabilizer for PVC wire and cable (60-105 C), wire coatings, flexible profiles, film sheeting, plastisols; wh. fine powd.; 99.5-100% thru 325 mesh; nil sol. in water; sp.gr. 3.8; 60.9% PbO.

Halstab M-1. [Halstab] Lead stabilizer/lubricant complex; stabilizer/lubricant for rigid PVC inj. molding; improves dynamic thermal stability; exc. regrind stability; wh. fine powd.; 99.5% thru 180 mesh; sp.gr. 4.0.

Halstab P-1. [Halstab] Blend; stabilizer/lubricant for rigid PVC pipe, conduit, sewer, DWV, rigid profiles; improves dynamic thermal stability, early color; wh. fine powd.; 99.5% thru 180 mesh; sp.gr. 3.02; 72% PbO.

Halstab P-2. [Halstab] Blend; stabilizer/lubricant for rigid PVC pipe, conduit, sewer, DWV, rigid profiles; improves dynamic thermal stability; wh. fine powd.; 99.5% thru 180 mesh; sp.gr. 2.9; 71% PbO.

Haltex™ 304. [Hitox] Alumina trihydrate; CAS 21645-51-2; smoke suppressive and flame retardant filler for plastics incl. cross-linked polymers, glass-reinforced unsat. polyesters, and many thermoplastics;

used for carpet backings, tub and shower stalls, wire and cable insulation, elec. uses, vinyl coated fabrics, rubber prods.; wh. fine powd., odorless; 4 μ median particle size; 90% < 10 μ; insol. in water; sp.gr. 2.42; bulk dens. 37 lb/ft^3 (loose); oil absorp. 27; brightness 91; m.p. 2015 ± 15 C; pH 8; 65.3% Al_2O_3; *Toxicology:* ACGIH TLV 10 mg/m^3 (dust); high exposure to dust may cause irritation to respiratory tract; skin and eye contact may cause mech. abrasion irritation; avoid breathing dust; *Precaution:* nonflamm.

Haltex™ 310. [Hitox] Alumina trihydrate; CAS 21645-51-2; smoke suppressive and flame retardant filler for plastics incl. cross-linked polymers, glass-reinforced unsat. polyesters, and many thermoplastics; used for carpet backings, tub and shower stalls, wire and cable insulation, elec. uses, vinyl coated fabrics, rubber prods.; wh. fine powd., odorless; 10 μ median particle size; 50% < 10 μ; 0.15% on 325 mesh; insol. in water; sp.gr. 2.42; bulk dens. 56 lb/ft^3 (loose); oil absorp. 19; brightness 91; m.p. 2015 ± 15 C; pH 8; 65.3% Al_2O_3; *Toxicology:* ACGIH TLV 10 mg/m^3 (dust); high exposure to dust may cause irritation to respiratory tract; skin and eye contact may cause mech. abrasion irritation; avoid breathing dust; *Precaution:* nonflamm.

Halthal. [Halstab] Lead phthalate, dibasic; CAS 69011-06-9; heat stabilizer for high temp. applics., PVC wire and cable (60-105 C), plastisols, profile extrusions, calendered sheet; wh. fine powd.; 99.5-100% thru 325 mesh; nil sol. in water; sp.gr. 4.6; 78.5% PbO.

Halthal EP. [Halstab] Coated dibasic lead phthalate; CAS 69011-06-9; heat stabilizer for high temp. applics., PVC wire and cable (60-105 C), plastisols, profile extrusions, calendered sheet; provides extra lubricity; wh. fine powd.; 99.5-100% thru 325 mesh; nil sol. in water; sp.gr. 3.9; 73.9% PbO.

Halthal S. [Halstab] Dibasic lead phthalate/binder; dust-free stabilizer for PVC; strands; sp.gr. 2.9 ± 0.1; 66 ± 1.5% PbO.

Hamp-Ene® 100. [Hampshire] Tetrasodium EDTA; CAS 64-02-8; EINECS 200-573-9; general purpose chelating agent; used as part of catalyst in SBR mfg.; pale straw clear liq.; water-misc.; m.w. 380.2; sp.gr. 1.26-1.28; dens. 10.6 lb/gal; chel. value 100 mg $CaCO_3$/g min. (@ pH 11); pH 11-12 (1%); 38% min. act.

Hamp-Ene® Na_3 Liq. [Hampshire] Trisodium EDTA; CAS 150-38-9; EINECS 205-758-8; chelating agent; used as part of catalyst in SBR mfg.; pale straw clear liq.; water-misc.; m.w. 358.2; sp.gr. 1.25-1.28; dens. 10.5 lb/gal; chel. value 100 mg $CaCO_3$/g min. (@ pH 11); pH 9.0-9.4 (1%); 35.8% min. act.

Hamp-Ene® NaFe Purified Grade. [Hampshire] Sodium ferric EDTA, trihydrate; chelating agent used in photographic developers, oxidation-reduction systems, and emulsion polymerization; m.w. 421.1; dens. 5.0 lb/gal; pH 4-6 (1%); 98% min. act.; Unverified

Hamposyl® AL-30. [Hampshire] Ammonium lauroyl sarcosinate; CAS 68003-46-3; EINECS 268-130-2; anionic; surfactant for shampoos, skin cleansers, bath gels; sec. emulsifier for emulsion polymerization; liq.; HLB 29.0; 30% conc.

Hamposyl® C. [Hampshire; Chemplex Chems.] Cocoyl sarcosine; CAS 68411-97-2; EINECS 270-156-4; anionic; detergent, wetting/foaming agent, foam stabilizer, emulsifier, anticorrosive agent, conditioner for personal care, cosmetics, household cleaners, industrial cleaners, biotechnology, corrosion inhibition, dispersants, froth flotation, leather, pesticides, textiles, petrol. and lubricant prods., metalworking; emulsifier for emulsion polymerization; mold release for RIM; stabilizer for polyols; antifog and antistat in plastics; pale yel. liq.; sol. in most org. solv.; m.w. 280; sp.gr. 0.97-0.99; m.p. 23-26 C; soften. pt. 18-22 C; HLB 10.0; 100% conc.

Hamposyl® C-30. [Hampshire; Chemplex Chems.] Sodium cocoyl sarcosinate; CAS 61791-59-1; EINECS 263-193-2; anionic; detergent, wetting/foaming agent, foam stabilizer, emulsifier, anticorrosive agent, conditioner for personal care, cosmetics, household cleaners, industrial cleaners, biotechnology, corrosion inhibition, dispersants, froth flotation, leather, pesticides, textiles, petrol. and lubricant prods., metalworking; emulsifier for emulsion polymerization; mold release for RIM; stabilizer for polyols; antifog and antistat in plastics; colorless to very pale yel. liq.; misc. in water; m.w. 301; sp.gr. 1.02-1.03; visc. 30 cps; f.p. -1 C; HLB 27.0; pH 7.5-8.5 (10%); surf. tens. 30 dynes/cm; 30% act.

Hamposyl® L. [Hampshire; Chemplex Chems.] Lauroyl sarcosine; CAS 97-78-9; EINECS 202-608-3; anionic; detergent, wetting/foaming agent, foam stabilizer, emulsifier, anticorrosive agent, conditioner for personal care, cosmetics, household cleaners, industrial cleaners, biotechnology, corrosion inhibition, dispersants, froth flotation, leather, pesticides, textiles, petrol. and lubricant prods., metalworking; emulsifier for emulsion polymerization; mold release for RIM; stabilizer for polyols; antifog and antistat in plastics; wh. waxy solid; sol. in most org. solvs.; m.w. 270; sp.gr. 0.97-0.99; m.p. 34-37 C; HLB 13.0; 94% act.

Hamposyl® L-30. [Hampshire; Chemplex Chems.] Sodium lauroyl sarcosinate; CAS 137-16-6; EINECS 205-281-5; anionic; detergent, wetting/foaming agent, foam stabilizer, emulsifier, anticorrosive agent, conditioner for personal care, cosmetics, household cleaners, industrial cleaners, biotechnology, corrosion inhibition, dispersants, froth flotation, leather, pesticides, textiles, petrol. and lubricant prods., metalworking; emulsifier for emulsion polymerization; mold release for RIM; stabilizer for polyols; antifog and antistat in plastics; colorless liq.; misc. in water; m.w. 292; sp.gr. 1.02-1.03; visc. 30 cps; f.p. -1 C; HLB 30.0; pH 7.5-8.5 (10%); surf. tens. 30 dynes/cm; 30% act.

Hamposyl® L-95. [Hampshire; Chemplex Chems.] Sodium lauroyl sarcosinate; CAS 137-16-6;

EINECS 205-281-5; anionic; detergent, wetting/foaming agent, foam stabilizer, emulsifier, anticorrosive agent, conditioner for personal care, cosmetics, household cleaners, industrial cleaners, biotechnology, corrosion inhibition, dispersants, froth flotation, leather, pesticides, textiles, petrol. and lubricant prods., metalworking; emulsifier for emulsion polymerization; mold release for RIM; stabilizer for polyols; antifog and antistat in plastics; wh. powd.; sol. in water; m.w. 292; dens. 25 lb/ft^3; HLB 30.0; pH 7.5-8.5 (10%); 94% act.

Hamposyl® M. [Hampshire; Chemplex Chems.] Myristoyl sarcosine; CAS 52558-73-3; EINECS 258-007-1; anionic; detergent, wetting/foaming agent, foam stabilizer, emulsifier, anticorrosive agent, conditioner for personal care, cosmetics, household cleaners, industrial cleaners, biotechnology, corrosion inhibition, dispersants, froth flotation, leather, pesticides, textiles, petrol. and lubricant prods., metalworking; emulsifier for emulsion polymerization; mold release for RIM; stabilizer for polyols; antifog and antistat in plastics; wh. waxy solid; sol. in most org. solv.; m.w. 298; sp.gr. 0.97-0.99; m.p. 48-53 C; 94% act.

Hamposyl® M-30. [Hampshire] Sodium myristoyl sarcosinate; CAS 30364-51-3; EINECS 250-151-3; anionic; detergent, wetting/foaming agent, foam stabilizer, emulsifier, anticorrosive agent, conditioner for personal care, cosmetics, household cleaners, industrial cleaners, biotechnology, corrosion inhibition, dispersants, froth flotation, leather, pesticides, textiles, petrol. and lubricant prods., metalworking; emulsifier for emulsion polymerization; mold release for RIM; stabilizer for polyols; antifog and antistat in plastics; colorless liq.; misc. in water; m.w. 320; sp.gr. 1.02-1.03; visc. 30 cps; f.p. -1 C; pH 7.5-8.5 (10%); surf. tens. 30 dynes/cm; 30% act.

Hamposyl® O. [Hampshire; Chemplex Chems.] Oleoyl sarcosine; CAS 110-25-8; EINECS 203-749-3; anionic; detergent, wetting/foaming agent, foam stabilizer, emulsifier, anticorrosive agent, conditioner for personal care, cosmetics, household cleaners, industrial cleaners, biotechnology, corrosion inhibition, dispersants, froth flotation, leather, pesticides, textiles, petrol. and lubricant prods., metalworking; emulsifier for emulsion polymerization; mold release for RIM; stabilizer for polyols; antifog and antistat in plastics; yel. liq.; sol. in most org. solv.; m.w. 349; sp.gr. 0.95-0.97; visc. 250 cps; HLB 10.0; 94% act.

Hamposyl® S. [Hampshire] Stearoyl sarcosine; CAS 142-48-3; EINECS 205-539-7; anionic; detergent, wetting/foaming agent, foam stabilizer, emulsifier, anticorrosive agent, conditioner for personal care, cosmetics, household cleaners, industrial cleaners, biotechnology, corrosion inhibition, dispersants, froth flotation, leather, pesticides, textiles, petrol. and lubricant prods., metalworking; emulsifier for emulsion polymerization; mold release for RIM; stabilizer for polyols; antifog and antistat in plastics; wh. waxy solid; sol. in most org. solv.; m.w. 338; sp.gr. 0.96-0.98; m.p. 53-58 C; 94% act.

Hampton Dry Colorants. [Hampton Colours Ltd] Dry colorants for thermoplastic inj. molding, blow molding, extrusion, and rotomolding.

Hampton Masterbatches. [Hampton Colours Ltd] Masterbatch for coloring plastics; intended for use at only 1% letdown in most clear or translucent polymer bases.

Hardener OZ. [Bayer AG] Latent curing agent based on urethane bisoxazolidine for use in one-pack PU systems.

Haro® Chem ALT. [Akcros] Aluminum tristearate; CAS 637-12-7; EINECS 211-279-5; Unverified

Haro® Chem BG. [Akcros] Barium stearate; CAS 6865-35-6; heat stabilizer for rigid PVC, ABS.

Haro® Chem BP-108X. [Akcros] Barium/lead soap; heat stabilizer for flexibile and rigid PVC; solid.

Haro® Chem BSG. [Akcros] Barium laurate; heat stabilizer for rigid PVC; solid.

Haro® Chem CBHG. [Akcros] Calcium behenate; CAS 3578-72-1; EINECS 222-700-7; heat stabilizer for PVC, PE, PP, PU, ABS, thermosets.

Haro® Chem CGD. [Akcros] Calcium stearate; CAS 1592-23-0; EINECS 216-472-8; Metal soap stabilizer for PVC, LDPE, LLDPE, M/HDPE, PP, ABS, rubbers, thermosets, paint, and fertilizers; solid.; Unverified

Haro® Chem CGL. [Akcros] Calcium stearate; CAS 1592-23-0; EINECS 216-472-8; Unverified

Haro® Chem CGN. [Akcros] Calcium stearate; CAS 1592-23-0; EINECS 216-472-8; heat stabilizer for PVC, PE, PP, PU, ABS, thermosets.

Haro® Chem CHG. [Akcros] Calcium 12-hydroxystearate; heat stabilizer for PVC; solid.

Haro® Chem CP-4. [Akcros] Calcium/zinc complex; Unverified

Haro® Chem CP-17. [Akcros] Calcium/zinc complex; heat stabilizer for flexible and semirigid PVC, plastisols; paste.

Haro® Chem CPR-2. [Akcros] Calcium stearate; CAS 1592-23-0; EINECS 216-472-8; Metal soap stabilizer for PVC, LDPE, LLDPE, M/HDPE, PP, ABS, rubbers, thermosets, cosmetics, paper, pharmaceuticals, paint, fertilizers; solid.; Unverified

Haro® Chem CPR-5. [Akcros] Calcium stearate; CAS 1592-23-0; EINECS 216-472-8; heat stabilizer for PVC, PE, PP, PU, ABS, thermosets.

Haro® Chem CSG. [Akcros] Calcium laurate; EINECS 225-166-3; heat stabilizer for PVC, PE, PP, PU, ABS, thermosets.

Haro® Chem CZ-37, CZ-40. [Akcros] Calcium-zinc soap; nontoxic heat stabilizers for rigid and plasticized PVC; powd.

Haro® Chem KB-214SA. [Akcros] Barium-cadmium soap; stabilizer for rigid PVC incl. extrusion, profile, extruded foam, and calendered film; solid.

Haro® Chem KB-219SA, KB-521SA. [Akcros] Mixed-metal soaps; stabilizer for rigid PVC incl. extrusion, profile, extruded foam, and calendered film; solid; Unverified

Haro® Chem KB-350X. [Akcros] Cadmium-barium soap; heat and lt. stabilizer for flexibile, semirigid, and rigid vinyls.

Haro® Chem KB-353A. [Akcros] Stabilizer for rigid and plasticized PVC applics.; solid; Unverified

Haro® Chem KB-353AS. [Akcros] Unverified

Haro® Chem KB-554A, ZZ-019. [Akcros] Solid metal soap stabilizer for plasticized PVC applics.; Unverified

Haro® Chem KPR. [Akcros] Cadmium stearate; CAS 2223-93-0; heat and lt. stabilizer for flexible and rigid vinyls.

Haro® Chem KS. [Akcros] Cadmium laurate; heat and lt. stabilizer for flexible and rigid PVC.

Haro® Chem LHG. [Akcros] Lithium 12-hydroxystearate; Metal soap stabilizer for thermosets, oils, fats, and lubricants; solid.; Unverified

Haro® Chem MF-2. [Akcros] Magnesium stearate; CAS 557-04-0; EINECS 209-150-3; stabilizer for ABS, thermosets, cosmetics, pharmaceuticals, fertilizers, and wire drawing; solid.

Haro® Chem MF-3. [Akcros] Magnesium stearate; CAS 557-04-0; EINECS 209-150-3; stabilizer for ABS, PS.

Haro® Chem NG. [Akcros] Sodium stearate; CAS 822-16-2; EINECS 212-490-5; Metal soap stabilizer for LDPE, cosmetics, oils, fats, and lubricants, pharmaceuticals, paint, and wire drawing; solid.; Unverified

Haro® Chem P28G. [Akcros] Normal lead stearate; CAS 1072-35-1; stabilizer for rigid and plasticized PVC applics.; solid.

Haro® Chem P51. [Akcros] Dibasic lead stearate; stabilizer for rigid and plasticized PVC applics. offers heat stability and good elec. properties; solid.

Haro® Chem PC. [Akcros] Basic lead carbonate; CAS 598-63-0; EINECS 215-290-6; stabilizer for plasticized PVC applics. offers heat stability and good elec. properties.

Haro® Chem PDF. [Akcros] Dibasic lead phosphite; CAS 1344-40-7; stabilizer for rigid and plasticized PVC applics. offers heat and light stability and good elec. properties.

Haro® Chem PDF-B. [Akcros] Dibasic lead phosphite; CAS 1344-40-7; PVC stabilizer; Unverified

Haro® Chem PDP-E. [Akcros] Dibasic lead phthalate; CAS 69011-06-9; stabilizer for plasticized PVC applics. offers heat stability and good elec. properties.

Haro® Chem PPCS-X. [Akcros] Polybasic lead sulfate complex; stabilizer for rigid and plasticized PVC applics. offers heat stability and good elec. properties.; Unverified

Haro® Chem PTS-E. [Akcros] Tribasic lead sulfate; stabilizer for rigid and plasticized PVC applics. offers heat stability and good elec. properties.

Haro® Chem ZGD. [Akcros] Zinc stearate; CAS 557-05-1; EINECS 209-151-9; heat stabilizer; Unverified

Haro® Chem ZGN. [Akcros] Zinc stearate; CAS 557-05-1; EINECS 209-151-9; heat stabilizer for PVC, LDPE, M/HDPE, PS, rubbers, thermosets, and paint; solid.

Haro® Chem ZGN-T. [Akcros] Zinc stearate; CAS 557-05-1; EINECS 209-151-9; heat stabilizer for PVC, PS, expanded PS, HIPS; solid.

Haro® Chem ZPR-2. [Akcros] Zinc stearate; CAS 557-05-1; EINECS 209-151-9; heat stabilizer for PVC, LDPE, M/HDPE, PS, rubbers, thermosets, cosmetics, paper, pharmaceuticals, paint; solid.

Haro® Chem ZSG. [Akcros] Zinc laurate; CAS 2452-01-9; EINECS 219-518-5; heat stabilizer for PVC and rubbers; solid.

Haro® Mix BF-202. [Akcros] One-pack lead heat and light stabilizers for rigid PVC applics., UPVC nonpressure pipe extrusion; Unverified

Haro® Mix BK-105, BK-107. [Akcros] Lead one-pack; one-pack lead heat and light stabilizer for rigid PVC applics.; Unverified

Haro® Mix CE-701. [Akcros] Lead one-pack system; one-pack lead heat stabilizer systems for rigid PVC applics., UPVC pressure and nonpressure pipes.

Haro® Mix CH-205. [Akcros] Lead one-pack system; stabilizer/lubricant for extrusion of UPVC corrugated land drainage pipe; Unverified

Haro® Mix CH-606. [Akcros] Lead one-pack; one-pack lead heat stabilizer for plasticized PVC applics. (cable); offers good elec. properties.; Unverified

Haro® Mix CK-203. [Akcros] Lead one-pack system; stabilizer/lubricant for extrusion of UPVC corrugated land drainage pipe.

Haro® Mix CK-213. [Akcros] Lead one-pack system; stabilizer/lubricant for extrusion of UPVC nonpressure pipes.

Haro® Mix CK-711. [Akcros] Lead one-pack system; stabilizer/lubricant for extrusion of UPVC pressure or nonpressure pipes; Unverified

Haro® Mix FK-102. [Akcros] Lead one-pack; stabilizer/lubricant for plasticized PVC cable extrusion.

Haro® Mix IC-217. [Akcros] Lead one-pack; one-pack lead heat stabilizer for rigid and plasticized PVC inj. moldings.; Unverified

Haro® Mix IC-238. [Akcros] Lead one-pack; one-pack lead heat stabilizer for rigid PVC inj. moldings.; Unverified

Haro® Mix IH-108. [Akcros] Lead one-pack; stabilizer/lubricant for plasticized PVC cable extrusion; Unverified

Haro® Mix LK-218. [Akcros] Lead one-pack; stabilizer/lubricant for UPVC general profile extrusion.

Haro® Mix LK-228. [Akcros] Lead one-pack; stabilizer/lubricant for UPVC general profile extrusion; Unverified

Haro® Mix MH-204. [Akcros] Lead one-pack system; stabilizer/lubricant for extrusion of UPVC pressure pipes; Unverified

Haro® Mix MK-107. [Akcros] Lead one-pack system; stabilizer/lubricant for extrusion of UPVC foam pipes; Unverified

Haro® Mix MK-220. [Akcros] Lead one-pack; stabilizer/lubricant for plasticized PVC cable extrusion.

Haro® Mix MK-620. [Akcros] Lead one-pack system; stabilizer/lubricant for extrusion of UPVC foam pipes.

Haro® Mix MK-744. [Akcros] Lead one-pack system;

stabilizer/lubricant for extrusion of UPVC corrugated land drainage pipe; Unverified

Haro® Mix SK-602. [Akcros] Lead one-pack; stabilizer/lubricant for UPVC inj. molding; Unverified

Haro® Mix UC-213. [Akcros] Lead one-pack; one-pack lead heat and light stabilizer for rigid PVC inj. moldings.; Unverified

Haro® Mix UK-121. [Akcros] Lead one-pack; stabilizer/lubricant for UPVC general profile extrusion; Unverified

Haro® Mix VC-501. [Akcros] Lead one-pack; stabilizer/lubricant for UPVC inj. molding; Unverified

Haro® Mix YC-502. [Akcros] Barium/cadmium/lead; heat and lt. stabilizer for PVC; Unverified

Haro® Mix YC-601. [Akcros] Barium/cadmium/lead one-pack; one-pack heat and light stabilizer for rigid PVC applics. (profiles); Unverified

Haro® Mix YE-301. [Akcros] Barium/cadmium/lead one-pack; one-pack heat and light stabilizer system for rigid PVC sheet; Unverified

Haro® Mix YK-110. [Akcros] Barium/cadmium/lead one-pack; stabilizer/lubricant for UPVC window profile extrusion; Unverified

Haro® Mix YK-113. [Akcros] Barium/cadmium/lead one-pack; stabilizer/lubricant for UPVC window profile extrusion; Unverified

Haro® Mix YK-307. [Akcros] Barium/cadmium/lead one-pack; stabilizer/lubricant for UPVC window profile extrusion.

Haro® Mix YK-603. [Akcros] Barium/cadmium/lead one-pack; stabilizer/lubricant for UPVC window profile extrusion; Unverified

Haro® Mix ZC-028, ZC-029, ZC-030. [Akcros] Calcium/zinc one-pack; nontoxic one-pack heat and light stabilizer for rigid PVC applics. (profiles); Unverified

Haro® Mix ZC-031, ZC-032. [Akcros] Calcium/zinc one-pack; nontoxic one-pack heat stabilizer for plasticized PVC applics. (cables); offers good elec. properties.; Unverified

Haro® Mix ZC-036. [Akcros] Calcium/zinc one-pack; nontoxic one-pack heat stabilizer for rigid PVC blow moldings (bottles); Unverified

Haro® Mix ZC-309. [Akcros] Calcium/zinc one-pack; nontoxic one-pack heat stabilizer for rigid PVC applics. (pipes); Unverified

Haro® Mix ZC-311. [Akcros] Calcium/zinc one-pack; stabilizer/lubricant for extrusion of UPVC pipes.

Haro® Mix ZC-902. [Akcros] Calcium/zinc; one-pack stabilizer/lubricant for PVC bottle blow molding.

Haro® Mix ZT-025. [Akcros] Tin one-pack; nontoxic one-pack heat stabilizer for rigid PVC applics. (pipes); Unverified

Haro® Mix ZT-026. [Akcros] Tin one-pack; nontoxic one-pack heat stabilizer for rigid PVC inj. moldings.; Unverified

Haro® Mix ZT-504. [Akcros] Tin one-pack; stabilizer/lubricant for UPVC inj. molding; Unverified

Haro® Mix ZT-508. [Akcros] Tin one-pack; stabilizer/lubricant for extrusion of UPVC pipes; Unverified

Haro® Mix ZT-514. [Akcros] Tin one-pack; stabilizer/lubricant for extrusion of UPVC pipes; Unverified

Haro® Mix ZT-905. [Akcros] Tin one-pack based on dioctyltin mercaptide; CAS 58229-88-2; nontoxic one-pack heat stabilizer for rigid PVC blow moldings (bottles); Unverified

Haro® Wax L01-56, L03-58, L04-73, L09-00, L18-78, L21-96, L23-58, L24-98. [Akcros] Lubricant for rigid and plasticized PVC; suitable for food contact applics.; Unverified

Haro® Wax L02-99, L05-51, L06-62, L07-47, L08-57, L10-58, L11-85, L12-43. [Akcros] Lubricant for rigid PVC applics.; suitable for food contact.; Unverified

Haro® Wax L13-89, L14-77, L15-75, L16-92, L17-98, L19-99, L22-99, L25-99. [Akcros] Lubricant for rigid PVC applics.; suitable for food contact.; Unverified

Haro® Wax L20-98. [Akcros] Lubricant for rigid PVC applics.; Unverified

Haro® Wax L333. [Akcros] Lubricant blend for rigid and plasticized PVC applics., incl. food contact applics.; Unverified

Haro® Wax L-344. [Akcros] Lubricant blend for rigid PVC applics., incl. food contact applics.; Unverified

Haro® Wax L433. [Akcros] Lubricant blend; Unverified

Hartomer 4900-25. [Huntsman] Sodium vinyl sulfonate; anionic; monomer for latex emulsion polymerization systems; liq.; 25% act.

Hartopol L64. [Huntsman] Polyoxyalkylene glycol; nonionic; dispersant and emulsifier in emulsion polymerization; wetting agent for pulp/paper industry; defoamer, detergent for many industries; liq.; HLB 15.0; cloud pt. 58-62 C (1% aq.); 100% act.

Harwick F-300. [Harwick] Fatty acid; internal lubricant for use in plastics at low concs.; solid.

HB-40®. [Monsanto] Partially hydrog. terphenyl; plasticizer extender, polymer modifier, resin solvator for vinyl sheeting, films, fabric or paper coatings, vinyl protective coatings, adhesives.

HC 01. [Hughes Ind.] Modified azodicarbonamide in EVA carrier resin; blowing agent for inj. molding and extrusion; pellets; dec. pt. 375-385 F.

HC 05. [Hughes Ind.] Endothermic blowing agent in EVA carrier resin; endothermic blowing agent for inj. molding and extrusion processes; pellets; dec. pt. 380-390 F.

Hefti MS-55-F. [Hefti Ltd.] Polysorbate 60; CAS 9005-67-8; nonionic; o/w emulsifier for cosmetics, pharmaceuticals, tech. applics.; latex stabilizer for emulsion polymerization; antistatic agent; liq./paste; HLB 15.0; 100% conc.

Hefti PGE-400-DS. [Hefti Ltd.] PEG-8 distearate; CAS 9005-08-7; nonionic; surfactant for pharmaceuticals, cosmetics, to incorporate fats and ester waxes in creams; homogenizer for suppository bases; plasticizer for plastics; solid; HLB 8.0; 100% conc.

Hefti PGE-600-DS. [Hefti Ltd.] PEG-12 distearate; CAS 9005-08-7; nonionic; o/w emulsifier for cosmetics, pharmaceuticals; plasticizer for plastics, various tech. applics.; flakes; HLB 10.5; 100% conc.

Hefti PGE-600-ML. [Hefti Ltd.] PEG-12 laurate; CAS 9004-81-3; nonionic; emulsifier, solubilizer for pharmaceuticals, cosmetics, textile oils, insecticides, pesticides; antistat for plastics; liq.; HLB 15.0; 100% conc.

Hefti TO-55-E. [Hefti Ltd.] PEG-18 sorbitan trioleate; CAS 9005-70-3; nonionic; o/w and w/o emulsifier for min. oils, veg. oils, train oils, waxes, etc.; for cattle feed, textiles, biocides, paints, varnishes, plastics, leather, fur, tech. applics., cosmetics, pharmaceuticals; liq.; HLB 10.5; 100% conc.

Hefti TO-55-EL. [Hefti Ltd.] PEG-17 sorbitan trioleate; CAS 9005-70-3; nonionic; o/w and w/o emulsifier for min. oils, veg. oils, train oils, waxes, etc.; for cattle feed, textiles, biocides, paints, varnishes, plastics, leather, fur, tech. applics., cosmetics, pharmaceuticals; liq.; HLB 10.0; 100% conc.

Heliogen®. [BASF AG] Phthalocyanine pigments; for superfast letterpress, offset, flexographic and gravure printing inks, paints, lacquers, for coloring plastics, for prod. of artists' chalks, paints and crayons.

Heloxy® 7. [Shell] Alkyl C8-C10 glycidyl ethers; visc. reducing modifier for epoxy formulations used in flooring, casting, tooling, laminating, potting, coatings, etc.; low volatility, but some color; Gardner 1 max. color; sp.gr. 0.88-0.91; dens. 7.6 lb/gal; visc. 3-5 cps; flash pt. 215 F; EEW 220-235.

Heloxy® 8. [Shell] Alkyl C12-C14 glycidyl ethers; visc. reducing modifier for epoxy formulations used in flooring, casting, tooling, laminating, potting, coatings, etc.; lower volatility and odor than Heloxy 7; FDA approval; Gardner 1 max. color; sp.gr. 0.88-0.91; dens. 7.5 lb/gal; visc. 6-9 cps; flash pt. 224 F; EEW 280-295.

Heloxy® 9. [Shell] Alkyl C12-C13 glycidyl ethers; visc. reducing modifier for epoxy formulations; low volatility; water-wh. color; dens. 7.5 lb/gal; visc. 6-9 cps; EEW 275-295.

Heloxy® 32. [Shell] Polyglycol diepoxide; reactive epoxy modifier, flexibilizer; maintains low temp. flexibility; low color; extends pot life; Gardner 2 max. color; dens. 8.9 lb/gal; visc. 45-90 cps; EEW 305-335.

Heloxy® 44. [Shell] Trimethylol ethane triglycidyl ether; trifunctional epoxy modifier which increases crosslink dens., maintains or increases reactivity, and provides moderate dilution efficiency; Gardner 4 max. color; dens. 9.9 lb/gal; visc. 200-330 cps; EEW 155-175.

Heloxy® 48. [Shell] Trimethylol propane triglycidyl ether; trifunctional epoxy modifier which increases crosslink dens., maintains or increases reactivity, and provides relatively efficient visc. reduction; Gardner 3 max. color; dens. 9.6 lb/gal; visc. 125-250 cps; EEW 140-160.

Heloxy® 56. [Shell] Dibromoneopentyl glycol diglycidyl ether; epoxy modifier for flame retardant applics.; low visc., low volatility; good compat. with aromatic brominated and nonbrominated resins; Gardner 5 max. color; dens. 12.4 lb/gal; visc. 275-500 cps; EEW 250-300.

Heloxy® 61. [Shell] Butyl glycidyl ether; CAS 2426-08-6; epoxy modifier; max. visc. reduction, min. loss of properties; increases impregnation of resin systems, and level of filler loading; used in elec., laminating, casting, tooling, flooring, and coatings; Gardner 1 max. color; sp.gr. 0.92-0.94; dens. 7.7 lb/gal; visc. 1-2 cps; EEW 145-155.

Heloxy® 62. [Shell] o-Cresyl glycidyl ether; CAS 2210-79-9; visc. reducing epoxy modifier; reactive diluent; increases level of filler loading in epoxy resins; used in flooring, low visc. casting, laminating, and decoupage; low volatility; Gardner 2 max. color; sp.gr. 1.07-1.09; dens. 9.0 lb/gal; visc. 5-10 cps; flash pt. 225 F; EEW 175-195.

Heloxy® 63. [Shell] Phenyl glycidyl ether; CAS 122-60-1; EINECS 204-557-2; visc. reducing modifier for epoxies; low volatility; Gardner 1 max. color; sp.gr. 1.07-1.09; dens. 9.2 lb/gal; visc. 4-7 cps; flash pt. 236 F; EEW 150-157; *Toxicology:* strong skin and eye irritant.

Heloxy® 64. [Shell] Nonyl phenyl glycidyl ether; visc. reducing modifier for epoxies; chem. intermediate; Gardner 2 max. color; dens. 8.2 lb/gal; visc. 90-130 cps; EEW 300-325.

Heloxy® 65. [Shell] p-tert-Butyl phenyl glycidyl ether; CAS 3101-60-8; visc. reducing modifier for epoxies; moderate reactive diluent; used in casting, tooling, laminating, and in flooring; good elec. props.; low volatility; Gardner 1 max. color; sp.gr. 1.01-1.03; dens. 8.5 lb/gal; visc. 20-30 cps; flash pt. 199 F; EEW 225-240.

Heloxy® 67. [Shell] Diglycidyl ether of 1,4 butanediol; CAS 2425-79-8; visc. reducing modifier, reactive diluent for liq. epoxies used in casting, laminating, tooling, potting, and elec. applics.; low volatility; useful for water-reducible systems; Gardner 1 max. color; some sol. in water; sp.gr. 1.09-1.12; dens. 9.2 lb/gal; visc. 13-18 cps; flash pt. 204 F; EEW 120-130.

Heloxy® 68. [Shell] Diglycidyl ether of neopentyl glycol; CAS 17557-23-2; visc. modifier for epoxies providing good visc. reduction, min. loss of properties; increases impregnation of resin systems, and level of filler loading; used in elec., laminating, casting, tooling, flooring, and coatings; Gardner 1 max. color; sp.gr. 1.04-1.08; dens. 8.9 lb/gal; visc. 13-18 cps; flash pt. 275 F; EEW 130-140.

Heloxy® 71. [Shell] Dimer acid diglycidyl ester; reactive epoxy modifier, flexibilizer; improves impact str. and toughness; Gardner 12 max. color; dens. 8.2 lb/gal; visc. 400-900 cps; EEW 390-490.

Heloxy® 84. [Shell] Polyglycidyl ether of an aliphatic polyol; reactive epoxy modifier, flexibilizer providing good visc. reduction, superior surf. wetting props.; Gardner 1 max. color; dens. 8.5 lb/gal; visc. 200-320 cps; EEW 620-680.

Heloxy® 107. [Shell] Diglycidyl ether of cyclohexanedimethanol; CAS 14228-73-0; modifier providing moderate visc. reduction of epoxy resins with min. loss of props.; used in casting, laminating, tooling, potting, elec., adhesive, and grouting applics.; good for applics. requiring min. deformation under load; Gardner 1 max. color; sp.gr. 1.09-1.11; dens. 9.1 lb/gal; visc. 55-75 cps; flash pt. 245 F; EEW

155-165.

Heloxy® 116. [Shell] 2-Ethylhexyl glycidyl ether; CAS 2461-15-6; EINECS 219-553-6; visc. reducing epoxy modifier with good dilution efficiency; nontoxic, nonirritating substitute for butyl glycidyl ether; used in casting, tooling, potting, laminating, and other low visc. applics.; Gardner 1 max. color; sp.gr. 0.90-0.92; dens. 7.6 lb/gal; visc. 2-4 cps; flash pt. 192 F; EEW 215-230.

Heloxy® 505. [Shell] Polyglycidyl ether of castor oil; reactive epoxy modifier, flexibilizer providing superior water resist., good exterior color retention and low-temp. flexibilization performance; low odor; nonvolatile; Gardner 8 max. color; dens. 8.5 lb/gal; visc. 300-500 cps; EEW 550-650.

Henley AZ EX 110. [Henley] Blowing agent in LDPE base resin; exothermic chemical blowing agent for structural foam molding of PP, HDPE; yel.; becomes wh. on dec.; dec. 375-425 F; 20% act.

Henley AZ EX 120. [Henley] Blowing agent in LDPE base resin; exothermic chemical blowing agent for molding and extrusion of all polyolefins; yel.; becomes wh. on dec.; dec. 365-410 F; 20% act.

Henley AZ EX 122. [Henley] Blowing agent in LDPE base resin; exothermic chemical blowing agent for extrusion of wire, cable, and profiles; recommended for LDPE extrusion at low temps. (< 340 F); wh.; dec. 270-310 F; 17% act.

Henley AZ EX 127. [Henley] Blowing agent in LDPE base resin; exothermic chemical blowing agent for molding and extrusion of all polyolefins, TPO, and TPE; yel.; becomes wh. on dec.; dec. 340-360 F; 21% act.

Henley AZ EX 210. [Henley] Blowing agent in PS base resin; exothermic chemical blowing agent for structural foam molding of PS, PPO resins; yel.; becomes wh. on dec.; dec. 375-425 F; 20% act.

Henley AZ EX 310. [Henley] Blowing agent in PVC base resin; exothermic chemical blowing agent for extrusion of PVC profiles, wire and cable, sheet; yel.; becomes wh. on dec.; dec. 310-350 F; 10% act.

Henley AZ LBA-30. [Henley] Azodicarbonamide in liq. carrier; CAS 123-77-3; EINECS 204-650-8; chemical blowing agent for molding or extrusion of thermoplastics (PE, PP, styrenics, PPO, etc.); bright yel. liq.; sp.gr. 1.13; dec. 150 C.

Henley AZ LBA-49. [Henley] Azodicarbonamide with activator in liq. carrier; CAS 123-77-3; EINECS 204-650-8; chemical blowing agent for rigid and flexible PVC processed by extrusion and inj. molding; suitable for foaming for wt. reduction for inj. molded or extruded parts; bright yel. liq.; sp.gr. 1.19; dec. 140 C.

Hercoflex® Plasticizer. [Aqualon] Pentaerythritol ester; plasticizer for PVC with outstanding heat resistance; wire insulation for high-temp. service and government specification cable construction and high-quality plastisol formulations; Pt-Co 150 max.; b.p. 143.3; sp.gr. 1.006-1.016; visc. 110 cps; pour pt. -51 C; sapon no. 390; ref. index 1.455; flash pt. 300 C (COC); fire pt. 325 C; vol. resist. 4.9×10^{11} ohm-cm.

Hercolite™ 240. [Hercules] Aromatic hydrocarbon resin; resin for use in polymer modification, adhesives, overprint lacquers, coatings; superior heat and uv stability; modifies flow in PVC, ABS, and block copolymers; processing aid for vinyl incl. flooring; good pigment and filler wetting for use in color additives and rubber and plastic compounding; tackifier for neoprene in contact adhesives for shoes; provides glossy nonyel. hard film in coatings; also for concrete curing compds., traffic paints; FDA 21CFR §175.105, 175.300, 175.320, 175.390, 176.170(b), 176.180, 177.1210; water-wh.; sol. in aromatic and chlorinated HC, ketones, ethers; insol. in aliphatic HC, alcohols, glycols; melt visc. 197 C (10 poise); soften. pt. (R&B) 122 C; 2% max. volatiles.

Hercolite™ 290. [Hercules] Aromatic hydrocarbon resin; resin for use in polymer modification, adhesives, overprint lacquers, coatings; superior heat stability; modifies flow in PVC, ABS, and block copolymers; processing aid for vinyl incl. flooring; good pigment and filler wetting for use in color additives and rubber and plastic compounding; tackifier for neoprene in contact adhesives for shoes; provides glossy nonyel. hard film in coatings; also for concrete curing compds., traffic paints; FDA 21CFR §175.105, 175.300, 175.320, 175.390, 176.170(b), 176.180, 177.1210; water-wh.; sol. in aromatic and chlorinated HC, ketones, ethers; insol. in aliphatic HC, alcohols, glycols; melt visc. 230 C (10 poise); soften. pt. (R&B) 142 C; 1% max. volatiles.

Hercolyn® D. [Hercules] Methyl hydrog. rosinate; CAS 8050-13-3; resinous plasticizer or tackifier in finished prods. such as lacquers, inks, adhesives, floor tiles, vinyl plastisols, artificial leather, and antifouling paints; fixative and carrier in perfumes and cosmetic preps.; lt. amber visc. liq.; low odor; sol. in ester, ketones, alcohols, ethers, coal tar, petrol. hydrocarbons, and veg. and min. oils; insol. in water; dens. 1.02 kg/l; visc. (G-H) Z2-Z3; b.p. 360-364 C; acid no. 7; sapon. no. 155; ref. index 1.52; flash pt. 183 C (COC).

Hercules® Ester Gum 10D. [Hercules] Glycerol ester of a partially dimerized rosin; thermoplastic resin used as a softener or plasticizer for elastomeric masticatory agents used in chewing gums; modifier for rubbers, film-formers, and waxes in adhesive and protective coating compositions; USDA Rosin M max. flakes; sol. in aromatic, aliphatic, and chlorinated hydrocarbons, esters, ketones; insol. in water; G-H visc. H; soften. pt. 116 C; acid no. 7.

Hetoxide NP-30. [Heterene] Nonoxynol-30; CAS 9016-45-9; nonionic; emulsifier for cosmetics, emulsion polymerization; Gardner 4 max. solid; sol. in water, IPA; HLB 17.0; hyd. no. 35-45; 99% conc.

HFVCD 11G. [Hughes Ind.] Azodicarbonamide; CAS 123-77-3; EINECS 204-650-8; blowing agent for calendered vinyl fabrics; provides high gas yield, fast gas generation, while retaining low temp. stability for even gauge prods. with good fabric adhesion; pre-dispersed, easily blended.

HFVCP 11G. [Hughes Ind.] Modified azodicarbonamide; CAS 123-77-3; EINECS 204-650-8; blowing agent for flexible vinyl processing; powd. or liq. disp.; dec. pt. 345-355 F.

HFVD 12CC. [Hughes Ind.] Modified azodicarbonamide in epoxidized soybean oil vehicle; blowing agent for flexible vinyl processing; designed to eliminate need to use heavy metal stabilizers, such as lead, for activation; liq. disp.; fineness (Hegman) 6.5 min.; dens. 9.75 ± 0.15 lb/gal; visc. 40,000 ± 10,000 cps (2.5 rpm); dec. pt. 385-395 F.

HFVP 03. [Hughes Ind.] Modified azodicarbonamide; CAS 123-77-3; EINECS 204-650-8; blowing agent for flexible vinyl processing; powd. or liq. disp.; dec. pt. 375-385 F.

HFVP 15P. [Hughes Ind.] Modified azodicarbonamide; CAS 123-77-3; EINECS 204-650-8; blowing agent for flexible vinyl processing; powd. or liq. disp.; dec. pt. 320-330 F.

HFVP 19B. [Hughes Ind.] Modified azodicarbonamide; CAS 123-77-3; EINECS 204-650-8; blowing agent for flexible vinyl processing; powd. or liq. disp.; dec. pt. 330-340 F.

High Purity MTBE. [Arco] Methyl t-butyl ether; CAS 1634-04-4; extraction solv., reaction medium in pharmaceuticals, for polymerizations, Grignard reactions; octane enhancer for gasoline; highly resist. to peroxide formation; Clear liq., terpene-like odor; sl. sol. in water; misc. with most org. solvs.; sp.gr. 0.737; f.p. -108 C; b.p. 55 C (760 mm Hg); ref. index 1.3694 (20 C); surf. tension 18.3 dynes/cm (32 C); visc. 0.350 cps (20 C); flash pt. -30 C; > 99.9% purity.

Hi-Pflex® 100. [Specialty Minerals] Calcium carbonate, surface-treated; filler improving impact str. and flex. mod.; high-performance reinforcing agent for plastics incl. flexible and rigid PVC (potable water pipe, wire and cable), PP (interior and exterior auto parts, toys, pallets, corrugated boxes); HDPE (pipe and other rigid applics., wire and cable and other flexible applics.), rubber; very wh. free-flowing powd.; 3.5 μ avg. particle size; sp.gr. 2.71; dens. 50 lb/ft^3; surf. area 3.1 m^2/g; oil absorp. 14 lb/100 lb; dry brightness 92; 96% $CaCO_3$.

Hi-Point® 90. [Witco/PAG] MEK peroxide in dimethyl phthalate; CAS 1338-23-4; catalyst/initiator for R.T. cures of polyesters; clear liq.; 9.0% act. oxygen.

Hi-Point® PD-1. [Witco/PAG] MEK peroxide in dimethyl phthalate; CAS 1338-23-4; catalyst/initiator for R.T. cures of polyesters; clear liq.; 5.25% act. oxygen.

Hi-Sil® 132. [PPG Industries] Precipitated silica; reinforcing agent for rubber incl. silicone rubbers; thickener and carrier; wh. powd.; 16 nm particle size; bulk dens. 10 lb/ft^3; surf. area 200 m^2/g; pH 7.

Hi-Sil® 135. [PPG Industries] Amorphous precipitated silica; reinforcing agent for elastomers (organic and silicone); Na_2SO_4 residual salt; wh. powd.; 18 nm particle size; bulk dens. 10 lb/ft^3; surf. area 170 m^2/g; pH 7 (5% slurry).

Hi-Sil® 210. [PPG Industries] Amorphous hydrated silica; CAS 112926-00-8; wh. pigment, reinforcing filler for rubber prods.; diluent, grinding aid, stabilizer, thixotrope; agric. chemicals; promotes adhesion in natural and syn. rubber-based adhesives; anticaking agent in lawn fertilizers, fungicides, grinding wheel abrasives, drain cleaners, laundry sour compds., hexamethylenetetramine for phenolic molding compds.; absorbent carrier and flow conditioner for solids and visc. control in liqs.; wh. nondusting pellets; 19 nm primary particle size; bulk dens. 15 lb/ft^3; surf. area 150 m^2/g; pH 7 (5% aq. susp.); 87.0% min. SiO_2 hydrate; *Toxicology:* OSHA TLV/TWA 6 mg/m^3 (total dust); skin contact with dust causes drying effect.

Hi-Sil® 233. [PPG Industries] Amorphous hydrated silica; CAS 112926-00-8; wh. pigment, reinforcing filler for rubber prods.; diluent, grinding aid, stabilizer, thixotrope; agric. chemicals; promotes adhesion in natural and syn. rubber-based adhesives; anticaking agent in lawn fertilizers, fungicides, grinding wheel abrasives, drain cleaners, laundry sour compds., hexamethylenetetramine for phenolic molding compds.; absorbent carrier and flow conditioner for solids and visc. control in liqs.; wh. powd., 0.5% max. retained on 325-mesh screen; 19 nm primary particle size; bulk dens. 10 lb/ft^3; surf. area 150 m^2/g; pH 7 (5% aq. susp.); 87.0% min. SiO_2 hydrate; *Toxicology:* OSHA TLV/TWA 6 mg/m^3 (total dust); skin contact with dust causes drying effect.

Hi-Sil® 243LD. [PPG Industries] Precipitated silica; CAS 112926-00-8; reinforcing agent for rubber (soling and footwear, tires, mech. goods); extender providing thixotropy and reinforcement, promoting adhesion in adhesives; wh. low-dusting nugget; 19 nm particle size; bulk dens. 17 lb/ft^3; surf. area 150 m^2/g; pH 7 (5% aq. susp.); 87.0% min. SiO_2 hydrate.

Hi-Sil® 250. [PPG Industries] Amorphous hydrated silica; wh. pigment, reinforcing filler for rubber prods.; diluent, grinding aid, stabilizer, thixotrope; agric. chemicals; promotes adhesion in natural and syn. rubber-based adhesives; anticaking agent in lawn fertilizers; fungicides, grinding wheel abrasives, drain cleaners, laundry sour compds., hexamethylenetetramine for phenolic molding compds.; absorbent carrier and flow conditioner for solids and visc. control in liqs.; wh. powd., 0.5% max. retained on 325-mesh screen; pH 6.5-7.3; 87.0% min. SiO_2 hydrate; *Toxicology:* OSHA TLV/TWA 6 mg/m^3 (total dust); skin contact with dust causes drying effect; Unverified

Hi-Sil® 255. [PPG Industries] Amorphous precipitated silica; reinforcing agent for rubbers; wh. gran.; surf. area 185 m^2/g.

Hi-Sil® 532EP. [PPG Industries] Amorphous precipitated silica; reinforcing agent for rubber; aids extrusion, resilience, adhesion, heat resist.; provides faster cure, improved dynamic props., higher filler loadings; carrier; replacement for carbon blk.; for tire carcass, wh. sidewalls, sporting goods, soft soling, cellular goods, resilient mountings and roll covers, adhesives, cements; wh. powd.; 0.04 μ ultimate particle size; bulk dens. 10 lb/ft^3; surf. area 60 m^2/g; pH 8 (5% aq. susp.); ref. index 1.44; 92% min. SiO_2 hydrate; *Toxicology:* suggested TLV/TWA 5 mg/m^3 (respirable dust); contact with powd. causes drying effect.

Hi-Sil® 752. [PPG Industries] Amorphous precipitated silica; reinforcing agent for elastomers; similar to Silene 732 D with exception of higher typical pH for elastomer cure rate adjustment; Na_2SO_4 residual salt; wh. powd.; pH 9.3.

Hi-Sil® 900. [PPG Industries] Amorphous precipitated silica; reinforcing agent for elastomers (organic and silicone); Na_2SO_4 residual salt; wh. powd.; 20 nm particle size; bulk dens. 8 lb/ft^3; surf. area 152 m^2/g; pH 7.

Hi-Sil® 915. [PPG Industries] Amorphous precipitated silica; reinforcing agent for elastomers (esp. silicone due to high surf. area, small gross particle size (typ. 4 μ), and low Na_2SO_4 salt content); wh. powd.; 16 nm particle size; bulk dens. 6 lb/ft^3; surf. area 220 m^2/g; pH 7 (5% slurry).

Hi-Sil® 2000. [PPG Industries] Amorphous precipitated silica; reinforcing agent intended specifically for use in dynamic elastomer applics.; wh. easy disp. gran. form; surf. area 250 m^2/g.

Hi-Sil® ABS. [PPG Industries] Precipitated silica; carrier to convert ester and liq. resin plasticizers and bonding agents to free-flowing powds. for introduction to rubber compds.; grinding aid, suspension aid, reinforcing filler, and anticaking agent; wh. powd.; 22 μm median particle diam.; sp.gr. 2.0; bulk dens. 8 lb/ft^3; surf. area 150 m^2/g; oil absorp. 305 cc/100 g; pH 7.0 (5% aq. slurry); 97.5% SiO_2.

Hi-Sil® EZ. [PPG Industries] Amorphous precipitated silica; reinforcing agent intended specifically for use in dynamic elastomeric applics.; wh. easy disp. gran. form; surf. area 170 m^2/g.

Hisolve DM. [Toho Chem. Industry] Diethylene glycol monomethyl ether; CAS 111-77-3; EINECS 203-906-6; solv. for lacquer thinners, inks, dyes, brake fluids, metal surf. finishing agents; liq.

Hisolve MC. [Toho Chem. Industry] Ethylene glycol monomethyl ether; CAS 109-86-4; EINECS 203-713-7; solv. for paints, inks, adhesives, plastics and plasticizers, cleaners, paint removers; liq.

Hisolve MDM. [Toho Chem. Industry] Diethylene glycol dimethyl ether; CAS 111-96-6; EINECS 203-924-4; reaction solv. in reduction, isomerization, substitution, alkylation condensation, polymerization, and metal including reaction; separation and extraction solv.; liq.

Hitox®. [Hitox] Rutile titanium dioxide; CAS 1317-80-2; buff-colored pigment developed as alternative to wh. titanium dioxide; used in coatings (alkyds, acrylic urethanes, high solids systems, water reducibles, water bases, powd. coatings), inks, adhesives, paper, foundry prods., building prods.; in plastics for PVC pipe and conduit, color concs., vinyl siding; beige fine powd., odorless; 1.5 μ avg. particle size; < 0.01% on 325 mesh; fineness (Hegman) 7; insol. in water; sp.gr. 4.1; dens. 34.22 lb/gal; bulk dens. 39.3 lb/ft^3 (pour), 65.5 lb/ft^3 (tapped); surf. area 1.124 m^2/g; oil absorp. 24-25; m.p. $\approx$ 1840 C; b.p. 2500-3000 C; pH 6.5-7.5; 95% rutile TiO_2; *Toxicology:* ACGIH TLV 10 mg/m^3 (resp.); dust can cause lung irritation; *Precaution:* nonflamm.

Hodag DOSS-70. [Calgene] Dioctyl sodium sulfosuccinate; CAS 577-11-7; EINECS 209-406-4; anionic; surfactant, wetting agent, surface tension reducer used in emulsion and suspension polymerization, dry cleaning, cleaning compds., industrial compds., paints, textiles, and cosmetics; colorless liq.; sol. in polar and nonpolar org. solvs.; water-disp.; sp.gr. 1.08; dens. 9.0 lb/gal; flash pt. (OC) 85 C; surf. tens. 26 dynes/cm (1% aq.); 70% act.

Hodag DOSS-75. [Calgene] Dioctyl sodium sulfosuccinate; CAS 577-11-7; EINECS 209-406-4; anionic; surfactant, wetting agent, surface tension reducer used in emulsion and suspension polymerization, dry cleaning, cleaning compds., industrial compds., paints, textiles, and cosmetics; colorless liq.; sol. in polar and nonpolar org. solvs.; disp. in water; sp.gr. 1.08; dens. 9.0 lb/gal; flash pt. (OC) 85 C; surf. tens. 26 dynes/cm (1% aq.); 75% act.

Hodag DTSS-70. [Calgene] Sodium ditridecyl sulfosuccinate; CAS 2673-22-5; EINECS 220-219-7; anionic; surfactant, wetting agent, surface tension reducer used in emulsion and suspension polymerization, dry cleaning, cleaning compds., industrial compds., paints, textiles, and cosmetics; pale liq.; sol. in polar and nonpolar org. solvs.; water-disp.; sp.gr. 1.02; dens. 8.5 lb/gal; flash pt. (OC) 91 C; surf. tens. 29 dynes/cm (1% aq.); 70% act.

Hodag PE-004. [Calgene] Phosphate ester; anionic; emulsifier, lubricant, antistat, corrosion inhibitor for pesticides, industrial alkaline detergents, drycleaning, textile wet processing, syn. fiber lubricants, emulsion polymerization, cosmetics; lt. amber clear visc. liq.; sol. in water, aromatic solvs.; sp.gr. 1.18-1.22; acid no. 140-165; pH 1.5-2.5 (10% aq.).

Hodag PE-104. [Calgene] Alkylaryl phosphate ester; anionic; emulsifier, EP lube additive, antistat, corrosion inhibitor, surfactant; for pesticides, industrial detergents, drycleaning, textile wet processing, lubricants, emulsion polymerization, cosmetics; lt. amber clear visc. liq.; sol. in aromatic and aliphatic solvs., disp. in water; sp.gr. 0.98-1.02; acid no. 110-130; pH 1.5-2.5 (10% aq.); 100% conc.

Hodag PE-106. [Calgene] Alkylaryl phosphate ester; anionic; emulsifier, EP lube additive, antistat, corrosion inhibitor, surfactant; for pesticides, industrial detergents, drycleaning, textile wet processing, lubricants, emulsion polymerization, cosmetics; lt. amber clear visc. liq.; sol. in aromatic and aliphatic solvs., disp. in water; sp.gr. 1.09-1.13; acid no. 100-115; pH 1.5-2.5 (10% aq.); 100% conc.

Hodag PE-109. [Calgene] Alkylaryl phosphate ester; anionic; emulsifier, EP lube additive, antistat, corrosion inhibitor, surfactant; for pesticides, industrial detergents, drycleaning, textile wet processing, lubricants, emulsion polymerization, cosmetics; lt. amber clear visc. liq.; sol. in water, aromatic and aliphatic solvs.; sp.gr. 1.07-1.11; acid no. 75-90; pH

1.5-2.5 (10% aq.); 100% conc.

Hodag PE-206. [Calgene] Alkylaryl phosphate ester; anionic; emulsifier, EP lube additive, antistat, corrosion inhibitor, surfactant; for pesticides, industrial detergents, drycleaning, textile wet processing, lubricants, emulsion polymerization, cosmetics; lt. amber sl. hazy visc. liq.; disp. in water; sol. in aromatic and aliphatic solvs.; sp.gr. 1.03-1.08; acid no. 85-100; pH 1.5-2.5 (10% aq.); 100% conc.

Hodag PE-209. [Calgene] Alkylaryl phosphate ester; anionic; emulsifier, EP lube additive, antistat, corrosion inhibitor, surfactant; for pesticides, industrial detergents, drycleaning, textile wet processing, lubricants, emulsion polymerization, cosmetics; lt. amber sl. hazy visc. liq.; disp. in water; sol. in aromatic and aliphatic solvs.; sp.gr. 1.04-1.08; acid no. 75-90; pH 1.5-2.5 (10% aq.); 100% conc.

Hodag PEG 200. [Calgene] PEG-4; CAS 25322-68-3; EINECS 203-989-9; cosmetics and pharmaceuticals formulation; plasticizer for adhesives; inks; resins and coatings; clear liq.; water-sol.; m.w. 190-210; pH 4.5-7.5 (5%).

Hodag PEG 300. [Calgene] PEG-6; CAS 25322-68-3; EINECS 220-045-1; cosmetics and pharmaceutical formulation; latex coagulating bath; plasticizer for adhesives, spray-on bandages; resins and coatings; clear liq.; water-sol.; m.w. 285-315; visc. 5.4-6.4 cSt (210 F); pH 4.5-7.5 (5%).

Hodag PEG 400. [Calgene] PEG-8; CAS 25322-68-3; EINECS 225-856-4; cosmetic and pharmaceutical formulation; humectant, coupler for lotions; release agent for rubber; latex coagulating bath; in PVAc paints; clear liq.; water-sol.; m.w. 380-420; visc. 6.8-8.0 cSt (210 F); pH 4.5-7.5 (5%).

Hodag PEG 540. [Calgene] PEG-6, PEG-32 (50/50 mixt.); cosmetic, pharmaceutical, and suppository formulation; humectant, plasticizer in adhesives; base for metal polishes; lubricant for paper sizes; inks; in alkyd resins and coatings; wh. waxy solid; water-sol.; m.w. 500-600; visc. 26-33 cSt (210 F); f.p. 38-41 C; pH 4.5-7.5 (5%).

Hodag PEG 3350. [Calgene] PEG-75; CAS 25322-68-3; cosmetic and pharmaceutical formulation; resins and coatings; humectant, plasticizer for adhesives; antistat for rubber conveyor belt; in shoe polish; lubricant for paper sizing; printing inks; tablet binder, lubricant; wh. waxy solid; water-sol.; m.w. 3015-3685; visc. 76-110 cSt (210 F); f.p. 53-56 C; pH 4.5-7.5 (5%).

Hodag PEG 8000. [Calgene] PEG-150; CAS 25322-68-3; cosmetic and pharmaceutical formulation; resins and coatings; antistat for rubber conveyor belting; in shoe polish; lubricant for paper size; printing inks; release agent for rubber; tablet binder/lubricant; wh. waxy solid; water-sol.; m.w. 7000-9000; visc. 470-900 cSt (210 F); f.p. 60-63 C; pH 4.5-7.5 (5%).

Hodag PPG-150. [Calgene] PPG (150); used for lubricants, metalworking compds., cosmetics, paints, urethane foams, hydraulic fluids, plasticizers, release agents, surfactant intermediates, textile lubricants, antifoam agents; straw clear liq.; water-sol.; m.w. 150; pour pt. -42 C; flash pt. (COC) 250 F.

Hodag PPG-400. [Calgene] PPG-9; CAS 25322-69-4; used for lubricants, metalworking compds., cosmetics, paints, urethane foams, hydraulic fluids, plasticizers, release agents, surfactant intermediates, textile lubricants, antifoam agents; straw clear liq.; water-sol.; m.w. 400; pour pt. -45 C; flash pt. (COC) 420 F.

Hodag PPG-1200. [Calgene] PPG-20; CAS 25322-69-4; used for lubricants, metalworking compds., cosmetics, paints, urethane foams, hydraulic fluids, plasticizers, release agents, surfactant intermediates, textile lubricants, antifoam agents; straw clear liq.; sol. 2% in water; m.w. 1200; pour pt. -40 C; flash pt. (COC) 450 F.

Hodag PPG-2000. [Calgene] PPG-26; CAS 25322-69-4; used for lubricants, metalworking compds., cosmetics, paints, urethane foams, hydraulic fluids, plasticizers, release agents, surfactant intermediates, textile lubricants, antifoam agents; straw clear liq.; sol. 0.2% in water; m.w. 2000; pour pt. -35 C; flash pt. (COC) 450 F.

Hodag PPG-4000. [Calgene] PPG-30; CAS 25322-69-4; used for lubricants, metalworking compds., cosmetics, paints, urethane foams, hydraulic fluids, plasticizers, release agents, surfactant intermediates, textile lubricants, antifoam agents; straw clear liq.; sol. 0.1% in water; m.w. 4000; pour pt. -29 C; flash pt. (COC) 440 F.

Hoechst Wax C. [Hoechst AG] Amide wax based on bis-stearoyl ethylenediamine; CAS 110-30-5; lubricant and release agent for polyolefins; improves processing in engineering thermoplastics such as polyamides and acetal; slip agent for printing inks and paints; increases m.p.; anti-adhesive coating; sanding aid for primer coats on furniture; almost wh. powd. (< 500 μm) or micropowd. (< 40 μm); insol. in water; dens. 0.99-1.01 g/cc (20 C); visc. ≈ 10 mm^2/s (150 C); drop pt. 139-144 C; acid no. < 8; sapon. no. < 8; *Toxicology:* LD50 (oral, mice) > 20,000 mg/kg; *Storage:* store in dry place @ R.T.

Hoechst Wax E. [Hoechst AG; Hoechst Celanese] Glycol/butylene glycol montanate; CAS 73138-45-1; internal/external lubricant and release agent for PVC, polyolefins, polyamide, PS, linear polyester, TPU, thermosets; for sheet, film and bottle applics., esp. for tin stabilized systems; also for emulsion polishes, solv.-based paste polishes, carbon paper; worldwide food approvals in plastics; pale ylsh. powd. (< 500 μm), fine powd. (< 125 μm), or flakes; insol. in water; dens. 1.01-1.03 g/cc (20 C); visc. ≈ 30 mm^2/s (100 C); drop pt. 79-85 C; acid no. 15-20; sapon. no. 130-160; *Toxicology:* LD50 (oral, mouse) > 20,000 mg/kg; *Storage:* store dry @ R.T.

Hoechst Wax KPS. [Hoechst Celanese; Hoechst AG] Glycol/butylene glycol montanate; CAS 73138-45-1; wax for polishes, esp. dry-bright emulsions; antiblocking agent for polymer dispersions; emulsions for citrus fruit coating; ylsh. flakes; insol. in water; dens. 1.00-1.02 g/cc (20 C); visc. ≈ 30 mm^2/s (100 C); drop pt. 80-85 C; acid no. 30-40; sapon.

no. 135-150.

Hoechst Wax LP. [Hoechst Celanese] Montan acid wax; CAS 68476-03-9; EINECS 270-664-6; lubricant for plastics; water repellent for textiles; also for emulsion polishes, color bases, scouring waxes, industrial wax emulsions for textiles, paper, wood, etc.; ylsh. flakes; insol. in water; dens. 1.00-1.02 g/cc (20 C); visc. ≈ 30 mm^2/s (100 C); drop pt. 82-89 C; acid no. 113-130; sapon. no. 140-160.

Hoechst Wax OP. [Hoechst AG; Hoechst Celanese] Butylene glycol montanate, calcium montanate; CAS 73138-44-0, 68308-22-5; internal/external lubricant and release agent for PVC, polyolefins, polyamide, PS, linear polyester, TPU, thermosets, etc.; solv.-based polishes esp. pastes, carbon paper; maintains clarity of finished prod.; worldwide food approvals for plastics; ylsh. flakes, powd. (< 500 μm) or fine powd. (< 125 μm); insol. in water; dens. 1.01-1.03 g/cc (20 C); visc. ≈ 300 mPa•s (120 C); drop pt. 98-104 C; acid no. 10-15; sapon. no. 100-120; *Toxicology:* LD50 (oral, mouse) > 20,000 mg/kg; *Storage:* store dry @ R.T.

Hoechst Wax PE 130. [Hoechst Celanese] Polyethylene; CAS 9002-88-4; EINECS 200-815-3; wax improving rub resist. of printing inks, solv.-based polishes, esp. pastes; external lubricant for rigid PVC pipe, profiles, flexible PVC cables; carrier for color concs.; also for lipsticks and solid stick antiperspirants; FDA approved; solid.

Hoechst Wax PE 190. [Hoechst AG; Hoechst Celanese] High dens., high m.w. linear polyethylene wax, nonpolar; CAS 9002-88-4; EINECS 200-815-3; wax for mfg. of transparent PVC articles with high surf. gloss; lubricant for plastics; increases m.p. of other waxes; wh. gran., powd. (< 500 μm), or micropowd. (< 40 μm); insol. in water; dens. 0.95-0.97 g/cc (20 C); visc. ≈ 25,000 mPa•s (140 C); soften. pt. ≈ 135 C; acid no. 0; sapon. no. 0; *Toxicology:* LD50 (oral, rat) > 15,000 mg/kg; *Storage:* store at R.T. in dry place.

Hoechst Wax PE 520. [Hoechst AG; Hoechst Celanese] Low dens. med. m.w. branched polyethylene wax, nonpolar; CAS 9002-88-4; EINECS 200-815-3; carrier for pigment and additive concs. for polyolefins; external lubricant for PVC; lubricant and tack regulator for rubber extrusion; flame-retardant ABS, PS, polyamide; solv.-based polishes, additive for hydrocarbon waxes; to improve rub resist. of printing inks, to modify the surf. props. of lacquers and coatings; wh. gran., fine grain (< 2000 μm) or powd. (< 500 μm); insol. in water; dens. 0.92-0.94 g/cc (20 C); visc. ≈ 650 mPa•s (140 C); drop pt. 117-122 C; acid no. 0; sapon. no. 0; *Toxicology:* LD50 (oral, rat) > 15,000 mg/kg; *Storage:* store at R.T. in dry place.

Hoechst Wax PED 191. [Hoechst AG] High m.w. linear oxidized polyethylene wax; CAS 68441-17-8; lubricant for plasticized and unplasticized PVC, highly transparent articles; processing aid for rigid PVC compds.; reduces sticking of tin-stabilized PVC melts to hot machine parts; emulsions for polishes; almost wh. flakes or powd. (< 500 μm); insol. in water; dens. 0.97-0.99 g/cc (20 C); visc. ≈ 4000 mPa•s (140 C); drop pt. 120-125 C; acid no. 15-17; sapon. no. 20-35; *Toxicology:* LD50 (oral, rat) > 15,000 mg/kg; *Storage:* store at R.T. in dry place.

Hoechst Wax PED 521. [Hoechst AG] Polyethylene wax, polar; CAS 9002-88-4; EINECS 200-815-3; external lubricant for plasticized and unplasticized PVC, PA, TPU, other thermoplastics; reduces sticking of tin-stabilized PVC melts to hot machine parts; emulsions for textiles, leather, paper, and polishes; BGA approval; flakes, fine grains.

Hoechst Wax PP 230. [Hoechst AG] Polypropylene wax; external lubricant for PVC; carrier for pigment concs.; additive to hydrocarbon waxes; BGA approval; gran., fine grain.

Hoechst Wax S. [Hoechst AG; Hoechst Celanese] Montan acid wax; CAS 68476-03-9; EINECS 270-664-6; lubricant and internal release agent for PVC, polyolefins, polyamide, PS, linear polyesters, TPU, thermosets, etc.; also for emulsion polishes, color bases, scouring waxes, industrial wax emulsions for textiles, paper, wood, etc.; worldwide food approvals for use in plastics; pale ylsh. powd. (< 500 μm), fine powd. (< 125 μm), or flakes; insol. in water; dens. 1.00-1.02 g/cc; visc. ≈ 30 mm^2/s (100 C); drop pt. 81-87 C; acid no. 130-150; sapon. no. 155-175; *Toxicology:* LD50 (oral, rat) > 15,000 mg/kg; *Storage:* store dry @ R.T.

Hoechst Wax SW. [Hoechst Celanese; Hoechst AG] Montan acid wax; CAS 68476-03-9; EINECS 270-664-6; wax for beauty cosmetic pastes, gels, lipstick; lubricant for plastics; almost wh. flakes; insol. in water; dens. 1.00-1.02 g/cc; visc. ≈ 30 mm^2/s (100 C); drop pt. 81-87 C; acid no. 115-135; sapon. no. 145-165.

Hoechst Wax UL. [Hoechst Celanese] Montan acid wax; CAS 68476-03-9; EINECS 270-664-6; lubricant for plastics; also for emulsion polishes, color bases, scouring waxes, industrial wax emulsions for textiles, paper, wood, etc.; ylsh. flakes; insol. in water; dens. 1.00-1.02 g/cc (20 C); visc. ≈ 30 mm^2/s (100 C); drop pt. 81-87 C; acid no. 100-115; sapon. no. 130-160.

Hordaflam® NK 70. [Hoechst AG] Highly chlorinated aliphatic hydrocarbon; flame retardant for plastics; 70% chlorine.

Hordaflam® NK 72. [Hoechst AG] Solid chlorinated paraffin; flame retardant for PS, elastomers, LDPE, unsat. polyester; 70% chlorine.; Unverified

Hordaflam® NL 70. [Hoechst AG] Highly chlorinated aliphatic hydrocarbon; flame retardant for plastics; 70% chlorine.

Hostaflam® AP 750. [Hoechst AG] Ammonium polyphosphate; intumescent halogen-free flame retardant for thermoplastics incl. PP, PE, EVA, polyolefin blends for extrusion and inj. molding; thermal stability to 260 C, low smoke evolution, environmentally friendly; also for hot-melt adhesives or coating compds.; wh. free-flowing powd.; dens. 1.8 g/cc; decomp. pt. > 260 C; 21% P, 12% N; *Toxicology:* LD50 (roal, rats) > 5000 mg/kg; mild skin irritant; *Storage:* store @ R.T. in dry place; reseal partially emptied containers.

Hostaflam® System F1912. [Hoechst AG] Flame retardant in LDPE carrier; flame retardant conc. for LDPE; may also be used in EVA.

Hostaflon® TF 9202. [Hoechst Celanese] PTFE wax; CAS 9002-84-0; internal/external lubricant for thermoplastics, thermosets, elastomers, coatings; solid.

Hostaflon® TF 9203. [Hoechst Celanese] PTFE wax; CAS 9002-84-0; internal/external lubricant for coatings, pkg.; FDA accepted; solid.

Hostalub® CAF 485. [Hoechst Celanese/Spec. Chem.] Calcium stearate, syn. paraffin wax, and oxidized polyethylene wax; lubricant for high line-speed extrusion of vinyl siding; Gardner 7 beads; drop pt. 100-108 C; acid no. 13-20; alkali no. 68-76.

Hostalub® FA 1. [Hoechst AG] Bis-stearoyl ethylene diamine-type amide wax; CAS 110-30-5; internal/ external lubricant for plasticized or unplasticized PVC and thermosets; carrier for pigments; also for PS, ABS; exc. heat stability; almost wh. powd. (< 500 μm) or micropowd. (< 40 μm); dens. 0.99-1.01 g/cc (20 C); visc. ≈ 10 mm²/s (150 C); drop pt. 139-144 C; acid no. < 15; sapon. no. < 15; *Toxicology:* LD50 (oral, mice) > 20,000 mg/kg; *Storage:* store in dry place @ R.T.

Hostalub® FE 71. [Hoechst Celanese/Waxes] Glycerol ester of sat. fatty acid; internal lubricant.

Hostalub® H 3. [Hoechst AG] Polyethylene wax; CAS 9002-88-4; EINECS 200-815-3; external lubricant for plastics; carrier for pigments.

Hostalub® H 4. [Hoechst AG] Modified hydrocarbon wax; external lubricant for PVC pipe and profile extrusion and for plasticized PVC; wh. fine grain (< 2000 μm); dens. 0.91-0.93 g/cc (20 C); visc. ≈ 10 mm²/s (120 C); drop pt. 108-113 C; acid no. 0; sapon. no. 0.

Hostalub® H 12. [Hoechst AG] Oxidized polyethylene wax; CAS 68441-17-8; lubricant, release agent, processing aid for plastics; reduces sticking of tin-stabilized PVC melts to hot machine parts; almost wh. fine grain (< 2000 μm); dens. 0.94-0.96 g/cc (20 C); visc. ≈ 350 mPa•s (120 C); drop pt. 100-108 C; acid no. 15-19; sapon. no. 30-45; *Toxicology:* LD50 (oral, rat) > 15,000 mg/kg; *Storage:* store at R.T. in dry place.

Hostalub® H 22. [Hoechst AG] Polyethylene wax, polar; CAS 9002-88-4; EINECS 200-815-3; lubricant for plastics; reduces sticking of tin-stabilized PVC melts to hot machine parts; almost wh. fine grain (< 2000 μm); dens. 0.95-0.97 g/cc (20 C); visc. ≈ 300 mPa•s (120 C); drop pt. 103-108 C; acid no. 22-28; sapon. no. 45-65; *Toxicology:* LD50 (oral, rat) > 15,000 mg/kg; *Storage:* store at R.T. in dry place.

Hostalub® VP Ca W 2. [Hoechst AG] Calcium montanate; CAS 68308-22-5; internal lubricant for PVC, polyamide, thermoplastic PU, other thermoplastics and thermosets.

Hostalub® WE4. [Hoechst AG; Hoechst Celanese] Glyceryl montanate; CAS 68476-38-0; lubricant and release agent for PVC, polyolefins, polyamide, PS, linear polyesters, TPU, thermosets; carriers for pigment concs.; worldwide food approvals for plastics; ylsh. powd. (< 500 μm) or flakes; insol. in water; dens. 1.00-1.02 g/cc (20 C); visc. ≈ 60 mm²/s (100 C); drop pt. 78-85 C; acid no. 20-30; sapon. no. 130-160; *Toxicology:* pract. nontoxic; *Storage:* store dry @ R.T.

Hostalub® WE40. [Hoechst AG] Montan ester wax; external lubricant in PVC; for calendering, blow molding, pipe and profile extrusion, inj. molding of unplasticized PVC.

Hostalub® XL 50. [Hoechst Celanese/Waxes] Blend of paraffin wax and calcium stearate; external lubricant for PVC inj. molding applics.; maintains fast fusion time; Gardner 12 max. beads; drop pt. 100-110 C; acid no. 15-22; alkali no. 57-65.

Hostalub® XL 165. [Hoechst Celanese/Waxes] Blend of paraffinic wax components; external lubricant for rigid PVC extrusion, wire and cable applics.; Gardner 6 max. prill or beads; drop pt. 72-79 C; acid no. < 1; ref. index 1.43-1.44 (80 C).

Hostalub® XL 165FR. [Hoechst Celanese/Waxes] Fully refined, single component paraffin wax; CAS 8002-74-2; EINECS 232-315-6; external lubricant for more demanding applics. such as rigid PVC profile or high line-speed extrusion; also for textile, waterproofing, paper coating, candles, as antiozonants for rubber; Gardner < 3 prill or bead; drop pt. 72-77 C; congeal pt. 68-74 C; ref. index 1.435-1.441 (80 C).

Hostalub® XL 165P. [Hoechst Celanese/Waxes] Blend of linear syn. paraffinic waxes having a sl. greater proportion of higher melting components than XL 165; CAS 8002-74-2; EINECS 232-315-6; external lubricant for rigid PVC, pipe; PPI and NSF accepted for pipe; Gardner 3 max. prill or bead form; dens. 0.91-0.92 g/cc (20 C); congeal pt. 65-76 C; acid no. < 1.

Hostalub® XL 355. [Hoechst Celanese/Waxes] Blend of Hostalub® XL 165 components with oxidized polyethylene wax; lubricant for PVC extrusion; Gardner 6 max. prill or bead; dens. 0.90-0.92 g/cc (20 C); drop pt. 90-105 C; acid no. 1-2.5.

Hostalub® XL 445. [Hoechst Celanese/Waxes] Blend of paraffin wax, oxidized polyethylene wax, and calcium stearate; lubricant for rigid PVC pipe extrusion; NSF and PPI compliances; Gardner 12 max. prill or bead; drop pt. 100-110 C; acid no. 15 max.; alkali no. 35-44.

Hostalub® System G1970. [Hoechst AG] Lubricant in LDPE carrier; lubricant conc. for LDPE, LLDPE, HDPE, EVA, PP; FDA and BGA approvals.

Hostalub® System G1971. [Hoechst AG] Lubricant in LDPE carrier; lubricant conc. for LDPE, LLDPE, HDPE, EVA, PP; FDA and BGA approvals.

Hostalub® System G1972. [Hoechst AG] Lubricant in LDPE carrier; lubricant conc. for LDPE, LLDPE, EVA; may also be used in HDPE, PP; BGA approvals.

Hostalub® System G1973. [Hoechst Celanese/Spec. Chem.] Fluroelastomer in LDPE carrier; processing/flow aid for LLDPE film extrusion, HDPE; pellet; dens. 0.90-0.93 g/cc (20 C); m.p. 110 C; 1.25-1.5% F.

Hostalux KCB. [Hoechst Celanese] Benzoxazole type; optical brightener, whitening agent used in all types of polymer processing; used in plastic films, press molding, and inj. molding material fibers and bristles, paints and lacquers; ylsh.-grn. powd.; sol. (mg/100 ml): 1020 mg in toluene; 950 mg in perchloroethylene; 860 mg in mesitylene; 800 mg in CCl_4; 760 mg in dimethyl formamide; 170 mg in acetone; m.p. 211 C.

Hostamont® CaV 102. [Hoechst AG] Calcium salt of long-chain carboxylic acids; nucleating agent for plastics.

Hostamont® NaS 102. [Hoechst AG] Mixt. of long-chain carboxylic acids and their sodium salts; nucleating agent for plastics.

Hostamont® NaV 101. [Hoechst AG] Sodium salt of long-chain carboxylic acids; nucleating agent for plastics.

Hostanox® O3. [Hoechst Celanese/Spec. Chem.; Hoechst AG] Benzene propanoic acid; CAS 501-52-0; EINECS 207-924-5; antioxidant, primary stabilizer for HDPE, LDPE, PP, PS, PU, polyester, PA, POM; radical scavenger; protects against oxidation, discoloration, degradation, crosslinking; low volatility, good extraction stability; food approvals in most countries; wh. cryst. powd.; sol. @ 20% in methanol, ethanol, IPA, MEK, cyclohexanone, diethyl ether, ethyl acetate, acetone; insol. in water; m.w. 795.08; dens. 1.1 g/cc (20 C); m.p. 170 C; *Toxicology:* LD50 (oral, rat) > 15 g/kg; *Storage:* store in dry place @ R.T.

Hostanox® OSP 1. [Hoechst Celanese/Spec. Chem.; Hoechst AG] Phenol, 4,4′-thiobis 2-(1,1-dimethylethyl) phosphite; metal deactivator inhibiting the autooxidation of copper and other metals; antioxidant for HDPE, PP, PMMA, PA, esp. polyolefins for wire and cable, polymers filled with metal powds. and asbestos, and PP fittings in contact with nonferrous metals; wh. amorphous powd.; sol. in more common aromatic solvs., aliphatic esters, ketones, chlorinated HC; insol. in water; m.w. 1103; dens. 1.1 g/cc (20 C); soften. pt. ≈ 110 C.

Hostanox® PAR 24. [Hoechst Celanese/Spec. Chem.; Hoechst AG] Tris-(2,4-di-t-butylphenyl) phosphite; CAS 31570-04-4; antioxidant for LLDPE, LDPE, HDPE, PP; color stabilizer in TPU; thermal stabilizer in nylon; also effective in PS, PC, polyesters; wh. gran. powd.; sol. (g/100 g): 40 g in chloroform, 35 g toluene, 13 g heptane, 7 g ethyl acetate, 1 g acetone, < 0.1 g water, methanol; m.w. 647; m.p. 185 C; acid no. ≈ 0.5; 4.5% P.

Hostanox® SE 1. [Hoechst Celanese/Spec. Chem.; Hoechst AG] Dilaurylthiodipropionate; CAS 123-28-4; EINECS 204-614-1; antioxidant for plastics.

Hostanox® SE 2. [Hoechst Celanese/Spec. Chem.; Hoechst AG] Distearyl thiodipropionate; CAS 693-36-7; EINECS 211-750-5; antioxidant for plastics; used in combination with Hostanox O3 to achieve long-term stability, esp. for food contact applics.; effective in reinforced and filled polyolefins; food approvals in most countries; wh. fine gran.; sol. (g/100 g): 20 g in chloroform, 9 g in toluene, 6 g in turpentine, < 1 g in acetone, ethyl acetate, heptane, hexane, IPA, methanol, < 0.01 g in water; dens. 0.99 g/cc (20 C); drop pt. 65 C; m.p. 170 C ≥ 4.5% S; *Toxicology:* LD50 (oral, rat) > 5 g/kg; *Storage:* store in dry place @ R.T.

Hostanox® SE 10. [Hoechst Celanese/Spec. Chem.; Hoechst AG] Dioctadecyl disulfide; antioxidant for HDPE, LDPE, PP, PMMA, PA; synergist; thermal stabilizer; costabilizer with phenolic antioxidants; strongly synergistic with Hostanox O3; peroxide decomposer; food approvals in several countries; wh. powd.; m.w. 571; sol. (g/100 g): 15 g in chloroform, turpentine, toluene, 7 g in heptane, hexane, 5 g in petrol. ether, < 1 g in acetone, ethyl acetate, IPA, methanol, < 0.01 g in water; dens. 0.85 g/cc (55 C); bulk dens. 0.50 g/cc; m.p. 53-58 C ≥ 9% S; *Toxicology:* LD50 (oral, rat) > 15 g/kg; *Storage:* store in dry place @ R.T.

Hostanox® ZnCS 1. [Hoechst Celanese/Spec. Chem.; Hoechst AG] Zinc salt of a thiocarbonic acid; antioxidant for plastics.

Hostanox® System A1961. [Hoechst Celanese/Spec. Chem.] Antioxidant blend (hindered phenol, thioester, phosphite) in LDPE carrier; antioxidant conc. to enhance the stability of molded or extruded polyolefins (LDPE, LLDPE, HDPE, EVA); FDA approved; unpigmented pellet.

Hostanox® System A4962. [Hoechst Celanese/Spec. Chem.] Hindered phenol and costabilizers in PP carrier; antioxidant conc. for PP; FDA approved; unpigmented pellet.

Hostanox® System A4965. [Hoechst Celanese/Spec. Chem.] Phenolic/costabilizer in PP carrier; antioxidant conc. for PP.

Hostanox® System P1961. [Hoechst Celanese/Spec. Chem.] Phenolic/costabilizer in PE carrier; antioxidant conc. for PE.

Hostapal BV Conc. [Hoechst Celanese/Colorants & Surf.; Hoechst AG] Alkylaryl polyglycol ether sulfate, sodium salt; anionic; dispersant for pigments, wetting agent, detergent for fibers, emulsifier for emulsion polymerization; yel. brn. gelatinous paste; misc. with water; dens. 9.10 lb/gal; 49-51% act.

Hostaprime® HC 5. [Hoechst Celanese/Spec. Chem.; Hoechst AG] Maleic anhydride grafted PP; coupling agent for glass-filled PP, PP/PA alloys; adhesion promoter improving incorporation of glass fibers in PP; FDA 21CFR §175.105; wh. powd., almost odorless; dens. 0.9 g/cc; apparent dens. 300 kg/m^3; m.p. 153-159 C; *Storage:* store @ R.T. in dry place.

Hostastab® SnS 10. [Hoechst AG] Dioctytin thioglycolate; stabilizer for PVC; food contact approval in Germany, France, UK, Italy, Japan; Gardner 4 max. color; dens. 1.08 g/cc (20 C); visc. 80 mPa•s (20 C); solid. pt. -50 C; ref. index 1.502 (20 C); *Toxicology:* LD50 (oral, rat) ≈ 2000 mg/kg; *Storage:* store dry @ R.T. in tightly sealed drums.

Hostastab® SnS 11. [Hoechst AG] Dioctytin thioglycolate contg. 25% glycerin fatty acid ester; stabilizer for PVC for use where internal lubrication is required; food contact approval in Germany, France, UK, Italy, Japan; Gardner 6 max. color; dens. 1.06 g/cc (20 C); visc. 120 mPa•s (20 C); solid. pt. 2 C; ref. index 1.494 (20 C); *Toxicology:* LD50 (oral, rat) ≈ 2000 mg/kg; *Storage:* store dry @ R.T. in tightly sealed drums.

Hostastab® SnS 15. [Hoechst AG] Dioctytin thioglycolate; stabilizer for PVC.

Hostastab® SnS 20. [Hoechst AG] Octyltin thioglycolate; stabilizer for PVC; FDA 21CFR §178.2650; food contact approval in Germany, France, UK, Italy, Japan; Gardner 6 max. color; dens. 1.09 g/cc (20 C); visc. 80 mPa•s (20 C); solid. pt. -50 C; ref. index 1.507 (20 C); *Toxicology:* LD50 (oral, rat) ≈ 2000 mg/kg; *Storage:* store dry @ R.T. in tightly sealed drums.

Hostastab® SnS 41. [Hoechst AG] Dioctytin thioglycolate; stabilizer for PVC; food contact approval in Germany, France, UK, Italy, Japan; Gardner 4 max. color; dens. 1.08 g/cc (20 C); visc. 50 mPa•s (20 C); solid. pt. -50 C; ref. index 1.502 (20 C); *Toxicology:* LD50 (oral, rat) ≈ 3000 mg/kg; *Storage:* store dry @ R.T. in tightly sealed drums.

Hostastab® SnS 44. [Hoechst AG] Dioctytin thioglycolate; stabilizer for plastics esp. S-PVC; food contact approval in Germany, France, UK, Italy, Japan; Gardner 4 max. color; dens. 1.09 g/cc (20 C); visc. 80 mPa•s (20 C); solid. pt. -50 C; ref. index 1.503 (20 C); *Toxicology:* LD50 (oral, rat) ≥ 2000 mg/kg; *Storage:* store dry @ R.T. in tightly sealed drums.

Hostastab® SnS 45. [Hoechst AG] Octyltin thioglycolate; stabilizer for PVC, esp. where good initial color is desired; food contact approval in Germany, France, UK, Italy, Japan; Gardner 4 max. color; dens. 1.10 g/cc (20 C); visc. 80 mPa•s (20 C); solid. pt. -50 C; ref. index 1.503 (20 C); *Toxicology:* LD50 (oral, rat) ≈ 2000 mg/kg; *Storage:* store dry @ R.T. in tightly sealed drums.

Hostastab® SnS 61. [Hoechst AG] Butyltin thioglycolate; stabilizer for rigid PVC sheet, profiles, pipes, inj. molded articles, films and tech. goods made from plasticized PVC (films, hoses, inj. molded parts); Gardner 6 max. color; dens. 1.10 g/cc (20 C); visc. 60 mPa•s (20 C); solid. pt. -50 C; ref. index 1.503 (20 C); *Toxicology:* LD50 (oral, rat) ≥ 200 mg/kg; skin and eye irritant; *Storage:* store dry @ R.T. in tightly sealed drums.

Hostastat® FA 14. [Hoechst Celanese/Spec. Chem.; Hoechst AG] Ethoxylated cocoamine; antistat for PE, PP, PS, ABS; FDA 21CFR §178.3130; ylsh. low-visc. liq.; dens. 0.88-0.91 g/cc (50 C); visc. 30 mPa•s (50 C); solid. pt. ≤ 5 C; pH 9-10 (1% aq.); 1% max. moisture; *Toxicology:* LD50 (oral, rat) 1720 mg/kg; strong skin and very strong mucous membrane irritant; *Storage:* hygroscopic; store dry @ R.T. in tightly closed containers; if flocculation occurs below solid. pt., warm and stir.

Hostastat® FA 15. [Hoechst Celanese/Spec. Chem.; Hoechst AG] Ethoxylated syn. alkylamine; antistat for PE, PP, PS, ABS, HIPS; processing aid for thin-gauge HDPE film and biaxially oriented PP film; FDA 21CFR §178.3130; Japan, France, Germany, Italy approvals; almost wh. clear low-visc. liq.; dens. 0.90 g/cc (50 C); visc. 30 mPa•s (50 C); solid. pt. ≤ 5 C; alkali no. 185-205; *Toxicology:* strong skin and very strong mucous membrane irritant; *Storage:* hygroscopic; store in closed containers @ R.T. in dry environment; if flocculation occurs below solid. pt., stir and warm.

Hostastat® FA 18. [Hoechst Celanese/Spec. Chem.; Hoechst AG] Ethoxylated tallowamine; antistat for plastics; FDA 21CFR §178.3130; ylsh. fine grain powd.; bulk dens. ≈ 500 g/l; drop pt. 85-100 C; alkali no. 160-180; *Toxicology:* strong skin and very strong mucous membrane irritant; *Storage:* hygroscopic; store dry in tightly closed containers @ R.T.

Hostastat® FA 38. [Hoechst Celanese/Spec. Chem.] Ethoxylated tallowalkylamine; antistat for thermoplastics, esp. polyolefins; FDA 21CFR §178.3130; nearly wh. free-flowing powd.; dens. 0.99 g/cc (20 C); drop pt. 95 C; acid no. 5; congeal pt. 60 C.

Hostastat® FE 2. [Hoechst Celanese/Spec. Chem.; Hoechst AG] Fatty acid ester; antistat for plastics; FDA 21CFR §175.300; almost wh. fine grain powd.; dens. 1.01-1.03 g/cc (20 C); bulk dens. ≈ 500 g/l; drop pt 55-62 C; acid no. ≤ 5; sapon. no. 160-180; *Toxicology:* LD50 (oral, mouse) > 5000 mg/kg; *Storage:* hygroscopic; store dry in tightly closed containers @ R.T.

Hostastat® FE 20. [Hoechst Celanese/Spec. Chem.; Hoechst AG] Fatty ester diglyceride; antistat for PP, PS, PVC, nylon, ABS, PVC; can be used in transparent tin-stabilized rigid PVC, semirigid PVC, and plasticized PVC; nearly colorless liq. (above 20 C); dens. 1.02 g/cc (23 C); visc. 250 mPa•s (50 C); solid. pt. < 25 C; acid no. < 5; sapon. no. 160-180; *Storage:* hygroscopic; store in tightly closed drums when not in use.

Hostastat® HS1. [Hoechst Celanese/Spec. Chem.; Hoechst AG] Sodium C10-18 alkyl sulfonate; antistat, processing aid for thermoplastics incl. PS, HIPS, ABS, PPO, PET, PVC, styrenic alloys, copolymers, and acrylic esters; can be incorporated before processing or applied externally to finished surfaces; improves pigment dispersibility, compat., gloss; FDA 21CFR §178.3130, 178.3400; nearly wh.; avail. in fine grain or masterbatch form; < 5 mm particle size; sol. in water, chlorinated HC, IPA; sp.gr. 1.05 (30% aq., 20 C); bulk dens. 450 g/l (fine grain); pH 7-8 (2% aq.); *Toxicology:* LD50 (oral, rat) 1850 mg/kg; *Storage:* hygroscopic; store dry in tightly closed containers @ R.T.

Hostastat® System E1902. [Hoechst AG] Antistat in LDPE carrier; antistat conc. for LDPE, LLDPE, EVA; BGA approval; *Toxicology:* irritant.

Hostastat® System E1906. [Hoechst AG] Antistat in LDPE carrier; antistat conc. for LDPE, LLDPE, EVA, HDPE, PP; FDA and BGA approvals.

Hostastat® System E1956. [Hoechst Celanese/Spec. Chem.] Laurylamide in LDPE carrier; antistat conc. for LDPE, LLDPE, HDPE, PP extrusion, molding; suited for electronics applics.; FDA 21CFR §178.3130; unpigmented pellets; *Storage:* hygroscopic; store in dry area; protect from moisture.

Hostastat® System E2903. [Hoechst Celanese/ Spec. Chem.] Ethoxylated alkylamine in HDPE carrier; antistat conc. for HDPE; FDA approved; unpigmented pellet; *Toxicology:* irritant.

Hostastat® System E3904. [Hoechst AG] Antistat in PS carrier; antistat conc. for PS; BGA approvals; *Toxicology:* harmful.

Hostastat® System E3952. [Hoechst Celanese/ Spec. Chem.] Sodium alkyl sulfonate predispersed in PS carrier; antistat systems for extruded or molded PS, HIPS, and styrene copolymer parts; also improves processability and pigment dispersion; FDA 21CFR §178.3120, 178.3400; unpigmented pellets; *Storage:* hygroscopic; store in dry, low humidity area.

Hostastat® System E3953. [Hoechst Celanese/ Spec. Chem.] Sodium alkyl sulfonate predispersed in PS carrier; antistat systems for extruded or molded PS, HIPS, and styrene copolymer parts; also improves processability and pigment dispersion; FDA 21CFR §178.3120, 178.3400; unpigmented pellets; *Storage:* hygroscopic; store in dry, low humidity area.

Hostastat® System E3954. [Hoechst Celanese/ Spec. Chem.] Sodium alkyl sulfonate predispersed in PS carrier; antistat systems for extruded or molded PS, HIPS, and styrene copolymer parts; also improves processability and pigment dispersion; FDA 21CFR §178.3120, 178.3400; unpigmented pellets; *Storage:* hygroscopic; store in dry, low humidity area.

Hostastat® System E4751. [Hoechst Celanese/ Spec. Chem.] Sodium alkyl sulfonate in ABS carrier; antistat conc. reducing surf. resist. and static decay in extruded or molded ABS parts; also improves processability and pigment dispersion; FDA 21CFR §178.3130, 178.3400; unpigmented pellets; *Storage:* hygroscopic; store in dry, low humidity area.

Hostastat® System E4905. [Hoechst AG] Antistat in PP carrier; antistat conc. for PP; FDA and BGA approvals; *Toxicology:* irritant.

Hostastat® System E5951. [Hoechst Celanese/ Spec. Chem.] Laurylamide in a PP carrier; antistat system for PP molding and extrusion; esp. suited for electronics applics.; FDA approved; unpigmented pellets.

Hostastat® System E6952. [Hoechst Celanese/ Spec. Chem.] Laurylamide in a LLDPE carrier; antistat system for polyolefins (LLDPE, HDPE film); reduces surf. resist. and static decay in extruded and molded parts; extra high tear str.; esp. suited for electronics applics.; FDA 21CFR §178.3130; unpigmented pellets; *Storage:* hygroscopic; store in dry area.

Hostatron® System P1931. [Hoechst AG] Endothermic chemical blowing agent in LDPE carrier; blowing agent conc. for LDPE, LLDPE, HDPE, PP; may also be used in EVA or PS; FDA and BGA approvals.

Hostatron® System P1933. [Hoechst Celanese/ Spec. Chem.] Bicarbonate/citric acid and proprietary additives in LDPE carrier; endothermic blowing agent conc. for LDPE, LLDPE, HDPE, PP, PS; FDA approved; unpigmented pellets.

Hostatron® System P1935. [Hoechst AG] Endothermic chemical blowing agent in LDPE carrier; blowing agent conc. for LDPE, LLDPE, HDPE, PP, PS; may also be used in EVA; FDA and BGA approvals.

Hostatron® System P1940. [Hoechst Celanese/ Spec. Chem.] Bicarbonate/citric acid and proprietary additives in LDPE carrier; endothermic blowing agent conc. for thin gauge sheet and film of LDPE, LLDPE, HDPE, PP, PS, ABS, e.g., thermoformed cups and trays, wrapping film, foam tape, insulation sheet, envelopes, foamed cable sheathing, blown film, inj. molding; FDA approved; unpigmented pellets.

Hostatron® System P1941. [Hoechst Celanese/ Spec. Chem.] Bicarbonate/citric acid and proprietary additives in LDPE carrier; endothermic blowing agent conc. for thin gauge sheet and film of LDPE, LLDPE, HDPE, PP, PS, ABS, e.g., thermoformed cups and trays, wrapping film, foam tape, insulation sheet, envelopes, foamed cable sheathing, blown film, inj. molding; FDA approved; unpigmented pellets.

Hostatron® System P9937. [Hoechst Celanese/ Spec. Chem.] Bicarbonate/citric acid and proprietary additives in proprietary PE carrier; endothermic blowing agent conc. for extruded articles and lower temp. moldings from PS, HIPS, ABS, PPO, PE, and PP; nucleating agent for foamed PS and PE, mech. foaming agents (air, freon, pentane); FDA approved; unpigmented pellets.

Hostatron® System P9947. [Hoechst Celanese/ Spec. Chem.] Bicarbonate/citric acid and proprietary additives in proprietary PZE carrier; endothermic blowing agent for inj. molded articles fro PS, HIPS, ABS, PPO, PE, and PP; nucleating agent for foamed PS and PE, mech. foaming agents (air, freon, pentane); FDA approved; unpigmented pellets.

Hostavin® ARO 8. [Hoechst Celanese/Spec. Chem.; Hoechst AG] Benzophenone-12; CAS 1843-05-6; EINECS 217-421-2; uv absorber for plastics, esp. LDPE, HDPE, PP, polyisobutylene, cellulosics, PC, EVA copolymers, plasticized PVC, thick-walled articles; transforms harmful uv radiation into heat energy; approved for food pkg. in various countries; pale yel. cryst. powd.; m.w. 326; sol. (g/100 g): 75 g acetone, 70 g ethyl acetate, 65 g toluene, 58 g chloroform, 25 g petrol. ether; sp.gr. 1.1; m.p. 48 C.

Hostavin® N 20. [Hoechst Celanese/Spec. Chem.; Hoechst AG] 2,2,6,6-Tetramethylpiperidine-4-(substituted) carbonamide; lt. stabilizer effective for thin-walled polyolefins, natural or dyed polymers incl. HDPE, PP, nylon, EVA, HIPS, PS, PA, ABS, PU, PC, and cellulosics; used in thick films and molded or extruded parts (sheet, drums, tanks, beverage crates, containers, bumpers, stadium seats, garden furniture, pkg. straps, netting); wh. cryst. powd.; sol. (g/100 g): 11 g in chloroform, 2 g IPA, turpentine oil, 1 g toluene, < 0.01 g water; sp.gr. 1.06; bulk dens. 0.46 g/cc; m.p. 225-227 C; *Storage:* store dry @ R.T.

Hostavin® N 24. [Hoechst Celanese/Spec. Chem.; Hoechst AG] Sterically HALS, low m.w.; uv stabilizer for PMMA, PE, PU, PA, PET.; *Toxicology:* irritant; Unverified

Hostavin® N 30. [Hoechst Celanese/Spec. Chem.; Hoechst AG] Oligomeric sterically HALS; uv stabilizer for LDPE, HDPE, PP, PE, EVA, PMMA, styrene copolymers, ABS, cellulosics, polyesters, PC, elastomers, POM, PU; useful for stabilizing colorless articles; exc. extraction resist.; contributes to antioxidative stability esp. of polyolefins; for films, fibers, tapes; wh. powd.; sol. (g/100 G): 40 g toluene, 30 g ethanol, 20 g acetone, < 1 g heptane, hexane, < 0.01 g water; m.w. > 1500; bulk dens. 0.45 kg/l; *Toxicology:* LD50 (oral, rat) > 2000 mg/kg; not a skin irritant; insignificant irritation by mucous membrane contact; *Storage:* store in dry place @ R.T.

Hostavin® VP NiCS 1. [Hoechst Celanese/Spec. Chem.] Nickel-containing; light stabilizer, antioxidant for polyolefins, esp. HDPE and LDPE; ideal for polyolefin films for agric. use; grn. powd.; m.w. 747; sol. (g/100 g): 35 g chloroform, 30 g toluene; m.p. 150 C; Unverified

Hostavin® System L1922. [Hoechst Celanese/Spec. Chem.] Polymeric HALS, uv absorber in LDPE carrier; uv stabilizer conc. for LDPE, LLDPE, HDPE; FDA approved; unpigmented pellets.

Hostavin® System L1923. [Hoechst AG] UV stabilizer in LDPE carrier; uv stabilizer conc. for LDPE; may also be used in LLDPE, EVA.

Hostavin® System L1924. [Hoechst Celanese/Spec. Chem.] Polymeric HALS in LDPE carrier; uv stabilizer conc. for LDPE, LLDPE, HDPE; unpigmented pellets.

Hostavin® System L1970. [Hoechst Celanese/Spec. Chem.] Monomeric HALS in LLDPE carrier; lt. stabilizer conc. for HDPE, LLDPE; unpigmented pellets.

Hostavin® System L1973. [Hoechst Celanese/Spec. Chem.] HALS in PE carrier; uv stabilizer conc. for PE, molding, film applics.

Hostavin® System L2925. [Hoechst Celanese/Spec. Chem.] Monomeric HALS, uv absorber in HDPE carrier; uv stabilizer conc. for HDPE for prod. of film and inj. molded articles; unpigmented pellets.

Hostavin® System L2970. [Hoechst Celanese/Spec. Chem.] Monomeric HALS in HDPE carrier; lt. stabilizer conc. for HDPE, LLDPE; unpigmented pellets.

Hostavin® System L2971. [Hoechst Celanese/Spec. Chem.] Polymeric HALS in HDPE carrier; lt. stabilizer conc. for HDPE, LLDPE; unpigmented pellets.

Hostavin® System L2972. [Hoechst Celanese/Spec. Chem.] Polymeric HALS in PP carrier; lt. stabilizer conc. for PP; unpigmented pellets.

Hostavin® System L4926. [Hoechst Celanese/Spec. Chem.] Monomeric HALS in PP carrier; uv stabilizer conc. for molded and extruded PP articles with wall thickness above 4 mil meeting highest longevity demands (e.g., outdoor furniture, stadium seats, crates, pipe, sheet); high extraction resist.; FDA approved; unpigmented pellets.

Hostavin® System L4927. [Hoechst Celanese/Spec. Chem.] Polymeric HALS in PP carrier; lt. stabilizer for molded and extruded PP articles, esp. for thinner gauge applics. meeting highest longevity demands (e.g., fiber, film, tape); FDA approved; pellets.

Hostavin® System L4928. [Hoechst Celanese/Spec. Chem.] Monomeric HALS, uv absorber in PP carrier; uv stabilizer conc. for PP; FDA approved; unpigmented pellets.

Hostavin® System L4962, L4963, L4964. [Hoechst Celanese/Spec. Chem.] HALS in PP carrier; uv stabilizer conc. for PP, molding, film applics.

HP. [Hughes Ind.] Azodicarbonamide; CAS 123-77-3; EINECS 204-650-8; general purpose blowing agents for prod. of cellular polymers @ 385-425 F; pale yel. powd., paste, liq. dispersion, or pellet conc.; 2-12 μ particle sizes; dens. 1.6 g/cc.

HPC-9. [Witco/PAG] Cumene hydroperoxide/MEK peroxide sol'n.; facilitates R.T. curing of vinyl ester or other unsat. polyester resins; combines good reactivity with low foaming and moderate exotherm chars.; clear to sl. yel. liq.; sol. in most org. solvs.; very sl. sol. in water; sp.gr. 1.07; flash pt. (Seta CC) 72 C; 8.9% act. oxygen; *Toxicology:* moder. skin irritant; severe eye irritant; *Precaution:* never mix directly with promoters or accelerators; may cause violent decomp. or explosion; once ignited, will burn intensely; *Storage:* store in closed original containers below 25 C, away from heat sources, flame, direct sunlight, contamination, strong acids, reducing agents, promoters, accelerators, heavy metals.

HRVP 01. [Hughes Ind.] Modified azodicarbonamide; CAS 123-77-3; EINECS 204-650-8; blowing agent for rigid vinyl processing incl. profile extrusion (baseboards, picture frames, door and window moldings), pipe extrusion (cellular core DWV, sewer), sheet extrusion (signage, siding); powd. or liq. disp.; dec. pt. 320-330 F.

HRVP 19. [Hughes Ind.] Modified azodicarbonamide; CAS 123-77-3; EINECS 204-650-8; blowing agent for rigid vinyl processing incl. profile extrusion (baseboards, picture frames, door and window moldings), pipe extrusion (cellular core DWV, sewer), sheet extrusion (signage, siding); powd. or liq. disp.; dec. pt. 375-385 F.

Huber 35. [J.M. Huber/Engineered Mins.] Hydrous kaolin clay; CAS 1332-58-7; EINECS 296-473-8; med. brightness, coarse particle size, low oil and water demand functional filler for adhesives, paint, plastics, and inks; fine powd.; 4 μm particle diam.; 0.1% max. on 325 mesh; fineness (Hegman) 4+; sp.gr. 2.60; bulking value 21.7 lb/gal; surf. area 6-11 m^2/g; oil absorp. 25-30; brightness 81-83.5; pH

6.0-7.5 (100 g clay/250 ml water); 1% max. moisture.

Huber 40C. [J.M. Huber/Engineered Mins.] Anhyd. kaolin clay; CAS 1332-58-7; EINECS 296-473-8; calcined functional filler with med. brightness and particle size, moderate oil and water demand; for paint and plastics applics.; fine powd.; 1.2 μm particle diam.; 0.1% max. on 325 mesh; fineness (Hegman) 5+; sp.gr. 2.5; bulking value 20.8 lb/gal; surf. area 10-12 m^2/g; oil absorp. 44-54; brightness 84.5-87; pH 4.5-6.0 (100 g clay/250 ml water); 0.5% max. moisture.

Huber 65A. [J.M. Huber/Engineered Mins.] Hyd. kaolin clay; CAS 1332-58-7; EINECS 296-473-8; air floated functional filler with med. brightness, med. particle size; for adhesives, paints, plastics; fine powd.; 0.9 μm particle diam.; 0.3% max. on 325 mesh; fineness (Hegman) 3+; sp.gr. 2.60; bulking value 21.7 lb/gal; surf. area 15-19 m^2/g; oil absorp. 30-35; brightness 79-81; pH 5.0-7.5 (100 g clay/250 ml water); 1% max. moisture.

Huber 70C. [J.M. Huber/Engineered Mins.] Anhyd. kaolin clay; CAS 1332-58-7; EINECS 296-473-8; calcined functional filler featuring high brightness, med. particle size, moderate oil and water demand; for paint and plastics; fine powd.; 1.4 μm particle diam.; 0.02% max. on 325 mesh; fineness (Hegman) 5+; sp.gr. 2.63; bulking value 21.9 lb/gal; surf. area 7-9 m^2/g; oil absorp. 46-56; brightness 90-92; pH 5-6 (100 g clay/250 ml water); 0.5% max. moisture.

Huber 80. [J.M. Huber/Engineered Mins.] Hyd. kaolin clay; CAS 1332-58-7; EINECS 296-473-8; water-washed functional filler with intermediate brightness and particle size, moderate oil and water demand; pulverized for easy dispersion; for adhesives, paints, plastics, inks; fine powd.; 0.7 μm particle diam.; 0.01% max. on 325 mesh; fineness (Hegman) 5+; sp.gr. 2.60; bulking value 21.7 lb/gal; surf. area 12-16 m^2/g; oil absorp. 40-45; brightness 86-87.5; pH 6-7 (100 g clay/250 ml water); 1% max. moisture.

Huber 80B. [J.M. Huber/Engineered Mins.] Hyd. kaolin clay; CAS 1332-58-7; EINECS 296-473-8; water-washed functional filler with intermediate brightness and particle size, moderate oil and water demand; for adhesives, paints, plastics, inks; spray-dried beads; 0.7 μm particle diam.; 0.01% max. on 325 mesh; fineness (Hegman) 5+; sp.gr. 2.60; bulking value 21.7 lb/gal; surf. area 12-16 m^2/g; oil absorp. 40-45; brightness 86-87.5; pH 6-7 (100 g clay/250 ml water); 1% max. moisture.

Huber 80C. [J.M. Huber/Engineered Mins.] Anhyd. kaolin clay; CAS 1332-58-7; EINECS 296-473-8; calcined functional filler featuring higher brightness, med. particle size, higher oil and water demand; for paint and plastics applics.; fine powd.; 1.4 μm particle diam.; 0.02% max. on 325 mesh; fineness (Hegman) 5+; sp.gr. 2.63; bulking value 21.9 lb/gal; surf. area 7-9 m^2/g; oil absorp. 48-58; brightness 91-93; pH 5-6 (100 g clay/250 ml water); 0.5% max. moisture.

Huber 90. [J.M. Huber/Engineered Mins.] Hyd. kaolin clay; CAS 1332-58-7; EINECS 296-473-8; water-washed functional filler with higher brightness, finer particle size, moderate oil and water demand; pulverized for easy dispersion; for adhesives, paints, plastics, inks; fine powd.; 0.6 μm particle diam.; 0.01% max. on 325 mesh; fineness (Hegman) 5+; sp.gr. 2.60; bulking value 21.7 lb/gal; surf. area 13-17 m^2/g; oil absorp. 40-45; brightness 87-88.5; pH 6-7.5 (100 g clay/250 ml water); 1% max. moisture.

Huber 90B. [J.M. Huber/Engineered Mins.] Hyd. kaolin clay; CAS 1332-58-7; EINECS 296-473-8; water-washed functional filler with higher brightness, finer particle size, moderate oil and water demand; for adhesives, paints, plastics, inks; spray-dried beads; 0.6 μm particle diam.; 0.01% max. on 325 mesh; fineness (Hegman) 5+; sp.gr. 2.60; bulking value 21.7 lb/gal; surf. area 13-17 m^2/g; oil absorp. 40-45; brightness 87-88.5; pH 6-7.5 (100 g clay/250 ml water); 1% max. moisture.

Huber 90C. [J.M. Huber/Engineered Mins.] Anhyd. kaolin clay; CAS 1332-58-7; EINECS 296-473-8; calcined functional filler featuring highest brightness, fine particle size, highest oil and water demand; for paint and plastics applics.; fine powd.; 0.7 μm particle diam.; 0.01% max. on 325 mesh; fineness (Hegman) 5+; sp.gr. 2.63; bulking value 21.9 lb/gal; surf. area 13-17 m^2/g; oil absorp. 75-85; brightness 92-94; pH 5-6 (100 g clay/250 ml water); 0.5% max. moisture.

Huber 95. [J.M. Huber/Engineered Mins.] Kaolin clay; CAS 1332-58-7; EINECS 296-473-8; water washed functional filler with med. brightness, fine particle size; for adhesives, paints, plastics, and inks; spray-dried beads or slurry; 0.3 μm particle diam.; 0.3% max. on 325 mesh; fineness (Hegman) 5+; sp.gr. 2.60; bulking value 21.7 lb/gal; surf. area 19-23m^2/g; oil absorp. 35-40; brightness 81-83; pH 6-8 (100 g clay/250 ml water); 1% max. moisture.

Huber ARO 60. [J.M. Huber/Engineered Carbons] Carbon black; CAS 1333-86-4; EINECS 215-609-9; for the rubber industry; 90 nm avg. particle diam.; 0.05% max 325-mesh residue; sp.gr. 1.8; pour dens. 32 lb/ft^3.

Huber HG. [J.M. Huber/Engineered Mins.] Hyd. kaolin clay; CAS 1332-58-7; EINECS 296-473-8; water-washed functional filler with higher brightness, very fine particle size, moderate oil and water demand; for adheisves, paints, plastics, and inks; spray-dried beads or slurry; 0.3 μm particle diam.; 0.01% max. on 325 mesh; fineness (Hegman) 5+; sp.gr. 2.60; bulking value 21.7 lb/gal; surf. area 20-24 m^2/g; oil absorp. 40-45; brightness 87-88.5; pH 6.0-7.5 (100 g clay/250 ml water); 1% max. moisture.

Huber HG90. [J.M. Huber/Engineered Mins.] Hyd. kaolin clay; CAS 1332-58-7; EINECS 296-473-8; water-washed functional filler with higher brightness, very fine particle size, moderate oil and water demand; for adhesives, paint, plastics, and inks; spray-dried beads or slurry; 0.2 μm particle diam.; 0.01% max. on 325 mesh; fineness (Hegman) 5+; sp.gr. 2.60; bulking value 21.7 lb/gal; surf. area 20-

24 m²/g; oil absorp. 40-45; brightness 90-92; pH 6.0-7.5 (100 g clay/250 ml water); 1% max. moisture.

Huber N110. [J.M. Huber/Engineered Carbons] Carbon black; CAS 1333-86-4; EINECS 215-609-9; for the rubber industry; 20 nm avg. particle diam.; 0.05% max 325-mesh residue; sp.gr. 1.8; pour dens. 22 lb/ft³.

Huber N220. [J.M. Huber/Engineered Carbons] Carbon black; CAS 1333-86-4; EINECS 215-609-9; for the rubber industry; 22 nm avg. particle diam.; 0.05% max 325-mesh residue; sp.gr. 1.8; pour dens. 22 lb/ft³.

Huber N234. [J.M. Huber/Engineered Carbons] Carbon black; CAS 1333-86-4; EINECS 215-609-9; for the rubber industry; 21 nm avg. particle diam.; 0.05% max 325-mesh residue; sp.gr. 1.8; pour dens. 21 lb/ft³.

Huber N299. [J.M. Huber/Engineered Carbons] Carbon black; CAS 1333-86-4; EINECS 215-609-9; for the rubber industry; 23 nm avg. particle diam.; 0.05% max 325-mesh residue; sp.gr. 1.8; pour dens. 21 lb/ft³.

Huber N326. [J.M. Huber/Engineered Carbons] Carbon black; CAS 1333-86-4; EINECS 215-609-9; for the rubber industry; 26 nm avg. particle diam.; 0.05% max 325-mesh residue; sp.gr. 1.8; pour dens. 28 lb/ft³.

Huber N330. [J.M. Huber/Engineered Carbons] Carbon black; CAS 1333-86-4; EINECS 215-609-9; for the rubber industry; 29 nm avg. particle diam.; 0.05% max 325-mesh residue; sp.gr. 1.8; pour dens. 23 lb/ft³.

Huber N339. [J.M. Huber/Engineered Carbons] Carbon black; CAS 1333-86-4; EINECS 215-609-9; for the rubber industry; 25 nm avg. particle diam.; 0.05% max 325-mesh residue; sp.gr. 1.8; pour dens. 21 lb/ft³.

Huber N343. [J.M. Huber/Engineered Carbons] Carbon black; CAS 1333-86-4; EINECS 215-609-9; for the rubber industry; 24 nm avg. particle diam.; 0.05% max 325-mesh residue; sp.gr. 1.8; pour dens. 21 lb/ft³.

Huber N347. [J.M. Huber/Engineered Carbons] Carbon black; CAS 1333-86-4; EINECS 215-609-9; for the rubber industry; 26 nm avg. particle diam.; 0.05% max 325-mesh residue; sp.gr. 1.8; pour dens. 21 lb/ft³.

Huber N351. [J.M. Huber/Engineered Carbons] Carbon black; CAS 1333-86-4; EINECS 215-609-9; for the rubber industry; 28 nm avg. particle diam.; 0.05% max 325-mesh residue; sp.gr. 1.8; pour dens. 21 lb/ft³.

Huber N375. [J.M. Huber/Engineered Carbons] Carbon black; CAS 1333-86-4; EINECS 215-609-9; for the rubber industry; 24 nm avg. particle diam.; 0.05% max 325-mesh residue; sp.gr. 1.8; pour dens. 21 lb/ft³.

Huber N539. [J.M. Huber/Engineered Carbons] Carbon black; CAS 1333-86-4; EINECS 215-609-9; for the rubber industry; 35 nm avg. particle diam.; 0.05% max 325-mesh residue; sp.gr. 1.8; pour dens. 24 lb/ft³.

Huber N550. [J.M. Huber/Engineered Carbons] Carbon black; CAS 1333-86-4; EINECS 215-609-9; for the rubber industry; 35 nm avg. particle diam.; 0.05% max 325-mesh residue; sp.gr. 1.8; pour dens. 22 lb/ft³.

Huber N650. [J.M. Huber/Engineered Carbons] Carbon black; CAS 1333-86-4; EINECS 215-609-9; for the rubber industry; 60 nm avg. particle diam.; 0.05% max 325-mesh residue; sp.gr. 1.8; pour dens. 23 lb/ft³.

Huber N660. [J.M. Huber/Engineered Carbons] Carbon black; CAS 1333-86-4; EINECS 215-609-9; for the rubber industry; 60 nm avg. particle diam.; 0.05% max 325-mesh residue; sp.gr. 1.8; pour dens. 27 lb/ft³.

Huber N683. [J.M. Huber/Engineered Carbons] Carbon black; CAS 1333-86-4; EINECS 215-609-9; for the rubber industry; 60 nm avg. particle diam.; 0.05% max 325-mesh residue; sp.gr. 1.8; pour dens. 21 lb/ft³.

Huber N762. [J.M. Huber/Engineered Carbons] Carbon black; CAS 1333-86-4; EINECS 215-609-9; for the rubber industry; 88 nm avg. particle diam.; 0.05% max 325-mesh residue; sp.gr. 1.8; pour dens. 32 lb/ft³.

Huber N774. [J.M. Huber/Engineered Carbons] Carbon black; CAS 1333-86-4; EINECS 215-609-9; for the rubber industry; 88 nm avg. particle diam.; 0.05% max 325-mesh residue; sp.gr. 1.8; pour dens. 30 lb/ft³.

Huber N787. [J.M. Huber/Engineered Carbons] Carbon black; CAS 1333-86-4; EINECS 215-609-9; for the rubber industry; 62 nm avg. particle diam.; 0.05% max 325-mesh residue; sp.gr. 1.8; pour dens. 28 lb/ft³.

Huber N990. [J.M. Huber/Engineered Carbons] Carbon black; CAS 1333-86-4; EINECS 215-609-9; for the rubber industry; 320 nm avg. particle diam.; 0.05% max 325-mesh residue; sp.gr. 1.8; pour dens. 40 lb/ft³.

Huber S212. [J.M. Huber/Engineered Carbons] Carbon black; CAS 1333-86-4; EINECS 215-609-9; for the rubber industry; 0.05% max 325-mesh residue; sp.gr. 1.8; pour dens. 27 lb/ft³.

Huber S315. [J.M. Huber/Engineered Carbons] Carbon black; CAS 1333-86-4; EINECS 215-609-9; for the rubber industry; 0.05% max 325-mesh residue; sp.gr. 1.8; pour dens. 28 lb/ft³.

Huber SM. [J.M. Huber/Engineered Mins.] Mica, surface-treated organo-functional muscovite; CAS 12001-26-2; filler for improved flexural, tensile, and impact properties in polyolefin plastics; see also Huber WG-1; wh. powd., flakes; 82.0-87.0% -325 mesh; negligible odor; negligible sol. in water; sp.gr. 2.80; bulk dens. 11.0-13.0 lb/ft³; m.p. ≈ 1000 C; pH 7.0-8.0 (28% solids); flash pt. none.

Huber WG-1. [J.M. Huber/Engineered Mins.] Mica, wet-ground muscovite; CAS 12001-26-2; filler for plastics (reinforcement in many polyolefins, nylons, thermoplastic and thermosetting polyesters, ABS, etc.), rubber, coatings, and pearlescent pigment applics.; off-wh.; 87.0-92.0% -325 mesh; negligible sol. in water; sp.gr. 2.80; bulk dens. 11.0-13.0 lb/ft^3; m.p. ≈ 1000 C; pH 7.0-8.0 (28% solids); flash pt. none.

Huber WG-2. [J.M. Huber/Engineered Mins.] Mica, wet-ground muscovite; CAS 12001-26-2; filler for plastics (reinforcement in many polyolefins, nylons, thermoplastic and thermosetting polyesters, ABS, etc.), rubber, coatings, and pearlescent pigment applics.; off-wh. powd., flake; 82.0-87.0% -325 mesh; negligible odor; negligible sol. in water; sp.gr. 2.80; bulk dens. 11.0-13.0 lb/ft^3; m.p. ≈ 1000 C; pH 7.0-8.0 (28% solids); flash pt. none.

Hubercarb® Q 1. [J.M. Huber/Engineered Mins.] Calcium carbonate; filler/extender for plastics, caulks/sealants, rubber, adhesives, glass, ceramics, paper, cleansers, paints/coatings, pesticides, asphalt, drilling mud, rice polishing, environmental cleanup; ultrafine powd.; 1 μ median particle size; 99.9% finer than 10 μ, 90% finer than 3 μ; fineness (Hegman) 5.5; bulk dens. 30 lb/ft^3 (loose); oil absorp. 20; brightness 90; 96.5% $CaCO_3$, 0.6% moisture.

Hubercarb® Q 1T. [J.M. Huber/Engineered Mins.] Calcium carbonate, surface modified with 1% stearates; filler/extender for plastics, caulks/sealants, rubber, adhesives, glass, ceramics, paper, cleansers, paints/coatings, pesticides, asphalt, drilling mud, rice polishing, environmental cleanup; surf. modification provides more rapid and complete dispersion in polymers, higher potential filler loadings, reduced abrasion to equip., less sensitivity to moisture; oil absorp. 21; 96.5% $CaCO_3$.

Hubercarb® Q 2. [J.M. Huber/Engineered Mins.] Calcium carbonate; filler/extender for plastics, caulks/sealants, rubber, adhesives, glass, ceramics, paper, cleansers, paints/coatings, pesticides, asphalt, drilling mud, rice polishing, environmental cleanup; fine powd.; 2 μ median particle size; 99.9% finer than 10 μ, 99% finer than 8 μ; fineness (Hegman) 6.5; bulk dens. 44 lb/ft^3 (loose); oil absorp. 18; brightness 90; 96.5% $CaCO_3$, 0.25% moisture.

Hubercarb® Q 2T. [J.M. Huber/Engineered Mins.] Calcium carbonate, surface modified with 1% stearates; filler/extender for plastics, caulks/sealants, rubber, adhesives, glass, ceramics, paper, cleansers, paints/coatings, pesticides, asphalt, drilling mud, rice polishing, environmental cleanup; surf. modification provides more rapid and complete dispersion in polymers, higher potential filler loadings, reduced abrasion to equip., less sensitivity to moisture; oil absorp. 16; 96.5% $CaCO_3$.

Hubercarb® Q 3. [J.M. Huber/Engineered Mins.] Calcium carbonate; filler/extender for plastics, caulks/sealants, rubber, adhesives, glass, ceramics, paper, cleansers, paints/coatings, pesticides, asphalt, drilling mud, rice polishing, environmental cleanup; fine powd.; 3 μ median particle size; 99.99% finer than 20 μ, 97% finer than 10 μ; fineness (Hegman) 6; bulk dens. 40 lb/ft^3 (loose); oil absorp. 18; brightness 89; 96.5% $CaCO_3$, 0.2% moisture.

Hubercarb® Q 3T. [J.M. Huber/Engineered Mins.] Calcium carbonate, surface modified with 1% stearates; filler/extender for plastics, caulks/sealants, rubber, adhesives, glass, ceramics, paper, cleansers, paints/coatings, pesticides, asphalt, drilling mud, rice polishing, environmental cleanup; surf. modification provides more rapid and complete dispersion in polymers, higher potential filler loadings, reduced abrasion to equip., less sensitivity to moisture; oil absorp. 15; 96.5% $CaCO_3$.

Hubercarb® Q 4. [J.M. Huber/Engineered Mins.] Calcium carbonate; filler/extender for plastics, caulks/sealants, rubber, adhesives, glass, ceramics, paper, cleansers, paints/coatings, pesticides, asphalt, drilling mud, rice polishing, environmental cleanup; fine powd.; 4 μ median particle size; 99.99% thru 325 mesh, 95% finer than 10 μ; fineness (Hegman) 6; bulk dens. 40 lb/ft^3 (loose); oil absorp. 17; brightness 88; 96.5% $CaCO_3$, 0.2% moisture.

Hubercarb® Q 6. [J.M. Huber/Engineered Mins.] Calcium carbonate; filler/extender for plastics, caulks/sealants, rubber, adhesives, glass, ceramics, paper, cleansers, paints/coatings, pesticides, asphalt, drilling mud, rice polishing, environmental cleanup; fine powd.; 6 μ median particle size; 99.99% thru 325 mesh, 95% finer than 20 μ; fineness (Hegman) 4; bulk dens. 45 lb/ft^3 (loose); oil absorp. 16; brightness 87; 96.5% $CaCO_3$, 0.15% moisture.

Hubercarb® Q 6-20. [J.M. Huber/Engineered Mins.] Calcium carbonate; filler/extender for plastics, caulks/sealants, rubber, adhesives, glass, ceramics, paper, cleansers, paints/coatings, pesticides, asphalt, drilling mud, rice polishing, environmental cleanup; gran.; 95% thru 8 mesh, 60% thru 12 mesh; bulk dens. 90 lb/ft^3 (loose); 96.5% $CaCO_3$, 0.05% moisture.

Hubercarb® Q 20-60. [J.M. Huber/Engineered Mins.] Calcium carbonate; filler/extender for plastics, caulks/sealants, rubber, adhesives, glass, ceramics, paper, cleansers, paints/coatings, pesticides, asphalt, drilling mud, rice polishing, environmental cleanup; gran.; 100% thru 12 mesh, 98% thru 20 mesh; bulk dens. 85 lb/ft^3 (loose); 96.5% $CaCO_3$, 0.05% moisture.

Hubercarb® Q 40-200. [J.M. Huber/Engineered Mins.] Calcium carbonate; filler/extender for plastics, caulks/sealants, rubber, adhesives, glass, ceramics, paper, cleansers, paints/coatings, pesticides, asphalt, drilling mud, rice polishing, environmental cleanup; gran.; 98% thru 40 mesh, 70% thru 60 mesh; bulk dens. 85 lb/ft^3 (loose); 96.5% $CaCO_3$, 0.05% moisture.

Hubercarb® Q 60. [J.M. Huber/Engineered Mins.] Calcium carbonate; filler/extender for plastics, caulks/sealants, rubber, adhesives, glass, ceramics, paper, cleansers, paints/coatings, pesticides, asphalt, drilling mud, rice polishing, environmental cleanup; med. fine powd.; 20 μ median particle size; 99.5% thru 40 mesh, 97% thru 100 mesh; bulk dens. 55 lb/ft^3 (loose); oil absorp. 12; brightness 80; 96.5% $CaCO_3$, 0.05% moisture.

Hubercarb® Q 100. [J.M. Huber/Engineered Mins.] Calcium carbonate; filler/extender for plastics, caulks/sealants, rubber, adhesives, glass, ceramics, paper, cleansers, paints/coatings, pesticides, asphalt, drilling mud, rice polishing, environmental cleanup; med. fine powd.; 24 μ median particle size; 99% thru 60 mesh, 98% thru 100 mesh; bulk dens. 55 lb/ft^3 (loose); oil absorp. 12; brightness 83; 96.5% $CaCO_3$, 0.05% moisture.

Hubercarb® Q 200. [J.M. Huber/Engineered Mins.] Calcium carbonate; filler/extender for plastics, caulks/sealants, rubber, adhesives, glass, ceramics, paper, cleansers, paints/coatings, pesticides, asphalt, drilling mud, rice polishing, environmental cleanup; med. fine powd.; 19 μ median particle size; 99.9% thru 60 mesh, 99% thru 100 mesh; bulk dens. 55 lb/ft^3 (loose); oil absorp. 12; brightness 84; 96.5% $CaCO_3$, 0.05% moisture.

Hubercarb® Q 200T. [J.M. Huber/Engineered Mins.] Calcium carbonate, surface modified with 1% stearates; filler/extender for plastics, caulks/sealants, rubber, adhesives, glass, ceramics, paper, cleansers, paints/coatings, pesticides, asphalt, drilling mud, rice polishing, environmental cleanup; surf. modification provides more rapid and complete dispersion in polymers, higher potential filler loadings, reduced abrasion to equip., less sensitivity to moisture; oil absorp. 12; 96.5% $CaCO_3$.

Hubercarb® Q 325. [J.M. Huber/Engineered Mins.] Calcium carbonate; filler/extender for plastics, caulks/sealants, rubber, adhesives, glass, ceramics, paper, cleansers, paints/coatings, pesticides, asphalt, drilling mud, rice polishing, environmental cleanup; med. fine powd.; 13 μ median particle size; 99.7% thru 200 mesh, 99.5% thru 325 mesh; bulk dens. 50 lb/ft^3 (loose); oil absorp. 14; brightness 86; 96.5% $CaCO_3$, 0.1% moisture.

Hubercarb® W 2. [J.M. Huber/Engineered Mins.] Calcium carbonate; filler and extender with low moisture pickup; ideal for moisture-sensitive applics. such as SMC, BMC, TMC, polyethylene film and single-component urethane caulks, and for paints and coatings, rubber, adhesives, ceramics, paper, nonabrasive cleaners; powd.; 2 μ median particle size; 100% finer than 10 μ; 89% finer than 5 μ; fineness (Hegman) 6.5; bulk dens. 50 lb/ft^3 (loose); oil absorp. 17; brightness 93; 99.3% $CaCO_3$; 0.07% moisture.

Hubercarb® W 2T. [J.M. Huber/Engineered Mins.] Calcium carbonate, surface modified with 1% stearates; filler and extender with low moisture pickup; ideal for moisture-sensitive applics. such as SMC, BMC, TMC, polyethylene film and single-component urethane caulks, and for paints and coatings, rubber, adhesives, ceramics, paper, nonabrasive cleaners; surface modification provides better dispersion in polymers, higher potential filler loadings, reduced abrasion to equip., less sensitivity to moisture; powd.; 2 μ median particle size; 100% finer than 10 μ, 89% finer than 5 μ; fineness (Hegman) 6.5; bulk dens. 50 lb/ft^3 (loose); oil absorp. 16; brightness 93; 99.3% $CaCO_3$; 0.07% moisture.

Hubercarb® W 3. [J.M. Huber/Engineered Mins.] Calcium carbonate; filler and extender with low moisture pickup; ideal for moisture-sensitive applics. such as SMC, BMC, TMC, polyethylene film and single-component urethane caulks, and for paints and coatings, rubber, adhesives, ceramics, paper, nonabrasive cleaners; powd.; 3 μ median particle size; 99.9% finer than 20 μ; 94% finer than 8 μ; fineness (Hegman) 6; bulk dens. 50 lb/ft^3 (loose); oil absorp. 15; brightness 93; 99.3% $CaCO_3$; 0.07% moisture.

Hubercarb® W 3T. [J.M. Huber/Engineered Mins.] Calcium carbonate, surface modified with 1% stearates; filler and extender with low moisture pickup; ideal for moisture-sensitive applics. such as SMC, BMC, TMC, polyethylene film and single-component urethane caulks, and for paints and coatings, rubber, adhesives, ceramics, paper, nonabrasive cleaners; surface modification provides better dispersion in polymers, higher potential filler loadings, reduced abrasion to equip., less sensitivity to moisture; powd.; 3 μ median particle size; 99.9% finer than 20 μ, 94% finer than 8 μ; fineness (Hegman) 6; bulk dens. 50 lb/ft^3 (loose); oil absorp. 15; brightness 93; 99.3% $CaCO_3$; 0.07% moisture.

Hubercarb® W 4. [J.M. Huber/Engineered Mins.] Calcium carbonate; filler and extender with low moisture pickup; ideal for moisture-sensitive applics. such as SMC, BMC, TMC, polyethylene film and single-component urethane caulks, and for paints and coatings, rubber, adhesives, ceramics, paper, nonabrasive cleaners; powd.; 4 μ median particle size; 99.5% fthr4u 325 mesh; 92% finer than 20 μ; fineness (Hegman) 6; bulk dens. 55 lb/ft^3 (loose); oil absorp. 13; brightness 92; 99.3% $CaCO_3$; 0.05% moisture.

Hubersil® 162. [J.M. Huber/Chems.] Silica, precipitated amorphous; reinforcing filler, carrier for rubber applics. incl. hose, belting, molded and extruded parts, tires, footwear, wire and cable, thermolastic elastomers, caulks and sealants; wh. powd.; 12 μ avg. particle size; dens. 2.0 gml; surf. area 160 m^2/g; oil absorp. 190 cc/100 g; ref. index 1.44; pH 7.0 (5%).

H-White. [J.M. Huber/Engineered Mins.] Calcium carbonate; functional filler extender; used in coatings, plastic and rubber fillers, building prods., ceramics flux, paper fillers, adhesives, cleaning compds., polishes; wh. irreg., uniaxial particles; particle size 3.0 μ 99.99% thru 500 mesh; water-sol.; sp.gr. 2.71; dens. 22.6 lb/solid gal, 45 lb/ft^3; oil absorp. 16 lb oil/100 lb; ref. index 1.6; hardness (Mohs) 3; 97.7% $CaCO_3$.

Hyamine® 1622 50%. [Lonza] Benzethonium chloride; CAS 121-54-0; EINECS 204-479-9; cationic; germicide, disinfectant for restaurant, veterinary, surgical, pharmaceutical topicals, industrial/household sanitizers; antistat, bacteriostat on fabrics; preservative for starch, glue, casein; cocatalyst for curing polyesters; deodorant; swimming pool algicide; lt. amber liq.; sol. in water, lower alcohols, glycols, ethoxyethanol, tetrachloroethane; misc. with ethylene dichloride, CCl_4; sp.gr. 1.03; dens. 8.56 lb/gal; pour pt. 25 F; flash pt. (Seta) 110 F; pH 8-10 (5%); surf. tens. 30 dynes/cm (0.1%); 50% act.; *Toxicology:* LD50 (oral, rat) 420 mg/kg, (dermal, rabbit) > 3 g/kg; may cause eye and skin irritation.

Hy Dense Calcium Stearate HP Gran. [Mallinckrodt] Calcium stearate; CAS 1592-23-0; EINECS 216-472-8; release agent in rigid PVC processing; FDA accepted; gran.

Hy Dense Calcium Stearate RSN Powd. [Mallinckrodt] Calcium stearate; CAS 1592-23-0; EINECS 216-472-8; internal/external lubricant, release agent in rigid and flexible PVC processing, PP, ABS, cellulosics, SMC/BMC polyesters, color concs.; FDA accepted; powd.

Hy Dense Zinc Stearate XM Powd. [Mallinckrodt] Zinc stearate; CAS 557-05-1; EINECS 209-151-9; internal/external lubricant for SMC/BMC polyester, phenolics, PVC; release agent for use where flow and green strength are critical and in intricate die cavities; FDA accepted; powd.

Hy Dense Zinc Stearate XM Ultra Fine. [Mallinckrodt] Zinc stearate; CAS 557-05-1; EINECS 209-151-9; internal/external lubricant for alkyd, melamine, color concs., phenolics, PVC; release agent used where sintered strength and dimensional stability are important; FDA accepted; powd.

Hydramax™ HM-B8, HM-B8-S. [Climax Performance] Magnesium hydroxide; CAS 1309-42-8; EINECS 215-170-3; fire retardant, smoke suppressant for engineering plastics, rubber, high-temp. adhesives, coatings, other high-temp. polymer systems; suitable for flexbile and rigid PVC, nitrile rubbers, neoprene, polyolefins, EPDM, SBR, EPR, urethanes, EVA copolymners, nylons, other engineering polymer systems; HM-B8-S preferred for high loadings in engineering resins; wh. free-flowing powd.; 1.1 μ median particle size; 99% < 8 μ; sp.gr. 2.38; surf. area 8 m^2/g; oil absorp. 30; ref. index 1.58; hardness (Mohs) 2.5; 0.5% moisture.

Hydramax™ HM-C9, HM-C9-S. [Climax Performance] Magnesium calcium carbonate hydroxide; fire retardant, smoke suppressant for engineering thermoplastics, coatings, other high-temp. polymer systems; suitable for cross-linked PE, PP, PBT, other engineering polymers; wh. free-flowing powd.; 0.9 μ median particle size; 99% < 9 μ; sp.gr. 2.60; surf. area 19 m^2/g; oil absorp. 40; ref. index 1.6; 0.5% moisture.

Hydrex®. [J.M. Huber/Chems.] Sodium aluminum silicate; high brightness filler; 4 μ avg. particle size; surf. area 75 m^2/g; oil absorp. 115 cc/100 g; pH 9.7.

Hydrex® R. [J.M. Huber/Chems.] Sodium magnesium aluminosilicate, precipitated amorphous; filler for the rubber industry; wh. powd.; 3 μ avg. particle size; 0.2% max. 325 mesh residue; dens. 2.1 g/ml; surf. area 70 m^2/g; oil absorp. 125 cc/100 g; brightness 98.5%; ref. index 1.55; pH 10.5 (20%).

Hydrite 121-S. [Dry Branch Kaolin] Kaolinite; CAS 1332-58-7; EINECS 296-473-8; extender pigment featuring intermediate particle size, intermediate water and oil demand; offers stain removal, hiding power and suspension, flow control; used in interior emulsions and primer paints and polyester premix plastics; slurry, particle size 1.5 μ; sp.gr. 2.58; dens. 21.66 lb/gal; visc. 285 cps; oil absorp. (Gardner Coleman) 39%; ref. index 1.56; pH 6.5-7.5; 38.38% Al_2O_3, 45.30% SiO_2.

Hydrobrite 200PO. [Witco/Petroleum Spec.] Wh. min. oil USP; binder, carrier, conditioner, deformer, dispersant, extender, heat transfer agent, lubricant, moisture barrier, plasticizer, protective agent and/or softener in adhesives, agric. chems., cleaning, pkg. plastics, textiles; FDA 21CFR §172.878, 178.3620(a); odorless, tasteless; sp.gr. 0.845-0.885; visc. 33.5-46 cst (40 C); distill. pt. 243 C min. (10 mm, 2.5%); pour pt. 15 F max.

Hydrobrite 300PO. [Witco/Petroleum Spec.] Wh. min. oil USP; binder, carrier, conditioner, deformer, dispersant, extender, heat transfer agent, lubricant, moisture barrier, plasticizer, protective agent and/or softener in adhesives, agric. chems., cleaning, pkg. plastics, textiles; FDA 21CFR §172.878, 178.3620(a); odorless, tasteless; sp.gr. 0.850-0.880; visc. 48-60 cSt (40 C); distill. pt. 260 C min. (10 mm, 2.5%); pour pt. 10 F.

Hydrobrite 380PO. [Witco/Petroleum Spec.] Wh. min. oil USP; binder, carrier, conditioner, deformer, dispersant, extender, heat transfer agent, lubricant, moisture barrier, plasticizer, protective agent and/or softener in adhesives, agric. chems., cleaning, pkg. plastics, textiles; FDA 21CFR §172.878, 178.3620(a); odorless, tasteless; m.w. 485; sp.gr. 0.858-0.873; visc. 69-82 cSt (40 C); distill. pt. 280 C min. (10 mm, 2.5%); pour pt. 20 F max.; flash pt. (COC) 495 F.

Hydrobrite 550PO. [Witco/Petroleum Spec.] Wh. min. oil USP; binder, carrier, conditioner, deformer, dispersant, extender, heat transfer agent, lubricant, moisture barrier, plasticizer, protective agent and/or softener in adhesives, agric. chems., cleaning, pkg. plastics, textiles; FDA 21CFR §172.878, 178.3620(a); odorless, tasteless; m.w. 540; sp.gr. 0.860-0.880; visc. 100-125 cSt (40 C); distill. pt. 295 C min. (10 mm, 2.5%); pour pt. 15 F max.; flash pt. (COC) 490 F min.; ref. index 1.470-1.478.

Hydrocerol® BIF. [Boehringer Ingelheim; Henley] Inorg. blowing system with special particle structure; endothermal blowing agent for foaming of thermoplastics in inj. molding and extrusion esp. extrusion of foamed profiles and plates of rigid PVC; German food and FDA compliances; wh. free-flowing fine powd.; dec. 120 C; *Storage:* exc. storage stability when kept in cool, dry place.

Hydrocerol® BIH. [Boehringer Ingelheim; Henley] Blowing agent for foaming of thermoplastics in inj. molding and extrusion; high gas yield at lower temps.; German food and FDA compliance; wh. free-flowing powd.; dec. 140 C; *Storage:* exc. storage stability when kept in cool, dry place.

Hydrocerol® BIH 10. [Boehringer Ingelheim; Henley] Multicomponent system in PE carrier; blowing agent for foaming of thermoplastics in inj. molding and extrusion; high gas yield at lower temps.; German food and FDA compliance; wh. cylindrical gran.; dec. 140 C; 10% act.

Hydrocerol® BIH 25. [Boehringer Ingelheim; Henley] Multicomponent system in PE carrier; blowing agent for foaming of thermoplastics in inj. molding and extrusion; high gas yield at lower temps.; German food and FDA compliance; wh. cylindrical gran.; dec. 140 C; 25% act.

Hydrocerol® BIH 40. [Boehringer Ingelheim; Henley] Multicomponent system in PE wax carrier; blowing agent for foaming of thermoplastics in inj. molding and extrusion; high gas yield at lower temps.; German food and FDA compliance; wh. cylindrical gran.; dec. 140 C; 40% act.

Hydrocerol® BIH 70. [Boehringer Ingelheim; Henley] Multicomponent system in PE wax carrier; blowing agent for foaming of thermoplastics in inj. molding and extrusion; high gas yield at lower temps.; German food and FDA compliance; wh. cylindrical gran.; dec. 140 C; 70% act.

Hydrocerol® BIN. [Henley] Blowing agent for foaming of thermoplastics in inj. molding and extrusion esp. extrusion of foamed profiles and plates from rigid PVC; German food and FDA compliance; wh. free-flowing powd.; dec. 120 C; *Storage:* exc. storage stability when kept in cool, dry place.

Hydrocerol® CF. [Boehringer Ingelheim; Henley] Multicomponent system; blowing and nucleating agent for fine-celled foaming of thermoplastics in inj. molding and extrusion, active cell regulation to achieve a uniform, fine-celled foam structure in direct gassed systems; German food and FDA compliance; wh. free-flowing powd.; dec. 150 C; *Storage:* exc. storage stability when kept in cool, dry place.

Hydrocerol® CF 5. [Boehringer Ingelheim; Henley] Multicomponent system in LDPE carrier; blowing and nucleating agent for fine-celled foaming of thermoplastics in extrusion, active cell regulation to achieve a uniform, fine-celled foam structure in direct gassed systems; German food and FDA compliance; wh. cylindrical gran.; dec. 150 C; 5% act.

Hydrocerol® CF 20E. [Boehringer Ingelheim; Henley] Multicomponent system in LDPE carrier; blowing and nucleating agent for fine-celled foaming of thermoplastics in inj. molding and extrusion, active cell regulation to achieve a uniform, fine-celled foam structure in direct gassed systems; German food and FDA compliance; wh. cylindrical gran.; dec. 150 C; 20% act.

Hydrocerol® CF 40E. [Boehringer Ingelheim; Henley] Multicomponent system in LDPE carrier; blowing and nucleating agent for fine-celled foaming of thermoplastics in inj. molding and extrusion, active cell regulation to achieve a uniform, fine-celled foam structure in direct gassed systems; German food and FDA compliance; wh. cylindrical gran.; dec. 150 C; 40% act.

Hydrocerol® CF 40S. [Boehringer Ingelheim; Henley] Multicomponent system in PS carrier; blowing and nucleating agent for fine-celled foaming of thermoplastics in inj. molding and extrusion, active cell regulation to achieve a uniform, fine-celled foam structure in direct gassed systems; esp. recommended for ABS, PS; lt. cylindrical gran.; dec. 150 C; 40% act.

Hydrocerol® CF-60V. [Henley] Multicomponent system in EVA carrier; chemical blowing and nucleating agent for fine-celled foaming of thermoplastics in inj. molding and extrusion, plus mold release for complicated plastic parts; active cell regulation; German food and FDA compliance; wh. cylindrical gran.; dec. 150 C; 60% act.; *Storage:* exc. storage stability when kept in cool, dry place.

Hydrocerol® CF 70. [Boehringer Ingelheim; Henley] Multicomponent system in PE wax carrier; blowing and nucleating agent for fine-celled foaming of thermoplastics in inj. molding and extrusion, active cell regulation to achieve a uniform, fine-celled foam structure in direct gassed systems; German food and FDA compliance; wh. cylindrical gran.; 70% act.

Hydrocerol® CLM 70. [Boehringer Ingelheim; Henley] Multicomponent system in PE wax carrier; blowing and nucleating agent for fine-celled foaming of thermoplastics in inj. molding and extrusion, active cell regulation to achieve a uniform, fine-celled foam structure; German food and FDA compliance; wh. cylindrical gran.; dec. 160 C; 70% act.

Hydrocerol® Compound. [Boehringer Ingelheim; Henley] Multicomponent system; blowing and nucleating agent for foaming of thermoplastics in inj. molding and extrusion, cell regulation to achieve a uniform, fine-celled foam structure; German food and FDA compliance; wh. free-flowing powd., odorless; sol. in water developing CO_2; dec. 160 C; *Storage:* exc. storage stability when kept in cool, dry place.

Hydrocerol® CP. [Boehringer Ingelheim; Henley] Multicomponent system; chemical blowing agent for fine-celled foaming of thermoplastics in inj. molding and extrusion; German food and FDA compliance; wh. free-flowing powd.; dec. 150 C; *Storage:* exc. storage stability when kept in cool, dry place.

Hydrocerol® CP 70. [Boehringer Ingelheim; Henley] Multicomponent system in PE wax carrier; chemical blowing agent for fine-celled foaming of thermoplastics in inj. molding and extrusion; German food and FDA compliance; wh. cylindrical gran.; dec. 150 C; 70% act.

Hydrocerol® CT-211. [Boehringer Ingelheim; Henley] Citric acid esters; chemical blowing agent for foaming of thermoplastics at higher processing temps.; esp. for PC; wh. pellets; dec. 210 C.

Hydrocerol® CT-219. [Boehringer Ingelheim; Henley] Blowing agent in PS carrier; chemical blowing agent for foaming of thermoplastics in inj. molding, preferably at processing temps. of 210-270 C;

German Plastics Commission, FDA 21CFR §177.1210, 178.3010; yel. cylindrical pellets; dec. 180 C; 70% act.

Hydrocerol® CT-232. [Henley] Azodicarbonamide and Hydrocerol system; blowing agent for foaming of thermoplastic resins at relatively low processing temps. when high gas yield and high wt. reduction are required; predominantly used with PVC and TPE; Germany compliance, FDA 21CFR § 177.1210; ylsh. free-flowing powd.; dec. 120-140 C; *Storage:* exc. storage stability when kept in cool, dry place.

Hydrocerol® CT-1001. [Henley] Multicomponent system in PE carrier; chemical blowing agent for foaming of thermoplastics in inj. molding and extrusion; provides high gas yield at lower temps.; German food and FDA compliance; wh. cylindrical gran.; dec. 140 C; 55% act.

Hydrocerol® HK. [Boehringer Ingelheim; Henley] Citric acid-based system; chemical blowing agent for foaming of thermoplastics in inj. molding and extrusion, preferably at processing temps. of 210-270 C; German food and FDA compliance; wh. free-flowing powd.; dec. 180 C; *Storage:* exc. storage stability when kept in cool, dry place.

Hydrocerol® HK 20. [Henley] Citric acid-based system in PE carrier; chemical blowing agent for foaming of thermoplastics in inj. molding and extrusion, preferably at processing temps. of 210-270 C; German food and FDA compliance; wh. pellets; dec. 180 C; 20% act.

Hydrocerol® HK 40. [Henley] Citric acid-based system in PE carrier; chemical blowing agent for foaming of thermoplastics in inj. molding and extrusion, preferably at processing temps. of 210-270 C; German food and FDA compliance; wh. pellets; dec. 180 C; 40% act.

Hydrocerol® HK 40 B. [Henley] Multicomponent system in special PE carrier with good nucleation props.; foaming and nucleating agent for thermoplastic resins, esp. PET sheets to achieve fine cells and high wt. reduction; German food and FDA compliance; wh. cylindrical gran.; dec. 180-190 C; 40% act.

Hydrocerol® HK 70. [Boehringer Ingelheim; Henley] Citric acid-based system in PE wax carrier; chemical blowing agent for foaming of thermoplastics in inj. molding and extrusion, preferably at processing temps. of 210-270 C; German food and FDA compliance; wh. cylindrical gran.; dec. 180 C; 70% act.

Hydrocerol® HP 20 P. [Henley] Multicomponent system in polymer carrier; chemical blowing agent for fine-celled foaming of thermoplastics in inj. molding and extrusion; esp. for foaming of PP ribbons and films; wh. gran.; dec. 130 C; 20% act.

Hydrocerol® HP 40 P. [Henley] Multicomponent system in polymer carrier; chemical blowing agent for fine-celled foaming of thermoplastics in inj. molding and extrusion; esp. for foaming of PP ribbons and films; wh. gran.; dec. 180 C; 40% act.

Hydrocerol® LBA-38. [Boehringer Ingelheim; Henley] Multicomponent system; endothermic chemical blowing and nucleating agent for thermoplastics (PP, HDPE, PS, ABS, HIPS, PPO, PPE) in extrusion, structural foam molding, and sink removal; nucleation in direct gas processes; food contact approvals in many countries; off-wh. oily liq.; sp.gr. 1.13; dec. 150 C.

Hydrocerol® LBA-40. [Boehringer Ingelheim; Henley] Citric acid deriv.; endothermic chemical blowing agent for PC for inj. molding or extrusion; also suitable for PPO, ABS, other engineering resins; suitable for wt. reduction of molded parts; clear transparent visc. liq.; sp.gr. 1.1; dec. 210 C.

Hydrocerol® LBA-47. [Henley] Multicomponent system; endothermic chemical blowing agent for PP, HDPE, HIPS for inj. molding or extrusion; primarily for wt. reduction of molded parts; also suitable for sink mark removal and nucleation in direct gassing; food contact approvals in many countries; off-wh. oily liq.; sp.gr. 1.13; dec. 140 C.

Hydrocerol® LC. [Boehringer Ingelheim; Henley] Citric acid esters; blowing agent for foaming of thermoplastics at higher processing temps.; esp. suitable for PC; colorless clear sol'n.; dec. 210 C.

Hydrocerol® LC 40 C. [Henley] Citric acid esters in polymer carrier; chemical blowing agent for foaming of thermoplastics at higher processing temps.; esp. for PC; lt. gray-brnsh. cylindrical gran.; dec. 210 C; 40% act.

Hydrocerol® SH. [Boehringer Ingelheim; Henley] Citric acid-based system; blowing agent for foaming of thermoplastics in inj. molding and extrusion, preferably at processing temps. of 230-270 C; German food and FDA compliance; wh. free-flowing powd.; dec. 180 C; *Storage:* exc. storage stability when kept in cool, dry place.

Hydrocerol® SH 70. [Boehringer Ingelheim; Henley] Citric acid-based system in PE wax carrier; blowing agent for foaming of thermoplastics in inj. molding and extrusion, preferably at processing temps. of 230-270 C; German food and FDA compliance; wh. cylindrical gran.; dec. 180 C; 70% act.

Hydrocerol® TAF. [Boehringer Ingelheim; Henley] Activated nucleating agent for foamed plastics; achieves uniform, fine-celled foam in prod. of foamed films and sheets using direct gassing process; German food and FDA compliance; wh. free-flowing powd.; dec. 150 C; *Storage:* exc. storage stability when kept in cool, dry place.

Hydrocerol® TAF 50. [Boehringer Ingelheim; Henley] Activated nucleating agent in PE wax carrier; nucleating agent for extrusion processes for direct gassing PS and polyolefins in prod. of foamed films and sheets; German food and FDA compliance; wh. cylindrical gran.; dec. 150 C; 50% act.

Hydrofol 1800. [Procter & Gamble] Stearic acid, 90%; CAS 57-11-4; EINECS 200-313-4; rubber grade; Gardner 12 color; m.w. 278; acid no. 193-210; iodine no. 10 max.; sapon. no. 193-215.

Hydromax 100. [Amspec] Alumina trihydrate; flame retardant for most polymers.

Hydrosulfite AWC. [Henkel/Functional Prods.] Sodium formaldehyde sulfoxylate; CAS 149-44-0; reducing agent for redox-catalyzed polymerization; powd.; 100% solids.

Hydro Zinc™. [C.P. Hall] Zinc stearate disp.; CAS 557-05-1; EINECS 209-151-9; provides antistick coating for rubber and other polymers, slabs or continuous sheets; cream-colored paste; disp. in water; sp.gr. 0.84 ± 0.1; pH 7 min. (10%); 28 ± 3% solids.

Hymocal. [Calumet Lubricants] Naphthenic process oil; for rubber industry, resin extending, PVC, textiles.

Hyonic GL 400. [Henkel/Functional Prods.] Polyethoxy alkyl phenol; nonionic; surfactant, coemulsifier for specialty polymerizations; liq.; HLB 18.0; 70% act.

Hyonic OP-70. [Henkel/Functional Prods.; Henkel/Organic Prods.] Octoxynol-7; CAS 9002-93-1; nonionic; surfactant, emulsifier, wetting agent, dispersant, detergent, intermediate for textile lubricants, agric. emulsifiable concs.; color development enhancer in latex paint; post-polymerization stabilizer; clear liq.; disp. in water; dens. 1.06 g/ml; HLB 12; pour pt. 11 C; cloud pt. 23 C (1% aq.); pH 7.0 (1% aq.); surf. tens. 27 dynes/cm (0.01%); > 99% act.

Hypermer 1083. [ICI Surf. Am.; ICI Surf. UK] Polymeric; nonionic; surfactant and dispersant for industrial applics.; emulsifier for polymerization; lt. brn. liq.; HLB 5.0.

Hypermer 1599A. [ICI Surf. UK] Polymeric; nonionic; surfactant and dispersant for industrial applics.; polymerization emulsifier; red-brn. liq.; HLB 5.5.

Hypermer 2296. [ICI Surf. Am.; ICI Surf. UK] Polymeric surfactant; nonionic; emulsifier for polymerization, industrial applics.; brn. liq.; HLB 5.0.

Hypermer 2524. [ICI Surf. Am.] Polymeric; nonionic; polymerization emulsifier; brn. liq.; HLB 5.0.

Hypermer A60. [ICI Surf. Am.; ICI Surf. UK] Modified polyester surfactant; nonionic; surfactant and dispersant for industrial applics.; emulsifier for metalworking and hydraulic fluids, polymerization; dk. brn. liq.; HLB 6.0.

Hypermer A65. [ICI Surf. UK] Modified polyester surfactant; nonionic; surfactant and dispersant for industrial applics.; emulsifier for metalworking and hydraulic fluids, polymerization; dk. brn. liq.; HLB 5.0.

Hypermer A95. [ICI Surf. Am.; ICI Surf. UK] Modified polyester surfactant; nonionic; surfactant and dispersant for industrial applics.; emulsifier for metalworking and hydraulic fluids, polymerization; yel. brn. liq.; HLB 6.0.

Hypermer A109. [ICI Surf. Am.; ICI Surf. UK] Modified polyester surfactant; nonionic; surfactant and dispersant for industrial applics.; emulsifier for metalworking and hydraulic fluids, polymerization; lt. brn. liq.; HLB 14.0.

Hypermer A200. [ICI Surf. Am.; ICI Surf. UK] Modified polyester surfactant; nonionic; surfactant and dispersant for industrial applics.; emulsifier for metalworking and hydraulic fluids, polymerization; yel. brn. liq.; HLB 7.0.

Hypermer A256. [ICI Surf. UK] Modified polyester surfactant; nonionic; surfactant and dispersant for industrial applics.; emulsifier for metalworking and hydraulic fluids, polymerization; dk. brn. liq.; HLB 9.0.

Hypermer A394. [ICI Surf. Am.; ICI Surf. UK] Modified polyester surfactant; nonionic; surfactant and dispersant for industrial applics.; emulsifier for metalworking and hydraulic fluids, polymerization; yel. brn. liq.; HLB 8.0.

Hypermer A409. [ICI Surf. Am.; ICI Surf. UK] Modified polyester surfactant; nonionic; surfactant and dispersant for industrial applics.; emulsifier for metalworking and hydraulic fluids, polymerization; yel. brn. liq.; HLB 9.0.

Hypermer B239. [ICI Surf. Am.; ICI Surf. UK] Block copolymer; nonionic; surfactant and dispersant for industrial applics.; emulsifier for polymerization; yel. brn. liq.; HLB 6.0.

Hypermer B246. [ICI Surf. Am.; ICI Surf. UK] Block copolymer; nonionic; surfactant and dispersant for industrial applics.; emulsifier for polymerization; red brn. waxy solid; HLB 6.0.

Hypermer B259. [ICI Surf. Am.; ICI Surf. UK] Block copolymer; nonionic; surfactant and dispersant for industrial applics.; emulsifier for polymerization; yel. brn. waxy solid; HLB 12.0.

Hypermer B261. [ICI Surf. Am.; ICI Surf. UK] Block copolymer; nonionic; surfactant and dispersant for industrial applics.; emulsifier for polymerization; red brn. waxy solid; HLB 8.0.

HyQual NF. [Mallinckrodt] Magnesium stearate; CAS 557-04-0; EINECS 209-150-3; internal/external lubricant for rigid vinyls (with calcium stearate), cellulose acetate, ABS; solid.

Hystrene® 3675. [Witco Corp.] 75% dimer acid; CAS 61788-89-4; corrosion inhibitor, intermediate; derivs. used as syn. lube components, corrosion inhibitors for petrol. processing, as extenders and crosslinking agents for high polymeric systems; mildness additive in detergents; Gardner 9 max. liq.; visc. 9000 cSt; acid no. 189-197; sapon. no. 189-199; 100% conc.

Hystrene® 3675C. [Witco Corp.] 75% Dimer acid, 3% monomer [dimer acid (CTFA)] CAS 61788-89-4; corrosion inhibitor, intermediate; derivs. used as syn. lube components, corrosion inhibitors for petrol. processing, as extenders and crosslinking agents for high polymeric systems; mildness additive in detergents; Gardner 9 max. liq.; visc. 7500 cSt; acid no. 189-197; sapon. no. 189-199; 100% conc.

Hystrene® 3680. [Witco Corp.] 80% Dimer acid; CAS 61788-89-4; corrosion inhibitor, intermediate; derivs. used as syn. lube components, corrosion inhibitors for petrol. processing, as extenders and crosslinking agents for high polymeric systems; mildness additive for detergents; Gardner 8 max. liq.; visc. 8500 cSt; acid no. 190-197; sapon. no. 190-199; 100% conc.

Hystrene® 3695. [Witco Corp.] 95% Dilinoleic acid; CAS 61788-89-4; corrosion inhibitor, intermediate;

derivs. used as syn. lube components, corrosion inhibitors for petrol. processing, as extenders and crosslinking agents for high polymeric systems; mildness additive in detergents; Gardner 6 max. liq.; visc. 7500 cSt; acid no. 190-196; sapon. no. 190-202.

Hystrene® 5016 NF. [Witco/H-I-P] Stearic acid NF, triple pressed; CAS 57-11-4; EINECS 200-313-4; stabilizer, lubricant, textile aux., emulsifier, plasticizer, intermediate, used in cosmetics, shampoos, pharmaceuticals, PVC pipe and sheeting; solid; acid no. 206-210; iodine no. 0.5 max.; sapon. no. 206-211; 100% conc.

Hystrene® 9718 NF. [Witco/H-I-P] Stearic acid NF(92%); CAS 57-11-4; EINECS 200-313-4; lubricant, textile aux., emulsifier, plasticizer, intermediate, used in cosmetics, shampoos, pharmaceuticals, PVC pipe and sheeting; solid; acid no. 196-201; iodine no. 0.8 max.; sapon. no. 196-202; 100% conc.

HyTech AX/603. [Mallinckrodt] Aluminum stearate; CAS 7047-84-9; EINECS 230-325-5; internal/external lubricant for nylon, rigid PVC; FDA approved; solid.

HyTech RSN 1-1. [Mallinckrodt] Magnesium stearate; CAS 557-04-0; EINECS 209-150-3; internal/external lubricant for ABS; solid.

HyTech RSN 11-4. [Mallinckrodt] Calcium stearate; CAS 1592-23-0; EINECS 216-472-8; internal/external lubricant for rigid and flexible PVC, PP, ABS, cellulosics, SMC/BMC polyesters, color concs.; FDA accepted; solid.

HyTech RSN 131 HS/Gran. [Mallinckrodt] Zinc stearate; CAS 557-05-1; EINECS 209-151-9; internal/external lubricant for PS, IPS, polyolefins; FDA accepted; solid.

HyTech RSN 248D. [Mallinckrodt] Calcium stearate; CAS 1592-23-0; EINECS 216-472-8; internal/external lubricant for PP, DAP, epoxy, alkyd resins; FDA accepted; solid.

HyTech T/351. [Mallinckrodt] Aluminum stearate; CAS 7047-84-9; EINECS 230-325-5; internal/external lubricant for nylon, rigid PVC, polyesters; FDA approved; solid.

HyTemp NPC-50. [Zeon] Quaternary ammonium compd.; nonpost cure agent used with all HyTemp 4050 series elastomers; sp.gr. 1.01; 50% act.

HyTemp NS-70. [Zeon] Sodium stearate disp.; CAS 822-16-2; EINECS 212-490-5; curative for sulfur-activated HyTemp 4050 and AR 70 series; sp.gr. 1.09; 70% act.

HyTemp SC-75. [Zeon] Amine cure pkg.; fast curative for HyTemp 4050 series; sp.gr. 1.05; 75% act.

HyTemp SO-40. [Zeon] Soap; fast curative for sulfur-activated HyTemp 4050 and AR70 series; liq.; sp.gr. 1.00; 40% act.

HyTemp SR-50. [Zeon] N,N′-Diphenylurea; CAS 102-07-8; nonpost cure retarder used with all HyTemp 4050 series; sp.gr. 1.04; 50% act.

HyTemp ZC-50. [Zeon] Triazine cure pkg.; fast curative for HyTemp AR70 series; sp.gr. 1.14; 50% act.

I

Ice # 2. [Van Den Bergh Foods] Glyceryl stearate and polysorbate 80; nonionic; stabilizer and emulsifier for the frozen desserts providing body, overrun, and dryness; lubricant for textiles and plastics; fabric softener; FDA 21CFR §182.4505, 172.840; ivory bead; HLB 5.2; m.p. 59-63 C; 32-38% alpha monoglyceride; *Storage:* store in cool, dry place; reseal drums between use; 3 mo life for optimum free-flowing props.

Iceberg®. [Burgess Pigment] Aluminum silicate, anhyd., calcined; pigment used in paints, paper, board, rubber and plastics; thin flat plate; 0.15% on 325 mesh; sp.gr. 2.63; oil absorp. 55; ref. index 1.62; pH 5.0-6.0; 51.0-52.4% silica, 42.1-44.3% alumina; 0.5% max. moisture.

Icecap® K. [Burgess Pigment] Aluminum silicate, anhyd., calcined; extender for TiO_2 in coatings where high Hegman grind is required; used in wire and cable, molded and extruded rubber and plastic prods., paper coatings, paints; good electricals, low compr. set, low water absorp.; thin flat plate; 0.009-0.03% on 325 mesh; sp.gr. 2.63; oil absorp. 55; ref. index 1.62pH 5.0-6.0; 51.0-52.4% silica, 42.1-44.3% alumina; 0.5% max. moisture.

Iconol NP-30. [BASF] Nonoxynol-30; CAS 9016-45-9; nonionic; wetting agent, penetrant, detergent, cleaning agent, emulsifier, latex stabilizer, dispersant for industrial, institutional and household cleaning prods.; emulsifier for emulsion polymerization, asphalt emulsions; wh. wax; water-sol.; m.w. 1535; sp.gr. 1.08 (50/25 C); visc. 30 cps (100 C); HLB 17.0; m.p. 41 C; cloud pt. > 100 C (1% aq.); pH 6.0-7.5 (5% aq.); surf. tens. 42 dynes/cm (0.1% aq.); 100% conc.

Iconol NP-30-70%. [BASF] Nonoxynol-30; CAS 9016-45-9; nonionic; wetting agent, penetrant, detergent, cleaning agent, emulsifier, latex stabilizer, dispersant for industrial, institutional and household cleaning prods.; emulsifier for emulsion polymerization, asphalt emulsions; APHA 100 max. liq.; water-sol.; m.w. 1535 (act.); sp.gr. 1.09; visc. 1100 cps; HLB 17.0; pour pt. 0 C; cloud pt. > 100 C (1% aq.); pH 6.0-7.5 (5% aq.); surf. tens. 42 dynes/cm (0.1% aq.); 28.5-31.5% water.

Iconol NP-40. [BASF] Nonoxynol-40; CAS 9016-45-9; nonionic; wetting agent, penetrant, detergent, cleaning agent, emulsifier, latex stabilizer, dispersant for industrial, institutional and household cleaning prods.; emulsifier for emulsion polymerization, asphalt emulsions; wh. wax; water-sol.; m.w. 1975; sp.gr. 1.09 (50/25 C); visc. 40 cps (100 C); HLB 18.0; m.p. 48 C; cloud pt. > 100 (1% aq.); pH 6.0-7.5 (5% aq.); 100% conc.

Iconol NP-40-70%. [BASF] Nonoxynol-40; CAS 9016-45-9; nonionic; wetting agent, penetrant, detergent, cleaning agent, emulsifier, latex stabilizer, dispersant for industrial, institutional and household cleaning prods.; emulsifier for emulsion polymerization, asphalt emulsions; water-sol.; m.w. 1975 (act.); sp.gr. 1.1; visc. 1400 cps; HLB 18.0; pour pt. 7 C; cloud pt. > 100 C (1% aq.); pH 6.0-7.5 (5% aq.); 28.5-31.5% water.

Iconol NP-50. [BASF] Nonoxynol-50; CAS 9016-45-9; nonionic; wetting agent, penetrant, detergent, cleaning agent, emulsifier, stabilizer, dispersant, coemulsifier; for industrial, institutional, and household cleaners; as primary emulsifiers for acrylic and vinyl emulsion polymerization and asphalt emulsions; wh. wax; water-sol.; m.w. 2415; sp.gr. 1.1 (50/25 C); visc. 60 cps (100 C); HLB 19.0; m.p. 49 C; cloud pt. > 100 C (1% aq.); pH 6.0-7.5 (5% aq.); 100% conc.

Iconol NP-50-70%. [BASF] Nonoxynol-50; CAS 9016-45-9; nonionic; wetting agent, penetrant, detergent, cleaning agent, emulsifier, latex stabilizer, dispersant for industrial, institutional and household cleaning prods.; emulsifier for emulsion polymerization, asphalt emulsions; liq.; HLB 19.0; 70% conc.

Iconol NP-70. [BASF] Nonoxynol-70; CAS 9016-45-9; nonionic; wetting agent, penetrant, detergent, cleaning agent, emulsifier, latex stabilizer, dispersant for industrial, institutional and household cleaning prods.; emulsifier for emulsion polymerization, asphalt emulsions; wh. wax; water-sol.; m.w. 3300; sp.gr. 1.1 (50/25 C); visc. 80 cps (100 C); HLB 18.6; pour pt. 51 C; cloud pt. > 100 C (1% aq.); pH 6.0-7.5 (5% aq.); 100% conc.

Iconol NP-70-70%. [BASF] Nonoxynol-70; CAS 9016-45-9; nonionic; wetting agent, penetrant, detergent, cleaning agent, emulsifier, latex stabilizer, dispersant for industrial, institutional and household cleaning prods.; emulsifier for emulsion polymerization, asphalt emulsions; APHA 100 max. liq.; water-sol.; m.w. 3300; sp.gr. 1.1; visc. 3600 cps; HLB 18.6; pour pt. 17 C; cloud pt. > 100 C (1% aq.);

pH 6.0-7.5 (5% aq.); 70% conc.

Iconol NP-100. [BASF] Nonoxynol-100; CAS 9016-45-9; 26027-38-3; nonionic; wetting agent, penetrant, detergent, cleaning agent, emulsifier, latex stabilizer, dispersant for industrial, institutional and household cleaning prods.; emulsifier for emulsion polymerization, asphalt emulsions; Gardner 1 max. flake, cast solid; water-sol.; m.w. 4315; sp.gr. 1.12 (50/25 C); visc. 150 cps (100 C); HLB 19.0; m.p. 52 C; cloud pt. > 100 C (1% aq.); pH 6.0-7.5 (5% aq.); 100% conc.

Iconol NP-100-70%. [BASF] Nonoxynol-100; CAS 9016-45-9; nonionic; wetting agent, penetrant, detergent, cleaning agent, emulsifier, latex stabilizer, dispersant for industrial, institutional and household cleaning prods.; emulsifier for emulsion polymerization, asphalt emulsions; liq.; HLB 19.0; cloud pt. > 100 C (1% aq.); 70% conc.

Iconol OP-10. [BASF] Octoxynol-10; CAS 9002-93-1; nonionic; wetting agent, penetrant, detergent, cleaning agent, emulsifier, latex stabilizer, dispersant for industrial, institutional and household cleaning prods.; emulsifier for emulsion polymerization, asphalt emulsions; APHA 100 max. liq.; water-sol.; m.w. 650; sp.gr. 1.06; visc. 250 cps; HLB 14.0; pour pt. 7 C; cloud pt. 63-67 C (1% aq.); surf. tens. 30 dynes/cm (0.1% aq.); 100% conc.

Iconol OP-30. [BASF] Octoxynol-30; CAS 9002-93-1; nonionic; emulsifier, wetting agent, dispersant, syn. latex stabilizer, detergent in formulating industrial, institutional, and household cleaning prods.; primary emulsifier for acrylic and vinyl emulsion polymerization and for asphalt emulsion systems; solid; HLB 17; cloud pt. > 100 C (1% aq.); surf. tens. 46.4 dynes/cm (0.1%); 100% conc.

Iconol OP-30-70%. [BASF] Octoxynol-30; CAS 9002-93-1; nonionic; emulsifier, wetting agent, dispersant, syn. latex stabilizer, detergent in formulating industrial, institutional, and household cleaning prods.; primary emulsifier for acrylic and vinyl emulsion polymerization and for asphalt emulsion systems; liq.; HLB 17; cloud pt. > 100 C (1% aq.); 70% conc.

Iconol OP-40. [BASF] Octoxynol-40; CAS 9002-93-1; nonionic; wetting agent, penetrant, detergent, cleaning agent, emulsifier, latex stabilizer, dispersant for industrial, institutional and household cleaning prods.; emulsifier for emulsion polymerization, asphalt emulsions; Gardner 1 max. cast solid; water-sol.; m.w. 1970; sp.gr. 1.09 (50/25 C); visc. 40 cps (100 C); HLB 18.0; m.p. 50 C; cloud pt. > 100 C (1% aq.); pH 6.0-7.5 (5% aq.); surf. tens. 42 dynes/cm (0.1% aq.); 100% conc.

Iconol OP-40-70%. [BASF] Octoxynol-40; CAS 9002-93-1; nonionic; wetting agent, penetrant, detergent, cleaning agent, emulsifier, latex stabilizer, dispersant for industrial, institutional and household cleaning prods.; emulsifier for emulsion polymerization, asphalt emulsions; APHA 100 max. liq.; water-sol.; m.w. 1970; sp.gr. 1.1; visc. 500 cps; HLB 18.0; pour pt. -4 C; cloud pt. > 100 C (1% aq.); pH 6.0-7.5 (5% aq.); 70% conc.

Idex™-400. [Idex Int'l.] 40% 3-Iodo-2-propynyl butyl carbamate in inert ingreds.; CAS 55406-53-6; EINECS 259-627-5; broad-spectrum industrial fungicide for aq. and solv. systems such as oleoresinous and latex paints, wood prods., cutting oils, textiles, paper coatings, inks, plastics, adhesives, canvas, and cordage; amber liq., char. odor; sol. in aromatics and alcohols; sp.gr. 1.13-1.16; dens. 9.4-9.7 lb/gal; 40% min. act.; *Storage:* store above 32 F; do not use or store near heat or open flame.

Idex™-1000. [Idex Int'l.] 97% 3-Iodo-2-propynyl butyl carbamate in inert ingreds.; CAS 55406-53-6; EINECS 259-627-5; broad-spectrum industrial fungicide for aq. and solv. systems such as oleoresinous and latex paints, wood prods., cutting oils, textiles, paper coatings, inks, plastics, adhesives, canvas, and cordage; off-wh. cryst. solid; sol. in most aromatic solvs. and alcohols; sp.gr. 1.74-1.77 (20 C); m.p. 65-68 C; 97% min. active, 1% max. moisture; *Storage:* store @ 0-32 C; do not use or store near heat or open flame.

Igepal® CA-877. [Rhone-Poulenc Surf. & Spec.] Nonoxynol-25; CAS 9016-45-9; nonionic; surfactant; primary emulsifier for vinyl acrylate polymerizations; post stabilizer for syn. latexes; dyeing assistant; emulsifier for fats and waxes; HLB 16.6; cloud pt. > 100 C (1% aq.); 70% act.

Igepal® CA-880. [Rhone-Poulenc Surf. & Spec.] Octoxynol-30; CAS 9002-93-1; nonionic; emulsifier for fats and oils, vinyl acetate and acrylate polymerization; post-stabilizer for syn. latices; dyeing assistant; FDA 21CFR §175.105, 176.170, 176.180; solid; HLB 17.4; surf. tens. 38 dynes/cm (at CMC); 100% conc.

Igepal® CA-887. [Rhone-Poulenc Surf. & Spec.] Octoxynol-30 aq. sol'n.; CAS 9002-93-1; nonionic; primary emulsifier for vinyl acrylate polymerizations; post stabilizer for syn. latexes; dyeing assistant; emulsifier for fats and waxes; FDA 21CFR §175.105, 176.170, 176.180; pale yel. liq.; aromatic odor; sp.gr. 1.10; HLB 17.4; cloud pt. > 100 C (1% aq.); flash pt. (PMCC) > 200 F; pour pt. 36 ± 2 F; surf. tens. 39 dynes/cm (0.01%); 70% act.

Igepal® CA-890. [Rhone-Poulenc Surf. & Spec.] Octoxynol-40; CAS 9002-93-1; nonionic; emulsifier for emulsion polymerization, stabilizer for plastics; dyeing assistant; FDA 21CFR §175.105, 176.170, 176.180; off-wh. wax; aromatic odor; sol. in butyl Cellosolve, ethanol, water; sp.gr. 1.08 (50 C); HLB 18.0; cloud pt. > 212 F (1% aq.); flash pt. > 200 F (PMCC); pour pt. 115 ± 2 F; surf. tens. 42 dynes/cm (0.01%); 100% act.

Igepal® CA-897. [Rhone-Poulenc Surf. & Spec.] Octoxynol-40; CAS 9002-93-1; nonionic; primary emulsifier for vinyl acrylate polymerizations; post stabilizer for syn. latexes; dyeing assistant; emulsifier for fats and waxes; FDA 21CFR §175.105, 176.180, 178.3400; pale yel. liq.; aromatic odor; sp.gr. 1.10; HLB 18.0; cloud pt. > 100 C (1% aq.); flash pt. (PMCC) > 200 F; pour pt. 25 ± 2 F; surf. tens. 48 dynes/cm (0.01%); 70% act.

Igepal® CO-430. [Rhone-Poulenc Surf. & Spec.]

Nonoxynol-4; CAS 9016-45-9; nonionic; coemulsifier, stabilizer, detergent, dispersant; plasticizer and antistat for PVAc; freeze/thaw stabilizer for latexes; also for petrol. oils, agric., personal care prods.; intermediate for anionic surfactants; ashless corrosion inhibitor for two-cycle engine oils; biodeg.; FDA, EPA compliance; pale yel. liq., aromatic odor; sol. in kerosene, Stod., naphtha, xylene, butyl Cellosolve, perchloroethylene, ethanol; sp.gr. 1.02; visc. 160-260 cps; solid. pt. -21 F; HLB 8.8; cloud pt. < 20 C (1% aq.); flash pt. (PMCC) > 200 F; pour pt. -16 ± 2 F; 100% act.; *Toxicology:* severe eye irritant; LD50 (oral, rat) 7.4 g/kg.

Igepal® CO-530. [Rhone-Poulenc Surf. & Spec.] Nonoxynol-6; CAS 9016-45-9; nonionic; emulsifier for silicones, detergent, dispersant for agric., petrol., paper, plastics, metalworking industries; de-icing fluid for jet aircraft fuels and automotive gasoline; biodeg.; FDA, EPA compliance; pale yel. liq., aromatic odor; sol. in Stod., naphtha, xylene, butyl Cellosolve, perchloroethylene, ethanol; sp.gr. 1.04; visc. 230-300 cps; HLB 10.8; flash pt. > 200 F (PMCC); pour pt. -26 ± 2 F; surf. tens. 28 dynes/cm (0.01%); Ross-Miles foam 15 mm (0.1% aq., initial); 100% act.; *Toxicology:* severe eye irritant; LD50 (oral, rat) 1.98 g/kg.

Igepal® CO-630. [Rhone-Poulenc Surf. & Spec.] Nonoxynol-9; CAS 9016-45-9; nonionic; detergent, wetting and rewetting agent, corrosion inhibitor, penetrant, emulsifier, dispersant for textile, paper, leather, household/industrial cleaners, agric., paints, metal processing, emulsion cleaning, emulsion polymerization; biodeg.; FDA 21CFR §175.105, 176.210, 178.3400, EPA compliance; almost colorless liq., aromatic odor; sol. in naphtha, xylene, butyl Cellosolve, perchloroethylene, ethanol, water; sp.gr. 1.06; visc. 225-300 cps; HLB 13.0; cloud pt. 126-133 F (1%); flash pt. > 200 F (PMCC); pour pt. 31 ± 2 F; surf. tens. 31 dynes/cm (0.01%); Ross-Miles foam 80 mm (0.1% aq., initial); 100% act.; *Toxicology:* severe eye irritant; LD50 (oral, rat) 3 g/kg.

Igepal® CO-710. [Rhone-Poulenc Surf. & Spec.] Nonoxynol-10.5; CAS 9016-45-9; nonionic; detergent, wetting and rewetting agent, corrosion inhibitor, penetrant, emulsifier for textile, paper, leather, household/industrial cleaners, agric., paints, metal processing, emulsion cleaning, emulsion polymerization; biodeg.; FDA, EPA compliance; pale yel. liq., aromatic odor; sol. in naphtha, xylene, butyl Cellosolve, perchloroethylene, ethanol, water; sp.gr. 1.06; visc. 240-300 cps; HLB 13.6; cloud pt. 158-165 F (1%); flash pt. > 200 F (PMCC); pour pt. 49 ± 2 F; surf. tens. 32 dynes/cm (0.01%); Ross-Miles foam 110 mm (0.1% aq., initial); 100% act.

Igepal® CO-850. [Rhone-Poulenc Surf. & Spec.] Nonoxynol-20; CAS 9016-45-9; nonionic; detergent, wetting agent, dispersant, emulsifier, for industrial cleaners, metalworking fluids, polyester resins; latex stabilizer; demulsifier for crude petrol. oil emulsions; glass mold release agent; in silicone emulsions; FDA 21CFR §175.105, 176.210, 178.3400; pale yel. wax; aromatic odor; sol. in naphtha, xylene, butyl Cellosolve, perchloroethylene, ethanol, water; sp.gr. 1.08 (50 C); HLB 16.0; cloud pt. 68-72 C (1% in 10% NaCl); flash pt. (PMCC) > 200 F; pour pt. 91 ± 2 F; surf. tens. 39 dynes/cm (0.01%); Ross-Miles foam 120 mm (0.1% aq., initial); 100% act.

Igepal® CO-880. [Rhone-Poulenc Surf. & Spec.] Nonoxynol-30; CAS 9016-45-9; nonionic; detergent, wetting agent, dispersant, emulsifier, for industrial cleaners, polyester resins, emulsion polymerization, agric.; latex post-stabilizer; demulsifier for crude petrol. oil emulsions; textile scouring; solubilizer for chlordane, toxaphene, kerosene, and essential oils; FDA 21CFR §175.105, 176.180, 178.3400, EPA compliance; pale yel. wax; aromatic odor; sol. in naphtha, xylene, butyl Cellosolve, perchloroethylene, ethanol, water; sp.gr. 1.08 (50 C); HLB 17.2; cloud pt. 74-78 C (1% in 10% NaCl); flash pt. (PMCC) > 200 F; pour pt. 109 ± 2 F; surf. tens. 43 dynes/cm (0.01%); Ross-Miles foam 120 mm (0.1% aq., initial); 100% act.; *Toxicology:* minimal eye irritant; LD50 (oral, rat) > 16 g/kg.

Igepal® CO-887. [Rhone-Poulenc Surf. & Spec.] Nonoxynol-30; CAS 9016-45-9; nonionic; detergent, wetting agent, dispersant, emulsifier, for industrial cleaners, polyester resins, emulsion polymerization; latex post-stabilizer; demulsifier for crude petrol. oil emulsions; textile scouring; solubilizer for chlordane, toxaphene, kerosene, essential oils; FDA 21CFR §175.105, 176.180, 178.3400, EPA compliance; pale yel. liq.; aromatic odor; sp.gr. 1.09; solid. pt. 28 F; HLB 17.2; cloud pt. 74-78 C (1% in 10% NaCl); flash pt. (PMCC) > 200 F; pour pt. 34 ± 2 F; 70% act.

Igepal® CO-890. [Rhone-Poulenc Surf. & Spec.] Nonoxynol-40; CAS 9016-45-9; nonionic; emulsifier, stabilizer, wetting agent, dyeing assistant for plastics, agric., latexes, floor polishes, etc.; emulsion polymerization surfactant; latex post-stabilizer; textile scouring; solubilizer for chlordane, toxaphene, kerosene, essential oils; FDA 21CFR §175.105, 178.3400, EPA compliance; off-wh. wax; aromatic odor; sol. in ethanol, ethylene dichloride, water; sp.gr. 1.09 (50 C); HLB 17.8; cloud pt. > 100 C (1% aq.); flash pt. (PMCC) > 200 F; pour pt. 112 ± 2 F; Ross-Miles foam 115 mm (0.1% aq., initial); 100% act.

Igepal® CO-897. [Rhone-Poulenc Surf. & Spec.] Nonoxynol-40; CAS 9016-45-9; nonionic; emulsifier, stabilizer, wetting agent, dyeing assistant for plastics, latexes, floor polishes, etc.; emulsion polymerization surfactant; latex post-stabilizer; textile scouring; solubilizer for chlordane, toxaphene, kerosene, essential oils; FDA 21CFR §178.3400, EPA compliance; pale yel. liq.; aromatic odor; sp.gr. 1.10; solid. pt. 40 F; HLB 17.8; cloud pt. > 100 C (1% aq.); flash pt. (PMCC) > 200 F; pour pt. 46 ± 2 F; 70% act.

Igepal® CO-970. [Rhone-Poulenc Surf. & Spec.] Nonoxynol-50; CAS 9016-45-9; nonionic; emulsifier, stabilizer, wetting agent for floor polishes, waxes; polymerization surfactant for vinyl acetate and acrylic emulsions; latex stabilizer; dyeing assistant for acid and azo dyes; effective at elevated temps. or in conc. electrolytes; FDA 21CFR §178.3400, EPA compliance; off-wh. wax; aromatic odor; sol. in ethanol, water, ethylene dichloride; sp.gr. 1.10 (50 C); HLB 18.2; cloud pt. > 100 C (1% aq.); flash pt. (PMCC) > 200 F; pour pt. 114 ± 2 F; Ross-Miles foam 100 mm (0.1% aq., initial); 100% conc.

Igepal® CO-977. [Rhone-Poulenc Surf. & Spec.] Nonoxynol-50; CAS 9016-45-9; nonionic; emulsifier, stabilizer, wetting agent for floor polishes, waxes; polymerization surfactant for vinyl acetate and acrylic emulsions; latex stabilizer; dyeing assistant for acid and azo dyes; effective at elevated temps. or in conc. electrolytes; FDA 21CFR §178.3400, EPA compliance; pale yel. liq.; aromatic odor; sol. in ethanol, water; sp.gr. 1.0 (50 C); HLB 18.2; pour pt. 52 ± 2 F; cloud pt. > 100 C (1% aq.); flash pt. (PMCC) > 200 F; surf. tens. 44 dynes/cm (@ CMC); 70% conc.

Igepal® CO-980. [Rhone-Poulenc Surf. & Spec.] Nonoxynol-70; CAS 9016-45-9; nonionic; stabilizer and dyeing assistant; polymerization emulsifier for vinyl acetate and acrylic emulsions; solid; water-sol.; HLB 18.7; cloud pt. > 212 F (1% aq.); 100% conc.

Igepal® CO-987. [Rhone-Poulenc Surf. & Spec.] Nonoxynol-70; CAS 9016-45-9; nonionic; polymerization emulsifier for vinyl acetate and acrylic emulsions; latex stabilizer; dyeing assistant for acid and azo dyes; emulsifier/stabilizer for floor waxes and polishes; effective at elevated temps. and in conc. electrolytes; off-wh. paste, aromatic odor; water-sol.; sp.gr. 1.10 (50 C); solid. pt. 58 F; HLB 18.6; pour pt. 63 F; cloud pt. > 100 C (1% aq.); flash pt. (PMCC) > 200 F; 70% conc.

Igepal® CO-990. [Rhone-Poulenc Surf. & Spec.] Nonoxynol-100; CAS 9016-45-9; nonionic; polymerization surfactant for vinyl acetate and acrylic emulsions; latex stabilizer; dyeing assistant for acid and azo dyes; emulsifier/stabilizer for floor waxes, polishes; agric. emulsions; effective at elevated temps. and in conc. electrolytes; off-wh. flake; aromatic odor; sol. in ethanol, water; sp.gr. 1.12 (50 C); HLB 19.0; pour pt. 122 ± 2 F; cloud pt. > 100 C (1% aq.); flash pt. (PMCC) > 200 F; surf. tens. 50 dynes/cm (@ CMC); 100% act.

Igepal® CO-997. [Rhone-Poulenc Surf. & Spec.] Nonoxynol-100; CAS 9016-45-9; nonionic; polymerization surfactant for vinyl acetate and acrylic emulsions; latex stabilizer; dyeing assistant for acid and azo dyes; emulsifier/stabilizer for floor waxes, polishes; effective at elevated temps. and in conc. electrolytes; EPA compliance; pale yel. paste; aromatic odor; sp.gr. 1.11; solid. pt. 65 F; HLB 19.0; pour pt. 68 ± 2 F; cloud pt. > 100 C (1% aq.); flash pt. (PMCC) > 200 F; surf. tens. 50 dynes/cm (@ CMC); 70% act. in water.

Igepal® CTA-639W. [Rhone-Poulenc Surf. & Spec.] Alkylphenol ethoxylated (9 EO); nonionic; detergent, wetting; emulsifier in rubber latex emulsion paints; surfactant; liq.; water-sol; surf. tens. 34 dynes/cm (@ CMC); 100% conc.

Igepal® DM-430. [Rhone-Poulenc Surf. & Spec.] Nonyl nonoxynol-7; CAS 9014-93-1; nonionic; emulsifier for agric., emulsion polymerization, leather fat liquoring; EPA compliance; yel. slightly hazy liq.; sol. in naphtha, xylene, perchloroethylene, ethanol; dens. 8.3 lb/gal; sp.gr. 0.995; visc. 13 cps (100 C); HLB 9.5; pour pt. -7 C; cloud pt. 5< 20 C (1% aq.); flash pt. (COC) 500-520 F; 100% act.

Igepal® DM-530. [Rhone-Poulenc Surf. & Spec.] Nonyl nonoxynol-9; CAS 9014-93-1; nonionic; emulsifier for acid cleaners, cutting oils, agric., textile finishing oils, dry cleaning soaps, inks, lacquers, paints, metalworking fluids, pesticides, cosmetic emulsions, emulsion polymerization, leather fat liquoring; EPA compliance; yel. slightly hazy liq.; sol. in naphtha, xylene, perchloroethylene, ethanol; partly sol. water; sp.gr. 1.010; dens. 8.43 lb/gal; visc. 15 cps (100 C); HLB 10.6; pour pt. 0 C; cloud pt. < 20 C (1% aq.); flash pt. (COC) 500-520 F; surf. tens. 29 dynes/cm (0.01%); 100% act.

Igepal® DM-710. [Rhone-Poulenc Surf. & Spec.] Nonyl noxoxynol-15; CAS 9014-93-1; nonionic; detergent, emulsifier, dispersant, wetting agent, antistat for industrial cleaners, textiles, leather, metal cleaners, paper, latex, pesticides; co-emulsifier for syn. latexes; EPA compliance; pale yel. opaque liq.; aromatic odor; sol. in naphtha, xylene, perchloroethylene, ethanol, water; sp.gr. 1.045; dens. 8.72 lb/gal; visc. 19 cps (100 C); HLB 13.0; pour pt. 18 C; cloud pt. 48-52 C (1% aq.); flash pt. (COC) 500-520 F; surf. tens. 29 dynes/cm (0.01%); 100% act.

Igepal® DM-730. [Rhone-Poulenc Surf. & Spec.] Nonyl nonoxynol-24; CAS 9014-93-1; nonionic; detergent, emulsifier for textiles, leather, metal cleaners, paper, latex, pesticides, emulsion polymerization, latex stabilization; pale yel. paste; aromatic odor; sol. in xylene, ethanol, water; sp.gr. 1.049; dens. 8.75 lb/gal; visc. 25 cps (100 C); HLB 15.0; pour pt. 25 C; cloud pt. > 100 C (1% aq.); flash pt. (COC) 500-520 F; surf. tens. 33 dynes/cm (0.01%); 100% act.

Igepal® DM-880. [Rhone-Poulenc Surf. & Spec.] Nonyl nonoxynol-49; CAS 9014-93-1; nonionic; emulsifier, solubilizer for emulsion polymerization, latex stabilization, pesticides, essential oils, cleaners; pale yel. wax; aromatic odor; sol. in xylene, ethanol, water; sp.gr. 1.050 (50 C); visc. 55 cps (100 C); HLB 17.2; pour pt. 47 C; cloud pt. > 100 C (1% aq.); flash pt. (COC) 500-520 F; surf. tens. 38 dynes/cm (0.01%); 100% act.

Igepal® OD-410. [Rhone-Poulenc Surf. & Spec.] Phenoxydiglycol; nonionic; solv. for vinyl, phenolic, polyester, alkyd, NC, and cellulose acetate resins; ingred. of metal cleaners, paint strippers, and cleaning compds. where solvs. and solv. boosters are required; used as ink vehicle; liq.; 100% act.

Imicure® AMI-2. [Air Prods./Perf. Chems.] 2-Methyl imidazole; CAS 693-98-1; EINECS 211-765-7; curing agent for printed circuit board laminates, powd. coatings, adhesives, encapsulation; accelerator for dicyandiamide and anhydrides; wh. powd.; dens. 8.4 lb/gal; m.p. 165 F.

Imicure® EMI-24. [Air Prods./Perf. Chems.] 2-Ethyl 4-methyl imidazole; CAS 931-36-2; EINECS 213-234-5; elevated temp. curing agent for structural adhesives, composites, filament winding, solder-resist. inks, potting compds., molding compds., elec. laminates; accelerator for dicyandiamide or anhydride curing agents; pale yel. liq.; m.w. 110; sp.gr. 0.990; dens. 8.2 lb/gal; visc. 6500 cps; heat distort. temp. 313 F; *Storage:* may solidify as result of thermal or mech. shock; gradual warming above 115 F returns material to stable liq.

Imsil® 1240. [Unimin Spec. Minerals] Microcryst. silica; filler for paints and coatings (traffic, interior/exterior architectural, protective maintenance, marine), buffing and polishing compds., rubber, adhesives and sealants, and plastics; antiblock for LDPE film; GRAS for indirect food contact; wh. powd.; 98.6% thru 325 mesh; 5.8 μ median particle size; sp.gr. 2.65; dens. 22.07 lb/gal; bulking value 0.04531; surf. area 1.08 m^2/g; oil absorp. 29; brightness 85.3; m.p. 1610 C; ref. index 1.54-1.55; pH 6.4; hardness (Mohs) 6.5; 99.075% SiO_2; *Toxicology:* prolonged inh. of dust may cause silicosis; may cause cancer.

Imsil® A-8. [Unimin Spec. Minerals] Microcryst. silica; filler used in finishes, enamels, maintenance paints, plastic film antiblocks, urethane rubber, polishes, adhesives and sealants; wh. fine powd.; 100% thru 325 mesh; 2.1 μ median particle size; fineness (Hegman) 7; sp.gr. 2.65; dens. 22.07 lb/gal; surf. area 2.0 m^2/g; oil absorp. 28; brightness 87.7; m.p. 1610 C; ref. index 1.54-1.55; pH 6.4; hardness (Mohs) 6.5; 99% SiO_2; *Toxicology:* prolonged inh. of dust may cause silicosis; may cause cancer.

Imsil® A-10. [Unimin Spec. Minerals] Microcryst. silica, amorphous; filler for coatings (primers, metal finishes, interior and exterior architectural coatings, marine paints, protective and maintenance coatings), plastic film antiblocks, silicone rubber; wh. fine powd.; 99.0% < 10 μ; 2.3 μ median particle size; fineness (Hegman) 6.75; sol. in hydrofluoric acid; sp.gr. 2.65; dens. 22.07 lb/gal; surf. area 1.9 m^2/g; oil absorp. 28; brightness 87.3; m.p. 1610 C; ref. index 1.54-1.55; pH 6.4; hardness (Mohs) 6.5; 99% SiO_2; *Toxicology:* prolonged inh. of dust may cause silicosis; may cause cancer.

Imsil® A-15. [Unimin Spec. Minerals] Microcryst. silica, amorphous; filler for coatings (interior and exterior architectural, protective, maintenance, marine), silicone and urethane rubbers, adhesives and sealants; wh. fine powd.; 99.0% < 15 μ; 2.5 μ median particle size; fineness (Hegman) 5.75; sol. in hydrofluoric acid; sp.gr. 2.65; dens. 22.07 lb/gal; surf. area 1.6 m^2/g; oil absorp. 29; brightness 86.6; m.p. 1610 C; ref. index 1.54-1.55; pH 6.4; hardness (Mohs) 6.5; 99% SiO_2; *Toxicology:* prolonged inh. of dust may cause silicosis; may cause cancer.

Imsil® A-25. [Unimin Spec. Minerals] Microcryst. silica, amorphous; filler/extender in coatings (interior and exterior architectural, protective, maintenance, marine), silicone rubber molding compds., adhesives and sealants; wh. fine powd.; 99.9+% thru 400mesh; 3.5 μ median particle size; fineness (Hegman) 5.0; sol. in hydrofluoric acid; sp.gr. 2.65; dens. 22.07 lb/gal; surf. area 1.5 m^2/g; oil absorp. 28; brightness 86.3; m.p. 1610 C; ref. index 1.54-1.55; pH 6.1; hardness (Mohs) 6.5; 99% SiO_2; *Toxicology:* prolonged inh. of dust may cause silicosis; may cause cancer.

Imsil® A-30. [Unimin Spec. Minerals] Microcryst. silica; filler, flatting agent for coatings (interior and exterior architectural, protective, maintenance, marine), plastics, adhesives and sealants, urethane rubber compds., buffing/polishing compds.; wh. coarse powd.; 99.6% thru 325 mesh; 5.4 μ median particle size; fineness (Hegman) 3.25; sp.gr. 2.65; dens. 22.07 lb/gal; surf. area 1.1 m^2/g; oil absorp. 28; brightness 85.9; ref. index 1.54-1.55; pH 6.5; hardness (Mohs) 6.5; 99% SiO_2; *Toxicology:* prolonged inh. of dust may cause silicosis; may cause cancer.

Incropol L-7. [Croda Inc.] Laureth-7; nonionic; wetting agent for hard surf. cleaners; emulsifier for oils, emulsion polymerization; liq.; water-sol.; HLB 11.8; Custom

Industrene® 105. [Witco Corp.] Oleic acid; CAS 112-80-1; EINECS 204-007-1; intermediate used in alkyd resins, rubber compding., water repellents, polishes, soaps, abrasives, cutting oils, candles, crayons, emulsifiers, personal care prods.; activator, plasticizer, softener in NR and SR; pigment carrier in dispersions and inks; Gardner 6 liq.; solid. pt. 145 C max.; acid no. 195-204; iodine no. 85-95; sapon. no. 195-205; 100% conc.

Industrene® 143. [Witco Corp.] Tallow acid; CAS 61790-37-2; EINECS 263-129-3; intermediate used in alkyd resins, rubber compding., water repellents, polishes, soaps, abrasives, cutting oils, candles, crayons, emulsifiers; FG grades as lubricant, release agent, binder, defoamer in foods, intermediate for food emulsifiers; Gardner 5 paste; solid. pt. 39-43 C; acid no. 202-206; iodine no. 50-65; sapon. no. 202-207; 100% conc.

Industrene® 205. [Witco Corp.] Oleic acid; CAS 112-80-1; EINECS 204-007-1; intermediate used in alkyd resins, rubber compding., water repellents, polishes, soaps, abrasives, cutting oils, candles, crayons, emulsifiers; FG grades as lubricant, release agent, binder, defoamer in foods, intermediate for food emulsifiers; wh. liq.; acid no. 195-204; iodine no. 85-95; sapon. no. 195-205; 100% conc.

Industrene® 206. [Witco Corp.] Oleic acid NF; CAS 112-80-1; EINECS 204-007-1; intermediate used in alkyd resins, rubber compding., water repellents, polishes, soaps, abrasives, cutting oils, candles, crayons, emulsifiers, personal care prods.; liq.; solid. pt. 6 C max.; acid no. 199-204; iodine no. 95 max.; sapon. no. 200-205; 100% conc.

Industrene® 225. [Witco Corp.] Soya acid, dist.; CAS

67701-08-0; EINECS 269-657-0; intermediate used in alkyd resins, rubber compding., water repellents, polishes, soaps, abrasives, cutting oils, candles, crayons, emulsifiers; FG grades as lubricant, release agent, binder, defoamer in foods, intermediate for food emulsifiers; Gardner 3-4 liq.; solid. pt. 25 C max.; acid no. 195-201; iodine no. 135-145; sapon. no. 197-204; 100% conc.

Industrene® 226. [Witco Corp.] Soya acid, dist.; CAS 67701-08-0; intermediate used in alkyd resins, rubber compding., water repellents, polishes, soaps, abrasives, cutting oils, candles, crayons, emulsifiers; Gardner 3-4 liq.; solid. pt. 26 C max.; acid no. 195-203; iodine no. 125-135; sapon. no. 197-204; 100% conc.

Industrene® 325. [Witco Corp.] Dist. coconut acid; CAS 61788-47-4; EINECS 262-978-7; intermediate used in alkyd resins, rubber compding., water repellents, polishes, soaps, abrasives, cutting oils, candles, crayons, emulsifiers, personal care prods.; paste; acid no. 265-277; iodine no. 6-15; sapon. no. 265-278; 100% conc.

Industrene® 328. [Witco Corp.] Stripped coconut acid; CAS 61788-47-4; EINECS 262-978-7; intermediate used in alkyd resins, rubber compding., water repellents, polishes, soaps, abrasives, cutting oils, candles, crayons, emulsifiers, personal care prods.; paste; acid no. 253-260; iodine no. 5-14; sapon. no. 253-260; 100% conc.

Industrene® 365. [Witco Corp.] Mixt. caprylic/capric acid; CAS 67762-36-1; intermediate used in alkyd resins, rubber compding., water repellents, polishes, soaps, abrasives, cutting oils, candles, crayons, emulsifiers; FG grades as lubricant, release agent, binder, defoamer in foods, intermediate for food emulsifiers; acid no. 355-369; iodine no. 1 max.; sapon. no. 355-374; 100% conc.

Industrene® 1224. [Witco/H-I-P; Sovereign] Stearic acid; CAS 57-11-4; EINECS 200-313-4; activator for rubber; tall oil based; lower cost than tallow based grades.

Industrene® 4518. [Witco Corp.] Single pressed stearic acid; CAS 57-11-4; EINECS 200-313-4; intermediate used in alkyd resins, rubber compding., water repellents, polishes, soaps, abrasives, cutting oils, candles, crayons, emulsifiers; FG grades as lubricant, release agent, binder, defoamer in foods, intermediate for food emulsifiers; Gardner 3 solid; acid no. 204-211; iodine no. 8-11; sapon. no. 204-212; 100% conc.

Industrene® 5016. [Witco; Sovereign] Double pressed stearic acid; CAS 57-11-4; EINECS 200-313-4; intermediate, activator used in alkyd resins, rubber compding., water repellents, polishes, soaps, abrasives, cutting oils, candles, crayons, emulsifiers; solid; acid no. 207-210; iodine no. 4-7; sapon. no. 207-211; 100% conc.

Industrene® 7018. [Witco Corp.] 70% Stearic acid; CAS 57-11-4; EINECS 200-313-4; intermediate used in alkyd resins, rubber compding., water repellents, polishes, soaps, abrasives, cutting oils, candles, crayons, emulsifiers; solid; acid no. 200-207; iodine no. 1 max.; sapon. no. 200-208; 100% conc.

Industrene® 9018. [Witco Corp.] Stearic acid (90%); CAS 57-11-4; EINECS 200-313-4; intermediate used in alkyd resins, rubber compding., water repellents, polishes, soaps, abrasives, cutting oils, candles, crayons, emulsifiers, personal care prods.; solid; acid no. 196-201; iodine no. 2 max.; sapon. no. 196-202; 100% conc.

Industrene® B. [Witco Corp.] Hydrog. stearic acid; CAS 57-11-4; intermediate used in alkyd resins, rubber compding., water repellents, polishes, soaps, abrasives, cutting oils, candles, crayons, emulsifiers; FG grades as lubricant, release agent, binder, defoamer in foods, intermediate for food emulsifiers; solid; acid no. 199-207; iodine no. 3 max.; sapon. no. 199-208; 100% conc.

Industrene® D. [Witco Corp.] 40% Dimer acid; intermediate for lubricants, corrosion inhibitors for petrol. industry, extenders and cross-linking agents for high polymeric systems, in hot-melt adhesives, epoxy curing agents; Gardner 8 color; visc. 600 cSt; acid no. 184-188; sapon. no. 184-197.

Industrene® R. [Witco; Sovereign] Hydrog. rubber grade stearic acid; CAS 57-11-4; intermediate, activator, process aid used in alkyd resins, rubber compding., water repellents, polishes, soaps, abrasives, cutting oils, candles, crayons, emulsifiers; Gardner 10 solid; acid no. 193-213; iodine no. 10 max.; sapon. no. 193-214; 100% conc.

Initiator BK. [Bayer/Fibers, Org., Rubbers] Silyl ether of benzopinacol in triethylphosphate (30%) and toluene (5%); initiator for radical polymerization processes with vinyl monomers, acrylic compds. or olefins; also suitable for BMC and SMC polyester prods.; safer processing than peroxide; yel.-grn. liq.; sp.gr. 1.14; visc. 6000 cps (23 C); 65% act.; *Storage:* storage stable.

Interlube™ 292. [C.P. Hall] Blend of a fatty acid salt with highly absorbent filler; internal process aid for natural and syn. rubber compds. with min. fillers and/or carbon blk.; no adverse effect on adhesion and bonding chars. or cure; improves pigment disp. by wetting out the polymer/filler interface; increases open mill mixing efficiency, flow in transfer, compr., and inj. molding and extrusion; tan lumpy powd.; sp.gr. 0.42-0.64; bulk dens. 0.53 ± 0.12 g/ml.

Interlube A. [Air Prods./Perf. Chems.] Blend of zinc fatty acids and lubricants; lubricant for rubber compds.; improves mold and mill roll release; straw paste; sp.gr. 0.81; m.p. 107 C; Unverified

Interlube P/DS. [Air Prods./Perf. Chems.] Blend of zinc fatty acids and lubricants on inert carrier; lubricant for rubber compds.; improves mold and mill roll release; cream to buff dust-suppressed powd.; sp.gr. 1.1; Unverified

Internal Lubricant D-148™ Dry. [C.P. Hall] Fatty acid compd.; internal lubricant improving polymer processing and pigment disp., eliminating sticking problems, permitting increased extrusion speeds, and improving flow in transfer, compr., and inj. moldings; neutral in most cure systems; tan lumpy powd.; sp.gr. 0.49 ± 0.8 g/ml.

Internal Lubricant D-148™ Wet. [C.P. Hall] Fatty acid compd.; internal lubricant improving polymer processing and pigment disp., eliminating sticking problems, permitting increased extrusion speeds, and improving flow in transfer, compr., and inj. moldings; neutral in most cure systems; yel.-brn. paste; sp.gr. 1.05 ± 0.03.

Interstab® BC-100S. [Akzo Nobel] Barium-cadmium; zinc-free stabilizer offering heat stability; used for flexible compds., esp. hose and profile, filled systems, film and sheet; liq.

Interstab® BC-103. [Akzo Nobel] Barium-cadmium-zinc; low zinc stabilizer offering heat and lt. stability; used for plastisol applics.; liq.

Interstab® BC-103A. [Akzo Nobel] Barium-cadmium-zinc; med. zinc stabilizer offering heat and lt. stability; nonsulfur staining; used for plastisol applics. and flexible compds.; liq.

Interstab® BC-103C. [Akzo Nobel] Barium-cadmium-zinc; med. zinc stabilizer offering heat and lt. stability; nonsulfur staining; used for plastisol applics., flexible compds. esp. hose, profile, filled systems.

Interstab® BC-103L. [Akzo Nobel] Barium-cadmium-zinc; low zinc stabilizer offering heat and lt. stability; used for plastisol applics. and flexible compds.; liq.

Interstab® BC-109. [Akzo Nobel] Barium-cadmium-zinc; high zinc stabilizer offering heat and lt. stability, non plate-out, plastisol air release, plastisol visc. control for plastisol applics.; liq.

Interstab® BC-110. [Akzo Nobel] Barium-cadmium-zinc; high zinc stabilizer offering heat and lt. stability, non plate-out, plastisol air release, and plastisol visc. control for plastisol applics.; liq.

Interstab® BC-4195. [Akzo Nobel] Barium-cadmium-zinc; heat and lt. stabilizer, lubricant for flexible and semirigid PVC for exterior automotive applics.; liq.

Interstab® BC-4377. [Akzo Nobel] Cadmium-zinc; heat and lt. stabilizer for vinyl foams; liq.

Interstab® BZ-4828A. [Akzo Nobel] Barium-zinc; heat and light stabilizer for flexible and semirigid PVC; good lubrication and clarity; liq.

Interstab® BZ-4836. [Akzo Nobel] Barium-zinc-phosphite complex; liq. metal soap heat and light stabilizer for flexible PVC.

Interstab® C-16. [Akzo Nobel] Aminocrotonate; stabilizer, lubricant for flexible, semirigid, and rigid PVC, bottles.

Interstab® CA-18-1. [Akzo Nobel] Calcium stearate; CAS 1592-23-0; EINECS 216-472-8; lubricant, release agent, and processing aid in extrusion, calendering, inj., or blow molding of rigid or flexible PVC compds.; lubricant or metal scavenger in PP prod.; used in plastics, rubber, coatings, cosmetic, and metallurgical fields; food contact applic.; fine particle size.

Interstab® CH-55. [Akzo Nobel] Organophosphites; chelating agents to improve clarity in PVC compd.; aux. heat and light stabilizers; food contact and medical applics.

Interstab® CH-55R. [Akzo Nobel] Organophosphite; chelating agents to improve clarity in PVC compd.; aux. heat and light stabilizers; food contact and medical applics.

Interstab® CH-90. [Akzo Nobel] Phosphite; nonstaining antioxidant for ABS, PP, PVC.

Interstab® CLB-747. [Akzo Nobel] Barium-zinc; heat and lt. stabilizer for flexible and semirigid PVC, plastisols, organosols, foams, fluorescent pigments; liq.

Interstab® CZ-10. [Akzo Nobel] Calcium-zinc soap; nontoxic, nonstaining heat and light stabilizer for vinyls (flexible, rigid, plastisols, organosols, foams); liq.

Interstab® CZ-11. [Akzo Nobel] Calcium-zinc soap; nontoxic, nonstaining light stabilizer, lubricant for flexible, semirigid, and rigid vinyl compds.; paste.

Interstab® CZ-11D. [Akzo Nobel] Calcium-zinc soap; nontoxic, nonstaining heat and light stabilizer for flexible, semirigid, rigid, plastisol, and organosol vinyls; paste.

Interstab® CZ-19A. [Akzo Nobel] Calcium-zinc soap; nontoxic, nonstaining heat and light stabilizer for flexible, semirigid, rigid, plastisol, and organosol vinyls; powd.

Interstab® CZ-22. [Akzo Nobel] Calcium-zinc soap; nontoxic, nonstaining heat and light stabilizer for flexible, semirigid, rigid, plastisol, and organosol vinyls; powd.

Interstab® CZ-23. [Akzo Nobel] Calcium-zinc soap; nontoxic, nonstaining heat and light stabilizer for flexible, semirigid, rigid, plastisol, and organosol vinyls; powd.

Interstab® F-402. [Akzo Nobel] Calcium-zinc soap; nontoxic stabilizer offering heat stability, clarity, nonplate-out, and nonstaining props.; for rigid vinyl, bottles; powd.

Interstab® G-140. [Akzo Nobel] Glycerol ester; syn. lubricant used in rigid calendering, extrusion, blow and inj. molding as internal lubricant; excellent clarity, sparkle, and surf. finish to compds.; food contact applic.

Interstab® G-8257. [Akzo Nobel] Ethylene bis-stearamide; lubricant; functions at low levels as internal lubricant in rigid PVC applic.; food contact applic. in PVC and polymers; wax.

Interstab® LF 3615. [Akzo Nobel] Lead coprecipitate; heat and lt. stabilizer for flexible PVC.

Interstab® LT-4805R. [Akzo Nobel] Barium-zinc; lubricating heat and light stabilizer for flexible and rigid PVC, plastisols, foams; liq.

Interstab® LT-4361 [Akzo Nobel] Barium-calcium-zinc; heat and lt. stabilizer, lubricant for flexible PVC, plastisols, organosols, foams; liq.

Interstab® MT981. [Akzo Nobel] Barium/cadmium complex; stabilizer for rigid PVC, window profiles; polyol-free version; useful for dark pigmentations; nondusting tablets.

Interstab® MT982. [Akzo Nobel] Barium/cadmium complex; polyol stabilizer for PVC window profiles; nondusting tablets.

Interstab® R-4048. [Akzo Nobel] Cadmium-free, low toxicity stabilizer for heat stability, nonplate-out, nonsulfide stain; for plastisol applics. (slush molding, rotational molding, dipping, spreading, film and sheet).

Interstab® R-4114. [Akzo Nobel] Barium-cadmium-zinc; low zinc stabilizer for heat stability; flexible compds., film and sheet applics.; liq.

Interstab® R-4150. [Akzo Nobel] Barium-cadmium-zinc; heat and lt. stabilizer, lubricant for flexible and semirigid PVC, plastisols, organosols; liq.

Interstab® ZN-18-1. [Akzo Nobel] Zinc stearate; CAS 557-05-1; EINECS 209-151-9; lubricant exhibiting internal and external properties used in plastics industry; excellent early color and clarity to clear PVC prods.; efficient color dispersant and lubricant in polyolefins and PS; food contact applic.; fine particle size.

Ionet DL-200. [Sanyo Chem. Industries] POE dilaurate; CAS 9005-02-1; nonionic; emulsifier for emulsion polymerization, metal processing lubricant and personal care prods.; liq.; HLB 6.6; 100% conc.

Ionet DO-200. [Sanyo Chem. Industries] POE dioleate; CAS 9005-07-6; nonionic; emulsifier for emulsion polymerization of vinyl resins, for metal processing lubricants, cosmetics; liq.; HLB 5.3; 100% conc.

Ionet DO-400. [Sanyo Chem. Industries] POE dioleate; CAS 9005-07-6; nonionic; emulsifier for emulsion polymerization of vinyl resins, for metal processing lubricants, cosmetics; liq.; HLB 8.4; 100% conc.

Ionet DO-600. [Sanyo Chem. Industries] POE dioleate; CAS 9005-07-6; nonionic; emulsifier for emulsion polymerization of vinyl resins, for metal processing lubricants, cosmetics; liq.; HLB 10.4; 100% conc.

Ionet DO-1000. [Sanyo Chem. Industries] POE dioleate; CAS 9005-07-6; nonionic; emulsifier for emulsion polymerization of vinyl resins, for metal processing lubricants, cosmetics; solid; HLB 12.9; 100% conc.

Ionet DS-300. [Sanyo Chem. Industries] POE distearate; CAS 9005-08-7; nonionic; emulsifier for emulsion polymerization of vinyl resins, for metal processing lubricants, cosmetics; solid; HLB 7.3; 100% conc.

Ionet DS-400. [Sanyo Chem. Industries] POE distearate; CAS 9005-08-7; nonionic; emulsifier for emulsion polymerization of vinyl resins, for metal processing lubricants, cosmetics; solid; HLB 8.5; 100% conc.

Ionol. [Shell] 2,6-Di-tert-butyl-4-methylphenol; antioxidant for petrol. prods.; rubber compds.; stabilizer for neoprene and plastic; Pt-Co 45 max.; sol. in wh. oil; solid. pt. 68.8 C min.; 98.0% min. purity; Unverified

Ionpure Type A. [U.S. Cosmetics] Silver borosilicate; amorphous water-sol. inorg. preservative for cosmetics and plastic pkg.; stable to 500 C; resist. to light; USA approved; *Toxicology:* nontoxic, nonsensitizing, nonmutagenic.

Ionpure Type B. [U.S. Cosmetics] Silver aluminum magnesium phosphate; amorphous water-sol. inorg. preservative for cosmetics and plastic pkg.; stable to 500 C; resist. to light; USA approved; *Toxicology:* nontoxic, nonsensitizing, nonmutagenic.

IPP. [PPG Industries] Diisopropyl peroxydicarbonate; CAS 105-64-6; initiator for polymerization of unsat. monomers; reduces time of batch runs; helps control polymerization; wh. solid; m.w. 206.18; sol. 0.04% in water; sp.gr. 1.080 (15.5/4C); m.p. 8-10 C; 98.5% min. conc.; 7.8% act. oxygen.

Ircogel® 900. [Lubrizol] Calcium org.; rheology control agent, thixotropic, antisag, and flow control agent for use in coating plastisols and organosols, cloth coating plastisols, PVC sealants, polysulfide sealants, polymercaptan sealants; exhibits wetting or dispersant effect on fillers; lt. brn. soft gel; sp.gr. 1.10 (15.6 C); dens. 9.00 lb/gal (15.6 C); flash pt. (COC) 204 C; 100% calcium org. gel.

Ircogel® 903. [Lubrizol] Calcium complex in diisononyl phthalate; thixotrope and visc. control agent for coating plastisols and organosols, fabric coatings, PVC sealants; lt. brn. liq.; dens. 8.75 lb/gal (15.6 C); flash pt. (COC) 204 C; 33.3% diisononyl phthalate.

Ircogel® 904. [Lubrizol] Calcium org. gel in Rule 66 Stoddard solv.; thixotropic, antisag, and flow control agent for use in coating organosols, PVC sealants; exhibits wetting or dispersant effect on fillers; lt. brn. liq.; sp.gr. 0.98 (15.6 C); dens. 8.25 lb/gal (15.6 C); flash pt. (PMCC) 38 C min.; 70% act., 30% Stoddard solv.

Irgafos® 168. [Ciba-Geigy/Additives] Tris(2,4-di-t-butylphenyl) phosphite; CAS 31570-04-4; antioxidant for polyolefins; suitable for food contact applics.; FDA 21CFR §177.1010, 177.1500, 177.1350, 177.1520(c), 177.1570, 177.1580, 177.2600, 177.1640, etc.

Irganox® 129. [Ciba-Geigy/Additives] 2,2′-Ethylidenebis (4,5-di-t-butylphenol); CAS 35958-30-6; antioxidant and thermal stabilizer for polymers; food pkg. applic.; used in PP, polyethylene, PVC, PS, ABS, hydrocarbon resins, EVA-modified compds.; FDA 21CFR §178.2010; wh. free-flowing cryst. powd.; sol. (g/100 g): > 100 g acetone, 44 g toluene, 18 g heptane, < 0.03 g water; m.w. 438; sp.gr. 1.01 ± 0.03; dens. 35 lb/ft^3; m.p. 161-163 C; b.p. > 550 F; flash pt. (COC) > 193 C; 99%+ purity; *Toxicology:* LD50 (oral, rat) > 10,000 mg/kg, (dermal, rabbit) > 2 g/kg; nontoxic; nonirritating.

Irganox® 245. [Ciba-Geigy/Additives] Ethylenebis (oxyethylene) bis (3-t-butyl-4-hydroxy-5-methylhydrocinnamate); CAS 36443-68-2; antioxidant/stabilizer for use in org. polymers (ABS, ASA, BR, SBR, NBR, HIPS, SAN, acetal, acrylic, PPO, polyamide, PU, PBT, rigid and flexible PVC); suitable for food contact applics.; FDA 21CFR §175.105, 177.1640, 177.2470, 177.2480, 178.2010; wh. to sl. ylsh. cryst. powd.; odorless; m.w. 586.8; sol. > 50% in acetone, > 40% in chloroform and methylene chloride, 37% in ethyl acetate, 18% in benzene, 12% in methanol, ≈ 6% in styrene, < 0.01% in water; m.w. 587; sp.gr. 1.14; m.p. 76-79 C; decomp. pt. >

220 C; flash pt. (PM) > 302 F.

Irganox® 259. [Ciba-Geigy/Additives] 1,6-Hexamethylene bis-(3,5-di-tert-butyl-4-hydroxyhydrocinnamate); CAS 35074-77-2; antioxidant; stabilizer for polyolefin, elastomer, styrenic, polyacetal, petrol. prods., and org. substrates; suitable for food contact applics.; FDA 21CFR §175.105, 175.125, 175.300(b), 175.320(b), 176.170(a)(b), 177.2470, 177.2480, 178.2010 etc.; off-wh. cryst. powd.; sol. in benzene, acetone, chloroform, min. oil, hexane, and water; m.w. 639; m.p. 103-108 C.

Irganox® 565. [Ciba-Geigy/Additives] 4-[[4,6-Bis(octylthio)-s-triazin-2-yl]amino]-2,6-di-t-butylphenol; CAS 991-84-4; antioxidant/stabilizer for polymers, adhesives, rubber articles (SBS footwear); suitable for food contact applics.; FDA 21CFR §175.105, 175.125, 175.300(b), 177.1640, 177.1810, 177.2600, 178.2010 etc.

Irganox® 1010. [Ciba-Geigy/Additives] Tetrakis [methylene (3,5-di-tert-butyl-4-hydroxyhydrocinnamate)] methane; CAS 6683-19-8; EINECS 229-722-6; nondiscoloring, nonstaining antioxidant and thermal stabilizer for org. and polymeric materials, polyolefin, elastomer, acetal, acrylic, PET, PU, PMMA, PVC, food pkg. and adhesive applic., petrol. prods.; U.S. patents 3,285,855 and 3,644,482; FDA 21CFR §175.105, 175.125, 175.300(b), 175.320(b), 176.170(a), 177.2470, 178.2010 etc.; wh. free-flowing cryst. powd.; sol. (g/100 g): 163 g methylene chloride, 144 g chloroform, 122 g benzene, 89 g acetone; m.w. 1178; sp.gr. 1.15; m.p. 110-125 C.

Irganox® 1019. [Ciba-Geigy/Additives] N,N′-1,3-Propanediylbis(3,5-di-t-butyl-4-hydroxyhydrocinnamamide); stabilizer for polymers, rubber; suitable for food contact applics.; FDA 21CFR §177.2600, 178.2010.

Irganox® 1035. [Ciba-Geigy/Additives] Thiodiethylene bis-(3,5-di-t-butyl-4-hydroxy) hydrocinnamate; CAS 41484-35-9; stabilizer and antioxidant for polymer, org. substrate, polyolefin, BR, SBR, NBR, SEBS, elastomer, petrol. prods., food pkg. applic.; FDA 21CFR §175.105, 175.125, 176.170, 176.180, 177.1210, 177.2600, 178.2010; wh. cryst. powd.; sol. in toluene, acetone, methanol, min. oil, and water; m.w. 642; sp.gr. 1.19; m.p. 63 C min.

Irganox® 1076. [Ciba-Geigy/Additives] Octadecyl 3,5-di-t-butyl-4-hydroxyhydrocinnamate; CAS 2082-79-3; EINECS 218-216-0; antioxidant and thermal stabilizer used in org. and polymeric material; stabilizer for polyolefins, styrenics, elastomers, PVC, urethane, acrylic coatings, adhesive, and petrol. prods.; food pkg. applic.; FDA 21CFR §175.105, 177.1010, 177.1350, 177.1520(c), 177.1580, 177.1640, 178.2010, 178.3790 etc.; wh. cryst. free-flowing powd., odorless; sol. (g/100 g): 133 g benzene, 61 g ethyl acetate, 47 g hexane, 23 g acetone, 12 g min. oil, < 0.01 g water; m.w. 531; sp.gr. 1.02; m.p. 50-55 C; *Toxicology:* may cause irritation esp. after prolonged/repeated exposure and in body areas of heavy perspiration.

Irganox® 1098. [Ciba-Geigy/Additives] N,N′-hexamethylene bis (3,5-di-tert-butyl-4-hydroxyhydrocinnamamide); CAS 23128-74-7; antioxidant for polymers, rubber, SBR, polyacetals, linear saturated polyesters, PVC, polyolefins; suitable for food contact applics.; FDA 21CFR §175.300(b), 177.1210, 177.1500, 177.2470, 177.2480, 177.2600, 178.2010; wh. cryst. powd.; m.w. 637; sol. in chloroform, methanol, acetone, ethyl acetate, water, benzene, and hexane; m.p. 156-161 C.

Irganox® 1330. [Ciba-Geigy/Additives] 1,3,5-Trimethyl-2,4,6-tris(3,5-di-t-butyl-4-hydroxybenzyl) benzene; CAS 1709-70-2; antioxidant for polymers; suitable for food contact applics.; FDA 21CFR §177.1500, 178.2010.

Irganox® 1425 WL. [Ciba-Geigy/Additives] Calcium bis[monoethyl(3,5-di-t-butyl-4-hydroxybenzyl) phosphonate) 50% with polyethylene wax; CAS 65140-91-2; antioxidant for polyolefins, BR, SBR, NBR; suitable for food contact applics.; FDA 21CFR §175.105, 175.125, 175.300(b), 175.320(b), 177.1520(c), 178.2010 etc.; 50% act.

Irganox® 1520. [Ciba-Geigy/Additives] 2,4-Bis [(octylthio) methyl]-o-cresol; CAS 110553-27-0; antioxidant for polymers; used in the base stabilization of elastomers and the compd. stabilization of adhesives; effective during elastomer dynamic processing without the use of phosphites; suitable for food contact applics.; FDA 21CFR §175.105, 175.300(b), 175.125 176.170, 176.180, 177.1210, 177.2600, 178.2010; pale yel. low-visc. liq.; sol. in methanol, ethanol, acetone, ethyl acetate, methylene chloride, chloroform, toluene, n-hexane; sol. < 0.01 g/100 g water; m.w. 424.7; dens. 0.9787 g/ml; flash pt. > 200 C.

Irganox® 3114. [Ciba-Geigy/Additives] 1,3,5-Tris(3,5-di-t-butyl-4-hydroxybenzyl)-s-triazine-2,4,6-(1H,3H,5H) trione; CAS 27676-62-6; antioxidant for polyolefins (PP fiber, filled PP); suitable for food contact applics.; FDA 21CFR §175.105, 177.1520, 178.2010.

Irganox® 3125. [Ciba-Geigy/Additives] 3,5-Di-t-butyl-4-hydroxyhydrocinnamic acid triester with 1,3,5-tris(2-hydroxyethyl)-s-triazine-2,4,6(1H,3H,5H)-trione; antioxidant for polymers; suitable for food contact applics.; FDA 21CFR §175.105, 177.1520, 178.2010.

Irganox® 5057. [Ciba-Geigy/Additives] N-Phenylbenzeneamine reaction prods. with 2,4,4-trimethylpentenes; stabilizer for polymers, rubber, adhesives; suitable for food contact applics.; FDA 21CFR §175.105, 177.2600.

Irganox® B 215. [Ciba-Geigy/Additives] Irganox 1010/Irgafos 168, 1:2 ratio; antioxidant-process stabilizer combination for PP, PE; food pkg. applic.; colorless powd.; sol. in chloroform, methylene chloride, benzene, hexane, ethyl acetate, acetone, methanol, and water; m.w. 647; m.p. 180-185 C.

Irganox® B 225. [Ciba-Geigy/Additives] Irganox 1010/Irgafos 168, 1:1 ratio; antioxidant-process stabilizer combination for PP, PE, ASA, BR, SBR, NBR, EPDM, SBS footwear, SEBS, wire and cable compds., PET, PVC bottles; food pkg. applic.; m.w.

647; sol. in chloroform, methylene chloride, benzene, hexane, ethyl acetate, acetone, methanol, and water; m.p. 180-185 C.

Irganox® B 501W. [Ciba-Geigy/Additives] Irganox 1425:Irgafos 168:polyethylene wax (1:2:1); antioxidant/costabilizer blend for PP fiber, film, and slit tape.

Irganox® B 561. [Ciba-Geigy/Additives] Irganox 1010:Irgafos 168 (1:4); antioxidant/costabilizer blend for LDPE, HDPE, PE film, rotomolding, inj. molding, blow molding.

Irganox® B 900. [Ciba-Geigy/Additives] Irganox 1076:Irgafos 168 (1:4); antioxidant/costabilizer blend for LDPE, HDPE, PE film, rotomolding, inj. and blow molding, ABS, SAN, PET, PC.

Irganox® B 921. [Ciba-Geigy/Additives] Irganox 1076/Irgafos 168 (1:8 ratio); antioxidant/costabilizer blend.

Irganox® B 1171. [Ciba-Geigy/Additives] Irganox 1098:Irgafos 168 (1:1); antioxidant/costabilizer blend for polyamides.

Irganox® B 1411. [Ciba-Geigy/Additives] Irganox 3114:Irgafos 168 (1:1); antioxidant/costabilizer blend for PP fiber, film, slit tape, filled PP.

Irganox® B 1412. [Ciba-Geigy/Additives] Irganox 3114/Irgafos 168 (1:2 ratio); antioxidant/costabilizer blend.

Irganox® MD 1024. [Ciba-Geigy/Additives] 1,2-Bis (3,5-di-t-butyl-4-hydroxyhydrocinnamoyl) hydrazine; CAS 32687-78-8; antioxidant, metal deactivator, stabilizer; used in extending the lifetime of PP, polyethylene, and certain thermoplastic elastomers, EPDM, ABS, ASA, peroxide, nylon, and polyacetal; suitable for food contact applics.; FDA 21CFR §175.105, 177.2470, 177.2480, 178.2010; wh. cryst. powd.; sol. in THF, acetone, toluene, water, and paraffin oil; sp.gr. 1.12; m.p. 227-229 C.

Irgastab® 2002. [Ciba-Geigy/Additives] Nickel bis[O-ethyl(3,5-di-tert-butyl-4-hydroxybenzyl)] phosphonate; uv lt. stabilizer and antioxidant for polyolefins; useful in pigmented or opaque fibers and in films; dyesite for nickel chelatable dyes for polyolefins; provides stability during processing of PP; imparts negligible odor to substrates; resistant to extraction; resists discoloration by nitrogen oxides; low volatility; sulfur-free; lt. tan powd.; m.w. 713.5; sol. 100g/100 ml or more in most common hydrocarbon solvs.; m.p. 180 C min.; 8.24% Ni.

Irgastab® T 265. [Ciba-Geigy/Additives] Mixed mono- and dioctyltin (C10-16) alkylthioglycolates; organotin heat stabilizer for food-grade PVC bottle and sheet applics.; for improved start-to-finish color protection; low volatility and extraction resistance in processing; almost colorless clear liq.; sp.gr. 1.04; flash pt. 302 F; pH 3 (@200 g/l water).

Irgastab® T 634. [Ciba-Geigy/Additives] Tin, butyl 2-ethylhexanoate laurate oxo complexes; heat and light stabilizer for vinyl systems, incl. rigid and plasticized PVC; slightly ylsh. low-visc. liq.; sol. 40 ppm in water; sp.gr. 1.08; b.p. > 250 C; m.p. < 20 C; ref. index 1.485; pH 4 (2g/10 ml water).

Isanol®. [BASF AG] 60% iso- and 40% n-butanol; CAS 71-36-3; EINECS 200-751-6; solv. for prod. of adhesives and surface coating resins.

Isatin. [BASF AG] Intermediate for prod. of pharmaceuticals and dyes, stabilizer for plastics industry.

Isobutyl Niclate®. [R.T. Vanderbilt] Nickel diisobutyldithiocarbamate; CAS 15317-78-9; nonstaining antioxidant/antiozonant for protection in epichlorohydrin rubber; for automotive and appliance molded goods, wire and cable applics.; improves heat aging; grnsh. powd.; 99.8% min thru 100 mesh; sol. in acetone, chloroform, toluene; pract. insol. in water; m.w. 467.47; dens. 1.27 ± 0.03 mg/m^3; m.p. 173-181 C; 11.5-13.5% Ni.

Isobutyl Tuads®. [R.T. Vanderbilt] Tetraisobutylthiuram disulfide; CAS 3064-73-1; ultra accelerator, vulcanizing agent, sulfur donor for natural and syn. rubber; generates a low level of a nonregulated, less toxic nitrosamine; for SBR or BR blends, NR, EPDM, NBR; activator of thiazole and sulfenamides in SBR or BR blends; lt. yel. powd.; negligible sol. in water; dens. 1.14 mg/m^3; m.p. 65-70 C; *Toxicology:* ingestion of alcoholic beverages before/after handling may cause antabuse reaction; *Precaution:* incompat. with strong acids, reducing agents; hazardous decomp. prods.: oxides of nitrogen, sulfur, carbon.

Isobutyl Zimate®. [R.T. Vanderbilt] Zinc diisobutyldithiocarbamate; CAS 36190-62-2; ultra accelerator for rubber; generates a low level of a nonregulated, less toxic nitrosamine; for NR, EPDM, NBR; for SBR or BR blends as activator of thiazole and sulfenamides; wh. to cream powd., odorless; 99.9% min. thru 100 mesh; negligible sol. in water; dens. 1.24 mg/m^3; m.p. 115 C; 13.5-14.5% Zn; *Toxicology:* dust or vapor may cause irritation to respiratory tract, skin, eyes; may cause antabuse reaction when alcohol is consumed; recommended OSHA PEL/TWA 15 mg/m^3 (total dust), 5 mg/m^3 (respirable dust); *Precaution:* may react with strong oxidizing agents; hazardous decomp. prods.: oxides of carbon, nitrogen, sulfur, hydrocarbons; *Storage:* keep container closed.

Isonate® 125M. [Dow] MDI; CAS 101-68-8; processing aid, intermediate for the prod. of cast, RIM, and thermoplastic PU elastomers, adhesives, binders, coatings, and sealants; solid; b.p. 200 C (5 mm Hg); dens. 1.18 g/ml (43 C); visc. 5 cps (43 C); decomp. pt. 230 C; isocyanate equiv. 125.5; flash pt. (COC) 199 C; 99% purity; 33.5% NCO content.

Isonate® 143L. [Dow] MDI; CAS 101-68-8; processing aid for PU industry for use alone or with prepolymers; used in cast and RIM processing, adhesives, binders, coatings, and sealants; liq.; b.p. 200 C (5 mm Hg); dens. 1.21 g/ml; visc. 35 cps; decomp. pt. 230 C; isocyanate equiv. 144.9; flash pt. (COC) 199 C; 29.0% NCO content.

Isonate® 181. [Dow] MDI; CAS 101-68-8; processing aid for PU industry; polyether quasi-prepolymer for use in formulating cast elastomers and microcellular rubber for shoe soles and related applics., adhesives, binders, coatings, and sealants; liq.; b.p. 200 C (5 mm Hg); dens. 1.21 g/ml; visc. 850

cps; decomp. pt. 230 C; isocyanate equiv. 183; COC flash pt. 199 C; 22.9% NCO content.

Isonate® 240. [Dow] MDI; CAS 101-68-8; processing aid for PU industry; polyester quasi-prepolymer yielding high physical property elastomers or microcellulars via the cast, low pressure or RIM dispensing process; for structural polymers, dynamic elastomers, adhesives, binders, coatings, and sealants; liq.; b.p. 200 C (5 mm Hg); dens. 1.22 g/ml; visc. 1100 cps; decomp. pt. 230 C; isocyanate equiv. 225; COC flash pt. 199 C; 18.7% NCO content.

Isonox® 132. [Schenectady] 2,6-Di-tbutyl-4-sec-butylphenol; CAS 17540-75-9; antioxidant, stabilizer used in polyols, PVC, polyolefins, rubber systems, and functional fluids; FDA 21CFR §175.105; pale to straw yel. clear liq.; m.w. 262.4; sp.gr. 0.902; dens. 7.5 lb/gal; visc. 75 cps; m.p. 18 C; b.p. 275 C (760 mm); flash pt. (Seta) > 94 C; 95% min. act.; *Toxicology:* not known to cause irritation to eyes, skin, or by inh.; however, may be irritant to sensitive individuals; irritant to GI tract; may cause CNS effects; *Precaution:* incompat. with strong oxidants; thermal decomp. may produce CO, CO_2; *Storage:* store in cool, dry area (40-80 F) in closed containers; may solidify if stored below 70 F for long periods; reheat to 100 F to liquefy.

Isonox® 232. [Schenectady] CAS 4306-88-1; low cost, low volatility antioxidant, stabilizer for polyols, lubes, oils, and greases; APHA 500 max. liq.; f.p. < -40 C; 90% min. act.; Custom

I.T. 3X [R.T. Vanderbilt] Industrial talc (hydrous magnesium calcium silicate); CAS 14807-96-6; EINECS 238-877-9; filler/extender in NR and syn. rubbers, paints, sealants, mastics, latex compds.; wh. powd.; 9.3 μm median particle size; 98% min. thru 325 mesh; dens. 2.85 ± 0.03 mg/m^3; bulk dens. 59 lb/ft^3 (compacted); oil absorp. 29; brightness 90; pH 9.4 (10% aq. slurry); 56.0% SiO_2, 30.7% MgO, 7.0% CaO.

I.T. 5X [R.T. Vanderbilt] Industrial talc (hydrous magnesium calcium silicate); CAS 14807-96-6; EINECS 238-877-9; filler/extender in NR and syn. rubbers, paints, sealants, mastics, latex compds.; wh. powd.; 7.6 μm median particle size; 99.25% min. thru 325 mesh; dens. 2.85 ± 0.03 mg/m^3; bulk dens. 50 lb/ft^3 (compacted); oil absorp. 30; brightness 90; pH 9.4 (10% aq. slurry); 56.0% SiO_2, 30.7% MgO, 7.0% CaO.

I.T. 325 [R.T. Vanderbilt] Industrial talc (hydrous magnesium calcium silicate); CAS 14807-96-6; EINECS 238-877-9; filler/extender in NR and syn. rubbers, paints, sealants, mastics, latex compds.; wh. powd.; 5.5 μm median particle size ; 99.9% min. thru 325 mesh; dens. 2.85 ± 0.03 mg/m^3; bulk dens. 55 lb/ft^3 (compacted); oil absorp. 29; brightness 90; pH 9.4 (10% aq. slurry); 56.0% SiO_2, 30.7% MgO, 7.0% CaO.

I.T. FT. [R.T. Vanderbilt] Industrial talc (hydrous magnesium calcium silicate); CAS 14807-96-6; EINECS 238-877-9; filler/extender in NR and syn. rubbers, paints, sealants, mastics, latex compds.; wh. powd.; 7.0 μm median particle size; 99.4% min. thru 325 mesh; dens. 2.85 ± 0.03 mg/m^3; bulk dens. 50 lb/ft^3 (compacted); oil absorp. 29; brightness 90; pH 9.4 (10% aq. slurry); 56.0% SiO_2, 30.7% MgO, 7.0% CaO.

I.T. X. [R.T. Vanderbilt] Industrial talc (hydrous magnesium calcium silicate); CAS 14807-96-6; EINECS 238-877-9; filler/extender in NR and syn. rubbers, paints, sealants, mastics, latex compds.; particulate; 10.2 μm median particle size; 98% < 325 mesh; dens. 2.85 mg/m^3; bulk dens. 64 lb/ft^3 (compacted); oil absorp. 23; brightness 90; pH 9.4 (10% aq. slurry); 56.0% SiO_2, 30.7% MgO, 7.0% CaO.

J

Jayflex® 77. [Exxon] Diisoheptyl phthalate; CAS 71888-89-6; plasticizer; Pt-Co < 25 color; sp.gr. 0.995 (20/20 C); dens. 8.28 lb/gal (20 C); visc. 51 cSt (20 C); vapor pressure 2.2 mm Hg (200 C); b.p. 220 C (5 mm); pour pt. -40 C; flash pt. (TCC) 390 F; surf. tens. 28 dynes/cm (20 C); 99.6% purity.

Jayflex® 210. [Exxon] Proprietary; CAS 64742-53-6; secondary plasticizer; Pt-Co < 1 color; sp.gr. 0.87 (20/20 C); dens. 7.25 lb/gal (20 C); visc. 12 cSt (40 C); vapor pressure < 0.01 mm Hg (200 C); b.p. 330 C (5 mm); pour pt. -45 C; flash pt. (TCC) 295 F.

Jayflex® 215. [Exxon] Proprietary; CAS 64742-14-9; secondary plasticizer; sp.gr. 0.762 (20/20 C); dens. 6.43 lb/gal (20 C); visc. 3 cSt; vapor pressure 5 mm Hg (20 C); b.p. 271 C (5 mm); pour pt. -48 C; flash pt. (TCC) 250 F.

Jayflex® DHP. [Exxon] Dihexyl phthalate; CAS 68515-50-4; EINECS 201-559-5; plasticizer; Pt-Co < 25 color; sp.gr. 1.008 (20/20 C); dens. 8.39 lb/gal (20 C); visc. 37 cSt (20 C); vapor pressure 3 mm Hg (200 C); b.p. 210 C (5 mm); pour pt. -33 C; flash pt. (TCC) 380 F; surf. tens. 26 dynes/cm (20 C); 99.6% purity.

Jayflex® DIDP. [Exxon] Diisodecyl phthalate; CAS 68515-49-1; EINECS 247-977-1; plasticizer; Pt-Co < 25 color; sp.gr. 0.967 (20/20 C); dens. 8.05 lb/gal (20 C); visc. 129 cSt (20 C); vapor pressure 0.35 mm Hg (200 C); b.p. 256 C (5 mm); pour pt. -50 C; flash pt. (TCC) 435 F; surf. tens. 33.5 dynes/cm (20 C); 99.6% purity.

Jayflex® DINA. [Exxon] Diisononyl adipate; CAS 33703-08-1; EINECS 251-646-7; plasticizer; Pt-Co < 25 color; sp.gr. 0.924 (20/20 C); dens. 7.68 lb/gal (20 C); visc. 19.6 cSt; vapor pressure 0.9 mm Hg (200 C); b.p. 233 C (5 mm); pour pt. -59 C; flash pt. (TCC) 390 F; surf. tens. 32 dynes/cm (20 C); 99% purity.

Jayflex® DINP. [Exxon] Diisononyl phthalate; CAS 68515-48-0; EINECS 249-079-5; plasticizer; Pt-Co < 25 color; sp.gr. 0.973 (20/20 C); dens. 8.10 lb/gal (20 C); visc. 102 cSt (20 C); vapor pressure 0.5 mm Hg (200 C); b.p. 252 C (5 mm); pour pt. -48 C; flash pt. (TCC) 415 F; surf. tens. 33 dynes/cm (20 C); 99.6% purity.

Jayflex® DIOP. [Exxon] Diisooctyl phthalate; CAS 27554-26-3; plasticizer; Pt-Co < 25 color; sp.gr. 0.985 (20/20 C); dens. 8.20 lb/gal (20 C); visc. 85 cSt (20 C); vapor pressure 1 mm Hg (200 C); b.p. 230 C (5 mm); pour pt. -46 C; flash pt. (TCC) 400 F; surf. tens. 30 dynes/cm (20 C); 99.6% purity.

Jayflex® DOA. [Exxon] Dioctyl adipate; CAS 103-23-1; EINECS 203-090-1; plasticizer; Pt-Co < 25 color; sp.gr. 0.927 (20/20 C); dens. 7.71 lb/gal (20 C); visc. 16 cSt (20 C); vapor pressure 2.5 mm Hg (200 C); b.p. 215 C (5 mm); pour pt. < -60 C; flash pt. (TCC) 380 F; 99% purity.

Jayflex® DTDP. [Exxon] Ditridecyl phthalate; CAS 68515-47-9; EINECS 204-294-3; plasticizer; Pt-Co < 75 color; sp.gr. 0.956 (20/20 C); dens. 7.94 lb/gal (20 C); visc. 322 cSt (20 C); vapor pressure 0.085 mm Hg (200 C); b.p. 286 C (5 mm); pour pt. -37 C; flash pt. (TCC) 445 F; 99.6% purity.

Jayflex® L9P. [Exxon] Linear phthalate; high permanence plasticizer for extrusion and molding film, sheet and coated fabric for automotive, weather stripping, pool liners, membranes, tarps, specialty wire and cable applics.; Pt-Co 25 color, clear appearance, mild char. odor; sp.gr. 0.967-0.973; visc. 54 cP (20 C); flash pt. 213 C; 99.6% act.

Jayflex® L11P. [Exxon] Diundecyl linear phthalate; CAS 3648-20-2; EINECS 222-884-9; plasticizer; Pt-Co < 75 color; sp.gr. 0.954 (20/20 C); dens. 7.94 lb/gal (20 C); visc. 57 cSt; vapor pressure < 0.5 mm Hg (200 C); b.p. 245 C (5 mm); pour pt. -15 C; flash pt. (TCC) 460 F; surf. tens. 33 dynes/cm (20 C); 99.6% purity.

Jayflex® L911P. [Exxon] Dinonyl undecyl linear phthalate; plasticizer; Pt-Co < 50 color; sp.gr. 0.961 (20/20 C); dens. 7.97 lb/gal (20 C); visc. 57 cSt; vapor pressure < 0.2 mm Hg (200 C); b.p. 245 C (5 mm); pour pt. -20 C; flash pt. (TCC) 392 F; 99.6% purity.

Jayflex® L7911P. [Exxon] Diheptyl, nonyl, undecyl linear phthalate; plasticizer; Pt-Co < 25 color; sp.gr. 0.972 (20/20 C); dens. 8.09 lb/gal (20 C); visc. 65 cSt (20 C); vapor pressure 0.5 mm Hg (200 C); b.p. 252 C (5 mm); pour pt. < -46 C; flash pt. (TCC) 400 F; 99.6% purity.

Jayflex® TINTM. [Exxon] Triisononyl trimellitate; CAS 53894-23-8; plasticizer; Pt-Co < 100 color; sp.gr. 0.979 (20/20 C); dens. 8.15 lb/gal (20 C); visc. 430 cSt (20 C); vapor pressure < 0.1 mm Hg (200 C); b.p. 311 C (5 mm); pour pt. -40 C; flash pt. (TCC) 465 F; 99% purity.

Jayflex® TOTM. [Exxon] Trioctyl trimellitate; CAS 3319-31-1; plasticizer; Pt-Co < 100 color; sp.gr.

0.992 (20/20 C); dens. 8.30 lb/gal (20 C); visc. 312 cSt (20 C); vapor pressure < 0.16 mm Hg (200 C); b.p. 300 C (5 mm); pour pt. -46 C; flash pt. (TCC) 430 F; 99% purity.

Jayflex® UDP. [Exxon] Undecyl dodecyl phthalate; CAS 68515-47-9; plasticizer; Pt-Co < 75 color; sp.gr. 0.959 (20/20 C); dens. 7.98 lb/gal (20 C); visc. 185 cSt; vapor pressure 0.15 mm Hg (200 C); b.p. 275 C (5 mm); pour pt. -40 C; flash pt. (TCC) 437 F; 99.6% purity.

Jeffamine® BuD-2000. [Huntsman] Urea condensate of POP polyamine; epoxy modifier; nonreactive additive used in conc. of 5-20 phr to provide enhancement of metal-to-metal adhesion, thermal shock properties; results in increased elongation, higher impact and tensile strength, and lowered modulus, while heat deflection values are only slightly affected; reactive with formaldehyde to produce polymeric materials; lt. yel. visc. liq.; visc. 22,000 cps; flash pt. (PMCC) 471 F.

Jeffamine® D-230. [Huntsman] POP polyamine; epoxy curing agent and modifier used in heat-cured sol'n. coatings, castings, adhesives, and laminates; colorless liq.; sol. in water, glycol ethers, esters, alcohols, ketones, and hydrocarbons; sp.gr. 0.9480; visc. 8.7 cs; ref. index 1.466; pH 11.3 (1% aq.); flash pt. (COC) 256 F; distort. temp 60-75 C (264 psi).

Jeffamine® D-400. [Huntsman] POP diamine; epoxy curing agent and modifier used in coatings, castings, adhesives; colorless liq.; sol. in water, glycol ethers, esters, alcohols, ketones, and hydrocarbons; sp.gr. 0.9702; visc. 22 cs; ref. index 1.4482; pH 11.3 (1% aq.); flash pt. (COC) 347 F; distort. temp. 42-45 C (264 psi).

Jeffamine® D-2000. [Huntsman] POP diamine; epoxy curing agent and modifier useable alone or in combination; pale yel. liq.; sol. in water, glycol ethers, esters, alcohols, ketones, and hydrocarbons; sl. sol. in water; sp.gr. 0.9964; visc. 265 cs; ref. index 1.4514; pH 10.1 (1% aq.); flash pt. (COC) 460 F.

Jeffamine® DU-700. [Huntsman] Urea condensate of POP polyamine; epoxy curing agent useable alone or in combination; pale yel. sl. visc. liq.; visc. 1500 cps; distort. temp. < 25 C (264 psi).

Jeffamine® EDR-148. [Huntsman] Polyoxyalkylene diamine; epoxy curing agent; also in polyamide, polyurea, modified urethane resins, in adhesives, elastomers, foam formulas; intermediate for textile and paper treatment chemicals; colorless to lt. yel. liq.; m.w. 148; sp.gr.1.0154 (25/4 C); visc. 8 cSt; flash pt. (PMCC) 265 F.

Jeffamine® T-403. [Huntsman] POP triamine; epoxy curing agent and modifier; colorless liq.; sol. in water, glycol ethers, esters, alcohols, ketones, and hydrocarbons; sp.gr. 0.9812; visc. 76.5 cs; ref. index 1.4606; pH 11.2 (1% aq.); flash pt. (COC) 380 F; distort. temp. 60-70 C (264 psi).

Jeffamine® T-3000. [Huntsman] Polyoxyalkylene triamine; epoxy curing agent; also in polyamide, polyurea, modified urethane resins, in adhesives, elastomers, foam formulas; intermediate for textile and paper treatment chemicals; colorless to lt. yel. hazy liq.; m.w. 3000; sp.gr.1.1203; visc. 467 cSt; flash pt. (PMCC) 455 F.

Jeffamine® T-5000. [Huntsman] Polyoxyalkylene triamine; epoxy curing agent; also in polyamide, polyurea, modified urethane resins, in adhesives, elastomers, foam formulas; intermediate for textile and paper treatment chemicals; colorless to lt. yel. hazy liq.; m.w. 5000; sp.gr. 0.9967 (25/4 C); visc. 829 cSt; flash pt. (PMCC) 410 F.

Jeffcat DD. [Huntsman] Tert. amine; CAS 34745-96-5; catalyst for flexible polyether urethane slabstock foams; colorless clear liq.; water-sol.; sp.gr. 0.848 (20/20 C); f.p. < -49 C; b.p. 205 C; flash pt. (TCC) 170 F; *Toxicology:* LD50 (oral, rat) 350 mg/kg, (dermal, rabbit) 430 mg/kg; extremely irritating to skin and eyes; toxic by ingestion and skin absorption; *Storage:* store under dry inert gas blanket; reduce fire hazards.

Jeffcat DM-70. [Huntsman] Catalyst for producing ester or ether flexible PU slabstock foam, molded high resilience foam, one-component foam; sp.gr. 0.99 (20/20 C); f.p. -32 C; i.b.p. 150 C; flash pt. (TCC) 102 F; *Toxicology:* extremely irritating to skin and eyes; sl. to mod. toxicity by ingestion and skin absorption.

Jeffcat DMCHA. [Huntsman] Tert. amine; catalyst for rigid and semirigid urethane foams; for insulating applics. incl. spray, slabstock, laminations, and refrigeration; lt. yel. clear liq.; sol. in most polyols and org. solvs.; insol. in water; sp.gr. 0.85 (20/20 C); visc. 2.4 cp (20 C); f.p. < -78 C; b.p. 160 C (756 mm); flash pt. (PMCC) 104 F; *Toxicology:* LD50 (oral, rat) 350-565 mg/kg, (dermal, rabbit) 500-1000 mg/kg; mod. toxic by oral and dermal exposure; extremely irritating to eyes, skin, mucous membranes; *Precaution:* alkaline corrosive liq. (DOT); *Storage:* store under dry inert gas blanket.

Jeffcat DMDEE. [Huntsman] 2,2′-Dimorpholinodiethyl ether; catalyst for polyurethane flexible and rigid foam, one-component foam, slabstock foam applics.; sp.gr. 1.1 (20/20 C); f.p. -28 C; b.p. 309 C; flash pt. (TCC) 295 F; *Toxicology:* extremely irritating to skin and eyes; sl. to mod. toxicity by ingestion and skin absorption.

Jeffcat DME. [Huntsman] N,N-Dimethylethanolamine; CAS 108-01-0; EINECS 203-542-8; catalyst for polyurethane rigid foam, flexible ether slabstock; sp.gr. 0.88 (20/20 C); f.p. -59 C; b.p. 130 C; flash pt. (TCC) 105 F; *Toxicology:* extremely irritating to skin and eyes; sl. to mod. toxicity by ingestion and skin absorption.

Jeffcat DMP. [Huntsman] N,N′-Dimethylpiperazine; CAS 106-58-1; catalyst for polyurethane molded high resilience flexible foam, ester slabstock; sp.gr. 0.85 (20/20 C); f.p. -1 C; b.p. 133 C; flash pt. (TCC) 72 F; *Toxicology:* extremely irritating to skin and eyes; sl. to mod. toxicity by ingestion and skin absorption.

Jeffcat DPA. [Huntsman] N,N-(Dimethyl)-N′,N′-diisopropanol-1,3-propanediamine; patented catalyst for flexible and rigid PU foam, low-dens. (0.5-0.6 pcf) pkg. foam; lt. yel. liq., low odor; sp.gr. 0.94 (20/20 C); visc. 86 cSt; ; f.p. < -26 C; i.b.p. 212 C; hyd. no. 505-530; flash pt. (PMCC) 215 F; *Toxicology:* LD50 (oral, rat) 2.2 g/kg, (dermal, rabbit) 3.3 g/kg; sl. to mod. toxicity by ingestion and skin absorption; extremely irritating to skin and eyes.

Jeffcat DPA-50. [Huntsman] N,N-(Dimethyl)-N′,N′-diisopropanol-1,3-propanediamine (50%) and N,N-dimethylethanolamine (50%); catalyst for PU pkg. foam; flash pt. (PMCC) 110 F; *Toxicology:* extremely irritating to skin and eyes; sl. to mod. toxicity by ingestion and skin absorption.

Jeffcat M-75. [Huntsman] Amine in solv.; catalyst for polyester PU flexible foams, polyester and polyether slabstock; Pt-Co 100 max. clear liq., low odor; sp.gr. 0.9014 (20/4 C); f.p. < -50 C; flash pt. (PMCC) 126 F; *Toxicology:* LD50 (oral, rat) 0.38 g/kg, (dermal, rabbit) 2.08 g/kg; toxic by ingestion; sl. toxic by skin absorption; extremely irritating to skin, eyes, respiratory system.

Jeffcat MM-70. [Huntsman] Catalyst for flexible PU ester slabstock foam; sp.gr. 0.94; f.p. -80 C; i.b.p. 157 C; flash pt. (TCC) 105 F; *Toxicology:* extremely irritating to skin and eyes; sl. to mod. toxicity by ingestion and skin absorption.

Jeffcat NEM. [Huntsman] N-Ethylmorpholine; CAS 100-74-3; EINECS 202-885-0; catalyst for flexible and rigid polyester and polyether PU foam, polyester slabstock, highly resilient flexible foams; colorless clear liq.; m.w. 115.2; sp.gr. 0.914 (20/20 C); dens. 7.6 lb/gal; ; f.p. -63 C; b.p. 138 C (760 mm); flash pt. (TCC) 90 F; *Toxicology:* LD50 (oral, rat) 1.8 g/kg; sl. toxic by ingestion; severely irritating to eyes and skin; TLV/TWA 20 ppm; *Precaution:* flamm.; avoid exposure to sparks and open flame; *Storage:* store under dry inert gas blanket.

Jeffcat NMM. [Huntsman] N-Methylmorpholine; CAS 109-02-4; EINECS 203-640-0; catalyst for flexible polyester urethane foam, rigid foam molding applics., high rise rigid foam panels; low-color clear liq., penetrating ammoniacal odor; m.w. 101.15; sp.gr. 0.9213 (20/20 C); vapor pressure 18 mm (20 C); f.p. -65.9 C; b.p. 115.6 C; flash pt. (TCC) 55 F; *Toxicology:* LD50 (oral, rat) 2.7 g/kg, (dermal, rabbit) 1.4 ml/kg; sl. toxicity by ingestion and skin absorption; extremely irritating to skin and eyes; vapor may cause hazy vision; *Precaution:* flamm. and corrosive; reduce fire hazards; combustion prods.: CO, CO_2, ammonia; *Storage:* hygroscopic; prod. will discolor when exposed to air; store under dry nitrogen gas blanket.

Jeffcat T-10. [Huntsman] Stannous octoate sol'n. in dioctyl phthalate; catalyst for polyether PU foams; controls gelation, open cell content; synergistic with tert. amine catalysts; pale yel. liq.; sp.gr. 1.107 (20/4 C); visc. ≤ 180 cp (20 C); flash pt. (COC) ≥ 157 C; ref. index 1.490; 50% act.; ≥ 14% total tin; *Toxicology:* may cause skin and eye irritation on prolonged exposure; *Precaution:* susceptible to oxidation; *Storage:* store under dry nitrogen blanket in cool, dry place; repad open containers.

Jeffcat T-12. [Huntsman] Dibutyltin dilaurate; CAS 77-58-7; EINECS 201-039-8; catalyst for crosslinking of the polyol-isocyanate reaction in urethane industry; for high resilient molded foams, rigid foams; yel. clear liq.; sol. in most polyether polyols, plasticizers, org. solvs.; sp.gr. 1.07 (10/4 C); visc. ≤ 80 cp; flash pt. (COC) ≥ 180 C; ref. index 1.479 (10 C); 18.5% tin content; *Toxicology:* LD50 (rat) 200-300 mg/kg; irritating to skin and eyes; *Precaution:* somewhat susceptible to oxidation; *Storage:* store under inert gas blanket in cool, dry place.

Jeffcat TD-20. [Huntsman] Specialty amine; catalyst for PU ether flexible slabstock and rigid foam; sp.gr. 0.91 (20/20 C); f.p. < -20 C; i.b.p. 138 C; flash pt. (PMCC) 108 F; *Toxicology:* extremely irritating to skin and eyes; sl. to mod. toxicity by ingestion and skin absorption.

Jeffcat TD-33. [Huntsman] Triethylenediamine in propylene glycol; general purpose catalyst for producing polyurethane foam systems (flexible and rigid); sp.gr. 1.0 (20/20 C); f.p. -13 C; b.p. 185 C; flash pt. (PMCC) 206 F; 33% act.; *Toxicology:* extremely irritating to skin and eyes; sl. to mod. toxicity by ingestion and skin absorption.

Jeffcat TD-33A. [Huntsman] Triethylenediamine in dipropylene glycol; general purpose catalyst for producing polyurethane foam systems (flexible and rigid); sp.gr. 1.0 (20/20 C); f.p. < -24 C; flash pt. (PMCC) 196 F; 33% act.; *Toxicology:* extremely irritating to skin and eyes; sl. to mod. toxicity by ingestion and skin absorption.

Jeffcat Z-65. [Huntsman] Blend of aliphatic tert. amines in glycol diluent; catalyst for mfg. of flexible polyether PU slabstock foams; recommended for use with nonconventional foam softening and blowing agents; accelerates the cure rate, gives improved processing latitude and max. rise potential; Pt-Co < 150 clear liq.; sp.gr. 0.9105 (20/20 C); visc. 4.17 cp; b.p. 130 C; flash pt. (PMCC) 88 F; *Toxicology:* mod. toxic by single oral or dermal exposures; extremely irritating to eyes; corrosive to skin; may cause allergic skin reactions; vapors may cause hazy vision; *Storage:* store under dry inert gas blanket.

Jeffcat ZF-22. [Huntsman] Bis-(2-dimethylaminoethyl)ether, dipropylene glycol; general purpose catalyst for flexible and rigid PU foam, polyether slabstock; Pt-Co 100 clear liq.; sp.gr. 0.9041 (20/20 C); visc. 4 cp (20 C); ; f.p. < -60 C; b.p. 188 C; flash pt. (TCC) 164 F; 70% act.; *Toxicology:* extremely irritating to skin and eyes; irritating to mucous membranes, respiratory tract; mod. toxic by ingestion and skin absorption; *Precaution:* corrosive (DOT).

Jeffcat ZF-24. [Huntsman] Bis (2-dimethylaminoethyl) ether in dipropylene glycol; general purpose catalyst for flexible and rigid PU molded foam, polyether slabstock foams; Pt-Co 100 clear liq.; sp.gr. 0.9845 (20/20 C); visc. 37 cSt; f.p. -40 C; b.p. 204 C; flash pt. (TCC) 200 F; 23% act.; *Toxicology:* LD50 (oral, rat) 0.5-2.0 g/kg, (dermal, rabbit) 1.0-3.0 g/kg; mod. toxic by ingestion, sl. toxic by single dermal exposure; sl. irritating to skin and eyes; vapors may cuase respiratory irritation.

Jeffcat ZF-26. [Huntsman] Bis-(2-dimethylaminoethyl) ether in dipropylene glycol; general purpose catalyst for flexible and rigid PU foam, polyether slabstock; Pt-Co 100 clear liq.; sp.gr. 1.008 (20/20 C); visc. 54 cSt; f.p. -30 C; b.p. 216 C; flash pt. (TCC) 210 F; 11% act.; *Toxicology:* LD50 (oral, rat) 2.0-5.0 g/kg, (dermal, rabbit) 1.0-3.0 g/kg; sl. toxic by single oral and dermal exposure; sl. irritating to skin and eyes.

Jeffcat ZF-51. [Huntsman] Blend of Texacat TD-33A, ZF-22, and DME in dipropylene glycol; catalyst for flexible and rigid PU foam, polyether slabstock; lt. yel. clear liq.; sp.gr. 0.9814 (20/20 C); visc. 27 cSt; f.p. -40 C; i.b.p. 152 C; flash pt. (PMCC) 148 F; *Toxicology:* sl. toxic by single oral dose; mod. toxic by skin absorption; mod. irritating to eyes and skin; vapors may produce hazy vision; *Storage:* store under dry inert gas blanket.

Jeffcat ZF-52. [Huntsman] Blend of Texacat ZF-22, DME, and dipropylene glycol partially neutralized with formic acid; catalyst for PU ether flexible foam slabstock; sp.gr. 1.0 (20/20 C); f.p. -38 C; i.b.p. 131 C; flash pt. (PMCC) 166 F; *Toxicology:* extremely irritating to skin and eyes; sl. to mod. toxicity by ingestion and skin absorption.

Jeffcat ZF-53. [Huntsman] Blend of Texacat TD-33A and ZF-22; catalyst for PU ether flexible foam slabstock; sp.gr. 1.0 (20/20 C); f.p. -39 C; flash pt. (PMCC) 178 F; *Toxicology:* extremely irritating to skin and eyes; sl. to mod. toxicity by ingestion and skin absorption.

Jeffcat ZF-54. [Huntsman] Texacat ZF-22 partially neutralized with formic acid; catalyst for PU molded high resilience flexible foam; sp.gr. 1.1 (20/20 C); f.p. -38 C; b.p. 127 C; flash pt. (PMCC) 164 F; *Toxicology:* extremely irritating to skin and eyes; sl. to mod. toxicity by ingestion and skin absorption.

Jeffcat ZR-50. [Huntsman] N,N-Bis-(3-dimethylaminopropyl)-N-isopropanolamine; catalyst for flexible and rigid PU foam, low-dens. pkg. foam, highly resilient molded foams; Pt-Co 100 max. clear liq.; low odor; sp.gr. 0.8895 (20/4 C); b.p. 290 C; flash pt. (PMCC) 285 F; *Toxicology:* LD50 (oral, rat) 1344 mg/kg, (dermal, rabbit) 3570 mg/kg; mod. toxic by ingestion; pract. nontoxic by skin absorption; corrosive to skin and eyes; *Precaution:* corrosive (DOT).

Jeffcat ZR-70. [Huntsman] 2-(2-Dimethylaminoethoxy)ethanol; CAS 1704-62-7; EINECS 216-940-1; catalyst for polyurethane molded high resilience flexible foam, ether slabstock, pkg. foam; *Toxicology:* extremely irritating to skin and eyes; sl. to mod. toxicity by ingestion and skin absorption.

Jeffox PPG-400. [Huntsman] PPG-400; intermediate yielding esters; useful as lubricants, defoaming agents in rubber and pharmaceuticals, solvs. and humectant modifiers for inks, plasticizers, and functional fluids; water-wh. visc. liq., faint ether-like odor; m.w. 400; dens. 8.40 lb/gal; visc. 150-175 SUS (100 F); flash pt. (PMCC) 320 F; pour pt. -35 F; pH 5-7.

Jeffox PPG-2000. [Huntsman] PPG-2000; intermediate yielding esters; useful as lubricants, defoaming agents in rubber and pharmaceuticals, solvs. and humectant modifiers for inks, plasticizers, and functional fluids; water-wh. visc. liq., faint ether-like odor; m.w. 2000; dens. 8.37 lb/gal; flash pt. (PMCC) 370 F; pour pt. -25 F; pH 5-7.

Jet Fil® 200. [R.T. Vanderbilt] Platy talc, hydrous magnesium silicate; CAS 14807-96-6; EINECS 238-877-9; filler for PP and other resin systems; wh. powd.; 9 μm mean particle size; dens. 48 lb/ft^3 (tapped); oil absorp. 36; pH 9.5.

Jet Fil® 350. [Luzenac Am.; R.T. Vanderbilt] Platy talc, hydrous magnesium silicate; CAS 14807-96-6; EINECS 238-877-9; filler for PP and other resin systems; wh. powd.; 7.5 μm median particle size; fineness (Hegman) 4.0; bulk dens. 22-26 lb/ft^3 (loose); oil absorp. 40; brightness 84; pH 9.5.

Jet Fil® 500. [Luzenac Am.; R.T. Vanderbilt] Platy talc, hydrous magnesium silicate; CAS 14807-96-6; EINECS 238-877-9; filler for PP and other resin systems; antiblock; wh. powd.; 5.5 μm median particle size; fineness (Hegman) 5.0; bulk dens. 9-13 lb/ft^3 (loose); oil absorp. 44; brightness 86; pH 9.5.

Jet Fil® 575C. [Luzenac Am.] Talc; CAS 14807-96-6; EINECS 238-877-9; filler for general industrial, compacted plastics applics.; powd.; 3.4 μ median particle size; fineness (Hegman) 5.5; bulk dens. 25-30 lb/ft^3 (loose); brightness 87.

Jet Fil® 625C. [Luzenac Am.] Talc; CAS 14807-96-6; EINECS 238-877-9; filler for good impact, compacted plastics; powd.; 2.2 μ median particle size; fineness (Hegman) 6.0; bulk dens. 15-20 lb/ft^3 (loose); brightness 88.

Jet Fil® 700C. [Luzenac Am.; R.T. Vanderbilt] Talc; CAS 14807-96-6; EINECS 238-877-9; filler for polyolefins, PVC, elastomers; raises heat deflection temps., flex. mod. and impact str. in polyolefins; nucleating agent for mfg. of plastic resins; pellet; 1.5 μ median particle size; 100% thru 325 mesh; fineness (Hegman) 6.0; bulk dens. 65-70 lb/ft^3 (loose); bulking value 0.044; surf. area 13.5; oil absorp. 56; brightness 88; 98-99% talc, 1-2% magnesite; 61% SiO_2, 32% MgO.

K

KA 301. [Kenrich Petrochemicals] Diisobutyl (oleyl) aceto acetyl aluminate; coupling agent; lt. yel. liq.; sol. in IPA, xylene, toluene, DOP, min. oil; sp.gr. 0.97; visc. < 1000 cps; flash pt. 72 F; pH 6.0.

KA 322. [Kenrich Petrochemicals] Diisopropyl (oleyl) aceto acetyl aluminate; coupling agent; pale grn. liq.; sol. in xylene, toluene, DOP, min. oil; sp.gr. 0.99; visc. < 1000 cps; flash pt. 70 F; pH 6.

Kadox®-215. [Zinc Corp. of Am.] French process zinc oxide; CAS 1314-13-2; EINECS 215-222-5; pelleted Kadox®-15; pellets; 0.12 μ mean particle size; sp.gr. 5.6; sp.vol. 0.18; pkg. dens. 60 lb/ft^3; surf. area 9 m^2/g; oil absorp. 14 lb oil/100 lb ZnO; 99.6% zinc oxide.; Unverified

Kadox®-272. [Zinc Corp. of Am.] French process zinc oxide; CAS 1314-13-2; EINECS 215-222-5; pelleted form of Kadox®-72; 0.18 μ mean particle size; 99.99% thru 325 mesh; sp.gr. 5.6; sp.vol. 0.18; pkg. dens. 60 lb/ft^3; surf. area 6 m^2/g; oil absorp. 13 lb oil/100 lb ZnO; 99.7% zinc oxide.; Unverified

Kadox® 720. [Zinc Corp. of Am.] French process zinc oxide; CAS 1314-13-2; EINECS 215-222-5; pigment for use as activator for both wh. and colored rubber compds.; 0.18 μ mean particle size; 99.97% thru 325 mesh; sp.gr. 5.6; sp.vol. 0.18; dens. 50 lb/ft^3 resp.; surf. area 6.0 m^2/g; oil absorp. 13 lb oil/100 lb ZnO; 98.8% zinc oxide.

Kadox®-911. [Zinc Corp. of Am.] French process zinc oxide; CAS 1314-13-2; EINECS 215-222-5; pigment providing highest activating power and reinforcement in rubber; also used in insulated wire, transparent rubber, mechanical goods, latex, lubricating oils; improves lubrication and inhibits corrosion; avail. in pelleted form as Kadox 911P and propionic coated form as Kadox 911C; 0.12 μ mean particle size; 99.99% thru 325 mesh; sp.gr. 5.6; sp.vol. 0.18; pkg. dens. 30 lb/ft^3; surf. area 9 m^2/g; 99.6% zinc oxide.

Kadox®-920. [Zinc Corp. of Am.] French process zinc oxide; CAS 1314-13-2; EINECS 215-222-5; pigment used in latex, mechanical goods, insulated wire, footwear, and tires where uniform activation and a degree of reinforcement desired; also in the prod. of resinates, ceramics, textiles, phosphate solutions; avail. in pelleted form as Kadox 920P and propionic coated form as Kadox 920C; 0.18 μ mean particle size; 99.99% thru 325 mesh; sp.gr. 5.6; sp.vol. 0.18; pkg. dens. 35 lb/ft^3; surf. area 6 m^2/g; oil absorp. 13 lb oil/100 lb ZnO; 99.7% zinc oxide.

Kadox® 930. [Zinc Corp. of Am.] Am. process zinc oxide; CAS 1314-13-2; EINECS 215-222-5; pigment for use as activator for both wh. and colored rubber compds.; avail. in pelleted form as Kadox 930P and propionic coated form as Kadox 930C; 0.36 μ mean particle size; 99.97% thru 325 mesh; sp.gr. 5.6; sp.vol. 0.18; dens. 35 lb/ft^3; surf. area 3.2 m^2/g; oil absorp. 13 lb oil/100 lb ZnO; 98.9% zinc oxide.

Kane Ace® B-11A. [Kaneka] Impact modifier for PVC providing high impact str., glossy surf., clarity, easy processing in blow molding, extrusion, inj. and calendering to rigid and semirigid prods., and high shear str., glossy surf., leather feel to plasticized compds.; suitable for food pkg.; provides exc. clarity to film and sheet applics.; non-crease whitening; FDA 21CFR §178.3790; wh. free-flowing low-dusting powd.; < 1% on 16 mesh; bulk dens. > 0.35 g/cc; *Precaution:* inflamm.; can trigger dust explosions under certain conditions.

Kane Ace® B-12. [Kaneka] Impact modifier for PVC providing high impact str., glossy surf., clarity, easy processing in blow molding, extrusion, inj. and calendering to rigid and semirigid prods., and high shear str., glossy surf., leather feel to plasticized compds.; suitable for food pkg.; general purpose modifier for bottle, film, and sheet, sl. fluorescent color, moderate impact str.; FDA 21CFR § 178.3790; wh. free-flowing low-dusting powd.; < 1% on 16 mesh; bulk dens. > 0.35 g/cc; *Precaution:* inflamm.; can trigger dust explosions under certain conditions.

Kane Ace® B-18A-1. [Kaneka] Impact modifier for PVC providing high impact str., glossy surf., clarity, easy processing in blow molding, extrusion, inj. and calendering to rigid and semirigid prods., and high shear str., glossy surf., leather feel to plasticized compds.; suitable for food pkg.; general purpose modifier for bottle, film, and sheet, natural color, med. impact str.; FDA 21CFR §178.3790; wh. free-flowing low-dusting powd.; < 1% on 16 mesh; bulk dens. > 0.35 g/cc; *Precaution:* inflamm.; can trigger dust explosions under certain conditions.

Kane Ace® B-22. [Kaneka] Impact modifier for PVC providing high impact str., glossy surf., clarity, easy processing in blow molding, extrusion, inj. and calendering to rigid and semirigid prods., and high shear str., glossy surf., leather feel to plasticized compds.; suitable for food pkg.; general purpose modifier for bottle, film, and sheet, sl. fluorescent color, good impact str.; FDA 21CFR §178.3790; wh. free-flowing low-dusting powd.; < 1% on 16 mesh; bulk dens. > 0.35 g/cc; *Precaution:* inflamm.; can trigger dust explosions under certain conditions.

Kane Ace® B-28. [Kaneka] Impact modifier for PVC providing high impact str., glossy surf., clarity, easy processing in blow molding, extrusion, inj. and calendering to rigid and semirigid prods., and high shear str., glossy surf., leather feel to plasticized compds.; suitable for food pkg.; general purpose modifier for profile, pipe, fitting, plate, bottle, film, and sheet, natural color, exc. impact str.; FDA 21CFR §178.3790; wh. free-flowing low-dusting powd.; < 1% on 16 mesh; bulk dens. > 0.35 g/cc; *Precaution:* inflamm.; can trigger dust explosions under certain conditions.

Kane Ace® B-28A. [Kaneka] Impact modifier for PVC providing high impact str., good clarity; good for Ca-Zn formulations; FDA 21CFR §178.3790; *Precaution:* inflamm.; may cause dust explosions under certain conditions.

Kane Ace® B-31. [Kaneka] Impact modifier for PVC providing high impact str., glossy surf., clarity, easy processing in blow molding, extrusion, inj. and calendering to rigid and semirigid prods., and high shear str., glossy surf., leather feel to plasticized compds.; suitable for food pkg.; provides exc. clarity to film and sheet applics.; non-crease whitening; FDA 21CFR §178.3790; wh. free-flowing low-dusting powd.; < 1% on 16 mesh; bulk dens. > 0.35 g/cc; *Precaution:* inflamm.; can trigger dust explosions under certain conditions.

Kane Ace® B-38A. [Kaneka] Impact modifier for PVC providing good impact str., clarity, and oil resist. to cooking oil bottles; FDA 21CFR §178.3790; *Precaution:* inflamm.; may cause dust explosions under certain conditions.

Kane Ace® B-51. [Kaneka] Impact modifier for PVC providing high impact str. at low temp., high clarity; non-crease whitening; for sheet and plate applics.; FDA 21CFR §178.3790; *Precaution:* inflamm.; may cause dust explosions under certain conditions.

Kane Ace® B-52. [Kaneka] Impact modifier for PVC providing high impact str. and good clarity to sheet, film, bottle, and plate applics.; FDA 21CFR §178.3790; *Precaution:* inflamm.; may cause dust explosions under certain conditions.

Kane Ace® B-56. [Kaneka] Impact modifier for PVC providing very high impact str. to sheet, film, bottles, plate, and pipe; opaque; FDA 21CFR §178.3790; *Precaution:* inflamm.; may cause dust explosions under certain conditions.

Kane Ace® B-58. [Kaneka] Impact modifier for PVC providing high impact str. and good clarity to bottles, plate, and pipe; FDA 21CFR §178.3790; *Precaution:* inflamm.; may cause dust explosions under certain conditions.

Kane Ace® B-58A. [Kaneka] Impact modifier for PVC providing high impact str., good clarity; good for Ca-Zn formulations; FDA 21CFR §178.3790; *Precaution:* inflamm.; may cause dust explosions under certain conditions.

Kane Ace® B-513. [Kaneka] Impact modifier for PVC providing med. impact str., exc. clarity; non-crease whitening; FDA 21CFR §178.3790; *Precaution:* inflamm.; may cause dust explosions under certain conditions.

Kane Ace® B-521. [Kaneka] Impact modifier for PVC providing high impact str. at low temps. and good clarity to sheet, film, and plate applics.; FDA 21CFR §178.3790; *Precaution:* inflamm.; may cause dust explosions under certain conditions.

Kane Ace® B-522. [Kaneka] Impact modifier for PVC providing high impact str. and good clarity; FDA 21CFR §178.3790; *Precaution:* inflamm.; may cause dust explosions under certain conditions.

Kane Ace® FM-10. [Kaneka] Weatherable impact modifier for PVC, PC, PBT, and their alloys; for building and industrial materials, PVC pipes, soft PVC; improves impact str. and durability; minimal weathering discoloration over long periods; reduces low-temp. brittleness in soft PVC; high impact grade; *Precaution:* inflamm.; may cause dust explosions under certain conditions.

Kane Ace® FM-20. [Kaneka] Weatherable impact modifier for PVC, PC, PBT, and their alloys; for building and industrial materials, PVC pipes, soft PVC; improves impact str. and durability; minimal weathering discoloration over long periods; reduces low-temp. brittleness in soft PVC; ultra high impact grade; *Precaution:* inflamm.; may cause dust explosions under certain conditions.

Kane Ace® FT-80. [Kaneka] Weatherable impact modifier for PVC improving weatherability, cold temp. impact str., and moldability; well suited for inj. molding applics.; minimal weathering discoloration over long periods for enduring outdoor use; *Precaution:* inflamm.; may cause dust explosions under certain conditions.

Kane Ace® M-511. [Kaneka] Impact modifier for engineering plastics providing ultra high impact esp. at low temp.; opaque; for pipe; FDA 21CFR §178.3790; *Precaution:* inflamm.; may cause dust explosions under certain conditions.

Kane Ace® M-521. [Kaneka] Impact modifier for engineering plastics providing ultra high impact str., good processability to pipe; opaque; FDA 21CFR §178.3790; *Precaution:* inflamm.; may cause dust explosions under certain conditions.

Kane Ace® PA-10. [Kaneka] Acrylic resin; processing aid for PVC providing good flow and air mark resist., low m.w.; FDA 21CFR §178.3790; *Precaution:* inflamm.; can trigger dust explosions under certain conditions.

Kane Ace® PA-20. [Kaneka] Acrylic modifier; processing aid for PVC; improves primary and sec. processability and prod. appearance; for extrusion, inj. molding, calendering, rolling, blow molding, expansion molding, vacuum and pressure forming;

increases tens. str. and elong., heat resist., gelation speed, melt visc.; also effective for CPVC, ABS-PVC, CPE-PVC; suitable for food pkg.; FDA 21CFR §178.3790; wh. free-flowing low-dusting powd.; 100% thru 16 mesh; bulk dens. > 0.35 g/cc; *Precaution:* inflamm.; can trigger dust explosions under certain conditions.

Kane Ace® PA-30. [Kaneka] Acrylic resin; processing aid for PVC providing improved gel acceleration, vacuum forming, draw-down, hot elong., high melt str., foam processability, high m.w.; FDA 21CFR §178.3790; *Precaution:* inflamm.; can trigger dust explosions under certain conditions.

Kane Ace® PA-50. [Kaneka] Acrylic resin; processing aid for flexible and semirigid PVC providing improved gel acceleration, vacuum forming, draw-down, flow mark, hot elong.; FDA 21CFR §178.3790; *Precaution:* inflamm.; can trigger dust explosions under certain conditions.

Kane Ace® PA-100. [Kaneka] High-polymer modifier; processing aid for PVC providing improved lubricating processability, peeling from metal; FDA 21CFR §178.3790; *Precaution:* inflamm.; can trigger dust explosions under certain conditions.

Kanevinyl® XEL A. [Kaneka] Gloss inhibitor for extrusion (plasticized PVC).

Kanevinyl® XEL B. [Kaneka] Gloss inhibitor for inj. molding and calendering (plasticized PVC).

Kanevinyl® XEL C. [Kaneka] Gloss inhibitor for inj. molding, calendering, and blow molding.

Kanevinyl® XEL D. [Kaneka] Gloss inhibitor for extrusion, blow molding, inj. molding (rigid PVC).

Kanevinyl® XEL E. [Kaneka] Gloss inhibitor for calendering (rigid PVC) and inj. molding (rigid or plasticized PVC).

Kanevinyl® XEL F. [Kaneka] Gloss inhibitor for extrusion (rigid PVC).

Kantstik™ 66. [Specialty Prods.] Natural org. fatty compd. in aerosol form; environmentally friendly nonsilicone paintable external release agent for most thermosets and thermoplastic molding resins incl. PE, PS, PP, nylon, polyesters, epoxy, phenolics, polysulfone, vinyl resins, PU foams and elastomers; food grade mold spray; non-ozone depleting; non-CFC; improves flow on the mold; mold temp. range 40-500 F; *Toxicology:* nontoxic; ACGIH TLV 1000 ppm; *Precaution:* nonflamm. aerosol; keep away from heat, sparks, open flames; do not puncture or incinerate container; do not expose to heat above 120 F; *Storage:* 1 yr min shelf life @ R.T.

Kantstik™ 79. [Specialty Prods.] Natural org. fatty compd. with high flash min. spirits carrier; environmentally friendly paintable external release agent for most thermosets and thermoplastic molding resins incl. PE, PS, PP, nylon, polyesters, epoxy, phenolics, polysulfone, vinyl resins, PU foams and elastomers; food grade mold release; non-ozone depleting; improves flow on the mold; *Precaution:* combustible; keep away from open flames, sources of heat and ignition; *Storage:* keep containers closed at all times when not in use; store in dry, cool, well-ventilaterd areas away from heat source.

Kantstik™ 94. [Specialty Prods.] Org. polymer; semi-permanent dry film-forming mold release agent for thermoplastics (PE, PP, ABS, PC), thermosets (polyester, phenolic, epoxy) for inj., rotational, or compr. molding, gel coat laminates, resin transfer, casting, filament winding; provides mutliple releases, exacting part detail, and good luster; clear liq.; *Precaution:* flamm.; keep away from open flames, heat; if cloudiness, milky appearance, or precipitate are noticed, material has been contaminated; *Storage:* store in dry, cool, well-ventilated area away from heat in closed containers.

Kantstik™ 325. [Specialty Prods.] Blend; external lubricant for filament winding, BMC, SMC and TMC polyester, epoxy, phenolic, rubber; liq.

Kantstik™ 504. [Specialty Prods.] Fatty alcohol; water-based external lubricant for polyester FRP composites; liq.

Kantstik™ BW. [Specialty Prods.] Water-based silicone-free; environmentally friendly external release agent for phenolic, epoxy, and rubber resins molded at 180 F or higher; for compr. and/or transfer molding; can be removed from molds with water; yields paintable surfs. on molded parts; *Precaution:* nonflamm.

Kantstik™ FX-7. [Specialty Prods.] Complex ester; internal lubricant for polyester, epoxy, PU; liq.

Kantstik™ FX-9. [Specialty Prods.] Complex ester; internal lubricant for polyester, vinyl esters, acetals, furan, phenolics, thermoplastics, PE, PP, nylon, PS; liq.

Kantstik™ LM Conc. B. [Specialty Prods.] Water-based polymer dispersion; silicone-free; semipermanent external release agent for most thermoset (polyester, epoxy, phenolic melamine, PU, furan) and thermoplastic (acrylic, PS, PC, nylon, PVA) resin systems; provides dry film lubrication for multiple releases; also as mold release for printed circuit boards; environmentall friendly; wh. liq.; *Toxicology:* nonhazardous; *Precaution:* nonflamm.; *Storage:* prod. will freeze at 32 F; warm to restore to liq. state.

Kantstik™ M-55. [Specialty Prods.] Blend; external lubricant, mold release agent for compr. molding prepregs, polyester, epoxy, phenolic, SMC; liq.

Kantstik™ M-56. [Specialty Prods.] Silicone aq. emulsion; environmentally friendly external release agent for inj. molding thermoplastics, thermosets (polyester, epoxy, phenolics), and rubber (NR, EPDM, polyacetate, Viton), compr. molding, prepregs or preforms incl. SMC, BMC, TMC; exc. temp. resist.; non-ozone depleting; complies with the Federal Clean Air Act Amendment; wh. liq., sweet mild odor; misc. with water; b.p. 212 F; pH 6.5-7.5; 35-40% act.; *Precaution:* nonflamm.; free of ozone depleting substances; *Storage:* store @ R.T.

Kantstik™ PE. [Specialty Prods.] Aq./alcohol sol'n. of complex high m.w. polymers; environmentally friendly external mold release for PUR casting applics.; provides dry film lubrication for multiple releases; also suitable for RIM, RRIM molding processes; wh. clear liq., sl. pleasant alcoholic odor; *Toxicology:* contains alcohol; avoid contact with eyes and skin, inh. of vapors; *Storage:* store in cool, dry, well-ventilated area away from open flames and heat sources in tightly sealed containers.

Kantstik™ Q Powd. [Specialty Prods.] Refined syn. wax ester; mold release, internal lubricant for inj. molding; improves plastic flow in hard-to-reach mold areas; used in butyrate, PP, nylon, glass-filled nylon, PC, SAN, styrene, polyethylene, ABS, HIPS, PVC, BMC/SMC; improves lustrous surf.; color dispersant for thermoplastics; sl. ylsh. free-flowing granular powd., wax-like odor; insol. in water; sp.gr. 1.01-1.03 (20 C); flash pt. 220 C; *Toxicology:* LD50 (oral, mouse) > 20,000 mg/kg; nuisance dust—ACGIH TLV 10 mg/m^3 (total), 5 mg/m^3 (respirable); nonirritating to skin; *Precaution:* extreme heat may cause decarboxylation; avoid open flames; thermal decomp. may produce CO and CO_2.

Kantstik™ RIM. [Specialty Prods.] Water-based mold release for PU resin systems, RIM and RRIM applics.; good multiple release.

Kantstik™ Sealer. [Specialty Prods.] Polymeric; sealer providing hard surf. coating to tooling molds; eliminates mold porosity; for use on metal and FRP molds; withstands temps. in excess of 700 F; produces high gloss parts with exc. reproducibility; clear liq.; *Precaution:* flamm.; keep away from open flames, heat; if cloudiness, milky appearance, or precipitate are noticed, material has been contaminated; *Storage:* store in dry, cool, well-ventilated area away from heat in closed containers.

Kantstik™ SPC. [Specialty Prods.] Org. polymer; semipermanent external mold release agent for use when fabricating polyester composite parts; provides dry film lubrication for multiple releases; for casting, gel coat laminates, resin transfer, lay-up; also sutiable for epoxy and phenolic compoosit; withstands temps. in excess of 700 F; produces high gloss parts with exc. reproducibility; clear liq.; *Precaution:* flamm.; keep away from open flames, heat; if cloudiness, milky appearance, or precipitate are noticed, material has been contaminated; *Storage:* store in dry, cool, well-ventilated area away from heat in closed containers.

Kantstik™ S Powd. [Specialty Prods.] Amide wax; internal/external lubricant for extrusion, calendering, blow and inj. molding of thermoplastics (ABS, PVC, PS, SAN, PE, PP) and thermosets (polyester BMC); powd.; sp.gr. 0.97; m.p. 138-140 C; acid no. 3-10; flash pt. 277 C min.; dielec. str. 233 V/mil; dielec. const. 2.28 (10^6 cycles); vol. resist. 0.99 x 10^{13} ohm-cm.

Kaopolite® 1147. [Kaopolite] Aluminum silicate, anhyd.; CAS 1327-36-2; EINECS 215-475-1; abrasive for auto, metals, and plastics polishes, dentifrices, cleaning compds.; speeds cleaning without damage and develops glossy surf.; larger particles, whiter than SF type; powd.; 1.8 μ median particle size; 0.01% on 325 mesh; sp.gr. 2.62; dens. 21.91 lb/solid gal; bulking value 0.046 gal/lb; oil absorp. 65; brightness 93; ref. index 1.62; pH 5.5 (20% solids); hardness (Mohs) 5.

Kaopolite® 1152. [Kaopolite] Aluminum silicate, anhyd.; mild polishing props. for auto polishes, plastics; gentle abrasive that speeds cleaning; antiblocking agent for plastic film; powd.; 0.8 μ median particle size; 0.01% on 325 mesh; disp. in aq. and nonaq. systems; sp.gr. 2.62; dens. 21.91 lb/solid gal; bulking value 0.046 gal/lb; oil absorp. 90; brightness 93; ref. index 1.62; pH 5.5 (20% solids); hardness (Mohs) 5.

Kaopolite® 1168. [Kaopolite] Aluminum silicate, anhyd.; CAS 1327-36-2; EINECS 215-475-1; fastest cutting abrasive for auto, metals, and plastics polishes, cleaning compds.; antiblocking agent for plastic film; powd.; 1.8 μ median particle size; 0.01% on 325 mesh; sp.gr. 2.62; dens. 21.91 lb/solid gal; bulking value 0.046 gal/lb; oil absorp. 55; brightness 91; ref. index 1.62; pH 5.0 (20% solids); hardness (Mohs) 7.

Kaopolite® SF. [Kaopolite] Aluminum silicate, anhyd.; CAS 1327-36-2; EINECS 215-475-1; gentle abrasive for auto, metal, and plastic polishes, household, dentifrices, cosmetics; antiblocking agent in plastic film; promotes good suspension in most liq. systems; easy passage in aerosol valves; fine dry powd.; 0.7 μ median particle size; 0.01% on 325 mesh; sp.gr. 2.62; dens. 21.91 lb/solid gal; bulking value 0.046 gal/lb; oil absorp. 50; brightness 90; ref. index 1.62; pH 4.7 (20% solids); hardness (Mohs) 7.

Kaopolite® SFO-NP. [Kaopolite] Aluminum silicate, anhyd., surface treated; CAS 1327-36-2; EINECS 215-475-1; gentle abrasive with enhanced oil wettability for auto, metals, and plastic polishes; ease of dispersion in solv. and oil-based systems; antiblocking agent in plastic film; powd.; 0.7 μ median particle size; sp.gr. 2.62; dens. 21.91 lb/solid gal; bulking value 0.046 gal/lb; oil absorp. 30; brightness 90; ref. index 1.62; hardness (Mohs) 7.

Kathon® LX. [Rohm & Haas] 5-Chloro-2-methyl-4-isothiazolin-3-one; CAS 26172-55-4; EINECS 247-500-7; antimicrobial, preservative for polymer emulsions.; Unverified

Kaydol®. [Witco/Petroleum Spec.] Wh. min. oil USP; emollient and lubricant in cosmetics and pharmaceuticals; also for food processing, agric., chemicals, pkg., textiles; catalyst carrier and plasticizer for polymers; FDA 21CFR §172.878, §178.3620a; water-wh., odorless, tasteless; sp.gr. 0.869-0.885; visc. 64-70 cSt (40 C); pour pt. -23 C; flash pt. 216 C.

Kaydol® S. [Witco/Petroleum Spec.] Wh. min. oil USP; binder, carrier, conditioner, deformer, dispersant, extender, heat transfer agent, lubricant, moisture barrier, plasticizer, protective agent and/or softener in adhesives, agric. chems., cleaning, pkg. plastics, textiles; FDA 21CFR §172.878, 178.3620(a); odorless, tasteless; sp.gr. 0.869-0.885; visc. 64.5-86 cSt (40 C); distill. pt. 243 C min. (10 mm, 2.5%); pour pt. 0 F; flash pt. (COC) 440 F.

Kayphobe-ABO. [Kaopolite] Kaolin, surface treated with 4% methyl hydrogen polysiloxane; ultrafine abrasive and polishing agent for industrial coatings, offset inks, plastic compds.; antiblocking agent for thin plastic film; free flowing agent; FDA 21CFR §186.1256, GRAS; wh. fine powd., odorless; disp. in org. systems; sp.gr. 2.63 (of kaolin); pH 4.5-7.0 (20% aq. susp.); > 96% act.; *Toxicology:* TLV 10 mg/m^3 (total dust); nuisance dust; *Precaution:* nonflamm., nonexplosive; aq. slurry is slippery.

KC 31. [Georgia Marble] Alumina trihydrate; flame retardant for most polymers.

Kemamide® B. [Witco Corp.] Behenamide; CAS 3061-75-4; EINECS 221-304-1; lubricant, slip, antiblock, and mold release agent for plastics, crayons, petrol. prods., asphalts, inks, metals, textiles; mold release agent for thermoplastic resins in inj. molding; defoamer and water repellent in industrial/household applic.; corrosion inhibitor; pigment grinding aid and dyestuff dispersant in paints, enamels, varnishes, and lacquers; intermediate for textile emulsifiers and softeners; foam stabilizer in household detergents; FDA compliance; Gardner 4 max. waxy solid, powd., and pellet; sol. (g/100 g solv. @ 60 C): > 10 g in IPA, MEK, 10 g in methanol and toluene; water-insol.; m.w. 312; dens. 0.807 g/ml (130 C); visc. 6.5 cP (130 C); m.p. 98-106 C; acid no. 4 max.; iodine no. 4 max.; flash pt. (COC) 257 C; 95% conc.

Kemamide® E. [Witco Corp.] Erucamide; CAS 112-84-5; EINECS 204-009-2; lubricant, slip, antiblock, and mold release agent for plastics, crayons, petrol. prods., asphalts, inks, metals, textiles; mold release agent for thermoplastic resins in inj. molding; defoamer and water repellent in industrial/household applic.; FDA compliance; Gardner 5 max solid; m.w. 335; sol. (g/100 g solv.) 25 g in chloroform, 8 g in IPA, 4 g in MEK, 3 g in methyl alcohol and toluene; water insol.; m.p. 76-86 C; acid no. 4 max.; iodine no. 70-80; 100% conc.

Kemamide® E-180. [Witco Corp.] Stearyl erucamide; CAS 10094-45-8; lubricant additive, friction modifier for high-temp. plastics applics.; FDA compliance; wh. powd.; sol. (g/100 g solv. @ 50 C) < 50 g in chloroform and toluene, 50 g in dichloroethane, 40 g in VM&P naphtha, 30 g in IPA, 25 g in MEK; dens. 0.8074 (110 C); m.p. 72-76 C; acid no. 3 max.; iodine no. 44; flash pt. (CC) 258 C.

Kemamide® E-221. [Witco Corp.] Erucyl erucamide; lubricant additive, friction modifier for high-temp. plastics applics.; wh. powd.; sol. (g/100 g solv. @ 40 C) < 50 g in dichloroethane and VM&P naphtha, 50 g in IPA, 30 g in MEK; dens. 0.8165 (110 C); m.p. 55-58 C; acid no. 4 max.; iodine no. 80; flash pt. (CC) 260 C.

Kemamide® O. [Witco Corp.] Oleamide, tech.; CAS 301-02-0; EINECS 206-103-9; lubricant, slip, antiblock, and mold release agent for plastics, crayons, petrol. prods., asphalts, inks, metals, textiles; mold release agent for thermoplastic resins in inj. molding; defoamer and water repellent in industrial/household applic.; release agent migrating from food pkg.; FDA 21CFR §175.105, 175.300, 175.380, 175.390, 178.3910, 179.45, 181.28; Gardner 7 max. solid; m.p. 68-78 C; acid no. 5 max.; iodine no. 72-90; 100% conc.

Kemamide® P-181. [Witco Corp.] Oleyl palmitamide; CAS 16260-09-6; lubricant additive, friction modifier for high-temp. plastics applics.; FDA compliance; wh. powd.; sol. (g/100 g solv @ 50 C) 50 g in MEK, < 50 g in dichlorethane, IPA, VM&P naphtha, toluene, 5 g in methyl alcohol; dens. 0.8076 g/ml (110 C); m.p. 69-72 C; acid no. 8 max.; iodine no. 43; flash pt. 262 C (CC).

Kemamide® S. [Witco Corp.] Stearamide; CAS 124-26-5; EINECS 204-693-2; lubricant, slip, antiblock, and mold release agent for plastics, crayons, petrol. prods., asphalts, inks, metals, textiles; mold release agent for thermoplastic resins in inj. molding; defoamer and water repellent in industrial/household applic.; release agent migrating from food pkg.; FDA 21CFR §175.105, 177.1210, 178.3860, 178.3910, 179.45, 181.28; Gardner 4 max. waxy solid, powd., and pellet; m.w. 278; sol. (g/100 g solv. @ 50 C) > 10 g in chloroform, 10 g in IPA, 4 g in MEK, 3 g in methyl alcohol, 2 g in toluene; dens. 0.809 g/ml (130 C); visc. 5.8 cP (130 C); m.p. 98-108 C; acid no. 4 max.; iodine no. 3 max.; flash pt. 246 C (COC); fire pt. 268 C (COC); 95% min. amide.

Kemamide® S-65. [Witco Corp.] Tallow amide.

Kemamide® S-180. [Witco Corp.] Stearyl stearamide; lubricant additive, friction modifier for high-temp. plastics applics.; wh. powd.; sol. (g/100 g solv. @ 60 C) 50 g in toluene, < 50 g in chloroform, 20 g in dichloroethane, 15 g in IPA, MEK, and VM&P naphtha; dens. 0.8042 g/ml (110 C); m.p. 92-95 C; acid no. 8 max.; iodine no. 1; flash pt. 246 C (CC).

Kemamide® S-221. [Witco Corp.] Erucyl stearamide; lubricant additive, friction modifier for high-temp. plastics applics.; wh. powd.; sol. (g/100 g solv. @ 50 C) < 50 g in chloroform and toluene, 20 g in VM&P naphtha, 14 g in dichloroethane and IPA, 7 g in MEK; dens. 0.7877 g/ml (110 C); m.p. 72-75 C; acid no. 8 max.; iodine no. 46; flash pt. 268 C (CC).

Kemamide® U. [Witco Corp.] Oleamide; CAS 301-02-0; EINECS 206-103-9; lubricant, slip, antiblock, and mold release agent for plastics, crayons, petrol. prods., asphalts, inks, metals, textiles; mold release agent for thermoplastic resins in inj. molding; defoamer and water repellent in industrial/household applic.; release agent migrating from food pkg.; FDA 21CFR §175.105, 175.300, 175.380, 175.390, 178.3910, 179.45, 181.28; Gardner 5 max. waxy solid, powd., and pellet; m.w. 275; sol. (g/100 g solv. @ 30 C) > 30 g in IPA, 28 g in methyl alcohol, 25 g in toluene, > 20 g in MEK; dens. 0.823 g/ml (130 C); visc. 5.5 cP (130 C); m.p. 68-78 C; acid no. 4 max.; iodine no. 72-90; flash pt. 245 C (COC); 95% min. amide.

Kemamide® W-20. [Witco Corp.] Ethylene dioleamide; CAS 110-31-6; EINECS 203-756-1; lubricant, slip, antiblock, and mold release agent for plastics, crayons, petrol. prods., asphalts, inks, metals, textiles; mold release agent for thermoplastic resins in inj. molding; defoamer and water repellent in industrial/household applic.; internal and external lubricants in ABS, PS, polyethylene, PP, PVC, nylon, cellulose acetate, PVAc, and phenolic resins; defoamer in paper industry blk. liquoring, fabric dyeing, latex systems; metal processing, asphalts; FDA compliance; Gardner 6 max. powd. and flake; sol. (g/100 g solv. @ 35 C) > 20 g in toluene, 12 g in IPA, 4 g in dichloroethane, 3 g in MEK; m.p. 120 C; acid no. 10 max.; flash pt. 296 C (COC); fire pt. 315 C (COC).

Kemamide® W-39. [Witco Corp.] Ethylene distearamide; lubricant, slip, antiblock, and mold release agent for plastics, crayons, petrol. prods., asphalts, inks, metals, textiles; mold release agent for thermoplastic resins in inj. molding; defoamer and water repellent in industrial/household applic.; release agent migrating from food pkg.; FDA 21CFR §175.105, 175.300, 175.380, 175.390, 176.170, 177.1200, 177.2470, 177.2480, 181.28; Gardner 18 max. flakes; sol. (g/100 g solv. @ 70 C) 2.0 g toluene, 1.6 g dichloroethane; 1.4 g IPA, 0.9 g MEK; m.p. 140 C; acid no. 10 max.; flash pt. 299 C (COC); fire pt. 315 C (COC).

Kemamide® W-40. [Witco Corp.] Ethylene distearamide; lubricant, slip, antiblock, and mold release agent for plastics, crayons, petrol. prods., asphalts, inks, metals, textiles; mold release agent for thermoplastic resins in inj. molding; defoamer and water repellent in industrial/household applic.; release agent migrating from food pkg.; FDA 21CFR §175.105, 175.300, 175.380, 175.390, 176.170, 177.1200, 177.2470, 177.2480, 181.28; Gardner 3 max. powd. and flake; sol. (g/100 g solv. @ 70 C) 2.0 g toluene, 1.6 g dichloroethane; 1.4 g IPA, 0.9 g MEK; m.p. 140 C; acid no. 10 max.; flash pt. 299 C (COC); fire pt. 315 C (COC).

Kemamide® W-40/300. [Witco Corp.] Ethylene distearamide; lubricant, slip, antiblock, and mold release agent for plastics, crayons, petrol. prods., asphalts, inks, metals, textiles; mold release agent for thermoplastic resins in inj. molding; defoamer and water repellent in industrial/household applic.; Gardner 3 max. atomized very fine powd.; m.p. 140 C; acid no. 10 max.

Kemamide® W-40DF. [Witco Corp.] Ethylene distearamide; defoamer grade lubricant, slip, antiblock, and mold release agent for plastics, crayons, petrol. prods., asphalts, inks, metals, textiles; mold release agent for thermoplastic resins in inj. molding; water repellent in industrial/household applics.; Gardner 3 max. powd. and flake; sol. (g/100 g solv. @ 70 C) 2.0 g toluene, 1.6 g dichloroethane; 1.4 g IPA, 0.9 g MEK; m.p. 140 C; acid no. 5 max.; flash pt. 299 C (COC); fire pt. 315 C (COC).

Kemamide® W-45. [Witco Corp.] Ethylene distearamide; lubricant, slip, antiblock, and mold release agent for plastics, crayons, petrol. prods., asphalts, inks, metals, textiles; mold release agent for thermoplastic resins in inj. molding; defoamer and water repellent in industrial/household applic.; release agent migrating from food pkg.; FDA 21CFR §175.105, 175.300, 175.380, 175.390, 176.170, 177.1200, 177.2470, 177.2480, 181.28; Gardner 3 max. powd. and flake; sol. (g/100 g solv. @ 70 C) 2.0 g toluene, 1.6 g dichloroethane; 1.4 g IPA, 0.9 g MEK; m.p. 145 C; acid no. 10 max.; flash pt. 304 C (COC); fire pt. 322 C.

Kemamine® AS-650. [Witco Corp.] Nitrogen derivs.; cationic; antistat for polyolefins, styrenics, and other plastics, esp. film applics.; lubricity aid, mold release aid, pigment dispersant; suitable for food pkg. applics.; FDA 21CFR §178.3130; pale straw clear liq.; sp.gr. 0.9058; m.p. 4 C; f.p. 40 F; 97% min. tert. amine; *Toxicology:* skin irritant; *Precaution:* corrosive (DOT).

Kemamine® AS-974. [Witco Corp.] Nitrogen derivs.; cationic; antistat for polyolefins, styrenics, and other plastics, esp. film applics.; lubricity aid, mold release aid, pigment dispersant; suitable for food pkg. materials; FDA 21CFR §178.3130; amber paste; sp.gr. 0.904; m.p. 25 C; 97% min. tert. amine; *Toxicology:* skin irritant.

Kemamine® AS-974/1. [Witco Corp.] Nitrogen derivs.; cationic; antistat for polyolefins, styrenics, and other plastics, esp. film applics.; lubricity aid, mold release aid, pigment dispersant; suitable for food pkg. materials; FDA 21CFR §178.3130; wh. free-flowing powd.; 60% min. tert. amine; *Toxicology:* skin irritant.

Kemamine® AS-989. [Witco Corp.] Nitrogen derivs.; cationic; antistat for polyolefins, styrenics, and other plastics, esp. film applics.; lubricity aid, mold release aid, pigment dispersant; suitable for food pkg. materials; FDA 21CFR §178.3130; pale straw clear liq.; sp.gr. 0.904; f.p. 31 F; m.p. 1 C; cloud pt. 55 F; 96.8% min. tert. amine; *Toxicology:* skin irritant; *Precaution:* corrosive (DOT).

Kemamine® AS-990. [Witco Corp.] Nitrogen derivs.; cationic; antistat for LDPE, LLDPE, and other plastics, esp. film applics.; lubricity aid, mold release aid, pigment dispersant; suitable for food pkg. materials; FDA 21CFR §178.3130; wh. free-flowing powd.; sp.gr. 0.8065 (100 C); m.p. 50-55 C; 97% min. tert. amine; *Toxicology:* skin irritant.

Kemamine® D-190. [Witco Corp.] Arachidyl-behenyl 1,3-propylenediamine; cationic; gasoline detergent, bactericide, corrosion inhibitor in petrol. prod., epoxy hardener, dispersant, asphalt emulsifier; Gardner 9 max. solid; iodine no. 10 max.; 88% conc.

Kemamine® D-650. [Witco Corp.] N-Coconut 1,3-propylenediamine; CAS 61791-63-7; EINECS 263-195-3; cationic; gasoline detergent, bactericide, corrosion inhibitor in petrol. prod., epoxy hardener, dispersant, asphalt emulsifier; Gardner 6 max. paste; 88% conc.

Kemamine® D-970. [Witco Corp.] N-hydrog. tallow 1,3-propylenediamine; CAS 68603-64-5; EINECS 271-696-6; cationic; gasoline detergent, bactericide, corrosion inhibitor in petrol. prod., epoxy hardener, dispersant, asphalt emulsifier; Gardner 6 max. solid; 88% conc.

Kemamine® D-974. [Witco Corp.] Tallowaminopropylamine; CAS 68439-73-6; EINECS 270-416-7; cationic; gasoline detergent, bactericide, corrosion inhibitor in petrol. prod., epoxy hardener, dispersant, asphalt emulsifier; Gardner 12 max. solid; 88% conc.

Kemamine® D-989. [Witco Corp.] N-Oleyl 1,3-propylenediamine; CAS 7173-62-8; EINECS 230-528-9; cationic; gasoline detergent, bactericide, corrosion inhibitor in petrol. prod., epoxy hardener, dispersant, asphalt emulsifier; Gardner 7 max. liq.; 88% conc.

Kemamine® D-999. [Witco Corp.] Soyaminopropylamine; gasoline detergent, bactericide, corrosion inhibitor in petrol. prod., epoxy hardener, dispersant, asphalt emulsifier; Gardner 8 max. paste; iodine no. 70 min.; 85% min. act.

Kemamine® DD-3680. [Witco Corp.] Dimer diamine; chemical intermediate, extender, crosslinking agent in polymeric systems; corrosion inhibitor; in epoxy systems; Gardner 14 max. color.

Kemamine® DP-3680. [Witco Corp.] Dimer diprimary amine; chemical intermediate, extender, crosslinking agent in polymeric systems; corrosion inhibitor; in epoxy systems; Gardner 14 max. color.

Kemamine® DP-3695. [Witco Corp.] Dimer diprimary amine; chemical intermediate, extender, crosslinking agent in polymeric systems; corrosion inhibitor; in epoxy systems; Garder 14 max. color; 93% conc.

Kemamine® P-190, P-190D. [Witco Corp.] 90% Arachidyl-behenyl primary amine (P-190D—dist.); cationic; emulsifier, flotation agent, dispersing and flushing agent, intermediate, used in metalworking oils, as fuel oil additive; mold release for rubber and plastics; lubricant and spinning aid in metalworking oils; Gardner 3 and 1 resp., solid; sol. in common org. solv.; 93, 97% conc.

Kemamine® P-650D. [Witco Corp.] Cocamine, dist.; CAS 61788-46-3; EINECS 262-977-1; cationic; emulsifier, flotation agent, dispersing and flushing agent, intermediate, used in metalworking oils, as fuel oil additive; mold release for rubber and plastics; lubricant and spinning aid in metalworking oils; Gardner 1 liq.; sol. in common org. solv.; iodine no. 12 max.; 97% conc.

Kemamine® P-970. [Witco Corp.] Hydrog. tallow amine (tech.); CAS 61788-45-2; EINECS 262-976-6; cationic; emulsifier, flotation agent, dispersing and flushing agent, intermediate, used in metalworking oils, as fuel oil additive; mold release for rubber and plastics; lubricant and spinning aid in metalworking oils; Garder 3 solid; sol. in common org. solvs.; iodine no. 3 max.; 93% conc.

Kemamine® P-970D. [Witco Corp.] Hydrog. tallow amine (dist.); CAS 61788-45-2; EINECS 262-976-6; cationic; emulsifier, flotation agent, dispersing and flushing agent, intermediate, used in metalworking oils, as fuel oil additive; mold release for rubber and plastics; lubricant and spinning aid in metalworking oils; Gardner 1 solid; sol. in common org. solv.; iodine no. 3 max.; 93, 97% conc.

Kemamine® P-974D. [Witco Corp.] Tallow amine (dist.); CAS 61790-33-8; EINECS 263-125-1; cationic; emulsifier, flotation agent, dispersing and flushing agent, intermediate, used in metalworking oils, as fuel oil additive; mold release for rubber and plastics; lubricant and spinning aid in metalworking oils; Gardner 1 paste; sol. in common org. solv.; iodine no. 38 min.; 97% conc.

Kemamine® P-989D. [Witco Corp.] Oleamine, dist.; CAS 112-90-3; EINECS 204-015-5; cationic; emulsifier, flotation agent, dispersing and flushing agent, intermediate, used in metalworking oils, as fuel oil additive; mold release for rubber and plastics; lubricant and spinning aid in metalworking oils; Gardner 1 liq.; sol. in common org. solv.; iodine no. 70 min.; 97% conc.

Kemamine® P-990, P-990D. [Witco Corp.] Stearamine (tech. and dist. resp.); CAS 124-30-1; EINECS 204-695-3; cationic; emulsifier, flotation agent, dispersing and flushing agent, intermediate, used in metalworking oils, as fuel oil additive; mold release for rubber and plastics; lubricant and spinning aid in metalworking oils; Gardner 3 and 1 resp., solid; sol. in common org. solv.; 93, 97% conc.

Kemamine® P-999. [Witco Corp.] Tech. oleic-linoleic amine; emulsifier, flotation agent, dispersing and flushing agent, intermediate, used in metalworking oils, as fuel oil additive; mold release for rubber and plastics; lubricant and spinning aid in metalworking oils; Gardner 5 max.; iodine no. 85 min.; 93% conc.

Kemamine® T-1902D. [Witco Corp.] Dist. dimethyl-90% arachidyl-behenyl tert. amine; cationic; chemical intermediate for quat. ammonium derivs. used for textiles; acid scavenger in petrol. prods.; epoxy hardener, catalyst in mfg. of flexible PU foams; Gardner 1 max. liq.; 95% conc.

Kemamine® T-6501. [Witco Corp.] Methyl dicoconut tert. amine; CAS 61788-62-3; EINECS 262-990-2; cationic; chemical intermediate for quat. ammonium derivs. used for textiles; acid scavenger in petrol. prods.; epoxy hardener, catalyst in mfg. of flexible PU foams; Gardner 3 max. liq.; 95% conc.

Kemamine® T-6502D. [Witco Corp.] Dist. dimethyl cocamine; CAS 61788-93-0; EINECS 263-020-0; cationic; chemical intermediate for quat. ammonium derivs. used for textiles; acid scavenger in petrol. prods.; epoxy hardener, catalyst in mfg. of flexible PU foams; Gardner 1 max.; 95% conc.

Kemamine® T-9701. [Witco Corp.] Dihydrog. tallow methylamine; CAS 61788-63-4; EINECS 262-991-8; cationic; chemical intermediate for quat. ammonium derivs. used for cosmetics and textiles; acid scavenger in petrol. prods.; epoxy hardener, catalyst in mfg. of flexible PU foams; corrosion inhibitor; gasoline additive; Gardner 3 max. liq.; 95% conc.

Kemamine® T-9702D. [Witco Corp.] Dist. dimethyl hydrog. tallow amine; CAS 61788-95-2; EINECS 263-022-1; cationic; chemical intermediate for quat. ammonium derivs. used for textiles; acid scavenger in petrol. prods.; epoxy hardener, catalyst in mfg. of flexible PU foams; Gardner 1 max. liq.; 95% conc.

Kemamine® T-9902D. [Witco Corp.] Dist. dimethyl

stearamine; CAS 124-28-7; EINECS 204-694-8; cationic; chemical intermediate for quat. ammonium derivs.; acid scavenger in petrol. prods.; epoxy hardener, catalyst in mfg. of flexible PU foams; corrosion inhibitor; gasoline additive; Gardner 1 max. liq.; 95% conc.

Kemamine® T-9972D. [Witco Corp.] Dist. dimethyl soyamine; CAS 61788-91-8; EINECS 263-017-4; cationic; chemical intermediate for quat. ammonium derivs. used for textiles; acid scavenger in petrol. prods.; epoxy hardener, catalyst in mfg. of flexible PU foams; Gardner 1 max. liq.; 95% conc.

Kemester® 143. [Witco Corp.] Methyl tallowate; intermediate in prod. of superamides, in metalworking lubricants, as solv.; lubricant, plasticizer for leather, rubber prods.; Gardner 8 max.; acid no. 4.0 max.; iodine no. 60 max.; sapon. no. 195 min.

Kemester® 4000. [Witco Corp.] Butyl oleate; CAS 142-77-8; EINECS 205-559-6; nonionic; emollient, wetting agent; plasticizer for textiles, leathers, elastomers, personal care prods.; Gardner 2 max. liq.; m.p. 2 C; acid no. 2 max.; iodine no. 72-81; sapon. no. 164-172; 100% conc.

Kemester® 5221SE. [Witco Corp.] PEG-2 stearate SE; CAS 9004-99-3; anionic; emollient, emulsifier, plasticizer, lubricant for cosmetics, rubber, textiles; Gardner 3 max. flake; m.p. 46-56 C; acid no. 95-105; iodine no. 3 max.; sapon. no. 163-178; 100% conc.

Kemester® 6000. [Witco Corp.] Glyceryl stearate; nonionic; cosmetic and industrial emulsifier, emollient; plasticizer for elastomers; Gardner 3 max. bead; m.p. 57-60 C; acid no. 3 max.; iodine no. 2 max.; sapon. no. 162-176; 100% conc.

Kemester® EGDS. [Witco/H-I-P] Glycol distearate; CAS 627-83-8; EINECS 211-014-3; intermediate in prod. of superamides, in metalworking lubricants, specialized solv.; industrial lubricant; opacifier, pearling additive, thickener for cosmetics and pharmaceuticals; internal/external lubricant for PVC; FDA accepted; Gardner 2 max. solid; m.p. 60-63 C; acid no. 6 max.; iodine no. 1 max.; sapon. no. 190-200; 100% conc.

Kemester® GMS (Powd.). [Witco/H-I-P] Glyceryl stearate; industrial lubricant; internal antistat for PE PP, PS, rigid PVC, BMC, SMC, epoxy, phenolic; FDA approved; wh. powd.; m.p. 58-59 C; acid no. 1.

KemGard 425. [Sherwin-Williams] Molybdate complex; flame retardant for thermoset polyester, PE, PP, PVC, PU.

KemGard 981. [Sherwin-Williams] Zinc phosphate complex; flame retardant for acrylics, PE, PP, PVC.

Kempore® 60/14FF. [Uniroyal] Azodicarbonamide-based; CAS 123-77-3; EINECS 204-650-8; chemical blowing agent for dynamic foaming processes, e.g., inj. molding, extrusion, and calendering; operating temp. 330-450 F; pale yel. powd.; sp.gr. 1.52 dens. 30 lb/ft^3; gas yield 200 ml/g.

Kencolor®. [Kenrich Petrochemicals] Pigment and catalyst dispersions for silicone rubber; low-visc. pastes or dry cuttable pastes.

Kencure™ C9P. [Kenrich Petrochemicals] Cumyl phenol; CAS 599-64-4; activator; low cost replacement for nonyl phenol, sl. less active; liq.

Kencure™ MPP. [Kenrich Petrochemicals] Curative activator and catalyst; sp.gr. 1.19.

Kencure™ MPPJ. [Kenrich Petrochemicals] Curative activator and catalyst; sp.gr. 1.13.

Kenflex® A. [Kenrich Petrochemicals] Aromatic hydrocarbon polymeric plasticizer; discoloring plasticizer, softener, processing aid for chlorosulfonated PE providing improved extrusion, exc. electricals, mold flow/release, high temp. stability, permanence, nonextractability, nonmigrance, nonvolatility; coupling for poly blends; improves ozone resist.; also for polysulfide, nitrile, neoprene and butyl rubber; lt. amber solid; m.w. 662; sp.gr. 1.08; m.p. 58-74 C; i.b.p. 204 C; acid no. 0.3; iodine no. 65; flash pt. (COC) 240 $\pm$ 20 C; *Toxicology:* LD50 (oral) > 20 g/kg (exceptionally low oral toxicity).

Kenflex® A-30. [Kenrich Petrochemicals] Plasticizer, softener for nitrile rubber; offers rapid incorporation; visc. oily liq.; sp.gr. 1.09.

Kenflex® N. [Kenrich Petrochemicals] Aromatic hydrocarbon polymeric plasticizer; discoloring plasticizer, softener, processing aid for chlorosulfonated PE providing superior processing, improved extrusion, exc. electricals, exc. mold flow and release, high temp. stability, permanence, nonextractability, nonmigrance, nonvolatility; coupling for poly blends; also for polysulfide, neoprene; yel. to lt. brn. liq.; m.w. 472 $\pm$ 15; sp.gr. 1.01; i.b.p. 176 C; flash pt. (COC) 190 $\pm$ 20 C.

Ken Kem® AA. [Kenrich Petrochemicals] Titanium acetylacetonate; sp.gr. 0.99.

Ken Kem® CP-45. [Kenrich Petrochemicals] Cumyl phenol; CAS 599-64-4; modifier for epoxy, furan, and phenolic resins; epoxy cure accelerator; chemical intermediate; Gardner > 18 liq.; sp.gr. 1.07; dist. range 340-730 F; pour pt. 30 F; hyd. no. 119; flash pt. (COC) 250 F min.; 45% act., 50% α-methyl styrene oligomer, 5% light end components.

Ken Kem® CP-99. [Kenrich Petrochemicals] Cumyl phenol; CAS 599-64-4; modifier for epoxy, furan, and phenolic resins; epoxy cure accelerator; chemical intermediate; cryst. powd.; sp.gr. 1.08; 99% act.

Kenlastic®. [Kenrich Petrochemicals] Polymeric dispersions of activators and accelerators for the polymer industry.

Kenlastic® K-6640. [Kenrich Petrochemicals] Red lead polymer disp. in EPM base rubber; processing aid for neoprene; red soft rubber-like slab; sp.gr. 4.56; 90% act.

Kenlastic® K-6641. [Kenrich Petrochemicals] Rubbermakers sulfur disp. in EPM base rubber; processing aid for rubber compounding; slab; sp.gr. 1.61; 80% act.

Kenlastic® K-6642. [Kenrich Petrochemicals] Fumed litharge polymer disp. in EPM base rubber; processing aid for chlorosulfonated PE; yel. soft rubber slab; sp.gr. 4.71; 90% act.

Kenlastic® K-9273. [Kenrich Petrochemicals] Zinc oxide polymer disp. in EPM base rubber; processing aid for rubber compds.; slab; sp.gr. 3.1; 85% act.

Ken-Mag®. [Kenrich Petrochemicals] Magnesium oxide in binder; CAS 1309-48-4; EINECS 215-171-9; processing aid for the polymer industry incl. neoprene and chlorosulfonated PE formulas; beige bar-form paste disp.; sp.gr. 2.02; dens. 126 lb/ft^3; 75% act.

Kenmix®. [Kenrich Petrochemicals] Paste dispersions of activators and accelerators for the polymer industry.

Kenmix® LItharge/P 5/1. [Kenrich Petrochemicals] Litharge paste disp. in aromatic hydrocarbon (Kenflex P); processing aid for chlorosulfonated PE; increases water resist., elec. props., weather resist., flexibility; yel. soft paste; sp.gr. 3.92; 83% act.

Kenmix® LItharge/P 9/1. [Kenrich Petrochemicals] Litharge paste disp. in aromatic hydrocarbon (Kenflex P); processing aid for chlorosulfonated PE; increases water resist., elec. props., weather resist., flexibility; yel. soft paste; sp.gr. 5.11; 90% act.

Kenmix® Red Lead/P 9.1. [Kenrich Petrochemicals] Red lead in 10% Kenflex P; processing aid for neoprene; red soft paste; sp.gr. 4.93; 90% act.

Kenplast® A-450. [Kenrich Petrochemicals] Plasticizer; sp.gr. 1.00.

Kenplast® BG. [Kenrich Petrochemicals] Plasticizer; sp.gr. 1.02.

Kenplast® ES-2. [Kenrich Petrochemicals] Cumylphenyl acetate; plasticizer for urethanes; high flash pt. reactive diluent for epoxy, reducing odor and irritation; ideal for polyamide-cured epoxy floorings and coal tar pipe coatings; comonomer and impact modifier for phenolics; improves adhesion of vinyl plastisols coating metals and polar plastics; Gardner 9 liq.; sp.gr. 1.08; dens. 9.0 lb/gal; visc. 325 cps; dist. range 321-355 C; flash pt. (COC) 160 C; .pour pt. -12 C; acid no. < 1; sapon. no. 154.

Kenplast® ESB. [Kenrich Petrochemicals] Cumylphenyl benzoate; primary plasticizer for PVC and PVC/nitrile; process aid for semirigid PVC extrudates; solvates conductive polyester and acrylic inks to improve impact; Gardner > 18 liq.; sp.gr. 1.13 dens. 9.4 lb/gal; visc. 15,000 cps; dist. range 250-427 C; pour pt. 16 C; acid no. < 1; sapon. no. 124; flash pt. (COC) 171 C; 70% act. in a-methyl styrene oligomer.

Kenplast® ESI. [Kenrich Petrochemicals] Biscumylphenyl trimellitate; plasticizer, process aid for extrusion of PVC, urethanes; lubricant and process aid for filled PS, PC; reactive diluent and flow promoter for epoxy powd. coatings; prevents oxidative depolymerization in nylons; Gardner > 18 powd.; sp.gr. 1.18; dens. 9.8 lb/gal; soften. pt. 80 C; acid no. < 1; sapon. no. 172; flash pt. (COC) 260 C; 90% act. in a-methyl styrene oligomer.

Kenplast® ESN. [Kenrich Petrochemicals] Cumylphenyl neodecanoate; primary plasticizer for PVC, PVC/nitrile; process aid for semirigid PVC extrusions; solvates and impact modifies conductive polyester and acrylic inks; Gardner > 18 liq.; sp.gr. 1.05; dens. 8.75 lb/gal; visc. 5800 cps; pour pt. 1.66 C; acid no. < 1; sapon. no. 130; flash pt. (COC) 196 C; 85% act. in a-methyl styrene oligomer.

Kenplast® G. [Kenrich Petrochemicals] Mixed alkylated phenanthrenes; nonreactive discoloring primary plasticizer for dark PVC, nitrile, and urethane compds.; flexibilizer and diluent for epoxy resins; amber color; sp.gr. 1.06; visc. 25 cps (100 F); i.b.p. 510 F; flash pt. (COC) 290 F; *Toxicology:* LD50 (oral, rat) ≈ 2.5 g/kg; nonmutagenic; *Storage:* will crystallize during storage below 50 F; warm before use to eliminate crystals; do not store above 70 F.

Kenplast® LG. [Kenrich Petrochemicals] Plasticizer; sp.gr. 1.08.

Kenplast® LT. [Kenrich Petrochemicals] Mixed dibasic ester; plasticizer for low temp. polymers incl. PVC, NBR, SBR; offers exc. heat stability and good electricals; coalescing agent, flow improver for latex alkyds and acrylics; Gardner 2 liq.; sp.gr. 0.9; visc. 20 cps; i.b.p. 320 F; sapon. no. 260; flash pt. (COC) 260 F.

Kenplast® PG. [Kenrich Petrochemicals] Phenyl glycol ether; plasticizer; nonreactive diluent and flexibilizer for epoxies; Gardner 3 liq.; sp.gr. 1.11; f.p. -50 F; b.p. 450 F; flash pt. (COC) 230 F.

Kenplast® PPE. [Kenrich Petrochemicals] Plasticizer; sp.gr. 1.20.

Kenplast® RD. [Kenrich Petrochemicals] Plasticizer; sp.gr. 0.992.

Kenplast® RDN. [Kenrich Petrochemicals] Plasticizer; sp.gr. 1.00.

Kenplast® RG. [Kenrich Petrochemicals] Plasticizer; sp.gr. 1.09.

Ken-React® 7 (KR 7). [Kenrich Petrochemicals] Isopropyl dimethacryl isostearoyl titanate (monoalkoxy), IPA; coupling agent which sometimes also acts as adhesion promoters, antioxidants, antistats, antifoaming agents, accelerators, blowing agent activators, catalysts, curatives, corrosion inhibitors, disp. aids, emulsifiers, flame retardants, foaming agents; grinding aids, hardeners, impact modifiers, internal lubes, process aids, release agents, stabilizers for thermoplastics, thermosets, elastomers; dk. red brn. liq.; sol. in IPA, xylene, toluene, DOP, min. oil; insol. in water; sp.gr. 1.02 (16 C); visc. 220 cps; b.p. 250 F (initial); pH 5 (sat. sol'n.); flash pt. (TCC) 110 F; 95+% solids in IPA.

Ken-React® 9S (KR 9S). [Kenrich Petrochemicals] Isopropyl tridodecylbenzenesulfonyl titanate (monoalkoxy), IPA; coupling agent which sometimes also acts as adhesion promoters, antioxidants, antistats, antifoaming agents, accelerators, blowing agent activators, catalysts, curatives, corrosion inhibitors, disp. aids, emulsifiers, flame retardants, foaming agents; grinding aids, hardeners, impact modifiers, internal lubes, process aids, release agents, stabilizers for thermoplastics, thermosets, elastomers; transparent redsh. br. liq.; sol. in IPA, xylene, toluene, min. oil; reacts slowly in DOP; insol. in water; sp.gr. 1.08 (16 C); visc. 8000 cps; b.p. 130 F (initial); pH 2 (sat. sol'n.); flash pt. (TCC) 97 F; 88+% solids in IPA; *Toxicology:* LD50 (oral, rat) 8 g/kg.

Ken-React® 12 (KR 12). [Kenrich Petrochemicals] Isopropyl tri(dioctylphosphato) titanate (monoalkoxy), IPA; coupling agent which sometimes also acts as adhesion promoters, antioxidants, antistats, antifoaming agents, accelerators, blowing agent activators, catalysts, curatives, corrosion inhibitors, disp. aids, emulsifiers, flame retardants, foaming agents; grinding aids, hardeners, impact modifiers, internal lubes, process aids, release agents, stabilizers for thermoplastics, thermosets, elastomers; translucent to translucent off-wh. liq.; sol. in IPA, xylene, toluene, min. oil; reacts slowly in DOP; insol. in water; sp.gr. 1.04 (16 C); visc. 1500 cps; b.p. 170 F (initial); pH 4.5 (sat. sol'n.); flash pt. (TCC) 150 F; 95+% solids in IPA; *Toxicology:* LD50 (oral, rat) 7 g/kg; severe skin irritant.

Ken-React® 26S (KR 26S). [Kenrich Petrochemicals] Isopropyl 4-aminobenzenesulfonyl di(dodecylbenzenesulfonyl) titanate (monoalkoxy), IPA; coupling agent which sometimes also acts as adhesion promoters, antioxidants, antistats, antifoaming agents, accelerators, blowing agent activators, catalysts, curatives, corrosion inhibitors, disp. aids, emulsifiers, flame retardants, foaming agents; grinding aids, hardeners, impact modifiers, internal lubes, process aids, release agents, stabilizers for thermoplastics, thermosets, elastomers; grey liq.; sol. in IPA, xylene, toluene; reacts slowly in DOP; insol. in min. oil, water; sp.gr. 1.12 (16 C); visc. 30,000 cps; b.p. 250 F (initial); pH 6.5 (sat. sol'n.); flash pt. (TCC) 75 F; 95+% solids in IPA; *Toxicology:* LD50 (oral, rat) 5 g/kg; severe skin irritant.

Ken-React® 33DS (KR 33DS). [Kenrich Petrochemicals] Alkoxy trimethacryl titanate (monoalkoxy), IPA; coupling agent which sometimes also acts as adhesion promoters, antioxidants, antistats, antifoaming agents, accelerators, blowing agent activators, catalysts, curatives, corrosion inhibitors, disp. aids, emulsifiers, flame retardants, foaming agents; grinding aids, hardeners, impact modifiers, internal lubes, process aids, release agents, stabilizers for thermoplastics, thermosets, elastomers; tan to red br. liq.; sol. in IPA, xylene, toluene; reacts slowly in DOP; insol. in min. oil, water; sp.gr. 1.11 (16 C); visc. 2500 cps; b.p. 170 F (initial); pH 3.5 (sat. sol'n.); flash pt. (TCC) 120 F; 78+% solids in IPA.

Ken-React® 38S (KR 38S). [Kenrich Petrochemicals] Isopropyl tri(dioctylpyrophosphato) titanate (monoalkoxy), IPA; coupling agent which sometimes also acts as adhesion promoters, antioxidants, antistats, antifoaming agents, accelerators, blowing agent activators, catalysts, curatives, corrosion inhibitors, disp. aids, emulsifiers, flame retardants, foaming agents; grinding aids, hardeners, impact modifiers, internal lubes, process aids, release agents, stabilizers for thermoplastics, thermosets, elastomers; yel. to amber liq.; sol. in IPA, xylene, toluene, min. oil; reacts slowly in DOP; insol. in water; sp.gr. 1.09 (16 C); visc. 1500 cps; b.p. 170 F (initial); pH 2 (sat. sol'n.); flash pt. (TCC) 100F; 99+% solids in IPA; *Toxicology:* LD50 (oral, rat) 5 g/kg; severe skin irritant.

Ken-React® 39DS (KR 39DS). [Kenrich Petrochemicals] Alkoxy triacryl titanate (monoalkoxy), IPA; coupling agent which sometimes also acts as adhesion promoters, antioxidants, antistats, antifoaming agents, accelerators, blowing agent activators, catalysts, curatives, corrosion inhibitors, disp. aids, emulsifiers, flame retardants, foaming agents; grinding aids, hardeners, impact modifiers, internal lubes, process aids, release agents, stabilizers for thermoplastics, thermosets, elastomers; tan to red br. liq.; sol. in IPA, xylene, toluene; reacts slowly in DOP; insol. in min. oil, water; sp.gr. 1.07 (16 C); visc. 100 cps; b.p. 220 F (initial); pH 3 (sat. sol'n.); flash pt. (TCC) 180 F; 49+% solids in IPA.

Ken-React® 41B (KR 41B). [Kenrich Petrochemicals] Tetraisopropyl di (dioctylphosphito) titanate (coordinate type), IPA; coupling agent which sometimes also acts as adhesion promoters, antioxidants, antistats, antifoaming agents, accelerators, blowing agent activators, catalysts, curatives, corrosion inhibitors, disp. aids, emulsifiers, flame retardants, foaming agents; grinding aids, hardeners, impact modifiers, internal lubes, process aids, release agents, stabilizers for thermoplastics, thermosets, elastomers; yel. liq.; sol. in IPA, xylene, toluene, DOP, min. oil; insol. in water; sp.gr. 0.96 (16 C); visc. 15 cps; b.p. 160 F (initial); pH 6 (sat. sol'n.); flash pt. (TCC) 130 F; 98+% solids in IPA; *Toxicology:* LD50 (oral, rat) 11.5 g/kg; nonsensitizing to skin.

Ken-React® 44 (KR 44). [Kenrich Petrochemicals] Isopropyl tri(N ethylamino-ethylamino) titanate (monoalkoxy), IPA; coupling agent which sometimes also acts as adhesion promoters, antioxidants, antistats, antifoaming agents, accelerators, blowing agent activators, catalysts, curatives, corrosion inhibitors, disp. aids, emulsifiers, flame retardants, foaming agents; grinding aids, hardeners, impact modifiers, internal lubes, process aids, release agents, stabilizers for thermoplastics, thermosets, elastomers; yel. br. liq.; sol. in IPA; reacts slowly in DOP; insol. in xylene, toluene, min. oil, water; sp.gr. 1.2 (16 C); visc. 36,000 cps; b.p. 180 F (initial); pH 10 (sat. sol'n.); flash pt. (TCC) 130F; 95+% solids in IPA.

Ken-React® 46B (KR 46B). [Kenrich Petrochemicals] Tetraoctyloxytitanium di (ditridecylphosphite) (coordinate type), IPA; coupling agent which sometimes also acts as adhesion promoters, antioxidants, antistats, antifoaming agents, accelerators, blowing agent activators, catalysts, curatives, corrosion inhibitors, disp. aids, emulsifiers, flame retardants, foaming agents; grinding aids, hardeners, impact modifiers, internal lubes, process aids, release agents, stabilizers for thermoplastics, thermosets, elastomers; pale yel. liq.; sol. in IPA, xylene, toluene, DOP, min. oil; insol. in water; sp.gr. 0.92 (16 C); visc. 50 cps; b.p. 160 F (initial); pH 6 (sat. sol'n.); flash pt. (TCC) 180 F; 98+% solids in IPA; *Toxicology:* LD50 (oral, rat) 5 g/kg; severe skin irritant.

Ken-React® 55 (KR 55). [Kenrich Petrochemicals] Tetra (2, diallyoxymethyl-1 butoxy titanium di (ditridecyl) phosphite (coordinate type), IPA; coupling agent which sometimes also acts as adhesion promoters, antioxidants, antistats, antifoaming agents, accelerators, blowing agent activators, catalysts, curatives, corrosion inhibitors, disp. aids, emulsifiers, flame retardants, foaming agents; grinding aids, hardeners, impact modifiers, internal lubes, process aids, release agents, stabilizers for thermoplastics, thermosets, elastomers; yel. liq.; sol. in IPA, xylene, toluene, DOP, min. oil; insol. in water; sp.gr. 0.97 (16 C); visc. 50 cps; b.p. 120 F (initial); pH 5 (sat. sol'n.); flash pt. (TCC) 200 F; 78+% solids in IPA; *Toxicology:* LD50 (oral, rat) 10.3 g/kg.

Ken-React® 133DS (KR 133DS). [Kenrich Petrochemicals] Titanium dimethacrylate oxyacetate (oxyacetate chelate type), IPA; coupling agent which sometimes also acts as adhesion promoters, antioxidants, antistats, antifoaming agents, accelerators, blowing agent activators, catalysts, curatives, corrosion inhibitors, disp. aids, emulsifiers, flame retardants, foaming agents; grinding aids, hardeners, impact modifiers, internal lubes, process aids, release agents, stabilizers for thermoplastics, thermosets, elastomers; amber liq.; limited sol. in IPA, xylene, toluene, min. oil; reacts slowly in DOP; insol. in water; sp.gr. 1.06 (16 C); visc. 100 cps; b.p. 272 F (initial); pH 2.5 (sat. sol'n.); flash pt. (TCC) 150 F; 45+% solids in IPA.

Ken-React® 134S (KR 134S). [Kenrich Petrochemicals] Titanium di (cumylphenylate) oxyacetate (oxyacetate chelate type), IPA; coupling agent which sometimes also acts as adhesion promoters, antioxidants, antistats, antifoaming agents, accelerators, blowing agent activators, catalysts, curatives, corrosion inhibitors, disp. aids, emulsifiers, flame retardants, foaming agents; grinding aids, hardeners, impact modifiers, internal lubes, process aids, release agents, stabilizers for thermoplastics, thermosets, elastomers; red liq.; sol. in xylene, toluene, DOP; insol. in IPA, min. oil, water; sp.gr. 1.14 (16 C); visc. 8000 cps; b.p. 180 F (initial); pH 5 (sat. sol'n.); flash pt. (TCC) 195 F; 95+% solids in IPA.

Ken-React® 138D (KR 138D). [Kenrich Petrochemicals] 2-(N,N-Dimethylamino methyl propanol adduct of KR 138S, IPA; coupling agent which sometimes also acts as adhesion promoters, antioxidants, antistats, antifoaming agents, accelerators, blowing agent activators, catalysts, curatives, corrosion inhibitors, disp. aids, emulsifiers, flame retardants, foaming agents; grinding aids, hardeners, impact modifiers, internal lubes, process aids, release agents, stabilizers for thermoplastics, thermosets, elastomers; tan brn. liq.; sol. in IPA, xylene, toluene, DOP, water; insol. in min. oil; sp.gr. 1.06 (16 C); visc. < 1000 cps; b.p. 180 F (initial); pH 8 (sat. sol'n.); flash pt. (TCC) 95 F; 94+% solids in IPA.

Ken-React® 138S (KR 138S). [Kenrich Petrochemicals] Titanium di (dioctylpyrophosphate) oxyacetate (oxyacetate chelate type), IPA; coupling agent which sometimes also acts as adhesion promoters, antioxidants, antistats, antifoaming agents, accelerators, blowing agent activators, catalysts, curatives, corrosion inhibitors, disp. aids, emulsifiers, flame retardants, foaming agents; grinding aids, hardeners, impact modifiers, internal lubes, process aids, release agents, stabilizers for thermoplastics, thermosets, elastomers; pale yel. liq.; sol. in IPA, xylene, toluene, DOP; insol. in min. oil, water; sp.gr. 1.12 (16 C); visc. 1000 cps; b.p. 160 F (initial); pH 3 (sat. sol'n.); flash pt. (TCC) 100 F; 99+% solids in IPA; *Toxicology:* LD50 (oral, rat) 5 g/kg; moderate skin irritant.

Ken-React® 158D (KR 158D). [Kenrich Petrochemicals] 2-(N,N-Dimethylamino) isubutanol adduct of KR 158, IPA; coupling agent which sometimes also acts as adhesion promoters, antioxidants, antistats, antifoaming agents, accelerators, blowing agent activators, catalysts, curatives, corrosion inhibitors, disp. aids, emulsifiers, flame retardants, foaming agents; grinding aids, hardeners, impact modifiers, internal lubes, process aids, release agents, stabilizers for thermoplastics, thermosets, elastomers; pale yel. liq.; sol. in IPA, water; limited sol. in xylene, toluene; reacts slowly in DOP; insol. in min. oil; sp.gr. 1.08 (16 C); visc. 400 cps; b.p. 162 F (initial); pH 7.2 (sat. sol'n.); flash pt. (TCC) 85 F; 92% solids in IPA.

Ken-React® 158FS (KR 158FS). [Kenrich Petrochemicals] Titanium di (butyl, octyl pyrophosphate) di (dioctyl, hydrogen phosphite) oxyacetate (oxyacetate chelate type), IPA; coupling agent which sometimes also acts as adhesion promoters, antioxidants, antistats, antifoaming agents, accelerators, blowing agent activators, catalysts, curatives, corrosion inhibitors, disp. aids, emulsifiers, flame retardants, foaming agents; grinding aids, hardeners, impact modifiers, internal lubes, process aids, release agents, stabilizers for thermoplastics, thermosets, elastomers; yel.-tan liq.; sol. in IPA, xylene, toluene, DOP; insol. in min. oil, water; sp.gr. 1.16 (16 C); visc. ≤ 200 cps; b.p. 170 F (initial); pH 3 (sat. sol'n.); flash pt. (TCC) 110; 80+% solids in IPA; *Toxicology:* LD50 (oral, rat) 7.8 g/kg; minimal skin irritant.

Ken-React® 212 (KR 212). [Kenrich Petrochemicals] Di (dioctylphosphato) ethylene titanate (A, B ethylene chelate type), IPA; coupling agent which sometimes also acts as adhesion promoters, antioxidants, antistats, antifoaming agents, accelerators, blowing agent activators, catalysts, curatives, corrosion inhibitors, disp. aids, emulsifiers, flame retardants, foaming agents; grinding aids, hardeners, impact modifiers, internal lubes, process aids, release agents, stabilizers for thermoplastics, thermosets, elastomers; pale orange-red liq.; sol. in IPA, xylene, toluene, DOP, min. oil; insol. in water; sp.gr. 0.98 (16 C); visc. 300 cps; b.p. 185 F (initial); pH 4.8 (sat. sol'n.); flash pt. (TCC) 70 F; 75+%

solids in IPA; *Toxicology:* LD50 (oral, rat) 5 g/kg; moderate skin irritant.

Ken-React® 238A (KR 238A). [Kenrich Petrochemicals] Acrylate functional amine adduct of KR 238S, IPA; coupling agent which sometimes also acts as adhesion promoters, antioxidants, antistats, antifoaming agents, accelerators, blowing agent activators, catalysts, curatives, corrosion inhibitors, disp. aids, emulsifiers, flame retardants, foaming agents; grinding aids, hardeners, impact modifiers, internal lubes, process aids, release agents, stabilizers for thermoplastics, thermosets, elastomers; yel. liq.; sol. in IPA, water; insol. in xylene, toluene, DOP, min. oil; sp.gr. 1.03 (16 C); visc. < 300 cps; b.p. 170 F (initial); pH 8 (saturated sol'n.); flash pt. (TCC) 95 F; 95+% solids in IPA.

Ken-React® 238J (KR 238J). [Kenrich Petrochemicals] Methacrylamide functional amine adduct of KR 238S, IPA; coupling agent which sometimes also acts as adhesion promoters, antioxidants, antistats, antifoaming agents, accelerators, blowing agent activators, catalysts, curatives, corrosion inhibitors, disp. aids, emulsifiers, flame retardants, foaming agents; grinding aids, hardeners, impact modifiers, internal lubes, process aids, release agents, stabilizers for thermoplastics, thermosets, elastomers; amber/red liq.; sol. in IPA, water; insol. in xylene, toluene, DOP, min. oil; sp.gr. 1.06 (16 C); visc. < 400 cps; b.p. 220 F (initial); pH 7 (saturated sol'n.); flash pt. (TCC) 85 F; 88+% solids in IPA; *Toxicology:* LD50 (oral, rat) 5.5 g/kg; nonsensitizing to skin.

Ken-React® 238M (KR 238M). [Kenrich Petrochemicals] Methacrylate functional amine adduct of KR 238S, IPA; coupling agent which sometimes also acts as adhesion promoters, antioxidants, antistats, antifoaming agents, accelerators, blowing agent activators, catalysts, curatives, corrosion inhibitors, disp. aids, emulsifiers, flame retardants, foaming agents; grinding aids, hardeners, impact modifiers, internal lubes, process aids, release agents, stabilizers for thermoplastics, thermosets, elastomers; yel. to amber liq.; sol. in IPA, xylene, toluene, DOP; insol. in min. oil, water; sp.gr. 1.04 (16 C); visc. < 400 cps; b.p. 170 F (initial); pH 5.5 (sat. sol'n.); flash pt. (TCC) 95 F; 94+% solids in IPA.

Ken-React® 238S (KR 238S). [Kenrich Petrochemicals] Di (dioctylpyrophosphato) ethylene titanate (A, B ethylene chelate type), IPA; coupling agent which sometimes also acts as adhesion promoters, antioxidants, antistats, antifoaming agents, accelerators, blowing agent activators, catalysts, curatives, corrosion inhibitors, disp. aids, emulsifiers, flame retardants, foaming agents; grinding aids, hardeners, impact modifiers, internal lubes, process aids, release agents, stabilizers for thermoplastics, thermosets, elastomers; pale orange-red liq.; sol. in IPA, xylene, toluene, DOP, min. oil; insol. in water; sp.gr. 1.08 (16 C); visc. 800 cps; b.p. 160 F (initial); pH 3 (sat. sol'n.); flash pt. (TCC) 70 F; 78+% solids in IPA; *Toxicology:* LD50 (oral, rat) 13.5 g/kg; nonsensitizing to skin.

Ken-React® 238T (KR 238T). [Kenrich Petrochemicals] Triethylamine adduct of KR 238S, IPA; coupling agent which sometimes also acts as adhesion promoters, antioxidants, antistats, antifoaming agents, accelerators, blowing agent activators, catalysts, curatives, corrosion inhibitors, disp. aids, emulsifiers, flame retardants, foaming agents; grinding aids, hardeners, impact modifiers, internal lubes, process aids, release agents, stabilizers for thermoplastics, thermosets, elastomers; pale yel. to orange/red liq.; sol. in IPA, water; insol. in xylene, toluene, DOP, min. oil; sp.gr. 1.02 (16 C); visc. < 200 cps; b.p. 170 F (initial); pH 7 (sat. sol'n.); flash pt. (TCC) 100 F; 95+% solids in IPA.

Ken-React® 262A (KR 262A). [Kenrich Petrochemicals] Acrylate functional amine adduct of KR 262A, IPA; coupling agent which sometimes also acts as adhesion promoters, antioxidants, antistats, antifoaming agents, accelerators, blowing agent activators, catalysts, curatives, corrosion inhibitors, disp. aids, emulsifiers, flame retardants, foaming agents; grinding aids, hardeners, impact modifiers, internal lubes, process aids, release agents, stabilizers for thermoplastics, thermosets, elastomers; pale yel. liq.; sol. in IPA, water; limited sol. in xylene, toluene, DOP; insol. in min. oil; sp.gr. 1.11 (16 C); visc. < 5000 cps; b.p. 170 F (initial); pH 6.5 (sat. sol'n.); flash pt. (TCC) 95 F; 88+% solids in IPA.

Ken-React® 262ES (KR 262ES). [Kenrich Petrochemicals] Di (butyl, methyl pyrophosphato) ethylene titanate di (dioctyl, hydrogen phosphite) (A, B ethylene chelate type), IPA; coupling agent which sometimes also acts as adhesion promoters, antioxidants, antistats, antifoaming agents, accelerators, blowing agent activators, catalysts, curatives, corrosion inhibitors, disp. aids, emulsifiers, flame retardants, foaming agents; grinding aids, hardeners, impact modifiers, internal lubes, process aids, release agents, stabilizers for thermoplastics, thermosets, elastomers; yel.-tan liq.; sol. in IPA, xylene, toluene, DOP, min. oil; insol. in water; sp.gr. 1.17 (16 C); visc. 100 cps; b.p. 170 F (initial); pH 2 (sat. sol'n.); flash pt. (TCC) 85 F; 82+% solids in IPA.

Ken-React® OPP2 (KR OPP2). [Kenrich Petrochemicals] Dicyclo (dioctyl) pyrophosphato titanate, IPA; coupling agent which sometimes also acts as adhesion promoters, antioxidants, antistats, antifoaming agents, accelerators, blowing agent activators, catalysts, curatives, corrosion inhibitors, disp. aids, emulsifiers, flame retardants, foaming agents; grinding aids, hardeners, impact modifiers, internal lubes, process aids, release agents, stabilizers for thermoplastics, thermosets, elastomers; ylsh.-br. liq.; sol. in xylene, toluene; limited sol. in min. oil; reacts slowly in IPA, DOP, water; sp.gr. 1.13 (16 C); visc. 7000 cps; b.p. 210 F (initial); pH 5 (sat. sol'n.); flash pt. (TCC) 140 F; 66% solids in IPA.

Ken-React® OPPR (KR OPPR). [Kenrich Petrochemicals] Cyclo (dioctyl) pyrophosphato dioctyl titanate, IPA; coupling agent which sometimes also acts as adhesion promoters, antioxidants, antistats, antifoaming agents, accelerators, blowing agent activators, catalysts, curatives, corrosion inhibitors, disp. aids, emulsifiers, flame retardants, foaming agents; grinding aids, hardeners, impact modifiers, internal lubes, process aids, release agents, stabilizers for thermoplastics, thermosets, elastomers; ylsh.-br. liq.; sol. in xylene, toluene; limited sol. in min. oil; reacts slowly in IPA, DOP, water; sp.gr. 1.06 (16 C); visc. 110 cps; b.p. 280 F (initial); pH 6 (sat. sol'n.); flash pt. (TCC) 160 F; 66% solids in IPA.

Ken-React® KR TTS. [Kenrich Petrochemicals] Isopropyl titanium triisostearate, IPA; CAS 61417-49-0; EINECS 262-774-8; coupling agent which sometimes also acts as adhesion promoters, antioxidants, antistats, antifoaming agents, accelerators, blowing agent activators, catalysts, curatives, corrosion inhibitors, disp. aids, emulsifiers, flame retardants, foaming agents; grinding aids, hardeners, impact modifiers, internal lubes, process aids, release agents, stabilizers for thermoplastics, thermosets, elastomers; reddish-brn. transparent liq.; sol. in IPA, xylene, toluene, DOP, min. oil; insol. i water; sp.gr. 0.95 (16 C); visc. 125 cps; b.p. 300 F (initial); pH 5.5 (sat. sol'n.); flash pt. (TCC) > 200 F; 95+% solids in IPA; *Toxicology:* LD50 (oral, rat) 30 g/kg; moderate skin irritant.

Ken-Stat™ KS N100. [Kenrich Petrochemicals] Antistat for ABS, acrylics, nylon, acetal, PC.

Ken-Stat™ KS Q100P. [Kenrich Petrochemicals] Internal antistat for cellulosics, PE, PP, PS, flexible and rigid PVC, polyester.

Ken-Stat™ ZZ-1441H. [Kenrich Petrochemicals] Neoalkoxy organometallic in 50% silica carrier; antistat for most polymers; powd.; 50% act.

Ken-Stat™ ZZ-1441L. [Kenrich Petrochemicals] Neoalkoxy organometallic in 80% LLDPE; antistat for polyolefins; pellets; 20% act.

Ken-Stat™ ZZ-1441R. [Kenrich Petrochemicals] Neoalkoxy organometallic in 80% Resin 18; antistat for PVC, PS; powd.; 20% act.

Ken-Zinc®. [Kenrich Petrochemicals] Zinc oxide disp. in 15% hydrocarbon; CAS 1314-13-2; EINECS 215-222-5; processing aid for neoprene providing improved tens. props. and good aging; also suitable for NR and SBR; bar-form paste disp.; sp.gr. 3.25; dens. 203 lb/ft^3; 85% solids.

KEP 010P, 020P, 070P. [Goldsmith & Eggleton] Ethylene propylene copolymer; CAS 9010-79-1; EPM rubber used as polyolefin modifier; also for wire and cable insulation compds., brake parts; peroxide-curable; wh. to light gray bales; sp.gr. 0.86; Mooney visc. 19, 24, and 70 resp. (ML1+4, 100 C); Unverified

Kessco® PEG 200 DL. [Stepan; Stepan Canada] PEG-4 dilaurate; CAS 9005-02-1; emulsifier, thickener, solubilizer, emollient, opacifier, spreading agent, wetting agent, dispersant for cosmetics (creams/lotions, makeup, bath oils, shampoos, conditioners, sunscreens), pharmaceuticals (ointments, suppositories), food, agric., plastics, etc.; lubricant, emulsifier, softener for textile and metal-working applics.; lt. yel. liq.; sol. in IPA, acetone, CCl_4, ethyl acetate, toluol, IPM, wh. oil; water-disp.; sp.gr. 0.951; dens. 7.9 lb/gal; f.p. < 9C; HLB 5.9; acid no. 10.0 max.; iodine no. 9.0; sapon. no. 176-186; flash pt. (COC) 460 F; fire pt. 510 F; 100% act.

Kessco® PEG 200 DO. [Stepan; Stepan Canada] PEG-4 dioleate; CAS 9005-07-6; nonionic; emulsifier, thickener, solubilizer, emollient, opacifier, spreading agent, wetting agent, dispersant for cosmetics (creams/lotions, makeup, bath oils, shampoos, conditioners, sunscreens), pharmaceuticals (ointments, suppositories), food, agric., plastics, etc.; lt. amber liq.; f.p. < -15 C; sol. in naptha, kerosene, IPA, acetone, CCl_4, ethyl acetate, toluol, IPM, peanut oil, wh. oil; water-disp.; sp.gr. 0.942; dens. 7.9 lb/gal; HLB 6.0; acid no. 10.0 max; sapon. no. 148-158; pH 5.0 (3%); flash pt. (COC) 545 F.

Kessco® PEG 200 DS. [Stepan; Stepan Canada] PEG-4 distearate; CAS 9005-08-7; nonionic; emulsifier, thickener, solubilizer, emollient, opacifier, spreading agent, wetting agent, dispersant for cosmetics (creams/lotions, makeup, bath oils, shampoos, conditioners, sunscreens), pharmaceuticals (ointments, suppositories), food, agric., plastics, etc.; wh. to cream soft solid; sol. in naptha, kerosene, IPA, acetone, CCl_4, ethyl acetate, toluol, IPM, peanut oil, wh. oil; water-disp.; sp.gr. 0.9060 (65 C); HLB 5.0; m.p. 34 C; acid no. 10.0 max.; sapon. no. 153-162; pH 5.0 (3%); flash pt. (COC) 475 F.

Kessco® PEG 200 ML. [Stepan; Stepan Canada] PEG-4 laurate; CAS 9004-81-3; nonionic; emulsifier, thickener, solubilizer, emollient, opacifier, spreading agent, wetting agent, dispersant for cosmetics (creams/lotions, makeup, bath oils, shampoos, conditioners, sunscreens), pharmaceuticals (ointments, suppositories), food, agric., plastics, etc.; lt. yel. liq.; f.p. < 5 C; sol. in IPA, acetone, CCl_4, ethyl acetate, toluol, IPM, water-disp.; sp.gr. 0.985; dens. 8.2 lb/gal; HLB 9.8; acid no. 5.0 max.; sapon. no. 132-142; pH 4.5; flash pt. (COC) 385 F.

Kessco® PEG 200 MO. [Stepan; Stepan Canada] PEG-4 oleate; CAS 9004-96-0; EINECS 233-293-0; nonionic; emulsifier, thickener, solubilizer, emollient, opacifier, spreading agent, wetting agent, dispersant for cosmetics (creams/lotions, makeup, bath oils, shampoos, conditioners, sunscreens), pharmaceuticals (ointments, suppositories), food, agric., plastics, etc.; lt. amber liq.; f.p. < -15 C; sol. in IPA, acetone, CCl_4, ethyl acetate, toluol, water-disp.; sp.gr. 0.973; dens. 8.1 lb/gal; HLB 8.0; acid no. 5.0 max.; sapon. no. 115-124; pH 5.0; flash pt. (COC) 395 F.

Kessco® PEG 200 MS. [Stepan; Stepan Canada] PEG-4 stearate; CAS 9004-99-3; EINECS 203-358-8; nonionic; emulsifier, thickener, solubilizer, emollient, opacifier, spreading agent, wetting agent, dispersant for cosmetics (creams/lotions, makeup, bath oils, shampoos, conditioners, sunscreens), pharmaceuticals (ointments, suppositories), food, agric., plastics, etc.; wh. to cream soft solid; sol. in IPA, acetone, CCl_4, ethyl acetate, toluol, IPM, peanut oil; water-disp.; sp.gr. 0.9360

(65 C); HLB 7.9; m.p. 31 C; acid no. 5.0 max.; sapon. no. 120-129; flash pt. (COC) 410 F.

Kessco® PEG 300 DL. [Stepan; Stepan Canada] PEG-6 dilaurate; CAS 9005-02-1; nonionic; emulsifier, thickener, solubilizer, emollient, opacifier, spreading agent, wetting agent, dispersant for cosmetics (creams/lotions, makeup, bath oils, shampoos, conditioners, sunscreens), pharmaceuticals (ointments, suppositories), food, agric., plastics, etc.; lt. yel. liq.; f.p. < 13 C; sol. in naptha, IPA, acetone, toluol, IPM; water-disp.; sp.gr. 0.975; dens. 8.1 lb/gal; HLB 9.8; acid no. 10.0 max.; sapon. no. 148-158; flash pt. (COC) 475 F.

Kessco® PEG 300 DO. [Stepan; Stepan Canada] PEG-6 dioleate; CAS 9005-07-6; nonionic; emulsifier, thickener, solubilizer, emollient, opacifier, spreading agent, wetting agent, dispersant for cosmetics (creams/lotions, makeup, bath oils, shampoos, conditioners, sunscreens), pharmaceuticals (ointments, suppositories), food, agric., plastics, etc.; lt. amber liq.; f.p. < -5 C; sol. in IPA, acetone, CCl_4, ethyl acetate, toluol, IPM, peanut oil; water-disp.; sp.gr. 0.962; dens. 8.0 lb/gal; HLB 7.2; acid no. 10.0 max.; sapon. no. 128-137; pH 5.0; flash pt. (COC) 510 F.

Kessco® PEG 300 DS. [Stepan; Stepan Canada] PEG-6 distearate; CAS 9005-08-7; nonionic; emulsifier, thickener, solubilizer, emollient, opacifier, spreading agent, wetting agent, dispersant for cosmetics (creams/lotions, makeup, bath oils, shampoos, conditioners, sunscreens), pharmaceuticals (ointments, suppositories), food, agric., plastics, etc.; wh. to cream soft solid; sol. in naptha, kerosene, IPA, acetone, CCl_4, ethyl acetate, toluol, IPM, peanut oil, wh. oil; water-disp.; HLB 6.5; m.p. 32 C; acid no. 10.0 max.; sapon. no. 130-139; pH 5.0.

Kessco® PEG 300 ML. [Stepan; Stepan Canada] PEG-6 laurate; CAS 9004-81-3; EINECS 219-136-9; nonionic; emulsifier, thickener, solubilizer, emollient, opacifier, spreading agent, wetting agent, dispersant for cosmetics (creams/lotions, makeup, bath oils, shampoos, conditioners, sunscreens), pharmaceuticals (ointments, suppositories), food, agric., plastics, etc.; lt. yel. liq.; f.p. < 8C; sol. in IPA, acetone, CCl_4, ethyl acetate, toluol; water-disp.; sp.gr. 1.011; dens. 8.4 lb/gal; HLB 11.4; acid no. 5.0 max.; sapon. no. 104-114; pH 4.5; flash pt. (COC) 445 F.

Kessco® PEG 300 MO. [Stepan; Stepan Canada] PEG-6 oleate; CAS 9004-96-0; nonionic; emulsifier, thickener, solubilizer, emollient, opacifier, spreading agent, wetting agent, dispersant for cosmetics (creams/lotions, makeup, bath oils, shampoos, conditioners, sunscreens), pharmaceuticals (ointments, suppositories), food, agric., plastics, etc.; lt. amber liq.; f.p. < -5 C; sol. in IPA, acetone, CCl_4, ethyl acetate, toluol; water-disp.; sp.gr. 0.998; dens. 8.3 lb/gal; HLB 9.6; acid no. 5.0 max.; sapon. no. 94-102; pH 5.0 (3% disp.); flash pt. (COC) 450 F.

Kessco® PEG 300 MS. [Stepan; Stepan Canada] PEG-6 stearate; CAS 9004-99-3; nonionic; emulsifier, thickener, solubilizer, emollient, opacifier, spreading agent, wetting agent, dispersant for cosmetics (creams/lotions, makeup, bath oils, shampoos, conditioners, sunscreens), pharmaceuticals (ointments, suppositories), food, agric., plastics, etc.; wh. to cream soft solid; sol. in IPA, acetone, CCl_4, ethyl acetate, toluol, IPM, peanut oil; water-disp.; sp.gr. 0.9660 (65 C); HLB 9.7; m.p. 28 C; acid no. 5.0 max.; sapon. no. 97-105; pH 5.0 (3% disp.); flash pt. (COC) 475 F.

Kessco® PEG 400 DL. [Stepan; Stepan Canada] PEG-8 dilaurate; CAS 9005-02-1; nonionic; emulsifier, thickener, solubilizer, emollient, opacifier, spreading agent, wetting agent, dispersant for cosmetics (creams/lotions, makeup, bath oils, shampoos, conditioners, sunscreens), pharmaceuticals (ointments, suppositories), food, agric., plastics, etc.; lt. yel. liq.; f.p. 18 C; sol. in naptha, IPA, acetone, CCl_4, ethyl acetate, toluol, IPM, peanut oil; water-disp.; sp.gr. 0.990; dens. 8.3 lb/gal; HLB 9.8; acid no. 10.0 max.; sapon. no. 127-137; flash pt. (COC) 480 F.

Kessco® PEG 400 DO. [Stepan; Stepan Canada] PEG-8 dioleate; CAS 9005-07-6; nonionic; emulsifier, thickener, solubilizer, emollient, opacifier, spreading agent, wetting agent, dispersant for cosmetics (creams/lotions, makeup, bath oils, shampoos, conditioners, sunscreens), pharmaceuticals (ointments, suppositories), food, agric., plastics, etc.; lt. amber liq.; f.p. < 7 C; sol. in naptha, IPA, acetone, CCl_4, ethyl acetate, toluol, IPM, peanut oil; water-disp.; sp.gr. 0.977; dens. 8.1 lb/gal; HLB 8.5; acid no. 10.0 max.; sapon. no. 113-122; pH 5.0; flash pt. (COC) 520 F.

Kessco® PEG 400 DS. [Stepan; Stepan Canada] PEG-8 distearate; CAS 9005-08-7; emulsifier, thickener, solubilizer, emollient, opacifier, spreading agent, wetting agent, dispersant for cosmetics (creams/lotions, makeup, bath oils, shampoos, conditioners, sunscreens), pharmaceuticals (ointments, suppositories), food, agric., plastics, etc.; lubricant, emulsifier, softener for textile and metalworking applics.; wh. to cream soft solid; sol. in naptha, IPA, acetone, CCl_4, ethyl acetate, toluol, IPM, peanut oil, wh. oil; water-disp; sp.gr. 0.9390 (65 C); HLB 8.5; m.p. 36 C; acid no. 10.0 max.; sapon. no. 115-124; pH 5.0 (3% disp.); flash pt. (COC) 500 F; 100% act.

Kessco® PEG 400 ML. [Stepan; Stepan Canada] PEG-8 laurate; CAS 9004-81-3; nonionic; emulsifier, thickener, solubilizer, emollient, opacifier, spreading agent, wetting agent, dispersant for cosmetics (creams/lotions, makeup, bath oils, shampoos, conditioners, sunscreens), pharmaceuticals (ointments, suppositories), food, agric., plastics, etc.; lt. yel. liq.; f.p. 12 C; sol. in water, IPA, acetone, CCl_4, ethyl acetate, toluol; sp.gr. 1.028; dens. 8.6 lb/gal; HLB 13.1; acid no. 5.0 max.; sapon. no. 86-96; flash pt. (COC) 475 F.

Kessco® PEG 400 MO. [Stepan; Stepan Canada] PEG-8 oleate; CAS 9004-96-0; nonionic; emulsifier, thickener, solubilizer, emollient, opacifier, spreading agent, wetting agent, dispersant for cosmetics (creams/lotions, makeup, bath oils, shampoos, conditioners, sunscreens), pharmaceuticals (ointments, suppositories), food, agric., plastics, etc.; lt. amber liq.; f.p. < 10 C; sol. in IPA, acetone, CCl_4, ethyl acetate, toluol; water-disp.; sp.gr. 1.013; dens. 8.4 lb/gal; HLB 11.4; acid no. 5.0 max.; sapon. no. 80-89; pH 5.0 (3% disp.); flash pt. (COC) 510 F.

Kessco® PEG 400 MS. [Stepan; Stepan Canada] PEG-8 stearate; CAS 9004-99-3; nonionic; emulsifier, thickener, solubilizer, emollient, opacifier, spreading agent, wetting agent, dispersant for cosmetics (creams/lotions, makeup, bath oils, shampoos, conditioners, sunscreens), pharmaceuticals (ointments, suppositories), food, agric., plastics, etc.; wh. to cream soft solid; sol. in IPA, acetone, CCl_4, ethyl acetate, toluol; water-disp.; sp.gr. 0.9780 (65 C); HLB 11.6; m.p. 32 C; acid no. 5.0 max.; sapon. no. 83-92; pH 5.0 (3% disp.); flash pt. (COC) 480 F.

Kessco® PEG 600 DL. [Stepan; Stepan Canada] PEG-12 dilaurate; CAS 9005-02-1; emulsifier, thickener, solubilizer, emollient, opacifier, spreading agent, wetting agent, dispersant for cosmetics (creams/lotions, makeup, bath oils, shampoos, conditioners, sunscreens), pharmaceuticals (ointments, suppositories), food, agric., plastics, etc.; lubricant, emulsifier, softener for textile and metalworking applics.; liq.; sol. in IPA, acetone, CCl_4, ethyl acetate, toluol, IPM; water-disp.; sp.gr. 0.9820 (65 C); f.p. 24 C; HLB 11.7; acid no. 10.0 max.; sapon. no. 102-112; flash pt. (COC) 465 F; 100% act.

Kessco® PEG 600 DO. [Stepan; Stepan Canada] PEG-12 dioleate; CAS 9005-07-6; nonionic; emulsifier, thickener, solubilizer, emollient, opacifier, spreading agent, wetting agent, dispersant for cosmetics (creams/lotions, makeup, bath oils, shampoos, conditioners, sunscreens), pharmaceuticals (ointments, suppositories), food, agric., plastics, etc.; lt. amber liq.; f.p. 19 C; sol. in IPA, acetone, CCl_4, ethyl acetate, toluol, IPM, peanut oil; water-disp.; sp.gr. 1.001; dens. 8.3 lb/gal; HLB 10.5; acid no. 10.0 max.; sapon. no. 92-102; pH 5.0 (3% disp.); flash pt. (COC) 495 F.

Kessco® PEG 600 DS. [Stepan; Stepan Canada] PEG-12 distearate; CAS 9005-08-7; emulsifier, thickener, solubilizer, emollient, opacifier, spreading agent, wetting agent, dispersant for cosmetics (creams/lotions, makeup, bath oils, shampoos, conditioners, sunscreens), pharmaceuticals (ointments, suppositories), food, agric., plastics, etc.; lubricant, emulsifier, softener for textile and metalworking applics.; wh. to cream soft solid; sol. in IPA, acetone, CCl_4, ethyl acetate, toluol, IPM, peanut oil; water-disp.; sp.gr. 0.9670 (65 C); HLB 10.7; m.p. 39 C; acid no. 10.0 max.; sapon. no. 93-102; pH 5.0 (3% disp.); flash pt. (COC) 490 F; 100% act.

Kessco® PEG 600 ML. [Stepan; Stepan Canada] PEG-12 laurate; CAS 9004-81-3; emulsifier, thickener, solubilizer, emollient, opacifier, spreading agent, wetting agent, dispersant for cosmetics (creams/lotions, makeup, bath oils, shampoos, conditioners, sunscreens), pharmaceuticals (ointments, suppositories), food, agric., plastics, etc.; lubricant, emulsifier, softener for textile and metalworking applics.; lt. yel. liq.; sol. in water, Na_2SO_4, IPA, acetone, CCl_4, ethyl acetate, toluol; sp.gr. 1.050; dens. 8.8 lb/gal; f.p. 23 C; HLB 14.6; acid no. 5.0 max.; sapon. no. 64-74; flash pt. (COC) 475 F; 100% act.

Kessco® PEG 600 MO. [Stepan; Stepan Canada] PEG-12 oleate; CAS 9004-96-0; nonionic; emulsifier, thickener, solubilizer, emollient, opacifier, spreading agent, wetting agent, dispersant for cosmetics (creams/lotions, makeup, bath oils, shampoos, conditioners, sunscreens), pharmaceuticals (ointments, suppositories), food, agric., plastics, etc.; lt. amber liq.; f.p. 23 C; sol. in water, IPA, acetone, CCl_4, ethyl acetate, toluol; sp.gr. 1.037; dens. 8.7 lb/gal; HLB 13.5; acid no. 5.0 max.; sapon. no. 60-69; pH 5.0 (3% disp.); flash pt. (COC) 525 F.

Kessco® PEG 600 MS. [Stepan; Stepan Canada] PEG-12 stearate; CAS 9004-99-3; nonionic; emulsifier, thickener, solubilizer, emollient, opacifier, spreading agent, wetting agent, dispersant for cosmetics (creams/lotions, makeup, bath oils, shampoos, conditioners, sunscreens), pharmaceuticals (ointments, suppositories), food, agric., plastics, etc.; wh. to cream soft solid; sol. in water, IPA, acetone, CCl_4, ethyl acetate, toluol; sp.gr. 1.000 (65 C); HLB 13.6; m.p. 37 C; acid no. 5.0 max.; sapon. no. 61-70; pH 5.0 (3% disp.); flash pt. (COC) 480 F.

Kessco® PEG 1000 DL. [Stepan; Stepan Canada] PEG-20 dilaurate; CAS 9005-02-1; nonionic; emulsifier, thickener, solubilizer, emollient, opacifier, spreading agent, wetting agent, dispersant for cosmetics (creams/lotions, makeup, bath oils, shampoos, conditioners, sunscreens), pharmaceuticals (ointments, suppositories), food, agric., plastics, etc.; cream soft solid; f.p. 38 C; sol. in water, IPA, acetone, CCl_4, ethyl acetate; toluol, IPM; sp.gr. 1.015 (65 C); HLB 14.5; acid no. 10.0 max.; sapon. no. 68-78; flash pt. (COC) 475 F.

Kessco® PEG 1000 DO. [Stepan; Stepan Canada] PEG-20 dioleate; CAS 9005-07-6; nonionic; emulsifier, thickener, solubilizer, emollient, opacifier, spreading agent, wetting agent, dispersant for cosmetics (creams/lotions, makeup, bath oils, shampoos, conditioners, sunscreens), pharmaceuticals (ointments, suppositories), food, agric., plastics, etc.; cream soft solid; f.p. 37 C; sol. in water, IPA, acetone, CCl_4, ethyl acetate, toluol; sp.gr. 1.005 (65 C); HLB 13.1; acid no. 10.0 max.; sapon. no. 64-74; pH 5.0 (3% disp.); flash pt. (COC) 505 F.

Kessco® PEG 1000 DS. [Stepan; Stepan Canada] PEG-20 distearate; CAS 9005-08-7; nonionic; emulsifier, thickener, solubilizer, emollient, opacifier, spreading agent, wetting agent, dispersant for cosmetics (creams/lotions, makeup, bath oils, shampoos, conditioners, sunscreens), pharmaceuticals (ointments, suppositories), food, agric., plastics, etc.; cream wax; sol. in water, IPA, acetone, CCl_4, ethyl acetate, toluol; sp.gr. 1.005 (65 C); HLB 12.3; m.p. 40 C; acid no. 10.0 max.; sapon. no. 65-74; pH 5.0 (3% disp.); flash pt. (COC) 485 F.

Kessco® PEG 1000 ML. [Stepan; Stepan Canada] PEG-20 laurate; CAS 9004-81-3; nonionic; emulsifier, thickener, solubilizer, emollient, opacifier, spreading agent, wetting agent, dispersant for cosmetics (creams/lotions, makeup, bath oils, shampoos, conditioners, sunscreens), pharmaceuticals (ointments, suppositories), food, agric., plastics, etc.; cream soft solid; f.p. 40 C; sol. in water, propylene glycol (hot), Na_2SO_4, IPA, acetone, CCl_4, ethyl acetate, toluol; sp.gr. 1.035 (65 C); HLB 16.5; acid no. 5.0 max.; sapon. no. 41-51; flash pt. (COC) 490.

Kessco® PEG 1000 MO. [Stepan; Stepan Canada] PEG-20 oleate; CAS 9004-96-0; nonionic; emulsifier, thickener, solubilizer, emollient, opacifier, spreading agent, wetting agent, dispersant for cosmetics (creams/lotions, makeup, bath oils, shampoos, conditioners, sunscreens), pharmaceuticals (ointments, suppositories), food, agric., plastics, etc.; cream soft solid; f.p. 39 C; sol. in water, Na_2SO_4 (5%), IPA, acetone, CCl_4, ethyl acetate, toluol; sp.gr. 1.035 (65 C); HLB 15.4; acid no. 5.0 max.; sapon. no. 40-49; pH 5.0; flash pt. (COC) 515 F.

Kessco® PEG 1000 MS. [Stepan; Stepan Canada] PEG-20 stearate; CAS 9004-99-3; nonionic; emulsifier, thickener, solubilizer, emollient, opacifier, spreading agent, wetting agent, dispersant for cosmetics (creams/lotions, makeup, bath oils, shampoos, conditioners, sunscreens), pharmaceuticals (ointments, suppositories), food, agric., plastics, etc.; cream wax; sol. in water, Na_2SO_4 (5%), IPA, acetone, CCl_4, ethyl acetate, toluol; sp.gr. 1.030 (65 C); HLB 15.6; m.p. 41 C; acid no. 5.0 max.; sapon. no. 40-48; pH 5.0 (3% disp.); flash pt. (COC) 475 F.

Kessco® PEG 1540 DL. [Stepan; Stepan Canada] PEG-32 dilaurate; CAS 9005-02-1; nonionic; emulsifier, thickener, solubilizer, emollient, opacifier, spreading agent, wetting agent, dispersant for cosmetics (creams/lotions, makeup, bath oils, shampoos, conditioners, sunscreens), pharmaceuticals (ointments, suppositories), food, agric., plastics, etc.; cream wax; f.p. 42 C; sol. in water, Na_2SO_4 (5%); hot in propylene glycol, IPA, acetone, CCl_4, ethyl acetate, toluol; sp.gr. 1.04 (65 C); HLB 15.7; acid no. 10.0 max.; sapon. no. 48-56; pH 4.5 (3% disp.); flash pt. (COC) 450 F.

Kessco® PEG 1540 DO. [Stepan; Stepan Canada] PEG-32 dioleate; CAS 9005-07-6; nonionic; emulsifier, thickener, solubilizer, emollient, opacifier, spreading agent, wetting agent, dispersant for cosmetics (creams/lotions, makeup, bath oils, shampoos, conditioners, sunscreens), pharmaceuticals (ointments, suppositories), food, agric., plastics, etc.; cream wax; f.p. 44 C; sol. in water, propylene glycol, Na_2SO_4; hot in IPA, acetone, CCl_4, ethyl acetate, toluol; sp.gr. 1.025 (65 C); HLB 15.0; acid no. 10.0 max.; sapon. no. 45-55; pH 5.0 (3% disp.); flash pt. (COC) 480 F.

Kessco® PEG 1540 DS. [Stepan; Stepan Canada] PEG-32 distearate; CAS 9005-08-7; nonionic; emulsifier, thickener, solubilizer, emollient, opacifier, spreading agent, wetting agent, dispersant for cosmetics (creams/lotions, makeup, bath oils, shampoos, conditioners, sunscreens), pharmaceuticals (ointments, suppositories), food, agric., plastics, etc.; cream wax; sol. in water, IPA, acetone, CCl_4, ethyl acetate, toluol; sp.gr. 1.015 (65 C); HLB 14,8; m.p. 45 C; acid no. 10.0 max.; sapon. no. 49-58; pH 5.0; flash pt. (COC) 490 F.

Kessco® PEG 1540 ML. [Stepan; Stepan Canada] PEG-32 laurate; CAS 9004-81-3; nonionic; emulsifier, thickener, solubilizer, emollient, opacifier, spreading agent, wetting agent, dispersant for cosmetics (creams/lotions, makeup, bath oils, shampoos, conditioners, sunscreens), pharmaceuticals (ointments, suppositories), food, agric., plastics, etc.; cream wax; f.p. 46 C; sol. in water, Na_2SO_4 (5%); sol. hot in propylene glycol, IPA, acetone, CCl_4, ethyl acetate, toluol; sp.gr. 1.06 (65 C); HLB 17.6; acid no. 5.0 max.; sapon. no. 26-36; pH 4.5 (3% disp.); flash pt. (COC) 445 F.

Kessco® PEG 1540 MO. [Stepan; Stepan Canada] PEG-32 oleate; CAS 9004-96-0; nonionic; emulsifier, thickener, solubilizer, emollient, opacifier, spreading agent, wetting agent, dispersant for cosmetics (creams/lotions, makeup, bath oils, shampoos, conditioners, sunscreens), pharmaceuticals (ointments, suppositories), food, agric., plastics, etc.; cream wax; f.p. 45 C; sol. in water, propylene glycol, Na_2SO_4; sol. hot in IPA, acetone, CCl_4, ethyl acetate, toluol; sp.gr. 1.050 (65 C); HLB 17.0; f.p. 47 C; acid no. 5.0 max.; sapon. no. 28-37; pH 5.0 (3% disp.); flash pt. (COC) 520 F.

Kessco® PEG 1540 MS. [Stepan; Stepan Canada] PEG-32 stearate; CAS 9004-99-3; nonionic; emulsifier, thickener, solubilizer, emollient, opacifier, spreading agent, wetting agent, dispersant for cosmetics (creams/lotions, makeup, bath oils, shampoos, conditioners, sunscreens), pharmaceuticals (ointments, suppositories), food, agric., plastics, etc.; cream wax; sol. in water, Na_2SO_4 (5%), IPA, acetone, CCl_4, ethyl acetate, toluol; sp.gr. 1.050 (65 C); HLB 17.3; m.p. 47 C; acid no. 5.0 max.; sapon. no. 27-36; pH 5.0; flash pt. (COC) 495 F.

Kessco® PEG 4000 DL. [Stepan; Stepan Canada] PEG-75 dilaurate; CAS 9005-02-1; nonionic; emulsifier, thickener, solubilizer, emollient, opacifier, spreading agent, wetting agent, dispersant for cosmetics (creams/lotions, makeup, bath oils, shampoos, conditioners, sunscreens), pharmaceuticals (ointments, suppositories), food, agric., plastics, etc.; cream wax; f.p. 52 C; sol. in water, Na_2SO_4 (5%); sol. hot in propylene glycol, IPA, acetone, CCl_4, ethyl acetate, toluol; sp.gr. 1.065 (65 C); HLB 17.6; acid no. 5.0 max.; sapon. no. 20-30; pH 4.5 (3% disp.); flash pt. (COC) 495 F.

Kessco® PEG 4000 DO. [Stepan; Stepan Canada] PEG-75 dioleate; CAS 9005-07-6; nonionic; emulsifier, thickener, solubilizer, emollient, opacifier, spreading agent, wetting agent, dispersant for cosmetics (creams/lotions, makeup, bath oils, shampoos, conditioners, sunscreens), pharmaceuticals (ointments, suppositories), food, agric., plastics, etc.; cream wax; f.p. 49 C; sol. in water, propylene glycol, Na_2SO_4; sol. hot in IPA, acetone, CCl_4, ethyl acetate, toluol; sp.gr. 1.060 (65 C); HLB 17.8; acid no. 5.0 max.; sapon. no. 19-27; pH 5.0; flash pt. (COC) 500 F.

Kessco® PEG 4000 DS. [Stepan; Stepan Canada] PEG-75 distearate; CAS 9005-08-7; nonionic; emulsifier, thickener, solubilizer, emollient, opacifier, spreading agent, wetting agent, dispersant for cosmetics (creams/lotions, makeup, bath oils, shampoos, conditioners, sunscreens), pharmaceuticals (ointments, suppositories), food, agric., plastics, etc.; cream wax; sol. in water, Na_2SO_4 (5%), IPA, acetone, CCl_4, ethyl acetate, toluol; sp.gr. 1.060 (65 C); HLB 17.3; m.p. 51 C; acid no. 5.0 max.; sapon. no. 19-27; pH 5.0 (3% disp.); flash pt. (COC) 515 F.

Kessco® PEG 4000 ML. [Stepan; Stepan Canada] PEG-75 laurate; CAS 9004-81-3; nonionic; emulsifier, thickener, solubilizer, emollient, opacifier, spreading agent, wetting agent, dispersant for cosmetics (creams/lotions, makeup, bath oils, shampoos, conditioners, sunscreens), pharmaceuticals (ointments, suppositories), food, agric., plastics, etc.; cream wax; f.p. 55 C; sol. in water, Na_2SO_4 (5%); sol. hot in propylene glycol, IPA, acetone, CCl_4, ethyl acetate, toluol; sp.gr. 1.075 (65 C); HLB 18.8; acid no. 5.0 max.; sapon. no. 9-18; pH 4.5; flash pt. (COC) 515 F.

Kessco® PEG 4000 MO. [Stepan; Stepan Canada] PEG-75 oleate; CAS 9004-96-0; nonionic; emulsifier, thickener, solubilizer, emollient, opacifier, spreading agent, wetting agent, dispersant for cosmetics (creams/lotions, makeup, bath oils, shampoos, conditioners, sunscreens), pharmaceuticals (ointments, suppositories), food, agric., plastics, etc.; cream wax; f.p. 55 C; sol. in water, propylene glycol, Na_2SO_4; sol. hot in IPA, acetone, CCl_4, ethyl acetate, toluol; sp.gr. 1.075 (65 C); HLB 18.3; acid no. 5.0 max.; sapon. no. 10-18; pH 5.0; flash pt. (COC) 495 F.

Kessco® PEG 4000 MS. [Stepan; Stepan Canada] PEG-75 stearate; CAS 9004-99-3; nonionic; emulsifier, thickener, solubilizer, emollient, opacifier, spreading agent, wetting agent, dispersant for cosmetics (creams/lotions, makeup, bath oils, shampoos, conditioners, sunscreens), pharmaceuticals (ointments, suppositories), food, agric., plastics, etc.; cream wax; sol. in water, Na_2SO_4 (5%), IPA, acetone, CCl_4, ethyl acetate, toluol; sp.gr. 1.075 (64 C); HLB 18.6; m.p. 56 C; acid no. 5.0 max.; sapon. no. 10-18; pH 5.0; flash pt. (COC) 465 F.

Kessco® PEG 6000 DL. [Stepan; Stepan Canada] PEG-150 dilaurate; CAS 9005-02-1; nonionic; emulsifier, thickener, solubilizer, emollient, opacifier, spreading agent, wetting agent, dispersant for cosmetics (creams/lotions, makeup, bath oils, shampoos, conditioners, sunscreens), pharmaceuticals (ointments, suppositories), food, agric., plastics, etc.; cream wax; f.p. 57 C; sol. in water, Na_2SO_4 (5%); sol. hot in propylene glycol, IPA, acetone, CCl_4, ethyl acetate, toluol; sp.gr. 1.077 (65 C); HLB 18.7; m.p. 56 C; acid no. 9.0 max.; sapon. no. 12-20; pH 4.5 (3% disp.); flash pt. (COC) 435 F.

Kessco® PEG 6000 DO. [Stepan; Stepan Canada] PEG-150 dioleate; CAS 9005-07-6; nonionic; emulsifier, thickener, solubilizer, emollient, opacifier, spreading agent, wetting agent, dispersant for cosmetics (creams/lotions, makeup, bath oils, shampoos, conditioners, sunscreens), pharmaceuticals (ointments, suppositories), food, agric., plastics, etc.; cream wax; f.p. 56 C; sol. in water, propylene glycol, Na_2SO_4 (5%); sol. hot in IPA, acetone, CCl_4, ethyl acetate, toluol; sp.gr. 1.070 (65 C); HLB 18.3; acid no. 9.0 max.; sapon. no. 13-21; pH 5.0 (3% disp.); flash pt. 500 F.

Kessco® PEG 6000 DS. [Stepan; Stepan Canada] PEG-150 distearate; CAS 9005-08-7; emulsifier, thickener, solubilizer, emollient, opacifier, spreading agent, wetting agent, dispersant for cosmetics (creams/lotions, makeup, bath oils, shampoos, conditioners, sunscreens), pharmaceuticals (ointments, suppositories), food, agric., plastics, etc.; lubricant, emulsifier, softener for textile and metalworking applics.; cream wax; sol. in propylene glycol, Na_2SO_4 (5%), IPA, acetone, CCl_4, ethyl acetate, toluol; sp.gr. 1.075 (65 C); HLB 18.4; m.p. 55 C; acid no. 9.0 max.; sapon. no. 14-20; pH 5.0 (3% disp.); flash pt. (COC) 475 F; 100% act.

Kessco® PEG 6000 ML. [Stepan; Stepan Canada] PEG-150 laurate; CAS 9004-81-3; nonionic; emulsifier, thickener, solubilizer, emollient, opacifier, spreading agent, wetting agent, dispersant for cosmetics (creams/lotions, makeup, bath oils, shampoos, conditioners, sunscreens), pharmaceuticals (ointments, suppositories), food, agric., plastics, etc.; cream wax; f.p. 61 C; sol. in water, Na_2SO_4 (5%); sol. hot in propylene glycol, IPA, acetone, CCl_4, ethyl acetate, toluol; sp.gr. 1.085 (65 C); HLB 19.2; acid no. 5.0 max.; sapon. no. 7-13; pH 4.5.

Kessco® PEG 6000 MO. [Stepan; Stepan Canada] PEG-150 oleate; CAS 9004-96-0; nonionic; emulsifier, thickener, solubilizer, emollient, opacifier, spreading agent, wetting agent, dispersant for cosmetics (creams/lotions, makeup, bath oils, shampoos, conditioners, sunscreens), pharmaceuticals (ointments, suppositories), food, agric., plastics, etc.; cream wax; f.p. 59 C; sol. in water, propylene glycol, Na_2SO_4 (5%); sol. hot in IPA, acetone, CCl_4, ethyl acetate, toluol; sp.gr. 1.085 (65 C); HLB 19.0; acid no. 5.0 max.; sapon. no. 7-13; pH 5.0; flash pt. 470 F.

Kessco® PEG 6000 MS. [Stepan; Stepan Canada] PEG-150 stearate; CAS 9004-99-3; nonionic; emulsifier, thickener, solubilizer, emollient, opacifier, spreading agent, wetting agent, dispersant for cosmetics (creams/lotions, makeup, bath oils, shampoos, conditioners, sunscreens), pharmaceuticals (ointments, suppositories), food, agric., plastics, etc.; cream wax; sol. in water, propylene glycol, Na_2SO_4 (5%), IPA, acetone, CCl_4, ethyl acetate, toluol; sp.gr. 1.080 (65 C); HLB 18.8; m.p. 61 C; acid no. 5.0 max.; sapon. no. 7-13; pH 5.0 (3% disp.); flash pt. (COC) 480 F.

Ketjenblack® EC-300 J. [Akzo Nobel] Carbon black; CAS 1333-86-4; EINECS 215-609-9; electroconductive carbon black for use as internal antistat for polyolefins, PVC, elastomers; pellets.

Ketjenblack® EC-310 NW. [Akzo Nobel] Carbon black aq. disp.; CAS 1333-86-4; EINECS 215-609-9; electroconductive carbon black for rubber compding.; aq. disp.; 10% act.

Ketjenblack® EC-350 J Spd. [Akzo Nobel] Carbon black disp. in paraffinic oil; electroconductive carbon black for rubber compding.; powd.; 50% act. in paraffinic oil.

Ketjenblack® EC-350 N DOP. [Akzo Nobel] Carbon black disp. in dioctyl phthalate; electroconductive carbon black for rubber compding.; powd.; 50% act. in DOP.

Ketjenblack® EC-600 JD. [Akzo Nobel] Carbon black; CAS 1333-86-4; EINECS 215-609-9; super electroconductive carbon black for use in rubber compounding for elec. conductivity; pellets.

Ketjenblack® ED-600 JD. [Akzo Nobel] Carbon black; CAS 1333-86-4; EINECS 215-609-9; electroconductive carbon black for rubber compding.; pellets.

Ketjenflex® 8. [Akzo Nobel] Ethyl toluene sulfonamide; plasticizer for resins, coatings, electroplating sol'ns., thermoplastics and thermosets, nitrocellulose lacquers; lt. yel. visc. liq.; dens. 1.2 g/ml; visc. 400 mPa•s; m.p. 0 C; b.p. > 340 C; flash pt. (COC) 224 C.

Ketjenflex® 9. [Akzo Nobel] Toluene sulfonamide; plasticizer for difficult resins, e.g., nylon, other polyamides, shellac, cellulose, cellulose-acetate, casein, and other protein materials; fungicide in protective coatings; brightener in electroplating; plasticizer in thermoplastics; component in fluorescent resins; wh. to lt. cream fine cryst. powd.; m.p. 102-107 C; flash pt. (COC) 180 C; 1.5% max. moisture.

Ketjensil® SM 405. [Akzo Nobel] Sodium-aluminum silicate, precipitated; reinforcing filler for silicone rubber applics.; filler in disp. paints for partial replacement of TiO_2; powd.; 5.5 μm avg. particle size; dens. 2.1 g/cc; bulk dens. 180 kg/m^3; surf. area 70-80 m^2/g; pH 10.5 (10% aq. disp.); 82% SiO2, 9% Al_2O_3.

Kevlar® 29, 49. [DuPont] Aramid; high str., lightweight, flexible material; used for aircraft/aerospace, boat hulls, prosthetics, footwear, sporting goods, ropes and cables, fiber optics, bullet-resist. vests, fabrics, brakes and other friction prods., in radial tires, as asbestos replacement, as reinforcement for mech. rubber goods; avail. as continuous filament yarn, staple, engineered short fiber, pulp, spun yarn, needlepunched felt, paper, woven fabrics, cord, narrow webbing; Kevlar 29, 49 yarn props.: dens. 1.44 g/cc; tens. str. 2930 MPa; tens. elong. 3.6 and 2.5% resp. (break); Kevlar 29, 49 fiber props.: dens. 1.44 g/cc; tens. str. 3620 MPa; tens. elong. 4.4 and 2.9% (break).

Keycide® X-10. [Witco Corp.] Tributyltin oxide, stabilized; CAS 56-35-9; EINECS 200-268-0; antimildew additive, antimicrobial for PVAc latex paints for packaged stability and mildew-resistant applied films; Gardner 1 clear liq.; disp. in water; dens. 7.85 lb/gal; 12% act.

Kisuma 5A. [Kyowa Chem. Industry] Noble magnesium hydroxide compd.; CAS 1309-42-8; EINECS 215-170-3; nontoxic fire retardant; compatibile with plastic to produce high quality composites; used with thermoplastic resins; eliminates toxic gas emissions and reduces smoke emissions; heat stabilizer for resins containing halogen; superfine particles; sp.gr. 0.69 m/100 g; dens. 2.36; ref. index 1.56; hardness (Moh) 2.5; 96.8% $Mg(OH)_2$.

Kisuma 5B. [Kyowa Chem. Industry] Noble magnesium hydroxide compd.; CAS 1309-42-8; EINECS 215-170-3; nontoxic fire retardant; compatibile with plastic to produce high quality composites; used with thermoplastic resins; eliminates toxic gas emissions and reduces smoke emissions; heat stabilizer for resins containing halogen; superfine particles; sp.gr. 0.58 m/100 g; dens. 2.36; ref. index 1.56; hardness (Moh) 2.5; 98.1% $Mg(OH)_2$.

Kisuma 5E. [Kyowa Chem. Industry] Noble magnesium hydroxide compd.; CAS 1309-42-8; EINECS 215-170-3; nontoxic fire retardant; compatibile with plastic to produce high quality composites; used with thermoplastic resins; eliminates toxic gas emissions and reduces smoke emissions; heat stabilizer for resins containing halogen; superfine particles; sp.gr. 0.56 m/100 g; dens. 2.36; ref. index 1.56; hardness (Moh) 2.5; 98.8% $Mg(OH)_2$.

Klearfac® AA040. [BASF] Phosphate ester; anionic; solubilizer for nonionics in high electrolyte sol'ns.; for nonionic surfactants and dedusters for powd. alkalies; emulsifier; antistat used with syn. fiber yarns; used in textile processing to improve effectiveness of kier boiling, scouring, bleaching, soaping, and print-washing operations; used in heavy-duty cleaning formulations, pesticides, herbicides, insecticides, liq. fertilizers, and plastics; liq.; sp.gr. 1.112; visc. 240 cps; pour pt. 12.8 C; cloud pt. 2 C (1% aq.); biodeg.; 60% min. act.

Kloro 3000. [Ferro/Keil] Chlorinated org.; flame retardant for ABS, acrylics, cellulosics, thermoset polyester, PE, PP, PS, PVAc, PVC, PU.

Kloro 6001. [Ferro/Keil] Chlorinated paraffin; flame retardant for ABS, acrylic, cellulose acetate, epoxy, ethyl cellulose, nitrile, thermoset polyester, PE, PP, PS, PVAc, PVC, flexible PU foam; very good heat stability and resist. to hydrolysis; used for sol. oils, synthetics and semi-synthetic.

Kodaflex® 240. [Eastman] Plasticizer; Pt-Co 10 max. color; sp.gr. 1.125 (20/20 C); dens. 9.36 lb/gal (20 C); flash pt. (COC) 160 C; 96.5% min. assay.

Kodaflex® DBP. [Eastman] Dibutyl phthalate; CAS 84-74-2; EINECS 201-557-4; primary plasticizer-solv. for nitrocellulose lacquers; for rubbers and CAB, ethyl cellulose, PVAc, and syn. resins; solv.

for oil-sol. dyes, insecticides, peroxides, and org. compds.; antifoamer and fiber lubricant in textile mfg.; cosmetics plasticizer; Pt-Co 15 ppm liq.; sol. 0.46% in water; m.w. 278; sp.gr. 1.048; dens. 8.72 lb/gal (20 C); visc. 15 cP; f.p. -35 C; b.p. 340 C (760 mm); ref. index 1.4920; flash pt. (COC) 190 C; fire pt. (COC) 202 C; dissip. factor 0.08 x 10^{-2} (1 MHz); dielec. const. 5.8 (1 MHz); vol. resist. 3.0 x 10^{9} ohm cm; 99% min. assay; *Precaution:* combustible.

Kodaflex® DEP. [Eastman] Diethyl phthalate; CAS 84-66-2; EINECS 201-550-6; plasticizer; wetting agent in grinding pigments; pigment-disp. medium in cellulose acetate sol'ns. and plastics, and solv. for natural resins and polymers; seldom used in PVC prods. due to relatively high volatility; also for cosmetics; Pt-Co 10 ppm liq.; sol. 0.12 g/l in water; m.w. 222; sp.gr. 1.120; dens. 9.32 lb/gal (20 C); visc. 9.5 cP; f.p. < -50 C; b.p. 298 C (760 mm); ref. index 1.4990; flash pt. (COC) 161 C; fire pt. (COC) 171 C; dissip. factor 0.10 x 10^{-2} (1 MHz); dielec const. 6.7 (1 MHz); vol. resist. 1.45 x 10^{9} ohm cm; 99% min. assay.

Kodaflex® DMP. [Eastman] Dimethyl phthalate; CAS 131-11-3; EINECS 205-011-6; plasticizer with high solv. power for cellulose acetate extrusion compds.; compatible with ethyl cellulose, CAB, PS, PVAc, polyvinyl butyral, and PVC; used in NC-based printing inks; also for cosmetics and personal care prods.; Pt-Co 5 ppm liq.; sol. 0.45 g/l in water; m.w. 194; sp.gr. 1.192 (20/20 C); dens. 9.93 lb/gal (20 C); visc. 11.0 cP; f.p. -1 C; b.p. 284 C; ref. index 1.513; flash pt. (COC) 157 C; fire pt. (COC) 168 C; dissip. factor 0.13 x 10^{-2} (1 mc); dielec. const. 7.5 (1 mc); vol. resist. 1.07 x 10^{9} ohm cm; 99% min. assay.

Kodaflex® DOA. [Eastman] Dioctyl adipate; CAS 103-23-1; EINECS 203-090-1; plasticizer providing flexibility at low temps. to vinyl prods.; used in unfilled garden hose, clear sheeting, elec. insulation; also for cosmetics and personal care prods.; Pt-Co 20 ppm liq.; sol. < 0.1 g/l in water; m.w. 370; sp.gr. 0.927; dens. 7.71 lb/gal (20 C); visc. 12 cP; f.p. < -70 C; b.p. 417 C; ref. index 1.4472; flash pt. (COC) 206 C; fire pt. (COC) 229 C; dissip. factor 0.04 x 10^{-2} (1 MHz); dielec. const. 3.9 (1 MHz); vol. resist. 9.3 x 10^{11} ohm cm; 99% min. assay.

Kodaflex® DOP. [Eastman] Dioctyl phthalate; CAS 117-81-7; EINECS 204-211-0; all-purpose plasticizer used with PVC resins incl. film and sheeting for upholstery, clothing, food pkg., paper coatings, molded vinyl prods., elec. wire insulation; compatible with PS, methyl methacrylate, chlorinated rubber, NC, and CAB; low odor, relatively low toxicity, and low volatility; Pt-Co 15 ppm liq.; sol. < 0.1 g/l in water; m.w. 390.57; sp.gr. 0.9852; dens. 8.20 lb/gal; visc. 56.5 cP; f.p. -50 C; b.p. 384 C; ref. index 1.4836; flash pt. (COC) 216 C; fire pt. (COC) 240 C; dissip. factor 0.64 x 10^{-2} (1 MHz); dielec. const. 4.8 (1 MHz); vol. resist 2.2 x 10^{11} ohm cm; 99% min. assay.

Kodaflex® DOTP. [Eastman] Dioctyl terephthalate; CAS 422-86-2; EINECS 225-091-6; primary plasticizer used with PVC resins, in PVC plastisols, rubber; applic. incl. wire coatings, automotive and furniture upholstery; compatible with acrylics, CAB, cellulose nitrate, polyvinyl butyral, styrene, oxidizing alkyds, nitrile rubber; also for cosmetics; Pt-Co 15 ppm liq.; sol. 0.004 g/l in water; m.w. 390.57; sp.gr. 0.9835; dens. 8.18 lb/gal (20 C); visc. 63 cP; f.p. 48 C; b.p. 400 C; ref. index 1.4867; flash pt. (COC) 238 C; fire pt. (COC) 266 C; dissip. factor 0.1 x 10^{-2} (1 MHz); dielec. const. 4.6 (1 MHz); vol. resist. 3.9 x 10^{12} ohm-cm; 99% min. assay.

Kodaflex® HS-3. [Eastman] Butyl octyl phthalate; EINECS 201-562-1; high-solvating primary plasticizer for polyvinyl homopolymer and copolymer resins, vinyl plastisols, expanded vinyl foams, and rotational molding and dip coating; Pt-Co 35 ppm liq.; sol. < 0.01 g/l in water; sp.gr. 0.988-0.996; dens. 8.26 lb/gal (20 C); visc. 42 cP; f.p. -48 C; b.p. 350 C; ref. index 1.4848; flash pt. (COC) 208 C; fire pt. (COC) 224 C.

Kodaflex® HS-4. [Eastman] 60% Dioctyl phthalate, 5% dibutyl phthalate; high-solvating primary plasticizer; used in formulating vinyl plastisols having processing chars. required to be mechanically frothed; for PVC formulations, rotational molding, slush molding, dip coating, and filament coating applic.; Pt-Co 35 ppm liq.; sol. < 0.01 g/l in water; sp.gr. 0.993; dens. 8.26 lb/gal (20 C); visc. 31.8 cP; f.p. -47 C; b.p. 337 C; ref. index 1.4798; flash pt. (COC) 182 C; fire pt. (COC) 200 C; dissip. factor 7.9 x 10^{-2} (1 MHz); dielec. const. 3.4 (1 MHz); vol. resist. 2.0 x 10^{14} ohm-cm.

Kodaflex® PA-6. [Eastman] Adhesion-promoting plasticizer; provides improved adhesion of vinyl dispersions to metals; eliminates need for PVC resins in automotive sealants; Gardner 1 clear liq.; sp.gr. 1.004 (20/20 C); dens. 8.35 lb/gal (20 C); visc. 800 cP; f.p. -22 C; b.p. 404 C; acid no. 8; flash pt. (COC) 201 C; ref. index 1.4810; *Precaution:* combustible.

Kodaflex® TEG-EH. [Eastman] Triethylene glycol di-2-ethylhexanoate; CAS 94-28-0; plasticizer; Pt-Co 50 max. color; sp.gr. 0.967 (20/20 C); dens. 8.065 lb/gal (20 C); b.p. 385 C; flash pt. (COC) 216 C.

Kodaflex® TOTM. [Eastman] Trioctyl trimellitate; CAS 3319-31-1; primary plasticizer used in vinyl film and vinyl-coated fabrics; also for cosmetics; Pt-Co 100 ppm liq.; sol. 0.006 g/l in water; m.w. 547; sp.gr. 0.989 (20/20 C); dens. 8.22 lb/gal (20 C); visc. 194 cP; f.p. -38 C; b.p. 414 C; ref. index 1.4832; flash pt. (COC) 263 C; fire pt. (COC) 297 C; dissip. factor 0.7 x 10^{-2} (1 kHz); dielec. const. 4.9 (1 kHz); vol. resist. 5 x 10^{11} ohm-cm; 99% min. assay.

Kodaflex® Triacetin. [Eastman] Glyceryl triacetate; CAS 102-76-1; EINECS 203-051-9; low-toxicity plasticizer for vinyl compds.; used in adhesives, resinous and polymeric coatings, paper, and paperboard for food contact; water-insol. hydroxyethyl cellulose films; Pt-Co 5 liq.; sl. fatty odor; moderate sol. in water; m.w. 218; sp.gr. 1.160; dens. 9.65 lb/gal (20 C); visc. 17 cP; f.p. 3.2 C (supercools to -70 C); b.p. 258 C; flash pt. (COC)

153 C; dissip. factor 0.18 x 10^{-2} (1 MHz); dielec. const. 6.2 (1 MHz); vol. resist. 8.3 x 10^{9} ohm-cm; 99% min. assay.

Kodaflex® Tripropionin. [Eastman] Glyceryl tripropionate; CAS 139-45-7; plasticizer; kosher certified; Pt-Co 25 max. color; sp.gr. 1.085 (20/20 C); dens. 9.07 lb/gal (20 C); f.p. -58 C; b.p. 282 C; flash pt. (COC) 157 C; 96.5% min. assay.

Kodaflex® TXIB. [Eastman] 2,2,4-Trimethyl-1,3-pentanediol diisobutyrate; CAS 6846-50-0; primary plasticizer used in surf. coatings, vinyl floorings, moldings, and vinyl prods.; compatible with film-forming vehicles used in lacquers for wood, paper, and metals; primary plasticizer for PVC plastisols for rotocasting and slush molding; used in PVC organosols processed by extrusion and inj. molding; also for cosmetics and personal care prods.; clear liq.; sl. fruity odor; sol. 0.42 g/l in water; m.w. 286.4; sp.gr. 0.945; dens. 7.86 lb/gal (20 C); visc. 9 cP; f.p. -70 C; b.p. 280 C; ref. index 1.4300; flash pt. 143 C (COC); fire pt. 152 C (COC); dissip. factor 0.13 x 10^{-2} (1 MHz); dielec. const. 4.5 (1 MHz); vol. resist. 1.5 x 10^{11} ohm-cm; 98% min. assay.

Koreforte®. [BASF AG] Reinforcing resins for natural and syn. rubbers.

Koresin®. [BASF AG] Butylphenol/acetylene condensation prod.; tackifier for natural and syn. rubber in compds. and cements; yel. to brn. pellets or powd.; sol. in hydrocarbons; sp.gr. 1.02-1.04; soften. pt. (B&R) 135-150 C; drop pt. 140-160 C; *Storage:* pract. unlimited storage stability.

Korestab®. [BASF AG] Antioxidant for natural and syn. rubbers.

Koretack®. [BASF AG] Tackifier for natural and syn. rubbers.

Koretack® 5193. [BASF AG] Thermoplastic octylphenol-formaldehyde resin; tackifier for natural and syn. rubbers incl. BR, CR, SBR, NBR, and NR; for compds. and cements; brn. pellets; sp.gr. 1.02 (20 C); soften. pt. (B&R) 90-105 C; acid no. 55-75.

Kosmos® 10. [Goldschmidt AG] Stannous octoate, dioctyl phthalate; tin-org. catalyst for the mfg. of polyether PU foam; accelerates the gel reaction and intensifies the activation of the blowing reaction; sl. yel. liq.; sp.gr. 1.067 ± 0.02; visc. < 200 mPa•s; solid. pt. < -60 C; flash pt. (COC) > 170 C; ref. index 1.4887 ± 0.009; 33% sol'n. in dioctyl phthalate; > 9.035% tin content; *Storage:* 9 mos storage stability in original sealed containers in dry place; protect against high temps. and humidity.

Kosmos® 15. [Goldschmidt AG] Stannous octoate, dioctyl phthalate; tin-org. catalyst for the mfg. of polyether PU foam; accelerates the gel reaction and intensifies the activation of the blowing reaction; sl. yel. liq.; sp.gr. 1.107 ± 0.02; visc. < 180 mPa•s; solid. pt. < -60 C; flash pt. (COC) > 157 C; ref. index 1.490 ± 0.009; 50% sol'n. in dioctyl phthalate; > 14.0% tin content; *Storage:* 9 mos storage stability in original sealed containers in dry place; protect against high temps. and humidity; Unverified

Kosmos® 16. [Goldschmidt AG] Stannous octoate; CAS 301-10-0; tin-org. catalyst for the mfg. of polyether PU foam; accelerates the gel reaction and intensifies the activation of the blowing reaction; slightly yel. liq.; sp.gr. 1.024 ± 0.007; visc. ≤ 150 mPas; solid. pt. < -10 C; flash pt. (COC) > 120 C; ref. index 1.48 ± 0.02; 50% sol'n. in min. oil; ≥ 13.9% tin content; Unverified

Kosmos® 19. [Goldschmidt AG] Di-n-butyltindilaurate; CAS 77-58-7; EINECS 201-039-8; catalyst for PU flexible, rigid, and semirigid foams, for cold cure high resiliency PU foams, and as cocatalyst; clear yel. liq.; sol. with polyols, plasticizers, and common org. solvs.; sp.gr. 1.07 ± 0.01; visc. < 80 mPas; solid. pt. < -10 C; flash pt. 200 C; ref. index 1.479 ± 0.009; 18.5 ± 0.5% tin content; Unverified

Kosmos® 21. [Goldschmidt AG] Dialkyltinmercaptide; catalyst for PU polymerization, for molded PU foams, e.g., microcellular foams, rigid foams, high resilient foams, and RIM; sl. yel. clear liq.; sp.gr. 1.020 ± 0.01; visc. 25 ± 7 mPa•s; f.p. < -2 C; flash pt. (COC) 260 C; ref. index 1.502 ± 0.007; 18.5 ± 0.5% tin content; *Storage:* 10 mos min. storage stability in original sealed containers; protect against high temps. and humidity; Unverified

Kosmos® 23. [Goldschmidt AG] Dioctyltinmercaptide; CAS 58229-88-2; catalyst for the prod. of PU plastics, esp. where extended pot life is required; clear to sl. yel. liq.; sol. in polyols; insol. in water; dens. 1.085 ± 0.01 g/cc; visc. 100 ± 50 mPas; flash pt. (OC) 130 C; ref. index 1.499 ± 0.008; Unverified

Kosmos® 24. [Goldschmidt AG] Organotin compd.; catalyst for the prod. of PU plastics; yel. liq.; slight odor; sol. in polyols; insol. in water; dens. 1.085 ± 0.005 g/cc; visc. 95 ± 10 mPas; flash pt. > 100 C; ref. index 1.499 ± 0.008; Unverified

Kosmos® 25. [Goldschmidt AG] Organotin (IV) catalyst; for the prod. of PU plastics; lt. yel. liq., sl. odor; sol. in polyols; insol. in water; dens. 1.028 g/cc; visc. 35 mPa•s; flash pt. (OC) 109 C; ref. index 1.499 ± 0.008; Unverified

Kosmos® 29. [Goldschmidt AG] Stannous octoate; CAS 301-10-0; catalyst for the gelling reaction during mfg. of polyether PU foams; pale liq.; sol. in polyols and most org. solvs.; insol. in water, alcohol; sp.gr. 1.25 ± 0.02; visc. < 500 mPa•s; solid. pt. < -25 C; ref. index 1.491 ± 0.008; 28 ± 0.5% tin content; Unverified

Kosmos® 64. [Goldschmidt AG] Potassium octoate and polyglycol; catalyst for the prod. of polyisocyanurate foams; yel. to sl. brn. clear liq.; sp.gr. 1.092 ± 0.02; visc. 400 ± 20 mPa•s; hyd. no. 420; flash pt. (COC) 200 C; 43.3 ± 0.5% potassium octoate content; Unverified

Koster Keunen Candelilla. [Koster Keunen] Candelilla wax; CAS 8006-44-8; EINECS 232-347-0; gellant, emollient used for lipsticks, creams, lotions, gel prods.; leather dressings; furniture and other polish; cements; varnishes; sealing wax; elec. insulating compositions; phonograph records; paper sizing; rubber; waterproofing and insectproofing; paint removers; soft wax stiffeners; lubricants; adhesives; candy and gum; FDA21CFR §172.615, 175.105, 175.320, 176.180; wax; sol. hot in alcohol, benzene, petroleum ether; sp.gr. 0.9820-0.9930; m.p. 68.5-72.5 C; acid no. 11-19; iodine no. 19-44; sapon. no. 44-66; ref. index 1.4555; dielec. const. 2.50-2.63.

Koster Keunen Ceresine. [Koster Keunen] Ceresine; CAS 8001-75-0; EINECS 232-290-1; used for cosmetics, creams, lotions, ointments, salves, pharmaceuticals; candles; shoe, floor, and leather polishes; antifouling paints; wood polishes/filler; incandescent gas mantles; paper sizing; waxed paper; lubricants; wax figures; toys; elec. insulation; impregnating/preserving agent; crayons; perfume pastes; pomades; rubber mixtures; waterproofing textiles; adhesives; FDA §175.105; wax; sp.gr. 0.88-0.92; m.p. various grades from 133-163 F; acid no. 0; sapon. no. < 1; ref. index 1.4416-1.4465; dielec. const. 2.15-2.33.

Koster Keunen Microcrystalline Waxes. [Koster Keunen] Microcrystalline wax; CAS 64742-42-3; used for cosmetics, pharmaceuticals; laminating of paper, cloth; waterproofing paper, fiberboard, textiles, wood; potting ocmpds. for condensers; polishes for floor, furniture, skis, leather; rust prevention compding. of rubber; pattern-making; printing inks; lubricants; records; FDA §172.886, 178.3710; wh. to amber wax; sp.gr. 0.90-0.94; visc. 50-100 (210 F); m.p. 140-190 F; acid no. 0-0.2; iodine no. 0.-1.5; sapon. no. 0-2; flash pt. > 425 F; fire pt. > 550 F.

Koster Keunen Ozokerite. [Koster Keunen] Ozokerite; CAS 8021-55-4; used for cosmetics, creams, lotions, pomades; paints, varnishes; polishes for leather, automobile, floor; printing inks; pharmaceutical ointments; crayons; waxed paper; linen/ cotton sizing; elec. insulating; lubricants and sealing wax compositions; process engraving, lithography; rubber filler; FDA approved; wax; sol. (g/100 g): 12.99 carbon bisulfide, 11.83 g petrol. ether (75 C), 6.06 g turpentine (160 C), 3.95 g xylene (137 C), 2.83 g toluene (109 C), 2.42 g chloroform; sp.gr. 0.85-0.95; m.p. various grades from 149-190 F; acid no. 0; iodine no. 7-9; sapon. no. 0; ref. index 1.440 (60 C); dielec. const. 2.37-2.55.

Koster Keunen Synthetic Candelilla. [Koster Keunen] Syn. candelilla wax; CAS 136097-95-5; gellant, thickener, stabilizer, and moisturizer used for lipsticks, creams, lotions, gel prods.; leather dressings; furniture and other polish; cements; varnishes; sealing wax; elec. insulating compositions; phonograph records; paper sizing; rubber; waterproofing and insectproofing; paint removers; soft wax stiffeners; lubricants; adhesives; candy and gum; wax; sol. hot in alcohol, benzene, petrleum ether; m.p. 68.5-72.5 C; acid no. 11-20; sapon. no. 44-66.

Kotamite®. [ECC Int'l.] Calcium carbonate, coated with < 2% stearic acid; CAS 1317-65-3, 57-11-4 resp.; coated pigment, filler, reinforcement with easy dispersion in plastic compds., e.g., polyolefins, rigid and flexible PVC; for wire and cable insulation compds., water sealant compds., improved impact props. in PP; hydrophobic; NSF compliance; wh. powd., odorless; 3 μ mean particle size; fineness (Hegman) 6.0; sp.gr. 2.71; dens. 50 lb/ft^3 (loose); surf. area 2.8 m^2/g; oil absorp. 13; pH 9.5 (5% slurry); nonflamm.; 97.6% $CaCO_3$, 0.2% max. moisture; *Toxicology:* TLV/TWA 10 mg/m^3, considered nuisance dust; *Precaution:* incompat. with acids.

KP-140®. [FMC] Tributoxyethyl phosphate; CAS 78-51-3; EINECS 201-122-9; plasticizer; leveling agent in floor polish formulations; flame retardant for plastics or syn. rubbers of lower flammability; imparts low temp. flexibility to plastics or acrylonitrile rubbers; reduces visc. in plastisols; APHA 75 max. liq.; mild butyl odor; sol. in org. liqs. and gasoline; m.w. 398; sp.gr. 1.016-1.023 (20/20 C); dens. 8.5 lb/gal (20 C); visc. 12.2 cp (20 C); vapor pressure < 0.1 mm Hg (150 C); f.p. < -70 C; b.p. 215-228 C (4 mm); pour pt. < -70 C; flash pt. (PMCC) 224 C; fire pt. 252 C; ref. index 1.434 ± 0.002; surf. tens. 30 dynes/cm (20 C).

Kronitex® 25. [FMC] Triaryl phosphate; CAS 68937-41-7; flame retardant plasticizer for PC/ABS, engineering resins, PVC, phenolic laminates, cellulosics; APHA 75 max. clear liq., odorless; m.w. 360; sp.gr. 1.190-1.200 (20/20 C); dens. 10 lb/gal (20 C); visc. 19-22 cp (100 F); pour pt. -29 C; flash pt. (COC) 245 C; fire pt. (COC) 320 C; ref. index 1.5574; 8.8% P, < 0.1% moisture.

Kronitex® 50. [FMC] Triaryl phosphate; CAS 68937-41-7; flame retardant plasticizer for PVC, flexible polyurethanes, syn. rubber, belting; APHA 75 max. clear liq., odorless; m.w. 375; sp.gr. 1.170-1.180 (20/20 C); dens. 9.8 lb/gal (20 C); visc. 70 cp (20 C); pour pt. -26 C; flash pt. (COC) 260 C; fire pt. (COC) 329 C; ref. index 1.553; 8.3% P, < 0.1% moisture.

Kronitex® 100. [FMC] Triaryl phosphate; CAS 68937-41-7; flame retardant plasticizer for PVC; aids fusion; in plastisols, visc. stability; compatibilizing plasticizer; catalyst carrier and pigment vehicle for PU; processing aid in rubber belting and mech. goods; flame retardant and processing aid in engineering resins; APHA 75 max. clear liq., odorless; m.w. 390; sp.gr. 1.150-1.165 (20/20 C); dens. 9.6 lb/gal; visc. 90 cp (20 C); f.p. -31 C; b.p. 220-270 C (4 mm); pour pt. -25 C; flash pt. (COC) 254 C; fire pt. (COC) 332 C; ref. index 1.552; 7.9% P, < 0.1% moisture.

Kronitex® 200. [FMC] Triaryl phosphate; CAS 68937-41-7; flame retardant plasticizer for cellulose polymers (NC lacquers and coatings), PVC (wire and cable sheathing, conveyor belting, carpet backing, plastisols, wall covering); low volatility, exc. resist. to extraction by solvs., superior elec. props.; APHA 75 max. clear liq., odorless; m.w. 420; sp.gr. 1.132-1.142 (20/20 C); dens. 9.5 lb/gal; visc. 46-52 cp (100 F); flash pt. (COC) 257 C; fire pt. (COC) 282 C; ref. index 1.552; 7.4% P, < 0.1% moisture.

Kronitex® 1840. [FMC] Triaryl phosphate; CAS 68937-41-7; thixotropic flame retardant gel for use in coating of fiberglass and other media requiring high loading and low volatility.

Kronitex® 1884. [FMC] Triaryl phosphate; CAS 68937-41-7; nonionic; emulsifiable flame retardant plasticizer for PVC and other polymers, rubber, aq. systems, latexes based on PVC, SBR, PVAc, acrylics, EVA, paints, dyes, nonwoven fabric binders, textile back coatings, carpet backings, asbestos paper bindings, adhesives, pigment carriers; water-disp.; sp.gr. 1.145-1.161 (20/20 C); acid no. 0.26.

Kronitex® 3600. [FMC] Triisopropylphenyl phosphate; plasticizer with improved low temp. flexibility, high flame retardance, and low smoke evolution; ideal for vinyl film and sheeting, wire and cable insulation, coated fabrics, plastisols; APHA 100 max. clear liq.; mild org. odor; sp.gr. 1.030-1.090 (20/20 C); dens. 8.94 lb/gal; visc. 28 cp; pour pt. -48 C; flash pt. (PMCC) 175 C; fire pt. (COC) 275 C; 7.4% P, < 0.1% moisture.

Kronitex® PB-460. [FMC] Brominated aromatic phosphate ester; flame retardant for engineering thermoplastics incl. PC, modified PPO, PBT, PET, ABS and blends; sol. (g/100 g): 43 g in methylene chloride, 25 g in toluene, 11 g in MEK, 0.3 g in methanol; m.p. 110 C.

Kronitex® PB-528. [FMC] Brominated aromatic phosphate ester; flame retardant for engineering thermoplastics incl. PC, modified PPO, PBT, PET, ABS and blends; sol. (g/100 g): 12 g in methylene chloride, 11 g in toluene, 1g in MEK; m.p. 225 C.

Kronitex® TBP. [FMC] Tributyl phosphate; CAS 126-73-8; EINECS 204-800-2; primary plasticizer for NC, cellulose acetate, chlorinated rubber, vinyls; high boiling solv. for lithographic inks; antifoam for paints, paper coating compds., aq. adhesives, inks, casein sol'ns., textile sizes, detergent sol'ns.; pigment dispersant; also for uranium extraction; APHA 50 max. color, char. odor; sol. in most org. liqs., min. oil, gasoline, 0.1% in water; m.w. 266; sp.gr. 0.977-0.983 (20/20 C); dens. 8.14 lb/gal (20 C); visc. 3.7 cp (20 C); vapor pressure 7.3 mm Hg (150 C); f.p. < -80 C; pour pt. < -80 C; b.p. 137-145 C; flash pt. (PMCC) 115 C; fire pt. 182 C; ref. index 1.423 ± 0.001; surf. tens. 29 dynes/cm (20 C); *Toxicology:* severe skin irritant.

Kronitex® TCP. [FMC] Tricresyl phosphate; CAS 68952-35-2; general purpose flame retardant for vinyl compds.; low air, oil, and water loss; processing aid by improving flux rate of compds. containing slow-solvating plasticizers; rapid gelation and fusion rate make it useful in plastisols; plasticizer for NC lacquers and coatings; plasticizer and processing aid for rubbers; flame retardant sheeting; APHA 75 max. clear liq., very sl. odor; m.w. 370; sp.gr. 1.160-1.175 (20/20 C); dens. 9.7 lb/gal; visc. 80 cp (20 C); vapor pressure < 0.02 mm Hg (150 C); b.p. 241-255 C (4 mm); pour pt. -28 C; flash pt. (COC) 252 C; fire pt. (COC) 338 C; ref. index 1.554; 8.4% P, < 0.1% moisture.

Kronitex® TOF. [FMC] Trioctyl phosphate; CAS 78-42-2; EINECS 201-116-6; low temp. plasticizer for use in PVC, rubber, paints, coatings; highly efficient solv.

Kronitex® TPP. [FMC] Triphenyl phosphate; CAS 115-86-6; EINECS 204-112-2; flame retardant plasticizer for engineering resins, syn. rubbers, vinyl resins, cellulose acetate film; low volatility, good permanence, resist. to extraction by org. solvs.; wh. flake, solid, sl. aromatic odor; sp.gr. 1.2 (60 C); vapor pressure 0.1 mm Hg (150 C); m.p. 48 C min.; b.p. 399 C (760 mm); flash pt. (CC) 220 C; 9.5% P, 0.1% max. moisture.

Kronitex® TXP. [FMC] Trixylenyl phosphate; CAS 25155-23-1; flame retardant with better milling action in filled PVC compds.; for superior elect. compds. (wire and cable applic.); APHA 100 max. clear liq.; very sl. odor; m.w. 400; sp.gr. 1.130-1.155 (20/20 C); dens. 9.6 lb/gal; visc. 190 cp (20 C); vapor pressure < 0.02 mm Hg (150 C); b.p. 248-265 C (4 mm); pour pt. -20 C; flash pt. (COC) 263 C; fire pt. (COC) 343 C; ref. index 1.553; 7.8% P, < 0.1% moisture.

Kronos® 1000. [Kronos] Anatase titanium dioxide; CAS 13463-67-7; EINECS 236-675-5; pigment for fibers, paper and coatings, elastomers, food pkg. materials; FDA 21CFR §175.300, 176.170, 176.180; EC, German approvals; dens. 3.8 g/cc; oil absorp. 20; 98% TiO_2; *Toxicology:* inert dust; TLV 6 mg/m^3 air.

Kronos® 2020. [Kronos] Rutile titanium dioxide, stabilized with Al; CAS 13463-67-7; EINECS 236-675-5; pigment for coatings (interior trade sales finishes, maintenance paints, industrial coatings, wh. exterior latex house paints, gel coats), plastisols, food pkg. materials, PVC pipe for potable water; NSF approved; FDA 21CFR §175.300, 176.170, 176.180; EC, German approvals; dens. 4.1 g/cc; oil absorp. 17; 94% TiO_2; *Toxicology:* inert dust; TLV 6 mg/m^3 air.

Kronos® 2073. [Kronos] Rutile titanium dioxide, stabilized with Al; CAS 13463-67-7; EINECS 236-675-5; pigment primarily for plastics (PC, PE, LLDPE, PS, plasticized and rigid interior or exterior PVC, ABS, other engineering thermoplastics, elastomers, linoleum, masterbatches, PVC pipe for potable water, food pkg. materials; NSF approved; FDA 21CFR §175.300, 176.170, 176.180; EC, German approvals; dens. 4.2 g/cc; oil absorp. 11; 97% TiO_2; *Toxicology:* inert dust; TLV 6 mg/m^3 air.

Kronos® 2081. [Kronos] Rutile titanium dioxide, stabilized with Al, Si; CAS 13463-67-7; EINECS 236-675-5; pigment primarily for paper laminates; also for urea-melamine resins, rigid PVC outdoor applics., food pkg. materials; highest lightfastness and greying resist.; FDA 21CFR §175.300, 176.170, 176.180; EC, German approvals; dens. 4.0 g/cc; oil absorp. 20; 90% TiO_2; *Toxicology:* inert dust; TLV 6 mg/m^3 air.

Kronos® 2101. [Kronos] Rutile titanium dioxide, stabilized with Al, Si; CAS 13463-67-7; EINECS 236-675-5; pigment for coatings (aq. and nonaq. exterior or interior trade sales finishes, weatherable maintenance and insutrial coatings), roofing, pastel rigid PVC siding, PVC pipe for potable water, food pkg. materials; NSF approved; FDA 21CFR §175.300, 176.170, 176.180; EC, German approvals; dens. 4.0 g/cc; oil absorp. 20; 92% TiO_2; *Toxicology:* inert dust; TLV 6 mg/m^3 air.

Kronos® 2160. [Kronos] Rutile titanium dioxide, stabilized with Al, Si; CAS 13463-67-7; EINECS 236-675-5; pigment for coatings where max. gloss retention and chalk resist. are essential, powd. coatings, automotive finishes; also suggested for plasticiced and rigid exterior PVC, plastisols, epoxy and polyester resins, food pkg. materials; PVC pipe for potable water; NSF approved; FDA 21CFR §175.300, 176.170, 176.180; EC, German approvals; dens. 3.9 g/cc; oil absorp. 18; 91% TiO_2; *Toxicology:* inert dust; TLV 6 mg/m^3 air.

Kronos® 2200. [Kronos] Rutile titanium dioxide, stabilized with Al; CAS 13463-67-7; EINECS 236-675-5; high str. blue-toned pigment primarily for plastics (PS and copolymers, polyolefins, plasticized PVC), elastomers, linoleum, masterbatches, food pkg. materials; FDA 21CFR §175.300, 176.170, 176.180; EC, German approvals; dens. 4.1 g/cc; oil absorp. 13; 96% TiO_2; *Toxicology:* inert dust; TLV 6 mg/m^3 air.

Kronos® 2210. [Kronos] Rutile titanium dioxide, stabilized with Al; CAS 13463-67-7; EINECS 236-675-5; pigment primarily for plastics (plasticized and rigid interior PVC, PC, polyolefins, PS and copolymers, plastisols, rigid exterior PVC), masterbatches, food pkg. materials; designed for ease of disp. and bluest tint tone; not recommended for very high temp. applics.; FDA 21CFR §175.300, 176.170, 176.180; EC, German approvals; dens. 4.1 g/cc; oil absorp. 13; 96% TiO_2; *Toxicology:* inert dust; TLV 6 mg/m^3 air.

Kronos® 2220. [Kronos] Rutile titanium dioxide, stabilized with Al, Si; CAS 13463-67-7; EINECS 236-675-5; pigment primarily for plastics (PC, polyolefin fibers, PS and copolymers, plasticized and rigid PVC, plastisols, other engineering thermoplastics), masterbatches, liq. colors, powd. coatings, food pkg. materials; moderate weather resist., max. ease o; FDA 21CFR §175.300, 176.170, 176.180; EC, German approvals; dens. 4.0 g/cc; oil absorp. 17; 93% TiO_2; *Toxicology:* inert dust; TLV 6 mg/m^3 air.

Kronos® 2230. [Kronos] Rutile titanium dioxide, stabilized with Al; CAS 13463-67-7; EINECS 236-675-5; pigment primarily for plastics (PC, PE, LLDPE, ABS, PS, plasticized and rigid PVC, other engineering thermoplastics), masterbatches, food pkg. materials; max. ease of disp.; FDA 21CFR §175.300, 176.170, 176.180; EC, German approvals; dens. 4.1 g/cc; oil absorp. 11; *Toxicology:* inert dust; TLV 6 mg/m^3 air.

Kronos® 2310. [Kronos] Rutile titanium dioxide, stabilized with Al, Si, Zr; CAS 13463-67-7; EINECS 236-675-5; pigment for coatings; also suitable for epoxy and thermoset polyester resins, food pkg. materials; FDA 21CFR §175.300, 176.170, 176.180; EC, German approvals; dens. 4.0 g/cc; oil absorp. 18; 93% TiO_2; *Toxicology:* inert dust; TLV 6 mg/m^3 air.

Krynac® 34.140. [Bayer/Fibers, Org., Rubbers] Butadiene-acrylonitrile copolymer, nonstaining, cold polymerized; CAS 9003-18-3; NBR used for plasticizer masterbatches, as visc. modifier, for o-rings, lip seals, gaskets; vulcanized with sulfur, sulfur donor, or peroxide; sp.gr. 0.98; Mooney visc. 140 (ML1+4, 100 C); 34% bound ACN.

Krynac® P30.49. [Bayer/Fibers, Org., Rubbers] NBR, calcium stearate dusting agent; modifier for thermoplastics; FDA 21CFR §177.2600; powd.; 0.8 mm avg. particle size; sp.gr. 0.97; Mooney visc. 42(ML1+4, 100 C, unmassed); 30% ACN.

Krynac® P34.52. [Bayer/Fibers, Org., Rubbers] NBR, calcium stearate dusting agent; modifier for thermoplastics; FDA 21CFR §177.2600; powd.; 0.8 mm avg. particle size; sp.gr. 0.98; Mooney visc. 46 (ML1+4, 100 C, unmassed); 34% ACN.

Krynac® P34.82. [Bayer/Fibers, Org., Rubbers] NBR; modifier for thermoplastics; powd.; 0.7 mm avg. particle size; sp.gr. 0.98; Mooney visc. 70 (ML1+4, 100 C, unmassed); 34% ACN.

Krynac® PXL 34.17. [Bayer/Fibers, Org., Rubbers] Butadiene-acrylonitrile copolymer, PVC dusting agent, nonstaining, hot polymerized; impact modifier for PVC; powd.; 0.45 mm avg. particle size; sp.gr. 0.98; Mooney visc. 48 (ML1+4, 100 C, unmassed); 34% ACN.

Krynac® PXL 38.20. [Bayer/Fibers, Org., Rubbers] Butadiene-acrylonitrile copolymer, sl. staining, hot polymerized; CAS 9003-18-3; impact modifier for phenolic resins; spray-dried powd.; 0.15 mm avg. particle size; sp.gr. 0.98; Mooney visc. 105 (ML1+4, 100 C, massed); 38% ACN.

Krynac® XL 31.25. [Bayer/Fibers, Org., Rubbers] Crosslinked butadiene-acrylonitrile copolymer, nonstaining, cold polymerized; CAS 9003-18-3; modifier for PVC and other thermoplastics; used with other elastomers to give processing base for extruded and calendered goods and as a nonmigratory plasticizer for PVC in crash pad covers, soling, inj. moldings; bales; sp.gr. 0.98; Mooney visc. 67 (ML1+4, 100 C, unmassed); 30.5% ACN.

K-Stay 21®. [King Industries; Struktol] Low visc. min. oil contg. a high m.w. sulfonic acid; plasticizer, softener, and mild processing aid for elastomers; amber liq.; sp.gr. 0.89 (15.6 C); dens. 7.4 lb/gal; visc. 45 SUS (37.8 C); acid no. 8.85; flash pt. (COC) 110 C min.

Kycerol 91. [Rit-Chem] Modified sodium bicarbonate; CAS 144-55-8; EINECS 205-633-8; endothermic blowing agent for inj. molding and extrusion of common thermoplastics (PE, PP, PS, ABS, PPO for film, profiles, sheet, casings, strings, ribbons, foamed prods. incl. containers, furniture, functional parts); nucleating agent in direct gassing of common thermoplastics; produces uniform fine-celled foam structure; suitable for food pkg., toys; FDA GRAS; wh. free-flowing fine powd., odorless; sol. in water; m.p. 150-160 C; gas vol. 115-125 ml/g in air;

95% min. assay; *Storage:* exc. safety in storage and handling under normal conditions.

Kycerol 92. [Rit-Chem] Modified sodium bicarbonate; CAS 144-55-8; EINECS 205-633-8; endothermic blowing agent for inj. molding and extrusion of common thermoplastics (PE, PP, PS, ABS, PPO for film, profiles, sheet, casings, strings, ribbons, foamed prods. incl. containers, furniture, functional parts); nucleating agent in direct gassing of common thermoplastics; produces uniform fine-celled foam structure; suitable for food pkg., toys; FDA GRAS; wh. free-flowing fine powd., odorless; sol. in water; m.p. 180-190 C; gas vol. 114-124 ml/g in air; 95% min. assay; *Storage:* exc. safety in storage and handling under normal conditions.

KZ 55. [Kenrich Petrochemicals] Tetra (2,2 diallyloxymethyl) butyl, di (ditridecyl) phosphito zirconate, IPA; coupling agent sometimes also acting as adhesion promoter, antioxidant, antistat, antifoam, accelerator, blowing agent activator, catalyst, curative, corrosion inhibitor, disp. aid, emulsifier, flame retardant, foaming agent, hardener, impact modifier, internal lube, process aid, release agent, retarder, stabilizer, surfactant, suspension aid, thixotrope, wetting agent for thermoplastics, thermosets, elastomers; lt. brn. liq.; sol. in IPA, xylene, toluene, DOP, min. oil; sp.gr. 1.00; visc. 100 cps; i.b.p. 380 F; flash pt. (TCC) > 200 F; pH 5.7; 90+% solids in IPA.

KZ OPPR. [Kenrich Petrochemicals] Cyclo (dioctyl) pyrophosphato dioctyl zirconate, methyl naphthalene solv.; coupling agent sometimes also acting as adhesion promoter, antioxidant, antistat, antifoam, accelerator, blowing agent activator, catalyst, curative, corrosion inhibitor, disp. aid, emulsifier, flame retardant, foaming agent, hardener, impact modifier, internal lube, process aid, release agent, retarder, stabilizer, surfactant, suspension aid, thixotrope, wetting agent for thermoplastics, thermosets, elastomers; pale yel. liq.; sol. in xylene, toluene, DOP, min. oil; sp.gr. 1.12; visc. 1000 cps; flash pt. (TCC) 150 F; 90% solids in methyl naphthalene solv.

KZ TPP. [Kenrich Petrochemicals] Cyclo [dineopentyl (diallyl)] pyrophosphato dineopentyl (diallyl) zirconate, IPA; coupling agent sometimes also acting as adhesion promoter, antioxidant, antistat, antifoam, accelerator, blowing agent activator, catalyst, curative, corrosion inhibitor, disp. aid, emulsifier, flame retardant, foaming agent, hardener, impact modifier, internal lube, process aid, release agent, retarder, stabilizer, surfactant, suspension aid, thixotrope, wetting agent for thermoplastics, thermosets, elastomers; ylsh. brn. liq.; sol. in IPA, xylene, toluene, DOP, min. oil; sp.gr. 1.18; visc. 3280 cps; i.b.p. 170 F; flash pt. (TCC) > 200 F; 95% solids in IPA.

KZ TPPJ. [Kenrich Petrochemicals] Cycloneopentyl, cyclo (dimethylaminoethyl) pyrophosphato zirconate, di mesyl salt; coupling agent sometimes also acting as adhesion promoter, antioxidant, antistat, antifoam, acclerator, blowing agent activator, catalyst, curative, corrosion inhibitor, disp. aid, emulsifier, flame retardant, foaming agent, hardener, impact modifier, internal lube, process aid, release agent, retarder, stabilizer, surfactant, suspension aid, thixotrope, wetting agent for thermoplastics, thermosets, elastomers; tan liq.; sol. in IPA, toluene, DOP, water; sp.gr. 1.08 (16 C).

L

L-55R®. [Reheis] Aluminum-magnesium hydroxy carbonate, co-dried; acid neutralizing agent used in prod. of polyolefin resins, esp. where high act. polymerization catalysts are used; avail. with sodium, calcium, or zinc stearate coatings; 90% min. thru 10 μ particle size; 27% MgO, 16% Al_2O_3.

Lamefix 680. [Grünau GmbH] Fatty acid polyglycol ether and PEG; nonionic; accelerator for HT-fixation of polyester and triacetate; liq.; 98% conc.; Unverified

Lamigen ES 30. [Dai-ichi Kogyo Seiyaku] PEG lanolin fatty acid ester; nonionic; softener and antistat for textile; oiling agent for leather; emulsifier for emulsion polymerization; additive to lubricating oil and water sol. paint; paste; HLB 11.0; 100% conc.

Lamigen ES 60. [Dai-ichi Kogyo Seiyaku] PEG lanolin fatty acid ester; nonionic; softener and antistat for textile; oiling agent for leather; emulsifier for emulsion polymerization; additive to lubricating oil and water sol. paint; solid; HLB 13.0; 100% conc.

Lamigen ES 100. [Dai-ichi Kogyo Seiyaku] PEG lanolin fatty acid ester; nonionic; softener and antistat for textile; oiling agent for leather; emulsifier for emulsion polymerization; additive to lubricating oil and water sol. paint; solid; HLB 14.5; 100% conc.

Lamigen ET 20. [Dai-ichi Kogyo Seiyaku] PEG lanolin fatty alcohol ether; nonionic; softener and antistat for textile; oiling agent for leather; emulsifier for emulsion polymerization; additive to lubricating oil and water sol. paint; paste; HLB 12.0; 100% conc.

Lamigen ET 70. [Dai-ichi Kogyo Seiyaku] PEG lanolin fatty acid ester; nonionic; softener and antistat for textile; oiling agent for leather; emulsifier for emulsion polymerization; additive to lubricating oil and water sol. paint; solid; HLB 14.0; 100% conc.

Lamigen ET 90. [Dai-ichi Kogyo Seiyaku] PEG lanolin fatty acid ester; nonionic; softener and antistat for textile; oiling agent for leather; emulsifier for emulsion polymerization; additive to lubricating oil and water sol. paint; solid; HLB 15.0; 100% conc.

Lamigen ET 180. [Dai-ichi Kogyo Seiyaku] PEG lanolin fatty acid ester; nonionic; softener and antistat for textile; oiling agent for leather; emulsifier for emulsion polymerization; additive to lubricating oil and water sol. paint; solid; HLB 16.0; 100% conc.

Langford® Clay. [R.T. Vanderbilt] Soft kaolin; CAS 1332-58-7; EINECS 296-473-8; inert filler for all elastomers, nonblack stocks; cream powd.; 1.3 μm median particle size; 99% min. thru 300 mesh; dens. 2.62 ± 0.03 mg/m^3; bulk dens. 44 lb/ft^3 (compacted); oil absorp. 36; brightness 69; pH 3.9 (10% aq. slurry); 44.5% SiO_2, 39.8% Al_2O_3.

Lankrocell® D15L. [Akcros] Blend of org. surfactants; mechanical foam promoter for PVC for carpet and floor backing applics.; amber liq.; misc. with DOP, BBP, chlorinated paraffin, phosphate plasticizers; sp.gr. 1.07.

Lankrocell® KLOP. [Akcros] Surfactant and Ba/Cd/Zn metal soap stabilizer; stabilizer and mechanical foam promoter for foamed PVC plastisols used for carpet and floor backing applics.; heat stabilizer for plastisol during gelation step; brn. liq.; sp.gr. 1.5.

Lankrocell® KLOP/CV. [Akcros] Surfactant and Ba/Zn metal soap stabilizer; stabilizer and mechanical foam promoter for foamed PVC plastisols used for carpet and floor backing applics.; heat stabilizer for plastisol during gelation step; brn. liq.; sp.gr. 1.0.

Lankroflex® ED 6. [Akcros] Epoxidized ester of mixed fatty acids; plasticizer/stabilizer for epoxy; Unverified

Lankroflex® GE. [Akcros] Epoxidized soya bean oil; CAS 8013-07-8; EINECS 232-391-0; plasticizer/stabilizer for PVC; suitable for food-contact applics.; Europe, FDA §181.27; Unverified

Lankroflex® L. [Akcros] Epoxidized linseed oil; plasticizer/stabilizer for PVC; suitable for food-contact applics.; Europe, FDA §178.3740; Unverified

Lankroflex® Series. [Akcros] Epoxidized ester and oil; plasticizer and co-stabilizer for thermoplastics, mainly PVC; liq.; Unverified

Lankromark® BL277. [Akcros] Butyltin carboxylate; Unverified

Lankromark® BLT105. [Akcros] Butylthiotin stabilizer; PVC stabilizer; Unverified

Lankromark® BM122. [Akcros] Butyltin carboxylate; Unverified

Lankromark® BM205. [Akcros] Butyltin carboxylate; Unverified

Lankromark® BM271. [Akcros] Butyltin carboxylate; transparent heat and light stabilizer for rigid and plasticized PVC applics.; Unverified

Lankromark® BM286. [Akcros] Butyltin carboxylate; Unverified

Lankromark® BM400. [Akcros] Butyltin carboxylate; Unverified

Lankromark® BT050. [Akcros] Butyl thiotin; transparent heat stabilizer for rigid and plasticized PVC applics.; Unverified

Lankromark® BT120. [Akcros] Butylthiotin stabilizer; PVC stabilizer; Unverified

Lankromark® BT120A. [Akcros] Dibutyltin mercaptide; stabilizer for PVC processing; exc. heat stability and clarity; for rigid sheeting, profiles, inj. molding; Unverified

Lankromark® BT190. [Akcros] Butylthiotin stabilizer; PVC stabilizer; Unverified

Lankromark® BT339. [Akcros] Butylthiotin stabilizer; PVC stabilizer; Unverified

Lankromark® DLTDP. [Akcros] Dilauryl thiodipropionate; CAS 123-28-4; EINECS 204-614-1; stabilizer for polyolefins and other polymers; synergist with phenolic antioxidant or organophosphite; Unverified

Lankromark® DP6404Z. [Akcros] Barium/zinc; self-lubricating stabilizer for suspension PVC for demanding semirigid formulations; Unverified

Lankromark® DP6452Z. [Akcros] Barium/zinc; Unverified

Lankromark® DSTDP. [Akcros] Distearyl thiodipropionate; CAS 693-36-7; EINECS 211-750-5; stabilizer for polyolefins and other polymers; synergist with phenolic antioxidant or organophosphite; Unverified

Lankromark® DTDTDP. [Akcros] Ditridecyl thiodipropionate; CAS 10595-72-9; Unverified

Lankromark® LC68. [Akcros] Barium/cadmium; nonlubricating stabilizers for suspension PVC resins; LC310 grade also for emulsion resins; liqs.; Unverified

Lankromark® LC90. [Akcros] Cadmium/zinc; stabilizer/activator for chemically blown PVC; fast action; liq.; Unverified

Lankromark® LC244. [Akcros] Barium/cadmium/zinc; Unverified

Lankromark® LC299. [Akcros] Barium/cadmium; Unverified

Lankromark® LC310. [Akcros] Barium/cadmium/zinc; Unverified

Lankromark® LC431. [Akcros] Barium/cadmium; Unverified

Lankromark® LC475. [Akcros] Cadmium-containing; stabilizer for suspension PVC resins; high efficiency, self-lubricating; Unverified

Lankromark® LC486. [Akcros] Barium/cadmium/zinc; Unverified

Lankromark® LC541. [Akcros] Cadmium-containing; stabilizer for suspension PVC resins; high efficiency, self-lubricating; Unverified

Lankromark® LC563. [Akcros] Cadmium-containing; stabilizer for suspension PVC resins; high efficiency, self-lubricating; Unverified

Lankromark® LC585. [Akcros] Cadmium-containing; stabilizer for suspension PVC resins; high efficiency, self-lubricating; Unverified

Lankromark® LC629. [Akcros] Cadmium-containing; stabilizer for suspension PVC resins; high efficiency, self-lubricating; Unverified

Lankromark® LC651. [Akcros] Cadmium-containing; Unverified

Lankromark® LC662. [Akcros] Barium/cadmium/zinc; Unverified

Lankromark® LD299. [Akcros] Barium/cadmium; Unverified

Lankromark® LE65. [Akcros] Triphenyl phosphite; CAS 101-02-0; EINECS 202-908-4; stabilizer for rigid and flexible PVC, PU; epoxy curing agent; Unverified

Lankromark® LE76. [Akcros] Dialkylaryl phosphite; stabilizer for rigid and flexible PVC; Unverified

Lankromark® LE87. [Akcros] Complex phosphite; stabilizer for flexible and nontoxic rigid PVC; Unverified

Lankromark® LE98. [Akcros] Alkyldiaryl phosphite; stabilizer for rigid and flexible PVC; antioxidant for polyolefins; Unverified

Lankromark® LE109. [Akcros] Tris (nonyl phenyl) phosphite; CAS 26523-78-4; EINECS 247-759-6; stabilizer, chelator for nontoxic rigid and flexible PVC; antioxidant for ABS/MBS, polyolefins, SBR; suitable for food-contact applics.; Europe, FDA §178.2010; Unverified

Lankromark® LE131. [Akcros] Alkyldiaryl phosphite; stabilizer for rigid and flexible PVC; antioxidant for polyolefins; Unverified

Lankromark® LE230. [Akcros] Substituted benzophenone; stabilizer for polyolefins, ABS, MBS, and SBR; Unverified

Lankromark® LE274. [Akcros] Benzotriazole derivs.; stabilizer for polyolefins, ABS, MBS, and SBR; Unverified

Lankromark® LE285. [Akcros] 2-Hydroxy-4-octoxybenzophenone; CAS 1843-05-6; EINECS 217-421-2; UV absorber for PVC and polyolefins; yel. powd.; m.p. 48-50 C; Unverified

Lankromark® LE296. [Akcros] 2-Hydroxy-4-methoxybenzophenone; CAS 131-57-7; EINECS 205-031-5; UV absorber for PVC and other polymers; Unverified

Lankromark® LE340. [Akcros] Benzotriazole deriv.; Unverified

Lankromark® LE373. [Akcros] Pentaerythrityl tetrakis [3-(3′,5′-di-t-butyl-4-hydroxyphenyl) propionate]; CAS 6683-19-8; antioxidant; Unverified

Lankromark® LE384. [Akcros] Octadecyl 3-(3′,5′-di-t-butyl-4-hydroxyphenyl)propionate; CAS 2082-79-3; EINECS 218-216-0; Unverified

Lankromark® LZ121. [Akcros] Barium/zinc; nonlubricating stabilizer for PVC emulsion resins; liqs.; Unverified

Lankromark® LZ187. [Akcros] Barium/zinc; stabilizer/activator for chemically blown PVC; slow action; liq.; Unverified

Lankromark® LZ242. [Akcros] Barium/zinc; Unverified

Lankromark® LZ440. [Akcros] Zinc; stabilizer/activator for chemically blown PVC; fast action; liq.; Unverified

Lankromark® LZ495. [Akcros] Calcium/zinc; nonlubricating stabilizer for PVC suspension and emulsion resins; liqs.; Unverified

Lankromark® LZ561. [Akcros] Potassium/zinc; stabilizer/activator for chemically blown PVC; fast action; liq.; Unverified

Lankromark® LZ616. [Akcros] Barium/zinc; high efficiency self-lubricating stabilizer for PVC suspension resins; LZ616 also for emulsion resins (rotational casting); liqs.; Unverified

Lankromark® LZ638. [Akcros] Lead/zinc; stabilizer/activator for chemically blown PVC; fast action; liq.; Unverified

Lankromark® LZ649. [Akcros] Calcium/zinc; nonlubricating stabilizer for PVC emulsion resins; liqs.; Unverified

Lankromark® LZ693. [Akcros] Barium/zinc; high efficiency nonlubricating stabilizers for PVC suspension and emulsion resins; liqs.; Unverified

Lankromark® LZ704. [Akcros] Barium/zinc; Unverified

Lankromark® LZ770. [Akcros] Barium/zinc; Unverified

Lankromark® LZ792. [Akcros] Barium/zinc; Unverified

Lankromark® LZ836. [Akcros] Barium/zinc; Unverified

Lankromark® LZ858. [Akcros] Barium/zinc; Unverified

Lankromark® LZ935. [Akcros] Calcium/zinc; Unverified

Lankromark® LZ968. [Akcros] Barium/zinc; Unverified

Lankromark® LZ1023. [Akcros] Barium/zinc; Unverified

Lankromark® LZ1034. [Akcros] Organic zinc/epoxy blend; plasticizer for stabilization of rigid PVC bottles; Unverified

Lankromark® LZ1045. [Akcros] Calcium/zinc; nontoxic stabilizer for PVC suspension and emulsion resins; liqs.; Unverified

Lankromark® LZ1056. [Akcros] Barium/zinc; Unverified

Lankromark® LZ1067. [Akcros] Barium/zinc; Unverified

Lankromark® LZ1144. [Akcros] Barium/zinc; Unverified

Lankromark® LZ1155. [Akcros] Barium/zinc; Unverified

Lankromark® LZ1166. [Akcros] Barium/zinc; Unverified

Lankromark® LZ1177. [Akcros] Calcium/zinc; Unverified

Lankromark® LZ1188. [Akcros] Calcium/zinc; Unverified

Lankromark® LZ1199. [Akcros] Potassium/zinc; Unverified

Lankromark® LZ1210. [Akcros] Calcium/zinc; Unverified

Lankromark® LZ1221. [Akcros] Potassium/zinc; Unverified

Lankromark® LZ1232. [Akcros] Potassium/zinc; stabilizer/activator for chemically blown PVC; med. action; liq.; Unverified

Lankromark® OT050, OT250. [Akcros] Octyl thiotin; transparent heat stabilizer for rigid PVC applics.; suitable for food contact applics.; Europe approvals; Unverified

Lankromark® OT052. [Akcros] Tin stabilizer; stabilizer for PVC; suitable for food contact applics.; Europe approvals; FDA approved to 3 phr max.; Unverified

Lankromark® OT252. [Akcros] Tin stabilizer; stabilizer for PVC; suitable for food contact applics.; Europe approvals; FDA approved to 3 phr max.; Unverified

Lankromark® OT450. [Akcros] Octyl thiotin; transparent heat stabilizer for rigid and plasticized PVC applics.; suitable for food contact applics.; Europe approvals; Unverified

Lankromark® OT452. [Akcros] Tin stabilizer; stabilizer for PVC; suitable for food contact applics.; Europe approvals; FDA approved to 3 phr max.; Unverified

Lankromark® OT650. [Akcros] Octyl thiotin; transparent heat stabilizer for rigid PVC applics.; suitable for food contact applics.; Europe approvals; Unverified

Lankromark® OT652. [Akcros] Tin stabilizer; stabilizer for PVC; suitable for food contact applics.; Europe approvals; FDA approved to 3 phr max.; Unverified

Lankroplast® L110. [Akcros] Liquid tackifier for flexible PVC; solid; Unverified

Lankroplast® L542. [Akcros] Tackifier for use in highly filled calendered flexible PVC formulations; liq.; Unverified

Lankroplast® L553. [Akcros] Glyceryl stearate; Unverified

Lankroplast® V2012. [Akcros] Visc. modifier for PVC plastisols; Unverified

Lankroplast® V2023. [Akcros] Visc. depressant for PVC plastisols; Unverified

Lankroplast® V2067. [Akcros] Visc. depressant for PVC plastisols; Unverified

Lankroplast® V2100. [Akcros] Visc. depressant for PVC plastisols; Unverified

Lankropol® ADF. [Harcros] Modified sulfosuccinamate; anionic; foaming agent used in foamed aq. polymer systems, textile, leather, upholstery industries; lt. br. clear liq.; char. odor; sol. in water; sp.gr. 1.078; visc. 11 cs; flash pt. > 200 F (COC); pour pt. -5 C; pH 6-8 (1% aq.); 35% act.; Unverified

Lankropol® ATE. [Harcros UK] Tetrasodium N-(1,2-dicarboxyethyl)-N-octadecyl sulfosuccinamate; anionic; primary emulsifier and mechanical stabilizer for emulsion polymers, aux. foaming agent, solubilizing agent; amber slightly hazy liq.; ethanolic odor; sp.gr. 1.119; visc. 36 cs; flash pt. 73 F (Abel CC); pour pt. 3 C; pH 7.0-8.5 (1% aq.); 35% act.; Unverified

Lankropol® KMA. [Harcros UK] Sodium dihexyl sulfosuccinate, ethanol; CAS 6001-97-4; anionic; emulsifier, wetting agent esp. in sol'ns. of electrolytes; solubilizer for soaps, emulsion polymerization aid; pale straw hazy liq.; ethanolic odor; sol. in water; sp.gr. 1.082; visc. 31 cs; flash pt. 91 F (Abel

CC); pour pt. < 0 C; pH 6.0-7.5 (1% aq.); surf. tens. 46 dynes/cm (0.1%); 60% act. in ethanol.

Lankropol® KN51. [Harcros UK] Sodium dicyclohexylalkoxide sulfosuccinate aq. sol'n.; anionic; low foam primary emulsifier for emulsion polymerization affording poor film rewettability; amber liq., char. odor; sol. in water; sp.gr. 1.063; visc. 200 cps; pour pt. -8 C; flash pt. (Abel CC) > 100 C; pH 7.0 (1% aq.); surf. tens. 42 dynes/cm (0.1%); 50% act.

Lankropol® KNB22. [Harcros UK] Monoalkyl sulfosuccinate; anionic; foaming agent for personal care prods.; primary emulsifier for latex prod.; cement foaming agent; liq.; 29% conc.

Lankropol® KO2. [Harcros UK] Sodium dioctyl sulfosuccinate, ethanol; CAS 577-11-7; anionic; wetting agent, emulsifier for emulsion polymerization; pale straw clear liq.; ethanolic odor; sol. up to 0.5% in water; sp.gr. 0.996; visc. 43 cs; flash pt. (Abel CC) 27 C; pour pt. < 0 C; pH 6.5 (1% aq.); surf. tens. 32 dynes/cm (0.1%); 60% act. in ethanol.

Lankropol® ODS/LS. [Harcros UK] Disodium octadecyl sulfosuccinamate; CAS 14481-60-8; EINECS 238-479-5; anionic; foaming agent for aq. polymer dispersions; lt. amber hazy liq., paste; mild odor; sol. in water; sp.gr. 1.066-1.082; visc. 12 cs; flash pt. > 200 F (COC); pour pt. -5 to 4 C; pH 7-9 (1% aq.); 35% act.

Lankropol® ODS/PT. [Harcros UK] Disodium N-octadecyl sulfosuccinamate; CAS 14481-60-8; EINECS 238-479-5; anionic; foaming agent for aq. polymer dispersions; lt. amber hazy liq., paste; mild odor; sol. in water; sp.gr. 1.066-1.082; visc. 5400 cs; flash pt. > 200 F (COC); pour pt. -5 to 4 C; pH 7-9 (1% aq.); 35% act.; Unverified

Lankropol® OPA. [Harcros UK] Potassium salt of a fatty acid sulfonate; anionic; surfactant used in metal industry and for household use; wetting agent, detergent, dispersant; emulsifier for emulsion polymerization; electrolyte stable; biodeg.; dk. amber clear liq.; mild fatty odor; sol. in water; sp.gr. 1.110; visc. 223 cs; flash pt. (COC) > 95 C; pour pt. < 0 C; pH 6.0 (1% aq.); surf. tens. 37 dynes/cm (0.1%); 50% act.

Lankrostat® 16. [Akcros] Antistats for plasticized PVC; Unverified

Lankrostat® 38. [Akcros] Antistat; Unverified

Lankrostat® 104. [Akcros] Antistats for polyolefins, PS; Unverified

Lankrostat® 0600. [Akcros] Antistat; Unverified

Lankrostat® CA2. [Akcros] Antistat for polyolefins, PS, ABS; Unverified

Lankrostat® LA3. [Akcros] Antistat; Unverified

Lankrostat® LDN. [Akcros] Antistat; Unverified

Lankrostat® LME. [Akcros] Antistat for polyolefins, PS, crystal PS; Unverified

Lankrostat® NP6. [Akcros] Antistat; Unverified

Lankrostat® QAT. [Akcros] Cationic; antistat for plasticized PVC, PS, and crystal PS; Unverified

Larostat® 60A. [PPG/Specialty Chem.] Internal antistat for nat. latex rubber; can be formulated into carpet cleaners for compliance with Stainmaster® requirements.

Larostat® 88. [PPG/Specialty Chem.] Modified soyadimethylethyl ammonium ethosulfate; CAS 68308-67-8; cationic; noncorrosive mold release agent; surface-active antistat for PE, PP; liq.; 10% act. in water.

Larostat® 96. [PPG/Specialty Chem.] External antistat for rigid PVC, textiles; liq.; 100% act.

Larostat® 143. [PPG/Specialty Chem.] Oleyldimethylethyl ammonium ethosulfate; cationic; general purpose, low toxicity surface active antistat; liq.; 100% act.

Larostat® 264 A. [PPG/Specialty Chem.] Modified soyadimethylethyl ammonium ethosulfate; CAS 68308-67-8; cationic; surface active internal/topical antistat for syn. fibers, fiberglass, plastics (nylon, acrylic, PS, rigid PVC, polyester, polyethylene), PVC floor finishes, dust control; yel. clear liq.; water-sol.; 35% act. in water.

Larostat® 264 A Anhyd. [PPG/Specialty Chem.] Modified soyadimethylethyl ammonium ethosulfate; CAS 68308-67-8; cationic; internal/external antistat for syn. fibers, fiberglass, plastic, polyethylene; surface active; gel-waxy solid; water- and solv.-sol.; 99% act.

Larostat® 264 A Conc. [PPG/Specialty Chem.] Modified soyadimethylethyl ammonium ethosulfate; CAS 68308-67-8; cationic; noncorrosive mold release agent; internal/external antistat; surface active; liq.; water-sol.; 90% act.

Larostat® 377 DPG. [PPG/Specialty Chem.] Lauric myristic dimethylethyl ammonium ethosulfate, dipropylene glycol; noncorrosive mold release agent; internal antistat; useful in flexible polyurethane foam; surface active; anhyd. liq.; water-sol.; 80% act.

Larostat® 377 FR. [PPG/Specialty Chem.] Quaternary; flame-retardant surface-active internal antistat for polyurethane foam; liq.; water-sol.; 50% act.

Larostat® 451 P. [PPG/Specialty Chem.] Stearyldimethylethyl ammonium ethosulfate; cationic; noncorrosive release agent forming a hard film; imparts gloss and external antistatic properties; post treatment in polystyrenes and fiberglass; surface active; liq.; water-sol.; 30% act.

Larostat® 477. [PPG/Specialty Chem.] Surface-active antistat with improved stability; for ABS, acrylic, cellulosics, nylon, polyacetal, PC, PE, PP, PS, PVC; liq.; water-sol.; 100% act.

Larostat® 519. [PPG/Specialty Chem.] Soyethyl dimonium ethosulfate on silica; general purpose internal antistat for plastics, syn. rubber, PS; powd.; water-disp.; 60% act.

Larostat® 902 A. [PPG/Specialty Chem.] Amide; non-amine internal/topical antistat for PE and PP film; does not affect film quality or clarity; nonreactive with PC; esp. for electronics pkg.; FDA 21CFR §175.105, 176.210, 177.1210; meets MIL spec. B-81705C; liq.; water-sol.; 99% act.

Larostat® 902 S. [PPG/Specialty Chem.] Non-amine antistat on silica; internal antistat for polyolefins; stable to 230 C; FDA 21CFR §175.105, 176.210, 177.1210; meets MIL spec. B-81705C; free-flowing powd.; water-disp.; 60% act.

Larostat® 903. [PPG/Specialty Chem.] Quaternary; internal antistat for flexible PVC; will not discolor the vinyl; meets MIL Spec. B-81705B; liq.; water-sol.; 60% act.

Larostat® 904. [PPG/Specialty Chem.] Antistat for engineering plastics to 300 C, high temp. fiberglass.

Larostat® 905. [PPG/Specialty Chem.] Antistat for flexible PVC, nonwovens, engineering plastics to 250 C.

Larostat® 906. [PPG/Specialty Chem.] Antistat for flexible PVC, engineering plastics to 240 C, high temp. fiberglass.

Larostat® 3001. [PPG/Specialty Chem.] anionic; external antistat for ABS, acrylic, cellulosics, nylon, acetal, PC, PE, PP, PS, flexible and rigid PVC.

Larostat® C-2. [PPG/Specialty Chem.] Ethoxylated cocamine; internal antistat for all types of polyolefins; FDA 21CFR §178.3130; liq.; insol. in water; 99% act.

Larostat® FPE. [PPG/Specialty Chem.] Amide; high performance internal antistat for food-grade polyolefin films, esp. PP; less corrosive than ethoxylated amines; does not impart greasy feel to film; suitable for electronics pkg.; FDA 21CFR § 175.105, 176.180, 177.2260, 177.2800, 178.3130; solid; water-disp.; 99% act.

Larostat® FPE-S. [PPG/Specialty Chem.] Amide antistat on silica; internal antistat for food-grade polyolefin films; FDA 21CFR §175.105, 176.180, 177.2260, 177.2800, 178.3130; free-flowing powd.; water-disp.; 60% act.

Larostat® GMOK. [PPG/Specialty Chem.] Glyceryl oleate; general purpose internal antistat; GRAS; semisolid; insol. in water; 99% act.

Larostat® GMSK. [PPG/Specialty Chem.] Glyceryl stearate; general purpose internal antistat; GRAS; solid; insol. in water; 99% act.

Larostat® HTS 904. [PPG/Specialty Chem.] Quaternary; heat-stable internal antistat for thermoplastics, Noryl nylon; stable to 300 C; liq.; partly sol. in water; 90% act.

Larostat® HTS 904 S. [PPG/Specialty Chem.] Quaternary (HTS 904 on silica); heat-stable internal antistat for thermoplastics, Noryl nylon; powd.; disp. in water; 60% act.

Larostat® HTS 905. [PPG/Specialty Chem.] Quaternary; heat-stable internal antistat for thermoplastics, HIPS, PP, nonwovens; stable to 250 C; liq.; sol. in water; 90% act.

Larostat® HTS 905 S. [PPG/Specialty Chem.] Quaternary (HTS 905 on silica); heat-stable internal antistat for thermoplastics, HIPS, PP, nonwovens; powd.; disp. in water; 60% act.

Larostat® HTS 906. [PPG/Specialty Chem.] Anionic; heat-stable internal antistat; stable to 240 C; paste; sol. in oil; insol. in water; 98% act.

Larostat® LTQ. [PPG/Specialty Chem.] Quaternary; low-toxicity antistat for topical applic. to acrylic, nylon, and polyester; compat. with dyes and finish baths; good thermal stability, wetting props.; onset of thermal degradation at 225 C; brn. liq.; water-sol.; visc. 100 cps max.; flash pt. > 200 C; pH 7.5-9.0; 99% act.

Larostat® PVC. [PPG/Specialty Chem.] Antistat for rigid and flexible PVC, clear PVC; MIL Spec B-81705C compliant.

Larostat® T-2. [PPG/Specialty Chem.] Ethoxylated tallowamine; antistat for all types of polyolefins incl. food-grade LDPE and HDPE; FDA 21CFR §178.3130.

Latekoll®. [BASF AG] Polyacrylic derivs.; thickener for polymer dispersions and latexes; for paints, adhesives, sealants, prod. of fiber webs.

Laural LS. [Ceca SA] Sodium laureth sulfate; CAS 9004-82-4; anionic; base for shampoo and foaming baths; detergent for household prods.; emulsifier for emulsion polymerization; APHA < 500 gel; pH 4-7 (10%); surf. tens. 32 dynes/cm; ≥ 22% act.

Laurex® 4526. [Albright & Wilson UK] Primary fatty alcohol blend; lubricant for rigid PVC for inj. molding processes; feedstock for ethoxylation; wh. waxy flake; dens. 0.45 g/cc; m.p. 48-53 C; flash pt. 202 C.

Laurex® CS [Albright & Wilson Am.; Albright & Wilson UK] Cetearyl alcohol BP; nonionic; mfg. of surfactants; raw material for ethoxylation, sulfation, etc.; stabilizer in emulsion polymerization; lubricant in rigid PVC, also for pharmaceutical creams, hand lotions, bath oils, shaving creams; wh. waxy flake; dens. 0.4 g/cc; m.p. 48-53 C; acid no. 0.5 max.; sapon. no. 2.0 max.; flash pt. 150 C; 100% act.

Laurex® NC. [Albright & Wilson UK] Lauryl alcohol; CAS 112-53-8; EINECS 203-982-0; raw material for ethoxylation, sulfation, etc.; stabilizer in emulsion polymerization; foam stabilizer for fire-fighting foams; superfatting agent for shampoos; wh. soft solid; dens. 0.84 g/cc; m.p. 20-25 C; acid no. 0.2 max.; sapon. no. 1.0 max.; flash pt. 132 C.

Laurox®. [Akzo Nobel] Dilauroyl peroxide; CAS 105-74-8; initiator for elevated-temp. polyester cures, for prod. of PVC resins, and acrylates; flakes; 98% assay; 3.93% active oxygen.

Laurox® W-25. [Akzo Nobel] Dilauroyl peroxide; CAS 105-74-8; initiator; suspension; 25% assay; 1% act. oxygen.

Laurox® W-40. [Akzo Nobel] Dilauroyl peroxide; CAS 105-74-8; efficient initiator for prod. of PVC resins; pumpable form; suspension; 40% assay; 1.61% act. oxygen; *Storage:* store @ 0-20 C.

Laurox® W-40-GD1. [Akzo Nobel] Dilauroyl peroxide; CAS 105-74-8; initiator for PVC prod.; suspension; 40% assay; 1.61% act. oxygen; *Storage:* store @ 0-20 C.

Leadstar. [Syn. Prods.] Lead stearate, normal; CAS 1072-35-1; internal/external lubricant and aux. heat stabilizer for PVC; solid.

Leadstar Envirostab. [Syn. Prods.] Lead stearate; CAS 1072-35-1; external lubricant for rigid and flexible PVC compds. and records; used in conjunction with Tribase prods.; efficient lubricant-stabilizer for rigid inj. molding.

Lectro 60. [Syn. Prods.] Lead chlorosilicate complex; economical lead stabilizer for vinyl elec. insulation and vinyl tapes; wh. fine powd.; sp.gr. 3.9; 47.5% PbO.

Lectro 78. [Syn. Prods.] Tetrabasic lead fumarate; vulcanizing and curing agent for chlorosulfonated and crosslinked PE; low levels used in phonograph records; gives tight cure and good water resist. in lt. colored compds. where yel. color and photosensitivity of litharge are objectionable; lt. cream powd.; sp.gr. 6.5; 90% PbO.

Lectro 90. [Syn. Prods.] Lead sulfophthalate complex; multipurpose heat stabilizer for all classes of vinyl wire insulation from 60 C to 105 C; good compat. with phthalates and trimellitates; sp.gr. 3.4; 59% PbO.

Leecure B-110. [Leepoxy Plastics] Boron trifluoride-based; CAS 7637-07-2; hardener, epoxy curing agent providing water, heat and chem. resist. compds. with high physical str. and elec. props.; used in electronic potting compds., elec. varnishes, adhesives, filament winding, fiberglass composites, prepregs; brn. liq.; visc. 15,000 cps; dens. 9 lb/gal; min. cure temp. 65 C; *Storage:* nonhygroscopic, but avoid prolonged exposure to humidity; keep containers tightly sealed; blanket with dry N.

Leecure B-550. [Leepoxy Plastics] Boron trifluoride-based; CAS 7637-07-2; hardener, epoxy curing agent providing water, heat and chem. resist. compds. with high physical str. and elec. props.; used in electronic potting compds., elec. varnishes, adhesives, filament winding, fiberglass composites, prepregs; brn. liq.; visc. 40,000 cps; dens. 10 lb/gal; min. cure temp. 110 C; *Storage:* nonhygroscopic, but avoid prolonged exposure to humidity; keep containers tightly sealed; blanket with dry N.

Leecure B-610. [Leepoxy Plastics] Boron trifluoride-based; CAS 7637-07-2; hardener, epoxy curing agent providing water, heat and chem. resist. compds. with high physical str. and elec. props.; used in electronic potting compds., elec. varnishes, adhesives, filament winding, fiberglass composites, prepregs; purple-brn. liq.; visc. 10,000 cps; dens. 9 lb/gal; *Storage:* nonhygroscopic, but avoid prolonged exposure to humidity; keep containers tightly sealed; blanket with dry N.

Leecure B-612. [Leepoxy Plastics] Boron trifluoride-based; CAS 7637-07-2; hardener, epoxy curing agent providing water, heat and chem. resist. compds. with high physical str. and elec. props.; used in electronic potting compds., elec. varnishes, adhesives, filament winding, fiberglass composites, prepregs; purple-amber liq.; visc. 10,000 cps; dens. 9 lb/gal; *Storage:* nonhygroscopic, but avoid prolonged exposure to humidity; keep containers tightly sealed; blanket with dry N.

Leecure B-614. [Leepoxy Plastics] Boron trifluoride-based; CAS 7637-07-2; hardener, epoxy curing agent providing water, heat and chem. resist. compds. with high physical str. and elec. props.; used in electronic potting compds., elec. varnishes, adhesives, filament winding, fiberglass composites, prepregs; amber liq.; visc. 11,000 cps; dens. 9 lb/gal; min. cure temp. 65 C; *Storage:* nonhygroscopic, but avoid prolonged exposure to humidity; keep containers tightly sealed; blanket with dry N.

Leecure B-950. [Leepoxy Plastics] Boron trifluoride-based; CAS 7637-07-2; hardener, epoxy curing agent providing water, heat and chem. resist. compds. with high physical str. and elec. props.; used in electronic potting compds., elec. varnishes, adhesives, filament winding, fiberglass composites, prepregs; red-brn. liq.; visc. 33,000 cps; dens. 10 lb/gal; min. cure temp. 80 C; *Storage:* nonhygroscopic, but avoid prolonged exposure to humidity; keep containers tightly sealed; blanket with dry N.

Leecure B-1310. [Leepoxy Plastics] Boron trifluoride-based; CAS 7637-07-2; hardener, epoxy curing agent providing water, heat and chem. resist. compds. with high physical str. and elec. props.; used in electronic potting compds., elec. varnishes, adhesives, filament winding, fiberglass composites, prepregs; amber liq.; visc. 12,000 cps; dens. 9 lb/gal; min. cure temp. 65 C; *Storage:* nonhygroscopic, but avoid prolonged exposure to humidity; keep containers tightly sealed; blanket with dry N.

Leecure B-1550. [Leepoxy Plastics] Boron trifluoride-based; CAS 7637-07-2; hardener, epoxy curing agent providing water, heat and chem. resist. compds. with high physical str. and elec. props.; used in electronic potting compds., elec. varnishes, adhesives, filament winding, fiberglass composites, prepregs; honey-colored liq.; visc. 15,000 cps; dens. 10 lb/gal; min. cure temp. 120 C; *Storage:* nonhygroscopic, but avoid prolonged exposure to humidity; keep containers tightly sealed; blanket with dry N.

Leecure B-1600. [Leepoxy Plastics] Boron trifluoride-based; CAS 7637-07-2; hardener, epoxy curing agent providing water, heat and chem. resist. compds. with high physical str. and elec. props.; used in electronic potting compds., elec. varnishes, adhesives, filament winding, fiberglass composites, prepregs; amber liq.; visc. 7000 cps; dens. 9 lb/gal; min. cure temp. 120 C; *Storage:* nonhygroscopic, but avoid prolonged exposure to humidity; keep containers tightly sealed; blanket with dry N.

Leecure B-1700. [Leepoxy Plastics] Boron trifluoride-based; CAS 7637-07-2; hardener, epoxy curing agent providing water, heat and chem. resist. compds. with high physical str. and elec. props.; used in electronic potting compds., elec. varnishes, adhesives, filament winding, fiberglass composites, prepregs; amber liq.; visc. 4000 cps; dens. 8.5 lb/gal; min. cure temp. 120 C; *Storage:* nonhygroscopic, but avoid prolonged exposure to humidity; keep containers tightly sealed; blanket with dry N.

Leocon 1070B. [Lion] Polyoxyalkylene glycol; nonionic; defoamer for polymerization processes; liq.; Unverified

Leomin AN. [Hoechst Celanese/Colorants & Surf.; Hoechst AG] Alkyl phosphonate; anionic; surfactant for textile processing; antistat for fiber mfg. and processing, plastics processing; sl. yel. clear liq.; visc. 200 cps; pH 8 (1% aq.); 87% act.

Leucophor KNR. [Sandoz] Fluorescent whitener for commerical/industrial laundry detergents, rug/upholstery cleaners, fabric softeners, laundry bleach, whitening soap, brightening polymers and plastics; esp. for cellulosics.

Leucopure EGM Powd. [Sandoz] CI fluorescent brightener 236; CAS 3333-62-8; Fluorescent whitener/brightener for commerical/industrial laundry detergents, rug/upholstery cleaners, fabric softeners, laundry bleach, whitening soap, polymers and plastics; esp. for synthetics and wool; enhances whiteness of TiO_2; masks yellowness in polymers; thermally stable to 350 C; FDA 21CFR §177.1520, 177.2800; BGA approved; greenish-yel. powd.; insol. in water; m.p. 253-254 C; *Toxicology:* LD50 (oral, rat) > 5000 mg/kg; virtually nontoxic; nonirritating to skin and eyes; nonsensitizing; *Precaution:* dusts may be explosive hazard; incompat. with strong oxidizing agents; thermal decomp. may produce oxides of carbon and nitrogen; *Storage:* keep containers closed.

Levapren 400. [Bayer AG; Bayer/Fibers, Org., Rubbers] Vinyl acetate/ethylene rubber; syn. rubber for tech. moldings and extrudates, lamp seals, cable sheathings and insulations, cellular rubber goods, footwear soles, waterproof sheeting, hot-melt and pressure-sensitive hot-melt adhesives, etc.; suited for cable jackets; impact modifier for PVC; admixt. to bituminous compositions; good low temp. props.; gran.; sol. in aromatic and chlorinated HC; sp.gr. 0.98; Mooney visc. 20 (ML1+4, 100 C); melt index 5 g/10 min; 40 ± 1.5% VA content.

Levapren 450HV. [Bayer AG; Bayer/Fibers, Org., Rubbers] Vinyl acetate/ethylene rubber; used in peroxide-cured rubber goods with good resist. to heat, ozone, and weathering and blended with natural or syn. rubbers to improve ozone and weather resist.; applics. incl. molded and extruded goods, cable insulation and jackets, proofed goods, and cellular rubber prods.; impact modifier for PVC; admixt. to bituminous compositions; gran.; sol. in aromatic and chlorinated HC; sp.gr. 0.99; Mooney visc. 27 (ML1+4, 100 C); melt index 5 g/10 min max.; 45% VA.

Levapren 450P. [Bayer AG; Bayer/Fibers, Org., Rubbers] EVA copolymer; CAS 24937-78-8; impact modifier for PVC; raises impact strength and notched impact strength of unplasticized PVC; applied mainly by graft polymerization of vinyl chloride onto Levapren; PVC modified with Levapren is easy to process and highly resistant to lt. and weathering, and useful in construction industry; lentil-shaped gran.; Unverified

Levapren 452. [Bayer AG; Bayer/Fibers, Org., Rubbers] Vinyl acetate/ethylene rubber; for tech. moldings and extrudates, cable sheathings, insulation, cellular rubber goods, footwear soles, waterproof sheeting, fabric proofings; very good resist. to weathering, ozone, light; exc. hot air resist.; impact modifier for PVC; gran.; sol. in aromatic and chlorinated HC, cyclic ethers; sp.gr. 0.99; Mooney visc. 11 (ML1+4, 100 C); melt index 11 g/10 min; soften. pt. (R&B) 134 C; tens. str. 5.4 MPa; tens. elong. 900% (break); Shore hardness A30; 45% VA.

Levapren 500HV. [Bayer/Fibers, Org., Rubbers] Vinylacetate/ethylene rubber; used in peroxide-cured rubber goods with good resist. to heat, ozone, and weathering and blended with natural or syn. rubbers to improve ozone and weather resist.; applics. incl. molded and extruded goods, cable insulation and jackets, proofed goods, and cellular rubber prods.; impact modifier for PVC; admixt. to bituminous compositions; gran.; sol. in aromatic and chlorinated HC; sp.gr. 1.00; Mooney visc. 27 (ML1+4, 100 C); melt index 5 g/10 min max.; 50% VA.

Levapren 700HV. [Bayer/Fibers, Org., Rubbers] Vinyl acetate/ethylene rubber; used in peroxide-cured rubber goods with good resist. to heat, ozone, and weathering and blended with natural or syn. rubbers to improve ozone and weather resist.; applics. incl. molded and extruded goods, cable insulation and jackets, proofed goods, and cellular rubber prods.; impact modifier for PVC; admixt. to bituminous compositions; gran.; sol. in ketones, esters, aromatic and chlorinated HC; sp.gr. 1.08; Mooney visc. 27 (ML1+4, 100 C); melt index 5 g/10 min max.; 70% VA.

Levapren KA 8385. [Bayer/Fibers, Org., Rubbers] Vinyl acetate/ethylene rubber; for tech. moldings and extrudates, cable sheathings, insulation, cellular rubber goods, footwear soles, waterproof sheeting, fabric proofings; very good resist. to weathering, ozone, light; exc. hot air resist.; impact modifier for PVC; gran.; sp.gr. 1.04; Mooney visc. 24 (ML1+4, 100 C); 60% VA.

Lexolube® B-109. [Inolex] Tridecyl stearate; CAS 31556-45-3; EINECS 250-696-7; drawing and heat setting lubricant for textile/industrial filament yarns, plastic extrusion, magnetic tapes, metalworking, coatings and inks (mar and slip agent); APHA 150 max. liq.; sp.gr. 0.857; visc. 25 cP; m.p. 4 C; acid no. 1 max.; iodine no. 1; sapon. no. 120-128; hyd. no. 1.5 max.; flash pt. 221 C; smoke pt. 144 C; 0.1% max. moisture.

Lexolube® BS-Tech. [Inolex] Butyl stearate; CAS 123-95-5; EINECS 204-666-5; fiber or coning oil lubricant for textile syn. and natural fibers; lubricant/plasticizer for metalworking, plastics, rubber; lubricant/slip agent for coatings and inks; APHA 400 max. liq.; sp.gr. 0.853; visc. 9 cP; m.p. 18 C; acid no. 3 max.; iodine no. 3; sapon. no. 170-180; hyd. no. 0.5 max.; flash pt. 180 C; smoke pt. 102 C; 0.1% max. moisture.

Lexolube® NBS. [Inolex] n-Butyl stearate; CAS 123-95-5; EINECS 204-666-5; fiber and coning oil lubricant for textile syn. and natural fibers; lubricant/plasticizer for metalworking, plastics, rubber industries; lubricant and slip agent for coatings and inks; FDA approved as metal can rolling lubricant; APHA 10 max. liq.; sp.gr. 0.853; visc. 9 cP; m.p. 20 C; acid no. 1 max.; iodine no. 1; sapon. no. 165-175; hyd. no. 0.7 max.; flash pt. 180 C; smoke pt. 115 C; 0.1% max. moisture.

Lexolube® T-110. [Inolex] 2-Ethylhexyl stearate; lubricant for textile, metalworking, plastics industries; APHA 300 max. liq.; sp.gr. 0.860; visc. 16 cP; m.p. -38 C; acid no. 3 max.; sapon. no. 148; hyd. no. 1 max.; flash pt. 218 C; smoke pt. 128 C; 0.2% max. moisture.

LICA 01. [Kenrich Petrochemicals] Neopentyl (diallyl)oxy trineodecanonyl titanate, IPA; coupling agents which also act as adhesion promoters, antioxidants, antistats, antifoaming agents, accelerators, blowing agent activators, catalysts, curatives, corrosion inhibitors, disp. aids, emulsifiers, flame retardants, foaming agents, hardeners, impact modifiers, internal lubes, process aids, release agents, retarders, stabilizers, surfactants, suspension aids, thixotropes, wetting agents for thermoplastics, thermosets, elastomers; brnsh. orange liq.; b.p. 320 F (initial); sol. in IPA, xylene, toluene, DOP, min. oil; insol. in water; sp.gr. 1.02 (16 C); visc. 850 cps; pH 5 (sat. sol'n.); flash pt. (TCC) 160 F; 95% solids in IPA; *Toxicology:* LD50 (oral, rat) 5 g/kg.

LICA 09. [Kenrich Petrochemicals] Neopentyl (diallyl)oxy, tri(dodecyl)benzenesulfonyl titanate, IPA; coupling agents which also act as adhesion promoters, antioxidants, antistats, antifoaming agents, accelerators, blowing agent activators, catalysts, curatives, corrosion inhibitors, disp. aids, emulsifiers, flame retardants, foaming agents, hardeners, impact modifiers, internal lubes, process aids, release agents, retarders, stabilizers, surfactants, suspension aids, thixotropes, wetting agents for thermoplastics, thermosets, elastomers; grnsh.-br. liq.; b.p. 170 F (initial); sol. in IPA, xylene, toluene, DOP, min. oil; insol. in water; sp.gr. 1.04 (16 C); visc. 2000 cps; pH 2 (sat. sol'n.); flash pt. (TCC) 180 F; 90% solids in IPA; *Toxicology:* LD50 (oral, rat) 1.7 g/kg.

LICA 12. [Kenrich Petrochemicals] Neopentyl (diallyl) oxy, tri (dioctyl)phosphato titanate, IPA; coupling agents which also act as adhesion promoters, antioxidants, antistats, antifoaming agents, accelerators, blowing agent activators, catalysts, curatives, corrosion inhibitors, disp. aids, emulsifiers, flame retardants, foaming agents, hardeners, impact modifiers, internal lubes, process aids, release agents, retarders, stabilizers, surfactants, suspension aids, thixotropes, wetting agents for thermoplastics, thermosets, elastomers; orange liq.; sol. in xylene, toluene, DOP, min. oil; limited sol. in IPA; insol. in water; sp.gr. 1.03 (16 C); visc. 300 cps; b.p. 160 F (initial); pH 5 (sat. sol'n.); flash pt. (TCC) 160 F; 95% solids in IPA; *Toxicology:* LD50 (oral, rat) 5 g/kg.

LICA 38. [Kenrich Petrochemicals] Neopentyl (diallyl) oxy, tri (dioctyl)pyrophosphato titanate, IPA; coupling agents which also act as adhesion promoters, antioxidants, antistats, antifoaming agents, accelerators, blowing agent activators, catalysts, curatives, corrosion inhibitors, disp. aids, emulsifiers, flame retardants, foaming agents, hardeners, impact modifiers, internal lubes, process aids, release agents, retarders, stabilizers, surfactants, suspension aids, thixotropes, wetting agents for thermoplastics, thermosets, elastomers; grnsh. br. liq.; sol. in IPA, xylene, toluene, DOP, min. oil; insol. in water; sp.gr. 1.13 (16 C); visc. 5000 cps; b.p. 160 F (initial); pH 3.5 (sat. sol'n.); flash pt. (TCC) 160 F; 95% solids in IPA; *Toxicology:* LD50 (oral, rat) 2.5-5 g/kg.

LICA 38A. [Kenrich Petrochemicals] Acrylate functional amine adduct; coupling agents which also act as adhesion promoters, antioxidants, antistats, antifoaming agents, accelerators, blowing agent activators, catalysts, curatives, corrosion inhibitors, disp. aids, emulsifiers, flame retardants, foaming agents, hardeners, impact modifiers, internal lubes, process aids, release agents, retarders, stabilizers, surfactants, suspension aids, thixotropes, wetting agents for thermoplastics, thermosets, elastomers; ylsh.-br. liq.; b.p. 235 F (initial); sol. in IPA; limited sol. in toluene, water; reacts slowly in DOP; insol. in xylene, min. oil; sp.gr. 1.08 (16 C); visc. 1100 cps; pH 7.5 (sat. sol'n.); flash pt. (TCC) 150 F; 95+% solids in IPA.

LICA 38J. [Kenrich Petrochemicals] Methacrylate functional amine adduct of LICA 38; coupling agents which also act as adhesion promoters, antioxidants, antistats, antifoaming agents, accelerators, blowing agent activators, catalysts, curatives, corrosion inhibitors, disp. aids, emulsifiers, flame retardants, foaming agents, hardeners, impact modifiers, internal lubes, process aids, release agents, retarders, stabilizers, surfactants, suspension aids, thixotropes, wetting agents for thermoplastics, thermosets, elastomers; ylsh.-red liq.; b.p. 220 F (initial); sol. in IPA, water; limited sol. in xylene, toluene, DOP; insol. in min. oil; sp.gr. 1.09 (16 C); visc. 5100 cps; pH 7.5 (sat. sol'n.); flash pt. (TCC) 160 F; 95+% solids in IPA.

LICA 44. [Kenrich Petrochemicals] Neopentyl (diallyl)oxy, tri (N-ethylenediamino) ethyl titanate, IPA; coupling agents which also act as adhesion promoters, antioxidants, antistats, antifoaming agents, accelerators, blowing agent activators, catalysts, curatives, corrosion inhibitors, disp. aids, emulsifiers, flame retardants, foaming agents, hardeners, impact modifiers, internal lubes, process aids, release agents, retarders, stabilizers, surfactants, suspension aids, thixotropes, wetting agents for thermoplastics, thermosets, elastomers; brnsh. orange liq.; sol. in IPA; reacts slowly in DOP; insol. in xylene, toluene, min. oil, water; sp.gr. 1.17 (16 C); visc. 10,000 cps; b.p. 250 F (initial); pH 11 (sat. sol'n.); flash pt. (TCC) 200 F; 95% solids in IPA; *Toxicology:* LD50 (oral, rat) 5 g/kg.

LICA 97. [Kenrich Petrochemicals] Neopentyl(diallyl) oxy, tri(m-amino)phenyl titanate, phenyl glycol ether solv.; coupling agents which also act as adhesion promoters, antioxidants, antistats, antifoaming agents, accelerators, blowing agent activators, catalysts, curatives, corrosion inhibitors, disp. aids, emulsifiers, flame retardants, foaming agents, hardeners, impact modifiers, internal lubes, process aids, release agents, retarders, stabilizers, surfactants, suspension aids, thixotropes, wetting agents for thermoplastics, thermosets, elastomers; brn. liq.; sol. in ether solvents; limited sol. in IPA; insol. in xylene, toluene, DOP, min. oil, water; sp.gr. 1.17 (16 C); visc. 2600 cps; b.p. 180 F (initial); pH 6 (sat. sol'n.); flash pt. (TCC) 160 F; 56% solids.

LICA 99. [Kenrich Petrochemicals] Neopentyl (diallyl)oxy, trihydroxy caproyl titanate; coupling agents which also act as adhesion promoters, antioxidants, antistats, antifoaming agents, accelerators, blowing agent activators, catalysts, curatives, corrosion inhibitors, disp. aids, emulsifiers, flame retardants, foaming agents, hardeners, impact modifiers, internal lubes, process aids, release agents, retarders, stabilizers, surfactants, suspension aids, thixotropes, wetting agents for thermoplastics, thermosets, elastomers; tan liq.; sol. in IPA, DOP; sp.gr. 1.03.

Lilamin AC-59 P. [Berol Nobel AB] Long-chain fatty nitrogen deriv. with C14-22 alkyl chain length; mold release agent, processing aid for mfg. of natural and syn. rubber prods.

Lindol XP Plus. [Akzo Nobel] Tricresyl phosphate; flame retardant for acrylic, cellulosics, epoxy, thermoset polyester, PS, PVAc, PVC, PU flexible foam.

Lipolan 1400. [Lion] α-Olefin sulfonate; anionic; emulsifier for cosmetics; emulsifier, dispersant for emulsion polymerization; powd.; 100% conc.; Unverified

Lipolan LB-440. [Lion] α-Olefin sulfonate; anionic; detergent base; emulsifier for cosmetics and emulsion polymerization; liq.; 37% conc.

Lipolan PJ-400. [Lion] α-Olefin sulfonate; anionic; emulsifier for cosmetics; dispersant for emulsion polymerization; powd.; 100% conc.

Lipomin LA. [Lion] Alanine type; external antistat for textiles and plastics; liq.; Unverified

Liponox NC 2Y. [Lion] Alkylphenol ether, ethoxylated; nonionic; emulsifier for emulsion polymerization; flakes; 100% conc.; Unverified

Liponox NC-500. [Lion] POE alkylphenol ether; nonionic; emulsifier for emulsion polymerization; flakes; HLB 18.2; 100% conc.

Liquax 488. [Astor Wax] Microcryst. wax, zinc stearate; hard wax; used in mold release compds. for PU and other plastics, liq. solv. polishes, corrosion-resistant compds.; wh. flakes; m.p. 82-84 C; penetration 4-6; acid no. 4-8; sapon. no. 5-10.

Liquazinc AQ-90. [Witco; Sovereign] Zinc stearate aq. emulsion; CAS 557-05-1; EINECS 209-151-9; partitioning agent, antitack agent, mold release, lubricant for rubber, liq. slab dips; release aid and lubricant for abrasive papers; special sub micron particle size prevents settling out; very low visc.; wh. disp.; fineness 99.9% thru 325 mesh; sp.gr. 1.02; visc. 1000 cps; 50% solids.

LiSIPA. [Eastman] Lithium salt of sulfonated isophthalic acid; specialty monomer for polymer applics.

Lite-R-Cobs®. [Andersons] Corncob meal; inert plastic extender and filler; replaces wood flour in wood particle molding with phenolic resins, in profile and sheet stock prod.; filler in glue, asphalt, caulking compds., and rubber; also used in industrial abrasives, as absorbent, as agric. chemical carriers, livestock feed roughage; tan gran.; essentially odorless; 20.7% sol. in 1% sodium hydroxide, 7.4% in hot water, 4.0% in alcohol, 2.4% in 10% sulfuric acid; 2.1% in acetone; sp.gr. 0.8; bulk dens. 8-15 lb/ft^3; oil absorp. 500%; water absorp. 727.0%; flash pt. (OC) 350 F; hardness (Mohs) 1.0; 35.7% cellulose.

Litharge 28. [Eagle-Picher] High-purity lead oxide; CAS 1317-36-8; EINECS 215-267-0; activator and vulcanizing agent in rubber; mfg. of dry colors, greases, high-pressure lubricants, brake linings, ceramics, glass, piezoelectric devices, various chemical processes; yel.; > 10 μ mean particle size; 99.9% -325 mesh; sp.gr. 9.4; apparent dens. 32-38 g/in.^{3}; 99.9% PbO.

Litharge 33. [Eagle-Picher] High-purity lead oxide; CAS 1317-36-8; EINECS 215-267-0; acid acceptor, activator, and vulcanizing agent in rubber compding; also in mfg. of dry colors, high-pressure lubricants, brake linings, ceramics, glass, piezoelectric devices, chemical processes; high surf. area and ease of dispersion; yel. powd.; < 2 μ median particle diam.; 99.9% -325 mesh; sp.gr. 9.4; apparent dens. 13-17 g/in.^{3}; bulking value 0.126 gal/lb; 99.9% PbO.

Lithene AH. [Revertex Ltd] Butadiene liq. telomer; contains cyclic polymer segments which impart toughness and thermal stability to cured polymers; peroxide reactivity, good dielec. chars. and high filler tolerance; suited for elec. insulating materials; as curing agents for EPM and EPDM rubber compositions; clear, colorless to pale-yel. liq.; m.w. 1800; dens. 0.93 g/ml; visc. 450 poise (35 C); COC flash pt. > 260 C; 95% polybutadiene; Unverified

Lithene AL. [Revertex Ltd] Butadiene liq. telomer; curing agent for EPM and EPDM rubber compositions; imparts toughness and thermal stability to cured polymers; lower visc. grade for incorporation into water-reducible systems via maleinization; clear, colorless to pale-yel. liq.; m.w. 1000; dens. 0.93 g/ml; visc. 40 poise; COC flash pt. 186 C; 90% polybutadiene; Unverified

Lithol® Pigments. [BASF AG] Azo dye lakes for letterpress, offset, flexographic and gravure inks, for paints, for coloring plastics.

Lithopone 30% DS. [Sachtleben Chemie GmbH] Lithopone, surf. treated, micronized; CAS 1345-05-7; wh. pigment used as TiO_2 substitute for thermoplastic masterbatches, powd. coatings in thermosets based on urea, melamine, and polyester, in natural and syn. elastomers; dens. 4.3 g/ml; surf. area 3 m^2/g; oil absorp. 8-9; pH 7-8; hardness (Mohs) 3.

Lithopone 30% L. [Sachtleben Chemie GmbH] Lithophone; CAS 1345-05-7; wh. pigment used as TiO_2 substitute for thermoplastic masterbatches, powd. coatings in thermosets based on urea, melamine, and polyester, in natural and syn. elastomers; wh.; dens. 4.3 g/ml; surf. area 3 m^2/g; oil absorp. 8-9; pH 7-8; hardness (Mohs) 3.

Lithopone D (Red Seal 30% ZnS). [Sachtleben Chemie GmbH] Lithopone coated with org. surfactant; CAS 1345-05-7; wh. pigment used as TiO_2 substitute for thermoplastic masterbatches, in thermosets based on urea, melamine, and polyester, in natural and syn. elastomers, and in paper applics.; dens. 4.3 g/cc; bulk dens. 1.70 l/kg (poured), 0.90 l/kg (tapped); surf. area 3 m^2/g; oil absorp. 9; pH 8; hardness (Mohs) 3.

Lithopone D (Silver Seal 60% ZnS). [Sachtleben Chemie GmbH] Lithopone coated with org. surfactant; CAS 1345-05-7; wh. pigment used as TiO_2 substitute for thermoplastic masterbatches, in thermosets based on urea, melamine, and polyester, in natural and syn. elastomers, and in paper applics.; dens. 4.2 g/cc; bulk dens. 1.80 l/kg (poured), 1.00 l/kg (tapped); surf. area 5 m^2/g; oil absorp. 10; pH 7; hardness (Mohs) 3; Unverified

Lithopone DS (Red Seal 30% ZnS). [Sachtleben Chemie GmbH] Lithopone; CAS 1345-05-7; wh. pigment used as TiO_2 substitute for thermoplastic masterbatches, powd. coatings in thermosets based on urea, melamine, and polyester, in natural and syn. elastomers; micronized form; dens. 4.3 g/ml; bulk dens. 1.60 l/kg (poured), 0.80 l/kg (tapped); surf. area 3 m^2/g; oil absorp. 8; pH 8; hardness (Mohs) 3.

Lithopone L (Red Seal 30% ZnS). [Sachtleben Chemie GmbH] Lithopone; CAS 1345-05-7; wh. pigment used as TiO_2 substitute for thermoplastic masterbatches, in thermosets based on urea, melamine, and polyester, in natural and syn. elastomers, and in paper applics.; dens. 4.3 g/cc; bulk dens. 1.80 l/kg (poured), 0.90 l/kg (tapped); surf. area 3 m^2/g; oil absorp. 9; pH 7; hardness (Mohs) 3; Unverified

Lithopone L (Silver Seal 60% ZnS). [Sachtleben Chemie GmbH] Lithopone; CAS 1345-05-7; wh. pigment used as TiO_2 substitute for thermoplastic masterbatches, in thermosets based on urea, melamine, and polyester, in natural and syn. elastomers, and in paper applics.; dens. 4.2 g/cc; bulk dens. 1.90 l/kg (poured), 1.00 l/kg (tapped); surf. area 5 m^2/g; oil absorp. 10; pH 7; hardness (Mohs) 3; Unverified

LK-221®. [Air Prods./Perf. Chems.] Nonsilicone surfactant; cell stabilizer for PU rigid and elastomeric applics.; liq.; insol. in water; sp.gr. 1.027-1.036; visc. 2000 cps (23 C); f.p. < -17 C; hyd. no. 41; flash pt. (COC) 187.7 C.

LK®-443. [Air Prods./Perf. Chems.] Nonsilicone surfactant; cell stabilizer for rigid PU applics.; liq.; insol. in water; sp.gr. 1.075-1.082; visc. 2500 cps (23 C); f.p. < -21 C; hyd. no. 44; flash pt. (PMCC) 116 C.

Lomar® HP. [Henkel/Functional Prods.] Condensed potassium naphthalene sulfonate; anionic; dispersant; sec. emulsifier for emulsion polymerization; powd.; 94% conc.

Lomar® LS. [Henkel/Functional Prods.; Henkel/Textiles] Condensed sodium naphthalene sulfonate; CAS 9084-06-4; anionic; dispersant, emulsifier for emulsion polymerization, dyestuff mfg., agric. formulations; leveling agent for dyeing fibers; low salt; tan powd.; sol. @ 5% in water; cloud pt. 2 C; flash pt. 92 F; pH 9.5 (10% aq.); 95% act.

Lomar® LS Liq. [Henkel/Functional Prods.] Condensed sodium naphthalene sulfonate; anionic; dispersant; primary emulsifier for emulsion polymerization; liq.; 46% solids.

Lomar® PW. [Henkel/Emery; Henkel/Functional Prods.; Henkel/Textile] Condensed sodium naphthalene sulfonate; CAS 9084-06-4; anionic; dispersant for pigments, extenders, and fillers in aq. media; used in dyeing syn. and natural fibers; emulsifier in emulsion polymerization; ceramics; gypsum board, for pigments, printing, rubber and wet milling; food pkg. applics.; agric. prods.; suspending agent, stabilizer for paint and paper industries; EPA-exempt; tan powd.; water-sol. @ 5%; dens. 0.66 g/cc; cloud pt. 5 C; flash pt. 87 F; pH 9.5 (10%); 87% act.

Lomar® PWA. [Henkel/Emery; Henkel/Functional Prods.] Condensed ammonium naphthalene sulfonate; anionic; visc. depressant; for molding and extruding operations in ceramics; dispersant for emulsion paints, agric. formulations; emulsifier for emulsion polymerization of syn. elastomers; visc. reducer for pigment slurries; stabilizer; EPA-exempt; lt. tan powd.; sol. in water; pH 3.5 (10%); 92% act.

Lomar® PWA Liq. [Henkel/Emery; Henkel/Functional Prods.] Condensed ammonium naphthalene sulfonate; anionic; dispersant; sec. emulsifier for emulsion polymerization, agric. formulations; EPA-exempt; dk. brn. liq.; pH 7.3 (10%); 44% act.

Lonza Coke PC 40. [Lonza G+T] Coke; CAS 50-36-2; plastics additive providing wear reduction in PTFE, creep resist., chemical inertia and stability, thermal conductivity and stability.

Lonza KS 44. [Lonza G+T] Graphite; CAS 7782-42-5; EINECS 231-955-3; relatively round particle shape, high elec. and thermal conductivity, good compressibility; for plastics, carbon brushes, batteries, electrochemistry, pencils, hard metals, lubricants, catalysts; fine powd.; bulk dens. 0.20 g/cc; surf. area 10 m^2/g; oil absorp. 109 g/100 g (in DBP) $\geq$ 99.9% C.

Lonza T 44. [Lonza G+T] Graphite; CAS 7782-42-5; EINECS 231-955-3; more flake-shaped particles, larger surf. area, greater hardness than KS types; exc. lubricating props.; for plastics, carbon brushes, lubricants, catalysts; fine powd.; bulk dens. 0.18 g/cc; surf. area 10 m^2/g; oil absorp. 121 g/100 g (in DBP) $\geq$ 99.9% C.

Lonzacure® M-CDEA. [Lonza Ltd] 4,4′-Methylenebis (3-chloro-2,6-diethylaniline); CAS 106246-33-7; chain extender for elastomer PU; curing agent for epoxides; precursor for polyimides; intermediate for org. synthesis; off-wh. cryst.; sol. in toluene, xylene, DMSO, DMF, n-butanol, aniline; sol. in triethylene glycol @ 80 C; insol. in water; m.w.

379.38; bulk dens. 0.61-0.65 kg/l; m.p. 88-90 C; 97% min. assay; *Toxicology:* LD50 (oral, rat) > 5000 mg/kg, (dermal, rat) > 2000 mg/kg; nonirritating to skin and eyes; nonmutagenic; *Storage:* 1 yr min. storage stability.

Lonzacure® M-DEA. [Lonza Ltd] 4,4′-Methylenebis 2,6-diethylaniline; chain extender for RIM PU; curing agent for epoxy laminates and composites; precursor for polyimides; wh. to brn. flakes; m.w. 310.49; m.p. 88 C min.; 99% min. assay; *Toxicology:* LD50 (oral, rat) 1901 mg/kg; nonirritating to skin; nonmutagenic.

Lonzacure® M-DIPA. [Lonza Ltd] 4,4′-Methylenebis 2,6-diisopropylaniline; chain extender for RIM PU; reddish brn. to red solidified melt; m.w. 366.60; solid. pt. 10-30 C; *Toxicology:* LD50 (oral, rat) 1110 mg/kg; nonirritating to skin; nonmutagenic.

Lonzacure® M-DMA. [Lonza Ltd] 4,4′-Methylenebis 2,6-dimethylaniline; starting material for polyimides; curing agent for epoxy resins; off-wh. cryst.; m.w. 254.38; m.p. 117.5 C min.; 99% min. assay; *Toxicology:* LD50 (oral, rat) 724 mg/kg; nonirritating to skin; nonmutagenic.

Lonzacure® M-MEA. [Lonza Ltd] 4,4′-Methylenebis 2-ethyl-6-methylaniline; precursor for polyimides, in photoresists; wh. to brn. flakes; m.w. 282.43; m.p. 85 C min.; 99% min. assay; *Toxicology:* LD50 (oral, rat) 1582 mg/kg; nonirritating to skin; nonmutagenic.

Lonzacure® M-MIPA. [Lonza Ltd] 4,4′-Methylenebis 2-isopropyl-6-methylaniline; chain extender for RIM PU; ylsh. brn. solidified melt; m.w. 310.49; solid. pt. 10-30 C; *Toxicology:* LD50 (oral, rat) 2015 mg/kg; nonirritating to skin; nonmutagenic.

Lonzaine® 16S. [Lonza] Cetyl betaine; CAS 693-33-4; EINECS 211-748-4; internal antistat for ABS, acrylic, nylon, rigid PVC; cosmetics surfactant.

Lonzaine® 18S. [Lonza] Stearyl betaine; CAS 820-66-6; EINECS 212-470-6; surfactant; internal antistat for ABS, acrylic, nylon, rigid PVC.

Lonzamon® AAEA. [Lonza Ltd] 2-(Acetoacetoxy) ethyl acrylate; CAS 21282-96-2; visc. reducer in polymer processing (varnishes, adhesives, polyesters); better radiation curing, better surf. curing in presence of atmospheric oxygen; 92% min. assay; *Toxicology:* LD50 (oral, rat) 1132 mg/kg, (dermal, rat) > 2000 mg/kg; mod. skin irritant.

Lonzamon® AAEMA. [Lonza Ltd] 2-(Acetoacetoxy) ethyl methacrylate; CAS 21282-97-3; visc. reducer in polymer processing (varnishes, adhesives, polyesters); m.w. 214.22; b.p. 100 C; acid no. 3-8; 95% min. assay; *Toxicology:* LD50 (oral, rat) > 5000 mg/kg; pract. nonirritating to skin; *Storage:* store in cool, dark place.

Loobwax 0597. [Astor Wax] Petroleum wax; plastics lubricant for pipe, profiles, wire and cable; meets PPI TR3/4 requirements.

Loobwax 0598. [Astor Wax] Ester wax; oxidized lubricant for internal lubrication of plastic resins, PVC bottles, calendering and extrusion of film and sheet; FDA accepted; solid.

Loobwax 0605. [Astor Wax] Syn. hydrocarbon and petroleum wax; plastics lubricant for high shear applics., profiles, siding.

Loobwax 0638. [Astor Wax] Syn. polyolefin blend; internal/external lubricant for calendering and extrusion for thin profiles; FDA accepted; solid.

Loobwax 0651. [Astor Wax] Polyethylene wax; CAS 9002-88-4; EINECS 200-815-3; oxidized lubricant for high-shear applics., pipes, profiles, clear films.

Loobwax 0740. [Astor Wax] Syn. hydrocarbon-based wax; plastics lubricant for single screw extrusion of pipe, profiles.

Loobwax 0750. [Astor Wax] Petroleum wax; high-temp. plastics lubricant with high external lubricant props.; for siding, profiles.

Loobwax 0761. [Astor Wax] Esterified, oxidized polyethylene wax; CAS 68441-17-8; plastics lubricant.

Loobwax 0782. [Astor Wax] Fatty acid ester; internal lubricant for rigid PVC; solid.

Lo-Vel® 27. [PPG Industries] Amorphous syn. silica; CAS 7631-86-9; thickener, flatting agent for controlling gloss in topcoat lacquers for vinyl fabrics and textiles (hunting and ski parkas), furniture lacquers, pigmented metal finishes; also in auto finishes, incandescent bulb coating, agric. chem. carrier, as antiblock for plastic films; wh. powd.; 1.7 μm median size; sp.gr. 2.1; bulking value 5.8 gal/100 lb; bulk dens. 4 lb/ft^3 (tapped); surf. area 170 m^2/g; oil adsorp. 220 linseed oil lb/100 lb; pH 6.5-7.3 (5%); ref. index 1.455; 87% min. SiO_2; *Toxicology:* OSHA TLV/TWA 6 mg/m^3 (total dust); contact with skin causes drying effect; *Precaution:* dust can create explosion hazard.

Lo-Vel® 28. [PPG Industries] Amorphous syn. silica; CAS 7631-86-9; thickener, flatting agent for coil coating; also in auto finishes, incandescent bulb coating, agric. chem. carrier, as antiblock for plastic films; wh. powd.; 4.2 μm median size; sp.gr. 2.1; bulking value 5.8 gal/100 lb; bulk dens. 7 lb/ft^3 (tapped); surf. area 170 m^2/g; oil adsorp. 220 linseed oil lb/100 lb; pH 7; ref. index 1.455; 87% min. SiO_2; *Toxicology:* OSHA TLV/TWA 6 mg/m^3 (total dust); contact with skin causes drying effect; *Precaution:* dust can create explosion hazard.

Lo-Vel® 29. [PPG Industries] Amorphous syn. silica; CAS 7631-86-9; thickener, flatting agent for coil coating; also in auto finishes, incandescent bulb coating, agric. chem. carrier, as antiblock for plastic films; wh. powd.; 4.8 μm median size; sp.gr. 2.1; bulking value 5.8 gal/100 lb; bulk dens. 7 lb/ft^3 (tapped); surf. area 170 m^2/g; oil adsorp. 220 linseed oil lb/100 lb; pH 7; ref. index 1.455; 87% min. SiO_2; *Toxicology:* OSHA TLV/TWA 6 mg/m^3 (total dust); contact with skin causes drying effect; *Precaution:* dust can create explosion hazard.

Lo-Vel® 39A. [PPG Industries] Amorphous syn. silica; CAS 7631-86-9; thickener, flatting agent for coatings, micro texture finish; also in auto finishes, incandescent bulb coating, agric. chem. carrier, as antiblock for plastic films; wh. powd.; 6.7 μm median size; sp.gr. 2.1; bulking value 5.8 gal/100 lb; bulk dens. 10 lb/ft^3 (tapped); surf. area 170 m^2/g; oil adsorp. 220 linseed oil lb/100 lb; pH 7; ref. index 1.455; 87% min. SiO_2; *Toxicology:* OSHA TLV/TWA 6 mg/m^3 (total dust); contact with skin causes drying effect; *Precaution:* dust can create explosion hazard.

Lo-Vel® 66. [PPG Industries] Amorphous syn. silica surf. treated with wax for easy resuspension; CAS 7631-86-9; flatting agent reducing gloss of lacquers, coatings, furniture lacquer; also in auto finishes, incandescent bulb coating, agric. chem. carrier, as antiblock for plastic films; wh. powd.; 2.4 μ median size; sp.gr. 2.1; bulking value 5.8 gal/100 lb; bulk dens. 4 lb/ft^3; pH 6.5-7.3 (5%); ref. index 1.455; 87% min. SiO_2; *Toxicology:* OSHA TLV/TWA 6 mg/m^3 (total dust); contact with skin causes drying effect; *Precaution:* dust can create explosion hazard.

Lo-Vel® 275. [PPG Industries] Amorphous syn. silica; CAS 7631-86-9; thickener, flatting agent for general purpose and coil coatings; also in auto finishes, incandescent bulb coating, agric. chem. carrier, as antiblock for plastic films; wh. powd.; 3.6 μm median size; sp.gr. 2.1; bulking value 5.8 gal/100 lb; bulk dens. 7 lb/ft^3 (tapped); surf. area 170 m^2/g; oil adsorp. 220 linseed oil lb/100 lb; pH 7; ref. index 1.455; 87% min. SiO_2; *Toxicology:* OSHA TLV/TWA 6 mg/m^3 (total dust); contact with skin causes drying effect; *Precaution:* dust can create explosion hazard.

Lo-Vel® HSF. [PPG Industries] Amorphous syn. silica surf. treated with wax for easy resuspension; CAS 7631-86-9; flatting agent reducing gloss of high-solids, low-VOC coatings with minimal visc. increase and high efficiency; suggested for camouflage coatings; also in auto finishes, incandescent bulb coating, agric. chem. carrier, as antiblock for plastic films; wh. powd.; 10 μ median size; sp.gr. 2.1; bulking value 5.8 gal/100 lb; bulk dens. 14 lb/ft^3; pH 6.5-7.3 (5%); ref. index 1.455; 87% min. SiO_2; *Toxicology:* OSHA TLV/TWA 6 mg/m^3 (total dust); contact with skin causes drying effect; *Precaution:* dust can create explosion hazard.

Lowilite® 22. [Great Lakes] Benzophenone-12; CAS 1843-05-6; EINECS 217-421-2; UV absorber for PP, HDPE, LDPE, flexible and rigid PVC, EVA, PC, PS, epoxies, cosmetics; exc. heat resist. at extrusion temps.; approved for food contact; cream to lt. yel. powd.; sol (g/100 g): 190 g in acetone, toluene, 16 g in ethyl acetate, 140 g in MEK, 52 g in DOS, 48 g in DOA, 44 g in DOP; m.w. 326.21; solid. pt. 47 C min.

Lowilite® 26. [Great Lakes] 2-(2′-Hydroxy-3′-t-butyl-5′-methylphenyl)-5-chlorobenzotriazole; CAS 3896-11-5; EINECS 223-445-4; UV stabilizer for polyolefins, PVC, unsat. polyester resins and coatings, other polymers; heat stabilizing effect in PVC; improves weathering resist.; approved for food contact; sl. yel. cryst. powd.; sol (g/100 ml): 13 g styrene, 5 g methyl methacrylate, 3 g MEK, 2 g ethyl acetate; m.p. 138 C min.

Lowilite® 55. [Great Lakes] 2-(2′-Hydroxy-5′-methylphenyl) benzotriazole; CAS 2440-22-4; EINECS 219-470-5; UV absorber (strong absorber below 400 nm) for polymers incl. polyester, PVC, PS, HIPS, rubber, PC, PMMA; approved for food contact; sl. ylsh. cryst. powd.; sol. (g/100 g): 11 g toluene, 10 g acetone, 5 g ethyl acetate, 1 g DOP, < 0.1 g water; m.p. 128 C min.; > 99% purity.

Lowilite® 62. [Great Lakes] α-Methylstyrene/N-(2,2,6,6-tetramethylpiperidinyl-4)maleimide/N-stearyl maleimide terpolymer; HALS protecting LDPE, HPDE, LLDPE, PP against degradation by oxidation and photo-oxidation; very high thermal resist. to 320 C; also for ethylene copolymers, PU, surf. coatings; brn. pellets or ylsh. to amber powd.; sol. (g/100 g): 45 g in toluene, benzene, xylene, 30 g in dichloromethane; soften. pt. 95-125 C; 0.5% max. water; *Toxicology:* nontoxic; LD50 (oral, rat) > 6000 mg/kg.

Lowilite® 77. [Great Lakes] Bis (2,2,6,6-tetramethyl-4-piperidinyl) sebacate; CAS 52829-07-0; EINECS 258-207-9; HALS for polyolefins, PS, HIPS, ABS, SAN, ASA; also suitable for PU, polyamides, acetal; wh. to sl. ylsh. cryst. powd.; sol. (g/100 g): 108 g in methanol, 98 g in trichloromethane, 46 g in ethyl acetate, 20 g in acetone, < 0.1 g in water; m.w. 481; m.p. 80-85 C; > 99% purity.

Lowinox® 22IB46. [Great Lakes] 2,2′-Isobutylidenebis(4,6-dimethylphenol); CAS 33145-10-7; antioxidant for rubber, latex; approved for food contact; solid.

Lowinox® 22M46. [Great Lakes] 2,2′-Methylenebis (4-methyl-6-t-butylphenol); CAS 119-47-1; antioxidant for rubber, latex, adhesives, ABS, polyacetate; approved for food contact; solid.

Lowinox® 44B25. [Great Lakes] 4,4′-Butylidenebis(2-t-butyl-5-methylphenol); antioxidant for rubber, latex, adhesives, ABS, polyamide; approved for food contact; solid.

Lowinox® 44S36. [Great Lakes] 4,4′-Thiobis (2-t-butyl-5-methylphenol); CAS 96-69-5; antioxidant for latex, adhesives, plastics, cross-linked polymers; approved for food contact; solid.

Lowinox® BHT. [Great Lakes] BHT; CAS 128-37-0; EINECS 204-881-4; antioxidant for rubber, adhesives, plastics, polyolefins, polystyrene; approved for food contact; solid.

Lowinox® BHT Food Grade. [Great Lakes] BHT; CAS 128-37-0; EINECS 204-881-4; food-grade nondiscoloring antioxidant/stabilizer for plastics (polystyrene, LDPE, etc.) protecting against degradation by heat and oxygen; BGA, FDA; wh. cryst.; sol. (g/100 g): 244 g in acetone, 212 g in ethyl acetate, 186 g in toluene, tetrachloromethane, 56 g in methanol; f.p. 69.2 C; 99% min. purity.

Lowinox® BHT-Superflakes. [Great Lakes] BHT; CAS 128-37-0; EINECS 204-881-4; food-grade antioxidant/stabilizer but with stronger, typ. odor and tendency to sl. discolor on extended storage; suggested for use in syn. rubbers and their latexes; also for polyols, PUR systems, PVC; BGA, FDA; wh.-ylsh. flakes; sol. (g/100 g): 244 g in acetone, 212 g in ethyl acetate, 186 g in toluene, tetrachloromethane, 56 g in methanol; f.p. 69.2 C;

99% min. purity.

Lowinox® CA 22. [Great Lakes] 1,1,3-Tris-(2-methyl-4-hydroxy-5-t-butylphenyl) butane; CAS 1843-03-4; antioxidant; approved for food contact; solid.

Lowinox® CPL. [Great Lakes] Sterically hindered polynuclear phenol; antioxidant for rubber, latex; approved for food contact; solid.

Lowinox® DLTDP. [Great Lakes] Dilauryl-3,3′-thiodipropionate; CAS 123-28-4; EINECS 204-614-1; antioxidant for plastics, polyolefins; approved for food contact; solid.

Lowinox® DSTDP. [Great Lakes] Distearyl-3,3′-thiodipropionate; CAS 693-36-7; EINECS 211-750-5; antioxidant for plastics, polyolefins; approved for food contact; solid.

Lowinox® PP35. [Great Lakes] Pentaerythrityl tetrakis-3-(3′,5′-di-t-butyl-4′-hydroxyphenyl) propionate; CAS 6683-19-8; antioxidant for adhesives, hot-melts, plastics, polyolefins, PS, PC, SAN; approved for food contact; solid.

Loxiol® G 10. [Henkel KGaA/Dehydag] Monoglycerol ester of unsat. fatty acid; internal lubricant used in clear or opaque applic., rigid and soft PVC, food applic.; pigment dispersing aux. in PVC paste processing; costabilizer in combination with tin; aids flow in pigmented and filled rigid PVC compds.; FDA 21CFR §175.300, 175.320, 176.170, 177.1210, 181.27, 182.1323, 182.4505, GRAS; BGA, Japan approvals; pale yel. liq.; f.p. -1 to -8 C; dens. 7.9 lb/gal; visc. 200-225 cps; acid no. 3.0 max.; ref. index 1.467 ± 0.005; flash pt. (CC) 435 F.

Loxiol® G10P. [Pulcra SA] Fatty acid glycerol ester; internal lubricant for rigid PVC extrusion, flexible PVC calendering, extrusion, inj. molding, PVC paste; yel. turbid liq., neutral odor; pract. insol. in water; dens. 0.95 g/cc; visc. 150-240 mPa•s; solid. pt. < 4 C; flash pt. > 180 C; *Toxicology:* LD50 (oral, rat) > 5 g/kg.

Loxiol® G 11. [Henkel KGaA/Dehydag] Fatty acid ester of polyol; internal lubricant for rigid and soft PVC, used in rigid PVC film and sheeting; FDA 21CFR §175.105, 176.170, 176.180, 176.210, 177.2800, 178.3120, 178.3130; BGA, Japan approvals; liq.; sp.gr. 0.970-0.985; visc. 500-610 mPa.s; acid no. < 1; ref. index 1.473-1.478; flash pt. > 220 C.

Loxiol® G 12. [Henkel KGaA/Dehydag] Monoglycerol ester of sat. fatty acid; internal lubricant for rigid and soft PVC, used in rigid PVC film and sheeting; FDA 21CFR §175.210, 175.300, 175.320, 177.1010, 182.1324, GRAS; BGA, Japan approvals; wh. beads, 20 mesh particle size; dens. 20-30 lb/ft^3; acid no. 2 max.; ref. index 1.441 ± 0.005 (80 C); flash pt. (OC) 428 F.

Loxiol® G 13. [Henkel KGaA/Dehydag] Fatty acid ester of polyol; lubricant for rigid PVC esp. in cable and inj. molding; FDA 21CFR §175.105, 176.170, 176.180, 176.210, 177.2800, 178.3120; Japan approvals; liq.; sp.gr. 0.905-0.915; visc. 35-40 mPa•s; acid no. < 1; ref. index 1.460-1.466; flash pt. > 190 C.

Loxiol® G 15. [Henkel/Process & Polymer Chems.] Fatty acid ester of polyol; internal lubricant for rigid PVC for use in inj. molding, extrusion, and calendering; FDA 21CFR §175.300, 176.170, 176.180, 177.1200, 177.1210, 177.1350, 177.1400, 177.2420, 177.2800, 178.3280; BGA, Japan approvals; pale tan flakes, off-wh. beads, 20 mesh particle size; dens. 20-30 lb/ft^3; acid no. 5 max.; ref. index 1.445 ± 0.005 (95 C); flash pt. (OC) > 300 C.

Loxiol® G 16. [Henkel/Process & Polymer Chems.] Glycerol ester of unsat. fatty acid; internal lubricant for rigid PVC film, calendered and rigid sheet, clear, opaque inj. molding and profiles; costabilizer in tin-stabilized rigid, semirigid, or plasticized systems; printing and laminating properties to articles; FDA 21CFR §175.105, 176.210, 178.3120, 182.90, 182.4505, GRAS; BGA, Japan approvals; pale yel. fluid; dens. 7.3 lb/gal; visc. 90-120 cP; f.p. < -5 C; acid no. 1.0 max.; flash pt. (CC) 460 F; ref. index 1.468 ± 0.005.

Loxiol® G 20. [Henkel/Process & Polymer Chems.] Stearic acid; CAS 57-11-4; EINECS 200-313-4; external lubricant for rigid and soft PVC, esp. for extrusions; FDA 21CFR §175.105, 175.300, 175.320, 176.170, 176.180, 176.210, 177.1010, 177.1200, 177.1210, 177.1350, 177.1400, 177.2260, 177.2420, 177.2600, 177.2800, 178.3120, 178.3570, 178.3910, 184.1090, GRAS; BGA, Japan approvals; solid; sp.gr. 0.840-0.850 (80 C); visc. 8-11 mPa•s (80 C); acid no. 207-210; flash pt. > 180 C; ref. index 1.436-1.437 (60 C).

Loxiol® G 21. [Henkel/Process & Polymer Chems.] Hydroxystearic acid; CAS 106-14-9; EINECS 203-366-1; external lubricant for rigid and soft PVC, esp. for extrusions; antiplate-out effect; FDA 21CFR §175.105, 176.210, 178.3120, 178.3570; BGA, Japan approvals; solid; sp.gr. 0.885-0.891 (80 C); visc. 30-35 mPa•s (80 C); acid no. 172-180; flash pt. > 210 C; ref. index 1.440-1.442 (80 C).

Loxiol® G 22. [Henkel KGaA/Dehydag] Hard paraffin; CAS 8002-74-2; EINECS 232-315-6; external lubricant for rigid PVC; FDA 21CFR §175.125, 175.300, 177.1200, 177.1210; BGA, Japan approvals; solid; visc. 10 mPa.s (120 C); acid no. < 0.1; flash pt. > 280 C.

Loxiol® G 23. [Henkel KGaA/Dehydag] Hard paraffin; CAS 8002-74-2; EINECS 232-315-6; lubricant; FDA 21CFR §175.320, 176.170, 176.180, 176.200, 177.1200, 177.1400, 177.2420, 177.2600, 177.2800, 178.3710, 179.45; BGA, Japan approvals.

Loxiol® G 32. [Henkel KGaA/Dehydag] Fatty acid ester; all purpose lubricant for rigid PVC extrusion; PC prod.; FDA 21CFR §178.3450; BGA, Japan approvals; solid; sp.gr. 0.816-0.819 (80 C); visc. 6-7 mPa.s (80 C); acid no. < 2; ref. index 1.430-1.435 (80 C); flash pt. > 230 C; Unverified

Loxiol® G 33. [Henkel/Process & Polymer Chems.] Simple ester; internal/external lubricant for rigid PVC, esp. extrusions; lubricant for PC; solid.

Loxiol® G 33 Bead. [Henkel KGaA/Dehydag] Simple fatty acid esters; internal and external lubricant used in rigid PVC inj. moldings, calendered or extruded sheet, and complicated interior and exterior profiles; superior flow and offers superior lt., heat, and moisture resistance suitable for outdoor applic.; partial substitute for calcium stearates due to excellent flow properties; resists plate-out; BGA, Japan approvals; dens. 20-25 lb/ft^3; acid no. 6.0 max.; ref. index 1.436 ± 0.005 (70 C); flash pt. > 280 F.

Loxiol® G 40. [Henkel/Process & Polymer Chems.] Simple ester from branched chain fatty alcohol and fatty acid; lubricant for calendering rigid PVC sheet, flexible PVC film, extrusion of exterior profiles and semirigid wire and cable compds.; intermediate lubricant in phenolic inj. molding compds. and nylon resins; BGA, Japan approvals; pale yel. liq.; dens. 7.2 lb/gal; visc. 25-35 cps; f.p. 8 C max.; acid no. 2.0 max.; flash pt. 401 F; ref. index 1.452 ± 0.005.

Loxiol® G 41. [Henkel KGaA/Dehydag] Fatty acid ester; compatible lubricant for rigid PVC in glass clear articles; lubricant for duroplasts; BGA, Japan approvals; solid; sp.gr. 0.820-0.827 (80 C); visc. 5-7 mPa.s (80 C); acid no. < 2; ref. index 1.430-1.435 (80 C); flash pt. > 220 C.

Loxiol® G 47. [Henkel/Process & Polymer Chems.] Fatty acid ester; externally act. lubricant for rigid PVC extrusion and PC prod.; BGA, Japan approvals; solid; sp.gr. 0.816-0.821 (80 C); visc. 7-10 mPa•s (80 C); acid no. < 2; flash pt. > 250 C; ref. index 1.433-1.437 (80 C).

Loxiol® G 52. [Henkel/Functional Prods.] C16-18 fatty alcohol; surfactant for polymerization; solid/flakes; m.p. 48-52 C.

Loxiol® G 53. [Henkel/Functional Prods.; Henkel KGaA/Dehydag] C16-18 fatty alcohol; surfactant for polymerizations; internal lubricant for rigid PVC; inj. molding; FDA 21CFR §175.105, 176.170, 176.180, 176.210, 177.1010, 177.2800, 178.3120, 178.3910; BGA, Japan approvals; solid; sp.gr. 0.790-0.802 (80 C); visc. 4-6 mPa.s (80 C); m.p. 48-51 C; acid no. < 0.2; ref. index 1.427-1.430 (80 C); flash pt. > 160 C; 100% conc.

Loxiol® G 60. [Henkel/Process & Polymer Chems.] Dicarboxylic acid ester of sat. aliphatic alcohols; universally applicable internal lubricant for rigid PVC; BGA, Japan approvals; solid; sp.gr. 0.878-0.884 (80 C); visc. 10-13 mPa•s (80 C); acid no. < 2; flash pt. > 230 C; ref. index 1.453-1.458 (80 C).

Loxiol® G 70. [Henkel/Process & Polymer Chems.] Polymeric complex ester of sat. fatty acids; internal/external lubricant with release props. for rigid PVC, esp. sheeting and film; replacement of montan esters on cost performance basis; good heat stability and results in superior clarity; wh. flakes; dens. 25-30 lb/ft^3; acid no. 17.5 max.; flash pt. > 280 F; ref. index 1.454 ± 0.005 (70 C).

Loxiol® G 70 S. [Henkel/Process & Polymer Chems.] High molecular complex ester; highly effective release agent, external lubricant for rigid PVC, esp. in calendered film; FDA 21CFR §178.3690; BGA, Japan approvals; solid.

Loxiol® G 71. [Henkel/Process & Polymer Chems.] Complex ester from unsat. fatty acids; external lubricant with release props. for rigid and flexible PVC processing; effective in semirigid sheet blown film, wire and cable compd.; prods. articles of sparkling clarity; does not adversely affect HDT and hot-fill temps.; lt. visc. liq.; dens. 7.86 lb/gal; visc. 500-700 cP; f.p. -10 to 0 C; acid no. 17.5 max.; flash pt. 530 F; ref. index 1.469 ± 0.005.

Loxiol® G 71 S. [Henkel/Process & Polymer Chems.] High molecular complex ester; external lubricant, release agent for rigid and flexible PVC; BGA, Japan approvals; liq.

Loxiol® G 72. [Henkel KGaA/Dehydag] High-molecular complex esters; compatible lubricant with release properties for rigid PVC; useful for rigid PVC bottles, film, and sheeting; BGA, Japan approvals; solid; sp.gr. 0.887-0.895 (80 C); visc. 18-27 mPa.s (80 C); acid no. < 7; flash pt. > 240 C.

Loxiol® HOB 7107. [Henkel/Process & Polymer Chems.] Polymeric ester of sat. fatty acids; external lubricant, release agent for rigid PVC, esp. calendered film; used in food contact applic.; good heat stability and resists water blush; components individually comply with FDA 21CFR §178.3280, 178.3690; BGA, Japan approvals; wh. beads; 20 mesh particle size; dens. 15-20 lb/ft^3; acid no. 17.5 max.; flash pt. 585 F; ref. index 1.444 ± 0.005 (95 C).

Loxiol® HOB 7111. [Henkel/Process & Polymer Chems.] Fatty acid ester of polyol; highly compt. internal lubricant for rigid PVC extrusion, inj. molding, calenders, clear blow-molded bottles, sheet; produces high-clarity articles; strong synergistic effect with stabilizers; FDA 21CFR §175.300, 176.170, 177.1200, 177.1210, 177.1350; BGA, Japan approvals; wh. flakes; dens. 15-20 lb/ft^3; acid no. 5 max.; flash pt. > 285 F; ref. index 1.438 ± 0.005 (95 C).

Loxiol® HOB 7119. [Henkel/Process & Polymer Chems.] Fatty acid ester of polyol; highly effective external lubricant for inj. molded PC, ABS, rigid PVC; solid.

Loxiol® HOB 7121. [Henkel/Process & Polymer Chems.] Fatty acid ester of polyol; internal lubricant/costabilizer for rigid PVC processing esp. clear rigid articles; superior flow in rigid PVC inj. molding and complicated profiles; extrusion films, calendered sheets, blow-molded bottles; good thermal and lt. stability; effective in high calcium stearate compds.; highly costabilizing in CaZn stabilized and lead stabilized PVC systems; resistant to plate-out; FDA 21CFR §175.105, 176.210, 178.2010, 178.3120; BGA, Japan approvals; off-wh. beads or flakes, 20 mesh bead; dens. 30-35 lb/ft^3 (bead), 15-25 lb/ft^3 (flake); acid no. 2.0 max.; flash pt. 535 F; ref. index 1.450 ± 0.004 (70 C).

Loxiol® HOB 7131. [Henkel KGaA/Dehydag] Monoglycerol ester of sat. fatty acid; internal lubricant for clear or opaque applic.; food applic.; mildly costabilizing in tin-stabilized systems; good flow in pigmented and filled rigid PVC compds.; FDA 21CFR §175.210, 175.300, 175.320, 177.1010, 182.1324, GRAS; BGA, Japan approvals; wh.

beads, 20 mesh bead; dens. 25-30 lb/ft^3; acid no. 3.0 max.; ref. index 1.445 ± 0.005 (70 C); flash pt. (CC) 428 F.

Loxiol® HOB 7138. [Henkel KGaA/Dehydag] Hard paraffin; CAS 8002-74-2; EINECS 232-315-6; lubricant; FDA 21CFR §172.878; BGA, Japan approvals; Unverified

Loxiol® HOB 7140. [Henkel/Process & Polymer Chems.] High m.w. complex ester; external lubricant, release agent for rigid PVC, esp. calendered film; solid.

Loxiol® HOB 7162. [Henkel/Process & Polymer Chems.] Simple ester; internal/external lubricant for rigid PVC, esp. extrusions; solid.

Loxiol® HOB 7169. [Henkel KGaA/Dehydag] Hard paraffin; CAS 8002-74-2; EINECS 232-315-6; processing aid for extruded rigid PVC; FDA 21CFR §175.320, 176.170, 176.180, 176.200, 177.1200, 177.1400, 177.2420, 177.2600, 177.2800, 178.3710, 179.45; BGA, Japan approvals; solid.

Loxiol® P 1141. [Henkel/Process & Polymer Chems.] Fatty acid ester of polyol; internal lubricant for rigid PVC film; liq.

Loxiol® P 1304. [Henkel/Process & Polymer Chems.] Fatty acid ester of polyol; internal lubricant for rigid PVC; pigment dispersing aux. for PVC paste processing; liq.

Loxiol® P 1404. [Henkel KGaA/Dehydag] BGA, Japan approvals.

Loxiol® P 1420. [Henkel/Functional Prods.] C16-18 fatty alcohol; surfactant for polymerization; solid/flakes; m.p. 48-52 C.

Loxiol® VGE 1523. [Henkel KGaA/Dehydag] Combination lubricant containing metal soaps; BGE, Japan approvals.

Loxiol® VGE 1727. [Henkel KGaA/Dehydag] Combination lubricant containing metal soaps; BGE, Japan approvals.

Loxiol® VGE 1728. [Henkel/Process & Polymer Chems.] Costabilizing lubricant system for rigid PVC extrusion; improves thermostability; suitable for calcium stearate/EBS type rheology; solid.

Loxiol® VGE 1837. [Henkel/Process & Polymer Chems.] Blend; highly effective internal/external lubricant for difficult extrusions of rigid PVC, foamed PVC, glass clear rigid PVC prods.; solid.

Loxiol® VGE 1875. [Henkel/Process & Polymer Chems.] Costabilizing lubricant system for rigid PVC foamed profiles; improves thermostability; highly compat.; solid.

Loxiol® VGE 1884. [Henkel/Process & Polymer Chems.] Costabilizing lubricant system for rigid PVC extrusion; improves thermostability; suitable for calcium stearate/paraffin type rheology; solid.

Loxiol® VGS 1460. [Henkel KGaA/Dehydag] Combination lubricant containing metal soaps; components individually comply with FDA 21CFR §178.2010, 178.3690; BGA, Japan approvals.

Loxiol® VGS 1877. [Henkel/Process & Polymer Chems.] Costabilizing internal lubricant for rigid PVC inj. molded articles; solid.

Loxiol® VGS 1878. [Henkel/Process & Polymer Chems.] Costabilizing internal/external lubricant for rigid PVC inj. molded articles; solid.

Loxiol® VPG 1354. [Henkel/Functional Prods.] Stearyl alcohol; CAS 112-92-5; EINECS 204-017-6; nonionic; surfactant for polymerization; flakes; m.p. 55-57 C; 100% conc.

Loxiol® VPG 1451. [Henkel/Functional Prods.] Behenyl alcohol; CAS 661-19-8; EINECS 211-546-6; surfactant for polymerization; solid/flakes; m.p. 63-65 C.

Loxiol® VPG 1496. [Henkel/Functional Prods.] C12-18 fatty alcohol; CAS 67762-25-8; EINECS 267-006-5; surfactant for polymerization; liq./solid; m.p. 18-23 C; 100% act.

Loxiol® VPG 1732. [Henkel/Process & Polymer Chems.] High molecular complex ester; external lubricant, release agent for rigid PVC and acetate-modified PVC pkg. articles; solid.

Loxiol® VPG 1743. [Henkel/Functional Prods.] Cetyl alcohol; CAS 36653-82-4; EINECS 253-149-0; nonionic; surfactant for polymerization; flakes; m.p. 46-49 C; 100% conc.

Loxiol® VPG 1781. [Henkel/Process & Polymer Chems.] Blend; highly effective internal/external lubricant for difficult extrusions of rigid PVC, foamed PVC, tin carboxylate systems; solid.

LP®-3. [Morton Int'l./Polymer Systems] Polysulfide polymer, mercaptan-terminated; CAS 9080-49-3; epoxy modifier; used for adhesives, potting and encapsulating compds., concrete adhesives and coatings; exc. resist. to most acids and bases, very good elec. resist.; cured using epoxy resins; amber liq., slight mercaptan odor; m.w. 1000; sp.gr. 1.27; visc. 10 poise; Of compd.: Shore hardness A15-60.

LP-100. [Eagle-Picher] Lead dioxide; CAS 1309-60-0; EINECS 215-174-5; catalyst/curing agent for polysulfide, low m.w. butyl and polyisoprene rubber; oxidizer in mfg. of dyes and to control burning rate of incendiary fuses or pyrotechnics; slow cure; brn.; 99.9% -325 mesh; sp.gr. 9.2; surf. area < 4.25 m^2/g; 90% PbO_2, 0.2% water.

LP-200. [Eagle-Picher] Lead dioxide; CAS 1309-60-0; EINECS 215-174-5; catalyst/curing agent for polysulfide, low m.w. butyl and polyisoprene rubber; oxidizer in mfg. of dyes and to control burning rate of incendiary fuses or pyrotechnics; med. cure; brn.; 99.9% -325 mesh; sp.gr. 9.2; surf. area 4.25-6.5 m^2/g; 90% PbO_2, 0.2% water.

LP-300. [Eagle-Picher] Lead dioxide; CAS 1309-60-0; EINECS 215-174-5; catalyst/curing agent for polysulfide, low m.w. butyl and polyisoprene rubber; oxidizer in mfg. of dyes and to control burning rate of incendiary fuses or pyrotechnics; fast cure; brn.; 99.9% -325 mesh; sp.gr. 9.2; surf. area 6.5-10 m^2/g; 90% PbO_2, 0.2% water.

LP-400. [Eagle-Picher] Lead dioxide; CAS 1309-60-0; EINECS 215-174-5; catalyst/curing agent for polysulfide, low m.w. butyl and polyisoprene rubber; oxidizer in mfg. of dyes and to control burning rate of incendiary fuses or pyrotechnics; very fast cure; brn.; 99.9% -325 mesh; sp.gr. 9.2; surf. area 10-14 m^2/g; 90% PbO_2, 0.2% water.

Lube 105. [CDC Int'l.] Alcohol ester; external lubricant for nylon monofilaments; liq.

Lube 106. [CDC Int'l.] Alcohol ester; external lubricant for polyester, epoxy, PE, phenolic, PS, nylon, cellulose acetate, cellulose acetate butyrate, diallyl phthalate, ethyl cellulose; liq.

Lubrazinc® W, Superfine. [Witco Corp.] Zinc stearate; CAS 557-05-1; EINECS 209-151-9; lubricant for fiber-reinforced plastics; dry lubricant for powd. metallurgy applic.; used with iron powd. to yield compacts of high uniform density; wh. powd.; sol. in hot turpentine, benzene, toluene, xylene, CCl_4, veg. and min. oils, waxes; insol. in water; sp.gr. 1.10; soften. pt. 120 C; 14.0% conc.

Lubricin 25. [CasChem] processing aid for rigid PVC during calendering, inj. molding of pipe, fittings, conduit, sheet and profiles and for rubber improving appearance of molded goods.

Lubrizol® 2106. [Lubrizol] Barium phenate; heat stabilizer for PVC coatings, films, and fabricated materials; liq.; 28% Ba.

Lubrizol® 2116. [Lubrizol] Barium carboxylate; heat stabilizer for PVC coatings, films, and fabricated materials; liq.; 34% Ba.

Lubrizol® 2117. [Lubrizol] Calcium carboxylate; heat stabilizer for PVC coatings, films, bottles, extruded pipe, and food contact applics.; FDA approved; liq.; 14% Ca.

Lubrizol® 2152. [Lubrizol] Calcium sulfonate; CAS 61789-86-4; anionic; pigment dispersant and wetting agent for color concs., paints, coatings, inks; pigment flushing aid for org. and inorg. pigments; visc. stabilizer/reducer in plastisols; amber liq.; sp.gr. 0.969-0.999; dens. 8.08-8.33 lb/gal; visc. 2000-6000 cps; flash pt. (PMCC) 155 C; 100% act.

Lubrizol® 2153. [Lubrizol] Succinimide; nonionic; pigment dispersant and wetting agent for color concs., inks, plastisols and organosols; amber liq.; sp.gr. 0.910-0.940; dens. 7.59-7.84 lb/gal; visc. 8000-24,000 cps; flash pt. (PMCC) 160 C; 100% act.

Lubrizol® 2155. [Lubrizol] Succinimide; nonionic; pigment dispersant for color concs., inks, plastisols and organosols; wetting agent for carbon fibers; amber liq.; sp.gr. 0.909-0.939; dens. 7.58-7.83 lb/gal; visc. 13,000-23,000 cps; flash pt. (PMCC) 160 C; 100% act.

Lubrol 90. [CNC Int'l. L.P.] Ester; nonoily press release agent for rubber roll release problems in pulp/paper industry; surface lubricant for tabulator card stock and paper board.

Luchem AS-946. [Elf Atochem N. Am./Org. Peroxides] Methyl methacrylate/allyl methacrylate copolymer; antishrink additive.

Luchem AS-946-25. [Elf Atochem N. Am./Org. Peroxides] Methyl methacrylate/allyl methacrylate copolymer; antishrink additive; 25% sol'n. in ADC.

Lucidol 70. [Elf Atochem N. Am./Org. Peroxides] Benzoyl peroxide; CAS 94-36-0; EINECS 202-327-6; initiator for bulk, sol'n., and suspension polymerization and high-temp. and R.T. cure of polyester resins; used in pharmaceutical applic.; granular wet solid; 70% act.; 4.36-4.52% act. oxygen.

Lucidol 75. [Elf Atochem N. Am./Org. Peroxides] Benzoyl peroxide with 25% water; CAS 94-36-0; EINECS 202-327-6; polymerization initiator and crosslinking agent in prod. of various plastics; catalyst for free radical polymerization in prod. of PS, PVC, methacrylates and for crosslinking of unsat. polyester; wh. gran. solid; sol. in org. solvs., e.g., acetone, various esters; sparingly sol. in alcohols; insol. in water; m.w. 242; bulk dens. 41.2 lb/ft^3; m.p. ≥ 100 C (rapid decomp.); 75 ± 2% assay, 4.83-5.09% act. oxygen; *Precaution:* avoid contact with heat, flames, ignition sources, contamination, strong acids, strong alkalies, strong oxidizers, amines, promoters, reducing agents; *Storage:* store in original containers @ 38 C max.

Lucidol 75FP. [Elf Atochem N. Am./Org. Peroxides] Benzoyl peroxide with 25% water; CAS 94-36-0; EINECS 202-327-6; active constituent in anti-acne creams and soaps; polymerization initiator for plastics prod.; catalyst for free radical polymerization for PS, PVC, polyester resins; wh. fine powd.; 5-15 μ particle size; sol. in org. solvs., e.g., acetone, various esters; sparingly sol. in water, alcohol; m.w. 242; bulk dens. 510-570 kg/m^3; 75 ± 2% assay, ≈ 6.5% act. oxygen; *Precaution:* avoid contact with heat, sparks, ignition sources, contamination, strong acids, strong alkalies, strong oxidizers, amines, promoters, reducing agents; *Storage:* store in original containers @ 38 C.

Lucidol 78. [Elf Atochem N. Am./Org. Peroxides] Benzoyl peroxide; CAS 94-36-0; EINECS 202-327-6; initiator for bulk, sol'n., and suspension polymerization, and high-temp. and R.T. cure of polyester resins; granular wet solid; 78% act.; 4.95-5.28% act. oxygen.

Lucidol 98. [Elf Atochem N. Am./Org. Peroxides] Benzoyl peroxide; CAS 94-36-0; EINECS 202-327-6; initiator for bulk, sol'n., and suspension polymerization, and high-temp. and R.T. cure of polyester resins; also used for curing elastomers, cure of acrylic syrup, and polymer modification, thermoplastic crosslinking; granular solid; 98% act.; 6.5% act. oxygen.

Luconyl®. [BASF AG] Concs. of org. and inorg. pigments using nonionic dispersants; for coloring polymer emulsion paints, inks, water-based wood glazes.

Luperco 101-P20. [Elf Atochem N. Am./Org. Peroxides] 20% disp. of 2,5-dimethyl-2,5-di (t-butylperoxy) hexane on a PP powd. carrier; crosslinking agent for elastomers and thermoplastic resins; wh. free-flowing powd.; m.w. 290.45; bulk dens. 32.8 lb/ft^3.

Luperco 101-XL. [Elf Atochem N. Am./Org. Peroxides] 2,5-Dimethyl-2,5-di(t-butylperoxy) hexane on inert filler ($CaCO_3$); initiator for curing elastomers, polymer modification thermoplastic crosslinking, and high-temp. cure of polyester resins; free-flowing powd.; m.w. 290.45; dens. 1.248 g/cc; 45% conc.; 4.96-5.29% act. oxygen.

Luperco 130-XL. [Elf Atochem N. Am./Org. Peroxides] 2,5-Dimethyl-2,5-di (t-butylperoxy) hexyne-3 on inert filler ($CaCO_3$); crosslinking agent, initiator for thermoplastic modification, curing elastomers, high temp. cure of polyesters, and cure of acrylic syrup; free-flowing powd.; m.w. 286.42; dens. 1.263 g/cc; 45% conc.; 5.03% act. oxygen.

Luperco 230-XL. [Elf Atochem N. Am./Org. Peroxides] n-Butyl-4,4-bis (t-butylperoxy) valerate on inert filler; initiator for curing elastomers, and for high-temp. cure of polyester resins; free-flowing powd.; m.w. 334.4; dens. 0.526 g/ml; ; 40% act.; 3.83% act. oxygen.

Luperco 231-SRL. [Elf Atochem N. Am./Org. Peroxides] 1,1-Di (t-butylperoxy) 2,2,5-trimethyl cyclohexane; CAS 6731-36-8; crosslinking agent for thermoplastic modification, curing elastomers such as polybutadiene and EPDM, and high temp. cure of polyesters; wh. powd.; bulk dens. 1.413 g/cc; 40% conc.; 4.23% act. oxygen.

Luperco 231-XL. [Elf Atochem N. Am./Org. Peroxides] 1,1-Di (t-butylperoxy) 3,3,5-trimethyl cyclohexane on inert filler (CaCO3); initiator for curing elastomers, crosslinking agent for thermoplastic modification; free-flowing powd.; m.w. 302.5; dens. 1.413 g/ml; 40% act.; 4.23% min. act. oxygen.

Luperco 233-XL. [Elf Atochem N. Am./Org. Peroxides] Ethyl-3,3-di (t-butylperoxy) butyrate on inert filler; initiator for curing elastomers and for polymer modification thermoplastic cross-linking; free-flowing powd.; m.w. 292.4; dens. 0.3551 g/ml; ; 40% act.; 4.39% act. oxygen.

Luperco 331-XL. [Elf Atochem N. Am./Org. Peroxides] 1,1-Di (t-butylperoxy) cyclohexane on inert filler; initiator for high-temp. cure of polyester resins and for curing elastomers; free-flowing powd.; m.w. 260.3; dens. 0.3277 g/ml; 40% act.; 4.92% act. oxygen.

Luperco 500-40C. [Elf Atochem N. Am./Org. Peroxides] Dicumyl peroxide on calcium carbonate; initiator for curing elastomers, polymer modification thermoplastic crosslinking, and high-temp. cure of polyester resins; free-flowing powd.; m.w. 270.37; dens. 1.611 g/cc; 40% conc.; 2.34-2.46% act. oxygen.

Luperco 500-40KE. [Elf Atochem N. Am./Org. Peroxides] Dicumyl peroxide on Burgess KE clay; initiator for curing elastomers, polymer modification thermoplastic crosslinking, and high-temp. cure of polyester resins; free-flowing powd.; m.w. 270.37; dens. 1.579 g/cc; 40% conc.; 2.34-2.46% act. oxygen.

Luperco 500-SRK. [Elf Atochem N. Am./Org. Peroxides] Dicumyl peroxide; CAS 80-43-3; crosslinking agent for thermoplastics, curing of elastomers, and high temp. cure of polyester resins; powd.; 40% conc.; 2.37% act. oxygen.

Luperco 801-XL. [Elf Atochem N. Am./Org. Peroxides] t-Butyl cumyl peroxide on inert filler; initiator for high-temp. cure of polyester resins, curing elastomers, and polymer modification thermoplastic cross-linking; solid; 40% act.; 3.07% act. oxygen.

Luperco 802-40KE. [Elf Atochem N. Am./Org. Peroxides] α-α-Bis (t-butylperoxy) diisopropylbenzene on Burgess clay; crosslinking agent for polymer modification, curing elastomers, high-temp. cure of polyester resins; 40% solids; 3.73-3.92% act. oxygen.

Luperco AA. [Elf Atochem N. Am./Org. Peroxides] Benzoyl peroxide blend with wheat starch; initiator for polymer modification, thermoplastic crosslinking, high-temp. and R.T. cure of polyester resins; powd.; 33% act.; 2.11-2.18% act. oxygen.

Luperco ACP. [Elf Atochem N. Am./Org. Peroxides] Benzoyl peroxide blend with inorg. phosphates; initiator for high-temp. and R.T. cure of polyester resins; powd.; 35% act.; 2.31% min. act. oxygen.

Luperco AFR-250. [Elf Atochem N. Am./Org. Peroxides] Benzoyl peroxide; CAS 94-36-0; EINECS 202-327-6; fire retardant initiator for high-temp. and R.T. cure of polyester resins; paste; 25% act.; 1.58-1.78% act. oxygen.

Luperco AFR-400. [Elf Atochem N. Am./Org. Peroxides] Benzoyl peroxide; CAS 94-36-0; EINECS 202-327-6; initiator for high-temp. and room-temp. cures of polyester resins; pourable paste; 40% act.; 2.64% act. oxygen.

Luperco AFR-500. [Elf Atochem N. Am./Org. Peroxides] Benzoyl peroxide; CAS 94-36-0; EINECS 202-327-6; fire retardant initiator for high-temp. and R.T. cure of polyester resins; paste; 50% act.; 3.3% min. act. oxygen.

Luperco ANS. [Elf Atochem N. Am./Org. Peroxides] Benzoyl peroxide with plasticizer; CAS 94-36-0; EINECS 202-327-6; initiator for polymer modification, thermoplastic crosslinking, high-temp. and R.T. cure of polyester resins, and cure of acrylic syrup; paste; 55% act.; 3.6% min. act. oxygen.

Luperco ANS-P. [Elf Atochem N. Am./Org. Peroxides] Benzoyl peroxide with plasticizer; CAS 94-36-0; EINECS 202-327-6; crosslinking agent for high temp. and R.T. cure of polyesters, cure of acrylic syrup; paste; 55% conc.; 3.63% act. oxygen.

Luperco AST. [Elf Atochem N. Am./Org. Peroxides] Benzoyl peroxide with silicone oil; initiator for curing elastomers; paste; 50% act.; 3.3% min. act. oxygen.

Luperco ATC. [Elf Atochem N. Am./Org. Peroxides] Benzoyl peroxide with tricresyl phosphate; initiator for polymer modification, thermoplastic crosslinking, high-temp. and R.T. cure of polyester resins; paste; 50% act.; 3.30-3.43% act. oxygen.

Luperfoam 40. [Elf Atochem N. Am./Org. Peroxides] Aq. mixt. of t-butylhydrazinium chloride and cupric chloride; chemical blowing agent for unsat. polyester resins.

Luperfoam 329. [Elf Atochem N. Am./Org. Peroxides] Aq. mixt. of t-butylhydrazinium chloride and ferric chloride; chemical blowing agent for R.T. foaming of unsat. polyester resins.

Luperox 2,5-2,5. [Elf Atochem N. Am./Org. Peroxides] 2,5-Dihydroperoxy-2,5-dimethylhexane; initiator for bulk, sol'n., emulsion, and suspension polymerization; solid; 70% conc. in water; 11.85-12.57% act. oxygen.

Luperox 118. [Elf Atochem N. Am./Org. Peroxides] 2,5-Dimethyl-2,5-bis-(benzoylperoxy)hexane; initiator for vinyl polymerization; solid; m.w. 386; sol. 16-20% in ethylene chloride, 11-15% in tetrahydro furan, 6-10% in benzene, 3-5% in ethyl acetate; insol. in water; dens. 24.5 lb/ft^3; m.p. 237 F; 92.5% conc.; 7.66% min. act. oxygen.

Luperox 500R. [Elf Atochem N. Am./Org. Peroxides] Dicumyl peroxide; CAS 80-43-3; initiator for bulk, sol'n., and suspension polymerization, polymer modification thermoplastic crosslinking, curing elastomers, high-temp. cure of polyester resins; cryst. solid; m.w. 270.37; sp.gr. 1.00 (40 C); 99% act.; 5.87% min. act. oxygen.

Luperox 500T. [Elf Atochem N. Am./Org. Peroxides] Dicumyl peroxide; CAS 80-43-3; initiator for bulk, sol'n., and suspension polymerization, polymer modification thermoplastic crosslinking, curing elastomers, high-temp. cure of polyester resins; semicryst. solid; m.w. 270.37; sp.gr. 0.997-1.009 (40 C); 91-93% act.; 5.40% min. act. oxygen.

Luperox 802. [Elf Atochem N. Am./Org. Peroxides] α-α′-Bis(t-butylperoxy) diisopropylbenzene; CAS 25155-25-3; crosslinking agent for thermoplastic modification, curing elastomers, and high temp. cure of polyesters; initiator for vinyl polymerization; semicryst. solid; sp.gr. 0.930; 96% conc.; 9.26% act. oxygen.

Lupersol 10. [Elf Atochem N. Am./Org. Peroxides] t-Butyl peroxyneodecanoate; CAS 26748-41-4; initiator for vinyl polymerizations; liq.; insol. in water; sp.gr. 0.90; flash pt. 1.45 F; m.w. 244; 95% min. assay; 6.2% min. act. oxygen.

Lupersol 10-M75. [Elf Atochem N. Am./Org. Peroxides] t-Butyl peroxyneodecanoate in odorless min. spirits; initiator for vinyl polymerizations; sol'n.; m.w. 244; insol. in water; sp.gr. 0.87; visc. 8.2 cps (32 F); ref. index 1.44; flash pt. > 100 F; 74-76% assay; 4.8-5.0% act. oxygen.

Lupersol 11. [Elf Atochem N. Am./Org. Peroxides] t-Butyl peroxypivalate in odorless min. spirits; initiator for vinyl polymerizations; sol'n.; insol. in water; m.w. 174; f.p. -2 F; sp.gr. 0.85; visc. 2.8 cps (32 F); ref. index 1.41; flash pt. 155 F; 74-76% assay; 6.8-7.0% act. oxygen.

Lupersol 70. [Elf Atochem N. Am./Org. Peroxides] t-Butyl peroxyacetate in odorless min. spirits; initiator for vinyl polymerizations; sol'ns.; insol. in water; m.w. 132; f.p. < -22 F; sp.gr. 0.89; visc. 1.4 cps; ref. index 1.40; flash pt. 140 F; 74-76% assay; 8.9-9.2% act. oxygen.

Lupersol 75-M. [Elf Atochem N. Am./Org. Peroxides] t-Butyl peroxyacetate in odorless min. spirits; initiator for vinyl polymerizations; sol'n.; insol. in water; m.w. 132; f.p. < -15 F; sp.gr. 0.83; visc. 1.4 cps; ref. index 1.40; flash pt. 130 F; 49-51% assay; 5.93-6.18% act. oxygen.

Lupersol 76-M. [Elf Atochem N. Am./Org. Peroxides] t-Butylperoxy acetate sol'n. in odorless min. spirits; initiator for bulk, sol'n., and suspension polymerization, for high-temp. cure of polyester resins, and for cure of acrylic syrup; sol'n.; 75% act.; 7.26% act. oxygen.

Lupersol 80. [Elf Atochem N. Am./Org. Peroxides] t-Butyl peroxyisobutyrate in odorless min. spirits; initiator for vinyl polymerizations; sol'n.; insol. in water; m.w. 160; f.p. < -40 F; sp.gr. 0.87; visc. 1.4 cps; ref. index 1.41; flash pt. 140 F; 74-76% assay; 7.4-7.6% act. oxygen.

Lupersol 101. [Elf Atochem N. Am./Org. Peroxides] 2,5-Dimethyl-2, 5-di(t-butylperoxy) hexane; CAS 78-63-7; initiator for bulk, sol'n., and suspension polymerization, polymer modification thermoplastic crosslinking, curing elastomers, high-temp. cure of polyester resins; liq.; m.w. 290.45; sp.gr. 0.8650; b.p. 115 C (10 mm); f.p. < 8 C; flash pt. (Seta) 43 C; ref. index 1.4160; 90% act.; 9.92% min. act. oxygen.

Lupersol 130. [Elf Atochem N. Am./Org. Peroxides] 2,5-Dimethyl-2 5-di(t-butylperoxy) hexyne-3; CAS 1068-27-5; initiator for bulk, sol'n., and suspension polymerization, high-temp. curing of polyester resins, and cure of acrylic syrup; liq.; m.w. 286.42; sp.gr. 0.886-0.890; b.p. 113 C (10 mm); fg.p. < 8 C; flash pt. (Seta) 87 C; ref. index 1.4260-1.4300; 90-95% act.; 10.05-10.61% act. oxygen.

Lupersol 188-M75. [Elf Atochem N. Am./Org. Peroxides] α-Cumylperoxy neodecanoate in odorless min. spirits; initiator for bulk, sol'n., emulsion, and suspension polymerization; sol'n.; 75% conc.; 3.86-3.97% act. oxygen.

Lupersol 219-M60. [Elf Atochem N. Am./Org. Peroxides] Diisononanoyl peroxide in odorless min. spirits; crosslinking agent for bulk, sol'n., and suspension polymerization; sol'n.; 60% conc.; 3.82% act. oxygen.

Lupersol 220-D50. [Elf Atochem N. Am./Org. Peroxides] 2,2-Di(t-butylperoxy) butane in DOP; initiator for bulk, sol'n., and suspension polymerization, high-temp. curing of polyester resins, and cure of acrylic syrup; liq.; m.w. 234.1; sp.gr. 0.924; f.p. -14 C; flash pt. (Seta) 53 C; 50% act.; 6.80-7.0% act. oxygen.

Lupersol 221. [Elf Atochem N. Am./Org. Peroxides] Di(n-propyl) peroxydicarbonate; CAS 16066-38-9; initiator for bulk, sol'n., emulsion, and suspension polymerization; liq.; 99% act.; 7.68% min. act. oxygen.

Lupersol 223. [Elf Atochem N. Am./Org. Peroxides] Di (2-ethylhexyl) peroxydicarbonate; CAS 16111-62-4; initiator for bulk, sol'n., emulsion, and suspension polymerization; liq.; 97% act.; 4.50% min. act. oxygen.

Lupersol 223-M40. [Elf Atochem N. Am./Org. Peroxides] Di (2-ethylhexyl) peroxydicarbonate sol'n. in odorless min. spirits; initiator for bulk, sol'n., and suspension polymerization; sol'n.; 40% act.; 1.86% act. oxygen.

Lupersol 223-M75. [Elf Atochem N. Am./Org. Peroxides] Di (2-ethylhexyl) peroxydicarbonate in odorless min. spirits; initiator for bulk, sol'n., emulsion, and suspension polymerization; sol'n.; 75% act.; 3.41-3.51% act. oxygen.

Lupersol 224. [Elf Atochem N. Am./Org. Peroxides]

2,4-Pentanedione peroxide; curing agent for unsat. polyester thermoset resins; sol'n.; water-misc.; sp.gr. 1.10; visc. 7.6 cps; ref. index 1.4260; flash pt. 214 F.

Lupersol 225. [Elf Atochem N. Am./Org. Peroxides] Di (s-butyl) peroxydicarbonate; CAS 19910-65-7; initiator for bulk, sol'n., and suspension polymerization and cure of acrylic syrup; liq.; 98% act.; 6.69% min. act. oxygen.

Lupersol 225-M60. [Elf Atochem N. Am./Org. Peroxides] Di (sec-butyl) peroxydicarbonate in odorless min. spirits; initiator for bulk, sol'n., and suspension polymerization and cure of acrylic syrup; sol'n.; 60% act.; 4.03-4.17% act. oxygen.

Lupersol 225-M75. [Elf Atochem N. Am./Org. Peroxides] Di (sec-butyl) peroxydicarbonate sol'n. in odorless min. spirits; initiator for bulk, sol'n., and suspension polymerization; sol'n.; 75% act.; 5.12% act. oxygen.

Lupersol 230. [Elf Atochem N. Am./Org. Peroxides] n-Butyl-4,4-bis (t-butylperoxy) valerate; CAS 995-33-5; initiator for bulk, sol'n., and suspension polymerization, for high-temp. cure of polyester resins, for curing elastomers, and for cure of acrylic syrup; liq.; m.w. 334.4; sp.gr. 0.955; f.p. -21 C; flash pt. (Seta) 64-94 C; 90% act.; 8.61% act. oxygen.

Lupersol 231. [Elf Atochem N. Am./Org. Peroxides] 1,1-Di (t-butylperoxy) 3,3,5-trimethyl cyclohexane; CAS 6731-36-8; initiator for bulk, sol'n., and suspension polymerization, curing elastomers, high-temp. cure of polyester resins, and cure of acrylic syrup; colorless liq.; m.w. 302.5; sp.gr. 0.9050; f.p. -40 C; flash pt. (Seta) 46 C; 92% act.; 9.73% min. act. oxygen.

Lupersol 231-P75. [Elf Atochem N. Am./Org. Peroxides] 1,1-Di (t-butylperoxy) 3,3,5-trimethyl cyclohexane sol'n. in dibutyl phthalate; initiator for high-temp. cure of polyester resins and for cure of acrylic syrup; liq.; m.w. 302.5; sp.gr. 0.9345; f.p. < -40 C; flash pt. (Seta) 86 C; 75% act.; 7.94% act. oxygen.

Lupersol 233-M75. [Elf Atochem N. Am./Org. Peroxides] Ethyl-3,3-di (t-butylperoxy) butyrate in odorless min. spirits; initiator for bulk, sol'n., and suspension polymerization, high-temp. cure of polyesters and acrylics; liq.; m.w. 292.4; sp.gr. 0.9505; f.p. 9 C; flash pt. (Seta) 61 C; 75% act.; 8.21% min. act. oxygen.

Lupersol 256. [Elf Atochem N. Am./Org. Peroxides] 2,5-Dimethyl-2,5-bis(2-ethylhexanoylperoxy) hexane; initiator for vinyl polymerizations; liq.; insol. in water; m.w. 431; sp.gr. 0.93; visc. 50.6 cps; f.p. < -4 F; ref. index 1.44; flash pt. 190 F; 90% conc.; 6.7% act. oxygen.

Lupersol 288-M75. [Elf Atochem N. Am./Org. Peroxides] α-Cumyl peroxyneoheptanoate in odorless min. spirits; initiator for bulk, sol'n. and suspension polymerization; sol'n.; 75% conc.; 4.48-4.60% act. oxygen.

Lupersol 331-80B. [Elf Atochem N. Am./Org. Peroxides] 1,1-Di (t-butylperoxy) cyclohexane in butylbenzyl phthalate; initiator for bulk, sol'n., and suspension polymerization, high-temp. curing of polyester resins, and cure of acrylic syrup; liq.; m.w. 260.3; sp.gr. 0.975; f.p. -56 C; flash pt. (Seta) 70 C; ref. index 1.4546; 80% conc.; 9.70-9.95% act. oxygen.

Lupersol 531-80B. [Elf Atochem N. Am./Org. Peroxides] 1,1-Di-(t-amylperoxy) cyclohexane in butyl benzyl phthalate; initiator for bulk, sol'n., and suspension polymerization, for high-temp. cure of polyester resins, and for cure of acrylic syrup; liq.; m.w. 288.4; sp.gr. 0.959; f.p. < -25 C; flash pt. (Seta) 78 C; 80% act.; 11.10% act. oxygen.

Lupersol 531-80M. [Elf Atochem N. Am./Org. Peroxides] 1,1-Di (t-amylperoxy) cyclohexane in odorless min. spirits; crosslinking agent for bulk, sol'n., and suspension polymerization and cure of acrylic syrup; liq.; 80% conc.; 8.88% act. oxygen.

Lupersol 533-M75. [Elf Atochem N. Am./Org. Peroxides] Ethyl 3,3-di(t-amylperoxy)butyrate in odorless min. spirits; CAS 67567-23-1; initiator for curing of acrylic syrup; crosslinking agent for bulk, sol'n., and suspension polymerization, thermoplastic modification; sol'n.; m.w. 320; sp.gr. 0.893; f.p. -55 C; flash pt. (Seta) 42 C; 75% conc.; 7.39-7.59% act. oxygen.

Lupersol 546-M75. [Elf Atochem N. Am./Org. Peroxides] t-Amylperoxy neodecanoate sol'n. in OMS; initiator for bulk, sol'n., and suspension polymerization, and cure of acrylic syrup; sol'n.; 75% act.; 4.64% act. oxygen.

Lupersol 553-M75. [Elf Atochem N. Am./Org. Peroxides] 2,2-Di (t-amylperoxy) propane in odorless min. spirits; CAS 3052-70-8; initiator for high temp. cure of polyesters, cure of acrylic syrup; crosslinking agent for bulk, sol'n., and suspension polymerization, thermoplastic modification; sol'n.; m.w. 248; sp.gr. 0.857; f.p. -27 C; flash pt. (Seta) 45 C; 75% conc.; 9.53-9.79% act. oxygen.

Lupersol 554-M50, 554-M75. [Elf Atochem N. Am./Org. Peroxides] t-Amylperoxy pivalate sol'ns. in OMS; initiators for bulk, sol'n., and suspension polymerization and for cure of acrylic syrup; sol'n.; 50 and 75% act. resp.; 4.25 and 6.38% act. oxygen resp.

Lupersol 555-M60. [Elf Atochem N. Am./Org. Peroxides] t-Amyl peroxyacetate in odorless min. spirits; crosslinking agent for bulk, sol'n., and suspension polymerization, high temp. cure of polyester, and cure of acrylic syrup; 60% sol'n.; 6.6% act. oxygen.

Lupersol 575. [Elf Atochem N. Am./Org. Peroxides] t-Amylperoxy-2-ethylhexanoate; CAS 686-31-7; initiator for bulk, sol'n., and suspension polymerization, for high-temp. cure of polyester resins, and for cure of acrylic syrup; liq.; 95% act.; 6.60% act. oxygen.

Lupersol 575-M75. [Elf Atochem N. Am./Org. Peroxides] t-Amylperoxy-2-ethylhexanoate sol'ns. in OMS; initiator for bulk, sol'n., and suspension polymerization, for high-temp. cure of polyester resins, and for cure of acrylic syrup; sol'n.; 75% act.; 5.21% act. oxygen.

Lupersol 575-P75. [Elf Atochem N. Am./Org. Peroxides] t-Amylperoxy-2-ethylhexanoate sol'n. in plasticizer; CAS 686-31-7; initiator for bulk and sol'n. polymerization, for high-temp. cure of polyester resins, and for cure of acrylic syrup; sol'n.; 75% act.; 5.14-5.28% act. oxygen.

Lupersol 610-M50. [Elf Atochem N. Am./Org. Peroxides] 3-Hydroxy-1,1-dimethylbutyl peroxyneodecanoate in odorless min. spirits; highly efficient low temp. initiator for polymerization of vinyl chloride; provides improved kinetics and no odorous decomp. prods.; colorless clear liq.; sol. in aromatic and aliphatic hydrocarbons, chlorinated hydrocarbons, ketones, esters; insol. in water; m.w. 288.4; sp.gr. 0.8465; flash pt. 122 F; 49-51% assay, 2.72-2.83% act. oxygen; *Precaution:* easily ignited, burns vigorously; avoid contamination esp. by strong oxidizing agents; *Storage:* store away from combustible materials in original containers.

Lupersol 665-M50. [Elf Atochem N. Am./Org. Peroxides] 1,1-Dimethyl-3-hydroxybutyl peroxy-2-ethylhexanoate; crosslinking agent for polymerization, cure of acrylic syrup.

Lupersol 665-T50. [Elf Atochem N. Am./Org. Peroxides] 1,1-Dimethyl-3-hydroxybutylperoxy-2-ethylhexanoate in toluene; crosslinking agent for emulsion, bulk, sol'n., and suspension polymerization, cure of acrylic syrup; 50% sol'n.; 3.07% act. oxygen.

Lupersol 688-M50. [Elf Atochem N. Am./Org. Peroxides] 1,1-Dimethyl-3-hydroxybutylperoxyneoheptanoate in odorless min. spirits; crosslinking agent for bulk, sol'n., emulsion, and suspension polymerization, and cure of acrylic syrup; sol'n.; 50% conc.; 3.25% act. oxygen.

Lupersol 688-T50. [Elf Atochem N. Am./Org. Peroxides] 1,1-Dimethyl-3-hydroxybutylperoxyneoheptanoate in toluene; crosslinking agent for bulk, sol'n., emulsion, and suspension polymerization, and cure of acrylic syrup; sol'n.; 50% conc.; 3.25% act. oxygen.

Lupersol 801. [Elf Atochem N. Am./Org. Peroxides] t-Butyl cumyl peroxide; CAS 3457-61-2; initiator for bulk, sol'n., and suspension polymerization; polymer modification, thermoplastic crosslinking; curing elastomers; high temp. cure of polyester resins; cure of acrylic syrup; liq.; 90% act., 6.91-7.30% act. oxygen.

Lupersol DDM-9. [Elf Atochem N. Am./Org. Peroxides] MEK peroxide; CAS 1338-23-4; EINECS 215-661-2; polymerization initiator for cure of promoted unsat. polyester resins and vinyl ester resins at ambient temps.; promoter, transition metal salt, activates decomposition of peroxide initiator; flexibility in useful conc. range and pot life; clear liq.; insol. in water; sp.gr. 1.0815; visc. 14.8 cps; f.p. < -30 C; ref. index 1.4615; flash pt. 58 C (SETA); 8.8 ± 0.1% act. oxygen.

Lupersol DDM-30. [Elf Atochem N. Am./Org. Peroxides] Ketone peroxide in dimethyl phthalate diluent; curing agent for unsat. polyester thermoset resins; filament winding applic.; sol'n.; sp.gr. 1.490; visc. 11.0 cps; flash pt. 143 F.

Lupersol Delta-3. [Elf Atochem N. Am./Org. Peroxides] MEK peroxide; CAS 1338-23-4; EINECS 215-661-2; initiator designed for high speed cures in GP resins; offers fast mold turnover at ambient temps. and improved cure rates at low ambient temps.; clear liq.; sol. in alcohols, ethers, ketones, esters, hydrocarbons; insol. in water; sp.gr. 1.1472; visc. 15.7 cps; flash pt. 68 C; 8.7-9.0% act. oxygen; *Storage:* store @ 18-29 C (38 C max.).

Lupersol Delta-X-9. [Elf Atochem N. Am./Org. Peroxides] MEK peroxide; CAS 1338-23-4; EINECS 215-661-2; polymerization initiator for cure of promoted unsat. polyester resins and vinyl ester resins at ambient temps.; promoter, transition metal salt, activates decomposition of peroxide initiator; flexibility in useful conc. range and pot life; clear liq.; insol. in water; sp.gr. 1.1471; visc. 15.8 cps; f.p. < -30 C; ref. index 1.4758; Setaflash pt. 68 C; 8.8 ± 0.1% act. oxygen.

Lupersol DFR. [Elf Atochem N. Am./Org. Peroxides] Ketone peroxide; initiator for R.T. cure of polyester resins; 8.7% min. act. oxygen.

Lupersol DHD-9. [Elf Atochem N. Am./Org. Peroxides] MEK peroxide; CAS 1338-23-4; EINECS 215-661-2; polymerization initiator for cure of promoted unsat. polyester resins and vinyl ester resins at ambient temps.; promoter, transition metal salt, activates decomposition of peroxide initiator; flexibility in useful conc. range and pot life; clear liq.; insol. in water; sp.gr. 1.1363; visc. 16.0 cps; f.p. < -20 C; ref. index 1.4777; flash pt. (Seta) 66 C; 8.8 ± 0.1% act. oxygen.

Lupersol DSW-9. [Elf Atochem N. Am./Org. Peroxides] Mixt. of peroxides and hydroperoxides; initiator for cure of acrylic syrup; 8.7% min. act. oxygen.

Lupersol KDB. [Elf Atochem N. Am./Org. Peroxides] Di-t-butyl diperoxyphthalate in DBP; initiator for vinyl polymerizations; sol'n.; insol. in water; m.w. 310; sp.gr. 1.06; visc. 31 cps; f.p. 41 F; ref. index 1.49; flash pt. 235 F; 50% min. assay; 5.2% min. act. oxygen.

Lupersol P-31. [Elf Atochem N. Am./Org. Peroxides] t-Butyl peroctoate and 1,1-di (t-butylperoxy) cyclohexane blend in min. oil; initiator for high-temp. cure of polyester resins; liq.; 7.17-7.49% act. oxygen.

Lupersol P-33. [Elf Atochem N. Am./Org. Peroxides] t-Butyl peroctoate and 1,1-di (t-butylperoxy) cyclohexane blend in min. oil; initiator for high-temp. cure of polyester resins; liq.; 6.70-6.97% act. oxygen.

Lupersol PDO. [Elf Atochem N. Am./Org. Peroxides] t-Butyl peroxy-2-ethylhexanoate in DOP; initiator for vinyl polymerizations; sol'n.; insol. in water; m.w. 216; sp.gr. 0.93; visc. 12.0 cps; f.p. < -22 F; ref. index 1.45; flash pt. 190 F; 50% min. assay; 3.7% min. act. oxygen.

Lupersol PMS. [Elf Atochem N. Am./Org. Peroxides] t-Butyl peroxy-2-ethylhexanoate in odorless min. spirits; initiator for vinyl polymerizations; sol'n.; insol. in water; m.w. 216; sp.gr. 0.82; visc. 2.2 cps; f.p. < -22 F; ref. index 1.42; flash pt. 155 F; 50% min. assay; 3.37% min. act. oxygen.

Lupersol TAEC. [Elf Atochem N. Am./Org. Peroxides] OO-t-amyl O-(2-ethylhexyl) monoperoxycarbonate; crosslinking agent for emulsion, bulk, sol'n., and suspension polymerization, high temp. cure of polyester, and cure of acrylic syrup; liq.; 95% conc.; 5.8% act. oxygen.

Lupersol TBEC. [Elf Atochem N. Am./Org. Peroxides] OO-t-butyl O-(2-ethylhexyl) monoperoxycarbonate; initiator for bulk, sol'n., and suspension polymerization, curing elastomers, high-temp. cure of polyester resins, and cure of acrylic syrup; liq.; 6.17% act. oxygen.

Lupersol TBIC-M75. [Elf Atochem N. Am./Org. Peroxides] OO-t-Butyl O-isopropyl monoperoxycarbonate in odorless min. spirits; initiator for vinyl polymerizations; liq.; m.w. 176; sp.gr. 0.87; visc. 2.05 cps; f.p. -65 F; ref. index 1.41; flash pt. 140 F; 74-76% assay; 6.72-6.90% act. oxygen.

Lustrabrite® S. [Telechemische] Toluene sulfonamide/epoxy resin; patented formaldehyde-free nail enamel resin providing superior clarity, brilliance, high gloss, adhesion; plasticizer for nitrocellulose.

Luviskol® K17. [BASF; BASF AG] PVP; CAS 9003-39-8; EINECS 201-800-4; film-forming agent, hair fixative, thickener, protective colloid, stabilizer, visc. modifier, suspending agent, and dispersant; used for emulsion and suspension polymerizations, cosmetics, adhesives, sealants, paints, coatings, paper, detergents, glass fibers, inks, ceramics, nonpharmaceutical tableting, photographic films; powd. or liq.; > 99% > 250 μm particle size (powd.); sol. in alcohols, esters, ether alcohols, ketones, lactams, chlorinated hydrocarbons, nitroparaffins, acids, amines; bulk dens. 0.4-0.5 g/cc (powd.); visc. 5.4 cps (5% in methanol); pH 3-7 (10%, powd.), 7-10 (10%, sol'n.); > 92% solids (powd.), 50% solids in water (sol'n.).

Luviskol® K30. [BASF; BASF AG] PVP; CAS 9003-39-8; EINECS 201-800-4; film-forming agent, hair fixative, thickener, protective colloid, stabilizer, visc. modifier, suspending agent, and dispersant; used for emulsion and suspension polymerizations, cosmetics, adhesives, sealants, paints, coatings, paper, detergents, glass fibers, inks, ceramics, nonpharmaceutical tableting, photographic films; powd. or liq.; particle size > 99% > 250 μm (powd.); sol. in alcohols, esters, ether alcohols, ketones, lactams, chlorinated hydrocarbons, nitroparaffins, acids, amines; bulk dens. 0.4-0.5 g/cc; visc. 7.0 cps (5% in methanol); pH 3-7 (10%, powd.), 7-10 (10%, sol'n.); > 95% solids (powd.), 30% solids in water (sol'n.).

Luviskol® K60. [BASF; BASF AG] PVP; CAS 9003-39-8; EINECS 201-800-4; film-forming agent, hair fixative, thickener, protective colloid, stabilizer, visc. modifier, suspending agent, and dispersant; used for emulsion and suspension polymerizations, cosmetics, adhesives, sealants, paints, coatings, paper, detergents, glass fibers, inks, ceramics, nonpharmaceutical tableting, photographic films; liq.; sol. in alcohols, esters, ether alcohols, ketones, lactams, chlorinated hydrocarbons, nitroparaffins, acids, amines; pH 7-10 (10%); 45% solids in water.

Luviskol® K90. [BASF; BASF AG] PVP; CAS 9003-39-8; EINECS 201-800-4; film-forming agent, hair fixative, thickener, protective colloid, stabilizer, visc. modifier, suspending agent, and dispersant; used for emulsion and suspension polymerizations, cosmetics, adhesives, sealants, paints, coatings, paper, detergents, glass fibers, inks, ceramics, nonpharmaceutical tableting, photographic films; powd. or liq.; particle size > 99% > 250 μm (powd.); sol. in alcohols, esters, ether alcohols, ketones, lactams, chlorinated hydrocarbons, nitroparaffins, acids, amines; bulk dens. 0.3-0.4 g/cc (powd.); visc. 30.8 cps (5% in methanol); pH 5-9 (10%, powd.), 7-10.5 (10%, sol'n.); > 95% solids (powd.), 15% or 20% solids in water (sol'n.).

Luwax® A. [BASF] LDPE homopolymer wax; CAS 9002-88-4; EINECS 200-815-3; additive for printing inks; flatting and antisettling agent in paints; dispersant and color enhancer; in floor polishes; lubricant for plastics processing; processing aid for plastics and natural and syn. rubber; aids flow and release; wh. powd.; dens. 0.92 g/cc; visc. 1250 mm^2/s (120 C); m.p. 103 C; drop pt. 106 C; acid no. 0; sapon. no. 0.

Luwax® AF 30. [BASF] Micronized polyethylene wax; CAS 9002-88-4; EINECS 200-815-3; additive for printing inks, as flattening and antisettling agent in paints, dispersant in color concs.; improves hardness in wax compds., black heel mark resist. in floor polishes; lubricant in plastics processing; powd.; 6 μ avg. particle size; dens. 0.94 g/cc; visc. 185 mm^2/s (120 C); m.p. 109 C; drop pt. 114 C.

Luwax® AH 3. [BASF] HDPE wax; additive for printing inks, as flattening and antisettling agent in paints, dispersant in color concs.; improves hardness in wax compds., black heel mark resist. in floor polishes; lubricant in plastics processing; powd.; dens. 0.945 g/cc; visc. 185 mm^2/s (120 C); m.p. 109 C; drop pt. 114 C; acid no. nil.

Luwax® AH 6. [BASF] HDPE homopolymer; CAS 9002-88-4; EINECS 200-815-3; additive for PVC processing, mfg. of printing inks, hot-melt adhesives, encapsulating compds., solv.-type polishes; gran., powd.; dens. 0.94 g/cc; visc. 1250 mm^2/s (120 C); m.p. 110 C; drop pt. 115 C; acid no. nil; penetration hardness 1 dmm.

Luwax® AL 3. [BASF] LDPE homopolymer; CAS 9002-88-4; EINECS 200-815-3; additive for prod. of color concs., wax dispersions, wax coatings and polishes, PVC processing; powd.; dens. 0.92 g/cc; visc. 185 mm^2/s (120 C); m.p. 100 C; drop pt. 101 C; acid no. nil; penetration hardness 2 dmm.

Luwax® AL 61. [BASF] LDPE homopolymer; CAS 9002-88-4; EINECS 200-815-3; additive for solv.-based polishes, printing inks, masterbatches, wax coatings, corrosion inhibitors, mold release agents; processing aid for plastics and rubber; powd.; dens. 0.93 g/cc; visc. 1250 mm^2/s (120 C); m.p. 107 C; drop pt. 113 C; acid no. nil; penetration hardness 1 dmm.

Luwax® EAS 1. [BASF] Ethylene/acrylic acid copolymer; CAS 9010-77-9; additive for printing inks, as flattening and antisettling agent in paints, dispersant in color concs.; improves hardness in wax compds., black heel mark resist. in floor polishes; lubricant in engineering plastics processing; gran.; dens. 0.935 g/cc; visc. 1450 mm^2/s (120 C); m.p. 107 C; drop pt. 104 C; acid no. 43.

Luwax® ES 9668. [BASF] Isotactic polypropylene wax with high crystallinity; wax additive for color concs., paints and coatings, toner, PP processing; visc. 800-1200 mm^2/s (200 C); m.p. 159-166 C; acid no. 0.

Luwax® OA 2. [BASF] Oxidized polyethylene wax; CAS 68441-17-8; additive for printing inks, as flattening and antisettling agent in paints, dispersant in color concs.; improves hardness in wax compds., black heel mark resist. in floor polishes; lubricant in plastics processing; gran., powd.; dens. 0.97 g/cc; visc. 360 mm^2/s (120 C); m.p. 104 C; drop pt. 109 C; acid no. 22; penetration hardness 1 dmm.

Luwax® OA 5. [BASF] Oxidized polyethylene wax; CAS 68441-17-8; high hardness wax for coating of fruits, paper, cellophane, textiles, and floor polishes; lubricant for plastics processing; gran. or powd.; dens. 0.97 g/cc; visc. 410 mm^2/s (120 C); m.p. 104 C; drop pt. 110 C; acid no. 16; penetration hardness < 1 dmm.

Luzenac 8170. [Luzenac Am.; R.T. Vanderbilt] Talc; CAS 14807-96-6; EINECS 238-877-9; ultrabright appearance grade for polyolefins, PVC, elastomers, most other plastics; raises heat deflection temps., flex. mod., and impact str. in polyolefins; good for compounding natural color PP, color matching in wh. and light colored prods.; 10 μ median particle size; 99% thru 325 mesh; bulk dens. 16 lb/ft^3 (loose); oil absorp. 28; brightness 90-91; 92% talc, 6% chlorite, 1% dolomite; 60% SiO_2, 31% MgO.

LV-31. [J.M. Huber/Engineered Mins.] Alumina trihydrate; filler with flame retarding and smoke suppressing properties; used as a resin extender in polyester, vinyls, PU, latex, neoprene foam systems, wire and cable insulation, vinyl wall and floor coverings, epoxies; also suitable for high loadings and high visc. systems like rubber compds.; used in polyester cast onyx; particulate, 60-70% thru 325 mesh; sp.gr. 2.42; dens. 1.0 g/cm^3; surf. area 0.50 m^2/g; ref. index 1.57; hardness (Moh) 2.5-3.5; 65% Al_2O_3; Unverified

Lyndcoat™ 10-RTU. [Rhone-Poulenc Rubber Spec.] Solv.-free, ready-to-use mold release coating that cures to clear durable finish; exc. release and lubricating props.; 13% solids.

Lyndcoat™ 15-RTU. [Rhone-Poulenc Rubber Spec.] Solv.-free, ready-to-use mold release coating that cures to clear durable finish; exc. release and lubricating props.; visc. 30-60 cps; pH 3.5-5.5; 11% solids.

Lyndcoat™ 20-RTU. [Rhone-Poulenc Rubber Spec.] Solv.-free, ready-to-use mold release coating that cures to clear durable finish; exc. release and lubricating props.; visc. 1000-2000 cps; pH 3.5-5.5; 15% solids.

Lyndcoat™ BR-18/M. [Rhone-Poulenc Rubber Spec.] Solv.-free silicone-based coating with high mica content; ready-to-use release agent for butyl rubber bladders; cures during vulcanization to a tight durable coating with exc. release and lubricity; promotes air bleed; reduces flamm., toxicity, and dusting problems; visc. 1000-2000 cps; pH 3.5-5.5; 36% solids.

Lyndcoat™ BR 500-RTU. [Rhone-Poulenc Rubber Spec.] Aq. silicone emulsion/acrylic latex emulsion/surfactant blend; bladder release agent; free of hydrogen, solvs., and formaldehyde; ready-to-use with no mica; visc. 550-850 cps; pH 3.5-5.5; 23.5% solids.

Lyndcoat™ BR 601-RTU. [Rhone-Poulenc Rubber Spec.] Solv.-free silicone-based coating with no mica; ready-to-use bladder release agent with superior lubricity and release; dries clear, glossy, and durable with superior lubricity and release; replaces lining cements; eliminates flamm., toxicity, and dust problems; visc. 550-850 cps; pH 3.5-5.5; 23.5% solids.

Lyndcoat™ BR 790-RTU. [Rhone-Poulenc Rubber Spec.] Solv.-free silicone-based coating with no mica; ready-to-use bladder release agent with superior lubricity and release; replaces lining cements; eliminates flamm., toxicity, and dust problems; visc. low; pH 3.5-5.5; 21.5% solids.

Lyndcoat™ BR 880-RTU. [Rhone-Poulenc Rubber Spec.] Solv.-free silicone-based coating with no mica; ready-to-use bladder release agent with superior lubricity and release; replaces lining cements; eliminates flamm., toxicity, and dust problems; visc. 1000-2000 cps; pH 3.4-5.5; 25.5% solids.

Lyndcoat™ M 755-RTU. [Rhone-Poulenc Rubber Spec.] Solv.-free silicone-based coating with low mica content; ready-to-use release agent for butyl rubber bladders; promotes air bleed; visc. 1000-2000 cps; pH 3.5-5.5; 28% solids.

Lyndcoat™ M 771-RTU. [Rhone-Poulenc Rubber Spec.] Solv.-free silicone-based coating with low mica content; ready-to-use release agent for butyl rubber bladders; promotes air bleed; visc. high; 32.5% solids.

M

MR 575. [Sovereign] Sulfonate/oil blend; plasticizer, peptizer for NR and SR.

Mackester™ EGDS. [McIntyre] Glycol distearate; CAS 627-83-8; EINECS 211-014-3; emulsifier, lubricant, antistat, defoamer for metalworking, textile lubricants, plastics, paper; emulsifier, pearlescent, emollient for cosmetics; flake; 100% conc.

Mackester™ EGMS. [McIntyre] Glycol stearate; CAS 111-60-4; EINECS 203-886-9; emulsifier, lubricant, antistat, defoamer for metalworking, textile lubricants, plastics, paper; emulsifier, pearlescent, emollient for cosmetics; flake; 100% conc.

Mackester™ IP. [McIntyre] Glycol stearate, other ingreds.; CAS 111-60-4; EINECS 203-886-9; emulsifier, lubricant, antistat, defoamer for metalworking, textile lubricants, plastics, paper; emulsifier, pearlescent, emollient for cosmetics; flake; 100% conc.

Mackester™ SP. [McIntyre] Glycol stearate and stearamide MEA; emulsifier, lubricant, antistat, defoamer for metalworking, textile lubricants, plastics, paper; emulsifier, pearlescent, emollient for cosmetics; flake; 100% conc.

Macol® 300. [PPG/Specialty Chem.] PPG-7 buteth-10; CAS 9038-95-3; nonionic; detergent for toilet cleaners, laundry detergents, emulsion polymerization, defoamers, metalworking fluids, hydraulic fluids; liq.; sol. @ 5% in water, alcohols, ketones; disp. in min. oil; sp.gr. 1.033; visc. 60 cst (100 F); pour pt. -40 C; flash pt. (PMCC) 340 F; ref. index 1.456; 100% conc.

Macol® 660. [PPG/Specialty Chem.] PPG-12 buteth-16; CAS 9038-95-3; nonionic; detergent for toilet bowl cleaners, laundry; defoamer, rubber lubricant, intermediate; hydraulic, heat transfer, and metal working fluids; mold release agent; emulsion polymerization; APHA 100 max. clear visc. liq.; sol. in acetone, propylene glycol, oleic acid, castor oil, water, toluene, IPA; sp.gr. 1.047; dens. 8.72 lb/gal; visc. 140 cst (100 F); acid no. 0-1; pour pt. -37 C; cloud pt. 61 C (1% aq.); flash pt. (COC) 430 F; ref. index 1.459; pH 5-7; 100% conc.

Macol® 3520. [PPG/Specialty Chem.] PPG-28 buteth-35; CAS 9038-95-3; nonionic; detergent for toilet bowl cleaners, laundry; defoamer, rubber lubricant, intermediate; hydraulic, heat transfer, and metal working fluids; mold release agent; emulsion polymerization; APHA 100 max. clear visc. liq.; sol. in acetone, propylene glycol, oleic acid, castor oil, water, toluene, IPA; sp.gr. 1.050; dens. 8.75 lb/gal; visc. 760 cst (100 F); pour pt. -28 C; acid no. 0-1; ref. index 1.461; pH 5-7; flash pt. (COC) 430 F; 100% conc.

Macol® 5100. [PPG/Specialty Chem.] PPG-33 buteth-45; CAS 9038-95-3; nonionic; detergent for toilet bowl cleaners, laundry; defoamer, rubber lubricant, intermediate; hydraulic, heat transfer, and metal working fluids; mold release agent; emulsion polymerization; food processing; FDA compliance; APHA 100 max. clear visc. liq.; sol. in acetone, propylene glycol, oleic acid, castor oil, water, toluene, IPA; sp.gr. 1.050; dens. 8.75 lb/gal; visc. 1100 cst (100 F); acid no. 0-1; pour pt. -28 C; cloud pt. 55 C (1% aq.); flash pt. (COC) 430 F; ref. index 1.462; pH 5-7; 100% conc.

Macol® OP-40(70). [PPG/Specialty Chem.] Octoxynol-40; CAS 9002-93-1; nonionic; emulsifier, detergent, wetting agent, dispersant, solubilizer, coupling agent for cosmetics, textile, metalworking, household, industrial and other applics.; emulsifier for vinyl acetate and acrylate polymerization; liq.; sol. @ 5% in water; sp.gr. 1.10; visc. 490 cps; HLB 17.9; pour pt. 4 C; cloud pt. > 100 C (1% aq.); flash pt. (PMCC) > 350 F; 70% act. in water.

Magala® 0.5E. [Akzo Nobel] Di-n-butyl magnesium/triethyl aluminum complex (0.5:1); CAS 61632-57-3; used in prod. of catalysts for polymerization of olefins or dienes and as an alkylating agent.

MagChem® 20M. [Martin Marietta Magnesia Spec.] Magnesium oxide; CAS 1309-48-4; EINECS 215-171-9; lightburned reactive grade for mfg. of magnesium chems., construction prods., lubrication oil additives, fuel additives, oil drilling chems., in neoprene and chlorinated polymers, sugar refining, uranium ore processing; powd.; 1 μ median particle size; 99% thru 325 mesh; bulk dens. 20 lb/ft^3; surf. area 10 m^2/g; 98.2% MgO.

MagChem® 30. [Martin Marietta Magnesia Spec.] Magnesium oxide; CAS 1309-48-4; EINECS 215-171-9; lightburned reactive grade for mfg. of magnesium chems., construction prods., lubrication oil additives, fuel additives, oil drilling chems., in neoprene and chlorinated polymers, sugar refining, uranium ore processing, acid neutralization; powd.; 4 μ median particle size; 98% thru 325 mesh; bulk dens. 22 lb/ft^3 (loose); surf. area 25 m^2/g;

98.2% MgO.

MagChem® 30G. [Martin Marietta Magnesia Spec.] Magnesium oxide; CAS 1309-48-4; EINECS 215-171-9; lightburned reactive grade for mfg. of magnesium chems., construction prods., lubrication oil additives, fuel additives, oil drilling chems., in neoprene and chlorinated polymers, sugar refining, uranium ore processing, acid neutralization, free-flowing gran.; 300 μ median particle size; 95% thru 40 mesh, 10% thru 100 mesh; bulk dens. 60 lb/ft^3 (loose); surf. area 30 m^2/g; 98.2% MgO.

MagChem® 35. [Martin Marietta Magnesia Spec.] Magnesium oxide; CAS 1309-48-4; EINECS 215-171-9; lightburned reactive grade for mfg. of magnesium chems., construction prods., lubrication oil additives, fuel additives, oil drilling chems., in neoprene and chlorinated polymers, sugar refining, uranium ore processing, acid neutralization; free-flowing powd.; 4 μ median particle size; 98% thru 325 mesh; bulk dens. 22 lb/ft^3 (loose); surf. area 35 m^2/g; 98.2% MgO.

MagChem® 35K. [Martin Marietta Magnesia Spec.] High purity magnesium oxide; CAS 1309-48-4; EINECS 215-171-9; moderately reactive curing and vulcanizing agent for chlorprene and fluoroelastomers; powd.; 5 μ mean particle size; 99.5% thru 325 mesh; bulk dens. 22 lb/ft^3 (loose); surf. area 30 m^2/g; 98% MgO; *Storage:* store in dry place; reseal open containers.

MagChem® 40. [Martin Marietta Magnesia Spec.] Magnesium oxide; CAS 1309-48-4; EINECS 215-171-9; lightburned reactive grade for mfg. of magnesium chems., construction prods., lubrication oil additives, fuel additives, oil drilling chems., in neoprene and chlorinated polymers, sugar refining, uranium ore processing, acid neutralization; curing/vulcanizing agent for chloroprene, fluoroelastomers; free-flowing powd.; 4 μ median particle size; 99% thru 325 mesh; bulk dens. 22 lb/ft^3 (loose); surf. area 50 m^2/g; 98.2% MgO.

MagChem® 50. [Martin Marietta Magnesia Spec.] Light burned tech. magnesium oxide; CAS 1309-48-4; EINECS 215-171-9; reactive grade for mfg. of magnesium chems., rubber compounding, plastic thickening, filtrate clarification, odor removal, selective adsorptions; curing/vulcanizing agent for chloroprene, fluoroelastomers; wh. fine free-flowing powd.; 5 μ median particle size; 99.5% thru 325 mesh; bulk dens. 18 lb/ft^3 (loose); surf. area 65 m^2/g; 98.2% MgO.

MagChem® 50Y. [Martin Marietta Magnesia Spec.] High purity magnesium oxide; CAS 1309-48-4; EINECS 215-171-9; med. reactivity curing and vulcanizing agent for chlorprene and fluoroelastomers; powd.; 5 μ mean particle size; 99.5% thru 325 mesh; bulk dens. 22 lb/ft^3 (loose); surf. area 65 m^2/g; 98% MgO; *Storage:* store in dry place; reseal open containers.

MagChem® 60. [Martin Marietta Magnesia Spec.] Light burned tech. magnesium oxide; CAS 1309-48-4; EINECS 215-171-9; magnesium source for mfg. of magnesium compds.; used in rubber compounding, plastic thickening, filtrate clarification, odor removal, and selective adsorptions; curing/vulcanizing agent for chloroprene, fluoroelastomers; offers high reactivity, low bulk dens., exc. flow props.; wh. fine powd.; 5 μ mean particle size; 100% thru 100 mesh, 99.5% thru 325 mesh; bulk dens. 18 lb/ft^3 (loose); surf. area 80 m^2/g; 98% MgO; *Storage:* store in dry place; exposure to moisture may cause caking.

MagChem® 125. [Martin Marietta Magnesia Spec.] High purity, highly reactive magnesium oxide; CAS 1309-48-4; EINECS 215-171-9; scorch retarder, vulcanizing and curing agent in polychloroprene, fluoroelastomers, halogenated elastomers, other rubber compds.; thickening agent in SMC, BMC; powd.; 5 μ median particle size; 99.8% thru 325 mesh; bulk dens. 22 lb/ft^3 (loose); surf. area 120 m^2/g; 98.1% MgO; *Storage:* store in dry place, quickly reseal opened containers; will readily absorb moisture and CO_2 on exposure to air.

MagChem® 200-AD. [Martin Marietta Magnesia Spec.] High purity magnesium oxide; CAS 1309-48-4; EINECS 215-171-9; highly reactive thickener for polyester SMC, BMC; curing agent for neoprene/phenolic adhesives; micronized powd.; 1 μ median particle size; 99.9% thru 325 mesh; bulk dens. 13 lb/ft^3 (loose); surf. area 160 m^2/g; 93% MgO; *Storage:* store in dry place; reseal open containers; will readily absorb moisture and CO_2 on exposure to air.

MagChem® 200-D. [Martin Marietta Magnesia Spec.] High purity, highly reactive magnesium oxide; CAS 1309-48-4; EINECS 215-171-9; scorch retarder in Neoprene G, Hypalon, Viton, and other elastomers; powd.; 3 μ median particle size; 99.9% thru 325 mesh; bulk dens. 23 lb/ft^3 (loose); surf. area 170-200 m^2/g; iodine no. 150; 93% MgO; *Storage:* store in dry place, quickly reseal opened containers; will readily absorb moisture and CO_2 on exposure to air.

Maglite® D. [Marine Magnesium] Magnesium oxide; CAS 1309-48-4; EINECS 215-171-9; used in cure of polychloroprene compds. to aid aging stability, absorbing HCl.

Magnifin® H5. [Martinswerk GmbH] Magnesium hydroxide; CAS 1309-42-8; EINECS 215-170-3; halogen-free general-purpose flame retardant/filler for plastics and rubber; dens. 2.4 g/cc; bulk dens. ≥ 300 g/l; surf. area 4-6 m^2/g; brightness > 96%; ref. index 1.56-1.58; hardness (Mohs) 2.5; conduct. ≤ 350 μS; > 99.8% $Mg(OH)_2$.

Magnifin® H5B. [Martinswerk GmbH] Magnesium hydroxide, aminosilane surf. treated; CAS 1309-42-8; EINECS 215-170-3; halogen-free flame retardant/filler for plastics and rubber, esp. polyamide, polyarylamide, T-EVA.

Magnifin® H5C. [Martinswerk GmbH] Magnesium hydroxide, fatty acid surf. treated; CAS 1309-42-8; EINECS 215-170-3; halogen-free flame retardant/filler for plastics and rubber, esp. PVC, EVA, EPDM, PE copolymers; environmentally friendly.

Magnifin® H5C/3. [Martinswerk GmbH] Magnesium hydroxide, increased levels of fatty acid surf. treatment; CAS 1309-42-8; EINECS 215-170-3; halogen-free flame retardant/filler for plastics and rubber, esp. PVC, EVA, EPDM, PE copolymers; improved rheology.

Magnifin® H5D. [Martinswerk GmbH] Magnesium hydroxide, proprietary hydrophobizing surf. treated; CAS 1309-42-8; EINECS 215-170-3; halogen-free flame retardant/filler for plastics and rubber, esp. EPRs, EVA, Kraton.

Magnifin® H-7. [Lonza; Martinswerk GmbH] Magnesium hydroxide; CAS 1309-42-8; EINECS 215-170-3; halogen-free general-purpose flame retardant/filler for PVC, PP, PE, nylon, PS; dens. 2.4 g/cc; bulk dens. ≥ 300 g/l; surf. area 7-9 m^2/g; brightness > 96%; ref. index 1.56-1.58; hardness (Mohs) 2.5; conduct. ≤ 350 μS.

Magnifin® H7A. [Martinswerk GmbH] Magnesium hydroxide; CAS 1309-42-8; EINECS 215-170-3; halogen-free flame retardant/filler for plastics and rubber, esp. EPDM, EPM, XL-PE, XL-EVA; environmentally friendly.

Magnifin® H7C. [Martinswerk GmbH] Magnesium hydroxide, fatty acid surf. treated; CAS 1309-42-8; EINECS 215-170-3; halogen-free flame retardant/filler for plastics and rubber, esp. PVC, EVA, EPDM, PE copolymers; environmentally friendly.

Magnifin® H7C/3. [Martinswerk GmbH] Magnesium hydroxide, increased levels of fatty acid surf. treatment; CAS 1309-42-8; EINECS 215-170-3; halogen-free flame retardant/filler for plastics and rubber, esp. PVC, EVA, EPDM, PE copolymers; improved rheology.

Magnifin® H7D. [Martinswerk GmbH] Magnesium hydroxide, proprietary hydrophobizing surf. treated; CAS 1309-42-8; EINECS 215-170-3; halogen-free flame retardant/filler for plastics and rubber, esp. EPRs, EVA, Kraton.

Magnifin® H-10. [Lonza; Martinswerk GmbH] Magnesium hydroxide; CAS 1309-42-8; EINECS 215-170-3; halogen-free flame general-purpose retardant/filler for PVC, PP, PE, ABS, nylon, PS, PU, rubber; environmentally friendly; dens. 2.4 g/cc; bulk dens. ≈ 350 g/l; surf. area 9-11 m^2/g; brightness > 96%; ref. index 1.56-1.58; hardness (Mohs) 2.5; conduct. ≤ 350 μS.

Magnifin® H10A. [Martinswerk GmbH] Magnesium hydroxide, vinylsilane surf. treated; CAS 1309-42-8; EINECS 215-170-3; halogen-free flame retardant/filler for plastics and rubber, esp. crosslinked rubber and PE.

Magnifin® H10B. [Martinswerk GmbH] Magnesium hydroxide, aminosilane surf. treated; CAS 1309-42-8; EINECS 215-170-3; halogen-free flame retardant/filler for plastics and rubber, esp. polyamide, polyarylamide, T-EVA.

Magnifin® H10C. [Martinswerk GmbH] Magnesium hydroxide, fatty acid surf. treated; CAS 1309-42-8; EINECS 215-170-3; halogen-free flame retardant/filler for plastics and rubber, esp. PVC, EVA, EPDM, PE copolymers; environmentally friendly.

Magnifin® H10D. [Martinswerk GmbH] Magnesium hydroxide, proprietary hydrophobizing surf. treated; CAS 1309-42-8; EINECS 215-170-3; halogen-free flame retardant/filler for plastics and rubber, esp. EPRs, EVA, Kraton.

Magnifin® H10F. [Martinswerk GmbH] Magnesium hydroxide, proprietary surf. treated; CAS 1309-42-8; EINECS 215-170-3; halogen-free flame retardant/filler for plastics, esp. polyolefins.

Magnum-White. [RMc Minerals] Magnesium hydroxide and calcium carbonate blend; fire retardant/smoke suppressant filler for PVC compds. and SBR latex formulations; powd.; 99.8% thru 325 mesh; 10-11 μ mean particle size; sp.gr. 2.51; bulk dens. 39.3 lb/ft^3 (loose); oil absorp. 20.3 ml/100 g; brightness 91-92; ref. index 1.157-1.59; pH 9.5 (10% aq.); hardness (Mohs) 2.7; 63.29% $CaCO_3$, 36.55% $Mg(OH)_2$.

Magocarb-33. [Kaopolite] Magnesium carbonate; CAS 546-93-0; EINECS 208-915-9; flame retardant filler in plastics.

Magoh-S. [Kaopolite] Magnesium hydroxide; CAS 1309-42-8; EINECS 215-170-3; fire retardant filler for plastics and rubber compds.; extender pigment for flame retardant coatings; fine powd. or 55% liq. slurry.

Magotex™. [Kaopolite] Fused magnesium oxide; CAS 1309-48-4; EINECS 215-171-9; high thermal conduct. coeff., low oil absorp., good elec. resistivity, low moisture; filler for plastics, thermal conductive molding compds., brake lining, elec. potting compds., special refractories, caulks, mastics, sealants, ceramics, coatings; 14 μ median particle size; 9% on 325 mesh; sp.gr. 3.6; dens. 30.10 lb/solid gal; bulking value 0.0332 gal/lb; oil absorp. 17; GE brightness 70%; ref. index 1.73; pH 11 (20% solids); hardness (Mohs) 5.5-6.0; resist. 12,000 ohm-cm; 95% MgO, 3% SiO_2, 1.8% CaO.

Magsil Diamond Talc 200 mesh. [Colin Stewart Minchem Ltd] Talc; CAS 14807-96-6; EINECS 238-877-9; filler; wh. free-flowing powd.; 95% min. thru 200 mesh; 99.99% thru 60 mesh; brightness 89; 59% SiO_2, 30% MgO.

Magsil Diamond Talc 350 mesh. [Colin Stewart Minchem Ltd] Talc; CAS 14807-96-6; EINECS 238-877-9; filler; wh. free-flowing powd.; 98% min. thru 350 mesh; 99.9% thru 60 mesh; brightness 89; 59% SiO_2, 30% MgO.

Magsil Star Talc 200 mesh. [Colin Stewart Minchem Ltd] Talc; CAS 14807-96-6; EINECS 238-877-9; filler; wh. free-flowing powd.; 98% min. thru 200 mesh; 100% thru 60 mesh; brightness 86; 58% SiO_2, 30% MgO.

Magsil Star Talc 350 mesh. [Colin Stewart Minchem Ltd] Talc; CAS 14807-96-6; EINECS 238-877-9; filler; wh. free-flowing powd.; 98% min. thru 350 mesh; 100% thru 60 mesh; brightness 86; 58% SiO_2, 30% MgO.

Manganese Violet. [Holliday Pigments] Manganese violet; CAS 101-66-3; EINECS 233-257-4; inorg. violet pigment used for coloration of plastics, powd. coatings, artists' colors, and cosmetics; exc. lightfastness; various grades differing in shade; alkali sensitive; acid stable; special grades avail.; FDA approval for cosmetics intended for use around the eyes; sp.gr. 2.70; heat stability 260 C; *Toxicology:* nontoxic.

Manobond™ 680-C. [Rhone-Poulenc Rubber Spec.] Cobalt boro acylate; cobalt adhesion promoter; promotes adhesion of rubber to steel; of primary importance in steel-belted radial tires or other steel-reinforced rubber prods.; 22.5% cobalt metal.

Manro BA Acid. [Hickson Manro Ltd.] Linear dodecylbenzene sulfonic acid; anionic; raw material used in emulsifiers, heavy and lt. duty detergents, hand cleaning gels, machine degreasers, tank cleaners; emulsion polymerization; catalyst; metalworking; dk. brn. visc. liq.; char. odor; sp.gr. 1.04; visc. 1500 cps; biodeg.; 96% conc.

Manro DL 28. [Hickson Manro Ltd.] Sodium lauryl sulfate, modified; CAS 151-21-3; anionic; foaming and wetting agent used in foamable precoats and no-gel foam compds. in latex industry, emulsion polymerization; biodeg.; pale yel. clear liq.; mild char. odor; sp.gr. 1.05; visc. 50 cps; cloud pt. 0 C; pH 7-8 (10% aq.); 28% min. act.

Manro DL 32. [Hickson Manro Ltd.] Sodium dodecylbenzene sulfonate; anionic; emulsion polymerization surfactant; liq.; 32% conc.

Manro FCM 90LV. [Hickson Manro Ltd.] Xylene sulfonic acid; CAS 25321-41-9; EINECS 246-839-8; catalyst in prod. of resin bound sand castings; curing agent in prod. of resins; dk. brnsh. liq.; slight aromatic odor; sp.gr. 1.28; visc. 1100 cps; 91% min. act.

Manro MA 35. [Hickson Manro Ltd.] Sodium tallow sulfosuccinamate; CAS 68988-69-2; anionic; foaming agent used in no-gel foam systems based on high solids, noncarboxylated SBR latex or natural rubber latex; lt. cream fluid; mild, char. odor; sp.gr. 1.05; pH 7.5-9.5 (10% aq.); 35% solids.

Manro PTSA/C. [Hickson Manro Ltd.] Para toluene sulfonic acid monohydrate; catalyst in prod. of esters; curing agent for coating resins; adhesive systems; mfg. of dyes and pigments; wh. to pink crystals; 97% min. act.

Manro PTSA/E. [Hickson Manro Ltd.] Toluene sulfonic acid; anionic; catalyst for prod. of resin bound sand castings, mfg. of esters, resin prod.; intermediate for hydrotrope prod.; amber clear liq.; slight odor; sp.gr. 1.24; visc. 10 cps; 65% act.

Manro PTSA/H. [Hickson Manro Ltd.] Toluene sulfonic acid; anionic; catalyst for prod. of resin bound sand castings, mfg. of esters, resin prod.; intermediate for hydrotrope prod.; liq.; 62.5% conc.

Manro PTSA/LS. [Hickson Manro Ltd.] Toluene sulfonic acid; anionic; catalyst for prod. of resin bound sand castings, mfg. of esters, resin prod.; intermediate for hydrotrope prod.; amber clear liq., sl. odor; sp.gr. 1.24; visc. 10 cps; 65% conc.

Manro SDBS 25/30. [Hickson Manro Ltd.] Sodium dodecylbenzene sulfonate; anionic; detergent, emulsifier, foaming agent, wetting agent used in liq. detergents and cleaning prods., emulsion polymerization, rubber, plastics, textiles, insecticides; biodeg.; amber clear visc. liq.; mild odor; sp.gr. 1.05 (30 C); visc. 1000 cps (30 C); pH 7-9 (10% aq.); 29% min. act.

Manro SLS 28. [Hickson Manro Ltd.] Sodium lauryl sulfate; CAS 151-21-3; EINECS 205-788-1; anionic; foaming and wetting agent used in cosmetics, toiletries, emulsion polymerization, plastics, rubber, foam rubber, carpet and upholstery shampoos, industrial cleaning; biodeg.; very pale yel. clear liq.; mild odor; sp.gr. 1.04; visc. 200-7000 cps; pH 7-8 (10% aq.); 28% min. act.

Manro TDBS 60. [Hickson Manro Ltd.] TEA dodecylbenzene sulfonate; CAS 27323-41-7; EINECS 248-406-9; anionic; mild detergent used in car shampoos, bubble baths, emulsion polymerization, household and industrial cleaners; amber clear visc. liq.; slight, alcoholic odor; sp.gr. 1.06; visc. 400 cps; pH 6.5-7.5 (10% aq.); 60% act.

Mapeg® CO-25. [PPG/Specialty Chem.] PEG-25 castor oil; CAS 61791-12-6; nonionic; surfactant; solubilizer, coupling agent for oils, solvs., waxes; for cosmetic, paper, metalworking fluid, and emulsion polymerization; liq.; sol. @ 5% in IPA; disp. in water, min. spirits, toluene; sp.gr. 1.040; HLB 10.8; acid no. 2 max.; sapon. no. 83; pour pt. 5 C; flash pt. (PMCC) > 350 F.

Mapeg® CO-25H. [PPG/Specialty Chem.] PEG-25 hydrog. castor oil; CAS 61788-85-0; nonionic; surfactant for formulation of gels; solubilizer, coupling agent; for cosmetic, paper, metalworking fluids, and emulsion polymerization; liq.; sol. @ 5% in IPA, toluene, min. oil; disp. in water, min. spirits; sp.gr. 1.040; HLB 10.8; acid no. 2 max.; sapon. no. 82; pour pt. 5 C; flash pt. (PMCC) > 350 F; 100% conc.

Maphos® 15. [PPG/Specialty Chem.] Aromatic phosphate ester; nonionic/anionic; surfactant, hydrotrope, detergency aid, antistat for drycleaning and lubricant systems, emulsion polymerization; sp.gr. 1.08; acid no. 55 (to pH 5.2); flash pt. (PMCC) > 300 F; pH 2 (1% aq.); 70% act., 2.8% phosphorus; *Toxicology:* skin irritant.

Maphos® 17. [PPG/Specialty Chem.] Aromatic phosphate ester; anionic; emulsifier for emulsion polymerization; solubilizer; yel. clear visc. liq.; sol. @ 5% in water @ pH 2.0 and 9.5; sp.gr. 1.11; acid no. 70 (to pH 5.2); pour pt. < 0 C; flash pt. (PMCC) > 300 F; pH 1.5-2.5 (1% aq.); 70% act., 3.8% phosphorus; *Toxicology:* skin irritant.

Maphos® 18. [PPG/Specialty Chem.] Aliphatic phosphate ester; nonionic/anionic; surfactant, hydrotrope, detergency aid, antistat for drycleaning and lubricant systems, emulsion polymerization; sp.gr. 1.16; acid no. 120 (to pH 5.2); flash pt. (PMCC) > 300 F; pH 2 (1% aq.); 99.5% act., 6.7% phosphorus; *Toxicology:* skin irritant.

Maphos® 30. [PPG/Specialty Chem.] Aliphatic phosphate ester; nonionic/anionic; surfactant, hydrotrope, detergency aid, antistat for drycleaning and lubricant systems, emulsion polymerization, metalworking; liq.; water-sol.; sp.gr. 1.10; acid no. 190 (to pH 5.2); flash pt. (PMCC) > 200 F; pH 2 (1% aq.); 99.5% act., 11.1% phosphorus; *Toxicology:* skin irritant.

Maphos® 41A. [PPG/Specialty Chem.] Aromatic phosphate ester; nonionic/anionic; emulsifier, lubricant with anticorrosive/antifrictional props. for oil and water-sol. lubricant systems, e.g., greases, syn. cutting oils, drawing compds., chain-belt lubricants, gear oils, rust preventatives; particle size modifier for emulsion polymerization; soft, waxy paste; sp.gr. 1.12; acid no. 73 (to pH 5.2); flash pt. (PMCC) > 300 F; pH 2 (1% aq.); 82% act., 4.6% phosphorus; *Toxicology:* skin irritant.

Maphos® 54. [PPG/Specialty Chem.] Aromatic phosphate ester; nonionic/anionic; surfactant, hydrotrope, detergency aid, antistat for drycleaning and lubricant systems, emulsion polymerization; liq.; oil-sol.; sp.gr. 1.08; acid no. 70 (to pH 5.2); flash pt. (PMCC) > 300 F; pH 2 (1% aq.); 99.5% act., 4.1% phosphorus; *Toxicology:* skin irritant.

Maphos® 55. [PPG/Specialty Chem.] Aromatic phosphate ester; nonionic/anionic; surfactant, hydrotrope, detergent, antistat for drycleaning, metalworking lubricants, emulsion polymerization; liq.; oil-sol.; sp.gr. 1.14; acid no. 110 (to pH 5.2); flash pt. (PMCC) > 300 F; pH 2 (1% aq.); 80% act., 8.2% phosphorus; *Toxicology:* skin irritant.

Maphos® 56. [PPG/Specialty Chem.] Aliphatic phosphate ester; nonionic/anionic; surfactant, hydrotrope, detergent, dispersant, wetting agent, antistat for drycleaning, metalworking lubricants, emulsion polymerization; liq.; oil-sol.; sp.gr. 1.16; acid no. 130 (to pH 5.2); flash pt. (PMCC) > 300 F; pH 2 (1% aq.); 90% act., 7.4% phosphorus.

Maphos® 76. [PPG/Specialty Chem.] Aromatic phosphate ester; nonionic/anionic; detergent, antistat, anticorrosive agent, emulsifer, solubilizer, hydrotrope used in drycleaning formulations, hard surface cleaners, emulsion polymerization for PVAc, acrylate polymers; yel. clear visc. liq.; sol. in dist. water @ pH 9.5 (@ 5%), in perchloroethylene (@ 10%); sp.gr. 1.09; acid no. 55 (to pH 5.2); pour pt. < 0 C; flash pt. (PMCC) > 300 F; pH 1.5-2.5 (1% aq.); 99.5% act., 3.1% phosphorus; *Toxicology:* skin irritant.

Maphos® 76 NA. [PPG/Specialty Chem.] Partial sodium salt of aromatic phosphate ester; anionic; detergent, antistat, anticorrosive agent, solubilizer used in drycleaning formulations; emulsifier for PVAc and acrylate polymerization; lt. yel. clear liq.; sol. in dist. water @ pH 2.0 and 9.5 (@ 5%), in perchloroethylene (@ 10%); sp.gr. 1.08; pour pt. 8 C; flash pt. (PMCC) > 300 F; pH 5.0-6.0 (1% aq.); 90% act., 2.6% phosphorus.

Maphos® 77. [PPG/Specialty Chem.] Aliphatic phosphate ester; nonionic/anionic; surfactant, hydrotrope, detergency aid, antistat for drycleaning and lubricant systems, emulsion polymerization; lt. yel. clear liq.; sp.gr. 1.010; acid no. 34 (to pH 5.2); flash pt. (PMCC) > 300 F; pH 2 (1% aq.); 99.5% act., 1.9% phosphorus; *Toxicology:* skin irritant.

Maphos® 151. [PPG/Specialty Chem.] Aromatic phosphate ester; nonionic/anionic; surfactant, hydrotrope, detergency aid, antistat for drycleaning and lubricant systems, emulsion polymerization; sp.gr. 1.09; acid no. 50 (to pH 5.2); flash pt. (PMCC) > 300 F; pH 2 (1% aq.); 99.5% act., 2.6% phosphorus; *Toxicology:* skin irritant.

Maphos® 236. [PPG/Specialty Chem.] Aliphatic phosphate ester; nonionic/anionic; surfactant, hydrotrope, detergency aid, antistat for drycleaning and lubricant systems, emulsion polymerization; sp.gr. 1.04; acid no. 95 (to pH 5.2); flash pt. (PMCC) > 300 F; pH 2 (1% aq.); 99.5% act., 5.4% phosphorus; *Toxicology:* skin irritant.

Maphos® FDEO. [PPG/Specialty Chem.] Phosphate ester; nonionic/anionic; surfactant, hydrotrope, detergency aid, antistat for drycleaning and lubricant systems, emulsion polymerization; sp.gr. 1.12; acid no. 105 (to pH 5.2); flash pt. (PMCC) > 200 F; pH 2 (1% aq.); 90% act., 5.7% phosphorus; *Toxicology:* skin irritant.

Maphos® L-6. [PPG/Specialty Chem.] Complex org. phosphate acid ester; anionic; o/w dispersant, emulsion polymerization, syn. cutting fluids; yel. liq.; 100% conc.

Maphos® L 13. [PPG/Specialty Chem.] Aliphatic phosphate ester; nonionic/anionic; surfactant, lubricant, anticorrosive, coupling agent for metalworking, dry cleaning, hard surf. cleaning, dedusting, lubrication, emulsion polymerization; lt. yel. amber clear liq.; sol. in dist. water @ pH 9.5 (@ 5%), in perchloroethylene, min. oil, and xylene (@ 10%), clear to boiling in 5% TSP (@ 1%); sp.gr. 1.015; acid no. 90 (to pH 5.2); pour pt. 3 C; flash pt. (PMCC) > 300 F; pH 1.5-2.5 (1% aq.); 99.5% act., 5.4% phosphorus; *Toxicology:* skin irritant.

Maranil CB-22. [Pulcra SA] TEA dodecylbenzene sulfonate; anionic; surfactant for foam baths, liq. detergents, carwashes, heavy duty detergents; emulsifier for carnauba wax and pine oil, emulsion polymerization; pigment dispersant; liq.; 50% conc.

Maranil Paste A 65. [Pulcra SA] Sodium dodecylbenzene sulfonate; anionic; scouring and wetting agent in the textile industry and emulsifier for emulsion polymerization in the rubber and plastic industry; paste; 65 ± 2% conc.

Maranil Powd. A. [Henkel/Functional Prods.; Henkel KGaA/Dehydag] Sodium dodecylbenzene sulfonate; anionic; base for mfg. detergents, dishwashes, cleaning agents; wetting agent; emulsifier for PVC copolymers, carboxylated S/B latexes; powd.; 86-90% act.

Marble Dust™. [ECC Int'l.] Calcium carbonate; CAS 471-34-1, 1317-65-3; coarse ground pigment, filler, reinforcement for putties, glazes, mild abrasive compds., rubber and latex compds.; wh. powd., odorless; 21 μ mean particle size; 18% on 325 mesh; negligible sol. in water; sp.gr. 2.71; dens. 70 lb/ft^3 (loose); pH 9.5 (5% slurry); nonflamm.; 96.6% $CaCO_3$, 0.2% max. moisture; *Toxicology:* TLV/TWA 10 mg/m^3, considered nuisance dust; *Precaution:* incompat. with acids.

Marblemite™. [ECC Int'l.] Calcium carbonate; CAS 1317-65-3; high brightness pigment, filler, reinforcement for high loading and min. black specks in cultured marble applics.; also in putties, caulks, floor tile, cast rigid or elastomeric compds., foamed compds.; wh. powd., odorless; 3% on 40 mesh, 45% thru 200 mesh; negligible sol. in water; sp.gr. 2.71; dens. 87 lb/ft^3 (loose); pH 9.5 (5% slurry); nonflamm.; 97.1% $CaCO_3$, 0.2% max. moisture; *Toxicology:* TLV/TWA 10 mg/m^3, considered nuisance dust; *Precaution:* incompat. with acids.

Marblewhite® MW200. [Specialty Minerals] Limestone; CAS 1317-65-3; filler for paints, polymers, caulks and putties, rubber, joint compds., building prods.; good brightness, controlled particle size; powd.; 20 μ median particle size; 11% on 325 mesh; sp.gr. 2.7; dens. 109.5 lb/ft^3 (tapped); bulk dens. 51 lb/ft^3; bulking value 22.5; oil absorp. 11 g/ 100 g; brightness 93; 99% $CaCO_3$, 1.2% $MgCO_3$.

Marblewhite® MW325. [Specialty Minerals] Limestone; CAS 1317-65-3; filler for paints, polymers, caulks and putties, rubber, building prods.; good brightness, controlled particle size; powd.; 13.8 μ median particle size; 1% on 325 mesh; sp.gr. 2.7; dens. 97.5 lb/ft^3 (tapped); bulk dens. 45.6 lb/ft^3; bulking value 22.5; oil absorp. 12 g/100 g; brightness 95; 98% $CaCO_3$, 1.2% $MgCO_3$.

Marblewhite® MW A 200. [Specialty Minerals] Limestone; CAS 1317-65-3; filler for paints, polymers, caulks and putties, joint compds., rubber, building prods.; high brightness, controlled particle size; powd.; 18.2 μ median particle size; 1% on 200 mesh; sp.gr. 2.7; dens. 100.6 lb/ft^3 (tapped); bulk dens. 47.5 lb/ft^3; brightness 88.5; 96.7% $CaCO_3$, 0.9% $MgCO_3$.

Marblewhite® MW A 325. [Specialty Minerals] Limestone; CAS 1317-65-3; filler for paints, polymers, caulks and putties, joint compds., rubber, building prods.; high brightness, controlled particle size; powd.; 11.5 μ median particle size; 1% on 325 mesh; sp.gr. 2.7; dens. 94.8 lb/ft^3 (tapped); bulk dens. 40.1 lb/ft^3; brightness 88.5; 96.7% $CaCO_3$, 0.9% $MgCO_3$.

Mark® 133. [Witco/PAG] Calcium-zinc; heat and lt. stabilizer for vinyl plastisols, organosols; high clarity; liq.

Mark® 135. [Witco/PAG] Nonstaining antioxidant for cellulosics, polyester, PC, polyethers.

Mark® 144. [Witco/PAG] Barium-zinc; heat stabilizer for highly filled compds., flexible and semirigid PVC, foams, tile; powd.

Mark® 152. [Witco/PAG] Calcium-zinc; stabilizer for food wrap. pkg. film, beverage tubing, bottles, and vacuum-formed sheet; nontoxic; paste.

Mark® 155. [Witco/PAG] Barium/zinc; heat stabilizer for highly filled compds., flexible, semirigid and foamed PVC; powd.

Mark® 158. [Witco/PAG] Phosphite complex; antioxidant for use in stabilization of polyolefin resins; used in extrusion and molding compds., monofilament and fiber applic.; pale yel. liq.; sol. in hexane, MEK, and benzene; f.p. < -20 C; sp.gr. 0.982; Gardner Z1 visc.; ref. index 1.488; flash pt. (COC) 169 C.

Mark® 224. [Witco/PAG] Epoxy; high effieicncy heat and lt. stabilizer for flexible and rigid PVC, plastisols, organosols, tile, chlorinated paraffins.

Mark® 260. [Witco/PAG] Phosphite; antioxidant for stabilization of polyolefins; pale yel. visc. liq.; m.w. 1238; sol. in hexane, MEK, and benzene; sp.gr. 0.943; ref. index 1.515; flash pt. 138 C.

Mark® 281B. [Witco/PAG] Cadmium-zinc; one-pkg. stabilizer/antifog systems for meat and produce wrap; stabilizer/activators and cell stabilizers for plastisol and calendered foam; nitrogenous flooring stabilizers, and other special compds.; liq.

Mark® 329. [Witco/PAG] Phosphite; nonstaining antioxidant for ABS, PP, PVC; FDA 21CFR §178.2010.

Mark® 366. [Witco/PAG] Phosphite stabilizer; for use as aux. stabilizers with Ba/Cd and Ca/Zn systems to improve color and clarity in flexibles and rigids for both general purpose and nontoxic end uses.

Mark® 398. [Witco/PAG] Barium-zinc; heat and lt. stabilizer for flexible and semirigid PVC, foams, tile, fluorescent pigments; low plateout; powd.

Mark® 462. [Witco/PAG] Barium/cadmium/phosphite; stabilizer for all clear and pigmented, flexible and semirigid compds., plastisols, organosols, and vinyl sol'ns.; liq.

Mark® 462A. [Witco/PAG] Barium cadmium zinc; high efficiency heat and light stabilizer for flexible and semirigid PVC, plastisols, organosols, and foams; high clarity, low plateout; liq.

Mark® 495. [Witco/PAG] Calcium-zinc; heat and lt. stabilizer for flexible and semirigid PVC, plastisols, organosols, bottles; nontoxic; FDA approved; powd.

Mark® 522. [Witco/PAG] Phosphite; antioxidant for stabilization of PP compds.; pale yel. visc. liq.; m.w. 1881; sol. in hexane, MEK, and benzene; sp.gr. 0.975; ref. index 1.502; flash pt. (COC) 193 C.

Mark® 550, 556. [Witco/PAG] Barium lead stabilizer; stabilizer for DWV pipe, elec., conduit, wire and cable, records, etc.

Mark® 565A. [Witco/PAG] Calcium-zinc; heat and lt. stabilizer for filled systems, vinyl plastisols and organosols; liq.

Mark® 577A. [Witco/PAG] Calcium-zinc; heat and lt. stabilizer for rigid PVC, bottles; nontoxic; FDA approved; powd.

Mark® 630. [Witco/PAG] Calcium-zinc; one-pkg. stabilizer/antifog systems for meat and produce wrap; stabilizer/activators and cell stabilizers for plastisol and calendered foam; nitrogenous flooring stabilizers, and other special compds.; paste.

Mark® 1014. [Witco/PAG] Barium cadmium zinc; heat and lt. stabilizer, lubricant for flexible and semirigid PVC, foams, tile, phosphate plasticizers; powd.

Mark® 1043A. [Witco/PAG] Stabilizer for food wrap. pkg. film, beverage tubing, bottles, and vacuum-formed sheet; nontoxic.

Mark® 1178. [Witco/PAG] Phosphite; nonstaining antioxidant, heat/color/process stabilizer for ABS, cellulosics, polyester, PE, PP, PS, PVC for food wrap. pkg. film, beverage tubing, bottles, and vacuum-formed sheet; nontoxic; FDA 21CFR §178.2010.

Mark® 1178B. [Witco/PAG] Phosphite; nonstaining antioxidant, heat/color/process stabilizer for ABS, cellulosics, polyester, PE, PP, PS, PVC; chelating agent used with stabilizers for polymers and plastics; food pkg., hot melt formulations; FDA 21CFR §178.2010; pale yel. visc. liq.; sp.gr. 0.985; flash pt. (COC) 210 C; ref. index 1.5250.

Mark® 1220. [Witco/PAG] Organo complex; antioxidant/stabilizer for styrene and ABS polymers; food applic.; stabilizer for ABS latex; pale yel. visc. liq.; sp.gr. 0.978; visc. 4125 cps; ref. index 1.5152; flash pt. 186 C.

Mark® 1409. [Witco/PAG] Phenolic-phosphite; antioxidant/stabilizer for polyolefins; inj. molding and extrusion of film and fiber; food additives; lt. brn. visc. liq.; sp.gr. 0.989; visc. 20,000; ref. index 1.5291; flash pt. 190 C.

Mark® 1413. [Witco/PAG] 2-Hydroxy-4-n-octoxybenzophenone; CAS 1843-05-6; EINECS 217-421-2; uv absorber in vinyls, polyolefins, and other polymers; lt. straw powd.; sp.gr. 1.16; m.p. 46-48 C.

Mark® 1500. [Witco/PAG] Phosphite; nonstaining antioixdant, heat/color/process stabilizer for ABS, PVC, food wrap. pkg. film, beverage tubing, bottles, and vacuum-formed sheet; nontoxic; FDA 21CFR §178.2010.

Mark® 1535. [Witco/PAG] Benzophenone; antioxidant/uv stabilizer for polyolefins; inj. molding and extrusion of film and fiber; food additives; liq.

Mark® 1589. [Witco/PAG] Phenolic complex; antioxidant used in polyolefins and elastomers; wh. powd.; sp.gr. 1.17; m.p. 208 C.

Mark® 1589B. [Witco/PAG] Phenolic-thio-complex; antioxidant/stabilizer for polyolefins; inj. molding and extrusion of film and fiber; food additives; wh. powd.; sp.gr. 1.06.

Mark® 1600. [Witco/PAG] Barium-zinc; stabilizer for applics. where the absence of cadmium is desired, without loss in dynamic heat stability; liq.

Mark® 1900. [Witco/PAG] Organotin; antioxidant for ABS, PVC; stabilizer for high impact and unmodified rigid compds., incl. the super tins, weather-resistant systems, etc., potable water pipe, vinyl siding, and rigid foam.

Mark® 1905. [Witco/PAG] Organotin; stabilizer for high impact and unmodified rigid compds., incl. the super tins, weather-resistant systems, etc., potable water pipe, vinyl siding, and rigid foam.

Mark® 1939. [Witco/PAG] Tin; economical heat stabilizer for rigid PVC pipe.

Mark® 1984. [Witco/PAG] Tin; heat stabilizer for rigid PVC bottles; FDA approved.

Mark® 2077. [Witco/PAG] Barium cadmium zinc; high efficiency heat and light stabilizer for flexible and semirigid PVC, plastisols, organosols, and foams; high clarity, low plateout; liq.

Mark® 2100, 2100A. [Witco/PAG] Sulfur-free organotins; heat stabilizers for rigid PVC compds.; additives to boost the heat and lt. stability of std. organotin mercaptide stabilized compds.; stabilizers for halogenated resins other than PVC; wh. to off-wh. powd.

Mark® 2112. [Witco/PAG] Substituted triphenyl phosphite; CAS 101-02-0; EINECS 202-908-4; nondiscoloring antioxidant for polymer processing (ABS, cellulosics, polyester, PE, PP, PS, PVC, PC); stabilizer for PP, HDPE, and LDPE; synergist; FDA 21CFR §178.2010; wh. powd.; sol. in org. solv.; insol. in water; sp.gr. 0.98; m.p. 186 C.

Mark® 2115. [Witco/PAG] Antimony mercaptide; stabilizer for rigid PVC pipe.

Mark® 2140. [Witco/PAG] Pentaerythrityl hexylthiopropionate; CAS 95823-35-1; stabilizer for use in polyolefins and other polymeric systems; synergistic with primary antioxidants; color which reduces or eliminates the need for phosphite; used at 0.1-0.3 phr in PP, 0.05-0.10 phr in HDPE, 0.25-0.5 phr in elastomers, and 0.5-1.0 phr in S.B. latex; very pale straw liq.; mild odor; m.w. 825; sol. in common org. solvs.; sp.gr. 1.0570; flash pt. (COC) 274 C.

Mark® 2180. [Witco/PAG] Organo-metallic compd.; stabilizer; nucleating agent for PP; reduces haze in PP when used at 0.1-0.3 phr; increases the crystallization temp. without affecting the heat aging of the compd.; wh. powd.; sol. in hot water; sp.gr. 1.24; m.p. decomp. @ 300 C.

Mark® 2326. [Witco/PAG] Calcium-zinc; heat and lt. stabilizer for flexible and rigid PVC, bottles; nontoxic; FDA approved; powd.

Mark® 4700. [Witco/PAG] Ba-Zn salt; non-Cd heat stabilizer for PVC filled and clear compds.; for extrusion and calendering; Gardner 7+ clear liq.; sp.gr. 1.100; flash pt. (COC) 103 C; ref. index 1.5225.

Mark® 4701. [Witco/PAG] Ba-Zn salt; non-Cd heat stabilizer for highly filled PVC plastisols; for extrusion, calendering; Gardner 4+ clear liq.; sp.gr. 0.930; flash pt. (COC) 53 C; ref. index 1.4665; *Toxicology:* may cause irritation; avoid contact with eyes, skin, clothing; *Storage:* store in tightly closed drums, out of contact with moist air and oxidizing agents; keep away from flame and excessive heat.

Mark® 4702. [Witco/PAG] Ba-Zn salt; non-Cd heat stabilizer for clear general purpose PVC applics.; for extrusion, calendering; Gardner 7 clear liq.; sp.gr. 1.102; flash pt. (COC) 100 C; ref. index 1.5145; *Toxicology:* may cause irritation; avoid contact with eyes, skin, clothing; *Storage:* store in tightly closed drums, out of contact with moist air and oxidizing agents; keep away from flame and excessive heat.

Mark® 4716. [Witco/PAG] Ba-Zn salt; non-Cd heat stabilizer for general purpose PVC filled and clear applics.; for extrusion, calendering, inj. molding, plastisols; Gardner 7 clear liq.; sp.gr. 1.1047; flash pt. (COC) 95 C; ref. index 1.5045; *Toxicology:* may cause irritation; avoid contact with eyes, skin, clothing; *Storage:* store in tightly closed drums, out of contact with moist air and oxidizing agents; keep away from flame and excessive heat.

Mark® 4722. [Witco/PAG] Ba-Zn salt; non-Cd heat stabilizer for darker colored PVC applics.; good long term stability; for extrusion, calendering; Gardner 5 clear liq.; sp.gr. 1.020; flash pt. (COC)

120 C; ref. index 1.4920; *Toxicology:* may cause irritation; avoid contact with eyes, skin, clothing; *Storage:* store in tightly closed drums, out of contact with moist air and oxidizing agents; keep away from flame and excessive heat.

Mark® 4723. [Witco/PAG] Ba-Zn salt; non-Cd heat stabilizer for low fogging automotive PVC applics.; for extrusion, calendering, inj. molding, and plastisols; Gardner 10 clear liq.; sp.gr. 1.108; flash pt. (COC) 100 C; ref. index 1.5083; *Toxicology:* may cause irritation; avoid contact with eyes, skin, clothing; *Storage:* store in tightly closed drums, out of contact with moist air and oxidizing agents; keep away from flame and excessive heat.

Mark® 4726. [Witco/PAG] K-Zn salt; non-Ba, non-Cd heat stabilizer for filled PVC compds.; for extrusion, calendering, inj. molding, and plastisols; Gardner 4+ clear liq.; sp.gr. 1.107; flash pt. (COC) 115 C; ref. index 1.4982; *Toxicology:* may cause irritation; avoid contact with eyes, skin, clothing; *Storage:* store in tightly closed drums, out of contact with moist air and oxidizing agents; keep away from flame and excessive heat.

Mark® 5060. [Witco/PAG] Distearyl pentaerythritol diphosphite; CAS 3806-34-6; EINECS 223-276-6; heat and lt. stabilizer/antioxidant for use in a wide range of polymers; heat/processing additive for molding, extrusion, fibers, and films with polymers; synergistic with hindered phenolic antioxidants, thioesters, UV absorbers, and/or both amine and nickel type lt. stabilizers; used where improved hydrolytic stability is required; wh. flakes; sp.gr. 1.0; 7.0-7.8% P.

Mark® 5082. [Witco/PAG] Phosphite; hydrolytically stable antioxidant for PE, PP, elastomers; FDA 21CFR §178.2010.

Mark® 5089. [Witco/PAG] Pentaerythritol alkyl thiodipropionate; CAS 96328-09-1; antioxidant/stabilizer for plastics.

Mark® 5095. [Witco/PAG] Lauryl/stearyl thiodipropionate; antioxidant for polyolefins and other polymeric systems incl. syn. rubber; also for pharmaceuticals, cosmetics, industrial oils, greases, lubricants; FDA regulated; wh. free-flowing powd.

Mark® 5111. [Witco/PAG] Phenol; nondiscoloring antioxidant for ABS, polyester, PE, PP, PS, HIPS, PU; FDA approved.

Mark® 6000. [Witco/PAG] Barium complex; heat stabilizer for vinyl records; powd.

Mark® 7118. [Witco/PAG] Barium cadmium; high efficiency heat and light stabilizer for flexible and rigid PVC; liq.

Mark® 7308. [Witco/PAG] Barium cadmium; high efficiency heat and light stabilizer for flexible and semirigid PVC, plastisols, organosols, and foams; liq.

Mark® C. [Witco/PAG] Phosphite stabilizer; for use as aux. stabilizers with Ba/Cd and Ca/Zn systems to improve color and clarity in flexible and rigid PVC for both general purpose and nontoxic end uses; also for ABS, PP.

Mark® DPDP. [Witco/PAG] Phosphite; heat and lt. stabilizer for PE, PP, ABS, PS, SAN.

Mark® DPP. [Witco/PAG] Phosphite; heat and lt. stabilizer for epoxies.

Mark® GS, RFD. [Witco/PAG] Zinc complex; heat and lt. stabilizer for flexible, semirigid, plastisol and organosol vinyls; generally used with BaCd.

Mark® LL. [Witco/PAG] Barium/cadmium/phosphite; stabilizer for all clear and pigmented, flexible and semirigid compds., plastisols, organosols, and vinyl sol'ns.; liq.

Mark® OHM. [Witco/PAG] Proprietary; heat and lt. stabilizer for flexible and semirigid PVC, elec. insulation.

Mark® OTM. [Witco/PAG] Octyltin; heat and lt. stabilizer for food wrap. pkg. film, beverage tubing, bottles, and vacuum-formed sheet; for flexible, rigid, plastisol and organosol vinyls; nontoxic.

Mark® PDDP. [Witco/PAG] Phosphite; heat and lt. stabilizer for PE, PP, ABS, PS, SAN.

Mark® QED. [Witco/PAG] Calcium-zinc; heat and lt. stabilizer for flexible and rigid PVC, bottles; nontoxic; FDA approved; powd.

Mark® RFD. [Witco/PAG] Zinc; stabilizer; liqs. and solids.

Mark® TDP. [Witco/PAG] Tridecyl phosphite; heat and lt. stabilizer for PE, PP, ABS, PS, SAN; water-wh. liq.; m.w. 502; sp.gr. 0.880-0.9050; flash pt. (COC) 140 C; 6.1% P.

Mark® TPP. [Witco/PAG] Triphenyl phosphite; CAS 101-02-0; EINECS 202-908-4; heat and lt. stabilizer for epoxies; water- wh. liq.; m.w. 310; sp.gr. 1.172-1.186; dens. 9.86 lb/gal; m.p. 25 C; flash pt. (COC) 155 C; 9.5% phosphorus.

Mark® TT. [Witco/PAG] Barium-cadmium; stabilizer for plasticized and rigid compds. for calendering, extrusion, inj. molding, powd. molding, slush and rotational molding, dip-and spread-coating, etc.; sulfide stain protection, plate-out resist., good weatherability; blowing agent catalyst; solid.

Mark® WS. [Witco/PAG] Barium-cadmium; stabilizer for plasticized and rigid compds. for calendering, extrusion, inj. molding, powd. molding, slush and rotational molding, dip-and spread-coating, etc.; sulfide stain protection, plate-out resist., good weatherability; solid.

Marklear AFL-23. [Witco/PAG] Antistat/antifog coating for all plastics; FDA approved.

Markpet. [Witco/PAG] Petrolatum; external lubricant for rigid PVC; FDA 21CFR §172.880; solid.

Markstat® AL-1. [Witco/PAG] Antistat for rigid PVC; FDA-sanctioned; liq.

Markstat® AL-12. [Witco/PAG] Quat. ammonium chloride deriv. of polyalkoxy tert. amines; cationic; antistat additive for PU films and thermoplastics; lt. amber oily liq.; sol. in IPA, acetone, MEK, ethanol, water; sp.gr. 1.01; visc. 1800 cps; 100% act.

Markstat® AL-13. [Witco/PAG] Nonionic; general purpose antistat for rigid PVC; liq.

Markstat® AL-14. [Witco/PAG] Antistat for plasticized PVC, surface treatments of all materials; high wetting; FDA-sanctioned; liq.

Markstat® AL-15. [Witco/PAG] Nonionic; general

purpose antistat for plasticized PVC; liq.

Markstat® AL-22. [Witco/PAG] Quat. ammonium chloride deriv., modified; low foam general purpose antistat used as surf. treatment for plastics; amber clear liq.; water-sol.; sp.gr. 1.042; visc. (Gardner) V-W; ref. index 1.4726; 100% act.

Markstat® AL-26. [Witco/PAG] Cationic; antistat for rigid PVC and acrylics; liq.

Markstat® AL-44. [Witco/PAG] General purpose antistat for plasticized PVC; liq.

Markstat® AS-7. [Witco/PAG] Antistat for polyolefins; FDA-sanctioned; solid.

Markstat® AS-16. [Witco/PAG] Antistat for polyolefins; FDA-sanctioned to 0.2 phr; paste.

Markstat® AS-18. [Witco/PAG] Antistat for polyolefins; FDA-sanctioned to 0.3 phr; solid.

Markstat® AS-20. [Witco/PAG] Antistat for modified styrene polymers; FDA-sanctioned; soft solid.

Marlican®. [Hüls AG] Straight-chain dodecylbenzene; CAS 67774-74-7; anionic; detergent intermediate, solubilizer; sec. plasticizer; reduces visc. of PVC pastes; solv. for carbonless copy papers; biodeg.; liq.; 100% act.

Marquat Pigments. [Degussa] Cadmium, cobalt and titanium pigments; for coatings industry, coloring plastics; stable to 500 C.

Martinal® 103 LE. [Lonza] Aluminum trihydrate; CAS 21645-51-2; EINECS 244-492-7; flame retardant filler for PVC.

Martinal® 104 LE. [Lonza] Aluminum trihydrate; CAS 21645-51-2; EINECS 244-492-7; flame retardant filler for PVC, PE, PU thermoplastic polyester, phenolics.

Martinal® 107 LE. [Lonza] Aluminum trihydrate; CAS 21645-51-2; EINECS 244-492-7; flame retardant filler for PVC, PE, PU, thermoplastic polyester, phenolics.

Martinal® OL-104. [Martinswerk GmbH] Aluminum hydroxide; CAS 21645-51-2; EINECS 244-492-7; filler, flame retardant for polyester resins (SMC, BMC, laminates), epoxy and acrylic resins, crosslinked elastomers (conveyor belts, cables, profiles, molded parts), latexes (wall paper, carpetbacking), thermoplastic and thermoplastic elastomers (PP, PE and copolymers, EVA), PVC (plasticized, rigid, foam), aq. disps., PU castings, lacquers, elastomers, flexible and rigid foam); powd.; 1.3-2.6 μm median particle size; dens. 2.4 g/cc; bulk dens. ≈ 320 kg/m^3; surf. area 3-5 m^2/g; oil absorp. 27-35 cc/100 g; brightness (TAPPI) > 94; ref. index 1.58; pH 9 ± 1; conduct. < 100 μS/cm; 99.6% act.

Martinal® OL-104/C. [Martinswerk GmbH] Aluminum hydroxide; CAS 21645-51-2; EINECS 244-492-7; filler, flame retardant for thermoplastics and thermoplastic elastomers (PP, PE and copolymers, EVA), PVC (plasticized, rigid, foam), crosslinked elastomer cable and conveyor belts; powd.; 1.3-2.6 μm median particle size; dens. 2.4 g/cc; bulk dens. ≈ 400 kg/m^3; surf. area 3-5 m^2/g; oil absorp. 25-30 cc/100 g; brightness (TAPPI) > 94; ref. index 1.58; pH 9 ± 1; conduct. < 100 μS/cm; 99.6% act.

Martinal® OL-104 LE. [Martinswerk GmbH] Aluminum hydroxide; CAS 21645-51-2; EINECS 244-492-7; finely precipitated filler, flame retardant; esp. suitable for demanding elec. applics., crosslinked elastomer cables, profiles, molded parts, thermoplastics and thermoplastic elastomers (PP, PE and copolymers, EVA); powd.; 1.3-2.6 μm median particle size; dens. 2.4 g/cc; bulk dens. ≈ 370 kg/m^3; surf. area 3-5 m^2/g; oil absorp. 30-40 cc/100 g; brightness (TAPPI) > 94; ref. index 1.58; pH 9 ± 1; conduct. < 60 μS/cm; 99.6% act.

Martinal® OL-104/S. [Martinswerk GmbH] Aluminum hydroxide; CAS 21645-51-2; EINECS 244-492-7; filler, flame retardant for acrylic resin composites, crosslinked elastomer cables, EVA; powd.; 1.3-2.6 μm median particle size; dens. 2.4 g/cc; bulk dens. ≈ 430 kg/m^3; surf. area 3-5 m^2/g; oil absorp. 27-35 cc/100 g; brightness (TAPPI) > 94; ref. index 1.58; pH 9 ± 1; conduct. < 100 μS/cm; 99.6% act.

Martinal® OL-107. [Martinswerk GmbH] Aluminum hydroxide; CAS 21645-51-2; EINECS 244-492-7; filler, flame retardant for polyester resins (SMC, BMC, laminates), epoxy and acrylic resins, crosslinked elastomers (conveyor belts, cables), thermoplastics and thermoplastic elastomers (PP, PE and copolymers, EVA), PVC (plasticized, rigid and foam), aq. disps.; powd.; 0.9-1.3 μm median particle size; dens. 2.4 g/cc; bulk dens. ≈ 200 kg/m^3; surf. area 6-8 m^2/g; oil absorp. 34-44 cc/100 g; brightness (TAPPI) > 95; ref. index 1.58; pH 9 ± 1; conduct. < 200 μS/cm; 99.6% act.

Martinal® OL-107/C. [Martinswerk GmbH] Aluminum hydroxide; CAS 21645-51-2; EINECS 244-492-7; filler, flame retardant for thermoplastics and thermoplastic elastomers (PP, PE and copolymers, EVA), PVC (plasticized, rigid, and foam), crosslinked elastomer cable; powd.; 0.9-1.3 μm median particle size; dens. 2.4 g/cc; bulk dens. ≈ 300 kg/m^3; surf. area 6-8 m^2/g; oil absorp. 30-39 cc/100 g; brightness (TAPPI) > 95; ref. index 1.58; pH 9 ± 1; conduct. < 200 μS/cm; 99.6% act.

Martinal® OL-107 LE. [Martinswerk GmbH] Aluminum hydroxide; CAS 21645-51-2; EINECS 244-492-7; finely precipitated filler, flame retardant; esp. suitable for demanding elec. applics., crosslinked elastomer cables, thermoplastics and thermoplastic elastomers (PP, PE and copolymers, EVA); powd.; 0.9-1.3 μm median particle size; dens. 2.4 g/cc; bulk dens. ≈ 200 kg/m^3; surf. area 6-8 m^2/g; oil absorp. 40-55 cc/100 g; brightness (TAPPI) > 95; ref. index 1.58; pH 9 ± 1; conduct. < 150 μS/cm; 99.6% act.

Martinal® OL-111 LE. [Martinswerk GmbH] Aluminum hydroxide; CAS 21645-51-2; EINECS 244-492-7; finely precipitated filler, flame retardant esp. suitable for demanding elec. applics.; also for crosslinked elastomer cables and conveyor belts, EVA elastomers, aq. dispersions, flexible PU foam; powd.; 0.7-0.9 μm median particle size; dens. 2.4 g/cc; bulk dens. ≈ 160 kg/m^3; surf. area 10-12 m^2/g; oil absorp. 45-60 cc/100 g; brightness (TAPPI) > 95.5; ref. index 1.58; pH 9 ± 1; conduct. < 200 μS/cm;

99.6% act.

Martinal® OL/Q-107. [Martinswerk GmbH] Aluminum hydroxide; CAS 21645-51-2; EINECS 244-492-7; filler, flame retardant for crosslinked elastomer cables and conveyor belts, aq. disps., rigid PU foam; powd.; 0.9-1.3 µm median particle size; dens. 2.4 g/cc; bulk dens. ≈ 220 kg/m^3; surf. area 6-8 m^2/g; oil absorp. 40-55 cc/100 g; brightness (TAPPI) > 95; ref. index 1.58; pH 9 ± 1; conduct. < 800 µS/cm; 99.6% act.

Martinal® OL/Q-111. [Martinswerk GmbH] Aluminum hydroxide; CAS 21645-51-2; EINECS 244-492-7; filler, flame retardant for crosslinked elastomer cables and conveyor belts, EVA elastomers, aq. disps.; powd.; 0.7-0.9 µm median particle size; dens. 2.4 g/cc; bulk dens. ≈ 160 kg/m^3; surf. area 10-12 m^2/g; oil absorp. 52-64 cc/100 g; brightness (TAPPI) > 95.5; ref. index 1.58; pH 9 ± 1; conduct. < 1000 µS/cm; 99.6% act.

Martinal® ON. [Martinswerk GmbH] Aluminum hydroxide; CAS 21645-51-2; EINECS 244-492-7; filler, flame retardant for thermoset plastics, polyester (concrete, artificial marble), epoxy casting resins, crosslinked elastomer cables, carpetbacking latexes, urea/phenolic chipboard; powd.; 50-70 µm median particle size; dens. 2.4 g/cc; bulk dens. ≈ 1100 kg/m^3; brightness (TAPPI) ≈ 85; ref. index 1.58; pH 9 ± 1; conduct. ≈ 80 µS/cm.

Martinal® ON-310. [Martinswerk GmbH] Aluminum hydroxide; CAS 21645-51-2; EINECS 244-492-7; filler, flame retardant for polyester resins (SMC, BMC, laminates, artificial marble), epoxy, acrylic, crosslinked elastomer (conveyor belts, profiles, molded parts, flooring), PVC (plasticized, rigid foam), latexes (carpetbacking, wallpaper), PU (flexible and rigid foam, castings, lacquer, elastomers), aq. disps.; powd.; 9-13 µm median particle size; dens. 2.4 g/cc; bulk dens. ≈ 640 kg/m^3; oil absorp. 24-28 cc/100 g; brightness (TAPPI) ≈ 93; ref. index 1.58; pH 9 ± 1; conduct. < 140 µS/cm; 99.6% act.

Martinal® ON-313. [Martinswerk GmbH] Aluminum hydroxide; CAS 21645-51-2; EINECS 244-492-7; filler, flame retardant for polyester resins (SMC, BMC, laminates, artificial marble), epoxy, acrylic, crosslinked elastomers (conveyor belts, cables, flooring), PU (flexible and rigid foam, castings), latexes (carpetbacking, wallpaper), aq. disps.; powd.; 11-15 µm median particle size; dens. 2.4 g/cc; bulk dens. ≈ 720 kg/m^3; oil absorp. 22-24 cc/100 g; brightness (TAPPI) ≈ 92; ref. index 1.58; pH 9 ± 1; conduct. < 130 µS/cm; 99.6% act.

Martinal® ON-320. [Martinswerk GmbH] Aluminum hydroxide; CAS 21645-51-2; EINECS 244-492-7; filler, flame retardant for polyester resins (SMC, BMC, laminates, artificial marble), epoxy, acrylic, PU (flexible and rigid foam, castings), latexes (carpetbacking), crosslinked elastomers (cables, flooring), aq. disps.; powd.; 15-25 µm median particle size; dens. 2.4 g/cc; bulk dens. ≈ 900 kg/m^3; oil absorp. 18-22 cc/100 g; brightness (TAPPI) ≈ 90; ref. index 1.58; pH 9 ± 1; conduct. < 120 µS/cm; 99.6% act.

Martinal® ON-920/V. [Martinswerk GmbH] Aluminum hydroxide; CAS 21645-51-2; EINECS 244-492-7; filler, flame retardant for polyester SMC and BMC applics.; powd.; 10-20 µm median particle size; dens. 2.4 g/cc; bulk dens. ≈ 600 kg/m^3; oil absorp. 8-13 cc/100 g; brightness (TAPPI) ≈ 91; ref. index 1.58; pH 9 ± 1; 99.6% act.

Martinal® ON-4608. [Martinswerk GmbH] Aluminum hydroxide; CAS 21645-51-2; EINECS 244-492-7; filler, flame retardant for thermoset plastics, polyester (SMC, BMC, laminates), epoxy, acrylic, crosslinked elastomer conveyor belts, profiles, molded parts, flooring, PVC (plasticized, rigid foam), latexes (carpetbacking, wallpaper), aq. disps.; powd.; 7-11 µm median particle size; dens. 2.4 g/cc; bulk dens. ≈ 550 kg/m^3; oil absorp. 25-29 cc/100 g; brightness (TAPPI) ≈ 94; ref. index 1.58; pH 9 ± 1; conduct. < 150 µS/cm.

Martinal® OS. [Martinswerk GmbH] Aluminum hydroxide; CAS 21645-51-2; EINECS 244-492-7; filler, flame retardant for use where coarse fillers are needed, e.g., artificial marble, polymer concrete, resin flooring, chipboard; powd.; 40-60 µm median particle size; dens. 2.4 g/cc; bulk dens. ≈ 950 kg/m^3; brightness (TAPPI) ≈ 86; ref. index 1.58; pH 9 ± 1; conduct. ≈ 40 µS/cm.

Martinal® OX. [Martinswerk GmbH] Aluminum hydroxide; CAS 21645-51-2; EINECS 244-492-7; filler, flame retardant for thermoset plastics (polyester artificial marble), carpetbacking latexes; powd.; 70-100 µm median particle size; dens. 2.4 g/cc; bulk dens. ≈ 1000 kg/m^3; brightness (TAPPI) ≈ 82; ref. index 1.58; pH 9 ± 1; conduct. ≈ 60 µS/cm.

Masil® 1066C. [PPG/Specialty Chem.] Dimethicone copolyol; provides lubricity and soft feel to cosmetic creams and lotions, hair conditioners; resin plasticizer for hair sprays and mousses; lubricant and antistat for plastics, textiles, metal processing; wetting and leveling char.; antifog for glass cleaners; liq.; water-sol.; sp.gr. 1.02; visc. 1800 cSt; cloud pt. 42 C (1% aq.); flash pt. (PMCC) > 300 F; pour pt. -50 C; surf. tens. 25.9 dynes/cm (1% aq.).

Masil® 1066D. [PPG/Specialty Chem.] Dimethicone copolyol; provides lubricity and soft feel to cosmetic creams and lotions, hair conditioners; resin plasticizer for hair sprays and mousses; lubricant and antistat for plastics, textiles, and metal processing; wetting props.; liq.; water-sol.; sp.gr. 1.03; visc. 1050 cSt; cloud pt. 39 C (1% aq.); flash pt. (PMCC) > 300 F; pour pt. -33 C; surf. tens. 25.2 dynes/cm (1% aq.).

Masil® 2132. [PPG/Specialty Chem.] Silicone glycol; antistat and wetting agent for personal care prods., textile, plastics, lubricants and formulations; solv. based coating, dispersant, and antifoam; liq.; sp.gr. 1.03; visc. 2000 cSt; cloud pt. 38 C (1% aq.); flash pt. (PMCC) > 300 F; pour pt. -40 C; surf. tens. 26.5 dynes/cm (1% aq.); 100% conc.

Masil® 2133. [PPG/Specialty Chem.] Silicone glycol; antistat and wetting agent for personal care prods., textile, plastics, lubricants and formulations; solv.

based coating, dispersant, and antifoam; liq.; sp.gr. 1.03; visc. 1050 cSt; cloud pt. 37 C (1% aq.); flash pt. (PMCC) > 300 F; pour pt. -33 C; surf. tens. 25.8 dynes/cm (1% aq.).

Masil® 2134. [PPG/Specialty Chem.] Silicone glycol; antistat and wetting agent for personal care prods., textile, plastics, lubricants and formulations; solv. based coating, dispersant, and antifoam; liq.; sp.gr. 1.02; visc. 1800 cSt; cloud pt. 39 C (1% aq.); flash pt. (PMCC) > 300 F; pour pt. -50 C; surf. tens. 32.0 dynes/cm (1% aq.).

Masil® EM 100. [PPG/Specialty Chem.] Dimethylpolysiloxane fluids aq. emulsion; nonionic; emulsifier, release aid in molding, extrusion, laminating, and casting for rubber, plastics, and metals; for leather, glass, and vinyl cleaners, polishes, textile softeners, textile/fiber lubricants; recommended for glass hard surf. cleaner applics., and imparts nonsmearing, low gloss, and ease of wipe properties to such formulations; emulsion; sp.gr. 0.99; visc. 100 cSt; 35% act.

Masil® EM 100 Conc. [PPG/Specialty Chem.] Dimethylpolysiloxane fluids aq. emulsion; nonionic; emulsifier, release aid in molding, extrusion, laminating, and casting for rubber, plastics, and metals; for leather, glass, and vinyl cleaners, polishes, textile softeners, textile/fiber lubricants; emulsion; sp.gr. 0.99; visc. 100 cSt; 60% act.

Masil® EM 100D. [PPG/Specialty Chem.] Dimethylpolysiloxane fluids aq. emulsion; nonionic; emulsifier, release aid in molding, extrusion, laminating, and casting for rubber, plastics, and metals; for leather, glass, and vinyl cleaners, polishes, textile softeners, textile/fiber lubricants; emulsion; sp.gr. 0.99; visc. 100 cSt; 30% act.

Masil® EM 100P. [PPG/Specialty Chem.] Dimethylpolysiloxane fluids aq. emulsions; nonionic; emulsifier, release aid in molding, extrusion, laminating, and casting for rubber, plastics, and metals; for leather, glass, and vinyl cleaners, polishes, textile softeners, textile/fiber lubricants; used in automotive and furniture polish formulations; quick breaking emulsion, ease of rub-out and uniform deposition of silicone; emulsion; disp. in water; sp.gr. 0.99; visc. 100 cSt; 60% act.

Masil® EM 250 Conc. [PPG/Specialty Chem.] Dimethylpolysiloxane fluids aq. emulsion; nonionic; emulsifier, release aid in molding, extrusion, laminating, and casting for rubber, plastics, and metals; for leather, glass, and vinyl cleaners, polishes, textile softeners, textile/fiber lubricants; recommended for glass hard surf. cleaners imparting nonsmearing, low gloss, and ease of wipe properties to such formulations; emulsion; sp.gr. 0.99; visc. 250 cSt; 60% act.

Masil® EM 266 (50). [PPG/Specialty Chem.] Mixed alkyl methyl polysiloxane; emulsifier, release aid in molding, extrusion, laminating, and casting for rubber, plastics, and metals; for leather, glass, and vinyl cleaners, polishes, textile softeners, textile/fiber lubricants; milky wh. emulsion; sp.gr. 0.95; visc. 100 cSt; flash pt. (PMCC) none; 50% act. silicone.

Masil® EM 350X. [PPG/Specialty Chem.] Dimethylpolysiloxane fluids aq. emulsion; nonionic; emulsifier, release aid in molding, extrusion, laminating, and casting for rubber, plastics, and metals; for leather, glass, and vinyl cleaners, polishes, textile softeners, textile/fiber lubricants; emulsion; sp.gr. 0.99; visc. 350 cSt; 35% act.

Masil® EM 350X Conc. [PPG/Specialty Chem.] Dimethylpolysiloxane fluids aq. emulsion; nonionic; emulsifier, release aid in molding, extrusion, laminating, and casting for rubber, plastics, and metals; for leather, glass, and vinyl cleaners, polishes, textile softeners, textile/fiber lubricants; recommended for glass, leather, and vinyl cleaners, in textile softening applics.; printing release agent minimizing ink smearing and scuffing; emulsion; sp.gr. 0.99; visc. 350 cSt; 60% act.

Masil® EM 1000. [PPG/Specialty Chem.] Dimethylpolysiloxane fluids aq. emulsion; nonionic; emulsifier, release aid in molding, extrusion, laminating, and casting for rubber, plastics, and metals; for leather, glass, and vinyl cleaners, polishes, textile softeners, textile/fiber lubricants; emulsion; sp.gr. 0.99; visc. 1000 cSt; 35% act.

Masil® EM 1000 Conc. [PPG/Specialty Chem.] Dimethylpolysiloxane fluids aq. emulsion; nonionic; emulsifier, release aid in molding, extrusion, laminating, and casting for rubber, plastics, and metals; for leather, glass, and vinyl cleaners, polishes, textile softeners, textile/fiber lubricants; emulsion; sp.gr. 0.99; visc. 1000 cSt; 60% act.

Masil® EM 1000P. [PPG/Specialty Chem.] Dimethylpolysiloxane fluids aq. emulsion; nonionic; emulsifier, release aid in molding, extrusion, laminating, and casting for rubber, plastics, and metals; for leather, glass, and vinyl cleaners, polishes, textile softeners, textile/fiber lubricants; emulsion; sp.gr. 0.99; visc. 1000 cSt; 60% act.

Masil® EM 10,000. [PPG/Specialty Chem.] Dimethylpolysiloxane fluids aq. emulsion; emulsifier, release aid in molding, extrusion, laminating, and casting for rubber, plastics, and metals; for leather, glass, and vinyl cleaners, polishes, textile softeners, textile/fiber lubricants; emulsion; sp.gr. 1.09; visc. 10,000 cSt; 35% act.

Masil® EM 10,000 Conc. [PPG/Specialty Chem.] Dimethylpolysiloxane fluids aq. emulsion; nonionic; emulsifier, release aid in molding, extrusion, laminating, and casting for rubber, plastics, and metals; for leather, glass, and vinyl cleaners, polishes, textile softeners, textile/fiber lubricants; emulsion; sp.gr. 0.99; visc. 10,000 cSt; 60% act.

Masil® EM 60,000. [PPG/Specialty Chem.] Dimethylpolysiloxane fluids aq. emulsion; nonionic; emulsifier, release aid in molding, extrusion, laminating, and casting for rubber, plastics, and metals; for leather, glass, and vinyl cleaners, polishes, textile softeners, textile/fiber lubricants; emulsion; sp.gr. 0.99; visc. 60,000 cSt; 35% act.

Masil® EM 100,000. [PPG/Specialty Chem.] Dimethylpolysiloxane fluids aq. emulsion; nonionic; emulsifier, release aid in molding, extrusion, laminating, and casting for rubber, plastics, and metals; for leather, glass, and vinyl cleaners, polishes, textile softeners, textile/fiber lubricants; emulsion; sp.gr. 1.00; visc. 100,000 cSt; 35% act.

Masil® SF 5. [PPG/Specialty Chem.] Dimethicone; release aid, defoamer for nonaq. processes, esp. in the petrol., foods, and printing inks industries; internal lubricant for plastics, rubber, and metal; also in furniture and auto-wax polishes, household and personal care prods.; textile lubricant; lower visc. fluids recommended as emollients and antiwhitening agents for cosmetic applications (antiperspirants, creams and lotions); water-wh. oily, clear liq.; odorless, tasteless; sp.gr. 0.916; visc. 5 cSt; pour pt. -65 C; flash pt. (PMCC) 280 F; ref. index 1.3970.

Masil® SF 10. [PPG/Specialty Chem.] Dimethicone; release aid, defoamer for nonaq. processes, esp. in the petrol., foods, and printing inks industries; internal lubricant for plastics, rubber, and metal; also in furniture and auto-wax polishes, household and personal care prods.; lower visc. fluids recommended as emollients and antiwhitening agents for cosmetic applications (antiperspirants, creams and lotions); sp.gr. 0.940; pour pt. -65 C; flash pt. (PMCC) 320 F; ref. index 1.3990.

Masil® SF 20. [PPG/Specialty Chem.] Dimethicone; release aid, defoamer for nonaq. processes, esp. in the petrol., foods, and printing inks industries; internal lubricant for plastics, rubber, and metal; also in furniture and auto-wax polishes, household and personal care prods.; lower visc. fluids recommended as emollients and antiwhitening agents for cosmetic applications (antiperspirants, creams and lotions); water-wh. oily, clear liq.; odorless, tasteless; sp.gr. 0.953; visc. 20 cSt; pour pt. -65 C; flash pt. (PMCC) 395 F; ref. index 1.4010.

Masil® SF 50. [PPG/Specialty Chem.] Dimethicone; release aid, defoamer for nonaq. processes, esp. in the petrol., foods, and printing inks industries; internal lubricant for plastics, rubber, and metal; also in furniture and auto-wax polishes, household and personal care prods.; elegant, nongreasy emollient for skin care props.; provides nontacky water barrier props. to eyeliners and eyeshadows; water-wh. oily, clear liq.; odorless, tasteless; sp.gr. 0.963; visc. 50 cps; pour pt. -55 C; flash pt. (PMCC) 460 F; ref. index 1.4020.

Masil® SF 100. [PPG/Specialty Chem.] Dimethicone; release aid, defoamer for nonaq. processes, esp. in the petrol., foods, and printing inks industries; internal lubricant for plastics, rubber, and metal; also in furniture and auto-wax polishes, household and personal care prods.; elegant, nongreasy emollient for skin care props.; provides nontacky water barrier props. to eyeliners and eyeshadows; water-wh. oily, clear liq.; odorless, tasteless; sp.gr. 0.968; visc. 100 cps; pour pt. -55 C; flash pt. (PMCC) 461 C; ref. index 1.4030.

Masil® SF 200. [PPG/Specialty Chem.] Dimethicone; release aid, defoamer for nonaq. processes, esp. in the petrol., foods, and printing inks industries; internal lubricant for plastics, rubber, and metal; also in furniture and auto-wax polishes, household and personal care prods.; water-wh. oily, clear liq.; odorless, tasteless; sp.gr. 0.972; visc. 200 cSt; pour pt. -50 C; flash pt. (PMCC) 460 F; ref. index 1.4031.

Masil® SF 350. [PPG/Specialty Chem.] Dimethicone; release aid, defoamer for nonaq. processes, esp. in the petrol., foods, and printing inks industries; internal lubricant for plastics, rubber, and metal; also in furniture and auto-wax polishes, household and personal care prods.; elegant, nongreasy emollient for skin care props.; provides nontacky water barrier props. to eyeliners and eyeshadows; water-wh. oily, clear liq.; odorless, tasteless; sp.gr. 0.973; visc. 350 cSt; pour pt. -50 C; flash pt. (PMCC) 500 F; ref. index 1.4032.

Masil® SF 350 FG. [PPG/Specialty Chem.] Dimethicone; CAS 63148-62-9; release aid, defoamer for nonaq. processes, esp. in the petrol., foods, and printing inks industries; internal lubricant for rubber, plastics, fibers, rubber, and metal; also in furniture and auto-wax polishes, household and personal care prods.; colorless clear liq., bland odor; misc. in aliphatic/aromatic/halogenated solvs.; insol. in water, lower alcohols; sp.gr. 0.966-0.972; visc. 332.5-367.5 cst; b.p. > 300 F; pour pt. -50 C; flash pt. (PMCC) 500 F; ref. index 1.4025-1.4045; *Toxicology:* temporary eye irritant; nonirritating to skin; *Storage:* store in well-ventilated area below 120 F.

Masil® SF 500. [PPG/Specialty Chem.] Dimethicone; release aid, defoamer for nonaq. processes, esp. in the petrol., foods, and printing inks industries; internal lubricant for plastics, rubber, and metal; also in furniture and auto-wax polishes, household and personal care prods.; elegant, nongreasy emollient for skin care props.; provides nontacky water barrier props. to eyeliners and eyeshadows; water-wh. oily, clear liq.; odorless, tasteless; sp.gr. 0.973; visc. 500 cSt; pour pt. -50 C; flash pt. (PMCC) 500 F; ref. index 1.4033.

Masil® SF 1000. [PPG/Specialty Chem.] Dimethicone; release aid, defoamer for nonaq. processes, esp. in the petrol., foods, and printing inks industries; internal lubricant for plastics, rubber, and metal; also in furniture and auto-wax polishes, household and personal care prods.; nonoily gloss and lubricity aid for hair conditioners, skin protectants, water-barrier props. in creams, lotions, foundations, and blushers; water-wh. oily, clear liq.; odorless, tasteless; sp.gr. 0.974; visc. 1000 cSt; pour pt. -50 C; flash pt. (PMCC) 500 F; ref. index 1.4035; CC flash pt. 260 C.

Masil® SF 5000. [PPG/Specialty Chem.] Dimethicone; release aid, defoamer for nonaq. processes, esp. in the petrol., foods, and printing inks industries; internal lubricant for plastics, rubber, and metal; also in furniture and auto-wax polishes, household and personal care prods.; water-wh. oily, clear liq.; odorless, tasteless; sp.gr. 0.975; visc. 5000 cSt; pour pt. -49 C; flash pt. (PMCC) 500 F; ref. index 1.4035.

Masil® SF 10,000. [PPG/Specialty Chem.] Dimethicone; release aid, defoamer for nonaq. processes, esp. in the petrol., foods, and printing inks industries; internal lubricant for plastics, rubber, and metal; also in furniture and auto-wax polishes, household and personal care prods.; nonoily gloss and lubricity aid for hair conditioners, skin protectants, water-barrier props. in creams, lotions, foundations, and blushers; water-wh. oily, clear liq.; odorless, tasteless; sp.gr. 0.975; visc. 10,000 cSt; pour pt. -47 C; flash pt. (PMCC) 500 F; ref. index 1.4035.

Masil® SF 12,500. [PPG/Specialty Chem.] Dimethicone; release aid, defoamer for nonaq. processes, esp. in the petrol., foods, and printing inks industries; internal lubricant for plastics, rubber, and metal; also in furniture and auto-wax polishes, household and personal care prods.; skin protectant, water repellent, emollient, softener for cuticle care prods., hair care prods.; water-wh. oily, clear liq.; odorless, tasteless; sp.gr. 0.975; visc. 12,500 cSt; pour pt. -47 C; flash pt. (PMCC) 500 F; ref. index 1.4035.

Masil® SF 30,000. [PPG/Specialty Chem.] Dimethicone; release aid, defoamer for nonaq. processes, esp. in the petrol., foods, and printing inks industries; internal lubricant for plastics, rubber, and metal; also in furniture and auto-wax polishes, household and personal care prods.; skin protectant, water repellent, emollient, softener for cuticle care prods., hair care prods.; water-wh. oily, clear liq.; odorless, tasteless; sp.gr. 0.976; visc. 30,000 cSt; pour pt. -46 C; flash pt. (PMCC) 500 F; ref. index 1.4035.

Masil® SF 60,000. [PPG/Specialty Chem.] Dimethicone; release aid, defoamer for nonaq. processes, esp. in the petrol., foods, and printing inks industries; internal lubricant for plastics, rubber, and metal; also in furniture and auto-wax polishes, household and personal care prods.; skin protectant, water repellent, emollient, softener for cuticle care prods., hair care prods.; water-wh. oily, clear liq.; odorless, tasteless; sp.gr. 0.977; visc. 60,000 cSt; pour pt. -44 C; flash pt. (PMCC) 500 F; ref. index 1.4035.

Masil® SF 100,000. [PPG/Specialty Chem.] Dimethicone; release aid, defoamer for nonaq. processes, esp. in the petrol., foods, and printing inks industries; internal lubricant for plastics, rubber, and metal; also in furniture and auto-wax polishes, household and personal care prods.; water-wh. oily, clear liq.; odorless, tasteless; sp.gr. 0.978; visc. 100,000 cSt; pour pt. -40 C; flash pt. (PMCC) 500 F; ref. index 1.4035.

Masil® SF 300,000. [PPG/Specialty Chem.] Dimethicone; release aid, defoamer for nonaq. processes, esp. in the petrol., foods, and printing inks industries; internal lubricant for plastics, rubber, and metal; also in furniture and auto-wax polishes, household and personal care prods.; water-wh. oily, clear liq.; odorless, tasteless; sp.gr. 0.978; visc. 300,000 cSt; pour pt. -40 C; flash pt. (PMCC) 500 F; ref. index 1.4035.

Masil® SF 500,000. [PPG/Specialty Chem.] Dimethicone; release aid, defoamer for nonaq. processes, esp. in the petrol., foods, and printing inks industries; internal lubricant for plastics, rubber, and metal; also in furniture and auto-wax polishes, household and personal care prods.; sp.gr. 0.978; pour pt. -40 C; flash pt. (PMCC) 500 F; ref. index 1.4035.

Masil® SF 600,000. [PPG/Specialty Chem.] Dimethicone; release aid, defoamer for nonaq. processes, esp. in the petrol., foods, and printing inks industries; internal lubricant for plastics, rubber, and metal; also in furniture and auto-wax polishes, household and personal care prods.; water-wh. oily, clear liq.; odorless, tasteless; sp.gr. 0.979; visc. 600,000 cSt; pour pt. -34 C; flash pt. (PMCC) 500 F; ref. index 1.4035.

Masil® SF 1,000,000. [PPG/Specialty Chem.] Dimethicone; release aid, defoamer for nonaq. processes, esp. in the petrol., foods, and printing inks industries; internal lubricant for plastics, rubber, and metal; also in furniture and auto-wax polishes, household and personal care prods.; sp.gr. 0.979; pour pt. -25 C; flash pt. (PMCC) 500 F; ref. index 1.4035.

Masil® SFR 70. [PPG/Specialty Chem.] Dimethiconol; CAS 31692-79-2; spreading agent and emollient in antiperspirant formulations; offers humidity resist. to hair spray formulas; reactive fluid; raw material in compding. silicone RTV systems, textile and paper coatings, plasticizer/processing aid for silicone elastomers, hydrophobizing silica, in water repellent formulations; water-wh. clear liq., odorless, tasteless; visc. 55-90 cst; 92.5% min. solids.

Mazol® SFR 100. [PPG/Specialty Chem.] Dimethiconol; CAS 31692-79-2; spreading agent and emollient in antiperspirant formulations; offers humidity resist. to hair spray formulas; plasticizer for silicone elastomers; hydrophobizing agent in treating silica; in water repllents, textile and paper coatings; raw material for RTV silicone systems; water-wh. clear liq., odorless, tasteless; visc. 90-150 cst; acid no. 0.05 max.

Mazol® SFR 750. [PPG/Specialty Chem.] Dimethiconol; CAS 31692-79-2; uniform spreading agent, softener, and nongreasy emollient for skin and nail care prods.; plasticizer for silicone elastomers; hydrophobizing agent in treating silica; in water repllents, textile and paper coatings; raw material for RTV silicone systems; water-wh. clear liq., odorless, tasteless; visc. 675-825 cps.

Mastertek® ABS 3710. [Campine Am.] Masterbatch of antimony oxide and halogenated flame retardants; flame retardant masterbatch for applics. requiring UL94V-0 classifications; max. processing temp. 230 C; wh.; sp.gr. 1.9; bulk dens. 1.0 g/cc; 80% act.

Mastertek® PA 3664. [Campine Am.] Masterbatch of antimony oxide and brominated flame retardant in polyamide 6; flame retardant masterbatch for polyamide 6 and 6/6 formulations with UL94V-0 classifications; max. processing temp. 300 C; wh.; sp.gr. 2.5; bulk dens. 1.3 g/cc; 80% act.

Mastertek® PBT 3659. [Campine Am.] Masterbatch of antimony oxide and oligomeric brominated flame retardant in PBT carrier resin; flame retardant masterbatch for PBT and glass fiber-reinforced PBT; max. processing temp. 320 C; wh.; sp.gr. 2.05; 80% act.

Mastertek® PC 3686. [Campine Am.] Masterbatch of antimony oxide and brominated flame retardant in PC carrier resin; flame retardant masterbatch for PC; max. processing temp. 300 C; wh.; sp.gr. 1.8; bulk dens. 1.0 g/cc; 80% act.

Mastertek® PE 3660. [Campine Am.] Masterbatch of antimony oxide and brominated flame retardant in LDPE carrier resin; flame retardant masterbatch for crosslinked LDPE formulations with UL94V-0 classification; max. processing temp. 280 C; ylsh.; melt flow 0.27 g/10 min; sp.gr. 1.98; bulk dens. 1.20 g/cc; 75% act.

Mastertek® PE 3668. [Campine Am.] Masterbatch of antimony oxide and brominated flame retardant in carrier resin; flame retardant masterbatch for LDPE films used in the contstruction industry or as a drape for equipment; max. processing temp. 280 C; wh.; melt flow 6 g/10 min; sp.gr. 1.5; bulk dens. 0.8 g/cc; 50% act.

Mastertek® PP 3658. [Campine Am.] Masterbatch of antimony oxide and brominated flame retardant in PP carrier resin; flame retardant masterbatch for PP homopolymer and copolymer formulations with UL94V-0 classifications; max. processing temp. 280 C; wh.; melt flow 1.1 g/10 min; sp.gr. 2.2; bulk dens. 1.1 g/cc; 80% act.

Mastertek® PP 5601. [Campine Am.] Masterbatch of antimony oxide and chlorinated flame retardant in PP carrier resin; flame retardant masterbatch for PP homopolymer and copolymer formulations with UL94V-2 classifications; max. processing temp. 210 C; wh.; melt flow > 1000 g/10 min; sp.gr. 1.57; 80% act.

Mastertek® PP 5604. [Campine Am.] Masterbatch of antimony oxide and brominated flame retardant in PP carrier resin; flame retardant masterbatch for PP homopolymer and copolymer formulations with UL94V-2 classifications; max. processing temp. 290 C; ylsh.; melt flow 7 g/10 min; sp.gr. 1.785; 75% act.

Mastertek® PS 3698. [Campine Am.] Masterbatch of antimony oxide and brominated flame retardant in carrier resin; flame retardant masterbatch for HIPS requiring UL94V-2 or VDE 0471 (glow wire) certification; max. processing temp. 240 C; wh.; melt flow > 15 g/10 min; sp.gr. 1.2; 30% act.

Mathe 6T. [Norac] Aluminum stearate; CAS 7047-84-9; EINECS 230-325-5; internal/external lubricant for nylon, rigid PVC, polyesters, color concs.; FDA approved; solid.

Mathe Barium Stearate. [Norac] Barium stearate; CAS 6865-35-6; heat stabilizer, internal/external lubricant for flexible and rigid vinyl and records, PVC rigid foams, PS foams; powd.

Mathe Calcium Stearate. [Norac] Calcium stearate; CAS 1592-23-0; EINECS 216-472-8; internal/external lubricant for ABS, PS, polyolefins, rigid and flexible PVC, SMC/TMC/BMC polyesters, cellulosics, DAP, epoxy, alkyd, phenolics, polyesters, color concs.; FDA accepted; solid.

Mathe Lithium Stearate. [Norac] Lithium stearate; CAS 4485-12-5; EINECS 224-772-5; internal/external lubricant for nylon; solid.

Mathe Magnesium Stearate. [Norac] Magnesium stearate; CAS 557-04-0; EINECS 209-150-3; internal/external lubricant for ABS, rigid vinyls (with calcium stearate), cellulose acetate; solid.

Mathe Sodium Stearate. [Norac] Sodium stearate; CAS 822-16-2; EINECS 212-490-5; internal/external lubricant for nylon, PP, PS, rigid vinyls (with calcium stearate); solid.

Mathe Zinc Stearate 25S. [Norac] Zinc stearate; CAS 557-05-1; EINECS 209-151-9; internal/external lubricant for polyolefins, ABS, PS, IPS, rigid/flexible PVC, SMC/TMC/BMC polyesters, phenolics, melamine, alkyd, U-F, polyesters, color concs.; good clarity; FDA accepted; solid.

Mathe Zinc Stearate S. [Norac] Zinc stearate; CAS 557-05-1; EINECS 209-151-9; internal/external lubricant for polyolefins, ABS, PS, IPS, rigid/flexible PVC, SMC/TMC/BMC polyesters, phenolics, melamine, alkyd, U-F, polyesters, color concs.; FDA accepted; solid.

Mayzo RA-0010A, RA-10B, RA-10C. [Mayzo] Polyurethane release coat in toluene; release coating for tape printing, pressure-sensitive labels, film, tapes such as PP, PE, polyester; clear liq., aromatic odor; flash pt. < 78 F; 1, 3, and 10% solids resp.; *Precaution:* highly flamm.

Mayzo RA-1315W. [Mayzo] Fluoroalkyl acrylate copolymer aq. emulsion with acetic acid, acetone, ethylene glycol, and emulsifier; mildly cationic; release coating for film tapes incl. pressure-sensitive tapes of PP, polyester, PE, cellophane, foil, paper; applied to non-adhesive side of the tape; prevents delamination or tearing; pale yel. emulsion; water-disp.; sp.gr. 1.05; pH 2-4; 15% act.; *Precaution:* contains flamm. solv.; *Storage:* 3 yr storage in sealed containers stored in cool area.

Mayzo RA-0095H. [Mayzo] Polyvinyl octadecyl carbamate; CAS 70892-21-6; release coating for pressure-sensitive labels, film, tapes such as PP, PE, polyester, cellophane, duct tapes, foil, paper; prevents delamination or tearing; sl. yel.-wh. dry powd.; insol. in water; m.w. 110,000-135,000; m.p. 87-93 C; cloud pt. < 30 C (5% in toluene); 99.5% min. assay; *Storage:* 2 yr storage in sealed containers stored in cool, dry area.

Mayzo RA-0095HS. [Mayzo] Polyvinyl octadecyl carbamate; CAS 70892-21-6; aging resist. release coating for pressure-sensitive and hot-melt pressure-sensitive tapes of PP, polyester, PE, cellophane, duct tapes, foil, paper; prevents delamination and tearing; wh. dry powd.; insol. in water; m.w. 420,000; m.p. 102-120 C; cloud pt. < 25 C (5% in toluene); 99.5% min. assay; *Storage:* 2 yr storage in sealed containers stored in cool, dry area.

Mazawet® 77. [PPG/Specialty Chem.] Alkyl polyoxyalkylene ether; nonionic; wetting agent, surfactant used in metalworking, textile processing, spray-dried detergents, emulsion polymerization, dry-cleaning systems, paints, inks, hard surface cleaners, deresination of sulfite pulp; Gardner 1 liq.; m.w. 426 avg.; sol. @ 5% in toluene, perchloroethylene, Stod., propylene glycol, water; sp.gr. 0.965; pour pt. 11 C; cloud pt. 45 C (1% aq.); flash pt. (PMCC) 340 F; pH 6.5 (3% aq.); surf. tens. 28.1 dynes/cm (0.1%); Draves wetting 4 s (0.1% aq.); 100% act.

Mazawet® DF. [PPG/Specialty Chem.] Alkyl polyoxyalkylene ether; nonionic; low foam wetting surfactant for metalworking fluids, paint, printing inks, polishes, floor waxes, hard surface cleaning, and emulsion polymerization; liq.; sp.gr. 1.06; cloud pt. 60 C (1% aq.); surf. tens. 27.3 dynes/cm (0.1% aq.); Draves wetting 6 s (0.1% aq.); 100% act.

Mazeen® C-2. [PPG/Specialty Chem.] PEG-2 cocamine; CAS 61791-14-8; cationic; emulsifier, rewetting agent, lubricant, coupler used in insecticides and herbicides, grease additives, textile lubricants, water-based inks, cosmetics; plastics antistat; visc. modifier and rust inhibitor in acid media for metalworking; Gardner 10 liq.; sol. in benzene, acetone, IPA, min. oil, toluene, min. spirits, forms gel in water; m.w. 285; sp.gr. 0.874; flash pt. (PMCC) > 350 F; surf. tens. 28 dynes/cm (0.1%); 100% conc.

Mazeen® C-5. [PPG/Specialty Chem.] PEG-5 cocamine; CAS 61791-14-8; cationic; emulsifier, rewetting agent, lubricant, coupler used in insecticides and herbicides, grease additives, textile lubricants, water-based inks, cosmetics; plastics antistat; rust inhibitor in acid media for metalworking; Gardner 10 liq.; sol. in water, benzene, acetone, IPA, min. oil, toluene; disp. in min. spirits; m.w. 425; sp.gr. 0.976; flash pt. (PMCC) > 350 F; surf. tens. 33 dynes/cm (0.1%); 100% conc.

Mazeen® C-10. [PPG/Specialty Chem.] PEG-10 cocamine; CAS 61791-14-8; cationic; emulsifier, rewetting agent, lubricant, coupler used in insecticides and herbicides, grease additives, textile lubricants, water-based inks, cosmetics; plastics antistat; Gardner 11 liq.; sol. in water, benzene, acetone, IPA, toluene; m.w. 645; sp.gr. 1.017; flash pt. (PMCC) > 350 F; surf. tens. 39 dynes/cm (0.1%); 100% conc.

Mazeen® C-15. [PPG/Specialty Chem.] PEG-15 cocamine; CAS 61791-14-8; cationic; emulsifier, rewetting agent, lubricant, coupler used in metalworking, insecticides and herbicides, grease additives, textile lubricants, water-based inks, cosmetics; plastics antistat; Gardner 9 liq.; sol. in water, benzene, acetone, IPA, toluene; m.w. 860; sp.gr. 1.042; flash pt. (PMCC) > 350 F; surf. tens. 41 dynes/cm (0.1%); 100% conc.

Mazeen® S-2. [PPG/Specialty Chem.] PEG-2 soyamine; CAS 61791-24-0; cationic; emulsifier, rewetting agent, lubricant, coupler used in insecticides and herbicides, grease additives, textile lubricants, water-based inks, cosmetics; plastics antistat; Gardner 14 liq.; sol. in benzene, acetone, IPA, min. oil; disp. in min. spirits, toluene; m.w. 350; sp.gr. 0.911; flash pt. (PMCC) > 350 F; surf. tens. 26 dynes/cm (0.1%); 100% conc.

Mazeen® S-5. [PPG/Specialty Chem.] PEG-5 soyamine; CAS 61791-24-0; cationic; emulsifier, rewetting agent, lubricant, coupler used in insecticides and herbicides, grease additives, textile lubricants, water-based inks, cosmetics; plastics antistat; Gardner 14 liq.;sol. in benzene, IPA; partly sol. acetone, min. oil; m.w. 480; sp.gr. 0.951; flash pt. (PMCC) > 350 F; surf. tens. 33 dynes/cm (0.1%); 100% conc.

Mazeen® S-10. [PPG/Specialty Chem.] PEG-10 soyamine; CAS 61791-24-0; cationic; emulsifier, rewetting agent, lubricant, coupler used in insecticides and herbicides, grease additives, textile lubricants, water-based inks, cosmetics; plastics antistat; Gardner 14 liq.; sol. in water, benzene, acetone, IPA, toluene; m.w. 710; sp.gr. 1.020; flash pt. (PMCC) > 350 F; surf. tens. 40 dynes/cm (0.1%); 100% conc.

Mazeen® S-15. [PPG/Specialty Chem.] PEG-15 soyamine; CAS 61791-24-0; cationic; emulsifier, rewetting agent, lubricant, coupler used in insecticides and herbicides, grease additives, textile lubricants, water-based inks, cosmetics; plastics antistat; Gardner 18 liq.; sol. in water, benzene, acetone, IPA, toluene; m.w. 930; sp.gr. 1.040; flash pt. (PMCC) > 350 F; surf. tens. 43 dynes/cm (0.1%); 100% conc.

Mazeen® T-2. [PPG/Specialty Chem.] PEG-2 tallow amine; CAS 61791-44-4; cationic; emulsifier, rewetting agent, lubricant, coupler used in insecticides and herbicides, grease additives, textile lubricants, water-based inks, cosmetics, metalworking; plastics antistat; visc. modifier, rust inhibitor in acid media; Gardner 11 liq.; sol. in benzene, acetone, IPA, min. oil; m.w. 350; sp.gr. 0.916; flash pt. (PMCC) > 350 F; surf. tens. 29 dynes/cm (0.1%); 100% conc.

Mazeen® T-5. [PPG/Specialty Chem.] PEG-5 tallow amine; CAS 61791-44-4; cationic; emulsifier, rewetting agent, lubricant, coupler used in insecticides and herbicides, grease additives, textile lubricants, water-based inks, cosmetics, metalworking; plastics antistat; Gardner 12 liq.; sol. in benzene, acetone, IPA, min. oil; gels in water; m.w. 480; sp.gr. 0.966; flash pt. (PMCC) > 350 F; surf. tens. 34 dynes/cm (0.1%); 100% conc.

Mazeen® T-10. [PPG/Specialty Chem.] PEG-10 tallow amine; cationic; emulsifier, rewetting agent, lubricant, coupler used in insecticides and herbicides, grease additives, textile lubricants, water-based inks, cosmetics; plastics antistat.

Mazeen® T-15. [PPG/Specialty Chem.] PEG-15 tallow amine; cationic; emulsifier, rewetting agent, lubricant, coupler used in insecticides and herbicides, grease additives, textile lubricants, water-based inks, cosmetics; plastics antistat; Gardner

18 liq.; sol. in water, benzene, acetone, IPA; m.w. 925; sp.gr. 1.028; flash pt. (PMCC) > 350 F; surf. tens. 41 dynes/cm (0.1%); 100% conc.

Mazol® 159. [PPG/Specialty Chem.] PEG-7 glyceryl cocoate; emulsifier for food prods.; emollient and solubilizer used in cosmetics, toiletries, pharmaceuticals, cleansing prods., bath oils, shower gels, skin fresheners, lubricants, mold release compds.; plasticizer in syn. fabrics and plastics; amber liq.; sol. in water, min. oil; disp. in min. spirits, toluene, IPA; HLB 13.0; acid no. 5 max.; sapon. no. 82-98; flash pt. (PMCC) > 350 F.

Mazol® GMS-90. [PPG/Specialty Chem.] Glyceryl stearate; emulsifier for food prods., cosmetics, toiletries, pharmaceuticals, lubricants, mold release compds., in plasticizers for syn. fabrics and plastics; tan flake; sol. in IPA; sol. hot in min. oil, toluene; HLB 3.9; acid no. 2 max.

Mazol® GMS-D. [PPG/Specialty Chem.] Glyceryl stearate SE; emulsifier for food prods.; emollient; used in cosmetics, toiletries, pharmaceuticals, lubricants, mold release compds., metalworking applics.; plasticizer in syn. fabrics and plastics; tan flake; sol. in IPA; sol. hot in min. oil, toluene; disp. in water; HLB 6.0; acid no. 3.5 max.; sapon. no. 145-160; flash pt. (PMCC) > 350 F.

Mazol® PGO-104. [PPG/Specialty Chem.] Decaglyceryl tetraoleate; CAS 34424-98-1; EINECS 252-011-7; emulsifier for food prods., cosmetics, toiletries, pharmaceuticals, lubricants, mold release compds., in plasticizers for syn. fabrics and plastics; FDA 21CFR §172.854; kosher; dk. liq.; sol. in IPA, min. oil, veg. oil, toluene; disp. in ethanol, propylene glycol, min. spirits; insol. in water; HLB 6.2; acid no. 8 max.; iodine no. 61; sapon. no. 125-145; flash pt. (PMCC) > 350 F.

Mazon® 1045A. [PPG/Specialty Chem.] POE sorbitol fatty acid ester; nonionic; emulsifier for pesticide, herbicide, metalworking, die-cast lubricant formulations, and emulsion polymerization; humectant, emollient; liq.; water-sol.; sp.gr. 1.02; visc. 260 cps; HLB 13.0; sapon. no. 90; flash pt. (PMCC) > 350 F; 100% conc.

Mazon® 1086. [PPG/Specialty Chem.] POE sorbitol ester; nonionic; emulsifier, coupling agent for agric. pesticide and herbicide formulations, emulsion polymerization, metalworking lubricants, die-cast lubricants; liq.; water-sol.; sp.gr. 1.02; visc. 200 cps; HLB 10.4; sapon. no. 97; flash pt. (PMCC) > 350 F; 100% conc.

Mazon® 1096. [PPG/Specialty Chem.] POE sorbitol ester; nonionic; emulsifier, coupling agent for agric. pesticide and herbicide formulations, emulsion polymerization, metalworking lubricants, die-cast lubricants; liq.; water-sol.; sp.gr. 1.02; visc. 240 cps; HLB 11.2; sapon. no. 85; flash pt. (PMCC) > 350 F; 100% conc.

Mazu® DF 200SX. [PPG/Specialty Chem.] Silicone compd.; defoamer for industrial use; used in adhesives, solv.-based inks and paints, insecticides, resin polymerization, and petrol. industry; liq.; sp.gr. 0.99; dens. 8.3 lb/gal; visc. 2000 cSt; flash pt. (PMCC) > 350 F; 100% silicone.

Mazu® DF 200SXSP. [PPG/Specialty Chem.] Silicone compd.; defoamer for industrial use; used in adhesives, solv.-based inks and paints, insecticides, resin polymerization, and petrol. industry; sp.gr. 0.99; dens. 8.3 lb/gal; visc. 2500 cSt; flash pt. (PMCC) > 350 F; 100% silicone.

Mazu® DF 210SXSP. [PPG/Specialty Chem.] Silicone emulsion; industrial defoamer for highly acidic and alkaline systems, adhesives, antifreeze, brines, hot aq. systems, inks, insecticides, paints, resin polymerization, starch processing, petrol., latex binders and emulsions, textile, paper; sp.gr. 1.00; flash pt. (PMCC) none; pH 7.0; 10% silicone.

McLube 1700. [McLube] Solv.-based PTFE; CAS 9002-84-0; hard, dry, bonded lubricant coating for leather, plastics, elastomers for gaskets, packings; cord, twine, rope and cable from natural or syn. fibers; machine parts and fittings; instruments, office machines; metalworking; musical instruments; household items (hinges, locks, zippers, etc.).

McLube 1700L. [McLube] Solv.-based hard, dry, bonded release coating for epoxies and thermosets, and natural, nitrile, SBR and silicone rubber; for use on cool (< 200 F) surfaces.

McLube 1704. [McLube] Solv.-based PTFE; CAS 9002-84-0; hard, dry, bonded lubricant coating for leather, plastics, elastomers for gaskets, packings; cord, twine, rope and cable from natural or syn. fibers; machine parts and fittings; instruments, office machines; metalworking; musical instruments; household items (hinges, locks, zippers, etc.); exc. stick-slip, low coeff. of friction.

McLube 1711L. [McLube] Solv.-based hard, dry, bonded release coating for epoxies and thermosets, and natural, nitrile, SBR and silicone rubber; for use on hot (> 150 F) mold surfaces.

McLube 1725L. [McLube] Solv.-based protective coating for molds in storage; all-purpose, bonded coating esp. useful for PU, epichlorohydrin, EPDM, fluoroelatomers, and thermoplastic rubbers.

McLube 1733L. [McLube] Solv.-based flexible, dry film, bonded release coating providing exc. release from nonmetallic molds; increases life of RTV molds; useful with polyacrylics and urethane compds.

McLube 1775. [McLube] Solv.-based PTFE; CAS 9002-84-0; flexible, dry film, bonded lubricant/release/antistick coating exhibiting low friction and exc. release for rubber and plastics; esp. effective for hose mfg. with flexible and rigid mandrels; lubricant for sliding metal surfs. exposed to moisture, reactive chems., active solvs.

McLube 1777. [McLube] TFE polymer disp.; CAS 9002-84-0; release coating for hot molds used to form rubber and plastic parts; antistick coating for cure of hose and other extrusions and on tools and process equip.

McLube 1777-1. [McLube] Conc. aq. disp. of PTFE particles; CAS 9002-84-0; water-based dry, bonded release coating for hot molds used to form rubber and plastic parts; antistick coating for pan or open-steam cure of hose and other extrusions; antistick coating to reduce buildup of compds. on rolls, cutting, and forming tools, and other process equip.

McLube 1779-1. [McLube] Water-based release coating esp. designed for peroxide-cured fluoroelastomer and similar hard-to-release compds.

McLube 1782. [McLube] TFE polymer disp.; CAS 9002-84-0; mold release, antistick coating for silicone rubber.

McLube 1782-1. [McLube] Water-based release coating esp. useful for silicone rubber.

McLube 1800. [McLube] Fluorochemical; water-based dry, bonded mold release for hot surfs.; exceptional dry, nondusting, antitack coating for uncured elastomers; antistick coating for process equip. such as dry cans, rolls, blades, oven flights, conveyor belts, etc.

McLube 1804L. [McLube] Solv.-based clear, dry, release coating, esp. useful in release of hose from rigid steel and aluminum mandrels; suitable for use with food grade hose; useful for PU.

McLube 1829. [McLube] Fluorochemical/resin mixt.; release coating for rubber molding processes; antitack coating for uncured rubber.

McLube 1849. [McLube] Fluorochemical emulsion; water-based release coating effective with a wide variety of elastomers and plastics.

McLube 2000 Series. [McLube] Solv.-based PTFE; CAS 9002-84-0; lubricant coating meeting MIL-L-60326; for leather, plastics, elastomers for gaskets, packings; cord, twine, rope and cable from natural or syn. fibers; machine parts and fittings; instruments, office machines; metalworking; musical instruments; household items (hinges, locks, zippers, etc.).

McNamee® Clay. [R.T. Vanderbilt] Kaolin; CAS 1332-58-7; EINECS 296-473-8; soft extender and reinforcing filler for elastomers, polyester, styrene, epoxies, phenol formaldehyde resins; preferred over calcium carbonate giving substantially better flexural modulus, and improved flow chars.; wh. to cream powd., 1.2 μm median particle size, 99.7% min. thru 325 mesh; dens. 2.62 mg/m^3; bulk dens. 35 lb/ft^3 (compacted); oil absorp. 35; brightness 75; pH 4.8 (10% aq. slurry); 44.46% SiO_2, 39.84% Al_2O_3.

MDI AB-925. [Modern Disp.] 25% Carbon black in SAN/ABS carrier; black conc. for pipe, molding applics.

MDI AC-725. [Modern Disp.] 5% Antistatic/static dissipative conc. in LDPE carrier resin; antistat/static dissipative conc. for polyethylene and PP film; does not affect clarity or colorability; disperses easily in many systems; complies with MIL-81705B; effective at low r.h.; imparts 10^9-10^{10} ohm/sq resist.; free-flowing pellet.

MDI EC-940. [Modern Disp.] 40% Carbon black in LDPE carrier; black conc. for extrusion coating applics.

MDI HD-650. [Modern Disp.] 50% Titanium dioxide in HDPE carrier; wh. conc. for film applics., sheet extrusion.

MDI HD-935. [Modern Disp.] 35% Carbon black in HDPE carrier; black conc. for sheet extrusion; uv stable.

MDI MB-650. [Modern Disp.] 50% Titanium dioxide in EVA carrier; wh. conc. for universal applics.

MDI NY-905. [Modern Disp.] 25% Carbon black in nylon carrier; jet black conc. featuring high color.

MDI PC-925. [Modern Disp.] 25% Carbon black in PC carrier; jet black conc. featuring high color.

MDI PE-135. [Modern Disp.] 35% Carbon black in LDPE carrier; black conc. for FDA applics.; FDA compliance.

MDI PE-200UVA. [Modern Disp.] 10% HALS, 2% antioxidant in LDPE carrier; light stabilizer/antioxidant conc. for film applics.

MDI PE-500. [Modern Disp.] 50% Carbon black in LDPE carrier; black conc. for thin-gauge film applics.

MDI PE-500-20F. [Modern Disp.] 40% Carbon black, 15% filler in LDPE carrier; black conc. for noncritical film applics.

MDI PE-500A. [Modern Disp.] 50% Carbon black in LDPE carrier; black conc. for pipe, molding, compounding applics.

MDI PE-500HD. [Modern Disp.] 50% Carbon black in HDPE carrier; black conc. for HDPE film applics.

MDI PE-500LL. [Modern Disp.] 50% Carbon black in LLDPE carrier; black conc. for thin-gauge film applics.

MDI PE-540. [Modern Disp.] 40% Carbon black in LDPE carrier; black conc. for blown film applics.; darker color.

MDI PE-650F. [Modern Disp.] 50% Titanium dioxide in LDPE carrier; wh. conc. for film applics.

MDI PE-650HF. [Modern Disp.] 50% Titanium dioxide in LDPE carrier; wh. conc. for high-flow molding applics.

MDI PE-650LL. [Modern Disp.] 50% Titanium dioxide in LLDPE carrier; wh. conc. for film applics.

MDI PE-670T. [Modern Disp.] 70% Titanium dioxide in LDPE carrier; wh. conc. for film and compounding applics.

MDI PE-675. [Modern Disp.] 75% Titanium dioxide in LDPE carrier; wh. conc. for molding and extrusion applics.

MDI PE-800FR. [Modern Disp.] 80% Flame retardant in LDPE carrier; flame retardant conc. for molding and extrusion.

MDI PE-907. [Modern Disp.] 40% Carbon black in LDPE carrier; black conc. for molding, compounding applics.

MDI PE-907HD. [Modern Disp.] 40% Carbon black in HDPE carrier; black conc. for HDPE molding applics.

MDI PE-931WC. [Modern Disp.] 31% Carbon black in LDPE carrier; black conc. for wire and cable applics.

MDI PE-940. [Modern Disp.] 40% Carbon black in LDPE carrier; black conc. for imparting max. uv stability.

MDI PP-130. [Modern Disp.] 30% Carbon black in PP

carrier; black conc. for FDA applics.; FDA compliance.

MDI PP-535. [Modern Disp.] 35% Carbon black in PP carrier; black conc. for molding, compounding applics.

MDI PP-940. [Modern Disp.] 40% Carbon black in PP carrier; black conc. providing uv stability.

MDI PS-125. [Modern Disp.] 25% Carbon black in PS carrier; black conc. for FDA applics.; FDA compliance.

MDI PS-650. [Modern Disp.] 50% Titanium dioxide in PS carrier; wh. conc. for molding, extrusion applics.

MDI PS-901. [Modern Disp.] 33% Carbon black in PS carrier; black conc. for molding, compounding applics.

MDI PS-903. [Modern Disp.] 30% Carbon black in PS carrier; black conc. for sheet extrusion.

MDI PV-940. [Modern Disp.] 40% Carbon black in PVC carrier; black conc. for extrusion, molding, and compounding applics.

MDI SC-305. [Modern Disp.] 5% Slip additive in LDPE carrier; slip conc. for film applics.

MDI SF-320. [Modern Disp.] 20% Antiblock additive in LDPE carrier; antiblock conc. for film applics.

MDI SF-325. [Modern Disp.] 5% Slip and 20% antiblock additive in LDPE carrier; slip/antiblock conc. for film applics.

MDI SN-948. [Modern Disp.] 30% Carbon black in SAN carrier; black conc. for compounding; gran.

MEA Commercial Grade. [Dow] Monoethanolamine; CAS 141-43-5; EINECS 205-483-3; used in surfactants, cosmetics/toiletries, pharmaceuticals, metalworking fluids, textile chemicals, gas conditioning chemicals, agric. intermediates, adhesives, coatings, petroleum, rubber, wood pulping, and cement grinding aids; sp.gr. 1.0113 (25/4 C); dens. 8.45 lb/gal; visc. 18.9 cps; f.p. 10 C; b.p. 171 C (760 mm Hg); flash pt. (Seta CC) 201 F; fire pt. 200 F; ref. index 1.4525.

MEA Low Freeze Grade. [Dow] Monoethanolamine; CAS 141-43-5; EINECS 205-483-3; used in surfactants, cosmetics/toiletries, pharmaceuticals, metalworking fluids, textile chemicals, gas conditioning chemicals, agric. intermediates, adhesives, coatings, petroleum, rubber, wood pulping, and cement grinding aids; sp.gr. 1.0113 (25/4 C); dens. 8.45 lb/gal; visc. 18.9 cps; f.p. 10 C; b.p. 171 C (760 mm Hg); flash pt. (Seta CC) 201 F; fire pt. 200 F; ref. index 1.4525.

MEA Low Iron Grade. [Dow] Monoethanolamine; CAS 141-43-5; EINECS 205-483-3; used in surfactants, cosmetics/toiletries, pharmaceuticals, metalworking fluids, textile chemicals, gas conditioning chemicals, agric. intermediates, adhesives, coatings, petroleum, rubber, wood pulping, and cement grinding aids; sp.gr. 1.0113 (25/4 C); dens. 8.45 lb/gal; visc. 18.9 cps; f.p. 10 C; b.p. 171 C (760 mm Hg); flash pt. (Seta CC) 201 F; fire pt. 200 F; ref. index 1.4525.

MEA Low Iron-Low Freeze Grade. [Dow] Monoethanolamine; CAS 141-43-5; EINECS 205-483-3; used in surfactants, cosmetics/toiletries, pharmaceuticals, metalworking fluids, textile chemicals, gas conditioning chemicals, agric. intermediates, adhesives, coatings, petroleum, rubber, wood pulping, and cement grinding aids; sp.gr. 1.0113 (25/4 C); dens. 8.45 lb/gal; visc. 18.9 cps; f.p. 10 C; b.p. 171 C (760 mm Hg); flash pt. (Seta CC) 201 F; fire pt. 200 F; ref. index 1.4525.

MEA NF Grade. [Dow] Monoethanolamine NF; CAS 141-43-5; EINECS 205-483-3; used in surfactants, cosmetics/toiletries, pharmaceuticals, metalworking fluids, textile chemicals, gas conditioning chemicals, agric. intermediates, adhesives, coatings, petroleum, rubber, wood pulping, and cement grinding aids; sp.gr. 1.0113 (25/4 C); dens. 8.45 lb/gal; visc. 18.9 cps; f.p. 10 C; b.p. 171 C (760 mm Hg); flash pt. (Seta CC) 201 F; fire pt. 200 F; ref. index 1.4525.

Mearlin® Card Gold. [Mearl] Mica coated with titanium dioxide and/or iron oxide; nonmetallic colors with metallic gold appearance with bright highlights; designed for coating, printing, or incorporation into credit cards or other plastic cards; also for plastics, surf. coatings, printing inks; powd.; 6-48 μ particle size; sp.gr. ≈ 3.0; bulk dens. ≈ 10 lb/ft^3.

Mearlin® Hi-Lite Super Blue. [Mearl] Mica coated with titanium dioxide and/or iron oxide; interference luster pigment providing blue by reflection, yel. color by transmission; very intense color for plastics, surf. coatings, and printing inks; powd.; 6-48 μ particle size; sp.gr. ≈ 3.0; bulk dens. ≈ 10 lb/ft^3.

Mearlin® Hi-Lite Super Gold. [Mearl] Mica coated with titanium dioxide and/or iron oxide; interference luster pigment providing yellow-gold by reflection, bluish color by transmission; very intense color for plastics, surf. coatings, and printing inks; powd.; 6-48 μ particle size; sp.gr. ≈ 3.0; bulk dens. ≈ 10 lb/ft^3.

Mearlin® Hi-Lite Super Green. [Mearl] Mica coated with titanium dioxide and/or iron oxide; interference luster pigment providing green by reflection, red color by transmission; very intense color for plastics, surf. coatings, and printing inks; powd.; 6-48 μ particle size; sp.gr. ≈ 3.0; bulk dens. ≈ 10 lb/ft^3.

Mearlin® Hi-Lite Super Orange. [Mearl] Mica coated with titanium dioxide and/or iron oxide; interference luster pigment providing orange by reflection, blue-grn. color by transmission; good color intensity for plastics, surf. coatings, and printing inks; powd.; 6-48 μ particle size; sp.gr. ≈ 3.0; bulk dens. ≈ 10 lb/ft^3.

Mearlin® Hi-Lite Super Red. [Mearl] Mica coated with titanium dioxide and/or iron oxide; interference luster pigment providing red by reflection, grn. color by transmission; very intense color for plastics, surf. coatings, and printing inks; powd.; 6-48 μ particle size; sp.gr. ≈ 3.0; bulk dens. ≈ 10 lb/ft^3.

Mearlin® Hi-Lite Super Violet. [Mearl] Mica coated with titanium dioxide and/or iron oxide; interference luster pigment providing violet by reflection, yel.-grn. color by transmission; good color intensity for plastics, surf. coatings, and printing inks; powd.; 6-48 μ particle size; sp.gr. ≈ 3.0; bulk dens. ≈ 10 lb/ft^3.

Mearlin® Inca Gold. [Mearl] Mica coated with titanium dioxide and/or iron oxide; nonmetallic pigment with gold color by reflection, yel. by transmission; intense gold effect for plastics, surf. coatings, printing inks; powd.; 6-75 μ particle size; sp.gr. ≈ 3.0; bulk dens. ≈ 10 lb/ft^3.

Mearlin® MagnaPearl 1000. [Mearl] Mica coated with titanium dioxide; lustrous wh. pearl pigment for plastics, surf. coatings, and printing inks based on water or solvs.; exceptional brilliance, cleaner luster; wh. powd.; 8-48 μ particle size, 20 μ avg. size; bulk dens. 10 lb/ft^3.

Mearlin® MagnaPearl 1100. [Mearl] 71% Mica coated with 28% rutile titanium dioxide and 1% stannic oxide; pigment with pearly reflection; for plastics; off-wh. free-flowing powd., odorless; 17.7-21.3 μm mean particle size; insol. in water; dens. 3.0 kg/l; bulk dens. 16 g/100 cm^3; m.p. > 1000 C; *Precaution:* avoid dust formation; hazardous decomp. prods.: CO, CO_2; *Storage:* store in dry, well ventilated area in securely closed original container.

Mearlin® MagnaPearl 1110. [Mearl] Mica coated with titanium dioxide with anti-yellowing surf. treatment; wh. pearlescent pigment treated to prevent yellowing in plastics; wh. powd.; 8-48 μ particle size.

Mearlin® MagnaPearl 2000. [Mearl] Mica coated with titanium dioxide; lustrous wh. pearl pigment for plastics, surf. coatings, and printing inks based on water or solvs.; smooth texture, exc. coverage; wh. powd.; 5-25 μ particle size, 10 μ avg. size; bulk dens. 9 lb/ft^3.

Mearlin® MagnaPearl 2110. [Mearl] MagnaPearl 2100 (mica coated with titanium dioxide) with anti-yellowing surf. treatment; wh. pearlescent pigment treated to prevent yellowing in plastics; wh. powd.; 5-25 μ particle size.

Mearlin® MagnaPearl 3000. [Mearl] Mica coated with anatase titanium dioxide; ultrafine luster pigment for plastics, surf. coatings, and printing inks; fine, bright wh. luster, higher opacity, exc. coverage; wh. powd.; 2-10 μ particle size, 5 μ avg. size; bulk dens. 9 lb/ft^3.

Mearlin® MagnaPearl 3100. [Mearl] Mica coated with rutile titanium dioxide; ultrafine luster pigment for plastics, surf. coatings, and printing inks; wh. powd.; 2-10 μ particle size, 5 μ avg. size; bulk dens. 9 lb/ft^3.

Mearlin® MagnaPearl 4000. [Mearl] Mica coated with anatase titanium dioxide; luster pigment for plastics, surf. coatings, and printing inks; exhibits silvery-sparkle effect and clean, bright wh. appearance; silvery-wh. sparkle powd.; 15-150 μ particle size, 53 μ avg. size; bulk dens. 12 lb/ft^3.

Mearlin® MagnaPearl 5000. [Mearl] Mica coated with titanium dioxide; luster pigment for plastics, surf. coatings, and printing inks; exhibits silvery-sparkle effect and clean, bright wh. appearance; more economical; silvery-wh. sparkle powd.; 14-95 μ particle size, 36 μ avg. size; bulk dens. 11 lb/ft^3.

Mearlin® Nu-Antique Silver. [Mearl] Mica coated with titanium dioxide and/or iron oxide; nonmetallic pigment with dark gunmetal color; can be blended with wh. Mearlins to pewter or to bright silvers; for plastics, surf. coatings, printing inks; powd.; 6-90 μ particle size; sp.gr. ≈ 3.0; bulk dens. ≈ 10 lb/ft^3.

Mearlin® Pearl White. [Mearl] Mica coated with titanium dioxide and/or iron oxide; pigment for plastics, surf. coatings, and printing inks with good luster; economical, widely used; wh. powd.; 6-90 μ particle size; sp.gr. ≈ 3.0; bulk dens. ≈ 10 lb/ft^3.

Mearlin® Satin White. [Mearl] Mica coated with titanium dioxide and/or iron oxide; luster pigment for plastics, surf. coatings, and printing inks; bright, wh., cleaner luster, exc. opacity and coverage; wh. powd.; 4-32 μ particle size; sp.gr. ≈ 3.0; bulk dens. ≈ 10 lb/ft^3.

Mearlin® Silk White. [Mearl] Mica coated with titanium dioxide and/or iron oxide; smooth, satin luster pigment for plastics, surf. coatings, and printing inks; wh. powd.; 4-75 μ particle size; sp.gr. ≈ 3.0; bulk dens. ≈ 10 lb/ft^3.

Mearlin® Sparkle. [Mearl] Mica coated with titanium dioxide and/or iron oxide; glittery, silver luster pigment for plastics, surf. coatings, and printing inks; quite transparent; wh. powd.; 10-110 μ particle size; sp.gr. ≈ 3.0; bulk dens. ≈ 10 lb/ft^3.

Mearlin® Supersparkle. [Mearl] Mica coated with titanium dioxide and/or iron oxide; luster pigment for plastics, surf. coatings, and printing inks; low proportions provide a high gloss wet look finish; higher concs. produce a fine sandpaper texture; wh. powd.; 10-150 μ particle size; sp.gr. ≈ 3.0; bulk dens. ≈ 10 lb/ft^3.

Mearlite® Ultra Bright UDQ. [Mearl] Bismuth oxychloride in alkyd/DBP vehicle; industrial pearlescent pigment with high brilliance for solv.-based spray coatings, liq. inks, gelcoats and casting resins; dens. 14.0 lb/gal; flash pt. (Seta CC) 79 F; VOC 3.22 lb/gal; 50% pigment, 77% total solids.

Mearlite® Ultra Bright UMS. [Mearl] Bismuth oxychloride in DBP vehicle; industrial pearlescent pigment with high brilliance for gelcoats and casting resins; also compat. with solv.-based spray coatings, liq. inks; dens. 10.0 lb/gal; flash pt. (Seta CC) 105 F; VOC 1.40 lb/gal; 15% pigment, 86% total solids.

Mearlite® Ultra Bright UTL. [Mearl] Bismuth oxychloride in acrylic vehicle; industrial pearlescent pigment with high brilliance for solv.-based spray coatings, liq. inks, gelcoats and casting resins; dens. 15.5 lb/gal; flash pt. (Seta CC) 45 F; VOC 4.96 lb/gal; 60% pigment, 68% total solids.

Mekon® White. [Petrolite] Hard microcryst. wax; CAS 63231-60-7; EINECS 264-038-1; release agent; used in hot-melt coatings and adhesives, paper coatings, printing inks, plastic modification (as lubricant and processing aid), lacquers, paints, varnishes, as binder in ceramics, for potting/filling in elec./electronic components, in investment casting, rubber and elastomers (plasticizer, antisunchecking, antiozonant), as emulsion wax size in papermaking, as fabric softener ingred., in emulsion and latex coatings, and in cosmetic hand creams and lipsticks; chewing gum base; incl. FDA §172.230, 172.615, 175.105, 175.300, 176.170, 176.180, 176.200, 177.1200, 178.3710, 179.45; wh. wax; very low sol. in org. solvs.; dens. 0.78 g/cc (99 C); visc. 15 cps (99 C); m.p. 94 C.

Merix Anti-Static #79 Conc. [Merix] Proprietary; internal/external antistat for ABS, acrylic, cellulose nitrate, polyacetal, PC, PP, PS, flexible and rigid PVC.

Merix Anti-Static #79-OL Super Conc. [Merix] Proprietary; internal/external antistat for ABS, acrylic, cellulosics, nylon, PE, PP, flexible and rigid PVC; FDA approved.

Merix Anti-Static #79 Super Conc. [Merix] Proprietary; internal/external antistat for all plastics.

Merix MCG Compd. [Merix] Proprietary; antifog, heat and lt. stabilizer for flexible and rigid PVC, plastisols, organosols, bottles, ABS, chlorinated PE, acrylics; good clarity.

Merix Mold-Ease Conc. PCR. [Merix] Proprietary; internal/external antistat for all plastics.

Merpol® A. [DuPont] Ethoxylated phosphate; nonionic; low foaming wetting agent, surf. tens. reducer for chemical mfg., cosmetics, metal processing, paper, petrol., inks, plastics, soaps, syn. fibers, textiles; stable to acids, bases, heat to 100 C, freezing; colorless to pale yel. liq., mild odor; sol. in polar solvs., 0.1-1% in nonpolar solvs.; disp. in water to 1%; sp.gr. 1.07 g/mL; dens. 8.9 lb/gal; visc. 104 cP; HLB 6.7; cloud pt. < 25 C (upper, 1%); flash pt. (TCC) 138 C; pH 5-7 (1% emulsion); surf. tens. 26 dynes/cm (0.1%); 100% act.

Merpol® HCS. [DuPont] Alcohol ethoxylate; nonionic; wetting agent, detergent, emulsifier, penetrant, antistat, leveling agent, dyeing assistant, stabilizer used in textiles, leather, paper, metal processing, rubber, emulsion polymerization, paints, inks, medicinal ointments, antiperspirants, cutting oils, polishes, cosmetics; pigment dispersant; lt. yel. clear, visc. liq.; mild fatty odor; 40% sol. in water, sol. in org. solvs. that are misc. with water; sp.gr. 1.03 g/mL; dens. 8.63 lb/gal; HLB 15.3; cloud pt. > 100 C (upper, 1%); flash pt. > 235 F; pH 6-8 (10%); surf. tens. 42.9 dynes/cm (0.1%); 60% act.

Mesamoll®. [Bayer AG; Bayer/Fibers, Org., Rubbers] Alkyl sulfonic phenyl ester; universal plasticizer used in PVC calendering, extrusion, inj. molding, dip coating, high-pressure foam, rotational and compression moldings, film for linings and food pkg., shower curtains, floorcoverings, imitation leather, tarpaulins, protective clothing, cable insulation, structure profiles, tubing, shoes, tech. articles, toys; also used with PS, joint sealants, natural, S/B, nitrile/butadiene, chlorinated and butyl rubber; cleansing agent for PU processing equip.; Hazen < 450; dens. 1.04-1.07 g/cc; visc. 100-125 mPa•s; acid no. < 0.1; flash pt. (OC) 210-240 C; ref. index 1.4970-1.5000; 70:30 PVC:plasticizer properties: tens. str. 22 mPa; tens. elong. 340%; hardness (Shore D) 30.

Mesamoll® II. [Bayer/Fibers, Org., Rubbers] Alkyl sulfonic phenyl ester; plasticizer for PVC and PU compds.; imparts reduced oil and fat extraction and lower volatiles along with exc. hydrolysis resist.; dens. 1.04-1.07 g/cm^3; visc. 110-140 cps (23 C).

Metablen® C-301. [Elf Atochem N. Am.] Polymeric resin; impact modifier for PVC blow molding compds.; improves oil resist. of bottles; used for olive oil bottles; FDA sanctioned; food approved in Europe; 2% max. on 16 mesh (1000 μ); sp.gr. 1.08; bulk dens. 0.25-0.40; 1% max. volatiles; *Toxicology:* prolonged exposure may cause skin irriation; may be eye irritant; vapors may be pungent and irritating to respiratory system.

Metacure® T-1. [Air Prods./Perf. Chems.] Dibutyl tin diacetate; CAS 1067-33-0; catalyst for use in prod. of PU coatings, adhesives, and sealants; fastest cure in series; straw yel. liq.; insol. in water; sp.gr. 1.32; dens. 11 lb/gal; flash pt. (COC) 290 F; 33.9% total tin; *Toxicology:* skin and eye irritant; vapors on heating may cause irritation.

Metacure® T-5. [Air Prods./Perf. Chems.] Tin-based; catalyst for prod. of PU coatings, adhesives, and sealants; increased pot life over Metacure T-12; lt. amber liq.; sp.gr. 1.41; dens. 11.76 lb/gal; PMCC flash pt. > 390 F; 44.9% total tin.

Metacure® T-9. [Air Prods./Perf. Chems.] Stannous octoate; CAS 301-10-0; catalyst for use in prod. of PU coatings, adhesives, and sealants; uniform activity and exc. stability; colorless to pale yel. clear liq.; sp.gr. 1.25; dens. 10.5 lb/gal; visc. 310 cps; pour pt. < -25 C; flash pt. (PM) 142 C; 28% total tin; *Toxicology:* eye and skin irritant.

Metacure® T-12. [Air Prods./Perf. Chems.] Dibutyl tin dilaurate; CAS 77-58-7; EINECS 201-039-8; curing catalyst for prod. of PU coatings, adhesives, and sealants; formulated to remain liq. > 18 C for easier handling; FDA 21CFR §175.105, 177.1680; oily liq., solid below R.T.; sp.gr. 1.05; visc. 43 cps; f.p. 18 C; decomp. pt. > 150 C; flash pt. (COC) 235 C; ref. index 1.4686; 18% total tin; *Toxicology:* LD50 (oral, rat) 3954 mg/kg, sl. toxic; sl. toxic by acute dermal exposure; severely irritating to skin and eyes; causes burns; irritant on inhalation of vapors.

Metacure® T-45. [Air Prods./Perf. Chems.] Potassium octoate; catalyst for PU; catalyzes the trimerization of isocyanates and polyol-isocyanate reaction; pale yel. clear visc. liq.; sol. in water and polar solvs., esp. alcohols, glycols; sp.gr. 1.09; visc. 20,000 cps; f.p. -5 C; TCC flash pt. 102 C; hyd. no. 86; 14% potassium content; *Toxicology:* sl. toxic by ingestion; mild to moderate eye and skin irritant on prolonged/repeated exposure; *Storage:* store in closed container with min. of air space or blanket with nitrogen or dry air; hygroscopic.

Metacure® T-120 [Air Prods./Perf. Chems.] Organotin; catalyst for one-part moisture cure and two-part isocyanate coatings; compatible with amine co-catalysts; pale amber liq.; sol. in higher alcohols, org. solvs., insol. in water; sp.gr. 0.995; visc. 19 cps; f.p. -20 C; decomp. pt. 320 C; flash pt. (PMCC) 121 C; 17.5% total tin; *Toxicology:* skin irritant on prolonged exposure; eye irritant; irritant on inhalation of vapors.

Metacure® T-125 [Air Prods./Perf. Chems.] Organotin; catalyst for PU coatings, adhesives, and sealants; suitable for one-part moisture cure and two-part reactions; lt. yel. liq.; sp.gr. 1.15; visc. 282 cps; f.p. -25 C; decomp. pt. 215 C; PMCC flash pt. 123 C; 16.5% total tin; *Toxicology:* skin and eye irritant; vapors when heated can cause irritation.

Metacure® T-131 [Air Prods./Perf. Chems.] Organotin; catalyst for PU coatings, adhesives, and sealants; provides delayed action catalysis of isocyanate/polyol reaction; lt. yel. liq.; sol. in alcohols, org. solvs., insol. in water; sp.gr. 1.11; visc. 33 cps; f.p. -23 C; decomp. pt. 255 C; flash pt. (PMCC) 130 C; 17.5% total tin; *Toxicology:* skin and eye irritant; vapors when heated can cause irritation.

Metallic 9500. [D.J. Enterprises] Fe, Mn, C, Si; conductive additive for coatings or composites; used in thermoplastics for EMI shielding and ESD protection for elec. prods.; noncombustible; gray/ blk. solid; < 44 μ particle size; sp.gr. > 7.6; m.p. 1371-1482 C.

Metasap® Barium Stearate. [Syn. Prods.] Barium stearate; CAS 6865-35-6; internal/external lubricant for vinyl compositions; solid.

Metasap® Cadmium Stearate. [Syn. Prods.] Cadmium stearate; CAS 2223-93-0; stabilizer, internal lubricant for PVC; solid.

Metasap® Zinc Stearate. [Syn. Prods.] Zinc stearate; CAS 557-05-1; EINECS 209-151-9; internal/external lubricant for melamine, phenolics, PVC; FDA accepted; solid.

Metazene® 60%. [Pestco] Mixt. of fatty alcohol esters of methyl methacrylic acid; tech. odor counteractant; clear liq., mild odor; insol. in water; sp.gr. 0.836; visc. 5 cps; vapor pressure 7 mm Hg (320 F); m.p. -7.6 F; f.p. -8 F; b.p. 417.8 F; flash pt. 186 F; < 1% volatiles; *Toxicology:* LD50 (oral, rat) > 5000 mg/kg, (dermal, rabbit) > 3000 mg/kg; mod. skin irritation; sl. eye irritation; inh. may cause nose and throat irritation; concs. may cause allergic reaction; *Precaution:* exposure to excessive heat may cause exothermic polymerizations; heating closed containers to ≥ 558 F may cause bursting; incompat. with strong oxidizing and reducing agents; avoid heat, contamination, oxygen-free atm., bright light; *Storage:* 12 mos shelf life at ambient temps.; additional 6 mos. if head space of containers is flushed with pure oxygen; store in cool area, keep from freezing; if freezing occurs, thaw at 18-40 C.

Metazene® 80%. [Pestco] Mixt. of fatty alcohol esters of methyl methacrylic acid; tech. odor counteractant; clear liq., mild odor; insol. in water; sp.gr. 0.852; visc. 5 cps; vapor pressure 7 mm Hg (320 F); m.p. -7.6 F; f.p. -8 F; b.p. 433.4 F; flash pt. 208 F; < 1% volatiles; *Toxicology:* LD50 (oral, rat) > 5000 mg/kg, (dermal, rabbit) > 3000 mg/kg; mod. skin irritation; sl. eye irritation; inh. may cause nose and throat irritation; concs. may cause allergic reaction; *Precaution:* exposure to excessive heat may cause exothermic polymerizations; heating closed containers to ≥ 558 F may cause bursting; incompat. with strong oxidizing and reducing agents; avoid heat, contamination, oxygen-free atm., bright light; *Storage:* 12 mos shelf life at ambient temps.; additional 6 mos. if head space of containers is flushed with pure oxygen; store in cool area, keep from freezing; if freezing occurs, thaw at 18-40 C.

Metazene® 99%. [Pestco] Mixt. of fatty alcohol esters of methyl methacrylic acid; tech. odor counteractant; clear liq., mild odor; negligible sol. in water; sp.gr. 0.868; visc. 5 cps; vapor pressure 7 mm Hg (320 F); m.p. -7.6 F; f.p. -8 F; b.p. 558 F; flash pt. 230 F; < 1% volatiles; *Toxicology:* LD50 (oral, rat) > 5000 mg/kg, (dermal, rabbit) > 3000 mg/kg; mod. skin irritation; sl. eye irritation; inh. may cause nose and throat irritation; concs. may cause allergic reaction; *Precaution:* exposure to excessive heat may cause exothermic polymerizations; heating closed containers to ≥ 558 F may cause bursting; incompat. with strong oxidizing and reducing agents; avoid heat, contamination, oxygen-free atm., bright light; *Storage:* 12 mos shelf life at ambient temps.; additional 6 mos. if head space of containers is flushed with pure oxygen; store in cool area, keep from freezing; if freezing occurs, thaw at 18-40 C.

Methasan®. [Monsanto] Zinc dimethyldithiocarbamate; CAS 137-30-4; sec. accelerator for thiazoles and sulfenamides; fast curing accelerator for NR, SBR, and latexes; for low or room temp. cure; blooms at high levels; pellets or powd.

Methocel® A. [Dow] Methylcellulose; CAS 9004-67-5; nonionic; plasticizer for ceramic and refractory shapes and furnace linings; visc. stabilizer for latex and emulsion paints; thickener; wh. to off-wh. powd., odorless; sp.gr. 1.39; dens. 11.6 lb/gal; f.p. 0 C (2%); pH 7; ref. index 1.336 (2%); surf. tens. 47-53 dynes/cm; 27.5-31.5% methoxyl.

Methocel® E. [Dow] Hydroxypropyl methylcellulose; CAS 9004-65-3; nonionic; plasticizer for ceramic and refractory shapes and furnace linings; visc. stabilizer for latex and emulsion paints; thickener; wh. to off-wh. powd., odorless; sp.gr. 1.39; dens. 11.6 lb/gal.

Methyl Cumate®. [R.T. Vanderbilt] Copper dimethyldithiocarbamate; CAS 137-29-1; accelerator for high-speed vulcanization of SBR, IIR, EPDM; dk. brn. powd.; sol. in acetone, toluene, chloroform; pract. insol. in water, alcohol, gasoline; m.w. 303.98; dens. 1.75 ± 0.03 mg/m^3; m.p. > 325 C; 18-20% Cu.

Methyl Cumate® Rodform. [R.T. Vanderbilt] Copper dimethyldithiocarbamate; CAS 137-29-1; rubber accelerator; dk. brn. rods; negligible sol. in water; dens. 1.74 mg/m^3; 98% act.; *Toxicology:* LD50 (oral, rat) > 5000 mg/kg, (dermal, rabbit) > 2000 mg/ kg); harmful by inh.; may cause eye irritation; dust may cause irregular breathing; contains min. oil: OSHA PEL/TWA 5 mg/m^3; *Precaution:* incompat. with acids; hazardous decomp. prods.: copper oxides, amines, CO_2, CO, CS_2.

Methyl Ledate. [R.T. Vanderbilt] Lead dimethyldithiocarbamate; CAS 19010-66-3; ultra accelerator;

NR, SBR, IIR, IR, BR ultra accelerator for high speed, high temp. vulcanization; effective under continuous curing conditions; generally used with thiazole modifiers; wh. powd., 99.9% thru 100 mesh; m.w. 447.65; dens. 2.43 ± 0.03 mg/m³; m.p. > 310 C; 45.5-47.5% lead content (powd.); 41.0-43.0% lead content (rods).

Methyl Namate®. [R.T. Vanderbilt] Sodium dimethyldithiocarbamate aq. sol'n.; rubber accelerator; water treatment chemical; readily forms water-insol. salts with heavy metals such as cadmium, copper, chromium, and nickel; clarification agent for wastewater from plating, photo finishing, ore beneficiation processes; yel. to lt. amber liq.; misc. with water; dens. 1.18 mg/m³; pH 10; *Toxicology:* LD50 (oral, rat) 1000 mg/kg; *Precaution:* hazardous decomp. prods.: oxides of sulfur, nitrogen, carbon, and sodium at combustion temps.

Methyl Niclate®. [R.T. Vanderbilt] Nickel dimethyldithiocarbamate; CAS 15521-65-0; nonstaining antioxidant for CR, CSM, ECO, NBR, SBR, peroxide-vulcanized elastomers; for hose and belting, automotive and appliance molded goods, wire and cable; improves heat aging; grnsh. powd.; 99.8% min. thru 100 mesh; insol. in water; m.w. 299.12; dens. 1.77 ± 0.03 mg/m³; m.p. > 290 C; 18-20% Ni.

Methyl Selenac®. [R.T. Vanderbilt] Selenium dimethyldithiocarbamate; CAS 144-34-4; ultra accelerator for NR, SBR, IIR; also vulcanizing agent; effective in low sulfur and sulfurless heat resistant compds.; nondiscoloring in lt. stocks; generally used with thiazoles to balance scorch and curing chars.; yel. powd., 99.9% thru 100 mesh; sol. in toluene, carbon disulfide, chloroform; m.w. 559.78; dens. 1.58 ± 0.03 mg/m³; m.p. 140-172 C; 13.0-15.0% selenium content (powd.); 12.5-14.0% selenium content (rods).

Methyl Tuads®. [R.T. Vanderbilt] Tetramethylthiuram disulfide; CAS 137-26-8; ultra accelerator; for NR and syn. rubbers (esp. IIR, CR); accelerator and vulcanizing agent; cure modifier for Neoprene (retards G types; accelerates vulcanization of W types); nondiscoloring in lt. stocks; wh. to cream powd., sl. aromatic odor; 99.9% thru 100 mesh; sol. in toluene, carbon disulfide, chloroform; m.w. 240.44; dens. 1.42 ± 0.03 mg/m³; m.p. 142-156 C; flash pt. (PMOC) 140 C; 13.3% avail. sulfur; *Toxicology:* LD50 (oral, female rat) 1800 mg/kg, sl. toxic; may cause nose, throat, skin irritation, allergic reaction; consumption of alcohol after repeated exposure may cause skin irritation; OSHA TWA 5 mg/m³; contains min. oil (as mist): OSHA TWA 5 mg/m³; *Precaution:* dust may form explosive mixt. with air; incompat. with strong acids, oxidizers; hazardous decomp. prods.: oxides of carbon, nitrogen, and sulfur upon combustion; *Storage:* store in cool, dry place.

Methyl Tuads® Rodform. [R.T. Vanderbilt] Tetramethylthiuram disulfide; CAS 137-26-8; ultra accelerator; for NR and syn. rubbers (esp. IIR, CR); accelerator and vulcanizing agent; cure modifier for Neoprene (retards G types; accelerates vulcanization of W types); nondiscoloring in lt. stocks; wh. to cream and blue rodform, sl. aromatic odor; sol. in toluene, carbon disulfide, chloroform; m.w. 240.44; dens. 1.42 ± 0.03 mg/m³; m.p. 142-156 C; flash pt. (PMOC) 140 C; 12% avail. sulfur.

Methyl Zimate®. [R.T. Vanderbilt] Zinc dimethyldithiocarbamate; CAS 137-30-4; ultra accelerator for NR and syn. rubbers; latex accelerator; act. over wide temp. range; generally requires thiazole modifier for safe processing and wide curing range; nondiscoloring in lt. stocks; wh. powd., 99.9% thru 100 mesh; pract. insol. in water; m.w. 305.82; dens. 1.71 ± 0.03 mg/m³; m.p. 242-257 C; 19.5-23.0% Zn (powd.), 19-21% Zn (rods).

Methyl Zimate® Slurry. [R.T. Vanderbilt] Zinc dimethyldithiocarbamate with sodium polynaphthalene sulfonate in aq. slurry; ultra accelerator for NR and syn. rubbers, latexes; act. over wide temp. range; generally requires thiazole modifier for safe processing and wide curing range; nondiscoloring in lt. stocks; wh. aq. slurry; negligible sol. in water; dens. 1.28 mg/m³; pH 9.0-11.5; 50% act.; *Toxicology:* may be fatal if vapor is inhaled; may cause eye, skin irritation, skin sensitization; antabuse effect with ingestion of alcoholic beverages; ziram: LD50 (oral, rat) 1400 mg/kg, toxic; (dermal, rabbit) > 2000 mg/kg, low toxicity; *Precaution:* not a fire hazard; exposed to flame, emits acrid fumes; incompat. with acids, oxidizers; hazardous decomp. prods.: oxides of C, S, N, and Zn; *Storage:* keep container closed; keep from freezing.

MFM-401. [Rohm Tech] Allyl methacrylate; CAS 96-05-9; EINECS 202-473-0; graft crosslinker for syn. resins and emulsions.

MFM-405. [Rohm Tech] 1,4-Butanediol dimethacrylate; CAS 2082-81-7; peroxide crosslinker for elastomers, PVC plastisols, dental materials.

MFM-407. [Rohm Tech] 1,3-Butanediol dimethacrylate; peroxide crosslinker for elastomers, PVC plastisols, dental materials.

MFM-413. [Rohm Tech] Triethylene glycol dimethacrylate; CAS 109-16-0; crosslinking agent for anaerobic adhesives, PVC plastisols, photoresists, photopolymer plates, dental materials.

MFM-415. [Rohm Tech] Trimethylolpropanetrimethacrylate with 250 ppm MEHQ; peroxide crosslinker for elastomers, photoresists, photopolymer plates, adhesion promter and hardener in PVC plastisols.

MFM-416. [Rohm Tech] Ethylene glycol dimethacrylate; CAS 97-90-5; EINECS 202-617-2; peroxide crosslinker for elastomers, anaerobic adhesives, dental materials.

MFM-418. [Rohm Tech] Diethylene glycol dimethacrylate; CAS 2358-84-1; peroxide crosslinker for elastomers, anaerobic adhesives, photopolymer plastes.

MFM-425. [Rohm Tech] Tetraethylene glycol dimethacrylate; CAS 25852-47-5; crosslinker for anaerobic adhesives, photoresists, syn. resins, unsat. polyesters.

MFM-786 V. [Rohm Tech] Trimethylolpropane trimethacrylate with 100 ppm MEHQ; peroxide crosslinker for elastomers, photoresists, photopolymer plates, adhesion promoter/hardener in PVC plastisols.

MGH-93. [RMc Minerals] Magnesium hydroxide; CAS 1309-42-8; EINECS 215-170-3; plastics additive permitting higher processing temps. for olefins, PP, nylons; absorbs more heat than hydrated alumina; does not generate poisonous gases during combustion; dilutes smoke produced by decomposing polymers; powd.; 100% thru 200 mesh, 99.7% thru 325 mesh; 10 μ mean particle size; sp.gr. 2.39; brightness 88 min.; pH 9.9; hardness (Mohs) 2.5; 93.99% $Mg(OH)_2$; *Toxicology:* nontoxic; does not generate poisonous gases during combustion; dilutes smoke produced by decomposing polymers.

Micawhite 200. [Franklin Industrial Minerals] Mica; CAS 12001-26-2; filler for building prods., stucco/joint cement applics., noncritical paint, plastics applics.; wh. platy powd.; 35 μ mean particle size; 98.5% thru 100mesh, 90% thru 200 mesh; sp.gr. 2.8; bulk dens. 16 ± 2 lb/ft^3 (loose); pH 7-8; 0.5% max. moisture.

Michel XO-24. [M. Michel] Modified amine; antistat additive, integral release agent for linear polyethylene and PVC; FDA approved for food contact applics.

Michel XO-85. [M. Michel] Modified amine; antistat additive, integral release agent for PP.

Michel XO-108. [M. Michel] Modified amine; antistat for PP, polyethylene, polyester film; water-insol.

Micral® 532. [J.M. Huber/Engineered Mins.] Alumina trihydrate; flame retardant, smoke suppressor, high loading filler for thermoset, thermoplastic, rubber , adhesives, paints, coatings, paper applics.; 9 μ median particle diam.; 99.9% thru 325 mesh; sp.gr. 2.42; bulk dens. 0.65 g/cc (loose); surf. area 4-5 m^2/g; oil absorp. 27-29; brightness (TAPPI) 89; ref. index 1.57; hardness (Moh) 2.5-3.5; 64.9% Al_2O_3.

Micral® 632. [J.M. Huber/Engineered Mins.] Alumina trihydrate; smoke suppressor/flame retardant for PVC and XLPE wire/cable insulation, SBR compds., epoxies, silicones, vinyl sheet flooring, adhesives, coatings, paints, polishes; powd.; 3.5 μ avg. particle diam.; sp.gr. 2.42; oil absorp. 32; brightness (TAPPI) 95; ref. index 1.57; hardness (Moh) 2.5-3.5; 64.9% Al_2O_3.

Micral® 855. [J.M. Huber/Engineered Mins.] Alumina trihydrate; flame retardant/smoke suppressant for wire and cable jacketing and insulation, inj. molded and extruded polyolefins, thermoplastic and crosslinked compds., epoxy elec. laminates,m rigid PVC, molded or calendered SBR, SBR and urethane elastomers; silicone rubber; powd.; 100% thru 325 mesh, 99.995% thru 500 mesh; 2 μ median particle size; sp.gr. 2.42; bulk dens. 0.3 g/cc (loose); surf. area 12 m^2/g; oil absorp. 38; brightness (TAPPI) 97+; 64.9% Al_2O_3.

Micral® 916. [J.M. Huber/Engineered Mins.] Alumina trihydrate; halogen-free smoke suppressor/flame retardant for wire and cable insulation, inj.-molded polyolefins, coatings, adhesives, rubber goods, paper filler and coating, PVC, EPDM, EPR, ABS, XLPE, and compr.-molded thermosets; powd.; 0.8 μ median particle diam.; 100% through 325 mesh; sp.gr. 2.42 g/cm^3; bulk dens. 0.15 g/cm^3 (loose), 0.35 g/cm^3 (packed); surf. area 13 m^2/g; oil absorp. 46 ml/100 g; brightness (Photovolt) 97+; 64.9% Al_2O_3; Unverified

Micral® 932. [J.M. Huber/Engineered Mins.] Alumina trihydrate; halogen-free smoke suppressor/flame retardant for wire/cable insulation, inj.-molded polyolefins, coatings, adhesives, rubber goods, paper filler and coating, epoxy elec. laminates, FRP elec. laminates, SMC/BMC, flexible/rigid PVC, plastisols, EPDM, EPR, ABS, XLPE, and compr.-molded thermosets; powd.; 2 μ median particle diam.; 100% thru 325 mesh; sp.gr. 2.42; bulk dens. 0.35 g/cc (loose), 0.5 g/cc (packed); surf. area 13 m^2/g; oil absorp. 38; brightness (TAPPI) 95; 64.9% Al_2O_3.

Micral® 1000. [J.M. Huber/Engineered Mins.] Alumina trihydrate; economical smoke suppressor/flame retardant with high brightness; for wire and cable insulation, inj. molded polyolefins, coatings, adhesives, rubber goods, PVC, EPDM, EPR, XLPE, EVA, compr. molded thermosets; powd.; 1.1 μ median particle diam.; 100% thru 325 mesh; sp.gr. 2.42; surf. area 12 m^2/g; oil absorp. 33-35; brightness (TAPPI) 97; hardness (Moh) 3.0; 64.9% Al_2O_3.

Micral® 1500. [J.M. Huber/Engineered Mins.] Alumina trihydrate; economical smoke suppressor/flame retardant with high brightness; for wire and cable insulation, inj. molded polyolefins, coatings, adhesives, rubber goods, PVC, EPDM, EPR, XLPE, EVA, compr. molded thermosets; powd.; 1.5 μ median particle diam.; 100% thru 325 mesh; sp.gr. 2.42; surf. area 10 m^2/g; oil absorp. 33-35; brightness (TAPPI) 96; hardness (Moh) 3.0; 64.9% Al_2O_3.

Micro P Extender. [D.J. Enterprises] Amorphous mineral silicate; lightweight resin extender, filler for aircraft, military, appliances, business equip., construction, consumer prods., elec./electronic, land transport, and marine applics.; high filler loading capability; water-resist.; nonhygroscopic; reduces shrinkage and exotherm temp.; free-flowing micro particle; 44-150 μ range; insol. in water/org. acids; bulk dens. 8-10 lb/ft^3; 71-75% SiO_2, 12.5-18% Al_2O_3.

Microbloc®. [Specialty Minerals] Surface-treated talc with proprietary coating; CAS 14807-96-6; EINECS 238-877-9; antiblock for plastic film industry, engineering thermoplastics; enhanced compat. with polyolefins for improved film clarity and low COF; 2.3 μ avg. particle size; < 0.05% 325 mesh residue; dry brightness 89 ± 1.5; sp.gr. 2.7; bulk dens. 13 lb/ft^3 (loose), 37 lb/ft^3 (tapped).

Micro-Cel® C. [Celite] Syn. calcium silicate; CAS 1344-95-2; EINECS 215-710-8; functional filler used to convert sticky, visc. liqs. to dry liqs. for use in rubber, agric., chemical, plastics, food processing, animal feed, and pharmaceutical industries; wh. fine powd.; 5% 325-mesh residue; sp.gr. 2.40; bulk dens. 8.5 lb/ft^3; surf. area 175 m^2/g; water absorp. 405%; oil absorp. 340%; ref. index 1.55; pH 10 (10% aq. slurry); 50% SiO_2, 29% CaO; 8% moisture.

Micro-Cel® E. [Celite] Calcium silicate; functional filler used as carriers, grinding aids, anticaking agent, and conditioner in agric. chemicals; toxicants; carriers for liq. seed inoculants; inert carriers to convert sticky visc. liq. to dry liqs.; in rubber goods with high oil loadings; also used as inert catalyst carriers, vanadium catalyst used in mfg. of sulfuric acid, and phosphoric acid catalyst of petrol. industry; off-wh. powd., 6.0% 325-mesh residue; sp.gr. 2.45; dens. 5.5 lb/ft^3; surf. area 120 m^2/g; oil absorp. 490%; ref. index 1.55; pH 8.4 (10% aq.); 58.6% SiO_2, 22% CaO.

Micro-Cel® T-21. [Celite] Calcium silicate; functional filler, absorbent, inert carrier, conditioner, anticaking agent, diluent for chemical, food, cosmetics, agric., paper, rubber, plastic, and other industrial applics.; converts liqs. to dry powds.; lt. gray particulate; 3% 325 mesh residue; sp.gr. 2.4; bulk dens. 12.8 lb/ft^3; surf. area 190 m^2/g; water absorp. 220%; oil absorp. 170%; ref. index 1.54; pH 7.6 (10% aq. slurry); 65% SiO_2, 14.9% MgO.

Micro-Cel® T-49. [Celite] Calcium silicate; functional filler, absorbent, inert carrier, conditioner, anticaking agent, diluent for chemical, food, cosmetics, agric., paper, rubber, plastic, and other industrial applics.; converts liqs. to dry powds.; off-wh. particulate; 10% 325 mesh residue; sp.gr. 2.1; bulk dens. 12 lb/ft^3; surf. area 105 m^2/g; water absorp. 240%; oil absorp. 190%; ref. index 1.53; pH 10 (10% aq. slurry); 36% SiO_2, 33% CaO; 8% moisture.

Micro-Chek® 11. [Ferro/Bedford] 2-n-Octyl-4-isothiazolin-3-one in plasticizer (epoxidized soybean oil); mildewcide, antimicrobial, fungicide, preservative for PVC, PU used for roofing membranes, automotive trim, awnings, pond liners, marine upholstery, outdoor furniture, interior applics.; dens. 8.3 lb/gal; 4% sol'n.

Micro-Chek® 11 P. [Ferro/Bedford] 2-n-Octyl-4-isothiazolin-3-one; CAS 26530-20-1; antimicrobial, mildewcide for PVC, polyurethane, other polymers for use in roofing membranes, exterior automotive trims, awnings, tarpaulins, pond liners, marine upholstery, shower curtains, outdoor furniture; EPA registered; off-wh. free-flowing powd.; dens. 9.2 lb/gal; soften. pt. 149 C; 4% act.; *Toxicology:* do not ingest; can cause skin irritations and possible eye damage; toxic to fish; *Storage:* store prod. in secure area.

MicroPflex 1200. [Specialty Minerals] Surface-modified microtalc; CAS 14807-96-6; EINECS 238-877-9; filler for engineering thermoplastics; powd.; 1.5 μ median particle size; nil retained on 325 mesh; sp.gr. 2.70; dens. 11.0 lb/ft^3 (loose), 21.5 lb/ft^3 (tapped); dry brightness 90.0; hardness (Mohs) 1; 60.5% SiO_2; 31.7% MgO.

Microtalc® MP10-52. [Specialty Minerals] High purity Montana talc; CAS 14807-96-6; EINECS 238-877-9; filler for solv. and water-based paints and coatings, semigloss paints, OEM/specialty areas, rigid vinyl, and polyolefins; powd.; 0.8 μ median particle size; fineness (Hegman) 6.5; sp.gr. 2.8; dens. 22 lb/ft^3 (tapped); bulk dens. 6.4 lb/ft^3; bulking value 23.3 lb/gal; oil absorp. 55 g/100 g; brightness 91; pH 8.8; 60% SiO_2, 33% MgO, 1% Al_2O_3.

Microtalc® MP12-50. [Specialty Minerals] High purity Montana talc; CAS 14807-96-6; EINECS 238-877-9; filler for solv. and water-based paints and coatings, semigloss paints, primers, OEM/specialty areas, rubber, polyolefins; powd.; 1.2 μ median particle size; fineness (Hegman) 6.0; sp.gr. 2.8; dens. 22.6 lb/ft^3 (tapped); bulk dens. 7.5 lb/ft^3; bulking value 23.3 lb/gal; oil absorp. 54 g/100 g; brightness 89.5; pH 8.8; 60% SiO_2, 33% MgO, 1% Al_2O_3.

Microtalc® MP15-38. [Specialty Minerals] High purity Montana talc; CAS 14807-96-6; EINECS 238-877-9; filler for solv. and water-based paints and coatings, semigloss paints, primers, OEM/specialty areas, and polyolefins; powd.; 2.0 μ median particle size; fineness (Hegman) 5.75; sp.gr. 2.8; dens. 33 lb/ft^3 (tapped); bulk dens. 12 lb/ft^3; bulking value 23.3 lb/gal; oil absorp. 40 g/100 g; brightness 90; pH 8.8; 60% SiO_2, 33% MgO, 1% Al_2O_3.

Microtalc® MP25-38. [Specialty Minerals] High purity Montana talc; CAS 14807-96-6; EINECS 238-877-9; filler for solv. and water-based paints and coatings, semigloss paints, primers, OEM/specialty areas, and polyolefins; powd.; 2.4 μ median particle size; trace on 325 mesh; fineness (Hegman) 5.5; sp.gr. 2.8; dens. 34 lb/ft^3 (tapped); bulk dens. 12.5 lb/ft^3; bulking value 23.3 lb/gal; oil absorp. 39 g/100 g; brightness 90; pH 8.8; 60% SiO_2, 33% MgO, 1% Al_2O_3.

Microtalc® MP44-26. [Specialty Minerals] Untreated talc; CAS 14807-96-6; EINECS 238-877-9; filler for polyolefins; powd.; 5 μ avg. particle size, 44 μ top size.

Microtuff® 325F. [Specialty Minerals] Treated talc; CAS 14807-96-6; EINECS 238-877-9; filler for polyolefins, engineering thermoplastics; powd.; 5 μ avg. particle size, 44 μ top size.

Microtuff® 1000. [Specialty Minerals] Surface-treated platy talc; CAS 14807-96-6; EINECS 238-877-9; filler for polyolefins; powd.; 1.5 μ avg. particle size; uncoated properties: sp.gr. 2.70; bulk dens. 6.7 lb/ft^3 (loose); dry brightness 90; oil absorp. 50; pH 8.8; 60.1% SiO_2; 33.1% MgO.

Microtuff® F. [Specialty Minerals] Surface-treated platy talc; CAS 14807-96-6; EINECS 238-877-9; filler for polyolefins for food pkg. applics. (thermoformed containers, blown film, inj. molded pkg.) and for engineering thermoplastics; powd.; 1.5 μ avg. particle size; sp.gr. 2.70; bulk dens. 7.0 lb/ft^3 (loose); pH 8.8; dry brightness 90; 60.1% SiO_2, 33.1% MgO.

Micro-White® 07 Slurry. [ECC Int'l.] Calcium carbonate; CAS 1317-65-3; ultrafine pigment, filler, reinforcement for paper, paint, polymers, food, caulks, etc.; high brightness pigment for coating high quality papers; offers low visc., low abrasion; slurry, odorless; negligible sol. in water; sp.gr. 2.71; dens.

15.87 lb/gal; visc. 250 cps; surf. area 12 m^2/g; pH 9-10; nonflamm.; 75% solids; *Precaution:* incompat. with acids.

Micro-White® 10. [ECC Int'l.] Calcium carbonate; CAS 1317-65-3; extra fine ground pigment providing improved opacity and gloss props. in paints, coatings, paper; improved physical props. and surf. gloss in plastics and elastomers; slurry; 1 μ mean particle size; oil absorp. 19-21.

Micro-White® 10 Slurry. [ECC Int'l.] Calcium carbonate; CAS 1317-65-3; EINECS 207-439-9; ultrafine pigment, filler, reinforcement with exc. optical props., low visc., low adhesive demand in coatings, good retention as filler; for paper, paint, polymers, food, caulks, etc.; slurry, odorless; negligible sol. in water; sp.gr. 2.71; dens. 15.87 lb/gal; visc. 250 cps; surf. area 8 m^2/g; pH 9-10; nonflamm.; 75% solids; *Precaution:* incompat. with acids.

Micro-White® 15. [ECC Int'l.] Calcium carbonate; CAS 1317-65-3; ultrafine wet-ground pigment, filler, reinforcement for paint and polymer industries; used in water and solv.-based coatings, rubber and plastics compds., caulks, sealants, adhesives, etc.; NSF compliance; slurry or wh. powd., odorless; 2.0 μm mean particle size; fineness (Hegman) 6.0; negligible sol. in water; sp.gr. 2.71; dens. 39 lb/ft^3 (loose); surf. area 5.0 m^2/g; oil absorp. 17; pH 9.5 (5% slurry); nonflamm.; 97.6% $CaCO_3$, 0.2% max. moisture; *Toxicology:* TLV/TWA 10 mg/m^3, considered nuisance dust; *Precaution:* incompat. with acids.

Micro-White® 15 SAM. [ECC Int'l.] Calcium carbonate, coated with < 2% stearic acid; CAS 1317-65-3, 57-11-4 resp.; ultrafine coated pigment, filler, reinforcement for applics. requiring superior physical props.; for PVC pipe and conduit, flexible sheeting and molding compds., filled PP and PE, wire and cable compds., rubber-based sealants and adhesives; hydrophobic; NSF compliance; wh. powd., odorless; 2.0 μm mean particle dize; fineness (Hegman) 6.0; sp.gr. 2.71; dens. 45 lb/ft^3 (loose); surf. area 5.0 m^2/g; oil absorp. 14; pH 9.5 (5% slurry); nonflamm.; 97.6% $CaCO_3$, 0.2% max. moisture; *Toxicology:* TLV/TWA 10 mg/m^3, considered nuisance dust; *Precaution:* incompat. with acids.

Micro-White® 15 Slurry. [ECC Int'l.] Calcium carbonate; CAS 1317-65-3; EINECS 207-439-9; fine, wet-ground filling pigment, reinforcement offering exc. retention in the sheet, exc. optical props.; for paper, paint, polymers, food, caulks, matte or dull coatings, as precoating for blade-applied topcoats; slurry, odorless; negligible sol. in water; sp.gr. 2.71; dens. 15.87 lb/gal; visc. 500 cps; surf. area 5 m^2/g; pH 9-10; nonflamm.; 75% solids; *Precaution:* incompat. with acids.

Micro-White® 25. [ECC Int'l.] Calcium carbonate; CAS 1317-65-3; easy dispersing pigment, filler, reinforcement for paint, rubber, plastics incl. PVC compds., filled polyolefins, water and solv.-based coatings, caulks, sealants, floor coverings, etc.; NSF compliance; wh. powd., odorless; 3.0 μ mean particle size; fineness (Hegman) 6.0; negligible sol. in water; sp.gr. 2.71; dens. 45 lb/ft^3 (loose); surf. area 2.8 m^2/g; oil absorp. 15; pH 9.5 (5% slurry); nonflamm.; 97.6% $CaCO_3$, 0.2% max. moisture; *Toxicology:* TLV/TWA 10 mg/m^3, considered nuisance dust; *Precaution:* incompat. with acids.

Micro-White® 25 SAM. [ECC Int'l.] Calcium carbonate, surface treated; coated pigment designed for easy dispersion in plastic compds., e.g., polyolefins, rigid and flexible PVC; hydrophobic for superior wire and cable insulation compds., improved impact resist. in PP; wh. powd., odorless; 3.0 μm mean particle size; sp.gr. 2.71.

Micro-White® 25 Slurry. [ECC Int'l.] Calcium carbonate; CAS 1317-65-3; EINECS 207-439-9; med.-fine wet-ground filler pigment, reinforcement for paper, paint, polymers, food, caulks, matte or dull coatings; slurry, odorless; negligible sol. in water; sp.gr. 2.71; dens. 15.32 lb/gal; visc. 500 cps; surf. area 3.5 m^2/g; pH 9-10; nonflamm.; 72% solids; *Precaution:* incompat. with acids.

Micro-White® 40. [ECC Int'l.] Calcium carbonate; CAS 1317-65-3; economic med.-fine ground pigment, filler, reinforcement for flexible PVC applics., caulks, adhesives, coatings; wh. powd., odorless; 4 μ mean particle size; fineness (Hegman) 5.0; negligible sol. in water; sp.gr. 2.71; dens. 40 lb/ft^3 (loose); oil absorp. 12; pH 9.5 (5% slurry); nonflamm.; 96% $CaCO_3$, 0.2% max. moisture; *Toxicology:* TLV/TWA 10 mg/m^3, considered nuisance dust; *Precaution:* incompat. with acids.

Micro-White® 100. [ECC Int'l.] Calcium carbonate; CAS 1317-65-3; wh. general-purpose pigment, filler, reinforcement for paints, rubber goods, caulks, putties, spackles, joint compds., rug and textile backings, mild abrasive compds.; wh. powd., odorless; 13.5 μ mean particle size; 7% on 325 mesh; negligible sol. in water; sp.gr. 2.71; dens. 60 lb/ft^3 (loose); pH 9.5 (5% slurry); nonflamm.; 95% $CaCO_3$, 0.2% max. moisture; *Toxicology:* TLV/TWA 10 mg/m^3, considered nuisance dust; *Precaution:* incompat. with acids.

Mikrofine ADC-EV. [High Polymer Labs] Modified azodicarbonamide; CAS 123-77-3; EINECS 204-650-8; chem. foaming agent for EVA, HDPE, LDPE; processing temps. 250-320 F.

Mikrofine AZDM. [High Polymer Labs] 2,2′-Azobisisobutyronitrile; CAS 78-67-1; EINECS 201-132-3; chem. foaming agent for silicone rubber, semirigid PVC; processing temps. 200-240 F; gas yield 125 cc/g.

Mikrofine BSH. [High Polymer Labs] Benzene sulfonyl hydrazide; CAS 80-17-1; chem. foaming agent for rubber cloth laminates; processing temps. 195-230 F; gas yield 90 cc/g.

Mikrofine OBSH. [High Polymer Labs] 4,4′-Oxybis (benzenesulfonyl) hydrazine; CAS 80-51-3; chem. foaming agent for EVA, LDPE, PS, flexible PVC; processing temps. 250-375 F; gas yield 120-125 cc/g.

Mikrofine TSH. [High Polymer Labs] p-Toluene sulfonyl hydrazide; CAS 877-66-7; chem. foaming agent for unsat. polyester, rubber cloth laminates; processing temps. 220-250 F; gas yield 115 cc/g.

Mikrofine TSSC. [High Polymer Labs] p-Toluene sulfonyl semicarbazide; chem. foaming agent for ABS, acetal, acrylic, EVA, HDPE, LDPE, PPO, PP, PS, HIPS, flexible and rigid PVC, TPE; processing temps. 390-455 F; gas yield 145 cc/g.

Miljac Barium Stearate. [Miljac] Barium stearate; CAS 6865-35-6; internal/external lubricant for vinyls, PVC, PS foams; solid.

Miljac Cadmium Stearate. [Miljac] Cadmium stearate; CAS 2223-93-0; internal/external lubricant for PVC; solid.

Miljac Calcium Stearate. [Norac] Calcium stearate; CAS 1592-23-0; EINECS 216-472-8; internal/external lubricant for ABS, PS, PVC, SMC/TMC/BMC polyesters; solid.

Miljac Magnesium Stearate. [Miljac] Magnesium stearate USP; CAS 557-04-0; EINECS 209-150-3; internal/external lubricant for ABS, rigid vinyls (with calcium stearate), cellulose acetate; FDA accepted; solid.

Miljac Zinc Stearate. [Miljac] Zinc stearate; CAS 557-05-1; EINECS 209-151-9; internal/external lubricant for polyolefins, ABS, PS, IPS, rigid/flexible PVC, SMC/TMC/BMC polyesters, phenolics, melamine, alkyd, U-F, polyesters, color concs.; FDA accepted; solid.

Millad® 3905. [Milliken] CAS 32647-67-9; additive to enhance clarity and aesthetics of polyolefins, esp. PP, LLDPE, some HDPE; improves gloss, surf. appearance, increases flex. mod., tens. str., heat distort. temp.; for inj. or blow molding, film or sheet extrusion processes; avail. as precompounded commercial resins from a variety of manufacturers; FDA 21CFR §178.3295; wh. powd.; bulk dens. 20 lb/ft^3; m.p. 220-225 C; 98% purity, < 1% moisture.

Millad® 3940. [Milliken] CAS 54686-97-4, 87826-41-3; additive to enhance clarity and aesthetics of polyolefins, esp. PP, LLDPE, some HDPE; improves gloss, surf. appearance, increases flex. mod., tens. str., heat distort. temp.; for inj. or blow molding, film or sheet extrusion processes; avail. as precompounded commercial resins from a variety of manufacturers; FDA 21CFR §178.3295; wh. powd.; bulk dens. 20 lb/ft^3; m.p. 245-250 C; 98% purity, < 1% moisture.

Millad® 3988. [Milliken] Sorbitol-based; CAS 135861-56-2; clarifying agent for polyolefins providing highest clarity, gloss, surf. smoothness, exceptional nucleation capability, enhanced physical props., faster processing; suitable for food contact or odor-sensitive applics.; FDA 21CFR §177.1520(c) (1.1),(3.1), (3.2); BGA accepted as indirect food additive; ELINCS accepted; Canadian approval pending; wh. powd.; dens. 0.37 g/cc; > 96% purity.

Millad® Conc. 5C41-10. [Milliken] 10% conc. of Millad 3905 in PP random copolymer; clarifying agent for use in PP homopolymers and random copolymers; FDA cleared.

Millad® Conc. 5L71-10. [Milliken] 10% conc. of Millad 3905 in LDPE; clarifying agent for use in LLDPE; FDA cleared.

Millad® Conc. 8C41-10. [Milliken] 10% conc. of Millad 3988 in PP random copolymer; clarifying agent for use in PP homopolymers and random copolymers; FDA cleared.

Millad® HBPA. [Milliken] Hydrog. bisphenol A; used in prep. of alkyd, polyester, and epoxy resins where good color stability and improved weatherability are important; for casting, laminating, coatings and fiber prod.; wh. flake; sol. in wide range of org. solvs.; m.w. 240; m.p. 150 C; hyd. no. 430 min.; 0.5% max. water; *Storage:* keep prod. sealed in storage.

Millamine® 5260. [Milliken] Cycloaliphatic diamine (93% 1,2-diaminocyclohexane, 6% methylpentamethylenediamine, 0.8% hexamethylenediamine); epoxy curing agent offering outstanding heat distort. temp., chem. resist. and physical props. in cured system; also as corrosion inhibitor, PU crosslinker, intermediate for polyamides and other chems.; colorless low visc. liq.; m.w. 114; sp.gr. 0.9408; b.p. 188 C; ref. index 1.4869; *Toxicology:* corrosive; *Storage:* avoid contact with the atmosphere to prevent yellowing.

Milldride® 5060. [Milliken] Cycloaliphatic dianhydride; epoxy curing agent, crosslinker giving high HDT and chem. resist. to polyesters, polyimides, polyamides; as polyimide intermediate; component for wire coatings; in specialty coatings such as uv-cured inks; improved sol. in epoxy resins, improved filler loading, longer pot life; fine powd.; 98% thru 200 mesh; m.w. 264; bulk dens. 0.3 g/cc; m.p. 70-90 C; sapon. no. 925; *Toxicology:* may cause skin and eye irritation, sensitization in susceptible individuals; produces irritating vapors when heated above its m.p.; *Storage:* limit exposure to moixture and humidity; reseal containers after removing the prod.

Milldride® DDSA. [Milliken] Dodecenyl succinic anhydride (C12-branched); curing agent for epoxy resins, corrosion inhibitor for nonaq. lubricating oils, intermediate for prep. of alkyd or unsat. polyester resins, platicizers, intermediate in chem. reactions; also for drying oils, lacquers, paints, surfactants, asphalt additives, waterproofing agents, textiles, leather, starch modifier, germicides, hydrocarbon fuel additives; lt. yel. clear visc. liq.; m.w. 266; sp.gr. 1.003-1.008; visc. 440 cps; b.p. 220 C (10 mm Hg); sapon. no. 418; neutral. no. 394-432; flash pt. 343-347 F.

Milldride® HDSA. [Milliken] Hexadecenyl succinic anhydride; curing agent for epoxy resins, corrosion inhibitor for nonaq. lubricating oils, intermediate for prep. of alkyd or unsat. polyester resins, intermediate in chem. reactions; m.w. 294; sp.gr. 0.95; b.p. 235 C (5 mm Hg); sapon. no. 348.

Milldride® HHPA. [Milliken] Hexahydrophthalic anhydride; CAS 85-42-7; epoxy curing agent; also for prep. of alkyd and polyester resins where good color stability is important; for casting, laminating, embedding, coating, and impregnating elec. components; wh. solid; misc. with most org. solvs.; m.w. 154; sp.gr. 1.18 (40 C); visc. 23 cps (60 C); m.p. 35

C; b.p. 110 C (5 mm); solid. pt. 35 C min.; neutral. no. 720 ± 10.

Milldride® MHHPA. [Milliken] 4-Methyl hexahydrophthalic anhydride and a small amt. of hexahydrophthalic anhydride in a eutectic blend; epoxy curing agent; for casting and impregnation applics. where low visc., light color and good heat resist. are required; colorless clear liq.; misc. with most org. solvs.; m.w. 164; sp.gr. 1.17; visc. 50-70 cps; m.p. -15 C; b.p. 127 C (5 mm Hg); flash pt. (OC) 160 C; *Toxicology:* LD50 (oral, rat) 3300 mg/kg; sl. to mod. skin irritation; *Storage:* hygroscopic; keep free of moisture during storage; undergoes hydrolysis in presence of moisture to form the dicarboxylic acid.

Milldride® nDDSA. [Milliken] n-Dodecenyl succinic anhydride (C12-linear); curing agent for epoxy resins, corrosion inhibitor for nonaq. lubricating oils, intermediate for prep. of alkyd or unsat. polyester resins, intermediate in chem. reactions; m.w. 266; sp.gr. 0.96; b.p. 175 C (1 mm Hg); solid. pt. 35 C; sapon. no. 421.

Milldride® nDSA. [Milliken] n-Decenyl succinic anhydride; curing agent for epoxy resins, corrosion inhibitor for nonaq. lubricating oils, intermediate for prep. of alkyd or unsat. polyester resins, intermediate in chem. reactions; m.w. 238; sp.gr. 1.005; b.p. 195 C (10 mm Hg); solid. pt. 16 C; sapon. no. 471.

Milldride® ODSA. [Milliken] Octadecenyl succinic anhydride; curing agent for epoxy resins, corrosion inhibitor for nonaq. lubricating oils, intermediate for prep. of alkyd or unsat. polyester resins, intermediate in chem. reactions; detergent dispersant for lubricating oil; detergent builder; pour pt. depressant, visc. index improver; paper sizing; water repellent compositions; yel.-wh. waxy solid; m.w. 350; sp.gr. 0.9428; visc. 175-250 SUS (210 F); b.p. 251 C (4 mm Hg); solid. pt. 69 C; sapon. no. 320; flash pt. 210 C; *Storage:* avoid moisture during storage; undergoes slow hydrolysis to form corresponding alkenylsuccinic acid.

Milldride® OSA. [Milliken] Octenyl succinic anhydride; CAS 26680-54-6; curing agent for epoxy resins, corrosion inhibitor for nonaq. lubricating oils, intermediate for prep. of alkyd or unsat. polyester resins, intermediate in chem. reactions; m.w. 210; sp.gr. 1.0; f.p. 50 F max.; b.p. 168 C (10 mm Hg); solid. pt. 10 C; sapon. no. 534; flash pt. (COC) 185 C.

Milldride® TDSA. [Milliken] Tetradecenyl succinic anhydride; curing agent for epoxy resins, corrosion inhibitor for nonaq. lubricating oils, intermediate for prep. of alkyd or unsat. polyester resins, intermediate in chem. reactions; APHA 300 color; m.w. 294; sp.gr. 0.95; b.p. 235 C (15 mm Hg); solid. pt. 45 C; sapon. no. 381.

Millithix® 925. [Milliken] Dibenzylidene sorbitol; CAS 32647-67-9; EINECS 251-136-4; thixotrope and gellant for use in unsat. polyester and vinyl ester resins; also as clear antiperspirant; solid wh. powd.; *Precaution:* a static charge may accumulate during handling; observe precautions when using around flamm. or explosive liqs.; *Storage:* store in closed container in dry environment away from sources of ignition.

Millrex™. [C.P. Hall] Processing and release aid eliminating sticking in compounded elastomeric stocks, esp. neoprene, soft sponge, friction, or other low durometer stocks; reduces milling, mixing, calendering, and extruding temps.; helps prevent scorching; processing aid for high reclaim content stocks; colorless clear oily liq.; sp.gr. 0.85 ± 0.01.

Minbloc™ 16. [Unimin Spec. Minerals] Nepheline syenite; CAS 37244-96-5; filler for clear vinyl, LDPE and LLDPE films incl. agric. film, theromset polyester compds., food pkg.; GRAS for indirect food contact; wh. powd., odorless; negligible sol. in water; sp.gr. 2.61; m.p. 1223 C.

Minbloc™ 20. [Unimin Spec. Minerals] Nepheline syenite; CAS 37244-96-5; filler for clear vinyl, LDPE and LLDPE films incl. agric. film, theromset polyester compds., food pkg.; GRAS for indirect food contact; wh. powd., odorless; negligible sol. in water; sp.gr. 2.61; m.p. 1223 C.

Minbloc™ 30. [Unimin Spec. Minerals] Nepheline syenite; CAS 37244-96-5; filler for clear vinyl, LDPE and LLDPE films incl. agric. film, theromset polyester compds., food pkg.; GRAS for indirect food contact; wh. powd., odorless; negligible sol. in water; sp.gr. 2.61; m.p. 1223 C.

Minex® 4. [Unimin Canada; Unimin Spec. Minerals] Nephylene syenite; CAS 37244-96-5; filler for architectural and traffic paints and coatings, thermoset plastics, adhesives, caulks, and sealants, friction prods., rubber compds., polyester BMC; wh. fine powd., odorless; 99.98% thru 325 mesh; 7.5 μ median particle size; fineness (Hegman) 4.5; sp.gr. 2.61; dens. 21.7 lb/gal; surf. area 0.7 m^2/g; oil absorp. 26; brightness 88.4; m.p. 1722 C; ref. index 1.53; pH 9.8; hardness (Mohs) 5.5-6; 60% SiO_2, 23.6% Al_2O_3, 10.6% Na_2O, 4.8% K_2O; *Toxicology:* ACGIH TLV 10 mg/m^3; prolonged inh. of dust may cause delayed lung injury.

Minex® 7. [Unimin Canada; Unimin Spec. Minerals] Nephylene syenite; CAS 37244-96-5; filler for coatings (architectural, traffic, powd., primers), polyester BMC, PVC plastics and plastisols, polyolefin films, adhesives and sealants, and rubber compds.; high brightness contributes blue undertones; wh. fine powd., odorless; 100% thru 325 mesh; 3.9 μ median particle size; fineness (Hegman) 5.5+; sp.gr. 2.61; dens. 21.7 lb/gal; surf. area 1.2 m^2/g; oil absorp. 31; brightness 89.8; m.p. 1722 C; ref. index 1.53; pH 10.0; hardness (Mohs) 6; 59.9% SiO_2, 23.5% Al_2O_3, 10.5% Na_2O, 4.9% K_2O; *Toxicology:* ACGIH TLV 10 mg/m^3; prolonged inh. of dust may cause delayed lung injury.

Minex® 10. [Unimin Canada; Unimin Spec. Minerals] Nephylene syenite; CAS 37244-96-5; filler for coatings (primers, interior/exterior architectural, powd.), thermoset plastics, PVC plastics and plastisols, polyolefin films, polyester gelcoats, adhesives and sealants, and inks; high brightness contributes blue undertones; wh. fine powd., odorless;

100% thru 325 mesh; 2.3 μ median particle size; fineness (Hegman) 6+; sp.gr. 2.61; dens. 21.7 lb/gal; surf. area 1.7 m^2/g; oil absorp. 33; brightness 90.4; m.p. 1722 C; ref. index 1.53; pH 10.1; hardness (Mohs) 6; 60% SiO_2, 23.4% Al_2O_3, 10.4% Na_2O, 4.8% K_2O; *Toxicology:* ACGIH TLV 10 mg/m^3; prolonged inh. of dust may cause delayed lung injury.

Min-U-Sil® 5. [Floridin; U.S. Silica] Natural cryst. silica; high purity filler and extender for architectural and traffic paints, powd. coatings, protective coatings, wire and cable coating compds., adhesives/sealants, ceramic, high-temp. insulation, epoxy castings; semireinforcing filler for silicone rubber; filler in thermoplastics and thermosets incl. polyester and epoxy; wh. powd.; fineness (Hegman) 7.5; sp.gr. 2.65; dens. 22.07 lb/gal; bulk dens. 35.7 lb/ft^3 (untapped); oil absorp. 42-44; pH 7-8; GE brightness 88.8-95.8; 99.14-99.25% SiO_2.

Min-U-Sil® 10. [Floridin; U.S. Silica] Natural cryst. silica; high purity filler and extender for architectural and traffic paints, powd. coatings, protective coatings, silicone rubber, wire and cable coating compds., adhesives/sealants, plastics, ceramic, high-temp. insulation, epoxy castings; semireinforcing filler for silicone rubber; filler in thermoplastics and thermosets incl. polyester and epoxy; wh. powd.; fineness (Hegman) 7.25-7.5; sp.gr. 2.65; dens. 22.07 lb/gal; bulk dens. 42.7 lb/ft^3 (untapped); oil absorp. 33.3-35.1; pH 7.3-7.7; GE brightness 91-94.5; 99.52-99.55% SiO_2.

Min-U-Sil® 15. [Floridin; U.S. Silica] Natural cryst. silica; high purity filler and extender for architectural and traffic paints, powd. coatings, protective coatings, silicone rubber, wire and cable coating compds., adhesives/sealants, plastics, ceramic, high-temp. insulation, epoxy castings; semireinforcing filler for silicone rubber; filler in thermoplastics and thermosets incl. polyester and epoxy; wh. powd.; fineness (Hegman) 7.25; sp.gr. 2.65; dens. 22.07 lb/gal; bulk dens. 46 lb/ft^3 (untapped); oil absorp. 33; pH 7.7; GE brightness 94.1; 99.58% SiO_2.

Min-U-Sil® 30. [Floridin; U.S. Silica] Natural cryst. silica; high purity filler and extender for architectural and traffic paints, powd. coatings, protective coatings, silicone rubber, wire and cable coating compds., adhesives/sealants, plastics, ceramic, high-temp. insulation, epoxy castings; semireinforcing filler for silicone rubber; filler in thermoplastics and thermosets incl. polyester and epoxy; wh. powd.; fineness (Hegman) 5.25-5.5; sp.gr. 2.65; dens. 22.07 lb/gal; bulk dens. 48.5 lb/ft^3 (untapped); oil absorp. 23-26.5; pH 7.0-7.7; GE brightness 89.3-89.9; 99.62-99.76% SiO_2.

Min-U-Sil® 40. [Floridin; U.S. Silica] Natural cryst. silica; high purity filler and extender for architectural and traffic paints, powd. coatings, protective coatings, silicone rubber, wire and cable coating compds., adhesives/sealants, plastics, ceramic, high-temp. insulation, epoxy castings; semireinforcing filler for silicone rubber; filler in thermoplastics and thermosets incl. polyester and epoxy; wh. powd.; fineness (Hegman) 4.5; sp.gr. 2.65; dens. 22.07 lb/gal; oil absorp. 22; pH 7.0; GE brightness 90.5; 99.76% SiO_2.

MIPA. [Dow] Monoisopropanolamine; used to produce soaps with good hard surf. detergency, shampoos, emulsifiers, textile specialties, agric. and polymer curing chemicals, adhesives, coatings, metalworking, petroleum, rubber processing, gas conditioning chemicals; sp.gr. 0.960 (20/4 C); dens. 7.95 lb/gal; visc. 23 cps; f.p. 3 C; b.p. 159 C (760 mm Hg); flash pt. (TCC) 173 F; ref. index 1.4456.

Miramine® C. [Rhone-Poulenc Surf. & Spec.] Coco hydroxyethyl imidazoline; CAS 67784-90-1; EINECS 263-170-7; cationic; emulsifier, corrosion inhibitor, softener, antistat, for textiles, asphalt, plastics, petrol. industry, cutting oils; water repellent treatment of cement, concrete, and plaster; antifungal agent for wood; tar emulsion breaker; slime control additive in paperboard; brn. paste; amine odor; sol. in aq. acid; disp. in water; dens. 0.96 g/ml; f.p. 20 C; 85% act.

Miranol® CM-SF Conc. [Rhone-Poulenc Surf. & Spec.] Sodium cocoamphopropionate; CAS 68919-41-5; amphoteric; mild foaming agent, conditioner, detergent, emulsifier, foam booster/stabilizer for shampoos, liq. soaps, facial cleansers, bath gels, bubble baths; coemulsifier for emulsion polymerization; emulsifier, wetting agent for industrial, institutional and household cleaners; biodeg.; lt. amber clear liq.; sol. in water and alcohol; pH 9.5-10.5; 36-38% solids.

Miranol® J2M Conc. [Rhone-Poulenc Surf. & Spec.] Disodium capryloamphodiacetate; CAS 68608-64-0; amphoteric; emulsifier, caustic soda wetting agent, food washing and peeling, industrial, institutional and household cleaners, wax stripper, emulsion polymerization of syn. rubbers and resins; biodeg.; clear liq.; pH 8.2-8.6; 48-50% solids.

Miranol® SM Conc. [Rhone-Poulenc Surf. & Spec.] Sodium caproamphoacetate; CAS 68608-61-7; amphoteric; wetting agent, foaming agent, detergent used in medicated and germicidal shampoos and hand soaps, rug and upholstery shampoos, in emulsion polymerization; biodeg.; clear liq.; pH 9.0-9.5; 40-42% solids.

MiRaSlip VRA-102. [M-R-S] Blend; internal lubricant for rigid and flexible PVC and plastisols; liq.

MiRaStab 113-K. [M-R-S] Calcium-zinc; heat and lt. stabilizer for flexible and rigid PVC, plastisols; nontoxic; paste.

MiRaStab 130-K. [M-R-S] Calcium-zinc; heat stabilizer, lubricant for flexible and rigid PVC, plastisols, organosols; nontoxic; powd.

MiRaStab 140-K. [M-R-S] Calcium-zinc; heat and lt. stabilizer, lubricant for flexible and rigid PVC, plastisols, organosols; nontoxic; powd.

MiRaStab 154-K. [M-R-S] Magnesium-zinc; heat and lt. stabilizer for flexible and rigid PVC, plastisols; nontoxic; powd.

MiRaStab 400-K. [M-R-S] Lead; heat and lt. stabilizer,

activator for flexible and rigid PVC, platisols, organosols, foams, tile.

MiRaStab 403-K. [M-R-S] Barium-zinc; heat stabilizer, activator for fleixble PVC, plastisols, organosols, foams; liq.

MiRaStab 704-K. [M-R-S] Phosphite; heat and lt. stabilizer for flexible and rigid PVC, plastisols, organosols, foams, bottles, tile; nontoxic.

MiRaStab 706-K. [M-R-S] Phosphite; antioxidant for flexible and rigid PVC, plastisols, organosols, foams, bottles, tile.

MiRaStab 709-ZK. [M-R-S] Phosphite fortified with zinc; heat and lt. stabilizer for flexible PVC, plastisols, organosols.

MiRaStab 800-K. [M-R-S] Epoxy; low temp. plasticizer, heat and lt. stabilizer for flexible and rigid PVC, plastisols, organosols, foams, bottles, tile.

MiRaStab 919-K. [M-R-S] Calcium-zinc; heat stabilizer, lubricant for flexible and rigid PVC, bottles; nontoxic; powd.

MiRaStab 922-K. [M-R-S] Calcium-tin-zinc; heat and lt. stabilizer for flexible and semirigid PVC, plastisols, organosols.

MiRaStab 950-K. [M-R-S] Calcium-zinc; heat and lt. stabilizer for filled systems, flexible PVC, plastisols, organosols; liq.

Mistron CB. [Luzenac Am.] Platy talc, surf.-treated; CAS 14807-96-6; EINECS 238-877-9; reinforcing filler offering exc. impact props. to polyolefins, good film props. to films, high modulus, superior processing, improved tear and aging props. in rubber; ultrafine powd.; 1.7 μ mean particle size; fineness (Hegman) 5.5; sp.gr. 2.8; bulk dens. 8 lb/ft^3 (loose); bulking value 23.3 lb/gal; surf. area 16 m^2/g; oil absorp. 42; brightness (TAPPI) 86; pH 9.5 (10% slurry); 98% talc; 63.3% SiO_2, 31.1% MgO; 0.5% moisture.

Mistron PXL®. [Luzenac Am.] Platy microcryst. talc, surf. treated with a coupling agent; CAS 14807-96-6; EINECS 238-877-9; reinforcing pigment used with peroxide-cured vulcanized rubber enhancing mech. and elec. props.; provides high modulus, low compr., superior rheology; exhibits exc. adhesion to rubber; permits higher loadings; ultrafine powd.; 1.7 μ mean particle size; fineness (Hegman) 5.5; sp.gr. 2.8; bulking value 23.3 lb/gal; surf. area 16 m^2/g; oil absorp. 42; brightness (GE) 87; dens. 8 lb/ft^3 (loose); 98% talc; 63.3% SiO_2, 31.1% MgO; 0.5% moisture.

Mistron Vapor. [Luzenac Am.] Talc; CAS 14807-96-6; EINECS 238-877-9; provides higher extrusion rates in rubber industry, pitch control in pulp and paper industry.

Mistron Vapor® R. [Luzenac Am.] Platy microcryst. talc; CAS 14807-96-6; EINECS 238-877-9; reinforcing filler, processing aid for elastomers, plastics; rheological control agent for resins, plastisols, adhesives; nucleating agent for plastic foams; ultrafine powd.; 1.7 μ mean particle size; fineness (Hegman) 5.5; sp.gr. 2.8; bulk dens. 8 lb/ft^3 (loose); bulking value 23.3 lb/gal; surf. area 16 m^2/g; oil absorp. 40; brightness (TAPPI) 87; pH 9.5 (10% slurry); 98% talc; 61% SiO_2, 31% MgO; 0.5% moisture.

Mistron ZSC. [Luzenac Am.] Talc, zinc stearate-coated; flatting agent in solv.-based coatings, rheological control in resins, plastisols and adhesives, nucleating agent in plastic foams, mold release agent, rubber dusting aid; powd.; 1.5 μ median diam.; fineness (Hegman) 6.0; sp.gr. 2.8; dens. 8 lb/ft^3 (loose); bulking value 23.3 lb/gal; surf. area 16 m^2/g; oil absorp. 43; 0.5% moisture.

Mitsubishi Carbon Black. [Mitsubishi Kasei] Carbon black; CAS 1333-86-4; EINECS 215-609-9; various grades for high color, med. color, regular color, long flow, oil pellet, conductive props. for use in inks, rubber, plastics, and paint.

Mixxim® AO-20. [Fairmount] Phenol; nondiscoloring, nonstaining antioxidant for PP, PE, EPDM; FDA 21CFR §177.1520; colorless, odorless.

Mixxim® AO-30. [Fairmount] Phenol; nondiscoloring, nonstaining antioxidant for ABS, cellulosics, nylon, polyester, polyolefins, PS, PVC, acetals, EPDM, PU; FDA 21CFR §175.105, 178.2010; colorless, odorless.

Mixxim® BB/100. [Fairmount] Bis[2-hydroxy-5-t-octyl-3-(benzotriazol-2-yl)phenyl]methane; CAS 103597-45-1; uv light absorber for processing engineering resins (nylon, PC, PET, PBT, PPO) and PVC, styrene, and acrylics; highly effective where long term permanent uv light stability is required; wh. flowable powd.; sol. (g/100 ml): 3 g in ethyl acetate, 2.5 g in xylene, < 1 g in ethanol, hexane, acetone; insol. in water; m.w. 658; m.p. 196 C; *Precaution:* avoid dusting, open flames or sparks; *Storage:* store in cool, dry area in closed containers to minimize moisture contact.

Mixxim® BB/200. [Fairmount] Bis[2-hydroxy-5-t-methyl-3-(benzotriazol-2-yl)phenyl]methane; CAS 30653-05-5; uv light absorber for processing engineering resins (nylon, PC, PPS, PET, PBT, PPO), esp. at high temps.; low volatility; broad uv absorp. range (290-390 nm); lt. yel. powd.; sol. (g/100 g): 0.1 g in DMF, pseudocumene; insol. in methanol, acetonitrile; m.w. 462; m.p. 285 C min.; 98% min. assay; *Precaution:* avoid dusting, open flames or sparks; *Storage:* store in cool, dry area in closed containers to minimize moisture contact.

Mixxim® DBM. [Fairmount] Beta diketone; heat stabilizer for rigid PVC; replacement for heavy metals.

Mod Acid. [Ruetgers-Nease] Modified toluene sulfonic acid prod.; hydrotrope, solv., intermediate; catalyst; crosslinking catalyst for phenolic adhesives, melamine-formaldehyde and acrylic polymers; curing catalyst in polymerization of furan for foundry molds or cores; vulcanization of rubber; prep. of acetals, esters; cryst. range 0-15 C; 92% min. assay; *Toxicology:* corrosive; avoid contact with skin and eyes.

Modaflow®. [Monsanto] Plasticizer.

Modarez APVC 8. [Protex] Acrylic polymer; modifier for rigid PVC for mfg. of bottles and hollow casings; improves surface and gloss; gives exc. transparency; for extrusion, glass-blowing extrusion, inj. molding, calendering; FDA 21CFR §178.3790; various European approvals regarding food contact; wh. free-flowing powd.; misc. with toluene, ethyl acetate; insol. in water; apparent dens. 0.40 ±0.10 g/cc; 99.5% min. dry material; *Toxicology:* only very sl. toxic; *Precaution:* incompat. with strong oxidizers; *Storage:* store in well-ventilated area away from exposure to direct light, humidity, excessive heat.

Modicol S. [Henkel/Functional Prods.] Sulfated fatty acid; anionic; stabilizer; in rubber industry to stabilize natural and syn. latexes during compding., storage, and applic; ensures mechanical and chemical stability; prevents premature coagulation during high-speed agitation, acidification, or pigmentation; dk. rdsh.-brn. liq.; fatty odor; sol. in water and sol. alcohols; hazy sol'n. in 4% sodium hydroxide, in toluol, and in min. and veg. oils; sp.gr. 1.1; pH 7.5 (2%); 55% act.

Mogul® L. [Cabot/Special Blacks Div] Carbon black, aftertreated; CAS 1333-86-4; EINECS 215-609-9; colorant, uv stabilizer for elec. resistivity in plastics, optimum flow in offset inks, carbon paper, and ribbons; good for dry toners; fluffy; 24 nm particle size; dens. 15 lb/ft^3; surf. area 138 m^2/g; pH 2.5-4.0; 5.0% volatile.

Mold-Ease. [Merix] Nonsilicone mold lubricant for plastics, ceramic, and rubber articles.

Molder's Edge ME-100. [Zip-Chem Prods.] Silicone spray; general purpose mold release and lubricant compat. with wide variety of polymers incl. ABS, acetal, acrylic, alkyd, cellulosics, PC, PP, PS, polysulfone, PVC, urethane elastomers and foams, invest waxes; nonflamm.; free of ozone-damaging chlorofluorocarbons; FDA/USDA compliance.

Molder's Edge ME-110. [Zip-Chem Prods.] Silicone spray; heavy-duty mold release and lubricant compat. with ABS, acetal, alkyd, amino, cellulosics, PC, PVC, urethane elastomers and foams; exc. on problem molds, plaster casts, plastic extrusions; nonflamm.; free of ozone-damaging chlorofluorocarbons; FDA compliance.

Molder's Edge ME-120. [Zip-Chem Prods.] Silicone spray; general purpose mold release and lubricant recommended for PC and plastics with cirtical surf. tolerance such as nylon, polyethylene, acetal; also suitable for ABS, acrylic, alkyd, amino, cellulosics, PS, polysulfone, PVC, urethane elastomers; nonflamm.; free of ozone-damaging chlorofluorocarbons; FDA compliance.

Molder's Edge ME-175. [Zip-Chem Prods.] Silicone spray; mold release and lubricant for molding and processing of medical, pharmaceutical, and parts intended for incidental food contact; esp. effective with Noryl® resins; also for ABS, acetal, acrylic, alkyd, cellulosics, PC, polyester, PE, PP, PS, PVC, PU; nonflamm.; free of ozone-damaging chlorofluorocarbons; FDA and USDA compliances.

Molder's Edge ME-211. [Zip-Chem Prods.] Copolymer; non-CFC paintable mold lubricant and release for most types of plastic and rubber processing incl. ABS, acetal, acrylic, cellulosics, nylon, PE, PP, PS, PVC, PU elastomers, investment waxes; not recommended for PC; *Toxicology:* nontoxic.

Molder's Edge ME-232. [Zip-Chem Prods.] Lecithin USP; non-CFC release agent for silicone rubber molds, investment waxes; reduces stress cracking and crazing; inert; FDA/USDA compliance; *Toxicology:* nontoxic.

Molder's Edge ME-263. [Zip-Chem Prods.] General purpose non-CFC paintable mold lubricant and release for molding structural foam parts, parts with deep draws; compat. with ABS, acetal, acrylic, alkyd, amino, cellulosics, nylon, PC, polyester, PE, PP, PS, PVC, rubber, urethane elastomers; recognized under UL-94 flamm. tests; *Toxicology:* nontoxic.

Molder's Edge ME-301. [Zip-Chem Prods.] Petrolatum USP; CAS 8027-32-5; non-CFC lubricant, pattern release for silicone rubber molds; will not cause cracking or crazing; FDA/USDA compliance.

Molder's Edge ME-304. [Zip-Chem Prods.] Mineral oil USP; CAS 8027-32-5; non-CFC lubricant, release agent for nylon, acetals, polyethylene, PS, and similar thermoplastics; reduced environmental stress cracking and crazing; inert; FDA/USDA compliance; *Toxicology:* nontoxic.

Molder's Edge ME-341. [Zip-Chem Prods.] Fluorotelemer; non-CFC dry lubricant, mold release for virtually all thermoplastic and most thermoset resins; useful on hot or cold molds; withstands high temps.; exc. for molding silicone rubber.

Molder's Edge ME-345. [Zip-Chem Prods.] Paintable non-CFC mold lubricant for thermoset applics.; general purpose release for processing thermoplastics, rubber, investment wax casting; compat. with ABS, acetal, acrylic, amino, nylon, PC, polyester, polyethylene, PP, PS, PVC, rubber, urethane elastomers; FDA compliance; *Toxicology:* nontoxic.

Molder's Edge ME-369. [Zip-Chem Prods.] Zinc stearate USP; CAS 557-05-1; EINECS 209-151-9; non-CFC high temp. lubricant for acrylics, PC, ABS, styrenes, polysulfone, PP, acetal, alkyd, amino, cellulosics, nylon, polyester, polyethylene, rubber.

Molder's Edge ME-448. [Zip-Chem Prods.] Non-CFC release agent for epoxy, phenolic, polyester, urethane elastomers; works on steel, aluminum, plaster, and stone molds; exc. for use in inj., transfer and compr. molding, lamination, encapsulation, pour casting, and pressure molding; *Toxicology:* low toxicity.

Molder's Edge ME-514. [Zip-Chem Prods.] Non-CFC paintable mold lubricant for processing of electronic, computer, optical, medical, and pharmaceutical parts; esp. useful with PC, but suitable for virtually all thermoplastic resins, rubber, and some epoxies and phenolics; FDA/USDA compliance; *Toxicology:* nontoxic.

Molder's Edge ME-515. [Zip-Chem Prods.] Silicone-based; non-CFC release agent for urethane elastomers and foams; inert; exc. for use in inj. molding, casting, and pour molding; *Toxicology:* nontoxic.

Molder's Edge ME-5440. [Zip-Chem Prods.] Similar to Teflon®; non-CFC heavy-duty, multipurpose dry lubricant stable to 500 F; contains a bonding agent that allows it to adhere to molds, reducing transfer to parts; ideal for cold molding; conforms to MIL specs; compat. with ABS, acetal, acrylic, alkyd, amino, cellulosics, epoxy, phenolics, nylon, PC, polyester, polyethylene, PP, PS, polysufone, rubber, urethane elastomers and foam.

Molder's Edge ME-7000. [Zip-Chem Prods.] Silicone blend spray; general purpoose non-CFC mold release and lubricant compat. with ABS, acetal, acrylic, alkyd, amino, cellulosics, PC, PP, PS, polysulfone, PVC, PU elastomers and foams, investment waxes; nonflamm.; free of ozone-damaging chlorofluorocarbons; FDA compliance.

Mold Lubricant™ 426. [C.P. Hall] Water-based nonsilicone mold release conc.; mold release agent, lubricant for molding of mech. goods, sponge prods., belts, and footwear; compat. with all elastomers; colorless clear visc. liq.; misc. with water in all proportions; sp.gr. 1.05 ± 0.03; pH 7.5 min. (10%).

MoldPro 613. [Witco/PAG] Mold release for epichlorohydrin, EPDM, fluoroelastomers, TPE, nitrile, acrylic; imparts drastic reduction in coeff. of friction on elastomeric surfs.; internal processing agent; powd.; dens. 0.91 mg/m^3; m.p. 178 F; *Toxicology:* not primary skin irritants, but may cause drying on prolonged contact; *Precaution:* will burn if ignited; dust may cause explosion hazard.

MoldPro 616. [Witco/PAG] Mold release for acrylic elastomers; mold flow agent for epichlorohydrin and EPDM; powd.; dens. 0.93 mg/m^3; m.p. 248 F; *Toxicology:* not primary skin irritants, but may cause drying on prolonged contact; *Precaution:* will burn if ignited; dust may cause explosion hazard.

MoldPro 619. [Witco/PAG] General processing and extrusion aid, mold flow agent for nitrile rubber; general processing aid for neoprene, butyl rubber; powd.; dens. 0.93 mg/m^3; m.p. 284 F; *Toxicology:* not primary skin irritants, but may cause drying on prolonged contact; *Precaution:* will burn if ignited; dust may cause explosion hazard.

MoldPro 759. [Witco/PAG] Internal mold release and flow agent for nylon 6/6, nylon 6.

MoldPro 830. [Witco/PAG] Internal mold release agent for styrene butadiene block copolymer.

Mold-Release 225. [Lilly] Blend; general-purpose mold parting agent, external lubricant for thermoplastic and thermosetting compds.; liq.

Mold-Release 605. [Lilly] Blend; external lubricant for polyester; liq.

Mold Wiz DCZ. [Axel Plastics Research Labs] Internal antistat for acrylics, cellulose acetate, PE, PP, PS, flexible PVC.

Mold Wiz Ext. 249. [Axel Plastics Research Labs] External lubricant for polyesters, epoxies, natural and syn. rubbers, preforms, filament winding, PVC; liq.

Mold Wiz Ext. 424/7. [Axel Plastics Research Labs] External lubricant for all PU foam, RIM processes, shoe soles; liq.

Mold Wiz Ext. AZN. [Axel Plastics Research Labs] General purpose external lubricant for sulfones, polyolefins, acetals, nylons, etc.; liq.

Mold Wiz Ext. F-57. [Axel Plastics Research Labs] External lubricant for polyesters, epoxies, elastomers, phenolics; liq.

Mold Wiz Ext. FFIH. [Axel Plastics Research Labs] Lubricant for rigid PU foam, RIM processes; liq.

Mold Wiz Ext. P. [Axel Plastics Research Labs] External lubricant for PU; liq.

Mold Wiz Int. 18-36. [Axel Plastics Research Labs] Internal lubricant for epoxies, phenolics, melamines, urea, formaldehyde; safe as indirect additive; liq.

Mold Wiz Int. 33PA. [Axel Plastics Research Labs] Internal lubricant for PVC, PE, PS, PP, ABS, nylon, TPU; safe as indirect additive; solid.

Mold Wiz Int. 33UDK. [Axel Plastics Research Labs] Internal lubricant for PC, clear and glass-filled; nylon; PP; PS; FDA accepted; solid.

Mold Wiz Int. 937. [Axel Plastics Research Labs] Internal lubricant for polyester, PU (soft and rigid foams), PS, engineering thermoplastics, thermosets; safe as indirect additive; liq.

Mold Wiz OY. [Axel Plastics Research Labs] External antistat for acrylics, cellulose acetate, PE, PP, PS.

Molgard. [Lilly] internal/external lubricant for polyester, esp. pultrusion; liq.

Monafax 785. [Mona Industries] Nonoxynol-9 phosphate; anionic; emulsifier, lubricant, antistat, detergent, corrosion inhibitor for emulsion polymerization, agric., metalworking lubricants, alkaline cleaners, industrial use; antisoil redeposition for dry cleaning; clear to slightly hazy visc. liq.; sol. in water, ethanol, perchlorethylene, xylene; sp.gr. 1.115; dens. 9.2 lb/gal; acid no. 70 ± 3; pH < 2.5 (10%); 100% act.

Monafax 1293. [Mona Industries] Org. phosphate ester; anionic; biodeg. hydrotrope, coupling agent for nonionics and other detergents which are only slightly sol. in high electrolyte concs.; surfactant, wetting agent, antistat, dispersant; for household and industrial hard surf. detergents, metal cleaners, agric., textile, paper/pulp, drycleaning, emulsion polymerization; Gardner 3 clear liq.; sol. in water, ethanol, trichlorethylene, kerosene, toluene, hexane; disp. in min. oil, min. spirits; sp.gr. 1.07; dens. 8.9 lb/gal; acid no. 111 (@ pH 4.5), 185 (@ pH 9.4); pH < 2 (10%); 98% act.

Monarch® 120. [Cabot/Special Blacks Div] Carbon black; CAS 1333-86-4; EINECS 215-609-9; blue toned colorant, uv stabilizer for plastics, inks, coatings, cement; low oil absorp., easy dispersion; exc. for coloring or tinting vs. lampblack; fluffy; 75 nm particle size; dens. 18 lb/ft^3; surf. area 25 m^2/g; pH 6-8; 0.5% volatile.

Monarch® 700. [Cabot/Special Blacks Div] Carbon black; CAS 1333-86-4; EINECS 215-609-9; med. jetness colorant, uv stabilizer for plastics and coatings; exc. elec. conductivity; fluffy; 18 nm particle size; dens. 9 lb/ft^3; surf. area 200 m^2/g; pH 6-8; 1.5% volatiles.

Monarch® 800. [Cabot/Special Blacks Div] Carbon black; CAS 1333-86-4; EINECS 215-609-9; med. jetness colorant, uv stabilizer for plastics and coatings; fluffy; 17 nm particle size; dens. 15 lb/ft^3; surf. area 210 m^2/g; pH 6-8; 1.5% volatiles.

Monarch® 880. [Cabot/Special Blacks Div] Carbon black; CAS 1333-86-4; EINECS 215-609-9; med. jetness colorant, uv stabilizer for with exc. dispersion chars. for plastics and coatings; fluffy; 16 nm particle size; dens. 8 lb/ft^3; surf. area 220 m^2/g; pH 6-8; 1.5% volatiles.

Monarch® 900. [Cabot/Special Blacks Div] Carbon black; CAS 1333-86-4; EINECS 215-609-9; med. high jetness colorant for plastic compds., lacquers, and enamels; fluffy; 15 nm particle size; dens. 14 lb/ft^3; surf. area 230 m^2/g; pH 6-8; 2.0% volatiles.

Monarch® 1000. [Cabot/Special Blacks Div] Carbon black, aftertreated; CAS 1333-86-4; EINECS 215-609-9; med. high jetness colorant with good stability for coatings and resistive plastics; fluffy; 16 nm particle size; dens. 11 lb/ft^3; surf. area 343 m^2/g; pH 2.5-4.0; 9.5% volatiles.

Monarch® 1100. [Cabot/Special Blacks Div] Carbon black; CAS 1333-86-4; EINECS 215-609-9; colorant; high jetness furnace black for plastics and coatings; fluffy; 14 nm particle size; dens. 15 lb/ft^3; surf. area 240 m^2/g; pH 6-8; 2.0% volatiles.

Monawet 1240. [Mona Industries] Disodium nonoxynol-10 sulfosuccinate; anionic; wetting agent for emulsion polymerization of vinyl acetate for use in adhesives, paints; FDA §175.105 compliance; APHA 25 clear liq.; sp.gr. 1.11; dens. 9.25 lb/gal; pH 7.0 (10%); Ross-Miles foam 275 mm (0.1%); 35% total solids.

Monawet MB-45. [Mona Industries] Diisobutyl sodium sulfosuccinate; CAS 127-39-9; EINECS 204-839-5; anionic; wetting, dispersing, emulsifying, penetrating and solubilizing agent used in emulsion polymerization of S/B for rug backing, paper coating, water treatment; EPA and FDA §178.3400 compliance; APHA 25 clear colorless liq.; sol. in water, fairly sol. in polar and nonpolar solvs.; m.w. 332; sp.gr. 1.12; dens. 9.3 lb/gal; cloud pt. 13 C; flash pt. 215 F (PMCC); surf. tens. 54 dynes/cm (0.1%); pH 6.0 (10%); Ross-Miles foam 30 mm (0.1%); 45% act. in water.

Monawet MM-80. [Mona Industries] Dihexyl sodium sulfosuccinate, 15% water, 5% IPA; anionic; wetting agent, detergent for emulsion and suspension polymerization, rug backing, paper coating, textiles, paint, agric., cosmetic, detergent, mining, water treatment, electroplating baths, and food industries; electrolyte tolerant; suitable for styrene, S/B, acrylonitrile/vinyl chloride systems; EPA and FDA §175.105, 176.170, 177.1210, 178.3400 compliances; colorless clear liq.; sol. 33 g/100 g water, in polar and nonpolar solvs.; m.w. 388; sp.gr. 1.10; dens. 9.2 lb/gal; cloud pt. < 0 C; flash pt. (PMCC) 110 F; pH 6 ± 1 (10%); surf. tens. 46 dynes/cm (0.1%); Ross-Miles foam 60 mm (0.1%); 80% act.; *Toxicology:* not a primary skin irritant.

Monawet MO-70. [Mona Industries] Dioctyl sodium sulfosuccinate, 20% water, 10% diethylene glycol butyl ether; anionic; wetting, dispersing, emulsifying, penetrating and solubilizing agent used in emulsion and suspension polymerization of vinyl acetate for adhesives, paints, textile, fertilizer, mining, water treatment, fire fighting, cosmetic, food industries; EPA and FDA §175.105, 175.300, 175.320, 176.170, 176.210, 177.1200, 177.2800, 178.3400 compliance; colorless clear liq.; sol. in polar and nonpolar solvs.; m.w. 444; sp.gr. 1.08; dens. 9.0 lb/gal; cloud pt. < -5 C; flash pt. (PMCC) 325 F; pH 6.0 (10%); surf. tens. 29 dynes/cm (0.1%); Ross-Miles foam 225 mm (0.1%); 70% act.

Monawet MO-70E. [Mona Industries] Dioctyl sodium sulfosuccinate, 19% water, 11% ethanol; anionic; wetting agent for industrial applics., emulsion polymerization of VA, adhesives, paints, textiles, agric., cosmetics, glass cleaners, mining, water treatment, wall paper removal, food pkg. plants; EPA and FDA §175.105, 175.300, 175.320, 176.170, 176.210, 177.1200, 177.2800, 178.3400 compliance; colorless clear liq.; sol. in polar and nonpolar solvs.; m.w. 444; sp.gr. 1.08; dens. 9.0 lb/gal; cloud pt. < -5 C; flash pt. (PMCC) 82 F; pH 6.0 (10%); surf. tens. 29 dynes/cm (0.1%); Ross-Miles foam 220 mm (0.1%); 70% act.

Monawet MT-70. [Mona Industries] Ditridecyl sodium sulfosuccinate, 12% water, 18% hexylene glycol; anionic; wetting, dispersing, emulsifying, penetrating and solubilizing agent used in emulsion and suspension polymerization of vinyl chloride and styrene for paints, coatings, indirect food additives; EPA and FDA §176.180, 178.3400 compliances; lt. straw clear liq.; sol. in polar and nonpolar solvs.; m.w. 584; sp.gr. 1.02; dens. 8.5 lb/gal; cloud pt. -2 C; flash pt. (PMCC) 230 F; pH 6.0 (10%); surf. tens. 29 dynes/cm (0.1%); 70% act.

Monawet SNO-35. [Mona Industries] Tetrasodium dicarboxyethyl stearyl sulfosuccinamate; CAS 3401-73-8; EINECS 222-273-7; anionic; wetting agent, solubilizer, emulsifier, dispersant, visc. depressant, mild detergent used in emulsion polymerization of styrene for paints, coatings, textile, cosmetic, agric. prods.; biodeg.; EPA and FDA §175.105, 176.170, 176.180, 178.3400 compliances; lt. amber clear liq.; m.w. 653; sol. in water, high electrolyte salt sol'ns.; sp.gr. 1.14; dens. 9.5 lb/gal; visc. 16-18 s (#2 Zahn cup); f.p. 45 ± 5 F; acid no. 2.0; iodine no. 0.5; pH 7.5; surf. tens. 43 dynes/cm (0.1%); Draves wetting 232 s (0.25%, 30 C); Ross-Miles foam 185 mm (0.1%); 35% solids; *Toxicology:* very low acute oral toxicity; mild eye irritant.

Mono-Coat® 65-RT. [Chem-Trend] Solv.-based release agent system formulated to cure and adhere to the mold surface at R.T.; for use on laminates and/or composite molding; also as primary release for epoxy and polyester resin systems with graphite, aramid, or fiberglass reinforcements; slightly hazy fluid; dens. 8.62 lb/gal; visc. < 100 cP (21 C); flash pt. none.

Mono-Coat® E76. [Chem-Trend] Solv.-based release system for all types of rubber molding and for mold conditioning; colorless water-thin, clear liq.; dens. 10.52 lb/gal; visc. < 10 cP (21 C); flash pt. none.

Mono-Coat® E91. [Chem-Trend] Solv.-based release system for molding operations where high degree of slip is required; used in rubber molding (compr., transfer, and inj. of peroxide-cured fluoroelastomers, dry EPDM compds., some epichlorohydrin compds.), and in composite/laminate molding; colorless clear fluid; dens. 10.71 lb/gal; visc. < 10 cP (21 C); flash pt. none.

Monolan® 2500 E/30. [Harcros UK] EO-PO block polymer; nonionic; detergent, wetting agent for emulsion polymerization, metal cleaning, resin plasticizers, latex stabilization, textile processing; coemulsifier for phosphate toxicants; colorless clear liq., faint odor; water-sol.; sp.gr. 1.043; visc. 650 cs; HLB 6.0; flash pt. (COC) > 200 C; pour pt. < 0 C; pH 7.0 (1% aq.); surf. tens. 43 dynes/cm (0.1%); 100% act.

Monolan® 3000 E/60, 8000 E/80. [Harcros UK] EO-PO block polymer; nonionic; detergent, wetting agent, antifoam for industrial and domestic detergents, emulsion polymerization, metal cleaning, resin plasticizers, latex stabilization, textile processing; wh. solid and wh. flake resp., faint odor; water-sol.; sp.gr. 1.052 and 1.070; visc. 355 cs and 1100 cs (60 C) resp.; flash pt. > 400 F (COC); pour pt. 31 and 49 C; pH 6-8 and 7-8 (1% aq.); 100% act.; Unverified

Monolan® 8000 E/80. [Harcros UK] EO/PO block polymer; CAS 9003-11-6; nonionic; low foam wetting agent and detergent, dispersant; primary emulsifier in emulsion polymerization; wh. flake, faint odor; water-sol.; sp.gr. 1.070; visc. 1100 cs (60 C); HLB 16.0; pour pt. 49 C; cloud pt. 100 C (1% aq.); flash pt. (COC) > 200 C; pH 7.0 (1% aq.); surf. tens. 52 dynes/cm (0.1%); 100% act.

Monolan® O Range. [Harcros UK] EO/PO copolymer; nonionic; low foam surfactants and defoamers for agric., textiles, emulsion polymerization, syn. lubricants.

Monolan® PPG440, PPG1100, PPG2200. [Harcros UK] PPG; nonionic; lubricant, antistat, plasticizer, cosolvs. in dyestuff, ink, resin, rubber industries, cosmetic preparations, intermediates in surfactants and plastic prod.; colorless liq.; faint odor; sol. in water (PPG440), insol. in water (PPG1100, 2200); m.w. 400, 1000, 2000 resp.; sp.gr. 1.010, 1.005, and 1.004; visc. 80, 180, and 450 cs; flash pt. > 450 F (COC); pour pt. < 0 C; 100% act.; Unverified

Monoplex® DDA. [C.P. Hall] Diisodecyl adipate; CAS 27178-16-1; EINECS 248-299-9; lubricant; plasticizer for PVC and other polymers; APHA 20 clear liq.; sol. in hexane, toluene, kerosene, ethanol, acetone, min. oil; insol. in water; m.w. 427; sp.gr. 0.916; visc. 30 cps; f.p. -43 C; acid no. 0.1; iodine no. nil; sapon. no. 263; flash pt. 227 C; ref. index 1.4501; 0.05% moisture.

Monoplex® DIOA. [C.P. Hall] Diisooctyl adipate; plasticizer for PVC and other polymers; APHA 50 clear liq.; sol. in hexane, toluene, kerosene, ethanol, acetone, min. oil; insol. in water; m.w. 373; sp.gr. 0.926; f.p. -65 C; acid no. 0.05; iodine no. 1; sapon. no. 303; flash pt. 193 C; ref. index 1.446; 0.05% moisture.

Monoplex® DOA. [C.P. Hall] Di-2-ethylhexyl adipate; CAS 103-23-1; EINECS 203-090-1; plasticizer for PVC food pkg. film; APHA 15 clear liq.; sol. in hexane, toluene, kerosene, ethanol, acetone, min. oil; insol. in water; m.w. 373; sp.gr. 0.925; visc. 18 cps; f.p. -65 C; acid no. 0.1; iodine no. nil; sapon. no. 302; flash pt. 199 C; ref. index 1.445; 0.04% moisture.

Monoplex® DOS. [C.P. Hall] Di-2-ethylhexyl sebacate; plasticizer for elec. PVC compds., low temp. sheet, film; APHA 70 clear liq.; sol. in hexane, toluene, kerosene, ethanol, acetone, min. oil; insol. in water; m.w. 430; sp.gr. 0.913; visc. 25 cps; f.p. < -65 C; acid no. 0.02; iodine no. nil; sapon. no. 262; flash pt. 216 C; ref. index 1.4488; 0.04% moisture.

Monoplex® NODA. [C.P. Hall] n-Octyl, n-decyl adipate; plasticizer for PVC and other polymers; APHA 20 clear liq.; sol. in hexane, toluene, kerosene, ethanol, acetone, min. oil; insol. in water; m.w. 400; sp.gr. 0.913; visc. 16 cps; f.p. 0 C; acid no. 0.1; iodine no. nil; sapon. no. 280; flash pt. 212 C; ref. index 1.446; 0.04% moisture.

Monoplex® S-73. [C.P. Hall] Epoxidized octyl tallate; CAS 61788-72-5; stabilizer, plasticizer for PVC, PS; Gardner 1 clear liq.; sol. in hexane, toluene, kerosene, ethanol, acetone, min. oil; insol. in water; m.w. 420; sp.gr. 0.928; visc. 40 cps; f.p. -7 C; acid no. 0.7; iodine no. 1.5; sapon. no. 140; flash pt. 224 C; ref. index 1.458; 0.04% moisture.

Monoplex® S-75. [C.P. Hall] Epoxidized glycol dioleate; stabilizer, plasticizer for PVC, PS; Gardner 1 clear liq.; sol. in hexane, toluene, kerosene, ethanol, acetone, min. oil; insol. in water; m.w. 420; sp.gr. 0.950; visc. 90 cps; f.p. 0 C; acid no. 0.6; iodine no. 1.3; sapon. no. 185; flash pt. 296 C; ref. index 1.461; 0.05% moisture.

Monostearyl Citrate (MSC). [Morflex] Monostearyl citrate; CAS 1337-33-3; EINECS 215-654-4; chelating agent; stabilizer for oils; synergistic with antioxidants; sequestrant in food and feed applics.; plasticizer or component in food pkg. materials; FDA 21CFR §175.300, 175.380, 175.390, 176.170, 177.1210, 178.3910, 181.27, 182.685, 582.6851; waxy solid; sol. ≈ 1% in olive oil, safflower oil, corn oil, peanut oil, veg. oil; sp.gr. 0.92 (85 C); m.p. 47 C min.; sapon. no. 140-170; neutral. no. 100-125; flash pt. (COC) > 400 F.

Monosulf. [Henkel/Organic Prods.] Sulfated castor oil; CAS 8002-33-3; EINECS 232-306-7; anionic; penetrant, emulsifier; textile dyeing assistant; fat liquor for suede leather; paper coating evener; plasticizer for starch, glues; emulsifier for latex; liq.; 68% conc.

Monothiurad®. [Monsanto] Tetramethyl thiuram monosulfide; CAS 97-74-5; general purpose primary accelerator for NBR, CR; pellets and powd.

Montaclere®. [Monsanto] Styrenated phenol; CAS

61788-44-1; nonstaining antioxidant for NR, synthetics, latexes; fiber lubricant stabilization, precursor for ethoxylated emulsifying agents, precursor for pigment/dye grinding medium; liq.

Montosol PB-25. [Pulcra SA] Sodium nonoxynol-25 sulfate; anionic; emulsifier for emulsion polymerization; liq.; 35% conc.; Unverified

Montovol RF-10. [Pulcra SA] Sodium lauryl sulfate; CAS 151-21-3; EINECS 205-788-1; anionic; emulsifier; additive for emulsion polymerization; used in shampoos and specialty cleaners; liq., paste; 29 ± 1% conc.; Unverified

Morfax®. [R.T. Vanderbilt] 4-Morpholinyl-2-benzothiazole disulfide; CAS 95-32-9; sulfenamide accelerator for NR and syn. rubbers; provides good curing activity with adequate processing safety; used for tires and mech. goods requiring max. strength and wearing quality; cream to lt. yel. friable powd.; m.w. 284.35; sol. in toluene, chloroform; practically insol. in water; dens. 1.51 ± 0.03 mg/m^3; m.p. 116-128 C.

Morflex® 190. [Morflex] n-Butyl phthalyl-n-butyl glycolate; CAS 85-70-1; EINECS 201-624-8; vinyl plasticizer.

Morflex® 560. [Morflex] Tri-n-hexyl trimellitate; CAS 1528-49-0; plasticizer with high efficiency, good permanence.

Morflex® 1129. [Morflex] Dimethyl isophthalate; CAS 1459-93-4; EINECS 215-951-9; plasticizer; chemical intermediate in prod. of syn. fibers in no-tear envelopes, wallpaper backing, and polyester bottles.

Mould Release Agent N 32. [Chemetall GmbH] Solvent-free mold release agent on a plastic base for the rubber and plastics industry.

MPC Channel Black TR 354. [Struktol] Carbon black; CAS 1333-86-4; EINECS 215-609-9; filler; powd.; 28 nm avg. particle size; 0.08% on 45 mm sieve; surf. area 92 m^2/g; pH 4.

MPDiol® Glycol. [Arco] 2-Methyl-1,3-propanediol; CAS 2163-42-0; intermediate used in prod. of solvs., urethanes, unsat. polyesters, gel coats, sat. polyester and alkyd coatings, polymeric plasticizers; produces unsat. polyester resins with light color, exc. styrene miscibility, exc. balance of tens. str., elong., and flexibility; colorless clear low-visc. liq.; sol. in water; m.w. 90.1; sp.gr. 1.015; dens. < 8.47 lb/gal; visc. 174 cps; vapor pressure 4.3 mm Hg (100 C); m.p. -91 C; b.p. 213 C; hyd. no. 1200; flash pt. (OC) > 230 F; 98.5% assay; *Precaution:* store away from strong min. acids, dehydrating agents, strong oxidizing agents; *Storage:* avoid excessive exposure to moisture; keep away from heat, sparks, and flame.

Mr 245. [Sovereign] Oxidized asphalt, treated with 1% limestone; min. rubber used as plasticizer, process aid, and low-cost filler in mech. goods, mats, floor treads, tire components, retreads, sealing compds., wire and cable; used in compds. with short mix cycles and low mix temps. where a low m.p. is required; powd.; m.p. 245 F.

Mr 285. [Sovereign] Oxidized asphalt; min. rubber used as plasticizer, process aid, and low-cost filler in mech. goods, mats, floor treads, tire components, retreads, sealing compds., wire and cable; used in compds. with short mix cycles and low mix temps. where a low m.p. is required; powd.; m.p. 285 F.

Mr 305. [Sovereign] Oxidized asphalt; min. rubber used as plasticizer, process aid, and low-cost filler in rubber goods; mainly for brake linings and other friction prods.; powd.; m.p. 305 F.

MR1085C. [Sovereign] Rosin oil; tackifier for tires; highly refined version of Alatac 100-10, providing better aged tack; liq.

MS-122N/CO$_2$. [Miller-Stephenson] Tetrafluoroethylene telomer (1%) with 1,1-dichloro-1-fluoroethane (92-95%), carbon dioxide (2-4%), isopropyl alcohol (2-4%); release agent, dry lubricant for use on cold molds, esp. for epoxy potting/encapsulating, PU, nylon, acrylics, PP, PC phenolics, PS, foams, rubber molding, glass, wood, paper, gears, belts, wire and cable, metals; wh. milky liq., faint ethereal odor; negligible sol. in water; dens. 1.23 g/cc; b.p. ≈ 40 C; pH neutral; flash pt. none; *Toxicology:* mild eye irritant; inh. of high concs. of vapor may be harmful and may cause heart irregularities, unconsciousness, or death; inh. of vapors in presence of tobacco prods. causes polymer fume fever; *Precaution:* avoid heat, sparks, flame; incompat. with strong acids and alkalies, finely powd. Al, Mg, Zn, strong oxidizing agents, some desiccants; decomp. prods. incl. CO, CO_2, HF, HCl; flamm. limits in air 7.4-15.5% by vol.; *Storage:* store in clean dry area; do not heat above 125 F; rotate stock to shelf life of 1 yr.

MS-136N/CO$_2$. [Miller-Stephenson] Tetrafluoroethylene telomer; release agent for hot molds, esp. for PU, nylon, ABS, polypropylene oxides, acetals, vinyls, PVC, cellulosics, phenolics, elastomers, TPE, melamines, etc.; solid.

Multifex™ FF. [Specialty Minerals] Precipitated calcium carbonate; filler for rubber, SMC/BMC applics.; powd.; 0.7 μ avg. particle size, 2 μ top size.

Multifex-MM®. [Specialty Minerals] Precipitated calcium carbonate; filler for rubber, PVC plastisols; fine powd.; 0.07 μ avg. particle size, 0.3 μ top size.

Multiflow®. [Monsanto] Plasticizer, resin modifier for nonaq. coatings; improves flow, substrate wetting in industrial powd. coatings.

Multiwax® 180-M. [Witco/Petroleum Spec.] Microcryst. wax NF; CAS 63231-60-7; EINECS 264-038-1; plasticizer or modifier for polymeric coatings and adhesives; hot melt adhesives and coatings, chewing gum base, protective coatings, cosmetics/pharmaceuticals; FDA §172.886, 178.3710; lt. yel.; misc. with petrol. prods., many essential oils, most animal and veg. fats, oils, and waxes; visc. 14.3-18.0 cSt (99 C); m.p. 82-88 C; flash pt. (COC) 277 C min.

Multiwax® W-835. [Witco/Petroleum Spec.] Microcryst. wax; CAS 63231-60-7; EINECS 264-038-1; plasticizer or modifier for polymeric coatings and adhesives; used in silk screen printing, cold creams, cleansing creams, hair pomades, pharmaceuticals, crayons, paste-up adhesive, chewing gum, dental prods., elec., sealants, etc.; FDA §172.886, 178.3710; wh.; visc. 14.3-18.0 cSt (99 C); m.p. 74-79 C; flash pt. (COC) 246 C min.

Multiwax® X-145A. [Witco/Petroleum Spec.] Microcryst. wax NF; CAS 63231-60-7; EINECS 264-038-1; plasticizer or modifier for polymeric coatings and adhesives; cheese coating, laminating of cellophane and plastic film, waterproofing/protective linings, cosmetics/pharmaceuticals; FDA §172.886, 178.3710; lt. yel.; misc. with petrol. prods., many essential oils, most animal and veg. fats, oils, and waxes; visc. 14.3-18.0 cSt (99 C); m.p. 66-71 C; flash pt. (COC) 260 C min.

Myrj® 45. [ICI Spec. Chem.; ICI Surf. Am.; ICI Surf. Belgium] PEG-8 stearate; CAS 9004-99-3; nonionic; general purpose o/w emulsifiers for cosmetics, pharmaceuticals, etc.; internal antistat for PE, EVA; Canada compliance; wh. cream-colored soft waxy solid; sol. in alcohol, disp. in water; sp.gr. 1.0; HLB 11.1; pour pt. 28 C; sapon. no. 82-95; 100% conc.

Myvaplex® 600P. [Eastman] Glyceryl monostearate (from hydrog. soybean oil); emulsifier for foods, cosmetics and personal care prods.; starch complexing agent, lubricant, and processing aid for extruded foods; used for pasta, cereals, instant potatoes, snack foods, pet foods, ice cream, frozen desserts; processing aid for expanded PS; internal lubricant in rigid PVC; FDA 21CFR §182.1324, 182.4505; powd.; sp.gr. 0.92 (80 C); m.p. 69 C; congeal pt. 70 C; clear pt. 78 C; acid no. 3 max.; iodine no. 5 max.; 90% min. monoester; *Storage:* 24 mo shelf life.

Myvaplex® 600PK. [Eastman] Glyceryl monostearate (from hydrog. soybean oil); nonionic; dispersant, mold release, processing aid, antistat, antifog, lubricant, antiblock for PS, polyolefins, PVC, PU; powd.; sp.gr. 0.92 (80 C); m.p. 69 C; acid no. 3 max.; iodine no. 5 max.; 90% min. monoester.

Myverol® 18-04. [Eastman] Dist. hydrog. palm oil glyceride; CAS 67784-87-6; nonionic; emulsifier for cosmetics, personal care prods., baked goods, confectionery prods., dehydrated potatoes, etc.; improves texture and reduces stickiness in toffees; emulsifier for infant formulas; emulsion stabilizer, antispattering aid in margarine; stabilizes oil, improves mouth-feel and flavor in peanut butter; dispersant, mold release, processing aid, antistat, antifog, lubricant, antiblock for PS, polyolefins, PVC, PU; FDA 21CFR §184.1505 (GRAS), EC E471; small beads; sp.gr. 0.94 (80 C); m.p. 66 C; HLB 3.8-4.0; acid no. 3 max.; iodine no. 5 max.; 90% min. monoester; *Storage:* 24 mo shelf life.

Myverol® 18-06. [Eastman] Hydrog. soy glyceride, dist.; CAS 61789-08-0; nonionic; emulsifier for foods, cosmetics; emulsion stabilizer, aerater in cake mixes, whipped toppings; improves texture, reduces stickiness in toffees; softener, plasticizer for chewing gum; infant formulas; gloss aid in confectionery coatings; antispattering aid in margarine; extrusion aid (pet foods); starch complexing (instant potatoes); dispersant, mold release, processing aid, antistat, antifog, lubricant, antiblock for PS, polyolefins, PVC, PU; FDA 21CFR §184.1505 (GRAS), EC E471; small beads; negligible sol. in water; sp.gr. 0.92 (80 C); m.p. 69 C; b.p. 460 C; HLB 3.8-4.0; acid no. 3 max.; iodine no. 5 max.; flash pt. (COC) 227 C; 90% min. monoester; *Toxicology:* LD50 (oral, rat) > 5 g/kg, (dermal, rat) > 2 g/kg; nonirritating to skin; sl. eye irritant; *Storage:* 24 mo shelf life.

Myverol® 18-07. [Eastman] Hydrog. cottonseed glyceride, dist.; CAS 61789-07-9; nonionic; emulsifier for foods, cosmetics; improves texture and reduces stickiness in toffees; improves gloss and crystallization in confectionery coatings; emulsion stabilizer, antispattering aid in margarine; aerator for whipped toppings; dispersant, mold release, processing aid, antistat, antifog, lubricant, antiblock for PS, polyolefins, PVC, PU; FDA 21CFR §184.1505 (GRAS), EC E471; small beads; sp.gr. 0.92 (80 C); m.p. 68 C; HLB 3.8-4.0; acid no. 3 max.; iodine no. 5 max.; 90% min. monoester; *Storage:* 24 mo shelf life.

Myverol® 18-99. [Eastman] Canola oil glyceride; nonionic; emulsifier for foods, cosmetics, personal care prods.; sustained release formulations and microspheres, topical permeation enhancement, solubilization; antifoam for food processing; emulsion stabilizer for diet spreads; emulsifier, aerator for icings; stabilizes oil, improves mouth-feel and flavor in peanut butter; dispersant, mold release, processing aid, antistat, antifog, lubricant, antiblock for PS, polyolefins, PVC, PU; FDA 21CFR §184.1505 (GRAS), EC E471; semiplastic; sp.gr. 0.93 (80 C); m.p. 35 C; HLB 3.8-4.0; acid no. 3 max.; iodine no. 90-95; 90% min. monoester; *Storage:* 6 mo shelf life.

N

Nacconol® 40G. [Stepan; Stepan Canada] Sodium dodecylbenzene sulfonate; anionic; foamer, dispersant, wetting agent, detergent for agric., cement, dyeing, emulsion polymerization, textile, metal cleaning, metalworking, mining, paper industries; biodeg.; cream gran. powd.; sol. 2 g/100 ml in water; dens. 35 lb/ft^3; pH 6.4-7.6 (1% aq.); 38-42% act.

Nacconol® 90G. [Stepan; Stepan Canada; Stepan Europe] Sodium dodecylbenzene sulfonate; anionic; foamer, dispersant, wetting agent, detergent for agric., cement, dyeing, emulsion and latex polymerization, textile, metal cleaning, laundry prods., metalworking, mining, paper industries; biodeg.; wh. to beige gran.; sol. 15 g/100 ml in water; dens. 30 lb/ft^3; pH 6.0-7.5 (1% aq.); 90% act.

Nafol® 10 D. [Vista] C10 alcohol (90% C10, 10% C8); intermediate for mfg. of toiletries, cosmetics, detergents, laundry softeners, lubricating oil additives, plasticizers, plastics additives, textile/leather additives, disinfectants, agrochem., paper defoamers, flotation agents; Hazen 10 max. color; m.w. 155-162; sp.gr. 0.829 (20/4 C); solid. pt. 3 C; b.p. 215-240 C; acid no. 0.05 max.; iodine no. 0.2 max.; hyd. no. 345-365; flash pt. 95 C; 99% min. act.

Nafol® 20+. [Vista] C20-24 alcohol (50% C20, 29% C22, 14% C24); EINECS 307-145-1; intermediate for mfg. of toiletries, cosmetics, detergents, laundry softeners, lubricating oil additives, plasticizers, plastics additives, textile/leather additives, disinfectants, agrochem., paper defoamers, flotation agents; Hazen 1800 max. color; sp.gr. 0.804 (80/4 C); solid. pt. 53-58 C; acid no. 1.0 max.; iodine no. 20 max.; hyd. no. 130-150; flash pt. 210 C; 80% act.

Nafol® 810 D. [Vista] C8-10 alcohol (43% C8, 55% C10); CAS 68603-15-5; EINECS 287-621-2; intermediate for mfg. of toiletries, cosmetics, detergents, laundry softeners, lubricating oil additives, plasticizers, plastics additives, textile/leather additives, disinfectants, agrochem., paper defoamers, flotation agents; Hazen 10 max. color; m.w. 143-148; sp.gr. 0.827 (20/4 C); b.p. 195-240 C; acid no. 0.05 max.; iodine no. 0.15 max.; hyd. no. 380-390; pour pt. 11 C; flash pt. 85 C; 99% min. act.

Nafol® 1014. [Vista] C10-14 alcohol (15% C10, 47% C12, 38% C14); EINECS 283-066-5; intermediate for mfg. of toiletries, cosmetics, detergents, laundry softeners, lubricating oil additives, plasticizers, plastics additives, textile/leather additives, disinfectants, agrochem., paper defoamers, flotation agents; Hazen 20 max. color; m.w. 187-193; sp.gr. 0.832 (20/4 C); solid. pt. 14-18 C; b.p. 230-285 C; acid no. 0.05 max.; iodine no. 0.2 max.; hyd. no. 290-300; flash pt. 118 C; 99% min. act.

Nafol® 1218. [Vista] C12-18 alcohol (40% C12, 30% C14, 18% C16, 10% C180; CAS 67762-25-8; EINECS 267-006-5; intermediate for mfg. of toiletries, cosmetics, detergents, laundry softeners, lubricating oil additives, plasticizers, plastics additives, textile/leather additives, disinfectants, agrochem., paper defoamers, flotation agents; Hazen 30 max. color; m.w. 204-216; sp.gr. 0.823 (40/4 C); solid. pt. 25-28 C; b.p. 270-335 C; acid no. 0.1 max.; iodine no. 0.4 max.; hyd. no. 260-275; flash pt. 145 C; 98.5% min. act.

Naftocit® Di 4. [Chemetall GmbH; Oakite Spec.] Zinc dimethyldithiocarbamate; CAS 137-30-4; vulcanizing accelerator for rubber and latex processing; powd.

Naftocit® Di 7. [Chemetall GmbH; Oakite Spec.] Zinc diethyldithiocarbamate; CAS 14323-55-1; vulcanizing accelerator for rubber and latex processing; powd.

Naftocit® Di 13. [Chemetall GmbH; Oakite Spec.] Zinc dibutyldithiocarbamate; CAS 136-23-2; vulcanizing accelerator for rubber and latex processing; powd.

Naftocit® DPG. [Chemetall GmbH; Oakite Spec.] N,N′-Diphenyl guanidine; CAS 102-06-7; vulcanizing accelerator for rubber and latex processing; powd.

Naftocit® MBT. [Chemetall GmbH; Oakite Spec.] 2-Mercaptobenzothiazole; CAS 149-30-4; EINECS 205-736-8; vulcanizing accelerator for rubber and latex processing; powd.

Naftocit® MBTS. [Chemetall GmbH; Oakite Spec.] 2,2′-Dibenzothiazyl disulfide; CAS 120-78-5; vulcanizing accelerator for rubber and latex processing; powd.

Naftocit® Mi 12. [Chemetall GmbH; Oakite Spec.] Ethylene thiourea; CAS 96-45-7; EINECS 202-506-9; vulcanizing accelerator for rubber and latex processing; powd.

Naftocit® NaDBC. [Chemetall GmbH; Oakite Spec.] Sodium dibutyldithiocarbamate; CAS 136-30-1; vulcanizing accelerator for rubber and latex processing; powd; liq.

Naftocit® NaDMC. [Chemetall GmbH; Oakite Spec.] Sodium dimethyldithiocarbamate; vulcanizing accelerator for rubber and latex processing; powd., liq.

Naftocit® Thiuram 16. [Chemetall GmbH; Oakite Spec.] Tetramethylthiuram disulfide; CAS 137-26-8; vulcanizing accelerator for rubber and latex processing; powd.

Naftocit® ZBEC. [Chemetall GmbH; Oakite Spec.] Zinc dibenzyldithiocarbamate; CAS 14726-36-4; vulcanizing ultra-accelerator for syn. rubber and latex processing, e.g., NR, SBR, EPDM, IIR, IR, NBR; activator for thiuram, mercapto, and sulfenamide accelerators; does not produce hazardous nitrosamines during vulcanization; for carpet backings, foam rubber, dipped prods., rubber threads, rubberized textiles, soft rubber prods., nontoxic rubber goods, footwear, adhesives, wire insulation, roof foils; wh. to cream-colored powd.; partly sol. in chlorinated hydrocarbons and acetone; insol. in water and petrol.; m.w. 610.2; sp.gr. 1.40 g/cc (20 C); m.p. 182 C min.; *Storage:* 2 yrs shelf life when avoiding extremely high temps. and humidity.

Naftocit® ZMBT. [Chemetall GmbH; Oakite Spec.] Zinc-2-mercaptobenzothiazole; CAS 155-04-4; vulcanizing accelerator for rubber and latex processing; powd.

Naftolen® H, NV, ZD, ZD103, ZD105, ZD106, ZM. [Chemetall GmbH] Aromatic plasticizer and extender oil; for natural and syn. rubber; sp.gr. 0.947, 0.976, 0.961, 0.971, 0.949, 0.981, 0.957 resp.; visc. 39, 260, 880, 135, 1250, 950, 500 mm^2/s resp. (40 C); Unverified

Naftolen® N400, N401, N402, N403, N404, N405, N406, N407, N408. [Chemetall GmbH] Naphthenic plasticizer and extender oil; for natural and syn. rubber; sp.gr. 0.874, 0.868, 0.868, 0.863, 0.857, 0.853, 0.848, 0.855, 0.866 resp.; visc. 20, 12.7, 31, 46, 68, 100, 185, 460, 21.7 mm^2/s resp. (40 C); Unverified

Naftolen® ND, P 603, P 611. [Chemetall GmbH] Relatively naphthenic plasticizer and extender oil; for natural and syn. rubber; sp.gr. 0.834, 0.827, 0.828 resp.; visc. 110, 24, 24 mm^2/s resp. (40 C); Unverified

Naftolen® P600, P604, P606, P612, P613, P614, P616. [Chemetall GmbH] Paraffinic plasticizer and extender oil; for natural and syn. rubber; sp.gr. 0.820, 0.817, 0.814, 0.815, 0.799, 0.806, and 0.806 resp.; visc. 120, 146, 77, 85.5, 570, 89, and 91 mm^2/s resp. (40 C); Unverified

Naftonox® 2246. [Chemetall GmbH; Oakite Spec.] 2,2′-Methylenebis (4-methyl-6-t-butylphenol); CAS 119-47-1; antioxidant for prod. and processing of natural and syn. rubbers and latexes, fuels and oils, and adhesives.

Naftonox® BBM. [Chemetall GmbH] 4,4′-Butylidene bis (2-t-butyl-5-methylphenol); antioxidant for solid rubber and latex processing, adhesives, plastics.; Unverified

Naftonox® BHT. [Chemetall GmbH] 2,6-Di-t-butyl-4-methylphenol; antioxidant for prod. and processing of natural and syn. rubbers and latexes, thermoplastics, adhesives, fuels and oils, foodstuffs and animal feeds.; Unverified

Naftonox® IMB . [Chemetall GmbH] 2,2′-Isobutylenebis (4,6-dimethylphenol); antioxidant for solid rubber and latex processing, adhesives.; Unverified

Naftonox® PA. [Chemetall GmbH] Sterically hindered high alkylated phenols; antioxidant for syn. rubber and latex prod. and processing.; Unverified

Naftonox® PS. [Chemetall GmbH] Sterically hindered styrenated phenols; antioxidant for syn. rubber and latex prod. and processing.; Unverified

Naftonox® TMQ. [Chemetall GmbH] Polymerized 2,2,4-trimethyl-1,2-dihydroquinoline; CAS 26780-96-1; antioxidant for solid rubber and latex processing, adhesives, plastics; flakes and powd.

Naftonox® ZMP. [Chemetall GmbH; Oakite Spec.] Polymerized sterically hindered polyphenol; antioxidant for solid rubber and latex processing, adhesives, plastics.

Naftopast® Antimontrioxid. [Chemetall GmbH] Antimony trioxide; CAS 1309-64-4; EINECS 215-175-0; processing aid in rubber compounding; solid disp.; 77% conc.; Unverified

Naftopast® Antimontrioxid-CP. [Chemetall GmbH] Antimony trioxide; CAS 1309-64-4; EINECS 215-175-0; processing aid in rubber compounding; solid disp.; 80% conc.; Unverified

Naftopast® Di7-P. [Chemetall GmbH] Zinc diethyldithiocarbamate; CAS 14323-55-1; processing aid in rubber compounding; solid disp.; 75% conc.; Unverified

Naftopast® Di13-P. [Chemetall GmbH] Zinc dibutyldithiocarbamate; CAS 136-23-2; processing aid in rubber compounding; solid disp.; 85% conc.; Unverified

Naftopast® GMF. [Chemetall GmbH] GMF; processing aid in rubber compounding; solid disp.; 60% conc.; Unverified

Naftopast® Litharge A. [Chemetall GmbH] Lead oxide; processing aid in rubber compounding; solid disp.; 90% conc.; Unverified

Naftopast® MBT-P. [Chemetall GmbH] 2-Mercaptobenzothiazole; CAS 149-30-4; EINECS 205-736-8; processing aid in rubber compounding; solid disp.; 70% conc.; Unverified

Naftopast® MBTS-A. [Chemetall GmbH] 2,2′-Dibenzothiazyl disulfide; CAS 120-78-5; processing aid in rubber compounding; solid disp.; 70 and 75% concs.; Unverified

Naftopast® MBTS-P. [Chemetall GmbH] 2,2′-Dibenzothiazyl disulfide; CAS 120-78-5; processing aid in rubber compounding; solid disp.; 70% conc.; Unverified

Naftopast® MgO-A. [Chemetall GmbH] MgO; CAS 1309-48-4; EINECS 215-171-9; processing aid in rubber compounding; solid disp.; 60 and 66.6% conc.; Unverified

Naftopast® Mi12-P. [Chemetall GmbH] Ethylenethiourea; CAS 96-45-7; EINECS 202-506-9; processing aid in rubber compounding; solid disp.; 80% conc.; Unverified

Naftopast® Red Lead A, P. [Chemetall GmbH] Red lead, Pb_3O_4; processing aid in rubber compounding; solid disp.; 90%conc.; Unverified

Naftopast® Schwefel-P. [Chemetall GmbH] Sulfur; CAS 7704-34-9; EINECS 231-722-6; processing aid in rubber compounding; solid disp.; 75% conc.; Unverified

Naftopast® Thiuram 16-P. [Chemetall GmbH] Tetramethylthiuram disulfide; CAS 137-26-8; processing aid in rubber compounding; solid disp.; 70% conc.; Unverified

Naftopast® TMTM-P. [Chemetall GmbH] TMTM; processing aid in rubber compounding; solid disp.; 70% conc.; Unverified

Naftopast® ZnO-A. [Chemetall GmbH] Zinc oxide; CAS 1314-13-2; EINECS 215-222-5; processing aid in rubber compounding; solid disp.; 80% conc.; Unverified

Naftozin® N, Spezial. [Chemetall GmbH] Stearic acid; CAS 57-11-4; EINECS 200-313-4; processing aid for solid rubber processing.; lubricant in PVC processing; flakes and pearls; iodine no. 8.0 and 3.0 max. resp.; Unverified

Nansa® 1042/P. [Albright & Wilson UK] Dodecylbenzene sulfonic acid; CAS 27176-87-0; anionic; intermediate for neutralization; sodium salts as emulsifiers in emulsion polymerization of plastics and syn. rubbers; dk. br. visc. liq.; sp.gr. 1.05 g/cc; visc. 1900 cs (20 C); 95% act.; Unverified

Nansa® 1106/P. [Albright & Wilson UK] Sodium dodecylbenzene sulfonate; surfactant for emulsion polymerization; golden yel. clear liq.; sp.gr. 1.0 g/cc; visc. 2500 cs (20 C); cloud pt. 11 C; pH 8.0 ± 1 (5% aq.); 30 ± 1% act.; Unverified

Nansa® 1169/P. [Albright & Wilson UK] Sodium dodecylbenzene sulfonate; surfactant for emulsion polymerization; golden yel. cloudy liq.; dens. 1.0 g/cc; visc. 3000 cs (20 C); cloud pt. 25 C; pH 8 ± 1 (5% aq.); 30% act.; Unverified

Nansa® EVM50. [Albright & Wilson UK] Calcium dodecylbenzene sulfonate in aromatic solv.; CAS 68953-96-8; EINECS 247-557-8; anionic; emulsifier, dispersant for agrochemicals, textiles, surf. coatings, polymerization, leather industries; brn. visc. liq.; flash pt. (CC) > 48 C; pH 5.5-7.5 (3%); 50 ± 1.5% act.

Nansa® EVM70/B. [Albright & Wilson UK] Calcium dodecylbenzene sulfonate in hexanol; anionic; emulsifier, dispersant for agrochemicals, textiles, surf. coatings, polymerization, leather industries; brn. visc. liq.; flash pt. (CC) > 28 C; pH 5.5-7.5 (3%); 68.5 ± 1.5% act.

Nansa® EVM70/E. [Albright & Wilson UK] Calcium dodecylbenzene sulfonate in isobutanol; anionic; emulsifier, dispersant for agrochemicals, textiles, surf. coatings, polymerization, leather industries; brn. visc. liq.; flash pt. (CC) > 57 C; pH 5.5-7.5 (3%); 67 ± 2% act.

Nansa® HS85/S. [Albright & Wilson UK] Sodium dodecylbenzene sulfonate; CAS 25155-30-0; anionic; formulation of detergents, hard surface and bottle cleaners; metal treatment and paper processing; scouring and wetting agent for textile industry; foamer and mortar plasticizer in building industry; rubber/plastics emulsifier; biodeg.; cream flake; dens. 0.5 g/cm^3; pH 9-11 (1%); 85 ± 3% act.

Nansa® SSA. [Albright & Wilson UK] Dodecylbenzene sulfonic acid; CAS 68584-22-5; anionic; detergent intermediate; in prep. of emulsifiers for emulsion polymerization; dk. br. visc. liq.; dens. 1.05 g/cc; visc. 1900 cs (20 C); 96% act.

Nansa® SSA/P. [Albright & Wilson UK] Dodecylbenzene sulfonic acid; intermediate for neutralization; sodium salts as emulsifiers in emulsion polymerization of plastics and syn. rubbers; dk. br. visc. liq.; sp.gr. 1.05 g/cc; visc. 1900 cs (20 C); 95% act.

NAO 105. [Sovereign] French process zinc oxide, untreated; CAS 1314-13-2; EINECS 215-222-5; general purpose activator for rubber; also used in oil-based paints and some chem. applics. where low reactivity is required; powd.; 0.21 μ mean particle size; surf. area 5.0 m^2/g.

NAO 115. [Sovereign] French process zinc oxide, treated; CAS 1314-13-2; EINECS 215-222-5; general purpose activator for rubber for use where fast dispersion is required; exhibits good flow in soft stocks; treated version of NAO 105.

NAO 125. [Sovereign] French process zinc oxide; CAS 1314-13-2; EINECS 215-222-5; activator for rubber for use in automatic handling equip. and applics. where min. dust is required; not recommended for use in soft compds.; pelleted version of NAO 105; pellets.

NAO 135. [Sovereign] French process zinc oxide; CAS 1314-13-2; EINECS 215-222-5; activator for rubber for use in automatic handling equip. and applics. where min. dust is required; not recommended for use in soft compds.; pelleted form of NAO 115; pellets.

Natrosol® Hydroxyethylcellulose. [Aqualon] Hydroxyethylcellulose; CAS 9004-62-0; thickener for hair care prods., creams and lotions, latex paints; protective colloid in emulsion polymerization; suspending aid in joint and tile cements; binder for welding rods; sol. in hot and cold water, DMSO; tolerates up to 70% polar org. solv. in water.

Naugard® 10. [Uniroyal/Spec. Chem.] Tetrakis [methylene (3,5-di-t-butyl-4-hydroxyhydrocinnamate)] methane; CAS 6683-19-8; EINECS 229-722-6; antioxidant effective against thermal oxidative degradation during long term heat aging; for polyolefins, styrenics, elastomers, adhesives, lubricants, and oils; FDA approved for PP, LDPE, LLDPE, HDPE, HIPS, ABS, polyester, EVA; FDA 21CFR §178.2010, 176.170(a)(5), 175.125, 175.300 (b)(2,3), 175.320(b)(3), 176.170(b)(2), 175.300, 176.180, 177.2470 (b), 178.3800, 178.3850; wh. free-flowing powd. or gran.; sol. (g/100 ml): 100 g acetone, chloroform, benzene; m.w. 1178; sp.gr. 1.05; m.p. 110-125 C; flash pt. 299 C; *Toxicology:* LD50 (oral, rat) > 5000 mg/kg; LD50 (dermal, rabbit) > 3160 mg/kg; nonirritating to eyes, skin; *Storage:* 1 yr storage life in sealed containers in cool, dry areas; avoid extended storage at elveated temps.

Naugard® 76. [Uniroyal/Spec. Chem.] Octadecyl 3,5-di-t-butyl-4-hydroxyhydrocinnamate; CAS 2082-79-3; EINECS 218-216-0; nondiscoloring antioxidant for stabilizing polymeric substances such as polyolefins, styrenics, EPDM, and PVC; provides good thermal and color stability; FDA approved for PP, LDPE, LLDPE, HDPE, HIPS, ABS, PVC, PC, EVA, polyamides; FDA 21CFR §178.2010; wh. free-flowing powd. or gran.; sol. (g/100 ml): 140 g chloroform, 104 g xylene, 65 g ethyl acetate, 52 g hexane; m.w. 531; sp.gr. 1.02; m.p. 50-55 C; flash pt. 273 C; *Toxicology:* LD50 (oral, rat) > 10,000 mg/kg (low oral toxicity); LD50 (dermal, rabbit) > 2000 mg/kg (irritating); *Storage:* 1 yr storage life in sealed containers in cool, dry areas; avoid extended storage above 120 F, direct heat.

Naugard® 431. [Uniroyal/Spec. Chem.] Hindered phenolic; nondiscoloring antioxidant, stabilizer used in hot melt adhesives applic.; FDA approved for PP, LDPE, LLDPE, HDPE, HIPS, EVA, polyamides; FDA approval.

Naugard® 445. [Uniroyal/Spec. Chem.] 4,4′-Bis (α,α-dimethylbenzyl) diphenylamine; CAS 10081-67-1; nondiscoloring antioxidant for polyolefins, styrenics, polyether polyols, hot-melt adhesives, lubricant additives, nylon and other polymers; FDA approved for polyester, EVA, polyamides; FDA §175.105, 175.300, 177.1590; wh. to off-wh. powd.; sol. (g/100 ml): 41.2 g MEK, 40.8 g acetone, 35.2 g benzene, 32.8 g toluene; m.w. 446; sp.gr. 1.14; m.p. 98-100 C; *Storage:* store in cool, dry location away from direct heat and light sources.

Naugard® 492. [Uniroyal/Spec. Chem.] Phenolic phosphite; nonstaining, nondiscoloring antioxidant with FDA approvals for PP, LDPE, LLDPE, HDPE, HIPS, EVA, polyamides; FDA approvals; liq.

Naugard® 524. [Uniroyal/Spec. Chem.] Tris (2,4-di-t-butyl phenyl) phosphite; CAS 31570-04-4; nondiscoloring antioxidant used in thermoplastic and thermoset polymers where color and processing stability are critical; FDA approved for PP, LDPE, LLDPE, HDPE, HIPS, ABS, nylon 6, PC, EVA, polyamides; exceptional hydrolytic stability; FDA 21CFR §178.2010; wh. free-flowing powd.; sol. in methylene chloride, chlorobenzene, hexane, hot ethanol; insol. in water; m.w. 647; sp.gr. 1.03; m.p. 180-186 C; flash pt. 225 C; *Toxicology:* LD50 (oral, rat) > 5000 mg/kg; low oral toxicity; nonirritating to skin and eyes; *Storage:* 1 yr life in sealed containers stored in cool, dry areas; avoid extended storage above 120 F, direct heat.

Naugard® 529. [Uniroyal/Spec. Chem.] Butylated styrenated p-cresol; antioxidant providing exc. color and processing stability; for PP, LDPE, LLDPE, HDPE, HIPS; lubricant for fuels, syn. lubricants, oils, and greases; FDA approved; liq.

Naugard® A. [Uniroyal/Spec. Chem.] Sec. amine; antioxidant protecting polymers from loss of physical props. on extended exposure to heat; FDA approved for EVA and polyamides; FDA approved.

Naugard® BHT. [Uniroyal/Spec. Chem.] Di-t-butyl-p-cresol; CAS 128-37-0; EINECS 204-881-4; antioxidant; stabilizer used in hot melt adhesives applics.; for PP, LDPE, LLDPE, HDPE, HIPS, ABS, PVC, polyester, EVA, polyamides; avail. as tech. or food grade; lubricant for fuels, gear oil, turbine oils, syn. lubricants, hydraulic oils, greases; FDA approvals.

Naugard® I-2. [Uniroyal/Spec. Chem.] N,N′-Bis(1,4-dimethylpentyl)-p-phenylenediamine; CAS 3081-14-9; styrene polymerization inhibitor; column antifoulant.

Naugard® I-3. [Uniroyal/Spec. Chem.] N-(1,4-Dimethylpentyl)-N′-phenyl-p-phenylenediamine; styrene polymerization inhibitor; column antifoulant.

Naugard® I-4. [Uniroyal/Spec. Chem.] N-Phenyl,N′-isopropyl-p-phenylenediamine; styrene polymerization inhibitor; column antifoulant.

Naugard® J. [Uniroyal/Spec. Chem.] N,N′-Diphenyl-p-phenylenediamine; CAS 74-31-7; acrylic polymerization inhibitor.

Naugard® NBC. [Uniroyal/Spec. Chem.] Nickel dibutyldithiocarbamate; CAS 13927-77-0; nickel chelating uv stabilizer for polyolefins; dk. grn. powd.; sol. (g/100 ml solv.): 101 g benzene, 58 g toluene, ethylene dichloride, 28 g CCl_4, 11 g acetone; m.w. 467.5; sp.gr. 1.26; m.p. 86-90 C; flash pt. (TOC) 500 F.

Naugard® P. [Uniroyal/Spec. Chem.] Tris(nonylphenyl) phosphite; CAS 26523-78-4; EINECS 247-759-6; antioxidant; processing and color stabilizer for substrates; hot melt adhesive applic.; for PP, LDPE, LLDPE, HDPE, HIPS, ABS, PVC, PC, EVA, polyamides; FDA 21CFR §178.2010.

Naugard® PHR. [Uniroyal/Spec. Chem.] Tris(nonylphenyl) phosphite with 1% triisopropanolamine; antioxidant; processing and color stabilizer for substrates; hot melt adhesive applic., FDA approved for PP, LDPE, LLDPE, HDPE, HIPS, ABS, PVC, PC, EVA, polyamides; FDA approvals.

Naugard® SP. [Uniroyal/Spec. Chem.] Phenolic/bisphenolic; nonstaining antioxidant and processing stabilizer; used in cellulose, nylon, PE, PP, PS, PVC, urethane, EVA and polyamide hot melt adhesives; FDA 21CFR §175.105; liq.

Naugard® Super Q. [Uniroyal/Spec. Chem.] Polymerized trimethyl dihydroquinoline; CAS 26780-96-1; antioxidant with exceptional processing and long term heat stabilization, esp. in carbon black-filled polymers; used in LDPE, LLDPE, HDPE, and EVA.

Naugard® XL-1. [Uniroyal/Spec. Chem.] 2,2′-Oxamido bis-[ethyl 3-(3,5-di-tert-butyl-4-hydroxyphenyl) propionate]; CAS 70331-94-1; antioxidant and metal deactivator; used in polymerization, processing, and in end use applics., wire and cable insulation, pipe and inj. parts for automobiles and appliances; processing stabilizer for polyolefins, film, sheet, and blow molded bottles; FDA approved for PP, LDPE, LLDPE, HDPEZ, HIPS, ABS, EVA, and polyamides; FDA 21CFR §178.2010, 177.1520(c), 177.1640, 175.105; wh. free-flowing powd.; sol. (g/100 ml): 35 g chloroform, 10 g acetone, 0.01 g water; m.w. 697; sp.gr. 1.12; m.p. 170-180 C; flash pt. (TOC) 260 C; *Toxicology:* LD50 (oral, rat) > 20 g/kg; low oral toxicity; sl. irritating to eyes; nonirritating to skin; *Storage:* 1 yr storage life in sealed containers stored in cool, dry areas; avoid extended storage at high temps.

Naugard® XL-517. [Uniroyal/Spec. Chem.] Phenolic phosphite; blend of high performance antioxidant/ metal deactivator with color stabilizer to provide optimal color and processing stability; FDA approved for PP, LDPE, LLDPE, HDPE, HIPS, ABS, EVA, and polyamides; FDA approvals; solid.

Naugawhite®. [Uniroyal] Hindered bisphenolic; nonstaining general-purpose antioxidant for thermoplastics; hot melt adhesives; FDA approved for ABS, EVA, polyamides; FDA approved.

Naxonac™ 510. [Ruetgers-Nease] Nonyl phenol ether phosphate; anionic; detergent, wetting agent, emulsifier, lubricant, hydrotrope for heavy-duty and household detergents, waterless hand cleaners, solv. degreasers, emulsion polymerization, paint and wax strippers, electrolytic cleaners; clear liq.; sol. and stable in aq. sol'ns. with high alkaline concs.; pH 2; 99% conc.

Naxonac™ 610. [Ruetgers-Nease] Nonyl phenol ether phosphate; anionic; detergent, emulsifier, wetting agent, stabilizer; solubilizer for nonionics; for household and industrial cleaners, electrolytic cleaning, dairy cleaning, acid cleaners, solv. cleaners; emulsion polymerization of vinyl acetate and highly carboxylated acrylic monomers/comonomers; clear visc. liq., mild odor; sol. and stable in aq. sol'ns. with high concs. of alkaline builders; sp.gr. 1.05; pH 1.5-2.5; 100% act.

Naxonac™ 690-70. [Ruetgers-Nease] Nonyl phenol ether phosphate; anionic; emulsifier imparting corrosion resistance; for emulsion polymerization of vinyl acetate and highly carboxylated acrylics; does not discolor when used to dedust alkaline powds.; colorless to pale yel. visc. liq.; sp.gr. 1.07; pH 1.7 (10%); 70% act.

NC-4. [Mitsui Toatsu] Bis (p-ethylbenzylidene) sorbitol; nucleating agent improving optical and physical props. of PP for container, pkg. film, inj. syringe and medical applics., food pkg.; FDA 21CFR § 178.3295; approved for food contact applics. in Japan, Europe, Australia; m.w. 414; m.p. 210-230 C; 99.5% min. purity; *Toxicology:* nontoxic.

Negomel AL-5. [StanChem] Internal antistat for flexible PVC.

Neoprene FB. [DuPont] Polychloroprene (2-chlorobutadiene 1,3); very low m.w. vulcanizing plasticizer for Neoprene, other syn. elastomers; used in cement applics. requiring high solids-to-visc. ratio; in preparation of caulks and sealants; dk. brn. wax-like solid.

Neutral Degras. [Fanning] Fatty esters of wool grease; CAS 68815-23-6; lubricant with EP and slip chars.; wire drawing compds., slushing and cutting oils, lubricants; rust preventative; plasticizer and lubricant for adhesives; textile lubricant; inhibits crystallization of wax components; ink formulations; dispersant for other waxes; used in leather in stuffing greases; waterproofing agent; sol. in chloroform, trichloroethylene, and 100 parts of boiling anhyd. alcohol; insol. in water.

Nevastain® 21. [Neville] Hindered styrenated phenol; CAS 61788-44-1; nonstaining antioxidant used in mastic adhesives, rubber goods such as mech. and molded goods, PE, PP, PS, caulking compds., cement, and antiskinning agents; FDA 21CFR §175.105; Gardner 7 visc. liq.; sol. in alcohols, chlorinated hydrocarbons, esters, ethers, ketones (except acetone), aromatic, naphthenic, and terpene hydrocarbons; m.w. 300; sp.gr. 1.075; hyd. no. 190; COC flash pt. 325 F.

Nevastain® A. [Neville] Hindered phenolic compd.; nonstaining antioxidant used in mastic adhesives, rubber goods such as mech. and molded goods, caulking compds., cement, and antiskinning agents; FDA 21CFR §175.105, 177.2600; Gardner 14 visc. liq.; sol. in alcohols, chlorinated hydrocarbons, esters, ethers, ketones (except acetone), aromatic, naphthenic, and terpene hydrocarbons; m.w. 300; sp.gr. 1.08; hyd. no. 210; flash pt. (COC) 380 F.

Nevastain® B. [Neville] Hindered phenolic compd.; nonstaining antioxidant used in mastic adhesives, rubber goods such as mech. and molded goods, caulking compds., cement, and antiskinning agents; FDA 21CFR §175.105, 177.2600; Gardner 15 flakes; sol. in alcohols, chlorinated hydrocarbons, esters, ethers, ketones (except acetone), aromatic, naphthenic, and terpene hydrocarbons; m.w. 490; sp.gr. 1.10; soften. pt. (R&B) 95 C; hyd. no. 120; flash pt. (COC) 525 F.

Nevchem® 70. [Neville] Alkylated petrol. hydrocarbon resin; resin used in adhesives (hot melt pressure sensitive, mastic); coatings (aluminum, emulsion, industrial, marine, paper, traffic, wire/cable), inks, rubber (cements, mechanical and molded goods, tires), concrete-curing compds., caulking compds.; processing aid in rubber compds.; FDA 21CFR §175.105, 177.2600, 178.3800; Gardner 12 (50% in toluene) solid; sol. in esters, ethers, ketones (except acetone), chlorinated, aromatic, naphthenic, and terpene hydrocarbons; m.w. 430; sp.gr. 1.02; soften. pt. (R&B) 72 C; iodine no. (Wijs) 70; flash pt. (COC) 415 F.

Nevchem® 100. [Neville] Alkylated petrol. hydrocarbon resin; resin used in adhesives (hot melt pressure sensitive, mastic); coatings (aluminum, emulsion, industrial, marine, paper, traffic, wire/cable), inks, rubber (cements, mechanical and molded goods, tires), concrete-curing compds., caulking compds.; processing aid in rubber compds.; FDA 21CFR §175.105, 177.2600, 178.3800; Gardner 12 (50% in toluene) solid, flakes; sol. in esters, ethers, ketones (except acetone), chlorinated, aromatic, naphthenic, and terpene hydrocarbons; m.w. 640; sp.gr. 1.063; soften. pt. (R&B) 103 C;

iodine no. (Wijs) 100; flash pt. (COC) 455 F.

Nevchem® 110. [Neville] Alkylated petrol. hydrocarbon resin; resin used in adhesives (hot melt pressure sensitive, mastic); coatings (aluminum, emulsion, industrial, marine, paper, traffic, wire/cable), inks, rubber (cements, mechanical and molded goods, tires), concrete-curing compds., caulking compds.; processing aid in rubber compds.; FDA 21CFR §175.105, 177.2600, 178.3800; Gardner 12 (50% in toluene) solid, flakes; sol. in esters, ethers, ketones (except acetone), chlorinated, aromatic, naphthenic, and terpene hydrocarbons; m.w. 715; sp.gr. 1.064; soften. pt. (R&B) 110 C; iodine no. (Wijs) 100; flash pt. (COC) 465 F.

Nevchem® 120. [Neville] Alkylated petrol. hydrocarbon resin; resin used in adhesives (hot melt pressure sensitive, mastic); coatings (aluminum, emulsion, industrial, marine, paper, traffic, wire/cable), inks, rubber (cements, mechanical and molded goods, tires), concrete-curing compds., caulking compds.; processing aid in rubber compds.; FDA 21CFR §175.105, 177.2600, 178.3800; Gardner 12 (50% in toluene) solid, flakes; sol. in esters, ethers, ketones (except acetone), chlorinated, aromatic, naphthenic, and terpene hydrocarbons; m.w. 880; sp.gr. 1.070; soften. pt. (R&B) 120 C; iodine no. (Wijs) 100; flash pt. (COC) 475 F.

Nevchem® 130. [Neville] Alkylated petrol. hydrocarbon resin; resin used in adhesives (hot melt pressure sensitive, mastic); coatings (aluminum, emulsion, industrial, marine, paper, traffic, wire/cable), inks, rubber (cements, mechanical and molded goods, tires), concrete-curing compds., caulking compds.; processing aid in rubber compds.; FDA 21CFR §175.105, 177.2600, 178.3800; Gardner 12 (50% in toluene) solid, flakes; sol. in esters, ethers, ketones (except acetone), chlorinated, aromatic, naphthenic, and terpene hydrocarbons; m.w. 1092; sp.gr. 1.076; soften. pt. (R&B) 130 C; iodine no. (Wijs) 107; flash pt. (COC) 510 F.

Nevchem® 140. [Neville] Alkylated petrol. hydrocarbon resin; resin used in adhesives (hot melt pressure sensitive, mastic); coatings (aluminum, emulsion, industrial, marine, paper, traffic, wire/cable), inks, rubber (cements, mechanical and molded goods, tires), concrete-curing compds., caulking compds.; processing aid in rubber compds.; FDA 21CFR §175.105, 177.2600, 178.3800; Gardner 12 (50% in toluene) solid, flakes; sol. in esters, ethers, ketones (except acetone), chlorinated, aromatic, naphthenic, and terpene hydrocarbons; m.w. 1110; sp.gr. 1.080; soften. pt. (R&B) 140 C; iodine no. (Wijs) 107; flash pt. (COC) 520 F.

Nevchem® 150. [Neville] Alkylated petrol. hydrocarbon resin; resin used in adhesives (hot melt pressure sensitive, mastic); coatings (aluminum, emulsion, industrial, marine, paper, traffic, wire/cable), inks, rubber (cements, mechanical and molded goods, tires), concrete-curing compds., caulking compds.; processing aid in rubber compds.; FDA 21CFR §175.105, 177.2600, 178.3800; Gardner 12 (50% in toluene) solid, flakes; sol. in esters, ethers, ketones (except acetone), chlorinated, aromatic, naphthenic, and terpene hydrocarbons; m.w. 1200; sp.gr. 1.09; soften. pt. (R&B) 150 C; iodine no. (Wijs) 107; flash pt. (COC) 550 F.

Nevex® 100. [Neville] Modified hydrocarbon resin; enhances compatibility of wax and ethylene-vinyl acetate systems; in adhesives (hot melt, hot melt pressure sensitive, mastic, pressure sensitive), coatings (can & drum, industrial, paper), rubber (cements, mechanical and molded goods, tires); FDA 21CFR §175.105, 175.300, 176.170, 176.180, 177.1210, 177.2600, 178.3800; Gardner 12 (50% in toluene) solid, flake; sol. in esters, ethers, ketones (except acetone), and chlorinated, aromatic, naphthenic, and terpene hydrocarbons; m.w. 640; sp.gr. 1.120; soften. pt. (R&B) 100 C; acid no. < 1; iodine no. (Wijs) 55; flash pt. (COC) 410 F.

Nevillac® 10° XL. [Neville] Hydroxy modified resin; plasticizer/tackifier in nitrile rubber; used in adhesives (hot melt, hot melt pressure sensitive, pressure sensitive), epoxy coatings, rubber cements, antiskinning agents; FDA 21CFR §175.105, 177.2600; Gardner 8 liq.; sol. in alcohols, esters, ethers, ketones (except acetone), and chlorinated, aromatic, naphthenic, and terpene hydrocarbons; m.w. 250; sp.gr. 1.091; soften. (R&B) pt. 12 C; iodine no. (Wijs) 80; hyd. no. 200; flash pt. (COC) 365 F.

Nevillac® Hard. [Neville] Hydroxy modified resin; plasticizer/tackifier for nitrile rubber; maintains oil and solv. resist. of the compounded nitrile; FDA 21CFR §175.105, 177.2600; Gardner 15 (50% in toluene) solid; sol. in alcohols, esters, ethers, ketones (except acetone), and chlorinated, aromatic, naphthenic, and terpene hydrocarbons; m.w. 430; sp.gr. 1.144; soften. pt. (R&B) 76 C; iodine no. (wijs) 140; hyd. no. 170; flash pt. (COC) 475 F.

Nevpene® 9500. [Neville] Modified hydrocarbon resin; aromatic producing an exceptional tackifier with modification; used in adhesives, coatings, rubber, and caulking compds.; FDA 21CFR §175.105, 175.300, 176.170, 176.180, 177.1210, 177.2600, 178.3800; Gardner 8 (50% in toluene) solid, flakes; sol. in esters, ethers, ketones (except acetone), and chlorinated, aromatic, naphthenic, and terpene hydrocarbons; m.w. 1040; sp.gr. 1.050; soften. pt. (R&B) 95 C; acid no. < 1; iodine no. (Wijs) 97; flash pt. (COC) 450 F.

Nevtac® 10°. [Neville] Syn. polyterpene resin; tackifier for PP, PE, NR, SIS block copolymers, EVA, polyisoprene, butyl rubber used in solv. and hot-melt adhesives, coatings, rubber prods., concrete-curing compds., and caulking compds.; FDA 21 CFR §175.125, 175.105, 175.300, 175.320, 176.170, 176.180, 177.1210, 177.2600; Gardner 6 (50% in toluene) liq.; sol. in esters, ethers, ketones (except acetone), aromatic, naphthenic, chlorinated, and terpene hydrocarbons; m.w. 363; sp.gr. 0.99; soften. pt. (R&B) 10 C; flash pt. (COC) 400 F.

Nevtac® 80. [Neville] Syn. polyterpene resin; tackifier for PP, PE, NR, SIS block copolymers, EVA, polyisoprene, butyl rubber used in solv. and hot-melt adhesives, coatings, rubber prods., concrete-curing compds., and caulking compds.; FDA 21 CFR §175.125, 175.105, 175.300, 175.320, 176.170, 176.180, 177.1210, 177.2600; Gardner 5 (50% in toluene) solid; sol. in esters, ethers, ketones (except acetone), and chlorinated, aromatic, naphthenic, and terpene hydrocarbons; m.w. 1070; sp.gr. 0.97; soften. pt. (R&B) 81 C; flash pt. (COC) 440 F.

Nevtac® 100. [Neville] Syn. polyterpene resin; tackifier for PP, PE, NR, SIS block copolymers, EVA, polyisoprene, butyl rubber used in solv. and hot-melt adhesives, coatings, rubber prods., concrete-curing compds., and caulking compds.; FDA 21 CFR §175.125, 175.105, 175.300, 175.320, 176.170, 176.180, 177.1210, 177.2600; Gardner 3+ (50% in toluene) solid, flakes; sol. in esters, ethers, ketones (except acetone), and chlorinated, aromatic, naphthenic, and terpene hydrocarbons; m.w. 1215; sp.gr. 0.97; soften. pt. (R&B) 101 C; flash pt. (COC) 450 F.

Nevtac® 115. [Neville] Syn. polyterpene resin; tackifier for PP, PE, NR, SIS block copolymers, EVA, polyisoprene, butyl rubber used in solv. and hot-melt adhesives, coatings, rubber prods., concrete-curing compds., and caulking compds.; FDA 21 CFR §175.125, 175.105, 175.300, 175.320, 176.170, 176.180, 177.1210, 177.2600; Gardner 6 (50% in toluene) solid, flakes; sol. in esters, ethers, ketones (except acetone), and chlorinated, aromatic, naphthenic, and terpene hydrocarbons; m.w. 1405; sp.gr. 0.99; soften. pt. (R&B) 114 C; flash pt. (COC) 460 F.

Newcol 180T. [Nippon Nyukazai] POE stearate; nonionic; surfactant used in emulsion polymerization; solid; 100% conc.

Newcol 261A, 271A. [Nippon Nyukazai] Sodium alkyl diphenyl ether disulfonate; anionic; detergent, emulsifier used in emulsion polymerization; liq.; 45% conc.

Newcol 506. [Nippon Nyukazai] POE nonylphenyl ether; nonionic; antifoam, mold lubricant, wetting agent, penetrant, spreading agent, dispersant, emulsifier for detergents, agric. chemicals, machine oils, emulsion polymerization; solid; 100% conc.

Newcol 508. [Nippon Nyukazai] POE nonylphenyl ether; nonionic; antifoam, mold lubricant, wetting agent, penetrant, spreading agent, dispersant, emulsifier for detergents, agric. chemicals, machine oils, emulsion polymerization; solid; 100% conc.

Newcol 560SF. [Nippon Nyukazai] Ammonium POE nonylphenyl ether sulfate; anionic; emulsifier, wetting agent, penetrant, detergent used in emulsion polymerization; liq.; 50% conc.

Newcol 560SN. [Nippon Nyukazai] Sodium POE nonylphenyl ether sulfate; anionic; emulsifier, wetting agent, penetrant, detergent used in emulsion polymerization; liq.; 30% conc.

Newcol 568. [Nippon Nyukazai] POE nonylphenyl ether; nonionic; low-foaming detergent; surfactant for emulsion polymerization; solid; 100% conc.

Newcol 607, 610, 614, 623. [Nippon Nyukazai] POE alkylaryl ether; nonionic; emulsifier, solubilizer, detergent used in emulsion polymerization; liqs. and solids; 100% conc.

Newcol 704, 707. [Nippon Nyukazai] POE alkylaryl ether; nonionic; emulsifier, solubilizer, detergent for emulsion polymerization; liq.; 100% conc.

Newcol 707SF. [Nippon Nyukazai] POE alkylaryl ether; CAS 104042-16-2; anionic; emulsifier, detergent for emulsion polymerization; liq.; 30% conc.

Newcol 710, 714, 723. [Nippon Nyukazai] POE alkylaryl ether; CAS 104042-16-2; anionic; emulsifier, solubilizer, detergent for emulsion polymerization; liq.; 30% conc.

Newcol 861S. [Nippon Nyukazai] Sodium octylphenoxyethoxyethyl sulfonate; anionic; emulsifier, wetting agent, penetrant, detergent used in emulsion polymerization; paste; 30% conc.

Newcol 1305SN, 1310SN. [Nippon Nyukazai] Sodium POE tridecyl ether sulfate; anionic; emulsifier, wetting agent, penetrant, detergent used in emulsion polymerization; liq.; 30% conc.

Newpol PE-61. [Sanyo Chem. Industries] EO/PO block copolymer; CAS 9003-11-6; nonionic; base material for household and industrial detergents; plasticizer, antistat for phenol resins; emulsifier for agric. pesticides and emulsion polymerization; pigment and pitch dispersant; liq.; HLB 5.8; 100% conc.

Newpol PE-62. [Sanyo Chem. Industries] EO/PO block copolymer; CAS 9003-11-6; nonionic; base material for household and industrial detergents; plasticizer, antistat for phenol resins; emulsifier for agric. pesticides and emulsion polymerization; pigment and pitch dispersant; liq.; HLB 6.3; 100% conc.

Newpol PE-64. [Sanyo Chem. Industries] EO/PO block copolymer; CAS 9003-11-6; nonionic; base material for household and industrial detergents; plasticizer, antistat for phenol resins; emulsifier for agric. pesticides and emulsion polymerization; pigment and pitch dispersant; liq.; HLB 10.1; 100% conc.

Newpol PE-68. [Sanyo Chem. Industries] EO/PO block copolymer; CAS 9003-11-6; nonionic; base material for household and industrial detergents; plasticizer, antistat for phenol resins; emulsifier for agric. pesticides and emulsion polymerization; pigment and pitch dispersant; solid; HLB 14.9; 100% conc.

Newpol PE-74. [Sanyo Chem. Industries] EO/PO block copolymer; CAS 9003-11-6; nonionic; base material for household and industrial detergents; plasticizer, antistat for phenol resins; emulsifier for agric. pesticides and emulsion polymerization; pigment and pitch dispersant; liq.; HLB 10.1; 100% conc.

Newpol PE-75. [Sanyo Chem. Industries] EO/PO block copolymer; CAS 9003-11-6; nonionic; base material for household and industrial detergents; plasticizer, antistat for phenol resins; emulsifier for agric. pesticides and emulsion polymerization; pigment and pitch dispersant; liq.; HLB 10.7; 100% conc.

Newpol PE-78. [Sanyo Chem. Industries] EO/PO block copolymer; CAS 9003-11-6; nonionic; base material for household and industrial detergents; plasticizer, antistat for phenol resins; emulsifier for agric. pesticides and emulsion polymerization; pigment and pitch dispersant; solid; HLB 14.8; 100% conc.

Newpol PE-88. [Sanyo Chem. Industries] EO/PO block copolymer; CAS 9003-11-6; nonionic; base material for household and industrial detergents; plasticizer, antistat for phenol resins; emulsifier for agric. pesticides and emulsion polymerization; pigment and pitch dispersant; solid; HLB 14.6; 100% conc.

Niaproof® Anionic Surfactant 4. [Niacet] Sodium tetradecyl sulfate; CAS 139-88-8; anionic; detergent, wetting agent, penetrant, emulsifier used in adhesives and sealants, coatings, photo chemicals, emulsion polymerization, metal processing, electrolytic cleaning, pickling baths, plating, pharmaceuticals, leather, textiles; FDA compliance; colorless liq., mild char. odor; misc. with water; sp.gr. 1.031; dens. 8.58 lb/gal; b.p. 92 C; COC flash pt. none; pH 8.5 (0.1% aq.); surf. tens. 47 dynes/cm (0.1% aq.); Draves wetting 20 s (0.26%); Ross-Miles foam 10 mm; 27% act. in water; *Toxicology:* moderate oral and skin toxicity; eye irritant; LD50 (oral, rats) 4.95 ml/kg; *Precaution:* corrosive, slippery.

Niaproof® Anionic Surfactant 08. [Niacet] Sodium 2-ethylhexyl sulfate; CAS 126-92-1; EINECS 204-812-8; anionic; detergent, wetting agent, penetrant, emulsifier used in textile mercerizing, metal cleaning, electroplating, photo chemicals, adhesives, emulsion polymerization, household and industrial cleaners, agric., pharmaceuticals; stable to high concs. of electrolytes; FDA compliance; colorless liq., mild char. odor; misc. with water; sp.gr. 1.109; dens. 9.23 lb/gal; b.p. 95 C; COC flash pt. none; pH 7.3 (0.1% aq.); surf. tens. 63 dynes/cm (0.1% aq.); Ross-Miles foam 10 mm (initial); 39% act.; *Toxicology:* moderate oral and skin toxicity; eye irritant; LD50 (oral, rats) 7.27 ml/kg.

Nicron 325. [Luzenac Am.] Platy talc; CAS 14807-96-6; EINECS 238-877-9; general purpose economical filler for industrial coatings and plastics that require max. filler loadings; 7.5 μ median particle size; Hegman grind 0-1; oil absorp. 26; dry brightness 85.

Nicron 400. [Luzenac Am.] Platy talc; CAS 14807-96-6; EINECS 238-877-9; general purpose stir-in filler with minimum binder demand; for all types of paints and plastics; 5.0 μ median particle size; Hegman grind 4 min.; oil absorp. 31; dry brightness 86.

Nikkol GO-430. [Nikko Chem. Co. Ltd.] PEG-30 sorbitan tetraoleate; nonionic; emulsifier, solubilizer, superfatting agent used in drugs and cosmetics, for emulsion polymerization, agric. chemicals, printing inks; pale yel. liq.; sol. in ethanol, ethyl acetate, xylene; partly sol. in water; sp.gr. 1.048; HLB 11.5; ref. index 1.4727; 100% conc.

Nikkol GO-440. [Nikko Chem. Co. Ltd.] PEG-40 sorbitan tetraoleate; CAS 9003-11-6; nonionic; emulsifier, solubilizer, superfatting agent used in drugs and cosmetics, for emulsion polymerization, agric. chemicals, printing inks; pale yel. liq.; sol. in ethanol, ethyl acetate, xylene; partly sol. in water, propylene glycol; sp.gr. 1.054; HLB 12.5; 100% conc.

Nikkol GO-460. [Nikko Chem. Co. Ltd.] PEG-60 sorbitan tetraoleate; nonionic; emulsifier, solubilizer, superfatting agent used in drugs and cosmetics, for emulsion polymerization, agric. chemicals, printing inks; pale yel. liq.; sol. in water, ethanol, ethyl acetate, xylene; partly sol. in propylene glycol; sp.gr. 1.060; HLB 14.0; 100% conc.

Niox KQ-34. [Pulcra SA] Oxo-alcohol, ethoxylated; nonionic; wetting agent; emulsifier for fats and oleins; aux. prod. for emulsion polymerization; solid; HLB 15.4; 100% conc.; Unverified

Nipol® 1000X88. [Zeon] Butadiene-acrylonitrile copolymer, slightly staining antioxidant; CAS 9003-18-3; used for molded and extruded prods., adhesives, plastic modifications; exc. oil and fuel resist.; vulcanized with sulfur-accelerator, sulfur donor, or peroxide systems; gum; sp.gr. 1.00; Mooney visc. 70-90 (ML4, 100 C); 43% ACN.

Nipol® 1001 CG. [Zeon] Butadiene-acrylonitrile copolymer, slightly staining antioxidant; CAS 9003-18-3; cement grade used for coated materials, sheet packing, molded and extruded goods, o-rings, oil seals, adhesives, plastic modification; exc. oil and fuel resist.; vulcanized with sulfur-accelerator, sulfur donor, or peroxide systems; gum; sp.gr. 1.00; Mooney visc. 75-100 (ML4, 100 C); 41% ACN.

Nipol® 1022X59. [Zeon] Precrosslinked butadiene-acrylonitrile copolymer; CAS 9003-18-3; used in plastics compding. to impart rubbery props. to thermoplastics; sp.gr. 0.98; Mooney visc. 53-68 (ML, 100 C); 33% ACN.

Nipol® 1042X82. [Zeon] Precrosslinked butadiene-acrylonitrile copolymer, nonstaining antioxidant; CAS 9003-18-3; used for coated materials, rubber-covered rolls, molded and extruded goods; exc. calendering stock; vulcanized with sulfur-accelerator, sulfur donor, or peroxide systems; may be blended with other polymers to improve extrusion; gum; sp.gr. 0.98; Mooney visc. 75-95 (ML4, 100 C); 33% ACN.

Nipol® 1053. [Zeon] Butadiene-acrylonitrile copolymer, nonstaining antioxidant; CAS 9003-18-3; used for flooring, hose, molded and extruded goods, o-rings, oil seals, plastic modification; superior processing; vulcanized with sulfur-accelerator, sulfur donor, or peroxide systems; gum; sp.gr. 0.97; Mooney visc. 45-60 (ML4, 100 C); 29% ACN.

Nipol® 1312. [Zeon] Butadiene-acrylonitrile copolymer, nonstaining antioxidant; CAS 9003-18-3; plasticizer used for nitrile, neoprene, and PVC compds.; improves knitting and flow; also for plastisols and phenolic resins; FDA 21CFR §175.105, 177.2600; liq.; sp.gr. 0.96; visc. 20,000-30,000 cps; 100% solids; 28% ACN.

Nipol® 1312 LV. [Zeon] Butadiene-acrylonitrile copolymer, nonstaining antioxidant; CAS 9003-18-3; plasticizer, softener used for nitrile, neoprene, phenolic, and PVC compds.; improves knitting and flow; esp. suitable for plastisol compounding; FDA 21CFR §175.105, 177.2600; liq.; sp.gr. 0.96; visc. 9000-16,000 cps; 100% solids; 26% ACN.

Nipol® 1411. [Zeon] Butadiene-acrylonitrile copolymer, slightly staining antioxidant with 5% talc partitioning agent; used for friction prods., phenolic resin modification; vulcanized with sulfur-accelerator systems; finely divided powd.; 0.1 mm median particle size; nonsol.; sp.gr. 1.00; 38% ACN.

Nipol® 1422. [Zeon] Butadiene-acrylonitrile copolymer, nonstaining antioxidant with 4% talc and 2% silica partitioning agent; nonmigratory, nonvolatile, nonextractable plasticizer for PVC compds. improving low-temp. flexibility, impact resist., resist. to oils and fuels; for automotive door panel sheet, furniture upholstery, luggage casings, footwear, wire and cable covers, tubing, hose, pads, mmounts, extruded profiles, sheet, general molding; finely divided powd.; 0.1 mm median particle size; nonsol.; sp.gr. 0.98; 33% ACN.

Nipol® 1422X14. [Zeon] Precrosslinked butadiene-acrylonitrile copolymer, nonstaining antioxidant, 6% talc partitioning agent; powd. form of Nipol 1022X59; nonextractable modifier for PVC, ABS to impart flexibility, impact resist., exc. uv and thermal resist.; coarse powd.; 1.0 mm median particle size; Mooney visc. 53-68 (base NBR); 33% ACN.

Nipol® 1452X8. [Zeon] Butadiene-acrylonitrile copolymer, nonstaining antioxidant, 7% talc partitioning agent; for plastics modification, adhesives, and coatings; improved thermal and uv stability; powd.; 9.5 mm median particle size; sp.gr. 0.98; Mooney visc. 45-60; 33% ACN.

Nipol® 1453 HM. [Zeon] Butadiene-acrylonitrile copolymer with 5% talc partitioning agent; plastics modifier (ABS, PVC); also used for adhesives; FDA compliances; crumb rubber; 9.5 mm median particle size; sp.gr. 0.97; Mooney visc. 60-75 (base NBR); 29% ACN.

Nipol® 1472. [Zeon] Carboxylated butadiene-acrylonitrile copolymer with 5% talc partitioning agent; used for adhesives, composites, epoxy and phenolic resin modification; crumb; 9.5 mm median particle size; sp.gr. 0.98; Mooney visc. 25-35 (base NBR); 27% ACN.

Nipol® DP5123P. [Zeon] Noncrosslinked NBR, nonstaining antioxidant with 9% PVC partitioning agent; nonmigratory, nonextractable, nonvolatile modifier for plasticized PVC for improved oil and chem. resist., low temp. and heat aging props., superior physical props., increased abrasion resist. and enhanced rubber-like feel; good thermal stability; free-flowing powd.; 0.5 mm median particle size; sp.gr. 0.98; Mooney visc. 25-40 (ML-4, 100 C); 33% ACN.

Nipol® DP5125P. [Zeon] Partially crosslinked NBR, nonstaining antioxidant with 9% nominal PVC partitioning agent; nonmigratory, nonextractable modifier for PVC for improved oil and chem. resist., low temp. and heat aging props., increased abrasion resist. and enhanced rubber-like feel; good thermal stability; free-flowing powd.; 0.5 mm median particle size; sp.gr. 0.98; Mooney visc. 45-65 (ML1+4, 100 C); 33% ACN; *Storage:* store below 75 F in low-humidity area to prevent premature agglomeration; use on first-in-first-out basis within 3-4 mos.

Nipol® DP5128P. [Zeon] Highly crosslinked NBR, nonstaining antioxidant with 9% PVC partitioning agent; nonmigratory, nonextractable, nonvolatile modifier for PVC for improved oil and chem. resist., low temp. and heat aging props., superior physical props., increased abrasion resist. and enhanced rubber-like feel; good thermal stability; extrusion aid; free-flowing powd.; 0.5 mm median particle size; sp.gr. 0.98; Mooney visc. 70-90 (ML1+4, 100 C); 33% ACN.

Nirez® 300. [Arizona] Terpene phenolic resin; tackifier for adhesives and pressure-sensitive adhesives; FDA 21CFR §175.105, 175.125; Gardner 4+ color; soften. pt. (R&B) 112 C.

Nirez® 2019. [Arizona] Terpene phenolic resin; tackifier for adhesives and pressure-sensitive adhesives; FDA 21CFR §175.105, 175.125; Gardner 5 color; soften. pt. (R&B) 123 C.

Nirez® V-2040. [Arizona] Terpene phenolic resin; tackifier for adhesives and pressure-sensitive adhesives; FDA 21CFR §175.105, 175.125; Gardner 4+ color; soften. pt. (R&B) 118 C.

Nirez® V-2040HM. [Arizona] Terpene phenolic resin; tackifier for adhesives and pressure-sensitive adhesives; FDA 21CFR §175.105, 175.125; Gardner 4+ color; soften. pt. (R&B) 125 C.

Nirez® V-2150. [Arizona] Terpene phenolic resin; tackifier for adhesives; Gardner 5 color; soften. pt. (R&B) 150 C.

Nissan Anon BF. [Nippon Oils & Fats] Dimethyl cocoalkyl betaine; amphoteric; plastic antistatic agent, softener, germicide in foods and cleaning industries, asphalt antistripping agent, textile treatment; lt. yel. liq.; 25% min act.

Nissan Elegan S-100. [Nippon Oils & Fats] Special n; antistat for ABS, PS, PP, and PE plastics; solid.; Unverified

Nissan New Elegan A. [Nippon Oils & Fats] Special c; antistat for soft PVC; paste.; Unverified

Nissan New Elegan ASK. [Nippon Oils & Fats] Special c; antistat for rigid PVC; powd.; Unverified

Nissan Nonion HS-204.5. [Nippon Oils & Fats] POE octylphenyl ether; nonionic; detergent, emulsifier for household and industrial cleaners and detergents, textile processing, paints, printing inks, agric., latex polymerization; APHA 120 liq.; sol. in xylene, kerosene, ether, methanol; disp. in water; HLB 9.8; cloud pt. 0 C (1% aq.); 100% conc.

Nissan Nonion HS-206. [Nippon Oils & Fats] POE octylphenyl ether; nonionic; detergent, emulsifier for household and industrial cleaners and detergents, textile processing, paints, printing inks, agric., latex polymerization; APHA 120 liq.; sol. in xylene, kerosene, ether, methanol; disp. in water; HLB 11.2; 100% conc.

Nissan Nonion HS-208. [Nippon Oils & Fats] POE octylphenyl ether; nonionic; detergent, emulsifier for household and industrial cleaners and detergents, textile processing, paints, printing inks, agric., latex polymerization; APHA 120 liq.; sol. in xylene, ether, methanol; disp. in water; HLB 12.6; 100% conc.

Nissan Nonion HS-210. [Nippon Oils & Fats] POE octylphenyl ether; nonionic; detergent, emulsifier for household and industrial cleaners and detergents, textile processing, paints, printing inks, agric., latex polymerization; APHA 120 liq.; sol. in water, xylene, ether, methanol; HLB 13.3; cloud pt. 70-78 C (1% aq.); 100% conc.

Nissan Nonion HS-215. [Nippon Oils & Fats] POE octylphenyl ether; nonionic; detergent, emulsifier for household and industrial cleaners and detergents, textile processing, paints, printing inks, agric., latex polymerization; APHA 120 liq.; sol. in water, xylene, ether, methanol; HLB 15.0; cloud pt. 91-99 C (1% aq.); 100% conc.

Nissan Nonion HS-220. [Nippon Oils & Fats] POE octylphenyl ether; nonionic; detergent, emulsifier for household and industrial cleaners and detergents, textile processing, paints, printing inks, agric., latex polymerization; APHA 120 liq.; sol. in water, methanol, cloudy in xylene, ether; HLB 16.2; cloud pt. 100 C min. (1% aq.); 100% conc.

Nissan Nonion HS-240. [Nippon Oils & Fats] PEG-40 octylphenyl ether; CAS 9002-93-1; nonionic; detergent, emulsifier for household and industrial cleaners and detergents, textile processing, paints, printing inks, agric., latex polymerization; APHA 120 liq.; sol. in water, methanol; HLB 17.9; cloud pt. 100 C min. (1% aq.); 100% conc.

Nissan Nonion HS-270. [Nippon Oils & Fats] PEG-70 octylphenol ether; CAS 9002-93-1; nonionic; detergent, emulsifier for household and industrial cleaners and detergents, textile processing, paints, printing inks, agric., latex polymerization; semisolid; HLB 18.7; 100% conc.

Nissan Nonion NS-202. [Nippon Oils & Fats] POE nonylphenyl ether; nonionic; detergent, emulsifier for household and industrial cleaners, textile processing, paints, printing inks, agric. preparations, and latex polymerization; APHA 120 max. liq.; sol. in xylene, kerosene, ether, methanol; disp. in water; HLB 5.7; cloud pt. 0 C (1% aq.); 100% conc.

Nissan Nonion NS-204.5. [Nippon Oils & Fats] POE nonylphenyl ether; nonionic; detergent, emulsifier for household and industrial cleaners, textile processing, paints, printing inks, agric. preparations, and latex polymerization; APHA 120 max. liq.; sol. in xylene, kerosene, ether, methanol; disp. in water; HLB 9.5; cloud pt. 0 C max. (1% aq.); 100% conc.

Nissan Nonion NS-206. [Nippon Oils & Fats] PEG-6 nonylphenyl ether; CAS 9016-45-9; nonionic; detergent, emulsifier for household and industrial cleaners, textile processing, paints, printing inks, agric. preparations, and latex polymerization; APHA 120 max. liq.; sol. in xylene, ether, methanol; disp. in water; HLB 10.9; cloud pt. 0 C max. (1% aq.); 100% conc.

Nissan Nonion NS-208.5. [Nippon Oils & Fats] POE nonylphenyl ether; nonionic; detergent, emulsifier for household and industrial cleaners, textile processing, paints, printing inks, agric. preparations, and latex polymerization; APHA 120 max. liq.; sol. in xylene, ether, methanol; disp. in water; HLB 12.6; cloud pt. 33-41 C (1% aq.); 100% conc.

Nissan Nonion NS-209. [Nippon Oils & Fats] POE nonylphenyl ether; nonionic; detergent, emulsifier for household and industrial cleaners, textile processing, paints, printing inks, agric. preparations, and latex polymerization; liq.; HLB 12.9; 100% conc.

Nissan Nonion NS-210. [Nippon Oils & Fats] POE nonylphenyl ether; nonionic; detergent, emulsifier for household and industrial cleaners, textile processing, paints, printing inks, agric. preparations, and latex polymerization; APHA 120 max. liq.; sol. in water, xylene, ether, methanol; HLB 13.3; cloud pt. 46-54 C (1% aq.); 100% conc.

Nissan Nonion NS-212. [Nippon Oils & Fats] PEG-12 nonylphenyl ether; CAS 9016-45-9; nonionic; detergent, emulsifier for household and industrial cleaners, textile processing, paints, printing inks, agric. preparations, and latex polymerization; APHA 120 max. liq.; sol. in water, xylene, ether, methanol; HLB 14.1; cloud pt. 73-84 C (1% aq.); 100% conc.

Nissan Nonion NS-215. [Nippon Oils & Fats] POE nonylphenyl ether; nonionic; detergent, emulsifier for household and industrial cleaners, textile processing, paints, printing inks, agric. preparations, and latex polymerization; APHA 120 max. liq.; sol. in water, xylene, ether, methanol; HLB 15; cloud pt. 89-97 C (1% aq.); 100% conc.

Nissan Nonion NS-220. [Nippon Oils & Fats] POE nonylphenyl ether; nonionic; detergent, emulsifier for household and industrial cleaners, textile processing, paints, printing inks, agric. preparations, and latex polymerization; APHA 120 max. semisolid; sol. in water, xylene, ether, methanol; HLB 16; cloud pt. 100 C min. (1% aq.); 100% conc.

Nissan Nonion NS-230. [Nippon Oils & Fats] POE nonylphenyl ether; nonionic; detergent, emulsifier for household and industrial cleaners, textile processing, paints, printing inks, agric. preparations, and latex polymerization; APHA 120 max. solid; sol. in water, xylene, methanol, cloudy in ether; HLB 17.1; cloud pt. 100 C min. (1% aq.); 100% conc.

Nissan Nonion NS-240. [Nippon Oils & Fats] POE nonylphenyl ether; nonionic; detergent, emulsifier for household and industrial cleaners, textile processing, paints, printing inks, agric. preparations, and latex polymerization; APHA 120 max. solid; sol. in water, methanol; HLB 17.8; cloud pt. 100 C min. (1% aq.); 100% conc.

Nissan Nonion NS-250. [Nippon Oils & Fats] POE nonylphenyl ether; nonionic; detergent, emulsifier for household and industrial cleaners, textile processing, paints, printing inks, agric. preparations, and latex polymerization; liq.; HLB 18.2; 100% conc.

Nissan Nonion NS-270. [Nippon Oils & Fats] POE nonylphenyl ether; nonionic; detergent, emulsifier for household and industrial cleaners, textile processing, paints, printing inks, agric. preparations, and latex polymerization; liq.; HLB 18.7; 100% conc.

Nissan Persoft EK. [Nippon Oils & Fats] Sulfated fatty alcohol ethoxylate, sodium salt; anionic; emulsifier for vinyl polymerization; textile detergent; dyeing assistant; degreaser; liq.; 30% conc.

Nissan Plonon 102. [Nippon Oils & Fats] POE-POP ether; nonionic; emulsifier, solubilizer, dispersant, detergent, antifoaming agent, wetting agent used in soaps, syn. resins, metal cleaning, emulsion polymerization, fermentation, paper industries; colorless liq.; odorless and tasteless; m.w. 1250; sol. in water, ethanol, acetone, benzol, dioxane; visc. 105 cs (37.8 C); HLB 7; cloud pt. 22 C (10% aq.); surf. tens. 41.1 dynes/cm (0.1%); 100% conc.

Nissan Plonon 104. [Nippon Oils & Fats] POE-POP ether; nonionic; emulsifier, solubilizer, dispersant, detergent, antifoaming agent, wetting agent used in soaps, syn. resins, metal cleaning, emulsion polymerization, fermentation, paper industries; colorless liq.; odorless and tasteless; m.w. 1650; sol. in water, ethanol, acetone, benzol, dioxane; visc. 105 cs (37.8 C); HLB 15.0; cloud pt. 64 C (10% aq.); surf. tens. 43.5 dynes/cm (0.1%); 100% conc.

Nissan Plonon 108. [Nippon Oils & Fats] POE-POP ether; nonionic; emulsifier, solubilizer, dispersant, detergent, antifoaming agent, wetting agent used in soaps, syn. resins, metal cleaning, emulsion polymerization, fermentation, paper industries; m.w. 4000; HLB 30.5; cloud pt. 100 C (10% aq.); 100% conc.; Unverified

Nissan Plonon 171. [Nippon Oils & Fats] POE-POP ether; nonionic; emulsifier, solubilizer, dispersant, detergent, antifoaming agent, wetting agent used in soaps, syn. resins, metal cleaning, emulsion polymerization, fermentation, paper industries; m.w. 2000; sol. in ethanol, acetone, benzol, dioxane; visc. 160 cs (37.8 C); HLB 3; cloud pt. 24 C (10% aq.); surf. tens. 39.5 dynes/cm (0.1%); 100% conc.; Unverified

Nissan Plonon 172. [Nippon Oils & Fats] POE-POP ether; nonionic; emulsifier, solubilizer, dispersant, detergent, antifoaming agent, wetting agent used in soaps, syn. resins, metal cleaning, emulsion polymerization, fermentation, paper industries; m.w. 2400; sol. in ethanol, acetone, benzol, dioxane; visc. 210 cs (37.8 C); HLB 7; cloud pt. 28 C (10% aq.); 100% conc.; Unverified

Nissan Plonon 201. [Nippon Oils & Fats] POE-POP ether; nonionic; emulsifier, solubilizer, dispersant, detergent, antifoaming agent, wetting agent used in soaps, syn. resins, metal cleaning, emulsion polymerization, fermentation, paper industries; colorless liq.; odorless and tasteless; m.w. 2200; sol. in ethanol, acetone, benzol, dioxane; visc. 190 cs (37.8 C); HLB 3; cloud pt. 21 C (10% aq.); surf. tens. 38 dynes/cm (0.1%); 100% conc.

Nissan Plonon 204. [Nippon Oils & Fats] POE-POP ether; nonionic; emulsifier, solubilizer, dispersant, detergent, antifoaming agent, wetting agent used in soaps, syn. resins, metal cleaning, emulsion polymerization, fermentation, paper industries; colorless paste; odorless and tasteless; m.w. 3400; sol. in water, ethanol; visc. 370 cs (37.8 C); HLB 13.5; cloud pt. 64 C (10% aq.); surf. tens. 39.8 dynes/cm (0.1%); 100% conc.

Nissan Plonon 208. [Nippon Oils & Fats] POE-POP ether; nonionic; emulsifier, solubilizer, dispersant, detergent, antifoaming agent, wetting agent used in soaps, syn. resins, metal cleaning, emulsion polymerization, fermentation, paper industries; colorless paste; odorless and tasteless; m.w. 8000; sol. in water; HLB 28; cloud pt. 100 C (10% aq.); surf. tens. 44.1 dynes/cm (0.1%); 100% conc.

Nissan Rapisol B-30, B-80, C-70. [Nippon Oils & Fats] Sodium dioctyl sulfosuccinate; anionic; wetting agent, polymerization agent for PVC, dyeing aux.; liq.; 30, 80, and 70% conc. resp.

Nissan Tert. Amine AB. [Nippon Oils & Fats] Octadecyl-dimethylamine; CAS 124-28-7; EINECS 204-694-8; intermediate for various surfactants, visc. index improver for lubricating oil, curing catalyst for epoxy resin, corrosion inhibitor, germicide; lt. yel. liq. or half-solid; 95% min. tert. amine; Unverified

Nissan Tert. Amine ABT. [Nippon Oils & Fats] Tallowhydrog. alkyl dimethylamine; CAS 61788-95-2; intermediate for various surfactants, visc. index improver for lubricating oil, curing catalyst for epoxy resin, corrosion inhibitor, germicide; lt. yel. liq. or half-solid; 95% min. tert. amine; Unverified

Nissan Tert. Amine BB. [Nippon Oils & Fats] Dodecyl-dimethylamine; intermediate for various surfactants, visc. index improver for lubricating oil, curing catalyst for epoxy resin, corrosion inhibitor, germicide; lt. yel. liq.; 95% min. tert. amine; Unverified

Nissan Tert. Amine FB. [Nippon Oils & Fats] Cocoalkyl dimethylamine; intermediate for various surfactants, visc. index improver for lubricating oil, curing catalyst for epoxy resin, corrosion inhibitor, germicide; lt. yel. liq.; 95% min. tert. amine; Unverified

Nissan Tert. Amine MB. [Nippon Oils & Fats] Tetradecyl dimethylamine; intermediate for various surfactants, visc. index improver for lubricating oil, curing catalyst for epoxy resin, corrosion inhibitor, germicide; lt. yel. liq. or half-solid; 95% min. tert. amine; Unverified

Nissan Tert. Amine PB. [Nippon Oils & Fats] Hexadecyl-dimethylamine; intermediate for various surfactants, visc. index improver for lubricating oil, curing catalyst for epoxy resin, corrosion inhibitor, germicide; lt. yel. liq. or half-solid; 95% min. tert.

amine; Unverified

Nissan Trax K-300. [Nippon Oils & Fats] POE alkyl ether sulfate; anionic; emulsifier for emulsion polymerization; liq.; 34% conc.

Nissan Trax N-300. [Nippon Oils & Fats] POE alkylaryl ether sulfate; anionic; emulsifier for emulsion polymerization; liq.; 34% conc.

Nitropore® ATA. [Uniroyal] Azodicarbonamide/DNPT blend; chemical blowing agent.

Nitropore® OBSH. [Olin] 4,4′-Oxybis (benzenesulfonhydrazide); CAS 80-51-3; nitrogen-releasing blowing agent for elastomers, thermoplastics, rubber-resin blends; applics. incl. cellular pipe insulation, athletic padding, molded gaskets, carpet underlayment, flotation prods., and coaxial cable; wh. powd.; no odor; sol. in acetone, dimethylformamide, and dimethylsulfoxide; insol. in benzene, hexane, and water; sp.gr. 1.54; dens. 27 lb/ft^3; gas yield 125 ml/g; decomp. pt. 165 C; Unverified

Nix Stix L195WF. [Dwight Prods.] Long-lasting nontransferable water-based mold release for rubber.

Nix Stix L478. [Dwight Prods.] Silicone emulsion; mold release for urethane and rubber prods.

Nix Stix L515. [Dwight Prods.] Silicone dispersion; mold release for urethane prods.; recommended for cast urethanes, both rigid and flexible types, and most rubber compds.; also for polyesters, epoxies, phenolics, most thermosetting plastics.

Nix Stix L529. [Dwight Prods.] Semipermanent nontransferable water-based release for rubber.

Nix Stix LO582. [Dwight Prods.] Silicone emulsion; mold release for rubber prods.

Nix Stix X9022. [Dwight Prods.] Aerosol release agent; nonflamm. silicone spray for rubber and urethane.

Nix Stix X9027. [Dwight Prods.] Aerosol release agent; silicone spray; passes the flame extension test; release for rubber and urethane.

Nix Stix X9028. [Dwight Prods.] Aerosol release agent; flamm. silicone spray for rubber and urethane.

NMP. [Arco] 2-Methyl-2-pyrrolidone; CAS 872-50-4; solv. and cosolv.; used as coatings solv., in stripping and cleaning of paints and varnishes, industrial cleaning, mold cleaning, petrochem. processing, agric. solv.; polymer solv. for PVAc, PVDF, PS, vinyl copolymers, nylon and aromatic polyamides and polyimides, polyesters, acrylics, PC, cellulose derivs., syn. elastomers, waxes; colorless mobile liq., mild amine-like odor; completely misc. with water; m.w. 99.13; sp.gr. 1.028; dens. 8.6 lb/gal (20 C); vapor press 0.2 mm Hg (20 C); f.p. -25 C; b.p. 202-205 C; flash pt. 93 C; ref. index 1.4680-1.4695; 99.9+% assay; *Toxicology:* relatively nontoxic; *Precaution:* avoid contact with strong acids or alkalies; *Storage:* air and moisture exclusion will prolong shelf life; hygroscopic; store in dry area, protect from weather.

NMP-Plus™. [ANGUS] 60% 2-Nitro-2-methyl-1-propanol (CAS 76-39-1) on silica (CAS 7631-86-9); methylene donor for crosslinking resorcinol resins to promote tire laminate adhesion; wh. gran. solid; m.w. 119.1 (of act.); bulk dens. 0.28 g/ml (loose), 0.44 g/ml (packaged); flash pt. (PMCC) > 200 F; 60% act.; *Toxicology:* toxic by ingestion; mod. irritating to eyes; sl. irritating to skin; severely irritating to eyes in sol'n.; excessive contact can cause drying of mucous membranes; SiO_2 PEL 6 mg/m^3; *Precaution:* high heat (> 400 F) may cause decomp. with evolution of gases (formaldehyde, oxides of C and N); incompat. with alkaline materials; *Storage:* keep containers in cool, dry place free of alkaline (amine) fumes.

No. 1 White™. [ECC Int'l.] Calcium carbonate; CAS 1317-65-3; general-purpose pigment, filler, reinforcement where fineness is of secondary importance; for paints, coatings, hand cleaners, rubber compds., latex foams, polishing compds., putties, caulks, adhesives; wh. powd., odorless; 13 μ mean particle size; 0.5% on 325 mesh; negligible sol. in water; sp.gr. 2.71; dens. 55 lb/ft^3 (loose); pH 9.5 (5% slurry); nonflamm.; 97% $CaCO_3$, 0.2% max. moisture; *Toxicology:* TLV/TWA 10 mg/m^3, considered nuisance dust; *Precaution:* incompat. with acids.

No. 3 White. [ECC Int'l.] Calcium carbonate; wh. general-purpose filler for low-cost paints, rubber goods, mild abrasive compds., and putties; wh. powd., odorless; 18 μ mean particle size; Unverified

No. 8 White. [Georgia Marble] Calcium carbonate; filler with controlled color for applics. where consistently whiter fillers are needed; wh. powd.; 25 ± 5%. on 325 mesh; brightness 90.

No. 9 NCS. [Georgia Marble] Calcium carbonate; noncolor critical filler for use where neither color uniformity or high brightness are required; wh. powd.; 9 ± 3% on 325 mesh; brightness > 88.

No. 9 White. [Georgia Marble] Calcium carbonate; filler with controlled color for applics. where consistently whiter fillers are needed; wh. powd.; 9 ± 3% on 325 mesh; brightness 92.

No. 10 White. [Georgia Marble] Calcium carbonate; filler with controlled color for applics. where consistently whiter fillers are needed; used in paints, wall sealants, caulks, floor and ceiling tile, plastics, putties, nonblocking agents, wall and floor mastics; additive for natural and syn. rubbers, foamed thermoplastic materials; wh. powd.; 16 μ mean particle size; 1% max. on 325 mesh; sp.gr. 2.71; brightness 92; hardness (Mohs) 3; 95% min. $CaCO_3$.

No. 22 Barytes™. [Cimbar Perf. Minerals] Natural unbleached barytes; CAS 7727-43-7; EINECS 231-784-4; additive in brake linings, linings for jar lids, acoustical compds., urethane foams, and other prods.; powd.; 9 μ mean particle size; 0.5% on 325 mesh; sp.gr. 4.35; bulk dens. 100 lb/ft^3 (loose); oil absorp. 9; pH 7; 95% $BaSO_4$, 0.15% moisture.

Noiox AK-41. [Pulcra SA] PEG oleate; nonionic; lubricant, solubilizer and emulsifier; wetting agent and pigment dispersant; antistat for plastics and aux. dispersant; liq.; 100% conc.; Unverified

Noiox KS-10, -12, -13, -14, -16. [Pulcra SA] PEG; nonionic; lubricant, solv. for dyestuffs and resins;

plasticizer for casein, gelatin compositions and printing inks; used in cosmetic and pharmaceutical items; intermediate prods.; liq., solid, paste; water-sol.; 100% conc.; Unverified

Nonanol N. [BASF AG] Mixt. of isomeric primary nonyl alcohols; starting material for prod. of plasticizers for PVC and vinyl chloride copolymers; low volatile leveling agent for baking finishes; antifoam.

Nonipol 20. [Sanyo Chem. Industries] POE nonyl phenyl ether; CAS 9016-45-9; nonionic; penetrant, wetting agent, spreader-sticker; base material for detergents; emulsifier for agric. pesticides and emulsion polymerization, org. solv., machine oils, liq. paraffins; liq.; HLB 5.7; 100% conc.

Nonipol 40. [Sanyo Chem. Industries] POE nonylphenyl ether; nonionic; penetrant, wetting agent, spreader-sticker; base material for detergents; emulsifier for agric. pesticides and emulsion polymerization, org. solv., machine oils, liq. paraffins; liq.; HLB 8.9; 100% conc.

Nonipol 55. [Sanyo Chem. Industries] POE nonylphenyl ether; nonionic; penetrant, wetting agent, spreader-sticker; base material for detergents; emulsifier for agric. pesticides and emulsion polymerization, org. solv., machine oils, liq. paraffins; liq.; HLB 10.5; 100% conc.

Nonipol 60. [Sanyo Chem. Industries] POE nonylphenyl ether; nonionic; penetrant, wetting agent, spreader-sticker; base material for detergents; emulsifier for agric. pesticides and emulsion polymerization, org. solv., machine oils, liq. paraffins; liq.; HLB 10.9; 100% conc.

Nonipol 70. [Sanyo Chem. Industries] POE nonylphenyl ether; nonionic; penetrant, wetting agent, spreader-sticker; base material for detergents; emulsifier for agric. pesticides and emulsion polymerization, org. solv., machine oils, liq. paraffins; liq.; HLB 11.7; 100% conc.

Nonipol 85. [Sanyo Chem. Industries] POE nonylphenyl ether; nonionic; penetrant, wetting agent, spreader-sticker; base material for detergents; emulsifier for agric. pesticides and emulsion polymerization, org. solv., machine oils, liq. paraffins; liq.; HLB 12.6; 100% conc.

Nonipol 95. [Sanyo Chem. Industries] POE nonyl phenyl ether; penetrant, wetting agent, spreader-sticker; base material for detergents; emulsifier for agric. pesticides and emulsion polymerization, org. solv., machine oils, liq. paraffins; liq.; HLB 13.1; 100% conc.

Nonipol 100. [Sanyo Chem. Industries] POE nonylphenyl ether; nonionic; penetrant, wetting agent, spreader-sticker; base material for detergents; emulsifier for agric. pesticides and emulsion polymerization, org. solv., machine oils, liq. paraffins; liq.; HLB 13.1; 100% conc.

Nonipol 110. [Sanyo Chem. Industries] POE nonylphenyl ether; nonionic; penetrant, wetting agent, spreader-sticker; base material for detergents; emulsifier for agric. pesticides and emulsion polymerization, org. solv., machine oils, liq. paraffins; liq.; HLB 13.8; 100% conc.

Nonipol 120. [Sanyo Chem. Industries] POE nonylphenyl ether; nonionic; penetrant, wetting agent, spreader-sticker; base material for detergents; emulsifier for agric. pesticides and emulsion polymerization, org. solv., machine oils, liq. paraffins; liq.; HLB 14.1; 100% conc.

Nonipol 130. [Sanyo Chem. Industries] POE nonylphenyl ether; nonionic; penetrant, wetting agent, spreader-sticker; base material for detergents; emulsifier for agric. pesticides and emulsion polymerization, org. solv., machine oils, liq. paraffins; liq.; HLB 14.5; 100% conc.

Nonipol 160. [Sanyo Chem. Industries] POE nonylphenyl ether; nonionic; penetrant, wetting agent, spreader-sticker; base material for detergents; emulsifier for agric. pesticides and emulsion polymerization, org. solv., machine oils, liq. paraffins; solid; HLB 15.2; 100% conc.

Nonipol 200. [Sanyo Chem. Industries] POE nonylphenyl ether; nonionic; penetrant, wetting agent, spreader-sticker; base material for detergents; emulsifier for agric. pesticides and emulsion polymerization, org. solv., machine oils, liq. paraffins; solid; HLB 16.0; 100% conc.

Nonipol 400. [Sanyo Chem. Industries] POE nonylphenyl ether; nonionic; penetrant, wetting agent, spreader-sticker; base material for detergents; emulsifier for agric. pesticides and emulsion polymerization, org. solv., machine oils, liq. paraffins; solid; HLB 17.8; 100% conc.

Nonox® WSP. [Akzo Nobel] 2,2′-Methylenebis [4-methyl-6-(1-methyl-cyclohexyl) phenol]; antioxidant for plastics, esp. polyolefin, PS; cryst. powd.

Nopalcol 1-L. [Henkel/Functional Prods.] PEG-2 laurate; CAS 9004-81-3; nonionic; general purpose emulsifier, plasticizer, lubricant, wetting agent, dispersant, binding and thickening agent for emulsion polymerization, dry cleaning, leather, min. oil emulsions, paper industry, wall-tile mastics, solv. emulsions; liq.; HLB 6.0; 98% act.

Nopalcol 1-TW. [Henkel/Functional Prods.] PEG-2 tallowate; nonionic; emulsifier, plasticizer, lubricant, wetting agent, defoamer, binding and thickening agent, used in emulsion polymerization, cosmetics, dry cleaning, leather, textile industries; solid; HLB 4.1; 99% act.

Nopalcol 2-DL. [Henkel/Functional Prods.] PEG-4 dilaurate; CAS 9005-02-1; nonionic; general purpose emulsifier, plasticizer, lubricant, wetting agent, dispersant, binding and thickening agent for emulsion polymerization, dry cleaning, leather, min. oil emulsions, paper industry, wall-tile mastics, solv. emulsions; liq.; HLB 5.9; 99% act.

Nopalcol 4-C. [Henkel/Functional Prods.] PEG-8 cocoate; nonionic; general purpose emulsifier, plasticizer, lubricant, wetting agent, dispersant, binding and thickening agent for emulsion polymerization, dry cleaning, leather, min. oil emulsions, paper industry, wall-tile mastics, solv. emulsions; liq.; HLB 13.9; 99% act.

Nopalcol 4-CH. [Henkel/Functional Prods.] PEG-8 cocoate (hyd.); nonionic; general purpose emulsifier, plasticizer, lubricant, wetting agent, dispersant, binding and thickening agent for emulsion polymerization, dry cleaning, leather, min. oil emulsions, paper industry, wall-tile mastics, solv. emulsions; liq.; HLB 13.9; 99% act.

Nopalcol 4-L. [Henkel/Functional Prods.] PEG-8 laurate; CAS 9004-81-3; nonionic; general purpose emulsifier, plasticizer, lubricant, wetting agent, dispersant, binding and thickening agent for emulsion polymerization, dry cleaning, leather, min. oil emulsions, paper industry, wall-tile mastics, solv. emulsions; liq.; HLB 13.1; 99% act.

Nopalcol 4-O. [Henkel/Functional Prods.] PEG-8 oleate; CAS 9004-96-0; nonionic; emulsifier; dispersant for leather pigments; paper coating defoamer, plasticizer and leveling agent; emulsion polymerization; liq.; HLB 11.7; 95% act.

Nopalcol 4-S. [Henkel/Functional Prods.] PEG-8 stearate; CAS 9004-99-3; nonionic; general purpose emulsifier, plasticizer, lubricant, wetting agent, dispersant, binding and thickening agent for emulsion polymerization, dry cleaning, leather, min. oil emulsions, paper industry, wall-tile mastics, solv. emulsions; paste; HLB 11.3; 99% act.

Nopalcol 6-DO. [Henkel/Functional Prods.] PEG-12 dioleate; CAS 9005-07-6; nonionic; general purpose emulsifier, plasticizer, lubricant, wetting agent, dispersant, binding and thickening agent for emulsion polymerization, dry cleaning, leather, min. oil emulsions, paper industry, wall-tile mastics, solv. emulsions; liq.; HLB 12.7; 98% act.

Nopalcol 6-DTW. [Henkel/Functional Prods.] PEG-12 ditallowate; nonionic; general purpose emulsifier, plasticizer, lubricant, wetting agent, dispersant, binding and thickening agent for emulsion polymerization, dry cleaning, leather, min. oil emulsions, paper industry, wall-tile mastics, solv. emulsions; paste; HLB 11.2; 99% act.

Nopalcol 6-L. [Henkel/Functional Prods.] PEG-12 laurate; CAS 9004-81-3; nonionic; general purpose emulsifier, plasticizer, lubricant, wetting agent, dispersant, binding and thickening agent for emulsion polymerization, dry cleaning, leather, min. oil emulsions, paper industry, wall-tile mastics, solv. emulsions; liq.; HLB 15.0; 99% act.

Nopalcol 6-R. [Henkel/Functional Prods.] PEG-12 ricinoleate; CAS 9004-97-1; nonionic; general purpose emulsifier, plasticizer, lubricant, wetting agent, dispersant, binding and thickening agent for emulsion polymerization, dry cleaning, leather, min. oil emulsions, paper industry, wall-tile mastics, solv. emulsions; liq.; HLB 13.9; 99% act.

Nopalcol 6-S. [Henkel/Functional Prods.] PEG-12 stearate; CAS 9004-99-3; nonionic; general purpose emulsifier, plasticizer, lubricant, wetting agent, dispersant, binding and thickening agent for emulsion polymerization, dry cleaning, leather, min. oil emulsions, paper industry, wall-tile mastics, solv. emulsions; paste; HLB 13.2; 99% act.

Nopalcol 10-COH. [Henkel/Functional Prods.] PEG-20 hydrog. castor oil; CAS 61788-85-0; nonionic; general purpose emulsifier, plasticizer, lubricant, wetting agent, dispersant, binding and thickening agent for emulsion polymerization, dry cleaning, leather, min. oil emulsions, paper industry, wall-tile mastics, solv. emulsions; liq.; HLB 10.3; 99% act.

Nopalcol 12-CO. [Henkel/Functional Prods.] PEG 1200 castor oil; nonionic; general purpose emulsifier, plasticizer, lubricant, wetting agent, dispersant, binding and thickening agent for emulsion polymerization, dry cleaning, leather, min. oil emulsions, paper industry, wall-tile mastics, solv. emulsions; liq.; HLB 11.3; 99% act.

Nopalcol 12-COH. [Henkel/Functional Prods.] PEG 1200 hydrog. castor oil; CAS 61788-85-0; nonionic; general purpose emulsifier, plasticizer, lubricant, wetting agent, dispersant, binding and thickening agent for emulsion polymerization, dry cleaning, leather, min. oil emulsions, paper industry, wall-tile mastics, solv. emulsions; liq.; HLB 11.3; 99% act.

Nopalcol 19-CO. [Henkel/Functional Prods.] PEG 1900 castor oil; nonionic; general purpose emulsifier, plasticizer, lubricant, wetting agent, dispersant, binding and thickening agent for emulsion polymerization, dry cleaning, leather, min. oil emulsions, paper industry, wall-tile mastics, solv. emulsions; liq.; HLB 14.9; 99% conc.

Nopalcol 30-TWH. [Henkel/Functional Prods.] PEG-60 hydrog. tallowate; nonionic; general purpose emulsifier, plasticizer, lubricant, wetting agent, dispersant, binding and thickening agent for emulsion polymerization, dry cleaning, leather, min. oil emulsions, paper industry, wall-tile mastics, solv. emulsions; solid; HLB 18.0; 99% act.

Nopalcol 200. [Henkel/Functional Prods.] PEG 200; CAS 25322-68-3; nonionic; general purpose emulsifier, plasticizer, lubricant, wetting agent, dispersant, binding and thickening agent for emulsion polymerization, dry cleaning, leather, min. oil emulsions, paper industry, wall-tile mastics, solv. emulsions; liq.; 99% act.

Nopalcol 400. [Henkel/Functional Prods.] PEG 400; CAS 25322-68-3; nonionic; general purpose emulsifier, plasticizer, lubricant, wetting agent, dispersant, binding and thickening agent for emulsion polymerization, dry cleaning, leather, min. oil emulsions, paper industry, wall-tile mastics, solv. emulsions; liq.; 99% act.

Nopalcol 600. [Henkel/Functional Prods.] PEG 600; CAS 25322-68-3; nonionic; general purpose emulsifier, plasticizer, lubricant, wetting agent, dispersant, binding and thickening agent for emulsion polymerization, dry cleaning, leather, min. oil emulsions, paper industry, wall-tile mastics, solv. emulsions; paste; 99% act.

Nopco® 2031. [Henkel/Functional Prods.; Henkel-Nopco] Sulfonated fatty prod., sodium neutralized; anionic; emulsifier for emulsion polymerization; liq.; 60% conc.

Nopco ESA120. [Pulcra SA] Sodium dioctyl sulfosuccinate; anionic; dispersant for pigments; wetting and penetrating agent; used in drycleaning detergents and emulsion polymerization; rewetting agent for textile and paper industries; liq.; sol. in water, alcohol; 70 ± 2% conc.

Nopco® NXZ. [Henkel/Cospha] Defoamer for syn. latex emulsions, paint and adhesives from SBR, PVAc, acrylic, water-sol. resins; amber hazy liq.; forms milky emulsion in water; sp.gr. 0.91; dens. 7.6 lb/gal; pH 7 (2%); flash pt. 171 C; 100% act.

Nopcosant. [Henkel/Process & Polymer Chems.] Sodium salt of condensed naphthalene sulfonate; dispersant for NR, SR latexes, pigments; used for paints, cements, sealants; tan powd.; water-sol.; sp.gr. 1.02; pH 7.5-8.0 (2%).

Nopcostat 092. [Henkel] Internal/external antistat for most plastics.

Nopcostat HS. [Henkel] Internal/external antistat for ABS, acetal, PE, PP, PS, flexible and rigid PVC.

Noram S. [Ceca SA] N-tallow amine; CAS 61790-33-8; EINECS 263-125-1; cationic; industrial detergent; synthesis intermediate; lubricant, textile, and oil industries; base for prod. of N-alkyl sulfosuccinamate, used to make carpetbacking latex foams; pasty.

Noram SH. [Ceca SA] N-Hydrog. tallow amine; CAS 61788-45-2; EINECS 262-976-6; cationic; industrial detergent; anti-caking agent, corrosion inhibitor; faciliates dispersion of fillers in natural rubber (0.5-1% of filler weight); eases moldstripping chiefly for hard rubbers; flakes.

Noramium CES 80. [Ceca SA] Surf. antistat for plastics and resins.

Noroplast 820X. [Ceca SA] Cationic; permanent antistat for PP.

Noroplast 2000. [Ceca SA] Cationic; permanent antistat for ABS, PS (impact and std. grades), and polyolefins.

Noroplast 8000. [Ceca SA] Cationic; permanent antistat for PE.

Noroplast 8500. [Ceca SA] Cationic; permanent antistat for PP.

Norox® MCP. [Norac] MEK peroxides and cumyl hydroperoxide in plasticizers; polymerization initiator for R.T. cure of unsat. polyester and vinyl ester resins; low peak exotherm, longer gel time; pale yel. liq.; sol. in oxygenated org. solvs.; sl. sol. in water; sp.gr. 1.07 (25/4 C); flash pt. (COC) 200 F min. fire pt. 200 F min.; 9% max. act. oxygen; *Precaution:* spilled peroxides can become decontaminated and decomp. in a hazardous, violent manner; ignite readily and burn vigorously; *Storage:* store ≤ 80 F in original containers away from sources of heat, sparks, or flames; out of direct sunlight; away from promoters, accelerators, reducing agents, strong acids/bases.

Norsolene 9090. [Total] Hydrocarbon resin; processing aid for rubber extrusion, molding, calendering; enhances green tack; for rubber tires, industrial goods (sealants, hoses, belts, conveyors, shoes); also in hot-melt, pressure-sensitive, and solv.-based adhesives, paints and varnishes, binders for lithographic inks, in road bitumen and asphalt, wood impregnation; Gardner 4-7 (50% in toluene) solid; sp.gr. 1.06-1.08 (20 C); soften. pt. 98-108 C; acid no. < 0.1; iodine no. < 25; sapon. no. < 3.

Norsolene 9110. [Total] Hydrocarbon resin; processing aid for rubber extrusion, molding, calendering; enhances green tack; for rubber tires, industrial goods (sealants, hoses, belts, conveyors, shoes); also in hot-melt, pressure-sensitive, and solv.-based adhesives, paints and varnishes, binders for lithographic inks, in road bitumen and asphalt, wood impregnation; Gardner 4-7 (50% in toluene) solid; sp.gr. 1.06-1.08 (20 C); soften. pt. 120-130 C; acid no. < 0.1; iodine no. < 25; sapon. no. < 3.

Norsolene A 90. [Total] Modified hydrocarbon resin; processing aid for rubber extrusion, molding, calendering; enhances green tack; for rubber tires, industrial goods (sealants, hoses, belts, conveyors, shoes); also in hot-melt, pressure-sensitive, and solv.-based adhesives, paints and varnishes, binders for lithographic inks, in road bitumen and asphalt, wood impregnation; Gardner 4-6 (50% in toluene) solid; sp.gr. 1.03-1.04 (20 C); soften. pt. 90-100 C; acid no. < 0.1; iodine no. 40; sapon. no. < 3.

Norsolene A 100. [Total] Modified hydrocarbon resin; processing aid for rubber extrusion, molding, calendering; enhances green tack; for rubber tires, industrial goods (sealants, hoses, belts, conveyors, shoes); also in hot-melt, pressure-sensitive, and solv.-based adhesives, paints and varnishes, binders for lithographic inks, in road bitumen and asphalt, wood impregnation; Gardner 4-6 (50% in toluene) solid; sp.gr. 1.03-1.04 (20 C); soften. pt. 101-106 C; acid no. < 0.1; iodine no. 45; sapon. no. < 3.

Norsolene A 110. [Total] Modified hydrocarbon resin; processing aid for rubber extrusion, molding, calendering; enhances green tack; for rubber tires, industrial goods (sealants, hoses, belts, conveyors, shoes); also in hot-melt, pressure-sensitive, and solv.-based adhesives, paints and varnishes, binders for lithographic inks, in road bitumen and asphalt, wood impregnation; Gardner 4-6 (50% in toluene) solid; sp.gr. 1.04-1.05 (20 C); soften. pt. 107-112 C; acid no. < 0.1; iodine no. 55; sapon. no. < 3.

Norsolene D 3005. [Total] Hydrocarbon resin; processing aid for rubber extrusion, molding, calendering; enhances green tack; for rubber tires, industrial goods (sealants, hoses, belts, conveyors, shoes); visc. reducer in epoxy systems; increases pot life, flexibility in floor coatings, etc.; also in hot-melt, pressure-sensitive, and solv.-based adhesives, paints and varnishes, binders for lithographic inks, in road bitumen and asphalt, wood impregnation; Gardner 3-5 liq.; sp.gr. 1.02-1.03 (20 C); visc. 4-10 poise; acid no. < 0.5; iodine no. 75; sapon. no. < 3.

Norsolene I 130. [Total] Modified hydrocarbon resin; processing aid for rubber extrusion, molding, calendering; enhances green tack; for rubber tires, industrial goods (sealants, hoses, belts, conveyors, shoes); also in hot-melt, pressure-sensitive, and solv.-based adhesives, paints and varnishes, binders for lithographic inks, in road bitumen and asphalt, wood impregnation; Gardner 4-7 (50% in toluene) solid; sp.gr. 1.09-1.11 (20 C); soften. pt. 123-133 C; acid no. < 0.1; iodine no. < 25; sapon. no. < 3.

Norsolene I 140. [Total] Modified hydrocarbon resin; processing aid for rubber extrusion, molding, calendering; enhances green tack; for rubber tires, industrial goods (sealants, hoses, belts, conveyors, shoes); also in hot-melt, pressure-sensitive, and solv.-based adhesives, paints and varnishes, binders for lithographic inks, in road bitumen and asphalt, wood impregnation; Gardner 7-10 (50% in toluene) solid; sp.gr. 1.06-1.07 (20 C); soften. pt. 135-142 C; acid no. < 0.1; iodine no. 25; sapon. no. < 3.

Norsolene L 2010. [Total] Hydrocarbon resin; processing aid for rubber extrusion, molding, calendering; enhances green tack; for rubber tires, industrial goods (sealants, hoses, belts, conveyors, shoes); visc. reducer in epoxy systems; increases pot life, flexibility in floor coatings, etc.; also in hot-melt, pressure-sensitive, and solv.-based adhesives, paints and varnishes, binders for lithographic inks, in road bitumen and asphalt, wood impregnation; Gardner 12-15 liq.; sp.gr. 1.02-1.03 (20 C); visc. 0.5-5.5 poise; acid no. < 0.1; iodine no. < 25; sapon. no. < 3.

Norsolene M 1080. [Total] Modified hydrocarbon resin; processing aid for rubber extrusion, molding, calendering; enhances green tack; for rubber tires, industrial goods (sealants, hoses, belts, conveyors, shoes); also in hot-melt, pressure-sensitive, and solv.-based adhesives, paints and varnishes, binders for lithographic inks, in road bitumen and asphalt, wood impregnation; Gardner 4-7 (50% in toluene) solid; sp.gr. 1.09-1.11 (20 C); soften. pt. 88-98 C; acid no. < 0.1; iodine no. 35; sapon. no. < 3.

Norsolene M 1090. [Total] Modified hydrocarbon resin; processing aid for rubber extrusion, molding, calendering; enhances green tack; for rubber tires, industrial goods (sealants, hoses, belts, conveyors, shoes); also in hot-melt, pressure-sensitive, and solv.-based adhesives, paints and varnishes, binders for lithographic inks, in road bitumen and asphalt, wood impregnation; Gardner 3-6 (50% in toluene) solid; sp.gr. 1.09-1.11 (20 C); soften. pt. 98-108 C; acid no. < 0.1; iodine no. 35; sapon. no. < 3.

Norsolene S 85. [Total] Hydrocarbon resin; processing aid for rubber extrusion, molding, calendering; enhances green tack; for rubber tires, industrial goods (sealants, hoses, belts, conveyors, shoes); also in hot-melt, pressure-sensitive, and solv.-based adhesives, paints and varnishes, binders for lithographic inks, in road bitumen and asphalt, wood impregnation; Gardner 4-9 (50% in toluene) solid; sp.gr. 1.06-1.08 (20 C); soften. pt. 78-88 C; acid no. < 0.1; iodine no. < 25; sapon. no. < 3.

Norsolene S 95. [Total] Hydrocarbon resin; processing aid for rubber extrusion, molding, calendering; enhances green tack; for rubber tires, industrial goods (sealants, hoses, belts, conveyors, shoes); also in hot-melt, pressure-sensitive, and solv.-based adhesives, paints and varnishes, binders for lithographic inks, in road bitumen and asphalt, wood impregnation; Gardner 4-9 (50% in toluene) solid; sp.gr. 1.06-1.08 (20 C); soften. pt. 88-99 C; acid no. < 0.1; iodine no. < 25; sapon. no. < 3.

Norsolene S 105. [Total] Hydrocarbon resin; processing aid for rubber extrusion, molding, calendering; enhances green tack; for rubber tires, industrial goods (sealants, hoses, belts, conveyors, shoes); also in hot-melt, pressure-sensitive, and solv.-based adhesives, paints and varnishes, binders for lithographic inks, in road bitumen and asphalt, wood impregnation; Gardner 4-9 (50% in toluene) solid; sp.gr. 1.06-1.08 (20 C); soften. pt. 99-109 C; acid no. < 0.1; iodine no. < 25; sapon. no. < 3.

Norsolene S 115. [Total] Hydrocarbon resin; processing aid for rubber extrusion, molding, calendering; enhances green tack; for rubber tires, industrial goods (sealants, hoses, belts, conveyors, shoes); also in hot-melt, pressure-sensitive, and solv.-based adhesives, paints and varnishes, binders for lithographic inks, in road bitumen and asphalt, wood impregnation; Gardner 4-8 (50% in toluene) solid; sp.gr. 1.06-1.08 (20 C); soften. pt. 110-120 C; acid no. < 0.1; iodine no. < 25; sapon. no. < 3.

Norsolene S 125. [Total] Hydrocarbon resin; processing aid for rubber extrusion, molding, calendering; enhances green tack; for rubber tires, industrial goods (sealants, hoses, belts, conveyors, shoes); also in hot-melt, pressure-sensitive, and solv.-based adhesives, paints and varnishes, binders for lithographic inks, in road bitumen and asphalt, wood impregnation; Gardner 4-8 (50% in toluene) solid; sp.gr. 1.06-1.08 (20 C); soften. pt. 120-130 C; acid no. < 0.1; iodine no. < 25; sapon. no. < 3.

Norsolene S 135. [Total] Hydrocarbon resin; processing aid for rubber extrusion, molding, calendering; enhances green tack; for rubber tires, industrial goods (sealants, hoses, belts, conveyors, shoes); also in hot-melt, pressure-sensitive, and solv.-based adhesives, paints and varnishes, binders for lithographic inks, in road bitumen and asphalt, wood impregnation; Gardner 4-8 (50% in toluene) solid; sp.gr. 1.06-1.08 (20 C); soften. pt. 130-140 C; acid no. < 0.1; iodine no. < 25; sapon. no. < 3.

Norsolene S 145. [Total] Hydrocarbon resin; processing aid for rubber extrusion, molding, calendering; enhances green tack; for rubber tires, industrial goods (sealants, hoses, belts, conveyors, shoes); also in hot-melt, pressure-sensitive, and solv.-based adhesives, paints and varnishes, binders for lithographic inks, in road bitumen and asphalt, wood impregnation; Gardner 3-8 (50% in toluene) solid; sp.gr. 1.06-1.08 (20 C); soften. pt. 140-150 C; acid no. < 0.1; iodine no. < 25; sapon. no. < 3.

Norsolene S 155. [Total] Hydrocarbon resin; processing aid for rubber extrusion, molding, calendering; enhances green tack; for rubber tires, industrial goods (sealants, hoses, belts, conveyors, shoes); also in hot-melt, pressure-sensitive, and solv.-based adhesives, paints and varnishes, binders for lithographic inks, in road bitumen and asphalt, wood impregnation; Gardner 3-8 (50% in toluene) solid; sp.gr. 1.06-1.08 (20 C); soften. pt. 150-160 C; acid no. < 0.1; iodine no. < 25; sapon. no. < 3.

No Stik 802. [Ross Chem.] Silicone emulsion; mold release for rubber, plastics, urethane; ingred. in polishes; disp. in water; Unverified

No Stik 806. [Ross Chem.] Silicone compd.; mold release agent for plastic, paper, and other molded articles in contact with foods; 35% act.

Nouryflex 520. [Akzo Nobel] Phthalate ester; plasticizer for lacquer and coatings, esp. in varnishes based on NC and alkyds; Gardner 1 max. color; s.gr. 0.988-1.000; visc. 50 mPa•s; b.p. 185-198 C; flash pt. 193 C; acid no. 2 max.; ref. index 1.485.

Novacarb. [Nova Polymers] Carbon black; CAS 1333-86-4; EINECS 215-609-9; black conc. and impact modifier for ABS and PVC; produces molded parts with exc. gloss; free-flowing spheroid particulate.

Novalar. [Nova Polymers] Impact modifier for ABS, PVC, vinyl chloride copolymers, PC, PU, epoxies, PBT, acrylics; improves impact str., ductility, low temp. props., eliminates brittleness, upgrades scrap.

Novalene. [Nova Polymers] Impact modifier for homopolymer PP and HDPE; upgrades scrap PP homopolymer to copolymer; improves ductility and low temp. props.; eliminates brittleness.

Novor 924. [Akrochem] Urethane vulcanizing system; accelerator for high-temp. curing of natural rubber; deep golden br. dustless powd.; sp.gr. 1.27; m.p. 220-230 C; 75% act. in naphthenic process oil.

Novor 950. [Akrochem; Rubber Consultants] Methylene bis(4-phenylisocyanate)-based; CAS 101-68-8; vulcanizing agent for natural and syn. rubbers giving exceptional reversion and high-temp. aging resist., reduced emission levels.

Noxamium S2-50. [Ceca SA] Ammonium quat. deriv. of ethoxylated tallow amines; cationic; antistat, bactericide, emulsifier; surf. antistat for plastics and resins; liq.; 50% conc.

NP-10. [Neville] Aromatic plasticizer; chemically inert plasticizer, nonsaponifiable grade, with low reactivity; used in adhesives (mastic, pressure sensitive), rubber (cements, mechanical and molded goods, tires), and caulking compds.; FDA 21CFR §175.105, 177.2600; Gardner 14 semisolid (50% in toluene); sol. in ethers and chlorinated, aromatic, naphthenic, and terpene hydrocarbons; m.w. 500; sp.gr. 0.99; soften. pt. (R&B) 10 C; iodine no. (Wijs) 40; flash pt. (COC) 400 F.

NP-25. [Neville] Aromatic plasticizer; chemically inert plasticizer, nonsaponifiable grade, with low reactivity; used in adhesives (mastic, pressure sensitive), rubber (cements, mechanical and molded goods, tires), and caulking compds.; FDA 21CFR §175.105, 177.2600; Gardner 14 semisolid (50% in toluene); sol. in ethers and chlorinated, aromatic, naphthenic, and terpene hydrocarbons; m.w. 520; sp.gr. 1.01; soften. pt. (R&B) 25 C; iodine no. (Wijs) 73; flash pt. (COC) 450 F.

NP-Dip®. [Struktol] Zinc stearate-free; antitack coating for highly plasticized or hydrophobic rubber compds.; incorporates release and partitioning agents that cover any newly exposed surf. area while rubber awaits further processing; 50% solids; *Toxicology:* nontoxic.

NS-20 [SVO Enterprises] Sunflower oil, refined, bleached, winterized, deodorized; CAS 8001-21-6; EINECS 232-273-9; food grade, environmentally friendly veg. oil for use as feedstocks/raw materials, in plastics, polymers, rubber, and pharmaceuticals; very high oxidative and thermal stability; readily biodeg.

Nuodex S-1421 Food Grade. [Syn. Prods.] Calcium stearate FCC; CAS 1592-23-0; EINECS 216-472-8; tablet mold release, powder flow aid, plastic additive, direct food additive; wh. free-flowing powd.; 100% thru 325 mesh; apparent dens. 0.20 g/cc; m.p. 154 C; 2.5% moisture.

Nuodex S-1520 Food Grade. [Syn. Prods.] Calcium stearate FCC; CAS 1592-23-0; EINECS 216-472-8; tablet mold release, powder flow aid, plastic additive, direct food additive; wh. free-flowing powd.; 95% thru 200 mesh; apparent dens. 0.35 g/cc; m.p. 154 C; 2.5% moisture.

Nuoplaz® 849. [Hüls Am.] Epoxidized soybean oil; CAS 8013-07-8; EINECS 232-391-0; plasticizer for PVC homo or copolymer resins; imparts synergistic boost to heat and lt. stability; provides high compat. allowing for increased usage in Ba/Zn stabilized systems where additional stability is required; for food contact, esp. film for pkg. red meat and chicken, and medical devices; biodeg.; FDA 21CFR §181.27; APHA 150 max. liq., low odor; m.w. 1000; sp.gr. 0.988-0.998; visc. 320 cps; pour pt. 32 F; acid no. 0.47 max.; flash pt. (COC) 550 F; ref. index 1.4720; 0.10% max. moisture; *Toxicology:* low acute and chronic toxicity.

Nuoplaz® 1046. [Hüls Am.] Proprietary; plasticizer for PVC; imparts exc. resist. to staining in floor and wall coverings; resist. to org. extraction; minimal migration; APHA 90 max. liq.; sp.gr. 1.029-1.036; acid no. 0.2 max.; ref. index 1.4930; 60% diester, 30% dibenzoate.

Nuoplaz® 6000. [Hüls Am.] Proprietary benzoate; plasticizer for PVC; imparts exc. resist. to staining in floor and wall coverings; resist. to org. extraction; minimal migration; APHA 125 max. liq.; sp.gr. 1.015-1.021; acid no. 0.3 max.; ref. index 1.4850; 55% diester, 20% dibenzoate.

Nuoplaz® 6159. [Hüls Am.] Proprietary benzoate; plasticizer for vinyl automotive sealants; provides visc. stability, adhesion, shape retention to uncred sealant; after cure, imparts high tens. str.; also for flooring and wall coverings, PU and epoxy formulations which require external plasticizers; APHA 200 max. color; sp.gr. 1.080 ± 0.003; acid no. 2 max.; hyd. no. 10 max.; flash pt. (COC) > 400 F; ref. index 1.532 ± 0.003; 0.1% max. moisture.

Nuoplaz® 6534. [Hüls Am.] Proprietary benzoate; plasticizer for PVC and PVC acetate copolymers, plastisols, organosols; exc. stain resist. in flooring applics.; exc. gloss and paintability for automotive sealers; biodeg.; APHA 150 max. color; sp.gr. 1.078-1.083; kinematic visc. 780 cs; acid no. 0.1 max.; flash pt. (COC) 415 F; ref. index 1.5270-1.5290; 0.10% max. moisture; *Toxicology:* low acute and chronic toxicity.

Nuoplaz® 6934. [Hüls Am.] Proprietary adipate; plasticizer for PVC combining low temp. props., low volatiles, and compat.; ideal for outdoor upholstery and apparel; biodeg.; APHA 20 max. clear sl. visc. liq.; sp.gr. 0.9195-0.9255; visc. 12.6 cs (100 C); acid no. 0.3 max.; flash pt. 445 F; ref. index 1.4519; 0.05% max. moisture; *Toxicology:* low acute and chronic toxicity.

Nuoplaz® 6938. [Hüls Am.] Butyl cyclohexyl phthalate; CAS 84-64-0; EINECS 201-548-5; plasticizer for PVC providing good low temp. props., exc. processing, resist. to staining; offers fast fusion in plastisols with exc. visc. props. for coating and spray applics.; good resist. to extraction by water and oil; biodeg.; clear liq., mild char. odor; misc. with common org. solvs.; sp.gr. 0.975; sapon. no. 279; *Toxicology:* low acute and chronic toxicity.

Nuoplaz® 6959. [Hüls Am.] Tri (2-ethylhexyl) trimellitate; CAS 3319-31-1; high purity plasticizer for PVC for medical applics. incl. blood bags, dialysis tubing, catheters; exc. clarity, low temp. flexibility, low extractability; APHA 100 max. clear liq.; m.w. 546; sp.gr. 0.980-0.990; visc. 190 cps; pour pt. -50 F; acid no. 0.1 max.; flash pt. (COC) 500 F; ref. index 1.4848; 99% act.

Nuoplaz® DIBP. [Hüls Am.] Diisobutyl phthalate; CAS 84-69-5; plasticizer for automotive (exterior trim, sealants, gasketing, caulks, interior surfs.), consumer items (shower curtains, produce wrap, toys, luggage, bookbinding, flooring, pool liners, wearing apparel); roofing, textiles, tubing, medical devices, tape, foam, wire insulation, agric. film, pond liners, dip coats; Gardner 3 max. clear; sp.gr. 1.038; pour pt. -50 F; acid no. 0.1 max.; hyd. no. 3 max.; flash pt. (COC) 350 F; ref. index 1.4881; 99% act.

Nuoplaz® DIDA. [Hüls Am.] Diisodecyl adipate; CAS 27178-16-1; EINECS 248-299-9; plasticizer for PVC and copolymer resins providing exc. low temp. props. and very low volatility; used in metal lubricant formulations; APHA 30 max. clear liq.; m.w. 426; sp.gr. 0.912-0.916; visc. 14.6-15.1 cps; pour pt. -80 F; acid no. 0.2 max.; flash pt. (COC) 445 F; ref. index 1.4495-1.4510; 99.5% act.

Nuoplaz® DIDP. [Hüls Am.] Diisodecyl phthalate; CAS 68515-49-1; EINECS 247-977-1; low volatility plasticizer for PVC, high temp. elec. insulation; low fogging chars. for automotive industry; low solvating for plastisols which exhibit low initial visc. and exc. visc. stability; APHA 20 max. liq.; m.w. 446; sp.gr. 0.962-0.978; visc. 76 cps; pour pt. -46 F; acid no. 0.1 max.; flash pt. (COC) 450 F; ref. index 1.4835; 99% act.; *Toxicology:* low acute and chronic toxicity.

Nuoplaz® DOA. [Hüls Am.] Di (2-ethylhexyl) adipate; CAS 103-23-1; EINECS 203-090-1; plasticizer imparting exc. low temp. props. to PVC; blends easily with polymeric plasticizers to impart desirable hand and drape; exc. wetting for plastisol use; ideal for meat wrap where its oxygen transmission props. enhance the color of red meats and its low temp. flexibility suits refrigerated environment; FDA 21CFR §178.3740; APHA 50 max. clear liq.; m.w. 320; sp.gr. 0.924; visc. 11 cps; pour pt. -85 F; acid no. 0.04; flash pt. (COC) 390 F; ref. index 1.4454; 99.9% act.

Nuoplaz® DOM. [Hüls Am.] Dioctyl maleate; conomoner used in polymerization with vinyl acetate, vinyl chloride, styrene and derivs. of acrylic and methacrylic acids; used in latex paints, textiles; as specialty plasticizer; APHA 50 max. liq.; m.w. 340; sp.gr. 0.939-0.944; visc. 9.1-9.5 cs (100 F); pour pt. -75 F; acid no. 0.10 max.; flash pt. (COC) 370 F; ref. index 1.452-1.454; 94% act.

Nuoplaz® DOP. [Hüls Am.] Di (2-ethylhexyl) phthalate; CAS 117-81-7; EINECS 204-211-0; plasticizer for PVC; offers compat., low volatility, low temp. flexibility, lt. stability, good heat stability, exc. elec. props.; biodeg.; APHA 20 max. liq.; m.w. 390; sp.gr. 0.980-0.985; visc. 50 cps; pour pt. -50 F; acid no. 0.1 max.; flash pt. (COC) 425 F; ref. index 1.4845; 99% act.; *Toxicology:* low acute and chronic toxicity.

Nuoplaz® DTDA. [Hüls Am.] Ditridecyl adipate; plasticizer for PVC; intermediate for high efficiency syn. lubricants, providing superior low temp. fluidity and high temp. resist.; APHA 50 max. clear liq.; m.w. 510; sp.gr. 0.910; visc. 26-31 cs (100 F); pour pt. -65 F; acid no. 0.10 max.; flash pt. 490 F; ref. index 1.4556; 0.05% max. moisture.

Nuoplaz® DTDP. [Hüls Am.] Ditridecyl phthalate; primary plasticizer for PVC, esp. high temp. insulated wire compds. and low fogging automotive applics.; stabilized to prevent chem. breakdown at elevated temps; biodeg.; APHA 50 max. liq.; m.w. 530; sp.gr. 0.938-0.947; visc. 195 cps; acid no. 0.1 max.; flash pt. (COC) 460 F; ref. index 1.4831; 99.5% act.; *Toxicology:* low acute and chronic toxicity.

Nuoplaz® TIDTM. [Hüls Am.] Triisodecyl trimellitate; plasticizer; APHA 200 max. clear liq.; sp.gr. 0.975 ± 0.005; acid no. 0.1 max.; 99% act.

Nuoplaz® TOTM. [Hüls Am.] Tri (2-ethylhexyl) trimellitate; CAS 3319-31-1; plasticizer featuring permanence, low volatility, and resist. to extraction by polar solvs. and soapy water, superior high temp. performance, exc. elec. props.; good for wire and elec. insulation formulations; APHA 75 max. clear liq.; m.w. 546; sp.gr. 0.987; visc. 240 cps; pour pt. -50 F; acid no. 0.1; flash pt. (COC) 500 F; ref. index 1.484; 99.3% act.

Nusolv™ ABP-62. [Ridge Tech.] 1,1′-Biphenyl (1-methylethyl); CAS 25640-78-2; solv. possessing exc. solvency, chem. and thermal stability, nonvolatility; for aerosols, adhesives, electronic parts mfg. and cleaning, industrial cleaning, inks and paper coatings, metal cleaning, paints, plastics, sealants, tile mfg.; FDA and EPA clearances; colorless clear liq., fragrant aromatic odor; sp.gr. 0.985; visc. 6.2 cst (100 F); i.b.p. 480 F; pour pt. -55

F; flash pt. (COC) 300 F; fire pt. 310 F; ref. index 1.583; dielec. str. > 45 KV; dielec. const. 2.60 (100 C); vol. resist. 5 x 10^{12} ohm-cm (100 C).

Nusolv™ ABP-74. [Ridge Tech.] Alkyl biphenyl mixt.; CAS 81846-81-3; solv. possessing exc. solvency, chem. and thermal stability, nonvolatility; for aerosols, adhesives, electronic parts mfg. and cleaning, industrial cleaning, inks and paper coatings, metal cleaning, paints, plastics, sealants, tile mfg.; FDA and EPA clearances; colorless clear liq., sl. pleasant odor; sp.gr. 0.975; dens. 8.10 lb/gal (60 F); visc. 7.4 cst (100 F); i.b.p. 570 F; pour pt. -50 F; flash pt. (COC) > 300 F.

Nusolv™ ABP-87. [Ridge Tech.] Alkyl biphenyl mixt.; solv. possessing exc. solvency, chem. and thermal stability, nonvolatility; for aerosols, adhesives, electronic parts mfg. and cleaning, industrial cleaning, inks and paper coatings, metal cleaning, paints, plastics, sealants, tile mfg.; FDA and EPA clearances; pale yel. clear liq., sl. mild odor; sp.gr. 0.971; visc. 8.7 cst (100 F); i.b.p. 564 F; f.p. -27.2 C; flash pt. (COC) > 300 F; ref. index 1.571.

Nusolv™ ABP-103. [Ridge Tech.] Bis(methylethyl)-1,1′-biphenyl; CAS 69009-90-1; solv. possessing exc. solvency, chem. and thermal stability, nonvolatility; for aerosols, adhesives, electronic parts mfg. and cleaning, industrial cleaning, inks and paper coatings, metal cleaning, paints, plastics, sealants, tile mfg.; non-V.O.C., high flash pt., sl. or no odor and color, chemically and thermally stable, nonconductive; FDA and EPA clearances; sl. yel. clear liq., odorless; negligible sol. in water; m.w. 238; sp.gr. 0.965; dens. 8.049 lb/gal (60 F); visc. 10.3 cst (100 F); vapor pressure < 0.01 mm Hg; m.p. -40 F; b.p. 313 C; pour pt. -40 F; flash pt. (COC) 320 F; ref. index 1.567; pH neutral; *Toxicology:* LD50 (oral, rat) 4650 mg/kg (sl. toxic); sl./moderate eye irritant; prolonged/repeated exposure may cause dermatitis; may be irritating to respiratory tract; *Precaution:* combustible; incompat. with high heat, sparks, open flame, strong oxidizers; partial burning produces fumes, smoke, CO.

Nusolv™ ABP-164. [Ridge Tech.] Alkyl biphenyl mixt.; CAS 81846-81-3; solv. possessing exc. solvency, chem. and thermal stability, nonvolatility; for aerosols, adhesives, electronic parts mfg. and cleaning, industrial cleaning, inks and paper coatings, metal cleaning, paints, plastics, sealants, tile mfg.; non-V.O.C.; FDA and EPA clearances; sl. yel. clear liq., sl. mild odor; sp.gr. 0.958; visc. 16.4 cst (100 F); i.b.p. 657 F; pour pt. -40 F; flash pt. (COC) 345 F.

Nyacol® A-1530. [PQ Corp.] Colloidal disp. of antimony pentoxide in water; CAS 1314-60-9; EINECS 215-237-7; anionic; flame retardant additive to latex emulsions; durable treatment for fabrics, nonwovens, fiberfill, paper, fiberglass, vinyls; suitable for FR adhesives; disp. particle size 15-30 mμ; sp.gr. 1.37; visc. 5 cps; 30% antimony pentoxide.

Nyacol® A-1540N. [PQ Corp.] Colloidal antimony pentoxide aq. disp.; CAS 1314-60-9; EINECS 215-237-7; flame retardant; used in latex compds. used for textile fabrics, nonwovens, adhesives; aq. disp.; 40% oxide.

Nyacol® A-1550. [PQ Corp.] Colloidal disp. of antimony pentoxide in water; CAS 1314-60-9; EINECS 215-237-7; anionic; flame retardant additive to latex emulsions; durable treatment for fabrics, nonwovens, fiberfill, paper, fiberglass, vinyls; suitable for FR adhesives; disp. particle size 15-30 mμ; sp.gr. 1.81; visc. 10 cps; 50% antimony pentoxide.

Nyacol® A-1588LP. [PQ Corp.] Colloidal antimony pentoxide powd.; CAS 1314-60-9; EINECS 215-237-7; flame retardant for epoxies; powd.; 10-40 μ particle range; 87% oxide.

Nyacol® AB40. [PQ Corp.] Colloidal antimony pentoxide disp. in polyester polyol; flame retardant for rigid foams, RIM urethane coatings and elastomers; nonaq. disp.; 40% oxide.

Nyacol® AP50. [PQ Corp.] Colloidal antimony pentoxide disp. in nonreactive high m.w. tert. amine; CAS 1314-60-9; EINECS 215-237-7; flame retardant for epoxy resins, ketone sol'ns.; nonaq. disp.; 50% oxide.

Nyacol® APE1540. [PQ Corp.] Colloidal antimony pentoxide disp. in unsat. polyester; flame retardant for halogenated polyester laminates; nonaq. disp.; 40% oxide.

Nyacol® N22. [PQ Corp.] Bromine/colloidal antimony pentoxide composition; flame retardant blendable with latices (i.e. SBR, acrylic); used for textile coating or impregnation; clear film when dried; nonsettling; fluid; pH 8; 68% solids (48% bromine; 22% colloidal antimony pentoxide).

Nyacol® N24. [PQ Corp.] Colloidal antimony pentoxide aq. disp.; CAS 1314-60-9; EINECS 215-237-7; flame retardant; used in latex compds. used for textile fabrics, nonwovens, adhesives; aq. disp.; 11.2% oxide; 33% halogen (chlorine and bromine); 73% solids.

Nyacol® ZTA. [PQ Corp.] Colloidal antimony pentoxide powd.; CAS 1314-60-9; EINECS 215-237-7; flame retardant for vinyls, ABS, HIPS, PP; powd.; 10-40 μ particle range; 80% oxide.

Nyad® 200, 325, 400. [Nyco Minerals] Milled wollastonite, untreated; CAS 13983-17-0; cost-effective functional fillers and additives for plastics, coating, friction, refractory, ceramic, construction, elastomer, sealant, and adhesive applics.; wh. acicular free-flowing powd., odorless; sol. 0.0095 g/100 cc in water; m.w. 116; sp.gr. 2.9; dens. 24.2 lb/solid gal; bulk value 0.0413 gal/lb; m.p. 1540 C; ref. index 1.63; pH 9.9 (10% slurry); hardness (Mohs) 4.5; 51% SiO_2, 47.5% CaO; *Toxicology:* nuisance dust, TLV/TWA 10 mg/m^3 (8 h, total dust); long-term cumulative inh. may cause restriction of airways; may cuase minor skin irritation; *Storage:* keep dry and cool in original shipping containers until use.

Nyad® 475, 1250. [Nyco Minerals] Fine particle size wollastonite, untreated; CAS 13983-17-0; cost-effective functional fillers and additives for plastics, coating, friction, refractory, ceramic, construction, elastomer, sealant, and adhesive applics.; wh. acicular free-flowing fine powd., odorless; sol. 0.0095 g/100 cc in water; m.w. 116; sp.gr. 2.9; dens. 24.2 lb/solid gal; bulk value 0.0413 gal/lb; m.p. 1540 C; ref. index 1.63; pH 9.9 (10% slurry); hardness (Mohs) 4.5; 51% SiO_2, 47.5% CaO; *Toxicology:* nuisance dust, TLV/TWA 10 mg/m^3 (8 h, total dust); long-term cumulative inh. may cause restriction of airways; may cuase minor skin irritation; *Storage:* keep dry and cool in original shipping containers until use.

Nyad® FP, G, G Special. [Nyco Minerals] High aspect ratio wollastonite, untreated; CAS 13983-17-0; cost-effective functional fillers and additives for plastics, coating, friction, refractory, ceramic, construction, elastomer, sealant, and adhesive applics.; wh. acicular free-flowing powd., odorless; sol. 0.0095 g/100 cc in water; m.w. 116; sp.gr. 2.9; dens. 24.2 lb/solid gal; bulk value 0.0413 gal/lb; m.p. 1540 C; ref. index 1.63; pH 9.9 (10% slurry); hardness (Mohs) 4.5; 51% SiO_2, 47.5% CaO; *Toxicology:* nuisance dust, TLV/TWA 10 mg/m^3 (8 h, total dust); long-term cumulative inh. may cause restriction of airways; may cuase minor skin irritation; *Storage:* keep dry and cool in original shipping containers until use.

Nyad G® Wollastocoat®. [Nyco Minerals] Surface-modified wollastonite; CAS 13983-17-0; cost-effective functional fillers and additives for plastics, coating, friction, refractory, ceramic, construction, elastomer, sealant, and adhesive applics.; surf. modifications can improve processing, bonding between resin and filler, mech. and phys. props., material handling and warehousing, etc.; wh. acicular powd., faint odor; sol. 0.01 g/100 cc in water; sp.gr. 2.9; dens. 24.2 lb/solid gal; bulk value 0.0413 gal/lb; m.p. 1540 C; ref. index 1.63; pH 9.9 (10% slurry); hardness (Mohs) 4.5; > 99% wollastonite, < 1% proprietary treatments; *Toxicology:* nuisance dust, TLV/TWA 10 mg/m^3 (8 h, total dust); long-term cumulative inh. can cause restriction of airways; may cause minro skin irritation; *Precaution:* surf. treatments may oxidize or decomp. at elevated temp.s; burning may produce sm. amts. of oxides of carbon, silicon, nitrogen, or sulfur; *Storage:* keep dry and cool in original shipping containers until use.

Nycor® R. [Nyco Minerals] High aspect ratio wollastonite, untreated; CAS 13983-17-0; cost-effective functional fillers and additives for plastics, coating, friction, refractory, ceramic, construction, elastomer, sealant, and adhesive applics.t; wh. acicular free-flowing powd., odorless; sol. 0.0095 g/100 cc in water; m.w. 116; sp.gr. 2.9; dens. 24.2 lb/solid gal; bulk value 0.0413 gal/lb; m.p. 1540 C; ref. index 1.63; pH 9.9 (10% slurry); hardness (Mohs) 4.5; 51% SiO_2, 47.5% CaO; *Toxicology:* nuisance dust, TLV/TWA 10 mg/m^3 (8 h, total dust); long-term cumulative inh. may cause restriction of airways; may cuase minor skin irritation; *Storage:* keep dry and cool in original shipping containers until use.

Nytal® 100. [R.T. Vanderbilt] Industrial talc (hydrous magnesium calcium silicate); CAS 14807-96-6; EINECS 238-877-9; reinforcing filler for elastomers, PVC, vinyl asbestos tile, polyester in match molded articles, body patching compds., PP, nylon, phenol formaldehyde, polyethylene, ceramic wall tile and artware; improves stiffness; aux. flux in vitreous ceramic bodies; wh. powd.; 17 μm median particle size; 90% < 325 mesh; dens. 2.85 mg/m^3; bulk dens. 61 lb/ft^3 (compacted); oil absorp. 21; brightness 90; pH 9.4 (10% aq. slurry); 56.20% SiO_2, 28.80% MgO, 8.58% CaO.

Nytal® 200. [R.T. Vanderbilt] Industrial talc (hydrous magnesium calcium silicate); CAS 14807-96-6; EINECS 238-877-9; reinforcing filler for elastomers, PVC, vinyl asbestos tile, polyester in match molded articles, body patching compds., PP, nylon, phenol formaldehyde, polyethylene, ceramic wall tile and artware; improves stiffness; wh. powd.; 10.2 μm median particle size; 97.5% < 325 mesh; dens. 2.85 mg/m^3; bulk dens. 60 lb/ft^3 (compacted); oil absorp. 23; brightness 90; pH 9.4 (10% aq. slurry).

Nytal® 300. [R.T. Vanderbilt] Industrial talc (hydrous magnesium calcium silicate); CAS 14807-96-6; EINECS 238-877-9; reinforcing filler for elastomers, PVC, vinyl asbestos tile, polyester in match molded articles, body patching compds., PP, nylon, phenol formaldehyde, polyethylene, ceramic wall tile and artware; improves stiffness; wh. powd.; 5.5 μm median particle size; 99.9% < 325 mesh; dens. 2.85 mg/m^3; bulk dens. 41 lb/ft^3 (compacted); oil absorp. 29; brightness 90; pH 9.4 (10% aq. slurry).

Nytal® 400. [R.T. Vanderbilt] Industrial talc (hydrous magnesium calcium silicate); CAS 14807-96-6; EINECS 238-877-9; reinforcing filler for elastomers, PVC, vinyl asbestos tile, polyester in match molded articles, body patching compds., PP, nylon, phenol formaldehyde, polyethylene, ceramic wall tile and artware; improves stiffness; wh. powd.; 3.3 μm median particle size; 99.9% < 325 mesh; dens. 2.85 mg/m^3; bulk dens. 25 lb/ft^3 (compacted); oil absorp. 39; brightness 90; pH 9.4 (10% aq. slurry).

NZ 01. [Kenrich Petrochemicals] Neopentyl (diallyl) oxy, trineodecanoyl zirconate, IPA; coupling agents which sometimes also act as adhesion promoters, antioxidants, antistats, antifoams, accelerators, blowing agent activators, catalysts, curatives, corrosion inhibitors, disp. aids, emulsifiers, flame retardants, foaming agents, hardeners, impact modifiers, internal lubes, process aids, release agents, retarders, stabilizers, surfactants, suspension aids, thixotropes, wetting agents for thermoplastics, thermosets, elastomers; amber liq.; sol. in xylene, toluene, DOP, min. oil; limited sol. in IPA; insol. in water; sp.gr. 0.96 (16 C); visc. 40 cps; b.p. 188 F (initial); pH 8 (sat. sol'n.); flash pt. (TCC) 195 F; 95% solids in IPA.

NZ 09. [Kenrich Petrochemicals] Neopentyl (diallyl) oxy, tri (dodecyl) benzene sulfonyl zirconate, IPA; coupling agents which sometimes also act as adhesion promoters, antioxidants, antistats, antifoams, accelerators, blowing agent activators, catalysts, curatives, corrosion inhibitors, disp. aids, emulsifiers, flame retardants, foaming agents, hardeners, impact modifiers, internal lubes, process aids, release agents, retarders, stabilizers, surfactants, suspension aids, thixotropes, wetting agents for thermoplastics, thermosets, elastomers; brn. liq.; sol. in xylene, toluene, DOP; limited sol. in IPA, min. oil; insol. in water; sp.gr. 1.09 (16 C); visc. 190 cps; b.p. 300 F (initial); pH 4 (sat. sol'n.); flash pt. (TCC) 150 F; 95% solids in IPA.

NZ 12. [Kenrich Petrochemicals] Neopentyl (diallyl) oxy, tri (dioctyl) phosphato zirconate, IPA; coupling agents which sometimes also act as adhesion promoters, antioxidants, antistats, antifoams, accelerators, blowing agent activators, catalysts, curatives, corrosion inhibitors, disp. aids, emulsifiers, flame retardants, foaming agents, hardeners, impact modifiers, internal lubes, process aids, release agents, retarders, stabilizers, surfactants, suspension aids, thixotropes, wetting agents for thermoplastics, thermosets, elastomers; orange/red liq.; sol. in xylene, toluene, DOP, min. oil; limited sol. in IPA; insol. in water; sp.gr. 1.06 (16 C); visc. 160 cps; b.p. 220 F (initial); pH 6 (sat. sol'n.); flash pt. (TCC) 170 F; 95% solids in IPA.

NZ 33. [Kenrich Petrochemicals] Neopentyl (diallyl) oxy, trimethacryl zirconate, IPA; coupling agents which sometimes also act as adhesion promoters, antioxidants, antistats, antifoams, accelerators, blowing agent activators, catalysts, curatives, corrosion inhibitors, disp. aids, emulsifiers, flame retardants, foaming agents, hardeners, impact modifiers, internal lubes, process aids, release agents, retarders, stabilizers, surfactants, suspension aids, thixotropes, wetting agents for thermoplastics, thermosets, elastomers; yel./orange liq.; sol. in xylene, toluene, DOP; sp.gr. 1.09 (16 C); visc. < 100 cps; i.b.p. > 300 F; flash pt. (TCC) 150 F; pH 4; 95+% solids in IPA.

NZ 38. [Kenrich Petrochemicals] Neopentyl (diallyl) oxy, tri(dioctyl) pyrophosphato zirconate, IPA; coupling agents which sometimes also act as adhesion promoters, antioxidants, antistats, antifoams, accelerators, blowing agent activators, catalysts, curatives, corrosion inhibitors, disp. aids, emulsifiers, flame retardants, foaming agents, hardeners, impact modifiers, internal lubes, process aids, release agents, retarders, stabilizers, surfactants, suspension aids, thixotropes, wetting agents for thermoplastics, thermosets, elastomers; reddsh. liq.; sol. in IPA, xylene, toluene, DOP; insol. in min. oil, water; sp.gr. 1.10 (16 C); visc. 360 cps; b.p. 345 F (initial); pH 6 (sat. sol'n.); flash pt. (TCC) 170 F; 95% solids in IPA.

NZ 39. [Kenrich Petrochemicals] Neopentyl (diallyl) oxy, triacryl zirconate, IPA; coupling agents which sometimes also act as adhesion promoters, antioxidants, antistats, antifoams, accelerators, blowing agent activators, catalysts, curatives, corrosion inhibitors, disp. aids, emulsifiers, flame retardants, foaming agents, hardeners, impact modifiers, internal lubes, process aids, release agents, retarders, stabilizers, surfactants, suspension aids, thixotropes, wetting agents for thermoplastics, thermosets, elastomers; yel./orange liq.; sol. in xylene, toluene, DOP, min. oil; insol. in water; sp.gr. 1.07 (16 C); visc. < 100 cps; i.b.p. > 300 F; flash pt. (TCC) < 200 F; pH 4 (sat. sol'n.); 95+% solids in IPA.

NZ 44. [Kenrich Petrochemicals] Neopentyl (diallyl) oxy, tri (N-ethylenediamino) ethyl zirconate, IPA; coupling agents which sometimes also act as adhesion promoters, antioxidants, antistats, antifoams, accelerators, blowing agent activators, catalysts, curatives, corrosion inhibitors, disp. aids, emulsifiers, flame retardants, foaming agents, hardeners, impact modifiers, internal lubes, process aids, release agents, retarders, stabilizers, surfactants, suspension aids, thixotropes, wetting agents for thermoplastics, thermosets, elastomers; yel./orange liq.; sol. in IPA; limited sol. in xylene, toluene; reacts slowly in DOP; insol. in min. oil, water; sp.gr. 1.17 (16 C); visc. > 50 cps; b.p. 300 F (initial); pH 11 (sat. sol'n.); flash pt. (TCC) > 220 F; 95% solids in IPA.

NZ 49. [Kenrich Petrochemicals] Neopentyl (diallyl) oxy, tri (9,10 epoxy stearoyl) zirconate, IPA; coupling agents which sometimes also act as adhesion promoters, antioxidants, antistats, antifoams, accelerators, blowing agent activators, catalysts, curatives, corrosion inhibitors, disp. aids, emulsifiers, flame retardants, foaming agents, hardeners, impact modifiers, internal lubes, process aids, release agents, retarders, stabilizers, surfactants, suspension aids, thixotropes, wetting agents for thermoplastics, thermosets, elastomers; amber waxy liq.; sol. in xylene, toluene, DOP, min. oil; insol. in water; sp.gr. 1.01 (16 C); 95% solids in IPA.

NZ 89. [Kenrich Petrochemicals] Neopentyl (diallyl) oxy, trimercapto-phenyl zirconate; coupling agents which sometimes also act as adhesion promoters, antioxidants, antistats, antifoams, accelerators, blowing agent activators, catalysts, curatives, corrosion inhibitors, disp. aids, emulsifiers, flame retardants, foaming agents, hardeners, impact modifiers, internal lubes, process aids, release agents, retarders, stabilizers, surfactants, suspension aids, thixotropes, wetting agents for thermoplastics, thermosets, elastomers; brn. waxy liq.; sol. in xylene, toluene; sp.gr. 1.06 (16 C).

NZ 97. [Kenrich Petrochemicals] Neopentyl (diallyl) oxy, tri(m-amino) phenyl zirconate, phenyl glycol ether solv.; coupling agents which sometimes also act as adhesion promoters, antioxidants, antistats, antifoams, accelerators, blowing agent activators, catalysts, curatives, corrosion inhibitors, disp. aids, emulsifiers, flame retardants, foaming agents, hardeners, impact modifiers, internal lubes, process aids, release agents, retarders, stabilizers, surfactants, suspension aids, thixotropes, wetting agents for thermoplastics, thermosets, elastomers; brn. liq.; sol. in ether solvents; limited sol. in IPA; reacts slowly in DOP; insol. in xylene, toluene, min. oil, water; sp.gr. 1.20 (16 C); visc. 4400 cps; b.p. 300 F (initial); pH 6 (sat. sol'n.); flash pt. (TCC) 190 F; 67% solids in phenyl glycol ether.

O

Octamine® Flake, Powd. [Uniroyal] Octylated diphenylamine; antioxidant.

Octoate® Z. [R.T. Vanderbilt] Zinc di-2-ethylhexoate; activator for natural and syn. rubbers; used in sol. cure systems in natural and polyisoprene; lt. amber liq.; m.w. 351.77; dens. 1.12 ± 0.02 mg/m^3; 80% min. act.; 17-19% Zn.

Octoate® Z Solid. [R.T. Vanderbilt] Zinc di-2-ethylhexoate; rubber activator; 72% act. on inert filler.

Octocure 456. [Tiarco] Sulfur aq. disp.; CAS 7704-34-9; EINECS 231-722-6; rubber accelerator; also for vulcanization processes in aq. latex compds.; 70% total solids.

Octocure 462. [Tiarco] French process zinc oxide aq. disp.; CAS 1314-13-2; EINECS 215-222-5; pigment and accelerator for vulcanization of rubber; well-suited for use in gelled latex compds.; extremely fine particle size; 60% conc.

Octocure ZBZ-50. [Tiarco] Zinc dibenzyldithiocarbamate aq. disp.; CAS 14726-36-4; anionic; latex and rubber accelerator; for SBR and natural latex compds.; used where the formation of volatile nitrosamines must be reduced; off-wh. liq. to paste, sl. typ. odor; < 5 μ avg. particle size; insol. in water; sp.gr. 1.07; visc. 1000-3000 cps; b.p. 212 F; flash pt. > 212 F; pH 9.0-10.0; 51-53% solids; *Toxicology:* eye irritant; *Precaution:* thermal decomp. may produce sulfur oxides, nitric oxides, carbon oxides; *Storage:* store in tightly closed containers; protect from freezing; stir prior to use.

Octocure ZDB-50. [Tiarco] Zinc dibutyldithiocarbamate; CAS 136-23-2; latex and rubber accelerator.

Octocure ZDE-50. [Tiarco] Zinc diethyldithiocarbamate; CAS 14323-55-1; latex and rubber accelerator.

Octocure ZDM-50. [Tiarco] Zinc dimethyldithiocarbamate; CAS 137-30-4; latex and rubber accelerator.

Octocure ZIX-50. [Tiarco] Zinc isopropyl xanthate aq. disp.; CAS 42590-53-4; anionic; latex and rubber accelerator; for SBR and natural latex compds.; wh. liq. to paste, char. odor; < 5 μ avg. particle size; insol. in water; sp.gr. 1.10; visc. 1000-3000 cps; b.p. 212 F; flash pt. > 212 F; pH 9-10; 51-53% solids; *Toxicology:* LD50 (oral, rat) 3200-4000 mg/kg; mod. toxic by ingestion; may cause headache; eye irritant; contains < 5% potassium hydroxide (TLV 2 mg/m^3 ACGIH); *Precaution:* reacts with acids liberating highly flamm. CS_2; thermal decomp. produces sulfur oxies, nitric oxides, carbon oxides, CS_2; incompat. with strong acids, bases, oxidizers; *Storage:* store in tightly closed containers; protect from freezing; stir prior to use; storage should be kept to < 3 mos.

Octocure ZMBT-50. [Tiarco] Zinc mercaptobenzothiazole; CAS 155-04-4; latex and rubber accelerator.

Octoguard FR-01. [Tiarco] Decabromodiphenyloxide aq. disp.; CAS 1163-19-5; EINECS 214-604-9; flame retardant for water-based polymer compds. such as latex adhesives, binders, coatings, and foams; 65% conc.

Octoguard FR-10. [Tiarco] Antimony trioxide; CAS 1309-64-4; EINECS 215-175-0; flame retardant for water-based polymer compds. such as latex adhesives, binders, coatings, and foams; 65% conc.

Octoguard FR-15. [Tiarco] Antimony trioxide/decabromodiphenyloxide aq. disp. (1:5 ratio); flame retardant for water-based polymer compds. such as latex adhesives, binders, coatings, and foams; 67.5% conc.

Octolite 561. [Tiarco] Bisphenol aq. dispersion; antioxidant suitable for all white or light-colored latex compds. where color is critical; also for rubber intended for repeated or continuous contact with food; FDA clearance; 50% conc.

Octolite AO-28. [Tiarco] Polymeric hindered phenol/thioester; antioxidant; emulsion; 63% conc.

Octopol NB-47. [Tiarco] Sodium dibutyldithiocarbamate; CAS 136-30-1; ultra accelerator for SBR and natural rubber latex compds.; in polymerization of chloroprene rubber; esp. for latex compds. where copper staining of zinc salt dithiocarbamates is a problem; liq.; 47% act.

Octopol SBZ-20. [Tiarco] Sodium dibenzyldithiocarbamate aq. sol'n.; CAS 55310-46-8; vulcanization accelerator; for latex applics. where reduction of volatile nitrosamines is desired; polymerization shortstop in copolymerization of butadiene, styrene, and neoprene; lt. grn./yel. clear liq., sl. amine odor; completely sol. in water; sp.gr. 1.04; b.p. 212 F; flash pt. > 212 F; pH 12.5-13.5; 19.5-20.5% act.; *Toxicology:* severe eye irritant; strong skin irritant; *Precaution:* corrosive (DOT); incompat. with strong acids, oxidizers; thermal decomp. produces toxic prods.; acid decomp. releases CS_2 and amine; *Storage:* keep containers closed; store in cool, well-ventilated area; avoid storage below 40 F to prevent crystallization of act. ingred.

Octopol SDB-50. [Tiarco] Sodium dibutyldithiocarbamate aq. sol'n.; CAS 136-30-1; polymerization ingred. in syn. rubber industry; peptizer in the mfg. of chloroprene rubber; lt. yel./grn. clear liq., sl. amine odor; completely sol. in water; sp.gr. 1.05-1.09; b.p. 212 F; flash pt. > 212 F; pH 12.5-13.5; 49-51% act.; *Toxicology:* severe eye irritant; strong skin irritant; *Precaution:* corrosive (DOT); incompat. with strong acids, oxidizing agents; thermal decomp. produces toxic prods.; acid decomp. releases CS_2 and amine; *Storage:* avoid storage below 40 F to prevent crystallization of the act. ingred.

Octopol SDE-25. [Tiarco] Sodium diethyldithiocarbamate; CAS 148-18-5; natural rubber latex preservative; precipitant for heavy metals in waste water treatment; liq.; 25% act.

Octopol SDM-40. [Tiarco] Sodium dimethyldithiocarbamate; polymerization shortstop in SBR rubber; precipitant for heavy metals in waste water treatment; liq.; 40% act.

Octosol 449. [Tiarco] Potassium oleate; CAS 143-18-0; EINECS 205-590-5; anionic; foaming agent, stabilizer, emulsifier, dispersant; primary frothing aid in gelled latex foam compds.; clear liq.; dens. 8.75 lb/gal; pH 9.8-10.2; 16.5-17.5% solids.

Octosol 496. [Tiarco] Proprietary surfactant; anionic; emulsifier, dispersant, stabilizer, foaming agent for frothed latex adhesive compds. for carpet backing and laminating; amber clear liq.; sol. in water; pH 7-10; 29-31% solids in water.

Octosol 562. [Tiarco] Lauryl trimethyl ammonium chloride; CAS 112-00-5; EINECS 203-927-0; cationic; gel sensitizer for latex foam rubber; 33% solids.

Octosol 571. [Tiarco] Lauryl trimethyl ammonium chloride; CAS 112-00-5; EINECS 203-927-0; cationic; emulsifier for cationic or blended ionic emulsion systems; frothing agent, gel sensitizer in latex foam rubber; Gardner 1 max. liq.; sol. in water; pH 7-9; 49-51% solids in aq. IPA.

Octosol A-1. [Tiarco] Disodium N-[3-(dodecyloxy) propyl] sulfosuccinamate; CAS 58353-68-7; anionic; emulsifier, dispersant, wetting agent, foaming agent for frothed latex compds. and adhesives; suspending agent in emulsion polymerization; textile softener; straw yel. to lt. amber liq.; sol. in warm water; pH 9-10; 33-35% solids.

Octosol A-18. [Tiarco] Disodium N-octadecyl sulfosuccinamate; CAS 14481-60-8; EINECS 238-479-5; anionic; emulsifier, dispersant, foaming agent for latex compds., cleaners; suspending agent in emulsion polymerization; stable in acid and alkaline aq. systems; wh. to cream-colored paste; sol. in warm water; pH 7-9; 34-36% solids in aq. disp.

Octosol A-18-A. [Tiarco] Diammonium N-octadecyl sulfosuccinamate; CAS 68128-59-6; anionic; emulsifier, stabilizer, foaming agent for acrylic latex frothed compds., formulations where reduced sodium ion content is desirable; suspending agent in emulsion polymerization; clear to creamy wh. liq.; sol. in warm water; pH 8-10; 34-36% solids.

Octosol HA-80. [Tiarco] Sodium dihexyl sulfosuccinate; anionic; emulsifier for latex emulsion polymerization; liq.; 80% conc.

Octosol IB-45. [Tiarco] Sodium diisobutyl sulfosuccinate; CAS 127-39-9; EINECS 204-839-5; emulsifier for latex emulsion polymerization; liq.; 45% conc.

Octosol SLS. [Tiarco] Sodium lauryl sulfate; CAS 151-21-3; anionic; stabilizer, frothing aid, emulsifier, dispersant for latex compds.; low cloud pt., low salt content; lt. color clear liq., low odor; sol. in water; sp.gr. 1.03-1.08; visc. < 500 cps; cloud pt. < 18 C; pH 7-9; 28-30% solids in water.

Octosol SLS-1. [Tiarco] Sodium lauryl sulfate; CAS 151-21-3; anionic; stabilizer, frothing aid for latex compds.; emulsifier, dispersant; low cloud pt., low salt content; clear liq.; sol. in water; sp.gr. 1.03-1.08; dens. 8.67 lb/gal; visc. < 800 cps; cloud pt. 40 F; pH 7.5-9.0; 28-31% solids in water.

Octosol TH-40. [Tiarco] Sodium dicyclohexyl sulfosuccinate; CAS 23386-52-9; EINECS 245-629-3; anionic; surfactant for emulsion polymerization, mfg. of carboxylated latexes; paste; 40% conc.

Octowax 321. [Tiarco] Refined paraffin wax aq. emulsion; CAS 8002-74-2; EINECS 232-315-6; heat-stable, nondiscoloring wax emulsion with 125-130 F melting range; for use in paper and wood coating, sizing, textile lubrication, as processing aid in latex foams; antiozonant chars.; improves water resist. and aging chars. in latex compds.; FDA 21CFR §175.105, 177.2600, 178.3710; wh. emulsion; visc. 200-1600 cps; m.p. 125-130 F; pH 8.5-9.5; 49-51% solids; *Storage:* store in tightly closed containers; protect from freezing; stir before use.

Octowax 518. [Tiarco] HDPE emulsion; CAS 9002-88-4; EINECS 200-815-3; nonionic; wax emulsion for use in paper and wood coating, sizing, textile lubrication, as processing aid in latex foams; antiozonant chars.; 25% conc.

Octowet 70PG. [Tiarco] Sodium dioctyl sulfosuccinate; anionic; high speed wetting agent, emulsifier, penetrant for textile processing, paints, coatings, polymerization, agric., mining, paper, printing, industrial applics. where high flash is important; colorless liq.; m.w. 444.63; dens. 9.08 lb/gal; visc. 200-400 cps; acid no. < 2; iodine no. < 0.15; flash pt. > 212 F; pH 5-7; Draves wetting < 5 s (0.075%); 69-71% act. in water, propylene glycol.

#1 Oil. [CasChem] Castor oil, tech.; tech. grade for industrial use; plasticizer, tackifier, penetrant for adhesives, inks, coatings, sealants; industrial lubricants; pigment dispersion in coatings; leather softener/preserver; solubilizer for detergents; plasticizer, mold release for rubber; waxes and polishes; textile processing; Gardner 2+ oil; sp.gr. 0.959; visc. 7.5 stokes; pour pt. -10 F; acid no. 2; iodine no. 86; sapon. no. 180; hyd. no. 158.

#15 Oil. [CasChem] Polymerized castor oil; pigment wetting/dispersing agent; plasticizer for resins, gums, polymers; lubricant, penetrant; coupling solv.; adhesion promoter; for cellulose lacquers, inks, adhesives, industrial lubricants, polishes, caulks, leather dressing, hydraulic fluids, rubber compding.; FDA approval; Gardner 11 color; sp.gr. 1.013; visc. 250 stokes; pour pt. 35 F; acid no. 14; iodine no. 64; sapon. no. 220; hyd. no. 137.

#30 Oil. [CasChem] Castor oil, polymerized; plasticizer and processing aid for rubber; nonvolatile, nonextrudable; extremely low vapor pressure; highly compat. with rubber, neoprene, most syn. elastomers; Gardner 13 visc. liq.; sp.gr. 1.019; visc. 500 stokes; pour pt. 45 F; acid no. 13; iodine no. 63; sapon. no. 220; hyd. no. 136.

Okstan M 15. [Bärlocher GmbH] Butyltin carboxylate; stabilizer for PVC coating through dip molding (tool handles, bicicyle handles, PVC boots and gloves, wire baskets); exc. transparency and color, very good photostability; liq.

Okstan M 69 S. [Bärlocher GmbH] Butyltin mercaptide; stabilizer for plastisol processing; provides superior stability and color props.; resist. to water staining; esp. low-odor; suitable for automotive dashboard panels, PVC coating through dip molding; liq., low odor.

Omacide® P-711-5. [Olin] 5% Zinc pyrithione, 67% C7-C9-C11 phthalate blend plasticizer; antimicrobial disp. for protection of PVC systems; EPA reg. no. 1258-1184, 1258-1183 (of act.); off-wh. powd., faint aromatic odor; sp.gr. ≈ 1; visc. 6000-15,000 cps; 5% act.; *Storage:* store in cool, well-ventilated place, away from sources of ignition.

Omacide® P-BBP-5. [Olin] 5% Zinc pyrithione, 80% butyl benzyl phthalate plasticizer; antimicrobial disp. for protection of PVC systems; EPA reg. no. 1258-1184, 1258-1183 (of act.); off-wh. powd., faint aromatic odor; sp.gr. ≈ 1; visc. 6000-15,000 cps; 5% act.; *Storage:* store in cool, well-ventilated place, away from sources of ignition.

Omacide® P-DIDP-5. [Olin] 5% Zinc pyrithione, 65% diisodecyl phthalate plasticizer; antimicrobial disp. for protection of PVC systems; EPA reg. no. 1258-1184, 1258-1183 (of act.); off-wh. powd., faint aromatic odor; sp.gr. ≈ 1; visc. 6000-15,000 cps; 5% act.; *Storage:* store in cool, well-ventilated place, away from sources of ignition.

Omacide® P-DOP-5. [Olin] 5% Zinc pyrithione, 77% dioctyl phthalate plasticizer; antimicrobial disp. for protection of PVC systems; EPA reg. no. 1258-1184, 1258-1183 (of act.); off-wh. powd., faint aromatic odor; sp.gr. ≈ 1; visc. 6000-15,000 cps; 5% act.; *Storage:* store in cool, well-ventilated place, away from sources of ignition.

Omacide® P-ESO-5. [Olin] 5% Zinc pyrithione, 75% epoxidized soybean oil plasticizer; antimicrobial disp. for protection of PVC systems; EPA reg. no. 1258-1184, 1258-1183 (of act.); off-wh. powd., faint aromatic odor; sp.gr. ≈ 1; visc. 6000-15,000 cps; 5% act.; *Storage:* store in cool, well-ventilated place, away from sources of ignition.

Ongard AZ11. [Anzon] Multimetal complex; flame retardant for thermoset polyester, PVC.

Onifine-C. [Otsuka] 20% Azodicarbonamide, 75% polymeric carrier, 5% barium-zinc complex stabilizer; blowing agent for prod. of cross-linked PVC foam of closed-cell structures; cushioning and absorbing materials; insulation, water flotation items; sealant in motor and construction industries; lt. yel. powd.; Unverified

Onifine-CC. [Otsuka] 40% Azodicarbonamide, modified, 50% polymeric carrier, 10% barium-zinc complex stabilizer; blowing agent for prod. of cross-linked PVC foam of closed-cell structures; cushioning and absorbing materials; insulation, water flotation items; sealant in motor and construction industries; lt. yel. powd.; Unverified

Onifine-CE. [Otsuka] 93.5% Azodicarbonamide, modified, 6.5% barium-zinc complex stabilizer; blowing agent for highly cross-linked PVC foaming prods.; esp. designed for extrusion; PVC expanded prods.; Unverified

Onyx Classica OC-1000. [Alcoa] Hydrated alumina; developed for syn. onyx for syn. marble industry; features high whiteness, purity, consistency, flame retardancy, compatibility.

Onyx Premier WP-31. [Alcan] Alumina trihydrate; filler for cultured onyx; provides highest whiteness, exc. translucency; flame retardant, smoke suppressant; compat. with polyester, acrylic, and other resin systems; very wh. free-flowing cryst. powd.; 19-26 μ avg. particle size; 20-40% on 325 mesh; sp.gr. 2.42; bulk dens. 1.0 g/cc (loose); ref. index 1.57; hardness (Mohs) 2.5-3.5; 65% Al_2O_3.

Opacicoat™. [ECC Int'l.] Calcium carbonate, coated with < 2% stearic acid; ultrafine coated pigment, filler, reinforcement with unique particle size distribution; permits higher loadings in polymer systems while still maintaining props.; hydrophobic; NSF compliance; wh. powd., odorless; 1.1 μ mean particle size; fineness (Hegman) 6-7; sp.gr. 2.71; dens. 45 lb/ft^3 (loose); pH 9.5 (5% slurry); nonflamm.; 97.6% $CaCO_3$, 0.2% max. moisture; *Toxicology:* TLV/TWA 10 mg/m^3, considered nuisance dust; *Precaution:* incompat. with acids.

Opex® 80. [Uniroyal] Dinitrosopentamethylenetetramine; CAS 101-25-7; chemical blowing agent for ABS, PU, silicone, NR, SBR; 80% act.

Oppasin®. [BASF AG] Inorg. and org. pigment concs.; for coloring rubber compds. and sol'ns.; optical brighteners for textile fibers.

Optiwhite®. [Burgess Pigment] Aluminum silicate, thermo-optic; pigment retaining whiteness and hiding when embedded in binders; for use in paper coating, paints, rubber, and plastics to extend TiO_2 or other costly pigments; exc. dispersion, good electricals, high hiding, improved film props.; U.S. patent 3,309,214; amorphous particles; 0.15% max. on 325 mesh; sp.gr. 2.2; oil absorp. 57; ref. index 1.62; pH 5.5-6.2; 0.5% max. moisture.

Optiwhite® P. [Burgess Pigment] Aluminum silicate, thermo-optic; pigment for processing rubber, plastics, and coatings with good dispersion chars.; in paints, permits reduction of TiO_2 levels and flatting agents; U.S. patent 3,021,195; amorphous particles; 0.15% max. on 325 mesh; sp.gr. 2.2; oil absorp. 65; ref. index 1.62; pH 4.0-4.5; 0.5% max.

moisture.

Ortegol® 204. [Goldschmidt AG] Delayed reaction crosslinking additive for prod. of high resilience PU slabstock foams; sol. in water and water/amine blends; sp.gr. 1.25 g/cc; visc. 80 ± 20 mPa•s; hyd. no. 860 ± 20; 25% water; *Storage:* store @ 10-40 C to avoid precipitation; Unverified

Ortegol® 300. [Goldschmidt AG] Additive for mfg. of flexible PU slabstock and molded foams of reduced hardness; unlimited sol. in water; dens. 1.09 ± 0.01 g/m^3; visc. 8 ± 1 mPa•s; 50% water; *Toxicology:* irritating to eyes, respiratory system, and skin; *Storage:* stir before use; store @ 0-50 C in closed containers; Unverified

Ortegol® 310. [Goldschmidt AG] Additive for mfg. of flexible PU slabstock and molded foams of reduced hardness; unlimited sol. in water; dens. 1.085 ± 0.010; visc. 10 ± 2 mPa•s; flash pt. > 100 C; pH 8 ± 1; 50% water; *Toxicology:* irritating to eyes, respiratory system, and skin; *Storage:* 6 mos storage stability @ 0-50 C in closed containers; Unverified

OSO® 440. [Hitox] Syn. yel. iron oxide; CAS 1309-37-1; EINECS 215-168-2; pigment with good suspension and stabilization props. for coatings, automotive finishes, appliance enamels, paints, plastics, building prods., rubbers, inks; med. yel. fine powd., odorless; < 0.01% on 325 mesh; fineness (Hegman) 6+; insol. in water; sp.gr. 4.06; dens. 33.90 lb/gal; oil absorp. 45-49; pH 5.0; 88+% Fe_2O_3; *Toxicology:* ACGIH TLV 5 mg/m^3; may cause irritation from abrasion on contact with skin; eye irritant; prolonged inh. produces siderosis; avoid breathing dust; *Precaution:* nonflamm.; incompat. with hydrazine, calcium hypochlorite, performic acid, bromine pentafluoride.

OSO® 1905. [Hitox] Syn. red iron oxide; CAS 1309-37-1; EINECS 215-168-2; pigment with good suspension and stabilization props. for coatings, automotive finishes, appliance enamels, paints, plastics, building prods., rubbers, inks; med. red fine powd., odorless; nil on 325 mesh; fineness (Hegman) 6+; insol. in water; sp.gr. 4.86; dens. 40.60 lb/gal; oil absorp. 19; pH 6.5; 96% Fe_2O_3; *Toxicology:* ACGIH TLV 5 mg/m^3; may cause irritation from abrasion on contact with skin; eye irritant; prolonged inh. produces siderosis; avoid breathing dust; *Precaution:* nonflamm.; incompat. with hydrazine, calcium hypochlorite, performic acid, bromine pentafluoride.

OSO® NR 830 M. [Hitox] Micronized natural red iron oxide; CAS 1309-37-1; EINECS 215-168-2; pigment for highly loaded coatings such as primers, industrial maintenance finishes, other exterior applics., building prods., rubber, and plastics; good for applics. requiring permanency of color; red fine powd., odorless; 0.8 μmedian particle size; < 0.001% on 325 mesh; fineness (Hegman) 6+; insol. in water; sp.gr. 4.31; dens. 36 lb/gal; bulk dens. 42.1 lb/ft^3 (loose); oil absorp. 21-23; m.p. 1840 C; pH 8.5; 76.6% Fe_2O_3; *Toxicology:* ACGIH TLV 5 mg/m^3; may cause irritation from abrasion on contact with skin; eye irritant; long-term exposure to silica causes silicosis; avoid breathing dust; *Precaution:* nonflamm.

OSO® NR 950. [Hitox] Natural red iron oxide; CAS 1309-37-1; EINECS 215-168-2; pigment for highly loaded coatings such as primers, industrial maintenance finishes, other exterior applics., building prods., rubber, and plastics; good for applics. requiring permanency of color; deep red fine powd., odorless; 3.2 μmedian particle size; 0.02% on 325 mesh; fineness (Hegman) 7; insol. in water; sp.gr. 5.0; dens. 41.7 lb/gal; oil absorp. 14; m.p. 1840 C; pH 6.5-7.5; 94.9% Fe_2O_3; *Toxicology:* ACGIH TLV 5 mg/m^3; may cause irritation from abrasion on contact with skin; eye irritant; long-term exposure to silica causes silicosis; avoid breathing dust; *Precaution:* nonflamm.

Ottalume 2100. [Ferro] Fluorescent zinc oxide; CAS 1314-13-2; EINECS 215-222-5; uv stabilizer in plastics and paints; opacifier in clear plastics where lt. transmittance is needed, and the lt. needs to be highly scattered; wh. daylight, fluorescent lt. blue, micron-sized powd.; Unverified

Oulu 356. [Veitsiluoto Oy] Polymerization emulsifier.

Oulumer 70. [Veitsiluoto Oy] Tall oil rosin; tackifier for pressure-sensitive and hot-melt adhesives; raw material for high softening pt. rosin esters; also in printing inks.

Oulupale XB 100. [Veitsiluoto Oy] Pale rosin; tackifier with exc. thermal stability for disposable hot melts in hygiene sector, pressure-sensitive adhesives, bookbinding adhesives; compat. with EVA, SB, SIS, and NR elastomers and waxes.

Oulutac 20 D. [Veitsiluoto Oy] Modified tall oil rosin aq. disp.; plasticizer for other resin disps.; gives very soft, water-resist. tacky film; good storage and mech. stability and resist. against sl. freezing; visc. 400 mPa•s (20 C); soften. pt. 15 C; pH 9.0; 52% solids.

Oulutac 80 D/HS. [Veitsiluoto Oy] Glyceryl rosinate aq. disp.; tackifier for natural and syn. rubber-based latex adhesives; good mech. stability and resist. to sl. freezing.

Oxi-Chek 114. [Ferro] Hindered phenol; primary antioxidant for rubber polymers, polyolefins, acetals; cream-colored powd.; Unverified

Oxi-Chek 116. [Ferro] Hindered phenol; antioxidant for polyolefins, styrenics, elastomers, acrylics, adhesives, PVC; wh. to off-wh. powd.; Unverified

Oxi-Chek 414. [Ferro] Hindered phenol; antioxidant for elastomers, latexes, plastics, adhesives; wh. powd.; Unverified

Oxitol. [Shell] Ethylene glycol monoethyl ether; CAS 110-80-5; EINECS 203-804-1; med. boiling solv. with toluene dilution ratio of 4.9; strong solv. action on NC and alkyd resins; also for phenolics, Epon resins, various oils and dyes; used in mfg. of surf. coatings, esp. lacquers where it improves gloss, and in printing and dyeing textiles; used in leather finishing; coupling agent for soap-hydrocarbon systems; used in varnish removers and cleaning sol'ns.; colorless liq.; mild odor; water-misc.; sp.gr. 0.9256-0.9286; 0.10% max. water; Unverified

Oxp. [Cardinal Carolina] Polyethylene wax; CAS

9002-88-4; EINECS 200-815-3; internal/external lubricant for PVC pipe extrusion, film, sheet, profiles; solid.

OZ. [Georgia Marble] Calcium carbonate; screen filler with tightly controlled range of particle sizes; powd.; 5% on 6 mesh, 5% thru 14 mesh.

P

P®-10 Acid. [CasChem] Ricinoleic acid; CAS 141-22-0; EINECS 205-470-2; chemical intermediate; imparts lubricity and rust-proofing to sol. cutting oils; basis for grease, soaps, resin plasticizers, and ethoxylated derivs.; modifier for coatings and adhesive polymers; used in inks, coatings, plastics, cosmetics and toilet goods; FDA approval; Gardner 5 liq.; sp.gr. 0.940; visc. 4 stokes; pour pt. 10 F; acid no. 180; iodine no. (Wijs) 89; sapon. no. 186.

PA-57. [Akrochem] Lightly cross-linked natural rubber extended with a lt.-colored, nonstaining oil; processing aid for extrusions, calendering, and open steam curing; crumb form of rubber compressed into bales; 57% cross-linked rubber.

PA-80. [Akrochem] Lightly cross-linked natural rubber extended with a lt.-colored, nonstaining oil; processing aid when blended with natural rubber, SBR, neoprene, or nitrile rubber; for extrusions, calendering, and open steam cure; crumb form of rubber compressed into bales; 57% cross-linked rubber.

PAG DLTDP. [Witco/PAG] Dilauryl thiodipropionate; CAS 123-28-4; EINECS 204-614-1; antioxidant used for polyoelfins, thermoplastic elastomers, syn. rubber; antioxidant for cosmetics and pharmaceuticals; FDA regulated; wh. free-flowing powd.

PAG DMTDP. [Witco/PAG] Dimyristyl thiodipropionate; CAS 16545-54-3; antioxidant for polyolefins and other polymeric systems; FDA regulated; wh. flakes.

PAG DSTDP. [Witco/PAG] Distearyl thiodipropionate; CAS 693-36-7; EINECS 211-750-5; antioxidant for polyolefins and other polymeric systems where long term heat stability is required; also for pharmaceutical and cosmetic prods., oils, greases, and lubricants; FDA regulated.

PAG DTDTDP. [Witco/PAG] Ditridecyl thiodipropionate; CAS 10595-72-9; antioxidant for polyolefins and thermoplastic elastomers, esp. latexes where a liq. antioxidant/stabilizer is dispersed more effectively; water-wh. liq.

PAG DXTDP. [Witco/PAG] Distearyl thiodipropionate; CAS 693-36-7; EINECS 211-750-5; antioxidant for polyolefins and other polymeric systems where max. compat. is desired; FDA regulated; wh. flakes.

Paintable Mist No. P112-A. [IMS] 3 ± 2% Fluid release agent (CAS 68037-77-4) with 50-60% HFC-152A, 30-40% dimethyl ether, 5-15% aliphatic petrol. distillate (CAS 64742-89-8); non-CFC, non-ozone depleting low-conc. mold release for use wherever parts are to be painted, plated, printed, or stamped; effective for ABS, epoxy, PMMA, PC, polyester, polyethylene, polysulfone; clear mist with sl. ethereal odor as dispensed from aerosol pkg.; sl. sol. in water; sp.gr. > 1; vapor pressure 60 ± 10 psig; flash pt. < 0 F; > 95% volatile; *Toxicology:* inh. may cause CNS depression with dizziness, headache; higher exposure may cause heart problems; gross overexposure may cause fatality; eye and skin irritation; direct contact with spray can cause frostbite; *Precaution:* flamm. limits 1-18%; at elevated temps. (> 130 F) aerosol containers may burst, vent, or rupture; incompat. with strong oxidizers, strong caustics, reactive metals (Na, K, Zn, Mg, Al); *Storage:* store in cool, dry area out of direct sunlight; do not puncture, incinerate, or store above 120 F.

Paintable Neutral Oil Spray No. N1616. [IMS] 10-20% Mineral oil (CAS 8042-47-5) with 50-60% HFC 152a, 25-30% dimethyl ether, and 25-35% HCFC 141b; non-CFC nonsilicone mold release for all polyolefins and PVC, ABS, acetal, epoxy, nylon, PMMA, PC, PS, PU, rubber; minimizes environmental stress cracking and crazing; suitable for food pkg.; FDA 21CFR §178.3620; clear to wh. mist with sl. ethereal odor as dispensed from aerosol pkg.; sl. sol. in water; sp.gr. > 1; vapor pressure 60 ± 10 psig; flash pt. < 0 F; > 75% volatile; *Toxicology:* inh. may cause CNS depression with dizziness, headache; higher exposure may cause heart problems; gross overexposure may cause fatality; eye and skin irritation; direct contact with spray can cause frostbite; *Precaution:* flamm. limits 1.2-18%; at elevated temps. (> 130 F) aerosol containers may burst, vent, or rupture; incompat. with strong oxidizers, strong caustics, reactive metals (Na, K, Zn, Mg, Al); *Storage:* store in cool, dry area out of direct sunlight; do not puncture, incinerate, or store above 120 F.

Paintable Organic Oil Spray No. O316. [IMS] 1-5% Lecithin (CAS 8029-76-3) with 35-45% HFC-152a, 25-30% dimethyl ether, 25-35% HCFC-141b, 5-15% aliphatic petrol. distillate (CAS 64742-89-8); non-CFC nonsilicone paintable mold release and parting agent which minimizes environmental stress cracking and crazing on polyethylene; also effective for PP, ABS, acetal, nylon, PS, PVC, rubber; suitable for food pkg.; FDA 21CFR §184.1400; clear to wh. mist with sl. ethereal odor as dispensed from aerosol pkg.; sl. sol. in water; sp.gr. > 1; vapor pressure 55 ± 10 psig; flash pt. < 0 F; > 90% volatile; *Toxicology:* inh. may cause CNS depression with dizziness, headache; higher exposure may cause heart problems; gross overexposure may cause fatality; eye and skin irritation; direct contact with spray can cause frostbite; *Precaution:* flamm. limits 1.2-18%; at elevated temps. (> 130 F) aerosol containers may burst, vent, or rupture; incompat. with strong oxidizers, strong caustics, reactive metals (Na, K, Zn, Mg, Al); *Storage:* store in cool, dry area out of direct sunlight; do not puncture, incinerate, or store above 120 F.

Palamoll®. [BASF AG] Plasticizers for plasticized PVC prods. resist. to oil, gasoline and bitumen with little migration tendency.

Palatinol® 11. [BASF AG] Undecyl phthalate; plasticizer for plasticized PVC.

Palatinol® 91P. [BASF] High m.w. linear phthalate; primary plasticizer for PVC homopolymer systems; used in applics. that require high-temp. processing and heat aging property retention for wire and cable; elec. properties; APHA 30 max.; mild, char. odor; sol. < 0.01% sol. in water; sp.gr. 0.953-0.963; visc. 48 cps; acid no. 0.1 max.; ref. index 1.4815; flash pt. (COC) 210 C; 50 phr in PVC film: tens. str. 3040 psi; tens. elong. 360%; hardness (Durometer A) 78; 99.0% min. ester content.

Palatinol® 711. [BASF AG] Phthalic acid ester consisting of C7-11 alcohols; plasticizer with low volatility for plasticized PVC prods. with low brittle temps.

Palatinol® DBP. [BASF] Di-n-butylphthalate; CAS 84-74-2; EINECS 201-557-4; plasticizer for PVAc-based coatings and adhesives; used to bond paper and in lamination of vinyl wall coverings to gypsum board; NC lacquers and cellophane coatings; mfg. of smokeless powds. for military use, processing aid for nitrile rubbers; carrier solv. for pigments and dyes, solv. for printing ink vehicles, fixative for perfumes, modifier of phenolic laminates, component of floor waxes; APHA 35 max.; mild, char. odor; m.w. 278.35; b.p. 182 C; sol. < 0.01% in water; sp.gr. 1.045-1.051; dens. 8.7 lb/gal; visc. 20.1 cps; pour pt. -40 C; acid no. 0.07 max.; ref. index 1.4926; flash pt. (COC) 171 C.

Palatinol® DIDP. [BASF] Diisodecylphthalate; primary monomeric plasticizer for vinyls; used in high-temp. processing in applic. such as wire and cable formulations, automotive formulations; APHA 35 max.; mild, char. odor; m.w. 446.7; b.p. 260 C; sol. < 0.01% in water; sp.gr. 0.963-0.969; visc. 114.4 cps; pour pt. -40 C; acid no. 0.07 max.; ref. index 1.4835; flash pt. (COC) 232 C; 50 phr in PVC film: tens. str. 3290 psi; tens. elong. 290%; hardness (Durometer A) 85.

Palatinol® DOA. [BASF] Di (2-ethylhexyl) adipate; CAS 103-23-1; EINECS 203-090-1; plasticizer used in PVC, NC, and rubber; food contact applics.; plastisols; dip coating formulations; APHA 25 max.; mild odor; m.w. 370.58; b.p. 214 C; sol. < 0.01% in water; sp.gr. 0.924-0.930; visc. 13.7 cps; pour pt. -75 C; acid no. 0.07 max.; ref. index 1.4466; flash pt. (COC) 196; 99.0% min. ester content.

Palatinol® DOP. [BASF] Di (2-ethylhexyl) phthalate; general-purpose plasticizer for vinyl compositions; solvating power for PVC; APHA 30 max.; mild, char. odor; m.w. 390.57; b.p. 231 C; sol. < 0.01% in water; sp.gr. 0.982-0.986; visc. 81.4 cps; pour pt. -46 C; acid no. 0.07 max.; ref. index 1.4859; flash pt. (COC) 218 C; 50 phr in PVC film: tens. str. 3210 psi; tens. elong. 276%; hardness (Durometer A) 79; 99.0% min. ester.

Palatinol® FF21. [BASF] Dialkyl phthalate; primary plasticizer for vinyl chloride homopolymer and copolymer resins; for molding and extrusion of vinyl compds., highly-filled systems, and dip, slush, and rotational molding of plastisol/organosol systems; APHA 35 max.; mild, char. odor; b.p. 224 C; sol. < 0.01% sol. in water; sp.gr. 0.990-0.998; visc. 57 cps; pour pt. -36 C; acid no. 0.1 max.; ref. index 1.4781; flash pt. (COC) 210 C; 50 phr in PVC film: tens. str. 2865 psi; tens. elong. 330%; hardness (Durometer A) 77; 99.0% min. ester.

Palatinol® FF31. [BASF] Dialkyl phthalate; primary plasticizer for vinyl chloride homopolymer and copolymer resins; for molding and extrusion of vinyl compds., highly-filled systems, and dip, slush, and rotational molding of plastisol/organosol systems; used in film, sheeting, garden hose, massback carpeting and flooring, chemically expanded construction; APHA 35 max.; mild, char. odor; b.p. 220 C; sol. < 0.01% in water; sp.gr. 0.992-1.000; visc. 55 cps; ref. index 1.4875; flash pt. (COC) 210 C; 50 phr in PVC film: tens. str. 2870 psi; tens. elong. 325%; hardness (Durometer A) 74; 99.0% min. ester.

Palatinol® FF41. [BASF] Dialkyl phthalate; primary plasticizer for vinyl chloride homopolymer and copolymer resins; for molding and extrusion of vinyl compds., highly-filled systems, and dip, slush, and rotational molding of plastisol/organosol systems; high solvating; provides rheological properties to plastisols and organosols at high shear rates; used in adhesives and sealants; APHA 35 max.; mild char. odor; b.p. 217 C; sol. < 0.01% in water; sp.gr. 0.995-1.003; visc. 51 cps; pour pt. -37 C; acid no. 0.1 max.; ref. index 1.4877; flash pt. (COC) 210 C; 50 phr in PVC film: tens. str. 2850 psi; tens. elong. 335%; hardness (Durometer A) 73; 99.0% min. ester.

Palatinol® M. [BASF AG] Dimethyl phthalate; CAS 131-11-3; EINECS 205-011-6; plasticizer for nail polish, deodorants; supplementary plasticizer for paints and varnishes.

Palatinol® N. [BASF] Diisononyl phthalate; primary plasticizer for PVC, plastisols; used for film for the building trade and engineering; APHA 30 clear liq.; mild, char. odor; m.w. 418; b.p. 237 C (5 mm Hg); < 0.01% sol. in water; sp.gr. 0.973-0.978; visc. 80 cps; pour pt. -49 C; acid no. 0.07 max.; ref. index

1.484-1.488; flash pt. (COC) > 200 C; 50 phr in PVC, 70-mil film properties: tens. str. 2410 psi; tens. elong. 311%; 100%. mod. 1470 psi; hardness 82 (15 s, Durometer A); brittle temp. -30 C; 99.0% min. ester.

Palatinol® TOTM. [BASF] Tri (2-ethylhexyl) trimellitate; primary, monomeric plasticizer for PVC homopolymer and copolymer resins; used where extreme low volatility of a plasticizer is required; used in vinyl compositions, in high temp. wire insulation and critical automotive applics.; APHA 150 max.; mild, char. odor; b.p. 283 C; sol. < 0.01% in water; sp.gr. 0.986-0.994; visc. 260 cps; pour pt. -46 C; acid no. 0.2 max.; ref. index 1.485; flash pt. (COC) 250 C; 50 phr in PVC film: tens. str. 3050 psi; tens. elong. 310%; hardness (Durometer A) 89; 99.0% min ester.

Pale 4. [CasChem] Polymerized castor oil; pigment wetting/dispersing agent; plasticizer for resins, gums, polymers; lubricant, penetrant; coupling solv.; adhesion promoter; for cellulose lacquers, inks, adhesives, industrial lubricants, polishes, caulks, leather dressing, hydraulic fluids, rubber compding., gasket cement; FDA approval; Gardner 4 color; sp.gr. 0.998; visc. 32 stokes; pour pt. 5 F; acid no. 16; iodine no. 70; sapon. no. 212; hyd. no. 158.

Pale 16. [CasChem] Polymerized castor oil; pigment wetting/dispersing agent; plasticizer for resins, gums, polymers; lubricant, penetrant; coupling solv.; adhesion promoter; for cellulose lacquers, inks, adhesives, industrial lubricants, polishes, caulks, leather dressing, hydraulic fluids, rubber compding., gasket cement; FDA approval; Gardner 9 color; sp.gr. 1.025; visc. 250 stokes; pour pt. 25 F; acid no. 24; iodine no. 56; sapon. no. 237; hyd.no. 136.

Pale 170. [CasChem] Polymerized castor oil; pigment wetting/dispersing agent; plasticizer for resins, gums, polymers; lubricant, penetrant; coupling solv.; adhesion promoter; for cellulose lacquers, inks, adhesives, industrial lubricants, polishes, caulks, leather dressing, hydraulic fluids, rubber compding., gasket cement; FDA approval; Gardner 2 color; sp.gr. 0.970; visc. 11 stokes; pour pt. -5 F; acid no. 4; iodine no. 80; sapon. no. 184; hyd. no. 160.

Pale 1000. [CasChem] Polymerized castor oil; pigment wetting/dispersing agent; plasticizer for resins, gums, polymers; lubricant, penetrant; coupling solv.; adhesion promoter; for cellulose lacquers, inks, adhesives, industrial lubricants, polishes, caulks, leather dressing, hydraulic fluids, rubber compding., gasket cement; FDA approval; Gardner 9 color; sp.gr. 1.018; visc. 120 stokes; pour pt. 25 F; acid no. 20; iodine no. 59; sapon. no. 230; hyd. no. 139.

Paliogen®. [BASF AG] Org. pigments; for high-quality paints, coloring plastics, tin plate inks, artists' paints and crayons.

Paliotol®. [BASF AG] Org. pigments; for special surf. coatings, for coloring plastics, for prod. of artists' paints and crayons, for tin plate inks.

Pamolyn® 125. [Climax Fluids Additives] Oleic acid; CAS 112-80-1; EINECS 204-007-1; for conversion to soaps and sulfonates for use as textile processing aids, automotive additives, mold lubricants, surfactants, agents for prod. of syn. rubber.

Parabis. [Dow] Purer para-para grade of bisphenol A; CAS 80-05-7; EINECS 201-245-8; resin intermediate used to produce PC and polysulfone engineering thermoplastic resins with toughness, clarity, and low dens.; high strength glazing, rugged appliance and tool housings, wheels, and high heat lighting enclosures; antioxidants for rubber, soap, and brake fluids, and fungicides, stabilizers, and dyeability enhancers; wh. flakes; m.w. 228; sol. (g/100 g solv.) 85 g acetone, 33 g ether, 210 g methanol; sp.gr. 1.195; dens. 37-41 lb/ft^3; b.p. 220 C; f.p. 156.5 C min.; b.p. 220 C (4 mm Hg); suspended dust in air can be flamm.; flash pt. (COC) 415 F.

Parabolix® 100. [Merix] Degreaser, destaticizer for electronics, parabolic light fixtures; uv stabilizer, antistatic coating for ABS, acrylic, cellulosics, nylon, acetal, PC, PE, PP, PS, PVC.

Para-Flux® 4156. [C.P. Hall] ASTM Type 101 aromatic process oil; plasticizer for CR compds.; Gardner 18 liq.; sp.gr. 0.993 (60 F); f.p. -15 C; acid no. 0.13; flash pt. 235 C.

Paraloid® BTA-702. [Rohm & Haas] Methacrylate/butadiene styrene; impact modifier; FDA 21CFR §178.3790; Unverified

Paraloid® BTA-715. [Rohm & Haas] Methacrylate/butadiene styrene; impact modifier; FDA 21CFR §178.3790; Unverified

Paraloid® BTA-730. [Rohm & Haas] Methacrylate-butadiene-styrene; impact modifier for transparent semirigid and rigid vinyl film and sheet; provides toughness for nonweatherable applics.; high impact resist. and exc. clarity with good resist. to crease whitening; used for blister packs, thermoformed containers, pharmaceutical and food contact pkg.; FDA 21CFR §178.3790; wh. fine powd.; *Toxicology:* airborne dust may be irritating; *Precaution:* airborne dust may constitute an explosion hazard.

Paraloid® BTA-733. [Rohm & Haas] Methacrylate-butadiene-styrene; impact modifier for vinyl; provides superior toughness for nonweatherable applics.; most efficient modifier for clear vinyl; FDA 21CFR §178.3790.

Paraloid® BTA-751. [Rohm & Haas] Methacrylate-butadiene-styrene; impact modifier for vinyl; provides superior toughness for nonweatherable applics.; max. toughness in lower m.w./high flow formulations.

Paraloid® BTA-753. [Rohm & Haas] Methacrylate-butadiene-styrene; impact modifier for vinyl; provides superior toughness for nonweatherable applics.; high impact resist. at ambient and low temps. for opaque vinyl; FDA 21CFR §178.3790.

Paraloid® BTA-III-N2. [Rohm & Haas] Methacrylate/butadiene styrene; impact modifier for PVC; FDA

21CFR §178.3790.

Paraloid® EXL-3330. [Rohm & Haas] Butyl acrylate-based; CAS 141-32-2; EINECS 205-480-7; acrylic toughener for engineering plastics incl. PC, PBT and PET; for weatherable applics. requiring good color retention because of its acrylic composition; good inherent thermal stability, resist. to UV degradation, and balance of toughness props.

Paraloid® EXL-3361. [Rohm & Haas] Butyl acrylate-based; CAS 141-32-2; EINECS 205-480-7; acrylic toughener for engineering plastics; for weatherable applics. requiring good color retention because of its acrylic composition; enhanced thermal stability and good balance of props.

Paraloid® EXL-3387. [Rohm & Haas] Reactive butyl acrylate-based; CAS 141-32-2; EINECS 205-480-7; acrylic toughener for engineering plastics; for weatherable applics. requiring good color retention because of its acrylic composition; esp. recommended for glass-reinforced nylon.

Paraloid® EXL-3611. [Rohm & Haas] Reactive methacrylate/butadiene styrene; toughener for engineering plastics; superior low-temp. impact resist.; reactive toughener for nylon.

Paraloid® EXL-3647. [Rohm & Haas] Methacrylate-butadiene-styrene; toughener for engineering plastics; superior low-temp. impact resist.; improved thermal stability.

Paraloid® EXL-3657. [Rohm & Haas] Methacrylate/butadiene styrene; toughener; Unverified

Paraloid® EXL-3691. [Rohm & Haas] Methacrylate-butadiene-styrene; toughener for engineering plastics; superior low-temp. impact resist.; general purpose toughener.

Paraloid® EXL-4151. [Rohm & Haas] Acrylic-imide copolymer; blending additive; Unverified

Paraloid® EXL-4261. [Rohm & Haas] Acrylic-imide copolymer; blending additive; Unverified

Paraloid® EXL-5137. [Rohm & Haas] Acrylic; gloss reducer; Unverified

Paraloid® EXL-5375. [Rohm & Haas] Acrylic conc.; specialty modifier used to improve impact str. and nucleation esp. in C-PET for dual-ovenable trays.

Paraloid® HT-100. [Rohm & Haas] Specialty modifier for heat distort. resist. in vinyl.

Paraloid® HT-510®. [Rohm & Haas] Polyglutarimide acrylic copolymer; heat distort. temp. modifier for PVC; FDA approved; Unverified

Paraloid® K-120N. [Rohm & Haas] Methyl methacrylate polymer; CAS 9011-14-7; acrylic processing aid for rigid and plasticized vinyl; produces rapid fusion, homogeneous melt, increased melt str., ease of thermoformability; FDA 21CFR §178.3790; wh. lower dusting fine free-flowing powd.; sol. in MEK, toluene, cyclohexanone, ethylene dichlroide; sp.gr. 1.18; bulk dens. 9.8 lb/gal; ref. index 1.49; 0.5% volatiles; *Toxicology:* airborne dust may be irritating to eyes and respiratory tract, and to skin after prolonged/repeated exposure; *Precaution:* airborne dust may present a severe explosive hazard; keep away from open flames, radiant heaters, etc.

Paraloid® K-120ND. [Rohm & Haas] Acrylic; processing aid for vinyl; produces rapid fusion, homogeneous melt, increased melt str., and ease of thermoformability; superior clarity and dispersibility; FDA 21CFR §178.3790.

Paraloid® K-125. [Rohm & Haas] Acrylic; processing aid for vinyl; produces rapid fusion, homogeneous melt, increased melt str., and ease of thermoformability; highest efficiency; FDA 21CFR §178.3790.

Paraloid® K-130. [Rohm & Haas] Acrylic; processing aid for vinyl; produces rapid fusion, homogeneous melt, increased melt str., and ease of thermoformability; best balance of clarity and efficiency; FDA 21CFR §178.3790.

Paraloid® K-147. [Rohm & Haas] Acrylic; processing aid for vinyl; produces rapid fusion, homogeneous melt, increased melt str., and ease of thermoformability; for low m.w. vinyl.

Paraloid® K-175. [Rohm & Haas] Acrylic; processing aid for vinyl; produces fapid fusion, homogeneous melt, increased melt str., and ease of thermoformability; also lubricates to reduce sticking and plate-out; FDA 21CFR §178.3790.

Paraloid® K-400. [Rohm & Haas] Foam modifier for cellular PVC; most efficient grade allowing substantial dens. reduction and exc. surf.; improves processability and durability; for foam core pipe, foam profiles, rigid foam sheet.

Paraloid® KF-710. [Rohm & Haas] Acrylic polymer; specialty gloss control modifier used for a matte appearance on vinyl surfs.; for opaque vinyl applics. in the building industry and clear and opaque calendered film in pkg. applics.

Paraloid® KM-318F. [Rohm & Haas] Acrylic; modifiers for cellular PVC; high surf. quality; FDA 21CFR §178.3790.

Paraloid® KM-334. [Rohm & Haas] Acrylic; impact modifier for vinyl; offers toughness, color retention and weatherability for long-term outdoor exposure; most efficient and weatherable; FDA 21CFR §178.3790.

Paraloid® KM-390. [Rohm & Haas] Acrylic; impact modifier with processing aid for vinyl; offers toughness, color retention, and weatherability for long-term outdoor exposure; FDA 21CFR §178.3790.

Paraloid® PM-800. [Rohm & Haas] Acrylic; modifier, processing aid for enhancing thermoformability of PP; wh. translucent prilled bead; 1-3 mm bead diam.; melt flow 0.4 dg/min (230 C, 2.16 kg); bulk dens. 33-36 lb/ft^3; 0.1% water, 0.4% V.O.C.

Paraplex® G-25 70%. [C.P. Hall] Polyester sebacate; plasticizer for epoxies; Gardner 1 visc. liq.; sol. in hexane, toluene, kerosene; partly sol. in ethanol, min. oil; insol. in water; m.w. 8000; sp.gr. 0.997; visc. 2300 cps; f.p. < -20 C; acid no. 1.0; iodine no. nil; sapon. no. 318; flash pt. 2 C; ref. index 1.478; 0.04% moisture.

Paraplex® G-25 100%. [C.P. Hall] High m.w. polyester sebacate; plasticizer for epoxies, elec. tapes, high temp. insulation, coaxial cable, upholstery, coated fabric; Gardner 2 visc. liq.; sol. in hexane, toluene; partly sol. in kerosene, ethanol, min. oil;

insol. in water; m.w. 8000; sp.gr. 1.057; visc. 160,000 cps; f.p. 12 C; acid no. 1.4 ; iodine no. nil; sapon. no. 455; flash pt. 304 C; ref. index 1.468; 100% act., 0.04% moisture.

Paraplex® G-30. [C.P. Hall] Mixed dibasic acid polyester; plasticizer for epoxies; APHA 100 visc. liq.; sol. in hexane, toluene, kerosene, ethanol, acetone, min. oil; insol. in water; m.w. 800; sp.gr. 1.085; visc. 1400 cps; f.p. -29 C; acid no. 1.0; iodine no. nil; sapon. no. 415; flash pt. 257 C; ref. index 1.499; 0.06% moisture.

Paraplex® G-31. [C.P. Hall] Mixed dibasic acid polyester; plasticizer for epoxies; APHA 160 visc. liq.; sol. in hexane, toluene, kerosene, ethanol, acetone, min. oil; insol. in water; m.w. 1000; sp.gr. 1.11; visc. 5500 cps; f.p. -20 C; acid no. 1.0; iodine no. nil; sapon. no. 440; flash pt. 274 C; ref. index 1.504; 0.05% moisture.

Paraplex® G-40. [C.P. Hall] High m.w. polyester adipate; plasticizer for PVC, epoxies, gasoline hose, flooring, tapes, aprons; resists migration into rubber, oils, solvs.; APHA 150 visc. liq.; sol. in toluene; partly sol. in kerosene, ethanol, acetone, min. oil; insol. in water; m.w. 6000; sp.gr. 1.15; visc. 150,000 cps; f.p. -20 C; acid no. 1.5; iodine no. nil; sapon. no. 595; flash pt. 293 C; ref. index 1.469; 0.05% moisture.

Paraplex® G-41. [C.P. Hall] High m.w. polyester adipate; plasticizer for PVC, epoxies, hoses, pkg., tape, liners, gaskets; APHA 125 visc. liq.; sol. in toluene; partly sol. in kerosene, ethanol, acetone, min. oil; insol. in water; m.w. 5000; sp.gr. 1.12; visc. 100,000 cps; f.p. -28 C; acid no. 1.5; iodine no. nil; sapon. no. 550; flash pt. 288 C; ref. index 1.469; 0.04% moisture.

Paraplex® G-50. [C.P. Hall] Polyester adipate; pigment grinding medium; plasticizer for PVC, epoxies; for insulation, upholstery, window channels, liners, gaskets, coated fabrics; Gardner 1 visc. liq.; sol. in toluene; partly sol. in kerosene, ethanol, acetone, min. oil; insol. in water; m.w. 2200; sp.gr. 1.10; visc. 2700 cps; f.p. -18 C; acid no. 0.5; iodine no. nil; sapon. no. 536; flash pt. 268 C; ref. index 1.465; 0.06% moisture.

Paraplex® G-51. [C.P. Hall] Polyester adipate; plasticizer for PVC, epoxies; for wall cover, insulation tape, gaskets, apparel; Gardner 1 visc. liq.; sol. in toluene; partly sol. in kerosene, ethanol, acetone, min. oil; insol. in water; m.w. 2200; sp.gr. 1.10; visc. 2800 cps; f.p. -24 C; acid no. 1.2; iodine no. nil; sapon. no. 542; flash pt. 268 C; ref. index 1.464; 0.04% moisture.

Paraplex® G-54. [C.P. Hall] Polyester adipate; plasticizer for PVC, epoxies; gaskets, insulation, upholstery, wall cover; APHA 100 visc. liq.; sol. in toluene; partly sol. in kerosene, ethanol, acetone, min. oil; insol. in water; m.w. 3300; sp.gr. 1.092; visc. 5600 cps; f.p. 0 C; acid no. 1.1; iodine no. nil; sapon. no. 510; flash pt. 288 C; ref. index 1.466; 0.05% moisture.

Paraplex® G-56. [C.P. Hall] Polyester adipate; plasticizer for epoxies; Gardner 2 visc. liq.; sol. in toluene; partly sol. in kerosene, ethanol, acetone, min. oil; insol. in water; m.w. 4200; sp.gr. 1.12; visc. 11,000 cps; f.p. -10 C; acid no. 0.8; iodine no. nil; sapon. no. 573; flash pt. 293 C; ref. index. 1.466; 0.06% moisture.

Paraplex® G-57. [C.P. Hall] Polyester adipate; plasticizer for epoxies; APHA 100 visc. liq.; sol. in toluene; partly sol. in kerosene, ethanol, acetone, min. oil; insol. in water; m.w. 3400; sp.gr. 1.10; visc. 6800 cps; f.p. -10 C; acid no. 1.0; iodine no. nil; sapon. no. 526; flash pt. 277 C; ref. index. 1.466; 0.05% moisture.

Paraplex® G-59. [C.P. Hall] Polyester adipate; plasticizer for PVC, epoxies, NC, CAP, chlorinated rubber; Gardner 4 visc. liq.; sol. in toluene; partly sol. in kerosene, ethanol, acetone, min. oil; insol. in water; m.w. 4900; sp.gr. 1.13; dens. 9.39 lb/gal; visc. 20,000 cps; f.p. 7 C; acid no. 0.7; iodine no. nil; sapon. no. 555; flash pt. 293 C; ref. index 1.471; 0.05% moisture.

Paraplex® G-60. [C.P. Hall] Epoxidized soybean oil; CAS 8103-07-8; EINECS 232-391-0; polymeric type plasticizer for coating formulations based on PVC and copolymers, PS, NC, chlorinated rubber and paraffin; permanence in surface coating films under severe exposure, high plasticizing efficiency; good flexibility at low temp.; stabilizes materials with acid-producing components; Gardner 1 clear liq.; sol. in hexane, toluene, kerosene, acetone, min. oil; partly sol. in ethanol; insol. in water; m.w. 10000; sp.gr. 0.988; dens. 8.2 lb/gal; visc. 375 cps; f.p. 5 C; acid no. 0.2; iodine no. 9.0; sapon. no. 182; flash pt. 310 C; ref. index 1.471; 100% solids, 0.04% moisture.

Paraplex® G-62. [C.P. Hall] Epoxidized soybean oil; CAS 8013-07-8; EINECS 232-391-0; plasticizer and stabilizer in flexible and semiflexible vinyl compds., PS, chlorinated rubbers; for food pkg., general use; APHA 100 clear liq., low odor and taste; sol. in toluene, ethanol, acetone; partly sol. in hexane, kerosene, min. oil; m.w. 1000; sp.gr. 0.993; dens. 8.3 lb/gal; visc. 450 cps; f.p. 5 C; acid no. 0.2; iodine no. 0.9; sapon. no. 183; flash pt. 310 C; ref. index 1.471; 0.04% moisture.

Paraplex® RG-2. [Rohm & Haas] Oil-modified sebacic acid-type plasticizing coating resin; compatible with ethyl cellulose, polyvinyl butyral; used in cable lacquers, artificial leather, coated fabric formulations requiring exterior durability; used on rubber; Gardner 4-7; dens. 8.0 lb/gal; visc. (G-H) U; acid no. 22-35; Film properties (35% Paraplex/65% NC): tens. str. 8420 psi; tens. elong 5.8% (ultimate); 60 ±2% solids in toluol; also avail. in 100% solid form.; Unverified

Paraplex® RG-8. [Rohm & Haas] Oil-modified sebacic acid; all-purpose grinding medium for NC lacquers offering pigment wetting and stability chars.; sole plasticizer in NC films or in combination with castor oil; Gardner 4-7; dens. 8.3 lb/gal; visc. (G-H) Z2-Z3; acid no. 0-3.5; 100% solids; Unverified

Par® Clay. [R.T. Vanderbilt] Hard kaolin; CAS 1332-58-7; EINECS 296-473-8; reinforcer and inert filler for elastomers, nonblack compds.; cream powd.; 0.2 μm median particle size; 99.5% min. thru 325 mesh; dens. 2.62 ± 0.03 mg/m^3; bulk dens. 26 lb/ft^3 (compacted); oil absorp. 40; brightness 68; pH 4.8 (10% aq. slurry); 44.9% SiO_2, 37.8% Al_2O_3.

Paricin® 1. [CasChem] Methyl hydroxystearate; CAS 141-23-1; EINECS 205-471-8; lubricant/processing aid for butyl rubber; wax firming agent in cosmetics and specialty inks; source of hydroxystearic acid for glycerin-free lithium greases; sp.gr. 1.02; m.p. 52 C; acid no. 5; iodine no. (Wijs) 3; sapon. no. 179; hyd. no. 164; penetration hardness 19.

Paricin® 6. [CasChem] Butyl acetoxystearate; oxidation-stable plasticizer for vinyls; used in plastisols for non-exudation, easy deaeration and maintenance of low visc.; lubricant penetrant; Gardner 2 color; sp.gr. 0.918; visc. 1.3 stokes; acid no. 4; iodine no. (Wijs) 2; sapon. no. 276; hyd. no. 5; pour pt. 20 F.

Paricin® 8. [CasChem] Glyceryl tri(acetoxystearate); CAS 139-43-5; EINECS 295-625-0; lubricant plasticizer for vinyls, esp. high-temp. wire jacketing; heat and oxidation-stable; grinding med. for pigment dispersions; plasticizer for nitrocellulose; high-temp. lubricant for metals; Gardner 1 color; sp.gr. 0.955; visc. 1.8 stokes (37.8 C); acid no. 1; iodine no. (Wijs) 2; sapon. no. 298; hyd. no. 5; pour pt. 30 F.

Paricin® 210. [CasChem] N-Stearyl 12-hydroxystearamide; lubricant/antistat for plastics, metals; mold release, antiblocking agent for textile coatings; slip agent for varnishes and lacquers; also for elec. potting compds., crayons, wax blends, high-temp. greases; release agent in contact with food; FDA approval; sp.gr. 0.98; m.p. 102 C; acid no. 1; iodine no. (Wijs) 5; sapon. no. 10; hyd. no. 93; penetration hardness 1.

Paricin® 220. [CasChem] N (2-Hydroxyethyl) 12-hydroxystearamide; internal mold release agent, lubricant, antistat for polyolefins, PVC, styrenics, metals; antiblocking agent for textile coatings; slip agent for varnishes and lacquers; also in elec. potting compds., plug valve lubricants, crayons, wax blends, high-temp. greases; FDA approval; solid wax; sp.gr. 1.05; m.p. 104 C; acid no. 1; iodine no. (Wijs) 4; sapon. no. 14; hyd. no. 300.

Paricin® 285. [CasChem] N,N′-Ethylene bis 12-hydroxystearamide; internal lubricant, mold release, slip additive for PVC; antiblocking agent for textile coatings; slip agent for varnishes, lacquers; also in elec. potting compds., plug valve lubricants, caryons, wax blends, high-temp. greases; FDA approval; solid wax; sp.gr. 1.02; m.p. 140 C; acid no. 4; iodine no. (Wijs) 4; sapon. no. 14; hyd. no. 165; penetration hardness 1.

Paroil® 140. [Dover] Chlorinated paraffin; plasticizer for polychloroprene and other polymers.

Parolite. [Henkel/Functional Prods.; Henkel/Textiles] Zinc formaldehyde sulfoxylate; CAS 24887-06-7; reducing agent for redox-catalyzed polymerization; stripping agent for removing dyes from wool and nylon; powd.; 100% solids.

Patcote® 512. [Am. Ingreds.] Silicone-containing; defoamer for use in urethane-modified resins; lt. amber liq.; sp.gr. 0.892 ± 0.012; dens. 7.43 ± 0.1 lb/gal; pour pt. < -30 F; PMCC flash pt. 143 F; 100% act.; *Toxicology:* eye or skin irritant on prolonged contact; do not take internally.

Patcote® 513. [Am. Ingreds.] Silicone-containing; defoamer for use in water-reducible acrylic coatings; wh. cloudy liq.; sp.gr. 0.830 ± 0.012; dens. 6.91 ± 0.1 lb/gal; pour pt. < -30 F; PMCC flash pt. 145 F; 100% act.; *Toxicology:* eye or skin irritant on prolonged contact; do not take internally.

Patcote® 519. [Am. Ingreds.] Silicone-containing; defoamer for use in trade sales and industrial acrylic lacquers; wh. cloudy liq.; sp.gr. 0.840 ± 0.012; dens. 7.00 ± 0.1 lb/gal; pour pt. < -30 F; PMCC flash pt. 148 F; 100% act.; *Toxicology:* eye or skin irritant on prolonged contact; do not take internally.

Patcote® 520. [Am. Ingreds.] Silicone-containing; defoamer for use in water-reducible alkyd, industrial acrylic, and urethane resins; wh. cloudy liq.; sp.gr. 0.839 ± 0.012; dens. 6.99 ± 0.1 lb/gal; pour pt. < -30 F; PMCC flash pt. 160 F; cloud pt. 77-85 F; 100% act.; *Toxicology:* eye or skin irritant on prolonged contact; do not take internally.

Patcote® 525. [Am. Ingreds.] Silicone-containing; defoamer for use in water-reducible alkyd and industrial acrylic systems; clear, slightly opalescent liq.; sp.gr. 0.822 ± 0.012; dens. 6.85 ± 0.1 lb/gal; pour pt. < -30 F; PMCC flash pt. 158 F; 100% act.; *Toxicology:* eye or skin irritant on prolonged contact; do not take internally.

Patcote® 531. [Am. Ingreds.] Silicone-containing; defoamer for water-reducible acrylic systems; wh. cloudy liq.; dens. 6.98 lb/gal; pour pt. < 30 F; flash pt. (PMCC) 158 F; 100% active.; *Toxicology:* eye or skin irritant on prolonged contact; do not take internally.

Patcote® 550. [Am. Ingreds.] Silicone-containing; defoamer for use in water-reducible alkyds and acrylics as well as acrylic latices for both industrial and trade sales formulations; wh. slightly cloudy liq.; sp.gr. 0.874 ± 0.012; dens. 7.28 ± 0.1 lb/gal; pour pt. < -30 F; PMCC flash pt. 150 F; 100% act.; *Toxicology:* eye or skin irritant on prolonged contact; do not take internally.

Patcote® 555. [Am. Ingreds.] Silicone-containing; defoamer for acrylic latex systems for trade sales; milky wh. liq.; sp.gr. 0.994; dens. 8.28 lb/gal; pour pt. 30 F; flash pt. (PMCC) < 300 F; 100% active.; *Toxicology:* eye or skin irritant on prolonged contact; do not take internally; *Storage:* may settle on aging, stir before use.

Patcote® 577. [Am. Ingreds.] Silicone-containing; defoamer for use in PVA resins for trade sales and water-reducible alkyds for industrial coatings; lt. amber cloudy liq.; sp.gr. 0.826 ± 0.012; dens. 6.88 ± 0.1 lb/gal; pour pt. < -30 F; PMCC flash pt. 140 F; 100% act.; *Toxicology:* eye or skin irritant on prolonged contact; do not take internally.

Patcote® 597. [Am. Ingreds.] Silicone-containing;

defoamer for water-reducible alkyds and acrylic emulsions; sl. cloudy liq.; sp.gr. 0.814; dens. 6.78 lb/gal; pour pt. < 30 F; flash pt. (PMCC) 155 F; 100% active.; *Toxicology:* eye or skin irritant on prolonged contact; do not take internally; *Precaution:* keep away from open flame.

Patcote® 598. [Am. Ingreds.] Silicone; defoamer for use in oil-modified urethanes and water-reducible alkyds; amber cloudy liq.; sp.gr. 0.810 ± 0.012; dens. 6.75 ± 0.1 lb/gal; pour pt. < -30 F; PMCC flash pt. 159 F; 100% act.

Patcote® 801. [Am. Ingreds.] Nonsilicone; defoamer for use in PVA-acrylic copolymers and terpolymer emulsions for trade sales; FDA clearance; wh. opaque liq.; sp.gr. 0.911 ± 0.012; dens. 7.59 ± 0.1 lb/gal; pour pt. -5 F; flash pt. (PMCC) 300 F; 100% act.; *Toxicology:* eye or skin irritant on prolonged contact; do not take internally.

Patcote® 802. [Am. Ingreds.] Nonsilicone; defoamer for use in acrylic and terpolymer emulsions in trade sales; whitish amber cloudy liq.; sp.gr. 0.905 ± 0.012; dens. 7.54 ± 0.1 lb/gal; pour pt. 5 F; flash pt. (PMCC) > 300 F; 100% act.

Patcote® 803. [Am. Ingreds.] Nonsilicone; defoamer for acrylic and terpolymer emulsions for trade sales; FDA compliance; wh. opaque liq.; sp.gr. 0.896; dens. 7.47 lb/gal; pour pt. 5 F; flash pt. (PMCC) > 300 F; 100% active.; *Toxicology:* eye or skin irritant on prolonged contact; do not take internally.

Patcote® 806. [Am. Ingreds.] Nonsilicone; defoamer for acrylic and terpolymer emulsions in trade sales and some industrial applics.; FDA clearance; FDA §175.105, 175.300, 176.170, 176.180; lt. amber liq.; sp.gr. 0.8787; dens. 7.32 lb/gal; pour pt. 5 F; flash pt. (PMCC) > 300 F; 100% active.; *Toxicology:* eye or skin irritant on prolonged contact; do not take internally; *Storage:* prod. settles, stir before use.

Patcote® 811. [Am. Ingreds.] Nonsilicone; defoamer for graphic arts water-based acrylic systems; creamy opaque; mild oily odor; sp.gr. 0.924; dens. 7.70 lb/gal; pour pt. 34-37 F; flash pt. (PMCC) > 300 F; 100% active.; *Toxicology:* eye or skin irritant on prolonged contact; do not take internally; *Storage:* may show separation, stir before use.

Patcote® 812. [Am. Ingreds.] Nonsilicone; defoamer for nonionic and anionic acrylic emulsion systems; creamy opaque; mild oily odor; sp.gr. 0.924; dens. 7.70 lb/gal; pour pt. 34-37 F; flash pt. (PMCC) > 300 F; 100% active.; *Toxicology:* eye or skin irritant on prolonged contact; do not take internally; *Storage:* may show separation, stir before use.

Patcote® 841M. [Am. Ingreds.] Nonsilicone; defoamer for acrylic, terpolymer, and PVC systems for trade sales paints, screen printing inks; amber opaque; sp.gr. 0.879; dens. 7.32 lb/gal; pour pt. -5 F; flash pt. (PMCC) > 300 F; *Toxicology:* eye or skin irritant on prolonged contact; do not take internally; *Storage:* may show sl. separation on extended storage, stir before use.

Patcote® 845. [Am. Ingreds.] Nonsilicone; defoamer for solvent, high solids and water-based acrylic systems; clear liq.; sp.gr. 0.8643; dens. 7.20 lb/gal; pour pt. < -30 F; flash pt. (SCC) 97 F; flamm.; 100% active.; *Toxicology:* eye or skin irritant on prolonged contact; do not take internally; *Precaution:* flamm.; *Storage:* will not settle on standing.

Patcote® 847. [Am. Ingreds.] Nonsilicone; defoamer for use in both solvent- and water-based alkyds and high solids systems; clear liq.; sp.gr. 0.864 ± 0.012; dens. 7.20 ± 0.1 lb/gal; pour pt. < -30 F; flash pt. (PMCC) 107 F; flamm.; 100% act.; *Toxicology:* eye or skin irritant on prolonged contact; do not take internally; *Precaution:* flamm.; keep away from open flame; *Storage:* will not settle on standing.

Patcote® 883. [Am. Ingreds.] Nonsilicone; defoamer for use in trade sales acrylic emulsions; amber cloudy liq.; sp.gr. 0.871 ± 0.012; dens. 7.26 ± 0.1 lb/gal; pour pt. 20 F; flash pt. (PMCC) 395 F; 100% act.; *Toxicology:* eye or skin irritant on prolonged contact; do not take internally; *Precaution:* keep away from open flame.

Pationic® 900. [Am. Ingreds./Patco] Glyceryl monostearate/oleate from animal sources; antistat for polyolefins; mold release for PP, PE, TPO, TPE; internal lubricant for PVC; enhances melt flow in polyolefins, vinyls, styrenics, TPE, TPO; color dispersant, pigment wetter in various polymers; FDA 21CFR §184.1324, GRAS; ivory wh. fine beads; m.p. 66 C; 96% monoglyceride.

Pationic® 901. [Am. Ingreds./Patco] Dist. glyceryl stearate derived from fully hydrog. veg. oil; antistat for polyolefins; mold release for PP; internal lubricant for PVC; enhances melt flow in polyolefins, vinyls, styrenics, TPE, TPO; color dispersant, pigment wetter in various polymers; FDA 21CFR §184.1324, GRAS; ivory wh. fine beads; m.p. 72 C; 96% monoglyceride.

Pationic® 902. [Am. Ingreds./Patco] Dist. glyceryl stearate derived from fully hydrog. animal fat; antistat and mold release agent for polyolefins; internal lubricant for rigid PVC; melt flow enhancer, pigment dispersion/wetting for polymer systems; FDA 21CFR §184.1324, GRAS; ivory wh. fine beads; m.p. 68 C; 96% monoglyceride.

Pationic® 905. [Am. Ingreds./Patco] Dist. glyceryl stearate; antistat in polyolefins; mold release in PP; internal lubricant in PVC (enhances dynamic heat stability in rigid compds.); melt flow enhancer in polyolefins, vinyls, styrenics, TPEs and TPOs; colorant dispersant, pigment wetting in polymers; FDA 21CFR §184.1324, GRAS; ivory wh. fine beads; m.p. 69 C; 96% monoglyceride.

Pationic® 907. [Am. Ingreds./Patco] Glyceryl oleate from veg. sources; internal lubricant for rigid PVC (bottles, clear sheet extrusion, calendered film, profile extrusion, inj. molding compds.; enhanced dynamic heat stability); antifog for PVC and polyolefins; antistat for polyolefins; FDA 21CFR §184.1505, GRAS; ivory wh. liq. @ 40-45 C, soft paste at ambient temps.; 96% monoglyceride.

Pationic® 909. [Am. Ingreds./Patco] Dist. glyceryl stearate derived from fully hydrog. veg. oil; additive for polymer systems and compds. requiring dry blending or surf. coating of the additive directly onto resin, pellets, powd., or beads; processing aid in EPS; color dispersant, lubricant, binder; color conc. distribution; FDA 21CFR §184.1324, GRAS; ivory wh. fine beads; 98% thru 100 mesh; m.p. 72 C; 96% monoglyceride.

Pationic® 914. [Am. Ingreds./Patco] Glyceryl mono/tristearate derived from fully hydrog. veg. oil; processing aid for expanded PS, providing lubrication and antistatic props., minimizing lumping and caking; colorant dispersant and lubricant; binder for dry colorant formulations; improves flow and distribution of color concs.; for dry blending or surface coating applic.; FDA 21CFR §182.70, 184.1324, GRAS; ivory wh. very fine beads; m.p. 65 C; 65% monoglyceride.

Pationic® 919. [Am. Ingreds./Patco] Glyceryl tristearate; CAS 555-43-1; EINECS 209-097-6; lubricant, anticaking agent, processing aid for expandable PS, dispersant for colorants; for dry blending or surface coating onto substrates; FDA 21CFR §182.70; ivory wh. fine beads; 98% thru 100 mesh; m.p. 65 C.

Pationic® 920. [Am. Ingreds./Patco] Sodium stearoyl lactylate; CAS 25383-99-7; EINECS 246-929-7; acid and catalyst scavenger/neutralizer, internal/external lubricant for flame retardant compds., filled and unfilled polymers, color concs.; for polyolefins, PVC; FDA 21CFR §172.846; off-wh. powd.; 99% thru 20 mesh; m.p. 47-53 C; sapon. no. 210-235; 3.5-5.0% Na.

Pationic® 925. [Am. Ingreds./Patco] Calcium/sodium stearoyl lactylate; acid neutralizer/acceptor/scavenger for polyolefins, flame retardants, pigments; lubricant for PVC, polyolefins, filled compds.; FDA 21CFR §172.844, 172.846; off-wh. powd.; 99% thru 20 mesh; m.p. 45-55 C; sapon. no. 195-230.

Pationic® 930. [Am. Ingreds./Patco] Calcium stearoyl-2-lactylate; CAS 5793-94-2; EINECS 227-335-7; acid neutralizer/acceptor/scavenger for polyolefins, flame retardants, pigments; lubricant for PVC, polyolefins, filled compds.; FDA 21CFR §172.844; off-wh. powd.; 99% thru 20 mesh; m.p. 45-55 C; sapon. no. 195-230; 4.2-5.2% Ca.

Pationic® 930K. [Am. Ingreds./Patco] Calcium stearoyl-2-lactylate; CAS 5793-94-2; EINECS 227-335-7; acid neutralizer/acceptor/scavenger for polyolefins, flame retardants, pigments; lubricant for PVC, polyolefins, filled compds.; direct food additive; FDA 21CFR §172.844; off-wh. powd.; m.p. 45-55 C; 4.5% Ca.

Pationic® 940. [Am. Ingreds./Patco] Calcium stearoyl-2-lactylate; CAS 5793-94-2; EINECS 227-335-7; acid/catalyst neutralizer, lubricant for polyolefins and other thermoplastic systems; neutralizes corrosive acids; melt flow enhancer and stabilizer; FDA 21CFR §172.844; ivory free-flowing powd.; basic pH; 12% Ca.

Pationic® 1019. [Am. Ingreds./Patco] Glyceryl tristearate; CAS 555-43-1; EINECS 209-097-6; internal/external lubricant for PVC (minimizes sticking, enhances processing); colorant dispersant and lubricant; binder for dry colorant formulations; color concs. distribute more readily; FDA 21CFR §182.70; ivory wh. very fine flakes; m.p. 65 C.

Pationic® 1042. [Am. Ingreds./Patco] Glyceryl stearate; antistat, mold release agent, flow modifier, internal lubricant for polyolefins, styrenics, PVC, TPEs, and TPOs; FDA 21CFR §184.1505, GRAS; ivory wh. very fine flakes; m.p. 60-63 C; 43% monoglyceride.

Pationic® 1042B. [Am. Ingreds./Patco] Glyceryl stearate; antistat, mold release agent, flow modifier, internal lubricant for polyolefins, styrenics, PVC, TPEs, and TPOs; FDA 21CFR §184.1505, GRAS; ivory wh. fine beads; m.p. 61 C; 43% α-monoglyceride.

Pationic® 1042K. [Am. Ingreds./Patco] Glyceryl stearate, kosher; antistat, mold release, flow modifier, internal lubricant for polyolefins, styrenics, PVC, TPEs, and TPOs; FDA 21CFR §184.1505, GRAS; ivory wh. very fine flakes; m.p. 62 C; 43% α-monoglyceride.

Pationic® 1042KB. [Am. Ingreds./Patco] Glyceryl stearate, kosher; antistat, mold release, flow modifier, internal lubricant for polyolefins, styrenics, PVC, TPEs, and TPOs; FDA 21CFR §184.1505, GRAS; ivory wh. fine beads; m.p. 62 C; 43% α-monoglyceride.

Pationic® 1052. [Am. Ingreds./Patco] Glyceryl stearate from animal sources; antistat, mold release agent, flow modifier, internal lubricant for polyolefins, styrenics, PVC, TPEs, and TPOs; FDA 21CFR §184.1505, GRAS; ivory wh. very fine flakes; m.p. 62 C; 53% α-monoglyceride.

Pationic® 1052B. [Am. Ingreds./Patco] Glyceryl stearate; antistat, mold release agent, flow modifier, internal lubricant for polyolefins, styrenics, PVC, TPEs, and TPOs; FDA 21CFR §184.1505, GRAS; ivory wh. fine beads; m.p. 62 C; 53% α-monoglyceride.

Pationic® 1052K. [Am. Ingreds./Patco] Glyceryl stearate from veg. sources, kosher; antistat, mold release, flow modifier, internal lubricant for polyolefins, styrenics, PVC, TPEs, and TPOs; FDA 21CFR §184.1505, GRAS; ivory wh. very fine flakes; m.p. 62 C; 53% α-monoglyceride.

Pationic® 1052KB. [Am. Ingreds./Patco] Glyceryl stearate, kosher; antistat, mold release, flow modifier, internal lubricant for polyolefins, styrenics, PVC, TPEs, and TPOs; FDA 21CFR §184.1505, GRAS; ivory wh. fine beads; m.p. 62 C; 53% α-monoglyceride.

Pationic® 1061. [Am. Ingreds./Patco] Glyceryl oleate from veg. sources; internal lubricant for PVC; helps maintain clarity and early color hold; in PVC antifogging formulations to promote compat.; antistat, flow modifier, dispersant in polyolefins; also additive to styrenics; FDA 21CFR §184.1555, GRAS; amber visc. liq. @ 19 C; m.p. 39-41 C; 44% α-monoglyceride.

Pationic® 1064. [Am. Ingreds./Patco] Glyceryl oleate from veg. sources; CAS 111-03-5; internal lubricant, antifog aid, antistat, flow modifier, dispersant for PVC, polyolefins, and styrenics; helps maintain clarity and early color hold; FDA 21CFR §184.1505, GRAS; ivory wh. semisolid, liq. above its m.p. 39-41 C; 44% α-monoglyceride.

Pationic® 1074. [Am. Ingreds./Patco] Glyceryl oleate from veg. sources; CAS 111-03-5; internal lubricant, antifog aid, antistat, flow modifier, additive dispersant for PVC, polyolefins, and styrenics; FDA 21CFR §184.1505, GRAS; ivory wh. semisolid, liq. above its m.p. 48-50 C; 43% α-monoglyceride.

Pationic® 1145. [Am. Ingreds./Patco] Mixt. of glycerol esters derived from hydrog. veg. oil; antistat, processing aid for EPS providing lubrication, minimizing lumping and caking; colorant dispersant, binder, and lubricant; for use in polymer systems requiring dry blending or surf. coating directly onto resin, compd., pellets, powd., or bead; FDA 21CFR §184.1324, GRAS; ivory wh. very fine beads; m.p. 61 C; 100% glycerides.

Pationic® 1230. [Am. Ingreds./Patco] Calcium lactate; CAS 814-80-2; EINECS 212-406-7; acid/catalyst neutralizer for polyolefins and other thermoplastic systems; neutralizes corrosive acids in polymers; melt flow stabilizer; minimizes polymer color development; FDA 21CFR §184.1207, GRAS; wh. free-flowing powd.; pH 7.1 (1%); 13% Ca.

Pationic® 1240. [Am. Ingreds./Patco] Calcium lactate; CAS 814-80-2; EINECS 212-406-7; acid/catalyst neutralizer for polyolefins and other thermoplastic systems; neutralizes corrosive acids in polymers; melt flow stabilizer; minimizes polymer color development; FDA 21CFR §184.1207, GRAS; wh. free-flowing powd.; pH basic; 19.5% Ca.

Pationic® 1250. [Am. Ingreds./Patco] Calcium lactate; CAS 814-80-2; EINECS 212-406-7; acid/catalyst neutralizer in polyolefins and other thermoplastics; provides corrosion resist. by neutralizing mildly and strongly acidic catalyst residues; provides improved melt flow stability, minimizes polymer color development; FDA 21CFR §172.844, 184.1207; wh. free-flowing powd.; pH basic; 15.5% Ca.

Pationic® AS38. [Am. Ingreds./Patco] Proprietary ester derived from natural oils; antistat for PVC resin to inhibit the generation of static charge; enhances heat stability of PVC resin; FDA 21CFR §174.5; avail. in kosher version; liq.; sp.gr. 0.95; dens. 7.89 lb/gal; surf. tens. 32.34 dynes/cm (0.1%); 100% conc.; *Toxicology:* LD50 (oral, rat) > 2.5 g/kg; nontoxic by ingestion; not a primary skin irritant; nonirritating to eyes.

PE-AB 07. [Additive Polymers Ltd] 10% Natural silica in LLDPE-based masterbatch; antiblock agent for plastics.

PE-AB 19. [Additive Polymers Ltd] 15% Natural silica in LLDPE-based masterbatch; antiblock agent for plastics.

PE-AB 22. [Additive Polymers Ltd] 40% Calcium carbonate in LLDPE-based masterbatch; antiblock agent for plastics.

PE-AB 30. [Additive Polymers Ltd] 30% Natural silica in LLDPE-based masterbatch; antiblock agent for plastics.

PE-AB 33. [Additive Polymers Ltd] 5% Precipitated silica in LLDPE-based masterbatch; antiblock agent for plastics.

PE-AB 48. [Additive Polymers Ltd] 50% Natural silica in LLDPE-based masterbatch; antiblock agent for plastics.

PE-AB 73. [Additive Polymers Ltd] 25% Natural silica in LLDPE-based masterbatch; antiblock agent for plastics.

PE-AO 51. [Additive Polymers Ltd] 10% Phenolic in LLDPE-based masterbatch; antioxidant for plastics.

PE-AO 91. [Additive Polymers Ltd] 8% Phosphite and 2% phenolic in LLDPE-based masterbatch; antioxidant for plastics.

PE-AS 05. [Additive Polymers Ltd] 2.5% Ethoxylated amine in LLDPE-based masterbatch; antistat for plastics.

PE-AS 74. [Additive Polymers Ltd] 5% Ethoxylated amine in LLDPE-based masterbatch; antistat for plastics.

PE-AS 109. [Additive Polymers Ltd] 5% Ethoxylated amine in LLDPE-based masterbatch; antistat for plastics.

PeFlu 727. [Astor Wax] PTFE wax; CAS 9002-84-0; internal/external lubricant for thermoplastics, thermosets, elastomers, coatings; solid.

Pegosperse® 50 MS. [Lonza] Glycol stearate; CAS 111-60-4; EINECS 203-886-9; nonionic; surfactant, opacifier, pearlescent, aux. emulsifier for cosmetics (hair and skin care prods.), bath and shower prods., household formulations; dispersant, emulsifier for o/w emulsions for industrial, textile, plastics, water treatment; wh. beads; sol. in methanol, ethanol, acetone, ethyl acetate, toluol, naphtha, min. and veg. oils; sp.gr. 0.96; m.p. 58 C; HLB 2; acid no. < 5; iodine no. < 1; sapon. no. 170-190; pH 5.0 (3% aq.); 100% conc.

Pegosperse® 100 L. [Lonza] PEG-2 laurate SE; nonionic; self-dispersing surfactant, w/o or o/w emulsifier for cosmetic creams and lotions, textile spin finishes and lubricants, solv. and solventless coatings, paper defoamer formulations; internal antistat for PP; FDA approved as indirect food additive; straw clear liq.; sol. in ethanol, toluol, naphtha, min. oil; disp. in water; sp.gr. 0.97; HLB 7; solid. pt. < 13 C; acid no. < 4; sapon. no. 160-170; pH 9 (5% aq.); 100% conc.

Pegosperse® 100-S. [Lonza] PEG-2 stearate SE; CAS 106-11-6; nonionic; self-emulsifying surfactant; thickener for cosmetic creams and lotions; dispersant and thickener for household polishes, waxes, pastes; antitack agent in rubber molding lubricants; off-wh. beads; sol. in methanol, ethanol, toluol, naphtha, min. and veg. oils; disp. in water; sp.gr. 0.96; m.p. 52 C; HLB 4; acid no. 95-105; sapon. no. 165-175; pH 7 (5%); 100% conc.

Pegosperse® 200 DL. [Lonza] PEG-4 dilaurate; CAS 9005-02-1; nonionic; surfactant, dispersant, emulsifier for cosmetic bath oils, lotions, household pastes, polishes; emulsifier, dispersant, opacifier, visc. control agent, defoamer for textiles, plastics, water treatment; yel. clear liq.; sol. in ethanol, min.

and veg. oil; disp. in water; sp.gr. 0.96; solid. pt. 3 C; HLB 7 ± 1; acid no. 5 max.; sapon. no. 170-185; 100% conc.

Pegosperse® 200 ML. [Lonza] PEG-4 laurate; CAS 9004-81-3; nonionic; dispersant, emulsifier, emollient, visc. modifier for cosmetics (bath oils, lotions, cream rinses, shampoos); visc. modifier for PVC plastisols; defoamer for aq. coatings; emulsifier for textile spin finish formulations; lt. yel. clear liq.; sol. in methanol, ethanol, acetone, ethyl acetate, toluol; misc. with water, sp.gr. 0.99; HLB 9; solid pt. < 5; acid no. 5 max.; sapon. no. 149-159; pH 5 (5% aq.); 100% conc.

Pegosperse® 400 DL. [Lonza] PEG-8 dilaurate; CAS 9005-02-1; nonionic; surfactant, emulsifier for cosmetic lotions, bath oils, formulations requiring clarity; provides low temp. flexibility and visc. modification for PVC plastisols; lubricant in textile spin finishes; yel. clear liq.; sol. in methanol, ethanol, acetone, ethyl acetate, toluol, naphtha, min. and veg. oils; sp.gr. 0.99; solid. pt. 5-12 C; HLB 15; acid no. < 5; iodine no. 7.5; sapon. no. 130-140; pH 4 (5% aq.); 100% conc.

Pegosperse® 400 DO. [Lonza] PEG-8 dioleate; CAS 9005-07-6; nonionic; surfactant for cosmetic lotions, bath oils; visc. modifier for PVC plastisols; yel. clear liq.; sol. in methanol, ethanol, ethyl acetate, toluol, naphtha, min. and veg. oils; disp. in water; sp.gr. 0.97; solid. pt. < 0 C; HLB 8.0 ± 0.5; acid no. < 10; iodine no. 52; sapon. no. 115-125; pH 5.5 (5% aq.); 100% conc.

Pegosperse® 400 ML. [Lonza] PEG-8 laurate; CAS 9004-81-3; nonionic; surfactant, emulsifier, dispersant for cosmetic lotions, hair conditioners, clear bath oils; visc. modifier in PVC plastisols; emulsifier, lubricant in textile spin finishes; emulsifier, leveling agent, pigment dispersant, and wetting agent for coatings; solubilizer for oils and solvs.; straw clear liq.; sol. in methanol, ethanol, acetone, ethyl acetate, toluol, veg. and min. oil; misc. with water; sp.gr. 1.03; HLB 14; solid. pt. < 7 C; acid no. < 3; sapon. no. 90-100; pH 5 (5% aq.); 100% conc.

Pegosperse® 400 MS. [Lonza] PEG-8 stearate; CAS 9004-99-3; nonionic; surfactant, o/w emulsifier, thickener, pigment grinding aid in hair, skin, and makeup prods., household polishes, cleaners, silicone-based lubricants; emulsifier, lubricant, and softener for textile processing oils and finishes; component of diamond abrasive pastes; antigel agent in starch sol'ns. for paper industry; internal antistat for PS; wh. soft solid; sol. in methanol, ethanol, acetone, ethyl acetate, toluol, naphtha, min. and veg. oils; disp. in water; ; sp.gr. 1.0; m.p. 30 C min.; HLB 11; acid no. 3 max.; sapon. no. 83-94; pH 5 (5% aq.); 100% conc.

Pegosperse® 1500 MS. [Lonza] PEG-6-32 stearate; CAS 9004-99-3; nonionic; emulsifier for cosmetic creams, skin care prods., silicone-based lubricants; internal antistat for PE, PP; wh. solid; sol. in ethanol, ethyl acetate, toluol, veg. oil, methanol, naphtha; misc. with water; sp.gr. 1.05; m.p. 27-31 C; HLB 14; acid no. 8.5 max.; sapon. no. 57-67; pH 4 (5% aq.); 100% conc.

Penacolite® B-18-S. [Indspec] Resorcinol-formaldehyde resin; dry bonding agent formulated for rubber-to-wire adhesion; for tires, industrial belts and retreads.

Penacolite® R-2170. [Indspec] Resorcinol-formaldehyde resin; CAS 24969-11-7; bonding agent for dipping formulas for bonding syn. industrial fabrics (aramid, polyester and glass) to rubber.

Penetron OT-30. [Hart Prods. Corp.] 2-Ethylhexyl sulfosuccinate; anionic; penetrant, wetting agent, surf. tens. depressant for textiles, paper, paint, plastic, rubber and metal industries; liq.; 25-30% conc.

Pennad 150. [Elf Atochem N. Am./Org. Chems.] Diethylaminoethanol; CAS 100-37-8; EINECS 202-845-2; intermediate, emulsifier, catalyst in urethane foams, curing agent, corrosion inhibitor; m.w. 117.2; sp.gr. 0.880-0.890; b.p. 158-163.5 C; flash pt. (CC) 52 C; 99.5% min. purity.

Pennstop® 1866. [Elf Atochem N. Am./Org. Chems.] N,N-Diethylhydroxylamine; CAS 3710-84-7; free radical scavenger used by the rubber industry as an emulsion polymerization inhibitor; vapor phase inhibitor for olefin or styrene monomer recovery systems; in-process inhibitor for prod. of styrene, divinyl benzene, butadiene, isoprene; intermediate in the synthesis of silicone rubber and photographic developers; APHA 250 max. color; m.w. 89.1; sp.gr. 0.902; flash pt. (TCC) 46 C; 85% min. purity.

Pennstop® 2049. [Elf Atochem N. Am./Org. Chems.] N,N-diethylhydroxylamine; CAS 3710-84-7; free radical scavenger used by the rubber industry as an emulsion polymerization inhibitor; vapor phase inhibitor for olefin or styrene monomer recovery systems; in-process inhibitor for prod. of styrene, divinyl benzene, butadiene, isoprene; intermediate in the synthesis of silicone rubber and photographic developers; colorless to lt. straw liq.; m.w. 89.1; sp.gr. 0.902; dens. 7.5 lb/gal; b.p. 125-133 C; f.p. -25 C; flash pt. (TCC) 46 C; 85% min. purity; Unverified

Pennstop® 2697. [Elf Atochem N. Am./Org. Chems.] N,N-diethylhydroxylamine; CAS 3710-84-7; free radical scavenger used by the rubber industry as an emulsion polymerization inhibitor; vapor phase inhibitor for olefin or styrene monomer recovery systems; in-process inhibitor for prod. of styrene, divinyl benzene, butadiene, isoprene; intermediate in the synthesis of silicone rubber and photographic developers; colorless to lt. yel. liq.; m.w. 89.1; sp.gr. 0.865; dens. 7.2 lb/gal; b.p. 125-130 C (dec.); f.p. -6 C; flash pt. (TCC) 50 C; 98% min. purity; Unverified

Penreco 1520. [Penreco] Petrolatum, tech.; CAS 8009-03-8; used in rubber processing aids, carbon papers, buffing and polishing compds., corrosion preventatives, general purpose lubricants, printing inks, solder pastes; dk. green; visc. 70-115 SUS (210 F); m.p. 115-135 F.

Penreco 3070. [Penreco] Petrolatum, tech.; CAS

8009-03-8; used for rubber processing; carbon paper; buffing/polishing compds.; corrosion prevention; general-purpose lubricant; shoe polish; printing inks; solder pastes; dk. green; visc. 70-95 SUS (210 F); m.p. 125-140 F.

Penreco Amber. [Penreco] Petrolatum USP; CAS 8027-32-5; EINECS 232-373-2; emollient, base for cosmetic and pharmaceutical preparations; waterproofing agent for butcher paper; lubricant, water repellent, moisture barrier for textile and paper; carrier for modeling clays, soldering paste and flux; pigment carrier for carbon paper; binder and conditioner for crayons; animal feed supplements; fruit/veg. coatings; food pkg. materials; external lube for PVC; rubber processing aid; FDA 21CFR 172.880, 178.3700, 573.720; visc. 68-82 SUS (210 F); m.p. 122-135 F; congeal pt. 123 F; solid. pt. 122 F.

Penreco Red. [Penreco] Tech. petrolatum; CAS 8009-03-8; tech.-grade base and binder for buffing compds., polishes, shoe polishes; rubber processing aids; carbon papers; corrosion preventatives; general-purpose lubricants; printing inks; solder pastes; red; visc. 70-82 SUS (210 F); m.p. 120-135 F.

Penreco Snow. [Penreco] Wh. petrolatum USP; CAS 8027-32-5; EINECS 232-373-2; emollient, base, solv., and carrier for cosmetic and pharmaceutical preparations (creams, lotions, petrol. jellies, ointments, dental adhesives); foods (animal feed supplement, fruit/veg. coatings, food pkg.); sanitary lubricant in food prod. machinery; rust-preventive coating in food processing equipment; carrier in adhesive tapes and compds.; lubricant for textile, paper, PVC; shoe polishes; FDA 21CFR 172.880, 178.3700, 573.720; visc. 64-75 SUS (210 F); m.p. 122-135 F; congeal pt. 123 F; solid. pt. 121 F.

PE-PR 37. [Additive Polymers Ltd] 2% Fluoroelastomer in LLDPE-based masterbatch; processing aid for plastics.

PE-PR 58. [Additive Polymers Ltd] 4% Fluoroelastomer in LLDPE-based masterbatch; processing aid for plastics.

Peptizer 566. [C.P. Hall] Naphthenic oil/sulfonate ester blend; processing aid for nonpolar elastomers, silicones; amber clear oily liq.; sol. in hexane, toluene, kerosene, min. oil; sp.gr. 0.922; visc. 75 cps; f.p. -35 C; acid no. 1.5; sapon. no. 1; flash pt. 179 C; ref. index 1.502; 0.40% moisture.

Peptizer 932. [C.P. Hall] Min. oil/sulfonate blend; processing aid for nonpolar elastomers, silicones; amber clear oily liq.; sol. in hexane, toluene, kerosene, min. oil; sp.gr. 0.843; visc. 43 cps; f.p. -5 C; acid no. 1.1; sapon. no. 1; flash pt. 135 C; ref. index 1.466; 0.06% moisture.

Peptizer 965. [C.P. Hall] Min. oil/sulfonate ester blend; processing aid for nonpolar elastomers, silicones; amber clear oily liq.; sol. in hexane, toluene, kerosene, min. oil; sp.gr. 0.866; visc. 15 cps; f.p. -5 C; acid no. 0.1; sapon. no. 10; flash pt. 135 C; ref. index 1.4655; 0.06% moisture.

Peptizer 7010. [C.P. Hall] Min. oil/sulfonate blend; processing aid for nonpolar elastomers, silicones; amber clear oily liq.; sol. in hexane, toluene, kerosene, min. oil; sp.gr. 0.850; visc. 18 cps; f.p. -5 C; acid no. 8.1; sapon. no. 8; flash pt. 74 C; ref. index 1.465; 0.20% moisture.

Perchem® 97. [Akzo Nobel] Modified refined montmorillonite clay; CAS 1318-93-0; EINECS 215-288-5; off-wh. free-flowing powd.; 95% < 75 μm.

Perkacit® CBS. [Akzo Nobel] N-Cyclohexyl-2-benzothiazole sulfenamide; CAS 95-33-0; accelerator for rubber; soft gran., powd., oil-coated powd.; 95% act.

Perkacit® CDMC. [Akzo Nobel] Copper dimethyldithiocarbamate; CAS 137-29-1; accelerator for rubber; soft gran., oil-coated powd.; 95% act.

Perkacit® DCBS. [Akzo Nobel] N,N′-Dicyclohexyl-2-benzothiazole sulfenamide; CAS 4979-32-2; accelerator for rubber; soft gran.; 95% act.

Perkacit® DOTG. [Akzo Nobel] N,N-Di-o-tolylguanidine; CAS 97-39-2; accelerator for rubber; oil-coated powd.; 96% act.

Perkacit® DPG. [Akzo Nobel] N,N-Di-phenylguanidine; CAS 102-06-7; accelerator for rubber; oil-coated powd., powd.; 96% act.

Perkacit® DPTT. [Akzo Nobel] Dipentamethylenethiuram tetrasulfide; CAS 120-54-7; accelerator for rubber; oil-coated powd., powd.; 95% act.

Perkacit® ETU. [Akzo Nobel] Ethylene thiourea; CAS 96-45-7; EINECS 202-506-9; accelerator for rubber; oil-coated powd.; 97% act.

Perkacit® MBS. [Akzo Nobel] 2-(Morpholinothio) benzothiazole; CAS 102-77-2; accelerator for rubber; soft gran.; 95% act.

Perkacit® MBT. [Akzo Nobel] 2-Mercaptobenzothiazole; CAS 149-30-4; EINECS 205-736-8; accelerator for rubber; oil-coated powd. and powd.; 95% act.

Perkacit® MBTS. [Akzo Nobel] Dibenzothiazole disulfide; CAS 120-78-5; accelerator for rubber; oil-coated powd. and powd.; 95% act.

Perkacit® NDBC. [Akzo Nobel] Nickel dibutyl dithiocarbamate; CAS 13927-77-0; accelerator for rubber; soft gran., oil-coated powd.; 96% act.

Perkacit® SDMC. [Akzo Nobel] Sodium dimethyldithiocarbamate; accelerator for rubber industry; shortstopper for polymer prod.; liq.; 40% aq. sol'n.

Perkacit® TBBS. [Akzo Nobel] Butyl 2-benzothiazole sulfenamide; CAS 95-31-8; accelerator for rubber; soft gran., dedusted gran.; 95% act.

Perkacit® TBzTD. [Akzo Nobel] Tetrabenzylthiuram disulfide; CAS 10591-85-2; accelerator for rubber; oil-coated powd., powd.

Perkacit® TDEC. [Akzo Nobel] Tellurium diethyl dithiocarbamate; CAS 20941-65-5; accelerator for rubber; oil-coated powd.

Perkacit® TETD. [Akzo Nobel] Tetraethylthiuram disulfide; CAS 97-77-8; accelerator for rubber; oil-coated powd., powd., crystals; 98% act.

Perkacit® TMTD. [Akzo Nobel] Tetramethylthiuram disulfide; CAS 137-26-8; accelerator for rubber; soft gran., microgran., oil-coated powd., powd.; 97% act.

Perkacit® TMTM. [Akzo Nobel] Tetramethylthiuram monosulfide; CAS 97-74-5; accelerator for rubber;

soft gran., powd.; 95% act.

Perkacit® ZBEC. [Akzo Nobel] Zinc dibenzyl dithiocarbamate; CAS 14726-36-4; accelerator for rubber; powd., oil-coated powd.; 99% act.

Perkacit® ZDBC. [Akzo Nobel] Zinc di-n-butyl dithiocarbamate; CAS 136-23-2; accelerator for rubber; oil-coated powd., powd.; 98% act.

Perkacit® ZDEC. [Akzo Nobel] Zinc diethyl dithiocarbamate; CAS 14323-55-1; accelerator for rubber; oil-coated powd., powd., wettable powd.; 98% act.

Perkacit® ZDMC. [Akzo Nobel] Zinc dimethyl dithiocarbamate; CAS 137-30-4; accelerator for rubber; oil-coated powd., powd.; 98% act.

Perkacit® ZMBT. [Akzo Nobel] Zinc-2-mercaptobenzothiazole; CAS 155-04-4; accelerator for rubber; powd.; 95% act.

Perkadox® 14. [Akzo Nobel] Di-(2-t-butylperoxyisopropyl) benzene; CAS 25155-25-3; low reactivity peroxide useful as a finishing initiator at high temps. for styrenics; synergist for some halogen-containing flame retardants; also for crosslinking of olefin copolymers, EPDM, SBR, Neoprene, Hypalon; solid; 90% assay; 8.51% act. oxygen; *Storage:* store @ 40 C max.

Perkadox® 14/40. [Akzo Nobel] a,a´-Bis(t-butylperoxy) diisopropylbenzene with inorg. phlegmatizer; CAS 25155-25-3; initiator for elevated-temp. polyester cures when long ambient temp. or high ambient temp. storage of molding compds. is necessary; crosslinking of olefin copolymers, EPDM, SBR, Neoprene, Hypalon; powd.; 40% act.

Perkadox® 14-40B-pd. [Akzo Nobel] Bis (t-butylperoxy) diisopropyl benzene on calcium carbonate carrier; CAS 25155-25-3; initiator where very long or high ambient temp. storage of molding compds. is necessary; also for crosslinking of olefin copolymers, EPDM, SBR, Neoprene, Hypalon; powd.; 40% act.

Perkadox® 14-40K-pd. [Akzo Nobel] Di-(2-t-butylperoxyisopropyl)benzene on clay carrier; CAS 25155-25-3; crosslinking agent for rubber; powd.; 40% assay; 3.78% act. oxygen.

Perkadox® 14-90. [Akzo Nobel] Di-(2-t-butylperoxyisopropyl)benzene; CAS 25155-25-3; crosslinking agent for rubber; semicryst. solid; 90% act.

Perkadox® 14S. [Akzo Nobel] Di-(2-t-butylperoxyisopropyl)benzene; CAS 25155-25-3; initiator for PP; crosslinking agent for rubber; solid; 96% assay; 9.07% act. oxygen; *Storage:* store @ 30 C max.

Perkadox® 14S-20PP-pd. [Akzo Nobel] Di-(2-t-butylperoxyisopropyl)benzene; CAS 25155-25-3; initiator for PP; powd.; 20% assay, 1.89% act. oxygen; *Storage:* store @ 20 C max.

Perkadox® 14S-fl. [Akzo Nobel] Di-(2-t-butylperoxyisopropyl)benzene; CAS 25155-25-3; initiator for PP; crosslinking agent for rubber; flakes; 96% assay; 9.07% act. oxygen; *Storage:* store @ 30 C max.

Perkadox® 16. [Akzo Nobel] Bis(4-t-butylcyclohexyl) peroxydicarbonate; CAS 15520-11-3; ultrafast initiator for polyester cure above 180 F; pultrusion, matched die molding; short ambient temp. compd. shelf life; powd.; 98% act.

Perkadox® 16/35. [Akzo Nobel] Bis-(4-t-butyl cyclohexyl) peroxydicarbonate; CAS 15520-11-3; ultrafast initiator for curing polyester; powd.; 35% on inorg. filler.

Perkadox® 16N. [Akzo Nobel] Bis (4-t-butyl cyclohexyl) peroxydicarbonate; CAS 15520-11-3; initiator for PVC prod., acrylic polymerization; liq.; 98% act.

Perkadox® 16S. [Akzo Nobel] Di-(4-t-butylcyclohexyl) peroxydicarbonate; CAS 15520-11-3; initiator for acrylates, PS, PVC; powd.; 95% assay; 3.81% act. oxygen; *Storage:* store @ 20 C max.

Perkadox® 16-W25-GB1. [Akzo Nobel] Di-(4-t-butylcyclohexyl)peroxydicarbonate; CAS 15520-11-3; suspension; 25% assay; 1.00% act. oxygen.

Perkadox® 16/W40. [Akzo Nobel] Bis(4-t-butylcyclohexyl) peroxydicarbonate; CAS 15520-11-3; initiator for PVC prod., acrylic polymerization; suspension; 40% act. in water; 1.61% act. oxygen; *Storage:* store @ 0-20 C.

Perkadox® 16-W40-GB2. [Akzo Nobel] Di-(4-t-butylcyclohexyl)peroxydicarbonate; CAS 15520-11-3; initiator for PS, PVC; suspension; 40% assay; 1.61% act. oxygen; *Storage:* store @ 0-20 C.

Perkadox® 16-W40-GB5. [Akzo Nobel] Di-(4-t-butylcyclohexyl) peroxydicarbonate; CAS 15520-11-3; initiator for polymerization of vinyl chloride, acrylates, methacrylates, etc.; recommended for mfg. of PVC for elec. applics.

Perkadox® 16/W70. [Akzo Nobel] Bis-(4-t-butylcyclohexyl) peroxydicarbonate; CAS 15520-11-3; ultrafast initiator for curing polyester and for PVC; powd.; 70% act. in water.

Perkadox® 18. [Akzo Nobel] Dicyclohexyl peroxydicarbonate; CAS 1561-49-5; initiator for acrylates, PVC; solid; m.w. 286.3; 92% assay, 5.14% act. oxygen; *Storage:* store @ 0 C max.

Perkadox® 20. [Akzo Nobel] Di(2-methylbenzoyl) peroxide; CAS 3034-79-5; powd.; 78% assay; 4.62% act. oxygen.

Perkadox® 20-W40. [Akzo Nobel] Di(2-methylbenzoyl) peroxide; CAS 3034-79-5; suspension; 40% assay; 2.36% active oxygen.

Perkadox® 24. [Akzo Nobel] Dicetyl peroxydicarbonate; CAS 26332-14-5; initiator for PVC; powd.; 90% assay, 2.52% act. oxygen; *Storage:* store @ 20 C max.

Perkadox® 24-fl. [Akzo Nobel] Dicetyl peroxydicarbonate; CAS 26332-14-5; initiator for PVC; flakes; 95% assay, 2.66% act. oxygen; *Storage:* store @ 20 C max.

Perkadox® 24-W40. [Akzo Nobel] Dicetyl peroxydicarbonate aq. susp.; CAS 26332-14-5; initiator for PVC; suspension; 40% assay, 1.12% act. oxygen; *Storage:* store @ 15 C max.

Perkadox® 26-fl. [Akzo Nobel] Dimyristyl peroxydicarbonate; CAS 53220-22-7; initiator for PVC; flakes; 96% assay; 1.98% act. oxygen; *Storage:* store @ 15 C max.

Perkadox® 26-W40. [Akzo Nobel] Dimyristyl peroxydicarbonate; CAS 53220-22-7; initiator for PVC;

suspension; 40% assay; 1.24% act. oxygen; *Storage:* store @ 0-15 C.

Perkadox® 30. [Akzo Nobel] 2,3-Dimethyl-2,3-diphenylbutane; CAS 1889-67-4; initiator for PS; liq.; *Storage:* store @ 30 C max.

Perkadox® 58. [Akzo Nobel] 3,4-Dimethyl-3,4-diphenylhexane; CAS 10192-93-5; initiator for PS; paste; m.w. 266.4; 95% assay; *Storage:* store @ 30 C max.

Perkadox® 64-10PP-pd. [Akzo Nobel] 1,4-Di(2-t-butylperoxyisopropyl)benzene; CAS 2781-00-2; initiator for PP; powd.; 10% assay, 0.94% act. oxygen; *Storage:* store @ 30 C max.

Perkadox® 64-20PP-pd. [Akzo Nobel] 1,4-Di(2-t-butylperoxyisopropyl)benzene; CAS 2781-00-2; initiator for PP; powd.; 20% assay, 1.89% act. oxygen; *Storage:* store @ 30 C max.

Perkadox® 64-40PP-pd. [Akzo Nobel] 1,4-Di(2-t-butylperoxyisopropyl)benzene; CAS 2781-00-2; initiator for PP; powd.; 40% assay, 3.78% act. oxygen; *Storage:* store @ 30 C max.

Perkadox® AIBN. [Akzo Nobel] 2,2′-Azobis (isobutyronitrile); CAS 78-67-1; EINECS 201-132-3; initiator for acrylates, PS; powd.; m.2. 164.2; 98% assay; *Storage:* store @ 25 C max.

Perkadox® AMBN. [Akzo Nobel] 2,2′-Azobis(2-methylbutyronitrile); CAS 13472-08-7; EINECS 236-740-8; initiator for acrylates, PS; powd.; m.w. 192.3; 96% assay; *Storage:* store @ 25 C max.

Perkadox® BC. [Akzo Nobel] Dicumyl peroxide; CAS 80-43-3; high-temp. initiator for acrylates, PE, PP, PS; used as a flame retardant synergist; also as crosslinking agent for a variety of natural and syn. rubbers and olefins; cryst. powd.; 98% assay; 5.80% act. oxygen; *Storage:* store @ 30 C max.

Perkadox® BC-40Bpd. [Akzo Nobel] Dicumyl peroxide on calcium carbonate carrier; crosslinking agent for a variety of natural and syn. rubbers as well as olefins; powd.; 40% act.

Perkadox® BC-40K-pd. [Akzo Nobel] Dicumyl peroxide on silane-treated clay carrier; crosslinking agent for a variety of natural and syn. rubbers as well as olefins; powd.; 40% act.

Perkadox® BC-40S. [Akzo Nobel] Dicumyl peroxide; CAS 80-43-3; crosslinking agent for rubber; paste; 40% act.; *Storage:* store @ 30 C max.

Perkadox® BC-FF. [Akzo Nobel] Dicumyl peroxide; CAS 80-43-3; initiator for acrylates, PE, PP, PS; free-flowing powd.; m.p. 39 C; 99% assay; 5.86% act. oxygen; *Storage:* store @ 30 C max.

Perkadox® BPSC. [Akzo Nobel] t-Butylperoxy stearyl carbonate; CAS 62476-60-6; initiator for acrylates, PS; solid; 95% assay, 3.93% act. oxygen; *Storage:* store @ 10 C max.

Perkadox® IPP-AT50. [Akzo Nobel] Di(isopropyl) peroxydicarbonate, toluene; initiator for use in PVC; liq.; 50% act. in toluene.

Perkadox® SE-8. [Akzo Nobel] Dioctanoyl peroxide; CAS 762-16-3; initiator for acrylates, PVC; flakes; m.w. 286.4; 98% assay; 5.47% active oxygen; *Storage:* store @ 5 C max.

Perkadox® SE-10. [Akzo Nobel] Didecanoyl peroxide; CAS 762-12-9; initiator for acrylates, PVC; flakes; m.w. 342.5; 98.5% assay; 4.60% active oxygen; *Storage:* store @ 10 C max.

Perkalink® 300. [Akzo Nobel] Triallyl cyanurate; CAS 101-37-1; EINECS 202-936-7; coagent to improve efficiency of peroxide-induced crosslinking of rubber; sensitizer for radiation-cured compds.; solid; 99% act.

Perkalink® 300-50D-pd. [Akzo Nobel] Triallyl cyanurate on silica carrier; co-agent to improve the efficiency of peroxide induced crosslinking in rubber compds.; powd.; 50% act.

Perkalink® 300-50DX. [Akzo Nobel] Triallyl cyanurate on silica/wax carrier; co-agent to improve the efficiency of peroxide induced crosslinking in rubber compds.; gran.; 50% act.

Perkalink® 301. [Akzo Nobel] Triallyl isocyanurate; CAS 1025-15-6; coagent to improve efficiency of peroxide-induced crosslinking of rubber; sensitizer for radiation-cured compds.; solid; 98% act.

Perkalink® 301-50D. [Akzo Nobel] Triallyl isocyanurate on silica carrier; CAS 1025-15-6; co-agent to improve efficiency of peroxide-induced crosslinking of rubber; recommended for fluoroelastomers; powd.; 50% act.

Perkalink® 400. [Akzo Nobel] Trimethylolpropane trimethacrylate; CAS 3290-92-4; co-agent to improve efficiency of peroxide-induced crosslinking of rubber; sensitizer for radiation-cured compds.; liq.; 99% act.

Perkalink® 400-50D. [Akzo Nobel] Trimethylolpropane trimethacrylate on silica carrier; CAS 3290-92-4; co-agent to improve efficiency of peroxide-induced crosslinking of rubber; sensitizer for radiation-cured compds.; powd.; 50% act.

Perkalink® 401. [Akzo Nobel] Ethylene glycol dimethacrylate; CAS 97-90-5; EINECS 202-617-2; co-agent to improve efficiency of peroxide-induced crosslinking of rubber; sensitizer for radiation-cured compds.; liq.; 99% act.

Perkalink® 900. [Akzo Nobel] 1,3-Bis(citraconimidomethyl) benzene; CAS 119462-56-5; antireversion agent for rubber; stops loss of props. that result from reversion of sulfur crosslinks; improves compd. heat resist while maintaining flex fatigue, tear, modulus, and tens. str.; for NR, IR, NR/IR blends with BR/SBR, truck and OTR compds., conveyor belt, large rubber goods, compds. subject to high service temps., aircraft/racing tire parts; off-wh. pastilles; sol. in water, xylene, dichloromethane; m.w. 324; dens. 1270 kg/m^3; bulk dens. 620-670 kg/m^3; m.p. 87 C; *Toxicology:* LD50 (oral, rat) > 2000 mg/kg; severe eye irritant; may cause allergic skin reactions; nuisance dust may cause irritation of respiratory tract; *Precaution:* may form flamm. dust-air mixts.; incompat. with strong oxidizers; hazardous decomp. prods.: CO, CO_2, NO_x; *Storage:* store in closed pkg. in cool place; protect against direct sunlight; 1 yr max. storage life.

Perkasil® KS 207. [Akzo Nobel] Aluminum-magnesium-sodium silicate; reinforcing filler for rubber industry; medium reinforcing props.; powd.

Perkasil® KS 300. [Akzo Nobel] Precipitated silica; highly dispersible reinforcing filler with exc. processing; for rubber industry; powd., gran.

Perkasil® KS 404. [Akzo Nobel] Precipitated silica; highly dispersible reinforcing filler with exc. transparency; for rubber industry; powd., gran.

Perkasil® VP 406. [Akzo Nobel] Precipitated silica; reinforcing filler for silicone rubber applics.; powd.

Perlankrol® ATL40. [Akcros] TEA lauryl sulfate; CAS 139-96-8; EINECS 205-388-7; anionic; biodeg. surfactant for prep. of high foaming shampoos and toiletries; emulsifier for emulsion polymerization; pale straw clear liq.; mild odor; water-sol.; sp.gr. 1.048; visc. 60 cs.; pour pt. < 0 C; flash pt. (COC) > 95 C; pH 7.5 (1% aq.); surf. tens. 32 dynes/cm (0.1%); 40% act.

Perlankrol® DGS. [Akcros] Sodium primary alcohol sulfate; anionic; solubilizer, textile aux.; detergent base in household and industrial cleaners; foaming agent in syn. latex industry; primary emulsifier in emulsion polymerization; wh. slurry; faint fatty alcohol odor; water-sol.; sp.gr. 1.041; visc. 220 cs (40 C); flash pt. > 200 F (COC); pour pt. 12 C; pH 7.0-8.5 (1% aq.); 28% act.; Unverified

Perlankrol® DSA. [Akcros] Sodium lauryl sulfate; CAS 151-21-3; EINECS 205-788-1; anionic; foaming agent for syn. latexes, emulsion polymerization aid; base for prep. of high foaming shampoos and toiletries; wetting agent; industrial detergent additive; wh. slurry; faint fatty alcohol odor; water-sol.; sp.gr. 1.041; visc. 220 cs (40 C); pour pt. 12 C; flash pt. (COC) > 95 C; pH 8.0 (1% aq.); surf. tens. 30 dynes/cm (0.1%); 28% act.

Perlankrol® EAD60. [Akcros] Ammonium alcohol ether sulfate, ethanol; anionic; stabilizer, detergent for industrial and household cleaners; foaming agent for fire fighting foam compds.; dispersant and emulsifier for chlorinated solvs.; emulsifier for emulsion polymerization; clear mobile straw gel; faint alcoholic odor; water-sol.; sp.gr. 1.033; visc. 240 cs; flash pt. (Abel CC) 85 F; pour pt. < 0 C; pH 7-9 (1% aq.); surf. tens. 34 dynes/cm (0.1%); 60% act. in ethanol.

Perlankrol® EP12. [Akcros] Fatty alcohol ether sulfate; anionic; emulsifier for emulsion polymerization; biodeg.; pale yel. clear to hazy liq.; surf. tens. 39 dynes/cm (0.1%); 20% act.

Perlankrol® EP24. [Akcros] Fatty alcohol ether sulfate; anionic; emulsifier for emulsion polymerization; biodeg.; straw clear liq.; surf. tens. 42 dynes/cm (0.1%); 28% act.

Perlankrol® EP36. [Akcros] Fatty alcohol ether sulfate; anionic; emulsifier for emulsion polymerization; biodeg.; straw clear liq.; surf. tens. 40 dynes/cm (0.1%); 28% act.

Perlankrol® ESD. [Akcros] Syn. primary alcohol ether sulfate, sodium salt; anionic; biodeg. high foam additive for liq. household and industrial detergents; emulsifier for emulsion polymerization; water-wh. clear liq.; faint odor; water-sol.; sp.gr. 1.043; visc. 26 cs; flash pt. (COC) > 95 C; pour pt. < 0 C; pH 8.0 (1% aq.); surf. tens. 40 dynes/cm (0.1%); 27% act. in water.

Perlankrol® ESD60. [Akcros] Sodium fatty alcohol ether sulfate; anionic; biodeg. foam booster for liq. detergent formulations, fire fighting foams; emulsifier for emulsion polymerization; water-wh. clear to hazy mobile gel; faint alcoholic odor; water-sol.; sp.gr. 1.046; visc. 210 cs; pour pt. < 0 C; flash pt. (COC) 24 C; pH 8.0 (1% aq.); surf. tens. 39 dynes/cm (0.1%); 60% act. in aq. alcohol.

Perlankrol® FB25. [Akcros] Alkyl phenol ether sulfate, modified, sodium salt; anionic; primary emulsifier for emulsion polymerization; clear straw liq.; faint alcoholic odor; water-sol.; sp.gr. 1.038; visc. 48 cs; flash pt. 86 F (Abel CC); pour pt. -3 C; pH 6.0-8.5 (1% aq.); 25% act.; Unverified

Perlankrol® FD63. [Akcros] Alkyl phenol ether sulfate, ammonium salt; anionic; primary emulsifier for emulsion polymerization; clear straw liq.; faint alcoholic odor; water-sol.; sp.gr. 1.068; visc. 132 cs; flash pt. (Abel CC) 29 C; pour pt. < -5 C; pH 7.5 (1% aq.); surf. tens. 29 dynes/cm (0.1%); 63% act. in aq. alcohol.

Perlankrol® FF. [Akcros] Alkyl phenol ether sulfate, ammonium salt; anionic; foam booster/stabilizer for industrial detergents; emulsifier for emulsion polymerization; amber hazy visc. liq.; faint ammoniacal odor; water-sol.; sp.gr. 1.123; visc. 4200 cs; flash pt. (Abel CC) 30 C; pour pt. 5 C; pH 7.5 (1% aq.); surf. tens. 29 dynes/cm (0.1%); 90% act. in aq. alcohol.

Perlankrol® FN65. [Akcros] Alkyl phenol ether sulfate, sodium salt; anionic; detergent base, foam stabilizer used in specialty detergent formulations; primary emulsifier in emulsion polymerization; clear straw liq.; faint alcoholic odor; water-sol.; sp.gr. 1.123; visc. 398 cs; flash pt. (Abel CC) 30 C; pour pt. < -5 C; pH 7.5 (1% aq.); surf. tens. 27 dynes/cm (0.1%); 65% act. in aq. ethanol.

Perlankrol® FT58. [Akcros] Ammonium alkyl phenol ether sulfate; anionic; primary emulsifier for emulsion polymerization; detergent base, foam stabilizer/booster; emulsifier for cresylic acid; amber clear/sl. hazy liq., faint alcoholic odor; water-sol.; sp.gr. 1.058; visc. 119 cs; pour pt. < -5 C; flash pt. (Abel CC) 32 C; pH 7.8 (1% aq.); 59% act. in aq. alcohol.

Perlankrol® FV70. [Akcros] Ammonium alkyl phenol ether sulfate; anionic; primary emulsifier in emulsion polymerization, latex compding., textile auxiliaries, oilfield applics.; straw clear/sl. hazy liq., faint alcoholic odor; water-sol.; sp.gr. 1.097; visc. 848 cs; pour pt. -15 C; flash pt. (Abel CC) 29 C; pH 7.5 (1% aq.); 70% act. in aq. alcohol.

Perlankrol® FX35. [Akcros] Alkylphenol ether sulfate, sodium salt; anionic; emulsifier for emulsion polymerization; pale yel. clear liq.; surf. tens. 33 dynes/cm (0.1%); 35% act.

Perlankrol® PA Conc. [Akcros] Alkyl phenol ether sulfate, ammonium salt; CAS 30416-77-4; anionic; foam booster/stabilizer, detergent base, frothing agent used in liq. detergent formulations; emulsifier for cresylic acid, emulsion polymerization; amber hazy visc. liq.; faint ammoniacal odor; water-sol.; sp.gr. 1.110; visc. 6000 cs; flash pt. (Abel CC) 30 C; pour pt. 7 C; pH 7.5 (1% aq.); surf. tens. 37 dynes/cm (0.1%); 90% act. in aq. alcohol.

Perlankrol® RN75. [Akcros] Alkyl phenol ether sulfate, sodium salt; anionic; speciality foaming and wetting agent in high electrolyte concs.; primary emulsifier in emulsion polymerization; straw clear liq.; faint alcoholic odor; water-sol. sp.gr. 1.118; visc. 347 cs; flash pt. (Abel CC) 31 C; pour pt. < -5 C; pH 7.5 (1% aq.); surf. tens. 32 dynes/cm (0.1%); 75% act. in aq. alcohol.

Perlankrol® SN. [Akcros] Alkyl phenol ether sulfate, sodium salt; anionic; foam booster and stabilizer, wetting agent, emulsifier in emulsion polymerization; industrial detergent formulations; straw clear liq.; faint odor; water-sol.; sp.gr. 1.060; visc. 116 cs; flash pt. (COC) > 95 C; pour pt. < -5 C; pH 7.5 (1% aq.); surf. tens. 30 dynes/cm (0.1%); 30% act. in water.

Permalease™ 2040. [George Mann] General-purpose semi-permanent release coating for all plastic and elastomer applics.; nontransfer coating leaves parts ready to finish or bond; used for silicone molds, rubber molds.

Permalease™ 2264. [George Mann] Semi-permanent release coating for urethane elastomers, flexible and rigid foams, and high resilience sealing foams; nontransferring; provides multiple releases.

Permalease™ 3000. [George Mann] Semi-permanent release coating for rubber mech. parts and polyester molding; forms a tough multi-release film with no transfer to molded parts.

Permalease™ 3207. [George Mann] Water emulsion form of Permalease 3000; semi-permanent release coating for natural and syn. rubber elastomers.

Permalease™ 3500. [George Mann] Heavy-duty semi-permanent release coating for EPDM seal compds., RTM composite epoxies, polyesters; leaves a matte finish.

Permalease™ 5000. [George Mann] R.T. rapid cure semi-permanent release coating for rubber, polyester, epoxy composites, and castings.

Permalease™ 5500. [George Mann] Provides improved release over Permalease 5000 and matte finish.

Permanax™ 6PPD. [Akzo Nobel] N-1,3-Dimethylbutyl-N′-phenyl-p-phenylene diamine; CAS 793-24-8; staining antiozonant for rubber; pastilles.

Permanax™ BL. [Akzo Nobel] Acetone/diphenylamine condensate; Staining antioxidant for rubber; visc. liq.

Permanax™ BLN. [Akzo Nobel] Acetone/diphenylamine condensate; staining antioxidant for rubber; visc. liq.

Permanax™ BLW. [Akzo Nobel] Acetone/diphenylamine condensate on inert carrier; staining antioxidant for rubber; powd.

Permanax™ CNS. [Akzo Nobel] nonstaining antioxidant for rubber; powd.

Permanax™ CR. [Akzo Nobel] staining antiozonant and antioxiant for chloroprene rubber; gran.

Permanax™ DPPD. [Akzo Nobel] Diphenyl-p-phenylene diamine; CAS 74-31-7; staining antiozonant for rubber; powd., flakes.

Permanax™ HD. [Akzo Nobel] Heptylated diphenylamine; staining antioxidant for rubber; visc. liq.

Permanax™ HD (SE). [Akzo Nobel] Heptylated diphenylamine, self-emulsifying; Self-emulsifying grade; staining antioxidant for rubber; visc. liq.

Permanax™ IPPD. [Akzo Nobel] N-Isopropyl-N′-phenyl-p-phenylene diamine; CAS 101-72-4; staining antiozonant for rubber; pastilles.

Permanax™ OD. [Akzo Nobel] Octylated diphenylamine; staining antioxidant for rubber; rods.

Permanax™ OZNS. [Akzo Nobel] Nonstaining antiozonant and antioxidant for rubber; liq., powd.

Permanax™ TQ. [Akzo Nobel] Polymerized 2,2,4-trimethyl-1,2-dihydroquinoline; CAS 26780-96-1; staining antioxidant for rubber; pastilles.

Permanax™ WSL. [Akzo Nobel] 2,4-Dimethyl-6-(1-methyl cyclohexyl) phenol; nonstaining antioxidant for rubber, ABS, PS; liq.

Permanax™ WSL Pdr. [Akzo Nobel] 2,4-Dimethyl-6-(1-methyl cyclohexyl) phenol on inert carrier; nonstaining antioxidant for rubber; powd.

Permanax™ WSO. [Akzo Nobel] High m.w. phenolic compd.; nonstaining antioxidant for rubber; powd.

Permanax™ WSP. [Akzo Nobel] 2,2′-Methylene-bis(6-(1-methylcyclohexyl)-p-cresol); nonstaining antioxidant for rubber, PE, PP; powd.

Permanax™ WSP (PQ). [Akzo Nobel] 2,2′-Methylene-bis(6-(1-methylcyclohexyl)-p-cresol); nonstaining antioxidant for use in polyethylene; powd.

Permethyl® 100 Epoxide. [Permethyl] Methyl branched isomers of pentadecane-2,3-epoxide (50%), methyl branched isomers of octadecane-2,3-epoxide (35%), and methyl branched isomers of eneicosane-2,3-epoxide (15%); reactive plasticizer/toughener/diluent for epoxy resin systems in sealants, casting and potting compds., coatings, lacquers, and adhesives applics.; modifier for other polymer systems (polyolefins, PVC, PU, polybutene epoxides); improves processing, weathering, uv, and soiling resist., visc. reduction; lt. yel. clear visc. liq., sl. odor; negligible sol. in water; sp.gr. 0.86 (15 C); visc. 15-20 cps (18 C); b.p. 226-273 C; acid no. 2.5; iodine no. 8; flash pt. 121 C; flash pt. (TOC) 121 C; epoxide value 0.39-0.40 mol epoxide/100 g; *Toxicology:* LD50 (oral, rat) > 5000 mg/kg; inh. of vapor, if heated, may cause respiratory tract irritation; eye and skin irritant; ingestion causes mouth and GI tract irritation; may cause allergic reactions; *Precaution:* incompat. with strong oxidizing or reducing agents; empty container may contain explosive vapors; *Storage:* keep container closed; keep away from heat, sparks, open flame.

Permyl® B 100. [Ferro/Bedford] Proprietary benzophenone; uv absorber for unsat. polyesters, fiberglass corrugated outdoor panels, PVC film for outdoor use, extruded garden hose, acrylic and epoxy coatings, cellulosic eyeglass frames; wh. to yel. cryst. powd.; very sol. in alcohols, ketones, esters; mod. sol. in aromatic hydrocarbons; low sol.

in aliphatic hydrocarbons; m.w. 214; sp.gr. 1.33; m.p. 117 C.

Peroxidol 780. [C.P. Hall; Lenape] Epoxidized soya; CAS 8013-07-8; EINECS 232-391-0; plasticizer offering heat and lt. stability, migration and extraction properties; does not mar ABS or PS; used in refrigerator gasketing and food wrap; Gardner 2; m.w. 1000; sp.gr. 0.990; dens. 8.2 lb/gal; pour pt. 2 F; ref. index 1.4710; Custom

Peroxidol 781. [C.P. Hall; Lenape] Epoxidized tall oil; sec. plasticizer stabilizer; reacts synergistically with metallic stabilizers to give heat and lt. stability; used in chlorinated rubber; Gardner 2; m.w. 420; sp.gr. 0.923; dens. 7.7 lb/gal; pour pt. 2 F; ref. index 1.4579; flash pt. > 500 F; Custom

Peroximon® DC 40 MF. [Akrochem] Dicumyl peroxide/EPM masterbatch; CAS 80-43-3; vulcanizing and crosslinking agent for ethylene-propylene elastomers, polyethylene, EVA copolymers, nitrile and PU elastomers, SBR, PVC, neoprene; for rubber car parts, inj. molded articles, elec. cable insulation, conveyor belts, shoe soles; sheets; sp.gr. 0.90; flash pt. (MCOC) > 115 C; 40% peroxide, 60% EPM, 2.36% act. oxygen; *Toxicology:* low toxicity; mildly irritating to eyes and skin; *Precaution:* inflamm.; oxidizing agent; *Storage:* store in original pkg. away from direct sunlight in cool, dry place @ < 30 C.

Peroximon® DC 40 MG. [Akrochem] Dicumyl peroxide/EPM masterbatch; vulcanizing and crosslinking agent for ethylene-propylene elastomers, polyethylene, EVA copolymers, nitrile and PU elastomers, SBR, PVC, neoprene; for rubber car parts, inj. molded articles, elec. cable insulation, conveyor belts, shoe soles; pellets; sp.gr. 0.90; bulk dens. 0.58 g/ml; flash pt. (MCOC) > 115 C; 40% peroxide, 60% EPM, 2.36% act. oxygen; *Toxicology:* low toxicity; mildly irritating to eyes and skin; *Precaution:* inflamm.; oxidizing agent; *Storage:* store in original pkg. away from direct sunlight in cool, dry place @ < 30 C.

Peroximon® S-164/40P. [Akrochem] 1,1-Di(t-butylperoxy)-3,3,5-trimethylcyclohexane; CAS 6731-36-8; vulcanizing agent for rubber.

PE-SA 21. [Additive Polymers Ltd] 5% Erucamide and 10% natural silica in LLDPE-based masterbatch; slip/antiblock agent for plastics.

PE-SA 24. [Additive Polymers Ltd] 2.5% Erucamide and 15% natural silica in LLDPE-based masterbatch; slip/antiblock agent for plastics.

PE-SA 40. [Additive Polymers Ltd] 3% Oleamide and 10% natural silica in LLDPE-based masterbatch; slip/antiblock agent for plastics.

PE-SA 49. [Additive Polymers Ltd] 10% Erucamide and 10% precipitated silica in LLDPE-based masterbatch; slip/antiblock agent for plastics.

PE-SA 60. [Additive Polymers Ltd] 4% Erucamide and 25% natural silica in LLDPE-based masterbatch; slip/antiblock agent for plastics.

PE-SA 72. [Additive Polymers Ltd] 2% Erucamide and 10% natural silica in LLDPE-based masterbatch; slip/antiblock agent for plastics.

PE-SA 92. [Additive Polymers Ltd] 5% Oleamide and 10% natural silica in LLDPE-based masterbatch; slip/antiblock agent for plastics.

PE-SL 04. [Additive Polymers Ltd] 3% Erucamide in LLDPE-based masterbatch; slip agent for plastics.

PE-SL 31. [Additive Polymers Ltd] 3% Oleamide in LLDPE-based masterbatch; slip agent for plastics.

PE-SL 98. [Additive Polymers Ltd] 7.5% Erucamide in LLDPE-based masterbatch; slip agent for plastics.

Petcat R-9. [Laurel Industries] Antimony trioxide; CAS 1309-64-4; EINECS 215-175-0; catalyst in PET polyester prod.; wh. fine cryst. powd., odorless; very sl. sol. in water; dissolves in conc. HCl, sulfuric acid, strong alkalies; m.w. 291.52; sp.gr. 5.3; m.p. 656 C; b.p. 1425 C; ref. index 2.1; 99.3-99.9% Sb_2O_3; *Toxicology:* avoid skin or eye contact, inhalation; use only with adequate ventilation; *Storage:* store in dry, well-ventilated area.

Petrac® 165. [Syn. Prods.] Paraffin wax; CAS 8002-74-2; EINECS 232-315-6; lubricant used in rigid PVC compds. to control external lubrication and fusion; FDA accepted; wh. small flakes; dens. 0.92 g/cm^2; m.p. 162 F; congeal pt. 158 F; acid no. 0.1; flash pt. (COC) 510 F.

Petrac® 215. [Syn. Prods.] Oxidized polyethylene wax; lubricant used to control external lubrication of rigid PVC compds. and provide high gloss surf.; lubricant/release agent in PVC compds. for film, sheet, and inj. molding; off-wh. powd.; soften. pt. 215 F; acid no. 12.

Petrac® 245. [Syn. Prods.] Dist. fatty acids; used in industrial prods. incl. syn. lubricants, bar soaps, cosmetics, rubber tires; Gardner 5; acid no. 204; iodine no. 68; sapon. no. 205; Unverified

Petrac® 250. [Syn. Prods.] Stearic acid, rubber grade; CAS 57-11-4; EINECS 200-313-4; activator, dispersant, plasticizer, lubricant, release agent, vulcanization accelerator in rubber compd. processing; thickener and gelling agent for greases; emulsifier in polishing/buffing compds.; wire drawing lubricant; FAC 5 color; acid no. 197; iodine no. 4; sapon. no. 199; 0.2% moisture.

Petrac® 270. [Syn. Prods.] Tech. stearic acid; CAS 57-11-4; EINECS 200-313-4; ext. lubricant, processing aid in flexible PVC processing; chemical intermediate for metallic stearates, esters, etc.; dispersant, plasticizer, activator, lubricant in rubber compounding; thicknener for greases; suitable for color-sensitive applics.; Lovibond 2Y/0.2R (5$^1/_4$ in.) color; acid no. 205; iodine no. 1; 0.2% moisture.

Petrac® Calcium Stearate CP-11. [Syn. Prods.] Calcium stearate; CAS 1592-23-0; EINECS 216-472-8; mold release agent, lubricant, pigment suspension aid, and flow control agent for plastics (PVC, phenolics), paints, inks, foundry, industrial cleaners, waterproofing of cement and clay tile, lubrication and glossing in paper coatings; wire drawing compds.; wh. powd., 97% thru 100 mesh; sp.gr. 1.03; dens. 26 lb/ft^3; m.p. 151 C; acid no. 205; 9.8% CaO.

Petrac® Calcium Stearate CP-11 LS. [Syn. Prods.] Calcium stearate; CAS 1592-23-0; EINECS 216-

472-8; mold release agent, lubricant, pigment suspension aid, and flow control agent for plastics (PVC, phenolics), paints, inks, foundry, industrial cleaners, waterproofing of cement and clay tile, lubrication and glossing in paper coatings; wire drawing compds.; powd., 90% thru 200 mesh; sp.gr. 1.03; dens. 26 lb/ft^3; m.p. 154 C; soften. pt. 150 C; 9.4% CaO.

Petrac® Calcium Stearate CP-11 LSG. [Syn. Prods.] Calcium stearate; CAS 1592-23-0; EINECS 216-472-8; mold release agent, lubricant, pigment suspension aid, and flow control agent for plastics (PVC, phenolics), paints, inks, foundry, industrial cleaners, waterproofing of cement and clay tile, lubrication and glossing in paper coatings; wire drawing; recommended for applic. where high purity and fine particle size are important; powd., 99.5% thru 325 mesh; sp.gr. 1.03; dens. 24 lb/ft^3; m.p. 154 C; soften. pt. 150 C; 9.4% CaO.

Petrac® Calcium Stearate CP-22G. [Syn. Prods.] Calcium stearate; CAS 1592-23-0; EINECS 216-472-8; lubricant for rigid PVC, polyolefins; emulsifier; FDA accepted; gran.; 100% conc.

Petrac® Calcium/Zinc Stearate CZ-81. [Syn. Prods.] Calcium/zinc stearate; lubricant and mold release agent in polyethylene/PP extrusion and molding operations; release agent in bulk and sheet molding compds. of unsat. polyester; particulate, 99% thru 200 mesh; dens. 24 lb/ft^3; m.p. 97-107 C; 0.3% moisture.

Petrac® Eramide®. [Syn. Prods.] Erucamide; CAS 112-84-5; EINECS 204-009-2; lubricant, slip, antitack, antistat, internal mold release agent; used in PP for extrusion of sheets, in inj. molding; in PVC film and sheeting; slip and antiblock agent in PE film; also in EVA copolymer, PPE, thermoplastic polyester, inks, coatings; internal mold release agent in molded prods.; lamination of polyethylene to cellophane and in polyethylene extrusion coatings; withstands high processing temps.; food contact applics.; APHA 10 dry powd., pellets, or flakes; char. mild odor; m.p. 80 C; 99.5% total amide.

Petrac® GMS. [Syn. Prods.] Glyceryl stearate; internal/external lubricant, mold release for rigid PVC inj. molding, film, sheet, bottle compds.; pigment suspension in paints and inks, waterproofing agent for cement, clay tile; lubricant, glossing aid in paper coatings; foundry resins; lubricant, antistat, and antifog in polyolefins; off-wh. flake; m.p. 140 F; acid no. 2; iodine no. 2.5; sapon. no. 167; 43% alpha monoglycerides.

Petrac® Magnesium Stearate MG-20 NF. [Syn. Prods.] Magnesium stearate NF; CAS 557-04-0; EINECS 209-150-3; dry lubricant and anticaking agent used in food prods., and filling of pharmaceutical capsules; antistick properties in tableting; improves stability, smoothness, texture of emulsions, creams, ointments; improves texture and water-repellency in baby and medical powds.; internal/external lubricant in ABS; FDA accepted; wh. fluffy powd., 95% thru 200 mesh; 85% thru 325 mesh; dens. 22 lb/ft^3; soften. pt. 140 C.

Petrac® PHTA. [Syn. Prods.] Partially hydrog. tallow fatty acids; CAS 61790-38-3; EINECS 263-130-9; primary emulsifier in polymerization of SBR, ABS, methyl methacrylate-butadiene-styrene polymers; also used in industrial prods. incl. syn. lubricants, bar soaps, cosmetics, rubber tires; Lovibond 7Y/0.7R ($5^1/_4$ in.) color; acid no. 205; iodine no. 38; sapon. no. 206; 0.2% moisture.

Petrac® Slip-Eze®. [Syn. Prods.] Oleamide; CAS 301-02-0; EINECS 206-103-9; slip, release, antitack, and/or internal mold release agent; polyethylene film and sheeting; inj. or extruded molded prods.; PVC film and plastisols; polyester SMC, EVA copolymers, inks, coatings, wax coatings, crayons, petrol. oils, greases; textile chemicals, detergents, food pkg. materials; APHA 10 dry powd., pellets, or flakes, char. mild bland odor; m.p. 73 C; 99.5% total amide.

Petrac® Slip-Quick®. [Syn. Prods.] Amide blend; slip, antiblock, mold release in PE, PP, PVC film; powd.

Petrac® Vyn-Eze®. [Syn. Prods.] Stearamide; CAS 124-26-5; EINECS 204-693-2; slip, antitack, antiblock, and/or internal mold release agent; PVC molded prods., film, and plastisol systems; polyethylene film; inks, coatings, food pkg. materials; pigment dispersant in rubber; intermediate for softeners and waterproofing agents for textiles; foam stabilizers, rinse aid release control in detergents; APHA 10 dry powd., pellets, or flakes; char. mild bland odor; m.p. 104 C; 99.5% total amide.

Petrac® Zinc Stearate ZN-41. [Syn. Prods.] Zinc stearate USP; CAS 557-05-1; EINECS 209-151-9; lubricant and mold release agent in PS, PVC, melamine, U-F, phenol-formaldehyde, and polyester molding resins; dusting agent for uncured rubber slabs; suspending and flattening agent in solv.- and water-based paints; anticaking agent used in extinguishers; lubricant in powd. metallurgy; powd.; 99.5% thru 325 mesh; dens. 19 lb/ft^3; m.p. 121 C; pH neutral; 0.3% moisture.

Petrac® Zinc Stearate ZN-42. [Syn. Prods.] Zinc stearate; CAS 557-05-1; EINECS 209-151-9; release agent in plastics bulk molding or sheet molding compds.; lubricant/stabilizer in vinyl compds.; rubber compd. where high bulk dens. is preferred; particulate, 99% thru 200 mesh; m.p. 121 C; 14% ZnO.; Unverified

Petrac® Zinc Stearate ZN-44 HS. [Syn. Prods.] Zinc stearate; CAS 557-05-1; EINECS 209-151-9; mold release agent and lubricant used in plastic applics. where minimal color development is desired; for polyolefins, PS; powd.; 99.5% thru 325 mesh; dens. 19 lb/ft^3; m.p. 121 C; 13.5% ZnO.

Petrac® Zinc Stearate ZW-45. [Syn. Prods.] Zinc stearate in aq. disp.; CAS 557-05-1; EINECS 209-151-9; lubricant and mold release agent for plastics industry, pigment suspension in paints and inks, waterproofing agent for cement and clay tile, lubricant and glossing aid in paper coatings; used in rigid PVC pipe, foundry resins; wh. disp.; visc. 400 cps; pH 8.7; 45% solids; Unverified

Petrarch® A0699. [United Chem. Tech.] N-(2-Aminoethyl)-3-aminopropylmethyldimethoxy silane; CAS 3069-29-2; coupling agent, chem. intermediate, blocking agent, release agent, lubricant, primer, reducing agent; liq.; m.w. 206.4; sp.gr. 0.975; b.p. 129-130 C (10 mm); ref. index 1.4447; 95% purity.

Petrarch® A0700. [United Chem. Tech.] N-2-Aminoethyl-3-aminopropyltrimethoxysilane; CAS 1760-24-3; EINECS 212-164-2; coupling agent, chem. intermediate, blocking agent, release agent, lubricant, primer, reducing agent; coupler for cellulosics, epoxy, phenolic, melamine, nylons, PVC, acrylics, polyolefins, polyamide/imide, PC, PPO, PPS, polyimide, polyvinyl butyral, PU, nitrile and SAN sealants, epichlorohydrin and silicone rubber; liq.; m.w. 222.4; sp.gr. 1.01; b.p. 140 C (15 mm); flash pt. 121 C; ref. index 1.442.

Petrarch® A0701. [United Chem. Tech.] Aminoethylamino propyltrimethoxy silane, tech. grade; CAS 1760-24-3; EINECS 212-164-2; coupler for epoxy, phenolic, melamine, nylons, PVC, acrylics, polyolefins, PU and nitrile rubber; liq.; water-sol.

Petrarch® A0742. [United Chem. Tech.] 3-Aminopropylmethyldiethoxy silane; CAS 3179-76-8; coupling agent, chem. intermediate, blocking agent, release agent, lubricant, primer, reducing agent; chemically bonds, strengthens thermoset furanand melamine composites; liq.; m.w. 191.4; sp.gr. 0.92 (20 C); b.p. 85-88 C (8 mm); flash pt. 125 C; ref. index 1.427; 97% purity.

Petrarch® A0743. [United Chem. Tech.] 3-Aminopropylmethyldiethoxy silane, tech. grade; CAS 3179-76-8; coupling agent, chem. intermediate, blocking agent, release agent, lubricant, primer, reducing agent; chemically bonds, strengthens thermoset furan composites; liq.; m.w. 191.4; sp.gr. 0.92 (20 C); b.p. 85-88 C (8 mm); flash pt. 125 C; ref. index 1.427; 97% purity.

Petrarch® A0750. [United Chem. Tech.] 3-Aminopropyltriethoxysilane; CAS 919-30-2; EINECS 213-048-4; coupling agent, chem. intermediate, blocking agent, release agent, lubricant, primer, reducing agent; chemical bonder/strengthener for resin/plastic composites; for DAP, epoxy, furan, phenol-formaldehyde, U-F, urethane, cellulosics, nylon 6, polyamide-imide, PBT, PC, PVAc, PPS, polysulfone, PES, PVB, PVC, sealants (polysulfide, nitrile, SAN), rubber (fluorocarbon, epichlorohydrin, silicone); liq.; m.w. 221.4; sp.gr. 0.95 (20 C); b.p. 122-123 C (30 mm); flash pt. 96 C; ref. index 1.423 (20 C).

Petrarch® A0800. [United Chem. Tech.] Aminopropyltrimethoxysilane; CAS 13822-56-3; EINECS 237-511-5; coupling agent, chem. intermediate, blocking agent, release agent, lubricant, primer, reducing agent; liq.; water-sol.; m.w. 179.2; sp.gr. 1.01 (20 C); b.p. 80 C (8 mm); flash pt. 104 C; ref. index 1.420; 98% purity.

Petrarch® B2500. [United Chem. Tech.] Bis(trimethylsilyl)acetamide; CAS 10416-59-8; coupling agent, chem. intermediate, blocking agent, release agent, lubricant, primer, reducing agent; liq.; m.w. 203.4; sp.gr. 0.83 (20 C); b.p. 71-73 C (35 mm); flash pt. 17 C; ref. index 1.418 (20 C); 95% purity.

Petrarch® C3292. [United Chem. Tech.] 3-Chloropropyltriethoxysilane; CAS 5089-70-3; EINECS 225-805-6; coupling agent, chem. intermediate, blocking agent, release agent, lubricant, primer, reducing agent; chemically bonds, strengthens PVC composites; liq.; m.w. 240.8; sp.gr. 1.009 (20 C); b.p. 98-102 C (10 mm); flash pt. 61 C; ref. index 1.420 (20 C); 98% purity.

Petrarch® C3300. [United Chem. Tech.] 3-Chloropropyltrimethoxysilane; CAS 2530-87-2; EINECS 219-787-9; coupling agent, chem. intermediate, blocking agent, release agent, lubricant, primer, reducing agent; liq.; m.w. 198.7; sp.gr. 1.081; b.p. 183 C; flash pt. 66 C; ref. index 1.464; 98% purity.

Petrarch® G6720. [United Chem. Tech.] 3-Glycidoxypropyltrimethoxysilane; CAS 2530-83-8; EINECS 219-784-2; coupling agent, chem. intermediate, blocking agent, release agent, lubricant, primer, reducing agent; chemically bonds, strengthens thermoset and thermoplastic composites; for epoxy, melamine, phenol-formaldehyde, PBT, PVAc, PS, polysulfide sealants, butyl rubber, polysaccharide polymers; liq.; m.w. 236.3; sp.gr. 1.070 (20 C); b.p. 120 C (2 mm); flash pt. 79 C; ref. index 1.428; 98% purity.

Petrarch® H7300. [United Chem. Tech.] Hexamethyldisilazane; CAS 999-97-3; EINECS 213-668-5; coupling agent, chem. intermediate, blocking agent, release agent, lubricant, primer, reducing agent; liq.; m.w. 161.4; sp.gr. 0.77 (20 C); b.p. 126-127 C; flash pt. 27 C; ref. index 1.408 (20 C); 98% purity.

Petrarch® I7810. [United Chem. Tech.] Isobutyltrimethoxysilane; CAS 18395-30-7; coupling agent, chem. intermediate, blocking agent, release agent, lubricant, primer, reducing agent; liq.; m.w. 178.3; sp.gr. 0.93 (20 C); b.p. 154-157 C; flash pt. 14 C; ref. index 1.396 (20 C); 98% purity.

Petrarch® M8550. [United Chem. Tech.] 3-Methacryloxypropyltrimethoxysilane; CAS 2530-85-0; EINECS 219-785-8; coupling agent, chem. intermediate, blocking agent, release agent, lubricant, primer, reducing agent; chemically bonds, strengthens thermoplastic and thermoset composites; for methyl methacrylate, unsat. polyester, polyacrylate, polyether ketone, PEEK, PP, PS, sealants (acrylic, SBR), rubbers (vinyl-terminated silicone); liq.; m.w. 248.4; sp.gr. 1.045; b.p. 190 C; flash pt. 92 C; ref. index 1.429; 97% purity.

Petrarch® M8450. [United Chem. Tech.] 3-Mercaptopropylmethyldimethoxysilane; CAS 31001-77-1; coupling agent, chem. intermediate, blocking agent, release agent, lubricant, primer, reducing agent; liq.; m.w. 180.4; sp.gr. 0.99 (20 C); b.p. 96 C (30 mm); flash pt. 97 C; ref. index 1.4502; 95% purity.

Petrarch® M8500. [United Chem. Tech.] 3-Mercaptopropyltrimethoxysilane; CAS 4420-74-0;

EINECS 224-588-5; coupling agent, chem. intermediate, blocking agent, release agent, lubricant, primer, reducing agent; chemically bonds, strengthens thermoset and thermoplastic composites; for epoxy, polysulfide sealants, neoprene, isoprene, and epichlorohydrin rubbers, gold and precious metals; liq.; m.w. 196.3; sp.gr. 1.05; b.p. 219-220 C; flash pt. 93 C; ref. index 1.440; 97% purity.

Petrarch® M9050. [United Chem. Tech.] Methyltriethoxysilane; CAS 2031-67-6; EINECS 217-983-9; coupling agent, chem. intermediate, blocking agent, release agent, lubricant, primer, reducing agent; liq.; m.w. 178.3; sp.gr. 0.90 (20 C); b.p. 141-143 C; flash pt. 38 C; ref. index 1.383 (20 C); 98% purity.

Petrarch® M9100. [United Chem. Tech.] Methyltrimethoxysilane; CAS 1185-55-3; EINECS 214-685-0; coupling agent, chem. intermediate, blocking agent, release agent, lubricant, primer, reducing agent; liq.; m.w. 136.2; sp.gr. 0.955 (20 C); b.p. 102-103 C; flash pt. 8 C; ref. index 1.3696 (20 C); 98% purity.

Petrarch® O9835. [United Chem. Tech.] Octyltriethoxysilane; CAS 2943-75-1; EINECS 220-941-2; coupling agent, chem. intermediate, blocking agent, release agent, lubricant, primer, reducing agent; liq.; m.w. 276.5; sp.gr. 0.88 (20 C); b.p. 98-99 C (2 mm); flash pt. 100 C; ref. index 1.415 (20 C); 98% purity.

Petrarch® T1807. [United Chem. Tech.] Tetraethoxysilane; CAS 78-10-4; coupling agent, chem. intermediate, blocking agent, release agent, lubricant, primer, reducing agent; liq.; m.w. 208.3; sp.gr. 0.934 (20 C); b.p. 169 C; flash pt. 60 C; ref. index 1.383 (20 C); 98% purity.

Petrarch® T1918. [United Chem. Tech.] Tetrakis (2-ethoxyethoxy) silane; CAS 18407-94-8; coupling agent, chem. intermediate, blocking agent, release agent, lubricant, primer, reducing agent; liq.; m.w. 384; sp.gr. 1.02 (20 C); b.p. 200 C (10 mm); flash pt. 131 C; 90% purity.

Petrarch® T1980. [United Chem. Tech.] Tetramethoxysilane; CAS 681-84-5; coupling agent, chem. intermediate, blocking agent, release agent, lubricant, primer, reducing agent; liq.; m.w. 152.2; sp.gr. 1.05 (20 C); b.p. 121-122 C; flash pt. 45 C; ref. index 1.368 (20 C); 99% purity.

Petrarch® T2090. [United Chem. Tech.] Tetrapropoxysilane; CAS 682-01-9; coupling agent, chem. intermediate, blocking agent, release agent, lubricant, primer, reducing agent; liq.; m.w. 264.4; sp.gr. 0.92 (20 C); b.p. 224-225 C; flash pt. 95 C; ref. index 1.401 (20 C); 98% purity.

Petrarch® T2503. [United Chem. Tech.] N-[3-(Triethoxysilyl)-propyl)-4,5-dihydroimidazole; CAS 58068-97-6; coupling agent, chem. intermediate, blocking agent, release agent, lubricant, primer, reducing agent; chemically bonds, strengthens thermoset phenol-formaldehyde and thermoplastic PVB composites; liq.; m.w. 274.1; sp.gr. 1.00 (20 C); b.p. 134 C (2 mm); flash pt. 78 C; ref. index 1.45; 90% purity.

Petrarch® T2910. [United Chem. Tech.] Trimethoxysilylpropyldiethylenetriamine; CAS 35141-30-1; coupling agent, chem. intermediate, blocking agent, release agent, lubricant, primer, reducing agent; chemically bonds, strengthens thermoset and thermoplastic composites; for polyimide, PPO, polysulfone, PES, copper, iron; liq.; m.w. 265.4; sp.gr. 1.03 (20 C); b.p. 114-168 C (2 mm); flash pt. 137 C; ref. index 1.463; 91% purity.

Petrarch® V4900. [United Chem. Tech.] Vinyltrichlorosilane; CAS 75-94-5; EINECS 200-917-8; coupling agent, chem. intermediate, blocking agent, release agent, lubricant, primer, reducing agent; liq.; m.w. 161.5; sp.gr. 1.24 (20 C); b.p. 92-93 C; flash pt. 21 C; ref. index 1.43 (20 C); 97% purity.

Petrarch® V4910. [United Chem. Tech.] Vinyltriethoxysilane; CAS 78-08-0; EINECS 201-081-7; coupling agent, chem. intermediate, blocking agent, release agent, lubricant, primer, reducing agent; liq.; m.w. 190.3; sp.gr. 0.90 (20 C); b.p. 160-161 C; flash pt. 34 C; ref. index 1.396 (20 C); 98% purity.

Petrarch® V4917. [United Chem. Tech.] Vinyltrimethoxysilane; CAS 2768-02-7; EINECS 220-449-8; coupling agent, chem. intermediate, blocking agent, release agent, lubricant, primer, reducing agent; chemically bonds, strengthens thermoset and thermoplastic composites; for methyl methacrylate, unsat. polyester, polyacrylate, PVAc, acrylic and SBR sealants, silicone (hydroxyl- or vinyl-terminated) rubbers; liq.; m.w. 148.2; sp.gr. 1.13 (20 C); b.p. 123 C; flash pt. 23 C; ref. index 1.393; 98% purity.

Petrarch® V5000. [United Chem. Tech.] Vinyltris (methoxyethoxy)silane; CAS 1067-53-4; EINECS 213-934-0; coupling agent, chem. intermediate, blocking agent, release agent, lubricant, primer, reducing agent; chemically bonds, strengthens thermoset methyl methacrylate composites; liq.; m.w. 280.4; sp.gr. 1.04 (20 C); b.p. 284-286 C; flash pt. 71 C; ref. index 1.427; 97% purity.

Petrolatum RPB. [Witco/Petroleum Spec.] Petrolatum tech.; binder, carrier, lubricant, moisture barrier, plasticizer, protective agent, softener for tech. applics., (elec., paper and pkg., plastics, rubber, textiles, industrial use); dk. brn.; m.p. 71-77 C.

Petrolite® C-700. [Petrolite] Hard microcryst. wax consisting of n-paraffinic, branched paraffinic, and naphthenic hydrocarbons; CAS 63231-60-7; EINECS 264-038-1; used in hot-melt coatings and adhesives, paper coatings, printing inks, plastic modification (as lubricant and processing aid), lacquers, paints, varnishes, as binder in ceramics, for potting, filling in elec./electronic components; in investment castings, as emulsion wax size in papermaking, as fabric softener ingred. in permanent-press fabrics, in emulsion and latex coatings, and in cosmetic hand creams and lipsticks; color 1.5 (D1500) wax; very low sol. in org. solvs.; sp.gr. 0.93; visc. 16 cps (99 C); m.p. 93 C.

Petrolite® C-1035. [Petrolite] Hard microcryst. wax consisting of n-paraffinic, branched paraffinic, and naphthenic hydrocarbons; CAS 63231-60-7; EINECS 264-038-1; used in hot-melt coatings and adhesives, paper coatings, printing inks, plastic modification (as lubricant and processing aid), lacquers, paints, varnishes, as binder in ceramics, for potting, filling in elec./electronic components; in investment castings, as emulsion wax size in papermaking, as fabric softener ingred. in permanent-press fabrics, in emulsion and latex coatings, in cosmetic hand creams and lipsticks, chewing gum base, microcapsule for flavoring substances; incl. FDA §172.230, 172.615, 175.105, 175.300, 176.170, 176.180, 176.200, 177.1200, 178.3710, 179.45; color 0.5 (D1500) wax; very low sol. in org. solvs.; sp.gr. 0.93; visc. 15 cps (99 C); m.p. 94 C.

Petrolite® C-3500. [Petrolite] Oxidized polymer; general purpose polymer for use in adhesives, textiles, and high-wax polish applics.; lubricant for PVC extrusion; FDA 21CFR §175.105, 176.200, 176.210, 177.2800; solid; visc. 152 SUS (116 C); m.p. 107 C; congeal pt. 93 C; acid no. 24; sapon. no. 42.

Petrolite® CP-7. [Petrolite] Ethylene/propylene compd.; CAS 9010-79-1; component for solv.-based PU mold release agents; additives for adhesives, cosmetics, paraffin modification, printing inks, investment casting; FDA §175.105; m.w. 650; dens. 0.94 g/cc; visc. 12 cps (99 C); m.p. 96 C.

Petrolite® CP-11. [Petrolite] Syn. copolymer; component for PU mold release, laminating and hot-melt adhesives, printing inks, cosmetics, investment casting wax; m.w. 1100; dens. 0.94 g/cc; visc. 13 cs (149 C); m.p. 110 C.

Petrolite® CP-12. [Petrolite] Syn. copolymer; component for PU mold release, laminating and hot-melt adhesives, printing inks, cosmetics, investment casting wax; m.w. 1200; dens. 0.94 g/cc; visc. 16 cs (149 C); m.p. 112 C.

Petrolite® E-2020. [Petrolite] Oxidized HDPE; CAS 68441-17-8; wax producing emulsions for formulating textile lubricants, mold release agents, lubricants, and antiblocking agents; useful in ceramic binders; processing aid in the fabrication of thermoplastic resins; exhibits external and internal lubricating properties; color 0.5 (D1500) wax; m.p. 115.5 C; acid no. 23; sapon. no. 42.

Petronauba® C. [Petrolite] Oxidized microcryst. wax; external lubricant, process aid, slip, antiblock, scuff, mar resist. aid in styrenics, polyolefins, TPOs; used in the formulation of emulsions, textile finish softeners, floor, auto, and liq. solv.-based polishes; carnauba substitute; color 1.5 (D1500) wax; m.p. 93 C; acid no. 26; sapon. no. 54.

Petro-Rez 100. [Akrochem] Thermoplastic, aromatic hydrocarbon resin; plasticizer, esp. for colored rubber articles; builds tack; aids processing of min.-filled elastomers; Gardner 10+ (50% in toluol) flake; sol. in benzene, toluol, xylene, hexane, and ketones; insol. in alcohols; sp.gr. 1.07; soften. pt. (B&R) 100 C; acid no. < 1; iodine no. 35.

Petro-Rez 103. [Akrochem] Thermoplastic, aromatic hydrocarbon resin; plasticizer, tack builder; aids processing of min.-filled elastomers; extender in black-loaded compds.; Gardner 11+ (50% in toluol) flake; sol. in benzene, toluol, xylene, hexane, and ketones; insol. in alcohols; sp.gr. 1.07; soften. Pt. (B&R) 100 C; acid no. < 1; iodine no. 90.

Petro-Rez 200. [Akrochem] Thermoplastic aromatic hydrocarbon resin; plasticizer, tackifier, processing aid; extender in black-loaded compds.; used in rubber compding. applics., e.g., molded mechanical goods, extrusions, shoe soling, elec. insulation, and flooring; Gardner 11+ to 13+ (50% in toluol) flake; sol. in aromatic and aliphatic solvs.; insol. in alcohols; sp.gr. 1.07; soften. pt. (B&R) 97-105 C; acid no. < 1.

Petro-Rez 215. [Akrochem] Aromatic-aliphatic hydrocarbon resin; plasticizer for rubber compding. applics. incl. molding-grade mechanical goods, extrusion compds., shoe soling, elec. insulation, and flooring prods.; reinforcing agent, processing aid; tackifier in adhesive in hot-melt, solv.-type, and emulsion adhesives; Gardner 11+ to 13+ (50% in toluene) flake; sol. in aromatic and aliphatic solvs., incl. low-KB ink oils; insol. in alcohols, esters, and ketones; soften. pt. 112-120 C; acid no. < 1.0; sapon. no. < 1.0.

Petro-Rez 801. [Akrochem] Cycloaliphatic hydrocarbon resin; tackifier for rubber compding. applics. incl. molded mech. goods, extrusions, shoe soling, elec. insulation, flooring; processing aid for min.-filled elastomers; extender in black loaded compds.; amber solid; sol. in aromatic and aliphatic solvs.; insol. in alcohols; sp.gr. 1.07; soften. pt. (B&R) 105 C; acid no. nil; *Storage:* store in cool area.

Petro-Rez PTH. [Akrochem] Low m.w. plasticizer polymerized from mixed petrol. streams; plasticizer/tackifier; modifies other resins and polymers in rubber compding., adhesive systems, and high-solids ink vehicles; Gardner 16-17 (50% in toluene); sp.gr. 1.02; melt visc. 3.6 poises (85 C); soften. pt. (B&R) 25 C; acid no. nil.

Petro-Rez PTL. [Akrochem] Low m.w. plasticizer polymerized from mixed petrol. stream; plasticizer/tackifier; modifies other resins and polymers in rubber compding., adhesive systems, and high-solids ink vehicles; Gardner 16-17 (50% in toluene); sp.gr. 1.02; melt visc. 1 poise (85 C); R&B soften. pt. 10 C; acid no. nil.

PE-UV 01. [Additive Polymers Ltd] 5% HALS, 5% benzophenone in LLDPE-based masterbatch; uv stabilizer for plastics.

PE-UV 11. [Additive Polymers Ltd] 10% HALS II in LLDPE-based masterbatch; uv stabilizer for plastics.

PE-UV 12. [Additive Polymers Ltd] 10% Benzophenone in LLDPE-based masterbatch; uv absorber for plastics.

PE-UV 13. [Additive Polymers Ltd] 10% HALS II and 10% benzophenone in LLDPE-based masterbatch; uv stabilizer for plastics.

PE-UV 18. [Additive Polymers Ltd] 10% Nickel complex and 10% benzophenone in LLDPE-based masterbatch; uv stabilizer for plastics.

PE-UV 23. [Additive Polymers Ltd] 10% HALS III and 10% benzophenone in LLDPE-based masterbatch; uv stabilizer for plastics.

PE-UV 25. [Additive Polymers Ltd] 10% Nickel complex and 5% benzophenone in LLDPE-based masterbatch; uv stabilizer for plastics.

PE-UV 32. [Additive Polymers Ltd] 7.5% HALS II and 7.5% benzophenone in LLDPE-based masterbatch; uv stabilizer for plastics.

PE-UV 41. [Additive Polymers Ltd] 7.5% HALS II in LLDPE-based masterbatch; uv stabilizer for plastics.

PE-UV 45. [Additive Polymers Ltd] 20% HALS II in LLDPE-based masterbatch; uv stabilizer for plastics.

PE-UV 57. [Additive Polymers Ltd] 10% HALS III in LLDPE-based masterbatch; uv stabilizer for plastics.

PE-UV 67. [Additive Polymers Ltd] 20% HALS III in LLDPE-based masterbatch; uv stabilizer for plastics.

PE-UV 84. [Additive Polymers Ltd] 10% HALS III and 5% benzophenone in LLDPE-based masterbatch; uv stabilizer for plastics.

PE-UV 99. [Additive Polymers Ltd] 10% HALS II, 10% benzophenone, and 2% anitoxidant in LLDPE-based masterbatch; uv stabilizer for plastics.

PE-UV 102. [Additive Polymers Ltd] 10% HALS III, 10% benzophenone, and 2% anitoxidant in LLDPE-based masterbatch; uv stabilizer for plastics.

PE-UV 118. [Additive Polymers Ltd] 20% Benzophenone in LLDPE-based masterbatch; uv absorber for plastics.

PE-UV 119. [Additive Polymers Ltd] 25% Microground titanium dioxide in LLDPE-based masterbatch; uv absorber for plastics.

Pfinyl® 402. [Specialty Minerals] Treated ground limestone; CAS 1317-65-3; filler for rubber, flexible vinyl, PVC plastisols; powd.; 5 µ avg. particle size, 25 µ top size.

Phos-Chek P/30. [Monsanto] Ammonium polyphosphate; flame retardant for ABS, PE, PP, PS, PU rigid and flexible foam, EVA, TPE, TPR.

Phosflex 71B. [Akzo Nobel] t-Butyl phenyl diphenyl phosphate; flame retardant, plasticizer for NBR elastomers, PVC, PVAc, cellulosics, PS, ABS, engineering resins, polyesters, alloys; PVAc emulsions.

Phosflex 362. [Akzo Nobel] 2-Ethylhexyl diphenyl phosphate; CAS 1241-94-7; plasticizer for PVC, PVAc, cellulosics, PS, ABS, engineering resins, polyesters, alloys; suitable for food film.

Phosflex 370. [Akzo Nobel] Alkyl/diaryl phosphate blend; plasticizer for NBR, NBR/PVC, CR, CPE elastomers, polyacrylate, fluoroelastomers, PVC, PVAc, cellulosics, PS, ABS, engineering resins, polyesters, alloys; flame retardant.

Phosflex 390. [Akzo Nobel] Isodecyl diphenyl phosphate; plasticizer for NBR elastomers, PVC, PVAc, cellulosics, PS, ABS, engineering resins, polyesters, alloys; esp. for PVC wallcovering.

Phosflex TBEP. [Akzo Nobel] Tributoxyethyl phosphate; CAS 78-51-3; EINECS 201-122-9; flame retardant for acrylics, cellulosics, epoxy, nitrile, PS, PVAc, PVC.

Phosphanol Series. [Toho Chem. Industry] Phosphate ester surfactants; emulsifier for textile spinning oils, emulsion polymerization of vinyl acetate and copolymers, acrylate, etc.; antistat for PVC film; anticorrosive agent; cosmetics solubilizer and emulsifier; liq., paste, solid.

Phospholan® PDB3. [Harcros UK] Fatty alcohol ethoxylate phosphate ester; anionic; wetting agent, detergent component, textiles, drycleaning; metal cleaning; emulsifier for phenols and in emulsion polymerization; agric.; waterless hand cleaners; metalworking lubricants with anticorrosive properties; iodophors; antistat; pale straw liq.; mild odor; disp. in water; sp.gr. 1.000; visc. 2400 cs; flash pt. (COC) > 150 C; pour pt. 10 C; pH 2.5 (1% aq.); surf. tens. 30 dynes/cm (0.1%); 100% act.; *Toxicology:* extremely irritating to eyes and skin; *Precaution:* corrosive org. acid.

Phospholan® PNP9. [Harcros UK] Nonylphenol ethoxylate phosphate ester, acid form; CAS 51811-79-1; anionic; electrolyte sol. wetting agent and emulsifier; for alkaline and industrial cleaners, textiles, dyestuff carrier, metal cleaning, emulsion polymerization, agric., waterless hand cleaners, metalworking lubes, iodophors; dk. amber visc. liq.; mild odor; water-sol.; sp.gr. 1.200; visc. 11,500 cs (40 C); flash pt. (COC) > 150 C; pour pt. 15 C; pH 2.5 (1% aq.); surf. tens. 36 dynes/cm; 100% act.; *Toxicology:* extremely irritating to eyes and skin; *Precaution:* corrosive org. acid.

Picco® 5070. [Hercules] Aromatic hydrocarbon resin; low m.w. nonpolar thermoplastic resin used as tackifier in elastomer-based solv., water, and hot-melt sealant and adhesive systems; as processing and reinforcing agent in rubber compd. and extruding applics.; exhibit exc. leafing and solv.-release props. in metallic cold-cut paints; food-pkg. and processing operations; Gardner 11 solid; sol. in aromatic, aliphatic, and chlorinated hydrocarbons, low-KB aliphatic ink oils, benzyl alcohol, cyclohexanol, MEK, butyl Carbitol acetate, and diethyl Carbitol; dens. 1.03 kg/l; G-H visc. I (70% in toluene); soften. pt. (R&B) 70 C; acid no. < 1; sapon. no. < 1; flash pt. (COC) 221 C.

Picco® 6070. [Hercules] Aromatic hydrocarbon resin; low m.w. nonpolar thermoplastic resin used as tackifiers in various elastomer-based solv., water, and hot-melt adhesive systems; as tackifiers, reinforcing agents, and extenders in rubber compd. and extruding applics.; in metallic cold-cut paints; suitable in heat set letterpress and web offset printing ink vehicles; FDA cleared for use in food-pkg. and processing operations; Gardner 12 (50% in toluene) solid; sol. in aromatic, aliphatic, and chlorinated hydrocarbons, low-KB aliphatic ink oils, benzyl alcohol, cyclohexanol, MEK, butyl Carbitol acetate, and diethyl Carbitol; dens. 1.02 kg/l; G-H visc. G (70% in toluene); soften. pt. (R&B) 70 C; acid no. < 1; sapon. no. < 1; flash pt. (COC) 237 C.

Picco® 6100. [Hercules] Aromatic hydrocarbon resin; tackifier, reinforcing agent, and extender in rubber compd. and extruding applics.; in metallic cold-cut paints; suitable in heat set letterpress and web offset printing ink vehicles; FDA cleared for use in food-pkg. and processing operations; Gardner 11 solid, flakes; sol. in aromatic, aliphatic, and chlorinated hydrocarbons, low-KB aliphatic ink oils, benzyl alcohol, cyclohexanol, MEK, butyl Carbitol acetate, and diethyl Carbitol; dens. 1.06 kg/l; G-H visc. T (70% in toluene); soften. pt. (R&B) 100 C; acid no. < 1; sapon. no. < 1; flash pt. (COC) 246 C.

Picco® 6115. [Hercules] Aromatic hydrocarbon resin; tackifier, reinforcing agent, and extender in rubber compd. and extruding applics.; in metallic cold-cut paints; suitable in heat set letterpress and web offset printing ink vehicles; FDA cleared for use in food-pkg. and processing operations; Gardner 11 solid, flakes; sol. in aromatic, aliphatic, and chlorinated hydrocarbons, low-KB aliphatic ink oils, benzyl alcohol, cyclohexanol, MEK, butyl Carbitol acetate, and diethyl Carbitol; dens. 1.07 kg/l; G-H visc. W (70% in toluene); soften. pt. (R&B) 115 C; acid no. < 1; sapon. no. < 1; flash pt. (COC) 260 C.

Piccodiene® 2215SF. [Hercules] Aliphatic hydrocarbon resin produced largely from petrol.-derived dicyclopentadiene monomer; low m.w. heat-reactive thermoplastic resin with acid-, alkali-, and water-resistance, fast solv. release, high gloss, and balanced tack, adhesive, and cohesive props. when blended with elastomers; used as tackifiers, stiffeners, processing aids in rubber compds., fire carcass stocks, SBR-based mastics/sealants; fluxing resins to aid filler and pigment disp.; in traffic paints; FDA cleared for used in food pkg./processing; Gardner 11 solid, flakes; sol. in aliphatic, aromatic, and chlorinated hydrocarbons; water-insol.; dens. 1.13 kg/l; G-H visc. K-L (60% in min. spirits); soften. pt. (R&B) 102 C; acid no. < 1; sapon. no. < 2; flash pt. (COC) 224 C.

Piccolastic® A5. [Hercules] Hydrocarbon resin derived from a pure styrene monomer; low m.w. nonpolar thermpolastic resin offering lt. color, and solubility, and wide compatibility; softener and primary plasticizer in hot-melt compositions, adhesives, coatings, and rubber compd.; food-pkg. and processing operations; Gardner 3 liq.; sol. in aromatic, aliphatic, and chlorinated hydrocarbons, ketones, pyridine, carbon bisulfide, ethyl and butyl acetates, and turpentine; insol. in water; dens. 1.04 kg/l; visc. 20 stokes (40 C); soften. pt. 5 C; acid no. < 1; sapon. no. < 1; ref. index 1.57; flash pt. (COC) 165 C.

Piccolastic® A75. [Hercules] Hydrocarbon resin derived from a pure styrene monomer; low m.w. nonpolar thermpolastic resin offering lt. color, and solubility, and wide compatibility; softener and primary plasticizer in hot-melt compositions, adhesives, coatings, and rubber compd.; food-pkg. and processing operations; Gardner 2 hard solid; sol. in aromatic, aliphatic, and chlorinated hydrocarbons, ketones, pyridine, carbon bisulfide, ethyl and butyl acetates, and turpentine; insol. in water; dens. 1.06 kg/l; soften. pt. 75 C; acid no. < 1; sapon. no. < 1; ref. index 1.60; flash pt. (COC) 260 C.

Piccolyte® A115. [Hercules] Hydrocarbon resin derived from the terpene monomer alphapinene; low m.w. pale, inert, thermoplastic resin characterized by lt. color, high gloss, broad compat. and sol., and balanced tack, adhesive, and cohesive props.; tackifiers in pressure sensitive and hot-melt systems based on natural and syn. rubbers; reinforcing resins in rubber compds.; gloss-promoters in paints, varnishes, and coatings; in sealants, caulks, and rubber; in lost-wax investment castings and in waterproofing agents; food contact use; Gardner 4 (50% in toluene) solid; sol. in aliphatic, aromatic, and chlorinated hydrocarbons, ketones, VM&P naphtha, and ethyl acetate; dens. 0.98 kg/l; melt visc. 200 C (1 poise); soften. pt. (R&B) 115 C; ref. index 1.53; flash pt. (COC) 216 C.

Piccolyte® A125. [Hercules] Hydrocarbon resin derived from the terpene monomer α-pinene; low m.w. pale, inert, thermoplastic resin characterized by lt. color, high gloss, broad compat. and sol., and balanced tack, adhesive, and cohesive props.; tackifiers in pressure sensitive and hot-melt systems based on natural and syn. rubbers; reinforcing resins in rubber compds.; gloss-promoters in paints, varnishes, and coatings; in sealants, caulks, and rubber; in lost-wax investment castings and in waterproofing agents; food contact use; Gardner 4 (50% in toluene) solid; sol. in aliphatic, aromatic, and chlorinated hydrocarbons, ketones, VM&P naphtha, and ethyl acetate; dens. 0.98 kg/l; melt visc. 210 C (1 poise); soften. pt. (R&B) 125 C; ref. index 1.53; flash pt. (COC) 219 C.

Piccolyte® A135. [Hercules] Hydrocarbon resin derived from the terpene monomer α-pinene; low m.w. pale, inert, thermoplastic resin characterized by lt. color, high gloss, broad compat. and sol., and balanced tack, adhesive, and cohesive props.; tackifiers in pressure sensitive and hot-melt systems based on natural and syn. rubbers; reinforcing resins in rubber compds.; gloss-promoters in paints, varnishes, and coatings; in sealants, caulks, and rubber; in lost-wax investment castings and in waterproofing agents; food contact use; Gardner 5 (50% in toluene) solid; sol. in aliphatic, aromatic, and chlorinated hydrocarbons, ketones, VM&P naphtha, and ethyl acetate; dens. 0.98 kg/l; melt visc. 220 C (1 poise); soften. pt. (R&B) 135 C; ref. index 1.53; flash pt. (COC) 238 C.

Piccolyte® C115. [Hercules] Polydipentene; CAS 9003-73-0; thermoplastic resin used as tackifier in adhesives, modifier resin in rubber compds., in hot-melt coatings, laminations, wax modification; also cleared under FDA regulations for use as masticatory agents in chewing gum compositions; Gardner 4 (50% in toluene) solid, flakes; sol. in aliphatic, aromatic, chlorinated hydrocarbons, min. oil, VM&P naphtha, turpentine, ether, amyl and butyl acetates, long-chain aliphatic alcohols; dens. 0.99 kg/l; melt visc. 204 C (1 poise); soften. pt. (R&B) 115 C; flash pt. (COC) 235 C.

Piccolyte® C125. [Hercules] Terpene hydrocarbon resin produced from d-limonene; thermoplastic resin used as tackifier in adhesives, modifier resin in rubber compds., in hot-melt coatings, laminations, wax modification; also cleared under FDA regulations for use as masticatory agents in chewing gum compositions; Gardner 4 (50% in toluene) solid, flakes; sol. in aliphatic, aromatic, and chlorinated hydrocarbons, min. oil, VM&P naphtha, turpentine, ether, amyl and butyl acetates, long-chain aliphatic alcohols; dens. 0.99 kg/l; melt visc. 214 C; soften. pt. (R&B) 125 C; flash pt. (COC) 241 C.

Piccolyte® C135. [Hercules] Terpene hydrocarbon resin produced from d-limonene; thermoplastic resin used as tackifier in adhesives, modifier resin in rubber compds., in hot-melt coatings, laminations, wax modification; also cleared under FDA regulations for use as masticatory agents in chewing gum compositions; Gardner 4 (50% in toluene) solid, flakes; sol. in aliphatic, aromatic, and chlorinated hydrocarbons, min. oil, VM&P naphtha, turpentine, ether, amyl and butyl acetates, long-chain aliphatic alcohols; dens. 0.99 kg/l; melt visc. 224 C; soften. pt. (R&B) 131 C; flash pt. (COC) 260 C.

Piccolyte® S115. [Hercules] Hydrocarbon resin produced from the monomeric terpene betapinene; Thermoplastic resin as tackifier for adhesives, tapes, hot melts, in rubber compds., medical goods; modifier for EVA resins and waxes; waterproofing resin for textile sizing; in sealants, printing inks; cleared under FDA regulations for use in closures with sealing gaskets for food containers, markings on fruits and vegs., chewing gum bases; Gardner 2 solid; sol. in aliphatic, aromatic, and chlorinated hydrocarbons, esters, ethers, long-chain alcohols, higher ketones; insol. in water; dens. 0.99 kg/l; melt visc. 220 C; soften pt. 115 C; flash pt. (COC) 234 C.

Piccolyte® S125. [Hercules] Hydrocarbon resin produced from the monomeric terpene betapinene; thermoplastic resin as tackifier for adhesives, tapes, hot melts, in rubber compds., medical goods; modifier for EVA resins and waxes; waterproofing resin for textile sizing; in sealants, printing inks; cleared under FDA regulations for use in closures with sealing gaskets for food containers, etc.; Gardner 2 (50% in toluene) solid; sol. in aliphatic, aromatic, and chlorinated hydrocarbons, esters, ethers, long-chain alcohols, higher ketones; insol. in water; dens. 0.99 kg/l; melt visc. 225 C (1 poise); soften. pt. (R&B) 125 C; flash pt. (COC) 238 C.

Piccolyte® S135. [Hercules] Hydrocarbon resin produced from the monomeric terpene betapinene; thermoplastic resin as tackifier for adhesives, tapes, hot melts, in rubber compds., medical goods; modifier for EVA resins and waxes; waterproofing resin for textile sizing; in sealants, printing inks; S135 offers additional clearances under FDA regulations for use in closures with sealing gaskets for food containers, chewing gum base; Gardner 2 (50% in toluene) solid, flakes; sol. in aliphatic, aromatic, and chlorinated hydrocarbons, esters, ethers, long-chain alcohols, higher ketones; insol. in water; dens. 0.99 kg/l; melt visc. 232 C (1 poise); soften. pt. (R&B) 135 C; flash pt. (COC) 259 C.

Picconol® A100. [Hercules] Aliphatic hydrocarbon resin emulsions; anionic; resin emulsion used in combination with other aq. thermoplastic and/or elastomeric systems to produce coatings, paints, and adhesives; tackifiers for natural and syn. rubber systems; waterproof finishes for paper, textiles, and textile backings; A100 esp. suitable for use in high-styrene SBR, EVA resin, and polyacrylic-based latex paints, and in adhesives for bonding films, fibers, and granular materials; food-pkg. and processing operations; Gardner 9 (50% in toluene) liq.; dens. 0.97 kg/l; visc. 3000 cps; soften. pt. (R&B) 63 C; pH 8.5; 50% aq. disp.

Piccopale® 100. [Hercules] Aliphatic hydrocarbon resins derived mainly from dienes and other reactive olefin monomers; tackifier, binder for rubber, adhesives, construction materials; saturant/waterproofing agent for paper, textiles; replacements for gloss oils in paints, varnishes, and coatings; dens. 0.97 kg/l (all); melt. visc. 230 C (1 poise); soften. pt. (R&B) 100 C; COC flash pt. 259 C; acid no. < 1; sapon. no. < 2.

Piccotac® 95BHT. [Hercules] Aliphatic hydrocarbon resin; narrow m.w. distribution resin for the adhesives industry; tackifier for hot-melt adhesive applics.; in rubber-based pressure sensitive adhesives and high-ethylene E/VA resin-based hot-melts; as a saturant and waterproofing agent in various paper, paperboard, and textile prods.; in hot-melt road-marking compds.; Gardner 5 (50% solids in toluene) solid, flakes; sol. in aliphatic and aromatic hydrocarbons, MIBK, ethyl ether, and long-chain alcohols; dens. 0.96 kg/l; melt visc. 205 C (1 poise); soften. pt. (R&B) 95 C; acid no. < 1; sapon. no. < 2; flash pt. (COC) 480 F.

Piccotac® B. [Hercules] Aliphatic hydrocarbon resin derived largely from mixed monomers of petrol. origin; low m.w. thermoplastic resin offering lt. color, balance of tack, adhesive, and cohesive props., heat resistance, and wide compat. and sol.; tackifier for natural rubber, polyisoprene, and SIS rubbers in pressure sensitive adhesives; as a modifier for waxes used in lost-wax investment casting operations and in EVA resin/wax/low m.w. polyethylene hot-melts; avail. in two stabilized grades (BBHT and BHM); Gardner 5+ (50% solids in toluene) solid, flakes; sol. in aliphatic and aromatic hydrocarbons, long-chain alcohols (ethyl-hexanol), ethylene dichloride, MIBK, and trichloroethylene; dens. 0.97 kg/l; visc. 360 cps (70% solids in toluene); soften. pt. (R&B) 100 C (BBHT) and 98 C (BHM); acid no. < 1; sapon. no. < 1; flash pt. (COC) 249 C.

Piccovar® AP10. [Hercules] Aromatic hydrocarbon resin; plasticizer, softener and tackifier in rubber compd., pressure sensitive applic., and adhesive systems; food processing operations; Gardner 13 soft solid; sol. in heptane, octane, and min. spirits, aromatic and chlorinated hydrocarbons, ethyl acetate, ethyl ether, and higher m.w. alcohols; insol. in water; dens. 1.02 kg/l; visc. 215 SUS (99 C); soften. pt. 11 C; acid no. < 1; sapon. no. < 2; flash pt. (COC) 113 C.

Piccovar® AP25. [Hercules] Aromatic hydrocarbon resin; plasticizer, softener and tackifier in rubber compd., pressure sensitive applic., and adhesive systems; food processing operations; Gardner 13 soft solid; sol. in heptane, octane, and min. spirits, aromatic and chlorinated hydrocarbons, ethyl acetate, ethyl ether, and higher m.w. alcohols; insol. in water; dens. 1.01 kg/l; visc. 100 SFS (99 C); soften. pt. 32 C; acid no. < 1; sapon. no. < 2; flash pt. (COC) 221 C.

Piccovar® L60. [Hercules] Dicylopentadiene alkylaryl hydrocarbon resins; thermoplastic plasticizer and tackifier in solution- and hot-melt applied coatings and adhesives; food pkg.; compatible with petrol. waxes, EVA copolymers, polyethylene, and high m.w. hydrocarbon polymers; Gardner 11 solid; sol. in aliphatic, aromatic, chlorinated, and mixed hydrocarbon solvs., ethers, esters, and ketones; insol. in water; dens. 1.05 kg/l; melt visc. 129 C; soften. pt. 58 C; acid no. < 1; sapon. no. < 2; flash pt. (COC) 243 C.

Plasadd® PE 9017. [Cabot Plastics Int'l.] Antistatic masterbatch based on LDPE; antistatic masterbatch for film extrusion, blow molding, inj. molding; compat. with LDPE, LLDPE, HDPE, EVA.

Plasadd® PE 9134. [Cabot Plastics Int'l.] HALS plus absorber combination in LDPE; uv stabilizer masterbatch for film extrusion, blow molding, inj. molding; compat. with LDPE, LLDPE, HDPE, EVA; recommended for critical sealing and printing applics.

Plasblak® EV 1755. [Cabot Plastics Int'l.] EVA masterbatch with 50% SRF carbon blk.; blk. masterbatch offering adequate weathering props., good opacity, easy processing for molding and extrusion applics.; compatible with LDPE, HDPE, PP, PS; pellets; dens. 1.23 mg/m^3; melt flow 40 g/10 min.

Plasblak® HD 3300. [Cabot Plastics Int'l.] Masterbatch with 25% carbon black in HDPE; specialty blk. masterbatch for very thin HDPE films; also for pipe applics.

Plasblak® LL 2590. [Cabot Plastics Int'l.] 40% Carbon blk. (small particle size) in LLDPE carrier with pentaerythrityl tetrakis (3,5-di-tert-butyl-4-hydroxyphenyl propionate) antioxidant; blk. masterbatch for cold potable water and cable sheathing applics.; max. protection against uv degradation; food contact use; pellets.

Plasblak® LL 2612. [Cabot Plastics Int'l.] Masterbatch with 40% special carbon black in LLDPE; stabilized masterbatch for critical HDPE and LLDPE film applics.; also compat. with LDPE; suitable for food contact use.

Plasblak® LL 3282. [Cabot Plastics Int'l.] 40% carbon blk. (avg. particle size 23 nm) in LLDPE copolymer carrier; blk. masterbatch for high-quality pkg. film and agric. film demanding max. resist. to weathering and approval for contact with foodstuffs and which also need to run at fast LLDPE extrusion rates; compatible with LDPE, HDPE, PP, and ethylene copolymers; pellets; dens. 1150 kg/m^3; melt flow 55 g/10 min.

Plasblak® PE 1371. [Cabot Plastics Int'l.] 40% carbon blk. (avg. particle size 20 nm) in LDPE carrier; blk. masterbatch for pipes, high-quality pkg. film and agric. film demanding max. resist. to weathering and approval for contact with foodstuffs; compatible with LDPE, LLDPE, HDPE, PP; pellets or micropulverized form; dens. 1140 kg/m^3; melt flow 15 g/10 min.

Plasblak® PE 1380. [Cabot Plastics Int'l.] 40% furnace blk. (large particle size) in polyethylene carrier; blk. masterbatch offering adequate weatherability, good opacity, easy processing; pellets; dens. 932 kg/m^3; melt flow 2.3 g/10 min; tens. str. 8.34 Mn/m^2; tens. elong. 514%; Vicat soften. pt. 95 C; Unverified

Plasblak® PE 1851. [Cabot Plastics Int'l.] 50% SRF carbon blk. (avg. particle size 60 nm) in LDPE carrier with calcium carbonate extender as antiblocking agent; blk. masterbatch for film, molding, and extrusion applics.; compatible with LDPE, HDPE, PP, and ethylene copolymers; pellets; dens. 1380 kg/m^3; melt flow 30 g/10 min.

Plasblak® PE 1873. [Cabot Plastics Int'l.] 40% SRF carbon blk. (avg. particle size 60 nm) in LDPE carrier with calcium carbonate extender as antiblocking agent; blk. masterbatch designed for film, molding, and extrusion applics., e.g., building film, agric. film, general purpose molding and extrusion; compatible with LDPE, HDPE, PP, and ethylene copolymers; pellets; dens. 1420 kg/m^3; melt flow 55 g/10 min.

Plasblak® PE 2249. [Cabot Plastics Int'l.] 50% Elftex 160 type carbon blk. (avg. particle size 50 nm) in LDPE carrier; blk. masterbatch for extrusion of general purpose film, pipe, sheathing, and for molding applics.; compatible with LDPE, HDPE, PP, and ethylene copolymers; pellets; dens. 1230 kg/m^3; melt flow 8 g/10 min.; Unverified

Plasblak® PE 2272. [Cabot Plastics Int'l.] 50% SRF carbon blk. (avg. particle size 60 nm) in LDPE carrier; blk. masterbatch for extrusion of general purpose film, pipe, sheathing, and for molding applics.; compatible with LDPE, HDPE, PP, and ethylene copolymers; pellets; dens. 1230 kg/m^3; melt flow 75 g/10 min.

Plasblak® PE 2377. [Cabot Plastics Int'l.] 40% SRF carbon blk. (avg. particle size 60 nm) in LDPE carrier with calcium carbonate extender as antiblocking agent; blk. masterbatch for film, molding, and extrusion applics., e.g., building film, general purpose agric. film, inj. molding, and pipe extrusion; compatible with LDPE, HDPE, PP, and ethylene copolymers; pellets; dens. 1510 kg/m^3; melt flow 52 g/10 min.

Plasblak® PE 2479. [Cabot Plastics Int'l.] 30% SRF carbon blk. (avg. particle size 60 nm) in LDPE carrier; blk. masterbatch for film, molding, and extrusion applics., e.g., building film, general purpose agric. film, inj. molding, and extrusion; compatible with LDPE, LDPE copolymers, LLDPE, HDPE, PP; pellets; dens. 1500 kg/m^3; melt flow 46 g/10 min.

Plasblak® PE 2515. [Cabot Plastics Int'l.] 40% carbon blk. (avg. particle size 27 nm) in LDPE carrier with calcium carbonate extender as antiblocking agent; blk. masterbatch for film, molding, and extrusion applics., e.g., building film, general purpose agric. film, inj. molding, and extrusion; compatible with LDPE, LDPE copolymers, LLDPE, HDPE, PP; pellets; dens. 1280 kg/m^3; melt flow 36 g/10 min.; Unverified

Plasblak® PE 2570. [Cabot Plastics Int'l.] 27% carbon blk. (avg. particle size 20 nm) in LDPE copolymer carrier with calcium carbonate extender as antiblocking agent; blk. masterbatch giving max. protection against uv degradation for film, molding, and extrusion applics., e.g., building film, general purpose agric. film, inj. molding, and extrusion; copatible with LDPE and LDPE copolymers; pellets; dens. 1300 kg/m^3; melt flow 30 g/10 min.; Unverified

Plasblak® PE 3006. [Cabot Plastics Int'l.] Masterbatch with 40% carbon black in LDPE; general-purpose pipe-grade masterbatch providing good uv protection; compat. with LDPE, LLDPE, HDPE.

Plasblak® PE 3168. [Cabot Plastics Int'l.] Masterbatch with 50% carbon black in LDPE; masterbatch for thin film and specialty film applics.; compat. with LDPE, LLDPE, HDPE; esp. for critical LDPE films requiring max. dispersion.

Plasblak® PE 4056. [Cabot Plastics Int'l.] Masterbatch with 25% med. color carbon black and filler in LDPE; low cost blk. masterbatch for budget molding and recycling applics.; compat. with LDPE, LLDPE, HDPE, PP.

Plasblak® PE 4135. [Cabot Plastics Int'l.] 40% carbon blk. (small particle size) in LDPE carrier; blk. masterbatch giving maximum protection against uv degradation for use in agric. films with good resist. to fertilizers, potable water applics., coatings for metal pipes, pipes for conveying liqs. or gases under pressure; pellets; dens. 1.130 kg/m^3; melt flow 60 g/10 min.; Unverified

Plasblak® PP 3393. [Cabot Plastics Int'l.] 30% carbon blk. (avg. particle size 27 nm) in PP homopolymer carrier; blk. masterbatch for general purpose PP molding and extrusion applics.; pellets; dens. 1030 kg/m^3; melt flow 16 g/10 min.; Unverified

Plasblak® PP 3583. [Cabot Plastics Int'l.] 40% Carbon blk. (avg. particle size 20 nm) in PP copolymer carrier; blk. masterbatch for general purpose PP molding and extrusion applics. demanding optimum resist. to uv weathering, e.g., reservoir linings; food contact use; pellets; dens. 1160 kg/m^3; melt flow 20 g/10 min.

Plasblak® PP 3585. [Cabot Plastics Int'l.] 40% carbon blk. (avg. particle size 20 nm) in high flow PP copolymer carrier; blk. masterbatch designed for fiber extrusion applics.; also suitable for general purpose extrusion and molding; gives maximum protection against uv degradation; food contact use; pellets; dens. 1220 kg/m^3; melt flow 100 g/10 min (21.6 kg, 230 C).

Plasblak® PP 4045. [Cabot Plastics Int'l.] 40% carbon blk. (avg. particle size 20 nm) in PP homopolymer carrier; blk. masterbatch for general purpose PP fibers, molding, and extrusion applics.; pellets; dens. 1110 kg/m^3; melt flow 55 g/10 min.; Unverified

Plasblak® PS 0469. [Cabot Plastics Int'l.] 30% Carbon blk. (avg. particle size 17 nm) in PS carrier; blk. masterbatch specially designed for coloring PS; used for applics. such as packing materials (boxes, lids), electronic equip. cases (casettes), radios, TVs, domestic appliances, food contact; compat. with PS, SAN, ABS; pellets; dens. 1200 kg/m^3; melt flow 55 g/10 min.

Plasblak® PS 1844. [Cabot Plastics Int'l.] 35% carbon blk. (avg. particle size 23 nm) in crystal PS carrier; blk. masterbatch specially designed for coloring PS; used for applics. such as packing materials (boxes, lids), electronic equip. cases (casettes), radios, TVs, domestic appliances, food contact; compat. with PS, SAN, ABS; pellets; dens. 1160 kg/m^3; melt flow 12 g/10 min.; Unverified

Plasblak® PS 3294. [Cabot Plastics Int'l.] 50% SRF carbon blk. (avg. particle size 60 nm) in PS carrier; blk. masterbatch designed for coloring PS in general purpose extrusion and molding applics.; compat. with PS, SAN, ABS; pellets; dens. 1330 kg/m^3; melt flow 12 g/10 min.

Plasblak® PS 4054. [Cabot Plastics Int'l.] 40% Carbon blk. (avg. particle size 27 nm) in crystal PS carrier; blk. masterbatch designed for coloring crystal and semi-impact PS; recommended for use in coloring thin section articles or where a lower addition rate is required; food contact use; compat. with PS, SAN, ABS; pellets; dens. 1100 kg/m^3; melt flow 5 g/10 min.

Plasblak® SA 3176. [Cabot Plastics Int'l.] 30% carbon blk. (avg. particle size 17 nm) in SAN carrier; blk. masterbatch designed for coloring ABS, SAN, and PS; used for molding articles for pkg. (boxes, lids), electronic equip. cases (cassettes), radios, TVs, domestic appliances; pellets; dens. 950 kg/m^3; melt flow 20 g/10 min.; Unverified

Plasblak® UN 2014. [Cabot Plastics Int'l.] 50% Carbon blk. (avg. particle size 17 nm) in LDPE copolymer carrier; blk. masterbatch designed for high gloss, jet color pigmentation of PS, ABS, SAN, and PVC for inj. molding; food contact use; pellets; also avail. in micropulverized form as UM 2014; dens. 1220 kg/m^3; melt flow 30 g/10 min.

Plasblak® UN 2016. [Cabot Plastics Int'l.] 50% Carbon blk. (avg. particle size 27 nm) in LDPE copolymer carrier; blk. masterbatch designed for regular color pigmentation of PS, ABS, SAN, and PVC for inj. molding; food contact use; pellets; also avail. in micropulverized form as UM 2016; dens. 1220 kg/m^3; melt flow 20 g/10 min; flamm. HB.

Plas-Chek® 775. [Ferro/Bedford] Epoxidized soybean oil; CAS 8013-07-8; EINECS 232-391-0; stabilizer, plasticizer for PVC compds.; uses incl. food wrap film, refrigerator gaskets, medical tubing and bags, chlorinated paraffin stabilizer, toys, wall coverings, halogen and acid scavenger, PVC foams, infant wear film and sheet, pigment disps., flooring, beverage tubing, ubholstery; Gardner 1 max.; negligible odor; m.w. 1000; sol. in most org. solvs.; sp.gr. 0.992 min.; visc. 5 stokes max.; flash pt. 600 F; pour pt. 25 F; acid pt. 0.5 max.; 7.0% min. oxirane oxygen.

Plas-Chek® 795. [Ferro/Bedford] Epoxidized linseed oil; aux. plasticizer for PVC compds.; sol. in most org. solvs.; Unverified

Plasdeg PE 9321. [Cabot Plastics Int'l.] LDPE/starch-based masterbatch; degradation promoter imparting controlled degradability to polyolefins incl. LDPE, LLDPE, ethylene copolymers, HDPE, and PP; mainly used in LDPE, LLDPE, and copolymer films and moldings; FDA compliance; pellets.; Unverified

Plastech EP 8126. [Cabot Plastics Int'l.] PP copolymer, rubber-modified; rubber masterbatch modifying impact resist. of PP; melt index 0.8 g/10 min.

Plasthall® 6-10P. [C.P. Hall] plasticizer for PVC providing superior low temp. performance, improved thermal oxidative stability, reduced flamm.

Plasthall® 8-10 TM. [C.P. Hall] Linear trimellitate; plasticizer; Gardner 1-2 clear liq.; sol. in hexane, toluene, kerosene, acetone, min. oil; insol. in water; m.w. 595; sp.gr. 0.971; visc. 107.5 cps; acid no. 0.1; iodine no. 1.0; sapon. no. 283; flash pt. 279 C; ref. index 1.482; 0.06% moisture.

Plasthall® 8-10 TM-E. [C.P. Hall] n-Octyl, n-decyl trimellitate; plasticizer for polychloroprene compds., acrylate elastomers.

Plasthall® 83SS. [C.P. Hall] Dibutoxyethoxyethyl sebacate substitute; plasticizer for nitrile rubber, PVAc; Gardner 18 clear liq.; sol. in hexane, toluene, kerosene, ethyl alcohol, acetone, min. oil, ASTM oil #1; m.w. 505; sp.gr. 0.989; visc. 30 cps; f.p. -10 C; acid no. 1.9; sapon. no. 202; flash pt. 152 C; ref. index 1.446; 0.05% moisture.

Plasthall® 100. [C.P. Hall] Isooctyl tallate; EINECS 269-788-3; plasticizer for SBR, CR; yel. to amber liq.; sol. in hexane, toluene, kerosene, min. oil; insol. in water; m.w. 400; sp.gr. 0.873; visc. 20 cps; f.p. -60 C; acid no. 0.9; sapon. no. 145; flash pt. 216 C; ref. index 1.4592; 0.10% moisture.

Plasthall® 200. [C.P. Hall] Dibutoxyethyl phthalate; CAS 117-83-9; plasticizer for PU; Gardner 2 clear liq.; sol. in toluene, kerosene, ethanol, acetone; insol. in water; m.w. 366; sp.gr. 1.060; visc. 46 cps; f.p. 50 C; acid no. 0.9; sapon. no. 307; flash pt. 225 C; ref. index 1.484; 0.05% moisture.

Plasthall® 201. [C.P. Hall] Dibutoxyethyl glutarate; plasticizer for PVAc; APHA 200 clear liq.; sol. in hexane, toluene, kerosene, ethanol, acetone; partly sol. in min. oil; insol. in water; m.w. 335; sp.gr. 1.002; visc. 17 cps; f.p. < -60 C; acid no. 0.8; iodine no. nil; sapon. no. 338; flash pt. 193 C; ref. index 1.440; 0.10% moisture.

Plasthall® 203. [C.P. Hall] Dibutoxyethyl adipate; plasticizer for PVAc, CR; APHA 40 clear liq.; sol. in hexane, toluene, kerosene, ethanol, acetone; partly sol. in min. oil; insol. in water; m.w. 346; sp.gr. 0.995; visc. 20 cps; f.p. -34 C; acid no. 0.7; iodine no. nil; sapon. no. 328; flash pt. 188 C; ref. index 1.441; 0.10% moisture.

Plasthall® 205. [C.P. Hall] Dibutoxyethyl azelate; plasticizer; Gardner 2 clear liq.; sol. in hexane, toluene, kerosene, ethanol, acetone, min. oil; insol. in water; m.w. 384; sp.gr. 0.975; visc. 20 cps; f.p. -25 C; acid no. 0.6; sapon. no. 292; flash pt. 210 C; ref. index 1.4444; 0.10% moisture.

Plasthall® 207. [C.P. Hall] Dibutoxyethyl sebacate; plasticizer for PVAc, CR; APHA 100 clear liq.; sol. in hexane, toluene, kerosene, ethanol, acetone, min. oil; insol. in water; m.w. 402; sp.gr. 0.969; visc. 25 cps; f.p. -20 C; acid no. 1.0; sapon. no. 280; flash pt. 238 C; ref. index 1.445; 0.05% moisture.

Plasthall® 220. [C.P. Hall] Dibutoxyethoxyethyl phthalate; plasticizer for PU; Gardner 3 clear liq.; sol. in hexane, toluene, kerosene, ethanol, acetone, min. oil; insol. in water; m.w. 454; sp.gr. 1.072; f.p. -41C; acid no. 0.71; sapon. no. 252; flash pt. 204 C; ref. index 1.4830; 0.04% moisture.

Plasthall® 224. [C.P. Hall] Dibutoxyethoxyethyl glutarate; plasticizer for PVAc; APHA 200 clear liq.; sol. in hexane, toluene, kerosene, ethanol, acetone; partly sol. in propylene glycol; insol. in water; m.w. 423; sp.gr. 1.016; visc. 22 cps; f.p. < -60 C; acid no. 0.9; iodine no. nil; sapon. no. 245; flash pt. 143 C; ref. index 1.4437; 0.10% moisture.

Plasthall® 226. [C.P. Hall] Dibutoxyethoxyethyl adipate; CAS 141-17-3; plasticizer for PVAc, CR; APHA 100 clear liq.; sol. in hexane, toluene, kerosene, ethanol, acetone; partly sol. in propylene glycol; insol. in water; m.w. 434; sp.gr. 1.01; visc. 25 cps; f.p. -25 C; acid no. 0.5; sapon. no. 232; flash pt. 152 C; ref. index 1.4445; 0.07% moisture.

Plasthall® 325. [C.P. Hall] Butoxyethyl oleate; plasticizer for NBR (20% ACN); Gardner 5 clear oily liq.; sol. in hexane, toluene, kerosene, ethanol, acetone, min. oil; insol. in water; m.w. 379; sp.gr. 0.888; visc. 15 cps; f.p. -35 C; acid no. 1.0; iodine no. 70; sapon. no. 148; flash pt. 202 C; ref. index 1.42; 0.05% moisture.

Plasthall® 503. [C.P. Hall] Butyl oleate; CAS 142-77-8; EINECS 205-559-6; textile surf. finisher, softener, thread lubricant and antistat; plasticizer for SBR, CR; Gardner 11 clear oily liq.; sol. in hexane, toluene, kerosene, ethyl alcohol, acetone, min. oil, ASTM oil #1; insol. in water; m.w. 388; sp.gr. 0.875; visc. 14 cps; f.p. -55 C; acid no. 1.5; iodine no. 127; sapon. no. 165; flash pt. 179 C; ref. index 1.457; 0.10% moisture.

Plasthall® 914. [C.P. Hall] n-Butyl oleate; CAS 142-77-8; EINECS 205-559-6; plasticizer for SBR, CR.

Plasthall® 4141. [C.P. Hall] PEG-3 caprate-caprylate; plasticizer for rubber goods, CR, PC; lubricant for aluminum can industry; Gardner 2 clear liq.; sol. in hexane, toluene, kerosene, ethanol, acetone, min. oil; partly sol. in propylene glycol; insol. in water; m.w. 430; sp.gr. 0.968; visc. 25 cps; f.p. -5 C; acid no. 0.5; sapon. no. 260; hyd. no. 15; flash pt. 213 C; ref. index 1.446; 0.08% moisture.

Plasthall® 7006. [C.P. Hall] Di (alkyl alkylether) adipate; monomeric plasticizer for nitrile rubber, PVAc; APHA 125 clear liq.; sol. in hexane, toluene, kerosene, ethanol, acetone, min. oil; m.w. 353; sp.gr. 0.973; visc. 18 cps; f.p. -65 C; flash pt. 188 C; acid no. 0.9; sapon. no. 318; ref. index 1.443; 0.08% moisture.

Plasthall® 7041. [C.P. Hall] Long chain alkyl alkylether diester; plasticizer for rubber, CR, PVAc; transfer aid on correctable ribbon; penetration and tack agent for computer ribbons, carbon paper; yel. liq.; sol. in hexane, toluene, kerosene, ethanol, acetone, min. oil; insol. in water; m.w. 492; sp.gr. 0.925; visc. 15 cps; f.p. -51 C; acid no. 0.5; iodine no. nil; sapon. no. 228; flash pt. 182 C; ref. index 1.449; 0.08% moisture.

Plasthall® 7045. [C.P. Hall] Alkyl alkylether diester; plasticizer for PVAc.

Plasthall® 7049. [C.P. Hall] Alkyl oleate; plasticizer for SBR; transfer aid on correctable ribbon; penetration and tack agent for computer ribbons, carbon paper; Gardner 5 clear oily liq.; sol. in hexane, toluene, kerosene, ethanol, acetone, min. oil; insol. in water; sp.gr. 0.874; visc. 25 cps; f.p. -55 C; acid no. 1.3; iodine no. 69; sapon. no. 145; flash pt. 191 C; ref. index 1.4590; 0.10% moisture.

Plasthall® 7050. [C.P. Hall] Dialkyl diether glutarate; monomeric plasticizer for adhesives applics., nitrile rubber; yel. liq.; sol. in toluene, ethyl alcohol, acetone, propylene glycol, min. oil, ASTM oil #1; partly sol. in water; m.w. 450; sp.gr. 1.068; visc. 40 cps; f.p. -60 C; acid no. 0.7; iodine no. nil; sapon. no. 249; flash pt. 193 C; ref. index. 1.447; 0.10% moisture.

Plasthall® 7059. [C.P. Hall] Isooctyl oleate; plasticizer for SBR, CR.

Plasthall® BSA. [C.P. Hall] N,N-Butyl benzene sulfonamide; CAS 3622-84-2; EINECS 222-823-6; plasticizer for nylon 6/6 and high polarity polymers, emulsion adhesives, pkg., caulk, printing ink, surf. coatings; APHA 35 clear liq.; sol. in toluene, ethanol, acetone, propylene glycol; insol. in water; m.w. 213; sp.gr. 1.146; visc. 110 cps; acid no. 0.3; flash pt. 194 C; ref. index 1.522.

Plasthall® CF. [C.P. Hall] Mixed acid diester; plasticizer for PVC, CR; APHA 50 clear liq.; sol. in hexane, toluene, kerosene, ethanol, acetone, min. oil; insol. in water; m.w. 429; sp.gr. 0.915; f.p. -54 C; acid no. 0.3; sapon. no. 262; flash pt. 222 C; ref. index 1.450.

Plasthall® DBEA. [C.P. Hall] Di (butoxyethyl) adipate; monomeric plasticizer for nitrile rubber; APHA 40 color; m.w. 346; sp.gr. 0.995; f.p. -34 C; acid no. 0.7; flash pt. 188 C; 0.10% moisture.

Plasthall® DBEEA. [C.P. Hall] Di (butoxyethoxyethyl) adipate; CAS 141-17-3; monomeric plasticizer for nitrile rubber; APHA 100 color; m.w. 494; sp.gr. 1.010; f.p. -25 C; acid no. 0.5; flash pt. 152 C; 0.07% moisture.

Plasthall® DBP. [C.P. Hall] Dibutyl phthalate; CAS 84-74-2; EINECS 201-557-4; plasticizer; APHA 10 clear liq.; sp.gr. 1.10465; f.p. -40 C; flash pt. 340 F; ref. index 1.4930; 99% act.

Plasthall® DBS. [C.P. Hall] Dibutyl sebacate; CAS 109-43-3; EINECS 203-672-5; plasticizer for PVC; APHA 50 clear liq.; sol. in hexane, toluene, kerosene, ethanol, acetone, min. oil; insol. in water; m.w. 316; sp.gr. 0.934; visc. 14 cps; f.p. -11 C; acid no. 0.2; sapon. no. 355; flash pt. 185 C; ref. index 1.440; 0.05% moisture.

Plasthall® DBZZ. [C.P. Hall] Dibenzyl azelate; plasticizer; yel. liq.; sol. in toluene, kerosene, ethanol, acetone; insol. in water; m.w. 372; sp.gr. 1.072; visc. 38 cps; f.p. 4 C; acid no. 0.6; iodine no. nil; sapon. no. 302; flash pt. 218 C; ref. index 1.5204; 0.20% moisture.

Plasthall® DIBA. [C.P. Hall] Diisobutyl adipate; CAS 141-04-8; EINECS 205-450-3; plasticizer for PVC; APHA 30 clear liq.; sol. in hexane, toluene, kerosene, ethanol, acetone, min. oil; insol. in water; m.w. 259; sp.gr. 0.950; visc. 13 cps; f.p. -17 C; acid no. 0.1; iodine no. nil; sapon. no. 433; flash pt. 204 C; ref. index 1.432; 0.06% moisture.

Plasthall® DIBZ. [C.P. Hall] Diisobutyl azelate; plasticizer; amber liq.; sol. in hexane, toluene, kerosene, ethanol, acetone, min. oil; insol. in water; m.w. 303; sp.gr. 0.932; visc. 15 cps; f.p. -30 C; acid no. 0.6; iodine no. nil; sapon. no. 369; flash pt. 160 C; ref. index 1.4355; 0.20% moisture.

Plasthall® DIDA. [C.P. Hall] Diisodecyl adipate; CAS 27178-16-1; EINECS 248-299-9; plasticizer for PVC, CR; APHA 50 clear liq.; sol. in hexane, toluene, kerosene, ethanol, acetone, min. oil; insol. in water; m.w. 427; sp.gr. 0.918; visc. 30 cps; f.p. -43 C; flash pt. 227 C; acid no. 0.1; iodine no. nil; sapon. no. 263; flash pt. 227 C; ref. index 1.450; 0.10% moisture.

Plasthall® DIDG. [C.P. Hall] Diisodecyl glutarate; plasticizer for PVC, CR; lubricant additive; APHA 50 clear liq.; sol. in hexane, toluene, kerosene, ethanol, acetone, min. oil; insol. in water; m.w. 416; sp.gr. 0.920; visc. 23 cps; f.p. -65 C; acid no. 0.3; iodine no. nil; sapon. no. 275; flash pt. 204 C; ref. index 1.450; 0.08% moisture.

Plasthall® DIDP-E. [C.P. Hall] Diisodecyl phthalate; CAS 68515-49-1; EINECS 247-977-1; plasticizer; APHA 15 clear liq.; sp.gr. 0.9650; f.p. -48 C; acid no. 0.005; sapon. no. 251; flash pt. 450 F; ref. index 1.4840; 99.8% act.

Plasthall® DINP. [C.P. Hall] Diisononyl phthalate; CAS 68515-48-0; EINECS 249-079-5; plasticizer; APHA 15 clear liq.; sp.gr. 0.9725; f.p. -45 F; flash pt. 435 F; ref. index 1.485; 99.6% min. act.

Plasthall® DIOA. [C.P. Hall] Diisooctyl adipate; plasticizer for PVC; APHA 20 clear liq.; sol. in hexane, toluene, kerosene, ethanol, acetone, min. oil; insol. in water; m.w. 373; sp.gr. 0.92; visc. 18 cps; f.p. < -60 C; acid no. 0.3; iodine no. nil; sapon. no. 301; flash pt. 179 C; ref. index 1.446; 0.05%

moisture.

Plasthall® DIODD. [C.P. Hall] Diisooctyl dodecanedioate; lubricant additive; plasticizer for PVC; APHA 50 clear liq.; sol. in hexane, toluene, kerosene, ethyl alcohol, acetone, min. oil, ASTM oil #1; insol. in water; m.w. 454; sp.gr. 0.909; f.p. -70 C; acid no. 0.1; sapon. no. 246; flash pt. 238 C; ref. index 1.450.

Plasthall® DIOP. [C.P. Hall] Diisooctyl phthalate; CAS 27554-26-3; plasticizer; APHA 15 clear liq., low odor; sp.gr. 0.9164; visc. 58 cps; acid no. 0.05 max.; flash pt. (COC) 216 C; ref. index 1.485; 99.6% min. act.

Plasthall® DOA. [C.P. Hall] Dioctyl adipate; CAS 103-23-1; EINECS 203-090-1; monomeric plasticizer for nitrile rubber, CR, PVC; APHA 45 clear liq.; sol. in hexane, toluene, kerosene, ethanol, acetone, min. oil; m.w. 373; sp.gr. 0.925; visc. 18 cps; f.p. -65 C; acid no. 0.3; iodine no. nil; sapon. no. 301; flash pt. 193 C; ref. index 1.445; 0.08% moisture.

Plasthall® DODD. [C.P. Hall] Dioctyl dodecanedioate dioate; lubricant additive; useful as textile surf. finishes, softeners, thread lubricants and/or antistats; plasticizer for PVC; APHA 50 clear liq.; sol. in hexane, toluene, kerosene, ethanol, acetone, min. oil; insol. in water; m.w. 454; sp.gr. 0.909; f.p. -45 C; acid no. 0.3; sapon. no. 247; flash pt. 222 C; ref. index 1.450; 0.05% moisture.

Plasthall® DOP. [C.P. Hall] Di-2-ethylhexyl phthalate; CAS 117-81-7; EINECS 204-211-0; monomeric plasticizer for nitrile rubber, CR.

Plasthall® DOS. [C.P. Hall] Dioctyl sebacate; plasticizer for PVC, CR; APHA 25 clear liq.; sol. in hexane, toluene, kerosene, ethanol, acetone, min. oil; insol. in water; m.w. 430; sp.gr. 0.913; visc. 25 cps; f.p. -65 C; acid no. 0.2; sapon. no. 261; flash pt. 216 C; ref. index 1.499; 0.05% moisture.

Plasthall® DOSS. [C.P. Hall] Dioctyl sebacate substitute; plasticizer for rubber industry, CR, PVC; amber liq.; sol. in hexane, toluene, kerosene, ethyl alcohol, acetone, min. oil, ASTM oil #1; insol. in water; m.w. 445; sp.gr. 0.918; visc. 51 cps; f.p. -55 C; acid no. 1.5; sapon. no. 264; flash pt. 232 C; ref. index 1.449; 0.05% moisture.

Plasthall® DOZ. [C.P. Hall] Dioctyl azelate; CAS 103-24-2; plasticizer for PVC, CR; APHA 60 clear liq.; sol. in hexane, toluene, kerosene, ethanol, acetone, min. oil; insol. in water; m.w. 412; sp.gr. 0.915; visc. 25 cps; f.p. -60 C; acid no. 0.3; iodine no. nil; sapon. no. 272; flash pt. 199 C; ref. index 1.448; 0.05% moisture.

Plasthall® ESO. [C.P. Hall] Epoxidized soybean oil; CAS 8013-07-8; EINECS 232-391-0; plasticizer for PS; APHA 100 clear liq.; sol. in toluene, ethanol, acetone; partly sol. in hexane, kerosene, min. oil; insol. in water; m.w. 1000; sp.gr. 0.993; visc. 450 cps; f.p. 5 C; acid no. 0.2; iodine no. 1.5; sapon. no. 183; flash pt. 310 C; ref. index 1.471; 0.04% moisture.

Plasthall® HA7A. [C.P. Hall] Polyester adipate; plasticizer for epoxy resins; APHA 150 clear visc. liq.; sol. in toluene; partly sol. in kerosene, ethanol, acetone, min. oil; insol. in water; m.w. 3000; sp.gr. 1.153; visc. 22,000 cps; f.p. -10 C; acid no. 1.5; iodine no. nil; sapon. no. 589; flash pt. 266 C; ref. index 1.4662; 0.08% moisture.

Plasthall® LTM. [C.P. Hall] Linear trimellitate; plasticizer; Gardner 1-2 clear liq.; sol. in hexane, toluene, kerosene, acetone, min. oil; insol. in water; m.w. 585; sp.gr. 0.972; visc. 120 cps; f.p. -55 C; acid no. 0.08; iodine no. < 1; sapon. no. 288; flash pt. 260 C; ref. index 1.482; 0.06% moisture.

Plasthall® NODA. [C.P. Hall] n-Octyl, n-decyl adipate; plasticizer for PVC; APHA 25 clear liq.; sol. in hexane, toluene, kerosene, ethanol, acetone, min. oil; insol. in water; m.w. 390; sp.gr. 0.916; visc. 13 cps; f.p. -4 C; acid no. 0.1; iodine no. nil; sapon. no. 288; flash pt. 198 C; ref. index 1.445; 0.04% moisture.

Plasthall® P-545. [C.P. Hall] Polyester; plasticizer; APHA 150 clear liq.; sp.gr. 1.059; visc. 4200 cps; acid no. 0.6; sapon. no. 436; flash pt. 520 F; ref. index 1.4750; 0.1% max. moisture.

Plasthall® P-550. [C.P. Hall] Polyester glutarate; plasticizer for PVC, epoxies; flexibilizing, permanent, nonmigrating; adhesive for film backing and varieties of tape; APHA 150 clear visc. liq.; sol. in toluene; partly sol. in kerosene, ethanol, acetone, min. oil; insol. in water; m.w. 2500; sp.gr. 1.063; visc. 4200 cps; f.p. -41 F; acid no. 0.6; iodine no. nil; sapon. no. 485; flash pt. 520 F; ref. index 1.4638; 0.05% moisture.

Plasthall® P-612. [C.P. Hall] Polyester adipate; plasticizer; APHA 100 clear liq.; sp.gr. 1.04; visc. 400 cps; f.p. -24 F; acid no. 1.5; sapon. no. 530; flash pt. 440 F; ref. index 1.4580; 0.1% max. moisture.

Plasthall® P-622. [C.P. Hall] Polyester adipate; plasticizer; APHA 100 clear liq.; sp.gr. 0.4746; visc. 6500 cps; acid no. 27.5; flash pt. 400 F; ref. index 1.4746; 0.05% moisture.

Plasthall® P-643. [C.P. Hall] Polyester adipate; plasticizer for PVC, epoxies; flexibilizing, permanent, nonmigrating; adhesive for film backing and varieties of tape; APHA 100 visc. liq.; sol. in toluene; partly sol. in kerosene, ethanol, acetone, min. oil; insol. in water; m.w. 2000; sp.gr. 1.08; visc. 2650 cps; f.p. -50 C; acid no. 1.0; iodine no. nil; sapon. no. 486; flash pt. 282 C; ref. index 1.4649; 0.05% moisture.

Plasthall® P-650. [C.P. Hall] Polyester adipate; plasticizer for epoxies; APHA 100 visc. liq.; sol. in toluene; partly sol. in kerosene, ethanol, acetone, min. oil; insol. in water; m.w. 1200; sp.gr. 1.05; visc. 3300 cps; f.p. -10 C; acid no. 0.5; iodine no. nil; sapon. no. 470; flash pt. 260 C; ref. index; 0.10% moisture.

Plasthall® P-670. [C.P. Hall] Polyester adipate; plasticizer for epoxies; APHA 100 visc. liq.; sol. in toluene; partly sol. in kerosene, ethanol, acetone, min. oil; insol. in water; m.w. 800; sp.gr. 1.08; visc. 1200 cps; f.p. -45 C; acid no. 1.0; iodine no. nil; sapon. no. 505; flash pt. 265 C; ref. index 1.464; 0.10% moisture.

Plasthall® P-1070. [C.P. Hall] Polyester sebacate; plasticizer for epoxies; Gardner 3 visc. liq.; sol. in hexane, toluene, ethanol, acetone; partly sol. in kerosene, min. oil; insol. in water; m.w. 2000; sp.gr. 1.069; visc. 5000 cps; f.p. -22 C; acid no. 1.0; sapon. no. 455; flash pt. 218 C; ref. index 1.465; 0.05% moisture.

Plasthall® P-7035. [C.P. Hall] Polyester glutarate; plasticizer for flexible PVC, epoxies; Gardner 7 visc. liq.; sol. in toluene; partly sol. in kerosene, ethanol, acetone, min. oil; insol. in water; m.w. 4500; sp.gr. 1.08; visc. 11,000 cps; f.p. -12 C; acid no. 1.0; iodine no. nil; sapon. no. 498; flash pt. 260 C; ref. index 1.467; 0.06% moisture.

Plasthall® P-7035M. [C.P. Hall] Polyester glutarate; plasticizer for epoxies; Gardner 4 visc. liq.; sol. in toluene; partly sol. in kerosene, ethanol, acetone, min. oil; insol. in water; m.w. 3600; sp.gr. 1.089; visc. 6300 cps; f.p. -12 C; acid no. 1.0; iodine no. nil; sapon. no. 510; flash pt. 260 C; ref. index 1.4653; 0.06% moisture.

Plasthall® P-7046. [C.P. Hall] Polyester glutarate; plasticizer for epoxies; Gardner 5 visc. liq.; sol. in toluene; partly sol. in kerosene, ethanol, acetone, min. oil; insol. in water; m.w. 4200; sp.gr. 1.11; visc. 12,000 cps; f.p. -25 C; acid no. 1.0; iodine no. nil; sapon. no. 530; flash pt. 266 C; ref. index 1.466; 0.08% moisture.

Plasthall® P-7068. [C.P. Hall] Polyester phthalate; plasticizer for epoxies, CR; Gardner 1 clear liq.; sol. in hexane, toluene, kerosene, ethanol, acetone, min. oil; insol. in water; m.w. 650; sp.gr. 1.03; visc. 900 cps; f.p. -35 C; acid no. 0.8; sapon. no. 310; flash pt. 243 C; ref. index 1.501; 0.10% moisture.

Plasthall® P-7092. [C.P. Hall] Polyester glutarate; plasticizer for epoxies; Gardner 9 visc. liq.; sol. in toluene; partly sol. in kerosene, ethanol, acetone, min. oil; insol. in water; m.w. 5000; sp.gr. 1.11; visc. 24,000 cps; f.p. -20 C; acid no. 0.8; iodine no. nil; sapon. no. 520; flash pt. 271 C; ref. index 1.467; 0.08% moisture.

Plasthall® P-7092D. [C.P. Hall] Polyester glutarate; plasticizer for epoxies; Gardner 5 visc. liq.; sol. in toluene; partly sol. in kerosene, ethanol, acetone, min. oil; insol. in water; m.w. 3500; sp.gr. 1.11; visc. 6500 cps; f.p. -25 C; acid no. 1.0; iodine no. nil; sapon. no. 530; flash pt. 271 C; ref. index 1.465; 0.05% moisture.

Plasthall® R-9. [C.P. Hall] Octyl tallate; transfer aid on correctable ribbon; penetration and tack agent for computer ribbons, carbon paper; plasticizer for SBR; yel. to amber liq.; sol. in hexane, toluene, kerosene, min. oil; insol. in water; m.w. 398; sp.gr. 0.873; visc. 17 cps; f.p. -10 C; acid no. 0.2; sapon. no. 141; flash pt. 213 C; ref. index 1.459; 0.10% moisture.

Plasthall® TIOTM. [C.P. Hall] Triisooctyl trimellitate; CAS 53894-23-8; plasticizer for acrylate elastomers; Gardner 1-2 clear liq.; sol. in hexane, toluene, kerosene, acetone, min. oil; insol. in water; m.w. 550; sp.gr. 0.990; visc. 270 cps; f.p. -52 C; acid no. 0.08; iodine no. < 1; sapon. no. 306; flash pt. 260 C; ref. index 1.484; 0.06% moisture.

Plasthall® TOTM. [C.P. Hall] Trioctyl trimellitate; CAS 3319-31-1; monomeric plasticizer for nitrile rubber, CR, acrylate elastomers; Gardner 1-2 clear liq.; sol. in hexane, toluene, kerosene, acetone, min. oil; insol. in water; m.w. 550; sp.gr. 0.990; visc. 260 cps; f.p. -54 C; acid no. 0.08; iodine no. < 1; sapon. no. 306; flash pt. 260 C; ref. index 1.484; 0.06% moisture.

Plasticizer 9. [BASF AG] Glycerol ether alcohol; plasticizer for naphtha-resist. and benzene-resist. nitro lacquers and plastics.

Plasticizer REO. [Akrochem] Mixt. of paraffinic/naphthenic process oils and petroleum sulfonates; plasticizer, process aid for syn. and natural rubber compds.; peptizer; reclaiming agent; improves filler dispersion in dry compds.; amber; sp.gr. 0.88; visc. (Gardner) A-D.

Plastiflow CW-2. [Syn. Prods.] Internal lubricant for semirigid and rigid PVC; solid.

Plastifoam. [Plastics & Chem.] Azodicarbonamide; CAS 123-77-3; EINECS 204-650-8; chem. foaming agent for ABS, acetal, acrylic, EVA, HDPE, LDPE, PPO, PP, PS, HIPS, flexible PVC, TPE; processing range 300-450 F; gas yield 220 cc/g.

Plastigel®. [Plasticolors] Chemical thickeners for polyester SMC, BMC, TMC.

Plastigel PG9033. [Plasticolors] Disp. of reactive magnesium oxide in an unsat., nonmonomer-containing polyester vehicle; thickener in the prod. of thickenable polyester molding compds. when a controlled visc. increase is required; liq.; neutral; dens. 12.5 ± 0.5 lb/gal; visc. 12,000-20,000 cps (RVT, #6 spindle, 20 rpm); 37-39% MgO.

Plastigel PG9037. [Plasticolors] Disp. of reactive magnesium hydroxide in an unsat., nonmonomer-containing polyester vehicle; thickener in the prod. of thickenable polyester molding compds. when a controlled visc. increase is required; neutral liq.; dens. 11.4 ± 0.5 lb/gal; visc. 24,000-26,000 cps; 38.4-40.4% magnesium hydroxide.; Unverified

Plastigel PG9068. [Plasticolors] Disp. of reactive calcium oxide in an unsat., nonmonomer-containing polyester vehicle; thickener in the prod. of thickenable polyester molding compds. when a controlled visc. increase is required; neutral liq.; dens. 13.8 ± 0.5 lb/gal; visc. 17,000-19,000 cps; 51.0-53.0% calcium oxide.; Unverified

Plastigel PG9089. [Plasticolors] Disp. of reactive magnesium hydroxide/calcium hydroxide in an unsat. monomer-containing vehicle, thickener in the prod. of thickenable polyester molding compds. when a controlled visc. increase is required; neutral liq.; visc. 10.500-18.000 cps; 27.5-29.5% magnesium hydroxide plus calcium hydroxide in a 3.75:1 ratio.; Unverified

Plastigel PG9104. [Plasticolors] Disp. of reactive calcium hydroxide in an unsat., monomer-containing vehicle; thickener in the prod. of thickenable polyester molding compds. when a controlled visc. increase is required; neutral liq.; dens. 9.6 ± 0.5 lb/gal; visc. 20,000-25,000 cps; 28.5-30.5% calcium hydroxide.; Unverified

Plastilease 250. [Lilly] Silicone; paintable release agent for molding of parts to be post-finished or bonded; effective in overcoming rubber-to-metal friction and vibrational squealing; dens. 6.1 lb/gal; Seta CC flash pt. 10 F; 98% volatiles; Unverified

Plastilease 512-B. [Lilly] PVAL resins; film-forming release agent providing fast drying rate and easy cleanup; spray applics. is preferred; deep grn.; dens. 7.7 lb/gal; Seta CC flash pt. 62 F; 92% volatiles; Unverified

Plastilease 512-CL. [Lilly] PVAL resins; film-forming release agent providing fast drying rate and easy cleanup; spray applics. is preferred; clear.; Unverified

Plastilease 514. [Lilly] Polyolefin milky disp.; release agent for processing resins in 200 F range; works well with preform release; textured mold surfs.; dens. 6.7 lb/gal; Seta CC flash pt. 17 F; 93% volatiles; Unverified

Plastilit® 3060. [BASF AG] Polypropylene glycol alkylphenol ether; plasticizer for polymer dispersions in the paint, construction, chemical, adhesive and sealant industries.

Plastisan B. [3V] Nonhalogenated organic; flame retardant for epoxy, PC, thermoset polyester, PU flexible and rigid foam, nylon, PET, PBT.

Plastisorb. [Plasticolors] UV and weather-resist. colorants and additives for polymer systems.

Plastisperse®. [Plasticolors] Low dusting, high loaded, dry concs. for thermoplastic elastomers, bulk molding compds.

Plastistab 2005. [OM Group] Calcium-zinc; heat and lt. stabilizer for flexible PVC, plastisols, organosols, foams; low flame spread; liq.

Plastistab 2020. [OM Group] Calcium-zinc; heat and lt. stabilizer for flexible PVC, plastisols, organosols, foams; low VOC; liq.

Plastistab 2160. [OM Group] Calcium-zinc; heat and lt. stabilizer for flexible and rigid PVC, plastisols, organosols, foams; FDA approval; powd.

Plastistab 2210. [OM Group] Calcium-zinc; heat and lt. stabilizer for flexible PVC, plastisols, organosols, foams; FDA approval; liq.

Plastistab 2310. [OM Group] Barium-zinc; heat and lt. stabilizer for flexible and semirigid PVC, plastisols, organosols, foams, automotive low-fog applics.; liq.

Plastistab 2320. [OM Group] Barium-zinc; heat and lt. stabilizer for flexible and semirigid PVC, plastisols, organosols, foams, high clarity applics.; liq.

Plastistab 2330. [OM Group] Barium-zinc; heat and lt. stabilizer for flexible and semirigid PVC, plastisols, organosols, foams; low plateout; liq.

Plastistab 2560. [OM Group] Barium-zinc; multipurpose heat and lt. stabilizer for flexible and semirigid PVC, tiles; powd.

Plastistab 2580. [OM Group] Barium-zinc; heat and lt. stabilizer for flexible and semirigid PVC, plastisols, wire and cable applics.; powd.

Plastistab 2800. [OM Group] Tin; heat and lt. stabilizer for flexible and rigid PVC, plastisols, organosols, bottles.

Plastistab 2900. [OM Group] Epoxy; heat and lt. stabilizer for flexible and rigid PVC, plastisols, organosols, foams, bottles, tile; FDA approved.

Plastistab 2931. [OM Group] Phosphite; heat and lt. stabilizer for flexible and rigid PVC, plastisols, organosols, foams, bottles.

Plastistab 2941. [OM Group] Phosphite; heat and lt. stabilizer for flexible and rigid PVC, plastisols, organosols, foams, bottles; FDA approved.

Plastoflex® 2307. [Akzo Nobel] Epoxidized unsat. soya bean oils; CAS 8013-07-8; EINECS 232-391-0; plasticizer/stabilizer for PVC; aux. plasticizer or stabilizer; uses incl. plastisols, calendered and extruded prods., wetting of pigments, chlorinated paraffins, alkyd resins, neoprene, ethyl/cellulose, PVC/PVA emulsions; Gardner 3 max. clear liq.; dens. 0.99 g/ml; visc. 3-7 poise; acid no. 1.0 max.; ref. index 1.473; 7.0% min. conc.

Plastogen. [R.T. Vanderbilt] Plasticizer and softener for elastomers, NR latex films; amber to mahogany liq.; dens. 0.82 $\pm$ 0.02 mg/m^3; acid no. 1.0-1.1.

Plastogen E®. [King Industries; Struktol] Low visc. min. oil contg. a high m.w. sulfonic acid; plasticizer, softener, and mild processing aid for elastomers; useful in sponge and low durometer stocks; amber liq.; sp.gr. 0.88 (15.6 C); dens. 7.3 lb/gal; visc. 41 SUS (37.8 C); acid no. 1.1; flash pt. (COC) 120 C.

Plastol®. [BASF; BASF AG] Low-visc. polymer plasticizers for PVC; for prods. which must have moderate resistance to chemicals.

Plastolein® 9048. [Henkel/Emery] Dibutyl azelate; CAS 2917-73-9; low temp. vinyl plasticizer esp. for cellulosic molding compds.; sp.gr. 0.94; dens. 7.8 lb/gal; solid. pt. $\leq$ -70 F; flash pt. 325 F; acid no. 0.15 max.

Plastolein® 9049. [Henkel/Emery] Low temp. plasticizer; sp.gr. 0.94; dens. 7.8 lb/gal; visc. 8 cSt (100 F); solid. pt. < -70 F; flash pt. 325 F; acid no. 0.15 max.

Plastolein® 9050. [Henkel/Emery] Plasticizer for use in food pkg. films; color 85/98% min. trans. (440/550 nm); sp.gr. 0.93; dens. 7.7 lb/gal; visc. 8.0 cSt (100 F); solid. pt. -9 F; acid no. 1.0 max.; hyd. no. 4.0 max.; ref. index 1.444; flash pt. 395 F; 45 phr in PVC, 75-mil sheet: tens. str. 3000 psi; elong. 360%; 100% mod. 1250 psi; hardness 84/80 (10 s Durometer A); brittle pt. -62 C.

Plastolein® 9051. [Henkel/Emery] Di-n-hexyl azelate; CAS 109-31-9; low temp. plasticizer for food pkg. films; sp.gr. 0.93; dens. 7.7 lb/gal; visc. 8 cSt (100 F); solid. pt. -9 F; flash pt. 395 F; acid no. 0.2 max.; ref. index 1.444.

Plastolein® 9058. [Henkel/Emery] Di-2-ethylhexyl azelate; CAS #103-24-2; CAS 103-24-2; low temp. vinyl plasticizer imparting exc. hand, drape, and softness; FDA approved; sp.gr. 0.92; dens. 7.6 lb/gal; visc. 10 cSt (100 F); solid. pt. -90 F; flash pt. 415 F; acid no. 1.0 max.; ref. index 1.446.

Plastolein® 9071. [Henkel/Emery] Low-temp. plasticizer for PVC compds.; sp.gr. 0.91; dens. 7.5 lb/gal; visc. 13.5 cSt (100 F); solid. pt. 10 F; acid no. 1.0 max.; hyd. no. 4.0 max.; ref. index 1.451; flash pt.

390 F; 50 phr in PVC, 75-mil sheet: tens. str. 3050 psi; elong. 325%; 100% mod. 1475 psi; hardness 89/86 (10 s Durometer A); brittle pt. -58 C.

Plastolein® 9091. [Henkel/Emery] Low-temp. plasticizer for PVC; dk. color; sp.gr. 0.92; dens. 7.6 lb/gal; visc. 11 cSt (100 F); solid. pt. -60 F; acid no. 1.0 max.; hyd. no. 4.0 max.; ref. index 1.447; flash pt. 410 F; 45 phr in PVC, 75-mil sheet: tens. str. 3225 psi; elong. 295%; 100% mod. 1575 psi; hardness 89/86 (10 s Durometer A); brittle pt. -57 C.

Plastolein® 9232. [Henkel/Emery] Epoxy soya; CAS 8013-07-8; EINECS 232-391-0; plasticizer; offers low extraction and low volatility; heat and lt. stability to PVC formulations by acting as HCl scavenger; Gardner 2; All properties 50:50 blend with DOP: sp.gr. 0.99; dens. 8.2 lb/gal; visc. 161 cSt (100 F); solid. pt. 19 F; acid no. 1.0 max.; ref. index 1.470; flash pt. 585 F; fire pt. 649 F; 54 phr in PVC sheet: tens. str. 2650 psi; tens. elong 345%; hardness 86/81 (Durometer A).

Plastolein® 9404. [Henkel/Emery] Triethylene glycol dipelargonate; CAS 106-06-9; low temp. plasticizer for syn. elastomers and natural rubbers; sp.gr. 0.965; dens. 8.1 lb/gal; visc. 59 cSt (100 F); solid. pt. < 0 F; flash pt. 410 F; ref. index 1.448.

Plastolein® 9717. [Henkel/Emery] Polymeric plasticizer; low visc. plasticizer providing permanence and low-temp. performance at min. cost; suitable for pigment grinding; Gardner 5 max.; sp.gr. 1.02; dens. 8.5 lb/gal; visc. 255 cSt (100 F); solid. pt. 18 F; acid no. 2.5 max.; ref. index 1.469; flash pt. 450 F; fire pt. 500 F; 56 phr in PVC sheet: tens. str. 3100 psi; tens. elong. 340%; hardness 87/82 (Durometer A).

Plastolein® 9720. [Henkel/Emery] Polymeric plasticizer; easily processed and handled, low volatility and resistance to oil extraction; used in plastisols and calendered fabrics; Gardner 5 max.; sp.gr. 1.03; dens. 8.6 lb/gal; visc. 207 cSt (100 F); solid. pt. 18 F; acid no. 3.0 max.; ref. index 1.462; flash pt. 500 F; fire pt. 560 F; 56 phr in PVC sheet: tens. str. 3125 psi; tens. elong. 355%; hardness 87/80 (Durometer A).

Plastolein® 9749. [Henkel/Emery] Med. m.w. polymeric general purpose plasticizer offering exc. balance of performance properties incl. resistance to humidity aging; Gardner 5 color; sp.gr. 1.06; dens. 8.9 lb/gal; visc. 857 cSt (100 F); solid. pt. 32 F; flash pt. 535 F; acid no. 3.0 max.; ref. index 1.477.

Plastolein® 9752. [Henkel/Emery] Low-to-med. visc. polymeric plasticizer with exc. processing properties, good permanence; used for appliance cords, gasketing, wall coverings, plastisols, calendered coated fabrics, automotive constructions; Gardner 3 color; sp.gr. 1.09; dens. 9.1 lb/gal; visc. 1130 cSt (100 F); solid. pt. 40 F; flash pt. 465 F; acid no. 3.0 max.; ref. index 1.496.

Plastolein® 9761. [Henkel/Emery] Polymeric plasticizer; suited for PVC compds. in contact with alkyd-type finishes, lacquers, modified and unmodified PS, and ABS; Gardner 2 max.; sp.gr. 1.06; dens. 8.9 lb/gal; visc. 1945 cSt (100 F); solid. pt. 10 F; acid no. 3.0 max.; ref. index 1.469; flash pt. 515 F; fire pt. 600 F; 62 phr in PVC sheet: tens. str. 3050 psi; tens. elong. 340%; hardness 86/80 (Durometer A).

Plastolein® 9762. [Henkel/Emery] Polymeric plasticizer for low-temp. applics.; used in vinyl tapes, elec. insulation, and upholstery; Gardner 2 max.; sp.gr. 1.08; dens. 9.0 lb/gal; visc. 457 cSt (100 F); solid. pt. 5 F; acid no. 3.0 max.; hyd. no. 20.0 max.; ref. index 1.466; flash pt. 560 F; 56 phr in PVC, 75-mil sheet: tens. str. 3080 psi; elong. 300%; 100% mod. 1450 psi; hardness 85/80 (10 s Durometer A); brittle pt. -22 C.

Plastolein® 9765. [Henkel/Emery] Polymeric plasticizer; med.-to-high visc. plasticizer with resistance to migration, extraction media, attack by fungus, and humidity aging; Gardner 6 max.; sp.gr. 1.08; dens. 9.0 lb/gal; visc. 2985 cSt (100 F); solid. pt. 35 F; acid no. 3.0 max.; ref. index 1.479; flash pt. 530 F; fire pt. 595 F; 68 phr in PVC sheet: tens. str. 3025 psi; tens. elong 340%; hardness 86/81 (Durometer A).

Plastolein® 9776. [Henkel/Emery] Polymeric plasticizer; med.-to-high visc. plasticizer used in appliance gasketing, wall coverings, appliance cords, and automotive constructions; Gardner 2 max.; sp.gr. 1.08; dens. 9.0 lb/gal; visc. 2572 cSt (100 F); solid. pt. -4 F; acid no. 2.0 max.; ref. index 1.466; flash pt. 575 F; fire pt. 600 F; 62 phr in PVC sheet: tens. str. 3200 psi; tens. elong. 335%; hardness 88/82 (Durometer A).

Plastolein® 9784. [Henkel/Emery] Polymeric plasticizer; for pressure-sensitive adhesive applics.; sp.gr. 1.07; dens. 8.9 lb/gal; visc. 1000-1250 cSt (100 F); solid. pt. -5 F; flash pt. 570 F; acid no. 3.0 max.; ref. index 1.469.

Plastolein® 9789. [Henkel/Emery] Polymeric plasticizer; formulations requiring max. permanence; calendering tracking aid and tackifier; Gardner 4 max.; sp.gr. 1.08; dens. 9.0 lb/gal; visc. 16,000 cSt (100 F); solid. pt. -20 F; acid no. 5.0 max.; ref. index 1.460; flash pt. 580 F; fire pt. 635 F; 67 phr in PVC sheet: tens. str. 2800 psi; tens. elong. 325%; hardness 84/79 (Durometer A).

Plastolein® 9790. [Henkel/Emery] Polymeric plasticizer; high m.w. plasticizer offering outstanding performance and permanence where lt. color is of minimal importance; Gardner 7 max.; sp.gr. 1.08; dens. 9.0 lb/gal; visc. 16,000 cSt (100 F); solid. pt. -20 F; acid no. 5.0 max.; ref. index 1.460; flash pt. 580 F; fire pt. 635 F; 67 phr in PVC sheet: tens. str. 2775 psi; tens. elong. 320%; hardness 85/79 (Durometer A).

Plastolube. [Plastics & Chem.] Calcium stearate; CAS 1592-23-0; EINECS 216-472-8; internal/external lubricant for vinyls, phenolics, PP, ABS, reinforced polyester; solid.

Plastomag® 170. [Morton Int'l./Plastics Additives] Oil-dispersed magnesium oxide; CAS 1309-48-4; EINECS 215-171-9; chemical thickener for polyester resins; anticaking agent; used in syn. rubber compding., adhesives, fuel oil additives, and as acid acceptor for specialty plastics; soft micropellets; 75% thru 40 mesh; 20% on 40 mesh; 5% on 20 mesh; sp.gr. 2.45; bulk dens. 50 lb/ft^3 (aerated); 65% Elastomag® 170; 35% naphthenic process oil; 98% MgO.

Plastomoll®. [BASF AG] Adipic acid esters; for low-temp. resist. and light-stable nitro lacquers; plasticizers for plasticized PVC prods. with low brittle temps.

Plaswite® LL 7014. [Cabot Plastics Int'l.] 70% Rutile titanium dioxide in LLDPE carrier; wh. masterbatch for high quality LLDPE film pigmentation for applics. requiring good dispersion, thermal and lt. stability, and which need to run at fast LLDPE extrusion rates; compatible with LDPE, LLDPE, HDPE, PP, EVA; pellets; dens. 2010 kg/m^3; melt flow 0.5 g/10 min.

Plaswite® LL 7015. [Cabot Plastics Int'l.] Masterbatch with 50% titanium dioxide plus extender in LLDPE; pigment/antiblock masterbatch for general film applics.; compat. with LDPE, LLDPE, HDPE, PP.

Plaswite® LL 7041. [Cabot Plastics Int'l.] Masterbatch with 70% titanium dioxide and lithopone, plus antistat in LLDPE; wh. pigmentary masterbatch for static-sensitive applics. (general film, blow molding); compat. with LDPE, LLDPE, HDPE, PP.

Plaswite® LL 7049. [Cabot Plastics Int'l.] 60% Rutile titanium dioxide in LLDPE carrier; wh. masterbatch for high quality LLDPE film pigmentation for applics. requiring good dispersion, thermal and lt. stability, and which need to run at fast LLDPE extrusion rates; compatible with LDPE, LLDPE, HDPE, PP, EVA; pellets; dens. 1710 kg/m^3; melt flow 2 g/10 min.; Unverified

Plaswite® LL 7057. [Cabot Plastics Int'l.] Masterbatch with 60% titanium dioxide in LLDPE; stabilized high performance masterbatch specially formulated for LLDPE and HDPE thin blown film applics.; also compat. with LDPE, PP.

Plaswite® LL 7105. [Cabot Plastics Int'l.] 70% Titanium dioxide/lithopone blend in LLDPE carrier; wh. masterbatch for high quality LLDPE film pigmentation for applics. requiring good dispersion, thermal stability, and which need to run at fast LLDPE extrusion rates; compatible with LDPE, LLDPE, HDPE, PP, EVA; pellets; dens. 2010 kg/m^3; melt flow 2 g/10 min.

Plaswite® LL 7108. [Cabot Plastics Int'l.] 75% Titanium dioxide/lithopone blend in LLDPE carrier; wh. masterbatch for high quality LLDPE film pigmentation for applics. requiring good dispersion, thermal stability, and which need to run at fast LLDPE extrusion rates; compatible with LDPE, LLDPE, HDPE, PP, EVA; pellets; dens. 2018 kg/m^3; melt flow 2-3 g/10 min.

Plaswite® LL 7112. [Cabot Plastics Int'l.] Masterbatch with 70% titanium dioxide in LLDPE; wh. weathering resist. masterbatch for agric. film; heat stabilized; compat. with LDPE, LLDPE, HDPE, PP.

Plaswite® PE 7000. [Cabot Plastics Int'l.] 75% Rutile titanium dioxide/calcium carbonate extender in LDPE carrier; wh. masterbatch for general purpose polyethylene film pigmentation where antiblocking props. are required for inj. molding; compatible with LDPE, LLDPE, HDPE, PP; pellets; dens. 2070 kg/m^3; melt flow 7 g/10 min.

Plaswite® PE 7001. [Cabot Plastics Int'l.] 75% Rutile titanium dioxide/calcium carbonate extender in LDPE carrier; wh. masterbatch for heavy-duty polyethylene film pigmentation where antiblocking props. are required, and for inj. molding; compatible with LDPE, LLDPE, HDPE, PP, EVA; pellets; dens. 1900 kg/m^3; melt flow 5 g/10 min.

Plaswite® PE 7002. [Cabot Plastics Int'l.] 75% Rutile titanium dioxide in LDPE carrier; wh. masterbatch for high quality, high gloss polyethylene film with an exc. surf. for printing; compatible with LDPE, LLDPE, HDPE, PP, EVA; pellets; dens. 2180 kg/m^3; melt flow 4 g/10 min.

Plaswite® PE 7003. [Cabot Plastics Int'l.] 65% Rutile titanium dioxide in LDPE carrier; wh. masterbatch for coloring various articles incl. thin-walled containers such as yogurt pots, disposable drinking cups, domestic appliances, toys, pkg. materials for food contact; compatible with LDPE, HDPE; pellets; dens. 1840 kg/m^3; melt flow 6 g/10 min.; Unverified

Plaswite® PE 7004. [Cabot Plastics Int'l.] 40% Rutile titanium dioxide/30% calcium carbonate extender in LDPE carrier; wh. masterbatch for polyethylene film pigmentation and inj. molding; compatible with LDPE; pellets; dens. 1.86 mg/m^3; melt flow 8 g/10 min.

Plaswite® PE 7005. [Cabot Plastics Int'l.] 70% Titanium dioxide/lithopone in LDPE carrier; wh. masterbatch for high-quality polyethylene film pigmentation and inj. molding; compatible with LDPE, LLDPE, HDPE, PP, EVA; pellets; dens. 2070 kg/m^3; melt flow 10 g/10 min.

Plaswite® PE 7006. [Cabot Plastics Int'l.] 75% Rutile titanium dioxide in LDPE carrier; wh. masterbatch for pigmentation of high quality polyethylene film with good aging props., inj. molding; compatible with LDPE, HDPE; pellets; dens. 2.18 mg/m^3; melt flow 4 g/10 min.; Unverified

Plaswite® PE 7024. [Cabot Plastics Int'l.] 60% Rutile titanium dioxide in LDPE carrier; wh. masterbatch for high quality, high gloss film with an exc. surf. for printing; compatible with LDPE, LLDPE, HDPE, PP, EVA; pellets; dens. 1710 kg/m^3; melt flow 8 g/10 min.

Plaswite® PE 7031. [Cabot Plastics Int'l.] 70% Rutile titanium dioxide with calcium carbonate extender as antiblocking agent in LDPE carrier; wh. masterbatch for high quality polyethylene film where antiblocking props. and good printability are required; compatible with LDPE, LLDPE, HDPE, PP, EVA; pellets; dens. 1950 kg/m^3; melt flow 15 g/10 min.

Plaswite® PE 7079. [Cabot Plastics Int'l.] 50% Rutile titanium dioxide in specified LDPE carrier; wh. masterbatch designed for high temp. performance for blown and cast film applics. such as thin film for diaper linings; compatible with LDPE, LLDPE, HDPE, PP, EVA; pellets; dens. 1500 kg/m^3; melt flow 3 g/10 min.

Plaswite® PE 7097. [Cabot Plastics Int'l.] 75% Rutile titanium dioxide/calcium carbonate extender in LDPE carrier; wh. masterbatch for general purpose polyethylene film pigmentation where antiblocking props. are required and for inj. molding; compatible with LDPE, LLDPE, HDPE, PP, EVA; pellets; dens. 2020 kg/m^3; melt flow 7 g/10 min.; Unverified

Plaswite® PE 7192. [Cabot Plastics Int'l.] 75% Rutile titanium dioxide/calcium carbonate extender in LDPE carrier; wh. masterbatch for heavy-duty polyethylene film pigmentation where antiblocking props. are required and for inj. molding; compatible with LDPE, LLDPE, HDPE, PP; pellets; dens. 1980 kg/m^3; melt flow 5 g/10 min.

Plaswite® PP 7161. [Cabot Plastics Int'l.] 50% Rutile titanium dioxide in PP homopolymer carrier, heat-stabilized; wh. masterbatch for mfg. of various articles incl. thin-walled containers such as yogurt pots, disposable drinking cups, domestic appliances, toys, pkg. materials for food contact; compat. with PP; specially designed for prod. of economic articles where reduced wall thickness is required; compatible with PP; pellets; dens. 1.480 mg/m^3; melt flow 28 g/10 min.

Plaswite® PP 7269. [Cabot Plastics Int'l.] 65% Rutile titanium dioxide in PP homopolymer carrier, heat-stabilized; wh. masterbatch for mfg. of various articles incl. thin-walled containers such as yogurt pots, disposable drinking cups, domestic appliances, toys, pkg. materials for food contact; compat. with PP; pellets; dens. 1830 kg/m^3; melt flow 31 g/10 min.

Plaswite® PS 7028. [Cabot Plastics Int'l.] 60% Titanium dioxide in crystal PS carrier; wh. masterbatch for mfg. of various articles incl. thin-walled containers such as yogurt pots, disposable drinking cups, domestic appliances, toys, pkg. materials for food contact; compatible with PS; pellets.

Plaswite® PS 7037. [Cabot Plastics Int'l.] Masterbatch with 70% titanium dioxide and lithopone in PS; low abrasion, easy-flow masterbatch for molding and thermoforming applics.; compat. with PS, SAN, ABS.

Plaswite® PS 7131. [Cabot Plastics Int'l.] Masterbatch with 15% titanium dioxide and extender in PS; cost-effective masterbatch for thick-sectioned moldings, thermoforming; compat. with PS, SAN, ABS.

Plaswite® PS 7174. [Cabot Plastics Int'l.] 50% Rutile titanium dioxide in crystal PS carrier, with additional toner; wh. masterbatch for mfg. of various articles incl. thin-walled containers such as yogurt pots, disposable drinking cups, domestic appliances, toys, pkg. materials; compatible with PS, SAN, ABS; pellets.

Plaswite® PS 7179. [Cabot Plastics Int'l.] Masterbatch with 40% titanium dioxide in PS; easy diluting masterbatch with positive effect on impact str. resist.; for inj. molding and thermoforming; compat. with PS, SAN, ABS.

Plaswite® PS 7231. [Cabot Plastics Int'l.] 70% Rutile titanium dioxide/lithopone extender in crystal PS carrier; wh. masterbatch for mfg. of various articles incl. thin-walled containers such as yogurt pots, disposable drinking cups, domestic appliances, toys, pkg. materials for food contact; compatible with PS; pellets; dens. 2180 kg/m^3; melt flow 13 g/10 min.

Pliabrac® 519. [Albright & Wilson Am.] Triaryl phosphate ester made from isopropyl phenol and phenol; CAS 68937-41-7; flame retardant plasticizer for NBR, CR, CPE elastomers, polyacrylate, fluoroelastomers, PVC, PVAc, cellulosics, PS, ABS, engineering resins, polyester, alloys, urea-formaldehyde resins, antifungal PVC, lubricants, hydraulic fluids; 8.3% P.

Pliabrac® 521. [Albright & Wilson Am.] Triaryl phosphate ester made from isopropyl phenol and phenol; CAS 68937-41-7; flame retardant plasticizer for NBR, NBR/PVC, CR, CPE elastomers, polyacrylate, fluoroelastomers, PVC, PVAc, cellulosics, PS, ABS, urea-formaldehyde resins, elastomers, engineering resins, polyester, alloys, inks, lubricants, hydraulic fluids; 8.1% P.

Pliabrac® 524. [Albright & Wilson Am.] Triaryl phosphate ester made from isopropyl phenol and phenol; CAS 68937-41-7; flame retardant plasticizer for NBR, NBR/PVC, CR, CPE elastomers, polyacrylate, fluoroelastomers, PVC, PVAc, cellulosics, PS, ABS, engineering resins, urea-formaldehyde, elastomers, polyester, alloys, lubricants, hydraulic fluids; exc. dipping visc. for PVC plastisol; 7.7% P.

Pliabrac® TCP. [Albright & Wilson Am.] Tricresyl phosphate; CAS 1330-78-5; EINECS 215-548-8; flame retardant/plasticizer for NBR, NBR/PVC, CR, CPE elastomers, PVC, PVAc, cellulosics, PS, ABS, urea-formaldehyde resins, elastomers, engineering resins, polyester, alloys, lubricants, hyddraulic fluids; 8.3% P.

Pliabrac® TXP. [Albright & Wilson Am.] Trixylenyl phosphate; CAS 25155-23-1; flame retardant/plasticizer for PVC, PVAc, cellulosics, PS, ABS, urea-formaldehyde resins, elastomers, PVC wire insulation, lubricants, hydraulic fluids; 7.6% P.

Pliolite® S-6B, S-6F. [Goodyear] S/B polymer; CAS 9003-55-8; reinforcers to increase hardness, stiffness, and abrasion resist. of rubber; wh. gran. solid, essentially odorless; dens. 1.04 mg/m^3; m.p. (R&B) 42-52 C; S/B ratio 82.5/17.5.

Pliovic® M-50. [Goodyear] Vinyl homopolymer resin; blending resin for modifying plastisol compds. resulting in lower visc., improved visc. stability, increased hardness, minimized surf. gloss; esp. suited for expanded vinyl foams, automotive caulks, sealants; also for rotogravure flooring, fabric coating, gasketing, glove coating, slush molded and rotocasted items; wh. fine powd.; 100% thru 80 mesh, 98.5% thru 200 mesh; inherent visc. 0.73; 100% PVC.

Pliovic® M-70. [Goodyear] Vinyl homopolymer resin; blending resin for modifying plastisol compds. resulting in lower visc., improved visc. stability, increased hardness, minimized surf. gloss; for rotogravure flooring, fabric coating, gasketing, glove coating, slush molded and rotocasted items, expanded and nonexpanded vinyl applic.; wh. fine powd.; 100% thru 80 mesh, 98.5% thru 200 mesh; inherent visc. 0.90; 100% PVC.

Pliovic® M-70SC. [Goodyear] Vinyl homopolymer resin; blending resin for modifying plastisol compds. resulting in lower visc., improved visc. stability, increased hardness, minimized surf. gloss; esp. suitable for strand coating, very thin knife or roll coating applics.; also for rotogravure flooring, fabric coating, gasketing, glove coating, slush molded and rotocasted items, expanded and nonexpanded vinyl applic.; wh. fine powd.; 100% thru 80 mesh, 99% thru 200 mesh; inherent visc. 0.90; 100% PVC.

Pliovic® M-90. [Goodyear] High m.w. vinyl homopolymer resin; blending resin for modifying plastisol compds. resulting in lower visc., improved visc. stability, increased hardness, minimized surf. gloss; good film-tear str. and visc. stability; for rotogravure flooring, fabric coating, gasketing, glove coating, slush molded and rotocasted items, expanded and nonexpanded vinyl applic.; wh. fine powd.; 100% thru 80 mesh, 99% thru 200 mesh; 100% PVC.

Pluracol® 355. [BASF] Amine-based polyol; cross-linking agent for semiflexible urethane foams, coatings, adhesives, and polymers; APHA 60 max. liq.; sp.gr. 1.03; visc. 3200 cps; pH 11.0.

Pluracol® 364. [BASF] POP deriv. of sucrose; cross-linking agent producing extra strong rigid foams; Gardner 10; sp.gr. 1.09; visc. 22,000 cps; pH 11.0.

Pluracol® 450. [BASF] POP deriv. of pentaerythritol; crosslinking agent for rigid urethane foams; m.w. 405; sp.gr. 1.08; visc. 2200 cps; acid no. 0.06; pH 6.5.

Pluracol® 550. [BASF] POP deriv. of pentaerythritol; cross-linking agent for rigid urethane foams; m.w. 500; sp.gr. 1.06; visc. 1300 cps; acid no. 0.06; pH 6.5.

Pluracol® 650. [BASF] POP deriv. of pentaerythritol; crosslinking agent for rigid urethane foams; m.w. 594; sp.gr. 1.05; visc. 1200 cps; acid no. 0.06; pH 6.5.

Pluracol® 669. [BASF] POP deriv. of sucrose; cross-linking agent producing extra strong rigid foams; Gardner 10; visc. 44,000 cps; pH 9.8.

Pluracol® E200. [BASF] PEG-4; CAS 25322-68-3; EINECS 203-989-9; intermediate for preparation of nonionic surfactants; binder, base, coating, stabilizer, solv., vehicle, extender, and coupling agent for pharmaceutical, cosmetic, and toiletries; lubricant for metal applics., rubber industry; wood treatment; textile conditioning, antistat, and sizing agent, softener; colorless clear liq.; m.w. 200; sol. in water and org. solvs. except aliphatic hydrocarbons; dens. 9.4 lb/gal; sp.gr. 1.12; visc. 4.36 cs (210 F); flash pt. 360 F; surf. tens. 57.2 dynes/cm (1%); pH 6.5 (5% aq.).

Pluracol® E300. [BASF] PEG-6; CAS 25322-68-3; EINECS 220-045-1; intermediate for preparation of nonionic surfactants; binder, base, coating, stabilizer, solv., vehicle, extender, and coupling agent for pharmaceutical, cosmetic, and toiletries; lubricant for metal applics., rubber industry; wood treatment; textile conditioning, antistat, and sizing agent, softener; dispersant in food tablets and preparations; plasticizer; colorless clear liq.; m.w. 300; sol. in water and org. solvs. except aliphatic hydrocarbons; dens. 9.4 lb/gal; sp.gr. 1.12; visc. 5.9 cs (99 C); flash pt. (COC) 210 C; pour pt. -13 C; surf. tens. 62.9 dynes/cm (1%); pH 5.7 (5% aq.).

Pluracol® E400. [BASF] PEG-8; CAS 25322-68-3; EINECS 225-856-4; intermediate for preparation of nonionic surfactants; binder, base, coating, stabilizer, solv., vehicle, extender, and coupling agent for pharmaceutical, cosmetic, and toiletries; lubricant for metal applics., rubber industry; wood treatment; textile conditioning, antistat, and sizing agent, softener; dispersant in food tablets and preparations; plasticizer; colorless clear liq.; sol. in water and org. solvs. except aliphatic hydrocarbons; dens. 9.4 lb/gal; sp.gr. 1.12; visc. 7.39 cs (210 F); flash pt. 460 F; surf. tens. 66.6 dynes/cm (1%); pH 6.2 (5% aq.).

Pluracol® E400 NF. [BASF] PEG-8; CAS 25322-68-3; EINECS 225-856-4; chemical intermediate, base, coupler, thickener, lubricant, mold release agent, defoamer, softener, conditioner, antistat, sizing agent, dispersant for pharmaceutical, cosmetic, and oral care preparations, in metal polishing and cleaning formulations, rubber prods., paper and wood prods., textile processing, ink formulations; liq.; m.w. 400; visc. 7.4 cs (99 C); pour pt. 5 C; flash pt. (COC) 182 C.

Pluracol® E600. [BASF] PEG-12; CAS 25322-68-3; EINECS 229-859-1; intermediate for preparation of nonionic surfactants; binder, base, coating, stabilizer, solv., vehicle, extender, and coupling agent for pharmaceutical, cosmetic, and toiletries; lubricant for metal applics., rubber industry; wood treatment; textile conditioning, antistat, and sizing agent, softener; dispersant in food tablets and preparations; plasticizer; colorless clear liq.; sol. in water and org. solvs. except aliphatic hydrocarbons; dens. 9.4 lb/gal; sp.gr. 1.12; visc. 10.83 cs (210 F); flash pt. 480 F; surf. tens. 65.2 dynes/cm (1%); pH 5.3 (5% aq.).

Pluracol® E600 NF. [BASF] PEG-12; CAS 25322-68-3; EINECS 229-859-1; chemical intermediate, base, coupler, thickener, lubricant, mold release agent, defoamer, softener, conditioner, antistat, sizing agent, dispersant for pharmaceutical, cosmetic, and oral care preparations, in metal polishing and cleaning formulations, rubber prods., paper and wood prods., textile processing, ink formulations; liq.; m.w. 600; visc. 10.8 cs (99 C); pour pt. 20 C; flash pt. (COC) 249 C.

Pluracol® E1000. [BASF] PEG-20; CAS 25322-68-3; chemical intermediate, base, coupler, thickener, lubricant, mold release agent, defoamer, softener, conditioner, antistat, sizing agent, dispersant for pharmaceutical, cosmetic, and oral care preparations, in metal polishing and cleaning formulations, rubber prods., paper and wood prods., textile processing, ink formulations; solid; m.w. 1000; visc. 17.5 cs (99 C); m.p. 38 C; flash pt. (COC) 255 C.

Pluracol® E1450. [BASF] PEG-32; CAS 25322-68-3; chemical intermediate, base, coupler, thickener, lubricant, mold release agent, defoamer, softener, conditioner, antistat, sizing agent, dispersant for pharmaceutical, cosmetic, and oral care preparations, in metal polishing and cleaning formulations, rubber prods., paper and wood prods., textile processing, ink formulations; solid; m.w. 1450; visc. 28.5 cs (99 C); m.p. 45 C; flash pt. (COC) 255 C.

Pluracol® E1450 NF. [BASF] PEG-32; CAS 25322-68-3; chemical intermediate, base, coupler, thickener, lubricant, mold release agent, defoamer, softener, conditioner, antistat, sizing agent, dispersant for pharmaceutical, cosmetic, and oral care preparations, in metal polishing and cleaning formulations, rubber prods., paper and wood prods., textile processing, ink formulations; solid; m.w. 600; visc. 28.5 cs (99 C); m.p. 45 C; flash pt. (COC) 255 C.

Pluracol® E1500. [BASF] PEG-6-32; intermediate for preparation of nonionic surfactants; binder, base, coating, stabilizer, solv., vehicle, extender, and coupling agent for pharmaceutical, cosmetic, and toiletries; lubricant for metal applics., rubber industry; wood treatment; textile conditioning, antistat, and sizing agent, softener; dispersant in food tablets and preparations; plasticizer; wh. waxy solid; sol. in water and org. solvs. except aliphatic hydrocarbons; dens. 10.0 lb/gal; sp.gr. 1.20; m.p. 46.0-47.5 C; flash pt. > 490 F; surf. tens. 62.8 dynes/cm (1%); pH 6.7 (5% aq.); Unverified

Pluracol® E2000. [BASF] PEG-40; CAS 25322-68-3; chemical intermediate, base, coupler, thickener, lubricant, mold release agent, defoamer, softener, conditioner, antistat, sizing agent, dispersant for pharmaceutical, cosmetic, and oral care preparations, in metal polishing and cleaning formulations, rubber prods., paper and wood prods., textile processing, ink formulations; solid; m.w. 2000; visc. 43.5 cs (99 C); m.p. 52 C; flash pt. (COC) > 260 C.

Pluracol® E4000. [BASF] PEG-75; CAS 25322-68-3; intermediate for preparation of nonionic surfactants; binder, base, coating, stabilizer, solv., vehicle, extender, and coupling agent for pharmaceutical, cosmetic, and toiletries; lubricant for metal applics., rubber industry; wood treatment; textile conditioning, antistat, and sizing agent, softener; dispersant in food tablets and preparations; plasticizer; wh. waxy solid; sol. in water and org. solvs. except aliphatic hydrocarbons; dens. 10.0 lb/gal; sp.gr. 1.20; m.p. 59.5 C; flash pt. > 490 F; surf. tens. 61.9 dynes/cm (1%); pH 6.7 (5% aq.).

Pluracol® E4500. [BASF] PEG; chemical intermediate, base, coupler, thickener, lubricant, mold release agent, defoamer, softener, conditioner, antistat, sizing agent, dispersant for pharmaceutical, cosmetic, and oral care preparations, in metal polishing and cleaning formulations, rubber prods., paper and wood prods., textile processing, ink formulations; solid; m.w. 4500; visc. 170 cs (99 C); m.p. 60 C; flash pt. (COC) > 260 C.

Pluracol® E6000. [BASF] PEG-150; CAS 25322-68-3; intermediate for preparation of nonionic surfactants; binder, base, coating, stabilizer, solv., vehicle, extender, and coupling agent for pharmaceutical, cosmetic, and toiletries; lubricant for metal applics., rubber industry; wood treatment; textile conditioning, antistat, and sizing agent, softener; wh. waxy solid; sol. in water and org. solvs. except aliphatic hydrocarbons; dens. 10.0 lb/gal; sp.gr. 1.21; m.p. 61.0 C; flash pt. > 490 F; surf. tens. 62.1 dynes/cm (1%); pH 6.7 (5% aq.); Unverified

Pluracol® E8000. [BASF] PEG-150; CAS 25322-68-3; chemical intermediate, base, coupler, thickener, lubricant, mold release agent, defoamer, softener, conditioner, antistat, sizing agent, dispersant for pharmaceutical, cosmetic, and oral care preparations, in metal polishing and cleaning formulations, rubber prods., paper and wood prods., textile processing, ink formulations; solid; m.w. 8000; visc. 750 cs (99 C); m.p. 61 C; flash pt. (COC) > 260 C.

Pluracol® P-410. [BASF] PPG-9; CAS 25322-69-4; chemical intermediate, antifoam agent in fermentation and in paint formulations, antiblooming agent for pentachlorophenol-treated wood; binder and lubricant for ceramics; plasticizer of resin-treated papers; preparation of PU foams; hydraulic and grinding fluids; ore flotation; water-wh. liq.; slight ether-like odor; water-sol.; m.w. 425; dens. 8.36 lb/gal; sp.gr. 1.005; visc. 35 cps (100 F); flash pt. > 400 F (PMCC); pour pt. -35 F; pH 6-7.

Pluracol® P-710. [BASF] PPG-12; CAS 25322-69-4; chemical intermediate, antifoam agent in fermentation and in paint formulations, antiblooming agent for pentachlorophenol-treated wood; binder and lubricant for ceramics; plasticizer of resin-treated papers; preparation of PU foams; hydraulic and grinding fluids; ore flotation; water-wh. liq.; slight ether-like odor; water-sol.; m.w. 775; dens. 8.35 lb/gal; sp.gr. 1.004; visc. 65 cps (100 F); flash pt. > 400 F (PMCC); pour pt. 35 F; pH 6-7.

Pluracol® P-1010. [BASF] PPG-17; CAS 25322-69-4; chemical intermediate, antifoam agent in fermentation and in paint formulations, antiblooming agent for pentachlorophenol-treated wood; binder and lubricant for ceramics; plasticizer of resin-treated papers; preparation of PU foams; hydraulic and grinding fluids; ore flotation; water-wh. liq.; slight ether-like odor; insol. in water; m.w. 1050; dens. 8.38 lb/gal; sp.gr. 1.007; visc. 80 cps (100 F); flash pt. > 400 F (PMCC); pour pt. -35 F; pH 6-7.

Pluracol® P-2010. [BASF] PPG-26; CAS 25322-69-4; chemical intermediate, antifoam agent in fermentation and in paint formulations, antiblooming agent for pentachlorophenol-treated wood; binder and lubricant for ceramics; plasticizer of resin-treated papers; preparation of PU foams; hydraulic and grinding fluids; ore flotation; water-wh. liq.; slight ether-like odor; insol. in water; m.w. 2000; dens. 8.34 lb/gal; sp.gr. 1.002; visc. 175 cps (100

F); flash pt. > 400 F (PMCC); pour pt. -35 F; pH 6-7.

Pluracol® P-3010. [BASF] PPG; chemical intermediate, antifoam agent in fermentation and in paint formulations, antiblooming agent for pentachlorophenol-treated wood; binder and lubricant for ceramics; plasticizer of resin-treated papers; preparation of PU foams; hydraulic and grinding fluids; ore flotation; water-wh. liq.; slight ether-like odor; insol. in water; m.w. 3000; dens. 8.32 lb/gal; sp.gr. 1.00; visc. 295 cps (100 F); flash pt. > 400 (PMCC); pour pt. -20 F; pH 6-7.

Pluracol® P-4010. [BASF] PPG-30; CAS 25322-69-4; chemical intermediate, antifoam agent in fermentation and in paint formulations, antiblooming agent for pentachlorophenol-treated wood; binder and lubricant for ceramics; plasticizer of resin-treated papers; preparation of PU foams; hydraulic and grinding fluids; ore flotation; water-wh. liq.; slight ether-like odor; insol. in water; m.w. 4000; dens. 8.33 lb/gal; sp.gr. 1.00; visc. 550 cps (100 F); flash pt. > 400 (PMCC); pour pt. -20 F; pH 6-7.

Pluracol® WD90K. [BASF] Polyalkoxylated polyether; chemical intermediate for halides, ethers, esters, urethane derivs.; defoamer for boiler water treatment; in compressor lubricants, hydraulic, brake, metalworking, and heat transfer fluids, rubber lubricants, demulsifiers, textile lubricants; liq.; m.w. 13,000; visc. 110,000 SUS (37.8 C); pour pt. 5 C; flash pt. (COC) 280 C.

Pluriol® E 200. [BASF AG] PEG; nonionic; solubilizer, impregnating agent, humectant, mold release agent; flow improver, thermal and hydraulic fluid, org. intermediate; detergent and cleaner; dye and pigment dispersant; inks; textile and coatings industry; coloring ceramics; softener in paper industry; plasticizer in adhesives industry and in prod. of cellulose film; ceramics and metalworking lubricant; clear colorless liq.; m.w. 200; water-sol.; dens. 1.12 g/cc; visc. 55-65 cs.; m.p. < -40 C; flash pt. > 150 C; pH 6.0-7.5 (1% aq.); 100% conc.

Pluriol® E 1500. [BASF AG] PEG; nonionic; solubilizer, humectant; binder and hardener in personal care prods.; dispersing dyes and pigments; inks; textile and coating industry; coloring ceramics; paper industry softener; plasticizer in adhesives industry and prod. of cellulose film; ceramics and metalworking lubricants; aux. for copper and nickel electroplating baths; electrolytic polishing of steel; wh. fine powd.; m.w. 1500; water-sol.; bulk dens. 0.6 kg/l; visc. 25-35 cs (99 C); m.p. 45 C; flash pt. > 230 C; pH 6.0-7.5 (1% aq.); 100% conc.

Pluriol® E 4000. [BASF AG] PEG; nonionic; solubilizer, humectant; binder and hardener in personal care prods.; dispersing dyes and pigments; inks; textile and coating industry; coloring ceramics; paper industry softener; plasticizer in adhesives industry and prod. of cellulose film; wh. fine powd.; m.w. 4000; water-sol.; bulk dens. 0.6 kg/l; visc. 80-130 cs (99 C); m.p. 55 C; flash pt. > 240 C; pH 6.0-7.5 (1% aq.); 100% conc.

Pluriol® E 6000. [BASF AG] PEG; nonionic; solubilizer, humectant; binder and hardener in personal care prods.; dispersing dyes and pigments; inks; textile and coating industry; coloring ceramics; paper industry softener; plasticizer in adhesives industry and prod. of cellulose film; wh. fine powd.; m.w. 6000; water-sol.; bulk dens. 0.6 kg/l; visc. 300 cs (99 C); m.p. 60 C; flash pt. > 240 C; pH 6.0-7.5 (1% aq.); 100% conc.

Pluriol® E 9000. [BASF AG] PEG; nonionic; solubilizer, humectant; binder/hardener in personal care prods.; dispersing dyes/pigments; inks; textiles; coatings; ceramics; paper industry softener; plasticizer in adhesives industry and prod. of cellulose film; mold release for rubbers; wh. fine powd.; m.w. 9000; water-sol.; bulk dens. 0.6 kg/l; visc. 1900 cs (99 C); m.p. 65 C; flash pt. > 240 C; pH 6.0-7.5 (1% aq.); 100% conc.

Pluriol® P 600. [BASF AG] PPG; nonionic; mold release agent, additive for oils and fluids; lubricant and antifoam for rubber; consistency improver and solubilizer; intermediate in industrial applics.; colorless clear liq.; misc. with water and oil; m.w. 600; dens. 1.0 g/cc; visc. 130 cs; flash pt. 216 C; pour pt. -43 C; pH 6.5-7.5 (1% aq.); 100% conc.

Pluriol® P 900. [BASF AG] PPG; nonionic; mold release agent, additive for oils and fluids; lubricant and antifoam for rubber; consistency improver and solubilizer; intermediate in industrial applics.;solv.-type cleaners; colorless clear liq.; m.w. 900; dens. 1.0 g/cc; visc. 180 cs; cloud pt. 33 C (1% aq.); flash pt. 220 C; pour pt. -38 C; pH 6.5-7.5 (1% aq.); 100% conc.

Pluriol® P 2000. [BASF AG] PPG; nonionic; mold release agent, additive for oils and fluids; lubricant and antifoam for rubber; consistency improver and solubilizer; intermediate in industrial applics.; colorless clear liq.; m.w. 2000; dens. 1.0 g/cc; visc. 440 cs; flash pt. 222 C; pour pt. -35 C; pH 6.5-7.5 (1% aq.); 100% conc.

Pluriol® PE 6200. [BASF AG] PO/EO block polymer; nonionic; wetting agent, emulsifier, demulsifier used in mechanical cleaning processes and splitting crude oil emulsions; household rinse aids; phosphatizing baths; emulsifier in polymerization processes; colorless clear liq.; sol. in water, 10% HCl, ethanol, 2-propanol, toluene and tetrachloroethylene; m.w. 2500; dens. 1.04 g/cc; visc. 500 cps; cloud pt. 32 C (1% aq.); pour pt. -14 C; ref. index 1.45; pH 7 (5% aq.); surf. tens. 44 dynes/cm; 100% conc.

Pluriol® PE 6400. [BASF AG] PO/EO block polymer; nonionic; detergent for dishwashing machines, dairy cleaners; coolant and lubricant for grinding, drilling and cutting; dispersant, emulsifier; emulsion polymerization; slightly dull liq.; sol. in water, 10% HCl, ethanol, 2-propanol, toluene and tetrachloroethylene; m.w. 3000; dens. 1.05 g/cc; visc. 850 cps; cloud pt. 59 C (1% aq.); pour pt. 16 C; ref. index 1.45; pH 7 (5% aq.); surf. tens. 44 dynes/cm; 100% conc.

Pluriol® PE 10500. [BASF AG] PO/EO block polymer; nonionic; emulsifier for agrochemicals; household and industrial cleaners; emulsion polymerization; wh. waxy; sol. in water, 10% HCl, ethanol, 2-propanol; m.w. 6500; dens. 1.03 g/cc; cloud pt. > 100 C (1% aq.); ref. index 1.45 (70 C); pH 7 (5% aq.); surf. tens. 36 dynes/cm; 100% conc.

Plysurf A207H. [Dai-ichi Kogyo Seiyaku] Complex org. phosphate ester, free acid; CAS 39464-64-7; anionic; antistat, emulsifier for agric., metal cleaning, emulsion polymerization; pigment dispersant; general purpose detergent; liq.; HLB 7.1; 99% conc.

Plysurf A208B. [Dai-ichi Kogyo Seiyaku] Complex org. phosphate ester, free acid; anionic; antistat, emulsifier for agric., metal cleaning, emulsion polymerization; pigment dispersant; general purpose detergent; liq.; HLB 6.6; 99% conc.

Plysurf A208S. [Dai-ichi Kogyo Seiyaku] Complex org. phosphate ester, free acid; anionic; antistat, emulsifier for agric., metal cleaning, emulsion polymerization; pigment dispersant; general purpose detergent; liq.; HLB 7.0; 99% conc.

Plysurf A210G. [Dai-ichi Kogyo Seiyaku] Complex org. phosphate ester, free acid; anionic; antistat, emulsifier for agric., metal cleaning, emulsion polymerization; pigment dispersant; general purpose detergent; liq.; HLB 9.6; 99% conc.

Plysurf A212C. [Dai-ichi Kogyo Seiyaku] Complex org. phosphate ester, free acid; CAS 9046-01-9; anionic; antistat, emulsifier for agric., metal cleaning, emulsion polymerization; pigment dispersant; general purpose detergent; paste; HLB 9.4; 99% conc.

Plysurf A215C. [Dai-ichi Kogyo Seiyaku] Complex org. phosphate ester, free acid; anionic; antistat, emulsifier for agric., metal cleaning, emulsion polymerization; pigment dispersant; general purpose detergent; liq.; HLB 11.5; 99% conc.

Plysurf A216B. [Dai-ichi Kogyo Seiyaku] Complex org. phosphate ester, free acid; anionic; antistat, emulsifier for agric., metal cleaning, emulsion polymerization; pigment dispersant; general purpose detergent; paste; HLB 14.4; 99% conc.

Plysurf A217E. [Dai-ichi Kogyo Seiyaku] Complex org. phosphate ester, free acid; anionic; antistat, emulsifier for agric., metal cleaning, emulsion polymerization; pigment dispersant; general purpose detergent; solid; HLB 14.9; 99% conc.

Plysurf A219B. [Dai-ichi Kogyo Seiyaku] Complex org. phosphate ester, free acid; anionic; antistat, emulsifier for agric., metal cleaning, emulsion polymerization; pigment dispersant; general purpose detergent; solid; HLB 16.2; 99% conc.

Plysurf AL. [Dai-ichi Kogyo Seiyaku] Complex org. phosphate ester, free acid; anionic; antistat, emulsifier for agric., metal cleaning, emulsion polymerization; pigment dispersant; general purpose detergent; liq.; HLB 5.6; 98.5% conc.

PMF® Fiber 204. [Pall Process Filtration] Calcium-alumino silicate; CAS 1327-39-5; filler-reinforcement; used as a partial replacement for milled and chopped glass fibers in plastics such as nylon, PP, and phenolics; used in combination with particulate fillers, provides friction and wear properties in nonasbestos brake and clutch materials; also used in nonasbestos industrial papers and felts; off-wh. short fibers; 200 μ avg. fiber length; odorless; nil sol. in water; sp.gr. 2.7; 99% min. wool fiber.

PMF® Fiber 204AX. [Pall Process Filtration] Calcium-alumino silicate surface-treated with an organosilane; CAS 1327-39-5; filler-reinforcement; treated to provide improved wetting and fiber-polymer bonding in a variety of polymers incl. phenolic, epoxy, melamine, nylon, polyimide, PBT, PVC, PC, and urethane.

PMF® Fiber 204BX. [Pall Process Filtration] Calcium-alumino silicate surface treated with an organosilane; CAS 1327-39-5; filler-reinforcement; treated to provide improved wetting and fiber-polymer bonding in a variety of polymers incl. phenolic, epoxy, melamine, nylon, polyimide, PBT, PVC, PC, and urethane.

PMF® Fiber 204CX. [Pall Process Filtration] Calcium-alumino silicate surface treated with a low level of a fatty acid-type material; CAS 1327-39-5; filler-reinforcement; treated for improved fiber disp. during dry or low-solvent blending.

PMF® Fiber 204EX. [Pall Process Filtration] Calcium-alumino silicate surface treated with an organosilane; CAS 1327-39-5; filler-reinforcement; treated to provide improved wetting and fiber-polymer bonding in DAP, polybutadiene, urethane, thermoset polyester, and polyolefin materials.

Pogol 300. [Huntsman] PEG; nonionic; solubilizer, dispersant, emulsifier; herbicides and pesticides; plasticizer for starch, paste and polyethylene sheet; clear liq.; m.w. 285-315; sp.gr. 1.13; 100% act.

Polacure® 740M. [Air Prods./Perf. Chems.] Diamine curative; curative for high-performance elastomers; FDA-approved; equiv. wt. 157.

Polamine® 250. [Air Prods./Perf. Chems.] Polytetramethylene ether glycol diamine; curative for high-performance R.T. cast and cured elastomers for cast prototypes, coatings, adhesives, sealants, and spray systems; epoxy flexibilizer; solid; visc. 460 poise (30 C); m.p. 56 C; equiv. wt. 235.

Polamine® 650. [Air Prods./Perf. Chems.] Polytetramethyleneoxide-di-p-aminobenzoate; CAS 54667-43-5; curative for high-performance R.T. cast and cured elastomers for cast prototypes, coatings, adhesives, sealants, and spray systems; epoxy flexibilizer; brown liq.; m.w. 710-950; sp.gr. 1.0-1.05 (40 C); visc. 2500 cps max. (40 C); m.p. -47 C; equiv. wt. 355-475.

Polamine® 1000. [Air Prods./Perf. Chems.] Polytetramethyleneoxide-di-p-aminobenzoate; CAS 54667-43-5; curative for high-performance R.T. cast and cured elastomers for cast prototypes, coatings, adhesives, sealants, and spray systems; epoxy flexibilizer; amber liq.; sol. in most org. solvs.; insol. in water; m.w. 1238; sp.gr. 1.01-1.06 (20 C); visc. 3000 cps max. (40 C); m.p. 18-21 C; equiv. wt. 575-625; *Toxicology:* LD50 > 5000 mg/kg; mild eye and skin irritant; nonmutagenic; *Storage:* indefinite shelf life when stored in closed container.

Polamine® 2000. [Air Prods./Perf. Chems.] Polytetramethyleneoxide-di-p-aminobenzoate; CAS 54667-43-5; curative for high-performance R.T. cast and cured elastomers for cast prototypes, coatings, adhesives, sealants, and spray systems; epoxy flexibilizer; amber waxy solid; m.w. 1820-2486; sp.gr. 0.99-1.04 (40 C); visc. < 1000 cps (85 C); m.p. 36 C; equiv. wt. 940-1243.

Polarite 420. [ECC Int'l.] Low-profile additive with filler, polymeric coating, and pigment; pigment additive for low-shrink DMC and BMC applics.; provides high gloss, improved physical props., exc. weathering, improved resin/filler bonding; free-flowing beads; 80 μ bead size; 3 μ mean particle size; sp.gr. 2.5 g/ml; 0.2% max. moisture.

Polarite 880E(W). [ECC Int'l.] Low profile additive with aluminum trihydrate and polymeric coating; low profile flame retardant/smoke suppressant additive for BMC moldings; also provides zero shrink, high gloss, uniform pigmentation, improved physical props.; wh. free-flowing beads; 80 μm bead size, 9.2 μm particle size before treatment; sp.gr. 2.35 g/ml; 0.5% surf. moisture.

Polarite 880G(B). [ECC Int'l.] Low profile additive with aluminum trihydrate, polymeric coating, and carbon black; low profile flame retardant/smoke suppressant additive for BMC moldings; also provides zero shrink, high gloss, uniform pigmentation, improved physical props.; blk. nondusty free-flowing beads; 300 μm bead size, 9.2 μm particle size before treatment; sp.gr. 2.3 g/ml; 0.5% surf. moisture.

Polirol 10. [Auschem SpA] Alkylaryl phosphate, acid form; CAS 51811-79-1; anionic; emulsifier for emulsion polymerization; liq.; 100% conc.

Polirol 1BS. [Auschem SpA] POE/POP alkyl ether; nonionic; emulsifier for emulsion polymerization; solid; 100% conc.

Polirol 215. [Auschem SpA] POE/POP alkyl ether; CAS 68002-96-0; nonionic; emulsifier for emulsion polymerization; solid; 100% conc.

Polirol 23. [Auschem SpA] Sodium alkyl ether sulfate; nonionic; emulsifier for emulsion polymerization; liq.; 25% conc.

Polirol 4, 6. [Auschem SpA] Ammonium alkylphenol ether sulfate; anionic; emulsifier for emulsion polymerization; liq.; 25% conc.

Polirol C5. [Auschem SpA] PEG-200 cocoate; CAS 67762-35-0; nonionic; emulsifier for emulsion polymerization; liq.; HLB 11.0; 100% conc.

Polirol DS. [Auschem SpA] Alkylaryl sulfonate; CAS 27176-87-0; anionic; emulsifier for emulsion polymerization; liq.; 100% conc.

Polirol L400. [Auschem SpA] Ethoxylated lauric acid; CAS 9004-81-3; nonionic; emulsifier for emulsion polymerization; liq.; HLB 14.0; 100% conc.

Polirol LS. [Auschem SpA] Sodium lauryl sulfate; CAS 151-21-3; anionic; emulsifier for emulsion polymerization; liq./paste; 28% conc.

Polirol NF80. [Auschem SpA] Nonylphenol polyglycol ether; CAS 9016-45-9; nonionic; emulsifier for emulsion polymerization; flakes; 100% conc.

Polirol O55. [Auschem SpA] Cetoleth-55; CAS 68920-66-1; nonionic; emulsifier for emulsion polymerization; flakes; 100% conc.

Polirol SE 301. [Auschem SpA] Disodium sulfosuccinate monoester; anionic; emulsifier for emulsion polymerization; liq.; 30% conc.

Polirol TR/LNA. [Auschem SpA] Ditridecyl sodium sulfosuccinate; CAS 2673-22-5; EINECS 220-219-7; anionic; emulsifier for emulsion polymerization; liq.; 60% conc.

Polu-U. [Climax Performance] Molybdic oxide, undensified; CAS 1313-27-5; EINECS 215-204-7; smoke suppressant for polymers, esp. PVC and PVC alloy; also for polyesters, thermoplastic elastomers in transportation, construction, and wire and cable applics. that must meet exacting smoke standards; pale blue-gray free-flowing powd.; 2.5 μ avg. particle size; 99.9% < 32 μ; sol. 0.68 g/100 ml water; sp.gr. 4.69; bulk dens. 23.7 lb/ft^3; oil absorp. (DOP) 41 g/100 g; ref. index > 2.1; pH 2.9 (10% slurry).

Poly TDP 2000. [Eastman] Polymeric thioester; visc. suppressant, antioxidant for polymers (ABS, cellulosics, PE, PP, PVC, polyolefins, EPDM, EVA).

Polyace™ 573. [AlliedSignal/Perf. Addit.] Ethylene-maleic anhydride copolymer; bonding/compatibilizing agent for substrates, polymers, and polymer blends for paper coatings, wood prods., adhesives/sealants, plastics (olefin blends and filled olefins); imparts abrasion resist., lubricity, antiblocking, barrier props., and surf. gloss; FDA 21CFR §175.105, 176.170, 176.180; wh. to off-wh. free-flowing prill or flake; sol. in nonpolar org. solvs., e.g., toluene, xylene; insol. in water; visc. 500 cps (140 C); drop pt. 105 C; sapon. no. < 6; hardness 4 dmm; *Precaution:* elevated temps. for prolonged periods in presence of oxygen may cause degradation; *Storage:* store in cool, dry place; avoid excess moisture; nitrogen blanketing and/or addition of antioxidants recommended for storage under extreme temp. conditions.

Polyace™ 804. [AlliedSignal/Perf. Addit.] Custom blend of polyethylene homopolymer and fully refined paraffin wax; additive enhancing props. of petrol. waxes in adhesives, candles, coatings, cosmetics, pkg., plastics, and rubber; enhances blocking resist.; quick melting; easy to disperse; FDA 21CFR §175.105, 176.180; free-flowing fine prills; 25% < 500 μm, +95% < 1700 μm; visc. 5-10 cps (210 F); drop pt. 75-79 C; congeal pt. 62-66 C; acid no. nil; *Precaution:* elevated temps. for prolonged periods in presence of oxygen may cause degradation; *Storage:* store in cool, dry place; nitrogen blanketing and/or addition of antioxidants recommended for storage under extreme temp. conditions.

Poly bd® 600. [Elf Atochem N. Am.] Epoxidized polybutadiene polymer; used as sole resin in an epoxy system or as modifiers for epoxy systems; in flexible and impact-resist. coatings and potting compds.; visc. 5500 mPa•s; EEW 460; 0.10% water.

Poly bd® 605. [Elf Atochem N. Am.] Epoxidized polybutadiene polymer; used as sole resin in an epoxy system or as modifiers for epoxy systems; in flexible and impact-resist. coatings and potting compds.; modifier in cationic uv-cured coatings; visc. 25,000 mPa•s; EEW 260; 0.10% water.

Polybond® 1000. [Uniroyal/Spec. Chem.] PP homopolymer, acrylic acid modified; thermoplastic for use as a chemical coupling agent, compatibilizing agent, metal adhesive and nucleating agent; melt flow 100; dens. 0.91 g/cc; m.p. 161 C; tens. str. 31 MPa (yield); tens. elong. 36% (ultimate); flex. mod. 1.3 GPa (secant); impact str. (Izod) 373.7 J/m (unnotched); water absorp. 0.1-0.3% (24 h); distort. temp. 56 C (264 psi); dissip. factor 2.1×10^{-4}; dielec. const. 1.7.

Polybond® 1001. [Uniroyal/Spec. Chem.] PP homopolymer, acrylic acid modified; thermoplastic for use as a chemical coupling agent, compatibilizing agent, metal adhesive and nucleating agent; melt flow 40; dens. 0.91 g/cc; m.p. 161 C; tens. str. 31 MPa (yield); tens. elong. 36% (ultimate); flex. mod. 1.3 GPa (secant); impact str. (Izod) 373.7 J/m (unnotched); water absorp. 0.1-0.3% (24 h); distort. temp. 56 C (264 psi); dissip. factor 2.1×10^{-4}; dielec. const. 1.7.

Polybond® 1002. [Uniroyal/Spec. Chem.] PP homopolymer, acrylic acid modified; thermoplastic for use as a chemical coupling agent, compatibilizing agent, metal adhesive and nucleating agent; melt flow 20; dens. 0.91 g/cc; m.p. 161 C; tens. str. 31 MPa (yield); tens. elong. 36% (ultimate); flex. mod. 1.3 GPa (secant); impact str. (Izod) 373.7 J/m (unnotched); water absorp. 0.1-0.3% (24 h); distort. temp. 56 C (264 psi); dissip. factor 2.1×10^{-4}; dielec. const. 1.7.

Polybond® 1009. [Uniroyal/Spec. Chem.] HDPE homopolymer, acrylic acid modified; thermoplastic with good adhesion to a wide variety of substrates, exc. compat. with fillers and reinforcers; impact modifier, compatibilizer; chemical coupling agent for glass and mica-reinforced polyethylene; reduces moisture sensitivity in polyamides; melt index 5-6 g/10 min; dens. 0.95 g/cc; tens. str. 22.4 MPa (break); tens. elong. 10% (ultimate); flex. mod. 0.85 GPa (secant); impact str. (Izod) 23 J/m (notched); hardness (Shore D) 60; distort. temp. 41.1 C (264 psi); dissip. factor 3.9×10^{-4}; dielec. const. 2.35; vol. resist. 2.0×10^{16} ohm-cm.

Polybond® 2015. [Uniroyal/Spec. Chem.] PP, acrylic acid; additive for adhesion to metals and polar polymers; film and sheet extrusion; melt flow 12; hardness (Shore D) 60.

Polybond® 3001. [Uniroyal/Spec. Chem.] PP homopolymer, maleic anhydride-modified; thermoplastic used as coupling and compatibilizing agent for reinforced PP, polymer alloys; stabilized for processing and long-term heat-aging resist.; can be processed at temps. exceeding 220 C; melt flow 5.0 g/10 min; dens. 0.91 g/cc; m.p. 157 C; tens. str. 32.3 MPa (break); tens. elong. 436% (ultimate); flex. mod. 1.4 GPa (secant); impact str. (Izod) 50.6 J/m (notched); distort. temp. 53 C (264 psi).

Polybond® 3002. [Uniroyal/Spec. Chem.] PP homopolymer, maleic anhydride-modified; additive in polymer alloys, reinforced PP; melt flow 7.

Polybond® 3009. [Uniroyal/Spec. Chem.] HDPE, maleic anhydride-modified; additive in polymer alloys, reinforced HDPE; melt flow 5.

Polycat® 5. [Air Prods./Perf. Chems.] N,N,N′,N′′,N′′-Pentamethyldiethylenetriamine; high-activity blowing catalyst for rigid PU foams and improved processing of flexible slabstock PU foam; liq.; insol. in water; sp.gr. 0.83; flash pt. 75 C.

Polycat® 8. [Air Prods./Perf. Chems.] N,N-Dimethyl-N-cyclohexylamine; gelling catalyst for rigid PU applics.; liq.; insol. in water; sp.gr. 0.8512 (20 C); visc. 2.4 cps (20 C); vapor pressure 9.77 mm Hg (38 C); f.p. < -78 C; flash pt. (CC) 40 C.

Polycat® 9. [Air Prods./Perf. Chems.] Tert. amine; catalyst for semiflexible and rigid PU applics.; reduced odor; liq.; sol. in water; sp.gr. 0.8487; visc. 10 cps (20 C); vapor pressure 8.89 mm Hg (38 C); f.p. < -78 C; flash pt. (CC) > 100 C.

Polycat® 12. [Air Prods./Perf. Chems.] Tert. amine; catalyst for semiflexible and rigid PU applics.; liq.; insol. in water; sp.gr. 0.9235; visc. 10 cps; vapor pressure 11.4 mm Hg (38 C); f.p. < -78 C; flash pt. (CC) 100 C.

Polycat® 15. [Air Prods./Perf. Chems.] Tert. amine with reactive sec. amine; catalyst for semiflexible and rigid PU applics.; liq.; sol. in water; sp.gr. 0.8437; visc. 7.3 cps (20 C); vapor pressure 2.74 mm Hg (38 C); f.p. < -78 C; hyd. no. 300; flash pt. (CC) 88.3 C.

Polycat® 17. [Air Prods./Perf. Chems.] N,N,N′,N′′,N′′-Pentamethyldipropylenetriamine; catalyst for semiflexible PU foam, flexible slabstock, rigid, and pkg. applics., esp. cold-cured high-resiliency automotive molded foams; liq.; sol. in water; sp.gr. 0.9041; visc. 12.5 cps (23 C); vapor pressure 6 mm Hg (38 C); f.p. < -78.4 C; hyd. no. 353; flash pt. (CC) 95 C.

Polycat® 33. [Air Prods./Perf. Chems.] Amine-based gelling catalyst for rigid PU applics.; reduced odor; liq.; insol. in water; sp.gr. 0.8512 (20 C); visc. 2.4 cps (20 C); vapor pressure 9.77 mm Hg (38 C); f.p. < -78 C; flash pt. (CC) 39 C.

Polycat® 41. [Air Prods./Perf. Chems.] Amine-based lamination co-catalyst, recommended for use in combination with Polycat 43; for rigid PU applics.; liq.; sol. in water; sp.gr. 0.92; visc. 26.5 cps (38 C); vapor pressure 0.1 mm Hg (60 C); f.p. 141 C; flash pt. (CC) > 104 C.

Polycat® 43. [Air Prods./Perf. Chems.] Amine-based back-end cure catalyst for lamination systems, rigid PU applics.; co-catalyst with Polycat 41; liq.; sol. in water; sp.gr. 1.0893; visc. 1420 cps (26 C); vapor pressure < 52 mm Hg (38 C); f.p. -19 C; hyd. no. 542; flash pt. (CC) 94 C.

Polycat® 46. [Air Prods./Perf. Chems.] Potassium acetate in ethylene glycol; metal-based catalyst for rigid PU applics.; liq.; sol. in water; visc. 200 cps; hyd. no. 1122; flash pt. > 110 C.

Polycat® 58. [Air Prods./Perf. Chems.] Amine-based catalyst for PU flexible molded foam; for improved surface cure and cold collapse protection; liq.; sol. in water; sp.gr. 0.886; visc. 7.1 cps; vapor pressure 16 mm Hg (21 C); hyd. no. 1033; flash pt. > 37.8 C.

Polycat® 70. [Air Prods./Perf. Chems.] Tert. amine blend; catalyst for PU flexible slabstock applics.; liq.; sol. in water; sp.gr. 0.867; visc. 5.2 cps; vapor pressure 15 mm Hg (38 C); f.p. < -78.4 C; hyd. no. 447; flash pt. (CC) 42.2 C.

Polycat® 77. [Air Prods./Perf. Chems.] Tert. amine; catalyst for PU flexible molded foam, semiflexible, flexible slabstock, and rigid applics.; promotes both gelling and blowing reactions; gives open cells in flexible foam; liq.; sol. in water; sp.gr. 0.828; visc. 2.5 cps (23 C); vapor pressure 4 mm Hg (38 C); f.p. < -78 C; flash pt. (CC) 92 C; 100% act.

Polycat® 79. [Air Prods./Perf. Chems.] Amine-based delayed-action blowing catalyst for flexible molded PU; liq.; sol. in water; sp.gr. 0.873; vapor pressure 4.0 mm Hg (21 C); f.p. -60 C; flash pt. (Seta) 102 C.

Polycat® 85. [Air Prods./Perf. Chems.] Tert. amine; catalyst for rigid PU applics., appliance foam; liq.; sol. in water; sp.gr. 0.8475; vapor pressure 9 mm Hg (21 C); flash pt. (CC) 35.6 C.

Polycat® 91. [Air Prods./Perf. Chems.] Amine-based catalyst for improved processing of flexible slabstock PU foam; gives delayed action.

Polycat® DBU. [Air Prods./Perf. Chems.] 1,8-Diazabicyclo (5.4.0) undecene-7; CAS 6674-22-2; EINECS 229-713-7; highly active catalyst for PU elastomers, shoe soles, coatings; provides R.T. cure.; liq.; sol. in water; sp.gr. 1.0192; visc. 14 cps; vapor pressure 5.3 mm Hg (38 C); f.p. < -78 C; flash pt. (CC) > 96 C.

Polycat® SA-1. [Air Prods./Perf. Chems.] Polycat DBU with an organic acid blocker; CAS 6674-22-2; EINECS 229-713-7; delayed-action catalyst for PU coatings, rigid, elastomeric, shoe sole applics.; liq.; sol. in water; sp.gr. 1.073; visc. 433 cps; vapor pressure 1.29 mm Hg (38 C); f.p. < -78 C; flash pt. (CC) > 94 C.

Polycat® SA-102. [Air Prods./Perf. Chems.] Polycat DBU with an acid blocker; CAS 6674-22-2; EINECS 229-713-7; delayed-action catalyst for PU coatings, rigid, elastomeric, shoe sole applics.; liq.; sol. in water; sp.gr. 1.0172; visc. 3749 cps; vapor pressure > 2.1 mm Hg (38 C); f.p. 1-2 C; flash pt. (CC) > 94 C.

Polycat® SA-610/50. [Air Prods./Perf. Chems.] Polycat DBU with an acid blocker; CAS 6674-22-2; EINECS 229-713-7; delayed-action catalyst for PU flexible slabstock, coatings, rigid, elastomeric, shoe sole applics.; liq.; sol. in water; sp.gr. 1.0810; visc. 244 cps; vapor pressure 4.0 mm Hg (38 C); f.p. < -78 C; hyd. no. 400; flash pt. (CC) > 94 C.

Polycone 1000. [Olin] Modified silicone emulsion; mold release agent for rubber; liq.; Unverified

Polycup® 172. [Hercules] Polyamide aq. sol'n.; cationic; crosslinking agent for carboxylated SBR latex, carboxymethylcellulose, PVA, and other water-sol. polymers; also gives water resist. to starch adhesives; liq.; visc. 50 cps; 12.5% total solids.

Polycup® 1884. [Hercules] Polyamide aq. sol'n.; cationic; crosslinking agent for carboxylated SBR latex, carboxymethylcellulose, PVA, and other water-sol. polymers; also gives water resist. to starch adhesives; liq.; visc. 325 cps; 35% total solids.

Polycup® 2002. [Hercules] Polyamide aq. sol'n.; cationic; crosslinking agent for carboxylated SBR latex, carboxymethylcellulose, PVA, and other water-sol. polymers; also gives water resist. to starch adhesives; liq.; visc. 85 cps; 27% total solids.

Polydis® TR 016. [Struktol] Blend of calcium fatty acid salt and amide with sites avail. for hydrogen bonding; processing aid; dispersant for highly filled systems; processing aid, internal lubricant for polyolefins, PVC, PS, ABS, PET, TPO, TPE systems; approved for food and drug contact; pale yel. pellets or microbeads, neutral odor; sp.gr. 1.0; m.p. 99 C; flash pt. > 290 C; *Toxicology:* harmless; *Storage:* > 2 yrs storage stability.

Polydis® TR 060. [Struktol] Mixt. of aliphatic resins with m.w. < 2000; processing aid for plastics; binder, tackifier at processing temps.; improves blending of TPO compds., flame retardant formulations, filled polymer systems; improves surf. appearance, flow during molding; nonblooming; lt. amber powd. or pellet; good sol. in aliphatic, aromatic, and chlorinated hydrocarbons; sp.gr. 0.95; soften. pt. 102 C; flash pt. > 230 C; *Toxicology:* harmless; *Storage:* 2 yrs min. stability under normal storage conditions.

Polydis® TR 121. [Struktol] Oleamide; CAS 301-02-0; EINECS 206-103-9; slip agent for polyethylene films; lubricant and mold release for inj. molding applics., processing thermoplastic resins, thermoplastic elastomers, and thermoset rubber systems; dispersant; FDA 21CFR §175.105, 175.300, 175.320, 176.180, 177.1210, 177.1350, 177.1400, 179.45; cream-colored pellets; 85% 2-4 mm; insol. in water; limited sol. in alcohols, ketones, aromatic solvs.; bulk dens. 37.5 lb/ft^3; m.p. 70-76 C; acid no. 1 max.; iodine no. 80-90; flash pt. (COC) 210 C; 98% amide; *Precaution:* nonflamm.; *Storage:* negligible corrosivity; storage at normal temps. will not affect prod., but extended storage should be avoided.

Polydis® TR 131. [Struktol] Erucamide; CAS 112-84-5; EINECS 204-009-2; slip agent for polyolefin films; release agent for polymer systems; dispersant in color concs. and printing inks; process aid/lubricant for thermoplastic elastomers, thermoplastic resins, thermoset rubber systems; FDA 21CFR §175.105, 175.300, 175.320, 176.180, 177.1200, 177.1210, 177.1350, 177.1400, 178.3860, 179.45; Gardner 4 max. color, faint odor; sp.gr. 0.93; bulk dens. 34.7 lb/ft^3 (beads), 37.5 lb/ft^3 (pellets); m.p. 79-85 C; acid no. 3 max.; iodine no. 72-78; flash pt. (COC) 220 C; 99% amide; *Precaution:* nonflamm.; keep away from strong oxidizing agents; rotate inventory; *Storage:* noncorrosive; normal storage will not affect prod., but extended storage above 175 C should be avoided.

Poly-Dispersion® AAD-75. [Rhein Chemie] Dibenzothiazyl disulfide encapsulated in NBR elastomer; accelerator for rubber industry; tan slabs; sp.gr. 1.34; 75% act.

Poly-Dispersion® A (SAN)D-65. [Rhein Chemie] N-Cyclohexyl-2-benzothiazole sulfenamide encapsulated in NBR elastomer; accelerator for rubber industry; beige to lt. tan slabs; sp.gr. 1.14; 65% act.

Poly-Dispersion® ASD-75. [Rhein Chemie] Sol. sulfur encapsulated in NBR elastomer; predispersed rubber vulcanization agent; yel. slabs; sp.gr. 1.62; 75% act.

Poly-Dispersion® A (TEF-1)D-80. [Rhein Chemie] Polytetrafluoroethylene encapsulated in NBR elastomer; predispersed rubber chem.; wh. slabs; sp.gr. 1.74; 80% act.

Poly-Dispersion® A (TEF-1)D-80P. [Rhein Chemie] Polytetrafluoroethylene encapsulated in NBR elastomer; predispersed rubber chem.; wh. pellets; sp.gr. 1.74; 80% act.

Poly-Dispersion® A (TI)D-80. [Rhein Chemie] Titanium dioxide encapsulated in NBR elastomer; predispersed rubber colorant; wh. slabs; sp.gr. 2.54; 80% act.

Poly-Dispersion® AV (TBS)D-80. [Rhein Chemie] Tribasic lead sulfate encapsulated in NBR/PVC elastomer; predispersed rubber chem.; heat stabilizer for elec. and vinyl compds. requiring high heat stability; wh. slabs; sp.gr. 3.17; 80% act.

Poly-Dispersion® AV (TBS)D-80P. [Rhein Chemie] Tribasic lead sulfate encapsulated in NBR/PVC elastomer; predispersed rubber chem.; heat stabilizer for elec. and vinyl compds. requiring high heat stability; wh. pellets; sp.gr. 3.17; 80% act.

Poly-Dispersion® A (Z-CN)D-85. [Rhein Chemie] Zinc oxide encapsulated in NBR elastomer; predispersed activator for carboxylated NBR; cream slabs; sp.gr. 3.26; 85% act.

Poly-Dispersion® AZFD-85. [Rhein Chemie] French process zinc oxide encapsulated in NBR elastomer; predispersed activator for rubber; wh. slabs; sp.gr. 3.26; 85% act.

Poly-Dispersion® A (ZMAM)D-666P. [Rhein Chemie] 4- and 5-methylmercaptobenzimidazole zinc salt/low temp. diphenylamineacetone reaction prod. encapsulated in NBR elastomer; antidegradant for rubber; tan pellets; sp.gr. 1.26; 44.4/22.2% act.

Poly-Dispersion® A (ZMTI)D-50. [Rhein Chemie] 4- and 5-methylmercaptobenzimidazole zinc salt encapsulated in NBR elastomer; antidegradant for rubber; off-wh. slabs; sp.gr. 1.24; 50% act.

Poly-Dispersion® DCP-60P. [Rhein Chemie] Dicumyl peroxide encapsulated in ACM elastomer; predispersed catalyst for plastics and rubber vulcanization; blue pellets; sp.gr. 1.08; 60% act.

Poly-Dispersion® EAD-75. [Rhein Chemie] Dibenzothiazyl disulfide encapsulated in EPM elastomer; accelerator for rubber industry; pale grn. slabs; sp.gr. 1.27; 75% act.

Poly-Dispersion® EAD-75P. [Rhein Chemie] Dibenzothiazyl disulfide encapsulated in EPM elastomer; accelerator for rubber industry; pale grn. pellets; sp.gr. 1.27; 75% act.

Poly-Dispersion® ECSD-70. [Rhein Chemie] Insol. sulfur encapsulated in EPM elastomer; predispersed rubber vulcanization agent; yel. slabs; sp.gr. 1.43; 70% act.

Poly-Dispersion® E (DIC)D-40. [Rhein Chemie] Dicumyl peroxide encapsulated in EPM elastomer; predispersed catalyst for plastics and rubber vulcanization; gray slabs; sp.gr. 1.13; 40% act.

Poly-Dispersion® E (DOTG)D-65P. [Rhein Chemie] Diorthotolyl guanidine encapsulated in EPM elastomer; accelerator for rubber industry; tan pellets; sp.gr. 1.06; 65% act.

Poly-Dispersion® E (DPS)D-80. [Rhein Chemie] Dibasic lead phosphite encapsulated in EPM elastomer; predispersed heat and lt. stabilizer for vinyl plastics and chlorinated polymers; wh. slabs; sp.gr. 2.75; 80% act.

Poly-Dispersion® E (DYT)D-80. [Rhein Chemie] Dibasic lead phthalate encapsulated in EPM elastomer; predispersed heat and lt. stabilizer for vinyl; wh. slabs; sp.gr. 2.44; 80% act.

Poly-Dispersion® EMD-75. [Rhein Chemie] 2-Mercaptobenzothiazole encapsulated in EPM elastomer; accelerator for rubber industry; gray-tan slabs; sp.gr. 1.25; 75% act.

Poly-Dispersion® E (MX)D-75. [Rhein Chemie] Tetramethylthiuram monosulfide encapsulated in EPM elastomer; accelerator for rubber industry; deep yel. slabs; sp.gr. 1.19; 75% act.

Poly-Dispersion® E (MZ)D-75. [Rhein Chemie] Zinc dimethyldithiocarbamate encapsulated in EPM elastomer; accelerator for rubber industry; wh. slabs; sp.gr. 1.35; 75% act.

Poly-Dispersion® E (NBC)D-70. [Rhein Chemie] Nickel dibutyldithiocarbamate encapsulated in EPM elastomer; accelerator for rubber industry; dk. grn. slabs; sp.gr. 1.14; 70% act.

Poly-Dispersion® E (NBC)D-70P. [Rhein Chemie] Nickel dibutyldithiocarbamate encapsulated in EPM elastomer; accelerator for rubber industry; dk. grn. pellets; sp.gr. 1.14; 70% act.

Poly-Dispersion® END-75. [Rhein Chemie] Ethylene thiourea encapsulated in EPM elastomer; accelerator for rubber industry; wh. slabs; sp.gr. 1.22; 75% act.

Poly-Dispersion® END-75P. [Rhein Chemie] Ethylene thiourea encapsulated in EPM elastomer; accelerator for rubber industry; wh. pellets; sp.gr. 1.22; 75% act.

Poly-Dispersion® ERD-90. [Rhein Chemie] Red lead (Pb_3O_4) encapsulated in EPM elastomer; predispersed rubber chem.; orange slabs; sp.gr. 4.57; 90% act.

Poly-Dispersion® E (SAN-NS)D-70. [Rhein Chemie] N-t-Butylbenzothiazole-2-sulfenamide encapsulated in EPM elastomer; accelerator for rubber industry; lt. tan slabs; sp.gr. 1.12; 70% act.

Poly-Dispersion® ESD-80. [Rhein Chemie] Sol. sulfur encapsulated in EPM elastomer; predispersed rubber vulcanization agent; yel. slabs;

sp.gr. 1.61; 80% act.

Poly-Dispersion® E (SR)D-75. [Rhein Chemie] 4,4′-Dithiodimorpholine encapsulated in EPM elastomer; accelerator for rubber industry; pink slabs; sp.gr. 1.18; 75% act.

Poly-Dispersion® E (TET)D-70. [Rhein Chemie] Dipentamethylene thiuram tetrasulfide encapsulated in EPM elastomer; accelerator for rubber industry; cream slabs; sp.gr. 1.22; 70% act.

Poly-Dispersion® E (TET)D-70P. [Rhein Chemie] Dipentamethylene thiuram tetrasulfide encapsulated in EPM elastomer; accelerator for rubber industry; cream pellets; sp.gr. 1.22; 70% act.

Poly-Dispersion® E (VC)D-40. [Rhein Chemie] α,α′-Bis(t-butylperoxy)diisopropyl benzene encapsulated in EPM elastomer; predispersed additive for plastics, rubber vulcanization; pink slabs; sp.gr. 1.15; 40% act.

Poly-Dispersion® EZFD-85. [Rhein Chemie] French process zinc oxide encapsulated in EPM elastomer; predispersed activator for rubber; wh. slabs; sp.gr. 3.05; 85% act.

Poly-Dispersion® G (BAC)D-80. [Rhein Chemie] Barium carbonate encapsulated in ECO elastomer; predispersed rubber chem.; off-wh. slabs; sp.gr. 2.71; 80% act.

Poly-Dispersion® G (DPS)D-80. [Rhein Chemie] Dibasic lead phosphite encapsulated in ECO elastomer; predispersed heat and lt. stabilizer for vinyl plastics and chlorinated polymers; wh. slabs; sp.gr. 3.14; 80% act.

Poly-Dispersion® G (DYT)D-80. [Rhein Chemie] Dibasic lead phthalate encapsulated in ECO elastomer; predispersed heat and lt. stabilizer for vinyl; wh. slabs; sp.gr. 2.76; 80% act.

Poly-Dispersion® GND-75. [Rhein Chemie] Ethylene thiourea encapsulated in ECO elastomer; accelerator for rubber industry; wh. slabs; sp.gr. 1.30; 75% act.

Poly-Dispersion® GRD-90. [Rhein Chemie] Red lead (Pb_3O_4) encapsulated in ECO elastomer; predispersed rubber chem.; orange slabs; sp.gr. 5.25; 90% act.

Poly-Dispersion® H (202)D-80. [Rhein Chemie] Lead silicate encapsulated in CM elastomer; predispersed rubber chem.; off-wh. slabs; sp.gr. 2.97; 80% act.

Poly-Dispersion® H (202)D-80P. [Rhein Chemie] Lead silicate encapsulated in CM elastomer; predispersed rubber chem.; off-wh. pellets; sp.gr. 2.97; 80% act.

Poly-Dispersion® H (DYT)D-80. [Rhein Chemie] Dibasic lead phthalate encapsulated in CM elastomer; predispersed heat and lt. stabilizer for vinyl; wh.-off-wh. slabs; sp.gr. 2.74; 80% act.

Poly-Dispersion® H (DYT)D-80P. [Rhein Chemie] Dibasic lead phthalate encapsulated in CM elastomer; predispersed heat and lt. stabilizer for vinyl; wh.-off-wh. pellets; sp.gr. 2.74; 80% act.

Poly-Dispersion® JZFD-90P. [Rhein Chemie] French process zinc oxide encapsulated in NR elastomer; predispersed activator for rubber; wh. pellets; sp.gr. 3.60; 88% act.

Poly-Dispersion® K (NMC)D-70. [Rhein Chemie] Nickel dimethyldithiocarbamate encapsulated in CM elastomer; antidegradant for rubber; dk. grn. slabs; sp.gr. 1.44; 70% act.

Poly-Dispersion® K (PX-17)D-30. [Rhein Chemie] n-Butyl 4,4-bis (t-butylperoxy) valerate encapsulated in CM elastomer; predispersed additive for plastics, rubber vulcanization; cream slabs; sp.gr. 1.36; 30% act.

Poly-Dispersion® PLD-90. [Rhein Chemie] Calcined litharge (PbO) encapsulated in PIB elastomer; predispersed rubber accelerator; stabilizer; lt. orange slabs; sp.gr. 4.89; 90% act.

Poly-Dispersion® PLD-90P. [Rhein Chemie] Calcined litharge (PbO) encapsulated in PIB elastomer; predispersed rubber accelerator; lt. orange pellets; sp.gr. 4.89; 90% act.

Poly-Dispersion® S (AX)D-70. [Rhein Chemie] p-(p-Toluenesulfonylamido)diphenylamine encapsulated in SBR elastomer; antidegradant for rubber; lt. gray slabs; sp.gr. 1.18; 70% act.

Poly-Dispersion® SCSD-70. [Rhein Chemie] Insol. sulfur encapsulated in SBR elastomer; predispersed rubber vulcanization agent; yel. slabs; sp.gr. 1.44; 70% act.

Poly-Dispersion® SHD-65. [Rhein Chemie] Hexamethylenetetramine encapsulated in SBR elastomer; accelerator for rubber industry; wh. slabs; sp.gr. 1.12; 65% act.

Poly-Dispersion® S (PVI)D-50P. [Rhein Chemie] N-Cyclohexylthiophthalimide encapsulated in SBR elastomer; predispersed rubber chem.; red pellets; sp.gr. 1.06; 50% act.

Poly-Dispersion® SR-517-50. [Rhein Chemie] Crosslinking agent encapsulated in ACM elastomer; co-agent for rubber processing; brn. slabs; sp.gr. 1.12; 50% act.

Poly-Dispersion® SSD-75. [Rhein Chemie] Sol. sulfur encapsulated in SBR elastomer; predispersed rubber vulcanization agent; yel. slabs; sp.gr. 1.56; 75% act.

Poly-Dispersion® S (UR)D-75. [Rhein Chemie] Urea encapsulated in SBR elastomer; predispersed rubber chem.; off-wh. slabs; sp.gr. 1.17; 75% act.

Poly-Dispersion® SZFD-85. [Rhein Chemie] French process zinc oxide encapsulated in SBR elastomer; predispersed activator for rubber; wh. slabs; sp.gr. 3.15; 85% act.

Poly-Dispersion® SZFND-825. [Rhein Chemie] French process zinc oxide/ETU encapsulated in SBR elastomer; predispersed activator for rubber; wh. slabs; sp.gr. 2.62; 75/7.5% act.

Poly-Dispersion® TAC-50. [Rhein Chemie] Triallylcyanurate encapsulated in ACM elastomer; co-agent for rubber processing; red slabs; sp.gr. 1.14; 50% act.

Poly-Dispersion® TATM-50. [Rhein Chemie] Triallyltrimellitate encapsulated in ACM elastomer; co-agent for rubber processing; yel. slabs; sp.gr. 1.16; 60% act.

Poly-Dispersion® T (AZ)D-75. [Rhein Chemie] Zinc

dibenzyldithiocarbamate encapsulated in EPDM elastomer; accelerator for rubber industry; off-wh. slabs; sp.gr. 1.22; 75% act.

Poly-Dispersion® T (BZ)D-75. [Rhein Chemie] Zinc dibutyldithiocarbamate encapsulated in EPDM elastomer; accelerator for rubber industry; off-wh. slabs; sp.gr. 1.21; 75% act.

Poly-Dispersion® T (DIC)D-40P. [Rhein Chemie] Dicumyl peroxide encapsulated in EPDM elastomer; predispersed catalyst for plastics and rubber vulcanization; blue pellets; sp.gr. 1.17; 40% act.

Poly-Dispersion® T (DPG)D-65. [Rhein Chemie] Diphenylguanidine encapsulated in EPDM elastomer; accelerator for rubber industry; lt. blue slabs; sp.gr. 1.05; 65% act.

Poly-Dispersion® T (DPG)D-65P. [Rhein Chemie] Diphenylguanidine encapsulated in EPDM elastomer; accelerator for rubber industry; lt. blue pellets; sp.gr. 1.05; 65% act.

Poly-Dispersion® T (DYT)D-80P. [Rhein Chemie] Dibasic lead phthalate encapsulated in EPDM elastomer; predispersed heat and lt. stabilizer for vinyl; wh. pellets; sp.gr. 2.44; 80% act.

Poly-Dispersion® T (HRL)D-90. [Rhein Chemie] Heat-resist. litharge (PbO) encapsulated in EPDM elastomer; predispersed rubber accelerator; lt. orange-off-yel. slabs; sp.gr. 4.66; 90% act.

Poly-Dispersion® T (HVA-2)D-70. [Rhein Chemie] N,N′-m-Phenylenedimaleimide encapsulated in EPDM elastomer; predispersed rubber chem.; off yel.-lt. grn. slabs; sp.gr. 1.21; 70% act.

Poly-Dispersion® T (HVA-2)D-70P. [Rhein Chemie] N,N′-m-Phenylenedimaleimide encapsulated in EPDM elastomer; predispersed rubber chem.; off yel.-lt. grn. pellets; sp.gr. 1.21; 70% act.

Poly-Dispersion® T (LC)D-90. [Rhein Chemie] Calcined litharge (PbO) encapsulated in EPDM elastomer; predispersed rubber accelerator; lt. orange slabs; sp.gr. 4.87; 90% act.

Poly-Dispersion® T (LC)D-90P. [Rhein Chemie] Calcined litharge (PbO) encapsulated in EPDM elastomer; predispersed rubber accelerator; lt. orange pellets; sp.gr. 4.87; 90% act.

Poly-Dispersion® TMC-50. [Rhein Chemie] 1,1-Di(t-butylperoxy)-3,3,5-trimethylcyclohexane encapsulated in ACM elastomer; predispersed additive for plastics, rubber vulcanization; yel. powd.; sp.gr. 1.02; 50% act.

Poly-Dispersion® TRD-90P. [Rhein Chemie] Red lead (Pb_3O_4) encapsulated in EPDM elastomer; predispersed rubber chem.; orange pellets; sp.gr. 4.55; 90% act.

Poly-Dispersion® TSD-80. [Rhein Chemie] Sol. sulfur encapsulated in EPDM elastomer; predispersed rubber vulcanization agent; yel. slabs; sp.gr. 1.62; 80% act.

Poly-Dispersion® TSD-80P. [Rhein Chemie] Sol. sulfur encapsulated in EPDM elastomer; predispersed rubber vulcanization agent; yel. pellets; sp.gr. 1.62; 80% act.

Poly-Dispersion® TTD-75. [Rhein Chemie] Tellurium diethyldithiocarbamate encapsulated in EPDM elastomer; accelerator for rubber industry; yel./orange slabs; sp.gr. 1.22; 75% act.

Poly-Dispersion® TTD-75P. [Rhein Chemie] Tellurium diethyldithiocarbamate encapsulated in EPDM elastomer; accelerator for rubber industry; yel./orange pellets; sp.gr. 1.22; 75% act.

Poly-Dispersion® T (VC)D-40P. [Rhein Chemie] a,a′-Bis(t-butylperoxy)diisopropyl benzene encapsulated in EPDM elastomer; predispersed additive for plastics, rubber vulcanization; pink pellets; sp.gr. 1.15; 40% act.

Poly-Dispersion® TVP-50. [Rhein Chemie] n-Butyl 4,4-bis (t-butylperoxy) valerate encapsulated in ACM elastomer; predispersed additive for plastics, rubber vulcanization; grn. powd.; sp.gr. 1.06; 50% act.

Poly-Dispersion® TZFD-88P. [Rhein Chemie] French process zinc oxide encapsulated in EPDM elastomer; predispersed activator for rubber; wh. pellets; sp.gr. 3.40; 88% act.

Poly-Dispersion® VC-60P. [Rhein Chemie] a,a′-Bis(t-butylperoxy)diisopropyl benzene encapsulated in ACM/EPDM elastomer; predispersed additive for plastics, rubber vulcanization; pink pellets; sp.gr. 1.02; 60% act.

Poly-Dispersion® V (MT)D-75. [Rhein Chemie] Tetramethylthiuram disulfide encapsulated in EAM elastomer; accelerator for rubber industry; pale blue slabs; sp.gr. 1.25; 75% act.

Poly-Dispersion® V (MT)D-75P. [Rhein Chemie] Tetramethylthiuram disulfide encapsulated in EAM elastomer; accelerator for rubber industry; pale blue pellets; sp.gr. 1.25; 75% act.

Poly-Dispersion® V(NOBS)M-65. [Rhein Chemie] 2-Morpholinothiobenzothiazole encapsulated in EAM elastomer; accelerator for rubber industry; tan to pink slabs; sp.gr. 1.13; 65% act.

Poly DNB®. [Lord] Poly-p-dinitrosobenzene wax disp.; CAS 9003-34-3; conditioner for butyl rubber to improve grn. str., processing safety, and elec. resistivity.

Polyfil 100. [Custom Grinders] Aluminatrihydrate; flame retardant for ABS, acrylics, epoxy, nitrile, phenolic, PC, thermoset polyester, PE, PP, PS, PVAc, PVC, PU.

Polyfil® WC. [J.M. Huber/Engineered Mins.] Anhyd. organofunctional pigment; pigment reducing water vapor transmission, yielding exc. wet and dry elec. props. and good long-term stability in EPR and crosslinked polyethylene; wh. powd.; 1.4 μ particle diam.; 0.01% max. on 325 mesh; sp.gr. 2.63; surf. area 7-9 m^2/g; oil absorp. 48-60; brightness 90-93; pH 5-6 (28%).

Polyfin. [Crowley Chem.] Polyethylene wax; CAS 9002-88-4; EINECS 200-815-3; internal/external lubricant for inj. molding, calendering, extrusion; solid.

Poly-G® 200. [Olin] PEG 200; CAS 25322-68-3; EINECS 203-989-9; chemical intermediate for prod. of surfactants for cleaners, textiles, paper, cosmetics; carrier for pharmaceuticals; also in cosmetics and personal care prods., textiles, rubber mold releases, printing inks and dyes, metalworking fluids, foods, paints, paper, wood prods., adhesives agric., ceramics, elec. equip., petrol. prods., photographic prods., resins; APHA 25 max. liq.; m.w. 200; sol. in water, acetone, ethanol, ethyl acetate, toluene; sp.gr. 1.125; dens. 9.38 lb/gal; visc. 4.3 cs (99 C); flash pt. (COC) 171 C.

Poly-G® 300. [Olin] PEG 300; CAS 25322-68-3; EINECS 220-045-1; chemical intermediate for prod. of surfactants for cleaners, textiles, paper, cosmetics; carrier for pharmaceuticals; also in cosmetics and personal care prods., textiles, rubber mold releases, printing inks and dyes, metalworking fluids; foods, paints, paper, wood prods., adhesives agric., ceramics, elec. equip., petrol. prods., photographic prods., resins; APHA 25 max. liq.; m.w. 300; sol. in water, acetone, ethanol, ethyl acetate, toluene; sp.gr. 1.125; dens. 9.38 lb/gal; visc. 5.8 cs (99 C); f.p. -15 to -8 C; flash pt. (COC) 196 C.

Poly-G® 400. [Olin] PEG 400; CAS 25322-68-3; EINECS 225-856-4; chemical intermediate for prod. of surfactants for cleaners, textiles, paper, cosmetics; carrier for pharmaceuticals; also in cosmetics and personal care prods., textiles, rubber mold releases, printing inks and dyes, metalworking fluids; foods, paints, paper, wood prods., adhesives agric., ceramics, elec. equip., petrol. prods., photographic prods., resins; APHA 25 max. liq.; m.w. 400; sol. in water, acetone, ethanol, ethyl acetate, toluene; sp.gr. 1.127; dens. 9.4 lb/gal; visc. 7.3 cs (99 C); f.p. 4-10 C; flash pt. (COC) 224 C; pour pt. 3-10 C.

Poly-G® 600. [Olin] PEG 600; CAS 25322-68-3; EINECS 229-859-1; chemical intermediate for prod. of surfactants for cleaners, textiles, paper, cosmetics; carrier for pharmaceuticals; also in cosmetics and personal care prods., textiles, rubber mold releases, printing inks and dyes, metalworking fluids; foods, paints, paper, wood prods., adhesives agric., ceramics, elec. equip., petrol. prods., photographic prods., resins; APHA 25 max. liq.; m.w. 600; sol. in water, acetone, ethanol, ethyl acetate, toluene; sp.gr. 1.127; dens. 9.4 lb/gal; visc. 10.5 cs (99 C); f.p. 20-25 C; flash pt. (COC) 246 C; pour pt. 19-24 C.

Poly-G® 1000. [Olin] PEG 1000; CAS 25322-68-3; chemical intermediate for prod. of surfactants for cleaners, textiles, paper, cosmetics; carrier for pharmaceuticals; also in cosmetics and personal care prods., textiles, rubber mold releases, printing inks and dyes, metalworking fluids; foods, paints, paper, wood prods., adhesives agric., ceramics, elec. equip., petrol. prods., photographic prods., resins; wh. waxy solid; m.w. 1000; somewhat less sol. in water than liq. glycols; sp.gr. 1.104 (50/20 C); dens. 9.20 lb/gal (50/20 C); visc. 17.4 cs (99 C); f.p. 38-41 C; flash pt. (COC) 260 C; pour pt. 40 C.

Poly-G® 1500. [Olin] PEG 1500; chemical intermediate for prod. of surfactants for cleaners, textiles, paper, cosmetics; carrier for pharmaceuticals; also in cosmetics and personal care prods., textiles, rubber mold releases, printing inks and dyes, metalworking fluids; foods, paints, paper, wood prods., adhesives agric., ceramics, elec. equip., petrol. prods., photographic prods., resins; wh. waxy solid; m.w. 1500; somewhat less sol. in water than liq. glycols; sp.gr. 1.104 (50/20 C); dens. 9.20 lb/gal (50/20 C); visc. 28 cs (99 C); f.p. 43-46 C; flash pt. (COC) 266 C; pour pt. 45 C.

Poly-G® 2000. [Olin] PEG 2000; CAS 25322-68-3; chemical intermediate for prod. of surfactants for cleaners, textiles, paper, cosmetics; carrier for pharmaceuticals; also in cosmetics and personal care prods., textiles, rubber mold releases, printing inks and dyes, metalworking fluids; foods, paints, paper, wood prods., adhesives agric., ceramics, elec. equip., petrol. prods., photographic prods., resins; APHA 25 max. liq.; m.w. 2000; somewhat less sol. in water than lower m.w. glycols; sp.gr. 1.113; dens. 9.26 lb/gal; visc. 11.7 cs (99 C); f.p. -20 C; 60% aq. sol'n.

Poly-G® B1530. [Olin] PEG 500-600; chemical intermediate for prod. of surfactants for cleaners, textiles, paper, cosmetics; carrier for pharmaceuticals; also in cosmetics and personal care prods., textiles, rubber mold releases, printing inks and dyes, metalworking fluids; foods, paints, paper, wood prods., adhesives agric., ceramics, elec. equip., petrol. prods., photographic prods., resins; wh. waxy solid; m.w. 900; somewhat less sol. in water than liq. glycols; sp.gr. 1.104 (50/20 C); dens. 9.20 lb/gal (50/20 C); visc. 15 cs (99 C); f.p. 38-41 C; flash pt. (COC) 254 C; pour pt. 38 C.

Poly-G® WS 280X. [Olin] Heat-transfer fluid and lubricant for plastic pipe extrusion and drawing; annealing and curing medium for thermosetting and thermoplastic resins; Gardner 10 max. clear liq.; sol. no clouding @ 50% by vol. in water; sp.gr. 1.042; visc. 53-63 cs (38 C); pour pt. -37 C; ref. index 1.4584; pH 5.5-7.5 (10% aq.); flash pt (COC) 280 C; sp. heat 0.503 cal/g/C.; Unverified

Polygard®. [Uniroyal] Tris (mixed mono and dinonyl phenyl) phosphite; antioxidant, processing and color stabilizer for PP, LDPE, LLDPE, HDPE, HIPS, ABS, PVC, PC, and EVA and polyamide hot-melt adhesives; FDA approvals.

Polygard® HR. [Uniroyal] Polygard with 1% triisopropanolamine; antioxidant with more resist. to hydrolysis than Polygard; FDA approved for PP, LDPE, LLDPE, HDPE, HIPS, ABS, PVC, PC, EVA, and polyamides; FDA approvals.

Polylite. [Uniroyal] Amine; discoloring antioxidant for ABS; liq.

Polylube J. [Custom Compounding] PTFE wax; CAS 9002-84-0; internal/external lubricant for thermoplastics, thermosets, elastomers; solid.

Polylube RE. [Huntsman] Blend of polyoxyalkylene derivs.; nonionic; lubricant, antistat for acetate, nylon; liq.; 85% act.

Polymel #7. [Frank B. Ross] Modified polyethylene wax; CAS 9002-88-4; EINECS 200-815-3; internal/external lubricant; low m.w. wax incorporated into rubber batches giving exc. mold release; wh.; m.p.

205-215 F; flash pt. 420 F min.; acid no. nil.

Polymica 200. [Franklin Industrial Minerals] Wet-processed muscovite mica; CAS 12001-26-2; filler for coating and performance polymer applics. where high brightness, color, particle size, and consistency are important; wh. platy powd.; 30 μ mean particle size; 98.5% min. thru 100 mesh, 90% max. thru 200 mesh; sp.gr. 2.8; bulk dens. 14±2 lb/ft^3 (loose); oil absorp. 55-65; pH 7-8; 0.5% max. moisture.

Polymica 325. [Franklin Industrial Minerals] Wet-processed muscovite mica; CAS 12001-26-2; filler for coating and performance polymer applics. where high brightness, color, particle size, and consistency are important; wh. platy powd.; 25 μ mean particle size; 100% thru 100 mesh, 98% min. thru 200 mesh; sp.gr. 2.8; bulk dens. 15 ± 2 lb/ft^3 (loose); oil absorp. 50-60; pH 7-8; 0.5% max. moisture.

Polymica 400. [Franklin Industrial Minerals] Wet-processed muscovite mica; CAS 12001-26-2; filler for coating and performance polymer applics. where high brightness, color, particle size, and consistency are important; wh. platy powd.; 18.5 μ mean particle size; 100% thru 200 mesh, 99% thru 325 mesh, 95% min. thru 400 mesh; sp.gr. 2.8; bulk dens. 11.5±2 lb/ft^3 (loose); oil absorp. 45-55; pH 7-8; 0.5% max. moisture.

Polymica 3105. [Franklin Industrial Minerals] Wet-processed muscovite mica; CAS 12001-26-2; filler for coating and performance polymer applics. where high brightness, color, particle size, and consistency are important; wh. platy powd.; 23 μ mean particle size; 99.9% min. thru 100 mesh, 88% min. thru 325 mesh; sp.gr. 2.8; bulk dens. 13±2 lb/ft^3 (loose); pH 7-8; 0.5% max. moisture.

Polymist® 284. [Ausimont] Polytetrafluoroethylene; CAS 9002-84-0; additive in elastomers or plastics where improved lubricity and/or wear resist. are required; also for paints, coatings, printing inks, oils and greases; improves wear chars., uv resist., nonstick/antifriction props., corrosion resist., thermal stability; wh. free-flowing powd.; < 10 μ avg. particle; sp.gr. 2.22; bulk dens. 450 g/l; surf. area 3 m^2/g; m.p. 327-333 C; coeff. of friction < 0.1; *Toxicology:* avoid inh. of fumes when heated above 450 F; *Storage:* indefinite shelf life when stored @ R.T.

Polymist® F-5. [Ausimont] Polytetrafluoroethylene; CAS 9002-84-0; additive in elastomers or plastics where improved lubricity and/or wear resist. are required; also for paints, coatings, printing inks, oils and greases; improves wear chars., uv resist., nonstick/antifriction props., corrosion resist., thermal stability; wh. free-flowing powd.; < 10 μ avg. particle; sp.gr. 2.28; bulk dens. 450 g/l; surf. area 3 m^2/g; m.p. 320-325 C; coeff. of friction < 0.1; *Toxicology:* avoid inh. of fumes when heated above 450 F; *Storage:* indefinite shelf life when stored @ R.T.

Polymist® F-5A. [Ausimont] Polytetrafluoroethylene; CAS 9002-84-0; additive in elastomers or plastics where improved lubricity and/or wear resist. are required; also for paints, coatings, printing inks, oils and greases; improves wear chars., uv resist., nonstick/antifriction props., corrosion resist., thermal stability; wh. free-flowing powd.; < 6 μ avg. particle; sp.gr. 2.28; bulk dens. 400 g/l; surf. area 3 m^2/g; m.p. 320-325 C; coeff. of friction < 0.1; *Toxicology:* avoid inh. of fumes when heated above 450 F; *Storage:* indefinite shelf life when stored @ R.T.

Polymist® F-5A EX. [Ausimont] Polytetrafluoroethylene; CAS 9002-84-0; additive in elastomers or plastics where improved lubricity and/or wear resist. are required; also for paints, coatings, printing inks, oils and greases; improves wear chars., uv resist., nonstick/antifriction props., corrosion resist., thermal stability; wh. free-flowing powd.; < 7 μ avg. particle; sp.gr. 2.28; bulk dens. 400 g/l; surf. area 3 m^2/g; m.p. 325-330 C; coeff. of friction < 0.1; *Toxicology:* avoid inh. of fumes when heated above 450 F; *Storage:* indefinite shelf life when stored @ R.T.

Polymist® F-510. [Ausimont] Polytetrafluoroethylene; CAS 9002-84-0; additive in elastomers or plastics where improved lubricity and/or wear resist. are required; also for paints, coatings, printing inks, oils and greases; improves wear chars., uv resist., nonstick/antifriction props., corrosion resist., thermal stability; wh. free-flowing powd.; < 25 μ avg. particle; sp.gr. 2.28; bulk dens. 475 g/l; surf. area 3 m^2/g; m.p. 325-330 C; coeff. of friction < 0.1; *Toxicology:* avoid inh. of fumes when heated above 450 F; *Storage:* indefinite shelf life when stored @ R.T.

Polymist® XPH-284. [Ausimont] PTFE wax; CAS 9002-84-0; internal/external lubricant for thermoplastics, thermosets, elastomers, coatings, lubricants; FDA accepted; solid.

Polyox® WSR 205. [Union Carbide] PEG-14M; CAS 25322-68-3; nonionic; water-sol. thermoplastic resin, thickener, lubricant, binder, flocculant, wet adhesive; for ceramics, papermaking; dispersant for vinyl polymerization; EPA 40CFR § 180.1001(d); FDA 21CFR §175.300, 175.380, 175.390, 176.170, 176.180, 177.1210, 177.1350; wh. gran. powd., mild ammoniacal odor; 100% thru 10 mesh, 96% thru 20 mesh; sol. in water, some chlorinated solvs., alcohols, aromatic hydrocarbons, ketones; m.w. 600,000; sp.gr. 1.15-1.26; bulk dens. 19-37 lb/ft^3; visc. 4500-8800 cps (5% aq.); m.p. 62-67 C; pH 8-10 (5% aq.); *Precaution:* slippery when wet; *Storage:* store in sealed containers below 25 C, away from heat; avoid dust buildup.

Polyox® WSR 301. [Union Carbide] PEG-90M; CAS 25322-68-3; nonionic; water-sol. thermoplastic resin, thickener, lubricant, binder, flocculant, wet adhesive; flow modifier, lubricant for construction, military, agric., mining; provides wet adhesion to personal care prods.; foam stabilizer in fermented malt beverages (300 ppm max.); dispersant for vinyl polymerization; EPA 40CFR §180.1001(d); FDA 21CFR §172.770, 175.300, 175.380, 175.390, 176.170, 176.180, 177.1210, 177.1350; wh. gran. powd., mild ammoniacal odor; 100% thru 10 mesh, 96% thru 20 mesh; sol. in water, some chlorinated solvs., alcohols, aromatic hydrocarbons, ketones; m.w. 4,000,000; bulk dens. 19-37 lb/ft^3; visc. 1650-5500 cps (1% aq.); m.p. 62-67 C; pH 8-10 (1% aq.); *Precaution:* slippery when wet; *Storage:* store in sealed containers below 25 C, away from heat; avoid dust buildup.

Polyox® WSR 303. [Union Carbide] Polyethylene oxide; CAS 25322-68-3; nonionic; water-sol. thermoplastic resin, thickener, lubricant, binder, flocculant, wet adhesive; for mining, papermaking industries; foam stabilizer in fermented malt beverages (300 ppm max.); dispersant for vinyl polymerization; EPA 40CFR §180.1001(d); FDA 21CFR §172.770, 175.300, 175.380, 175.390, 176.170, 176.180, 177.1210, 177.1350; wh. gran. powd., mild ammoniacal odor; 100% thru 10 mesh, 96% thru 20 mesh; sol. in water, some chlorinated solvs., alcohols, aromatic hydrocarbons, ketones; m.w. 7,000,000; bulk dens. 19-37 lb/ft^3; visc. 7500-10,000 cps (1% aq.); m.p. 62-67 C; pH 8-10 (sol'n.); *Precaution:* slippery when wet; *Storage:* store in sealed containers below 25 C, away from heat; avoid dust buildup.

Polyox® WSR 308. [Union Carbide] Polyethylene oxide; CAS 25322-68-3; nonionic; water-sol. thermoplastic resin, thickener, lubricant, binder, flocculant, wet adhesive; foam stabilizer in fermented malt beverages (300 ppm max.); dispersant for vinyl polymerization; EPA 40CFR §180.1001(d); FDA 21CFR §172.770, 175.300, 175.380, 175.390, 176.170, 176.180, 177.1210, 177.1350; wh. gran. powd., mild ammoniacal odor; 100% thru 10 mesh, 96% thru 20 mesh; sol. in water, some chlorinated solvs., alcohols, aromatic hydrocarbons, ketones; m.w. 8,000,000; sp.gr. 1.15-1.26; bulk dens. 19-37 lb/ft^3; visc. 10,000-15,000 cps (1% aq.); m.p. 62-67 C; pH 8-10 (sol'n.); *Precaution:* slippery when wet; *Storage:* store in sealed containers below 25 C, away from heat; avoid dust buildup.

Polyox® WSR 1105. [Union Carbide] PEG-20M; CAS 25322-68-3; nonionic; water-sol. thermoplastic resin, thickener, lubricant, binder, flocculant, wet adhesive; dispersant for vinyl polymerization; EPA 40CFR §180.1001(d); FDA 21CFR §175.300, 175.380, 175.390, 176.170, 176.180, 177.1210, 177.1350; wh. gran. powd., mild ammoniacal odor; 100% thru 10 mesh, 96% thru 20 mesh; sol. in water, some chlorinated solvs., alcohols, aromatic hydrocarbons, ketones; m.w. 900,000; sp.gr. 1.15-1.26; bulk dens. 19-37 lb/ft^3; visc. 8800-17,600 cps (5% aq.); m.p. 62-67 C; pH 8-10 (5% aq.); *Precaution:* slippery when wet; *Storage:* store in sealed containers below 25 C, away from heat; avoid dust buildup.

Polyox® WSR 3333. [Union Carbide] PEG-9M; CAS 25322-68-3; nonionic; water-sol. thermoplastic resin, thickener, lubricant, binder, flocculant; for wet adhesion in papermaking; dispersant for vinyl polymerization; EPA 40CFR §180.1001(d); FDA 21CFR §175.300, 175.380, 175.390, 176.170, 176.180, 177.1210, 177.1350; wh. gran. powd., mild ammoniacal odor; 100% thru 10 mesh, 96% thru 20 mesh; sol. in water, some chlorinated solvs., alcohols, aromatic hydrocarbons, ketones; m.w. 400,000; sp.gr. 1.15-1.26; bulk dens. 19-37 lb/ft^3; visc. 2250-3350 cps (5% aq.); m.p. 62-67 C; pH 8-10 (5% aq.); *Precaution:* slippery when wet; *Storage:* store in sealed containers below 25 C, away from heat; avoid dust buildup.

Polyox® WSR N-10. [Union Carbide] PEG-2M; CAS 25322-68-3; nonionic; water-sol. thermoplastic resin, thickener, lubricant, binder, flocculant, wet adhesive; dispersant for vinyl polymerization; EPA 40CFR §180.1001(d); FDA 21CFR §175.300, 175.380, 175.390, 176.170, 176.180, 177.1210, 177.1350; wh. gran. powd., mild ammoniacal odor; 100% thru 10 mesh, 96% thru 20 mesh; sol. in water, some chlorinated solvs., alcohols, aromatic hydrocarbons, ketones; m.w. 100,000; sp.gr. 1.15-1.26; bulk dens. 19-37 lb/ft^3; visc. 12-50 cps (5% aq.); m.p. 62-67 C; pH 8-10 (5% aq.); *Precaution:* slippery when wet; *Storage:* store in sealed containers below 25 C, away from heat; avoid dust buildup.

Polyox® WSR N-12K. [Union Carbide] PEG-23M; CAS 25322-68-3; nonionic; water-sol. thermoplastic resin, thickener, lubricant, binder, flocculant, wet adhesive; dispersant for vinyl polymerization; EPA 40CFR §180.1001(d); FDA 21CFR §175.300, 175.380, 175.390, 176.170, 176.180, 177.1210, 177.1350; wh. gran. powd., mild ammoniacal odor; 100% thru 10 mesh, 96% thru 20 mesh; sol. in water, some chlorinated solvs., alcohols, aromatic hydrocarbons, ketones; m.w. 1,000,000; sp.gr. 1.15-1.26; bulk dens. 19-37 lb/ft^3; visc. 400-800 cps (2% aq.); m.p. 62-67 C; pH 8-10 (2% aq.); *Precaution:* slippery when wet; *Storage:* store in sealed containers below 25 C, away from heat; avoid dust buildup.

Polyox® WSR N-60K. [Union Carbide] PEG-45M; CAS 25322-68-3; nonionic; water-sol. thermoplastic resin, thickener, lubricant, binder, flocculant, wet adhesive; lubricant/emollient in personal care prods.; dispersant for vinyl polymerization; EPA 40CFR §180.1001(d); FDA 21CFR §175.300, 175.380, 175.390, 176.170, 176.180, 177.1210, 177.1350; wh. gran. powd., mild ammoniacal odor; 100% thru 10 mesh, 96% thru 20 mesh; sol. in water, some chlorinated solvs., alcohols, aromatic hydrocarbons, ketones; m.w. 2,000,000; sp.gr. 1.15-1.26; bulk dens. 19-37 lb/ft^3; visc. 2000-4000 cps (2% aq.); m.p. 62-67 C; pH 8-10 (2% aq.); *Precaution:* slippery when wet; *Storage:* store in sealed containers below 25 C, away from heat; avoid dust buildup.

Polyox® WSR N-80. [Union Carbide] PEG-5M; CAS 25322-68-3; nonionic; water-sol. thermoplastic resin, thickener, lubricant, binder, flocculant, wet adhesive; for agric., ceramics applics.; dispersant for vinyl polymerization; EPA 40CFR § 180.1001(d); FDA 21CFR §175.300, 175.380, 175.390, 176.170, 176.180, 177.1210, 177.1350; wh. gran. powd., mild ammoniacal odor; 100% thru 10 mesh, 96% thru 20 mesh; sol. in water, some chlorinated solvs., alcohols, aromatic hydrocarbons, ketones; m.w. 200,000; sp.gr. 1.15-1.26; bulk dens. 19-37 lb/ft^3; visc. 65-115 cps (5% aq.); m.p. 62-67 C; pH 8-10 (5% aq.); *Precaution:* slippery when wet; *Storage:* store in sealed containers below 25 C, away from heat; avoid dust buildup.

Polyox® WSR N-750. [Union Carbide] PEG-7M; CAS 25322-68-3; nonionic; water-sol. thermoplastic resin, thickener, lubricant, binder, flocculant, wet adhesive for agric., elec., papermaking industries; emollient for personal care prods.; dispersant for vinyl polymerization; EPA 40CFR §180.1001(d); FDA 21CFR §175.300, 175.380, 175.390, 176.170, 176.180, 177.1210, 177.1350; wh. gran. powd., mild ammoniacal odor; 100% thru 10 mesh, 96% thru 20 mesh; sol. in water, some chlorinated solvs., alcohols, aromatic hydrocarbons, ketones; m.w. 300,000; sp.gr. 1.15-1.26; bulk dens. 19-37 lb/ft^3; visc. 600-1200 cps (5% aq.); m.p. 62-67 C; pH 8-10 (5% aq.); *Precaution:* slippery when wet; *Storage:* store in sealed containers below 25 C, away from heat; avoid dust buildup.

Polyox® WSR N-3000. [Union Carbide] PEG-14M; CAS 25322-68-3; nonionic; water-sol. thermoplastic resin, thickener, lubricant, binder, flocculant, wet adhesive; binder/coating in elec.; emollient in personal care prods.; dispersant for vinyl polymerization; EPA 40CFR §180.1001(d); FDA 21CFR §175.300, 175.380, 175.390, 176.170, 176.180, 177.1210, 177.1350; wh. gran. powd., mild ammoniacal odor; 100% thru 10 mesh, 96% thru 20 mesh; sol. in water, some chlorinated solvs., alcohols, aromatic hydrocarbons, ketones; m.w. 400,000; sp.gr. 1.15-1.26; bulk dens. 19-37 lb/ft^3; visc. 2250-4500 cps (5% aq.); m.p. 62-67 C; pH 8-10 (5% aq.); *Precaution:* slippery when wet; *Storage:* store in sealed containers below 25 C, away from heat; avoid dust buildup.

Polyox® Coagulant. [Union Carbide] PEG-115M; CAS 25322-68-3; nonionic; water-sol. thermoplastic resin; reduces friction in shave prods.; feel modifier providing soft feel; film-former, thickener for aq. systems; flocculant for mining, papermaking; foam stabilizer in fermented malt beverages (300 ppm max.); dispersant for vinyl polymerization; EPA 40CFR §180.1001(d); FDA 21CFR §172.770, 175.300, 175.380, 175.390, 176.170, 176.180, 177.1210, 177.1350; wh. gran. powd., mild ammoniacal odor; 100% thru 10 mesh, 96% thru 20 mesh; sol. in water, some chlorinated solvs., alcohols, aromatic hydrocarbons, ketones; m.w. 5,000,000; sp.gr. 1.15-1.26; bulk dens. 19-37 lb/ft^3; visc. 5500-7500 cps (1% aq.); m.p. 62-67 C; pH 8-10 (sol'n.); *Precaution:* slippery when wet; *Storage:* store in sealed containers below 25 C, away from heat; avoid dust buildup.

Poly-Pale® Ester 10. [Hercules] Glyceryl rosinate; pale thermoplastic resin for lacquers, varnishes, adhesives, and wax modification; tackifying resin in pressure-sensitive rubber-based adhesives, in solv. and emulsion types; in E/VA resin wax hot-melt adhesives and coatings; in varnishes to contribute hardness, rapid drying, and resistance to water and alkali; improves clarity as a wax modifier; USDA Rosin N flakes; sol. in esters, ketones, aromatic and aliphatic hydrocarbons, and chlorinated solvs.; sp.gr. 1.08; dens. 1.08 kg/l; G-H visc. F (60% solids in min. spirits); Hercules drop soften. pt. 114 C; acid no. 7.

Polyplate 90. [J.M. Huber/Engineered Mins.] Hyd. kaolin clay; CAS 1332-58-7; EINECS 296-473-8; delaminated functional filler with high brightness, finer particle size; for adhesives, paints, plastics, and inks; spray-dried beads, slurry; 0.6 μm particle diam.; 0.01% max. on 325 mesh; fineness (Hegman) 5+; sp.gr. 2.60; bulking value 21.7 lb/gal; surf. area 13-17 m^2/g; oil absorp. 40-45; brightness 90-91; pH 6.0-7.5 (100 g clay/250 ml water); 1% max. moisture.

Polyplate 852. [J.M. Huber/Engineered Mins.] Hyd. kaolin clay; CAS 1332-58-7; EINECS 296-473-8; delaminated washed functional filler with high brightness and moderate particle size; for adhesives, paint, plastics, inks; spray-dried beads, slurry; 0.9 μm particle diam.; 0.01% max. on 325 mesh; fineness (Hegman) 4+; sp.gr. 2.60; bulking value 21.7 lb/gal; surf. area 12-16 m^2/g; oil absorp. 40-45; brightness 87.5-89; pH 6.0-7.5 (100 g clay/250 ml water); 1% max. moisture.

Polyplate P. [J.M. Huber/Engineered Mins.] Hyd. kaolin clay; CAS 1332-58-7; EINECS 296-473-8; delaminated washed functional filler with high brightness and high aspect ratio; for adhesives, paints, plastics, and inks; large platey particles, spray-dried beads, slurry; 1 μm particle diam.; 0.01% max. on 325 mesh; fineness (Hegman) 4+; sp.gr. 2.60; bulking value 21.7 lb/gal; surf. area 11-15 m^2/g; oil absorp. 40-45; brightness 87.5-89; pH 6.0-7.5 (100 g clay/250 ml water); 1% max. moisture.

Polyplate P01. [J.M. Huber/Engineered Mins.] Hyd. kaolin clay; CAS 1332-58-7; EINECS 296-473-8; delaminated washed functional filler with high brightness, high aspect ratio; for adhesives, paint, plastics, inks; pulverized; fine powd.; 1 μm particle diam.; 0.01% max. on 325 mesh; fineness (Hegman) 4+; sp.gr. 2.60; bulking value 21.7 lb/gal; surf. area 11-15 m^2/g; oil absorp. 40-45; brightness 87.5-89; pH 6.0-7.5 (100 g clay/250 ml water); 1% max. moisture.

Polypol 19. [Crowley Chem.] Amorphous polypropylene; internal lubricant for adhesives, elastomers, PE extrusion; solid.

Polysar EPDM 227. [Bayer/Fibers, Org., Rubbers] EPDM terpolymer; modifier for polyolefin and TPEs; used alone to promote heat resist. or in blends to increase hardness while decreasing compd. visc.; for inj. molding and peroxide-cured heat-resist. compds.; slow cure rate; lt. colored bales, pellets; sp.gr. 0.86; Mooney visc. 25 (ML1+8, 100 C); 75/25 E/P ratio, 3% ENB.

Polysar EPDM 847XP. [Bayer/Fibers, Org., Rubbers] EPDM terpolymer; modifier for polyolefin and TPEs; used in hose, profiles, other extruded goods requiring exc. dimensional stability, tires, sponge prods., wire/cable insulation, molded goods (gaskets, seals), membranes and other calendered prods.; weatherstripping, roller covers; pellet; sp.gr. 0.86 g/cc; Mooney visc. 55 (ML1+8, 125 C); 74/26 E/P ratio, 4% ENB.

Polysar EPDM 6463. [Bayer/Fibers, Org., Rubbers] EPDM, with 50 phr nonstaining paraffinic oil; oil-extended rubber for optimum heat stability and weather resistance; std. cure rate; used for heat-resistant applics., hose, elec. insulation, roofing membranes, and molded, extruded, and calendered goods; modifier for polyolefins and TPEs; bale; sp.gr. 0.87; Mooney visc. 37 (ML1+8, 150 C); 68/32 E/P ratio; 4% ENB.

Polysar EPM 306. [Bayer/Fibers, Org., Rubbers] EPM; CAS 9010-79-1; rubber used principally for modifying PP, TPEs, other plastics; also for inj. molded and extruded goods (elec. components, wire insulation, O-rings, brake components); pellet and bales; sp.gr. 0.86; Mooney visc. 36 (ML1+8,100 C); 68/32 ethylene/propylene ratio; 60% ethylene content.

Polysar S-1018. [Bayer/Fibers, Org., Rubbers] Divinyl crosslinked emulsion SBR; CAS 9003-55-8; extrusion aid in SBR compds. to control die swell; in solv.-based adhesive applics.; sp.gr. 0.94; 23.5% bound styrene.

Polysar SS 260. [Bayer/Fibers, Org., Rubbers] High styrene self-reinforcing emulsion SBR; CAS 9003-55-8; visc. and hardness modifier for SBR and NBR compds.; stiffness modifier for SBR/NBR closed-cell sponge and microcellular footwear; sp.gr. 1.00; 64% bound styrene.

Polystat Agent #5033. [Polymer Research Corp. of Am.] Phosphoric acid, partial ester; antistat used in plastics.; Unverified

Polystay AA-1. [Goodyear] Anilino-phenyl methacrylamide; antioxidant in emulsion polymers, NBR, SBR, BR, ABS, CR; masterbatch for compounding polymer alloys and blends; cryst. powd.

Polystay AA-1R. [Goodyear] Anilino-phenyl methacrylamide; refined, decolorized version of Polystay AA-1; for more color critical applics.; cryst. powd.

Polystep® A-4. [Stepan; Stepan Canada] Sodium linear alkyl (C12) benzene sulfonate; anionic; emulsifier for emulsion polymerization, S/B, vinyl chloride, and vinylidene chloride latexes; FDA 21CFR §176.210, 177.2600, 178.3130, 178.3400; yel. turbid liq.; 50% act.

Polystep® A-7. [Stepan; Stepan Canada] Sodium dodecylbenzene sulfonate, linear; anionic; emulsifier for emulsion polymerization, S/B, vinyl chloride, vinylidene chloride latexes; FDA 21CFR §176.210, 177.2600, 178.3130, 178.3400; yel. slurry; 39% conc.

Polystep® A-11 [Stepan; Stepan Canada; Stepan Europe] Isopropylamine dodecylbenzene sulfonate, branched; anionic; emulsifier, pigment dispersant; emulsion polymerization (S/B, vinyl chloride, vinylidene chloride latexes); FDA 21CFR §176.210, 177.2600, 178.3130, 178.3400; pale clear visc. liq.; oil-sol.; 88% act.

Polystep® A-13. [Stepan; Stepan Canada] Linear dodecylbenzene sulfonic acid; emulsion polymerization surfactant; catalyst in acid catalyzed reactions; FDA 21CFR §176.210, 178.3400; dk. visc. liq.; 97% act.

Polystep® A-15. [Stepan; Stepan Canada] Sodium linear dodecylbenzene sulfonate; anionic; emulsifier for emulsion and styrene polymerization; FDA 21CFR §176.210, 177.2600, 178.3130, 178.3400; pale clear to hazy liq.; 22% conc.

Polystep® A-15-30K. [Stepan; Stepan Canada] Potassium dodecylbenzene sulfonate, linear; CAS 27177-77-1; EINECS 248-296-2; anionic; surfactant for styrene-butadiene, vinyl chloride, and vinylidene chloride latexes; thermal and hydrolytic stability; FDA 21CFR §176.210, 177.2600, 178.3130, 178.3400; hazy slurry; 30% conc.

Polystep® A-16. [Stepan; Stepan Canada] Sodium dodecylbenzene sulfonate, branched; anionic; emulsifier for styrene polymerization; FDA 21CFR §176.210, 177.2600, 178.3130, 178.3400; pale turbid liq.; 30% conc.

Polystep® A-16-22. [Stepan; Stepan Canada] Sodium dodecylbenzene sulfonate, branched; anionic; emulsifier for styrene, vinyl chloride polymerization; FDA 21CFR §176.210, 177.2600, 178.3130, 178.3400; pale turbid liq.; 22% conc.

Polystep® A-17. [Stepan; Stepan Canada] Dodecylbenzene sulfonic acid, branched; anionic; intermediate for the prod. of sodium salts; neutralized acid as emulsifier; salts as emulsifiers for SBR latex; emulsion polymerization; catalyst for acid catalyzed reactions; FDA 21CFR §176.210, 178.3400; dk. visc. liq.; water-sol.; dens. 9.1 lb/gal; sp.gr. 1.09; visc. 9000 cps (30 C); 97.4% act.

Polystep® A-18. [Stepan; Stepan Canada; Stepan Europe] Sodium alpha olefin (C14, C16) sulfonate; CAS 68439-57-6; EINECS 270-407-8; anionic; surfactant for vinyl and vinylidene chloride, acrylic, styrene-acrylaic, SBR polymerization; FDA 21CFR §175.105, 178.3400; pale yel. clear liq.; 40% act.

Polystep® B-1. [Stepan; Stepan Canada] Ammonium nonoxynol-4 sulfate; anionic; emulsifier for emulsion polymerization; FDA 21CFR §178.3400; yel. visc. liq.; 60% act.

Polystep® B-3. [Stepan; Stepan Canada; Stepan Europe] Sodium lauryl sulfate; CAS 151-21-3; anionic; emulsifier for emulsion polymerization of S/B, vinyl chloride, sol. acrylics; FDA 21CFR §175.105, 175.300, 176.170, 176.210, 177.1200, 177.1210, 177.2600, 177.2800, 178.3400; USP approved; wh. powd.; 97.5% act.

Polystep® B-5. [Stepan; Stepan Canada; Stepan Europe] Sodium lauryl sulfate; CAS 151-21-3; anionic; emulsifier for emulsion polymers, incl. vinyl chloride, S/B, acrylic; FDA 21CFR §175.105, 175.300, 176.170, 176.210, 177.1200, 177.1210, 177.2600, 177.2800, 178.3400; pale yel. clear liq.; 30% act.

Polystep® B-7. [Stepan; Stepan Canada] Ammonium lauryl sulfate; anionic; emulsion polymerization surfactant; latex foaming agent; water resistance in coatings; FDA 21CFR §175.105, 175.300, 176.170, 176.210, 177.1200, 177.1210, 177.2600, 177.2800, 178.3400; pale clear liq.; 30% act.

Polystep® B-11. [Stepan; Stepan Canada; Stepan Europe] Ammonium laureth sulfate (4 EO); anionic; emulsifier for emulsion polymerization (acrylics, styrene-acrylic, vinyl acrylics, S/B); FDA 21CFR §175.105, 176.210, 178.3400; pale clear liq.; 59% act.

Polystep® B-12. [Stepan; Stepan Canada; Stepan Europe] Sodium laureth sulfate (4 EO); CAS 9004-82-4; anionic; emulsifier for polymerization (acrylics, styrene-acrylics, vinyl acrylics, S/B); FDA 21CFR §175.105, 176.210; pale clear liq.; 60% act.

Polystep® B-19. [Stepan; Stepan Canada] Sodium laureth sulfate (30 EO); anionic; emulsifier for emulsion polymerization (acrylics, styrene-acrylics, vinyl acrylics), floor polish, finish polymer and latexes; good water resistance; FDA 21CFR §175.105, 176.210; pale clear liq.; 26% act.

Polystep® B-20. [Stepan; Stepan Canada] Ammonium laureth sulfate (30 EO); anionic; emulsifier for emulsion polymerization (acrylics, styrene-acrylics, vinyl acrylics), floor finish latexes; good water resistance; FDA 21CFR §175.105, 176.210, 178.3400; pale clear liq.; 30% act.

Polystep® B-22. [Stepan; Stepan Canada] Ammonium laureth sulfate (12 EO); anionic; emulsifier for acrylic copolymers; FDA 21CFR §175.105, 176.210, 178.3400; pale clear liq.; 30% act.

Polystep® B-23. [Stepan; Stepan Canada; Stepan Europe] Sodium laureth-12 sulfate; anionic; high active emulsifier for emulsion polymerization (acrylics, styrene-acrylics, vinyl acrylics); somewhat monomer sol.; FDA 21CFR §175.105, 176.210; amber hazy liq.; oil-sol.; 60% act.

Polystep® B-24. [Stepan; Stepan Canada; Stepan Europe] Sodium lauryl sulfate; CAS 151-21-3; anionic; emulsifier with low cloud pt.; for emulsion polymerization (S/B, vinyl chloride, acrylic, styrene acrylic, PVC); FDA 21CFR §175.105, 175.300, 176.170, 176.210, 177.1200, 177.1210, 177.2600, 177.2800, 178.3400; pale clear liq.; 30% act.

Polystep® B-25. [Stepan; Stepan Canada; Stepan Europe] Sodium decyl sulfate; anionic; emulsifier for emulsion polymerization (S/B, vinyl chloride, acrylic), high surf. tens. latex; hydrophilic; FDA 21CFR §175.105, 177.1210; pale clear liq.; 38% act.

Polystep® B-27. [Stepan; Stepan Canada; Stepan Europe] Sodium nonoxynol-4 sulfate; anionic; emulsifier for acrylics, SBR, vinyl chloride, butyl rubber; FDA 21CFR §175.105, 178.3400; pale yel. clear liq.; 30% act.

Polystep® B-29. [Stepan; Stepan Canada; Stepan Europe] Sodium octyl sulfate; anionic; low-foaming emulsifier for vinyl chloride systems; FDA 21CFR §176.210; water-wh. to pale yel. clear liq.; 33% act.

Polystep® B-LCP. [Stepan; Stepan Canada] Sodium alkyl sulfate; anionic; surfactant for vinyl chloride systems; FDA 21CFR §175.105, 177.1210; pale clear liq.; 30% conc.

Polystep® CM 4 S. [Stepan Europe] Sodium alkylphenol ether sulfate; anionic; emulsifier for various systems; gives latex with large particles; visc. disp.; 30% conc.

Polystep® C-OP3S. [Stepan; Stepan Canada; Stepan Europe] Sodium octoxynol-3 sulfate; anionic; emulsifier for controlling particle size in vinyl acetate specialty copolymers; FDA 21CFR §176.170; wh. visc. disp.; 20% act.

Polystep® F-1. [Stepan; Stepan Canada; Stepan Europe] Nonoxynol-4; CAS 9016-45-9; nonionic; nonfoaming pigment dispersant; emulsifier for emulsion polymerization (styrene acrylic, acrylic, vinyl acrylic, S/B); FDA 21CFR §175.105, 176.180, 178.3400; water wh. to pale yel. clear liq.; oil-sol.; 100% act.

Polystep® F-2. [Stepan; Stepan Canada; Stepan Europe] Nonoxynol-6; CAS 9016-45-9; nonionic; pigment dispersant; contributes mech. and freeze/thaw stability to latexes; allows control of particle size; FDA 21CFR §175.105, 176.180, 178.340; pale clear liq.; oil-sol.; 100% act.

Polystep® F-3. [Stepan; Stepan Canada; Stepan Europe] Nonoxynol-8; CAS 9016-45-9; nonionic; pigment dispersant, emulsifier for polymerization (styrene acrylic, acrylic, vinyl acrylic, S/B); contributes mech. and freeze/thaw stability to latexes; allows control of particle size; FDA 21CFR §175.105, 176.180, 178.340; water-wh. to pale yel. clear liq.; water-sol.; 100% act.

Polystep® F-4. [Stepan; Stepan Canada; Stepan Europe] Nonoxynol-10; CAS 9016-45-9; nonionic; emulsifier for emulsion polymerization (styrene acrylic, acrylic, vinyl acrylic, S/B); FDA 21CFR §175.105, 176.180, 178.340; water-wh. to pale yel. clear liq.; 100% act.

Polystep® F-5. [Stepan; Stepan Canada; Stepan Europe] Nonoxynol-12; CAS 9016-45-9; nonionic; emulsifier and stabilizer for emulsion polymerization (styrene acrylic, acrylic, vinyl acrylic, S/B); allows control of particle size; FDA 21CFR §175.105, 176.180, 178.340; water-wh. to pale yel. clear to turbid liq.; 100% act.

Polystep® F-6. [Stepan; Stepan Canada; Stepan Europe] Nonoxynol-14; CAS 9016-45-9; nonionic; emulsifier for emulsion polymerization (styrene acrylic, acrylic, vinyl acrylic, S/B); FDA 21CFR §175.105, 176.180, 178.340; water-wh. to pale yel. clear to turbid liq.; 100% act.

Polystep® F-9. [Stepan; Stepan Canada; Stepan Europe] Nonoxynol-30; CAS 9016-45-9; nonionic; emulsifier for acrylic systems; FDA 21CFR §175.105, 176.180, 178.340; water-wh. to pale yel. clear to hazy liq. or gel; 70% act.

Polystep® F-10. [Stepan; Stepan Canada; Stepan Europe] Nonoxynol-40; CAS 9016-45-9; nonionic; emulsifier for acrylics and vinyl acetate; FDA 21CFR §175.105, 176.180, 178.340; clear to hazy liq. or gel; 70% conc.

Polystep® F-95B. [Stepan; Stepan Canada; Stepan Europe] Nonoxynol-34; CAS 9016-45-9; nonionic;

emulsifier for acrylics and vinyl acetate; blended with other surfactants to increase the latex particle size; FDA 21CFR §175.105, 176.180, 178.340; clear liq.; 70% act.

Polysurf A212E. [Dai-ichi Kogyo Seiyaku] Complex org. phosphate ester, free acid; anionic; antistat, emulsifier for agric., metal cleaning, emulsion polymerization; pigment dispersant; general purpose detergent; liq.; HLB 10.3; 99% conc.

Polytac. [Crowley Chem.] Amorphous polypropylene; internal lubricant for adhesives, elastomers, PE extrusion; solid.

Polytalc 262. [Specialty Minerals] Treated talc; CAS 14807-96-6; EINECS 238-877-9; filler for polyolefins, engineering thermoplastics; powd.; 2 μ avg. particle size, 25 μ top size.

PolyTalc™ 445. [Specialty Minerals] Surface-modified platy talc; CAS 14807-96-6; EINECS 238-877-9; filler for polymer applics. where color is critical; provides enhanced long term heat stability; for polyolefins, engineering thermoplastics; powd.; 6 μ avg. particle size; 99% finer than 30 μ, 42% finer than 5 μ; sp.gr. 2.70; apparent dens. 21 lb/ft^3; pH 8.8; dry brightness 87 min.; 58-61% SiO_2, 31-34% MgO.

Poly-Tergent® 2EP. [Olin] Sodium dodecyl diphenyl ether disulfonate; anionic; surfactant for emulsion polymerization; yel. to brn. liq.; sol. in water, alcohols; sp.gr. 1.16; dens. 9.65 lb/gal; visc. 150 cp; surf. tens. 31.7 dynes/cm (0.1%); Draves wetting 152 s (0.5%); Ross-Miles foam 140 mm (1%, initial); 45% act.; *Toxicology:* LD50 (oral, rats) 2 g/kg; severe eye irritant; possible skin burns.

Polytrend® 850. [Hüls Am.] Pigment dispersions in hydroxyl functional isophthalic polyester resin; industrial colorants for polyester gel coat, sheet molding compd., cast polyester, potting compds., polyester PU, high-solids baking enamels, and RIM applics.

Polyvel A2010. [Polyvel] 20% Erucamide and 10% antiblock conc.; slip additive for PP blown and cast film.

Polyvel AI40. [Polyvel] Antifog agent conc.; antifog for LDPE and LLDPE blown and cast food contact film applics.; exc. thermal stability; FDA 21CFR §164.1505, 174.5; off-wh. pellet; melt flow 210 g/10 min (190 C, 2160 g load); soften. pt. 65 C; 40% act.

Polyvel AI509. [Polyvel] Glyceryl stearate conc.; antistat additive for polyolefin film, foam and inj. molding; esp. effective in PP; suitable for clear film applics., food pkg. applics.; wh. pellet; melt index 320 g/10 min; soften. pt. 70 C; 50% act.

Polyvel AI1645, AI1685, AI2645, AI2685. [Polyvel] Antifog agent conc.; antifog for LDPE, LLDPE, HDPE, and PP food pkg. applics.

Polyvel AI2163. [Polyvel] Amine-based antistat conc.; antistat for polyolefin film, extrusion, and inj. molding applics.

Polyvel AI2902. [Polyvel] Non-amine-based antistat conc.; antistat for polyolefin film, extrusion, and inj. molding applics., electronic and food pkg. applics.

Polyvel AI3129, AI5129. [Polyvel] GMS-based antistat conc.; antistat for LDPE, LLDPE, HDPE, and PP.

Polyvel AI3184. [Polyvel] Antifog agent conc.; antifog for agric. film.

Polyvel BA20, BA40, BA60. [Polyvel] Azodicarbonamide conc.; blowing agent conc. for PE, PP, HIPS, PPO, and elastomers for process temps. of 330-420 F.

Polyvel BA20T, BA40T. [Polyvel] Blowing agent conc.; blowing agent for LDPE, EVA, elastomers, and PU for process temps. of 250-340 F.

Polyvel BA205PT. [Polyvel] Blowing agent conc.; blowing agent for PP, PET, ABS, PPO, PBT, nylon, and PC for process temps. of 450-500 F.

Polyvel BA2015. [Polyvel] Blowing agent conc.; blowing agent for PPS, PBT, nylon, TFE, and PC for process temps. of 600-680 F.

Polyvel BA2575. [Polyvel] Blowing agent conc.; blowing agent for PPS, nylon, and TFE for process temps. of 700-750 F.

Polyvel C200. [Polyvel] Fragrance oil conc.; fragrance for K-resin for clear fragranced applics., air fresheners, Christmas decorations, toys, home prods., jewelry, pkg., etc.; clear or translucent dry pellet; melt index 6-12 g/10 min; soften. pt. 115 C.

Polyvel CA-P20. [Polyvel] Clarifying agent conc. in PP carrier; nucleating agent in PP homopolymer and random copolymers; provides improved clarity, better gloss, improved tens. and flex. str. and dimensional stability; FDA 21CFR §177.1520(c); wh. free-flowing pellet; melt flow 25 g/10 min; soften. pt. 160 C; for inj. molding, inj. blow molding, blown film, bottles, and sheet.

Polyvel CR-5. [Polyvel] 2,5-Dimethyl-2,5-di(t-butylperoxy) hexane; CAS 78-63-7; flow modifier, rheology and processing aid for PP; pellet; 5% act.

Polyvel CR-5F. [Polyvel] 2,5-Dimethyl-2,5-di(t-butylperoxy) hexane; CAS 78-63-7; flow modifier and processing aid for PP; FDA approved for food contact; 5% act.

Polyvel CR-5P. [Polyvel] Bis(t-butylperoxyisopropyl) benzene; flow modifier and processing aid for PP; 5% act.

Polyvel CR-5T. [Polyvel] t-Butyl perbenzoate; CAS 614-45-9; additive for grafting and extrusion reactions, esp. for nylon; pellet; 5% act.

Polyvel CR-10. [Polyvel] 2,5-Dimethyl-2,5-di(t-butylperoxy) hexane; CAS 78-63-7; flow modifier, rheology and processing aid for PP; pellet; 10% act.

Polyvel CR-10P. [Polyvel] Bis(t-butylperoxyisopropyl) benzene; flow modifier and processing aid for PP; 10% act.

Polyvel CR-10T. [Polyvel] t-Butyl perbenzoate; CAS 614-45-9; additive for grafting and extrusion reactions; 10% act.

Polyvel CR-20P. [Polyvel] Bis(t-butylperoxyisopropyl) benzene; flow modifier and processing aid for PP; pellet; 20% act.

Polyvel CR-25. [Polyvel] 2,5-Dimethyl-2,5-di(t-butylperoxy) hexane; CAS 78-63-7; flow modifier, rheology and processing aid for PP; 25% act.

Polyvel CR-L10. [Polyvel] 2,5-Dimethyl-2,5-di(t-butylperoxy)hexyne-3; CAS 1068-27-5; additive for roto molding and crosslinking; 10% act.

Polyvel CR-L20. [Polyvel] 2,5-Dimethyl-2,5-di(t-butylperoxy)hexyne-3; CAS 1068-27-5; additive for roto molding and controlled crosslinking of EVA and polyethylene to increase hot str. and physical props.; 20% act.

Polyvel E40. [Polyvel] Fragrance oil conc.; fragrance for PVC dry blends for simulated leather articles, shower curtains, baby toys, household articles; free-flowing powd.; 40% act.

Polyvel ER40. [Polyvel] Fragrance or flavor oil conc.; fragrance of flavor concs. for PVC plastisol applics. for roto molding and coating; fine particle size powd.

Polyvel P10. [Polyvel] Fragrance oil conc.; fragrance for HDPE film applics., e.g., pkg., novelty bags, garbage bags; dry pellet; melt index 0.7 g/10 min; soften. pt. 160 C; 10% fragrance.

Polyvel P15. [Polyvel] Fragrance oil conc.; fragrance for LDPE or LLDPE film applics., e.g., pkg., novelty bags, garbage bags; dry pellet; melt index 8 g/10 min; soften. pt. 112 C; 15% fragrance.

Polyvel P25. [Polyvel] Fragrance oil conc. for EVA, LDPE, LLDPE, HDPE, and PP; pellet; 25% act.

Polyvel PA10FP, FA10FP2. [Polyvel] Fluoroelastomer conc. in LLDPE carrier; processing aid for LLDPE, HDPE blow film processing; 10% act.

Polyvel PCL-10. [Polyvel] Dicumyl peoxide; CAS 80-43-3; additive for controlled crosslinking of polyethylene, EP, EPDM rubber, and EVA to increase hot str. and physical props.; pellet; 10% act.

Polyvel PCL-20. [Polyvel] Dicumyl peoxide; CAS 80-43-3; additive for controlled crosslinking of polyethylene, EP, EPDM rubber, and EVA to increase hot str. and physical props.; pellet; 20% act.

Polyvel PV15. [Polyvel] Fragrance or flavor oil conc.; fragrance or flavor conc. for flexible PVC inj. molding and extrusion applics.; pellet.

Polyvel PV30. [Polyvel] Fragrance oil conc.; fragrance for pelletized flexible PVC compd., extrusion, and inj. molding; 30% act.

Polyvel RE25. [Polyvel] Erucamide conc.; slip and release additive for LDPE, LLDPE, HDPE, and PP films, Kraton rubber, elastomers; off-wh. pellet; melt flow 42 g/10 min; soften. pt. 76 C; 25% act.

Polyvel RE40. [Polyvel] Erucamide conc.; slip and release additive for HDPE, PP, Kraton rubber, elastomers; 40% act.

Polyvel RM40. [Polyvel] Molybdenum disulfide conc.; additive for nylon 6, sheet, profile and tubing extrusion to increase heat distort. temp. and improve bearing and wear resist. props.; 40% MoS_2.

Polyvel RM40Z. [Polyvel] Molybdenum disulfide conc.; additive for nylon 6/6 sheet, profile and tubing extrusion to increase heat distort. temp. and improve bearing and wear resist. props.; 40% MoS_2.

Polyvel RM70. [Polyvel] Molybdenum disulfide conc.; additive for all nylon grades used in compounding and inj. molding to increase heat distort. temp. and improve bearing and wear resistance props.; 70% MoS_2.

Polyvel RO25. [Polyvel] Oleamide conc.; CAS 301-02-0; EINECS 206-103-9; slip and release additive for LDPE, EVA film and inj. molding applics.; off-wh. pellet; melt flow 42 g/10 min; soften. pt. 76 C; 25% act.

Polyvel RO40. [Polyvel] Oleamide conc.; CAS 301-02-0; EINECS 206-103-9; slip and release additive for LDPE, EVA film and inj. molding applics.; 40% act.

Polyvel RO252. [Polyvel] 25% Oleamide and 20% antiblock conc.; slip additive for LDPE and LLDPE film.

Polyvel RZ40. [Polyvel] Zinc stearate conc.; CAS 557-05-1; EINECS 209-151-9; slip and release additive for HIPS, ABS, and nylon; 40% act.

Polyvel UV25AG2. [Polyvel] UV stabilizer conc.; uv stabilizer for LDPE, LLDPE outdoor film, sheet and molding applics.; 25% act.

Polywax® 500. [Petrolite] Polyethylene homopolymer; CAS 9002-88-4; EINECS 200-815-3; lubricant, flow modifier, release agent, antiblock for plastics processing; nucleating agent for expandable PS; slip aid for printing inks; leveling/slip agent for powd. coatings; visc. modifier for hot-melt adhesives; cosmetic ingreds.; wax modifier; chewing gum base; incl. FDA §172.888, 175.105, 175.300, 176.170, 176.180, 176.200, 177.1200, 178.3720, 179.45; prilled; low sol. in org. solvs., esp. at R.T.; sol. in CCl_4, benzene, xylene, toluene, turpentine; m.w. 500; melt index > 5000 g/10 min; dens. 0.93 g/cc; visc. 6.6 cps (99 C); m.p. 88 C; soften. pt. (R&B) 88 C.

Polywax® 655. [Petrolite] Polyethylene homopolymer; CAS 9002-88-4; EINECS 200-815-3; lubricant, flow modifier, release agent, antiblock for plastics processing; nucleating agent for expandable PS; slip aid for printing inks; leveling/slip agent for powd. coatings; visc. modifier for hot-melt adhesives; cosmetic ingreds.; wax modifier; chewing gum base; incl. FDA §172.888, 175.105, 175.300, 176.170, 176.180, 176.200, 177.1200, 178.3720, 179.45; prilled; low sol. in org. solvs., esp. at R.T.; sol. in CCl_4, benzene, xylene, toluene, turpentine; m.w. 655; melt index > 5000 g/10 min; dens. 0.94 g/cc; visc. 5 cps (149 C); m.p. 99 C; soften. pt. (R&B) 99 C.

Polywax® 850. [Petrolite] Polyethylene homopolymer; CAS 9002-88-4; EINECS 200-815-3; lubricant, flow modifier, release agent, antiblock for plastics processing; nucleating agent for expandable PS; slip aid for printing inks; leveling/slip agent for powd. coatings; visc. modifier for hot-melt adhesives; cosmetic ingreds.; wax modifier; incl. FDA §172.888, 175.105, 175.300, 176.170, 176.180, 176.200, 177.1200, 178.3720, 179.45; m.w. 850; melt index > 5000 g/10 min; dens. 0.96 g/cc; visc. 8.5 cs (149 C); m.p. 107 C; soften. pt (R&B) 107 C.

Polywax® 1000. [Petrolite] HDPE; CAS 9002-88-4; EINECS 200-815-3; lubricant, flow modifier, release agent, antiblock for plastics processing; nucleating agent for expandable PS; slip aid for printing inks; leveling/slip agent for powd. coatings; visc. modifier for hot-melt adhesives; cosmetic ingreds.; wax modifier; FDA 21CFR §172.888, 175.105, 175.125, 175.300, 175.320, 176.170, 176.180, 176.200, 176.210, 177.1200, 177.1210, 177.2600, 177.2800, 178.3720, 178.3850, 179.45; prilled; low sol. in org. solvs., esp. at R.T.; sol. in CCl_4, benzene, xylene, toluene, turpentine; m.w. 1000; melt index > 5000 g/10 min; dens. 0.96 g/cc; visc. 12 cps (149 C); m.p. 113 C; soften. pt. (R&B) 113 C.

Polywax® 2000. [Petrolite] HDPE; CAS 9002-88-4; EINECS 200-815-3; lubricant, flow modifier, release agent, antiblock for plastics processing; nucleating agent for expandable PS; slip aid for printing inks; leveling/slip agent for powd. coatings; visc. modifier for hot-melt adhesives; cosmetic ingreds.; wax modifier; food-contact applics.; FDA 21CFR §173.20, 175.105, 175.125, 175.300, 175.320, 176.170, 176.180, 176.200, 176.210, 177.1200, 177.1210, 177.1520(c) (2.1)-(2.3), 177.2600, 178.3570, 178.3850, 179.45; prilled; low sol. in org. solvs., esp. at R.T.; sol. in CCl_4, benzene, xylene, toluene, turpentine; m.w. 2000; melt index > 5000 g/10 min; dens. 0.97 g/cc; visc. 48 cps (149 C); m.p. 126 C; soften. pt. (R&B) 126 C.

Polywax® 3000. [Petrolite] Polyethylene homopolymer; CAS 9002-88-4; EINECS 200-815-3; lubricant, flow modifier, release agent, antiblock for plastics processing; nucleating agent for expandable PS; slip aid for printing inks; leveling/slip agent for powd. coatings; visc. modifier for hot-melt adhesives; cosmetic ingreds.; wax modifier; FDA 21CFR §173.20, 175.105, 175.125, 175.300, 175.320, 176.170, 176.180, 176.200, 176.210, 177.1200, 177.1210, 177.1520(c) (2.1)-(2.3), 177.2600, 178.3570, 178.3850, 179.45; m.w. 3000; melt index > 5000 g/10 min; dens. 0.98 g/cc; visc. 130 cps (149 C); m.p. 129 C; soften. pt. (R&B) 129 C.

Poly-Zole® AZDN. [Uniroyal] 2,2′-Azobisisobutyronitrile; CAS 78-67-1; EINECS 201-132-3; initiator for addition-polymerization of vinyls, acrylics, and other monomers, in bulk, sol'n., and suspension; wh.or off-wh. powd.; m.w. 164.2; decomp. pt. 103 C; 99.5% assay.

Porofor® ADC/E. [Bayer AG; Bayer/Fibers, Org., Rubbers] Azodicarbonamide; CAS 123-77-3; EINECS 204-650-8; chemical blowing agent for prod. of plastic foams; used in plasticized and unplasticized PVC, polyolefins, PS, styrene copolymers, expanded UPVC pipe and sections; pale yel. to orange powd., 23 μm mean particle size; sol. in dimethyl sulfoxide and dimethylformamide; sp.gr. 1.65; dec. pt. 205-215 C; gas yield 220 cm^3/g.

Porofor® ADC/F. [Bayer AG; Bayer/Fibers, Org., Rubbers] Azodicarbonamide; CAS 123-77-3; EINECS 204-650-8; chemical blowing agent for prod. of plastic foams; used in plasticized and unplasticized PVC, polyolefins, PS, styrene copolymers, wallpaper, and extrusion of thermoplastics; pale yel. to orange powd.; 15 μm mean particle size; sp.gr. 1.65; dec. pt. 205-215 C; gas yield 220 cm^3/g.

Porofor® ADC/K. [Bayer AG] Azodicarbonamide, activator; CAS 123-77-3; EINECS 204-650-8; blowing agent for rubber goods; yel. powd.; dens. 1.6 g/cm^3; Unverified

Porofor® ADC/M. [Bayer/Fibers, Org., Rubbers] Azodicarbonamide; CAS 123-77-3; EINECS 204-650-8; chemical blowing agent for prod. of plastic foams and inj. molding; used in plasticized and unplasticized PVC, polyolefins, PS, styrene copolymers; lt. yel. to orange cryst. powd.; 5 μm mean particle size; m.w. 116.1; sp.gr. 1.65; dec. pt. 205-215 C; gas yield 220 cm^3/g.

Porofor® ADC/S. [Bayer AG] Azodicarbonamide; CAS 123-77-3; EINECS 204-650-8; chemical blowing agent for prod. of plastic foams; PVC expanded artificial leather and wallpaper; powd., particle size 7 μm.; Unverified

Porofor® B 13/CP 50. [Bayer AG] Benzene-1,3-disulfonyl hydrazide, chlorinated paraffin, pbw ratio 50:50; blowing agent for rubber goods; pale gray crumbly to pasty substance; dens. 1.5 g/cm^3; Unverified

Porofor® BSH Paste. [Bayer/Fibers, Org., Rubbers] Benzene sulfonyl hydrazide, paraffin oil, pbw ratio 75:25; blowing agent for rubber goods; pale gray paste; dens. 1.26 g/cm^3.

Porofor® BSH Paste M. [Bayer AG] Benzene sulfonyl hydrazide, paraffin oil, pbw ratio 75:25; blowing agent for rubber goods; pale gray paste; dens. 1.26 g/cm^3; Unverified

Porofor® BSH Powder. [Bayer AG] Benzene sulfonyl hydrazide; CAS 80-17-1; blowing agent for rubber goods; wh. powd.; dens. 1.48 g/cm^3; Unverified

Porofor® DNO/F. [Bayer AG] Dinitrosopentamethylene tetramine, desensitizing agent, pbw ratio 80:20; CAS 101-25-7; blowing agent for rubber goods; ylsh. powd.; dens. 1.43 g/cm^3; Unverified

Porofor® N. [Bayer AG] Azoisobutyric dinitrile; blowing agent for prod. of PVC; used in floats, buoys, net floats, bath mats; wh. to pale gray cryst. powd.; sol. in toluene, xylene, dichloroethane, ethyl acetate, chloroform, and most vinyl monomers; insol. in water; Unverified

Porofor® S 44. [Bayer AG] Diphenylene-oxide-4,4′-disulfohydrazide; blowing agent for prod. of foamed plastics.; Unverified

PP-AB 104. [Additive Polymers Ltd] 5% Precipitated silica in PP-based masterbatch; antiblock agent for plastics.

PP-AO 27. [Additive Polymers Ltd] 15% Phenolic and 15% phosphite in PP-based masterbatch; antioxidant for plastics.

PP-AO 47. [Additive Polymers Ltd] 10% Phenolic in PP-based masterbatch; antioxidant for plastics.

PP-AS 79. [Additive Polymers Ltd] 5% Ethoxylated amine in PP-based masterbatch; antistat for plastics.

PP-SA 46. [Additive Polymers Ltd] 4% Erucamide and 4% precipitated silica in PP-based masterbatch; slip/antiblock agent for plastics.

PP-SA 97. [Additive Polymers Ltd] 4% Erucamide and

7.5% precipitated silica in PP-based masterbatch; slip/antiblock agent for plastics.

PP-SA 128. [Additive Polymers Ltd] 5% Erucamide and 7.5% precipitated silica in PP-based masterbatch; slip/antiblock agent for plastics.

PP-SL 82. [Additive Polymers Ltd] 16% Erucamide in PP-based masterbatch; slip agent for plastics.

PP-SL 98. [Additive Polymers Ltd] 7.5% Erucamide in PP-based masterbatch; slip agent for plastics.

PP-SL 101. [Additive Polymers Ltd] 4% Erucamide in PP-based masterbatch; slip agent for plastics.

PP-UV 38. [Additive Polymers Ltd] 20% HALS I in PP-based masterbatch; uv stabilizer for plastics.

PP-UV 52. [Additive Polymers Ltd] 25% HALS II and 25% HALS III in PP-based masterbatch; uv stabilizer for plastics.

PP-UV 53. [Additive Polymers Ltd] 20% HALS IV in PP-based masterbatch; uv stabilizer for plastics.

PP-UV 54. [Additive Polymers Ltd] 20% HALS II in PP-based masterbatch; uv stabilizer for plastics.

PP-UV 61. [Additive Polymers Ltd] 20% HALS III in PP-based masterbatch; uv stabilizer for plastics.

PP-UV 62. [Additive Polymers Ltd] 10% HALS I in PP-based masterbatch; uv stabilizer for plastics.

PP-UV 63. [Additive Polymers Ltd] 10% HALS II in PP-based masterbatch; uv stabilizer for plastics.

PP-UV 64. [Additive Polymers Ltd] 10% HALS III in PP-based masterbatch; uv stabilizer for plastics.

PP-UV 66. [Additive Polymers Ltd] 10% HALS II and 10% HALS III in PP-based masterbatch; uv stabilizer for plastics.

PP-UV 75. [Additive Polymers Ltd] 10% HALS I and 10% HALS III in PP-based masterbatch; uv stabilizer for plastics.

PP-UV 77. [Additive Polymers Ltd] 50% HALS III in PP-based masterbatch; uv stabilizer for plastics.

PP-UV 87. [Additive Polymers Ltd] 10% HALS IV in PP-based masterbatch; uv stabilizer for plastics.

PP-UV 88. [Additive Polymers Ltd] 50% HALS IV in PP-based masterbatch; uv stabilizer for plastics.

PP-UV 113. [Additive Polymers Ltd] 20% HALS I and 20% HALS III in PP-based masterbatch; uv stabilizer for plastics.

Prespersion PAB-262. [Syn. Prods.] 70% Ethylene thiourea in petrol. oil; accelerator disp.; paste.

Prespersion PAB-866. [Syn. Prods.] 83.3% Zinc oxide in petrol. oil; accelerator disp.; paste.

Prespersion PAB-1196. [Syn. Prods.] 83.3% Sodium bicarbonate in petrol. oil; paste.

Prespersion PAB-1724. [Syn. Prods.] 80% Sulfur in petrol. oil; paste.

Prespersion PAB-4493. [Syn. Prods.] 80% Calcium oxide in petrol. oil; paste.

Prespersion PAB-8912. [Syn. Prods.] 67% Dibasic lead phthalate in DIDP; stabilizer for PVC plastisols.

Prespersion PAB-9541. [Syn. Prods.] 66.6% Titanium dioxide in DOP; colorant/filler disp.; paste.

Prespersion PAC-332. [Syn. Prods.] 72.5% Tribasic lead sulfate in DIDP; stabilizer for PVC plastisols.

Prespersion PAC-363. [Syn. Prods.] Dibasic lead phosphite in DIDP; stabilizer for PVC plastisols.

Prespersion PAC-1451. [Syn. Prods.] 70% Cadmium stearate in DOP; stabilizer disp.; paste.

Prespersion PAC-1712. [Syn. Prods.] 80% Antimony oxide in petrol. oil; flame retardant disp.; paste.

Prespersion PAC-2451. [Syn. Prods.] 80% Agerite® HP-S in petrol. oil; antioxidant disp.; paste.

Prespersion PAC-2941. [Syn. Prods.] 75% Basic lead carbonate in DIDP; stabilizer for PVC plastisols.

Prespersion PAC-3656. [Syn. Prods.] 75% Zinc oxide in DOP; accelerator disp.; paste.

Prespersion PAC-3936. [Syn. Prods.] 80% Dibasic lead phosphite in epoxidized soybean oil; stabilizer for PVC plastisols.

Prespersion PAC-4893. [Syn. Prods.] 80% Tribasic lead sulfate in epoxidized soybean oil; stabilizer for PVC plastisols.

Prespersion PAC-4911. [Syn. Prods.] 75% N-Nitroso diphenyl amide in DOP; paste.

Prespersion SIB-7211. [Syn. Prods.] 70% Titanium dioxide in silicone base; wh. colorant conc.

Prespersion SIB-7324. [Syn. Prods.] 2,5-Dimethyl-2,5-di(t-butylperoxy)hexane in silicone base; catalyst disp.; liq.

Prespersion SIC-3233. [Syn. Prods.] 70% Dicup R (dicumyl peroxide) in silicone base; catalyst disp.; paste.

Prespersion SIC-3342. [Syn. Prods.] 66.6% Red iron oxide in silicone base; colorant conc.

Prespersion SIC-3536. [Syn. Prods.] 50% Carbon black (N-765) in silicone base; colorant conc.

Prifac 5902, 5904, 5905. [Unichema Int'l.] Partially hydrog. tallow and palm oil fatty acids; emulsifier for emulsion polymerization of SBR.

Prifrac 2920. [Unichema] Lauric acid; CAS 143-07-7; EINECS 205-582-1; emulsifier for hot emulsion polymerization of NBR, NR latex stabilization; acid no. 277-282; iodine no. 0.5 max.; 92-94% conc.

Prifrac 2922. [Unichema] Lauric acid; CAS 143-07-7; EINECS 205-582-1; emulsifier for mfg. of nitrile rubbers, stabilization of NR latex; acid no. 278-282; iodine no. 0.2 max.; 98-100% conc.

Primax UH-1060. [Air Prods./Perf. Chems.] UHMWPE, modified; CAS 9002-88-4; EINECS 200-815-3; surface modified particles producing cast elastomer composites with high abrasion resist. and low coeff. of friction; imparts improved abrasion, corrosion and chem. resist. in coatings; mesh size 60.

Primax UH-1080. [Air Prods./Perf. Chems.] UHMWPE, modified; CAS 9002-88-4; EINECS 200-815-3; surface modified particles producing cast elastomer composites with high abrasion resist. and low coeff. of friction; imparts improved abrasion, corrosion and chem. resist. in coatings; mesh size 80.

Primax UH-1250. [Air Prods./Perf. Chems.] UHMWPE, modified; CAS 9002-88-4; EINECS 200-815-3; surface modified particles producing cast elastomer composites with high abrasion resist. and low coeff. of friction; imparts improved abrasion, corrosion and chem. resist. in coatings;

mesh size 250.

Printex P. [Degussa] Furnace black; CAS 1333-86-4; EINECS 215-609-9; for plastics formulations requiring highest uv absorp.

Priolene 6901. [Unichema] Oleic acid; CAS 112-80-1; EINECS 204-007-1; emulsifier for cold emulsion polymerization of SBR latex; foaming agent for NR, SBR latex, NBR, CR rubbers.

Priolene 6907. [Unichema Int'l.] Oleic acid; CAS 112-80-1; EINECS 204-007-1; emulsifier for cold emulsion polymerization of SBR latex.

Priolene 6911. [Unichema Int'l.] Primary foaming agent for NR, SBR latex, NBR, CR rubbers.

Priolene 6930. [Unichema Int'l.] Oleic acid; CAS 112-80-1; EINECS 204-007-1; emulsifier for cold emulsion polymerization of SBR latex.

Priolube 1405. [Unichema Int'l.] n-Butyl oleate; CAS 142-77-8; EINECS 205-559-6; plasticizer for chloroprene rubber.

Priolube 1407. [Unichema Int'l.] Glyceryl oleate; lubricant for PVC processing; antifog agent for PVC film.

Priolube 1408. [Unichema] Glyceryl oleate; antifog for PVC film.

Priolube 1409. [Unichema Int'l.] Glyceryl dioleate; CAS 25637-84-7; EINECS 247-144-2; lubricant for PVC processing; antifog for PVC film.

Priolube 1414. [Unichema Int'l.] Isobutyl oleate; CAS 84988-79-4; EINECS 284-868-8; additive to improve ozone resist. of chloroprene rubber.

Priolube 1447. [Unichema] Polyethylene glycol dioleate; lubricant for PVC processing.

Priolube 1451. [Unichema Int'l.] n-Butyl stearate.; CAS 123-95-5; EINECS 204-666-5; plasticizer for butyl rubber; lubricant for PVC processing.

Priolube 1458. [Unichema Int'l.] Isooctyl stearate; CAS 91031-48-0; EINECS 292-951-5; lubricant for PVC processing.

Priplast 1431. [Unichema Int'l.] Epoxidized oleate; plasticizer for PVC.

Priplast 1562. [Unichema Int'l.] Plasticizer for nitrile rubber to be used for petrol. hose connections.

Priplast 3013. [Unichema Int'l.] Di-n-hexyl azelate; CAS 109-31-9; plasticizer for PVC.

Priplast 3018. [Unichema] Di-2-ethylhexyl azelate; CAS 103-24-2; plasticizer for chloroprene and nitrile rubbers; Gardner 1 max. color; sp.gr. 0.92; visc. 16 mPa•s; acid no. 0.5 max.; hyd. no. 3 max.; flash pt. (COC) 215 C; ref. index 1.448.

Priplast 3114. [Unichema] Polymeric; plasticizer; Gardner 3 max. color; sp.gr. 1.09; visc. 3000-3500 mPa•s; acid no. 3 max.; hyd. no. 35 max.; flash pt. (COC) 280 C; ref. index 1.465.

Priplast 3124. [Unichema] Polymeric; plasticizer; Gardner 2 max. color; sp.gr. 1.06; visc. 700-900 mPa•s; acid no. 3 max.; hyd. no. 20 max.; flash pt. (COC) 240 C; ref. index 1.476.

Priplast 3149. [Unichema] Polymeric; plasticizer; Gardner 7 max. color; sp.gr. 1.08; visc. 35,000-45,000 mPa•s; acid no. 3 max.; hyd. no. 20 max.; flash pt. (COC) 320 C; ref. index 1.468.

Priplast 3155. [Unichema] Polymeric; plasticizer; Gardner 5 max. color; sp.gr. 1.05; visc. 8000-10,000 mPa•s; acid no. 3 max.; hyd. no. 10-20; flash pt. (COC) 275 C; ref. index 1.486.

Priplast 3157. [Unichema] Polymeric; plasticizer for rubber with exc. heat stability and good solv. extraction resist.; Gardner 2 max. color; sp.gr. 1.07; visc. 4200-5000 mPa•s; acid no. 2 max.; hyd. no. 22 max.; flash pt. (COC) 280 C; ref. index 1.466.

Priplast 3159. [Unichema] Polymeric; plasticizer; Gardner 2 max. color; sp.gr. 1.10; visc. 35,000-45,000 mPa•s; acid no. 2 max.; hyd. no. 33 max.; flash pt. (COC) 305 C; ref. index 1.471.

Priplast 3183. [Unichema] Aromatic polyester polyol; polymer building block for CRC-free PU foams with improved flame retardant props.; high functionality, low visc.; visc. 4500-8000 mPa•s; acid no. 2 max.; hyd. no. 460-510.

Priplast 3184. [Unichema] Aromatic polyester polyol; polymer building block for CRC-free PU foams with improved flame retardant props.; visc. 4500-7000 mPa•s; acid no. 2 max.; hyd. no. 380-420.

Priplast 3185. [Unichema] Aliphatic polyester polyol; polymer building block for CRC-free PU foams with improved flame retardant props.; sp.gr. 1.05; visc. 6000-10,000 mPa•s; acid no. 2 max.; hyd. no. 330-360.

Pripol 1004. [Unichema] C44 dimer acid, hydrog.; polymer building block; modifier providing flexibility, toughness, improved impact resist. and low temp. props.; modifier for nylon, PBT, PET, polyester fibers, polyamides for hot melts, urethane elastomers, spin finishes; reasonably biodeg.; APHA 100 max. liq., faint odor; insol. in water; sp.gr. 0.92; visc. 4100 mPa•s; b.p. > 200 C; acid no. 159-164; iodine no. 7; sapon. no. 160-166; pour pt. -10 C; flash pt. (COC) 325 C; ref. index 1.475; *Precaution:* incompat. with strong oxidizing agents.

Pripol 1009. [Unichema] Hydrog. dimer acid; CAS 68783-41-5; building block/modifier permitting prod. of higher m.w. condensation polymers with flexibility, toughness, improved impact resist., low moisture absorp., hydrolytic resist.; used in polyester fibers, polyamides, urethane elastomers; reasonably biodeg.; virtually colorless liq., faint odor; insol. in water; sp.gr. 0.92; dens. 950 kg/m^3; visc. 8000 mPa•s; vapor pressure < 1 mbar (20 C); b.p. > 200 C; pour pt. 0 C; acid no. 194-198; iodine no. 7; sapon. no. 196-200; flash pt. (COC) 305 C; fire pt. 340 C; 98% dimer; *Toxicology:* LD50 (oral, rat) > 5 g/kg; nontoxic by ingestion; sl. irritating to skin; mildly irritating to eyes; *Precaution:* incompat. with strong oxidizing agents.

Pripol 1013. [Unichema] Dimer acid; polymer building block; Gardner 5 max. liq.; sp.gr. 0.93; visc. 7100 mPa•s; acid no. 194-198; sapon. no. 197-201; flash pt. (COC) 285 C; fire pt. 320 C; 93-98% dimer, 2-4% trimer.

Pripol 1017. [Unichema] Dimer acid; polymer building block; aux. emulsifier for chloroprene rubbers; Gardner 8 max. liq.; sp.gr. 0.95; visc. 8000 mPa•s; acid no. 192-197; sapon. no. 195-202; flash pt. (COC) 275 C; fire pt. 300 C; 75-82% dimer, 18-22%

trimer.

Pripol 1022. [Unichema] Dimer acid; polymer building block; aux. emulsifier for chloroprene rubbers; Gardner 8 max. liq.; sp.gr. 0.94; visc. 5800 mPa•s; acid no. 192-196; sapon. no. 197-202; flash pt. (COC) 275 C; fire pt. 300 C; 75-82% dimer, 18-21% trimer.

Pripol 1025. [Unichema] Hydrog. dimer acid; polymer building block; for hot melt adhesives; polyamide for thermographic inks; Gardner 3 max. visc. liq.; sp.gr. 0.95; visc. 8900 mPa•s; acid no. 192-197; sapon. no. 195-202; flash pt. (COC) 275 C; fire pt. 300 C; 75-82% dimer, 18-21% trimer.

Pripol 1040. [Unichema] Trimer acid; polymer building block; used in polyamino-amides for PVC plastisols to be used as car underbody coatings and in water-sol. alkyd resins; Gardner 16 liq.; sp.gr. 1.00; visc. 45,000 mPa•s; acid no. 184-194; sapon. no. 195-205; flash pt. (COC) 320 C; fire pt. 340 C; 22% dimer, 78% trimer.

Pripol 2033. [Unichema] Dimer diol; polymer building block; cosmetics ingred.; APHA 50 max. liq.; visc. 2500 mPa•s; acid no. 0.2 max.; iodine no. 10; sapon. no. 2 max.; hyd. no. 195-205; 97.5% dimer.

Pristerene 4900. [Unichema] Stearic acid; CAS 57-11-4; EINECS 200-313-4; lubricant for PVC processing.

Proaid® 9802. [Akrochem] Nonstaining, nondiscoloring processing and dispersing aid for polychloroprene and EPDM; lt. br. pellets; sp.gr. 0.91.

Proaid® 9810. [Akrochem] Zinc salt of fatty acids; nonstaining, nonblooming processing aid for unsat. polymers; peptizing agent for natural and polyisoprene rubber (and blends); aids disp. of fillers; imparts some mold release benefits; tan pellets; sp.gr. 1.05; m.p. 172 F.

Proaid® 9814. [Akrochem] Homogenizing agent and softening resin for use in most elastomers; improves processing; dk. br. flakes; sp.gr. 1.19; soften. pt. 75-85 C.

Proaid® 9826. [Akrochem] Homogenizing agent and softening resin for use in most elastomers; improves processing; for lt.-colored compds.; lt. br. pellets; sp.gr. 1.00; soften. pt. 85 C.

Proaid® 9831. [Akrochem] Zinc salt of fatty acids; nonstaining, nonblooming processing aid for unsat. polymers; peptizing agent for natural and polyisoprene rubber (and blends); aids disp. of fillers; imparts some mold release benefits; sp.gr. 1.1; m.p. 165-185 F.

Proaid® 9904. [Akrochem] Specialty processing aid for use in CM (CPE) and CSM (Hypalon) polymers; gives some mold release properties; stabilizes hot air aging properties; wh. flakes; sp.gr. 1.02.

Proaid® FILL. [Akrochem] Mixt. of partially oxidized hydrocarbons; processing aid for inj. molding of syn. and natural rubber; lt. colored pastilles; sp.gr. 0.85; m.p. 131-144 F.

Proaid® FLOW. [Akrochem] Mixt. of fatty acids; processing aid for extrusion of syn. and natural rubber; lt. colored pastilles; sp.gr. 0.90; m.p. 122-131 F.

Proaid® PEP. [Akrochem] Zinc salts of high m.w. fatty acids; peptizer, processing aid for natural and syn. rubber; ivory-colored pastilles; sol. in aromatic solvs.; insol. in acetone, ethanol; sp.gr. 1.05; mp. 205-216 F.

Prodox® 120. [PMC Specialties] Styrenated phenol (46-50% distyrenated phenol, 34-40% tristyrenated phenol, 12.5-16% monostyrenated phenol); CAS 61788-44-1; antioxidant for rubber used in shoe soles; as ethoxylate for use in other prods.; pale yel. liq.; sp.gr. 1.06-1.10 (60/60 F); visc. 200 poise max.

proFLOW 1000. [PolyVisions] Low m.w. PP homopolymer; flow modifier for polymers (polyolefins, polyesters, elastomers); processing lubricant, mold release for engineering thermoplastics (polyester, polyarylate, nylon, PC); nucleating agent for crystallizabale PET; adhesion to glass and metals; oil retention; delayed fusion additive for CPVC, PVDC; pellet or powd.; m.w. 46,900 (peak); melt index ≈ 2600 g/10 min; melt visc. 240 poise (190 C); m.p. 161 C; impact str. (Izod) 0.14 ft-lb/in. notched.

proFLOW 3000. [PolyVisions] Low m.w. ethylene/propylene copolymer; CAS 9010-79-1; flow modifier for polymers (polyolefins, polyesters, elastomers); processing lubricant, mold release for engineering thermoplastics (polyester, polyarylate, nylon, PC); adhesion to glass and metals; oil retention; delayed fusion additive for PVC; carrier resin for cellulosic composites; pellet or powd.; m.w. 42,600 (peak); melt index [a]e 2600 g/10 min; melt visc. 330 poise (190 C); m.p. 1420C; impact str. (Izod) 0.32 ft-lb/in. notched.

Promotor 301. [Akzo Nobel] Metal compd., hydrocarbon solv.; accelerator used in combination with peresters to achieve an optimal cure in hot press molding and a fast cure at R.T. and elevated temps.; sol'n.

Proplast 015. [Aquatec Quimica SA] Fatty ester; lubricant for thermoplastics; org. pigments dispersant; water-insol.; Unverified

Proplast 050, 075. [Aquatec Quimica SA] Glycerol ester; lubricant for thermoplastics; liq.; Unverified

Proplast 058. [Aquatec Quimica SA] Glycerol ester; antistat for thermoplastics; flakes.; Unverified

Proplast 060. [Aquatec Quimica SA] Glycerol ester; lubricant for PVC; flakes.; Unverified

Proplast 290. [Aquatec Quimica SA] Diamide of fatty acid; lubricant for thermoplastics; powd.; Unverified

Propyl Zithate®. [R.T. Vanderbilt] Zinc isopropyl xanthate (CAS 1000-90-4) with mixed petrol. process oil (CAS 64741-96-4 and 64741-97-5); sec. accelerator for natural and syn. rubbers, cements, adhesives; accelerator in latexes to prevent copper discoloration; wh. to lt. yel. powd.; sl. sol. in alcohol, chloroform, petrol. ether, toluene; insol. in water; m.w. 335.83; dens. 1.56 ± 0.03 mg/m^3; m.p. 140 C min. with dec.; 18.4-20.6% Zn; *Toxicology:* may cause irritation to skin or mucous membranes; zinc isopropyl xanthate: LD50 (oral, rat) > 2000 mg/kg; petrol. process oil (as mist): LD50 (oral, rat) > 5000 mg/kg; minute amts. in lungs may cause mild-severe pulmonary injury; OSHA PEL 5 mg/m^3; *Precaution:* unstable thermally; avoid temps. above 140 C; hazardous decomp. prods.: ZnO, CO_2, CS_2, hydrogen sulfide; *Storage:* 1 yr storage life.

Prosil® 196. [PCR] γ-Mercaptopropyl trimethoxy silane; CAS 4420-74-0; EINECS 224-588-5; coupling agent having both org. and inorg. reactivity; for acrylic, epichlorohydrin, nitrile, polysulfone, PS, PVC, urethane thermoplastics; thermoset acrylic, epoxy, nitrile/phenolic, phenolic, polybutadiene; and elastomerics; lt. straw clear liq.; m.w. 196.3; b.p. 220 C (760 mm Hg); sp.gr. 1.05; ref. index 1.440; flash pt. (TCC) 170 F; 97% purity; Unverified

Prosil® 220. [PCR] γ-Aminopropyl triethoxy silane, tech. grade; CAS 919-30-2; EINECS 213-048-4; coupling agent enhancing and promoting chemical bonding between inorg. and org. molecules; for acetal, acrylic, epichlorohydrin, nitrile, NC, polyamide, PC, polyethylene, polyimide, polymethacrylate, PP, polysulfone, PS, PVC, urethane, and vinyl thermoplastics; thermoset acrylic (thermoset and latex), alkyd, epoxy, furan, melamine, nitrile/ phenolic, phenolic, polyester, vinyl butyral/phenolic; and elastomers; lt. straw clear liq.; m.w. 221.3; b.p. 217 C (760 mm Hg); sp.gr. 0.950; flash pt. (COC) 113 F; 85% purity.

Prosil® 221. [PCR] γ-Aminopropyl triethoxy silane, high purity grade; CAS 919-30-2; EINECS 213-048-4; coupling agent enhancing and promoting chemical bonding between inorg. and org. molecules; for acetal, acrylic, epichlorohydrin, nitrile, NC, polyamide, PC, polyethylene, polyimide, polymethacrylate, PP, polysulfone, PS, PVC, urethane, and vinyl thermoplastics; thermoset acrylic (thermoset and latex), alkyd, epoxy, furan, melamine, nitrile/ phenolic, phenolic, polyester, vinyl butyral/ phenolic; and elastomers; Pt-Co APHA 20 liq.; m.w. 221.3; b.p. 217 C (760 mm Hg); sp.gr. 0.946; ref. index 1.420; flash pt. (COC) 182 F; 98% purity.

Prosil® 248. [PCR] γ-Methacryloxypropyl trimethoxy silane; CAS 2530-85-0; EINECS 219-785-8; coupling agent having reactive methacrylate and trimethoxysilyl groups; improves adhesion of org. thermoset resins to inorg. materials such as fiberglass, clay, quartz, and other siliceous surfaces; for ABS, acrylic, polyethylene, polyimide, polymethacrylate, PP, PS, silicone, SAN, and urethane thermoplastics; alkyd, DAP, epoxy, polybutadiene, polyester, and crosslinked polyethylene thermosets; and elastomers; Pt-Co APHA 30 clear liq.; m.w. 248.1; b.p. 255 C (760 mm Hg); sp.gr. 1.04; ref. index 1.43; flash pt. (COC) 190 F; 97.5% purity.

Prosil® 3128. [PCR] n-(2-Aminoethyl)-3-aminopropyl trimethoxysilane; CAS 1760-24-3; EINECS 212-164-2; coupling agent for epoxies, phenolics, melamines, nylons, PVC, urethanes, acrylics.

Prosil® 5136. [PCR] 3-Glycidoxypropyl trimethoxysilane; CAS 2530-83-8; EINECS 219-784-2; coupling agent for epoxies, urethanes, acrylics, polysulfides.

Prosil® HMDS. [PCR] Hexamethyldisilazane; coupling agent; silica treatment for silicone elastomers, novolac photoresist adhesion promoter.

Prote-pon P-2 EHA-02-K30. [Protex] Alkyl ether potassium phosphate; wetting agent, detergent, hydrotrope, emulsifier, rust inhibitor, EP lubricant for alkaline detergents, metal cleaners/lubricants, hard surf. cleaners, textile scours/lubricants, emulsion polymerization, agric., and drycleaning formulations; liq.; pH 7.2 ± 0.3.

Prote-pon P 2 EHA-02-Z. [Protex] Alkyl ether phosphoric acid; wetting agent, detergent, hydrotrope, emulsifier, rust inhibitor, EP lubricant for alkaline detergents, metal cleaners/lubricants, hard surf. cleaners, textile scours/lubricants, emulsion polymerization, agric., and drycleaning formulations; liq.

Prote-pon P 2 EHA-Z. [Protex] Alkyl ether phosphoric acid; wetting agent, detergent, hydrotrope, emulsifier, rust inhibitor, EP lubricant for alkaline detergents, metal cleaners/lubricants, hard surf. cleaners, textile scours/lubricants, emulsion polymerization, agric., and drycleaning formulations; liq.

Prote-pon P-0101-02-Z. [Protex] Alkyl ether phosphoric acid; wetting agent, detergent, hydrotrope, emulsifier, rust inhibitor, EP lubricant for alkaline detergents, metal cleaners/lubricants, hard surf. cleaners, textile scours/lubricants, emulsion polymerization, agric., and drycleaning formulations; liq.

Prote-pon P-L 201-02-K30. [Protex] Alkyl ether potassium phosphate; wetting agent, detergent, hydrotrope, emulsifier, rust inhibitor, EP lubricant for alkaline detergents, metal cleaners/lubricants, hard surf. cleaners, textile scours/lubricants, emulsion polymerization, agric., and drycleaning formulations; emulsion; pH 7.0 ± 0.5.

Prote-pon P-L 201-02-Z. [Protex] Alkyl ether phosphoric acid; wetting agent, detergent, hydrotrope, emulsifier, rust inhibitor, EP lubricant for alkaline detergents, metal cleaners/lubricants, hard surf. cleaners, textile scours/lubricants, emulsion polymerization, agric., and drycleaning formulations; wax.

Prote-pon P-NP-06-K30. [Protex] Alkyl ether potassium phosphate; wetting agent, detergent, hydrotrope, emulsifier, rust inhibitor, EP lubricant for alkaline detergents, metal cleaners/lubricants, hard surf. cleaners, textile scours/lubricants, emulsion polymerization, agric., and drycleaning formulations; liq.; pH 7.2 ± 0.3.

Prote-pon P-NP-06-Z. [Protex] Alkyl ether phosphoric acid; wetting agent, detergent, hydrotrope, emulsifier, rust inhibitor, EP lubricant for alkaline detergents, metal cleaners/lubricants, hard surf. cleaners, textile scours/lubricants, emulsion polymerization, agric., and drycleaning formulations; liq.

Prote-pon P-NP-10-K30. [Protex] Alkyl ether potassium phosphate; wetting agent, detergent, hydrotrope, emulsifier, rust inhibitor, EP lubricant for alkaline detergents, metal cleaners/lubricants, hard surf. cleaners, textile scours/lubricants, emulsion polymerization, agric., and drycleaning formulations; liq.; pH 7.2 ± 0.3.

Prote-pon P-NP-10-MZ. [Protex] Alkyl ether phosphoric acid; wetting agent, detergent, hydrotrope, emulsifier, rust inhibitor, EP lubricant for alkaline detergents, metal cleaners/lubricants, hard surf. cleaners, textile scours/lubricants, emulsion polymerization, agric., and drycleaning formulations; liq.

Prote-pon P-NP-10-Z. [Protex] Alkyl ether phosphoric acid; wetting agent, detergent, hydrotrope, emulsifier, rust inhibitor, EP lubricant for alkaline detergents, metal cleaners/lubricants, hard surf. cleaners, textile scours/lubricants, emulsion polymerization, agric., and drycleaning formulations; liq.

Prote-pon P-OX 101-02-K75. [Protex] Alkyl ether potassium phosphate; wetting agent, detergent, hydrotrope, emulsifier, rust inhibitor, EP lubricant for alkaline detergents, metal cleaners/lubricants, hard surf. cleaners, textile scours/lubricants, emulsion polymerization, agric., and drycleaning formulations; liq.; pH 6.5 ± 0.3.

Prote-pon P-TD-06-K13. [Protex] Alkyl ether potassium phosphate; wetting agent, detergent, hydrotrope, emulsifier, rust inhibitor, EP lubricant for alkaline detergents, metal cleaners/lubricants, hard surf. cleaners, textile scours/lubricants, emulsion polymerization, agric., and drycleaning formulations; liq.; pH 7.5 ± 0.5.

Prote-pon P-TD-06-K30. [Protex] Alkyl ether potassium phosphate; wetting agent, detergent, hydrotrope, emulsifier, rust inhibitor, EP lubricant for alkaline detergents, metal cleaners/lubricants, hard surf. cleaners, textile scours/lubricants, emulsion polymerization, agric., and drycleaning formulations; liq.; pH 7.2 ± 0.3.

Prote-pon P-TD 06-K60. [Protex] Alkyl ether potassium phosphate; wetting agent, detergent, hydrotrope, emulsifier, rust inhibitor, EP lubricant for alkaline detergents, metal cleaners/lubricants, hard surf. cleaners, textile scours/lubricants, emulsion polymerization, agric., and drycleaning formulations; gel; pH 7.5 ± 0.5.

Prote-pon P-TD-06-Z. [Protex] Alkyl ether phosphoric acid; wetting agent, detergent, hydrotrope, emulsifier, rust inhibitor, EP lubricant for alkaline detergents, metal cleaners/lubricants, hard surf. cleaners, textile scours/lubricants, emulsion polymerization, agric., and drycleaning formulations; liq.

Prote-pon P-TD-09-Z. [Protex] Alkyl ether phosphoric acid; wetting agent, detergent, hydrotrope, emulsifier, rust inhibitor, EP lubricant for alkaline detergents, metal cleaners/lubricants, hard surf. cleaners, textile scours/lubricants, emulsion polymerization, agric., and drycleaning formulations; paste.

Prote-pon P-TD-12-Z. [Protex] Alkyl ether phosphoric acid; wetting agent, detergent, hydrotrope, emulsifier, rust inhibitor, EP lubricant for alkaline detergents, metal cleaners/lubricants, hard surf. cleaners, textile scours/lubricants, emulsion polymerization, agric., and drycleaning formulations; paste.

Prote-pon TD-09-K30. [Protex] Alkyl ether potassium phosphate; wetting agent, detergent, hydrotrope, emulsifier, rust inhibitor, EP lubricant for alkaline detergents, metal cleaners/lubricants, hard surf. cleaners, textile scours/lubricants, emulsion polymerization, agric., and drycleaning formulations; liq.; pH 7.0 ± 0.3.

Protint. [PMS Consolidated] Transparent color concs. for clarified PP; offers lightfastness, high chroma and clarity, bleed resist., nonextractability, good heat stability.

Protopet® White 1S. [Witco/Petroleum Spec.] Petrolatum USP; CAS 8027-32-5; EINECS 232-373-2; med. consistency and m.p. petrolatum functioning as carrier, lubricant, emollient, moisture barrier, protective agent, softener for cosmetics, pharmaceutical ointment, industrial applics.; for confectionery lubricants, meat packing, release agents and lubricants for food industry; catalyst carrier, extrusion aid, plasticizer for polymers; FDA 21CFR §172.880; Lovibond 1.5Y color, odorless; visc. 10-16 cSt (100 C); m.p. 54-60 C.

Protopet® Yellow 2A. [Witco/Petroleum Spec.] Petrolatum USP; CAS 8027-32-5; EINECS 232-373-2; med. consistency and m.p. petrolatum functioning as carrier, lubricant, emollient, moisture barrier, protective agent, softener for cosmetics, pharmaceutical ointment, industrial applics.; for confectionery lubricants, meat packing, release agents and lubricants for food industry; catalyst carrier, extrusion aid, plasticizer for polymers; FDA 21CFR §172.880; Lovibond 30Y/2.5R color, odorless; visc. 10-16 cSt (100 C); m.p. 54-60 C.

Prox-onic CSA-1/04. [Protex] Ceteareth-4; CAS 68439-49-6; emulsifier, emulsion stabilizer, detergent, wetting agent, dispersant, solubilizer, defoamer, dye assistant, leveling agent for cosmetics, textiles, metal cleaners, industrial, institutional and household cleaners, emulsion polymerization; solid; HLB 8.0.

Prox-onic CSA-1/06. [Protex] Ceteareth-6; CAS 68439-49-6; emulsifier, emulsion stabilizer, detergent, wetting agent, dispersant, solubilizer, defoamer, dye assistant, leveling agent for cosmetics, textiles, metal cleaners, industrial, institutional and household cleaners, emulsion polymerization; solid; HLB 10.1.

Prox-onic CSA-1/010. [Protex] Ceteareth-10; CAS 68439-49-6; emulsifier, emulsion stabilizer, detergent, wetting agent, dispersant, solubilizer, defoamer, dye assistant, leveling agent for cosmetics, textiles, metal cleaners, industrial, institutional and household cleaners, emulsion polymerization; solid; HLB 12.4; cloud pt. 73 C (1% aq.).

Prox-onic CSA-1/015. [Protex] Ceteareth-15; CAS 68439-49-6; emulsifier, emulsion stabilizer, detergent, wetting agent, dispersant, solubilizer, defoamer, dye assistant, leveling agent for cosmetics, textiles, metal cleaners, industrial, institutional and household cleaners, emulsion polymerization; paste; HLB 14.3; cloud pt. 95-100 C (1% aq.).

Prox-onic CSA-1/020. [Protex] Ceteareth-20; CAS 68439-49-6; emulsifier, emulsion stabilizer, detergent, wetting agent, dispersant, solubilizer, defoamer, dye assistant, leveling agent for cosmetics, textiles, metal cleaners, industrial, institutional and household cleaners, emulsion polymerization; solid; cloud pt. > 100 C (1% aq.).

Prox-onic CSA-1/030. [Protex] Ceteareth-30; CAS 68439-49-6; emulsifier, emulsion stabilizer, detergent, wetting agent, dispersant, solubilizer, defoamer, dye assistant, leveling agent for cosmetics, textiles, metal cleaners, industrial, institutional and household cleaners, emulsion polymerization; solid; cloud pt. > 100 C (1% aq.).

Prox-onic CSA-1/050. [Protex] Ceteareth-50; CAS 68439-49-6; emulsifier, emulsion stabilizer, detergent, wetting agent, dispersant, solubilizer, defoamer, dye assistant, leveling agent for cosmetics, textiles, metal cleaners, industrial, institutional and household cleaners, emulsion polymerization; solid; HLB 16.9; cloud pt. < 100 C (1% aq.).

Prox-onic DT-03. [Protex] PEG-3 tallow diamine; cationic; surfactant, emulsifier, lubricant additive, antistat, detergent for textile, metal, plastics, dyeing assistants, degreasers, corrosion inhibitor, agric.; intermediate for quats.; liq.; m.w. 475; HLB 5.5.

Prox-onic DT-015. [Protex] PEG-15 tallow diamine; cationic; surfactant, emulsifier, lubricant additive, antistat, detergent for textile, metal, plastics, dyeing assistants, degreasers, corrosion inhibitor, agric.; intermediate for quats.; liq.; m.w. 1020; HLB 13.0.

Prox-onic DT-030. [Protex] PEG-30 tallow diamine; cationic; surfactant, emulsifier, lubricant additive, antistat, detergent for textile, metal, plastics, dyeing assistants, degreasers, corrosion inhibitor, agric.; intermediate for quats.; liq.; m.w. 1665; HLB 15.9.

Prox-onic EP 1090-1. [Protex] Difunctional block polymer ending in primary hydroxyl groups; nonionic; defoamer for metalworking, cosmetic, pharmaceuticals, paper, textiles; base for low foaming surfactants, antifoams, dishwash, dispersing and wetting agents for paints, drilling muds, emulsifiers, petrol. demulsifiers, emulsion polymerization; component for sugar beet and yeast defoamers; liq.; m.w. 2000; HLB 3.0; cloud pt. 24 C (1% aq.); 100% act.

Prox-onic EP 1090-2. [Protex] Difunctional block polymer ending in primary hydroxyl groups; nonionic; defoamer for metalworking, cosmetic, paper, textiles; base for low foaming surfactants, antifoams, dishwash, dispersing and wetting agents for paints, drilling muds, emulsifiers, petrol. demulsifiers, emulsion polymerization; component for sugar beet and yeast defoamers; liq.; m.w. 2600; HLB 6.5; cloud pt. 28 C (1% aq.); 100% act.

Prox-onic EP 2080-1. [Protex] Difunctional block polymer ending in primary hydroxyl groups; nonionic; defoamer for metalworking, cosmetic, paper, textiles; base for low foaming surfactants, antifoams, dishwash, dispersing and wetting agents for paints, drilling muds, emulsifiers, petrol. demulsifiers, emulsion polymerization; component for sugar beet and yeast defoamers; liq.; m.w. 2500; HLB 7.0; cloud pt. 30 C (1% aq.); 100% act.

Prox-onic EP 4060-1. [Protex] Difunctional block polymer ending in primary hydroxyl groups; nonionic; defoamer for metalworking, cosmetic, paper, textiles; base for low foaming surfactants, antifoams, dishwash, dispersing and wetting agents for paints, drilling muds, emulsifiers, petrol. demulsifiers, emulsion polymerization; component for sugar beet and yeast defoamers; liq.; m.w. 3000; HLB 1.0; cloud pt. 16 C (1% aq.); 100% act.

Prox-onic HR-05. [Protex] PEG-5 castor oil; CAS 61791-12-6; nonionic; emulsifier, pigment dispersant, leveling agent, softener, rewetting agent, degreaser, antistat, emulsion stabilizer, lubricant for leather, paint, paper, plastics, textile, and cosmetics industries; solubilizer for perfumes; liq.; HLB 3.8; sapon. no. 145.

Prox-onic HR-016. [Protex] PEG-16 castor oil; CAS 61791-12-6; nonionic; emulsifier, pigment dispersant, leveling agent, softener, rewetting agent, degreaser, antistat, emulsion stabilizer, lubricant for leather, paint, paper, plastics, textile, and cosmetics industries; solubilizer for perfumes; liq.; HLB 8.6; sapon. no. 100.

Prox-onic HR-025. [Protex] PEG-25 castor oil; CAS 61791-12-6; nonionic; emulsifier, pigment dispersant, leveling agent, softener, rewetting agent, degreaser, antistat, emulsion stabilizer, lubricant for leather, paint, paper, plastics, textile, and cosmetics industries; solubilizer for perfumes; liq.; HLB 10.8; sapon. no. 80.

Prox-onic HR-030. [Protex] PEG-30 castor oil; CAS 61791-12-6; nonionic; emulsifier, pigment dispersant, leveling agent, softener, rewetting agent, degreaser, antistat, emulsion stabilizer, lubricant for leather, paint, paper, plastics, textile, and cosmetics industries; solubilizer for perfumes; liq.; HLB 11.7; sapon. no. 73.

Prox-onic HR-036. [Protex] PEG-36 castor oil; CAS 61791-12-6; nonionic; emulsifier, pigment dispersant, leveling agent, softener, rewetting agent, degreaser, antistat, emulsion stabilizer, lubricant for leather, paint, paper, plastics, textile, and cosmetics industries; solubilizer for perfumes; liq.; HLB 12.6; sapon. no. 68.

Prox-onic HR-040. [Protex] PEG-40 castor oil; CAS 61791-12-6; nonionic; emulsifier, pigment dispersant, leveling agent, softener, rewetting agent, degreaser, antistat, emulsion stabilizer, lubricant for leather, paint, paper, plastics, textile, and cosmetics industries; solubilizer for perfumes; liq.; HLB 12.9; sapon. no. 61.

Prox-onic HR-080. [Protex] PEG-80 castor oil; CAS 61791-12-6; nonionic; emulsifier, pigment dispersant, leveling agent, softener, rewetting agent, degreaser, antistat, emulsion stabilizer, lubricant for leather, paint, paper, plastics, textile, and cosmetics industries; solubilizer for perfumes; solid; HLB 15.8; sapon. no. 34.

Prox-onic HR-0200. [Protex] PEG-200 castor oil; CAS 61791-12-6; nonionic; emulsifier, pigment dispersant, leveling agent, softener, rewetting agent, degreaser, antistat, emulsion stabilizer, lubricant for leather, paint, paper, plastics, textile, and cosmetics industries; solubilizer for perfumes; liq.; HLB 18.1; sapon. no. 16.

Prox-onic HR-0200/50. [Protex] PEG-200 castor oil; CAS 61791-12-6; nonionic; emulsifier, pigment dispersant, leveling agent, softener, rewetting agent, degreaser, antistat, emulsion stabilizer, lubricant for leather, paint, paper, plastics, textile, and cosmetics industries; solubilizer for perfumes; liq.; HLB 18.1; sapon. no. 16; 50% act.

Prox-onic HRH-05. [Protex] PEG-5 hydrogenated castor oil; CAS 61788-85-0; nonionic; emulsifier, pigment dispersant, leveling agent, softener, rewetting agent, degreaser, antistat, emulsion stabilizer, lubricant for leather, paint, paper, plastics, textile, and cosmetics industries; solubilizer for perfumes; liq.; HLB 3.8; sapon. no. 142.

Prox-onic HRH-016. [Protex] PEG-16 hydrogenated castor oil; CAS 61788-85-0; nonionic; emulsifier, pigment dispersant, leveling agent, softener, rewetting agent, degreaser, antistat, emulsion stabilizer, lubricant for leather, paint, paper, plastics, textile, and cosmetics industries; solubilizer for perfumes; liq.; HLB 8.6; sapon. no. 100.

Prox-onic HRH-025. [Protex] PEG-25 hydrogenated castor oil; CAS 61788-85-0; nonionic; emulsifier, pigment dispersant, leveling agent, softener, rewetting agent, degreaser, antistat, emulsion stabilizer, lubricant for leather, paint, paper, plastics, textile, and cosmetics industries; solubilizer for perfumes; liq.; HLB 10.8; sapon. no. 80.

Prox-onic HRH-0200. [Protex] PEG-200 hydrogenated castor oil; CAS 61788-85-0; nonionic; emulsifier, pigment dispersant, leveling agent, softener, rewetting agent, degreaser, antistat, emulsion stabilizer, lubricant for leather, paint, paper, plastics, textile, and cosmetics industries; solubilizer for perfumes; liq.; HLB 18.1; sapon. no. 17.

Prox-onic HRH-0200/50. [Protex] PEG-200 hydrogenated castor oil; CAS 61788-85-0; nonionic; emulsifier, pigment dispersant, leveling agent, softener, rewetting agent, degreaser, antistat, emulsion stabilizer, lubricant for leather, paint, paper, plastics, textile, and cosmetics industries; solubilizer for perfumes; liq.; HLB 18.1; sapon. no. 17; 50% act.

Prox-onic L 081-05. [Protex] POE (5) linear alcohol ether; biodeg. low foam detergent, wetting agent, emulsifier for household, agric. and industrial cleaners; coupling agent and solubilizer for perfumes and org. additives; cosmetic and pharmaceutical emulsions, shampoos, gels, shaving creams, antiperspirant; emulsion polymerization; modifies plastisol visc.

Prox-onic L 101-05. [Protex] POE (5) linear alcohol ether; biodeg. low foam detergent, wetting agent, emulsifier for household, agric. and industrial cleaners; coupling agent and solubilizer for perfumes and org. additives; cosmetic and pharmaceutical emulsions, shampoos, gels, shaving creams, antiperspirant; emulsion polymerization; modifies plastisol visc.

Prox-onic L 102-02. [Protex] POE (2) linear alcohol ether; biodeg. low foam detergent, wetting agent, emulsifier for household, agric. and industrial cleaners; coupling agent and solubilizer for perfumes and org. additives; cosmetic and pharmaceutical emulsions, shampoos, gels, shaving creams, antiperspirant; emulsion polymerization; modifies plastisol visc.

Prox-onic L 121-09. [Protex] POE (9) linear alcohol ether; biodeg. low foam detergent, wetting agent, emulsifier for household, agric. and industrial cleaners; coupling agent and solubilizer for perfumes and org. additives; cosmetic and pharmaceutical emulsions, shampoos, gels, shaving creams, antiperspirant; emulsion polymerization; modifies plastisol visc.

Prox-onic L 161-05. [Protex] POE (5) linear alcohol ether; biodeg. low foam detergent, wetting agent, emulsifier for household, agric. and industrial cleaners; coupling agent and solubilizer for perfumes and org. additives; cosmetic and pharmaceutical emulsions, shampoos, gels, shaving creams, antiperspirant; emulsion polymerization; modifies plastisol visc.

Prox-onic L 181-05. [Protex] POE (5) linear alcohol ether; biodeg. low foam detergent, wetting agent, emulsifier for household, agric. and industrial cleaners; coupling agent and solubilizer for perfumes and org. additives; cosmetic and pharmaceutical emulsions, shampoos, gels, shaving creams, antiperspirant; emulsion polymerization; modifies plastisol visc.

Prox-onic L 201-02. [Protex] POE (2.5) linear alcohol ether; biodeg. low foam detergent, wetting agent, emulsifier for household, agric. and industrial cleaners; coupling agent and solubilizer for perfumes and org. additives; cosmetic and pharmaceutical emulsions, shampoos, gels, shaving creams, antiperspirant; emulsion polymerization; modifies plastisol visc.

Prox-onic LA-1/02. [Protex] Laureth-2; CAS 3055-93-4; EINECS 221-279-7; coupling agent, solubilizer, emulsion stabilizer for cosmetic and hair care prods.; with anionic surfactants for emulsion polymerization; in coning and textile spin finishes; liq.; HLB 6.4.

Prox-onic LA-1/04. [Protex] Laureth-4; CAS 5274-68-0; EINECS 226-097-1; coupling agent, solubilizer, emulsion stabilizer for cosmetic and hair care prods.; with anionic surfactants for emulsion polymerization; in coning and textile spin finishes; liq.; HLB 9.2; cloud pt. 52 C (1% aq.).

Prox-onic LA-1/09. [Protex] Laureth-9; CAS 3055-99-0; EINECS 221-284-4; coupling agent, solubilizer, emulsion stabilizer for cosmetic and hair care prods.; with anionic surfactants for emulsion polymerization; in coning and textile spin finishes; liq.; HLB 13.3; cloud pt. 73-76 C (1% aq.).

Prox-onic LA-1/012. [Protex] Laureth-12; CAS 3056-00-6; EINECS 221-286-5; coupling agent, solubilizer, emulsion stabilizer for cosmetic and hair care prods.; with anionic surfactants for emulsion polymerization; in coning and textile spin finishes; solid; HLB 14.5; cloud pt. < 100 C (1% aq.).

Prox-onic LA-1/023. [Protex] Laureth-23; CAS 9002-92-0; coupling agent, solubilizer, emulsion stabilizer for cosmetic and hair care prods.; with anionic surfactants for emulsion polymerization; in coning and textile spin finishes; solid; HLB 16.7.

Prox-onic MC-02. [Protex] PEG-2 cocamine; CAS 61791-14-8; cationic; surfactant, emulsifier, lubricant additive, antistat, detergent for textile, metal, plastics, dyeing assistants, degreasers, corrosion inhibitor, agric.; intermediate for quats.; liq.; m.w. 290; HLB 6.1.

Prox-onic MC-05. [Protex] PEG-5 cocamine; CAS 61791-14-8; cationic; surfactant, emulsifier, lubricant additive, antistat, detergent for textile, metal, plastics, dyeing assistants, degreasers, corrosion inhibitor, agric.; intermediate for quats.; liq.; m.w. 425; HLB 10.4.

Prox-onic MC-015. [Protex] PEG-15 cocamine; CAS 61791-14-8; cationic; surfactant, emulsifier, lubricant additive, antistat, detergent for textile, metal, plastics, dyeing assistants, degreasers, corrosion inhibitor, agric.; intermediate for quats.; liq.; m.w. 890; HLB 15.0.

Prox-onic MHT-05. [Protex] PEG-5 hydrog. tallow amine; cationic; surfactant, emulsifier, lubricant additive, antistat, detergent for textile, metal, plastics, dyeing assistants, degreasers, corrosion inhibitor, agric.; intermediate for quats.; paste; m.w. 495; HLB 9.0.

Prox-onic MHT-015. [Protex] PEG-15 hydrog. tallow amine; cationic; surfactant, emulsifier, lubricant additive, antistat, detergent for textile, metal, plastics, dyeing assistants, degreasers, corrosion inhibitor, agric.; intermediate for quats.; liq.; m.w. 925; HLB 14.3.

Prox-onic MO-02. [Protex] PEG-2 oleamine; cationic; surfactant, emulsifier, lubricant additive, antistat, detergent for textile, metal, plastics, dyeing assistants, degreasers, corrosion inhibitor, agric.; intermediate for quats.; liq.; m.w. 358; HLB 4.1.

Prox-onic MO-015. [Protex] PEG-15 oleamine; cationic; surfactant, emulsifier, lubricant additive, antistat, detergent for textile, metal, plastics, dyeing assistants, degreasers, corrosion inhibitor, agric.; intermediate for quats.; liq.; m.w. 930; HLB 14.2.

Prox-onic MO-030. [Protex] PEG-30 oleamine; cationic; surfactant, emulsifier, lubricant additive, antistat, detergent for textile, metal, plastics, dyeing assistants, degreasers, corrosion inhibitor, agric.; intermediate for quats.; liq.; m.w. 1600; HLB 16.5.

Prox-onic MO-030-80. [Protex] PEG-30 oleamine; cationic; surfactant, emulsifier, lubricant additive, antistat, detergent for textile, metal, plastics, dyeing assistants, degreasers, corrosion inhibitor, agric.; intermediate for quats.; 80% act.

Prox-onic MS-02. [Protex] PEG-2 stearamine; CAS 10213-78-2; EINECS 233-520-3; cationic; surfactant, emulsifier, lubricant additive, antistat, detergent for textile, metal, plastics, dyeing assistants, degreasers, corrosion inhibitor, agric.; intermediate for quats.; solid; m.w. 353; HLB 6.9.

Prox-onic MS-05. [Protex] PEG-5 stearamine; CAS 26635-92-7; cationic; surfactant, emulsifier, lubricant additive, antistat, detergent for textile, metal, plastics, dyeing assistants, degreasers, corrosion inhibitor, agric.; intermediate for quats.; solid; m.w. 485; HLB 9.0.

Prox-onic MS-011. [Protex] PEG-11 stearamine; CAS 26635-92-7; cationic; surfactant, emulsifier, lubricant additive, antistat, detergent for textile, metal, plastics, dyeing assistants, degreasers, corrosion inhibitor, agric.; intermediate for quats.; liq.; m.w. 750; HLB 12.9.

Prox-onic MS-050. [Protex] PEG-50 stearamine; CAS 26635-92-7; cationic; surfactant, emulsifier, lubricant additive, antistat, detergent for textile, metal, plastics, dyeing assistants, degreasers, corrosion inhibitor, agric.; intermediate for quats.; solid; m.w. 2465; HLB 17.8.

Prox-onic MT-02. [Protex] PEG-2 tallow amine; CAS 61791-44-4; cationic; surfactant, emulsifier, lubricant additive, antistat, detergent for textile, metal, plastics, dyeing assistants, degreasers, corrosion inhibitor, agric.; intermediate for quats.; paste; m.w. 350; HLB 5.0.

Prox-onic MT-05. [Protex] PEG-5 tallow amine; cationic; surfactant, emulsifier, lubricant additive, antistat, detergent for textile, metal, plastics, dyeing assistants, degreasers, corrosion inhibitor, agric.; intermediate for quats.; liq.; m.w. 490; HLB 9.0.

Prox-onic MT-015. [Protex] PEG-15 tallow amine; cationic; surfactant, emulsifier, lubricant additive, antistat, detergent for textile, metal, plastics, dyeing assistants, degreasers, corrosion inhibitor, agric.; intermediate for quats.; liq.; m.w. 930; HLB 14.3.

Prox-onic MT-020. [Protex] PEG-20 tallow amine; cationic; surfactant, emulsifier, lubricant additive, antistat, detergent for textile, metal, plastics, dyeing assistants, degreasers, corrosion inhibitor, agric.; intermediate for quats.; liq.; m.w. 1150; HLB 15.7.

Prox-onic NP-1.5. [Protex] Nonoxynol-1; EINECS 248-762-5; nonionic; emulsifier, detergent, wetting agent, dispersant, solubilizer, coupling agent, defoamer for emulsion polymerization, latex carpet, textiles, metalworking, household, industrial, agric., paper, paint, and cosmetics industries; liq.; oil-sol.; HLB 4.6.

Prox-onic NP-04. [Protex] Nonoxynol-4; CAS 9016-45-9; nonionic; emulsifier, detergent, wetting agent, dispersant, solubilizer, defoamer, coupling agent for emulsion polymers, latex carpet, textiles, metalworking, household, industrial, agric., paper, paint, and cosmetics industries; liq.; HLB 8.9.

Prox-onic NP-06. [Protex] Nonoxynol-6; CAS 9016-45-9; nonionic; emulsifier, detergent, wetting agent, dispersant, solubilizer, coupling agent, defoamer for emulsion polymerization, latex carpet, textiles, metalworking, household, industrial, agric., paper, paint, and cosmetics industries; liq.; HLB 10.9; cloud pt. < 25 C (1% aq.).

Prox-onic NP-09. [Protex] Nonoxynol-9; CAS 9016-45-9; nonionic; emulsifier, detergent, wetting agent, dispersant, solubilizer, coupling agent, defoamer for emulsion polymerization, latex carpet, textiles, metalworking, household, industrial, agric., paper, paint, and cosmetics industries; liq.; HLB 13.0; cloud pt. 54 C (1% aq.).

Prox-onic NP-010. [Protex] Nonoxynol-10; CAS 9016-45-9; EINECS 248-294-1; nonionic; emulsifier, detergent, wetting agent, dispersant, solubilizer, coupling agent, defoamer for emulsion polymerization, latex carpet, textiles, metalworking, household, industrial, agric., paper, paint, and cosmetics industries; liq.; HLB 13.5; cloud pt. 72 C (1% aq.).

Prox-onic NP-015. [Protex] Nonoxynol-15; CAS 9016-45-9; nonionic; emulsifier, detergent, wetting agent, dispersant, solubilizer, coupling agent, defoamer for emulsion polymerization, latex carpet, textiles, metalworking, household, industrial, agric., paper, paint, and cosmetics industries; paste; HLB 15.0; cloud pt. 96 C (1% aq.).

Prox-onic NP-020. [Protex] Nonoxynol-20; CAS 9016-45-9; nonionic; emulsifier, detergent, wetting agent, dispersant, solubilizer, coupling agent, defoamer for emulsion polymerization, latex carpet, textiles, metalworking, household, industrial, agric., paper, paint, and cosmetics industries; solid; HLB 16.0; cloud pt. > 100 C (1% aq.).

Prox-onic NP-030. [Protex] Nonoxynol-30; CAS 9016-45-9; nonionic; emulsifier for fats, oils, and waxes; detergent, wetting agent, dispersant, solubilizer, coupling agent for emulsion polymers, textiles, metalworking, household, industrial, agric., paper, paint, and cosmetics industries; solid; HLB 17.1; cloud pt. > 100 C (1% aq.).

Prox-onic NP-030/70. [Protex] Nonoxynol-30; CAS 9016-45-9; nonionic; emulsifier for fats, oils, and waxes; detergent, wetting agent, dispersant, solubilizer, coupling agent for emulsion polymers, textiles, metalworking, household, industrial, agric., paper, paint, and cosmetics industries; liq.; HLB 17.1; cloud pt. > 100 C (1% aq.); 70% act.

Prox-onic NP-040. [Protex] Nonoxynol-40; CAS 9016-45-9; nonionic; emulsifier, detergent, wetting agent, dispersant, solubilizer, coupling agent for emulsion polymers, latex carpet, textiles, metalworking, household, industrial, agric., paper, paint, and cosmetics industries; for high temps. and electrolyte use; solid; HLB 17.8; cloud pt. > 100 C (1% aq.).

Prox-onic NP-040/70. [Protex] Nonoxynol-40; CAS 9016-45-9; nonionic; emulsifier, detergent, wetting agent, dispersant, solubilizer, coupling agent for emulsion polymers, latex carpet, textiles, metalworking, household, industrial, agric., paper, paint, and cosmetics industries; for high temps. and electrolyte use; liq.; HLB 17.8; cloud pt. > 100 C (1% aq.); 70% act.

Prox-onic NP-050. [Protex] Nonoxynol-50; CAS 9016-45-9; nonionic; emulsifier, detergent, wetting agent, dispersant, solubilizer, coupling agent for emulsion polymers, latex carpet, textiles, metalworking, household, industrial, agric., paper, paint, and cosmetics industries; for high temps. and electrolyte use; solid; HLB 18.2; cloud pt. > 100 C (1% aq.).

Prox-onic NP-050/70. [Protex] Nonoxynol-50; CAS 9016-45-9; nonionic; emulsifier, detergent, wetting agent, dispersant, solubilizer, coupling agent for emulsion polymers, latex carpet, textiles, metalworking, household, industrial, agric., paper, paint, and cosmetics industries; for high temps. and electrolyte use; liq.; HLB 18.2; cloud pt. > 100 C (1% aq.); 70% act.

Prox-onic NP-0100. [Protex] Nonoxynol-100; CAS 9016-45-9; nonionic; emulsifier, detergent, wetting agent, dispersant, solubilizer, coupling agent for emulsion polymers, latex carpet, textiles, metalworking, household, industrial, agric., paper, paint, and cosmetics industries; for high temps. and electrolyte use; solid; HLB 19.0; cloud pt. > 100 C (1% aq.).

Prox-onic NP-0100/70. [Protex] Nonoxynol-100; CAS 9016-45-9; nonionic; emulsifier, detergent, wetting agent, dispersant, solubilizer, coupling agent for emulsion polymers, latex carpet, textiles, metalworking, household, industrial, agric., paper, paint, and cosmetics industries; for high temps. and electrolyte use; liq.; HLB 19.0; cloud pt. > 100 C (1% aq.); 70% act.

Prox-onic OA-1/04. [Protex] Oleth-4; CAS 9004-98-2; nonionic; coupling agent, solubilizer, emulsion stabilizer for cosmetic and hair care prods.; with anionic surfactants for emulsion polymerization; in coning and textile spin finishes; liq.; HLB 7.9.

Prox-onic OA-1/09. [Protex] Oleth-9; CAS 9004-98-2; nonionic; coupling agent, solubilizer, emulsion stabilizer for cosmetic and hair care prods.; with anionic surfactants for emulsion polymerization; in coning and textile spin finishes; liq.; HLB 11.9; cloud pt. 52 C (1% aq.).

Prox-onic OA-1/020. [Protex] Oleth-20; CAS 9004-98-2; nonionic; emulsifier for min. oils, fatty acids, waxes; for polishes, cosmetics, polyethylene aq. disps.; stabilizer for rubber latex; solubilizer/emulsifier for essential oils, pharmaceuticals; wetting agent in metal cleaners; dyeing assistant; dyestuff dispersant for leather; solid; HLB 15.3; cloud pt. > 100 C (1% aq.).

Prox-onic OA-2/020. [Protex] Oleth-20; CAS 9004-98-2; nonionic; emulsifier for min. oils, fatty acids, waxes; for polishes, cosmetics, polyethylene aq. disps.; stabilizer for rubber latex; solubilizer/emulsifier for essential oils, pharmaceuticals; wetting agent in metal cleaners; dyeing assistant; dyestuff dispersant for leather; liq.; HLB 15.3; cloud pt. > 100 C (1% aq.).

Prox-onic OP-09. [Protex] Octoxynol-9; CAS 9002-93-1; nonionic; emulsifier for metal, textile processing, household and industrial cleaners, vinyl and acrylic polymerization; liq.; HLB 13.5; cloud pt. 65 C (1% aq.).

Prox-onic OP-030/70. [Protex] Octoxynol-30; CAS 9002-93-1; nonionic; coemulsifier for vinyl and acrylic polymerization; dye assistant; liq.; HLB 17.3; cloud pt. > 100 C (1% aq.); 70% act.

Prox-onic OP-040/70. [Protex] Octoxynol-40; CAS

9002-93-1; nonionic; coemulsifier for vinyl and acrylic polymerization; dye assistant; liq.; HLB 17.9; cloud pt. > 100 C (1% aq.); 70% act.

Prox-onic PEG-2000. [Protex] PEG-2M; CAS 25322-68-3; low foam wetting in paper pulping; emulsifier for metal degreasing; bottle cleaner defoamer; binder for tobacco; improves hydrophility and elasticity of PU; mold release agent; agric.; cosmetic/pharmaceutical emulsions; m.w. 2000; solid. pt. 48-52 C; hyd. no. 51-62.

Prox-onic PEG-20,000. [Protex] PEG-20M; CAS 25322-68-3; low foam wetting in paper pulping; emulsifier for metal degreasing; bottle cleaner defoamer; plasticizer for cement; ceramics binder; mold release for rubber; agric.; cosmetic/pharmaceutical emulsions; m.w. 20,000; solid. pt. 60 C; hyd. no. 7-11.

PS040. [Hüls Am.] Dimethicone; used in brake fluids, polishes, coatings, greases and oils, antifoams, hand creams; as heat transfer media; mold release for rubber, plastic and glass parts; lubricant; toner for photocopiers; in hydraulic systems; as dielec. fluid in electronics; sol. in methylene chloride, chlorofluorocarbons, ethyl ether, xylene, MEK; m.w. 3780; sp.gr. 0.960; visc. 50 cSt; pour pt. -65 C; flash pt. 285 C; ref. index 1.4015; surf. tens. 20.8 dynes/cm; dielec. str. 400 V/mil; dielec. const. 2.75.

PS041. [Hüls Am.] Dimethicone; used in brake fluids, polishes, coatings, greases and oils, antifoams, hand creams; as heat transfer media; mold release for rubber, plastic and glass parts; lubricant; toner for photocopiers; in hydraulic systems; as dielec. fluid in electronics; sol. in methylene chloride, chlorofluorocarbons, ethyl ether, xylene, MEK; m.w. 5970; sp.gr. 0.966; visc. 100 cSt; pour pt. -65 C; flash pt. 315 C; ref. index 1.4025; surf. tens. 20.9 dynes/cm; dielec. str. 400 V/mil; dielec. const. 2.75.

PS041.2. [Hüls Am.] Dimethicone; used in brake fluids, polishes, coatings, greases and oils, antifoams, hand creams; as heat transfer media; mold release for rubber, plastic and glass parts; lubricant; toner for photocopiers; in hydraulic systems; as dielec. fluid in electronics; sol. in methylene chloride, chlorofluorocarbons, ethyl ether, xylene, MEK; m.w. 9430; sp.gr. 0.968; visc. 200 cSt; pour pt. -60 C; flash pt. 315 C; ref. index 1.4030; surf. tens. 21 dynes/cm; dielec. str. 400 V/mil; dielec. const. 2.75.

PS041.5. [Hüls Am.] Dimethicone; used in brake fluids, polishes, coatings, greases and oils, antifoams, hand creams; as heat transfer media; mold release for rubber, plastic and glass parts; lubricant; toner for photocopiers; in hydraulic systems; as dielec. fluid in electronics; sol. in methylene chloride, chlorofluorocarbons, ethyl ether, xylene, MEK; m.w. 13,650; sp.gr. 0.970; visc. 350 cSt; pour pt. -60 C; flash pt. 315 C; ref. index 1.4031; surf. tens. 21.1 dynes/cm; dielec. str. 400 V/mil; dielec. const. 2.75.

PS042. [Hüls Am.] Dimethicone; used in brake fluids, polishes, coatings, greases and oils, antifoams, hand creams; as heat transfer media; mold release for rubber, plastic and glass parts; lubricant; toner for photocopiers; in hydraulic systems; as dielec. fluid in electronics; sol. in methylene chloride, chlorofluorocarbons, ethyl ether, xylene, MEK; m.w. 17,250; sp.gr. 0.971; visc. 500 cSt; pour pt. -55 C; flash pt. 315 C; ref. index 1.4033; surf. tens. 21.1 dynes/cm; dielec. str. 400 V/mil; dielec. const. 2.75.

PS043. [Hüls Am.] Dimethicone; used in brake fluids, polishes, coatings, greases and oils, antifoams, hand creams; as heat transfer media; mold release for rubber, plastic and glass parts; lubricant; toner for photocopiers; in hydraulic systems; as dielec. fluid in electronics; sol. in methylene chloride, chlorofluorocarbons, ethyl ether, xylene, MEK; m.w. 28,000; sp.gr. 0.971; visc. 1000 cSt; pour pt. -50 C; flash pt. 315 C; ref. index 1.4034; surf. tens. 21.2 dynes/cm; dielec. str. 400 V/mil; dielec. const. 2.75.

PS044. [Hüls Am.] Dimethicone; internal lubricant and process aid for thermoplastics and die casting, in rubber industry; as damping fluid; paint additive; sol. in methylene chloride, chlorofluorocarbons, ethyl ether, xylene, MEK; m.w. 49,350; sp.gr. 0.973; visc. 5000 cSt; pour pt. -48 C; flash pt. 315 C; ref. index 1.4035; surf. tens. 21.3 dynes/cm; dielec. str. 400 V/mil; dielec. const. 2.75.

PS045. [Hüls Am.] Dimethicone; internal lubricant and process aid for thermoplastics and die casting, in rubber industry; as damping fluid; paint additive; sol. in methylene chloride, chlorofluorocarbons, ethyl ether, xylene, MEK; m.w. 62,700; sp.gr. 0.974; visc. 10,000 cSt; pour pt. -48 C; flash pt. 315 C; ref. index 1.4035; surf. tens. 21.5 dynes/cm; dielec. str. 400 V/mil; dielec. const. 2.75.

PS046. [Hüls Am.] Dimethicone; internal lubricant and process aid for thermoplastics and die casting, in rubber industry; as damping fluid; paint additive; sol. in methylene chloride, chlorofluorocarbons, ethyl ether, xylene, MEK; m.w. 67,700; sp.gr. 0.974; visc. 12,500 cSt; pour pt. -46 C; flash pt. 315 C; ref. index 1.4035; surf. tens. 21.5 dynes/cm; dielec. str. 400 V/mil; dielec. const. 2.75.

PS047. [Hüls Am.] Dimethicone; internal lubricant and process aid for thermoplastics and die casting, in rubber industry; as damping fluid; paint additive; sol. in methylene chloride, chlorofluorocarbons, ethyl ether, xylene, MEK; m.w. 91,700; sp.gr. 0.976; visc. 30,000 cSt; pour pt. -43 C; flash pt. 315 C; ref. index 1.4035; surf. tens. 21.5 dynes/cm; dielec. str. 400 V/mil; dielec. const. 2.75.

PS047.5. [Hüls Am.] Dimethicone; internal lubricant and process aid for thermoplastics and die casting, in rubber industry; as damping fluid; paint additive; sol. in methylene chloride, chlorofluorocarbons, ethyl ether, xylene, MEK; m.w. 116,500; sp.gr. 0.976; visc. 60,000 cSt; pour pt. -42 C; flash pt. 315 C; ref. index 1.4035; surf. tens. 21.5 dynes/cm; dielec. str. 400 V/mil; dielec. const. 2.75.

PS048. [Hüls Am.] Dimethicone; internal lubricant and process aid for thermoplastics and die casting, in rubber industry; as damping fluid; paint additive; sol. in methylene chloride, chlorofluorocarbons, ethyl ether, xylene, MEK; m.w. 139,000; sp.gr.

0.977; visc. 100,000 cSt; pour pt. -41 C; flash pt. 321 C; ref. index 1.4035; surf. tens. 21.5 dynes/cm; dielec. str. 400 V/mil; dielec. const. 2.75.

PS048.5. [Hüls Am.] Dimethicone; internal lubricant and process aid for thermoplastics and die casting, in rubber industry; as damping fluid; paint additive; sol. in methylene chloride, chlorofluorocarbons, ethyl ether, xylene, MEK; m.w. 204,000; sp.gr. 0.977; visc. 300,000 cSt; pour pt. -41 C; flash pt. 321 C; ref. index 1.4035; surf. tens. 21.5 dynes/cm; dielec. str. 400 V/mil; dielec. const. 2.75.

PS049. [Hüls Am.] Dimethicone; internal lubricant and process aid for thermoplastics and die casting, in rubber industry; as damping fluid; paint additive; sol. in methylene chloride, chlorofluorocarbons, ethyl ether, xylene, MEK; m.w. 260,000; sp.gr. 0.978; visc. 600,000 cSt; pour pt. -41 C; flash pt. 321 C; ref. index 1.4035; surf. tens. 21.6 dynes/cm; dielec. str. 400 V/mil; dielec. const. 2.75.

PS049.5. [Hüls Am.] Dimethicone; internal lubricant and process aid for thermoplastics and die casting, in rubber industry; as damping fluid; paint additive; impact modifier for thermoplastics; mold release; sol. in methylene chloride, chlorofluorocarbons, ethyl ether, xylene, MEK; m.w. 308,000; sp.gr. 0.978; visc. 1,000,000 cSt; pour pt. -39 C; flash pt. 321 C; ref. index 1.4035; surf. tens. 21.6 dynes/cm; dielec. str. 400 V/mil; dielec. const. 2.75.

PS050. [Hüls Am.] Dimethicone; internal lubricant and process aid for thermoplastics and die casting, in rubber industry; as damping fluid; paint additive; impact modifier for thermoplastics; mold release; sol. in methylene chloride, chlorofluorocarbons, ethyl ether, xylene, MEK; m.w. 423,000; sp.gr. 0.978; visc. 2,500,000 cSt; pour pt. -38 C; flash pt. 321 C; ref. index 1.4035; surf. tens. 21.6 dynes/cm; dielec. str. 400 V/mil; dielec. const. 2.75.

PS071. [Hüls Am.] Ethylene oxide-modified polydimethylsiloxane; surfactant, wetting agent for photographic plates; antifog for glass and plastic optics; slip agent for flexographic and gravure inks; sol. in water; sp.gr. 1.00; visc. 20 cSt; m.p. 0 C; flash pt. 115 C; ref. index 1.4416; surf. tens. 21 dynes/cm (1% aq.); 75% nonsiloxane.

PS072. [Hüls Am.] Ethylene oxide/propylene oxide-modified polydimethylsiloxane; surfactant, lubricant for fibers and plastics, metal-to-plastic wear interfaces; antitack and mar resist. aid for urethane coatings; slip agent in flexographic and gravure inks; sp.gr. 1.02; visc. 1800 cSt; m.p. -50 C; flash pt. 101 C; ref. index 1.4456; surf. tens. 33 dynes/cm (1% aq.).

PS136.8. [Hüls Am.] (30-35%) Phenethylmethylsiloxane/(65-70%) dimethylsiloxane copolymer; mold release agent for rubber, plastics, die casting; sp.gr. 1.03; visc. 200-600 cSt; pour pt. -64 C; ref. index 1.469; surf. tens. 22 dynes/cm.

PS137. [Hüls Am.] (48-52%) Phenethylmethylsiloxane/(48-52%) hexylmethylsiloxane copolymer; mold release agent for rubber, plastics, die casting; sp.gr. 1.04; visc. 1500 cSt; pour pt. -46 C; ref. index 1.493.

PS138. [Hüls Am.] (α-Methylphenethyl) methyl (48-52%) dimethylsiloxane copolymer; mold release agent for rubber, plastics, die casting; sp.gr. 1.02; visc. 1000 cSt; ref. index 1.480.

PS140. [Hüls Am.] Polyoctylmethylsiloxane; CAS 68440-90-4; lubricant for soft metals; rubber and plastic lubricant esp. when mated against steel or aluminum; for aluminum machining operations; process aid and plasticizer in polyolefin rubbers; sp.gr. 0.91; visc. 600-1000 cSt; pour pt. -50 C; ref. index 1.445; surf. tens. 30.4 dynes/cm.

PSA. [Huntsman] Phenol sulfonic acid; anionic; electrolyte for tin plating; catalyst for resins; liq.; 65-68% act.

PTAL. [Mitsubishi Gas] p-Tolualdehyde; CAS 104-87-0; EINECS 203-246-9; additive for resins; intermediate for pharmaceuticals; fragrance; 86% biodeg.; colorless liq., aromatic odor; sol. 0.27% in water @ 40 C; m.w. 120.2; b.p. 205.9 C; m.p. -5.6 C; acid no. 0.2; flash pt. (COC) 88 C; 96.4% purity; *Toxicology:* LD50 (oral, rat) 1000 mg/kg; eye and skin irritant.

PTSA 70. [Huntsman] p-Toluene sulfonic acid; anionic; catalyst for resins; liq.; 68% act.

PTZ® Chemical Grade. [ICI Polymer Addit.] Phenothiazine; CAS 92-84-2; antioxidant, monomer stabilizer, polymerization inhihbitor, anti-aging heat, heat stabilizer, catalyst inhibitor for plastics, petrol., rubber, and asphalt industries; stabilizes thermoset polyester and vinyl esters; FDA 21CFR §175.105, 176.170, 176.180, 177.2600; pale yel. coarse powd. (80-100 mesh), gran. (8-35 mesh), micropulverized (10-12 μ); sol. 21% in acetone, 15% in ethyl amyl ketone, 11% in ethyl acetate; negligible sol. in water; m.w. 199.26; m.p. 184 C min.; b.p. 371 C; 99.6% min. purity; *Toxicology:* dust may produce irritation to eyes, nose, throat, or skin; *Storage:* nonhygroscopic, but store in closed containers; minimize accumulation of fine dust.

PTZ® Industrial Grade. [ICI Polymer Addit.] Phenothiazine; CAS 92-84-2; antioxidant, monomer stabilizer, polymerization inhihbitor, anti-aging heat, heat stabilizer, catalyst inhibitor for plastics, petrol., rubber, and asphalt industries; stabilizes thermoset polyester and vinyl esters; FDA 21CFR §175.105, 176.170, 176.180, 177.2600; lt. green coarse powd. (80-100 mesh), gran. (8-35 mesh), micropulverized (6-10 μ); sol. 21% in acetone, 15% in ethyl amyl ketone, 11% in ethyl acetate; negligible sol. in water; m.w. 199.26; m.p. 179 C min.; b.p. 371 C; 94% min. purity; *Toxicology:* dust may produce irritation to eyes, nose, throat, or skin; *Storage:* nonhygroscopic, but store in closed containers; minimize accumulation of fine dust.

PTZ® Purified Flake. [Zeneca Spec.] Phenothiazine; CAS 92-84-2; stabilizer for vinyl ester resins, ethylene acrylic acid polymers; yel. flakes; > 500 μ particle size; sol. (g/100 ml): 19 g acetone, 15 g ethyl amyl ketone, 10 g ethyl acetate, 6.5 g methyl methacrylate; negligible sol. in water; bulk dens. 0.87 g/ml; m.p. 184 C min.; b.p. 371 C; flash pt. (SCC) > 230 F; 99.6% min. purity; *Toxicology:* LD50 (oral, rat) > 5 g/kg; sl. toxic by ingestion; photosensitivity reactions possible; dust may produce irritation to eyes, nose, throat, or skin; ingestion can cause toxic hepatitis, GI irritation, kidney injury, etc.; OSHA PEL 5 mg/m^3 (skin); *Precaution:* may form explosive dust clouds in air; incompat. with strong acids; hazardous combustion prods.: CO_2, CO, NO_x, ammonia, sulfur oxides; *Storage:* nonhygroscopic, but store in closed containers; minimize accumulation of fine dust; do not reuse containers.

PTZ® Purified Powd. [Zeneca Spec.] Phenothiazine; CAS 92-84-2; stabilizer for vinyl ester resins, ethylene acrylic acid polymers; yel. powd.; < 700 μ particle size; sol. (g/100 ml): 19 g acetone, 15 g ethyl amyl ketone, 10 g ethyl acetate, 6.5 g methyl methacrylate; negligible sol. in water; bulk dens. 0.75 g/ml; m.p. 184 C min.; b.p. 371 C; flash pt. (SCC) > 230 F; 99.6% min. purity; *Toxicology:* LD50 (oral, rat) > 5 g/kg; sl. toxic by ingestion; photosensitivity reactions possible; dust may produce irritation to eyes, nose, throat, or skin; ingestion can cause toxic hepatitis, GI irritation, kidney injury, etc.; OSHA PEL 5 mg/m^3 (skin); *Precaution:* may form explosive dust clouds in air; incompat. with strong acids; hazardous combustion prods.: CO_2, CO, NO_x, ammonia, sulfur oxides; *Storage:* nonhygroscopic, but store in closed containers; minimize accumulation of fine dust; do not reuse containers.

Pura WBC608. [Air Prods./Perf. Chems.] Water-based release agent contg. silicone; release agent for PU cast elatomers providing improved surf. quality, reduced VOC emissions; effective at elevated temps.; wh. milky liq., mild odor; misc. with water in all proportions; sp.gr. 0.9886; vapor pressure 19 mm Hg (21 C); b.p. 100 C; flash pt. (Seta) 98 C; 24% solids; *Toxicology:* mild skin and eye irritant; *Storage:* avoid freezing temps. during storage.

Puxol CB-22. [Pulcra SA] TEA dodecylbenzene sulfonate; CAS 27323-41-7; EINECS 248-406-9; anionic; used for formulation of liq. detergents, personal care prods., car washing shampoos; emulsifier for emulsion polymerization reactions; pigment dispersant; liq.; 50 ± 1% conc.; Unverified

PVC Deodorant #5417, OS. [Andrea Aromatics] Mixture of fragrance materials; masks odors during processing and in finished prod.; Unverified

PVP K-60. [ISP] PVP; CAS 9003-39-8; EINECS 201-800-4; used in adhesives to bond glass, metal, and plastics; as film-former in pressure-sensitive, water-rewettable types; as visc. modifier in polymer based adhesives; m.w. 160,000; dens. 9.3 lb/gal; 45% min. active.

PX-104. [Aristech] Dibutyl phthalate; CAS 84-74-2; EINECS 201-557-4; reagent grade plasticizer; APHA 25 color; sp.gr. 1.045; visc. 16 cps; f.p. -40 C.

PX-111. [Aristech] Diundecyl phthalate; CAS 3648-20-2; EINECS 222-884-9; reagent grade plasticizer; APHA 50 color; sp.gr. 0.952; visc. 52 cps; pour pt. 9 C.

PX-120. [Aristech] Diisodecyl phthalate; CAS 68515-49-1; EINECS 247-977-1; reagent grade plasticizer; APHA 25 color; sp.gr. 0.965; visc. 86 cps; pour pt. -48 C.

PX-126. [Aristech] Ditridecyl phthalate; EINECS 204-294-3; reagent grade plasticizer; APHA 75 color; sp.gr. 0.953; visc. 215 cps; pour pt. -26 C.

PX-138. [Aristech] Dioctyl phthalate; CAS 117-81-7; EINECS 204-211-0; reagent grade plasticizer; APHA 25 color; sp.gr. 0.983; visc. 57 cps; pour pt. -50 C.

PX-139. [Aristech] Diisononyl phthalate; CAS 68515-48-0; EINECS 249-079-5; reagent grade plasticizer; APHA 25 color; sp.gr. 0.974; visc. 70 cps; pour pt. -45 C.

PX-212. [Aristech] Mixed normal alkyl adipate; reagent grade plasticizer; APHA 25 color; sp.gr. 0.917; visc. 13 cps; f.p. -4 C.

PX-238. [Aristech] Dioctyl adipate; CAS 103-23-1; EINECS 203-090-1; reagent grade plasticizer; APHA 25 color; sp.gr. 0.925; visc. 11 cps; f.p. -42 C.

PX-239. [Aristech] Diisononyl adipate; CAS 33703-08-1; EINECS 251-646-7; reagent grade plasticizer; APHA 25 color; sp.gr. 0.921; visc. 18 cps; f.p. -65 C.

PX-316. [Aristech] Mixed normal alkyl phthalate; reagent grade plasticizer; APHA 25 color; sp.gr. 0.970; visc. 34 cps; pour pt. -20 C.

PX-318. [Aristech] Mixed n-octyl n-decyl phthalate; CAS 119-07-3; EINECS 204-295-9; primary plasticizer for PVC and other highly polar polymers; for improved volatility while maintaining superior low temp. performance and easy processability; for mfg. of sheet, film, extrusion, and molding compds.; exhibits low initial visc. and exc. storage stabiltiy in plastisols; FDA 21CFR §175.10S, 177.2600, indirect food contact; char. odor; m.w. 421; sp.gr. 0.967; visc. 36 cps; f.p. -19 C; b.p. 255 C (5 mm); flash pt. (COC) 460 F; ref. index 1.482.

PX-336. [Aristech] Mixed normal alkyl trimellitate; reagent grade plasticizer; APHA 150 color; sp.gr. 0.975; visc. 103 cps; f.p. -17 C.

PX-338. [Aristech] Trioctyl trimellitate; CAS 3319-31-1; reagent grade plasticizer; APHA 150 color; sp.gr. 0.988; visc. 216 cps; pour pt. -46 C.

PX-339. [Aristech] Triisononyl trimellitate; CAS 53894-23-8; reagent grade plasticizer; Unverified

PX-504. [Aristech] Dibutyl maleate; CAS 105-76-0; EINECS 203-328-4; reagent grade plasticizer; APHA 50 color; sp.gr. 0.993; visc. 5 cps; pour pt. -65 C.

PX-538. [Aristech] Dioctyl maleate; CAS 142-16-5; EINECS 205-524-5; reagent grade plasticizer; APHA 50 color; sp.gr. 0.942; visc. 14 cps; pour pt. -85 C.

PX-800. [Aristech] Epoxidized soya oil; CAS 8013-07-8; EINECS 232-391-0; reagent grade plasticizer; Unverified

PX-911. [Aristech] Mixed alkyl phthalate; reagent grade plasticizer; APHA 25 color; sp.gr. 0.959; visc. 44 cps; pour pt. -20 C.

PX-914. [Aristech] Butyl octyl phthalate; EINECS 201-562-1; reagent grade plasticizer; APHA 25 color; sp.gr. 0.993; visc. 46 cps; pour pt. -35 C.

Pycal 94. [ICI Am.] POE aryl ether; emulsifier, wetting agent, plasticizer for industrial applics.; pale yel.

liq.; sol. in water, ether, alcohol, ketone, lower aliphatic ester, and aromatic hydrocarbon solv.; sp.gr. 1.1; visc. approx 50 cps; acid no. 2 max.; flash pt. > 300 F; ref. index 1.502; ; pH 4-6 (50% aq.).

Pyrax® A. [R.T. Vanderbilt] Pyrophyllite (hydrated aluminum silicate); CAS 12269-78-2; inert filler/extender for NR and syn. rubbers and latexes; off-wh. to tan powd.; 10 μm median particle size; 99.5% min. thru 100 mesh, 97% min. thru 200 mesh; dens. 2.80 ± 0.05 mg/m^3; bulk dens. 68 lb/ft^3 (compacted); oil absorp. 24; brightness 65; pH 6.5 (10% aq. slurry); 77.8% SiO_2, 16% Al_2O_3.

Pyrax® ABB. [R.T. Vanderbilt] Pyrophyllite (hydrated aluminum silicate); CAS 12269-78-2; diluent or carrier for agric. toxicants; used in cosmetics, pharmaceuticals, water paints, and dry wall plasters; filler for plastics, rubber; color additive for drugs and cosmetics; anticaking/blending/pelleting agent or carrier in animal feeds; variable powd., 11 μ diam.; pH 6.5 (10% solids).

Pyrax® B. [R.T. Vanderbilt] Pyrophyllite (hydrated aluminum silicate); CAS 12269-78-2; diluent or carrier for agric. toxicants; used in cosmetics, pharmaceuticals, water paints, and dry wall plasters; filler for plastics, rubber; color additive for drugs and cosmetics; anticaking/blending/pelleting agent or carrier in animal feeds; wh. powd.; 10 μ median particle size; 95% thru 325 mesh; dens. 2.80 mg/m^3; bulk dens. 52 lb/ft^3 (compacted); oil absorp. 26; brightness 80; ; pH 6.9 (10% aq. slurry).

Pyrax® WA. [R.T. Vanderbilt] Pyrophyllite (hydrated aluminum silicate); CAS 12269-78-2; diluent or carrier for agric. toxicants; used in cosmetics, pharmaceuticals, water paints, and dry wall plasters; filler for plastics, rubber; color additive for drugs and cosmetics; anticaking/blending/pelleting agent or carrier in animal feeds; wh. powd., 13 μ median particle size; 87% thru 325 mesh; dens. 2.80 mg/m^3; bulk dens. 61 lb/ft^3 (compacted); oil absorp. 24; brightness 78; pH 6.6 (10% aq. slurry).

Pyridine 1°. [Nepera] Pyridine; CAS 110-86-1; EINECS 203-809-9; chem. intermediate; can undergo electrophilic and nucleophilic substitution, act as an acid scavenger, catalyze reactions; used in agric., pharmaceuticals, photography, coatings, curing agents, rubber, plastics, antidandruff shampoos, textiles; water-wh. liq., char. odor; >99.5% assay; Unverified

Pyrobloc SAP. [Anzon] Sodium antimonate; flame retardant for ABS, acrylics, epoxy, thermoset polyester, PS, PE, PP, PVC, PU.

Pyro-Chek® 60PB. [Ferro/Keil] Brominated polystyrene; melt-blendable flame retardant for higher temp. engineering resins; melts at lower themp. than Pyro-Chek 68PB; coarse powd.; 20 μ median particle size; sp.gr. 2.0; 60% Br, 1% Cl.

Pyro-Chek® 60PBC. [Ferro/Keil] Brominated polystyrene, compacted, low dusting grade; flame retardant for engineering thermoplastics, polyamides, thermoplastic polyesters; high thermal stability; irreg. gran.; sp.gr. 1.9; bulk dens. 980 kg/m^3; 58% min. Br; *Toxicology:* nontoxic.

Pyro-Chek® 68PB. [Ferro/Keil] Brominated polystyrene; melt-blendable flame retardant for engineering thermoplastics, nylon, thermoplastic polyesters; high thermal stability; off-wh. coarse powd.; 20 μ median particle size; sp.gr. 2.1; bulk dens. 22 lb/ft^3; soften. pt. 215-225 C; 66% min. Br, 1% Cl; *Toxicology:* nontoxic.

Pyro-Chek® 68PBC. [Ferro/Keil] Brominated polystyrene, compacted, low dusting grade; flame retardant for engineering thermoplastics, nylon, thermoplastic polyesters; high thermal stability; irregular gran.; sp.gr. 2.1; bulk dens. 1050 kg/m^3; 66% min. Br, 1.5% max. Cl; *Toxicology:* nontoxic.

Pyro-Chek® 68PBG. [Ferro/Keil] Brominated polystyrene; flame retardant for polyolefins, thermoplastic elastomers, other low-temp. polymers; thermally stable; stable to uv light; wh. powd.; 7.5 μ avg. particle size; sp.gr. 2.1; 66% min. Br; *Toxicology:* nontoxic.

Pyro-Chek® 77B. [Ferro/Keil] Bis (pentabromophenoxy) ethane; flame retardant exhibiting resistance to discoloration on exposure to lt.; used in ABS, HIPS, styrene-maleic anhydride, and polyolefins; synergist with antimony oxide; sp.gr. 2.8; dens. 1.5 cc/g; m.p. 322 C; 77% bromine.; Unverified

Pyro-Chek® C60PB. [Ferro/Keil] Brominated polystyrene, compacted, low dusting grade; melt-blendable flame retardant for engineering thermoplastics, polyamides, and thermoplastic polyesters; high thermal stability; off-wh. irreg. gran.; sp.gr. 1.85; 58% min. Br; *Toxicology:* nontoxic.

Pyro-Chek® C68PB. [Ferro/Keil] Brominated polystyrene, compacted, low dusting grade; melt-blendable flame retardant for engineering thermoplastics, nylon, thermoplastic polyesters; high thermal stability; irreg. gran.; sp.gr. 2.1; bulk dens. 66 lb/ft^3; 66% min. Br; *Toxicology:* nontoxic.

Pyro-Chek® LM. [Ferro/Keil] Brominated polystyrene; melt-blendable flame retardant for styrenics, esp. HIPS, ABS; also for HDPE, nylon, PBT, etc.; high thermal stability; beige/tan powd.; sp.gr. 2.1; 66% min. Br; *Toxicology:* not considered hazardous.

2-Pyrol®. [ISP] 2-Pyrrolidone; CAS 616-45-5; EINECS 204-648-7; monomer for polypyrrolidone; solv. for polymers, insecticides, polyhydroxylic alcohols, sugars, iodine, specialty inks; used in petrol. processing and acrylonitrile mfg.; plasticizer and coalescing agent for acrylic latices; and acrylic-styrene copolymers in emulsion type floor coatings; intermediate for N-methylol derivs., alkaloids related to peganine, amino-butyric acid derivs.; liq.; f.p. 25 C; b.p. 245 C; flash pt. (OC) 129 C.

Q

Q-1300. [Wako Pure Chem. Ind.] N-Nitrosophenylhydroxylamine ammonium salt; CAS 135-20-6; EINECS 205-183-2; polymerization inhibitor for vinyl, butadiene, isoprene, chloroprene, divinyl benzene, acrylic monomers; storage stabilizer for precured resins (unsat. polyester); chelating agent; antioxidant for acrolein, furfural; germicides, fungicides, agric. chems.; corrosion inhibitor for metals; heat stabilizer for chlorosulfonated polyethylene; raw material for dye synthesis; analytical reagent; wh. or sl. yel. cryst. powd.; sol. in water, alcohol, and org. solvs.; m.w. 155.2; dec. 141-144 C.

Q-1301. [Wako Pure Chem. Ind.; Wako Chem. USA] N-Nitrosophenylhydroxylamine aluminum salt; CAS 15305-07-4; EINECS 239-341-7; polymerization inhibitor; ideal for uv ink stabilizer; pale brnsh. yel. powd.; sol. in THF; m.w. 438.34; m.p. 160 C min.; 98% min. assay.

Q-Cel® 300. [PQ Corp.] Inorg. silicate microspheres; extender/filler for plastics, fiberglass-reinforced plastics, cultured marble, cast polyester furniture and decorative parts, bowling ball cores, cast urethane and epoxy systems, autobody repair fillers, marine putties, PVC plastisol compds., slurry explosives; wh. particulate; particle size 75 μ; dens. 7.0 lb/ft^3; 70% solids.

Q-Cel® 636. [PQ Corp.] Hollow borosilicate glass microspheres; high-str. extender/filler for applics. with moderate pressure and shear; for thermoset and PVC plastisol formulations, low-dens. SMC formulations; provides max. filler loadings and low compd. densities; wh. spheres; dens. 0.36 g/cc; *Toxicology:* moderately alkaline; prolonged exposure may cause respiratory irritation.

Q-Cel® 640. [PQ Corp.] Hollow borosilicate glass microspheres; high-str. extender/filler for applics. with moderate pressure and shear; for thermoset and PVC plastisol formulations, low-dens. SMC formulations; provides max. filler loadings and low compd. densities; wh. spheres; dens. 0.42 g/cc; *Toxicology:* moderately alkaline; prolonged exposure may cause respiratory irritation.

Q-Cel® 650. [PQ Corp.] Hollow borosilicate glass microspheres; high-str. extender/filler for applics. with moderate pressure and shear; for thermoset and PVC plastisol formulations, low-dens. SMC formulations; provides max. filler loadings and low compd. densities; wh. spheres; dens. 0.48 g/cc; *Toxicology:* moderately alkaline; prolonged exposure may cause respiratory irritation.

Q-Cel® 2106. [PQ Corp.] Hollow borosilicate glass microspheres; lightweight extender/filler for plastics, fiberglass-reinforced plastics, urethane foams, putties, cultured marble, adhesives; wh. spheres; dens. 0.20 g/cc; *Toxicology:* moderately alkaline; prolonged exposure may cause respiratory irritation.

Q-Cel® 2116. [PQ Corp.] Hollow borosilicate glass microspheres; extender/filler for plastics, casting applics. with polyesters, urethanes, PVC, and other systems where low visc. and low shear processing used; sensitizer for bulk and general purpose packaged explosives; wh. spheres; dens. 0.13 g/cc; *Toxicology:* moderately alkaline; prolonged exposure may cause respiratory irritation.

Q-Cel® 2135. [PQ Corp.] Hollow borosilicate glass microspheres; lightweight extender/filler for plastics; as sensitizer in industrial explosives; wh. spheres; dens. 0.28 g/cc; *Toxicology:* moderately alkaline; prolonged exposure may cause respiratory irritation.

Q-Cel® 6717. [PQ Corp.] Hollow borosilicate glass microspheres; lightweight extender/filler for plastics, automotive repair compds., moldings, thin films as low as 4 mils; produces very smooth surf. parts; wh. spheres; dens. 0.20 g/cc; *Toxicology:* moderately alkaline; prolonged exposure may cause respiratory irritation.

Q-Cel® 6832. [PQ Corp.] Hollow borosilicate glass microspheres; lightweight extender/filler for alkali-sensitive resins incl. thermosets (polyesters, epoxies, urethanes), sprayable PVC plastisol underbody sealants, fiberglass-reinforced plastics; as sensitizer in emulsion explosives; wh. spherical powd.; 45 μ mean particle size; dens. 0.32 g/cc; bulk dens. 0.19 g/cc (untamped); *Toxicology:* moderately alkaline; prolonged exposure may cause respiratory irritation.

Q-Cel® 6835. [PQ Corp.] Hollow borosilicate glass microspheres; lightweight extender/filler for alkali-sensitive resins incl. thermosets (polyesters, epoxies, urethanes), sprayable PVC plastisol underbody sealants, fiberglass-reinforced plastics; as sensitizer in emulsion explosives; wh. spherical powd.; 40 μ mean particle size; dens. 0.35 g/cc; bulk dens. 0.20 g/cc (untamped); *Toxicology:* moderately alkaline; prolonged exposure may cause respiratory irritation.

Q-Cel® 6920. [PQ Corp.] Hollow borosilicate glass microspheres; lightweight extender/filler for plastics, fiberglass-reinforced plastics, urethane foams, putties, cultured marble, adhesives; provides smaller top end particles to improve surf. appearance of finished prods.; wh. spheres; dens. 0.20 g/cc; *Toxicology:* moderately alkaline; prolonged exposure may cause respiratory irritation.

Q-Cel® Ultralight. [PQ Corp.] Hollow sodium borosilicate glass microspheres; extender/filler for thermoset casting applics. with polyesters, urethanes, PVC, and other systems where low visc. and low shear processing used; provides high filler loadings and low compd. densities; wh. spherical powd.; 75 μ mean particle size; dens. 0.10 g/cc; *Toxicology:* moderately alkaline; prolonged exposure may cause respiratory irritation.

QDO®. [Lord] p-Quinone dioxime; CAS 105-11-3; nonsulfur vulcanizing agent for syn. elastomers; used for rapid curing of butyl stocks requiring heat resist.

QO® Furan. [QO Chem.] Furan; chemical intermediate in the mfg. of herbicides, pharmaceuticals, plastics, and fine chemicals; colorless; sol. in most org. liqs.

QO® Furfural. [QO Chem.] 2-Furaldehyde; CAS 98-01-1; EINECS 202-627-7; chemical intermediate for mfg. of derivatives (furan and THF); solv. for separating sat. from unsat. compds. in petrol. lubricating oil, gas oil, and diesel fuel; extractive distillation of C4 and C5 hydrocarbons for the mfg. of syn. rubber; decolorizing agent for wood rosin; solv. and processing aid for anthracene; ingred. in resins; reactive solv. and wetting agent in abrasive wheels and brake linings; colorless when freshly distilled; industrial furfural is lt. yel. to brn. liq.; pungent almond-like odor; m.w. 96.08; dens. 1.1545; visc. 1.49 cps; b.p. 161.7 C (760 mm); f.p. -36.5 C; TCC flash pt. 61.7 C; surf. tens. 40.7 dynes/cm (29.9 C); ref. index 1.5235.

QO® Furfuryl Alcohol (FA®). [QO Chem.] Furfuryl alcohol; CAS 98-00-0; EINECS 202-626-1; used in the prod. of foundry sand binders and corrosion-resistant resins; intermediate for esterification and etherification; impregnating sol'n. and carbon binder; wood adhesive component; solv. and temporary plasticizer for phenolic resins in the mfg. of cold-molded abrasive wheels; visc. reducer, cure promoter, and carrier in amine-cured epoxy resins; pale yel. liq.; sol. in water and many common org. solvs.

QO® Tetrahydrofurfuryl Alcohol (THFA®). [QO Chem.] Tetrahydrofurfuryl alcohol; CAS 97-99-4; EINECS 202-625-6; High boiling solv. and carrier for pesticides; FDA approved for use in paper processing; chemical intermediate; also in industrial and consumer cleaners, leather and textile dyeing, epoxies, coatings, inks, paints, and adhesives; plasticizer and vinyl stabilizer carrier; colorless liq.; misc. with water; biodeg.

Quadrilan® SK. [Harcros UK] Quat. fatty amine ethoxylate; cationic; antistat used in polymers; liq.; 100% conc.; Unverified

Quadrol®. [BASF] Tetra (2-hydroxypropyl) ethylenediamine; CAS 102-60-3; EINECS 203-041-4; polyol; chelating agent; intermediate used in resins, emulsifiers, surfactants, pharmaceuticals, herbicides, fungicides, insecticides, adhesives, and plasticizers; liq.; m.w. 292; sol. in water, ethyl alcohol, toluene, ethylene glycol, perchloroethylene; sp.gr. 1.03; ref. index 1.478; 99.2% tert. amine.

Quickset® Extra. [Witco/PAG] MEK peroxide; CAS 1338-23-4; EINECS 215-661-2; high purity initiator for R.T. curing of polyester resins; increased reactivity with lower peroxide conc.; liq.; 9.0% act. oxygen.

Quickset® Super. [Witco/PAG] MEK peroxide; CAS 1338-23-4; EINECS 215-661-2; high purity initiator for R.T. curing of polyester resins; increased reactivity with lower conc.; liq.; 9.0% act. oxygen.

Quikote™. [C.P. Hall] Zinc stearate disp.; CAS 557-05-1; EINECS 209-151-9; provides waterproof antistick coating to rubber stock; no interference with molding operations; can be used on water-cooled takeaway systems before cooling; wh. uniform paste; sp.gr. 1.04 ± 0.6; pH 6 min. (10%); 30 ± 12.5% solids.

Quikote™ C. [C.P. Hall] Calcium stearate disp.; CAS 1592-23-0; EINECS 216-472-8; provides waterproof antistick coating to rubber stock; no interference with molding operations; can be used on water-cooled takeaway systems before cooling; wh. uniform paste; sp.gr. 0.85 ± 0.11; pH 7 min. (10%); 18-23% solids.

Quikote™ C-LD. [C.P. Hall] Low dusting calcium stearate disp.; CAS 1592-23-0; EINECS 216-472-8; provides waterproof antistick coating to rubber sheet stock and preforms; no interference with molding operations; enhances release from mold; wh. uniform paste; sp.gr. 0.88 ± 0.14; pH 9 min. (10%); 17-22% solids.

Quikote™ C-LD-A. [C.P. Hall] Low dusting calcium stearate disp.; CAS 1592-23-0; EINECS 216-472-8; provides waterproof antistick coating to rubber sheet stock and preforms, good short and long-term release props. to all polymers; no interference with molding operations; enhances release from mold; designed for applics. where heavy metal ions or dust cannot be tolerated; all of release coating remains on rubber stock and does not flake or dust off; wh. uniform paste; sp.gr. 0.88 ± 0.14; pH 9 min. (10%); 17-22% solids.

Quikote™ C-LM. [C.P. Hall] Low melting calcium stearate disp.; CAS 1592-23-0; EINECS 216-472-8; provides waterproof antistick coating to rubber sheet stock and preforms; no interference with molding operations; enhances release from mold; esp. for water-cooled takeaways; wh. uniform paste; sp.gr. 0.9 ± 0.1; pH 7 min.; 26 ± 3% solids.

Quikote™ C-LMLD. [C.P. Hall] Low melting, low dusting calcium stearate disp.; CAS 1592-23-0;

EINECS 216-472-8; provides waterproof antistick coating to rubber sheet stock and preforms; no interference with molding operations; enhances release from mold; esp. for water-cooled takeaways; wh. uniform paste; pH 7.5 min.; 25±3% solids.

Quikote™ M. [C.P. Hall] Zinc stearate disp.; CAS 557-05-1; EINECS 209-151-9; provides waterproof antistick coating to rubber stock; no interference with molding operations; can be used for water-cooled takeaways before cooling; wh. uniform paste; sp.gr. 1.04±0.6; pH 6 min. (10%); 30±2.5% solids.

Quimipol EA 2503. [CPB] Syn. primary alcohol ethoxylate; nonionic; emulsifier for min. oils and aliphatic solvs.; visc. depressant for PVC plastisols; liq.; HLB 7.8; 100% conc.

Quimipol ENF 200. [CPB] Nonylphenol ethoxylate; nonionic; emulsifier for emulsion polymerization; wetting agent and detergent at high temps. and high electrolyte concs.; solid; HLB 16.0; 100% conc.

Quimipol ENF 230. [CPB] Nonylphenol ethoxylate; nonionic; emulsifier for emulsion polymerization; wetting agent and detergent at high temps. and high electrolyte concs.; solid; HLB 16.4; 100% conc.

Quimipol ENF 300. [CPB] Nonylphenol ethoxylate; nonionic; emulsifier for emulsion polymerization; wetting agent and detergent at high temps. and high electrolyte concs.; solid; HLB 17.1; 100% conc.

Quindo Magenta RV-6863. [Bayer/Fibers, Org., Rubbers] Pigment Red 202; quinacridone magenta pigment toner for plastics, polyolefins, PVC, PS, ABS, powd. coatings, industrial and architectural finishes, printing inks, textile inks, artists' colors; very good to exc. lightfastness; easy to disperse; relatively high opacity; sp.gr. 1.59; dens. 3.53 lb/gal (tapped); oil absorp. 51 g/100 g.

R

1R70. [AlliedSignal/Ind. Fibers] Nylon; reinforcing material for in-rubber composites providing thermal stability, high tenacity; tenacity 8.7 g/d; shrinkage 8% (350 F); toughness 0.7 g/d.

R7234. [Sovereign] Hexamethoxymethylmelamine; CAS 3089-11-0; adhesion promoter for rubber, esp. tires where it improves adhesion of NR and syn. rubber to steel or textile cord; powd.

R7500E. [Sovereign] Octylphenol formaldehyde resin; thermoreactive resin used to cure butyl and other elastomers with low unsaturation; for butyl bladders, air bags, seals, pharmaceutical parts, masking tapes, prods. subject to high heat and steam; also for TPR elastomers; flake; soften. pt. (R&B) 80-95 C.

R7510. [Sovereign] Octylphenol formaldehyde resin; tackifier for most tire applics.; gives better aged tack than hydrocarbon or rosin-based tackifiers; flakes or pastilles; soften. pt. (R&B) 90-105 C.

R7521. [Sovereign] Octylphenol formaldehyde resin; tackifier for most tire applics.; gives better aged tack than hydrocarbon or rosin-based tackifiers; flakes or pastilles; soften. pt. (R&B) 85-95 C.

R7530E. [Sovereign] Octylphenol formaldehyde resin; thermoreactive resin used to cure butyl and other elastomers with low unsaturation; for butyl bladdes, air bags, seals, pharmaceutical parts, masking tapes, prods. subject to high heat and steam; also for TPR elastomers; gives superior life cycle to butyl prods.; flake; soften. pt. (R&B) 80-90 C.

R7557P. [Sovereign] Modified phenol-formaldehyde novolak; cashew-modified reinforcing resin for use in tire beads, apex strips, sidewalls, shoe soles, and coextruded window profiles; pastille; soften. pr. (R&B) 95-105 C.

R7559. [Sovereign] Modified phenol-formaldehyde; modified reinforcing resin for use in tire beads, apex strips, sidewalls, shoe soles, and coextruded window profiles; flake or pastille.

R7578. [Sovereign] p-Octylphenol formaldehyde resin; high performance tackifier for race tires, tread cements, splice cements, belts, and hoses; best for initial tack and aged tack even after several days storage under adverse conditions; flakes or pastilles; soften. pt. (R&B) 118-132 C.

RA-061. [Himont] Ethylene-propylene copolymer; CAS 9010-79-1; EPM rubber used as impact modifier and raw material for TPO; flex. mod. 55,000 psi; Izod impact no break notched (40 C); Unverified

Rakusol®. [BASF AG] Org. and inorg. pigments in paraffin oil and glycerol ester; used for coloring plastics; less suitable for PVC.

Ralox® 02. [Raschig] 4,4′-Methylene-bis-(2,6-di-t-butylphenol); CAS 118-82-1; nonstaining antioxidant; FDA approved.

Ralox® 35. [Raschig] 3-(3,5-Di-t-butyl-4-hydroxyphenyl) propionic methyl ester; CAS 6886-38-5; antioxidant for min. oil, plastics, rubber.

Ralox® 46. [Raschig] 2,2′-Methylene-bis-(6-t-butyl-4-methylphenol); CAS 119-47-1; EINECS 204-327-1; nonstaining antioxidant/stabilizer for plastics (polyacetal, ABS), sol'n. and emulsion polymers, rubber, nat. and syn. latex, adhesives, hot melts, petrochem. prods., cables; anti-aging agent for lt. colored rubber goods; with synergists, as processing and long-term stabilizer for thermoplastics incl. glass-reinforced grades; FDA and Germany approvals for food contact; cream-colored free-flowing powd., odorless; sol. 800 g/l in ethanol; insol. in water; m.w. 340.5; sp.gr. 1.07 g/cc (20 C); bulk dens. 300 kg/m^3; m.p. 125-132 C; flash pt. (PM) 195 C min.; pH 4-7 (10%, 20 C); *Toxicology:* LD50 (oral, rat) 4880 mg/kg; irritating to eyes and respiratory system; nonirritating to skin; *Precaution:* keep away from ignition sources; dust can form explosive mixts. with air; incompat. with strong acids, bases, oxidizers; hazardous decomp. prods.: CO, hydrocarbons; *Storage:* 12 mos storage life in unopened original pkg. in dry place; sl. pinking may occur after 6 mos.

Ralox® 240. [Raschig] 2-t-Butyl-4-methylphenol; CAS 2409-55-4; antioxidant for min. oil, plastics, rubber.

Ralox® 530. [Raschig] Octadecyl-3-(3,5-di-t-butyl-4-hydroxyphenyl) propionate; CAS 2082-79-3; EINECS 218-216-0; nonstaining antioxidant for plastics (PE, PP, PS, styrene copolymers, polydienes, polyesters, PVC), adhesives, hot melts, waxes, lubricating oils, cables, syn. rubbers (EPDM, EPM); protects against heat degradation, light-aging; partially biodeg.; FDA, BGA approval; wh. cryst., odorless; sol. 330 g/l in acetone; difficult to dissolve in water; m.w. 530.9; dens. 1.05 g/cc; apparent dens. 0.4 g/ml; bulk dens. 400 kg/m; m.p. 50-53 C; flash pt. 270 C; *Toxicology:* LD50 (oral, rat) > 5000 mg/kg, (dermal, rabbit) > 2000 mg/kg;

nonirritating to skin, eyes; nonsensitizing; *Precaution:* keep away from ignition sources; dust can form explosive mixts. in air; incompat. with strong acids, bases, oxidizing agents; hazardous decomp. prods.: CO, hydrocarbons; *Storage:* store in original container.

Ralox® 630. [Raschig] Tetrakis [methylene (3,5-di-t-butyl-4-hydroxyhydrocinnamate)] methane; CAS 6683-19-8; EINECS 229-722-6; nonstaining antioxidant for plastics (PE, PP, polybuytene, polyacetals, polyamides, PU, polyester, PVC, PS, ABS), adhesives, hot melts, waxes, greases, lubricants, cables, syn. rubbers (EPDM, EPM); provides light-aging stability; FDA, BGA approval; wh. cryst. powd., odorless; sol. 890 g/l in acetone; insol. in water; m.w. 1177.7; apparent dens. 0.6 kg/l; dens. 1.45 g/ml (20 C); bulk dens. ≈ 900 kg/m^3; m.p. 110-125 C; *Toxicology:* LD50 (oral, rat) > 5000 mg/kg; nonirritating to eyes, skin; nonsensitizing; *Precaution:* keep away from ignition sources; dust can form explosive mixt. in air; incompat. with strong oxidizing agents, acids, bases; hazardous decomp. prods.: CO, hydrocarbons; *Storage:* 12 mos storage life kept in tightly closed original container (dry, 25 C).

Ralox® BHT food grade. [Raschig] BHT; CAS 128-37-0; EINECS 204-881-4; nonstaining antioxidant for plastics in contact with food, polyols, rubber, latex, waxes, adhesives, hot melts, min. oil processing (lubricants), petrol., feedstuffs; FDA, BGA approval; wh. cryst., almost odorless; sol. 330 g/l in ethanol; insol. in water; m.w. 220.3; dens. 1.05 g/cc (20 C); bulk dens. 450 kg/m^3; vapor pressure 0.02 mbar (20 C); solid. pt. 69.2 C min.; m.p. 69.5 C; b.p. 265 C; flash pt. 127 C; 99.5% min. purity; *Toxicology:* LD50 (oral, rat) > 2000 mg/kg; *Precaution:* keep away from ignition sources; dust can form explosive mixt. with air; incompat. with strong acids bases, oxidizing agents; hazardous decomp. prods.: flamm. gases; *Storage:* protect from light; store below 30 C.

Ralox® BHT Tech. [Raschig] BHT; CAS 128-37-0; EINECS 204-881-4; nonstaining antioxidant for min. oil, plastics, rubber.

Ralox® LC. [Raschig] Butylated reaction prod. of p-cresol and dicyclopentadiene; CAS 68610-5-15; EINECS 271-867-2; low-staining antioxidant for rubber threads, latex dipping articles, latex foam coatings, high temp. applics.; stabilizer for hot melts; FDA approved; wh. to beige powd., brn. pastilles or flakes, odorless; insol. in water; sp.gr. 1.1 g/ml; bulk dens. 300-400 kg/m^3; m.p. 105-120 C; pH 7-8 (10%); *Toxicology:* LD50 (oral, rat) > 5000 mg/kg, (dermal, rabbit) > 5000 mg/kg; *Precaution:* keep away from ignition sources; dust can form explosive mixts. in air; incompat. with strong acids, bases, oxidizing agents; hazardous decomp. prods.: CO, gaseous hydrocarbons; *Storage:* store in original container tightly closed.

Ralox® TMQ-G. [Raschig] 2,2,4-Trimethyl-1,2-dihydroquinoline polymer; CAS 26780-96-1; staining antioxidant for rubber processing; heat antioxidant for NR, IR, SBR, BR, and NBR vulcanized goods, thermoplastic rubbers and their blends; provides some protection against rubber deterioration by heavy metals; recommended for stabilization of rubber mixts. for tech. articles; amber to brn. pastilles, perceptible odor; sol. 3000 g/l in acetone; insol. in water; dens. 1.08 g/cc (20 C); bulk dens. 400-700 kg/m^3; soften. pt. (R&B) 71-91 C; *Toxicology:* LD50 (oral, rat) 3200-7900 mg/kg; repeated/prolonged exposure may produce allergic skin reaction; *Precaution:* dust can form explosive mixts. with air; incompat. with strong oxidizing agents; hazardous decomp. prods.: NO_x, CO; *Storage:* store in cool, dry place up to 30 C max. in original container.

Ralox® TMQ-H. [Raschig] 2,2,4-Trimethyl-1,2-dihydroquinoline polymer; CAS 26780-96-1; staining antioxidant for stabilization of latex foams for carpet back coating; also for peroxide cross-linked PE (e.g., for cable coverings); lt. yel. to ochreous powd., amber to brn. pastilles, perceptible odor; sol. 3000 g/l in acetone; insol. in water; dens. 1.08 g/cc (20 C); bulk dens. 400-700 kg/m^3; m.p. 100 C min.; *Toxicology:* LD50 (oral, rat) 3200-7900 mg/kg; repeated/prolonged exposure may cause allergic skin reaction; avoid dust formation; *Precaution:* hazardous reaction with strong oxidizing agents; dust can form explosive mixts. with air; hazardous combustion gases: NO_x, CO; *Storage:* store in cool, dry place up to 30 C max. in original container.

Ralox® TMQ-R. [Raschig] 2,2,4-Trimethyl-1,2-dihydroquinoline polymer; CAS 26780-96-1; staining antioxidant for rubber processing; heat antioxidant for NR, IR, BR, and NBR vulcanized goods, thermoplastic rubbers and their blends; provides some protection against rubber deterioration by heavy metals; provides best sol. for rubber mixts., useful where short mixing times are required; stabilizes rubber mixts. for tires; amber to brn. pastilles, perceptible odor; sol. 3000 g/l in acetone; insol. in water; dens. 1.08 g/cc (20 C); bulk dens. 400-700 kg/m^3; soften. pt. (R&B) 71-81 C; *Toxicology:* LD50 (oral, rat) 3200-7900 mg/kg; repeated/prolonged exposure may produce allergic skin reaction; *Precaution:* dust can form explosive mixts. with air; incompat. with strong oxidizing agents; hazardous decomp. prods.: NO_x, CO; *Storage:* store in cool, dry place up to 30 C max. in original container.

Ralox® TMQ-T. [Raschig] 2,2,4-Trimethyl-1,2-dihydroquinoline polymer; CAS 26780-96-1; staining antioxidant for stabilization of latex foams for carpet back coating; also for peroxide cross-linked PE; lt. yel. to ochreous powd., amber to brn. pastilles, perceptible odor; sol. 3000 g/l in acetone; insol. in water; dens. 1.08 g/cc (20 C); bulk dens. 400-700 kg/m^3; m.p. 83 C min.; *Toxicology:* LD50 (oral, rat) 3200-7900 mg/kg; repeated/prolonged exposure may cause allergic skin reaction; avoid dust formation; *Precaution:* hazardous reaction with strong oxidizing agents; dust can form explosive mixts. with air; hazardous combustion gases: NO_x, CO; *Storage:* store in cool, dry place up to 30

C max. in original container.

Rapidblend 1793. [Air Prods./Perf. Chems.] Disp. of zinc oxide, magnesium oxide, and liq. alkylated diphenylamine antioxidant (5:4:2 ratio); vulcanizing/antidegradant system for polychloroprene compds.; rods of stiff, thermoplastic putty consistency; sp.gr. 2.0; 100% act.; Unverified

RapidPurge 2. [RapidPurge] Blend of inert minerals, inorg. salts, org. salts, and thermoplastic polyolefins; nonabrasive purging compd. effective on all thermoplastics; for cleaning inj. molding, hot runner tools, extrusion, multi-layer dies, blow molding equip.; gray-wh. very fine powd. and pellets, possible sl. ammonia odor; < 3% sol. in water; sp.gr. 1.40; soften. pt. > 200 F; flash pt. > 650 F (polyethylene component); *Toxicology:* nuisance dust 10 mg/m^3 (total dust); fumes of ammonia may cause tearing of eyes; ing. may cause constipation; *Precaution:* incompat. with strong oxidizers, calcium, hypochlorite bleaches; hazardous decomp. prods.: CO; *Storage:* store in cool, dry area.

RC 7. [Releasomers] Fluorocarbon; mold release agent and lubricant for silicone rubber molding operations, thermoset plastic molding; high service temp. (260 C); solv. sol'n.; *Toxicology:* avoid prolonged/repeated skin contact, breathing of vapors; *Precaution:* keep away from heat, open flame, sparks; *Storage:* store in closed containers when not in use.

RD-1. [Ciba-Geigy/Plastics] Butyl glycidyl ether; CAS 2426-08-6; diluent for epoxy; Gardner 2 max. color; visc. 1-5 cp; EEW 130-149; Unverified

RD-2. [Ciba-Geigy/Plastics] 1,4-Butanediol diglycidyl ether; CAS 2425-79-8; diluent for epoxy; Gardner 2 max. color; visc. 1-5 cp; EEW 130-149; Unverified

Reactint®. [Milliken] Polymeric colorants for flexible urethane foams.

Reactint® Black 57AB. [Milliken] Reactive polymeric urethane colorant offering high color str. at low conc.; for coloring RIM, RRIM, microcellular, and cast elastomers; not recommended for slabstock urethane foam or other large block applics. where high temps. are involved; blk.; sol. in cold water, polyols, and most other urethane components; dens. 9.2 lb/gal; visc. 3000 cps; hyd. no. 175.

Reactint® Black X40LV. [Milliken] Reactive polymeric urethane colorant offering high color str. at low conc.; for coloring flexible urethane foams, rigid and semirigid urethane foams, RIM, RRIM applics.; blk.; sol. in cold water; dens. 9.2 lb/gal; visc. 3000 cps; hyd. no. 154.

Reactint® Blue 17AB. [Milliken] Reactive polymeric urethane colorant offering high color str. at low conc.; for coloring RIM, RRIM, microcellular and cast elastomer; not recommended for slabstock urethane foam or other large block applic. where high temps. involved; blue; sol. in cold water, polyol, most other urethane components; dens. 10 lb/gal; visc. 3500 cps; hyd. no. 210.

Reactint® Blue X3LV. [Milliken] Reactive polymeric urethane colorant offering high color str. at low conc.; for coloring flexible urethane foams, rigid and semirigid urethane foams, RIM, RRIM applics.; blue (aqua); sol. in cold water; dens. 9.2 lb/gal; visc. 2500 cps; hyd. no. 205.

Reactint® Blue X19. [Milliken] Bright colorant providing dual compat. in ether and ester foams; for coloring flexible or rigid urethane foams, RIM, RRIM applics.; blue; sol. in cold water; dens. 9.75 lb/gal; visc. 2000 cps; hyd. no. 260.

Reactint® Orange X38. [Milliken] Reactive polymeric urethane colorant offering high color str. at low conc.; for coloring flexible urethane foams, rigid and semirigid urethane foams, RIM, RRIM applics.; orange; sol. in cold water; dens. 9.5 lb/gal; visc. 3000 cps; hyd. no. 105.

Reactint® Red X52. [Milliken] Reactive polymeric urethane colorant offering high color str. at low conc.; for coloring flexible urethane foams, rigid and semirigid urethane foams, RIM, RRIM applics.; red; sol. in cold water; dens. 9.5 lb/gal; visc. 1000 cps; hyd. no. 150.

Reactint® Violet X80LT. [Milliken] Reactive polymeric urethane colorant offering high color str. at low conc.; for coloring flexible urethane foams, rigid and semirigid urethane foams, RIM, RRIM applics.; violet; sol. in cold water; dens. 9.1 lb/gal; visc. 2000 cps; hyd. no. 80.

Reactint® Yellow X15. [Milliken] Reactive polymeric urethane colorant offering high color str. at low conc.; for coloring flexible urethane foams, rigid and semirigid urethane foams, RIM, RRIM applics.; lemon yel.; sol. in cold water; dens. 9.2 lb/gal; visc. 2500 cps; hyd. no. 84.

Reed C-ABS-7526. [Reed Spectrum] ABS-based wh. conc.; wh. color conc. for ABS resins; FDA sanctioned.

Reed C-ABS-17415. [Reed Spectrum] ABS-based blk. conc. (25:1 ratio); wh. color conc. for general purpose applics.

Reed C-BK-7. [Reed Spectrum] 40% Carbon black EVA-based conc.; blk. color conc. for olefin resins and other materials.

Reed C-EPES-1331. [Reed Spectrum] 21% Carbon blk. polyester-based conc.; blk. color conc. for thermoplastic elastomers incl. Hytrel®, Lomod®, and Riteflex®.

Reed C-NY-261. [Reed Spectrum] Nylon-based color conc.; blk. color conc. for use in nylon 6 and 6/6.

Reed C-NY-4892. [Reed Spectrum] Nylon-based color conc. with 20% carbon blk.; blk. color conc. for use in nylon 6 and 6/6 esp. for filled and toughened prods.; superior hiding power.

Reed C-PBT-1338. [Reed Spectrum] PBT polyester-based color conc., 20% carbon blk.; blk. color conc. for use in PBT polyester.

Reed C-PET-1333. [Reed Spectrum] 21% Carbon blk. polyester-based conc.; blk. color conc. for PET resins incl. unfilled (strapping) and filled (elec. connector) applics.

Reed C-PPR-2096. [Reed Spectrum] 60% Titanium dioxide PP-based conc.; wh. color conc. for homopolymer and copolymer PP; FDA approved.

Reed C-PS-4230. [Reed Spectrum] 60% Titanium

dioxide PS-based conc.; wh. color conc. for PS, MIPS, and HIPS; ideal for food and medical applics.; FDA approved.

Reed C-PUR-1051. [Reed Spectrum] Blk. PU-based conc.; blk. color conc. for ester-based PU applics.; suitable for inj. molding and extrusion.

Reed C-PY-1232. [Reed Spectrum] PC-based blk. conc.; blk. color conc. for use in clear, transparent, or mixed color prods.

Reed PWC-1. [Reed Spectrum] 60% Titanium dioxide conc.; wh. color conc. for inj. molded PE and PP food and medical applics.; FDA sanctioned.

ReedLite® C-NY. [Reed Spectrum] Heavy metal-free nylon color concs.; for automotive, electronic and mech. applics.

ReedLite® CPC. [Reed Spectrum] Heavy metal-free PC color concs.; for elec., mech., automotive, appliance and business machine applics.

Regal® 300. [Cabot/N. Am. Rubber Black] N326 Carbon black; CAS 1333-86-4; EINECS 215-609-9; carbon black for rubbers featuring high elong. and good fatigue resist.; 0.1% max. on 325 mesh; pour dens. 465 kg/m^3; iodine no. 82.

Regal® 300R. [Cabot/Special Blacks Div] Carbon black; CAS 1333-86-4; EINECS 215-609-9; colorant providing high tinting str., good jetness, and good dispersion for inks, coatings, plastics, paper; fluffy; 27 nm particle size; dens. 13 lb/ft^3; surf. area 80 m^2/g; 1.0% volatile.

Regal® 330. [Cabot/Special Blacks Div] Carbon black; CAS 1333-86-4; EINECS 215-609-9; colorant, uv stabilizer providing high tinting str., good jetness, and good dispersion for inks, coatings, plastics, paper; pellets; 25 nm particle size; dens. 28 lb/ft^3; surf. area 94 m^2/g; 1.0% volatile.

Regal® 330R. [Cabot/Special Blacks Div] Carbon black; CAS 1333-86-4; EINECS 215-609-9; colorant, uv stabilizer providing high tinting str., good jetness, and good dispersion for inks, coatings, plastics, paper; fluffy; 25 nm particle size; dens. 19 lb/ft^3; surf. area 94 m^2/g; 1.0% volatile.

Regal® 400. [Cabot/Special Blacks Div] Carbon black; CAS 1333-86-4; EINECS 215-609-9; colorant providing exc. dispersion in gloss printing and carbon paper inks, stable tinting of enamels, resistive plastics; pellets; 25 nm particle size; dens. 30 lb/ft^3; surf. area 96 m^2/g; 3.5% volatile.

Regal® 400R. [Cabot/Special Blacks Div] Carbon black, aftertreated; CAS 1333-86-4; EINECS 215-609-9; colorant providing exc. dispersion in gloss printing and carbon paper inks, stable tinting of enamels, resistive plastics (higher dielec. PVC cable compds.); fluffy; 25 nm particle size; dens. 14 lb/ft^3; surf. area 96 m^2/g; pH 2.5-4.0; 3.5% volatile.

Regal® 660. [Cabot/Special Blacks Div] Carbon black; CAS 1333-86-4; EINECS 215-609-9; colorant providing high tint str., good jetness, low oil absorp. for inks, coatings, plastics, toners; pellets; 24 nm particle size; dens. 31 lb/ft^3; surf. area 112 m^2/g; 1.0% volatile.

Regal® 660R. [Cabot/Special Blacks Div] Carbon black; CAS 1333-86-4; EINECS 215-609-9; colorant providing high tint str., good jetness, low oil absorp. for inks, coatings, plastics, toners; fluffy; 24 nm particle size; dens. 15 lb/ft^3; surf. area 112 m^2/g; 1.0% volatile.

Regulator ZL. [Bayer AG] Used to regulate the pot life of solv.-free PU systems through dealkalization.

Release Agent E-155. [Wacker Silicones] Silicone polymer contg. polydimethylsiloxane (CAS 63148-62-9) and alkylene diaminofunctional polydimethylsiloxane; release agent, lubricant providing multiple releases when molding urethane or epoxy plastic parts; also for metal cables to enhance longevity; for use as is or diluted with a chlorinated or hydrocarbon solv.; clear liq., sl. odor; negligible sol. in water; sp.gr. 0.968; dens. 8.1 lb/gal; visc. 225 cSt; vapor pressure 0.1 mm Hg (68 F); flash pt. (PMCC) 138 C; 100% act.; *Toxicology:* eye irritant; methanol (hydrolysis by-prod.): causes optic neuropathy, metabolic acidosis, respiratory depression, ingestion may cause blindness or death; can generate formaldehyde vapors @ 150 C (potential cancer hazard, irritant, toxic); *Precaution:* burns with difficulty, but will support combustion; hazardous/thermal decomp. prods.: SiO_2, CO, CO_2, formaldehyde, various hydrocarbon fragments; *Storage:* 1 yr min. shelf life stored in closed container; store in cool, dry, well-ventilated area; containers may contain flamm. vapors due to hydrolysis.

Release Agent NL-1. [Akzo Nobel] Wax mixture, hydrocarbon solv.; release agent for molded prods.; used as a first thin layer on the mold surface; paste.

Release Agent NL-2. [Akzo Nobel] Polyvinyl alcohol sol'n., alcohol water solv.; film-forming release agent; sol'n.

Release Agent NL-10. [Akzo Nobel] Wax mixture, paraffin oil solv.; release agent for molded prods.; used as a first thin layer on the mold surface; paste.

Release Gel 1765 G. [Goldschmidt] Sprayable gelatinous release agent for prod. of radiator hoses; 100% act.; Unverified

Remcopal 229. [Ceca SA] Ceteareth-25; CAS 68439-49-6; nonionic; emulsifier for emulsion polymerization, washing detergents; solid; HLB 15.2; 100% conc.

Remcopal 33820. [Ceca SA] Nonoxynol-20; CAS 9016-45-9; nonionic; emulsifier for emulsion polymerization; paste; HLB 15.5; 100% conc.

Renacit® 7. [Bayer/Fibers, Org., Rubbers] Pentachlorothiophenol with activating and dispersing additives; CAS 133-49-3; peptizing agent facilitating open mill and internal mixer mastication in rubber industry; lt. gray powd.; substantially insol. in common solvs.; sp.gr. 2.3; m.p. partly fusible; 46.5% act.

Renacit® 7/WG. [Bayer/Fibers, Org., Rubbers] Renacit 7 (pentachlorothiophenol) containing stearic acid and paraffin wax; peptizing agent facilitating open mill and internal mixer mastication in rubber industry; lt. gray gran.; sp.gr. 2.10; m.p. partly fusible.

Renacit® 8/LG. [Bayer/Fibers, Org., Rubbers] Activated zinc salts of unsat. fatty acids; peptizing agent facilitating open mill and internal mixer mastication; esp. suitable for NR; brn. lentil-shaped gran.; dens. 1.1 g/cm^3; m.p. ≥ 75 C.

Reogen E®. [King Industries; Struktol] High m.w. sulfonic acid and low visc. min. oil; plasticizer and processing aid for elastomers; aids compounding and mold flow; retards scorch with min. loss of props.; mahogany liq.; sp.gr. 0.89 (25.6 C); dens. 7.4 lb/gal; visc. 45 SUS (37.8 C); acid no. 8.1; flash pt. (COC) 120 C.

Research Curing Agent RSC-1246. [Shell] Aromatic amine; curing agent for high performance composites; long gel time; improved hot/wet performance; solid; m.p. 210-215 F.

Research Curing Agent RSC-2215. [Shell] Aliphatic amine, accelerated; fast R.T. curing agent for epoxy composites, structural adhesives, civil engineering, coatings, tooling; good mech. props. and chem. resist., exc. toughness; Gardner < 1 low-visc. liq.; dens. 8.175 lb/gal; visc. 175 cps; *Toxicology:* may cause skin irritation, serious systemic effects; *Precaution:* corrosive liq. (DOT); *Storage:* stable at R.T. in unopened container; becomes darker on exposure to the atmosphere.

Resimene® 3520. [Monsanto] Hexamethylmelamine; CAS 68002-20-0; with a methylene acceptor; reinforces rubber compds., latex dips and enhances rubber adhesion to textiles, brass plated or zinc plated steel cord; visc. liq. or 72% act. powd.

Resin 731D. [Hercules] Modified dehydrogenated (disproportionated) rosin; pale, oxidation-resistant, thermoplastic resin used in hot-melt-applied adhesives and coatings for paper and paperboard substrates, as tackifier and processing aid for rubber-based adhesives and molding compds.; for use in contact with food; USDA Rosin N solid, flakes; sol. in alcohols, esters, ketones, min. spirits, and aromatic hydrocarbons; dens. 1.058 kg/l; R&B soften. pt. 73 C; COC flash pt. 209 C; acid no. 154; sapon. no. 159.

Resin 3072. [Hercules] Mixed emulsifiers containing both disproportionated fatty acid and rosin acid; used in emulsion polymerization of elastomers calling for mixed emulsifier system.

Resinall 70R. [Sovereign] Aromatic hydrocarbon resin; tackifier for rubber providing exc. heat, chem., UV, and oxidation resist.; no effect on cure rate due to low acid no.; flake; soften. pt. (R&B) 100 C.

Resinall 153. [Resinall] Zinc resinate of modified rosin; high m.p. resin with exc. sol. in low KB aliphatic solvs.; used in adhesives and rubber compding.; also as modifier for phenolic resins, printing inks and gloss oil; amber clear solid, mild odor; negligible sol. in water; sp.gr. 1.140; dens. 9.4 lb/gal; visc. (Gardner) C-L (60% toluene); m.p. (R&B) 155-165 C; acid no. < 10; flash pt. (COC) 305 C; 7.5 ± 0.5% metallic zinc; *Toxicology:* minimal toxicity by oral ingestion; dust may be mildly irritating to respiratory tract; *Precaution:* incompat. with strong oxidants; explosive dust/air mixts. are possible but not expected.

Resinall 203. [Sovereign] Modified rosin; tackifier and retarder for NR compds.; used in large farm and OTR tires where long slow cures require extra scorch safety; tightly controlled high acid no.; flake; soften. pt. (R&B) 95 C.

Resinall 219. [Sovereign] Modified rosin; low-cost process aid and tackifier; promotes shear and aids dispersion; gives good release from mixer; flake; soften. pt. (R&B) 100 C.

Resinall 286. [Sovereign] Modified rosin; low-cost process aid and tackifier for rubber; flake.

Resinall 605. [Sovereign] Glyceryl rosinate; process aid for rubber; flake; soften. pt. (R&B) 80 C.

Resinall 610. [Sovereign] Pentaerythrityl rosinate; CAS 8050-26-8; EINECS 232-479-9; tackifier for rubber, hot-melt adhesives, mastics, contact cements, pressure-sensitive adhesives; good tack, light color, very low acid no.; exc. compat. with solvs., elastomers, rubber, and waxes; can be emulsified for use in SBR, NR, and CR latex adhesives; flake; soften. pt. (R&B) 100 C.

Resinall 711. [Sovereign] Aromatic hydrocarbon resin; tackifier for rubber; lighter color and lower iodine no. than 70R; flake; soften. pt. (R&B) 100 C.

Resinall 725. [Sovereign] Aliphatic/aromatic hydrocarbon resin; tackifier for rubber; soften. pt. (R&B) 100 C.

Resinall 766. [Sovereign] Aliphatic hydrocarbon resin; tackifier for rubber with good heat and chem. stability; lower cost than aromatics; flake; soften. pt. (R&B) 105 C.

Resinall 767. [Sovereign] Aliphatic hydrocarbon resin; tackifier for rubber; provides improved processing, reduced scorching, and increases plasticity while providing good tack; good sol., wide compat. with most polymers; lighter color than Resinall 766; flake; soften. pt. (R&B) 105 C.

Resinall 792. [Sovereign] Aliphatic rosin-modified hydrocarbon resin; economical tackifier for rubber; darker than Resinall 766; flake; soften. pt. (R&B) 100 C.

Retarder AK. [Akrochem] Modified phthalic anhydride; CAS 85-44-9; EINECS 201-607-5; nondiscoloring retarding agent to reduce scorching of rubber compds. at processing temps.; also acts as an activator for certain blowing agents; wh. nondusting powd.; 1.0% max. retained 100 mesh; practically odorless; sol. in alcohol, benzene, and acetone; slightly sol. in water; sp.gr. 1.45-1.51; m.p. 123-132 C.

Retarder BA, BAX. [Akrochem] Predominantly benzoic acid; CAS 65-85-0; EINECS 200-618-2; retarding agent for natural and syn. rubbers and latexes; nonstaining; acts as an activator for certain blowing agents; processing aid with certain cis-polybutadiene rubbers; BAX is oil-treated; wh. flakes; sp.gr. 1.30; m.p. 122 C.

Retarder PX. [Akrochem] Phthalic anhydride, oil treated; CAS 85-44-9; EINECS 201-607-5; nondiscoloring retarding agent to reduce scorching of rubber compds. at processing temps.; off-wh. powd.; pract. odorless; sl. sol. in water; sol. in alcohol; sp.gr. 1.52; m.p. 129-134 C.

Retarder SAFE. [Akrochem] Treated aromatic sulfonamide; nonstaining, nondiscoloring retarder for nat. and syn. rubber compds.; also useful for replasticizing of sl. scorched stocks by cold mill mixing; wh. to tan powd., sl. typ. odor; 1% m ax. on 100 mesh; 95% sol. in benzene, CCl_4; insol. in water; sp.gr. 1.68; m.p. 108-113 C (for sulfonamide content); *Toxicology:* not considered hazardous for normal industrial use; *Storage:* good storage stability.

Retarder SAX. [Akrochem] Tech. salicylic acid (90%) and lt. process oil treatment (10%); retarder; vulcanization inhibitor for SBR and natural rubber compds.; also as accelerator for W types of Neoprene; blowing agent activator in sponge rubber compds.; off-wh. cryst. powd.; practically odorless; easily disperses in dry polymers; sp.gr. 1.31.

Retilox® F 40 MF. [Akrochem] α,α′-bis(t-butylperoxy)-m/p-diisopropylbenzene/EPM masterbatch; curing and crosslinking agents for EPM, EPDM, polyethylene, silicone rubbers, NBR, EVA copolymers, SBR, chlorosulfonated polyethylene, PVC, PU rubbers, polybutadiene rubbers, neoprene, natural rubber, chlorinated polyethylene; for mfg. of rubber articles, membranes for waterproofing, elec. cable insulation, conveyor belts, shoe soles; sheets; sp.gr. 0.90; flash pt. (OC) > 100 C; 40% peroxide, 60% EPM, 3.78% act. oxygen; *Toxicology:* nontoxic, but may cause mild irritation; *Precaution:* will burn in contact with naked flame; *Storage:* store in cool, dry place below 30 C in original containers.

Retilox® F 40 MG. [Akrochem] α,α′-bis(t-butylperoxy)-m/p-diisopropylbenzene/EPM masterbatch; curing and crosslinking agents for EPM, EPDM, polyethylene, silicone rubbers, NBR, EVA copolymers, SBR, chlorosulfonated polyethylene, PVC, PU rubbers, polybutadiene rubbers, neoprene, natural rubber, chlorinated polyethylene; for mfg. of rubber articles, membranes for waterproofing, elec. cable insulation, conveyor belts, shoe soles; cubic pellets; sp.gr. 0.90; bulk dens. 0.55 g/ml; 40% peroxide, 60% EPM, 3.78% act. oxygen; *Toxicology:* nontoxic, but may cause mild irritation; *Precaution:* will burn in contact with naked flame; *Storage:* store in cool, dry place below 30 C in original containers.

Retrocure® G. [Akrochem] Blend of 4- and 5-methylbenzotriazole (tolyltriazole); CAS 136-85-6; prevulcanization retarder for sulfur-modified (G type) polychloroprene rubbers; also for NBR systems where MBTS/sulfur cure systems are used; lt. beige gran.; sp.gr. 1.2; m.p. 83 C; 98% min. act.

Rewoderm® S 1333. [Rewo GmbH] Disodium ricinoleamido MEA-sulfosuccinate; anionic; detergent for very mild and skin-friendly shampoos, intimate preps., dermatological lotions, skin protection prods., washing up liqs.; decreases irritancy of alkylbenzene sulfonate and other surfactants; emulsifier for emulsion polymerization; amber liq.; pH 6.5-7.5 (5%); surf. tens. 28 mN/m; 40% act.

Rewomat B 2003. [Rewo GmbH] Tetrasodium (1,2-dicarboxyethyl)-N-alkyl sulfosuccinamide; anionic; detergent, foaming agent and stabilizer; emulsion polymerization additive; industrial, metal, household and all-purpose cleaner; solubilizer for detergent raw materials; soldering aid; pigment dispersant; emulsifier for wax and oil; cosmetics; amber clear liq.; sol. in water, alkali and electrolytes; pH 7-9 (10% solids); 35% solids; Unverified

Rewomat TMS. [Rewo GmbH] Disodium alkyl sulfosuccinamide; anionic; foaming agent, stabilizer, emulsifier, foam additive for natural and syn. latexes; emulsion polymerization; wetting agent for latex impregnation; soft creamy paste or liq.; pH 9-10 (10% solids); 35% solids; Unverified

Rewopal® HV 5. [Rewo GmbH] Nonoxynol-5; CAS 9016-45-9; nonionic; emulsifier, solubilizer used in emulsion polymerization and insecticides, raw material for rinsing, washing, and cleaning agents; wetting agent in the paper and cellulose industry, degreaser used in pickling and alkaline immersion-bath cleaners; BGA and FDA compliance; Gardner 2 max. liq.; sol. in min. oil, alcohol, ketone, toluene, xylene, chlorinated hydrocarbons; cloud pt. 59-62 C; pH 5-7 (1%); surf. tens. 30 dynes/cm; 100% act.

Rewopal® HV 9. [Rewo GmbH] Nonoxynol-9; CAS 9016-45-9; nonionic; emulsifier, solubilizer used in emulsion polymerization and insecticides, raw material for rinsing, washing, and cleaning agents; wetting agent in the paper and cellulose industry, degreaser used in pickling and alkaline immersion-bath cleaners; Gardner 2 max. liq.; sol. in min. oil, alcohol, ketone, toluene, xylene, chlorinated hydrocarbons; cloud pt. 59-62 C (2%); pH 5-7 (1% solids).

Rewopal® HV 10. [Rewo GmbH] Nonoxynol-10; CAS 9016-45-9; nonionic; emulsifier, solubilizer used in emulsion polymerization and insecticides, raw material for rinsing, washing, and cleaning agents; wetting agent in the paper and cellulose industry, degreaser used in pickling and alkaline immersion-bath cleaners; BGA and FDA compliance; Gardner 2 max. liq.; sol. in min. oil, alcohol, ketone, toluene, xylene, chlorinated hydrocarbons; cloud pt. 70-73 C (2%); pH 5-7 (1% solids); surf. tens. 34 dynes/cm; 100% act.

Rewopal® HV 25. [Rewo GmbH] Nonoxynol-25; CAS 9016-45-9; nonionic; emulsifier, solubilizer used in emulsion polymerization, insecticides, detergents with high electrolyte content (e.g., electrolating); raw material for rinsing, washing, and cleaning agents; wetting agent in the paper and cellulose industry; degreaser used in pickling and alkaline immersion-bath cleaners; BGA and FDA compliance; wh. wax; sol. in min. oil, alcohol, ketone, toluene, xylene, chlorinated hydrocarbons; cloud pt. 73-76 C (2%); pH 5-7 (1% solids); surf. tens. 42 dynes/cm; 100% act.

Rewopal® HV 50. [Rewo GmbH] Nonoxynol-50; CAS 9016-45-9; nonionic; emulsifier for emulsion polymerization; wax; 100% conc.

Rewophat E 1027. [Rewo GmbH] Alkylphenol polyglycol ether phosphate; anionic; corrosion inhibitor, emulsifier for min. oils, dispersant, wetting agent; for emulsion polymerization; metalworking fluids; antistat; high pressure additive; textile antistat; yel. liq.; sol. in org. solvs.; sol. < 5 g/1000 ml water; pH 2-4 (1%); surf. tens. 32 dynes/cm; 100% act.

Rewophat NP 90. [Rewo GmbH] Nonylphenol polyglycol ether phosphate; anionic; emulsifier for emulsion polymerization; textile auxiliaries; antistat, raw material for industrial cleaners; visc. liq.; sol. in water, org. solvs.; pH 2-4 (1%); surf. tens. 34 dynes/cm; 100% conc.

Rewopol® 15/L. [Rewo GmbH] Sodium lauryl sulfate; CAS 151-21-3; anionic; surfactant for emulsion polymerization; liq.; 15% conc.; Unverified

Rewopol® B 1003. [Rewo GmbH] Disodium tallow sulfosuccinamate; CAS 90268-48-7; EINECS 290-850-0; anionic; foaming and antigelling agent for latex foam backings and coatings; emulsifier for emulsion polymerization; flotation agent; FDA compliance; paste; sol. 35 g/1000 ml water (40 C); pH 8-10 (1%); surf. tens. 42 dynes/cm; 35% conc.

Rewopol® B 2003. [Rewo GmbH] Tetrasodium dicarboxyethyl stearyl sulfosuccinamate; anionic; flotation reagent; emulsifier for emulsion polymerization; foaming agent for latex emulsion (carpet backing); antigelling agent, cleaning agent for paper mill felts; FDA compliance; low visc. liq.; sol. in water; partly sol. in org. solvs.; pH 7-8 (1%); surf. tens. 40 dynes/cm; 35% conc.

Rewopol® NL 2-28. [Rewo GmbH] Sodium laureth sulfate; CAS 9004-82-4; anionic; surfactant for shampoos, shower gels, foam baths, liq. soaps, dishwashing liqs., emulsion polymerization, air entrainment agent, textile auxiliaries; liq.; 28% conc.

Rewopol® NL 3. [Rewo GmbH] Sodium laureth sulfate; CAS 9004-82-4; anionic; detergent, foamer; cleaning formulations; personal care products; emulsifier for emulsion polymerization; BGA compliance; colorless clear liq.; low odor; visc. 50-150 cps; pH 6.5-7.5 (1%); surf. tens. 38 mN/m; 28% act.

Rewopol® NL 3-28. [Rewo GmbH] Sodium laureth sulfate; CAS 9004-82-4; anionic; surfactant for shampoos, shower gels, foam baths, liq. soaps, dishwashing liqs., emulsion polymerization, air entrainment agent, textile auxiliaries; liq.; 28% conc.

Rewopol® NL 3-70. [Rewo GmbH] Sodium laureth sulfate; CAS 9004-82-4; anionic; surfactant for shampoos, shower gels, foam baths, liq. soaps, dishwashing liqs., emulsion polymerization, air entrainment agent, textile auxiliaries; paste; HLB 18.0; 70% conc.

Rewopol® NLS 15 L. [Rewo GmbH] Sodium lauryl sulfate; CAS 151-21-3; anionic; emulsifier for emulsion polymerization; BGA and FDA compliance; liq.; water-sol.; pH 9-9.5 (1%); surf. tens. 30 mN/m; 15% conc.

Rewopol® NLS 28. [Rewo GmbH] Sodium lauryl sulfate; CAS 151-21-3; EINECS 205-788-1; anionic; detergent, emulsifier in emulsion polymerization; raw material for lt. duty detergents, body cleaning agents, shampoos, detergent pastes, cosmetics; air entrainment agents; BGA and FDA compliance; lt. yel. low visc. liq.; water-sol.; m.w. 300; pH 7-8 (1%); surf. tens. 30 mN/m; 28% act.

Rewopol® NLS 30 L. [Rewo GmbH] Sodium lauryl sulfate; CAS 151-21-3; anionic; emulsifier for emulsion polymerization; BGA and FDA compliance; liq.; water-sol.; pH 9-9.5 (1%); surf. tens. 30 mN/m; 30% conc.

Rewopol® NOS 5. [Rewo GmbH] Nonyl phenol polyglycol ether sulfate; anionic; emulsion polymerization surfactant esp. for styrene; BGA and FDA compliance; liq.; sol. in water; pH 6-8 (1%); surf. tens. 29 mN/m; 30% conc.

Rewopol® NOS 8. [Rewo GmbH] Nonyl phenol polyglycol ether sulfate; anionic; emulsifier for styrene polymerization and all other monomers; BGA and FDA compliance; liq.; water-sol.; pH 7-8 (1%); surf. tens. 35 mN/m; 33% conc.

Rewopol® NOS 10. [Rewo GmbH] Nonyl phenol polyglycol ether sulfate; anionic; emulsifier for styrene polymerization and all other monomers; BGA and FDA compliance; liq.; water-sol.; pH 7-8 (1%); surf. tens. 36 mN/m; 35% conc.

Rewopol® NOS 25. [Rewo GmbH] Nonyl phenol polyglycol ether sulfate; anionic; emulsifier for styrene polymerization and all other monomers; BGA and FDA compliance; liq.; water-sol.; pH 7-8 (1%); surf. tens. 40 mN/m; 35% conc.

Rewopol® SBDB 45. [Rewo GmbH] Diisobutyl sodium sulfosuccinate; CAS 127-39-9; EINECS 204-839-5; anionic; emulsifier for emulsion polymerization; stabilizer for disps.; pigment dispersant; BGA and FDA compliance; liq.; sol. 240 g/1000 ml water; pH 5-7 (1%); surf. tens. 49 dynes/cm; 45% conc.

Rewopol® SBDC 40. [Rewo GmbH] Dicyclohexyl sodium sulfosuccinate; CAS 23386-52-9; EINECS 245-629-3; anionic; emulsifier for emulsion polymerization; stabilizer for disps.; pigment dispersant; BGA and FDA compliance; paste; sol. 10 g/1000 ml water; sol. hot in org. solvs.; pH 5-7 (1%); surf. tens. 40 dynes/cm; 40% conc.

Rewopol® SBDO 75. [Rewo GmbH] Dioctyl sodium sulfosuccinate; CAS 577-11-7; EINECS 209-406-4; anionic; wetting agent, solubilizer; emulsion polymerization; BGA and FDA compliance; visc. liq.; sol. 10 g/1000 ml water; very sol. in org. solvs.; pH 5-7 (1%); surf. tens. 27 dynes/cm; 75% conc.

Rewopol® SBFA 50. [Rewo GmbH] Fatty alcohol polyglycol ether sulfosuccinate; anionic; emulsion polymerization emulsifier, post-stabilizer, dispersant for pigments; best for acrylates and vinyl acetates; FDA compliance; low visc. liq.; HLB 14.0; pH 5-7 (1%); surf. tens. 30 mN/m; 30% conc.

Rewopol® SBMB 80. [Rewo GmbH] Diisohexyl sulfosuccinate; anionic; emulsion polymerization surfactant; stabilizer for disps.; pigment dispersant; BGA and FDA compliance; visc. liq.; sol. 400 g/1000 ml water; sol. in org. solvs.; pH 5-7 (1%); surf. tens. 30 dynes/cm; 80% conc.

Rewopol® SMS 35. [Rewo GmbH] Alkyl disodium sulfosuccinamate; anionic; emulsifier for emulsion polymerization; foaming agent for latex emulsion; antigelling and cleaning agent for paper mill felts; paste; 35% conc.; Unverified

Rewopol® TMSF. [Rewo GmbH] Alkyl sulfosuccinamate; anionic; emulsifier for emulsion polymerization; foaming agent for latex emulsions; antigelling agent; liq.; 31% conc.

Rewoquat CPEM. [Rewo GmbH] PEG-5 cocomonium methosulfate; CAS 68989-03-7; cationic; conditioner for antistatic hair care prods., emulsifier in emulsion polymerization; visc. liq.; 100% conc.

Reworyl® B 70. [Rewo GmbH] Benzene sulfonic acid; CAS 98-11-3; EINECS 202-638-7; anionic; catalyst for esterification, polymerization, and polycondensation in foundry resins; straw clear liq.; 70% conc.; Unverified

Reworyl® C 65. [Rewo GmbH] Cumene sulfonic acid; CAS 28631-63-2; anionic; catalyst for esterification, polymerization, and polycondensation in foundry resins; liq.; 65% conc.; Unverified

Reworyl® NKS 50. [Rewo GmbH] Sodium dodecylbenzene sulfonate; anionic; biodeg. raw material for detergents and cleaners; textile industry; emulsifier for styrene polymerization; BGA and FDA compliance; paste; sol. 250 g/1000 ml water; pH 7-8 (1%); surf. tens. 38 mN/m; 50% conc.

Reworyl® T 65. [Rewo GmbH] p-Toluene sulfonic acid; anionic; catalyst for prep. of esters, acid hydrolysis, polymerization reactions; industrial applics.; intermediate; leather tanning agent; toxic crop-dusting agents; prod. of syn. fibers; pale clear liq.; char. odor; 65% act.; Unverified

Reworyl® X 65. [Rewo GmbH] Xylene sulfonic acid; CAS 25321-41-9; EINECS 246-839-8; anionic; catalyst for esterification, polymerization, and polycondensation in foundry resins; liq.; 65% conc.; Unverified

Rexanol™. [C.P. Hall] Soap-type lubricant; lubricant providing nondusting, antistick surf. to latex and rubber slabs or continuous sheets; wh. opalescent cream; sp.gr. 0.95 ± 0.15; pH 9 min. (10%); 20.8 ± 3.2% solids.

Rexol 25/4. [Huntsman] Nonoxynol-4; CAS 9016-45-9; nonionic; detergent, dispersant, stabilizer; low foaming emulsifier for oils and petrol. solvs., emulsion polymerization; intermediate for anionic sulfonates; pigment dispersant; clear liq.; oil-sol.; sp.gr. 1.02; visc. 200 cps; HLB 8.6; ref. index 1.5000; 100% act.

Rexol 25/7. [Huntsman] Nonoxynol-7; CAS 9016-45-9; nonionic; emulsifier for herbicides and pesticides; intermediate for anionic sulfates; antistatic plasticizer for plastics; clear liq.; HLB 11.6; cloud pt. 76-78 C (10% in 25% in diethylene glycol butyl ether); 100% act.

Rexol 25/8. [Huntsman] Nonoxynol-8; CAS 9016-45-9; nonionic; detergent, wetting agent, emulsifier, dispersant; base surfactant for household and industrial detergents; used for leather, paint, textile, pesticides, pulp and paper industries, emulsion polymerization; clear liq.; water-sol.; HLB 12.4; cloud pt. 23-27 C (1% aq.); 100% act.

Rexol 25/10. [Huntsman] Nonoxynol-10; CAS 9016-45-9; nonionic; detergent, dispersant, emulsifier, wetting agent; for paint, textiles, pulp/paper industries, emulsion polymerization; degreaser for leather; scouring agent for raw wool; liq.; water-sol.; HLB 13.4; cloud pt. 63-66 C (1% aq.); 99% act.

Rexol 25/12. [Huntsman] nonionic; emulsifier, wetting agent, stabilizer for emulsion polymerization; liq.; 100% act.

Rexol 25/40. [Huntsman] Nonoxynol-40; CAS 9016-45-9; nonionic; emulsifier used in polymer emulsification for paints and coatings; solid; HLB 17.8; cloud pt. 74-76 C (1% in 10% NaCl); 100% act.

Rexol 25/307. [Huntsman] Nonoxynol-30; CAS 9016-45-9; nonionic; solubilizer for essential oils and pesticides; surfactant for latex emulsion polymerization; liq.; HLB 17.2; cloud pt. 74-76 C (1% in 10% NaCl); 70% act. in water.

Rexol 25/407. [Huntsman] Nonoxynol-40; CAS 9016-45-9; nonionic; detergent, wetting agent; emulsifier for vinyl acetate and acrylate polymerization; demulsifier of petrol. oils; textile scouring; stabilizer for syn. latexes; yel. liq.; water-sol.; sp.gr. 1.10; HLB 17.8; cloud pt. 74-76 C (1% in 10% NaCl); 70% act. in water.

Rexol 45/307. [Huntsman] Octoxynol-30; CAS 9002-93-1; nonionic; emulsifier for vinyl acetate and acrylate emulsion polymerization; detergent, dispersant; liq.; HLB 17.4; cloud pt. 73-77 C (1% in 10% NaCl); 70% act.

Rexol 45/407. [Huntsman] Octoxynol-40; CAS 9002-93-1; nonionic; emulsifier for vinyl acetate and acrylate emulsion polymerization; liq.; HLB 18.0; cloud pt. 73-77 C (1% in 10% NaCl); 70% act.

Rezol® 4393. [Witco/New Markets] Polyfunctional specialty polyether; used as semirigid foam crosslinkers, rigid foam additives, precursors for reactive diluents and chain extenders in urethane synthesis; APHA 80 color; m.w. 292; hyd. no. 770.

Rez-O-Sperse® 3. [Dover] Chlorinated paraffin (Paroil 170-HV) aq. emulsion; nonionic; flame retardant for plastics, rubbers, adhesives, paints, fabric coatings, inks, carpet backings, paper coatings; plasticizer, tackifier; cream wh. liq.; sp.gr. 1.540; dens. 11.3 lb/gal; visc. 150-300 poise (#6 spindle, 10 rpm); pH 7; 66.5% solids, 45% Cl.

Rez-O-Sperse® A-1. [Dover] Resinous chlorinated paraffin aq. disp.; nonionic; flame retardant for plastics, rubbers, adhesives, paints, fabric coatings, inks, carpet backings, paper coatings; used where increased hardness is desired; cream wh. liq.; sp.gr. 1.600; dens. 11 lb/gal; visc. 48 poise (#5 spindle, 20 rpm); pH 7.5; 65% solids, 45% Cl.

Rhenoblend® N 6011. [Rhein Chemie] NBR/PVC blend (60:40); used for rubber articles requiring high resist. to oil, weathering, and aging; increases impact resist. of PVC, makes PVC foils more flexible; lt. brn. translucent sheets; sp.gr. 1.3; Mooney visc. 55 (MS1+4, 100 C).

Rhenocure® ADT. [Rhein Chemie] Aminedialkyldithiophosphate coated with min. oil; accelerator for EPDM rubber, tech. molded and extruded goods; wh. cryst. powd.; sp.gr. 1.0.

Rhenocure® CA. [Rhein Chemie] N,N′-Diphenylthiourea; CAS 102-08-9; accelerator for tech. molded and extruded rubber goods, cable coverings, sheeting; fast vulcanization, high degree of crosslinking when using low sulfur or sulfur donors; wh. to ylsh. powd.; sp.gr. 1.3.

Rhenocure® CMT. [Rhein Chemie] Accelerator blend contg. dithiocarbamates, thiazoles, thiurams; accelerator for the cross-linking of EPDM, tech. rubber goods esp. molded articles (transfer and inj. molding); fast vulcanization using sulfur; scorch-resist.; ylsh. crumbling coated powd.; sp.gr. 1.4.

Rhenocure® CMU. [Rhein Chemie] Accelerator blend contg. dithioicarbamates, thiazoles, thiurams, and thioureas; accelerator for the cross-linking of EPDM, tech. rubber goods, extruded articles such as profiles, seals, sheeting; fast vulcanization when using sulfur; scorch-resist.; ylsh. crumbling coated powd.; sp.gr. 1.3.

Rhenocure® CUT. [Rhein Chemie] Copper-dialkyldithiophosphate, coated with min. oil; accelerator for EPDM and other diene rubbers, tech. molded and extruded goods, e.g., profiles, foils, buffers, dock fenders, etc.; fast vulcanization, high degree of crosslinking; scorch-resist.; sl. staining; yel.-grn. paste; sp.gr. 1.1.

Rhenocure® Diuron. [Rhein Chemie] 3-(3,4-Dichlorophenyl)-1,1-dimethylurea bound to ACM; cross-linking agent; for oil-resistant seals based on ACM for automotive applics.; wh.-ylsh. gran.; sp.gr. 1.4.

Rhenocure® EPC. [Rhein Chemie] Accelerator blend contg. dithioicarbamates, thiazoles, thiurams, and thioureas; accelerator for the crosslinking of EPDM, tech. rubber goods, extruded articles such as profiles, seals, sheeting; fast vulcanization when using sulfur; scorch-resist.; yel.-brn. elastomer-bound gran.; sp.gr. 1.3; 80% act.

Rhenocure® IS 60. [Rhein Chemie] 60% Insol. sulfur, 40% sol. sulfur; curing agent for NR and SR compds. where blooming of sulfur should be avoided; used in tires, conveyor belts, repair material, assembled goods; yel. powd.; sp.gr. 1.95.

Rhenocure® IS 60-5. [Rhein Chemie] 95% Rhenocure IS 60, 5% oil; curing agent for NR and SR compds. where blooming of sulfur should be avoided; used in tires, conveyor belts, repair material, assembled goods; yel. powd.; sp.gr. 1.8.

Rhenocure® IS 60/G. [Rhein Chemie] 80% Rhenocure IS 60, 20% elastomer binder; curing agent for NR and SR compds. where blooming of sulfur should be avoided; used in tires, conveyor belts, repair material, assembled goods; yel. gran.; sp.gr. 1.6.

Rhenocure® IS 90-20. [Rhein Chemie] 80% Sulfur, 20% oil; curing agent for NR and SR compds. where blooming of sulfur should be avoided; used in tires, conveyor belts, repair material, assembled goods; yel. powd.; sp.gr. 1.6.

Rhenocure® IS 90-33. [Rhein Chemie] 67% sulfur, 33% oil and inorg. dispersant; curing agent for NR and SR compds. where blooming of sulfur should be avoided; used in tires, conveyor belts, repair material, assembled goods; yel. powd.; sp.gr. 1.5; Unverified

Rhenocure® IS 90-40. [Rhein Chemie] 60% sulfur, 40% oil and inorg. dispersant; curing agent for NR and SR compds. where blooming of sulfur should be avoided; used in tires, conveyor belts, repair material, assembled goods; yel. powd.; sp.gr. 1.4; Unverified

Rhenocure® IS 90/G. [Rhein Chemie] 70% Sulfur, 30% elastomer binder; curing agent for NR and SR compds. where blooming of sulfur should be avoided; used in tires, conveyor belts, repair material, assembled goods; yel. gran.; sp.gr. 1.5.

Rhenocure® M. [Rhein Chemie] 4,4′-Dithiodimorpholine; CAS 103-34-4; vulcanizing agent for reversion resist. goods based on NR, e.g., tire carcasses, ageing resist. goods based on SBR, NBR, and EPDM, e.g., mech. rubber goods; wh. powd.; sp.gr. 1.3.

Rhenocure® M/G. [Rhein Chemie] 80% 4,4′-Dithiodimorpholine and 20% elastomer binder; CAS 103-34-4; vulcanizing agent for reversion resist. goods based on NR, e.g., tire carcasses, ageing resist. goods based on SBR, NBR, and EPDM, e.g., mech. rubber goods; gray elastomer-bound gran.; sp.gr. 1.25.

Rhenocure® S/G. [Rhein Chemie] 80% Dithiodicaprolactam and 20% elastomer binder; sulfur donor; vulcanizing agent for NR and SR, reversion resistant good, e.g., tire carcasses, with thiuram sec. accelerator for nitrile rubber seals; ageing resist. goods based on SBR, NBR, EPDM; wh.-ylsh. to gray elastomer-bound gran.; dens. 1.25 g/cc; 80% act.

Rhenocure® TDD. [Rhein Chemie] Thiadiazole deriv. bound to CM; vulcanizing agent for peroxide-free cross-linking of CM and other sat., halogen-contg. elastomers; pale gray gran.; sp.gr. 1.4.

Rhenocure® TP/G. [Rhein Chemie] 50% Zinc dialkyldithiophosphate with 50% elastomer binder; accelerator for EPDM and other diene rubbers, tech. molded and extruded goods, e.g., profiles, hoses, sheeting, coatings; scorch-resist.; beige gran.; sp.gr. 1.25.

Rhenocure® TP/S. [Rhein Chemie] Zinc-dialkyldithiophosphate bound to silica; accelerator for EPDM and other diene rubbers, tech. molded and extruded goods, e.g., profiles, hoses, sheeting, coatings; scorch-resist.; wh. crumbling powd.; sp.gr. 1.3.

Rhenocure® ZAT. [Rhein Chemie] Zinc-amine-dithiophosphate complex, coated with min. oil; accelerator for EPDM and other diene rubbers, tech. molded and extruded goods, e.g., hoses, dock fenders, sheeting; provides fast vulcanization, high degree of crosslinking; wh.-gray paste; sp.gr. 1.1.

Rhenodiv® 20. [Rhein Chemie] Silicone-free mold release agent for rubber; colorless to ylsh. clear liq.; sp.gr. 1.05; pH 7 (10%); 3-10% conc.

Rhenodiv® 30. [Rhein Chemie] Alkali salt of a fatty acid deriv.; mold release for molded goods, tires, ebonite, soling slabs, goods which are bonded together; suitable for subsequent lacquering, welding, or cementing together of the vulcanizates; whitish thixotropic disp.; dens. 0.9 g/cc; pH 7-9 (aq.); 0.5-1.5% conc. in water.

Rhenodiv® A. [Rhein Chemie] Release agent for rubber compounding, extruding; ylsh. liq.; sp.gr. 1.0; pH 9-10 (10%); 2-5% conc.

Rhenodiv® F. [Rhein Chemie] Surfactant-polyvalent alcohols aq. sol'n.; release agent for rubber compounding and extruding; ylsh.-brn. low-visc. liq.; sp.gr. 1.1; pH 10.5-11 (10%); 5-15% conc.

Rhenodiv® KS. [Rhein Chemie] Release agent for rubber compounding; yel. to grn. powd.; sp.gr. 1.3; pH 11 (1%); 3-5% conc.

Rhenodiv® LE. [Rhein Chemie] Fatty acid salts aq. paste; release agent for rubber compounding and extruding; ylsh.-wh. paste; pH 10-11 (10%); 5-20% conc.

Rhenodiv® LL. [Rhein Chemie] Fatty acid salt aq. sol'n.; release agent for rubber compounding and extruding; ylsh.-wh. visc. liq.; sp.gr. 1.0; pH 9.5-10 (10%); 5-20% conc.

Rhenodiv® LS. [Rhein Chemie] Fatty acid salts with film-formers bound to silica; release agent for rubber compounding; wh. to yel. powd.; sp.gr. 1.1; pH 9 (1%); 5-20% conc.

Rhenodiv® PV. [Rhein Chemie] Release agent for rubber compounding, esp. for EPDM; yel. to grn. powd.; sp.gr. 1.35; pH 11 (1%); 3-5% conc.

Rhenodiv® S. [Rhein Chemie] Release agent for rubber compounding, extruding; whitish paste; pH 9.5-10.5 (10%); 3-5% conc.

Rhenodiv® ZB. [Rhein Chemie] Zinc stearate aq. disp.; CAS 557-05-1; EINECS 209-151-9; release agent for rubber compounding; wh. low-visc. stable suspension; 3-5% conc.

Rhenofit® 1600. [Rhein Chemie] Urea derivs./dispersants in aq. paste form; blow promoter for cellular and microcellular rubber compds.; increases plasticity of uncured compd., accelerates vulcanization, delays scorch; lt.-colored paste; sp.gr. 1.25.

Rhenofit® 1987. [Rhein Chemie] Urea/surfactant blend bound to silica; filler activator for rubbers containing lt.-colored reinforcing fillers, e.g., molded soles, rubber mats, floor coverings, household and bathing goods, cellular and microcellular rubber compds.; wh. fine crumbling powd.; sp.gr. 1.5.

Rhenofit® 2009. [Rhein Chemie] Urea deriv./surfactant blend bound to silica; filler activator for rubbers containing lt.-colored reinforcing fillers, e.g., molded soles, rubber mats, floor coverings, household and bathing goods, cellular and microcellular rubber compds.; wh. fine crumbling powd.; sp.gr. 1.8.

Rhenofit® 2642. [Rhein Chemie] Urea deriv. bound to silica; filler activator for rubbers containing lt.-colored reinforcing fillers, e.g., molded soles, rubber mats, floor coverings, household and bathing goods, cellular and microcellular rubber compds.; wh. fine crumbling powd., hygroscopic; sp.gr. 1.5.

Rhenofit® 3555. [Rhein Chemie] Amine derivs. bound to silica; filler activator for rubbers requiring rapid vulcanization and fast-curing molded goods, e.g., soles, heels, tiles, floor coverings; wh. powd.; sp.gr. 1.4.

Rhenofit® B. [Rhein Chemie] Sec. amine; filler activator for colored articles based on NR and syn. rubbers, e.g., soles, heels, bicycle tires, cable coverings, mech. rubber goods; ylsh. to grn. liq.; sp.gr. 0.9.

Rhenofit® BDMA/S. [Rhein Chemie] 70% 1,4-Butanediol dimethacrylate, 30% silica; cross-linking activator for peroxide vulcanization of tech. molded and extruded goods based on EPDM, EPM, NBR, CM, etc.; for seals, sleeves, cables, profiles, etc.; wh. nondusting powd.; sp.gr. 1.2.

Rhenofit® CF. [Rhein Chemie] Treated very finely divided calcium hydroxide; CAS 1305-62-0; EINECS 215-137-3; cross-linking activator for fluoroelastomers, tech. rubber goods; wh. powd.; sp.gr. 2.2.

Rhenofit® EDMA/S. [Rhein Chemie] 70% Ethyleneglycol dimethacrylate, 30% silica; cross-linking activator for peroxide vulcanization of tech. molded and extruded goods based on EPDM, EPM, NBR, CM, etc.; for seals, sleeves, cables, profiles, etc.; wh. nondusting powd.; sp.gr. 1.25.

Rhenofit® NC. [Rhein Chemie] Fatty acid amide-amine; activating accelerator for the cross-linking of CM; ylsh. pellets; sp.gr. .095.

Rhenofit® TAC/S. [Rhein Chemie] 70% Triallylcyanurate, 30% silica; cross-linking activator for peroxide vulcanization of tech. molded and extruded goods based on EPDM, EPM, NBR, CM, etc.; for seals, sleeves, cables, profiles, etc.; wh. fine crumbling powd.; sp.gr. 1.25.

Rhenofit® TRIM/S. [Rhein Chemie] 70% Trimethylolpropane trimethacrylate, 30% silica; cross-linking activator for peroxide vulcanization of tech. molded and extruded goods based on EPDM, EPM, NBR, CM, etc.; for seals, sleeves, cables, profiles, etc.; wh. nondusting powd.; sp.gr. 1.25.

Rhenofit® UE. [Rhein Chemie] Activators bound to silica; activating accelerator for the cross-linking of CO and ECO, tech. rubber goods requiring high ageing resist., oil resist. and cold flexibility; wh., slightly hygroscopic powd.; sp.gr. 1.5.

Rhenogran® BPH-80. [Rhein Chemie] 2,2′-Methylenebis-(4-methyl-6-t-butylphenol) encapsulated in EPDM/EVA elastomer; predispersed antioxidant for rubber; gray-wh. gran.; dens. 1.04 g/cc; 80% act.

Rhenogran® CaO-50/FPM. [Rhein Chemie] Fine particle sized calcium oxide encapsulated in FPM elastomer; predispersed drying agent for rubber; gray-brn. slabs; dens. 2.15 g/cc; 50% act.

Rhenogran® $Ca(OH)_2$-50/FPM. [Rhein Chemie] Fine particle sized calcium hydroxide encapsulated in FPM elastomer; predispersed stabilizer/activator for fluorocarbon rubbers; gray-brn. slabs; dens.

1.95 g/cc; 50% act.

Rhenogran® CHM21-40/FPM. [Rhein Chemie] Fine particle sized calcium hydroxide and magnesium oxide (2:1) encapsulated in FPM elastomer; predispersed activator for fluorocarbons; brnsh. slabs; dens. 1.85 g/cc; 40% act.

Rhenogran® DCBS-80. [Rhein Chemie] N,N′-Dicyclohexyl-2-benzothiazyl sulfenamide encapsulated in EPDM/EVA elastomer; predispersed accelerator for rubber; beige gran.; dens. 1.15 g/cc; 80% act.

Rhenogran® Fe-gelb-50. [Rhein Chemie] Hydrated iron (III) oxide encapsulated in EPDM/EVA elastomer; predispersed pigments for rubber; yel.-brn. gran.; dens. 2.0 g/cc; 50% act.

Rhenogran® Fe-rot-70. [Rhein Chemie] Iron (III) oxide encapsulated in EPDM/EVA elastomer; predispersed pigments for rubber; dk. red-brn. gran.; dens. 2.1 g/cc; 70% act.

Rhenogran® IPPD-80. [Rhein Chemie] N-Isopropyl-N′-phenyl-p-phenylene diamine encapsulated in EPDM/EVA elastomer; predispersed antioxidant for rubber; gray-blk. gran.; dens. 1.10 g/cc; 80% act.

Rhenogran® MBI-80. [Rhein Chemie] 2-Mercaptobenzimidazole encapsulated in EPDM/EVA elastomer; predispersed antioxidant for rubber; wh.-gray gran.; dens. 1.28 g/cc; 80% act.

Rhenogran® MgO-40/FPM. [Rhein Chemie] Magnesium oxide encapsulated in FPM elastomer; predispersed stabilizer/activator for chlorinated rubber; brn.-gray slabs; dens. 1.95 g/cc; 40% act.

Rhenogran® MMBI-70. [Rhein Chemie] Methylmercaptobenzimidazole encapsulated in EPDM/EVA elastomer; predispersed antioxidant for rubber; beige gran.; dens. 1.16 g/cc; 70% act.

Rhenogran® MPTD-80. [Rhein Chemie] Dimethyl diphenyl thiuram disulfide encapsulated in EPDM/EVA elastomer; predispersed accelerator for rubber; beige gran.; dens. 1.2 g/cc; 80% act.

Rhenogran® MTT-80. [Rhein Chemie] 3-Methyl thiazolidine-thione-2 encapsulated in EPDM/EVA elastomer; predispersed accelerator for rubber; gray-beige gran.; dens. 1.25 g/cc; 80% act.

Rhenogran® NDBC-70/ECO. [Rhein Chemie] Nickel dibutyldithiocarbamate encapsulated in ECO elastomer; predispersed antioxidant for rubber; dk. grn. gran.; dens. 1.21 g/cc; 70% act.

Rhenogran® NKF-80. [Rhein Chemie] Sterically hindered bisphenol encapsulated in EPDM/EVA elastomer; predispersed antioxidant for rubber; ylsh. gran.; dens. 1.07 g/cc; 80% act.

Rhenogran® OTBG-50. [Rhein Chemie] o-Tolylbiguanide encapsulated in EPDM/EVA elastomer; predispersed accelerator for rubber; wh.-gray gran.; dens. 1.25 g/cc; 50% act.

Rhenogran® OTBG-75. [Rhein Chemie] o-Tolylbiguanide encapsulated in EPDM/EVA elastomer; predispersed accelerator for rubber; wh.-gray gran.; dens. 1.14 g/cc; 75% act.

Rhenogran® P-50/EVA. [Rhein Chemie] Polycarbodiimide encapsulated in EVA elastomer; predispersed antidegradant for rubber; brn.-beige gran.; dens. 1.01 g/cc; 50% act.

Rhenogran® PBN-80. [Rhein Chemie] Phenyl-β-naphthylamine encapsulated in EPDM/EVA elastomer; predispersed antioxidant for rubber; lt. gray-pink gran.; dens. 1.15 g/cc; 80% act.

Rhenogran® PbO-80/FPM. [Rhein Chemie] Lead (II) oxide encapsulated in FPM elastomer; predispersed stabilizer, vulcanization aid for rubber; yel.-orange slabs; dens. 4.9 g/cc; 80% act.

Rhenogran® Resorcin-80. [Rhein Chemie] Resorcinol encapsulated in EPDM/EVA elastomer; predispersed adhesive aid for rubber-to-textiles; whitish gran.; dens. 1.19 g/cc; 80% act.

Rhenogran® Resorcin-80/SBR. [Rhein Chemie] Resorcinol encapsulated in SBR elastomer; predispersed adhesive aid for rubber-to-textiles; brnsh. gran.; dens. 1.21 g/cc; 80% act.

Rhenogran® Vulk. E-80. [Rhein Chemie] Sulfonamide deriv. encapsulated in EPDM/EVA elastomer; predispersed retarder for sulfur cures of rubber; ochre gran.; dens. 1.39 g/cc; 80% act.

Rhenomag® G1. [Rhein Chemie] Magnesium oxide in polymeric binder; CAS 1309-48-4; EINECS 215-171-9; acid acceptor and vulcanization activator for rubber goods; tan pellets; sp.gr. 1.65; 60% act.; *Storage:* stable when stored under normal storage conditions (< 35 C and 50% r.h.); use within 1 yr.

Rhenomag® G3. [Rhein Chemie] Magnesium oxide in polymer-bound gran. form; CAS 1309-48-4; EINECS 215-171-9; acid acceptor and vulcanization activator for rubber goods, tech. molded and extruded articles, adhesives based on CR, erasers, chlorinated paraffins; wh.-gray gran.; sp.gr. 2.25.

Rhenopor 1843. [Rhein Chemie] Sodium hydrogen carbonate with dispersants (50:50); inorg. blowing agent for cellular and microcellular rubber articles; wh. powd.; sp.gr. 2.0.

Rhenosin® 140. [Rhein Chemie] Thermoplastic copolymeric resin based on selected hydrocarbon fractions; softening agent and homogenizer for dark colored NR and SR rubbers, blends, tires, conveyor belts; dk. brn.-blk. pellets; sp.gr. 1.1; soften. pt. 85 C.

Rhenosin® 143. [Rhein Chemie] Thermoplastic copolymeric resin based on selected hydrocarbon fractions; homogenizer for dk.-colored NR, SR, and blends, tires, conveyor belts; dk. brn.-blk. pellets; sp.gr. 1.1; soften. pt. 90 C.

Rhenosin® 145. [Rhein Chemie] Thermoplastic copolymeric resin based on selected hydrocarbon fractions; homogenizer for dk.-colored NR, SR, and blends, tires, conveyor belts; dk. brn.-blk. pellets; sp.gr. 1.1; soften. pt. 95 C.

Rhenosin® 260. [Rhein Chemie] Thermoplastic aromatic hydrocarbon resin; softening agent and homogenizer for lt. colored NR and SR and blends, tech. molded and extruded goods; lt. brn. pellets; sp.gr. 1.1; soften. pt. 85 C.

Rhenosol®. [Rhein Chemie] Chemicals encapsulated by selected fatty acid derivs.; predispersed rubber chems.

Rhenosorb® C. [Rhein Chemie] Coated calcium oxide; CAS 1305-78-8; EINECS 215-138-9; desiccant for extruded rubber goods (profiles, seals), V-belts, straps, conveyor belts, roll coverings, cable jackets, rubber floorings; gray dusting powd.; sp.gr. 2.8; 90% act.

Rhenosorb® C/GW. [Rhein Chemie] Calcium oxide in elastomer carrier; CAS 1305-78-8; EINECS 215-138-9; desiccant for extruded rubber goods (profiles, seals), V-belts, straps, conveyor belts, roll coverings, cable jackets, rubber floorings; pale gray elastomer-bound gran.; sp.gr. 2.25; 80% act.

Rhenosorb® F. [Rhein Chemie] Finely divided calcium oxide, treated; CAS 1305-78-8; EINECS 215-138-9; desiccant for seals, molded goods based on fluoro rubber; wh. powd.; sp.gr. 3.0; 96% act.

Rhenovin® CBS-70. [Rhein Chemie] 70% N-cyclohexyl-2-benzothiazyl sulfenamide, 30% silica and dispersants; vulcanization accelerator; gray powd.; sp.gr. 1.1; Unverified

Rhenovin® DDA-70. [Rhein Chemie] 70% Diphenylamine deriv., 30% silica filler; antioxidant; brn. powd.; sp.gr. 1.3.

Rhenovin® FH-70. [Rhein Chemie] 70% Aromatic polyether (Vulkanol FH), 30% silica filler; syn. plasticizer; wh. powd.; sp.gr. 1.25.

Rhenovin® MBT-70. [Rhein Chemie] 70% 2-Mercaptobenzothiazole, 30% silica and dispersants; vulcanization accelerator; beige powd.; sp.gr. 1.25; Unverified

Rhenovin® MBTS-70. [Rhein Chemie] 70% Dibenzothiazyl disulfide, 30% silica and dispersants; vulcanization accelerator; yel.-beige powd.; sp.gr. 1.25; Unverified

Rhenovin® Na-stearat-80. [Rhein Chemie] 80% Sodium stearate, 20% inorg. dispersants; crosslinking activator for ACM; wh. powd.; sp.gr. 1.2.

Rhenovin® S-90. [Rhein Chemie] 90% Sulfur, 10% dispersants; vulcanizing agent; yel. powd.; sp.gr. 1.9; Unverified

Rhenovin® S-stearat-80. [Rhein Chemie] 80% Potassium stearate, 20% inorg. dispersants; crosslinking activator for ACM; wh. powd.; sp.gr. 1.1.

Rhenovin® TMTD-70. [Rhein Chemie] 70% Tetramethylthiuram disulfide, 30% silica and dispersants; vulcanization accelerator; whitish powd.; sp.gr. 1.1; Unverified

Rhenovin® TMTM-70. [Rhein Chemie] 70% Tetramethylthiuram monosulfide, 30% silica and dispersants; vulcanization accelerator; yel. powd.; sp.gr. 1.2; Unverified

Rhenovin® ZnO-90. [Rhein Chemie] 90% Zinc oxide, 10% dispersants; vulcanization accelerator; yel.-wh. powd.; sp.gr. 4.0; Unverified

Rhodacal® 301-10. [Rhone-Poulenc Surf. & Spec.] Sodium C14-16 olefin sulfonate; CAS 68439-57-6; EINECS 270-407-8; anionic; emulsion polymerization surfactant; liq.; surf. tens. 29 dynes/cm (@ CMC); 40% solids.

Rhodacal® 301-10P. [Rhone-Poulenc Surf. & Spec.] Sodium C14-16 olefin sulfonate; CAS 68439-57-6; EINECS 270-407-8; anionic; emulsion polymerization surfactant.

Rhodacal® 330. [Rhone-Poulenc Surf. & Spec.] Isopropylamine dodecylbenzene sulfonate; CAS 26264-05-1; EINECS 247-556-2; anionic; emulsifier, wetting agent, grease/pigment dispersant, lubricant, solubilizer, solv., penetrant, high foaming base for shampoos and cleaners, metalworking fluids, agric. pesticides, emulsion polymerization, drycleaning, latex paints, metal cleaning; amber liq.; oil-sol.; sp.gr. 1.03; dens. 8.497 lb/gal; visc. 6500 cps; HLB 11.7; flash pt. > 200 C; pH 3-6 (5%); 90% act.

Rhodacal® A-246L. [Rhone-Poulenc Surf. & Spec.] Sodium C14-16 olefin sulfonate; CAS 68439-57-6; EINECS 270-407-8; anionic; detergent, foaming agent, emulsifier, wetting agent, dispersant for hair shampoos, liq. soaps, skin cleansers, industrial cleaners, emulsion polymerization; liq.; surf. tens. 29 dynes/cm (@ CMC); 40% act.

Rhodacal® BA-77. [Rhone-Poulenc Surf. & Spec.] Sodium isopropylnaphthalene sulfonate; CAS 1322-93-6; anionic; emulsifier, wetting agent, dispersant without detergent props.; for industrial cleaners, leather dyeing, textile processing, paints, inks, pesticides, syn. latex emulsions, plastics; latex stabilizer; prevents coagulation of syn. rubbers; FDA compliance; tan powd.; m.w 314; pH 8-10 (5%); 75% act.

Rhodacal® BX-78. [Rhone-Poulenc Surf. & Spec.] Sodium dibutylnaphthalene sulfonate; CAS 25417-20-3; anionic; emulsifier, wetting agent, penetrant, dispersant for industrial cleaning, textiles, dyeing, leather, insecticides, herbicides, paper, dyes/pigments, wallpaper pastes, rubber, latex polymerization; cream to tan powd.; m.w. 326; water-sol.; pH 6-8 (5%); surf. tens. 36 dynes/cm (@ CMC); 75% solids.

Rhodacal® DOV. [Rhone-Poulenc Surf. & Spec.] TEA dodecylbenzene sulfonate; anionic; emulsion polymerization surfactant; liq.; 40% solids.

Rhodacal® DS-10. [Rhone-Poulenc Surf. & Spec.] Sodium dodecylbenzene sulfonate; CAS 25155-30-0; anionic; emulsifier for emulsion polymerization; emulsifier, dispersant for agric. formulations; FDA, EPA compliance; flake; surf. tens. 32 dynes/cm (@ CMC); 98% conc.

Rhodacal® DSB. [Rhone-Poulenc Surf. & Spec.] Sodium dodecyl diphenyloxide disulfonate; CAS 28519-02-0; anionic; detergent, foamer, emulsifier, textile dye leveling agent, coupling agent, solubilizer; dispersant in metal and other industrial cleaners; coemulsifier in emulsion polymerization; wetting agent, dispersant for agric. formulations; alkaline and chlorine stable; FDA, EPA compliance; liq.; surf. tens. 32 dynes/cm (@ CMC); 45% solids.

Rhodacal® LDS-22. [Rhone-Poulenc Surf. & Spec.] Linear sodium dodecylbenzene sulfonate; CAS 25155-30-0; anionic; emulsion polymerization surfactant; FDA compliance; liq.; surf. tens. 32 dynes/cm (@ CMC); 23% solids.

Rhodafac® BX-660. [Rhone-Poulenc Surf. & Spec.] Aromatic phosphate ester; anionic; emulsifier, stabilizer for emulsion polymerization; emulsifier, wetting agent, hydrotrope for use in alkaline cleaning sol'ns.; liq.; 80% act.

Rhodafac® BX-760. [Rhone-Poulenc Surf. & Spec.] Phosphate ester; anionic; hydrotrope for use in alkaline cleaning sol'ns.; emulsifier and stabilizer for emulsion polymerization; solubilizer for nonionic surfactants in highly alkaline aq. sol'ns.; liq.; 90% act.

Rhodafac® MC-470. [Rhone-Poulenc Surf. & Spec.] Sodium laureth-4 phosphate; CAS 42612-52-2; anionic; detergent, emulsifier, visc. builder for creams and lotions, facial cleansers, polymerization and stabilization of latexes; fatliquoring of leathers; metalworking fluids; textile antistat, lubricant, softener; emulsifier for min. oils; clear visc. liq.; sol. in min. oil, kerosene, xylene, perchloroethylene, disp. in water; sp.gr. 1.02-1.04; dens. 8.580 lb/gal; pour pt. 0 C; flash pt. > 200 C; pH 5.0-6.5 (10%); 95% act.

Rhodafac® PE-510. [Rhone-Poulenc Surf. & Spec.] Nonoxynol-6 phosphate; CAS 68412-53-3; anionic; detergent for drycleaning, waterless hand cleaners; intermediate for textile lubricants; emulsifier for emulsion polymerization (PVAc and acrylic films); emulsifier, dispersant for agric. formulations; corrosion inhibitor; good electrolyte tolerance; FDA, EPA compliance; lt. colored clear to sl. hazy visc. liq.; sol. in kerosene, Stod., xylene, butyl Cellosolve, perchloroethylene, ethanol; disp. in water; sp.gr. 1.08-1.09; dens. 9.0 lb/gal; pour pt. 20 C; pH < 2.5 (10%); 100% act.

Rhodafac® PS-17. [Rhone-Poulenc Surf. & Spec.] Complex org. phosphate ester, potassium salt; anionic; detergent, wetting agent, dispersant, foamer, foam stabilizer for heavy-duty liq. detergents, paraffin emulsions, emulsion polymerization, textile wet processing; stable to electrolytes; limited acid compat.; visc. liq.; sol. in water, ethanol, some aromatic and aliphatic solvs. and trichloroethylene; 80% conc.

Rhodafac® PS-19. [Rhone-Poulenc Surf. & Spec.] Complex org. phosphate ester, potassium salt; anionic; detergent, wetting agent, dispersant, foamer, foam stabilizer for emulsion polymerization, textile wet processing; alkaline compat.; limited acid compat.; visc. liq.; 80% conc.

Rhodafac® RE-610. [Rhone-Poulenc Surf. & Spec.] Nonoxynol-9 phosphate; CAS 68412-53-3; anionic; detergent, emulsifier, wetting agent, dispersant, antistat, lubricant, dedusting agent for drycleaning, pesticides, emulsion polymerization, textile wet processing, metals, household and industrial detergents; FDA, EPA compliance; slightly hazy visc. liq.; sol. in xylene, butyl Cellosolve, perchloroethylene, ethanol, water; dens. 9.2 lb/gal; sp.gr. 1.1-1.12; pour pt. < 0 C; surf. tens. 34 dynes/cm (@ CMC); 100% act.

Rhodafac® RE-877. [Rhone-Poulenc Surf. & Spec.] Aromatic phosphate ester; anionic; emulsifier, stabilizer, solubilizer, detergent, dispersant, wetting agent, antistat, lubricant for pesticides, textile wet processing, fiber and metal lubricants, emulsion polymerization, cosmetics; clear visc. liq.; sol. in xylene, butyl Cellosolve, ethanol, water; sol. hazy in perchloroethylene; sp.gr. 1.155; dens. 9.6 lb/gal; pour pt. 2 C; acid no. 60-74; pH < 2.5 (10%); 75% act.

Rhodafac® RE-960. [Rhone-Poulenc Surf. & Spec.] Aromatic phosphate ester of ethoxylated nonylphenol, free acid; CAS 51811-79-1; anionic; emulsifier, wetting agent, dispersant for industrial cleaners, emulsion polymerization, pesticides, drycleaning, textile wet processing, metals; promotes freeze/thaw and mech. stability in PVAc and acrylic latexes; FDA compliance; soft waxy paste; sol. in butyl Cellosolve, ethanol, water; dens. 9.8 lb/gal; sp.gr. 1.17-1.18 (50 C); pour pt. 20 C; pH < 2.5 (10%); surf. tens. 43 dynes/cm (@ CMC); 90% act.

Rhodafac® RS-410. [Rhone-Poulenc Surf. & Spec.] Trideceth-3 phosphate; CAS 9046-01-9; anionic; emulsifier, detergent, wetting agent, dispersant for industrial, institutional and household cleaners, pesticides, drycleaning, textiles, metal treatment; antistat for plastics; FDA, EPA compliance; hazy visc. liq.; sol. in min. oil, kerosene, Stod., xylene, butyl Cellosolve, perchloroethylene, ethanol, disp. in water; sp.gr. 1.03-1.04; dens. 8.6 lb/gal; pour pt. -15 C; pH < 2.5 (10%); 100% act.

Rhodafac® RS-610. [Rhone-Poulenc Surf. & Spec.] Trideceth-6 phosphate; CAS 9046-01-9; anionic; emulsifier for emulsion polymerization, waterless hand cleaners, pesticides; detergent for drycleaning formulations, textile wetting agent, lubricant for fiber and metal treatment, paper-mill felt washing; antistat for aerosols, plastics; FDA, EPA compliance; hazy, visc. liq.; sol. in kerosene, Stod., xylene, butyl Cellosolve, perchloroethylene, ethanol; disp. in water; sp.gr. 1.04-1.06; dens. 8.7 lb/gal; pour pt. < 0 C; pH < 2.5 (10%); 100% act.

Rhodafac® RS-710. [Rhone-Poulenc Surf. & Spec.] Trideceth-10 phosphate; CAS 9046-01-9; anionic; detergent, wetting agent, emulsifier, dispersant for textile wet processing, industrial cleaners, pesticides, drycleaning, metal treatment; antistat for plastics; FDA, EPA compliance; opaque visc. liq.; sol. in Stod., xylene, butyl Cellosolve, perchloroethylene, ethanol, water; sp.gr. 1.04-1.06; dens. 8.7 lb/gal; pour pt. 18 C; pH < 2.5 (10%); 100% act.

Rhodamox® CAPO. [Rhone-Poulenc Surf. & Spec.] Cocamidopropyl dimethylamine oxide; CAS 68155-09-9; EINECS 268-938-5; nonionic/cationic; foaming agent/stabilizer, thickener, emollient for shampoos, bath prods., dishwash, rug shampoos, fine fabric detergents, shaving creams, lotions, foam rubber, electroplating, paper coatings; used in toiletries for mildness; colorless clear liq.; faint odor; water sol.; disp. in min. oil; dens. 1.0 g/ml; f.p. < 0 C; 30% conc.

Rhodamox® LO. [Rhone-Poulenc Surf. & Spec.] Lauryl dimethylamine oxide; CAS 1643-20-5; EINECS 216-700-6; cationic; foaming agent/stabilizer, thickener, emollient for shampoos, bath prods., dishwash, fine fabric detergents, shaving creams, lotions, textile softeners, foam rubber, in electroplating, paper coatings; used in toiletries for mildness; colorless clear liq.; faint odor; water sol.; disp. in min. oil; dens. 1.0 g/ml; f.p. < 0 C; 30% conc.

Rhodapex® AB-20. [Rhone-Poulenc Surf. & Spec.] Ammonium laureth sulfate; anionic; emulsifier for emulsion polymerization and min. oils; detergent; liq.; 30% conc.

Rhodapex® CO-433. [Rhone-Poulenc Surf. & Spec.] Sodium nonoxynol-4 sulfate; CAS 68891-39-4; anionic; high foaming detergent, wetting agent, dispersant for dishwashing, scrub soaps, car washes, rug and hair shampoos; emulsifier for emulsion polymerization, petrol. waxes; antistat for plastics and syn. fibers; lime soap dispersant; FDA compliance; Varnish 4 max. clear liq.; mild aromatic odor; water-sol.; sp.gr. 1.065; dens. 8.9 lb/gal; visc. 2500 cps; surf. tens. 32 dynes/cm (1%); 28% act.

Rhodapex® CO-436. [Rhone-Poulenc Surf. & Spec.] Ammonium nonoxynol-4 sulfate; CAS 9051-57-4; anionic; high foaming detergent, wetting agent, dispersant for dishwashing, scrub soaps, car washes, rug and hair shampoos; emulsifier for emulsion polymerization, petrol. waxes; antistat for plastics and syn. fibers; lime soap dispersant; FDA, EPA compliance; Varnish 5 max clear liq.; alcoholic odor; water-sol.; sp.gr. 1.065; dens. 8.9 lb/gal; visc. 100 cps; surf. tens. 34 dynes/cm (1%); 58% act.

Rhodapex® ES. [Rhone-Poulenc Surf. & Spec.] Sodium laureth (3) sulfate; CAS 9004-82-4; anionic; emulsifier, high foaming base for shampoos, lt. duty detergents, bubble baths; polymerization surfactant; liq.; 27% conc.

Rhodapex® EST-30. [Rhone-Poulenc Surf. & Spec.] Sodium trideth sulfate; CAS 25446-78-0; anionic; emulsifier, wetting agent, dispersant for baby shampoo, other personal care prods., household, industrial, institutional and industrial formulations, emulsion polymerization of styrene systems, textile scouring, dishwash; FDA compliance; Gardner 2 max. liq.; cloud pt. 14 C max.; pH 7.5-8.5 (10%); surf. tens. 33 dynes/cm (@ CMC); 29-30% act.

Rhodapon® CAV. [Rhone-Poulenc Surf. & Spec.] Sodium isodecyl sulfate; CAS 68299-17-2; anionic; wetting agent, emulsifier, detergent, foamer, rinse aid, visc. control agent; post-stabilizer in latex paints; metal treatment; textile and plywood mfg.; fruit/veg. washing; hard surface cleaners; emulsion polymerization of vinyl, acrylic, SBR, PVC; clear liq.; dens. 8.95 lb/gal; visc. 150 cps; cloud pt. < 10 C: pH 9.5-10.5 (10%); surf. tens. 34 dynes/cm (@ CMC); 39-40% act.

Rhodapon® L-22. [Rhone-Poulenc Surf. & Spec.] Ammonium lauryl sulfate; CAS 2235-54-3; EINECS 218-793-9; anionic; high foaming detergent, emulsifier for shampoo, bubble bath, pet shampoos, industrial and institutional cleaners, wool scouring, fire fighting foams, assistant for pigment dispersion; emulsion polymerization aid; clear visc. liq.; visc. 1000 cps; HLB 31.0; cloud pt. 14 C; pH 6.8 (10%); 28% act.

Rhodapon® L-22/C. [Rhone-Poulenc Surf. & Spec.] Ammonium lauryl sulfate; CAS 2235-54-3; EINECS 218-793-9; anionic; high foaming detergent, emulsifier for shampoo, bubble bath, pet shampoos, industrial and institutional cleaners, wool scouring, fire fighting foams, assistant for pigment dispersion; emulsion polymerization aid; clear visc. liq.; visc. 1000 cps; HLB 31.0; cloud pt. 14 C; pH 6.8 (10%); 28% act.

Rhodapon® L-22HNC. [Rhone-Poulenc Surf. & Spec.] Ammonium lauryl sulfate; CAS 2235-54-3; EINECS 218-793-9; surfactant for personal care prods. and SBR latex froth applics.; liq.; 30% act.

Rhodapon® LCP. [Rhone-Poulenc Surf. & Spec.] Sodium lauryl sulfate; CAS 151-21-3; anionic; low cloud pt. emulsion polymerization surfactant; FDA compliance; liq.; surf. tens. 30 dynes/cm (@ CMC); 30% solids.

Rhodapon® OLS. [Rhone-Poulenc Surf. & Spec.] Sodium octyl sulfate; CAS 142-31-4; anionic; wetting agent; rinse aid for industrial, institutional and household cleaners; mercerizing agent for cotton goods; surfactant in electrolyte baths for metal cleaning; hard surface cleaning; neoprene dispersant; emulsifier for emulsion polymerization; clear liq.; visc. 100 cps; pH 8 (10%); surf. tens. 33 dynes/cm (@ CMC); 33% act.

Rhodapon® OS. [Rhone-Poulenc Surf. & Spec.] Sodium oleyl sulfate; CAS 1847-55-8; EINECS 217-430-1; anionic; specialty emulsifier for emulsion polymerization; FDA compliance; liq.; surf. tens. 34 dynes/cm (@ CMC); 26% solids.

Rhodapon® SB. [Rhone-Poulenc Surf. & Spec.] Sodium lauryl sulfate; CAS 151-21-3; anionic; emulsifier for emulsion polymerization; liq.; HLB 40.0; surf. tens. 31 dynes/cm (@ CMC); 29% conc.

Rhodapon® TDS. [Rhone-Poulenc Surf. & Spec.] Sodium tridecyl sulfate; CAS 3026-63-9; EINECS 221-188-2; anionic; emulsifier, wetting agent, emulsion polymerization of vinyl chloride, styrene, and styrene/acrylic monomers; detergent formulations; base for shampoos, foaming bath oils, cosmetic emulsions; FDA compliance; colorless liq.; visc. 100 cps; cloud pt. 10 C; pH 8.5 (10%); surf. tens. 34 dynes/cm (@ CMC); 25% act.

Rhodapon® UB. [Rhone-Poulenc Surf. & Spec.] Sodium lauryl sulfate; CAS 151-21-3; anionic; emulsifier for sensitive emulsion polymerizations; FDA compliance; liq.; HLB 40.0; surf. tens. 29 dynes/cm (@ CMC); 30% solids.

Rhodasurf® A 24. [Rhone-Poulenc Surf. & Spec.] Laureth-3; CAS 9002-92-0; nonionic; wetting agent, emulsifier, dispersant, detergent, coupling agent, solubilizer; emulsion stabilizer for cosmetics and hair care systems; also in coning oils, textile spin finishes; with anionics in emulsion polymerization; solid; HLB 8.0; cloud pt. > 20 C (1% aq.); 100% conc.

Rhodasurf® BC-737. [Rhone-Poulenc Surf. & Spec.] Trideceth-14; CAS 78330-21-9; nonionic; surfactant for high temp. processes; stabilizer/emulsifier for syn. latexes; also for textile wetting, scouring, degreasing; solubilizer, emulsifier for essential oils, solvs., fats, waxes; HLB 15.1; cloud pt. < 100 C (1% aq.).

Rhodasurf® BC-840. [Rhone-Poulenc Surf. & Spec.] Trideceth-15; CAS 78330-21-9; nonionic; foam builder, detergent; solubilizer for alkylaryl sulfonates; for lt. and heavy duty high foaming detergent, agric. formulations; stabilizer/emulsifier for syn. latexes; for high temp. processes; EPA compliance; paste; sp.gr. 1.01-1.04 (40 C); HLB 15.4; pour pt. 33-40 C; cloud pt. < 100 C (1% aq.); flash pt. (PMCC) > 200 F; pH 6-8 (10%); 100% act.

Rhodasurf® E 400. [Rhone-Poulenc Surf. & Spec.] PEG-8; CAS 25322-68-3; EINECS 225-856-4; nonionic; surfactant intermediate; binder and lubricant in compressed tablets; softener for paper, plasticizer for starch pastes and polyethylene films; coupling agent for skin care lotions; liq.; pour pt. 6 C; 100% act.

Rhodasurf® E 600. [Rhone-Poulenc Surf. & Spec.] PEG-12; CAS 25322-68-3; nonionic; surfactant intermediate; binder and lubricant in compressed tablets; color stabilizer for fuel oils; softener for paper; plasticizer for starch paste and polyethylene film; coupling agent in skin care lotions; liq.; pour pt. 22 C; 100% act.

Rhodasurf® L-4. [Rhone-Poulenc Surf. & Spec.] Laureth-4; CAS 68002-97-1; EINECS 226-097-1; nonionic; emulsifier, thickener, wetting agent, pigment dispersant, lubricant, solubilizer for cosmetic and industrial emulsions; textile scouring agent, emulsion polymerization, metal cleaning, monomer systems, floor waxes, paper finishes, rubber; emollient for pharmaceuticals; liq.; HLB 9.7; cloud pt. 0-5; pH 6.5 (1%); 100% conc.

Rhodasurf® L-25. [Rhone-Poulenc Surf. & Spec.] Laureth-23; CAS 9002-92-0; nonionic; surfactant for cosmetic and industrial emulsions; wetting agent, pigment dispersant, lubricant, solubilizer, textile scouring agent, emulsion polymerization, metal cleaning, rubber and monomer systems; stabilizer for emulsion polymers; floor waxes, paper finishes, emollient for pharmaceuticals; wax; HLB 16.9; cloud pt. > 95 C (1%); pH 6.5 (1%); 100% act.

Rhodasurf® L-790. [Rhone-Poulenc Surf. & Spec.] Laureth-7; CAS 9002-92-0; EINECS 221-283-9; nonionic; surfactant for cosmetic and industrial emulsions; wetting agent, pigment dispersant, lubricant, solubilizer, textile scouring agent, emulsion polymerization, metal cleaning; liq.; HLB 12.1; cloud pt. 53-63 C (10%); 90% act.

Rhodasurf® ON-870. [Rhone-Poulenc Surf. & Spec.] Oleth-20; CAS 9004-98-2; nonionic; high foaming emulsifier, stabilizer, dispersant, wetting agent, solubilizer for min. oils, fatty acids, waxes; for industrial cleaners, metal cleaners, agric., paints, dyes, adhesives, textile, leather, cosmetic, pharmaceutical industries; polyethylene dispersant; stabilizer for natural and syn. latexes; FDA, EPA compliance; wh. solid wax; sol. in water, xylene, ethanol, ethylene glycol, butyl Cellosolve; sp.gr. 1.04; HLB 15.4; pour pt. 46 C; cloud pt. < 100 C (1% aq.); flash pt. (PMCC) 93 C; surf. tens. 37 dynes/cm (0.1%); 100% act.

Rhodasurf® ON-877. [Rhone-Poulenc Surf. & Spec.] Oleth-20; CAS 9004-98-2; nonionic; stabilizer for natural and syn. rubber latex emulsions; acid degreaser, dyeing assistant for textiles; dyestuff dispersant, tanning assistant, fat liquor, degreaser for leathers; solubilizer and emulsifier for essential oils and pharmaceuticals; emulsifier for aq. dispersions of polyethylene; liq.; HLB 15.4; cloud pt. < 100 C (1% aq.); 70% conc.

Rhodasurf® PEG 3350. [Rhone-Poulenc Surf. & Spec.] PEG-75; CAS 25322-68-3; mold release and antistat for rubber prods.; binder; waxy flakes; 100% conc.

Rhodasurf® PEG 8000. [Rhone-Poulenc Surf. & Spec.] PEG 8000; mold release and antistat for rubber; binder; waxy flakes; 100% conc.

Rhodialux Q84. [Great Lakes] [2,2´-Thiobis (4-t-octyl phenolato)]-n-butylamine nickel; uv stabilizer for ABS, PE, PP; powd.; m.p. 258-281 C.

Rhodianox MBPS. [Great Lakes] Phenol; nondiscoloring, nonstaining, nonblooming antioxidant for ABS, cellulosics, nylon, polyester, PE, PP, PS, PVC, PU, polychloroprene, polyisoprene; FDA 21CFR §175.105, 177.206.

Rhodianox MO14. [Great Lakes] Phenol; nonstaining, nonmigrating antioxidant for ABS, cellulosics, polyester, PE, PP, PS, PVC, EVA; suitable for food pkg. adhesives, food wrap for nonfatty foods; FDA approved.

Rhodianox TBM6 TP. [Great Lakes] Phenol; nondiscoloring, nonstaining antioxidant for ABS, cellulosics, nylon, polyester, polyolefins, PS, PVC.

Rhodorsil® AF 422. [Rhone-Poulenc Silicones] Dimethylpolysiloxane emulsion; nonionic; antifoam for aq. systems, agric., textile, chemical, rubber, metallurgy industries; EPA compliance; wh. to wh.-gray milky liq.; faint odor; sp.gr. 1.0; pH 5-6; 10% act. in water.

RIA CS. [Olin] Modified urea; CAS 57-13-6; EINECS 200-315-5; cure accelerator and activator for compds. expanded with Opex blowing agent; powd.; Unverified

Ricaccel. [Ricon Resins] Cure rate accelerator for chloroprene elastomers; imparts good storage life and gives outstanding cured props.; designed as 1:1 replacement for ETU; paste.

Ricobond 1031, 1731, 1756. [Ricon Resins] Polymeric; reactive adhesive promoter for compding. with elastomers to give increased adhesion to metal, elastomer, plastic, mineral, fabric, fiber, etc.

Ricon 100. [Ricon Resins] SBR random copolymer; CAS 9003-55-8; thermosetting liq. resin system, outstanding elec. properties, thermal stability, moisture and age resistance, adhesion, and chem. resistance; Ricon 100 used for elec. potting and impregnation of transformers, capacitors, motors laminates, molding compds. and castings, rubber modifiers, mica paper binder, nuclear heat shield; for harder cure zinc-rich coatings, molding compds.; hazy to clear amber very visc. liq.; low odor; m.w. 2400; negligible sol. in water; sp.gr.

0.89; dens. 7.4 lb/gal; visc. 40,000 cps (45 C); b.p. > 300 F; flash pt. 200 C; 80% butadiene, 20% styrene, 65% 1,2-vinyl; 98.5% min. NV.

Ricon 130. [Ricon Resins] 1,2 Polybutadiene homopolymer resin; low visc. rubber and plastics additive; m.w. 2600; visc. 700 cps; 24% 1,2-vinyl; 48% 1,4-trans, 26% 1,4-cis.

Ricon 131. [Ricon Resins] 1,2 Polybutadiene homopolymer resin; plasticizer, impact modifier; coatings; m.w. 5500; visc. 3500 cps; 16-20% 1,2-vinyl; 44% 1,4-trans, 36% 1,4-cis.

Ricon 142. [Ricon Resins] 1,2 Polybutadiene homopolymer resin; PVC plasticizer; impact modifier for plastics; tackifier resin; coatings; m.w. 4500; visc. 9,500 cps; 55% 1,2-vinyl; 26% 1,4-trans, 19% 1,4-cis.

Ricon 150. [Ricon Resins] High vinyl 1,2 polybutadiene homopolymer resin; thermosetting resin used in elec. potting and impregnation of transformers, capacitors, motors laminates, molding compds. and castings, rubber modifiers, mica paper binder, nuclear heat shield; for food pkg. and contact; impact modifier for PE or PP; coagent for NR, EPR, EPDM, CR, NBR, and SBR for peroxide cures, EPR and EPDM for sulfur cure; lowers Mooney visc. allowing increased filler loading; amber visc. liq.; low odor; m.w. 2400; sp.gr. 0.89; dens. 7.4 lb/gal; visc. 40,000 cps (45 C); flash pt. (TCC) > 300 F; 70% 1,2-vinyl, 20-22% trans-1,4, 6-8% cis-1,4; 100% resin.

Ricon 152. [Ricon Resins] 1,2 Polybutadiene homopolymer; thermosetting resin; coagent for NR, EPR, EPDM, CR, NBR, and SBR; process aid for rubber and plastics; elec. cable crosslinker; amber visc. liq.; low odor; m.w. 1600; sp.gr. 0.89; dens. 7.4 lb/gal; visc. 20,000 cps (45 C); flash pt. 200 C; acid no. nil; 80% 1,2 vinyl; 14-16% trans-1,4; 6-8% cis-1,4; 98% NV.

Ricon 153. [Ricon Resins] 1,2 Polybutadiene homopolymer resin; thermosetting resin for wire coating, EPDM peroxide-cured modifier, coatings, processing aid; crosslinker and coagent for rubber; modifier for rubber, plastics, mastics, adhesives; oil well cables; heat-resist. rubber; amber very visc. liq.; m.w. 3200; sp.gr. 0.89; dens. 7.4 lb/gal; visc. 60,000 cps (45 C); flash pt. 200 C; 80% 1,2 vinyl; 14-16% trans-1,4; 6-8% cis-1,4; 98.5% min. NV.

Ricon 157. [Ricon Resins] 1,2 Polybutadiene homopolymer resin; thermosetting resin for thin film coatings, molded elec. parts, coatings, motor wire varnish, transformer impregnating; improves die swell in rubber extrusions; process aid, plasticizer; clear visc. liq.; m.w. 1025 ± 75; sp.gr. 0.89; dens. 7.4 lb/gal; visc. 70 ± 20 poise; flash pt. (TCC) > 300 F; acid no. nil; 70% 1,2-vinyl; 20-% 1,4-trans, 10% 1,4-cis; 98.5% min. NV.

Ricon 181. [Ricon Resins] 1,2 polybutadiene-styrene copolymer resin (random); thermosetting resin used in coatings, liq. rubber; rubber and ink additive; liq.; m.w. 3000; visc. 10,000 cps; 100% solids.

Ricon P30/Dispersion. [Ricon Resins] 1,2-Vinylpolybutadiene disp. in Micro-Cel E; additive for carbon-reinforced natural rubber; nonextractable plasticizer and processing aid resulting in improved carbon black loading; crosslinkable peptizer; dry powd.; low odor; b.p. > 300 F; nil sol. in water; dens. 5.4 lb/gal; flash pt. > 300 F; 65% act. resin.

Ridacto®. [Kenrich Petrochemicals] Ethanol, 2,2′-oxybis reaction prods. with ammonia, morpholine derivs. and residues; CAS 68909-77-3; activator for prevention of reversion in natural/syn. rubber blends; dk. liq., ammoniacal odor; sol. in water; sp.gr. 1.045; vapor pressure < 1 mm Hg (20 C); b.p. 480 F; flash pt. (PMCC) 305 F; 100% act.; *Toxicology:* irritant to skin, eyes, and respiratory tract; harmful if swallowed; *Precaution:* incompat. with nitrosating agents; avoid excessive heating, ignition sources; decomp. produces ammonia, CO_x; *Storage:* store containers closed in cool, dry place.

Rim Wiz. [Axel Plastics Research Labs] external lubricant for RIM, PU molding; liq.

Rit-Cizer #8. [Rit-Chem] N-Ethyl o/p toluenesulfonamide; plasticizer for adhesives, paints, printing inks, epoxy resins, polyamide resins, phenolics, melamine resins, nitrocellulose lacquers, PVAc, EVA, cellulose acetate; imparts improved adhesion to adhesives and coatings, increased flexibility at lower temps., improved gloss, oil resist., stability at higher temps.; lt. yel. visc. liq., sl. char. odor; sol. in alcohol, aromatic solvs.; pract. insol. in water, petrol. hydrocarbons; m.w. 199.2; sp.gr. 1.180-1.190 (50 C); f.p. 0 C max.; b.p. 340 C (760 mm); flash pt. (COC) 175 C.

Rit-Cizer #9. [Rit-Chem] o/p Toluene sulfonamide; plasticizer for thermosetting resins (melamine, urea, phenolics), nylon, casein, PVAc; imparts good gloss, improves wetting action; FDA approved for food pkg. applics.; wh. fine gran. solid, essentially odorless; sol. 1% in water @ 34 C; sp.gr. 1.353; m.p. 115-118 C; decomp. pt. 214 C; flash pt. 420 F; pH 6.5-7.5; 0.5% max. moisture; *Toxicology:* only sl. toxic when large doses are ingested.

Rit-Cizer 10-EHFA. [Rit-Chem] Non-petroleum based; plasticizer designed to prevent migration and blooming; plasticizer/extender for PVC, rubber compds., neoprene, nitrile-butadiene, EPDM, SBR, polybutadiene, PU elastomers, PP; lt. yel. liq.; sol. in alcohols, aromatics, aliphatics; insol. in water; sp.gr. 0.87; visc. 6.5 cps; acid no. 0.3; iodine no. (Wijs) 97; flash pt. (Seta CC) 200 F; 100% purity.

Rit-Cizer 10-EM. [Rit-Chem] Non-petroleum based; plasticizer, softener, extender for PVC, nitrile rubber, Hypalon, neoprene, polybutadiene, polyacrylates, and PU elastomers; Gardner 3+ color; sol. in alcohols, aromatics and aliphatics; insol. in water; sp.gr. < 1; visc. 40 cps; m.p. < 25 C; b.p. > 300 F; acid no. 1.5; flash pt. (Seta CC) > 200 F; 100% purity.

Rit-Cizer 10-ETA. [Rit-Chem] Non-petroleum based; plasticizer offering stable low temp., flexible props. without loss of abrasion resist.; no loss of plasticity in Neoprenes and Hypalons after prolonged high temps.; extender/plasticizer for polybutadienes,

PVC, PP, EPDM, polyacrylates, PU monomers; Gardner 2 color; sol. in alcohols, aromatics and aliphatics; insol. in water; sp.gr. 0.870-0.873; dens. 7.23 lb/gal; visc. 6-8 cps; b.p. > 300 F; acid no. 0.5; cloud pt. < 4 C; flash pt. (CC) > 200 F; 100% purity.

Rit-Cizer 10-MME. [Rit-Chem] Non-petroleum based; economical plasticizer/extender for use where color is not a factor; for PVC, EPDM, SBR, butyl, and nitrile-butadiene; amber low visc. liq.; sol. in alcohols, aromatics, aliphatics; insol. in water; sp.gr. 0.89; dens. 7.44 lb/gal; visc. 10-11 cps; acid no. 5; iodine no. (Wijs) 72; flash pt. (Seta CC) > 400 F; 100% purity.

Rit-O-Lite MS-80. [Rit-Chem] Promotes, modifies, and improves adhesion, gloss retention, solv. release, and lowers visc. of other resin systems; compat. with most commercial resins, esp. nitrocellulose, vinyls, acrylics, and urethanes; FDA 21CFR §175.105, 177.1200; Gardner 3 max. clear visc. sol'n.; sp.gr. 1.171; visc. (Gardner) Z2-Z4; pH 6.8-7.2; 80 ± 1% resin solids.

RL-90. [Sovereign] Rutile titanium dioxide; CAS 13463-67-7; EINECS 236-675-5; pigment with easy processing, high whiteness, blue undertone, high opacity; for rubber, thermoplastic elastomers, and rigid and flexible PVC.

RO-1. [Rogers Anti-Static] Internal/external antistat for all plastics; FDA approved.

RO-40. [Georgia Marble] Ground calcium carbonate; filler for asphalt, putty, ceramic material, foamed compds.; 73 ± 3% min. thru 200 mesh.

Rokon. [R.T. Vanderbilt] 2-Mercaptobenzothiazole; CAS 149-30-4; EINECS 205-736-8; metal deactivator, copper corrosion inhibitor in fuels, industrial lubricants, automotive chemicals, and industrial cleaners; rubber accelerator; lt. yel. to lt. buff powd.; sol. 25% in Cellosolve, 20% in butyl Cellosolve, 30% in Carbitol, 1.5% in toluene, and < 5% in highly aromatic oils; slightly sol. in water; dens. 1.50 mg/m^3; m.p. 164-175 C.

Ross Carnauba Wax. [Frank B. Ross] Carnauba wax; CAS 8015-86-9; EINECS 232-399-4; avail. as Prime or No. 1 Yel., Med. or No. 2 Yel., No. 2 or No. 3 North Country grades; used for polishes, leather finishes, cosmetic creams and lipsticks, casting, lubricants, buffing compds., glazing of candies, gum, pills, and paper, inks; protective coatings, candles, medicinal ointments and salves; lubricant for melamine, phenolics; FDA 21CFR §182.1978; flakes or powd.; sp.gr. 0.996-0.998; m.p. 181.4 F min.; acid no. 2-10; iodine no. 7-14; sapon. no. 78-88; flash pt. 570 F min.; ref. index 1.4540.

Ross Ceresine Wax. [Frank B. Ross] Ceresine wax; CAS 8001-75-0; EINECS 232-290-1; avail. in various grades; used for adhesives, textile waterproofing/mildewproofing, polishes, sunchecking in rubbers, cosmetic pomades and perfumes, crayons, medicinal ointments and slaves, lubricants, mold releases, paper impregnants/sizes, paints, casting; FDA 21CFR §175.105; wh., yel., tan, or orange grades; sp.gr. 0.880-0.935; m.p. 53.3-87.8 C; acid no. nil; sapon. no. 2 max.; ref. index 1.425-1.435.

Ross Montan Wax. [Frank B. Ross] Montan wax; CAS 8002-53-7; EINECS 232-313-5; wax for use as carnauba substitutes, defoaming agents, in carbon paper and printing inks, polishes, rubber prods., plastics, records, insulation, sizing, adhesives; FDA 21CFR §176.210, 175.105, 177.2600; German crude: pellets, powd.; m.p. 83-89 C; acid no. 31-38; sapon. no. 87-104; Domestic crude/refined: dk. brn., brn., tan flakes, powd.; m.p. 79.4-89 C; acid no. 24-55; sapon. no. 75-135.

Ross Ouricury Wax Replacement. [Frank B. Ross] Wax used as carnauba replacement where color is not important; for shoe polishes, inks for carbon paper, lubricants, mold release, roll leaf, stamping; dk. color; sp.gr. 0.997-1.010; m.p. 180-188 F; acid no. 2-7; ester no. 76-88; sapon. no. 78-95; flash pt. 289.9 C min.

Ross Wax #100. [Frank B. Ross] Synthetic wax; CAS 8002-74-2; mold release in rubber and plastics; external lubricant for PVC; increases m.p., opacity, gloss in wax blends, floor waxes, textiles, coatings, candles, hot melts, paints, inks, asphalt; FDA §172.886, 172.888; wh. flakes; sol. hot in most solvs.; visc. 80 SUS (120 C); congeal pt. 200-210 F; acid no. nil; sapon. no. nil.

Ross Wax #140. [Frank B. Ross] Synthetic wax; CAS 110-30-5; high m.p. wax used as ingredient, finish or processing aid in adhesives, ammunition, asphalt, explosives, paints, paper, pyrotechnics, lubricants, mold releases, PVC, textile finishes, powd. metallurgy, etc.; FDA 21CFR §175.300, 176.170, 176.180, 175.105, 177.1200; wh. flakes, powd., or atomized; sol. in hot naphtha, xylene, Carbitol, IPA, trichlorethylene; sp.gr. 0.97; m.p. 138-140 C; flash pt. 530 F min.; acid no. 3-10; dielec. str. 233 V/mil; vol. resist. 0.99 x 10^{13} ohm-cm.

Ross Wax #145. [Frank B. Ross] High m.p. fully refined petroleum wax; CAS 8002-74-2; lubricant for formulating PVC and elec. wire and cable compds.; FDA 21CFR §178.3710; wh. flakes; sol. in benzene; sp.gr. 0.930-0.936; m.p. 140-150 F.

Ross Wax #160. [Frank B. Ross] Synthetic wax; CAS 123237-14-9; high m.p. wax used as ingredient, finish or processing aid in adhesives, ammunition, explosives, hot-melt coatings, lubricants, paints, paper, pyrotechnics, release agent, propellants, textile finishes, powd. metallurgy, varnish, plastic film, etc.; tan med. hard wax in lumps, flakes, powd., or atomized; sol. in hot naphtha, xylene, Carbitol, IPA, trichlorethylene, toluene; sp.gr. 1.0232; m.p. 157-162 C; flash pt. 590 F min.; acid no. 10 max.; iodine no. 7.5.

Ross Wax #165. [Frank B. Ross] High m.p. fully refined petroleum wax; CAS 8002-74-2; lubricant for formulating PVC and elec. wire and cable compds.; FDA 21CFR 178.3710; wh. chips; insol. in acetone and water (#165); sp.gr. 0.930-0.936; m.p. 158-165 F.

Rotax®. [R.T. Vanderbilt] 2-Mercaptobenzothiazole; CAS 149-30-4; EINECS 205-736-8; accelerator; primary accelerator for natural and syn. rubbers;

nonstaining and nondiscoloring; used in proofing compds. where lowest odor is desired; corrosion inhibitor in automotive chems., industrial cleaners; protects silverware from sulfur blackening; pale yel. powd., 99.9% thru 100 mesh; very sol. in dilute caustic, toluene, carbon disulfide, chloroform, alcohol; pract. insol. in water; m.w. 167.25; dens. 1.52 ± 0.03 mg/m^3; m.p. 169-180 C.

Royalene 301-T. [Uniroyal] EPDM terpolymer; used as an antiozonant in tire sidewalls and coverstrip and in blends with butyl rubber to improve heat and ozone resist. in inner tubes; vulcanizable with sulfur, peroxide, or radiation; lt. amber bales; sp.gr. 0.87; Mooney visc. 40 (ML1+4, 125 C); 68/32 ethylene/propylene ratio.

Royalene 7565. [Uniroyal] EPDM terpolymer/HDPE blend (65/35 ratio); impact modifier for PP; free-flowing pellets; 52/48 ethylene/propylene ratio.; Unverified

Royaltuf 372. [Uniroyal] EPDM/SAN graft polymer (50/50 ratio); impact modifier for PC, polyester/PC alloys, PVC; sp.gr. 0.97; Unverified

Royaltuf 465A. [Uniroyal] Maleated EPDM; impact modifier for nylon; pellets; sp.gr. 0.90; Mooney visc. 60 (ML1+4, 125 C); 75/25 ethylene/propylene ratio.; Unverified

RPP. [Aster] Reprocessed paint polymer mixt.; tackifier, plasticizer, extender, and binder in PVC sealers, automotive adhesives and sealers, rubber, plastics, pressure sensitives, hot melts, asphalt coatings, sound deadeners, sealant tapes, caulks; impact modifier for plastics; solid, rubber masterbatch, paste, powd., aq. emulsion, or solv. cutback; Solid: purple or bluish-gray; sp.gr. 1.1; melt index 2-20 g/min (150 F); > 98% solids.

RR 5. [Releasomers] Semipermanent mold release agent for most thermosetting rubber and plastic materials; avail. in cold and hot formulations for applic. to mold surfs. below 140 F and above 200 F; solv. sol'n.; *Toxicology:* avoid prolonged/repeated skin contact, breathing of vapors; *Precaution:* keep away from heat, spark, flames; *Storage:* store in closed containers when not in use.

RR Zinc Oxide (Untreated). [Akrochem] American process zinc oxide; CAS 1314-13-2; EINECS 215-222-5; activator for rubber; powd.; 0.27 μ mean particle size; sp.gr. 5.6; dens. 30 lb/ft^3; 99.2% ZnO.

RR Zinc Oxide-Coated. [Akrochem] American process zinc oxide coated with propionic acid; activator for rubber; powd.; 0.27 μ mean particle size; sp.gr. 5.6; dens. 40 lb/ft^3; 98.8% ZnO.

RS-80. [SVO Enterprises] High oleic rapeseed oil, refined, bleached, deodorized; CAS 8002-13-9; EINECS 232-299-0; lubricant and slip agent for plastics, polymers, rubber; biodeg. base oil for industrial lubricants; improved high temp. oxidative stability; food grade; oil; sp.gr. 0.919 (60 F); visc. 37 cSt (40 C); iodine no. 100; pour pt. -21 C; flash pt. 240 C; 70% min. oleic acid.

Rubber Lubricant™. [Jet-Lube] Water-based silicone prod.; lubricant for rubber and plastic parts, providing a temporary, wet, slippery surf. which either dries to a thin protective film or is absorbed into natural or syn. rubber; mold release agent for metal, plastic, and rubber castings (acrylic, cellulosics, ABS, PS, melamines); heat stable to 500 F; milky-wh. opaque fluid, odorless, tasteless; disp. in water; sp.gr. 1.00; vapor pressure < 0.01 mm Hg; m.p. 25 F; b.p. 212 F; *Toxicology:* repeated skin contact may cause redness and irritation in sensitive individuals; *Precaution:* incompat. with strong oxidizing materials; *Storage:* store @ R.T.

Rubbermakers Sulfur. [Akrochem] Sulfur; CAS 7704-34-9; EINECS 231-722-6; for rubber compounding; powd.; 99.5% min. thru 100 mesh, 90-96% min. thru 200 mesh; sp.gr. 2.07; 99.5% min. purity.

Rubber Shield. [Zyvax] Release in aliphatic and aromatic naphtha carrier; general purpose release coating for use with heated molds and tools; releases most rubber polymers incl. NR, SBR, neoprene, EPDM, nitrile, epichlorohdyrin, and halogenated butyl; provides multiple releases with fast cure, high gloss, high slip; thermal stability to > 700 F; clear thin liq., char. solv. odor; dens. 7.29 lb/gal; flash pt. (CC) > 100 F; *Precaution:* flamm.; *Storage:* 1 yr min shelf life stored in cool, dry area (< 90 F); prod. is moisture-sensitive; store in tightly closed container when not in use.

Rubichem Cal-Dip®. [Struktol] Stearate-based; antitack for rubber industry; yields uniform low-dusting antitack coating which will disperse in elastomers with cure temps. as low as 120 C; easy to disperse, environmentally safe; 25% solids.

Rubichem Cal-Dip® HS. [Struktol] Stearate-based; antitack for rubber industry; yields uniform low-dusting antitack coating which will disperse in elastomers with cure temps. as low as 120 C; easy to disperse, environmentally safe; 50% solids.

Rubichem Cal-Dip® W. [Struktol] Antitack for rubber industry; yields uniform low-dusting antitack coating that covers any newly exposed surf. area while the rubber awaits further processing.

Rubichem Clay-Dip®. [Struktol] Zinc stearate-free slurry; antitack dipping compd. formulated to uniformly coat highly plasticized or hydrophobic rubber compds.; environmentally safe; minimal dusting; exc. nonsettling; 50% solids; *Toxicology:* nontoxic.

Rubichem Dry-Dip® A. [Struktol] Zinc stearate-free; antitack coating for highly plasticized or hydrophobic rubber compds.; environmentally safe; nondusting; powd.; disp. in cold water; *Toxicology:* nontoxic.

Rubichem Dry-Dip® C. [Struktol] Zinc stearate-free; antitack coating for highly plasticized or hydrophobic rubber compds.; environmentally safe; nondusting; powd.; disp. in cold water; *Toxicology:* nontoxic.

Rubichem Dry-Dip® NP. [Struktol] Zinc stearate-free; antitack coating for highly plasticized or hydrophobic rubber compds.; environmentally safe; nondusting; powd.; disp. in cold water; *Toxicology:* nontoxic.

Rubichem Zinc-Dip®. [Struktol] Zinc stearate-based slurry; CAS 557-05-1; EINECS 209-151-9; antitack for rubber industry; uniformly coats highly plasticized or hydrophobic rubber compds.; disperses in elastomers with cure temps. as low as 120 C; 25% solids.

Rubichem Zinc-Dip® HS. [Struktol] Zinc stearate-based slurry; CAS 557-05-1; EINECS 209-151-9; antitack for rubber industry; uniformly coats highly plasticized or hydrophobic rubber compds.; disperses in elastomers with cure temps. as low as 120 C; 40% solids.

Rubichem Zinc-Dip® W. [Struktol] Zinc stearate-based; CAS 557-05-1; EINECS 209-151-9; antitack for rubber industry; uniformly coats highly plasticized or hydrophobic rubber compds.; covers newly exposed surfs. while the rubber awaits further processing; easy to disperse.

Rudol®. [Witco/Petroleum Spec.] Lt. min. oil NF; white oil functioning as binder, carrier, conditioner, defoamer, dispersant, extender, heat transfer agent, lubricant, moisture barrier, plasticizer, protective agent, and/or softener in adhesives, agric., chemicals, cleaning, cosmetics, food, pkg., plastics, and textiles industries; FDA 21CFR §172.878, §178.3620a; water-wh., odorless, tasteless; sp.gr. 0.852-0.870; visc. 28-30 cSt (40 C); pour pt. -7 C; flash pt. 188 C.

RX-13117. [C.P. Hall] Monomeric plasticizer for CR; APHA 250 liq.; m.w. 391; sp.gr. 0.865; f.p. -50 C; acid no. 0.12; flash pt. 213 C.

RX-13154. [C.P. Hall] Monomeric plasticizer for CR; APHA 30 liq.; m.w. 500; sp.gr. 0.910; f.p. -53 C; acid no. 0.18; flash pt. 216 C.

RX-13411. [C.P. Hall] Polymeric plasticizer for PVC and PS providing improved low-temp. brittleness performance, less tendency to migrate to PS substrates, less extractability after long-term humidity aging; APHA 250 liq.; m.w. 391; sp.gr. 0.865; f.p. -50 C; acid no. 0.12; flash pt. 213 C.

RX-13412, -13413. [C.P. Hall] Polymeric plasticizer for PVC, ABS providing improved low-temp. brittleness performance and greater compound softening efficiency; APHA 250 liq.; m.w. 391; sp.gr. 0.865; f.p. -50 C; acid no. 0.12; flash pt. 213 C.

RX-13414, -13415. [C.P. Hall] Polymeric plasticizer for PVC; APHA 250 liq.; m.w. 391; sp.gr. 0.865; f.p. -50 C; acid no. 0.12; flash pt. 213 C.

Rylex® 30. [Ferro/Bedford] Alkyl phenol polysulfide composed primarily of p-t-butyl phenol; CAS 60303-68-6; vulcanizing agent and antioxidant for rubbers; aux. cross-linking agent for halobutyl rubber, NR and SBR adhesives, chlorobutyl/natural rubber blends; covulcanizer for neoprene/phenolic resin adhesives; antioxidant in tire sidewalls; amber flakes, tack-free, mild sulfur odor; sol. in aromatic and chlorinated solvs., partially sol. in aliphatic solvs., insol. in water; sp.gr. 1.15; soften. pt. 95 C min.; 29.5% min. S; *Storage:* good storage stability under normal conditions.

Rylex® 3010. [Ferro/Bedford] Alkyl phenol polysulfide/stearic acid blend; vulcanizing agent and antioxidant for rubbers; aux. cross-linking agent for halobutyl rubber, NR and SBR adhesives, chlorobutyl/natural rubber blends; covulcanizer for neoprene/phenolic resin adhesives; antioxidant in tire sidewalls; yel. flakes, tack-free, mild sulfur odor; sol. in aromatic and chlorinated solvs., partially sol. in aliphatic solvs., insol. in water; sp.gr. 1.08; soften. pt. 80 C min.; 26.4% min. S; *Storage:* good storage stability under normal conditions.

S

S160 Beads and Powd. [Degussa] Carbon black; CAS 1333-86-4; EINECS 215-609-9; easy processing blk. for med. quality PVC and calendered systems, PVC cable compds., PP, flexible, rigid ABS, PS.

S.A. 100. [Amspec] Sodium antimonate; flame retardant for ABS, acrylics, epoxy, thermoset polyester, PS, PE, PP, PVC, PU.

Sable™ 3. [J.M. Huber/Engineered Carbons] Carbon black; CAS 1333-86-4; EINECS 215-609-9; carbon black for demanding rubber applics.; 29 nm particle size; 5 ppm on 60 mesh, 30 ppm on 325 mesh; bulk dens. 23.5 lb/ft^3 (loose); surf. area 75 m^2/g; pH 7.0.

Sable™ 5. [J.M. Huber/Engineered Carbons] Carbon black; CAS 1333-86-4; EINECS 215-609-9; carbon black for demanding rubber applics.; 35 nm particle size; 5 ppm on 60 mesh, 30 ppm on 325 mesh; bulk dens. 22 lb/ft^3 (loose); surf. area 42 m^2/g; pH 7.0.

Sable™ 15. [J.M. Huber/Engineered Carbons] Carbon black; CAS 1333-86-4; EINECS 215-609-9; carbon black for demanding rubber applics.; 320 nm particle size; 1 ppm on 60 mesh, 10 ppm on 325 mesh; bulk dens. 39.9 lb/ft^3 (loose); surf. area 8.2 m^2/g; pH 9.4.

Sable™ 17. [J.M. Huber/Engineered Carbons] Carbon black; CAS 1333-86-4; EINECS 215-609-9; carbon black for demanding rubber applics.; 26 nm particle size; 5 ppm on 60 mesh, 30 ppm on 325 mesh; bulk dens. 29 lb/ft^3 (loose); surf. area 78 m^2/g; pH 8.0.

Sachtolith® HD. [Sachtleben Chemie GmbH] Zinc sulfide, coated; CAS 1314-98-3; EINECS 215-251-3; wh. pigment, flame retardant used in thermoplastics (HDPE, PP), thermosets (melamine, urea, polyester), flame-resistant plastics, glass-reinforced plastics, pigment concs. and masterbatches, elastomers, textile fibers, paper, sealants, lubricants; opacifier; HD grade is coated with a small amt. of org. surfactant and have improved wetting-out and disp. char. in aq. and org. systems; FDA compliance; 0.35 μm mean particle size; dens. 4.0 g/ml; surf. area 8 m^2/g; oil absorp. 14; ref. index 2.37; pH 7; hardness (Mohs) 3; 98% ZnS content.

Sachtolith® HD-S. [Sachtleben Chemie GmbH] Zinc sulfide, coated; CAS 1314-98-3; EINECS 215-251-3; wh. pigment, flame retardant used in thermoplastics (HDPE, PP), thermosets (melamine, urea, polyester), flame-resistant plastics, glass-reinforced plastics, pigment concs. and masterbatches, elastomers, textile fibers, paper, sealants, lubricants; opacifier; HD-S grade is coated with a small amt. of org. surfactant and have improved wetting-out and disp. char. in aq. and org. systems; FDA compliance; 0.30 μm mean particle size; dens. 4.0 g/ml; surf. area 8 m^2/g; oil absorp. 13; ref. index 2.37; pH 7; hardness (Mohs) 3; 98% ZnS.

Sachtolith® L. [Sachtleben Chemie GmbH] Zinc sulfide; CAS 1314-98-3; EINECS 215-251-3; wh. pigment, flame retardant used in thermoplastics (HDPE, PP), thermosets (melamine, urea, polyester), flame-resistant plastics, glass-reinforced plastics, pigment concs. and masterbatches, elastomers, textile fibers, paper, sealants, lubricants; opacifier; L is uncoated std. grade; FDA compliance; 0.35 μm mean particle size; dens. 4.0 g/ml; surf. area 8 m^2/g; oil absorp. 14; ref. index 2.37; pH 6; hardness (Mohs) 3; 98% ZnS content.

Safoam® AP-40. [Reedy Int'l.] Safoam plus an exothermic agent in PS and rubber carrier; chemical nucleating and blowing agent for all major resins and advanced engineering polymers such as Noryl, ABS, rigid PVC; for inj. molding and extrusion for very fine cell size for large, thick wall parts, e.g., electronic enclosures, cabinets, industrial structural elements; decomp. pt. 165-204 C; 40% act.

Safoam® FP. [Reedy Int'l.] Chemical nucleating and blowing agent for all major resins such as ABS, PS, PVC, EVA, PE, HDPE, LDPE, PPO, PPE, etc.; for foam sheet extrusion, general purpose inj. molding; provides greater flexibility, finer texture, and opacity; 7 μ particle powd.; decomp. pt. 165-204 C; 100% act.

Safoam® FP-20. [Reedy Int'l.] 20% Safoam FP in low m.w. PS carrier; endothermic chemical nucleating and blowing agent for resins benefitting from dispersion, melt flow, and phys. props. such as ABS, PS, PVC, TPE, and TPU; for inj. molding, extrusion, very fine cell size for structural foam extrusions and profiles; decomp. pt. 165-204 C; 20% act.

Safoam® FP-40. [Reedy Int'l.] 40% Safoam FP in low m.w. PS carrier; endothermic chemical nucleating and blowing agent for resins benefitting from dispersion, melt flow, and phys. props. such as ABS, PS, PVC, TPE, and TPU; for general purpose inj. molding and extrusion, for very fine cell size for uniform structural foam extrusions and profiles; decomp. pt. 165-204 C; 40% act.

Safoam® FPE-20. [Reedy Int'l.] 20% Safoam FP in low m.w. PE carrier; endothermic chemical nucleating and blowing agent for polyolefin-based resins benefitting from improved melt flow and dispersion maximizing wt. reduction and cell distribution; for inj. molding and extrusion with fine cell size; provides uniform dosing for improved surf. features and elimination of sink marks; decomp. pt. 165-204 C; 20% act.

Safoam® FPE-50. [Reedy Int'l.] 50% Safoam FP in low m.w. PE carrier; endothermic chemical nucleating and blowing agent for polyolefin-based resins benefitting from improved melt flow and dispersion maximizing wt. reduction and cell distribution; for inj. molding and extrusion with fine cell size; provides exc. dens. reductions with improved physical props.; decomp. pt. 165-204 C; 50% act.

Safoam® P. [Reedy Int'l.] Endothermic chemical nucleating and blowing agent for all major resins such as ABS, PS, PVC, EVA, PE, HDPE, LDPE, PPO, PPE, etc.; for foam sheet extrusion, general purpose inj. molding; provides greater rigidity, texture, and impact str.; 35 μ particle powd.; decomp. pt. 165-204 C; 100% act.

Safoam® P-20. [Reedy Int'l.] 20% Safoam P in low m.w. PS carrier; endothermic chemical nucleating and blowing agent for resins benefitting from dispersion, melt flow, and phys. props. such as ABS, PS, PVC, TPE, and TPU; for general purpose extrusion and inj. molding; provides improved cycle time, plasticity, and eliminates sink marks; decomp. pt. 165-204 C; 20% act.

Safoam® P-50. [Reedy Int'l.] 50% Safoam P in low m.w. PS carrier; endothermic chemical nucleating and blowing agent for resins benefitting from dispersion, melt flow, and phys. props. such as ABS, PS, PVC, TPE, and TPU; for general purpose inj. molding and foam sheet extrusions esp. for thermoformed food pkg.; decomp. pt. 165-204 C; 50% act.

Safoam® PCE-40. [Reedy Int'l.] Safoam in EEA carrier; endothermic chemical nucleating and blowing agent for moisture-sensitive, high-temp. resins such as acetal, nylon, PC, and PC/ABS blends; for inj. molding and extrusion for fine cell size; provides exc. dens. reductions with improved physical props.; decomp. pt. 182-316 C; 40% act.

Safoam® PE-20. [Reedy Int'l.] 20% Safoam P in low m.w. PE carrier; endothermic chemical nucleating and blowing agent for polyolefin-based resins benefitting from improved melt flow and dispersion maximizing wt. reduction and cell distribution; for inj. molding and extrusion with standard cell size; provides uniform dosing for improved surf. features and elimination of sink marks; decomp. pt. 165-204 C; 20% act.

Safoam® PE-50. [Reedy Int'l.] 50% Safoam P in low m.w. PE carrier; endothermic chemical nucleating and blowing agent for polyolefin-based resins benefitting from improved melt flow and dispersion maximizing wt. reduction and cell distribution; for inj. molding and extrusion with standard cell size; provides significant dens. reduction without loss of physical props.; decomp. pt. 165-204 C; 50% act.

Safoam® RIC. [Reedy Int'l.] Chemical nucleating and blowing agent for all major resins but esp. compat. with flexible and rigid PVC; for general purpose extrusion and inj. molding at lower temps.; provides greater rigidity, texture, and impact str.; 30 μ particle powd.; decomp. pt. 137-204 C; 100% act.

Safoam® RIC-25. [Reedy Int'l.] Safoam in PE carrier; endothermic chemical nucleating and blowing agent for polyolefin-based resins benefitting from improved melt flow and dispersion maximizing wt. reduction and cell distribution; for inj. molding and extrusion at reduced temps.; provides exc. dens. reductions with improved physical props.; 30 μ particle powd.; decomp. pt. 137-204 C; 25% act.

Safoam® RIC-30. [Reedy Int'l.] Safoam in PE carrier; endothermic chemical nucleating and blowing agent for all major resins, esp. compat. with flexible and rigid PVC and PE; for inj. molding and extrusion at reduced temps.; provides microcellular nucleation for PBA systems; 30 μ particle powd.; decomp. pt. 137-204 C; 30% act.

Safoam® RPC. [Reedy Int'l.] Chemical nucleating and blowing agent for moisture-sensitive resins such as acetal, nylon, PET, PC, and ABS blends; for inj. molding and extrusion of low dens. foam sheet for thermoformed food pkg.; 2 μ particle powd.; decomp. pt. 182-316 C; 100% act.

Safoam® RPC-40. [Reedy Int'l.] PS conc.; endothermic chemical nucleating and blowing agent for moisture-sensitive, high-temp. resins such as acetal, nylon, PET, PC, and ABS blends; for inj. molding and extrusion for fine cell size; provides exc. dens. reductions with an increase in physical props.; decomp. pt. 182-316 C; 40% act.

SAFR-C70. [SAFR Compds.] Ethylenediamine phosphate; nonhalogen flame retardant for PP and PE resin systems; at 1:1 ratio, achieves UL94 V0 rating (1/16 in.); produces nonblooming, non-moisture-sensitive compds. with relatively low dens.; sl. off-wh. pellet.

SAFR-CELL 40. [SAFR Compds.] Endothermic and exothermic blowing agents in LLDPE carrier; chemical blowing agent conc. for thermoplastics processed at 360-430 F incl. polyolefins, styrenics, and ABS; efficient for finer cell development; suitable for extrusion and inj. molding; lt. yel. pellet, sl. odor; sp.gr. 0.98 ± 0.03; decomp. pt. 320 F, full decomp. @ 375 F.

SAFR-CELL 40 LT. [SAFR Compds.] Endothermic and exothermic blowing agents in EVA carrier; low temp. chemical blowing agent conc. for thermoplastics esp. flexible PVC; efficient in finer cell development; suitable for extrusion and inj. molding; lt. yel. pellet, sl. odor; sp.gr. 0.98 ± 0.03; decomp. pt. 265 F, full decomp. @ 320 F.

SAFR-EPS K. [SAFR Compds.] Ethylene-propylene-styrene terpolymer; impact modifier for rigid PVC; resin compatibilizer for PVC, polyolefin, styrenics;

process aid; improves compd. uniformity and lubrication; in finished prod., acts as intergranular shock and stress absorber; wh. rubber pellet; sp.gr. 0.91 ± 0.02; soften. pt. 105-110 C.

SAFR-PURGE. [SAFR Compds.] Recyclable thermoplastic purge compd. compat. with most resin systems processed below 550 F; effective surf. active cleaning; nonabrasive and nonhazardous; for extrusion and inj. molding applics.; process temp. range 285-550 F; wh. pellet; sp.gr. 0.96 ± 0.01.

SAFR-T-5K. [SAFR Compds.] 20% conc. of titanium IV tetra tridecyl, bis (tris tridecyl) phosphite in SAFR-EPS K resin carrier; strongly hydrophobic coupling agent conc. capable of superplasticizing elastomers, flexible vinyl and polyolefins; wets pigment and filler surfs. for improved adhesion and increased solids loading; internal and external lubricant in PVC; improves processing; process temp. range 245-500 F; lt. pinkish/tan pellet, mild odor; sp.gr. 0.93 ± 0.03; 20% conc.

SAFR-T-12K. [SAFR Compds.] 20% conc. of oxy bis titanium IV tris (bis tridecyl) phosphate in SAFR-EPS K resin carrier; hydrophobic coupling agent conc. capable of improving mech. and chem. props. of some PVC alloys, some styrenics, and polyolefins; wets pigment and filler surfs. for improved adhesion and increased solids loading; improves str. and impact resist.; improves processing; for PTFE, caclium carbonate, talc, and some pigments dispersed in PP, PE, PVC alloys, ABS and unfilled resins; process temp. range 245-500 F; lt. pinkish/tan pellet, mild odor; sp.gr. 0.93 ± 0.03; 20% conc.

SAFR-Z-12K. [SAFR Compds.] 20% conc. of oxy bis zirconium IV tris (bis tridecyl) phosphate in SAFR-EPS K resin carrier; coupling agent conc.; wets and bonds polyolefin resins, and metal, metal oxides, or other alkaline surfs. to nonpolar resins; for use with calcium carbonate-filled, pigment-filled, conductive metal-filled polyolefins; process temp. range 245-650 F; off-wh./pinkish pellet, sl. odor; sp.gr. 0.93 ± 0.03; 20% conc.

SAFR-Z-12SK. [SAFR Compds.] 20% conc. of oxy bis zirconium IV tris (bis ethoxylated butyl) phosphate in SAFR-EPS K resin carrier; coupling agent conc.; wets and bonds polar plastics and metal, metal oxides, or other alkaline surfs. to semipolar resins; for use with mica-filled, glass-filled, conductive metal-filled, and flame retardant-filled polyolefins; improves processing in filled and unfilled styrenics; process temp. range 245-650 F; lt. brn. pellet, sl. odor; sp.gr. 0.93 ± 0.03; 20% conc.

SAFR-Z-22E K. [SAFR Compds.] 20% conc. of zirconium IV bis (mixed) alcoholato cyclo bis alcoholato diphsophate adduct 2 moles trid tridecyl phosphite in SAFR-EPS K resin carrier; coupling agent conc. designed for rigid PVC applics.; wets and bonds calcium carbonate and titanium dioxide to the resin; improves hot melt flow, prod. str.; reduces or eliminates need for thermal stabilizers; process temp. range 245-650 F; yel./lt. brn. pellet, mild odor; sp.gr. 0.93 ± 0.03; 20% conc.

SAFR-Z-44 K. [SAFR Compds.] 20% conc. of zirconium IV propyl (tris-aminoethyl) ethanolamine in SAFR-EPS K resin carrier; moderately hydrophobic coupling agent conc. capable of crosslinking epoxies, polyesters, polyamides, and polyurethanes, efficiently wetting pigment and filler surfs. for improved adhesion to polar plastic and metallic surfs.; increases filler loading; improves str.; aids wetting and dispersion of fillers and pigments; increases elong. and sl. increases tens. str. in unfilled resins; aids processing; process temp. range 245-700 F; dk. brn. pellet, sl. odor; sp.gr. 0.93 ± 0.03; 20% conc.

Sandet 60. [Sanyo Chem. Industries] Sodium dodecylbenzene sulfonate; CAS 25155-30-0; anionic; detergent base and emulsifier for emulsion polymerization of vinyl resin; liq.

Sandostab 4030. [Sandoz] Bis(4-(1,1-dimethylethyl) benzoato-o)hydroxy aluminum; CAS 13170-05-3; stabilizer; wh. fine free-flowing powd., odorless; insol. in water; sp.gr. 1.14; pH 5-6 (10% aq.); *Toxicology:* LD50 (oral, rat) > 5000 mg/kg; nonirritating to eyes, mildly irritating to skin; inh. of fine aluminum powd. may cause pulmonary fibrosis, may be implicated in Alzheimers; *Precaution:* dusts may be explosive with spark or flame initiation; thermal decomp. may produce oxides of carbon and aluminum; *Storage:* keep containers closed.

Sandostab P-EPQ. [Sandoz] Phosphorus trichloride, reaction prods. with 1,1′-biphenyl and 2,4-bis (1,1-dimethylethyl) phenol; CAS 119345-01-6; effective processing stabilizer, sec. antioxidant for polymers incl. filled polyolefins, ABS, PS, polybutylene, PBT, PC, thermoplastic polyester, nitrile barrier resins; peroxide decomposer for plastics mfg.; prevents polymer yellowing; synergistic wsith phenolic antioxidants; reduces equipment corrosion in flame-retardant applics.; FDA 21CFR §178.2010; off-wh. to sl. ylsh. powd.; sol. > 100 g/l in hexane, toluene, ethyl acetate, acetone, benzene, cyclohexane @ 50 C; insol. in water; sp.gr. 1.045; bulk dens. 38 g/100 ml; m.p. 75 C; flash pt. 350 C; 5.4-5.9% P; *Toxicology:* LD50 (oral, rat) > 5000 mg/kg (dermal, rat) > 5000 mg/kg; nonirritating to eyes; mildly irritating to skin; *Precaution:* incompat. with strong oxidizing agents; thermal decomp. may produce oxides of carbon; *Storage:* keep containers closed.

Sandoz Phosphorester 510. [Sandoz] Complex phosphate ester; anionic; detergent, emulsifier in waterless hand cleaners, dry cleaning detergents; antistat, corrosion inhibitor in textile fiber lubricants; primary emulsifier in emulsion polymerization of PVAc, acrylic; clear visc. liq.; water-disp.; sol. in conc. electrolyte systems; pH 1.5-2.5 (10%); 99% solids.

Sandoz Phosphorester 610. [Sandoz] Complex phosphate ester; emulsifier for aromatic and chlorinated solvs.; solubilizer for nonionic surfactants; emulsion polymerization surfactant used in vinyl acetate and acrylic systems; clear visc. liq.; sol. in conc. electrolyte systems; pH 1.5-2.5 (10%); 99% solids.

Sandoz Phosphorester 690. [Sandoz] Complex

phosphate ester; emulsifier for emulsion polymerization; household and industrial cleaners; soft waxy paste; pH 1.5-2.5; 95% solids.

Sandoz Sulfonate 2A1. [Sandoz] Sodium dodecyl diphenyl ether sulfonate; anionic; solubilizer; household detergents, industrial, disinfectant, and metal cleaners, emulsifier in emulsion polymerization; textile aux.; also avail in acid form; yel./brn. liq.; sol. in electrolyte sol'ns., caustic, HCl, TKPP; sp.gr. 1.16; visc. 145 cps; 45% act.

Sanduvor® 3050. [Sandoz] Sterically hindered amine lt. stabilizer; lt. stabilizer for two-coat metallic systems, automotive paints, plastics; stabilizes enamel against gloss reduction and cracking; visc. liq. to waxy paste, mild odor; sol. > 2000 g/l in toluene, xylene, wh. spirit, ethyl acetate, acetone, n-butanol, < 0.02 g/l in water; sp.gr. 0.978; m.p. 86-113 F; flash pt. 105 C; *Toxicology:* LD50 (oral, rat) > 5000 mg/kg; nonirritating to eyes; irritating to skin; contains o-xylene which can cause severe eye irritation, mod. skin irritation, respiratory irritation, nausea, vomiting, diarrhea; *Precaution:* incompat. with acids, alkalies, strong oxidizing agents; thermal decomp. may produce oxides of carbon and nitrogen; avoid dust formation; *Storage:* unlimited storage stability when stored at R.T. in closed containers.

Sanduvor® 3051. [Sandoz] Radical scavenger/sterically hindered amine lt. stabilizer; lt. stabilizer for powd. coatings for exterior applics., plastics; aids gloss retention during weathering, resists cracking; for clear coats, thin film systems, pigmented one-coat systems; wh. powd., odorless; sol. 1.25 g/l in n-butanol, < 1 g/l in toluene, xylene, wh. spirit, ethyl acetate, acetone; insol. in water; sp.gr. 1.06 (20 C); m.p. > 225 C; *Toxicology:* LD50 (oral, rat) 2800 mg/kg; nonirritating to skin; moderate eye irritant; *Precaution:* dusts may be explosive hazard; incompat. with strong oxidizing agents; thermal decomp. may produce oxides of carbon and nitrogen; *Storage:* unlimited storage stability in closed containers @ R.T.

Sanduvor® 3052. [Sandoz] Sterically hindered amine lt. stabilizer; lt. stabilizer for plastics and clear or pigmented coatings incl. automotive urethanes, industrial maintenance, uv-cured coatings, solv.-free polyesters, air-drying systems, baking enamels; improves gloss retention, decreases crack formation; pract. colorless low-visc. liq., odorless; sol. > 20000 g/l in toluene, xylene, wh. spirit, ethyl acetate, acetone, n-butanol, > 0.05 g/l in water; sp.gr. 0.9125; b.p. > 280 C; solid. pt. < -20 C; flash pt. (CC) 134 C; pH 10-10.5 (10% aq.); 100% act.; *Toxicology:* LD50 (oral, rat) 732 mg/kg; irritating to eyes and skin; *Precaution:* incompat. with strong oxidizing agents; thermal decomp. may produce oxides of carbon and nitrogen; *Storage:* unlimited storage stability in closed containers @ R.T.

Sanduvor® VSU. [Sandoz] 2-Ethyl, 2′-ethoxy-oxalanilide; CAS 23949-66-8; uv absorber for acrylates, acrylic melamine, alkyd/melamine, two-component PU resins, unsat. polyesters; improves light and weathering fastness of pigments, dyes, and topcoats; stabilizes two-coat metallic auto paints; protects against blistering; wh. powd., mild odor; sol. (g/l): 60 g in toluene, MEK, 40 g in ethyl acetate, 30 g in xylene; insol. in water; sp.gr. 1.209; m.p. 123-127 C; flash pt. > 200 F; *Toxicology:* LD50 (oral, rat) > 5000 mg/kg; virtually nontoxic; nonirritating to skin and eyes; *Precaution:* dusts may be explosive with spark or flame initiation; incompat. with strong oxidizing agents; thermal decomp. may produce CO_x and NO_x; *Storage:* unlimited storage stability when stored in a dry place @ R.T.; keep container closed.

Sanitized® LX 91-01 RF. [Sanitized AG] Halogenated phenoxy compd. and isothiazolinone derivs.; antimicrobial for treatment of natural or syn. rubber, foam rubber, latex dispersions, sealing compds. based on polysulfides; for rubber sheeting, molded goods, dipped goods, rubber thread, coatings, and foams; brn. liq.; sol. in org. solvs., water, latex emulsions; sp.gr. 1.05 ± 0.02 (20 C); flash pt. (Abel-Pensky CC) > 100 C; *Storage:* 12 mos storage stability in original closed containers.

Sanstat 2012-A. [Sanyo Chem. Industries] Sodium dioctyl sulfosuccinate; cationic; external antistat for plastics incl. polyolefins, PS, polyamide, PVC, ABS; liq.; Unverified

Santicizer 97. [Monsanto] Dialkyl adipate; low temp. flex plasticizer for PVC film, sheeting, and coatings; for film and sheet processing, coated fabrics for rainwear, film for luggage and accessories, coated industrial fabrics exposed to refrigeration; used in PVC plastisols, rubber formulations; APHA 50 max. clear oily liq.; m.w. 370; b.p. 224 C; insol. in water; sp.gr. 0.916-0.924; dens. 7.7 lb/gal; visc. 12.8 cSt; ref. index 1.441-1.447; flash pt. 400 F (COC); fire pt. 450 F (COC).

Santicizer 141. [Monsanto] 2-Ethylhexyl diphenyl phosphate; CAS 1241-94-7; flame-retardant low-temp. plasticizer for PVC, cellulose nitrate, CAB, ethyl cellulose, polymethyl methacrylate, PS, buna N rubber; used in vinyl film and sheeting, textile coatings, plastisols, organosols, adhesives, and pkg. materials; APHA 40 max. clear oily liq.; odorless; m.w. 362; b.p. 239 C; sol. 0.003% in water; sp.gr. 1.088-1.093; dens. 9.1 lb/gal; visc. 16.4 cSt; pour pt. -54 C; ref. index 1.507-1.510; flash pt. 435 F (COC); fire pt. 460 F (COC); 43 phr in PVC sheet: tens. str. 2930 psi; tens. elong. 320%; hardness (Shore A) 84.

Santicizer 143. [Monsanto] Proprietary, modified triaryl phosphate ester; flame-retardant plasticizer contributing clarity and toughness to end prod. for PVC and vinyl nitrile elastomers, emulsions, cellulosics; APHA 75 max. clear oily liq.; b.p. 232 C; sol. < 0.001% in water; sp.gr. 1.152; dens. 9.58 lb/gal; visc. 44.2 cSt; ref. index 1.542; flash pt. 475 F; fire pt. 525 F; 50 phr in PVC film: tens. str. 3090 psi; tens. elong. 326%; hardness (Shore A) 84.0.

Santicizer 148. [Monsanto] Isodecyl diphenyl phosphate; flame-retardant low-temp. plasticizer for PVC and copolymers, PVAc, acrylics; high solvating; for finished film or coated fabric applics., vinyl plastisols; also for PVC adhesives, ethyl cellulose, NC, SBR and butyl rubbers; APHA clear oily liq.; odorless; m.w. 390; b.p. 245 C; insol. in water; sp.gr. 1.069-1.079; dens. 8.94 lb/gal; visc. 22.5 cSt; pour pt. < -50 C; ref. index 1.503-1.509; flash pt. 465 F (COC); fire pt. 500 F (COC); 43 phr in PVC sheet: tens. str. 2930 psi; tens. elong. 1940 psi; hardness (Shore A) 83.

Santicizer 154. [Monsanto] t-Butylphenyl diphenyl phosphate; flame retardant plasticizer used in PVC, vinyl and vinyl nitrile foams, PVAc emulsions, cellulosic resins; APHA 60 max. clear mobile liq.; odorless; m.w. 368; sol. < 0.001% in water; sp.gr. 1.175-1.185; dens. 9.8 lb/gal; visc. 58 cSt; pour pt. -25 C; ref. index 1.5535-1.5565; flash pt. 505 F (COC); fire pt. 590 F (COC).

Santicizer 160. [Monsanto] Butylbenzyl phthalate; CAS 85-68-7; EINECS 201-622-7; high-solvating, stain-resisting, general-purpose plasticizer used in PVC, NC lacquers and films, PVAc, acrylic coatings, ethyl cellulose; plasticizes difficult resins, chlorinated rubber, cellulose propionate, polyamide; APHA clear oily liq.; slight char. odor; m.w. 312; b.p. 240 C; sol. 0.0003% in water; sp.gr. 1.115-1.123; dens. 9.3 lb/gal; visc. 41.5 ± 1.5 cSt; pour pt. -45 C; ref. index 1.535-1.540; flash pt. 390 F (COC); fire pt. 450 F (COC); 43 phr in PVC sheet: tens. str. 3090 psi; tens. elong. 350%; hardness (Shore A) 86.

Santicizer 261. [Monsanto] Alkyl benzyl phthalate; monomeric plasticizer offering permanence; used in film and sheeting, coated fabrics, plastisols, organosols, vinyl films, and acrylic lacquers, calendering, extrusions and vinyl foams; APHA 75 max. clear oily liq.; slight char. odor; m.w. 368; b.p. 252; sol. < -0.01% in water; sp.gr. 1.065-1.074; dens. 8.9 lb/gal; visc. 53 cSt; pour pt. -45 C; ref. index 1.523-1.529; flash pt. 445 F (COC); hardness (Shore A) 89 (43 phr in PVC).

Santicizer 278. [Monsanto] Benzyl phthalate; high solvating plasticizer offering permanence; used in PVC, paints and acrylic coatings, caulks and sealants based on chlorinated rubber, butyl rubber, polysulfides, or PU; APHA clear oily liq.; slight char. odor; m.w. 455; b.p. 243 C; insol. in water; sp.gr. 1.094-1.101; dens. 9.1 lb/gal; visc. 860 cSt; pour pt. -6.5 C; ref. index 1.517-1.521; flash pt. 440 F (COC); fire pt. 535 F (COC); tens. str. 3180 psi; tens. elong. 258%; hardness (Shore A) 97.

Santocure®. [Monsanto] N-Cyclohexyl-2-benzothiazole sulfenamide; CAS 95-33-0; accelerator for sulfur curable elastomers, esp. EPDM; fast cure rate; 7 mm pellets or powd.

Santocure® DCBS. [Monsanto] N,N-Dicyclohexyl-2-benzothiazole sulfenamide; CAS 4979-32-2; slow cure, low modulus accelerator for rubber; very long scorch delay; millipellets.

Santocure® IPS. [Monsanto] N,N-diisopropyl-2-benzothiazole sulfenamide; accelerator for sulfur-curable elastomers; slow cure rate for thick sections; 7 mm pellets.

Santocure® MOR. [Monsanto] 2-(Morpholinothio) benzothiazole; CAS 102-77-2; accelerator for sulfur-curable elastomers; slow cure rate; millipellet.

Santocure® NS. [Monsanto] N-t-Butyl-2-benzothiazole sulfenamide; CAS 95-31-8; accelerator for sulfur-curable elastomers; fast cure rate; 7 mm pellets or dust-suppressed powd.

Santocure® TBSI. [Monsanto] N-t-Butyl-2-benzothiazole sulfenamide; CAS 3741-08-8; primary amine derived accelerator providing long scorch safety and slow cure rate; reversion resist.; very stable in storage; useful in steel skim stock; powd.

Santoflex® 13. [Monsanto] N-(1,3-Dimethylbutyl) N′-phenyl-p-phenylenediamine; CAS 793-24-8; antiozonant/antiflex cracking agent; SBR stabilizer; monomer polymerization inhibitor; refinery catalyst antifouling agent; 7 mm pastiles and molten liq.

Santoflex® 77. [Monsanto] N,N′-Bis (1,4-dimethylpentyl)-p-phenylenediamine; CAS 3081-14-9; antiozonant for static applics.; oil and fuel additives, oil refinery process additives, polymerization stop, catalyst antifouling agents; liq.

Santoflex® 134. [Monsanto] Alkylaryl p-phenylenediamine blend; antiozonant/antiflex agent; polymerization inhibitor for monomers; SBR stabilizer; liq.

Santoflex® 715. [Monsanto] Alkylaryl and dialkyl p-phenylenediamine blend; antiozonant for static/dynamic applics.; liq.

Santoflex® AW. [Monsanto] 6-Ethoxy-1,2-dihydro-2,2,4-trimethylquinoline and related prod.; CAS 91-53-2; EINECS 202-075-7; antiozonant and antiflex-cracking agent; staining; liq.

Santoflex® IP. [Monsanto] N-Isopropyl-N′-phenyl-p-phenylenediamine; CAS 101-72-4; antioxidant/antiozonant for elastomers; deactivator for metal catalyzed degradation; staining; flakes; low sol. in general purpose elastomers.

Santogard® PVI. [Monsanto] N-(Cyclohexylthio) phthalimide; CAS 17796-82-6; prevulcanization inhibitor; provides predicatble scorch control in most sulfur vulcanizates without affecting cure rate or cured props.; improves cost performance through marginal stock recovery, higher processing temps.; may bloom at concs. above 0.5 phr; powd. or 80% wax pellets.

Santone® 3-1-SH. [Van Den Bergh Foods] Polyglyceryl-3 oleate; CAS 9007-48-1; nonionic; emulsifier and aerating agent used in food industry, cakes; replacement for polysorbates in icings and icing shortenings; textile and plastic lubricant; color dispersant; FDA 21CFR §172.854; kosher; wh. plastic; m.p. 84-92 F; HLB 7.2; acid no. 8 max.; iodine no. 36-48; sapon no. 125-135; hyd. no. 345-390; 100% conc.; *Storage:* store sealed in cool, dry place away from odor-producing substances; 1 yr storage life at 40-95 F.

Santonox®. [Monsanto] 4,4′-Thiobis (6-t-butyl-m-cresol); CAS 96-69-5; highly effective nonstaining antioxidant for polyolefins; forestalls degradation caused by processing heat or UV radiation; paper film coatings, polyolefin staiblization, pulp/paper stabilization, thermal print stabilization; fiber lubricant stabilization; powd.

Santovar® A. [Monsanto] 2,5-Di (t-amyl) hydroquinone; CAS 79-74-3; staining antioxidant in noncuring applics., adhesives; NBR stabilizer; polymerization arrestor; powd.

Santoweb® D. [Monsanto] Treated cellulose fibers; CAS 9004-34-6; EINECS 232-674-9; short fiber reinforcing agent for NR, SBR, BR, and CR; contains methylene acceptors; blk. fibrous aggregate.

Santoweb® DX. [Monsanto] Treated cellulose fibers; CAS 9004-34-6; EINECS 232-674-9; short fiber reinforcing agent for NR, SBR, BR, and CR; blk. fibrous aggregate.

Santoweb® H. [Monsanto] Treated cellulose fibers; CAS 9004-34-6; EINECS 232-674-9; short fiber reinforcing agent for EPDM and IIR elastomers; blk. fibrous aggregate.

Santoweb® W. [Monsanto] Treated cellulose fibers; CAS 9004-34-6; EINECS 232-674-9; short fiber reinforcing agent for PVC, NBR, and nonblack rubber compds.; cream-colored fibrous aggregate.

Santowhite® Crystals. [Monsanto] 4,4´-Thiobis-(6-t-butyl-m-cresol); CAS 96-69-5; nonstaining thermal antioxidant for NR, syns., and latexes; syn. polymer stabilizer; powd.

Santowhite® ML. [Monsanto] Butylated polymer of p-cresol dicyclopentadiene; CAS 68610-51-5; EINECS 271-867-2; nonstaining, nonmigrating antioxidant for nonblack rubber compds. and latexes, polyolefin stabilization, crosslinked PE stabilization; powd.

Santowhite® PC. [Monsanto] 2,2´-Methylenebis-(4-methyl-6-t-butyl-phenol); CAS 119-47-1; high activity nonstaining antioxidant for NR, syns., latexes; syn. polymer stabilizer; antidegradant for polyolefins and coatings; powd.

Santowhite® Powd. [Monsanto] 4,4´-Butylidenebis-(6-t-butyl-m-cresol); CAS 85-60-9; nonstaining thermal antioxidant and UV discoloration protectant for NR, syns., latexes; syn. polymer stabilizer; antidegradant for polyolefins and thermal coatings; paper film coatings, thermal print stabilization; powd.

Saret® 500. [Sartomer] Nitroso contg.; trifunctional crosslinking agents, scorch retarder; offers processing flexibility when used with the peroxide cure of elastomers; suitable for inj., transfer, and compr. molding; for elastomers, plastisols, sealants, TPE; Gardner 17-18 dk. liq., mild odor; sp.gr. 1.064; dens. 9 lb/gal; visc. 46 cps; b.p. > 200 C (1 mm); flash pt. (PMCC) 212 F; ref. index 1.4751.

Saret® 515. [Sartomer] Nitroso-contg.; trifunctional crosslinking agent, scorch retarder; offers processing flexibility when used with the peroxide cure of elastomers; suitable for inj., transfer, and compr. molding; for elastomers, plastisols, sealants, TPE; Gardner 17-18 dk. liq., mild odor; sp.gr. 1.066; dens. 9 lb/gal; visc. 47 cps; b.p. > 200 C (1 mm); flash pt. (PMCC) 155 F; 95+% act.

Saret® 516. [Sartomer] Non-nitriso; difunctional crosslinking agent, scorch retarder, coagent; incorporated into peroxide-cured natural and syn. rubber systems; for elastomers, plastisols, sealants, TPE; low staining; Gardner 3 clear liq., mild odor; sp.gr. 1.008; visc. 7 cps; 98% reactive esters.

Saret® 517. [Sartomer] Non-nitroso; trifunctional crosslinking agent, scorch retarder, coagent; incorporated into peroxide-cured natural and syn. rubber systems; for elastomers, plastisols, sealants, TPE; low staining; Gardner 4-5 clear liq., mild odor; sp.gr. 1.058; visc. 46 cps; 99% reactive esters.

Saret® 518. [Sartomer] Non-nitroso; tetrafunctional crosslinking agent, scorch retarder, coagent; incorporated into peroxide-cured natural and syn. rubber systems; for elastomers, sealants, TPE; low staining; Gardner 14-15 clear liq., mild odor; sp.gr. 1.112; visc. 624 cps; 99+% reactive esters.

Saret® 519. [Sartomer] Non-nitroso triacrylate; trifunctional crosslinking agent, scorch retarder for elastomers, plastisols, sealants, TPE; for use in rubber rolls and hoses where compr. set and resist. to moisture are important; low staining; Gardner 15 liq., pungent odor; sp.gr. 1.108; visc. 60-80 cps; 97+ % reactive ester.

Saret® 633. [Sartomer] Zinc diacrylate; CAS 14643-87-9; scorch retarder, coagent in peroxide cured stocks; used with hydrog. nitrile rubber, polybutadiene, ethylene propylene diene rubber, silicone, SBR, natural rubber; also for adhesives; fast cure rate; high crosslink dens.; off-wh. free-flowing powd.; 99% thru 200 mesh; m.w. 239; sp.gr. 1.608; dens. 1.6 mg/m^3; m.p. > 250 C; 90% reactive esters; *Storage:* store in cool, preferably air-conditioned area below 80 F in tightly closed containers; hygroscopic; use within 6 mos for optimum results.

Saret® 634. [Sartomer] Zinc dimethacrylate; scorch retarder, coagent in peroxide cured stocks; used with hydrog. nitrile rubber, polybutadiene, ethylene propylene diene rubber, silicone, SBR, natural rubber; also for adhesives; fast cure rate, high crosslink dens.; off-wh. free-flowing powd.; 99% thru 200 mesh; m.w. 235; sp.gr. 1.478; dens. 1.5 mg/m^3; m.p. 240 C; 97% reactive esters; *Storage:* store in cool, preferably air-conditioned area below 80 F in tightly closed containers; hygroscopic; use within 6 mos for optimum results.

Satintone® SP-33®. [Engelhard] Calcined aluminum silicate; reinforcing extender for rubber and polymer systems; designed as an extender for PVC elec. insulation compds.; very wh. highly pulverized powd.; 1.4 μm avg. particle size; 0.03% max. residue +325 mesh; sp.gr. 2.50 g/cc; bulk dens. 16 lb/ft^3 (loose), 30 lb/ft^3 (tamped); bulking value 0.048 gal/lb; oil absorp. 50-60 lb/100 lb; ref. index 1.62; pH 5-6; brightness (GE) 84-86; 52.2% SiO_2; 44.3% Al_2O_3.

Satintone® Whitetex®. [Engelhard] Calcined aluminum silicate; general-purpose reinforcing extender used in elastomers, PVC, polyamides; very wh. highly pulverized powd.; 1.4 μm avg. particle size; 0.02% max. residue +325 mesh; sp.gr. 2.63 g/cc; bulk dens. 18 lb/ft^3 (loose), 33 lb/ft^3 (tamped); bulking value 0.046 gal/lb; oil absorp. 50-60 lb/100 lb; ref. index 1.62; pH 5-6; brightness (GE) 90-92;

52.4% SiO_2; 44.5% Al_2O_3.

Saytex® 102E. [Albemarle] Decabromodiphenyl oxide; CAS 1163-19-5; EINECS 214-604-9; flame retardant for PS in TV and VCR cabinets, thermoplastic polyesters, nylon, PC, PP, PE, TPE, thermoplastic PU, silicone, PVC, EPDM, syn. rubber, phenolics, textiles, paints, hot-melt adheisves; high-purity elec. grade for wire and cable; wh. powd.; 5 μ avg. particle size; sol. 0.5% in toluene, < 0.1% in water, acetone, methanol; m.w. 959.2; sp.gr. 3.25; bulk dens. 44 lb/ft^3 (loose); m.p. 304 C; 97.5% asaay, 83% Br; *Toxicology:* avoid prolonged/repeated skin contact, inh. of dust, contact with eyes.

Saytex® 111. [Albemarle] Octabromodiphenyl oxide; CAS 32536-52-0; EINECS 251-087-9; flame retardant for ABS, nylon, adhesives, and coatings; semiplasticizing additive for styrenics; off-wh. powd.; sol. 35.3% in toluene, 12.2% in acetone; somewhat sol. in chlorinated and aromatic solvs.; insol. in water; sp.gr. 2.63; bulk dens. 53 lb/ft^3 (loose); m.p. 79-87 C; acid no. 0.02; 79% Br; *Toxicology:* avoid prolonged/repeated skin contact, inh. of dust, contact with eyes.

Saytex® 120. [Albemarle] Tetradecabromodiphenoxy benzene; CAS 58965-66-5; EINECS 261-526-6; flame retardant for nylon, engineering resins and alloys, styrenic polymers, PE, PP, HIPS, ABS, PBT, TPE, XLPE, epoxy, wire and cable compds., adhesives, paints; exc. thermal stability, good uv resist.; wh. powd.; 7 μ max. avg. particle size; insol. in water and common inorgs.; m.w. 1366.9; sp.gr. 3.25; bulk dens. 44 lb/ft^3 (loose); m.p. 375 C; 81.8% Br; *Toxicology:* avoid prolonged/repeated skin contact, inh. of dust, contact with eyes.

Saytex® 8010. [Albemarle] Flame retardant for HIPS, other styrenics, thermoplastic polyolefins; electronic applics., wire and cable applics. with iginition resist. at low loadings, computer enclosures which meet demanding performance requirements with lower cost resins.

Saytex® BCL-462. [Albemarle] Dibromoethyldibromocyclohexane; CAS 3322-93-8; EINECS 222-036-8; flame retardant for expandable, cryst. and high-impact PS, SAN resins, thermoplastic PU, PVC wire and cable, adhesives, coatings, textile treatment; good thermal stability; wh. cryst. powd.; sol. 54.2% in styrene, 52.4% in toluene, 35% in acetone, 0.06% in water; m.w. 427.8; sp.gr. 2.38; bulk dens. 47 lb/ft^3 (loose); m.p. 70-76 C; 74.7% Br; *Toxicology:* avoid prolonged/repeated skin contact, inh. of dust, contact with eyes.

Saytex® BN-451. [Albemarle] Ethylenebis dibromonorbornane dicarboximide; CAS 52907-07-0; EINECS 258-250-3; thermally stable flame retardant for PP, polyamides 610 and 612, PU elastomers and coatings, textiles; outstanding uv stability; off-wh. powd.; sol. ≤ 0.1% in water, methanol, toluene; m.w. 672.0; sp.gr. 2.05; bulk dens. 23 lb/ft^3 (loose); m.p. 297-304 C; acid no. 0.28; 47.6% Br; *Toxicology:* avoid prolonged/repeated skin contact, inh. of dust, contact with eyes.

Saytex® BT-93®. [Albemarle] Ethylenebistetrabromophthalimide; CAS 32588-76-4; EINECS 251-118-6; flame retardant for HIPS, ABS, PE, PP, PBT, EPDM, PC, XLPE, epoxy, ionomer resins, textile treatment, elec. parts, underhood automotive, wire and cable applics. with ignition resist.; exc. thermal and uv stability; lt. yel. powd.; < 16 μ particle size; insol. in water and org. solvs.; m.w. 951.5; sp.gr. 2.77; bulk dens. 33 lb/ft^3 (loose); m.p. 450-455 C; acid no. 1.0 max.; 67.2% Br; *Toxicology:* avoid prolonged/repeated skin contact, inh. of dust, contact with eyes.

Saytex® BT-93W. [Albemarle] Ethylenebistetrabromophthalimide; CAS 32588-76-4; EINECS 251-118-6; flame retardant for elec. parts, switches, connectors, computer housings, film, wire and cable, underhood automotive, and composites for PBT, HIPS, ABS, PE, PP, epoxy, and elastomers; wh. powd.; 99% < 16 μ particle size; sol. < 0.1% in water, acetone, methanol, toluene; m.w. 951.5; sp.gr. 2.77; bulk dens. 33 lb/ft^3 (loose); m.p. 450-455 C; acid no. 1.0 max.; 67.2% Br; *Toxicology:* avoid prolonged/repeated skin contact, inh. of dust, contact with eyes.

Saytex® FR-1138. [Albemarle] 84-87% Dibromoneopentyl glycol, 10-15% tribromoneopentyl alcohol, and 1-3% monobromopentaerythritol; reactive flame retardant for unsat. polyesters, rigid and flexible PU foam, PU elastomers; wh. sm. flakes; sol. 56.4% in methanol, 55.7% in acetone, 42.0% in propylene glycol, 27.2% in ethylene glycol, 2% in water; m.w. 261.9; sp.gr. 2.04; bulk dens. 53.92 lb/ft^3 (loose); m.p. 85-105 C; hyd. no. 395-425; 61% Br; *Toxicology:* avoid prolonged/repeated skin contact, inh. of dust, contact with eyes.

Saytex® HBCD-LM. [Albemarle] Hexabromocyclododecane; CAS 3194-55-6; EINECS 221-695-9; low-melting flame retardant for low-dens. expandable and extruded PS foam, crystal and high-impact PS, PP, PVC wire and cable, textiles, hot-melt adhesives, coatings; good thermal stability; wh. powd. or gran.; sol. in common solvs.; sol. 12.3% in styrene, 10.2% in acetone, 9.7% in toluene, < 0.01% in water; m.w. 641.7; sp.gr. 2.36; bulk dens. 69 lb/ft^3 (gran., loose), 60 lb/ft^3 (powd., loose); m.p. 178-188 C; 74.7% Br; *Toxicology:* avoid prolonged/repeated skin contact, inh. of dust, contact with eyes.

Saytex® RB-49. [Albemarle] Tetrabromophthalic anhydride; CAS 632-79-1; EINECS 211-185-4; reactive flame retardant filler for unsat. polyester, epoxy; reactive intermediate for preparation of polyols, esters, and imides; textile applics.; exc. thermal stability; wh. powd.; sol. in dimethyl formamide, nitrobenzene, 8.1% in styrene, < 0.1% in water; m.w. 463.7; sp.gr. 2.91; bulk dens. 71 lb/ft^3 (loose); m.p. 276-282 C; 68.9% Br; *Toxicology:* avoid prolonged/repeated skin contact, inh. of dust, contact with eyes.

Saytex® RB-79. [Albemarle] Tetrabromophthalate diol; CAS 20566-35-2; EINECS 243-885-0; reactive flame-retardant for rigid urethane and urethane-modified isocyanurate foams for UL and factory mutual class #1 and 2 specs; slab stock, pour-in-place, and spray foams; also for urethane coatings and elastomers; amber liq.; sol. > 100% in acetone, toluene, < 0.1% in water, methanol, CFC-11; m.w. 628.0; sp.gr. 1.8; dens. 15.02 lb/gal; visc. 80,000-120,000 cps; acid no. 0.5 min.; hyd. no. 200-235; 45% Br; *Toxicology:* avoid prolonged/ repeated skin contact, inh. of dust, contact with eyes.

Saytex® RB-100. [Albemarle] Tetrabromobisphenol A; CAS 79-94-7; EINECS 201-236-9; reactive or additive source of bromine for flame retardancy for ABS, PC, epoxy, phenolics, unsat. polyester, paints; reactive intermediate for preparation of brominated epoxy resins, PC, and unsat. polyesters; intermediate for other flame retardants; for circuit board mfg.; wh. powd.; sol. in alcohols and ketones; sl. sol. in aromatic and halogenated solvs.; insol. in water; m.w. 543.9; sp.gr. 2.18; bulk dens. 67 lb/ft^3 (loose); f.p. 181 C; 98.5% min. assay; 58.7% Br; *Toxicology:* avoid prolonged/repeated skin contact, inh. of dust, contact with eyes.

Saytex® VBR. [Albemarle] Monomeric vinyl bromide; CAS 593-60-2; flame retardant; intermediate in org. synthesis and in the mfg. of flame retardants, polymers, copolymers, pharmaceuticals, fumigants, and other chemicals; also used in modified acrylic fibers, textiles, adhesives, coatings, photographic plates and; off-wh. low-boiling liq., char. pungent odor; m.w. 106.96; f.p. -139.2 C; b.p. 15.8 C (760 mm Hg); dens. 12.7 lb/gal (20 C); flash pt. (COC) none; 99.8% vinyl bromide; 74.5% Br.

SB-30. [J.M. Huber/Engineered Mins.] Alumina trihydrate; flame retardant, smoke suppressor, high loading filler, resin extender in polyester, vinyls, PU, latex, neoprene foam systems, wire and cable insulation, vinyl wall and floor coverings, epoxies; used in pharmaceuticals, epoxy elec. laminates, epoxy encapsulating/casting, FRP elec. laminates, SMC/BMC phenolic compr. molding, in plastics, rubber, and cellulosics; off-wh. unground particulate; 50 μ median particle diam.; 10% on 100 mesh, 95% on 325 mesh; sp.gr. 2.42; bulk dens. 1.2 g/cc (loose); surf. area 0.1 m^2/g; brightness (TAPPI) 80; ref. index 1.57; hardness (Moh) 2.5-3.5; 64.9% Al_2O_3.

SB-31. [J.M. Huber/Engineered Mins.] Alumina trihydrate; flame retardant, smoke suppressor, high loading filler, resin extender in polyester, vinyls, PU, latex, neoprene foam systems, wire and cable insulation, vinyl wall and floor coverings, epoxies; off-wh. particulate; median particle diam. 35 μ, 35-70% thru 325 mesh; sp.gr. 2.42; dens. 1.0 g/cc; surf. area 0.5 m^2/g; 65.0% Al_2O_3; Unverified

SB-31C. [J.M. Huber/Engineered Mins.] Alumina trihydrate; flame retardant, smoke suppressor, high loading filler, resin extender in polyester, vinyls, PU, latex, neoprene foam systems, wire and cable insulation, vinyl wall and floor coverings, epoxies; off-wh. particulate; median particle diam. 90 μ, 3-15% thru 325 mesh; sp.gr. 2.42; dens. 1.2 g/cc; surf. area 0.15 m^2/g; 65.0% Al_2O_3; Unverified

SB-136. [J.M. Huber/Engineered Mins.] Alumina trihydrate; flame retardant, smoke suppressor, high loading filler, resin extender in polyester, flexible PVC, PU, latex, neoprene foam systems, wire and cable insulation, vinyl wall and floor coverings, epoxy encapsulating/potting; off-wh. ground particulate; 18.5 μ median particle diam.; 85% thru 325 mesh; sp.gr. 2.42; bulk dens. 0.8 g/cc (loose); surf. area 2-3 m^2/g; oil absorp. 20-22; brightness (TAPPI) 86; ref. index 1.57; hardness (Moh) 2.5-3.5; 64.9% Al_2O_3.

SB-331. [J.M. Huber/Engineered Mins.] Alumina trihydrate; flame retardant, smoke suppressor, high loading filler, resin extender in polyester, vinyls, PU, latex, neoprene foam systems, wire and cable insulation, vinyl wall and floor coverings, epoxies; also builds visc. fast, thus limiting load factor; good wet out, blending, and suspension properties; SMC/BMC compression molding, epoxies, polishes, adhesives, and coatings; particulate; median particle diam. 10-12 μ, 98% thru 325 mesh; sp.gr. 242; dens. 0.6-0.8 g/cc; surf. area 3-4 m^2/g; oil absorp. 27-28; ref. index 1.57; hardness 2.5-3.5 Moh; 65.0% Al_2O_3; Unverified

SB-332. [J.M. Huber/Engineered Mins.] Alumina trihydrate; high loading med. grind filler, flame retardant, smoke suppressant for thermoplastics, thermosets, adhesives, paints, coatings, paper, elastomers, and rubber prods.; esp. for adhesives/ sealants; ground particulate; 11 μ median particle diam.; 98% thru 325 mesh; sp.gr. 2.42; bulk dens. 0.7 g/cc (loose); surf. area 3-4 m^2/g; oil absorp. 27-28; brightness (TAPPI) 88; ref. index 1.57; hardness (Moh) 2.5-3.5; 64.9% Al_2O_3.

SB-335. [J.M. Huber/Engineered Mins.] Alumina trihydrate; flame retardant, smoke suppressor, high loading filler for thermoplastics, thermosets, adhesives, paints, coatings, paper, elastomers, rubber prods.; for spray-up or land lay-up FRP applic., filament winding, panel prod., resin injection, SMC/ BMC/acrylic sheet rigidizing, and cast polyester parts; particulate; median particle diam. 14 μ, 90-94% thru 325 mesh; sp.gr. 2.42; dens. 0.7 g/cc; surf. area 2-3 m^2/g; oil absorp. 25-26 ml/100 g; ref. index 1.57; 65.0% Al_2O_3; Unverified

SB-336. [J.M. Huber/Engineered Mins.] Alumina trihydrate; high loading filler, flame retardant, smoke suppressor; for spray-up or land lay-up FRP applic., filament winding, panel prod., resin injection, SMC/BMC, acrylic sheet rigidizing, cast polyester parts, epoxy encapsulating/potting, powd. coatings; ground particulate; 15.5 μ median particle diam.; 95% thru 325 mesh; sp.gr. 2.42; bulk dens. 0.75 g/cc (loose); oil absorp. 22-24; brightness (TAPPI) 87; ref. index 1.57; hardness (Moh) 2.5-3.5; 64.9% Al_2O_3.

SB-431. [J.M. Huber/Engineered Mins.] Alumina trihydrate; high loading filler, flame retardant, smoke suppressor; used for SMC, BMC, and resin inj.; provides optimum in visc., flame, elec., and molding properties; rapid disp. in resin; mold flow and wet-out chars; yield exc. surf. profile, minimal porosity, pigmentation, filler and reinforcement distribution; particulate; wh. median particle diam. 8-10 μ, 99.5% thru 325 mesh; sp.gr. 2.42; dens. 0.6-0.7 g/cc; surf. area 4-5 m^2/g; oil absorp. 27-29; ref. index 1.57; hardness (Mohs) 2.5-3.5; 64.9% Al_2O_3; Unverified

SB-432. [J.M. Huber/Engineered Mins.] Alumina trihydrate; high loading filler, flame retardant, smoke suppressor for adhesives/sealants, FRP contact molding, polyester thermoset molding compds., SMC/BMC, phenolic compr. molding, flexible PVC, plastisols; off-wh. ground particulate; 9 μ median particle diam.; 99.9% thru 325 mesh; sp.gr. 2.42; bulk dens. 0.65 g/cc (loose); surf. area 4-5 m^2/g; oil absorp. 27-29; brightness (TAPPI) 89; ref. index 1.57; hardness (Moh) 2.5-3.5; 64.9% Al_2O_3.

SB-632. [J.M. Huber/Engineered Mins.] Alumina trihydrate; filler, flame retardant, smoke suppressor for plastic and rubber applic.; suitable where high surf. area and high loading levels are required; used in adhesives/sealants, , neoprene foam, urethane flexible foam, silicone rubber, rigid/flexible PVC, plastisols, wire and cable insulation, SBR belting, EPDM, EPR, coatings, ABS, polyethylene, and XPLE; particulate; 3.5 μ median particle diam.; 99.99% thru 325 mesh; sp.gr. 2.42; bulk dens. 0.45 g/cc (loose); surf. area 5-6 m^2/g; oil absorp. 32; brightness (TAPPI) 90-95; ref. index 1.57; hardness (Moh) 2.5-3.5; 64.9% Al_2O_3.

SB-805. [J.M. Huber/Engineered Mins.] Alumina trihydrate; high loading filler, flame retardant, smoke suppressor commonly used where a high degree of active surf. area is required and high loading levels must be maintained; for neoprene foam, thermoplastics, thermosets, adhesives, paints, coatings, paper; superfine powd.; 2.6 μ median particle diam.; 99.99% thru 325 mesh; sp.gr. 2.42; bulk dens. 0.4 g/cc (loose), 0.7 g/cc (packed); surf. area 5-6 m^2/g; oil absorp. 34; brightness (TAPPI) 90-95; ref. index 1.57; hardness (Moh) 2.5-3.5; 64.9% Al_2O_3.

SB-932. [J.M. Huber/Engineered Mins.] Alumina trihydrate; high loading filler, flame retardant, smoke suppressor; used for applic. requiring high surf. area prod.; improved physical and handling properties providing low cost approach to flame retardance and smoke suppression; in coatings, adhesives, vinyls, wire and cable insulation, EPDM, EPR, ABS, polyolefins, and XLPE; particulate; median particle diam. 1.5 μ, 100% thru 325 mesh; sp.gr. 2.42; dens. 0.3 g/cc; surf. area 6.5 m^2/g; oil absorp. 38; ref. index 1.57; hardness (Mohs) 2.5-3.5; 64.9% Al_2O_3; Unverified

SC-53. [ECC Int'l.] Calcium carbonate; CAS 471-34-1, 1317-65-3; coarse ground pigment, filler, reinforcement for paper, paint, polymers, food, esp. polyolefins, carpet backing, caulks, sealants, putties, as mild abrasives in cleaners; wh. powd., odorless; 53% thru 325 mesh; negligible sol. in water; sp.gr. 2.71; dens. 22.57 lb/gal; bulk dens. 85 lb/ft^3 (loose); bulking value 0.044; ref. index 1.59; pH 9.5; nonflamm.; *Toxicology:* TLV/TWA 10 mg/m^3, considered nuisance dust; *Precaution:* incompat. with acids.

Schercophos NP-6. [Scher] Nonoxynol-6 phosphate; anionic; emulsifier for polar and nonpolar solvs., emulsion polymerization; corrosion inhibitor; liq.; 100% conc.

Schercowet DOS-70. [Scher] Sodium dioctyl sulfosuccinate; anionic; emulsifier for emulsion polymerization; wetting and rewetting agent for textile wet processing; liq.; 70% act.

Scorchguard '0'. [Rhein Chemie] 74% Activated magnesia in 26% org. binder; predispersed rubber chem.; buff/gray bars; sp.gr. 1.94; 74% act.

Scotchlite™. [3M/Industrial Spec.] Glass bubbles; engineered fillers, resin extenders for thermoplastics with high durability and chem. resist.; for transportation, aerospace, automotive, construction, explosives, and sporting goods industries.

Secolat. [Stepan Europe] Alkyl disodium sulfosuccinamate; anionic; foamer for latex emulsions; emulsifier for emulsion polymerization; liq.; 35% conc.; Unverified

Secosol® ALL40. [Stepan Europe] Sodium laureth sulfosuccinate; anionic; mild foamer for shampoos, bubble baths, liq. soaps, shower gels, bath salts; emulsifier for emulsion polymerization; water-wh. to pale yel. liq.; 31% act.

Secosol® DOS 70. [Stepan Europe] Sodium dioctyl sulfosuccinate; anionic; emulsifier, dispersant and wetting/rewetting agent for household/industrial cleaners, emulsion polymerization, paints, inks, oilfield prod.; textile additive; water-wh. to pale yel. liq.; 70% act.

Secoster® SDG. [Stepan Europe] Glyceryl stearate; nonionic; antistat and lubricant for polyolefins; wh. solid; 100% act.

Seenox 412S. [Witco/PAG] Pentaerythrityl tetrakis (β-laurylthiopropionate); CAS 29598-76-3; antioxidant for polyolefins, thermoplastic elastomers, engineering thermoplastics; synergistic with primary antioxidants; outstanding long-term heat aging performance; wh. cryst. powd.

Select-A-Sorb Powdered, Compacted. [R.T. Vanderbilt] Industrial talc (hydrous magnesium silicate); CAS 14807-96-6; EINECS 238-877-9; reinforcing filler for NR and syn. rubbers, wire and cable stocks; high hydrophobicity; mixes quickly into elastomers, gives low moisture pickup in vulcanizates; FDA 21CFR §175.300, 177.1210, 182.90; wh. powd.; 2.4 μm median particle size; 99.9% < 325 mesh; dens. 2.75 mg/m^3; bulk dens. 22 lb/ft^3 (compacted); oil absorp. 56 ; brightness 85; pH 9.6 (10% aq. slurry).

Sellogen HR. [Henkel/Organic Prods.; Henkel/Textiles; Henkel Canada; Henkel-Nopco] Sodium dialkyl naphthalene sulfonate; anionic; wetting agent, dispersant; heavy duty household and industrial cleaners; fire control, paint strippers, agric. chemical formulations, emulsion and suspension polymerization aids, latex paints, textile dyeing on cellulosics; stable in high concs. of electrolytes; EPA-exempt; lt. tan powd.; water-sol.; dens. 0.40 g/cc; surf. tens. 44.1 dynes/cm (0.25%); pH 9.0 (5%); 75% act.

SEM-135. [Harcros] Silicone emulsion; lubricant and release agent for rubber, plastics, metalworking; water-disp.; 35% act.; Unverified

Semtol® 40. [Witco/Petroleum Spec.] Wh. min. oil, tech.; white oils functioning as binder, carrier, conditioner, defoamer, dispersant, extender, heat transfer agent, lubricant, moisture barrier, plasticizer, protective agent, and/or softener in adhesives, agric., chemicals, cleaning, pkg., plastics, and textiles industries; FDA 21CFR §178.3620b; sp.gr. 0.804-0.820; visc. 4-5 cSt (40 C); flash pt. 135 C min.; pour pt. 2 C max.

Semtol® 70. [Witco/Petroleum Spec.] Wh. min. oil, tech.; white oils functioning as binder, carrier, conditioner, defoamer, dispersant, extender, heat transfer agent, lubricant, moisture barrier, plasticizer, protective agent, and/or softener in adhesives, agric., chemicals, cleaning, pkg., plastics, textiles; FDA 21CFR §178.3620b; sp.gr. 0.837-0.853; visc. 11-14 cSt (40 C); pour pt. -7 C max.; flash pt. 177 C min.

Semtol® 85. [Witco/Petroleum Spec.] Wh. min. oil, tech.; white oils functioning as binder, carrier, conditioner, defoamer, dispersant, extender, heat transfer agent, lubricant, moisture barrier, plasticizer, protective agent, and/or softener in adhesives, agric., chemicals, cleaning, pkg., plastics, textiles; FDA 21CFR §178.3620b; sp.gr. 0.839-0.855; visc. 14-17 cSt (40 C); pour pt. -7 C max.; flash pt. 179 C min.

Semtol® 100. [Witco/Petroleum Spec.] Wh. min. oil, tech.; white oils functioning as binder, carrier, conditioner, defoamer, dispersant, extender, heat transfer agent, lubricant, moisture barrier, plasticizer, protective agent, and/or softener in adhesives, agric., chemicals, cleaning, pkg., plastics, textiles; FDA 21CFR §178.3620b; sp.gr. 0.839-0.855; visc. 18-20 cSt (40 C); pour pt. -7 C max.; flash pt. 182 C min.

Semtol® 350. [Witco/Petroleum Spec.] Wh. min. oil, tech.; white oils functioning as binder, carrier, conditioner, defoamer, dispersant, extender, heat transfer agent, lubricant, moisture barrier, plasticizer, protective agent, and/or softener in adhesives, agric., chemicals, cleaning, pkg., plastics, textiles; FDA 21CFR §178.3620b; sp.gr. 0.850-0.890; visc. 64-90 cSt (40 C); pour pt. -12 C max.; flash pt. 216 C min.

Sequestrene® 30A. [Ciba-Geigy/Dyestuffs] Tetrasodium EDTA; CAS 64-02-8; EINECS 200-573-9; chelating agent for water softening, enhanced oil recovery, liq. soaps, detergents, chemical cleaning, scale removal, beerstone removal, cosmetics, photography, textile, paper, leather, metal treatment, syn. rubber, plastics; straw to pale yel. liq.; misc. with water; sp.gr. 1.29-1.31; chel. value 100; pH 11.5-12.5 (10% aq.); 39% act. in water.

Sequestrene® NAFe 13% Fe. [Ciba-Geigy/Dyestuffs] Monosodium ferric EDTA; chelating agent; micronutrient; animal feeds, photographic uses, polymerization catalyst for syn. rubber; powd.

Serdet DFK 40. [Servo Delden BV] Sodium lauryl sulfate; CAS 151-21-3; EINECS 205-788-1; anionic; personal care and dishwashing formulations; emulsifier in emulsion polymerization; paste; biodeg.; 41% conc.; Unverified

Serdet DJK 30. [Servo Delden BV] Sodium decyl sulfate; anionic; emulsifier for emulsion polymerization; liq.; 28% conc.

Serdet DM. [Servo Delden BV] Dodecylbenzene sulfonic acid; anionic; used in scouring powds. and liq. detergents; emulsifier for emulsion polymerization; liq.; 97% conc.; Unverified

Serdet DNK 30. [Servo Delden BV] Sodium nonoxynol-4 sulfate; anionic; detergent base; emulsifier for emulsion polymerization; liq.; 31% conc.; Unverified

Serdet DSK 40. [Servo Delden BV] Sodium 2-ethylhexyl sulfate; anionic; wetting agent; latex stabilizer; spreading, dispersing agent, detergent; emulsion polymerization; liq.; water-sol.; 40% conc.; Unverified

Serdet Perle Conc. [Servo Delden BV] Lauryl ether sulfate and alkylolamide; anionic; wetting agent; latex stabilizer; spreading, dispersing agent, detergent; emulsion polymerization; liq.; 34% conc.; Unverified

Serdox NNP 25. [Servo Delden BV] Nonoxynol-25; CAS 9016-45-9; nonionic; emulsifier for emulsion polymerization; solid; HLB 16.5; 100% conc.; Unverified

Serdox NNP 30/70. [Servo Delden BV] Nonoxynol-30; CAS 9016-45-9; nonionic; dyeing assistant; lime soap dispersant; emulsifier and stabilizer for emulsion polymerization; liq.; HLB 17.0; 70% conc.; Unverified

Serdox NOP 30/70. [Servo Delden BV] Octoxynol-30; CAS 9002-93-1; nonionic; emulsifier and stabilizer for emulsion polymerization; liq.; HLB 17.5; 70% conc.; Unverified

Serdox NOP 40/70. [Servo Delden BV] Octoxynol-40; CAS 9002-93-1; nonionic; emulsifier and stabilizer for emulsion polymerization; liq.; HLB 18.0; 70% conc.; Unverified

Serdox NSG 400. [Servo Delden BV] PEG-9 stearate; CAS 9004-99-3; EINECS 226-312-9; nonionic; biodeg. surfactant for textile processing and cosmetic emulsions; antistat for plastics; solid; water-disp.; 100% conc.

Sermul EA 30. [Servo Delden BV] Sodium laureth-3 sulfate; CAS 9004-82-4; 68891-38-3; anionic; emulsifier for emulsion polymerization; liq.; 27% conc.

Sermul EA 54. [Servo Delden BV] Sodium nonoxynol-4 sulfate; CAS 9014-90-8; anionic; emulsifier for emulsion polymerization; liq.; 31% conc.

Sermul EA 129. [Servo Delden BV] Ammonium lauryl sulfate; CAS 2235-54-3; anionic; emulsifier for emulsion polymerization; biodeg.; liq.; 27% conc.

Sermul EA 136. [Servo Delden BV] Nonoxynol-15

phosphate; anionic; emulsifier for emulsion polymerization; liq.; 100% conc.

Sermul EA 139. [Servo Delden BV] 2-Ethylhexyl polyglycol ether (3 EO) phosphate, acid form; anionic; emulsifier for emulsion polymerization; biodeg.; liq.; 100% conc.

Sermul EA 146. [Servo Delden BV] Sodium nonylphenol polyglycol ether (15 EO) sulfate; CAS 9014-90-8; anionic; emulsifier for emulsion polymerization; liq.; 35% conc.

Sermul EA 150. [Servo Delden BV] Sodium lauryl sulfate; CAS 151-21-3; anionic; emulsifier for emulsion polymerization; biodeg.; paste; 41% conc.

Sermul EA 151. [Servo Delden BV] Sodium nonoxynol-10 sulfate; CAS 9014-90-8; anionic; emulsifier for emulsion polymerization; liq.; 35% conc.

Sermul EA 152. [Servo Delden BV] Ammonium nonoxynol-10 sulfate; anionic; emulsifier for emulsion polymerization; liq.; 35% conc.

Sermul EA 176. [Servo Delden BV] Sodium nonoxynol-10 sulfosuccinate; anionic; emulsifier for emulsion polymerization; biodeg.; liq.; 35% conc.

Sermul EA 188. [Servo Delden BV] Nonoxynol-10 phosphate, acid form; anionic; emulsifier for emulsion polymerization; liq.; 100% conc.

Sermul EA 205. [Servo Delden BV] Nonoxynol-50 phosphate, acid form; anionic; emulsifier for emulsion polymerization; liq.; 100% conc.

Sermul EA 211. [Servo Delden BV] Nonoxynol-6 phosphate; anionic; emulsifier for emulsion polymerization; liq.; 100% conc.

Sermul EA 214. [Servo Delden BV] Sodium alpha olefin sulfonate; CAS 68439-57-6; anionic; emulsifier for emulsion polymerization; liq.; 37% conc.

Sermul EA 221. [Servo Delden BV] Sodium C12-15 pareth-3 sulfosuccinate; anionic; emulsifier for emulsion polymerization; liq.; 40% conc.

Sermul EA 224. [Servo Delden BV] Oleyl polyglycol ether phosphate; anionic; emulsifier for emulsion polymerization; liq.; 100% conc.

Sermul EA 242. [Servo Delden BV] Dodecylbenzene sulfonate; anionic; emulsifier for emulsion polymerization; liq.; 100% conc.

Sermul EA 370. [Servo Delden BV] Sodium laureth sulfate; CAS 9004-82-4; anionic; emulsifier for emulsion polymerization; liq.; 28% conc.

Sermul EK 330. [Servo Delden BV] N-Talloil aminopropyl N,N-dimethylamine; cationic; emulsifier for natural resins, min. and natural oils; intermediate; liq.; 100% conc.

Sermul EN 15. [Servo Delden BV] Nonoxynol-15; CAS 9016-45-9; nonionic; emulsifier for natural resins; liq.; 100% conc.

Sermul EN 20/70. [Servo Delden BV] Nonoxynol-20; CAS 9016-45-9; nonionic; post stabilizer for emulsion polymerization; solid; 70% conc.

Sermul EN 30/70. [Servo Delden BV] Nonoxynol-30; CAS 9016-45-9; nonionic; post stabilizer for emulsion polymerization; solid; 70% conc.; Unverified

Sermul EN 145. [Servo Delden BV] Nonoxynol-30; CAS 9016-45-9; nonionic; post-stabilizer for emulsion polymerization; liq.; 70% conc.

Sermul EN 155. [Servo Delden BV] Alkyl polyglycol ether; nonionic; emulsifier for emulsion polymerization; liq.; 100% conc.

Sermul EN 229. [Servo Delden BV] Nonoxynol-12; CAS 9016-45-9; nonionic; emulsifier for natural resins; liq.; 100% conc.

Sermul EN 237. [Servo Delden BV] Alkyl polyglycol ether; nonionic; emulsifier for emulsion polymerization; liq.; 100% conc.

Sermul EN 312. [Servo Delden BV] Alkyl polyglycol ether; nonionic; emulsifier for emulsion polymerization; liq.; 100% conc.

Servoxyl VPGZ 7/100. [Servo Delden BV] Cetyl oleyl ether phosphate, acid form (7 EO); anionic; detergent in drycleaning, emulsifier in formulation of metal cleaners in emulsion polymerization; pesticide and cosmetic preparations; liq.; 100% conc.; Unverified

Servoxyl VPIZ 100. [Servo Delden BV] Acid butyl phosphate; detergent in drycleaning, emulsifier in formulation of metal cleaners in emulsion polymerization; pesticide and cosmetic preparations; liq.; 100% conc.; Unverified

Servoxyl VPNZ 10/100. [Servo Delden BV] Nonylphenol ether phosphate, acid form (10 EO); CAS 51609-41-7; anionic; detergent in drycleaning, emulsifier in formulation of metal cleaners in emulsion polymerization; pesticide and cosmetic preparations; liq.; 100% conc.; Unverified

Servoxyl VPQZ 9/100. [Servo Delden BV] Dinonylphenol polyglycol ether phosphate, acid form (9 EO); CAS 66172-82-5; EINECS 266-218-5; anionic; detergent in drycleaning, emulsifier in formulation of metal cleaners, in emulsion polymerization; pesticide and cosmetic preparations; liq.; 100% conc.; Unverified

Servoxyl VPTZ 3/100. [Servo Delden BV] 2-Ethylhexyl polyglycol ether phosphate, acid form; anionic; detergent in drycleaning, emulsifier in formulation of metal cleaners in emulsion polymerization; pesticide and cosmetic preparations; liq.; 100% conc.; Unverified

Servoxyl VPTZ 100. [Servo Delden BV] 2-Ethylhexyl phosphate, acid form; CAS 39407-03-9; anionic; detergent in drycleaning, emulsifier in formulation of metal cleaners, in emulsion polymerization; pesticide and cosmetic preparations; liq.; 100% conc.; Unverified

Servoxyl VPUZ. [Servo Delden BV] Acid methyl phosphate; detergent in drycleaning, emulsifier in formulation of metal cleaners in emulsion polymerization; pesticide and cosmetic preparations; liq.; 100% conc.; Unverified

Servoxyl VPYZ 500. [Servo Delden BV] Acid phosphate ester, modified; anionic; detergent in drycleaning, emulsifier in formulation of metal cleaners in emulsion polymerization; pesticide and cosmetic preparations; liq.; 100% conc.; Unverified

Setsit® 5. [R.T. Vanderbilt] Activated dithiocarbamate; accelerator for NR, SBR latexes; primary accelerator for latex; dilute 1:1 with water before adding; sec. accelerator for thiazole activation;

redsh. brn. liq.; sol. in alcohol, acetone; dens. 1.01 ± 0.02 mg/m^3; flash pt. 55 C.

Setsit® 9. [R.T. Vanderbilt] Activated dithiocarbamate blend; accelerator for NR, SBR, and CR latexes; activator for NR, SBR, and NR/SBR latex foam; amber to br. liq.; mod. sol. in toluene, gasoline, chloroform; misc. with water, alcohol; dens. 1.00 ± 0.02 mg/m^3; flash pt. (PM) 55 C.

Setsit® 104. [R.T. Vanderbilt] Activated dithiocarbamate; ultrafast accelerator for NR, SBR, and NR/SBR latex foam and film applics.; little or no precure will occur if zinc oxide is withheld from compd.; should be diluted 1:2 with water; amber liq.; water-misc.; dens. 1.18 ± 0.02 mg/m^3.

SF18-350. [GE Silicones] Dimethicone fluid; lubricant, antifoam, mold release; rubber and plastic lubricant; base fluid for grease; mold release for rubber, plastic, and food applics.; antifoam in food applic. (fermentation, corn oil mfg., deep-fat frying) and aq. defoaming formulations; FDA compliance; water-wh. clear oily liq., tasteless, odorless; disp. in org. solvs.; sp.gr. 0.973; dens. 8.0 lb/gal; visc. 350 cs; pour pt. -58 F; flash pt. (PMCC) 204 C; ref. index 1.4030.

SF81-50. [GE Silicones] Dimethyl silicone; defoamers, release agents, in cosmetics, polishes, paint additives, and mechanical devices; lubricant in rubber or plastic-to-metal applics.; for damping and heat transfer in mechanical/elec. applics.; water-wh. clear, oily fluid; sp.gr. 0.972; visc. 50 cstk; pour pt. -120 F; ref. index 1.4030; flash pt. 460 F; surf. tens. 21.0 dynes/cm; conduct. 0.087 Btu/h-°F ft^2/ft; sp. heat 0.36 Btu/lb/F; dissip. factor 0.0001; dielec. str. 35.0 kV; dielec. const. 2.74; vol. resist. 1 x 10^{14} ohm-cm.

SF96®. [GE Silicones] Polydimethylsiloxane; nonionic; emollient, lubricant for polishes, antifoams, textiles, chemical specialties; plastic and rubber lubrication; dampening or heat transfer fluids; oil defoamer, paint additives; mold release for tires, rubber, plastics; textile softener/modifier; water-wh. liq.; sol. in aliphatic, aromatic, and chlorinated hydrocarbons, alcohols, and ketones, higher hydrocarbons; sp.gr. 0.916-0.974; dens. 8.0 lb/gal; visc. 50-1000 cs; pour pt. -120 to -58 F; flash pt. (PMCC) 204 C; ref. index 1.3970-1.4035; 100% act.

SF96® (100 cst) [GE Silicones] Dimethicone; used for damping, heat transfer, hydraulic fluids, rubber/plastic, base fluid for grease, polishes, cosmetics/toiletries, mold release for tires, rubber, plastics, aq. defoaming prods., thread/fiber lubricants; paint additives for flow control, mar resist., gloss; sp.gr. 0.968; visc. 100 cSt; pour pt. -67 F; flash pt. (PMCC) 575 F; ref. index 1.4030; surf. tens. 20.9 dynes/cm; sp. heat 0.36 Btu/lb/F; dissip. factor 0.0001; dielec. str. 35 kV; dielec. const. 2.74; vol. resist. 1 x 10^{14} ohm-cm.

SF96® (350 cst) [GE Silicones] Dimethicone; used for damping, heat transfer, hydraulic fluids, rubber/plastic, base fluid for grease, polishes, cosmetics/toiletries, mold release for tires, rubber, plastics, aq. defoaming prods., thread/fiber lubricants; paint additives for flow control, mar resist., gloss; sp.gr. 0.973; visc. 350 cSt; pour pt. -58 F; flash pt. (PMCC) 637 F; ref. index 1.4032; surf. tens. 21.1 dynes/cm; sp. heat 0.36 Btu/lb/F; dissip. factor 0.0001; dielec. str. 35 kV; dielec. const. 2.75; vol. resist. 1 x 10^{14} ohm-cm.

SF96® (500 cst) [GE Silicones] Dimethicone; used for damping, heat transfer, hydraulic fluids, rubber/plastic, base fluid for grease, polishes, cosmetics/toiletries, mold release for tires, rubber, plastics, aq. defoaming prods., thread/fiber lubricants; paint additives for flow control, mar resist., gloss; sp.gr. 0.973; visc. 350 cSt; pour pt. -58 F; flash pt. (PMCC) 662 F; ref. index 1.4033; surf. tens. 21.1 dynes/cm; sp. heat 0.36 Btu/lb/F; dissip. factor 0.0001; dielec. str. 35 kV; dielec. const. 2.76; vol. resist. 1 x 10^{14} ohm-cm.

SF96® (1000 cst) [GE Silicones] Dimethicone; used for damping, heat transfer, hydraulic fluids, rubber/plastic, base fluid for grease, polishes, cosmetics/toiletries, mold release for tires, rubber, plastics, aq. defoaming prods., thread/fiber lubricants; paint additives for flow control, mar resist., gloss; sp.gr. 0.974; visc. 350 cSt; pour pt. -58 F; flash pt. (PMCC) 658 F; ref. index 1.4035; surf. tens. 21.1 dynes/cm; sp. heat 0.36 Btu/lb/F; dissip. factor 0.0001; dielec. str. 35 kV; dielec. const. 2.77; vol. resist. 1 x 10^{14} ohm-cm.

SF1080. [GE Silicones] Methyl alkyl polysiloxane silicone fluid; nonionic; mold release agent, lubricant; used in rubber, plastic, and metal industries, hair care prods.; internal mold release agent in vinyl slush molding; aluminum die cast mold release agent; lt. yel. liq.; sol. in aliphatic, aromatic, and chlorinated hydrocarbons, higher alcohols, and higher ketones; sp.gr. 1.035; visc. 1500 cs; pour pt. -50 F; ref. index 1.4930; flash pt. 400 F (CC); 100% act.

SF1188. [GE Silicones] Dimethicone copolyol; emollient, lubricant, and release agent for cosmetics and toiletries, paint, plastic mold release, and rubber lubricants; textile softener/modifier and thread/fiber lubricants; amber clear fluid; sol. in water below 43 C; sol. in acetone, toluene, lower alcohols, and some hydrocarbons; sp.gr. 1.04; dens. 8.65 lb/gal; visc. 1000 cps; flash pt. (PMCC) 200 F; ref. index 1.4470; surf. tens. 25.5 dynes/cm; 100% silicone.

SF1265. [GE Silicones] Polydimethyl diphenyl siloxane fluid; for heat transfer applics., base fluid in high temp. greases, plastics additive, dielec. coolant, high temp. ultrasonic coupler, high temp. bath and oxide protector for solder baths; lt. straw liq.; sp.gr. 1.10; visc. 450 cstk; ref. index 1.5250; flash pt. 600 F; surf. tens. 25.0 dynes/cm.

SFR 100. [GE Silicones] Silicone fluid; flame retardant for polyolefins; excellent lubrication during processing; improves impact resistance of PP; easy mold fill and lower temp. processing; colorless clear high visc. liq.; sp.gr. 1.0; dens. 1.04 g/cc; visc. 300,000-900,000 cps; flash pt. 202 C (PM); 100% silicone conc.

Si 69. [Degussa] Bis(3-triethoxysilylpropyl) tetrasulfane; CAS 40372-72-3; reinforcing agent, crosslinking/coupling agent for rubber compds. contg. silicic acid fillers (silica, clay, and talc); couples nonblk. pigments; curing agent for good heat aging; provides cure equilibrium for reversion resist.; for footwear, rolls, mech. molded goods, hose, solid tires, belts, articles which undergo dynamic stress; ylsh. brn. liq.; m.w. 537.4; sp.gr. 1.074 (20 C); visc. 7-8 cps; set pt. 80 C; b.p. > 250 C (dec.); flash pt. 218 C; 22% min. total sulfur, 9.5% min. polysulfidic sulfur.

Si 264. [Degussa] 3-Thiocyanatopropyltriethoxy silane; reinforcing agent improving props. of fillers (silicas, silicates, clays and whitings) in unsat. polymers with double bonds (NR, IR, SBR, BR, NBR, EPDM); not a dangerous substance as defined by German transport regs.; amber liq., lt. typical nonoffensive odor; sol. in all common org. solvs.; insol. in water; m.w. 249.3; dens. 1 g/cc; visc. 3 mPa•s; b.p. > 280 C (dec.); flash pt. (PM) 125 C; *Toxicology:* sl. toxic; *Storage:* 12 mos storage life in original closed container @ R.T.

Sico®. [BASF AG] Predominantly azo pigments; for baking and air-drying finishes, for letterpress, offset, flexographic and special gravure inks, for coloring plastics.

Sicolen®. [BASF AG] Org. and inorg. pigments in polyethylene; for coloring polyethylene hollow articles, inj. molding and extrusion articles.

Sicolub® DSP. [BASF AG] Distearyl phthalate; CAS 90193-76-3; EINECS 290-580-3; lubricant for PVC; powd.

Sicolub® E. [BASF AG] Montan wax deriv.; CAS 8002-53-7; EINECS 232-313-5; lubricant for PVC; powd.

Sicolub® EDS. [BASF AG] Ethylene diamine distearyl amide; lubricant for PVC and PS; powd.

Sicolub® LB 1, LB 2, LBK, LK 3. [BASF AG] Syn. wax and montan wax-based; lubricant for PVC; powd.

Sicolub® OA 2, OA 4. [BASF AG] Oxidized polyethylene wax; lubricant for PVC; powd.

Sicolub® OP. [BASF AG] Partly saponified montan wax; CAS 8002-53-7; EINECS 232-313-5; lubricant for PVC; powd.

Sicolub® TDS. [BASF AG] Isotridecyl stearate; CAS 31565-37-4; lubricant for PVC; liq.

Sicomin®. [BASF AG] Chrome yellow and molybdate orange pigments; for paints and surface coatings, for coloring plastics, for flexographic and gravure inks for pkg. materials, laminated paper coloring.

Sicopal®. [BASF AG] Inorg. pigments having a spinel structure based on various metal oxides; for industrial finishes with exc. fastness and for coloring plastics.

Sicoplast®. [BASF AG] Predispersed pigment mixts. for mass coloring of thermoplastics.

Sicostab®. [BASF AG] Heat stabilizers for rigid and plasticized PVC, for internal and external applics., PVC foams.

Sicostyren®. [BASF AG] Org. and inorg. pigment concs. in polystyrene; for coloring polystyrene and styrene copolymers.

Sicotan®. [BASF AG] Inorg. mixed-phase pigments with structure of rutile titanium dioxide and other metal oxides; in combination with org. pigments for luminous hues for surface coatings and plastics with high processing temps.

Sicotrans®. [BASF AG] Transparent inorg. pigments of extremely fine particle size; for high-quality paint systems, esp. metallics, and for coloring plastics.

Sicoversal®. [BASF AG] Org. and inorg. pigment concs.; for coloring thermoplastics.

Silacto®. [Kenrich Petrochemicals] Ethanol, 2,2′-oxybis-, reaction prods. with ammonia, morpholine derivs., residues; CAS 68909-77-3; activator for silica-loaded elastomeric compds.; dk. liq., ammoniacal odor; appreciable sol. in water; sp.gr. 1.10; vapor press < 1 mm Hg (20 C); b.p. 480 F; flash pt. (PMCC) 305 F; 100% act.; *Toxicology:* LD50 (oral, rat) 8.75 g/kg; severe eye irritant, skin irritant; harmful if swallowed or inhaled; *Precaution:* incompat. with acids; avoid excess heat; decomp. produces ammonia, CO, CO_2; *Storage:* store containers closed in cool, dry place.

Silcat® R. [Union Carbide] Vinylsilane; crosslinking agent for modification of polyethylene or its copolymers to yield a moisture crosslinkable system; lt. straw clear liq.; sp.gr. 0.9762; visc. 0.7 cSt; flash pt. (TCC) 20 C; Unverified

Silcogum®. [Gayson] Pigment dispersions in reactive silicone gums; color dispersions for rubber, paint, ink, and plastics industries; handles like a rubber masterbatch; FDA compliances; avail. in wh., blk., gray, red, blue, yel., brn., and grn. colors based on titanium dioxide, red oxide, cadmium red, etc.; 50-80% pigment conc.

Silcopas®. [Gayson] Pigment dispersions in silicone fluids; color dispersions for rubber, paint, ink, and plastics industries; for RTV and liq. inj. molding systems; FDA compliances; avail. in wh., blk., gray, red, blue, yel., orange, brn., and grn. colors based on titanium dioxide, red oxide, cadmium red, ultra blue, chromium oxide, etc.

Sil-Co-Sil® 40. [U.S. Silica] Ground silica; filler for paints, plastics, rubber, polishes, cleansers, ceramic frits and glazes, fiberglass, precision castings, etc.; wh. powd.; 3.7-5.8 μ avg. particle size; 0.8-1.0% +325 mesh; fineness (Hegman) 4; bulk dens. 84-87 lb/ft^3 (tapped); surf. area 3700-6100 cm^2/g; oil absorp. 23-33; brightness 84-88; pH 7.2-7.8; 99.6-99.8% SiO_2.

Sil-Co-Sil® 45. [U.S. Silica] Ground silica; filler for paints, plastics, rubber, polishes, cleansers, ceramic frits and glazes, fiberglass, precision castings, etc.; wh. powd.; 4.0 μ avg. particle size; 1.6% +325 mesh; fineness (Hegman) 3.5; bulk dens. 87 lb/ft^3 (tapped); surf. area 5700 cm^2/g; oil absorp. 27; brightness 85; pH 7.5; 99.8% SiO_2.

Sil-Co-Sil® 47. [U.S. Silica] Ground silica; filler for paints, plastics, rubber, polishes, cleansers, ceramic frits and glazes, fiberglass, precision castings, etc.; wh. powd.; 6.0 μ avg. particle size; 2.2% +325 mesh; fineness (Hegman) 3.5; bulk dens. 88

lb/ft^3 (tapped); surf. area 3800 cm^2/g; oil absorp. 26; brightness 84; pH 7.0; 99.6% SiO_2.

Sil-Co-Sil® 49. [U.S. Silica] Ground silica; filler for paints, plastics, rubber, polishes, cleansers, ceramic frits and glazes, fiberglass, precision castings, etc.; wh. powd.; 5.0-5.5 μ avg. particle size; 2.5% +325 mesh; fineness (Hegman) 3.5; bulk dens. 88-89 lb/ft^3 (tapped); surf. area 4100-4500 cm^2/g; oil absorp. 22-24; brightness 84-88; pH 6.6-7.2; 99.5-99.7% SiO_2.

Sil-Co-Sil® 51. [U.S. Silica] Ground silica; filler for paints, plastics, rubber, polishes, cleansers, ceramic frits and glazes, fiberglass, precision castings, etc.; wh. powd.; 5.0 μ avg. particle size; 4.0% +325 mesh; fineness (Hegman) 3; bulk dens. 89 lb/ft^3 (tapped); surf. area 4500 cm^2/g; oil absorp. 21; brightness 87; pH 7.2; 99.7% SiO_2.

Sil-Co-Sil® 52. [U.S. Silica] Ground silica; filler for paints, plastics, rubber, polishes, cleansers, ceramic frits and glazes, fiberglass, precision castings, etc.; wh. powd.; 5.0-5.6 μ avg. particle size; 2.6-4.0% +325 mesh; fineness (Hegman) 3; bulk dens. 88-89 lb/ft^3 (tapped); surf. area 4000-4500 cm^2/g; oil absorp. 23-26; brightness 84-85; pH 6.6-7.0; 99.5-99.6% SiO_2.

Sil-Co-Sil® 53. [U.S. Silica] Ground silica; filler for paints, plastics, rubber, polishes, cleansers, ceramic frits and glazes, fiberglass, precision castings, etc.; wh. powd.; 4.9-6.5 μ avg. particle size; 4.5-7.0% +325 mesh; fineness (Hegman) 3; bulk dens. 89-90 lb/ft^3 (tapped); surf. area 3500-4600 cm^2/g; oil absorp. 21-26; brightness 83-87; pH 6.8-7.5; 99.6-99.8% SiO_2.

Sil-Co-Sil® 63. [U.S. Silica] Ground silica; filler for paints, plastics, rubber, polishes, cleansers, ceramic frits and glazes, fiberglass, precision castings, etc.; wh. powd.; 5.5-6.6 μ avg. particle size; 6-10% +325 mesh; fineness (Hegman) 2-2.5; bulk dens. 90-92 lb/ft^3 (tapped); surf. area 3400-4200 cm^2/g; oil absorp. 21-25; brightness 80-84; pH 6.8-7.3; 99.6-99.8% SiO_2.

Sil-Co-Sil® 75. [U.S. Silica] Ground silica; filler for paints, plastics, rubber, polishes, cleansers, ceramic frits and glazes, fiberglass, precision castings, etc.; wh. powd.; 6.0-6.7 μ avg. particle size; 12-13% +325 mesh; fineness (Hegman) 1; bulk dens. 93-94 lb/ft^3 (tapped); surf. area 3400-3800 cm^2/g; oil absorp. 20-24; brightness 82-86; pH 6.8-7.2; 99.5-99.7% SiO_2.

Sil-Co-Sil® 90. [U.S. Silica] Ground silica; filler for paints, plastics, rubber, polishes, cleansers, ceramic frits and glazes, fiberglass, precision castings, etc.; wh. powd.; 6.5-7.0 μ avg. particle size; 16-21% +325 mesh; fineness (Hegman) 0; bulk dens. 95 lb/ft^3 (tapped); surf. area 3200-3500 cm^2/g; oil absorp. 20-24; brightness 82-85; pH 6.8-7.0; 99.6-99.8% SiO_2.

Sil-Co-Sil® 106. [U.S. Silica] Ground silica; filler for paints, plastics, rubber, polishes, cleansers, ceramic frits and glazes, fiberglass, precision castings, etc.; wh. powd.; 6.2-7.0 μ avg. particle size; 22-29% +325 mesh; bulk dens. 96 lb/ft^3 (tapped); surf. area 3200-3700 cm^2/g; oil absorp. 19-23 ; brightness 83-86; pH 6.8-7.0; 99.7-99.8% SiO_2.

Sil-Co-Sil® 125. [U.S. Silica] Ground silica; filler for paints, plastics, rubber, polishes, cleansers, ceramic frits and glazes, fiberglass, precision castings, etc.; wh. powd.; 6.5-7.5 μ avg. particle size; 26-35% +325 mesh; bulk dens. 96 lb/ft^3 (tapped); surf. area 3000-3500 cm^2/g; oil absorp. 18.5-20.5; brightness 83-85; pH 6.8; 99.6-99.7% SiO_2.

Sil-Co-Sil® 250. [U.S. Silica] Ground silica; filler for paints, plastics, rubber, polishes, cleansers, ceramic frits and glazes, fiberglass, precision castings, etc.; wh. powd.; 9.5 μ avg. particle size; 47% +325 mesh; bulk dens. 99 lb/ft^3 (tapped); surf. area 2400 cm^2/g; oil absorp. 20; brightness 80; pH 6.8; 99.8% SiO_2.

Silene® 732D. [PPG Industries] Amorphous precipitated silica; CAS 112926-00-8; semi-reinforcing agent for rubber; aids extrusion, resilience; provides smooth appearance, low sp.gr., improved dimensional stability; for colored EPDM rubber, other nat. and syn. rubbers incl. SBR, neoprene, nitrile, Hypalon®; for soling, footwear, mech. rubber goods, tire sidewalls; wh. powd.; 0.08 μ ultimate particle size; sp.gr. 2.0 (in rubber); dens. 10 lb/ft^3; surf. area 35 m^2/g; pH 9.0 (5% aq. susp.); ref. index 1.45; 88% SiO_2 hydrate; *Toxicology:* TLV 6 mg/m^3 (OSHA); contact with skin causes drying effect.

Silene® D. [PPG Industries] Amorphous hydrated silica; reinforcing pigment used in syn. and natural rubber compds.; fast, smooth calendering and extrusion; used in soling, footwear, mechanical rubber goods, tire sidewalls; wh. powd.; sp.gr. 1.93; ref. index 1.45; pH 9.6 (5% aq.); 82% SiO_2 hydrate.; Unverified

Silfin® Z. [Hoffmann Min.] After-treated Sillitin Z86 with higher brightness; filler, min. additive for thermoplastics, thermosets, thermoplastic PU, paints, lacquers incl. electrophoresis; partial replacement for wh. pigments; whiter alternative to Sillitin; almost entirely free from active Fe ions resulting in greatly improved aging props. for certain props.; loose structures; surf. area 16 m^2/g; brightness (Hunter) 91.

Silicone Emulsion 350, 1M, 10M, 60M. [Akrochem] Formulated emulsions of dimethyl silicone fluids in water; nonionic e; release agent for most tire and mechanical goods operations, wire and cable, and plastics operations; wh. emulsion; sp.gr. 0.98-1.02; dens. 8.25 lb/gal; visc. in centistokes indicated by number in grade; pH 7.0; 35% silicone content.

Silicone Emulsion E-131. [Wacker Silicones] Silicone emulsion contg. polydimethylsiloxane (CAS 63148-62-9) and alkyl phenol ethoxylate; mold release for rubber and plastic parts; also for auto/furniture polish emulsions, cosmetic hand creams and lotions, textile thread lubricants and release, and aerosol window cleaners; wh. liq., mild char. odor; completely sol. in water; sp.gr. 0.99; dens. 8.3 lb/gal; visc. 350 cSt (of silicone); flash pt. (PM) > 160 F; 60% solids in water; *Toxicology:* eye and skin irritant; *Precaution:* hazardous/thermal decomp.

prods.: SiO_2, CO, CO_2, formaldehyde, various hydrocarbon fragments; *Storage:* 6 mos storage stability in unopened containers stored @ 0-32 C; store in cool, dry well-ventilated area; protect from freezing.

Silicone Emulsion E-133. [Wacker Silicones] Silicone emulsion; mold release for rubber and plastic parts; also for auto/furniture polish emulsions, cosmetic hand creams and lotions, textile release and softeners; wh. liq.; dens. 8.3 lb/gal; visc. 1000 cSt (of silicone); 60% solids in water; *Storage:* 6 mos storage stability in unopened containers stored @ 0-32 C.

Silicone Fluid 350, 1M, 5M, 10M, 30M, 60M. [Akrochem] Dimethylpolysiloxane; release and slip agents for tire and mechanical goods (e.g., fan belts, O rings, floor mats, hose, toys, shoe heels, floor tile, bath mats, wire and cable applics.); additives in rubber and plastics for water-repellent treatments; water-wh. clear fluids; misc. with nonpolar liqs. (hydrocarbons, ethers, etc.); immiscible with water, glycerin, alcohols, and other polar liqs.; sp.gr. 0.95-0.98; dens. 7.8-8.0 lb/gal; visc. 20-60,000 cSt (visc. in centistokes indicated by number in grade).

Silicone Mist No. S116-A. [IMS] Silicone release agent (CAS 63148-62-9) with HFC-152A, dimethyl ether, and aliphatic petrol. distillate (CAS 64742-89-8); non-CFC, non-ozone depleting low conc. mold release for use where low residue is critical; for ABS, acetal, acrylic, nylon, PC, polyethylene, PS, rubber; suitable for food pkg.; FDA 21CFR §181.28; clear mist with sl. ethereal odor as dispensed from aerosol pkg.; sl. sol. in water; sp.gr. > 1; vapor pressure 60 ± 10 psig; flash pt. < 0 F; > 95% volatile; *Toxicology:* inh. may cause CNS depression with dizziness, headache; higher exposure may cause heart problems; gross overexposure may cause fatality; eye and skin irritaiton; *Precaution:* flamm. limits 1.2-18%; at elevated temps. (> 130 F) aerosol containers may burst, vent, or rupture; incompat. with strong oxidizers, strong caustics, reactive metals (Na, K, Zn, Mg, Al); contains methylpolysiloxanes which can generate formaldehyde; *Storage:* store in cool, dry area out of direct sunlight; do not puncture, incinerate, or store above 120 F.

Silicone Release Agent #5038. [Polymer Research Corp. of Am.] Dimethyl silicone fluid emulsion; release agent for rubber and plastics; applied by spraying, brushing, or wiping; milky wh. liq. emulsion; dilutable with water; dens. 8.0 lb/gal; 35% act.; Unverified

Silicone Spray No. S312-A. [IMS] 3% Silicone release agent (CAS 63148-62-9) with HFC-152A, dimethyl ether, and aliphatic petrol. distillate (CAS 64742-89-8); non-CFC, non-ozone depleting general purpose mold release forming long-lasting film on molds for ABS, acetal, acrylic, nylon, PMMA, PC, polyethylene, PS, polysulfone, rubber; mold release for food pkg.; FDA 21CFR §181.28; clear mist with sl. ethereal odor as dispensed from aerosol pkg.; sl. sol. in water; sp.gr. > 1; vapor pressure 60 ± 10 psig; flash pt. < 0 F; > 95% volatile; *Toxicology:* inh. may cause CNS depression with dizziness, headache; higher exposure may cause heart problems; gross overexposure may cause fatality; eye and skin irritaiton; *Precaution:* flamm. limits 1.2-18%; at elevated temps. (> 130 F) aerosol containers may burst, vent, or rupture; incompat. with strong oxidizers, strong caustics, reactive metals (Na, K, Zn, Mg, Al); contains methylpolysiloxanes which can generate formaldehyde; *Storage:* store in cool, dry area out of direct sunlight; do not puncture, incinerate, or store above 120 F.

Sillikolloid P 87. [Hoffmann Min.] 71% Quartz, 29% kaolinite; CAS 14808-60-7, 1318-74-7; EINECS 310-127-6; filler for rubber, lacquer and paint, polish, microcoating, industrial protection applics.; < 70 mg/kg residue > 40 µm; 0.6% water sol.; dens. 2.6 g/cc; bulk dens. 0.24 g/cc; visc. 51 Pas (in DOP); surf. area 15.7 m^2/g; oil absorp. 51; brightness (Hunter) 87; ref. index 1.55; pH 8.2; vol. resist. 10^{12} ohm-cm; 81.7% SiO_2, 11.4% Al_2O_3; 0.8% max. moisture.

Sillitin N 82. [Hoffmann Min.] 80% Quartz, 20% kaolinite; CAS 14808-60-7, 1318-74-7; EINECS 310-127-6; filler for rubber, lacquer and paint, polish, microcoating, industrial protection applics.; < 100 mg/kg residue > 40 µm; 0.5% water sol.; dens. 2.6 g/cc; bulk dens. 0.34 g/cc; visc. 17 Pas (in DOP); surf. area 14.7 m^2/g; oil absorp. 43; brightness (Hunter) 82; ref. index 1.55; pH 8.0; vol. resist. 10^{12} ohm-cm; 86.2% SiO_2, 8% Al_2O_3; 0.8% max. moisture.

Sillitin N 85. [Hoffmann Min.] 80% Quartz, 20% kaolinite; CAS 14808-60-7, 1318-74-7; EINECS 310-127-6; filler for rubber, lacquer and paint, polish, microcoating, industrial protection applics.; < 100 mg/kg residue > 40 µm; 0.5% water sol.; dens. 2.6 g/cc; bulk dens. 0.33 g/cc; visc. 17 Pas (in DOP); surf. area 12.2 m^2/g; oil absorp. 43; brightness (Hunter) 85; ref. index 1.55; pH 8; vol. resist. 10^{12} ohm-cm; 87.3% SiO_2, 7.7% Al_2O_3; 0.8% max. moisture.

Sillitin V 85. [Hoffmann Min.] 86% Quartz, 14% kaolinite; CAS 14808-60-7, 1318-74-7; EINECS 310-127-6; filler for rubber, lacquer and paint, polish, microcoating, industrial protection applics.; < 100 mg/kg residue > 40 µm; 0.4% water sol.; dens. 2.6 g/cc; bulk dens. 0.39 g/cc; visc. 10 Pas (in DOP); surf. area 11.5 m^2/g; oil absorp. 42; brightness (Hunter) 85; ref. index 1.55; pH 8; vol. resist. 10^{12} ohm-cm; 90.8% SiO_2, 5.4% Al_2O_3; 0.8% max. moisture.

Sillitin Z 86. [Hoffmann Min.] 74% Quartz, 26% kaolinite; CAS 14808-60-7, 1318-74-7; EINECS 310-127-6; filler for rubber, lacquer and paint, polish, microcoating, industrial protection applics.; < 100 mg/kg residue > 40 µm; 0.5% water sol.; dens. 2.6 g/cc; bulk dens. 0.27 g/cc; visc. 35 Pas (in DOP); surf. area 14.2 m^2/g; oil absorp. 48; brightness (Hunter) 86; ref. index 1.55; pH 8.2; vol. resist. 10^{12} ohm-cm; 83.2% SiO_2, 10.4% Al_2O_3; 0.8% max.

moisture.

Sillitin Z 89. [Hoffmann Min.] 73% Quartz, 27% kaolinite; CAS 14808-60-7, 1318-74-7; EINECS 310-127-6; filler for rubber, lacquer and paint, polish, microcoating, industrial protection applics.; < 100 mg/kg residue > 40 μm; 0.4% water sol.; dens. 2.6 g/cc; bulk dens. 0.26 g/cc; visc. 46 Pas (in DOP); surf. area 14.3 m^2/g; oil absorp. 50; brightness (Hunter) 89; ref. index 1.55; pH 8.5; vol. resist. 10^{12} ohm-cm; 83.4% SiO_2, 10.6% Al_2O_3; 0.8% max. moisture.

Sillum-200. [D.J. Enterprises] Semicalcined silica-alumina; filler and flame retardant for the plastic and epoxy industries; beige; 36.90% 200 mesh; 39.50% 325 mesh; 61.97% silica, 21.25% alumina.

Sillum-200 Q/P. [D.J. Enterprises] Alumina silicate; filler with exc. suspension and dispersion qualities; high filler loading capability 40-60%; water-resist.; reduces shrinkage, improves hardness, reduces exotherm temp.; nonhygroscopic; free-flowing bead; 44-75 μ screen; sp.gr. 2.35; bulk dens. 50-60 lb/ft^3; m.p. > 2600 F; pH 6-7; 62% SiO_2, 37% Al_2O_3.

Sillum-PL 200. [D.J. Enterprises] Aluminum and silica oxides; flame retardant additive for plastics; sp.gr. 2.6; hardness (Mohs) 3.0; 70.4% SiO_2, 24.8% Al_2O_3.

Siloxan Tego® 457. [Goldschmidt] Organic modified polysiloxane; release agent for rubber molding; water-sol.; 100% act.; Unverified

Siltech® CE-2000. [Siltech] Tri (octyldodecyl) citrate; CAS 126121-35-5; nonionic; patented; for polycarbonate mold release; amber clear to hazy liq., mild odor; insol. in water; sp.gr. 0.97; b.p. > 100 C (760 mm); flash pt. (PMCC) > 200 C; 100% act.; *Toxicology:* may cause skin irritation, moderate eye irritation; ingestion may cause abdominal discomfort, nausea, vomiting, diarrhea; *Storage:* keep container tightly closed.

Siltech® CE-2000 F3.5. [Siltech] Fluoro-Guerbet citrate; patented; for polycarbonate lubrication; amber clear to hazy liq., mild odor; insol. in water; sp.gr. 0.97; b.p. > 100 C (760 mm); flash pt. (PMCC) > 200 C; 100% act.; *Toxicology:* may cause skin irritation, moderate eye irritation; ingestion may cause abdominal discomfort, nausea, vomiting, diarrhea; *Storage:* keep container tightly closed.

Siltech® E-2140. [Siltech] Water-based dimethylpolysiloxane emulsion (50-70% dimethicone (CAS 63148-62-9), 2-10% alcohol ethoxylates (CAS 61702-78-1)); nonionic; nonpaintable general purpose mold release agent for metal foundry, rubber and plastic molding applics., vinyl and tire treatments; lubricant and softener for textiles; milky wh. liq., mild odor; disp. in water; sp.gr. 1.0; b.p. > 100 C (760 mm); flash pt. (PMCC) none; *Toxicology:* may cause skin irritation, moderate to severe eye irritation; ingestion may cause throad irritation, nausea, vomiting, diarrhea; *Storage:* keep container tightly closed.

Siltech® E-2170. [Siltech] Water-based reactive dimethylpolysiloxane emulsion; durable mold release agent; provides softness and hand on textiles; 60% act.

Siltech® MFF-3000. [Siltech] Water-based alkyl silicone emulsion (30-50% polyphenylmethylsiloxane (CAS 2116-84-9), 2-10% alcohol ethoxylates (CAS 61702-78-1)); mold release agent recommended where paintability is desired; milky wh. liq., mild odor; disp. in water; sp.gr. 1.0; b.p. > 100 C (760 mm); flash pt. (PMCC) none; *Toxicology:* may cause skin irritation, moderate to severe eye irritation; ingestion may cause throad irritation, abdominal discomfort, nausea, vomiting, diarrhea; *Storage:* keep container tightly closed.

Siltech® PF. [Siltech] Solv.-based; mold release agent where paintability is desired.

Siltech® STD-100. [Siltech] Solv.-based; durable mold release.

Siltex™. [Kaopolite] Fused silica; high performance extender pigment, filler, abrasive for industrial coatings, polymer systems, epoxy compds., elec. molding and potting compds., translucent/transparent sealants, caulks, mastics, cleaning compds., metal polish; outstanding elec. props., low ref. index and thermal expansion.

Siltex™ 50+ 100 mesh. [Kaopolite] Fused silica, free of cryst. silica; abrasive for cleaning compds., cosmetics (hand and body cleaners), metal polish, plastic compds.

Siltex™ 44. [Kaopolite] Fused silica; extender pigment, filler for coatings (high peformance corrosion-resist., translucent or transparent, flat interior or exterior, marine, insulting elec., moisture-sensitive), cleaners and polishes; polymers (low shrinkage, high dimensional stability engineered thermoplastics, moisture-cured urethanes, moisture-sensitive CIP systems, exc. elec. props.); powd. 8.8 μ median particle size; 2.5% max. on 325 mesh; sp.gr. 2.17; dens. 18.08 lb/solid gal; bulking value 0.0553 gal/lb; oil absorp. 20-25; GE brightness 90-92%; ref. index 1.46; pH 6.5-7.5 (20% solids); hardness (Mohs) 5.5-6.0; sp.resist. 150,000-200,000 ohm/cm; 99.50% min. SiO_2.

Siltex™ 44C. [Kaopolite] Fused silica; extender pigment, filler for coatings (high peformance corrosion-resist., translucent or transparent, flat interior or exterior, marine, insulting elec., moisture-sensitive), cleaners and polishes; polymers (low shrinkage, high dimensional stability engineered thermoplastics, moisture-cured urethanes, moisture-sensitive CIP systems, exc. elec. props.); powd. 7.0 μ median particle size; 5% max. on 325 mesh; sp.gr. 2.17; dens. 18.08 lb/solid gal; bulking value 0.0553 gal/lb; oil absorp. 18-23; GE brightness 84-86%; ref. index 1.46; pH 6.5-7.5 (20% solids); hardness (Mohs) 5.5-6.0; sp.resist. 150,000-200,000 ohm/cm; 99.50% min. SiO_2.

Silver Bond 45. [Unimin Spec. Minerals] Cryst. silica; filler, flatting agent for traffic paint, exterior block fillers, mastics and adhesives, buffing compds., elec. epoxy compds.; wh. coarse powd.; 13.3 μ median particle size; 96.8% thru 325 mesh; sp.gr. 2.65; dens. 22.07 lb/gal; surf. area 0.5 m^2/g; oil absorp. 23.4; brightness 87.2; m.p. 1610 C; ref.

index 1.54-1.55; pH 6.9; hardness (Mohs) 6.5; 99.6% SiO_2; *Toxicology:* prolonged inh. of dust may cause delayed lung injury (silicosis); may produce cancer.

Silver Bond B. [Unimin Spec. Minerals] Cryst. silica; filler, flatting agent for traffic paint, exterior block fillers, mastics and adhesives, buffing compds., elec. epoxy compds.; wh. coarse powd.; 9.9 μ median particle size; 99.4% thru 325 mesh; sp.gr. 2.65; dens. 22.07 lb/gal; surf. area 0.6 m^2/g; oil absorp. 23.7; brightness 87.3; m.p. 1610 C; ref. index 1.54-1.55; pH 6.7; hardness (Mohs) 6.5; 99.23% SiO_2; *Toxicology:* prolonged inh. of dust may cause delayed lung injury (silicosis); may produce cancer.

Silverline 400. [Luzenac Am.] Platy talc; CAS 14807-96-6; EINECS 238-877-9; general purpose stir-in filler with minimum binder demand; for all types of paints and plastics; 5.0 μ median particle size; Hegman grind 4.0 min.; oil absorp. 31; dry brightness 76-82.

Silverline 665. [Luzenac Am.] Platy talc; CAS 14806-96-6; EINECS 238-877-9; ultrafine, high purity talc with exc. reinforcing and dielec. props.; for extrusion of wire and cable; additive for gloss control in paints and coatings; 1.5 μ median particle size; Hegman grind 6.0+; oil absorp. 48; dry brightness 84 ± 2.

Silwax® C. [Siltech] Dimethiconol hydroxystearate; CAS 133448-13-2; patented, highly lubricious wax for personal care applics., polishes, textile lubrication and softening, laundry prods., dryer sheet softeners, syn. lubricants, plastics lubricants; off-wh. waxy paste, mild odor; sol. @ 5% in IPA; disp. in silicone fluid; insol. in water; sp.gr. 0.99; b.p. > 100 C (760 mm); flash pt. (PMCC) > 200 C; 100% act.; *Toxicology:* may cause skin irritation, moderate eye irritation; ingestion may cause abdominal discomfort, nausea, vomiting, diarrhea.

Silwax® F. [Siltech] Dimethiconol fluoroalcohol dilinoleic acid; patented prod. forming films which are hydrophobic, nonocclusive, and resist. to many chemical agents; highly lubricious, chem. resist. paintable mold release; additive for barrier creams and other skin care prods.; off-wh. to brn. waxy paste, mild odor; water-insol.; sp.gr. 1.07; m.p. 120 F; b.p. > 100 C (760 mm); flash pt. (PMCC) > 200 C; pH 5-7 (10%); 100% solids, 8% F; *Toxicology:* may cause skin irritation, moderate eye irritation; ingestion may cause abdominal discomfort, nausea, vomiting, diarrhea; *Storage:* keep container tightly closed.

Silwax® S. [Siltech] Dimethiconol stearate; CAS 130169-63-0; nonionic; patented, highly lubricious wax for personal care applics., polishes, textile lubrication and softening, laundry prods., dryer sheet softeners, syn. lubricants, plastics lubricants; off-wh. waxy paste, mild odor; sol. @ 5% in IPA, min. spirits; disp. in min. oil, oleic acid; insol. in water; sp.gr. 0.99; b.p. > 100 C (760 mm); flash pt. (PMCC) > 200 C; 100% act.; *Toxicology:* may cause skin irritation, moderate eye irritation; ingestion may cause abdominal discomfort, nausea, vomiting, diarrhea; *Storage:* keep container tightly closed.

Silwax® WD-IS. [Siltech] Dimethicone copolyol isostearate; CAS 133448-16-5; nonionic; patented, highly lubricious wax for personal care conditioning, polishes, textile lubrication and softening, laundry prods., dryer sheet softeners, syn. lubricants, plastics lubricants; forms microemulsions in water without added surfactants; amber liq., mild odor; sol. @ 5% in IPA, glyceryl trioleate, oleic acid; disp. in water; sp.gr. 1.05; b.p. > 100 C (760 mm); flash pt. (PMCC) > 200 C; 100% act.; *Toxicology:* may cause skin irritation, moderate eye irritation; ingestion may cause abdominal discomfort, nausea, vomiting, diarrhea; *Storage:* keep container tightly closed.

Silwax® WD-S. [Siltech] Polydimethylsiloxane wax; CAS 133448-15-4; patented, highly lubricious wax for personal care applics., polishes, textile lubrication and softening, laundry prods., dryer sheet softeners, syn. lubricants, plastics lubricants; more dispersible than Silwax S, more substantive than WD-IS; amber liq., mild odor; sol. @ 5% in IPA, glyceryl trioleate, oleic acid; initially disp. in water, but creams upon standing; sp.gr. 1.06; b.p. > 100 C (760 mm); flash pt. (PMCC) > 200 C; 100% act.; *Toxicology:* may cause skin irritation, moderate eye irritation; ingestion may cause abdominal discomfort, nausea, vomiting, diarrhea; *Precaution:* keep container tightly closed.

Silwax® WS. [Siltech] Dimethicone copolyol wax; nonionic; patented, highly lubricious wax for personal care applics., polishes, textile lubrication and softening, laundry prods., dryer sheet softeners, syn. lubricants, plastics lubricants; produces high gloss on a variety of substrates; clear waxy solid, mild odor; sol. @ 5% in water, IPA, PEG 400, glyceryl trioleate, oleic acid; insol. in min. oil; sp.gr. 1.07; b.p. > 100 C (760 mm); flash pt. (PMCC) > 200 C; 100% act.; *Toxicology:* may cause skin irritation, moderate eye irritation; ingestion may cause abdominal discomfort, nausea, vomiting, diarrhea; *Storage:* keep container tightly closed.

Silwet® L-7001. [OSi Specialties] Dimethicone copolyol; CAS 67762-85-0; nonionic; surfactant, dispersant, emulsifier, leveling/flow control agent, antifog, lubricant, antiblock, slip additive for adhesives, agric., automotive, coatings, printing inks, textiles, household specialties, cutting fluids, petrol. extraction, paper, personal care prods., plastics and rubber; pale yel. clear liq.; m.w. 20,000; sol. in water, methanol, IPA, acetone, xylene, methylene chloride; sp.gr. 1.023; dens. 8.50 lb/gal; visc. 1700 cSt; HLB 9-12; cloud pt. 39 C (1%); flash pt. (PMCC) 97 C; pour pt. -48 C; surf. tens. 30.5 dyne/cm (0.1% aq.); Draves wetting > 300 s (0.25%); Ross-Miles foam 46 mm (0.1%, initial); 75% act.

Silwet® L-7500. [OSi Specialties] Dimethicone copolyol; CAS 68440-66-4; nonionic; surfactant, antifoam, dispersant, emulsifier, leveling and flow control agent, lubricant, slip additive for adhesives, automotive, chemical processing, coatings, petrol. extraction, paper, personal care prods., plastics and rubber, pharmaceutical, textile applics.; lt. yel. clear liq.; sol. in methanol, IPA, acetone, xylene, hexanes, methylene chloride; insol. in water; m.w. 3000; sp.gr. 0.982; dens. 8.16 lb/gal; visc.140 cSt; HLB 5-8; pour pt. -43 C; flash pt. (PMCC) 121 C; 100% act.

Silwet® L-7600. [OSi Specialties] Dimethicone copolyol; CAS 68938-54-5; nonionic; surfactant, wetting agent for adhesives, window cleaners, textiles, personal care prods.; internal lubricant for plastics and rubber; lt. amber clear liq.; sol. in water, methanol, IPA, acetone, xylene, methylene chloride; m.w. 4000; sp.gr. 1.066; dens. 8.86 lb/gal; visc. 110 cSt; HLB 13-17; pour pt. 2 C; cloud pt. 64 C (1% aq.); flash pt. (PMCC) 74 C; surf. tens. 25.1 dyne/cm (1.1% aq.); Draves wetting 131 s (0.1%); Ross-Miles foam 94 mm (0.1%, initial); 100% act.

Silwet® L-7602. [OSi Specialties] Dimethicone copolyol; CAS 68938-54-5; nonionic; surfactant, defoamer, dispersant, emulsifier, leveling and flow control agent, gloss agent; lubricant, release agent, antiblock and slip additive for plastics and rubber, wetting agent for adhesives, agric., automotive specialties, chem. processing; coatings, petrol. extraction, skin care prods., urethane bubble release, pharmaceutical, printing inks, textiles; pale yel. clear liq.; sol. in methanol, IPA, acetone, xylene, methylene chloride; disp. in water; m.w. 3000; sp.gr. 1.027; dens. 8.54 lb/gal; visc. 100 cSt; HLB 5-8; flash pt. (PMCC) 127 C; pour pt. -15 C; surf. tens. 26.6 dyne/cm (1.1% aq.); 100% act.

Sinonate 960SF. [Sino-Japan] Ethoxylated alkylphenol sulfate; anionic; detergent, foaming agent, wetting agent, penetrant; for emulsion polymerization; paste; 50% conc.

Sinonate 960SN. [Sino-Japan] Ethoxylated alkylphenol sulfate; anionic; detergent, foaming agent, wetting agent, penetrant; for emulsion polymerization; liq.; 30% conc.

Sinopol 610. [Sino-Japan] Surfactant; nonionic; low foaming emulsifier, solubilizer, detergent; for emulsion polymerization; liq.; HLB 13.6; 100% conc.

Sinopol 623. [Sino-Japan] Surfactant; nonionic; low foaming emulsifier, solubilizer, detergent; for emulsion polymerization; liq.; HLB 16.6; 100% conc.

Sinopol 707. [Sino-Japan] Surfactant; nonionic; low foaming emulsifier for emulsion polymerization; detergent, cleaning agent; liq.; HLB 12.0; 100% conc.

Sinopol 714. [Sino-Japan] Surfactant; nonionic; low foaming emulsifier for emulsion polymerization; detergent, cleaning agent; paste; HLB 15.0; 100% conc.

Sinopol 806. [Sino-Japan] POE octylphenyl ether; nonionic; stabilizer, dispersant, suspending agent; for emulsion polymerization; solid; HLB 17.3; 100% conc.

Sinopol 808. [Sino-Japan] POE octylphenyl ether; nonionic; stabilizer, dispersant, suspending agent; for emulsion polymerization; solid; HLB 17.9; 100% conc.

Sinopol 908. [Sino-Japan] POE nonylphenyl ether; nonionic; stabilizer, dispersant, suspending agent; for emulsion polymerization; flake; HLB 17.8; 100% conc.

Sinopol 910. [Sino-Japan] POE nonylphenyl ether; nonionic; stabilizer, dispersant, suspending agent; for emulsion polymerization; flake; HLB 18.2; 100% conc.

Sinoponic PE 61. [Sino-Japan] POP/POE ether; nonionic; low foaming detergent, wetting agent for phosphatizing bath; emulsifier for acrylic polymers; liq.; HLB 3.0; 100% conc.

Sinoponic PE 62. [Sino-Japan] POP/POE ether; nonionic; low foaming detergent, wetting agent for phosphatizing bath; emulsifier for acrylic polymers; liq.; HLB 7.0; 100% conc.

Sinoponic PE 64. [Sino-Japan] POP/POE ether; nonionic; low foaming detergent, wetting agent for phosphatizing bath; emulsifier for acrylic polymers; liq.; HLB 15.0; 100% conc.

Sinoponic PE 68. [Sino-Japan] POP/POE ether; nonionic; low foaming detergent, wetting agent for phosphatizing bath; emulsifier for acrylic polymers; flake; HLB 29.0; 100% conc.

Sipernat® 22. [Degussa] Syn. amorphous precipitated silica; CAS 112926-00-8; hydrophilic adsorbent, anticaking and free-flow agents; used as aid to convert liqs. into powds.; processing aid; hydrophilic; for adhesives, sealants, detergents, foods (table salt, powd. sweetener), plastics, cosmetics industries; 21CFR §172.480, 173.340, 175.105, 175.300, 176.200, 176.210, 177.1200, 177.2600, 178.3210; wh. fluffy powd., 18 nm avg. particle size; dens. 270 g/l (tapped); surf. area 190 m^2/g; pH 6.3 (5% aq. susp.); 98% assay.

Sipernat® 22LS. [Degussa] Precipitated silica; thickening agent for cosmetics, pharmaceuticals, elec., insulation, paper, film, pesticides, plastics industries; for thickening and thixotropy in filled polyester gel coats, laminating resins; antiblock in PE and PP blown films; FDA approved; tapped dens. 80 g/l; surf. area 190 m^2/g; pH 6.2; 98% SiO_2.

Sipernat® 22S. [Degussa] Syn. amorphous precipitated silica; CAS 112926-00-8; hydrophilic free-flow/anticaking agent for powd. detergents, TPEs, sealants, foodstuffs (spices, guar gum, dried soup, powd. cheese), pharmaceuticals, fire extinguishers; converts liqs. to powds., e.g., PVC stabilizers; 21CFR §172.480, 173.340, 175.105, 175.300, 176.200, 176.210, 177.1200, 177.2600, 178.3210; wh. loose powd.; 18 nm avg. particle size; dens. 120 g/l (tapped); surf. area 190 m^2/g; pH 6.3 (5% aq. susp.); 98% assay.

Sipernat® 44. [Degussa] Aluminum silicate; hydrophilic antiblocking agent in PE and PP blown films; reinforcing and antisedimentation filler for RTV silicone rubber casting and molding compds.; powd.; dens. 500 g/l (tapped); pH 11.8 (5% aq. susp.); 42% SiO_2, 36% Al_2O_3, 22% Na_2O.

Sipernat® 50LS. [Degussa] Precipitated silica; antiblocking agent for PE and PP blown films.

Sipernat® 50S. [Degussa] Syn. amorphous precipitated silica; CAS 112926-00-8; free-flow, anticaking and defoaming agent for food use (cocoa powd., dried soup, cake mix); indirect food additive for adhesives, paper, coatings, rubber, etc.; for prevention of plate out in PVC calendered films; anticaking for pre-expanded PS foam grans.; improves storability of phenolic and melamine resins; 21CFR §172.480, 173.340, 175.105, 176.200, 176.210, 175.300, 177.1200, 177.2600, 178.3210; wh. loose powd.; 7 μm avg. agglomerate size; tapped dens. 100 g/l; surf. area 480 m^2/g; pH 6.7 (5%); 99% SiO_2, 5% moisture; *Toxicology:* inhalation may cause respiratory tract irritation.

Sipernat® D11. [Degussa] Silica and dimethicone; free-flow agent for fire extinguishers; filler for thickening, reinforcing and improving weatherability of polysulfide sealants.

Sipernat® D17. [Degussa] Silica silylate; hydrophobic anticaking and free-flow agent for PS, fire extinguishers, pesticides; EPA approved; wh. fluffy powd., 17 nm avg. particle size; dens. 150 g/l (tapped); surf. area 100 m^2/g; pH 8; 99.5% assay.

Sipomer® AAE. [Rhone-Poulenc Surf. & Spec.] Allyl alcohol ethoxylate; nonionic; reactive intermediate for silylation; bound protective colloid; copolymerizable stabilizer; m.w. 500; sp.gr. 1.08; visc. 50 cps; flash pt. > 150 C; 99% assay.

Sipomer® AM. [Rhone-Poulenc Surf. & Spec.] Allyl methacrylate; CAS 96-05-9; EINECS 202-473-0; monomer, modifier for coatings, elastomers, adhesives, intermediates; contributes hardness and scratch resist.; crosslinker/hardener; APHA 15 color; m.w. 126; sp.gr. 0.94; visc. 1 cps; b.p. 42 C (15 mm); flash pt. (TCC) 37 C; 99% act.

Sipomer® BEM. [Rhone-Poulenc Surf. & Spec.] Behenyl polyethoxy methacrylate; copolymerizable surfactant for acrylic latexes with sol'n. viscs. that are shear thickening; produces latexes with associative thickening props.; m.w. 1500; 50% act., 25% methacrylic acid.

Sipomer® β-CEA. [Rhone-Poulenc Surf. & Spec.] β-Carboxyethyl acrylate; CAS 79-10-7; monomer for emulsion polymerization, adhesives and coatings; improves latex props., adhesion; APHA 15 color; m.w. 144; sp.gr. 1.2; visc. 75 cps; b.p. 103 C (19 mm); flash pt. 88 C; 98% act.

Sipomer® COPS 1. [Rhone-Poulenc Surf. & Spec.] Sodium allyloxy hydroxypropyl sulfonate; anionic; monomer for emulsion polymerization; copolymerizable stabilizer; provides latex stability at low surfactant levels; reduces latex foaming; provides improved film props.; APHA 50 color; m.w. 218; sp.gr. 1.17; visc. 10 cps; f.p. -12 C; b.p. 100 C (760 mm); flash pt. > 100 C; 40% act.

Sipomer® DCPA. [Rhone-Poulenc Surf. & Spec.] Dicyclopentenyl acrylate; CAS 12542-30-2; monomer for air-drying adhesives, coatings, caulks, sealants, elastomers; crosslinks acrylics, unsat. polyesters and alkyds; Gardner 5 color; m.w. 204; sp.gr. 1.08; visc. 18 cps; f.p. -40 C; b.p. 110 C (3.8 mm); flash pt. (PMCC) 128 C; 98% assay.

Sipomer® DCPM. [Rhone-Poulenc Surf. & Spec.] Dicyclopentenyl methacrylate; CAS 51178-59-7; monomer for concrete resurfacing, air-drying adhesives and coatings, elastomers; improves cohesive str. through crosslinking; amber; m.w. 218; sp.gr. 1.05; visc. 25 cps; b.p. 137 C (13 mm); flash pt. (TCC) > 110 C; 98% assay.

Sipomer® HEM-5. [Rhone-Poulenc Surf. & Spec.] PEG-5 monomethacrylate; CAS 25736-86-1; monomer for emulsion polymerization; thickener, dispersant, suspending agent; copolymerizable stabilizer; bound protective colloid; modifying comonomer/crosslinker; APHA 50 color; m.w. 306; sp.gr. 1.1; visc. 35 cps; 88% assay.

Sipomer® HEM-10. [Rhone-Poulenc Surf. & Spec.] PEG-10 monomethacrylate; nonionic; monomer for emulsion polymerization; thickener, dispersant, suspending agent; copolymerizable stabilizer; bound protective colloid; modifying comonomer/crosslinker; APHA 50 color; m.w. 526; sp.gr. 1.1; visc. 65 cps; 83% assay.

Sipomer® HEM-D. [Rhone-Poulenc Surf. & Spec.] 2-Hydroxyethyl methacrylate; CAS 868-77-9; EINECS 212-782-2; monomer for creating and modifying a wide range of polymers; for paints, industrial coatings, textile finishes, adhesives, textile coatings, biomedical applics.; increases moisture adsorption, improves adhesion; APHA 25 color; m.w. 130; sp.gr. 1.07; visc. 200 cps; f.p. -20 C; b.p. 68 C (1 mm); flash pt. (TCC) 106 C; 96.5% act.

Sipomer® IBOA. [Rhone-Poulenc Surf. & Spec.] Isobornyl acrylate; CAS 588-33-5; monomer for creating and modifying a wide range of polymers; for paints, textile finishes, adhesives, industrial and textile coatings; resists uv degradation; APHA 25 color; m.w. 208; sp.gr. 0.98; visc. 9.5 cps; f.p. 5 C; b.p. 275 C (760 mm); flash pt. 116 C; 99.5% act.

Sipomer® IBOA HP. [Rhone-Poulenc Surf. & Spec.] Isobornyl acrylate, high purity; CAS 588-33-5; monomer for creating and modifying a wide range of polymers; for paints, textile finishes, adhesives, industrial and textile coatings; m.w. 208; 99.5% act.

Sipomer® IBOMA. [Rhone-Poulenc Surf. & Spec.] Isobornyl methacrylate; CAS 7534-94-3; monomer for creating and modifying a wide range of polymers; for high-performance coatings, paints, textile finishes, adhesives, industrial and textile coatings; resists uv degradation; contributes hardness and heat resist.; APHA 25 color; m.w. 222; sp.gr. 0.98; visc. 6 cps; f.p. -59 C; b.p. 140 C (20 mm); flash pt. 101 C; 99% act.

Sipomer® IBOMA HP. [Rhone-Poulenc Surf. & Spec.] Isobornyl methacrylate, high purity; CAS 7534-94-3; monomer for creating and modifying a wide range of polymers; for paints, textile finishes, adhesives, industrial and textile coatings; m.w. 222; 99% act.

Sipomer® IDM. [Rhone-Poulenc Surf. & Spec.] Isodecyl methacrylate; CAS 29964-84-9; soft monomer for coatings, adhesives, potting compds., leather treatments; resists uv degradation; Hazen 150 color; m.w. 226; sp.gr. 0.87; visc. 3 cps; f.p. -41 C; flash pt. > 100 C; 98.5% assay.

Sipomer® TMPTMA. [Rhone-Poulenc Surf. & Spec.] Trimethylolpropane trimethacrylate; CAS 3290-92-4; crosslinker/hardener for PVC wire and cable coatings; modifying comonomer; crosslinker for anaerobic adhesives; for coatings, adhesives, caulks, sealants; APHA 100 color; m.w. 338; sp.gr. 1.05; visc. 90 cps; f.p. -15 C; 99% assay.

Sipomer® WAM. [Rhone-Poulenc Surf. & Spec.] Allyl/ureido monomer; CAS 85356-84-9; monomer for emulsion polymerization of acrylic and vinyl-acrylic, latex paints, adhesives, caulks, sealants; wet adhesion promoter; improves paint film scrub resist.; m.w. 345; Gardner 6 color; m.w. 345; sp.gr. 1.15; visc. 500 cps; f.p. -10 C; b.p. 100 C (760 mm); flash pt. > 100 C; 90% act.

Sipomer® WAM II. [Rhone-Poulenc Surf. & Spec.] Proprietary; monomer for emulsion polymerization of acrylic and vinyl-acrylic latex paints with improved wet adhesion chars.; m.w. 197; 70% act.

SM 70 Glass Coating Emulsion. [GE Silicones] Reactive polydimethylsiloxane aq. emulsion; anionic; for treatment of glass, ceramics, metals, and plastics; protects glass against scratching; imparts water repellency to food containers; FDA compliance; wh. emulsion; 50% silicone; *Storage:* 12 mos storage stability.

SM 2059. [GE Silicones] Amodimethicone; cationic; silicone emulsion which cures to a durable, detergent-resistant film for mold release, particle treatment, textile finishes, vinyl and auto polish applics., catalysts for reactive polymers; off-wh. emulsion; water-disp.; dens. 8.25 lb/gal; visc. 20 cps; 35-41% total solids, 35% silicone; *Storage:* 12 mos storage stability.

SM 2061. [GE Silicones] Silicone dimethyl o/w aq. emulsion; nonionic; release agent for mech. rubber goods (automobile floor mats, shock mounts, fan belts, o-rings, footwear, floor tile), plastics, foundry release, softener/modifier for textiles, in polishes; wh. emulsion; dens. 8.2 lb/gal; visc. 60,000 cst; 36.5-38.5% total solids, 35% silicone; *Storage:* 12 mos storage stability.

SM 2068A. [GE Silicones] Dimethyl silicone aq. emulsion; anionic; release agent for mech. rubber goods (automobile floor mats, shock mounts, fan belts, o-rings, footwear, floor tile), plastics, foundry release, softener/modifier for textiles, in polishes; wh. emulsion; dens. 8.2 lb/gal; visc. > 100,000 cst; 34-38% total solids, 35% silicone; *Storage:* 12 mos storage stability.

SM 2079. [GE Silicones] Aq. emulsion of a high-visc. silicone polymer; nonionic; curable silicone emulsion forms durable, detergent-resistant film for mold release, particle treatment, textile finishes, fabric softener, hand modifier, vinyl and auto polish applics. (requires catalyst); resin extender; wh. emulsion; water-disp.; dens. 8.30 lb/gal; visc. 10,000 cst; 60-64% total solids, 60% silicone; *Storage:* 12 mos storage stability.

SM 2128. [GE Silicones] Silicone aq. emulsion based on SF 18 (350); nonionic; release agent for mech. rubber goods (automobile floor mats, shock mounts, fan belts, o-rings, footwear, floor tile), plastics, paper prods., indirect food contact; latex dip additive; FDA compliance; wh. emulsion; dens. 8.2 lb/gal; visc. 350 cst; 37-41% total solids, 35% silicone; *Storage:* 12 mos storage stability.

SM 2140. [GE Silicones] Dimethyl polysiloxane aq. emulsion; nonionic; release agent for mech. rubber goods (automobile floor mats, shock mounts, fan belts, o-rings, footwear, floor tile), plastics, wire and cable, foundry release, softener/modifier for textiles, in polishes; wh. emulsion; dens. 8.2 lb/gal; visc. 10,000 cst; 51.5-54.5% total solids, 50% silicone; *Storage:* 12 mos storage stability.

SM 2154. [GE Silicones] Alkyl-modified silicone emulsion based on SF 1080; nonionic; paintable release agent for mech. rubber goods (automobile floor mats, shock mounts, fan belts, o-rings, footwear, floor tile), plastics, metals, foundry release, softener/modifier for textiles, in polishes; wh. to lt. yel. emulsion; easily diluted with water; dens. 8.2 lb/gal; visc. 1500 cst; 53-56% total solids, 50% silicone; *Precaution:* light sensitive.

SM 2155. [GE Silicones] Dimethyl silicone aq. emulsion based on SF 96 (1000 cs); nonionic; release agent for mech. rubber goods (automobile floor mats, shock mounts, fan belts, o-rings, footwear, floor tile), plastics, wire and cable, foundry release, softener/modifier for textiles, in polishes; wh. emulsion; dens. 8.2 lb/gal; visc. 1000 cst; 51.5-54.5% total solids, 50% silicone; *Storage:* 12 mos storage stability.

SM 2162. [GE Silicones] Dimethyl silicone aq. emulsion based on SF 96; nonionic; release agent for mech. rubber goods (automobile floor mats, shock mounts, fan belts, o-rings, footwear, floor tile), plastics, wire and cable, foundry release, softener/modifier for textiles, in polishes; wh. liq., mild odor; dilutable with water; sp.gr. 1.04; dens. 8.25 lb/gal; visc. 350 cst; b.p. 212 F; flash pt. > 575 F; 52.5-55.5% total solids, 50% silicone; *Storage:* 12 mos storage stability.

SM 2163. [GE Silicones] Dimethyl silicone aq. emulsion based on SF 96; nonionic; release agent for mech. rubber goods (automobile floor mats, shock mounts, fan belts, o-rings, footwear, floor tile), plastics, foundry release, softener/modifier for textiles, in polishes, printing processes; wh. liq.; water-dilutable; dens. 8.25 lb/gal; visc. 350 cst; 60-64% total solids, 60% silicone; *Storage:* 12 mos storage stability.

SM 2164. [GE Silicones] Dimethyl silicone aq. emulsion based on SF 96; anionic; latex foam surfactant, latex thread lubricant; wh. emulsion; water-dilutable; dens. 8.2 lb/gal; visc. 100 cst; 50-53% total solids, 50% silicone; *Storage:* 12 mos storage stability.

SMA® 1000. [Elf Atochem N. Am.] 1:1 Styrene/maleic anhydride copolymer; CAS 9011-13-6; anionic; soil release agent; used in ammoniacal water sol'n. in carpet shampoos, paints, inks, paper coatings, commercial laundries; in emulsion polymerization, temporary coatings, oven cleaners; as leveling resin for floor polishes; dispersant; powd.; 99% act.

SMA® 2000. [Elf Atochem N. Am.] Styrene/maleic anhydride copolymer; CAS 9011-13-6; anionic; soil release agent; used in ammoniacal water sol'n. in carpet shampoos, paints, inks, paper coatings, commercial laundries; in emulsion polymerization, temporary coatings, oven cleaners; as leveling resin for floor polishes; dispersant; powd.; 99% act.

S-Maz® 80. [PPG/Specialty Chem.] Sorbitan oleate; CAS 1338-43-8; EINECS 215-665-4; nonionic; lubricant, antistat, textile softener, process defoamer, opacifier, coemulsifier, solubilizer, dispersant, suspending agent, coupler; prepares exc. w/o emulsions; binder in eye shadow, pressed powd. makeup; emulsifier for cleansing and night creams; lubricant, rust inhibitor, penetrant in metalworking formulations; dedusting agent for plastics; amber liq.; sol. in ethanol, naphtha, min. oils, min. spirits, toluene, veg. oil; disp. in water; sp.gr. 1.0; visc. 1000 cps; HLB 4.6; acid no. 7.5 max.; sapon. no. 149-160; hyd. no. 193-209; flash pt. (PMCC) > 350 F; 100% conc.

Smokebloc 11, 12. [Anzon] Multimetal complex; flame retardant for thermoset polyester, PVC.

Snowflake P.E.™ [ECC Int'l.] Calcium carbonate; CAS 1317-65-3; med. ground pigment, filler, reinforcement for max. loading in highly filled systems, e.g., unsat. polyester resins by SMC, BMC, TMC, XMC, transfer molding, pultrusion, wet molding, and spray up methods; wh. powd., odorless; 5.5 μ mean particle size; 0.1% on 325 mesh; negligible sol. in water; sp.gr. 2.71; dens. 48 lb/ft^3 (loose); oil absorp. 9; pH 9.5 (5% slurry); nonflamm.; 97.6% $CaCO_3$, 0.1% max. moisture; *Toxicology:* TLV/TWA 10 mg/m^3, considered nuisance dust; *Precaution:* incompat. with acids.

Snowflake White™. [ECC Int'l.] Calcium carbonate; CAS 1317-65-3; general-purpose easy dispersing pigment, extender, filler, reinforcement for protective coatings, water- and solv.-based paints, rubber, plastics, caulks, sealants, glazing compds., mastics; wh. powd., odorless; 5.5 μ mean particle size; fineness (Hegman) 5.0; negligible sol. in water; sp.gr. 2.71; dens. 45 lb/ft^3 (loose); oil absorp. 11; pH 9.5 (5% slurry); nonflamm.; 97.6% $CaCO_3$, 0.2% max. moisture; *Toxicology:* TLV/TWA 10 mg/m^3, considered nuisance dust; *Precaution:* incompat. with acids.

Sodium Bicarbonate T.F.F. Treated Free Flowing. [Rhone-Poulenc Food Ingreds.] Sodium bicarbonate, tricalcium phosphate FCC; used for foods (self-rising four and corn meal), industrial applics. (fire extinguishers, blowing agent for sponge rubber and some plastics); wh. free-flowing cryst., odorless; 2% min. on 80 mesh, 60% max. on 325 mesh; bulk dens. 60 lb/ft^3; 99% min. assay.

Sodium Bicarbonate USP No. 1 Powd. [Rhone-Poulenc Food Ingreds.] Sodium bicarbonate USP; CAS 144-55-8; EINECS 205-633-8; used for foods (colors, conditioners, starches, candies), pharmaceuticals (antibiotic mfg.), industrial applics. (mild alkaline buffering, fire extinguishers, neutralizing agent in mfg. of polyesters); wh. free-flowing cryst., odorless; 2% min. on 80 mesh, 20% max. on 200 mesh; bulk dens. 60 lb/ft^3; 99% min. assay.

Sodium Bicarbonate USP No. 3 Extra Fine Powd. [Rhone-Poulenc Food Ingreds.] Sodium bicarbonate USP; CAS 144-55-8; EINECS 205-633-8; used for pharmaceuticals (deodorant powds.), industrial applics. (fire extinguishers, dusting agent in pkg. plastic film sheets); wh. free-flowing cryst., odorless; 40% min. on 60 mesh, 5% max. on 200 mesh; bulk dens. 42 lb/ft^3; 99% min. assay.

Soft Detergent 95. [Lion] α-Olefin sulfonate; anionic; emulsifier and dispersant for emulsion polymerization; powd.; 95% conc.; Unverified

Softenol® 3107. [Hüls Am.] C7 fatty acid triglyceride; lubricant for machinery for confectionary and food industries; mold release agent in processing of plastics; additive to cutting oils, lacquer and varnish systems; antiblocking agent for artificial skins; liq.; neutral odor and taste; dens. 0.930 g/cc; visc. 20.6 mm^2/s; sapon. no. 385-395; biodeg.

Softenol® 3119. [Hüls Am.] Triglyceride; release agent for prod. and processing of PS polymers; microfine powd.

Softenol® 3408. [Hüls Am.] C8-10 fatty acid-1,2-propanediol ester; additive in mfg. of textile and glass fibers; lubricant for cutting devices; mold release agent in processing of plastics; liq.; neutral odor and taste; dens. 0.920 g/cc; visc. 10.9 mm^2/s; flash pt. 193 C; sapon. no. 320-340; biodeg.

Softenol® 3701. [Hüls Am.] Partial glyceride; release agent for prod. and processing of PS polymers; paste.

Softenol® 3900. [Hüls Am.] Glyceryl stearate; nonionic; emulsifier, lubricant, antistat, demolding and antiblocking agent, greaser, plasticizer, transparency and pigment dispersant used in mfg. and processing of plastics, textiles, leather, oils and polishing materials; powd.; neutral odor and taste; dens. 0.893 g/cc (80 C); visc. 22.9 mm^2/s (80 C); m.p. 58-65 C; sapon. no. 155-175; biodeg.

Softenol® 3925, 3945, 3960. [Hüls Am.] Glyceryl stearate; release agent for prod. and processing of PS polymers; powds. except 3945 (microcryst. powd.).

Softenol® 3991. [Hüls Am.] Glyceryl stearate; nonionic; lubricant, emulsifier, antistat, pigment dispersant, plasticizer, antiblocking agent for plastics, textile and leather processing aids, drilling and cutting oils, polishes, rubber; biodeg.; powd.; neutral odor and taste; dens. 0.914 g/cc (80 C); visc. 34.5 mm^2/s (80 C); m.p. 65 C; sapon. no. 155-165; biodeg.; 90% 1-monoester.

Solef® 11012/1001. [Solvay Polymers] PVDF copolymer; CAS 24937-79-9; processing aid to enhance extrusion of polyolefins; high thermal stability; for HDPE and LLDPE extruded blown film, PE and PP pipe extrusion, sheet and profile extrusion, HDPE and HMW HDPE blow molding.

Sole Terge 8. [Calgene] Disodium oleamido MIPA-sulfosuccinate; CAS 43154-85-4; EINECS 256-120-0; high foaming base for bubble baths, shampoos, etc.; frother in specialized ore flotation; emulsifier and stabilizer in emulsion polymerization; stable at all concs. at pH 4-7; amber liq.; sol. in water, ethanol, glycols, acetone; sp.gr. 1.09-1.12; dens. 9.1 lb/gal; pH 6-7 (1% aq.); surf. tens. 35 dynes/cm (0.1%); 32-35% act.; *Toxicology:* mild eye irritant.

Solricin® 135. [CasChem] Potassium ricinoleate; CAS 7492-30-0; EINECS 231-314-8; anionic; detergent, emulsifier, mild germicide, glycerized rubber lubricant, foam stabilizer in foamed rubber; making of cutting and sol. oils, household and cosmetic prods.; FDA approval; Gardner 2 liq.; sp.gr. 1.034; dens. 8.6 lb/gal; visc. 0.9 stokes; 32% act.; Unverified

Solricin® 235. [CasChem] Potassium castor soap sol'n. in water; CAS 8013-05-6; EINECS 232-388-4; emulsifier, dispersant, mild germicide; glycerized rubber lubricant; emulsifier, foam stabilizer for foamed rubber; Gardner 4 liq.; water-sol.; sp.gr. 1.034; visc. 0.9 stokes; 35% solids.

Solricin® 435. [CasChem] Sodium ricinoleate aq. sol'n.; CAS 5323-95-5; EINECS 226-191-2; emulsifier, stabilizer, defoamer for emulsion polymerization of resins (PVC, PVAc); FDA approval; Gardner 4 liq.; water-sol.; sp.gr. 1.022; visc. 2 stokes; 35% solids.

Solricin® 535. [CasChem] Sodium ricinoleate sol'n. in water; CAS 5323-95-5; EINECS 226-191-2; mild germicide; glycerized rubber lubricant; emulsifier and foam stabilizer for foamed rubber; FDA approval; Gardner 4 liq.; water-sol.; sp.gr. 1.025; visc. 1 stokes; 35% solids.

Soprophor® NPF/10. [Rhone-Poulenc Surf. & Spec.] Acid nonylphenyl ether phosphate; anionic; detergent in industrial cleaners; household detergents; paint stripping agent; dedusting; emulsifier for acrylic, PVAc emulsion polymerization; liq.; 100% conc.

Sorbax HO-40. [Chemax] POE sorbitol ester; emulsifier for agric. pesticide/herbicide, emulsion polymerization, metalworking lubricants, die-cast lubricants; liq.; sapon. no. 97; HLB 10.4.

Sorbax HO-50. [Chemax] POE sorbitol ester; o/w emulsifier for solvs., veg. and petrol. oils used in the textile, agric., emulsion polymerization, and metal lubricant industries; liq.; sapon. no. 85; HLB 11.4.

Sotex 3CW. [Morton Int'l./Specialty Chem.] Fatty acid ester, long chain; nonionic; dispersant for paints and coatings, ink vehicle; wetting agent for dyes; internal/external antistat for nylon, PP, PS, flexible and rigid PVC; visc. liq.; sol. in aliphatic and aromatic hydrocarbons; sp.gr. 0.930-0.960; b.p. > 200 C; 100% act.

SP-25. [Schenectady] Alkyl-phenol formaldehyde resin; tackifier for NBR adhesives and sealants; nonheat reactive; yel. to amber lumps; sol. in aromatic and chlorinated hydrocarbons, ketones, esters; sp.gr. 1.07; R&B soften. pt. 100-110 C.

SP-553. [Schenectady] Terpene phenolic resin; tackifier for SBR, NBR, NR, and CR compds. and adhesives; lengthens bonding range of Neoprene adhesives; yel. flakes; sol. in aliphatic and aromatic hydrocarbons, ketones, esters, and chlorinated hydrocarbons; sp.gr. 1.00; R&B soften. pt. 110-120 C.

SP-560. [Schenectady] Terpene phenolic resin; tackifier for SBR, NR, NBR, CR, and BR compds. and adhesives; lengthens bonding range of Neoprene adhesives with min. effect on heat resistance; lt. amber flakes; sol. in aliphatic and aromatic hydrocarbons, ketones, esters, and chlorinated hydrocarbons; sp.gr. 1.10; R&B soften. pt. 146-156 C.

SP-1044. [Schenectady] Alkyl phenol formaldehyde resin; crosslinking agent for butyl rubber, esp. curing bladders for automobile tires; requires addition of Lewis acid (Neoprene W or stannous chloride); lump resin; sol. in aromatic, aliphatic, and chlorinated hydrocarbons, ketones, and esters; sp.gr. 1.03-1.07; m.p. 60-70 C; 7.0-9.5% methylol content.

SP-1045. [Schenectady] Alkyl phenol-formaldehyde resin; crosslinking agent for butyl rubber, esp. curing bladders for automobile tires; requires addition of Lewis acid (Neoprene W or stannous chloride); amber lumps; sp.gr. 1.05; m.p. 60-66 C; 8-11% methylol content.

SP-1055. [Schenectady] Bromomethylated and methylolated alkyl phenol formaldehyde resin; crosslinking agent for unsat. elastomers (butyl or EPDM); used in vulcanizing butyl rubber curing bladders for automobile tires; requires no additional source of labile halogen; rdsh. lumps; sp.gr. 1.10; m.p. 57-66 C; 9.0-12.5% methylol content; 4% max. bromine content.

SP-1056. [Schenectady] Bromomethylated and methylolated alkyl phenol formaldehyde resin; crosslinking agent for unsat. elastomers (butyl, EPDM, and natural rubber); used to partially crosslink elastomer in pressure-sensitive tapes; requires no additional source of labile halogen; rdsh. lumps; sp.gr. 1.10; m.p. 57-66 C; 7.5-11.0% methylol content; 6% min. bromine content.

SP-1068. [Schenectady] Alkyl phenol formaldehyde resin; tackifier for BR, CR, SBR, NBR, NR compds. and in cements; nonheat reactive; lt. amber flakes; sol. in aromatic and aliphatic hydrocarbons, ketones, and esters; sp.gr. 1.02; R&B soften. pt. 85-95 C.

SP-1077. [Schenectady] Alkyl phenol formaldehyde resin; tackifier for BR, CR, SBR, NBR, NR, and EPDM elastomers; allows lowest loading of resin component to achieve desired tack level; lt. amber flakes; sol. in aliphatic and aromatic hydrocarbons, ketones, and esters; sp.gr. 1.02; R&B soften. pt. 92-103 C.

SP-6700. [Schenectady] Oil-modified phenol formaldehyde resin; resin in NBR, SBR, NR, and CR as plasticizer during processing; acts as hardener with 8% hexa after cure; use where scorching problems preclude use of hexa-containing materials such as SP-6600; brn. flakes; sol. in ketones, alcohols, and esters; sp.gr. 1.16; R&B soften. pt. 95 C.

SP-6701. [Schenectady] Oil-modified phenol formaldehyde resin; resin in SBR as plasticizer during processing and hardener after cure with 8% hexa; use where scorch problems preclude use of hexa-containing materials such as SP-6601; amber flakes; sol. in alchols, ketones, and esters; sp.gr. 1.16; R&B soften. pt. 95 C.

Specialty 101. [Specialty Prods.] Internal lubricant for thermoplastic polyester, PE, urea-formaldehyde; metal molds.

Specialty 102. [Specialty Prods.] Internal lubricant for polyester, phenolics, phenolformaldehyde, vinyl ester, DAP, epoxy, metal and FRP molds; liq.

Specialty 103. [Specialty Prods.] Internal lubricant for epoxy pultrusion, polyester; hot press, metal molds; liq.

Specialty 105. [Specialty Prods.] Internal lubricant for polyester, polyester gel coats; FRP, aluminum, epoxy molds; liq.

Spectraflo®. [Ferro/Plastic Colorants] Liq. colorants for engineering resins, high temp. molding.

SpectraPurge™. [Ferro/Plastic Colorants] Pourable liq. purging compd. for inj. directly into the feed throat of plastic processing equip.; 40-50% avg. savings in downtime changing from color to color or resin to resin; compat. with all resins; nontoxic; formulated with FDA listed materials; processing temps. 350-550 F; wh. liq.; dens. 16.9 lb/gal; visc. 10-12,000 cps; < 3% volatiles by vol.

Spectratech® CM 00524. [Quantum/USI] 30% Silver color conc. in LDPE carrier; silver color additive for general purpose film, trash bags, duct tape; FDA approved for food pkg.

Spectratech® CM 10540. [Quantum/USI] Dual uv inhibitor (Cyasorb UV 1084/531) for extrusion molding; 10% loading.

Spectratech® CM 10608. [Quantum/USI] Economy grade antiblock conc. for film applics.; FDA approved for food pkg.; 20% loading; Unverified

Spectratech® CM 10777. [Quantum/USI] Antistat conc. for film and sheet, inj. molding, blow molding; FDA approved for food pkg.; 4.5% loading.

Spectratech® CM 10778. [Quantum/USI] Antiblock conc. for high clarity film applics.; FDA approved for food pkg.; 50% loading.

Spectratech® CM 10779. [Quantum/USI] Antiblock conc. for colored film applics.; FDA approved for food pkg.; 50% loading.

Spectratech® CM 11013. [Quantum/USI] Antiblock conc. for film applics.; FDA compliance for food pkg.; 20% loading.

Spectratech® CM 11014. [Quantum/USI] Fast slip conc. for film applics.; FDA compliance for food pkg.; 5% loading.

Spectratech® CM 11045. [Quantum/USI] Antistat conc. for film and sheet, inj. molding, blow molding; limited FDA approval; 5% loading.

Spectratech® CM 11053. [Quantum/USI] Flame retardant conc. for LDPE film; max. processing temp. 380 F.

Spectratech® CM 11126. [Quantum/USI] Slow slip conc. for film applics.; FDA compliance for food pkg.; 5% loading.

Spectratech® CM 11172, 11174. [Quantum/USI] Slip concs. for film applics.; fast and slow resp.

Spectratech® CM 11194. [Quantum/USI] Slow slip conc. for film applics.; FDA compliance for food pkg.; 1% loading.

Spectratech® CM 11246. [Quantum/USI] UV absorber (Cyasorb UV 531) for pkg., agric. films, molded items; limited FDA approval; 10% loading.

Spectratech® CM 11340. [Quantum/USI] BHT; CAS 128-37-0; EINECS 204-881-4; antioxidant conc. for HDPE cereal liners; limited FDA approval; 5% loading.

Spectratech® CM 11357. [Quantum/USI] UV inhibitor (Cyasorb UV 1084) for agric. film, pool coverings, greenhouse film; 10% loading.

Spectratech® CM 11367. [Quantum/USI] UV inhibitor (Cyasorb UV 1084) for agric. film, pool coverings, greenhouse film; 20% loading.

Spectratech® CM 11489. [Quantum/USI] Flame retardant conc. for HDPE film.

Spectratech® CM 11513. [Quantum/USI] Slip/antiblock conc. for film applics.; limited FDA approval; 2.5% slip, 13.3% antiblock.

Spectratech® CM 11524. [Quantum/USI] Primary/secondary blend; antioxidant conc. for polyolefin films and molded items; limited FDA approval; 0.84% primary, 4.16% secondary.

Spectratech® CM 11591. [Quantum/USI] Flame retardant conc. for LDPE, HDPE, PP film and wire and cable; max. processing temp. 600 F.

Spectratech® CM 11593. [Quantum/USI] Primary/secondary blend; antioxidant conc. for polyolefin films and molded items; limited FDA approval; 1.33% primary, 6.65% secondary.

Spectratech® CM 11615. [Quantum/USI] UV inhibitor (Tinuvin 770) for PP slit tape, HDPE, and PP molded goods; 10% loading.

Spectratech® CM 11616. [Quantum/USI] UV inhibitor and antioxidant for blk. PE and PP mulch film.; Unverified

Spectratech® CM 11638. [Quantum/USI] Antistat conc. for film and sheeting for electrical pkg.; 20% loading.

Spectratech® CM 11681. [Quantum/USI] UV inhibitor for greenhouse film, PP slit tape, molded goods.; Unverified

Spectratech® CM 11698. [Quantum/USI] Optical brightener for film applics.

Spectratech® CM 11704. [Quantum/USI] UV inhibitor for HDPE slit tape, greenhouse and construction film.; Unverified

Spectratech® CM 11715. [Quantum/USI] UV inhibitor (Tinuvin 622) for HDPE slit tape, pkg., and molded goods.; limited FDA approval; 12.5% loading.

Spectratech® CM 11720. [Quantum/USI] UV inhibitor (Chimassorb 944LD) for HDPE slit tape, greenhouse and construction film; limited FDA approval; 10% loading.

Spectratech® CM 11812. [Quantum/USI] Optical brightener for film applics.; limited FDA approval.
Spectratech® CM 22801. [Quantum/USI] 50% Gray color conc. in LDPE carrier; gray color additive for general purpose blown film.
Spectratech® CM 30757. [Quantum/USI] 46% Blue color conc. in LDPE carrier with zinc stearate; lt. blue color additive for general purpose industrial liners; FDA approved for food pkg.
Spectratech® CM 34286. [Quantum/USI] 40% Blue color conc. in LDPE carrier with zinc stearate; med. blue color additive for general purpose blown films, medical and personal care films, food film; FDA approved for food pkg.
Spectratech® CM 45288. [Quantum/USI] 50% Greem color conc. in LDPE carrier with zinc stearate; forest green color additive for general purpose, films, trash bags and industrial liners.
Spectratech® CM 56483. [Quantum/USI] 50% Yel. color conc. in LDPE carrier with zinc stearate; yel. color additive for grocery sacks, general purpose films, merchandise bags; formulated for heat stability.
Spectratech® CM 63745. [Quantum/USI] 26% Brn. color conc. in LDPE carrier with zinc stearate; dk. brn. color additive for trash bags, industrial liners.
Spectratech® CM 77242. [Quantum/USI] Antistat conc. for film and sheeting for elec. pkg.; 20% loading.
Spectratech® CM 80206. [Quantum/USI] 50% Wh. color conc. in LDPE carrier; color additive for general-purpose blown film; high dispersion quality; for blow molding for general purpose, household, chemical, toiletry, and cosmetic bottles; for inj. molding of containers; FDA approved for food pkg.
Spectratech® CM 88100. [Quantum/USI] 50% Wh. color conc. in LDPE carrier; color additive for extrusion coating; premium grade for laminations; snack food pkg.; exc. disp. and processability; FDA approved for food pkg.
Spectratech® CM 88200. [Quantum/USI] 50% Wh. color conc. in LDPE carrier with zinc stearate; color additive for blown and cast film; critical let-down ratios; optimum dispersion quality; FDA approved for food pkg.
Spectratech® CM 92049. [Quantum/USI] 50% Blk. color conc. in LDPE carrier; blk. color additive for blown film applics.; critical let-down ratios; optimum dispersion quality.
Spectratech® CM 92101. [Quantum/USI] 50% Blk. color conc. in LDPE carrier; blk. color additive for general-purpose film, extrusion, and molding applics.
Spectratech® CM 92473. [Quantum/USI] 50% Blk. color conc. in LDPE carrier; economical blk. color additive for thick blown film, inj. molding, sheet, or pipe applics.
Spectratech® HM 82192. [Quantum/USI] 50% Wh. color conc. in EAA carrier; color additive for extrusion coating, laminations, condiment pkg.; FDA approved for food pkg.
Spectratech® HM 99506. [Quantum/USI] 55% Blk. color conc. in EVA carrier with zinc stearate; blk. color additive for general purpose molding and compounding.
Spectratech® IM 88763. [Quantum/USI] 70% Wh. color conc. in EMA carrier; color additive for extrusion coating, laminations, condiment pkg., cookie pkg.; FDA approved for food pkg.
Spectratech® KM 11264. [Quantum/USI] BHT; CAS 128-37-0; EINECS 204-881-4; antioxidant conc. for HDPE cereal liners; limited FDA approval; 10% loading.
Spectratech® KM 80531. [Quantum/USI] 50% Wh. color conc. in HDPE carrier; color additive for general-purpose film applics.; optimum dispersion quality; also for inj. molding of general purpose containers; FDA approved for food pkg.
Spectratech® PM 10634. [Quantum/USI] Slip/antiblock conc. for film applics.; limited FDA approval; 2.5% slip, 13.3% antiblock.
Spectratech® PM 11607. [Quantum/USI] Processing aid for LLDPE and HDPE extrusion; limited FDA approval; 1.7% loading.
Spectratech® PM 11725. [Quantum/USI] Processing aid for LLDPE and HDPE extrusion; limited FDA approval; 3% loading.
Spectratech® PM 80230. [Quantum/USI] 50% Wh. color conc. in LLDPE carrier; color additive for general-purpose blown film; exc. dispersion quality; FDA approved for food pkg.
Spectratech® PM 82215. [Quantum/USI] 50% Filled wh. color conc. in LLDPE carrier; color additive for blow molding for general purpose, household, chemical, toiletry, motor oil, and cosmetic bottles; FDA approved for food pkg.
Spectratech® PM 92742. [Quantum/USI] 40% N330 Carbon blk. color conc. in LLDPE carrier; blk. color additive for geo membranes; outstanding dispersion; improved uv stability.
Spectratech® RM 88964. [Quantum/USI] 40% Wh. color conc. in PP carrier with 21.3% Polywax 520 additive; blue-wh. color additive for inj. molding and extrusion of general-purpose containers; FDA approved for food pkg.
Spectratech® SM 88008. [Quantum/USI] 30% Wh. color conc. in HIPS carrier with zinc stearate additive; blue-wh. color additive for inj. molding and extrusion of general-purpose containers, plastic utensils; enhances rigidty; FDA approved for food pkg.
Spectratech® SM 92042. [Quantum/USI] 30% Blk. color conc. in PS carrier; blk. color additive for general purpose molding applics.
Spectratech® YM 80751. [Quantum/USI] 80% Wh. color conc. in EVA carrier; general-purpose high loaded color additive for film extrusion and molding applics.; FDA approved for food pkg.
Spectratech® YR 92363. [Quantum/USI] 25% Blk. color conc. in proprietary carrier resin with antioxidant and calcium stearate; blk. color additive for semiconductive films, electronic and explosive pkg., medical films.
Sphericel® 110P8. [Potters Industries] Hollow boro-

silicate glass spheres; CAS 65997-17-3; tough spheres which withstand molding pressures and reduce weight in engineering grade plastic compds. and molded parts; improve dimensional stability and flow props.; wh. spherical powd.; 8 μ mean size; insol. in water; sp.gr. .1-1.5 g/cc; dens. 1.1 g/cc; soften. pt. > 600 C; *Toxicology:* ACGIH TLV 10 mg/m^3 nuisance dust, 5 mg/m^3 (respirable dust); dust may be irritating to respiratory tract, eyes; *Precaution:* nonflamm.; incompat. with hydrofluoric acid; spillages may be slippery.

Spheriglass® Coated. [Potters Industries] Solid borosilicate glass spheres; ultrafine inorg. reinforcement for resins; developed to provide smaller spheres to impart antiblock props. in polyolefin films; 3, 2, and 4 μ avg. particle size resp.

Spheriglass® 1922. [Potters Industries] Solid soda-lime glass spheres; inorganic reinforcement for thermoplastic and thermoset resins; provides improved flow props., high resin displacement, low shrinkage and warpage, better molded parts, dimensional stability; 219 μ mean particle size; sp.gr. 2.5; bulk dens. 98 lb/ft^3 (tapped); oil absorp. 18 g oil/100 g spheres; soften. pt. 704 C; ref. index 1.51; hardness (Moh) 6.0; dielec. const. 6.9 (10^6 Hz, 22 C); vol. resist. 6.5 x 10^{12} ohm-cm; 72.5% SiO_2, 13.7% Na_2O, 9.8% CaO, 3.3% MgO; *Toxicology:* nontoxic.

Spheriglass® 2530. [Potters Industries] Solid soda-lime glass spheres; inorganic reinforcement for thermoplastic and thermoset resins; provides improved flow props., high resin displacement, low shrinkage and warpage, better molded parts, dimensional stability; 66 μ mean particle size; sp.gr. 2.5; bulk dens. 98 lb/ft^3 (tapped); oil absorp. 18 g oil/100 g spheres; soften. pt. 704 C; ref. index 1.51; hardness (Moh) 6.0; dielec. const. 6.9 (10^6 Hz, 22 C); vol. resist. 6.5 x 10^{12} ohm-cm; 72.5% SiO_2, 13.7% Na_2O, 9.8% CaO, 3.3% MgO; *Toxicology:* nontoxic.

Spheriglass® 3000E. [Potters Industries] Solid borosilicate glass spheres; inorg. reinforcement for thermoplastic and thermoset resins; provides improved flow props., high resin displacement, low shrinkage and warpage, better molded parts, dimensional stability; 26 μmean particle size; sp.gr. 2.54; bulk dens. 101 lb/ft^3 (tapped); oil absorp. 17 g oil/100 g spheres; soften. pt. 846 C; ref. index 1.55; hardness (Moh) 6.5; dielec. const. 5.8 (10^6 Hz, 22 C); vol. resist. 10^{13}-10^{16} ohm-cm; 52.5% SiO_2, 22.5% CaO, 14.5% Al_2O_3, 8.6% B_2O_3.

Spheriglass® 4000E. [Potters Industries] Solid borosilicate glass spheres; inorg. reinforcement for thermoplastic and thermoset resins; provides improved flow props., high resin displacement, low shrinkage and warpage, better molded parts, dimensional stability; 18 μmean particle size; sp.gr. 2.54; bulk dens. 83 lb/ft^3 (tapped); oil absorp. 14 g oil/100 g spheres; soften. pt. 846 C; ref. index 1.55; hardness (Moh) 6.5; dielec. const. 5.8 (10^6 Hz, 22 C); vol. resist. 10^{13}-10^{16} ohm-cm; 52.5% SiO_2, 22.5% CaO, 14.5% Al_2O_3, 8.6% B_2O_3.

Spheriglass® 5000. [Potters Industries] Solid soda-lime glass spheres; inorganic reinforcement for thermoplastic and thermoset resins; provides improved flow props., high resin displacement, low shrinkage and warpage, better molded parts, dimensional stability; 11 μ mean particle size; sp.gr. 2.5; bulk dens. 77 lb/ft^3 (tapped); oil absorp. 17 g oil/100 g spheres; soften. pt. 704 C; ref. index 1.51; hardness (Moh) 6.0; dielec. const. 6.9 (10^6 Hz, 22 C); vol. resist. 6.5 x 10^{12} ohm-cm; 72.5% SiO_2, 13.7% Na_2O, 9.8% CaO, 3.3% MgO; *Toxicology:* nontoxic.

Spheriglass® 5000E. [Potters Industries] Solid borosilicate glass spheres; inorg. reinforcement for thermoplastic and thermoset resins; provides improved flow props., high resin displacement, low shrinkage and warpage, better molded parts, dimensional stability; 11 μmean particle size; sp.gr. 2.54; bulk dens. 99 lb/ft^3 (tapped); oil absorp. 19 g oil/100 g spheres; soften. pt. 846 C; ref. index 1.55; hardness (Moh) 6.5; dielec. const. 5.8 (10^6 Hz, 22 C); vol. resist. 10^{13}-10^{16} ohm-cm; 52.5% SiO_2, 22.5% CaO, 14.5% Al_2O_3, 8.6% B_2O_3.

Spheriglass® 6000. [Potters Industries] Solid soda-lime glass spheres; inorganic reinforcement for thermoplastic and thermoset resins; provides improved flow props., high resin displacement, low shrinkage and warpage, better molded parts, dimensional stability; 4 μ mean particle size; sp.gr. 2.5; bulk dens. 90 lb/ft^3 (tapped); oil absorp. 20 g oil/100 g spheres; soften. pt. 704 C; ref. index 1.51; hardness (Moh) 6.0; dielec. const. 6.9 (10^6 Hz, 22 C); vol. resist. 6.5 x 10^{12} ohm-cm; 72.5% SiO_2, 13.7% Na_2O, 9.8% CaO, 3.3% MgO; *Toxicology:* nontoxic.

Spheriglass® Coated. [Potters Industries] Solid glass spheres with coupling agent coatings; inorg. reinforcement for thermoplastic and thermoset resins; coupling agent coatings reduce mfg. cost and optimize flex. and tens. str. props.

Spin-Flam MF-82. [Himont] Nonhalogenated organic; flame retardant for ABS, PE, PP, PS, PU rigid and flexible foam, EVA, TPE, TPR.

Spinomar NaSS. [Tosoh] Sodium p-styrenesulfonate; emulsifier for soapless emulsion of acrylic, vinyl acetate, and SBR latex; dye assistant; flocculant/scale inhibitor; hair fixative, cosmetic dispersant; artificial bio-membranes; antistat for paper, fibers, plastics; pharmaceuticals; metal plating; wh. powd., nil odor; sol. in water; insol. in aliphatic, halogenated, or high alcohol solvs.; m.w. 206.20; apparent sp. dens. 0.5; decomp. pt. 330 C; flash pt. nil; 81.5% act.; *Toxicology:* LD 50 (oral, mouse) 16 g/kg (as 40% olive oil susp.); *Storage:* store in airtight containers in dark place; if dried, subject to slow oxidation and/or polymerization.

Sprayon 803. [Sprayon Prods.] Silicone; food grade, fast-evaporating, non-ozone depleting, nonstaining mold release for wax, rubber, TFE, cermaics, and plastics; performs at mold temps. from 40-500 F; used for investment castings, compr. molds, plastic inj. molds, blow molds, candle molds, extruders, and in foundries; FDA approved at 10 ppm max.

Sprayon 805. [Sprayon Prods.] Silicone; heavy-duty, non-ozone depleting mold release for rubber, plastic, elastomeric materials, PS and PU foams; heat stable to 500 F; minimizes crazing, spotting, sticking, and other causes for rejects; FDA approved at 1 ppm max.

Sprayon 807. [Sprayon Prods.] Non-ozone depleting, heavy-duty paintable mold release for plastic, rubber, wax, moldable resins, elastomeric compds. for inj. molding, hot or cold molding processes; good for large intracte or difficult-to-release parts; for appliances, housewares, furniture, sporting goods, tires (green carcass), automotive, electronic components.

Sprayon 811 Dry Film Vydax. [Sprayon Prods.] Non-ozone depleting mold release providing exc. antistick props. for multiple releases of most moldable resins and elastomeric compds., urethanes, styrenes; for low and high temp. molds; for PS, nylon, foams, filament winding, epoxy potting/encapsulation, PP, rubber, metals, phenolic resins.

Sprayon 812. [Sprayon Prods.] Zinc stearate; CAS 557-05-1; EINECS 209-151-9; dust-repelling, non-ozone depleting mold release for thermosetting and thermoplastic applics., clear styrenes and acrylics, reflex lenses, PP, PC, PS, polysulfonates, plastic molds, molding ceramics; effective to 450 F.

Sprayon 815. [Sprayon Prods.] Silicone; non-ozone depleting, noncorrosive mold release for PS, elastomeric compds., plastics, PU; minimizes crazing, spotting, sticking, and other causes of rejects; FDA approved @ 1 ppm max.

Sprayon 829. [Sprayon Prods.] Non-ozone depleting paintable mold release compat. with plastic, elastomeric compds., and most moldable resins; used for appliances, housewares, furniture, automotive, electronic, sporting goods; ideal for applics. where a min. of silicone is required.

Sprayset® MEKP. [Witco/PAG] MEK peroxide in anhyd. ethyl acetate; catalyst for spray gun applics. in the curing of polyesters; clear low-visc. liq.; 5.25% act. oxygen; *Precaution:* flamm.

SR 111. [Sartomer] Metallic diacrylate; monomer, coagent offering high crosslink dens. to adhesives, elastomers, golf balls; improves processability and cure performance; wh. powd.; sol. in water; m.w. 207; sp.gr. 1.670; 99% reactive esters.

SR-201. [Sartomer] Allyl methacrylate with 100 ppm MEHQ inhibitor; monomeric acrylic ester capable of polymerizing to hard, infusible resin that is water wh., clear, and glass-like; as comonomer with vinyl-type monomers to produce crosslinked polymers; reactive diluent; unsymmetrical crosslinking agent where a two-stage polymerizing or drying action is desired; for adhesives, coatings, dental polymers, paints, plastics, chem. intermediates; APHA 10 clear liq., pungent, disagreeable odor; m.w. 126; sp.gr. 0.930; dens. 7.7 lb/gal; visc. 2 cps; b.p. 150 C; flash pt. (PMCC) 88 F; ref. index 1.4343-1.4350; 99% reactive esters.

SR-203. [Sartomer] Tetrahydrofurfuryl methacrylate; with 100 ppm hydroquinone, 900 ppm MEHQ; high boiling, low-visc., low-volatility monomer as reactive diluent; polymerization initiated by conventional methods (peroxide catalysts, thermally, ionizing radiation, and uv radiation); for adhesives (anaerobic), cosmetics, elastomers, plastics, photopolymers, electronics (photoresists); straw APHA 65 clear liq., mild, pleasant odor; sol. in alcohol, ether, ketones, esters, and low m.w. aliphatic hydrocarbons; low sol. in water; m.w. 170; sp.gr. 1.041; dens. 8.75 lb/gal; visc. 5 cps; b.p. 265 C; flash pt. (PMCC) 135 F; sapon. no. 172; ref. index 1.4554; 99% reactive esters.

SR-205. [Sartomer] Triethylene glycol dimethacrylat; with 80 ppm hydroquinone inhibitor; noncorrosive, low visc., high boiling crosslinking monomeric ester; in vinyl plastisols reduces initial visc. and oil extractability, and improves ultimate hardness, heat distort., hot tear strength, and stain resistance; in cast acrylic sheet and rod, contact lens, elastomers, ion exchange resins, dental polymers, adhesives, coatings (paper, plastic), cosmetics, paints, sealants, photopolymers, electronics (photoresists, solder masks); colorless clear liq., mild, pleasant odor; sol. in alcohol, ether, ketones, esters, and low m.w. aliphatic hydrocarbons; m.w. 286; sp.gr. 1.070; dens. 8.96 lb/gal; visc. 11 cps; flash pt. (PMCC) 146 F; ref. index 1.4595; 99% reactive esters.

SR-206. [Sartomer] Ethylene glycol dimethacrylate with 70 ppm hydroquinone inhibitor; crosslinking agent used in emulsion polymerization, cast acrylic sheet, fiberglass-reinforced polyesters, ion exchange resins, rubber compds., adhesives, coatings (plastic), cosmetics, dental polymers, electronics; colorless clear liq., mild musty odor; sol. in alcohols, ethers, ketones, esters, aromatic and aliphatic hydrocarbons; insol. in water; m.w. 198.21; sp.gr. 1.049; dens. 8.75 lb/gal; visc. 6 cps; b.p. 260 C; flash pt. (PMCC) 155 F; ref. index 1.4522-1.4532; 99% reactive esters.

SR-209. [Sartomer] Tetraethylene glycol dimethacrylate with 75 ppm hydroquinone inhibitor; crosslinking agent used in castings, plastisols, coatings, fibers, papers, adhesives (anaerobic), concrete polymers and sealants, dental polymers, elastomers, electronics (photoresists, solder masks), photopolymers, sealants; APHA 30 clear liq., mild pleasant odor; sol. in alcohols, esters, ethers, ketones, and aromatic hydrocarbons; insol. in water; m.w. 330.37; sp,.gr. 1.077; dens. 9.0 lb/gal; visc. 14 cps; b.p. 220 C; acid no. 0.5; flash pt. (PMCC) 161 F; ref. index 1.4605-1.4610; 99% reactive esters.

SR-210. [Sartomer] PEG dimethacrylate with 75 ppm hydroquinone inhibitor; difunctional monomer, crosslinking agent for radiation and peroxide cure systems incl. adhesives (anaerobic), concrete polymers and sealants, cosmetics, dental polymers, elastomers, plastics, plastisols, sealants; colorless clear liq., mild, pleasant odor; sol. in alcohols, ethers, ketones, esters, and aromatic hydrocarbons; m.w. 330.37; sp.gr. 1.080; dens. 9.0 lb/gal; visc. 15 cps; b.p. > 200 C; acid no. 0.5; flash pt. (PMCC) 166 F; ref. index 1.4615; 99% reactive esters.

SR-212. [Sartomer] 1,3-Butylene glycol diacrylate with 500 ppm hydroquinone inhibitor; low visc. monomer; curing agent; polymerizes to hard, insol., infusible, thermoset resin; exothermic polymerization reaction, initiated thermally or by common free radical initiators, i.e., high energy and uv radiation, peroxide compds.; for acrylics, coatings, plastics; high stain resist.; straw clear liq., musty odor; sol. in aromatics, esters, and ketones; limited sol. in aliphatics; m.w. 198; sp.gr. 1.051; dens. 8.7 lb/gal; visc. 9 cps; acid no. 1.4; flash pt. (PMCC) 133 F; 99% reactive esters.

SR-213. [Sartomer] 1,4-Butanediol diacrylate with 75 ppm hydroquinone inhibitor; difunctional monomer, crosslinking agent for radiation and peroxide cure systems incl. elastomers, plastics; provides high solvency, weatherability, chem., heat, abrasion, and water resist.; APHA 50 clear liq., pleasant odor; m.w. 198; sp.gr. 1.054; dens. 8.9 lb/gal; visc. 8 cps; flash pt. (PMCC) 147 F; 99% reactive esters.

SR-214. [Sartomer] 1,4-Butanediol dimethacrylate with 240 ppm MEHQ inhibitor; difunctional monomer, crosslinking agent for radiation and peroxide cure systems incl. concrete polymers, elastomers, photopolymers, plastics, electronics (photoresists); offers high solvency, weatherability, chem., heat, abrasion, and water resist.; APHA 40 clear liq.; m.w. 226; sp.gr. 1.019; visc. 7 cps; ref. index 1.4548; 98% reactive esters.

SR-239. [Sartomer] 1,6-Hexanediol methacrylate with 95 ppm hydroquinone inhibitor; difunctional monomer, crosslinking agent used in casting compds., glass fiber-reinforced plastics, adhesives, coatings, ion-exchange resins, textile prods., plastisols, dental polymers, elastomers, acrylics, electronics (encapsulants, photoresisists); sealants; APHA 30 clear liq., bland odor; sol. in org. solv.; insol. in water; m.w. 254; sp.gr. 0.982; dens. 8.31 lb/gal; visc. 8 cps; flash pt. (PMCC) 157 F; ref. index 1.4565; 99% reactive esters.

SR-244. [Sartomer] 2-Methoxyethyl acrylate with 100 ppm MEHQ inhibitor; low-visc. monomer as reactive diluent for radiation and peroxide cure applics. incl. elastomers, chem. intermediates; APHA 50 clear liq.; sl. disp. in water; m.w. 130; sp.gr. 1.012; visc. 13 cps; 98% reactive esters.

SR-252. [Sartomer] PEG-12 dimethacrylate with 1000 ppm MEHQ inhibitor; difunctional monomer, crosslinking agent for radiation and peroxide cure systems incl. adhesives, coatings, concrete sealants, elastomers, electronics (photoresists, solder masks), photopolymers, plastics; produces soft, flexible films; APHA 35 clear liq.; disp. in water; m.w. 736; sp.gr. 1.101; visc. 67 cps; ref. index 1.466; 99% reactive esters; *Toxicology:* low skin irritation.

SR-290. [Sartomer] Tris (2-hydroxyethyl) isocyanurate trimethacrylate with 80 ppm MEHQ; trifunctional monomer, crosslinking agent for radiation and peroxide cure systems incl. coatings (metal, optical, plastic, wood), elastomers, plastisols; heat resist., fast cure response; wh. cryst. solid; m.w. 437; sp.gr. 1.292; visc. 20,000 cps; 96% reactive esters.

SR-295. [Sartomer] Pentaerythritol tetraacrylate with 340 ppm MEHQ inhibitor; crosslinking agent in adhesives (binder), coatings (glass, metal, optical, paper, PVC floor, sood), elastomers, electronics (solder masks), inks (flexo, gravure, offset, screen), paints, sealants; APHA 35 clear liq., musty odor; m.w. 352; sp.gr. 1.179; dens. 9.9 lb/gal; visc. 342 cps (38 C); m.p. 15-18 C; flash pt. (PMCC) 158 F; ref. index 1.4823; 99% reactive ester.

SR-297. [Sartomer] 1,3-Butylene glycol dimethacrylate with 200 ppm MEHQ; low visc. difunctional monomer; polymerizes to hard, insol., infusible, thermoset resin; mildly exothermic polymerization initiated by common free radical initiators; used in adhesives, coatings, ion-exchange resins, textile prods., plastisols, dentures, glass fiber-reinforced plastics, casting compds., and rubber compding.; improves resistance to scratching, attack by oils and solvs., heat deformation, and hardness; straw clear liq., mildly disagreeable odor; sol. in alcohols, ethers, ketones, esters, aromatic and aliphatic hydrocarbons; m.w. 226.26; sp.gr. 1.011; dens. 8.4 lb/gal; visc. 7 cps; b.p. 290 C; acid no. 0.5; flash pt. (PMCC) 137 F; ref. index 1.449; 99% reactive esters.

SR-313. [Sartomer] Lauryl methacrylate with 400 ppm MEHQ inhibitor; low-visc. monomer as reactive diluent in radiation and peroxide cure applics. incl. acrylics, adhesives, chem. intermediates, coatings (metal, textile), elastomers, plastics, sealants, photopolymers, electronics (encapsulants, photoresists); APHA 30 clear liq.; m.w. 254; sp.gr. 0.872; visc. 6 cps; ref. index 1.4420; 99% reactive esters.

SR-344. [Sartomer] PEG 400 diacrylate with 490 ppm MEHQ inhibitor; difunctional monomer, curing agent for radiation and peroxide cure systems incl. adhesives, coatings (glass, metal, optical, paper, plastic, PVC floor, wood), concrete polymers, inks (flexo, gravure, litho, offset, screen), photopolymers; produces soft, flexible films; APHA 35 clear liq., mild odor; water-sol.; m.w. 58; sp.gr. 1.1165; visc. 57 cps; flash pt. (PMCC) 51 F; 99% reactive esters; *Toxicology:* low skin irritation.

SR-348. [Sartomer] Ethoxylated bisphenol A dimethacrylate with 340 ppm MEHQ inhibitor; crosslinking agent used as additive in peroxide-cured rubber compds., irradiated coatings, polyesters, adhesives (anaerobic), concrete polymers, dental polymers, electronics (photoresists), inks (flexo, gravure), photopolymers, plastisols, sealants; very low volatility, fast cure response; Gardner 2-3 clear liq., musty odor; m.w. 452; sp.gr. 1.119; dens. 9.2 lb/gal; visc. 1082 cps; acid no. < 0.5; flash pt. (PMCC) 168 F; ref. index 1.5389; 99% reactive esters.

SR-349. [Sartomer] Ethoxylated bisphenol A diacrylate with 750 ppm hydroquinone inhibitor;

difunctional monomer, curing agent for radiation and peroxide cure systems incl. adhesives, coatings, elastomers, inks (flexo, gravure, screen); very low volatility; Gardner 2-3 clear liq., mild odor; m.w. 424; sp.gr. 1.145; dens. 9.4 lb/gal; visc. 1600 cps; flash pt. (PMCC) > 150 F; ref. index 1.5424; 99% reactive esters.

SR-350. [Sartomer] Trimethylolpropane trimethacrylate with 65 ppm hydroquinone inhibitor; trifunctional monomer, crosslinking agent for radiation and peroxide cure systems incl. acrylics, adhesives, coatings (glass, metal, PVC floor, wood), concrete polymers and sealants, dental polymers, elastomers, inks, paints, plastics, plastisols; low volatility, fast cure response; APHA 40 clear liq., mild odor; sol. in alcohols, ethers, ketones, esters, aliphatic hydrocarbons; insol. in water; m.w. 338.39; sp.gr. 1.061; dens. 8.8 lb/gal; visc. 44 cps; b.p. > 200 C; acid no. 0.5; flash pt. (PMCC) 149 F; ref. index 1.4700; 99% reactive esters.

SR-365. [Sartomer] Zinc dimethacrylate; monomer, coagent offering low compr. set props. to adhesives, elastomers, golf balls; improves processability and cure performance; off-wh. powd.; sol. in warm acetic acid; sl. sol. in water; m.w. 235; sp.gr. 1.480; 99% reactive esters.

SR-368. [Sartomer] Tris (2-hydroxyethyl) isocyanurate triacrylate with 100 ppm MEHQ inhibitor; trifunctional monomer, crosslinking agent for radiation and peroxide cure systems incl. coatings (metal, optical, plastic, wood), elastomers, electronics (photoresists), inks (screen), plastics; adhesion promoter; fast cure response; offers abrasion resist.; Gardner 1 cryst. solid; m.w. 423; sp.gr. 1.300; m.p. 52-54 C; 99% reactive esters.

SR-376. [Sartomer] Zinc methacrylate; coagent improving processability and cure performance in elastomers, golf balls, adhesives; wh. powd.; sl. sol. in water; m.w. 168; sp.gr. 1.800; 99% reactive esters.

SR-379. [Sartomer] Glycidyl methacrylate with 60 ppm hydroquinone inhibitor; monofunctional monomer as reactive diluent; polymerized by applics. of heat, heat and peroxidic catalysts, and irradiation by uv, beta, gamma, or x-ray; in hydrogels for contact lenses and membranes, molding and casting compds.; impregnating paper, concrete, and wood, coatings (metal, powd.), printing inks, adhesives and sealants, elastomers, plastics, electronics (photoresists), chem. intermediates; clear liq., pungent odor; sol. in most organic solvs.; m.w. 142.15; sp.gr. 1.073; dens. 8.9 lb/gal; visc. 5 cps; b.p. 75 C; flash pt. (PMCC) 76 C; ref. index 1.4470; 99% reactive esters.

SR-399. [Sartomer] Dipentaerythritol pentaacrylate with 270 ppm MEHQ inhibitor; high m.w., multifunctional acrylate monomer susceptible to polymerization by radiation and uv initiation; curing agent; for acrylics, adhesives (binder, structural), coatings (glass, metal, paper, plastic, PVC floor, wood), electronics; inks (flexo, gravure, litho, offset, screen), paints, photopolymers, plastics, plastisols, sealants; APHA 50 clear liq., mild odor; sol. in esters, ketones; m.w. 524.51; sp.gr. 1.192; dens. 9.5 lb/gal; visc. 13,600 cps; acid no. 1.4; flash pt. (PMCC) 240 F; ref. index 1.4889; 99% reactive esters; *Toxicology:* low skin irritation.

SR-416. [Sartomer] Zinc diacrylate; CAS 14643-87-9; coagent improving processability and cure performance in elastomers, adhesives, golf balls; good dispersion, high crosslink dens.; wh. powd.; m.w. 239; sol. in warm acetic acid; sl. sol. in water; sp.gr. 1.608; 92% reactive esters.

SR-423. [Sartomer] Isobornyl methacrylate with 150 ppm MEHQ inhibitor; monofunctional monomer as reactive diluent for radiation and peroxide cure applics. incl. adhesives (structural), chem. intermediates, coatings (metal), concrete polymers, plastics; APHA 20 clear liq.; m.w. 222; sp.gr. 0.979; visc. 11 cps; ref. index 1.4738; 99% reactive esters.

SR-444. [Sartomer] Pentaerythritol triacrylate with 350 ppm MEHQ inhibitor; trifunctional monomer, crosslinking agent used in adhesives (binder), coatings (glass, metal, optical, paper, PVC floor, wood), elastomers, electronics (solder masks), inks (flexo, gravure, offset, screen), sealants; fast cure response; APHA 50 clear liq., musty odor; m.w. 298; sp.gr. 1.162; dens. 9.9 lb/gal; visc. 520 cps; flash pt. (PMCC) 174 F; ref. index 1.4790; 99% reactive esters; *Toxicology:* low skin irritation.

SR-454. [Sartomer] Ethoxylated trimethylolpropane triacrylate with 255 ppm MEHQ inhibitor; trifunctional monomer, curing agent for radiation and peroxide cure systems incl. acrylics, adhesives (pressure sensitive, structural), coatings (glass, metal, optical, paper, PVC floor, release, textile, wood), concrete polymers; electronics (conformal, photoresists, solder masks), inks (flexo, gravure, litho, screen), paints, photopolymers; APHA 55 clear liq., mild odor; m.w. 428; sp.gr. 1.103; dens. 9.25 lb/gal; visc. 60 cps; flash pt. (PMCC) 188 F; ref. index 1.4686; 99% reactive esters; *Toxicology:* low skin irritation.

SR-494. [Sartomer] Ethoxylated pentaerythrityl tetraacrylate with 190 ppm MEHQ; multifunctional monomer, crosslinking agent for radiation and peroxide cure systems incl. adhesives (binder), chem. intermediates, coatings (glass, metal, paper, plastic, PVC floor, wood), elastomers; electronics (solder masks), inks (flexo, gravure, offset, screen), sealants; fast cure response; APHA 80 clear liq.; sp.gr. 1.128; visc. 150 cps; 99% reactive ester.

SR-506. [Sartomer] Isobornyl acrylate with 170 ppm MEHQ inhibitor; monomer, curing agent, reactive diluent for radiation and peroxide cure applics. incl. adhesives, coatings (glass, metal, optical, paper, PVC floor, release), concrete polymers, electronics (photoresists, solder masks), inks (screen), photopolymers; APHA 20 clear liq.; m.w. 208; sp.gr. 0.987; dens. 8.2 lb/gal; visc. 9 cps; flash pt. (PMCC) 240 F; ref. index 1.4722; 99% reactive esters.

SR-527. [Sartomer] Modified metallic diacrylate; scorch retarder, coagent improving processability and cure performance in elastomers, adhesives;

exc. dispersion, high crosslink dens.; wh. powd.; m.w. 239; sol. in warm acetic acid; sl. sol. in water; sp.gr. 1.600; 92% reactive esters.

SR-7475. [Firestone Syn. Rubber] Stereospecific butadiene/styrene copolymer; CAS 9003-55-8; blendable modifier for thermoplastic resins and asphalt; APHA 5 small pellet; melt index 12; sp.gr. 0.96; ref. index 1.5495; tens. str. 1000 psi; Shore hardness A 85; 43% bound styrene.

SR-9010. [Sartomer] Methacrylate ester with enhanced inhibitor pkg. (60 ppm hydroquinone, 200 ppm MEHQ); trifunctional adhesion-promoting monomer for coatings (glass, metal, plastic), elastomers, plastisols, sealants; fast cure response; extended shelf life; APHA 430 clear liq.; sp.gr. 1.075; visc. 32 cps; ref. index 1.4723; 99% reactive esters.

SR-9011. [Sartomer] Methacrylate ester with 90 ppm hydroquinone; trifunctional adhesion-promoting monomer for coatings (glass, metal, plastic), elastomers, plastisols, sealants; fast cure response; APHA 140 clear liq.; sp.gr. 1.078; visc. 35 cps; 99% reactive esters.

SR-9041. [Sartomer] Pentaacrylate ester with 265 ppm MEHQ inhibitor; multifunctional monomer, crosslinking agent for radiation and peroxide cure systems incl. acrylics, adhesives (binder, structural), coatings (glass, metal, paper, plastic, PVC floor, wood), electronics, inks (flexo, gravure, litho, offset, screen); paints, photopolymers, plastics, plastisols, sealants; fast cure response; APHA 65 clear liq.; sp.gr. 1.192; visc. 15,195 cps; ref. index 1.4887; *Toxicology:* low skin irritation.

SRF-1501. [Schenectady] Resorcinol formaldehyde preformed resin sol'n.; bonding agent used in rubber compds. to improve adhesion; red to brnsh. flakes; sol. in alcohols, ketones, esters, and water; sp.gr. 1.37; soften. pt. 100-110 C.

SS 4177. [GE Silicones] Silicone polymer blend; release agent for rubber, plastics, and resinous materials, esp. thermoset materials (epoxy resins, polyester resins, urethane elastomers); also used in polishes; clear liq.; dens. 7.9 lb/gal; visc. 20 cps; PMCC flash pt. 34 C; 50% silicone in mixed aliphatic/aromatic solvs.

SSIPA. [Eastman] Sodium salt of sulfonated isophthalic acid; specialty monomer for polymer applics.

Stabaxol P. [Rhein Chemie] Hydrolysis agent for millable urethane; heat-resist. stabilizer for curable EVA.

Stabilite 470. [Arizona] Phenolic; antioxidant for mech. rubber goods, NR, reclaimed rubber, SBR, butyl, polyisoprene, nitrile, EVA, PVC, epoxy resins, ABS, PE, PP, PS; FDA 21CFR §175.105.

Stabilizer 1097. [Bayer AG; Bayer/Fibers, Org., Rubbers] Org. acid chloride sol'n. in butyl acetate; stabilizer used to extend the pot life of the bonding agent system/plastisol mixt.; dens. ≈ 0.95 g/cm^3; flash pt. 27 C; 20% solids.

Stabilizer 2013-P®. [TSE Industries] Carbodiimide; CAS 622-16-2; activator; improves resistance to hydrolysis by acids, bases, and hot water in vulcanizates based on millable PU; dk. amber liq.; aromatic odor; sp.gr. 1.05; reacts slowly with water; b.p. 75 C; flash pt. 176 F (TOC).

Stabilizer 7000. [Raschig] Bis (2,6-diisopropylphenyl) carbodiimide; CAS 2162-74-5; EINECS 218-487-5; stabilizer; process and post long-term protective agent against the effects of hydrolysis in polymers; recommended for PET exposed to high temps. and extreme environments, e.g., monofilaments for use in dryer felts for paper industry, TPU casting elastomers, EVA, PBT; wh. cryst., perceptible odor; sol. in acetone, insol. in water; m.w. 362.56; dens. 0.97 g/cc (20 C); bulk dens. 600-700 kg/m^3; m.p. 48-50 C; *Toxicology:* LD50 (oral, rat) 2395 mg/kg; nonirritating to skin and eyes; may cause sensitization, allergic skin reaction on repeated/prolonged exposure; *Precaution:* incompat. with acids, alkalies, oxidizing agents; dust can form explosive mixts. with air; hazardous decomp. prods.: NO_x, isocyanates; keep away from sources of ignition; *Storage:* store in cool place below 30 C in tightly closed original containers for 2 yr storage life.

Stabilizer C. [Bayer AG] Diphenyl thiourea; CAS 102-08-9; org. stabilizer for PVC; highly effective in concs. as low as 0.3-1.0%; normally restricted to emulsion PVC; used with calcium/zinc compds., suitable for the stabilization of suspension or bulk polymerized PVC food pkg. goods; used in the prod. of unplasticized PVC food pkg. goods e.g., film and bottles; calendered food pkg. film from unplasticized PVC; tarpaulins, industrial gloves, soft floor coverings, unplasticized PVC pipe, sections and sheet; wh. powd.

Stabiol CZ 1336/4E. [Pulcra SA] Zinc- and magnesium-based in epoxidized soybean oil; PVC stabilizer; yel. liq., neutral odor; pract. insol. in water; dens. 0.98 g/cc; visc. 1200 mPa•s max.; solid. pt. < -5 C; flash pt. > 240 C.

Stabiol CZ 1616/6F. [Pulcra SA] Calcium and zinc soaps in epoxidized soybean oil; PVC stabilizer; U.S. FDA, Belgium, French, Italian, German approvals; ylsh. paste, fatty odor; pract. insol. in water; dens. 1.00 g/cc (20 C); flash pt. > 280 C.

Stabiol CZ 2001/1. [Henkel] Calcium-zinc; heat stabilizer for flexible and semirigid PVC, plastisols; for flexible opaque applics.; powd.

Stabiol VCA 1733. [Henkel] Calcium-zinc; heat stabilizer for rigid PVC, bottles; good clarity, self-lubricating; FDA approvals; powd.

Stabiol VZn 1603. [Pulcra SA] Zinc-based in epoxidized soybean oil; PVC stabilizer; yel. liq., pract. odorless; pract. insol. in water; dens. 1.00-1.02 g/cc; visc. ≈ 650 mPa•s; solid. pt. < 0 C; flash pt. > 300 C; 2.4-2.6% Zn.

Stabiol VZN 1783. [Henkel] Calcium-zinc; heat stabilizer for rigid PVC, bottles; good clarity; FDA approvals; powd.

Sta-Flame 88/99. [Miljac] Pentaerythritol; CAS 115-77-5; EINECS 204-104-9; flame retardant for ABS, PE, PS, PVC, PU flexible and rigid foam.

Staflex 550. [C.P. Hall; Lenape] Butylene glycol

polyester; plasticizer for general vinyl applics. and offering permanence, heat stability, low oil extraction; elec. properties and humidity stability in refrigerators and elec. applics.; APHA 175; m.w. 2500; sp.gr. 1.077; dens. 9.0 lb/gal; visc. 2500 cSt; pour pt. 20 F; ref. index 1.4658; flash pt. > 500 F; cloud pt. 20 F; Custom

Staflex 802. [C.P. Hall; Lenape] Butylene glycol polyester; plasticizer with elec. properties; useful in wire and cable industry; APHA 100; m.w. 1900; sp.gr. 1.084; dens. 9.0 lb/gal; visc. 3620 cSt; pour pt. 5 F; ref. index 1.4808; flash pt. 485 F; cloud pt. 5 F; Custom

Staflex 804. [C.P. Hall; Lenape] Butylene glycol polyester; plasticizer with low volatility and oil extraction resistance and elec. properties and rubber migration resistance; used in wire and cable industries; APHA 100; m.w. 1900; sp.gr. 1.070; dens. 8.9 lb/gal; visc. 2270 cSt; pour pt. 0 F; ref. index 1.4724; flash pt. > 500 F; cloud pt. 0 F; Custom

Staflex 809. [C.P. Hall; Lenape] Butylene glycol polyester; plasticizer offering low odor and odor; offers high resistance to PS migration, humidity stability and elec. properties; used in refrigerator gasketing; APHA 100; m.w. 2100; sp.gr. 1.081; dens. 9.0 lb/gal; visc. 3100 cSt (0 C); pour pt. 10 F; ref. index 1.4748; flash pt. 510 F; cloud pt. 10 F; Custom

Staflex BDP. [C.P. Hall; Lenape] Phthalate ester; plasticizer useful in vinyl and rubbers; applic. incl. elec. insulation, automotive vinyl and rubber applic., adhesives, roofing, pool and pond liners, consumer goods, pkg., cosmetics, and wall coverings; APHA 25; m.w. 363; sp.gr. 0.985; dens. 8.2 lb/gal; visc. 350 cSt (0 C); pour pt. -60 F; ref. index 1.4850; flash pt. 425 F; cloud pt. < -60 F; Custom

Staflex BOP. [C.P. Hall; Lenape] Butyl octyl phthalate; EINECS 201-562-1; plasticizer; useful where increased solvating action is desired in vinyl; APHA 30; m.w. 368; sp.gr. 0.996; dens. 8.3 lb/gal; visc. 287 cSt (0 C); pour pt. -52 F; ref. index 1.4862; flash pt. 415 F; cloud pt. -22 F; Custom

Staflex DBEA. [C.P. Hall; Lenape] Dibutoxy ethyl adipate; plasticizer similar to DOA in most properties with slightly more efficiency and greater heat and lt. stability in vinyls; compatibility and low-temp. properties in rubber.

Staflex DBEP. [C.P. Hall; Lenape] Dibutoxy ethyl phthalate; CAS 117-83-9; plasticizer; rapid fluxing by high solvating in vinyl; better rolling bank in calendering; better permanence than DBP in NC.; Custom

Staflex DBF. [C.P. Hall; Lenape] Dibutyl fumarate; CAS 105-76-9; plasticizer; paint bases; chemical intermediate; internal plasticizer for copolymerization with vinyl acetate, vinyl chloride, acrylates, and styrene; APHA 50; m.w. 228; sp.gr. 0.986; dens. 8.2 lb/gal; visc. 3.8 cSt (100 F); pour pt. -40 F; ref. index 1.4546; flash pt. 285 F; cloud pt. 0 F.

Staflex DBP. [C.P. Hall; Lenape] Dibutyl phthalate; CAS 84-74-2; EINECS 201-557-4; general purpose plasticizer for ethylcellulose, rubbers, oil-sol. dyes, printing inks, emulsions; APHA 20; m.w. 278; sp.gr. 1.050; dens. 8.7 lb/gal; visc. 75 cSt (0 C); pour pt. < -80 F; ref. index 1.4910; flash pt. 360 F; cloud pt. < -80 F; Custom

Staflex DIBA. [C.P. Hall; Lenape] Adipate ester; plasticizer; aliphatic; often used in blends with phthalate plasticizers for improved low-temp. performance and lower volatility; specialty lubricant applic.; APHA 10; m.w. 258; sp.gr. 0.952; dens. 7.9 lb/gal; visc. 15 cSt (0 C); pour pt. -18 F; ref. index 1.4304; flash pt. 310 F; cloud pt. -18 F.

Staflex DIBCM. [C.P. Hall; Lenape] Bis diisobutyl maleate (diisobutyl Carbitol maleate); plasticizer used for special applics. as reactive monomer, formulation of specialty surfactants.

Staflex DIDA. [C.P. Hall; Lenape] Diisodecyl adipate; CAS 27178-16-1; EINECS 248-299-9; lacquer marring resistance of low temp. plasticizer; heat and lt. stability; APHA 10; m.w. 427; sp.gr. 0.918; dens. 7.6 lb/gal; visc. 50 cSt (0 C); pour pt. < -80 F; ref. index 1.4513; flash pt. 445 F; cloud pt. < -80 F; Custom

Staflex DIDP. [C.P. Hall; Lenape] Phthalate ester; plasticizer useful in vinyl and rubbers; applic. incl. elec. insulation, automotive vinyl and rubber applic., adhesives, roofing, pool and pond liners, consumer goods, pkg., cosmetics, and wall coverings; APHA 25; m.w. 446; sp.gr. 0.964; dens. 8.0 lb/gal; visc. 700 cSt (0 C); pour pt. -40 F; ref. index 1.4836; flash pt. 465 F; cloud pt. -8 F; Custom

Staflex DINA. [C.P. Hall; Lenape] Diisononyl adipate; CAS 33703-08-1; EINECS 251-646-7; plasticizer offering heat and lt. stability, low temp. properties, and permanence in vinyls; food pkg. film, upholstery, elec. insulation, gasketing; APHA 10; m.w. 398; sp.gr. 0.919; dens. 7.6 lb/gal; visc. 66 cSt (0 C); pour pt. -59 F; ref. index 1.4502; flash pt. 444 F; cloud < -80 F.

Staflex DINM. [C.P. Hall; Lenape] Diisononyl maleate; plasticizer, intermediate used in specialty applic.; lower volatility than DIOM.

Staflex DIOA. [C.P. Hall; Lenape] Diisooctyl adipate; plasticizer giving heat and lt. stability, low temp. in vinyls; used in syn. elastomers; provides low initial visc. in plastisols.; Custom

Staflex DIOM. [C.P. Hall; Lenape] Diisooctyl maleate; CAS 1330-76-3; stable, high purity plasticizer; comonomer with other monomers such as vinyl acetate; liq.

Staflex DIOP. [C.P. Hall; Lenape] Diisooctyl phthalate; CAS 27554-26-3; plasticizer; provides compatibility in vinyl; low rate of solvation in ethylcellulose and rubbers; APHA 25; m.w. 390; sp.gr. 0.985; dens. 8.2 lb/gal; visc. 280 cSt (0 C); pour pt. -50 F; ref. index 1.4850; flash pt. 425 F; cloud pt. -36 F; Custom

Staflex DMAM. [C.P. Hall; Lenape] Dimethyl amyl maleate; CAS 105-52-2; plasticizer compatible with syn. and natural resins and rubbers; preparation of corresponding sodium sulfosuccinate as ingred. in detergents; sol. in org. solvs.

Staflex DMP. [C.P. Hall; Lenape] Dimethyl phthalate;

CAS 131-11-3; EINECS 205-011-6; plasticizer gives molding chars. in molding cellulose acetate; quick tack and low temp. emulsion coalescence in PVA emulsions; peroxide diluent.

Staflex DOA. [C.P. Hall; Lenape] Dioctyl adipate; CAS 103-23-1; EINECS 203-090-1; plasticizer used in food wrap; heat and lt. stability in vinyl; low initial visc. in plastisols; APHA 10; m.w. 371; sp.gr. 0.926; dens. 7.7 lb/gal; visc. 30 cSt (0 C); pour pt. < -80 F; ref. index 1.4462; flash pt. 410 F; cloud pt. < -80 F; Custom

Staflex DOF. [C.P. Hall; Lenape] Di-2-ethylhexyl fumarate; CAS 141-02-6; EINECS 220-835-6; internal plasticizer for copolymerization with vinyl acetate, vinyl chloride, acrylates, and styrene; exterior paint formulations; APHA 50; m.w. 340; sp.gr. 0.942; dens. 7.8 lb/gal; visc. 16 cSt (100 F); pour pt. < -80 F; ref. index 1.4570; flash pt. 370 F; cloud pt. < -80 F; Custom

Staflex DOM. [C.P. Hall; Lenape] Di-2-ethylhexyl maleate; plasticizer; latex paint formulation; internal plasticizer for copolymerization with vinyl acetate, vinyl chloride, acrylates, and styrene; chemical intermediate for prod. of surfactant, sodium dioctyl sulfosuccinate; APHA 50; m.w. 340; sp.gr. 0.943; dens. 7.8 lb/gal; visc. 15 cSt (100 F); pour pt. < -70 F; ref. index 1.4540; flash pt. 370 F; cloud pt. -50 F.

Staflex DOP. [C.P. Hall; Lenape] Di (2-ethylhexyl) phthalate; plasticizer for vinyl; uses incl. film sheeting, textile coatings, pkg. materials, adhesives; APHA 35; m.w. 390; sp.gr. 0.985; dens. 8.2 lb/gal; visc. 375 cSt (0 C); pour pt. -60 F; ref. index 1.4851; flash pt. 420 F; cloud pt. < -60 F; Custom

Staflex DOZ. [C.P. Hall; Lenape] Dioctyl azelate; CAS 103-24-2; plasticizer; in vinyl, provides low volatility, compatibility, low water extraction; resistance to heat and lt.; used in upholstery, raincoats, shower curtains, and outerwear; APHA 45; m.w. 412; sp.gr. 0.920; dens. 7.6 lb/gal; visc. 47 cSt (0 C); pour pt. < -80 F; ref. index 1.4500; flash pt. 455 F; cloud pt. -60 F; Custom

Staflex DTDM. [C.P. Hall; Lenape] Ditridecyl maleate; high purity, low odor, and low color plasticizer and intermediate; used in fields of copolymerization and surfactants, the latter via the bisulfite addition process.

Staflex DTDP. [C.P. Hall; Lenape] Ditridecyl phthalate; plasticizer with elec. properties, permanence, good compatibility in vinyls; APHA 40; m.w. 531; sp.gr. 0.952; dens. 7.9 lb/gal; visc. 1900 cSt (0 C); pour pt. -48 F; ref. index 1.4833; flash pt. 500 F; cloud pt. < -48 F; Custom

Staflex NODA. [C.P. Hall; Lenape] n-Octyl, n-decyl adipate; plasticizer with elec. properties for outdoor wire and cable applic.; heat and lt. stability with high compatibility in vinyls; APHA 10; m.w. 399; sp.gr. 0.918; dens. 7.6 lb/gal; visc. 60 cSt (0 C); pour pt. 26 F; ref. index 1.4458; flash pt. 430 F; cloud pt. < 26 F.

Staflex NONDTM. [C.P. Hall; Lenape] n-Octyl, n-decyl trimellitate; plasticizer used in high-temp. wire and cable applics.; low temp. flexibility, permanence, and ease of processing in vinyl applic.; APHA 80; m.w. 585; sp.gr. 0.978; dens. 8.1 lb/gal; visc. 125 cps; pour pt. -30 F; ref. index 1.4821; flash pt. > 500 F; cloud pt. -30 F.

Staflex ODA. [C.P. Hall; Lenape] Adipate ester; plasticizer with low temp. flexibility and limited compatibility than that of phthalates; aliphatic in nature; blended with phthalate plasticizers for improved low-temp. performance and lower volatility; specialty lubricant applics.; APHA 10; m.w. 399; sp.gr. 0.922; dens. 7.7 lb/gal; visc. 38 cSt (0 C); pour pt. < -80 F; ref. index 1.4488; flash pt. 430 F; cloud pt. < -80 F; Custom

Staflex ODP. [C.P. Hall; Lenape] Octyldecyl phthalate; CAS 119-07-3; EINECS 204-295-9; plasticizer for vinyl applics.; lower in volatility than DOP; APHA 20; m.w. 418; sp.gr. 0.973; dens. 8.1 lb/gal; visc. 500 cSt (0 C); pour pt. -60 F; ref. index 1.4838; flash pt. 440 F; cloud pt. -60 F; Custom

Staflex TIOTM. [C.P. Hall; Lenape] Triisooctyl trimellitate; CAS 53894-23-8; plasticizer used in high-temp. wire and cable applics.; APHA 100; m.w. 546; sp.gr. 0.986; dens. 8.2 lb/gal; visc. 1400 cSt (0 C); pour pt. -40 F; ref. index 1.4830; flash pt. 490 F; cloud pt. -40 F.

Staflex TOTM. [C.P. Hall; Lenape] Trioctyl trimellitate; plasticizer used in high-temp. wire and cable applics.; used in vinyls where good low temp. properties must be combined with low volatility; APHA 100; m.w. 546; sp.gr. 0.987; dens. 8.2 lb/gal; visc. 1800 cSt (0 C); pour pt. -30 F; ref. index 1.4838; flash pt. > 500 F; cloud pt. -30 F.

Stanclere® A-121, A-121 C, A-221. [Akzo Nobel] Antimony mercaptide; stabilizer offering heat stability, nonplate-out, plastisol air release, and plastisol visc. control; used in rigid compds. esp. pipe and profile applics; liq.

Stanclere® C26. [Akzo Nobel] Urea; CAS 57-13-6; EINECS 200-315-5; stabilizer for flexible and rigid PVC bottles.

Stanclere® T-55. [Akzo Nobel] Tin carboxylate; stabilizer for PVC window profiles; outstanding weatherability.

Stanclere® T-57, T-80, T-81. [Akzo Nobel] Alkyl-tin-carboxylate; stabilizer for PVC.

Stanclere® T-85, TL. [Akzo Nobel] Butyl tin carboxylate; stabilizer offering lt. stability, clarity, nonplate-out, plastisol air release, and plastisol visc. control; used in flexible compds. esp. hose and profile; liq.

Stanclere® T-94 C. [Akzo Nobel] Butyl tin mercaptide; stabilizer offering heat stability, clarity, nonplate-out, plastisol air release, and plastisol visc. liq.

Stanclere® T-126. [Akzo Nobel] Butyl tin mercaptide; stabilizer offering lubricity, clarity, plastisol air release, and plastisol visc. control, used in rigid compds. esp. blow molding and inj. molding; liq.

Stanclere® T-160. [Akzo Nobel] Butyltin; stabilizer for profiles.

Stanclere® T-161, T-163, T-164, T-165, T-174, T-182, T-183. [Akzo Nobel] Butylmercaptide; stabilizer for PVC.

Stanclere® T-184. [Akzo Nobel] Tin mercaptide; stabilizer for PVC window profiles; good light stability.

Stanclere® T-186. [Akzo Nobel] Dibutyltin β-mercaptopropionate; stabilizer for PVC.

Stanclere® T-192, T-193, T-194. [Akzo Nobel] Butylmercaptide; stabilizer for PVC.

Stanclere® T-197. [Akzo Nobel] Ba-tin complex; stabilizer for PVC window profiles; exc. light and weather stability; powd.

Stanclere® T-208. [Akzo Nobel] Estertin mercaptide; stabilizer offering heat and lt. stability, clarity, nonplate-out, low odor, plastisol air release, and plastisol visc. control; used for flexible compds. (film, sheet, hose, profile, inj. molding, filled systems) and plastisol applics. (slush molding, rotational molding, dipping, and spreading); liq.

Stanclere® T-222, T-233. [Akzo Nobel] Estertin mercaptide; stabilizers offering heat and lt. stability, clarity, nonplate-out, plastisol air release, and plastisol visc. control; used for rigid compds. (pipe, profile, film, sheet, blow molding, inj. molding, siding); liq.

Stanclere® T-250, T-250SD. [Akzo Nobel] Estertin mercaptide; stabilizers offering lower processing odor, low migration, and improved lt. stability compared to alkyl-tin mercaptides; T-250 and T-250SD grades feature heat and lt. stability, lubricity, clarity, nonplate-out, plastisol air release, and plastisol visc. control; applics. incl. rigid compds. (pipe, profile, foam profile, siding); liq.

Stanclere® T-470, T-473, T-482, T-483, T-484, T-582. [Akzo Nobel] Octylmercaptide; stabilizer for PVC.

Stanclere® T-801. [Akzo Nobel] Butyl tin mercaptide; stabilizer offering heat and lt. stability, lubricity, clarity, nonplate-out, plastisol air release, and plastisol visc. control; used in rigid compds., esp. pipe, profile, sheet, film, foam profile, siding; liq.

Stanclere® T-876. [Akzo Nobel] Butyltin carboxylate; stabilizer offering lt. stability, clarity, nonplate-out, plastisol air release, and plastisol visc. control; used in plastisol applics. (slush molding, rotational molding, dipping, spreading); liq.

Stanclere® T-877. [Akzo Nobel] Butyltin carboxylate; stabilizer offering lt. stability, clarity, nonplate-out, plastisol air release, and plastisol visc. control; used in flexible compds., esp. hose and profile applics.; liq.

Stanclere® T-878, T-55. [Akzo Nobel] Butyltin carboxylate; stabilizer offering lt. stability, clarity, nonplate-out; plastisol air release, and plastisol visc. control; used in flexible compds., esp. hose and profile; liq.

Stanclere® T-883. [Akzo Nobel] Sulfur-containing octyltin; stabilizer for semirigid and rigid PVC bottles and sheet.

Stanclere® T-4356. [Akzo Nobel] Sulfur-containing organotin stabilizer; used in filled or pigmented rigid inj. molding applics., siding, and profile market; offers improved uv stability, impact strength, and heat distort. temps. while not sacrificing heat stability chars.; usage levels: 1.0-2.0 phr; Gardner 2 liq.; sp.gr. 1.210; visc. 50 cps.

Stanclere® T-4654. [Akzo Nobel] Tin; stabilizer for rigid PVC exterior profiles, foams, bottles.

Stanclere® T-4704. [Akzo Nobel] Tin; stabilizer for rigid PVC bottles; liq.

Stanclere® T-4817. [Akzo Nobel] Sulfur-containing butyltin stabilizer.

Stanclere® T-5507. [Akzo Nobel] Tin; stabilizer for semirigid and rigid PVC exterior profiles and siding, organosols, records, bottles.

Stanclere® TL. [Akzo Nobel] Butyltin carboxylate; lubricant/stabilizer for vinyls and PU offering lt. stability, clarity, nonplate-out, plastisol air release, and plastisol visc. control; used in flexible compds. esp. hose and profile; liq.

Stanclere® TM. [Akzo Nobel] Dibutyl-tin-maleate; CAS 15535-69-0; stabilizer for PVC.

Stangard 500. [Harwick] Amine; high temp. antioxidant for NBR.

Stangard PC. [Harwick] Phenol; nondiscoloring, nonstaining antioxidant for ABS, PE, PP, EPDM, EVA; FDA approved for adhesives for polyolefins (0.1% max.); off-wh. powd.

Stangard SP/SPL. [Harwick] Phenol; nondiscoloring, nonstaining antioxidant for PE, PP, PS, PVC.

Stan Wax. [Harwick] Polyethylene wax; CAS 9002-88-4; EINECS 200-815-3; internal lubricant for PE, PP, PVC; FDA approved; solid.

Star Stran 748. [Schuller] Continuous filament elec. grade borosilicate glass strand coated with silane sizing treatment; reinforcement for PP, PPS, PEI; enhances mech. props.

Starwax® 100. [Petrolite] Hard microcryst. wax consisting of n-paraffinic, branched paraffinic, and naphthenic hydrocarbons; CAS 63231-60-7; EINECS 264-038-1; wax used in hot-melt coatings and adhesives, paper coatings, printing inks, plastic modification (as lubricant and processing aid), lacquers, paints, and varnishes, as binder in ceramics, for potting/impregnant in elec./electronic components, rubber and elastomers (plasticizer, antisunchecking, antiozonant), as emulsion wax size in papermaking, as fabric softener ingred., in emulsion and latex coatings, and in cosmetic hand creams and lipsticks; chewing gum base; incl. FDA §172.230, 172.615, 175.105, 175.300, 176.170, 176.180, 176.200, 177.1200, 178.3710, 179.45; color 1.0 (D1500) wax; very low sol. in org. solvs.; sp.gr. 0.93; visc. 15 cps (99 C); m.p. 88 C.

Sta-Tac®. [Arizona] C_5 hydrocarbon resin; tackifier for pressure-sensitive adhesive and hot-melt applics.; FDA 21CRR §175.105; Gardner 4 color; soften. pt. (R&B) 100 C.

Sta-Tac® B. [Arizona] C_5 hydrocarbon resin; tackifier for pressure-sensitive adhesive, hot-melt, and laminating adhesives; FDA 21CFR §175.105, 175.300, 176.170, 176.180; Gardner 3+ color; soften. pt. (R&B) 100 C.

Sta-Tac® T. [Arizona] C_5 hydrocarbon resin; tackifier for pressure-sensitive hot-melts and laminating adhesives; FDA 21CFR §175.105; Gardner color 4+ color; soften. pt. (R&B) 102 C.

Statexan K1. [Bayer AG; Bayer/Fibers, Org., Rubbers] Sulfonated aliphatic hydrocarbon; internal antistatic additive or external coating for rigid PVC, PS, HIPS, ABS, etc.; lt.-colored wax-like flakes; hygroscopic; sol. in water, methanol, and ethanol; pH 7 min. (10 g/L water); 94% assay, 1.5% max. water.

Staticide®. [ACL] Water-based topical antistat for spray, dip, or wipe-on applic. to any material; for industrial, commercial and institutional facilities, clean room environments; biodeg.

Statik-Blok® FDA-3. [Amstat Industries] Polyether type; nonionic; surface-active antistatic sol'n. for food contact surfaces; exc. on PVC, polystyrene, nylon, and other plastics; FDA §178.3400; clear, colorless liq., pract. odorless; sol. in water, IPA; pH 7.

Statik-Blok® J-2. [Amstat Industries] Antistatic coating for most plastics.

Stat-Kon® AS-F. [LNP] HDPE; CAS 9002-88-4; EINECS 200-815-3; antistatic grade composite for protection of mod. sensitive elec. components from low voltages; internal/external antistat; sp.gr. 0.94; tens. str. 3500 psi; tens. elong. 250%; flex. str. 3000 psi; Izod impact 4.5 ft-lb/in. notched; distort. temp. 130 F (264 psi); flamm. HB.

Stat-Kon® AS-FE. [LNP] HDPE; CAS 9002-88-4; EINECS 200-815-3; antistatic grade composite for protection of mod. sensitive elec. components from low voltages; suitable for extrusion; internal/external antistat; sp.gr. 0.94; tens. str. 3300 psi; tens. elong. 300%; flex. str. 2800 psi; Izod impact 5.0 ft-lb/in. notched; distort. temp. 130 F (264 psi); flamm. HB.

Stavinor ATA-1056. [Elf Atochem N. Am.] Barium-zinc; heat and lt. stabilizer for flexible and semirigid PVC; good clarity; liq.

Stavinor CA PSE. [Elf Atochem N. Am.] Zinc complex; heat and lt. stabilizer for flexible and rigid PVC, plastisols, organosols, foams, records, bottles, tiles, PU; NSF approved.

Staybelite®. [Hercules] Acidic resin made by partially hydrogenating wood rosin; thermoplastic resin used as plasticizer, tackifier, or processing aid for reclaimed, natural, and syn. rubbers; as component in sealing wax, optical lens pitch, elec. cable sats., waxes for paper coating, syn. resins, metal resinates, soldering fluxes, and ceramic ink vehicles; pale solid; soften. pt. (Drop) 75 C; acid no. 160.

Staybelite® Ester 5. [Hercules] Glyceryl hydrog. rosinate; thermoplastic syn. resin; as softener/plasticizer for the masticatory agent in chewing gum bases; as resin modifier for film-formers, elastomers, and waxes in adhesive and protective coating compositions; where low odor and resistance to aging are important; used in food-pkg. and -processing operations; USDA Rosin solid, flakes; low odor; sol. in esters, ketones, higher alcohols, glycol ethers, and aliphatic, aromatic, and chlorinated hydrocarbons; dens. 1.06 kg/l; Hercules drop soften. pt. 81 C; acid no. 5.

Staybelite® Ester 10. [Hercules] Glyceryl hydrogenated rosinate; thermoplastic syn. resin in solv., hot-melt, and emulsion adhesives based on thermoplastic and elastomeric materials; in emulsion form as tackifier for natural rubber, SBR, and neoprene latexes; as tackifier in industrial tape masses; in lamination of metal, foil, paper, etc.; in hot-melt-applied barrier coatings, chlorinated rubber finishes, etc.; as modifier for paraffin wax; used in food-pkg. and -processing operations; USDA Rosin WG solid, flakes; sol. in esters, ketones, higher alcohols, glycol ethers, and aliphatic, aromatic, and chlorinated hydrocarbons; sp.gr. 1.07; dens. 1.07 kg/l; G-H visc. I-M (75% solids in toluene); Hercules drop soften. pt. 83 C; acid no. 8.

Stellar 500. [Luzenac Am.] Talc; CAS 14087-96-6; EINECS 238-877-9; for plastics applics. where good color and brightness are required; antiblocking agent in polyolefin films; powd.; 4.4 μ median particle size; fineness (Hegman) 5.0; sp.gr. 2.8; dens. 11 lb/ft^3; bulking value 23.3 lb/gal; surf. area 7.8 m^2/g; oil absorp. 43; pH 9.0 (10% slurry).

Stellar 510. [Luzenac Am.] Talc; CAS 14087-96-6; EINECS 238-877-9; for plastics applics. where good color and heat aging are required; easily compounded at levels from 20-40%; powd.; 4.4 μ median particle size; fineness (Hegman) 5.5; sp.gr. 2.8; dens. 15 lb/ft^3 (loose); bulking value 23.3 lb/gal; surf. area 6.2 m^2/g; oil absorp. 44; pH 9.0 (10% slurry).

Steol® 4N. [Stepan; Stepan Canada] Sodium laureth sulfate; CAS 9004-82-4; anionic; detergent, emulsifier, foamer, wetting agent used in personal care prods.; car wash, dishwash; textile mill applics.; emulsion polymerization; water-wh. liq.; water-sol.; sp.gr. 1.045; visc. 31 cps; cloud pt. -4 C; pH 8.0; 28% act.

Steol® CA-460. [Stepan; Stepan Canada] Ammonium laureth sulfate; CAS 32612-48-9; anionic; detergent, emulsifier, foamer, dispersant, visc. modifier, and wetting agent used in shampoos, bath and cleansing preps., dishwashers, car washers, textile mill applics., emulsion polymerization; pale yel. liq.; water-sol.; sp.gr. 1.016; visc. 67 cps; cloud pt. 19 C; pH 7.0; 60% conc.

Steol® COS 433. [Stepan Canada] Sodium nonoxynol-4 sulfate; anionic; emulsifier for acrylics, SBR, vinyl chloride, butyl rubber; pale yel. clear liq.; 30% act.

Stepanol® WA-100. [Stepan; Stepan Canada; Stepan Europe] Sodium lauryl sulfate USP/NF; CAS 151-21-3; EINECS 205-788-1; anionic; detergent, foamer, wetting and suspending agent; dentrifrice formulations where minimal taste contribution is important; pharmaceuticals; emulsion polymerization; clay dispersions; wh. powd.; water-sol.; pH 7.5-10 (10%); 98.5% min. act.

Stepfac® 8170. [Stepan; Stepan Canada] Ethoxylated nonylphenol phosphate; anionic; hydrotrope for nonionics; emulsifier for agric., emulsion polymerization, oils, metalworking lubricants, corrosion inhibitors, pigment dispersants; heavy-duty industrial/household alkali cleaners; compatibility agent for liq. fertilizers; amber clear liq.; sol. in xylene, water; pH 2.0 (1%); 100% act.

Stereon® 205. [Firestone Syn. Rubber] Sol'n. polymerized styrene butadiene copolymer; CAS 9003-55-8; rubber processing aid; contributes unique props. in adhesives and asphalt modification; rubbery solid, little, if any, color or odor; nil sol. in water; sp.gr. 0.94; visc. 8.8 cP (5.43% in toluene); Mooney visc. 48; 25% bound styrene; *Precaution:* combustion byprods.: oxides of carbon; *Storage:* store below 150 F.

Stereon® 210. [Firestone Syn. Rubber] Sol'n. polymerized styrene butadiene copolymer with antiblocking agent to prevent re-agglomeration after grinding; CAS 9003-55-8; rubber processing aid; contributes unique props. in adhesives and asphalt modification; rubbery solid, little, if any, color or odor; nil sol. in water; sp.gr. 0.94; visc. 8.8 cP (5.43% in toluene); Mooney visc. 48 (ML4, 212 F); 25% bound styrene; *Precaution:* combustion byprods.: oxides of carbon; *Storage:* store below 150 F.

Stereon® 721A. [Firestone Syn. Rubber] Low visc. sol'n. polymerized S/B copolymer rubber; CAS 9003-55-8; graftable impact modifier for ABS and other engineering resins; APHA 10 max. rubbery solid, low or no odor; nil sol. in water; sp.gr. 0.91; visc. 29 cP (5.43% in toluene); Mooney visc. 35 (ML4, 212 F); ref. index 1.5245; 10% bound styrene; *Precaution:* combustion byprods.: oxides of carbon; *Storage:* store below 150 F.

Stereon® 730A. [Firestone Syn. Rubber] Low visc. sol'n. polymerized S/B copolymer rubber; CAS 9003-55-8; graftable impact modifier for PS, ABS, and other engineering resins; blendable impact modifier for HIPS, flame retardant HIPS; APHA 10 max. rubbery solid, low or no odor; nil sol. in water; sp.gr. 0.94; visc. 25 cP (5.43% in toluene); ref. index 1.5391; 30% bound styrene; *Precaution:* combustion byprods.: oxides of carbon; *Storage:* store below 150 F.

Stereon® 840A. [Firestone Syn. Rubber] Stereospecific SBR block copolymer; CAS 9003-55-8; rubber modifier for thermoplastic resins, esp. for HIPS, flame-retardant HIPS, PS, PP, polyethylene films; gives alloyed prods. improved impact resist. and toughness; for hot-melt and solv.-based adhesives, nonwoven elastic/feminine pad attachment, pressure sensitives, tapes, labels, graphic arts binding, assembly/laminating for bonding plastics, wood, metals, glass, caulks, sealants, coatings, asphalt; FDA 21CFR §177.1520, 177.1640; APHA 14 translucent pellets, low or no odor; nil sol. in water; melt index 12; sp.gr. 0.96; inherent visc. 0.8; ref. index 1.5495; tens. str. 1900 psi; hardness (Shore A) 85; 44.5% bound styrene; B/S ratio 57/43; *Precaution:* combustion byprods.: oxides of carbon; *Storage:* store below 150 F.

Stereon® 841A. [Firestone Syn. Rubber] Stereospecific butadiene/styrene copolymer; CAS 9003-55-8; blendable modifier for HIPS, flame retardant HIPS, polyolefins, adhesives, asphalt; APHA 5 translucent pellet, low or no odor; nil sol. in water; melt index 11.5; sp.gr. 0.96; inherent visc. 0.8; ref. index 1.5495; tens. str. > 1000 psi; hardness (Shore A) 85; 44.5% bound styrene; *Precaution:* combustion byprods.: oxides of carbon; *Storage:* store below 150 F.

Sterling® 1120. [Cabot/Industrial Rubber Blacks] Carbon black; CAS 1333-86-4; EINECS 215-609-9; additive for molded rubber prods. that require exc. flex, durability, and heat dissipation with ease of processing; suitable for engine mounts and other dynamic applics., shock absorber bushings, high-durometer molded goods, down-hole oil field parts.

Sterling® 2320. [Cabot/Industrial Rubber Blacks] Carbon black; CAS 1333-86-4; EINECS 215-609-9; additive for sealing, dynamic, and hose rubber prods. requiring superior hot tear str., compr. set, and resilience; for seals, o-rings, engine mount components, high-durometer molded goods, thin-walled/intricate parts, hose applics.

Sterling® 3420. [Cabot/Industrial Rubber Blacks] Carbon black; CAS 1333-86-4; EINECS 215-609-9; easy extruding carbon blk. for extruded and inj. molded rubber prods. where low scrap rates are critical; cleaner than std. semireinforcing blks.; for windshield wiper blades, o-rings, hose covers, seals, dynamic parts.

Sterling® 4620. [Cabot/Industrial Rubber Blacks] Carbon black; CAS 1333-86-4; EINECS 215-609-9; additive for superior surf. appearance and ease of processing in extruded, hose, and molded rubber prods., esp. automotive and industrial hose, automotive and architectural weatherstripping, inner tubes, rubber diaphragms.

Sterling® 5550. [Cabot/Industrial Rubber Blacks] Carbon black; CAS 1333-86-4; EINECS 215-609-9; easy-dispersing carbon blk. for improved performance in inj. molded dynamic rubber applics., incl. ducting, bushings, mounts, belts.

Sterling® 5630. [Cabot/Industrial Rubber Blacks] Carbon black; CAS 1333-86-4; EINECS 215-609-9; cleaner carbon blk. with well balanced dispersion and reinforcing props. for extruded and molded industrial rubber applics. incl. weatherstripping, window channeling, roofing membranes, hoses, belts.

Sterling® 6630. [Cabot/Industrial Rubber Blacks] Carbon black; CAS 1333-86-4; EINECS 215-609-9; carbon blk. featuring greater processing safety and improved cleanliness for extruded and molded rubber parts incl. seals, weatherstripping, hoses, belts, extruded profiles, mounts and bushings.

Sterling® 6640. [Cabot/Industrial Rubber Blacks] Carbon black; CAS 1333-86-4; EINECS 215-609-9; carbon blk. featuring high microwave receptivity for optimized performance in microwave cured extruded sponge applics.; for automotive and architectural sponge weatherstripping, industrial spogne, dual durometer profiles.

Sterling® 6740. [Cabot/Industrial Rubber Blacks] Carbon black; CAS 1333-86-4; EINECS 215-609-9; additive for superior tear str., durability, surf.

appearance, and cost-effectiveness in molded and extruded rubber prods., general-purpose molding compds., belts, microwave-cured extruded sponge profile, pulley lagging, casters.

Sterling® 8860. [Cabot/Industrial Rubber Blacks] Carbon black; CAS 1333-86-4; EINECS 215-609-9; improved carbon blk. providing exc. abrasion resist. and damping for wear-resist. rubber surfs. and other reinforcing applics.; for solid tires, floor mats, shoe soles, casters, mounts, and bushings.

Sterling® C. [Cabot/Industrial Rubber Blacks] Carbon black; CAS 1333-86-4; EINECS 215-609-9; cost-effective additive for conductive and static-dissipative rubber compd. applics. incl. wire and cable jacketing, static-dissipative hoses and belts, general-purpose static-dissipative parts.

Sterling® NS. [Cabot/N. Am. Rubber Black] N774 Carbon black; CAS 1333-86-4; EINECS 215-609-9; semi-reinforcing carbon black for rubbers; similar to N762 but smoother extrusion in most compds.; 0.1% max. on 325 mesh; pour dens. 496 kg/m^3; iodine no. 28.

Sterling® NS-1. [Cabot/N. Am. Rubber Black] N762 Carbon black; CAS 1333-86-4; EINECS 215-609-9; semi-reinforcing carbon black for rubbers featuring low hysteresis, high loading capacity; 0.1% max. on 325 mesh; pour dens. 521 kg/m^3; iodine no. 28.

Sterling® SO. [Cabot/N. Am. Rubber Black] N550 Carbon black; CAS 1333-86-4; EINECS 215-609-9; smooth extruding semi-reinforcing carbon black for rubbers; widely used in tires and extruded prods.; 0.1% max. on 325 mesh; pour dens. 352 kg/m^3; iodine no. 43.

Sterling® V. [Cabot/N. Am. Rubber Black] N660 Carbon black; CAS 1333-86-4; EINECS 215-609-9; semi-reinforcing carbon black widely used in tire body and industrial rubber compds.; 0.1% max. on 325 mesh; pour dens. 424 kg/m^3; iodine no. 36.

Sterling® VH. [Cabot/N. Am. Rubber Black] N650 Carbon black; CAS 1333-86-4; EINECS 215-609-9; semi-reinforcing carbon black for rubbers; processing props. similar to N550 with marginally lower reinforcement; 0.1% max. on 325 mesh; pour dens. 360 kg/m^3; iodine no. 36.

Sterling® XC-72. [Cabot/Industrial Rubber Blacks] Carbon black; CAS 1333-86-4; EINECS 215-609-9; additive for high-conductive and static-dissipative rubber compd. applics., e.g., wire and cable jacketing, static-dissipative flooring, hoses, and belts, conductive roll coverings, power distribution parts, high-impact conductive plastics.

Stoner E206. [Stoner] Proprietary silicone release blend; mold release for release of plastics, rubber, and waxes; for mold temps. of 40 to 500 F; nonflamm. aerosol; *Toxicology:* ACGIH TLV 1000 ppm; *Storage:* 1 yr min. shelf life @ R.T.

Stoner E302. [Stoner] Proprietary release blend; food-grade paintable mold release for release of plastics; for mold temps. of 40 to 500 F; nonflamm. aerosol; *Toxicology:* ACGIH TLV 1000 ppm; *Storage:* 1 yr min. shelf life @ R.T.

Stoner E313. [Stoner] Proprietary release blend; paintable mold release for release of plastics; for mold temps. of 40 to 500 F; nonflamm. aerosol; *Toxicology:* ACGIH TLV 1000 ppm.

Stoner E408. [Stoner] Dry film release blend; paintable dry film mold release for plastics, rubber, and composites; for mold temps. of 40 to 500 F; nonflamm. aerosol; *Toxicology:* ACGIH TLV 1000 ppm; *Storage:* 1 yr min. shelf life @ R.T.

Stoner E965. [Stoner] Paintable mold release agent for plastic, rubber, wax, and ceramic molding incl. inj., compr., transfer, vacuum form, pour cast, die cast, extrusion, and foam molding operations; antistick agent; allows molded parts to be painted, plated, decorated; clear colorless liq.; nonflam.; Unverified

Struktol® 40 MS. [Struktol] Mixt. of aromatic hydrocarbon resins; resin plasticizer; homogenizing agent for elastomer blend compds.; improves extrusion, calendering, green tack; not suitable for lt. colored compds.; recommended for butyl, halobutyl, chloroprene, EPDM, NR, BR, SBR rubbers; dk. brn. nonhardening resin; good sol. in aromatic and chlorinated hydrocarbons; sp.gr. 1.0; soften. pt. 55 C; *Storage:* 2 yr min. storage stability under normal storage conditions.

Struktol® 40 MS Flakes. [Struktol] Modified mixture of rubber-compatible nonhardening syn. resins; resin plasticizer; homogenizing agent for elastomer blend compds.; improves extrusion, calendering, green tack; not suitable for lt. colored compds.; dk. brn. pastilles; good sol. in aromatic and chlorinated hydrocarbons; sp.gr. 1.1; bulk dens. 650 g/l; soften. pt. 100 C; *Storage:* 2 yrs min storage stability under normal storage conditions; store below 95 F; do not double-stack; use on first-in first-out basis.

Struktol® 60 NS. [Struktol] Mixt. of aliphatic hydrocarbon resins; resin plasticizer; homogenizing agent for lt.-colored elastomer compds.; improves extrusion, calendering, green tack; recommended for butyl, halobutyl, CR, EPDM, NR, BR, SBR; amber nonhardening resin; good sol. in aromatic and chlorinated hydrocarbons; sp.gr. 0.95; soften. pt. 50 C; *Storage:* 2 yrs min storage stability under normal storage conditions.

Struktol® 60 NS Flakes. [Struktol] Mixture of rubber-compatible lt.-colored aliphatic resins; resin plasticizer; homogenizing agent for lt.-colored elastomer compds.; improves extrusion, calendering, green tack; amber flakes; good sol. in aromatic and chlorinated hydrocarbons; sp.gr. 0.97; bulk dens. 600 g/l; soften. pt. 100 C; *Storage:* 2 yrs min storage stability under normal storage conditions; store below 95 F; do not double-stack; use on first-in first-out basis.

Struktol® A 50. [Struktol] Zinc soaps of unsat. fatty acids; processing aid, peptizer for natural and syn. rubber incl. butyl, epichlorohydrin, BR, sol'n. SBR; beige flakes or pearls; sp.gr. 1.1; bulk dens. 580 g/l; drop pt. 100 C; 10.5% Zn; *Storage:* 18 mos min. storage stability under normal storage conditions.

Struktol® A 60. [Struktol] Mixt. of zinc soaps of high

m.w. fatty acids; peptizer for natural rubber; processing aid for natural and syn. rubber esp. butyl, epichlorohydrin, BR; beige flakes; rubber-sol.; sp.gr. 1.15; bulk dens. 610 g/l; drop pt. 90 C; *Storage:* 18 mos min. storage stability under normal storage conditions.

Struktol® A 61. [Struktol] Mixt. of zinc soaps of high m.w. fatty acids; peptizer for natural rubber; processing aid for natural and syn. rubber; brn. flakes; rubber-sol.; sp.gr. 1.20; bulk dens. 610 g/l; drop pt. 100 C; *Storage:* 18 mos min. storage stability under normal storage conditions.

Struktol® A 80. [Struktol] Blend of organo-metal complexes, other peptizing agents, org. and inorg. dispersants; peptizer and processing aid for natural and syn. rubber esp. SBR, CR; bluish-gray pastilles; easily dispersed in rubber; sp.gr. 1.10; m.p. 80 C; *Storage:* 1 yr min. storage stability under normal storage conditions.

Struktol® A 82. [Struktol] Blend of organo-metal complexes, other peptizing agents, org. and inorg. dispersants; peptizer and processing aid for natural and syn. rubber esp. SBR; bluish-gray pastilles; easily dispersed in rubber; sp.gr. 1.40; bulk dens. 710 g/l; m.p. 50 C; *Storage:* 1 yr min. storage stability under normal storage conditions.

Struktol® A 86. [Struktol] Blend of organo-metal complexes, other peptizing agents, org. and inorg. dispersants; peptizer and processing aid for natural and syn. rubber esp. SBR; higher actives than A 80 and A 82; greenish-blue pastilles; sp.gr. 1.30; bulk dens. 620 g/l; m.p. 50 C; *Storage:* 1 yr min. storage stability under normal storage conditions.

Struktol® AW-1. [Struktol] Syn. ester; antistatic plasticizer for min.-filled NBR, SBR, and NR compds.; ylsh. liq.; insol. in aliphatic hydrocarbons, oils, and greases; sp.gr. 1.10; visc. 140 mPa•s (20 C); pour pt. 10 C; flash pt. 215 C; *Storage:* 2 yr min. storage stability under normal storage conditions.

Struktol® Activator 73. [Struktol] Mixt. of zinc salts of aliphatic and aromatic carboxylic acids; activator for the sulfur vulcanization of diene rubbers, esp. natural rubber; provides prolonged reversion time, higher modulus, less heat build-up under dynamic load, reduced compr. set at elevated temps.; peptizer for mastication of natural rubber; lt. gray pastilles; sp.gr. 1.2; drop pt. 100 C; 18% Zn; *Storage:* 2 yrs min. storage stability under normal storage conditions.

Struktol® EP 52. [Struktol] Blend of rubber-compat. nonhardening syn. resins and fatty acid soaps; processing aid esp. for highly loaded EPDM compds.; also for butyl and halobutyl rubber compds.; improves extrusion uniformity, flow and smoothness of molded and calendered prods.; dk. brn. flakes; sp.gr. 1.10; bulk dens. 610 g/l; soften. pt. 87 C; *Storage:* 1 yr min. storage stability under normal storage conditions.

Struktol® FA 541. [Struktol] Blend of activators and dispersants bound to a highly active filler; activator for lt. reinforcing fillers; shortens cure time without reducing flow time, avoids premature scorch; deodorizing effect in cellular and sponge rubber; suitable for all sulfur-cured rubber compds.; lt. free-flowing powd.; sp.gr. 1.30; bulk dens. 770 g/l; reaction acidic; *Storage:* 2 yrs min. storage stability under normal storage conditions in closed bags in cool, dry area.

Struktol® HP 55. [Struktol] Mixt. of aromatic hydrocarbon resins and fatty acid derivs.; high performance process aid for rubber, esp. tires based on SBR and NR; also for conveyor belts, motor mounts, hose; improves mixing and milling chars., visc. reduction, flow props.; dk. brn. pastilles; sp.gr. 1.04; bulk dens. 620 g/l; drop pt. 99 C; *Storage:* 2 yrs min. storage stability under normal storage conditions.

Struktol® HPS 11. [Struktol] Blend of fatty acid derivs.; processing aid for rubber; increases flow, promotes release; lt. tan pearls, fatty acid odor; sp.gr. 0.98; bulk dens. 603 g/l; drop pt. 86 ± 6 C; *Storage:* 2 yr min. storage stability under normal storage conditions.

Struktol® IB 531. [Struktol] Amine-based salts bound to highly active fillers; activator for accelerators; regulates crosslinking and avoids premature reversion; filler activator for compds. contg. highly active min. fillers; accelerates cure; for inj. and transfer molding, continuous cure; lt. free-flowing powd.; sp.gr. 1.50; bulk dens. 600 g/l; reaction basic; *Storage:* 2 yrs min. storage stability under normal storage conditions in closed bags in cool, dry area.

Struktol® KW 400. [Struktol] Syn. ester; plasticizer for NBR, and other common rubbers; improves low temp. flexibility; lt. yel. liq.; sol. in aliphatic and aromatic hydrocarbons; sp.gr. 1.00; visc. 15 mPa•s (20 C); pour pt. -60 C; flash pt. 195 C; *Storage:* 2 yr min. storage stability under normal storage conditions.

Struktol® KW 500. [Struktol] Syn. ester; plasticizer for NBR, and other common rubbers; improves low temp. flexibility and heat resist.; lt. liq.; sol. in aliphatic and aromatic hydrocarbons; sp.gr. 1.00; visc. 50 mPa•s (20 C); pour pt. -20 C; flash pt. 235 C; *Storage:* 2 yr min. storage stability under normal storage conditions.

Struktol® PE H-100. [Struktol] Low m.w. polyethylene homopolymer wax; CAS 9002-88-4; EINECS 200-815-3; improves flow and processability in natural and syn. elastomers; provides release from equip. with improved pigment dispersion and finish to molded articles; wh. waxy gran. or powd.; sp.gr. 0.91; visc. 180 cps (140 C); drop pt. 102 ± 5 C; acid no. nil.

Struktol® SU 50. [Struktol] Insoluble sulfur prep.; used for rubber compounding; offers fast incorporation and optimum sulfur dispersion to reduce sulfur blooming; suitable for most critical applics.; nondusting friable powd.; sp.gr. 1.25; bulk dens. 550 g/l; 50% sulfur, 45% insol. sulfur, 50% org. dispersant; *Storage:* 1 yr min. storage stability under normal storage conditions in closed bags in cool area.

Struktol® SU 95. [Struktol] Coated soluble sulfur contg. dispersants and wetting agents; CAS 7704-

34-9; EINECS 231-722-6; used for any sulfur-cured natural or syn. rubber compds.; nondusting powd.; 5% max. on 325 mesh; sp.gr. 1.9; bulk dens. 550 g/l; 95% sulfur, 5% org. dispersant; *Storage:* 2 yrs min. storage stability under normal storage conditions.

Struktol® SU 105. [Struktol] Coated soluble sulfur contg. dispersants and wetting agents; CAS 7704-34-9; EINECS 231-722-6; used for any sulfur-cured natural or syn. rubber compds.; paste; 5% max. on 325 mesh; sp.gr. 1.6; 50% sulfur, 25% org. dispersant, 25% inorg. dispersant; *Storage:* 2 yrs min. storage stability under normal storage conditions.

Struktol® SU 109. [Struktol] Coated insoluble sulfur; used for rubber compounding; offers fast incorporation and optimum sulfur dispersion to reduce sulfur blooming; suitable where low content of dispersing agent is desirable; nondusting friable powd.; sp.gr. 1.50; bulk dens. 600 g/l; 75% sulfur, 67.5% insol. sulfur, 24% org. dispersant, 1% inorg. dispersant; *Storage:* 1 yr min. storage stability under normal storage conditions in closed bags in cool area.

Struktol® SU 120. [Struktol] Coated soluble sulfur contg. dispersants and wetting agents; CAS 7704-34-9; EINECS 231-722-6; used for any sulfur-cured natural or syn. rubber compds.; nondusting powd.; 5% max. on 325 mesh; sp.gr. 1.65; bulk dens. 700 g/l; 83% sulfur, 16% org. dispersant, 1% inorg. dispersant; *Storage:* 2 yrs min. storage stability under normal storage conditions.

Struktol® SU 135. [Struktol] Insoluble sulfur prep.; used for rubber compounding; offers fast incorporation and optimum sulfur dispersion to reduce sulfur blooming; nondusting friable powd.; sp.gr. 1.50; bulk dens. 800 g/l; 75% sulfur, 34% insol. sulfur, 24% org. dispersant, 1% inorg. dispersant; *Storage:* 1 yr min. storage stability under normal storage conditions in closed bags in cool area.

Struktol® TH 10 Flakes. [Struktol] Mixt. of aliphatic and aromatic resins; tackifying and homogenizing agent, processing aid; provides optimum dispersion of ingreds. in raw compd.; maintains green tack for several weeks; improves extrudability and storage stability; suitable for lt.-colored compds.; amber flakes; sol. in aliphatic, aromatic, and chlorinated hydrocarbons; sp.gr. 1.00; soften. pt. 100 C; *Storage:* 2 yrs min. storage stability under normal storage conditions.

Struktol® TH 20 Flakes. [Struktol] Mixt. of aliphatic and aromatic resins; tackifying and homogenizing agent, processing aid; provides optimum dispersion of ingreds. in raw compd.; maintains green tack for several weeks; improves extrudability and storage stability; restricted to dk.-colored compds.; dk. brn. flakes; sol. in aliphatic, aromatic, and chlorinated hydrocarbons; sp.gr. 1.10; soften. pt. 105 C; *Storage:* 2 yrs min. storage stability under normal storage conditions.

Struktol® T.M.Q. [Struktol] Polymerized 2,2,4-trimethyl-1,2-dihydroquinoline; CAS 26780-96-1; antioxidant in rubber tires, belts, mech. goods, sponge and retreading; minimally discoloring; sp.gr. 1.06; soften. pt. 90 ± 5 C.

Struktol® T.M.Q. Powd. [Struktol] Polymerized 2,2,4-trimethyl-1,2-dihydroquinoline; CAS 26780-96-1; antioxidant in rubber tires, belts, mech. goods, sponge and retreading; minimally discoloring; amber powd.; 99.5% thru 80 mesh; sp.gr. 1.06; soften. pt. 90 ± 5 C.

Struktol® TR 041. [Struktol] Fatty acid based additive; maximizes wetting to polar fillers and nonpolar polymers allowing improved incorporation of additives into the polymer matrix; decreases melt temps. and visc.; for HIPS; lt. tan, neutral odor; sp.gr. 0.95; bulk dens. 37.6 lb/ft^3; m.p. 84 C; acid no. < 20; flash pt. > 260 C; *Storage:* > 2 yrs storage stability.

Struktol® TR 042. [Struktol] Processing aid for plastics; dispersant for powd. materials; improves filler incorporation; prevents sticking of elastomers and compds. to rotors and rolls; plasticizing action, improved mold release; lt. sm. beads; sp.gr. 1.10; bulk dens. 680 g/l; drop pt 57 C; acid no. 5; *Storage:* 2 yrs min. stability under normal storage conditions.

Struktol® TR 044. [Struktol] Fatty acid ester blend; processing aid for polymers; dispersant for powd. materials, providing more uniform filler incorporation; reduces decomp. risk in highly loaded compds.; improves filler dispersion and flow props. and release during processing; FDA 21CFR §175.105, 176.170, 176.210, 177.1200, 178.2010; lt. powd.; sp.gr. 0.95; bulk dens. 32 lb/ft^3; drop pt. 57 C; acid no. < 10; *Storage:* 2 yrs min. stability under normal storage conditions.

Struktol® TR 065. [Struktol] Blend of med. m.w. resins; homogenizer, high temp. processing aid and blending agent for plastics; effective in binding filler materials to the polymer system; stable to 700 F processing temps.; compat. with polar and nonpolar polymers; used for highly filled polyesters, nylon, nitriles, PVC, and alloys or blends of dissimilar materials; FDA 21CFR §172.280, 172.615, 175.105, 175.125, 177.1200, 177.2600; lt. tan color, resinous odor; sp.gr. 1.01; drop pt. 108 C; *Storage:* 2 yrs min. stability under normal storage conditions.

Struktol® TR 071. [Struktol] Blend of fatty acid zinc salt and amide; blending aid; physical coupling agent between polymer and filler systems; wets polar and nonpolar fillers; internal lubricant at process temps.; esp. effective with flame retarded HIPS, selected brominated compds.; lt. tan color, neutral odor; m.p. 117 C; flash pt. > 230 C; *Storage:* 2 yrs min. stability under normal storage conditions.

Struktol® TR 077. [Struktol] Zinc fatty acid soap; low melting processing aid; zinc stearate replacement in PS, concs., thermosets, some olefins; improved mech. handling and blending; beige pellets; sp.gr. 1.1; bulk dens. 580 g/l; drop pt. 100 C; 10.5% Zn; *Storage:* 18 mos min. stability under normal storage conditions.

Struktol® TR 141. [Struktol] Sat. primary amide; processing aid; blooms in an invisible monolayer for surf. appearance and mold release; also provides visc. reduction; cream-colored microbeads; relatively insol. in water; limited sol. in alcohol, ketones, aromatic solvs.; bulk dens. 0.6 lb/ft^3; m.p. 98-104 C; acid no. 9; iodine no. 1 max.; flash pt. 210 C; 94% amide.

Struktol® TR 251. [Struktol] EBS replacement for use in ABS and filled PP; neutral color; 4 mm pellets or microbeads; drop pt. 119 C.

Struktol® TR PE(H)-100. [Struktol] Polyethylene homopolymer wax; CAS 9002-88-4; EINECS 200-815-3; processing aid for natural and syn. elastomers; improves flow, provides good release from equip. with improved pigment dispersion and finish to molded articles; wh. waxy gran. or powd.; sp.gr. 0.91; visc. 180 cps (140 C); drop pt. 102 ± 5 C; acid no. nil.

Struktol® W 34 Flakes. [Struktol] Mixt. of high m.w. natural fatty acid esters and fatty acid soaps, bound to cheically indifferent fillers; dispersing and processing aid for rubber; improves plasticity of elastomers which leads to rapid incorporation of fillers; dispersant in highly filled compds.; reduces sticking to process equip.; improves extrusion, release; good for CR, NR, nitrile, SBR; lt. brn. pastilles; sp.gr. 1.2; bulk dens. 620 g/l; drop pt. 90 C; acid no. < 30; *Storage:* 2 yr min. storage stability under normal storage conditions.

Struktol® W 80. [Struktol] Mixt. of lubricants and fatty acid derivs.; processing aid for rubber; aids dispersion, plasticity providing easier incorporation of fillers; prevents agglomeration of fillers in highly loaded compds.; reduces sticking to process equip.; improves extrusion, release; lt. brn. paste; sp.gr. 0.89; bulk dens. 890 g/l; drop pt. 42 C; *Storage:* 1 yr min. storage stability under normal storage conditions.

Struktol® WA 48. [Struktol] Mixt. of esters and zinc soaps of natural fatty acids; processing aid for special polymers, esp. epichlorohydrin rubber to improve flow and release; improves extrusion of NBR/PVC; amber flakes; sp.gr. 1.00; bulk dens. 530 g/l; m.p. 80 C; iodine no. 60; *Storage:* 2 yr min. storage stability under normal storage conditions.

Struktol® WB 16. [Struktol] Mixt. of fatty acid soaps, predominantly calcium; lubricant for rubber; improves flow props., mold release; reduces friction; eliminates sticking; sl. activating effect on cure rate of sulfur-contg. compds.; aids stability in polychloroprene; recommended for butyl, halobutyl, EPDM, carboxy-nitrile, polyacrylate, TPE; lt. colored pearl; sp.gr. 0.98; bulk dens. 600 g/l; drop. pt. 104 C; *Storage:* 2 yr min. storage stability under normal storage conditions.

Struktol® WB 212. [Struktol] Blend of high m.w. aliphatic fatty acid esters and condensation prods., bound to inert fillers; processing aid for rubber; dispersant for powd. materials; promotes faster filler incorporation; prevents sticking; reduces scorching esp. in highly loaded compds.; plasticizer; mold release aid; sl. activation in chlorosulfonated polyethylenes; recommended for CPE, SBR; good for CR, ECO, NR, nitrile, polyacrylate, sol'n. SBR; lt. sm. beads; sp.gr. 1.10; bulk dens. 680 g/l; drop pt. 57 C; acid no. 5; reaction neutral; 10% water; *Storage:* 2 yr min. storage stability under normal storage conditions.

Struktol® WB 222. [Struktol] Water-free blend of high molecular aliphatic fatty acid esters and condensation prods.; rubber processing aid; dispersant for powd. materials; promotes faster filler incorporation; prevents sticking; reduces scorching esp. in highly loaded compds.; plasticizer; mold release aid; improves flow props., release; aids demolding in nitriles; recommended for CR, CPE, Hypalon, nitrile, carboxy-nitrile, SBR; wh. sm. beads; sp.gr. 0.95; bulk dens. 510 g/l; drop pt. 57 C; acid no. 5; reaction neutral; *Storage:* 2 yr min. storage stability under normal storage conditions.

Struktol® WB 300. [Struktol] Syn. ester; plasticizer for NBR, epichlorohydrin, and ACM rubber compds.; improves oil and fuel resist.; highly compat. with phenolic resin modified with NBR and NBR/PVC; end uses incl. petrol hoses, printing rollers, oil seals, etc.; processing aid; ylsh. liq.; insol. with aliphatic hydrocarbons, oils, and greases; sp.gr. 1.10; visc. 400 mPa•s (20 C); pour pt. -32 C; flash pt. 250 C; *Storage:* 2 yr min. storage stability under normal storage conditions.

Struktol® WS 280 Paste. [Struktol] Organo-silicone compd. on inorganic carrier (25%); processing aid for special polymers; offers lubrication effects of silicones without exudation or bloom problems; improves mill handling of sticky fluoro-silicone compds.; paste; sp.gr. 1.00; *Storage:* 2 yr min. storage stability under normal storage conditions.

Struktol® WS 280 Powd. [Struktol] Organo-silicone compd. on inorganic carrier (25%); processing aid for special polymers; offers lubrication effects of silicones without exudation or bloom problems; improves mill handling of sticky fluoro-silicone compds.; crumbly powd.; sp.gr. 1.20; bulk dens. 470 g/l; *Storage:* 2 yr min. storage stability under normal storage conditions.

Struktol® ZEH. [Struktol] Zinc 2-ethylhexanoate; activator for natural rubber; offers heat stability (reversion resist.) in NR compds. contg. normal levels of sulfur, esp. with thiazole accelerators; replacement for stearic acid in syn. rubber; highly visc. liq.; sp.gr. 1.20; visc. 5000 mPa•s; 23% Zin; *Storage:* 2 yr min. storage stability under normal storage conditions.

Struktol® ZEH-DL. [Struktol] 66.7% Zinc 2-ethylhexanoate, 33.3% silica; activator for natural rubber; replacement for stearic acid in syn. rubber; also for masticating NR; lt. powd.; sp.gr. 1.40; bulk dens. 690 g/l; 15% Zn; *Storage:* 2 yr min. storage stability under normal storage conditions.

Struktol® ZP 1014. [Struktol] Zinc peroxide with inorg. and organic dispersants; curing agent in carboxylated NBR; improved scorch resist. and storage stability over zinc oxide; easier handling and improved dispersion in the rubber mix; lt. gray nondusting powd., sl. pleasant odor; insol. in water;

sp.gr. 2.3; bulk dens. 1000 g/l; flash pt. 160 C (org. dispersant); 50% act. (26% ZnO_2, 24% ZnO), 30% inorg. dispersant, 20% org. dispersant; *Toxicology:* contains a mild eye irritant; *Precaution:* oxidizer; contact with water, acids, heavy metal salts can cause decomp. (generation of oxygen); flamm.; keep away from ignition sources and heat; *Storage:* 6 mos min. storage stability under normal storage conditions; store in cool, dry place.

Stygene R-2. [Chemfax] Polynuclear aromatic polymer; plasticizer and softener for natural and syn. rubber, and vinyl polymers; low volatility and good plasticizing action useful in tread and camelback and as a tackifier for carcasses, mech. goods, and footwear stocks; lowest m.w. member of series; as sat. for wood, paper, fabrics, cord, rope, and felt; liq.; sp.gr. 1.06; dens. 8.9 lb/gal; visc. 90 SSU (210 F); COC flash pt. 250 F.

Stygene Series. [Chemfax] Polynuclear aromatic polymers derived from specially prepared petrol. stream; resins in rubber compding., joint cements, plastic compds., protective coatings, fiber board, inks, epoxy potting compds., adhesives, insecticides, briquettes, floor tile, etc.; soft grades as rubber plasticizers; hard grades as rubber extenders; carrier for insecticidal toxicants; in calendered and extruded goods, mech. goods; Barrett 22 liq. to hard solids; sol. in aromatics, terpenes, chlorinated hydrocarbons, higher ketones and esters; acid no. 0 (neutral); flamm. self-extinguishing when flame source is removed.

Styrid. [Specialty Prods.] Styrene suppressant for fiberglass molding; reduces emissions and odors without affecting interlaminar bonding; meets Calif. Rule 1162 in polyester resin; USDA approved; wh. visc. liq., aromatic odor; negligible sol. in water; sp.gr. 0.785-0.788; dens.6.55-6.58 lb/gal; b.p. 305-396 F; flash pt. (TCC) 38 C; 55% volatile; *Toxicology:* contains 55% min. spirits (OSHA PEL 100 ppm; LD50 (oral, rat) > 25 ml/kg, (dermal, rat) 4 ml/kg); irritating to respiratory tract at high vapor concs.; may cause headaches, dizziness; prolonged/repeated skin contact may cause irritation, dermatitis; *Precaution:* flamm.; flamm. limits 0.9-7.0%; do not handle near heat, sparks, flame, or strong oxidants; thermal decomp. may produce fumes, smoke, CO, oxides of sulfur; *Storage:* material will thicken to paste in colder temps.; heat and/or mix @ R.T. to reliquefy; indefinite shelf life if kept in original containers away from extreme heat; keep closed when not in use.

SU 3. [Releasomers] Copolymer in solv. sol'n.; release agent for release of cast urethane incl. hard-to-release MDI types; minimal mold buildup; also effective with epoxies and polyesters; provides immediate release (no cure time needed); for hot or cold molds; *Toxicology:* avoid prolonged/repeated skin contact, breathing vapors; *Precaution:* keep away from heat, flame, sparks; *Storage:* keep container closed when not in use.

Suconox-4®. [Zeeland] N-Butyryl-p-aminophenol; processing aid for thermoplastics; Custom

Suconox-9®. [Zeeland] N-Pelargonoyl-p-aminophenol; processing aid for thermoplastics; Custom

Suconox-12®. [Zeeland] N-Lauroyl-p-aminophenol; processing aid for thermoplastics; Custom

Suconox-18®. [Zeeland] N-Stearoyl-p-amino phenol; CAS 103-99-1; processing aid and antioxidant in thermoplastics; m.p. 130-134 C; > 98% assay.

Sulfads®. [R.T. Vanderbilt] Essentially dipentamethylene thiuram hexasulfide; CAS 120-54-7; ultra accelerator for NR and syn. rubbers; vulcanizing agent; lt. yel. to buff powd.; 99.9% thru 100 mesh; sol. in chloroform, toluene, acetone; practically insol. in water; m.w. 448.82; dens. 1.50 ± 0.03 mg/m^3; m.p. 115 C min.; 35.0% avail. sulfur (powd.).

Sulfads® Rodform. [R.T. Vanderbilt] Dipentamethylenethiuram tetrasulfide; CAS 120-54-7; rubber accelerator; lt. yel. to lt. buff rods; negligible sol. in water; dens. 1.48 mg/m^3; 32.5% avail. sulfur; *Toxicology:* may cause mild skin irritation; *Precaution:* incompat. with strong oxidizing agents; hazardous decomp. prods.: oxides of N, S, C at combustion temps.

Sulfasan® R. [Monsanto] 4,4′-Dithiodimorpholine; CAS 103-34-4; vulcanizing agent, crosslinking agent for elastomers; sulfur donor for EV and semi-EV cure systems; avail. reg. grind, special grind, wax pellets; 80% act.

Sulfochem 436. [Chemron] Ammonium nonoxynol-4 sulfate; anionic; high-foaming detergent for dishwashing, carwash, and carpet shampoo formulations; emulsion polymerization surfactant for SBR, acrylic, vinyl acrylic systems; pale yel. liq.; 58-60% act.

Sulfochem 437. [Chemron] Ammonium nonylphenol ether sulfate; anionic; emulsion polymerization surfactant for SBR, acrylic, vinyl acrylic, PVAc, and copolymer systems; pale yel. liq.; 30% act.

Sulfochem 438. [Chemron] Ammonium nonylphenol ether sulfate; anionic; emulsion polymerization surfactant for SBR, acrylic, vinyl acrylic, PVAc, and copolymer systems; pale yel. liq.; 30% act.

Sulfolane W. [Shell] Tetramethylene sulfone; CAS 126-33-0; EINECS 204-783-1; highly polar compd. with outstanding solv. properties and chemical and thermal stability; for extraction of benzene, toluene, other aromatic hydrocarbons from oil refinery streams; used in Sulfinol process to remove acid gases; used for separation of low boiling alcohols, min. oils, tars; plasticizer; polymerization solv.; dielec. in elec. equipment; solv. in surf. coatings; component in hydraulic fluids; sp.gr. 1.2600-1.2615; 2.6-3.0% water.

Sulfopon® 101/POL. [Pulcra SA] Sodium lauryl sulfate; CAS 151-21-3; anionic; surfactant for emulsion polymerization; liq.; m.w. 302; visc. < 500 mPa•s; pH 7.5-8.5 (10%); 28-30% act.

Sulfopon® 101 Special. [Henkel KGaA/Cospha; Pulcra SA] Sodium lauryl sulfate; CAS 151-21-3; EINECS 205-788-1; anionic; surfactant for shampoos, specialty cleaners, and lt.-duty detergents; emulsifier for emulsion polymerization; low freezing pt.; paste; m.w. 302; pH 7.5-8.5 (101%); 30% conc.

Sulfopon® 102. [Henkel KGaA] Sodium lauryl sulfate C10-C16; CAS 151-21-3; anionic; detergent, emulsifier for emulsion polymerization of syn. rubber, PVC, and other polymers; liq.; 30% conc.

Sulfopon® P-40. [Pulcra SA] Sodium lauryl sulfate; CAS 151-21-3; EINECS 205-788-1; anionic; dispersant and emulsifier for acrylates, styrene acrylic, vinyl chloride, vinyl acetate copolymers; also for cream shampoos, specialty cleaners, rug shampoos; paste; m.w. 302; pH 6.0-7.0 (10%); 38-42% act.

Sulfotex DOS. [Henkel Canada] Disodium oleamido PEG sulfosuccinate; anionic; detergent, emulsifier, lathering agent for skin cleansers; foamer for carpet backing and fire fighting; emulsifier for polymerization; liq.; 28% conc.

Sumine® 2015. [Zeeland] Tertiary catalyst; epoxy curative.

Sunnol DOS. [Lion] POE alkylphenyl ether sulfate; anionic; emulsion polymerization agent; liq.; 30% conc.; Unverified

Sunnol DP-2630. [Lion] POE alkylphenyl ether sulfate; anionic; emulsion polymerization agent; liq.; 30% conc.

Sunolite® 100. [Witco; Sovereign] Microcrystalline wax/paraffin wax blend; antisunchecking agent and antiozonant used in rubber prods.; sole protectant; provides static ozone protection thru surf. migration; lt. yel. flakes; sp.gr. 0.92; m.p. 63-67 C; flash pt. 243 C.

Sunolite® 127. [Witco; Sovereign] Microcrystalline wax/paraffin wax blend; antisunchecking agent and antiozonant used in rubber prods.; sole protectant; provides static ozone protection thru surf. migration; microcryst. structure provides more controlled rate of bloom than straight paraffins; yel. flakes; sp.gr. 0.92; m.p. 64-68 C; flash pt. 247 C.

Sunolite® 130. [Witco; Sovereign] Paraffin wax; CAS 8002-74-2; EINECS 232-315-6; processing aid for rubber; provides some static ozone protection; pastille; m.p. 130 F.

Sunolite® 160. [Witco; Sovereign] Paraffin wax; CAS 8002-74-2; EINECS 232-315-6; external lubricant and processing aid in rubber and in rigid PVC formulations for extrusion and inj. molding processes; imparts smoothness and high gloss to PVC prods. as pipes and fittings for potable water, natural gas, DWV, sewage, drainage, and irrigation; elec. conduits and profiles for construction industry; food pkg. applic.; wh. flakes or powd.; nil odor and taste; sp.gr. 0.922; visc. 50 SUS; m.p. 70 C; flash pt. (COC) 268 C.

Sunolite® 240. [Witco; Sovereign] Microcrystalline/paraffin blend; antisunchecking agent and antiozonant used in rubber prods.; sole protectant; provides static ozone protection thru surf. migration; provides broadest range of protection, in both low and high temp. applics.; lt. yel. flakes; sp.gr. 0.92; m.p. 66-71 C; flash pt. 248 C.

Sunolite® 240TG. [Witco; Sovereign] Microcrystalline/paraffin blend; tire grade antisunchecking agent and antiozonant; nonstaining and nondiscoloring in most compds.; tan pastille.

Sunolite® 666. [Witco; Sovereign] Microcryst./paraffin blend; antiozonant used in rubber goods; sole protectant; provides static ozone protection thru surf. migration; provides best accelerated ozone resist. to prods. exposed to 75 F conditions; lt. yel. flakes; sp.gr. 0.92; m.p. 63-67 C; flash pt. 249 C.

Sunproofing Wax 1343. [Frank B. Ross] Complex hydrocarbon mixt.; antiozonant, anticracking and sunchecking wax for the rubber industry; FDA 21CFR §177.2600; wh. solid, odorless; insol. in water; sp.gr. 0.917-0.939; m.p. 70-74 C; acid no. nil; sapon. no. 1.0 max.; flash pt. (COC) 400 F min.; *Toxicology:* may irritate eyes, skin; molten wax can cause burns; fumes from decomp. may cause respiratory irritation; dust may cause irritation of mucous membranes and respiratory tract; TLV/TWA 5 mg/m^3 (respirable dust); *Precaution:* incompat. with strong oxidizing agents.

Sunyl® 80. [SVO Enterprises] High oleic sunflower oil; CAS 8001-21-6; EINECS 232-273-9; engineering plastic raw material feedstock; impact modifier; plasticizer; stabilizer; internal/external lubricant; rheological agent; biodeg. base oil for industrial lubricants; exc. high temp. oxidative stability; pharmaceutical diluent, carrier, emulsifier, surfactant, emollient, nutrient, binder in tablets; sp.gr. 0.915 (60 F); visc. 40 cSt (40 C); pour pt. -12 C; iodine no. 86; flash pt. 240 C.

Sunyl® 80 ES. [SVO Enterprises] Low polyunsat. glyceryl trioleate with additional stabilizers; engineering plastic raw material feedstock; impact modifier; plasticizer; stabilizer; internal/external lubricant; rheological agent; in pharmaceuticals as substitute for min. oils and high price esters; sp.gr. 0.917 (60 F); visc. 45 cSt (40 C); iodine no. 75; sapon. no. 190; drop pt. 68 F; pour pt. 0 C; cloud pt. 80 F; flash pt. 240 C; ref. index 1.4584 (48 C).

Sunyl® 80 RBD. [SVO Enterprises] High oleic sunflower oil, refined, bleached, deodorized; CAS 8001-21-6; EINECS 232-273-9; food grade environmentally friendly veg. oil for use as feedstocks/raw materials, in plastics, polymers, rubber, and pharmaceuticals; oxidative and thermal stability; readily biodeg.; 77-82% oleic acid.

Sunyl® 80 RBD ES. [SVO Enterprises] High oleic sunflower oil, refined, bleached, deodorized; CAS 8001-21-6; EINECS 232-273-9; extended stability, food grade, environmentally friendly veg. oil for use as feedstocks/raw materials, in plastics, polymers, rubber, and pharmaceuticals; oxidative and thermal stability; readily biodeg.; 77-82% oleic acid.

Sunyl® 80 RBWD. [SVO Enterprises] High oleic sunflower oil, refined, bleached, winterized, deodorized; CAS 8001-21-6; EINECS 232-273-9; food grade, environmentally friendly veg. oil for use as feedstocks/raw materials, in plastics, polymers, rubber, and pharmaceuticals; oxidative and thermal stability; readily biodeg.; 77-82% oleic acid.

Sunyl® 80 RBWD ES. [SVO Enterprises] High oleic sunflower oil, refined, bleached, winterized, deodorized; CAS 8001-21-6; EINECS 232-273-9;

extended stability, food grade, environmentally friendly veg. oil for use as feedstocks/raw materials, in plastics, polymers, rubber, and pharmaceuticals; oxidative and thermal stability; readily biodeg.; 77-82% oleic acid.

Sunyl® 90. [SVO Enterprises] Very high oleic sunflower oil; CAS 8001-21-6; EINECS 232-273-9; engineering plastic raw material feedstock; impact modifier; plasticizer; stabilizer; internal/external lubricant; rheological agent; biodeg. base oil for industrial lubricants; exc. high temp. oxidative stability; pharmaceutical diluent, carrier, emulsifier, surfactant, emollient, nutrient, binder in tablets; sp.gr. 0.916 (60 F); visc. 40 cSt (40 C); pour pt. -18 C; iodine no. 82; flash pt. 240 C.

Sunyl® 90 RBD. [SVO Enterprises] Very high oleic sunflower oil, refined, bleached, deodorized; CAS 8001-21-6; EINECS 232-273-9; food grade, environmentally friendly veg. oil for use as feedstocks/ raw materials, in plastics, polymers, rubber, and pharmaceuticals; oxidative and thermal stability; readily biodeg.; 85% min. oleic acid.

Sunyl® 90 RBWD. [SVO Enterprises] Very high oleic sunflower oil, refined, bleached, winterized, deodorized; CAS 8001-21-6; EINECS 232-273-9; food grade, environmentally friendly veg. oil for use as feedstocks/raw materials, in plastics, polymers, rubber, and pharmaceuticals; oxidative and thermal stability; readily biodeg.; 85% min. oleic acid.

Sunyl® 90E RBWD. [SVO Enterprises] Very high oleic sunflower oil, refined, bleached, winterized, deodorized; CAS 8001-21-6; EINECS 232-273-9; food grade, environmentally friendly veg. oil for use as feedstocks/raw materials, in plastics, polymers, rubber, and pharmaceuticals; oxidative and thermal stability; readily biodeg.; expeller grade; 85% min. oleic acid.

Sunyl® 90E RBWD ES 1016. [SVO Enterprises] Very high oleic sunflower oil, refined, bleached, winterized, deodorized; CAS 8001-21-6; EINECS 232-273-9; extended stability, food grade, environmentally friendly veg. oil for use as feedstocks/raw materials, in plastics, polymers, rubber, and pharmaceuticals; oxidative and thermal stability; readily biodeg.; expeller grade; low color; 87% min. oleic acid.

Sunyl® HS 500. [SVO Enterprises] High oleic sunflower oil, refined, bleached, deodorized, partially hydrog.; CAS 8001-21-6; EINECS 232-273-9; extended stability, food grade, environmentally friendly veg. oil for use as feedstocks/raw materials, in plastics, polymers, rubber, and pharmaceuticals; very high oxidative and thermal stability; readily biodeg.; 77-82% oleic acid.

Supercoat®. [ECC Int'l.] Calcium carbonate coated with < 2% stearic acid; CAS 1317-65-3, 57-11-4 resp.; coated ultrafine ground pigment offering easy incorporation and dispersion in plastics with improved physical props. esp. impact resist.; for rigid and flexible PVC, filled PP and PE, nonaq. sealant systems, filled plastic sheet; hydrophobic; NSF compliance; wh. powd., odorless; 1 μ mean particle size; fineness (Hegman) 7.0; sp.gr. 2.71; dens. 45 lb/ft^3 (loose); surf. area 7.2 m^2/g; oil absorp. 16; pH 9.5 (5% slurry); nonflamm.; 97.6% $CaCO_3$, 0.2% max. water; *Toxicology:* TLV/TWA 10 mg/m^3, considered nuisance dust; *Precaution:* incompat. with acids.

Super-Fil®. [Specialty Minerals] Untreated ground limestone; CAS 1317-65-3; filler for rigid vinyl, blown film applics.; powd.; 2 μ avg. particle size, 10 μ top size.

Super Microtuff F. [Specialty Minerals] Treated talc; CAS 14807-96-6; EINECS 238-877-9; filler for polyolefins, engineering thermoplastics; powd.; 1 μ avg. particle size, 10 μ top size.

Supermite®. [ECC Int'l.] Calcium carbonate; CAS 1317-65-3; ultrafine pigment, filler, reinforcement developing superior physical props., opacity, and surf. gloss in plastics (filled polyolefin and PVC compds.), elastomers, sealants, paints, inks, and coatings; NSF compliance; wh. powd., odorless; 1 μ mean particle size; fineness (Hegman) 7.0; negligible sol. in water; sp.gr. 2.71; dens. 32 lb/ft^3 (loose); surf. area 7.2 m^2/g; oil absorp. 19; pH 9.5 (5% slurry); nonflamm.; 97.6% $CaCO_3$, 0.2% max. moisture; *Toxicology:* TLV/TWA 10 mg/m^3, considered nuisance dust; *Precaution:* incompat. with acids.

Super Nevtac® 99. [Neville] Syn. polyterpene resin; tackifier for PP, PE, NR, SIS block copolymers, EVA, polyisoprene, butyl rubber used in solv. and hot-melt adhesives, coatings, rubber prods., concrete-curing compds., and caulking compds.; FDA 21 CFR §175.125, 175.105, 175.300, 175.320, 176.170, 176.180, 177.1210, 177.2600; Gardner 5 (50% in toluene) solid, flakes; sol. in esters, ethers, ketones (except acetone), and chlorinated, aromatic, naphthenic, and terpene hydrocarbons; m.w. 1040; sp.gr. 0.98; soften. pt. (R&B) 99 C; flash pt. (COC) 450 F.

Superox® 46-709. [Reichhold] Standard general purpose polymerization initiator for curing of unsat. polyester and vinyl ester resins.

Superox® 46-747. [Reichhold] MEK peroxide in plasticizer; CAS 1338-23-4; EINECS 215-661-2; crosslinking polymerization initiator with unsat. polyester resins or vinyl ester resins; gives faster gel times and cure times than other MEK peroxide formulations; gives higher exotherms, esp. in thick laminate sections; colorless clear liq.; sl. sol. in water; sp.gr. 1.10; flash pt. (COC) 200 F min.; fire pt. 200 F min.; 9% max. act. oxygen; *Toxicology:* may cause blindness; *Precaution:* flamm.; avoid direct mixing with metal soaps, amine, or other polymerization accelerator/promoter as violent decomp. will result; strong oxidizing material; never dilute with acetone; *Storage:* 1 yr shelf life stored in closed, original containers in cool area below 75 F (3 mos shelf life @ 86-100 F); store away from flamm., heat sources, sunlight, strong oxidizing/ reducing agents/promoters.

Superox® 46-748. [Reichhold] MEK peroxide in plasticizer; CAS 1338-23-4; EINECS 215-661-2;

polymerization curing agent for prepromoted vinyl ester resins or unsat. polyester resins; minimizes foaming in vinyl esters; optimum crosslinking for vinyl esters; colorless clear liq.; sl. sol. in water; sp.gr. 1.10; flash pt. (COC) 200 F min.; fire pt. 200 F min.; 9% max. act. oxygen; *Toxicology:* may cause blindness; *Precaution:* flamm.; avoid direct mixing with metal soaps, amine, or other polymerization accelerator/promoter as violent decomp. will result; strong oxidizing material; never dilute with acetone; *Storage:* 1 yr shelf life stored in closed, original containers in cool area below 75 F (3 mos shelf life @ 86-100 F); store away from flamm., heat sources, sunlight, strong oxidizing/reducing agents/promoters.

Superox® 46-753-00. [Reichhold] Cumyl hydroperoxide and MEK peroxide in plasticizer; polymerization initiator for R.T. curing of unsat. polyester and vinyl ester resins; imparts longer gel times and lower peak exotherm, improves cure development over MEK peroxides; pale yel. clear liq.; sol. in oxygenated org. solvs.; sl. sol. in water; sp.gr. 1.07; flash pt. (COC) 200 F min.; fire pt. 200 F min.; 9% max. act. oxygen; *Toxicology:* may cause blindness; *Precaution:* flamm.; avoid direct mixing with metal soaps, amine, or other polymerization accelerator/promoter as violent decomp. will result; strong oxidizing material; never dilute with acetone; *Storage:* 1 yr shelf life stored in closed, original containers in cool area below 75 F (3 mos shelf life @ 86-100 F); store away from flamm., heat sources, sunlight, strong oxidizing/reducing agents/promoters.

Superox® 702. [Reichhold] MEK peroxide in a plasticizer sol'n.; CAS 1338-23-4; EINECS 215-661-2; catalyst for R.T. curing of unsat. polyester, vinyl ester, and isophthalic resins; used in gel coats to eliminate or reduce porosity; colorless clear liq.; sol. in polar org. compds. and polyester resins; insol. in water; sp.gr. 1.16; visc. 14 cps; f.p. < -22 F; flash pt. > 140 F; ref. index 1.48; 8.9 ± 0.1% act. oxygen; *Toxicology:* LD50 (oral, rat) 500-5000 mg/kg; TLV 0.2 ppm (in workroom air); may cause blindness; *Precaution:* flamm.; strong oxidizing material; protect from flamm., heat sources, sunlight, strong oxidizing and reducing agents, promoters; never dilute with acetone; *Storage:* 1 yr shelf life stored in closed, original containers in cool area below 75 F (3 mos shelf life @ 86-100 F).

Superox® 709. [Reichhold] MEK peroxide in sol'n.; CAS 1338-23-4; EINECS 215-661-2; general purpose, high efficiency catalyst for curing of almost all resin systems; colorless clear liq.; misc. in polyester resins, dimethyl phthalate, MEK, ethyl acetate, polar org. compds.; insol. in water; sp.gr. 1.15; visc. 11 cps; f.p. < -35 C; flash pt. (Seta) > 60 C; ref. index 1.4714; 8.9 ± 0.1% act. oxygen; *Toxicology:* LD50 (oral, rat) 500-5000 mg/kg; TLV 0.2 ppm (in workroom air); may cause blindness; *Precaution:* flamm.; strong oxidizing material; protect from flamm., heat sources, sunlight, strong oxidizing and reducing agents, promoters; never dilute with acetone; *Storage:* 1 yr shelf life stored in closed, original containers in cool area below 75 F (3 mos shelf life @ 86-100 F).

Superox® 710. [Reichhold] MEK peroxide in sol'n.; CAS 1338-23-4; EINECS 215-661-2; catalyst for cure of unsat. polyester and vinyl ester resins; gives faster cure in many resin systems; colorless clear liq.; misc. in polyester resins, dimethyl phthalate, MEK, ethyl acetate, polar org. compds.; insol. in water; sp.gr. 1.145; visc. 9.5 cps; f.p. < -35 C; flash pt. (Seta) > 60 C; ref. index 1.4640; 8.9 ± 0.1% act. oxygen; *Toxicology:* LD50 (oral, rat) 500-5000 mg/kg; TLV 0.2 ppm (in workroom air); may cause blindness; *Precaution:* flamm.; strong oxidizing material; protect from flamm., heat sources, sunlight, strong oxidizing and reducing agents, promoters; never dilute with acetone; *Storage:* 1 yr shelf life stored in closed, original containers in cool area below 75 F (3 mos shelf life @ 86-100 F).

Superox® 711. [Reichhold] MEK peroxide in sol'n.; CAS 1338-23-4; EINECS 215-661-2; catalyst for cure of unsat. polyester and vinyl ester resins; faster in many resin systems; primarily used for polyester foaming and curing clay pipe seals; colorless clear liq.; misc. in polyester resins, dimethyl phthalate, MEK, ethyl acetate, polar org. compds.; insol. in water; sp.gr. 1.121; visc. 12 cps; f.p. < -35 C; flash pt. (Seta) > 60 C; ref. index 1.4466; 8.9 ± 0.1% act. oxygen; *Toxicology:* LD50 (oral, rat) 500-5000 mg/kg; TLV 0.2 ppm (in workroom air); may cause blindness; *Precaution:* flamm.; strong oxidizing material; protect from flamm., heat sources, sunlight, strong oxidizing and reducing agents, promoters; never dilute with acetone; *Storage:* 1 yr shelf life stored in closed, original containers in cool area below 75 F (3 mos shelf life @ 86-100 F).

Superox® 712. [Reichhold] MEK peroxide in sol'n.; CAS 1338-23-4; EINECS 215-661-2; general purpose, high efficiency catalyst for curing of almost all resin systems; identical to Superox 709 except for color; red clear liq.; misc. in polyester resins, dimethyl phthalate, MEK, ethyl acetate, polar org. compds.; insol. in water; sp.gr. 1.15; visc. 11 cps; f.p. < -35 C; flash pt. (Seta) > 60 C; ref. index 1.4714; 8.9 ± 0.1% act. oxygen; *Toxicology:* LD50 (oral, rat) 500-5000 mg/kg; TLV 0.2 ppm (in workroom air); may cause blindness; *Precaution:* flamm.; strong oxidizing material; protect from flamm., heat sources, sunlight, strong oxidizing and reducing agents, promoters; never dilute with acetone; *Storage:* 1 yr shelf life stored in closed, original containers in cool area below 75 F (3 mos shelf life @ 86-100 F).

Superox® 713. [Reichhold] MEK peroxide in sol'n.; CAS 1338-23-4; EINECS 215-661-2; catalyst for cure of unsat. polyester and vinyl ester resins; low visc. diluted version of Superox 709; useful in spray-up applics. and high ambient temp. conditions; colorless clear liq.; misc. in polyester resins, dimethyl phthalate, MEK, ethyl acetate, polar org. compds.; insol. in water; sp.gr. 1.118; visc. 4.8 cps; f.p. < -35 C; flash pt. (Seta) > 60 C; ref. index 1.4490;

5.5 ± 0.05% act. oxygen; *Toxicology:* LD50 (oral, rat) 500-5000 mg/kg; TLV 0.2 ppm (in workroom air); may cause blindness; *Precaution:* flamm.; strong oxidizing material; protect from flamm., heat sources, sunlight, strong oxidizing and reducing agents, promoters; never dilute with acetone; *Storage:* 1 yr shelf life stored in closed, original containers in cool area below 75 F (3 mos shelf life @ 86-100 F).

Superox® 730. [Reichhold] MEK peroxide in sol'n.; CAS 1338-23-4; EINECS 215-661-2; catalyst for cure of unsat. polyester and vinyl ester resins; std. visc. diluted version of Superox 709; useful in spray-up applics. and high ambient temp. conditions; colorless clear liq.; misc. in polyester resins, dimethyl phthalate, MEK, ethyl acetate, polar org. compds.; insol. in water; sp.gr. 1.158; visc. 9.6 cps; f.p. < -35 C; flash pt. (Seta) > 60 C; ref. index 1.4830; 5.5 ± 0.05% act. oxygen; *Toxicology:* LD50 (oral, rat) 500-5000 mg/kg; TLV 0.2 ppm (in workroom air); may cause blindness; *Precaution:* flamm.; strong oxidizing material; protect from flamm., heat sources, sunlight, strong oxidizing and reducing agents, promoters; never dilute with acetone; *Storage:* 1 yr shelf life stored in closed, original containers in cool area below 75 F (3 mos shelf life @ 86-100 F).

Superox® 732. [Reichhold] MEK peroxide in a low visc. plasticizer sol'n.; CAS 1338-23-4; EINECS 215-661-2; catalyst for cure of unsat. polyester and vinyl ester resins; low visc. diluted version of Sueprox 702; virtually eliminates porosity in gel coats; useful in spray-up applics. and high ambient temp. conditions; colorless clear liq.; misc. in polyester resins, dimethyl phthalate, MEK, ethyl acetate, polar org. compds.; insol. in water; sp.gr. 1.126; visc. 5.4 cps; f.p. < -35 C; flash pt. (Seta) > 60 C; ref. index 1.4551; 5.5 ± 0.05% act. oxygen; *Toxicology:* LD50 (oral, rat) 500-5000 mg/kg; TLV 0.2 ppm (in workroom air); may cause blindness; *Precaution:* flamm.; strong oxidizing material; protect from flamm., heat sources, sunlight, strong oxidizing and reducing agents, promoters; never dilute with acetone; *Storage:* 1 yr shelf life stored in closed, original containers in cool area below 75 F (3 mos shelf life @ 86-100 F).

Superox® 739. [Reichhold] MEK peroxide in a low visc. plasticizer sol'n.; CAS 1338-23-4; EINECS 215-661-2; catalyst for cure of unsat. polyester and vinyl ester resins; useful in spray-up applics. and high ambient temp. conditions; identical to Superox 730 except for color; red clear liq.; misc. in polyester resins, dimethyl phthalate, MEK, ethyl acetate, polar org. compds.; insol. in water; sp.gr. 1.138; visc. 9.6 cps; f.p. < -35 C; flash pt. (Seta) > 60 C; ref. index 1.4830; 5.5 ± 0.05% act. oxygen; *Toxicology:* LD50 (oral, rat) 500-5000 mg/kg; TLV 0.2 ppm (in workroom air); may cause blindness; *Precaution:* flamm.; strong oxidizing material; protect from flamm., heat sources, sunlight, strong oxidizing and reducing agents, promoters; never dilute with acetone; *Storage:* 1 yr shelf life stored in closed, original containers in cool area below 75 F (3 mos shelf life @ 86-100 F).

Superox® 744. [Reichhold] Benzoyl peroxide in a nonvolatile plasticizer disp.; polymerization initiator or catalyst for polyester and vinyl ester resins, vinyl monomers; wh. liq. disp.; sol. in org. solvs., unsat. polyester resins, and monomers; sp.gr. 1.15; dens. 9.6 lb/gal; visc. 5000 cps; flash pt. (Seta) 140 F.

Super-Pflex® 100. [Specialty Minerals] Surface-modified precipitated calcium carbonate; filler; provides exc. extrusion processing, impact str. retention, and enhanced color stability on outdoor exposure in extruded rigid PVC siding and profiles, small diam. conduit and pipe; NSF compliance; powd.; 0.7 μ particulate; 0.03% +325 mesh; sp.gr. 2.7; dens. 50 lb/ft^3 (tapped); bulk dens. 25 lb/ft^3; surf. area 7 m^2/g; oil absorp. 26 g/100 g; dry brightness 97; 98% $CaCO_3$.

Super-Pflex® 200. [Specialty Minerals] Surface-modified precipitated calcium carbonate; reinforcing agent for inj. molding of rigid PVC; used in elec. conduit, DWV pipe, pressure pipe, PVC fittings, rainwater gutters and downspouts, house siding, and various profiles; powd.; 0.7 μ avg. particle size; 0.03% +325 mesh; sp.gr. 2.7; dens. 50 lb/ft^3 (tapped); bulk dens. 25 lb/ft^3; surf. area 7 m^2/g; dry brightness 97; 98% $CaCO_3$.

Super 33 Silicone Spray No. S3312-A. [IMS] 33% Silicone release agent (CAS 63148-62-9) with 50-60% HFC-152A and 30-40% dimethyl ether; non-CFC, non-ozone depleting high conc. mold release for tough molding jobs for ABS, acetal, acrylic, nylon, PMMA, PC, polyester, polyethylene, PS, rubber; suitable for food pkg.; FDA 21CFR §181.28; clear mist with sl. ethereal odor as dispensed from aerosol pkg.; sl. sol. in water; sp.gr. > 1; vapor pressure 60 ± 10 psig; flash pt. < 0 F; > 95% volatile; *Toxicology:* inh. may cause CNS depression with dizziness, headache; higher exposure may cause heart problems; gross overexposure may cause fatality; eye and skin irritation; direct contact with spray can cause frostbite; *Precaution:* flamm. limits 1-18%; at elevated temps. (> 130 F) aerosol containers may burst, vent, or rupture; incompat. with strong oxidizers, strong caustics, reactive metals (Na, K, Zn, Mg, Al); contains methylpolysiloxanes which can generate formaldehyde; *Storage:* store in cool, dry area out of direct sunlight; do not puncture, incinerate, or store above 120 F.

Super Sta-Tac® 80. [Arizona] Specialty hydrocarbon based on mixed olefins; for pressure-sensitive adhesive, hot-melt, and sealants; tackifier for S-I-S and S-B-S polymers; FDA 21CFR §175.105; Gardner 4+ color; soften. pt. (R&B) 82 C.

Super Sta-Tac® 100. [Arizona] Specialty hydrocarbon based on mixed olefins; tackifier for pressure-sensitive adhesive, hot-melt, and sealants; FDA 21CFR §175.105; Gardner 4+ color; soften. pt. (R&B) 95 C.

Supra EF. [Luzenac Am.] Platy talc; CAS 14807-96-6; EINECS 238-877-9; for cosmetic applics. incl. dusting and pressed powds., creams and lotions, antiperspirants, bath and loose powds.; exc. slip; also for most stringent rubber dusting applics.; powd.; 8.2 μ median particle size; 98% thru 200 mesh; sp.gr. 2.8; bulk dens. 25.6 lb/ft^3 (loose); surf. area 5 m^2/g; oil absorp. 42; brightness (GE) 86; pH 8.5 (10% slurry); 60.4% SiO_2, 32% MgO.

Supra EF A. [Luzenac Am.] Platy talc USP; CAS 14807-96-6; EINECS 238-877-9; extrafine, extremely platy talc with exc. brightness, purity, surface passivity, and slip; used for dusting and pressed powds., antiperspirants, creams, lotions, bath and loose powds. and as detackifying agent for fine rubber prods.; powd.; 99.6% thru 200 mesh; median diam. 8 μ; tapped dens. 59 lb/ft^3; pH 9 (10% slurry).

Supragil® NK. [Rhone-Poulenc Surf. & Spec.] Sodium dibutyl naphthalene sulfonate; CAS 25417-20-3; anionic; wetting agent for pesticides; emulsifier, wetting and dispersing agent, foamer for dyes, pigments, emulsion polymerization, industrial use; EPA compliance; cream to tan powd.; m.w. 326; water-sol.; pH 6-8 (5%); surf. tens. 36 dynes/cm (@ CMC); 65% conc.

Supralate® WAQ. [Witco/New Markets] Sodium lauryl sulfate; CAS 151-21-3; anionic; emulsifier, detergent for emulsion polymerization, general detergency; liq.

Supralate® WAQE. [Witco/New Markets] Sodium lauryl sulfate, tech.; CAS 151-21-3; anionic; emulsifier for emulsion polymerization; pale yel. liq.; m.w. 302; water sol., act. ingred. is sol. in polar solv. with some electrolyte precipitation; dens. 8.6 lb/gal; visc. 500 cps; cloud pt. 65 F; 29-30.5% act.

Suprmix DBEEA. [C.P. Hall] Dibutoxyethoxyethyl adipate; CAS 141-17-3; alternative to conventional dry liq. concs. based on calcium silicate; eliminates handling problems in rubber processing.

Surfac® 554. [Shell] PVC adhesion promoter for low temp. plastisol bakes; provides films with superior color props.; Gardner 8 color; sp.gr. 1.057; dens. 8.82 lb/gal; visc. 8400 cps; amine no. 190.

Surfac® 555. [Shell] PVC adhesion promoter for low temp. cure; develops superior PVC plastisol adhesions to metal substrates at low temps.; Gardner 12 color; dens. 8.1 lb/gal; visc. 17,000 cps (43.3 C); amine no. 380.

Surfac® 580. [Shell] PVC adhesion promoter developed for improved paintable low temp. cures; Gardner 12 color; dens. 7.94 lb/gal; visc. 10,950 cps; amine no. 190.

Surfac® 600. [Shell] PVC adhesion promoter developed to give exc. adhesion of PVC plastisols to various metallic and coated surfs. at very low bake temps.; esp. for compatibility with wet-on-wet (high solid) paint; Gardner 12 max. color; dens. 8.4 lb/gal; visc. 3000 cps; acid no. 1.5 max.; amine no. 75.

Surfac® 610. [Shell] PVC adhesion promoter developed to give exc. adhesion of PVC plastisols to various metallic and coated surfs. at very low bake temps.; exhibits improved odor and adhesive bond; gives wet-on-wet (high solid) paint compatibility to properly formulated PVC plastisols; Gardner 12 max. color; dens. 9.1 lb/gal; visc. 60 poise; acid no. 1.5 max.; amine no. 75.

Surfagene FAD 105. [Chem-Y GmbH] Nonyl phenol deriv.; anionic; emulsifier for emulsion polymerization; Gardner 3 clear liq.; sp.gr. 1.01; visc. 10,000 mPa•s; pH 2 (10%); 99% act.

Surfagene FAZ 109. [Chem-Y GmbH] Nonoxynol-9 phosphate; anionic; emulsifier for emulsion polymerization; Gardner 3 clear liq.; sp.gr. 1.1; visc. 10,000 mPa•s; pH 2 (10%); 99% act.

Surfagene FAZ 109 NV. [Chem-Y GmbH] Nonoxynol-9 phosphate; anionic; emulsifier for emulsion polymerization; liq.; 20% act.

Surfax 215. [Aquatec Quimica SA] Ammonium lauryl sulfate; CAS 2235-54-3; anionic; emulsion and suspension polymerization of polystyrene; liq.; 29% conc.

Surfax 218. [Aquatec Quimica SA] Sodium laureth sulfate; CAS 9004-82-4; 3088-31-1; anionic; foaming agent for rubber systems; liq.; 27% conc.

Surfax 220. [Aquatec Quimica SA] Sodium alkyl sulfate; anionic; emulsifier for emulsion polymerization of styrene butadiene, vinyl acetate, vinyl chloride, acrylic copolymers; liq.; 28% conc.

Surfax 495. [Aquatec Quimica SA] Sodium alkylaryl sulfonate; anionic; emulsifier for emulsion polymerization of vinyl acetate; liq.; 95% conc.

Surfax 502. [Aquatec Quimica SA] Disodium ethoxylated alcohol half ester of sulfosuccinic acid; anionic; emulsifier for emulsion polymerization of vinyl acetate and acrylates; 34% conc.

Surfax 536. [Aquatec Quimica SA] Disodium ethoxylated alcohol half ester of sulfosuccinic acid; anionic; emulsifier for emulsion polymerization of vinyl acetate and acrylates; 33% conc.

Surfax 539. [Aquatec Quimica SA] Disodium ethoxylated alcohol half ester of sulfosuccinic acid; anionic; emulsifier for emulsion polymerization of vinyl acetate and acrylates; 34% conc.

Surfax 550. [Aquatec Quimica SA] Sodium dioctyl sulfosuccinate; anionic; for emulsion polymerization of vinyl acetate; wetting agent for pigments; liq.; 68% conc.

Surfax 560. [Aquatec Quimica SA] Sodium dialkyl sulfosuccinate; anionic; for emulsion polymerization of modified styrene-butadiene; powd.; 83% conc.

Surfynol® 82S. [Air Prods./Perf. Chems.] 3,6-Dimethyl-4-octyne-3,6-diol on amorphous silica carrier; nonionic; defoamer/wetting agent in pesticide wettable powds., electroplating baths, cement, plastics, coatings; solubilizer and clarifier in shampoos; EPA compliance; wh. free-flowing powd.; disp. in water; sp.gr. 0.47 g/ml; dens. 3.9 lb/gal; pH 7.6 (5% aq.); Draves wetting 4 s (4.35%); 46% conc.

Surfynol® 104A. [Air Prods./Perf. Chems.] Tetramethyl decynediol and 2-ethyl hexanol; CAS 126-86-3; nonionic; defoamer, wetting agent for PVC prod., emulsion polymerization, pesticides, coatings, dyestuffs, aq. systems; lubricity additive for metalworking formulations; EPA approved; lt.

yel. liq.; sp.gr. 0.869; dens. 7.2 lb/gal; HLB 4.0; m.p. < 0 C; surf. tens. 33.0 (0.1% aq.); 50% conc.

Surfynol® 104E. [Air Prods./Perf. Chems.] Tetramethyl decynediol and ethylene glycol; nonionic; wetting agent, defoamer, dispersant, visc. stabilizer for emulsion polymerization, PVC prod., latex dipping, foundry, adhesives, dyes, pigments, oilfield, hard surf. cleaners, metalworking fluids, textiles, cement, electroplating, EPA approved; lt. yel. clear liq.; sp.gr. 1.001; dens. 8.3 lb/gal; HLB 4.0; m.p. < 0 C; surf. tens. 36.2 (0.1% aq.); 50% conc.

Surfynol® 104PA. [Air Prods./Perf. Chems.] Tetramethyl decynediol in IPA; nonionic; wetting and defoaming agent in water-based systems, e.g., coatings, adhesives, inks, cements, metalworking fluids, latex dipping and paper coatings, emulsion polymerization, PVC prod.; lt. yel. liq.; sp.gr. 0.839 (25 C); dens. 8.1 lb/gal; HLB 4.0; m.p. -16 C; 50% conc.

Surfynol® 465. [Air Prods./Perf. Chems.] PEG-10 tetramethyl decynediol; CAS 9014-85-1; nonionic; wetting agent, defoamer for aq. coatings, inks, adhesives; surfactant for emulsion polymerization; electroplating additive; FDA 21 CFR §175.105, 175.300, 176.170, 176.180, 176.200, 177.210, EPA compliance; straw clear liq.; sol. in water, CCl_4, xylene, ethylene glycol, min. oil; sp.gr. 1.038; visc. < 200 cps; HLB 13.0; pour pt. 44 F; cloud pt. 63 C (5%); pH 6-8 (1% aq.); surf. tens. 33.2 (0.1% aq.); 100% conc.

Surfynol® 485. [Air Prods./Perf. Chems.] PEG-30 tetramethyl decynediol; CAS 9014-85-1; nonionic; wetting agent, defoamer for aq. coatings, inks, adhesives, agric., electroplating, oilfield chems., paper coatings, emulsion polymerization; FDA 21CFR §175.105, 175.300, 176.170, 176.180, 176.200, 177.210, EPA compliance; straw clear liq.; sol. in water, CCl_4, xylene, ethylene glycol, min. oil; sp.gr. 1.080; visc. < 350 cps; HLB 17.0; pour pt. 85 F; cloud pt. > 100 C (5%); pH 6-8 (1% aq.); surf. tens. 40.1 (0.1% aq.); 100% conc.

Surfynol® D-101. [Air Prods./Perf. Chems.; Air Prods. Nederland BV] Nonsilicone; defoamer/antifoamer for aq. systems, coatings, adhesives; effective for PVAc, ethylene-vinyl acetate systems; FDA compliance; yel. clear liq.; emulsifiable in water; sp.gr. 0.92; b.p. 171 C (109 mm); cloud pt. < 25 F; flash pt. (PMCC) 165 F; pH 5 (5% aq.); 100% act.

Surfynol® D-201. [Air Prods./Perf. Chems.; Air Prods. Nederland BV] Nonsilicone; defoamer/antifoamer for aq. systems, coatings, adhesives; effective for PVAc, ethylene-vinyl acetate systems; FDA compliance; yel. clear liq.; emulsifiable in water; sp.gr. 0.93; b.p. 204 C (17 mm); flash pt. (PMCC) > 25 F; pH 5 (5% aq.); 100% act.

Surfynol® DF-110D. [Air Prods./Perf. Chems.] Higher m.w. acetylenic glycol in dipropylene glycol; nonionic; defoamer, de-air entrainment aid for water-based coatings, inks, adhesives, and highly pigmented systems (concrete, paper coatings, grouts, ceramics), PVC prod., emulsion polymerization; liq.; HLB 3.0.

Surfynol® DF-110L. [Air Prods./Perf. Chems.; Air Prods. Nederland BV] Higher m.w. acetylenic glycol in mixed glycols; nonionic; defoamer, de-air entrainment aid for water-based coatings, inks, adhesives, and highly pigmented systems (concrete, paper coatings, grouts, ceramics), PVC prod., emulsion polymerization; liq.; HLB 3.0.

Sur-Wet® R. [Air Prods./Perf. Chems.] Modified aliphatic amine; R.T. epoxy curing agent which adheres and cures well when applied under water; requires acceleration with Ancamine 2205, 2072, or 2280 to give harder faster cures; water accelerates cure; for underwater coatings, adhesives, splash-zone compds.; Gardner 4 liq.; sp.gr. 0.980; visc. 6500 cps; amine no. 195; equiv. wt. 222; heat distort. temp. 113 F (264 psi).

Sustane® BHA. [UOP Food Antioxidants] BHA; CAS 25013-16-5; EINECS 246-563-8; antioxidant, stabilizer for fats, oils, other foods, flavors, vitamins, waxes, tallow, sausage, chewing gum base, shortening, lard, food pkg. materials, potatoes, and cereals; in cosmetic creams, lotions, hair dressings; antioxidant for cellulosics, PE, PP, PS, PVC; FDA 21CFR §172.110, 172.115, 182.3169; USDA 9CFR §318.7, 381.147; wh. solid tablets; sol. in propylene glycol, ethanol, glyceryl oleate, soybean and cottonseed oils, acetone; insol. in water; m.w. 180.2; sp.gr. 1.020; visc. 3.3 cSt (100 C); m.p. 57 C; b.p. 270 C (760 mm Hg); flash pt. (CC) 130 C; 98.5% min. purity; *Toxicology:* may cause irritation of skin, nostrils; *Precaution:* dusts may present an explosion hazard.

Sustane® BHT. [UOP Food Antioxidants] BHT; CAS 128-37-0; EINECS 204-881-4; preservative and antioxidant for foods, food pkg. materials, tallow, animal feeds, soaps, cosmetic creams, lotions, hair dressings; inhibits oxidation of fats and oils and retards rancidity and off-flavors caused by oxidation; nonstaining, nondiscoloring antioxidant for plastics (ABS, cellulosics, nylon, PVC, PE, PP, PS, polyester); FDA 21CFR §172.115, 182.3173; USDA 9CFR §318.7, 381.147; wh. cryst.; sol. (g/100 solv.): 48 g in lard (50 C), 40 g in benzene, 30 g in min. oil, 28 g in linseed oils, 20 g in methanol, nil in water; m.w. 220.3; sp.gr. 1.01 (20/4 C); dens. 37.5 lb/ft^3; visc. 3.5 cSt (80 C); m.p. 70 C; b.p. 265 C (760 mm Hg); flash pt. (CC) 118 C; ref. index 1.486; 100% act.; *Toxicology:* may cause irritation of skin, nostrils; *Precaution:* dusts may present an explosion hazard.

Sustane® TBHQ. [UOP Food Antioxidants] TBHQ; CAS 1948-33-0; EINECS 217-752-2; food-grade preservative and antioxidant for foodstuffs and meat prods., esp. in preservation of shortenings and oils derived from cottonseed, soybean, safflower, etc.; color stable and useful as substitute for reactive antioxidants that tend to form purple complexes with iron or copper; antioxidant for polyester; FDA 21CFR §172.185, 177.2420; USDA 9CFR §318.7, 381.147; off-wh. cryst. powd.; sol. (g/100 g): 100 g in methanol, 30 g in propylene glycol, 10 g in corn oil, 5 g in lard (50 C), nil in water; m.w.

166.2; dens. 27 lb/ft^3; m.p. 126.5-128.5 C; b.p. 295 C (760 mm Hg); flash pt. (CC) 171 C; 99.0% min purity; *Toxicology:* may cause irritation of skin, nostrils; *Precaution:* dusts may present an explosion hazard.

Swanox BHEB. [Harwick] Phenol; nondiscoloring, nonstaining antioxidant for PE.

S-Wax-1. [Cardinal Carolina] internal/external lubricant for PVC extrusions, multiscrew inj. molding; solid.

S-Wax-3. [Cardinal Carolina] General-purpose internal/external lubricant for rigid PVC, single-screw and multiscrew extrusions and inj. molding; solid.

SY-AB 70. [Additive Polymers Ltd] Antiblock additive in K-Resin®-based masterbatch; antiblock for plastics.

SY-AO 68. [Additive Polymers Ltd] Antioxidant blend in K-Resin®-based masterbatch; antioxidant for plastics.

SY-AS 140. [Additive Polymers Ltd] Antistatic system in K-Resin®-based masterbatch; antistat for plastics.

Sylvaros® 20. [Arizona] Tall oil rosin; printing ink binder as resin or salt, paper sizing agent, emulsifier for SBR polymerization as soap, tackifier resin in adhesives, imidazoline modifier in corrosion inhibitors, elastomer modifier in emulsion polymerization, dust control additive; film former/plasticizer in lacquers and varnishes; Gardner 6- color; sp.gr. 1.01; soften. pt. (R&B) 75 C; acid no. 176; flash pt. (OC) 226 C.

Sylvaros® 80. [Arizona] Tall oil rosin; printing ink binder as resin or salt, paper sizing agent, emulsifier for SBR polymerization as soap, tackifier resin in adhesives, imidazoline modifier in corrosion inhibitors, elastomer modifier in emulsion polymerization, dust control additive; film former/plasticizer in lacquers and varnishes; Gardner 6 color; sp.gr. 1.01; soften. pt. (R&B) 75 C; acid no. 171; flash pt. (OC) 226 C.

Sylvaros® 315. [Arizona] Lower acid number tall oil rosins; for mfg. of paper size, intermediate in rosin deriv. prod., printing ink binders as resins, tackifier resin in sealants and mastics, as starting point rosin for resin esters; Gardner 13 color; sp.gr. 1.03; soften. pt. (R&B) 60 C; acid no. 80-115; flash pt. (OC) 210 C; 55% rosin acids.

Sylvaros® R. [Arizona] Polymerized rosin; tackifier for pressure-sensitive adhesives, construction adhesives, pick-up gums for labeling, adhesion promoter for difficult-to-bond substrates; Gardner 8 color; soften. pt. (R&B) 77 C; acid no. 157.

Sylvaros® RB. [Arizona] Lower acid number tall oil rosins; for mfg. of paper size, intermediate in rosin deriv. prod., printing ink binders as resins, tackifier resin in sealants and mastics, as starting point rosin for resin esters; Gardner 15 color; sp.gr. 1.03; soften. pt. (R&B) 55 C; acid no. 60-80; flash pt. (OC) 201 C; 35% rosin acids.

Sylvatac® 5N. [Arizona] Rosin ester; tackifier for adhesives; FDA 21CFR §175.105, 176.210; 175.300, 176.170, 176.180, only as reactant in oil-based or fatty acid-based alkyd resins; Gardner 10 max. liq.; acid no. 60-66.

Sylvatac® 40N. [Arizona] Rosin ester; tackifier for adhesives; FDA 21CFR §175.105, 175.300, 176.170, 176.180, 176.210; Gardner 7+ max. color; soften. pt. (R&B) 42 C; acid no. 3-11.

Sylvatac® 80N. [Arizona] Rosin ester; tackifier for adhesives; FDA 21CFR §175.105, 175.300, 176.170, 176.180, 176.210; Gardner 7 max. color; soften. pt. (R&B) 81 C; acid no. 3-11.

Sylvatac® 95. [Arizona] Polymerized rosin; tackifier for adhesives; FDA 21CFR §175.105, 175.300, 176.170, 176.180, 176.210; Gardner 8 color; soften. pt. (R&B) 96 C; acid no. 149-157.

Sylvatac® 100NS. [Arizona] Rosin ester; tackifier for adhesives; FDA 21CFR §175.105, 175.300, 176.170, 176.180, 176.210; Gardner 9 max. color; soften. pt. (R&B) 100 C; acid no. 5-13.

Sylvatac® 105NS. [Arizona] Rosin ester; tackifier for adhesives; FDA 21CFR §175.105, 175.300, 176.170, 176.180, 176.210; Gardner 8 max. color; soften. pt. (R&B) 105 C; acid no. 5-13.

Sylvatac® 140. [Arizona] Polymerized rosin; tackifier for pressure-sensitive adhesives, construction adhesives, pick-up gums for labeling, adhesion promoter for difficult-to-bond substrates; FDA 21CFR §175.105, 175.300, 176.170, 176.180, 176.210; Gardner 11 color; soften. pt. (R&B) 139 C; acid no. 133-147.

Sylvatac® 295. [Arizona] Polymerized rosin; tackifier for pressure-sensitive adhesives, construction adhesives, pick-up gums for labeling, adhesion promoter for difficult-to-bond substrates; FDA 21CFR §175.105, 175.300, 176.170, 176.180, 176.210; Gardner 7 color; visc. 9000-13,000 cps (250 F); soften. pt. (R&B) 97 C; acid no. 158-165.

Sylvatac® 1085. [Arizona] Rosin ester; tackifier for adhesives; FDA 21CFR §175.105, 175.300, 176.170, 176.180, 178.3870; Gardner 5 max. color; soften. pt. (R&B) 85 C; acid no. 3-11.

Sylvatac® 1100. [Arizona] Rosin ester; tackifier for adhesives; FDA 21CFR §175.105, 175.300, 176.170, 176.180, 176.210; Gardner 5 max. color; soften. pt. (R&B) 100 C; acid no. 5-13.

Sylvatac® 1103. [Arizona] Rosin ester; tackifier for adhesives; FDA 21CFR §175.105, 175.300, 176.170, 176.180, 176.210; Gardner 5 max. color; soften. pt. (R&B) 102 C; acid no. 5-13.

Sylvatac® 2100. [Arizona] Rosin ester; tackifier for adhesives; FDA 21CFR §175.105, 175.300, 176.170, 176.180, 176.210; Gardner 5 max. color; soften. pt. (R&B) 97 C; acid no. 10 max.

Sylvatac® 2103. [Arizona] Rosin ester; tackifier for adhesives; FDA 21CFR §175.105, 175.300, 176.170, 176.180, 176.210; Gardner 5 max. color; soften. pt. (R&B) 102 C; acid no. 13 max.

Sylvatac® 2110. [Arizona] Rosin ester; tackifier for adhesives; FDA 21CFR §175.105, 175.300, 176.170, 176.180, 176.210; Gardner 5 max. color; soften. pt. (R&B) 108 C; acid no. 9 max.

Sylvatac® AC. [Arizona] Modified tall oil rosin; for mfg. of paper size, intermediate in rosin deriv. prod., printing ink binders as resins, tackifier resin in sealants and mastics, as starting point rosin for resin esters; Gardner 6 color; sp.gr. 1.03; soften. pt. (R&B) 85-92 C; acid no. 194; flash pt. (OC) > 177 C.

Sylvatac® ACF. [Arizona] Modified tall oil rosin; used in mfg. of paper size, intermediate in rosin deriv. prod., printing ink binders, tackifier resin in sealants and mastics, starting point rosin for resin esters; Gardner 9+ color; sp.gr. 1.03; soften. pt. (R&B) 75-82 C; acid no. 164; flash pt. (OC) > 177 C.

Sylvatac® R85. [Arizona] Polymerized rosin; tackifier for pressure-sensitive adhesives, construction adhesives, pick-up gums for labeling, adhesion promoter for difficult-to-bond substrates; Gardner 6 color; visc. 9440 cps (225 F); soften. pt. (R&B) 84 C; acid no. 155-162.

Sylvatac® RX. [Arizona] Polymerized rosin; tackifier for pressure-sensitive adhesives, construction adhesives, pick-up gums for labeling, adhesion promoter for difficult-to-bond substrates; Gardner 7-color; soften. pt. (R&B) 74 C; visc. 5150 cps (225 F); acid no. 141-149.

Syncrolube 2528. [Croda Chem. Ltd.] Complex ester; external lubricant for PVC and PC; FDA accepted; solid.

Syncrolube 3532. [Croda Chem. Ltd.] Alcohol ester; internal/external lubricant and processing aid for rigid and flexible PVC; improves melt flow, imparts high gloss; liq.

Syncrolube 3780. [Croda Chem. Ltd.] Alcohol ester; internal/external lubricant and processing aid for polymers esp. general-purpose rigid PVC; improves inj. melt flow; FDA 21CFR §178.3450; solid.

Syncrolube 3795. [Croda Chem. Ltd.] Alcohol ester; internal lubricant and processing aid for rigid and flexible PVC; improves melt flow; solid.

Syn Fac® 8026. [Milliken] Aromatic polyether polyol; intermediate in epoxy and polyester resins, as urethane crosslinker; liq.; 100% act.

Syn Fac® 8031. [Milliken] Aromatic polyether polyol; nonionic; reactive diluent, dispersant for solv. coatings and other applics.; liq.; 100% act.

Synox 5LT. [Great Lakes] 2,2′-Methylene bis (4-methyl-6-t-butylphenol); CAS 119-47-1; nondiscoloring, nonstaining, nonblooming antioxidant for natural, S/B, BR, CR, and polyisoprene rubbers, cellulosics, ABS, nylon, polyester, PE, PP, PS, PVC; FDA 21CFR §175.105, 177.206; sol. in alcohols, aromatics, esters, ketones.

Synperonic OP16. [ICI Am.] Octoxynol-16; CAS 9002-93-1; nonionic; emulsifier for emulsion polymerization, cosmetic creams, lotions; solubilizer for topical pharmaceuticals and shampoos; liq.; HLB 15.6; 70% conc.; Unverified

Synperonic OP30. [ICI Am.; ICI plc] Octoxynol-30; CAS 9002-93-1; nonionic; surfactant for emulsion polymerization; emulsifier for creams, lotions; solubilizer for topical pharmaceuticals and shampoos; pale yel. solid; HLB 17.2; 100% conc.

Synperonic PE30/80. [ICI plc] PO/EO block copolymer; nonionic; emulsifier, dispersant, stabilizer; emulsion polymerization; used in pigments in emulsion paints, inks; for household, industrial, and agrochem. applics.; flake; m.w. 8500; m.p. 50 C; HLB 29; cloud pt. > 100 C (1% aq.); surf. tens. 48 dynes/cm; pH 6-8 (2.5%); 99% min. act.; Unverified

Synpro® Aluminum Stearate 303. [Syn. Prods.] Aluminum stearate; CAS 7047-84-9; EINECS 230-325-5; internal/external lubricant for nylon, rigid PVC, polyesters; FDA accepted; solid.

Synpro® Barium Stearate. [Syn. Prods.] Barium stearate; CAS 6865-35-6; process aid, internal/external lubricant for PVC records and rigid foam, polyolefins, PS, ABS; solid.

Synpro® Cadmium Stearate. [Syn. Prods.] Cadmium stearate; CAS 2223-93-0; process aid, internal stabilizer/lubricant for PVC, polyolefins, PS, ABS; solid.

Synpro® Calcium Pelargonate. [Syn. Prods.] Calcium pelargonate; process aid, lubricant for polyolefins; 99% thru 200 mesh; soften. pt. 195 C; 3.0% moisture.

Synpro® Calcium Stearate 12B. [Syn. Prods.] Calcium stearate; CAS 1592-23-0; EINECS 216-472-8; process aid, lubricant for PVC; 95% thru 200 mesh; soften. pt. 155 C; 3.0% moisture.

Synpro® Calcium Stearate 15 [Syn. Prods.] Calcium stearate; CAS 1592-23-0; EINECS 216-472-8; process aid, lubricant for PVC, phenolics, polyesters; in color concs., foundry, paint flattening, industrial cleaners, wire drawing compds., masonry/cement, and waterproofing applics.; fineness 99% thru 200 mesh; soften. pt. 155 C; 3.0% moisture.

Synpro® Calcium Stearate 15F. [Syn. Prods.] Calcium stearate; CAS 1592-23-0; EINECS 216-472-8; process aid, lubricant for PVC; fineness 60% thru 40 mesh; soften. pt. 155 C; 3.0% moisture.

Synpro® Calcium Stearate 114-36. [Syn. Prods.] Calcium stearate; CAS 1592-23-0; EINECS 216-472-8; process aid, lubricant for polyolefins; 95% thru 200 mesh; soften. pt. 155 C; 3.0% moisture.

Synprolam 35. [ICI plc] C13-15 alkyl primary amine; CAS 68155-27-1; cationic; anticaking agent for fertilizer; flotation agent, corrosion inhibitor; metal cleaning formulations, pigment dispersions, rubber processing auxs., bitumen emulsifier; intermediate for prod. of amine salts, ethoxylates and sulfosuccinates; Hazen 50 clear liq.; sp.gr. 0.8; visc. 4.5 cps; flash pt. 127 C (PMCC); 99% primary amine.

Synprolam 35DM. [ICI plc] Syn. C13-15 dimethyl tert. fatty amine; CAS 68391-04-8; cationic; chemical intermediate for prod. of quat. salts and amine oxides used in cosmetics; catalyst for PU foam; corrosion inhibitor; Hazen 50 clear liq.; sp.gr. 0.79; visc. 3.9 cps; flash pt. 143 C (PMCC); 97% tert. amine.

Synprolam 35 Lauramide. [ICI plc] Difatty amides derived from Synprolam 35; nonionic; low melting lubricant for plastics industry; water repellent; flakes; 100% conc.; Unverified

Synprolam 35MX1. [ICI plc] PEG-1 C13-15 alkyl methyl amine; CAS 92112-62-4; cationic; emulsifier, textile processing aid, wetting agent; antistat for polyolefin and PVC; liq.; 100% conc.

Synprolam 35MX3. [ICI plc] PEG-3 C13-15 alkyl methyl amine; cationic; antistat for polyolefin and PVC; liq.; 100% conc.

Synprolam 35MX5. [ICI plc] PEG-5 C13-15 alkyl methyl amines; cationic; antistat for polyolefin and PVC; liq.; 100% conc.; Unverified

Synprolam 35 Stearamide. [ICI plc] Difatty amides derived from Synprolam 35; nonionic; low melting lubricant for plastics industry; water repellent; flakes; 100% conc.; Unverified

Synprolam 35TMQS. [ICI plc] C13-15 alkyl trimethyl ammonium methosulfate in 1,4-butanediol; plastics antistat; esp. for PU compositions; liq.; 80% act.; Unverified

Synprolam 35X2. [ICI plc] PEG-2 C13-15 alkyl amine; cationic; emulsifier, PU catalyst, textile and plastic processing aid, textile dyeing aux., antistat; liq.; 100% conc.

Synprolam 35X2QS, 35X5QS, 35X10QS. [ICI plc] PEG-2, -5, and -10 C13-15 alkyl methyl ammonium methosulfate; cationic; antistat for PVC and PU; hydrotrope, wetting agent, conditioner, textile processing aid; alkali-stable; liq.; 90-100% conc.

Synprolam 35X5. [ICI plc] PEG-5 C13-15 alkyl amine; cationic; emulsifier, PU catalyst, textile and plastic processing aid, textile dyeing aux., antistat; 100% conc.

Synprolam 35X10. [ICI plc] PEG-10 C13-15 alkyl amine; cationic; emulsifier, PU catalyst, textile and plastic processing aid, textile dyeing aux., antistat; 100% conc.

Synprolam 35X15. [ICI plc] PEG-15 C13-15 alkyl amine; cationic; emulsifier, PU catalyst, textile and plastic processing aid, textile dyeing aux., antistat; 100% conc.

Synprolam 35X20. [ICI plc] PEG-20 C13-15 alkyl amine; cationic; emulsifier, PU catalyst, textile and plastic processing aid, textile dyeing aux., antistat; 100% conc.

Synprolam 35X25. [ICI plc] PEG-25 C13-15 alkyl amine; cationic; emulsifier, PU catalyst, textile and plastic processing aid, textile dyeing aux., antistat; 100% conc.

Synprolam 35X35. [ICI plc] PEG-35 C13-15 alkyl amine; cationic; emulsifier, PU catalyst, textile and plastic processing aid, textile dyeing aux., antistat; 100% conc.

Synprolam 35X50. [ICI plc] PEG-50 C13-15 alkyl amine; cationic; emulsifier, PU catalyst, textile and plastic processing aid, textile dyeing auxs., antistat; 100% conc.

Synpro® Lead Stearate. [Syn. Prods.] Lead stearate; CAS 1072-35-1; internal/external lubricant and aux. heat stabilizer for PVC; solid.

Synpro® Lithium Stearate. [Syn. Prods.] Lithium stearate; CAS 4485-12-5; EINECS 224-772-5; lubricant for nylon; solid.

Synpro® Magnesium Stearate 90. [Syn. Prods.] Magnesium stearate; CAS 557-04-0; EINECS 209-150-3; processing aid, lubricant for ABS; 99% thru 200 mesh; soften. pt. 147 C; 3.0% moisture.

Synpron 231. [Syn. Prods.] Zinc phosphite; PVC stabilizer for applics. requiring exc. clarity; FDA accepted; liq.

Synpron 239. [Syn. Prods.] Sn/Ca/Zn; patented general purpose PVC stabilizer providing outstanding clarity, exc. early and intermediate color hold; FDA accepted; powd.

Synpron 241. [Syn. Prods.] Phosphite chelator; aux. PVC heat stabilizer providing crisp initial color and long term stability to rigid and plasticized nontoxic formulations; FDA accepted; liq.

Synpron 350. [Syn. Prods.] Barium-zinc; stabilizer for PVC wire and cable jacketing and insulating compds.; replacement for lead-based stabilizers; gives exc. color, long term heat stability, good elec. props.; wh. powd.; sp.gr. 1.5.

Synpron 351. [Syn. Prods.] Noncadmium; stabilizer for flexible PVC applics. for calendered, extruded, and inj. molded compds.; booster stabilizer for semirigid and flexible compds. where the primary stabilization is obtained by cadmium-free liqs.; off-wh. powd.; 99% thru 100 mesh; sp.gr. 1.14; apparent dens. 0.3 g/cc.

Synpron 352. [Syn. Prods.] Lead-free; non heavy metal stabilizer for extrusion of wire jacketing; reduces auto fogging and maintains color specs after aging; exc. compat.; off-wh. fine powd., odorless; < 2% on 200 mesh; sp.gr. 1.52.

Synpron 353. [Syn. Prods.] Lead-free; non heavy metal stabilizer for extrusion of 60-80 C wire insulation; exc. compat.; off-wh. fine powd., odorless; < 2% on 200 mesh; sp.gr. 1.54.

Synpron 354. [Syn. Prods.] Lead-free; non heavy metal stabilizer for extrusion of 90-105 C wire insulation; exc. compat.; off-wh. fine powd., odorless; < 2% on 200 mesh; sp.gr. 1.50.

Synpron 355. [Syn. Prods.] Lead-free; stabilizer for extrusion of 90-105 C wire insulation; exc. compat.; off-wh. fine powd., odorless; < 2% on 200 mesh; sp.gr. 1.59.

Synpron 357. [Syn. Prods.] Barium/cadmium; PVC stabilizer for calendered or extruded clear applics.; liq.

Synpron 431. [Syn. Prods.] Phosphite chelator; aux. PVC heat and lt. stabilizer for automotive interior materials requiring low fog props.; low volatility; liq.

Synpron 687. [Syn. Prods.] Nonmetallic; aux. heat and lt. stabilizer enhancing long term processing stability and improving lt. stability; recommended for flexible vinyl/CPE blends; powd.

Synpron 755. [Syn. Prods.] Calcium/zinc; stabilizer for plastisol applics. requiring long term, low temp. heat aging performance, e.g., for exterior applics.; liq.

Synpron 940. [Syn. Prods.] High barium/zinc; general purpose plastisol stabilizer providing exc. high heat stability and clarity; liq.

Synpron 1002. [Syn. Prods.] Modified butyltin contng. sulfur; PVC heat stabilizer providing long term stability in bottle compds., PVC pipe, and inj. molded fittings; liq.

Synpron 1004. [Syn. Prods.] Non-sulfur-contg. organotin; PVC heat stabilizer for flooring and other applics. requiring exc. color and good lt. stability; liq.

Synpron 1007. [Syn. Prods.] Mixed metal/sulfur barium tin; patented PVC heat stabilizer for twin and multiscrew extrusions of PVC pipe and conduit, in base stock of capped PVC siding; NSF accepted for potable water pipe; dk. amber liq., char. odor.

Synpron 1009. [Syn. Prods.] Dibutyltin dilaurate; CAS 77-58-7; EINECS 201-039-8; stabilizer providing exc. uv stability in calendered and extrusion applics.; also as catalyst in urethane foam and RIM applics.; liq.

Synpron 1011. [Syn. Prods.] Sulfur-contg. organotin; PVC heat stabilizer for bottle and sheet compds. due to exc. color and process heat stability; liq.

Synpron 1027. [Syn. Prods.] Antimony mercaptide; patented PVC heat stabilizer esp. for pipe and conduit applics.; recommended for use with calcium stearate for optimum stabilization; may be used in NSF potable water pipe at 1.0 phr max.; clear straw liq., mild mercaptide odor; sp.gr. 1.2; Gardner visc. B.

Synpron 1034. [Syn. Prods.] Antimony mercaptide; patented PVC heat stabilizer esp. for pipe and conduit applics.; recommended for use with calcium stearate for optimum stabilization; may be used in NSF potable water pipe at 1.0 phr max.; clear liq., mild mercaptide odor; Gardner visc. B.

Synpron 1048. [Syn. Prods.] Antimony mercaptide; PVC heat stabilizer for calendered or extruded rigid sheeting such as for credit card stock or building panels; also for inj. molding of fittings, in the base for capstock of siding compds.; clear liq., mild mercaptide odor.

Synpron 1108. [Syn. Prods.] Barium/cadmium/high zinc; stabilizer for plastisol or heavily filled calendering applics.; liq.

Synpron 1112. [Syn. Prods.] Zinc; stabilizer/activator for PVC plastisol foams; aux. stabilizer for traditional Ba/Cd systems; liq.

Synpron 1135. [Syn. Prods.] Barium/cadmium/mod.-high zinc; economical general purpose plastisol stabilizer; liq.

Synpron 1236. [Syn. Prods.] Zinc; stabilizer/activator for PVC plastisol foams; aux. stabilizer for traditional Ba/Cd systems; liq.

Synpron 1321. [Syn. Prods.] Zinc/calcium; PVC stabilizer providing extended processing time, early color hold and clarity, minimal color change during processing; useful for blown film; FDA accepted; paste.

Synpron 1363. [Syn. Prods.] Barium/zinc; stabilizer for plastisol flooring applics. where max. clarity, processing, and aging chars. are required; liq.

Synpron 1384. [Syn. Prods.] Barium/cadmium/low zinc; PVC stabilizer for extruded clear applics.; liq.

Synpron 1410. [Syn. Prods.] Phosphite chelator; aux. PVC stabilizer for applics. where low phenolic chars. are desired for improved heat and lt. stability; liq.

Synpron 1419. [Syn. Prods.] Barium/cadmium/mod. zinc; economical PVC calendering stabilizer for clear or filled applics.; liq.

Synpron 1427. [Syn. Prods.] Barium/cadmium/mod. zinc; PVC stabilizer for calendered or extruded clear, filled, or pigmented applics.; liq.

Synpron 1428. [Syn. Prods.] Barium/cadmium/low zinc; economical PVC extrusion stabilizer for filled and pigmented applics.; liq.

Synpron 1456. [Syn. Prods.] Phosphite chelator; stabilizer, chelator for rigid and flexible PVC compds. that will be calendered, extruded, or inj. molded; enhances outdoor weathering of PVC; improved clarity and initial color; liq.

Synpron 1470. [Syn. Prods.] High cadmium/barium; lubricating booster stabilizer for calendering operations to give extended color hold with min. plate out during processing; powd.

Synpron 1522. [Syn. Prods.] Calcium/zinc; stabilizer for plastisol wear layer and top coatings in flooring; exc. moderate temp. (190-200 F) and lt. stability; liq.

Synpron 1528. [Syn. Prods.] Barium/cadmium/high zinc; stabilizer for plastisol when bright color hold is required; liq.

Synpron 1536. [Syn. Prods.] High barium/cadmium; lubricating stabilizer for wire and cable applics. where retention of physical props. under aging conditions is required; powd.

Synpron 1538. [Syn. Prods.] Zinc soap; PVC stabilizer for low odor and taste applics., e.g., refrigerator gasketing; FDA accepted; powd.

Synpron 1558. [Syn. Prods.] Barium/cadmium/high zinc; high efficiency PVC stabilizer for exterior applics.; liq.

Synpron 1566. [Syn. Prods.] Zinc/calcium; PVC stabilizer providing extended processing time, early color hold and clarity, minimal color change during processing; useful for blown film; FDA accepted; powd.

Synpron 1578. [Syn. Prods.] Barium/cadmium; PVC stabilizer for calendered or extruded clear applics. where exceptional color hold is required; liq.

Synpron 1585. [Syn. Prods.] High cadmium/barium; lubricating booster stabilizer for calendering operations to give extended color hold with min. plate out during processing; nondusting powd.

Synpron 1618. [Syn. Prods.] Barium/cadmium/zinc; high efficiency general purpose lubricating stabilizer for exterior applics.; liq.

Synpron 1642. [Syn. Prods.] Zinc-potassium; stabilizer/activator for PVC plastisol foams; aux. stabilizer for traditional Ba/Cd systems; liq.

Synpron 1652. [Syn. Prods.] Barium/zinc; stabilizer designed for vinyl chloride-vinyl acetate copolymer resin systems; general purpose record stabilizer; imparts lubrication and surf. bloom resist.; powd.

Synpron 1678. [Syn. Prods.] Barium/high cadmium/mod. zinc; high efficiency PVC stabilizer for extrusion or calendering; designed for exterior clear applics. requiring exc. UV resist.; liq.

Synpron 1693. [Syn. Prods.] Zinc/calcium; PVC stabilizer for food wrap and medical tubing; FDA accepted; paste.

Synpron 1699. [Syn. Prods.] Zinc/calcium; PVC stabilizer providing extending processing time with exc. early color hold; FDA accepted; powd.

Synpron 1749. [Syn. Prods.] Zinc/calcium/phosphite; PVC stabilizer for food wrap and medical tubing; FDA accepted; liq.

Synpron 1750. [Syn. Prods.] Zinc soap; PVC stabilizer; aids dispersibility; FDA accepted; paste.

Synpron 1774. [Syn. Prods.] Barium/cadmium/mod. zinc; high efficiency lubricating stabilizer for PVC calendering or extrusion; recommended for exterior applics.; liq.

Synpron 1800. [Syn. Prods.] Barium/cadmium; high efficiency PVC extrusion stabilizer for clear applics.; liq.

Synpron 1810. [Syn. Prods.] Barium/cadmium/mod. zinc; high efficiency PVC stabilizer for calendering or extrusion for clear or filled pigmented applics.; liq.

Synpron 1811, 1812. [Syn. Prods.] Barium/cadmium/ mod. zinc; high efficiency PVC stabilizer for extrusion; provides exc. plate out resist. and color hold; calendering stabilizer for filled and pigmented applics.; liq.

Synpron 1820. [Syn. Prods.] Barium/cadmium/high zinc; high efficiency PVC plastisol stabilizer providng exc. visc. control; liq.

Synpron 1831. [Syn. Prods.] Mod. barium/high zinc; plastisol stabilizer for exterior applics. where exceptional color hold during processing and through field exposure is required; liq.

Synpron 1840. [Syn. Prods.] Zinc; general purpose plastisol stabilizer providing exceptional color hold, clarity, and accelerated aging chars.; liq.

Synpron 1850. [Syn. Prods.] High barium/cadmium; lubricating booster stabilizer; extends processing time or as primary stabilizer for applics. requiring low smoking, low fogging, or low volatility; powd.

Synpron 1852. [Syn. Prods.] High barium/cadmium; lubricating booster stabilizer; extends processing time or as primary stabilizer for applics. requiring low smoking, low fogging, or low volatility; nondusting powd.

Synpron 1860. [Syn. Prods.] High cadmium/barium; lubricating booster stabilizer for calendering and extrusion where extended color hold during processing, accelerated aging, and UV exposure is required; powd.

Synpron 1861. [Syn. Prods.] High cadmium/barium; booster stabilizer formulated to give exc. color hold and compat. when used in printed or heat sealed applics.; powd.

Synpron 1870. [Syn. Prods.] Barium/cadmium/zinc; lubricating primary stabilizer for automotive applics.; provides low fogging while maintaining color hold during processing and aging; powd.

Synpron 1871. [Syn. Prods.] Barium/cadmium/zinc; lubricating primary stabilizer for automotive applics.; provides extended color hold during processing; powd.

Synpron 1881. [Syn. Prods.] Barium/zinc; lubricating stabilizer giving exc. color hold during processing and field exposure; recommended for pigmented exterior calendered or plastisol applics.; powd.

Synpron 1890. [Syn. Prods.] Zinc; lubricating stabilizer for calendering and extrusion where long term, low temp. heat stability is required; powd.

Synpron 1900. [Syn. Prods.] Calcium/zinc; PVC stabilizer providing exc. lubrication and processing time; FDA accepted; powd.

Synpron 1913. [Syn. Prods.] Calcium/zinc; PVC stabilizer providing exc. lubrication and processing time; FDA accepted; paste.

Synpro® Sodium Stearate ST. [Syn. Prods.] Sodium stearate; CAS 822-16-2; EINECS 212-490-5; internal/external lubricant for nylon, PP, PS; solid.

Synpro® Stannous Stearate. [Syn. Prods.] Stannous stearate; internal/external lubricant, antistat, stabilizer for plastics; FDA accepted; solid.

Synpro® Zinc Dispersion Zincloid. [Syn. Prods.] Zinc stearate; CAS 557-05-1; EINECS 209-151-9; slab dip for rubber; powd.; 99% thru 200 mesh; 80% moisture.

Synpro® Zinc Stearate 8. [Syn. Prods.] Zinc stearate; CAS 557-05-1; EINECS 209-151-9; process aid, lubricant for polyolefins, PS; also for color concs.; 99% thru 200 mesh; soften. pt. 120 C; 0.5% moisture.

Synpro® Zinc Stearate ACF. [Syn. Prods.] Zinc stearate; CAS 557-05-1; EINECS 209-151-9; dusting agent for rubbers; slab dip; fineness 99% thru 200 mesh; soften. pt. 120 C; 0.5% moisture.

Synpro® Zinc Stearate D. [Syn. Prods.] Zinc stearate; CAS 557-05-1; EINECS 209-151-9; process aid, lubricant for polyesters; 99% thru 325 mesh; soften. pt. 120 C; 0.5% moisture.

Synpro® Zinc Stearate GP. [Syn. Prods.] Zinc stearate; CAS 557-05-1; EINECS 209-151-9; process aid, lubricant for PVC, melamine, rubber compounding; 99% thru 200 mesh; soften. pt. 120 C; 0.5% moisture.

Synpro® Zinc Stearate GP Flake. [Syn. Prods.] Zinc stearate; CAS 557-05-1; EINECS 209-151-9; process aid, lubricant for PVC, melamine, rubber compounding; flake; soften. pt. 120 C; 0.5% moisture.

Synpro® Zinc Stearate HSF. [Syn. Prods.] Zinc stearate; CAS 557-05-1; EINECS 209-151-9; process aid, lubricant for polyolefins, PS; flake; soften. pt. 120 C; 0.5% moisture.

Syntase® 62. [Great Lakes] Benzophenone-3; CAS 131-57-7; EINECS 205-031-5; uv absorber for OTC sunscreen lotions and oil-based cosmetics and for protection of PS, PVC, methacrylate polymers, and polyesters against uv degradation over prolonged exposure; useful in protecting clear varnishes and lacquers, linseed oil-based alkyds and phenolic coatings intended for use on uv-sensitive surfaces; FDA 21CFR §177.1010; pale cream powd.; sol. 75% in ethyl acetate, 68% in toluene, 65% in acetone, 60% in MEK; insol. in water; m.w. 228; sp.gr. 1.339; m.p. 62 C min.; pH 7.2 (10% aq. slurry); 99.5% min. purity.

Syntex® 3981. [Shell] accelerator for use with urea formaldehyde or melamine thermosetting resin systems; provides good package stability, reduction of bake times; Unverified

Synthetic Carbon 910-44M. [D.J. Enterprises] Derived from high purity metallurgical carbon; CAS 7440-44-0; lower-cost replacement for carbon black. petrol.-based prods in rubber compds., coatings, and plastic composites; uniform dispersion; < 44 μ particle size; sp.gr. 1.9-2.2; 99% fixed C, 1% S, < 1% moisture.

SY-SA 69. [Additive Polymers Ltd] Slip/antiblock blend in K-Resin®-based masterbatch; slip/antiblock for plastics.

SY-SL 131. [Additive Polymers Ltd] Amide slip additive in K-Resin®-based masterbatch; slip/mold release agent for plastics.

SY-SY 132. [Additive Polymers Ltd] Stearate system in K-Resin®-based masterbatch; lubricant for plastics.

SY-UV 34. [Additive Polymers Ltd] UV absorber system in K-Resin®-based masterbatch; uv absorber for plastics.

T

T-600. [Olin] PPG triol; plasticizer, chemical intermediate; used in brake fluids and in the mfg. of resins; produces rigid and flexible urethane foams, and nonfoam urethane coatings, adhesives, elastomers, and sealants and caulks; m.w. 600; f.p. -29 C; sp.gr. 1.037; dens. 8.64 lb/gal; Unverified

Tactix H41. [Dow Plastics] Aromatic diamine blend; epoxy curing agent, hardener; sp.gr. 1.009; visc. 80-120 cps; flash pt. 250 F; 16-17% N.

Tafigel PUR 40. [Münzing Chemie GmbH] Polyurethane thickener for gloss emulsion paints, wood preservative stains, adhesives.

TAHP-80. [Witco/PAG] 80% t-Amyl hydroperoxide sol'n. in t-amyl alcohol/water; initiator; water-wh. clear liq., sharp odor; sol. < 5% in water; m.w. 104.15; sp.gr. 0.904 (22 C); b.p. 26 C (3 mm); flash pt. (Seta CC) 46 C; 80% conc.; 12.3% act. oxygen; *Toxicology:* severe eye and skin irritant; may cause blindness; harmful or fatal if swallowed; *Precaution:* incompat. with strong acids, bases, promoters, accelerators, readily oxidizables, metal salts; hazardous decomp. prods.: CO, CO_2 from burning; once ignited, will burn vigorously; potential explosion hazard with promoters; *Storage:* store in closed original containers below 38 C, away from acids, heat, flame, direct sunlight.

Taipol BR 015H. [Goldsmith & Eggleton] Sol'n. polybutadiene, cobalt Ziegler catalyst, nonstaining; CAS 9003-17-2; imparts exc. gloss and low temp. props.; modifier for HIPS and other plastics; sp.gr. 0.91; visc. 45-75 cP (5% styrene sol'n.); Mooney visc. 40 (ML4, 100 C); 96% cis; Unverified

Taktene 1202. [Bayer/Fibers, Org., Rubbers] Sol'n. polybutadiene, nonstaining; CAS 9003-17-2; used in passenger car and truck tires for improved treadwear in blends with SBR/NR; exc. low temp. props.; also for golf centers, footwear, belting, hose, floor tile, molded and extruded goods; mill processable; vulcanized with sulfur-accelerator systems or with peroxide for max. resilience; PS impact modifier; FDA compliance; lt. colored bales; sp.gr. 0.92; visc. 55-80 mPa•s (5% in styrene); Mooney visc. 37 (ML1+4, 100 C); very high cis content.

Taktene 1203. [Bayer/Fibers, Org., Rubbers] Sol'n. polybutadiene, nonstaining; CAS 9003-17-2; impact modifier for PS and ABS produced by in situ polymerization; FDA compliance; sp.gr. 0.91; Mooney visc. 42 (ML1+4, 100 C); very high cis content.

Taktene 1203G1. [Bayer/Fibers, Org., Rubbers] Sol'n. polybutadiene, nonstaining; CAS 9003-17-2; impact modifier for PS and ABS produced by in situ polymerization; FDA compliance; sp.gr. 0.91; Mooney visc. 40.5 (ML1+4, 100 C); very high cis content.

Tamol® NH 9103. [BASF AG] Sodium naphthalene sulfonate; dispersant for pigments; stabilizer for dispersions and emulsions; plasticizer for concrete, mortar, and gypsum; powd.; water-sol.; 90% conc.; Unverified

Tamol® NMC 9301. [BASF AG] Calcium naphthalene sulfonate; dispersant for pigments; stabilizer for dispersions and emulsions; plasticizer for concrete, mortar, and gypsum; powd.; water-sol.; 95% conc.; Unverified

Tamol® NN 4501. [BASF AG] Sodium naphthalene sulfonate; dispersant for pigments, carbon blk., dyestuffs; stabilizer for dispersions and emulsions; dispersant for rubber syntheses, processing of rubber latexes; liq.; water-misc.; 45% conc.

Tamsil 8. [Unimin Spec. Minerals] Microcryst. silica; filler for coatings (primers, metal finishes, enamels, protective, maintenance), plastic film antiblocks, urethane rubber, polishes; wh. fine powd.; 100% thru 325 mesh; 2.0 μ median particle size; fineness (Hegman) 6.75; sp.gr. 2.65; dens. 22.07 lb/gal; surf. area 2.1 m^2/g; oil absorp. 28; brightness 87.5; m.p. 1610 C; ref. index 1.54-1.55; pH 6.3; hardness (Mohs) 6.5; 98.81% SiO_2; *Toxicology:* prolonged inh. of dust can cause delayed lung injury (silicosis); may produce cancer.

Tamsil 10. [Unimin Spec. Minerals] Microcryst. silica; filler for coatings (metal finishes, primers, interior/exterior architectural, marine, protective, maintenance), plastic film antiblocks, silicone rubber; wh. fine powd.; 100% thru 325 mesh; 2.1 μ median particle size; sp.gr. 2.65; dens. 22.07 lb/gal; surf. area 2.0 m^2/g; oil absorp. 28; brightness 86.4; m.p. 1610 C; ref. index 1.54-1.55; pH 6.3; hardness (Mohs) 6.5; 98.9% SiO_2; *Toxicology:* prolonged inh. of dust can cause delayed lung injury (silicosis); may produce cancer.

Tamsil 15. [Unimin Spec. Minerals] Microcryst. silica; filler for coatings (interior/exterior architectural, protective, maintenance, marine), silicone and urethane rubbers; wh. fine powd.; trace thru 325 mesh; 2.5 μ median particle size; sp.gr. 2.65; dens. 22.07 lb/gal; surf. area 1.6 m^2/g; oil absorp. 29.9; brightness 86; m.p. 1610 C; ref. index 1.54-1.55; pH 6.3; hardness (Mohs) 6.5; 98.92% SiO_2; *Toxicology:* prolonged inh. of dust can cause delayed lung injury (silicosis); may produce cancer.

Tamsil 25. [Unimin Spec. Minerals] Microcryst. silica; filler for coatings (interior/exterior architectural, protective, maintenance, marine), silicone rubber molding compds.; wh. fine powd.; trace thru 325 mesh; 3.5 μ median particle size; sp.gr. 2.65; dens. 22.07 lb/gal; surf. area 1.5 m^2/g; oil absorp. 28.5; brightness 85.6; m.p. 1610 C; ref. index 1.54-1.55; pH 5.9; hardness (Mohs) 6.5; 99% SiO_2; *Toxicology:* prolonged inh. of dust can cause delayed lung injury (silicosis); may produce cancer.

Tamsil 30. [Unimin Spec. Minerals] Microcryst. silica; filler for coatings (interior/exterior architectural, protective, maintenance, marine), urethane rubber, buffing compds.; wh. coarse powd.; 99.6% thru 325 mesh; 5.8 μ median particle size; sp.gr. 2.65; dens. 22.07 lb/gal; surf. area 1.0 m^2/g; oil absorp. 29; brightness 85.2; m.p. 1610 C; ref. index 1.54-1.55; pH 6.2; hardness (Mohs) 6.5; 99.12% SiO_2; *Toxicology:* prolonged inh. of dust can cause delayed lung injury (silicosis); may produce cancer.

Tamsil Gold Bond. [Unimin Spec. Minerals] Microcryst. silica; filler for traffic paint, interior/exterior architectural coatings, protective, maintenance and marine coatings, mastics and adhesives, elec. epoxy compds., buffing compds.; wh. coarse powd.; 98.6% thru 325 mesh; 8.4 μ median particle size; sp.gr. 2.65; dens. 22.07 lb/gal; surf. area 1.1 m^2/g; oil absorp. 29; brightness 85.2; m.p. 1610 C; ref. index 1.54-1.55; pH 6.1; hardness (Mohs) 6.5; 99.075% SiO_2; *Toxicology:* prolonged inh. of dust can cause delayed lung injury (silicosis); may produce cancer.

Tartac 20. [Sovereign] Pine tar; plasticizer, tackifier, and process aid for rubber; replacement for oil; retards cure due to high acidity; low-visc. liq.

Tartac 30. [Sovereign] Pine tar; plasticizer, tackifier, and process aid for rubber; replacement for oil; retards cure due to high acidity; med.-visc. liq.

Tartac 40. [Sovereign] Pine tar; plasticizer, tackifier, and process aid for rubber; replacement for oil; retards cure due to high acidity; high-visc. liq.

TBEP. [Rhone-Poulenc Surf. & Spec.] Tributoxyethylphosphate; CAS 78-51-3; EINECS 201-122-9; nonfoaming emulsifier, wetting agent, plasticizer, dispersant, leveling agent for acrylic floor finishes; lowers visc. of PVC plastisols; plasticizer for acrylonitrile and other rubbers; acid and electrolyte stable; liq.; 100% act.

TBHP-70. [Witco/PAG] t-Butyl hydroperoxide sol'n. with di-t-butyl peroxide and sm. amts. of t-butyl alcohol and water as diluents; initiator; colorless liq.; 70% act.; 12.4% act. oxygen.

TBP. [FMC] Tributyl phosphate; CAS 126-73-8; EINECS 204-800-2; antifoam agent used in paper coating compds., water adhesives, inks, casein sol'ns., textile sizes, and detergent sol'ns.; plasticizer and solv. for NC, cellulose acetate, chlorinated rubber, and vinyls; solv. for natural gums and syn. resins; component in extraction of rare earth metal salts; Pt-Co 50 max. liq.; char. odor; m.w. 266; f.p. < -80 C; b.p. 137-145 C; sol. in org. liqs., min. oil, and gasoline; sp.gr. 0.978 ± 0.003; dens. 975 kg/m^3; visc. 3.7 cp; pour pt. < -80 C; ref. index 1.423 ± 0.001; flash pt. 166 C; fire pt. 182 C; Unverified

T-Det® BP-1. [Harcros] PO/EO block copolymer; nonionic; emulsifier, detergent, dispersant and polymerization agent; pesticide formulations; FDA compliance; wh. solid; water-sol.; sp.gr. 1.04 (105 F); dens. 8.7 lb/gal (105 F); HLB 15.7; pour pt. 85 F; cloud pt. 154 F (1%); flash pt. (PMCC) > 400 F; pH 4-7 (1% aq.); 99.5% act.

T-Det® C-40. [Harcros] PEG-40 castor oil; CAS 61791-12-6; nonionic; emulsifier, solubilizer, degreaser, lubricant, dispersant, penetrant used in leather, paper, textile and metal processing, rubber, paint; dispersant for pigment slurries; leveling agent, defoamer, stabilizer; wax and polish preparations; FDA compliance; yel. liq.; mild, oily odor; sol. in acetone, butyl Cellosolve, CCl_4, ethanol, ether, ethylene glycol, methanol, veg. oil, water, xylene; sp.gr. 1.05; dens. 8.7 lb/gal; HLB 14.2; pour pt. 60 F; flash pt. (PMCC) > 200 F; pH 6.0 (1%); 99.5% min. act.

T-Det® N-20. [Harcros] Nonoxynol-20; CAS 9016-45-9; nonionic; surfactant for high temp. detergents, emulsion polymerization, petrol. processing; emulsifier; FDA compliance; opaque solid; water-sol.; sp.gr. 1.08 (105 F); dens. 9.0 lb/gal (105 F); HLB 16.0; pour pt. 90 F; flash pt. (PMCC) > 500 F; cloud pt. > 212 F (1%); pH 5-7 (1% aq.); 99.5% min. act.

T-Det® N-40. [Harcros] Nonoxynol-40; CAS 9016-45-9; nonionic; emulsifier for emulsion polymerization, asphalt, elevated temp. applics.; FDA compliance; pale yel. waxy solid; mild aromatic odor; sol. in water; sp.gr. 1.07 (132 F); HLB 17.7; cloud pt. 212 F (1%); flash pt. > 500 F; pour pt. 120 F; pH 6.5-8.0 (1%); 99.5% min. act.

T-Det® N-50. [Harcros] Nonoxynol-50; CAS 9016-45-9; nonionic; detergent, emulsifier for emulsion polymerization, asphalt, textiles, agric., elevated temp. applics.; FDA compliance; wh. solid; water-sol.; sp.gr. 1.07 (132 F); dens. 8.9 lb/gal (132 F); HLB 18.0; pour pt. 120 F; cloud pt. > 212 F (1%); flash pt. (PMCC) > 500 F; pH 5-7 (1% aq.); 99.5% min. act.

T-Det® N-70. [Harcros] Nonoxynol-70; CAS 9016-45-9; nonionic; high temp. detergent, emulsifier for oils, fats and waxes; stabilizer in latex and alkyd emulsions; emulsion polymerization; asphalt; scouring textile operations; FDA compliance; pale yel. waxy solid; mild aromatic odor; dens. 9.0 lb/gal (135 F); sp.gr. 1.08 (135 F); HLB 18.7; cloud pt. 212 F (1%); flash pt. > 500 F; pour pt. 125 F; pH 6.0-7.0 (1%); 99.5% min. act.

T-Det® N-100. [Harcros] Nonoxynol-100; CAS 9016-45-9; nonionic; emulsifier for asphalt, emulsion polymerization, textiles, toilet block prods.; high temp. and high electrolyte applics.; FDA compliance; wh. waxy solid; mild aromatic solid; sol. in ethylene dichloride, ethanol and water; sp.gr. 1.08 (135 F); dens. 9.0 lb/gal (132 F); HLB 19.0; cloud pt. 212 F (1%); flash pt. (PMCC) > 500 F; pour pt. 127 F; pH 6-7 (1%); 99.5% min. act.

T-Det® N-307. [Harcros] Nonoxynol-30; CAS 9016-45-9; nonionic; high temp. detergent, emulsifier for emulsion polymerization, asphalt, textiles; stabilizer; FDA compliance; clear liq.; water-sol.; sp.gr. 1.10; dens. 9.2 lb/gal; visc. 806 cps; HLB 17.0; cloud pt. > 100 C (1%); flash pt. > 212 F (TOC); pour pt. 30 F; 70 ± 0.5% act.

T-Det® N-407. [Harcros] Nonoxynol-40; CAS 9016-45-9; nonionic; emulsifier for emulsion polymerization, asphalt, paint systems, elevated temp. applics.; FDA compliance; clear liq.; dens. 9.0 lb/gal; sp.gr. 1.08; visc. 900 cps; HLB 17.7; cloud pt. > 212 F (1%); flash pt. > 212 F (TOC); pour pt. 30 F; 70 ± 0.5% act.

T-Det® N-507. [Harcros] Nonoxynol-50; CAS 9016-45-9; nonionic; detergent, emulsifier for emulsion polymerization, asphalt, paint systems, textiles, agric., elevated temp. applics.; FDA compliance; APHA 100 opaque liq.; sp.gr. 1.09; dens. 9.1 lb/gal; visc. 760 cps; HLB 18.0; cloud pt. > 100 C (1%); flash pt. > 212 F (TOC); pour pt. 40 F; 70 ± 0.5% act.

T-Det® N-705. [Harcros] Nonoxynol-70; CAS 9016-45-9; nonionic; high temp. detergent, emulsifier for oils, fats and waxes; stabilizer in latex and alkyd emulsions; emulsion polymerization; asphalt; scouring textile operations; clear to hazy liq.; mild aromatic odor; dens. 9.0 lb/gal; sp.gr. 1.08; cloud pt. 212 F (1%); flash pt. > 500 F; pour pt. 40 F; pH 5.0-6.5 (1%); 50% min. act.

T-Det® N-707. [Harcros] Nonoxynol-70; CAS 9016-45-9; nonionic; detergent, emulsifier for oils, fats and waxes; stabilizer in latex and alkyd emulsions; emulsion polymerization; scouring textile operations; FDA compliance; clear to hazy liq.; mild aromatic odor; dens. 9.0 lb/gal; sp.gr. 1.08; HLB 18.7; cloud pt. 212 F (1%); flash pt. > 500 F; pour pt. 40 F; pH 5.0-6.5 (1%); 50% min. act.

T-Det® N-1007. [Harcros] Nonoxynol-100; CAS 9016-45-9; nonionic; emulsifier for asphalt, emulsion polymerization, textiles, high temp. and high electrolyte applics.; FDA compliance; clear-opaque visc. liq.; mild aromatic odor; water-sol.; sp.gr. 1.1; dens. 9.2 lb/gal; HLB 19.0; cloud pt. 212 F (1%); flash pt. (PMCC) > 500 F; pour pt. 62 F; pH 6-7 (1%); 70 ± 0.5% act.

T-Det® O-40. [Harcros] Octoxynol-40; CAS 9002-93-1; nonionic; emulsifier, stabilizer for agric. formulations, detergents, textiles, emulsion polymerization; FDA compliance; wh. solid; water-sol.; sp.gr. 1.06 (150 F); dens. 8.8 lb/gal (150 F); HLB 17.9; pour pt. 120 F; flash pt. (PMCC) > 500 F; cloud pt. > 212 F (1%); 99.5% min. act.

T-Det® O-307. [Harcros] Octoxynol-30; CAS 9002-93-1; nonionic; emulsifier, stabilizer for emulsion polymerization, industrial intermediate and agric. applics.; liq.; HLB 17.3; 70% conc.

TEA 85. [Dow] Triethanolamine (85%), diethanolamine (15%); used in surfactants, cosmetics/toiletries, metalworking, textile chemicals, gas conditioning chemicals, agric. intermediates, adhesives, antistats, cotings, corrosion inhibition, electroplating, petroleum, polymers, rubber, cement grinding aids; sp.gr. 1.1179; dens. 9.34 lb/gal; visc. 590.5 cps; f.p. 17 C; b.p. 325 C (760 mm Hg); flash pt. (PMCC) 354 F; fire pt. 410 F; ref. index 1.4836.

TEA 85 Low Freeze Grade. [Dow] Triethanolamine (85%), water (15%); CAS 102-71-6; EINECS 203-049-8; used in surfactants, cosmetics/toiletries, metalworking, textile chemicals, gas conditioning chemicals, agric. intermediates, adhesives, antistats, coatings, corrosion inhibition, electroplating, petroleum, polymers, rubber, cement grinding aids; sp.gr. 1.1179; dens. 9.34 lb/gal; visc. 590.5 cps; f.p. 17 C; b.p. 325 C (760 mm Hg); flash pt. (PMCC) 354 F; fire pt. 410 F; ref. index 1.4836.

TEA 99 Low Freeze Grade [Dow] Triethanolamine; CAS 102-71-6; EINECS 203-049-8; used in surfactants, cosmetics/toiletries, metalworking, textile chemicals, gas conditioning chemicals, agric. intermediates, adhesives, antistats, cotings, corrosion inhibition, electroplating, petroleum, polymers, rubber, cement grinding aids; sp.gr. 1.1205; dens. 9.35 lb/gal; visc. 600.7 cps; f.p. 21 C; b.p. 340 C (760 mm Hg); flash pt. (COC) 350 F; fire pt. 420 F; ref. index 1.4839; 15% water.

TEA 99 Standard Grade. [Dow] Triethanolamine; CAS 102-71-6; EINECS 203-049-8; used in surfactants, cosmetics/toiletries, metalworking, textile chemicals, gas conditioning chemicals, agric. intermediates, adhesives, antistats, cotings, corrosion inhibition, electroplating, petroleum, polymers, rubber, cement grinding aids; sp.gr. 1.1205; dens. 9.35 lb/gal; visc. 600.7 cps; f.p. 21 C; b.p. 340 C (760 mm Hg); flash pt. (COC) 350 F; fire pt. 420 F; ref. index 1.4839.

Tebol™. [Arco] Tert. butyl alcohol; CAS 75-65-0; solv., cosolvent, compatibilizer, coupling agent, processing aid for pharmaceuticals, personal care prods., aq. coatings and adhesives, agric. formulations, polymer processing, cleaners/disinfectants; chlorinated hydrocarbon stabilizer; sol. 100% in water; m.w. 74.12; sp.gr. 0.80 (20/20 C); dens. 6.7 lb/gal (20 C); vapor pressure 30 mm Hg (20 C); b.p. 175 F; flash pt. 52 F.

Tebol™ 99. [Arco] High purity tert. butyl alcohol; CAS 75-65-0; solv., cosolvent, compatibilizer, coupling agent, processing aid for pharmaceuticals, personal care prods., aq. coatings and adhesives, agric. formulations, polymer processing, cleaners/disinfectants; chlorinated hydrocarbon stabilizer; Pt-Co ≤ 10 color; sol. in water; misc. with most org. solvs.; sp.gr. 0.78; dens. 6.5 lb/gal; visc. 3.3 cps (30 C); i.b.p. 81.5 C (760 mm Hg); f.p. 24.5 C; flash pt. (CC) 11 C; 99.3% act.

Techfil 7599. [Luzenac Am.] Platy talc; CAS 14807-96-6; EINECS 238-877-9; additive in PP for appliance and automotive grades; improves vapor and liq. water resist., shelf life; imparts a smoother surf.

for paint prods.; low-cust functional fuller for roofing prods., caulks, and sealants; powd.; 12 μ median particle size; 99.5% thru 325 mesh; fineness (Hegman) 3.5; bulk dens. 26 lb/ft^3 (loose), 60 lb/ft^3 (tapped); surf. area 3.0 m^2/g; oil absorp. 42; brightness 79; 75% talc, 15% magnesite, 10% dolomite; 46% SiO_2, 29% MgO.

Tech Pet F™. [Witco/Petroleum Spec.] Petrolatum tech.; binder, carrier, lubricant, moisture barrier, plasticizer, protective agent, softener for tech. applics., (elec., paper and pkg., plastics, rubber, textiles, industrial use); dk. brn.; m.p. 50-65 C.

Tech Pet P. [Witco/Petroleum Spec.] Petrolatum tech.; catalyst carrier, extrusion aid, plasticizer for polymers; industrial applics. incl. metal polishes, modeling clay, printing ink, wire rope lubricant, soldering paste, rust preventatives; brn.; m.p. 74-82 C.

Techpolymer MB-4, MB-8. [Sekisui Plastics] Spherical fine polymer powd. with PMMA base material; matting agent for paints and inks; resin modifier, light diffusion agent for PMMA, PC, PET, PVC, etc.; film additive providing light diffusion, delustering, antiblocking; filler for toiletries and cosmetics; lubrication improver for foundation; pore-forming agent in porous ceramics; in dental materials; powd. 4 and 8 μm mean particle diam. resp.

Techpolymer MBX-5 to MBX-50. [Sekisui Plastics] Spherical fine polymer powd. with crosslinked PMMA base material; matting agent for paints and inks; resin modifier, light diffusion agent for PMMA, PC, PET, PVC, etc.; film additive providing light diffusion, delustering, antiblocking; filler for toiletries and cosmetics; lubrication improver for foundation; pore-forming agent in porous ceramics; in dental materials; suitable for applics. requiring heat, solv., and weather resist.; powd. 5-50 μm mean particle diam.; sp.gr. 1.20; bulk dens. 0.55 (MBX-8); ref. index 1.49.

Techpolymer SBX-6, -8, -17. [Sekisui Plastics] Spherical fine polymer powd. with crosslinked PS base material; matting agent for paints and inks; resin modifier, light diffusion agent for PMMA, PC, PET, PVC, etc.; film additive providing light diffusion, delustering, antiblocking; filler for toiletries and cosmetics; lubrication improver for foundation; pore-forming agent in porous ceramics; in dental materials; suitable for applics. requiring heat and solv. resist.; powd. 6, 8, and 17 μm mean particle diam. resp.; sp.gr. 1.06; ref. index 1.59.

TEDA. [Tosoh] Triethylenediamine; CAS 280-57-9; general purpose gelling catalyst for flexible slabstock and molded PU, semirigid and rigid applics., elastomer shoe soles; wh. fine powd., ammoniacal odor; f.p. 159.8 C; b.p. 174 C; 99.95% purity; *Toxicology:* LD50 1870 mg/kg; mod. toxic; may cause eye and skin inflammation; *Storage:* store in cool, dry, dark area; hygroscopic; keep container closed; may yel. over prolonged storage.

TEDA-D007. [Tosoh] Stannous octoate; CAS 301-10-0; std. catalyst for PU flexible foam; also for slabstock and elastomer shoe soles; sp.gr. 1.260 (20/20 C); visc. 400 cps; flash pt. (PMCC) 110 C.

TEDA-L25B. [Tosoh] Triethylenediamine in 1,4-butanediol; gelling catalyst for semirigid PU and elastomer shoe sole applics.; ylsh. clear liq.; sol. in water; sp.gr. 1.030 (20/20 C); visc. 123 cps; f.p. < -2 C; b.p. 199-211 C; hyd. no. 934; flash pt. (COC) 102 C; 25% act.; *Toxicology:* LD50 (rat) 1970 mg/kg; *Storage:* store in cool, dry, dark area.

TEDA-L33. [Tosoh] Triethylenediamine in dipropylene glycol; gelling catalyst for PU slabstock and molded flexible, rigid, and elastomer shoe shoe applics.; ylsh. clear liq.; sol. in water; sp.gr. 1.033 (20/20 C); visc. 114 cps; f.p. < -20 C; b.p. 194-204 C; hyd. no. 558; flash pt. (COC) 104 C; 33% act.; *Toxicology:* LD50 (rat) 4480 mg/kg; *Storage:* store in cool, dry, dark area.

TEDA-L33E. [Tosoh] Triethylenediamine in ethylene glycol; gelling catalyst for semirigid PU and elastomer shoe sole applics.; ylsh. clear liq.; sp.gr. 1.098 (20/20 C); visc. 61 cps; f.p. < -20 C; b.p. 184-196 C; hyd. no. 1205; flash pt. (COC) 96 C; 33% act.; *Toxicology:* LD50 (rat) 3900 mg/kg; *Storage:* store in cool, dry, dark area.

TEDA-T401. [Tosoh] Tetravalent tin; catalyst for flexible molded PU, rigid PU, elastomer shoe sole applics.; sp.gr. 1.300 (20/20 C); visc. < 20 cps; f.p. 5-8 C; flash pt. (COC) 140 C.

TEDA-T411. [Tosoh] Dibutyltin dillaurate; CAS 77-58-7; EINECS 201-039-8; std. catalyst for PU elastomer applics.; also for slabstock, semirigid, and rigid applics.; sp.gr. 1.050 (20/20 C); visc. < 40 cps; f.p. 10-13 C; flash pt. (COC) 250 C.

TEDA-T811. [Tosoh] Tetravalent tin; catalyst for flexible molded PU, rigid PU, elastomer shoe sole applics.; sp.gr. 1.010 (20/20 C); visc. < 60 cps; flash pt. (COC) 230 C.

Teflon® MP 1000. [DuPont/Polymer Prods.] PTFE; CAS 9002-84-0; fluoroadditive for modifying thermoplastics, elastomers, ink additives, coatings, lubricants, and sealants; general-purpose lubricant additive to improve surf. wear and friction; FDA food contact and EEC food approvals; wh. powd.; 8-15 μ avg. particle size; 90% > 3 μ; sp.gr. 2.2-2.3; bulk dens. 450-600 g/L; surf. area 7-10 m^2/g; melting peak temp. 325 ± 10 C.

Teflon® MP 1100. [DuPont/Polymer Prods.] PTFE; CAS 9002-84-0; fluoroadditive for modifying ink additives, coatings, lubricants, and sealants; lubricant for thermoplastics, thermosets, elastomers; wh. powd.; 1.8-4 μ avg. particle size; 90% > 0.3 μ; sp.gr. 2.2-2.3; bulk dens. 200-425 g/L; surf. area 5-10 m^2/g; melting peak temp. 320 ± 10 C.

Teflon® MP 1200. [DuPont/Polymer Prods.] PTFE; CAS 9002-84-0; fluoroadditive improving performance of printing inks and other coatings; nonagglomerated; lubricant for thermoplastics, thermosets, elastomers; wh. powd.; 2.5-4.5 μ avg. particle size; 90% > 1 μ; sp.gr. 2.2-2.3; bulk dens. 375-525 g/L; surf. area 2.3-4.5 m^2/g; melting peak temp. 320 ± 10 C.

Teflon® MP 1300. [DuPont/Polymer Prods.] PTFE; CAS 9002-84-0; fluoroadditive for modifying thermoplastics, ink additives, coatings, lubricants, and sealants; lubricant additive improving surf. wear and friction; nonagglomerated; FDA food contact and EEC food approvals; wh. powd.; 8-15 μ avg. particle size; 90% > 3 μ; sp.gr. 2.2-2.3; bulk dens. 350-500 g/L; surf. area 2.3-4.5 m^2/g; melting peak temp. 325 ± 10 C.

Teflon® MP 1400. [DuPont/Polymer Prods.] PTFE; CAS 9002-84-0; fluoroadditive for modifying thermoplastics, elastomers, ink additives, coatings, lubricants, and sealants; lubricant additive improving surf. wear and friction; easily disp. in thermoplastics, inks, and coatings; nonagglomerated; broad FDA approvals incl. 21CFR §177.1550 and EEC food approvals; wh. powd.; 7-12 μ avg. particle size; 90% > 3 μ; sp.gr. 2.2-2.3; bulk dens. 300-500 g/L; surf. area 2.3-4.5 m^2/g; melting peak temp. 325 ± 10 C.

Teflon® MP 1500. [DuPont/Polymer Prods.] PTFE; CAS 9002-84-0; fluoroadditive for modifying elastomers, ink additives, coatings, lubricants, and sealants; improve surf. lubricity and tear str. in elastomers without loss of elastic behavior; wh. powd.; 20 μ avg. particle size; 90% > 10 μ; sp.gr. 2.20-2.23; bulk dens. > 300 g/L; surf. area 8-12 m^2/g; melting peak temp. 325 ± 10 C.

Teflon® MP 1600. [DuPont/Polymer Prods.] PTFE; CAS 9002-84-0; fluoroadditive for modifying elastomers, ink additives, coatings, lubricants, and sealants; esp. for composite coating and coil coating applics.; thickener for greases; broad FDA approvals incl. 21CFR §177.1550 and EEC food approvals; wh. powd.; 4-12 μ avg. particle size; sp.gr. 2.2-2.3; bulk dens. 250-500 g/L; surf. area 8-12 m^2/g; melting peak temp. 325 ± 10 C.

Tegiloxan®. [Goldschmidt] Methylsilicone oils; antifoams for min. oils, release agent in rubber and plastics industry; lubricant for tire prod.; additive for polishes; 100% act.; Unverified

TegMeR® 703. [C.P. Hall] PEG-3 diheptanoate; plasticizer for PC and other polymers; lubricant for aluminum can industry; APHA 50 clear liq.; sol. in hexane, toluene, kerosene, ethanol, acetone, min. oil; partly sol. in propylene glycol; insol. in water; m.w. 388; sp.gr. 0.990; visc. 28 cps; acid no. 0.8; sapon. no. 289; flash pt. 210 C; ref. index 1.444; 0.10% moisture.

TegMeR® 704. [C.P. Hall] PEG-4 diheptanoate; plasticizer for PC and other polymers; lubricant for aluminum can industry; APHA 25 clear liq.; sol. in hexane, toluene, kerosene, ethanol, acetone; partly sol. in propylene glycol; insol. in water; m.w. 418; sp.gr. 0.997; visc. 26 cps; f.p. -20 C; acid no. 0.1; sapon. no. 265; flash pt. 229 C; ref. index 1.446; 0.3% moisture.

TegMeR® 803. [C.P. Hall] Triethylene glycol di-2-ethylhexoate; plasticizer for adhesives, nitrile rubber, PC, other polymers; APHA 50 clear liq.; sol. in hexane, toluene, kerosene, ethyl alcohol, acetone, min. oil; partly sol. in propylene glycol; insol. in water; m.w. 374; sp.gr. 0.968; visc. 21 cps; f.p. -73 C; acid no. 0.3; sapon. no. 248; flash pt. 204 C; ref. index 1.4425; 0.05% moisture.

TegMeR® 804. [C.P. Hall] Tetraethylene glycol di-2-ethylhexoate; plasticizer for rubber industry, CR, PC, other polymers; textile surf. finishes, softeners, thread lubricants, and/or antistats; Gardner 5 clear liq.; sol. in hexane, toluene, kerosene, ethyl alcohol, acetone; partly sol. in propylene glycol; insol. in water; m.w. 449; sp.gr. 0.984; visc. 18 cps; f.p. -65 C; acid no. 0.45; sapon. no. 240; hyd. no. 10; flash pt. 204 C; ref. index 1.445; 0.06% moisture.

TegMeR® 804 Special. [C.P. Hall] PEG-4 di-2-ethylhexoate; plasticizer for PVC and other polymers; lubricant for aluminum can industry; APHA 60 clear liq.; sol. in hexane, toluene, kerosene, ethanol, acetone; partly sol. in propylene glycol; insol. in water; m.w. 449; sp.gr. 0.981; visc. 20 cps; f.p. -62 C; acid no. 0.4; sapon. no. 242; hyd. no. 5; flash pt. 218 C; ref. index 1.4450; 0.06% moisture.

TegMeR® 809. [C.P. Hall] PEG 400 di-2-ethylhexoate; plasticizer for PC and other polymers; lubricant for aluminum can industry; Gardner 2 clear liq.; sol. in hexane, toluene, kerosene, ethanol, acetone; insol. in water; m.w. 652; sp.gr. 1.02; visc. 50 cps; f.p. -48 C; acid no. 0.5; sapon. no. 170; flash pt. 262 C; ref. index 1.452; 0.05% moisture.

TegMeR® 903. [C.P. Hall] PEG-3 dipelargonate; CAS 106-06-9; lubricant for aluminum can industry; plasticizer for PC and other polymers; APHA 50 clear liq.; sol. in hexane, toluene, kerosene, ethanol, acetone, min. oil; partly sol. in propylene glycol; insol. in water; m.w. 420; sp.gr. 0.97; visc. 24 cps; acid no. 0.2; sapon. no. 254; hyd. no. 15; flash pt. 224 C; ref. index 1.447; 0.05% moisture.

Tego® IMR® 412 T. [Goldschmidt] Internal mold release agent for demolding microcellular PU systems, e.g. RIM formulations, and elastomers; Unverified

Tego® IMR® 830. [Goldschmidt AG] Internal release agent for RIM formulations; silicone-free; Unverified

Tego® IMR 918. [Goldschmidt AG] Polysiloxane polyoxyalkylene block copolymer; additive improving flowability in RIM systems and high density integral skin foams; sol. in polyol; insol. in water; dens. 0.990 g/cc; visc. 50 mPa•s; flash pt. > 80 C.

Tego® Airex 975. [Tego] Silicone modified org. polymer; deaerator for low- to med.-polar solv.-based systems, esp. chlorinated rubber paints, sol'n. vinyls (vehicles for tank linings, pool paints, pond liners); sp.gr. 0.95; dens. 7.9 lb/gal; flash pt. 25 C; 25% act. in xylene; Unverified

Tegoamin® 33. [Goldschmidt] Triethylene diamine; CAS 280-57-9; catalyst for the mfg. of PU foams and elastomers, esp. hot cured and high resilience block foams; sol'n.; sol. in water; sp.gr. 1.033 ± 0.005; solid. pt. < -20 C; b.p. 206-227 C; flash pt. 90 C; 33% sol'n. in dipropylene glycol; Unverified

Tegoamin® 100. [Goldschmidt] Triethylene diamine; CAS 280-57-9; catalyst for the mfg. of PU foams and elastomers; 100% act.; Unverified

Tegoamin® BDE. [Goldschmidt] Bisdimethylaminoethyl ether; highly reactive catalyst for the prod. of flexible polyether PU foams; colorless clear liq., typ. amine odor; sol. in water; sp.gr. 0.902; visc. 4.1 mPas; solid. pt. < -80 C; b.p. 186-226 C; flash pt. 75 C; pH 12.8 (33% aq.); Unverified

Tegoamin® CPE. [Goldschmidt] Tert. amine; activator for the mfg. of polyester PU foams of reduced odor; sol. in water; sp.gr. 0.958 ± 0.005; solid. pt. -9 C; b.p. 127-324 C; flash pt. 33 C; Unverified

Tegoamin® CPU 33. [Goldschmidt] Tert. amine; catalyst for the prod. of conventional flexible PU foam; for slabstock and molded foams; liq.; sol. in water; dens. 0.945 g/cc; visc. 23 mPa•s; solid. pt. < -50 C; b.p. 122-231 C; flash pt. 42 C; Unverified

Tegoamin® DMEA. [Goldschmidt] Dimethylethanolamine; CAS 108-01-0; EINECS 203-542-8; catalyst for the prod. of flexible PU foams; sp.gr. 0.887; solid. pt. < -20 C; b.p. 133-135 C; flash pt. 40 C; Unverified

Tegoamin® EPS. [Goldschmidt] Tert. amine; catalyst in the mfg. of polyester PU slabstock foams; sol. in water; sp.gr. 0.988 ± 0.005; solid. pt. > -31 C; flash pt. 29 C; Unverified

Tegoamin® PDD 60. [Goldschmidt] Tert. amine; catalyst used as an activator for the mfg. of flexible PU foam; sol. in water; dens. 0.983 ± 0.005 g/cc; f.p. < -20 C; b.p. 188-230 C; flash pt. 92 C; Unverified

Tegoamin® PMD. [Goldschmidt] Tert. amine; catalyst for the mfg. of flexible PU foams; for slabstock and molded foams; clear colorless liq.; sol. in water; dens. 0.870 ± 0.005 g/cc; f.p. < -20 C; flash pt. 43 C; Unverified

Tegoamin® PTA. [Goldschmidt] Tert. amine; activator for the mfg. of flexible hot-cured and high resilience PU foams; catalyst for the blowing and gelling reactions; sol. in water; sp.gr. 0.962 ± 0.005; solid. pt. < -20 C; b.p. 138-229 C; flash pt. 61 C; Unverified

Tegoamin® SMP. [Goldschmidt] Tert. amine; catalyst for the mfg. of conventional flexible PU slabstock and molded foams (hot-cured foam); colorless to sl. yel. clear liq.; sol. in water; sp.gr. 0.941 ± 0.008; visc. 9 ± 2 mPas; solid. pt. < -50 C; b.p. 128-224 C; flash pt. 34 C; Unverified

Tego® Antiflamm® N. [Goldschmidt] Flame retardant for PU foams; Unverified

Tegocolor®. [Goldschmidt] Pigment disps. in a polyether polyol; heat-stable color pastes for polyether PU foams; Unverified

Tego® Dispers 700. [Tego GmbH] Salt of a high m.w. fatty acid deriv.; wetting and dispersing additive for solv.-based paint systems; deflocculant, antiflooding agent; for alkyd, epoxide, chlorinated rubber, bitumen, polyurethane, and polyester systems; sp.gr. 0.93 g/cc; flash pt. 25 C; 50% act. in xylene; Unverified

Tego® Effect L 104. [Tego GmbH] Polyurethane-based sol'n. in butyl diglycol; thickener for aq. polymeric emulsions based on polyacrylates, PU, PVA/PE or polysiloxanes; improves sliding, leveling props.; high visc. pourable liq.; dens. 1.01 g/cc (20 C); pH 8 ± 1; 25% act.; *Storage:* 1 yr storage stability in closed containers @ 20 C; Unverified

Tego® Emulsion 240. [Goldschmidt AG] Plasticized silicone emulsion; aq. release agent for thermosetting and thermoplastic materials; provides semipermanent mold release allowing several operations before reapplication; 40% act.

Tego® Emulsion 3454. [Goldschmidt AG] Silicone emulsion; release agent for prod. of food pkg.

Tego® Emulsion 3529. [Goldschmidt AG] Silicone-free emulsion; aq. release agent for natural and syn. rubbers; 35% act.

Tego® Emulsion ASL. [Goldschmidt AG] Silicone emulsion; release agent for nat. and syn. rubbers and the foundry industry; 35% act.

Tego® Emulsion DF 19. [Goldschmidt AG] Emulsion of a special liq. silicone; aq. release agent for thermosetting and thermoplastic materials; provides semipermanent mold release allowing several operations before reapplication; 40% act.

Tego® Emulsion DT. [Goldschmidt AG] Emulsion of a special silicone; cationic; aq. release agent for thermosetting and thermoplastic materials; provides semipermanent mold release allowing several operations before reapplication; 40% act.

Tego® Emulsion N. [Goldschmidt AG] Silicone emulsion with org. release agents; aq. release agent for rubber molding; 20% act.

Tego® Emulsion S 3. [Goldschmidt AG] Emulsion of Getren® S 3 Conc. 100; aq. release agent for demolding of plastic and rubber articles; 35% act.

Tego® Foamex 800. [Tego GmbH] Polysiloxane-polyether copolymer o/w emulsion; CAS 68937-54-2; nonionic; defoamer for water-based emulsion paints and water-thinnable systems, PU paints, PU-acrylics, furniture paint, wood varnish, adhesives, dispersion paints, building paints, polymer dispersions; sp.gr. 1.0 g/cc; 20% act. in water; *Storage:* 6 mos storage stability in closed containers under frost-free conditions; Unverified

Tego® Foamex 805. [Tego GmbH] Polysiloxane-polyether copolymer o/w emulsion; CAS 68937-54-2; nonionic; defoamer for water-based emulsion paints and water-thinnable systems, PU paints, PU-acrylics, furniture paint, wood varnish, adhesives, dispersion paints, polymer dispersions; sp.gr. 1.0 g/cc; 20% act. in water; *Storage:* 6 mos storage stability in closed containers under frost-free conditions; Unverified

Tego® Foamex 1488. [Tego GmbH] Dimethicone copolyol emulsion; nonionic/anionic; defoamer for water-based emulsion paints and water-thinnable systems, building protection coatings (acrylic, styrene-acrylic, PVA), wood/furniture varnishes, anticorrosive coatings/industrial paints (alkyd, polyester, acrylic), polymer emulsions; BGA approved; sp.gr. 1.0 g/cc; 20% act. in water; *Storage:* 6 mos storage stability in closed containers under frost-free conditions; Unverified

Tego® Foamex 7447. [Tego GmbH] Dimethicone copolyol emulsion; nonionic; defoamer for water-based emulsion paints and water-thinnable systems, building protection coatings (acrylic, styrene-acrylic, PVAc), wood/furniture varnishes, anticorrosion/industrial paints (alkyd, polyester, acrylic), printing inks, polymer disps.; BGA approved; sp.gr. 1.0 g/cc; 20% act. in water; *Storage:* 6 mos storage stability in closed containers under frost-free conditions; Unverified

Tego® Release Agent M 379. [Goldschmidt AG] Org. modified polysiloxane; internal release agent for thermosetting resins; 100% act.

Tegosil®. [Goldschmidt] Silicone-based; aerosol for release of inj. molded plastic and rubber moldings; Unverified

Tego® Silicone Paste A. [Goldschmidt AG] Lubricant with low load bearing props. for screw threads, rubber and plastic seals and gaskets.

Tegostab® B 1048. [Goldschmidt AG] Silicone; foam stabilizer PU figid foams; for continuously laminated PU boardstock, refrigeration and pipe insulation, rigid block foams; cell-regulating effect; influences dimensional stability and heat conductivity; dens. 1.04 ± 0.1 g/cc; visc. 800 ± 150 mPa•s; solid. pt. < -25 C; cloud pt. 38 ± 2 C (4% aq.); pH 10 ± 0.5 (4% aq.); *Precaution:* incompat. with tin-based catalysts; *Storage:* 6 mos min. storage stability; Unverified

Tegostab® B 1400A. [Goldschmidt AG] Silicone; foam stabilizer for mfg. of rigid PU foams; cell-regulating effect; modified dimensional stability and heat conductivity; esp. suitable for refrigerator systems and for panels; sp.gr. 1.04 ± 0.01 g/cc; visc. 800 ± 150 mPa•s; turbidity pt. 36-40 C (4% aq.); pH 10 ± 0.4 (4% aq.); *Storage:* 6 mos storage stability; Unverified

Tegostab® B 1605. [Goldschmidt AG] Polysiloxane-polyoxyalkylene copolymer; stabilizer in mfg. of flexible polyether PU foams, primarily slabstock foams, but also suitable for molded foams; broad processing range; sp.gr. 1.045 ± 0.007 g/cc; visc. 1100 ± 150 mPa•s; f.p. < -40 C; turbidity pt. 38 ± 2 C (4% aq.); pH 10 ± 0.5 (4% aq.); *Storage:* 6 mos min. storage stability; Unverified

Tegostab® B 1648. [Goldschmidt AG] Polysiloxane-polyoxyalkylene copolymer; stabilizer for mfg. of rigid PU foams based on crude MID or TDI; produces rigid foams with fine cell structure, high closed cell content, good dimensional stability; sol. in rigid foam polyols, polyol/fluorocarbon blends; sp.gr. 1.075 ± 0.007; visc. 270 ± 70 mPa•s; solid. pt. < -15 C; hyd. no. 95 ± 7; cloud pt. 60 ± 3 C (4% aq.); pH 10 ± 0.5 (4% aq.); *Storage:* 6 mos min. storage stability; Unverified

Tegostab® B 1903. [Goldschmidt AG] Silicone; foam stabilizer for mfg. of rigid PU foam and systems based on sucrose polyols; improves dimensional stability and heat conductivity; dens. 1.045 ± 0.01; visc. 1.050 ± 150 mPa•s; cloud pt. 44 ± 2 C (4% aq.); pH 10 ± 0.4 (4% aq.); *Storage:* 6 mos min. storage stability; Unverified

Tegostab® B 2173. [Goldschmidt AG] Silicone; foam stabilizer for prod. of molded flexible polyether foam; provides extremely open cell structure; also for slabstock prod. under certain conditions; sp.gr. 1.030 ± 0.007; visc. 950 ± 170 cps; solid. pt. < -25 C; turbidity pt. 35-39 C (4% aq.); pH 10 ± 0.4 (4% aq.); Unverified

Tegostab® B 2219. [Goldschmidt AG] Polysiloxane-polyoxyalkylene copolymer; stabilizer for mfg. of rigid PU foams, continuously laminated PU boardstock, rigid block foams; improves dimensional stability and insulating efficacy; dens. 1.03 ± 0.01 g/cc; visc. 1100 ± 150 mPa•s; solid. pt. < 5 C; cloud pt. 45 ± 2 C (4% aq.); pH 10 ± 0.5 (4% aq.); *Storage:* 6 mos min. storage stability; Unverified

Tegostab® B 3136. [Goldschmidt AG] Surfactant, stabilizer for use in flexible PU foam formulations based on highly reactive polyethers; dens. 1.03 ± 0.01 g/cc; visc. 470 ± 100 mPa•s; solid. pt. < -25 C; cloud pt. 32 ± 2 C (4% aq.); pH 10 ± 0.5 (4% aq.); *Storage:* 6 mos min. storage stability; Unverified

Tegostab® B 3640. [Goldschmidt AG] Silicone surfactant; used for prod. of flame retarded flexible polyether PU foam; synergistic with flame retardants; sp.gr. 1.04 ± 0.01 g/cc; visc. 750 ± 150 mPa•s; solid. pt. < -25 C; cloud pt. 38 ± 2 C (4% aq.); pH 10 ± 0.5 (4% aq.); *Storage:* 6 mos storage stability in original sealed containers, protected against air, light, and raised temps.; Unverified

Tegostab® B 3752. [Goldschmidt AG] Foam stabilizer for hot-cured molded PU foam; synergistic with flame retardant; Unverified

Tegostab® B 4113. [Goldschmidt AG] Silicone surfactant; cell-regulating stabilizer for high resilience cold-cured PU foam; liq.; sol. in polyols, Freon, org. solvs.; insol. in water; sp.gr. 0.983 ± 0.01 g/cc; visc. 135 ± 15 mPa•s; hyd. no. 39 ± 3; flamm. (PM) > 100 C; *Storage:* 12 mos storage stability; Unverified

Tegostab® B 4351. [Goldschmidt AG] Stabilizer and cell-regulating surfactant for the mfg. of high resilient PU foams; sl. yel. clear liq.; sol. in polyether polyols; insol. in water; sp.gr. 0.99 ± 0.02 g/cc; visc. 120 ± 30 mPa•s; hyd. no. 35 ± 3; *Storage:* 12 mos storage stability; Unverified

Tegostab® B 4380. [Goldschmidt AG] Silicone; foam stabilizer for high resiliency PU foams; cell-regulating additive; clear to sl. opaque liq.; sol. in polyols, Freon, complete resin mixts.; insol. in water; sp.gr. 0.98 ± 0.02 g/cc; visc. 130 ± 20 mPa•s; hyd. no. 37 ± 4; flash pt. > 100 C; *Storage:* 6 mos min. storage stability; Unverified

Tegostab® B 4617. [Goldschmidt AG] Silicone; foam stabilizer for high resilience PU foams; cell-regulating additive; sl. yel. clear liq.; sol. in polyether polyols; insol. in water; sp.gr. 0.98 ± 0.01 g/cc; visc. 122 ± 14 mPa•s; hyd. no. 36 ± 4; *Storage:* 12 mos min. storage stability; Unverified

Tegostab® B 4690. [Goldschmidt AG] Stabilizer for high resilience cold-cured PU foam; exhibits strong cell regulating effect and weak stabilizing effect; insol. in water; dens. 0.99 ± 0.02 g/cc; visc. 137 ± 12 mPa•s; hyd. no. 36 ± 4; flash pt. > 100 C; < 0.1% water; *Storage:* 6 mos min. storage stability; Unverified

Tegostab® B 4900. [Goldschmidt AG] Polysiloxane-polyoxyalkylene block copolymer; stabilizer in mfg.

of flexible polyether PU foams; dens. 1.04 ± 0.01 g/cc; visc. 1050 ± 150 mPa•s; solid. pt. < -10 C; cloud pt. 40 ± 2 C (4% aq.); pH 10.1 ± 0.5; *Storage:* 6 mos min. storage stability; Unverified

Tegostab® B 4970. [Goldschmidt AG] Polysiloxane-polyoxyalkylene copolymer; stabilizer for mfg. of flexible polyether PU foams; broad processing range; primarily for slabstock foams, but also suitable for molded foams; sp.gr. 1.03 ± 0.01; visc. 1250 ± 200 cps; f.p. < -25 C; cloud pt. 39 ± 2 C; pH 10 ± 0.05 (4% aq.); *Storage:* 6 mos min. storage stability; Unverified

Tegostab® B 5055. [Goldschmidt AG] Stabilizer for polyester PU foams; sp.gr. 1.07 ± 0.01 g/cc; visc. 470 ± 100 mPa•s; solid. pt. < -20 C; pH 10 ± 1 (4% aq.); *Storage:* 6 mos storage stability; Unverified

Tegostab® B 8002. [Goldschmidt AG] Polysiloxane-polyoxyalkylene block copolymer; stabilizer in mfg. of flexible polyether PU foams; dens. 1.03 ± 0.01 g/cc; visc. 500 ± 150 mPa•s; solid. pt. < -25 C; cloud pt. 40 ± 2 C (4% aq.); pH 10 ± 0.5 (4% aq.); *Storage:* 6 mos min. storage stability; Unverified

Tegostab® B 8014. [Goldschmidt AG] Polysiloxane-polyoxyalkylene block copolymer; foam stabilizer in mfg. of flexible hot-cured polyether PU foams; sp.gr. 1.03 ± 0.01 g/cc; visc. 1600 ± 200 mPa•s; cloud pt. 38 ± 2 C (4% aq.); pH 10 ± 0.5 (4% aq.); *Storage:* 6 mos min. storage stability; Unverified

Tegostab® B 8017. [Goldschmidt AG] Polysiloxane-polyoxyalkylene block copolymer; stabilizer in mfg. of flexible polyether PU foams; sp.gr. 1.04 ± 0.01 g/cc; visc. 1100 ± 150 mPa•s; f.p. < -25 C; turbidity pt. 38 ± 2 C (4% aq.); pH 10 ± 0.5 (4% aq.); *Storage:* 6 mos min. storage stability; Unverified

Tegostab® B 8021. [Goldschmidt AG] Polysiloxane-polyoxyalkylene block copolymer; foam stabilizer in mfg. of flexible polyether PU foams; sp.gr. 1.03 ± 0.1 g/cc; visc. 1200 ± 200 mPa•s; solid. pt. < -25 C; cloud pt. 37 ± 2 C (4% aq.); pH 10 ± 0.5 (4% aq.); *Storage:* 6 mos min. storage stability; Unverified

Tegostab® B 8030. [Goldschmidt AG] Polysiloxane-polyoxyalkylene block copolymer; foam stabilizer in mfg. of flexible polyether PU foams; dens. 1.03 ± 0.1 g/cc; visc. 1700 ± 200 mPa•s; cloud pt. 38 ± 2 C (4% aq.); pH 10 ± 0.4 (4% aq.); *Storage:* 6 mos min. storage stability; Unverified

Tegostab® B 8100. [Goldschmidt AG] Polysiloxane-polyoxyalkylene block copolymer; foam stabilizer in mfg. of flexible hot-cured polyether PU foams; dens. 1.03 ± 0.01 g/cc; visc. 900 ± 300 mPa•s; solid. pt. < -25 C; cloud pt. 33 ± 3 C (4% aq.); pH 7 ± 1.5 (4% aq.); *Storage:* 6 mos min. storage stability; Unverified

Tegostab® B 8200. [Goldschmidt AG] Stabilizer for flexible flame-retarded PU block foam; Unverified

Tegostab® B 8202. [Goldschmidt AG] Silicone surfactant; foam stabilizer in mfg. of flexible polyether PU foams; wide processing latitude; synergistic with flame retardants; suitable for flame retarded and normal hot cured PU foam; sp.gr. 1.04 ± 0.01 g/cc; visc. 1950 ± 250 mPa•s; solid. pt. < -25 C; cloud pt. 40 ± 2 C (4% aq.); pH 10 ± 0.5 (4% aq.); *Storage:* 6 mos min. storage stability; Unverified

Tegostab® B 8231. [Goldschmidt AG] Silicone surfactant; foam stabilizer in mfg. of flexible polyether PU foams; wide processing latitude; synergistic with flame retardants; for flame retarded and normal hot-cured PU foams; sl. yel. liq.; dens. 1.03 ± 0.01 g/cc; visc. 1400 ± 200 mPa•s; solid. pt. < -25 C; cloud pt. 40 ± 2 C (4% aq.); pH 10.1 ± 0.5 (4% aq.); Unverified

Tegostab® B 8300. [Goldschmidt AG] Stabilizer for prod. of polyester PU foams; produces fine-celled foam without addition of cell regulators or emulsifiers; sp.gr. 1.073 ± 0.01 g/cc; visc. 400 ± 100 mPa•s; solid. pt. < -20 C; flash pt. > 65 C; < 3% water; *Storage:* 12 mos storage stability; Unverified

Tegostab® B 8301. [Goldschmidt AG] Stabilizer for mfg. of die-cuttable polyester PU foams with an open and regular cell structure; no need for additional cell regulating additives; dens. 1.058 ± 0.05 g/cc; visc. 500 ± 100 mPa•s; solid. pt. 1 C; flash pt. 155 C; < 0.5% water; *Storage:* 12 mos storage stability; Unverified

Tegostab® B 8404. [Goldschmidt AG] Polysiloxane-polyether copolymer surfactant; foam stabilizer for mfg. of rigid PU foams, continuously laminated PU boardstock, pour-in-place foams, rigid block foams; clear visc. liq.; dens. 1.055 ± 0.01 g/cc; visc. 450 ± 120 mPa•s; solid. pt. < -15 C; hyd. no. 120 ± 12; cloud pt. 64 ± 3 C (4% aq.); pH 6.5 ± 1.5 (4% aq.); *Storage:* 12 mos storage stability; Unverified

Tegostab® B 8406. [Goldschmidt AG] Polysiloxane-polyether copolymer surfactant; stabilizer for mfg. of rigid PU and polyisocyanurate foams; superior emulsifying props.; clear visc. liq.; dens. 1.07 ± 0.01 g/cc; visc. 620 ± 110 mPa•s; solid. pt. < 13 C; hyd. no. 135 ± 10; cloud pt. 80 ± 4 C (4% aq.); pH 6.5 ± 1.5 (4% aq.); *Storage:* 12 mos storage stability; Unverified

Tegostab® B 8407. [Goldschmidt AG] Polysiloxane-polyether copolymer surfactant; stabilizer for rigid PU refrigeration and pipe insulation, polyisocyanurate foams; clear visc. liq.; dens. 1.055 ± 0.01 g/cc; visc. 270 ± 80 mPa•s; solid. pt. < -9 C; hyd. no. < 5; cloud pt. 56 ± 3 C (4% aq.); pH 6.5 ± 1.5 (4% aq.); *Storage:* 12 mos storage stability; Unverified

Tegostab® B 8408. [Goldschmidt AG] Polysiloxane-polyether copolymer surfactant; foam stabilizer for rigid PU foams, continuously laminated PU boardstock, modified polyisocyanurate foams, spray foams; synergistic with flame retardants; clear visc. liq.; sp.gr. 1.072 ± 0.01 g/cc; visc. 730 ± 150 mPa•s; f.p. < 10 C; hyd. no. 115 ± 5; cloud pt. 84 ± 3 C (4% aq.); flash pt. > 85 C; pH 6.5 ± 1.5 (4% aq.); *Storage:* 12 mos storage stability; Unverified

Tegostab® B 8409. [Goldschmidt AG] Polysiloxane-polyether copolymer surfactant; foam stabilizer for rigid PU and polyisocyanurate foams, pour-in-place PU foams, refrigeration and pipe insulation; clear visc. liq.; dens. 1.055 ± 0.009 g/cc; visc. 750 ± 170 mPa•s; solid. pt. < -5 C; hyd. no. 118 ± 10; cloud pt. 58 ± 3 C (4% aq.); pH 7 ± 1.5 (4% aq.); *Storage:* 12 mos storage stability; Unverified

Tegostab® B 8418. [Goldschmidt AG] Polysiloxane polyether copolymer; foam stabilizer for mfg. of rigid PU foams, high-dens. molded foams; colorless to sl. yel. clear liq.; dens. 1.045 ± 0.01 g/cc; visc. 1100 ± 200 mPa•s; solid. pt. < -10 C; cloud pt. 51 ± 3 C (4% aq.); pH 6.5 ± 1.5 (4% aq.); *Storage:* 6 mos min. storage stability; avoid prolonged exposure to humidity; Unverified

Tegostab® B 8423. [Goldschmidt AG] Polysiloxane-polyether block copolymer; foam stabilizer and emulsifier for rigid PU and polyisocyanurate foams with max. demand on flow properties; esp. for refrigeration and pipe insulation; clear visc. liq.; sol. in common rigid foam polyols and resin preblends; dens. 1.055 ± 0.01 g/cc; visc. 750 ± 170 mPa•s; solid. pt. < -8 C; hyd. no. 100 ± 10; cloud pt. 70 ± 3 C (4% aq.); pH 6.5 ± 1.5 (4% aq.); *Storage:* 12 mos storage stability; Unverified

Tegostab® B 8425. [Goldschmidt AG] Polysiloxane polyether copolymer; surfactant, stabilizer for rigid PU foams for refrigeration and pipe insulation; sl. yel. clear liq.; dens. 1.055 ± 0.01 g/cc; visc. 300 ± 60 mPa•s; solid. pt. < 0 C; hyd. no. 116 ± 12; cloud pt. 70 ± 2 C (4% aq.); pH 6.5 ± 1.5 (4% aq.); *Storage:* 12 mos storage stability; Unverified

Tegostab® B 8427. [Goldschmidt AG] Polysiloxane-polyether copolymer; stabilizer for mfg. of rigid PU foams; clear visc. liq.; dens. 1.06 ± 0.1 g/cc; visc. 1130 ± 200 mPa•s; solid. pt. < -5 C; hyd. no. 110 ± 10; cloud pt. 71 ± 2 C (4% aq.); flash pt. > 80 C; pH 6.5 ± 1.5 (4% aq.); *Storage:* 12 mos storage stability; Unverified

Tegostab® B 8432. [Goldschmidt AG] Silicone surfactant; foam stabilizer for mfg. of rigid PU foams, continuous laminated PU boardstock, pour-in-place applics.; colorless or lt. yel. clear visc. liq.; dens. 1.06 ± 0.01 g/cc; visc. 380 ± 100 mPa•s; solid. pt. < 4 C; hyd. no. 105 ± 10; cloud pt. 63 ± 3 C (4% aq.); pH 6.0 ± 1.0 (4% aq.); *Storage:* 12 mos storage stability; Unverified

Tegostab® B 8433. [Goldschmidt AG] Polysiloxane-polyether copolymer; foam stabilizing surfactant for mfg. of rigid PU foams, esp. in mfg. of appliances; sl. amber clear liq.; sp.gr. 1.06 ± 0.01 g/cc; visc. 1000 ± 200 mPa•s; pour pt. < -2 C; cloud pt. 71 ± 2 C (4% aq.); flash pt. > 80 C; pH 6.5 ± 1.5 (4% aq.); Unverified

Tegostab® B 8435. [Goldschmidt AG] stabilizer; pale clear liq.; sp.gr. 1.07 ± 0.01 g/cc; visc. 500 ± 120 mPa•s; solid. pt. < -10 C; hyd. no. 160 ± 20; cloud pt. 50 ± 3 C (4% aq.); flash pt. > 65 C; pH 8 ± 1 (4% aq.); *Storage:* 6 mos min. storage stability; Unverified

Tegostab® B 8444. [Goldschmidt AG] Stabilizer for high density molded rigid PU foams; Unverified

Tegostab® B 8450. [Goldschmidt AG] Silicone; stabilizer for mfg. of polyester PU shoe soles; sol. in water and in polyester polyol components; dens. 1.077 ± 0.1 g/cc; visc. 320 ± 60 mPa•s; solid. pt. 13 ± 2 C; cloud pt. 91 ± 3 C; flash pt. > 100 C; Unverified

Tegostab® B 8622. [Goldschmidt AG] Polysiloxane-polyalkylene copolymer surfactant; stabilizer for mfg. of high resilient PU foams; colorless to sl. yel. clear liq.; sol. in polyols; insol. in water; sp.gr. 0.98 ± 0.01 g/cc; visc. 43 ± 10 mPa•s; hyd. no. 104 ± 5; *Storage:* 12 mos storage stability; Unverified

Tegostab® B 8631. [Goldschmidt AG] Surfactant for prod. of cold-curing PU foams; colorless to lt. yel. liq.; sol. in polyol components; insol. in water; sp.gr. 0.965 ± 0.01 g/cc; visc. 25 ± 7 mPa•s; hyd. no. 130 ± 15; flash pt. > 100 C; *Storage:* 12 mos storage stability; Unverified

Tegostab® B 8636. [Goldschmidt AG] Silicone surfactant; foam stabilizer for prod. of combustion-modified high-resilience PU slabstock foams, esp. foams contg. melamine as fire protecting agent; colorless clear liq.; sp.gr. 0.99 ± 0.01 g/cc; visc. 41 ± 4 mPa•s; hyd. no. 110; flash pt. 121 C; *Storage:* 12 mos storage stability; avoid pollution by tin compds.; Unverified

Tegostab® B 8680. [Goldschmidt AG] Silicone; stabilizer for high resiliency cold-cured PU foam where high degree of stabilization is required; colorless clear liq.; sol. in polyols; insol. in water; sp.gr. 0.955 ± 0.005 g/cc; visc. 12 ± 1 mPa•s; hyd. no. 116; flash pt. 75 C; *Storage:* 12 mos storage stability; Unverified

Tegostab® B 8681. [Goldschmidt AG] Silicone based on organo-modified siloxane; stabilizer for high resiliency slabstock foams and molded cold-cure PU foams; colorless to sl. yel. clear liq.; sol. in polyols; insol. in water; dens. 0.95 ± 0.01 g/cc; visc. 13 ± 3 mPa•s; hyd. no. 115 ± 10; flash pt. 65 C; *Storage:* 12 mos storage stability; Unverified

Tegostab® B 8694. [Goldschmidt AG] Silicone; stabilizer with active cell opening prop. for prod. of high resiliency molded foam based on MDI; colorless clear liq.; sol. in polyols and ready-for-use polyol blends; insol. in water; dens. 0.960 ± 0.01 g/cc; visc. 15 ± 2 mPa•s; f.p. -18 C; hyd. no. 120 ± 10; cloud pt. 10 C; flash pt. > 114 C; *Storage:* 12 mos storage stability; Unverified

Tegostab® B 8863 T. [Goldschmidt AG] Silicone-based; stabilizer for low-dens. PU foams used in pkg.; also provides open cell structure; Unverified

Tegostab® B 8896. [Goldschmidt AG] Silicone surfactant; foam stabilizer in mfg. of phenolic resin foams; sl. turbid liq.; sp.gr. 1.02 ± 0.02 g/cc; visc. 1200 ± 200 mPa•s; solid. pt. < 0 C; flash pt. > 200 C; pH 7 ± 1; *Storage:* 12 mos storage stability; Unverified

Tegostab® B 8901. [Goldschmidt AG] Silicone stabilizer; stabilizer for polyester PU foams with integral skins, in shoe sole systems; sol. in water and in polyester-polyol components; dens. 1.035 ± 0.007 g/cc; visc. 2600 ± 400 mPa•s; solid. pt. < -20 C; cloud pt. 45 ± 4 C; flash pt. > 80 C; Unverified

Tegostab® B 8905. [Goldschmidt AG] Silicone; stabilizer for prod. of microcellular polyether urethane foams (integral foams); improves cell structure and use props. of manufactured molded parts; colorless to lt. yel. liq.; sol. in water; dens. 1.065 ± 0.01 g/cc; visc. 650 ± 150 mPa•s; hyd. no. 169 ± 20;

pour pt. -12 ± 3 C; cloud pt. 56 ± 2 C (4% aq.); flash pt. 108 C; pH 7.2 ± 1 (4% aq.); *Storage:* 12 mos storage stability; Unverified

Tegostab® B 8906, 8910. [Goldschmidt AG] Org. stabilizers providing a smooth surf. for PU spray foams; Unverified

Tegostab® BF 2270. [Goldschmidt AG] Polysiloxane-polyoxyalkylene block copolymer; foam stabilizer for hot-cured polyether PU foams; sp.gr. 1.03 ± 0.1 g/cc; visc. 1200 ± 150 mPa•s; solid. pt. < -25 C; cloud pt. 38 ± 2 C (4% aq.); pH 10 ± 0.5 (4% aq.); *Storage:* 6 mos min. storage stability; Unverified

Tegostab® BF 2370. [Goldschmidt AG] Polysiloxane-polyoxyalkylene block copolymer; foam stabilizer in mfg. of flexible polyether PU slabstock and molded foam; wide processing latitude for hot-cured PU foam; sp.gr. 1.03 ± 0.01 g/cc; visc. 850 ± 150 mPa•s; solid. pt. < -25 C; cloud pt. 37 ± 2 C (4% aq.); pH 10 ± 0.5 (4% aq.); *Storage:* 6 mos min. storage stability; Unverified

Tegotens 4100. [Goldschmidt AG] Palmitic/stearic acid mono/diglycerides; nonionic; surface coating for expanded polystyrene beads containing a propellant; antistat for PE/PP; fabric softener; coemulsifier; powd.

Tegotrenn® LH 157 A. [Goldschmidt] Release agent for demolding of flexible hot-cured PU foam; solv.-based disp.; dilutable with org. solvs. in 1:1 or 1:2 ratio; Unverified

Tegotrenn® LH 525. [Goldschmidt AG] Release agent for demolding of flexible hot-cured PU foam; aq. emulsion; dilutable with demineralized water in rat 1:5-1:9; 42% act.; Unverified

Tegotrenn® LI 197. [Goldschmidt] Release agent for demolding of integral skinned PU foam; conc. version; Unverified

Tegotrenn® LI 237. [Goldschmidt] Silicone-contg. release agent for demolding of integral skinned PU foam, esp. pigmented systems; Unverified

Tegotrenn® LI 344. [Goldschmidt AG] Release agent for demolding of integral skinned PU foam, esp. those which are to be painted; Unverified

Tegotrenn® LI 448. [Goldschmidt] Release agent for PU RIM systems; Unverified

Tegotrenn® LI 537 W. [Goldschmidt] Release agent suitable for water carrier for PU RIM applics.; Unverified

Tegotrenn® LI 747. [Goldschmidt] Release agent for PU RIM applics.; silicone-containing; Unverified

Tegotrenn® LI 748. [Goldschmidt] Release agent for PU RIM applics.; silicone-free; Unverified

Tegotrenn® LI 766. [Goldschmidt] Release agent for demolding of integral skinned PU foam, esp. for thin film applics.; Unverified

Tegotrenn® LI 841-237. [Goldschmidt AG] Silicone-contg.; release agent for demolding of flexible integral skinned PU foam with sl. shining surfaces; Unverified

Tegotrenn® LI 850-344. [Goldschmidt AG] Release agent for demolding of flexible integral skinned PU foam with matt surfaces and for parts which are painted, or when the in-mold coating tech. is applied; silicone-free; Unverified

Tegotrenn® LI 901. [Goldschmidt] Release agent for demolding of integral skinned PU foam, RIM applics.; Unverified

Tegotrenn® LI 903. [Goldschmidt AG] Release agent for demolding of rigid high density PU foam, e.g, electronic housings and for RIM applics.; silicone-free; Unverified

Tegotrenn® LI 904. [Goldschmidt AG] Release agent for high mold temp. RIM applics.; Unverified

Tegotrenn® LK 104. [Goldschmidt] Release agent for demolding of cold cured (high resilience) PU foam; paste; Unverified

Tegotrenn® LK 260, 498/F, 755, 859, 860. [Goldschmidt] Release agents for demolding of cold cured (high resilience) PU foam; Unverified

Tegotrenn® LK 870. [Goldschmidt] Release agent for demolding of cold cured (high resilience) PU foam, esp. for sound deadening foam mats; Unverified

Tegotrenn® LK 873. [Goldschmidt] Release agent for demolding of cold cured (high resilience) PU foam, esp. for furniture seat cushions, car seats, neck and arm rests; Unverified

Tegotrenn® LK 876. [Goldschmidt] Release agent for demolding of cold cured (high resilience) PU foam, esp. for pure MDI systems; Unverified

Tegotrenn® LR 201/F, 236. [Goldschmidt] Release agents for demolding or rigid, semirigid, and integral skinned PU foam; Unverified

Tegotrenn® LR 309 F. [Goldschmidt] Release agents for demolding of rigid PU pour-in-place foams; Unverified

Tegotrenn® LR 468 F. [Goldschmidt] Release agents for demolding of rigid PU pipe insulation; high releasing and sliding props.; Unverified

Tegotrenn® LR 486. [Goldschmidt] Release agent for PU pipe insulation; Unverified

Tegotrenn® LR 560. [Goldschmidt] Release agent for demolding high dens. rigid PU foams with integral skin; Unverified

Tegotrenn® LR 808-877. [Goldschmidt AG] Release agent for demolding of rigid PU foam; paste; Unverified

Tegotrenn® LR 869. [Goldschmidt AG] Release agent for demolding of rigid and semirigid PU foam; silicone-free; Unverified

Tegotrenn® LS 584. [Goldschmidt] Disp. of release waxes in org. solvs.; release agent for polyether PU shoe sole foams with integral skins; Unverified

Tegotrenn® LS 809/F, 815. [Goldschmidt] Release agent for demolding of PU shoe soles; Unverified

Tegotrenn® LS 814 F. [Goldschmidt] Silicone contg. release agent for demolding of polyether PU shoe sole foams with integral skins; Unverified

Tegotrenn® LS 828. [Goldschmidt AG] Silicone conc. for prod. of ready-to-use release agents for demolding of ester PU shoe soles; Unverified

Tegotrenn® LS 829 F. [Goldschmidt] Release agent for demolding of polyester PU shoe soles; Unverified

Tegotrenn® LS 951-156. [Goldschmidt AG] Release agent for demolding of PU elastomers; dilutable

1:2-1:5 with demineralized water; Unverified

Tegotrenn® LS 952. [Goldschmidt] Water-based release agent for demolding microcellular PU foams, RIM parts; Unverified

Tegotrenn® LS 960. [Goldschmidt AG] Sprayable release agent for demolding of polyether and polyester PU shoe soles; solv.-free; Unverified

Tegotrenn® LS 961. [Goldschmidt AG] Sprayable release agent for demolding of PU two-color shoe soles; good compat. with lacquers; Unverified

Tegotrenn® LS 969. [Goldschmidt AG] Conc. for prod. of ready-to-use release agents for demolding of mid soles; Unverified

Tegotrenn® S 2100. [Goldschmidt] Silicone conc.; release agent for demolding of ether PU shoe soles; exc. wettability; Unverified

Tegotrenn® S 2200. [Goldschmidt AG] Silicone conc.; for prod. of ready-to-use release agents for demolding of ether PU shoe soles; Unverified

Teknor Color Crystals. [Teknor Color] Very highly loaded specialty conc. for use with PE, PP, impact PS, ABS, TPE, EVA, and flexible PVC; suitable for inj. molding, blow molding, sheet extrusion, and profile extrusion; very cost effective for use with recycled resins.

Telalloy® A-10. [Kaneka] ABS modifier; modifier providing improved heat resist., impact str., heat elong. of PVC resins; for rigid sheet, plate, pipe; *Precaution:* inflamm.; may cause dust explosions under certain conditions.

Telalloy® A-15. [Kaneka] ABS modifier; modifier providing improved heat resist., impact str., heat elong. in PVC resins; for rigid sheet, plate, pipe; *Precaution:* inflamm.; may cause dust explosions under certain conditions.

Telalloy® A-50B. [Kaneka] ABS modifier; modifier providing improved heat resist., transparency, heat elong. in PVC resins; in soft and semirigid leather, enhances the stability of embossing in thermoforming; FDA 21CFR §178.3790; *Precaution:* inflamm.; may cause dust explosions under certain conditions.

Telloy. [R.T. Vanderbilt] Tellurium; vulcanizing agent; sec. vulcanizing agent used in NR, IR, SBR; increases state of cure and improves heat aging of sulfurless and low sulfur stocks; reduces curing time in NR hard rubber and allows use of lower sulfur for increase flexibility; steel to dk. gray metallic powd., 100% thru 200 mesh; m.w. 127.61; practically insol. in water; dens. 6.26 ± 0.03 mg/m^3; m.p. > 450 C.

Temex BA-1. [Cookson Spec. Addit.] Barium-lead; heat stabilizer, lubricant for flexible and rigid vinyls; powd.

Tenox® BHA. [Eastman] BHA; CAS 25013-16-5; EINECS 246-563-8; antioxidant for foods, beverage and dessert mixes, chewing gum, flavorings, citrus oils, min. oil, nuts, vitamins, waxes, yeast, food pkg., plastics, rubber, and cosmetics; carry through effect in cooked foods, providing extended shelf life; stabilizer for fats, oils, vitamins, pet foods, etc.; FDA 21CFR §161.175, 164.110, 165.175, 166.110, 172.110, 172.515, 182.3169; USDA 9CFR 318.7, 381.147; kosher; EC listed; wh. waxy tablets or flakes, sl. odor; sol. in ethanol, diisobutyl adipate, propylene glycol, glyceryl oleate, soya oil, lard, yel. grease, and paraffin; insol. in water; m.w. 180.25; b.p. 264-270 C (733 mm); m.p. 48-63 C; *Storage:* 6 mo shelf life.

Tenox® PG. [Eastman] Propyl gallate; CAS 121-79-9; EINECS 204-498-2; nondiscoloring antioxidant, stabilizer for cellulosics, CO, ECO, EPDM, cosmetics, fats and oils incl. veg. oils, poultry fat; improves storage life of butter oils, poultry fat, etc.; often in combination with citric acid, BHA, and/or BHT; suggested for use in veg. oils in countries where TBHQ not permitted; FDA 21CFR §164.110, 165.175, 166.110, 172.615, 184.1660; USDA 9CFR 318.7, 381.147; kosher; EC listed; wh. cryst. powd., very sl. odor; sol. in ethanol, propylene glycol, glycerol, < 1% in water; m.w. 212.20; m.p. 146-150 C; b.p. 148 C dec.; *Storage:* 6 mo shelf life.

Tenox® TBHQ. [Eastman] t-Butyl hydroquinone; CAS 1948-33-0; EINECS 217-752-2; antioxidant and stabilizer used in plastics, cosmetics and in oils and fats for food applics. incl. edible fats, fish oils and prods., frying oils, margarine, nuts, poultry fat, veg. oils, waxes; used in combination with BHA and/or BHT in meat and poultry prods. to provide max. protection with USDA compliance; FDA 21CFR §164, 110, 165.175, 166.110, 172.185; USDA 9CFR 318.7, 381.147; kosher; wh. to lt. tan cryst., very sl. odor; sol. in ethanol, ethyl acetate, propylene glycol, veg. oils; < 1% in water; m.w. 166.22; m.p. 126.5-128.5 C; b.p. 300 C (760 mm); *Storage:* 6 mo shelf life.

Tensopol A 79. [Hickson Manro Ltd.] Sodium fatty alcohol sulfate; CAS 73296-89-6; anionic; surfactant for toiletries, hair care and bath prods., household detergents, industrial cleaning, emulsion polymerization, pigment dispersion; needles; 90% act.

Tensopol A 795. [Hickson Manro Ltd.] Sodium fatty alcohol sulfate; CAS 73296-89-6; anionic; surfactant for toiletries, hair care and bath prods., household detergents, industrial cleaning, emulsion polymerization, pigment dispersion; needles; 95% act.

Tensopol ACL 79. [Hickson Manro Ltd.] Broadcut sodium fatty alcohol sulfate; CAS 68955-19-1; anionic; surfactant for toiletries, hair care and bath prods., household detergents, industrial cleaning, emulsion polymerization, pigment dispersion; needles; 90% act.

Tensopol S 30 LS. [ICI plc] Sodium fatty alcohol sulfate; CAS 73296-89-6; anionic; used in latex foam, emulsion polymerization; paste; 29-31% conc.

TEPA. [Air Prods./Perf. Chems.] Tetraethylenepentamine; CAS 112-57-2; EINECS 203-986-2; R.T. epoxy curing agent; intermediates for amidoamines, imidazolines, epoxy adducts, and other addition compds.; occasionally used for floorings, sol'n. coatings, and encapsulation; FDA 21CFR §175.105, 175.300, 176.170; Gardner 5 liq.; sp.gr. 0.996; visc. 80 cps; amine no. 1340; equiv. wt. 34;

heat distort. temp. 127 F (264 psi).

Tergitol® 15-S-3. [Union Carbide] C11-15 pareth-3; CAS 68131-40-8; nonionic; biodeg. detergent, emulsifier, wetter, defoamer for aq. systems, intermediate used in textiles, solv. cleaners, drycleaning, metalworking fluids, water treatment, oilfield chems., pulp/paper deinking, latex emulsions, plastics antistat, agric.; FDA, EPA compliance; clear liq.; sol. in oil, chlorinated solvs., most org. solvs.; m.w. 336; sp.gr. 0.930; dens. 7.74 lb/gal; visc. 26 cs; HLB 8.3; hyd. no. 167; pour pt. -46 C; cloud pt. < 0 (1% aq.); flash pt. (PMCC) 174 C; pH 4.0 (1% in aq. IPA); 100% act.

Tergitol® 15-S-5. [Union Carbide] C11-15 pareth-5; CAS 68131-40-8; nonionic; biodeg. detergent, emulsifier, wetting agent, intermediate for household and industrial detergents, textiles, drycleaning, water treatment, metalworking/cleaning, oilfield chems., pulp/paper deinking, agric.; latex emulsion stabilizer; plastics antistat; defoamer for aq. systems; fuel de-icing; FDA, EPA compliance; clear liq.; sol. in oils, chlorinated solvs., most org. solvs.; m.w. 415; sp.gr. 0.965; dens. 8.03 lb/gal; visc. 35 cP; HLB 10.6; hyd. no. 135; cloud pt. < 0 C (1% aq.); flash pt. (PMCC) 178 C; pour pt. -24 C; pH 6-8 (1%); 100% act.

Tergitol® 15-S-30. [Union Carbide] C11-15 pareth-30; CAS 68131-40-8; nonionic; emulsifier, detergent, wetting agent for use at elevated temps., in presence of strong electrolytes; for alkaline industrial/household cleaners/degreasers, textile scouring, dye carriers, vinyl acetate and acrylate polymerization; stabilizer for syn. latexes; demulsifier for petrol. oil emulsions; FDA, EPA compliance; wh. waxy solid; sol. in water, chlorinated solvs., polar org. solvs.; m.w. 1558; sp.gr. 1.055; dens. 8.78 lb/gal (55 C); visc. 92 cP (50 C); HLB 17.5; hyd. no. 36; pour pt. 39 C; cloud pt. > 100 C (1% aq.); flash pt. (PMCC) 249 C; pH 6.5 (1% aq.); surf. tens. 39 dynes/cm (0.1% aq.); 100% act.

Tergitol® 15-S-40. [Union Carbide] C11-15 pareth-40; CAS 68131-40-8; nonionic; emulsifier, detergent, wetting agent for use at elevated temps. in presence of strong electrolytes; for alkaline industrial/household cleaners/degreasers, textile scouring, dye carriers, vinyl acetate and acrylate polymerization; stabilizer for syn. latexes; demulsifier for petrol. oil emulsions; FDA, EPA compliance; wh. waxy solid; sol. in water, chlorinated solvs., polar org. solvs.; m.w. 2004; sp.gr. 1.061; dens. 8.83 lb/gal (55 C); visc. 166 cP (50 C); HLB 18.0; hyd. no. 28; pour pt. 44 C; cloud pt. > 100 C (1% aq.); flash pt. (PMCC) 252 C; pH 7.0 (1% aq.); surf. tens. 42 dynes/cm (0.1% aq.); 100% act.

Tergitol® D-683. [Union Carbide] Alkoxylated alkylphenol; CAS 37251-69-7; nonionic; emulsifier for fiber finishing operations; dispersant for pigments in resins, plastics and for abrasives in hard surf. cleaners; improves wetting of oil-based materials in coatings and adhesives; clear liq.; sol. in water, chlorinated solvs., most org. solvs.; m.w. 1004; sp.gr. 1.019 (30/20 C); dens. 8.48 lb/gal (50 C); visc. 320 cP; hyd. no. 56; pour pt. -1 C; cloud pt. 21 C (1% aq.); flash pt. (PMCC) 210 C; pH 6.4 (1% aq.); surf. tens. 32 dynes/cm (0.1% aq.); Ross-Miles foam 38 mm (0.1%, initial); 100% act.

Tergitol® NP-40. [Union Carbide] Nonoxynol-40; CAS 9016-45-9; nonionic; detergent, wetting agent, asphalt emulsifier, emulsion polymerization, leveling agent; wh. solid; mild char. odor; m.w. 1980; water-sol.; sp.gr. 1.080; dens. 8.97 lb/gal (55 C); HLB 17.8; m.p. 48 C; b.p. > 250 C; flash pt. 525 F(COC); cloud pt. 100 (0.5%); pH 4.0-8.0 (10% aq.); surf. tens. 45 dynes/cm (0.1% aq.); 100% act.

Tergitol® NP-40 (70% Aq.). [Union Carbide] Nonoxynol-40; CAS 9016-45-9; nonionic; surfactant for emulsion polymerization; leveling agent; clear liq.; mild odor; water-sol.; sp.gr. 1.104; dens. 9.19 lb/gal; visc. 533 cs; solid. pt. -2 C; 70% act.

Tergitol® XD. [Union Carbide] PPG-24-buteth-27; CAS 9038-95-3; nonionic; emulsifier, dispersant, stabilizer for agric. insecticides/herbicides, latex polymerization, iodophor mfg. for germicidal cleaning, latex paints, dye pigments, leather; emulsifier for silicone oils, diacyl peroxides; FDA, EPA compliance; wh. waxy solid; sol. in chlorinated solvs., many polar org. solvs.; m.w. 3117; sp.gr. 1.041; dens. 8.66 lb/gal (40 C); visc. 251 cP (50 C); solid. pt. 33 C; hyd. no. 18; pour pt. 35 C; cloud pt. 76 C (1% aq.); flash pt. (PMCC) 208 C; pH 6.5 (20% aq.); surf. tens. 38 dynes/cm (0.1% aq.); Ross-Miles foam 76 mm (0.1%, initial); 100% act.

Tergitol® XH. [Union Carbide] EO/PO copolymer; CAS 9038-95-3; nonionic; emulsifier, dispersant for agric., latex polymerization, iodophor mfg. for germicidal cleaning, latex paints, dye pigments, leather finishes, toilet bowl cleaners; emulsifier for silicone oils, diacyl peroxides; FDA, EPA compliance; wh. waxy solid; sol. in chlorinated solvs., many polar org. solvs.; m.w. 3740; sp.gr. 1.048; dens. 8.72 lb/gal (50 C); visc. 319 cP (50 C); solid. pt. 41 C; hyd. no. 15; pour pt. 44 C; cloud pt. 99 C (1% aq.); flash pt. (PMCC) 144 C; pH 5.4 (10% aq.); surf. tens. 39 dynes/cm (0.1% aq.); Ross-Miles foam 58 mm (0.1% aq., initial); 100% act.

Tergitol® XJ. [Union Carbide] EO/PO copolymer; CAS 9038-95-3; nonionic; emulsifier, dispersant, stabilizer for latex polymerization, agric., latex paints, iodophors, dye pigments, leather finishes; emulsifier for silicone oils; FDA, EPA compliance; wh. waxy solid; sol. in chlorinated solvs., many polar org. solvs.; m.w. 2550; sp.gr. 1.023; dens. 8.51 lb/gal (45 C); visc. 149 cP (50 C); hyd. no. 22; pour pt. 26 C; cloud pt. 50 C (1% aq.); flash pt. (PMCC) 216 C; pH 6.5 (1% aq.); surf. tens. 36 dynes/cm (0.1% aq.); Ross-Miles foam 50 mm (0.1% aq., initial); 100% act.

Teric 16A22. [ICI Australia] Ceteareth-22; CAS 68439-49-6; nonionic; emulsifier for agric.; enzyme coating; laundry detergents; metal soaking; rubber latex stabilization; Hazen 100 flakes; sol. in water, benzene, ethyl acetate, ethyl Icinol, perchlorethylene, ethanol; sp.gr. 1.01 (50 C); visc. 92 cps (50 C); m.p. 40 ± 2 C; HLB 15.8; cloud pt. > 100 C; surf.

tens. 41.3 dynes/cm; pH 6-8 (1% aq.); biodeg.; 100% act.

Teric 18M5. [ICI Australia] PEG-5 stearamine; CAS 26635-92-7; nonionic; wetting agent, dispersant; emulsion stabilizer; emulsifier used in agric. toxicants and processing of textiles, paper, leather and bldg. board; corrosion inhibitor in lubricants and greases; softener and antistat in solv. cleaning of textiles and compd. of plastics; solid; sol. in benzene, ethyl acetate, ethyl Icinol, perchlorethylene, ethanol, kerosene, min., paraffin and veg. oil, olein; disp. in water; sp.gr. 0.935 (50 C); visc. 46 cps (50 C); m.p. 35 ± 2 C; HLB 9.5; pH 8-10 (1% aq.); 100% act.

Teric 18M10. [ICI Australia] PEG-10 stearamine; CAS 26635-92-7; nonionic; wetting agent, dispersant; emulsion stabilizer; emulsifier used in agric. toxicants, textiles, paper, leather and bldg. board; corrosion inhibitor in lubricants and greases; softener, antistat for plastics, textiles; liq.; sol. in water, benzene, ethyl acetate, ethyl Icinol, perchlorethylene, ethanol, kerosene, veg. oil, olein; sp.gr. 0.997; visc. 255 cps; m.p. 18 ± 2 C; HLB 11.2; cloud pt. > 100 C; surf. tens. 34.0 dynes/cm; pH 8-10 (1% aq.); 100% act.

Teric C12. [ICI Australia] PEG-12 castor oil; CAS 61791-12-6; nonionic; detergent, emulsifier and coemulsifier in formulation of sol. oils, solv. cleaners and temp. protective coatings; corrosion resistant, used in industrial and institutional cleaning preparations; die lubricant in metal forming operations; mold release agent in plastics; fiber lubricant in textile processing; Hazen > 250 liq.; sol. in water, benzene, ethyl acetate, ethyl Icinol, ethanol, veg. oil, olein; sp.gr. 1.048; visc. 5000 cps; m.p. 3 ± 2 C; HLB 12.7; cloud pt. 67-69 C; surf. tens. 42.0 dynes/cm; pH 6-8 (1% aq.); 100% act.

Teric N8. [ICI Australia] Nonoxynol-8 (8.5 EO); CAS 9016-45-9; nonionic; wetting agent, dispersant, emulsifier, solubilizer, detergent; used in concrete mfg., agric. sprays, solv. cleaners, paints; detergents, emulsion polymerization; Hazen 100 liq.; sol. in water, benzene, ethyl acetate, ethyl Icinol, perchlorethylene, ethanol, veg. oil, olein; sp.gr. 1.056; visc. 350 cps; m.p. 0 ± 2 C; HLB 12.3; cloud pt. 30-34 C; surf. tens. 29.4 dynes/cm; pH 6-8 (1% aq.); 100% act.

Teric N9. [ICI Australia] Nonoxynol-9; CAS 9016-45-9; nonionic; wetting agent, dispersant, emulsifier, solubilizer, detergent; used in concrete mfg., agric. sprays, solv. cleaners, paints; detergents, emulsion polymerization; Hazen 100 liq.; sol. in water, benzene, ethyl acetate, ethyl Icinol, perchlorethylene, veg. oil, olein; sp.gr. 1.060; visc. 330 cps; m.p. 0 ± 2 C; HLB 12.8; cloud pt. 50-55 C; surf. tens. 30.6 dynes/cm; pH 6-8 (1% aq.); 100% act.

Teric N10. [ICI Australia] Nonoxynol-10; CAS 9016-45-9; nonionic; wetting agent, dispersant, emulsifier, solubilizer, detergent; used in concrete mfg., agric. sprays, solv. cleaners, paints; detergents, emulsion polymerization; Hazen 100 liq.; sol. in water, benzene, ethyl acetate, ethyl Icinol, perchlorethylene, veg. oil, olein; sp.gr. 1.063; visc. 360 cps; m.p. 5 ± 2 C; HLB 13.3; cloud pt. 67 ± 2 C; surf. tens. 30.6 dynes/cm; pH 6-8 (1% aq.); 100% act.

Teric N12. [ICI Australia] Nonoxynol-12; CAS 9016-45-9; nonionic; wetting agent, dispersant, emulsifier, solubilizer, detergent; used in concrete mfg., agric. sprays, solv. cleaners, paints; detergents, emulsion polymerization; Hazen 100 liq.; sol. in water, benzene, ethyl acetate, ethyl Icinol, perchlorethylene, ethanol, olein; sp.gr. 1.045 (50 C); visc. 67 cps (50 C); m.p. 11 ± 2 C; HLB 13.9; cloud pt. 82 ± 2 C; surf. tens. 35.2 dyens/cm; pH 6-8 (1% aq.); 100% act.

Teric N13. [ICI Australia] Nonoxynol-13; CAS 9016-45-9; nonionic; wetting agent, dispersant, emulsifier, solubilizer, detergent; used in concrete mfg., agric. sprays, solv. cleaners, paints; detergents, emulsion polymerization; Hazen 100 liq.; sol. in water, benzene, ethyl acetate, ethyl Icinol, perchlorethylene, ethanol, olein; sp.gr. 1.049 (50 C); visc. 75 cps (50 C); m.p. 14+2 C; HLB 14.4; cloud pt. 89 ± 2 C; surf. tens. 34.9 dynes/cm; pH 6-8 (1% aq.); 100% act.

Teric N15. [ICI Australia] Nonoxynol-15; CAS 9016-45-9; nonionic; wetting agent, coemulsifier; used in paints, perfumes, detergents and sanitizers, emulsion polymerization; Hazen 100 paste; sol. in water, benzene, ethyl acetate, ethyl Icinol, perchlorethylene, ethanol, olein; sp.gr. 1.051 (50 C); visc. 82 cps (50 C); m.p. 21 ± 2 C; HLB 15.0; cloud pt. 95 ± 2 C; surf. tens. 33.4 dynes/cm; pH 6-8 (1% aq.); 100% act.

Teric N20. [ICI Australia] Nonoxynol-20; CAS 9016-45-9; nonionic; wetting agent, coemulsifier; used in paints, perfumes, detergents and sanitizers, emulsion polymerization, latex stabilization; Hazen 100 solid; sol. in water, benzene, ethyl acetate, ethyl Icinol, perchlorethylene, ethanol, olein; sp.gr. 1.061 (50 C); visc. 100 cps (50 C); m.p. 30 ± 2 C; HLB 16.0; hyd. no. 42-57; cloud pt. > 100 C; surf. tens. 41.7 dynes/cm; pH 6-8 (1% aq.); 100% act.

Teric N30. [ICI Australia] Nonoxynol-30; CAS 9016-45-9; nonionic; wetting agent, coemulsifier; used in paints, perfumes, detergents and sanitizers, emulsion polymerization; Hazen 150 flakes; sol. in water, benzene, ethyl acetate, ethyl Icinol, perchlorethylene, ethanol; sp.gr. 1.066 (50 C); visc. 150 cps (50 C); m.p. 40 ± 2 C; HLB 17.2; hyd. no. 31-36; cloud pt. > 100 C; surf. tens. 42.8 dynes/cm; pH 6-8 (1% aq.); 100% act.

Teric N40. [ICI Australia] Nonoxynol-40; CAS 9016-45-9; nonionic; wetting agent, coemulsifier; used in paints, perfumes, detergents and sanitizers, emulsion polymerization; Hazen 200 flakes; sol. in water, benzene, ethyl acetate, ethyl Icinol, perchlorethylene, ethanol; sp.gr. 1.080 (50 C); visc. 250 cps (50 C); m.p. 45 ± 2 C; HLB 18.0; hyd. no. 25-32; cloud pt. > 100 C; surf. tens. 41.0 dynes/cm; pH 6-8 (1% aq.); 100% act.

Teric PE68. [ICI Australia] POP + 150 EO; nonionic; wetting agent, dispersant, emulsifier, detergent, intermediate; stabilizer in mfg. of latex paints, paper coatings, dairy cleaners and santizers; plasticizer for resins; Hazen 100 flakes; sol. in water, benzene, ethyl acetate, ethyl Icinol, perchlorethylene, ethanol, olein; m.p. 50 ± 2 C; HLB 29; cloud pt. > 100 C; surf. tens. 49.0 dynes/cm; pH 6 (1% aq.); 100% act.

Teric PEG 300. [ICI Australia] PEG 300; CAS 25322-68-3; EINECS 220-045-1; pesticide solubilizer/carrier; visc. modifier for brake fluids; intermediate for PEG esters, PU foams; plasticizer/solv.; cosmetics; metalworking lubricants; paints/resins; paper/film; pharmaceuticals; printing inks; rubber release; textile aux.; biodeg.; APHA 25 color; sol. in water and most polar org. solvs.; m.w. 285-315; sp.gr. 1.128; visc. 5.8 cst (99 C); pour pt. -12 C; flash pt. (OC) > 190 C; ref. index 1.463; pH 4.5-7.5; surf. tens. 44.6 dynes/cm; 99% act.

Teric PEG 400. [ICI Australia] PEG 400; CAS 25322-68-3; EINECS 225-856-4; biodeg.; emulsifier, antistat for textiles; rubber; inks; cosmetics; pharmaceuticals; paper/film; pesticide solubilizer/carrier; intermediate for PEG esters, PU foams; plasticizer/solv. for cork; metalworking lubricants; paints/resins; APHA 25 color; sol. in water and most polar org. solvs.; m.w. 380-420; sp.gr. 1.130; visc. 7.3 cst (99 C); pour pt. 6 C; flash pt. (OC) > 215 C; ref. index 1.465; pH 4.5-7.5; surf. tens. 44.6 dynes/cm; 99% act.

Teric PEG 600. [ICI Australia] PEG 600; CAS 25322-68-3; EINECS 229-859-1; biodeg.; emulsifier for textiles; pesticide solubilizer/carrier; binder for ceramics; intermediate for PEG esters, PU foams; cosmetics; pharmaceuticals; metalworking lubricants; resins; paper/film; inks; rubber release; APHA 10 color (25% aq.); sol. in water and most polar org. solvs.; m.w. 560-630; sp.gr. 1.127; visc. 10.4 cst (99 C); pour pt. 19 C; flash pt. (OC) > 230 C; ref. index 1.454; pH 4.5-7.5; surf. tens. 44.6 dynes/cm; 99% act.

Teric PEG 1500. [ICI Australia] PEG 1500; biodeg.; textile finishing and sizing; wood processing; latex lubricant; printing inks; pharmaceuticals; paper/film; APHA 30 color (25% aq.); sol. in water and most polar org. solvs.; m.w. 1430-1570; sp.gr. 1.208; visc. 28.4 cst (99 C); pour pt. 46 C; flash pt. (OC) > 260 C; ref. index 1.456; pH 4.5-7.5; surf. tens. 53.1 dynes/cm (50% aq.); 100% act.

Teric PEG 4000. [ICI Australia] PEG 4000; CAS 25322-68-3; biodeg.; binder/plasticizer for ceramics; intermediate for copolymers, PEG esters; cosmetics; pharmaceticals; metalworking lubricants and electropolishes; resins; paper/film; printing inks; rubber antistat, release, compding. aid; textile aux.; APHA 50 color (25% aq.); sol. in water and most polar org. solvs.; m.w. 3300-4000; sp.gr. 1.217; visc. 130-180 cst (99 C); pour pt. 56 C; flash pt. (OC) > 260 C; ref. index 1.456; pH 4.5-7.5; surf. tens. 54.4 dynes/cm (50% aq.); 100% act.

Teric PEG 8000. [ICI Australia] PEG 8000; CAS 25322-68-3; biodeg.; intermediate for copolymers, PEG esters; cosmetics; pharmaceuticals; thickener for inks; release for rubber molding; sol. in water and most polar org. solvs.; m.w. 7000-9000; sp.gr. 1.2; visc. 700-900 cst (99 C); pour pt. 55-60 C; flash pt. (OC) > 260 C; ref. index 1.456; pH 4.5-7.5; 100% act.

Teric PPG 1000. [ICI Australia] PPG 1000; biodeg.; intermediate for surfactants, ethers, esters; antifoam for ceramics, rubber; hydraulic brake fluids; dyeing; metalworking lubricants; plasticizer for plastics; latex coagulant; textile lubricant; APHA 100 color; sol. in polar org. solvs.; sp.gr. 1.006; visc. 196 cps; hyd. no. 106-118; pour pt. -36 C; flash pt. (OC) 227 C; ref. index 1.448; surf. tens. 31.3 dynes/cm; 99% act.

Teric PPG 1650. [ICI Australia] PPG 1650; biodeg.; intermediate for surfactants, ethers, esters; antifoam for ceramics, rubber; lubricant/softener for leather; metalworking lubricant; demulsifier for petrol. industry; plasticizer for plastics; solv. for cosmetics; sol. in polar org. solvs.; sp.gr. 1.006; visc. 330 cps; hyd. no. 67-71; pour pt. -32 C; flash pt. (OC) 230 C; ref. index 1.449; pH 5-8 (1% aq. disp.); surf. tens. 31.8 dynes/cm; 99% act.

Teric PPG 2250. [ICI Australia] PPG 2250; biodeg.; intermediate for surfactants, ethers, esters; hydraulic brake fluids; cosmetics; dyeing; lubricant/softener for leather; metalworking lubricant; demulsifier for petrol.; latex coagulant; rubber release agent; solv. for veg. oils; APHA 50 color; sol. in polar org. solvs.; sp.gr. 1.005; visc. 482 cps; hyd. no. 47-54; pour pt. -29 C; flash pt. (OC) 240 C; ref. index 1.449; pH 4.0-7.5 (10% in aq. methanol); surf. tens. 32.1 dynes/cm; 99% act.

Teric PPG 4000. [ICI Australia] PPG 4000; biodeg.; intermediate for surfactants, ethers, esters; cosmetics; lubricant/softener for leather; metalworking lubricants; demulsifier for petrol. industry; mold release for rubber; APHA 250 color; sol. in polar org. solvs.; sp.gr. 1.004; visc. 1232 cps; hyd. no. 26-30; pour pt. -30 C; flash pt. (OC) 240 C; ref. index 1.450; pH 5-8 (10% in aq. methanol); surf. tens. 32.2 dynes/cm; 99% act.

TETA. [Air Prods./Perf. Chems.] Triethylenetetramine; CAS 112-24-3; EINECS 203-950-6; R.T. epoxy curing agent; intermediates for amidoamines, imidazolines, epoxy adducts, and other addition compds.; occasionally used for floorings, sol'n. coatings, and encapsulation; FDA 21CFR §175.300, 176.170; Gardner 1 liq.; sp.gr. 0.984; visc. 20 cps; amine no. 1540; equiv. wt. 27; heat distort. temp. 126 F (264 psi).

Tetralin. [DuPont] 1,2,3,4-Tetrahydronaphthalene; CAS 119-64-2; EINECS 204-340-2; solv. for oils, resins, waxes, rubber, and asphalt; colorless to pale yel. clear liq.; m.w. 132.2; f.p. -31 C; b.p. 207 C (760 mm Hg); negligible sol. in water; sp.gr. 0.970; dens. 8.1 lb/gal; flash pt. 82 C (TCC); 97.0% min. tetrahydronaphthalene.

Tetrathal. [Monsanto] Tetrachlorophthalic anhydride; flame retardant for epoxy, phenolic, thermoset polyester, rigid PU foam.

Tetronic® 90R4. [BASF] EO/PO ethylene diamine block copolymer; CAS 107397-59-1; nonionic; surfactant series functioning as emulsion stabilizers, solubilizers, dispersants, wetting agents, antistats, penetrants, plasticizers, defoaming agents, demulsifiers in the petrol., paint, paper, cement, ink, cosmetic, drug, plastic, detergent, and metalworking industries; rubber activator; R series for low foaming applics.; liq.; m.w. 7240; visc. 3870 cps; HLB 1-7; pour pt. 12 C; cloud pt. 43 C (1% aq.); surf. tens. 43 dynes/cm (0.1%); 100% act.; *Toxicology:* minimal skin and minimal to mild eye irritation.

Tetronic® 150R1. [BASF] EO/PO ethylene diamine block copolymer; CAS 107397-59-1; nonionic; surfactant series functioning as emulsion stabilizers, solubilizers, dispersants, wetting agents, antistats, penetrants, plasticizers, defoaming agents, demulsifiers in the petrol., paint, paper, cement, ink, cosmetic, drug, plastic, detergent, and metalworking industries; rubber activator; R series for low foaming applics.; liq.; m.w. 8000; visc. 1840 cps; HLB 1-7; pour pt. -17 C; cloud pt. 20 C (1% aq.); surf. tens. insol. (0.1%); 100% act.; *Toxicology:* minimal skin and minimal to mild eye irritation.

Tetronic® 304. [BASF] Poloxamine 304; CAS 11111-34-5; nonionic; emulsifier, thickener, wetting agent, dispersant, solubilizer, stabilizer for cosmetics and pharmaceuticals; demulsifier in petrol. industry; detergent ingred.; antistat for polyethylene and resin molding powds.; metal treatment; emulsion polymerization; used in latex-based paints, aq.-based syn. cutting fluids and vulcanization of rubber; colorless liq.; m.w. 1650; ref. index 1.4649; water-sol.; sp.gr. 1.06; visc. 450 cps; HLB 16; cloud pt. 94 C (1%); pour pt. -11 C; surf. tens. 53.0 dynes/cm (0.1%); 100% act.; *Toxicology:* none to mild eye and minimal to moderate skin irritation.

Tetronic® 701. [BASF] Poloxamine 701; CAS 11111-34-5; nonionic; emulsifier, thickener, wetting agent, dispersant, solubilizer, stabilizer for cosmetics and pharmaceuticals; demulsifier in petrol. industry; detergent ingred.; antistat for polyethylene and resin molding powds.; metal treatment; emulsion polymerization; used in latex-based paints, aq.-based syn. cutting fluids and vulcanization of rubber; colorless liq.; m.w. 3400; ref. index. 1.4553; sp.gr. 1.02; visc. 575 cps; HLB 3; cloud pt. 22 C (1%); pour pt. -21 C; surf. tens. 36.1 dynes/cm; 100% act.; *Toxicology:* none to mild eye and minimal to moderate skin irritation.

Tetronic® 704. [BASF] Poloxamine 704; CAS 11111-34-5; nonionic; emulsifier, thickener, wetting agent, dispersant, solubilizer, stabilizer for cosmetics and pharmaceuticals; demulsifier in petrol. industry; detergent ingred.; antistat for polyethylene and resin molding powds.; metal treatment; emulsion polymerization; used in latex-based paints, aq.-based syn. cutting fluids and vulcanization of rubber; colorless liq.; m.w. 5500; ref. index 1.4613; water-sol.; sp.gr. 1.04; visc. 850 cps; HLB 15; cloud pt. 65 C (1%); pour pt. 18 C; surf. tens. 40.3 dynes/cm (0.1%); 100% act.; *Toxicology:* none to mild eye and minimal to moderate skin irritation.

Tetronic® 901. [BASF] Poloxamine 901; CAS 11111-34-5; nonionic; emulsifier, thickener, wetting agent, dispersant, solubilizer, stabilizer for cosmetics and pharmaceuticals; demulsifier in petrol. industry; detergent ingred.; antistat for polyethylene and resin molding powds.; metal treatment; emulsion polymerization; used in latex-based paints, aq.-based syn. cutting fluids and vulcanization of rubber; colorless liq.; m.w. 4750; ref. index 1.4545; sp.gr. 1.02; visc. 700 cps; HLB 2.5; cloud pt. 20 C (1%); pour pt. -23 C; 100% act.; *Toxicology:* none to mild eye and minimal to moderate skin irritation.

Tetronic® 904. [BASF] Poloxamine 904; CAS 11111-34-5; nonionic; emulsifier, thickener, wetting agent, dispersant, solubilizer, stabilizer for cosmetics and pharmaceuticals; demulsifier in petrol. industry; detergent ingred.; antistat for polyethylene and resin molding powds.; metal treatment; emulsion polymerization; used in latex-based paints, aq.-based syn. cutting fluids and vulcanization of rubber; colorless liq.; m.w. 7500; ref. index 1.4604; water-sol.; sp.gr. 1.04; visc. 6000 cps; HLB 14.5; cloud pt. 64 C (1%); pour pt. 29 C; surf. tens. 35.4 dynes; 100% act.; *Toxicology:* none to mild eye and minimal to moderate skin irritation.

Tetronic® 908. [BASF] Poloxamine 908; CAS 11111-34-5; nonionic; emulsifier, thickener, wetting agent, dispersant, solubilizer, stabilizer for cosmetics and pharmaceuticals; demulsifier in petrol. industry; detergent ingred.; antistat for polyethylene and resin molding powds.; metal treatment; emulsion polymerization; used in latex-based paints, aq.-based syn. cutting fluids and vulcanization of rubber; flakes; m.w. 27,000; water-sol.; m.p. 58 C; HLB 30.5; cloud pt. > 100 C (1%); surf. tens. 45.7 dynes/cm (0.1%); 100% act.; *Toxicology:* none to mild eye and minimal to moderate skin irritation.

Tetronic® 1307. [BASF] Poloxamine 1307; CAS 11111-34-5; nonionic; emulsifier, thickener, wetting agent, dispersant, solubilizer, stabilizer for cosmetics and pharmaceuticals; demulsifier in petrol. industry; detergent ingred.; antistat for polyethylene and resin molding powds.; metal treatment; emulsion polymerization; used in latex-based paints, aq.-based syn. cutting fluids and vulcanization of rubber; solid; m.w. 18,600; water-sol.; m.p. 54 C; HLB 23.5; cloud pt. > 100 C (1%); surf. tens. 43.8 dynes/cm (0.1%); 100% act.; *Toxicology:* none to mild eye and minimal to moderate skin irritation.

Texacar® EC. [Texaco] Ethylene carbonate; CAS 96-49-1; solv. for org. and inorg. materials; Rule 66 exempt; also used as reactant and plasticizer in fibers and textiles, plastics and resins, aromatic hydrocarbon extraction, electrolytes, hydraulic brake fluids; wh. cake solid, mild odor; melts to clear liq. > 95 F; m.w. 88.06; sp.gr. 1.3218 (39/4 C); dens. 10.87 lb/gal (40 C); visc. 1.9 cs (40 C); f.p. 35.5 C min.; b.p. 248.2 C (760 mm); flash pt. 305 F; ref. index 1.4158; pH 6.5-7.5 (10% aq.); 99% min. assay.

Texacar® PC. [Texaco] Propylene carbonate; CAS 108-32-7; EINECS 203-572-1; solv. for org. and inorg. materials; Rule 66 exempt; also used as reactant and plasticizer in fibers and textiles, hydraulic fluids, plastics and resins, gas treating, aromatic hydrocarbon extraction, metal extraction, surf. coatings, foundry sand binders, lubricants, electrolytes, personal care prods.; gellant for clays in greases and cosmetics; clear mobile liq., pract. odorless; m.w. 102.09; sp.gr. 1.2057 (20/4 C); dens. 10.05 lb/gal; visc. 2.8 cs; b.p. 241.9 C (760 mm); f.p. -49.2 C; flash pt. (PMCC) 275 F; pour pt. -100 F; ref. index 1.4210; pH 6.5-7.5 (10% aq.); 99.4% min. assay.

Texacure EA-20. [Huntsman] Diethylenetriamine; CAS 111-40-0; EINECS 203-865-4; epoxy curing agent featuring rapid R.T. cure, good adhesive and mech. props., good elec. and insulating props., good chem. and corrosion resist.; water-wh. clear liq., strong amine odor; hygroscopic; Pt-Co 30 max. clear liq.; m.w. 103.2; sp.gr. 0.954 (20/20 C); visc. 4 cs (100 F); vapor pressure 0.4 mm Hg (20 C); f.p. -39 C; b.p. 207 C; flash pt. (TCC) 204 F; 98.5% min. act.; *Toxicology:* causes severe eye and skin burns; may cause allergic skin and respiratory reactions; *Storage:* store under dry nitrogen blanket; prod. will discolor on exposure to air.

Texacure EA-24. [Huntsman] High boiling mixt. of triethylenetetramine components; CAS 112-23-3; EINECS 203-950-6; epoxy curing agent featuring rapid R.T. cure, good adhesive and mech. props., good elec. and insulating props., good chem. resist.; Pt-Co 100 max. clear liq.; m.w. 146; dens. 8.02 lb/gal; visc. 12 cp; 95% min. act.; *Toxicology:* causes severe skin and eye burns; may cause allergic skin and respiratory reactions; *Storage:* discolors on exposure to air; hygroscopic; store under dry nitrogen gas blanket.

Texadril 8780. [Henkel/Functional Prods.] EO/PO block copolymer; nonionic; stabilizer, emulsifier esp. for vinyl acetate homo- and copolymers; solid; 100% act.

Texanol® Ester-Alcohol. [Eastman] 2,2,4-Trimethyl-1,3-pentanediol monoisobutyrate; CAS 25265-77-4; slow-evaporating solv. used as coalescing agent in latex finishes and water-base inks, PVAc latices, PVAc-acrylic latices, and acrylic, EVA, and B/S latices; used as solv. in electrodeposition coatings, lacquer coatings; defoamer in waterborne systems and drilling muds for oil industry; chemical intermediate for prod. of esters. used as stain-resistance plasticizers, lubricant base stocks, solvs., syn. detergents, and herbicides; Pt-Co 20 max. clear liq.; mild char. odor; m.w. 216.3; f.p. -50 C; b.p. 244 C min. (initial, 760 mm); water-insol.; sp.gr. 0.950; dens. 0.95 kg/l; pour pt. -57 C; ref. index 1.4423; flash pt. 120 C (COC); fire pt. 132 C; > 99% volatiles by vol.

Texapon® K-12. [Henkel/Cospha; Henkel/Functional Prods.] Sodium lauryl sulfate; CAS 151-21-3; anionic; emulsifier for emulsion polymerization; detergent for high-solids cleansers; also for dispersants, wettable powds.; wh. powd.; > 90% act.

Texapon® K-1296. [Henkel/Cospha; Henkel/Functional Prods.] Sodium lauryl sulfate; CAS 151-21-3; anionic; wetting agent and detergent for cleaning formulations; additive for mech. latex foaming; wh. powd.; > 96% act.

Texapon® NSE. [Pulcra SA] Sodium laureth sulfate; CAS 9004-82-4; anionic; surfactant for emulsion polymerization; liq.; m.w. 405; visc. < 350 mPa•s; pH 6.5-7.5 (10%); 26-28% act.

Texapon® OT Highly Conc. Needles. [Henkel/Functional Prods.; Henkel KGaA/Cospha] Sodium lauryl sulfate C12-C18; CAS 151-21-3; EINECS 205-788-1; anionic; detergent, shampoo base; for bubble bath, soaps; emulsifier for emulsion polymerization, additive for mech. latex foaming, carpet and upholstery cleaners; needles; 91-92% act.

Texapon® VHC Needles. [Henkel/Cospha; Henkel/Functional Prods; Henkel Canada; Henkel KGaA/Cospha] Sodium lauryl sulfate; CAS 151-21-3; EINECS 205-788-1; anionic; wetting agent, foamer, emulsifier, detergent and cosmetic base for personal care prods., scouring agents, pigment dispersions, emulsion polymerization, PVC processing; wh. fine needles; pH 7-8 (1% aq.); 86-90% act.

Texapon® ZHC Powder. [Henkel/Cospha; Henkel/Functional Prods.] Sodium lauryl sulfate C12-C18; CAS 151-21-3; EINECS 205-788-1; foaming dispersion and wetting agent for bubble baths, cosmetic cleansing creams and emulsions, emulsion polymerization, mech. latex foaming, carpet and upholstery cleaners; wh. fine powd.; pH 7.5-8.5 (0.25% aq.); 88% act.

Texicryl® 13-300, 13-302. [Scott Bader] Carboxylated acrylic copolymer aq. emulsions; emulsions used as thickening agents for natural and syn. latices; low visc., high m.w.; useful in formulating chemical resistant, high PVC sheen, and semigloss emulsion paints; 13-300 is more effective in bridging across latex particles; for thickening most syn. rubber latices and their compds.; both suitable in thickening acrylic emulsions used as print bonding adhesives in the prod. of nonwoven fabrics; emulsion; 0.2 μ particle size; sp.gr. 1.08; visc. 0.05-0.3 poise; pH 2-4; 40 and 35 ± 1% solids resp.; Unverified

Texlin® 300. [Huntsman] Triethylenetetramine; CAS 112-24-3; EINECS 203-950-6; improves yields and reduces processing time for many applics.; intermediate in asphalt additives, corrosion inhibitors, epoxy curing agents, surfactants in fabric softener and textile additives; in paper industry, petrol. prods.; Pt-Co 30 max. clear liq.; m.w. 146.2; sp.gr. 0.9918 (20/20 C); dens. 8.26 lb/gal; visc. 21 cSt; vapor pressure < 0.01 mm Hg (20 C); f.p. -41.9 C; b.p. 261 C; amine no. 1475; flash pt. (TCC) 310 F; pH 11 (25% aq.); 98.1% min. purity; *Precaution:* corrosive liq. (DOT).

Tex-Wet 1104. [Calgon] Phosphate, acid form; anionic; detergent, wetting agent, emulsifier for heavy-duty all-purpose liqs., pesticides, polymerization, drycleaning; liq.; 100% conc.

Tex-Wet 1143. [Calgon] Phosphate ester, acid form; anionic; detergent, wetting agent, emulsifier for pesticides, polymerization, drycleaning, dye carriers; liq.; 100% conc.

Thanecure®. [TSE Industries; Sovereign] Zinc chloride with benzothiazyl disulfide; CAS 14239-75-9; vulcanization activator for sulfur-curable millable urethane elastomer; cream colored powd., amine odor; sl. sol. in water; mod. sol. in benzene; sp.gr. 1.85; m.p. 290 ; flash pt. (OC) > 204 C.

Thermact® 1. [Olin] Amine; catalyst developed to allow the use of systems based on Thermolin RF-230 in two-component processing for foam; very sol. in ethanol, acetone, benzene, chloroform; sol. in water; slightly sol. in dioctyl phthalate; insol. in ether; sp.gr. 1.096 (24 C); visc. 207 cp; hyd. no. 520; pH 7.2 (10% aq.); flash pt. (COC) 98-99 C; Unverified

Thermacure®. [Cook Composites & Polymers] Dicumyl peroxide; CAS 80-43-3; vulcanizing agent or polymerizing catalyst in rubber and plastics industries, esp. for unsat. polyester resins, bulk polymerization of vinyl monomers such as styrene; wh. cryst. powd.; m.w. 270.37; sol. in org. solvs., syn. and natural rubbers, polyester resins; insol. in water; sp.gr. 1.012; flash pt. (Seta) 121 C; 99% min. assay; 5.87% act. oxygen; Unverified

Thermax® Floform® N-990. [Cancarb; R.T. Vanderbilt] Med. thermal carbon blk.; CAS 1333-86-4; EINECS 215-609-9; functional filler in rubber prods. with high loadings of carbon blk; reducing agent in metallurgy; uv protectant in wire and cable insulation jacket; also for refractories, adhesives, advanced ceramics, audio/video magnetic tape, carbon electrodes and seals, glass prod., high-temp. insulation; soft pellets; 270 nm mean particle diam.; 0.0015% 325-mesh residue; sp.gr. 1.84; pour dens. 40 lb/ft^3; surf. area 7-12 m^2/g; pH 9-11; 99.5% min. carbon content.

Thermax® Floform® Ultra Pure N-990. [Cancarb; R.T. Vanderbilt] Med. thermal carbon blk. with highest carbon content and extremely low in ash and sulfur; CAS 1333-86-4; EINECS 215-609-9; used where low impurities are required; functional filler in fluoroelastomers; reducing agent/carburizer in metallurgy; also in advanced ceramics, metal carbides, high-temp. insulation, specialty applics.; soft pellets; 270 nm mean particle diam.; 0.002% 325 mesh residue; sp.gr. 1.84; pour dens. 40 lb/ft^3; surf. area 7-12 m^2/g; pH 4-8; 99.6% min. carbon content.

Thermax® Powder N-991. [Cancarb] Med. thermal carbon blk.; CAS 1333-86-4; EINECS 215-609-9; used in low shear force mixing systems; functional filler in rubber goods; also for advanced ceramics, metal carbides, fluoroelastomers, metallurgical reducing agent, wire and cable jacket insulation, refractories, adhesives, ceramics, magnetic tape, glass prod.; powd.; 270 nm mean particle diam.; 0.025% 325-mesh residue; sp.gr. 1.84; pour dens. 22 lb/ft^3; surf. area 7-12 m^2/g; pH 9-11; 99.5% min. carbon content.

Thermax® Powder Ultra Pure N-991. [Cancarb; R.T. Vanderbilt] Med. thermal carbon blk.; CAS 1333-86-4; EINECS 215-609-9; for use in low shear force mixing systems; functional filler in fluoroelastomers; metallurgical reducing agent; also for advanced ceramics, metal carbides, high-temp. insulation, specialty applics.; powd.; 270 nm mean particle diam.; 0.0250% 325-mesh residue; sp.gr. 1.8; surf. area 7-12 m^2/g; pH 4-8; 99.6% min. carbon content.

Thermax® Stainless. [R.T. Vanderbilt] Med. thermal carbon blk.; CAS 1333-86-4; EINECS 215-609-9; nonstaining reinforcer for all elastomers, black stocks; dk. gray powd. and pellets; 0.1% max. residue on 325 mesh; dens. 1.80 ± 0.03 mg/m^3 (in rubber); iodine no. 5-13.

Thermax® Stainless Floform® N-907. [Cancarb; R.T. Vanderbilt] Med. thermal carbon blk.; CAS 1333-86-4; EINECS 215-609-9; nonstaining; functional filler in rubber goods (fluoroelastomers, gaskets, o-rings, profiles, seals, sponge); high-temp. insulation; soft pellets; 240 nm mean particle diam.; 0.0015% 325-mesh residue; sp.gr. 1.84; pour dens. 41 lb/ft^3; surf. area 7-12 m^2/g; pH 9-11; 99.6% min. carbon content.

Thermax® Stainless Powder N-908. [Cancarb] Med. thermal carbon blk.; CAS 1333-86-4; EINECS 215-609-9; nonstaining; functional filler in rubber goods (fluoroelastomers, gaskets, o-rings, profiles, seals, sponge); high-temp. insulation; powd.; 240 nm mean particle diam.; 0.025% 325-mesh residue; sp.gr. 1.84; pour dens. 22 lb/ft^3; surf. area 7-12 m^2/g; pH 9-11; 99.6% min. carbon content.

Thermax® Stainless Powder Ultra Pure N-908. [Cancarb] Med. thermal carbon blk.; CAS 1333-86-4; EINECS 215-609-9; nonstaining; used for critical applics. where low oil and high purity are required; for advanced ceramics, metal carbides, fluoroelastomers, high-temp. insulation, metallurgical reducing agent; powd.; 240 nm mean particle diam.; 0.0250% 325-mesh residue; sp.gr. 1.84; pour dens. 22 lb/ft^3; surf. area 7-12 m^2/g; pH 4-8.

Therm-Chek® 6-V-6A. [Ferro/Bedford] Barium-cadmium-zinc; lt. and heat stabilizer; primarily for use in plastisols and organisols; in rigidsol disps. containing < 30 phr of plasticizer; provides good clarity and early color retention, good air release and visc. control; lt. amber liq.; sp.gr. 1.04.

Therm-Chek® 130. [Ferro/Bedford] Barium-zinc; lt. and heat stabilizer; for use in plastisols; useful in organosols and rigidsol disp. containing < 30 phr plasticizer; good air release and visc. control; lt. amber liq.; dens. 8.8 lb.gal; visc. (Gardner) A2.

Therm-Chek® 135. [Ferro/Bedford] Strontium-zinc; heat and lt. stabilizer for flexible and semirigid PVC, plastisols, organosols.

Therm-Chek® 170. [Ferro/Bedford] Barium organic; heat stabilizer for flexible and semirigid PVC, plastisols.

Therm-Chek® 344. [Ferro/Bedford] Calcium-zinc stabilizer; lt. and heat stabilizer; for use in plastisols; useful in organosols and rigidsol disp. containing < 30 phr plasticizer; good air release and visc. control; lt. yel. liq.; dens. 8.5 lb/gal; visc. (Gardner) A1.

Therm-Chek® 659. [Ferro/Bedford] Calcium-zinc; heat and lt. stabilizer for flexible PVC, plastisols; nontoxic; liq.

Therm-Chek® 707-X. [Ferro/Bedford] Zinc, epoxy stabilizer; nontoxic stabilizer which produces very clear vinyl prods. having good initial color compared with other nontoxic systems; wh. to lt. yel paste; sp.gr. 0.99.

Therm-Chek® 714. [Ferro/Bedford] Calcium-zinc stabilizer; high efficiency, low odor stabilizer for all types of nontoxic vinyl formulations incl. rigids, but works best in calendered and extruded prods.; wh. powd.; dens. 2.3 lb/gal.

Therm-Chek® 763. [Ferro/Bedford] Calcium-zinc; heat and lt. stabilizer for flexible and rigid PVC, plastisols, organosols, bottles; FDA approved; paste.

Therm-Chek® 837. [Ferro/Bedford] Dibutyltin maleate; CAS 15535-69-0; nonsulfur-type heat stabilizer for rigid vinyl resin compds.; provides lt. stability; low level usage will optimize impact properties of a rigid vinyl; will not cross stain during compding. in the presence of lead, cadmium, or other metals that form colored sulfides; does not water blush; suitable for flexible vinyl formulations; Gardner 4 clear pale liq.; dens. 11.2 lb/gal; visc. (Gardner) G; 23.2% tin content.

Therm-Chek® 840. [Ferro/Bedford] Dibutyltin bis isooctyl thioglycolate; heat stabilizer for clear and pigmented rigid vinyl resin extrusions; used for rigid PVC pipe and siding, sheet, bottles, and profiles; suitable for flexible vinyl formulations; combination with epoxy compds.; nonsulfur staining, nonlubricating; Gardner 3 clear pale liq.; dens. 9.4 lb/gal; visc. (Gardner) A; 18.0% tin; 9.6% sulfur.

Therm-Chek® 900. [Ferro/Bedford] Epoxy; heat and lt. stabilizer for flexible and rigid PVC, plastisols, organosols, foams, tile.

Therm-Chek® 904. [Ferro/Bedford] Org. inhibitor; heat and light stabilizer stabilizer; produces a high degree of clarity; improves initial color and heat stability in both clear and pigmented stocks; can be used effectively in formulations containing phosphate plasticizers; also for calendered, molded, and extruded stocks, and for plastisols, organosols, foams, tiles; wh. to pale yel. liq.; sp.gr. 1.03.

Therm-Chek® 1238. [Ferro/Bedford] Barium-cadmium-zinc; heat and lt. stabilizer for flexible and semirigid PVC, plastisols, organosols, foams, tile; liq.

Therm-Chek® 1720. [Ferro/Bedford] Barium-cadmium-zinc; heat and lt. stabilizer for vinyl plastisols, organosols, foams; liq.

Therm-Chek® 1825. [Ferro/Bedford] Mixed cadmium-barium fatty acid soap with org. inhibitor; stabilizer for rigid formulations and calendered, molded, and extruded plasticized compositions; useful in clear, pigmented, and filled stocks; provides color in calendered expanded vinyls; offers long-term heat stabilization at moderate conc. levels, early color retention in pigmented stocks, compatibility; self-lubricating; slows dissipation of blowing agent in foams; lt. cream powd.; sp.gr. 1.21.

Therm-Chek® 1827. [Ferro/Bedford] Barium-cadmium; heat and lt. stabilizer, lubricant for flexible and rigid PVC, tile; powd.

Therm-Chek® 1864. [Ferro/Bedford] Barium-cadmium-zinc; heat and lt. stabilizer, lubricant for flexible, semirigid, and rigid PVC; powd.

Therm-Chek® 5469. [Ferro/Bedford] Barium-cadmium stabilizer with org. inhibitor; high potency, highly compatible stabilizer for general calendering, extruding, fluid bed, and plastisol applics. requiring max. clarity; produces bright, true colors in pigmented applics.; nonlubricating; recommended in areas where roll or mold plate-out is critical, for use with stearic acid to obtain optimum heat stability and to add lubrication; may be used with an epoxy plasticizer as synergist; lt. amber liq.; sp.gr. 1.07.

Therm-Chek® 5590. [Ferro/Bedford] Barium-cadmium; heat and lt. stabilizer, lubricant for flexible and rigid PVC, tile; low plateout; powd.

Therm-Chek® 5649. [Ferro/Bedford] Barium-cadmium-zinc; heat and lt. stabilizer for flexible and semirigid PVC, plastisols, organosols; good clarity; liq.

Therm-Chek® 5776. [Ferro/Bedford] Barium-cadmium; heat and lt. stabilizer for flexible and rigid PVC, plastisols, organosols; low lubricant; liq.

Therm-Chek® 5930. [Ferro/Bedford] Barium-cadmium-zinc; heat and lt. stabilizer for flexible PVC, plastisols, organosols, foams; good weatherability; liq.

Therm-Chek® 6160. [Ferro/Bedford] Barium-calcium-zinc; heat and lt. stabilizer, lubricant for flexible and semirigid PVC, plastisols, organosols, foam; liq.

Therm-Chek® 6253, 6254. [Ferro/Bedford] Calcium/zinc; vinyl heat stabilizer for 90 C and some 105 C wire and cable jacketing; powd.

Therm-Chek® 6258. [Ferro/Bedford] Bariumzinc; vinyl heat stabilizer for co-use with Therm-Chek 139 in filled and pigmented extrusion and calendering formulations; as cadmium replacement; powd.

Therm-Chek® BJ-2. [Ferro/Bedford] Barium-cadmium; heat stabilizer, lubricant for vinyl records; powd.

Therm-Chek® F-694. [Ferro/Bedford] Bariumzinc; vinyl heat stabilizer for general purpose clear and filled plastisol applics.; exc. balance of early color hold and process heat stability; liq.

Thermocel 1002. [Elf Atochem N. Am.] Tin; heat and lt. stabilizer for semirigid and rigid PVC, foam sheet, profile, siding.

Thermocel 2365. [Elf Atochem N. Am.] Tin; heat and lt. stabilizer for semirigid and rigid PVC, foam pipe.

Thermoguard® 210. [Elf Atochem N. Am./Plastics Addit.] Brominated polymeric epoxy resin; flame retardant additive for various thermosets such as unsat. polyester, phenolics, polyamides, and epoxy resins; lt. straw solid; soften. pt. 60-70 C; EEW 510-580 g/eq; 48-50% Br.

Thermoguard® 212. [Elf Atochem N. Am./Plastics

Addit.] Bromine, chlorine, and phosphorus blend; flame retardant additive for thermosets, esp. transparent unsat. polyesters; lt. straw clear liq.; sp.gr. 1.56; visc. 4750 cps; 32.5% Br, 7.4% Cl, 4.6% P; *Toxicology:* may be irritating to skin, eyes, and respiratory tract on repeated/prolonged contact.

Thermoguard® 213. [Elf Atochem N. Am./Plastics Addit.] Proprietary formulation based on brominated epoxy resin, antimony oxide, and styrene; efficient flame retardant additive for use in unsat. polyester resins; wh. opaque liq. (stable suspension); sp.gr. 1.7; visc. 3500 cps; 30% Br, 20% antimony oxide; *Toxicology:* may be irritating to skin, eyes, and respiratory tract on repeated/prolonged contact.

Thermoguard® 215. [Elf Atochem N. Am./Plastics Addit.] Brominated epoxy resin and antimony oxide; flame retardant additive for unsat. polyesters, epoxy, phenolic resins, PU, polyamides, and other specialty applics.; wh. powd.; bulk dens. 0.8 g/cc; epoxy equiv. 680 $\pm$ 35 g/eq.; 40% bromine; 20% antimony oxide.

Thermoguard® 220. [Elf Atochem N. Am./Plastics Addit.] Brominated polymeric epoxy resin; low m.w. flame retardant additive which can be blended with other resins (solids or liqs.) for use in potting, wet lay-up, pre-preg laminates, adhesives, molding compds., and coatings; Gardner 12 max. semisolid; soften. pt. 51-62 C; epoxy equiv. wt. 350-400 g/eq; 45-48% bromine.

Thermoguard® 230. [Elf Atochem N. Am./Plastics Addit.] Brominated polymeric epoxy resin; flame retardant additive for resin systems esp. nylon; off-wh. powd.; soften. pt. 90-110 C; epoxy equiv. wt. 1700-2100 g/eq; 49-51% bromine.

Thermoguard® 240. [Elf Atochem N. Am./Plastics Addit.] Brominated polymeric epoxy resin; flame retardant additive for thermosets and thermoplastics esp. in engineering resins such as polyester and polyamides; off-wh. solid; soften. pt. 186 C; epoxy equiv. 20,000 g/eq.; 49-51% bromine.

Thermoguard® 505. [Elf Atochem N. Am./Plastics Addit.] Decabromodiphenyloxide; CAS 1163-19-5; EINECS 214-604-9; flame retardant for thermoplastic and thermoset polymers, esp. in high performance applics.; suitable for PS, PE, epoxies, etc.; creamy wh. powd., odorless; negligible sol. in water; m.w. 959.2; sp.gr. 3.0; vapor pressure negligible; m.p. 300-310 C; 97% min. purity, 83.3% Br; *Toxicology:* may be irritating to respiratory system and eyes.

Thermoguard® 8218. [Elf Atochem N. Am./Plastics Addit.] 81-83% Decabromodiphenyloxide (Thermoguard 505) conc. in 17-19% LDPE; flame retardant in convenient pellet form to improve processing while reducing hazards and waste; wh. pellets, 1/8 in. avg. diam.; 68 $\pm$ 1% Br; *Toxicology:* may be irritating to respiratory system and eyes.

Thermoguard® 9010. [Elf Atochem N. Am./Plastics Addit.] 90% Antimony oxide (Thermoguard S) in LDPE; flame retardant in convenient pellet form to improve processing and reduce hazards and waste; pellets; 1/8 in. avg. diam.; *Toxicology:* may be irritating to respiratory system and eyes.

Thermoguard® CPA. [Elf Atochem N. Am./Plastics Addit.] Antimony; flame retardant for use as replacement for antimony oxide in many formulated plastics esp. in flexible PVC applics. such as wire and cable insulation and jacketing; wh. free-flowing powd.; 99.8% thru 325 mesh; sp.gr. 5.0; bulk dens. 0.5-0.7 g/cc; *Toxicology:* relatively nonhazardous; may be irritating to eyes and skin on prolonged/repeated exposure; inh. of dust may be irritating to respiratory tract.

Thermoguard® DM-9406. [Elf Atochem N. Am./Plastics Addit.] 80% Antimony oxide (Thermoguard S) in LDPE; flame retardant in convenient pellet form to improve processing and reduce hazards and waste; pellets; 1.8 in. avg. diam.

Thermoguard® FR. [Elf Atochem N. Am./Plastics Addit.] Sodium antimonate; flame retardant for vinyls and other plastics; features low opacity, low tinting strength; wh. fine powd.; 99.7% thru 200 mesh; sol. in conc. H_2SO_4, 20% HCl and ammonium salts; insol. in water, acetic acid, and tartaric acid; sp.gr. 4.80; 61.7% Sb; *Toxicology:* may be irritating to respiratory system and eyes.

Thermoguard® FR/PE 80/20. [Elf Atochem N. Am./Plastics Addit.] 80% Antimony (Thermoguard FR) in 20% LDPE; flame retardant for use in water and acid sensitive polymers such as PET; wh. free-flowing pellets; 1/8-1/4 in. size; 59.8-61.5% Sb; *Toxicology:* dust is irritating to upper respiratory tract when inhaled; irritating to cornea and mucous membranes of nose and throat; skin contact may result in dermatitis.

Thermoguard® L. [Elf Atochem N. Am./Plastics Addit.] Antimony oxide; CAS 1309-64-4; EINECS 215-175-0; flame retardant for use with a halogen-containing compd.; flame retardant pigment for PVC; also for use with chlorinated organics for producing flame-retardant polyesters and polyethylene compds.; low tinctorial strength; wh. fine powd.; 3.0 μ avg. particle size; 99.9% thru 325 mesh; insol. in common org. solvs.; very sl. sol. in water, aq. HCl, potassium hydroxide, and tartaric acid; sp.gr. 5.7; m.p. 656 C; ref. index 2.087; 83.0% antimony; 99.5% antimony oxide; *Toxicology:* may be irritating to respiratory system and eyes; avoid breathing dust, skin and eye contact.

Thermoguard® S. [Elf Atochem N. Am./Plastics Addit.] Antimony oxide; CAS 1309-64-4; EINECS 215-175-0; flame retardant for use in plastics, paper, textiles, and paints; flame retardant pigment for PVC; also for use with chlorinated organics for producing flame-retardant polyesters, polyethylene compds., mfg. of flame-retardant films, sheets, textiles, paper, paints; must be used with a halogen-containing compd. for flame retardant effect; high tinctorial str.; wh. superfine powd., odorless; 1.3 μ avg. particle size; 99.9% thru 325 mesh; insol. in common org. solvs.; very sl. sol. in water; sol. in aq. HCl, potassium hydroxide, and tartaric acid; m.w. 291.52; sp.gr. 5.67; vapor pressure 0 mm (20

C); m.p. 656 C; b.p. 1435 C; ref. index 2.087; 99.5% Sb_2O_3; 83.0% antimony; *Toxicology:* may be irritating to respiratory system, eyes, skin; ing. may cause severe GI symptoms; antimony oxide has caused lung tumors in test animals; ACGIH TLV 0.5 mg/m^3 (as Sb); contains < 0.4% arsenic and < 0.4% lead; *Precaution:* incompat. with strong oxidizers, acids, halogenated acids; hazardous decomp. prods.: toxic antimony fumes.

Thermoguard® UF. [Elf Atochem N. Am./Plastics Addit.] Antimony oxide; CAS 1309-64-4; EINECS 215-175-0; superfine grade flame retardant with high tinctorial str. for plastic, textiles, paper, and paint applics. in conjunction with halogenated flame retarders; wh. free-flowing powd.; 0.1% on 325 mesh; 50% 0.2-0.5 μ; sp.gr. 5.67; bulk dens. 25 lb/ft^3; m.p. 656 C; ref. index 2.1; 99.5% Sb_2O_3; *Toxicology:* may be irritating to respiratory system and eyes; avoid breathing dust and eye and skin contact.

Thermoguard® XS-70-T. [Elf Atochem N. Am./Plastics Addit.] Chlorinated paraffin; heat-stabilized additive for use in polymers where processing temps. are below 240-250 C.

Thermolin® 101. [Olin] Tetrakis (2-chloroethyl) ethylene diphosphate; nonreactive flame retardant additive for use in flexible PU foams for the transportation, bedding, and furniture industries, and in thermoplastic and thermoset resins such as acrylates, polyolefins, PAN, styrene, ABS, polyesters, epoxies, and PET; sol. in many common org. solvs., incl. simple alcohols, ketones, ethers, esters, and aromatic hydrocarbons; immiscible and virtually insol. in dry cleaning solvs.; sol. in TDI and in most polyether polyols for use in urethane foams; sp.gr. 1.45; dens. 12.1 lb/gal; visc. 260 cp; pour pt. -62 C; flash pt. 142 C; 30% chlorine; Unverified

Thermolin® RF-230. [Olin] Highly chlorinated polyol preblended with fluorocarbon-11; flame retardant for rigid foam boardstock in the construction industry; dk. amber; sp.gr. 1.5; dens. 12.5 lb/gal; visc. 20,000- 22,0-00 cps; acid no. 1.5 max.; hyd. no. 340.0 ± 10; pH 4.5-7.0 (in 10/6 IPA/water); flash pt. 145 C; 93% act. polyol; Unverified

Thermolite 12. [Elf Atochem N. Am./Plastics Addit.] Tin; heat and lt. stabilizer for flexible and rigid vinyl and vinyl copolymers, plastisols, organosols, foams, PU; good lubricant.

Thermolite 31. [Elf Atochem N. Am./Plastics Addit.] Tin; heat and lt. stabilizer with good processability for semirigid and rigid vinyl and vinyl copolymers, plastisols, organosols, records, bottles, tile, ABS.

Thermolite 35. [Elf Atochem N. Am./Plastics Addit.] Tin; stabilizer providing high heat stability to semirigid and rigid PVC and vinyl copolymers, ABS, bottles.

Thermolite 73. [Elf Atochem N. Am./Plastics Addit.] Tin; heat stabilizer providing good early color to semirigid and rigid PVC and vinyl copolymers, plastisols, records, and bottles.

Thermolite 139. [Elf Atochem N. Am./Plastics Addit.] Sulfur-contg. organotin; heat stabilizer for extrusion of weatherable PVC siding, window profiles, other rigid PVC applics. requiring weathering and color hold props.; sl. yel. clear oily liq.; sp.gr. 0.99; visc. 30 cs; ref. index 1.49.

Thermolite 176. [Elf Atochem N. Am./Plastics Addit.] Butyltin compd.; heat stabilizer for rigid PVC extrusion, pipe, conduit, and profiles; exc. color hold; low use level with exc. whiteness; NSF approved for potable water applics.; off-wh. to yel. clear oily liq.; insol. in water; sp.gr. 0.968; visc. 28 cs; pour pt. < -40 C; flash pt. 99 C; ref. index 1.49; *Toxicology:* may cause immediate or delayed skin and severe eye irritation; respiratory tract irritant; overexposure by inh. may cause coughing, headache, nausea; *Precaution:* incompat. with acids and oxidizers; direct sunlight causes deterioration to an inorg. tin salt; *Storage:* store in cool, dry area under normal warehouse conditions.

Thermolite 187. [Elf Atochem N. Am./Plastics Addit.] Phosphite; heat and lt. stabilizer for flexible and rigid PVC, plastisols, organosols, foams, records, bottles, tiles, vinyl copolymers; low plate-out antioxidant for ABS, PP, PVC; synergistic; clear.

Thermolite 340. [Elf Atochem N. Am./Plastics Addit.] Substituted bis(butyltin 2-ethylhexyl mercaptoacetate); stabilizer for PVC profile and siding applics. where color retention and weatherability are crucial; provides low lubrication and exc. dynamic processing stability in extrusions; superior performance in high TiO_2-loaded formulations; lt. yel. clear oily liq., char. odor; insol. in water; sp.gr. 1.2; visc. 77 cs; f.p. < -20 C; flash pt. (TCC) 129 C; ref. index 1.5243; 100% act.; *Toxicology:* ACGIH TLV 0.1 mg/m^3 (as Sn); may cause immediate or delayed skin irritation or severe eye irritation; inh. may cause respiratory tract irritation; overexposure may cause coughing, headache, nausea; *Precaution:* may produce irritating fumes, org. acid vapors on exposure to elevated temps. or flame; incompat. with acids, oxidizers; direct sunlight causes degradation to an inorg. tin salt; *Storage:* store in cool, dry area under normal warehouse conditions.

Thermolite 380. [Elf Atochem N. Am./Plastics Addit.] Sulfur-contg. organotin; heat stabilizer for extrusion of weatherable PVC siding, window profiles, other rigid PVC applics. requiring weathering and color hold props.; pale yel. clear oily liq.; sp.gr. 1.21; visc. 226 cs; pour pt. -23 C; ref. index 1.55.

Thermolite 390. [Elf Atochem N. Am./Plastics Addit.] Tin; heat and lt. stabilizer for semirigid and rigid PVC calendered sheet and film.

Thermolite 395. [Elf Atochem N. Am./Plastics Addit.] Tin; heat and lt. stabilizer for semirigid and rigid PVC extruded sheet, film, and bottles.

Thermolite 525. [Elf Atochem N. Am./Plastics Addit.] > 50% Dibutyltin dilaurate, < 30% dibutyltin bis(2-ethylhexyl mercaptoacetate), < 10% dibutyltin oxide; stabilizer; insol. in water; sp.gr. 1.075; f.p. 11 C; flash pt. > 200 F; *Toxicology:* may cause immediate or delayed severe eye irritation, skin irritation; inh. of vapors may irritate respiratory tract; harmful if swallowed; *Precaution:* incompat. with acids, oxidizers; avoid direct sunlight, elevated temps., open flame; *Storage:* store in tightly closed container in cool, dry area.

Thermolite 813. [Elf Atochem N. Am./Plastics Addit.] Tin; heat and lt. stabilizer for flexible and rigid PVC, plastisols, organosols, foams, bottles, nitrile polymers; sulfur-free; FDA approved; solid.

Thermolite 890S. [Elf Atochem N. Am./Plastics Addit.] Tin; heat stabilizer for semirigid and rigid PVC, plastisols, organosols, foams, bottles; FDA approved; very low odor.

Thermoplast®. [BASF AG] Special dyes sol. in plastics; for mass dyeing of thermoplastic and thermosetting plastics, e.g., styrene polymers (PS, S/B, SAN, ABS), rigid PVC, polymethacrylate, cellulose derivs., polycarbonates, unsat. polyester, etc.

THF. [Arco] Tetrahydrofuran; CAS 109-99-9; EINECS 203-726-8; solv. for vinyls; reaction solv.; for PTMEG, magnetic tape paint, adhesives, PU coatings, PVC cements, vinyl films and cellophane; colorless clear low-visc. liq.; m.w. 72.1; sol. in water; sp.gr. 0.89 (20/20 C); dens. 7.4 lb/gal (20 C); vapor pressure 155 mm Hg (20 C); b.p. 149 F; flash pt. 1 F.

Thiate® EF-2. [R.T. Vanderbilt] Trimethylthiourea; CAS 2489-77-2; accelerator for CR; used in press cures, inj. molding, LCM extrusion stocks, sponge; wh. cryst. powd.; mod. sol. in toluene, acetone, chloroform; insol. in water; m.w. 118.20; dens. 1.23 $\pm$ 0.03 mg/m^3; m.p. 80-90 C.

Thiate® H. [R.T. Vanderbilt] 1,3-Diethylthiourea; CAS 105-55-5; accelerator for CR, EPDM, and chlorobutyl; for fast initial set-up, continuous cured sponge, LCM extrusions, and hot air cures; wh. to lt. yel. flakes; sol. in methanol, chloroform, and acetone; pract. insol. in water; m.w. 132.33; dens. 1.11 $\pm$ 0.03 mg/m^3; m.p. 68-77 C.

Thiate® U. [R.T. Vanderbilt] 1,3-Dibutylthiourea; CAS 109-46-6; accelerator with same applics. as Thiate H but with slightly slower set-up and better scorch; wh. to lt. tan course powd.; sol. in methanol, chloroform, acetone, toluene; pract. insol. in water; m.w. 188.34; dens. 1.03 $\pm$ 0.03 mg/m^3; m.p. 56-65 C.

Thinner PU. [Bayer AG] Reactive thinner for use in solv.-free DD coatings.

Thiofide®. [Monsanto] 2,2′-Dithiobis (benzothiazole); CAS 120-78-5; primary accelerator for sulfur-curable elastomers; nonstaining, nondiscoloring in wh. stocks; plasticizer/retarder in CR; 2 mm pellets or dust-suppressed powd.

Thiotax®. [Monsanto] 2-Mercaptobenzothiazole; CAS 149-30-4; EINECS 205-736-8; primary accelerator for sulfur-curable elastomers, e.g., EPDM, IIR, low-temp. NR cures; powd. or dust-suppressed powd.

Thiovanol®. [Evans Chemetics] Thioglycerin; CAS 96-27-5; EINECS 202-495-0; stabilizer for acrylonitrile polymers; crosslinking agent for hard high-gloss coatings; accelerator for epoxy-amine condensation reactions; reducing agent; used in hair waving and straightening, hair dyes, depilatories, textiles, furs, pharmaceuticals, surfactants, foam stabilizing additives for detergents, shampoos, insecticides, pesticides, fungicides, and dessicants; practically clear and colorless sol'n.; mild char. odor; m.w. 108.2; b.p. 118 C (5 mm, anhyd.); misc. in all proportions with water and alcohol; insol. in ether; 90% aq. sol'n.

Thiurad®. [Monsanto] Tetramethylthiuram disulfide; CAS 137-26-8; sulfur donor, sec. accelerator for EV and semi-EV cure systems; may bloom above 1.0 phr; pellets or powd.

THPE-BZT. [Hoechst Celanese/Bulk Pharm.] 1,1,1-Tris(hydroxyphenylethane)benzotriazole; uv light stabilizer for plastics and coatings; reactive comonomer with polysulfones, PC, and other systems becoming part of polymer backbone; better thermal stability for higher temp. processing; nonmigrating; wh. free-flowing powd.; sol. (g/100 g): 9.7 g PM acetate, 5.7 g MIBK, 4.7 g 2-heptanone, 2.4 g ethyl acetate, 2.2 g butyl acetate, 2.6 g butanol; m.w. 423; m.p. 266-268 C.

THQ. [Eastman] Toluhydroquinone; CAS 95-71-6; EINECS 202-443-7; antioxidant for unsat. polyesters.

T-Hydro®. [Arco] t-Butyl hydroperoxide; CAS 75-91-2; EINECS 200-915-7; free radical polymerization initiator; co-initiator in polyethylene processes; finishing catalyst in emulsion polymerizations and unsat. polyester resin thermosets; intermediate for peroxygen derivs.; epoxidation reagent; oxidizing reagent; source of act. oxygen for pulp and paper bleaching, metal sulfide oxidations, etc.; clear sol'n., pungent odor; m.w. 90.12; sp.gr. 0.935; dens. 7.8 lb/gal (20 C); visc. 4.1 cps; vapor pressure 23 mm Hg (20 C); f.p. -2.8 C; b.p. 96 C (760 mm Hg); flash pt. (TCC) 43 C; ref. index 1.3246; pH 4.3; 70% aq. sol'n., 12% act. oxygen.

Tinuvin® 123. [Ciba-Geigy/Additives] Bis-(1-octyloxy-2,2,6,6,tetramethyl-4-piperidinyl) sebacate; CAS 129757-67-1; high-performance light stabilizer for PP fibers and tapes contg. flame retardants, coatings industry; non-interacting with acid catalysts, metal driers and pigments; good resist. to gas yellowing; pale yel. liq.; misc. with min. spirits, xylene, MIBK, butyl acetate, ethyl acetate, butanol; nil in water; m.w. 737.2; sp.gr. 0.97 (20 C).

Tinuvin® 144. [Ciba-Geigy/Additives] Bis(1,2,2,6,6-pentamethyl-4-piperidinyl)(3,5-di-t-butyl-4-hydroxybenzyl)butylpropanedioate; CAS 63843-89-0; uv lt. stabilizer giving protection against thermal oxidation; resistant to extraction in aq. environments, to volatilization during high-temp. processing; used in PP fiber, acrylic, coatings systems; off-wh. cryst. powd.; sol. (g/100 g): sol. 50 g in chloroform; 45 g in methylene chloride; 29 g in benzene; 8 g in ethyl acetate; < 0.01 g in water; m.w. 685; sp.gr. 1.07; m.p. 146-150 C.

Tinuvin® 213. [Ciba-Geigy/Additives] Methyl 3-[3-(2H-benzotriazol-2-yl)-5-t-butyl-4-hydroxyphenyl]proportionate and PEG 300; CAS 104810-47-1; lt. stabilizer for PVC, ABS, ASA, SBS footwear; lt. yel. liq.; m.w. > 600; sp.gr. 1.17 (20 C).

Tinuvin® 234. [Ciba-Geigy/Additives] 2-[2-Hydroxy-

3,5-di(1,1-dimethylbenzyl)phenyl]2H-benzotriazole; CAS 70321-86-7; uv absorber for PP, SBS footwear, SEBS, acetal, acrylic, polyamide, PET, PMMA, PBT, PC; recommended for thin section and fiber applics.; suitable for food contact applics.; FDA 21CFR §177.1580, 177.1630, 178.2010; lt. yel. powd.; m.w. 448; sp.gr. 1.22 (20 C); m.p. 135-141 C.

Tinuvin® 326. [Ciba-Geigy/Additives] 2-(3′-t-Butyl-2′-hydroxy-5′-methylphenyl)-5-chlorobenzotriazole; CAS 3896-11-5; uv lt. absorber for polyolefins, coatings; offers greater absorp. of longer wavelengths than Tinuvin P, better compatibility with polyolefins, lower volatility, less likely to discolor with metals; less effect on metal driers and catalyts; used in PP, PE, polybutylene, and related polymers; exhibits antioxidant properties; suitable for food contact applics.; FDA 21CFR § 177.1520(c), 178.2010; lt. yel. cryst. powd.; sol. (g/100 ml): 12.6 g in styrene; 4.9 g in methyl methacrylate; 2.9 g in MEK; 2.5 g in ethyl acetate; 1.8 g in petrol. ether; 0 g in water; m.w. 316; sp.gr. 1.32 (20 C); m.p. 140-141 C.

Tinuvin® 327. [Ciba-Geigy/Additives] 2-(3′,5′-Di-t-butyl-2′-hydroxyphenyl)-5-chlorobenzotriazole; CAS 2864-99-1; EINECS 223-383-8; uv lt. absorber for plastics, coatings; stable to heat and lt., low volatility, good color, superior wash fastness, resist. to gas fading; for polyolefins, EPDM, SAN, acrylic, flexible PVC, cold-cured polyesters, PU, dyes; pale yel. powd.; sol. (g/100 ml solv.): 16 g in toluene,15 g in xylene, 14 g in styrene and methyl methacrylate , 8 g in cyclohexane, 7 g in n-butyl acetate, 6 g ethyl acetate, 5 g in min. spirits, 4 g in MEK,3 g in dioctyl phthalate and hexane, 0.3 g in ethyl alcohol; m.w. 358; sp.gr. 1.26 (20 C); m.p. 154-158 C.

Tinuvin® 328. [Ciba-Geigy/Additives] 2-2′-Hydroxy-3,5′-di-t-amylphenyl) benzotriazole; CAS 25973-55-1; uv lt. absorber offering high solubility and desirable balance of props.; used for stabilizing coatings, automotive coatings, styrenics, polyolefins, acrylic, other substrates requiring uv lt. protection; suitable for food contact applics.; FDA 21CFR §175.105; off-wh. powd.; sol. (g/100 g solv.): 44 g in xylene; 28 g in n-butyl acetate; 24 g in MEK; 20 g in ethyl acetate; 14 g in min. spirits; nil sol. in water; m.w. 351.5; sp.gr. 1.17 (20 C); m.p. 79-87 C.

Tinuvin® 329. [Ciba-Geigy/Additives] 2-(2H-Benzotriazol-2-yl)-4-(1,1,3,3-tetramethylbutyl) phenol; light stabilizer for polymers; suitable for food contact applics.; FDA 21CFR §177.1580, 178.2010.

Tinuvin® 571. [Ciba-Geigy/Additives] Phenol, 2-(2H-benzotriazole-2-yl)-4-methyl-6-dodecyl; CAS 23328-53-2; UV absorber for plastics incl. thermoplastic PU, rigid and plasticized PVC, PVB, PMMA, PC, PVDC, EVOH, EVA, spin finishes for PA, PET, PU, PP fibers, latexes, waxes, adhesives, PS, styrenic copolymers, elastomers, and polyolefins; pale yel. clear liq.; misc. > 50% with acetone, chloroform, ethyl acetate, n-hexane, methylene chloride, toluene, styrene, petrol. ether; misc. < 0.01% with water; m.w. 394; sp.gr. 1.02 (20 C); ref. index 1.58 (20 C).

Tinuvin® 622. [Ciba-Geigy/Additives] Dimethyl succinate polymer with 4-hydroxy-2,2,6,6-tetramethyl-1-piperidine ethanol; CAS 65447-77-0; HALS; lt. stabilizer for polyolefin substrates; very low volatility, low migration tendency, no negative influence on polymer color; highly effective in carbon blk. systems; for PP, PE, acetal, flexible PVC; FDA 21CFR §178.2010; U.K., Japan, Germany compliance; off-wh. powd.; sol. (g/100 g): > 40 g in chloroform and methylene chloride; > 30 g in benzene; 5 g in ethyl acetate; 2 g in acetone; 0.1 g in methanol; < 0.01 g in hexane and water; m.w. > 2000; sp.gr. 1.18 (20 C); m.p. 55-70 C.

Tinuvin® 622LD. [Ciba-Geigy/Additives] Dimethyl succinate polymer with tetramethyl hydroxy-1-hydroxyethyl piperidine; CAS 65447-77-0; lt. and heat stabilizer for polyolefins, PP inj. molded bars, HDPE, linear LDPE plaques, PP tape, LDPE film, PP multifilament, ABS polymer systems, thermoplastic polyolefins, flexible PVC, food pkg.; FDA 21CFR §177.1350, 177.1520, 178.2010; off-wh. powd.; sol. (g/100 g): > 67 g in chloroform and methylene chloride, 18 g in toluene, 8 g in xylene, 0.01 g hexane and water; m.w. > 2500; sp.gr. 1.18 (20 C); m.p. 55-70 C; *Toxicology:* LD50 (oral, rat) > 15,000 mg/kg; may cause skin irritation and irritation of nasal and respiratory tract.

Tinuvin® 765. [Ciba-Geigy/Additives] Bis(1,2,2,6,6-pentamethyl-4-piperidinyl)sebacate; CAS 41556-26-7; HALS; lt. stabilizer providing superior performance in thick-section PP; functioning primarily as a free radical scavenger; uv stabilizer for PU; also for ABS, ASA, BR, SBR, NBR, SAN, flexible PVC; sl. yel. clear liq.; sol. in most components of urethane polymers; m.w. 509; sp.gr. 0.99 (20 C).

Tinuvin® 770. [Ciba-Geigy/Additives] Bis (2,2,6,6-tetramethyl-4-piperidinyl) sebacate; CAS 52829-07-0; EINECS 258-207-9; lt. stabilizer for polyolefins, incl. natural and pigmented PP multifilament slit film, polyethylene, ethlene-propylene copolymer and terpolymer, PU, ABS, impact PS, and other styrenic polymers and copolymers; synergistic with Tinuvin P in ABS, impact PS, and PU; useful with costabilizers such as high-performance phenolic antioxidants; wh. to sl. ylsh. cryst. powd.; sol. (g/100 ml): 56 g in methylene chloride; 46 g in benzene; 45 g in chloroform; 38 g in methanol; 19 g in acetone; 5 gin hexane; < 0.01 g in water; m.w. 481; sp.gr. 1.05 (20 C); vapor pressure 10^{-10} mm Hg; m.p. 81-85 C; *Toxicology:* eye irritant; may cause skin and respiratory tract irritation; may leave bitter metallic taste in mouth if inhaled or ingested.

Tinuvin® 783. [Ciba-Geigy/Additives] Poly[[6-[(1,1,3,3-tetramethylbutyl)amino]-s-triazine-2,4-diyl][[(2,2,6,6-tetramethyl-4-piperidyl) imino-hexamethylene [(2,2,6,6-tetramethyl-4-piperidyl) imino]] and dimethyl succinate polymer with 4-hydroxy-2,2,6,6-tetramethyl-1-piperidine ethanol; HALS; lt. stabilizer for polyolefins; effective in rotomolded parts, tape, film, and fiber; low volatility; FDA 21CFR §178.2010.

Tinuvin® P. [Ciba-Geigy/Additives] 2-(2′-Hydroxy-5′-methylphenyl) benzotriazole; CAS 2440-22-4; EINECS 219-470-5; uv lt. absorber (300-400 nm) offering protection to a wide range of plastics; dissipates photochemical energy as thermal energy; high thermal stability; does not discolor on lt. or heat aging; used in conventional heat-cured and flame-retardant, halogen-containing polyesters, plasticized and rigid PVC, PVC food pkg., PS homopolymer, impact PS, rubber, PC, acrylics, polyolefins, coatings, colorants, adhesives; FDA 21CFR §177.1010, 177.1315, 177.1580, 177.1630, 177.1640, 177.1980, 178.2010; wh. cryst. powd.; sol. (g/100 ml): 7.2 g in styrene, 6.0 g in toluene, 5.0 g in methyl methacrylate, 3.5 g in ethyl acetate, 2.5 g in acetone and dioctyl phthalate, 1.7 g in methyl Cellosolve, 1.5 g in min. spirits, nil sol. in water; m.w. 225; sp.gr. 1.51; m.p. 128-132 C; b.p. 225 C (10 mm).

Tioga Adhesion Promoter 30-0-100. [Tioga Coatings] Water-based nonconductive adhesion promoter for polypropylene, TPO, and TPR; 0 V.O.C.; clear; visc. 100-300 cps; pH 8.5-9.5; 15% solids.

Tioga Adhesion Promoter 30-6-600. [Tioga Coatings] Conductive water-based adhesion promoter for polypropylene, TPO, and TPR; 0 V.O.C.; blk.; visc. 200-500 cps; pH 8.5-9.5; 15% solids.

Tiona® RCL-4. [SCM] Rutile titanium dioxide, alumina surf. treatment; plastics grade pigment additive for faster processing, superior dispersion, higher tint str., outstanding resist. to PE yellowing; provides max. brightness; for polyolefins (blown and cast film, extrusion coatings); engineering resins (ABS, PPO/PPE, PC, nylon, acrylics, high-temp. resins), PS, HIPS, rigid vinyl siding, PVC profiles and pipe; exc. dispersion, high tint str. blue tint tone; NSF-listed; wh. powd.; sp.gr. 4.2; oil absorp. 12-14; 97% min. act.

Tiona® RCL-6. [SCM] Rutile titanium dioxide, alumina/silica surf. treatment; durable pigment for exterior applics., top coat rigid vinyl siding, acrylic applics., polyester; max. chalk resist. and color retention, good uv resist.; wh. powd.; 0.01% max. on 325 mesh; sp.gr. 4.0; dens. 33.3 lb/gal (solids); bulking value 0.030 gal/lb (solids); oil absorp. 14-24; pH 6.5-8.5 (10% slurry); 88% min. act.

Tiona® RCL-9. [SCM] Rutile titanium dioxide, alumina surf. treatment; pigment for thermosets and thermoplastics; easy dispersing; for polyester gel coats, plastisols, rigid and flexible PVC and pipe, acrylics; high gloss, med. exterior durability; NSF-listed for plastic pipe; wh. powd.; 0.01% max. on 325 mesh; sp.gr. 4.1 dens. 34.4 lb/gal (solids); bulking value 0.029 gal/lb (solids); oil absorp. 12-19; pH 6.5-8.5 (10% slurry); 94% min. act.

Tiona® RCL-69. [SCM] Rutile titanium dioxide, alumina surf. treatment; pigment for engineering plastics (ABS, PPO/PPE, nylon, high-temp. resins, PC, acrylics), PVC (rigid, flexible, pipe), PS, HIPS, polyolefins (blown and cast film, extrusion coatings); exc. dispersion, high tint str., bluest tint tone; NSF-listed for plastic pipe applics.; wh. powd.; sp.gr. 4.2; oil absorp. 13-15; 97% min. act.

TIPA 99. [Dow] Triisopropanolamine; CAS 122-20-3; EINECS 204-528-4; used to produce soaps with good hard surf. detergency, shampoos, pharmaceuticals, emulsifiers, textile specialties, agric. and polymer curing chemicals, adhesives, antistats, coatings, metalworking, rubber, gas conditioning chemicals; sp.gr. 0.988 (70/4 C); dens. 8.24 lb/gal (70 C); visc. 100 cps (60 C); f.p. 44 C; b.p. 306 C (760 mm Hg); flash pt. (COC) 320 F; ref. index 1.4595 (30 C).

TIPA 101. [Dow] Triisopropanolamine; CAS 122-20-3; EINECS 204-528-4; used to produce soaps with good hard surf. detergency, shampoos, emulsifiers, textile specialties, agric. and polymer curing chemicals, gas conditioning chemicals; sp.gr. 0.988 (70/4 C); dens. 8.24 lb/gal (70 C); visc. 100 cps (60 C); f.p. 44 C; b.p. 306 C (760 mm Hg); flash pt. (COC) 320 F; ref. index 1.4595 (30 C); 10% water.

TIPA Low Freeze Grade. [Dow] Triisopropanolamine; CAS 122-20-3; EINECS 204-528-4; used to produce soaps with good hard surf. detergency, shampoos, emulsifiers, textile specialties, agric. and polymer curing chemicals, gas conditioning chemicals; sp.gr. 0.988 (70/4 C); dens. 8.24 lb/gal (70 C); visc. 100 cps (60 C); f.p. 44 C; b.p. 306 C (760 mm Hg); flash pt. (COC) 320 F; ref. index 1.4595 (30 C); 15% water.

Ti-Pure® R-100. [DuPont] Titanium dioxide; CAS 13463-67-7; EINECS 236-675-5; high tinting str. wh. pigment with cream undertone for melt-processed plastics; provides bluest undertone in transmitted light applics.; exc. dry blend dispersibility; for high-temp. PE films, PE/PP moldings, flexible and rigid PVC, ABS, PS, PC, polyamide; NSF listed; wh. powd.; 0.32 μm median particle size; sp.gr. 4.2; oil absorp. 11 g/100 g; pH 8.0; 97% TiO_2.

Ti-Pure® R-101. [DuPont] Titanium dioxide; CAS 13463-67-7; EINECS 236-675-5; high tinting str. wh. pigment with neutral undertone for melt-processed plastics; exc. dry blend dispersibility; for high-temp. PE films, PE/PP moldings, flexible and rigid PVC, ABS, PS, PC, polyamide; NSF listed; wh. powd.; 0.29 μm median particle size; sp.gr. 4.2; oil absorp. 11 g/100 g; pH 8.5; 97% TiO_2.

Ti-Pure® R-102. [DuPont] Titanium dioxide; CAS 13463-67-7; EINECS 236-675-5; high tinting str. wh. pigment with blue undertone for plastics; exc. dispersibility in plasticized systems; exc. flocculation resist. in liq. colorants, plastisols; for rigid and flexible vinyls; also for PE/PP film and moldings, ABS, PS, PC, polyamide; NSF listed; wh. powd.; 0.25 μm median particle size; sp.gr. 4.1; oil absorp. 14 g/100 g; pH 7.5; 96% TiO_2.

Ti-Pure® R-103. [DuPont] Titanium dioxide; CAS 13463-67-7; EINECS 236-675-5; high str. wh. pigment with max. blueness; exc. discoloration resist. in polyolefins; adhesive grade with improved rheology; exc. for PE/PP molding and film, flexible and rigid PVC, ABS, PS, PC, polyamide; wh. powd.; 0.23 μm median particle size; sp.gr. 4.1; oil absorp. 14 g/100 g; pH 6.5; 96% TiO_2.

Ti-Pure® R-900. [DuPont] Titanium dioxide; CAS 13463-67-7; EINECS 236-675-5; wh. pigment for plastics with exc. dispersibility in plasticizers and flexible vinyls; exc. flocculation resist.; provides cream undertone for plastisols; for liq. colorants, PVC plastisol, flooring, sheet, and pipe applics.; NSF listed; wh. powd.; 0.31 μm median particle size; sp.gr. 4.0; oil absorp. 16 g/100 g; pH 8.5; 94% TiO_2.

Ti-Pure® R-902. [DuPont] Titanium dioxide; CAS 13463-67-7; EINECS 236-675-5; wh. pigment for plastics with good weathering resist. in all systems and only a small sacrifice in tinting str.; good dispersibility in plasticizers and flexible vinyls; for exterior, durable polyolefin and PVC applics.; wh. powd.; 0.32 μm median particle size; sp.gr. 4.0; oil absorp. 17 g/100 g; pH 8.7; 91% TiO_2.

Ti-Pure® R-960. [DuPont] Titanium dioxide; CAS 13463-67-7; EINECS 236-675-5; wh. pigment for plastics with good weathering resist. in all systems and only a small sacrifice in tinting str.; good dispersibility in plasticizers and flexible vinyls; for exterior, durable polyolefin and PVC applics.; also for PVC plastisol, lead-stabilized systems, ABS, PS, PC, polyamide; wh. powd.; 0.35 μm median particle size; sp.gr. 3.9; oil absorp. 18 g/100 g; pH 7.4; 89% TiO_2.

Tisyn®. [Burgess Pigment] Aluminum silicate, thermo-optic; high hiding pigment for paint formulations, paper coatings, plastics, rubber, water and solv. systems; U.S. patent 3,021,195; thin flat plate; 0.15% max. on 325 mesh; sp.gr. 2.2; oil absorp. 73; ref. index 1.62; pH 4.0-4.5; 0.5% max. moisture.

Titanium Dioxide P25. [Degussa] Titanium dioxide; CAS 13463-67-7; EINECS 236-675-5; catalyst carrier for fixed bed catalyst; heat stabilizer for HCR silicone rubber and flame retardant; uv absorber for sunscreen lotions; fluffy wh. powd., odorless, hydrophilic; 21 nm avg. particle size; insol. in water; dens. ≈ 3.7 g/ml; dens. ≈ 130 g/l (tapped); surf. area 50 ± 15 m^2/g; m.p. > 1800 C; pH 3.5-4.5 (4% aq. disp.); > 99.5% assay; *Toxicology:* TLV 10 mg/m^3 total dust; LD50 (oral, rat) > 5000 mg/kg; dust may cause respiratory irritation.

Tixosil 311. [Rhone-Poulenc] Hydrated silica, amorphous; CAS 7631-86-9; flatting agents providing gloss and sheen control in coatings, esp. high bake industrial enamels (grades 311, 321); rheology control agent for solv. and water-based systems (grades 331 and 375); for coatings, adhesives, plastics, paper, food, cosmetics, toothpaste, and pharmaceutical applics.; wh. powd.; 4 μ avg. particle size; < 0.1% retained 325 mesh; Hegman grind 6; sp.gr. 2.05; dens. 17.1 lb/gal; surf. area 250 m^2/g; oil absorp. 300%; ref. index 1.45; pH 7; 96% SiO_2.

Tixosil 321. [Rhone-Poulenc] Precipitated silica, amorphous; flatting agents providing gloss and sheen control in coatings, esp. high bake industrial enamels (grades 311, 321); rheology control agent for solv. and water-based systems (grades 331 and 375); for coatings, adhesives, plastics, paper, food, cosmetics, toothpaste, and pharmaceutical applics.; wh. powd.; 3 μ avg. particle size; < 0.1% retained 325 mesh; Hegman grind > 6; dens. 17.1 lb/gal; surf. area 250 m^2/g; oil absorp. 300%; ref. index 1.45; pH 7; 96% SiO_2; Unverified

Tixosil 331. [Rhone-Poulenc] Hydrated silica, amorphous; flatting agents providing gloss and sheen control in coatings; rheology control agent for solv. and water-based systems; for coatings, adhesives, plastics, paper, food, cosmetics, toothpaste, and pharmaceutical applics.; wh. powd.; 3 μ avg. particle size; < 0.1% retained 325 mesh; Hegman grind 7; dens. 17.1 lb/gal; surf. area 310 m^2/g; oil absorp. 320%; ref. index 1.45; pH 7; 96% SiO_2.

Tixosil 333, 343. [Rhone-Poulenc] Hydrated silica; rheology control agent for solv. and water-based systems; for coatings, adhesives, plastics, paper, food, cosmetics, toothpaste, and pharmaceutical applics.

Tixosil 375. [Rhone-Poulenc] Precipitated silica, amorphous; flatting agents providing gloss and sheen control in coatings, esp. high bake industrial enamels (grades 311, 321); rheology control agent for solv. and water-based systems (grades 331 and 375); for coatings, adhesives, plastics, paper, food, cosmetics, toothpaste, and pharmaceutical applics.; wh. powd.; 2 μ avg. particle size; < 0.1% retained 325 mesh; Hegman grind 7; dens. 17.5 lb/gal; surf. area 180 m^2/g; oil absorp. 290%; ref. index 1.45; pH 6.6; 98% SiO_2; Unverified

TL-115. [ICI Advanced Materials] PTFE; CAS 9002-84-0; general purpose lubricant powd. used in wide media; excellent bearing chars. when used as additive in polymeric systems; improves PV, static, and dynamic coefficients of friction, and wear factor; dry lubricant additive to thermoset, elastomeric, and rubber compds.; used in rubber industry as release agent when mixed with silicones; belting compds., seals, gaskets; improves tear resistance when added to silicone rubber; powd., 6-25 μ particle size; sp.gr. 2.15-2.25; dens. 475 g/l; Unverified

T-Maz® 20. [PPG/Specialty Chem.] Polysorbate 20; CAS 9005-64-5; nonionic; emulsifier, solubilizer, wetting agent, antistat, stabilizer, dispersant, visc. modifier, suspending agent used in the food, cosmetic, drug, textile and metalworking industries; very mild ingred. for no-tear shampoos, baby baths; solubilizer for emollients and fragrances into bath prods.; emulsifier for cleansing prods., skin care emulsions; dedusting agent for plastics; kosher; yel. liq.; sol. in water, ethanol, acetone, toluene, veg. oil, propylene glycol; sp.gr. 1.1; visc. 400 cps; HLB 16.7; acid no. 2 max.; sapon. no. 40-50; hyd. no. 96-108; flash pt. (PMCC) > 350 F; 97% act.

T-Mulz® 598. [Harcros] Phosphate ester, free acid; anionic; detergent, emulsifier, hydrotrope for highly alkaline sol'ns., dry cleaning, general purpose cleaners, agric. formulations, down hole scale inhibitor, emulsion polymerization; FDA compliance;

amber visc. liq.; sol. in ethanol, ethylene, xylene; disp. in min. oil, kerosene; sp.gr. 1.10; dens. 9.2 lb/gal; acid no. 65 (pH 5.5); pour pt. 23 F; flash pt. (PMCC) > 350 F; pH 2-3 (1% aq.); surf. tens. 34.4 dynes/cm (0.05% aq.); 99.5% min. act.

T-Mulz® 734-2. [Harcros] Phosphate ester, free acid; anionic; emulsifier, hydrotrope, detergent for agric. formulations, solvs., detergents, emulsion polymerization, alkaline cleaners; amber clear visc. liq.; sol. in ethanol, ethylene, xylene; sp.gr. 1.18 (68 F); dens. 9.8 lb/gal (68 F); acid no. 107 (pH 5.5); pour pt. 50 F; flash pt. > 350 F (PMCC); pH 2-3 (1% aq.); surf. tens. 38.4 dynes/cm (0.05% aq.); 99.5% min. act.

TOF. [Merrand] Tri 2-ethylhexyl phosphate; CAS 78-42-2; EINECS 201-116-6; low temp. plasticizer for NBR, NBR/PVC, CR, CPE elastomers, PVC, PVAc, cellulosics, PS, ABS; some flame retardance; Unverified

TOF™. [Rhone-Poulenc Surf. & Spec.] Trioctyl phosphate; CAS 78-42-2; EINECS 201-116-6; plasticizer for vinyls and syn. rubbers; reduces tackiness at high temp.; provides uniform plasticizing action over broad temp. range; very compat. with PVC; 100% act.

Toho Master Batch-100. [Toho Chem. Industry] Antistatic agent for PP pellet.

Toho Me-PEG Series. [Toho Chem. Industry] Methoxy polyethylene glycol (m.w. 225, 350, 550, 705, 1000); base material for surfactant, syn. resin, plasticizer, lubricating industries; wetting, softening, penetrating, lubricating and cleaning agent for textile, paper, ink, pigments, etc.; liq./solid.

Toho PEG Series. [Toho Chem. Industry] Polyethylene glycol (m.w. 200, 300, 400, 600, 1000, 1500, 1540, 2000, 4000); base material for surfactants, syn. resin, plasticizer, lubricating industries; wetting, softening, penetrating, lubricating and cleaning agent for textile, paper, ink, pigments, cosmetics and toiletries, etc.; liq./paste/solid.

Tolplaz TBEP. [Albright & Wilson Am.] Tributoxyethyl phosphate; CAS 78-51-3; EINECS 201-122-9; low temp. plasticizer for NBR, NBR/PVC, CR, CPE elastomers, polyacrylate, fluoroelastomers, PVAc, cellulosics, PS, ABS; lubricant additive; leveling agent in floor polish.

Tomah DA-14. [Exxon/Tomah] N-Isodecyloxypropyl-1,3-diaminopropane; CAS 72162-46-0; cationic; intermediate for textile foaming agents, surfactants, ethoxylates, agric. chemicals; corrosion inhibitor for metalworking fluids; additive for fuels, lubricants, petrol. refining; crosslinking agent for epoxy resins; bactericidal props.; lt. amber liq.; sol. 1.8 g/100 g in water; m.w. 295; sp.gr. 0.86; pour pt. -50 F; amine no. 375-395; flash pt. (COC) 108 C; 100% act.; 90% min. diamine.

Tomah DA-16. [Exxon/Tomah] N-Isododecyloxypropyl-1,3-diaminopropane; cationic; intermediate for textile foaming agents, surfactants, ethoxylates, agric. chemicals; corrosion inhibitor for metalworking fluids; additive for fuels, lubricants, petrol. refining; crosslinking agent for epoxy resins; liq.; m.w. 328; sp.gr. 0.87; pour pt. -30 F; amine no. 335-355; flash pt. (COC) 160 C; 100% conc.

Tomah DA-17. [Exxon/Tomah] N-Isotridecyloxypropyl-1,3-diaminopropane; cationic; intermediate for textile foaming agents, surfactants, ethoxylates, agric. chemicals; corrosion inhibitor for metalworking fluids; additive for fuels, lubricants, petrol. refining; crosslinking agent for epoxy resins; amber liq.; sol. 2.1 g/100 g in water; m.w. 340; sp.gr. 0.87; pour pt. -30 F; amine no. 325-350; flash pt. (COC) 160 C; 95% min. amine.

Tomah PA-10. [Exxon/Tomah] Hexyloxypropylamine; corrosion inhibitor for metalworking fluids; antistat; flotation collector; additive for fuel, lubricant, petrol. refining; intermediate for surfactants, textile foaming agents, ethoxylates and agric. chem.; crosslinking agent for epoxy resins; lt. amber liq.; m.w. 165; water-sol.; sp.gr. 0.84; pour pt. -30 F; amine no. 325-340; flash pt. (COC) 84 C; 95% min. amine.

Tomah PA-14. [Exxon/Tomah] Isodecyloxypropylamine; CAS 7617-78-9; cationic; emulsifier; corrosion inhibitor for metalworking fluids; antistat; flotation collector; additive for fuel, lubricant, petrol. refining; intermediate for surfactants, textile foamers, ethoxylates and agric. chem.; crosslinking agent for epoxies; colorless to lt. amber liq.; m.w. 229; sp.gr. 0.84; pour pt. -50 F; amine no. 240-255; flash pt. (COC) 104 C; 95% min. amine.

Tomah PA-16. [Exxon/Tomah] Isododecyloxypropylamine; corrosion inhibitor for metalworking fluids; antistat; flotation collector; additive for fuel, lubricant, petrol. refining; intermediate for surfactants, textile foamers, ethoxylates and agric. chem.; crosslinking agent for epoxies; liq.; m.w. 253; pour pt. -30 F; amine no. 215-230; flash pt. (COC) 120 C; 100% conc.

Tomah PA-19. [Exxon/Tomah] Linear C12-15 alkyloxypropylamine; CAS 68610-26-4; corrosion inhibitor for metalworking fluids; antistat; flotation collector; additive for fuel, lubricant, petrol. refining; intermediate for surfactants, textile foamers, ethoxylates and agric. chem.; crosslinking agent for epoxies; clear liq., ammoniacal odor; insol. in water; m.w. 272; sp.gr. 0.85; visc. 6 cSt; f.p. 50 f; b.p. 400 F min.; pour pt. 50 F; amine no. 195-215; flash pt. (PMCC) 201 F; *Toxicology:* eye and skin corrosive.

Tomah PA-1214. [Exxon/Tomah] Octyl/decyloxypropylamine; corrosion inhibitor for metalworking fluids; antistat; flotation collector; additive for fuel, lubricant, petrol. refining; intermediate for surfactants, textile foaming agents, ethoxylates and agric. chem.; crosslinking agent for epoxy resins; sp.gr. 0.85.

Tomah Q-14-2. [Exxon/Tomah] Isodecyloxypropyl dihydroxyethyl methyl ammonium chloride; CAS 125740-36-5; cationic; quat. used as acid corrosion inhibitor, plastics and textile antistat, and emulsifier; bactericidal props.; lt. amber liq.; water-sol.; sp.gr. 0.97; HLB 19.5; pH 6-9 (1% aq.); 74% act.; *Precaution:* flamm.

Tomah Q-18-15. [Exxon/Tomah] Stearyl PEG-15

methyl ammonium chloride; cationic; quat. used as acid corrosion inhibitor, plastics and textile antistat, and emulsifier; amber liq.; sp.gr. 1.06; HLB 16.3-18.0; pH 6-9 (5% aq.); 100% act.; *Precaution:* flamm.

Tomah Q-D-T. [Exxon/Tomah] Tallow dimethyl trimethyl propylene diammonium chloride, IPA; cationic; quat. used as acid corrosion inhibitor, plastics and textile antistat, and emulsifier; amber liq.; sp.gr. 0.95; HLB 15.0; pH 6-9 (5% aq.); 50% act. in IPA-water; *Precaution:* flamm.

Tonox®. [Uniroyal] p,p′-Diaminodiphenylmethane; CAS 101-77-9; curing agent for epoxies and urethanes offering exc. physical and elec. props., heat stability, chem. resist.; FDA 21CFR §175.300, 177.1680, 177.2280, 177.2600 (5% max.); tan to brn. waxy lumps; sol. (g/100 ml): 27.3 g in acetone, 18.7 g in methanol, 8.1 g in IPA, 0.1 g in water; sp.gr. 1.15; visc. 50-100 cps (80 C); m.p. 70-80 C; b.p. 253 C (20 mm); flash pt. (COC) 228 C; fire pt. (COC) 267 C; pH 9-10 (10% aq. slurry; *Toxicology:* LD50 (oral, rat) 830 mg/kg; may cause fever, vomiting, jaundice, irritation to skin, eyes, mucous membranes; may stain skin yel.; may cause cancer, damage to liver, kidneys; *Storage:* store in cool, dry area in closed containers.

Tonox® 22. [Uniroyal] Crude methylene dianiline; CAS 101-77-9; curing agent for epoxy resins; red br. visc. liq.; sp.gr. 1.15; visc. > 250,000 cps; hyd. equiv. wt. 52.4.

Tonox® 60/40. [Uniroyal] 60% Methylene dianiline and 40% m-phenylene diamine; curing agent for epoxies and urethanes offering elex. phys. props., chem. resist.; FDA 21CFR §177.2280; brn. visc. liq.; sol. (g/100 ml): 12.3 g in methanol, 12.1 g in benzene, 10.4 g in acetone, 2.3 g in water, 1.2 g in CCl_4; sp.gr. 1.14; visc. 1500 cps; flash pt. (COC) 115 C; fire pt. (COC) 196 C; *Toxicology:* MDA: LD50 (oral, rat) 830 mg/kg; may cause fever, vomiting, jaundice, irritation to skin, eyes, mucous membranes; may stain skin yel.; may cause cancer, damage to liver, kidneys; *Storage:* store in cool, dry area in closed containers.

Tonox® LC. [Uniroyal] Aromatic polyamine; curing agent for epoxies and urethanes offering good chem. resist., extended pot life over Tonox and Tonox 60/40; relatively nonstaining and nonvolatile; brn. visc. liq.; sol. in polar and aromatic solvs.; sl. sol. in aliphatic solvs.; insol. in water; sp.gr. 1.12; visc. 25,000-150,000 cps; vapor pressure 20 mm Hg (250 C); flash pt. 182 C; pot life (150 g mass) 100 min @ 70 C, 40 min @ 100 C; 100% act.; *Storage:* store in cool, dry area in closed containers.

Topanex 100BT. [ICI Polymer Addit.] 2-(2′-Hydroxy-5′-methylphenyl) benzotriazole; CAS 2440-22-4; EINECS 219-470-5; uv absorber, lt. stabilizer, antioxidant protecting plastics (PVC, styrenics, acrylics, unsat. polyesters), lacquers; effective at 290-380 nm; pale yel. cryst. powd.; sol. (g/100 ml): 7 g in styrene, 6 g in toluene; m.w. 225; m.p. 128-132 C.

Topanex 500H. [ICI Polymer Addit.] 1,5-Dioxaspiro[5.5]undecane 3,3-dicarboxylic acid, bis (2,2,6,6-tetramethyl-4-piperidinyl)ester; uv stabilizer for ABS, PE, PP; wh. powd.; sol. (g/100 g): 32.8 g methyl chloride, 21 g methanol, 10.3 g toluene; m.w. 522; bulk dens. 0.472 g/ml; m.p. 160-162 C.

Topanol® 205. [Zeneca Spec.] 2,2-Bis[4-(2-(3,5-di-t-butyl-4-hydroxyhydrocinnamoyloxy)) ethoxyphenyl] propane; high m.w. hindered phenolic antioxidant for retardation of thermal and oxidative degradation in polyolefins, styrenics, and engineering polymers; nondiscoloring; wh. free-flowing powd.; m.w. 836; sol. in common org. solvs.; sp.gr. 1.09; m.p. 100-102 C.

Topanol® CA. [Seal Sands Chems. Ltd.] > 85% 1,1,3-Tris (2-methyl-4-hydroxy-5-t-butyl phenyl) butane, 15% max. toluene; antioxidant for polyolefin and styrenic polymers, plasticizers, hot melt adhesives, SBR latexes, polyamides, polyesters; FDA approved; wh. cryst. powd., toluene odor; sol. in acetone, ethanol, diethyl ether, ethyl acetate; nil sol. in water; sp.gr. 1.08; bulk dens. 0.5 g/ml (powd.); m.p. 181 C min.; flash pt. (Seta CC) 72.8 C; *Toxicology:* LD50 (oral, rat) > 5000 mg/kg; nonirritating to skin and eyes; toluene: OSHA TWA 200 ppm; on heating to 150 F, toluene vapors may be released which can cause liver, kidney, CNS effects, cardiac irregularities, mucous membrane irritation; *Precaution:* flamm. solid; readily burns in a fire releasing toxic vapors; incompat. with strong oxidizing agents; hazardous decomp. prods.: CO, CO_2, dense smoke; *Storage:* keep away from heat and flame; keep container closed when not in use.

Topanol® CA-SF. [Seal Sands Chems. Ltd.] 1,1,3-Tris (2-methyl-4-hydroxy-5-t-butylphenyl) butane; CAS 1843-03-4; antioxidant; wh. free-flowing cryst. powd., odorless; sol. (g/100 g): 96 g acetone, 70 g ethanol, 63 g diethyl ether, 61 g ethyl acetate, 0.03 g water; sp.gr. 1.08; bulk dens. 0.5 g/ml (powd.); m.p. 181 C; 100% act.; *Precaution:* can burn in fire releasing toxic vapors; incompat. with strong oxidizing agents; hazardous decomp. prods.: CO, CO_2, dense smoke; *Storage:* keep container closed when not in use.

Toyocat®-B2. [Tosoh] Tert. amine blend; balanced blowing/gelling catalyst for flexible slabstock and rigid PU; exc. flowability and improved dimensional stabillty in rigid board applics.; ylsh. clear liq.; sol. in water; sp.gr. 0.808 (20/20 C); visc. 2 cps; f.p. < -20 C; b.p. 199-206 C; flash pt. (PMCC) 72 C; < 0.5% water; *Toxicology:* LD50 (oral, rat) 865 mg/kg; low toxicity; causes eye and skin inflammation on direct contact; *Storage:* prod. darkens on prolonged storage.

Toyocat®-B6. [Tosoh] Tert. amine blend; catalyst for flexible, semirigid, and rigid PU foams; exc. flowability and improved dimensional stability in rigid board applics.; ylsh. clear liq.; sol. in water; sp.gr. 1.017 (20/20 C); visc. 40 cps; f.p. < 0 C; flash pt. (COC) 72 C; < 0.5% water; *Toxicology:* LD50 (oral, rat) 2681 mg/kg; low toxicity; causes eye and skin inflammation on direct contact; *Storage:* prod. darkens on prolonged storage.

Toyocat®-B7. [Tosoh] Tert. amine blend; catalyst for flexible, semirigid, and rigid PU foams; exc. flowability and improved dimensional stability in rigid board applics.; ylsh. clear liq.; sol. in water; sp.gr. 0.963 (20/20 C); visc. 17 cps; f.p. < -5 C; flash pt. (COC) 74 C; < 0.5% water; *Toxicology:* LD50 (oral, rat) 2184 mg/kg; low toxicity; causes eye and skin inflammation on direct contact; *Storage:* prod. darkens on prolonged storage.

Toyocat®-B20. [Tosoh] Amine blend; delayed-action catalyst for rigid PU applics.; improves flowability and k-factor; amber liq.; sol. in water; sp.gr. 1.076 (20/20 C); visc. 61 cps; f.p. < -20 C; b.p. 165-186 C; flash pt. (TCC) 77 C; < 1.5% water; *Toxicology:* LD50 (oral, rat) 1943 mg/kg; low toxicity; causes eye and skin inflammation on direct contact; *Storage:* prod. darkens on prolonged storage.

Toyocat®-B54. [Tosoh] Amine blend; delayed-action catalyst for rigid PU applics.; improves flowability and k-factor; amber liq.; sol. in water; sp.gr. 1.048 (20/20 C); visc. 51 cps; f.p. < -10 C; b.p. 171-191 C; flash pt. (COC) 82 C; < 1% water; *Toxicology:* LD50 (oral, rat) 2508 mg/kg; low toxicity; causes eye and skin inflammation on direct contact; *Storage:* prod. darkens on prolonged storage.

Toyocat®-B61. [Tosoh] Amine blend; balanced blowing/gelling catalyst for flexible slabstock and rigid PU; sp.gr. 1.017 (20/20 C); visc. 40 cps; f.p. < 0 C; b.p. 188-201 C; flash pt. (TCC) 77 C.

Toyocat®-B71. [Tosoh] Amine blend; balanced blowing/gelling catalyst for rigid PU applics.; sp.gr. 0.977 (20/20 C); visc. 17 cps; f.p. < -20 C; b.p. 191-203 C; flash pt. (TCC) 75 C.

Toyocat®-DMA. [Tosoh] N,N′-Dimethylethanolamine; CAS 108-01-0; EINECS 203-542-8; catalyst for PU slabstock and molded flexible, semirigid, and rigid applics.; sp.gr. 0.888 (20/20 C); visc. 3.8 cps; f.p. -59 C; b.p. 135 C; flash pt. (TCC) 42 C.

Toyocat®-DMCH. [Tosoh] N,N′-Dimethylcyclohexylamine; catalyst for rigid PU applics.; sp.gr. 0.850 (20/20 C); visc. 2 cps; f.p. < -78 C; b.p. 162-165 C; flash pt. (TCC) 40 C.

Toyocat®-DT. [Tosoh] N,N,N′,N″, N″-Pentamethyldiethylenetriamine; blowing catalyst for PU slabstock and molded flexible and rigid applics.; improves foam flow and dimensional stability in low-dens. rigid foam systems, esp. when applied to highly complex molded prods.; colorless clear liq., low odor; sol. in water; sp.gr. 0.828 (20/20 C); visc. 2 cps; f.p. < -20 C; b.p. 201 C; flash pt. (COC) 82 C; < 0.5% water; *Toxicology:* LD50 (oral, rat) 1630 mg/kg; low toxicity; causes eye and skin inflammation on direct contact; *Storage:* prod. darkens on prolonged storage.

Toyocat®-ET. [Tosoh] Bis(dimethylaminoethyl) ether in dipropyleneglycol; blowing catalyst for PU slabstock and molded flexible and rigid applics.; good for isocyanate-water reactions; produces foam with fine and uniform cell structure; colorless clear liq.; sol. in water; sp.gr. 0.902 (20/20 C); visc. 5 cps; f.p. < -80 C; b.p. 170 C; flash pt. (COC) 80 C; < 1% water; *Toxicology:* LD50 (oral, rat) 1814 mg/kg; low toxicity; causes eye and skin inflammation on direct contact; *Storage:* prod. darkens on prolonged storage.

Toyocat®-ETF. [Tosoh] Amine; acid-blocked, delayed-action blowing catalyst for flexible slabstock and molded PU, semirigid applics.; in complex large molds, exhibits exc. foam flowability and fast cure; amber liq.; sol. in water; sp.gr. 1.046 (20/20 C); visc. 63 cps; f.p. < -20 C; b.p. 170-180 C; flash pt. (COC) 81 C; *Toxicology:* LD50 (oral, rat) 1260 mg/kg; low toxicity; causes eye and skin inflammation on direct contact; *Storage:* prod. darkens on prolonged storage.

Toyocat®-F2. [Tosoh] Amine blend; catalyst for flexible slabstock and molded PU applics.; optimized for water-blown flexible foams; special blend for CFC reduction; sp.gr. 1.038 (20/20 C); visc. 3 cps; f.p. < -5 C; b.p. 195-205 C; flash pt. (COC) 104 C.

Toyocat®-F4. [Tosoh] Amine blend; catalyst for flexible slabstock and molded PU applics.; optimized for water-blown flexible foams; special blend for CFC reduction; sp.gr. 0.995 (20/20 C); visc. 3 cps; f.p. < -7 C; b.p. 195-205 C; flash pt. (TCC) 76 C.

Toyocat®-F83. [Tosoh] Amine blend; strong gelling catalyst for rigid PU and elastomer shoe soles; special blend for CFC reduction; imrpoves friability in rigid foams; ylsh. clear liq.; sol. in water; sp.gr. 1.001 (20/20 C); visc. 15 cps; f.p. < -10 C; b.p. 199-209 C; flash pt. (COC) 97 C; *Toxicology:* LD50 (oral, rat) 1098 mg/kg; causes eye and skin inflammation on direct contact; *Storage:* prod. will darken on prolonged storage.

Toyocat®-F94. [Tosoh] Amine blend; catalyst for rigid PU and elastomer shoe soles; special blend for CFC reduction; improves friability in rigid foams; ylsh. clear liq.; sol. in water; sp.gr. 0.995 (20/20 C); visc. 3 cps; f.p. < -7 C; b.p. 195-205 C; flash pt. (TCC) 76 C; *Toxicology:* LD50 (oral, rat) 1090 mg/kg; causes eye and skin inflammation on direct contact; *Storage:* prod. will darken on prolonged storage.

Toyocat®-HP. [Tosoh] Amine; reactive thermosensitive catalyst for PU flexible and semirigid foams; reduces PVC discoloration for automotive semirigid foam systems; suitable for isocyanate-polyol reactions; nonmigratory; amber liq.; sol. in water; sp.gr. 0.994 (20/20 C); visc. 30 cps (35 C); f.p. < 30 C; flash pt. (COC) 114 C; < 0.5% water; *Toxicology:* LD50 (oral, rat) 6040 mg/kg; low toxicity; causes eye and skin inflammation on direct contact; *Storage:* prod. darkens on prolonged storage.

Toyocat®-HPW. [Tosoh] N-Methyl-N′-hydroxyethylpiperazine, 10% water; reactive catalyst for PU slabstock and molded flexible, semirigid applics.; sp.gr. 0.992 (20/20 C); visc. 30 cps; f.p. < 5 C; b.p. 219-231 C; flash pt. (PMCC) 98 C.

Toyocat®-HX4W. [Tosoh] Reactive tert. amine blend; catalyst for semirigid PU applics.; improves vinyl staining; sl. gelling; amber liq.; sol. in water; sp.gr.

0.987 (20/20 C); visc. 110 cps; f.p. < 5 C; b.p. 217-228 C; flash pt. (COC) 131 C; 9.7-10.3% water; *Toxicology:* LD50 (oral, rat) 6050 mg/kg; low toxicity; causes eye and skin inflammation on direct contact; *Storage:* prod. darkens on prolonged storage.

Toyocat®-HX15W. [Tosoh] Reactive tert. amine blend; catalyst for semirigid PU applics.; improves vinyl staining; amber liq.; sol. in water; sp.gr. 0.996 (20/20 C); visc. 5 cps; f.p. < -10 C; b.p. 210-220 C; flash pt. (COC) 130 C; 9.7-10.3% water; *Toxicology:* LD50 (oral, rat) 3410 mg/kg; low toxicity; causes eye and skin inflammation on direct contact; *Storage:* prod. darkens on prolonged storage.

Toyocat®-HX35W. [Tosoh] Reactive tert. amine blend; catalyst for semirigid PU applics.; improves vinyl staining; sl. blowing; provides exceptional flow and reactivity in semirigid systems; nonmigratory; amber liq., odorless; sol. in water; sp.gr. 0.962 (20/20 C); visc. 12 cps; f.p. < -10 C; b.p. 209-221 C; flash pt. (COC) 101 C; < 0.5% water; *Toxicology:* LD50 (oral, rat) 2460 mg/kg; low toxicity; causes eye and skin inflammation on direct contact; *Storage:* prod. darkens on prolonged storage.

Toyocat®-HX63. [Tosoh] Reactive tert. amine blend; catalyst for semirigid PU applics.; improves vinyl staining and moldability; sp.gr. 0.938 (20/20 C); visc. 7.5 cps; f.p. < -10 C; b.p. 130-230 C; flash pt. (TCC) 35 C.

Toyocat®-HX70. [Tosoh] Reactive tert. amine blend; catalyst for semirigid PU applics.; improves vinyl staining and moldability; sp.gr. 0.965 (20/20 C); visc. 50 cps; f.p. < -10 C; b.p. 150-230 C; flash pt. (COC) 107 C.

Toyocat®-L20M. [Tosoh] Amine blend; catalyst for rigid PU appliance foams; sp.gr. 0.913 (20/20 C); visc. 5 cps; f.p. < -18 C; b.p. 133-144 C; flash pt. (PMCC) 47 C.

Toyocat®-L75D. [Tosoh] Amine blend; catalyst for flexible PU slabstock foam; sp.gr. 0.980 (20/20 C); visc. 32 cps; f.p. < -20 C; b.p. 175-190 C; flash pt. (COC) 65 C.

Toyocat®-LE. [Tosoh] Amine blend; std. catalyst for flexible PU slabstock and molded foam; sp.gr. 1.006 (20/20 C); visc. 48 cps; f.p. < -20 C; b.p. 193-206 C; flash pt. (PMCC) 99 C.

Toyocat®-M30. [Tosoh] Amine blend; catalyst for flexible molded and semirigid PU; improves moldability and surf. cure; sp.gr. 0.849 (20/20 C); visc. 3.5 cps; f.p. < -5 C; b.p. 192-212 C; flash pt. (COC) 86 C.

Toyocat®-M50. [Tosoh] Amine blend; catalyst for flexible molded and semirigid PU; improves moldability and surf. cure; sp.gr. 0.902 (20/20 C); visc. 7 cps; f.p. < -10 C; b.p. 190-210 C; flash pt. (COC) 90 C.

Toyocat®-MR. [Tosoh] N,N,N′,N′-Tetramethylhexanediamine; non-thermosensitive strong gelling catalyst for PU slabstock and molded flexible, semirigid and rigid applics.; improves flow, dimensional stability and cure speed; colorless clear liq.; sol. in water; sp.gr. 0.800 (20/20 C); visc. 1 cps; f.p. < -70 C; b.p. 198 C; flash pt. (COC) 85 C; *Toxicology:* LD50 (oral, rat) 720 mg/kg; low toxicity; causes eye and skin inflammation on direct contact; *Storage:* prod. darkens on prolonged storage.

Toyocat®-N31. [Tosoh] Amine blend; catalyst for flexible slabstock and molded PU, rigid applics., elastomer shoe soles; improves cure; sp.gr. 0.954 (20/20 C); visc. 19 cps; f.p. < -20 C; b.p. 194-204 C; flash pt. (PMCC) 80 C.

Toyocat®-N81. [Tosoh] Amine blend; catalyst for flexible slabstock and molded PU, rigid applics., elastomer shoe soles; improves cure; sp.gr. 0.911 (20/20 C); visc. 5 cps; f.p. < -5 C; b.p. 192-203 C; flash pt. (COC) 73 C.

Toyocat®-NEM. [Tosoh] N-Ethylmorpholine; CAS 100-74-3; EINECS 202-885-0; catalyst for PU molded flexible, semirigid, and rigid applics.; sp.gr. 0.914 (20/20 C); visc. 2.5 cps; f.p. -60 C; b.p. 139 C; flash pt. (TCC) 29 C.

Toyocat®-NP. [Tosoh] N,N′,N′-Trimethylaminoethyl piperazine; thermosensitive catalyst for PU slabstock and molded flexbile, rigid, and elastomer shoe sole applics.; delays cream time; exc. flowability; fast curing; synergistic with org. tin catalysts; colorless clear liq.; sol. in water; sp.gr. 0.886 (20/20 C); visc. 3 cps; f.p. < -20 C; b.p. 212 C; flash pt. (COC) 93 C; *Toxicology:* LD50 (oral, rat) 1420 mg/kg; low toxicity; causes eye and skin inflammation on direct contact; *Storage:* prod. darkens on prolonged storage.

Toyocat®-SC2. [Tosoh] Amine blend; catalyst for flexible PU slabstock foam; sp.gr. 0.922 (20/20 C); f.p. < -20 C; b.p. 150-160 C; flash pt. (COC) 40 C.

Toyocat®-SPF. [Tosoh] Amine blend; acid-blocked delayed-action catalyst for flexible slabstock and molded PU, high resiliency formulations; improves moldability and cure; amber liq.; sol. in water; sp.gr. 1.019 (20/20 C); visc. 53 cps; f.p. < -10 C; b.p. 144-154 C; flash pt. (PMCC) 97 C; *Toxicology:* LD50 (oral, rat) 1940 mg/kg; low toxicity; causes eye and skin inflammation on direct contact; *Storage:* prod. darkens on prolonged storage.

Toyocat®-SPF2. [Tosoh] Amine blend; acid-blocked delayed-action catalyst for flexible slabstock and molded PU, high resiliency formulations; improves moldability and cure; amber liq.; sol. in water; sp.gr. 1.076 (20/20 C); visc. 61 cps; f.p. < -20 C; b.p. 165-186 C; flash pt. (PMCC) 77 C; *Toxicology:* LD50 (oral, rat) 1943 mg/kg; low toxicity; causes eye and skin inflammation on direct contact; *Storage:* prod. darkens on prolonged storage.

Toyocat®-SPF3. [Tosoh] Amine blend; acid-blocked delayed-action catalyst for flexible hot and semi-hot molded PU foam prods.; amber liq.; sol. in water; sp.gr. 1.006 (20/20 C); visc. 16 cps; f.p. < -20 C; flash pt. (PMCC) 35 C; *Toxicology:* LD50 (oral, rat) 1821 mg/kg; low toxicity; causes eye and skin inflammation on direct contact; *Storage:* prod. darkens on prolonged storage.

Toyocat®-TE. [Tosoh] N,N,N′,N′-Tetramethylethylenediamine; CAS 110-18-9; EINECS 203-744-6; balanced blowing/gelling catalyst for PU molded flexible, semirigid, and rigid applics.; ylsh. clear liq.; sol. in water; sp.gr. 0.776 (20/20 C); visc. 1 cps; f.p. < -55 C; b.p. 120 C; flash pt. (TCC) 21 C; *Toxicology:* LD50 (oral, rat) 1580 mg/kg; low toxicity; causes eye and skin inflammation on direct contact; *Storage:* prod. darkens on prolonged storage.

Toyocat®-TF. [Tosoh] Triethylenediamine-based blend; CAS 280-57-9; acid-blocked, delayed-action gelling catalyst for flexible slabstock and molded PU, semirigid and rigid PU, elastomer shoe soles; ylsh. clear liq.; sol. in water; sp.gr. 1.062 (20/20 C); visc. 186 cps; f.p. < -5 C; b.p. 147-159 C; flash pt. (COC) 111 C; < 1% water; *Toxicology:* LD50 (oral, rat) 3110 mg/kg; low toxicity; causes eye and skin inflammation on direct contact; *Storage:* prod. darkens on prolonged storage.

Toyocat®-THN. [Tosoh] Amine blend; acid-blocked catalyst for rigid PU applics.; improves flowability and cure; sp.gr. 0.975 (20/20 C); visc. 40 cps; f.p. < -20 C; b.p. 193-205 C; flash pt. (COC) 111 C.

Toyocat®-TMF. [Tosoh] Mixt. of tert. amines and ethylene glycol; delayed action strong blowing catalyst for PU foam; improves foam flowability, dimensions, stability in rigid systems; produces fine-celled foam with low k-factor; also for flexible and semirigid foams; ylsh. clear liq., ammoniacal odor; sol. in water, dipropylene glycol, glycol, methanol, ethanol, acetone; sp.gr. 1.06 (20/20 C); visc. 96 cps; vapor pressure < 13 Pa (20 C); f.p. < -20 C; b.p. 101-198 C; flash pt. (COC) 81 C; pH 8-9 (2% aq.); *Toxicology:* LD50 (oral, rat) 1976 mg/kg; harmful by ingestion and skin contact; causes burns; may cause skin sensitization; may be irritating to mouth, GI tract, respiratory tract; severe eye irritant; contains glycol (TLV/STEL 50 ppm CL vapor); *Precaution:* incompat. with acids, oxidizers, chlorinated org. compds., aldehydes, copper; hazardous combustion prods.: toxic and corrosive fumes of NO_x; decomp. prods.: CO, CO_2, nitrogen compds.; *Storage:* store in dry, cool, well-ventilated area in tightly closed containers; prod. darkens on prolonged storage.

Toyocat®-TRC. [Tosoh] Amine; special trimerization catalyst for polyisocyanurate foam, rigid PU applics.; sp.gr. 0.920 (20/20 C); visc. 26 cps; f.p. < -70 C; b.p. 140 C (1 mm); flash pt. (PMCC) 104 C.

TP 87/21. [Hoffmann Min.] Sillitin N 85 (quartz-kaolinite) coated with epoxy silane; filler, min. additive for thermoplastic PU.

TP 89/50. [Hoffmann Min.] Sillitin Z 86 (quartz-kaolinite) fluidized by a special coating; filler, min. additive for thermoplastic PU.

TP 90/127. [Hoffmann Min.] Sillitin N 85 (quartz-kaolinite), coated with amino silane; filler, min. additive for thermoplastic PU.

TP 91/01. [Hoffmann Min.] Silfin Z (quartz-kaolinite) fluidized by a special coating; filler, min. additive for thermoplastic PU.

Translink® 37. [Engelhard] Calcined aluminum silicate with vinyl functional surface treatment; reinforcing extender; used in crosslinked polyethylene and the ethylene-propylene rubbers used in elec. compds.; wh. highly pulverized powd.; 1.4 μm avg. particle size (pretreatment); 0.02% max. residue +325 mesh; sp.gr. 2.63 g/cc; bulk dens. 16 lb/ft³ (loose), 35 lb/ft³ (tamped); bulking value 0.046 gal/lb; oil absorp. 45-55 lb/100 lb; ref. index 1.62; brightness (GE) 90-92; 52.4% SiO_2; 44.5% Al_2O_3.

Translink® 77. [Engelhard] Calcined aluminum silicate with vinyl functional surface treatment; reinforcing extender; used in dielectric paper and EPDM med. and high-voltage wire and cable insulation; wh. highly pulverized powd.; 0.8 μm avg. particle size (pretreatment); 0.02% max. residue +325 mesh; sp.gr. 2.63 g/cc; bulk dens. 11 lb/ft³ (loose), 21 lb/ft³ (tamped); bulking value 0.046 gal/lb; oil absorp. 80-90 lb/100 lb; ref. index 1.62; brightness (GE) 90-92; 52.4% SiO_2; 44.5% Al_2O_3.

Translink® 445. [Engelhard] Calcined aluminum silicate with amino-silane surface treatment; reinforcing extender for polyamides, polyesters, and other polar polymers; wh. highly pulverized powd.; 1.4 μm avg. particle size (pretreatment); 0.02% max. residue +325 mesh; sp.gr. 2.63 g/cc; bulk dens. 16 lb/ft³ (loose), 35 lb/ft³ (tamped); bulking value 0.046 gal/lb; oil absorp. 45-55 lb/100 lb; ref. index 1.62; pH 8.5-9.5; brightness (GE) 90-92; 52.4% SiO_2; 44.5% Al_2O_3.

Translink® 555. [Engelhard] Calcined aluminum silicate with amino-silane surface treatment; reinforcing extender for polyamides, polyesters, and other polar polymers; maintains high impact str. while providing exc. balance of tens., flex., heat deflection, and dimensional stability props.; wh. highly pulverized powd.; 0.8 μm avg. particle size (pretreatment); 0.02% max. residue +325 mesh; sp.gr. 2.63 g/cc; bulk dens. 16 lb/ft³ (loose), 35 lb/ft³ (tamped); bulking value 0.046 gal/lb; oil absorp. 80-90 lb/100 lb; ref. index 1.62; pH 8.5-9.5; brightness (GE) 90-92; 52.4% SiO_2; 44.5% Al_2O_3.

Translink® HF-900. [Engelhard] Calcined aluminum silicate, amino-silane surface-modified; reinforcing extender for plastics (acrylics, epoxy, nylon, phenolics, PC, polyester, PVC, urethane) and rubbers (butyl, EPR, neoprene, nitrile, SBR, urethane); wh. highly pulverized powd.; 1.8 μm avg. particle size (pretreatment); 0.04% max. residue +325 mesh; sp.gr. 2.63 g/cc; bulk dens. 14 lb/ft³ (loose), 25 lb/ft³ (tamped); bulking value 0.046 gal/lb; oil absorp. 70-80 lb/100 lb; ref. index 1.62; pH 8.5-9.5; brightness (GE) 91-94; 52.4% SiO_2; 44.5% Al_2O_3.

TR-Calcium Stearate. [Struktol] Calcium stearate; CAS 1592-23-0; EINECS 216-472-8; high purity clear melt nonwettable processing aid; powd.; 7.5% max. on 200 mesh, 1% max. on 100 mesh; bulk dens. 20-22 lb/ft³; m.p. 148-150 C; 11.2% max. CaO; *Storage:* unlimited storage stability in cool, dry area.

Trenbest 500. [Lucas Meyer] Trimethyl hydroxyethyl ammonium ester of a carboxyglycerol phosphoric acid; mold release agent for the plastic and caoutchouc industry; used for molding and extrusion powds., laminated plastics, inj. molding powds., ABS; low-visc. liq.; sp.gr. 0.96; flash pt. < 14 C; 50%

solids in isopropyl acetate.

Trennspray Tego®. [Goldschmidt] Silicone-free aerosol for release of inj. molded plastic and rubber moldings; Unverified

Tribase. [Syn. Prods.] Tribasic lead sulfate; heat stabilizer for flexible and rigid PVC; for vinyl elec. insulation thru 90 C classification; effective with phthalates, adipates, and azelates; stabilizer of choice for rigid PVC extrusion, calendering, and inj. molding; activator for azodicarbonamide blowing agents; wh. fine powd.; sp.gr. 6.4 (25/4 C); 89.1% PbO.

Tribase EXL Special. [Syn. Prods.] Modified tribasic lead sulfate; stabilizer for wire and cable compds. up to 75 C classification; sp.gr. 4.0; 67% PbO.

Tribase XL. [Syn. Prods.] Tribasic lead sulfate; heat stabilizer for flexible and rigid PVC; for vinyl elec. insulation thru 90 C classification; effective with phthalates, adipates, and azelates; stabilizer of choice for rigid PVC extrusion, calendering, and inj. molding; activator for azodicarbonamide blowing agents; wh. coated fine powd.; sp.gr. 4.6 (25/4 C); 80.9% PbO.

Triglyme. [Ferro/Grant] Triethylene glycol dimethyl ether; CAS 112-49-2; solv. which tends to solvate cations; used in electrochemistry, polymer/boron chemistry; gas absorp., extraction, stabilization; used in industrial prods. such as fuels, lubricants, textiles, pharmaceuticals, pesticides; colorless clear; mild, nonresidual odor; m.w. 178.22; f.p. -45 C; b.p. 216 C (760 mm); water-sol.; sp.gr. 0.9862; dens. 8.23 lb/gal; visc. 3.8 cP; ref. index 1.4224; pH neutral; flash pt. 111 C (CC); surf. tens. 29.4 dynes/cm; 98.0% min. purity.

Trigonal® 14. [Akzo Nobel] Benzoin ether; initiator used in u.v. lt. curing of polyester resins; long shelf life in polyester formulation; UV sensitizer for lacquers and FRP laminates; liq.

Trigonal® 121. [Akzo Nobel] Ketone mixture; UV initiator used in combination with an amine accelerator for UV curing of FRP laminates; powd.

Trigonox® 17. [Akzo Nobel] n-Butyl 4,4-di(t-butylperoxy)valerate; CAS 995-33-5; initiator for cure of unsaturated polyester resins, acrylates; liq.; 95% peroxide; 9.1% act. oxygen; *Storage:* store @ 30 C max.

Trigonox® 17/40. [Akzo Nobel] n-Butyl-4,4-bis(t-butylperoxy) valerate; CAS 995-33-5; crosslinking agent for olefin copolymers, EPDM, SBR, Neoprene, Hypalon; powd.; 40% act. with inorg. phlegmatizer.

Trigonox® 17-40B-pd. [Akzo Nobel] 4,4-Di-t-butylperoxy-n-butyl valerate on calcium carbonate carrier; crosslinking agent for olefin copolymers, esp. chlorinated polyethylene; useful for crosslinking EPDM, SBR, Neoprene, Hypalon; powd.; 40% act.

Trigonox® 21. [Akzo Nobel] t-Butyl peroctoate; CAS 3006-82-4; initiator for acrylates, styrenics, and LDPE polymerization; used where presence of water is objectionable; liq.; 97% assay, 7.18% act. oxygen; *Storage:* store @ -30 to 10 C.

Trigonox® 21-C50. [Akzo Nobel] t-Butyl peroctoate in min. spirits; CAS 3006-82-4; used where a diluent can be tolerated; initiator for acrylates; does not require refrigerated storage and transportation; sol'n.; 50% assay; 3.70% act. oxygen; *Storage:* store @ -30 to 10 C.

Trigonox® 21 LS. [Akzo Nobel] Butylperoctoate; CAS 3006-82-4; initiator for cure of unsaturated polyester resins; sol'n.; 87% peroxide; 6.4% act. oxygen.

Trigonox® 21-OP50. [Akzo Nobel] t-Butyl-peroctoate in DOP; initiator for elevated-temp. polyester cures; can be substituted where a liq. is desired; used in pultrusion, SMC, and BMC were fast cure is needed; sol'n.; 50% assay; 3.70% act. oxygen.

Trigonox® 22-BB80. [Akzo Nobel] 1,1-Di-t-butylperoxy cyclohexane in butyl benzyl phthalate; CAS 3006-86-8; initiator for SMC and BMC formulations; sol'n.; 80% assay; 9.53% act. oxygen.

Trigonox® 22-C50. [Akzo Nobel] 1,1-Di(t-butylperoxy)cyclohexane in odorless min. spirits; initiator for acrylates; sol'n.; 50% assay; 6.14% act. oxygen; *Storage:* store @ 25 C max.

Trigonox® 22-E50. [Akzo Nobel] 1,1-Di(t-butylperoxy)cyclohexane in min. oil; initiator for acrylates; sol'n.; 50% assay; 6.14% act. oxygen; *Storage:* store @ 25 C max.

Trigonox® 23. [Akzo Nobel] t-Butyl peroxyneodecanoate; CAS 26748-41-4; initiator for acrylates, PE, PVC; liq.; 95% assay; 6.22% act. oxygen; *Storage:* store @ -30 to -10 C.

Trigonox® 23-C75. [Akzo Nobel] t-Butylperoxyneodecanoate in odorless min. spirits; CAS 26748-41-4; highly react. liq. used in polyester pultrusion and continuous sheet curing; initiator for LDPE polymerization, for styrenics, for acrylates, and for PVC polymerization; sol'n.; 75% assay, 4.91% act. oxygen; *Storage:* store @ -25 to -10 C.

Trigonox® 23-W40. [Akzo Nobel] t-Butyl peroxyneodecanoate aq. emulsion; CAS 26748-41-4; initiator for PVC; emulsion; 40% assay; 2.62% act. oxygen; *Storage:* store @ -20 to -10 C.

Trigonox® 25-C75. [Akzo Nobel] t-Butyl peroxypivalate in odorless min. spirits; CAS 927-07-1; initiator for LDPE polymerization, styrenics, and PVC; sol'n.; 75% assay; 6.89% active oxygen; *Storage:* store @ -25 to -5 C.

Trigonox® 27. [Akzo Nobel] t-Butyl peroxydiethylacetate; CAS 2550-33-6; initiator for acrylates, PE; liq.; 96% assay, 8.16% act. oxygen; *Storage:* store @ 15 C max.

Trigonox® 29. [Akzo Nobel] 1,1-Di(t-butylperoxy)-3,5,5-trimethylcyclohexane; CAS 6731-36-8; initiator for acrylates, PE, PS; liq.; 95% assay; 10.05% act. oxygen; *Storage:* store @ 25 C max.

Trigonox® 29-40B-pd. [Akzo Nobel] 1,1-Di-t-butylperoxy-3,3,5-trimethylcyclohexane on calcium carbonate carrier; CAS 6731-36-8; crosslinking agent for olefin copolymers; useful for crosslinking EPDM, SBR, Neoprene, Hypalon; powd.; 40% act.

Trigonox® 29-B75. [Akzo Nobel] 1,1-Di-t-butylperoxy-3,3,5-trimethylcyclohexane in DBP; initiator for elevated-temp. polyester cures, esp. SMC (conventional and low profile), BMC where longer flow time is required, automotive SMC processes; also for acrylates, PE, PS; sol'n.; 75% assay; 7.94% act. oxygen; *Storage:* store @ 25 C max.

Trigonox® 29-C75. [Akzo Nobel] 1,1-Di-t-butylperoxy-3,3,5-trimethylcyclohexane, odorless min. spirits; initiator for styrenics; liq.; 75% in odorless min. spirits.

Trigonox® 36-C60. [Akzo Nobel] Di(3,5,5-trimethylhexanoyl) peroxide in odorless min. spirits; CAS 3851-87-4; initiator for acrylates, PE, PVC; sol'n.; 60% assay; 3.05% act. oxygen; *Storage:* store @ -10 to 0 C.

Trigonox® 36-C75. [Akzo Nobel] Bis (3,5,5-trimethylhexanoyl) peroxide in odorless min. spirits; CAS 3851-87-4; initiator for LDPE polymerization, styrenics, acrylates, PVC polymerization; sol'n.; 75% assay; 3.82% active oxygen; *Storage:* store @ -10 to 0 C.

Trigonox® 36-W40. [Akzo Nobel] Di(3,5,5-trimethylhexanoyl) peroxide; CAS 3851-87-4; initiator; emulsion; 40% assay; 2.04% act. oxygen.

Trigonox® 40. [Akzo Nobel] 2,4-Pentanedione peroxide; initiator for ambient-temp. polyester cures; rapid hardness development in hand and spray layup; fire resistant; liq.; 4.0% act. oxygen.

Trigonox® 41-C50. [Akzo Nobel] t-Butyl peroxyisobutyrate in odorless min. spirits; CAS 109-13-7; initiator for acrylates, PE; sol'n.; 50% assay; 5.00% act. oxygen; *Storage:* store @ -15 to 10 C.

Trigonox® 41-C75. [Akzo Nobel] t-Butyl peroxyisobutyrate in odorless min. spirits; CAS 109-13-7; initiator acrylates, PE; sol'n.; 75% assay; 7.49% act. oxygen; *Storage:* store @ -15 to 10 C.

Trigonox® 42. [Akzo Nobel] t-Butyl peroxy-3,5,5-trimethyl hexanoate; CAS 13122-18-4; initiator for cure of unsaturated polyester resins; liq.; 97% peroxide; 6.7% act. oxygen.

Trigonox® 42 PR. [Akzo Nobel] t-Butyl peroxy-3,5,5-trimethyl hexanoate; CAS 13122-18-4; initiator for cure of unsaturated polyester resins; sol'n.; 87% peroxide; 6.1% act. oxygen.

Trigonox® 42S. [Akzo Nobel] t-Butyl peroxy-3,5,5-trimethylhexanoate; CAS 13122-18-4; initiator for acrylates, PE, PS; liq.; 97% assay; 6.73% act. oxygen; *Storage:* store @ -20 to 25 C.

Trigonox® 44P. [Akzo Nobel] 2,4-Pentanedione peroxide; initiator for ambient-temp. polyester cures; nonseparating paste form; applic.; incl. autobody repair putties and mine bolt cartridges; paste.

Trigonox® 61. [Akzo Nobel] Ketone peroxide mixture; initiator for cure of unsaturated polyester resins; sol'n.; 33% peroxide; 7.7% act. oxygen.

Trigonox® 63. [Akzo Nobel] Ketone peroxide mixture; initiator for cure of unsaturated polyester resins; sol'n.; 33% peroxide; 6.5% act. oxygen.

Trigonox® 67. [Akzo Nobel] Ketone peroxide mixture; initiator for cure of unsaturated polyester resins; sol'n.; 20% peroxide; 2.8% act. oxygen.

Trigonox® 93. [Akzo Nobel] t-Butyl peroxybenzoate; CAS 614-45-9; initiator for cure of unsaturated polyester resins; sol'n.; 77% peroxide; 6.4% act. oxygen.

Trigonox® 97. [Akzo Nobel] t-Butyl peroxy-2-methylbenzoate; CAS 22313-62-8; initiator for styrenics, acrylates; liq.; 98% assay, 7.52% act. oxygen; *Storage:* store @ 20 C max.

Trigonox® 97-C75. [Akzo Nobel] t-Butyl peroxy-2-methylbenzoate in odorless min. spirits; CAS 22313-62-8; initiator for styrenics, acrylates; sol'n.; 75% assay; 5.76% act. oxygen; *Storage:* store @ 20 C max.

Trigonox® 99-B75. [Akzo Nobel] Cumyl peroxyneodecanoate; CAS 26748-47-0; initiator; sol'n.; 75% assay; 3.92% act. oxygen.

Trigonox® 99-C75. [Akzo Nobel] α-Cumyl peroxyneodecanoate (CAS 26748-47-0) in odorless min. spirits; low-temp. initiator for all types of PVC prod.; liq.; m.w. 306.4 (of act.); 75% assay; 3.92% act. oxygen; *Storage:* store @ -30 to -20 C.

Trigonox® 99-W40. [Akzo Nobel] Cumyl peroxyneodecanoate aq. emulsion; CAS 26748-47-0; initiator for PVC; emulsion; 40% assay; 2.09% act. oxygen; *Storage:* store @ -30 to -20 C.

Trigonox® 101. [Akzo Nobel] 2,5-Dimethyl-2,5-di (t-butyl peroxy hexane); CAS 78-63-7; initiator for polyester when molding at elevated temps.; also for crosslinking of olefin copolymers, chlorinated polyethylene, EPDM, SBR, and vinyls; initiator for free radical polymerizations of styrene; crosslinking agent for olefin copolymers, chlorinated polyethylene, EPDM, SBR, and vinyls; liq.; f.p. 10 C; 92% assay; 10.14% act. oxygen; *Storage:* store @ 40 C max.

Trigonox® 101-7.5PP-pd. [Akzo Nobel] 2,5-Dimethyl-2,5-di(t-butylperoxy)hexane with PP; initiator for PP; powd.; 7.5% assay; 0.83% act. oxygen; *Storage:* store @ 30 C max.

Trigonox® 101/45Bpd. [Akzo Nobel] 2,5-Dimethyl-2,5-di (t-butyl peroxy hexane) on calcium carbonate carrier; CAS 78-63-7; initiator for polyester when molding at elevated temps.; also for crosslinking of olefin copolymers, chlorinated polyethylene, EPDM, SBR, and vinyls; initiator for free radical polymerizations of styrene; powd.; 45% act.

Trigonox® 101-45S-ps. [Akzo Nobel] 2,5-Dimethyl-2,5-di (t-butyl peroxy hexane) with silicone oil carrier; CAS 78-63-7; crosslinking agent for rubber; paste; 45% act.

Trigonox® 101-50PP-pd. [Akzo Nobel] 2,5-Dimethyl-2,5-di(t-butylperoxy)hexane with PP; initiator; powd.; 50% assay; 5.51% act. oxygen.

Trigonox® 101-E50. [Akzo Nobel] 2,5-Dimethyl-2,5-di(t-butylperoxy)hexane in min. oil; initiator for PP; sol'n.; f.p. 10 C; 50% assay; 5.51% act. oxygen; *Storage:* store @ 30 C max.

Trigonox® 107. [Akzo Nobel] Ethyl-o-benzoyl laurohydroximate; high-temp. crosslinking agent; liq.; 90% act.

Trigonox® 111-B40. [Akzo Nobel] Di-t-butyl peroxyphthalate in dibutyl phthalate; CAS 2155-71-7; initiator for acrylates; sol'n.; 40% assay; 4.12% act. oxygen; *Storage:* store @ 30 C max.

Trigonox® 117. [Akzo Nobel] t-Butylperoxy 2-ethylhexyl carbonate; CAS 34443-12-4; initiator for acrylates, PS; liq.; 98% assay, 6.36% act. oxygen.

Trigonox® 121. [Akzo Nobel] t-Amyl peroxy-2-ethylhexanoate; CAS 686-31-7; initiator for acrylates, PE, PS; liq.; 95% assay; 6.60% act. oxygen; *Storage:* store @ -20 to 10 C.

Trigonox® 121-BB75. [Akzo Nobel] t-Amyl peroxy-2-ethylhexanoate; CAS 686-31-7; initiator for cure of unsaturated polyester resins; sol'n.; 75% peroxide; 5.2% act. oxygen.

Trigonox® 123-C75. [Akzo Nobel] t-Amyl peroxyneodecanoate (CAS 68229-16-1) in odorless min. spirits; initiator for PVC, PE; liq.; 75% assay, 4.64% act. oxygen; *Storage:* store @ -25 to -15 C.

Trigonox® 125-C50. [Akzo Nobel] t-Amyl peroxypivalate in odorless min. spirits; CAS 29240-17-3; initiator for PE, PVC; sol'n.; 50% assay; 4.25% act. oxygen; *Storage:* store @ -30 to -10 C.

Trigonox® 125-C75. [Akzo Nobel] t-Amyl peroxypivalate in odorless min. spirits; CAS 29240-17-3; initiator used to provide faster curing of BMC and SMC; initiator for LDPE and PVC polymerization; sol'n.; 75% assay; 6.37% active oxygen; *Storage:* store @ -30 to -10 C.

Trigonox® 127. [Akzo Nobel] t-Amyl peroxybenzoate; CAS 4511-39-1; initiator for acrylates, PE; liq.; 94% assay; 7.22% act. oxygen; *Storage:* store @ 20 C max.

Trigonox® 131. [Akzo Nobel] t-Amylperoxy 2-ethylhexyl carbonate; CAS 70833-40-8; initiator for acrylates, PS; liq.; 98% assay, 6.01% act. oxygen; *Storage:* store @ 30 C max.

Trigonox® 133-C60. [Akzo Nobel] t-Amyl peroxyacetate in odorless min. spirits; initiator for acrylates, PE; sol'n.; 60% assay, 6.56% act. oxygen; *Storage:* store @ 30 C max.

Trigonox® 141. [Akzo Nobel] 2,5-Dimethyl-2,5-(2-ethylhexanoylperoxy)hexane; CAS 13052-09-0; initiator for acrylates, PE, PS; liq.; 90% assay; 6.91% act. oxygen; *Storage:* store @ -20 to 15 C.

Trigonox® 145. [Akzo Nobel] 2,5-Dimethyl-2,5-di(t-butylperoxy)-hexyne-3; CAS 1068-27-5; initiator, crosslinking agent; liq.; 92% assay; 10.05% act. oxygen; *Storage:* store @ 30 C max.

Trigonox® 145-45B. [Akzo Nobel] 2,5-Dimethyl-2,5-di(t-butylperoxy)-hexyne-3; CAS 1068-27-5; initiator, crosslinking agent; powd.; 45% peroxide.

Trigonox® 145-E85. [Akzo Nobel] 2,5-Dimethyl-2,5-di(t-butylperoxy)-hexyne-3 in min. oil; CAS 1068-27-5; initiator for acrylates, PP; liq.; 85% assay, 9.49% act. oxygen; *Storage:* store @ 30 C max.

Trigonox® 151-C50. [Akzo Nobel] 2,4,4-Trimethylpentyl-2-peroxyneodecanoate (CAS 51240-95-0) in odorless min. spirits; initiator for acrylates, PE, PVC; sol'n.; 50% assay; 2.66% act. oxygen; *Storage:* store @ -15 C max.

Trigonox® 151-C70. [Akzo Nobel] 2,4,4-Trimethylpentyl-2-peroxyneodecanoate (CAS 51240-95-0) in odorless min. spirits; initiator for acrylates, PE, PVC; sol'n.; 70% assay; 3.72% act. oxygen; *Storage:* store @ -15 C max.

Trigonox® 161. [Akzo Nobel] Mixture; initiator for cure of unsaturated polyester resins; sol'n.; 65% peroxide; 6.7% act. oxygen.

Trigonox® 169-NA50. [Akzo Nobel] Di-t-butyl diperoxyazelate in acetyl tributyl citrate; CAS 16580-06-6; initiator for acrylates, PE, PS; sol'n.; 50% assay; 4.81% act. oxygen; *Storage:* store @ -20 to 30 C.

Trigonox® 169-OP50. [Akzo Nobel] Di-t-butyl peroxyazelate; CAS 16580-06-6; initiator; sol'n.; 50% assay; 4.81% act. oxygen.

Trigonox® 171. [Akzo Nobel] t-Butyl 3-isopropenylcumyl peroxide; CAS 96319-55-0; initiator for acrylates; liq.; 90% assay, 8.50% act. oxygen.

Trigonox® 187-C30. [Akzo Nobel] Diisobutyryl peroxide (CAS 3437-84-1) in odorless min. spirits; initiator for PE, PVC; sol'n.; m.w. 174.2 (of act.); 30% assay, 2.75% act. oxygen.

Trigonox® 201. [Akzo Nobel] Di-t-amyl peroxide; CAS 10508-09-5; initiator for acrylates, PP; liq.; m.w. 174.3; 90% assay, 8.26% act. oxygen; *Storage:* store @ 40 C max.

Trigonox® 239A. [Akzo Nobel] Cumene hydroperoxide; CAS 80-15-9; initiator for cure of unsaturated polyester resins; sol'n.; 4.6% act. oxygen.

Trigonox® A-80. [Akzo Nobel] t-Butyl hydroperoxide; CAS 75-91-2; initiator for acrylates, PS; sol'n.; 80% assay, 15.60% act. oxygen; *Storage:* store @ 0-40 C.

Trigonox® A-W70. [Akzo Nobel] t-Butyl hydroperoxide in water; initiator for high-temp. acrylate and styrenic polymerizations; suited for use in redox polymerization of acrylic emulsion polymers; sol'n.; 70% assay; 12.43% act. oxygen; *Storage:* store @ 0-40 C.

Trigonox® ACS-M28. [Akzo Nobel] Acetyl cyclohexane sulfonyl peroxide in DMP; CAS 3179-56-4; reactive initiator avail. for use in producing PVC; liq.; 29% conc.

Trigonox® ADC. [Akzo Nobel] Peroxydicarbonate; CAS 19910-65-7; initiator for low- and med.-density polyethylene; initiator for acrylates, PVC polymerization; liq.; 98% assay; 7.1% act. oxygen; *Storage:* store @ -30 to -20 C.

Trigonox® ADC-NS60. [Akzo Nobel] Peroxydicarbonate mixture; CAS 78350-78-4; initiator; sol'n.; 60% assay; 4.3% act. oxygen.

Trigonox® B. [Akzo Nobel] Di-t-butyl peroxide; CAS 110-05-4; EINECS 203-733-6; initiator for LDPE polymerization; lower m.w. at higher temps.; high-temp. peroxide used as finishing initiator for styrenics; crosslinking agent for rubber, olefin copolymers; volatile liq. best suited when used in injected extruder applic.; liq.; 99% assay; 10.83% act. oxygen; *Storage:* store @ -30 to 40 C.

Trigonox® BPIC. [Akzo Nobel] t-Butyl peroxy isopropyl carbonate in odorless min. spirits; CAS 2372-21-6; initiator for BMC and SMC molding; initiator for free radical polymerizations of styrene and acrylates; crosslinking agent for elastomers incl. SBR, urethanes, EPR, EPDM, and nitrile rubbers; sol'n.; 75% assay; 6.81% act. oxygen; *Storage:* store @ -20 to 25 C.

Trigonox® C. [Akzo Nobel] t-Butyl perbenzoate; CAS 614-45-9; initiator for elevated-temp. polyester cures, for LDPE polymerization, and high-temp. polymerization of acrylic emulsion polymers; used in BMC and SMC molding in temp. range 275-325 F; crosslinking agent for rubber; liq.; 98% assay; 8.07% act. oxygen; *Storage:* store @ 8-25 C.

Trigonox® C-C75. [Akzo Nobel] t-Butyl perbenzoate; CAS 614-45-9; initiator; sol'n.; 75% assay; 6.18% act. oxygen.

Trigonox® D-C50. [Akzo Nobel] 2,2-Di-(t-butylperoxy)butane in odorless min. spirits; CAS 2167-23-9; initiator for acrylates, PE, PS; sol'n.; 50% assay; 6.83% act. oxygen; *Storage:* store @ 30 C max.

Trigonox® D-E50. [Akzo Nobel] 2,2-Di-(t-butylperoxy)butane in min. oil; CAS 2167-23-9; initiator for acrylates, PE, PS; sol'n.; 50% assay; 6.83% act. oxygen; *Storage:* store @ -20 to 30 C.

Trigonox® EHP. [Akzo Nobel] Bis(2-ethylhexyl) peroxydicarbonate; CAS 16111-62-9; versatile, highly reactive initiator for PE, PVC prod.; high m.w. reduces reactor fouling; liq.; 98% assay; 4.52% act. oxygen; *Storage:* store @ -20 C max.

Trigonox® EHP-AT70. [Akzo Nobel] Bis(2-ethylhexyl) peroxydicarbonate in toluene; versatile, highly reactive initiator for PVC prod.; high m.w. reduces reactor fouling; liq.; 70% act.

Trigonox® EHP-B75. [Akzo Nobel] Di-(2-ethylhexyl)peroxydicarbonate; CAS 16111-62-9; initiator; sol'n.; 75% assay; 3.46% act. oxygen.

Trigonox® EHP-C40. [Akzo Nobel] Bis(2-ethylhexyl) peroxydicarbonate in odorless min. spirits; CAS 16111-62-9; highly react. initiator for PE, PVC prod.; sol'n.; 40% assay, 1.85% act. oxygen; *Storage:* store @ -15 C.

Trigonox® EHP-C70. [Akzo Nobel] Di-(2-ethylhexyl)peroxydicarbonate in odorless min. spirits; CAS 16111-62-9; initiator for PE, PVC; sol'n.; 70% assay; 3.23% act. oxygen; *Storage:* store @ -15 C max.

Trigonox® EHP-C75. [Akzo Nobel] Bis(2-ethylhexyl) peroxydicarbonate in odorless min. spirits; CAS 16111-62-9; versatile, highly reactive initiator for PE, PVC prod.; high m.w. reduces reactor fouling; liq.; 75% assay; 3.46% act. oxygen; *Storage:* store @ -15 C max.

Trigonox® EHPS. [Akzo Nobel] Bis(2-ethylhexyl) peroxydicarbonate, stabilized; CAS 16111-62-9; initiator for PVC prod.; liq.; 98% assay, 4.52% act. oxygen; *Storage:* store @ -20 C.

Trigonox® EHPS-C75. [Akzo Nobel] Stabilized bis(2-ethylhexyl) peroxydicarbonate in odorless min. spirits; CAS 16111-62-9; initiator for PVC prod.; sol'n.; 75% assay; 3.46% act. oxygen; *Storage:* store @ -15 C max.

Trigonox® EHP-W40. [Akzo Nobel] Di-(2-ethylhexyl)peroxydicarbonate aq. emulsion; CAS 16111-62-9; initiator for PVC prod.; emulsion; 40% assay; 1.85% act. oxygen; *Storage:* store @ -15 C max.

Trigonox® EHP-W40S. [Akzo Nobel] Di-(2-ethylhexyl)peroxydicarbonate; CAS 16111-62-9; initiator for PVC prod.; flakes; 40% assay; 1.85% act. oxygen; *Storage:* store @ -15 C max.

Trigonox® EHP-W50. [Akzo Nobel] Di-(2-ethylhexyl)peroxydicarbonate aq. emulsion; CAS 16111-62-9; initiator for PVC prod.; emulsion; 50% assay; 2.31% act. oxygen; *Storage:* store @ -15 C max.

Trigonox® F-C50. [Akzo Nobel] t-Butyl peracetate in odorless min. spirits; CAS 107-71-1; initiator for low- and med.-dens. polyethylene, styrenics, acrylates; sol'n.; 50% assay; 6.05% act. oxygen; *Storage:* store @ -15 to 10 C.

Trigonox® HM. [Akzo Nobel] MIBK peroxide; CAS 37206-20-5; initiator for cure of unsaturated polyester resins; sol'n.; 45% peroxide; 8.7% act. oxygen.

Trigonox® K-80. [Akzo Nobel] Cumene hydroperoxide; CAS 80-15-9; initiator for acrylates; sol'n.; 80% assay; 8.51% act. oxygen; *Storage:* store @ -30 to 40 C.

Trigonox® K-95. [Akzo Nobel] Cumene hydroperoxide; CAS 80-15-9; liq.; 95% assay; 9.99% act. oxygen.

Trigonox® KSM. [Akzo Nobel] t-Butyl peroctoate and 1,1-bis (t-butylperoxy)-3,3,5-trimethylcyclohexane; SMC/BMC initiator offering good shelf life in compd. and molding temp. range of 230-290 F; polyester cures; also for acrylates; sol'n.; 75% assay; 6.10% act. oxygen; *Storage:* store @ 20 C max.

Trigonox® M-50. [Akzo Nobel] Diisopropylbenzene monohydroperoxide; CAS 26762-93-6; initiator for acrylates, PE; sol'n.; 50% assay; 4.10% act. oxygen; *Storage:* store @ -25 to 30 C.

Trigonox® NBP-C50. [Akzo Nobel] Di-n-butyl peroxydicarbonate in odorless min. spirits; CAS 16215-49-9; initiator for PE, PVC; sol'n.; 50% assay; 3.41% act. oxygen; *Storage:* store @ -30 to -20 C.

Trigonox® SBP. [Akzo Nobel] Di(s-butyl) peroxydicarbonate; CAS 19910-65-7; initiator for LDPE polymerization; fastest conversion to polymer; used for prod. of PVC; liq.; m.w. 234.2; 98% assay; 6.69% act. oxygen; *Storage:* store @ -20 C max.

Trigonox® SBP-AX30. [Akzo Nobel] Di(s-butyl) peroxydicarbonate in xylene; CAS 19910-65-7; efficient initiator give fast cycle times for acrylate polymerization; liq.; 30% act.

Trigonox® SBP-C30. [Akzo Nobel] Di-(s-butyl)peroxydicarbonate (CAS 19910-65-7) in odorless min. spirits; initiator for PVC; sol'n.; 30% assay; 2.05% act. oxygen; *Storage:* store @ -30 to -15 C.

Trigonox® SBP-C50. [Akzo Nobel] Di-(s-butyl)peroxydicarbonate; CAS 19910-65-7; initiator; sol'n.; 50% assay; 3.42% act. oxygen.

Trigonox® SBP-C60. [Akzo Nobel] Di (s-butyl) peroxydicarbonate) (CAS 19910-65-7) in odorless min. spirits; initiator for PVC giving fast cycle times for polymerization; liq.; 60% assay, 4.09% act. oxygen; *Storage:* store @ -30 to -20 C.

Trigonox® SBP-C75. [Akzo Nobel] Di(s-butyl)peroxydicarbonate in odorless min. spirits; CAS 19910-65-7; initiator giving fast cycle times for PVC polymerization; liq.; 75% act.

Trigonox® SBPS. [Akzo Nobel] Di(s-butyl) peroxydicarbonate, stabilized; CAS 19910-65-7; initiator for PVC prod.; liq.; m.w. 234.2; 98% assay; 6.69% act. oxygen; *Storage:* store @ -20 C max.

Trigonox® T. [Akzo Nobel] t-Butyl cumyl peroxide; CAS 3457-61-2; crosslinking agent for EPDM, SBR, Neoprene, Hypalon; low reactivity peroxide useful as a finishing initiator at high temps. for styrenics; a synergist for some halogen-containing flame retardants; liq.; 90% assay; 6.91% act. oxygen; *Storage:* store @ -20 to 40 C.

Trigonox® T-94. [Akzo Nobel] t-Butyl cumyl peroxide; CAS 3457-61-2; crosslinking agent for rubber; liq.; 95% act.

Trigonox® TAHP-W85. [Akzo Nobel] t-Amyl hydroperoxide in water; CAS 3425-61-4; initiator for acrylates, PE; sol'n.; 85% assay; 13.06% act. oxygen; *Storage:* store @ 30 C max.

Trigonox® TMPH. [Akzo Nobel] 2,4,4-Trimethylpentyl-2-hydroperoxide; CAS 5809-08-5; initiator for acrylates, PE; sol'n.; 80% assay; 9.91% act. oxygen; *Storage:* store @ -30 to 20 C.

Tri-Mal. [Syn. Prods.] Tribasic lead maleate; vulcanizing agent for chlorosulfonated polyethylene; gives tight cure and good water resist. in lt. colored compds.; cream powd.; sp.gr. 6.3; 89% PbO.

Trimet® Liquid. [IMC/Americhem] Trimethylolethane; CAS 77-85-0; raw material for alkyd and polyester resins for paints, syn. lubricants, plasticizers, stabilizers for plastics, coating agents for pigments; nitrate ester in explosives and propellants; colorless sol'n.; dens. 9.04 lb/gal (80 C); m.p. 199-200 C (solids); 80% solids aq. sol'n.; 41% hydroxyl content; *Toxicology:* essentially nontoxic; mildly irritating to abraded skin, nonirritating to eyes.

Trimet® Nitration Grade. [IMC/Americhem] Trimethylolethane; CAS 77-85-0; raw material for alkyd and polyester resins for paints, syn. lubricants, plasticizers, stabilizers for plastics, coating agents for pigments; nitration grade esp. for use in explosives; solid; dens. 6.6 and 6.5 lb/gal (briquets, gran.); m.p. 199-203 C; 41.75% hydroxyl content; *Toxicology:* essentially nontoxic; mildly irritating to abraded skin, nonirritating to eyes.

Trimet® Pure. [IMC/Americhem] Trimethylolethane; CAS 77-85-0; raw material for alkyd and polyester resins for paints, syn. lubricants, plasticizers, stabilizers for plastics, coating agents for pigments; nitrate ester in explosives and propellants; crystalline; dens. 6.6 and 6.5 lb/gal (briquets, gran.); m.p. 199-203 C; 41.75% hydroxyl content; *Toxicology:* essentially nontoxic; mildly irritating to abraded skin, nonirritating to eyes.

Trimet® Tech. [IMC/Americhem] Trimethylolethane; CAS 77-85-0; raw material for alkyd and polyester resins for paints, syn. lubricants, plasticizers, stabilizers for plastics, coating agents for pigments; nitrate ester in explosives and propellants; wh. briquets or gran.; sol. 140 g/100g water, 75.2 g/100 g methanol, 27.9 g/100 g ethanol; dens. 6.3 and 6.2 lb/gal (briquets, gran.); m.p. 190-200 C; flash pt. (COC) 320 F; 41% hydroxyl content; *Toxicology:* essentially nontoxic; mildly irritating to abraded skin, nonirritating to eyes.

Triton® 770 Conc. [Union Carbide; Union Carbide Europe] Sodium alkylaryl polyether sulfate; anionic; detergent, emulsifier, wetting agent, penetrant used in detergents, cleaners, textile and leather processing, emulsion polymerization; degreasing agent on skins prior to tanning; amber clear liq.; sol. in water, acid and alkaline sol'ns.; sp.gr. 0.98; dens. 8.3 lb/gal; visc. 15 cps; flash pt. 78 F (PMCC); pour pt. -20 F; pH 7.5 (5% aq.); surf. tens. 28 dynes/cm (1%); Ross-Miles foam 165 mm (0.1%, initial, 120 F); 30% act., 23% IPA.

Triton® N-60. [Union Carbide; Union Carbide Europe] Nonoxynol-6; nonionic; antifogging agent in plasticized PVC; corrosion inhibitor; dispersant for petrol. oils; liq.; 100% conc.

Triton® W-30 Conc. [Union Carbide; Union Carbide Europe] Sodium alkylaryl ether sulfate; anionic; detergent, wetting agent, penetrant for sizing, dyeing, desizing, textile processing; emulsifier in emulsion polymerization; amber clear liq.; sol. in water, acid and alkaline sol'ns.; sp.gr. 0.98; dens. 8.2 lb/gal; visc. 9 cps; flash pt. 74 F (PMCC); pour pt. -15 F; pH 7.5 (5% aq.); surf. tens. 29 dynes/cm (1%); Ross-Miles foam 110 mm (0.1%, initial, 120 F); 27% act., 27% IPA.

Triton® X-15. [Union Carbide; Union Carbide Europe] Octoxynol-1; CAS 9002-93-1; nonionic; surfactant, coupling agent, emulsifier for industrial/household cleaners, emulsion polymerization, agric., latex stabilizer; FDA 21CFR §172.710, 175.105, 176.210, EPA compliance; APHA 250 liq.; sol. in aliphatic hydrocarbons; misc. with aromatics, alcohols, glycols, ethers, ketones; m.w. 250; sp.gr. 0.985; dens. 8.2 lb/gal; visc. 790 cps; HLB 3.6; flash pt. > 300 F (TOC); pour pt. 15 F; 100% act.

Triton® X-35. [Union Carbide; Union Carbide Europe] Octoxynol-3; CAS 9002-93-1; nonionic; surfactant, coupling agent, emulsifier for industrial/household cleaners, emulsion polymerization, agric., latex stabilizer; FDA, EPA compliance; APHA 125 liq.; sol. in aliphatic hydrocarbons; misc. with aromatic hydrocarbons, alcohols, glycols, ethers, ketones; m.w. 338; sp.gr. 1.023; dens. 8.5 lb/gal; visc. 370 cps; HLB 7.8; flash pt. > 300 F (TOC); pour pt. -10 F; surf. tens. 29 dynes/cm (0.01%); Ross-Miles foam 5 mm (0.1%, initial, 120 F); 100% act.

Triton® X-165-70%. [Union Carbide; Union Carbide Europe] Octoxynol-16; CAS 9002-93-1; nonionic; detergent, emulsifier, wetting agent for industrial/household cleaners, emulsion polymerization, latex stabilizer; FDA compliance; APHA 125 aq. sol'n.; sol. in inorg. salt sol'ns., aq. min. acids; misc. with water, alcohols, glycols, ethers, ketones; m.w. 910; sp.gr. 1.080; dens. 9.0 lb/gal; visc. 540 cps; HLB 15.8; cloud pt. > 100 C (1%); flash pt. > 300 F (TOC); pour pt. 55 F; surf. tens. 35 dynes/cm (0.1%); Ross-Miles foam 145 mm (0.1%, initial, 120 F); 70% act.

Triton® X-200. [Union Carbide; Union Carbide Europe] Sodium octoxynol-2 ethane sulfonate; anionic; detergent for metal cleaning, pickling and plating baths, household cleaners; polymerization emulsifier, latex post-stabilizer; wetting agent in strong alkaline baths; dispersant for fulling in lime soap; dye leveling agent for acid dyestuffs; wh. liq.; dens. 8.9 lb/gal; visc. 7000 cps; flash pt. (TOC) > 300 F; pour pt. 25 F; surf. tens. 29 dynes/cm (1%); 28% conc.

Triton® X-301. [Union Carbide; Union Carbide Europe] Sodium alkylaryl polyether sulfate; anionic; wetting agent, emulsifier, penetrant, high-foam detergent for household and industrial cleaners, emulsion polymerization; wh. paste; sol. in water and acid and alkaline sol'ns.; sp.gr. 1.05; dens. 8.8 lb/gal; visc. 4200 cps; flash pt. > 300 F (TOC); pour pt. 30 F; pH 7.5 (5% aq.); surf. tens. 28 dynes/cm (0.1%); Ross-Miles foam 145 mm (0.1%, 120 F, initial); 20% act. in water.

Triton® X-305-70%. [Union Carbide; Union Carbide Europe] Octoxynol-30; CAS 9002-93-1; nonionic; detergent, emulsifier, wetting agent for household/ industrial cleaners, emulsion polymerization, agric., latex stabilizer; FDA 21CFR §172.710, 175.105, 176.210, 178.3400, EPA compliance; APHA 150 liq.; sol. in inorg. salt sol'ns., aq. min. acids; misc. with water, alcohols, glycols, ethers, ketones; m.w. 1526; sp.gr. 1.095; dens. 9.1 lb/gal; visc. 470 cps; HLB 17.3; pour pt. 35 F;cloud pt. > 100 C (1%); flash pt. > 300 F (TOC); surf. tens. 37 dynes/cm (0.1%); Ross-Miles foam 150 mm (0.1%, initial, 120 F); 70% act.

Triton® X-405-70%. [Union Carbide; Union Carbide Europe] Octoxynol-40; CAS 9002-93-1; nonionic; detergent, emulsifier, wetting agent for household/ industrial cleaners, emulsion polymerization, agric.; latex stabilizer; coupling agent in naphthol dyeing; FDA 21CFR §172.710, 175.105, 176.210, 178.3400, EPA compliance; APHA 250 liq.; sol. in inorg. salt sol'ns., aq. min. acids; misc. with water, alcohols, glycols, ethers, ketones; m.w. 1966; sp.gr. 1.102; dens. 9.2 lb/gal; visc. 490 cps; HLB 17.9; cloud pt. > 100 C (1%); flash pt. > 212 F (TOC); pour pt. 25 F; surf. tens. 37 dynes/cm (0.1%); Ross-Miles foam 126 mm (0.1%, initial, 120 F); 70% act.

Triton® X-705-70%. [Union Carbide; Union Carbide Europe] Octoxynol-70; CAS 9002-93-1; nonionic; emulsifier for emulsion polymerization; FDA 21CFR §172.710, 175.105, EPA compliance; APHA 500 liq.; sol. in inorg. salt sol'ns., aq. min. acids; misc. with water, alcohols, glycols, ethers, ketones; sp.gr. 1.1; dens. 9.2 lb/gal; visc. 505 cP; HLB 18.7; pour pt. 43 F; cloud pt. > 100 C (1% aq.); flash pt. (Seta CC) > 230 F; surf. tens. 39 dynes/cm (0.1%); Ross-Miles foam 95 mm (0.1%, initial, 120 F); 70% act.

Trycol® 6954. [Henkel/Emery] Nonoxynol-150; CAS 9016-45-9; nonionic; wetting agent, dispersant; emulsifier for emulsion polymerization; Gardner 1 solid; sol. in water; HLB 19.3; m.p. 60 C; cloud pt. 69 C (10% saline); flash pt. 500 F.

Trycol® 6957. [Henkel/Emery; Henkel/Textile] Nonoxynol-40; CAS 9016-45-9; nonionic; detergent, wetting agent, dispersant, coemulsifier, stabilizer; for emulsion polymerization, agric. formulations, textile processing; EPA-exempt; Gardner 1 solid; sol. in water, xylene; HLB 17.8; cloud pt. 90 C (5% saline); flash pt. 560 F; 100% conc.

Trycol® 6967. [Henkel/Emery; Henkel/Textile] Nonoxynol-20; CAS 9016-45-9; nonionic; mod. high foaming detergent, wetting agent, emulsfier for emulsion polymerization, textile processing; Gardner 1 solid; sol. in water, xylene; HLB 16.0; m.p. 34 C; cloud pt. 88 C (5% saline); flash pt. 515 F; 100% conc.

Trycol® 6968. [Henkel/Emery; Henkel/Textile] Nonoxynol-30; CAS 9016-45-9; nonionic; detergent, wetting agent; emulsifier in emulsion polymerization, for fats, oils, and waxes, agric. formulation, textile processing; EPA-exempt; Gardner 1 solid; sol. in water, xylene; HLB 17.1; m.p. 43 C; cloud pt. 93 C (5% saline); flash pt. 515 F; 100% conc.

Trycol® 6969. [Henkel/Emery; Henkel/Textile] Nonoxynol-30; CAS 9016-45-9; nonionic; detergent, wetting agent; emulsifier in emulsion polymerization, for fats, oils, and waxes, agric. formulations, textile processing; EPA-exempt; Gardner 1 liq.; sol. in water; dens. 9.0 lb/gal; visc. 260 cSt (100 F); HLB 17.1; cloud pt. 92 C (5% saline); pour pt. 5 C; 70% act. in water.

Trycol® 6970. [Henkel/Emery; Henkel/Textile] Nonoxynol-40; CAS 9016-45-9; nonionic; detergent, wetting agent, dispersant, coemulsifier, stabilizer; for emulsion polymerization, agric. formulations, textile processing; EPA-exempt; Gardner 1 liq.; sol. in water; dens. 9.2 lb/gal; visc. 385 cSt (100 F); HLB 17.8; cloud pt. 90 C (5% saline); pour pt. 7 C; 70% act. in water.

Trycol® 6971. [Henkel/Emery] Nonoxynol-50; CAS 9016-45-9; nonionic; moderately high foaming detergent, emulsifier; for emulsion polymerization, agric. formulations; EPA-exempt; Gardner 1 solid; sol. in water, xylene; HLB 18.2; m.p. 54 C; cloud pt. 76 C (10% saline); flash pt. 520 F; 100% conc.

Trycol® 6972. [Henkel/Emery] Nonoxynol-50; CAS 9016-45-9; nonionic; detergent, wetting agent, dispersant, coemulsifier, stabilizer; for emulsion polymerization; Gardner 1 liq.; sol. in water; dens. 9.0 lb/ gal; visc. 440 cSt (100 F); HLB 18.2; cloud pt. 76 C (10% saline); pour pt. -4 C; 70% act.

Trycol® 6981. [Henkel/Emery] Nonoxynol-100; CAS 9016-45-9; nonionic; moderately high foaming detergent, emulsifier; for emulsion polymerization; Gardner 1 liq.; sol. in water; dens. 9.2 lb/gal; visc. 564 cSt (100 F); HLB 19.0; cloud pt. 72 C (10% saline); pour pt. 18 C; 70% act.

Trycol® 6984. [Henkel/Emery; Henkel/Textile] Octoxynol-40; CAS 9002-93-1; nonionic; dispersant, wetting agent; emulsifier for emulsion polymerization of acrylic and vinyl monomers; textile processing; agric. formulations; EPA-exempt; Gardner 1 liq.; sol. in water; dens. 9.0 lb/gal; visc. 220 cSt (100 F); HLB 17.9; cloud pt. 74 C (10% saline); pour pt.

13 C; 70% act. in water.

Trycol® 6985. [Henkel/Emery; Henkel/Textile] Nonyl nonoxynol-8; CAS 9014-93-1; nonionic; emulsifier in textile dye carrier applics., insecticides, wax emulsions; foam control agent; spreading agent in pigment printing; post-stabilizer in emulsion polymerization; intermediate; EPA-exempt; Gardner 2 liq.; sol. in xylene, glycerol trioleate; disp. in water, min. oil; dens. 8.4 lb/gal; visc. 173 cSt (100 F); HLB 10.4; cloud pt. < 25 C; flash pt. 525 F; pour pt. 9 C; 100% conc.

Trycol® DNP-8. [Henkel/Emery] Nonyl nonoxynol-8; CAS 9014-93-1; nonionic; foam control agent; emulsifier for polar and nonpolar solvs. and textile jet dye carrier applics.; spreading agent in pigment printing; post-stabilizer in emulsion polymerization; intermediate in mfg. of anionics; surfactant component in acidic cleaners, aerosols, insecticides and wax emulsions; Gardner 2 liq.; disp. in water; dens. 8.4 lb/gal; visc. 410 cs; HLB 10.4; cloud pt. < 25 C; 100% act.; Unverified

Trycol® NP-20. [Henkel/Emery] Nonoxynol-20; CAS 9016-45-9; nonionic; surfactant used as detergent, wetting agent and emulsifier for emulsion polymerization; Gardner 1 solid; HLB 16.0; m.p. 34 C; flash pt. 515 F; cloud pt. 88 C; Unverified

Trycol® NP-30. [Henkel/Emery] Nonoxynol-30; CAS 9016-45-9; nonionic; detergent, wetting agent; emulsifier for fats, oils, and waxes and emulsion polymerization; Gardner 1 solid; m.p. 40 C; HLB 17.1; cloud pt. > 100 C; 100% act.; Unverified

Trycol® NP-307. [Henkel/Emery] Nonoxynol-30; CAS 9016-45-9; nonionic; detergent and wetting agent; emulsifier in emulsion polymerization and for fats, oils and waxes; Gardner 1 liq.; dens. 9.0 lb/gal; visc. 260 cSt (100 F); HLB 17.1; pour pt. 5 C; cloud pt. 92 C; Unverified

Trycol® OP-407. [Henkel/Emery] Octoxynol-40; CAS 9002-93-1; nonionic; hydrophilic dispersant and wetting agent; primary emulsifier in emulsion of polymerization of acrylic and vinyl monomers; Gardner 1 liq.; dens. 9.0 lb/gal; visc. 220 cSt (100 F); HLB 17.9; pour pt. 13 C; cloud pt. 74 C; Unverified

Trydet SA-50/30. [Henkel/Emery] PEG-50 stearate; CAS 9004-99-3; nonionic; hydrophilic emulsifier for preparing solubilized oils; visc. modifier, softener or plasticizer in acrylic or vinyl resin emulsions; Gardner 1 liq.; dens. 8.5 lb/gal; visc. 671 cSt (100 F); HLB 17.8; pour pt. 0 C; flash pt. 540 F; cloud pt. 81 C; Unverified

Tryfac® 5556. [Henkel/Emery; Henkel/Textile] Complex phosphate ester, free acid form; CAS 51811-79-1; anionic; wetting agent, dispersant, antistat in textile processing; solv. emulsifier for textile scours, detergents, pesticides; drycleaning detergent; also used in emulsion polymerization; EPA-exempt; Gardner 2 liq.; sol. in water, xylene; dens. 9.3 lb/gal; visc. 1700 cSt (100 F); cloud pt. 68 C (5% saline); flash pt. 450 F; pour pt. 5 C.

Trymeen® 6617. [Henkel/Emery; Henkel/Textile] PEG-50 stearyl amine; CAS 26635-92-7; cationic; emulsifier, antistat for metal buffing compds., latex rubber compding., agric. formulations; anticoagulant; lubricant, leveling agent for textile applics.; EPA-exempt; Gardner 4 solid; sol. in water; dens. 8.5 lb/gal (40 C); HLB 17.8; m.p. 35 C; cloud pt. 82 C (10% saline); flash pt. 540 F; 100% conc.

TR-Zinc Stearate. [Struktol] Zinc stearate; CAS 557-05-1; EINECS 209-151-9; high purity clear melt processing aid; powd.; 0.1% max. on 100 mesh, 1% max. on 325 mesh; bulk dens. 20-25 lb/ft^3; m.p. 117-121 C; 15% max. ZnO; *Storage:* unlimited storage stability in cool, dry area.

TSA-70. [Henkel Canada] Toluene sulfonic acid; anionic; catalyst in resin prod.; salts as hydrotropes in household and industrial cleaners, metal cleaning compds.; liq.; 70% conc.

TSS. [Dong Jin] p-Toluene sulfonyl semicarbazide; chem. foaming agent for ABS, acetal, acrylic, EVA, HDPE, LDPE, PPO, PP, PS, HIPS, flexible and rigid PVC, TPE; processing temps. 390-455 F; gas yield 145 cc/g.

Tuf Lube Hi-Temp NOD. [Specialty Prods.] PTFE wax; CAS 9002-84-0; external lubricant for non-CFC high-temp. thermosets and thermoplastics; liq.

Tullanox GT. [Tulco] Made from neither fumed nor precipitated silica base; has less structure than other Tullanox prods.; extremely hydrophobic.

Tullanox HM-100. [Tulco] Precipitated silica; same base as Tullanox HM-250, but with different rheology.

Tullanox HM-150. [Tulco] Precipitated silica; different base than Tullanox HM-250; offers ref. index which can provide clarity to adhesives, sealants, etc.

Tullanox HM-250. [Tulco] Hydrophobic precipitated silica, modified by org. silazane compd.; high surf. area, high water repellency; provides reinforcement, rheology control, corrosion resist., anticaking, thickening to silicone sealants, coatings, powds., polyester resins, liq. systems, elastomers, elec. insulation, defoamers; carrier for catalysts; filler/additive for plastics, paints, coatings, inks, pharmaceuticals, cosmetics, fertilizers, metals, adhesives, toners; wh. powd., extremely fine particle size; 0.3 μ particle diam.; sp.gr. 2.2; bulk dens. 5-6 lb/ft^3; surf. area 125 $\pm$ 20 m^2/g; ref. index 1.45-1.46; pH > 9 (4% in 50/50 IPA/water); *Toxicology:* may cause eye irritation.

Tylose® C, CB Series. [Hoechst Celanese/Colorants & Surf.] Sodium CMC; CAS 9004-32-4; anionic; binder in pencil leads; thickener in batteries, rubber industry, cosmetics, foodstuffs, pharmaceuticals, tobacco and textile industry; dispersant, emulsifier for insecticidal, fungicidal and herbicidal prods.; plasticizer in ceramics; surface sizing in paper industry; press aid and lubricant in welding electrodes; gran., powd.; 100% act.

Tylose® CBR Grades. [Hoechst Celanese/Colorants & Surf.; Hoechst AG] Sodium CMC; CAS 9004-32-4; binder for coatings; sedimenting aid in mining; gelling agent/binder/thickener in chemical tech. and rubber industry; sizing for paper and textile industry; plasticizer/filler in soaps and hand cleaning pastes; graying inhibitor for detergents; gran.

Tylose® CR. [Hoechst Celanese/Colorants & Surf.] Sodium CMC, tech.; CAS 9004-32-4; binder for coatings; sedimenting aid in mining; gelling agent/ binder/thickener in chemical tech. and rubber industry; sizing for paper and textile industry; plasticizer/filler in soaps and hand cleaning pastes; gran.; Unverified

Tylose® H Series. [Hoechst Celanese/Colorants & Surf.; Hoechst AG] Hydroxyethylcellulose; CAS 9004-62-0; nonionic; binder, thickener, plasticizer, visc. control agent, protective colloid in ceramics, emulsion polymerization, tobacco and textile industry, agric., cosmetics, soaps, and hand cleaning pastes; gran.; water-sol.; 100% act.

Tylose® MH Grades. [Hoechst Celanese/Colorants & Surf.; Hoechst AG] Methyl hydroxyethylcellulose; CAS 9032-42-2; nonionic; binder, thickener, pigment, foam, and filler stabilizer, dispersant, emulsifier, plasticizer, visc. control and sedimenting aid, and protective colloid used in coatings, paints, resins, mining, batteries, insecticides, fungicides, herbicides; rubber, textile, and leather industry; ceramics, suspension polymerization, and pharmaceuticals; gran.; water-sol.; 100% act.

Tylose® MHB. [Hoechst Celanese/Colorants & Surf.] Methyl hydroxyethylcellulose; CAS 9032-42-2; nonionic; binder, thickener, pigment, foam, and filler stabilizer, dispersant, emulsifier, plasticizer, visc. control and sedimenting aid, and protective colloid used in coatings, paints, resins, mining, batteries, insecticides, fungicides, herbicides; rubber, textile, and leather industry; ceramics, suspension polymerization, and pharmaceuticals; gran.; water-sol.

Tyzor AA. [DuPont] Acetylacetonate chelate; catalyst for esterification and olefin polymerization; crosslinking agent for automotive prods., coatings, elastomers, films/paints, graphic arts, plastics; surf. modifier for coatings, cosmetics, electronics, films, glass, graphic arts, metals, petrol. prods., plastics; red liq.; m.w. 364; sol. in IPA, IPA-water, ethyl acetate, IPA; sp.gr. 0.99; visc. 15 cP; flash pt. (PMCC) 12 C; 75% in IPA.

Tyzor DC. [DuPont] Ethylacetoacetate chelate; CAS 141-97-9; EINECS 205-516-1; catalyst for esterification and olefin polymerization; resin crosslinking agent for automotive prods., coatings, elastomers, films/paints, graphic arts, plastics; yel. to amber liq.; m.w. 424; sol. in IPA, acetone, ethanol, Freon TF solv.; sp.gr. 1.05; visc. 50 cP; flash pt. (PMCC) 27 C; 100% conc.

Tyzor LA. [DuPont] Lactic acid chelate, ammonium salt; catalyst for esterification and olefin polymerization; crosslinking agent for automotive prods., coatings, elastomers, films/paints, graphic arts, plastics; surf. modifier for coatings, cosmetics, electronics, films, glass, graphic arts, metals, petrol. prods., plastics; lt. amber liq.; m.w. 294 (of solids); water-misc.; sp.gr. 1.21; visc. 8 cP; 50% in water.

Tyzor TBT. [DuPont] Tetrabutyl titanate; CAS 5593-70-4; catalyst for esterification and olefin polymerization; crosslinking agent for automotive prods., coatings, elastomers, films/paints, graphic arts, plastics; surf. modifier for coatings, cosmetics, electronics, films, glass, graphic arts, metals, petrol. prods., plastics; pale yel. liq.; m.w. 340 (of solids); sol. in n-butanol, n-heptane; sp.gr. 0.99; visc. 65 cP; flash pt. (PMCC) 47 C; 100% conc.

Tyzor TE. [DuPont] TEA chelate; catalyst for esterification and olefin polymerization; crosslinking agent for automotive prods., coatings, elastomers, films/ paints, graphic arts, plastics; surf. modifier for coatings, cosmetics, electronics, films, glass, graphic arts, metals, petrol. prods., plastics; pale yel. liq.; m.w. 462 (of solids); sol. in water, IPA; sp.gr. 1.06; visc. 90 cP; flash pt. (PMCC) 16 C; 80% in IPA.

Tyzor TOT. [DuPont] Tetrakis (2-ethylhexyl) titanate; catalyst for esterification and olefin polymerization; crosslinking agent for automotive prods., coatings, elastomers, films/paints, graphic arts, plastics; surf. modifier for coatings, cosmetics, electronics, films, glass, graphic arts, metals, petrol. prods., plastics; pale yel. liq.; m.w. 565 (of solids); sol. in n-heptane, IPA; sp.gr. 0.91; visc. 140 cP; flash pt. (PMCC) 60 C; 100% conc.

Tyzor TPT. [DuPont] Tetraisopropyl titanate; CAS 546-68-9; catalyst for esterification and olefin polymerization; crosslinking agent for automotive prods., coatings, elastomers, films/paints, graphic arts, plastics; surf. modifier for coatings, cosmetics, electronics, films, glass, graphic arts, metals, petrol. prods., plastics; pale yel. liq.; m.w. 284 (of solids); sol. in IPA, n-heptane; sp.gr. 0.95; visc. 3 cP; flash pt. (PMCC) 21 C; 100% conc.

U

Ucarfloc® Polymer 300. [Union Carbide] Polyethylene oxide; CAS 25322-68-3; nonionic; water-sol. thermoplastic resin, thickener, lubricant, binder, flocculant, wet adhesive for mining; foam stabilizer in fermented malt beverages (300 ppm max.); dispersant for vinyl polymerization; EPA 40CFR §180.1001(d); FDA 21CFR §172.770, 175.300, 175.380, 175.390, 176.170, 176.180, 177.1210, 177.1350; wh. gran. powd., mild ammoniacal odor; 100% thru 10 mesh, 96% thru 20 mesh; sol. in water, some chlorinated solvs., alcohols, aromatic hydrocarbons, ketones; m.w. 4,000,000; sp.gr. 1.15-1.26; bulk dens. 19-37 lb/ft^3; visc. 1650-5500 cps (1% aq.); m.p. 62-67 C; pH 8-10 (sol'n.); *Precaution:* slippery when wet; *Storage:* store in sealed containers below 25 C, away from heat; avoid dust buildup.

Ucarfloc® Polymer 302. [Union Carbide] Polyethylene oxide; CAS 25322-68-3; nonionic; water-sol. thermoplastic resin, thickener, lubricant, binder, flocculant, wet adhesive for mining; foam stabilizer in fermented malt beverages (300 ppm max.); dispersant for vinyl polymerization; EPA 40CFR §180.1001(d); FDA 21CFR §172.770, 175.300, 175.380, 175.390, 176.170, 176.180, 177.1210, 177.1350; wh. gran. powd., mild ammoniacal odor; 100% thru 10 mesh, 96% thru 20 mesh; sol. in water, some chlorinated solvs., alcohols, aromatic hydrocarbons, ketones; m.w. 5,000,000; sp.gr. 1.15-1.26; bulk dens. 19-37 lb/ft^3; visc. 5500-7500 cps (1% aq.); m.p. 62-67 C; pH 8-10 (sol'n.); *Precaution:* slippery when wet; *Storage:* store in sealed containers below 25 C, away from heat; avoid dust buildup.

Ucarfloc® Polymer 304. [Union Carbide] Polyethylene oxide; CAS 25322-68-3; nonionic; water-sol. thermoplastic resin, thickener, lubricant, binder, flocculant, wet adhesive for mining; foam stabilizer in fermented malt beverages (300 ppm max.); dispersant for vinyl polymerization; EPA 40CFR §180.1001(d); FDA 21CFR §172.770, 175.300, 175.380, 175.390, 176.170, 176.180, 177.1210, 177.1350; wh. gran. powd., mild ammoniacal odor; 100% thru 10 mesh, 96% thru 20 mesh; sol. in water, some chlorinated solvs., alcohols, aromatic hydrocarbons, ketones; m.w. 7,000,000; sp.gr. 1.15-1.26; bulk dens. 19-37 lb/ft^3; visc. 7500-10,000 cps (1% aq.); m.p. 62-67 C; pH 8-10 (sol'n.); *Precaution:* slippery when wet; *Storage:* store in sealed containers below 25 C, away from heat; avoid dust buildup.

Ucarfloc® Polymer 309. [Union Carbide] Polyethylene oxide; CAS 25322-68-3; nonionic; water-sol. thermoplastic resin, thickener, lubricant, binder, flocculant, wet adhesive for mining; foam stabilizer in fermented malt beverages (300 ppm max.); dispersant for vinyl polymerization; EPA 40CFR §180.1001(d); FDA 21CFR §172.770, 175.300, 175.380, 175.390, 176.170, 176.180, 177.1210, 177.1350; wh. gran. powd., mild ammoniacal odor; 100% thru 10 mesh, 96% thru 20 mesh; sol. in water, some chlorinated solvs., alcohols, aromatic hydrocarbons, ketones; m.w. 8,000,000; sp.gr. 1.15-1.26; bulk dens. 19-37 lb/ft^3; visc. 10,000-15,000 cps (1% aq.); m.p. 62-67 C; pH 8-10 (sol'n.); *Precaution:* slippery when wet; *Storage:* store in sealed containers below 25 C, away from heat; avoid dust buildup.

Ucarsil® FR-1A, FR-1B. [OSi Specialties] Organosilicon chemical; additives allowing the processing of up to 70% alumina trihydrate into polyolefins.

Ufapol. [Unger Fabrikker AS] Surfactant blend; anionic; surfactant for emulsion polymerization; liq.

Ufasan 35. [Unger Fabrikker AS] Linear sodium dodecylbenzene sulfonate; CAS 25155-30-0; EINECS 246-680-4; anionic; biodeg. detergent, wetting agent, foaming agent for liq. detergents, dishwash, hair and car shampoos and in plastics, metal, agric., polish, textiles, mining, oil and cement industries; stable in hard water and acids; golden visc. liq., weak char. odor; m.w. 343-345; cloud pt. -7 C; pH 7.5-8.5; 34-36% act.

Ufasan 50. [Unger Fabrikker AS] Linear sodium alkylbenzene sulfonate; anionic; surfactant, emulsifier for detergents and in plastics, metal, agric., polish, textiles, mining, oil and cement industries; pumpable liq./paste; m.w. 344; 50% act.

Ufasan 60A. [Unger Fabrikker AS] Linear sodium dodecylbenzene sulfonate; anionic; biodeg., highly sol. surfactant for use in high active liq. detergents, dishwash, industrial cleaners, car shampoos, laundry washes and in plastics, metal, agric., polish, textiles, mining, oil and cement industries; stable in hard water, acid, alkalis; very pale mobile paste, weak odor; m.w. 337-339; pH 8-10 (1%); 59-61% act.

Ufasan 62B. [Unger Fabrikker AS] Sodium alkylbenzene sulfonate; partially biodeg. detergent, wetting and foaming agent for detergent powds., and in plastics, metal, agric., polish, textiles, mining, oil and cement industries; stable in hard water, acids, alkalis; wh. paste, weak odor; m.w. 343-345; pH 8-10 (1%); 61-63% act.

Ufasan 65. [Unger Fabrikker AS] Linear sodium dodecylbenzene sulfonate; anionic; biodeg. detergent, wetting and foaming agent for liq. detergents, dishwash, industrial detergents, shampoos, laundry powds. and in plastics, metal, agric., polish, textiles, mining, oil and cement industries; stable in hard water, acids, alkalis; wh. paste, weak odor; m.w. 343-345; pH 8-10 (1%); 64-66% act.

Ufasan IPA. [Unger Fabrikker AS] Isopropylamine alkylbenzene sulfonate; biodeg. raw material for detergents, dishwash, cleaning liqs., car shampoos, industrial cleaners and in plastics, metal, agric., polish, textiles, mining, oil and cement industries; emulsifier in solv.-based waterless hand and industrial cleaners; stable in hard water, diluted acids and alkalis; golden visc. liq., weak odor; sol. in water; m.w. 381; visc. 20,000-30,000 cp; cloud pt. < -10 C; pH 6.5-8.0 (1%); 94-96% act.

Ufasan TEA. [Unger Fabrikker AS] TEA dodecylbenzene sulfonate, linear; anionic; biodeg. surfactant for liq. detergents, dishwash, hair and car shampoos and in plastics, metal, agric., polish, textiles, mining, oil and cement industries; stable in hard water, diluted acids and alkalis; golden visc. liq., weak odor; sol. in water; m.w. 469-473; flash pt. (Abel-Penksy) > 63 C; 49-51% act.

UltraFine® II. [Laurel Industries] Antimony trioxide; CAS 1309-64-4; EINECS 215-175-0; flame retardant for liq. thermoplastic and thermoset systems, e.g., unsat. polyesters, epoxies, PU, phenolics, textile treatments; produces minimum loss of physical props. in ABS; catalyst in PET mfg.; wh. fine cryst. powd., odorless; 0.2-0.4 μ avg. particle size; 0.1% max. on 325 mesh; very sl. sol. in water; dissolves in conc. HCl, sulfuric acid, strong alkalies; m.w. 291.52; sp.gr. 5.3; bulk dens. 35 lb/ft^3; m.p. 656 C; b.p. 1425 C; ref. index 2.1; 99.3-99.9% Sb_2O_3; *Toxicology:* avoid skin or eye contact, inhalation; use only with adequate ventilation; *Storage:* store in dry, well-ventilated area.

Ultraflex®. [Petrolite] Microcryst. wax; CAS 63231-60-7; EINECS 264-038-1; plastic wax offering high ductility, flexibility at very low temps., provides protective barrier properties against moisture vapor and gases; uses incl. hot-melt laminating adhesives for papers, films, and foils; hot-melt coatings; in antisunchecking agents in rubber goods, elec. insulating agents, leather treating agents, water repellents for textiles, rustproof coatings, cosmetic ingreds., and as plasticizer for waxes used in crayons, dental compds., chewing gum base, candles; incl. FDA §172.230, 172.615, 175.105, 175.300, 176.170, 176.180, 176.200, 177.1200, 178.3710, 179.45; amber wax; also avail. in wh.; dens. 0.93 g/cc; visc. 13 cps (99 C); m.p. 69 C; flash pt. 293 C.

Ultramarine Blue. [Holliday Pigments] Sodium alumino sulfo silicate, ultramarine blue; inorg. bright red-shade blue pigment used for coloring plastics, rubber, printing ink, paint, powd. coatings, artists' colors, cosmetics, and detergents; exc. lightfastness; various grades differing in particle size, color str. and shade; special grades avail. (acid resist., low-moisture, non-dusting), violet and pink derivs. avail.; acid sensitive; alkali resist.; FDA approval for sensitive applics. (food-contact plastics, cosmetics for use around eyes); sp.gr. 2.35; heat stability 400 C; *Toxicology:* nontoxic.

Ultramoll® I. [Bayer AG] Polyadipate; polymeric plasticizer; good gelation, resistance to oils, fats, greases, and normal-grade petrol; used in calendering, extrusion, inj. molding; food pkg., oil tank linings, self-adherent film, crash pad film (antifogging), protective clothing, conveyor belts, tubing for milking machines and oil, shoe soles; used in VC copolymers, NC, cellulose acetobutyrate, and in natural, S/B, nitrile/butadiene, chlorinated, and butyl rubber; Hazen $\leq$ 150; dens. 1.075-1.090 g/cc; visc. 2000-3000 mPa•s; flash pt. 280-300 C; acid no. $\leq$ 1.0; sapon. no. 490-510; ref. index 1.472-1.473; 70:30 PVC:plasticizer suspension: tens. str. 23 MPa; tens. elong. 330% (break); hardness (Shore D) 41.

Ultramoll® II. [Bayer AG] Polyadipate; polymeric plasticizer; resistance to oils, fats, greases, bitumen, and normal-grade petrol; used for calendering, coating, extrusion, and inj. molding; pkg., foils, oil tank linings, self-adhesive film, crash pad film (antifogging), protective clothing; tablecloths, insulating sleeves, and strands for the radio and television industry, oil tubes, shoe soles; used in VC copolymers, PVAc, NC, cellulose acetobutyrate, and natural, S/B, nitrile/butadiene, chlorinated, and butyl rubbers; Hazen $\leq$ 150; dens. 1.100-1.115 g/cc; visc. 2000-3000 mPa•s; flash pt. 280-300 C; acid no. $\leq$ 1.0; sapon. no. 510-540; ref. index 1.472-1.473; 70:30 PVC:plasticizer suspension: tens. str. 24 MPa; tens. elong. 320% (break); hardness (Shore D) 40.

Ultramoll® III. [Bayer AG; Bayer/Fibers, Org., Rubbers] Polyadipate; polymeric plasticizer for PVC; resistance to oils, fats, greases, bitumen, and normal-grade petrol; used in coating, calendering, extrusion, inj. molding, and dip-coating for tarpaulins, protective clothing, imitation and expanded imitaion leather, pkg., foils, self-adhesive film, cable sheathing, sealing profiles, boots, gloves; used in VC copolymers, PVAc, NC, cellulose acetobutyrate, and in natural, S/B, chlorinated, and butyl rubber; Hazen $\leq$ 150; dens. 1.100-1.110 g/cc; visc. 1000-1300 mPa•s (50 C); acid no. $\leq$ 1.0; sapon. no. 510-530; flash. pt. 270-290 C; ref. index 1.469-1.470; 70:30 PVC:plasticizer suspension: tens. str. 23 MPa; tens. elong. 330% (break); hardness (Shore D) 44.

Ultramoll® M. [Bayer AG] Polyadipate; polymeric plasticizer for use in formulation of flame retardant pastes for polyester-based Moltopren®.

Ultramoll® PP. [Bayer AG] Polyphthalate; polymeric plasticizer; better resistance to oils, fats, greases than monomeric plasticizers; used in calendering, coating, extrusion, inj. molding, dip-coating, and rotational molding for tarpaulins, protective clothing, wallpapers, imitation leather, elec. insulating film, automotive films, protective foils, cables with resistance to high temps., profiles, tubing, conveyor belts, shoes, soles, gaskets, technical inj. moldings, industrial gloves, bellows; Hazen ≤ 100; dens. 1.035-1.045 g/cc; visc. 1200-1500 mPa•s; flash pt. 230-265 C; acid no. ≤ 0.5; sapon. no. 200-320; ref. index. 1.502-1.504; 70:30 PVC:plasticizer suspension: tens. str. 25 MPa; tens. elong. 245% (break); hardness (Shore D) 38.

Ultramoll® PU. [Bayer AG] PU; polymeric plasticizer; used in melt roll coating, extrusion, inj. molding, calender coatings for the clothing, upholstery, bag, and shoe industries, films (for garments to withstand chemical cleaning, plasters, optical industry, lining split leather and food pkg.), computer cables, cable sheathing, tubing, structural profiles, gaskets, bellows, shoe soles, boots for the oil and fisheries industry; powd.; dens. 1.14 g/cc; 70:30 PVC:plasticizer suspension: tens. str. 24 MPa; tens. elong. 275%; hardness (Shore D) 59; Unverified

Ultramoll® TGN. [Bayer AG] Polyphthalate; polymeric plasticizer; good gelation, resistance to oils, fats, greases, bitumen, and normal-grade petrol, comparatively low visc., and good spreading properties; used in pastes for spreading and dip-coating, calendering, extrusion, inj. molding for floor coverings (top layers and inlays), imitation leather, protective clothing and foils, tarpaulins, boots, gloves, self-adhesive films and tapes; used as desensitizing agents for peroxides; Hazen ≤ 100; dens. 1.090-1.100 g/cc; visc. 2000-2500 mPa.s; flash pt. 220-240 C; acid no. ≤ 1.0; sapon. no. 300-320; ref. index 1.503-1.505; 70:30 PVC:plasticizer suspension: tens. str. 30 MPa; tens. elong. 210% (break); hardness (Shore D) 55.

Ultranox® 210. [GE Specialty] Tetrakis [methylene(3,5-di-t-butyl-4-hydroxyhydrocinnamate)] methane; CAS 6683-19-8; EINECS 229-722-6; primary antioxidant for plastics; provides processing stability and long-term heat aging to styrenics, polyolefins, PVC, urethanes, acrylics, adhesives, and elastomers; FDA 21CFR §178.2010; Japan, Europe, Canada, and other approvals; wh. to off-wh. free-flowing powd.; sol. (g/100 g): 163 g in methylene chloride, 144 g in chloroform, 122 g in benzene, 89 g in acetone, < 0.01 g in water; sp.gr. 1.15; m.p. 110-125 C; flash pt. (COC) 273 C; *Toxicology:* LD50 (oral, rat) > 5 g/kg.

Ultranox® 226. [GE Specialty] 2,6 Di-t-butyl-4-methylphenol; CAS 128-37-0; EINECS 204-881-4; antioxidant for use in plastics incl. polyolefins, styrenics, PU, rubber, and food-pkg. materials; antioxidant for oils, greases, and lubricants; stabilizer in fuels, lubricating and min. oils; wh. cryst. solid; sol. in acetone, benzene, CCl_4, ethyl acetate, toluene; m.w. 220; sp.gr. 1.03; dens. 8.6 lb/gal; m.p. 69 C; b.p. 265 C (760 mm); flash pt. (COC) 127 C; 99% min. purity; Unverified

Ultranox® 254. [GE Specialty] Polymeric 2,2,4-trimethyl-1,2-dihydroquinoline; CAS 26780-96-1; antioxidant for natural and syn. rubber and their vulcanizates, e.g., in wire and cable applics; cream powd.; m.w. 173; sol. in acetone, ethyl acetate, methylene chloride, CCl_4, benzene, chloroform, ethanol; sp.gr. 1.08; dens. 9 lb/gal; m.p. 75 C; flash pt. (COC) 235 C min.; Unverified

Ultranox® 276. [GE Specialty] Octadecyl 3,5-di-t-butyl-4-hydroxyhydrocinnamate; CAS 2082-79-3; EINECS 218-216-0; high m.w. antioxidant/stabilizer for styrenics, polyolefins, PVC, urethane and acrylic coatings, adhesives, and elastomers; effective replacement for BHT in polyolefins; FDA 21CFR §178.2010; wh. cryst. powd.; sol. in most aprotic org. solvs. such as benzene, xylene, and acetone; m.w. 531; sp.gr. 1.02 (20 C); bulk dens. 22-23 lb/ft^3; m.p. 55 C; flash pt. (COC) 273 C; *Toxicology:* may cause irritation; avoid skin and eye contact, repeated/prolonged inhalation; *Precaution:* airborne dust constitutes an explosion hazard; *Storage:* store in cool, dry place away from direct sunlight, heat, spark, or flame.

Ultranox® 626. [GE Specialty] Bis (2,4-di-t-butylphenyl) pentaerythritol diphosphite, 0.5-1.2% triisopropanolamine; CAS 26741-53-7; EINECS 247-952-5; antioxidant for polyolefin, PVC, PET, styrenics, ABS, PP, min.-filled PP, HMWHDPE, and PC polymers; stabilizer; FDA 21CFR §175.105, 178.2010; approved in Canada, Germany, France, UK, Japan, and other countries; wh. cryst. powd.; 40% > 60 mesh, 15% 60-200 mesh; sol. in methylene chloride, THF, toluene; m.w. 604; dens. 0.43 g/ml; bulk dens. 24-28 lb/ft^3 (20 C); m.p. 160-175 C; flash pt. (PMCC) 168 C; acid no. 2.5 max.; 10.0-10.9% P; *Toxicology:* may cause irritation; avoid skin and eye contact, repeated/prolonged inhalation; *Precaution:* airborne dust constitutes an explosion hazard; *Storage:* store in cool, dry place away from direct sunlight, heat, spark, or flame.

Ultranox® 626A. [GE Specialty] Bis (2,4-di-t-butylphenyl) pentaerythritol diphosphite, ≤ 1.0% triisopropanolamine; CAS 26741-53-7; EINECS 247-952-5; larger particle size version of Ultranox 626; fewer fines for easier flow in some additive feeding equip.; 95% greater than 60 mesh; sol. (g/100 g): 43 g in methylene chloride, 35.7 g in toluene, 35 g in THF, 15 g in perchloroethylene, 10 g in min. oil, 8.5 g in acetone.

Ultranox® 627A. [GE Specialty] Bis (2,4-di-t-butylphenyl) pentaerythritol diphosphite with < 1% triisopropanolamine; CAS 26741-53-7; EINECS 247-952-5; antioxidant providing color stability, hydrolytic stability, reduction in polymer degradation during processing of ABS, PVC, PC; FDA 21CFR §178.2010; wh. free-flowing gran.; 95% > 250 μ; bulk dens. 32 lb/ft^3 (20 C); m.p. 150 C min.; acid no. 2.5 max.; flash pt. (PMCC) > 100 C;

9.1% min. P.

Ultranox® 815A Blend. [GE Specialty] Bis (2,4-di-t-butylphenyl) pentaerythritol diphosphite and 47-53% tetrakis [methylene(3,5-di-t-butyl-4-hydroxyhydrocinnamate)]methane (1:1); two-part antioxidant system providing improved stabilization of engineering polyester, HDPE, high-performance alloys, PP, thermoplastic olefins, adhesives during compounding and end-use applics.; FDA 21CFR §178.2010; Japan, Europe, Canada, and other approvals; wh. to off-wh. gran.; < 5% < 60 mesh; m.p. 104 C; acid no. 1.6 max.

Ultranox® 817A Blend. [GE Specialty] Bis (2,4-di-t-butylphenyl) pentaerythritol diphosphite and 30.3-36.3% tetrakis [methylene(3,5-di-t-butyl-4-hydroxyhydrocinnamate)]methane (2:1); two-part antioxidant system providing improved stabilization of engineering polyester, HDPE, high-performance alloys, PP, thermoplastic polyolefins during compounding and end-use applics.; FDA 21CFR §178.2010; Japan, Europe, Canada, and other approvals; wh. to off-wh. gran.; < 5% < 60 mesh; m.p. 105 C; acid no. 1.9 max.

Ultranox® 875A Blend. [GE Specialty] Bis (2,4-di-t-butylphenyl) pentaerythritol diphosphite and 47-53% octadecyl 3,5-di-t-butyl-4-hydroxyhydrocinnamate (1:1); two-part antioxidant system providing improved stabilization of ABS, HDPE, HIPS, PC, PP, thermoplastic olefins, adhesives, and elastomers during compounding and end-use applics.; FDA 21CFR §178.2010; Japan, Europe, Canada, and other approvals; wh. to off-wh. gran.; < 5% < 60 mesh; m.p. 46 C; acid no. 1.5 max.

Ultranox® 877A Blend. [GE Specialty] Bis (2,4-di-t-butylphenyl) pentaerythritol diphosphite and 30.3-36.3% octadecyl 3,5-di-t-butyl-4-hydroxyhydrocinnamate (2:1); two-part antioxidant system providing improved stabilization of ABS, HDPE, HIPS, PC, PP, thermoplastic olefins, and elastomers during compounding and end-use applics.; FDA 21CFR §178.2010; Japan, Europe, Canada, and other approvals; wh. to off-wh. gran.; < 5% < 60 mesh; m.p. 46 C; acid no. 1.8 max.

Ultranox® 2714A. [GE Specialty] 44.4% Bis(2,4-di-t-butylphenyl) pentaerythritol diphosphite, 44.4% tetrakis[methylene(3,5-di-t-butyl-4-hydroxyhydrocinnamate)]methane, 11.2% magnesium aluminum hydrotalcite; primary/sec. antioxidant blend for stabilization during compding. and in end-use applics.; inhibits degradation of m.w. and prevents discoloration; for adhesives, engineering polyester, PE, high-performance alloys, PP, thermoplastic olefins; FDA 21CFR §178.2010; Japan, Europe, Canada and other approvals; wh. to off-wh. gran.; < 5% < 60 mesh; m.p. 104 C; acid no. 1.6 max.

Ultra-Pflex®. [Specialty Minerals] Ultrafine surface-treated precipitated calcium carbonate; impact modifier in rigid PVC and plastisols; improves weatherability; produces inj. molded and extruded parts of high smoothness and gloss; thixotrope and flow control agent in liq. systems; provides exc. nucleation sites in foamed systems; powd.; 0.07 μ avg. particle size; sp.gr. 2.7; dens. 30 lb/ft^3 (tapped); bulk dens. 15 lb/ft^3; surf. area 19 m^2/g; dry brightness 98; 98% $CaCO_3$.

Ultrasil® VN3SP. [Degussa] High purity silica; catalyst carrier; high surf. area reinforcement for rubber compding.

Ultra Sulfate SL-1. [Witco Corp.] Sodium lauryl sulfate; CAS 151-21-3; EINECS 205-788-1; wetting agent, penetrant, lubricant, emulsifier, dye dispersant, scouring aid, antistat; base for personal care products; detergent for specialty household and industrial cleaners; textile surfactant; emulsion polymerization and latex stabilization; liq.; 30% conc.

Ultratalc™ 609. [Specialty Minerals] Untreated talc; CAS 14807-96-6; EINECS 238-877-9; filler for rubber, rigid vinyl, polyolefins; powd.; 0.9 μ avg. particle size, 6 μ top size.

Unads® Pellets. [R.T. Vanderbilt] Tetramethylthiuram monosulfide; CAS 97-74-5; rubber accelerator; yel. pellets; insol. in water; dens. 1.38 mg/m^3; flash pt. (CC) 156 C; *Toxicology:* LD50 (oral, rat) 1250-1390 mg/kg; harmful if swallowed; may cause skin sensitization upon repeated exposure; antabuse effect with alcohol consumption; *Precaution:* incompat. with strong acids and oxidizing agents; hazardous decomp. prods.: oxides of C, N, S at combustion temps.; toxic fumes may be evolved; *Storage:* store in cool, dry area.

Unads® Powd. [R.T. Vanderbilt] Tetramethylthiuram monosulfide with 1-2% antidusting oil; CAS 97-74-5; accelerator for NR, SR, esp. type W neoprenes; nondiscoloring in lt. stocks; for use alone or with thiazoles; yel. powd.; 99.9% min. thru 100 mesh; sol. in acetone, chloroform, toluene; sl. sol. in gasoline, hexane, water; dens. 1.37 mg/m^3; m.p. 103-114 C; flash pt. (CC) 156 C; 95% min. assay; *Toxicology:* LD50 (oral, rat) 1250-1390 mg/kg; harmful if swallowed; may cause skin sensitization upon repeated exposure; ingestion may cause unpleasant symptoms if alcohol is consumed (antabuse effect); *Precaution:* incompat. with strong acids and oxidizing agents; hazardous decomp. prods.: oxides of carbon, nitrogen, and sulfur; may evolve toxic fumes; *Storage:* store in cool, dry area.

Ungerol LS, LSN. [Unger Fabrikker AS] Sodium lauryl sulfate; CAS 151-21-3; anionic; carpet shampoos, furniture cleaners, polymerization emulsifier; paste; 29-41% conc.; Unverified

Unicell 5PT. [Dong Jin] 5-Phenyltetrazole; CAS 3999-10-8; chem. foaming agent for PPO, thermoplastic polyester, PC, polysulfone, nylon, PEI; processing range 450-550 F; gas yield 200 cc/g.

Unicell D. [Dong Jin] Azodicarbonamide; CAS 123-77-3; EINECS 204-650-8; chem. foaming agent for ABS, acetal, acrylic, EVA, HDPE, LDPE, PPO, PP, PS, HIPS, flexible PVC, TPE; processing range 300-450 F; gas yield 220 cc/g.

Unicell OH. [Dong Jin] 4,4′-Oxybis (benzenesulfonyl) hydrazine; CAS 80-51-3; chem. foaming agent for

EVA, LDPE, PS, flexible PVC; processing temps. 250-375 F; gas yield 120-125 cc/g.

Unicell TS. [Dong Jin] p-Toluene sulfonyl semicarbazide; chem. foaming agent for ABS, acetal, acrylic, EVA, HDPE, LDPE, PPO, PP, PS, HIPS, flexible and rigid PVC, TPE; processing temps. 390-455 F; gas yield 145 cc/g.

Unifine P-2. [Biddle Sawyer] Proprietary; endothermic chem. foaming agent for ABS, HDPE, PPO, HIPS, nylon, PC, PET; processing range 350-450 F; gas yield 100-105 cc/g.

Unifine P-3. [Biddle Sawyer] Proprietary; endothermic chem. foaming agent for PPO, nylon, PC, PET; processing range 400-480 F; gas yield 95-100 cc/g.

Unifine P-4. [Biddle Sawyer] Proprietary; endothermic chem. foaming agent for ABS, HDPE, PPO, rigid PVC; processing range 350-430 F; gas yield 125-135 cc/g.

Unifine P-5. [Biddle Sawyer] Proprietary; endothermic chem. foaming agent for ABS, LDPE, PS, rigid PVC; processing range 300-400 F; gas yield 115-125 cc/g.

Uniflex® 192. [Union Camp] Octyl tallate; plasticizer for syn. rubbers; metalworking; yel. liq.; sol. in min. oils, hexane, IPA, ethanol; insol. in glycerol, propylene glycol; sp.gr. 0.872; acid no. 0.5; flash pt. 227 C; fire pt. 240 C; 0.05% moisture.

Uniflex® 300. [Union Camp] Polymeric plasticizer; plasticizer for PVC; high temp. elec. applics.; high grade vinyl upholstery; adhesives for elec. tapes or wall coverings; gasketing and tubing; low migration; lt. colored liq.; sol. in esters, ketones, aromatic/chlorinated hydrocarbons; sp.gr. 1.099; visc. 3300 cSt; acid no. 2 max.; sapon. no. 520; flash pt. (COC) 293 C; fire pt. (COC) 327 C; ref. index 1.466.

Uniflex® 312. [Union Camp] Polymeric plasticizer; low visc. plasticizer showing good efficiency, flux rate, and low temp. flexibility in PVC compds. and low migration into rubbers; for vinyl tapes, elec. insulation, truck upholstery and other vinyl applics. which experience extremes in climate; compat. with PVC, PVC/PVA, nitrocellulose, cellulose acetate butyrate, chlorinated rubber; lt. colored liq.; sol. in esters, ketones, aromatic hydrocarbons, chlorinated solvs., alcohols; limited or insol. in water, glycerin, hexane; sp.gr. 1.08; visc. 980 cSt; acid no. 0.5; flash pt. (COC) 290 C; fire pt. (COC) 313 C; ref. index 1.46.

Uniflex® 314. [Union Camp] Polymeric plasticizer; plasticizer showing good compat. with the more polar resins, e.g., PVA, nitrocellulose, cellulose acetate, nylon; suggested for adhesives, heat-resist. lacquers, barrier coatings, binders, size for glass fibers, paints, sealants; lt. colored liq.; sol. in esters, most ketones, glycol ether solvs.; insol. in water; sp.gr. 1.192; visc. 5000 cSt; acid no. 33; flash pt. (COC) 315 C; fire pt. (COC) 338 C; ref. index 1.479.

Uniflex® 315. [Union Camp] Polymeric plasticizer; plasticizer with exc. permanence; compat. with PVC, PVC/PVA, nitrocellulose, cellulose acetate butyrate and propionate, chlorinated rubber; for PVC compds. (refrigerator gaskets), heavy-duty viny upholstery, elec. tapes, high temp. wire insulation; lt. colored liq.; sol. in esters, ketones, aromatic hydrocarbons, chlorinated solvs., alcohol; sp.gr. 1.11; visc. 209 cSt (99 C); acid no. 2 max.; flash pt. (COC) 279 C; fire pt. (COC) 315 C; ref. index 1.465.

Uniflex® 320. [Union Camp] Polymeric plasticizer; plasticizer with good permanence and ease of processing; resist. to extraction by oils and hydrocarbons; lower migration into rubber, enamels, and lacquer; compat. with PVC, PVC/PVA, NC, CAB, CAP, chlorinated rubber; suggested for vinyl film and sheet, plastisols; pigment grinding medium due to visc. and wetting props.; straw-colored liq.; sol. in esters, ketones, aromatic and chlorinated hydrocarbons; insol. in ethanol, hexane, water; sp.gr. 1.084; visc. 2500 cSt; acid no. 2 max.; flash pt. (COC) 279 C; fire pt. (COC) 307 C; ref. index 1.466.

Uniflex® 327. [Union Camp] Polymeric plasticizer; Gardner 3; sp.gr. 1.06; visc. 2800 cSt; acid no. 1.0; ref. index 1.469; flash pt. 263 C (COC); fire pt. 285 C (COC); Unverified

Uniflex® 330. [Union Camp] Polymeric plasticizer; plasticizer with good compat. with PVC, PVC/PVA, NC, CAB, CAP, chlorinated rubber; low odor transfer, low volatility, resist. to hydrocarbons, soap, detergents; for refrigerator gaskets, elec. tapes, wall coverings, vinyl upholstery sheeting, baby wear, hospital sheeting; lt. colored liq.; sol. in esters, ketones, chlorinated hydrocarbons, methanol, higher alcohols; sp.gr. 1.088; visc. 5300 cSt; acid no. 2 max.; flash pt. (COC) 290 C; fire pt. (COC) 318 C; ref. index 1.466.

Uniflex® 338. [Union Camp] Oil-modified sebacic acid alkyd polymeric plasticizer; permanent plasticizer with exc. pigment wetting and stability props.; grinding medium for NC lacquers; plasticizer for PVB, NC films, fabric coatings; ethyl cellulose cable, paper, wood, and metal lacquers; yel. visc. liq.; sol. in aromatic hydrocarbons and in mixts. with ethanol, butanol, and ethyl acetate; sp.gr. 0.998; visc. (G-H) Z2; acid no. 3.5 max.; flash pt. 288 C; fire pt. 313 C.

Uniflex® BYO. [Union Camp] Butyl oleate; CAS 142-77-8; EINECS 205-559-6; monomeric plasticizer and processing aid for plastics, rubber; textile lubricant, mold lubricant, in waterproofing agents, polishes, and metalworking lubricants; compat. with NR, most syn. rubbers incl. chlorinated, PS, cellulose nitrate, ethyl cellulose; FDA 21CFR § 177.2600, 177.2800; APHA 70; sp.gr. 0.868; visc. 7 cps; m.p. -10 C; acid no. 0.5; sapon. no. 170; flash pt. (COC) 199 C; fire pt. (COC) 218 C; ref. index 1.451.

Uniflex® BYS-CP. [Union Camp] Butyl stearate; CAS 123-95-5; EINECS 204-666-5; Unverified

Uniflex® BYS Special. [Union Camp] Butyl stearate; CAS 123-95-5; EINECS 204-666-5; Unverified

Uniflex® BYS-Tech. [Union Camp] n-Butyl stearate; CAS 123-95-5; EINECS 204-666-5; ester with low visc., good color, low odor for plasticizers, textile fiber lubricants, metalworking oils; compat. with nat. and syn. rubbers, chlorinated rubber, cellulose ethers, PS, PVB; APHA 40 color, low odor; sol. in alcohols, ketones, hydrocarbons, 0.2% in water; sp.gr. 0.86; visc. 9 cst; f.p. 20 C; acid no. 0.8; iodine no. 3; sapon. no. 175; flash pt. 201 C; fire pt. 221 C; ref. index 1.442.

Uniflex® DBS. [Union Camp] Dibutyl sebacate; CAS 109-43-3; EINECS 203-672-5; monomeric plasticizer for plastics and rubber (cellulosics, PVC food wraps, nitrile and neoprene rubbers); FDA 21CFR §175.105, 175.300, 175.320, 176.170, 176.180, 177.2600, 177.2800, 181.27; colorless oily liq.; sp.gr. 0.937; visc. 7 cps; f.p. -11 C; acid no. < 0.1; flash pt. (COC) 193 C; fire pt. (COC) 213 C; ref. index 1.4401; *Toxicology:* LD50 (oral, mouse) > 31 g/kg; nontoxic; nonirritating to eyes and skin.

Uniflex® DCA. [Union Camp] Dicapryl adipate; CAS 105-97-5; EINECS 203-349-9; monomeric plasticizer for plastics and rubber; exhibits exc. low temp. flexibility and good heat and lt. stability in vinyl film and sheeting, low initial visc. in plastisols; also compat. with nitrocellulose, PS, ethyl cellulose, most syn. rubbers; FDA 21CFR §177.2600; colorless oily liq.; sp.gr. 0.914; visc. 12 cps; f.p. -32 C; acid no. < 0.1; flash pt. (COC) 207 C; fire pt. (COC) 224 C; fef. index 1.439; *Toxicology:* LD50 (oral, rat) > 58 g/kg; nontoxic; nonirritating to eyes and skin.

Uniflex® DCP. [Union Camp] Dicapryl phthalate; monomeric plasticizer for plastics and rubber; gives low visc. in plastisols and organosols; compat. with PVC, nitrocellulose, chlorinated rubber; lt. colored liq.; sp.gr. 0.974; visc. 55 cps; f.p. -60 C; acid no. < 0.1; flash pt. (COC) 202 C; fire pt. (COC) 232 C; ref. index 1.479.

Uniflex® DOA. [Union Camp] Dioctyl adipate; CAS 103-23-1; EINECS 203-090-1; monomeric plasticizer for plastics and rubber; provides exc. low temp. props. to PVC compds., good low temp. fluidity and high temp. stability for syn. lubricants; FDA 21CFR §175.105, 177.1200,177.1210, 177.1400, 177.2600, 178.3740; colorless liq.; sp.gr. 0.927; dens. 7.7 lb/gal; visc. 12.5 cSt; f.p. < -70 C; acid no. < 0.1; flash pt. (COC) 207 C; fire pt. (COC) 230 C; ref. index 1.447.

Uniflex® DOS. [Union Camp] Dioctyl sebacate; monomeric plasticizer for plastics esp. PVC compds. requiring extreme low temp. flexibility; exc. heat, light, and weathering stability for use in tropics and arctic regions; syn. lubricant basestock providing good heat stability and low pour pt. in metalworking oils, multipurpose greases; FDA 21 CFR §175.105, 175.300, 176.170, 176.180, 177.1210, 177.1350, 177.2600, 178.3910; lt. colored liq.; sp.gr. 0.915; dens. 7.61 lb/gal; visc. 15.9 cSt; acid no. < 0.1; pour pt. -62 C; cloud pt. -55 C; flash pt. (COC) 227 C; fire pt. (COC) 260 C; ref. index 1.448.

Uniflex® EHT. [Union Camp] Octyl tallate; plasticizer for syn. rubbers; in metalworking; yel. liq.; sol. in min. oils, n-hexane, IPA, ethanol; insol. in glycerol, propylene glycol; sp.gr. 0.872; acid no. 1 max.; flash pt. 227 C; fire pt. 240 C.

Uniflex® TCTM. [Union Camp] Tricapryl trimellitate; monomeric plasticizer for plastics and rubber; APHA 80; sp.gr. 0.978; visc. 230 cSt; pour pt. -40 C; acid no. 0.5; ref. index 1.478; flash pt. 246 C (COC); fire pt. 268 C (COC); Unverified

Uniflex® TEG-810. [Union Camp] Triethylene glycol capric/caprylate; plasticizer for nitrile rubber; Unverified

Uniflex® TOTM. [Union Camp] Trioctyl trimellitate; monomeric plasticizer for plastics, rubber, and lubricants; APHA 100; sp.gr. 0.990; visc. 215 cSt; pour pt. -45 C; acid no. < 0.1; ref. index 1.485; flash pt. 260 C (COC); fire pt. 285 C (COC); Unverified

Unilin® 350 Alcohol. [Petrolite] C20-40 alcohols; chemical intermediate for oxidation, ethoxylation, sulfation, amination, esterification; coemulsifier and direct additive in coatings; plastics additive (processing aid, lubricant, dispersant); thickener, moisturizer, pigment dispersant, oil binder, stabilizer in cosmetic creams, lotions, lipsticks, antiperspirants, and soaps; FDA §175.105; slab; sol. in aliphatic and aromatic solvs.; m.w. 375; sp.gr. 0.78 (121 C); dens. 0.985 g/cc; visc. 5.9 cp (99 C); m.p. 78 C; hyd. no. 127.

Unilin® 425 Alcohol. [Petrolite] C20-40 alcohols; functional polymer for modification of PP, PVC, polyethylene, PS, and high-performance engineering resins; acts as antioxidant, heat stabilizer, uv stabilizer, or visc. depressant; promotes emollient protective films onto the skin in cosmetic creams and lotions; used in hot-melt and solv.-based coatings; textile/leather lubricants and finishes; chemical intermediate; defoamer for pulp/paper processing; FDA 21CFR §175.105; off-wh. prills; sol. in aliphatic and aromatic solvs.; m.w. 460; sp.gr. 0.78 (121 C); dens. 0.985 g/cc; visc. 7.8 cps (99 C); m.p. 91 C; hyd. no. 105.

Unilin® 550 Alcohol. [Petrolite] C30-50 alcohols; chemical intermediate for oxidation, ethoxylation, sulfation, amination, esterification; coemulsifier and direct additive in coatings; plastics additive (processing aid, lubricant, dispersant); thickener, moisturizer, pigment dispersant, oil binder, stabilizer in cosmetic creams, lotions, lipsticks, antiperspirants, and soaps; FDA 21CFR §175.105; off-wh. prills; sol. in toluene, MIBK, VM&P naphtha, hexane; m.w. 550; sp.gr. 0.78 (121 C); dens. 0.985 g/cc; visc. 5.5 cps (149 C); m.p. 99 C; hyd. no. 83.

Unilin® 700 Alcohol. [Petrolite] C40-60 alcohols; chemical intermediate for oxidation, ethoxylation, sulfation, amination, esterification; coemulsifier and direct additive in coatings; plastics additive (processing aid, lubricant, dispersant); thickener, moisturizer, pigment dispersant, oil binder, stabilizer in cosmetic creams, lotions, lipsticks, antiperspirants, and soaps; FDA 21CFR §175.105; off-wh. prills; sol. in toluene, MIBK, VM&P naphtha; m.w. 700; sp.gr. 0.79 g/cc (121 C); dens. 0.985 g/cc; visc. 7.9 cps (149 C); m.p. 105 C; hyd. no. 65.

Unilink® 4100. [UOP] Aromatic diamine; chain extender for PU elastomers; dk. reddish liq.; m.w. 220; sp.gr. 0.94 (60 F); dens. 7.8 lb/gal; visc. 38.6 cps (60 F); flash pt. (PMCC) 240 F.

Unilink® 4102. [UOP] Aromatic diamine; chain extender for PU applics.; dk. liq.; sp.gr. 0.94; dens. 7.8 lb/gal; visc. 36 cps (60 F); flash pt. (PMCC) 265 F.

Unilink® 4200. [UOP] Aromatic diamine; chain extender in PU elastomers; dk. amber liq.; m.w. 310; sp.gr. 0.996 (60 F); dens. 8.3 lb/gal; visc. 115 cps (100 F); flash pt. (PMCC) 300 F.

Unimoll® 66. [Bayer AG] Dicyclohexyl phthalate; CAS 84-61-7; EINECS 201-545-9; plasticizer for use in coatings, NC lacquers; reduces processing temp. of rigid PVC; increases tack and lowers visc. in hot-melt adhesives; good light stability, resist. to migration.

Unimoll® 66 M. [Bayer AG; Bayer/Fibers, Org., Rubbers] Dicyclohexyl phthalate; CAS 84-61-7; EINECS 201-545-9; plasticizer for use in formulation of delayed tack heat sealable coatings and adhesives; PVC processing aid to reduce paste visc. in automotive underbody sealants; wh. fine powd.; 99.9% < 40 μ; sol. in org. solvs.; insol. in water; sp.gr. 1.15; dens. 400 g/l; m.p. 63-65 C; b.p. 248 C; flash pt. 205-215 C.

Unimoll® BB. [Bayer AG] Benzylbutyl phthalate; CAS 85-68-7; EINECS 201-622-7; plasticizer used for PVC articles with resistance to bitumen, coating, calendering, extrusion, inj. molding, imitation leather, carpet backing, floor coverings, film, structural profiles, joint masking profiles; used in VC copolymer, PVAc, PS, NC, ethyl cellulose, CAB, acrylic surf. coatings, alkyd resins, polymethacrylate, natural, S/B, N/B, and chlorinated rubber; Hazen < 40; dens. 1.120-1.125 g/cc; visc. 61-63 mPa.s; acid no. < 0.1; sapon. no. 350-370; ref. index 1.5396-1.5412; flash pt. (OC) 205-215 C; tens. str. 21 MPa; tens. elong. 260%; hardness (Shore D) 30.

Unimoll® DB. [Bayer AG] Dibutyl phthalate; CAS 84-74-2; EINECS 201-557-4; plasticizer used only to assist gelling in combination with other plasticizers of lower gelling capacity; used in PVC pastes, calendering, extrusion, inj. molding, VC copolymers, PVAc, surf. coatings, adhesives, rubber, PS, polyacrylates, and alkyd resins; Hazen < 30; dens. 1.043-1.049 g/cc; visc. 19-21 MPa.s; acid no. < 0.1; sapon. no. 400-410; ref. index 1.4915-1.4930; flash pt. 173-177 C (OC); 30% in PVC: tens. str. 17 MPa; tens. elong. 265%; hardness (Shore D) 28; Unverified

Union Carbide® LE-45. [Union Carbide] Dimethylpolysiloxane emulsion; nonionic; lubricant in cosmetics; softener, release agent in textile nonwovens, plastics, glass molding; water-disp.; Unverified

Union Carbide® Y-11343. [Union Carbide] Low m.w. silicone/aminofunctional silane coupling agent; crosslinker, adhesion promoter for one- and two-pack RTV silicone sealant systems; amber to lt. gray clear to hazy liq.; visc. 70-120 cSt; flash pt. > 71 C.

Union Carbide® Y-11542. [Union Carbide] Ureidosilane; adhesion promoter between fillers or reinforcements and polymers; better pot life in reactive polymer systems such as phenolic, epoxy, urea-melamine, or PU; low VOC; for glass fiber sizes and finishes, wool insulation resin binders, primers, foundry sand binders, adhesives and sealants, phenolic brake shoes, abrasive grinding wheel binders; lt. straw clear liq.; sol. in methanol, ethanol, acetone, toluene, water; sp.gr. 1.15; b.p. 217-250 C (760 mm); flash pt. (TCC) 99 C; ref. index 1.386; 100% act.

Union Carbide® Y-11597. [Union Carbide] Organofunctional silane; adhesion promoter and crosslinker for substrate/matrix resin combinations incl. silicone RTV adhesives; compat. with a range of resins; yel. liq.; sp.gr. 1.17; visc. 350 cSt; flash pt. (PMCC) 216 F.

Union Carbide® Y-11602. [Union Carbide] Tert. alkyl carbamate-functional silane yielding an aminofunctional silane on decomp.; adhesion promoter; latent aminosilane reactive with wide range of thermoset and thermoplastic resins; for glass fiber sizes, binders for glass fiber insulation, hot-melt adhesives, foundry resin additives, sealants, caulks, min.-filled elastomers, thermoset, and thermoplastic resins, binders for abrasives, additives for coatings; lt. straw clear liq.; sp.gr. 1.035; flash pt. (Seta CC) 60 C; 100% act.

Uniplex 80. [Unitex] Triethyl citrate; CAS 77-93-0; EINECS 201-070-7; plasticizer for food pkg.; produces resins with improved light-fastness and low toxicity; FDA approved; essentially odorless; sol. 6.5 g/100 ml water; m.w. 276.3; sp.gr. 1.135-1.139dens. 9.48 lb/gal; visc. 35.2 cps; b.p. 127 C (1 mm Hg); pour pt. -50 F; flash pt. (COC) 155 C; ref. index 1.440; 99% min. assay.

Uniplex 82. [Unitex] Acetyl triethyl citrate; CAS 77-89-4; EINECS 201-066-5; plasticizer for plastic food wraps and in aerosol hair sprays and bandages; produces resins with exc. heat stability and low toxicity; FDA approved; APHA 50 max. color, essentially odorless; sol. 0.72 g/100 ml water; m.w. 318.3; sp.gr. 1.135-1.139; dens. 9.47 lb/gal; visc. 53.7 cps; b.p. 132 C (1 mm Hg); pour pt. -45 F; flash pt. (COC) 188 C; ref. index 1.438; 99% min. assay.

Uniplex 84. [Unitex] Acetyl tributyl citrate; CAS 77-90-7; EINECS 201-067-0; plasticizer for indirect and direct food contact applics.; milling lubricant for aluminum foil or sheet steel for use in cans for beverage and food prods.; in PVC toys, cellulose nitrate films, aerosol hair sprays, dairy prod. cartons; drink bottle caps, food jar caps, sol'n. coatings for foil and paper; produces resins with exc. heat stability and low toxicity; FDA 21CFR §172.515, 175.105, 175.300, 175.320, 175.380, 175.390, 176.170, 176.180, 177.1200, 177.1210, 178.3910, 181.27; APHA 30 max. color, essentially odorless; insol. in water; m.w. 402.5; sp.gr. 1.045-1.055; dens. 8.74 lb/gal; visc. 33 cps; b.p. 173 C (1 mm Hg); pour pt. -75 C; flash pt. (COC) 204 Cref. index 1.441; 99% min. assay.

Uniplex 108. [Unitex] N-Ethyl o/p-toluene sulfonamide; CAS 1077-56-1, 80-39-7; plasticizer for nylon, shellac, cellulose acetate, protein materials, PVAc adhesives (metal-to-rubber joins, bookbinding, shoe), nitrocellulose lacquers; suitable for food contact applics.; makes compds. resist. to oils, solvs., and greases; FDA 21CFR §175.105, 175.300, 175.320, 176.170, 176.180, 177.1200, 177.1210; APHA 200 max. clear oily liq., char. odor; m.w. 199; sp.gr. 1.184-1.187; cryst. pt. 40 C; solid. pt. 0 C; b.p. 196 C (10 mm Hg); flash pt. (COC) 345 F; ref. index 1.535-1.545; 0.1% max. moisture.

Uniplex 110. [Unitex] Dimethyl phthalate; CAS 131-11-3; EINECS 205-011-6; solv. and plasticizer for cellulose acetate butyrate compositions; solv. for org. catalysts; plasticizer and solv. for aerosol hair sprays; APHA 10 max. clear liq., essentially odorless; sp.gr. 1.188-1.192; visc. 14-15 cps; ref. index 1.512-1.514; 99% min. assay.

Uniplex 150. [Unitex] Dibutyl phthalate; CAS 84-74-2; EINECS 201-557-4; solv./plasticizer in fingernail polish, nail polish remover, hair sprays, org. peroxide catalysts, adhesives, coatings; compat. with cellulosics, methacrylate, PS, PVB, vinyl chloride, urea-formaldehyde, melamine-formaldehye, phenolics; suitable for food contact use; FDA 21CFR §175.105, 175.300, 175.320, 176.170, 177.1200, 177.2430; APHA 25 max. clear liq., mild odor; pract. insol. in water; m.w. 278; sp.gr. 1.042; dens. 8.72 lb/gal; visc. 59 cps (32 F); f.p. -35 C; b.p. 340 C (760 mm Hg); ref. index 1.4905; 99% min. assay.

Uniplex 155. [Unitex] Diisobutyl phthalate; CAS 84-69-5; plasticizer for thermoplastic and thermoset resins; solv./plasticizer in cellophane, resin coated sand for foundry casting, org. peroxides; suitable for food contact use; FDA 21CFR §175.105, 175.300, 175.320, 176.170, 177.1200, 177.2510, 177.2550, 177.2600, 177.2800, 177.2910; APHA 25 max. clear liq., mild odor; pract. insol. in water; m.w. 278; sp.gr. 1.038-1.040; dens. 8.65 lb/gal; b.p. 327 C (760 mm Hg); 99% min. assay.

Uniplex 171. [Unitex] o,p-Toluene sulfonamide; CAS 70-55-3, 88-19-7; plasticizer for thermoplastic and thermoset resins; imparts gloss and wetting to melamine, urea and phenolic resins; increases hardness of urea resin moldings; increases stability and compat. of melamine resins; improves flow of phenolics; increases flexibility in nylons; suitable for food contact use; FDA §175.105; wh. to lt. cream fine solid particles, essentially odorless; m.w. 171; flash pt. (COC) 420 F; 1% max. moisture.

Uniplex 173. [Unitex] p-Toluene sulfonamide; CAS 70-55-3; EINECS 200-741-1; plasticizer for thermoplastic and thermoset resins; imparts gloss and wetting to melamine, urea and phenolic resins; increases hardness of urea resin moldings; increases stability and compat. of melamines; improves flow of phenolics; increases flexibility in nylons; suitable for food contact use; FDA §175.105; wh. to lt. cream fine solid particles, essentially odorless; m.w. 171; flash pt. (COC) 420 F; 1% max. moisture.

Uniplex 214. [Unitex] N-Butylbenzene sulfonamide; CAS 3622-84-2; EINECS 222-823-6; plasticizer for polyamide resins, enhancing low temp. flexibility; for fishing line, parts, adhesives for nonwoven interlinings, auto fuel line, coil air hoses, other high performance applics.; APHA 50 max. clear liq., mild odor; sp.gr. 1.145-1.152; visc. 150-180 cps; ref. index 1.522-1.526; 99% min. assay.

Uniplex 225. [Unitex] N-(2-Hydroxypropyl) benzenesulfonamide; CAS 35325-02-1; plasticizer for polyamide, polyurethane, polyacrylic, cellulose ester; exc. antistatic props. and resist. to extraction by water and drycleaning solvs.; for paints, lacquers; stabilizer for pigmented unsat. polyester resins; APHA 50 max. clear visc. liq.; becomes reversibly hazy below 0 C; ref. index 1.529-1.530; pH 7-8; 93-95% solids.

Uniplex 250. [Unitex] Dicyclohexyl phthalate; CAS 84-61-7; EINECS 201-545-9; heat-activated plasticizer for heat seal applics. such as food wrappers/labels, pharmaceutical labels and other applics. where delayed heat activated adhesive is required; for printing ink formulations for paper, vinyl, textiles, other substrates; FDA 21CFR §175.105, 176.170, 177.1200; wh. cryst. solid lumps or gran., moderate char. odor; pract. insol. in water; m.w. 330; sp.gr. 1.29; m.p. 63-65 C; b.p. 231 C (10 mm Hg); pour pt. 60 C; 99% min. assay.

Uniplex 260. [Unitex] Glyceryl tribenzoate; CAS 614-33-5; polymer modifier; plasticizer; for heat seal applics., lacquers, films, in PVAc-based adhesives, cellophane coatings, nitrocellulose coatings, nail lacquer formulations, printing inks, polishes; extrusion and inj. molding processing aid; suitable for food contact applics.; FDA 21CFR §174.5, 175.105, 175.365, 175.380, 175.390, 176.170, 176.180, 176.1210; wh. cryst. lumps or gran.; m.w. 404; sp.gr. 1.2619 (30 C); m.p. 70-73 C; ref. index 1.565-1.570 (molten, 25 C); *Toxicology:* LD50 (rat) > 11.7 g/kg.

Uniplex 270. [Unitex] Dimethylisophthalate; CAS 1459-93-4; modifies clarity and melting pt. of PET resins used in films, blow-molded bottles, etc.

Uniplex 310. [Unitex] Cyclohexyl isooctyl phthalate; CAS 71486-48-1; EINECS 275-521-1; plasticizer for nylon and polyester; APHA 100 max. clear liq., mild odor; pract. insol. in water; m.w. 360; sp.gr. 1.031-1.033; dens. 8.46 lb/gal; flash pt. (COC) 440 F; ref. index 1.4980-1.4989; 99% min. assay.

Uniplex 413. [Unitex] Substituted benzene sulfonamide; CAS 98-10-2; EINECS 202-637-1; plasticizer and nucleating agent in PET resins; enhances impact resist.; improves mold release; reduces processing times; ylsh. soft particles, clear liq. when molten; sp.gr. 0.920-0.925 (83 C); m.p. 79-82 C.

Uniplex 552. [Unitex] Pentaerythritol tetrabenzoate; CAS 4196-86-5; plasticizer for adhesives intended for heat seal applics.; suitable for food contact use; FDA 21CFR §175.105; off-wh. cryst. lumps or gran., mild char. odor; m.w. 552; bulk dens. 34 lb/ft^3; m.p. 99 C min.; acid no. 0.28 max.; ref. index 1.569-

1.572 (molten, 50 C).

Uniplex 600. [Unitex] Toluenesulfonamide-formaldehyde resin; CAS 1338-51-8; modifier and adhesion promoter for syn. and natural resins used in adhesives and coatings applics.; extender in polyamide resins; suitable for food pkg. applics.; FDA 21CFR §177.1200, 176.170, 175.105; pract. colorless hard solid particles, faint formaldehyde odor; sol. in usual lacquer solvs.; insol. in water, aliphatic hydrocarbons, veg. oils; sp.gr. 1.35; soften. pt. 62 C; ; ref. index 1.4275-1.4325 (25 g/75 g butyl acetate).

Uniplex 680. [Unitex] o,p-Toluenesulfonamide-formaldehyde resin in butyl acetate; CAS 1338-51-8; used in nail polish formulations in conjunction with nitrocellulose to improve durability and gloss; also in PVAc adhesives to impart quick tack and green str., in formulations to bond cellophane to itself or paper, or aluminum foil to itself; 85% act.

Uniplex 809. [Unitex] PEG di-2-ethylhexoate; CAS 9004-93-7; plasticizer with low volatility and exc. heat resist.; for polyester and polyamide engineering plastics; improves mold release; 85% act.

Uni-Rez® 1502. [Union Camp] Polyamide resin in xylene sol'n.; curing agent; provides good adhesion to metals, concrete, many plastics, good film resiliency, toughness, and chem. resist.; used in epoxy coatings formulated to meet air pollution regulations; Unverified

Uni-Rez® 2100-P75. [Union Camp] Reactive polyamide; resin used as curing agent for epoxy sol'n. coatings meeting the solv. restrictions of Rules 66/3 meeting air pollution control requirements; Gardner 7- sol'n.; dens. 7.6 lb/gal; visc. 9500 cps; flash pt. (TCC) 59 C; 75 ± 1% solids in n-propanol.; Unverified

Uni-Rez® 2341. [Union Camp] Reactive polyamide; epoxy curing agent for adhesives, coatings, grouts, castings, and floor toppings; Gardner 17; visc. 2400 cps; 100% solids; Unverified

Uni-Rez® 2355. [Union Camp] Reactive polyamide; epoxy curing agent for adhesives, coatings, and floor toppings; Gardner 8; visc. 800 cps; 100% solids.

Uni-Rez® 3020. [Union Camp] Pentaerythrityl rosinate; CAS 8050-26-8; EINECS 232-479-9; modifier for adhesives (water and solv.-based, construction, pressure-sensitive, hot-melts), sealants, rubber compounding, coatings; offers good tack and stability; compat. with SBR, NR, butyl, chlorinated, neoprene rubbers, EVA, SBS, PE, PP, acrylic, phthalate and polyester plasticizers, alkyds, hydrocarbon resins; FDA 21CFR §175.105; Gardner 9 flakes; sol. in aromatic and aliphatic hydrocarbons, higher monoalcohols; dens. 9 lb/gal; visc. 150-260 cps (170 C); soften. pt. (R&B) 95-100; acid no. 13 max.; flash pt. 238 C; *Toxicology:* nonhazardous; *Storage:* flaked forms are prone to slow oxidation which can affect performance after prolonged storage; Unverified

Unislip 1753. [Unichema] Erucamide; CAS 112-84-5; EINECS 204-009-2; slip and antiblock agent; mold release for PP; Gardner 2 max. pastilles, beads; 2-4 mm (pastilles), 85% < 710 μm (beads); bulk dens. 0.60 g/ml; visc. 12 cSt (100 C); m.p. 79-85 C; acid no. 1 max.; iodine no. 72-79; flash pt. (COC) 230 C; 99% amide.

Unislip 1757. [Unichema] Oleamide (lubricant grade); CAS 301-02-0; EINECS 206-103-9; lubricant, slip and antiblock agent; Gardner 8 max. pastilles (2-4 mm); bulk dens. 0.60 g/ml; visc. 11 cSt (100 C); m.p. 66-72 C; acid no. 7 max.; iodine no. 85-95; flash pt. (COC) 210 C; 94% amide.

Unislip 1759. [Unichema] Oleamide; CAS 301-02-0; EINECS 206-103-9; lubricant, slip and antiblock agent; Gardner 2 max. pastilles, beads, 2-4 mm (pastilles), 85% < 710 μm (beads); bulk dens. 0.60 g/ml; visc. 11 cSt (100 C); m.p. 70-76 C; acid no. 1 max.; iodine no. 80-90; flash pt. (COC) 210 C; 99% amide.

Unistab D-33. [Union Carbide] 1,3,5-Trimethyl-2,4,6-tris(3,5-di-t-butyl-4-hydroxybenzyl) benzene (40%) in polymeric organosilicon; antioxidant for thermoplastics processing incl. PP, HDPE, LLDPE, nylon; suitable for film, fiber, stretched tape, and molded parts; esp. good for high processing temps.; FDA 21CFR §177.1520(c); wh. liq.; insol. in water; sp.gr. 1.02; visc. 8900 cP; Unverified

Uni-Tac® 68. [Union Camp] Disproportionated rosin; tackifier for adhesives; Gardner 9 color; soften. pt. (R&B) 68 C; acid no. 118.

Uni-Tac® 70. [Union Camp] Noncrystallizing stabilized rosin; tackifier for SBR, natural rubber, butyl rubber, ethylene-vinyl acetate and other polymers; for water and solv.-based construction adhesives, pressure-sensitive, sealant, hot melt and rubber compding.; FDA 21CFR §175.105, 176.210, 177.2600; Gardner 9 flakes, solid, or sol'n.; sol. in aromatic, aliphatic, and chlorinated solvs.; sol. @ ≥ 50% in ethanol, propanol, IPA; insol. in water; dens. 8.8 lb/gal; visc. (Gardner) Z (75% in min. spirits); soften. pt. (R&B) 80 C; acid no. 140; flas.

Uni-Tac® 72. [Union Camp] Noncrystallizing stabilized rosin in min. spirits; tackifier for SBR, NR, butyl rubber, EVA, other polymers; for solv. and water-based adhesives (construction, pressure-sensitive), sealant, and rubber compounding applics.; FDA 21CFR §175.105; Gardner 8 liq.; dens. 8.2 lb/gal; soften. pt. (R&B) 80 C; acid no. 105; flash pt. (PM) 41 C; 75% solids; *Precaution:* combustible.

Uni-Tac® 72M70. [Union Camp] Stabilized rosin sol'n. in Rule 66/3 min. spirits; tackifier for SBR, NR, butyl rubber, EVA, and other polymers for pressure-sensitive, sealant, and rubber compounding applics. incl. water- and solv.-based construction adhesives for carpeting, flooring, framing, and tile applics.; FDA 21CFR §175.105, 176.210; Gardner 15 max. liq.; dens. 8.1 lb/gal; visc. (G-H) T-W; soften. pt. (R&B) 80 C (solid resin); acid no. 94-102; 68-72% solids; *Precaution:* combustible.

Uni-Tac® 99-70. [Union Camp] Stabilized rosin ester; tackifier with heat and aging stability for construction and other adhesives, rubber compounding, sealants, coatings; high acid no.; compat. with EVA, EEA, SIS, SBS, SBR, NR, butyl rubber, neoprene, acrylic, LMWPE, amorphous PP, phthalate and polyester plasticizers, alkyds, hydrocarbon and terpene resins; FDA 21CFR §175.105; Gardner 8 solid or flakes; sol. in aromatic/aliphatic hydrocarbon solvs., esteers, ketones, chlorinated solvs.; insol. in water, alcohol; dens. 9.0 lb/gal; soften. pt. (R&B) 90 C; acid no. 53.

Uni-Tac® R40. [Union Camp] Rosin ester; tackifier/plasticizer for adhesive systems (pressure-sensitive, hot-melt); plasticizer/coupling agent for lacquers, high-solids coatings; good low temp. props.; compat. with SBR, NR, butyl, chlorinated, neoprene rubber, EVA, SBS, PE, PP, acrylic, phthalate and polyester plasticizers, alkyd and hydrocarbon resins; FDA 21CFR §175.105; Gardner 6 semi-solid; sol. in aromatic and aliphatic hydrocarbon solvs.; dens. 8.8 lb/gal; visc. 400 cps (100 C); soften. pt. (R&B) 35 C; acid no. 13 ; flash pt. 200 C.

Uni-Tac® R85. [Union Camp] Glyceryl rosinate; tackifier for use in EVA, SBS, SIS and other hot melts, pressure-sensitive adhesives, in rubber compding., sealants, coatings; stabilized to provide good heat and aging stability; FDA 21CFR §175.105; Gardner 6 flakes or solid; sol. in aromatic and aliphatic hydrocarbon solvs., esters, ketones, chlorinated solvs.; insol. in alcohol, water; dens. 9 lb/gal; visc. 1690 cps (125 C); soften. pt. (R&B) 82 C; acid no. 12; *Storage:* avoid prolonged storage of flaked form.

Uni-Tac® R85-Light. [Union Camp] Glyceryl rosinate; tackifier for EVA and other hot melts, pressure-sensitive adhesives, rubber compounding, sealants, coatings; offers exc. stability, light color, low odor and low volatility at elevated temps.; FDA 21CFR §175.105; Gardner 4+ flakes or solid; sol. in aromatic and aliphatic hydrocarbons, esters, ketones, chlorinated solvs.; insol. in alcohols, water; dens. 9 lb/gal; visc. 4100 cps (125 C); soften. pt. (R&B) 85 C; acid no. 12; flash pt. > 230 C; *Storage:* avoid prolonged storage of flaked form.

Uni-Tac® R86. [Union Camp] Stabilized polyol ester of rosin; tackifier for adhesives, esp. EVA and other hot melts, pressure-sensitive adhesives; also for rubber compounding, sealants, coatings; compat. with EVA, SIS, SBS, EEA, SBR, NR, butyl rubber, neoprene, acrylic, LMWPE, amorphous PP, phthalate and polyester plasticizers, alkyds, hydrocarbon and terpene resins; FDA 21CFR §175.105; Gardner 5 solid or flakes; sol. in aromatic/aliphatic/chlorinated solvs., esters, ketones; insol. in water, alcohol; dens. 9 lb/gal; soften. pt. (R&B) 85 C; acid no. 13; flash pt. > 230 C.

Uni-Tac® R99. [Union Camp] Pentaerythrityl rosinate; CAS 8050-26-8; EINECS 232-479-9; tackifier for EVA, SBS, SIS, other hot melts, construction adhesives, in rubber compding., sealants, coatings; stabilized to provide good heat and aging stability; FDA 21CFR §175.105; Gardner 6 flakes or solid; sol. in aromatic and aliphatic hydrocarbons, esters, ketones, chlorinated solvs.; insol. in alcohol, water; dens. 9 lb/gal; visc. 2400 cps (140 C); soften. pt. (R&B) 97 C; acid no. 13; *Storage:* avoid prolonged storage of flaked forms.

Uni-Tac® R100. [Union Camp] Pentaerythrityl rosinate; CAS 8050-26-8; EINECS 232-479-9; tackifier for EVA, SBS, SIS and other hot-melts, pressure-sensitive adhesives, construction adhesives, rubber compounding, sealants, coatings; stabilized to provide high heat and aging stability; low volatiliy; FDA 21CFR §175.105; Gardner 6 flakes or solid, low odor; sol. in aromatic and aliphatic hydrocarbons, esters, ketones, chlorinated solvs.; insol. in alcohols, water; dens. 9 lb/gal; visc. 4100 cps (140 C); soften. pt. (R&B) 99 C; acid no. 13; *Storage:* avoid prolonged storage of flaked form.

Uni-Tac® R100-Light. [Union Camp] Pentaerythrityl rosinate; CAS 8050-26-8; EINECS 232-479-9; tackifier for EVA and other hot melts, pressure-sensitive adhesives, rubber compounding, sealants, coatings; exc. stability, low volatility at elevated temps.; FDA 21CFR §175.105; Gardner 4 flakes or solid, low odor; sol. in aromatic and aliphatic hydrocarbons, esters, ketones, chlorinated solvs.; insol. in alcohols, water; dens. 9 lb/gal; visc. 23,000 cps (125 C); soften. pt. (R&B) 99 C; acid no. 13; *Storage:* avoid prolonged storage of flaked form.

Uni-Tac® R100RM. [Union Camp] Stabilized pentaerythrityl rosinate; CAS 8050-26-8; EINECS 232-479-9; binder in thermoplastic road marking compds. due to oxidation resist. and heat stability at elevated temps.; also for bitumen road surf. dressings to promote adhesion and antislip props.; compat. with EVA, SIS, SBS, EEA, SBR, NR, butyl rubber, neoprene, acrylic, LMWPE, amorphous PP, phthalate and polyester plasticizers, alkyds, hydrocarbon and terpene resins; Gardner 7 pellet or bulk molten, low odor; sol. in aromatic/aliphatic/chlorinated solvs., esters, ketones; insol. in water, alcohol; sp.gr. 1.08; melt visc. 90 cps (200 C); soften. pt. (R&B) 100 C; acid no. 12; flash pt. 238 C.

Uni-Tac® R101. [Union Camp] Pentaerythrityl rosinate, deodorized; CAS 8050-26-8; EINECS 232-479-9; tackifier for hot-melt and pressure-sensitive adhesives, construction adhesives, rubber compounding, sealants, coatings; exc. stability, low volatility, lt. color; compat. with EVA, SIS, SBS, EEA, SBR, NR, butyl rubber, neoprene, acrylic, LMWPE, amorphous PP, phthalate and polyester plasticizers, alkyds, hydrocarbon and terpene resins; FDA 21CFR §175.105; Gardner 4 solid or flakes, low odor; sol. in aromatic/aliphatic hydrocarbon solvs., esters, ketones, chlorinated solvs.; insol. in alcohol, water; dens. 9 lb/gal; melt visc. 4100 cps (140 C); soften. pt. (R&B) 100 C; acid no. 10.

Uni-Tac® R102. [Union Camp] Stabilized pentaerythrityl rosinate; CAS 8050-26-8; EINECS 232-479-9; tackifier for hot-melt and pressure-sensitive adhesives, rubber compounding, sealants, coatings, and where improved heat resist. is required; compat. with EVA, SIS, SBS, EEA, SBR, NR, butyl rubber, neoprene, acrylic, LMWPE, amorphous

PP, phthalate and polyester plasticizers, alkyds, hydrocarbon and terpene resins; FDA 21CFR §175.105; Gardner 4 solid or flakes, low odor; sol. in aromatic/aliphatic hydrocarbon solvs., esters, ketones, chlorinated solvs.; insol. in water, alcohol; dens. 9 lb/gal; soften. pt. (R&B) 102 C; acid no. 12.

Uni-Tac® R106. [Union Camp] Stabilized pentaerythrityl rosinate; CAS 8050-26-8; EINECS 232-479-9; tackifier for adhesives; provides good skinning resist. and color stability to EVA hot melts; also for pressure-sensitive adhesives, rubber compounding, sealants, coatings; compat. with EVA, styrene block copolymers, SBR, acrylics, NR, hydrocarbon and terpene resins; FDA 21CFR §175.105; Gardner 4 pellet, molten bulk, low odor; sol. in aromatic/aliphatic hydrocarbon solvs., esters, ketones, chlorinated solvs.; insol. in water, alcohol; melt visc. 500 cps (180 C); soften. pt. (R&B) 106 C; acid no. 10; flash pt. 238 C.

Uni-Tac® R112. [Union Camp] Stabilized pentaerythrityl rosinate; CAS 8050-26-8; EINECS 232-479-9; tackifier for adhesives, esp. EVA hot melts requiring extra heat resist.; also for rubber compounding, sealants, coatings; compat. with EVA, styrene block copolymers, SBR, acrylics, NR, hydrocarbon and terpene resins; FDA 21CFR §175.105; Gardner 4 pellets or molten bulk, low odor; sol. in aromatic/aliphatic/chlorinated solvs., esters, ketones; insol. in water, alcohol; dens. 9 lb/gal; melt visc. 500 cps (180 C); soften. pt. (R&B) 112 C; acid no. 10; flash pt. 238 C.

Unite. [Aristech] Chemically modified polyolefins containing maleic anhydride functionality; compatibilizers in polymer blends and alloys (incl. recycled plastics), as polymeric coupling agent in reinforced and filled polymers (esp. PP), as adhesive agent for bonding polyolefins to various substrates.

Unithox® 420. [Petrolite] C20-40 pareth-3; nonionic; o/w emulsifier for cosmetics, vehicle for inert, difficult-to-disperse colorants, oils, and waxes; provides silky lubricating feel, superior film-forming chars.; for coatings, metalworking fluids, pulp/paper processing, textiles; mold release for plastics; sol. in min. oil, butyl Cellosolve, xylene, toluene, MEK, perchloroethylene; sol. hazy in Stod., VM&P naphtha, kerosene; disp. in water, ethanol, IPA; m.w. 560; m.p. 91 C; HLB 4; hyd. no. 85; flash pt. 204 C; Ross-Miles foam 20 mm (initial, 50 C, 0.1%).

Unithox® 450. [Petrolite] C20-40 pareth-10; nonionic; component for water-based PU mold release agents; metalworking additive; o/w emulsifier for cosmetics, vehicle for inert, difficult-to-disperse colorants, oils, and waxes; provides silky lubricating feel, superior film-forming chars.; also for coatings, pulp/paper processing, textiles; mold release for plastics; solid; sol. in min. oil, cyclohexane, butyl Cellosolve, xylene, toluene, MEK, perchloroethylene; sol. hazy in Stod., VM&P naphtha, kerosene, ethanol, IPA; disp. in water, ethylene glycol; m.w. 900; m.p. 91 C; HLB 10; hyd. no. 55; flash pt. 218 C; surf. tens. 45 dynes/cm (0.01% aq.); Ross-Miles foam 25 mm (initial, 50 C, 0.1%); 50% EO.

Unithox® 480. [Petrolite] C20-40 pareth-40; nonionic; o/w emulsifier for cosmetics, vehicle for inert, difficult-to-disperse colorants, oils, and waxes; provides silky lubricating feel, superior film-forming chars.; also for coatings, metalworking fluids, pulp/paper processing, textiles; mold release for plastics; sol. in min. oil, cyclohexane, ethylene glycol, butyl Cellosolve, xylene, toluene, MEK, perchloroethylene, ethyl acetate; sol. cloudy in Stod., VM&P naphtha, kerosene, ethanol, IPA; m.w. 2250; m.p. 86 C; HLB 16; hyd. no. 22; flash pt. 218 C; surf. tens. 52 dynes/cm (0.01% aq.); Ross-Miles foam 28 mm (initial, 50 C, 0.1%).

Unithox® 550. [Petrolite] C30-50 pareth-10; nonionic; o/w emulsifier for cosmetics, vehicle for inert, difficult-to-disperse colorants, oils, and waxes; provides silky lubricating feel, superior film-forming chars.; also for coatings, metalworking fluids, pulp/paper processing, textiles; mold release for plastics; m.w. 1100; m.p. 99 C; HLB 10; hyd. no. 41; flash pt. 232 C; surf. tens. 47 dynes/cm (0.01% aq.); Ross-Miles foam 26 mm (initial, 50 C, 0.1%).

Unithox® 750. [Petrolite] C40-60 pareth-10; nonionic; o/w emulsifier for cosmetics, vehicle for inert, difficult-to-disperse colorants, oils, and waxes; provides silky lubricating feel, superior film-forming chars.; also for coatings, metalworking fluids, pulp/paper processing, textiles; mold release for plastics; m.w. 1400; m.p. 106 C; HLB 10; hyd. no. 33; flash pt. 232 C; surf. tens. 53 dynes/cm (0.01% aq.); Ross-Miles foam 22 mm (initial, 50 C, 0.1%).

Uniwax 1750. [Unichema] Stearamide; CAS 124-26-5; EINECS 204-693-2; antiblock additive for polyolefin films; lubricant, release agent for rubber compding., PVC processing; Gardner 5 max. color; 85% < 710 μm; bulk dens. 0.5 g/ml; visc. 6 cSt (120 C); m.p. 98-104 C; acid no. 9 max.; iodine no. 1 max.; flash pt. (COC) 210 C; 94% amide.

Uniwax 1760. [Unichema] Ethylene bisstearamide; lubricant, slip and antiblock agent, release agent for polyolefins, chloroprene rubber, PVC; Gardner 4 max. beads, pastilles, 2-4 mm (pastilles), 85% < 710 μm (beads); bulk dens. 0.55 g/ml; visc. 10 cSt (160 C); m.p. 143-149 C; acid no. 7 max.; iodine no. 1 max.; amine no. 2 max.; flash pt. (COC) 290 C.

Uniwax 1763. [Unichema] Ethylene bisstearamide; defoamer grade polymer additive; Gardner 5 max. color; 85% < 710 μm; bulk dens. 0.55 g/ml; visc. 10 cSt (160 C); m.p. 143-149 C; acid no. 4 max.; iodine no. 1 max.; amine no. 4 max.; flash pt. (COC) 290 C.

Uniwax AW-1050. [Astor Wax] Cryst. Ziegler-type polyethylene wax; CAS 9002-88-4; EINECS 200-815-3; hard, high-melting lubricant wax; produces high-gloss bright-drying nonionic emulsions (floor care, shoe polish) and silicone emulsions (furniture, auto polishes); pale yel. pellets; m.p. 85-90 C; congeal pt. 67-74 C; penetration 5-8; acid no. 16-20; sapon. no. 40-50; Unverified

Uralloy® Hybrid Polymer LP-2035. [Olin] PU polymer in styrene monomer; low profile additive designed for use in fiber-reinforced unsat. polyester composites; controls shrinkage, provides automotive Class A surf. appearance, offers easier processing; liq.; sp.gr. 1.00; visc. 1000-1400 cps; f.p. -35 C; acid no. 2.6; flash pt. (CC) 31 C; 60% styrene.

USP®-90MD. [Witco/PAG] 80% sol'n. of 1,1-di (t-amyl peroxy) cyclohexane in odorless min. spirits; CAS 15667-10-4; initiator for heat-cured polyester resin cures and acrylates; sol'n.; 80% peroxide, 8.9% act. oxygen.

USP®-240. [Witco/PAG] Ketone peroxide sol'n. from acetylacetone; CAS 37187-22-7; EINECS 215-661-2; fire-resist. peroxide used for rapidly curing polyester resins with low levels of cobalt accelerators; DOT org. peroxide label is not required; sol'n.; 4.0% act. oxygen.

USP®-245. [Witco/PAG] 2,5-Dimethyl-2,5-di(2-ethyl hexanoyl peroxy) hexane; CAS 13052-09-0; catalyst for heated curing of polyester resin systems; features rapid cures and outstanding surf. finishes; liq.; 90% min. peroxide; 6.7% act. oxygen.

USP®-355M. [Witco/PAG] 75% sol'n. of 3,5,5-trimethyl hexanoyl peroxide in odorless min. spirits; CAS 3851-87-4; initiator for polymers; sol'n.; 75% peroxide, 3.8% act. oxygen.

USP®-400P. [Witco/PAG] 80% sol'n. of 1,1-di (t-butyl peroxy) cyclohexane (CAS 3006-86-8) in butyl benzyl phthalate; initiator useful in heat-cured polyester resin systems where improved flow and pot life are critical; sol'n.; 80% peroxide, 9.8% act. oxygen.

USP®-690. [Witco/PAG] t-Butyl peroxy 2-ethyl hexanoate (CAS 3006-82-4) and 1,1-di (t-butyl peroxy) cyclohexane (CAS 3006-86-8) in a phthalate plasticizer; single initiator for elevated temp. curing of unsat. polyester resins and compds. over a broad temp. range; sol'n.; 6.9% act. oxygen.

USP®-800. [Witco/PAG] 70% t-butyl hydroperoxide with water as major diluent; CAS 75-91-2; initiator; colorless liq.; 70% peroxide, 12.4% act. oxygen.

UV-149. [M.A. Polymers] Hindered amine; uv stabilizer for PE; pellets.

UV-201. [M.A. Polymers] Proprietary; uv stabilizer for ABS, PS; pellets.

UV-247. [M.A. Polymers] Hindered amine; uv stabilizer for PE, PP; pellets.

UV-349. [M.A. Polymers] Proprietary; uv stabilizer for polyesters; pellets.

UV-822. [M.A. Polymers] Proprietary; uv stabilizer for PP; pellets.

UV-Absorber 325. [Rhein Chemie] 2-Hydroxy-4-methoxybenzophenone; CAS 131-57-7; EINECS 205-031-5; uv lt. absorber for the protection of plastics and surface coating material, except polyolefins; in cellulose derivs.; useful in plasticized and unplasticized PVC, PS, polymethacrylates, PVDC, unsat. polyester resins, and paint binders; sltly. ylsh powd.; freely sol. in org. solv., paint binders, and plasticizers.

UV-Absorber 340. [Rhein Chemie] Cyanacrylic ester deriv.; uv lt. absorber for the protection of plastics and surface coating materials; effective in PU and plasticized and unplasticized PVC, PS, polymethacrylates, NC, alkyd resins, and polyester resins; unsuitable for polyolefins.

UV-Absorber KL 3-2500. [Bayer AG] Benzophenone deriv.; uv lt. absorber for the protection of polyolefins; Unverified

Uvasil PSL6. [Great Lakes] HALS alkoxy deriv.; uv stabilizer and coupling agent for titanium dioxide-filled polyethylene.

Uvasorb 2 OH. [3V] Benzophenone-1; CAS 131-56-6; EINECS 205-029-4; uv B absorber with max. absorption at wavelengths 290-320; improves lightfastness of dyes used in cosmetics; also protects colored, liq. detergents; uv stabilizer for PE, PP, rigid PVC; pale yel. powd.; sol. in polar solvs. (> 10% in acetone, 40% in ethanol, 40% in methanol); insol. in water; m.w. 214; m.p. 144-146; 99% min. act.

Uvasorb 3C. [3V] Benzophenone-12; CAS 1843-05-6; EINECS 217-421-2; uv stabilizer for ABS, cellulosics, epoxy, polyesters, PE, PP, PS, flexible and rigid PVC, VDC; FDA cleared for polyolefins; powd.; m.p. 48-49 C.

Uvasorb HA-88. [3V] Hindered amine; uv stabilizer for PE, PP; powd.; m.p. 130-150 C.

Uvasorb MET. [3V] Benzophenone-3; CAS 131-57-7; EINECS 205-031-5; uv B filter with max. absorp. at wavelengths 280-320; for cosmetic prods., skin creams, sunscreens; uv stabilizer for ABS, cellulosics, polyesters, PS, PVC, VDC; pale yel. powd.; m.w. 228; sol. 20% in chloroform, 10% in flax oil, 4% in ethanol, 3% in methanol; insol. in water; m.p. 62-64 C; 99.5% act.

Uvasorb NI. [3V] [2,2′-Thiobis (4-t-octyl phenolato)]-n-butylamine nickel; uv stabilizer for ABS, PE, PP; powd.; m.p. 258-281 C.

Uvasorb SV. [3V] Drometrizole; CAS 2440-22-4; EINECS 219-470-5; uv stabilizer for ABS, cellulosics, epoxy, polyesters, PS, flexible and rigid PVC, VDC; powd.; m.p. 128-132 C.

UV-Chek® AM-101. [Ferro/Bedford] Nickel bis (octyl phenol sulfide); uv absorber for polyolefin; flakes; sol. in most org. solvs.; m.p. 130 C.

UV-Chek® AM-104. [Ferro/Bedford] Nickel dibutyldithiocarbamate; CAS 13927-77-0; lt. stabilizer for polyolefins, PP stretched tape, woven pkg. applics., HDPE pipe and pipe wrap, outdoor carpeting and syn. turf, thermoplastic elastomers; antioxidant/antiozonant for SBR, NBR, CR, CO, IIR; improves heat resist. of EPDM and CSM; dk. gr. brittle flakes; very sol. in aromatic or chlorinated solvs.; moderate sol. in ketones; low sol. in aliphatic solvents; insol. in water; m.w. 468; sp.gr. 1.22; m.p. 88 C; 12.5% Ni.

UV-Chek® AM-105. [Ferro/Bedford] Nickel bis (octyl phenol sulfide); lt. stabilizer for polymers, syn. elastomers; used in syn. PP turf, indoor-outdoor carpets, in synergistic combination with ZnO for stabilization of polyolefins; grn. brittle flakes; very sol. in aromatic or chlorinated solvs.; moderate sol. in aliphatic hydrocarbons and ketones; insol. in water; m.w. 1140; sp.gr. 1.11; m.p. 130 C; 7.7% Ni.

UV-Chek® AM-205. [Ferro/Bedford] Thiobis-2,2′-(4(1,1,3,3-tetramethylbutyl)phenol)-nickel-2-

ethylhexanoate; CAS 38727-83-2, 7580-31-6; lt. stabilizer for polymers, thermoplastic elastomers; used in polyethylene film for agric., construction, pkg. and greenhouse use, syn. PP turf, indoor-outdoor carpets, with ZnO for stabilization of polyolefins and elastomers; grn. brittle flakes; very sol. in aromatic or chlorinated solvs.; moderate sol. in aliphatic hydrocarbons and ketones; insol. in water; m.w. 1630; sp.gr. 1.19; m.p. 140 C; 13% Ni.

UV-Chek® AM-300. [Ferro/Bedford] 2-Hydroxy-4-n-octoxybenzophenone; CAS 1843-05-6; EINECS 217-421-2; uv absorber for plastics, flexible and semirigid PVC applics. (automotive body side molding, instrument panels, glove boxes), surf. coatings incl. automotive applics.; costabilizer for polyolefin mulch or greenhouse film; also for acetals, acrylics, ABS, alkyds, cellulosics, epoxies, PC, thermoplastic and thermoset polyester, LDPE, LLDPE, HDPE, PP, PS (crystal and impact), PVDC, SAN, surf. coatings, thermoplastic elastomers; lt. yel. cryst. powd.; very sol. in aromatic, aliphatic, and ketone solvs.; low sol. in alcohols; insol. in water; m.w. 326; sp.gr. 1.17; m.p. 48 C.

UV-Chek® AM-301. [Ferro/Bedford] 2-Hydroxy-4-n-octyloxybenzophenone; CAS 1843-05-6; EINECS 217-421-2; lt. stabilizer for polymers incl. acetals, acrylics, ABS, alkyds, cellulosics, epoxies, PC, thermoplastic and thermoset polyester, polyethylene, PP, PS, SAN, and surface coatings; costabilizer for polyethylene film in agric. or greenhouse applics., PVC flexible and semirigid applics. such as automotive body side molding, for automotive surface coatings; acts as uv absorber; lt. yel. free-flowing cryst. powd.; very sol. in aromatic, aliphatic, and ketone solvs.; low sol. in alcohols; insol. in water; m.w. 326; sp.gr. 1.17; m.p. 48 C.

UV-Chek® AM-320. [Ferro/Bedford] 4-Dodecyloxy-2-hydroxy benzophenone; uv stabilizer for ABS, cellulosics, epoxy, polyesters, PE, PP, PS, flexible and rigid PVC, VDC; powd.; m.p. 52 C.

UV-Chek® AM-340. [Ferro/Bedford] 2,4-Di-t-butylphenyl 3,5-di-t-butyl-4-hydroxybenzoate; CAS 4221-80-1; lt. stabilizer for polyolefins, nylon, ABS, PC, thermoplastic polyester, PS (crystal and impact), SAN, thermoplastic elastomers, esp. pigmented applics.; typical uses incl. molded containers, stadium seats, car bumpers, carpets, turf; imparts the least color to host resins of commonly used uv stabilizers; synergist/coadditive with other uv stabilizers, benzophenones, benzotriazoles, HALS, nickles; may be used in PP pkg. materials which contact dry or aq. foods; wh. to cream cryst. powd.; very sol. in aromatic hydrocarbons, chlorinated solvs.; mod. sol. in ketones, esters; low sol. in aliphatic hydrocarbons, alcohols; insol. in water; m.w. 438; sp.gr. 1.08; dens. 4.3 lb/gal; m.p. 196 C.

UV-Chek® AM-595. [Ferro/Bedford] Proprietary org./inorg. complex containing Ba, Na, and P; uv stabilizer for rigid or flexible PVC, e.g., rigid profiles, auto roofs, flexible sheeting; heat and lt. stabilizer for CPE, CPVC, PCDC, single-ply roofing or pond liners of CPE; acid scavenger in flame-retarded plastics; hygroscopic; wh. free-flowing fine powd.; 30 μ avg. particle size; disperses but does not melt or dissolve completely in polymers or solvs.; sp.gr. 2.12.

UV-Chek® AM-806. [Ferro/Bedford] 2,2,6,6-Tetramethylpiperidin-4-yl acrylate/methyl methacrylate copolymer; CAS 115340-81-3; lt. stabilizer for PP, ABS, PS, nylon, LDPE, LLDPE, HDPE, acrylics, PC, thermoplastic polyester, SAN, thermoplastic elastomers, surf. coatings; esp. effective in polyolefins for outstanding heat and processing stability; suitable for thin films, fibers, or molded parts; use with UV-Chek AM-340; lt. amber micro-pastille; very sol. in aromatic hydrocarbons, ketones, esters, some alcohols; very low sol. in aliphatic hydrocarbons and water; misc. with methylene, methylene chloride; m.w. > 2400; sp.gr. 1.03; soften. pt. 100 C min.; *Storage:* good storage stability if stored below 40 C.

Uvinox 770. [PMC Specialties] Phenol; antioxidant for plastics; wh. cryst. powd.

Uvi-Nox® 1494. [Rhone-Poulenc Surf. & Spec.] Diisobutyl nonyl phenol; CAS 4306-88-1; primary antioxidant; provides better thermal stabilization of polyolefins and lower vapor pressure than common antioxidants; also for syn. rug-backing, latex paints, rosin, ester gums, in gasoline and aviation fuels, insulating oils, paraffin wax; Gardner 8 visc. liq.; m.w. 280; sol. in abs. ethanol, benzene, MEK; pour pt. 3.3 C; flash pt. 152 C.

Uvinul® 408. [BASF] 2-Hydroxy-4-n-octoxybenzophenone; CAS 1843-05-6; EINECS 217-421-2; uv stabilizer for ABS, cellulosics, epoxy, polyesters, PE, PP, PS, flexible and semirigid PVC, VDC; FDA approved for polyolefins; powd.; m.p. 48-49 C.

Uvinul® 3000. [BASF; BASF AG] Benzophenone-1; CAS 131-56-6; EINECS 205-029-4; uv-A and B absorber; antiphoto-oxidant used for polyester, acrylics, PS, in outdoor paints and coatings, varnishes, colored liq. toiletries and cleaning agents, filters for photographic color films and prints, and rubber-based adhesives; Japan approval; ylsh. powd.; sol. in oils, alcohols, ether alcohols, propylene glycol, cyclic ethers, ketones, and esters; m.w. 214; m.p. ≥ 144 C; absorp. max. 288 nm; > 98% act.

Uvinul® 3008. [BASF] 2-Hydroxy-4-n-octoxy-benzophenone; CAS 1843-05-6; EINECS 217-421-2; uv absorber and stabilizer for polyethylene, PP, plasticized and rigid PVC, and other polymers; offers good compatibility, max. protection, and min. color and as a low order of toxicity; lt. yel.; m.w. 326; sol. (g/100 g solv. @ 30 C): 333 g in toluene; 332 g in acetone; 205 g in ethyl acetate; 185 g in MEK; 160 g in methyl pyrrolidone; 36 g in hexane; 5.8 g in ethanol; 3.1 g in methanol; sp.gr. 1.064; dens. (bulk) 0.96 l/kg (MEK); m.p. 48-49 C; 98.5% min. active.

Uvinul® 3035. [BASF] Etocrylene; CAS 5232-99-5; EINECS 226-029-0; noncolor contributing uv absorber; does not contain aromatic hydroxyl groups; effective under varying pH conditions; for NC lacquers and PVC; used in alkaline systems such as urea-formaldehyde and epoxyamine formulations, and in cosmetics; wh. powd.; m.w. 277; m.p. 94-96 C; > 98% act.

Uvinul® 3039. [BASF] Octocrylene; CAS 6197-30-4; EINECS 228-250-8; uv-B absorber for cosmetics esp. water-resist. formulations, plastics, coatings; in flexible and rigid PVC; used in NC lacquers, varnishes, vinyl flooring, and oil-based paints; in aerosol and oil-based suntan lotions; nonreative with metallic driers; USA (7-10%) approval; yel. clear visc. liq.; sol. in nonpolar plastics; misc. with oils, min. spirits, ethanol, IPM; m.w. 361; f.p. -10 C; absorp. max. 302 nm; > 98% act.

Uvinul® 3040. [BASF; BASF AG] Benzophenone-3; CAS 131-57-7; EINECS 205-031-5; uv-A/B absorber for cosmetics, plastics, and coatings, suncare preps., day creams; good weather resistance in resins and plastics; stabilizes PVC and polyesters against uv-lt. degradation; used in NC lacquers, varnishes, oil-based paints; USA (2-6%), EC (10%), and Japan (5%) approvals; lt. yel. powd.; sol. in oils, 12% in IPM, 6% in ethanol; m.w. 228; m.p. ≥ 62 C; absorp. max. 287 nm; > 98% act.

Uvinul® 3049. [BASF] Benzophenone-6; CAS 131-54-4; EINECS 205-027-3; economical uv-A and B absorber for cosmetics, plastics, coatings, textiles; greater heat stability and more sol. (in chlorinated and aromatic solvs.); gives broad protection to PVC, chlorinated polyesters, epoxies, acrylics, urethanes, cellulosics, oil-based paints and varnishes, and cosmetics; Japan approval; yel. powd.; sol. in oils; m.w. 274; m.p. ≥ 130 C; absorp. max. 339 nm; > 98% act.

Uvinul® 3050. [BASF; BASF AG] Benzophenone-2; CAS 131-55-5; EINECS 205-028-9; uv-A and B absorber with the broadest uv absorp. spectrum; retards fading of colorants, pigments, and dyestuffs; improves stability of fragrances to oxidation; prolongs the life of polymeric materials; photostabilizes cosmetic formulations and minimizes discoloration of syn. rubber of plastic latices; Japan approval; yel. powd.; sol. in oils, ethanol, propylene glycol, 8% in IPM; m.w. 246; m.p. 195 C; absorp. max. 345 nm; > 98% act.

Uvinul® 3093. [BASF] Benzophenone-11; CAS 1341-54-4; uv absorber; used in NC lacquer, fluorescent paint, inks, and for protecting furniture woods, colored liq. toiletries and cleaning agents, isocyanate systems, and butyrate metal lacquers; ylsh. powd.; > 98% act.

Uvinul® D-49. [BASF] 2,2′-Dihydroxy-4,4′-dimethoxy benzophenone; CAS 131-54-4; EINECS 205-027-3; uv stabilizer for epoxy, polyesters; powd.; m.p. 125 C.

Uvinul® D-50. [BASF] 2,2′,4,4′-Tetrahydroxy benzophenone; uv stabilizer for epoxy, polyesters; powd.; m.p. 195 C.

Uvinul® M-40. [BASF] 2-Hydroxy-4-methoxy benzophenone; CAS 131-57-7; EINECS 205-031-5; uv stabilizer for ABS, cellulosics, polyesters, PS, flexible and rigid PVC, VDC; powd.; m.p. 62-63.5 C.

Uvinul® N-35. [BASF] Ethyl-2-cyano-3,3-diphenyl acrylate; uv stabilizer for rigid PVC, VDC; powd.; m.p. 96 C.

Uvinul® N-539. [BASF] 2-Ethylhexyl-2-cyano-3,3-diphenyl acrylate; uv stabilizer for flexible and semi-rigid PVC, VDC; liq.; m.p. -10 C.

Uvitex® OB. [Ciba-Geigy/Additives] 2,2′-(2,5-Thiophenediyl)bis(5-t-butylbenzoxazole); CAS 7128-64-5; fluorescent whitener for the optical brightening of polymers at all stages of processing; offers brilliant, bluish wh. effects, good lt. fastness, excellent resistance to heat, high chemical stability, and easy sol. in a wide range of org. solvs.; used in thermoplastics, coatings, printing inks, syn. fibers; useful for safeguarding against forgeries, in waxes, fats, and oils, in artificial and expanded artificial leather, sheeting, household articles, elec. goods, food pkg. material, etc.; 21CFR §175.105, 177.1520(c), 177.1580, 177.1640, 177.2480, 178.3297; yel. cryst. powd.; sol. in waxes, paraffins, fats, and min. and veg. oils, and may be mixed with these substances in an org. solv.; m.w. 431; sp.gr. 1.26; m.p. 197-203 C.

V

V-30. [Wako Pure Chem. Ind.; Wako Chem. USA] 1-[(1-Cyano-1-methylethyl)azo]formamide; CAS 10288-28-5; polymerization initiator; pale yel. cryst.; sol. in ethanol, methanol, water; m.w. 140.14; m.p. 76-78 C.

V-40. [Wako Pure Chem. Ind.; Wako Chem. USA] 1,1′-Azobis(cyclohexane-1-carbonitrile); CAS 2094-98-6; EINECS 218-254-8; polymerization initiator; wh. cryst.; insol. in water; m.w. 244.34; m.p. 113-115 C.

V-50. [Wako Chem. USA] 2,2′-Azobis(2-amidinopropane) dihydrochloride; CAS 2997-92-4; EINECS 221-070-0; cationic; highly efficient polymerization initiator esp. useful for polymerization of water-sol. vinyl monomers; decomp. to give cationic free radicals in aq. sol'n. at relatively low temp.; also for acrylamide, acrylonitrile, cationic acrylic monomers, emulsion polymerization of styrene, VA, acryl esters, encapsulation or graft polymerization, photopolymerization; wh. to off-wh. gran., almost odorless; freely sol. in water; sol. in toluene, hexane; sparingly sol. in methanol, ethanol, acetone, dioxane, DMSO; m.w. 271.19; m.p. 160-169 C (dec.); *Toxicology:* LD50 (oral, rat) 410 mg/kg, (dermal, rat) > 5.9 g/kg; eye and skin irritant; allergic reactions; harmful if inhaled or ingested; may cause headaches, anorexia, vomiting, fever; prolonged exposure may cause shock, narcosis, death; *Precaution:* unstable above 40 C; causes exothermal decomp. @ 143 C; dust explosion hazard; incompat. with strong oxidizers and acids, persulfates; hazardous decomp. prods.: CO, NO_x, nitrogen gas, ammonium gas, hydrogen chloride; *Storage:* store away from sunlight in cool (< 40 C), well-ventilated area; container should not be tightly sealed.

V-59. [Wako Pure Chem. Ind.; Wako Chem. USA] 2,2′-Azobis(2-methylbutyronitrile); CAS 13472-08-7; EINECS 236-740-8; polymerization initiator; wh. cryst. powd.; sol. in toluene, ethanol, methanol; insol. in water; m.w. 192.26; m.p. 55-57 C; dec. 84-87 C.

V-60. [Wako Pure Chem. Ind.; Wako Chem. USA] 2,2′-Azobis(2-methylpropionitrile); CAS 78-67-1; EINECS 201-132-3; polymerization initiator; wh. cryst. powd.; sol. (g/100 g): 7.5 g methanol, 7 g toluene, 3 g ethanol; insol. in water; m.w. 164.21; m.p. 100-103 C.

V-65. [Wako Pure Chem. Ind.; Wako Chem. USA] 2,2′-Azobis(2,4-dimethylvaleronitrile); CAS 4419-11-8; EINECS 224-583-8; polymerization initiator; wh. cryst.; sol. (g/100 g): 72 g toluene, 22 g methanol, 20.5 g ethanol; insol. in water; m.w. 248.37; m.p. 45-70 C.

V-70. [Wako Pure Chem. Ind.; Wako Chem. USA] 2,2′-Azobis(4-methoxy-2,4-dimethylvaleronitrile); CAS 15545-97-8; EINECS 239-593-8; polymerization initiator; wh. cryst. powd.; insol. in water; m.w. 308.42; m.p. 50-96 C (dec.).

V-501. [Wako Pure Chem. Ind.; Wako Chem. USA] 4,4′-Azobis (4-cyanopentanoic acid); CAS 2638-94-0; EINECS 220-135-0; anionic; polymerization initiator; wh. cryst. powd., almost odorless; sol. in ethanol; sl. sol. in water; m.w. 280.28; m.p. 120-123 C (dec.); *Toxicology:* LD50 (oral, mouse) > 5000 mg/kg; harmful if inhaled or ingested; absorbed thru skin; dust irritant; ing. may cause anorexia, headaches, vomiting, fever; prolonged exposure may cause shock, narcosis, death; *Precaution:* unstable above 20 C; incompat. with oxidizers, strong acids; hazardous decomp. prods.: CO, NO_x, hydrogen cyanide gas and cyanide compds.; *Storage:* store in cool (< 20 C), dry, well-ventilated area away from sunlight; container should not be tightly closed.

V-601. [Wako Pure Chem. Ind.; Wako Chem. USA] Dimethyl 2,2′-azobis (2-methylpropionate); CAS 2589-57-3; EINECS 219-976-6; initiator for polymerization of vinyl acetate, acrylonitrile, styrene, methyl methacrylate, bulk polymerization; pale yel. cryst. or wax; sol. in toluene, hexane, ethanol, methanol, benzene, DMF, dioxane, DMSO; insol. in water; m.w. 230.26; sp.gr. 1.013 (30/4 C); m.p. 22-28 C; dec. 85-87 C; *Toxicology:* LD50 (oral, rat) 527 mg/kg; *Storage:* store below 15 C to prevent decomp.; store in cool, dry, dark place away from heat sources.

VA-044. [Wako Pure Chem. Ind.; Wako Chem. USA] 2,2′-Azobis[2-(2-imidazolin-2-yl)propane] dihydrochloride; CAS 27776-21-2; EINECS 248-655-3; cationic; polymerization initiator esp. useful for polymerization of water-sol. vinyl monomers; produces polymers with high linearity and m.w.; also for acrylamide, acrylic acid, styrene; wh. to pale yel. cryst. or cryst. powd.; sol. in toluene, hexane, 35.2 g/100 g water; m.w. 323.27; m.p. 188-193 C (dec.); pH 3.8 (1% aq.); *Toxicology:* LD50 (oral, rat) 3.0 g/

kg; *Precaution:* produces nitrogen gas when decomp.; *Storage:* 6 mos stability stored @ 40 C; store in cool, dry, dark place away from heat sources; do not store in tightly closed containers.

VA-061. [Wako Chem. USA] 2,2′-Azobis(N,N′-dimethylene isobutyramidine); CAS 20858-12-2; EINECS 244-085-4; cationic; polymerization initiator; wh. to sl. yel. cryst. powd., sl. char. odor; sol. in toluene; sl. sol. in water; m.w. 250.35; sp.gr. 0.455; m.p. 115-125 C (dec.); *Toxicology:* eye and skin irritant; allergic reactions; harmful if inhaled and ingested; may cause headaches, anorexia, nausa, vomiting, fever, etc.; *Precaution:* unstable above 10 C; causes violent exothermal decomp. @ 76 C; may create dust explosion; incompat. with oxidizers, strong acids, persulfates; hazardous decomp. prods.: CO, NO_x, N gas; *Storage:* store away from sunlight in cool (< 10 C), well-ventilated, dry area; container should not be tightly sealed.

VA-082. [Wako Pure Chem. Ind.; Wako Chem. USA] 2,2′-Azobis2-methyl-N-[1,1-bis(hydroxymethyl) ethyl]propionamide; CAS 104222-30-2; polymerization initiator; pale yel. cryst. powd.; sol. in toluene, hexane, 0.6 g/100 g water; m.w. 376.45; m.p. 156-161 C (dec.).

VA-086. [Wako Pure Chem. Ind.; Wako Chem. USA] 2,2′-Azobis[2-methyl-N-(2-hydroxyethyl)-propionamide] CAS 61551-69-7; nonionic; polymerization initiator; pale yel. cryst. powd.; sol. in toluene, hexane, 2.4 g/100 g water; m.w. 288.35; m.p. 140-145 C (dec.).

Valu-Fil. [Plastic Filler Sales] Calcium sulfate dihydrate; CAS 10101-41-4; flame retardant for thermoset polyester.

Vanax® 552. [R.T. Vanderbilt] Piperidinium pentamethylene dithiocarbamate; CAS 98-77-1; accelerator for NR, SR, cements, and latexes; peptizer for sulfur-modified G-type neoprenes; creamy wh. to lt. yel. powd.; m.w. 246.47; very sol. in chloroform; mod. sol. in acetone, toluene, alcohol; disp. in water; dens. 1.20 ± 0.03 mg/m^3; m.p. 167 C (decomp.).

Vanax® 808. [R.T. Vanderbilt] Butyraldehyde-aniline condensation prod.; CAS 34562-31-7; accelerator for NR, SBR, CR, IIR, and latexes; activator for acidic accelerators; also for reclaims, hard rubber stocks, CR cements contng. litharge; dk. amber liq.; dens. 0.98 ± 0.02 mg/m^3; flash pt. (PM) 135 C.

Vanax® 829. [R.T. Vanderbilt] Substituted 1,3,4-thiadiazole contg. process oil; rubber chemical; yel. powd.; negligible sol. in water; dens. 2.09 mg/m^3; *Toxicology:* LD50 (oral, rat) 6480 mg/kg, (dermal, rabbit) > 2000 mg/kg; may cause eye and skin irritation; process oil: OSHA TLV 5 mg/m^3 (as mist); *Precaution:* hazardous decomp. prods.: oxides of N, S, C at combustion temps.

Vanax® 833. [R.T. Vanderbilt] Butyraldehyde-monobutylamine condensation prod.; CAS 68411-19-8; accelerator for NR, SR, latexes, and reclaim; also in self-curing CR cements; red-amber liq.; dens. 0.86 ± 0.02 mg/m^3; flash pt. (PM) 96 C.

Vanax® A. [R.T. Vanderbilt] 4,4′-Dithiodimorpholine in paraffin wax; vulcanizing agent; sulfur donor for NR and syn. rubbers; functions as primary accelerator for NR, IR, SBR, NBR, IIR elastomers, and as primary and sec. accelerator in EPDM; powd.; wh. to lt. gray powd., pungent amine-like odor; also avail. in thread grade and rods (80% act.); 99% thru 100 mesh (powd. and thread grade); sol. in toluene, xylene, and CCl_4; m.w. 236.36; dens. 1.35 ± 0.03 mg/m^3 (rods); m.p. 123-125 C; flash pt. (COC) 154 C; pH 11; 26-29% sulfur (powd. and thread grade); 20.5-23% sulfur (rods); *Toxicology:* LD50 (oral, rat) 5600 mg/kg, (dermal, rabbit) > 5010 mg/kg; causes irritation to eyes, skin, respiratory tract; may cause allergic skin reaction; may liberate irritating free amine; may react with nitrosating agents to form an animal carcinogen; *Precaution:* may form explosive dust-air mixts.; incompat. with acids, reducing agents, nitrosating agents; hazardous decomp. prods.: morpholine vapors, oxides of N, S, and C at combustion temps.; *Storage:* store in cool, dry, well-ventilated area below 30 C, away from sulfur, acids, foodstuffs, heat, flame; do not reuse container.

Vanax® CPA. [R.T. Vanderbilt] Dimethylammonium hydrogen isophthalate; CAS 71172-17-3; accelerator for W and T-type neoprenes; used in press cured, inj. molded, and LCM stocks; wh. to lt. ivory powd.; 99.9% thru 100 mesh; very sol. in dimethylformamide, dimethyl sulfoxide; mod. sol. in acetone, chloroform; sl. sol. in alcohol, water; m.w. 211.24; dens. 1.35±0.03 mg/m^3; m.p. 190 C (dec.).

Vanax® DOTG. [R.T. Vanderbilt] N,N′-Di-o-tolylguanidine; CAS 97-39-2; accelerator for NR and SR; sec. accelerator; wh. to pink powd.; 99.9% thru 100 mesh; very sol. in alcohol, acetone; sl. sol. in toluene, water; m.w. 238.34; dens. 1.20 ± 0.03 mg/m^3; m.p. 173-179 C.

Vanax® DPG. [R.T. Vanderbilt] N,N′-diphenyl guanidine; CAS 102-06-7; accelerator for NR and SR; sec. accelerator; wh. to lt. yel. with pink tinge powd.; 99.9% thru 100 mesh; m.w. 210.28; very sol. in acetone, alcohol; slightly sol. in toluene, water; dens. 1.20 ± 0.03 mg/m^3; m.p. 144-149 C.

Vanax® DPG Pellets. [R.T. Vanderbilt] N,N′-diphenyl guanidine; CAS 102-06-7; accelerator for NR and SR; pellets.

Vanax® MBM. [R.T. Vanderbilt] m-Phenylenedimaleimide; CAS 3006-93-7; accelerator; coagent in peroxide-cured polymers; yel. to lt. brn. powd., rods; 99.9% min. thru 100 mesh (powd.); sol. in acetone; mod. sol. in chloroform; pract. insol. in water; dens. 1.44 ± 0.03 mg/m^3.

Vanax® NS. [R.T. Vanderbilt] N-t-Butyl-2-benzothiazolesulfenamide; CAS 95-31-8; accelerator; delayed action accelerator for natural and syn. rubbers; lower scorch and faster cure than Durax; used in tires, mechanical and extruded goods; lt. buff to tan powd., 99.9% min. thru 20 mesh; rods; pract. insol. in water; m.w. 238.37; dens. 1.28 ± 0.03 mg/m^3; m.p. 105 C min.

Vanax® PML. [R.T. Vanderbilt] Di-o-tolylguanidine salt of dicatechol borate; CAS 16971-82-7; accelerator for NR and SR stocks, CR, neoprenes for wire and cable and mechanical goods; activator and mild antioxidant in NR and SBR; tan to lt. brn. powd.; 99.9% thru 100 mesh; very sol. in acetone, chloroform; mod. sol. in alcohol, toluene; insol. in water; m.w. 467.37; dens. 1.25 ± 0.03 mg/m³; m.p. 165 C min.

Vanax® PY. [R.T. Vanderbilt] 23% poly-p-dinitrosobenzene polymer in wax; CAS 9003-34-3; chemical conditioner for IIR, cements; dk. brn. pellets; 99.9% thru 1/2 in. screen; sol. in toluene, chloroform, acetone; mod. sol. in hexane, gasoline; insol. in water; dens. 0.96 ± 0.03 mg/m³.

Vanax® TBSI. [R.T. Vanderbilt] N-t-Butyl 2-benzothiazole sulfenamide; CAS 95-31-8; rubber accelerator.

Vanchem® HM-50, HM-4346. [R.T. Vanderbilt] Aromatic polyisocyanate in monochlorobenzene and toluene resp.; CAS 108-90-7, 108-88-3; adhesion promoter in adhesives; primer or adhesive additive for adhering elastomeric coatings to syn. fiber fabrics coated with syn. rubbers; crosslinking agent for elastomers and plastics; dk. brn. liqs.; dens. 1.16 ± 0.02 and 1.05 ± 0.02 mg/m³ resp.; flash pt. (PM) 30 C and 7 C min. resp.

Vancide® 51Z Disp. [R.T. Vanderbilt] Dispersed form of Vancide 51Z; fungicide for addition to latex or for aq. applics.; misc. with water; sp.gr. 1.26 ± 0.02; 50% act. in water.

Vancide® 89. [R.T. Vanderbilt] Captan; CAS 133-06-2; EINECS 205-087-0; fungicide for natural and syn. rubber compds. containing susceptible plasticizers; industrial preservative for vinyl, polyethylene, paint, lacquer, soap, wallpaper flour paste; lt. tan powd.; m.w. 300.6; sol. in tetrachloroethane, chloroform, xylene, cyclohexanone, ethylene dichloride; sp.gr. 1.69 ± 0.03; m.p. 168-174 C; 90% min. assay.

Vancide® MZ-96. [R.T. Vanderbilt] Zinc dimethyldithiocarbamate; CAS 137-30-4; antimicrobial, preservative for starch and syn. latex adhesives, food pkg. adhesives; wh. powd.; m.w. 305.82; mod. sol. in dilute caustic, toluene, carbon disulfide, chloroform; water-disp.; dens. 1.71 ± 0.03 mg/m³; m.p. 252-260 C; 96% assay.

Vancide® TH. [R.T. Vanderbilt] Hexahydro-1,3,5-triethyl-s-triazine; CAS 7779-27-3; fungicide for use as industrial preservative for latex, adhesives, cutting fluids, marine lubricants; colorless to lt. yel. liq.; m.w. 171.29; sol. in acetone, ethanol, ether, water; mod. sol. in hydrocarbon solvs.; dens. 0.89 ± 0.02 mg/m³; b.p. 205-210 C; flash pt. (TCC) 66 C.

Vandex. [R.T. Vanderbilt] Selenium; sec. vulcanizing agent for NR, IR, SBR; increases state of cure and improves heat aging of sulfurless and low sulfur stocks; reduces curing time in NR hard rubber and allows use of lower sulfur for increased flexibility; dk. gray metallic powd., 99% thru 200 mesh; m.w. 78.96; sol. in carbon disulfide; practically insol. in water; dens. 4.80 ± 0.03 mg/m³; m.p. > 217 C.

Vanfre® AP-2. [R.T. Vanderbilt] Proprietary; polymer processing lubricant for NR, nitrile, SBR, sulfur-cured and peroxide-cured EPDM, butyl rubber, epichlorohydrin; reduces mixing time, aids dispersion of rubber additives, reduces compd. plasticity, improves mold flow; FDA 21CFR §175.105, 175.300, 177.1210, 177.2600, 178.3910; off-wh. to lt. tan friable flakes, fatty odor; negligible sol. in water; dens. 0.98 ± 0.03 mg/m³; m.p. 98 C min.; flash pt. (COC) 202 C; 1.8-2.5% Zn; *Toxicology:* low oral and dermal toxicity; prolonged exposure may cause transient irritation to eyes and skin; contains a reported carcinogen; lubricant: LD50 (oral, rat) 31,000 mg/kg; OSHA TLV/TWA 5 mg/m³; *Precaution:* incompat. with strong oxidizing agents; hazardous decomp. prods.: oxides of zinc and carbon.

Vanfre® AP-2 Special. [R.T. Vanderbilt] Polymer processing lubricant for NR, nitrile, SBR, sulfur-cured and peroxide-cured EPDM, butyl rubber, epichlorohydrin; reduces mixing time, aids dispersion of rubber additives, reduces compd. plasticity, improves mold flow; recommended for rubber articles in contact with food; contains no ingreds. with reported carcinogenic effects; FDA 21CFR § 175.105, 175.300, 177.1210, 177.2600, 178.3910; off-wh. to lt. tan friable flakes, fatty odor; negligible sol. in water; dens. 0.98 ± 0.03 mg/m³; m.p. 98 C min.; flash pt. (COC) 202 C; 1.8-2.5% Zn; *Toxicology:* low oral and dermal toxicity; prolonged exposure may cause transient irritation to eyes and skin; nuisance dust may be generated; lubricant: OSHA PEL/TWA 10 mg/m³ (total dust), 5 mg/m³ (respirable dust); *Precaution:* incompat. with strong oxidizing agents; hazardous decomp. prods.: oxides of zinc and carbon.

Vanfre® DFL. [R.T. Vanderbilt] Processing aid for mold lubrication; corrosion inhibitor; clear to straw liq.; dens. 1.08 ± 0.02 mg/m³; visc. 1200 cps min.; pH 7.2-8.5 (10%).

Vanfre® HYP. [R.T. Vanderbilt] Processing aid for Hypalon; reduces plasticity, improves mold flow and release; off-wh. to lt. tan friable flake; very sol. in toluene, chloroform; mod. sol. in acetone, hexane, gasoline; insol. water; dens. 0.97 ± 0.03 mg/m³; m.p. 81-97 C.

Vanfre® IL-1. [R.T. Vanderbilt] Sodium alkyl sulfates; internal lubricant for CR, NBR, CSM, and NR; improves flow chars. of highly loaded compds.; improves release from mill rolls; provides smoother extrusions; disperses easily; wh. to creamy wh. powd.; dens. 1.16 ± 0.03 mg/m³.

Vanfre® IL-2. [R.T. Vanderbilt] Fatty alcohols on inert carrier; internal lubricant; improves flow and molding chars. of highly loaded elastomer compds.; does not reduce tensile strength or appreciably affect water swell of vulcanizates; gray powd., 99.9% min. 100 mesh; mild fatty odor; dens. 1.17 ± 0.03 mg/m³.

Vanfre® M. [R.T. Vanderbilt] Processing aid for natural and syn. rubbers; reduces plasticity, improves mold flow; amber pellets; dens. 0.87 ± 0.03 mg/m³; m.p. 80 C min.

Vanfre® UN. [R.T. Vanderbilt] Lubricant, processing aid for ethylene/acrylic elastomers to improve release; lt. straw liq.; dens. 0.98 ± 0.02 mg/m³; visc. 160-180 cps; flash pt. (PM) 93 C; pH 2.0-2.5 (1%).

Vanfre® VAM. [R.T. Vanderbilt] Org. phosphate ester free acid; processing aid for VAMAC, ethylene/acrylic elastomer, other solv.-resistant polymers; tan to lt. brn. waxy solid; dens. 0.97 ± 0.03 mg/m³ (60 C); acid no. 98-110; pH 3.0-4.0 (1%).

Vanox® 3C. [R.T. Vanderbilt] N-Isopropyl-N′-phenyl-p-phenylenediamine; CAS 101-72-4; antioxidant, antiozonant for dk. colored NR and SR; purple-bk. flakes; very sol. in alcohol; mod. sol. in toluene; insol. in water; m.w. 226.4; dens. 1.10 ± 0.03 mg/m³; m.p. 70-79 C.

Vanox® 6H. [R.T. Vanderbilt] N-Cyclohexyl-N′-phenyl-p-phenylenediamine; CAS 101-87-1; antioxidant, antiozonant for dk. colored NR and SR; very sol. in acetone; mod. sol. in toluene; insol. in water; gray-violet powd.; m.w. 264.4; dens. 1.18 ± 0.03 mg/m³; m.p. 110-120 C.

Vanox® 12. [R.T. Vanderbilt] p,p′-Dioctyldiphenylamine; CAS 101-67-7; nonstaining, nonblooming antioxidant for NR, SR, CR elastomers used in adhesives, hot melts, latex, hose, belting, wire and cable, footwear applics.; inhibits oxygen attack, improves heat aging; off-wh. gran.; sol. in alcohol, toluene, gasoline; insol. in water; m.w. 393.66; dens. 1.01 ± 0.03 mg/m³; m.p. 94-100 C.

Vanox® 13. [R.T. Vanderbilt] Modified polyalkyl phosphited polyphenols; nonstaining, nonblooming antioxidant for NR, SR, TPR for adhesives, cements, hot melts; gel inhibitor polymer stabilizer; lt. amber liq.; sol. in chloroform, toluene, xylene, and petrol. hydrocarbons; pract. insol. in water; dens. 0.93 ± 0.02 mg/m³.

Vanox® 100. [R.T. Vanderbilt] Phenolic-type; nonstaining, nonblooming antioxidant for NR, SR, TPR, and peroxide-crosslinked chlorinated polyethylene; for footwear, adhesives, cements, hot melts; inhibits oxygen attack, improves heat aging; amber liq.; sol. in toluene, acetone, chloroform; pract. insol. in water; dens. 1.03 ± 0.02 mg/m³.

Vanox® 102. [R.T. Vanderbilt] Styrenated phenol-type; CAS 61788-44-1; nonstaining, nonblooming antioxidant for NR, SR for adheisves, cements, hot melts, latex applics.; gel inhibitor polymer stabilizer; lt. straw liq.; sol. in petrol. hydrocarbons, alcohols, esters; pract. insol. in water; dens. 1.08 ± 0.02 mg/m³.

Vanox® 200. [R.T. Vanderbilt] Amine blend; nonblooming antioxidant for NR, SR for tires, automotive and appliance molded goods; improves heat aging, inhibits environmental flex and stress cracking; dk. brn. liq.; negligible sol. in water; dens. 1.13 mg/m³; flash pt. (COC) 171 C; *Toxicology:* mists or vapors may be harmful; may cause irritation to skin and eyes; residual diphenylamine: OSHA TLV/TWA 10 mg/m³, systemic poison, irritant; residual triphenylphosphite; *Precaution:* incompat. with strong oxidizing agents, acids, water; hazardous decomp. prods.: oxides of N and C, aromatic/aliphatic hydrocarbons; *Storage:* store below 66 C.

Vanox® 1001. [R.T. Vanderbilt] Alkylated diphenylamines; CAS 68608-77-5; rubber chemical; amber liq.; negligible sol. in water; dens. 0.935 mg/m³; flash pt. (COC) 229 C; *Toxicology:* LD50 (oral, rat) 34,000 mg/kg, (dermal, rabbit) 3000 mg/kg; mild skin irritant; *Precaution:* incompat. with strong oxidizing agents; hazardous decomp. prods.: oxides of C and N, sm. amts. of aromatic and aliphatic hydrocarbons; *Storage:* store @ R.T. to prevent crystallization; keep container closed.

Vanox® 1005. [R.T. Vanderbilt] Alkylated-arylated bisphenolic phosphite; antioxidant for PS; gel inhibitor during polymerization; polymer stabilizer in syn. rubber mfg.; amber visc. liq.; sol. in benzene, chloroform, gasoline; pract. insol. in water; dens. 0.955-0.975 mg/m³; visc. 200-700 cps (60 C).

Vanox® 1030. [R.T. Vanderbilt] Phenolic-thio-ester; antioxidant for PP; wh. powd.; dens. 1.04 mg/m³; m.p. 61 C min.

Vanox® 1081. [R.T. Vanderbilt] p,p′-Dioctyldiphenylamine; CAS 101-67-7; antioxidant for hot melts; off-wh. gran.; sol. in oils, gasoline, acetone, alcohol, and toluene; pract. insol. in water; dens. 1.02 ± 0.03 mg/m³; m.p. 94-100 C.

Vanox® 1290. [R.T. Vanderbilt] 2,2′-Ethylidenebis (4,6-di-t-butylphenol); CAS 35958-30-6; nonstaining, nonblooming antioxidant for NR, SR; oxidative inhibitor for polymers; process stabilizer for polyolefins; stabilizer for PU and PS; latex applics.; wh. cryst. powd.; sol. in acetone, heptane, toluene, water; dens. 1.01 mg/m³; m.p. 161-164 C; flash pt. (COC) +380 F; 99% assay.

Vanox® 1320. [R.T. Vanderbilt] 2,6-Di-t-butyl-4-s-butyl phenol; CAS 17540-75-9; nonstaining, nonblooming antioxidant for NR, SR; oxidation inhibitor and scorch preventer for mfg. and storage of bun stock; stabilizer for PU foam; gel inhibitor polymer stabilizer; latex applics.; straw yel. liq.; sol. in oil, acetone, alcohol, heptane, toluene; insol. in water; dens. 0.93 ± 0.02 mg/m³; visc. 3.30 cSt (40 C); flash pt. (TCC) 94 C min.

Vanox® 2246. [R.T. Vanderbilt] 2,2′-Methylenebis (4-methyl-6-t-butylphenol); CAS 119-47-1; nonstaining, nonblooming antioxidant for NR, SR for hose and belting, footwear, adhesives, cements, hot melts, latex applics.; gel inhibitor polymer stabilizer; powd.; m.p. 125-130 C.

Vanox® AM. [R.T. Vanderbilt] Diphenylamine and acetone low-temp. reaction prod.; CAS 9003-79-6; antioxidant for NR and SR; lt. grn. tan powd.; 99.0% min. thru 100 mesh; very sol. in acetone, toluene; mod. sol. in alcohol; insol. in water; dens. 1.15 ± 0.03 mg/m³; soften. pt. (B&R) 85-97 C.

Vanox® AT. [R.T. Vanderbilt] Butyraldehyde-aniline condensation prod.; CAS 68411-20-1; nonstaining antioxidant for CR, NR, SBR, EPDM stocks, adhesives, cements, latexes; activator for thiuram and thiazoles; inhibits oxygen attack, environmental flex and stress cracking, improves heat aging; amber liq.; dens. 1.02 ± 0.02 mg/m³; flash pt. (PM) 80 C.

Vanox® GT. [R.T. Vanderbilt] Tris(3,5-di-t-butyl-4-

hydroxy benzyl) isocyanurate; CAS 27676-62-6; nonstaining, nonblooming antioxidant for NR, SR, TPR, PP, polyethylene, PU; for hot melts, latex applics.; improves heat aging; wh. powd.; sol. in toluene, acetone, chloroform, dimethylformamide; pract. insol. in water; m.w. > 500; dens. 1.03 mg/m^3; m.p. 217-225 C; flash pt. (COC) 553 F.

Vanox® HT. [R.T. Vanderbilt] Antioxidant blend contg. zinc; antioxidant for rubber; lt. gray powd., char. odor; insol. in water; dens. 1.54 mg/m^3; *Toxicology:* mild transient eye irritant; *Precaution:* incompat. with strong oxidizing agents; hazardous decomp. prods.: oxides of nitrogen, sulfur, carbon, and zinc upon combustion; may form explosive dust-air mixts.

Vanox® MBPC. [R.T. Vanderbilt] 2,2′-Methylene-bis(6-t-butyl-4-methylphenol); CAS 119-47-1; EINECS 204-327-1; rubber antioxidant; wh. to cream powd.; negligible sol. in water; dens. 1.08 mg/m^3; m.p. 125 C; *Toxicology:* LD50 (oral, rat) 5000 mg/kg; may cause eye irritation; *Precaution:* dust can be explosive if mixed with air in presence of ignition source; incompat. with strong oxidizing agents; hazardous decomp. prods.: CO_x at combustion temps.

Vanox® MTI. [R.T. Vanderbilt] 2-Mercaptotoluimidazole; CAS 53988-10-6; nonstaining, nondiscoloring, nonvolatile antioxidant for NR, SBR, NBR, EPDM; synergistic with Agerite antioxidants; outstanding for high temp. resist. and flex fatigue resist.; for hose and belting, automotive and appliance molded goods, wire and cable, latex applics.; lt. tan powd.; 99.9% min. thru 100 mesh; mod. sol. in acetone, alcohol; insol. in water; m.w. 165.25; dens. 1.33 ± 0.03 mg/m^3; m.p. 250 C min.

Vanox® NBC. [R.T. Vanderbilt] Nickel di-n-butyldithiocarbamate; CAS 13927-77-0; nonstaining antioxidant-antiozonant for SBR, NBR, CR, CSM, and ECO for hose and belting, automotive and appliance molded goods, wire and cable applics.; improves heat aging; dk. grn. cryst. powd.; 99.9% thru 20 mesh; very sol. in chloroform, toluene; mod. sol. in acetone; insol. water; m.w. 467.47; dens. 1.26 ± 0.03 mg/m^3; m.p. 86-90 C; 11.8-12.8% Ni.

Vanox® SKT. [R.T. Vanderbilt] 3,5-Di-t-butyl-4-hydroxyhydrocinnamic acid triester of 1,3,5-tris (2-hydroxyethyl)-s-triazine-2,4,6-(1H,3H,5H)-trione; CAS 34137-09-2; nonstaining, nonblooming antioxidant for NR, SR, TPR for adhesives, cements, hot melts, latex; stabilizer for polyolefins; food pkg. applics.; wh. cryst. powd.; sol. in aromatics; m.w. 1042; dens. 0.921 mg/m^3; m.p. 123-131 C; flash pt. (COC) 512 F; fire pt. (COC) 554 F.

Vanox® SWP. [R.T. Vanderbilt] 4,4′-Butylidene-bis-(6-t-butyl-m-cresol); CAS 85-60-9; antioxidant for natural and syn. latexes; nonstaining; wh. fine cryst.; very sol. in alcohol; mod. sol. in toluene; insol. in water; m.w. 382.5; dens. 1.03 ± 0.03 mg/m^3; m.p. 210-218 C.

Vanox® ZMTI. [R.T. Vanderbilt] Zinc 2-mercapto-toluimidazole; CAS 61617-00-3; nonstaining, nondiscoloring, nonvolatile antioxidant for NR, neoprene, SR, EPDM, nitrile stock; synergist with amines or phenols; outstanding for high temp. resist. and flex fatigue resist.; lt. tan powd.; 99.0% min. thru 100 mesh; sol. in methanol, ethanol; pract. insol. in water; m.w. 393.85; dens. 1.69 ± 0.03 mg/m^3; m.p. > 300 C; 15.5% min. Zn.

Vanox® ZS. [R.T. Vanderbilt] Fortified hindered phenol; nonstaining, nonblooming antioxidant for NR, SBR, IIR, NBR and their latexes; gel inhibitor during SBR processing; for tires, hose and belting, adhesives, cements, latex; wh. to pale yel. powd.; 99.0% thru 100 mesh; sl. sol. in most org. solvs.; insol. in water; dens. 1.30 ± 0.03 mg/m^3.

Vansil® W-10. [R.T. Vanderbilt] Wollastonite; CAS 13983-17-0; extender pigment/filler for solvent-thinned and latex paints, rubber applics.; wh. powd.; 18.5 μm median particle size; 97.3% < 200 mesh; dens. 2.90 mg/m^3; bulk dens. 80 lb/ft^3 (compacted); oil absorp. 19; brightness 89; pH 9.8 (10% aq. slurry); 49.83% SiO_2, 44.03% CaO.

Vansil® W-20. [R.T. Vanderbilt] Wollastonite; CAS 13983-17-0; extender pigment/filler for solvent-thinned and latex paints and rubber applics.; wh. powd.; 9.6 μm median particle size; 98% < 325 mesh; dens. 2.90 mg/m^3; bulk dens. 74 lb/ft^3 (compacted); oil absorp. 20; brightness 89; pH 9.8 (10% aq. slurry); 49.83% SiO_2, 44.03% CaO.

Vansil® W-30. [R.T. Vanderbilt] Wollastonite; CAS 13983-17-0; extender pigment/filler for solvent-thinned and latex paints, rubber applics.; wh. powd.; 6.1 μm median particle size; 99.9% < 325 mesh; dens. 2.90 mg/m^3; bulk dens. 62 lb/ft^3 (compacted; oil absorp. 21; brightness 89; pH 9.8 (10% aq. slurry); 49.83% SiO_2, 44.03% CaO.

Vanstay SC. [R.T. Vanderbilt] Phosphite org. complex; chelator/stabilizer which significantly improves clarity and early color; can be used with all primary Ba/Cd and Ba/Cd/Zn stabilizers for clear semirigid and plasticized PVC formulations which are calendered and extruded; water-wh. liq.

Vanstay SD. [R.T. Vanderbilt] Org. phosphite; low cost chelator/stabilizer/synergist; used with Ba/Cd/Zn and Ba/Cd stabilizers in flexible PVC; water-wh. liq.

Vanstay SG. [R.T. Vanderbilt] Org. phosphite; chelator/stabilizer/synergist to improve initial long term, low-temp. aging; used with Ba/Cd and Ba/Cd/Zn stabilizers; water-wh. liq.

Vanstay SH. [R.T. Vanderbilt] Org. phosphite; low cost, general purpose chelator/stabilizer/synergist for use with all Ba/Cd and Ba/Cd/Zn stabilizers; water-wh. liq.

Vantalc® 6H. [R.T. Vanderbilt] Talc (hydrous magnesium silicate); CAS 14807-96-6; EINECS 238-877-9; high purity filler for rubber applics., pigment for coatings; high hydrophobicity; mixes quickly into elastomers, gives low moisture pickup in vulcanizates; FDA 21CFR §175.300, 177.1210, 182.90; wh. powd.; 2.0 μm median particle size; 99.9% < 325 mesh; dens. 2.7 mg/m^3; bulk dens. 14 lb/ft^3 (compacted); oil absorp. 52; brightness 87; pH

9.6 (10% aq. slurry); 61.8% SiO_2; 31.3% MgO.

Vantard® PVI. [R.T. Vanderbilt] N-(Cyclohexylthio)phthalimide; CAS 17796-82-6; prevulcanization inhibitor; lt. tan to wh. cryst. and pellets, sl. mercaptan odor; insol. in water; dens. 1.3 mg/m^3; m.p. 90-95 C; flash pt. (PMOC) 165 C; *Toxicology:* LD50 (oral, rat) 2600 mg/kg, (dermal, rabbit) > 5010 mg/kg; eye and skin irritant, allergic sensitizer; inh. of odor after skin contact may cause nausea; *Precaution:* incompat. with acids; slowly degrades above 156 C; hazardous decomp. prods.: oxides of carbon, sulfur, and nitrogen upon combustion; *Storage:* keep container closed; store in cool, dry place away from foodstuffs and acids; 1 yr max. storage life.

Vanwax® H. [R.T. Vanderbilt] Protective wax for sunchecking inhibition of elastomerics; cream to lt. yel. flakes; mod. sol. in toluene, chloroform; pract. insol. in alcohol; insol. water; dens. 0.90 ± 0.03 mg/m^3; drop pt. 70-80 C.

Vanwax® H Special. [R.T. Vanderbilt] Petrol. wax; CAS 8002-74-2; protectant for elastomers exposed to sunlight and/or ozone under static conditions; wh. flakes; mod. sol. in toluene, chloroform; pract. insol. alcohol; insol. water; dens. 0.93 ± 0.03 mg/m^3; m.p. 65 C min.

Vanwax® OZ. [R.T. Vanderbilt] Wax blend; sunchecking inhibitor for elastomerics; wh. to lt. cream flakes; sol. in chloroform, toluene, other aromatics; insol. water and alcohol; dens. 0.90 ± 0.03 mg/m^3; drop pt. 58-63 C.

Varcum® 29008. [Occidental/Durez] Thermoplastic phenolic novolac; nonheat reactive tackifier for chlorobutyl and other syn. rubbers; flake; sol. in aromatics, esters, ketones; partially sol. in aliphatics, alcohols; soften. pt. (R&B) 125-140 C.

Varcum® 29427. [Occidental/Durez] Thermoplastic phenolic novolac; nonheat reactive tackifier for SBR and butyl rubbers; flake; sol. in aromatics, aliphatics, esters, ketones; partially sol. in alcohols; soften. pt. (R&B) 95-105 C.

Varcum® 29726. [Occidental/Durez] Epoxy resin; powdered epoxy resin for use as modifier to soften phenolic resins; shellac substitute; coarse particles; soften. pt. (Cap.) 60-70 C.

Varcum® 29421. [Occidental/Durez] Phenolic novolac resin; thermoplastic, nonheat reactive tackifier and processing aid for SBR, butyl, and chlorobutyl rubber; flaked; sol. (equal parts solv. and resin) in aromatics, aliphatics, esters, and ketones; soften. pt. (R&B) 85-105 C.

Varox® 130. [R.T. Vanderbilt] 2,5-Dimethyl-2,5-di(t-butylperoxy)-3-hexyne; CAS 1068-27-5; vulcanizing agent for EPDM, EPM, and CPE; crosslinking agent for PE and EVA; lt. yel. liq.; m.w. 286.5; dens. 0.89 ± 0.02 mg/m^3; 90-95% assay; 10.05-10.65% act. oxygen.

Varox® 130-XL. [R.T. Vanderbilt] 2,5-Dimethyl-2,5-di(t-butylperoxy)-3-hexyne; CAS 1068-27-5; vulcanizing agent for EPDM, EPM, and CPE; crosslinking agent for PE and EVA; wh. powd., 99% min. thru 100 mesh; m.w. 286.5; dens. 1.26 ± 0.03 mg/m^3; 45-55% assay; 4.52-5.83% act. oxygen.

Varox® 230-XL. [R.T. Vanderbilt] n-Butyl-4,4-bis (t-butylperoxy) valerate; CAS 995-33-5; crosslinking agent for elastomers, e.g., polybutadiene, CPE, EPDM; wh. powd.; 99% min. thru 100 mesh; m.w. 334.46; dens. 0.95 ± 0.03 mg/m^3; 39-41% assay; 3.83% min. act. oxygen.

Varox® 231. [R.T. Vanderbilt] 1,1-Bis (t-butylperoxy)-3,3,5-trimethylcyclohexane; CAS 6731-36-8; vulcanizing agent for SBR, NBR, EPDM, EPM, CPE, and silicones; crosslinking agent for PE and EVA; lt. yel. liq.; m.w. 302.46; dens. 0.91 ± 0.02 mg/m^3; 90% min. assay; 9.52% min. act. oxygen.

Varox® 231-XL. [R.T. Vanderbilt] 1,1-Bis (t-butylperoxy)-3,3,5-trimethylcyclohexane; CAS 6731-36-8; vulcanizing agent for SBR, NBR, EPDM, EPM, CPE, and silicones; crosslinking agent for PE and EVA; wh. powd.; 99% min. thru 100 mesh; m.w. 302.46; dens. 1.41 ± 0.03 mg/m^3; 39-41% assay; 3.8% min. act. oxygen.

Varox® 802-40KE. [R.T. Vanderbilt] 40% 2,2′-Bis(t-butylperoxy) diisopropyl benzene on KE clay; nonsulfur curing agent for NR and SR; crosslinking agent for PE, EPDM, EPM, EPR, silicone, EVA, CPA, VAE; provides exc. compr. set; wh. to cream powd., odorless; 1% max. on 100 mesh; insol. in water; dens. 1.50 mg/m^3; bulk dens. 31.4 lb/ft^3; flash pt. (Seta CC) 129 C; 40.5 ± 1% assay; 3.74-3.93% act. oxygen; *Toxicology:* LD50 (oral, rat) > 23,000 mg/kg; may irritate eyes, skin, respiratory system; dust inh. may cause sl. to mod. hyperemia in lungs, necropsy; clay: nuisance dust std. ACGIH TLV/TWA 30 mg/m^3; *Precaution:* org. peroxide; heat or contamination may cause gaseous decomp.; dust explosion hazard; thermally unstable; incompat. with strong acids, alkalies, oxidizers, copper, lead, iron; decomp. prods. are flamm. (methane, ethane, acetone, alcohols, ketone).

Varox® 802-40MB. [R.T. Vanderbilt] 40% 1,3- and 1,4-Di(t-butylperoxy) isopropyl benzenes in EPM rubber; nonsulfur curing agent, crosslinking agent; nonvolatile; yel. pellets, sl. odor; negligible sol. in water; dens. 0.92 mg/m^3; bulk dens. 57.4 lb/ft^3; *Toxicology:* peroxide: LD50 (oral, rat) > 23.1 g/kg; may cause sl. eye and skin irritation; *Precaution:* combustible; org. peroxide; thermally unstable; incompat. with strong acids, oxidizers, reducing agents; hazardous decomp. prods.: acetone, methane, ethane, t-butanol, CO_2; *Storage:* keep container closed; keep away from heat, sparks, flame; store below 38 C; do not reuse container.

Varox® ANS. [R.T. Vanderbilt] Benzoyl peroxide; CAS 94-36-0; EINECS 202-327-6; polymerization initiator.

Varox® DBPH. [R.T. Vanderbilt] 2,5-Dimethyl-2,5-di(t-butylperoxy) hexane; CAS 78-63-7; vulcanizing and crosslinking agent for elastomers and polyolefins, e.g., EPDM, EPM, PE, and NBR; water-wh. to lt. yel. liq.; m.w. 290.45; dens. 0.87 ± 0.02 mg/m^3; 90% min. assay; 9.92% min. act. oxygen.

Varox® DBPH-50. [R.T. Vanderbilt] 2,5-Dimethyl-2,5-di(t-butylperoxy) hexane on inert min. carrier; CAS

78-63-7; vulcanizing and crosslinking agent for elastomers and polyolefins, e.g., EPDM, EPM, PE, and NBR; wh. powd.; 99.9% thru 100 mesh; m.w. 290.45; dens. 1.45 ± 0.03 mg/m^3; 45% min. assay; 4.96% min. act. oxygen.

Varox® DCP-40C. [R.T. Vanderbilt] Dicumyl peroxide on calcium carbonate; cross-linking agent; wh. powd.; dens. 1.60 ± 0.03 mg/m^3; 39-41% assay.

Varox® DCP-40KE. [R.T. Vanderbilt] Dicumyl peroxide on KE clay; cross-linking agent; wh. powd.; dens. 1.61 ± 0.03 mg/m^3; 39-41% assay.

Varox® DCP-40MB. [R.T. Vanderbilt] Dicumyl peroxide on polymeric binder; cross-linking agent; 40% act.

Varox® DCP-R. [R.T. Vanderbilt] Dicumyl peroxide; CAS 80-43-3; cross-linking agent for PE, EPR, silicone, EVA, and VAE; nonsulfur curing agent for NR, IR, SBR, BR, NBR, EPDM, and CR; pale yel. recryst. cryst. solid; m.w. 270.37; dens. 1.001 mg/m^3; 99% min. assay.

Varox® DCP-T. [R.T. Vanderbilt] Dicumyl peroxide; CAS 80-43-3; cross-linking agent for PE, EPR, silicone, EVA, and VAE; nonsulfur curing agent for NR, IR, SBR, BR, NBR, EPDM, and CR; pale yel. semicryst. solid; m.w. 270.37; dens. 0.977-1.009 mg/m^3 (40 C); 91% min. assay.

Varox® TBPB. [R.T. Vanderbilt] t-Butyl perbenzoate; CAS 614-45-9; polymerization initiator.

Varsulf® NOS-25. [Witco/H-I-P] Sodium alkylphenol polyglycol ether sulfate; anionic; emulsifier for emulsion polymerization; skin care prods., hair conditioners; lt. yel. clear liq.; pH 7.0 (1%); 35% conc.

Vazo 52. [DuPont] 2,2′-Azobis(2,4-dimethylvaleronitrile); CAS 4419-11-8; EINECS 224-583-8; initiator for bulk, sol'n., emulsion, and suspension polymerization of vinyl monomers; polymerization of unsat. polyesters and copolymerization of vinyl compds.; source of radicals for vinyl polymerizations and chain reactions; decomposes in solvs. to give free radicals with no evidence of induced chain decomposition; used in pigmented or dyed systems that may be susceptible to oxidative degration; wh. cryst. solid; m.w. 248.37; sol. in functional compds. and aromatic hydrocarbons; insol. in water; dens. 400 kg/m^3; 98% min. act.

Versamag® DC. [Morton Int'l./Plastics Additives] Magnesium hydroxide; CAS 1309-42-8; EINECS 215-170-3; inorg. filler providing flame retardancy and smoke suppression to elastomers, plastics, and thermosets incl. EPDM, PP, PE, PVC; used in wire and cable compds., conduit/tubing, film and sheet; micropellets; sp.gr. 2.36; bulk dens. 25 lb/ft^3 (tamped); surf. area 18 m^2/g; ref. index 1.57; 96% assay.

Versamag® Tech. [Morton Int'l./Plastics Additives] Magnesium hydroxide; CAS 1309-42-8; EINECS 215-170-3; inorg. filler providing flame retardancy and smoke suppression to elastomers, plastics, and thermosets incl. EPDM, PP, PE, PVC; used in wire and cable compds., conduit/tubing, film and sheet; wh. powd.; 2 μ mean particle size; 0.5% retained on 325 mesh; sp.gr. 2.36; bulk dens. 14 lb/ft^3 (tamped); surf. area 18 m^2/g; ref. index 1.57; 96% assay.

Versamag® UF. [Morton Int'l./Plastics Additives] Magnesium hydroxide; CAS 1309-42-8; EINECS 215-170-3; fire retardant/smoke suppressant filler for plastics and elastomers incl. flexible and rigid PVC; wh. ultra fine powd.; 70% < 2 μ; 1 μ mean particle size; sp.gr. 2.36; bulk dens. 28.1 lb/ft^3 (tamped); surf. area 22 m^2/g; ref. index 1.57; 98.5% assay.

Versamid® 125. [Henkel/Coatings & Inks] Polyamide resin; epoxy curing agent offering chem. and corrosion resist., strong adhesion to a variety of substrates; for coatings, castings, potting, laminating, adhesives; Gardner 8 max. color; sp.gr. 0.97; dens. 8.1 lb/gal; visc. 6.5-9.5 poise (75 C); amine no. 330-360; tens. str. 8000 psi; tens. elong. 8%; distort. temp. 77 C; gel time 1 h 40 min; thru cure 24 h @ R.T.; 100% solids.

Versamid® 140. [Henkel/Coatings & Inks] Polyamide resin; epoxy curing agent offering chem. and corrosion resist., strong adhesion to a variety of substrates; for coatings, castings, potting, laminating, adhesives; Gardner 8 max. color; sp.gr. 0.97; dens. 8.1 lb/gal; visc. 80-120 poise; amine no. 370-400; tens. str. 8000 psi; tens. elong. 10%; distort. temp. 85 C; gel time 2 h 20 min; thru cure 24 h @ R.T.; 100% solids.

Versamid® 150. [Henkel/Coatings & Inks] Polyamide resin; epoxy curing agent offering chem. and corrosion resist., strong adhesion to a variety of substrates; used for coatings, castings, potting, laminating, adheisves; Gardner 8 max. color; sp.gr. 0.96; dens. 8.0 lb/gal; visc. 20-40 poise; amine no. 370-400; tens. str. 6000 psi; tens. elong. 10%; distort. temp. 85 C; gel time 1 h 50 min; thru cure 24 h @ R.T.; 100% solids.

Versamid® 674. [Henkel/Coatings & Inks] Polyamide resin; epoxy curing agent offering chem. and corrosion resist., strong adhesion to a variety of substrates; used for coatings, castings, potting, laminating, and adhesives; Gardner 9 max. color; sp.gr. 0.97; dens. 8.1 lb/gal; visc. 3-6 poise (75 C); amine no. 330-360; tens. str. 8000 psi; tens. elong. 8%; distort. temp. 72 C; gel time 2 h; thru cure 24-36 h @ R.T.; 100% solids; Unverified

Versamine® 900. [Henkel/Coatings & Inks] Modified aliphatic amine; epoxy curing agent offering faster cure times, lower reaction temps.; used for laminates, fast-set adhesives, castings, and tooling compds.; blendable with Genamid grades to provide high impact, flexible, high-build coatings and adhesives; Gardner 2 max. color; sp.gr. 1.02; dens. 8.5 lb/gal; visc. 2.9-8 poise; amine no. 900-1100; tens. str. 10,400 psi; tens. elong. 11%; distort. temp. 105 C; gel time 24 min; thru cure 10 h @ R.T.; 100% solids.

Versamine® 908. [Henkel/Coatings & Inks] Modified aliphatic amine; epoxy curing agent offering faster cure times, lower reaction temps.; for laminates, fast-set adhesives, castings, and tooling compds.;

used to accelerate Genamid 250, 490, 747; Gardner 4 max. color; sp.gr. 1.02; dens. 8.5 lb/gal; visc. 0.8-1.4 poise; amine no. 950-1200; tens. str. 9000 psi; tens. elong. 6%; distort. temp. 114 C; gel time 12 min; thru cure 8 h @ R.T.; 100% solids.

Versamine® 911. [Henkel/Coatings & Inks] Modified aliphatic amine; epoxy curing agent offering faster cure times, lower reaction temps.; med. blush epoxy hardener for fast-set and good flexibility; Gardner 3 max. color; sp.gr. 0.97; dens. 8.1 lb/gal; visc. 40-85 poise; amine no. 300-335; tens. str. 2500 psi; tens. elong. 15%; distort. temp. 40 C; gel time 7 min; thru cure 7 h @ R.T.; 100% solids.

Versamine® 1200. [Henkel/Coatings & Inks] Modified aliphatic amine; epoxy curing agent offering faster cure times, lower reaction temps.; for decorative epoxy flooring, fast-setting epoxy grouts, mortars, and thin-film high-solid industrial tooling; Gardner 4 max. color; sp.gr. 0.96; dens. 8.0 lb/gal; visc. 3.5-6.5 poise; amine no. 300-340; distort. temp. 60 C; gel time 14 min; thru cure 6 h @ R.T.; 100% solids; Unverified

Versamine® A50. [Henkel/Coatings & Inks] Aliphatic amine adduct; epoxy curing agent offering faster cure times, lower reaction temps.; for laminates, adhesives, castings, and tooling compds.; Gardner 3 max. color; sp.gr. 1.09; dens. 9.1 lb/gal; visc. 35-135 poise; amine no. 795-895; tens. str. 10,000 psi; tens. elong. 2.4%; distort. temp. 102 C; gel time 20 min; thru cure 15 h @ R.T.; 100% solids; Unverified

Versamine® A51. [Henkel/Coatings & Inks] Aliphatic amine adduct; epoxy curing agent offering faster cure times, lower reaction temps.; for solv.-based coatings with solid epoxy resins; Gardner 4 max. color; sp.gr. 1.00; dens. 8.3 lb/gal; visc. 6-9 poise; amine no. 166-240; thru cure 16-18 h @ R.T.; 48% solids; Unverified

Versamine® A52. [Henkel/Coatings & Inks] Aliphatic amine adduct; epoxy curing agent offering faster cure times, lower reaction temps.; for solv.-based coatings with solid epoxy resins; Gardner 4 max. color; sp.gr. 0.98; dens. 8.2 lb/gal; visc. 4-7 poise; amine no. 166-240; thru cure 16-18 h @ R.T.; 48% solids.

Versamine® A54. [Henkel/Coatings & Inks] Aliphatic amine adduct; epoxy curing agent for chem.-resist. coatings, castings, adhesives, and tooling compds.; fast cure, exc. chem. resist.; Gardner 7 max. color; dens. 8.9 lb/gal; visc. 32-45 poise; amine no. 800-900; After 7 days 25 C cure: tens. str. 12,400 psi; elong. 6%; flex. str. 15,800 psi; 100% solids.

Versamine® A55. [Henkel/Coatings & Inks] Modified aliphatic amine adduct; epoxy curing agent for chem.-resist. coatings, castings; lt. color, blush resist., exc. chem. resist.; Gardner 4 max. color; dens. 8.8 lb/gal; visc. 11-17 poise; amine no. 616-693; After 7 days 25 C cure: tens. str. 7400 psi; elong. 4%; flex. str. 20,100 psi; 87.4% solids.

Versamine® EH-30. [Henkel/Coatings & Inks] Tert. amine; accelerator for amine-cured epoxies; catalyst for accelerated mercaptan-cured epoxies; seldom used as sole curing agent; Gardner 8 max. color; sp.gr. 0.97; dens. 8.1 lb/gal; visc. 1.8-3.2 poise; amine no. 590-630; gel time 40 min (10 phr with 190 EEW Epoxy); 100% solids.

Versamine® EH-50. [Henkel/Coatings & Inks] Tert. amine; lt. color, low odor, formaldehyde-free accelerator for amine- or mercaptan-cured epoxies; Gardner 2 max. color; dens. 8.0 lb/gal; visc. 3-3.5 poise; amine no. 480-520; 100% solids.

Versamine® I-70. [Henkel/Coatings & Inks] Isolated amine adduct; epoxy curing agent offering fast cure, exc. chem. resist., lightfastness for coatings, castings, adhesives, and tooling compds.; Gardner 5 max. color; sp.gr. 1.07; dens. 8.95 lb/gal; visc. 30-50 poise; amine no. 755-835; tens. str. 10,200 psi; tens. elong. 18-21%; distort. temp. 75-97 C; gel time 15 min; thru cure 15 h @ R.T.; 100% solids.

Versene 100. [Dow] Tetrasodium EDTA; CAS 64-02-8; EINECS 200-573-9; chelating agent controlling trace metal ions to improve lathering and stability of personal care prods., pharmaceuticals, in water treatment, textiles, soaps/detergents, electroless copperplating, polymer prod., disinfectants, pulp/paper, enhanced oil recovery, metal cleaning and protection; for control of common heavy metal ions to pH 12, iron to pH 8, water hardness ions above pH 6; lt. straw-colored liq.; m.w. 1.290-1.325; dens. 10.9 lb/gal; pH 11.0-11.8 (1% aq.); chel. value 102; 39% act.

Versene 100 EP. [Dow] Tetrasodium EDTA; CAS 64-02-8; EINECS 200-573-9; chelating agent controlling trace metal ions to improve lathering and stability of personal care prods., pharmaceuticals, in water treatment, textiles, soaps/detergents, electroless copperplating, foods, polymer prod., disinfectants, pulp/paper, enhanced oil recovery, metal cleaning and protection; high purity version of Versene 100; liq.

Versene 100 LS. [Dow] Tetrasodium EDTA; CAS 64-02-8; EINECS 200-573-9; chelating agent controlling trace metal ions to improve lathering and stability of personal care prods., pharmaceuticals, in water treatment, textiles, soaps/detergents, electroless copperplating, foods, polymer prod., disinfectants, pulp/paper, enhanced oil recovery, metal cleaning and protection; low solids grade used in aerosols; liq.

Versene 100 SRG. [Dow] Tetrasodium EDTA; CAS 64-02-8; EINECS 200-573-9; chelating agent controlling trace metal ions to improve lathering and stability of personal care prods., pharmaceuticals, in water treatment, textiles, soaps/detergents, electroless copperplating, foods, polymer prod., disinfectants, pulp/paper, enhanced oil recovery, metal cleaning and protection; pH adjusted; liq.

Versene 100 XL. [Dow] Tetrasodium EDTA; CAS 64-02-8; EINECS 200-573-9; chelating agent controlling trace metal ions to improve lathering and stability of personal care prods., pharmaceuticals, in water treatment, textiles, soaps/detergents, electroless copperplating, foods, polymer prod., disinfectants, pulp/paper, enhanced oil recovery, metal cleaning and protection; low NTA version of Versene 100; liq.

Versene 220. [Dow] Tetrasodium EDTA tetrahydrate; CAS 64-02-8; EINECS 200-573-9; chelating agent controlling trace metal ions to improve lathering and stability of personal care prods., pharmaceuticals, in water treatment, textiles, soaps/detergents, electroless copperplating, polymer prod., disinfectants, pulp/paper, enhanced oil recovery, metal cleaning and protection; high purity, crystal form of Versene 100; wh. cryst.; m.w. 452; dens. 45 lb/ft^3; pH 10.5-11.5; chel. value 219; 99% act.

Versene Acid. [Dow] EDTA; CAS 60-00-4; EINECS 200-449-4; chelating agent controlling trace metal ions to improve lathering and stability of personal care prods., pharmaceuticals, in water treatment, textiles, soaps/detergents, electroless copperplating, polymer prod., disinfectants, pulp/paper, enhanced oil recovery, metal cleaning and protection; intermediate for prep. of other salt forms of EDTA; wh. powd.; m.w. 292; dens. 54 lb/ft^3; pH 2.5-3.0 (sat. aq. sol'n.); chel. value 339; 99% act.

Versene CA. [Dow] Calcium-disodium EDTA; CAS 62-33-9; EINECS 200-529-9; chelating agent controlling trace metal ions to improve lathering and stability of personal care prods., pharmaceuticals, in water treatment, textiles, soaps/detergents, electroless copperplating, foods, polymer prod., disinfectants, pulp/paper, enhanced oil recovery, metal cleaning and protection; high purity direct food additive preventing metal catalyzed oxidative breakdown; also for medical and pharmaceutical preps.; FCC, USP, and kosher compliance; wh. powd.; m.w. 410; dens. 40 lb/ft^3; pH 6.5-7.5 (1%); 97-102% act.

Versene Diammonium EDTA. [Dow] Diammonium EDTA; CAS 20824-56-0; EINECS 244-063-4; chelating agent controlling trace metal ions to improve lathering and stability of personal care prods., pharmaceuticals, in water treatment, textiles, soaps/detergents, electroless copperplating, polymer prod., disinfectants, pulp/paper, enhanced oil recovery, metal cleaning and protection; used when very sol. chelates are required or where sodium ions are undesirable; lt. straw-colored liq.; m.w. 326; sp.gr. 1.19-1.22; dens. 10.0 lb/gal; pH 4.6-5.2; chel. value 137; 44.6% act.

Versene NA. [Dow] Disodium EDTA dihydrate USP, FCC; CAS 139-33-3; EINECS 205-358-3; chelating agent controlling trace metal ions to improve lathering and stability of personal care prods., pharmaceuticals, in water treatment, textiles, soaps/detergents, electroless copperplating, foods, polymer prod., disinfectants, pulp/paper, enhanced oil recovery, metal cleaning and protection; high purity direct food additive preventing metal catalyzed oxidative breakdown; also for medical and pharmaceutical preps.; FCC, USP, and Kosher compliance; wh. cryst.; m.w. 372; dens. 67 lb/ft^3; pH 4.3-4.7 (1%); 99% act.

Versene Na$_2$. [Dow] Disodium EDTA dihydrate USP, FCC; CAS 139-33-3; EINECS 205-358-3; chelating agent used primarily in cosmetics and toiletries; crystals.

Versene Tetraammonium EDTA. [Dow] Tetraammonium EDTA; chelating agent controlling trace metal ions to improve lathering and stability of personal care prods., pharmaceuticals, in water treatment, textiles, soaps/detergents, electroless copperplating, polymer prod., disinfectants, pulp/paper, enhanced oil recovery, metal cleaning and protection; used when very sol. chelates are required or where sodium ions are undesirable; for chelation of iron to pH 9.5; lt. straw-colored liq.; m.w. 360; sp.gr. 1.16-1.19; dens. 9.7 lb/gal; pH 9.0-9.5; chel. value 130; 46.8% act.

Vertal 92. [Luzenac Am.] Platy talc; CAS 14807-96-6; EINECS 238-877-9; reinforcement in blk. PP compds.; dusting/parting agent for rubbers (tires, mech. goods); filler in autobody compds., caulks, putties, and sealants; powd.; 15 μ mean particle size; 92% thru 325 mesh; sp.gr. 2.8; bulk dens. 35 lb/ft^3 (loose); oil absorp. 33; brightness (GE) 72; pH 9.5 (10% slurry); 50-60% talc, 40-50% carbonates; 35.6% SiO_2, 32.9% MgO, 6.5% Fe_2O_3; 0.5% moisture.

Vertal 97. [Luzenac Am.] Talc; CAS 14807-96-6; EINECS 238-877-9; low cost plastics additive for general industrial and impact applics.; powd.; 10-14 μ median particle size; bulk dens. 27-37 lb/ft^3 (loose); brightness 70-78.

Vertal 200. [Luzenac Am.] Talc; CAS 14807-96-6; EINECS 238-877-9; plastics additive for general industrial, high impact applics.; powd.; 2.0 μ median particle size; fineness (Hegman) 6.0; bulk dens. 9-13 lb/ft^3 (loose); brightness 87.

Vertal 310. [Luzenac Am.] Talc; CAS 14807-96-6; EINECS 238-877-9; plastics additive for general industrial applics.; antiblock; powd.; 3.5 μ median particle size; fineness (Hegman) 5.5; bulk dens. 9-13 lb/ft^3 (loose); brightness 88.

Vertal 710. [Luzenac Am.] Talc; CAS 14807-96-6; EINECS 238-877-9; additive in filled PP to increase stiffness and modulus, reduce shrinkage, improve dimensional stability; dusting agent in rubber compds.; provides good gloss control and antisag props. in paint, exc. barrier props. in corrosion-resist. metal primers; powd.; 6.8 μ median diam.; fineness (Hegman) 4.0; sp.gr. 2.8; bulk dens. 20-24 lb/ft^3 (loose), 46 lb/ft^3 (tapped); oil absorp. 42; brightness 85; pH 9.0 (10% slurry); 95% talc; 58.7% SiO_2, 31.8% MgO.

Vertal C-2. [Luzenac Am.] Talc; CAS 14807-96-6; EINECS 238-877-9; dusting talc for elastomeric threads and rubber gloves; powd.; 10 μ mean particle size; brightness (GE) 82.

Vestamin® A139. [Hüls Am.] Blocked diamine; humidity-activated crosslinking agent for accelerating the hardening of isocyanate-contg. prepolymers; low-visc. liq.; dens. 0.86 g/cc.

Vestamin® IPD. [Hüls Am.] Isophorone diamine; CAS 2855-13-2; epoxy curative for coatings, adhesives, castings and composites; clear low visc. liq.; dens. 0.920-0.925 g/cc (20 C).

Vestamin® TMD. [Hüls Am.] Trimethylhexamethylenediamine; CAS 25620-58-0; epoxy curative for flexible coatings, adhesives, castings and composites; clear low-visc. liq.; dens. 0.865-0.870 g/cc (20 C).

Vestamin® V214. [Hüls Am.] Amine adduct; hardener for solv.-free epoxy systems requiring low visc.; colorless low-visc. liq.; dens. 0.92 g/cc (20 C).

Vestenamer® 6213. [Hüls Am.; Hüls AG] Polyoctenamer; nonstaining stabilizer; used as a blend component for other rubbers to improve plasticity, enhance filler incorporation and dispersion, improve flowability in extrusion, inj. and compr. molding, increase surf. smoothness in calendering operations; also for hose prod., tires, covulcanization; lt.-colored opaque gran.; dens. 0.89 g/cc; m.p. 33 C; < 62% trans-octenamer.

Vestenamer® 8012. [Hüls Am.; Hüls AG] Polyoctenamer; nonstaining stabilizer; blend component for other rubbers to improve plasticity, enhance filler incorporation and dispersion, improve flowability in extrusion, inj. and compr. molding, increase surf. smoothness in calendering operations; hose prod., tires, covulcanization; higher crystallinity than 6213; yields vulcanizates with good tens. str. and elasticity, high modulus and hardness, exc. abrasion resist. and low-temp. props., good resist. to aging and ozone; lt.-colored opaque gran.; weak odor; sol. in HC and halogenated HC; dens. 0.91 g/ml; m.p. 51 C; decomp. pt. 250 C; 80% trans-octenamer.

Vestolit® E 7012. [Hüls AG] Emulsion PVC; CAS 9002-86-2; emulsion PVC for paste processing; special prod. for chemical blowing; apparent bulk dens. 0.55 g/ml; visc. 125 ml/g; Unverified

Vestowax A-217. [Astor Wax] Syn. cryst. polyethylene wax from the Huels/Veba modification of the Ziegler synthesis; CAS 9002-88-4; EINECS 200-815-3; hard chemically inert lubricant for mold release compds., solv. and liq. polishes, solid wax compds., hot-melts, automobile rustproofing compds., solv.-dispersed agric. coatings, paper coating, inks, crayons, oils, elec., construction; processing aid, lubricant, mold release, aging protectant for plastics/rubber; decreases resin melt visc. in LDPE, PP; improves pigment disp. in color concs., filler and stabilizer disp.; improves mold flow and release in elastomers; FDA 21CFR §172.105, 172.125, 172.300, 172.320, 172.615, 172.888, 175.170, 175.180, 175.200, 175.210, 177.1200, 177.1210, 177.1320, 177.1520, 177.2600, 178.3720, 178.3850; wh. to ivory pellets; m.w. 1600; sp.gr. 0.92; visc. 100-160 cps (150 C); m.p. 115-125 C; congeal pt. 104-108 C; penetration 2-4; acid no. 0; sapon. no. 0; dielec. const. 2.1 (1 kc, 125 C); vol. resist. 5×10^{12} (300 V, 125 C).

Vestowax A-227. [Astor Wax] Syn. cryst. polyethylene wax from the Huels/Veba modification of the Ziegler synthesis; CAS 9002-88-4; EINECS 200-815-3; hard chemically inert lubricant for mold release compds., solv. and liq. polishes, solid wax compds., hot-melts, automobile rustproofing compds., solv.-dispersed agric. coatings, paper coating, inks, crayons, oils, elec., construction; processing aid, lubricant, mold release, aging protectant for plastics/rubber; decreases resin melt visc. in LDPE, PP; improves pigment disp. in color concs., filler and stabilizer disp.; improves mold flow and release in elastomers; wh. to ivory pellets, powd.; m.w. 2700; sp.gr. 0.92; visc. 180-240 cps (150 C); m.p. 102-110 C; congeal pt. 92-96 C; penetration 2-3; acid no. 0; sapon. no. 0; dielec. const. 2.0 (1 kc, 125 C); vol. resist. 1×10^{13} (300 V, 125 C).

Vestowax A-235. [Astor Wax] Syn. cryst. polyethylene wax from the Huels/Veba modification of the Ziegler synthesis; CAS 9002-88-4; EINECS 200-815-3; hard chemically inert lubricant for mold release compds., solv. and liq. polishes, solid wax compds., hot-melts, automobile rustproofing compds., solv.-dispersed agric. coatings, paper coating, inks, crayons, oils, elec., construction; processing aid, lubricant, mold release, aging protectant for plastics/rubber; decreases resin melt visc. in LDPE, PP; improves pigment disp. in color concs., filler and stabilizer disp.; improves mold flow and release in elastomers; wh. to ivory pellets; m.w. 3500; sp.gr. 0.92; visc. 300-380 cps (150 C0; m.p. 102-110 C; congeal pt. 91-95 C; penetration 1-2; acid no. 0; sapon. no. 0; dielec. const. 2.1 (1 kc, 125 C); vol. resist. 2×10^{13} (300 V, 125 C).

Vestowax A-415. [Astor Wax] Syn. cryst. polyethylene wax from the Huels/Veba modification of the Ziegler synthesis; CAS 9002-88-4; EINECS 200-815-3; hard chemically inert lubricant for mold release compds., solv. and liq. polishes, solid wax compds., hot-melts, automobile rustproofing compds., solv.-dispersed agric. coatings, paper coating, inks, crayons, oils, elec., construction; processing aid, lubricant, mold release, aging protectant for plastics/rubber; decreases resin melt visc. in LDPE, PP; improves pigment disp. in color concs., filler and stabilizer disp.; improves mold flow and release in elastomers; wh. to ivory pellets; m.w. 1500; sp.gr. 0.94; visc. 60-100 cps (150 C); m.p. 117-122 C; congeal pt. 104-108 C; penetration 1-2; acid no. 0; sapon. no. 0; dielec. const. 2.0 (1 kc, 125 C); vol. resist. 6×10^{13} (300 V, 125 C).

Vestowax A-616. [Astor Wax] Syn. cryst. polyethylene wax from the Huels/Veba modification of the Ziegler synthesis; CAS 9002-88-4; EINECS 200-815-3; hard chemically inert lubricant for mold release compds., solv. and liq. polishes, solid wax compds., hot-melts, automobile rustproofing compds., solv.-dispersed agric. coatings, paper coating, inks, crayons, oils, elec., construction; processing aid, lubricant, mold release, aging protectant for plastics/rubber; decreases resin melt visc. in LDPE, PP; improves pigment disp. in color concs., filler and stabilizer disp.; improves mold flow and release in elastomers; wh. to ivory pellets; m.w. 1600; sp.gr. 0.96; visc. 100-130 cps (150 C); m.p. 118-128 C; congeal pt. 109-113 C; penetration 1; acid no. 0; sapon. no. 0; dielec. const. 2.0 (1 kc,

125 C); vol. resist. 4 x 10^{13} (300 V, 125 C).

Vestowax AO-1539. [Astor Wax] Syn. cryst. oxidized Ziegler-type polyethylene wax; CAS 9002-88-4; EINECS 200-815-3; hard, high-melting lubricant wax; for anionic dry-bright emulsions; additive for polymer finishes; in powd. form as PVC lubricant component; wh. pellets, powd.; m.p. 105-110 C; congeal pt. 92-97 C; penetration 1-2; acid no. 20-25; sapon. no. 38-45.

Vestowax AS-1550. [Astor Wax] Oxidized polyethylene wax containing Ca soap; internal plastic lubricant; used for films, profiles, inj. molding, in dry-bright wax emulsions for floor maintenance and leather finishes; AS-1550P (powd. grade) as lubricant for clear, rigid PVC sheeting; yel. pellets; m.p. 102-107 C; congeal pt. 90-95 C; penetration 1-2; acid no. 22-26; sapon. no. 47-52.

Vestowax C-60. [Astor Wax] Oxidized wax; external plastic lubricant; used in pipe, sheet, profiles, and inj. molding PET bottles; produces smooth glossy surfaces; wh. powd.; congeal pt. 88-90 C; penetration 3-6; acid no. 27-32; sapon. no. 28-60.

Vestowax FT-150. [Astor Wax] Syn. wax; Fischer-Tropsch sat., syn. hard paraffins; CAS 8002-74-2; wax used in plastics processing for PVC lubrication, mold release compds. for PU and polyesters; slip agent in printing inks; in elec. and electronics for insulating compds., potting agents, nuclear reactor shielding; in hot-melt adhesives in paper coatings; in corrosion protection compds., polishes and maintenance compds., scratch-resistant paints, lacquers, varnishes, candle hardener and opacifier; FDA 21CFR §172.615; wh. pellets; m.w. 700; sol. @ elevated temps. in aliphatic and aromatic hydrocarbons, e.g., min. spirits, kerosene, benzol, turpentine, and all chlorinated solvs.; kinetic visc. ≈ 12 cstk (120 C); congeal. pt. 94-96 C; drop pt. 106-110 C; acid no. nil; iodine no. 0.1; sapon. no. nil; hardness (ball pressure) 305/265 kg/cm^2; dielec. const. 2.0 (125 C, 1 kc); vol. resist. 3 x 10^{13} (125 C, 300 V).

Vestowax FT-150P. [Astor Wax] Syn. wax; Fischer-Tropsch saturated, syn. hard paraffins; CAS 8002-74-2; wax used in plastics processing for PVC lubrication, mold release compds. for PU and polyesters; slip agent in printing inks; in elec. and electronics for insulating compds., potting agents, nuclear reactor shielding; in hot-melt adhesives in paper coatings; in corrosion protection compds., polishes and maintenance compds., scratch-resistant paints, lacquers, varnishes, candle hardener and opacifier; FDA 21CFR §172.615; wh. powd., waxy odor; sol. @ elevated temps. in aliphatic and aromatic hydrocarbons; m.w. 700; b.p. 500 F; sp.gr. 0.947; kinetic visc. ≈ 12 cstk (120 C); congeal. pt. 94-96 C; drop pt. 106-110 C; acid no. nil; iodine no. 0.1; sapon. no. nil; flash pt. (COC) ≈ 570 F; hardness (ball pressure) 305/265 kg/cm^2; dielec. const. 2.0 (125 C, 1 kc); vol. resist. 3 x 10^{13} (125 C, 300 V).

Vestowax FT-200. [Astor Wax] Syn. wax; Fischer-Tropsch saturated, syn. hard paraffins; CAS 8002-74-2; wax used in plastics processing for PVC lubrication, mold release compds. for PU and polyesters; slip agent in printing inks; in elec. and electronics for insulating compds., potting agents, nuclear reactor shielding; in hot-melt adhesives in paper coatings; in corrosion protection compds., polishes and maintenance compds., scratch-resistant paints, lacquers, varnishes, candle hardener and opacifier; FDA 21CFR §172.615; wh. pellets; sol. @ elevated temps. in aliphatic and aromatic hydrocarbons; m.w. 600; kinetic visc. ≈ 11 cstk (120 C); congeal. pt. 94-95 C; drop pt. 104-110 C; acid no. nil; iodine no. 0.1; sapon. no. nil; hardness (ball pressure) 190/150 kg; dielec. const. 2.0 (125 C, 1 kc); vol. resist. 3 x 10^{13} (125 C, 300 V).

Vestowax FT-300. [Astor Wax] Syn. wax; Fischer-Tropsch saturated, syn. hard paraffins; CAS 8002-74-2; wax used in plastics processing for PVC lubrication, mold release compds. for PU and polyesters; slip agent in printing inks; in elec. and electronics for insulating compds., potting agents, nuclear reactor shielding; in hot-melt adhesives in paper coatings; in corrosion protection compds., polishes and maintenance compds., scratch-resistant paints, lacquers, varnishes, candle hardener and opacifier; FDA 21CFR §172.615; wh. pellets or powd.; also avail. in micronized form (Vestofine SF-105, Eftofine FT-600); sol. @ elevated temps. in aliphatic and aromatic hydrocarbons; m.w. 730; kinetic visc. ≈ 12 cstk (120 C); congeal. pt. 96-98 C; drop pt. 107-111 C; acid no. nil; iodine no. 0.1; sapon. no. nil; hardness (ball pressure) 355/344 kg/cm^2; dielec. const. 2.0 (125 C, 1 kc); vol. resist. 3 x 10^{13} (125 C, 300 V).

Vicron® 10-20. [Specialty Minerals] Limestone; CAS 1317-65-3; filler with controlled particle size, low surf. area, high brightness; extender and property enhancer in paints, coatings, rigid vinyl, polyolefins, building prods.; powd.; 2.0 μ avg. particle size; 0.001% +325 mesh; sp.gr. 2.7; dens. 68 lb/ft^3 (tapped); bulk dens. 28 lb/ft^3; surf. area 4.1 m^2/g; oil absorp. 22 g/100 g oil; brightness 97; 97% $CaCO_3$.

Vicron® 15-15. [Specialty Minerals] Limestone; CAS 1317-65-3; filler with controlled particle size, low surf. area, high brightness, and low oil absorp.; extender and property enhancer in paints, coatings, rubber, flexible and rigid vinyl, PVC plastisols, polyolefins, building prods.; powd.; 3.5 μ avg. particle size; 0.004% +325 mesh; sp.gr. 2.7; dens. 62 lb/ft^3 (tapped); bulk dens. 30 lb/ft^3; surf. area 3.1 m^2/g; oil absorp. 20 g/100 g oil; brightness 96; 97% $CaCO_3$.

Vicron® 25-11. [Specialty Minerals] Limestone; CAS 1317-65-3; filler with controlled particle size, low surf. area, high brightness, and low oil absorp.; extender and property enhancer in paints, coatings, rubber, flexible and rigid vinyl, PVC plastisols, polyolefins, SMC/BMC, building prods.; powd.; 5.5 μ avg. particle size; 0.004% +325 mesh; sp.gr. 2.7; dens. 74 lb/ft^3 (tapped); bulk dens. 32 lb/ft^3; surf. area 1.8 m^2/g; oil absorp. 18 g/100 g oil; brightness 96; 97% $CaCO_3$.

Vicron® 31-6. [Specialty Minerals] Limestone; CAS

1317-65-3; filler with controlled particle size, low surf. area, high brightness, and low oil absorp.; extender and property enhancer in paints, coatings, rubber, flexible vinyl, SMC/BMC, building prods.; powd.; 6.5 μ avg. particle size; 0.1% +325 mesh; sp.gr. 2.7; dens. 78 lb/ft^3 (tapped); bulk dens. 36 lb/ft^3; surf. area 1.5 m^2/g; oil absorp. 16 g/100 g oil; brightness 95; 97% $CaCO_3$.

Vicron® 41-8. [Specialty Minerals] Limestone; CAS 1317-65-3; filler with controlled particle size, low surf. area, high brightness, and low oil absorp.; extender and property enhancer in paints, coatings, rubber, building prods.; powd.; 8.0 μ avg. particle size; 0.1% +325 mesh; sp.gr. 2.7; dens. 86 lb/ft^3 (tapped); bulk dens. 38 lb/ft^3; surf. area 1.3 m^2/g; oil absorp. 16 g/100 g oil; brightness 95; 97% $CaCO_3$.

Vicron® 45-3. [Specialty Minerals] Limestone; CAS 1317-65-3; filler with controlled particle size, low surf. area, high brightness; extender and property enhancer in paints, coatings, thermoplastic and thermoset polymers, building prods.; powd.; 13.2 μ avg. particle size; 0.1% +325 mesh; sp.gr. 2.7; dens. 102 lb/ft^3 (tapped); bulk dens. 43 lb/ft^3; surf. area 1.0 m^2/g; oil absorp. 12.5 g/100 g oil; brightness 95; 97% $CaCO_3$.

Victastab HMP. [Akzo Nobel] Di (polyoxyethylene) hydromethyl phosphonate; reactive flame retardant for cellulose nitrate, phenolic, thermoset polyester, PU applics.

Victawet® 12. [Akzo Nobel] EO reaction prod. with 2-ethyl hexanol and P_2O_5; nonionic; wetting agent, penetrant, dispersant, stabilizer; for pkg. dyeing of nylon, acid-type cleaners, emulsion polymerization, starch coatings; APHA 500 max. liq.; sol. in alcohol, acetone, toluene; disp. in water; dens. 9.34 lb/gal; sp.gr. 1.121; visc. 91.2 cs (100 F); b.p. 325 F; flash pt. (COC) 250 F; pour pt. -48 F; surf. tens. 29.4 dynes/cm^2 (0.2%); pH 7.0-7.4 (0.5%); 100% act.

Victawet® 85X. [Akzo Nobel] Buffered salt of Victawet 58B; anionic; stabilizer for vinyl resins, dispersant; powd.; 94% act.

Victory®. [Petrolite] Microcryst. wax; CAS 63231-60-7; EINECS 264-038-1; plastic wax offering high ductility, flexibility at very low temps.; provides protective barrier properties against moisture vapor and gases; used in hot-melt adhesives; hot-melt coatings; in antisunchecking agents in rubber goods, elec. insulating agents, leather treating agents, water repellents for textiles, rustproof coatings, cosmetic ingreds., and as plasticizer for other petrol. waxes used in crayons, dental compds., chewing gum base, and candles; incl. FDA §172.230, 172.615, 175.105, 175.300, 176.170, 176.180, 176.200, 177.1200, 178.3710, 179.45; wh., amber wax; dens. 0.93 g/cc; visc. 13 cps (99 C); m.p. 79 C; flash pt. 293 C.

Vicure® 55. [Akzo Nobel] Methyl phenylglyoxalate; CAS 15206-55-0; high purity photoinitiator for uv curable systems, esp. acrylate-based formulations.

Vikoflex. [Elf Atochem N. Am.] Dual-function epoxidized vegetable oils; epoxy plasticizers for flexible PVC prod.

Vikol® PX-15. [Vikon] Bis (tri-n-butyltin) oxide, stabilized; CAS 56-35-9; antimicrobial for vinyl to control bacterial growth and mildew; sol. in vinyl plasticizers and in vinyl polymer.; Unverified

Vinnol® K 704. [Wacker-Chemie GmbH] PVC/polyacrylate graft polymer; impact modifier for rigid PVC; produces flexible, unplasticized PVC components; spherical coarse particles; < 2000 μm particle size; sp.gr. 1.16; bulk dens. 0.64 g/cc; tens. str. 20 N/mm^2 (ultimate); tens. elong. 200% (break); hardness (Shore A) 95.

Vinuran®. [BASF AG] Polymers for modifying PVC; improves PVC props. incl. impact str., resist. to heat deformation and decomposition, processing, and deep-drawing behavior; used for hot water pipes, impact-resist. profiles, films, hollow articles, panels.

Vinylube® 36. [Lonza] Ester wax; internal lubricant for unplasticized clear vinyl extrusions; FDA accepted; cream bead; m.p. 59 C; flash pt. 210 C.

Vinylube® 38. [Lonza] Ester wax; external lubricant for unplasticized clear vinyl extrusions; FDA accepted; cream bead; m.p. 60 C; flash pt. 220 C.

Vinyzene® BP-5-2. [Morton Int'l./Plastics Additives] 10,10′-Oxybisphenoxarsine; CAS 58-36-6; antimicrobial sol'n. for PVC, PU, other plastics and syn. rubbers; for film, sheeting, extruded profiles, plastisols, molded goods, organosols, fabric coatings; end-prods. incl. shower curtains, floor/wall coverings, ditch liners, vinyl molding, marine upholstery, auto tops, tarps, awnings, refrigerator gaskets, weatherstripping; EPA reg. no. 2829-96; m.w. 502.2 (OBPA); *Toxicology:* skin irritant in undiluted form; toxic to fish and wildlife.

Vinyzene® BP-5-2 DIDP. [Morton Int'l./Plastics Additives] 10,10′-Oxybisphenoxarsine in diisodecyl phthalate plasticizer carrier; antimicrobial sol'n. for PVC, PU, other plastics and syn. rubbers; for film, sheeting, extruded profiles, plastisols, molded goods, organosols, fabric coatings; end-prods. incl. shower curtains, floor/wall coverings, ditch liners, vinyl molding, marine upholstery, auto tops, tarps, awnings, refrigerator gaskets, weatherstripping; EPA reg. no. 2829-96; m.w. 502.2 (OBPA); *Toxicology:* skin irritant in undiluted form; toxic to fish and wildlife.

Vinyzene® BP-5-2 DOP. [Morton Int'l./Plastics Additives] 10,10′-Oxybisphenoxarsine in dioctyl phthalate plasticizer carrier; antimicrobial sol'n. for PVC, PU, other plastics and syn. rubbers; for film, sheeting, extruded profiles, plastisols, molded goods, organosols, fabric coatings; end-prods. incl. shower curtains, floor/wall coverings, ditch liners, vinyl molding, marine upholstery, auto tops, tarps, awnings, refrigerator gaskets, weatherstripping; EPA reg. no. 2829-96; m.w. 502.2 (OBPA); *Toxicology:* skin irritant in undiluted form; toxic to fish and wildlife.

Vinyzene® BP-5-2 PG. [Morton Int'l./Plastics Additives] 10,10´-Oxybisphenoxarsine in plasticizer carrier; antimicrobial sol'n. for PVC, PU, other plastics and syn. rubbers; for film, sheeting, extruded profiles, plastisols, molded goods, organosols, fabric coatings; end-prods. incl. shower curtains, floor/wall coverings, ditch liners, vinyl molding, marine upholstery, auto tops, tarps, awnings, refrigerator gaskets, weatherstripping; EPA reg. no. 2829-96; m.w. 502.2 (OBPA); *Toxicology:* skin irritant in undiluted form; toxic to fish and wildlife.

Vinyzene® BP-5-2 S-160. [Morton Int'l./Plastics Additives] 10,10´-Oxybisphenoxarsine in plasticizer carrier; antimicrobial sol'n. for PVC, PU, other plastics and syn. rubbers; for film, sheeting, extruded profiles, plastisols, molded goods, organosols, fabric coatings; end-prods. incl. shower curtains, floor/wall coverings, ditch liners, vinyl molding, marine upholstery, auto tops, tarps, awnings, refrigerator gaskets, weatherstripping; EPA reg. no. 2829-96; m.w. 502.2 (OBPA); *Toxicology:* skin irritant in undiluted form; toxic to fish and wildlife.

Vinyzene® BP-5-2 U. [Morton Int'l./Plastics Additives] 10,10´-Oxybisphenoxarsine in plasticizer carrier; antimicrobial sol'n. for PVC, PU, other plastics and syn. rubbers; for film, sheeting, extruded profiles, plastisols, molded goods, organosols, fabric coatings; end-prods. incl. shower curtains, floor/wall coverings, ditch liners, vinyl molding, marine upholstery, auto tops, tarps, awnings, refrigerator gaskets, weatherstripping; EPA reg. no. 2829-96; m.w. 502.2 (OBPA); *Toxicology:* skin irritant in undiluted form; toxic to fish and wildlife.

Vinyzene® BP-505. [Morton Int'l./Plastics Additives] 5% 10,10´-Oxybisphenoxarsine in epoxidized soybean oil; antimicrobial, bacteriostat, fungistat for PVC, PU, other plastics, syn. rubber; recommended for film and sheet, extruded profiles, plastisols, molded goods, organosols, fabric coatings; m.w. 502.2 (OBPA); sp.gr. 1.049; flash pt. (COC) 205 F.

Vinyzene® BP-505 DIDP. [Morton Int'l./Plastics Additives] 5% 10,10´-Oxybisphenoxarsine in diisodecyl phthalate; antimicrobial, bacteriostat, fungistat for PVC, PU, other plastics, syn. rubber; recommended for film and sheet, extruded profiles, plastisols, molded goods, organosols, fabric coatings; end prods. incl. shower curtains, wall coverings, marine upholstery, urethane outsoles, auto tops, awnings, weatherstripping, foam gaskets, ditch liners, carpet underlay, leisure furniture, refrigerator gaskets, waterbed liners, swimming pool liners; EPA reg. no. 2829-125; m.w. 502.2 (OBPA); sp.gr. 0.993; flash pt. (COC) 205 F; *Toxicology:* skin irritant in undiluted form; toxic to fish and wildlife.

Vinyzene® BP-505 DOP. [Morton Int'l./Plastics Additives] 5% 10,10´-Oxybisphenoxarsine in di 2-ethylhexyl phthalate; antimicrobial, bacteriostat, fungistat for PVC, PU, other plastics, syn. rubber; recommended for film and sheet, extruded profiles, plastisols, molded goods, organosols, fabric coatings; end prods. incl. shower curtains, wall coverings, marine upholstery, urethane outsoles, auto tops, awnings, weatherstripping, foam gaskets, ditch liners, carpet underlay, leisure furniture, refrigerator gaskets, waterbed liners, swimming pool liners; EPA reg. no. 2829-125; m.w. 502.2 (OBPA); sp.gr. 1.007; flash pt. (COC) 215 F; *Toxicology:* skin irritant in undiluted form; toxic to fish and wildlife.

Vinyzene® BP-505 PG. [Morton Int'l./Plastics Additives] 5% 10,10´-Oxybisphenoxarsine in polypropylene glycol; antimicrobial, bacteriostat, fungistat for PVC, PU, other plastics, syn. rubber; recommended for film and sheet, extruded profiles, plastisols, molded goods, organosols, fabric coatings; end prods. incl. shower curtains, wall coverings, marine upholstery, urethane outsoles, auto tops, awnings, weatherstripping, foam gaskets, ditch liners, carpet underlay, leisure furniture, refrigerator gaskets, waterbed liners, swimming pool liners; EPA reg. no. 2829-125; m.w. 502.2 (OBPA); sp.gr. 1.031; flash pt. (COC) 215 F; *Toxicology:* skin irritant in undiluted form; toxic to fish and wildlife.

Vinyzene® BP-505 S160. [Morton Int'l./Plastics Additives] 5% 10,10´-Oxybisphenoxarsine in butylbenzyl phthalate; antimicrobial, bacteriostat, fungistat for PVC, PU, other plastics, syn. rubber; recommended for film and sheet, extruded profiles, plastisols, molded goods, organosols, fabric coatings; m.w. 502.2 (OBPA); sp.gr. 1.013; flash pt. (COC) 215 F.

Vinyzene® IT-3000 DIDP. [Morton Int'l./Plastics Additives] 2-n-Octyl-4-isothiazolin-3-one in diisodecyl phthalate carrier; fungicide for plastics (PVC, PU); recommended for vinyl film, sheeting, extruded profiles, plastisols, molded goods, organosols, fabric coatings, urethane shoe soles and foams; lt. straw liq., mild odor; sp.gr. 0.97; dens. 8.1 lb/gal; flash pt. (PMCC) 210 F; 4% act., 96% inert carrier; *Toxicology:* end prods. incl. shower curtains, wall coverings, marine upholstery, urethane outsoles, auto tops, awnings, weatherstripping, foam gaskets, ditch liners, carpet underlay, leisure furniture, refrigerator gaskets, waterbed liners, swimming pool liners.

Vinyzene® IT-3025 DIDP. [Morton Int'l./Plastics Additives] 2-n-Octyl-4-isothiazolin-3-one in diisodecyl phthalate carrier; industrial fungicide, mildewcide for plastics (PVC, PU, etc.); EPA reg. no. 707-208-2829; lt. straw clear liq., mild odor; sp.gr. 0.98; dens. 8.1 lb/gal; 25% act., 75% inert carrier; *Toxicology:* toxic to fish and wildlife; *Precaution:* corrosive to mild steel; do not contaminate with strong oxidizing or reducing agents.

Vinyzene® SB-1. [Morton Int'l./Plastics Additives] 10,10´-Oxybisphenoxarsine in PVC/PVA carrier; patented antimicrobial for PVC and other plastics; provides long-term protection against fungal and bacterial degradation, permanent staining, embrittlement; for interior applics. (PVC floor/wall coverings, coated upholstery fabrics, interior auto parts, refrigerator gaskets, shower curtains), exterior applics. (auto tops, exterior trim, tarps, awnings, ditch/pool liners, marine upholstery); EPA reg. no. 2829-115; clear lt. straw pellet; 1/16 x 1/16 in.; sol. in most common coating and printing inks solvs.; bulk dens. 50-53 lb/ft^3; glass transition temp. 70 C; 5% act., 95% carrier.

Vinyzene® SB-1 EAA. [Morton Int'l./Plastics Additives] 10,10′-Oxybisphenoxarsine in ethylene-acrylic acid copolymer carrier; patented fungicide, bactericide for polyolefins; for interior automotive parts, exterior drainage hose, underground cable jacketing, buoyant floats; EPA reg. no. 2829-115; off-wh. pellet, 3.5 mm x 2.5 mm; 5% act., 95% carrier.

Vinyzene® SB-1 PR. [Morton Int'l./Plastics Additives] 10,10′-Oxybisphenoxarsine in PP resin carrier; antimicrobial, fungicide, bacteristat for polyolefins; for interior applics. (interior auto parts, waste receptacles, insulation facing), exterior applics. (drainage hose, evaporation barriers, underground cable jacketing), carpet fibers and backing; water-wh. pellet, 3.5 x 2.5 mm; 5% act., 95% carrier.

Vinyzene® SB-1 PS. [Morton Int'l./Plastics Additives] 5% 10,10′-Oxybisphenoxarsine in a polymeric resin carrier; patented antimicrobial, fungicide, bacteristat for hot-melt adhesives, polyolefins, PU, PS, CPE; for interior applics. (polyolefin air conditioner cases, PVC floor coverings, coated upholstery fabrics, interior auto parts, refrigerator gaskets); exterior applics. (PE pipe wrap, underground cable jacketing, exterior auto parts, tarps, awnings, ditch/swimming pool liners, marine upholstery); EPA reg. no. 2829-115; clear pellets, 1/16 x 1/16 in.; 5% act., 95% carrier.

Vinyzene® SB-1 U. [Morton Int'l./Plastics Additives] 10,10′-Oxybisphenoxarsine in aliphatic polyester urethane carrier; patented antimicrobial, fungicide, bacteristat for thermoplastic PU and other plastics; for interior applics. (shoe soles, boots, hospital mattress coverings, cable), exterior applics. (underground cable jacketing, exterior auto parts, tarps, awnings, marine upholstery and coatings); EPA reg. no. 2829-115; straw-colored pellet; 5% act., 95% carrier.

Viosorb 520. [Aceto] Drometrizole; CAS 2440-22-4; EINECS 219-470-5; uv stabilizer for ABS, cellulosics, epoxy, polyesters, PS, flexible and rigid PVC, VDC; powd.; m.p. 128-132 C.

Viplex 525. [Crowley Chem.] Highly aromatic petrol. fraction; plasticizer/extender for PU, primer coatings; provides visc. reduction permitting higher filler loadings, good wetting of pigments, fillers, fiberglass; aids leveling, penetration; sp.gr. 1.10 (60 F); visc. 123 SSU (100 F); i.b.p. 538 F; flash pt. (COC) 335 F; pour pt. 0 F; Unverified

Viplex 680-P. [Crowley Chem.] Heavy paraffinic distillate solvent extract; CAS 64742-04-7; sec. vinyl plasticizer; used @ 40-60% replacement level; dk. brn. liq., mild odor; negligible sol. in water; sp.gr. 0.998 (60 F); dens. 8.28 lb/gal; visc. 107 cs (40 C); dist. range 710 F (5%); pour pt. 16 C; flash pt. (COC) 420 F; *Toxicology:* TLV 5.00 mg/m^3 (oil mist); sl. toxic orally; sl. eye irritant; moderate skin irritant; aromatic oils may cause skin cancer on repeated/prolonged contact; avoid breathing mists; *Precaution:* incompat. with strong oxidizers; *Storage:* store in cool area; maintain headspace gase concs. below flamm. limits; keep away from ignition sources.

Viplex 885. [Crowley Chem.] Highly aromatic (99%) petrol. distillate; CAS 68477-29-2; plasticizer/extender for epoxy systems, coatings, potting/encapsulation; sec. vinyl plasticizer; promotes low system visc., better wetting of fillers and pigments, higher loading levels; modifier for polyesters; lt. yel. liq., may form yel. cryst. when cooled or aged, oily aromatic odor; sp.gr. 1.11 (60 F); dens. 9.25 lb/gal; visc. 147 SSU (100 F); i.b.p. 660 F; pour pt. 0 F; flash pt. (COC) 365 F; 97% aromatic content; *Toxicology:* may cause severe eye irritation, moderate skin irritation on prolonged/repeated contact; ingestion may cause gastrointestinal irritation, nausea, vomiting, diarrhea; aspiration into lungs may case pneumonitis which may be fatal; *Precaution:* incompat. with strong oxidizing agents.

Viplex 895-BL. [Crowley Chem.] Polynuclear aromatic hydrocarbon; CAS 64741-81-7; sec. vinyl plasticizer esp. for dark and colored formulations @ 50-70% replacement levels of phthalate plasticizers; increases elec. props. in Hypalon, prevents crystallinity in Neoprene; increases ozone resist. in buty rubber; compat. with polymeric plasticizers in nitrile rubber; brnsh.-green liq., cracked hydrocarbon odor; insol. in water; sp.gr. 1.10 (60 F); dens. 9.16 lb/gal; visc. 250 SSU (100 F); vapor pressure nil; b.p. 500-1000 F; pour pt. 5 F; flash pt. (COC) 350 F; *Toxicology:* may cause skin tumors on repeated/prolonged contact; avoid contact with eyes and skin; hydrogen sulfide may be generated on heating (irritating to eyes, skin, lungs; may be fatal at higher concs.); *Precaution:* incompat. with strong oxidizing materials, heat, flame.

Vircol® 82. [Albright & Wilson Am.] Phosphorus-based polyol; flame retardant for use in rigid urethane foam prepolymers and premixes; enhances compat. of urethane pkg. systems; readily reacts with isocyanates; hydrolyzes slowly in water; liq.; 11.3% P.

Viscasil® (10,000 cSt). [GE Silicones] Dimethicone; used for damping, power transmission, rubber/plastic lubrication, base fluid for grease, polishes, cosmetics/toiletries, mold release for tires, rubber, plastics, petrol. antifoam, textile thread/fiber lubricants; sp.gr. 0.975; visc. 10,000 cSt; pour pt. -53 F; flash pt. (PMOC) 601 F; ref. index 1.4035; surf. tens. 21.3 dynes/cm; sp. heat 0.36 Btu/lb/F; dissip. factor 0.0001; dielec. str. 35 kV; dielec. const. 2.75; vol. resist. 1 x 10^{14} ohm-cm.

Viscasil® (12,500 cSt). [GE Silicones] Dimethicone; used for damping, rubber/plastic lubrication, polishes, cosmetics/toiletries, mold release for tires, rubber, plastics, petrol. antifoam, textile thread/fiber lubricants; sp.gr. 0.975; visc. 12,500 cSt; pour pt. -53 F; flash pt. (PMCC) 500 F; ref. index 1.4035; surf. tens. 21.3 dynes/cm; sp. heat 0.36 Btu/lb/F; dissip. factor 0.0001; dielec. str. 35 kV; dielec. const. 2.75; vol. resist. 1 x 10^{14} ohm-cm.

Viscobyk-4010. [Byk-Chemie USA] Aliphatic hydrocarbons with wetting agents; visc. depressant for PVC plastisols; for artificial leather, carpetbacking, casting, coil coating, conveyor belts, dipping, flooring, rotational molding, inks, tarps, wallpapers; improves wetting and disp. of inorg. fillers allowing higher filler loading; sl. ylsh. clear liq., hydrocarbon odor; insol. in water; sp.gr. 0.79; dens. 6.58 lb/gal; b.p. 271-285 C; acid no. 10; flash pt. (PMCC) 127 C; ref. index 1.444; *Toxicology:* LD50 (oral, rat) > 20,000 mg/kg; sl. eye irritant; nonirritating to skin; inh. of high concs. of vapor may cause respiratory tract irritation; repeated ingestion may irritate digestive tract; *Precaution:* incompat. with strong oxidizing agents; hazardous combustion prods.: CO, CO_2, phosphorous oxides; *Storage:* may become turbid below 5 C or solidify at lower temps.; warm to reverse effects; keep container tightly closed when not in use.

Viscobyk-4013. [Byk-Chemie USA] Aliphatic hydrocarbons with wetting agents; visc. depressant for filled, unfilled, and highly pigmented PVC plastisols; for artificial leather, carpetbacking, casting, coil coating, conveyor belts, dipping, flooring, rotational molding, inks, tarps, wallpapers; improves wetting and disp. of inorg. fillers allowing higher filler loading; ylsh. clear to sl. cloudy liq., hydrocarbon-like odor; insol. in water; sp.gr. 0.83; dens. 6.92 lb/gal; b.p. 113-125 C; acid no. 11.5; flash pt. (Seta CC) 108 C; ref. index 1.449; *Toxicology:* LD50 (oral, rat) > 10,000 mg/kg; sl. eye irritant; nonirritating to skin; inh. of vapor may cause respiratory tract irritation, headaches, nausea, vomiting, CNS depression; chronic absorp. may injur kidney; *Precaution:* flamm. limits 0.6-7.0% vol. in air; incompat. with strong oxidizing agents; hazardous combustion prods.: CO, CO_2; *Storage:* keep container tightly closed when not in use.

Viscobyk-4015. [Byk-Chemie USA] Aliphatic hydrocarbons with polysiloxane copolymers; visc. depressant with deaeration effect for unfilled PVC plastisols; for artificial leather, carpetbacking, casting, coil coating, conveyor belts, dipping, flooring, rotational molding, inks, tarps, wallpapers; colorless clear liq., paraffinic odor; insol. in water; sp.gr. 0.76; dens. 6.33 lb/gal; b.p. 204-235 C; flash pt. (Seta CC) 75 C; ref. index 1.430; *Toxicology:* LD50 (oral, rat) > 20,000 mg/kg; sl. eye irritant; nonirritating to skin; inh. of vapor may cause respiratory tract irritation, CNS depression; *Precaution:* incompat. with strong oxidizing agents; hazardous combustion prods.: CO, CO_2, silicon compds., formaldehyde; *Storage:* keep container tightly closed when not in use; avoid heat, flame, ignition sources.

Viscobyk-5025. [Byk-Chemie USA] Carboxylic acid derivs. with dispersing props.; low emission visc. depressant for filled PVC plastisols; for artificial leather, carpetbacking, casting, coil coating, conveyor belts, dipping, flooring, rotational molding, inks, tarps, wallpapers; lt. yel. liq., alcoholic odor; insol. in water; sp.gr. 0.88; dens. 7.33 lb/gal; acid no. 10; flash pt. (PMCC) 180 C; ref. index 1.452; *Toxicology:* LD50 (oral, rat) 10,000 mg/kg; nonirritating to skin/eyes; repeated ingestion may cause digestive tract irritation; *Precaution:* incompat. with strong oxidizing agents; hazardous combustion prods.: CO, CO_2; *Storage:* keep container tightly closed when not in use.

Viscobyk-5050. [Byk-Chemie USA] Carboxylic acid derivs. with dispersing props.; nonionic; low emission visc. depressant for unfilled PVC plastisols; for artificial leather, carpetbacking, casting, coil coating, conveyor belts, dipping, flooring, rotational molding, inks, tarps, wallpapers; lt. yel. liq., odorless; insol. in water; sp.gr. 0.88 (20 C0; dens. 7.33 lb/gal; flash pt. (PMCC) 174 C; ref. index 1.455; *Toxicology:* LD50 (oral, rat) > 10,000 mg/kg; nonirritating to skin/eyes; repeated ingestion may cause digestive tract irritation; *Precaution:* incompat. with strong oxidizing agents; hazardous combustion prods.: CO, CO_2; *Storage:* may become turbid below -5 C; warm and homogenize to reverse effect; keep container tightly closed when not in use.

Vistalon 503. [Exxon] Ethylene/propylene copolymer; CAS 9010-79-1; EPM rubber for modifying polyolefins; peroxide-curable; sp.gr. 0.86; Mooney visc. 28-38 (ML1±4, 257 F); Unverified

Vistalon 719. [Exxon] Ethylene/propylene copolymer; CAS 9010-79-1; EPM rubber for modifying polyolefins; free-flowing crumb or pellet; sp.gr. 0.89; Mooney visc. 49-59 (ML1±4, 257 F); Unverified

Vistalon 6505. [Exxon] EPDM terpolymer; can be blended and covulcanized with most common elastomers; exc. compr. set, ozone resist, good heat resist., exc. adhesion to brass and PU coatings; used for sponge, gaskets, molded goods, as antiozonant for high unsat. rubbers; sulfur-curable; very fast curing; wh. bales; sp.gr. 0.86; Mooney visc. 49-59 (ML1+4, 257 F); Unverified

Vistanex® MML-80, MML-100, MML-120, MML-140. [Exxon] Polyisobutylene (CAS 9003-27-4); syn. rubber used as polymeric additive for rubbers to impart stability, inertness, ozone and chem. resist., elec. inertness; also useful in molded and extruded prods. and coated materials; wh. to pale yel. bales; m.w. 90,000, 125,000, 166,000, and 211,000 resp.; sp.gr. 0.92; Unverified

Viton® Curative No. 20. [DuPont] 33% Organophosphonium salt and 67% fluoroelastomer; used in combination with Viton Curative No. 30 or No. 40 as a curing system for Viton fluoroelastomer; wh. to off-wh. pellets; sp.gr. 1.37.

Viton® Curative No. 30. [DuPont] 50% Dihydroxy aromatic compd. and 50% fluoroelastomer; used in combination with Viton Curative No. 20 as a curing system for Viton fluoroelastomers; tan to gray brn. pellets; sp.gr. 1.50.

Viton® Curative No. 40. [DuPont] 33% Benzophenone compd. and 67% fluoroelastomer; used in combination with Viton Curative No. 20 as a curing system for Viton fluoroelastomers; yel. pellets; sp.gr. 1.44.

Viton® Curative No. 50. [DuPont] Bisphenol-based;

accelerator/crosslinking and curative system for Viton® fluoroelastomer; provides reduced mold fouling, faster cure without significant loss in scorch safety over Viton Curative No. 20 or No. 30; for applics. incl. o-rings, seals, molded shapes, diaphragms, and tubing; pink to purple free-flowing prills, odorless; sol. in low m.w. polar solvs.; sp.gr. 1.38; m.p. 60-100 C; *Toxicology:* may cause irritation to skin and eyes and be harmful if inhaled; *Storage:* exc. storage stability.

Viton® Free Flow HD. [DuPont] 50% fluoropolymer/ 50% HDPE blend; processing additive for improved performance in HDPE resins and other polyolefins; effective in HDPE formulations contg. high levels of acid scavengers; FDA 21CFR §177.1520; wh. free-flowing powd.; 1% max. on 16 mesh, 50% min. on 80 mesh; sp.gr. 1.24 ± 0.05; 50 ± 2% fluoropolymer.

Viton® Free Flow TA. [DuPont] 50% fluoropolymer/ 50% talc partitioning agent blend; processing additive designed for reduced interactions with talc antiblocks in all polyolefins; FDA 21CFR §177.1520; wh. free-flowing powd.; 1% max. on 16 mesh, 50% min. on 80 mesh; sp.gr. 2.17 ± 0.05; 50 ± 2% fluoropolymer.

Vocol®, S-75. [Monsanto] Zinc-O,O-di-n-butylphosphorodithioate; CAS 6990-43-8; accelerator sulfur donor for EPDM cures; sec. accelerator sulfur donor for thiazoles and sulfenamides; 100% act. liq. and 75% act. powd.

Voidox 100%. [Guardian Chem.] Phenol; heat and lt. stabilizer, antioxidant, lubricant for PE, PP, ABS.

Vorite 105. [CasChem] Polymerized castor oil; pigment wetting/dispersing agent; plasticizer for resins, gums, polymers; lubricant, penetrant; coupling solv.; adhesion promoter; for cellulose lacquers, inks, adhesives, industrial lubricants, polishes, caulks, leather dressing, hydraulic fluids, rubber compding., gasket cement; Gardner 4 color; sp.gr. 0.975; visc. 26 stokes; pour pt. -5 F; acid no. 2; iodine no. 85; sapon. no. 170; hyd. no. 130.

Vorite 110. [CasChem] Polymerized castor oil; pigment wetting/dispersing agent; plasticizer for resins, gums, polymers; lubricant, penetrant; coupling solv.; adhesion promoter; for cellulose lacquers, inks, adhesives, industrial lubricants, polishes, caulks, leather dressing, hydraulic fluids, rubber compding., gasket cement; Gardner 3 oil; sp.gr. 0.990; visc. 115 stokes; pour pt. 15 F; acid no. 2; iodine no. 82; sapon. no. 166; hyd. no. 102.

Vorite 115. [CasChem] Polymerized castor oil; pigment wetting/dispersing agent; plasticizer for resins, gums, polymers; lubricant, penetrant; coupling solv.; adhesion promoter; for cellulose lacquers, inks, adhesives, industrial lubricants, polishes, caulks, leather dressing, hydraulic fluids, rubber compding., gasket cement; Gardner 3 oil; sp.gr. 0.995; visc. 192 stokes; pour pt. 20 F; acid no. 2; iodine no. 85; sapon. no. 165; hyd. no. 93.

Vorite 120. [CasChem] Polymerized castor oil; pigment wetting/dispersing agent; plasticizer for resins, gums, polymers; lubricant, penetrant; coupling solv.; adhesion promoter; for cellulose lacquers, inks, adhesives, industrial lubricants, polishes, caulks, leather dressing, hydraulic fluids, rubber compding., gasket cement; Gardner 4 oil; sp.gr. 1.001; visc. 700 stokes; pour pt. 50 F; acid no. 2; iodine no. 82; sapon. no. 160; hyd. no. 78.

Vorite 125. [CasChem] Polymerized castor oil; pigment wetting/dispersing agent; plasticizer for resins, gums, polymers; lubricant, penetrant; coupling solv.; adhesion promoter; for cellulose lacquers, inks, adhesives, industrial lubricants, polishes, caulks, leather dressing, hydraulic fluids, rubber compding., gasket cement; Gardner 4 oil; sp.gr. 1.007; visc. 900 stokes; pour pt. 55 F; acid no. 2; iodine no. 84; sapon. no. 157; hyd. no. 72.

VPA No. 1. [DuPont/Polymer Prods.] Wax/aromatic sulfur compd.; general purpose processing aid for all types of molded goods made from Viton®; provides smooth preforms, sl. faster cure rates and exc. mill and mold release chars.; cream to wh. powd.; sp.gr. 1.20; m.p. 150 C; *Toxicology:* can cause irritation of eyes and skin; *Storage:* keep containers closed to avoid contamination; store in cool, dry place.

VPA No. 2. [DuPont/Polymer Prods.] Vegetable wax; processing and extrusion aid allowing Viton® compds. to extrude faster, cooler, and smoother; has some adverse effect on high temp. compr. set; yel. flakes; sp.gr. 0.99; m.p. 80 C; *Toxicology:* no known toxicological hazards; *Storage:* keep containers closed to avoid contamination; store in cool, dry place.

VPA No. 3. [DuPont/Polymer Prods.] 75% Sulfur compd. on inert carrier; processing aid providing improved mold release for Viton® fluoroelastomers cured with bisphenol or peroxide systems; for use for o-rings, seals, molded shapes, diaphragms, and tubing; tan to lt. brn. free-flowing powd., musty odor; sp.gr. 1.45; 75% act.; *Toxicology:* may cause irritation to skin and eyes, be harmful if inhaled, be toxic if ingested; *Storage:* hygroscopic—loss of compd. scorch safety can occur if exposed to high humidity; keep containers closed to avoid contamination; store in cool, dry place.

V-Pyrol. [ISP] N-Vinyl-2-pyrrolidone; stabilized with N,N′-di-sec-butyl-p-phenylenediamine; reaction rate accelerator; copolymerizes readily with most other vinyl monomers; modifies hydrophilic properties in systems incl. adhesives, coatings, cosmetics, textiles, syn. fibers, textile sizes, protective colloids, lube oil additives; reactive diluent in UV- and electron beam-curable systems; aids pigment disp. (inks); intermediate for modified phenolic resins of interest as plasticizers, dye intermediates, textile assistants; liq.; f.p. 13.5 C; b.p. 193 C (400 mm); flash pt. 98 C (OC).

Vul-Cup 40KE. [Hercules] Bisperoxide supported on Burgess KE clay; efficient, scorch-resistant, low-odor vulcanizing agent and polymerization catalyst for elastomers and plastics; esp. suitable for transfer, inj., and rotational molding applics.; compds. can be extruded at higher rates because of higher extrusion temps. permissible.; wh. to off-wh. free-flowing powd.; m.w. 338; sol. in aliphatic, aromatic, and ketone solvs.; water-insol.; sp.gr. 1.50; 39.5-41.5% act. peroxide.

Vul-Cup R. [Hercules] Bisperoxide; efficient, scorch-resistant, low-odor vulcanizing agent and polymerization catalyst for elastomers and plastics; esp. suitable for transfer, inj., and rotational molding applics.; compds. can be extruded at higher rates because of higher extrusion temps. permissible.

Vulcabond® C 10. [Akzo Nobel] rubber to metal adhesion promoter; specially designed to improve adhesion between rubber and steel cord in radial tires; flakes.

Vulcabond® E. [Akzo Nobel] Polymeric phenolic compd. in aq. ammonia; textile bonding agent for rubber to polyester bonding; liq.; 20% sol'n.

Vulcabond® N 15. [Akzo Nobel] Rubber-to-metal adhesion promoter; flakes.

Vulcabond® TX. [Akzo Nobel] Isocyanate compd. in xylene; rubber to textile bonding agent; liq.; 50% sol'n.

Vulcabond® VP. [Akzo Nobel] 25% sol'n. of an isocyanurate-type polymer of TDI in dibutyl phthalate; bonding agent added to PVC plastisols to improve adhesion to a reinforcing substrate, e.g., nylon and polyester textiles; yel. visc., sl. cloudy liq.; sp.gr. 1.05; visc. 20-30 poise (20 C); flash pt. (PMCC) 146 C; 3.5-3.8% NCO content; *Toxicology:* reacts with body tissues; avoid prolonged/repeated contact; contains < 0.5% TDI (TLV 0.02 ppm); DBP TLV/TWA 5 mg/m^3; *Precaution:* not highly flamm.; if involved in a fire, may burn or decomp. and may produce flamm. and noxious vapors; *Storage:* 12 mos storage life; protect from moisture.

Vulcan® 2H. [Cabot/N. Am. Rubber Black] Carbon black; CAS 1333-86-4; EINECS 215-609-9; high structure reinforcing carbon black for rubbers; 0.1% max. on 325 mesh; pour dens. 344 kg/m^3; iodine no. 80.

Vulcan® 3. [Cabot/N. Am. Rubber Black] N330 Carbon black; CAS 1333-86-4; EINECS 215-609-9; reinforcing carbon black for rubbers featuring good abrasion resist. with moderate hysteresis props.; 0.1% max. on 325 mesh; pour dens. 368 kg/m^3; iodine no. 82.

Vulcan® 3H. [Cabot/N. Am. Rubber Black] N347 Carbon black; CAS 1333-86-4; EINECS 215-609-9; reinforcing carbon black for rubbers featuring lower abrasion resist. and lower hysteresis than Vulcan M; 0.1% max. on 325 mesh; pour dens. 336 kg/m^3; iodine no. 90.

Vulcan® 4H. [Cabot/N. Am. Rubber Black] N343 Carbon black; CAS 1333-86-4; EINECS 215-609-9; high structure, smooth extruding reinforcing carbon black for rubbers giving reinforcement between N339 and N234; 0.1% max. on 325 mesh; pour dens. 336 kg/m^3; iodine no. 92.

Vulcan® 6. [Cabot/N. Am. Rubber Black] N220 Carbon black; CAS 1333-86-4; EINECS 215-609-9; traditional reinforcing carbon black for tire treads and high quality IR prods.; 0.1% max. on 325 mesh; pour dens. 352 kg/m^3; iodine no. 121.

Vulcan® 6LM. [Cabot/N. Am. Rubber Black] N231 Carbon black; CAS 1333-86-4; EINECS 215-609-9; reinforcing carbon black for OTR tire treads; features high tearing and cutting resist.; 0.1% max. on 325 mesh; pour dens. 392 kg/m^3; iodine no. 121.

Vulcan® 7H. [Cabot/N. Am. Rubber Black] N234 Carbon black; CAS 1333-86-4; EINECS 215-609-9; reinforcing carbon black for rubbers featuring exceptionally high abrasion resist.; 0.1% max. on 325 mesh; pour dens. 329 kg/m^3; iodine no. 120.

Vulcan® 8H. [Cabot/N. Am. Rubber Black] Carbon black; CAS 1333-86-4; EINECS 215-609-9; reinforcing carbon black for rubbers featuring exceptionally high abrasion resist., smooth extrusion; 0.1% max. on 325 mesh; pour dens. 336 kg/m^3; iodine no. 120.

Vulcan® 9. [Cabot/N. Am. Rubber Black] N110 Carbon black; CAS 1333-86-4; EINECS 215-609-9; reinforcing carbon black for rubber processing featuring high tens. and tear props.; 0.1% max. on 325 mesh; pour dens. 352 kg/m^3; iodine no. 145.

Vulcan® 9A32. [Cabot/Special Blacks Div] Carbon black; CAS 1333-86-4; EINECS 215-609-9; colorant, uv stabilizer for polyolefin cable sheathing; pellets; 19 nm particle size; dens. 22 lb/ft^3; surf. area 140 m^2/g; 1.5% volatile.

Vulcan® 10H. [Cabot/N. Am. Rubber Black] Carbon black; CAS 1333-86-4; EINECS 215-609-9; reinforcing carbon black for rubber processing featuring max. abrasion resist.; 0.1% max. on 325 mesh; pour dens. 312 kg/m^3; iodine no. 142.

Vulcan® C. [Cabot/N. Am. Rubber Black] N293 Carbon black; CAS 1333-86-4; EINECS 215-609-9; conductive carbon black for rubbers combining high reinforcement with high elec. conductivity; 0.1% max. on 325 mesh; pour dens. 376 kg/m^3; iodine no. 145.

Vulcan® J. [Cabot/N. Am. Rubber Black] N375 Carbon black; CAS 1333-86-4; EINECS 215-609-9; reinforcing carbon black for rubbers giving props. similar to N220 at lower cost; 0.1% max. on 325 mesh; pour dens. 360 kg/m^3; iodine no. 90.

Vulcan® K. [Cabot/N. Am. Rubber Black] N351 Carbon black; CAS 1333-86-4; EINECS 215-609-9; reinforcing carbon black for use in tires and IR prods.; reinforcing yet dispersible in one-pass mixes; 0.1% max. on 325 mesh; pour dens. 352 kg/m^3; iodine no. 68.

Vulcan® M. [Cabot/N. Am. Rubber Black] N339 Carbon black; CAS 1333-86-4; EINECS 215-609-9; reinforcing carbon black for rubbers featuring high abrasion resist. and exc. extrusion props.; 0.1% max. on 325 mesh; pour dens. 344 kg/m^3; iodine no. 90.

Vulcan® P. [Cabot/Special Blacks Div] Carbon black; CAS 1333-86-4; EINECS 215-609-9; uv stabilizer for conductive and antistatic plastics applics.; pellets; 20 nm particle size; dens. 21 lb/ft^3; surf. area 140 m^2/g; 1.4% volatiles.

Vulcan® XC72. [Cabot/Special Blacks Div] N472 Carbon black; CAS 1333-86-4; EINECS 215-609-

9; conductive carbon black for rubbers featuring very high elec. conductivity with smooth extrusion and moderate reinforcement; also for conductive and antistatic plastic applics.; pellets; 30 nm particle size; 0.1% max. on 325 mesh; dens. 17 lb/ft^3; pour dens. 264 kg/m^3; surf. area 254 m^2/g; iodine no. 250; 1.5% volatile.

Vulcan® XC72R. [Cabot/Special Blacks Div] Carbon black; CAS 1333-86-4; EINECS 215-609-9; for coloring conductive and antistatic plastics; fluffy; 20 nm particle size; surf. area 254 m^2/g; 1.5% volatiles.

Vulcastab® EFA. [Akzo Nobel] Formaldehyde, ammonia, ethyl chloride condensate; sec. gelling agent and stabilizer for latex foams; visc. liq.

Vulcastab® LW. [Akzo Nobel] Ethylene oxide and fatty alcohol condensate; nonionic; stabilizer for latex rubber processing; flakes.

Vulcastab® T. [Akzo Nobel] Ammonium polymethacrylate sol'n.; stabilizer and thickener for latex rubber processing; visc. liq.

Vulcuren 2. [Bayer AG] 2-(4-Morpholinyldithio) benzothiazole; curing and vulcanizing agent; sulfur donor; heat resistant rubber applic.; yel. to beige powd.; dens. 1.48 g/cm^3; m.p. 1.27; Unverified

Vulcuren 2/EGC. [Bayer/Fibers, Org., Rubbers] 2-(4-Morpholinyl-dithio) benzothiazole; sulfur donor or vulcanizing agent with a much delayed onset of cure and high efficiency in the vulcanization of natural and syn. rubber goods that must withstand heat; used in aircraft and racing car tires, heat-resistant goods, technical goods, esp. seals and conveyor belting, and bulky goods; ylsh. to beige gran.; dens. 1.48 g/cm^3; m.p. 127 C.

Vulkacit 576. [Bayer/Fibers, Org., Rubbers] Condensation prod. of butyraldehyde with aromatic bases; semi-ultra accelerator for highly elastic goods that are subject to heavy dynamic stresses, goods that are cured in presses, steam, or hot air, when the mix contains acidic ingreds. or large amts. of reclaim or scrap, and for ebonite; reddish brn. liq.; sp.gr. 0.99.

Vulkacit CRV/LG. [Bayer/Fibers, Org., Rubbers] 3-Methyl-thiazolidinethione-2; CAS 1908-87-8; accelerator for polychloroprene elastomers, providing exc. heat resist. and resist. to reversion; used for mech. and tech. goods, insulation, sheathings, proofings, rubber sol'ns.; beige to brn. lentil-shaped gran.; sol. in methylene chloride, acetone, dimethyl formamide, methanol, toluene, ethyl acetate; m.w. 133; sp.gr. 1.39; m.p. ≥ 64 C; 98% min. act.

Vulkacit CZ/EG. [Bayer/Fibers, Org., Rubbers] Benzothiazyl-2-cyclohexyl sulfenamide; CAS 95-33-0; fast, scorch-safe accelerator giving delayed onset of cure; used for tires, dynamically stressed technical goods, technical moldings and extrudates; gran.; dens. 1.30 g/cm^3; m.p. ≥ 96 C.

Vulkacit CZ/EGC. [Bayer/Fibers, Org., Rubbers] Benzothiazyl-2-cyclohexyl sulfenamide; CAS 95-33-0; fast accelerator giving delayed onset of cure; used for tires, dynamically stressed technical goods, technical moldings and extrudates; coated gran.; dens. 1.30 g/cm^3; m.p. ≥ 96 C.

Vulkacit CZ/MGC. [Bayer/Fibers, Org., Rubbers] Benzothiazyl-2-cyclohexyl sulfenamide, min. oil coated; fast and scorch-safe accelerator for NR, IR, BR, SBR, and NBR, giving a much delayed onset of cure; used mainly for tires; suitable for dynamically stressed tech. goods (buffers and conveyor belting); tech. moldings and extrudates in general (seals, hoses, strip, footwear, cable sheathings, and insulations); lt. gray coated microgran.; sol. in benzene, CCl_4, methylene chloride, ethyl acetate, acetone; m.w. 264.1; sp.gr. 1.30; m.p. ≥ 96 C; 95% min. act.

Vulkacit D/C. [Bayer/Fibers, Org., Rubbers] Diphenyl guanidine; CAS 102-06-7; accelerator for NR, IR, BR, SBR, NBR, CR, with very slow onset of cure; used alone for bulky goods, e.g., tires, buffers, roll covers; sec. accelerator with mercaptos for mech. goods, extrudates and moldings, footwear, fabric proofings, cable sheathings and insulation; wh. to beige powd.; sol. in acetone, ethanol, ethyl acetate, methylene chloride, benzene; m.w. 211.1; sp.gr. 1.19; m.p. ≥ 145 C; 96% min. act.

Vulkacit D/EGC. [Bayer/Fibers, Org., Rubbers] Diphenyl guanidine; CAS 102-06-7; accelerator for NR, IR, BR, SBR, NBR, CR, with slow onset of cure; used alone for bulky goods, e.g., tires, buffers, roll covers; sec. accelerator with mercaptos for mech. goods, extrudates and moldings, footwear, fabric proofings, cable sheathings and insulation; wh. to beige coated gran.; sol. in acetone, ethanol, ethyl acetate, methylene chloride, benzene; m.w. 211.1; sp.gr. 1.19; m.p. ≥ 145 C; 96% min. act.

Vulkacit DM/C. [Bayer/Fibers, Org., Rubbers] Dibenzothiazyl disulfide, min. oil coated; delayed action, semi-ultra accelerator for NR, IR, BR, SBR, NBR, CR, IIR, chlorosulfonated polyethylene; used for mech. goods, tires, conveyor belts, cables, hoses, rubber footwear, expanded rubber goods; proofed fabrics, intricate molded goods; ylsh. coated powd.; sl. sol. in benzene, CCl_4, methylene chloride, ethanol, acetone; m.w. 332.2; sp.gr. 1.51; m.p. ≥ 168 C; 94% min. act.

Vulkacit DM/MGC. [Bayer/Fibers, Org., Rubbers] Dibenzothiazyl disulfide, min. oil coated; delayed action, semi-ultra accelerator for NR, IR, BR, SBR, NBR, CR, IIR, chlorosulfonated polyethylene; used for mech. goods, tires, conveyor belts, cables, hoses, rubber footwear, expanded rubber goods, proofed fabrics, intricate molded goods; ylsh. coated microgran.; sl. sol. in benzene, CCl_4, methylene chloride, ethanol, acetone; m.w. 332.2; sp.gr. 1.51; m.p. ≥ 168 C; 94% min. act.

Vulkacit DZ/EGC. [Bayer/Fibers, Org., Rubbers] Benzothiazyl-2-dicyclohexyl sulfenamide; CAS 95-33-0; accelerator with delayed onset of cure for dynamically stressed rubber, e.g., NR, IR, BR, SBR, NBR; used for large tires, conveyor belts, shock absorbers, intricate molded goods with long flow periods; cream to lt. brn. coated gran.; sol. in benzene, CCl_4, methylene chloride, ethyl acetate, acetone, ethanol, aliphatic hydrocarbons; m.w. 346.1; sp.gr. 1.20; m.p. ≥ 96 C; 95% min. act.

Vulkacit Merkapto/C. [Bayer/Fibers, Org., Rubbers]

2-Mercaptobenzothiazole, min. oil coated; semi-ultra accelerator for NR, IR, BR, SBR, NBR, IIR, EPDM; used alone in bulky goods or in combination for molded and extruded goods, hoses, conveyor belts, tires, footwear, cables, expanded rubber goods; proofed fabrics, latex articles; ylsh. coated powd.; sol. in ethyl acetate, acetone, methylene chloride, ethanol; m.w. 167.1; sp.gr. 1.52; m.p. ≥ 172 C; 93% min. act.

Vulkacit Merkapto/MGC. [Bayer/Fibers, Org., Rubbers] 2-Mercaptobenzothiazole, min. oil coated; semi-ultra accelerator for NR, IR, BR, SBR, NBR, IIR, EPDM; used alone in bulky goods or in combination for molded and extruded goods, hoses, conveyor belts, tires, footwear, cables, expanded rubber goods; proofed fabrics, latex articles; ylsh. coated microgran.; sol. in ethyl acetate, acetone, methylene chloride, ethanol; m.w. 167.1; sp.gr. 1.52; m.p. ≥ 174 C; 93% min. act.

Vulkacit MOZ/LG. [Bayer/Fibers, Org., Rubbers] Benzothiazyl-2-sulfene morpholide; CAS 102-77-2; accelerator with pronounced delayed onset of cure; used for dynamically stressed rubber goods, NR, IR, BR, SBR, and NBR; used for tires, conveyor belts, shock absorbers, mountings, seals, hoses, cables, footwear, intricate molded goods; yel. lentil-shaped gran.; sol. in acetone, methylene chloride, CCl_4, benzene, ethyl acetate, methanol; m.w. 252.1; sp.gr. 1.35; m.p. ≥ 78 C; 95% min. act.

Vulkacit MOZ/SG. [Bayer/Fibers, Org., Rubbers] Benzothiazyl-2-sulfene morpholide; CAS 102-77-2; fast, scorch-safe accelerator with delayed onset of cure; used for tires, dynamically stressed tech. goods, tech. moldings and extrudates; cylindrical gran.; sp.gr. 1.35; m.p. ≥ 78 C.

Vulkacit NZ/EG. [Bayer/Fibers, Org., Rubbers] Benzothiazyl-2-t-butyl sulfenamide; CAS 95-31-8; fast, scorch-safe accelerator giving delayed onset of cure; used for tires, dynamically stressed technical goods, technical moldings and extrudates; gran.; sp.gr. 1.29; m.p. ≥ 104 C.

Vulkacit ZBEC. [Bayer/Fibers, Org., Rubbers] Zinc salt of dibenzyl dithiocarbamate; CAS 14726-36-4; fast accelerator used in transparent, wh. and brilliantly colored tech. rubber goods and latex articles, incl. mech. goods, footwear, cables, fabric proofings, self-curing rubber mixes and sol'ns.; based on safe amine not subject to formation of traces of harmful volatile nitrosamines during vulcanization; slower curing and less scorchy than ZDMC, ZDEC, ZDBC; wh. to ylsh. wh. powd.; sp.gr. 1.40; m.p. ≥ 180 C.

Vulkacit ZM. [Bayer/Fibers, Org., Rubbers] Zinc salt of 2-mercaptobenzothiazole; CAS 155-04-4; semi-ultra accelerator for for rubber footwear and other hot air-cured goods, molding and tech. goods (roll covers, conveyor belting, transmission belting, footwear soles/heels, hoses, cables, bicycle and car tires, cellular rubber goods); esp. suitable as sec. accelerator for latex goods; sensitizing agent for natural and syn. latex mixes; ylsh. powd.; sp.gr. 1.70; m.p. > 310 C.

Vulkadur® A. [Bayer AG; Bayer/Fibers, Org., Rubbers] Phenol-formaldehyde resin plus hardener; reinforcing and hardening agent used mainly for goods based on NBR, e.g., in the mfg. of brake and clutch linings; also used as an ingred. of adhesives; wh. to yel. powd.; dens. 1.3 g/cc; soften. pt. 75-90 C.

Vulkadur® RA. [Bayer AG] Reinforcing resin for NR, SBR, BR, CR, NBR, and EPDM rubbers.

Vulkadur® RB. [Bayer AG] Reinforcing resin for NR, SBR, BR, CR, NBR, and EPDM rubbers.

Vulkadur® T, 40%. [Bayer AG] Precondensed resorcinol formaldehyde resin; bonding agent for latex dips for textiles, tire cord, and rubber goods; rdsh. brn. liq.; dens. 1.17 g/cc.

Vulkalent B/C. [Bayer/Fibers, Org., Rubbers] Phthalic anhydride; CAS 85-44-9; EINECS 201-607-5; retarder; increasing additions improve safety of mixes undergoing processing and prolong the flow times at curing temps.; not suitable for mixes containing Vulkacit Thiuram and without sulfur; wh. cryst. powd.; dens. 1.51 g/cm^3; m.p. 124 C.

Vulkalent E/C. [Bayer/Fibers, Org., Rubbers] Aromatic sulfonamide deriv. plus 5% calcium carbonate and 1-2% min. oil; vulcanization retarder to control the processing safety and flow time of sulfur-curing natural and syn. rubbers; useful for replasticization of slightly scorched compds.; beige-wh. fine coated powd.; sol. in ethyl acetate, benzene, methylene chloride, CCl_4; sp.gr. 1.68; m.p. 110 C (act. ingred.); 90% min. act.

Vulkalent TM. [Bayer/Fibers, Org., Rubbers] Tolyltriazole; CAS 29385-43-1; prevulcanization retarder for sulfur-modified CR and halobutyl rubbers; also effective in NBR when Vulkacit DM/sulfur curing systems are used; applics. incl. conveyor belting, hose, molded goods; lt. beige gran.; sol. in acetone, ethyl acetate, ethanol, MEK, methylene chloride; m.w. 133.2; sp.gr. 1.2; m.p. 83 C; 98% min. act.

Vulkanol® 80. [Bayer AG] Polyglycol ester of a mixt. of monocarboxylic acids; plasticizer to improve low-temp. flexibility and elastic behavior of vulcanizates; bright yel. clear liq.; dens. 0.97-0.99 g/cc.

Vulkanol® 81. [Bayer/Fibers, Org., Rubbers] Mixt. of a thioester and a carboxylic ester; syn. plasticizer used in tech. goods based on natural and syn. rubber, e.g., hoses and seals for vehicles, aircraft and machinery, moldings and extrudates, conveyor and transmission belting, cable sheathings, proofings; yel. clear liq.; b.p. cannot be distilled without decomp.; dens. 0.96-0.99 g/cm^3.

Vulkanol® 85. [Bayer/Fibers, Org., Rubbers] Ether thioether; plasticizer used to impart antistatic properties to flexible PVC; low-temp. strength in plasticized PVC; applic. incl. conveyor belts, mine air ducts, floorcoverings, shoe soles, protective work clothes, films; yel. liq., unpleasant odor; sol. in org. solvs.; insol. in water; m.w. 650; dens. 1.02-1.07 g/cm^3; visc. 40-80 MPa•s; flash pt. (OC) 190 C; ref. index. 1.468-1.473.

Vulkanol® 88. [Bayer AG; Bayer/Fibers, Org., Rubbers] Dibutyl methylene bisthioglycolate; plasticizer to improve low-temp. flexibility and elastic behavior of vulcanizates; used for nitrile and chloroprene rubber; yel. to brn. liq.; dens. 1.08-1.12 g/cc; b.p. 180-227 C; ref. index 1.485-1.495 (20 C).

Vulkanol® 90. [Bayer/Fibers, Org., Rubbers] Di-2-ethylhexyl ester thiodiglycolate; plasticizer for natural and syn. rubber incl. NBR, NBR/PVC blends, CR, SBR, BR, chlorosulfonated polyethylene, and chlorinated polyethylene; used for hoses, seals, shock absorbers, roll covers, conveyor belting; brnsh. clear liq.; sol. in aliphatic hydrocarbons, toluene, ethanol, acetone; dens. 0.97-0.99 g/cm^3; visc. 18 cps; b.p. cannot be distilled without decomp.; solid. pt. -60 C; flash pt. (OC) 200 C; ref. index 1.460-1.470 (20 C).

Vulkanol® BA. [Bayer AG] Dibenzyl ether; plasticizer to improve low-temp. flexibility and elastic behavior of vulcanizates; colorless liq.; b.p. 174-185 C; dens. 1.04 g/cc; Unverified

Vulkanol® FH. [Bayer AG] Xylene/formaldehyde condensation prod.; resinous plasticizer and tackifier for rubber mixes based on NBR, CR, SBR; gives high bldg. tack; pale yel. highly visc. liq.; misc. with aromatic and chlorinated hydrocarbons; partly sol. in ethers, esters, and aliphatic hydrocarbons; sp.gr. 1.08; visc. 12,000 cps (50 C); flash pt. (OC) 185 C; pour pt. 20 C; Unverified

Vulkanol® KA. [Bayer/Fibers, Org., Rubbers] Polyglycol ether; syn. plasticizer used in tech. goods based on natural and syn. rubber, e.g., hoses and seals for vehicles, aircraft and machinery, moldings and extrudates, conveyor and transmission belting, cable sheathings, proofings; also offers antistatic properties; brn. visc. liq.; misc. with org. solvs. other than aliphatic hydrocarbons and with water; sp.gr. 1.12; visc. 675 cps; b.p. cannot be distilled without decomp.; flash pt. (OC) 205 C; pour pt. 1 C.

Vulkanol® OT. [Bayer AG; Bayer/Fibers, Org., Rubbers] Ether thioether; plasticizer to improve low-temp. flexibility and elastic behavior of vulcanizates; used for nitrile and chloroprene rubber; applics. incl. hoses, seals, mech. goods; clear to yel. liq.; misc. with most org. solvs.; dens. 0.94-0.98 g/cm^3; visc. 40 cps; pour pt. -60 C; flash pt. (OC) 200 C; ref. index 1.471-1.477 (20 C).

Vulkanol® SF. [Bayer AG] Alkyl sulfonic acid alkyl phenyl ester; plasticizer to improve low-temp. flexibility and elastic behavior of vulcanizates; ylsh. transparent liq.; b.p. > 350 C; dens. 0.995-1.06 g/cc; Unverified

Vulkanox® 3100. [Bayer/Fibers, Org., Rubbers] Blend of phenyl- and tolyl-p-phenylene diamines; staining antioxidant and antiflexcracking agent for most diene rubbers (NR, IR, SBR, NBR); moderate antiozonant properties; used for tires, buffers, engine mounts, conveyor belts, mech. goods; brn. flakes; sol. in acetone, ethyl acetate, toluene, methylene chloride, ethanol, benzene, CCl_4; sp.gr. 1.20; solid. pt. ≥ 95 C.

Vulkanox® 4010 NA. [Bayer/Fibers, Org., Rubbers] N-isopropyl-N′-phenyl-p-phenylene diamine; CAS 101-72-4; staining and discoloring antioxidant/antiozonant for protection of rubber from ozone attack, oxidation, heat aging, flexcracking, and rubber poisons; suitable for natural and syn. rubbers; used for dynamically stressed goods, tech. goods, spring components, conveyor and transmission belting, seals, insulation; brn. to dk. violet gran.; sol. in benzene, methylene chloride, CCl_4, ethyl acetate, acetone, ethanol; m.w. 226; sp.gr. 1.07; m.p. ≥ 76 C; 95% min. act.

Vulkanox® 4020. [Bayer/Fibers, Org., Rubbers] N-(1,3-Dimethylbutyl)-N′-phenyl-p-phenylene diamine, staining stabilizer; CAS 793-24-8; antioxidant/antiozonant for cold- and hot-polymerized emulsion and sol'n. SBR and BR; protects vulcanizates against heat, oxidation, and flexcracking; brn. to brn. violet flakes or fused; sol. in benzene, methylene chloride, CCl_4, ethyl acetate, acetone, ethanol, aliphatic hydrocarbons; m.w. 268; sp.gr. 1.02; m.p. ≥ 46 C; 97% min. act.

Vulkanox® 4022. [Bayer/Fibers, Org., Rubbers] Blend of alkylaryl-p-phenylenediamines; staining and discoloring antioxidant offering protection to natural and syn. rubbers from dynamic stresses, oxidation, heat, ozone, and rubber poisons; used in tires, tech. goods, spring components, conveyer belting, hoses, transmission belting and seals, cable sheathings and insulations, inner tubes, roll covers; also offers antiflexcracking effects; dk. purple liq.; sol. in benzene, methylene chloride, aliphatic hydrocarbons, acetone; sp.gr. 1.01; visc. 80 SUS (75 C).

Vulkanox® 4023. [Bayer/Fibers, Org., Rubbers] Blend of alkylaryl-p-phenylenediamines and dialkyl-p-phenylenediamines; staining and discoloring antiozonant and antioxidant protecting natural and syn. rubbers against ozone attack, oxidation, heat aging, flexcracking, and rubber poisons; used in dynamically stressed goods, tires, conveyor belts, spring components; dk. purple liq.; sol. in benzene, methylene chloride, aliphatic hydrocarbons, acetone; sp.gr.1.00; visc. 70 SUS (75 C).

Vulkanox® 4025. [Bayer/Fibers, Org., Rubbers] Blend of alkylaryl-p-phenylenediamines and dialkyl-p-phenylenediamines; staining and discoloring antioxidant offering protection to natural and syn. rubbers from dynamic stresses, oxidation, heat, ozone, and rubber poisons; used in tires, tech. goods, spring components, conveyer belting, hoses, transmission belting and seals, cable sheathings and insulations, inner tubes, roll covers; also offers antiflexcracking effects; dk. purple liq.; dens. 0.97 g/cm^3.

Vulkanox® 4027. [Bayer/Fibers, Org., Rubbers] Blend of alkylaryl p-phenylene diamines; staining and discoloring antiozonant/antioxidant which protects natural and syn. rubber against ozone attack, oxidation, heat aging, flexcracking, and rubber poisons; used in dynamically stressed goods, tires, conveyor belts, spring components; dk. purple liq.; sol. in benzene, methylene chloride, aliphatic hydrocarbons, acetone; sp.gr. 0.96; visc. 80 SUS (75 C).

Vulkanox® 4030. [Bayer/Fibers, Org., Rubbers] N,N′-di-(1,4-dimethylpentyl)-p-phenylene diamine; CAS 3081-14-9; staining and discoloring antiozonant/antioxidant protecting natural rubber and syn. rubbers except CR from ozone attack, oxidation, heat aging, and flexcracking; used for dynamically stressed goods, tires, conveyor belts, hoses, spring components, cables and seals; dk. red low-visc. liq.; sol. in aliphatic hydrocarbons, benzene, methylene chloride, CCl_4, ethyl acetate, acetone, ethanol; sp.gr. 0.91; b.p. 237 C (14 Torr); 93% min. act.

Vulkanox® AFD. [Bayer/Fibers, Org., Rubbers] Unsat. aromatic ether; nonstaining antiozonant for NR, SBR, BR, NBR/PVC, and CR; liq.; dens. 1.02 g/cm^3.

Vulkanox® BKF. [Bayer/Fibers, Org., Rubbers] 2,2′-Methylene-bis(4-methyl-6-t-butylphenol); CAS 119-47-1; nonstaining antioxidant for natural and syn. rubbers; used in transparent, bathing, surgical tech., and latex goods, fabric proofings; stabilizer for hot- and cold-polymerized emulsion-SBR, BR, and NBR, and sol'n.-BR, IR, and ABS; wh. cryst. powd.; sol. in ethanol, acetone, ethyl acetate, methylene chloride, CCl_4, benzene; m.w. 340; sp.gr. 1.04; m.p. $\geq$ 125 C.

Vulkanox® CS. [Bayer/Fibers, Org., Rubbers] Sterically hindered bisphenol absorbed on clay; nonstaining antioxidant imparting resistance to oxygen and heat; wh. to beige powd.; dens. 1.91 g/cm^3; m.p. $\geq$ 150 C (act. substance); 25 $\pm$ 2% bisphenol.

Vulkanox® DDA. [Bayer/Fibers, Org., Rubbers] Styrenated diphenyl amine; staining and discoloring antioxidant for natural and syn. rubbers, exp. CR, giving protection against oxidation, heat, flexcracking, and rubber poisons; used for inner tubes, sponge rubber, seals/gaskets, soles, latex goods, heat-resistant articles; stabilizer for cold and hot-polymerized NBR and NBR or SBR latexes; orange to brn. visc. liq.; sol. in aliphatic hydrocarbons, benzene, CCl_4, methylene chloride, acetone, ethyl acetate; sp.gr. 1.08; visc. 15,000 cps; b.p. > 300 C.

Vulkanox® DS. [Bayer/Fibers, Org., Rubbers] Alkyl and aralkyl-substituted phenols; antioxidant for latex, transparent, dipped, bathing, and tech. goods, fabric proofings; yel. to rdsh. clear visc. liq.; b.p. 140 C; dens. 0.92 $\pm$ 0.05 g/cm^3.

Vulkanox® DS/F. [Bayer/Fibers, Org., Rubbers] Alkyl and aralkyl-substituted phenols; antioxidant for latex, transparent, dipped, bathing, and tech. goods, fabric proofings; wh. to beige-yel. powd.; b.p. > 140 C; dens. 1.30 $\pm$ 0.05 g/cm^3.

Vulkanox® HS/LG. [Bayer/Fibers, Org., Rubbers] 2,2,4-Trimethyl-1,2-dihydroquinoline, polymerized; CAS 26780-96-1; staining antioxidant for tech. goods and rubber prods.; affords very good heat protection; improves heat resist. of EPDM; also used in latex; yel. to amber lentil-shaped gran.; dens. 1.07 $\pm$ 0.02 g/cm^3; soften. pt. > 85 C.

Vulkanox® HS/Pellets. [Bayer/Fibers, Org., Rubbers] 2,2,4-Trimethyl-1,2-dihydroquinoline, polymerized; CAS 26780-96-1; weakly staining and discoloring antioxidant for NR, IR, BR, SBR, NBR%, EPDM; used for heat-resistant goods, tire, conveyor belts, hoses, seals, footwear, cables; yel. to tan pellets; m.w. $(172)_n$; sol. in acetone, ethyl acetate, ethanol, benzene, methylene chloride, CCl_4; sp.gr. 1.07; 95% min. act.

Vulkanox® HS/Powd. [Bayer/Fibers, Org., Rubbers] 2,2,4-Trimethyl-1,2-dihydroquinoline, polymerized; CAS 26780-96-1; weakly staining and discoloring antioxidant for NR, IR, BR, SBR, NBR%, EPDM; used for heat-resistant goods, tire, conveyor belts, hoses, seals, footwear, cables; yel. to tan powd.; m.w. $(172)_n$; sol. in acetone, ethyl acetate, ethanol, benzene, methylene chloride, CCl_4; sp.gr. 1.07; soften. pt. > 75 C.

Vulkanox® KB. [Bayer/Fibers, Org., Rubbers] 2,6-Di-t-butyl-p-cresol; CAS 128-37-0; EINECS 204-881-4; nondiscoloring/nonstaining antioxidant for lt. colored and transparent natural and syn. rubber goods; used in fabric proofings, toys, bathing, latex, and dipped goods; stabilizer in emulsion and sol'n.-polymerized elastomers (SBR, NBR, BR, IR); co-stabilizer for ABS; wh. to pink powd.; sol. in aliphatic hydrocarbons, benzene, CCl_4, methylene chloride, acetone, ethyl acetate, ethanol; m.w. 220; sp.gr. 1.03; m.p. $\geq$ 69 C; 99.8% min. act.

Vulkanox® MB-2/MGC. [Bayer/Fibers, Org., Rubbers] 4- and 5-Methylmercaptobenzimidazole, min. oil coated; nondiscoloring/nonstaining antioxidant for natural and syn. rubbers; synergist; protects against rubber poisons; sensitizing agent for latex; ylsh.-wh. coated microgran.; sol. in dimethyl formamide, acetone, ethanol; m.w. 164; sp.gr. 1.25; 96% min. act., 2% oil.

Vulkanox® NKF. [Bayer/Fibers, Org., Rubbers] 2,2′-Isobutylidene-bis(4,6-dimethyl phenol); CAS 33145-10-7; nonstaining/nondiscoloring antioxidant for natural and syn. elastomers; used for proofed and dipped goods, foam rubber, hoses, footwear, floor coverings, latex goods, carpet backing, other tech. goods; stabilizer for hot- and cold-polymerized emulsion-SBR, BR, and NBR, and sol'n.-BR, SBR, and IR; also effective in ABS; wh. cryst. powd.; m.w. 298; sol. in acetone, ethyl acetate, ethanol, methylene chloride, CCl_4; sp.gr. 1.12; m.p. > 156 C.

Vulkanox® OCD. [Bayer/Fibers, Org., Rubbers] Octylated diphenyl amine; antioxidant for tech. goods and rubber prods.; used in latex; pale gray brn. gran.; dens. 1.0 $\pm$ 0.05 g/cm^3; m.p. > 88 C.

Vulkanox® OCD/SG. [Bayer/Fibers, Org., Rubbers] Octylated diphenyl amine; weakly staining and discoloring antioxidant for CR, NR, IR, SBR, BR, NBR, protecting against rubber poisons; used for dynamically stressed goods, tires, air hoses, belts, sponge rubber, footwear, profiles; pale grayish-brn. cylindrical gran.; sol. in acetone, ethyl acetate, benzene, methylene chloride, CCl_4, aliphatic hydrocarbons; sp.gr. 1.0; m.p. $\geq$ 88 C.

Vulkanox® PAN. [Bayer/Fibers, Org., Rubbers]

Phenyl-α-naphthyl amine; CAS 90-30-3; antioxidant for tech. goods and rubber prods.; used in latex; also antiflexcracking agent on NR and IR; lt. brn. to lt. violet gran.; dens. 1.11 ± 0.03 g/cm³; solid. pt. > 52 C.

Vulkanox® SP. [Bayer/Fibers, Org., Rubbers] Styrenated phenol; CAS 61788-44-1; nonstaining antioxidant; yel. to brnsh. visc. liq.; dens. 1.08 g/cm³; 25 ± 2% bisphenol.

Vulkanox® ZKF. [Bayer/Fibers, Org., Rubbers] 2,2′-Methylene-bis(4-methyl-6-cyclohexyl phenol); nonstaining/nondiscoloring antioxidant for natural and syn. rubbers; used in transparent, bathing, surgical tech., and latex goods, proofed and dipped goods, foam rubber, hoses, footwear, floor coverings, carpet backing; wh. powd.; dens. 1.08 g/cm³; m.p. > 125 C.

Vulkanox® ZMB2/C5. [Bayer/Fibers, Org., Rubbers] Zinc salt of 4- and 5-methylmercaptobenzimidazole; nonstaining and nondiscoloring antioxidant for natural and syn. rubber; synergist; used mainly for heat-resistant thiuram-cured goods, wh. and colored and transparent goods, and latex foam; sensitizer for latexes; wh. to lt. beige coated powd.; sl. sol. in aliphatic hydrocarbons, benzene, CCl_4; m.w. 391.5; sp.gr. 1.75; m.p. ≥ 300 C; 15.5% zinc, 5% oil.

Vulkanox® ZMB-2/G. [Bayer/Fibers, Org., Rubbers] 4- and 5-methylmercaptobenzimidazole; nondiscoloring/nonstaining antioxidant for natural and syn. rubbers; synergist; protects against rubber poisons; sensitizing agent for latex; wh. to pale beige coated powd.; dens. 1.75 g/cm³; m.p. > 300 C.

Vulkasil® A1. [Bayer AG] Precipitated sodium aluminum silicate; med. reinforcing effect; filler used in all rubbers except silicone rubber; wh. amorphous powd.; dens. 2.0 g/cm³; surf. area (BET) 65 m²/g; pH 10-12.

Vulkasil® C. [Bayer AG] Precipitated silica with a small amt. of calcium silicate; med. reinforcing effect; filler used in all rubbers except silicone rubber; wh. amorphous powd.; dens. 2.0 g/cm³; surf. area (BET) 60 m²/g; pH 8.5-9.5; Unverified

Vulkasil® N. [Bayer AG] Reinforcing precipitated silica; filler for all rubbers except silicone rubber; ingred. of Cohedur or Cofill bonding systems; wh. amorphous powd. and gran.; dens. 2.0 g/cm³; surf. area (BET) 130 m²/g; pH 6.5-7.3; Unverified

Vulkasil® N/GR-S. [Bayer AG] Reinforcing precipitated silica; filler for all rubbers except silicone rubber; ingred. of Cohedur or Cofill bonding systems; wh. gran.; dens. 2.0 g/cm³; surf. area (BET) 130 m²/g; pH 6.5-7.3; Unverified

Vulkasil® S. [Bayer AG] Reinforcing precipitated silica; filler for all rubbers except silicone rubber; ingred. of Cohedur or Cofill bonding systems; wh. amorphous powd. and gran.; dens. 2.0 g/cm³; surf. area (BET) 170 m²/g; pH 5.5-7.0; Unverified

Vulkazon® AFD. [Bayer/Fibers, Org., Rubbers] Unsat. aromatic ether; nonstaining and nondiscoloring antiozonant for CR, and for NR, BR, SBR in conjunction with ozone protective waxes; used for extruded profiles, hoses, cables, and proofed fabrics; colorless to pale yel. liq.; bitter almond odor; sol. in acetone, ethanol, ethyl acetate, toluene, methylene chloride, aliphatic hydrocarbons; sp.gr. 1.02; b.p. 135 C (4.5 Torr).

Vulkazon® AFS-50. [Bayer/Fibers, Org., Rubbers] Bis-(1,2,3,6-tetrahydrobenzaldehyde) pentaerythritol acetal absorbed on silica, min. oil coated; nonstaining antiozonant for CR, IIR, CIIR, BIIR; used for hoses, extruded profiles, cables, sheeting, proofed goods; beige to gray powd.; m.w. 310; sol. in acetone, ethyl acetate, toluene, methylene chloride, ethanol; sp.gr. 1.34; m.p. 85 C; 50% act.

Vulkazon® AFS/LG. [Bayer/Fibers, Org., Rubbers] Bis (1,2,3,6-tetrahydrobenzaldehyde) pentaerythritol acetal; nonstaining antiozonant for CR, IIR, CIIR, BIIR, lt.-colored and carbon blk.-contg. vulcanizates which must be nonstaining; used for hoses, extruded profiles, cables, sheeting, proofed goods; brnsh. lentil-shaped gran.; sol. in acetone, ethyl acetate, toluene, methylene chloride, ethanol; m.w. 310; sp.gr. 1.06; m.p. 85 C min.

Vultac® 2. [Elf Atochem N. Am./Org. Chems.] Alkyl phenol disulfide; CAS 68555-98-6; NR, NBR, SBR vulcanizer; plasticizer; NBR, SBR tackifier; accelerates SBR; br. to straw tacky, resinous solid; sp.gr. 1.1-1.2; soften. pt. 50-60 C; flash pt. (COC) 235 C; 21.8-23.8% sulfur.

Vultac® 3. [Elf Atochem N. Am./Org. Chems.] Alkyl phenol disulfide; CAS 68555-98-6; vulcanizing agent for rubber and adhesives industries; used in chlorobutyl rubber, tires; brn. solid; sp.gr. 1.20; soften. pt. 78-93 C; flash pt. (COC) 171 C; 27-29% sulfur.

Vultac® 4. [Elf Atochem N. Am./Org. Chems.] Alkyl phenol disulfide and stearic acid; NR, NBR, SBR vulcanizer; plasticizer; tackifier; br. tacky, resinous solid; char. odor; sp.gr. 1.05-1.15; soften. pt. 52-60 C; flash pt. (COC) 231 C; 15.3-16.7% sulfur.

Vultac® 5. [Elf Atochem N. Am./Org. Chems.] Alkyl phenol disulfide on inert carrier; NR, NBR, SBR vulcanizer; accelerates SBR, NBR cures; br. lt. powd.; sp.gr. 1.435; flash pt. (COC) 171 C; 18.5-21% sulfur.

Vultac® 7. [Elf Atochem N. Am./Org. Chems.] Alkyl phenol disulfide; CAS 68555-98-6; vulcanizing agent for rubber and adhesives industries; used in chlorobutyl rubber, tires; dk. brn. flakes; sp.gr. 1.20; soften. pt. 105-125 C; flash pt. (COC) 204 C; 29.35-31.75% sulfur.

Vultac® 710. [Elf Atochem N. Am./Org. Chems.] Alkyl phenol disulfide, 10% stearic acid; vulcanizing agent for rubbers, incl. CR, NR, EPDM, chlorinated rubbers, and blends of SBR with reclaim or natural rubber; replacement for sulfur or thiuram; used in tires; dk. brn. flakes; soften. pt. 75-90 C; flash pt. (COC) 204 C; 26.4-28.4% S.

Vultamol®. [BASF AG] Sodium salt of condensed naphthalene sulfonic acid; dispersant for rubber syntheses and processing of rubber latexes; powd.; water-sol.; 91% conc.

Vultamol® SA Liq. [BASF AG] Sodium salt of condensed naphthalene sulfonic acid; dispersant for rubber syntheses and processing of rubber latexes; liq.; water-misc.; 45% conc.

Vybar® 103. [Petrolite] Syn. wax; ethylene-derived hydrocarbon polymer; CAS 8002-74-2; lubricant, anticaking agent, modifier; used in paraffin; used in candles to replace stearic acid; opacifies the candle and imparts resistance to thermal shock; pigment disp. to wet inorg. pigments and fillers at high loading levels; suitable for hot melt inks, mold release compds., plastic lubricants, protective coatings, polishes, slip and antimar additives; emollient, gloss, lubricant for cosmetics; FDA 21CFR §175.105, 176.170, 176.180, 178.3850; solid; limited sol. in org. solvs.; m.w. 2800; dens. 0.92 g/cc; visc. 310 cps (99 C); m.p. 72 C; iodine no. 14.

Vybar® 253. [Petrolite] Syn. wax; CAS 8002-74-2; plasticizer for rubber, PU mold release component, leveling agent for printing inks, additive for casting/pattern waxes, coatings, floor polishes, candles; increases hardness and opacity of paraffin without increasing cloud pt. or visc.; dens. 0.92; visc. 6 cps (99 C); soften. pt. 67 C.

Vybar® 260. [Petrolite] Syn. wax; olefin-derived hydrocarbon polymer; CAS 8002-74-2; paraffin wax modifier; hard, low-melting polymer with low melt visc; additive producing blends with paraffin which are harder and more opaque; upgrades properties of lower-grade waxes; useful in polishes, industrial coatings, hot-melt inks, mold-release compds., plastic lubricants, and slip and antimar additives; emollient, gloss, lubricant for cosmetics; FDA 21CFR §175.105, 176.170, 176.180, 178.3850; solid; sol. in min. spirits, xylene, and trichloroethylene; m.w. 2600; dens. 0.90 g/cc; visc. 330 cps (99 C); m.p. 51 C; iodine no. 15.

Vybar® 825. [Petrolite] Syn. wax; ethylene-derived hydrocarbon polymer; CAS 8002-74-2; plasticizer, lubricant with good slip chars.; additive in specialty lubricants, polishes, cutting oils, ski wax, and release agents; ozone barrier for elastomers; increases gloss, lubricity, and film deposition on skin for cosmetics applics.; liq.; sol. in org. solvs., min. oil, chlorinated solvs., and selected silicones; m.w. 1500; dens. 0.86 g/cc; visc. 52 cps (99 C); pour pt. -34 C; iodine no. 30.

Vycel. [Crowley Chem.] Petrol.-derived surfactant; surfactant for use in polyurethane foams; patent #5,104,904; lt. yel.; visc. 194 SSU (100 F); pour pt. 13 F; flash pt. (COC) 365 F.

Vydax® 1000. [DuPont] PTFE wax; CAS 9002-84-0; internal/external lubricant for thermosets, thermoplastics, elastomers; liq.

Vydax® AR. [DuPont] PTFE wax; CAS 9002-84-0; internal/external lubricant for thermosets, thermoplastics, elastomers; liq.

Vydax® Spray No. V312-A. [IMS] < 2% PTFE telomer mix release agent (CAS 79070-11-4, 65530-85-0) with 35-45% HFC-152a, 25-30% dimethyl ether, 25-35% HCFC-141b, < 6% IPA; non-CFC paintable multipurpose mold release and parting agent for thermosets and thermoplastics incl. acrylics and other engineering materials; clear to wh. mist with sl. ethereal odor as dispensed from aerosol pkg.; sl. sol. in water; sp.gr. < 1; vapor pressure 55 ± 10 psig; flash pt. < 0 F; > 90% volatile; *Toxicology:* inh. may cause CNS depression with dizziness, headache; higher exposure may cause heart problems; gross overexposure may cause fatality; eye and skin irritation; direct contact with spray can cause frostbite; may cause polymer fume fever; *Precaution:* flamm. limits 1.2-18%; at elevated temps. (> 130 F) aerosol containers may burst, vent, or rupture; incompat. with strong oxidizers, strong caustics, reactive metals (Na, K, Zn, Mg, Al); *Storage:* store in cool, dry area out of direct sunlight; do not puncture, incinerate, or store above 120 F.

Vydax® WD. [DuPont] Fluorotelomer disp.; mold lubricant and release agent where applic. to hot pressing surfs. is required; specially compded. to permit water dilution; excellent antistick properties; release coatings, water-based paints and inks; dry lubricant films; 5 μ particle size disp.; dens. 1.56 g/ml; visc. 585 cps.; Unverified

Vynate® 2-EH. [Union Carbide] Vinyl 2-ethylhexanoate; vinyl ester monomer; reactive intermediate in emulsion, suspension, and sol'n. polymerization for copolymer latexes, thermosetting coatings, water-sol. polymers; end-uses incl. coatings (interior/exterior, industrial, uv-curable, powd., electrodeposition, wood, cement, metal, paper), adhesives, caulks, sealants, concrete additives, printing inks, textile sizing, binders for nonwovens; sol. in acetone; insol. in water; m.w. 170.25; sp.gr. 0.874 (20 C); dens. 7.30 lb/gal; vapor pressure 0.42 mm Hg (20 C); f.p. < -90 C; b.p. 185.6 C (760 mm); flash pt. (CC) 65 C; *Precaution:* avoid contact with copper; *Storage:* avoid exposure to light.

Vynate® L-3. [Union Carbide] Vinyl propionate; vinyl ester monomer; reactive intermediate in emulsion, suspension, and sol'n. polymerization for copolymer latexes, thermosetting coatings, water-sol. polymers; end-uses incl. coatings (interior/exterior, industrial, uv-curable, powd., electrodeposition, wood, cement, metal, paper), adhesives, caulks, sealants, concrete additives, printing inks, textile sizing, binders for nonwovens; sol. 0.6% in water; m.w. 100.1; sp.gr. 0.917 (20 C); dens. 7.63 lb/gal; vapor pressure 33.9 mm Hg (20 C); f.p. -81 C; b.p. 95 C (760 mm); flash pt. (CC) 6 C; *Precaution:* flamm.; avoid contact with copper; *Storage:* inhibit before storage; 6 mos shelf life protected against heat and light; optimum storage below 25 C; blanket with nitrogen.

Vynate® Neo-5. [Union Carbide] Vinyl pivalate; vinyl ester monomer; reactive intermediate in emulsion, suspension, and sol'n. polymerization for copolymer latexes, thermosetting coatings, water-sol. polymers; end-uses incl. coatings (interior/exterior, industrial, uv-curable, powd., electrodeposition, wood, cement, metal, paper), adhesives, caulks, sealants, concrete additives, printing inks, textile sizing, binders for nonwovens; sol. in acetone, ethanol; insol. in water; m.w. 128.17; sp.gr. 0.871 (20 C); dens. 7.28 lb/gal; vapor pressure 18.1 mm Hg (20 C); f.p. < -75 C; b.p. 112 C (760 mm); flash pt. (CC) 14 C; *Precaution:* avoid contact with copper; *Storage:* avoid exposure to light; do not store in polyethylene.

Vynate® Neo-9. [Union Carbide] Vinyl neononanoate; vinyl ester monomer; reactive intermediate in emulsion, suspension, and sol'n. polymerization for copolymer latexes, thermosetting coatings, water-sol. polymers; end-uses incl. coatings (interior/exterior, industrial, uv-curable, powd., electrodeposition, wood, cement, metal, paper), adhesives, caulks, sealants, concrete additives, printing inks, textile sizing, binders for nonwovens; insol. in water; m.w. 184.28; sp.gr. 0.887 (20 C); dens. 7.38 lb/gal; vapor pressure 0.96 mm Hg (20 C); f.p. < -59 C; b.p. 188 C (760 mm); flash pt. (CC) 65 C; *Precaution:* avoid contact with copper; *Storage:* preserve MEHQ inhibitor content; avoid exposure to light; avoid storage in plastics; 1 yr max. storage @ 15-25 C, 2 mos @ 50 C.

Vynate® Neo-10. [Union Carbide] Vinyl neodecanoate with 5 ppm MEHQ inhibitor; vinyl ester monomer; reactive intermediate in emulsion, suspension, and sol'n. polymerization for copolymer latexes, thermosetting coatings, water-sol. polymers; end-uses incl. coatings (interior/exterior, industrial, uv-curable, powd., electrodeposition, wood, cement, metal, paper), adhesives, caulks, sealants, concrete additives, printing inks, textile sizing, binders for nonwovens; insol. in water; m.w. 198.3; sp.gr. 0.880 (20 C); dens. 7.32 lb/gal; vapor pressure 0.002 mm Hg (20 C); f.p. < -59 C; b.p. 84-86 C (10 mm); flash pt. (CC) 83 C; *Precaution:* avoid contact with copper; *Storage:* preserve MEHQ inhibitor content; avoid exposure to light; avoid storage in plastics; 1 yr max. storage @ 15-25 C, 2 mos @ 50 C.

W

1W70, 1W73. [AlliedSignal/Ind. Fibers] Polyester; reinforcing material for in-rubber composites providing high str. and low elong.; 1W73 has a special finish to make it adhesive active for exc. adhesion to rubber using a single-dip adhesive system; tenacity 8.9 g/d; shrinkage 8.5% (350 F); toughness 0.7 g/d.

1W74. [AlliedSignal/Ind. Fibers] Polyester; reinforcing material for in-rubber composites providing low shrinkage and special adhesive active finish; its toughness and exc. impact resist. make it ideal for conveyor belt applics.; tenacity 8.3 g/d; shrinkage 2% (350 F); toughness 0.9 g/d.

Wacker HDK® H15. [Wacker-Chemie GmbH] Fumed silica; CAS 112945-52-5; hydrophobic thickener, thixotrope, excipient, free-flow aid for nat. and syn. rubber, high temp. curing silicone rubber, cold curing elastomers (silicone, polysulfide), paints (epoxy, alkyd, zinc-rich), printing inks, powd. fire extinguisher; polymer powds.; wh. powd.; apparent dens. 40 g/l; surf. area 120 ± 20 m^2/g; ref. index 1.45; pH 3.8-4.5 (4% in 1:1 water/methanol disp.); > 99.8% SiO_2.

Wacker HDK® H20. [Wacker-Chemie GmbH; Wacker Silicones] Fumed silica; hydrophobic thickener, thixotrope, excipient, free-flow aid for paints/coatings, pharmaceuticals/cosmetic powds., sealing and jointing compds., printing inks, PVC dry-blending, rubbers; wh. powd.; apparent dens. 40 g/l; surf. area 170 ± 30 m^2/g; ref. index 1.45; pH 3.8-4.5 (4% in 1:1 water/methanol disp.).

Wacker HDK® N20. [Wacker-Chemie GmbH; Wacker Silicones] Fumed silica; hydrophilic thickener, thixotrope for paints/coatings, pharmaceuticals/ cosmetics (toothpaste, tablets, powds., aerosols, suspensions, ointments, creams), diazo paper, printing inks, PVC plastisols, rubbers, adhesives; wh. powd.; apparent dens. 40 g/l; surf. area 200 ± 30 m^2/g; ref. index 1.45; pH 3.6-4.3 (4% aq.).

Wacker HDK® S13. [Wacker-Chemie GmbH; Wacker Silicones] Fumed silica; hydrophilic thickener, thixotrope for high-temp. and cold curing silicone elastomers; wh. powd.; apparent dens. 50 g/l; surf. area 125 ± 15 m^2/g; ref. index 1.45; pH 3.6-4.3 (4% aq.); > 99.8% SiO_2.

Wacker HDK® T30. [Wacker-Chemie GmbH; Wacker Silicones] Fumed silica; hydrophilic thickener, thixotrope for high-temp. curing silicone rubber, PVC plastisols, organosols, plasticized PVC, cable compds., films, polyester gel coats, paints (polyester, alkyd, acrylic), printing inks; wh. powd.; apparent dens. 40 g/l; surf. area 300 ± 30 m^2/g; ref. index 1.45; pH 3.6-4.3 (4% aq.); > 99.8% SiO_2.

Wacker HDK® T40. [Wacker-Chemie GmbH; Wacker Silicones] Fumed silica; hydrophilic thickener, thixotrope for high-temp. curing silicone rubber, PVC plastisols, organosols, cable compds., paints (polyester, alkyd, acrylic), printing inks; wh. powd.; apparent dens. 40 g/l; surf. area 400 ± 40 m^2/g; ref. index 1.45; pH 3.6-4.3 (4% aq.); > 99.8% SiO_2.

Wacker HDK® V15. [Wacker-Chemie GmbH; Wacker Silicones] Fumed silica; hydrophilic thickener, thixotrope for pharmaceuticals/cosmetics (suspensions, ointments, creams), silicone rubber, polyurethane and polysulfide elastomers, acrylates; wh. powd.; apparent dens. 50 g/l; surf. area 150 ± 30 m^2/g; ref. index 1.45; pH 3.6-4.3 (4% aq.).

Wacker Processing Aid 772. [Wacker Silicones] Formulated prod.; processing aid which inhibits the tendency to crepe-harden and reduces initial plasticity of silicone compds.; soft opaque paste, char. odor; negligible sol. in water; sp.gr. 1.03; *Toxicology:* may release airborne contaminants during heat cure; avoid inh. exposure; *Precaution:* will burn with a lazy smoldering-type flame; hazardous/ thermal decomp. prods.: SiO_2, CO, CO_2, formaldehyde, various hydrcarbon fragments; *Storage:* store in cool, dry, well-ventilated area below 32 C; avoid excessive humidity and heat.

Wacker Silicone Antifoam Emulsion SE 9. [Wacker-Chemie GmbH] Simethicone; CAS 8050-81-5; antifoam, processing aid for foods, cosmetics, pharmaceuticals, fermentation, in mfg. of plastics in contact with food, for degassing of monomers (e.g., PVC, latex stripping); FDA §173.340, BGA II, VII; milky wh. med. visc. o/w emulsion; sp.gr. 1.0; pH 6-8; 15-16% solids in water.

Wacker Stabilizer R. [Wacker-Chemie GmbH; Wacker Silicones] Functional polydimethylsiloxane mixt. with calcium hydroxide; additive for Elastosil® R compds. catalyzed with Wacker Catalyst E; improves oil resist., prevents surf. bloom; reduces compr. set of silicone compds. catalyzed with 2,4-dichlorobenzoyl peroxide; off-wh. paste, sl. odor; negligible sol. in water; sp.gr. 1.58; flash pt. > 300 C; *Toxicology:* eye irritant; may cause permanent eye damage; calcium hydroxide: OSHA TLV 5 mg/m^3; *Precaution:* incompat. with strong oxidants; hazardous decomp. prods.: SiO_2, CO_2, CO, incompletely burned hydrocarbons, lime dust (at combustion), formaldehyde (at 150 C); *Storage:* store in cool, well-ventilated place, below 32 C, in closed containers; avoid excessive humidity and heat.

Wareflex® 650. [Sartomer] Dibutoxyethoxyethyl adipate; CAS 141-17-3; plasticizer for elastomers, TPE; enhances low temp. cure props., reduces processing visc.; APHA 5 clear liq.; sp.gr. 1.019; 98% reactive esters.

Water Shield. [Zyvax] Release agent in water and nonionic emulsifiers; nonionic; high performance, semipermanent release coating for multiple release of epoxies, phenolics, polyester resins, polyimides, polybismaleimides, rubbers; environmentally safe; no solvs. CFC, VOCs, ODCs; themally stable; high gloss, slip, lubricity; opaque liq., bland odor; dens. 8.0 lb/gal; flash pt. (CC) none; pH 9.2-9.5; *Storage:* 12 os shelf life stored in cool, dry area (< 100 F) in original unopened container; do not store at elevated or freezing temps.

Wayfos A. [Olin] Phosphate acid, acid form; anionic; antistat, emulsifier, wetting agent, detergent, coupling agent, solubilizer, lubricant, corrosion inhibitor; for alkaline built cleaners, textiles, plastics, metal, emulsion polymerization, agric. applics.; Gardner 2-5 liq.; sol. in alcohols, glycol ethers, aromatic and aliphatic hydrocarbons, chlorinated solvs.; disp. in water; sp.gr. 1.08; dens. 8.94 lb/gal; pH 1.5-2.0 (5% aq.); 100% act.; *Toxicology:* LD50 (oral, rats) < 5 g/kg; corrosive to eyes; possible skin corrosivity.

Wayfos D-10N. [Olin] Poly(oxy-1,2-ethanediyl), -nonylphenyl-hydroxy phosphate; anionic; antistat, dispersant, corrosion inhibitor, emulsifier, stabilizer, solubilizer for textile processing, alkaline and acid soak cleaners, hard surf. cleaners, dishwashes, wax removers, emulsion polymerization, agric., paper processing; Gardner 1-2 visc. liq.; sol. in water, alcohols, glycol ethers, aromatic and aliphatic hydrocarbons, chlorinated solvs.; sp.gr. 1.10; dens. 9.15 lb/gal; pH 1.7-2.0 (5% aq.); Draves wetting 44 s (0.25%); Ross-Miles foam 133; 100% act.; *Toxicology:* LD50 (oral, rats) > 5 g/kg; skin irritant; corrosive to eyes.

Wayfos M-60. [Olin] Aromatic phosphate ester, free acid; anionic; corrosion inhibitor, hydrotrope; drycleaning detergent; emulsifier for pesticides, emulsion polymerization; lt. amber visc. liq.; sol. in alcohols, glycol ethers, aromatic and aliphatic hydrocarbons, chlorinated solvs.; disp. in water; sp.gr. 1.12; dens. 9.32 lb/gal; pH 2.0-2.5 (1% aq.); 100% act.; *Toxicology:* LD50 (oral, rats) > 5 g/kg; irritating to skin and eyes.

Weston® 399. [GE Specialty] Trisnonylphenyl phosphite with 0.75% wt. triisopropanolamine; CAS 26523-78-4; EINECS 247-759-6; stabilizer for wide variety of polymers; improves color and processing stability during compounding, processing and enduse; synergistic with hindered phenols and other stabilizers; FDA 21CFR §175.105, 178.2010; approved in Canada, Germany, France, UK, Japan and other countries; APHA 100 max. clear liq.; m.w. 688; sp.gr. 0.980-0.992 (25/15.5 C); dens. 0.98 g/ml; visc. 250.0 cps (60 C); acid no. 0.50 max.; flash pt. (PMCC) 196 C; ref. index 1.5250-1.5280; 4.2% P; *Toxicology:* contact with skin or eyes may cause irritation; *Storage:* 1 yr storage life when stored in closed containers away from excess humidity; blanket under dry nitrogen.

Weston® 399B. [GE Specialty] Trisnonylphenyl phosphite (contains 1.0% wt. triisopropanolamine); CAS 26523-78-4; EINECS 247-759-6; stabilizer used in epoxies, hot-melt adhesives, PU, polyester, SBR, PP; in molding and extrusion of PP, HDPE, LDPE, HIPS, PC, ABS, PVC, polyesters, in calendering of ABS, PVC; in film applics. of PP, PE, PVC; fiber applics. of PP, polyesters; Gardner 3 max. clear liq.; m.w. 688; sp.gr. 0.980-0.992 (25/15.5 C); dens. 0.98 g/ml; visc. 250.0 cps (60 C); acid no. 0.50 max.; flash pt. (PM) 196 C; ref. index 1.5250-1.5280; 4.3% P; Custom

Weston® 430. [GE Specialty] Tris(dipropyleneglycol) phosphite; CAS 36788-39-3; EINECS 253-211-7; heat, color, and visc. stabilizer for a variety of polymers incl. polyester fibers and PU; APHA 50 max. clear liq.; m.w. 430; sp.gr. 1.088-1.098 (25/15.5 C); dens. 1.09 g/ml; visc. 42.2 cps (38 C); acid no. 0.20 max.; flash pt. (PMCC) 118 C; ref. index 1.4600-1.4635; 7.2% P; *Toxicology:* contact with skin or eyes may cause irritation; *Storage:* 1 yr storage life when stored in closed containers away from excess humidity; blanket under dry nitrogen.

Weston® 439. [GE Specialty] Poly 4,4′ isopropylidenediphenol neodol 25 alcohol phosphite; CAS 96152-48-6; EINECS 306-120-2; chelating agent in conjunction with metallic soaps to improve color and lt. stability and clarity of PVC formulations; improves color and light stability and clarity of PVC formulations; APHA 100 max. clear liq.; sol. in most common aprotic org. solvs.; sp.gr. 0.965-0.990 (25/15.5 C); b.p. 165 C (5 mm Hg); acid no. 0.2 max.; hyd. no. 20-40; flash pt. (PMCC) 177 C; ref. index 1.5040-1.5095; 5.4% P; *Toxicology:* contact with skin or eyes may cause irritation; *Storage:* 1 yr storage life when stored in closed containers away from excess humidity; blanket under dry nitrogen.

Weston® 474. [GE Specialty] Tris Neodol-25 phosphite; CAS 68610-62-8; stabilizer for hot-melt adhesives, PU, polyesters; used in molding, extrusion, and film applics. in PP, HDPE, LDPE, PVC, and polyesters; also useful for PP fiber applics. and calendering of PVC; APHA 50 max. clear liq.; m.w. 670; sp.gr. 0.875-0.885 (25/15.5 C); dens. 0.88 g/ml; visc. 21.0 cps (38 C); acid no. 0.05 max.; flash pt. (COC) 196 C; ref. index 1.4580-1.4610; 4.6% P; Custom

Weston® 491. [GE Specialty] Diphenyl didecyl (2,2,4-trimethyl-1,3-pentanediol) diphosphite; CAS 80584-87-8; stabilizer for PU, hot melt adhesives; used in molding and extrusion of PP, HDPE, LDPE, PC, ABS, PVC; used in film applics. for HDPE,

LDPE, and PP; used in calendering of PVC, molding of PU, and fiber PP applics.; APHA 50 max. clear liq.; m.w. 706; sp.gr. 0.985-0.998 (25/15.5 C); dens. 0.99 g/ml; visc. 11.5 cps (38 C); acid no. 0.20 max.; flash pt. (PM) 168 C; ref. index 1.4895-1.4925; 8.8% P; Custom

Weston® 494. [GE Specialty] Diisooctyl octylphenyl phosphite; CAS 68133-13-1; EINECS 268-665-1; heat and color stabilizer for variety of polymers; color and lt. stabilizer for PVC; APHA 50 max. clear liq.; m.w. 494; sp.gr. 0.920-0.930 (25/15.5 C); dens. 0.92 g/ml; visc. 24.5 cps (38 C); acid no. 0.05 max.; flash pt. (PMCC) 152 C; ref. index 1.4750-1.4800; 6.3% P; *Toxicology:* contact with skin or eyes may cause irritation; *Storage:* 1 yr storage life when stored in closed containers away from excess humidity; blanket under dry nitrogen.

Weston® 600. [GE Specialty] Diisodecyl pentaerythritol diphosphite; CAS 26544-27-4; EINECS 247-779-5; stabilizer for a variety of polymer systems incl. PC; antioxidant or corrosion inhibitor in lubricants; APHA 100 max. clear liq. (100 C); m.w. 508; sp.gr. 1.020-1.040 (25/15.5 C); dens. 1.04 g/ml; visc. 110.0 cps (38 C); acid no. 0.20 max.; flash pt. (PMCC) 152 C; ref. index 1.4720-1.4760; 12.1% P; *Toxicology:* contact with skin or eyes may cause irritation; *Storage:* 1 yr storage life when stored in closed containers away from excess humidity; blanket under dry nitrogen.

Weston® 618F. [GE Specialty] Distearyl pentaerythritol diphosphite; CAS 3806-34-6; EINECS 223-276-6; color and m.w. stabilizer for polyolefins, polyesters, elastomers, styrenics, engineering thermoplastics, adhesives; contribute thermal and uv stability; synergistic with lt. stabilizers such as benzophenones, benzotriazoles, HALS; FDA 21CFR §175.105, 178.2010; approved in Canada, Germany, France, UK, Japan, and other countries; wh. waxy flakes; m.w. 732; sp.gr. 0.920-0.935 (60/15.5 C); bulk dens. 27 lb/ft^3; m.p. 37-46 C; acid no. 1.0 max.; flash pt. (PMCC) 185 C; ref. index 1.4560-1.4590 (60 C); 7.9% P; *Toxicology:* contact with skin or eyes may cause irritation.

Weston® 619F. [GE Specialty] Distearyl pentaerythritol diphosphite with 1.0% wt. triisopropanolamine; CAS 3806-34-6; EINECS 223-276-6; color and m.w. stabilizer for polyolefins, polyesters, elastomers, styrenics, engineering thermoplastics, adhesives; contribute thermal and uv stability; synergistic with lt. stabilizers such as benzophenones, benzotriazoles, HALS; FDA 21CFR §175.105, 178.2010; approved in Canada, Germany, France, UK, Japan, and other countries; wh. waxy flakes; m.w. 732; sp.gr. 0.920-0.935 (60/15.5 C); bulk dens. 27 lb/ft^3; m.p. 37-46 C; acid no. 1.0 max.; flash pt. (PMCC) 188 C; ref. index 1.4560-1.4590 (60 C); 7.9% P; *Toxicology:* contact with skin or eyes may cause irritation.

Weston® DHOP. [GE Specialty] Poly (dipropyleneglycol)phenyl phosphite; CAS 80584-86-7; EINECS 279-499-4; sec. stabilizer for PVC; APHA 50 max. clear liq.; m.w. 2102; sp.gr. 1.168-1.180 (25/15.5 C); dens. 1.09 g/ml; visc. 250 cps (38 C); acid no. 0.10 max.; flash pt. (PMCC) 179 C; ref. index 1.5340-1.5380; 11.8% P; *Toxicology:* contact with skin or eyes may cause irritation; *Storage:* 1 yr storage life when stored in closed containers away from excess humidity; blanket under dry nitrogen.

Weston® DLP. [GE Specialty] Dilauryl phosphite; CAS 21302-09-0; stabilizer; APHA 50 max. clear liq.; m.w. 418; sp.gr. 0.898-0.906 (25/15.5 C); dens. 0.91 g/ml; visc. 8.5 cps (60 C); acid no. 4.00 max.; flash pt. (COC) 210 C; ref. index 1.4500-1.4530; 7.4% P; Custom

Weston® DOPI. [GE Specialty] Diisooctyl phosphite; CAS 36116-84-4; EINECS 252-287-4; lubricant additive improving antifriction and antiwear chars.; color and m.w. stabilizer for PP; APHA 70 max. clear liq.; m.w. 306; sp.gr. 0.925-0.933 (25/15.5 C); dens. 0.93 g/ml; visc. 8.0 cps (38 C); b.p. 129 C (5 mm); acid no. 3.0 max.; flash pt. (PMCC) 146 C; ref. index 1.4415-1.4430; 8.8-10.6% P; *Toxicology:* contact with skin or eyes may cause irritation; *Storage:* 1 yr storage life when stored in closed containers away from excess humidity; blanket under dry nitrogen.

Weston® DPDP. [GE Specialty] Diphenyl isodecyl phosphite; CAS 26544-23-0; EINECS 247-777-4; color and processing stabilizer for PC, PU, ABS, coatings; sec. stabilizer for flexible and rigid PVC; reacts principally by chelation of metallic chlorides during PVC compding.; for use in conjunction with primary heat stabilizers; APHA 50 max. clear liq.; sol. in most common aprotic org. solvs.; insol. in water; m.w. 374; sp.gr. 1.022-1.032 (25/15.5 C); dens. 8.6 lb/gal; visc. 10.3 cps (38 C); b.p. 190 C (5 mm); acid no. 0.05 max.; flash pt. (PM) 154 C; ref. index 1.5160-1.519; 8.3% P; *Toxicology:* LD50 (oral, rat) 3.99 g/kg; contact with skin or eyes may cause irritation; *Storage:* 1 yr storage life when stored in closed containers away from excess humidity; blanket under dry nitrogen.

Weston® DPP. [GE Specialty] Diphenyl phosphite; CAS 4712-55-4; EINECS 225-202-8; stabilizer for polyolefins; improves color of unsat. polyesters; sec. stabilizer improving lt. and color stability in PVC; APHA 50 max. clear liq.; m.w. 234; sp.gr. 1.200-1.220 (25/15.5 C); dens. 1.21 g/ml; visc. 10.5 cps (38 C); b.p. 140 C (5 mm); acid no. 15.0 max.; flash pt. (PMCC) 143 C; ref. index 1.5550-1.5575; 13.3% P; *Toxicology:* contact with skin or eyes may cause irritation; *Storage:* 1 yr storage life when stored in closed containers away from excess humidity; blanket under dry nitrogen.

Weston® DTDP. [GE Specialty] Di-tridecyl phosphite; CAS 36432-46-9; stabilizer; APHA 50 max. clear liq.; m.w. 446; sp.gr. 0.906-0.917 (25/15.5 C); dens. 0.91 g/ml; visc. 27.5 cps (38 C); acid no. 3.00 max.; flash pt. (COC) 224 C; ref. index 1.4530-1.4560; 7.0% P; Custom

Weston® EGTPP. [GE Specialty] Triphenyl phosphite with 0.5% triisopropanolamine; CAS 101-02-0; reactive diluent for epoxy applics. incl. adhesives, coatings, laminates, potting and soldering compds., tooling; visc. reducer and color stabilizer in polyesters; APHA 50 max. clear liq.; m.w. 310; sp.gr. 1.180-1.186 (25/15.5 C); dens. 9.8 lb/gal; visc. 12.0 cps (38 C); m.p. 22-25 C; b.p. 205 C (5 mm); acid no. 0.50 max.; flash pt. (PMCC) 154 C; ref. index 1.5860-1.5890; 10.0% P; *Toxicology:* contact with skin or eyes may cause irritation; *Storage:* 1 yr storage life when stored in closed containers away from excess humidity; blanket under dry nitrogen.

Weston® ODPP. [GE Specialty] Diphenyl isooctyl phosphite; CAS 26401-27-4; EINECS 247-658-7; color and processing stabilizer in ABS, PC, PU, coatings, PET fiber; sec. stabilizer improving color and heat stability in PVC; APHA 50 max. clear liq.; m.w. 346; sp.gr. 1.040-1.047 (25/15.5 C); dens. 1.04 g/ml; visc. 8.2 cps (38 C); b.p. 190 C (5 mm); acid no. 0.05 max.; flash pt. (PMCC) 182 C; ref. index 1.5210-1.5230; 9.0% P; *Toxicology:* contact with skin or eyes may cause irritation; *Storage:* 1 yr storage life when stored in closed containers away from excess humidity; blanket under dry nitrogen.

Weston® PDDP. [GE Specialty] Phenyl diisodecyl phosphite; CAS 25550-98-5; EINECS 247-098-3; color and processing stabilizer for ABS, PC, PU, coatings, PET fiber; sec. stabilizer improving color and heat stability in PVC; APHA 50 max. clear liq.; sol. in most common aprotic org. solvs.; insol. in water; m.w. 438; sp.gr. 0.938-0.947 (25/15.5 C); dens. 0.94 g/ml; visc. 12.0 cps (38 C); b.p. 190 C; acid no. 0.05 max.; flash pt. (PM) 160 C; ref. index 1.4780-1.4810; 7.1% P; *Toxicology:* LD50 (oral, rat) 9.4 g/kg; contact with skin or eyes may cause irritation; *Storage:* 1 yr storage life when stored in closed containers away from excess humidity; blanket under dry nitrogen.

Weston® PNPG. [GE Specialty] Phenyl neopentylene glycol phosphite; CAS 3057-08-7; EINECS 221-291-2; stabilizer for PU, hot melt adhesives; used in molding and extrusion of PP, HDPE, LDPE, PC, ABS, PVC; used in film applics. for HDPE, LDPE, and PP; used in calendering of PVC, molding of PU, and fiber PP applics.; APHA 35 max. clear liq.; m.w. 227; sp.gr. 1.130-1.139 (25/15.5 C); dens. 1.14 g/ml; visc. 13.0 cps (38 C); b.p. 130 C (5 mm); acid no. 0.14 max.; flash pt. (PMCC) 121 C; ref. index 1.5140-1.5188; 13.7% P; *Toxicology:* contact with skin or eyes may cause irritation; *Storage:* 1 yr storage life when stored in closed containers away from excess humidity; blanket under dry nitrogen.

Weston® PTP. [GE Specialty] Heptakis (dipropyleneglycol) triphosphite; CAS 13474-96-9; EINECS 236-753-9; stabilizer controlling color development and preventing bun scorching in foamed PU; 1% as flame retardant for flame lamination PU applics.; at 10% will flame retard PU foams; APHA 50 max. clear liq.; m.w. 1022; sp.gr. 1.105-1.120 (25/15.5 C); dens. 1.11 g/ml; visc. 130.0 cps (38 C); acid no. 0.20 max.; flash pt. (PMCC) 135 C; ref. index 1.4660-1.4710; 9.1% P; *Toxicology:* contact with skin or eyes may cause irritation; *Storage:* 1 yr storage life when stored in closed containers away from excess humidity; blanket under dry nitrogen.

Weston® TDP. [GE Specialty] Triisodecyl phosphite; CAS 25448-25-3; EINECS 246-998-3; sec. stabilizer for ABS, PET, PET fibers, coatings, lubricants, PVC, PU, polyolefins; APHA 50 max. clear liq.; m.w. 502; sp.gr. 0.884-0.904 (25/15.5 C); dens. 0.89 g/ml; visc. 11.6 cps (38 C); b.p. 180 C (5 mm); acid no. 0.05 max.; flash pt. (PMCC) 160 C; ref. index 1.4530-1.4610; 6.2% P; *Toxicology:* contact with skin or eyes may cause irritation; *Storage:* 1 yr storage life when stored in closed containers away from excess humidity; blanket under dry nitrogen.

Weston® THOP. [GE Specialty] Tetraphenyl dipropyleneglycol diphosphite; CAS 80584-85-6; EINECS 279-498-9; sec. stabilizer for many polymers, incl. PVC; APHA 50 max. clear liq.; m.w. 566; sp.gr. 1.164-1.188 (25/15.5 C); dens. 1.17 g/ml; visc. 39.0 cps (38 C); acid no. 0.20 max.; flash pt. (PMCC) > 177 C; ref. index 1.5570-1.5620; 10.9% P; *Toxicology:* contact with skin or eyes may cause irritation; *Storage:* 1 yr storage life when stored in closed containers away from excess humidity; blanket under dry nitrogen.

Weston® TIOP. [GE Specialty] Triisooctyl phosphite; CAS 25103-12-2; EINECS 246-614-4; color and m.w. stabilizer for acrylics, nylon, unsat. polyester, PVC; sulfur deactivator in lubricants; also improves antifriction and antiwear props.; APHA 50 max. clear liq.; m.w. 418; sp.gr. 0.886-0.900 (25/15.5 C); dens. 0.90 g/ml; visc. 14.1 cps (38 C); b.p. 180 C (5 mm); acid no. 0.05 max.; flash pt. (PMCC) 146 C; ref. index 1.4470-1.4520; 7.4% P; *Toxicology:* contact with skin or eyes may cause irritation; *Storage:* 1 yr storage life when stored in closed containers away from excess humidity; blanket under dry nitrogen.

Weston® TLP. [GE Specialty] Trilauryl phosphite; CAS 3076-63-9; EINECS 221-356-5; lubricant additive improving antifriction and antiwear props.; stabilizer in polyester fibers, PP; APHA 50 max. clear liq.; m.w. 586; sp.gr. 0.866-0.881 (25/15.5 C); dens. 0.88 g/ml; visc. 16.0 cps (38 C); b.p. 190 C (5 mm); acid no. 0.05 max.; flash pt. (COC) 232 C; ref. index 1.4530-1.4580; 5.3% P; *Toxicology:* contact with skin or eyes may cause irritation; *Storage:* 1 yr storage life when stored in closed containers away from excess humidity; blanket under dry nitrogen.

Weston® TLTTP. [GE Specialty] Trilauryl trithio phosphite; CAS 1656-63-9; EINECS 216-751-4; heat stabilizer for variety of polymers, esp. polyolefins where it improves processing and uv stability; antioxidant and sulfur deactivator in lubricants while improving antifriction and antiwear props.; APHA 50 max. clear liq. (@ 100 C); m.w. 634; sp.gr. 0.900-0.935 (25/15.5 C); visc. 25 cps (38 C); b.p. 200 C (5 mm); acid no. 0.50 max.; flash pt. (PMCC) 218 C; ref. index 1.4985-1.5025; 4.9% P; *Toxicology:* contact with skin or eyes may cause irritation; *Storage:* 1 yr storage life when stored in closed containers away from excess humidity; blanket under dry nitrogen.

Weston® TNPP. [GE Specialty] Trisnonylphenyl

phosphite; CAS 26523-78-4; EINECS 247-759-6; stabilizer for wide variety of polymers; improves color and processing stability during compounding, processing and end-use; synergistic with hindered phenols and other stabilizers; FDA 21CFR §175.105, 175.300, 175.390, 177.1210, 177.2600, 178.2010; approved in Canada, Germany, France, UK, Japan and other countries; APHA 100 max. clear liq.; m.w. 688; sp.gr. 0.980-0.992 (25/15.5 C); dens. 0.98 g/ml; visc. 250.0 cps (60 C); acid no. 0.10 max.; flash pt. (PMCC) 207 C; ref. index 1.5255-1.5280; 4.2% P; *Toxicology:* contact with skin or eyes may cause irritation; *Storage:* 1 yr storage life when stored in closed containers away from excess humidity; blanket under dry nitrogen.

Weston® TPP. [GE Specialty] Triphenyl phosphite; CAS 101-02-0; EINECS 202-908-4; stabilizer for styrenics, engineering thermoplastics, polyesters (to regulate visc., improve color stability), polyolefins (as catalyst adjuvant), PU (to prevent socrching during curing and improve color stability), epoxies, PVC, coatings, adhesives; APHA 50 max. clear liq.; m.w. 310; sp.gr. 1.180-1.186 (25/15.5 C); dens. 1.18 g/ml; visc. 12.0 cps (38 C); b.p. 205 C (5 mm); acid no. 0.50 max.; flash pt. (PMCC) 146 C; ref. index 1.5880-1.5900; 10.0% P; *Toxicology:* contact with skin or eyes may cause irritation; *Storage:* 1 yr storage life when stored in closed containers away from excess humidity; blanket under dry nitrogen.

Westvaco® 1480. [Westvaco] Tall oil acid; CAS 61790-12-3; EINECS 263-107-3; anionic; emulsifier for emulsion polymerization and post stabilization; liq.; iodine no. 80; 100% conc.

Westvaco® M-30. [Westvaco] Stabilized tall oil rosin and fatty acids; CAS 68152-92-1; anionic; emulsifier for polymerization of styrene and butadiene for syn. tire rubber; liq.; 100% conc.

Westvaco® M-40. [Westvaco] Stabilized tall oil rosin and fatty acids; CAS 68152-92-1; anionic; emulsifier for polymerization of styrene and butadiene for syn. tire rubber; liq.; 100% conc.

Westvaco® M-70. [Westvaco] Stabilized tall oil rosin and fatty acids; CAS 68152-92-1; anionic; emulsifier for polymerization of styrene and butadiene; liq.; 100% conc.

Westvaco® Resin 90. [Westvaco] Stabilized tall oil rosin; CAS 8050-09-7; anionic; emulsifier for syn. rubber emulsion polymerization; tackifier, stabilizer, plasticizer; Gardner 6 max. liq.; dens. 8.6 lb/gal; visc. 620 cps (200 F); pour pt. 130 F; acid no. 168-178; sapon. no. 175; 100% conc.

Westvaco® Resin 95. [Westvaco] Stabilized tall oil rosin; CAS 8050-09-7; anionic; emulsifier for SBR polymerization; tackifier, stabilizer, plasticizer; Gardner 11 max. liq.; dens. 8.6 lb/gal; soften. pt. (R&B) 112 F; acid no. 158-166; 100% conc.

Westvaco® Resin 790. [Westvaco] Sodium rosinate; CAS 61790-51-0; EINECS 263-144-5; anionic; emulsifier for emulsion polymerization; liq.; 70% conc.

Westvaco® Resin 795. [Westvaco] Sodium rosinate; CAS 61790-51-0; EINECS 263-144-5; anionic; emulsifier for emulsion polymerization; liq.; 70% conc.

Westvaco® Resin 895. [Westvaco] Potassium rosinate; CAS 61790-50-9; anionic; emulsifier for emulsion polymerization; liq.; 80% conc.

Westvaco® Rosin T. [Westvaco] Tall oil rosin; CAS 8052-10-6; CAS 8052-10-6; raw material for rubber processing oils, ink resins, dk. unfortified paper size, other applics. not sensitive to color; dens. 8.2 lb/gal; soften. pt. (R&B) 130 F; acid no. 100; sapon. no. 135.

Wet Zinc™ . [C.P. Hall] Zinc stearate disp.; CAS 557-05-1; EINECS 209-151-9; provides uniform antistick coating to polymers and rubber slabs or continuous sheet from takeaway mills, extruders, or calenders; uniform wh. paste; sp.gr. 0.85 ± 0.15; pH 7.8 ± 0.6; 26.2 ± 2% solids.

Wet Zinc™ P. [C.P. Hall] Zinc stearate disp.; CAS 557-05-1; EINECS 209-151-9; provides nonstick surf. to rubber slabs or continuous sheet from takeaway mills or extruders; high lubricating effect; wh. visc. liq.; sp.gr. 0.98 ± 0.08; pH 10.3 ± 0.8; 31 ± 3% solids.

WF-15. [Sovereign] Coated urea; CAS 57-13-6; EINECS 200-315-5; activator for nitrogen-type blowing agents.

WF-16. [Sovereign] Oil-treated urea; CAS 57-13-6; EINECS 200-315-5; activator for nitrogen-type blowing agents.

W.G.S. Hydrogenated Fish Glyceride 117. [Werner G. Smith] Triester of long chain fatty acids and glycerin; CAS 68424-59-9; softener, plasticizer used in wax compds., textile softeners and sizes, yarn lubricants, grease sticks, polishes, crayons, candles, leather stuffing, wire drawing compds., paper coatings, and plastics; off-wh. flakes, faint waxy odor; negligible sol. in water; sp.gr. 0.891 (60 F); m.p. 50 C; acid no. 5 max.; iodine no. 22-35; sapon. no. 188-195; flash pt. 500 F; *Toxicology:* irritating to eyes, skin, respiratory system, gastrointestinal tract; *Precaution:* incompat. with oxidizing agents; hazardous decomp. prods.: oxides of carbon, smoke; keep away from heat, sources of ignition; *Storage:* keep container closed when not in use.

W.G.S. Synaceti 116 NF/USP. [Werner G. Smith] Cetyl esters (syn. spermaceti); CAS 8002-23-1; EINECS 241-640-2; emollient, bodying agent for cosmetic emulsions, pharmaceuticals; slip aid for inks; gloss/slip aid for varnish; processing aid and lubricant for plastics; in binder formulations for pencils; extremely stable; FDA 21CFR §175.105, 175.300; wh. cryst. flakes, typ. waxy odor; negligible sol. in water; ; sp.gr. 0.82-0.84 (50 C); m.p. 43-47 C; acid no. 2 max.; iodine no. 1 max.; sapon. no. 109-120; flash pt. (COC) 450 F; *Toxicology:* passes LD50 toxicity and rabbit eye and skin irritation tests; inh. may cause respiratory irritation; ingestion may cause gastro-intestinal irritation; *Precaution:* incompat. with oxidizing agents; hazardous decomp. prods.: oxides of carbon, smoke; *Storage:* keep away from heat, sources of ignition; keep container closed when not in use.

WH-31. [Alcan] Alumina trihydrate; flame retardant, smoke suppressant for plastics and rubbers; in catalysts, pigments, frits, glazes, enamels, glass, ceramics, high purity chemicals; very wh. free-flowing cryst. powd.; 15-50% on 325 mesh, 1-8% on 200 mesh; sp.gr. 2.42; bulk dens. 1.0 g/cc (loose); ref. index 1.57; 65% Al_2O_3.

Whitcon® TL-5. [ICI Fluoropolymers] PTFE; CAS 9002-84-0; lubricant powd. suitable for compounding in rubber and plastics or dispersing in liqs.; also as additive to greases and thread sealants; wh. powd.; 2.2 μ particle size; sp.gr. 2.2; bulk dens. 475-550 g/l.

Whitcon® TL-6. [ICI Fluoropolymers] Finely milled PTFE; CAS 9002-84-0; lubricant esp. suitable for thermoplastic applics. where plateout and pearlescence are an issue; may also be used for compounding rubber and thermoset prods.; 11 μ particle size; bulk dens. 400 g/l.

Whitcon® TL-102-2. [ICI Fluoropolymers] PTFE; CAS 9002-84-0; general purpose lubricant for oils, greases, plastics, and elastomers; for aerosol formulations and solv. dispersion; exc. film-forming props.; widely used as EP additive in oils and greases; wh. powd.; 2 μ particle size; bulk dens. 315 g/l.

Whitcon® TL-155. [ICI Fluoropolymers] Finely milled PTFE; CAS 9002-84-0; lubricant powd. esp. suitable for thermoplastic applics. where plateout and pearlesence are an issue; also for compounding rubber and thermoset prods.; FDA compliant for food contact use; 14 μ particle size; bulk dens. 410 g/l.

Whitcon® TL-169. [ICI Fluoropolymers] PTFE; CAS 9002-84-0; lubricant powd. suitable for compounding in rubber and plastics or dispersing in liqs.; also as additive to greases and thread sealants; wh. powd.; 2-3 μ particle size; sp.gr. 2.2; bulk dens. 375-450 g/l.; Unverified

White Lead. [Syn. Prods.] Lead complex; heat and lt. stabilizer for flexible and rigid PVC, plastisols, organosols, foams, records, elec. insulation.

Wiltrol P. [Olin] Surface-treated phthalic anhydride; CAS 85-44-9; EINECS 201-607-5; scorch inhibitor for rubber stocks; powd.; Unverified

Wingdale White. [Georgia Marble] Calcium carbonate; general use filler; 6.5 μ median particle size; 0.200% on 325 mesh; fineness (Hegman) 3; oil absorp. 10; brightness 94.

Wingstay® 29. [Goodyear; R.T. Vanderbilt] P-oriented styrenated diphenylamines; stabilizer for SBR, CR, NBR; exc. for blk. loaded compds. using diene polymer systems, esp. for those requiring high temp. resist.; dk. tan liq. or tan powd.

Wingstay® 100. [Goodyear; R.T. Vanderbilt] Mixed diaryl p-phenylenediamines; CAS 68953-84-4; antioxidant/antiozonant for tires, camelback, mech. goods; stabilizer for SBR, polybutadiene , neoprene; blue-brn. flakes, amine odor; 100% min. thru 3/8 in. screen; insol. in water; dens. 1.18 mg/m^3; m.p. 87-97 C; 80% min. act.; *Toxicology:* LD50 (oral, rat) > 4000 mg/kg; may cause respiratory tract irritation, transient eye irritation, skin irritation or sensitization; OSHA PEL/TWA 5 mg/m^3 (respirable dust), 15 mg/m^3 (total dust) recommended; contains < 0.1% o-toluidine, aniline; *Precaution:* hazardous decomp. prods.: oxides of carbon and nitrogen upon combustion; under dusty conditions, static electricitymay cause an explosion; use grounded equip.

Wingstay® 100AZ. [Goodyear; R.T. Vanderbilt] Mixed diaryl p-phenylenediamine; CAS 68478-45-5; antioxidant/antiozonant for blk. neoprene prods. where resist. to weather and performance are prime requirements; exc. bin cure props.; blue-brn. flakes, amine odor; insol. in water; dens. 1.18 mg/m^3; 76% min. act.; *Toxicology:* LD50 (oral, rat) > 7500 mg/kg; may cause upper respiratory tract irritation, transient eye irritation, skin irritation or sensitization; OSHA PEL/TWA 5 mg/m^3 (respirable dust), 15 mg/m^3 (total dust) recommended; contains < 0.1% o-toluidine; *Precaution:* hazardous decomp. prods.: oxides of carbon and nitrogen upon combustion; under dusty conditions, static electricity may cause an explosion; ground all equip.

Wingstay® 200. [Goodyear; R.T. Vanderbilt] Mixed diaryl p-phenylenediamine; antioxidant, antiozonant, stabilizer for staining SBR and polybutadiene polymers; brn.-blk. semisolid.

Wingstay® Antioxidants. [Goodyear; R.T. Vanderbilt] Hindered phenols; antioxidants used in rubber compds. to resist deterioration from oxidation in presence of heat and lt.; liq, powd., and flakes.

Wingstay® C. [Goodyear; R.T. Vanderbilt] Butylated d:(dimethylbenzyl) phenol; stabilizer for nonstaining SBR and nitrile polymers; amber liq.

Wingstay® K. [Goodyear; R.T. Vanderbilt] Reaction prod. of p-nonyl phenol, dodecanethiol, and formaldehyde; highly act. self-synergized nonstaining and nondiscoloring antioxidant for raw polymer stabilization, esp. for SBR, NBR, and polybutadiene; water-wh. to amber liq.; sp.gr. 0.90-0.95; visc. 500-600 cP.

Wingstay® L. [Goodyear; R.T. Vanderbilt] 4-Methyl phenol reaction prods. with dicyclopentadiene and isobutylene; CAS 68610-51-5; nonstaining, nondiscoloring antioxidant for protection of natural rubber and carboxylated SBR latex compds., foam rubber, carpet backing, pressure-sensitive and hot-melt adhesives, latex, ABS compds., rubber thread, wh. sidewalls; lt. tan flakes or wh. powd.

Wingstay® L HLS. [R.T. Vanderbilt] Butylated reaction prod. of p-cresol and dicylopentadiene; CAS 68610-51-5; EINECS 271-867-2; rubber antioxidant.

Wingstay® S. [Goodyear; R.T. Vanderbilt] Styrenated phenol (CAS 61788-44-10 on hydrated amorphous silica (CAS 63231-67-4); nonstaining, nondiscoloring antioxidant for foam rubber, wh. sidewalls, kitchen and drug sundries, garden hose, shoe soles; stabilizer for SBR and BR; powd., phenolic odor; insol. in water; dens. 1.36 mg/m^3; *Toxicology:*

LD50 (oral) 2500 mg/kg; may irritate eyes, skin, respiratory tract; silica: ACGIH TWA/TLV 10 mg/m³ (total dust); *Precaution:* emits acrid fumes when exposed to flame; hazardous decomp. prods.: oxides of carbon at combustion temps.

Wingstay® SN-1. [Goodyear; R.T. Vanderbilt] β-Alkylthioester propionate; nonstaining, nondiscoloring antioxidant for lt. colored or blk. compds., carpet backing foams, laminating adhesives, SBR latex stabilization, rubber-based adhesives, footwear compds., mech. goods; lt. yel. cryst. solid or lt. tan powd.

Wingstay® T. [Goodyear; R.T. Vanderbilt] Hindered phenol; antioxidant, polymer stabilizer; amber liq.

Wingtack® 10. [Goodyear] C5 hydrocarbon resin; plasticizer/tackifier for pressure sensitives, hot melts and sealants; processing aid in rubber milling; compat. with SBR, polyisoprene, NR, butyl, SIS and SBS copolymers, EPDM, EVA, PE, PP; lt. yel. liq.; sp.gr. 0.88; visc. 20,000-40,000 cps; soften. pt. (R&B) 10-15 C.

Wingtack® 95. [Goodyear] C5 hydrocarbon resin; tackifier for hot melt coatings, adhesives, pressure-sensitive adhesives esp. based on SIS block polymer (surgical tape, masking tape, elec. tape, labels); tackifier and processing aid in rubber compounding (EPDM tires, hoses, belts, mech. goods, NR, SBR, polyisoprene); FDA approved; lt. yel. flake; sp.gr. 0.93; soften. pt. (R&B) 95-101 C.

Winnofil S, SP. [ICI Am.] Precipitated coated calcium carbonate; filler and impact modifier for thermoplastics and paints.; Unverified

Witco® 960. [Witco/H-I-P] Alkylphenol/EO adduct; pigment wetting and grinding aid for aq. paint systems; emulsifier for polymerization.; Unverified

Witco® 1298. [Witco/H-I-P] Dodecylbenzene sulfonic acid; anionic; detergent intermediate, o/w emulsifier, solubilizer, wetting agent, and detergent for household prods., metal cleaning; emulsion polymerization surfactant for latex stabilization and pigment dispersion; liq.; sol. in oil and water; sp.gr. 1.05; 97% act.

Witco® MRC. [Witco/PAG] Org. phosphate ester; mold release agent for unsat. polyester plastics.

Witco® Acid B. [Witco/H-I-P] Dodecylbenzene sulfonic acid; anionic; detergent intermediate; detergent, dispersant, o/w emulsifier, wetting agent for industrial and metal cleaning; solubilizer, emulsion polymerization surfactant for latex stabilization and pigment dispersion; liq.; sol. in oil and water; 100% conc.

Witco® Aluminum Stearate 18. [Witco/H-I-P] Aluminum stearate; CAS 7047-84-9; EINECS 230-325-5; thickener for hydrocarbon fluids, personal care applics.; internal/external lubricant for nylon, rigid PVC, polyesters; FDA accepted; solid; sol. hot with gelation on cooling in aliphatic and aromatic solvs., oils.

Witco® Aluminum Stearate 22. [Witco/H-I-P] Aluminum distearate; CAS 300-92-5; EINECS 206-101-8; pigment suspending agent for oil-based paints; thickener for hydrocarbon fluids; lubricant for wire drawing, stamping, metalworking, nylon, rigid PVC, polyesters; solid; sol. hot in aromatic and aliphatic solvs. and oils; water-insol.

Witco® Calcium Stearate A. [Witco/PAG] Calcium stearate; CAS 1592-23-0; EINECS 216-472-8; lubricant in processing of PP, ABS plastics; sol. in hot aromatic and aliphatic solvs. and oils; insol. in water, alcohol.

Witco® Calcium Stearate F. [Witco/PAG] Calcium stearate; CAS 1592-23-0; EINECS 216-472-8; water repellent additive for concrete mixes; lubricant in processing of PVC; sol. in hot aromatic and aliphatic solvs. and oils; insol. in water.

Witco® Calcium Stearate FP. [Witco/H-I-P] Calcium stearate; CAS 1592-23-0; EINECS 216-472-8; lubricant for flexible and rigid PVC, plastisols, organosols, PP, PU; personal care applics.

Witco® Heat Stable Sodium Stearate. [Witco/PAG] Sodium stearate; CAS 822-16-2; EINECS 212-490-5; lubricant for impact-modified PS; sol. in hot water, alcohol.

Witcolate™ 1276. [Witco/New Markets] Ammonium alcohol ether sulfate; anionic; wetting agent, penetrant, lubricant, emulsifier, dye dispersant, scouring aid, antistat; foamer for wallboard mfg. and petrol. industry; detergent for metal cleaning; textile surfactant; emulsion polymerization and latex stabilization; liq.; sol. in water, IPA; sp.gr. 1.05; dens. 8.8 lb/gal; visc. 30 cps; pour pt. < 10 F; pH 7.8; 60% solids.

Witcolate™ A Powder. [Witco/H-I-P] Sodium lauryl sulfate; CAS 151-21-3; EINECS 205-788-1; anionic; detergent, wetting agent, foamer used in personal care prods., wool detergents; polymerization emulsifier; powd.; water-sol.; sp.gr. 0.35; 93% act.

Witcolate™ D51-51. [Witco/New Markets] Nonoxynol-4 sulfate; anionic; antistat for syn. fibers and polymer prods.; surfactant, wetting agent and dispersant; emulsifier for polymerization of acrylics, vinyl acetate, vinyl acrylics, styrene, SAN, styrene acrylic, vinyl chloride; personal care formulations; FDA compliance; clear liq.; sol. in water, ethanol; sp.gr. 1.06; visc. 2500 cps; pH 8.0 (5% aq.); flash pt. (PMCC) > 93.3 C; surf. tens. 30.0 dynes/cm (1%); Ross-Miles foam 173 mm (initial, 1%, 49 C); 34% solids.

Witcolate™ D51-51EP. [Witco/New Markets] Sodium nonoxynol-4 sulfate; CAS 9014-90-8; emulsifier for emulsion polymerization of acrylic, vinyl acetate, vinyl acrylic, vinyl chloride; personal care formulations; FDA compliance; Gardner 4 color; visc. 400 cps; pH 8 (10% aq.); surf. tens. 30 dynes/cm (1%); 30% act.

Witcolate™ D51-52. [Witco/New Markets] Sodium alkylaryl polyether sulfate; anionic; emulsifier for polymerization, personal care formulations; liq.

Witcolate™ D51-53. [Witco/New Markets] Nonoxynol-10 sulfate; emulsifier for acrylic, vinyl acetate, vinyl acrylic, styrene, SAN, styrene acrylic, and vinyl chloride polymerization; personal care formulations; visc. 250 cps; surf. tens. 40 dynes/cm (1%);

Ross-Miles foam 218 mm (initial, 1%, 49 C); 34% solids.

Witcolate™ D51-53HA. [Witco/New Markets] Nonoxynol-10 sulfate; emulsifier for acrylic, vinyl acetate, vinyl acrylic, styrene, SAN, styrene acrylic, and vinyl chloride polymerization; personal care formulations; visc. 50 cps; surf. tens. 41.2 dynes/cm (1%); Ross-Miles foam 224 mm (initial, 1%, 49 C); 30% solids.

Witcolate™ D51-60. [Witco/New Markets] Nonoxynol-30 sulfate; emulsifier for acrylic, vinyl acetate, vinyl acrylic, styrene, SAN, styrene acrylic, and vinyl chloride polymerization; personal care formulations; visc. 300 cps; surf. tens. 43.3 dynes/cm (1%); Ross-Miles foam 234 mm (initial, 1%, 49 C); 40% solids.

Witcolate™ D-510. [Witco/New Markets] Sodium 2-ethylhexyl sulfate; CAS 126-92-1; EINECS 204-812-8; anionic; detergent, wetting agent and penetrant for industrial use and polymerization reactions; dispersant for bleaching powds.; lime soap and grease dispersant; personal care formulations; clear liq.; sol. (5%) in water, IPA; sp.gr. 1.12; pH 10.5 (10% aq.); flash pt. > 93 C; 39% act.

Witcolate™ SL-1. [Witco/PAG] Sodium lauryl sulfate; CAS 151-21-3; EINECS 205-788-1; emulsifier for acrylics, acrylonitrile, carboxylated SBR, chloroprene, styrene, vinyl chloride, vinyl acetate; personal care formulations; FDA compliance; visc. 50 cps; HLB 40; surf. tens. 31.8 dynes/cm (1%); Ross-Miles foam 180 mm (initial, 1%, 49 C); 29% solids.

Witco® Lithium Stearate 306. [Witco/H-I-P] Lithium stearate; CAS 4485-12-5; EINECS 224-772-5; internal/external lubricant for nylon; personal care applics.; solid.

Witco® Magnesium Stearate D. [Witco/PAG] Magnesium stearate; CAS 557-04-0; EINECS 209-150-3; internal/external lubricant for ABS, rigid vinyls, cellulose acetate; solid; sol. in hot aromatic and aliphatic solvs. and oils; insol. in water, alcohol.

Witconate™ 90. [Witco/New Markets] Linear alkylaryl sodium sulfonate; anionic; detergent, lubricant, emulsifier, antistat, wetting agent for household and industrial specialty compds., metal cleaning, personal care prods., textiles; emulsion polymerization and latex stabilization; flakes; 91% conc.

Witconate™ 1223H. [Witco/New Markets] Sodium branched dodecylbenzene sulfonate; emulsifier for ABS, SAN, SBR, styrene, vinyl acrylic, vinyl chloride polymerization; wetting agent for latexes; FDA compliance; visc. 500 cps; surf. tens. 31.7 dynes/cm (1%); Ross-Miles foam 175 mm (initial, 1%, 49 C); 23% solids.

Witconate™ 1230. [Witco/New Markets] Sodium dodecylbenzenesulfonate; anionic; base for liq. detergents; emulsion polymerization; liq.; 30% conc.

Witconate™ 1250. [Witco/New Markets] Sodium dodecylbenzene sulfonate; CAS 25155-30-0; EINECS 246-680-4; anionic; detergent, wetting agent, foaming agent, base for personal care, household, and industrial detergents, emulsion polymerization; slurry; water-sol.; sp.gr. 1.08; pH 7.5; 53% act.

Witconate™ 1255. [Witco/New Markets] Sodium dodecylbenzenesulfonate; CAS 25155-30-0; anionic; base for liq. detergents; emulsion polymerization; paste; 55% conc.

Witconate™ 1298. [Witco/New Markets] Linear alkylaryl sulfonic acid; anionic; penetrant, lubricant, dispersant, detergent, antistat, intermediate for wetting agents, emulsifiers, and detergents for household and industrial specialties, personal care prods.; textile surfactant; emulsion polymerization; liq.; 98% conc.

Witconate™ 3009-15. [Witco/New Markets] Sodium alkylaryl ether sulfate; latex emulsifier and stabilizer for polymerizations.

Witconate™ Acide TPB. [Witco/New Markets] Branched chain dodecylbenzene sulfonic acid.; CAS 68411-32-5; anionic; intermediate for prod. of branched sulfonates for polymerization and pesticides; liq.; 97% conc.

Witconate™ AOS. [Witco/New Markets] Sodium C14-16 olefin sulfonate; CAS 68439-57-6; EINECS 270-407-8; anionic; detergent, foamer, wetting agent, solubilizer for cosmetics and toiletries, shampoos, industrial detergents, textiles, petrol. industry; emulsifier for plastics/rubber polymerization; FDA compliance; liq.; sol. in water; disp. in IPA; sp.gr. 1.05; dens. 8.8 lb/gal; visc. 70 cps; HLB 11.8; pour pt. 32 F; pH 7.7; surf. tens. 35.0 dynes/cm (1%); Ross-Miles foam 168 mm (initial, 1%, 49 C); 39% act.

Witconate™ AOS-EP. [Witco/New Markets] Sodium alpha-olefin sulfonate; CAS 68439-57-6; EINECS 270-407-8; emulsifier for emulsion polymerization, latex paints, adhesives, binders, personal care formulations; lt. clear liq.; dens. 8.92 lb/gal; visc. 200 cps; flash pt. (PMCC) > 200 F; pH 7.5-8.5 (5%); 38-40% act.

Witconate™ D51-51. [Witco/New Markets] Sodium alkylaryl polyether sulfonate; anionic; emulsion polymerization surfactant; liq.; water-sol.

Witconol™ 1206. [Witco/H-I-P] Alkyl POE glycol ether; nonionic; wetting agent, detergent, o/w emulsifier for industrial uses; visc. and flow control agent for polymerization reactions; Gardner 1 liq.; water-sol.; sp.gr. 1.00; pH 7.0.

Witconol™ MST. [Witco/H-I-P] Glyceryl stearate; nonionic; emulsifier for cosmetic, pharmaceutical, aerosol formulations; internal lubricant, plasticizer, and emulsifier in industrial applics.; flow control agent for polymerization reactions; dispersant; flake; disp. in oil; sp.gr. 0.93; HLB 3.9; m.p. 58 C; 100% act.

Witconol™ NP-100. [Witco/New Markets] Nonoxynol-10; CAS 9016-45-9; nonionic; pigment dispersant, emulsifier, latex stabilizer, and leveling agent for polymerization reactions; visc. 350 cps; Ross-Miles foam 158 mm (initial, 1%, 49 C); 100% solids.

Witconol™ NP-300. [Witco/New Markets] Nonoxynol-30; CAS 9016-45-9; nonionic; personal care surfactant; paint industry emulsifier for emulsion polymerization; pigment wetting and grinding agent for aq. systems; spreading agent; solid, liq.; water-sol.

Witconol™ NP-330. [Witco/New Markets] Nonyl phenol ethoxy/propoxy (30); emulsifier for ABS, SAN, SBR, styrene, vinyl acrylic, vinyl chloride polymerization; plasticizer; visc. 600 cps; surf. tens. 41.1 dynes/cm (1%); Ross-Miles foam 237 mm (initial, 1%, 49 C); 100% solids.

Witco® Sodium Stearate T-1. [Witco Corp.] Sodium stearate; CAS 822-16-2; EINECS 212-490-5; internal/external lubricant for rigid vinyls (with calcium stearate), PS, nylon; solid.

Witco® Zinc Stearate 11. [Witco/PAG] Zinc stearate; CAS 557-05-1; EINECS 209-151-9; lubricant for fiber-reinforced and PS plastics; wh. powd.; 99.9% thru 325 mesh; sol. in hot turpentine, benzene, toluene, xylene, CCl_4, veg. and min. oils, waxes; sp.gr. 1.09; soften. pt. 119 C.

Witco® Zinc Stearate 42. [Witco/PAG] Zinc stearate; CAS 557-05-1; EINECS 209-151-9; lubricant for PE.

Witco® Zinc Stearate 44. [Witco/PAG] Zinc stearate; CAS 557-05-1; EINECS 209-151-9; lubricant for PS rubber; wh. powd.; 99.9% thru 325 mesh; sol. in hot turpentine, benzene, toluene, xylene, CCl_4, veg. and min. oils, waxes; sp.gr. 1.09; soften. pt. 120 C.

Witco® Zinc Stearate Disperso. [Witco/PAG] Zinc stearate; CAS 557-05-1; EINECS 209-151-9; lubricant and antitack agent for rubber; wh. disp.; sol. in hot turpentine, benzene, toluene, xylene, CCl_4, veg. and min. oils, waxes; sp.gr. 1.09; soften. pt. 120 C.

Witco® Zinc Stearate Heat-Stable. [Witco/PAG] Zinc stearate; CAS 557-05-1; EINECS 209-151-9; lubricant for PS plastics; wh. powd.; 99.9% thru 325 mesh; sol. in hot turpentine, benzene, toluene, xylene, CCl_4, veg. and min. oils, waxes; sp.gr. 1.09; soften. pt. 122 C.

Witco® Zinc Stearate LV. [Witco/PAG] Zinc stearate; CAS 557-05-1; EINECS 209-151-9; lubricant for fiber-reinforced plastics, SMC, BMC; wh. powd.; 99.7% thru 325 mesh; sol. in hot turpentine, benzene, toluene, xylene, CCl_4, veg. and min. oils, waxes; sp.gr. 1.09; soften. pt. 120 C.

Witco® Zinc Stearate NW. [Witco/PAG] Zinc stearate; CAS 557-05-1; EINECS 209-151-9; lubricant and antitack agent for rubber; wh. powd.; 99.9% thru 325 mesh; sol. in hot turpentine, benzene, toluene, xylene, CCl_4, veg. and min. oils, waxes; sp.gr. 1.09; soften. pt. 120 C.

Witco® Zinc Stearate Polymer Grade. [Witco/PAG] Zinc stearate; CAS 557-05-1; EINECS 209-151-9; internal/external lubricant for PS; wh. powd.; 99.9% thru 325 mesh; sol. in hot turpentine, benzene, toluene, xylene, CCl_4, veg. and min. oils, waxes; sp.gr. 1.09; soften. pt. 120 C.

Witco Zinc Stearate Regular. [Witco/PAG] Zinc stearate; CAS 557-05-1; EINECS 209-151-9; lubricant for urea/formaldehyde, melamine plastics; antitack dusting agent and lubricant for surfs. of uncured rubber slabs; flatting agent for solvent-based paints; wh. powd.; 99.9% thru 325 mesh; sol. in hot turpentine, benzene, toluene, xylene, CCl_4, veg. and min. oils, waxes; sp.gr. 1.09; soften. pt. 120 C.

Witco® Zinc Stearate REP. [Witco/PAG] Zinc stearate; CAS 557-05-1; EINECS 209-151-9; lubricant for fiber-reinforced polyester plastics; sol. in hot aromatic and aliphatic solvs. and oils; water-insol.

10, 325, 400 Wollastocoat®. [Nyco Minerals] Surface-modified wollastonite; CAS 13983-17-0; cost-effective functional fillers and additives for plastics, coating, friction, refractory, ceramic, construction, elastomer, sealant, and adhesive applics.; for nylon, phenolic molding compds., epoxy, polyester, PU/polyurea; surf. modifications can improve processing, bonding between resin and filler, mech. and phys. props., material handling and warehousing, etc.; wh. acicular powd., faint odor; sol. 0.01 g/100 cc in water; sp.gr. 2.9; dens. 24.2 lb/solid gal; bulk value 0.0413 gal/lb; m.p. 1540 C; ref. index 1.63; pH 9.9 (10% slurry); hardness (Mohs) 4.5; > 99% wollastonite, < 1% proprietary treatments; *Toxicology:* nuisance dust, TLV/TWA 10 mg/m^3 (8 h, total dust); long-term cumulative inh. can cause restriction of airways; may cause minro skin irritation; *Precaution:* surf. treatments may oxidize or decomp. at elevated temp.s; burning may produce sm. amts. of oxides of carbon, silicon, nitrogen, or sulfur; *Storage:* keep dry and cool in original shipping containers until use.

Wyfire® H-A-85. [Rhein Chemie] Antimony oxide encapsulated in CM elastomer; predispersed flame retardant for rubber; wh. slabs; sp.gr. 3.54.

Wyfire® H-A-85P. [Rhein Chemie] Antimony oxide encapsulated in CM elastomer; predispersed flame retardant for rubber; wh. pellets; sp.gr. 3.54.

Wyfire® S-A-51. [Rhein Chemie] Antimony oxide encapsulated in SBR elastomer; predispersed flame retardant for rubber; wh. slabs; sp.gr. 2.72.

Wyfire® S-BA-88P. [Rhein Chemie] Saytex BT-93/antimony oxide (60:20) encapsulated in SBR elastomer; predispersed flame retardant for rubber; lt. yel. pellets; sp.gr. 2.14.

Wyfire® Y-A-85. [Rhein Chemie] Antimony oxide encapsulated in EPDM elastomer; predispersed flame retardant for rubber; wh. slabs; sp.gr. 3.30.

Wyfire® Y-B-57. [Rhein Chemie] Saytex BT-93 encapsulated in EPDM elastomer; predispersed flame retardant for rubber; lt. yel. slabs; sp.gr. 1.88.

Wyfire® YP-BA-47P. [Rhein Chemie] Saytex BT-93/antimony oxide (59.9:23.1) encapsulated in EPDM elastomer; predispersed flame retardant for rubber; yel. pellets; sp.gr. 1.09.

Wytox® 312. [Uniroyal] Tris (nonylphenyl) phosphite; CAS 26523-78-4; EINECS 247-759-6; antioxidant for PE, PP, PS, PVC, ABS, nylon, food pkg.; processing and color stabilizer; FDA approvals; sol. in MEK, kerosene, amyl alcohol.

XY

1X30. [AlliedSignal/Ind. Fibers] Polyester; reinforcing material for in-rubber composites providing lower elongation and lower shrinkage than 1W70 and superior part-load and high dimensional stability; tenacity 7.8 g/d; shrinkage 4.5% (350 F); toughness 0.39 g/d.

1X33. [AlliedSignal/Ind. Fibers] Polyester; reinforcing material for in-rubber composites providing lower elong. and lower shrinkage than 1W70 and superior part-load and high dimensional stability; special adhesive active finish provides exc. adhesion to rubber using a single-dip system; tenacity 7.8 g/d; shrinkage 4.5% (350 F); toughness 0.39 g/d.

1X40, 1X43. [AlliedSignal/Ind. Fibers] Polyester; reinforcing material for in-rubber composites; designed as rayon replacement; offers highest dimensional stability at cost-effective replacement factor; special adhesive active finish provides exc. adhesion to rubber using a single-dip system; tenacity 7.8 g/d; shrinkage 4.5% (350 F); toughness 0.39 g/d.

X50-S®. [Degussa] Si69 (bistriethoxysilylpropyltetrasulfane), N330 carbon blk. (1:1); coupling agent, reinforcing filler for rubber compds. contg. silicic acid fillers (silica, clay, and talc); blk. pearls; sp.gr. 1.344; bulk dens. 880 g/l; pour dens. 770 g/l.

X230-S®. [Degussa] 1:1 Blend of N330 carbon blk. and silane; reinforcing agent for compds. contg. halogenated rubber and silicic acid fillers (silica, clay, and talc); blk. powd.; sp.gr. 1.30; pour dens. 550 g/l; 7% Cl.

X-743. [Neville] Aromatic plasticizer; aromatic, chemically inert, nonsaponifiable plasticizer exhibiting low reactivity; used in adhesives (mastic, pressure sensitive), rubber (cements, mechanical and molded goods, tires), and caulking compds.; FDA 21CFR §175.105, 177.2600; dk. brn. liq.; sol. in ethers and chlorinated, aromatic, naphthenic, and terpene hydrocarbons; m.w. 450; sp.gr. 1.10; flash pt. (COC) 215 F.

XC 9 Hot. [Releasomers] Water-based nonsilicone disp.; semipermanent mold release agent for most thermosetting rubber and plastic materials; designed for applics. to hot molds (above 225 F); *Toxicology:* avoid prolonged breathing of vapors, repeated skin contact; *Precaution:* keep away from heat, flame, sparks; *Storage:* keep closed when not in use.

XK 22. [Releasomers] Nonsilicone semipermanent mold release agent for the composite industry; offers fast cure, high chemical and abrasion resistance, and gives a film with exc. adhesion to the mold; operates easily above 850 F; solv. sol'n.; *Toxicology:* avoid prolonged/repeated skin contact, breathing of vapors; *Precaution:* keep away from heat, flame, sparks; *Storage:* keep closed when not in use.

XO White. [Georgia Marble] Ground calcium carbonate; screen-controlled filler for neutralization of acids, syn. marble, aggregate for cement finishes, vinyl asbestos tile, welding rods, polyester and epoxy floor tiles; 1% max. retained on 16 mesh; 15% min. thru 40 mesh; sp.gr. 2.71; pH 9.0-9.5; hardness (Moh) 3; 95.0% min. total carbonates.

XR 7. [Releasomers] Semipermanent mold release agent; forms a tough, durable film with exc. adhesion to mold surf.; gives easy and multiple releases with most thermosetting rubber and plastic materials; solv. sol'n.; *Toxicology:* avoid prolonged/repeated skin contact, breathing of vapors; *Precaution:* keep away from heat, flame, sparks; *Storage:* keep can closed when not in use.

XSA 80. [Huntsman] Xylene sulfonic acid; CAS 25321-41-9; EINECS 246-839-8; anionic; catalyst for resins; liq.; 80% act.

XSA 90. [Huntsman] Xylene sulfonic acid; CAS 25321-41-9; EINECS 246-839-8; anionic; catalyst for resins; liq.; 90% act.

XT 66. [Releasomers] Mold sealer for tool sealing in the composite industry; higher solids allow 1 coat sealing of most porosity and minor surf. imperfections; operates easily above 850 F; water-wh. clear sol'n.; sp.gr. 0.88 ± 0.03; flash pt. (TCC) 45 F; 5% solids in toluene; *Toxicology:* avoid prolonged/repeated skin contact, breathing vapors; *Precaution:* flamm.; keep away from heat, flame, sparks; *Storage:* 6 mos shelf life; keep container closed when not in use.

XUS 15615.00. [Dow Plastics] Static control additive providing inherent static control to flexible PU foams for electronics pkg. and other specialty pkg. applics.; clear to pale yel. liq., mild ammoniacal odor; partly sol. in water; sp.gr. 1.064; visc. 220 cst; hyd. no. 70; *Toxicology:* nonirritating to skin; may cause eye irritation; low single dose oral toxicity; *Precaution:* may decomp. in heat or fire releasing hydrogen fluoride; incompat. with strong min. acids and oxidizing agents; *Storage:* store in ventilated area; protect from moisture and extreme temps.

YSE-Cure B-001. [Ajinomoto] Amine adduct; epoxy curing agent for adhesives and coatings; flexible; APHA 200 colorless clear liq.; sp.gr. 1.09 (20 C); visc. 65-125 ps (20 C); amine no. 266-286; flash pt. 244 C; heat distort. temp. 62 C; *Toxicology:* vapor nonirritating to skin; *Storage:* store in tightly closed container; may precipitate wh. crystals in winter—heat to 60 C before use.

YSE-Cure B-002. [Ajinomoto] Amine adduct; epoxy curing agent for coating, flooring, casting; APHA 200 colorless clear liq.; sp.gr. 1.10 (20 C); visc. 20-60 ps (20 C); amine no. 317-337; flash pt. 210 C; heat distort. temp. 78 C; *Toxicology:* low toxicity; vapor nonirritating to skin; *Storage:* store in tightly closed container; may precipitate wh. crystals in winter—heat to 60 C before use.

YSE-Cure B-002W. [Ajinomoto] Amine adduct; epoxy curing agent for coating, flooring, casting; anti-crystallization version of B-002 suitable for winter use; APHA 200 colorless clear liq.; sp.gr. 1.10 (20 C); visc. 20-60 ps (20 C); amine no. 317-337; flash pt. 232 C; heat distort. temp. 73 C; *Toxicology:* vapor nonirritating to skin; *Storage:* store in tightly closed container.

YSE-Cure B-003. [Ajinomoto] Amine adduct; epoxy curing agent for coating, flooring, casting; clear colorless version of B-002; APHA 70 colorless clear liq.; sp.gr. 1.10 (20 C); visc. 20-60 ps (20 C); amine no. 317-337; flash pt. 210 C; heat distort. temp. 73 C; *Toxicology:* vapor nonirritating to skin; *Storage:* store in tightly closed container; may precipitate wh. crystals in winter—heat to 60 C before use.

YSE-Cure C-002. [Ajinomoto] Amine adduct; epoxy curing agent for lining and adhesives; acid resist.; APHA 300 colorless clear liq.; sp.gr. 1.10 (20 C); visc. 70-130 ps (20 C); amine no. 293-313; flash pt. 220 C; heat distort. temp. 80 C; *Toxicology:* vapor nonirritating to skin; *Storage:* store in tightly closed container; may precipitate wh. crystals in winter—heat to 60 C before use.

YSE-Cure F-100. [Ajinomoto] Amine; raw material for epoxy curing agents and imide compds.; APHA 100 paraffin-like solid; sp.gr. 1.11 (40 C); flash pt. 235 C; *Storage:* store in tightly closed containers.

YSE-Cure LX-1N. [Ajinomoto] Modified amine; epoxy curing agent for lining, potting; heat and chem. resist.; APHA 200 colorless clear liq.; sp.gr. 1.10 (20 C); visc. 1-3 ps (20 C); amine no. 480-520; flash pt. 141 C; heat distort. temp. 95 C; *Storage:* store in tightly closed container; may precipitate wh. crystals in winter—heat to 60 C before use.

YSE-Cure LX-2S. [Ajinomoto] Modified amine; epoxy curing agent for adhesives, grout and coating; flexible; Gardner 14 amber liq.; sp.gr. 0.98 (20 C); visc. 6-15 ps (20 C); amine no. 265-305; flash pt. 144 C; *Storage:* store in tightly closed container; may precipitate wh. crystals in winter—heat to 60 C before use.

YSE-Cure N-001. [Ajinomoto] Amine adduct; epoxy curing agent for elec. potting applics.; APHA 300 colorless clear liq.; sp.gr. 1.09 (20 C); visc. 25-41 ps (20 C); amine no. 330-350; flash pt. 220 C; heat distort. temp. 69 C; *Toxicology:* low toxicity; vapor nonirritating to skin; *Storage:* store in tightly closed container; may precipitate wh. crystals in winter—heat to 60 C before use.

YSE-Cure N-002. [Ajinomoto] Amine adduct; epoxy curing agent for elec. potting applics.; APHA 200 colorless clear liq.; sp.gr. 1.10 (20 C); visc. 7-27 ps (20 C); amine no. 358-378; flash pt. 222 C; heat distort. temp. 81 C; *Toxicology:* low toxicity; vapor nonirritating to skin; *Storage:* store in tightly closed container; may precipitate wh. crystals in winter—heat to 60 C before use.

YSE-Cure PX-3. [Ajinomoto] Modified amine; epoxy curing agent for grout and lining (tar-epoxy); inexpensive; Gardner 18 amber liq.; sp.gr. 1.09 (20 C); visc. 5-35 ps (20 C); amine no. 250-350; flash pt. 116 C; heat distort. temp. 64 C.

YSE-Cure QX-2. [Ajinomoto] Thiourea condensate; epoxy curing agent for lining, adhesives, and putty; fast curing; Gardner 10 amber liq.; sp.gr. 1.17 (20 C); visc. 70-180 ps (40 C); amine no. 265-305; flash pt. 185 C; heat distort. temp. 77 C; *Storage:* store in tightly closed container.

YSE-Cure QX-3. [Ajinomoto] Thiourea condensate; epoxy curing agent for lining, adhesives, and putty; fast curing; Gardner 10 amber liq.; sp.gr. 1.14 (20 C); visc. 25-65 ps (20 C); amine no. 375-415; flash pt. 166 C; *Storage:* store in tightly closed container.

YSE-Cure RX-2. [Ajinomoto] Accelerated modified amine; epoxy curing agent for lining and adhesives; fast curing; Gardner 18 amber liq.; sp.gr. 1.08 (20 C); visc. 10-40 ps (20 C); amine no. 240-260; flash pt. 82 C; heat distort. temp. 54 C; *Toxicology:* vapor nonirritating to skin; *Storage:* store in tightly closed container.

YSE-Cure RX-3. [Ajinomoto] Accelerated modified amine; epoxy curing agent for lining and flooring; fast curing; Gardner 6 reddish to brn. liq.; sp.gr. 1.09 (20 C); visc. 5-12 ps (20 C); amine no. 335-355; flash pt. 136 C; heat distort. temp. 58 C; *Storage:* store in tightly closed container.

YSE-Cure S-002. [Ajinomoto] Amine adduct sol'n.; epoxy curing agent; primer for metal, concrete, mortar, and wood; exc. adhesion; APHA 200 colorless clear liq.; sp.gr. 1.03 (20 C); visc. 40-120 ps (20 C); amine no. 81-93; flash pt. 6 C; heat distort. temp. 58 C.

Z

Zb™-112. [Climax Performance] Zinc borate; flame retardant, smoke suppressant; synergist in PVC and halogenated polyester formulations, and in elastomers, thermoplastic elastomers, polyamides, and polyolefins; strong char promoter; applics. incl. plastisols, coatings for cellulosics, textiles, and adhesives; wh. free-flowing powd., nonhygroscopic; 6 μ avg. particle size; 99.9% < 30 μ; sol. 0.5 g/100 ml water; sp.gr. 2.50; bulk dens. 18.1 lb/ft³; oil absorp. 39; ref. index 1.48.

Zb™-223. [Climax Performance] Zinc borate; flame retardant, smoke suppressant; synergist in PVC and halogenated polyester formulations, and in elastomers, thermoplastic elastomers, polyamides, and polyolefins; strong char promoter; applics. incl. calendered sheet, compounded thermoplastics and elastomers, thermoset applics.; useful for applics. requiring compounding or thermoforming @ elevated temps.; wh. free-flowing powd., nonhygroscopic; 4 μ avg. particle size; 99.9% < 30 μ; sol. 0.04 g/100 ml water; sp.gr. 2.83; bulk dens. 18.7 lb/ft³; oil absorp. 31; ref. index 1.57.

Zb™-237. [Climax Performance] Zinc borate; flame retardant, smoke suppressant; synergist in PVC and halogenated polyester formulations, and in elastomers, thermoplastic elastomers, polyamides, and polyolefins; strong char promoter; applics. incl. plastisols, coatings for textiles, adhesives; wh. free-flowing powd., nonhygroscopic; 2.6 μ avg. particle size; 99.9% < 16 μ; sol. 0.59 g/100 ml water; sp.gr. 2.31; bulk dens. 17.5 lb/ft³; oil absorp. 50; ref. index 1.47.

Zb™-325. [Climax Performance] Zinc borate; flame retardant, smoke suppressant; synergist in PVC and halogenated polyester formulations, and in elastomers, thermoplastic elastomers, polyamides, and polyolefins; strong char promoter; applics. incl. plastisols, coatings for textiles, adhesives; wh. free-flowing powd., nonhygroscopic; 2 μ avg. particle size; 99.9% < 28 μ; sol. 0.30 g/100 ml water; sp.gr. 2.66; bulk dens. 20 lb/ft³; oil absorp. 44; ref. index 1.53.

Zb™-467. [Climax Performance] Zinc borate; flame retardant, smoke suppressant; synergist in PVC and halogenated polyester formulations, and in elastomers, thermoplastic elastomers, polyamides, and polyolefins; strong char promoter; applics. incl. calendered sheet, compounded thermoplastics and elastomers, thermosets, applics. requiring compounding or thermoforming @ elevated temps.; wh. free-flowing powd., nonhygroscopic; 5 μ avg. particle size; 99.9% < 25 μ; sol. 0.1 g/100 ml water; sp.gr. 2.74; bulk dens. 29.3 lb/ft³; oil absorp. 24.2; ref. index 1.59.

Zeeospheres® 200. [Zeelan] Silica-alumina ceramic; strong, hard, inert, thick-walled hollow spheres for use as filler for a variety of plastic resins in inj. molding, extrusion, SMC, BMC, RTM, compression molding, potting/encapsulating, adhesives, tooling, casting, flooring, grouting, sealants, mastics, coatings, films, and other applics.; lt. gray spheres; 1.3 μ avg. particle size; 7+ Hegman grind; 0.01% retained on 325 mesh; sp.gr. 2.3 g/cc (D153); bulking value 19.2 lb/gal; soften. pt. 1200 C; pH 4.0-7.0; compr. str. > 60,000 psi; hardness 7 (Mohs); 99.4% spheres.

Zeeospheres® 400. [Zeelan] Silica-alumina ceramic; strong, hard, inert, thick-walled hollow spheres for use as filler for a variety of plastic resins in inj. molding, extrusion, SMC, BMC, RTM, compression molding, potting/encapsulating, adhesives, tooling, casting, flooring, grouting, sealants, mastics, coatings, films, and other applics.; lt. gray spheres; 1.6 μ avg. particle size; 67+ Hegman grind; 0.015% retained on 325 mesh; sp.gr. 2.2 g/cc; bulking value 18.3 lb/gal; soften. pt. 1200 C; pH 4.0-7.0; compr. str. > 60,000 psi; hardness 7 (Mohs); 99.4% spheres.

Zeeospheres® 600. [Zeelan] Silica-alumina ceramic; strong, hard, inert, thick-walled hollow spheres for use as filler for a variety of plastic resins in inj. molding, extrusion, SMC, BMC, RTM, compression molding, potting/encapsulating, adhesives, tooling, casting, flooring, grouting, sealants, mastics, coatings, films, and other applics.; lt. gray spheres; 1.8 μ avg. particle size; 3+ Hegman grind; 0.15% retained on 200 mesh; sp.gr. 2.1 g/cc; bulking value 17.5 lb/gal; soften. pt. 1200 C; pH 4.0-7.0; compr. str. > 60,000 psi; hardness 7 (Mohs); 99.4% spheres.

Zeeospheres® 800. [Zeelan] Silica-alumina ceramic; strong, hard, inert, thick-walled hollow spheres for use as filler for a variety of plastic resins in inj. molding, extrusion, SMC, BMC, RTM, compression molding, potting/encapsulating, adhesives, tooling, casting, flooring, grouting, sealants, mastics, coatings, films, and other applics.; lt. gray spheres; 3.0 μ avg. particle size; 7.0% retained on 200 mesh; sp.gr. 2.0 g/cc; bulking value 16.7 lb/gal; soften. pt. 1200 C; pH 4.0-7.0; compr. str. > 60,000 psi; hardness 7 (Mohs); 99.4% spheres.

Zeeospheres® 850. [Zeelan] Silica-alumina ceramic; strong, hard, inert, thick-walled hollow spheres for use as filler for a variety of plastic resins in inj. molding, extrusion, SMC, BMC, RTM, compression molding, potting/encapsulating, adhesives, tooling, casting, flooring, grouting, sealants, mastics, coatings, films, and other applics.; lt. gray spheres; 17.0 μ avg. particle size; 18.0% retained on 200 mesh; sp.gr. 2.0 g/cc; bulking value 16.7 lb/gal; soften. pt. 1200 C; pH 4.0-7.0; compr. str. > 60,000 psi; hardness 7 (Mohs); 99.4% spheres.

Zelec NE. [DuPont] Alcohol phosphate, neutralized; anionic; antistat for textiles, plastics, and films; internal mold release agent for unsat. thermosetting polyester resins; recommended where external heat is applied during curing cycle; protects ferrous metal molds from corrosion during storage; improves surf. appearance of molded parts; eliminates adhesion problems during subsequent applic. of finishes, printing inks, or adhesives to molded part; yel. visc. paste; mild fatty alcohol odor; sol. in water, polar solvs., glycerol, Freon TF, IPA, styrene, toluene; sp.gr. 1.10; dens. 1.10 g/ml; visc. > 100,000 cP; pH 7.0-7.5; flash pt. > 99 C (PMCC); 100% act.

Zelec NK. [DuPont] Fatty alcohol phosphate; anionic; antistat for textiles, plastics, and films; lt. yel. visc. paste; mild fatty alcohol odor; sol. in ethanol, IPA, ethylene glycol, benzene, toluene, or Freon TF solv.; disp. in water; sp.gr. 1.05; dens. 8.7 lb/gal; pH 7.0-7.5 (10% disp.); flash pt. > 210 F (PMCC); 100% act.

Zelec TY. [DuPont] Alcohol phosphate; anionic; antistat for polyolefin fibers, plastics, and films; lt. amber liq.; mild odor; water-sol.; dens. 9.8 lb/gal; pH 6.7-7.3 (10%).

Zelec UN. [DuPont] Alcohol phosphate, unneutralized; anionic; antistat; additive for oils to improve lubricity and provide corrosion protection; formulation of textile lubricants and finishes; release agent in plastic molding operations; pale yel. liq.; mild fatty alcohol odor; self-disp. in water; sol. in Freon TF, IPA, styrene, and toluene; sp.gr. 0.98; dens. 0.98 g/ml; visc. 165 cP; pH 2-3; flash pt. (PMCC) > 99 C; 100% act.

Zeolex® 23. [J.M. Huber/Chems.] Sodium silicoaluminate; reinforcing filler for rubber; powd.; 5 μ avg. particle size; 0.2% max. 325 mesh residue; dens. 2.1 g/ml; surf. area 73 m^2/g; oil absorp. 115 cc/100 g; ref. index 1.55; pH 10.2 (20%).

Zeolex® 80. [J.M. Huber/Chems.] Sodium silicoaluminate; titanium dioxide extender for rubber applics.; powd.; 4 μ avg. particle size; 0.2% max. 325 mesh residue; dens. 2.1 g/ml; surf. area 115 m^2/g; oil absorp. 115 cc/100 g; ref. index 1.51; pH 7.0 (20%).

Zeonet® A. [Zeon] Isocyanuric acid; CAS 108-80-5; EINECS 203-618-0; curing agent for Nipol AR (except 70 series), polyacrylate elastomers with epoxy cure sites; sp.gr. 2.50.

Zeonet® B. [Zeon] Trimethyl octadecyl ammonium bromide; accelerator for Nipol AR (except 70 series), polyacrylate elastomers with epoxy cure sites; sp.gr. 1.14.

Zeonet® U. [Zeon] N,N′-Diphenylurea; CAS 102-07-8; retarder to slow down the cure of Nipol AR and HyTemp 4050 series elastomers; sp.gr. 1.24.

Zeosil 45. [Sovereign] Precipitated silica; very high surf. area filler for reinforcement of shoe prods. and other applics. requiring high transparency and good dispersion; powd.

Zeosil 85. [Sovereign] Precipitated silica; low surf. area reinforcing filler for bromobutyl liners; gives good fatigue resist., adhesion, dispersion, easy processing; also used in sponge rubber, floor tiles, V-belts, and cables; micropearl.

Zeosil 125. [Sovereign] Precipitated silica; med. surf. area filler for rubber applics.; improves tear resist., heat buildup, and compr. set; gran., powd., or micropearl.

Zeosil 165G. [Sovereign] Precipitated silica; high surf. area reinforcing filler for rubber applics.; provides higher tear and tens. str. than std. silicas; lowest dust; used in applics. where low dust or automatic weighing are required; gran.

Zeosil 175MP. [Sovereign] Precipitated silica; high surf. area reinforcing filler for rubber applics.; provides higher tear, abrasion resist., and tens. str. than std. silicas; micropearl.

Zeosil 1165. [Sovereign] Precipitated silica; high surf. area reinforcing filler for rubber applics.; best suited for mech. goods, esp. belts and hoses; powd. form can replace fumed silica in some applics.; powd. or micropearl.

Zerogen® 10. [J.M. Huber/Engineered Mins.] Halogen-free magnesium hydroxide composition; CAS 1309-42-8; EINECS 215-170-3; flame retardant/smoke suppressant for thermoplastics and elastomeric formulations; stable to 332 C; 2.2-2.8 μ avg. particle size; sp.gr. 2.38; bulk dens. 0.39 g/cc (loose); surf. area 7-8 m^2/g; oil absorp. 27.

Zerogen® 11. [J.M. Huber/Engineered Mins.] Inorg. halogen-free flame retardant/smoke suppressor for thermoplastics such as EVA, polyethylene, PP, PBT, and nylon; acid scavenger; 2 μ avg. particle size; 0.01% retained on 325 mesh; sp.gr. 2.4; bulk dens. 0.27 g/cc (loose), 0.63 g/cc (packed); surf. area 15 m^2/g; oil absorp. 56.

Zerogen® 15. [J.M. Huber/Engineered Mins.] Halogen-free magnesium hydroxide composition; CAS 1309-42-8; EINECS 215-170-3; flame retardant and smoke suppressor for thermoplastic and elastomeric formulations incl. styrenics, polyamides, polyolefins, and thermoplastic polyesters; stable to 332 C; modified for ease of dispersion; powd.; 2.2-2.8 μ avg. particle size; 0.01% retained 325 mesh; sp.gr. 2.38; bulk dens. 0.34 g/cc (loose), 0.55 g/cc (packed); surf. area 7-8 m^2/g; oil absorp. 30.

Zerogen® 30. [J.M. Huber/Engineered Mins.] Halogen-free magnesium hydroxide composition; CAS

1309-42-8; EINECS 215-170-3; flame retardant and smoke suppressor for thermoplastic and elastomeric formulations; stable to 332 C; powd.; 0.8-1.2 μ avg. particle size; sp.gr. 2.38; bulk dens. 0.34 g/cc (loose), 0.57 g/cc (packed); surf. area 9-10 m^2/g; oil absorp. 34.

Zerogen® 33. [J.M. Huber/Engineered Mins.] Halogen-free proprietary blend; flame retardant and smoke suppressor for thermoplastics and elastomers, incl. styrenics, polyamides, polyolefins, and thermoplastic polyesters; 1.1 μ avg. particle size; 0.01% retained 325 mesh; sp.gr. 2.38 g/cc; bulk dens. 0.3 g/cc (loose), 0.6 g/cc (packed); surf. area 15 m^2/g; oil absorp. 56 cc/100 g.

Zerogen® 35. [J.M. Huber/Engineered Mins.] Halogen-free magnesium hydroxide composition; CAS 1309-42-8; EINECS 215-170-3; flame retardant and smoke suppressor for thermoplastics and elastomers, incl. styrenics, polyamides, polyolefins, and thermoplastic polyesters; offers improved disp. and flow properties; powd.; 0.8-1.2 μ avg. particle size; 0.01% retained 325 mesh; sp.gr. 2.38; bulk dens. 0.3 g/cc (loose), 0.49 g/cc (packed); surf. area 9-10 m^2/g; oil absorp. 31.

Zerogen® 60. [J.M. Huber/Engineered Mins.] Halogen-free magnesium hydroxide composition; CAS 1309-42-8; EINECS 215-170-3; med. particle sized flame retardant/smoke suppressant for thermoplastics and elastomeric formulations; 3.5 μ avg. particle size; sp.gr. 2.38; bulk dens. 0.4 g/cc (loose); surf. area 5.5 m^2/g; oil absorp. 32.

Zetax®. [R.T. Vanderbilt] Zinc 2-mercaptobenzothiazole; CAS 155-04-4; sec. accelerator in latex foam curing systems; pale yel. powd., 99.9% thru 100 mesh; pract. insol. in water; m.w. 397.86; dens. 1.70 ± 0.03 mg/m^3; m.p. > 300 C; 15-18% Zn.

Zetax® Dispersion. [R.T. Vanderbilt] Zinc 2-mercaptobenzothiazole aq. disp.; CAS 155-04-4; sec. accelerator in latex foam curing systems; lt. yel. liq.; misc. with water; dens. 1.35 ± 0.03 mg/m^3; pH 7.0-8.5; *Toxicology:* LD50 (oral, rat) 540 mg/kg; may irritate or sensitize skin, irritate eyes; avoid nitrosating agents to prevent formation of suspect carcinogens; *Precaution:* emits acrid fumes when exposed to flame; hazardous decomp. prods.: oxides of C, N, S, and Zn upon combustion; *Storage:* store above 32 F; keep from freezing.

Zewakol. [LignoTech] Modified lignosulfonates; extenders for U/F resins in particle board mfg.

Zic Stick '85'. [Rhein Chemie] 90% Zinc oxide in 10% org. binder; predispersed rubber chem.; wh. bars; sp.gr. 3.62; 90% act.

Zinc Borate 9506. [Joseph Storey] Zinc borate; synergist with halogens, antimony trioxide, alumina trihydrate in rubber and polymer flame retardant applics.; provides smoke reduction, afterglow suppression; for PVC, unsat. polyesters, nylon 6 and 6/6, cable (EVA, PE/EVA, EPDM), latex back coating; char forming and anticorrosive effect in paints.

Zinc Borate B9355. [Joseph Storey] Zinc borate ($2ZnO \cdot 3B_2O_3 \cdot 3.5 H_2O$); CAS 138265-88-0; synergist with halogens, antimony trioxide, alumina trihydrate in rubber and polymer flame retardant applics.; provides smoke reduction, afterglow suppression; for PVC, unsat. polyesters, nylon 6 and 6/6, cable (EVA, PE/EVA, EPDM), latex back coating; char forming and anticorrosive effect in paints; for higher temp. applics.; wh. amorphous powd., odorless; 0.2% max. on 200 mesh; sol. in acids and bases; < 1% sol. in water; m.w. 434.69; bulk dens. 0.3-0.5 g/cc; pH 7-8 (20% aq.); 99% act.; 37% min. ZnO, 47% min. B_2O_3; *Toxicology:* LD50 (oral, rat) > 10 g/kg, (dermal, rabbit) > 10 g/kg; TLV 10 mg/m^3; nuisance dust—may cause irritation to respiratory tract, eye irritation, drying of skin; *Storage:* store in cool, dry, well-ventilated area.

Zinc Omadine® 48% Fine Particle Disp. [Olin] Zinc pyrithione; CAS 13463-41-7; EINECS 236-671-3; antimicrobial inhibiting growth of gram-negative and gram-positive bacterial, fungi, mold and yeast; antidandruff agent for shampoos; preservation of cosmetics, surgical scrubs, acne preps., topical antibacterial prods.; metal coolant and cutting fluids, plastics (PVC, polyolefins, urethanes, nat. and syn. latex, SBR, EPDM), fabric coatings, paints, carpets, wire and cable insulation, adhesives, caulks, vinyl/urethane flooring; EPA reg. no. 1258-840, 1258-841; off-wh. disp. (particle size 90% < 1 μ), mild odor; m.w. 317.7 (zinc pyrithione); dens. 10 lb/gal; pH 6.5-8.5 (5% aq.); 48-50% act.; *Storage:* store in dry place at 10-54 C; keep containers tightly closed when not in use; do not store with strong oxidizing agents; shake or stir before use.

Zinc Omadine® 48% Std. Disp. [Olin] Zinc pyrithione; CAS 13463-41-7; EINECS 236-671-3; antimicrobial inhibiting growth of gram-negative and gram-positive bacterial, fungi, mold and yeast; antidandruff agent for shampoos; preservation of cosmetics, surgical scrubs, acne preps., topical antibacterial prods.; metal coolant and cutting fluids, plastics (PVC, polyolefins, urethanes, nat. and syn. latex, SBR, EPDM), fabric coatings, paints, carpets, wire and cable insulation, adhesives, caulks, vinyl/urethane flooring; EPA reg. no. 1258-840, 1258-841; off-wh. disp. (particle size 90% < 5 μ), mild odor; m.w. 317.7 (zinc pyrithione); dens. 10 lb/gal; pH 6.5-8.5 (5% aq.); 48-50% act.; *Storage:* store in dry place at 10-54 C; keep containers tightly closed when not in use; do not store with strong oxidizing agents; shake or stir before use.

Zinc Omadine® Powd. [Olin] Zinc pyrithione; CAS 13463-41-7; EINECS 236-671-3; antimicrobial inhibiting growth of gram-negative and gram-positive bacterial, fungi, mold and yeast; antidandruff agent for shampoos; preservation of cosmetics, surgical scrubs, acne preps., topical antibacterial prods.; metal coolant and cutting fluids, plastics (PVC, polyolefins, urethanes, nat. and syn. latex, SBR, EPDM), fabric coatings, paints, carpets, wire and cable insulation, adhesives, caulks, vinyl/urethane flooring; EPA reg. no. 1258-840, 1258-841; off-wh. powd., mild odor; 70 < 25 μ particle size; insol. in water; m.w. 317.7 (zinc pyrithione); sp.gr. 1.782; bulk dens. 0.35 g/ml; m.p. ≈ 240 C (dec.); pH 6.5-

8.5 (5% aq.); 95-99% assay; *Storage:* store in dry place at 10-54 C; keep containers tightly closed when not in use; do not store with strong oxidizing agents.

Zincote™. [C.P. Hall] Zinc stearate formulation; CAS 557-05-1; EINECS 209-151-9; provides uniform antistick coating to polymers and rubber slabs or continuous sheet from takeaway mills, extruders, or calenders; uniform wh. paste; sp.gr. 0.87 ± 0.16; pH 9.6 ± 0.7 (10%); 29 ± 3% solids.

Zinc Oxide No. 185. [Eagle Zinc] Tert. zinc oxide; CAS 1314-13-2; EINECS 215-222-5; accelerator activator, pigment, reinforcing agent for rubber and applics. not requiring a high degree of purity; also for prod. of zinc compds.; 99.6% -325 mesh; sp.gr. 5.6; bulk dens. 60 lb/ft^3; 96.5% ZnO.

Zinc Oxide No. 318. [Eagle Zinc] Amer. process zinc oxide, lead-free; CAS 1314-13-2; EINECS 215-222-5; slow curing fine particle size pigment and reinforcing agent for rubber industry; also for mfg. of various zinc compds.; 0.38 μ median particle diam.; 99.93% -325 mesh; sp.gr. 5.65; bulk dens. 38 lb/ft^3; bulking value 0.214 gal/lb; 99.2% ZnO.

Zinc Stearate 42. [Sovereign] Zinc stearate; CAS 557-05-1; EINECS 209-151-9; general purpose grade, partitioning agent in rubber; flatting agent in paint.

Zinc Stearate Spray No. Z212-B. [IMS] < 5% Zinc stearate release agent with 35-45% HFC-152a, 25-30% dimethyl ether, 25-35% HCFC-141b; non-CFC dry powd. paintable mold release for the food industry; effective for ABS, acrylic, PC, PPO and most other materials; FDA 21CFR §182.5994; clear to wh. mist with sl. ethereal odor as dispensed from aerosol pkg.; sl. sol. in water; sp.gr. < 1; vapor pressure 60 ± 10 psig; flash pt. < 0 F; > 90% volatile; *Toxicology:* inh. may cause CNS depression with dizziness, headache; higher exposure may cause heart problems; gross overexposure may cause fatality; eye and skin irritation; direct contact with spray can cause frostbite; may cause polymer fume fever; *Precaution:* flamm. limits 1.2-18%; at elevated temps. (> 130 F) aerosol containers may burst, vent, or rupture; incompat. with strong oxidizers, strong caustics, reactive metals (Na, K, Zn, Mg, Al); *Storage:* store in cool, dry area out of direct sunlight; do not puncture, incinerate, or store above 120 F.

Zink Oxide AT. [Bayer AG] Zinc oxide; CAS 1314-13-2; EINECS 215-222-5; activator for rubber processing.

Zink Oxide Transparent. [Bayer AG; Bayer/Fibers, Org., Rubbers] Highly disperse precipitated zinc oxide; CAS 1314-13-2; EINECS 215-222-5; vulcanization accelerator activator for transparent rubber goods, in natural and syn. rubbers, food-contacting goods, latex; acid acceptor in polychloroprene adhesives; lt. colored reinforcing filler; used for vulcanizates needing high elasticity or transparency, vulcanizates cured in hot air or with sulfenamides, vulcanizates cured with metal oxides and without sulfur; wh. to sl. ylsh. grn. powd.; sp.gr. 3.5; surf. area (BET) 45 m^2/g; pH 9-10; 70-73% ZnO; < 0.002% PbO.

Zinkoxyd Activ®. [Bayer AG; Bayer/Fibers, Org., Rubbers] Highly disperse precipitated zinc oxide; CAS 1314-13-2; EINECS 215-222-5; vulcanization accelerator activator for rubber goods based on natural and syn. elastomers and latex applics.; suitable at high levels for dynamically stressed articles, e.g., buffers and rollers, and in low concs. in transparent and translucent goods; lt. colored reinforcing filler; also for food-contacting goods; wh. to sl. ylsh. grn. powd.; sp.gr. 5.6; surf. area (BET) 50 m^2/g; pH 10-11; 93-96% ZnO.

Zisnet® F-PT. [Zeon] 2,4,6-Trimercapto-s-triazine; CAS 638-16-4; curing agent for epichlorohydrin and polyacrylate rubber; used in place of ethylene thiourea and red lead; gives improved heat resist., less mold fouling, reduced toxicity; oil treated to reduce dusting; sp.gr. 1.56.

ZN-18-1. [Akzo Nobel] Zinc complex; lubricant for flexible and rigid PVC, plastisols, organosols, foams, records, bottles, tiles, PE, PU.

Zoldine® ZT-65. [ANGUS] Hydroxymethyl dioxoazabicyclooctane; CAS 6542-37-6; EINECS 229-457-6; cross-linking agent for resorcinol phenol-formaldehyde or protein-based resin systems; raw material for synthesis; m.w. 145.1; water-sol.; f.p. -25 C; b.p. decomp.; flash pt. (TCC) > 200 F; pH 7.0 (0.1M aq. sol'n.).

Zonarez® 1040. [Arizona] Polyterpene resin; tackifier for adhesives; Gardner 4 color; soften. pt. (R&B) 40 C.

Zonarez® 1085. [Arizona] Polyterpene resin; tackifier for adhesives; Gardner 2 color; soften. pt. (R&B) 85 C.

Zonarez® 1100. [Arizona] Polyterpene resin; tackifier for adhesives; Gardner 1+ color; soften. pt. (R&B) 100 C.

Zonarez® 1115. [Arizona] Polyterpene resin; tackifier for adhesives; FDA 21CFR §175.105, 175.300, 176.170, 176.180; Gardner 1+ color; soften. pt. (R&B) 115 C.

Zonarez® 1115T. [Arizona] Polyterpene resin; tackifier for adhesives; FDA 21CFR §175.105, 175.300, 176.170, 176.180; Gardner 2 color; soften. pt. (R&B) 115 C.

Zonarez® 1125. [Arizona] Polyterpene resin; tackifier for adhesives; FDA 21CFR §175.105, 175.300, 176.170, 176.180; Gardner 1+ color; soften. pt. (R&B) 125 C.

Zonarez® 1135. [Arizona] Polyterpene resin; tackifier for adhesives; FDA 21CFR §175.105, 175.300, 176.170, 176.180; Gardner 1+ color; soften. pt. (R&B) 135 C.

Zonarez® 7085. [Arizona] Polydipentene; CAS 9003-73-0; low m.w., thermoplastic polymer which imparts high levels of tack and adhesion to many elastomeric and polymeric materials; stability in adhesive systems, resistance to attack by acids, alkalis, salts, and water, aging properties; used in the mfg. of pressure-sensitive adhesives, rubber cements, emulsion adhesives, hot melt adhesives/ coatings, can sealants, caulking and general sealants., ink, paints, concrete waterproofing agents, varnishes, chewing and bubble gum bases; Gardner 3 (1963, 50% in heptane) clear, bright; sol. in 2-ethyl hexyl alcohol, aliphatic hydrocarbons, aromatic hydrocarbons, chlorinated hydrocarbons, butyl and amyl acetates, terpenes, ethyl ether, and petrol. ether; sp.gr. 0.97; R&B soften. pt. 85 C; acid no. < 1; sapon. no. < 1.

Zonarez® 7115. [Arizona] Polyterpene resin produced from dipinene; bright, clear, pale-colored, low m.w. polymer imparting high levels of tack and adhesion to elastomeric and polymeric materials; in mfg. of pressure-sensitive adhesives, rubber cements, hot-melt adhesives/coatings, can sealants, caulking; inks, paints, concrete waterproofing agents, varnishes, chewing and bubble gum bases, moisture-resist. soft gelatin capsules and powds. of ascorbic acid and its salts; LITE resins feature water-wh. color; FDA compliance; FDA 21CFR §175.105, 175.300, 176.170, 176.180; Gardner 4 color; sp.gr. 0.99; soften. pt. (R&B) 115 C; flash pt. (CC) > 400 F.

Zonarez® 7115 LITE. [Arizona] Polyterpene resin produced from dipinene; bright, clear, pale-colored, low m.w. polymer imparting high levels of tack and adhesion to elastomeric and polymeric materials; in mfg. of pressure-sensitive adhesives, rubber cements, hot-melt adhesives/coatings, can sealants, caulking; inks, paints, concrete waterproofing agents, varnishes, chewing and bubble gum bases, moisture-resist. soft gelatin capsules and powds. of ascorbic acid and its salts; LITE resins feature water-wh. color; FDA compliance; FDA 21CFR §175.105, 175.300, 176.170, 176.180; Gardner 2+ color; sp.gr. 0.99; soften. pt. (R&B) 115 C; flash pt. (CC) > 400 F.

Zonarez® 7125. [Arizona] Polyterpene resin produced from dipinene; bright, clear, pale-colored, low m.w. polymer imparting high levels of tack and adhesion to elastomeric and polymeric materials; in mfg. of pressure-sensitive adhesives, rubber cements, hot-melt adhesives/coatings, can sealants, caulking; inks, paints, concrete waterproofing agents, varnishes, chewing and bubble gum bases, moisture-resist. soft gelatin capsules and powds. of ascorbic acid and its salts; LITE resins feature water-wh. color; FDA compliance; FDA 21CFR §175.105, 175.300, 176.170, 176.180; Gardner 4 color; sp.gr. 0.99; soften. pt. (R&B) 125 C; flash pt. (CC) > 400 F.

Zonarez® 7125 LITE. [Arizona] Polyterpene resin produced from dipinene; bright, clear, pale-colored, low m.w. polymer imparting high levels of tack and adhesion to elastomeric and polymeric materials; in mfg. of pressure-sensitive adhesives, rubber cements, hot-melt adhesives/coatings, can sealants, caulking; inks, paints, concrete waterproofing agents, varnishes, chewing and bubble gum bases, moisture-resist. soft gelatin capsules and powds. of ascorbic acid and its salts; LITE resins feature water-wh. color; FDA compliance; FDA 21CFR §175.105, 175.300, 176.170, 176.180; Gardner 3-color; sp.gr. 0.99; soften. pt. (R&B) 125 C; flash pt. (CC) > 400 F.

Zonarez® Alpha 25. [Arizona] Polyterpene resin based on α-pinene; clear, light-colored thermoplastic polymers imparting tack and str. to adhesives; increases hard resin loading; co-tackifier for the mfg. of pressure-sensitive adhesives, solv.-based adhesives, emulsion adhesives, and hot-melt adhesives/coatings; esp. designed as co-tackifier for mfg. of adhesive systems based on thermoplastic block copolymer such as Kraton®; plasticizing and softening props.; FDA compliance; FDA 21CFR §175.105, 175.300, 176.170, 176.180, 176.210; Gardner 4 color; sol. in aliphatic, aromatic, and chlorinated HC, esters, terpenes, ethyl ether, some alcohols; sp.gr. 0.90; soften. pt. (R&B) 25 C; flash pt. (CC) > 400 F.

Zonarez® B-10. [Arizona] Polyterpene resin (produced by the polymerization of betapinene); low m.w., thermoplastic polymer which imparts high levels of tack and adhesion to many elastomeric and polymeric materials; stability in adhesive systems, resistance to attack by acids, alkalis, salts, and water, aging properties; used in the mfg. of pressure-sensitive adhesives, rubber cements, emulsion adhesives, hot melt adhesives/coatings, can sealants, caulking and general sealants., ink, paints, concrete waterproofing agents, varnishes, chewing and bubble gum bases; Gardner 2+ (1963, 50% in heptane) clear, bright; sp.gr. 0.92; R&B soften. pt. 10 C; acid no. < 1; sapon. no. < 1.

Zonarez® B-115. [Arizona] Polyterpene resin; low m.w., thermoplastic polymer which imparts high levels of tack and adhesion to many elastomeric and polymeric materials; stability in adhesive systems, resistance to attack by acids, alkalis, salts, and water, aging properties; used in the mfg. of pressure-sensitive adhesives, rubber cements, emulsion adhesives, hot melt adhesives/coatings, can sealants, caulking and general sealants., ink, paints, concrete waterproofing agents, varnishes, chewing and bubble gum bases; FDA 21CFR §175.105, 1715.300, 176.170, 176.180; Gardner 3+ clear, bright color; sol. in 2-ethyl hexyl alcohol, aliphatic hydrocarbons, aromatic hydrocarbons, chlorinated hydrocarbons, ester, terpenes, ethyl ether, and petrol. ether; sp.gr. 0.98; soften. pt. (R&B) 115 C; acid no. < 1; sapon. no. < 1.

Zonarez® B-125. [Arizona] Polyterpene resin; low m.w., thermoplastic polymer which imparts high levels of tack and adhesion to many elastomeric and polymeric materials; stability in adhesive systems, resistance to attack by acids, alkalis, salts, and water, aging properties; used in the mfg. of pressure-sensitive adhesives, rubber cements, emulsion adhesives, hot melt adhesives/coatings, can sealants, caulking and general sealants., ink, paints, concrete waterproofing agents, varnishes, chewing and bubble gum bases; FDA 21CFR §175.105, 1715.300, 176.170, 176.180; Gardner 3+ clear, bright color; sol. in 2-ethyl hexyl alcohol, aliphatic hydrocarbons, aromatic hydrocarbons, chlorinated hydrocarbons, ester, terpenes, ethyl ether, and petrol. ether; sp.gr. 0.97; soften. pt. (R&B) 125 C; acid no. < 1; sapon. no. < 1.

Zonarez® M-1115. [Arizona] Polyterpene resin produced from β-pinene; oxidation-resist., low m.w. thermoplastic polymer developed as tackifier for adhesive systems based on the thermoplastic block copolymer Kraton® 1107; also works well with natural rubber-based adhesives; tackifier resin for mfg. of pressure-sensitive adhesives used in tapes and lables, solv.-based adhesives, emulsion adhesives, hot-melt adhesives and coatings; FDA compliance; FDA 21CFR §175.105, 1715.300, 176.170, 176.180; water-wh. bright clear color; sp.gr. 0.98; sol. in aliphatic, aromatic, and chlorinated HC, esters, terpenes, drying oils, rosin, and rosin derivs.; soften. pt. (R&B) 115 C; flash pt. (CC) > 400 F.

Zonarez® T-4115. [Arizona] Polyterpene resin; tackifier for adhesives; FDA 21CFR §175.105, 175.300, 176.170, 176.180; Gardner 2 color; soften. pt. (R&B) 115 C.

Zonatac® 85 LITE. [Arizona] Modified terpene; tackifier for adhesives; Gardner 3- color; soften. pt. (R&B) 89 C.

Zonatac® 105. [Arizona] Modified terpene hydrocarbon resin; light-colored, low odor thermoplastic resins for use as tackifier in hot-melt adhesives and coatings, hot-melt pressure-sensitive adhesives, and solv.-based thermoplastic elastomer pressure-sensitive adhesive systems; FDA 21CFR § 175.105; Gardner 3+ color; sp.gr. 1.02; soften. pt. (R&B) 105 C; flash pt. (CC) 480 F.

Zonatac® 105 LITE. [Arizona] Modified terpene hydrocarbon resin; light-colored, low odor thermoplastic resins for use as tackifier in hot-melt adhesives and coatings, hot-melt pressure-sensitive adhesives, and solv.-based thermoplastic elastomer pressure-sensitive adhesive systems; FDA 21CFR §175.105; water-wh.; sp.gr. 1.02; soften. pt. (R&B) 105 C; flash pt. (CC) 480 F.

Zonatac® 115 LITE. [Arizona] Modified terpene hydrocarbon resin; light-colored, low odor thermoplastic resins for use as tackifier in hot-melt adhesives and coatings, hot-melt pressure-sensitive adhesives, and solv.-based thermoplastic elastomer pressure-sensitive adhesive systems; FDA 21CFR §175.105; water-wh.; sp.gr. 1.02; soften. pt. (R&B) 115 C; flash pt. (CC) 480 F.

Zonatac® M-105. [Arizona] Modified terpene; tackifier for adhesives; FDA 21CFR §175.105; Gardner 1+ color; soften. pt. (R&B) 105 C.

Zonester® 25. [Arizona] Rosin ester; tackifier resin for adhesives; FDA 21CFR §175.105, 175.300, 176.170, 176.180, 176.210, 177.1210; Gardner 7 max. color; soften. pt. (R&B) 26 C; acid no. 20-24.

Zonester® 85. [Arizona] Rosin ester; tackifier resin for adhesives; FDA 21CFR §175.105, 175.125, 175.300, 176.170, 176.180, 177.1210, 178.3870; Gardner 8 max. color; soften. pt. (R&B) 82 C; acid no. 10 max.

Zonester® 100. [Arizona] Pentaerythritol ester of tall oil rosin; CAS 8050-26-8; EINECS 232-479-9; high-melting resin ester used as tackifier in mfg. of hot-melt adhesives, mastic adhesives, contact cements, and pressure-sensitive adhesives; in emulsion form used in SBR, natural rubber, and neoprene latex adhesives; FDA 21CFR §175.105, 175.300, 176.170, 176.180, 176.210; amber; sp.gr. 1.06; soften. pt. (R&B) 94 C; flash pt. (CC) > 400 F; acid no. 14 max.

Zonyl® FSA. [DuPont] Fluorochemical surfactant; anionic; wetting agent, emulsifier, dispersant, corrosion inhibitor, leveling agent for adhesives, agric., polishes, polymerization, pigment grinding, cleaners, coatings and paints, fire fighting, paper, ink, oil, plastics, and textile industries; mold release and calcium sulfate scale removal in polymer processing; liq.; sol > 2% in water, methanol; sp.gr. 1.03 mg/m^3; dens. 8.6 lb/gal; flash pt. (PMCC) 21 C; surf. tens. 18 dynes/cm (0.1%); 25% solids in water/IPA (1:1); *Precaution:* flamm.

Zonyl® FSB. [DuPont] Fluorochemical surfactant; amphoteric; wetting agent, emulsifier, dispersant, corrosion inhibitor, leveling agent for adhesives, agric., polishes, polymerization, pigment grinding, cleaners, coatings and paints, fire fighting, paper, ink, oil, plastics, and textile industries; foamer, froth flotation agent for solv. cleaners, elastomers; liq.; sol. > 2% in water, IPA, methanol; dens. 8.8 lb/gal; flash pt. 67 F (PMCC); surf. tens. 17 dynes/cm (0.1%); 40% solids in IPA/water.

Zonyl® FSC. [DuPont] Fluorochemical surfactant; cationic; wetting agent, emulsifier, dispersant, corrosion inhibitor, leveling agent for adhesives, agric., polishes, polymerization, pigment grinding, cleaners, coatings and paints, fire fighting, paper, ink, oil, plastics, and textile industries; foamer, froth flotation agent for solv. cleaners, elastomers; liq.; sol. > 2% in water, IPA, methanol, acetone; sp.gr. 1.16 mg/m^3; dens. 9.7 lb/gal; flash pt. (PMCC) 21 C; surf. tens. 19 dynes/cm (0.1% aq.); 50% solids in water/IPA (25:25); *Precaution:* flamm.

Zonyl® FSN. [DuPont] Fluorochemical surfactant; nonionic; wetting agent, emulsifier, dispersant, corrosion inhibitor, leveling agent for adhesives, agric., polishes, polymerization, pigment grinding, cleaners, coatings and paints, fire fighting, paper, ink, oil, plastics, and textile industries; antifogging props. in polymer processing; liq.; sol. > 2% in water, IPA, methanol, acetone, ethyl acetate, THF; sp.gr. 1.06 mg/m^3; dens. 8.8 lb/gal; flash pt. (PMCC) 22 C; surf. tens. 23 dynes/cm (0.1%); 40% solids in water/IPA (30:30); *Precaution:* flamm.

Zonyl® FSP. [DuPont] Fluorochemical surfactant; anionic; wetting agent, emulsifier, dispersant, corrosion inhibitor, leveling agent, antifoam for adhesives, agric., polishes, polymerization, pigment grinding, cleaners, coatings and paints, fire fighting, paper, ink, oil, plastics, and textile industries; lubricant, mold release, and calcium sulfate scale removal in polymer processing; liq.; sol. > 2% in water, methanol; sp.gr. 1.15 mg/m^3; dens. 9.6 lb/gal; flash pt. (PMCC) 24 C; surf. tens. 21 dynes/cm

(0.1%); 35% solids in water/IPA (45:20); *Precaution:* flamm.

Zonyl® TBS. [DuPont] Fluorochemical surfactant; anionic; leveling agent for emulsions; pigment dispersant; wetting and foaming agent for corrosive media; improves physical props. and increases polymerization rate in polymer technology; slurry; sp.gr. 1.20 mg/m^3; dens. 10.0 lb/gal; flash pt. (PMCC) > 93 C; surf. tens. 24 dynes/cm (0.1% aq.); 33% solids in water/acetic acid (64:3).

Zorapol LS-30. [Zohar Detergent Factory] Modified sodium lauryl sulfate; CAS 151-21-3; anionic; foaming agent for syn. latexes, emulsion polymerization aid, esp. in mfg. of polyacrylate emulsion; liq. to paste; 32% conc.

Zorapol SN-9. [Zohar Detergent Factory] Sodium nonoxynol-9 sulfate; CAS 9014-90-8; anionic/nonionic; emulsifier for emulsion polymerization of polyacrylates; in prep. of household detergents; liq.; 32% conc.

Zylac 235. [Petrolite] Polyethylene wax; CAS 9002-88-4; EINECS 200-815-3; external lubricant for PVC pipe extrusion; solid.

Part II
Chemical Dictionary/
Cross-Reference

Chemical Dictionary/Cross-Reference

This section includes both chemicals with their trade name cross-references and those that are used as additives for plastics and elastomers. Chemicals that do not have direct additive functions for either plastics or elastomers are included if they are a secondary constituent of a trade name product from Part I.

ABS. *See* Acrylonitrile-butadiene-styrene
Absolute alcohol. *See* Alcohol
Aceite de ricino. *See* Castor oil
Acenaphthylene. *See* Bromoacenaphthylene

Acetic acid
CAS 64-19-7; EINECS 200-580-7
Synonyms: Ethanoic acid; Vinegar acid; Methanecarboxylic acid; Ethyllic acid; Pyroligneus acid
Classification: Organic acid
Empirical: $C_2H_4O_2$
Formula: CH_3COOH
Properties: Clear colorless liq., pungent odor; misc. with water, alcohol, glycerol, ether; insol. in carbon disulfide; m.w. 60.03; dens. 1.0492 (20/4 C); m.p. 16.63 C; b.p. 118 C (765 mm); visc. 1.22 cps (20 C); flash pt. (OC) 43 C; ref. index 1.3715 (20 C)
Precaution: Combustible; moderate fire risk; DOT: corrosive, flamm. liq.
Toxicology: Pure acetic acid: moderately toxic by ingestion, inhalation; dilute approved FDA for food use; strong irritant to skin and tissue; TLV 10 ppm in air; LD50 (oral, rat) 3310 mg/kg
Uses: Mfg. of acetic anhydride, cellulose acetate, vinyl acetate monomer; acetic esters; prod. of plastics, pharmaceuticals, dyes, insecticides, photographic chemicals, food additives; solvent reagent; acidifier; flavoring agent
Regulatory: FDA 21CFR §73.85, 133, 172.814, 178.1010, 184.1005, GRAS; USDA 9CFR §318.7; Europe listed; UK approved
Manuf./Distrib.: Air Prods.; BASF; BP Chem.; General Chem.; Hoechst Celanese; Janssen Chimica; PMC; Quantum/USI; Saurefabrik; Schweizerhall

Acetic acid, butyl ester. *See* n-Butyl acetate
Acetic acid, ethenyl ester, polymer with ethene. *See* Ethylene/VA copolymer
Acetic acid, ethyl ester. *See* Ethyl acetate
Acetic acid, mercapto- isooctyl ester. *See* Isooctyl thioglycolate
Acetic ether. *See* Ethyl acetate
Acetin. *See* Glyceryl acetate, Triacetin
Acetoacetic acid ethyl ester. *See* Ethylacetoacetate

2-(Acetoacetoxy) ethyl acrylate
CAS 21282-96-2
Toxicology: LD50 (oral, rat) 1132 mg/kg, (dermal, rat) > 2000 mg/kg; mod. skin irritant
Uses: Visc. reducer in polymerization (varnishes, adhesives, polyesters)
Trade names: Lonzamon® AAEA

2-(Acetoacetoxy) ethyl methacrylate
CAS 21282-97-3
Synonyms: 2 (Methacryloyloxy) ethyl acetoacetate
Empirical: $C_{10}H_{10}O_5$
Formula: $CH_3COCH_2CO_2CH_2CH_2O_2CC(CH_3)=CH_2$

Acetone/diphenylamine condensate

Properties: Liq.; m.w. 214.22; sp.gr. 1.12 (20/20 C); b.p. 100 C; flash pt. 134 C
Toxicology: LD50 (oral, rat) > 5000 mg/kg; pract. nonirritating to skin
Uses: Visc. reducer in polymerization (varnishes, adhesives, polyesters); resin intermediate
Trade names: Lonzamon® AAEMA

Acetone/diphenylamine condensate
Uses: Antioxidant for rubbers
Trade names: BLE® 25; Permanax™ BL; Permanax™ BLN; Permanax™ BLW

Acetylacetonate
Uses: Catalyst for esterification and olefin polymerization; used in coatings; resin crosslinking agent for automotive prods., coatings, elastomers, films/paints, graphic arts, plastics
Manuf./Distrib.: Monomer-Polymer & Dajac Labs; Penta Mfg.; Union Carbide; Wacker
Trade names: Tyzor AA

Acetylacetone peroxide
CAS 37187-22-7; EINECS 215-661-2
Synonyms: 2,4-Pentanedione peroxide
Classification: Ketone peroxide
Toxicology: Hazardous
Uses: Initiator, curing agent for thermoset polyesters, esp. with cobalt accelerators
Trade names: Aztec® AAP-LA-M2; Aztec® AAP-NA-2; Aztec® AAP-SA-3; Lupersol 224; Trigonox® 40; Trigonox® 44P; USP®-240
Trade names containing: Esperfoam® FR

Acetylated lard glyceride
CAS 8029-92-3
Synonyms: Glycerides, lard mono-, acetates
Classification: Acetyl ester
Uses: Emulsifier, lubricant, emollient, deaerator; for food applics.
Regulatory: FDA 21CFR §172.828, 175.230

Acetylated palm kernel glycerides
Synonyms: Glycerides, palm kernel oil mono-, di-, and tri-, acetates
Classification: Acetyl ester
Uses: Lubricant and plasticizer for plastics and coatings; cosolv. for polar additives
Trade names: Cetodan® 95 CO

Acetyl cyclohexanesulfonyl peroxide
CAS 3179-56-4
Classification: Organic peroxide
Properties: M.w. 222.3
Uses: Polymerization initiator
Trade names: Aztec® ACSP-28-FT; Aztec® ACSP-55-W1
Trade names containing: Trigonox® ACS-M28

Acetylene black. *See* Carbon black
Acetylenicglycol. *See* Tetramethyl decynediol
2-(Acetyloxy)-1,2,3-propanetricarboxylic acid, tributyl ester. *See* Acetyl tributyl citrate
2-(Acetyloxy)-1,2,3-propanetricarboxylic acid, triethyl ester. *See* Acetyl triethyl citrate

Acetyl paracumyl phenol
Properties: Sp.gr. 1.08; m.p. 0 C; flash pt. (COC) 182 C
Uses: Plasticizer for PVC, VCA
Manuf./Distrib.: Kenrich Petrochem.

Acetyl tributyl citrate
CAS 77-90-7; EINECS 201-067-0
Synonyms: ATBC; 2-(Acetyloxy)-1,2,3-propanetricarboxylic acid, tributyl ester; Tributyl acetyl citrate
Classification: Aliphatic ester
Empirical: $C_{20}H_{34}O_8$
Formula: $CH_3COOC_3H_4(COOC_4H_9)_3$
Properties: Colorless sl. visc. liq., sweet herbaceous odor; sol. in alcohol; insol. in water; m.w. 402.49; dens.

1.14; b.p. > 300 C; flash pt. 204 C
Toxicology: Heated to decomp., emits acrid smoke and irritating fumes
Uses: Plasticizer for vinyl resins
Regulatory: FDA 21CFR §172.515, 175.105, 175.300, 175.320, 178.3910, 181.22, 181.27
Manuf./Distrib.: Morflex; Pfizer Spec.; Unitex
Trade names: Citroflex® A-4; Citroflex® A-4 Special; Estaflex® ATC; Uniplex 84
Trade names containing: Trigonox® 169-NA50

Acetyl triethyl citrate
CAS 77-89-4; EINECS 201-066-5
Synonyms: ATEC; 2-(Acetyloxy)-1,2,3-propanetricarboxylic acid, triethyl ester; Tricarballylic acid-β-acetoxytributyl ester; Triethyl acetylcitrate
Classification: Aliphatic ester
Empirical: $C_{14}H_{22}O_8$
Formula: $CH_3COOC_3H_4(COOC_2H_5)_3$
Properties: Colorless, odorless liq.; sl. sol. in water; m.w. 318.36; dens. 1.135 (25 C); flash pt. 187 C
Precaution: Combustible
Toxicology: Moderate toxicity by intraperitoneal route; mild toxicity by ingestion
Uses: Plasticizer for cellulosics, primarily ethyl cellulose
Regulatory: FDA 21CFR §175.105, 175.300, 175.320, 178.3910, 181.22, 181.27
Manuf./Distrib.: Morflex; Unitex
Trade names: Citroflex® A-2; Uniplex 82

Acetyl tri-n-hexyl citrate
CAS 24817-92-3
Synonyms: ATHC
Formula: $CH_3CO_2C[CO_2(CH_2)_5CH_3][CH_2CO_2(CH_2)_5CH_3]_2$
Properties: M.w. 468.65; sp.gr. 1.0046; flash pt. (COC) 240 C; ref. index. 1.447
Uses: Plasticizer for PVC; for polymeric medical articles
Manuf./Distrib.: Morflex; Unitex
Trade names: Citroflex® A-6

Acid calcium phosphate. *See* Calcium phosphate monobasic monohydrate
Acids, coconut. *See* Coconut acid
Acids, soy. *See* Soy acid
Acids, tall oil. *See* Tall oil acid
Acids, tallow. *See* Tallow acid
Acids, tallow, hydrogenated. *See* Hydrogenated tallow acid

ACM
Classification: Elastomer
Trade names containing: Poly-Dispersion® DCP-60P; Poly-Dispersion® TAC-50; Poly-Dispersion® TATM-50; Poly-Dispersion® TMC-50; Poly-Dispersion® TVP-50; Poly-Dispersion® VC-60P; Rhenocure® Diuron

Acroleic acid. *See* Acrylic acid

Acrylates copolymer
Synonyms: Acrylic/acrylate copolymer
Definition: Polymers of two or more monomers consisting of acrylic acid, methacrylic acid, or their simple esters
Uses: Surface coatings, emulsions, paints, paper and leather finishes
Regulatory: FDA 21CFR §175.105, 175.210, 175.300, 175.320, 176.170, 176.180, 177.1010, 178.3790
Trade names: Texicryl® 13-300, 13-302

Acrylic acid
CAS 79-10-7; EINECS 201-177-9
Synonyms: Acroleic acid; Propenoic acid; Ethylene carboxylic acid
Formula: $H_2C{:}CHCOOH$
Properties: Colorless liq., acrid odor; misc. with water, alcohol, ether; dens. 1.052 (20/20 C); b.p. 140.9 C; m.p. 12.1 C; flash pt. (OC) 54.5 C; ref. index 1.4224 (20 C); polymerizes readily
Precaution: May polymerize explosively; combustible
Toxicology: Irritant to skin; toxic by inhalation; TLV 10 ppm in air
Uses: Monomer for polyacrylic and polymethacrylic acids, other acrylic acids, acrylic polymers

Acrylic acid, hexyl ester

Manuf./Distrib.: BASF; Hoechst-Celanese; Hüls UK; Penta Mfg.; Rohm & Haas; Union Carbide
Trade names containing: Polybond® 1000; Polybond® 1001; Polybond® 1002; Polybond® 1009; Polybond® 2015

Acrylic acid, hexyl ester. *See* n-Hexyl acrylate
Acrylic acid polymers. *See* Polyacrylic acid
Acrylic/acrylate copolymer. *See* Acrylates copolymer
Acrylic fiber. *See* Acrylic resin

Acrylic-imide copolymer
Uses: Amorphous thermoplastic injection molding resin for lighting, automotive, optical, pkg., medical, and appliance applics.; blending additive
Trade names: Paraloid® EXL-4151; Paraloid® EXL-4261

Acrylic polymer. *See* Acrylic resin

Acrylic resin
CAS 9003-01-4
Synonyms: Acrylic polymer; Poly(acrylic acid); Acrylic fiber; Nitrile rubber
Definition: Thermoplastic polymer or copolymer of acrylic acid, methacrylic acid, esters of these acids, or acrylonitrile
Formula: $[—CH_2CH(CO_2H)—]$
Properties: Varies from hard to brittle solids to fibrous elastomeric structures, to viscous liqs.; able to transmit light for sheet and rod forms
Uses: Hard, shatterproof transparent or colored material, glass substitute in medical, dental, and industrial applics.; impact modifier, processing aid for PVC
Manuf./Distrib.: Air Prods.; Akzo Nobel; Alco; Am. Cyanamid; DSM UK; BFGoodrich; ICI Resins; Miles; Reichhold; Rhone-Poulenc; Rohm & Haas; Sybron; Synthron; Thibaut & Walker; UCB SA; Union Carbide
Trade names: Kane Ace® PA-10; Kane Ace® PA-30; Kane Ace® PA-50; Paraloid® EXL-5137; Paraloid® EXL-5375; Paraloid® K-120ND; Paraloid® K-125; Paraloid® K-130; Paraloid® K-147; Paraloid® K-175; Paraloid® KF-710; Paraloid® KM-334; Paraloid® KM-390; Paraloid® PM-800
See also Polyacrylate, Polyacrylic acid

Acrylic-styrene-acrylonitrile terpolymer
Synonyms: ASA terpolymer
Uses: Outdoor-weatherable injection molding resin for interior and exterior automotive/truck/RV trim and parts, building and construction applics., etc.; modifier for polymers
Trade names: Baymoflex A KU3-2069.A; Blendex® 975
Trade names containing: Blendex® 977

Acrylite. *See* Methyl methacrylate polymer
Acrylonitrile-butadiene copolymer. *See* Acrylonitrile-butadiene rubber

Acrylonitrile-butadiene rubber
Synonyms: NBR; Acrylonitrile rubber; Acrylonitrile-butadiene copolymer; Nitrile rubber
Classification: Syn. rubber
Formula: $[CH_2CH{=}CHCH_2CH_2CH(CN)]_x$
Properties: Dens. 0.98; tens. str. 1000-3000 psi; elong. 100-700%
Precaution: Combustible
Uses: Oil well parts, general-purpose oil-resistant applics., gaskets, grommets, o-rings; modifier for PVC and ABS resins
Trade names: Chemigum® N8; Krynac® P34.82
Trade names containing: Baymod® A 63 A; Baymod® A 72 A/74 A; Chem-Master R-102; Chem-Master R-215; Chem-Master R-936; Chem-Master RD-50; Krynac® P30.49; Krynac® P34.52; Nipol® DP5123P; Nipol® DP5125P; Nipol® DP5128P; Poly-Dispersion® AAD-75; Poly-Dispersion® A (SAN)D-65; Poly-Dispersion® ASD-75; Poly-Dispersion® A (TEF-1)D-80; Poly-Dispersion® A (TEF-1)D-80P; Poly-Dispersion® A (TI)D-80; Poly-Dispersion® A (Z-CN)D-85; Poly-Dispersion® AZFD-85; Poly-Dispersion® A (ZMTI)D-50

Acrylonitrile-butadiene rubber/PVC
Synonyms: NBR/PVC
Uses: Extrusion and molding composite for footwear, sponge, mechanical goods, thermoformed packaging, cable jacketing, embossed sheet, hose, flooring, and gaskets

Trade names: Chemigum® P83; Rhenoblend® N 6011
Trade names containing: Poly-Dispersion® AV (TBS)D-80; Poly-Dispersion® AV (TBS)D-80P

Acrylonitrile-butadiene-styrene
CAS 9003-56-9
Synonyms: ABS; Acrylonitrile polymer with 1,3-butadiene
Definition: Thermoplastic resin grafted from the 3 monomers from which its name is derived
Properties: Dens. ≈ 1.04; tens. str. ≈ 6500 psi; flex. str. ≈ 10,000 psi
Precaution: Combustible
Uses: Automotive body parts, telephones, pkg., shower stalls, etc.; impact modifier for PVC
Regulatory: FDA 21CFR §181.32
Manuf./Distrib.: Aiscondel SA; Bamberger Polymers; BASF; LNP; Mitsui Toatsu Chem.; Monsanto; Reichhold
Trade names: Baymod® A 50; Baymod® A 90/92; Baymod® A KU3-2086; Blendex® 131; Blendex® 336; Blendex® 338; Blendex® 467
Trade names containing: Baymod® A 63 A; Baymod® A 72 A/74 A; MDI AB-925

Acrylonitrile polymer with 1,3-butadiene. *See* Acrylonitrile-butadiene-styrene
Acrylonitrile rubber. *See* Acrylonitrile-butadiene rubber

Acrylonitrile/styrene/acrylate
Synonyms: ASA
Uses: Inj. molding or extrusion; esp. suitable for structural parts subject to outdoor exposure, e.g., letterboxes, parts for garden appliances, road signs, boat hulls, windsurfers, vehicle external parts, hot water drainage pipes, patio furniture, toy; modifier for PVC
Trade names: Baymod® A KA 8572

4-(2-Acryloyloxyethoxy)-2-hydroxybenzophenone polymer
Empirical: $(C_{17}H_{16}O_5)_n$
Properties: M.w. 50,000
Toxicology: LD50 (rat, oral) > 10 g/kg; LD50 (rabbit, dermal) > 5 g/kg
Uses: Light stabilizer, uv absorber for use in polymers (automotive, greenhouse, home siding, solar films and plastics)
Trade names: Cyasorb® UV 2126

ACS. *See* Styrene-acrylonitrile copolymer
Activated alumina. *See* Alumina
ADA. *See* Azodicarbonamide

Adipic acid
CAS 124-04-9; EINECS 204-673-3
Synonyms: Dicarboxylic acid C_6; Hexanedioic acid; 1,4-Butanedicarboxylic acid
Classification: Organic dicarboxylic acid
Empirical: $C_6H_{10}O_4$
Formula: $HOOC(CH_2)_4COOH$
Properties: Wh. monoclinic prisms; very sol. in alcohol; sol. in acetone; m.w. 146.16; dens. 1.360 (25/4 C); m.p. 152 C; b.p. 337.5 C; flash pt. (CC) 385 F
Toxicology: LD50 (oral, mouse) 1900 mg/kg
Uses: Manufacture of nylon and polyurethane foams; preparation of esters for use as plasticizers and lubricants; food additive (acidulant); baking powders; buffer, neutralizing agent; adhesives
Regulatory: FDA 21CFR §172.515, 184.1009, GRAS; GRAS (FEMA); USDA 9CFR §318.7; Japan approved; Europe listed (ADI 0-5 mg/kg, free acid basis); UK approved
Manuf./Distrib.: AlliedSignal; Asahi Chem Industry Co Ltd; Du Pont; Monsanto; Penta Mfg.; Rhone-Poulenc; UCB SA
Trade names: Adi-pure®

AEEA. *See* Aminoethylethanolamine
AEP. *See* Aminoethylpiperazine
Agalmatolite. *See* Pyrophyllite
AGE. *See* Allyl glycidyl ether
Agricultural limestone. *See* Limestone
AIBN. *See* 2,2′-Azobisisobutyronitrile
Alabaster. *See* Calcium sulfate dihydrate

Alcohol
CAS 64-17-5; EINECS 200-578-6
Synonyms: EtOH; Ethyl alcohol, undenatured; Ethanol; Ethanol, undenatured; Distilled spirits; Absolute alcohol
Definition: Undenatured ethyl alcohol
Empirical: C_2H_5OH
Formula: CH_3CH_2OH
Properties: Colorless limpid, volatile liq., vinous odor, pungent taste; misc. with water, methanol, ether, chloroform, 95% acetone; m.w. 46.08; dens. 0.816 (15.5 C); b.p. 78.3 C; f.p. -117.3 C; flash pt. (CC) 12.7 C; ref. index 1.365 (15 C)
Precaution: Flamm. liq.; can react vigorously with oxidizers; reacts violently with many chemicals
Toxicology: Depressant drug; TLV 1000 ppm in air; moderately toxic by ingestion; experimental tumorigen, teratogen
Uses: Solvent; extraction medium; mfg. of acetaldehyde, denatured alcohol, pharmaceuticals (tonics, colognes), perfumery, organic synthesis; octane booster in gasoline; topical antiseptic; active ingred. in OTC drug prods.
Regulatory: FDA 21CFR §169.3, 169.175, 169.176, 169.177, 169.178, 169.180, 169.181, 172.340, 172.560, 175.105, 176.200, 176.210, 177.1440, 184.1293, GRAS; 27CFR §2.5, 2.12; Japan restricted
Manuf./Distrib.: BP Chem.; Coyne; Eastman; Georgia-Pacific Resins; Gist-Brocades SpA; Grain Processing; Great Western; Quantum/USI; Union Carbide; Vista
Trade names containing: CB2408; Ethoquad® T/12; Lankropol® KMA; Lankropol® KO2; Monawet MO-70E

Alcohol C-6. *See* Hexyl alcohol
Alcohol C-8. *See* Caprylic alcohol, 2-Ethylhexanol
Alcohol C-10. *See* n-Decyl alcohol
Alcohol C-12. *See* Lauryl alcohol
Alcohol C-22. *See* Behenyl alcohol
Alcohols, C10-14. *See* C10-14 alcohols
Alcohols, C12-14. *See* C12-14 alcohols
Alcohols, C20-24. *See* C20-24 alcohols
Alkane C-6. *See* Hexane
Alkane C-7. *See* Heptane

Alkoxy triacryl titanate
Trade names containing: Ken-React® 33DS (KR 33DS); Ken-React® 39DS (KR 39DS)

Alkyd. *See* Alkyd resin

Alkyd resin
CAS 63148-69-6, 68459-31-4
Synonyms: Alkyd
Classification: Thermosetting coating polymer
Uses: Vehicles in exterior house paints, marine paints, and baking enamels; elec. components, encapsulation
Manuf./Distrib.: Bayer AG; Croda Resins Ltd; DSM UK; PPG Industries SA; Reichhold; Scott Bader
Trade names containing: Mearlite® Ultra Bright UDQ

Alkyl C10 bisphenol A phosphite. *See* Tetrakis isodecyl 4,4′-isopropylidene diphosphite

Alkyl C12-15 bisphenol A phosphite
CAS 93356-94-6
Uses: Stabilizer for PVC and vinyl copolymers
Trade names: Doverphos® 613

Alkyl phenol disulfide
CAS 86555-98-6
Uses: NR, NBR, SBR vulcanizer, plasticizer, tackifier; accelerates SBR; adhesives
Trade names: Vultac® 2; Vultac® 3; Vultac® 7
Trade names containing: Vultac® 4; Vultac® 5; Vultac® 710

Allyl-2,3-epoxypropyl ether. *See* Allyl glycidyl ether

Allyl glycidyl ether
CAS 106-92-3; EINECS 203-442-4
Synonyms: AGE; [(2-Propenyloxy)methyl]oxirane; Allyl-2,3-epoxypropyl ether; 1-Allyloxy-2,3-epoxypropane
Classification: Glycidyl ether monomer
Empirical: $C_6H_{10}O_2$
Formula: $CH_2CHCH_2OCH_2{=}CH_2CHO$
Properties: Colorless liq., pleasant odor; sol. in methanol, toluene; partly sol. in water; m.w. 114.15; dens. 0.9698 (20 C); f.p. -100 C; b.p. 153.9 C; flash pt. (OC) 135 F; ref. index 1.434
Precaution: Incompat. with strong acids, bases, oxidizing agents, amines; hazardous decomp. prods.: CO, hydrocarbons
Toxicology: LD50 (oral, rat) 922 mg/kg, (dermal, rabbit) 2550 mg/kg; poison by ingestion; moderately toxic by inhalation and skin contact; strong skin and eye irritant; TLV 5 ppm; may cause cancer
Uses: Modifier for elastomers, adhesives, fibers, reactive diluent for epoxies, reactive intermediate for coatings, sizing/finishing agent for fiberglass, silane intermediate in elec. coatings; stabilizer of chlorinated compds., vinyl resins, rubber; defoamer
Manuf./Distrib.: Alemark; Amber Syn.; CPS; Fluka; Monomer-Polymer & Dajac; Raschig; Spectrum Chem. Mfg.
Trade names: Ageflex AGE

Allyl methacrylate
CAS 96-05-9; EINECS 202-473-0
Empirical: $C_7H_{10}O_2$
Formula: $CH_2{:}C(CH_3)COOCH_2CH{:}CH_2$
Properties: M.w. 126.16; b.p. 59-60 C (35 mm); flash pt. 33 C; ref. index 1.4360 (20 C)
Uses: Silane monomer intermediate, crosslinker, polymer modifier for high impact plastics, adhesives, acrylic elastomers, optical polymers
Manuf./Distrib.: Aldrich; CPS; Monomer-Polymer & Dajac; Polysciences; Rhone-Poulenc Spec.; Richman; Rohm Tech; San Esters
Trade names: Ageflex AMA; MFM-401; Sipomer® AM
Trade names containing: SR-201

1-Allyloxy-2,3-epoxypropane. *See* Allyl glycidyl ether
Allyl 2,4,6-tribromophenyl ether. *See* Tribromophenyl allyl ether

Alumina
CAS 1344-28-1; EINECS 215-691-6
Synonyms: Aluminum oxide; Tabular alumina; Alumina, tabular; Calcined alumina; Alumina, calcined; Activated alumina; Alumina, activated; Alumite
Classification: Inorganic compd.
Empirical: Al_2O_3
Properties: Wh. powd., balls, or lumps; insol. in water; very sl. sol. in min. acids; m.w. 101.96; dens. 3-4; m.p. 2030 C
Precaution: Incombustible
Toxicology: Toxic by inhalation of dust; TLV:TWA 10 mg/m^3 (dust)
Uses: Prod. of aluminum, polyethylene, abrasives, refractories, ceramics, elec. insulators, catalysts and catalyst supports, paper, spark plugs, crucibles and lab ware, adsorbent for gases/water vapors, chromatographic analysis, heat-resist. fibers; food additives
Regulatory: Exempt from certification
Manuf./Distrib.: Air Prods.; Alcan; Alcoa; Aldrich; Atomergic Chemetals; BA Chem. Ltd.; Degussa; Ferro/ Transelco; Hüls Am.; Lonza Sarl; Nissan Chem. Ind.; Norton Chem. Process Prods.; Rhone-Poulenc; San Yuan; Zircar
Trade names: A-2; A-201; Aluminum Oxide C
Trade names containing: Aerosil® COK 84

Alumina, activated. *See* Alumina
Alumina, calcined. *See* Alumina

Alumina hydrate
CAS 1333-84-2
Synonyms: Alumina trihydrate; Aluminum hydroxide; Aluminum trihydroxide
Classification: Inorganic compd.
Empirical: $Al_2O_3 \cdot 3H_2O$
Properties: White powd., balls, or lumps; insol. in water, strong alkalis; m.w. 156.01; dens. 3.4-4.0; m.p.

2030 C; releases water on heating
Uses: Prod. of aluminum, abrasives, refractories, ceramics, elec. insulators, catalysts and catalyst supports, paper, spark plugs, crucibles and lab ware, adsorbent for gases/water vapors, chromatographic analysis, heat-resist. fibers, food additives; flame retardant for plastics
Regulatory: FDA 21CFR §173.1010, 177.1460
Manuf./Distrib.: Alcan; Alcoa; Atomergic Chemetals; R.E. Carroll; Climax Performance; Croxton & Garry Ltd; Franklin Ind. Min.; Nyco Minerals; Reheis; Solem
Trade names: Akrochem® Hydrated Alumina; Alcan BW53, BW103, BW153; Alcan FRF 5; Alcan FRF 10; Alcan FRF 20; Alcan FRF 30; Alcan FRF 40; Alcan FRF 60; Alcan FRF 80; Alcan FRF 80 S5; Alcan FRF LV1 S5; Alcan FRF LV2, LV4, LV5, LV6, LV7, LV8, LV9; Alcan FRF LV2 S5; Alcan H-10; Alcan Superfine 4; Alcan Superfine 7; Alcan Superfine 11; Alcan Ultrafine UF7; Alcan Ultrafine UF11; Alcan Ultrafine UF15; Alcan Ultrafine UF25; Alcan Ultrafine UF35; Baco DH101; 431-G; FRE; H-36; H-46; Haltex™ 304; Haltex™ 310; Hydromax 100; KC 31; LV-31; Micral® 532; Micral® 632; Micral® 855; Micral® 916; Micral® 932; Micral® 1000; Micral® 1500; 136-G; Onyx Classica OC-1000; Onyx Premier WP-31; Polyfil 100; SB-30; SB-31; SB-31C; SB-136; SB-331; SB-332; SB-335; SB-336; SB-431; SB-432; SB-632; SB-805; SB-932; 331-G; WH-31

Alumina silicate
Uses: Filler for plastics, coatings, foams
Trade names: Sillum-200 Q/P

Alumina, tabular. *See* Alumina
Alumina trihydrate. *See* Alumina hydrate

Alumino-silicate
Definition: A compd. of aluminum silicate with metal oxides or other radicals
Uses: Catalyst in refining petroleum, to soften water, detergents
Trade names: Extendospheres® XOL-50; Extendospheres® XOL-200

Aluminum calcium silicate. *See* Calcium aluminosilicate
Aluminum, dihydroxy (octadecanoato-o-). *See* Aluminum stearate

Aluminum distearate
CAS 300-92-5; EINECS 206-101-8
Definition: Aluminum salt of stearic acid
Empirical: $C_{36}H_{71}O_5Al$
Formula: $[CH_3(CH_2)_{16}COO]_2Al(OH)$
Properties: White powd.; insol. in water, alcohol, ether; forms gel w/ aliphatic and aromatic hydrocarbons; dens. 1.009; m.p. 145 C
Uses: Thickener in paints, inks, greases; water repellent; lubricant in plastics and cordage, in cement prod.
Regulatory: FDA 21CFR §172.863, 173.340, 175.105, 175.300, 175.320, 176.170, 176.200, 176.210, 177.1200, 177.1460, 177.2260, 178.3910, 179.45, 181.22, 181.29
Trade names: Witco® Aluminum Stearate 22

Aluminum hydrate. *See* Aluminum hydroxide

Aluminum hydroxide
CAS 21645-51-2; EINECS 244-492-7
Synonyms: Aluminum oxide trihydrate; Aluminum trihydrate; Aluminum hydrate; Hydrated alumina
Classification: Inorganic compd.
Empirical: AlH_3O_3
Formula: $Al(OH)_3$
Properties: White cryst. powd.; insol. in water; sol. in min. acids, caustic soda; m.w. 78.01; dens. 2.42; m.p. loses water @ 300 C
Toxicology: TLV:TWA 2 mg (Al)/m^3; poison by intraperitoneal route
Uses: Dyes, paints, textile finishing; flame retardant
Regulatory: FDA 21CFR §175.300, 177.1200, 177.2600, 182.90
Manuf./Distrib.: Alcan; Alcoa; Atomergic Chemetals; BA Chem Ltd; Nyco Minerals; Reheis; Rhone-Poulenc; Seimi Chem.; Solem; Vista; Whittaker, Clark & Daniels
Trade names: H-100; H-105; H-109; H-600; H-800; H-900; H-990; Martinal® 103 LE; Martinal® 104 LE; Martinal® 107 LE; Martinal® OL-104; Martinal® OL-104/C; Martinal® OL-104 LE; Martinal® OL-104/S; Martinal® OL-107; Martinal® OL-107/C; Martinal® OL-107 LE; Martinal® OL-111 LE; Martinal® OL/Q-107; Martinal® OL/Q-111; Martinal® ON; Martinal® ON-310; Martinal® ON-313; Martinal® ON-320;

Martinal® ON-920/V; Martinal® ON-4608; Martinal® O; Martinal® OX
Trade names containing: Polarite 880E(W); Polarite 880G(B)
See also Alumina hydrate

Aluminum-magnesium hydroxy carbonate
Uses: Acid neutralizing agent used in prod. of polyolefin resins
Trade names: L-55R®

Aluminum-magnesium-sodium silicate
Uses: Reinforcing filler for rubber industry
Trade names: Perkasil® KS 207

Aluminum nitride
CAS 24304-00-5; EINECS 246-140-8
Formula: AlN
Properties: Crystalline solid; dens. 3.10; m.p. 2150 C; Mohs hardness 9+
Uses: As semiconductor in electronics, nitriding of steel; filler for elastomeric, epoxy, polymeric systems
Manuf./Distrib.: Aldrich; Atomergic Chemetals; Carborundum; Dow; Mandoval Ltd
Trade names: AlNel A-100; AlNel A-200; AlNel AG

Aluminum oxide. *See* Alumina
Aluminum oxide trihydrate. *See* Aluminum hydroxide

Aluminum silicate
CAS 1327-36-2; 12141-46-7; EINECS 215-475-1
Synonyms: Pyrophyllite; CI 77004
Classification: Complex inorganic salt
Definition: Complex inorganic salt with 1 mole alumina, 1-3 moles silica
Formula: Al_2O_5Si
Properties: Varying proportions of Al_2O_3 and SiO_2; crystals or whiskers; high str.; m.w. 162.05
Uses: Clay used in reinforced plastics, dental cements, glass industry, paint filler, mfg. of precious stones, enamels
Regulatory: FDA 21CFR §175.300, 177.1200, 177.1460, 177.2600, 184.1155
Manuf./Distrib.: R.E. Carroll; CE Minerals; Dry Branch Kaolin; ECC Int'l.; Kaopolite; Solvay; Thiele Kaolin; R.T. Vanderbilt
Trade names: Akrodip™ Dry; Burgess 30; Burgess 30-P; Burgess 2211; Burgess 5178; Burgess CB; Burgess KE; Iceberg®; Icecap® K; Kaopolite® 1147; Kaopolite® 1152; Kaopolite® 1168; Kaopolite® SF; Kaopolite® SFO-NP; Optiwhite®; Optiwhite® P; Sipernat® 44; Tisyn®

Aluminum silicate, calcined
Uses: Reinforcing extender, pigment for rubber and polymer systems, paints, wire and cable insulation
Trade names: Satintone® SP-33®; Satintone® Whitetex®; Translink® 37; Translink® 77; Translink® 445; Translink® 555; Translink® HF-900

Aluminum silicate, hydrous
Uses: Reinforcing extender, pigment in rubber and polymer systems, paints; blocking agent for adhesives
Trade names: ASP® 072; ASP® 100; ASP® 102; ASP® 170; ASP® 200; ASP® 400; ASP® 400P; ASP® 600; ASP® 602; ASP® 900; ASP® NC; Buca®; Catalpo®; Catalpo® X-1
Trade names containing: ASP® 101

Aluminum sodium silicate. *See* Sodium silicoaluminate

Aluminum stearate
CAS 7047-84-9; EINECS 230-325-5
Synonyms: Aluminum, dihydroxy (octadecanoato-o-); Stearic acid aluminum dihydroxide salt
Definition: Aluminum salt of stearic acid
Empirical: $C_{18}H_{37}AlO_4$
Formula: $CH_3(CH_2)_{16}COOAl(OH)_2$
Properties: White powd.; insol. in water, alcohol, ether; sol. in alkali, petrol., turpentine oil; m.w. 344.48; dens. 1.070; m.p. 115 C
Toxicology: TLV:TWA 2 mg(Al)/m^3; heated to decomp., emits acrid smoke and irritating fumes
Uses: Paint, varnish drier, greases, waterproofing agent, cement additive, plastics lubricants, cutting compds., flatting agent, cosmetics, pharmaceuticals, and defoaming agent

Regulatory: FDA §121.1099, 172.863, 173.340, 181.29
Manuf./Distrib.: Elf Atochem N. Am./Wire Mill; Ferro/Grant; Magnesia GmbH; Norac; Syn. Prods.; Witco
Trade names: Cometals Aluminum Stearate; HyTech AX/603; HyTech T/351; Mathe 6T; Synpro® Aluminum Stearate 303; Witco® Aluminum Stearate 18

Aluminum trihydrate. *See* Aluminum hydroxide
Aluminum trihydroxide. *See* Alumina hydrate
Aluminum, tris(N-hydroxy-N-nitrosobenzenaminato-O,O´)-. *See* Nitrosophenylhydroxylamine aluminum salt

Aluminum tristearate
CAS 637-12-7; EINECS 211-279-5
Definition: Aluminum salt of stearic acid
Empirical: $C_{54}H_{105}O_6 \cdot Al$
Formula: $[CH_3(CH_2)_{16}COO]_3Al$
Properties: Hard material; sol. in alcohol, benzene, oil turpentine, min. oils; pract. insol. in water; m.w. 877.35; m.p. 117-120 C
Uses: Paint and varnish drier, greases, waterproofinga gent, cement additive, lubricants, cutting compds., flatting agent, cosmetics and pharmaceuticals, defoaming agent in beet sugar and yeast processing
Regulatory: FDA 21CFR §172.863, 173.340, 175.105, 175.300, 175.320, 176.170, 176.200, 176.210, 177.1200, 177.1460, 177.260, 178.3910, 179.45, 181.22, 181.29
Trade names: Haro® Chem ALT

Alumite. *See* Alumina
Amide C-18. *See* Stearamide
Amides, coconut oil, with sarcosine, sodium salts. *See* Sodium cocoyl sarcosinate
Amides, tallow. *See* Tallow amide
Amides, tallow, hydrogenated. *See* Hydrogenated tallow amide

Amidoamine resin
Uses: Hydrophobic component for formulating fabric finishes; softener, lubricant; epoxy curing agent
Manuf./Distrib.: Ashland; Chemron; Croda; Witco/Oleo-Surf.
Trade names: Epotuf Hardener 37-620; Genamid® 151; Genamid® 250; Genamid® 490; Genamid® 491; Genamid® 747; Genamid® 2000

Amines, coco alkyl dimethyl. *See* Dimethyl cocamine
Amines, hydrogenated tallow alkyl. *See* Hydrogenated tallowamine
Amines, tallow alkyl. *See* Tallow amine
3-Aminoaniline. *See* m-Phenylenediamine
3-Aminobutanoic acid, n-coco alkyl derivatives. *See* Cocaminobutyric acid
Aminocyclohexane. *See* Cyclohexylamine
4-Aminodiphenylamine. *See* N-Phenyl-p-phenylenediamine
p-Aminodiphenylamine. *See* N-Phenyl-p-phenylenediamine
2-Aminoethanol. *See* Ethanolamine

2-(2-Aminoethoxy) ethanol
CAS 929-06-6; EINECS 213-195-4
Formula: $NH_2CH_2CH_2OCH_2CH_2OH$
Properties: Colorless, slightly viscous liq., amine odor; misc. in water, alcohols; dens. 1.0572 (20/20 C); b.p. 221 C; f.p. -12.5 C; flash pt. 126.6 C
Toxicology: Strong irritant to tissue
Uses: Removal of acid components from gases; intermediate; prep. of foam stabilizers, wetting agents, emulsifiers, condensation polymers
Trade names: Diglycolamine® Agent (DGA®)

2-Aminoethyl alcohol. *See* Ethanolamine

N-(2-Aminoethyl)-3-aminopropyl methyldimethoxy silane
CAS 3069-29-2
Uses: Coupling agent, release agent, lubricant, blocking agent, chemical intermediate
Trade names: Dynasylan® 1411; Petrarch® A0699

N-2-Aminoethyl-3-aminopropyl trimethoxysilane
CAS 1760-24-3; EINECS 212-164-2

Synonyms: N-β-(-Aminoethyl) γ-aminopropyltrimethoxysilane; N-[3-(Trimethoxysilyl) propyl] ethylenediamine
Classification: Diamino silane
Empirical: $C_8H_{22}N_2O_3Si$
Formula: $NH_2CH_2CH_2NH(CH_2)_3Si(OCH_3)_3$
Properties: Liq.; m.w. 222.41; dens. 1.028 (20/4 C); b.p. 261-263 C; ref. index 1.442 (25 C); flash pt. 121 C
Toxicology: Poison by intravenous route, mild toxicity by ingestion and skin contact; severe eye irritant
Uses: Coupling agent for epoxies, phenolics, melamines, nylons, PVC, acrylics, urethanes, nitrile rubbers
Trade names: Dow Corning® Z-6020; Dynasylan® DAMO; Dynasylan® DAMO-P; Dynasylan® DAMO-T; Petrarch® A0700; Petrarch® A0701; Prosil® 3128

N-β-(-Aminoethyl) γ-aminopropyltrimethoxysilane. *See* N-2-Aminoethyl-3-aminopropyl trimethoxysilane
Aminoethylethandiamine. *See* Diethylenetriamine

Aminoethylethanolamine
CAS 111-41-1; EINECS 203-867-5
Synonyms: AEEA; Hydroxyethylethylenediamine
Empirical: $C_4H_{12}N_2O$
Formula: $NH_2CH_2CH_2NHCH_2CH_2OH$
Properties: M.w. 104.15; dens. 1.030; b.p. 238-240 C (752 mm); flash pt. > 110 C; ref. index 1.4861 (20 C)
Uses: Intermediate for textile finishing compounds (antifuming agents, dyestuffs, cationic surfactants), resins, rubber, insecticides, medicinals, surfactants, corrosion inhibitors
Manuf./Distrib.: Akzo Nobel; Allchem Ind.; Ashland; BASF; Coyne; Dow; Nippon Nyukazai; Schweizerhall; Union Carbide

Aminoethylpiperazine
CAS 140-31-8; EINECS 205-411-0
Synonyms: AEP; 1-(2-Aminoethyl) piperazine; 2-Piperazinoethylamione
Empirical: $C_6H_{15}N_3$
Formula: $H_2NC_2H_4NCH_2CH_2NHCH_2CH_2$
Properties: Liq.; sol. in water; m.w. 129.24; dens. 0.9837; b.p. 222.0 C; f.p. 17.6 C; flash pt. 200 F
Toxicology: Strong irritant to tissue; poison by intraperitoneal route, toxic by ingestion and skin contact
Uses: Epoxy curing agent, intermediate for pharmaceuticals, anthelmintics, surface-active agents, synthetic fibers, corrosion inhibitors, asphalt additives, emulsion breakers, emulsifiers, textile chems.
Manuf./Distrib.: Akzo Nobel; Dow; Fabrichem; Texaco; Tosoh; Union Carbide
Trade names: AEP; D.E.H. 39

1-(2-Aminoethyl) piperazine. *See* Aminoethylpiperazine
Aminoform. *See* Hexamethylene tetramine
Aminohexahydrobenzene. *See* Cyclohexylamine
α-(Aminomethyl)benzenemethanol. *See* Phenylethanolamine
Aminomethyl propanol. *See* 2-Amino-2-methyl-1-propanol

2-Amino-2-methyl-1-propanol
CAS 124-68-5; EINECS 204-709-8
Synonyms: AMP; Isobutanolamine; Aminomethyl propanol; Isobutanol-2 amine
Classification: Substituted aliphatic alcohol
Empirical: $C_4H_{11}NO$
Formula: $CH_3(CH_3)(NH_2)CH_2OH$
Properties: Solid or visc. liq.; misc. with water; sol. in alcohol; m.w. 89.1; dens. 0.93 (20/4 C); m.p. 30 C; b.p. 165 C (760 mm); flash pt. (TOC) 153 F
Precaution: Flamm. exposed to heat or flame
Toxicology: Moderately toxic by ingestion
Uses: Multifunctional additive for latex paint; pigment codispersant; solubilizer for resins; emulsifier; neutralizer for boiler water treatment; corrosion inhibitor; catalyst (as acid salt); absorbent for acidic gases, organic synthesis, cosmetics
Regulatory: FDA 21CFR §175.105, 176.170
Manuf./Distrib.: Allchem Ind.; ANGUS; Ashland; Janssen Chimica

2-Amino-1-phenylethanol. *See* Phenylethanolamine

N-(3-Aminopropyl)diethanolamine
Synonyms: APDEA

3-Aminopropylmethyldiethoxy silane

Properties: Clear liq.; sp.gr. 1.07; b.p. 167 C (760 mm); flash pt. (COC) 340 F; ref. index 1.4965
Uses: Intermediate in prep. of surfactants, fabric softeners, skin care prods., lube/fuel additives such as dispersants and detergents, corrosion inhibitors for metalworking, drugs, dyes, flocculants for paper mfg. and waste water treatment; integral skin PU, EPDM rubbers to improve blending and tear resist., buffers, chelating agents
Manuf./Distrib.: Texaco

3-Aminopropylmethyldiethoxy silane
CAS 3179-76-8; EINECS 221-660-8
Synonyms: 3-(Diethoxymethylsilyl) propylamine
Empirical: $C_8H_{21}NO_2Si$
Properties: M.w. 191.39; dens. 0.916; b.p. 85-88 C (8 mm); flash pt. 75 C; ref. index 1.4260
Toxicology: Poison by intraperitoneal route, moderate toxicity by ingestion and skin contact; skin irritant
Uses: Coupling agent, release agent, lubricant, blocking agent, chemical intermediate
Trade names: Dynasylan® 1505; Petrarch® A0742; Petrarch® A0743

Aminopropyltriethoxysilane
CAS 919-30-2; EINECS 213-048-4
Synonyms: γ-Aminopropyl triethoxysilane; 3-(Triethoxysilyl)-1-propanamine
Formula: $NH_2(CH_2)_3Si(OC_2H_5)_3$
Properties: Liq.; m.w. 221.37; dens. 0.942; b.p. 217 C; flash pt. 104 C
Toxicology: Poison by intraperitoneal route; moderate toxicity by ingestion and skin contact; skin and eye irritant
Uses: Sizing of glass fibers for making laminates; filler for thermosets, thermoplastics.
Manuf./Distrib.: Gelest; Hüls Am.; PCR
Trade names: Dynasylan® AMEO; Dynasylan® AMEO-P; Dynasylan® AMEO-T; Petrarch® A0750; Prosil® 220; Prosil® 221
Trade names containing: Aktisil® AM

γ-Aminopropyl triethoxysilane. *See* Aminopropyltriethoxysilane

Aminopropyltrimethoxysilane
CAS 13822-56-3; EINECS 237-511-5
Synonyms: 3-(Trimethoxysilyl)propylamine
Formula: $NH_2(CH_2)_3Si(OCH_3)_3$
Properties: Water-white liq.; m.w. 179.29; dens. 1.01; b.p. 91-92 C (15 mm); flash pt. 83 C; ref. index 1.42
Uses: Class fabric sizing, binder, adhesion promoter, release agent, lubricant
Trade names: Dynasylan® AMMO; Petrarch® A0800

Ammonium acetate
CAS 631-61-8; EINECS 211-162-9
Empirical: $C_2H_7NO_2$
Formula: CH_3COONH_4
Uses: Reagent in analytical chemistry, drugs, textile dyeing, preserving meats, foam rubbers, vinyl plastics, and explosives
Manuf./Distrib.: Aldrich; Allchem Ind.; Am. Biorganics; General Chem.; Great Western; Honig; Jarchem; Johnson Matthey; Magnablend; Spectrum Chem. Mfg.; Verdugt BV

Ammonium biborate
CAS 12228-87-4
Synonyms: Ammonium tetraborate; Diammonium tetraborate tetrahydrate (CAS 12007-58-8)
Formula: $(NH_4)_2B_4O_7 \cdot 4H_2O$
Properties: Wh. cryst. solid; partly sol. in water; m.w. 263.38; sp.gr. 1.63
Precaution: may release ammonia on heating; incompat. with alkaline conditions
Toxicology: LD50 (oral, mouse) > 4.2 g/kg; sl. transient eye irritant; sl. irritant to damaged skin; may cause nausea, vomiting, diarrhea, erythema, skin rash, dizziness
Uses: Neutralizing agent in mfg. of urea-formaldehyde resins; ingred. in flame-proofing formulations
Manuf./Distrib.: U.S. Borax

Ammonium bromide
CAS 12124-97-9; EINECS 235-183-8
Empirical: BrH_4N

Formula: NH_4Br
Properties: Colorless cryst. or ylsh. wh. powd.; sol. in water and alcohol; m.p. sublimes; noncombustible
Uses: Flame retardant for plastics, textiles, wood, chipboard, plywood; anticorrosive agents
Manuf./Distrib.: Aldrich; AmeriBrom; Cerac; Charkit; D&O; Great Lakes; Johnson Matthey SA; Spectrum Chem. Mfg.

Ammonium lactate
Synonyms: Lactic acid, ammonium salt
Formula: $NH_4C_3H_5O_3$
Properties: Colorless to yel. syrupy liq.; sol. in waer, alcohol; m.w. 107.08; dens. 1.19-1.21 (15 C)
Uses: Catalyst for esterification and olefin polymerization; used in coatings; resin crosslinking agent for automotive prods., coatings, elastomers, films/paints, graphic arts, plastics
Manuf./Distrib.: Magnablend; Purac Am.
Trade names: Tyzor LA

Ammonium laureth sulfate
CAS 32612-48-9 (generic); 67762-19-0
Synonyms: Ammonium lauryl ether sulfate
Definition: Ammonium salt of ethoxylated lauryl sulfate
Formula: $(C_2H_4O)_n \cdot C_{12}H_{26}O_4S \cdot H_3N$, n = 1-4
Toxicology: Moderate toxicity by ingestion; skin and eye irritant
Uses: Surfactant for emulsion polymerization
Regulatory: FDA 21CFR §175.105
Manuf./Distrib.: Aquatec Quimica SA; Ashland; Great Western; Lonza; Pilot; Rhone-Poulenc Spec.; Sandoz; Sea-Land; Witco/Oleo-Surf.
Trade names: Avirol® AE 3003; Polystep® B-11; Rhodapex® AB-20; Steol® CA-460

Ammonium laureth-9 sulfate
CAS 32612-48-9 (generic)
Synonyms: PEG-9 lauryl ether sulfate, ammonium salt; Ammonium PEG (450) lauryl ether sulfate; POE (9) lauryl ether sulfate, ammonium salt
Definition: Ammonium salt of ethoxylated lauryl sulfate
Formula: $CH_3(CH_2)_{10}(OCH_2CH_2)_nOSO_3NH_4$, avg. n = 9
Uses: Surfactant for emulsion polymerization
Trade names: Geropon® AB/20

Ammonium laureth-12 sulfate
CAS 32612-48-9 (generic)
Definition: Ammonium salt of ethoxylated lauryl sulfate
Formula: $(C_2H_4O)_n \cdot C_{12}H_{26}O_4S \cdot H_3N$, n = 12
Uses: Surfactant for acrylic copolymers
Regulatory: FDA 21CFR §175.105
Trade names: Polystep® B-22

Ammonium laureth-30 sulfate
Formula: $(C_2H_4O)_n \cdot C_{12}H_{26}O_4S \cdot H_3N$, n = 30
Uses: Emulsifier for emulsion polymerization
Trade names: Polystep® B-20

Ammonium lauroyl sarcosinate
CAS 68003-46-3; EINECS 268-130-2
Synonyms: N-Methyl-N-(1-oxododecyl) glycine, ammonium salt
Definition: Ammonium salt of lauroyl sarcosine
Formula: $C_{15}H_{29}NO_3 \cdot H_3N$
Uses: Sec. emulsifier for emulsion polymerization
Trade names: Hamposyl® AL-30

Ammonium lauryl ether sulfate. *See* Ammonium laureth sulfate

Ammonium lauryl sulfate
CAS 2235-54-3; 68081-96-9; EINECS 218-793-9
Synonyms: Sulfuric acid, monododecyl ester, ammonium salt
Definition: Ammonium salt of lauryl sulfate

Ammonium lignin sulfonate

Formula: $C_{12}H_{26}O_4S \cdot H_3N$
Properties: M.w. 283.48
Toxicology: Skin and eye irritant
Uses: Detergent, emulsifier, foaming agent, dispersant, wetting agent; for personal care prods., carpet shampoos, firefighting, dry wall mfg., dyes, chemical specialties, emulsion polymerization
Regulatory: FDA 21CFR §175.105, 175.210, 176.170, 177.1200
Manuf./Distrib.: Aquatec Quimica SA; Ashland; Lonza; Pilot; Rhone-Poulenc Spec.; Sandoz; Sea-Land; Stepan; Witco/Oleo-Surf.
Trade names: Akyposal ALS 33; Avirol® A; Calfoam NLS-30; Carsonol® ALS; Carsonol® ALS-S; Carsonol® ALS Special; Polystep® B-7; Rhodapon® L-22; Rhodapon® L-22/C; Rhodapon® L-22HNC; Sermul EA 129; Surfax 215

Ammonium lignin sulfonate. *See* Ammonium lignosulfonate

Ammonium lignosulfonate
CAS 8061-53-8
Synonyms: Ammonium lignin sulfonate
Uses: Wetting agent, emulsifier, dispersant, binder, leather tanning agent; for gypsum boards, metallurgy, briquetting, adhesives, resins, etc.; extender for U-F resins
Manuf./Distrib.: Borregaard LignoTech; Wesco Tech. Ltd; Westvaco

Ammonium manganese pyrophosphate. *See* Manganese violet

Ammonium naphthalene-formaldehyde sulfonate
Uses: Emulsifier for polymerization, rubber; pigment dispersant
Trade names: Dehscofix 930

Ammonium naphthalene sulfonate
Uses: Dispersant for pigments, carbon black, dyestuffs, in ceramics, paints; viscosity depressant; emulsion polymerization
Trade names: Lomar® PWA; Lomar® PWA Liq.

Ammonium nonoxynol-4 sulfate
CAS 9051-57-4; 31691-97-1 (generic); 63351-73-5
Definition: Ammonium salt of sulfated nonoxynol-4
Uses: High foaming surfactant, wetting agent, dispersant, emulsifier for shampoos, skin cleansers, lt. duty cleaners, emulsion polymerization
Regulatory: FDA 21CFR §178.3400
Trade names: Abex® EP-100; Polystep® B-1; Rhodapex® CO-436; Sulfochem 436

Ammonium nonoxynol-9 sulfate
Uses: Emulsfier, stabilizer for emulsion polymerization
Trade names: Abex® EP-110

Ammonium nonoxynol-10 sulfate
Uses: Emulsifier for emulsion polymerization
Trade names: Sermul EA 152

Ammonium nonoxynol-20 sulfate
Uses: Emulsifier for polymerization
Trade names: Abex® EP-115

Ammonium nonoxynol-30 sulfate
Uses: Emulsifier for emulsion polymerization
Trade names: Abex® EP-120

Ammonium nonoxynol-77 sulfate
Uses: Emulsifier for emulsion polymerization
Trade names: Abex® EP-227

Ammonium octamolybdate
Formula: $(NH_4)_4Mo_8O_{26}$
Uses: Smoke suppressant in polymers
Trade names: AOM

Ammonium PEG (450) lauryl ether sulfate. *See* Ammonium laureth-9 sulfate
Ammonium pentadecafluorooctanoate. *See* Ammonium perfluorooctanoate
Ammonium perfluorocaprylate. *See* Ammonium perfluorooctanoate

Ammonium perfluorooctanoate
CAS 3825-26-1
Synonyms: Ammonium perfluorocaprylate; Ammonium pentadecafluorooctanoate
Classification: Anionic fluorochemical surfactant
Formula: $C_8F_{15}COONH_4$
Properties: Wh. powd.; m.w. 431.13; dens. 0.6-0.7 g/cc; b.p. 125 C
Toxicology: TLV 0.1 mg/m^3 in air; toxic by inhalation and skin contact
Uses: Surfactant for emulsion polymerization of fluorinated monomers
Trade names: Fluorad® FC-118

Ammonium polyacrylate
CAS 9003-03-6
Synonyms: Poly(acrylic acid), ammonium salt; 2-Propenoic acid, homopolymer, ammonium salt
Definition: Ammonium salt of polyacrylic acid
Empirical: $(C_3H_4O_2)_x \cdot xH_3N$
Uses: Dispersant for paints and coatings; thickening and stabilizing agent for syn. latices; used in coatings, adhesives, dipped, cast, and molded goods, cements for rug backing, spraying, spreading, brushing, and extruding compds.
Regulatory: FDA 21CFR §175.105

Ammonium polymethacrylate
Uses: Stabilizer and thickener in latex rubber processing
Trade names: Daxad® 34A9; Vulcastab® T

Ammonium polyphosphate sol'n.
CAS 14728-39-3
Uses: Flame retardant for plastics, adhesives, elastomers, paints
Manuf./Distrib.: Albright & Wilson Am.; Fabrichem; Hoechst Celanese; Monsanto
Trade names: Amgard® MC; Amgard® PI; Exolit® 422; Exolit® 462; FRCROS 480; Hostaflam® AP 750; Phos-Chek P/30
Trade names containing: Exolit® IFR-10; Exolit® IFR-15; Exolit® IFR-23

Ammonium stearate
CAS 1002-89-7; EINECS 213-695-2
Synonyms: Octadecanoic acid, ammonium salt
Definition: Ammonium salt of stearic acid
Empirical: $C_{18}H_{36}O_2 \cdot H_3N$
Formula: $C_{17}H_{35}COONH_4$
Properties: Tan wax-like solid, free from ammonia odor; sol. in boiling water, hot toluene; partly sol. hot in butyl acetate, ethanol; m.w. 301.50; dens. 0.89 (22 C); m.p. 73-75 C
Uses: Vanishing creams, brushless shaving creams, other cosmetic prods., waterproofing of cements, concrete, stucco, paper, textiles, frothing aid and foam stabilizer in acrylic and SBR latex systems; foam rubber additive
Regulatory: FDA 21CFR §175.105, 175.320, 176.170, 176.200, 176.210, 177.1200, 177.2260
Manuf./Distrib.: Magnesia GmbH; Northern Prods.; Original Bradford Soap Works; ProChem

Ammonium tetraborate. *See* Ammonium biborate

Amodimethicone
Definition: Silicone polymer end blocked with amino functional groups
Formula: $HO[Si(CH_3)_2O][SiOH(CH_2)_3NHCH_2CH_2NH_2—O]H$
Uses: Silicone polymer curing to a durable, detergent-resistant film for mold release, particle treatment, textile finishes, and polish applics.
Trade names: GP-4 Silicone Fluid; SM 2059

AMP. *See* 2-Amino-2-methyl-1-propanol

t-Amyl alcohol
CAS 75-85-4; EINECS 200-908-9

Amylcarbinol

Synonyms: 2-Methyl-2-butanol; t-Pentyl alcohol; t-Pentanol; Dimethyl ethyl carbinol; Amylene hydrate
Empirical: $C_5H_{12}O$
Formula: $CH_3CH_2C(CH_3)_2OH$
Properties: Volatile liq., char. odor, burning taste; sol. in 8 parts water; misc. with alcohol, ether, benzene, chloroform, glycerin, oils; m.w. 88.15; dens. 0.808 (20/4 C); m.p. -9 C; b.p. 100-103 C; ref. index 1.405
Precaution: Keep tightly closed; protect from light
Toxicology: LD50 (oral, rat) 1 g/kg; mod. irritating to human mucous membranes; narcotic in high concs.
Uses: Solvent, flotation agent, organic synthesis, medicine (sedative)
Manuf./Distrib.: Allchem Ind.; BASF
Trade names containing: TAHP-80

Amylcarbinol. *See* Hexyl alcohol
Amylene hydrate. *See* t-Amyl alcohol

OO-t-Amyl O-(2-ethylhexyl) monoperoxycarbonate
Uses: Crosslinking agent for emulsion, bulk, sol'n., and suspension polymerization, high temp. cure of polyester, and cure of acrylic syrup
Trade names: Lupersol TAEC

t-Amyl hydroperoxide
CAS 3425-61-4
Classification: Organic peroxide
Empirical: $C_5H_{12}O_2$
Properties: Clear water-wh. liq., sharp odor; m.w. 104.15; sp.gr. 0.904 (22 C); b.p. 26 C (3 mm); flash pt. (Seta CC) 46 C
Precaution: Once ignited, burns vigorously; explosion hazard mixed with any promoter
Toxicology: Severe eye and skin irritation; may cause blindness; harmful or fatal if swallowed
Uses: Polymerization initiator
Manuf./Distrib.: Elf Atochem N. Am.; Witco/PAG
Trade names: Aztec® TAHP-80-AL; Trigonox® TAHP-W85
Trade names containing: TAHP-80

t-Amyl perbenzoate. *See* t-Amyl peroxybenzoate

t-Amyl peroxyacetate
CAS 690-83-5
Empirical: $C_7H_{14}O_3$
Formula: $CH_3COOOC(CH_3)_2C_2H_5$
Properties: M.w. 146.2
Uses: Crosslinking agent, initiator
Trade names containing: Lupersol 555-M60; Trigonox® 133-C60

t-Amyl peroxybenzoate
CAS 4511-39-1
Synonyms: t-Amyl perbenzoate
Classification: Organic peroxide
Empirical: $C_{12}H_{16}O_3$
Properties: Liq.; m.w. 208.2; sp.gr. 1.01; flash pt. (Seta CC) 49 C
Uses: Initiator for polymerization of ethylene, styrene, acrylates, and curing of unsat. polyester resins
Manuf./Distrib.: Elf Atochem N. Am.
Trade names: Aztec® TAPB; Esperox® 5100; Trigonox® 127
Trade names containing: Aztec® TAPB-90-OMS

t-Amylperoxy 2-ethylhexanoate
CAS 686-31-7
Classification: Organic peroxide
Properties: M.w. 230.3
Uses: Initiator for bulk, sol'n., and suspension polymerization, for cure of polyester resins and acrylic syrup
Manuf./Distrib.: Akzo Nobel; Aztec Peroxides; Witco/PAG
Trade names: Aztec® TAPEH; Esperox® 570; Lupersol 575; Trigonox® 121; Trigonox® 121-BB75
Trade names containing: Aztec® TAPEH-75-OMS; Aztec® TAPEH-90-OMS; Esperox® 570P; Lupersol 575-M75; Lupersol 575-P75

t-Amylperoxy 2-ethylhexyl carbonate
CAS 70833-40-8
Classification: Organic peroxide
Empirical: $C_{14}H_{28}O_4$
Formula: $C_4H_9CHC_2H_5CH_2OCOOOC(CH_3)_2C_2H_5$
Properties: M.w. 260.4
Uses: Initiator for polymerizations
Trade names: Trigonox® 131

t-Amylperoxyneodecanoate
CAS 68299-16-1
Uses: Initiator for polymerizations
Manuf./Distrib.: Akzo Nobel; Witco/PAG
Trade names: Aztec® TAPND-75-AL
Trade names containing: Esperox® 545M; Lupersol 546-M75; Trigonox® 123-C75

t-Amylperoxyneoheptanoate
Uses: Initiator for polymerizations
Trade names containing: Esperox® 747M

t-Amylperoxypivalate
CAS 29240-17-3
Classification: Organic peroxide
Empirical: $C_{10}H_{20}O_3$
Formula: $CH_3CH_2C(CH_3)_2OOCOC(CH_3)_3$
Properties: M.w. 188.3
Precaution: Flamm. oxidizing liq.
Toxicology: LD50 (rat, oral) 4.27 g/kg; LD50 (rabbit, dermal) > 2.0 g/kg
Uses: Low-temp. initiator for radical catalyzed polymerization of vinyl monomers
Manuf./Distrib.: Akzo Nobel; Witco/PAG
Trade names containing: Aztec® TAPPI-75-OMS; Esperox® 551M; Lupersol 554-M50, 554-M75; Trigonox® 125-C50; Trigonox® 125-C75

4-t-Amylphenol. *See* p-t-Amylphenol

p-t-Amylphenol
CAS 80-46-6
Synonyms: PTAP; 4-t-Amylphenol; Pentaphen, 2-methyl-2-p-hydroxyphenyl-butane
Formula: $C_2H_5C(CH_3)_2C_6H_4OH$
Properties: Colorless needles; m.w. 164.25; b.p. 255 C; m.p. 88-89 C; flash pt. (OC) 232 F
Toxicology: Moderate toxicity by ingestion and skin contact; severe eye irritant
Uses: Intermediate for chemical specialties; in germicidal formulations; also in mfg. of photographic chemicals, oil demulsifiers, phenolic resins, agric. surfactants, antiskinning agents
Manuf./Distrib.: Elf Atochem N. Am.

Amyltrichlorosilane
CAS 107-72-2
Synonyms: Pentyltrichlorosilane
Empirical: $C_5H_{11}Cl_3Si$
Properties: Colorless to yel. liq.; m.w. 205.60; dens. 1.132 (25/25 C); b.p. 168 C; flash pt. (COC) 62.8 C; ref. index 1.4152 (20 C)
Toxicology: Moderate toxicity by ingestion, skin contact; mild toxicity by inhalation; irritant to eyes, mucous membranes
Uses: Intermediate for silicones; coupling agent, blocking agent, release agent, lubricant
Manuf./Distrib.: Hüls Am.
Trade names: Dynasylan® AMTC

Anchoic acid. *See* Azelaic acid
Anhydrite (natural form). *See* Calcium sulfate
1,4-Anhydro-D-glucitol, 6-hexadecanoate. *See* Sorbitan palmitate
1,4-Anhydro-D-glucitol, 6-isooctadecanoate. *See* Sorbitan isostearate
Anhydrohexitol diisostearate. *See* Sorbitan diisostearate
Anhydrohexitol sesquioleate. *See* Sorbitan sesquioleate

Anhydrosorbitol monoisostearate. *See* Sorbitan isostearate
Anhydrosorbitol monolaurate. *See* Sorbitan laurate
Anhydrosorbitol monooleate. *See* Sorbitan oleate
Anhydrosorbitol monostearate. *See* Sorbitan stearate
Anhydrosorbitol sesquioleate. *See* Sorbitan sesquioleate
Anhydrosorbitol trioleate. *See* Sorbitan trioleate
Anhydrosorbitol tristearate. *See* Sorbitan tristearate
Anhydrous gypsum. *See* Calcium sulfate
Anhydrous lanolin. *See* Lanolin

Anilino-phenyl methacrylamide
Uses: Antioxidant for NBR, SBR, BR, ABS, CR
Trade names: Polystay AA-1; Polystay AA-1R

Antimonic acid. *See* Antimony pentoxide
Antimonic anhydride. *See* Antimony pentoxide

Antimony
CAS 7440-36-0; EINECS 231-146-5
Formula: Sb
Properties: Silver-white solid; insol. in water; sol. hot conc. H_2SO_4; m.w. 121.75; dens. 6.684 (25 C); m.p. 630.5 C; b.p. 1635 C; low thermal conductivity
Toxicology: Dermatitis, keratitis, conjunctivitis by contact, fumes, dust; TLV:TWA 0.5 mg (Sb)/m^3; adequate ventilation required
Uses: Hardening alloy for lead, bearing metal, type metal, solder, collapsible tubes and foil, sheet and pipe, semiconductor technology, pyrotechnics; flame retardant in plastics
Manuf./Distrib.: Aldrich; Amspec; Atomergic Chemetals
Trade names: Thermoguard® CPA

Antimony mercaptide
Uses: Stabilizer for PVC, pipe and profile applics.
Trade names: Mark® 2115; Stanclere® A-121, A-121 C, A-221; Synpron 1027; Synpron 1034; Synpron 1048

Antimony oxide. *See* Antimony trioxide
Antimony (III) oxide. *See* Antimony trioxide

Antimony pentoxide
CAS 1314-60-9; EINECS 215-237-7 (Sb_2O_5)
Synonyms: Antimonic anhydride; Antimonic acid; Stibic anhydride
Formula: O_5Sb_2
Properties: Yellowish powd., cube; sl. sol. in water; pract. insol. in HNO_3; sl. sol. in warm KOH or HCl; m.w. 323.52
Toxicology: LD50 (rats) 4 g/kg; toxic by intraperitoneal route
Uses: Flame/fire retardant for clothing, latex, resins
Manuf./Distrib.: All Chemie Ltd; Atomergic Chemetals; Cerac; Great Western; Laurel Ind.; Noah; PQ
Trade names: Nyacol® A-1530; Nyacol® A-1540N; Nyacol® A-1550; Nyacol® A-1588LP; Nyacol® AP50; Nyacol® N24; Nyacol® ZTA
Trade names containing: Nyacol® AB40; Nyacol® APE1540

Antimony peroxide. *See* Antimony trioxide

Antimony trioxide
CAS 1309-64-4; EINECS 215-175-0
Synonyms: Antimony white; Antimony (III) oxide; Antimony oxide; Antimony peroxide
Empirical: Sb_2O_3
Formula: O_3Sb_2
Properties: White crystalline powd., odorless; very sl. sol. in water; sol. in KOH, HCl and sulfuric acids, strong alkalies; m.w. 291.52; dens. 5.67; m.p. 655 C; b.p. 1425 C
Toxicology: LD50 (rats) > 20 g/kg; TLV:TWA 500 μg/m^3; poison by intravenous route; suspected human carcinogen, moderate toxicity otherwise
Uses: Flameproofing of textiles, paper, plastics; as paint pigment; ceramic opacifier; catalyst; intermediate staining iron and copper; phosphorus; mordant; glass decolorizer
Manuf./Distrib.: Aldrich; Amspec; Asarco; Atomergic Chemetals; Chemisphere Ltd; Chemson Ltd.; Elf

Atochem N. Am.; Hoechst-Celanese; Holtrachem; Laurel Ind.; Miljac; Nihon Kagaku Sangyo; Noah; Spectrum Chem. Mfg.

Trade names: Amsperse; AO; FireShield® H; FireShield® HPM; FireShield® HPM-UF; FireShield® L; Fyraway; Fyrebloc; Naftopast® Antimontrioxid; Naftopast® Antimontrioxid-CP; Octoguard FR-10; Petcat R-9; Thermoguard® L; Thermoguard® S; Thermoguard® UF; UltraFine® II

Trade names containing: Chem-Master R-135; Chem-Master R-321; Chem-Master R-8414; F-2430; Mastertek® ABS 3710; Mastertek® PA 3664; Mastertek® PBT 3659; Mastertek® PC 3686; Mastertek® PE 3660; Mastertek® PE 3668; Mastertek® PP 3658; Mastertek® PP 5601; Mastertek® PP 5604; Mastertek® PS 3698; Octoguard FR-15; Prespersion PAC-1712; Thermoguard® 215; Thermoguard® 9010; Thermoguard® DM-9406; Wyfire® H-A-85; Wyfire® H-A-85P; Wyfire® S-A-51; Wyfire® S-BA-88P; Wyfire® Y-A-85; Wyfire® YP-BA-47P

Antimony white. *See* Antimony trioxide
APDEA. *See* N-(3-Aminopropyl)diethanolamine

Arachidyl behenyl amine

Uses: Emulsifier, flotation agent, dispersing and flushing agent, intermediate, used in metalworking oils, as fuel oil additive; mold release for rubber and plastics; lubricant and spinning aid in metalworking oils

Trade names: Kemamine® P-190, P-190D

Arachidyl-behenyl 1,3-propylene diamine

Uses: Gasoline detergent, bactericide, corrosion inhibitor in petrol. prod., epoxy hardener

Trade names: Kemamine® D-190

Aramid

Definition: Generic name for class of highly aromatic polyamide fibers which are characterized by flame retardant properties

Uses: Used in protective clothing, dust-filter bags, tire cord, bullet-resistant structures

Trade names: Kevlar® 29, 49

Artificial oil of ants. *See* Furfural
ASA. *See* Acrylonitrile/styrene/acrylate
ASA terpolymer. *See* Acrylic-styrene-acrylonitrile terpolymer

Asphalt, oxidized

CAS 64742-93-4

Uses: Min. rubber used as plasticizer, process aid, and low-cost filler in mech. goods, mats, floor treads, tire components, retreads, sealing compds., wire and cable, etc.

Trade names: Mr 285; Mr 305

Trade names containing: Mr 245

A-stage resin, one-stage resin, resole. *See* Phenolic resin
ATBC. *See* Acetyl tributyl citrate
ATEC. *See* Acetyl triethyl citrate
ATHC. *See* Acetyl tri-n-hexyl citrate
Azacycloheptane. *See* Hexamethyleneimine

Azelaic acid

CAS 123-99-9; EINECS 204-669-1

Synonyms: Nonanedoic acid; Dicarboxylic acid C9; 1,7-Heptanedicarboxylic acid; Anchoic acid

Empirical: $C_9H_{16}O_4$

Formula: $HOOC(CH_2)_7COOH$

Properties: Yel. to wh. cryst. powd.; sol. in hot water, alcohol, org. solvs.; m.w. 188.23; dens. 1.029 (20/4 C); m.p. 106 C; b.p. 365 C (dec.); flash pt. 210 C

Uses: Organic synthesis, lacquers, alkyd resins, polyamides, polyester, adhesives, low temp. plasticizers, urethane elastomers

Manuf./Distrib.: Allchem Ind.; Charkit; Fluka; Henkel/Emery

1H-Azepine, hexahydro. *See* Hexamethyleneimine

2,2-Azobis(2-amidinopropane)dihydrochloride

CAS 2997-92-4; EINECS 221-070-0

Synonyms: 2,2-Azobis(2-methylpropionamidine)dihydrochloride

2,2′-Azobis[2-[N-(4-chlorophenyl)amidino]propane] dihydrochloride

Empirical: $C_8H_{18}N_6$ • 2HCl
Properties: Wh. or off-wh. gran., almost odorless; sol. in toluene, hexane, water; m.w. 271.19; m.p. 160-169 C (dec.)
Uses: Polymerization initiator
Trade names: V-50

2,2′-Azobis[2-[N-(4-chlorophenyl)amidino]propane] dihydrochloride. *See* 2,2′-Azobis[N-(4-chlorophenyl)-2-methylpropionamidine]dihydrochloride

2,2′-Azobis[N-(4-chlorophenyl)-2-methylpropionamidine]dihydrochloride
CAS 124960-38-9
Synonyms: 2,2′-Azobis[2-[N-(4-chlorophenyl)amidino]propane] dihydrochloride
Uses: Polymerization initiator

4,4′-Azobis (4-cyanopentanoic acid)
CAS 2638-94-0; EINECS 220-135-0
Synonyms: 4,4′-Azobis (4-cyanovaleric acid)
Empirical: $C_{12}H_{16}N_4O_4$
Properties: Wh. cryst. powd., almost odorless; sol. in ethanol; sl. sol. in water; m.w. 280.28; m.p. 120-123 C (dec.)
Precaution: Unstable above 20 C; incompat. with oxidizers, strong acids; hazardous decomp. prods.: CO, NO_x, hydrogen cyanide gas and cyanide compds.; container should not be tightly closed
Toxicology: LD50 (oral, mouse) > 5000 mg/kg; harmful if inhaled or ingested; absorbed thru skin; dust irritant; ing. may cause anorexia, headaches, vomiting, fever; prolonged exposure may cause shock, narcosis, death
Uses: Polymerization initiator
Trade names: V-501

4,4′-Azobis (4-cyanovaleric acid). *See* 4,4′-Azobis (4-cyanopentanoic acid)
Azobis (cyclohexanecarbonitrile). *See* 1,1′-Azobis(cyclohexane-1-carbonitrile)

1,1′-Azobis(cyclohexane-1-carbonitrile)
CAS 2094-98-6; EINECS 218-254-8
Synonyms: Azobis (cyclohexanecarbonitrile)
Empirical: $C_{14}H_{20}N_4$
Formula: $NCC_6H_{10}N{=}NC_6H_{10}CN$
Properties: Wh. cryst.; insol. in water; m.w. 244.34; m.p. 113-115 C
Precaution: Flamm.
Toxicology: Irritant
Uses: Polymerization initiator
Manuf./Distrib.: Aldrich
Trade names: V-40

2,2′-Azobis (N,N′-dimethyleneisobutyramidine)
CAS 20858-12-2; EINECS 244-085-4
Synonyms: 2,2′-Azobis[2-(2-imidazolin-2-yl)propane]
Empirical: $C_8H_{14}N_6$
Properties: Pale yel. powd.; sol. in toluene; sl. sol. in water; m.w. 250.35; m.p. 115-125 C (dec.)
Precaution: Unstable above 10 C; causes violent exothermal decomp. @ 76 C; may create dust explosion; incompat. with oxidizers, strong acids, persulfates; hazardous decomp. prods.: CO, NO_x, N gas
Toxicology: Eye and skin irritant; allergic reactions; harmful if inhaled and ingested; may cause headaches, anorexia, nausa, vomiting, fever, etc.
Uses: Polymerization initiator
Manuf./Distrib.: Spectrum Chem. Mfg.
Trade names: VA-061

2,2′-Azobis (N,N′-dimethyleneisobutyramidine) dihydrochloride. *See* 2,2′-Azobis[2-(2-imidazolin-2-yl) propane] dihydrochloride

2,2′-Azobis (2,4-dimethylvaleronitrile)
CAS 4419-11-8; EINECS 224-583-8
Empirical: $C_{14}H_{24}N_4$
Properties: Wh. cryst. solid; sol. in toluene, methanol, ethanol; insol. in water; m.w. 248.37; m.p. 45-70 C

Uses: Initiator for suspension polymerization of vinyl chlorides, solution polymerization of various monomers
Manuf./Distrib.: Janssen Chimica; Spectrum Chem. Mfg.
Trade names: V-65; Vazo 52

Azobisformamide. *See* Azodicarbonamide
1,1´-Azobisformamide. *See* Azodicarbonamide

2,2´-Azobis[2-[1-(2-hydroxyethyl)-2-imidazolin-2-yl] propane] dihydrochloride
CAS 118585-13-0
Uses: Polymerization initiator

2,2´-Azobis[2-(hydroxymethyl) propionitrile]
CAS 19706-80-0
Uses: Polymerization initiator

2,2´-Azobis[2-(2-imidazolin-2-yl)propane]. *See* 2,2´-Azobis (N,N´-dimethyleneisobutyramidine)

2,2´-Azobis[2-(2-imidazolin-2-yl) propane] dihydrochloride
CAS 27776-21-2; EINECS 248-655-3
Synonyms: 2,2´-Azobis (N,N´-dimethyleneisobutyramidine) dihydrochloride
Empirical: $C_7H_{14}N_6 \cdot 2HCl$
Properties: Wh. to pale yel. cryst. or cryst. powd.; sol. in toluene, hexane, 35.2 g/100 g water; m.w. 323.27; m.p. 188-193 C (dec.)
Uses: Polymerization initiator
Manuf./Distrib.: Wako Chem. USA
Trade names: VA-044

2,2´-Azobis(isobutyramide) dihydrate. *See* 2,2´-Azobis(2-methylpropionamide) dihydrate

2,2´-Azobisisobutyronitrile
CAS 78-67-1; EINECS 201-132-3
Synonyms: AIBN; 2,2´-Azobis(2-methylpropionitrile); Azodiisobutyronitrile; 2,2´-Azobis(2-methylpropane-nitrile)
Empirical: $C_8H_{12}N_4$
Formula: $(CH_3)_2C(CN)NNC(CN)(CH_3)_2$
Properties: Wh. cryst. powd.; insol. in water, sol. in many org. solvs. and vinyl monomers; m.w. 164.24; m.p. 105 C; gas yield 125 cc/g
Precaution: Explodes when heated with heptane
Toxicology: Toxic by ingestion; poison by intraperitoneal route; LD50 (mice, oral) 0.7 g/kg
Uses: Catalyst/initiator for vinyl polymerizations and for curing unsaturated polyester resins; blowing agent for plastics esp. for PVC prod. of floats, buoys
Manuf./Distrib.: Charkit; Fabrichem; Pfaltz & Bauer
Trade names: Ficel® AZDN-LF; Ficel® AZDN-LMC; Mikrofine AZDM; Perkadox® AIBN; Poly-Zole® AZDN; V-60

2,2´-Azobis(4-methoxy-2,4-dimethylvaleronitrile)
CAS 15545-97-8; EINECS 239-593-8
Empirical: $C_{16}H_{28}N_4O_2$
Properties: wh. cryst. powd.; insol. in water; m.w. 308.42; m.p. 50-96 C (dec.)
Uses: Polymerization initiator
Trade names: V-70

2,2´-Azobis{2-methyl-N-[1,1-bis(hydroxymethyl)ethyl] propionamide}
CAS 61551-69-7
Empirical: $C_{16}H_{32}N_4O_6$
Properties: Pale yel. cryst. powd.; sol. in toluene, hexane, 0.6 g/100 g water; m.w. 376.45; m.p. 156-161 C (dec.)
Uses: Polymerization initiator
Trade names: VA-082

2,2´-Azobis[2-methyl-N-[1,1-bis(hydroxymethyl)-2-hydroxyethyl] propionamide]
Uses: Polymerization initiator

2,2´-Azobis(2-methylbutyronitrile)
CAS 13472-08-7; EINECS 236-740-8
Empirical: $C_{10}H_{16}N_4$
Properties: Wh. cryst. powd.; sol. in toluene, ethanol, methanol; insol. in water; m.w. 192.26; m.p. 55-57 C; dec. 84-87 C
Uses: Polymerization initiator
Manuf./Distrib.: Wako Chem. USA
Trade names: Perkadox® AMBN; V-59

2,2´-Azobis[2-methyl-N-(2-hydroxyethyl)-propionamide]
CAS 61551-69-7
Empirical: $C_{12}H_{24}N_4O_4$
Properties: Pale yel. cryst. powd.; sol. in toluene, hexane, 2.4 g/100 g water; m.w. 288.35; m.p. 140-145 C (dec.)
Uses: Polymerization initiator
Trade names: VA-086

2,2´-Azobis(2-methyl-N-phenylpropionamidine) dihydrochloride
CAS 88684-42-8
Synonyms: 2,2´-Azobis[2-(N-phenylamidino) propane] dihydrochloride
Uses: Polymerization initiator

2,2´-Azobis(2-methylpropanenitrile). *See* 2,2´-Azobisisobutyronitrile

2,2´-Azobis(2-methylpropionamide) dihydrate
CAS 3682-94-8
Synonyms: 2,2´-Azobis(isobutyramide) dihydrate
Uses: Polymerization initiator

2,2-Azobis(2-methylpropionamidine)dihydrochloride. *See* 2,2-Azobis(2-amidinopropane)dihydrochloride
2,2´-Azobis(2-methylpropionitrile). *See* 2,2´-Azobisisobutyronitrile
2,2´-Azobis[2-(N-phenylamidino) propane] dihydrochloride. *See* 2,2´-Azobis(2-methyl-N-phenylpropionamidine) dihydrochloride

2,2´-Azobis[2-(3,4,5,6-tetrahydropyrimidin-2-yl) propane] dihydrochloride
CAS 102843-39-0
Uses: Polymerization initiator

2,2´-Azobis(2,4,4-trimethylpentane)
CAS 39198-34-0
Synonyms: Azodi-t-octane
Uses: Polymerization initiator
Manuf./Distrib.: Wako Chem. USA

Azodicarbonamide
CAS 123-77-3; EINECS 204-650-8
Synonyms: ADA; 1,1´-Azobisformamide; Azoformamide; Azobisformamide; Diazenedicarboxamide; Azodicarboxamide
Empirical: $C_2H_4N_4O_2$
Formula: $H_2NCON=NCONH_2$
Properties: Orange-red crystals; sol. in hot water; insol. in cold water, alcohol; m.w. 116.08; m.p. 225 C (dec.)
Precaution: Flamm.
Toxicology: Heated to decomp., emits toxic fumes of NO_x
Uses: Blowing and foaming agent for plastics; bleaching agent in cereal flour, maturing agent for flour
Regulatory: FDA 21CFR §172.806, 178.3010; Europe listed; UK approved
Manuf./Distrib.: Aldrich; Allchem Ind.; Charkit; Elf Atochem SA; Fabrichem; Gist-Brocades Food Ingreds.; Olin; Plastics & Chems.; Rit-Chem; Uniroyal; Witco/PAG
Trade names: Azo D; Azo DX72; Azofoam; Azofoam DS-1, DS-2; Azofoam M-1, M-2, M-3; Azofoam UC; Blo-Foam® 715; Blo-Foam® 754; Blo-Foam® 765; Blo-Foam® ADC 150; Blo-Foam® ADC 150FF; Blo-Foam® ADC 300; Blo-Foam® ADC 300FF; Blo-Foam® ADC 300WS; Blo-Foam® ADC 450; Blo-Foam® ADC 450FFNP; Blo-Foam® ADC 450WS; Blo-Foam® ADC 550; Blo-Foam® ADC 550FFNP; Blo-Foam® ADC 800; Blo-Foam® ADC 1200; Blo-Foam® ADC-MP; Blo-Foam® ADC-NP; Blo-Foam® KL9; Blo-Foam® KL10; Cellcom; Celogen® AZ 120; Celogen® AZ 130; Celogen® AZ 150; Celogen® AZ 180; Celogen®

AZ 199; Celogen® AZ 754; Celogen® AZ-760-A; Celogen® AZ-2100; Celogen® AZ-2990; Celogen® AZ 3990; Celogen® AZ-5100; Celogen® AZNP 130; Celogen® AZNP 199; Celogen® AZRV; Ficel® AC1; Ficel® AC2; Ficel® AC2M; Ficel® AC3; Ficel® AC3F; Ficel® AC4; Ficel® ACSP4; Ficel® ACSP5; Ficel® EPA; Ficel® EPB; Ficel® EPC; Ficel® EPD; Ficel® EPE; Ficel® LE; Ficel® SCE; Henley AZ LBA-30; Henley AZ LBA-49; HFVCD 11G; HFVCP 11G; HFVP 03; HFVP 15P; HFVP 19B; HP; HRVP 01; HRVP 19; Kempore® 60/14FF; Mikrofine ADC-EV; Plastifoam; Porofor® ADC/E; Porofor® ADC/F; Porofor® ADC/K; Porofor® ADC/M; Porofor® ADC/S; Unicell D

Trade names containing: Ampacet 10123; HC 01; HFVD 12CC; Hydrocerol® CT-232; Nitropore® ATA; Onifine-C; Onifine-CC; Onifine-CE; Polyvel BA20, BA40, BA60

Azodicarboxamide. *See* Azodicarbonamide
Azodiisobutyronitrile. *See* 2,2′-Azobisisobutyronitrile
Azodi-t-octane. *See* 2,2′-Azobis(2,4,4-trimethylpentane)
Azoformamide. *See* Azodicarbonamide

Azoisobutyrodinitrile
Uses: Blowing agent for prod. of PVC; used in floats, buoys, net floats, bath mats
Trade names: Porofor® N

Baking soda. *See* Sodium bicarbonate
BAPP. *See* Bis(aminopropyl)piperazine

Barium-cadmium complex
Uses: Stabilizer for flexible and rigid PVC compds., calendering, extrusion, inj. molding, powd. molding, slush and rotational molding, hose and profile, filled systems, film and sheet
Trade names: Haro® Chem KB-214SA; Interstab® BC-100S; Interstab® MT981; Interstab® MT982; Lankromark® LC68; Lankromark® LC299; Lankromark® LC431; Lankromark® LD299; Mark® 7118; Mark® 7308; Mark® TT; Mark® WS; Synpron 357; Synpron 1536; Synpron 1578; Synpron 1800; Synpron 1850; Synpron 1852; Therm-Chek® 1827; Therm-Chek® 5469; Therm-Chek® 5590; Therm-Chek® 5776; Therm-Chek® BJ-2

Barium-cadmium-lead complex
Uses: Stabilizer/lubricant for PVC profiles, sheet
Trade names: Haro® Mix YC-502; Haro® Mix YC-601; Haro® Mix YE-301; Haro® Mix YK-110; Haro® Mix YK-113; Haro® Mix YK-307; Haro® Mix YK-603

Barium-cadmium-phosphite complex
Uses: Stabilizer for flexible and semirigid compds., plastisols, organosols, and vinyl sol'ns.
Trade names: Mark® 462; Mark® LL

Barium-cadmium-zinc complex
Uses: Heat stabilizer for flexible vinyls processed by calendering, extrusion, inj. molding, or plastisol applics.; hose, profile, filled systems, etc.
Trade names: ADK STAB 189E; ADK STAB AC-133; ADK STAB AC-133N; ADK STAB AC-135; Ferro 1288; Interstab® BC-103; Interstab® BC-103A; Interstab® BC-103C; Interstab® BC-103L; Interstab® BC-109; Interstab® BC-110; Interstab® BC-4195; Interstab® R-4114; Interstab® R-4150; Lankromark® LC244; Lankromark® LC310; Lankromark® LC486; Lankromark® LC662; Mark® 462A; Mark® 1014; Mark® 2077; Synpron 1108; Synpron 1135; Synpron 1384; Synpron 1419; Synpron 1427; Synpron 1428; Synpron 1528; Synpron 1558; Synpron 1618; Synpron 1678; Synpron 1774; Synpron 1810; Synpron 1811, 1812; Synpron 1820; Synpron 1870; Synpron 1871; Therm-Chek® 6-V-6A; Therm-Chek® 1238; Therm-Chek® 1720; Therm-Chek® 1864; Therm-Chek® 5649; Therm-Chek® 5930
Trade names containing: Lankrocell® KLOP

Barium-calcium-zinc complex
Uses: Heat and light stabilizer for plastics
Trade names: Interstab® LT-4361; Therm-Chek® 6160

Barium carbonate
CAS 513-77-9; EINECS 208-167-3
Empirical: $CBaO_3$
Formula: $BaCO_3$
Properties: M.w. 197.37
Toxicology: A poison

Barium laurate

Uses: Treatment of brines in chlorine-alkali cells to remove sulfates, rodenticide, production of barium salts, ceramic flux, optical glass, case-hardening baths, ferrites, in radiation-resistant glass for color television tubes
Manuf./Distrib.: All Chemie Ltd; Barium & Chems.; Cerac; Bernardy Chimie SA; Great Western; Mallinckrodt; Osram Sylvania; Poly Research; San Yuan; Solvay GmbH; Spectrum Chem. Mfg.; US Petrochem. Ind.
Trade names containing: Chem-Master R-354; Poly-Dispersion® G (BAC)D-80

Barium laurate
EINECS 225-167-9
Synonyms: Dodecanoic acid, barium salt
Formula: $Ba(C_{12}H_{23}O_2)_2$
Properties: White leaf; sl. sol. in water and alcohol; m.w. 535.97; m.p. 260 C
Uses: Heat stabilizer for PVC
Trade names: Haro® Chem BSG

Barium-lead complex
Uses: Stabilizer for DWV pipe, elec., conduit, wire and cable, records, etc.
Trade names: Haro® Chem BP-108X; Mark® 550, 556; Temex BA-1

Barium-magnesium-zinc complex
Uses: PVC stabilizer
Trade names: ADK STAB AC-111; ADK STAB CPS-1; ADK STAB CPS-896; ADK STAB CPS-900

Barium metaborate
CAS 13701-59-2
Formula: $BaOB_2O_3 \cdot H_2O$
Properties: Wh. powd.; sol. 3-4 g/l water; m.p. 1000 C
Uses: Flame retardant for plastics
Manuf./Distrib.: Joseph Storey
Trade names: Busan® 11-M1

Barium phenate
Uses: Heat stabilizer for PVC coatings, films, and fabricated materials
Trade names: Lubrizol® 2106

Barium stannate. *See* Barium-tin complex

Barium stearate
CAS 6865-35-6
Formula: $Ba(C_{18}H_{35}O_2)_2$
Properties: White cryst. solid; insol. in water or alcohol; m.w. 706.42; dens. 1.145; m.p. 160 C
Precaution: Combustible
Toxicology: Moderately toxic by ingestion
Uses: Waterproofing agent, lubricant in metalworking, plastics, and rubber; wax compounding; prep. of greases; heat and light stabilizer in plastics
Manuf./Distrib.: Adeka Fine Chem.; Barium & Chems.; Cookson; Norac; Reagens SpA; Syn. Prods.; Witco/Oleo-Surf.
Trade names: Cometals Barium Stearate; Haro® Chem BG; Mathe Barium Stearate; Mathe Barium Stearate; Metasap® Barium Stearate; Miljac Barium Stearate; Synpro® Barium Stearate

Barium sulfate
CAS 7727-43-7; EINECS 231-784-4
Synonyms: Barytes (natural); Blanc fixe (artificial, precipitated); Basofor
Classification: Inorganic salt
Formula: $Ba \cdot H_2O_4S$
Properties: White or yellowish powd., odorless, tasteless; sol. in conc. sulfuric acid; insol. in water, dilute acids, alcohol; m.w. 233.40; dens. 4.25-4.5; m.p. 1580 C
Toxicology: TLV:TWA 10/mg^3 (total dust)
Uses: Weighting mud in oil drilling, paper coating, paints; filler and delustrant for textiles, rubber, plastics, and lithographic inks; radiation shield
Manuf./Distrib.: Am. Biorganics; Barium & Chems.; Barker Ind.; R.E. Carroll; Cyprus Ind. Min.; EM Ind.; J.M. Huber; Mallinckrodt; Ore & Chem.; Sachtleben Chemie GmbH; San Yuan; Spectrum Chem. Mfg.
Trade names: Bara-200C; Bara-200N; Bara-325C; Bara-325N; Barafoam 200; Barafoam 325; Barimite™;

Barimite™ 200; Barimite™ 1525P; Barimite™ 4009P; Barimite™ UF; Barimite™ XF; Bartex® 10; Bartex® 65; Bartex® 80; Bartex® FG-2; Bartex® FG-10; Bartex® OWT; Barytes Microsupreme 2410/M; Barytes Supreme; Blanc fixe F; Blanc fixe micro®; Blanc fixe N; Cimbar™ 325; Cimbar™ 1025P; Cimbar™ 1536P; Cimbar™ 3508P; Cimbar™ CF; Cimbar™ UF; Cimbar™ XF; G-50 Barytes™; No. 22 Barytes™

Barium-tin complex
Synonyms: Barium stannate
Uses: Stabilizer for PVC window profiles
Trade names: Stanclere® T-197; Synpron 1007

Barium-zinc complex
Uses: Heat and light stabilizer for flexible and rigid PVC, plastisols, topcoats, flooring, sheet and film; stabilizer/activator for chemically blown PVC
Trade names: ADK STAB 666; ADK STAB AC-118; ADK STAB AC-122; ADK STAB AC-141; ADK STAB AC-143; ADK STAB AC-153; ADK STAB AC-158; ADK STAB AC-167; ADK STAB AC-168; ADK STAB AC-169; ADK STAB AC-173; ADK STAB AC-181; ADK STAB AC-182; ADK STAB AC-183; ADK STAB AC-186; ADK STAB AC-190; ADK STAB AC-212; ADK STAB AC-216; ADK STAB AC-229; ADK STAB AC-230; ADK STAB AC-235; ADK STAB AC-243; ADK STAB AC-244; ADK STAB AC-248; ADK STAB AC-250; ADK STAB AC-252; ADK STAB AC-303; ADK STAB AC-307; ADK STAB AC-309; ADK STAB AC-310; ADK STAB AP-536; ADK STAB AP-539; ADK STAB AP-540; ADK STAB AP-543; ADK STAB AP-548; ADK STAB AP-550; ADK STAB AP-551; ADK STAB FL-43; ADK STAB FL-44; ADK STAB FL-45; ADK STAB FL-54; ADK STAB FL-55; ADK STAB FL-64; ADK STAB OF-14; ADK STAB OF-15; ADK STAB OF-19; ADK STAB OF-23; ADK STAB OF-31; ADK STAB RUP-9; ADK STAB RUP-10; ADK STAB RUP-11; ADK STAB RUP-14; B‰rostab® UBZ 76 BX; B‰rostab® UBZ 614; B‰rostab® UBZ 630; B‰rostab® UBZ 632; B‰rostab® UBZ 637; B‰rostab® UBZ 655 X; B‰rostab® UBZ 667; B‰rostab® UBZ 671; B‰rostab® UBZ 776 X; B‰rostab® UBZ 791; B‰rostab® UBZ 793; B‰rostab® UBZ 820 KA; Interstab® BZ-4828A; Interstab® CLB-747; Interstab® LT-4805R; Lankromark® DP6404Z; Lankromark® DP6452Z; Lankromark® LZ121; Lankromark® LZ187; Lankromark® LZ242; Lankromark® LZ616; Lankromark® LZ693; Lankromark® LZ704; Lankromark® LZ770; Lankromark® LZ792; Lankromark® LZ836; Lankromark® LZ858; Lankromark® LZ968; Lankromark® LZ1023; Lankromark® LZ1056; Lankromark® LZ1067; Lankromark® LZ1144; Lankromark® LZ1155; Lankromark® LZ1166; Mark® 144; Mark® 155; Mark® 398; Mark® 1600; Mark® 4700; Mark® 4701; Mark® 4702; Mark® 4716; Mark® 4722; Mark® 4723; MiRaStab 403-K; Plastistab 2310; Plastistab 2320; Plastistab 2330; Plastistab 2560; Plastistab 2580; Stavinor ATA-1056; Synpron 350; Synpron 940; Synpron 1363; Synpron 1652; Synpron 1831; Synpron 1881; Therm-Chek® 130; Therm-Chek® 6258; Therm-Chek® F-694
Trade names containing: Lankrocell® KLOP/CV; Onifine-C; Onifine-CC; Onifine-CE

Barium-zinc-magnesium-aluminum complex
Uses: PVC stabilizer
Trade names: ADK STAB CPS-2R; ADK STAB CPS-55R

Barium-zinc phosphite
Uses: Heat and light stabilizer for flexible PVC
Trade names: Interstab® BZ-4836

Barytes (natural). *See* Barium sulfate
Basic bismuth chloride. *See* Bismuth oxychloride
Basic magnesium carbonate. *See* Magnesium carbonate hydroxide
Basofor. *See* Barium sulfate
Battery acid. *See* Sulfuric acid
BBP. *See* Butyl benzyl phthalate
BDO. *See* 1,4-Butanediol

Behenamide
CAS 3061-75-4; EINECS 221-304-1
Synonyms: Behenic acid amide; Docosanamide
Classification: Aliphatic amide
Empirical: $C_{22}H_{45}NO$
Formula: $CH_3(CH_2)_{20}CONH_2$
Properties: m.p. 111-112 C
Uses: Antifoam in mfg. of detergents, in floor polishes, dripless candles; lubricant, slip, antiblock, and mold release in plastics
Regulatory: FDA 21CFR 117.1200

Behenic acid amide

Manuf./Distrib.: Croda Universal
Trade names: Kemamide® B

Behenic acid amide. *See* Behenamide

Behenyl alcohol
CAS 661-19-8; EINECS 211-546-6
Synonyms: 1-Docosanol; Alcohol C_{22}
Definition: Mixture of fatty alcohols chiefly of n-docosanol
Empirical: $C_{22}H_{46}O$
Formula: $CH_3(CH_2)_{20}CH_2OH$
Properties: Colorless waxy solid; insol. in water, sol. in ethanol, chloroform; m.w. 326.61; m.p. 71 C; b.p. 180 C (0.22 mm)
Uses: Synthetic fibers, lubricants, evaporation retardant on water surfaces; surfactant for polymerization
Regulatory: FDA 21CFR §178.3910
Manuf./Distrib.: Brown; Fluka; M. Michel; Schweizerhall; Vista
Trade names: Loxiol® VPG 1451

Behenyl dimethyl amine. *See* Dimethyl behenamine
Benylate. *See* Benzyl benzoate
Benzenamine, 4,4´-methylenebis (3-chloro-2,6-diethyl-(9Cl)). *See* 4,4´-Methylenebis (3-chloro-2,6-diethylaniline)
Benzenecarboxylic acid. *See* Benzoic acid
1,3-Benzenediamine. *See* m-Phenylenediamine
1,2-Benzenedicarboxylic acid bi (2-methoxyethyl) ester. *See* Dimethoxyethyl phthalate
Benzene dicarboxylic acid bis n-octyl n-decyl ester. *See* n-Octyl, n-decyl phthalate
1,2-Benzenedicarboxylic acid, 2-butoxy-2-oxoethyl, butyl ester. *See* n-Butyl phthalyl-n-butyl glycolate
1,2-Benzenedicarboxylic acid, butyl cyclohexyl ester. *See* Butyl cyclohexyl phthalate
1,2-Benzenedicarboxylic acid, butyl octyl ester. *See* Butyl octyl phthalate
1,2-Benzenedicarboxylic acid, butyl phenylmethyl ester. *See* Butyl benzyl phthalate
1,2-Benzenedicarboxylic acid, cyclohexyl isooctyl ester. *See* Cyclohexyl isooctyl phthalate
1,2-Benzenedicarboxylic acid, decyl octyl ester. *See* n-Octyl, n-decyl phthalate
1,2-Benzenedicarboxylic acid, dibutyl ester. *See* Dibutyl phthalate
1,2-Benzenedicarboxylic acid, dicyclohexyl ester. *See* Dicyclohexyl phthalate
1,2-Benzenedicarboxylic acid, dihexyl ester. *See* Dihexyl phthalate
1,2-Benzenedicarboxylic acid, diisodecyl ester. *See* Diisodecyl phthalate
1,2-Benzenedicarboxylic acid, diisononyl ester. *See* Diisononyl phthalate
1,2-Benzenedicarboxylic acid, diisotridecyl ester. *See* Diisotridecyl phthalate
1,2-Benzenedicarboxylic acid dimethyl ester. *See* Dimethyl phthalate
1,3-Benzenedicarboxylic acid, dimethyl ester. *See* Dimethyl isophthalate
1,2-Benzenedicarboxylic acid dioctyl ester. *See* Dioctyl phthalate
1,4-Benzenedicarboxylic acid, dioctyl ester. *See* Dioctyl terephthalate
1,2-Benzenedicarboxylic acid, ditridecyl ester. *See* Ditridecyl phthalate
1,2-Benzenedicarboxylic acid, diundecyl ester. *See* Diundecyl phthalate
1,2-Benzenedicarboxylic acid, lead (2+) salt, basic. *See* Lead phthalate, basic
1,3-Benzenediol. *See* Resorcinol
1,4-Benzenediol. *See* Hydroquinone

Benzene-1,3-disulfonyl hydrazide
Uses: Blowing agent for rubber goods
Trade names containing: Porofor® B 13/CP 50

Benzene, ethenyl-, homopolymer. *See* Polystyrene
Benzene, 1,1´-(1-methylethylidene)bis(3,5-dibromo-4-(2,3-dibromopropoxy)-(9Cl). *See* Tetrabromobisphenol A, bis (2,3-dibromopropyl ether)
Benzene monosulfonic acid. *See* Benzenesulfonic acid

Benzene propanoic acid
CAS 501-52-0; EINECS 207-924-5
Synonyms: Bis-[3,3-bis-(4´-hydroxy-3´-t-butylphenyl butanoic acid]-glycol ester; Butyric acid, 3,3-bis(3-t-butyl-4-hydroxyphenyl) ethylene ester
Empirical: $C_{50}H_{66}O_8$
Properties: M.w. 795.08

Uses: Antioxidant, stabilizer in plastics
Trade names: Hostanox® O3

Benzenepropanoic acid, 3,5-bis(1,1-dimethylethyl)-4-hydroxy-, thiodi-2,1-, ethanediyl ester. *See* Thiodiethylene bis (3,5-di-t-butyl-4-hydroxy) hydrocinnamate

Benzenesulfonamide
CAS 98-10-2; EINECS 202-637-1
Empirical: $C_6H_7NO_2S$
Formula: $C_6H_5SO_2NH_2$
Properties: M.w. 157.19; m.p. 150-152 C
Uses: Plasticizer and nucleating agent in PET resins
Manuf./Distrib.: Unitex
Trade names: Uniplex 413

Benzenesulfonic acid
CAS 98-11-3; 42615-29-2; EINECS 202-638-7
Synonyms: Benzene monosulfonic acid; Phenylsulfonic acid
Empirical: $C_6H_6O_3S$
Formula: $C_6H_5SO_3H$
Properties: Sol. in water, alcohol; sl. sol. in benzene; m.w. 158.18; m.p. 43-44 C
Precaution: Oxidizer
Toxicology: Highly toxic
Uses: Catalyst for esterification, polymerization, and polycondensation in foundry resins
Manuf./Distrib.: Howard Hall; Sloss Ind.
Trade names: Reworyl® B 70

Benzenesulfonic acid butyl amide. *See* N,N-Butyl benzene sulfonamide
Benzenesulfonic acid, dimethyl-. *See* Xylene sulfonic acid
Benzenesulfonic acid, tridecyl-. *See* Tridecylbenzene sulfonic acid

Benzenesulfonmethylamide
Properties: Sp.gr. 1.26 (20 C); flash pt. (COC) 180 C
Uses: Plasticizer for cellulosics
Manuf./Distrib.: Unitex

Benzenesulfonohydrazide. *See* Benzene sulfonyl hydrazide

Benzene sulfonyl hydrazide
CAS 80-17-1; EINECS 201-255-2
Synonyms: BSH; Benzenesulfonohydrazide
Empirical: $C_6H_8N_2O_2S$
Formula: $C_6H_5SO_2NHNH_2$
Properties: M.w. 172.21; m.p. 101-103 C; gas yield 90 cc/g
Uses: Blowing agent for rubber goods
Trade names: Mikrofine BSH; Porofor® BSH Powder
Trade names containing: Porofor® BSH Paste; Porofor® BSH Paste M

Benzene-1,2,4,5-tetracarboxylic dianhydride. *See* Pyromellitic dianhydride
1,2,4,5-Benzenetetracarboxylic dianhydride. *See* Pyromellitic dianhydride
1,2,4,5-Benzenetetracarboxylic 1,2:4,5 dianhydride. *See* Pyromellitic dianhydride

Benzethonium chloride
CAS 121-54-0; EINECS 204-479-9
Synonyms: N,N-Dimethyl-N-[2-[2-[4-(1,1,3,3-tetramethylbutyl) phenoxy] ethoxy] ethyl] benzene-methanaminium chloride; Diisobutylphenoxyethoxyethyl dimethyl benzyl ammonium Cl
Classification: Quaternary ammonium salt
Empirical: $C_{27}H_{42}ClNO_2$
Properties: Colorless, odorless plates, bitter; sol. in water, alcohol, acetone, chloroform; m.w. 448.15; m.p. 164-166 C
Toxicology: Poison by oral, subcutaneous, intraperitoneal, intravenous routes; severe eye irritant; LD50 (rats) 420 mg/kg
Uses: Cationic detergent; antiseptic; topical anti-infective; cocatalyst for curing polyesters

Regulatory: FDA 21CFR §175.105, EPA registered, Japan approved, Europe listed
Manuf./Distrib.: Lonza
Trade names: Hyamine® 1622 50%

2-Benzimidazolethiol. *See* 2-Mercaptobenzimidazole
Benzin. *See* Naphtha
Benzoate of soda. *See* Sodium benzoate
Benzoate sodium. *See* Sodium benzoate
1H,3H-Benzo(1,2-c;4,5-c′)difuran-1,3,5,7-tetrone. *See* Pyromellitic dianhydride

Benzoic acid
CAS 65-85-0; EINECS 200-618-2
Synonyms: Benzenecarboxylic acid; Phenylformic acid; Dracylic acid
Classification: Aromatic acid
Empirical: $C_7H_6O_2$
Properties: White scales, needles, crystals, benzoin odor; sol. in alcohol, ether, chloroform, benzene, carbon disulfide; sl. sol. in water; m.w. 122.13; dens. 1.2659; m.p. 121.25 C; b.p. 249.2 C; flash pt. 121.1 C
Precaution: Combustible when exposed to het or flame; reactive with oxidizing materials
Toxicology: LD50 (oral, rat) 2530 mg/kg; mod. toxic by ingestion, IP routes; poison by subcut. route; severe eye/skin irritant; may cause human intolerance reaction, asthma,hyperactivity in children; heated to decomp., emits acrid smoke and irritating fumes
Uses: Sodium and butyl benzoates, plasticizers, benzoyl chlorides, food preservatives, antimicrobial agent, flavors, perfumes, antifungal agent , retarder for rubbers and latexes
Regulatory: FDA 21CFR §150.141, 150.161, 166.40, 166.110, 175.300, 184.1021, GRAS 0.1% max. in foods; USA EPA registered; Japan 0.2% max.; Europe listed 0.5% max.; GRAS (FEMA); cleared by MID to retard flavor reversion in oleomargarine at 0.1%; Japan approved with limitations; Europe listed; UK approved
Manuf./Distrib.: Aldrich; Ashland; Jan Dekker; Elf Atochem SA; Fluka; R.W. Greeff; Int'l. Sourcing; Mallinckrodt; E. Merck; Napp; Penta Mfg.; Velsicol
Trade names: Retarder BA, BAX

Benzoic acid, benzyl ester. *See* Benzyl benzoate
Benzoic acid sodium salt. *See* Sodium benzoate

Benzoin ether
Uses: Initiator used in UV light curing of polyester resins; UV sensitizer for lacquers and FRP laminates
Trade names: Trigonal® 14

Benzoperoxide. *See* Benzoyl peroxide

Benzophenone
CAS 119-61-9; EINECS 204-337-6
Synonyms: Benzoylbenzene; Diphenyl ketone; Diphenylmethanone
Classification: Organic compd.
Empirical: $C_{13}H_{10}O$
Properties: Wh. rhombic cryst., persistent rose-like odor; sol. in fixed oils; sl. sol. in propylene glycol; m.w. 182.23; sp.gr. 1.0976 (α, 50/50 C), 1.108 (β, 23/40 C); m.p. 49 C (α), 26 C (β), 47 C (γ); b.p. 305 C
Precaution: Combustible when heated; incompat. with oxidizers
Toxicology: LD50 (oral, mouse) 2895 mg/kg; moderately toxic by ingestion and intraperitoneal routes; heated to decomp., emits acrid and irritating fumes
Uses: UV absorber for sunscreens; uv protective agents in polymers, petrol. waxes, etc.; photosensitizers in printing inks, wood and metal finishes; intermediate for mfg. of antihistamines, insecticides; flavoring agent
Regulatory: FDA 21CFR §172.515
Manuf./Distrib.: Aldrich; Allchem Ind.; Berje; Elf Atochem N. Am.; Fluka; R.W. Greeff; Penta Mfg.; Plastics & Chems.; Reedy Int'l.; Schweizerhall; Spectrum Chem. Mfg.; 3V; Velsicol
Trade names containing: PE-UV 12; PE-UV 118

Benzophenone-1
CAS 131-56-6; EINECS 205-029-4
Synonyms: 2,4-Dihydroxybenzophenone; Benzoresorcinol; 4-Benzoyl resorcinol
Classification: Organic benzophenone deriv.
Formula: $C_6H_5COC_6H_3(OH)_2$
Properties: Lt. yel. crystalline solid; insol. in water; sol. in ethanol, methanol, MEK, ethyl acetate; m.p. 142 C; b.p. 194 C (1 mm)

Uses: UV absorber in polymers
Manuf./Distrib.: Aceto; EM Industries; Ferro/Bedford; R.W. Greeff; Haarmann & Reimer; Hoechst Celanese; Quest Int'l.; Sartomer
Trade names: Uvasorb 2 OH; Uvinul® 3000

Benzophenone-2
CAS 131-55-5; EINECS 205-028-9
Synonyms: 2,2′,4,4′-Tetrahydroxy benzophenone
Classification: Organic benzophenone deriv.
Empirical: $C_{13}H_{10}O_5$
Properties: Powd.; m.w. 246.23; m.p. 200-203 C
Toxicology: Moderate toxicity by ingestion; eye irritant
Uses: UV absorber/stabilizer for epoxy, polyesters
Manuf./Distrib.: Aceto; EM Industries; Ferro/Bedford; R.W. Greeff; Haarmann & Reimer; Hoechst Celanese; Quest Int'l.; Sartomer
Trade names: Uvinul® 3050; Uvinul® D-50

Benzophenone-3
CAS 131-57-7; EINECS 205-031-5
Synonyms: 2-Hydroxy-4-methoxybenzophenone; (2-Hydroxy-4-methoxyphenyl) phenylmethanone; Oxybenzone
Classification: Organic benzophenone deriv.
Empirical: $C_{14}H_{12}O_3$
Properties: Yellowish cryst.; sol. in peanut oil, ethanol, PEG-8, oleyl alcohol, castor oil; insol. in water; m.w. 228.26; m.p. 62-63.5 C
Toxicology: Poison by intraperitoneal route; mild toxicity by ingestion; LD50 (rat, oral) > 10 g/kg; LD50 (rabbit, dermal) > 16 g/kg
Uses: Sunscreen; uv absorber for plastics (ABS, cellulosics, polyester, PS, flexible and rigid PVC, VDC)
Regulatory: FDA 21CFR §177.1010
Manuf./Distrib.: Aceto; EM Industries; Ferro/Bedford; R.W. Greeff; Haarmann & Reimer; Hoechst Celanese; Quest Int'l.; Sartomer
Trade names: Acetorb A; Cyasorb® UV 9; Lankromark® LE296; Syntase® 62; UV-Absorber 325; Uvasorb MET; Uvinul® 3040; Uvinul® M-40

Benzophenone-4
CAS 4065-45-6; EINECS 223-772-2
Synonyms: 2-Hydroxy-4-methoxybenzophenone-5-sulfonic acid; Sulisobenzone; 2-Hydroxy-4-methoxy-5-sulfo benzophenone
Classification: Organic benzophenone deriv.
Empirical: $C_{14}H_{12}O_6S$
Uses: UV absorber in sunscreen prods. and in hair sprays and shampoos for dyed and tinted hair; for leather and textile fibers; plastics
Manuf./Distrib.: Aceto; EM Industries; Ferro/Bedford; R.W. Greeff; Haarmann & Reimer; Hoechst Celanese; Quest Int'l.; Sartomer

Benzophenone-6
CAS 131-54-4; EINECS 205-027-3
Synonyms: 2,2′-Dihydroxy-4,4′-dimethoxybenzophenone; Bis(2-hydroxy-4-methoxyphenyl) methanone
Classification: Organic benzophenone deriv.
Empirical: $C_{15}H_{14}O_5$
Properties: Crystals; m.w. 274.26; m.p. 139-140 C
Uses: UV light absorber, esp. in paints, plastics (epoxy, polyester)
Manuf./Distrib.: Aceto; EM Industries; Ferro/Bedford; R.W. Greeff; Haarmann & Reimer; Hoechst Celanese; Quest Int'l.; Sartomer
Trade names: Uvinul® 3049; Uvinul® D-49

Benzophenone-8
CAS 131-53-3; EINECS 205-026-8
Synonyms: 2,2′-Dihydroxy-4-methoxybenzophenone; Dioxybenzone
Classification: Organic benzophenone deriv.
Empirical: $C_{14}H_{12}O_4$
Properties: Powd.; m.p. 68 C
Toxicology: Practically nontoxic in single oral doses; LD50 (rat, oral) > 10 g/kg; nonirritating to rabbit eye or

skin

Uses: UV absorber, light stabilizer for flexible PVC

Manuf./Distrib.: EM Industries; Ferro/Bedford; R.W. Greeff; Haarmann & Reimer; Hoechst Celanese; Quest Int'l.; Sartomer

Trade names: Cyasorb® UV 24

Benzophenone-11

CAS 1341-54-4

Classification: Organic benzophenone deriv.

Definition: Mixture of benzophenone-6 and -2 and other tetra-substituted benzophenone materials

Uses: UV absorber used in NC lacquer, fluorescent paint, inks, and for protecting furniture woods, colored liq. toiletries and cleaning agents, isocyanate systems, and butyrate metal lacquers

Manuf./Distrib.: EM Industries; Ferro/Bedford; R.W. Greeff; Haarmann & Reimer; Hoechst Celanese; Quest Int'l.; Sartomer

Trade names: Uvinul® 3093

Benzophenone-12

CAS 1843-05-6; EINECS 217-421-2

Synonyms: 2-Hydroxy-4-n-octoxybenzophenone; 2-Hydroxy-4-(octyloxy) benzophenone; [2-Hydroxy-4-(octyloxy)phenyl]phenylmethanone; Octabenzone

Classification: Organic benzophenone deriv.

Empirical: $C_{21}H_{26}O_3$

Properties: Crystals; m.w. 326.42; m.p. 45-46 C

Toxicology: Very low acute toxicity for animals; LD50 (rat, oral) > 10 g/kg; nonirritating to rabbit skin or eyes

Uses: UV stabilizer for polyethylene

Regulatory: FDA 21CFR §178.2010

Manuf./Distrib.: EM Industries; Enterprise; Ferro/Bedford; Great Lakes; R.W. Greeff; Haarmann & Reimer; Hoechst Celanese; Quest Int'l.; Sartomer; 3V

Trade names: BLS™ 531; Cyasorb® UV 531; Hostavin® ARO 8; Lankromark® LE285; Lowilite® 22; Mark® 1413; Uvasorb 3C; UV-Chek® AM-300; UV-Chek® AM-301; Uvinul® 408; Uvinul® 3008

3,3′,4,4′-Benzophenone tetracarboxylic dianhydride

CAS 2421-28-5

Synonyms: BTDA; 5,5′-Carbonylbis(1,3-isobenzofurandione; 4,4′-Carbonylbis(phthalic anhydride); 4,4′-Carbonyldiphthalic acid anhydride

Empirical: $C_{17}H_6O_7$

Properties: Amber to lt. tan powd. or flakes, odorless, tasteless; sol. 0.26 g/100 g water; m.w. 322.22; sp.gr. 1.57 (30 C); m.p. 221-225 C; b.p. 380-400 C; pH 2.4 (2.1 g/L)

Precaution: Incompat. with alkali metal hydroxides, amines, ammonia, alcohols, thiols, strong oxidizing/reducing agents

Toxicology: Fumes from molten BTDA are highly toxic; solid considered nontoxic; LD50 (oral, rats) > 12,800 mg/kg (as tetra-acid); may cause respiratory irritation/allergic response on overexposure

Uses: Epoxy curing agent; monomer in unsat. and saturated polyester resins; ester derivs. as intermediates in prod. of polyimides, lubricants, plasticizers; intermediate for alkyd, PU resins, polypyrrones; end-uses incl. electronic transfer molding compds., wire coatings, aerospace adhesives, advanced structural composites, uv-cured inks, insulating coatings and foams, flame retardant prods., release agents, molded engineering parts, refinery catalysts

Manuf./Distrib.: Allco; Chemie Linz N. Am.

Trade names: Allco BTDA

Benzophenone, 2,3,4-trihydroxy-. *See* 2,3,4-Trihydroxy benzophenone

Benzopinacol silyl ether

Trade names containing: Initiator BK

p-Benzoquinone dioxime. *See* p-Quinone dioxime
Benzoresorcinol. *See* Benzophenone-1
Benzothiazole disulfide. *See* Benzothiazyl disulfide
2-Benzothiazolesulfenamide, N-cyclohexyl-. *See* N-Cyclohexyl-2-benzothiazolesulfenamide
2-Benzothiazolethiol. *See* 2-Mercaptobenzothiazole
2(3H)-Benzothiazolethione. *See* 2-Mercaptobenzothiazole
2(3H)-Benzothiazolethione, zinc salt. *See* Zinc 2-mercaptobenzothiazole
2-Benzothiazolyl disulfide. *See* Benzothiazyl disulfide

2-Benzothiazolyl morpholinodisulfide. *See* 4-Morpholinyl-2-benzothiazole disulfide
2-Benzothiazolyl-n-morpholinosulfide. *See* N-Oxydiethylene benzothiazole-2-sulfenamide
Benzothiazyl-2-t-butyl sulfenamide. *See* Butyl 2-benzothiazole sulfenamide
Benzothiazyl-2-cyclohexyl sulfenamide. *See* N-Cyclohexyl-2-benzothiazolesulfenamide
Benzothiazyl 1-2-dicyclohexyl sulfenamide. *See* Dicyclohexylbenzothiazyl-2-sulfenamide
Benzothiazyl-2-dicyclohexyl sulfenamide. *See* Dicyclohexylbenzothiazyl-2-sulfenamide

Benzothiazyl disulfide
CAS 120-78-5
Synonyms: MBTS; Benzothiazole disulfide; 2,2′-Dithiobisbenzothiazole; 2-Benzothiazolyl disulfide; 2,2′-Dibenzothiazyl disulfide; 2-Mercaptobenzothiazole disulfide; Mercaptobenzthiazyl ether
Formula: $(C_6H_4SCN)_2S_2$
Properties: Gray-wh. to cream powd. or pellets, sl. odor; sparingly sol. in org. solvs.; insol. in water; m.w. 78.12; dens. 1.50 mg/m^3; m.p. 168 C; flash pt. (COC) 271 C
Precaution: Exposed to flame, emits acrid fumes; dust is sensitive to ignition
Toxicology: TLV:TWA 10 ppm; suspected human carcinogen; poison by inhalation, skin contact, IP, IV; mod. toxicity by ingestion, SQ; severe eye irritant
Uses: Accelerator for nat. and syn. rubbers (nitrile, SBR); plasticizer; vulcanization retarder; cure modifier
Manuf./Distrib.: Monsanto
Trade names: Akrochem® MBTS; Akrochem® MBTS Pellets; Ekaland MBTS; Naftocit® MBTS; Naftopast® MBTS-A; Naftopast® MBTS-P; Perkacit® MBTS; Thiofide®
Trade names containing: Altax®; Poly-Dispersion® AAD-75; Poly-Dispersion® EAD-75; Poly-Dispersion® EAD-75P; Rhenovin® MBTS-70; Thanecure®; Vulkacit DM/C; Vulkacit DM/MGC

Benzothiazyl-2-sulfene morpholide. *See* N-Oxydiethylene benzothiazole-2-sulfenamide
1H-Benzotriazole, methyl. *See* Tolyltriazole
2-(2H-Benzotriazol-2-yl)-4,6-bis(1,1-dimethylpropyl)phenol. *See* 2-(2′-Hydroxy-3,5′-di-t-amylphenyl) benzotriazole
2-(2H-Benzotriazol-2-yl)-4,6-bis(1-methyl-1-phenylethyl) phenol. *See* 2-[2-Hydroxy-3,5-di-(1,1-dimethylbenzyl)phenyl]-2H-benzotriazole
2-(2H-Benzotriazol-2-yl)-4-methylphenol. *See* Drometrizole
2-(2H-Benzotriazol-2-yl)-4-(1,1,3,3-tetramethylbutyl) phenol. See 2-(2-Hydroxy-5-t-octylphenyl) benzotriazole

Benzoxazole
CAS 273-53-0
Synonyms: 1-Oxa-3-azaindene
Properties: M.w. 119.12; m.p. 27-30 C; b.p. 182 C; flash pt. 58 C
Toxicology: Poison by intraperitoneal, intravenous routes; moderately toxic by ingestion
Uses: Optical brightener, whitener for polyesters and blends
Trade names: Ecco White® OP

Benzoylbenzene. *See* Benzophenone

Benzoyl peroxide
CAS 94-36-0; EINECS 202-327-6
Synonyms: Dibenzoyl peroxide; Benzoperoxide; Benzoyl superoxide
Classification: Organic peroxide
Empirical: $C_{14}H_{10}O_4$
Formula: $[C_6H_5C(O)]_2O_2$
Properties: Colorless to wh. gran., crust. solid, faint odor of benzaldehyde, tasteless; sl. sol. in alcohols, veg. oils, water; sol. in benzene, chloroform, ether; m.w. 242.23; dens. 1.3340 (25 C); m.p. 103-105 C
Precaution: Flamm. oxidizing liq.; explosion hazard
Toxicology: LD50 (oral, rat) 7710 mg/kg; highly toxic; TLV 5 mg/m^3; poison by ingestion, intraperitoneal routes; allergen; eye irritant
Uses: Oxidizing agent in bleaching oils, flours; catalyst for plastics; initiator in polymerization; bleaching agent
Regulatory: FDA 21CFR §184.1157, GRAS; Japan approved with limitations
Manuf./Distrib.: Abco Ind. Ltd; Akzo Chemie SA; Aztec Peroxides; Elf Atochem N. Am.; Fluka; Great Western; Norac
Trade names: Abcure S-40-25; Aztec® BP-05-FT; Aztec® BP-20-GY; Aztec® BP-40-S; Aztec® BP-50-FT; Aztec® BP-50-P1; Aztec® BP-50-PSI; Aztec® BP-50-SAQ; Aztec® BP-60-PCL; Aztec® BP-70-W; Aztec® BP-77-W; Benox® L-40LV; Cadet® BPO-70; Cadet® BPO-70W; Cadet® BPO-75W; Cadet® BPO-78; Cadet® BPO-78W; Cadox® 40E; Cadox® BFF-50L; Cadox® BPO-W40; Cadox® BT-50;

Cadox® BTW-50; Cadox® BTW-55; Lucidol 70; Lucidol 75; Lucidol 75FP; Lucidol 78; Lucidol 98; Luperco AFR-250; Luperco AFR-400; Luperco AFR-500; Luperco ANS; Luperco ANS-P; Varox® ANS

Trade names containing: Cadox® BCP; Cadox® BEP-50; Cadox® BFF-50; Cadox® BFF-60W; Cadox® BP-55; Cadox® BS; Cadox® BTA; Luperco AA; Luperco ACP; Luperco AST; Luperco ATC; Superox® 744

4-Benzoyl resorcinol. *See* Benzophenone-1
Benzoyl superoxide. *See* Benzoyl peroxide

Benzyl benzoate

CAS 120-51-4; EINECS 204-402-9
Synonyms: Benylate; Phenylmethyl benzoate; Benzoic acid, benzyl ester
Definition: Ester of benzyl alcohol and benzoic acid
Empirical: $C_{14}H_{12}O_2$
Formula: $C_6H_5COOCH_2C_6H_5$
Properties: Colorless oily liq., sl. aromatic odor; misc. with alcohol, chloroform, ether; insol. in water, glycerin; m.w. 212.26; sp.gr. 1.116; m.p. 21 C; b.p. 324 C; flash pt. (CC) 298 F; ref. index 1.568
Precaution: Combustible liq.; reactive with oxidizing materials
Toxicology: LD50 (oral, rat) 500 mg/kg; mod. toxic by ingestion, skin contact; heated to decomp., emits acrid and irritating fumes and smoke
Uses: Fixative and solvent for musk in perfumes and flavors; flavoring agent; external medicine; plasticizer for nitrocellulose and cellulose acetate; miticide
Regulatory: FDA 21CFR §172.515, 175.105
Manuf./Distrib.: Aldrich; Berje; Fluka; Haarmann & Reimer; Janssen Chimica; Kalama; Morflex; Novachem; Penta Mfg.; Schweizerhall

Benzyl butyl phthalate. *See* Butyl benzyl phthalate

N-Benzyldimethylamine

CAS 103-83-3; EINECS 203-149-1
Synonyms: DMBA; N-(Phenylmethyl) dimethylamine; N,N-Dimethylbenzylamine
Empirical: $C_9H_{13}N$
Formula: $C_6H_5CH_2N(CH_3)_2$
Properties: Colorless to lt. yel. liq.; m.w. 135.23; dens. 0.894 (27 C); b.p. 180-182 C
Toxicology: Poison by ingestion; moderately toxic by inhalation, skin contact; corrosive severe skin and eye irritant
Uses: Intermediate, dehydrohalogenating catalyst, corrosion inhibitor, acid neutralizer, potting compounds, adhesives, cellulose modifier; catalyst for rigid PU foam
Manuf./Distrib.: Aceto; Anhydrides & Chems.; R.W. Greeff; Penta Mfg.; PMC Specialties; Schweizerhall; Synthron; Zeeland
Trade names: Dabco® BDMA

Benzyldimethyldodecylammonium chloride. *See* Lauralkonium chloride

Benzyldimethyl ketal

Uses: Photoinitiator for wh. coatings, adhesives, inks, photopolymers, electronic photoresists, polyester-styrene wood filler composites
Manuf./Distrib.: Alemark; Amber Syn.
Trade names: Escacure® KB1; Escacure® KB60

Benzyloctyl adipate

Uses: Plasticizer for PVC, VC copolymers, NC, ethyl cellulose, PS, natural, S/B, N/B, chlorinated, and butyl rubbers
Trade names: Adimoll® BO

Benzyl phthalate

Uses: Plasticizer for PVC, paints and acrylic coatings, caulks and sealants based on chlorinated rubber, butyl rubber, polysulfides, or PU
Trade names: Santicizer 278

BGE. *See* Butyl glycidyl ether

BHA

CAS 25013-16-5; EINECS 204-442-7, 246-563-8

Synonyms: Butylated hydroxyanisole; (1,1-Dimethylethyl)-4-methoxyphenol; 3-t-Butyl-4-hydroxyanisole
Definition: Mixture of isomers of tertiary butyl-substituted 4-methoxyphenols
Empirical: $C_{11}H_{16}O_2$
Properties: Wh. waxy solid, faint char. odor; insol. in water; sol. in petrol. ether, 50% or higher alcohol, propylene glycol, fats, oils; m.w. 180.27; m.p. 48-55 C; b.p. 264-270 C (733 mm)
Precaution: Combustible
Toxicology: LD50 (oral, mouse) 2000 mg/kg; suspected carcinogen; moderate toxicity by ingestion, intraperitoneal routes; may cause rashes, hyperactivity; heated to decomp., emits acrid and irritating fumes
Uses: Antioxidant, preservative for foods, cosmetics, plastics, etc.
Regulatory: FDA 21CFR §166.110, 172.110, 172.515, 172.615, 173.340, 175.105, 175.125, 175.300, 175.380, 175.390, 176.170, 176.210, 177.1010, 177.1210, 177.1350, 178.3120, 178.3570, 179.45, 181.22, 181.24 (0.005% migrating from food pkg.), 182.3169 (0.02% max. of fat or oil), GRAS; GRAS (FEMA); USDA 9CFR 318.7, 381.147; Japan approved 0.2-1 g/kg; Europe approved; UK approved
Manuf./Distrib.: Aceto; Allchem Ind.; Eastman; Fluka; Int'l. Chem.; Penta Mfg.; Spectrum Chem. Mfg.; UOP
Trade names: Sustane® BHA; Tenox® BHA

BHEB. *See* 2,6-Di-t-butyl-4-ethyl phenol
BHMT. *See* Bis-hexamethylenetriamine

BHT
CAS 128-37-0; EINECS 204-881-4
Synonyms: DBPC; Butylated hydroxytoluene; 2,6-Di-t-butyl-4-methylphenol; 2,6-Di-t-butyl-p-cresol; 2,6-Bis (1,1-dimethylethyl)-4-methylphenol
Classification: Substituted toluene
Empirical: $C_{15}H_{24}O$
Formula: $[C(CH_3)_3]_2CH_3C_6H_2OH$
Properties: White cryst. solid; insol. in water; sol. in toluene, alcohols, MEK, acetone, Cellosolve, petrol. ether, benzene, most HC solvs.; m.w. 220.39; sp.gr. 1.048 (20/4 C); m.p. 68 C; b.p. 265 C; flash pt. (TOC) 260 F
Precaution: Combustible exposed to heat or flame; reactive with oxidizing materials
Toxicology: TLV: 10 mg/m^3; LD50 (oral, rat) 890 mg/kg; moderately toxic by ingestion; poison by IP, IV routes; suspected carcinogen; human skin irritant; eye irritant; may cause rashes, hyperactivity; heated to decomp., emits acrid smoke and fumes
Uses: Antioxidant for foods, animal feed, petrol. prods., syn. rubbers, plastics, soaps; antiskinning agent in paints and inks
Regulatory: FDA 21CFR §137.350, 166.110, 172.115, 172.615 (0.1% max.), 173.340 (0.1% of defoamer), 175.105, 175.125, 175.300, 175.380, 175.390, 176.170, 176.210, 177.1010, 177.1210, 177.1350, 177.2260, 177.2600, 178.3120, 178.3570, 179.45, 181.22; 181.24 (0.005% migrating from food pkg.), 182.3173 (0.02% max. of fat/oil), GRAS; USDA 9CFR §318.7 (0.003% in dry sausage, 0.01% in rendered animal fat, 0.02% in margarine), 381.147 (0.01% on fat in poultry); Japan 0.2-1 g/kg; Europe, UK approved
Manuf./Distrib.: Aceto; Aldrich; Ashland; Fluka; Great Lakes; Int'l. Chem.; Penta Mfg.; PMC Specialties; Raschig; Uniroyal
Trade names: Akrochem® Antioxidant BHT; CAO®-3; Ionol; Lowinox® BHT; Lowinox® BHT Food Grade; Lowinox® BHT-Superflakes; Naftonox® BHT; Naugard® BHT; Ralox® BHT food grade; Ralox® BHT Tech.; Spectratech® CM 11340; Spectratech® KM 11264; Sustane® BHT; Topanol® OC; Topanol® OF; Ultranox® 226; Vulkanox® KB
Trade names containing: Aquastab PA 58; Dimodan LS Kosher

Bicarbonate of soda. *See* Sodium bicarbonate

Bis (p-aminocyclohexyl) methane
CAS 1761-71-3
Synonyms: Methylene di(cyclohexylamine)
Formula: $CH_2(CH_6H_{10}NH_2)_2$
Uses: Epoxy curing agent
Manuf./Distrib.: Air Prods.; Aldrich; BASF
Trade names: Amicure® PACM

Bis(2-aminoethyl) amine. *See* Diethylenetriamine
N,N′-Bis(2-aminoethyl)-1,2-ethanediamine. *See* Triethylenetetramine

Bis(aminopropyl)piperazine
Synonyms: BAPP

Bis-[3,3-bis-(4´-hydroxy-3´-t-butylphenyl butanoic acid]-glycol ester

Properties: Clear liq.; sp.gr. 0.97; b.p. 295 C; flash pt. (COC) 325 F; ref. index 1.5001
Uses: Intermediate for prep. of textile and paper treating resins, dyestuffs, epoxy curing agents, pharmaceuticals, surfactants, polyamides
Manuf./Distrib.: Texaco

Bis-[3,3-bis-(4´-hydroxy-3´-t-butylphenyl butanoic acid]-glycol ester. *See* Benzene propanoic acid
2,2-Bis(bromomethyl)-3-bromo-1-propanol. *See* Tribromoneopentyl alcohol
2,2-Bis(bromomethyl)-1,3-propanediol. *See* Dibromoneopentyl glycol
2,5-Bis (5-t-butyl-2-benzoxazolyl) thiophene. *See* 2,2´-(2,5-Thiophenediyl)bis[5-t-butylbenzoxazole]
2,2´-Bis(6-t-butyl-p-cresyl)methane. *See* 2,2´-Methylenebis (6-t-butyl-4-methylphenol)
Bis (4-t-butylcyclohexyl) peroxydicarbonate. *See* Di (4-t-butylcyclohexyl) peroxydicarbonate
1,1-Bis(t-butylperoxy) cyclohexane. *See* 1,1-Di (t-butylperoxy) cyclohexane
α,α´-Bis (t-butylperoxy) diisopropyl benzene. *See* Di-(2-t-butylperoxyisopropyl)benzene
Bis (t-butylperoxy) diisopropyl benzene. *See* Di-(2-t-butylperoxyisopropyl)benzene
Bis (t-butylperoxy isopropyl) benzene. *See* Di-(2-t-butylperoxyisopropyl)benzene
1,1-Bis (t-butylperoxy)-3,3,5-trimethyl cyclohexane. *See* 1,1-Di (t-butylperoxy) 3,3,5-trimethyl cyclohexane

Bis(butyltin 2-ethylhexyl mercaptoacetate)
Uses: Stabilizer for PVC
Trade names: Thermolite 340

Bis (o-chlorobenzoyl) peroxide
Uses: Crosslinking agent used for curing silicone rubbers
Trade names: Cadox® OS

Bis (p-chlorobenzoyl) peroxide. *See* p,p-Dichlorobenzoyl peroxide

Bis (β-chloroethyl) vinyl phosphonate
Synonyms: Bis (beta chloroethyl) vinyl phosphonate
Uses: Reactive flame retardant for plastics
Trade names: Fyrol® Bis Beta

Bis (beta chloroethyl) vinyl phosphonate. *See* Bis (β-chloroethyl) vinyl phosphonate

1,3-Bis(citraconimidomethyl)benzene
CAS 119462-56-5
Properties: Off-wh. solid; sol. in water, xylene, dichloromethane; m.w. 324; dens. 1270 kg/m^3; m.p. 87 C
Precaution: May form flamm. dust-air mixts.; incompat. with strong oxidizers; hazardous decomp. prods.: CO, CO_2, NO_x
Toxicology: LD50 (oral, rat) > 2000 mg/kg; severe eye irritant; may cause allergic skin reactions; nuisance dust may cause irritation of respiratory tract
Uses: Antireversion agent for rubber vulcanization
Trade names: Perkalink® 900

Biscumylphenyl isophthalate
Properties: Sp.gr. 1.18; m.p. 85 C; flash pt. (COC) 500 C
Uses: Plasticizer for PVC, PS, some cellulosics
Manuf./Distrib.: Kenrich Petrochem.

Biscumylphenyl trimellitate
Uses: Plasticizer, process aid for extrusion of PVC, urethanes; lubricant and process aid for filled PS, PC; reactive diluent and flow promoter for epoxy powd. coatings
Trade names: Kenplast® ESI

1,2-Bis (3,5-di-t-butyl-4-hydroxyhydrocinnamoyl) hydrazine
CAS 32687-78-8
Uses: Antioxidant for polyolefins
Trade names: Irganox® MD 1024

2,2-Bis[4-(2-(3,5-di-t-butyl-4-hydroxyhydrocinnamoyloxy)ethoxyphenyl]propane
Empirical: $C_{53}H_{72}O_8$
Properties: White powd.; sol. in common org. solvs.; m.w. 836; sp.gr. 1.09; m.p. 100-102 C
Uses: Antioxidant for polymers

Trade names: Topanol® 205

Bis (2,4-di-t-butylphenyl) pentaerythritol diphosphite
CAS 26741-53-7; EINECS 247-952-5
Synonyms: Bis (2,4-di-t-butylphenyl) pentaerythrityl diphosphite
Formula: $C_5H_{18}[O_2POC_6H_3(C[CH_3]_3)2]_2$
Properties: M.w. 604
Uses: Antioxidant; impact modifier for PVC
Trade names: Blendex® 340
Trade names containing: Ultranox® 626; Ultranox® 626A; Ultranox® 627A; Ultranox® 815A Blend; Ultranox® 817A Blend; Ultranox® 875A Blend; Ultranox® 877A Blend; Ultranox® 2714A

Bis (2,4-di-t-butylphenyl) pentaerythrityl diphosphite. *See* Bis (2,4-di-t-butylphenyl) pentaerythritol diphosphite
Bis (dibutylthiocarbamoyl) disulfide. *See* Tetrabutyl thiuram disulfide

Bis (2,4-dichlorobenzoyl) peroxide
CAS 133-14-2
Synonyms: Di-2,4–διχηλοροβενζοψλ περοξιδε
Formula: $C_{18}H_{22}O_2$
Properties: M.w. 270.40
Toxicology: Mild toxicity by ingestion and others
Uses: Crosslinking agent used for curing silicone rubbers
Trade names: Aztec® DCLBP-50-PSI
Trade names containing: Cadox® TS-50

Bis(diethylthiocarbamyl) disulfide. *See* Tetraethylthiuram disulfide

Bisdiisobutyl maleate
Uses: Plasticizer used for special applics. as reactive monomer, formulation of specialty surfactants
Trade names: Staflex DIBCM

Bis (2,6-diisopropylphenyl) carbodiimide
CAS 2162-74-5; EINECS 218-487-5
Empirical: $C_{25}H_{34}N_2$
Properties: Wh. to lt. ylsh. cryst.; sol. in acetone, insol. in water; m.w. 362.56; m.p. 48 C
Precaution: Incompat. with acids, alkalies, oxidizing agents; dust can form explosive mixts. with air; keep away from sources of ignition; hazardous decomp. prods.: NO_x, isocyanates
Toxicology: LD50 (oral, rat) 2395 mg/kg; nonirritant to skin or eyes; may cause sensitization, allergic skin reaction on repeated/prolonged exposure
Uses: Stabilizer protecting polymers against hydrolysis in processing and long-term aging
Trade names: Stabilizer 7000

Bis-(2-dimethylaminoethyl) ether
Empirical: $C_8H_{20}N_2O$
Formula: $(CH_3)_2N(CH_2)_2O(CH_2)_2N(CH_3)_2$
Uses: Blowing catalyst for flexible PU
Trade names: Dabco® BL-16; Tegoamin® BDE
Trade names containing: Dabco® BL-11; Dabco® BL-17; Dabco® BLV; Dabco® RC; Jeffcat ZF-22; Jeffcat ZF-24; Jeffcat ZF-26; Jeffcat ZF-52; Jeffcat ZF-53; Toyocat®-ET

N,N-Bis (3-dimethylaminopropyl)-N-isopropanolamine
Formula: $[(CH_3)_2NCH_2CH_2CH_2]_2NCH_2CHCH_3OH$
Uses: Catalyst for flexible and rigid PU foam
Trade names: Jeffcat ZR-50

1,2-Bis (dimethylamono) ethane. *See* N,N,N´,N´-Tetramethylethylenediamine

4,4´-Bis (α,α-dimethylbenzyl) diphenylamine
CAS 10081-67-1
Classification: Amine antioxidant
Properties: Wh. to off-wh. powd.; m.w. 446; sp.gr. 1.14; m.p. 98-100 C
Uses: Antioxidant for polyolefins, styrenics, polyether polyols, hot-melt adhesives, lubricant additives, nylons
Trade names: Naugard® 445

Bis(dimethylcarbamodithioato,S,S′) zinc. *See* Zinc dimethyldithiocarbamate
2,5-Bis (1,1-dimethylethyl)-1,4-benzenediol. *See* Di-t-butylhydroquinone

Bis(4-(1,1-dimethylethyl)benzoato-o)hydroxy aluminum
CAS 13170-05-3
Empirical: $C_{22}H_{27}AlO_5$
Properties: Wh. fine free-flowing powd., odorless; insol. in water; sp.gr. 1.14; pH 5-6 (10% aq.)
Precaution: Dusts may be explosive with spark or flame initiation; thermal decomp. may produce oxides of carbon and aluminum
Toxicology: LD50 (oral, rat) > 5000 mg/kg; nonirritating to eyes, mildly irritating to skin; inh. of fine aluminum powd. may cause pulmonary fibrosis, may be implicated in Alzheimer's
Uses: Stabilizer
Trade names: Sandostab 4030

2,6-Bis (1,1-dimethylethyl)-4-methylphenol. *See* BHT
Bis (1,1-dimethylethyl) peroxide. *See* Di-t-butyl peroxide
2,4-Bis (1,1-dimethylethyl)phenyl-phosphite (3:1). *See* Tris (2,4-di-t-butylphenyl) phosphite
N,N′-Bis (1,4-dimethylpentyl)-p-phenylenediamine. *See* N,N′-Di-(1,4-dimethylpentyl)p-phenylenediamine

2-[4,6-Bis(2,4-dimethylphenyl)-1,3,5-triazin-2-yl]-5-(octyloxy)phenol
Empirical: $C_{33}H_{39}N_3O_2$
Properties: Powd.; m.w. 510; m.p. 87-89 C
Uses: Light absorber for plastics
Trade names: Cyasorb® UV 1164

2,5-Bis (1,1-dimethylpropyl)-1,4-benzenediol. *See* Diamylhydroquinone
Bis(dimethylthiocarbamoyl)sulfide. *See* Tetramethylthiuram monosulfide
Bis(dimethylthiocarbamyl) disulfide. *See* Tetramethylthiuram disulfide
Bis(dodecyloxycarbonylethyl) sulfide. *See* Dilauryl thiodipropionate
1,3-Bis(2,3-epoxypropoxy)-2,2-dimethyl propane. *See* Neopentyl glycol diglycidyl ether

Bis (p-ethylbenzylidene) sorbitol
Empirical: $C_{24}H_{30}O_6$
Properties: M.w. 414; m.p. 210-230 C
Uses: Nucleating agent for PP
Trade names: NC-4

Bis(ethylhexanoate) triethylene glycol. *See* Triethylene glycol di-2-ethylhexanoate
Bis (2-ethylhexyl) azelate. *See* Dioctyl azelate
Bis (2-ethylhexyl) decanedioate. *See* Dioctyl sebacate
Bis (2-ethylhexyl) ester, peroxydicarbonic acid. *See* Bis (2-ethylhexyl) peroxydicarbonate
Bis(2-ethylhexyl) hexanedioate. *See* Dioctyl adipate
Bis (2-ethylhexyl) maleate. *See* Dioctyl maleate

Bis (2-ethylhexyl) peroxydicarbonate
CAS 16111-62-9
Synonyms: Dioctyl peroxydicarbonate; Di(2-ethylhexyl)peroxydicarbonate; Bis (2-ethylhexyl) ester, peroxydicarbonic acid
Classification: Organic peroxide
Empirical: $C_{18}H_{34}O_6$
Properties: M.w. 346.4
Uses: Initiator for PVC polymerization
Manuf./Distrib.: Akzo Nobel; Elf Atochem N. Am.; Witco/PAG
Trade names: Aztec® EHPC-40-EAQ; Aztec® EHPC-40-ENF; Aztec® EHPC-65-AL; Aztec® EHPC-75-AL; Espercarb® 840; Lupersol 223; Trigonox® EHP; Trigonox® EHP-B75; Trigonox® EHPS; Trigonox® EHP-W40; Trigonox® EHP-W40S; Trigonox® EHP-W50
Trade names containing: Espercarb® 840M; Espercarb® 840M-40; Espercarb® 840M-70; Lupersol 223-M40; Lupersol 223-M75; Trigonox® EHP-AT70; Trigonox® EHP-C40; Trigonox® EHP-C70; Trigonox® EHP-C75; Trigonox® EHPS-C75

m-Bis(glycidyloxy) benzene. *See* Resorcinol diglycidyl ether

Bis-hexamethylenetriamine
CAS 68411-90-5
Synonyms: BHMT
Classification: Amine
Empirical: $C_{12}H_{29}N_3$
Formula: $NH_2(CH_2)_6NH(CH_2)_6NH_2$
Uses: Asphalt antistripping agents, cationic emulsifiers, ore flotation collectors, chelating agent, corrosion/ scale inhibitor, curing agent for epoxy resins and urethanes, flocculating agents, wet str. paper resins
Manuf./Distrib.: BASF; Monsanto
Trade names: BHMT Amine

Bis (2-hydroxy-3-t-butyl-5-ethylphenyl) methane. *See* 2,2′-Methylenebis (4-ethyl-6-t-butylphenol)

Bis (hydroxyethyl) aminopropyltriethoxy silane
CAS 7538-44-5
Uses: Coupling agent, release agent, lubricant, blocking agent, chemical intermediate
Trade names containing: CB2408

N,N-Bis(2-hydroxyethyl)aniline. *See* Phenyldiethanolamine
Bis (2-hydroxyethyl) cocamine. *See* PEG-2 cocamine
N,N (Bis (2-hydroxyethyl) cocamine oxide. *See* Dihydroxyethyl cocamine oxide
N,N-Bis (2-hydroxyethyl) coco amides. *See* Cocamide DEA
Bis (2-hydroxyethyl) coco amine. *See* PEG-2 cocamine
1,3-Bis(2-hydroxyethyl)-5,5-dimethyl-2,4-imidazolidinedione. *See* DEDM hydantoin
N,N-Bis(2-hydroxyethyl)dodecanamide. *See* Lauramide DEA

N,N-Bis (2-hydroxyethyl)-N-(3-dodecyloxy-2-hydroxypropyl) methyl ammonium methosulfate
CAS 18602-17-0
Empirical: $C_{21}H_{47}O_8NS$
Properties: M.w. 473.6
Uses: Antistat for PVC phonograph records, specialty packaging prods.
Trade names containing: Cyastat® 609

N,N-Bis (β-hydroxyethyl)-3-methylaniline. *See* m-Tolyl diethanolamine

Bis (2-hydroxyethyl) octyl methyl ammonium p-toluene sulfonate
CAS 58767-50-3
Synonyms: 1-Octanaminium, N,N-bis(2-hydroxyethyl)-N-methyl-, salt with 4-methylbenzenesulfonic acid (1:1)
Uses: Antistat for plastics
Trade names: Chemstat® 106G/60DC; Chemstat® 106G/90

Bis-2-hydroxyethyl oleamine. *See* PEG-2 oleamine
Bis (2-hydroxyethyl) soya amine. *See* PEG-2 soyamine
Bis-2-hydroxyethyl stearamine. *See* PEG-2 stearamine
N,N-Bis (2-hydroxyethyl) stearyl amine. *See* PEG-2 stearamine
Bis (2-hydroxyethyl) tallow amine. *See* PEG-2 tallowamine
Bis(hydroxymethanesulfinato-O,O′) zinc. *See* Zinc formaldehyde sulfoxylate
Bis(2-hydroxy-4-methoxyphenyl) methanone. *See* Benzophenone-6

Bis[2-hydroxy-5-methyl-3-(benzotriazol-2-yl)phenyl]methane
CAS 30653-05-5
Empirical: $C_{27}H_{22}N_6O_2$
Properties: Powd.; m.w. 462; m.p. 275 C
Uses: UV light absorber for engineering resins, ABS, cellulosics, epoxy, polyester, PS, flexible/rigid PVC, VDC
Trade names: Mixxim® BB/200

1,4-Bis(hydroxymethyl)cyclohexane (cis and trans). *See* 1,4-Cyclohexanedimethanol
2,2-Bis(hydroxymethyl)-1,3-propanediol. *See* Pentaerythritol
2,2-Bis(hydroxymethyl)propionic acid. *See* Dimethylolpropionic acid

Bis [2-hydroxy-5-t-octyl-3-(benzotriazol-2-yl) phenyl] methane
CAS 103597-45-1

2,2-Bis (4-hydroxyphenol) propane

Synonyms: 2,2´-Methylenebis[6-(2H-benzotriazol-2-yl)-4-(1,1,3,3-tetramethylbutyl)phenol
Empirical: $C_{41}H_{50}N_6O_2$
Properties: Powd.; m.w. 658; m.p. 196 C
Uses: UV light absorber for epoxy, polyesters, PS, flexible/rigid PVC
Trade names: Mixxim® BB/100

2,2-Bis (4-hydroxyphenol) propane. *See* Bisphenol A
Bis[1-hydroxy-2(1H)-pyridinethinato-O,S]-(T-4) zinc. *See* Zinc pyrithione
1,3-Bismaleimidobenzene. *See* m-Phenylenedimaleimide
1,2-Bis(methacryloyoxy) ethane. *See* Ethylene glycol dimethacrylate
Bis (2-methoxyethyl) ether. *See* Diethylene glycol dimethyl ether

Bis-(N-methylbenzamide) ethoxymethyl silane
CAS 16230-35-6
Uses: Coupling agent, release agent, lubricant, blocking agent, chemical intermediate
Trade names: Dynasylan® 0116

Bis(methylethyl)-1,1´-biphenyl
CAS 69009-90-1
Empirical: $C_{18}H_{22}$
Uses: Solvent
Trade names: Nusolv™ ABP-103

Bis(1-methylethyl) decanedioate. *See* Diisopropyl sebacate
Bis(1-methylethyl)hexanedioate. *See* Diisopropyl adipate
Bis(2-methylpropyl) hexanedioate. *See* Diisobutyl adipate
1,4-Bis(2-methylpropyl) sulfobutanedioate, sodium salt. *See* Diisobutyl sodium sulfosuccinate
N,N´-Bismorpholine disulfide. *See* 4,4´-Dithiodimorpholine
Bismuth chloride oxide. *See* Bismuth oxychloride

Bismuth dimethyl dithiocarbamate
CAS 21260-46-8
Synonyms: Tris (dimethyldithiocarbamate) bismuth
Formula: $Bi[(CH_3)_2NC(S)S]_3$
Properties: Lemon yel. powd.; sol. in chloroform; sl. sol. in benzene, carbon disulfide; insol. in water; m.w. 569.64; dens. 2.04; m.p. > 230 C
Toxicology: Suspected tumorigen
Uses: Accelerator for natural rubber and SBR
Trade names: Bismate® Powd.; Bismate® Rodform®; Bismet

Bismuth oxychloride
CAS 7787-59-9; EINECS 232-122-7
Synonyms: Bismuth chloride oxide; Basic bismuth chloride; Bismuth subchloride
Classification: Inorganic pigment
Formula: BiOCl
Properties: White crystalline powd.; sol. in acids; insol. in water; m.w. 260.48; dens. 7.717
Uses: Cosmetics, pigment, dry cell cathodes, artificial pearls
Regulatory: FDA 21CFR §73.1162, 73.2162
Manuf./Distrib.: Atomergic Chemetals; Great Western; ISP Van Dyk; Mallinckrodt; Mearl; Spectrum Chem. Mfg.
Trade names containing: Mearlite® Ultra Bright UDQ; Mearlite® Ultra Bright UMS; Mearlite® Ultra Bright UTL

Bismuth subchloride. *See* Bismuth oxychloride

Bis-(1-octyloxy-2,2,6,6,tetramethyl-4-piperidinyl) sebacate
CAS 129757-67-1
Uses: Light stabilizer for PP
Trade names: Tinuvin® 123

2,4-Bis [(octylthio) methyl]-o-cresol
CAS 110553-27-0
Synonyms: 2-Methyl-4,6-bis [(octylthio) methyl] phenol

Uses: Antioxidant for polymers
Trade names: Irganox® 1520

4-[[4,6-Bis(octylthio)-s-triazin-2-yl]amino]-2,6-di-t-butylphenol
CAS 991-84-4
Uses: Antioxidant/stabilizer for polymers, adhesives, rubber
Trade names: Irganox® 565

Bis (1-oxododecyl) peroxide. *See* Lauroyl peroxide

Bis (pentabromophenoxy) ethane
Uses: Flame retardant for ABS, HIPS, styrene-maleic anhydride, and polyolefins; synergist with antimony oxide
Trade names: Pyro-Chek® 77B

Bis(pentamethylenethiuram) tetrasulfide. *See* Dipentamethylene thiuram tetrasulfide

Bis (1,2,2,6,6-Pentamethyl-4-piperidinyl) (3,5-di-t-butyl-4-hydroxybenzyl) butyl propanedioate
CAS 63843-89-0
Uses: UV stabilizer used in coatings systems
Trade names: Tinuvin® 144

Bis(1,2,2,6,6-pentamethyl-4-piperidinyl)sebacate
CAS 41556-26-7
Uses: Light stabilizer for PVC
Trade names: Tinuvin® 765

Bisperoxide
Uses: Vulcanizing agent and polymerization catalyst for elastomers and plastics
Trade names: Vul-Cup R
Trade names containing: Vul-Cup 40KE

Bisphenol A
CAS 80-05-7; EINECS 201-245-8
Synonyms: 4,4|-Isopropylidenediphenol (CTFA); 2,2-Bis (4-hydroxyphenol) propane; 4,4|-(1-Methylethylidene) bisphenol
Empirical: $C_{15}H_{16}O_2$
Formula: $(CH_3)_2C(C_6H_4OH)_2$
Properties: White flakes, phenolic odor; insol. in water; sol. in alcohol; sl. sol. in CCl_4; m.w. 228.31; dens. 1.195 (25/25 C); m.p. 153 C; b.p. 220 C (4 mm); flash pt. 79.4 C
Toxicology: Poison by intraperitoneal route; moderate toxicity by ingestion and skin contact
Uses: Intermediate in mfg. of epoxy, polycarbonate, phenoxy, polysulfone, polyester resins, flame retardant, rubber chemicals, fungicide
Manuf./Distrib.: Aristech; ICC Ind.; Mitsui Petrochem. Ind.; Mitsui Toatsu; Shell
Trade names: Parabis

Bisphenol A dicyanate monomer/prepolymer
Uses: Co-reacts with and cures epoxy resins; for pultrusion, filament winding applics., for alloying with thermoplastic tougheners, in laminates, powd. coatings, molding powders, printed wiring boards, fiber-reinforcing prepregs, adhesives
Trade names: AroCy® B-10

Bisphenol A, hydrogenated
Empirical: $C_{16}H_{28}O_2$
Properties: M.w. 236.30; m.p. 150 C
Uses: In prep. of alkyd, polyester, and epoxy resins
Manuf./Distrib.: Allchem Ind.
Trade names: Millad® HBPA

Bis(phenoxarsin-10-yl)ether. *See* 10,10′-Oxybisphenoxyarsine
Bis(10-phenoxyarsinyl)oxide. *See* 10,10′-Oxybisphenoxyarsine
1,4-Bis(phenylamino)benzene. *See* N,N′-Diphenyl-p-phenylenediamine
Bis-o-(phenylmethylene)-D-glucitol. *See* Dibenzylidene sorbitol

N,N-Bis-stearoylethylenediamide. See Ethylene distearamide

Bis (1,2,3,6-tetrahydrobenzaldehyde) pentaerythritol acetal
Uses: Antiozonant for rubber
Trade names: Vulkazon® AFS/LG
Trade names containing: Vulkazon® AFS-50

Bis (2,2,6,6-tetramethyl-4-piperidinyl) decanedioate. *See* Bis (2,2,6,6-tetramethyl-4-piperidinyl) sebacate
Bis (2,2,6,6-tetramethylpiperidinyl) ester of cyclospiroketal. *See* 1,5-Dioxaspiro[5.5]undecane 3,3-dicarboxylic acid, bis(2,2,6,6-tetramethyl-4-piperidinyl)ester

Bis (2,2,6,6-tetramethyl-4-piperidinyl) sebacate
CAS 52829-07-0; EINECS 258-207-9
Synonyms: Decanedioic acid, bis (2,2,6,6-tetramethyl-4-piperidinyl) ester; Bis (2,2,6,6-tetramethyl-4-piperidyl) sebacate; Bis (2,2,6,6-tetramethyl-4-piperidinyl) decanedioate
Properties: White to ylsh. powd.; sol. (g/100 g): 108 g in methanol, 98 g in trichloromethane, 46 g in ethyl acetate, 20 g in acetone; m.w. 481; m.p. 80-85 C
Uses: Light stabilizer for polyolefins, PS, HIPS, ABS, SAN, ASA
Trade names: BLS™ 1770; Lowilite® 77; Tinuvin® 770
Trade names containing: Aquastab PC 45M

Bis (2,2,6,6-tetramethyl-4-piperidyl) sebacate. *See* Bis (2,2,6,6-tetramethyl-4-piperidinyl) sebacate

Bis (tribromophenoxy) ethane
Properties: White crystalline powd.
Uses: Flame retardant for many thermoplastic and thermoset systems
Manuf./Distrib.: Great Lakes
Trade names: Great Lakes FF-680™

Bis(tributyltin) oxide. *See* Tributyltin oxide
Bis (tri-n-butyltin) oxide. *See* Tributyltin oxide
Bis(tridecyl) phosphite. *See* Ditridecyl phosphite

Bis-(3-(triethoxysilyl)propyl)tetrasulfane
CAS 40372-72-3; EINECS 254-896-5
Synonyms: Bis (3-triethoxysilylpropyl) tetrasulfide
Empirical: $C_{18}H_{42}S_4Si_2O_6$
Properties: M.w. 537.4; m.p. 250 C
Uses: Reinforcing filler for sulfur-cured systems; crosslinking/coupling agent for rubber compds.
Trade names: Si 69
Trade names containing: Aktisil® PF 216; Coupsil VP 6109; Coupsil VP 8113; X50-S®

Bis (3-triethoxysilylpropyl) tetrasulfide. *See* Bis-(3-(triethoxysilyl)propyl)tetrasulfane

Bis (3,5,5-trimethylhexanoyl) peroxide
CAS 3851-87-4
Synonyms: Di (3,5,5-trimethylhexanoyl) peroxide
Classification: Organic peroxide
Uses: Initiator for LDPE polymerization, styrenics, acrylates, PVC polymerization
Trade names: Aztec® INP-37-AL; Aztec® INP-75-AL; Trigonox® 36-W40
Trade names containing: Trigonox® 36-C60; Trigonox® 36-C75

Bis (trimethylsilyl) acetamide
CAS 10416-59-8; EINECS 233-892-7
Synonyms: BSA; N,O-Bis(trimethylsilyl)acetamide
Empirical: $C_8H_{21}NOSi_2$
Formula: $CH_3C[=NSi(CH_3)_3]OSi(CH_3)_3$
Properties: Liq.; m.w. 203.43; dens. 0.832; b.p. 71-73 (35 mm); ref. index 1.418; flash pt. 11 C)
Toxicology: Moderately toxic by intraperitoneal route
Uses: Silylation agent in prep. of antibiotics and penicillins; coupling agent, blocking agent, release agent, lubricant
Manuf./Distrib.: Fluka; Howard Hall; Hüls AG; Hüls Am.
Trade names: Dynasylan® BSA; Petrarch® B2500

N,O-Bis(trimethylsilyl)acetamide. *See* Bis (trimethylsilyl) acetamide
Bis (trimethylsilyl) amine. *See* Hexamethyldisilazane
BKF. *See* 2,2′-Methylenebis (6-t-butyl-4-methylphenol)
Black lead. *See* Graphite
Blanc fixe (artificial, precipitated). *See* Barium sulfate
BLO. *See* Butyrolactone
Bolus alba. *See* Kaolin
BOP. *See* Butyl octyl phthalate
Borax. *See* Sodium borate decahydrate
Boric acid disodium salt. *See* Sodium borate decahydrate
Boric acid, zinc salt. *See* Zinc borate
Boron fluoride. *See* Boron trifluoride

Boron trifluoride
CAS 7637-07-2
Synonyms: Boron fluoride
Empirical: BF_3
Properties: M.w. 67.81; m.p. -127 C; b.p. -100 C
Precaution: Nonflamm. gas
Toxicology: Highly toxic
Uses: Catalyst in organic synthesis, production of diborane, instruments for measuring neutron intensity, soldering fluxes, gas brazing
Manuf./Distrib.: Air Prods.; Akzo Nobel; Aldrich; AlliedSignal; Atomergic Chemetals; Eagle-Picher; Elf Atochem N. Am.
Trade names: Leecure B-110; Leecure B-550; Leecure B-610; Leecure B-612; Leecure B-614; Leecure B-950; Leecure B-1310; Leecure B-1550; Leecure B-1600; Leecure B-1700

Boron trifluoride-amine complex
Uses: Epoxy curing agent
Trade names: Anchor® 1040; Anchor® 1115; Anchor® 1170; Anchor® 1171; Anchor® 1222

Borosilicate glass
Definition: A soda-lime glass contg. approx. 5% boric oxide; heat-resistant glass
Uses: Extender/filler for plastics, etc.
Trade names: Q-Cel® 636; Q-Cel® 640; Q-Cel® 650; Q-Cel® 2106; Q-Cel® 2116; Q-Cel® 2135; Q-Cel® 6717; Q-Cel® 6832; Q-Cel® 6835; Q-Cel® 6920; Sphericel® 110P8; Spheriglass® 3000E; Spheriglass® 4000E; Spheriglass® 5000E

BPIC. *See* t-Butyl peroxy isopropyl carbonate
BR. *See* Polybutadiene
Brassica campestris oil. *See* Rapeseed oil
Brazil wax. *See* Carnauba
Brimstone. *See* Sulfur

Bromoacenaphthylene
CAS 208-96-8
Synonyms: Cyclopenta(de)naphthalene; Acenaphthylene
Empirical: $C_{12}H_8$
Properties: m.w 152.20; m.p. 90-92 C; b.p. 280 C
Toxicology: Mutagenic data; irritant
Uses: Flame retardant imparting radiation resistance to rubbers and resins; used for insulated wires and cables (EPDM, CSM, crosslinked PE), protectors from x-ray, glove boxes, etc.
Manuf./Distrib.: Aldrich

Bromoethylene. *See* Vinyl bromide
Bromol. *See* 2,4,6-Tribromophenol

Bromopentaerythritol
Trade names containing: Saytex® FR-1138

Bromphthal. *See* Tetrabromophthalic anhydride
B/S. *See* Butadiene-styrene copolymer
BSA. *See* Bis (trimethylsilyl) acetamide

BSH. *See* Benzene sulfonyl hydrazide
BTDA. *See* 3,3′,4,4′-Benzophenone tetracarboxylic dianhydride
BTHC. *See* n-Butyryl tri-n-hexyl citrate

Burgess clay
Trade names containing: Luperco 802-40KE

Burnt lime. *See* Calcium oxide

1,3-Butadiene
CAS 106-99-0; EINECS 203-450-8
Formula: $H_2CCHCHCH_2$
Properties: Colorless gas @ R.T.; colorless liq. at lower temps. or higher pressures; sol. in ether, benzene, CCl_4, chloroform; sl. sol. in water, methanol, ethanol; m.w. 54.09; sp.gr. 0.627 (60 F); f.p. -164 F (760 mm); b.p. 24 F (760 mm)
Precaution: Flamm. gas
Toxicology: Suspected carcinogen
Uses: Monomer for SBR, polybutadiene rubber, general purpose rubber, nylon 66, ABS plastics; reactive chem. intermediate
Manuf./Distrib.: Aldrich; BP Chem.; Fluka; Miles; Phillips; Texaco; Texas Petrochem.

Butadiene/acrylonitrile copolymer
CAS 9003-18-3
Synonyms: 1,3-Butadiene, polymer with 2-propenenitrile; 2-Propenenitrile, polymer with 1,3-butadiene
Empirical: $(C_4H_6 \cdot C_3H_3N)_x$
Formula: $—[CH_2CH{=}CHCH_2]_x—[CHCNCH_2]_y$
Properties: Dens. 0.980
Toxicology: Suspected carcinogen
Uses: Elastomer for low temp. oil well specialties, belt covers, idler rolls, o-rings, seals, gaskets, hydraulic hose, footwear, packings, fuel hose, printer's blankets and roll covers, etc.; in plasticizer masterbatches, as impact modifier for thermoplastics, as visc. modifier
Regulatory: FDA 21CFR §175.125, 175.300, 175.320, 177.1200, 177.2600
Manuf./Distrib.: Aldrich
Trade names: Krynac® 34.140; Krynac® PXL 38.20; Krynac® XL 31.25; Nipol® 1000X88; Nipol® 1001 CG; Nipol® 1022X59; Nipol® 1042X82; Nipol® 1053; Nipol® 1312; Nipol® 1312 LV
Trade names containing: Krynac® PXL 34.17; Nipol® 1411; Nipol® 1422; Nipol® 1422X14; Nipol® 1452X8; Nipol® 1453 HM

Butadiene/acrylonitrile copolymer carboxylated
Trade names containing: Nipol® 1472

1,3-Butadiene, polymer with 2-propenenitrile. *See* Butadiene/acrylonitrile copolymer
Butadiene rubber. *See* Polybutadiene, Polybutadiene rubber

Butadiene-styrene copolymer
CAS 9003-55-8
Synonyms: B/S; Styrene polymer with 1,3-butadiene; Butadiene-styrene resin; Butadiene-styrene rubber
Toxicology: Eye irritant; heated to decomp., emits acrid smoke and irritating fumes
Uses: Mechanical rubber goods, flooring tile, tire tread, rug backing, adhesives, asphalt modification, inj. molded items, medical devices, containers, toys, food containers; blendable modifier, processing and extrusion aid; masticatory substance in chewing gum
Regulatory: FDA 21CFR §172.615, 181.30
Manuf./Distrib.: BASF; Firestone Syn. Rubber & Latex; Goodyear Tire & Rubber; Reichhold; Shell
Trade names: Duradene® 710; SR-7475

1,3-Butadiene-styrene copolymer. *See* Styrene-butadiene polymer
Butadiene-styrene resin. *See* Butadiene-styrene copolymer, Styrene-butadiene polymer
Butadiene-styrene rubber. *See* Butadiene-styrene copolymer
1,4-Butanedicarboxylic acid. *See* Adipic acid
Butanedioic acid, sulfo-, 1,4-dihexyl ester, sodium salt. *See* Dihexyl sodium sulfosuccinate
Butanediol. *See* 1,4-Butanediol

1,4-Butanediol
CAS 110-63-4; EINECS 203-786-5
Synonyms: BDO; 1,4-Butylene glycol; Tetramethylene glycol; Butanediol
Formula: $HO(CH_2)_4OH$
Properties: M.w. 90.12; dens. 1.017; m.p. 16 C; b.p. 230, 120-122 C (10 mm); flash pt. > 110 C
Uses: Intermediate; used in polyurethane formulation in the hard segment as a curative; plasticizer
Manuf./Distrib.: Allchem Ind.; Arco; Ashland; BASF; DuPont; Hüls UK; ISP
Trade names: BDO; Dabco® BDO
Trade names containing: Dabco® S-25; Synprolam 35TMQS; TEDA-L25B

1,4-Butanediol diacrylate
Uses: Crosslinking agent used in inks, adhesives, textile prod. modifiers, photoresists, modifiers for castings, polyesters, fiberglass, or radiation cured prods.
Manuf./Distrib.: Monomer-Polymer & Dajac
Trade names containing: SR-213

1,4-Butanediol diglycidyl ether
CAS 2425-79-8; EINECS 219-371-7
Synonyms: Diglycidyl ether of 1,4-butanediol
Empirical: $C_{10}H_{18}O_4$
Properties: M.w. 202.25; dens. 1.049 (20/4 C); b.p. 155-160 (11 mm)
Uses: Reactive epoxy diluent for laminates, tooling, potting, elec. and civil engineering applics.
Manuf./Distrib.: Fluka
Trade names: Epodil® 750; Heloxy® 67; RD-2

1,3-Butanediol dimethacrylate
Uses: Peroxide crosslinker for elastomers, PVC plastisols, syn. resins
Trade names: MFM-407

1,4-Butanediol dimethacrylate
CAS 2082-81-7
Formula: $[H_2C=C(CH_3)CO_2CH_2CH_2]_2$
Properties: M.w. 226.28; dens. 1.023; b.p. 132-134 C (4 mm); flash pt. > 110 C; ref. index 1.4560 (20 C)
Toxicology: Irritant
Uses: Crosslinking activator for peroxide vulcanization of tech. molded and extruded goods based on EPDM, EPM, NBR, CM, etc.
Manuf./Distrib.: Aldrich
Trade names: MFM-405
Trade names containing: Rhenofit® BDMA/S; SR-214

Butanoic acid, 3-amino-, N-coco alkyl derivatives. *See* Cocaminobutyric acid
1-Butanol. *See* Butyl alcohol

2-Butanol
CAS 15892-23-6; EINECS 240-029-8
Synonyms: s-Butyl alcohol; Methyl ethyl carbinol; Butylene hydrate; 2-Hydroxybutane
Empirical: $C_4H_{10}O$
Formula: $CH_3CH_2CH(OH)CH_3$
Properties: Liq.; m.w. 74.12; dens. 0.807 (20/4 C); b.p. 99-100 C; flash pt. 23 C; ref. index 1.397 (20 C)
Precaution: Flamm.
Regulatory: FDA 21CFR §172.515
Manuf./Distrib.: Ashland; Chemcentral; Shell
Trade names containing: Ancamide 400-BX-60

4-Butanolide. *See* Butyrolactone
2-Butanone. *See* Methyl ethyl ketone
2-Butanone peroxide. *See* Methyl ethyl ketone peroxide

1-Butene
CAS 106-98-9; EINECS 203-449-2
Synonyms: C-4 alpha olefin; Butene-1
Empirical: C_4H_8

Butene-1

Formula: CH_2:$CHCH_2CH_3$
Properties: Colorless gas; m.w. 56.11
Precaution: Flamm. gas
Toxicology: Handling hazards incl. fire, suffocation, and frostbite
Uses: Chemical intermediate (polyethylene and other polymers; plasticizers; syn. lubricants; gasoline additives; paper sizing; PVC lubricant)
Manuf./Distrib.: Albemarle; BP Chem.; Chevron; Coyne; Fluka; Phillips; Shell; Texas Petrochem.
Trade names: Gulftene® 4

Butene-1. *See* 1-Butene
2-Butenedioic acid bis (2-ethylhexyl) ester. *See* Dioctyl fumarate
2-Butenedioic acid, dibutyl ester. *See* Dibutyl maleate
cis-Butenedioic anhydride. *See* Maleic anhydride
1-Butene, homopolymer. *See* Polybutene

Butoxydiglycol
CAS 112-34-5; EINECS 203-961-6
Synonyms: PEG-2 butyl ether; Diethylene glycol butyl ether; 2-(2-Butoxyethoxy) ethanol
Formula: $C_4H_9OCH_2CH_2OCH_2CH_2OH$
Properties: Colorless liq., faint butyl odor; sol. in oils and water; m.w. 162.26; dens. 0.9553 (20/4 C); m.p. 68.1 C; b.p. 230.6 C; flash pt. 172 F; ref. index 1.4316 (20 C)
Precaution: Combustible; heated to dec., emits acrid smoke and irritating fumes
Toxicology: Moderately toxic by ingestion, intraperitoneal; mildly toxic by skin contact; severe eye irritant
Uses: Solvent for nitrocellulose, oils, dyes, gums, soaps, polymers; plasticizer intermediate
Regulatory: FDA 21CFR §175.105, 176.180
Manuf./Distrib.: Arco
Trade names: Butyl Dioxitol
Trade names containing: Monawet MO-70

1-t-Butoxy-2,3-epoxypropane. *See* t-Butyl glycidyl ether

Butoxyethanol
CAS 111-76-2; EINECS 203-905-0
Synonyms: 2-Butoxyethanol; Ethylene glycol monobutyl ether; Glycol butyl ether; Ethylene glycol butyl ether; Butyl glycol
Classification: Ether alcohol
Empirical: $C_6H_{14}O_2$
Formula: $HOCH_2CH_2OC_4H_9$
Properties: Colorless liq.; mild pleasant odor; m.w. 118.20; dens. 0.9019 (20/20 C); m.p. -74.8 C; b.p. 171.2 C; flash pt. (COC) 160 F
Precaution: Flamm. liq. exposed to heat or flame; incompat. with oxidizers, heat, flame
Toxicology: TLV 25 ppm in air; LD50 (IP, rat) 20 mg/kg; poison by ing., skin contact, IP, IV routes; mod. toxic by inh., subcut.; human systemic effects; experimental reproductive effects; skin/eye irritant; heated to dec., emits acrid smoke, irritating fumes
Uses: Solvent for nitrocellulose resins, spray lacquers, cosolvent, gas chromatography
Regulatory: FDA 21CFR §173.315, 175.105, 176.210, 177.1650, 178.1010, 178.3297, 178.3570
Manuf./Distrib.: Arco
Trade names: Butyl Oxitol; Ektasolve® EB

2-Butoxyethanol. *See* Butoxyethanol
2-(2-Butoxyethoxy) ethanol. *See* Butoxydiglycol

Butoxyethyl oleate
Synonyms: Ethylene glycol monobutyl ether oleate
Uses: Plasticizer for NBR
Trade names: Plasthall® 325

Butoxymethyl oxirane. *See* Butyl glycidyl ether
(t-Butoxymethyl)oxirane. *See* t-Butyl glycidyl ether
Butyl acetate. *See* n-Butyl acetate

n-Butyl acetate
CAS 123-86-4; EINECS 204-658-1

Synonyms: Acetic acid, butyl ester; Butyl acetate
Definition: Ester of butyl alcohol and acetic acid
Empirical: $C_6H_{12}O_2$
Formula: $CH_3COOC_4H_9$
Properties: Colorless liq., fruity odor; sol. in alcohol, ether, hydrocarbons; sl. sol. in water; m.w. 116.18; dens. 0.8826 (20/20 C); b.p. 126.3 C; f.p. -75 C; flash pt. (TOC) 36.6 C); ref. index 1.2951 (20 C)
Precaution: Flamm. liq.; mod. explosive when exposed to flame
Toxicology: LD50 (oral, rat) 14.13 g/kg; mildly toxic by inhalation, ingestion; moderately toxic by intraperitoneal route; skin irritant; severe eye irritant; mild allergen; TLV 150 ppm in air; heated to decomp., emits acrid and irritating fumes
Uses: Manufacture of lacquers, artificial leather, photographic films, plastics, safety glass, flavoring agent
Regulatory: FDA 21CFR §172.515, 175.105, 175.320, 177.1200; GRAS (FEMA); Japan approved as flavoring
Manuf./Distrib.: Allchem Ind.; BASF; BP Chem Ltd; Chisso Am.; Coyne; Eastman; General Chem.; Great Western; Hoechst AG; Hüls AG; Janssen Chimica; Penta Mfg.; Spectrum Chem. Mfg.; Union Carbide
Trade names containing: Bonding Agent 2005; Uniplex 680

Butyl acetoxystearate
Formula: $CH_3(CH_2)_5CH(CH_3COO)(CH_2)_{10}COOC_4H_9$
Properties: Sp.gr. 0.922; m.p. -7 C; flash pt. (COC) 207 C; ref. index 1.4480
Uses: Plasticizer for vinyls, cellulosics; textile oil, adhesives
Manuf./Distrib.: CasChem
Trade names: Paricin® 6

Butyl 12-(acetyloxy)-9-octadecenoate. *See* Butyl acetyl ricinoleate

Butyl acetyl ricinoleate
CAS 140-04-5; EINECS 205-393-4
Synonyms: Butyl 12-(acetyloxy)-9-octadecenoate
Definition: Butyl ester of acetyl ricinoleic acid
Empirical: $C_{24}H_{44}O_4$
Properties: Yel. oily liq., mild odor; misc. with org. solvs.; pract. insol. in water; dens. 0.940 (20/20 C); f.p. cloudy -32 C; solid. pt. -65 C; flash pt. 110 C; ref. index 1.4614 (20 C)
Uses: Plasticizer for vinyls, emulsifier, lubricant, detergent, protective coatings, special cleaning compds., quick-breaking emulsions
Regulatory: FDA 21 CFR §175.105, 177.2600, 177.2800
Manuf./Distrib.: CasChem
Trade names: Flexricin® P-6

Butyl acrylate
CAS 141-32-2; EINECS 205-480-7
Synonyms: Butyl-2-propenoate
Empirical: $C_7H_{12}O_2$
Formula: CH_2:$CHCOOC_4H_9$
Properties: Water-white, extremely reactive monomer; insol. in water; m.w. 128.19; dens. 0.89 (25/25 C); b.p. 69 C (50 mm); f.p. -64.6 C; flash pt. (OC) 120 F
Precaution: Flamm.
Toxicology: TLV:TWA 10 ppm (air); moderately toxic by ingestion, inhalation, skin contact, intraperitoneal route; skin and eye irritant
Uses: Intermediate in organic synthesis, polymers and copolymers for solvent coatings, adhesives, paints, binders, emulsifiers
Manuf./Distrib.: Allchem Ind.; Ashland; BASF; Hoechst-Celanese; ICD Group; Monomer-Polymer & Dajac; Union Carbide
Trade names: Paraloid® EXL-3330; Paraloid® EXL-3361; Paraloid® EXL-3387

Butyl alcohol
CAS 71-36-3; EINECS 200-751-6
Synonyms: n-Butyl alcohol; 1-Butanol; Propyl carbinol
Classification: Aliphatic alcohol
Empirical: $C_4H_{10}O$
Formula: $CH_3(CH_2)_2CH_2OH$
Properties: Colorless liq., vinous odor; sol. in water; misc. with alcohol, ether; m.w. 74.14; dens. 0.8109 (20/20 C); m.p. -90 C; f.p. -89.0 C; b.p. 117.7 C; flash pt. 35 C; ref. index 1.3993 (20 C)
Precaution: Flamm.; mod. explosive exposed to flame; incompat. with Al, oxidizing materials

n-Butyl alcohol

Toxicology: TLV:CL 50 ppm in air; LD50 (oral, rat) 790 mg/kg; poison by IV route; moderately toxic by skin contact, ingestion, subcutaneous, intraperitoneal routes; skin and severe eye irritant; heated to decomp., emits acrid smoke and fumes
Uses: Preparation of esters, solv. for resins, plasticizers; flavoring agent
Regulatory: FDA 21CFR §73.1, 172.515, 172.560, 175.105, 175.320, 176.200, 177.1200, 177.1440, 177.1650; 27CFR §21.99; GRAS (FEMA)
Manuf./Distrib.: Ashland; BASF; BP Chem Ltd; Eastman; Fluka; Hoechst Celanese; Shell; Union Carbide; Vista
Trade names: Isanol®
Trade names containing: Albrite® PA-75; Ancamide 700-B-75; GP-197 Resin Sol'n.; TBHP-70

n-Butyl alcohol. *See* Butyl alcohol
s-Butyl alcohol. *See* 2-Butanol

t-Butyl alcohol
CAS 75-65-0; EINECS 200-889-7
Synonyms: TBA; 2-Methyl-2-propanol; Trimethyl carbinol; 2-Propanol, 2-methyl-
Empirical: $C_4H_{10}O$
Formula: $(CH_3)_3COH$
Properties: Liq.; m.w. 74.12; dens. 0.775; m.p. 25-26 C; b.p. 83 C; flash pt. 11 C; ref. index 1.3870 (20 C)
Precaution: Flamm. liq.
Toxicology: Irritant
Uses: Solvent, coupling agent, processing aid for pharmaceuticals, personal care prods., aq. coatings and adhesives, agric. formulations, polymer processing, cleaners/disinfectants
Regulatory: FDA 21CFR §176.200, 178.3910, 27CFR §21.100
Manuf./Distrib.: AC Ind.; Aldrich; Allchem Ind.; Arco; Ashland; Fluka
Trade names: Tebol™; Tebol™ 99

Butylated d: (dimethylbenzyl) phenol
Uses: Stabilizer for SBR and nitrile polymers
Trade names: Wingstay® C

Butylated hydroxyanisole. *See* BHA
Butylated hydroxytoluene. *See* BHT
Butylated polyvinylpyrrolidone. *See* Butylated PVP

Butylated PVP
Synonyms: Butylated polyvinylpyrrolidone
Definition: Polymer of butylated vinylpyrrolidone
Uses: Moisture barrier, adhesive, protective colloid, and microencapsulating resin in cosmetics; pigment dispersant; as solubilizer for dyes; in petroleum industry as sludge and detergent dispersant; protective colloid in coatings; suspending aid in polymerization; dyeing assistant; antiredeposition agent in drycleaning; esp. as dispersant in aq. agric. chemicals or pigmented skin care prods.
Trade names: Ganex® P-904

Butylated triaryl phosphate
Uses: Plasticizer for NBR elastomers, PVC, PVAc, cellulosics, PS, ABS, engineering resins, polyesters, alloys; PVAc emulsions

N,N-Butyl benzene sulfonamide
CAS 3622-84-2; EINECS 222-823-6
Synonyms: Benzenesulfonic acid butyl amide
Empirical: $C_{20}H_{19}NO_2S$
Formula: $C_6H_5SO_2NHC_3H_9$
Properties: Amber to straw liq., pleasant odor; m.w. 337.46; dens. 1.148; b.p. 189-190 C (4.5 mm); ref. index 1.5235
Toxicology: Toxic by ingestion
Uses: Plasticizer for some cellulose compds., emulsion adhesives, pkg., caulk, printing ink, surf. coatings; synthesis of dyes, pharmaceuticals, other organic chemicals; in resin mfg.
Manuf./Distrib.: Allchem Ind.; Hardwicke; Unitex
Trade names: Dellatol® BBS; Plasthall® BSA; Uniplex 214

Butyl 2-benzothiazole sulfenamide
CAS 95-31-8, 3741-08-8
Synonyms: TBBS; N-t-Butyl-2-benzothiazole sulfenamide; Benzothiazyl-2-t-butyl sulfenamide
Formula: $C_6H_4NCS(SNH)C_4H_9$
Properties: Lt. buff powd. or flakes; sol. in most org. solvs.; m.w. 266.33; dens. 1.29; m.p. 104 C
Precaution: Combustible; emits very toxic fumes when heated
Toxicology: Poison by intravenous route
Uses: Rubber accelerator
Trade names: Akrochem® BBTS; Delac® NS; Perkacit® TBBS; Santocure® NS; Santocure® TBSI; Vanax® NS; Vanax® TBSI; Vulkacit NZ/EG
Trade names containing: Poly-Dispersion® E (SAN-NS)D-70

N-t-Butyl-2-benzothiazole sulfenamide. *See* Butyl 2-benzothiazole sulfenamide

Butyl benzyl phthalate
CAS 85-68-7; EINECS 201-622-7
Synonyms: BBP; 1,2-Benzenedicarboxylic acid, butyl phenylmethyl ester; Benzyl butyl phthalate
Classification: Aromatic ester
Formula: $C_4H_9OOCC_6H_4COOC_7H_7$
Properties: Clear oily liq., sl. odor; m.w. 312.39; dens. 1.116 (25/25 C); m.p. < -35 C; b.p. 370 C; flash pt. 390 F; ref. index 1.535-1.540
Toxicology: LD50 (oral, rat) 20,400 mg/kg; moderate toxicity by ingestion, intraperitoneal routes
Uses: Plasticizer for polyvinyl and cellulosic resins, organic intermediate
Regulatory: FDA 21CFR §175.105, 176.170, 176.180, 177.2420, 178.3740
Manuf./Distrib.: Allchem Ind.; Ashland; Miles; Monsanto
Trade names: Santicizer 160; Unimoll® BB
Trade names containing: Aztec® 1,1-BIS-80-BBP; Esperox® 28PD; Esperox® 570P; Lupersol 331-80B; Lupersol 531-80B; Omacide® P-BBP-5; Trigonox® 22-BB80; USP®-400P; Vinyzene® BP-505 S160

n-Butyl-4,4-bis (t-butylperoxy) valerate
CAS 995-33-5
Synonyms: 4,4-Di-t-butylperoxy-n-butyl valerate
Classification: Perketal
Empirical: $C_{17}H_{22}O_6$
Properties: M.w. 334.5
Uses: Initiator for polymerizations, for cure of polyester resins, acrylic syrup, and elastomers; crosslinking agent for olefin copolymers, esp. chlorinated PE, elastomers (EPDM, SBR, Neoprene, Hypalon)
Trade names: Aztec® NBV-40-G; Aztec® NBV-40-IC; Lupersol 230; Trigonox® 17; Trigonox® 17/40; Varox® 230-XL
Trade names containing: Luperco 230-XL; Poly-Dispersion® K (PX-17)D-30; Poly-Dispersion® TVP-50; Trigonox® 17-40B-pd

Butyl carbobutoxymethyl phthalate. *See* n-Butyl phthalyl-n-butyl glycolate
Butyl citrate. *See* Tributyl citrate

2-t-Butyl-p-cresol
CAS 2409-55-4; EINECS 219-314-6
Synonyms: MBPC; Monobutylated paracresol; 2-t-Butyl-4-methylphenol
Empirical: $C_{11}H_{16}O$
Properties: Clear liq.; sol. in org. solvs., aq. potassium hydroxide; m.w. 164.27; dens. 0.922; f.p. 23.1 C; m.p. 50-52 C; b.p. 244 C; flash pt. 47 C
Toxicology: Corrosive; poison by intraperitoneal, intravenous routes; moderately toxic by ingestion and skin contact; severe skin and eye irritant
Uses: Intermediate for antioxidants, uv absorbers, leather, perfume; antioxidant for min. oil, plastics, rubber
Manuf./Distrib.: Aldrich; Allchem Ind.; Biddle Sawyer; Fluka; Great Western
Trade names: Ralox® 240

6-t-Butyl-m-cresol
CAS 88-60-8
Synonyms: MBMC; 2-t-Butyl-5-methylphenol
Formula: $(CH_3)_3CC_6H_3(CH_3)OH$
Properties: Clear liq., solidifies sl. below R.T.; m.w. 164.25

6-t-Butyl-o-cresol

Precaution: Flamm.
Toxicology: Irritant
Uses: Germicide, synthesis of antioxidants and rubber processing chems., additives to lubricating oils, syn. resins
Manuf./Distrib.: Aldrich; PMC

6-t-Butyl-o-cresol
CAS 2219-82-1; EINECS 218-734-7
Synonyms: OBOC; Butyl-6-methylphenol; 2-t-Butyl-6-methylphenol
Formula: $(CH_3)_3CC_6H_3(CH_3)OH$
Properties: M.w. 164.25; dens. 0.964; m.p. 30-32 C; b.p. 230 C; flash pt. 107 C; ref. index 1.5190 (20 C)
Toxicology: Irritant
Manuf./Distrib.: Aldrich; Fluka

t-Butyl cumyl peroxide
CAS 3457-61-2
Classification: Organic peroxide
Properties: M.w. 208.3
Uses: Crosslinking agent for elastomers; initiator for curing polyester resins; finishing initiator for styrenics; a synergist for some halogen-containing flame retardants; polymer modifier
Trade names: Aztec® BCUP; Lupersol 801; Trigonox® T; Trigonox® T-94
Trade names containing: Luperco 801-XL

Butyl cyclohexyl phthalate
CAS 84-64-0; EINECS 201-548-5
Synonyms: 1,2-Benzenedicarboxylic acid, butyl cyclohexyl ester
Classification: Ester
Formula: $C_4H_9OOCC_6H_4COOC_6H_{11}$
Properties: Clear liq., very mild odor; misc. with most org. solvs.; m.w. 304.39; dens. 1.078
Precaution: Combustible
Uses: Plasticizer for PVC, plastisols, automotive undercoatings, plastics, flooring, processing aid for molded ethylcellulose prods.
Manuf./Distrib.: CPS; Hüls; Unitex
Trade names: Nuoplaz® 6938

n-Butyl 4,4-di(t-butylperoxy)valerate. *See* n-Butyl-4,4-bis (t-butylperoxy) valerate
1,4-Butylene glycol. *See* 1,4-Butanediol

1,3-Butylene glycol diacrylate
CAS 19485-03-1
Synonyms: Acrylic acid-1-methyltrimethylene ester; 1,3-Butanediol diacrylate; 1,3-Butylene diacrylate; 2-Propenoic acid-1-methyl-13-propanediyl ester
Empirical $C_{10}H_{14}O_4$
Properties: m.w. 198.24
Uses: Curing agent; polymerizes to hard, insol., infusible, thermoset resin
Toxicity: LD50 (oral, rat) 3540 mg/kg; mod. toxic by ingestion and skin contact; heated to decomp., emits acrid smoke and irritating fumes
Manuf./Distrib.: Monomer-Polymer & Dajac
Trade names containing: SR-212

1,3-Butylene glycol dimethacrylate
CAS 11890-98-8
Properties: M.w. 226.26
Uses: Crosslinker for plastisols, hard rubber rolls, cast acrylic sheet and rods, coagent for rubber compding., impregnant for metal and wood composites, adhesives, and glass-reinforced plastics
Manuf./Distrib.: CPS; Monomer-Polymer & Dajac
Trade names: Ageflex 1,3 BGDMA
Trade names containing: SR-297

Butylene glycol montanate
CAS 73138-44-0
Synonyms: Montan acid, butylene glycol ester
Definition: Ester of butylene glycol and montan acid wax
Trade names containing: Hoechst Wax OP

Butylene hydrate. *See* 2-Butanol
Butyl ethylene. *See* 1-Hexene

OO-t-Butyl 1-(2-ethylhexyl)monoperoxycarbonate
Trade names: Lupersol TBEC

OO-t-Butyl O-(2-ethylhexyl) monoperoxycarbonate
Uses: Initiator for polymerization, curing elastomers, high-temp. cure of polyester resins, and cure of acrylic syrup
Trade names: Lupersol TBEC

Butyl ethylhexyl phthalate. *See* Butyl octyl phthalate

Butyl glycidyl ether
CAS 2426-08-6; EINECS 219-376-4
Synonyms: BGE; Glycidyl butyl ether; Butoxymethyl oxirane
Classification: Glycidyl ether monomer
Empirical: $C_7H_{14}O_2$
Properties: M.w. 130.19; dens. 0.910; b.p. 164-166 C; flash pt. 55 C
Toxicology: TLV:TWA 25 ppm; moderately toxic by ingestion, skin contact, intraperitoneal route; mildly toxic by inhalation
Uses: Reactive diluent in epoxy resins, laminating, flooring, elec. casting and encapsulants
Manuf./Distrib.: Alemark; Amber Syn.; Ashland; CPS; Fluka; Monomer-Polymer & Dajac
Trade names: Ageflex BGE; Epodil® 741; Heloxy® 61; RD-1

t-Butyl glycidyl ether
CAS 7665-72-7; EINECS 231-640-0
Synonyms: TBGE; Oxirane, [1,1-dimethylethoxy)methyl]-; 1-t-Butoxy-2,3-epoxypropane; (t-Butoxymethyl) oxirane
Classification: Glycidyl ether monomer
Empirical: $C_7H_{14}O_2$
Properties: Colorless to ylsh. liq.; sol. in ethanol, org. solvs.; partly sol. in water; m.w. 130.19; dens. 0.917; m.p. -70 C; b.p. 152 C; flash pt. 44 C; ref. index 1.4170 (20 C)
Precaution: Flamm.; may form explosive peroxides; incompat. with strong acids, bases, oxidizing agents, amines; hazardous decomp. prods.: CO, hydrocarbons
Toxicology: LD50 (oral, rat) 2000 mg/kg; harmful by inh., ingestion, skin contact; irritating to eyes, respiratory system, skin; may cause sensitization by inh. and skin contct
Uses: Reactive diluent in epoxy resins, corrosion inhibitor in some solvs., modifier for amines, acids, and thiols; heat stabilizer for PVC; stabilizer for chlorinated hydrocarbons; polymerized to improve the antistatic props. of polyester fibers
Manuf./Distrib.: Aldrich; Fluka; Raschig
Trade names: Ageflex TBGE

Butyl glycol. *See* Butoxyethanol
n-Butyl glycol phthalate. *See* Dibutoxyethyl phthalate

t-Butylhydrazinium chloride
Uses: Chemical blowing agent for unsat. polyester resins
Trade names containing: Luperfoam 40; Luperfoam 329

t-Butyl hydroperoxide
CAS 75-91-2; EINECS 200-915-7
Synonyms: TBHP; 1,1-Dimethylethylhydroperoxide
Classification: Organic peroxide
Empirical: $C_4H_{10}O_2$
Formula: $(CH_3)_3COOH$
Properties: Water-white liq.; sl. sol. in water; sol. in org. solvs.; very sol. in esters and alcohols; m.w. 90.12; dens. 0.896 (20/4 C); m.p. -8 C; b.p. 35 C; flash pt. 62 C; ref. index 1.4007
Precaution: Flamm. oxidizing liq.
Toxicology: Poison by ingestion and inhalation; severe skin and eye irritant; LD50 (rat, oral) 560 mg/kg; LD50 (rat, dermal) 790 mg/kg
Uses: Catalyst in polymerization reactions; oxidative membrane in RBC suspensions; introduces peroxy group into organic molecules

t-Butyl hydroquinone

Manuf./Distrib.: Akzo Nobel; Aldrich; Allchem Ind.; Arco; Elf Atochem N. Am.; SAF; Witco/PAG
Trade names: Aztec® TBHP-22-OL1; Aztec® TBHP-70; T-Hydro®; Trigonox® A-80; Trigonox® A-W70; USP®-800
Trade names containing: TBHP-70

t-Butyl hydroquinone
CAS 1948-33-0; EINECS 217-752-2
Synonyms: TBHQ; Mono-t-butyl hydroquinone; 2-(1,1-Dimethylethyl)-1,4-benzenediol
Classification: Aromatic organic compd.
Empirical: $C_{10}H_{14}O_2$
Properties: Wh. to lt. tan cryst. solid; sol. in ethyl alcohol, ethyl acetate, acetone, ether; sl. sol. in water; m.w. 166.24; m.p. 126.5-128.5 C; flash pt. (COC) 171 C
Toxicology: LD50 (oral, rat) 700 mg/kg; poison by intraperitoneal route; moderately toxic by ingestion; irritant; heated to decomp., emits acrid smoke and irritating fumes
Uses: Polymerization inhibitor, industrial and food-grade antioxidant; food additive
Regulatory: FDA 21CFR 172.185 (limitation 0.02% of oil), 177.2420; USDA 9CFR §318.7 (limitation 0.003% in dry sausage, 0.006% with BHA/BHT, 0.01% in rendered animal fat, 0.02% with BHA and/or BHT, 0.02% in margarine), 381.147 (limitation 0.01% on fat in poultry)
Manuf./Distrib.: AC Ind.; Aceto; Allchem Ind.; Charkit; Eastman; Penta Mfg.; Schweizerhall; Showa Denko; UOP
Trade names: Eastman® MTBHQ; Sustane® TBHQ; Tenox® TBHQ

3-t-Butyl-4-hydroxyanisole. *See* BHA
2-(3´-t-Butyl-2´-hydroxy-5´-methylphenyl)-5-chlorobenzotriazole. *See* 2-(2´-Hydroxy-3´-t-butyl-5´-methylphenyl)-5-chlorobenzotriazole

4,4´-Butylidenebis (6-t-butyl-m-cresol)
CAS 85-60-9
Synonyms: 4,4´-Butylidenebis (3-methyl-6-t-butylphenol)
Formula: $C_{26}H_{38}O_2$
Properties: M.w. 382.64
Toxicology: Mildly toxic by ingestion
Uses: Antioxidant for NR, synthetics, latexes; polymer stabilizer
Trade names: Santowhite® Powd.; Vanox® SWP

4,4´-Butylidene bis (2-t-butyl-5-methylphenol)
Uses: Antioxidant for rubber, latex, adhesives, plastics
Trade names: Lowinox® 44B25; Naftonox® BBM

4,4´-Butylidenebis (3-methyl-6-t-butylphenol). *See* 4,4´-Butylidenebis (6-t-butyl-m-cresol)
Butyl-3-iodo-2-propynylcarbamate. *See* Iodopropynyl butylcarbamate

t-Butyl 3-isopropenylcumyl peroxide
CAS 96319-55-0
Properties: M.w. 248.4
Uses: Initiator for acrylates
Trade names: Trigonox® 171

OO-t-Butyl O-isopropyl monoperoxycarbonate
Classification: Peroxyester
Uses: Initiator for vinyl polymerizations
Trade names containing: Lupersol TBIC-M75

Butyl maleate
Synonyms: Monobutyl maleate
Properties: Sp.gr. 1.090-1.105; ref. index 1.4540-1.4561
Uses: Plasticizer
Manuf./Distrib.: Unitex

Butylmercaptide
Uses: Stabilizer for PVC
Trade names: Stanclere® T-161, T-163, T-164, T-165, T-174, T-182, T-183; Stanclere® T-192, T-193, T-194

t-Butyl methyl ether. *See* Methyl t-butyl ether
2-t-Butyl-4-methylphenol. *See* 2-t-Butyl-p-cresol
2-t-Butyl-5-methylphenol. *See* 6-t-Butyl-m-cresol
2-t-Butyl-6-methylphenol. *See* 6-t-Butyl-o-cresol
Butyl-6-methylphenol. *See* 6-t-Butyl-o-cresol
t-Butyl monoperoxy maleate. *See* t-Butyl peroxymaleic acid
Butyl octadecanoate. *See* Butyl stearate
n-Butyl octadecanoate. *See* Butyl stearate
Butyl 9-octadecenoate. *See* Butyl oleate

Butyl octyl phthalate
EINECS 201-562-1
Synonyms: BOP; 1,2-Benzenedicarboxylic acid, butyl octyl ester; Butyl ethylhexyl phthalate
Properties: Liq.; dens. 0.988-0.996 (20/20 C); f.p. -48 C; b.p. 350 C (760 mm); flash pt. (COC) 208 C; ref. index 1.4868
Precaution: Combustible
Uses: Plasticizer for PVC, PS, PVB, PVAc, cellulosics
Manuf./Distrib.: Aristech; Ashland; BASF; BP Oil; Eastman; Harwick; Unitex
Trade names: Kodaflex® HS-3; PX-914; Staflex BOP

Butyl oleate
CAS 142-77-8; EINECS 205-559-6
Synonyms: Butyl 9-octadecenoate
Definition: Ester of butyl alcohol and oleic acid
Empirical: $C_{22}H_{42}O_2$
Formula: $CH_3(CH_2)_7CH:CH(CH_2)_7COOC_4H_9$
Properties: M.w. 338; dens. 0.873; flash pt. 204 C
Uses: Plasticizer for PVC, PVB, PVAc, PS, cellulosics
Regulatory: FDA 21CFR §176.210, 177.2600, 177.2800
Manuf./Distrib.: Ferro/Keil; C.P. Hall; Harwick; Inolex; Sea-Land; Sybron; Union Camp; Velsicol; Witco/Oleo-Surf.
Trade names: Butyl Oleate C-914; Kemester® 4000; Plasthall® 503; Plasthall® 914; Priolube 1405; Uniflex® BYO

t-Butyl peracetate
CAS 107-71-1
Synonyms: t-Butyl peroxyacetate
Classification: Organic peroxide
Empirical: $C_6H_{12}O_3$
Formula: $(CH_3)_3COOCOCH_3$
Properties: Clear colorless; sol. in org. solvs., insol. in water; m.w. 132.18; dens. 0.923; flash pt. (COC) < 80 F
Precaution: Combustible oxidizing liq.; keep away from heat, flames, sparks
Toxicology: Moderately toxic by ingestion; mildly toxic by inhalation
Uses: Initiator
Manuf./Distrib.: Akzo Nobel; Aztec Peroxides; Elf Atochem N. Am.; Witco/PAG
Trade names containing: Aztec® TBPA-50-OMS; Aztec® TBPA-75-OMS; Esperox® 12MD; Lupersol 70; Lupersol 75-M; Lupersol 76-M; Trigonox® F-C50

t-Butyl perbenzoate
CAS 614-45-9
Synonyms: t-Butyl peroxybenzoate
Classification: Organic peroxide
Empirical: $C_{11}H_{14}O_3$
Formula: $(CH_3)_3COOCOC_6H_5$
Properties: Colorless to sl. yel. liq., mild aromatic odor; insol. in water; sol. in org. solvs.; m.w. 194.25; dens. 1.0; m.p. 8 C; b.p. 112 C (dec.); flash pt. 19 C
Precaution: Combustible oxidizing liq.
Toxicology: LD50 (rat, dermal) 3817 mg/kg; LD50 (rat, oral) 4838 mg/kg; sl. skin and eye irritant
Uses: Initiator for polymerization and/or crosslinking of monomers and unsaturated polymers
Manuf./Distrib.: Akzo Nobel; Aztec Peroxides; Elf Atochem N. Am.; Monomer-Polymer & Dajac; Norac; Witco/PAG
Trade names: Aztec® TBPB; Aztec® TBPB-50-FT; Aztec® TBPB-50-IC; Aztec® TBPB-SA-M2; Esperox® 10; Polyvel CR-5T; Polyvel CR-10T; Trigonox® 93; Trigonox® C; Trigonox® C-C75; Varox® TBPB

t-Butyl perisobutyrate. *See* t-Butyl peroxyisobutyrate
t-Butyl permaleaic acid. *See* t-Butyl peroxymaleic acid

t-Butyl peroctoate
CAS 3006-82-4; 62695-55-0
Synonyms: t-Butyl peroxy 2-ethylhexanoate
Classification: Organic peroxide
Empirical: $C_{12}H_{24}O_3$
Formula: $(CH_3)_3COOCOCHCH_2CH_3CH_2CH_2CH_2CH_3$
Properties: M.w. 216.3
Precaution: Combustible oxidizing liq.
Toxicology: Mild skin irritant, minimal eye irritant; LD50 (rat, oral) > 10,000 mg/kg; LD50 (rabbit, dermal) 16,818 mg/kg
Uses: Med.-temp. initiator for polymerization of vinyl monomers and the curing of styrene-unsaturated polyester resin compositions; catalyst in premix polyester BMC, SMC
Manuf./Distrib.: Akzo Nobel; Aztec Peroxides; Elf Atochem N. Am.; Witco/PAG
Trade names: Aztec® TBPEH; Aztec® TBPEH-30-AL; Esperox® 28; Trigonox® 21; Trigonox® 21 LS
Trade names containing: Aztec® TBPEH-50-DOP; Aztec® TBPEH-50-OMS; Esperox® 28MD; Esperox® 28PD; Lupersol P-31; Lupersol P-33; Lupersol PDO; Lupersol PMS; Trigonox® 21-C50; Trigonox® 21-OP50; Trigonox® KSM; USP®-690

t-Butyl peroxide. *See* Di-t-butyl peroxide
t-Butyl peroxyacetate. *See* t-Butyl peracetate
t-Butyl peroxybenzoate. *See* t-Butyl perbenzoate

t-Butyl peroxycrotonate
CAS 23474-91-1
Uses: Initiator for polymerization applics.
Trade names containing: Esperox® 13M

t-Butyl peroxydiethylacetate
CAS 2550-33-6
Classification: Organic peroxide
Empirical: $C_{10}H_{20}O_3$
Formula: $(C_2H_5)_2CHCOOOC(CH_3)_3$
Properties: M.w. 188.3
Uses: Initiator for acrylates, PE
Trade names: Trigonox® 27

t-Butyl peroxy 2-ethylhexanoate. *See* t-Butyl peroctoate

t-Butyl peroxy 2-ethylhexyl carbonate
CAS 34443-12-4
Empirical: $C_{13}H_{26}O_4$
Properties: Clear liq.; m.w. 246.34; sp.gr. 0.93 (20 C); f.p. < -75 C; flash pt. > 93 C
Uses: Initiator for polymerization of vinyl monomers, styrene, acrylates, and unsat. polyester resins
Trade names: Aztec® TBPEHC; Esperox® C-496; Trigonox® 117

t-Butyl peroxyisobutyrate
CAS 109-13-7
Synonyms: t-Butyl perisobutyrate
Classification: Organic peroxide
Empirical: $C_8H_{16}O_3$
Formula: $(CH_3)_3COOCOCH(CH_3)_2$
Properties: M.w. 160.2
Precaution: Combustible oxidizing liq.
Uses: Initiator in polymerization of vinyl monomers
Trade names: Aztec® TBPIB-75-AL
Trade names containing: Lupersol 80; Trigonox® 41-C50; Trigonox® 41-C75

t-Butyl peroxy isopropyl carbonate
CAS 2372-21-6
Synonyms: BPIC

Classification: Organic peroxide
Empirical: $C_8H_{16}O_4$
Formula: $(CH_3)_2CHOCOO_2C(CH_3)_3$
Properties: M.w. 176.2
Uses: Initiator for BMC and SMC molding; initiator for free radical polymerizations of styrene and acrylates; crosslinking agent for elastomers incl. SBR, urethanes, EPR, EPDM, and nitrile rubbers
Manuf./Distrib.: Akzo Nobel; Elf Atochem N. Am.
Trade names containing: Trigonox® BPIC

t-Butyl peroxy maleate. *See* t-Butyl peroxymaleic acid

t-Butyl peroxymaleic acid
CAS 1931-62-0
Synonyms: t-Butyl permaleaic acid; t-Butyl monoperoxy maleate; t-Butyl peroxy maleate
Formula: $(CH_3)_3COOCO \cdot CH{=}CHCO{\cdot}OH$
Properties: M.w. 188.20
Toxicology: Poison by intraperitoneal route
Uses: Polymerization initiator, curing agent for polymerization of various ethylenically unsaturated resins and monomers
Manuf./Distrib.: Witco/PAG
Trade names: Aztec® TBPM-50-FT; Aztec® TBPM-50-P; Esperox® 41-25; Esperox® 41-25A

t-Butyl peroxy 2-methyl benzoate
CAS 22313-62-8
Classification: Organic peroxide
Properties: M.w. 208.2
Uses: Initiator for polymerization of ethylene and styrene, high temp. molding of polyester resin systems, and vulcanization of silicon rubber
Trade names: Trigonox® 97
Trade names containing: Esperox® 497M; Trigonox® 97-C75

t-Butyl peroxyneodecanoate
CAS 26748-41-4
Classification: Organic peroxide
Empirical: $C_{14}H_{28}O_3$
Formula: $(CH_3)_3COOCOC_9H_{19}$-t (mixed isomers)
Properties: M.w. 244.4
Precaution: Combustible oxidizing liq.
Toxicology: LD50 (rat, oral) > 12,918 mg/kg; LD50 (rabbit, dermal) > 8000 mg/kg; non-eye irritant
Uses: Initiator for radical catalyzed polymerization of vinyl monomers
Manuf./Distrib.: Akzo Nobel; Witco/PAG
Trade names: Aztec® TBPND; Aztec® TBPND-40-EAQ; Aztec® TBPND-75-OMS; Lupersol 10; Trigonox® 23; Trigonox® 23-W40
Trade names containing: Esperox® 33M; Lupersol 10-M75; Trigonox® 23-C75

t-Butyl peroxyneoheptanoate
CAS 26748-38-9
Uses: Initiator
Trade names containing: Esperox® 750M

t-Butyl peroxypivalate
CAS 927-07-1
Classification: Organic peroxide
Empirical: $C_9H_{18}O_3$
Formula: $(CH_3)_3COOCOC(CH_3)_3$
Properties: Colorless liq.; insol. in water and ethylene glycol; sol. in most org. solvs.; m.w. 174.27; dens. 0.854 (25/25 C); m.p. < 19 C; flash pt. (OC) > 155 F
Precaution: Combustible oxidizing liq.
Toxicology: Mildly toxic by ingestion; LD50 (rabbit, dermal) 2500 mg/kg; LD50 (rat, oral) 4169 mg/kg
Uses: Low-temp. initiator for radical catalyzed polymerization of vinyl monomers
Manuf./Distrib.: Akzo Nobel; Aztec Peroxides; Elf Atochem N. Am.; Witco/PAG
Trade names: Aztec® TBPPI-25-AL
Trade names containing: Aztec® TBPPI-50-OMS; Aztec® TBPPI-75-OMS; Esperox® 31M; Lupersol 11; Trigonox® 25-C75

t-Butyl peroxy stearyl carbonate
CAS 62476-60-6
Empirical: $C_{23}H_{46}O_4$
Formula: $CH_3(CH_2)_{17}OCOOOC(CH_3)_3$
Properties: M.w. 386.6
Uses: Initiator for acrylates, PS
Trade names: Perkadox® BPSC

t-Butyl peroxy-3,5,5-trimethylhexanoate
CAS 13122-18-4
Classification: Organic peroxide
Empirical: $C_{13}H_{26}O_3$
Properties: M.w. 230.4
Uses: Initiator for cure of unsaturated polyester resins
Trade names: Aztec® TBPIN; Aztec® TBPIN-30-AL; Trigonox® 42; Trigonox® 42 PR; Trigonox® 42S

2-t-Butylphenol
CAS 88-18-6; EINECS 201-807-2
Synonyms: OTBP; o-t-Butylphenol
Empirical: $C_{10}H_{14}O$
Formula: $(CH_3)_3CC_6H_4OH$
Properties: Lt. yel. liq.; sol. in toluene, ethanol, isopentane; insol. in water; m.w. 150.22; dens. 0.982 (20/4 C); f.p. -7 C; b.p. 224 C; flash pt. (OC) 110 C; ref. index 1.523
Precaution: Combustible
Toxicology: Toxic by ingestion, irritating to respiratory system, skin, eyesf
Uses: Intermediate for syn. resins, plasticizers, surfactants, perfumes; antioxidant for aviation gasoline
Manuf./Distrib.: Biddle Sawyer

4-t-Butylphenol
CAS 98-54-4; EINECS 202-679-0
Synonyms: PTBP; p-t-Butylphenol
Empirical: $C_{10}H_{14}O$
Formula: $(CH_3)_3CC_6H_4OH$
Properties: Wh. cryst.; sol. in alcohol, ether; insol. in water; m.w. 150.24; dens. 1.03; f.p. -7 C; m.p. 98-99 C; b.p. 239 C; flash pt. (OC) 230 F
Precaution: Combustible
Toxicology: Irritant to skin and eyes
Uses: Plasticizer for cellulose acetate; intermediate for antioxidants, starches, phenolic resins; pour pt. depressor and emulsion breaker for petrol. oils, some plastics; syn. lubricants; insecticides; motor-oil additives
Manuf./Distrib.: AC Ind.; Fabrichem; Janssen Chimica; PMC Specialties; Schenectady

o-s-Butylphenol
CAS 89-72-5
Empirical: $C_{10}H_{14}O$
Formula: $C_2H_5CH(CH_3)C_6H_4OH$
Properties: Liq., sl. volatile; insol. in water; sl. sol. in alcohol, ether, alkalies; m.w. 150.2; sp.gr. 0.981; flash pt. (Seta) > 93 C
Precaution: Combustible
Uses: Herbicide, insecticide, polymerization inhibitor, stabilizer intermediate; intermediate in resin prep.
Manuf./Distrib.: Albemarle; Janssen Chimica; Schenectady

o-t-Butylphenol. *See* 2-t-Butylphenol
p-t-Butylphenol. *See* 4-t-Butylphenol

Butylphenol formaldehyde
Uses: Tackifier for rubbers, adhesives, cements
Trade names: Dyphene® 9474

t-Butylphenyl diphenyl phosphate
CAS 56803-37-3
Properties: Sp.gr. 1.15-1.18 (20 C); flash pt. (COC) 246 C; ref. index 1.553
Uses: Flame retardant plasticizer in PVC, vinyl and vinyl nitrile foams, PVAc emulsions, cellulosic resins

Manuf./Distrib.: Akzo Nobel; Ashland; FMC; C.P. Hall; Monsanto
Trade names: Phosflex 71B; Santicizer 154

4-t-Butylphenyl-2,3-epoxypropyl ether. *See* p-t-Butyl phenyl glycidyl ether

p-t-Butyl phenyl glycidyl ether
CAS 3101-60-8
Synonyms: 4-t-Butylphenyl-2,3-epoxypropyl ether
Properties: M.w. 206.29; dens. 1.038; b.p. 165-170 (14 mm); flash pt. 101 C
Uses: Reactive epoxy diluent for tooling, elec. applics., flooring, casting, laminating
Trade names: Epodil® 745; Heloxy® 65

Butyl phthalate. *See* Dibutyl phthalate

n-Butyl phthalyl-n-butyl glycolate
CAS 85-70-1; EINECS 201-624-8
Synonyms: 1,2-Benzenedicarboxylic acid, 2-butoxy-2-oxoethyl, butyl ester; Butyl carbobutoxymethyl phthalate; Dibutyl-o-carboxybenzoyloxyacetate; Dibutyl-o-(o-carboxybenzoyl) glycolate
Classification: Aromatic ester
Empirical: $C_{18}H_{24}O_6$
Properties: Colorless liq., odorless; insol. in water; m.w. 336.42; dens. 1.093-1.103 (25/25 C); b.p. 219 C (5 mm); flash pt. 390 F; ref. index 1.4880
Precaution: Combustible
Toxicology: LD50 (oral, rat) 7 g/kg; mildly toxic by IP; experimental teratogen, reproductive effects; mutagenic data; eye irritant; heated to decomp., emits acrid and irritating fumes
Uses: Plasticizer for PVC
Regulatory: FDA 21CFR §175.105, 175.300, 175.320, 181.22, 181.27
Manuf./Distrib.: Fabrichem; Morflex
Trade names: Morflex® 190

Butyl-2-propenoate. *See* Butyl acrylate

Butyl ricinoleate
Formula: $C_{17}H_{32}(OH)COOC_4H_9$
Properties: Yel. to colorless oleaginous liq.; sol. in alcohol, ether; insol. in water; sp.gr. 0.917; m.p. -10 C; flash pt. (COC) 207 C; ref. index 1.462
Uses: Plasticizer, lubricant for PVAc, PVB, cellulosics
Manuf./Distrib.: CasChem
Trade names: Flexricin® P-3

Butyl stearate
CAS 123-95-5; EINECS 204-666-5
Synonyms: Butyl octadecanoate; n-Butyl octadecanoate; Octadecanoic acid butyl ester
Definition: Ester of butyl alcohol and stearic acid
Empirical: $C_{22}H_{44}O_2$
Formula: $CH_3(CH_2)_{16}COO(CH_2)_3CH_3$
Properties: Crystals; sl. sol. in water; sol. in alcohol, ether; m.w. 340.60; dens. 0.86 (20/4 C); m.p. 17-22 C; b.p. 343 C; flash pt. (CC) 160 C; ref. index 1.4430 (20 C)
Toxicology: Heated to decomp., emits acrid smoke and irritating fumes
Uses: Solv., spreading and softening agent in plastics, textiles, cosmetics, rubbers; flavoring agent; plasticizer for food pkg. materials
Regulatory: FDA 21CFR §172.515, 173.340, 175.105, 175.300, 175.320, 176.200, 176.210, 177.2600, 177.2800, 178.3910, 181.22, 181.27
Manuf./Distrib.: Amerchol; Aquatec Quimica SA; Ashland; C.P. Hall; Henkel/Organic Prods.; Inolex; Kenrich Petrochem.; Mosselman NV; Penta Mfg.; Sea-Land; Stepan; Union Camp; Unitex; Witco/Oleo-Surf.
Trade names: Ablumol BS; ADK STAB LS-8; Butyl Stearate C-895; Lexolube® BS-Tech.; Lexolube® NBS; Priolube 1451; Uniflex® BYS-CP; Uniflex® BYS Special; Uniflex® BYS-Tech.

Butyl thiotin
Uses: Heat stabilizer for rigid and flexible PVC
Trade names: Lankromark® BLT105; Lankromark® BT050; Lankromark® BT120; Lankromark® BT190; Lankromark® BT339

Butyltin mercaptide
Uses: Heat stabilizer for PVC; catalyst for polyurethane coatings
Trade names: Dabco® T-120; Okstan M 69 S; Stanclere® T-94 C; Stanclere® T-126; Stanclere® T-801

Butyltin thioglycolate
Uses: Stabilizer for PVC
Trade names: Hostastab® SnS 61

Butyl titanate. *See* Tetrabutyl titanate

6-t-Butyl-2,4-xylenol
CAS 1879-09-0
Synonyms: 6TB24X
Properties: Liq.; sp.gr. 0.958 (60 F)
Manuf./Distrib.: Howard Hall

Butyraldehyde-aniline condensation product
CAS 68411-20-1; 34562-31-7
Uses: Accelerator for NR, SBR, CR, IIR, and latexes; activator for acidic accelerators; antioxidant for CR, NR, SBR, EPDM
Trade names: Vanax® 808; Vanox® AT

Butyraldehyde-monobutylamine condensation prod.
CAS 68411-19-8
Uses: Accelerator for NR, SR, latexes, and reclaim; also in self-curing CR cements
Trade names: Vanax® 833

Butyric acid, 3,3-bis(3-t-butyl-4-hydroxyphenyl) ethylene ester. *See* Benzene propanoic acid
Butyrolactam. *See* 2-Pyrrolidone

Butyrolactone
CAS 96-48-0; EINECS 202-509-5
Synonyms: BLO; Dihydro-2(3H)-furanone; γ-Butyrolactone; 4-Butanolide
Classification: Lactone
Empirical: $C_4H_6O_2$
Properties: Colorless oily liq., mild caramel odor; misc. with water; sol. in methanol, ethanol, acetone, ether, benzene; m.w. 86.09; dens. 1.120; m.p. -45 C; b.p. 204-205 C; flash pt. 98 C; ref. index 1.4348
Precaution: Combustible when exposed to heat or flame; reactive with oxidizing materials
Toxicology: LD50 (oral, rat) 1800 mg/kg; mod. toxic by ingestion, IV, intraperitoneal routes; suspected tumorigen; heated to decomp., emits acrid and irritating fumes
Uses: Intermediate for synthesis of polyvinylpyrrolidone, piperidine, phenylbutyric acid, etc.; solv. for polyacrylonitrile, cellulose acetate, methyl methacrylate polymers, polystyrene; constituent of paint removers, textile aids, drilling oils; flavoring agent for candy, soy milk; solv. for PAN, PS, fluorinated hydrocarbons, cellulose triacetate, shellac; in paint removers, petrol. processing, inks
Regulatory: GRAS
Manuf./Distrib.: Aldrich; Allchem Ind.; BASF; Gelest; Great Western; ISP; Janssen Chimica; Schweizerhall; Spectrum Chem. Mfg.; UCB SA
Trade names: BLO®

γ-Butyrolactone. *See* Butyrolactone
n-Butyroyl tri-n-hexyl citrate. *See* n-Butyryl tri-n-hexyl citrate

N-Butyryl-p-aminophenol
Trade names: Suconox-4®

n-Butyryl tri-n-hexyl citrate
CAS 82469-79-2
Synonyms: BTHC; n-Butyroyl tri-n-hexyl citrate
Classification: Organic compd.
Empirical: $C_{28}H_{50}O_8$
Properties: Sp.gr. 0.9910; flash pt. (COC) 204 C; ref. index 1.445
Uses: Vinyl plasticizer for medical applics.
Manuf./Distrib.: Morflex; Unitex
Trade names: Citroflex® B-6

C. *See* Carbon

C8-10 acid-1,2-propanediol ester
Uses: Additive in mfg. of textile and glass fibers; lubricant for cutting devices; mold release agent in processing of plastics
Trade names: Softenol® 3408

C7 acid triglyceride
Uses: Lubricant for machinery for confectionary and food industries; mold release agent in processing of plastics; additive to cutting oils, lacquer and varnish systems; antiblocking agent for artificial skins

C8-10 acid triglyceride
Uses: Lubricant for machinery used in food and confectionary industries and medical/technical equipment; antiblocking agent for artificial skins; oil in mfg. of aluminum foils; mold release agent for plastics; additive to cutting oils

Cadmium-barium complex
Uses: Heat and light stabilizer for PVC
Trade names: Haro® Chem KB-350X; Synpron 1470; Synpron 1585; Synpron 1860; Synpron 1861; Therm-Chek® 1825

Cadmium diamyldithiocarbamate
Uses: Accelerator for rubbers

Cadmium diethyldithiocarbamate
CAS 14239-68-0
Formula: $Cd[SC(S)N(C_2H_5)_2]_2$
Properties: White to cream colored rod; sol. in benzene, carbon disulfide, chloroform; insol. in water and gasoline; m.w. 408.96; dens. 1.39; m.p. 68-76 C
Toxicology: TLV 0.005 mg/m^3 in air; TLV:TWA 0.05 mg (Cd)/m^3
Uses: Accelerator for rubbers
Trade names: Ethyl Cadmate®

Cadmium laurate
CAS 2605-44-9; EINECS 220-017-9
Synonyms: Dodecanoic acid, cadmium salt
Uses: Heat and light stabilizer for PVC
Trade names: Haro® Chem KS

Cadmium oxide
CAS 1306-19-0; EINECS 215-146-2
Synonyms: Cadmium oxide brown
Empirical: CdO
Properties: M.w. 128.40; dens. 8.150
Toxicology: Highly toxic; suspected carcinogen
Uses: Cadmium plating baths, electrodes for storage batteries, cadmium salts, catalyst, ceramic glazes, phosphors, nematocide
Manuf./Distrib.: All Chemie Ltd; Aldrich; Alloychem; Asarco; Atomergic Chemetals; Cerac; Chemisphere Ltd; Great Western; Mallinckrodt; Nihon Kagaku Sangyo; Noah; Spectrum Chem. Mfg.
Trade names containing: Chem-Master R-215; Chem-Master R-303

Cadmium oxide brown. *See* Cadmium oxide

Cadmium stearate
CAS 2223-93-0
Empirical: $C_{36}H_{62}O_4{\cdot}Cd$
Properties: M.w. 681.48
Precaution: Flamm. in powder form
Toxicology: Toxic by inhalation of dust or fume; carcinogen; TLV 0.05 mg/m^3
Uses: Lubricant and stabilizer in plastics
Manuf./Distrib.: AC Ind.; Cometals; Cookson Spec.; Miljac; U.S. Chems.
Trade names: Cometals Cadmium Stearate; Haro® Chem KPR; Metasap® Cadmium Stearate; Miljac

Cadmium-zinc complex

Cadmium Stearate; Synpro® Cadmium Stearate
Trade names containing: Chem-Master R-482A; Prespersion PAC-1451

Cadmium-zinc complex
Uses: Stabiizer/activator for chemically blown PVC
Trade names: ADK STAB GR-16; ADK STAB GR-18; ADK STAB GR-22; Interstab® BC-4377; Lankromark® LC90; Mark® 281B

Calcined alumina. *See* Alumina
Calcined litharge. *See* Lead (II) oxide
Calcined magnesia. *See* Magnesium oxide

Calcium aluminosilicate
CAS 1327-39-5
Synonyms: Aluminum calcium silicate
Definition: Occurs in nature as the minerals: anorthite, didymolite, lawsonite, etc.
Empirical: Common forms: $CaAl_2Si_2O_8$ and $Ca_2Al_2SiO_7$
Uses: Constituent of cement; in refractories; filler-reinforcement in plastics
Trade names: PMF® Fiber 204; PMF® Fiber 204AX; PMF® Fiber 204BX; PMF® Fiber 204CX; PMF® Fiber 204EX

Calcium behenate
CAS 3578-72-1; EINECS 222-700-7
Synonyms: Calcium docosanoate; Docosanoic acid, calcium salt
Definition: Calcium salt of behenic acid
Empirical: $C_{22}H_{44}O_2 \cdot {}^1/_2Ca$
Uses: Emulsifier, emollient, opacifier, suspending agent for anhyd. systems; gellant, thickener for oils and waxes; heat stabilizer for thermoplastics and thermosets
Trade names: Haro® Chem CBHG

Calcium biphosphate. *See* Calcium phosphate monobasic monohydrate
Calcium bis(dihydrogenphsophate) monohydrate. *See* Calcium phosphate monobasic monohydrate

Calcium bis[monoethyl(3,5-di-t-butyl-4-hydroxybenzyl)phosphonate)
CAS 65140-91-2
Synonyms: Calcium monoethyl [[3,5-bis (1,1-dimethylethyl)-4-hydroxyphenyl]methyl]phosphonate (1:2)
Uses: Antioxidant for polyolefins
Trade names containing: Irganox® 1425 WL; Irganox® B 501W

Calcium carbonate
CAS 471-34-1; 1317-65-3; EINECS 207-439-9
Synonyms: Carbonic acid calcium salt (1:1); CI 77220; Precipitated calcium carbonate; Precipitated chalk
Classification: Inorganic salt
Empirical: $CaCO_3$
Properties: White powd. or colorless crystals, odorless, tasteless; very sl. sol. in water; sol. in acids with CO_2; m.w. 100.09; dens. 2.7-2.95; m.p. 825 C (dec.)
Precaution: Ignites on contact with F_2; incompat. with acids, alum, ammonium salts
Toxicology: TLV 5 mg/m^3 of air; LD50 (oral, rat) 6450 mg/kg; severe eye and moderate skin irritant
Uses: Filler; alkali; source of lime; neutralizing agent; opacifying agent in paper; putty; tooth powds.; antacid; whitewash; portland cement; paint; rubber; plastics; insecticides; in chemical analysis
Regulatory: FDA 21CFR §73.1070, 137.105, 137.155, 137.160, 137.165, 137.170, 137.175, 137.180, 137.185, 137.350, 169.115, 175.300, 177.1460, 181.22, 181.29, 182.5191, 184.1191, 184.1409, GRAS; BATF 27CFR §240.1051, limitation 30 lb/1000 gal of wine; Japan approved (1-2%); Europe listed; UK approved
Manuf./Distrib.: AluChem; Am. Ingreds.; BASF; R.E. Carroll; Cerac; ECC Int'l.; EM Industries; Genstar Stone Prods.; Georgia Marble; J.M. Huber; Mallinckrodt; Nichia Kagaku Kogyo; Pfizer; Whittaker, Clark & Daniels
Trade names: Albacar® 5970; Albafil®; Albaglos®; Amical® 85; Amical® 95; Amical® 101; Amical® 202; Amical® 303; Amical® SC2; Amical® SC3; Amical® SC5; Amical® SC7; Atomite®; Calwhite®; Calwhite® II; Camel-CAL®; Camel-CAL® ST; Camel-CAL® Slurry; Camel-CARB®; Camel-FIL™; Camel-FINE; Camel-FINE ST; Camel-TEX®; Camel-WITE®; Camel-WITE® Slurry; Camel-WITE® ST; CC™-101; CC™-103; CC™-105; CP Filler™; Drikalite®; Duramite®; Gamaco®; Gamaco® II; Gamaco® Slurry; Gama-Fil® 40; Gama-Fil® 55; Gama-Fil® 90; Gama-Fil® D-1; Gama-Fil® D-1T; Gama-Fil® D-2; Gama-

Fil® D-2T; Gama-Plas®; Gama-Sperse® 2; Gama-Sperse® 5; Gama-Sperse® 80; Gama-Sperse® 140; Gama-Sperse® 255; Gama-Sperse® 6451; Gama-Sperse® 6532; Gama-Sperse® 6532 NSF; G-White; Hallcote® 573; Hi-Pflex® 100; Hubercarb® Q 1; Hubercarb® Q 2; Hubercarb® Q 3; Hubercarb® Q 4; Hubercarb® Q 6; Hubercarb® Q 6-20; Hubercarb® Q 20-60; Hubercarb® Q 40-200; Hubercarb® Q 60; Hubercarb® Q 100; Hubercarb® Q 200; Hubercarb® Q 325; Hubercarb® W 2; Hubercarb® W 3; Hubercarb® W 4; H-White; Macromite; Marble Dust™; Marblemite™; Micro-White® 07 Slurry; Micro-White® 10; Micro-White® 10 Slurry; Micro-White® 15; Micro-White® 15 Slurry; Micro-White® 25; Micro-White® 25 Slurry; Micro-White® 40; Micro-White® 100; Multifex™ FF; Multifex-MM®; No. 1 White™; No. 3 White; No. 8 White; No. 9 NCS; No. 9 White; No. 10 White; Opacimite™; OZ; RO-40; SC-53; Snowflake P.E.™; Snowflake White™; Supercoat®; Supermite®; Super-Pflex® 100; Super-Pflex® 200; Ultra-Pflex®; Wingdale White; Winnofil S, SP; XO White

Trade names containing: Akrochem® DCP-40C; Akrochem® VC-40C; Aztec® 2,5-DI-50-C; Cabot® PE 9006; Cabot® PE 9026; Cabot® PE 9229; Colloids N 54/14/01; Di-Cup 40C; Dualite® M6001AE; Gama-Sperse® CS-11; Hallcote® DPD-524, DPD-547; Hubercarb® Q 1T; Hubercarb® Q 2T; Hubercarb® Q 3T; Hubercarb® Q 200T; Hubercarb® W 2T; Hubercarb® W 3T; Kotamite®; Luperco 101-XL; Luperco 130-XL; Luperco 231-XL; Luperco 500-40C; Magnum-White; Micro-White® 15 SAM; Micro-White® 25 SAM; Opacicoat™; PE-AB 22; Perkadox® 14-40B-pd; Perkadox® BC-40Bpd; Plasblak® PE 1851; Plasblak® PE 2377; Plasblak® PE 2515; Plasblak® PE 2570; Plaswite® PE 7000; Plaswite® PE 7001; Plaswite® PE 7004; Plaswite® PE 7031; Plaswite® PE 7097; Plaswite® PE 7192; Trigonox® 17-40B-pd; Trigonox® 29-40B-pd; Trigonox® 101/45Bpd; Varox® DCP-40C

See also Limestone

Calcium carbonate, natural. *See* Limestone
Calcium dihydrogen phosphate. *See* Calcium phosphate monobasic monohydrate
Calcium disodium edetate. *See* Calcium disodium EDTA

Calcium disodium EDTA

CAS 62-33-9; EINECS 200-529-9

Synonyms: Calcium disodium ethylenediamine tetraacetic acid; Edetate calcium disodium; Calcium disodium edetate

Classification: Substituted diamine

Empirical: $C_{10}H_{12}CaN_2Na_2O_8 \cdot 2H_2O$

Formula: $CaNa_2C_{10}H_{12}N_2O_8 \cdot xHOH$

Properties: White cryst. powd., odorless, faint salt taste; sol. in water; insol. in org. solvs.; m.w. 410.30

Toxicology: Possible link to liver damage in test animals

Uses: Medicine, preservative (foods), sequestrant, hog scald agent; in cosmetics, pharmaceuticals, in water treatment, textiles, soaps/detergents, electroless copperplating, foods, polymer prod., disinfectants, pulp/paper

Regulatory: FDA 21CFR §172.120; USDA 9CFR §318.7; Japan approved (0.035 g/kg max.); Europe listed; permitted in UK only in canned fish, shellfish

Manuf./Distrib.: Akzo Nobel; Am. Biorganics; Hampshire; Hickson Danchem

Trade names: Versene CA

Calcium disodium ethylenediamine tetraacetic acid. *See* Calcium disodium EDTA
Calcium docosanoate. *See* Calcium behenate

Calcium dodecylbenzene sulfonate

CAS 26264-06-2; 68953-96-8; EINECS 247-557-8

Uses: Emulsifier for herbicides/pesticides, polymerization, industrial use; dispersant for textiles, coatings, leather applics.

Trade names: Nansa® EVM50

Trade names containing: Nansa® EVM70/B; Nansa® EVM70/E

Calcium hydrate. *See* Calcium hydroxide
Calcium hydrogen orthophosphate. *See* Calcium phosphate dibasic
Calcium hydrogen phosphate anhydrous. *See* Calcium phosphate dibasic

Calcium hydroxide

CAS 1305-62-0; EINECS 215-137-3

Synonyms: Calcium hydrate; Hydrated lime; Slaked lime; Lime water

Classification: Inorganic base

Empirical: CaH_2O_2

Formula: $Ca(OH)_2$

Calcium hydroxide phosphate

Properties: Soft, white crystalline powd. with alkaline, sl. bitter taste; sl. sol. in water; sol. in glycerol, syrup, acid insol. in alcohols; m.w. 74.10; dens. 2.34; m.p. loses water at 580 C
Precaution: Violent reaction with maleic anhydride, nitroethane, nitromethane, nitroparaffins, etc.
Toxicology: TLV 5 mg/m^3 in air; LD50 (oral, rat) 7.34 g/kg; mildly toxic by ingestion; severe eye irritant; skin, mucous membrane irritant; dust is industrial hazard
Uses: Mortar, plasters, cements, calcium salts, disinfectant, lubricants, pesticides, mfg. of paper pulp, in SBR vulcanization, in water treatment; food additives, buffer, neutralizing agent, firming agent
Regulatory: FDA 21CFR §135.110, 184.1205, GRAS; USDA 9CFR §318.7; Japan restricted (1% max. as calcium); Europe listed; UK approved
Manuf./Distrib.: Am. Biorganics; EM Industries; Int'l. Chem.; Janssen Chimica; Mallinckrodt; Pfizer; Smith Lime Flour; Spec. Mins.; U.S. Gypsum
Trade names: Rhenofit® CF
Trade names containing: Cri-Spersion CRI-ACT-45; Cri-Spersion CRI-ACT-45-1/1; Cri-Spersion CRI-ACT-45-LV; Plastigel PG9089; Plastigel PG9104; Rhenogran® $Ca(OH)_2$ -50/FPM; Rhenogran® CHM21-40/FPM; Wacker Stabilizer R

Calcium hydroxide phosphate. *See* Calcium phosphate tribasic
Calcium-L-2-hydroxypropionate. *See* Calcium lactate

Calcium hydroxystearate
Uses: Heat stabilizer for PVC
Trade names: Haro® Chem CHG

Calcium lactate
CAS 814-80-2 (anhyd.); EINECS 212-406-7
Synonyms: 2-Hydroxypropanoic acid, calcium salt; Calcium-L-2-hydroxypropionate
Empirical: $C_6H_{10}CaO_6 \cdot xH_2O$, x < 5
Formula: $[CH_3CH(OH)COO]_2Ca \cdot xH_2O$
Properties: Wh. to cream-colored cryst. powd. or granules containing up to 5 moles of water of crystallization, almost odorless; sol. in water; pract. insol. in alcohol; m.w. 218.22 (anhyd.)
Toxicology: Heated to decomp., emits acrid smoke and irritating fumes
Uses: Buffer, dough conditioner, yeast food, firming agent, flavor enhancer, flavoring agent, leavening agent, nutrient supplement, stabilizer, thickener; acid/catalyst neutralizer for polyolefins and other thermoplastics
Regulatory: FDA 21CFR §184.1207, GRAS except for infant foods/formulas; USDA 9CFR §318.7 (0.6% max.); Japan approved (1% max. as calcium); Europe listed; UK approved
Manuf./Distrib.: Am. Biorganics; Atomergic Chemetals; EM Ind.; Int'l. Chem.; Patco; Spectrum Chem. Mfg.; Wilke Int'l.
Trade names: Pationic® 1230; Pationic® 1240; Pationic® 1250

Calcium laurate
EINECS 225-166-3
Synonyms: Dodecanoic acid, calcium salt
Formula: $Ca(C_{12}H_{23}O_2)_2 \cdot H_2O$
Properties: White needles; sl. sol. in water; sol. in alcohol; m.w. 456.71; m.p. 182 C
Uses: Heat stabilizer for PVC, PE, PP, PU, ABS, thermosets
Trade names: Haro® Chem CSG

Calcium metasilicate. *See* Wollastonite
Calcium monoethyl [[3,5-bis (1,1-dimethylethyl)-4-hydroxyphenyl]methyl]phosphonate (1:2). *See* Calcium bis[monoethyl(3,5-di-t-butyl-4-hydroxybenzyl)phosphonate)

Calcium montanate
CAS 68308-22-5
Synonyms: Montan acid, calcium salt
Definition: Calcium salt of the acids derived from montan acid wax
Uses: Lubricant for PVC
Trade names: Hostalub® VP Ca W 2
Trade names containing: Hoechst Wax OP

Calcium naphthalene sulfonate
Uses: Dispersant for pigments; stabilizer for dispersions and emulsions; plasticizer for concrete, mortar, and gypsum
Trade names: Tamol® NMC 9301

Calcium octadecanoate. *See* Calcium stearate
Calcium orthophosphate. *See* Calcium phosphate tribasic

Calcium oxide
CAS 1305-78-8; EINECS 215-138-9
Synonyms: Lime; Quicklime; Calx; Burnt lime
Classification: Inorganic oxide
Empirical: CaO
Properties: White or gray cryst. or powd., odorless; sol. in acids, glycerol, sugar sol'n.; sol. in water forming $Ca(OH)_2$ and generating heat; pract. insol. in alcohol; m.w. 56.08; dens. 3.40; m.p. 2570 C; b.p. 2850 C
Precaution: Noncombustible; powd. may react explosively with water; mixts. with ethanol may ignite if heated
Toxicology: TLV 2 mg/m^3; strong caustic; may cause severe irritation to skin, mucous membranes
Uses: Refractory, sewage treatment, insecticides, fungicides, mfg. of steel and aluminum; flotation of nonferrous ores; mfg. of glass, paper, Ca salts; in drilling fluids, lubricants; laboratory; desiccant for extruded rubber goods; alkali; nutrient, dietary supplement, dough conditioner, yeast food, hog/poultry scald agent
Regulatory: FDA 21CFR §182.1210, 182.5210, 184.1210, GRAS; USDA 9CFR §318.7, 381.147; Europe listed; UK approved
Manuf./Distrib.: AluChem; Ash Grove Cement; Cerac; GE; Hüls Am.; Mallinckrodt; Pfizer; Spec. Mins.; Smith Lime Flour; U.S. Gypsum
Trade names: Rhenosorb® C; Rhenosorb® C/GW; Rhenosorb® F
Trade names containing: BYK®-2615; Plastigel PG9068; Prespersion PAB-4493; Rhenogran® CaO-50/FPM

Calcium pelargonate
Uses: Process aid, lubricant for polyolefins
Trade names: Synpro® Calcium Pelargonate

Calcium phosphate dibasic
CAS 7757-93-9; EINECS 231-826-1
Synonyms: DCP-0; Dicalcium phosphate; Dicalcium orthophosphate anhyd. (E341); Calcium hydrogen phosphate anhydrous; Calcium hydrogen orthophosphate; Phosphoric acid calcium salt (1:1)
Classification: Inorganic salt
Empirical: $CaHPO_4$
Properties: Wh. cryst. powd., odorless, tasteless; sol. in dilute HCl, nitric, and acetic acids; insol. in alcohol; sl. sol. in water; m.w. 136.07; dens. 2.306; loses water at 109 C
Toxicology: Skin and eye irritant; nuisance dust
Uses: Foods, pharmaceuticals, dentifrice, medicine, glass, fertilizer, stabilizer for plastics, dough conditioner, yeast food, nutrient, dietary supplement
Regulatory: FDA 21CFR §181.29, 182.1217, 182.5217, 182.8217, GRAS; Japan approved (1% max. as calcium); Europe listed; UK approved
Manuf./Distrib.: Albright & Wilson; EM Industries; FMC; GE; Janssen Chimica; Mallinckrodt; OxyChem; Rhone-Poulenc Basic

Calcium phosphate monobasic monohydrate
CAS 10031-30-8; EINECS 231-837-1
Synonyms: MCP; Monocalcium phosphate monohydrate; Calcium dihydrogen phosphate; Acid calcium phosphate; Calcium bis(dihydrogenphsophate) monohydrate; Calcium biphosphate
Definition: Phosphoric acid calcium salt (2:1)
Empirical: $CaH_4O_8P_2 \cdot H_2O$
Formula: $CaH_4(PO_4)_2 \cdot H_2O$
Properties: Colorless pearly scales or powd., strong acid taste; deliq. in air; sol. in water and acids; m.w. 252.08; dens. 2.22 (18/4 C); m.p. loses water at 100 C, dec. 200 C
Precaution: Hygroscopic
Toxicology: Nuisance dust
Uses: Leavening acid in food prods., buffer, dough conditioner, firming agent, nutrient, dietary/mineral supplement, yeast food, sequestrant, fertilizer, stabilizer for plastics, to control pH in malt, glass mfg.
Regulatory: FDA 21CFR §136.110, 136.115, 136.130, 136.160, 136.165, 136.180, 137.80, 137.165, 137.175, 137.270, 150.141, 150.161, 155.200, 175.300, 181.29, 182.1217, 182.5217, 182.6215, 182.8217, GRAS; Japan approved
Manuf./Distrib.: Albright & Wilson; FMC; Kemira Kemi AB; Mallinckrodt; Monsanto; OxyChem; Nichia Kagaku Kogyo; Rhone-Poulenc Basic

Calcium phosphate tertiary. *See* Calcium phosphate tribasic

Calcium phosphate tribasic

CAS 7758-87-4, 12167-74-7; EINECS 231-840-8

Synonyms: TCP; Tricalcium phosphate; Tribasic calcium phosphate; Calcium hydroxide phosphate; Tricalcium orthophosphate; Calcium orthophosphate; Calcium phosphate tertiary; Precipitated calcium phosphate

Empirical: $Ca_3(PO_4)_2$

Formula: $3Ca_3(PO_4)_2 \cdot Ca(OH)_2$

Properties: Wh. cryst. powd.; odorless, tasteless; sol. in acid; insol. in water, alcohol, acetic acid; m.w. 1004.64; dens. 3.18; m.p. 1670 C; ref. index 1.63

Toxicology: Skin and eye irritant; nuisance dust

Uses: Foods, pharmaceuticals, polystyrene, ceramics, mordant, fertilizers, dentifrices, stabilizer for plastics, in meat tenderizer, anticaking agent, nutrient supplement, buffer; dessicant for insect pests; suspension polymerization agent

Regulatory: FDA 21CFR §137.105, 137.155, 137.160, 137.165, 137.170, 137.175, 137.180, 137.185, 169.179, 169.182, 175.300, 181.29, 182.1217, 182.5217, 182.8217, GRAS; USDA 9CFR §318.7; Japan approved (1% max. as calcium), restricted; Europe listed; UK approved

Manuf./Distrib.: Albright & Wilson; FMC; Mallinckrodt; Monsanto; Rhone-Poulenc Basic

Trade names containing: Sodium Bicarbonate T.F.F. Treated Free Flowing

Calcium ricinoleate

Definition: Derived from castor oil

Formula: $Ca[CH_3(CH_2)_5CHOHCH_2CHCH(CH_2)_7CO_2]_2$

Properties: Wh. powd., sl. odor of fatty acids; sol. in alcohols, glycols, ether-alcohols; dens. 1.04; m.p. 85 C

Precaution: Combustible

Toxicology: Heated to decomp., emits acrid smoke and irritating fumes

Uses: Stabilizer/lubricant for vinyls; thickener for plastisols

Regulatory: FDA 21CFR §181.29

Manuf./Distrib.: CasChem; Norman, Fox

Calcium silicate

CAS 1344-95-2, 10101-39-0; EINECS 215-710-8

Synonyms: Silicic acid, calcium salt

Definition: Hydrous or anhydrous silicate with varying proportions of calcium oxide and silica

Empirical: Common forms: $CaSiO_3$, Ca_2SiO_4, Ca_3SiO_5

Properties: White or cream-colored powd.; pract. insol. in water; dens. 2.10; bulk dens. 15-16 lb/ft^3; absorp. power 600% (water); surf. area 95-175 m^2/g

Toxicology: Nuisance dust

Uses: Constituent of lime glass, portland cement; reinforcing filler in elastomers and plastics; absorbent for liqs., gases, vapors; suspending agent; pigment and pigment extender; binder for refractories; in chromatography, road building; anticaking agent, filter aid for foods (baking powd., table salt)

Regulatory: FDA 21CFR §172.410, 175.300, 177.1460, 182.2227, GRAS (limitation 2% in table salt, 5% in baking powd.); Europe listed; UK approved

Manuf./Distrib.: Celite; Crosfield; Degussa; Great Western; J.M. Huber; Kraft; R.T. Vanderbilt

Trade names: Extrusil; Micro-Cel® C; Micro-Cel® E; Micro-Cel® T-21; Micro-Cel® T-49

Trade names containing: Vulkasil® C

Calcium/sodium stearoyl lactylate

Definition: Calcium/sodium salt derived from stearic and lactic acids

Uses: Acid neutralizer/acceptor/scavenger for polyolefins; lubricant for PVC, polyolefins, filled compds.

Trade names: Pationic® 925

Calcium-sodium-zinc complex

Uses: Stabilizer for dispersion PVC

Trade names: ADK STAB FL-22

Calcium stearate

CAS 1592-23-0; EINECS 216-472-8

Synonyms: Calcium octadecanoate; Stearic acid calcium salt

Definition: Calcium salt of stearic acid

Empirical: $C_{18}H_{35}O_2 \cdot {}^1/_2Ca$

Formula: $Ca(C_{18}H_{35}O_2)_2$

Properties: White powd.; insol. in water; sl. sol. in hot alcohol, hot veg. and min. oils; m.w. 707.00; bulk dens. 20 lb/ft^3; m.p. 149 C

Toxicology: Heated to decomp., emits acrid smoke and irritating fumes
Uses: Water repellent; flatting agent in paints, emulsions; release agent for plastic molding powds.; stabilizer for PVC resins; lubricant; in pencils and crayons; food grade as conditioning agent in foods and pharmaceuticals, anticaking agent, binder, emulsifier, lubricant, release agent, stabilizer, thickener
Regulatory: FDA 21CFR §169.179, 169.182, 172.863, 173.340; must conform to FDA specs for fats or fatty acids derived from edible oils, 175.105, 175.300, 175.320, 176.170, 176.200, 176.210, 177.1200, 177.2260, 177.2410, 177.2600, 178.2010, 179.45; 181.22, 181.29, 184.1229, GRAS
Manuf./Distrib.: Adeka Fine Chem.; Cometals; Eka Nobel Ltd; Elf Atochem N. Am./Wire Mill; Ferro/Grant; Henkel/Organic Prods.; Mallinckrodt; Miljac; PPG Industries; R.T. Vanderbilt; Witco
Trade names: ADK STAB CA-ST; Afco-Chem CS; Afco-Chem CS-1; Afco-Chem CS-S; Akrochem® Calcium Stearate; Akrochem® P 4000; Akrochem® P 4100; Akrodip™ CS-400; Cecavon CA 30; Coad 10; Cometals Calcium Stearate; Hallcote® CSD; Hallcote® CaSt 50; Haro® Chem CGD; Haro® Chem CGL; Haro® Chem CGN; Haro® Chem CPR-2; Haro® Chem CPR-5; Hy Dense Calcium Stearate HP Gran.; Hy Dense Calcium Stearate RSN Powd.; HyTech RSN 11-4; HyTech RSN 248D; Interstab® CA-18-1; Mathe Calcium Stearate; Miljac Calcium Stearate; Nuodex S-1421 Food Grade; Nuodex S-1520 Food Grade; Petrac® Calcium Stearate CP-11; Petrac® Calcium Stearate CP-11 LS; Petrac® Calcium Stearate CP-11 LSG; Petrac® Calcium Stearate CP-22G; Plastolube; Quikote™ C; Quikote™ C-LD; Quikote™ C-LD-A; Quikote™ C-LM; Quikote™ C-LMLD; Synpro® Calcium Stearate 12B; Synpro® Calcium Stearate 15; Synpro® Calcium Stearate 15F; Synpro® Calcium Stearate 114-36; TR-Calcium Stearate; Witco® Calcium Stearate A; Witco® Calcium Stearate F; Witco® Calcium Stearate FP
Trade names containing: Afco-Chem EX-95; Aquastab PD 31M; Gama-Sperse® CS-11; Hostalub® CAF 485; Hostalub® XL 50; Hostalub® XL 445; Krynac® P30.49; Krynac® P34.52

Calcium stearoyl lactylate
CAS 5793-94-2; EINECS 227-335-7
Synonyms: Calcium stearoyl-2-lactylate; Calcium stearyl-2-lactylate; Calcium stelate
Definition: Calcium salt of stearic acid ester of lactyl lactate
Empirical: $C_{24}H_{44}O_6 \cdot {}^1/_2Ca$
Properties: Cream-colored nonhygroscopic powd., caramel odor; sparingly sol. in water; m.w. 895.30; HLB 5-6; acid no. 50-86
Toxicology: When heated to decomp., emits acrid smoke and irritating fumes
Uses: Dough conditioner for bakery prods.; stabilizer; whipping agent; emulsifier in cosmetics and pharmaceuticals
Regulatory: FDA 21CFR §172.844; must conform to FDA specs for fats or fatty acids derived from edible oils; Japan approved with restrictions; Europe listed; UK approved
Trade names: Pationic® 930; Pationic® 930K; Pationic® 940

Calcium stearoyl-2-lactylate. *See* Calcium stearoyl lactylate
Calcium stearyl-2-lactylate. *See* Calcium stearoyl lactylate
Calcium stelate. *See* Calcium stearoyl lactylate

Calcium sulfate
CAS 7778-18-9 (anhyd.), 10101-41-4 (dihydrate); EINECS 231-900-3
Synonyms: Anhydrite (natural form); Calcium sulfonate; Gypsum; Anhydrous gypsum; Plaster of Paris
Classification: Inorganic salt
Empirical: CaO_4S
Formula: $Ca \cdot H_2O_4S$
Properties: Wh. to sl. yel.-wh. powd. or crystals, odorless; sl. sol. in water; m.w. 136.14; dens. 2.964; m.p. 1450 C
Precaution: Reacts violently with aluminum when heated; mixts. with phosphorus ignite at high temps.
Toxicology: Irritant; heated to decomp., emits toxic fumes of SO_x
Uses: Insol. anhydride: in cement formulations, as paper filler; Sol. anhydride: drying agent, desiccant; nutrient, dietary supplement, yeast food, dough conditioner/strengthener, firming agent, sequestrant, anticaking agent, coloring agent, leavening agent, pH control agent, processing aid, stabilizer, texturizer, thickener
Regulatory: FDA 21CFR §133, 133.102, 133.106, 133.111, 133.141, 133.165, 133.181, 133.195, 137.105, 137.155, 137.160, 137.165, 137.170, 137.175, 137.180, 137.185, 150.141, 150.161, 155.200, 175.300, 177.1460, 184.1230, GRAS; BATF 27CFR §240.1051, limitation 16.69 lb/1000 gal; Japan approved with restrictions (1% max.); Europe listed; UK approved
Manuf./Distrib.: Am. Ingreds.; EM Ind.; Fluka; Kemira Kemi AB; U.S. Gypsum
Trade names containing: Cadox® BTA

Calcium sulfate dihydrate
CAS 10101-41-4; EINECS 231-900-3

Calcium sulfonate

Synonyms: Native calcium sulfate; Precipitated calcium sulfate; Gypsum; Alabaster
Formula: CaO_4S
Properties: Lumps or powd.; sol. in water; very slowly sol. in glycerol; pract. insol. in most org. solvs.; m.w. 172.10; dens. 2.32
Uses: Mfg. of portland cement, plaster of Paris, artificial marble, sulfuric acid; as white pigment and filler in paints, pharmaceuticals, paper, insecticide dusts, yeast, polishing powds.; soil and water treatment; plaster casts (pharmaceutical); flame retardant for plastics
Manuf./Distrib.: R.E. Carroll; EM Ind.; Fluka; Franklin Ind. Mins.
Trade names: Valu-Fil

Calcium sulfonate
CAS 61789-86-4
Uses: Rust preventive, detergent for diesels; pigment dispersant for thermoplastic and epoxy color concs.; visc. stabilizer
Manuf./Distrib.: Ashland; King Ind.; Lubrizol; Stepan; Witco/Petrol. Spec.
Trade names: Lubrizol® 2152
See also Calcium sulfate

Calcium-tin complex
Uses: Heat and light stabilizer for PVC

Calcium-tin-zinc complex
Uses: Heat and light stabilizer for vinyls
Trade names: MiRaStab 922-K

Calcium-zinc complex
Uses: Heat stabilizer for rigid and plasticized PVC, profiles, bottles, pipe, plastisols
Trade names: ADK STAB 36; ADK STAB 593; ADK STAB AC-116; ADK STAB AC-311; ADK STAB SC-12; ADK STAB SC-24; ADK STAB SC-26; ADK STAB SC-32; ADK STAB SC-34; ADK STAB SP-55; ADK STAB SP-69; ADK STAB SP-76; ADK STAB SP-86; B‰ropan MC 8046 SP; B‰rostab® CT 901; B‰rostab® CT 926 HV; B‰rostab® CT 9140; B‰rostab® NT 210; B‰rostab® NT 211; B‰rostab® NT 212; B‰rostab® NT 224; Haro® Chem CP-4; Haro® Chem CP-17; Haro® Chem CZ-37, CZ-40; Haro® Mix ZC-028, ZC-029, ZC-030; Haro® Mix ZC-031, ZC-032; Haro® Mix ZC-036; Haro® Mix ZC-309; Haro® Mix ZC-311; Haro® Mix ZC-902; Interstab® CZ-10; Interstab® CZ-11; Interstab® CZ-11D; Interstab® CZ-19A; Interstab® CZ-22; Interstab® CZ-23; Interstab® F-402; Lankromark® LZ495; Lankromark® LZ649; Lankromark® LZ935; Lankromark® LZ1045; Lankromark® LZ1177; Lankromark® LZ1188; Lankromark® LZ1210; Mark® 133; Mark® 152; Mark® 495; Mark® 565A; Mark® 577A; Mark® 630; Mark® 2326; Mark® QED; MiRaStab 113-K; MiRaStab 130-K; MiRaStab 140-K; MiRaStab 919-K; MiRaStab 950-K; Plastistab 2005; Plastistab 2020; Plastistab 2160; Plastistab 2210; Stabiol CZ 2001/1; Stabiol VCA 1733; Stabiol VZN 1783; Synpron 755; Synpron 1522; Synpron 1900; Synpron 1913; Therm-Chek® 344; Therm-Chek® 659; Therm-Chek® 714; Therm-Chek® 763; Therm-Chek® 6253, 6254
Trade names containing: Stabiol CZ 1616/6F

Calcium-zinc stearate
Uses: Lubricant and mold release agent in polyethylene/PP extrusion and molding operations; release agent in bulk and sheet molding compds. of unsat. polyester
Trade names: Petrac® Calcium/Zinc Stearate CZ-81

C-8 alcohols. *See* Caprylic alcohol

C8-10 alcohols
CAS 68603-15-5; EINECS 287-621-2
Uses: Chemical intermediate; also for lube oil additives, plasticizers, surfactant feedstocks
Trade names: Nafol® 810 D

C10 alcohols
Uses: Intermediate for mfg. of toiletries, cosmetics, detergents, laundry softeners, lubricating oil additives, plasticizers, plastics additives, textile/leather additives, disinfectants, agrochem., paper defoamers, flotation agents
Trade names: Nafol® 10 D

C10-12 alcohols
EINECS 288-117-5

Uses: Intermediate for surfactants; lubricants in polymer processing, metal rolling rolls
Trade names: Alfol® 1012 HA

C10-14 alcohols
EINECS 283-066-5
Synonyms: Alcohols, C10-14
Uses: Lubricant in polymer processing, metal rolling, emollients/lubricants for cosmetics, defoamers, intermediate
Trade names: Alfol® 1014 CDC; Nafol® 1014

C12-14 alcohols
EINECS 279-420-3
Synonyms: Alcohols, C12-14
Uses: Lubricant in polymer processing, metal rolling; emollient in cosmetics; intermediate
Trade names: Alfol® 1214; Alfol® 1214 GC; Alfol® 1216 CO; Alfol® 1412

C12-16 alcohols
CAS 68855-56-1; EINECS 272-490-6
Uses: Intermediate for surfactants, plastics; lubricant and metal rolling rolls; emollient, opacifier, thickener
Trade names: Alfol® 1216

C12-18 alcohols
CAS 67762-25-8; EINECS 267-006-5
Uses: Surfactant intermediate; polymerization; lubricant in polymer processing, metal rolling; emollient for cosmetics
Trade names: Alfol® 1218 DCBA; Loxiol® VPG 1496; Nafol® 1218

C14-16 alcohols
EINECS 269-790-4
Uses: Lubricant in polymer processing, metal rolling; emollient for cosmetics; intermediate
Trade names: Alfol® 1416 GC

C14-18 alcohols
EINECS 267-009-1
Uses: Chemical intermediate; also for lube oil additives, plasticizers, surfactant feedstocks
Trade names: Alfol® 1418 DDB; Alfol® 1418 GBA

C16-18 alcohols. *See* Cetearyl alcohol

C20-24 alcohols
EINECS 307-145-1
Synonyms: Alcohols, C20-24
Uses: Intermediate for mfg. of toiletries, cosmetics, detergents, laundry softeners, lubricating oil additives, plasticizers, plastics additives, textile/leather additives, disinfectants, agrochem., paper defoamers, flotation agents
Trade names: Nafol® 20+

C20-40 alcohols
Definition: Mixture of syn. fatty alcohols with 20-40 carbons in the alkyl chain
Uses: Antioxidant, heat stabilizer, uv stabilizer, visc. depressant; promotes emollient protective films for skin care prods.
Trade names: Unilin® 350 Alcohol; Unilin® 425 Alcohol

C30-50 alcohols
Definition: Mixture of syn. fatty alcohols with 30-50 carbons in the alkyl chain
Uses: Surfactants; chemical intermediate; in coatings; plastics additive (processing aid, lubricant, dispersant); cosmetics
Trade names: Unilin® 550 Alcohol

C40-60 alcohols
Definition: Mixture of syn. fatty alcohols with 30-50 carbons in the alkyl chain
Uses: Functional polymer for modification of PP, PVC, polyethylene, PS, and high-performance engineering resins; acts as antioxidant, heat stabilizer, uv stabilizer, or visc. depressant; promotes emollient protective

films onto skin
Trade names: Unilin® 700 Alcohol

C8 alkyl dimethylamine. *See* Dimethyl octylamine
C10 alkyl dimethylamine. *See* Dimethyl decylamine
Calx. *See* Calcium oxide

C13-15 amine
CAS 68155-27-1
Uses: Anticaking agent for fertilizer; flotation agent, corrosion inhibitor; metal cleaning formulations, pigment dispersions, rubber processing auxs., bitumen emulsifier; intermediate for prod. of amine salts, ethoxylates and sulfosuccinates
Trade names: Synprolam 35

Candelilla synthetic
CAS 136097-95-5
Uses: Gellant, thickener, stabilizer, and moisturizer cosmetics, leather dressings, furniture and other polish; cements; varnishes; sealing wax; elec. insulating compositions; phonograph records; paper sizing; rubber; lubricants; adhesives; waterproofing
Trade names: Koster Keunen Synthetic Candelilla

Candelilla wax
CAS 8006-44-8; EINECS 232-347-0
Definition: Wax from various *Euphorbiaceae* species
Properties: Yel.-brown to translucent solid; pract. insol. in water; sparingly sol. in alcohol; sol. in acetone, benzene, carbon disulfide, hot petrol. ether, gasoline, oils, turpentine, CCl_4; dens. 0.983; m.p. 67-68 C; ref. index 1.4555; sapon. no. 50-65
Precaution: Combustible
Toxicology: Heated to decomp., emits acrid smoke and irritating fumes
Uses: Mfg. of cosmetics, rubber substitutes, polishes, candles, sealing waxes, varnishes, leather, creams; for waterproofing; elec. insulation; inks; molding compositions; sizing paper; hardening other waxes; protective coating for citrus; masticatory substance in chewing gum base
Regulatory: FDA 21CFR §175.105, 175.320, 176.180, 184.1976, GRAS; Japan approved
Manuf./Distrib.: Koster Keunen; Penta Mfg.; Frank B. Ross; Stevenson Cooper; Strahl & Pitsch
Trade names: Koster Keunen Candelilla

Canola oil glyceride
Definition: Monoglyceride derived from canola oil
Uses: Emulsifier for foods, cosmetics, personal care prods; emulsion stabilizer; antifoam; mold release, processing aid, antistat, antifog, lubricant, antiblock for plastics
Trade names: Myverol® 18-99

Canola oil (low erucic acid rapeseed oil). *See* Rapeseed oil
Capric acid chloride. *See* Decanoyl chloride
Caproamphoglycinate. *See* Sodium caproamphoacetate
Capryl alcohol. *See* Caprylic alcohol

Caprylic alcohol
CAS 111-87-5; EINECS 203-917-6
Synonyms: n-Octyl alcohol; Heptyl carbinol; 1-Octanol; n-Octanol; C-8 alcohols; Alcohol C-8; Capryl alcohol
Classification: Fatty alcohol
Empirical: $C_8H_{18}O$
Formula: $CH_3(CH_2)_6CH_2OH$
Properties: Colorless liq., fresh orange-rose odor, sl. herbaceous taste; sol. in water; misc. in alcohol, ether, and chloroform; m.w. 130.26; dens. 0.827; m.p. -16.7 C; b.p. 194.5 C; flash pt. 178 F; ref. index 1.429 (20 C)
Precaution: Combustible liq. when exposed to heat or flame; can react with oxidizers
Toxicology: LD50 (oral, mouse) 1790 mg/kg; poison by intravenous route; moderately toxic by ingestion; skin irritant
Uses: Wetting agent, emulsifier, surfactant intermediate, raw material for plasticizers; emulsion polymerization; tobacco-offshoots controller; detergent intermediate; flavoring agent, intermediate, solv.
Regulatory: FDA 21CFR §172.230, 172.515 (only for encapsulating lemon, dist. lime, orange, peppermint, and

spearmint oils), 172.864, 173.280, 175.105, 175.300, 176.210, 177.1010, 177.1200, 177.2800, 178.3480; GRAS (FEMA)
Manuf./Distrib.: Albemarle; Aldrich; M. Michel; Penta Mfg.; Schweizerhall; Vista
Trade names: Epal® 8

Caprylic/capric acid
CAS 67762-36-1
Uses: Used in alkyd resins, rubber compding., water repellents, polishes, soaps, abrasives, cutting oils, candles, crayons, emulsifiers; food grades as lubricant, release agent, binder, defoamer in foods, intermediate for food emulsifiers
Trade names: Industrene® 365

Capryloamphocarboxyglycinate. *See* Disodium capryloamphodiacetate
Capryloamphodiacetate. *See* Disodium capryloamphodiacetate

Captan
CAS 133-06-2; EINECS 205-087-0
Synonyms: N-Trichloromethylthiotetrahydrophthalimide; N-Trichloromethylthio-4-cyclohexene-1,2-dicarboximide
Classification: Organic compd.
Empirical: $C_9H_8Cl_3NO_2S$
Properties: Crystal, odorless; pract. insol. in water; partly sol. in benzene, chloroform, tetrachloroethane; m.w. 300.59; dens. 1.74; m.p. 178 C
Toxicology: Moderately toxic by inhalation; irritant; TLV 5 mg/m^3 of air; LD50 (rat, oral) 9000 mg/kg
Uses: Seed treatment, fungicide and bacteriostat in plants, plastics, leather, fabrics, fruit preservation; gas odorant
Manuf./Distrib.: Industrias Quimicas del Valles SA; R.T. Vanderbilt
Trade names: Vancide® 89

Carbamic acid, [4-[(4-aminocyclohexyl)methyl]cyclohexyl]-. *See* 4,4′-Methylenebis (cyclohexylamine) carbamate
Carbamic acid, butyl-3-iodo-2-propynyl ester. *See* Iodopropynyl butylcarbamate
Carbamide. *See* Urea
Carbamidic acid. *See* Urea
2-(Carbamoylazo) isobutyronitrile. *See* 1-[(1-Cyano-1-methylethyl)azo]formamide
Carbanilide. *See* N,N′-Diphenylurea
Carbinol. *See* Methyl alcohol
'Carbitol'. *See* Ethoxydiglycol

Carbodiimide
CAS 420-04-2; 622-16-2
Synonyms: Cyanamude; Cyanogenamide
Formula: HN:C:NH
Properties: Deliquescent crystals; very sol. in water, alcohol, ether, phenols, ketone; dens. 1.08; m.p. 43 C
Toxicology: Strong irritant to skin and mucous membranes; TLV 2 mg/m^3 of air
Uses: Activator for vulcanizates based on millable PU
Trade names: Stabilizer 2013-P®

Carbon
CAS 7440-44-0
Synonyms: C
Classification: Nonmetallic element
Empirical: C
Properties: Blk. cryst., powd. or diamond form; at.wt. 12.01; dens. 1.8-2.1 (amorphous); m.p. 3652-3697 C; b.p. $\approx$ 4200 C
Toxicology: Moderately toxic by intravenous route
Uses: Reducing agent, purifying metals, electronics; in rubber compds., coatings, and plastics
Trade names: Synthetic Carbon 910-44M

Carbonate magnesium. *See* Magnesium carbonate

Carbon black
CAS 1333-86-4; EINECS 215-609-9

Carbon black

Synonyms: Thermal black; Charcoal; Vegetable carbon; Channel black; Furnace black; Acetylene black; Lamp black; CI 77266

Definition: Finely divided particles of elemental carbon obtained by incomplete combustion of hydrocarbons (channel or impingement process)

Empirical: C

Properties: Insol. in water, ethanol, veg. oil

Toxicology: TLV 3.5 mg/m^3; low toxicity by ingestion, inhalation, and skin contact; nuisance dust in high concs.; suspected carcinogen

Uses: Inorganic colorant, filler, reinforcement; tire treads, belt covers, other abrasion-resistant rubber prods.; UV light absorber; colorant for printing inks; color additive, purification agent in food processing

Regulatory: Banned by U.S. FDA; may be used in foods for the European community; UK approved

Manuf./Distrib.: Akrochem; Akzo Nobel; Cabot; Degussa; Exxon; J.M. Huber; San Yuan; Spectrum Chem. Mfg.; R.T. Vanderbilt; Witco/Concarb

Trade names: Aquathene™ CM 92832; Black Pearls® 120; Black Pearls® 130; Black Pearls® 160; Black Pearls® 170; Black Pearls® 280; Black Pearls® 450; Black Pearls® 460; Black Pearls® 470; Black Pearls® 480; Black Pearls® 490; Black Pearls® 520; Black Pearls® 570; Black Pearls® 700; Black Pearls® 800; Black Pearls® 880; Black Pearls® 900; Black Pearls® 1000; Black Pearls® 1100; Black Pearls® 3700; Black Pearls® L; Continex® LH-10; Continex® LH-20; Continex® LH-30; Continex® LH-35; Continex® N110; Continex® N220; Continex® N234; Continex® N299; Continex® N326; Continex® N330; Continex® N339; Continex® N343; Continex® N351; Continex® N550; Continex® N650; Continex® N660; Continex® N683; Continex® N762; Continex® N774; Diablack® A; Diablack® E; Diablack® G; Diablack® H; Diablack® HA; Diablack® I; Diablack® II; Diablack® LH; Diablack® LI; Diablack® LR; Diablack® M; Diablack® N220M; Diablack® N234; Diablack® N339; Diablack® N550M; Diablack® N760M; Diablack® R; Diablack® SF; Diablack® SH; Diablack® SHA; Elftex® 120; Elftex® 125; Elftex® 160; Elftex® 280; Elftex® 285; Elftex® 415; Elftex® 430; Elftex® 435; Elftex® 460; Elftex® 465; Elftex® 570; Elftex® 670; Elftex® 675; FW 18; FW 200 Beads and Powd.; Huber ARO 60; Huber N110; Huber N220; Huber N234; Huber N299; Huber N326; Huber N330; Huber N339; Huber N343; Huber N347; Huber N351; Huber N375; Huber N539; Huber N550; Huber N650; Huber N660; Huber N683; Huber N762; Huber N774; Huber N787; Huber N990; Huber S212; Huber S315; Ketjenblack® EC-300 J; Ketjenblack® EC-310 NW; Ketjenblack® EC-600 JD; Ketjenblack® ED-600 JD; Mitsubishi Carbon Black; Mogul® L; Monarch® 120; Monarch® 700; Monarch® 800; Monarch® 880; Monarch® 900; Monarch® 1000; Monarch® 1100; MPC Channel Black TR 354; Novacarb; Printex P; Regal® 300; Regal® 300R; Regal® 330; Regal® 330R; Regal® 400; Regal® 400R; Regal® 660; Regal® 660R; S160 Beads and Powd.; Sable™ 3; Sable™ 5; Sable™ 15; Sable™ 17; Sterling® 1120; Sterling® 2320; Sterling® 3420; Sterling® 4620; Sterling® 5550; Sterling® 5630; Sterling® 6630; Sterling® 6640; Sterling® 6740; Sterling® 8860; Sterling® C; Sterling® NS; Sterling® NS-1; Sterling® SO; Sterling® V; Sterling® VH; Sterling® XC-72; Thermax® Floform® N-990; Thermax® Floform® Ultra Pure N-990; Thermax® Powder N-991; Thermax® Powder Ultra Pure N-991; Thermax® Stainless; Thermax® Stainless Floform® N-907; Thermax® Stainless Powder N-908; Thermax® Stainless Powder Ultra Pure N-908; Vulcan® 2H; Vulcan® 3; Vulcan® 3H; Vulcan® 4H; Vulcan® 6; Vulcan® 6LM; Vulcan® 7H; Vulcan® 8H; Vulcan® 9; Vulcan® 9A32; Vulcan® 10H; Vulcan® C; Vulcan® J; Vulcan® K; Vulcan® M; Vulcan® P; Vulcan® XC72; Vulcan® XC72R

Trade names containing: Ampacet 19153; Ampacet 19153-S; Ampacet 19200; Ampacet 19238; Ampacet 19258; Ampacet 19270; Ampacet 19278; Ampacet 19400; Ampacet 19458; Ampacet 19475-S; Ampacet 19492; Ampacet 19500; Ampacet 19552; Ampacet 19584; Ampacet 19614; Ampacet 19673; Ampacet 19699; Ampacet 19709; Ampacet 19716; Ampacet 19717; Ampacet 19786; Ampacet 19832; Ampacet 19873; Ampacet 19897; Ampacet 19975; Ampacet 19991; Ampacet 19999; Ampacet 39270; Ampacet 49270; Ampacet 49315; Ampacet 49370; Ampacet 49419; Ampacet 49842; Ampacet 49882; Ampacet 49920; Ampacet 49984; Ampacet 190014; Ampacet 190015; Black Out® Black; Colloids PE 48/32 C; Colloids PE 48/32 F; Colloids PE 48/58; Colloids PE 48/61 A; Colloids PE 48/61 F; Colloids PE 48/61/30; Colloids PE 48/62 A; Colloids PE 48/67; Colloids PE 48/72; Colloids PE 48/73; Colloids PE 48/76; Colloids PE 48/80; Colloids PE 48/85; Colloids PP 50/5; Colloids PP 50/7; Colloids PP 50/9; Colloids PS 47/31; Colloids PS 47/32; Colloids PS 47/38; Colloids SAN 52/8; Colloids UN 53/9D; Colloids UN 53/10; Colloids UN 53/12; Colloids UN 53/15; Colloids UN 53/17; Colloids UN 53/18; Diapol® WMB® 1808; Diapol® WMB® 1833; Diapol® WMB® 1847; Diapol® WMB® 1848; Diapol® WMB® 1849; Diapol® WMB® 1939; Diapol® WMB® S900; Diapol® WMB® S903; Diapol® WMB® S910; Diapol® WMB® S920; Diapol® WMB® S960; Ketjenblack® EC-350 J Spd; Ketjenblack® EC-350 N DOP; MDI AB-925; MDI EC-940; MDI HD-935; MDI NY-905; MDI PC-925; MDI PE-135; MDI PE-500; MDI PE-500-20F; MDI PE-500A; MDI PE-500HD; MDI PE-500LL; MDI PE-540; MDI PE-907; MDI PE-907HD; MDI PE-931WC; MDI PE-940; MDI PP-130; MDI PP-535; MDI PP-940; MDI PS-125; MDI PS-901; MDI PS-903; MDI PV-940; MDI SN-948; Plasblak® EV 1755; Plasblak® HD 3300; Plasblak® LL 2590; Plasblak® LL 2612; Plasblak® LL 3282; Plasblak® PE 1371; Plasblak® PE 1380; Plasblak® PE 1851; Plasblak® PE 1873; Plasblak® PE 2249; Plasblak® PE 2272; Plasblak® PE 2377; Plasblak® PE 2479; Plasblak® PE 2515; Plasblak® PE 2570;

Plasblak® PE 3006; Plasblak® PE 3168; Plasblak® PE 4056; Plasblak® PE 4135; Plasblak® PP 3393; Plasblak® PP 3583; Plasblak® PP 3585; Plasblak® PP 4045; Plasblak® PS 0469; Plasblak® PS 1844; Plasblak® PS 3294; Plasblak® PS 4054; Plasblak® SA 3176; Plasblak® UN 2014; Plasblak® UN 2016; Polarite 880G(B); Prespersion SIC-3536; Reed C-BK-7; Reed C-EPES-1331; Reed C-PBT-1338; Reed C-PET-1333; Spectratech® PM 92742; X50-S®; X230-S®

Carbon dioxide

CAS 124-38-9; EINECS 204-696-9
Synonyms: Carbonic acid gas; Carbonic anhydride
Classification: Gas
Empirical: CO_2
Properties: Colorless gas, odorless; m.w. 44.01; m.p. subl. @ -78.5 C
Precaution: Noncombustible; various dusts explode in CO_2 atmospheres
Toxicology: TLV 5000 ppm; asphyxiant at > 10%; experimental teratogen; skin contact can cause burns
Uses: Refrigeration, aerosol propellant, chemical intermediate, fire extinguishing, inert atmospheres, municipal water treatment, medicine, oil wells, mining, blowing agent; Aerating agent, carbonation, cooling agent, leavening agent, pH control agent, processing aid, propellant, carbonated beverages
Regulatory: FDA 21CFR §169.115, 169.140, 169.150, 184.120, GRAS, 193.45 (modified atm. for pest control); USDA 9CFR §318.7, 381.147; BATF 27CFR 240.1051; Japan approved; Europe listed; UK approved
Manuf./Distrib.: Air Prods.; BOC Gases; Carbonic Ind.; Coastal; Great Western; Nissan Chem. Ind.; Norsk Hydro AS; Scott Spec. Gases; Showa Denko
Trade names containing: MS-122N/CO_2

Carbon fiber

Synonyms: Graphite fiber; PAN carbon fiber; Pitch carbon fiber
Definition: Made from rayon, polyacrylonitrile, or petrol. pitch
Properties: Filament, yarn, fabric, whiskers
Uses: Reinforcement for polyester and epoxy composites for jet engine components, cements, metal, rubber; flameproof textile prods.; heating pads, protective clothing
Trade names: Dialead®

Carbonic acid calcium salt (1:1). *See* Calcium carbonate
Carbonic acid gas. *See* Carbon dioxide
Carbonic acid monosodium salt. *See* Sodium bicarbonate
Carbonic acid, 1,2-propylene glycol ester. *See* Propylene carbonate
Carbonic acid, zinc salt (1:1). *See* Zinc carbonate
Carbonic anhydride. *See* Carbon dioxide
5,5′-Carbonylbis 1,3-isobenzofurandione. *See* 3,3′,4,4′-Benzophenone tetracarboxylic dianhydride
4,4′-Carbonylbis(phthalic anhydride). *See* 3,3′,4,4′-Benzophenone tetracarboxylic dianhydride
Carbonyldiamide. *See* Urea
4,4′-Carbonyldiphthalic acid anhydride. *See* 3,3′,4,4′-Benzophenone tetracarboxylic dianhydride

β-Carboxyethyl acrylate

CAS 24615-84-7
Uses: Reactive monomer for emulsion polymerization, adhesion promoter; latexes; adhesives
Manuf./Distrib.: UCB Radcure
Trade names: Sipomer® β-CEA

N-(2-Carboxyethyl)-N-dodecyl-β-alanine, disodium salt. *See* Disodium lauriminodipropionate
N-(2-Carboxyethyl)-N-dodecyl-β-alanine, monosodium salt. *See* Sodium lauriminodipropionate
N-(2-Carboxyethyl)-N-(tallow acyl)-β-alanine. *See* Disodium tallowiminodipropionate
Carboxylic acid C_{18}. *See* Stearic acid

Carboxymethylcellulose sodium

CAS 9004-32-4
Synonyms: CMC; Cellulose gum; Sodium carboxymethylcellulose; Sodium CMC
Definition: Sodium salt of the carboxylic acid R-O-CH_2COONa
Formula: $[(C_6H_7O_2(OH)_2OCH_2COOH]_n$
Properties: Colorless or white powd. or granules, odorless; water sol. depends on degree of substitution; insol. in org. liqs.; m.w. 21,000-500,000; visc. 25 cps (2% aq.); m.p. > 300 C
Toxicology: Nontoxic; LD50 (oral, rat) 27,000 mg/kg
Uses: In drilling muds; in detergents as soil-suspending agent; in emulsion paints, adhesives, inks, textile sizes; as protective colloid; cosmetics; in pharmaceuticals as suspending agent, excipient, viscosity modifier;

food stabilizer, binder, thickener, extender, boiler water additive; in rubber industry
Regulatory: FDA 21CFR §133.134, 133.178, 133.179, 150.141, 150.161, 173.310, 175.105, 175.300, 182.70, 182.1745, GRAS; USDA 9CFR 318.7, limitation 1.5%, must be added dry; 9CFR 381.147; Japan restricted (2% max.); Europe listed; UK approved
Manuf./Distrib.: Aldrich; Aqualon; Courtaulds Water Soluble Polymers; J.W.S. Delavau; FMC; Hercules
Trade names: Tylose® C, CB Series; Tylose® CBR Grades; Tylose® CR

N-(Carboxymethyl)-N,N-dimethyl-1-decanaminium hydroxide, inner salt. *See* Decyl betaine
N-(Carboxymethyl)-N,N-dimethyl-1-hexadecanaminium hydroxide, inner salt. *See* Cetyl betaine
N-(Carboxymethyl)-N,N-dimethyl-1-octadecanaminium hydroxide, inner salt. *See* Stearyl betaine

Cardanol
Uses: Accelerator for epoxy systems; diluent; flexibilizer for phenolic systems
Trade names: Cardolite® NC-700

Cardanol acetate
Uses: Diluent/accelerator for epoxy mfg.; used to prepare adducts, in elec. applics.

Carnauba
CAS 8015-86-9; EINECS 232-399-4
Synonyms: Brazil wax; Carnauba wax
Definition: Exudate from leaves of Brazilian wax palm tree *Copernicia prunifera*
Properties: Yel. greenish brown lumps, solid, characteristic odor; sol. in ether, boiling alcohol and alkalies; insol. in water; dens. 0.995 (15/15 C); m.p. 82-85.5 C; sapon. no. 78-89; ref. index 1.4500
Toxicology: Heated to decomp., emits acrid smoke and irritating fumes
Uses: Shoe polishes, leather finishes, varnishes, waterproofing, furniture and floor polishes; hardening candles; substitute for beeswax; plasticizer for dental compds.; purified form in cosmetics, pharmaceuticals; anticaking agent, candy glaze/polish, formulation aid, lubricant, release agent, surface-finishing agent
Regulatory: FDA 21CFR §175.320, 184.1978, GRAS; Japan approved; Europe listed (permitted only in chocolate prods.); UK approved
Manuf./Distrib.: Penta Mfg.; Stevenson Cooper; Strahl & Pitsch
Trade names: Ross Carnauba Wax

Carnauba wax. *See* Carnauba

Castor oil
CAS 1323-38-2; 8001-79-4; EINECS 232-293-8
Synonyms: Ricinus oil; Aceite de ricino; Oil of Palma Christi; Tangantangan oil
Definition: Fixed oil obtained from seeds of *Ricinus communis*
Properties: Colorless to pale yel. viscous liq., characteristic odor; sol. in alcohol; misc. with glac. acetic acid, chloroform, ether; dens. 0.961; m.p. -12 C; b.p. 313 C; flash pt. (CC) 445 F; ref.index 1.478; sapon. no. 176-187
Precaution: Combustible when exposed to heat; spontaneous heating may occur
Toxicology: Moderately toxic by ingestion; allergen; eye irritant; purgative, laxative in large doses
Uses: Plasticizer in lacquers and nitrocellulose; polyurethane coatings; hydraulic fluids; industrial lubricants; used in mfg. of Turkey red oil, cosmetics, leather treatment; antisticking agent, release agent for foods; component of protective coatings; drying oil
Regulatory: FDA 21CFR §73.1, 172.510, 172.876, 175.300, 176.210, 177.2600, 177.2800, 178.3120, 178.3570, 178.3910, 181.22, 181.26, 181.28; GRAS (FEMA)
Manuf./Distrib.: Arista Industries; Ashland; CasChem; Climax Performance; Fanning; Harcros; Lipo; Norman, Fox
Trade names: AA Standard; #1 Oil

Castor oil, acetylated and hydrogenated. *See* Glyceryl (triacetoxystearate)
Castor oil acid, methyl ester. *See* Methyl ricinoleate
Castor oil, hydrogenated. *See* Hydrogenated castor oil

Castor oil, polymerized
Definition: Rubber-like polymer resulting from combination of castor oil with sulfur or diisocyanates
Uses: Pigment wetting/dispersing; plasticizer, processing aid for resins, gums, polymers, rubber; lubricant, penetrant, coupling solv.; adhesion promoter; for lacquers, inks, adhesives, lubricants, polishes, hydraulic fluids

Trade names: #15 Oil; #30 Oil; Pale 4; Pale 16; Pale 170; Pale 1000; Vorite 105; Vorite 110; Vorite 115; Vorite 120; Vorite 125

Castor oil, potassium salt. *See* Potassium castorate
Castor oil sulfated. *See* Sulfated castor oil

Castor oil, vulcanized
Uses: Extender, processing aid for rubbers
Trade names: Faktogel® Asolvan
Trade names containing: Faktogel® Asolvan T

Castorwax. *See* Hydrogenated castor oil
CBTS. *See* N-Cyclohexyl-2-benzothiazolesulfenamide

C44 dimer acid
Uses: Polymer building block; modifier for plastics, adhesives, elastomers
Trade names: Pripol 1004

'Cellosolve'. *See* Ethoxyethanol

Cellulose
CAS 9004-34-6, 65996-61-4; EINECS 232-674-9
Synonyms: Wood pulp, bleached; Cotton fiber; Cellulose powder; Cellulose gel
Definition: Natural polysaccharide derived from plant fibers
Properties: Colorless solid; insol. in water and org. solvs.; m.w. 160,000-560,000; dens. ≈ 1.5
Toxicology: Cannot be digested by humans; nuisance dust; heated to decomp., emits acrid smoke and irritating fumes
Uses: Fibers: reinforcing filler for textile, paper, rubber, plastics; for mfg. of nitrocellulose, cellulose acetate, etc.; in chromatography; as ion exchange material; microcrystalline: as binder-disintegrants in tableting; colloidal: stabilizer, emulsifier, thickener, dispersant, anticaking agent, binding agent in foods
Regulatory: FDA 21CFR §177.2260; GRAS; Europe listed; use in baby foods not permitted in UK
Manuf./Distrib.: Degussa; Eastman; FMC Int'l.; Hercules Ltd; Edw. Mendell
Trade names: CF 1500; CF 32,500T Coarse; Fibra-Cel® BH-70; Fibra-Cel® BH-200; Fibra-Cel® CBR-18; Fibra-Cel® CBR-40; Fibra-Cel® CBR-41; Fibra-Cel® CBR-50; Fibra-Cel® CC-20; Fibra-Cel® SW-8; Santoweb® D; Santoweb® DX; Santoweb® H; Santoweb® W

Cellulose gel. *See* Cellulose
Cellulose gum. *See* Carboxymethylcellulose sodium
Cellulose, 2-hydroxyethyl ether. *See* Hydroxyethylcellulose
Cellulose 2-hydroxypropyl methyl ether. *See* Hydroxypropyl methylcellulose
Cellulose methyl ether. *See* Methylcellulose
Cellulose powder. *See* Cellulose

Ceresin
CAS 8001-75-0; EINECS 232-290-1
Synonyms: White ozokerite wax; Ceresin wax; Earth wax; Mineral wax
Definition: Waxy mixture of hydrocarbons obtained by purification of ozokerite
Properties: White or yel. waxy cake; sol. in alcohol, benzene, chloroform, naphtha, hot oils; insol. in water; dens. 0.92-0.94; m.p. 68-72 C
Precaution: Combustible
Uses: Candles, sizing; bottles for hydrofluoric acid; shoe and leather polishes; antifouling paints; cosmetics; waterproofing textiles; substitute for beeswax; dental wax compds.; rubber compounding
Regulatory: FDA 21CFR §175.105
Manuf./Distrib.: M. Argueso & Co.; Astor Wax; Jonk BV; Koster Keunen; Frank B. Ross; Scheel; Stevenson Cooper; Strahl & Pitsch
Trade names: Koster Keunen Ceresine; Ross Ceresine Wax

Ceresin wax. *See* Ceresin

Ceteareth-4
CAS 68439-49-6 (generic)
Synonyms: PEG-4 cetyl/stearyl ether; POE (4) cetyl/stearyl ether
Definition: PEG ether of cetearyl alcohol

Ceteareth-6

Formula: $R(OCH_2CH_2)_nOH$, R = blend of cetyl and stearyl radicals, avg. n = 4
Uses: Emulsifier, lubricant, emollient, and conditioner for cosmetics, textiles, metal cleaners, industrial, institutional and household cleaners, emulsion polymerization.; dyeing assistant, leveling aid
Trade names: Prox-onic CSA-1/04

Ceteareth-6
CAS 68439-49-6 (generic)
Synonyms: PEG-6 cetyl/stearyl ether; POE (6) cetyl/stearyl ether
Definition: PEG ether of cetearyl alcohol
Formula: $R(OCH_2CH_2)_nOH$, R = blend of cetyl and stearyl radicals, avg. n = 6
Uses: Emulsifier, lubricant, emollient, and conditioner for cosmetics, textiles, metal cleaners, industrial, institutional and household cleaners, emulsion polymerization; dyeing assistant, leveling agent
Trade names: Prox-onic CSA-1/06

Ceteareth-10
CAS 68439-49-6 (generic)
Synonyms: PEG-10 cetyl/stearyl ether; POE (10) cetyl/stearyl ether
Definition: PEG ether of cetearyl alcohol
Formula: $R(OCH_2CH_2)_nOH$, R = blend of cetyl and stearyl radicals, avg. n = 10
Uses: Rubber emulsifier, lubricant, dye and fulling assistant; conditioner and emollient for personal care prods.; emulsion polymerization; floor waxes, paper finishes, pharmaceuticals
Regulatory: FDA 21CFR §177.2800
Trade names: Prox-onic CSA-1/010

Ceteareth-15
CAS 68439-49-6 (generic)
Synonyms: PEG-15 cetyl/stearyl ether; POE (15) cetyl/stearyl ether
Definition: PEG ether of cetearyl alcohol
Formula: $R(OCH_2CH_2)_nOH$, R = blend of cetyl and stearyl radicals, avg. n = 15
Uses: Detergent, emulsifier, leveling agent, intermediate; used in personal care prods., wax, oil and textiles, scouring agents, dyes, household formulations, silicone emulsification, surfactants
Regulatory: FDA 21CFR §177.2800
Trade names: Prox-onic CSA-1/015

Ceteareth-20
CAS 68439-49-6 (generic)
Synonyms: PEG-20 cetyl/stearyl ether; POE (20) cetyl/stearyl ether
Definition: PEG ether of cetearyl alcohol
Formula: $R(OCH_2CH_2)_nOH$, R = blend of cetyl and stearyl radicals, avg. n = 20
Uses: Emulsifier, dispersant, solubilizer, wetting agent, detergent, solubilizer, visc. control agent for cosmetics, pharmaceuticals, and industrial applics.; emulsion polymerization
Regulatory: FDA 21CFR §177.2800
Trade names: Prox-onic CSA-1/020

Ceteareth-22
CAS 68439-49-6 (generic)
Synonyms: PEG-22 cetyl/stearyl ether; POE (22) cetyl/stearyl ether
Definition: PEG ether of cetearyl alcohol
Formula: $R(OCH_2CH_2)_nOH$, R = blend of cetyl and stearyl radicals, avg. n = 22
Uses: Wetting agent, emulsifier, detergent, solubilizer, dispersant in hydrophobic conditions; dye and pigment carriers; textile applics.; mfg. of wax emulsions and polishes for household and industrial use; antistat; rubber latex stabilization
Trade names: Teric 16A22

Ceteareth-25
CAS 68439-49-6 (generic)
Synonyms: PEG-25 cetyl/stearyl ether; POE (25) cetyl/stearyl ether
Definition: PEG ether of cetearyl alcohol
Formula: $R(OCH_2CH_2)_nOH$, R = blend of cetyl and stearyl radicals, avg. n = 25
Uses: Emulsifier for cosmetic and pharmaceutical preparations, emulsion polymerization, washing detergents
Regulatory: FDA 21CFR §177.2800
Trade names: Remcopal 229

Ceteareth-30
CAS 68439-49-6 (generic)
Synonyms: PEG-30 cetyl/stearyl ether; POE (30) cetyl/stearyl ether
Definition: PEG ether of cetearyl alcohol
Formula: $R(OCH_2CH_2)_nOH$, R = blend of cetyl and stearyl radicals, avg. n = 30
Uses: Wetting agent, dispersant, detergent, emulsifier, leveling agent, intermediate, used for cosmetics, household formulations, silicone emulsification, textile processing, paints, rubber, floor waxes
Regulatory: FDA 21CFR §177.2800
Trade names: Prox-onic CSA-1/030

Ceteareth-50
CAS 68439-49-6 (generic)
Synonyms: PEG-50 cetyl/stearyl ether; POE (50) cetyl/stearyl ether
Definition: PEG ether of cetearyl alcohol
Formula: $R(OCH_2CH_2)_nOH$, R = blend of cetyl and stearyl radicals, avg. n = 50
Uses: Wetting agent, detergent for industrial/domestic applics.; emulsifier, foam control agent in syn. heavy duty detergents; emulsion polymerization
Regulatory: FDA 21CFR §177.2800
Trade names: Prox-onic CSA-1/050

Cetearyl alcohol
CAS 8005-44-5; 67762-30-5
Synonyms: Cetostearyl alcohol; Cetyl/stearyl alcohol; C16-18 alcohols
Definition: Mixture of fatty alcohols, predominantly cetyl and stearyl alcohols
Uses: Surfactant intermediate; raw material for ethoxylation, sulfation, etc.; stabilizer in emulsion polymerization; lubricant in rigid PVC; emollient, consistency factor for pharmaceutical creams, hand lotions, bath oils, shaving creams
Regulatory: FDA 21CFR §175.105
Trade names: Adol® 63; Alfol® 1618; Alfol® 1618 CG; Alfol® 1618 GC; Laurex® CS; Loxiol® G 52; Loxiol® G 53; Loxiol® P 1420

Cetearyl methacrylate
Synonyms: Cetyl-stearyl methacrylate
Uses: Monomer for mfg. of polymers used as oil additives, visc. index improvers, pour pt. depressants; internal plasticizer for adhesives, uv-curable resins, coatings
Trade names containing: Empicryl® 6030

Cetearyl stearate
CAS 93820-97-4, 136097-82-0
Synonyms: Cetostearyl stearate
Definition: Ester of cetearyl alcohol and stearic acid
Uses: Emollient for cosmetics; lubricant for PVC processing
Trade names: Estol 1481

Cetene. *See* Hexadecene-1

Cetethyl morpholinium ethosulfate
CAS 78-21-7; EINECS 201-094-8
Synonyms: Cetyl ethyl morpholinium ethosulfate; Quaternium-25; Morpholinium, 4-ethyl-4-hexadecyl, ethyl sulfate
Classification: Quaternary salt
Empirical: $C_{22}H_{46}NO \cdot C_2H_5O_4S$
Uses: Antistatic agent; textile specialties
Trade names: Barquat® CME-35

Cetoleth-5
Uses: Emulsifier for emulsion polymerization; wetting agent
Trade names: Disponil O 5

Cetoleth-6
Synonyms: PEG-6 cetyl/oleyl ether; POE (6) cetyl/oleyl ether
Definition: Polyethylene glycol ether of cetyl alcohol and oleyl alcohol

Cetoleth-10

Uses: Emulsifier, mfg. of textile lubricants, solv. and waterless hand cleaners; coemulsifier; detergent additive in petrol. oils; intermediate for anionic surfactants
Trade names: Ethylan® ME

Cetoleth-10
Synonyms: PEG-10 cetyl/oleyl ether; POE (10) cetyl/oleyl ether
Definition: Polyethylene glycol ether of cetyl alcohol and oleyl alcohol
Uses: Wetting agent, dispersant, emulsifier; used in processing of yarns and fabrics; textile dyeing; horticultural sprays; removal of oil slicks; solubilizer for sanitation chemicals; emulsion polymerization
Trade names: Disponil O 10

Cetoleth-13
Synonyms: PEG-13 cetyl/oleyl ether; PEG (13) cetyl/oleyl ether
Definition: Polyethylene glycol ether of cetyl alcohol and oleyl alcohol
Uses: Wetting agent, dispersant, emulsifier; used in processing of yarns and fabrics; textile dyeing; horticultural sprays; removal of oil slicks; solubilizer for sanitation chemicals; emulsion polymerization
Trade names: Ethylan® OE

Cetoleth-19
Uses: Emulsifier for cosmetics, emulsion polymerization; latex stabilizer; polish mfg.; dye leveling agent
Trade names: Ethylan® R

Cetoleth-20
Synonyms: PEG-20 cetyl/oleyl ether; POE (20) cetyl/oleyl ether
Definition: Polyethylene glycol ether of cetyl alcohol and oleyl alcohol
Uses: Emulsifier for emulsion polymerization; solubilizer, dispersant, latex stabilizer
Trade names: Disponil O 20

Cetoleth-25
Synonyms: PEG-25 cetyl/oleyl ether; POE (25) cetyl/oleyl ether
Definition: PEG ether of cetyl alcohol and oleyl alcohol
Formula: $R(OCH_2CH_2)_nOH$, R = blend of cetyl and oleyl radicals, avg. n = 25
Uses: Wetting agent, dispersant, emulsifier; emulsion polymerization; used in processing of yarns and fabrics; textile dyeing; horticultural sprays; removal of oil slicks; solubilizer for sanitation chemicals
Trade names: Disponil O 250

Cetoleth-55
CAS 68920-66-1
Uses: Emulsifier for emulsion polymerization
Trade names: Polirol O55

Cetostearyl alcohol. *See* Cetearyl alcohol
Cetostearyl stearate. *See* Cetearyl stearate

Cetrimonium chloride
CAS 112-02-7; EINECS 203-928-6
Synonyms: Cetyl trimethyl ammonium chloride; Palmityl trimethyl ammonium chloride; Hexadecyl trimethyl ammonium chloride
Classification: Quaternary ammonium salt
Formula: $C_{16}H_{33}(CH_3)_3NCl$
Uses: Emulsifier, dispersant, emollient, surfactant, softener, conditioner, bactericide, fungicide, and odor inhibitor in personal care prods.; antistat for hair, fibers, plastics; gel sensitizer in latex foam prod.; emulsion polymerization
Manuf./Distrib.: Witco/Oleo-Surf.
Trade names: Chemquat 16-50; Dehyquart® A
Trade names containing: Arquad® 16-50

Cetyl alcohol
CAS 36653-82-4; EINECS 253-149-0
Synonyms: Palmityl alcohol; C16 linear primary alcohol; 1-Hexadecanol
Classification: Fatty alcohol
Empirical: $C_{16}H_{34}O$
Formula: $CH_3(CH_2)_{14}CH_2OH$

Properties: White waxy solid; partially sol. in alcohol and ether; insol. in water; m.w. 242.27; dens. 0.8176 (49.5 C); m.p. 49.3 C; b.p. 344 C; ref. index 1.4283
Precaution: Flamm. when exposed to heat or flame; can react with oxidizing materials
Toxicology: LD50 (oral, rat) 6400 mg/kg; mod. toxic by ingestion, intraperitoneal routes; eye and human skin irritant; heated to decomp., emits acrid smoke and fumes
Uses: Perfumery; emulsifier, emollient, coupling agent; foam stabilizer in detergents; opacifier; thickener; chemical intermediate; cosmetics, pharmaceuticals; external mold release for acrylics; flavoring agent, color diluent, intermediate for confectionery, food supplements in tablet form, gum
Regulatory: FDA 21CFR §73.1, 73.1001, 172.515, 172.864, 175.105, 175.300, 176.200, 177.1010, 177.1200, 177.2800, 178.3480, 178.3910; GRAS (FEMA)
Manuf./Distrib.: Aarhus Oliefabrik A/S; Albemarle; Aldrich; Amerchol; Chemron; Croda; Lipo; Lonza; M. Michel; Norman, Fox; Procter & Gamble; Stepan; Vista
Trade names: Adol® 52; Alfol® 16; Epal® 16NF; Loxiol® VPG 1743
Trade names containing: Epal® 1214

Cetyl amine. *See* Palmitamine

Cetyl betaine
CAS 693-33-4; EINECS 211-748-4
Synonyms: N-(Carboxymethyl)-N,N-dimethyl-1-hexadecanaminium hydroxide, inner salt
Classification: A zwitterion (inner salt)
Empirical: $C_{20}H_{41}NO_2$
Uses: Wetting agent, detergent, emulsifier, dispersant, surfactant, conditioner; dyeing applics.; softener for textiles; leveling and rewetting agent in the paper industry; dyeing assistant and degreaser in the leather industry; antistat on plastic films
Trade names: Lonzaine® 16S
Trade names containing: Darvan® NS

Cetyl dimethylamine. *See* Dimethyl palmitamine

Cetyl esters
CAS 8002-23-1, 17661-50-6, 136097-97-7; EINECS 241-640-2
Synonyms: Synthetic spermaceti wax; Cetyl esters wax
Classification: Synthetic wax
Uses: Synthetic spermaceti NF; emollient and visc. builder for cosmetic and pharmaceutical preps.; processing aid and lubricant for plastics; gloss/sip aid for varnish
Manuf./Distrib.: Koster Keunen; Robeco; Werner G. Smith; Witco/Oleo-Surf.
Trade names: W.G.S. Synaceti 116 NF/USP

Cetyl esters wax. *See* Cetyl esters
Cetyl ethyl morpholinium ethosulfate. *See* Cetethyl morpholinium ethosulfate

Cetyl oleth-7 phosphate
Uses: Emulsifier for emulsion polymerization, metal cleaning, drycleaning detergent, pesticide and cosmetic preps.
Trade names: Servoxyl VPGZ 7/100

Cetyl-stearyl alcohol. *See* Cetearyl alcohol
Cetyl-stearyl methacrylate. *See* Cetearyl methacrylate
Cetyl trimethyl ammonium chloride. *See* Cetrimonium chloride
Channel black. *See* Carbon black
Charcoal. *See* Carbon black
CHDA. *See* 1,4-Cyclohexanedicarboxylic acid
CHDM. *See* 1,4-Cyclohexanedimethanol
China clay. *See* Kaolin
Chinese white. *See* Zinc oxide
Chlorallyl methenamine chloride. *See* Quaternium-15
Chlorcosane. *See* Paraffin, chlorinated

Chlorendic anhydride
CAS 115-27-5; EINECS 204-077-3
Synonyms: 4,5,6,7,8,8-Hexachloro-3a,4,7,7a-tetrahydro-4,7-methanoisobenzofuran-1,3-dione; 1,4,5,6,7,7-Hexachloro-5-norbornene-2,3-dicarboxylic anhydride; Hexachloroendomethylenetetrahydrophthalic an-

hydride
Empirical: $C_9H_2Cl_6O_3$
Properties: Fine white free-flowing crystals; sol. in acetone, benzene, toluene; sl. sol. in water, CCl_4; m.w. 370.86; dens. 1.73; m.p. 239-240 C
Uses: In flame-resistant polyester resins; chemical intermediate; source of chlorendic acid
Manuf./Distrib.: U.S. Chems.; Velsicol

Chlorinated dioctyl terephthalate. *See* Dioctyl terephthalate chlorinated
Chlorinated paraffin. *See* Paraffin, chlorinated
N-(3-Chloroallyl)hexaminium chloride. *See* Quaternium-15
1-(3-Chloroallyl)-3,5,7-triaza-1-azoniaadamantane chloride. *See* Quaternium-15
2-(5-Chloro-2H-benzotriazole-2-yl)-6-(1,1-dimethylethyl)-4-methylphenol. *See* 2-(2´-Hydroxy-3´-t-butyl-5´-methylphenyl)-5-chlorobenzotriazole
2-(5-Chloro-2H-benzotriazol-2-yl)-6-(1,1-dimethylethyl)-4-methylphenol. *See* 2-(2´-Hydroxy-3´-t-butyl-5´-methylphenyl)-5-chlorobenzotriazole
2-Chlorobutadiene 1,3. *See* Polychloroprene
Chlorocosane. *See* Paraffin, chlorinated
Chloroethene homopolymer. *See* Polyvinyl chloride
Chloroethylene polymer. *See* Polyvinyl chloride
3-Chloro-2-hydroxypropyltrimethyl ammonium chloride. *See* Chloro-2-hydroxypropyl trimonium chloride

Chloro-2-hydroxypropyl trimonium chloride
CAS 3327-22-8; EINECS 222-048-3
Synonyms: 3-Chloro-2-hydroxypropyltrimethyl ammonium chloride
Formula: $C_6H_{15}Cl_2NO$
Properties: M.w. 188.10; m.p. 189-190 C
Uses: Starch modifier for textiles; cationizing reagent for natural and syn. polymers
Manuf./Distrib.: Biddle Sawyer; Degussa; Fluka; Jarchem Ind.
Trade names: CHPTA 65%

5-Chloro-2-methyl-4-isothiazolin-3-one. *See* Methylchloroisothiazolinone
Chloroprene rubber. *See* Polychloroprene
Chloropropylene oxide elastomer. *See* Epichlorohydrin elastomer

3-Chloropropyltriethoxysilane
CAS 5089-70-3; EINECS 225-805-6
Empirical: $C_9H_{21}ClO_3Si$
Properties: Liq.; m.w. 240.80; dens. 1.004 (20/4 C); b.p. 98-102 C (10 mm); ref. index 1.420; flash pt. 61 C
Toxicology: Irritating to eyes, respiratory system, skin
Uses: Coupling agent, release agent, lubricant, blocking agent, chemical intermediate
Manuf./Distrib.: Howard Hall; Hüls Am.; OSi Spec.
Trade names: Dynasylan® CPTEO; Petrarch® C3292

3-Chloropropyltrimethoxysilane
CAS 2530-87-2; EINECS 219-787-9
Empirical: $C_6H_{15}ClO_3Si$
Properties: Liq.; m.w. 198.72; dens. 1.077; b.p. 195-196 C; ref. index 1.464 (25 C); flash pt. 78 C
Toxicology: Irritating to eyes, respiratory system
Uses: Coupling agent for epoxies, nylons, urethanes; release agent, lubricant
Manuf./Distrib.: Dow Corning; Howard Hall; Hüls Am.; OSi Spec.; PCR
Trade names: Dow Corning® Z-6076; Dynasylan® CPTMO; Petrarch® C3300

Chlorotrifluoroethene, homopolymer. *See* Polychlorotrifluoroethylene
CHP. *See* Cumyl hydroperoxide

C5 hydrocarbon resin
Uses: Tackifier, processing aid for adhesives, caulks, sealants, elastomers, polymers
Trade names: Eastotac™ H-100; Eastotac™ H-100E; Eastotac™ H-100L; Eastotac™ H-100R; Eastotac™ H-100W; Eastotac™ H-115E; Eastotac™ H-115L; Eastotac™ H-115R; Eastotac™ H-115W; Eastotac™ H-130E; Eastotac™ H-130L; Eastotac™ H-130R; Eastotac™ H-130W; Eastotac™ H-142R; Sta-Tac®; Sta-Tac® B; Sta-Tac® T; Wingtack® 10; Wingtack® 95

CI 77004. *See* Aluminum silicate

CI 77007. *See* Ultramarine blue
CI 77007. *See* Ultramarine violet
CI 77019. *See* Talc
CI 77220. *See* Calcium carbonate
CI 77266. *See* Carbon black
CI 77742. *See* Manganese violet
CI 77788. *See* Pigment yellow 53
CI 77891. *See* Titanium dioxide
CI 77947. *See* Zinc oxide

CI Fluorescent Brightener 236
CAS 3333-62-8
Classification: Coumarin fluorescent brightener
Empirical: $C_{25}H_{15}N_3O_2$
Properties: Green-yel. powd.; insol. in water; m.p. 482-486 F
Precaution: Dusts may be explosive with spark or flame initiation; incompat. with strong oxidizing agents; thermal decomp. may produce CO_x and NO_x
Toxicology: LD50 (oral, rat) > 5000 mg/kg; virtually nontoxic; nonirritating to skin and eyes; nonsensitizing
Uses: Fluorescent whitener/brightener for laundry detergents, soap, polymers and plastics
Regulatory: FDA 21CFR §177.1520, 177.2800
Trade names: Leucopure EGM Powd.

Citric acid
CAS 77-92-9 (anhyd.), 5949-29-1 (monohydrate); EINECS 201-069-1
Synonyms: 2-Hydroxy-1,2,3-propanetricarboxylic acid; β-Hydroxytricarballylic acid
Classification: Organic acid
Empirical: $C_6H_8O_7$
Formula: $HOC(COOH)(CH_2COOH)_2$
Properties: Colorless translucent crystals or powd.; odorless, tart taste; very sol. in water and alcohol; sol. in ether; m.w. 192.43; dens. 1.542; m.p. 153 C
Precaution: Combustible; potentially explosive reaction with metal nitrates
Toxicology: LD50 (oral, rat) 6730 mg/kg; poison by IV route; mod. toxic by subcutaneous and intraperitoneal routes; mildly toxic by ingestion; severe eye, mod. skin irritant; some allergenic props.; heated to decomp., emits acrid smoke and fumes
Uses: Preparation of citrates, acidifier, flavoring extracts, confections, soft drinks; antioxidant in foods; sequestering agent; dispersant; detergent builder; metal cleaner; curing accelerator
Regulatory: FDA 21CFR §131.111, 131.112, 131.136, 131.138, 131.144, 131.146, 133, 145.131, 145.145, 146.187, 150.141, 150.161, 155.130, 161.190, 166.40, 166.110, 169.115, 169.140, 169.150, 172.755, 173.160, 173.165, 173.280, 182.1033, 182.6033, GRAS; USDA 9CFR §318.7, 381.147; BATF 27CFR §240.1051, limitation 5.8 lb/1000 gal; Japan approved; Europe listed; UK approved
Manuf./Distrib.: Cargill; R.W. Greeff; Haarmann & Reimer; Hoffmann-La Roche; Penta Mfg.; PMC; Schweizerhall; U.S. Petrochem. Ind.
Trade names containing: Dimodan LS Kosher; Dur-Em® 117

Citric acid, octadecyl ester. *See* Stearyl citrate
C18 linear alcohol. *See* Stearyl alcohol
C6 linear alpha olefin. *See* 1-Hexene
C8 linear alpha olefin. *See* Octene-1
C14 linear alpha olefin. *See* Tetradecene-1
C12 linear primary alcohol. *See* Lauryl alcohol
C16 linear primary alcohol. *See* Cetyl alcohol
CM. *See* Polyethylene elastomer, chlorinated
CMBT. *See* Copper 2-mercaptobenzothiazolate
CMC. *See* Carboxymethylcellulose sodium
Coal tar naphtha. *See* Naphtha

Cobalt boro acylate
Uses: Cobalt adhesion promoter; promotes adhesion of rubber to steel, e.g., in tires
Trade names: Manobond™ 680-C

Cobalt-2-ethylhexoate. *See* Cobalt octoate

Cobalt octoate
CAS 136-52-7
Synonyms: Cobalt-2-ethylhexoate
Formula: $C_4H_9CH(C_2H_5)COOH$
Properties: Blue liq.; dens. 1.013 (25 C)
Uses: Accelerator, paint drier, whitener, catalyst
Manuf./Distrib.: OM Group; Shepherd; Witco/Oleo-Surf.
Trade names containing: Accelerator 55028; Accelerator NL-6; Accelerator NL-12; Accelerator NL49P; Accelerator NL51P; Accelerator NL53

Cocamide DEA
CAS 8051-30-7; 61791-31-9; 68603-42-9; EINECS 263-163-9
Synonyms: Coconut diethanolamide; Cocoyl diethanolamide; N,N-Bis (2-hydroxyethyl) coco amides
Definition: Ethanolamides of coconut acid
Formula: $RCO—N(CH_2CH_2OH)_2$, RCO represents the coconut acid radical
Uses: Detergent, thickener, emulsifier, foam booster/stabilizer, wetting agent, solubilizer for cosmetics, industrial and household cleaners, textiles, etc.; plastics antistat
Regulatory: FDA 21CFR §172.710, 173.322 (0.2% max.), 175.105, 176.210, 177.1200, 177.2260, 177.2800
Manuf./Distrib.: Aquatec Quimica SA
Trade names: Empilan® CDE; Empilan® CDX; Ethylan® LD; Ethylan® LDA-37; Ethylan® LDS

Cocamidopropylamine oxide
CAS 68155-09-9; EINECS 268-938-5
Synonyms: Cocamidopropyl dimethylamine oxide; Coco amides, N-[3-(dimethylamino)propyl], N-oxide; N-[3-(Dimethylamino)propyl]coco amides-N-oxide
Classification: Tertiary amine oxide
Formula: $RCO—NH(CH_2)_3N(CH_3)_3O$, RCO- represents the coconut fatty acids
Uses: Detergent, wetting agent, emulsifier, softener, conditioner, foam booster/stabilizer for rug shampoos, laundry detergents, dishwashing, shampoos, cleaners, antistatic softeners; foam stabilizer in foam rubber, electroplating, paper coatings
Trade names: Rhodamox® CAPO

Cocamidopropyl dimethylamine oxide. *See* Cocamidopropylamine oxide

Cocamine
CAS 61788-46-3; EINECS 262-977-1
Synonyms: Coconut amine
Classification: Primary aliphatic amine
Formula: RNH_2, R represents the coconut radical
Uses: Emulsifier, flotation agent, corrosion inhibitor, stripping agent for paints; mold release for rubber and plastics
Trade names: Kemamine® P-650D

Cocaminobutyric acid
CAS 68649-05-8; EINECS 272-021-5
Synonyms: Butanoic acid, 3-amino-, N-coco alkyl derivatives; 3-Aminobutanoic acid, n-coco alkyl derivatives
Classification: Substituted amino acid
Formula: $R—NH—CHCH_2COOHCH_3$, R represents the coconut radical
Uses: Pigment softening, dispersing agent; antifogging agent, foam booster, stabilizer, wetting agent in alkaline paint strippers, latex emulsions, latex rubber reclamation, inks, plastic films, cosmetics; cooling tower corrosion inhibitor
Trade names: Armeen® Z

Coco alkyl-2,2′-iminobisethanol
CAS 61791-31-9
Uses: Antistat for PP, PS, ABS, SAN resins
Trade names: Armostat® 410

Coco amides, N-[3-(dimethylamino)propyl], N-oxide. *See* Cocamidopropylamine oxide
Cocoamphopropionate. *See* Sodium cocoamphopropionate

Coco-betaine
CAS 68424-94-2, 85409-25-2; EINECS 270-329-4

Synonyms: Quat. ammonium compds., carboxymethyl (coco alkyl) dimethyl hydroxides, inner salts; Coconut betaine; Coco dimethyl betaine; Coco dimethyl glycine
Classification: Zwitterion (inner salt)
Uses: Detergent, wetting agent, emulsifier, foaming agent, solubilizer, bactericide, conditioner, visc. builder; industrial, household and cosmetic uses; plastics antistat
Manuf./Distrib.: Aquatec Quimica SA
Trade names: Nissan Anon BF

Coco diamino propane. *See* Cocopropylenediamine
Coco di(hydroxyethyl) amine oxide. *See* Dihydroxyethyl cocamine oxide
Coco dimethyl amine. *See* Dimethyl cocamine
Coco dimethyl betaine. *See* Coco-betaine
Coco dimethyl glycine. *See* Coco-betaine
Coco fatty acid. *See* Coconut acid

Coco morpholine
Uses: Catalyst in PU foams
Trade names: Armeen® N-CMD; Dabco® NCM

Coconut acid
CAS 61788-47-4; 67701-05-7, 68937-85-9; EINECS 262-978-7
Synonyms: Coco fatty acid; Coconut oil acids; Coconut fatty acids; Acids, coconut
Definition: Mixtures of fatty acids
Uses: Lubricant, intermediate used in alkyd resins, rubber compding., water repellents, polishes, soaps, abrasives, cutting oils, candles, crayons, emulsifiers, personal care prods.
Regulatory: FDA 21CFR §175.105, 175.320, 176.200, 176.210, 177.1010, 177.2260, 177.2600, 177.2800, 178.3570, 178.3910
Manuf./Distrib.: Akzo Nobel; Karlshamns; Norman, Fox; Procter & Gamble; Witco/Oleo-Surf.
Trade names: Industrene® 325; Industrene® 328
Trade names containing: Geropon® AS-200

Coconut amine. *See* Cocamine
Coconut betaine. *See* Coco-betaine
Coconut diethanolamide. *See* Cocamide DEA
Coconut fatty acid amidoethyl-N-2-hydroxyethylaminopropionate. *See* Sodium cocoamphopropionate
Coconut fatty acids. *See* Coconut acid
Coconut hydroxyethyl imidazoline. *See* Cocoyl hydroxyethyl imidazoline
Coconut oil acids. *See* Coconut acid
Coconut oil propane diamine. *See* Cocopropylenediamine
Coconut pentaethoxy methyl ammonium methyl sulfate. *See* PEG-5 cocomonium methosulfate
Coconut propylene diamine. *See* Cocopropylenediamine
Coconut trimethyl ammonium chloride. *See* Cocotrimonium chloride

Cocopropylenediamine
CAS 61791-63-7; EINECS 263-195-3
Synonyms: Coco diamino propane; Coconut oil propane diamine; Coconut propylene diamine
Uses: Chemical intermediate, emulsifier, gasoline detergent, bactericide, corrosion inhibitor in petrol. prod., epoxy hardener, flotation agent; used in metals, textiles, plastics, herbicides.
Trade names: Dinoram C; Duomeen® C; Kemamine® D-650

Cocotrimonium chloride
CAS 61789-18-2; EINECS 263-038-9
Synonyms: Coconut trimethyl ammonium chloride; Cocoyl trimethyl ammonium chloride; Quaternary ammonium compds., coco alkyl trimethyl, chlorides
Classification: Quaternary ammonium salt
Formula: $[RN(CH_3)_3]^+Cl^-$, R rep. alkyl groups from coconut oil
Uses: Emulsifier, dispersant used in corrosion inhibitor formulations for oil-field brines and HCl acidizing systems; antistat and lubricant for syn. fibers, plastics; bactericide, fungicide, disinfectant, sanitizer
Trade names: Chemquat C/33W
Trade names containing: Arquad® C-50

Cocoyl diethanolamide. *See* Cocamide DEA

Cocoyl hydroxyethyl imidazoline
CAS 61791-38-6; EINECS 263-170-7
Synonyms: 1H-Imidazole-1-ethanol, 4,5-dihydro-2-norcocoyl-; Cocoyl imidazoline; Coconut hydroxyethyl imidazoline
Classification: Heterocyclic compd.
Formula: R=N—N—$(CH_2)_2OH$, R is derived from coconut fatty radical
Uses: Emulsifier, antistat, corrosion inhibitor, softener for textiles, plastics, asphalt, tar emulsion breakers, paints, printing inks; water repellent treatment of cement, concrete, and plaster; fungicide
Trade names: Miramine® C

Cocoyl imidazoline. *See* Cocoyl hydroxyethyl imidazoline
N-Cocoyl-N-methyl glycine. *See* Cocoyl sarcosine

Cocoyl sarcosine
CAS 68411-97-2; EINECS 270-156-4
Synonyms: N-Cocoyl-N-methyl glycine; N-Methyl-N-(1-coconut alkyl) glycine
Definition: N-cocoyl deriv. of sarcosine
Uses: Detergent, wetting and foaming agent for hair and rug shampoos, cosmetics, industrial cleaners, pesticides, petrol. prods., textiles, metalworking; emulsifier for emulsion polymerization; plastics antistat, antifog; mold release for RIM; stabilizer for polyols
Regulatory: FDA 21CFR §178.3130
Manuf./Distrib.: Chemplex; Hampshire; R.T. Vanderbilt
Trade names: Hamposyl® C

Cocoyl trimethyl ammonium chloride. *See* Cocotrimonium chloride

Coke
CAS 50-36-2
Uses: Plastics additive providing wear reduction in PTFE, creep resist., chemical inertia and stability, thermal conductivity and stability
Manuf./Distrib.: Monsanto
Trade names: Lonza Coke PC 40

C4 alpha olefin. *See* 1-Butene
C12 alpha olefin. *See* Dodecene-1

C20-24 alpha olefin
CAS 64743-02-8
Uses: Intermediate for surfactants for personal care, specialty industrial chems. (polyethylene and other polymers; plasticizers; syn. lubricants; gasoline additives; paper sizing; PVC lubricants)
Trade names: Gulftene® 20-24

C24-28 alpha olefin
Uses: Intermediate for biodeg. surfactants for personal care and laundry, and specialty industrial chemicals (polyethylene and other polymers; plasticizers; syn. lubricants; gasoline additives; paper sizing; PVC lubricants)
Trade names: Gulftene® 24-28

C30 alpha olefin
Uses: Intermediate for biodeg. surfactants for personal care and laundry, and specialty industrial chemicals (polyethylene and other polymers; plasticizers; syn. lubricants; gasoline additives; paper sizing; PVC lubricants)
Trade names: Gulftene® 30+

Colloidal selenium. *See* Selenium
Colloidal silicon dioxide. *See* Silica
Cologel. *See* Methylcellulose
Colophony. *See* Rosin
Columbian spirits. *See* Methyl alcohol
Colza oil. *See* Rapeseed oil

Copolyether elastomer, thermoplastic
Uses: Offers performance of thermoset rubber with thermoplastic processing; used for automotive, elec./

electronic, industrial, consumer durables, and sporting goods; softening agent and homogenizer for rubbers
Trade names: Rhenosin® 140

Copper, bis(dimethyldithiocarbamate)-. *See* Copper dimethyldithiocarbamate
Copper chloride. *See* Cupric chloride

Copper dimethyldithiocarbamate
CAS 137-29-1
Synonyms: Dimethyldithiocarbamic acid copper salt; Wolfen; Copper, bis(dimethyldithiocarbamate)-
Empirical: $C_6H_{12}N_2S_4 \cdot Cu$
Formula: $[(CH_3)_2NC(S)S]_2$
Properties: Dark brown powd.; moderately sol. in acetone, benzene, chloroform; insol. in water, alcohol, gasoline; m.w. 303.98; dens. 1.75; m.p. > 325 C
Toxicology: Poison by intraperitoneal route
Uses: In SBR, primary accelerator, thiazole secondary accelerator; for molded and extruded goods
Trade names: Akrochem® Cu.D.D.; Akrochem® Cu.D.D.-PM; Methyl Cumate®; Methyl Cumate® Rodform; Perkacit® CDMC

Copper 2-mercaptobenzothiazolate
Synonyms: CMBT
Uses: Accelerator for EPDM
Trade names: Ekaland CMBT

Corn cob meal
Definition: Milled powder prepared from cobs of Zea mays
Properties: Powd.
Uses: Inert plastic extender and filler; replaces wood flour in wood particle molding with phenolic resins, in profile and sheet stock prod.; filler in glue, asphalt, caulking compds., and rubber; in industrial abrasives
Trade names: Grit-O'Cobs®; Lite-R-Cobs®

Cosmetic talc. *See* Talc
Cotton fiber. *See* Cellulose

Coumarone-indene resin
Classification: Thermosetting resin
Properties: Soft and sticky at R.T.; hardens on heating to solid; soften. pt. 126 C; ref. index 1.63-1.64
Toxicology: Heated to decomp., emits acrid smoke and irritating fumes
Uses: Extender in epoxy systems, adhesives, printing inks, floor tile binder; friction tape; paints; varnishes, enamels; protective coating on grapefruit, lemons, limes, oranges, tangerines; in chewing gum
Regulatory: FDA 21CFR §172.215
Manuf./Distrib.: Allchem Ind.; Natrochem; Neville

C11-15 pareth-3
CAS 68131-40-8 (generic)
Synonyms: Pareth-15-3
Definition: PEG ether of a mixture of syn. C11-15 fatty alcohols with avg. 3 moles of ethylene oxide
Uses: Detergent, emulsifier, wetting agent used in textile processing applics.; textile lubricant; latex emulsions; plastics antistat
Trade names: Tergitol® 15-S-3

C11-15 pareth-5
CAS 68131-40-8 (generic)
Synonyms: Pareth-15-5
Definition: PEG ether of a mixture of syn. C11-15 fatty alcohols with avg. 5 moles ethylene oxide
Uses: Detergent, emulsifier, wetting agent used in textile processing applics.; textile lubricant; latex emulsion stabilizer; plastics antistat
Trade names: Tergitol® 15-S-5

C11-15 pareth-30
CAS 68131-40-8 (generic)
Synonyms: Pareth-15-30
Definition: PEG ether of a mixture of syn. C11-15 fatty alcohols with avg. 30 moles of ethylene oxide

C11-15 pareth-40

Uses: Detergent, emulsifier, wetting agent used in textile processing applics.; textile lubricant; polymerization
Trade names: Tergitol® 15-S-30

C11-15 pareth-40
CAS 68131-40-8 (generic)
Synonyms: Pareth-15-40
Definition: PEG ether of a mixture of syn. C11-15 fatty alcohols with avg. 40 moles of ethylene oxide
Uses: Detergent, emulsifier, wetting agent used in textile processing applics.; textile lubricant; polymerization
Trade names: Tergitol® 15-S-40

C20-40 pareth-3
Definition: PEG ether of a mixture of syn. C20-40 alcohols with avg. 3 moles ethylene oxide
Uses: Emulsifier for cosmetics; vehicle; coatings, metalworking fluids, pulp/paper processing, textiles; mold release for plastics
Trade names: Unithox® 420

C20-40 pareth-10
Definition: PEG ether of a mixture of syn. C20-40 alcohols with avg. 10 moles ethylene oxide
Uses: Component for mold release agents; metalworking additive
Trade names: Unithox® 450

C20-40 pareth-40
Definition: PEG ether of a mixture of syn. C20-40 alcohols with avg. 40 moles ethylene oxide
Uses: Emulsifier for cosmetics; vehicle; coatings, metalworking fluids, pulp/paper processing, textiles; mold release for plastics
Trade names: Unithox® 480

C30-50 pareth-10
Definition: PEG ether of a mixture of syn. C30-50 alcohols with avg. 10 moles ethylene oxide
Uses: Emulsifier for cosmetics; vehicle; coatings, metalworking fluids, pulp/paper processing, textiles; mold release for plastics
Trade names: Unithox® 550

C40-60 pareth-10
Definition: PEG ether of a mixture of syn. C40-60 alcohols with avg. 10 moles ethylene oxide
Uses: Emulsifier for cosmetics; vehicle; coatings, metalworking fluids, pulp/paper processing, textiles; mold release for plastics
Trade names: Unithox® 750

C7-C9-C11 phthalate
Trade names containing: Omacide® P-711-5

CR. *See* Polychloroprene
p-Cresol dicyclopentadiene butylated polymer. *See* p-Cresol/dicyclopentadiene butylated reaction product

p-Cresol/dicyclopentadiene butylated reaction product
CAS 68610-51-5; EINECS 271-867-2
Synonyms: p-Cresol dicyclopentadiene butylated polymer
Properties: Powd.; sol. in acetone, ethyl acetate, ethanol, methylene chloride, liq. phosphites; insol. in water; m.w. 600-700; dens. 1.10; flash pt. (COC) 215 C
Uses: Antioxidant for rubber goods esp. latex applics.; also for EVA hot melts, thermoplastic rubber, styrenics
Trade names: Akrochem® Antioxidant 12; Ralox® LC; Santowhite® ML; Wingstay® L HLS

Cresyl diphenyl phosphate. *See* Diphenylcresyl phosphate

o-Cresyl glycidyl ether
CAS 2210-79-9
Synonyms: Glycidyl 2-methylphenyl ether
Empirical: $C_{10}H_{12}O_2$
Properties: M.w. 164.20
Toxicology: Irritant
Uses: Reactive epoxy diluent for tooling, elec. applics, coatings, flooring, casting, laminating, and decoupage

Manuf./Distrib.: Ashland; Raschig
Trade names: DY 023; Epodil® 742; Heloxy® 62

C13-15 trimethyl ammonium methosulfate
Uses: Antistat for plastics
Trade names containing: Synprolam 35TMQS

Cumene hydroperoxide
CAS 80-15-9
Synonyms: α,α-Dimethylbenzyl hydroperoxide
Classification: Organic peroxide
Empirical: $C_9H_{12}O_2$
Formula: $C_6H_5C(CH_3)_2OOH$
Properties: Colorless to pale yel. liq.; sl. sol. in water; sol. in alcohol, acetone, esters, hydrocarbons; m.w. 152.2; flash pt. 175 F
Precaution: Combustible; avoid contact with strong min. acids, other oxidizers, reducing agents, accelerators
Toxicology: Toxic by inhalation and skin absorption
Uses: Prod. of acetone and phenol; polymerization catalyst; curing agent for polyester resins
Manuf./Distrib.: AC Ind.; Elf Atochem N. Am.; JLM Marketing; Monomer-Polymer & Dajac; Witco/PAG
Trade names: Aztec® CHP-80; CHP-5; CHP-158; Trigonox® 239A; Trigonox® K-80; Trigonox® K-95
Trade names containing: HPC-9; Norox® MCP

Cumene peroxide. *See* Dicumyl peroxide

Cumene sulfonic acid
CAS 28631-63-2
Uses: Catalyst for foundry resins; descaling agent for metal cleaning; antistress additive and plating aid in electroplating bath; curing aid in the plastics industry; raw material in the mfg. of dyes and pigments; detergents industry
Trade names: Eltesol® CA 65; Eltesol® CA 96; Reworyl® C 65

Cumyl hydroperoxide
Synonyms: CHP
Manuf./Distrib.: Hüls AG
Trade names containing: Superox® 46-753-00

Cumyl peroxyneodecanoate
CAS 26748-47-0
Classification: Organic peroxide
Properties: M.w. 306.4
Uses: Initiator for PVC prod.
Trade names: Aztec® CUPND-75-AL; Trigonox® 99-B75; Trigonox® 99-W40
Trade names containing: Esperox® 939M; Lupersol 188-M75; Lupersol 288-M75; Trigonox® 99-C75

α-Cumylperoxyneoheptanoate
CAS 130097-36-8
Trade names containing: Esperox® 740M

Cumyl phenol
CAS 599-64-4
Synonyms: 4-Cumylphenol; 4-(2-Phenylisopropyl)phenol; p-Cumylphenol
Formula: $C_6H_5C(CH_3)_2C_6H_4OH$
Properties: White to tan crystals, phenol odor; m.w. 212.29; dens. 1.115 g/ml (25 C); m.p. 76-76 C; b.p. 335 C; f.p. 72 C; flash pt. 320 F
Precaution: Combustible
Uses: Intermediate for resins, insecticides, lubricants, surfactants; polycarbonate chain terminator; modifier for epoxy, furan, and phenolic resins
Manuf./Distrib.: AC Ind.; Hüls AG; ICI Specialties; PMC Specialties; Schenectady
Trade names: Ken Kem® CP-45; Ken Kem® CP-99

4-Cumylphenol. *See* Cumyl phenol
p-Cumylphenol. *See* Cumyl phenol

Cumylphenyl acetate
Properties: Sp.gr. 1.03; m.p. < 0 C; flash pt. (COC) 250 C
Uses: Plasticizer for urethanes; high flash pt. reactive diluent for epoxy, for polyamide-cured epoxy floorings and coal tar pipe coatings; comonomer and impact modifier for phenolics
Manuf./Distrib.: Kenrich Petrochem.
Trade names: Kenplast® ES-2

Cumylphenyl benzoate
Properties: Sp.gr. 1.10; m.p. 48-52 C; flash pt. (COC) 350 C
Uses: Primary plasticizer for PVC and PVC/nitrile; process aid for semirigid PVC extrudates; solvates conductive polyester and acrylic inks to improve impact
Manuf./Distrib.: Kenrich Petrochem.
Trade names: Kenplast® ESB

Cumylphenyl neodecanoate
Uses: Primary plasticizer for PVC, PVC/nitrile; process aid for semirigid PVC extrusions; solvates and impact modifies conductive polyester and acrylic inks
Trade names: Kenplast® ESN

Cupferron. *See* Nitrosophenylhydroxylamine ammonium salt

Cupric chloride
CAS 1344-67-8; EINECS 231-210-2
Synonyms: Copper chloride
Empirical: Cl_2Cu
Properties: Yel. to brown powd.; sol. in water, alcohol, acetone; m.w. 134.45; dens. 3.39 (25/4 C); partially dec. above 300 C; m.p. 498 C
Uses: Catalyst; deodorizing, desulfurizing agent in petrol. industry; mordant for dyeing and printing textiles; oxidizing agent for aniline dyes; in metallurgy, plating baths, photography, pyrotechnics; in mfg. of acrylonitrile; in pigments
Manuf./Distrib.: Am. Biorganics; Barker Ind.; Cerac; Cuproquim; Great Western; Hoechst Celanese; Spectrum Chem. Mfg.; Triple-S
Trade names containing: Luperfoam 40

Cyanamude. *See* Carbodiimide
Cyanogenamide. *See* Carbodiimide
Cyanoguanidine. *See* Dicyandiamide

1-[(1-Cyano-1-methylethyl)azo]formamide
CAS 10288-28-5
Synonyms: 2-(Carbamoylazo) isobutyronitrile
Empirical: $C_5H_8N_4O$
Properties: Pale yel. cryst.; sol. in ethanol, methanol, water; m.w. 140.14; m.p. 76-78 C
Uses: Polymerization initiator
Trade names: V-30

Cyanuric acid. *See* Isocyanuric acid
Cyclic ethylene carbonate. *See* Ethylene carbonate

Cyclic neopentanetetrayl bis(octadecyl phosphite)
Uses: Antioxidant for polyolefins

Cycloamylose. *See* Cyclodextrin
β-**Cycloamylose.** *See* Cyclodextrin

Cyclodextrin
CAS 7585-39-9, 10016-20-3; EINECS 231-493-2, 233-007-4
Synonyms: α-Cyclodextrin; Cyclomaltohexaose; Cycloamylose; β-Cycloamylose; Cyclomaltoheptaose
Definition: Cyclic polysaccharide comprised of six to eight glucopyranose units
Properties: M.w. 972.86; m.p. 278 (dec.)
Uses: Complex hosting guest molecules; increases the sol. and bioavailability of other substances; masks flavor, odor, or coloration; stabilizes against light, oxidation, heat, and hydrolysis; turns liqs. or volatiles into stable solid powds.; for use in pharmaceuticals, cosmetics, toiletries, foods, tobacco, pesticides, textiles,

paints, plastics, synthesis, polymers; Japan approved; not permitted in certain foods
Manuf./Distrib.: Am. Maize Prods.; Janssen Chimica; Pfanstiehl Labs; U.S. Biochemical; Wacker
Trade names: Beta W 7; Beta W7 P; Gamma W8

α-Cyclodextrin. *See* Cyclodextrin

Cyclo [dineopentyl (diallyl)] pyrophosphato dineopentyl (diallyl) zirconate
Uses: Coupling agent
Trade names containing: KZ TPP

Cyclo (dioctyl) pyrophosphato dioctyl titanate
Trade names containing: Ken-React® OPPR (KR OPPR)

Cyclo (dioctyl) pyrophosphato dioctyl zirconate
Uses: Coupling agent
Trade names containing: KZ OPPR

Cyclohexamethyleneimine. *See* Hexamethyleneimine
Cyclohexanamine. *See* Cyclohexylamine
1,2-Cyclohexanediamine. *See* 1,2-Diaminocyclohexane

1,4-Cyclohexanedicarboxylic acid
Synonyms: CHDA
Uses: Monomer for polymer applics.
Trade names: CHDA

1,2-Cyclohexanedicarboxylic anhydride. *See* Hexahydrophthalic anhydride

1,4-Cyclohexanedimethanol
CAS 105-08-8; EINECS 203-268-9
Synonyms: CHDM; 1,4-Bis(hydroxymethyl)cyclohexane (cis and trans)
Formula: $C_6H_{10}(CH_2OH)_2$
Properties: M.w. 144.21; b.p. 283 C; flash pt. 161 C
Precaution: Combustible
Uses: Monomer for polymer applics.
Manuf./Distrib.: Eastman; Fluka
Trade names: CHDM

1,4-Cyclohexanedimethanol dibenzoate
Uses: Plasticizer, modifier for hot-melt adhesives (EVA, block copolymer, polyester, polyamide, PU), delayed tack latex and hot-melts
Manuf./Distrib.: Velsicol

Cyclohexanedimethanol diglycidyl ether
CAS 14228-73-0
Properties: M.w. 256.35; dens. 1.100; flash pt. > 110 C; ref. index 1.4810 (20 C)
Toxicology: Irritant
Uses: Reactive epoxy diluent for casting, laminating, tooling, potting, elec., adhesive, civil engineering applics.
Trade names: Epodil® 757; Heloxy® 107

Cyclohexanone
CAS 108-94-1; EINECS 203-631-1
Synonyms: Ketohexamethylene; Pimelic ketone
Empirical: $C_6H_{10}O$
Formula: $CO(CH_2)_4CH_2$
Properties: Oily liq., peppermint-acetone odor; sl. sol. in water; sol. in alcohol, ether, common org. solvs.; m.w. 98.08; dens. 0.947 (20/4 C); m.p. -32.1 C; b.p. 156.7 C; flash pt. (CC) 63 C; ref. index 1.4507
Toxicology: LD50 (rat, oral) 1.62 ml/kg
Uses: Used as paint and varnish remover; solvent for cellulose acetate, nitrocellulose, natural resins, vinyl resins, rubber, waxes, fats; in prod. of adipic acid for nylon, cyclohexanone resins
Manuf./Distrib.: Aldrich; Allchem Ind.; AlliedSignal; Ashland; BASF; Coyne; Spectrum Chem. Mfg.; Union Carbide

Cyclohexanone peroxide
CAS 12262-58-7; EINECS 235-527-7
Synonyms: 1-Hydroperoxycyclohexyl 1-hydroxycyclohexyl peroxide
Formula: $C_6H_{10}(OOH)OOC_6H_{10}OH$
Properties: Grayish paste; insol. in water; sol. in most org. solvs.; dangerous fire risk
Uses: Initiator for ambient-temp. polyester cures; for automotive body putty, hobby and automotive kits
Trade names: Aztec® CHP-50-P1; Aztec® CHP-90-W1; Aztec® CHP-HA-1; Cyclonox® BT-50

Cyclohexylamine
CAS 108-91-8; EINECS 203-629-0
Synonyms: Aminocyclohexane; Aminohexahydrobenzene; Hexahydrobenzenamine; Cyclohexanamine; Hexahydroaniline
Empirical: $C_6H_{13}N$
Properties: Liq., strong fishy odor; m.w. 99.20; dens. 0.865; m.p. -17.7 C; b.p. 134.5 C; flash pt. 69.8 F
Precaution: Flamm. liq.; dangerous fire hazard exposed to heat, flame, or oxidizers; corrosive; heated to dec., emits toxic fumes of NO_x
Toxicology: TLV:TWA 10 ppm (skin); LD50 (oral, rat) 156 mg/kg; poison by ingestion, skin contact, IP; mod. toxic by subcut.; experimental teratogen, reproductive effects; severe human skin irritant; can cause dermatitis; human mutagenic data
Uses: Boiler water treatment, rubber accelerator, intermediate in organic synthesis
Regulatory: FDA 21CFR §173.310 (10 ppm max. in steam)
Manuf./Distrib.: Air Prods.; Allchem Ind.; Ashland; BASF; Elf Atochem N. Am.; Great Western; ICI Polyurethanes; Miles; PMC Specialties

Cyclohexylamine acetate
CAS 58695-41-3
Uses: Coagulant used in dipped goods from natural latex
Trade names: Coagulant CHA

N-Cyclohexyl-2-benzothiazolesulfenamide
CAS 95-33-0
Synonyms: CBTS; Benzothiazyl-2-cyclohexyl sulfenamide; 2-Benzothiazolesulfenamide, N-cyclohexyl-
Formula: $C_6H_4SNCSNHC_6H_{11}$
Properties: Cream-colored powd.; insol. in water; sol. in benzene, chloroform; dens. 1.27 (25 C); m.p. 93-100 C
Precaution: Dust may form explosive concs. in air; stable to 218 C; avoid reducing agents, acids
Toxicology: Irritating to eyes, skin and respiratory tract; may cause allergic skin reaction; LD50 (rat, oral) 5300 mg/kg (pract. nontoxic)
Uses: Rubber accelerator
Trade names: Akrochem® CBTS; Durax®; Durax® Rodform; Ekaland CBS; Perkacit® CBS; Santocure®; Vulkacit CZ/EG; Vulkacit CZ/EGC; Vulkacit DZ/EGC
Trade names containing: Poly-Dispersion® A (SAN)D-65; Rhenovin® CBS-70; Vulkacit CZ/MGC

Cyclohexyldimethylamine. *See* N,N-Dimethyl-N-cyclohexylamine

Cyclohexyl isooctyl phthalate
CAS 71486-48-1; EINECS 275-521-1
Synonyms: 1,2-Benzenedicarboxylic acid, cyclohexyl isooctyl ester
Uses: Plasticizer for nylon and polyester
Trade names: Uniplex 310

Cyclohexyl methacrylate
CAS 101-43-9; EINECS 202-943-5
Classification: Monomer
Formula: $H_2C{:}C(CH_3)COOC_6H_{11}$
Properties: Colorless monomeric liq.; insol. in water; m.w. 168.24; dens. 0.9626 (20/20 C); b.p. 210 C
Precaution: Combustible
Uses: Polymer modifier for optical lens systems, adhesives, floor polishes, vinyl polymerization, anaerobic adhesives; dental resins; encapsulation of electronic assemblies
Manuf./Distrib.: CPS; Fluka; Monomer-Polymer & Dajac; Polysciences; Rohm Tech
Trade names: Ageflex CHMA

Cyclohexyl-N′-phenyl-p-phenylenediamine
CAS 101-87-1

Uses: Antioxidant, antiozonant for dk. colored NR and SR
Trade names: Vanox® 6H

N-(Cyclohexylthio)phthalimide
CAS 17796-82-6
Synonyms: 1H-Isoindole-1,3-(2H)-dione, 2-(cyclohexylthio)
Properties: Cryst. or pellets, sl. mercaptan odor; insol. in water; dens. 1.3 mg/m^3; m.p. 90-95 C; flash pt. (PMOC) 165 C
Precaution: Fire produces toxic fumes; avoid contact with acids
Toxicology: Irritating to skin and eyes; may cause allergic skin reaction; LD50 (rat, oral) 2600 mg/kg, (rabbit, dermal) > 5010 mg/kg
Uses: Prevulcanization inhibitor; delays onset of accelerated sulfur vulcanization
Trade names: Santogard® PVI; Vantard® PVI
Trade names containing: Poly-Dispersion® S (PVI)D-50P

Cyclomaltoheptaose. *See* Cyclodextrin
Cyclomaltohexaose. *See* Cyclodextrin

Cycloneopentyl, cyclo (dimethlaminoethyl) pyrophosphato zirconate, di mesyl salt
Uses: Coupling agent, adhesion promoter, antioxidant, antistat, antifoam, accelerator, blowing agent activator, catalyst, curative, corrosion inhibitor, disp. aid, emulsifier, flame retardant, foaming agent, hardener, impact modifier, internal lube; process aid, release agent, retarder, stabilizer, surfactant, suspension aid, thixotrope, wetting agent for thermoplastics, thermosets, elastomers
Trade names: KZ TPPJ

Cyclopenta(de)naphthalene. *See* Bromoacenaphthylene
DACH. *See* 1,2-Diaminocyclohexane
DAHQ. *See* Diamylhydroquinone
DAMP. *See* 1,3-Pentanediamine
24DAMSP. *See* 2,4-Di-α-methylstyrylphenol
DAP. *See* Diallyl phthalate
Dapsone. *See* Sulfolane
DBDD. *See* 2,2′-Dibenzamidodiphenyl disulfide
DBEA. *See* Dibutoxyethyl adipate
DBEEA. *See* Dibutoxyethoxyethyl adipate
DBEEG. *See* Dibutoxyethoxyethyl glutarate
DBEEP. *See* Dibutoxyethoxyethyl phthalate
DBEG. *See* Dibutoxyethyl glutarate
DBEP. *See* Dibutoxyethyl phthalate
DBES. *See* Dibutoxyethyl sebacate
DBM. *See* Dibutyl maleate, Dibutyltin maleate
DBNPG. *See* Dibromoneopentyl glycol
DBP. *See* 2,4-Dibromophenol, Dibutyl phthalate
DBPC. *See* BHT
DBS. *See* Dibutyl sebacate
DBT 2-EHMA. *See* Dibutyltin bis(2-ethylhexyl mercaptoacetate)
DBTL. *See* Dibutyltin dilaurate
DBU. *See* Diazabicycloundecene
DCHP. *See* Dicyclohexyl phthalate
DCP. *See* Dicapryl phthalate
DCP-0. *See* Calcium phosphate dibasic
DDBSA. *See* Dodecylbenzene sulfonic acid
DDM. *See* n-Dodecyl mercaptan
DDSA. *See* Dodecenyl succinic anhydride
DEA. *See* Diethanolamine
DEAE. *See* Diethylaminoethanol

DEA-lauryl sulfate
CAS 143-00-0, 68585-44-4; EINECS 205-577-4
Synonyms: Sulfuric acid, monododecyl ester, compd. with 2,2′-iminodiethanol (1:1); Diethanolamine lauryl sulfate
Formula: $C_{12}H_{26}O_4S \cdot C_4H_{11}NO_2$
Uses: Detergent, foaming agent, wetting agent, emulsifier, visc. modifier for cosmetics, chemical specialties,

automobile cleaners; emulsion polymerization
Trade names: Carsonol® DLS

Deanol. *See* Dimethylethanolamine
DECA. *See* Decabromodiphenyl oxide
Decabromobiphenyl oxide. *See* Decabromodiphenyl oxide

Decabromodiphenyl oxide
CAS 1163-19-5; EINECS 214-604-9
Synonyms: DECA; Decabromobiphenyl oxide
Empirical: $C_{12}Br_{10}O$
Properties: Wh. powd., odorless; m.w. 959.2; m.p. 300-310 C
Uses: Flame retarant for HIPS, thermoset and thermoplastic polyesters, polypropylene, crosslinked polyethylene, elastomers, wire and cable insulation, adhesives, coatings, textile coatings
Manuf./Distrib.: Albemarle; Allchem Ind.; Great Lakes
Trade names: FR-1210; Great Lakes DE-83™; Great Lakes DE-83R™; Octoguard FR-01; Saytex® 102E; Thermoguard® 505
Trade names containing: Octoguard FR-15; Thermoguard® 8218

Decaglycerol tetraoleate. *See* Polyglyceryl-10 tetraoleate
Decaglyceryl tetraoleate. *See* Polyglyceryl-10 tetraoleate

Decahydronaphthalene
CAS 91-17-8; EINECS 202-046-9
Synonyms: Perhydronaphthalene; Dekalin; Naphthalane
Definition: Mixture of cis and trans
Empirical: $C_{10}H_{18}$
Properties: Insol. in water; sol. in alcohol; m.w. 138.14; dens. 0.8963; m.p. -43.26 C; b.p. 185.5 C
Uses: Solvent and stabilizer for shoe creams and floor waxes; solv. in paint and lacquers, oils, resins, rubber, and asphalt
Manuf./Distrib.: Fluka
Trade names: Decalin

Decanedioic acid, bis(1-methylethyl) ester. *See* Diisopropyl sebacate
Decanedioic acid, bis (2,2,6,6-tetramethyl-4-piperidinyl) ester. *See* Bis (2,2,6,6-tetramethyl-4-piperidinyl) sebacate
Decanedioic acid, dibutyl ester. *See* Dibutyl sebacate
Decanedioic acid, dimethyl ester. *See* Dimethyl sebacate
1-Decanol. *See* n-Decyl alcohol

Decanoyl chloride
CAS 112-13-0; EINECS 203-938-0
Synonyms: Capric acid chloride
Empirical: $C_{10}H_{19}ClO$
Formula: $CH_3(CH_2)_8COCl$
Properties: M.w. 190.71; dens. 0.919; b.p. 94-96 C (5 mm); ref. index 1.4410 (20 C)
Toxicology: Corrosive lachrymator
Uses: Intermediate, polymerization initiator
Manuf./Distrib.: Elf Atochem N. Am.; Janssen Chimica

Decanoyl peroxide
CAS 762-12-9
Classification: Organic peroxide
Empirical: $C_{20}H_{38}O_4$
Formula: $CH_3(CH_2)_8COOOCO(CH_2)_8CH_3$
Properties: M.w. 342.5; m.p. 38 C
Uses: Initator for polymerization, curing elastomers and polyester resins
Manuf./Distrib.: Elf Atochem N. Am.
Trade names: Aztec® DP; Decanox-F; Perkadox® SE-10

1-Decene. *See* Decene-1

Decene-1

CAS 872-05-9; EINECS 212-819-2
Synonyms: Linear C10 alpha olefin; Decylene; 1-Decene
Empirical: $C_{10}H_{20}$
Formula: $H_2C:CH(CH_2)_7CH_3$
Properties: Colorless liq., mild hydrocarbon odor; sol. in alcohol; sl. sol. in water; m.w. 140.27; dens. 0.741 (20/4 C); f.p. -66.3 C; b.p. 166-171 C; flash pt. (Seta) 114 F; ref. index 1.421 (20 C)
Precaution: Combustible
Toxicology: Irritating to skin and eyes; low acute inhalation toxicity; sl. toxic by ingestion
Uses: Intermediate for surfactants and specialty industrial chemicals (flavors, perfumes, pharmaceuticals, dyes, oils, resins)
Manuf./Distrib.: Albemarle; Chevron; Monomer-Polymer & Dajac; Shell
Trade names: Gulftene® 10

n-Decenyl succinic anhydride

Uses: Curing agent for epoxy resins, corrosion inhibitor for nonaq. lubricating oils, intermediate for prep. of alkyd or unsat. polyester resins, intermediate in chem. reactions
Trade names: Milldride® nDSA

Deceth-4 phosphate

CAS 52019-36-0 (generic); 68130-47-2; 9004-80-2 (generic)
Synonyms: PEG-4 decyl ether phosphate; PEG 200 decyl ether phosphate
Definition: Complex mixture of phosphoric acid esters of deceth-4
Uses: Detergent, foamer, dispersant, wetting agent for household, industrial, textile wet processing; coupler used in liq. alkali detergents; lubricant, antistat, corrosion inhibitor; emulsion polymerization
Regulatory: FDA 21CFR §176.210
Trade names: Cedephos® FA600

n-Decyl alcohol

CAS 112-30-1; 68526-85-2; EINECS 203-956-9
Synonyms: Alcohol C-10; Noncarbinol; Nonylcarbinol; 1-Decanol; Decylic alcohol
Classification: Fatty alcohol
Empirical: $C_{10}H_{22}O$
Formula: $CH_3(CH_2)_8CH_2OH$
Properties: Mod. visc. liq., sweet odor; sol. in alcohol, ether; insol. in water; m.w. 158.32; dens. 0.8297 (20/4 C); m.p. 7 C; b.p. 232.9 C; flash pt. (OC) 180 F; ref. index 1.43587
Precaution: Flamm. when exposed to heat or flame
Toxicology: LD50 (oral, rat) 4720 mg/kg; moderately toxic by skin contact; irritating to eyes, skin, respiratory system; heated to decomp., emits acrid smoke and irritating fumes
Uses: In mfg. of plasticizers, lubricants, petrol. additives, herbicides, surfactants, solvents; emulsion polymerization; moderate antifoaming capacity; flavoring agent
Regulatory: FDA 21CFR §172.515, 172.864, 175.300, 176.170, 178.3480, 178.3910; Japan approved as flavoring
Manuf./Distrib.: Albemarle; Aldrich; Brown; M. Michel; Penta Mfg.; Schweizerhall; Vista
Trade names: Epal® 10

Decyl betaine

CAS 2644-45-3; EINECS 220-152-3
Synonyms: N-(Carboxymethyl)-N,N-dimethyl-1-decanaminium hydroxide, inner salt; Decyl dimethyl glycine
Classification: Zwitterion (inner salt)
Formula: $C_{14}H_{29}NO_2$
Uses: Stabilizer for low alkaline latex systems; latex foam modifier, wetting agent
Trade names containing: Darvan® NS

Decyl dimethylamine. *See* Dimethyl decylamine
Decyl dimethyl glycine. *See* Decyl betaine

Decyl diphenyl phosphite

Uses: Stabilizer

Decylene. *See* Decene-1
Decylic alcohol. *See* n-Decyl alcohol
Decyl octyl alcohol. *See* Stearyl alcohol

Decyltridecyl phthalate
Properties: Sp.gr. 0.955; ref. index 1.4830
Uses: Plasticizer for PVC, PS, cellulosics
Manuf./Distrib.: Hüls

DEDM hydantoin
CAS 26850-24-8; EINECS 248-052-5
Synonyms: Di-(2-hydroxyethyl)-5,5-dimethyl hydantoin; 1,3-Bis(2-hydroxyethyl)-5,5-dimethyl-2,4-imidazolidinedione; Diethylol dimethyl hydantoin
Classification: Organic compd.
Empirical: $C_9H_{16}N_2O_4$
Uses: Intermediate for epoxies, urethane resins, and antistatic lubricants for the textile and plastics industries; crosslinker
Trade names: Dantocol® DHE

DEDM hydantoin dilaurate
Synonyms: Di-(2-hydroxyethyl)-5,5-dimethyl hydantoin dilaurate; Diethylol dimethyl hydantoin dilaurate
Definition: Diester of DEDM hydantoin
Empirical: $C_{33}H_{64}N_2O_6$
Uses: Intermediate for epoxies, urethane resins, and antistatic lubricants for the textiles industry
Trade names: Dantoest® DHE DL

DEG. *See* Diethylene glycol
DEHP. *See* Dioctyl phthalate
Dekalin. *See* Decahydronaphthalene
DEP. *See* Diethyl phthalate
DETA. *See* Diethylenetriamine
DETDA. *See* Diethyl toluene diamine
DETU. *See* 1,3-Diethylthiourea
Diacetin. *See* Glyceryl diacetate

Diacetone alcohol
CAS 123-42-2; EINECS 204-626-7
Synonyms: 4-Hydroxy-4-methyl-2-pentanone
Classification: Ketone
Empirical: $C_6H_{12}O_2$
Formula: $(CH_3)_2C(OH)CH_2COCH_3$
Uses: Solvent for nitrocellulose, cellulose acetate, oils, resins, waxes, fats, dyes, tars, lacquers, dopes, coatings, wood preservatives, rayon, artificial leather, metal cleaning; laboratory reagent; hydraulic fluids; textile stripping agent
Regulatory: FDA 21CFR §175.105
Manuf./Distrib.: Allchem Industries; Ashland; BP Chem. Ltd; Coyne; Elf Atochem N. Am.; Fabrichem; Great Western; Hoechst Celanese; Shell; Union Carbide

1,2-Diacetoxyethane. *See* Ethylene glycol diacetate

Dialkyl dimethyl ammonium chloride
Synonyms: Quaternium 31; Dimethyl dialkyl ammonium chloride
Toxicology: Heated to decomp., emits acrid smoke and irritating fumes
Uses: Softener
Regulatory: FDA 21CFR §173.400
Trade names containing: Adogen® 432

Diallyl phthalate
CAS 131-17-9; EINECS 205-016-3
Synonyms: DAP
Empirical: $C_{14}H_{14}O_4$
Formula: $C_6H_4(COOCH_2CH:CH_2)_2$
Properties: Colorless oily liq.; insol. in water; m.w. 246.27; dens. 1.120 (20/20 C); b.p. 158-165 (4 mm); f.p. -70 C; flash pt. 165.5 C
Precaution: Combustible
Toxicology: Toxic by ingestion
Uses: Primary plasticizer; intermediate

Manuf./Distrib.: Allchem Industries; Arco; Ashland; BP Chem. Ltd; C.P. Hall; Monomer-Polymer & Dajac; OxyChem; Rogers
Trade names: Ftalidap®
Trade names containing: Cadox® M-30; Cadox® MDA-30

1,2-Diaminocyclohexane
CAS 694-83-7; EINECS 211-776-7
Synonyms: DACH; 1,2-Cyclohexanediamine
Empirical: $C_6H_{14}N_2$
Properties: Misc. with water; m.w. 114; sp.gr. 0.94; b.p. 183 C (760 mm)
Precaution: Corrosive liq.
Uses: Epoxy curing agent; chelating agent for oilfield, textile, water treatment, detergent fields; herbicide intermediate; polyamide resins for adhesives, films, plastics, inks; PU for extenders; catalysts; scale/corrosion inhibitors
Manuf./Distrib.: Air Prods.; Aldrich; Du Pont; Fluka; Johnson Matthey; Milliken
Trade names: Ancamine® 1770; DCH-99
Trade names containing: Millamine® 5260

p,p´-Diaminodiphenylmethane. *See* 4,4´-Methylene dianiline
1,6-Diaminohexane. *See* Hexamethylenediamine
1,3-Diaminopentane. *See* 1,3-Pentanediamine

Diammonium cocoyl sulfosuccinate
Uses: Foaming agent for rubber latex compds.; emulsifier for emulsion polymerization
Trade names: Empimin® MSS

Diammonium edetate. *See* Diammonium EDTA

Diammonium EDTA
CAS 20824-56-0; EINECS 244-063-4
Synonyms: Diammonium edetate; Diammonium ethylene diamine tetraacetate
Classification: Substituted diamine
Empirical: $C_{10}H_{16}N_2O_8 \cdot 2H_3N$
Formula: $NCH_2CH_2N(CH_2COOH)_2(CH_2COONH_4)_2$
Uses: Chelating agent; photographic developer; for cosmetics, pharmaceuticals, in water treatment, textiles, soaps/detergents, electroless copperplating, polymer prod., disinfectants, pulp/paper, enhanced oil recovery, metal cleaning/protection
Trade names: Versene Diammonium EDTA

Diammonium ethylene diamine tetraacetate. *See* Diammonium EDTA
Diammonium octadecyl sulfosuccinamate. *See* Diammonium stearyl sulfosuccinamate

Diammonium stearyl sulfosuccinamate
CAS 68128-59-6
Synonyms: Diammonium octadecyl sulfosuccinamate
Uses: Foaming agent, emulsifier, stabilizer for acrylic latex; suspending agent in emulsion polymerization
Trade names: Octosol A-18-A

Diammonium tetraborate tetrahydrate. *See* Ammonium biborate

Diamylhydroquinone
CAS 79-74-3; EINECS 201-222-2
Synonyms: DAHQ; 2,5-Di (t-amyl) hydroquinone; 2,5-Di-t-pentylhydroquinone; 2,5-Bis (1,1-dimethylpropyl)-1,4-benzenediol
Empirical: $C_{16}H_{26}O_2$
Formula: $(C_5H_{11})_2(C_6H_2OH)_2$
Properties: Buff powd.; sl. sol. in water; sol. in alcohol, benzene; dens. 1.05 (25 C); m.p. 176 C
Uses: Antioxidant for uncured rubber; polymer inhibitor; food pkg.
Regulatory: FDA 21CFR §175.105
Trade names: Santovar® A

2,5-Di (t-amyl) hydroquinone. *See* Diamylhydroquinone

Di-t-amyl peroxide
CAS 10508-09-5
Empirical: $C_{10}H_{22}O_2$
Formula: $C_2H_5C(CH_3)_2OOC(CH_3)_2C_2H_5$
Properties: M.w. 174.3
Uses: Polymerization initiator
Trade names: Aztec® DTAP; Trigonox® 201

1,1-Di (t-amylperoxy) cyclohexane
CAS 15667-10-4
Uses: Initiator for polymerization, for high-temp. cure of polyester resins, and for cure of acrylic syrup
Trade names containing: Lupersol 531-80B; Lupersol 531-80M; USP®-90MD

2,2-Di-(t-amylperoxy) propane
CAS 3052-70-8
Classification: Peroxyketal
Uses: Initiator for high temp. cure of polyesters, cure of acrylic syrups; crosslinking agent for polymerization
Trade names containing: Lupersol 553-M75

2,4-Di-t-amylphenol
CAS 120-95-6
Synonyms: 24DTAP
Formula: $[C_2H_5C(CH_3)_2]_2C_6H_3OH$
Properties: M.w. 234.38; dens. 0.930; m.p. 25 C; b.p. 169-170 C (22 mm); flash pt. > 110 C
Toxicology: Toxic
Manuf./Distrib.: Elf Atochem N. Am.

Diamyl sodium sulfosuccinate
CAS 922-80-5; EINECS 213-085-6
Synonyms: Sodium diamyl sulfosuccinate; Sulfobutanedioic acid 1,4-dipentyl ester sodium salt; Sulfosuccinic acid dipentyl ester sodium salt
Definition: Sodium salt of the diester of an amyl alcohol and sulfosuccinic acid
Empirical: $C_{14}H_{26}O_7S \cdot Na$
Properties: White powd.; sol. in water, acetone, CCl_4, glycerol, hot kerosene, hot olive oil; m.w. 360.41
Uses: Emulsifier in emulsion polymerization; wetting agent
Trade names: Aerosol® AY-65

1,4-Dianilinobenzene. *See* N,N′-Diphenyl-p-phenylenediamine

Diaryl p-phenylene diamine
CAS 68478-45-5; 68953-84-4
Uses: Antioxidant/antiozonant for rubber goods
Trade names: Wingstay® 100; Wingstay® 100AZ; Wingstay® 200

Diatomaceous earth
CAS 7631-86-9, 68855-54-9; EINECS 231-545-4
Synonyms: Kieselguhr; Diatomite; Diatomaceous silica; Infusorial earth
Definition: Mineral material consisting chiefly of the siliceous frustules and fragments of various species of diatoms
Properties: Soft bulky solid; insol. in acids except HF; sol. in strong alkalies; dens. 1.9-2.35; bulk dens. 8-15 lb/ft^3; oil absorp. 135-185%; 88% silica; noncombustible
Toxicology: TLV:TWA 10 mg/m^3 (dust); poison by inhalation and ingestion; dust may cause fibrosis of the lungs
Uses: Filtration, clarifying, decolorizing; insulation absorbent; mild abrasive; catalyst carrier; anticaking agents in fertilizer; conditioning agent for dusts, soil; filler, processing aid for rubber
Regulatory: FDA 21CFR §73.1, 133.146, 160.105, 160.185, 172.230, 172.480, 173.340, 175.300, 177.1460, 177.2410, 182.90, 193.135, 561.145, 573.940; USDA 9CFR §318.7; Japan restricted (0.5% max. residual)
Manuf./Distrib.: Ashland; Celite; Coyne; CR Mins.; Great Western; L.A. Salomon; Seefast (Europe) Ltd; Spectrum Chem. Mfg.
Trade names: Celite® 270; Celite® 292; Celite® 350; Celite® Super Fine Super Floss; Celite® Super Floss; Celite® White Mist
Trade names containing: Ampacet 10063

Diatomaceous silica. *See* Diatomaceous earth

Diatomite. *See* Diatomaceous earth

Diazabicycloundecene
CAS 6674-22-2; EINECS 229-713-7
Synonyms: DBU; 1,8-Diazabicyclo[5.4.0]undec-7-ene (1,5-5); [2,3,4,6,7,8,9,10-Octapyrimidol[1,2-a]azepine]
Empirical: $C_9H_{16}N_2$
Uses: Accelerator for phenolic novolac and other epoxy cures
Manuf./Distrib.: Air Prods.; Aldrich; BASF; Fluka; SAF Bulk Chems.; Schweizerhall
Trade names: Amicure® DBU-E; Polycat® DBU; Polycat® SA-1; Polycat® SA-102; Polycat® SA-610/50

1,8-Diazabicyclo[5.4.0]undec-7-ene (1,5-5). *See* Diazabicycloundecene

Diazabicycloundecene, 2-ethylhexanoic acid salt
Uses: Epoxy curing agent
Trade names: Amicure® SA-102

Diazenedicarboxamide. *See* Azodicarbonamide
Diazobicyclooctane. *See* Triethylene diamine
1,4-Diazobicyclo [2.2.2] octane. *See* Triethylene diamine
DIBA. *See* Diisobutyl adipate
Dibasic lead carbonate. *See* Lead carbonate (basic)

2,2´-Dibenzamidodiphenyl disulfide
CAS 135-57-9
Synonyms: DBDD
Trade names containing: Akrochem® Peptizer 66

Dibenzo G-M-F. *See* Dibenzoyl-p-quinone dioxime
2,2´-Dibenzothiazyl disulfide. *See* Benzothiazyl disulfide
Dibenzoyl peroxide. *See* Benzoyl peroxide

Dibenzoyl-p-quinone dioxime
Synonyms: Dibenzo G-M-F
Formula: $(C_6H_5COON)_2C_6H_4$
Properties: Brownish-gray powd.; insol. in acetone, benzene, gasoline, ethylene dichloride, water; dens. 1.37; dec. above 200 C
Uses: Nonsulfur vulcanizing agent; in tire-curing bags, gaskets and wire insulation to impart heat resistance
Manuf./Distrib.: Lord
Trade names: DBQDO®

Dibenzyl azelate
Uses: Plasticizer
Trade names: Plasthall® DBZZ

Dibenzylidene sorbitol
CAS 32647-67-9; EINECS 251-136-4
Synonyms: Sorbitol acetal; Bis-o-(phenylmethylene)-D-glucitol; D-Glucitol, bis-o-(phenylmethylene)-
Classification: Substituted sorbitol
Empirical: $C_{20}H_{22}O_6$
Uses: Thixotrope and gellant for org. systems, unsat. polyester and vinyl ester resins
Regulatory: FDA 21CFR §178.3295
Trade names: Millithix® 925

DIBK. *See* Diisobutyl ketone
DIBP. *See* Diisobutyl phthalate

Dibromoethyldibromocyclohexane
CAS 3322-93-8; EINECS 222-036-8
Uses: Flame retardant for expandable PS, crystal and high impact PS, SAN, adhesives, coatings, textile treatment, polyurethanes
Manuf./Distrib.: Albemarle
Trade names: Saytex® BCL-462

Dibromoneopentyl glycol
CAS 3296-90-0; EINECS 221-967-7
Synonyms: DBNPG; Dibromopentaerythritol; 2,2-Bis(bromomethyl)-1,3-propanediol
Empirical: $C_5H_{10}Br_2O_2$
Formula: $(CH_2OH)_2C(CH_2Br)_2$
Properties: Off-white cryst. powd.; sol. in acetone, isopropanol, methanol; m.w. 261.94; dens. 2.23
Toxicology: LD50 (oral, rat) 1880 mg/kg; mild irritant to eyes, effects reversible; nonirritant to intact skin
Uses: Flame retardant for unsaturated polyesters and polyurethane rigid foams; flame retardant intermediate
Manuf./Distrib.: Albemarle
Trade names: FR-521; FR-522; FR-1138
Trade names containing: Saytex® FR-1138

Dibromoneopentyl glycol diglycidyl ether
Uses: Epoxy modifier for flame retardant applics.
Trade names: Heloxy® 56

Dibromopentaerythritol. *See* Dibromoneopentyl glycol

2,4-Dibromophenol
CAS 615-58-7; EINECS 210-436-5
Synonyms: DBP
Empirical: $C_6H_4Br_2O$
Formula: $Br_2C_6H_3OH$
Properties: Colorless needles; sol. in alcohol, acetone, benzene, CCl_4; m.w. 251.86; dens. 1.04; m.p. 33-36 C
Toxicology: Irritant to eyes; nonirritant to skin; LD50 (oral, mouse) 2780 mg/kg
Uses: Flame retardant for epoxy, phenolic, and polyester resins; flame retardant intermediate
Manuf./Distrib.: Esprit; Fluka; Great Lakes
Trade names: FR-612

Dibromostyrene
Uses: Reactive flame retardant for plastics
Manuf./Distrib.: Great Lakes
Trade names: Great Lakes DBS™

Dibromo/tribromo salicylanilide
Uses: Antimicrobial for resins, latex emulsions, plastics

Di-t-butoxydiacetoxysilane
CAS 13170-23-5
Formula: $C_{12}H_{24}SiO_6$
Properties: Liq.; m.w. 292.4; dens. 1.0196; b.p. 102 C (5 mm); ref. index 1.4040; flash pt. 95 C
Uses: Coupling agent, release agent, lubricant, blocking agent, chemical intermediate
Trade names: Dynasylan® BDAC

Dibutoxyethoxyethyl adipate
CAS 141-17-3
Synonyms: DBEEA
Properties: Sp.gr. 1.010-1.020; flash pt. (COC) 208 C; ref. index 1.444-1.446
Uses: Plasticizer for PVAc, PVB, some cellulosics
Manuf./Distrib.: C.P. Hall; Harwick; Inolex; Morton Int'l.; Novachem; Sartomer
Trade names: Plasthall® 226; Plasthall® DBEEA; Suprmix DBEEA; Wareflex® 650

Dibutoxyethoxyethyl formal
Properties: Sp.gr. 0.96-0.98; flash pt. (COC) 163 C; ref. index 1.435-1.45
Uses: Plasticizer for PVC, PVB, cellulose nitrate
Manuf./Distrib.: Sartomer
Trade names: Cryoflex® 660

Dibutoxyethoxyethyl glutarate
Synonyms: DBEEG
Properties: Sp.gr. 1.016; m.p. < -60 C; flash pt. (COC) 143 C; ref. index 1.4437
Uses: Plasticizer for PVC, PS, PVAc, cellulosics

Manuf./Distrib.: C.P. Hall
Trade names: Plasthall® 224

Dibutoxyethoxyethyl phthalate
Synonyms: DBEEP
Uses: Plasticizer for PU
Trade names: Plasthall® 220

Dibutoxyethyl adipate
CAS 141-18-4
Synonyms: DBEA
Formula: $(C_2H_4COOC_2H_4OC_4H_9)_2$
Properties: Colorless oily liq., mild butyl odor; dens. 0.997 (20/20 C); b.p. 203-215 (4 mm); f.p. -34 C; flash pt. 370 F
Precaution: Combustible
Uses: Primary plasticizer for most resins
Trade names: Plasthall® 203; Plasthall® DBEA; Staflex DBEA

Dibutoxyethyl azelate
Properties: Sp.gr. 0.975; m.p. -25 C; flash pt. (COC) 210 C; ref. index 1.444
Uses: Plasticizer for PVC, PVB, VCA, CAB
Manuf./Distrib.: C.P. Hall
Trade names: Plasthall® 205

2,2′-Dibutoxyethyl ether. *See* Diethylene glycol dibutyl ether

Dibutoxyethyl glutarate
Synonyms: DBEG
Properties: Sp.gr. 1.002; m.p. < -60 C; flash pt. (COC) 193 C; ref. index 1.440
Uses: Plasticizer for polymers
Manuf./Distrib.: C.P. Hall
Trade names: Plasthall® 201

Dibutoxyethyl phthalate
CAS 117-83-9
Synonyms: DBEP; n-Butyl glycol phthalate; Ethyldibutoxy phthalate
Formula: $C_6H_4(COOC_2H_4OC_4H_9)_2$
Properties: Colorless liq.; sol. in org. solvs.; dens. 1.06 (20 C); b.p. 270 C; f.p. -55 C; flash pt. 407 F
Uses: Plasticizer for PVC, polyvinyl acetate, and other resins
Manuf./Distrib.: Ashland; C.P. Hall; Unitex
Trade names: Plasthall® 200; Staflex DBEP

Dibutoxyethyl sebacate
Synonyms: DBES
Properties: Sp.gr. 0.97; m.p. -20 C; flash pt. (COC) 238 C; ref. index 1.445
Uses: Plasticizer for polymers
Manuf./Distrib.: C.P. Hall
Trade names: Plasthall® 207

Dibutyl adipate
CAS 105-99-7; EINECS 203-350-4
Synonyms: Hexanedioic acid, dibutyl ester
Definition: Diester of butyl alcohol and adipic acid
Empirical: $C_{14}H_{26}O_4$
Formula: $[—CH_2CH_2CO_2(CH_2)_3CH_3]_2$
Properties: M.w. 258.36; dens. 0.962; b.p. 305 C; flash pt. > 110 C
Uses: Emollient; oily component for day creams; plasticizer for cellulosics, PVAc, PVC, VCA
Manuf./Distrib.: Aquatec Quimica SA; Novachem

Dibutylammonium oleate
Formula: $(C_4H_9)_2NH_2COOC_{17}H_{33}$
Properties: Translucent lt. brown liq.
Precaution: Combustible

Dibutyl azelate

Uses: Activator for accelerators; processing aid for rubber and synthetic rubber
Trade names: Activator 1102

Dibutyl azelate
CAS 2917-73-9
Uses: Vinyl plasticizer esp. for cellulosic molding compds.
Trade names: Plastolein® 9048

Dibutyl-1,2-benzene dicarboxylate. *See* Dibutyl phthalate

2,6-Di-t-butyl-4-s-butyl phenol
CAS 17540-75-9
Formula: $C_2H_5CH(CH_3)C_6H_2[C(CH_3)_3]_2OH$
Properties: M.w. 262.44
Toxicology: Irritant
Uses: Antioxidant used in polyols and rubber systems
Trade names: Isonox® 132; Vanox® 1320

Dibutyl 'Carbitol.' *See* Diethylene glycol dibutyl ether
Dibutyl-o-(o-carboxybenzoyl) glycolate. *See* n-Butyl phthalyl-n-butyl glycolate
Dibutyl-o-carboxybenzoyloxyacetate. *See* n-Butyl phthalyl-n-butyl glycolate
2,6-Di-t-butyl-p-cresol. *See* BHT

Di-(4-t-butylcyclohexyl) peroxydicarbonate
CAS 15520-11-3
Synonyms: Bis (4-t-butylcyclohexyl) peroxydicarbonate
Classification: Organic peroxide
Properties: M.w. 398.5
Uses: Initiator for polymerization and copolymerization of vinyl chloride, polyester, acrylates, methacrylates
Trade names: Aztec® BCHPC; Aztec® BCHPC-40-SAQ; Aztec® BCHPC-75-W; Perkadox® 16; Perkadox® 16/35; Perkadox® 16N; Perkadox® 16S; Perkadox® 16-W25-GB1; Perkadox® 16/W40; Perkadox® 16-W40-GB2; Perkadox® 16-W40-GB5; Perkadox® 16/W70

Dibutyl decanedioate. *See* Dibutyl sebacate
Di-n-butyldiacetoxystannane. *See* Dibutyltin diacetate
2,6-Di-t-butyl-α-dimethylamino-p-cresol. *See* 2,6-Di-t-butyl-N,N-dimethylamino-p-cresol

2,6-Di-t-butyl-N,N-dimethylamino-p-cresol
CAS 88-27-7
Synonyms: 2,6-Di-t-butyl-α-dimethylamino-p-cresol
Formula: $(C_4H_9)_2C_6H_2OH[CH_2N(CH_3)_2]$
Properties: Lt. yel. crystalline solid; insol. in water and 10% NaOH; sol. in org. solvs.; m.w. 263.4; m.p. 93.9 C; flash pt. (OC) 280 F
Precaution: Combustible
Uses: Antioxidant in gasoline and oils including jet engine oils
Trade names: Ethanox® 703

Di-t-butyl diperoxyazelate
CAS 16580-06-6
Classification: Organic peroxide
Properties: M.w. 332.4
Trade names containing: Trigonox® 169-NA50

Di-t-butyl diperoxyphthalate. *See* Di-t-butyl peroxyphthalate
Dibutyldithiocarbamic acid sodium salt. *See* Sodium di-n-butyl dithiocarbamate

2,6-Di-t-butyl-4-ethyl phenol
Synonyms: BHEB
Uses: Plastics antioxidant
Manuf./Distrib.: Aceto

Dibutyl fumarate
CAS 105-76-9

Synonyms: Fumaric acid, dibutyl ester
Formula: $(C_4H_9)OOCCH{:}CHCOO(C_4H_9)$
Properties: Colorless liq.; insol. in water; dens. 0.9873 (20 C); b.p. 285.2 C; f.p. -15.6 C
Precaution: Combustible
Uses: Monomeric plasticizer for PVC, PVAc; copolymers, intermediate
Manuf./Distrib.: AC Ind.; Monomer-Polymer & Dajac; Penta Mfg.; Unitex
Trade names: Staflex DBF

2,5-Di-t-butyl hydroquinone. *See* Di-t-butylhydroquinone

Di-t-butylhydroquinone
CAS 88-58-4; EINECS 201-841-8
Synonyms: 2,5-Di-t-butyl hydroquinone; 2,5-Bis (1,1-dimethylethyl)-1,4-benzenediol
Empirical: $C_{14}H_{22}O_2$
Formula: $[C(CH_3)_3]_2C_6H_2(OH)_2$
Properties: White powd.; sol. in acetone, alcohol, benzene; insol. in water, aq. alkali; m.w. 222.23; m.p. 210-212 C; flash pt. (COC) 216 C
Uses: Polymerization inhibitor, antioxidant, stabilizer against UV deterioration of rubber
Regulatory: FDA 21CFR §175.105, 176.170, 176.180, 176.210, 177.2260, 177.2420, 177.2800
Manuf./Distrib.: Eastman; Fluka
Trade names: Eastman® DTBHQ

3,5-Di-t-butyl-4-hydroxybenzoic acid. *See* 3,5-Di-t-butyl-p-hydroxybenzoic acid

3,5-Di-t-butyl-p-hydroxybenzoic acid
CAS 1421-49-4; EINECS 215-823-2
Synonyms: 3,5-Di-t-butyl-4-hydroxybenzoic acid
Empirical: $C_{15}H_{22}O_3$
Properties: Powd.; m.w. 250.34; m.p. 200-210 C
Uses: UV stabilizer for PE, PP
Manuf./Distrib.: Fluka
Trade names: Aceto

3,5-Di-t-butyl-4-hydroxybenzoic acid, n-hexadecyl ester
Synonyms: n-Hexadecyl 3,5-di-t-butyl-4-hydroxybenzoate
Empirical: $C_{31}H_{54}O_3$
Properties: Wh. cryst., odorless; m.w. 475; m.p. 60 C
Toxicology: LD50 (rat, oral) > 5 g/kg; LD50 (rabbit, dermal) > 5 g/kg
Uses: Light stabilizer for polyolefins, polyesters, ABS; free radical scavenger
Trade names: Cyasorb® UV 2908

Dibutyl hydroxy hydrocinnamic acid triester
Trade names containing: Irganox® 3125

3,5-Di-t-butyl-4-hydroxyhydrocinnamic acid/1,3,5-tris(2-hydroxyethyl)-s-triazine-2,4,6-(1H,3H,5H)-trione triester
CAS 34137-09-2
Uses: Stabilizer/antioxidant for polyolefins, hot melt and food pkg.
Trade names: Vanox® SKT

2-(3′,5′-Di-t-butyl-2′-hydroxyphenyl)-5-chlorobenzotriazole
CAS 3864-99-1; EINECS 223-383-8
Uses: Light stabilizer for PVC
Trade names: Tinuvin® 327

3-(3,5-Di-t-butyl-4-hydroxyphenyl)propionic methyl ester
CAS 6886-38-5
Empirical: $C_{18}H_{28}O_3$
Uses: Antioxidant for min. oil, plastics, rubber
Trade names: Ralox® 35

Dibutyl maleate
CAS 105-76-0; EINECS 203-328-4
Synonyms: DBM; 2-Butenedioic acid, dibutyl ester
Empirical: $C_{12}H_{20}O_4$
Formula: $C_4H_9OOCCH:CHCOOC_4H_9$
Properties: Colorless oily liq.; insol. in water; m.w. 228.29; dens. 0.9964 (20/20 C); b.p. 274-277flash pt. (OC) 285 F
Precaution: Combustible
Uses: Copolymers, plasticizer for PVAc, intermediate
Manuf./Distrib.: AC Ind.; Allchem Ind.; Aristech; Ashland; Hoechst AG; Monomer-Polymer & Dajac; Pentagon Chem. Ltd; Penta Mfg.; Unitex
Trade names: PX-504

Dibutyl methylene bisthioglycolate
Uses: Plasticizer to improve low-temp. flexibility and elastic behavior of nitrile and chloroprene rubber
Trade names: Vulkanol® 88

2,6-Di-t-butyl-4-methylphenol. *See* BHT

Di (butyl, methyl pyrophosphato) ethylene titanate di (dioctyl, hydrogen phosphite)
Uses: Coupling agent, adhesion promoter, antioxidant, antistat, antifoam, accelerator, blowing agent activator, catalyst, curative, corrosion inhibitor, dispersion aid, emulsifier, flame retardant, foaming agent, grinding and process aid
Trade names containing: Ken-React® 262ES (KR 262ES)

Di-t-butyl peroxide
CAS 110-05-4; EINECS 203-733-6
Synonyms: DTBP; t-Butyl peroxide; Bis (1,1-dimethylethyl) peroxide
Classification: Organic peroxide
Empirical: $C_8H_{18}O_2$
Formula: $(CH_3)_3COOC(CH_3)_3$
Properties: Clear water-white liq.; sol. in styrene, ketones, most aliphatic and aromatic hydrocarbons; insol. in water; m.w. 146.23; dens. 0.791 (25/25 C); f.p. -40 C; b.p. 111 C; flash pt. (CC) 65 F
Precaution: Flamm. oxidizing liq.; dangerous fire hazard
Toxicology: LD50 (rat, oral) > 25,000 mg/kg
Uses: Polymerization catalyst for resins (e.g., olefins, styrene, styrenated alkyds, silicones); ignition accelerator for diesel fuel; organic synthesis; intermediate
Manuf./Distrib.: Akzo Nobel; Amber Syn.; Elf Atochem N. Am.; Witco/PAG
Trade names: Aztec® DTBP; Trigonox® B
Trade names containing: TBHP-70

Di-t-butyl peroxyazelate
CAS 16580-06-6
Trade names: Trigonox® 169-OP50

2,2-Di (t-butylperoxy) butane
CAS 2167-23-9
Classification: Organic peroxide
Properties: M.w. 234.3
Uses: Initiator for polymerization, high-temp. curing of polyester resins, and cure of acrylic syrup
Trade names: Aztec® BU-50-AL; Aztec® BU-50-WO
Trade names containing: Lupersol 220-D50; Trigonox® D-C50; Trigonox® D-E50

4,4-Di-t-butylperoxy-n-butyl valerate. *See* n-Butyl-4,4-bis(t-butylperoxy)valerate

1,1-Di (t-butylperoxy) cyclohexane
CAS 3006-86-8
Synonyms: 1,1-Bis(t-butylperoxy) cyclohexane
Classification: Organic peroxide
Empirical: $C_{14}H_{28}O_4$
Properties: M.w. 260.4
Precaution: Combustible oxidizing liq.
Uses: Crosslinking agent for unsat. polyester resins

Trade names: Aztec® 1,1-BIS-50-AL; Aztec® 1,1-BIS-50-WO
Trade names containing: Aztec® 1,1-BIS-80-BBP; Luperco 331-XL; Lupersol 331-80B; Lupersol P-31; Lupersol P-33; Trigonox® 22-BB80; Trigonox® 22-C50; Trigonox® 22-E50; USP®-400P; USP®-690

Di-n-butyl peroxydicarbonate
CAS 16215-49-9
Classification: Organic peroxide
Empirical: $C_{10}H_{18}O_6$
Formula: $C_4H_9OCOO_2CO_2C_4H_9$
Properties: M.w. 234.2
Uses: Initiator
Trade names containing: Espercarb® 438M-60; Lupersol 225-M60; Lupersol 225-M75; Trigonox® NBP-C50; Trigonox® SBP-AX30; Trigonox® SBP-C30; Trigonox® SBP-C60; Trigonox® SBP-C75

Di-(s-butyl)peroxydicarbonate
CAS 19910-65-7
Classification: Organic peroxide
Properties: M.w. 234.2
Uses: Initiator, crosslinking agent for polymers incl. vinyl, acrylic
Trade names: Lupersol 225; Trigonox® SBP; Trigonox® SBPS
Trade names containing: Trigonox® SBP-C50

Di-(2-t-butylperoxyisopropyl)benzene
CAS 25155-25-3
Synonyms: Bis (t-butylperoxy) diisopropyl benzene; α,α'-Bis (t-butylperoxy) diisopropyl benzene; Bis (t-butylperoxyisopropyl) benzene
Classification: Organic peroxide
Properties: M.w. 338.5
Uses: Initiator for styrenics; synergist for some halogen-containing flame retardants; crosslinking agent for olefin copolymers, EPDM, SBR, Neoprene, Hypalon, high temp. cure of polyesters; accelerator
Manuf./Distrib.: Aceto
Trade names: Aztec® DIPP-2; Aztec® DIPP-40-G; Aztec® DIPP-40-IC; Luperox 802; Perkadox® 14; Perkadox® 14/40; Perkadox® 14-90; Perkadox® 14S; Perkadox® 14S-20PP-pd; Perkadox® 14S-fl; Polyvel CR-5P; Polyvel CR-10P; Polyvel CR-20P
Trade names containing: Akrochem® VC-40C; Akrochem® VC-40K; Akroform® VC-40 EPMB; Akrosperse® VC-40 EPMB; Aztec® DIPP-40-IC1; Aztec® DIPP-40-IC5; Luperco 802-40KE; Perkadox® 14-40B-pd; Perkadox® 14-40K-pdPoly-Dispersion® E (VC)D-40; Poly-Dispersion® T (VC)D-40P; Poly-Dispersion® VC-60P; Retilox® F 40 MF; Retilox® F 40 MG; Varox® 802-40KE; Varox® 802-40MB

1,4-Di-(2-t-butylperoxyisopropyl) benzene
CAS 2781-00-2
Properties: M.w. 338.5
Uses: Initiator for PP
Trade names: Perkadox® 64-10PP-pd; Perkadox® 64-20PP-pd; Perkadox® 64-40PP-pd

Di-t-butyl peroxyphthalate
CAS 2155-71-7
Synonyms: Di-t-butyl diperoxyphthalate
Classification: Organic peroxide
Empirical: $C_{16}H_{22}O_6$
Formula: $C_6H_4[CO{\cdot}OOC(CH_3)_3]_2$
Properties: M.w. 310.4
Precaution: Shock-sensitive explosive; on dec., emits acrid smoke and fumes
Uses: Initiator for vinyl polymerizations
Trade names containing: Trigonox® 111-B40

1,1-Di (t-butylperoxy) 3,3,5-trimethyl cyclohexane
CAS 6731-36-8
Synonyms: 1,1-Bis (t-butylperoxy)-3,3,5-trimethyl cyclohexane
Classification: Organic peroxide
Empirical: $C_{17}H_{34}O_4$
Properties: M.w. 302.5
Precaution: Combustible oxidizing liq.

Toxicology: Nonirritating to eyes; LD50 (rat, oral) > 12,918 mg/kg; LD50 (rabbit, dermal) > 8000 mg/kg
Uses: Vulcanizing agent for SBR, NBR, EPDM, EPM, CPE, and silicones; crosslinking agent for unsat. PE and EVA
Trade names: Aztec® 3,3,5-TRI; Aztec® 3,3,5-TRI-40-G; Aztec® 3,3,5-TRI-40-IC; Aztec® 3,3,5-TRI-50-AL; Aztec® 3,3,5-TRI-50-FT; Aztec® 3,3,5-TRI-55-AL; Luperco 231-SRL; Lupersol 231; Peroximon® S-164/40P; Trigonox® 29; Varox® 231; Varox® 231-XL
Trade names containing: Aztec® 3,3,5-TRI-28-IC3; Aztec® 3,3,5-TRI-75-DBP; Luperco 231-XL; Lupersol 231-P75; Poly-Dispersion® TMC-50; Trigonox® 29-40B-pd; Trigonox® 29-B75; Trigonox® 29-C75; Trigonox® KSM

2,4-Di-t-butylphenol
CAS 96-76-4; EINECS 202-532-0
Synonyms: 24DTBP
Empirical: $C_{14}H_{22}O$
Properties: M.w. 206.33; m.p. 54-56 C; flash pt. 115 C
Uses: Antioxidant
Manuf./Distrib.: Albemarle; Biddle Sawyer; Great Lakes; Spectrum Chem. Mfg.

2,6-Di-t-butylphenol
CAS 128-39-2; EINECS 204-884-0
Synonyms: 26DTBP
Classification: Orthoalkylated aromatic
Empirical: $C_{14}H_{22}O$
Formula: $[(CH_3)_3C]_2C_6H_3OH$
Properties: Lt. straw cryst. solid; sol. in alcohol and benzene; insol. in water; m.w. 206.33; dens. 0.914 (20 C); m.p. 37 C; b.p. 253 C; flash pt. 245 F
Precaution: Combustible
Toxicology: Toxic by ingestion, inhalation, and skin absorption; strong irritant to tissue
Uses: Intermediate; antioxidant, stabilizer
Manuf./Distrib.: Albemarle; Allchem Ind.; Great Lakes; Penta Mfg.; PMC Specialties; Schenectady

2,4-Di-t-butylphenyl 3,5-di-t-butyl-4-hydroxybenzoate
CAS 4221-80-1
Properties: Powd.; m.p. 196 C
Uses: Initiator for styrenics; synergist for some halogen-containing flame retardants; also for crosslinking of olefin copolymers, EPDM, SBR, Neoprene, Hypalon; uv stabilizer
Trade names: UV-Chek® AM-340
Trade names containing: Aquastab PB 59

N,N´-Di-s-butyl-p-phenylenediamine
CAS 101-96-2
Empirical: $C_{14}H_{24}N_2$
Formula: $CH_3CH_2CH(CH_3)NHC_6H_4NHCH(CH_3)CH_2CH_3$
Properties: Amber to red liq.; sol. in gasoline, abs. alcohol, benzene; insol. in water or caustic sol'ns.; dens. 0.94 (15.5/15.5 C); f.p. 20 C; flash pt. (COC) 290 F
Precaution: Combustible
Toxicology: Toxic by ingestion, inhalation, skin absorption; strong irritant to tissue; causes skin burns; heated to dec., emits toxic fumes of NO_x
Uses: Antioxidant, stabilizer and metal deactivator for crude oil distillates, gasoline
Trade names containing: V-Pyrol

Dibutyl phthalate
CAS 84-74-2; EINECS 201-557-4
Synonyms: DBP; 1,2-Benzenedicarboxylic acid, dibutyl ester; Dibutyl-1,2-benzene dicarboxylate; Butyl phthalate
Definition: Aromatic diester of butyl alcohol and phthalic acid
Empirical: $C_{16}H_{22}O_4$
Formula: $C_6H_4(COOC_4H_9)_2$
Properties: Colorless stable oily liq., odorless; misc. with common org. solvs.; insol. in water; m.w. 278.17; dens. 1.0484 (20/20 C); b.p. 340 C; f.p. -35 C; flash pt. (COC) 340 F
Toxicology: Toxic; TLV 5 mg/m^3 of air
Uses: Plasticizer in nitrocellulose, lacquers, elastomers, explosives, nail polishes; solvent for perfumes, oils; perfume fixative; textile lubricating agent

Regulatory: FDA 21CFR §175.105, 175.300, 176.170, 176.300, 177.1200, 177-2420, 177.2600
Manuf./Distrib.: Aldrich; Allchem Ind.; Aristech; Ashland; BP Chem. Ltd; Chisso Am.; Coyne; Daihachi Chem. Ind.; Eastman; Great Western; C.P. Hall; Mitsubishi Gas; Novachem; Unitex
Trade names: Diplast® A; Kodaflex® DBP; Palatinol® DBP; Plasthall® DBP; PX-104; Staflex DBP; Unimoll® DB; Uniplex 150
Trade names containing: Aztec® 3,3,5-TRI-75-DBP; Bonding Agent 2001; Cadox® TDP; Cadox® TS-50; Kodaflex® HS-4; Lupersol 231-P75; Lupersol KDB; Mearlite® Ultra Bright UDQ; Mearlite® Ultra Bright UMS; Trigonox® 29-B75; Trigonox® 111-B40; Vulcabond® VP

Dibutyl sebacate
CAS 109-43-3; EINECS 203-672-5
Synonyms: DBS; Dibutyl decanedioate; Di-n-butyl sebacate; Decanedioic acid, dibutyl ester
Definition: Diester of butyl alcohol and sebacic acid
Empirical: $C_{18}H_{34}O_4$
Formula: $C_4H_9OCO(CH_2)_8OCOC_4H_9$
Properties: Clear colorless odorless liq.; insol. in water; m.w. 314.47; dens. 0.936 (20/20 C); b.p. 349 C (760 mm); f.p. -11 C; flash pt. 350 F; ref. index 1.4433 (15 C); stable, nonvolatile
Precaution: Combustible when exposed to heat or flame; can react with oxidizing materials
Toxicology: LD50 (oral, rat) 16 g/kg; mildly toxic by ingestion; experimental reproductive effects; heated to decomp., emits acrid smoke and fumes
Uses: Plasticizer, rubber softener, dielectric liquid, cosmetics and perfumes, flavoring agent
Regulatory: FDA 21CFR §172.515, 175.105, 175.300, 175.320, 176.170, 177.2600, 178.3910, 181.22, 181.27
Manuf./Distrib.: C.P. Hall; Harwick; Novachem; Richman; Union Camp; Unitex; Velsicol
Trade names: Plasthall® DBS; Uniflex® DBS

Di-n-butyl sebacate. *See* Dibutyl sebacate

1,3-Dibutylthiourea
CAS 109-46-6; EINECS 203-674-6
Synonyms: 1,3-Dibutyl-2-thiourea
Formula: $C_4H_9NHCSNHC_4H_9$
Properties: White to lt. tan powd.; sl. sol. in water; sol. in methanol, ether, acetone, benzene, ethyl acetate; insol. in gasoline; m.w. 188.34; m.p. 59-69 C
Uses: Rubber accelerator; corrosion inhibitor for pickling cast iron or carbon steel; reducing corrosion of ferrous metals and aluminum alloys in brine; intermediate
Manuf./Distrib.: AC Ind.; Faesy & Besthoff; Fluka
Trade names: Akroform® DBTU PM; Ekaland DBTU; Thiate® U

1,3-Dibutyl-2-thiourea. *See* 1,3-Dibutylthiourea

Dibutyltin bis(2-ethylhexyl mercaptoacetate)
CAS 10584-98-2
Synonyms: DBT 2-EHMA
Toxicology: TLV/ACGIH 0.1 mg/m^3; LD50 (oral, rat) 920 mg/kg (sl. toxic), (dermal, rabbit) 2000 mg/kg (moderately to sl. toxic)
Uses: Stabilizer
Trade names containing: Thermolite 525

Dibutyltin bisisooctyl thioglycolate
Properties: Yel. liq.; sol. in acetone, toluene
Uses: Catalyst
Trade names: Therm-Chek® 840

Dibutyltin (bis) mercaptide
Uses: Catalyst for PU coatings
Trade names: Dabco® T-131

Dibutyltin diacetate
CAS 1067-33-0; EINECS 213-928-8
Synonyms: Di-n-butyldiacetoxystannane
Formula: $(C_4H_9)_2Sn(C_2H_3O_2)_2$
Properties: Clear yel. liq.; sol. in water and most org. solvs.; m.w. 351.01; b.p. 130 C (2 mm); f.p. < 12 C; flash pt. 290 F

Dibutyltin dilaurate

Precaution: Combustible
Toxicology: TLV 0.1 mg/m^3 of air; toxic by skin absorption
Uses: Stabilizer for chlorinated organics; catalyst for PU coatings, condensation reactions
Manuf./Distrib.: Fluka
Trade names: Dabco® T-1; Fomrez® SUL-3; Metacure® T-1

Dibutyltin dilaurate
CAS 77-58-7; EINECS 201-039-8
Synonyms: DBTL
Empirical: $C_{32}H_{64}O_4Sn$
Formula: $(C_4H_9)_2Sn(OCOC_{10}H_{20}CH_3)_2$
Properties: Clear pale yel. liq.; sol. in acetone and benzene; insol. in water; m.w. 631.56; dens. 1.066; m.p. 23 C; f.p. 8 C; flash pt. 440 F
Precaution: Combustible
Toxicology: TLV 0.1 mg/m^3 of air; LD50 (oral, rat), 3954 mg/kg (sl. toxic), (dermal, rabbit) > 2000 mg/kg (sl. toxic)
Uses: Stabilizer for vinyl resins, lacquers, elastomers; catalyst for urethane and silicones
Manuf./Distrib.: Air Prods.; Cardinal Stabilizers; Elf Atochem N. Am.; Ferro/Bedford; Johnson Matthey; KMZ Chem. Ltd; OM Group; Witco/Oleo-Surf.
Trade names: ADK STAB BT-11; ADK STAB BT-18; ADK STAB BT-23; ADK STAB BT-25; ADK STAB BT-27; Cata-Chek® 820; Dabco® T-12; Fomrez® SUL-4; Jeffcat T-12; Kosmos® 19; Metacure® T-12; Synpron 1009; TEDA-T411
Trade names containing: Thermolite 525

Dibutyltin disulfide
Uses: Catalyst for PU coatings
Trade names: Dabco® T-5

Dibutyltin maleate
CAS 15535-69-0
Synonyms: DBM
Formula: $[(C_4H_9)_2Sn(OOCCH)_2]_x$
Properties: White powd.; insol. in water; sol. in benzene, organic esters; m.p. 110 C; flash pt. 400 F
Precaution: Combustible
Toxicology: TLV 0.1 mg/m^3 of air; toxic by skin absorption
Uses: Stabilizer for PVC resins; condensation catalyst
Manuf./Distrib.: Cardinal Stabilizers; Elf Atochem N. Am.; Ferro/Bedford; Gelest
Trade names: ADK STAB BT-31; ADK STAB BT-52; ADK STAB BT-53A; ADK STAB LS-2; Stanclere® TM; Therm-Chek® 837

Dibutyltin mercaptide
Uses: Stabilizer for PVC
Trade names: ADK STAB 1292; ADK STAB BT-83; Lankromark® BT120A

Dibutyltin β-mercaptopropionate
Properties: White crystals
Uses: Solvent; stabilizer for PVC
Trade names: Stanclere® T-186

Dibutyltin oxide
CAS 818-08-6
Formula: $[CH_3(CH_2)_3]_2$
Properties: M.w. 248.92
Toxicology: Highly toxic irritant
Uses: Condensation catalyst; intermediate for other organotins
Manuf./Distrib.: Aceto; Aldrich; Cardinal Stabilizers; Elf Atochem N. Am.
Trade names containing: Thermolite 525

Dicalcium orthophosphate anhyd.. *See* Calcium phosphate dibasic
Dicalcium phosphate. *See* Calcium phosphate dibasic

Dicapryl adipate
CAS 105-97-5; EINECS 203-349-9
Synonyms: Didecyl hexanedioate; Hexanedioic acid, didecyl ester

Definition: Diester of capryl alcohol and adipic acid
Empirical: $C_{26}H_{50}O_4$
Formula: $C_8H_{17}OOC(CH_2)_4COOC_8H_{17}$
Properties: Almost water-white liq.; b.p. 213-216 C (4 mm); flash pt. 352 F
Precaution: Combustible
Uses: Plasticizer for vinyl resins and cellulose ethers
Regulatory: FDA 21CFR §177.2600, 178.3740
Manuf./Distrib.: Inolex; Union Camp
Trade names: Uniflex® DCA

Dicapryl maleate
Properties: Sp.gr. 0.927; m.p. -50 C; flash pt. (COC) 196 C; ref. index 1.4485
Uses: Plasticizer
Manuf./Distrib.: Union Camp

Dicapryl phthalate
Synonyms: DCP; Di-(2-octyl) phthalate
Classification: Capryl compd.
Formula: $(C_8H_{17}OOC)_2C_6H_4$
Properties: Colorless viscous liq.; insol. in water; m.w. 390; dens. 0.965 (25 C); b.p. 227-234 C (4.5 mm); f.p. -60 C; flash pt. 295 F
Precaution: Combustible
Uses: Monomeric plasticizer for vinyl and cellulosic resins
Manuf./Distrib.: Union Camp
Trade names: Uniflex® DCP

Dicarboxylic acid C6. *See* Adipic acid
Dicarboxylic acid C9. *See* Azelaic acid

Dicatechol borate, di-o-tolyl guanidine salt
CAS 16971-82-7
Uses: Accelerator for NR and SR stocks, CR, neoprenes for wire and cable and mechanical goods; activator and mild antioxidant in NR and SBR
Trade names: Vanax® PML

Dicetyl peroxydicarbonate
CAS 26332-14-5
Empirical: $C_{34}H_{66}O_6$
Formula: $CH_3(CH_2)_{15}OCOOOOCOO(CH_2)_{15}CH_3$
Properties: M.w. 570.9
Uses: Polymerization initiator
Trade names: Aztec® CEPC; Aztec® CEPC-40-SAQ; Perkadox® 24; Perkadox® 24-fl.; Perkadox® 24-W40

2,4-Dichlorobenzoyl peroxide. *See* p,p-Dichlorobenzoyl peroxide
Di-(4-chlorobenzoyl) peroxide. *See* p,p-Dichlorobenzoyl peroxide

p,p-Dichlorobenzoyl peroxide
CAS 94-17-7
Synonyms: Bis (p-chlorobenzoyl) peroxide; 2,4-Dichlorobenzoyl peroxide; Di-(4-chlorobenzoyl) peroxide
Empirical: $C_{14}H_8Cl_2O_4$
Properties: Wh. granules; insol. in water; sol. in org. solvs.; m.w. 311.12
Precaution: Heat or contact with certain fumes can cause it to explode
Toxicology: Poison by ingestion and intraperitoneal route; irritant to skin and mucous membranes
Uses: Crosslinking agent for curing silicone rubbers; co-vulcanizing agent
Manuf./Distrib.: Akzo Nobel
Trade names: Cadox® PS; Cadox® TS-50S
Trade names containing: Cadox® TDP

Dichlorofluoroethane
CAS 1717-00-6
Synonyms: HCFC 141b; 1,1-Dichloro-1-fluoroethane
Empirical: CCl_2FCH_3
Properties: M.w. 116.95

1,1-Dichloro-1-fluoroethane

Uses: Blowing agent in rigid board, foam systems, flexible foam
Manuf./Distrib.: Spectrum Chem. Mfg.
Trade names: Genetron® 141b
Trade names containing: MS-122N/CO_2; Paintable Organic Oil Spray No. O316; Vydax® Spray No. V312-A; Zinc Stearate Spray No. Z212-B

1,1-Dichloro-1-fluoroethane. *See* Dichlorofluoroethane

3-(3,4-Dichlorophenyl)-1,1-dimethylurea
CAS 330-54-1
Synonyms: Diuron
Formula: $C_6H_3Cl_2NHCON(CH_3)_2$
Properties: White crystalline solid; low sol. in hydrocarbon solvs.; m.p. 159 C; dec. @ 180 C; stable towards oxidation and moisture
Toxicology: Toxic; TLV 10 mg/m^3 of air
Uses: Pre-emergence herbicide, sugar cane flowering suppressant
Trade names containing: Rhenocure® Diuron

Dichlorotrifluoroethane
CAS 306-83-2
Synonyms: 2,2-Dichloro-1,1,1-trifluoroethane
Empirical: $CHCl_2CF_3$
Properties: M.w. 152.91; dens. 1.462; b.p. 28.7 C
Precaution: Flamm.
Toxicology: Irritant; poison by inhalation
Uses: Blowing agent for rigid board, foam system insulation; refrigerant; in specialized solvent applics.
Manuf./Distrib.: Halocarbon Prods.; PCR

2,2-Dichloro-1,1,1-trifluoroethane. *See* Dichlorotrifluoroethane
Dicocoalkyl methylamine. *See* Dicoco methylamine
Dicoco dimethyl ammonium chloride. *See* Dicocodimonium chloride

Dicocodimonium chloride
CAS 61789-77-3; EINECS 263-087-6
Synonyms: Dicoco dimethyl ammonium chloride; Quaternium-34
Classification: Quaternary ammonium salt
Formula: $[RN(CH_3)_2R]^+Cl^-$ where R rep. the coconut radical
Uses: Emulsifier, corrosion inhibitor, coupling agent, bactericide, textile softener, asphalt emulsifier, hair conditioner; for car spray waxes, dust control oil, spot removal, petrol. processing, detergents, textiles, fabric softeners
Regulatory: FDA 21CFR §172.710, 177.1200
Trade names containing: Arquad® 2C-75

Dicoco methylamine
CAS 61788-62-3; EINECS 262-990-2
Synonyms: Dicocoalkyl methylamine
Classification: Dialkyl methylamine
Uses: Chemical intermediate for quat. ammonium derivs., acid scavenger in petrol. prods.; epoxy hardener, catalyst in mfg. of flexible PU foams
Trade names: Kemamine® T-6501

Dicumyl peroxide
CAS 80-43-3
Synonyms: Di-α-cumyl peroxide; Cumene peroxide; Diisopropylbenzene peroxide
Classification: Organic peroxide
Formula: $[C_6H_5C(CH_3)_2O]_2$
Properties: White crystals; insol. in water; sol. in acetone, aromatic solvs.; m.w. 270.4
Precaution: Strong oxidizer; may ignite org. materials on contact
Uses: Polymerization catalyst and vulcanizing agent; crosslinking agent for olefinic polymers
Manuf./Distrib.: AC Ind.; Akzo Nobel; Elf Atochem N. Am.; Hercules; Mitsui Petrochem. Ind.; Monomer-Polymer & Dajac; R.T. Vanderbilt; Witco/PAG
Trade names: Aztec® DCP-40-G; Aztec® DCP-40-IC; Aztec® DCP-40-IC1; Aztec® DCP-R; Di-Cup R, T; Esperal® 115RG; Luperco 500-SRK; Luperox 500R; Luperox 500T; Perkadox® BC; Perkadox® BC-40S;

Perkadox® BC-FF; Polyvel PCL-10; Polyvel PCL-20; Thermacure®; Varox® DCP-R; Varox® DCP-T
Trade names containing: Akrochem® DCP-40C; Akrochem® DCP-40K; Akroform® DCP-40 EPMB; Akrosperse® DCP-40 EPMB; Di-Cup 40C; Di-Cup 40KE; Luperco 500-40C; Luperco 500-40KE; Perkadox® BC-40Bpd; Perkadox® BC-40K-pd; Peroximon® DC 40 MF; Peroximon® DC 40 MG; Poly-Dispersion® DCP-60P; Poly-Dispersion® E (DIC)D-40; Poly-Dispersion® T (DIC)D-40P; Prespersion SIC-3233; Varox® DCP-40C; Varox® DCP-40KE; Varox® DCP-40MB

Di-α-cumyl peroxide. *See* Dicumyl peroxide

Dicyandiamide
CAS 461-58-5; EINECS 207-312-8
Synonyms: Cyanoguanidine
Empirical: $C_2H_4N_4$
Formula: $NH_2C(NH)(NHCN)$
Properties: Pure white crystals; sol. in liq. ammonia; partly sol. in hot water; m.w. 84.08; dens. 1.4 (25 C); m.p. 207-209 C; stable when dry
Precaution: Nonflamm.
Uses: Fertilizers; nitrocellulose stabilizer; organic synthesis, esp. of melamine, barbituric acid, and guanidine salts; explosives; curing agent, catalyst for epoxy resin; pharmaceuticals; fireproofing compds.; stabilizer in detergents; modifier for starch prod.
Manuf./Distrib.: Allchem Ind.; Andrulex Trading Ltd; CVC Spec.; San Yuan; SKW
Trade names: Amicure® CG-325; Amicure® CG-1200; Amicure® CG-1400; Amicure® CG-NA; Dicyanex® 200X; Dicyanex® 325; Dicyanex® 1200

Dicyclo (dioctyl) pyrophosphato titanate
Uses: Coupling agent, adhesion promoter, antioxidant, antistat, antifoam, accelerator, blowing agent activator, catalyst, curative, corrosion inhibitor, dispersion aid, emulsifier, flame retardant, foaming agent, grinding and process aid
Trade names containing: Ken-React® OPP2 (KR OPP2)

N,N′-Dicyclohexyl-2-benzothiazole sulfenamide. *See* Dicyclohexylbenzothiazyl-2-sulfenamide

Dicyclohexylbenzothiazyl-2-sulfenamide
CAS 4979-32-2
Synonyms: Benzothiazyl-2-dicyclohexyl sulfenamide; Benzothiazyl 1-2-dicyclohexyl sulfenamide; N,N′-Dicyclohexyl-2-benzothiazole sulfenamide
Uses: Accelerator for rubber
Trade names: Akrochem® DCBS; Perkacit® DCBS; Santocure® DCBS
Trade names containing: Rhenogran® DCBS-80

Dicyclohexyl peroxydicarbonate
CAS 1561-49-5
Classification: Organic peroxide
Properties: M.w. 286.3
Uses: Polymerization initiator
Manuf./Distrib.: Akzo Nobel
Trade names: Aztec® CHPC; Aztec® CHPC-90-W; Perkadox® 18

Dicyclohexyl phthalate
CAS 84-61-7; EINECS 201-545-9
Synonyms: DCHP; 1,2-Benzenedicarboxylic acid, dicyclohexyl ester
Formula: $C_6H_4(COOC_6H_{11})_2$
Properties: White granular solid; mildly aromatic odor; sol. in most org. solvs.; insol. in water; m.w. 330.43; dens. 1.20 (25/25 C); m.p. 62-65 C; flash pt. 405 F; nonvolatile
Precaution: Combustible
Uses: Plasticizer for nitrocellulose, ethylcellulose, chlorinated rubber, PVAc, PVC, and other polymers
Manuf./Distrib.: Miles; Morflex; Novachem; Schweizerhall; Unitex
Trade names: Unimoll® 66; Unimoll® 66 M; Uniplex 250
Trade names containing: Cadox® BFF-50; Cadox® BFF-60W

Dicyclohexyl sodium sulfosuccinate
CAS 23386-52-9; EINECS 245-629-3
Synonyms: Sodium dicyclohexyl sulfosuccinate; Succinic acid, sulfo-, 1,4-dicyclohexyl ester, sodium salt;

Dicyclopentenyl acrylate

Sodium 1,4-dicyclohexyl sulfobutanedioic acid
Definition: Diester of cyclohexyl alcohol and sulfosuccinic acid
Empirical: $C_{16}H_{26}O_7S \cdot Na$
Uses: Dispersant, surfactant, emulsifier for modified S/B; post additive to stabilize latex and promote adhesion
Regulatory: FDA 21CFR §178.3400
Trade names: Aerosol® A-196-40; Gemtex 691-40; Octosol TH-40; Rewopol® SBDC 40

Dicyclopentenyl acrylate
CAS 12542-30-2
Uses: Monomer for air-drying adhesives, coatings, caulks, sealants, elastomers; crosslinks acrylics, unsat. polyesters and alkyd
Trade names: Sipomer® DCPA

Dicyclopentenyl methacrylate
CAS 51178-59-7
Uses: Monomer for concrete resurfacing, air-drying adhesives and coatings, elastomers; improves cohesive str. through crosslinking
Trade names: Sipomer® DCPM

DIDA. *See* Diisodecyl adipate
Didecanoyl peroxide. *See* Decanoyl peroxide

Didecyl glutarate
Properties: Sp.gr. 0.918; flash pt. (COC) 227 C; ref. index 1.449
Uses: Plasticizer for PVC, PS, cellulosics
Manuf./Distrib.: C.P. Hall

Didecyl hexanedioate. *See* Dicapryl adipate

Didecyl hydrogen phosphite
Uses: Stabilizer

Didecyl methylamine
CAS 7396-58-9; EINECS 230-990-1
Synonyms: Methyl decyl-1-amino decane; N-Methyldidecylamine
Empirical: $C_{21}H_{45}N$
Formula: $[CH_3(CH_2)_9]_2NCH_3$
Properties: M.w. 311.60
Uses: Intermediate for mfg. of quaternary ammonium compds. for biocides, textile chemicals, oil field chemicals, amine oxides, betaines, polyurethane foam catalysis, epoxy curing agent; in fabric softeners, disinfectants, laundry detergents
Manuf./Distrib.: Fluka
Trade names: Armeen® M2-10D; Dama® 1010

Didecyl phenyl phosphite
CAS 1254-78-0
Uses: Stabilizer

Di (2,4-dichloro benzoyl) peroxide. *See* Bis (2,4-dichlorobenzoyl peroxide)
Di-2,4-dichlorobenzoyl peroxide. *See* p,p-Dichlorobenzoyl peroxide

N,N′-Di-(1,4-dimethylpentyl)-p-phenylene diamine
CAS 3081-14-9
Synonyms: N,N′-Bis (1,4-dimethylpentyl)-p-phenylenediamine; Diheptyl-p-phenylenediamine
Empirical: $C_{20}H_{36}N_2$
Formula: $C_6H_4(CNH_7H_{15})_2$
Properties: Amber to red liq.; m.w. 304.58; dens. 0.90; f.p. 7.2 C
Toxicology: Moderately toxic by ingestion, intraperitoneal routes; heated to dec., emits toxic fumes of NO_x
Uses: Antiozonant/antioxidant protecting natural rubber and syn. rubbers; for dynamically stressed goods, tires; styrene polymerization inhibitor; gasoline antioxidant and sweetener
Trade names: Naugard® I-2; Santoflex® 77; Vulkanox® 4030

Di (dioctylphosphato) ethylene titanate
Uses: Coupling agent, adhesion promoter, antioxidant, antistat, antifoam, accelerator, blowing agent activator, catalyst, curative, corrosion inhibitor, dispersion aid, emulsifier, flame retardant, foaming agent, grinding and process aid
Trade names containing: Ken-React® 212 (KR 212)

Di (dioctylpyrophosphato) ethylene titanate
Uses: Coupling agent, adhesion promoter, antioxidant, antistat, antifoam, accelerator, blowing agent activator, catalyst, curative, corrosion inhibitor, dispersion aid, emulsifier, flame retardant, foaming agent, grinding and process aid
Trade names containing: Ken-React® 238S (KR 238S)

Didodecyl 3,3´-thiodipropionate. *See* Dilauryl thiodipropionate
DIDP. *See* Diisodecyl phthalate

Diethanolamine
CAS 111-42-2; EINECS 203-868-0
Synonyms: DEA; 2,2´-Iminobisethanol; Di(2-hydroxyethyl) amine; 2,2´-Iminodiethanol
Classification: Aliphatic amine
Empirical: $C_4H_{11}NO_2$
Formula: $(HOCH_2CH_2)_2NH$
Properties: Colorless crystals or liq., mild ammoniacal odor; very sol. in water and alcohol; misc. with acetone; insol. in ether, benzene; m.w. 105.09; dens. 1.0881 (30/4 C); m.p. 28 C; b.p. 268 C; flash pt. 300 F
Precaution: Combustible
Toxicology: LD50 (rat, oral) 12.76 g/kg; TLV 3 ppm in air
Uses: Gas scrubbing; rubber chemicals intermediate; mfg. of surfactants for textiles, herbicides, petrol. demulsifier; emulsifier and dispersant in agric., cosmetics, pharmaceuticals; textile lubricants; humectant; softening agent; in organic synthesis; crosslinker for flexible PU foam slabstock
Regulatory: FDA 21CFR §175.105, 176.170, 176.180, 176.210, 177.2600
Manuf./Distrib.: Akzo Nobel; Allchem Ind.; Ashland; Coyne; Great Western; Hüls AG; Oxiteno; Union Carbide
Trade names: Dabco® DEOA-LF; DEA Commercial Grade; DEA Low Freeze Grade
Trade names containing: Empilan® LDX; TEA 85

Diethanolamine lauric acid amide. *See* Lauramide DEA
Diethanolamine lauryl sulfate. *See* DEA-lauryl sulfate
Diethanol-m-toluidine. *See* m-Tolyl diethanolamine
1,2-Diethoxyethane. *See* Ethylene glycol diethyl ether
3-(Diethoxymethylsilyl) propylamine. *See* 3-Aminopropylmethyldiethoxy silane

Diethylamine
CAS 109-89-7; EINECS 203-716-3
Empirical: $C_4H_{11}N$
Formula: $(C_2H_5)_2NH$
Properties: Liq.; m.w. 73.14; dens. 0.707; m.p. -50 C; b.p. 55 C; flash pt. -28 C; ref. index 1.3850 (20 C)
Precaution: Flamm. liq.
Toxicology: Corrosive
Uses: Rubber chemicals, textile specialties, selective solvent, dyes, flotation agents, resins, pesticides, polymerization inhibitor, pharmaceuticals, petroleum chemicals, electroplating, corrosion inhibitors
Manuf./Distrib.: AC Ind.; Air Prods.; Aldrich; Allchem Ind.; Ashland; BASF; Coyne; Elf Atochem N. Am.; Union Carbide

(Diethylamino) ethane. *See* Triethylamine

Diethylaminoethanol
CAS 100-37-8; EINECS 202-845-2
Synonyms: DEAE; Diethylethanolamine; N,N-Diethyl-2-aminoethanol; 2-Hydroxytriethylamine; β-Diethylaminoethyl alcohol
Empirical: $C_6H_{15}NO$
Formula: $(C_2H_5)_2NCH_2CH_2OH$
Properties: Colorless hygroscopic liq. base having props. of amines and alcohols; sol. in water, alcohol, ether, benzene; m.w. 117.19; dens. 0.88-0.89 (20/20 C); b.p. 161 C; flash pt. (OC) 140 F; f.p. -70 C; ref. index 1.4389
Precaution: Combustible; moderate fire risk; reactive with oxidizing materials

Toxicology: TLV 10 ppm in air; LD50 (oral, rat) 1300 mg/kg; poison by IP, IV routes; mod. toxic by ingestion, skin contact, subcut. routes; human systemic effects; skin, severe eye irritant; corrosive; heated to decomp., emits toxic fumes of NO_x
Uses: Water-sol. salts, textile softeners, pharmaceuticals, antirust compositions, emulsifying agents in acid media; curing agents for resins
Regulatory: FDA 21CFR §173.310 (15 ppm max. in steam)
Manuf./Distrib.: Ashland; BASF; Elf Atochem N. Am.; Union Carbide
Trade names: Pennad 150

N,N-Diethyl-2-aminoethanol. *See* Diethylaminoethanol

Diethylaminoethyl acrylate
CAS 2426-54-2
Classification: Amine monomer
Formula: $CH_2=CHCOOCH_2CH_2N(C_2H_5)_2$
Properties: M.w. 171.24; dens. 0.925 (20/4 C); b.p. 202 C
Toxicology: Corrosive; severe skin and eye irritant; harmful if swallowed or inhaled
Uses: Industrial and automotive coatings, dye and lube oil additives, cationic quat. monomers, intermediate for water treatment chemicals, etc.
Manuf./Distrib.: CPS; Monomer-Polymer & Dajac; Rhone-Poulenc Spec.
Trade names: Ageflex FA-2

β-Diethylaminoethyl alcohol. *See* Diethylaminoethanol

Diethylaminoethyl methacrylate
CAS 105-16-8
Classification: Amine monomer
Formula: $CH_2=C(CH_3)COOCH_2CH_2N(C_2H_5)_2$
Properties: M.w. 185.29
Precaution: Combustible
Toxicology: Severe eye and skin irritant; harmful if swallowed
Uses: Industrial and automotive coatings, dye additives, intermediate for water treatment and oil field chemicals, stabilizer for fuel oils
Manuf./Distrib.: CPS; Monomer-Polymer & Dajac; Polysciences; Rhone-Poulenc Spec.
Trade names: Ageflex FM-2

Diethyl 1,2-benzenedicarboxylate. *See* Diethyl phthalate

Diethyl N,N-bis(2-hydroxyethyl)aminomethylphosphonate
Classification: Organic
Properties: M.w. 255
Uses: Flame retardant for rigid PU foam
Trade names: Fyrol® 6

Di(2-ethylbutyl) phthalate. *See* Dihexyl phthalate

3,5-Diethyl-2,4-diaminotoluene
Uses: Chain extender for PU; hardener for epoxy resins
Trade names containing: DETDA 80

3,5-Diethyl-2,6-diaminotoluene
Uses: Chain extender for PU; hardener for epoxy resins
Trade names containing: DETDA 80

Diethylene diamine. *See* Piperazine
Diethylene dioxide. *See* 1,4-Dioxane
1,4-Diethylene dioxide. *See* 1,4-Dioxane
Diethylene ether. *See* 1,4-Dioxane

Diethylene glycol
CAS 111-46-6; EINECS 203-872-2
Synonyms: DEG; Dihydroxydiethyl ether; Diglycol; 2,2′-Oxybisethanol

Classification: Aliphatic diol
Empirical: $C_4H_{10}O_3$
Formula: $CH_2OHCH_2OCH_2CH_2OH$
Properties: Colorless syrupy liq., pract. odorless, sweetish taste; misc. with water, ethanol, acetone, ether, ethylene glycol; m.w. 106.12; dens. 1.1184 (20/20 C); b.p. 245 C; f.p. -80 C; flash pt. 255 F; extremely hygroscopic; lowers f.p. of water
Precaution: Combustible
Toxicology: Hazardous for household use in conc. of 10% or more
Uses: Prod. of polyurethane and unsat. polyester resins, triethylene glycol; textile softener; solvent for nitrocellulose, dyes and oils; dehydration of natural gas, elasticizers, and surfactants; humectant for tobacco, casein, and synthetic sponges
Regulatory: FDA 21CFR §175.105, 177.2420
Manuf./Distrib.: Allchem Ind.; Ashland; BASF; BP Chem. Ltd; Coyne; Du Pont; Eastman; Hoechst Celanese; Itochu Spec.; Mitsui Petrochem. Ind.; Mobil; Olin; OxyChem; Shell; Spectrum Chem. Mfg.; Texaco; Union Carbide
Trade names containing: Dabco® K-15

Diethylene glycol butyl ether. *See* Butoxydiglycol

Diethylene glycol di(aminopropyl) ether
Uses: Epoxy curing agent
Trade names: Ancamine® 1922

Diethylene glycol dibenzoate
CAS 120-55-8; EINECS 204-407-6
Synonyms: PEG 100 dibenzoate; POE (2) dibenzoate; PEG-2 dibenzoate
Definition: PEG diester of benzoic acid
Properties: Sp.gr. 1.1765-1.178 (20 C); m.p. 16 C; flash pt. (PMCC) 232 C; ref. index 1.5424-1.5449 (20 C)
Uses: Plasticizer for polymers, adhesives
Regulatory: FDA 21CFR §175.105, 176.170
Manuf./Distrib.: Hüls; Kalama; Unitex; Velsicol
Trade names: Benzoflex® 2-45
Trade names containing: Benzoflex® 50

Diethylene glycol dibutyl ether
CAS 112-73-2; EINECS 204-001-9
Synonyms: 2,2´-Dibutoxyethyl ether; Dibutyl Carbitol
Formula: $C_4H_9O(C_2H_4O)_2C_4H_9$
Properties: Almost colorless liq.; sl. sol. in water; m.w. 218.38; dens. 0.8853 (20/20 C); f.p. -60.2 C; b.p. 256 C; flash pt. 118 C
Precaution: Combustible
Toxicology: Moderately toxic by ingestion; mildly toxic by skin contact; skin and eye irritant; heated to dec., emits acrid smoke and irritating fumes
Uses: Solvent used in electrochemistry, polymer and boron chemistry; physical processes such as gas absorption, extraction, stabilization; coatings and inks; diluent in VC dispersions
Manuf./Distrib.: Brand-Nu Labs; Ferro/Grant; Fluka; Great Western; Hoechst AG
Trade names: Ethyl Diglyme

Diethylene glycol dimethacrylate
CAS 2358-84-1
Formula: $[H_2C{=}C(CH_3)CO_2CH_2CH_2]_2O$
Properties: M.w. 242.27
Uses: Crosslinker for rubber vulcanization, moisture barrier films and coatings, photopolymer printing plates and letterpress inks, conversion coatings and adhesives
Manuf./Distrib.: Aldrich; CPS; Monomer-Polymer & Dajac; Polysciences; Rohm Tech; Sartomer
Trade names: Ageflex DEGDMA; MFM-418

Diethylene glycol dimethyl ether
CAS 111-96-6; EINECS 203-924-4
Synonyms: Diglycol methyl ether; Dimethyldiglycol; Diglyme; 2-Methoxyethyl ether; Bis (2-methoxyethyl) ether
Formula: $CH_3(OCH_2CH_2)_2OCH_3$

Diethylene glycol dioleate

Properties: Colorless liq., mild odor; misc. with water and hydrocarbons; m.w. 134.18; dens. 0.9451 (20/20 C); b.p. 162 C; f.p. -68 C; visc. 1.089 cP (20 C)
Precaution: Combustible
Uses: Solvent, anhydrous reaction medium for organo-metallic synthesis; reaction solv. in reduction, isomerization, alkylation condensation, polymerization; extract and separation solv.
Manuf./Distrib.: Brand-Nu Labs; Ferro/Grant; Great Western; Hoechst AG
Trade names: Diglyme; Hisolve MDM

Diethylene glycol dioleate. *See* PEG-2 dioleate
Diethylene glycol distearate. *See* PEG-2 distearate
Diethylene glycol dodecyl ether. *See* Laureth-2
Diethylene glycol ethyl ether. *See* Ethoxydiglycol
Diethylene glycol laurate. *See* PEG-2 laurate
Diethylene glycol methyl ether. *See* Methoxydiglycol
Diethylene glycol monoethyl ether. *See* Ethoxydiglycol
Diethylene glycol monolaurate self-emulsifying. *See* PEG-2 laurate SE
Diethylene glycol monomethyl ether. *See* Methoxydiglycol
Diethylene glycol monooleate. *See* PEG-2 oleate
Diethylene glycol monooleate self-emulsifying. *See* PEG-2 oleate SE
Diethylene glycol monopropyl ether. *See* Diethylene glycol propyl ether
Diethylene glycol monostearate self-emulsifying. *See* PEG-2 stearate SE
Diethylene glycol monotallowate. *See* PEG-2 tallowate

Diethylene glycol propyl ether
CAS 6881-94-3
Synonyms: Diethylene glycol monopropyl ether
Empirical: $C_7H_{16}O_3$
Formula: $CH_2OHCH_2OCH_2CH_2OCH_2CH_2CH_3$
Properties: Liq.; sol. in water; m.w. 148.2; dens. 0.963 (20 C); b.p. 202 C (760 mm); f.p. < -90 C; flash pt. (TCC) 93 C
Uses: Solvent
Manuf./Distrib.: Ashland; Great Western
Trade names: Ektasolve® DP

Diethylene glycol stearate. *See* PEG-2 stearate
Diethyleneimide oxide. *See* Morpholine
Diethylene oxide. *See* 1,4-Dioxane
Diethylene oximide. *See* Morpholine

Diethylenetriamine
CAS 111-40-0; EINECS 203-865-4
Synonyms: DETA; Aminoethylethandiamine; 2,2′-Iminodiethylamine; Bis(2-aminoethyl) amine
Empirical: $C_4H_{13}N_3$
Formula: $NH_2C_2H_4NHC_2H_4NH_2$
Properties: Yel. liq., ammoniacal odor; sol. in water and hydrocarbons; m.w. 103.17; dens. 0.9542 (20/20 C); b.p. 206.7 C; f.p. -39 C; flash pt. 215 F; strongly alkaline; hygroscopic; corrosive to copper and its alloys
Toxicology: Toxic by ingestion, inhalation and skin absorption; strong irritant to eyes and skin; TLV 1 ppm in air
Uses: Solvent for sulfur, acid gases, various resins, dyes; saponification agent for acidic materials; fuel component; epoxy curing agent; corrosion inhibitors; surfactants; intermediate for textile finishes
Manuf./Distrib.: Akzo Nobel; Allchem Ind.; Ashland; Coyne; Fabrichem; Janssen Chimica; Nayler Chem. Ltd; Tosoh; Union Carbide
Trade names: D.E.H. 20; D.E.H. 52; D.E.H. 58; DETA; Texacure EA-20

N,N-Diethylethanamine. *See* Triethylamine
Diethylethanolamine. *See* Diethylaminoethanol
Di(2-ethylhexyl) adipate. *See* Dioctyl adipate
Di(2-ethylhexyl) azelate. *See* Dioctyl azelate
Di (2-ethylhexyl) fumarate. *See* Dioctyl fumarate
Di (2-ethylhexyl) isophthalate. *See* Dioctyl isophthalate
Di-(2-ethylhexyl) maleate. *See* Dioctyl maleate
Di(2-ethylhexyl)peroxydicarbonate. *See* Bis (2-ethylhexyl) peroxydicarbonate
Di(2-ethylhexyl) phthalate. *See* Dioctyl phthalate

Di-2-ethylhexyl sebacate. *See* Dioctyl sebacate
Di-2-ethylhexyl terephthalate. *See* Dioctyl terephthalate

Diethylhydroxylamine
CAS 3710-84-7; EINECS 223-055-4
Formula: $(C_2H_5)_2NOH$
Properties: Yel. liq.; m.w. 89.14; dens. 0.867; m.p. -25 C; b.p. 47 C (3 mm); flash pt. 45 C
Precaution: Combustible
Toxicology: Poison by skin contact; moderately toxic by ingestion, intraperitoneal route
Uses: Photographic developer, antioxidant, corrosive inhibitor; free radical scavenger used by the rubber industry as an emulsion polymerization inhibitor
Manuf./Distrib.: Elf Atochem N. Am.; Fluka; Great Western
Trade names: Pennstop® 1866; Pennstop® 2049; Pennstop® 2697

Diethylol dimethyl hydantoin. *See* DEDM hydantoin
Diethylol dimethyl hydantoin dilaurate. *See* DEDM hydantoin dilaurate

Diethyl phthalate
CAS 84-66-2; EINECS 201-550-6
Synonyms: DEP; Ethyl phthalate; Phthalic acid, diethyl ester; Diethyl 1,2-benzenedicarboxylate
Definition: Aromatic diester of ethyl alcohol and phthalic acid
Empirical: $C_{12}H_{14}O_4$
Formula: $C_6H_4(CO_2C_2H_5)_2$
Properties: Water-white liq., odorless, bitter taste; misc. with alcohols, ketones, esters, aromatic hydrocarbons; insol. in water; m.w. 222.24; dens. 1.120 (25/25 C); f.p. -40.5 C; b.p. 298 C; flash pt. (OC) 325 F; visc. 31.3 cs (0 C); stable
Precaution: Combustible when exposed to heat or flame; heated to decomp., emits acrid smoke and irritating fumes
Toxicology: TLV 5 mg/m^3 of air; LD50 (oral, rat) 8600 mg/kg; poison by IV route; mod. toxic by ingestion, subcut., IP routes; human systemic effects; strong irritant to eyes and mucous membranes; narcotic in high concs.; experimental reproductive effects
Uses: Solvent for nitrocellulose and cellulose acetate; plasticizer, wetting agent, insecticidal preps.; in perfumery as solvent and fixative; plasticizer in solid rocket propellants
Regulatory: FDA 21CFR §175.105, 175.300, 175.320, 178.3910, 181.22, 181.27, 212.177; 27CFR §21.105
Manuf./Distrib.: Allan; Allchem Ind.; Berje; BP Chem. Ltd; Daihachi Chem. Ind.; Eastman; Hüls Am.; Morflex; Penta Mfg.; Spectrum Chem. Mfg.; Unitex
Trade names: Kodaflex® DEP

N,N´-Diethylthiocarbamide. *See* 1,3-Diethylthiourea

1,3-Diethylthiourea
CAS 105-55-5; EINECS 203-308-5
Synonyms: DETU; N,N´-Diethylthiourea; N,N´-Diethylthiocarbamide
Formula: $(C_2H_5NH)_2CS$
Properties: Off-wh. solid, sl. typ. odor; sl. sol. in water; sol. in methanol, ether, acetone, benzene, ethyl acetate; insol. in gasoline; m.w. 132.2; dens. 1.1; m.p. 74 C
Toxicology: Carcinogen in animals; heated to dec., emits very toxic fumes of NO_x and SO_x, may cause severe eye irritation; poison by ingestion
Uses: Accelerator/activator in elastomers; corrosion inhibitor in metal pickling sol'ns.
Manuf./Distrib.: Faesy & Besthoff; Fluka; Howard Hall; Van Waters & Rogers
Trade names: Akrochem® D.E.T.U. Accelerator; Akroform® DETU PM; Ekaland DETU; Thiate® H

N,N´-Diethylthiourea. *See* 1,3-Diethylthiourea

Diethyl toluene diamine
CAS 68479-98-1
Synonyms: DETDA
Classification: Orthoalkylated aromatic
Uses: Intermediate for mfg. of pesticides, miticides, dyestuffs, antioxidants, pharmaceuticals, synthetic resins, fragrances; curing agent for polyurethane and epoxy resins
Trade names: Ethacure® 100

1,1-Difluoroethane
CAS 75-37-6
Synonyms: Ethylidene fluoride; HFC-152A
Formula: CH_3CHF_2
Properties: Colorless gas, odorless; insol. in water; m.w. 66.05; dens. 1.004 (-25 C); b.p. -24.7 C; f.p. -117 C
Precaution: Flamm., dangerous fire risk; flamm. limits in air 3.7-18%
Toxicology: Narcotic in high concs.
Uses: Intermediate
Manuf./Distrib.: PCR
Trade names containing: Blue Label Silicone Spray No. SB412-A; Paintable Organic Oil Spray No. 0316; Silicone Mist No. S116-A; Silicone Spray No. S312-A; Super 33 Silicone Spray No. S3312-A; Vydax® Spray No. V312-A; Zinc Stearate Spray No. Z212-B

1,1-Difluoroethene homopolymer. *See* Polyvinylidene fluoride
Diglycidyl ether of 1,4-butanediol. *See* 1,4-Butanediol diglycidyl ether
Diglycol. *See* Diethylene glycol
Diglycol laurate. *See* PEG-2 laurate
Diglycol methyl ether. *See* Diethylene glycol dimethyl ether
Diglycol oleate. *See* PEG-2 oleate
Diglycol stearate. *See* PEG-2 stearate
Diglyme. *See* Diethylene glycol dimethyl ether

Diheptyl, nonyl, undecyl linear phthalate
Uses: Plasticizer
Trade names: Jayflex® L7911P

Diheptyl-p-phenylenediamine. *See* N,N´-Di-(1,4-dimethylpentyl)-p-phenylenediamine

Di-n-heptyl phthalate
CAS 3648-21-3
Properties: M.w. 362; sp.gr. 0.991-0.993
Uses: Plasticizer
Manuf./Distrib.: Chisso Am.; Frinton Labs
Trade names: Diplast® E

Di-n-hexyl azelate
CAS 109-31-9
Properties: Sp.gr. 0.93; m.p. -23 C; flash pt. (COC) 204 C; ref. index 1.444
Uses: Plasticizer for food pkg. films
Manuf./Distrib.: Henkel
Trade names: Plastolein® 9051; Priplast 3013

Dihexyl maleate. *See* Dimethyl amyl maleate

Dihexyl phthalate
CAS 68515-50-4; 84-75-3; EINECS 201-559-5
Synonyms: Di(2-ethylbutyl) phthalate; 1,2-Benzenedicarboxylic acid, dihexyl ester
Formula: $C_6H_4(COOC_6H_{13})_2$
Properties: Oily, sl. aromatic liq.; m.w. 334.50; sp.gr. 1.008; b.p. 350 C (735 mm); f.p. -50 C; flash pt. (TCC) 380 F
Precaution: Combustible
Toxicology: Very mildly toxic by ingestion and skin contact; eye irritant
Uses: Plasticizer for cellulose ester and vinyl plastics
Manuf./Distrib.: Exxon; Frinton Labs; Unitex
Trade names: Jayflex® DHP

Dihexyl sodium sulfosuccinate
CAS 3006-15-3; 2373-38-8; 6001-97-4; EINECS 221-109-1
Synonyms: Sodium dihexyl sulfosuccinate; Butanedioic acid, sulfo-, 1,4-dihexyl ester, sodium salt
Definition: Sodium salt of the diester of 1-methylamyl alcohol and sulfosuccinic acid
Formula: $C_{16}H_{30}O_7S \cdot Na$
Uses: Dispersant, textile wetting agent, emulsifier, solubilizer; used for emulsion polymerization, battery

separators, electroplating, ore leaching
Regulatory: FDA 21CFR §175.105, 176.170, 177.1210, 178.3400
Manuf./Distrib.: Aquatec Quimica SA
Trade names: Aerosol® MA-80I; Astrowet H-80; Empimin® MA; Gemtex 680; Octosol HA-80
Trade names containing: Lankropol® KMA; Monawet MM-80

1,2-Dihydro-6-ethoxy-2,2,4-trimethylquinoline. *See* 6-Ethoxy-1,2-dihydro-2,2,4-trimethylquinoline
Dihydro-2(3H)-furanone. *See* Butyrolactone
Dihydrogenated tallow dimethyl ammonium chloride. *See* Quaternium-18

Dihydrogenated tallow methylamine
CAS 61788-63-4; 67700-99-6; EINECS 262-991-8
Classification: Tertiary aliphatic amine
Formula: $R_2N—CH_3$, R represents tallow radical
Uses: Chemical intermediate for quat. ammonium derivs., acid scavenger in petrol. prods.; epoxy hardener, catalyst in mfg. of flexible PU foams
Trade names: Kemamine® T-9701

2,5-Dihydroperoxy-2,5-dimethylhexane
Uses: Initiator for polymerization
Trade names: Luperox 2,5-2,5

1,4-Di (2-hydroperoxy isopropyl) benzene
Uses: Initiator
Trade names: Aztec® DIHP-55-W

4,5-Dihydro-1-[3-(triethoxysilyl)propyl]imidazole. *See* N-[3-(Triethoxysilyl)-propyl] 4,5-dihydroimidazole
1,2-Dihydro-2,2,4-trimethylquinoline homopolymer. *See* 2,2,4-Trimethyl-1,2-dihydroquinoline polymer
1,2-Dihydro-2,2,4-trimethyl quinoline polymer. See 2,2,4-Trimethyl-1,2-dihydroquinoline polymer

m-Dihydroxybenzene. *See* Resorcinol
p-Dihydroxybenzene. *See* Hydroquinone
2,4-Dihydroxybenzophenone. *See* Benzophenone-1
Dihydroxydiethyl ether. *See* Diethylene glycol
2,2´-Dihydroxy-4,4´-dimethoxybenzophenone. *See* Benzophenone-6
Di(2-hydroxyethyl) amine. *See* Diethanolamine

Dihydroxyethyl cocamine oxide
CAS 61791-47-7; EINECS 263-180-1
Synonyms: N,N (bis (2-hydroxyethyl) cocamine oxide; Coco di(hdyroxyethyl) amine oxide; Ethanol, 2,2´-iminobis, N-coco alkyl, N-oxide
Classification: Tertiary amine oxide
Formula: $R—N(CH_2CH_2OH)_2O$, R represents the coconut radical
Uses: Softener, conditioner, emulsifier, wetting agent, visc. builder, stabilizer, antistat for cosmetics; emollient, lubricant, and slip agent for shave creams; for textiles, petrol. additives, paper, plastics, rubber; gel sensitizer for latex foams
Trade names: Aromox® C/12-W
Trade names containing: Aromox® C/12

Di-(2-hydroxyethyl)-5,5-dimethyl hydantoin. *See* DEDM hydantoin
Di-(2-hydroxyethyl)-5,5-dimethyl hydantoin dilaurate. *See* DEDM hydantoin dilaurate
2,2´-Dihydroxy-4-methoxybenzophenone. *See* Benzophenone-8
1,2-Dihydroxypropane. *See* Propylene glycol
2,3-Dihydroxypropyl octacosanoic acid. *See* Glyceryl montanate
2,3-Dihydroxypropyl octadecanoate. *See* Glyceryl stearate

Diisobutyl adipate
CAS 141-04-8; EINECS 205-450-3
Synonyms: DIBA; Bis(2-methylpropyl) hexanedioate; Diisobutyl hexanedioate; Hexanedioic acid diisobutyl ester
Definition: Diester of isobutyl alcohol and adipic acid
Empirical: $C_{14}H_{26}O_4$
Formula: $[C_2H_4COOCH_2CH(CH_3)_2]_2$

Diisobutyl azelate

Properties: Colorless liq., odorless; sol. in most org. solvs.; insol. in water; m.w. 258.40; dens. 0.950 (25 C); b.p. 278-280 C; f.p. -20 C
Precaution: Combustible
Toxicology: LD50 (IP, rat) 5950 mg/kg; mod. toxic by IP route; mildly toxic by ingestion; experimental teratogen, reproductive effects; heated to decomp., emits acrid smoke and irritating fumes
Uses: Plasticizer for some cellulosics, PVAc, PVC, VCA
Regulatory: FDA 21CFR §175.105, 181.22, 181.27
Manuf./Distrib.: Aceto
Trade names: DIBA; Plasthall® DIBA

Diisobutyl azelate
Uses: Plasticizer
Trade names: Plasthall® DIBZ

Diisobutylene
CAS 25167-70-8; EINECS 246-690-9
Definition: Group of isomers (C_6H_{16}) of which 2,4,4-trimethylpentene-1 and 2,4,4-trimethylpentene-2 are most important
Empirical: C_8H_{16}
Properties: Colorless liq.; m.w. 112.22; dens. 0.7227 (15.5 C); b.p. 101-104 C; flash pt. -6.6 C
Precaution: Fire risk
Toxicology: Narcotic in high concs.
Uses: Alkylation, intermediate, antioxidants, surfactants, lube additive, plasticizers, rubber chemicals
Manuf./Distrib.: Fluka
Trade names containing: Agerite® Superlite® Emulcon®

Diisobutyl hexanedioate. *See* Diisobutyl adipate

Diisobutyl ketone
CAS 108-83-8; EINECS 203-620-1
Synonyms: DIBK; 2,6-Dimethyl-4-heptanone; Isovalerone
Empirical: $C_9H_{18}O$
Formula: $(CH_3)_2CHCH_2COCH_2CH(CH_3)_2$
Properties: M.w. 142.24; sp.gr. 0.8076; b.p. 169 C; flash pt. 45 C; ref. index 1.4130 (20 C)
Precaution: Flamm.
Toxicology: LD50 (oral, rat) 4300 mg/kg; ACGIH TLV 100 ppm, STEL 150 ppm; OSHA PEL 100 ppm; irritant to respiratory system
Uses: Solvent for nitrocellulose, rubber, synthetic resins; lacquers, coatings, organic synthesis, roll-coating inks, stains
Manuf./Distrib.: Aldrich; Allchem Ind.; Ashland; Coyne; Eastman; Fluka; Great Western; Hüls AG; Union Carbide
Trade names containing: BYK®-S 706

Diisobutyl nonyl phenol
CAS 4306-88-1
Trade names: Uvi-Nox® 1494

Diisobutyl (oleyl) aceto acetyl aluminate
Uses: Coupling agent
Trade names: KA 301

Diisobutylphenoxyethoxyethyl dimethyl benzyl ammonium Cl. *See* Benzethonium chloride

Diisobutyl phthalate
CAS 84-69-5
Synonyms: DIBP
Properties: Sp.gr. 1.040 (20 C); f.p. -50 C; flash pt. (PMCC) 185 C; ref. index 1.490
Uses: Plasticizer for PVC, PVB, PS, cellulosics
Manuf./Distrib.: Allchem Ind.; Hüls; Unitex
Trade names: Diplast® B; Nuoplaz® DIBP; Uniplex 155

Diisobutyl sodium sulfosuccinate
CAS 127-39-9; EINECS 204-839-5

Synonyms: Sodium diisobutyl sulfosuccinate; 1,4-Bis(2-methylpropyl) sulfobutanedioate, sodium salt; Sulfosuccinic acid diisobutyl ester sodium salt
Definition: Sodium salt of the diester of isobutyl alcohol and sulfosuccinic acid
Empirical: $C_{12}H_{22}NaO_7S$
Properties: White. powd.; sol. 760 g/l in water (25 C); sol. in glycerol, pine oil, oleic acid; m.w. 332.35
Toxicology: Irritating to eyes, mucous membranes
Uses: Wetting agent; emulsion polymerization of styrene, butadiene and copolymers
Regulatory: FDA 21CFR §178.3400
Trade names: Aerosol® IB-45; Astrowet B-45; Gemtex 445; Geropon® CYA/45; Geropon® CYA/DEP; Monawet MB-45; Octosol IB-45; Rewopol® SBDB 45

Diisobutyryl peroxide
CAS 3437-84-1
Empirical: $C_8H_{14}O_4$
Formula: $(CH_3)_2CHCOO_2COCH(CH_3)_2$
Properties: M.w. 174.2
Uses: Initiator for polymers
Trade names containing: Trigonox® 187-C30

2,4-Diisocyanatotoluene. *See* Toluene diisocyanate

Diisodecyl adipate
CAS 27178-16-1; EINECS 248-299-9
Synonyms: DIDA; Hexanedioic acid, diisodecyl ester
Definition: Diester of branched chain decyl alcohols and adipic acid
Empirical: $C_{26}H_{50}O_4$
Formula: $C_{10}H_{21}OOC(CH_2)_4COOC_{10}H_{21}$
Properties: Lt. colored oily liq., mild odor; insol. in glycerols, glycols, and some amines; sol. in most other org. solvs.; m.w. 426; dens. 0.918 (20/20 C); m.p. -70 C; b.p. 239-246 C (4 mm); f.p. -71 C; flash pt. 225 F
Precaution: Combustible
Uses: Primary plasticizer for polymers
Regulatory: FDA 21CFR §175.105, 177.2600
Manuf./Distrib.: C.P. Hall; Hatco; Henkel; Hüls; ICI Am.; Inolex; Werner G. Smith
Trade names: Monoplex® DDA; Nuoplaz® DIDA; Plasthall® DIDA; Staflex DIDA

Diisodecyl glutarate
Properties: Sp.gr. 0.920; m.p. -65 C; flash pt. (COC) 204 C; ref. index 1.450
Uses: Lubricant additive; plasticizer for PVC, VCA, cellulosics
Manuf./Distrib.: C.P. Hall
Trade names: Plasthall® DIDG

Diisodecyl pentaerythritol diphosphite
CAS 26544-27-4; EINECS 247-779-5
Formula: $C_{10}H_{21}OP(OCH_2)_2C(CH_2O)_2POC_{10}H_{21}$
Properties: Liq. @ 100 C; m.w. 508; dens. 1.020-1.040; ref. index 1.4710-1.4750; flash pt. (PMCC) 152 C
Uses: Stabilizer for polymers incl. PC; antioxidant, corrosion inhibitor in lubricants
Trade names: Weston® 600

Diisodecyl phthalate
CAS 68515-49-1, 26761-40-0; EINECS 247-977-1
Synonyms: DIDP; 1,2-Benzenedicarboxylic acid, diisodecyl ester
Formula: $C_6H_4(COOC_{10}H_{21})_2$
Properties: Clear liq., mild odor; insol. in glycerol, glycols, some amines; sol. in most other organics; m.w. 746; dens. 0.966 (20/20 C); f.p. -50 C; b.p. 250-257 C (4 mm); visc. 108 cp (20 C); flash pt. 450 F
Precaution: Combustible
Toxicology: Irritant
Uses: Plasticizer for PVC, PS, cellulosics
Manuf./Distrib.: Allchem Ind.; Aristech; Ashland; BASF; Coyne; Exxon; C.P. Hall; Harwick; Hatco; Hoechst AG; Hüls; ICC Ind.; OxyChem; Velsicol
Trade names: Diplast® R; Jayflex® DIDP; Nuoplaz® DIDP; Palatinol® DIDP; Plasthall® DIDP-E; PX-120
Trade names containing: Omacide® P-DIDP-5; Prespersion PAB-8912; Prespersion PAC-332; Prespersion PAC-363; Prespersion PAC-2941; Vinyzene® BP-5-2 DIDP; Vinyzene® BP-505 DIDP; Vinyzene® IT-3000 DIDP; Vinyzene® IT-3025 DIDP

Diisoheptyl phthalate
CAS 71888-89-6
Formula: $C_6H_{4-1},2\text{-}(CO_2C_7H_{15})_2$
Properties: Sp.gr. 0.995; flash pt. (TCC) 390 F
Toxicology: Suspected carcinogen
Uses: Plasticizer for PVC, PS, cellulosics
Manuf./Distrib.: Aldrich; Exxon
Trade names: Jayflex® 77

Diisohexyl sulfosuccinate
Uses: Emulsion polymerization surfactant
Trade names: Rewopol® SBMB 80

Diisononanoyl peroxide
Classification: Diacyl peroxide
Uses: Polymerization
Trade names containing: Lupersol 219-M60

Diisononanoyl peroxide. *See* 3,5,5-Trimethyl hexanoyl peroxide

Diisononyl adipate
CAS 33703-08-1; EINECS 251-646-7
Synonyms: Hexanedioic acid, diisononyl ester; Diisononyl hexanedioate
Definition: Diester of isononyl alcohol and adipic acid
Empirical: $C_{24}H_{46}O_4$
Formula: $C_9H_{19}OOC(CH_2)_4COOC_9H_{19}$
Properties: Sp.gr. 0.924; flash pt. (TCC) 390 F
Precaution: Combustible
Uses: Plasticizer for PVC, PS, cellulosics
Regulatory: FDA 21CFR §178.3740
Manuf./Distrib.: Aristech; Ashland; Exxon; Miles; Unitex
Trade names: Adimoll® DN; Jayflex® DINA; PX-239; Staflex DINA

Diisononyl hexanedioate. *See* Diisononyl adipate

Diisononyl maleate
Uses: Plasticizer, intermediate used in specialty applic.
Trade names: Staflex DINM

Diisononyl phthalate
CAS 14103-61-8, 68515-48-0; EINECS 249-079-5
Synonyms: DINP; 1,2-Benzenedicarboxylic acid, diisononyl ester
Properties: Sp.gr. 0.973; flash pt. (TCC) 415 F
Uses: Plasticizer for PVC, PS, cellulosics
Manuf./Distrib.: Allchem Ind.; Aristech; Ashland; BASF; Chemisphere Ltd; Chisso Am.; Exxon; C.P. Hall; Harwick; Henkel; ICC Ind.; Unitex
Trade names: Diplast® N; Jayflex® DINP; Palatinol® N; Plasthall® DINP; PX-139
Trade names containing: Dabco® T-16; Dabco® T-96

Diisooctyl adipate
CAS 1330-86-5
Synonyms: DIOA
Formula: $C_8H_{17}OOC(CH_2)_4COOC_8H_{17}$
Properties: Lt. straw colored liq., mild odor; dens. 0.924 (25 C); b.p. 214-226 (4 mm); f.p. > 75 C; flash pt. 370 F
Precaution: Combustible
Uses: Plasticizer for PVC
Manuf./Distrib.: Exxon; C.P. Hall; Henkel; Inolex; Miles; Unitex
Trade names: Monoplex® DIOA; Plasthall® DIOA; Staflex DIOA

Diisooctyl dodecanedioate
Properties: Sp.gr. 0.91; m.p. -70 C; flash pt. (COC) 238 C; ref. index 1.450
Uses: Lubricant additive; plasticizer for PVC, VCA, PS, some cellulosics

Manuf./Distrib.: C.P. Hall
Trade names: Plasthall® DIODD

Diisooctyl isophthalate
Properties: Sp.gr. 0.983 (23 C); flash pt. (COC) 232 C; ref. index 1.487 (23 C)
Uses: Plasticizer for vinyls, PS, cellulosics
Manuf./Distrib.: Unitex

Diisooctyl maleate
CAS 1330-76-3
Uses: Plasticizer; comonomer with other monomers such as vinyl acetate
Manuf./Distrib.: Hoechst AG; Ruger
Trade names: Staflex DIOM

Diisooctyl octylphenyl phosphite
CAS 68133-13-1; EINECS 268-665-1
Properties: Liq.; m.w. 494; dens. 0.920-0.930; ref. index 1.4750-1.4800; flash pt. (PMCC) 152 C
Uses: Heat and color stabilizer for polymers incl. PVC
Trade names: Weston® 494

Diisooctyl phosphite
CAS 36116-84-4; EINECS 252-287-4
Formula: $(C_8H_{17}O)_2POH$
Properties: Liq.; m.w. 306; dens. 0.925-0.933; b.p. 129 C (5 mm); ref. index 1.4415-1.4430; flash pt. (PMCC) 146 C
Uses: Stabilizer for PVC, PP; lubricant additive
Trade names: Weston® DOPI

Diisooctyl phthalate
CAS 27554-26-3
Synonyms: DIOP; Isooctyl phthalate
Classification: Isomeric esters
Empirical: $C_{24}H_{38}O_4$
Formula: $(C_8H_{17}COO)_2C_6H_4$
Properties: Nearly colorless visc. liq., mild odor; insol. in water; m.w. 390.62; dens. 0.980-0.983 (20/20 C); m.p. -50 C; b.p. 370 C; flash pt. 450 F
Precaution: Combustible
Toxicology: LD50 (oral, rat) 22 g/kg; mod. toxic by ingestion; mildly toxic by skin contact; skin irritant; heated to decomp, emits acrid smoke and irritating fumes
Uses: Plasticizer for vinyl, cellulosic, and acrylate resins and synthetic rubber
Regulatory: FDA 21CFR §181.27
Manuf./Distrib.: Exxon; C.P. Hall; Hüls; OxyChem; Unitex
Trade names: Jayflex® DIOP; Plasthall® DIOP; Staflex DIOP

Diisooctyl sebacate
Properties: Sp.gr. 0.912-0.916 (20 C); m.p. -50 to -42 C; flash pt. (COC) 235-246 C; ref. index 1.447
Uses: Plasticizer for PVC, PS, ethyl cellulose
Manuf./Distrib.: Hatco

Diisopropanolamine
CAS 110-97-4; EINECS 203-820-9
Synonyms: DIPA; 1,1′-Iminobis-2-propenol
Classification: Aliphatic amine
Empirical: $C_6H_{15}NO_2$
Formula: $HN(CH_2CHOHCH_3)_2$
Uses: Emulsifying agents for polishes, textile specialties, leather compounds, insecticides, cutting oils, aq. paints
Regulatory: FDA 21CFR § 175.105, 176.210
Manuf./Distrib.: Ashland; BASF; Coyne; Ruger; Van Waters & Rogers
Trade names: DIPA Commercial Grade; DIPA Low Freeze Grade 85; DIPA Low Freeze Grade 90; DIPA NF Grade

Diisopropyl adipate
CAS 6938-94-9; EINECS 248-299-9
Synonyms: Bis(1-methylethyl)hexanedioate; Hexanedioic acid, bis (1-methylethyl) ester
Definition: Diester of isopropyl alcohol and adipic acid
Empirical: $C_{12}H_{22}O_4$
Formula: $(CH_3)_2CHOCO(CH_2)_4COOCH(CH_3)_2$
Properties: Clear liq.; sol. in min. oil, ethanol, propylene glycol, IPM, oleyl alcohol; insol. in water, glycerin; sp.gr. 0.950-0.962; ref. index 1.4216-1.4245
Toxicology: LD50 (rat, oral) > 20 ± 3 ml/kg; nonirritating to eyes and skin
Uses: Emollient, coupling agent in aq. alcoholic systems, shave lotions, hair tonics; imparts spreadability to bath oils; plasticizer in hair sprays
Manuf./Distrib.: Inolex; ISP Van Dyk; Union Camp

Diisopropylbenzene monohydroperoxide
CAS 3736-26-3; 26762-93-6
Classification: Organic peroxide
Properties: M.w. 194.3
Uses: Initiator for acrylates, PE
Trade names: Trigonox® M-50

Diisopropylbenzene peroxide. *See* Dicumyl peroxide

Diisopropyl benzothiazole sulfenamide
Formula: $C_6H_4NC[SN(C_3H_7)_2]S$
Uses: Accelerator for sulfur-curable elastomers
Trade names: BIBBS; Santocure® IPS

Diisopropyl (oleyl) aceto acetyl aluminate
Uses: Coupling agent
Trade names: KA 322

Diisopropyl perdicarbonate
CAS 105-64-6
Synonyms: Diisopropyl peroxydicarbonate; Isopropyl percarbonate; Isopropyl peroxydicarbonate
Uses: Initiator for polymerization of unsat. monomers, PVC
Manuf./Distrib.: PPG Ind.
Trade names: IPP
Trade names containing: Perkadox® IPP-AT50

Diisopropyl peroxydicarbonate. *See* Diisopropyl perdicarbonate

Diisopropyl sebacate
CAS 7491-02-3; EINECS 231-306-4
Synonyms: Bis(1-methylethyl) decanedioate; Decanedioic acid, bis(1-methylethyl) ester
Definition: Diester of isopropyl alcohol and sebacic acid
Empirical: $C_{16}H_{30}O_4$
Formula: $(CH_3)_2CHOCO(CH_2)_8COOCH(CH_3)_2$
Properties: Sp.gr. 0.936; flash pt. (COC) 190 C; ref. index 1.4310
Uses: Emollient, solubilizer, coupling agent in creams, lotions, bath oils, men's grooming aids; plasticizer
Manuf./Distrib.: Union Camp

Diisotridecyl phthalate
EINECS 248-368-3
Synonyms: 1,2-Benzenedicarboxylic acid, diisotridecyl ester
Uses: Plasticizer for NC, vinyl, and other coatings

Dilaurin. *See* Glyceryl dilaurate
Dilauroyl peroxide. *See* Lauroyl peroxide

Dilauryl phosphite
CAS 21302-09-0
Formula: $(C_{12}H_{25}O)_2PHO$
Properties: Water-white liq.; m.w. 418; dens. 0.898-0.906 (30/15.5 C); ref. index 1.4500-1.4530; flash pt.

(PMCC) 138 C
Uses: Synthesis of organophosphorous compds. for extreme pressure lubricants, adhesives; catalyst in polymerization of unsaturated compds.; stabilizer, lubricant additive
Manuf./Distrib.: Albright & Wilson
Trade names: Weston® DLP

Dilauryl thiodipropionate
CAS 123-28-4; EINECS 204-614-1
Synonyms: Didodecyl 3,3´-thiodipropionate; Thiobis(dodecyl propionate); Thiodipropionic acid dilauryl ester; Bis(dodecyloxycarbonylethyl) sulfide
Classification: Diester
Definition: Diester of lauryl alcohol and 3,3´-thiodipropionic acid
Empirical: $C_{30}H_{58}O_4S$
Formula: $(C_{12}H_{25}OOCCH_2CH_2)_2S$
Properties: White flakes, sweetish odor; insol. in water; sol. in benzene, toluene, acetone, ether, chloroform; sl. sol. in alcohols, ethyl acetate; m.w. 514.94; dens. 0.975; m.p. 40 C; b.p. 240 C (1 mm); acid no. < 1
Toxicology: LD50 (oral, rat) > 10.3 g/kg; eye irritant; heated to decomp., emits toxic fumes
Uses: Additive for high-pressure lubricants and greases, plasticizer and softening agent, antioxidant for edible fats and oils, in cosmetics, pharmaceuticals; stabilizer for elastomers and plastics
Regulatory: FDA 21CFR §175.300, 181.22, 181.24 (0.005% migrating from food pkg.), 182.3280 (0.02% max. fat/oil)
Manuf./Distrib.: Cytec; Evans Chemetics; Morton Int'l.; Witco/PAG
Trade names: Aflux® 32; Carstab® DLTDP; Cyanox® LTDP; Evanstab® 12; Lankromark® DLTDP; Lowinox® DLTDP; PAG DLTDP
Trade names containing: Aquastab PA 48

Dilinoleic acid
CAS 6144-28-1; 61788-89-4
Synonyms: 9,12-Octadecadienoic acid, dimer; Dimer acid
Definition: 36-Carbon dicarboxylic acid formed by the catalytic dimerization of linoleic acid
Empirical: $C_{36}H_{64}O_4$
Properties: Liq., Gardner 6-9 color
Uses: Lubricant, corrosion inhibitor, mildness additive in household detergents, plastics, and protective coatings
Regulatory: FDA 21CFR §176.200, 176.210, 178.3910
Manuf./Distrib.: Arizona; Henkel/Emery; Union Camp; Witco
Trade names: Empol® 1016; Empol® 1018; Empol® 1022; Hystrene® 3695

Dimelamine phosphate
Uses: Flame retardant for polymers
Manuf./Distrib.: DSM Melamine Am.
Trade names: Amgard® ND

Dimer acid
CAS 68783-41-5
Definition: Dibasic acid usually contg. 36 carbons
Properties: Visc. liq.
Uses: Polymer building block; polymerizes with alcohols and polyols to yield plasticizers, lubricating oils, hydraulic fluids
Trade names: Empol® 1004; Empol® 1008; Empol® 1020; Empol® 1026; Hystrene® 3675; Hystrene® 3675C; Hystrene® 3680; Industrene® D; Pripol 1009; Pripol 1013; Pripol 1017, 1022; Pripol 1022; Pripol 1025
See also Dilinoleic acid

Dimer diamine
Uses: Chemical intermediate, extender, crosslinking agent in polymeric systems; corrosion inhibitor; in epoxy systems
Trade names: Kemamine® DD-3680

Dimethicone
CAS 9006-65-9; 9016-00-6; 63148-62-9; 68037-74-1 (branched)
Synonyms: PDMS; Dimethylpolysiloxane; Dimethyl silicone; Polydimethylsiloxane
Definition: Silicone oil consisting of dimethylsiloxane polymers

Dimethicone copolyol

Empirical: $(C_2H_6OSi)_xC_4H_{12}Si$
Formula: $(CH_3)_3SiO[SI(CH_3)_2O]_nSi(CH_3)_3$
Properties: Colorless visc. oil; sol. in hydrocarbon solvs.; misc. in chloroform, ether; insol. in water; m.w. 340-250,000; dens. 0.96-0.97; visc. 300-1050 cst; ref. index 1.400-1.404
Toxicology: Suspected neoplastigen; heated to dec., emits acrid smoke and irritating fumes
Uses: Ointment and topical drug ingredient, skin protectant, foam preventative, surface active agent, release material
Regulatory: FDA 21CFR §145.180, 145.181, 146.185, 173.340, 175.105, 175.300, 176.170, 176.200, 176.210, 177.2260, 177.2600, 177.2800, 178.3570, 178.3910, 181.22, 181.28; USDA 9CFR §318.7, 381.147; Europe listed; UK approved
Manuf./Distrib.: Dow Corning
Trade names: Akrochem® Silicone Emulsion 1M; Akrochem® Silicone Emulsion 10M; Akrochem® Silicone Emulsion 60M; Akrochem® Silicone Emulsion 100; Akrochem® Silicone Emulsion 350; Akrochem® Silicone Emulsion 350 Conc.; Akrochem® Silicone Fluid 350; Dabco® DC1630; Dow Corning® 200 Fluid; GP-50-A Modified Silicone Emulsion; GP-51-E Dimethyl Silicone Emulsion; GP-52-E Dimethyl Silicone Emulsion; GP-53-E Dimethyl Silicone Emulsion; GP-54-E Dimethyl Silicone Emulsion; GP-60-E Silicone Emulsion; GP-80-AE Silicone Emulsion; GP-7101 Silicone Wax; GP-7102 Silicone Wax; Masil® EM 100; Masil® EM 100 Conc.; Masil® EM 100D; Masil® EM 100P; Masil® EM 250 Conc.; Masil® EM 350X; Masil® EM 350X Conc.; Masil® EM 1000; Masil® EM 1000 Conc.; Masil® EM 1000P; Masil® EM 10,000; Masil® EM 10,000 Conc.; Masil® EM 60,000; Masil® EM 100,000; Masil® SF 5; Masil® SF 10; Masil® SF 20; Masil® SF 50; Masil® SF 100; Masil® SF 200; Masil® SF 350; Masil® SF 350 FG; Masil® SF 500; Masil® SF 1000; Masil® SF 5000; Masil® SF 10,000; Masil® SF 12,500; Masil® SF 30,000; Masil® SF 60,000; Masil® SF 100,000; Masil® SF 300,000; Masil® SF 500,000; Masil® SF 600,000; Masil® SF 1,000,000; PS040; PS041; PS041.2; PS041.5; PS042; PS043; PS044; PS045; PS046; PS047; PS047.5; PS048; PS048.5; PS049; PS049.5; PS050; Rhodorsil® AF 422; SF18-350; SF81-50; SF96®; SF96® (100 cst); SF96® (350 cst); SF96® (500 cst); SF96® (1000 cst); Silicone Emulsion 350, 1M, 10M, 60M; Silicone Fluid 350, 1M, 5M, 10M, 30M, 60M; Silicone Release Agent #5038; Siltech® E-2170; Silwax® WD-S; SM 70 Glass Coating Emulsion; SM 2061; SM 2068A; SM 2140; SM 2155; SM 2162; SM 2163; SM 2164; Union Carbide® LE-45; Viscasil® (10,000 cSt); Viscasil® (12,500 cSt)
Trade names containing: BYK®-LP W 6246; Cab-O-Sil® TS-720; Release Agent E-155; Silicone Emulsion E-131; Siltech® E-2140; Sipernat® D11; Wacker Stabilizer R

Dimethicone copolyol
CAS 63148-55-0; 64365-23-7; 67762-96-3
Synonyms: Dimethylsiloxane-glycol copolymer
Classification: Polymer
Uses: Silicone surfactant, dispersant, emulsifier, leveling and flow control agent, antifogging agent, lubricant, antiblock, slip additive for adhesives, agric., automotive, coatings, printing inks, textiles, household specialties, cutting fluids; in flexible slabstock PU foam
Trade names: Dow Corning® 190 Surfactant; Dow Corning® 193 Surfactant; GP-209 Silicone Polyol Copolymer; GP-215 Silicone Polyol Copolymer; Masil® 1066C; Masil® 1066D; SF1188; Silwax® WS; Silwet® L-7001; Silwet® L-7500; Silwet® L-7600; Silwet® L-7602; Tego® Foamex 1488; Tego® Foamex 7447
See also Polysiloxane polyether copolymer

Dimethicone copolyol isostearate
CAS 133448-16-5
Uses: Lubricious wax for personal care conditioning, polishes, textile lubrication and softening, laundry prods., dryer sheet softeners, syn. lubricants, plastics lubricants
Trade names: Silwax® WD-IS

Dimethicone/mercaptopropyl methicone copolymer
Uses: Release agent for plastics and rubber; internal lubricant and release agent for sulfur and peroxide cure rubber; coreactant in vinyl polymerization; synthesis of org./silicone copolymers; heat stabilizer for dimethyl silicone fluids
Trade names: GP-71-SS Mercapto Modified Silicone Fluid

Dimethiconol
CAS 31692-79-2
Synonyms: Poly[oxy(dimethylsilylene)], α-hydro-ω-hydroxy-
Definition: Dimethyl silicone terminated with hydroxyl groups
Uses: Silicone surfactant, textile softener, etc.; raw material for silicone RTV systems, textile/paper coatings; plasticizer/processing aid for silicone elastomers, in water repelllents

Trade names: Masil® SFR 70; Mazol® SFR 100; Mazol® SFR 750

Dimethiconol fluoroalcohol dilinoleic acid
Uses: Lubricious, chem. resist. paintable mold release; additive for barrier creams and other skin care prods.
Trade names: Silwax® F

Dimethiconol hydroxystearate
CAS 133448-13-2
Uses: Lubricious wax for personal care applics., polishes, textile lubrication and softening, laundry prods., dryer sheet softeners, syn. lubricants, plastics lubricants
Trade names: Silwax® C

Dimethiconol stearate
CAS 130169-63-0
Uses: Lubricious wax for personal care applics., polishes, textile lubrication and softening, laundry prods., dryer sheet softeners, syn. lubricants, plastics lubricants
Trade names: Silwax® S

Dimethoxyethyl phthalate
CAS 117-82-8
Synonyms: DMEP; 1,2-Benzenedicarboxylic acid bi (2-methoxyethyl) ester
Formula: $C_6H_4(COOCH_2CH_2OCH_3)_2$
Properties: Oily liq., mild odor; m.w. 282.32; dens. 1.172 (20/20 C); m.p. -40 C; b.p. 340 C; f.p. -45 C; flash pt. 194 C; combustible when exposed to heat or flame
Toxicology: Moderately toxic by ingestion, inhalation, and intraperitoneal routes; reacted to dec., emits acrid smoke and fumes
Uses: Plasticizer, solvent
Manuf./Distrib.: Ashland

Dimethyl adipate
CAS 627-93-0; EINECS 211-020-6
Synonyms: Dimethyl hexanedioate; Methyl adipate
Definition: Diester of methyl alcohol and adipic acid
Empirical: $C_8H_{14}O_4$
Formula: $CH_3O_2C(CH_2)_4CO_2CH_3$
Properties: Colorless liq.; m.w. 174.22; dens. 1.062 (20/4 C); m.p. 9-11 C; b.p. 115C (13 mm)
Toxicology: Moderately toxic by intraperitoneal route; suspected teratogen and reproductive effects; heated to dec., emits acrid smoke and irritating fumes
Uses: Solv. for coatings, cleaners, inks, textile lubricants, urethane prod.; plasticizer for flexible thermoset polyester; polymer intermediate for polyester polyols for urethanes, wet-str. paper resins, polyester resins; specialty chemical intermediate
Manuf./Distrib.: Ashland; Du Pont; Eastern Chem.; Morflex; UCB SA
Trade names: DBE-6
Trade names containing: DBE; DBE-2, -2SPG; DBE-3; DBE-9

Dimethylaminocyclohexane. *See* N,N-Dimethyl-N-cyclohexylamine
2-Dimethylaminoethanol. *See* Dimethylethanolamine

2-(2-Dimethylaminoethoxy) ethanol
CAS 1704-62-7; EINECS 216-940-1
Empirical: $C_6H_{15}NO_2$
Formula: $(CH_3)_2NCH_2CH_2OCH_2CH_2OH$
Properties: M.w. 133.22; dens. 0.95 (20/20 C); b.p. 201 C; f.p. < -50 C; flash pt. (TCC) 199 F
Toxicology: Poison by parenteral route, moderately toxic by ingestion and skin contact
Uses: Catalyst for PU
Trade names: Jeffcat ZR-70

Dimethylaminoethyl acrylate
CAS 2439-35-2
Empirical: $C_7H_{13}O_2N$
Formula: $CH_2=CHCOOCH_2CH_2N(CH_3)_2$
Properties: M.w. 143.19

Dimethylaminoethyl methacrylate

Toxicology: Corrosive; severe skin and eye irritant; harmful if swallowed or inhaled
Uses: Adhesion promoter in UV and EB cured coatings for metal, plastic, paper, and wood; catalyst; intermediate for water treatment chemicals, etc.
Manuf./Distrib.: CPS; Monomer-Polymer & Dajac
Trade names: Ageflex FA-1

Dimethylaminoethyl methacrylate
CAS 2867-47-2; EINECS 220-688-8
Synonyms: 2-(Dimethylamino)ethyl 2-methyl-2-propenoate
Classification: Amine monomer; aliphatic ester amine
Empirical: $C_8H_{15}NO_2$
Formula: $CH_2=C(CH_3)COOCH_2CH_2N(CH_3)_2$
Properties: Liq.; sol. in water and org. solvs.; m.w. 157.21; dens. 0.933 (25 C); b.p. 182-190 C; flash pt. 73.9 C
Precaution: Combustible
Toxicology: Poisonous; harmful if swallowed; causes severe eye burns, skin and mucous membrane irritation; strong lachrymator
Uses: Visc. index improvers, automotive and industrial coatings, dye additives, lube oil additives, detergent and sludge dispersant in lubricant oils, flocculant for waste water treatment, retention aid for paper making, acrylic floor polish; resin and rubber modifier; acid scavenger in PU foams; corrosion inhibitor
Regulatory: FDA 21CFR §178.3520
Manuf./Distrib.: CPS; Monomer-Polymer & Dajac; Polysciences; Rhone-Poulenc Spec.; Rohm Tech
Trade names: Ageflex FM-1

2-(Dimethylamino)ethyl 2-methyl-2-propenoate. *See* Dimethylaminoethyl methacrylate

Dimethylaminomethyl phenol
CAS 25338-55-0
Synonyms: m-Hydroxy-N,N-dimethyl aniline
Formula: $C_6H_4OHCH_2N(CH_3)_2$
Properties: Dk. red liq., phenolic odor; sol. in org. solvs.; moderately sol. in water; m.w. 137.09; dens. 1.010 (25/25 C); m.p. 85 C; b.p. 265-268 C
Uses: Antioxidants, stabilizers, catalysts, intermediates
Trade names: Ancamine® 1110

2-Dimethylaminomethyl propanol. *See* 2-Dimethylamino-2-methyl-1-propanol

2-Dimethylamino-2-methyl-1-propanol
CAS 7005-47-2; EINECS 230-279-6
Synonyms: 2-Dimethylaminomethyl propanol
Classification: Amino alcohol
Formula: $(CH_3)_2C(CH_2OH)N(CH_3)_2$
Properties: M.w. 117.22; dens. 0.910 (20/4 C); b.p. 158-160 C
Toxicology: Poison by intraperitoneal route; heated to dec., emits toxic NO_x
Uses: Solubilizer for resins in aq. coatings; emulsifier for waxes; vapor-phase corrosion inhibitor; urethane catalyst; titanate solubilizer; raw material for synthesis
Manuf./Distrib.: ANGUS; Fluka

Dimethylaminopropylamine
Synonyms: DMAPA
Properties: Clear liq.; dens. 6.8 lb/gal; f.p. < -56 C; b.p. 135 C (760 mm); flash pt. (COC) 110 F; ref. index 1.4350
Uses: Intermediate for hair prods., betaine mfg., gasoline additives, antistats, agric. emulsifiers, fabric softeners, asphalt antistripping agents, dyes; epoxy curing agent
Manuf./Distrib.: Elf Atochem N. Am.; Texaco

N-[3-(Dimethylamino)propyl]coco amides-N-oxide. *See* Cocamidopropylamine oxide
N-(3-Dimethylaminopropyl)-N,N-diisopropanolamine. *See* N,N-Dimethyl)-N′,N′-diisopropanol-1,3-propanediamine

Dimethylammonium hydrogen isophthalate
CAS 71172-17-3
Uses: Accelerator for neoprenes
Trade names: Vanax® CPA

Dimethyl amyl maleate
CAS 105-52-2
Synonyms: Di-4-methyl-2-amyl maleate; Dihexyl maleate
Empirical: $C_{16}H_{28}O_4$
Properties: Liq.; sol. < 0.01% in water at 20 C; m.w. 284.44; dens. 0.9602 (20/20 C); b.p. 179 C (10 mm); f.p. -70 C; flash pt. (OC) 290 F; combustible when exposed to heat
Toxicology: Mildly toxic by ingestion and skin contact; skin and eye irritant
Uses: Plasticizer; detergent intermediate
Trade names: Staflex DMAM

Di-4-methyl-2-amyl maleate. *See* Dimethyl amyl maleate

Dimethyl arachidyl-behenyl amine
Classification: Teritiary amine
Uses: Chemical intermediate for quat. ammonium derivs., acid scavenger in petrol. prods.; epoxy hardener, catalyst in mfg. of flexible PU foams
Trade names: Kemamine® T-1902D

Dimethyl 2,2´-azobisisobutyrate. *See* Dimethyl 2,2´-azobis (2-methylpropionate)

Dimethyl 2,2´-azobis (2-methylpropionate)
CAS 2589-57-3; EINECS 219-976-6
Synonyms: Dimethyl 2,2´-azobisisobutyrate
Empirical: $C_{10}H_{18}N_2O_4$
Properties: Pale yel. wax; sol. in toluene, hexane, ethanol, methanol; insol. in water; m.w. 230.26; m.p. 22-28 C; dec. 85-87 C
Uses: Polymerization initiator
Manuf./Distrib.: Wako Chem. USA
Trade names: V-601

Dimethyl behenamine
CAS 215-42-9; 21542-96-1
Synonyms: Behenyl dimethyl amine; N,N-Dimethyl-1-docosanamine
Classification: Tertiary aliphatic amine
Empirical: $C_{24}H_{51}N$
Formula: $CH_3(CH_2)_{20}CH_2N(CH_3)_2$
Uses: Neutralizer, conditioner, coemulsifier; in herbicides, ore flotation, pigment dispersion; aux. for textiles, leather, rubber, plastics, and metal industries
Trade names: Crodamine 3.ABD

Dimethylbenzene. *See* Xylene
Dimethyl 1,2-benzenedicarboxylate. *See* Dimethyl phthalate
Dimethylbenzenesulfonic acid. *See* Xylene sulfonic acid

Di-(2-methylbenzoyl) peroxide
CAS 3034-79-5
Classification: Organic peroxide
Properties: M.w. 270.3
Uses: Crosslinking agent
Trade names: Perkadox® 20; Perkadox® 20-W40

Di (4-methylbenzoyl) peroxide
Uses: Crosslinking agent
Trade names: Aztec® PMBP-50-PSI

N,N-Dimethylbenzylamine. *See* N-Benzyldimethylamine
α,α-Dimethylbenzyl hydroperoxide. *See* Cumene hydroperoxide

2,5-Dimethyl-2,5-bis (benzoylperoxy) hexane
Uses: Initiator for vinyl polymerization
Trade names: Luperox 118

2,5-Dimethyl-2,5-bis (t-butylperoxy) hexane. *See* 2,5-Dimethyl-2,5-di (t-butylperoxy) hexane

2,5-Dimethyl-2,5-bis (2-ethylhexanoylperoxy) hexane
Uses: Initiator for vinyl polymerizations
Trade names: Lupersol 256

Dimethyl butanedioate. *See* Dimethyl succinate

N-1,3-Dimethylbutyl-N′-phenyl-p-phenylene diamine
CAS 793-24-8
Uses: Antiozonant for rubber
Trade names: Akrochem® Antiozonant PD-2; Antozite® 67P; Permanax™ 6PPD; Santoflex® 13; Vulkanox® 4020

Dimethyl carbinol. *See* Isopropyl alcohol

Dimethyl cocamine
CAS 61788-93-0; EINECS 263-020-0
Synonyms: Coco dimethyl amine; Amines, coco alkyl dimethyl; Dimethyl coconut amine
Classification: Tertiary aliphatic amine
Formula: $R—N(CH_3)_2$, R represents the coconut radical
Uses: Chemical intermediate, raw material for surfactants
Trade names: Kemamine® T-6502D

Dimethyl coconut amine. *See* Dimethyl cocamine

N,N-Dimethyl-N-cyclohexylamine
CAS 98-94-2; EINECS 202-715-5
Synonyms: Cyclohexyldimethylamine; Dimethylaminocyclohexane
Empirical: $C_8H_{17}N$
Formula: $C_6H_{11}N(CH_3)_2$
Properties: M.w. 127.23; dens. 0.849 (20/4 C); b.p. 49-50 C (10 mm); flash pt. 39 C; ref. index 1.454 (20 C)
Precaution: Flamm.
Toxicology: Irritating to eyes, skin, respiratory system
Uses: Accelerator for NR, SBR, or latexes; catalyst for PU
Manuf./Distrib.: Air Prods.; BASF
Trade names: Polycat® 8; Toyocat®-DMCH

Dimethyl cyclohexyl ammonium dibutyl dithiocarbamate
Uses: Accelerator for NR, SBR, or latexes
Trade names: Akrochem® CZ-1

Dimethyl decylamine
CAS 1120-24-7; EINECS 214-302-7
Synonyms: Decyl dimethylamine; C10 alkyl dimethylamine; N,N-Dimethyl decylamine
Classification: Tertiary amine
Empirical: $C_{12}H_{27}N$
Formula: $CH_3(CH_2)_9N(CH_3)_2$
Properties: M.w. 185.36; dens. 0.778 (20/4 C); b.p. 234 C
Uses: Intermediate for quat. ammonium compds., amine oxides, betaines; for household prods., disinfectants, sanitizers, industrial hand cleaners, cosmetics, bubble baths, deodorants, polymer additives, PU foam catalysis, epoxy curing agent
Manuf./Distrib.: Fluka
Trade names: Adma® 10
Trade names containing: Adma® WC

N,N-Dimethyl decylamine. *See* Dimethyl decylamine
Dimethyl dialkyl ammonium chloride. *See* Dialkyl dimethyl ammonium chloride

2,5-Dimethyl-2,5-di(benzoylperoxy)hexane
CAS 2618-77-1
Uses: Polymerization initiator
Trade names: Aztec® DHPBZ-75-W

2,5-Dimethyl-2,5-di (t-butylperoxy) hexane
CAS 78-63-7
Synonyms: 2,5-Dimethyl-2,5-bis (t-butylperoxy) hexane
Classification: Organic peroxide
Empirical: $C_{16}H_{34}O_4$
Formula: $C_4H_9OOC(CH_3)_2CH_2CH_2C(CH_3)_2OOC_4H_9$
Properties: Colorless liq.; sol. in alcohol, many org. solvs.; insol. in water; m.w. 290.45; dens. 0.85; b.p. 50 C; f.p. 8 C; flash pt. 85 C; stable; flamm.
Precaution: Combustible oxidizing liq.
Toxicology: Moderately toxic by intraperitoneal route; LD50 (rat, oral) > 10,000 mg/kg; LD50 (rabbit, dermal) 4.1 ± 1.3 g/kg; minimal eye irritant
Uses: Catalyst in polyethylene crosslinking, styrene polymerization, polyester resins
Trade names: Aztec® 2,5-DI; Aztec® 2,5-DI-45-G; Aztec® 2,5-DI-45-IC; Aztec® 2,5-DI-45-IC1; Aztec® 2,5-DI-45-PSI; Esperal® 120; Lupersol 101; Polyvel CR-5; Polyvel CR-5F; Polyvel CR-10; Polyvel CR-25; Trigonox® 101; Varox® DBPH-50; Varox® DBPH
Trade names containing: Aztec® 2,5-DI-50-C; Aztec® 2,5-DI-70-S; Luperco 101-P20; Luperco 101-XL; Prespersion SIB-7324; Trigonox® 101-7.5PP-pd; Trigonox® 101/45Bpd; Trigonox® 101-45S-ps; Trigonox® 101-50PP-pd; Trigonox® 101-E50

2,5-Dimethyl-2,5-di-(t-butylperoxy)3-hexyne. *See* 2,5-Dimethyl-2,5-di (t-butylperoxy) hexyne-3

2,5-Dimethyl-2,5-di (t-butylperoxy) hexyne-3
CAS 1068-27-5
Synonyms: 2,5-Dimethyl-2,5-di-(t-butylperoxy)3-hexyne
Classification: Organic peroxide
Empirical: $C_{16}H_{30}O_4$
Formula: $(CH_3)_3COOC(CH_3)_2CCC(CH_3)_2OOC(CH_3)_3$
Properties: Pale yel. liq.; m.w. 286.4; sp.gr. 0.89 (20 C); f.p. < 8 C; flash pt. (Seta CC) 65 C
Precaution: Combustible oxidizing liq.
Toxicology: Nontoxic by skin contact; LD50 (mice, i.p.) 1850 mg/kg
Uses: Crosslinking agent for polyolefins

Dimethyldichlorosilane
CAS 75-78-5; 68611-44-9; EINECS 200-901-0
Formula: $(CH_3)_2SiCl_2$
Properties: Colorless liq.; sol. in benzene and ether; m.w. 129.0; dens. 1.062 (20 C); m.p. -76 C; b.p. 70 C; f.p.-86 C; flash pt. -8.9 C; ref. index 1.4023 (25 C)
Precaution: Flamm.
Toxicology: Irritating to eyes, respiratory system, and skin; PEL 6 mg/m^3
Uses: Intermediate for silicone prods.
Manuf./Distrib.: Dow Corning; Fluka; Hüls AG; Hüls Am.; PCR
Trade names containing: Cab-O-Sil® TS-610

2,5-Dimethyl-2,5-di(2-ethylhexanoyl peroxy) hexane
CAS 13052-09-0
Uses: Catalyst for heated curing of polyester resin systems
Trade names: Aztec® DHPEH; USP®-245

Dimethyldiglycol. *See* Diethylene glycol dimethyl ether
Dimethyl di(hydrogenated tallow)ammonium chloride. *See* Quaternium-18

N,N-Dimethyl)-N′,N′-diisopropanol-1,3-propanediamine
Synonyms: N-(3-Dimethylaminopropyl)-N,N-diisopropanolamine
Uses: Catalyst for PU foam
Trade names: Jeffcat DPA
Trade names containing: Jeffcat DPA-50

2,3-Dimethyl-2,3-diphenylbutane
CAS 1889-67-4
Properties: M.w. 238.4
Uses: Nonperoxide initiator for PS
Trade names: Perkadox® 30

3,4-Dimethyl-3,4-diphenylhexane
CAS 10192-93-5
Properties: M.w. 266.4
Uses: Nonperoxide initiator for PS
Trade names: Perkadox® 58

Dimethyl diphenyl thiuram disulfide
Uses: Accelerator for rubber
Trade names containing: Rhenogran® MPTD-80

Dimethyldithiocarbamic acid copper salt. *See* Copper dimethyldithiocarbamate
Dimethyldithiocarbamic acid sodium salt. *See* Sodium dimethyldithiocarbamate
N,N-Dimethyl-1-docosanamine. *See* Dimethyl behenamine
N,N-Dimethyl-1-dodecanamine-N-oxide. *See* Lauramine oxide
N,N-Dimethyl-N-dodecylbenzenemethanaminium chloride. *See* Lauralkonium chloride

Dimethyl erucylamine
Classification: Tertiary amine
Uses: Emulsifier for herbicides, ore flotation, pigment dispersion; aux. for textiles, leather, rubber, plastics, and metal industries
Trade names: Crodamine 3.AED

Dimethylethanolamine
CAS 108-01-0; EINECS 203-542-8
Synonyms: DMAE; 2-Dimethylaminoethanol; Deanol
Empirical: $C_4H_{11}NO$
Formula: $(CH_3)_2NCH_2CH_2OH$
Properties: Colorless liq., ammoniacal odor; m.w. 89.14; dens. 0.8866 (20/4 C); b.p. 133 C; flash pt. (OC) 105 F; ref. index 1.430flamm.
Toxicology: Moderately toxic by ingestion, inhalation, skin contact, intraperitoneal, subcutaneous routes; skin and eye irritant; central nervous system stimulant
Uses: Synthesis of dyestuffs, pharmaceuticals, textile auxiliaries; catalyst for PU foam and slabstock; epoxy curing agents
Manuf./Distrib.: Air Prods.; Allchem Ind.; Ashland; BASF; Elf Atochem N. Am.; Great Western; Nippon Nyukazai; Pelron; Texaco; Union Carbide
Trade names: Dabco® DMEA; Jeffcat DME; Tegoamin® DMEA; Toyocat®-DMA
Trade names containing: Dabco® R-8020®; Jeffcat DPA-50; Jeffcat ZF-51; Jeffcat ZF-52

Dimethyl ether
CAS 115-10-6; EINECS 204-065-8
Synonyms: Methane, oxybis-; Oxybismethane; Methyl ether
Classification: Organic compd.
Empirical: C_2H_6O
Formula: CH_3OCH_3
Uses: Aerosol propellant
Trade names containing: Blue Label Silicone Spray No. SB412-A; Paintable Organic Oil Spray No. O316; Silicone Mist No. S116-A; Silicone Spray No. S312-A; Super 33 Silicone Spray No. S3312-A; Vydax® Spray No. V312-A; Zinc Stearate Spray No. Z212-B

2-(1,1-Dimethylethyl)-1,4-benzenediol. *See* t-Butyl hydroquinone
Dimethyl ethyl carbinol. *See* t-Amyl alcohol

2,5-Dimethyl-2,5-(2-ethylhexanoylperoxy)hexane
CAS 13052-09-0
Classification: Organic peroxide
Empirical: $C_{24}H_{46}O_6$
Properties: M.w. 430.6
Uses: Initiator for acrylates, PE, PS
Trade names: Trigonox® 141

1,1-Dimethylethylhydroperoxide. *See* t-Butyl hydroperoxide
(1,1-Dimethylethyl)-4-methoxyphenol. *See* BHA

Dimethyl formamide
CAS 68-12-2; EINECS 200-679-5
Synonyms: DMF
Empirical: C_3H_7NO
Formula: $HCON(CH_3)_2$
Properties: Water-wh. liq.; a dipolar aprotic solv.; misc. with water, most org. solvs.; m.w. 73.10; dens. 0.953-0.954 (15.6/15.6 C); b.p. 152.8 C; f.p. -61 C; flash pt. 57.7 C
Precaution: Combustible; moderate fire risk
Toxicology: Toxic by skin absorption; strong irritant to skin and tissue; TLV 10 ppm in air
Uses: Solvent in vinyl resins and acetylene, butadiene, acid gases; polyacrylic fibers; catalyst in carboxylation reactions, organic synthesis, carrier for gases
Manuf./Distrib.: Aceto; Air Prods.; Ashland; BASF; Brown; Coyne; Du Pont; Great Western; ICI Spec.; Mallinckrodt; Mitsubishi Gas; Nissan Chem. Ind.; Spectrum Chem. Mfg.; UCB SA

Dimethyl glutarate
CAS 1119-40-0; EINECS 214-277-2
Synonyms: Dimethyl pentanedioate
Empirical: $C_7H_{12}O_4$
Formula: $CH_3OCO(CH_2)_3COOCH_3$
Properties: M.w. 160.17; dens. 1.087 (20/4 C); b.p. 96-103 C
Uses: Solv. for coatings, cleaners, inks, textile lubricants, urethane prod.; plasticizer for flexible thermoset polyester; polymer intermediate for polyester polyols for urethanes, wet-str. paper resins, polyester resins; specialty chemical intermediate
Manuf./Distrib.: Ashland
Trade names: DBE-5
Trade names containing: DBE; DBE-2, -2SPG; DBE-3; DBE-9

2,6-Dimethyl-4-heptanone. *See* Diisobutyl ketone

N,N´-Di-3(5-methylheptyl)-p-phenylenediamine
Trade names: Antozite® 2

Dimethyl hexanedioate. *See* Dimethyl adipate

Dimethyl hydrogenated tallow amine
CAS 61788-95-2; EINECS 263-022-1
Synonyms: Hydrogenated tallow dimethylamine; Hydrogenated tallowalkyl dimethylamine
Classification: Tertiary aliphatic amine
Formula: $R—N(CH_3)_2$, R represents hydrogenated tallow radical
Uses: Chemical intermediate, raw material for surfactants; in textiles; acid scavenger in petrol. prods.; epoxy hardener, catalyst in mfg. of flexible PU foams
Trade names: Kemamine® T-9702D; Nissan Tert. Amine ABT

1,1-Dimethyl-3-hydroxybutylperoxy-2-ethylhexanoate
Uses: Crosslinking agent for polymerization, cure of acrylic syrup
Trade names: Lupersol 665-M50
Trade names containing: Lupersol 665-T50

1,1-Dimethyl-3-hydroxybutylperoxyneoheptanoate
Uses: Crosslinking agent for polymerization, cure of acrylic syrup
Trade names containing: Lupersol 688-M50; Lupersol 688-T50

Dimethyl isophthalate
CAS 1459-93-4; EINECS 215-951-9
Synonyms: 1,3-Benzenedicarboxylic acid, dimethyl ester
Empirical: $C_{10}H_{10}O_4$
Formula: $C_6H_4(COOCH_3)_2$
Properties: M.w. 194.19; m.p. 66-67 C; flash pt. (COC) 143 C
Uses: Plasticizer for PVC, PVB, PS, cellulosics
Manuf./Distrib.: Fluka; Morflex; Unitex
Trade names: Morflex® 1129; Uniplex 270

Dimethyl lauramine
CAS 112-18-5; 67700-98-5; EINECS 203-943-8
Synonyms: Lauryl dimethylamine; Dodecyldimethylamine
Classification: Tertiary aliphatic amine
Empirical: $C_{14}H_{31}N$
Formula: $CH_3(CH_2)_{10}CH_2N(CH_3)_2$
Properties: M.w. 213.46
Toxicology: Moderately toxic by ingestion; severe skin and eye irritant; heated to dec., emits toxic NO_x
Uses: Liq. cationic detergent; corrosion inhibitor; acid-stable emulsifier; intermediate for quat. ammonium compds., amine oxides, betaines; for household prods., disinfectants, sanitizers, industrial hand cleaners, cosmetics; polymer additives, PU foam catalysis, epoxy curing agent
Regulatory: FDA 21CFR §177.1680
Manuf./Distrib.: Albemarle; Lonza; Mason
Trade names: Adma® 12; Nissan Tert. Amine BB
Trade names containing: Adma® 246-451; Adma® 246-621; Adma® 1214; Adma® 1416; Adma® WC

2,4-Dimethyl-6-(1-methyl cyclohexyl) phenol
Uses: Antioxidant for rubber
Trade names: Permanax™ WSL; Permanax™ WSL Pdr

3,3´-Dimethylmethylenedi(cyclohexylamine)
Uses: Epoxy curing agent
Trade names: Ancamine® 2049

Dimethyl methylphosphonate
CAS 756-79-6; EINECS 212-052-3
Synonyms: Methylphosphonic acid, dimethyl ester
Formula: $CH_3P(O)(OCH_3)_2$
Properties: M.w. 124.09; dens. 1.145; b.p. 181 C; f.p. 68 C
Toxicology: Mutagenic data, suspected carcinogen, experimental nerve gas stimulant
Uses: Flame retardant
Manuf./Distrib.: Aceto; Akzo Nobel; Albright & Wilson Am.; Fluka; Focus
Trade names: Antiblaze® DMMP; Fyrol® DMMP

Dimethyl myristamine
CAS 112-75-4; 68439-70-3; EINECS 204-002-4
Synonyms: Dimethyl myristylamine; Myristyl dimethylamine; Tetradecyl dimethylamine
Classification: Tertiary aliphatic amine
Empirical: $C_{16}H_{35}N$
Formula: $CH_3(CH_2)_{12}CH_2N(CH_3)_2$
Uses: Corrosion inhibitor; liq. cationic detergent; acid-stable; intermediate for quat. ammonium compds., amine oxides, betaines; for household prods., disinfectants, sanitizers, industrial hand cleaners, cosmetics, polymer additives; PU foam catalysis, epoxy curing agent
Trade names: Adma® 14; Nissan Tert. Amine MB
Trade names containing: Adma® 246-451; Adma® 246-621; Adma® 1214; Adma® 1416; Adma® WC

Dimethyl myristylamine. *See* Dimethyl myristamine
N,N-Dimethyl-1-octadecanamine. *See* Dimethyl stearamine

Dimethyl octylamine
CAS 7378-99-6; EINECS 230-939-3
Synonyms: DMOA; C8 alkyl dimethylamine; Octyl dimethylamine; N,N-Dimethyloctylamine
Formula: $CH_3(CH_2)_7N(CH_3)_2$
Properties: M.w. 157.30; dens. 0.765 (20/4 C); m.p. -57 C; b.p. 63-65 C; f.p. 65 C
Uses: Intermediate for quat. ammonium compds., amine oxides, betaines; for household prods., disinfectants, sanitizers, industrial hand cleaners, cosmetics, bubble baths, deodorants, polymer additives, PU foam catalysis, epoxy curing agent
Manuf./Distrib.: Fluka
Trade names: Adma® 8
Trade names containing: Adma® WC

N,N-Dimethyloctylamine. *See* Dimethyl octylamine

Dimethyl octynediol

CAS 1321-87-5; 78-66-0
Synonyms: 3,6-Dimethyl-4-octyne-3,6-diol
Classification: Aliphatic alcohol
Formula: $C_2H_5(CH_3)COHC:CCOH(CH_3)C_2H_5$
Properties: M.w. 170.35; m.p. 55 C; b.p. 222 C; f.p. > 110 C
Uses: Defoamer, wetting agent used in pesticide concs.; solubilizer and clarifier in shampoos; developer compds.; electroplating baths
Regulatory: FDA 21CFR §175.105
Trade names containing: Surfynol® 82S

3,6-Dimethyl-4-octyne-3,6-diol. *See* Dimethyl octynediol
Dimethylol dimethyl hydantoin. *See* DMDM hydantoin
1,3-Dimethylol-5,5-dimethyl hydantoin. *See* DMDM hydantoin

Dimethyl oleamine

CAS 14727-68-5; 28061-69-0
Synonyms: Oleyl dimethylamine
Classification: Tertiary amine
Uses: Surfactant intermediate; acid scavenger in petroleum prods.; emulsifier for herbicides, ore flotation, pigment dispersion; aux. for textiles, leather, rubber, plastics, and metal industries
Trade names: Crodamine 3.AOD

Dimethylolpropionic acid

CAS 4767-03-7
Synonyms: DMPA; 2,2-Bis(hydroxymethyl)propionic acid
Empirical: $C_5H_{10}O_4$
Properties: Off-wh. cryst. solid; sol. in water and methanol; sl. sol. in acetone; insol. in benzene; m.w. 134; m.p. 192-194 C
Toxicology: Essentially nontoxic; LD50 (mouse, oral) > 5000 mg/kg; sl. irritating to abraded skin; moderately irritating to eyes
Uses: In prep. of water-sol. alkyd resins, polyester resins, surfactants, chemical intermediates, syn. lubricants, plasticizers, pharmaceuticals, cosmetics
Manuf./Distrib.: Allchem Ind.; Fabrichem; Hoechst AG
Trade names: DMPA®

Dimethyl palmitamine

CAS 112-69-6; 68037-93-4; EINECS 203-997-2
Synonyms: Palmityl dimethylamine; Hexadecyl dimethylamine; Cetyl dimethylamine
Classification: Tertiary aliphatic amine
Empirical: $C_{18}H_{39}N$
Formula: $CH_3(CH_2)_{14}CH_2N(CH_3)_2$
Uses: Chemical intermediate, raw material for surfactants; for household prods., disinfectants, sanitizers, industrial hand cleaners, cosmetics, bubble baths, deodorants, polymer additives, PU foam catalysis, epoxy curing agent
Trade names: Adma® 16; Crodamine 3.A16D; Dabco® B-16; Nissan Tert. Amine PB
Trade names containing: Adma® 246-451; Adma® 246-621; Adma® 1416; Adma® WC

Dimethyl pentanedioate. *See* Dimethyl glutarate

N-(1,4-Dimethylpentyl)-N′-phenyl-p-phenylenediamine

Uses: Styrene polymerization inhibitor; column antifoulant
Trade names: Naugard® I-3

1,1-Dimethyl-3-phenylurea. *See* 1-Phenyl-3,3-dimethyl urea

Dimethyl phthalate

CAS 131-11-3; EINECS 205-011-6
Synonyms: DMP; Dimethyl 1,2-benzenedicarboxylate; 1,2-Benzenedicarboxylic acid dimethyl ester; Phthalic acid dimethyl ester
Definition: Diester of methyl alcohol and phthalic acid
Empirical: $C_{10}H_{10}O_4$
Formula: $C_6H_4(CO_2CH_3)_2$

1,4-Dimethylpiperazine

Properties: Colorless oily liq., sl. aromatic odor; misc. with alcohol, ether; insol. in water, paraffinic hydrocarbons; sl. sol. in min. oil; m.w. 194.08; dens. 1.189 (25/25 C); visc. 17.2 cps; m.p. 5.5 C; b.p. 282.0 C; flash pt. 149 C; ref. index 1.5138 (25 C)
Toxicology: TLV 5 mg/m^3 of air (TWA); moderately toxic by ingestion and intraperitoneal routes; midly toxic by inhalation; LD50 (rat, oral) 6.9 ml/kg
Uses: Solvent for resin, plasticizer for cellulose acetate and nitrocellulose lacquers; plastics, rubber; coating agents; safety glass
Regulatory: FDA 21CFR §175.105, 177.1010, 177.2420
Manuf./Distrib.: Allan; Allchem Ind.; Ashland; BASF; BP Chem.; Daihachi Chem. Ind.; Eastman; Great Western; Hüls Am.; Morflex; UCB SA; Unitex
Trade names: Kodaflex® DMP; Palatinol® M; Staflex DMP; Uniplex 110
Trade names containing: Cadox® M-30; Cadox® M-50; Hi-Point® 90; Hi-Point® PD-1; Trigonox® ACS-M28

1,4-Dimethylpiperazine. *See* N,N´-Dimethylpiperazine

N,N´-Dimethylpiperazine
CAS 106-58-1
Synonyms: 1,4-Dimethylpiperazine
Formula: $(CH_3)_2C_4H_8N_2$
Properties: Colorless mobile liq.; m.w. 114.13; dens. 0.8565 (20/4 C); b.p. 131 C; f.p. -1 C; flash pt. (TOC) 80 C
Precaution: Combustible
Uses: Curing agent for polyurethane foams, intermediate for cationic surfactants
Trade names: Jeffcat DMP

Dimethylpolysiloxane. *See* Dimethicone

Dimethyl sebacate
CAS 106-79-6; EINECS 203-431-4
Synonyms: Decanedioic acid, dimethyl ester
Formula: $[(CH_2)_4COOCH_3]_2$
Properties: Water-white liq.; m.w. 230.29; dens. 0.9896 (20/20 C); m.p. 24.5 C; b.p. ≈ 294 C; flash pt. 145 C
Precaution: Combustible
Uses: Solvent or plasticizer for nitrocellulose, vinyl resins; intermediate
Manuf./Distrib.: Aceto; Henkel; Hüls AG; Penta Mfg.; Richman; Union Camp; Unitex

Dimethyl silicone. *See* Dimethicone
Dimethylsiloxane-glycol copolymer. *See* Dimethicone copolyol

Dimethyl soyamine
CAS 61788-91-8; EINECS 263-017-4
Synonyms: Soya dimethyl amine
Classification: Tertiary aliphatic amine
Formula: $R—N(CH_3)_2$, R represents the soya radical
Uses: Surfactant intermediate; for textiles; acid scavenger in petrol. prods.; epoxy hardener, catalyst in mfg. of flexible PU foams
Trade names: Kemamine® T-9972D

Dimethyl stearamine
CAS 124-28-7; EINECS 204-694-8
Synonyms: Stearyl dimethyl amine; Octadecyl dimethylamine; N,N-Dimethyl-1-octadecanamine
Classification: Tertiary aliphatic amine
Formula: $CH_3(CH_2)_{16}CH_2N(CH_3)_2$
Uses: Chemical intermediate, raw material for surfactants; biocides, textile chems., oilfield chems., amine oxides, betaines, polyurethane foam catalysts, epoxy curing agents
Trade names: Adma® 18; Crodamine 3.A18D; Kemamine® T-9902D; Nissan Tert. Amine AB
Trade names containing: Adma® 1416; Adma® WC

2,4-Di-α-methylstyrylphenol
CAS 2772-45-4
Synonyms: 24DAMSP
Formula: $[C_6H_5C(CH_3)_2]_2C_6H_3OH$
Properties: M.w. 330.47

Manuf./Distrib.: Aldrich

Dimethyl succinate
CAS 106-65-0; EINECS 203-419-9
Synonyms: Dimethyl butanedioate; Methyl succinate
Empirical: $C_6H_{10}O_4$
Formula: $CH_3OCOCH_2CH_2COOCH_3$
Properties: Colorless liq., ethereal winey odor; m.w. 146.14; dens. 1.119 (20/4 C); m.p. 16-19 C; b.p. 190-193 C; flash pt. 90 C; ref. index 1.419
Uses: Light and heat stabilizer for polyolefins, ABS polymer systems, flexible PVC, food pkg.
Regulatory: FDA 21CFR §172.515
Manuf./Distrib.: Ashland; Chemie Linz N. Am.; Du Pont; Penta Mfg.; Schweizerhall
Trade names: DBE-4
Trade names containing: DBE; DBE-2, -2SPG; DBE-3; DBE-9

Dimethylsuccinate and tetramethyl hydroxy-1-hydroxyethyl piperidine polymer
CAS 65447-77-0
Uses: Light stabilizer for PVC, polyolefin substrates
Trade names: Tinuvin® 622; Tinuvin® 622LD

Dimethyl sulfoxide
CAS 67-68-5; EINECS 200-664-3
Synonyms: DMSO; Methyl sulfoxide
Empirical: C_2H_6OS
Formula: $(CH_3)_2SO$
Properties: M.w. 78.13; dens. 1.101; m.p. 18.4 C; b.p. 189 C; flash pt. 95 C; ref. index 1.4790 (20 C)
Precaution: Hygroscopic
Toxicology: Irritant
Uses: Solvent for polymerization; analytical reagent; industrial cleaners, pesticides, paint stripping, hydraulic fluids, medicine (anti-inflammatory), veterinary medicine, plant pathology and nutrition, pharmaceuticals, spinning syn. fibers
Manuf./Distrib.: Aldrich; Allchem Ind.; Elf Atochem N. Am.; Gaylord; Itochu Spec.; Monomer-Polymer & Dajac; Spectrum Chem. Mfg.
Trade names: Decap

N,N-Dimethyl-N-[2-[2-[4-(1,1,3,3-tetramethylbutyl) phenoxy] ethoxy] ethyl] benzenemethanaminium chloride. *See* Benzethonium chloride

N,N-Dimethyl-m-toluidine
CAS 121-72-2; EINECS 204-495-6
Empirical: $C_9H_{13}N$
Properties: M.w. 135.21; b.p. 212 C (760 mm)
Toxicology: poisonous
Uses: Cure accelerator for unsat. polyester and acrylate polymerizations
Manuf./Distrib.: First; Janssen Chimica; Spectrum Chem. Mfg.
Trade names: Firstcure™ DMMT

N,N-Dimethyl-p-toluidine
CAS 99-97-8; EINECS 202-805-4
Empirical: $C_9H_{13}N$
Properties: M.w. 135.23; b.p. 210-211 C (760 mm)
Uses: Accelerator for unsat. polyester and acrylate polymerizations
Manuf./Distrib.: Fabrichem; First; Monomer-Polymer & Dajac; Polysciences; R S A
Trade names: Firstcure™ DMPT

Dimorpholine N,N´-disulfide. *See* 4,4´-Dithiodimorpholine

2,2´-Dimorpholino diethylether
Uses: Catalyst for PU foam prod.
Trade names: Jeffcat DMDEE

Dimyristyl peroxydicarbonate
CAS 53220-22-7

Dimyristyl thiodipropionate

Classification: Organic peroxide
Properties: M.w. 514.8
Uses: Polymerization initiator, curing agent
Trade names: Aztec® MYPC; Aztec® MYPC-40-SAQ; Perkadox® 26-fl.; Perkadox® 26-W40

Dimyristyl thiodipropionate
CAS 16545-54-3; EINECS 240-613-2
Synonyms: Ditetradecyl 3,3′-thiobispropanoate; Propanoic acid, 3,3′-thiobis-, ditetradecyl ester
Classification: Diester
Empirical: $C_{34}H_{66}O_4S$
Formula: $S(CH_2CH_2COOC_{14}H_{29})_2$
Properties: M.w. 570; m.p. 48-50 C; acid no. 1.0 max.
Uses: Secondary antioxidant for polyolefins; stabilizer for elastomers, plastics; antioxidant for cosmetics, pharmaceuticals
Regulatory: FDA 21CFR §178.2010
Manuf./Distrib.: Evans Chemetics; Witco/PAG
Trade names: Cyanox® MTDP; Evanstab® 14; PAG DMTDP

N,N′-Di-2-naphthalenyl-1,4-benzenediamine. *See* N,N′-Di-β-naphthyl-p-phenylenediamine

N,N′-Di-β-naphthyl-p-phenylenediamine
CAS 93-46-9
Synonyms: DNPD; sym-Di-β-naphthyl-p-phenylenediamine; N,N′-Di-2-naphthalenyl-1,4-benzenediamine; 2-Naphthyl-p-phenylene diamine
Empirical: $C_{26}H_{20}N_2$
Formula: $C_6H_4(NHC_{10}H_7)_2$
Properties: Lt. tan to gray powd., rods, pellets, sl. amine odor; insol. in water; sl. sol. in acetone, chlorobenzene; m.w. 360.48; dens. 1.22-1.28
Precaution: Strong explosion hazard; avoid oxidizing agents
Toxicology: LD50 (rat, oral) 4500 mg/kg; human skin irritant; suspected tumorigen, mutagenic data; suspected skin and eye irritant
Uses: Antioxidant, stabilizer, polymerization inhibitor, intermediate in organic synthesis
Trade names: Agerite® White; Agerite® White White; Anchor® DNPD

sym-Di-β-naphthyl-p-phenylenediamine. *See* N,N′-Di-β-naphthyl-p-phenylenediamine

1,4-Di(2-neodecanoyl peroxy isopropyl) benzene
Trade names: Aztec® DIPND-50-AL

Dinitrosopentamethylene tetramine
CAS 101-25-7
Synonyms: DNPT; N,N-Dinitrosopentamethylenetetramine; NDPMT
Empirical: $C_5H_{10}N_6O_2$
Properties: M.w. 186.21; gas yield 240 cc/g
Toxicology: Poison by intravenous, subcutaneous, intraperitoneal routes; moderately toxic by ingestion; mutagenic data
Uses: Blowing agent for rubber goods, ABS
Trade names: Blo-Foam® DNPT 80; Blo-Foam® DNPT 93; Blo-Foam® DNPT 100; Opex® 80; Porofor® DNO/F

N,N-Dinitrosopentamethylenetetramine. *See* Dinitrosopentamethylene tetramine

Dinonyl undecyl phthalate
Uses: Plasticizer
Trade names: Jayflex® L911P

DINP. *See* Diisononyl phthalate
DIOA. *See* Diisooctyl adipate
Dioctadecyl dimethyl ammonium chloride. *See* Distearyldimonium chloride

Dioctadecyl disulfide
Synonyms: Distearyl disulfide
Empirical: $C_{36}H_{74}S_2$

Properties: M.w. 571
Uses: Antioxidant, costabilizer with phenolic antioxidants; for polyolefins
Trade names: Hostanox® SE 10

3,3´-Dioctadecyl thiodipropionate. *See* Distearyl thiodipropionate

Dioctanoyl peroxide
CAS 762-1603
Classification: Organic peroxide
Empirical: $C_{16}H_{30}O_4$
Formula: $CH_3(CH_2)_6COOOCO(CH_2)_6CH_3$
Properties: M.w. 286.4
Uses: Initiator for acrylates, PVC
Trade names: Perkadox® SE-8

Dioctyl adipate
CAS 103-23-1; EINECS 203-090-1
Synonyms: DOA; Bis(2-ethylhexyl) hexanedioate; Di(2-ethylhexyl) adipate; Hexanedioic acid, bis (2-ethylhexyl) ester
Definition: Diester of 2-ethylhexyl alcohol and adipic acid
Empirical: $C_{22}H_{42}O_4$
Formula: $[CH_2CH_2COOCH_2CH(C_2H_5)C_4H_9]_2$
Properties: Lt.-colored oily liq.; insol. in water; m.w. 370.64; dens. 0.9268 (20/20 C); b.p. 417 C; flash pt. 196 C; ref. index 1.4472
Toxicology: Suspected carcinogen and tetratogen; moderately toxic by IV route; mildly toxic by ingestion and skin contact; mutagenic data; eye and skin irritant
Uses: Plasticizer, commonly blended with general purpose plasticizers (DOP, DIOP); solvent, aircraft lubricants
Regulatory: FDA 21CFR §175.105, 175.300, 177.1200, 177.1210, 177.1400, 177.2600, 178.3740
Manuf./Distrib.: Allchem Ind.; Ashland; BASF; Chisso Am.; Coyne; Eastman; Esprit; Hüls AG; Inolex; Monsanto
Trade names: Adimoll® DO; Diplast® DOA; Jayflex® DOA; Kodaflex® DOA; Monoplex® DOA; Nuoplaz® DOA; Palatinol® DOA; Plasthall® DOA; PX-238; Staflex DOA; Uniflex® DOA

Dioctylated diphenylamine. *See* p,p´-Dioctyldiphenylamine

Dioctyl azelate
CAS 103-24-2
Synonyms: DOZ; Di(2-ethylhexyl) azelate; Bis (2-ethylhexyl) azelate
Empirical: $C_{25}H_{48}O_4$
Properties: Colorless liq., odorless; m.w. 412.73; dens. 0.919 (20/20 C); b.p. 376 C; flash pt. 221 C; ref. index 1.4472
Toxicology: Moderately toxic by intravenous route; mildly toxic by ingestion, skin contact; skin irritant
Uses: Plasticizer for vinyls (low-temp. plasticizer); base for synthetic lubricants
Manuf./Distrib.: C.P. Hall; Harwick; Henkel
Trade names: Plasthall® DOZ; Plastolein® 9058; Priplast 3018; Staflex DOZ

4,4´-Dioctyldiphenylamine. *See* p,p´-Dioctyldiphenylamine
Di-n-octyl diphenylamine. *See* p,p´-Dioctyldiphenylamine

p,p´-Dioctyldiphenylamine
CAS 101-67-7
Synonyms: Dioctylated diphenylamine; Di-n-octyl diphenylamine; 4,4´-Dioctyldiphenylamine
Formula: $C_8H_{17}C_6H_4NHC_6H_4C_8H_{17}$
Properties: Lt. tan powd.; sol. in benzene, gasoline, acetone, ethylene dichloride; insol. in water; m.w. 393.72; dens. 0.99; m.p. 80-90 C
Toxicology: Mildly toxic by ingestion
Uses: Antioxidant for petrol.-based and synthetic lubricants and plastics
Manuf./Distrib.: R.T. Vanderbilt
Trade names: Vanox® 12; Vanox® 1081
Trade names containing: Agerite® Hipar T; Prespersion PAC-2451

Dioctyl dodecanedioate
Properties: Sp.gr. 0.909; m.p. -45 C; flash pt. (COC) 222 C; ref. index 1.450
Uses: Plasticizer for PVC, VCA, PS, some cellulosics
Manuf./Distrib.: C.P. Hall

Dioctyl dodecanedioate dioate
Uses: Lubricant additive; as textile surf. finishes, softeners, thread lubricants and/or antistats
Trade names: Plasthall® DODD

Dioctyl fumarate
CAS 141-02-6; EINECS 220-835-6
Synonyms: DOF; Di (2-ethylhexyl) fumarate; 2-Butenedioic acid bis (2-ethylhexyl) ester
Formula: $C_{20}H_{36}O_4$
Properties: Clear mobile liq., mild odor; m.w. 340.56; dens. 0.942 (20/20 C); b.p. 211-220 C; flash pt. (COC) 365 F
Precaution: Combustible to heat and flame
Toxicology: Poison by intraperitoneal route; eye and severe skin irritant
Uses: Monomer for polymerization and copolymerization
Manuf./Distrib.: Monomer-Polymer & Dajac
Trade names: Staflex DOF

Dioctyl isophthalate
CAS 137-89-3
Synonyms: Di (2-ethylhexyl) isophthalate
Properties: Sp.gr. 0.9840 (20 C); pour pt. -46 C; flash pt. (COC) 232 C; ref. index 1.4875 (20 C)
Uses: Plasticizer for vinyls, PS, cellulosics
Manuf./Distrib.: Akzo Nobel; BP Chem.; JLM Marketing; Unitex

Dioctyl maleate
CAS 142-16-5; 2915-53-9; EINECS 205-524-5
Synonyms: Bis (2-ethylhexyl) maleate; Di-N-octyl maleate; Di-(2-ethylhexyl) maleate
Definition: Diester of 2-ethylhexyl alcohol and maleic acid
Empirical: $C_{20}H_{38}O_5$
Formula: $C_8H_{17}OCOCH=CHCOOC_8H_{17}$
Properties: Liq., char. mild odor; m.w. 358.52; sp.gr. 0.960-0.970; m.p. -50 to -85 C; flash pt. (COC) 182 C; ref. index 1.4480-1.4500
Toxicology: Mildly toxic by ingestion; heated to dec., emits acrid smoke and fumes
Uses: Emollient, fragrance coupler, solubilizer for benzophenone-3, antitackifier in antisperspirants, conditioner for hair prods.; in copolymerization of PVC and vinyl acetates; plasticizer for vinyl resins; also in latex paints
Manuf./Distrib.: Allchem Ind.; Aristech; BASF; Finetex; Hoechst AG; Hüls; Monomer-Polymer & Dajac; Novachem; Unitex
Trade names: Nuoplaz® DOM; PX-538; Staflex DOM

Di-N-octyl maleate. *See* Dioctyl maleate
Dioctyl peroxydicarbonate. *See* Bis (2-ethylhexyl) peroxydicarbonate

N,N´-Di (2-octyl)-p-phenylenediamine
CAS 103-96-8
Formula: $C_6H_4(NHC_8H_{17})_2$
Properties: Colorless liq.; misc. in methanol, pentane, benzene; dens. 0.912 (15 C); b.p. ≈ 390 C; pour pt. -4 C; flash pt. (PM) 201 C; ref. index 1.5129 (20 C)
Precaution: Combustible
Uses: Antioxidant, antiozonant for gasoline, mercaptans, synthetic rubber (NR, IR, BR, SBR)
Trade names: Antozite® 1

Dioctyl phthalate
CAS 117-81-7, 117-84-0; EINECS 204-211-0
Synonyms: DOP; DEHP; Di(2-ethylhexyl) phthalate; Di-s-octyl phthalate; 1,2-Benzenedicarboxylic acid dioctyl ester
Definition: Diester of 2-ethylhexyl alcohol and phthalic acid
Empirical: $C_{24}H_{38}O_4$
Formula: $C_6H_4[COOCH_2CH(C_2H_5)C_4H_9]_2$

Properties: Lt.-colored liq., odorless; insol. in water; misc. with min. oil; m.w. 390.62; dens. 0.9861 (20/20 C); b.p. 231 C (5 mm); flash pt. 218 C; ref. index 1.4836
Precaution: Combustible
Toxicology: TLV 5 mg/m³; STEL 10 mg/m³; LD50 (oral, rat) 30,600 mg/kg; poison by IV; mildly toxic by ingestion; skin and severe eye irritant; suspected human carcinogen; experimental teratogen; affects human GI tract; heated to decomp., emits acrid smoke
Uses: Plasticizer for many resins and elastomers
Regulatory: FDA 21CFR §175.105, 175.300, 175.310, 715.380, 175.390, 176.170, 176.210, 176.1210, 177.1010, 177.1200, 177.1210, 177.1400, 177.2600, 178.3120, 178.3910, 181.22, 181.27
Manuf./Distrib.: Allchem Ind.; Aristech; BASF; Chemisphere Ltd; Chisso Am.; Coyne; Daihachi Chem. Ind.; Eastman; Great Western; C.P. Hall; Hoechst AG; Hüls AG; Mitsubishi Gas; UCB SA
Trade names: Diplast® L8; Diplast® O; Kodaflex® DOP; Nuoplaz® DOP; Palatinol® DOP; Plasthall® DOP; PX-138; Staflex DOP
Trade names containing: Aztec® TBPEH-50-DOP; bioMeT 14; BYK®-2615; Dabco® T-10; Dabco® T-95; Fomrez® C-4; Jeffcat T-10; Ketjenblack® EC-350 N DOP; Kodaflex® HS-4; Kosmos® 10; Kosmos® 15; Lupersol 220-D50; Lupersol PDO; Omacide® P-DOP-5; Prespersion PAB-9541; Prespersion PAC-1451; Prespersion PAC-3656; Prespersion PAC-4911; Trigonox® 21-OP50; Vinyzene® BP-5-2 DOP; Vinyzene® BP-505 DOP

Di-(2-octyl) phthalate. *See* Dicapryl phthalate
Di-s-octyl phthalate. *See* Dioctyl phthalate

Dioctyl sebacate
CAS 122-62-3, 2432-87-3; EINECS 204-558-8
Synonyms: Di-2-ethylhexyl sebacate; Bis (2-ethylhexyl) decanedioate
Empirical: $C_{26}H_{50}O_4$
Properties: Pale straw-colored liq.; insol. in water; dens. 0.91 (25 C); b.p. 248 C (4 mm); f.p. -55 C; flash pt. (COC) 210 C; ref. index 1.447 (28 C)
Uses: Plasticizer for PVC, PS, cellulosics
Regulatory: FDA 21CFR §175.105, 175.300, 177.1210, 177.2600, 178.3910
Manuf./Distrib.: Allchem Ind.; Ashland; C.P. Hall; Harwick; Hatco; Novachem; Penta Mfg.; Union Camp; Unitex; Velsicol
Trade names: Monoplex® DOS; Plasthall® DOS; Uniflex® DOS

Dioctyl sodium sulfosuccinate
CAS 577-11-7; 1369-66-3; EINECS 209-406-4
Synonyms: DSS; Sodium dioctyl sulfosuccinate; Sodium di(2-ethylhexyl) sulfosuccinate; Docusate sodium
Definition: Sodium salt of the diester of 2-ethylhexyl alcohol and sulfosuccinic acid
Empirical: $C_{20}H_{38}O_7S \cdot Na$
Formula: $C_8H_{17}OOCCH_2CH(SO_3Na)COOC_8H_{17}$
Properties: White wax-like solid, octyl alchohol odor; slowly sol. in water; freely sol. in alcohol, glycerol, CCl_4, acetone, xylene; m.w. 445.63; m.p. 173-179 C
Toxicology: LD50 (oral, rat) 1900 mg/kg; moderately toxic by ingestion, intraperitoneal routes; poison by intravenous route; skin, severe eye irritant; heated to decomp., emits toxic fumes of SO_x and Na_2O
Uses: Food additive, emulsifier, wetting agent (processing aid in sugar industry); stabilizer for hydrophilic colloids; wetting agent, dispersant, emulsifier in cosmetic, pharmaceutical, emulsion polymerization, and industrial applics.; adjuvant in tablet
Regulatory: FDA 21CFR §73.1, with cocoa, 131.130, 131.132, 133.124, 133.133, 133.134, 133.162, 133.178, 133.179, 163.114, 163.117, 169.115, 169.150, 172.520, 172.808, 172.810, 175.105, 175.300, 175.320, 176.170, 176.210, 177.1200, 177.2800, 178.1010, 178.3400; USDA 9 CFR §318.7, 381.147
Manuf./Distrib.: Alco; Aquatec Quimica SA; Brotherton Ltd; Calgene; Cytec; Eastern Color & Chem.; EM Industries; Finetex; Hart Prod.; Henkel/Organic Prods.; McIntyre; Mona; Novachem; Witco/Oleo-Surf.
Trade names: Ablusol C-78; Aerosol® OT-75%; Aerosol® OT-100%; Aerosol® OT-S; Avirol® SO 70P; Chemax DOSS/70E; Chemax DOSS/70HFP; Chemax DOSS-75E; Disponil SUS IC 8; Drewfax® 0007; Emcol® DOSS; Empimin® OP70; Empimin® OT; Empimin® OT75; Gemtex PAX-60; Gemtex SC-75E; Geropon® CYA/60; Geropon® SDS; Geropon® SS-O-75; Geropon® WT-27; Hodag DOSS-70; Hodag DOSS-75; Nissan Rapisol B-30, B-80, C-70; Nopco ESA120; Octowet 70PG; Rewopol® SBDO 75; Sanstat 2012-A; Schercowet DOS-70; Secosol® DOS 70; Surfax 550
Trade names containing: Aerosol® OT-B; Gemtex PA-75; Gemtex PA-85P; Geropon® 99; Lankropol® KO2; Monawet MO-70; Monawet MO-70E

2,2′-[(Dioctylstannylene)bis(acetic acid)], diisooctyl ester
CAS 24601-97-8
Uses: Heat stabilizer for PVC

Dioctyl terephthalate
CAS 422-86-2; EINECS 225-091-6
Synonyms: DOTP; Di-2-ethylhexyl terephthalate; 1,4-Benzenedicarboxylic acid, dioctyl ester
Empirical: $C_{24}H_{38}O_4$
Formula: $C_6H_4(COOCH_2CH[C_2H_5]C_4H_9)_2$
Properties: Liq.; m.w. 390.57; sp.gr. 0.984; b.p. 400 C; f.p. -48 C; flash pt. (COC) 238 C; ref. index 1.489
Precaution: Combustible
Uses: Plasticizer for PVC, PS, cellulosics
Manuf./Distrib.: Ashland; Eastman; C.P. Hall; Harwick
Trade names: Kodaflex® DOTP

Dioctyl terephthalate chlorinated
Synonyms: Chlorinated dioctyl terephthalate
Properties: Sp.gr. 1.21; flash pt. (COC) none
Uses: Plasticizer for PVC, VCA, PVAc, PS
Manuf./Distrib.: Ferro/Keil

Dioctyl thiodiglycolate
Uses: Plasticizer for natural and syn. rubber incl. NBR, NBR/PVC blends, CR, SBR, BR, chlorosulfonated polyethylene, and chlorinated polyethylene; used for hoses, seals, shock absorbers, roll covers, conveyor belting
Trade names: Vulkanol® 90

Dioctyltin dilaurate
CAS 3648-18-8
Uses: PVC stabilizer
Trade names: ADK STAB OT-1

Dioctyltin maleate
Uses: PVC stabilizer
Trade names: ADK STAB OT-9

Dioctyltin mercaptide
CAS 58229-88-2
Synonyms: Di-n-octyltin mercaptide
Toxicology: TLV:TWA 0.1 mg(Sn)/m^3 (skin); moderately toxic by ingestion
Uses: Catalyst for the prod. of PU plastics; heat stabilizer for PVC bottles
Trade names: ADK STAB 465; ADK STAB 465E, 465L; Haro® Mix ZT-905; Kosmos® 23

Di-n-octyltin mercaptide. *See* Dioctyltin mercaptide

Dioctyltin oxide
CAS 870-08-6
Uses: Catalyst/polymerization regulator for plastics
Manuf./Distrib.: Aceto; Gelest

Dioctyltin thioglycolate
Uses: PVC stabilizer
Trade names: Hostastab® SnS 10; Hostastab® SnS 15; Hostastab® SnS 41; Hostastab® SnS 44
Trade names containing: Hostastab® SnS 11

N,N′-Dioleoylethylenediamine. *See* Ethylene dioleamide
DIOP. *See* Diisooctyl phthalate

1,4-Dioxane
CAS 123-91-1; EINECS 204-661-8
Synonyms: Diethylene ether; 1,4-Diethylene dioxide; Diethylene dioxide; Diethylene oxide
Classification: Ether

Formula: $OCH_2CH_2OCH_2CH_2$
Properties: Colorless liq., ethereal odor; stable; misc. with water, most org. solvs.; m.w. 88.11; dens. 1.0356 (20/20 C); b.p. 101.3 C; f.p. 10-12 C; flash pt. 18.3 C
Precaution: Flamm.
Toxicology: Carcinogen; toxic by inhalation, absorbed by skin; TLV 25 ppm in air; OSHA 100 ppm min. air
Uses: Stabilizer for chlorinated hydrocarbons; solv. for adhesives, dyes, cellulose, lacquer, wax, pharmaceuticals, coatings, natural and syn. rubbers, dry cleaning, metal surf. finishes, chem. reaction and extraction
Manuf./Distrib.: AC Ind.; Alemark; Allchem Ind.; Amber Syn.; Ashland; BASF; CPS; Ferro/Grant; Fluka; Mallinckrodt; Spectrum Chem. Mfg.; Toho; Union Carbide

1,5-Dioxaspiro[5.5]undecane 3,3-dicarboxylic acid, bis(2,2,6,6-tetramethyl-4-piperidinyl)ester
Synonyms: Bis (2,2,6,6-tetramethylpiperidinyl) ester of cyclospiroketal
Classification: Hindered amine light stabilizer
Empirical: $C_{29}H_{50}N_2O_6$
Properties: White powd.; sol. (g/100 g solv.): 32.8 g methyl chloride, 21 g methanol, 10.3 g toluene; m.w. 522; m.p. 160-162 C
Uses: UV stabilizer for PP, PE, ABS
Trade names: Cyasorb® UV 500; Topanex 500H

1,3-Dioxolan-2-one. *See* Ethylene carbonate
1,3-Dioxolan-2-one, 4-methyl. *See* Propylene carbonate
Dioxolone-2. *See* Ethylene carbonate
Dioxybenzone. *See* Benzophenone-8
DIPA. *See* Diisopropanolamine

Dipentaerythrityl pentaacrylate
Manuf./Distrib.: Monomer-Polymer & Dajac
Trade names containing: SR-399

Dipentamethylene hexasulfide
CAS 971-15-3
Uses: Accelerator for rubber

Dipentamethylene thiuram hexasulfide
Uses: Accelerator for NR and syn. rubbers; vulcanizing agent for heat-resistant latex
Trade names: Akrochem® DPTT; DPTT-S; Sulfads®

Dipentamethylene thiuram tetradisulfide
Synonyms: DPTT
Uses: Sec. accelerator for NR, SBR, EPDM, NBR
Trade names: Ekaland DPTT

Dipentamethylene thiuram tetrasulfide
CAS 120-54-7
Synonyms: DPTT; Tetrone A; Bis(pentamethylenethiuram) tetrasulfide; Piperidine, 1,1´-(tetrathiodicarbonothioyl)bis-
Formula: $[CH_2(CH_2)_4NCS]_2S_4$
Properties: Lt. gray powd.; sol. in chloroform, benzene, acetone; insol. in water; m.w. 384.70; dens. 1.53; m.p. 110 C min.
Toxicology: Poison by intraperitoneal route; heated to dec., emits toxic NO_x and SO_x
Uses: Ultra-accelerator for rubber
Trade names: Perkacit® DPTT; Sulfads® Rodform
Trade names containing: Poly-Dispersion® E (TET)D-70; Poly-Dispersion® E (TET)D-70P

2,5-Di-t-pentylhydroquinone. *See* Diamylhydroquinone

Diphenylamineacetone
CAS 9003-79-6
Uses: Antioxidant for NR and SR
Trade names: Agerite® Superflex; Agerite® Superflex Solid G; Aminox®; BXA Flake; Vanox® AM
Trade names containing: Agerite® Hipar T

N,N-Diphenyl-1,4-benzenediamine. *See* N,N´-Diphenyl-p-phenylenediamine

Diphenylcresyl phosphate
CAS 26444-49-5
Synonyms: Cresyl diphenyl phosphate; Tolyl diphenyl phosphate
Formula: $C_{19}H_{17}O_4P$
Properties: M.w. 340.33; sp.gr. 1.204-1.208; f.p. -38 C; flash pt. (COC) 233-237 C; ref. index 1.560
Toxicology: Mildly toxic by ingestion; heated to dec., emits PO_x
Uses: Plasticizer for PVC, PVB, PS, cellulosics
Manuf./Distrib.: FMC; Miles; Velsicol
Trade names: Disflamoll® DPK

Diphenyldidecyl (2,2,4-trimethyl-1,3-pentanediol) diphosphite
CAS 80584-87-8
Properties: Liq.; m.w. 706; dens. 0.985-0.998; b.p. 180 C (5 mm); ref. index 1.4895-1.4925; flash pt. (PMCC) 168 C
Uses: Heat and color stabilizer for polymers
Trade names: Weston® 491

Diphenylene oxide-4,4´-disulfohydrazide
Uses: Blowing agent for prod. of foamed plastics
Trade names: Porofor® S 44

Diphenyl-2-ethylhexyl phosphate. *See* Diphenyl octyl phosphate

Diphenylguanidine
CAS 102-06-7; EINECS 203-002-1
Synonyms: N,N´-Diphenylguanidine; 1,3-Diphenylguanidine; sym-Diphenylguanidine
Formula: $HN:C(NHC_6H_5)_2$
Properties: White powd., bitter taste, sl. odor; sol. in ethanol, CCl_4, chloroform, hot benzene, hot toluene; sl. sol. in water; m.w. 211.29; dens. 1.13; m.p. 147 C; dec. > 170 C
Toxicology: Poison by ingestion, intraperitoneal and other routes; experimental teratogen and reproductive effects; mutagenic data
Uses: Basic rubber accelerator, primary standard for acids
Manuf./Distrib.: Cytec; Fluka; Monsanto
Trade names: Akrochem® DPG; Anchor® DPG; DPG; Naftocit® DPG; Perkacit® DPG; Vanax® DPG; Vanax® DPG Pellets; Vulkacit D/C; Vulkacit D/EGC
Trade names containing: Poly-Dispersion® T (DPG)D-65; Poly-Dispersion® T (DPG)D-65P

1,3-Diphenylguanidine. *See* Diphenylguanidine
N,N´-Diphenylguanidine. *See* Diphenylguanidine
sym-Diphenylguanidine. *See* Diphenylguanidine

Diphenyl isodecyl phosphite
CAS 26544-23-0; EINECS 247-777-4
Synonyms: DPDP
Empirical: $C_{22}H_{31}O_3P$
Formula: $(C_6H_5O)_2POC_{10}H_{21}$
Properties: Liq.; m.w. 374; dens. 1.022-1.032; b.p. 190 C (5 mm); flash pt. (PMCC) 154 C; ref. index 1.5160-1.5190
Uses: Chelating agent with metal carboxylates as polymer additives esp. for chlorinated polymers; color, heat, and light stabilizer for PC, polyurethanes, ABS polymers, coatings; sec. stabilizer for PVC
Manuf./Distrib.: Akzo Nobel; Dover
Trade names: Doverphos® 8; Weston® DPDP

Diphenyl isooctyl phosphite
CAS 26401-27-4; EINECS 247-658-7
Synonyms: DPIOP
Formula: $(C_6H_5O)_2POC_8H_{17}$
Properties: Liq.; m.w. 346; dens. 1.040-1.047; b.p. 190 C (5 mm); ref. index 1.5210-1.5230; flash pt. (PMCC) 182 C
Uses: Color and processing stabilizer for ABS, PC, polyurethane, coatings, PET fiber; sec. stabilizer for PVC
Trade names: Doverphos® 9; Weston® ODPP

Diphenyl ketone. *See* Benzophenone

Diphenylmethane-4,4′-diisocyanate. *See* MDI
Diphenylmethanone. *See* Benzophenone
Diphenylnitrosamine. *See* Nitrosodiphenylamine

Diphenyl octyl phosphate
CAS 1241-94-7
Synonyms: Diphenyl-2-ethylhexyl phosphate; 2-Ethylhexyl diphenyl ester phosphoric acid; 2-Ethylhexyl diphenyl phosphate
Empirical: $C_{20}H_{27}O_4P$
Properties: M.w. 362.44; sp.gr. 1.088-1.093; flash pt. (COC) 224 C; ref. index 1.507-1.510
Toxicology: Poison by IV route; heated to decomp., emits toxic fumes of PO_x
Uses: Flame retardant plasticizer for typ. PVC applics., dip, rotationally, extruded and inj. molded parts, mechanical foam
Regulatory: FDA 21CFR §181.27
Manuf./Distrib.: Akzo Nobel; Ashland; Harwick; Miles; Monsanto
Trade names: Disflamoll® DPO; Phosflex 362; Santicizer 141

N,N′-Diphenyl-p-phenylenediamine
CAS 74-31-7
Synonyms: DPPD; 1,4-Bis(phenylamino)benzene; N,N-Diphenyl-1,4-benzenediamine; 1,4-Dianilinobenzene
Formula: $(C_6H_5NH)_2C_6H_4$
Properties: Gray powd.; insol. in water; sol. in acetone, benzene, monochlorobenzene, isopropyl acetate, DMF, ether, chloroform, ethyl acetate, glacial acetic acid; m.w. 260.36; dens. 1.28; m.p. 145-152 C
Precaution: Combustible; emits toxic fumes of NO_x
Toxicology: Poison by intravenous, intraperitoneal routes; moderately toxic by ingestion; eye irritant
Uses: Flex-resistant antioxidant, antiozonant in rubbers; stabilizer; polymerization inhibitor; retards copper degradation; intermediate for dyes, drugs, plastics, detergents
Trade names: Agerite® DPPD; Naugard® J; Permanax™ DPPD
Trade names containing: Agerite® Hipar T; Agerite® HP-S; Agerite® HP-S Rodform; Prespersion PAC-2451

Diphenyl phosphite
CAS 4712-55-4; EINECS 225-202-8
Synonyms: DPP
Formula: $(C_6H_5O)_2PHO$
Properties: Clear straw-colored liq.; m.w. 250.19; dens. 1.221 (25/15 C); m.p. 12 C; flash pt. 176 C; ref. index 1.557 (25 C)
Uses: Synthesis of organophosphorous compds.; color stabilizer for unsat. polyesters, polyolefins; sec. stabilizer for PVC
Manuf./Distrib.: Janssen Chimica; Spectrum Chem. Mfg.
Trade names: Doverphos® 213; Weston® DPP

Diphenylstibine 2-ethylhexoate
CAS 5035-58-5
Uses: Antimicrobial for protection of PVC systems
Trade names containing: bioMeT 14

N,N′-Diphenylthiocarbamide. *See* N,N′-Diphenylthiourea
1,2-Diphenyl-2-thiourea. *See* N,N′-Diphenylthiourea

N,N′-Diphenylthiourea
CAS 102-08-9; EINECS 203-004-2
Synonyms: sym-Diphenylthiourea; 1,2-Diphenyl-2-thiourea; Thiocarbanilide; N,N′-Diphenylthiocarbamide
Empirical: $C_{13}H_{12}N_2S$
Formula: $CS(NHC_6H_5)_2$
Properties: White to faint gray powd., bitter taste; pract. insol. in water; sol. in alcohol, ether, chloroform; m.w. 228.33; dens. 1.32 (25 C); m.p. 153-154 C; b.p. dec.
Precaution: Combustible
Toxicology: Moderately toxic by ingestion and intraperitoneal routes; MLD (rabbit, oral) 1.5 g/kg; heated to dec., emits highly toxic fumes of SO_x and NO_x
Uses: Accelerator for CR latex, NR latex, and cements, EPDM sponge compds.; activates thiazole accelerators; flotation agent; acid inhibitor; synthetic organic pharmaceuticals; sulfur dyes
Manuf./Distrib.: Charkit; Fluka; Novachem
Trade names: A-2; Akrochem® Thio No. 1; Rhenocure® CA; Stabilizer C

sym-Diphenylthiourea. *See* N,N´-Diphenylthiourea
asym-Diphenylurea. *See* N,N´-Diphenylurea

N,N´-Diphenylurea
CAS 102-07-8; EINECS 203-003-7
Synonyms: Carbanilide; asym-Diphenylurea
Empirical: $C_{13}H_{12}N_2O$
Formula: $(NHC_6H_5)CO(NHC_6H_5)$
Properties: Colorless prisms; sol. in alcohol, ether; very sl. sol. in water; m.w. 212.27; dens. 1.239; m.p. 235 C; b.p. 260 C
Toxicology: Poison by intraperitoneal route; moderately toxic by ingestion; heated to dec., emits toxic fumes of NO_x
Uses: Organic synthesis; retarder for elastomers; stabilizer for smokeless powders
Manuf./Distrib.: Fluka
Trade names: HyTemp SR-50; Zeonet® U

Diphosphoric acid, ammonium manganese (3+) salt (1:1:1). *See* Manganese violet
Diphosphoric acid tetrapotassium salt. *See* Tetrapotassium pyrophosphate

Dipropylene glycol
CAS 110-98-5; 25265-71-8; EINECS 203-821-4
Synonyms: DPG; Di-1,2-propylene glycol; 1,1´-Oxybis-2-propanol; Methyl-2(methyl-2) oxybispropanol
Classification: Mixture of diols
Empirical: $C_6H_{14}O_3$
Formula: $CH_3CHOHCH_2OCH_2CHOHCH_3$
Properties: Colorless, sl. visc. liq.; hygroscopic; sol. in toluene, water; m.w. 134.18; dens. 1.023; b.p. 233 C; flash pt. 280 F; ref. index 1.4410
Toxicology: Irritant; mildly toxic by ingestion
Uses: Solvent in hydraulic brake fluids, cutting oils, textile lubricants, inks; polyester and alkyd resins, reinforced plastics, plasticizers, solvents, fuel additives, in paints, cosmetics
Regulatory: FDA 21CFR §175.105, 176.170, 176.200, 178.3910
Manuf./Distrib.: Aldrich; Allchem Ind.; Arco; Ashland; Berje; Brown; Coyne; Great Western; Olin; PMC Specialties; Texaco
Trade names: Adeka Dipropylene Glycol
Trade names containing: Amicure® 33-LV; Chem-Master R-936; Dabco® 33-LV®; Dabco® BL-11; Dabco® BL-17; Dabco® BLV; Dabco® RC; Jeffcat TD-33A; Jeffcat ZF-22; Jeffcat ZF-24; Jeffcat ZF-26; Jeffcat ZF-51; Jeffcat ZF-52; Larostat® 377 DPG; Surfynol® DF-110D; TEDA-L33; Toyocat®-ET

Di-1,2-propylene glycol. *See* Dipropylene glycol

Dipropylene glycol dibenzoate
CAS 94-51-9, 27138-31-4; EINECS 202-340-7
Synonyms: 3,3´-Oxydyl-1-propanol dibenzoate
Empirical: $C_{20}H_{22}O_5$
Properties: Lt.-colored liq.; insol. in water; dens. 1.1271 (20/20 C); m.p. 200 C; b.p. 250 C (10 mm)
Uses: Plasticizer for cellulosics, PVC, PS, PVB, PVAc, VCA
Regulatory: FDA 21CFR §175.105, 176.170
Manuf./Distrib.: Ashland; Kalama; Unitex; Velsicol
Trade names: Benzoflex® 9-88; Benzoflex® 9-88 SG
Trade names containing: Benzoflex® 50

Dipropylene glycol methyl ether. *See* PPG-2 methyl ether
Dipropylene glycol methyl ether acetate. *See* PPG-2 methyl ether acetate
Dipropylene glycol monooleate. *See* PPG-2 oleate
Dipropylene glycol monostearate. *See* PPG-2 stearate

Di-n-propyl peroxydicarbonate
CAS 16066-38-9
Synonyms: n-Propyl percarbonate
Empirical: $C_8H_{14}O_6$
Properties: M.w. 206.22
Toxicology: Moderately toxic by ingestion and skin contact
Uses: Initiator for polymerization of unsat. monomers

Manuf./Distrib.: Elf Atochem N. Am.
Trade names: Lupersol 221

Disodium capryloamphodiacetate
CAS 7702-01-4; 68608-64-0; EINECS 231-721-0; 271-792-5
Synonyms: 1H-Imidazolium, 1-[2-(carboxymethoxy)ethyl]-1-(carboxymethyl)-2-heptyl-4,5-dihydro-, hydroxide, disodium salt; Capryloamphocarboxyglycinate; Capryloamphodiacetate
Classification: Amphoteric organic compd.
Empirical: $C_{16}H_{31}N_2O_6 \cdot 2Na$
Uses: Wetting agent, detergent, alkaline cleaners, household and industrial cleaners; in emulsion polymerization of syn. rubbers and resins
Trade names: Miranol® J2M Conc.

Disodium cetearyl sulfosuccinamate
Uses: Foaming agent for rubber latex compds.; emulsifier for emulsion polymerization
Trade names: Empimin® MKK/L

Disodium cetearyl sulfosuccinate
Synonyms: Disodium cetyl-stearyl sulfosuccinate; Sulfobutanedioic acid, cetyl/stearyl ester, disodium salt
Definition: Disodium salt of a cetearyl alcohol half ester of sulfosuccinic acid
Formula: $ROCOCHSO_3NaCH_2COONa$, R rep. alkyl groups from cetearyl alcohol
Uses: Foaming agent for rubber latex compds.; emulsifier for emulsion polymerization
Trade names: Empimin® MKK98

Disodium cetyl-stearyl sulfosuccinate. *See* Disodium cetearyl sulfosuccinate

Disodium cocoyl sulfosuccinate
Uses: Foaming agent for rubber latex compds.; emulsifier in the mfg. of polymers by emulsion polymerization
Trade names: Empimin® MHH

Disodium C12-15 pareth sulfosuccinate
CAS 39354-47-5
Synonyms: Disodium pareth-25 sulfosuccinate; Sulfobutanedioic acid, C12-15 pareth ester, disodium salt
Definition: Disodium salt of an ethoxylated, partially esterified sulfosuccinic acid
Formula: $R(OCH_2CH_2)_nOCOCHCH_2COONaSO_3Na$, R= C12-15 alkyl group, n = 1-4
Uses: Dispersant, wetting, foaming, detergent, emulsifying agent for cosmetics; emulsifier for acrylic, vinyl acetate, vinyl acrylic polymerization
Regulatory: FDA 21CFR §175.105
Trade names: Emcol® 4300

Disodium deceth-6 sulfosuccinate
CAS 68311-03-5 (generic), 68311-03-5
Synonyms: Sulfobutanedioic acid, deceth-6 ester, disodium salt
Definition: Disodium salt of the half ester of an ethoxylated decyl alcohol and sulfosuccinic acid
Uses: Emulsifier, solubilizer, foamer, dispersant, surfactant, wetting agent; used in emulsion polymerization of PVAc/acrylics; textile industry in emulsions, cosmetics, and pad-bath additive
Trade names: Aerosol® A-102

Disodium 3,3′-(dodecylimino) dipropionate. *See* Disodium lauriminodipropionate

Disodium dodecyloxy propyl sulfosuccinamate
CAS 58353-68-7
Uses: Emulsifier, dispersant, wetting agent, foaming agent for frothed latex compds. and adhesives; suspending agent in emulsion polymerization; textile softener
Trade names: Octosol A-1

Disodium edetate. *See* Disodium EDTA

Disodium EDTA
CAS 139-33-3; EINECS 205-358-3
Synonyms: Disodium edetate; Disodium ethyenediamine tetraacetate; Ethylenediaminetetraacetic acid, disodium salt; Edetate disodium
Classification: Substituted diamine

Disodium ethyenediamine tetraacetate

Empirical: $C_{10}H_{16}N_2O_8 \cdot 2Na$
Properties: White crystalline powd.; freely sol. in water; m.w. 336.24; m.p. 252 (dec.)
Toxicology: LD50 (oral, rat) 2 g/kg; poison by intravenous route; moderately toxic by ingestion; experimental teratogen, reproductive effects; mutagenic data; heated to decomp., emits toxic fumes of NO_x and Na_2O
Uses: Chelating agent in cosmetics, pharmaceuticals, water treatment, textiles, soaps/detergents, electroless copperplating, polymer prod., disinfectants, pulp/paper, enhanced oil recovery, metal cleaning/protection; food preservative, stabilizer, chelating and sequestering agent; promotes color retention in canned peas; hog scald agent; anticoagulant; pharmaceutic aid
Regulatory: FDA 21CFR §155.200, 169.115, 169.140, 169.150, 172.135, 175.105, 176.150, 176.170, 177.1200, 177.2800, 178.3570, 178.3910, 573.360; USDA 9CFR §318.7; Japan approved (0.25 g/kg max. as calcium disodium EDTA)
Manuf./Distrib.: Aldrich; R.W. Greeff; Hampshire; Int'l. Sourcing
Trade names: Versene NA

Disodium ethyenediamine tetraacetate. *See* Disodium EDTA
Disodium ethylene bisdithiocarbamate. *See* Nabam
Disodium ethylene-1,2-bisdithiocarbamate. *See* Nabam

Disodium hexamethylene bisthiosulfate
CAS 5719-73-3
Synonyms: Hexamethylene bisthiosulfate disodium salt
Uses: Post vulcanization stabilizer for sulfur cures of NR, IR, SBR, and NBR; used in tire treads, sidewalls, and general industrial prods. incl. belting and inj. molded goods; bonding promoter for rubber-based steel adhesion
Trade names: Duralink® HTS

Disodium isodecyl sulfosuccinate
CAS 37294-49-8; EINECS 253-452-8
Synonyms: Sulfobutanedioic acid, 4-isodecyl ester, disodium salt
Definition: Disodium salt of a half ester of a branched chain decyl alcohol and sulfosuccinic acid
Empirical: $C_{14}H_{26}O_7S \cdot 2Na$
Uses: Surfactant, sole emulsifier for PVC latexes-vinyl, vinylidene chloride, acrylics, surf. tens. depressant, solubilizer
Trade names: Aerosol® A-268

Disodium laureth sulfosuccinate
CAS 39354-45-5 (generic); 40754-59-4; 42016-08-0; 58450-52-5; EINECS 255-062-3
Synonyms: Sulfobutanedioic acid, 4-[2-[2-[2-(dodecyloxy)ethoxy]ethoxy]ethyl]ester, disodium salt; Disodium lauryl ether sulfosuccinate
Definition: Disodium salt of an ethoxylated lauryl alcohol half ester of sulfosuccinic acid
Formula: $(C_2H_4O)_xC_{16}H_{30}O_7S \cdot 2Na$
Uses: Emulsifier for emulsion polymerization; emulsifier, detergent, wetter for shampoo and bubble bath base, skin cleansers
Trade names: Geronol ACR/4; Geropon® ACR/4

Disodium lauriminodipropionate
CAS 3655-00-3; EINECS 222-899-0
Synonyms: Disodium N-lauryl-β-iminodipropionate; Disodium 3,3′-(dodecylimino) dipropionate; N-(2-Carboxyethyl)-N-dodecyl-β-alanine, disodium salt
Definition: Disodium salt of a substituted propionic acid
Formula: $CH_3(CH_2)_{10}CH_2—N(CH_2CH_2COONa)_2$
Uses: Detergent, lubricant, antistat, corrosion inhibitor, emulsifier, and solubilizer used in detergent and industrial formulations, personal care prods., leather, and textile fibers; emulsion polymerization
Trade names: Deriphat® 160

Disodium lauryl ether sulfosuccinate. *See* Disodium laureth sulfosuccinate
Disodium N-lauryl-β-iminodipropionate. *See* Disodium lauriminodipropionate
Disodium monooleamido MIPA-sulfosuccinate. *See* Disodium oleamido MIPA-sulfosuccinate

Disodium nonoxynol-10 sulfosuccinate
CAS 67999-57-9 (generic); 9040-38-4
Synonyms: Sulfobutanedioic acid, nonoxynol-10 ester, disodium salt
Definition: Disodium salt of the half ester of nonoxynol-10 and sulfosuccinic acid

Uses: Emulsifier, solubilizer, wetting agent, surfactant, dispersant, foamer, foam stabilizer, surf. tens. depressant; used in PVAc acrylic emulsions; textile emulsions, pad-bath additive, textile wetting
Regulatory: FDA 21CFR §175.105
Trade names: Aerosol® A-103; Geronol ACR/9; Geropon® ACR/9; Monawet 1240

Disodium octadecyl sulfosuccinamate. *See* Disodium stearyl sulfosuccinamate

Disodium oleamido MIPA-sulfosuccinate
CAS 43154-85-4; EINECS 256-120-0
Synonyms: Sulfobutanedioic acid, 4-[1-methyl-2-[(1-oxo-9-octadecenyl)amino]ethyl]ester, disodium salt; Disodium oleoyl isopropanolamide sulfosuccinate; Disodium monooleamido MIPA-sulfosuccinate
Definition: Disodium salt of a substituted isopropanolamide half ester of sulfosuccinic acid
Empirical: $C_{25}H_{45}NO_8S•2Na$
Uses: Dispersant, wetting, foaming, detergent, and emulsifying agent for bubble bath, shampoos, cleansers for cosmetics and toiletries; emulsion polymerization
Regulatory: FDA 21CFR §175.105, 176.170
Trade names: Emcol® K8300; Sole Terge 8

Disodium oleoyl isopropanolamide sulfosuccinate. *See* Disodium oleamido MIPA-sulfosuccinate

Disodium N-oleyl sulfosuccinamate
Uses: Foaming agent for rubber latices
Trade names: Empimin® MTT

Disodium oleyl sulfosuccinate
EINECS 303-773-5
Synonyms: Sulfobutanedioic acid, 1-(9-octadecenyl) ester, disodium salt
Definition: Disodium salt of an oleyl alcohol half ester of sulfosuccinic acid
Empirical: $C_{22}H_{40}O_7S•2Na$
Uses: Foaming agent for rubber latex compds.; emulsifier for emulsion polymerization
Trade names: Empimin® MTT/A

Disodium pareth-25 sulfosuccinate. *See* Disodium C12-15 pareth sulfosuccinate

Disodium ricinoleamido MEA-sulfosuccinate
CAS 40754-60-7; 65277-54-5; 67893-42-9; EINECS 267-617-7; 265-672-1
Synonyms: Sulfobutanedioic acid, 1-[2-[(12-hydroxy-1-oxo-9-octadecenyl)amino]ethyl]ester, disodium salt; Disodium ricinoleyl monoethanolamide sulfosuccinate
Definition: Disodium salt of a substituted ethanolamide half ester of sulfosuccinic acid
Empirical: $C_{24}H_{43}NO_9S•2Na$
Uses: Detergent used in cosmetics, personal care prods., household goods; emulsifier for emulsion polymerization
Trade names: Rewoderm® S 1333

Disodium ricinoleyl monoethanolamide sulfosuccinate. *See* Disodium ricinoleamido MEA-sulfosuccinate

Disodium stearyl sulfosuccinamate
CAS 14481-60-8; EINECS 238-479-5
Synonyms: Sulfobutanedioic acid, monooctadecyl ester, disodium salt; Disodium octadecyl sulfosuccinamate
Definition: Disodium salt of a stearyl amide of sulfosuccinic acid
Empirical: $C_{22}H_{43}NO_6S•2Na$
Uses: Emulsifier, dispersant, foamer, detergent, solubilizer for soaps and surfactants; alkaline cleaner formulations, brick and tile cleaners, emulsion polymerization of vinyl chloride and SBR; foaming agent for rubber latexes
Trade names: Aerosol® 18; Lankropol® ODS/LS; Lankropol® ODS/PT; Octosol A-18

Disodium tallowiminodipropionate
CAS 61791-56-8; EINECS 263-190-6
Synonyms: Disodium N-tallow-β iminodipropionate; N-(2-Carboxyethyl)-N-(tallow acyl)-β-alanine
Definition: Disodium salt of a substituted propionic acid
Formula: $R—N(CH_2CH_2COONa)_2$, R represents the alkyl group derived from tallow
Uses: Detergent, solubilizer, hydrotrope, moderate foaming surfactant used in textile, leather, emulsion polymerization, industrial and personal care prods.

Disodium N-tallow-β iminodipropionate

Trade names: Deriphat® 154

Disodium N-tallow-β iminodipropionate. *See* Disodium tallowiminodipropionate

Disodium tallow sulfosuccinamate
CAS 90268-48-7; EINECS 290-850-0
Synonyms: Sulfobutanedioic acid, tallow ester, disodium salt
Classification: Organic compd.
Formula: $RNHCOCH_2CHSO_3NaCOONa$, R = tallow alkyl groups
Uses: Foaming and antigelling agent for latex foam backings and coatings; emulsion polymerization; flotation agent
Trade names: Rewopol® B 1003

Distearyl dimethyl ammonium chloride. *See* Distearyldimonium chloride

Distearyldimonium chloride
CAS 107-64-2; EINECS 203-508-2
Synonyms: Distearyl dimethyl ammonium chloride; Quaternium-5; Dioctadecyl dimethyl ammonium chloride
Classification: Quaternary ammonium salt
Empirical: $C_{38}H_{80}N \cdot Cl$
Uses: Fabric softener, conditioner, antistat for commercial and institutional laundries; conditioner, antistat for hair prods.; foam and visc. builder in personal care prods.
Regulatory: FDA 21CFR §172.712, 177.1200
Trade names: Arosurf® TA-100; Dehyquart® DAM

Distearyl disulfide. *See* Dioctadecyl disulfide

Distearyl dithiodipropionate
CAS 6729-96-0
Synonyms: Dithio-bis(stearyl propionate)
Formula: $S_2(CH_2CH_2COOC_{18}H_{37})_2$
Properties: M.w. 698; m.p. 52-54 C
Uses: Antioxidant and stabilizer to prevent and retard degradation
Trade names: Evangard® DTB

Distearyl pentaerythritol diphosphite
CAS 3806-34-6; EINECS 223-276-6
Empirical: $C_{41}H_{82}O_6P_2$
Properties: Solid; m.w. 732; dens. 0.920-0.935; m.p. 40-70 C; ref. index 1.4560-1.4590; flash pt. (PMCC) 185 C
Uses: Color stabilizer, melt flow aid for polymer processing
Trade names: Doverphos® S-680; Doverphos® S-686, S-687; Mark® 5060; Weston® 618F
Trade names containing: Doverphos® S-682; Weston® 619F

Distearyl phosphite
CAS 19047-85-9; EINECS 242-784-9
Formula: $(C_{18}H_{37}O)_2PHO$
Properties: Wh. solid; m.w. 586; dens. 0.860-0.880 (60/15.5 C); ref. index 1.4450-1.4480 (60 C); flash pt. (PMCC) 185 C
Uses: Lubricant additive for antifriction and antiwear props.; color and m.w. stabilizer for PP

Distearyl phthalate
CAS 90193-76-3; EINECS 290-580-3
Uses: Lubricant for PVC
Trade names: Sicolub® DSP

Distearyl thiodipropionate
CAS 693-36-7; EINECS 211-750-5
Synonyms: DSTDP; 3,3′-Thiobispropanoic acid, dioctadecyl ester; 3,3′-Dioctadecyl thiodipropionate; Thiodipropionic acid, distearyl ester
Definition: Diester of stearyl alcohol and 3,3′-thiodipropionic acid
Empirical: $C_{42}H_{82}O_4S$
Formula: $(C_{18}H_{37}OOCCH_2CH_2)_2S$

Properties: White flakes; insol. in water; sol. in benzene, toluene, chloroform, and olefin polymers; m.w. 683; m.p. 58-62 C; b.p. 360 C (dec.)
Toxicology: Nonhazardous; heated to decomp., emits toxic fumes of SO_x
Uses: Antioxidant (cosmetics, pharmaceuticals), plasticizer, softening agent, stabilizer for plastics and elastomers (ABS, acrylonitrile, syn. rubbers, polyolefins)
Regulatory: FDA 21CFR §175.105, 175.300, 181.22, 181.24 (0.005% migrating from food pkg.)
Manuf./Distrib.: Cytec; Evans Chemetics; Hampshire; Morton Int'l.; Witco/PAG
Trade names: Carstab® DSTDP; Cyanox® STDP; Evanstab® 18; Hostanox® SE 2; Lankromark® DSTDP; Lowinox® DSTDP; PAG DSTDP; PAG DXTDP

Distilled spirits. *See* Alcohol

Distyryl phenyl
Uses: Optical brightener for nylon and blends
Trade names: Ecco White® FW-5

Disuccinic acid peroxide. *See* Succinic acid peroxide
Disulfide, bis(dibenzylthiocarbamoyl). *See* N,N,N′,N′-Tetrabenzylthiuram disulfide
Disulfide, bis(dibutylthiocarbamoyl). *See* Tetrabutyl thiuram disulfide
Disulfiram. *See* Tetraethylthiuram disulfide
Disulfur dichloride. *See* Sulfur chloride
Ditallowalkonium chloride. *See* Quaternium-18
Ditetradecyl 3,3′-thiobispropanoate. *See* Dimyristyl thiodipropionate
Dithane A-40. *See* Nabam
2,2′-Dithiobisbenzothiazole. *See* Benzothiazyl disulfide
4,4′-Dithiobis (morpholine). *See* 4,4′-Dithiodimorpholine
Dithio-bis(stearyl propionate). *See* Distearyl dithiodipropionate

Dithiodicaprolactam
Uses: Sulfur donor for NR and SR, ageing resistant articles based on SBR, NBR, EPDM
Trade names: Rhenocure® S/G

4,4′-Dithiodimorpholine
CAS 103-34-4
Synonyms: DTDM; 4,4′-Dithiobis (morpholine); Dimorpholine N,N′-disulfide; N,N′-Bismorpholine disulfide
Formula: $C_4H_8ONSSNOC_4H_8$
Properties: Gray to tan powd.; m.w. 236.38; dens. 1.36 (25 C); m.p. 122 C min.
Toxicology: Moderately toxic by ingestion, inhalation, and skin absorption; poison by intraperitoneal, intravenous routes
Uses: Rubber accelerator; fungicide; staining protector for rubber
Trade names: Akrochem® Accelerator R; Rhenocure® M; Rhenocure® M/G; Sulfasan® R
Trade names containing: Poly-Dispersion® E (SR)D-75; Vanax® A

4,4′-Dithiomorpholine
Uses: Sulfur donor for NR and SR; scorch retarder

N,N′-Di-o-tolyl guanidine. *See* Di-o-tolyl guanidine

Di-o-tolyl guanidine
CAS 97-39-2
Synonyms: DOTG; N,N′-Di-o-tolyl guanidine
Empirical: $C_{15}H_{17}N_3$
Formula: $(CH_3C_6H_4NH)_2CNH$
Properties: White crystals, nonhygroscopic; very sl. sol. in waer; sol. in warm alcohol; m.w. 239.35; dens. 1.10 (20/4 C); m.p.179 C
Toxicology: Toxic by ingestion, intraperitoneal route
Uses: Rubber accelerator
Manuf./Distrib.: Cytec
Trade names: Akrochem® DOTG; Anchor® DOTG; Ekaland DOTG; Perkacit® DOTG; Vanax® DOTG
Trade names containing: Poly-Dispersion® E (DOTG)D-65P

Di- tri-calcium phosphate
Trade names containing: Cadox® BCP

Ditridecyl adipate
CAS 26401-35-4; 16958-92-2; EINECS 247-660-8
Synonyms: Ditridecyl hexanedioate; Hexanedioic acid, ditridecyl ester
Definition: Diester of tridecyl alcohol and adipic acid
Empirical: $C_{32}H_{62}O_4$
Uses: Base stock for crankcase and compressor oils; fiber lubricant in yarns
Trade names: Nuoplaz® DTDA

Ditridecyl hexanedioate. *See* Ditridecyl adipate

Ditridecyl maleate
Uses: Plasticizer and intermediate; used in fields of copolymerization and surfactants
Trade names: Staflex DTDM

Ditridecyl phosphite
CAS 36432-46-9
Synonyms: Bis(tridecyl) phosphite
Formula: $(C_{13}H_{27}O)_2PHO$
Properties: Liq.; m.w. 446; dens. 0.906-0.917; b.p. 200 C (5 mm); ref. index 1.4530-1.4560; flash pt. (PMCC) 168 C
Uses: Lubricant additive to improve antifriction and antiwear chars.; stabilizer for PP; sec. stabilizer for PVC
Manuf./Distrib.: Albright & Wilson
Trade names: Weston® DTDP

Ditridecyl phthalate
CAS 119-06-2; 68515-47-9; EINECS 204-294-3
Synonyms: DTDP; 1,2-Benzenedicarboxylic acid, ditridecyl ester
Formula: $C_{34}H_{58}O_4$
Properties: M.w. 530.92; dens. 0.951 (20/20 C); b.p. > 285 C (5 mm); flash pt. (OC) 470 F; ref. index 1.484 (20 C)
Precaution: Combustible
Toxicology: Skin irritant
Uses: Plasticizer for PVC, PS, cellulosics
Manuf./Distrib.: Aristech; Ashland; Exxon; Hatco; Hüls; Egon Meyer
Trade names: Jayflex® DTDP; Nuoplaz® DTDP; PX-126; Staflex DTDP

Ditridecyl sodium sulfosuccinate
CAS 2673-22-5; EINECS 220-219-7
Synonyms: Sodium bistridecyl sulfosuccinate; Sodium ditridecyl sulfosuccinate; Sulfobutanedioic acid, 1,4-ditridecyl ester, sodium salt
Definition: Sodium salt of the diester of tridecyl alcohol and sulfosuccinic acid
Empirical: $C_{30}H_{58}O_7S \cdot Na$
Uses: Emulsifier, surfactant, detergent, foam modifier, wetting agent, dispersant; emulsion polymerization; dispersant, processing aid for resins, pigments, polymers, and dyes
Regulatory: FDA 21CFR §175.105, 176.180, 178.3400
Trade names: Aerosol® TR-70; Geropon® BIS/SODICO-2; Hodag DTSS-70; Polirol TR/LNA
Trade names containing: Monawet MT-70

Ditridecyl thiodipropionate
CAS 10595-72-9; EINECS 234-206-9
Synonyms: 3,3′-Thiobispropanoic acid, ditridecyl ester; Di(tridecyl) thiodipropionate
Definition: Diester of tridecyl alcohol and 3,3′-thiodipropionic acid
Empirical: $C_{32}H_{62}O_4S$
Formula: $S(CH_2CH_2COOC_{13}H_{27})_2$
Properties: M.w. 542.91; sol. in most org. solvs.; sl. sol. in methanol; b.p. 265 C (0.25 mm)
Toxicology: LD50 (rat, oral) > 10 ml/kg; LD50 (rabbit, dermal) > 5 ml/kg
Uses: Secondary antioxidant for polyolefins, ABS, petrol. lubricants, SBR latex, cosmetics, pharmaceuticals
Manuf./Distrib.: Cytec; Hampshire; Witco/PAG
Trade names: Cyanox® 711; Evanstab® 13; Lankromark® DTDTDP; PAG DTDTDP

Di(tridecyl) thiodipropionate. *See* Ditridecyl thiodipropionate
Di (3,5,5-trimethylhexanoyl) peroxide. *See* Bis (3,5,5-trimethylhexanoyl) peroxide

Diundecyl phthalate
CAS 3648-20-2; EINECS 222-884-9
Synonyms: 1,2-Benzenedicarboxylic acid, diundecyl ester
Empirical: $C_{30}H_{50}O_2$
Properties: M.w. 442.80; sp.gr. 0.954; f.p. 2-7 C; flash pt. (TCC) 460 F; ref. index 1.481 (20 C)
Precaution: Heated to dec., emits acrid smoke and irritating fumes
Toxicology: Eye irritant
Uses: Plasticizer for PVC, PS, cellulosics
Manuf./Distrib.: Aristech; Ashland; BASF; Exxon; Monsanto
Trade names: Jayflex® L11P; PX-111

Diuron. *See* 3-(3,4-Dichlorophenyl)-1,1-dimethylurea

Dixylyldisulfides
Uses: Reclaiming agent for scrap rubber
Trade names: Aktiplast®6 N

DMAE. *See* Dimethylethanolamine
DMAPA. *See* Dimethylaminopropylamine
DMBA. *See* N-Benzyldimethylamine

DMDM hydantoin
CAS 6440-58-0; EINECS 229-222-8
Synonyms: 1,3-Dimethylol-5,5-dimethyl hydantoin; 2,4-Imidazolidinedione, 1,3-bis(hydroxymethyl)-5,5-dimethyl-; Dimethylol dimethyl hydantoin
Classification: Organic compd.
Empirical: $C_7H_{12}N_2O_4$
Uses: Preservative
Regulatory: USA CIR approved, EPA registered; Europe listed 0.6% max.

DMEP. *See* Dimethoxyethyl phthalate
DMF. *See* Dimethyl formamide
DMOA. *See* Dimethyl octylamine
DMP. *See* Dimethyl phthalate, 2,4,6-Tris (dimethylaminomethyl) phenol
DMPA. *See* Dimethylolpropionic acid
DMSO. *See* Dimethyl sulfoxide
DNPD. *See* N,N´-Di-β-naphthyl-p-phenylenediamine
DNPT. *See* Dinitrosopentamethylene tetramine
DOA. *See* Dioctyl adipate
Docosanamide. *See* Behenamide
Docosanoic acid, calcium salt. *See* Calcium behenate
1-Docosanol. *See* Behenyl alcohol
13-Docosenamide. *See* Erucamide
cis 13-Docosenamide. *See* Erucamide
13-Docosenamide, N-octadecyl-. *See* Stearyl erucamide
Docusate sodium. *See* Dioctyl sodium sulfosuccinate

Dodecachloro dodecahydro dimethanodibenzo cyclooctene
CAS 13560-89-9
Empirical: $C_{18}H_{12}Cl_{12}$
Properties: Wh. cryst.; m.w. 654
Uses: Flame-retardant additive for polymeric systems
Trade names: Dechlorane® Plus 25; Dechlorane® Plus 35; Dechlorane® Plus 515

Dodecanamide. *See* Lauramide
1-Dodecanamine. *See* Lauramine
1-Dodecanethiol. *See* n-Dodecyl mercaptan
Dodecanoic acid. *See* Lauric acid
n-Dodecanoic acid. *See* Lauric acid
Dodecanoic acid, barium salt. *See* Barium laurate
Dodecanoic acid, cadmium salt. *See* Cadmium laurate
Dodecanoic acid, diester with 1,2,3-propanetriol. *See* Glyceryl dilaurate, Glyceryl dilaurate SE
Dodecanoic acid, 2,3-dihydroxypropyl ester. *See* Glyceryl laurate

Dodecanoic acid, ester with 1,2-ethanediol. *See* Glycol laurate SE
Dodecanoic acid 1,2-ethanediyl ester. *See* Glycol dilaurate
Dodecanoic acid, 2-hydroxypropyl ester. *See* Propylene glycol laurate
Dodecanoic acid methyl ester. *See* Methyl laurate
Dodecanoic acid, monoester with 1,2-propanediol. *See* Propylene glycol laurate
Dodecanoic acid, monoester with 1,2,3-propanetriol. *See* Glyceryl laurate
Dodecanoic acid, zinc salt. *See* Zinc laurate
1-Dodecanol. *See* Lauryl alcohol
Dodecanoyl peroxide. *See* Lauroyl peroxide
1-Dodecene. *See* Dodecene-1

Dodecene-1
CAS 112-41-4; 6842-15-5; EINECS 203-968-4
Synonyms: C12 alpha olefin; 1-Dodecene; α-Dodecylene; Tetrapropylene
Empirical: $C_{12}H_{24}$
Formula: $H_2C{:}CH(CH_2)_9CH_3$
Properties: Colorless liq.; insol. in water; sol. in alcohol, acetone, ether, petrol., coal tar solvs.; m.w. 168.32; dens. 0.764; m.p. -31.5 C; b.p. 213-215 C; flash pt. (Seta) 168 F; ref. index 1.430
Toxicology: Irritating to eyes and skin; low acute inhalation toxicity; sl. toxic by ingestion; narcotic in high concs.
Uses: Intermediate for surfactants and specialty industrial chemicals (polyethylene and other polymers; plasticizers; syn. lubricants; gasoline additives; paper sizing; PVC lubricants); flavors, perfumes, medicine, oils, dyes, resins
Manuf./Distrib.: Albemarle; Chevron; Shell
Trade names: Gulftene® 12

Dodecenyl succinic anhydride
CAS 25377-73-5; 26544-38-7; EINECS 246-917-1
Synonyms: DDSA
Empirical: $C_{16}H_{26}O_3$
Properties: Clear lt. yel. visc. liq.; m.w. 266; dens. 1.003-1.008 (60 F); b.p. 150 C (3 mm); flash pt. 343-347 F
Toxicology: Poison by intraperitoneal route; irritant and sensitizer
Uses: Epoxy curing agent; corrosion inhibitor for nonaq. lubricating oils, intermediate for prep. of alkyd or unsat. polyester resins, platicizers, intermediate in chem. reactions
Manuf./Distrib.: Anhydrides & Chems.; Buffalo Color; Dixie; Fluka; Humphrey; Milliken
Trade names: Milldride® DDSA; Milldride® nDDSA

Dodecoic acid. *See* Lauric acid
Dodecyl alcohol. *See* Lauryl alcohol
Dodecylamine. *See* Lauramine
Dodecylbenzene sodium sulfonate. *See* Sodium dodecylbenzenesulfonate

Dodecylbenzene sulfonic acid
CAS 27176-87-0; 68411-32-5; 68584-22-5; 68608-88-8; 85536-14-7; EINECS 248-289-4
Synonyms: DDBSA
Classification: Substituted aromatic acid
Empirical: $C_{18}H_{30}O_3S$
Properties: White to lt. yel. flakes, granules, and powd.; biodegradable
Uses: Anionic detergent raw material used in emulsifiers, heavy and lt. duty detergents, hand cleaning gels, machine degreasers, tank cleaners; emulsion polymerization; catalyst; metalworking
Regulatory: FDA 21CFR §176.210, 178.1010
Manuf./Distrib.: Allchem Ind.; Ashland; Biddle Sawyer; Pilot; Tradig; Witco/Oleo-Surf.
Trade names: Manro BA Acid; Nansa® 1042/P; Nansa® SSA; Nansa® SSA/P; Polystep® A-13; Polystep® A-17; Serdet DM; Witco® 1298; Witco® Acid B; Witconate™ Acide TPB

Dodecylbenzenesulfonic acid, compd. with 2,2′,2′′-nitrilotris[ethanol] (1:1). *See* TEA-dodecylbenzenesulfonate
Dodecylbenzenesulfonic acid, comp. with 2-propanamine (1:1). *See* Isopropylamine dodecylbenzenesulfonate
Dodecylbenzenesulfonic acid, potassium salt. *See* Potassium dodecylbenzene sulfonate
Dodecylbenzenesulfonic acid sodium salt. *See* Sodium dodecylbenzenesulfonate
Dodecyldimethylamine. *See* Dimethyl lauramine
Dodecyl dimethyl benzyl ammonium chloride. *See* Lauralkonium chloride
α-Dodecylene. *See* Dodecene-1

n-Dodecyl mercaptan
CAS 112-55-0; EINECS 203-984-1
Synonyms: DDM; m-Lauryl mercaptan; 1-Dodecanethiol; Mercaptan C_{12}
Formula: n-$C_{12}H_{25}SH$
Properties: Colorless liq.; sol. in ether, alcohol; insol. in water; m.w. 202.44; dens. 0.849 (15.5/15.5 C); m.p. -7 C; b.p. 143 C; flash pt. (OC) 127 C; ref. index 1.4589
Precaution: Combustible
Toxicology: Mutagenic data
Uses: Modifier in polymerization reactions, esp. for SBR; reducing initiator and chain transfer agent
Manuf./Distrib.: Elf Atochem N. Am.; Great Western; Phillips

Dodecyl methacrylate. *See* Lauryl methacrylate
Dodecyl 2-methyl-2-propenoate. *See* Lauryl methacrylate
2-[2-(Dodecyloxy)ethoxy]ethanol. *See* Laureth-2
2-[2-[2-(Dodecyloxy)ethoxy]ethoxy]ethanol. *See* Laureth-3

4-Dodecyloxy-2-hydroxybenzophenone
Formula: $C_{12}H_{25}OC_6H_3(OH)C(O)C_6H_5$
Properties: Pale yel. flakes; sol. in polar and nonpolar org. solvs.; m.p. 52 C
Uses: UV light inhibitor in plastics (ABS, cellulosics, epoxies, polyesters, PE, PP, PS, flexible and rigid PVC, VDC)
Manuf./Distrib.: Eastman
Trade names: UV-Chek® AM-320

Dodecylsulfate sodium salt. *See* Sodium lauryl sulfate
Dodecyl trimethyl ammonium chloride. *See* Laurtrimonium chloride
DOF. *See* Dioctyl fumarate
DOP. *See* Dioctyl phthalate
DOTG. *See* Di-o-tolyl guanidine
DOTP. *See* Dioctyl terephthalate
DOZ. *See* Dioctyl azelate
DPDP. *See* Diphenyl isodecyl phosphite
DPG. *See* Dipropylene glycol
DPIOP. *See* Diphenyl isooctyl phosphite
DPM. *See* PPG-2 methyl ether
DPP. *See* Diphenyl phosphite
DPPD. *See* N,N′-Diphenyl-p-phenylenediamine
DPTT. *See* Dipentamethylene thiuram tetradisulfide, Dipentamethylene thiuram tetrasulfide
Dracylic acid. *See* Benzoic acid

Drometrizole
CAS 2440-22-4; EINECS 219-470-5
Synonyms: 2-(2′-Hydroxy-5′-methylphenyl) benzotriazole; 2-(2H-Benzotriazol-2-yl)-4-methylphenol
Classification: Benzotriazole deriv.
Empirical: $C_{13}H_{11}N_3O$
Properties: Powd.; m.w. 225.27; m.p. 128 C
Toxicology: Mildly toxic by ingestion; eye irritant
Uses: UV light stabilizer for polymers (ABS, cellulosics, epoxy, polyester, PS, flexible and rigid PVC, VDC
Regulatory: FDA 21CFR §178.2010
Manuf./Distrib.: 3V
Trade names: Lowilite® 55; Tinuvin® P; Topanex 100BT; Uvasorb SV; Viosorb 520
Trade names containing: Aquastab PB 503

DSE. *See* Nabam
DSS. *See* Dioctyl sodium sulfosuccinate
DSTDP. *See* Distearyl thiodipropionate
24DTAP. *See* 2,4-Di-t-amylphenol
DTBP. *See* Di-t-butyl peroxide
24DTBP. *See* 2,4-Di-t-butylphenol
26DTBP. *See* 2,6-Di-t-butylphenol
DTDM. *See* 4,4′-Dithiodimorpholine
DTDP. *See* Ditridecyl phthalate
DTPANa$_5$. *See* Pentasodium pentetate

EAM

Classification: Elastomer

Trade names containing: Poly-Dispersion® V (MT)D-75; Poly-Dispersion® V (MT)D-75P; Poly-Dispersion® V(NOBS)M-65

Earth wax. *See* Ceresin
ECO. *See* Epichlorohydrin elastomer
Edathamil. *See* Edetic acid
Edetate calcium disodium. *See* Calcium disodium EDTA
Edetate disodium. *See* Disodium EDTA
Edetate sodium. *See* Tetrasodium EDTA
Edetate trisodium. *See* Trisodium EDTA

Edetic acid

CAS 60-00-4; EINECS 200-449-4

Synonyms: EDTA; N,N´-1,2-Ethanediylbis[N-(carboxymethyl) glycine]; Ethylene diamine tetraacetic acid; Edathamil

Classification: Substituted diamine

Empirical: $C_{10}H_{16}N_2O_8$

Formula: $(HOOCCH_2)_2NCH_2CH_2N(CH_2COOH)_2$

Properties: Colorless crystals; sl. sol. in water; insol. in common org. solvs.; m.w. 292.28; dec. 240 C

Toxicology: Irritant; poison by intraperitoneal route; mutagenic data

Uses: Chelating agent; stabilizer for cosmetics, pharmaceuticals, in water treatment, textiles, soaps/detergents, electroless copperplating, polymer prod., disinfectants, pulp/paper; antioxidant in foods

Regulatory: FDA 21CFR §175.105, 176.170

Manuf./Distrib.: Akzo Nobel; Aldrich; Allchem Ind.; Allied Colloids; Chemplex; Hampshire; Protex SA; Showa Denko

Trade names: Versene Acid

EDTA. *See* Edetic acid
EDTA Na_4. *See* Tetrasodium EDTA
EE. *See* Ethoxyethanol
EEA. *See* Ethylacetoacetate
EGDS. *See* Glycol distearate
EGMS. *See* Glycol stearate
2-EH. *See* 2-Ethylhexanol
EHGE. *See* 2-Ethylhexyl glycidyl ether
Elainic acid. *See* Oleic acid
Electrolyte acid. *See* Sulfuric acid
EMQ. *See* 6-Ethoxy-1,2-dihydro-2,2,4-trimethylquinoline
Entsufon sodium. *See* Sodium octoxynol-2 ethane sulfonate
Enzactin. *See* Triacetin
EPDM. *See* EPDM rubber

EPDM rubber

Synonyms: EPT; EPDM; Ethylene-propylene-diene terpolymer; Ethylene propylene terpolymer

Uses: Rubber for hose, molded and extruded goods, sponge, roofing; modifier for polyolefins

Trade names: Buna AP 147; Buna AP 331; Buna AP 437; Polysar EPDM 227; Polysar EPDM 847XP; Royalene 301-T; Vistalon 6505

Trade names containing: Chem-Master R-51; Chem-Master R-301; Chem-Master R-303; Chem-Master R-321; Poly-Dispersion® T (AZ)D-75; Poly-Dispersion® T (BZ)D-75; Poly-Dispersion® T (DIC)D-40P; Poly-Dispersion® T (DPG)D-65; Poly-Dispersion® T (DPG)D-65P; Poly-Dispersion® T (DYT)D-80P; Poly-Dispersion® T (HRL)D-90; Poly-Dispersion® T (HVA-2)D-70; Poly-Dispersion® T (HVA-2)D-70P; Poly-Dispersion® T (LC)D-90; Poly-Dispersion® T (LC)D-90P; Poly-Dispersion® T (VC)D-40P; Poly-Dispersion® TRD-90P; Poly-Dispersion® TSD-80; Poly-Dispersion® TSD-80P; Poly-Dispersion® TTD-75; Poly-Dispersion® TTD-75P; Poly-Dispersion® TZFD-88P; Poly-Dispersion® VC-60P; Polysar EPDM 6463; Rhenogran® BPH-80; Rhenogran® DCBS-80; Rhenogran® Fe-gelb-50; Rhenogran® Fe-rot-70; Rhenogran® IPPD-80; Rhenogran® MBI-80; Rhenogran® MMBI-70; Rhenogran® MPTD-80; Rhenogran® MTT-80; Rhenogran® OTBG-50; Rhenogran® OTBG-75; Rhenogran® PBN-80; Rhenogran® Resorcin-80; Royalene 7565; Wyfire® Y-A-85; Wyfire® Y-B-57; Wyfire® YP-BA-47P

EPDM/SAN graft polymer
Uses: Impact modifier for PC, polyester/PC alloys, PVC

Epichlorohydrin elastomer
Synonyms: ECO; Chloropropylene oxide elastomer
Classification: Epichlorohydrin elastomer
Formula: $(CHCH_2ClCH_2O)_x$
Uses: Elastomer used for fuel pump diaphragms, pipe gaskets, hose for fuel, oil, and gas, vibration isolators, motor mounts, rolls, adhesives, sponge goods, air conditioning hose and seals
Manuf./Distrib.: Ciba-Geigy; Conap; Hardman; Key Polymer; Morton Int'l.; Reichhold; Rhone-Poulenc/Perf. Resins; Sartomer; Shell; Union Carbide
Trade names containing: Chem-Master R-74; Chem-Master R-354; Poly-Dispersion® G (BAC)D-80; Poly-Dispersion® G (DPS)D-80; Poly-Dispersion® G (DYT)D-80; Poly-Dispersion® GND-75; Poly-Dispersion® GRD-90; Rhenogran® NDBC-70/ECO

EPM. *See* EPM rubber

EPM/EPDM copolymer
Uses: Elastomer for extruded, inj., transfer, and compression molded goods, foamed profiled articles; suitable for blending with SBR, NR, modifying polyolefin plastics
Trade names: Dutral® CO-059
Trade names containing: Dutral® PM-06PLE

EPM rubber
CAS 9010-79-1
Synonyms: EPR; EPM; Ethylene/propylene copolymer; EPR rubber; Ethene, polymer with 1-propene
Properties: Dens. 0.860
Uses: Rubber for inj. molded and extruded goods (elec. components, wire insulation, o-rings, brake components); modifying PP and other plastics
Regulatory: FDA 21CFR §177.1210
Manuf./Distrib.: Aldrich; Hüls AG
Trade names: Exxelor PE 805; Exxelor PE 808; Exxelor VA 1801; Exxelor VA 1803; Exxelor VA 1810; Exxelor VA 1820; KEP 010P, 020P, 070P; Petrolite® CP-7; Polysar EPM 306; proFLOW 3000; RA-061; Vistalon 719; Vistalon 503
Trade names containing: Akroform® DCP-40 EPMB; Akroform® VC-40 EPMB; Akrosperse® DCP-40 EPMB; Akrosperse® VC-40 EPMB; Chem-Master R-11; Chem-Master R-13; Chem-Master R-15; Chem-Master R-482A; Chem-Master R-486; Chem-Master R-524; Chem-Master R-8414; Kenlastic® K-6641; Kenlastic® K-6642; Kenlastic® K-9273; Peroximon® DC 40 MF; Peroximon® DC 40 MG; Poly-Dispersion® EAD-75; Poly-Dispersion® EAD-75P; Poly-Dispersion® ECSD-70; Poly-Dispersion® E (DIC)D-40; Poly-Dispersion® E (DOTG)D-65P; Poly-Dispersion® E (DPS)D-80; Poly-Dispersion® E (DYT)D-80; Poly-Dispersion® EMD-75; Poly-Dispersion® E (MX)D-75; Poly-Dispersion® E (MZ)D-75; Poly-Dispersion® E (NBC)D-70; Poly-Dispersion® E (NBC)D-70P; Poly-Dispersion® END-75; Poly-Dispersion® END-75P; Poly-Dispersion® ERD-90; Poly-Dispersion® E (SAN-NS)D-70; Poly-Dispersion® ESD-80; Poly-Dispersion® E (SR)D-75; Poly-Dispersion® E (TET)D-70; Poly-Dispersion® E (TET)D-70P; Poly-Dispersion® E (VC)D-40; Poly-Dispersion® EZFD-85; Retilox® F 40 MF; Retilox® F 40 MG; Varox® 802-40MB

Epoxidized flaxseed oil. *See* Epoxidized linseed oil

Epoxidized glycol dioleate
Uses: PVC stabilizer, plasticizer
Trade names: Monoplex® S-75

Epoxidized linseed oil
Synonyms: Epoxidized flaxseed oil
Properties: Clear pale yel. liq., low odor; sol. < 0.1% in water; sp.gr. 1.03; dec. 550 F; flash pt. (CC) 435 F
Precaution: Avoid oxidizing agents, strong acids, bases and amines; may produce CO on burning
Toxicology: LD50 (oral, rat) 30 gma/kg; nontoxic orally; no eye or skin irritation
Uses: Plasticizer/stabilizer for epoxy, PVC; food pkg.
Manuf./Distrib.: Elf Atochem N. Am.; Union Carbide; Witco/PAG
Trade names: ADK CIZER O-180A; Drapex® 10.4; Epoxol 9-5; Lankroflex® L; Plas-Chek® 795

Epoxidized linseed oil, butyl ester
Uses: Plasticizer; acid scavenger; stabilizer for PVC compds.; food pkg. materials

Epoxidized octyl stearate
CAS 106-84-3
Synonyms: 9,10-Octyl epoxy stearate
Uses: Heat stabilizer and plasticizer for plasticized and paste PVC formulations
Trade names: ADK CIZER D-32

Epoxidized octyl tallate
CAS 61788-72-5
Synonyms: Octyl epoxy tallate; 2-Ethylhexyl epoxy tallate
Properties: Sp.gr. 0.923 (20 C); pour pt. -8.5 C; flash pt. (COC) 235 C; ref. index 1.4581 (20 C)
Uses: Stabilizer, plasticizer for PVC, PS
Manuf./Distrib.: Elf Atochem N. Am.; Harwick; Witco/PAG
Trade names: Drapex® 4.4; Flexol® Plasticizer EP-8; Monoplex® S-73

Epoxidized oleate
Uses: Plasticizer for PVC
Trade names: Priplast 1431

Epoxidized 1,2-polybutadiene
Uses: As sole resin in an epoxy system or as modifiers for epoxy systems; in flexible and impact-resist. coatings and potting compds.
Trade names: ADK CIZER BF-1000; Poly bd® 600; Poly bd® 605

Epoxidized soybean oil
CAS 8013-07-8; EINECS 232-391-0
Synonyms: Soybean oil, epoxidized
Definition: Modified oil obtained from soybean oil by epoxidation
Properties: Clear pale yel. liq., low odor; sol. < 0.1% in water; sp.gr. 0.99; dec. 550 F; m.p. 25 F; iodine no. 6 max.; flash pt. (CC) 430 F
Precaution: Avoid oxidizing agents, strong acids, bases and amines
Toxicology: Nonhazardous; LD50 (oral, rat) 30 gm/kg; no eye or skin irritation (rabbit); heated to decomp., emits acrid smoke and irritating fumes
Uses: Plasticizer, stabilizer for PVC, epoxy, chlorinated rubber, coatings, inks; acid scavenger; food pkg.
Regulatory: FDA 21CFR §175.105, 177.1650, 178.3910, 181.22, 181.27
Manuf./Distrib.: Ashland; Elf Atochem N. Am.; FMC; Ferro; C.P. Hall; Henkel; Hüls; Union Carbide; Witco/PAG
Trade names: ADK CIZER O-130P; Drapex® 6.8; Epoxol 7-4; Estabex® 138-A; Estabex® 2307; Estabex® 2307 DEOD; Flexol® Plasticizer EPO; Lankroflex® GE; Nuoplaz® 849; Paraplex® G-60; Paraplex® G-62; Peroxidol 780; Plas-Chek® 775; Plasthall® ESO; Plastoflex® 2307; Plastolein® 9232; PX-800
Trade names containing: HFVD 12CC; Micro-Chek® 11; Omacide® P-ESO-5; Prespersion PAC-3936; Prespersion PAC-4893; Stabiol CZ 1616/6F; Vinyzene® BP-505

Epoxidized tallate
Properties: Sp.gr. 0.920; m.p. -22 C; flash pt. (COC) 220 C; ref. index 1.4580
Uses: Plasticizer for cellulosics, PVC
Manuf./Distrib.: Ashland; Elf Atochem N. Am.; Ferro/Keil; Henkel; Union Carbide; Witco/PAG

Epoxidized tall oil
Uses: Plasticizer stabilizer; used in chlorinated rubber
Trade names: Peroxidol 781

Epoxy acrylate
Uses: Curing agent; for paper clear coatings, wood top coatings, screen inks, litho inks, polyethylene coatings, metal decorative coatings, adhesive papers, wood fillers, solder masks and photoresists
Manuf./Distrib.: Monomer-Polymer & Dajac

Epoxy, bisphenol A
Uses: Heat stabilizer for PVC; exc. halogen capture for flame retardant resins
Trade names: ADK CIZER EP-13; ADK CIZER EP-17

1,2-Epoxy-3-phenoxypropane. *See* Phenyl glycidyl ether

2,3-Epoxypropyl methacrylate. *See* Glycidyl methacrylate

Epoxy resin
CAS 25928-94-3
Definition: Derived from epichlorohydrin and diethylene glycol
Formula: —OCH_2CHOCH_2
Toxicology: Strong skin irritant in uncured state; poison by inhalation; moderately toxic by ingestion
Uses: Surface coatings, adhesive, casting metal-forming tools and dies; encapsulation of elec. parts; stabilizer, modifier for other resins
Manuf./Distrib.: Asahi Chem Industry Co Ltd; Ciba-Geigy; Conap; Hardman; Key Polymer; Morton Int'l.; Reichhold; Rhone-Poulenc/Perf. Resins; Sartomer; Shell; Union Carbide
Trade names: Araldite® PY 306; Araldite® XD 897; CMD 5185; D.E.R. 337; D.E.R. 732; Epi-Tex® 611-Q; Varcum® 29726

Epoxy resin, brominated
Uses: Flame-retardant resin for elec. potting, encapsulation, and casting, high-pressure laminating, filament winding, and coatings; flame retardant for thermoplastics
Trade names: DX-5114; DX-5119; Epikote® 5050; Epikote® 5051 H; Epikote® 5057; Epikote® 5201; Epikote® 5201S; Epikote® 5203; Epikote® 5205; Epon® Resin 1183; F-2000 FR; F-2001; F-2016; F-2200; F-2300; F-2300H; F-2400; F-2400E; Thermoguard® 210; Thermoguard® 220; Thermoguard® 230; Thermoguard® 240
Trade names containing: F-2430; Thermoguard® 215

EPR. *See* EPM rubber
EPR rubber. *See* EPM rubber
EPT. *See* EPDM rubber

Erucamide
CAS 112-84-5; EINECS 204-009-2
Synonyms: Erucic acid amide; 13-Docosenamide; cis 13-Docosenamide
Classification: Aliphatic amide
Empirical: $C_{22}H_{43}ON$
Formula: $CH_3(CH_2)_7CH=CH(CH_2)_{11}CONH_2$
Properties: Solid; sol. in isopropanol; sl. sol. in alcohol, acetone; dens. 0.888; m.p. 75-80 C
Uses: Foam stabilizer; solvent for waxes and resins, emulsions; slip/antiblock agent for polyethylene; lubricant, mold release for rubber and plastics
Regulatory: FDA 21CFR §175.105, 177.1200, 177.1210, 178.3860
Manuf./Distrib.: Akzo Nobel; Chemax; Cookson Spec.; Croda Universal; Syn. Prods.; Witco/Oleo-Surf.
Trade names: Armid® E; Armoslip® EXP; Chemstat® HTSA #22; Crodamide E; Crodamide ER; Kemamide® E; Petrac® Eramide®; Polydis® TR 131; Unislip 1753
Trade names containing: Cabot® PE 9020; Cabot® PE 9166; Cabot® PE 9172; Colloids PE 48/10/13; PE-SA 21; PE-SA 24; PE-SA 49; PE-SA 60; PE-SA 72; PE-SL 04; PE-SL 98; Polyvel A2010; Polyvel RE25; Polyvel RE40; PP-SA 46; PP-SA 97; PP-SA 128; PP-SL 82; PP-SL 98; PP-SL 101

Erucic acid amide. *See* Erucamide

Erucyl erucamide
Uses: Lubricant, slip, antiblock, and mold release agent for plastics, crayons, petrol. prods., asphalts, inks, metals, textiles; mold release agent for thermoplastic resins in inj. molding; defoamer and water repellent in industrial/household applic.
Trade names: Kemamide® E-221

Erucyl stearamide
Definition: Ester of erucyl alcohol and stearic acid
Uses: Lubricant, slip, antiblock, and mold release agent for plastics, crayons, petrol. prods., asphalts, inks, metals, textiles; mold release agent for thermoplastic resins in inj. molding; defoamer and water repellent in industrial/household applic.
Trade names: Kemamide® S-221

Ethanediamine, N-(2-ethoxyphenyl)-N´-(2-ethylphenyl)-. *See* 2-Ethyl, 2´-ethoxy-oxalanilide
1,2-Ethanediol. *See* Glycol
1,2-Ethanediol dimethacrylate. *See* Ethylene glycol dimethacrylate
N,N´-1,2-Ethanediylbis[N-(carboxymethyl) glycine]. *See* Edetic acid

N,N´-1,2-Ethanediylbisoctadecanamide. *See* Ethylene distearamide
N,N´-1,2-Ethanediylbis-9-octadecenamide. *See* Ethylene dioleamide
2,2´-[1,2-Ethanediylbis(oxy)]bisethanol. *See* Triethylene glycol
Ethane, 1,1,1-trichloro-. *See* Trichloroethane
Ethanoic acid. *See* Acetic acid
Ethanol. *See* Alcohol

Ethanolamine
CAS 141-43-5; EINECS 205-483-3
Synonyms: MEA; 2-Aminoethanol; 2-Aminoethyl alcohol; Monoethanolamine; Glycinol; 2-Hydroxyethylamine
Classification: Monoamine
Empirical: C_2H_7NO
Formula: $NH_2CH_2CH_2OH$
Properties: Colorless liq., ammoniacal odor; hygroscopic; misc. with water, alcohol; sol. in chloroform; sl. sol. in benzene; m.w. 61.10; dens. 1.012; m.p. 10.5 C; b.p. 170 C; flash pt. 93 C
Precaution: Corrosive; flamm. exposed to heat or flame; powerful reactive base
Toxicology: TLV:TWA 3 ppm; LD50 (oral, rat) 214 mg/kg, (skin, rat) 1500 mg/kg; poison by IP route; mod. toxic by ingestion, skin contact, subcut., IV routes; corrosive irritant to eyes, skin, mucous membranes; heated to decomp., emits toxic fumes of NO_x
Uses: Scrubbing acid gases, esp. in synthesis of ammonia; nonionic detergents for dry cleaning wool treatment, emulsion paints, polishes, agricultural sprays; chemical intermediate; pharmaceuticals; corrosion inhibitor; chem. intermediate for mfg. rubber, vulcanization accelerators
Regulatory: FDA 21CFR §173.315, 175.105, 176.210, 176.300, 178.3120; not permitted for use in foods intended for babies and young infants in UK
Manuf./Distrib.: BP Chem. Ltd; OxyChem; Texaco; Union Carbide
Trade names: MEA Commercial Grade; MEA Low Freeze Grade; MEA Low Iron Grade; MEA Low Iron-Low Freeze Grade; MEA NF Grade

Ethanol, 2-butoxy-, phosphate (3:1). *See* Tributoxyethyl phosphate
Ethanol, 2,2´-iminobis, N-coco alkyl, N-oxide. *See* Dihydroxyethyl cocamine oxide
Ethanol, 2,2´-iminobis-, N-tallow alkyl derivs.. *See* N-Tallow alkyl-2,2´-iminobisethanol
Ethanol, 2-(2-methoxyethoxy)-. *See* Methoxydiglycol
Ethanol, undenatured. *See* Alcohol

2,2´-(1,2-Ethenediyldi-4,1-phenylene) bisbenzoxazole
CAS 1533-45-5
Empirical: $C_{28}H_{18}N_2O_2$
Properties: Yel. solid; sp.gr. 1.39
Toxicology: Irritant
Uses: Optical brightener for polymers
Trade names: Eastobrite® OB-1

Ethene, homopolymer. *See* Polyethylene
Ethene, homopolymer, oxidized. *See* Polyethylene, oxidized
Ethene, polymer with 1-propene. *See* EPM rubber
Ethenol homopolymer. *See* Polyvinyl alcohol
Ethenylbenzene, homopolymer. *See* Polystyrene
1-Ethenyl-2-pyrrolidinone. *See* N-Vinyl-2-pyrrolidone
1-Ethenyl-2-pyrrolidinone homopolymer. *See* PVP

N-(p-Ethoxycarbonylphenyl)-N´-ethyl-N´-phenylformamidine
CAS 65816-20-8
Empirical: $C_{18}H_{20}N_2O_2$
Properties: White to pale yel. powd.; sol. > 50 g/100 g in ethanol, methanol, IPA, butyl acetate; insol. in water; m.w. 296.4; dens. 1.077 (65 C); b.p. 215 C (2 mm); m.p. 60-65 C
Uses: UV absorber for cellulosics
Trade names: Givsorb® UV-2

N^2-(4-Ethoxycarbonylphenyl)-N´-methyl-N´-phenylformamidine
CAS 57834-33-0
Empirical: $C_{17}H_{18}N_2O_2$
Properties: Sl. yel. liq.; sol. > 50 g/100 g in ethanol, methanol, IPA, butyl acetate; insol. in water; m.w. 282.3; dens. 1.127 (20/4 C); b.p. 205 C (2 mm); ref. index 1.64

Uses: UV absorber for cellulosics
Trade names: Givsorb® UV-1

Ethoxydiglycol
CAS 111-90-0; EINECS 203-919-7
Synonyms: Diethylene glycol monoethyl ether; Diethylene glycol ethyl ether; Carbitol; 2-(2-Ethoxyethoxy) ethanol
Classification: Ether alcohol
Formula: $CH_2OHCH_2OCH_2CH_2OC_2H_5$
Properties: Colorless liq., mild pleasant odor; misc. with water, common org. solvs.; m.w. 134.20; dens. 1.0272 (20/20 C); b.p. 195-202 C; flash pt. 96.1 C; ref. index 1.425 (25 C)
Precaution: Combustible
Uses: Solvent for dyes, nitrocellulose, resins; textiles, textile printing, soaps; lacquers; organic synthesis; brake fluid diluent; mfg. of plasticizers
Regulatory: FDA 21CFR §175.105, 176.180
Manuf./Distrib.: Allchem Ind.; Ashland; Eastman; Great Western; Oxiteno; Union Carbide
Trade names: Dioxitol-Low Gravity
Trade names containing: Dioxitol-High Gravity; FC-520

6-Ethoxy-1,2-dihydro-2,2,4-trimethylquinoline
CAS 91-53-2; EINECS 202-075-7
Synonyms: EMQ; Ethoxyquin; 1,2-Dihydro-6-ethoxy-2,2,4-trimethylquinoline; Santoquine
Empirical: $C_{14}H_{19}NO$
Properties: Yel. liq.; m.w. 217.34; dens. 1.029-1.031 (25 C); m.p. $\approx$ 0 C; b.p. 125 C (2 mm); ref. index 1.569-1.572 (25 C)
Precaution: Combustible when exposed to heat or flame; can react with oxidizing materials
Toxicology: LD50 (oral, rat) 800 mg/kg; moderately toxic by ingestion; poison by intraperitoneal route; mutagenic data; heated to decomp., emits toxic fumes of NO_x
Uses: Insecticide; antioxidant for fats and rendered prods.; antiozonant, flex-cracking inhibitor; post-harvest preservation additive for food (human or animal); antioxidant for apples and pears; scald inhibitor; stabilizer, preservative
Regulatory: FDA §172.140, limitation 100 ppm in chili powd., 100 ppm in paprika, 5 ppm in uncooked meat fat, 3 ppm in uncooked poultry fat, 0.5 ppm in eggs, zero tolerance in milk
Trade names: Santoflex® AW

Ethoxyethanol
CAS 110-80-5; EINECS 203-804-1
Synonyms: EE; Ethylene glycol ethyl ether; 2-Ethoxyethanol; Cellosolve
Classification: Ether alcohol
Formula: $CH_3CH_2OCH_2CH_2OH$
Properties: Colorless liq., pract. odorless; misc. with hydrocarbons, alcohol, ether, acetone, liq. esters, water; m.w. 90.14; dens. 0.9311 (20/20 C); m.p. -70 C; b.p. 135.6 C; flash pt. 48.9 C; ref. index 1.4060 (25 C)
Toxicology: TLV 5 ppm (skin); LD50 (rat, oral) 3 g/kg; toxic by skin absorption; moderately toxic by ingestion, IV, IP routes; mildly toxic by inhalation, subcutaneous routes
Uses: Solvent for nitrocellulose, alkyd resins, lacquer, lacquer thinners, dyeing and printing textiles, varnish removers, cleaning solutions
Regulatory: FDA 21CFR §73.1, 175.105, 177.2600
Manuf./Distrib.: Arco; Allchem Ind.; Ashland; Brown; Great Western; Oxiteno; OxyChem; Union Carbide
Trade names: Oxitol

2-Ethoxyethanol. *See* Ethoxyethanol
2-(2-Ethoxyethoxy) ethanol. *See* Ethoxydiglycol
Ethoxylated 1-aminooctadecane. *See* PEG-50 stearamine
N-(2-Ethoxyphenyl)-N′-(2-ethylphenyl) ethanediamide. *See* 2-Ethyl, 2′-ethoxy-oxalanilide
Ethoxyquin. *See* 6-Ethoxy-1,2-dihydro-2,2,4-trimethylquinoline

Ethyl acetate
CAS 141-78-6; EINECS 205-500-4
Synonyms: Acetic ether; Acetic acid, ethyl ester; Vinegar naphtha
Definition: Ester of ethyl alcohol and acetic acid
Empirical: $C_4H_8O_2$
Formula: $CH_3COOC_2H_5$
Properties: Colorless liq., fragrant; sol. in chloroform, alcohol, ether; sl. sol. in water; m.w. 88.12; dens. 0.902

(20/4 C); bulk dens. 0.8945 g/ml (25 C); b.p. 77 C; f.p. -83.6 C; flash pt. -4.4 C; ref. index 1.3723
Precaution: Flamm.; very dangerous fire hazard exposed to heat or flame; can react vigorously with oxidizers
Toxicology: TLV 400 ppm in air; LD50 (oral, rat) 5620 mg/kg; poison by inhalation; mildly toxic by ingestion; irritant to eyes, skin, mucous membranes; mutagenic data; mildly narcotic; heated to decomp., emits acrid smoke and irritating fumes
Uses: General solvent in coatings and plastics, organic synthesis, smokeless powders, artificial leather, photographic films and plates, pharmaceuticals, synthetic fruit essences, flavoring agent; cleaning textiles; color diluent, flavoring agent, solv. for food use
Regulatory: FDA 21CFR § 73.1, 172.560, 173.228, 175.320, 177.1200, 182.60, GRAS; 27CFR §21.106; Japan approved with restrictions
Manuf./Distrib.: Allchem Industries; Berje; BP Chem. Ltd; Brown; Chisso Am.; Eastman; Hoechst Celanese; Hüls AG; Lonza AG; Mallinckrodt; Monsanto; Penta Mfg.; Union Carbide
Trade names containing: Desmodur® RC; Desmodur® RE; Desmodur® RFE; Sprayset® MEKP

Ethylacetoacetate
CAS 141-97-9; EINECS 205-516-1
Synonyms: EEA; 3-Oxobutanoic acid ethyl ester; Acetoacetic acid ethyl ester; Ethyl 3-oxobutanoate
Empirical: $C_6H_{10}O_3$
Formula: $CH_3COCH_2COOC_2H_5$
Properties: Colorless liq., fruity odor; sol. in ≈ 35 parts water; misc. with common org. solvs.; m.w. 130.14; dens. 1.0213 (25/4 C); m.p. -45 C; b.p. 180.8 C (760 mm); flash pt. (CC) 184 F; ref. index 1.4180-1.4195
Precaution: Combustible liq. when exposed to heat or flame; can react with oxidizing materials
Toxicology: LD50 (oral, rat) 3.98 g/kg; mod. toxic by ingestion; mod. irritating to skin, mucous membranes, eyes; heated to decomp., emits acrid smoke and irritating fumes
Uses: Catalyst for esterification and olefin polymerization; resin crosslinking agent for automotive prods., coatings, elastomers, films/paints, graphic arts, plastics; flavoring agent
Regulatory: FDA 21CFR §172.515; GRAS (FEMA); Japan approved as flavoring
Manuf./Distrib.: Aceto; Aldrich; Berje; Eastman; Lonza; Penta Mfg.; Spectrum Chem. Mfg.
Trade names: Tyzor DC

Ethyl alcohol, undenatured. *See* Alcohol

p-Ethylbenzaldehyde
CAS 4748-78-1; EINECS 225-268-8
Empirical: $C_9H_{10}O$
Properties: M.w. 134.2; sol. in ethanol, ether, toluene; insol. in water; dens. 0.979; b.p. 221 C; ref. index 1.5390 (20 C)
Toxicology: Irritant
Uses: Additive for resins; pharmaceutical intermediate; fragrance
Manuf./Distrib.: Esprit; Jonas; Penta Mfg.
Trade names: Ebal

Ethyl-o-benzoyl laurohydroximate
Uses: Crosslinking agent
Trade names: Trigonox® 107

Ethyl cadmate
Trade names containing: Chem-Master R-524

Ethyl citrate. *See* Triethyl citrate

Ethyl-2-cyano-3,3-diphenyl acrylate. ***See* Etocrylene**
Trade names: Uvinul® N-35

Ethyl 2-cyano-3,3-diphenyl-2-propenoate. *See* Etocrylene

Ethyl 3,3-di (t-amylperoxy) butyrate
CAS 67567-23-1
Uses: Initiator for curing of acrylic syrup; crosslinking agent for polymerization, thermoplastic modification
Trade names containing: Lupersol 533-M75

Ethyldibutoxy phthalate. *See* Dibutoxyethyl phthalate

Ethyldibutylperoxybutyrate
Uses: Initiator for curing elastomers and for polymer modification; thermoplastic crosslinking
Trade names: Aztec® EBU-40-G; Aztec® EBU-40-IC; Luperco 233-XL
Trade names containing: Lupersol 233-M75

Ethylene/acrylic acid copolymer
CAS 9010-77-9
Definition: Copolymer of ethylene and acrylic acid monomers
Properties: Dens. 0.960
Uses: Binder for nonwoven fibers; lubricant and processing aid for plastics; pigment dispersant
Regulatory: FDA 21CFR §177.1310, 178.1005
Trade names: A-C® 540, 540A; A-C® 580; A-C® 5120; A-C® 5180; Luwax® EAS 1
Trade names containing: Vinyzene® SB-1 EAA

Ethylene alcohol. *See* Glycol

Ethylenebis dibromonorbornane dicarboximide
CAS 52907-07-0; EINECS 258-250-3
Synonyms: Ethylenebis (5,6-dibromonorbornane-2,3-dicarboximide)
Uses: Flame retardant for PP, polyesters, polyamides, polyurethane elastomers, castables, coatings
Manuf./Distrib.: Albemarle
Trade names: Saytex® BN-451

Ethylenebis (5,6-dibromonorbornane-2,3-dicarboximide). *See* Ethylenebis dibromonorbornane dicarboximide
Ethylenebis (dithiocarbamate), disodium salt. *See* Nabam

N,N´-Ethylenebis 12-hydroxystearamide
Uses: Internal lubricant, mold release, slip additive for PVC
Trade names: Paricin® 285

Ethylenebis (oxyethylene) bis (3-t-butyl-4-hydroxy-5-methylhydrocinnamate)
CAS 36443-68-2
Properties: White powd.; m.w. 587
Uses: Stabilizer for polymers
Trade names: Irganox® 245

N,N´-Ethylene bis-ricinoleamide
Uses: Lubricant/antistat for plastics, metals; mold release, antiblocking agent for textile coatings; slip agent for varnishes and lacquers; also for elec. potting compds., crayons, wax blends, high-temp. greases
Trade names: Flexricin® 185

N,N´-Ethylene bisstearamide. *See* Ethylene distearamide

Ethylene bis-tetrabromophthalimide
CAS 32588-76-4; EINECS 251-118-6
Uses: Flame retardant for PBT, HIPS, ABS, polyethylene, PP, thermoplastic polyester, EPDM and other elastomers, ethylene copolymners, PC, ionomer resins, epoxies, polyimides, textile treatments
Trade names: Saytex® BT-93®; Saytex® BT-93W
Trade names containing: Wyfire® S-BA-88P; Wyfire® Y-B-57; Wyfire® YP-BA-47P

Ethylene/calcium acrylate copolymer
CAS 26445-96-5
Uses: Processing and performance additive; improves dispersion of additives in plastics; adhesion to variety of substrates; encapsulant for personal care prods.
Regulatory: FDA 21CFR §175.105
Trade names: AClyn® 201A

Ethylene carbonate
CAS 96-49-1; EINECS 202-510-0
Synonyms: Glycol carbonate; Dioxolone-2; 1,3-Dioxolan-2-one; Cyclic ethylene carbonate
Empirical: $C_3H_4O_3$

Formula: $(—CH_2O)_2CO$
Properties: Colorless solid or liq., odorless; misc. (40%) in water, alcohol, ethyl acetate, benzene, chloroform; sol. in ether, n-butanol, CCl_4; m.w. 88.07; dens. 1.3218 (39/4 C); m.p. 36.4 C; b.p. 248 C; flash pt. 143 C; ref. index 1.4158 (50 C)
Toxicology: Moderately toxic by intraperitoneal route; mildly toxic by ingestion; skin and eye irritant
Uses: Solvent for polymers and resins, plasticizer, intermediate for pharmaceuticals, hydroxyethylation reactions, rubber chemicals, textile finishing agent
Manuf./Distrib.: Fluka; Hüls AG; Monomer-Polymer & Dajac
Trade names: Texacar® EC

Ethylene carboxylic acid. *See* Acrylic acid
Ethylene diacetate. *See* Ethylene glycol diacetate

Ethylene diamine phosphate
Uses: Flame retardant for PP, PE, PVAc
Trade names: Amgard® NK; Amgard® NP; SAFR-C70

Ethylene diamine tetraacetic acid. *See* Edetic acid
Ethylene diamine tetraacetic acid, disodium salt. *See* Disodium EDTA
Ethylene diamine tetraacetic acid, sodium salt. *See* Tetrasodium EDTA
Ethylene dimethacrylate. *See* Ethylene glycol dimethacrylate
Ethylenedinitrilotetra-2-propanol. *See* Tetrahydroxypropyl ethylenediamine

Ethylene dioleamide
CAS 110-31-6; EINECS 203-756-1
Synonyms: N,N´-1,2-Ethanediylbis-9-octadecenamide; 9-Octadecenamide, N,N´-1,2-ethanediylbis-; N,N´-Dioleoylethylenediamine
Classification: Diamide
Empirical: $C_{38}H_{72}N_2O_2$
Uses: Syn. wax used as plastics processing lubricant, release agent, antiblock, antislip, antistat, m.p. modifier for waxes, industrial asphalts and tar, pigment dispersing agent for resin systems
Trade names: Advawax® 240; Glycolube® VL; Kemamide® W-20

Ethylene distearamide
CAS 110-30-5, 68955-45-3; EINECS 203-755-6, 273-277-0
Synonyms: N,N´-Ethylene bisstearamide; N,N´-1,2-Ethanediylbisoctadecanamide; N,N-Bis-stearoylethylenediamide
Classification: Diamide
Empirical: $C_{39}H_{76}O_2$
Formula: $CCH_3(CH_2)_{16}CONH(CH_2)_2NHCO(CH_2)_{16}CH_3$
Uses: Lubricant, processing aid for PVC, PS, ABS, nylon; peptizing and dispersing agents for tech. NR articles
Manuf./Distrib.: Aquatec Quimica SA; Chemax
Trade names: Abluwax EBS; Acrawax® C; Acrawax® C SG; Advawax® 280; Advawax® 290; Aktiplast® AS; Alkamide® STEDA; Chemstat® 327; Glycowax® 765; Hoechst Wax C; Hostalub® FA 1; Interstab® G-8257; Kemamide® W-39; Kemamide® W-40; Kemamide® W-40/300; Kemamide® W-40DF; Kemamide® W-45; Uniwax 1760; Uniwax 1763

Ethylene glycol. *See* Glycol
Ethylene glycol bisthioglycolate. *See* Glycol dimercaptoacetate
Ethylene glycol butyl ether. *See* Butoxyethanol

Ethylene glycol diacetate
CAS 111-55-7; EINECS 203-881-1
Synonyms: Glycol diacetate; Ethylene diacetate; 1,2-Diacetoxyethane
Classification: Ester
Formula: $CH_3COOCH_2CH_2OOCCH_3$
Properties: Colorless liq., faint odor; sol. in alcohol, ether, benzene; sl. sol. in water (10%); m.w. 146.14; dens. 1.1063 (20/20 C); b.p. 190.5 C; f.p. -31 C; flash pt. 96 C; ref. index 1.416
Precaution: Combustible
Toxicology: Moderately toxic by intraperitoneal route; mildly toxic by ingestion and skin contact; eye irritant
Uses: Extraction solv., foundry resins, perfume fixative, solv. for coatings; plasticizer for cellulosics, PVAc
Manuf./Distrib.: Aceto; Ashland; BP Chem.; Chemoxy Int'l. plc; CPS; Eastman
Trade names: Estol 1574

Ethylene glycol diethyl ether
CAS 629-14-1; EINECS 211-076-1
Synonyms: 1,2-Diethoxyethane; Ethyl glyme
Formula: $C_2H_5OCH_2CH_2OC_2H_5$
Properties: Colorless liq., sl. odor; immiscible in water; m.w. 118.20; dens. 0.8417 (20/20 C); f.p. -74 C; b.p. 121.4 C; flash pt. 35 C
Precaution: Flamm.
Toxicology: Moderately toxic by ingestion; mildly toxic by inhalation; eye irritant
Uses: Organic synthesis (reaction medium); solvent and diluent for detergents
Manuf./Distrib.: Brand-Nu Labs; Eastern Chem.; Ferro/Grant; Fluka
Trade names: Ethyl Glyme

Ethylene glycol dilaurate. *See* Glycol dilaurate

Ethylene glycol dimethacrylate
CAS 97-90-5; EINECS 202-617-2
Synonyms: 1,2-Ethanediol dimethacrylate; 1,2-Bis(methacryloyoxy) ethane; Ethylene dimethacrylate
Formula: $CH_2{:}C(CH_3)COOCH_2CH_2OCOC(CH_3){:}CH_2$
Properties: M.w. 198.1; dens. 1.053 (20/4 C); b.p. 66-68 C
Toxicology: Moderately toxic by ingestion, intraperitoneal routes
Uses: Crosslinker and modifier for ABS, acrylic sheet and rods, PVC, ion exchange resins, glaze coatings, dental polymers, paper processing aids, rubber modifier, adhesives, optical polymers, leather finishing, moisture barrier films
Manuf./Distrib.: Akzo Nobel; CPS; Hampford Research; Monomer-Polymer & Dajac; Rohm Tech; Sartomer
Trade names: Ageflex EGDMA; MFM-416; Perkalink® 401
Trade names containing: Rhenofit® EDMA/S; SR-206

Ethylene glycol dioleate. *See* Glycol dioleate
Ethylene glycol distearate. *See* Glycol distearate
Ethylene glycol ethyl ether. *See* Ethoxyethanol
Ethylene glycol methyl ether. *See* Methoxyethanol
Ethylene glycol monobutyl ether. *See* Butoxyethanol
Ethylene glycol monobutyl ether oleate. *See* Butoxyethyl oleate
Ethylene glycol monolaurate. *See* Glycol laurate
Ethylene glycol monomethyl ether. *See* Methoxyethanol
Ethylene glycol monooleate. *See* Glycol oleate
Ethylene glycol monophenyl ether. *See* Phenoxyethanol
Ethylene glycol monopropyl ether. *See* Ethylene glycol propyl ether
Ethylene glycol monoricinoleate. *See* Glycol ricinoleate
Ethylene glycol monostearate. *See* Glycol stearate
Ethylene glycol monostearate SE. *See* Glycol stearate SE
Ethylene glycol nonyl phenyl ether. *See* Nonoxynol-1
Ethylene glycol octyl phenyl ether. *See* Octoxynol-1

Ethylene glycol propyl ether
CAS 2807-30-9; EINECS 220-548-6
Synonyms: Propyl 'Cellosolve'; 2-Propoxyethanol; Ethylene glycol monopropyl ether
Empirical: $C_5H_{12}O_2$
Formula: $CH_2OHCH_2OCH_2CH_2CH_3$
Properties: Liq.; sol. in water; m.w. 104.17; dens. 0.913 (20/20 C); b.p. 149.5 C (760 mm); f.p. -90 C; flash pt. (TCC) 49 C
Precaution: Combustible
Toxicology: Moderately toxic by ingestion and skin contact; mild toxicity by inhalation; skin/eye irritant; heated to dec., emits acrid smoke, irritating fumes
Uses: Solvent for coatings, cosmetics, resins; coupling solv. for resin/water systems
Manuf./Distrib.: Ashland; Eastman
Trade names: Ektasolve® EP

Ethylene homopolymer. *See* Polyethylene
Ethylene/MA copolymer. *See* Ethylene-maleic anhydride copolymer

Ethylene/magnesium acrylate copolymer
Uses: Processing and performance additive; improves dispersion of additives in plastics; adhesion to variety of substrates; encapsulant for personal care prods.
Trade names: AClyn® 246A

Ethylene-maleic anhydride copolymer
CAS 9006-26-2
Synonyms: Ethylene/MA copolymer; 2,5-Furandione, polymer with ethene
Definition: Polymer of ethylene and maleic anhydride monomers
Uses: Bonding/compatibilizing agent for substrates, polymers, and polymer blends for paper coatings, wood prods., adhesives/sealants, plastics (olefin blends and filled olefins)
Regulatory: FDA 21CFR §175.105
Trade names: Polyace™ 573

Ethylene-methyl acrylate. *See* Ethylene-methyl acrylate copolymer

Ethylene-methyl acrylate-acrylic acid terpolymer
Uses: Impact modifier for engineering thermoplastics; adhesion modifier for rubber compds., TPOs, adhesives
Trade names: Escor® ATX-310; Escor® ATX-320; Escor® ATX-325

Ethylene-methyl acrylate copolymer
Synonyms: Ethylene-methyl acrylate
Trade names containing: Ampacet 11187; Ampacet 11200; Ampacet 11737; Ampacet 19238; Ampacet 19492

Ethylene/propylene copolymer. *See* EPM rubber
Ethylene-propylene-diene terpolymer. *See* EPDM rubber

Ethylene-propylene-styrene terpolymer
Uses: Impact modifier for rigid PVC; resin compatibilizer for PVC, polyolefin, styrenics; process aid; improves compd. uniformity and lubrication
Trade names: SAFR-EPS K

Ethylene propylene terpolymer. *See* EPDM rubber

Ethylene thiourea
CAS 96-45-7; EINECS 202-506-9
Synonyms: ETU; 2-Imidazolidinethione; Imidazoline-2-thiol; 2-Mercaptoimidazoline
Empirical: $C_3H_6N_2S$
Formula: $NHCH_2CH_2NHCS$
Properties: White to pale green crystals, faint amine odor; sl. sol. in cold water; very sol. in hot water; sl. sol. in R.T. methanol, ethanol, acetic acid, naphtha; m.w. 102.17; m.p. 199-204 C
Toxicology: Poison by ingestion, intraperitoneal routes; suspected carcinogen; mutagenic data; eye irritant; LD50 (rat, oral) 1832 mg/kg
Uses: Electroplating baths; intermediate for antioxidants, insecticides, fungicides, synthetic resins, vulcanization accelerators and rubber processing aid, dyes
Manuf./Distrib.: Faesy & Besthoff; Miljac; Ore & Chem. Corp.
Trade names: Ekaland ETU; Naftocit® Mi 12; Naftopast® Mi12-P; Perkacit® ETU
Trade names containing: Akroform® ETU-22 PM; Chem-Master R-15; Chem-Master R-74; Chem-Master R-301; Poly-Dispersion® END-75; Poly-Dispersion® END-75P; Poly-Dispersion® GND-75; Poly-Dispersion® SZFND-825; Prespersion PAB-262

Ethylene/VA copolymer
CAS 24937-78-8
Synonyms: EVA; EVA copolymer; Ethylene vinyl acetate; Ethylene/vinyl acetate copolymer; Acetic acid, ethenyl ester, polymer with ethene
Definition: Polymer of ethylene and vinyl acetate monomers
Formula: $(C_4H_6O_2 \cdot C_2H_4)_x$
Uses: Syn. rubber for tech. moldings and extrudates, lamp seals, cable sheathings and insulations, cellular rubber goods, footwear soles, waterproof sheeting, hot-melt and pressure-sensitive hot-melt adhesives, coatings, inks, lacquers, sealants; impact modifier for PVC
Regulatory: FDA 21CFR §175.300, 177.1200, 177.1210, 177.1350, 178.1005
Trade names: A-C® 400; A-C® 400A; A-C® 405M; A-C® 405S; A-C® 405T; A-C® 430; Baymod® L450 N; Baymod® L450 P; Baymod® L2450; Fusabond® MC-189D; Fusabond® MC-190D; Fusabond® MC-

197D; Levapren 450P

Trade names containing: Ampacet 19258; Ampacet 19278; Ampacet 19400; Ampacet 19500; Ampacet 19552; Ampacet 19786; HC 01; MDI MB-650; Plasblak® EV 1755; Reed C-BK-7; Rhenogran® BPH-80; Rhenogran® DCBS-80; Rhenogran® Fe-gelb-50; Rhenogran® Fe-rot-70; Rhenogran® IPPD-80; Rhenogran® MBI-80; Rhenogran® MMBI-70; Rhenogran® MPTD-80; Rhenogran® MTT-80; Rhenogran® OTBG-50; Rhenogran® OTBG-75; Rhenogran® P-50/EVA; Rhenogran® PBN-80; Rhenogran® Resorcin-80

Ethylene vinyl acetate. *See* Ethylene/VA copolymer
Ethylene/vinyl acetate copolymer. *See* Ethylene/VA copolymer

2-Ethyl, 2′-ethoxy-oxalanilide

CAS 23946-66-8

Synonyms: Ethanediamine, N-(2-ethoxyphenyl)-N′-(2-ethylphenyl)-; N-(2-Ethoxyphenyl)-N′-(2-ethylphenyl) ethanediamide

Classification: Oxalic anilide

Empirical: $C_{18}H_{20}N_2O_3$

Formula: $OC_2H_5NHCOCOHNC_2H_5$

Properties: Wh. powd., mild odor; insol. in water; sp.gr. 1.209; m.p. 126 C; flash pt. > 200 F

Precaution: Dusts may be explosive with spark or flame initiation; incompat. with strong oxidizing agents; thermal decomp. may produce CO_x and NO_x

Toxicology: LD50 (oral, rat) > 5000 mg/kg; virtually nontoxic; nonirritating to skin and eyes

Uses: UV absorber for acrylates, alkyd/melamines, PU, thermosetting polyester

Trade names: Sanduvor® VSU

Ethyl 3-ethoxypropionate

CAS 763-69-9

Empirical: $C_7H_{14}O_3$

Formula: $C_2H_5OC_3H_5O_2C_2H_5$

Properties: Liq.; sl. sol. in water; m.w. 146.19; dens. 0.9496 (20/20 C); b.p. 165-172 C; f.p. < -50 C; flash pt. (Seta) 58 C

Toxicology: Mildly toxic by ingestion and skin contact; skin and eye irritant

Uses: Solvent for polymerization; retarder solv. for coatings; intermediate for vitamin B_1

Manuf./Distrib.: Ashland; Eastman; Union Carbide

Trade names: Ektapro® EEP Solvent

Ethylformic acid. *See* Propionic acid
Ethyl glyme. *See* Ethylene glycol diethyl ether
2-Ethylhexanoic acid, 1-methyl-1,2-ethanediyl ester. *See* Propylene glycol dioctanoate

2-Ethylhexanol

CAS 104-76-7; EINECS 203-234-3

Synonyms: 2-EH; 2-Ethylhexyl alcohol; Octyl alcohol; Alcohol C_8

Formula: $CH_3(CH_2)_3CHC_2H_5CH_2OH$

Properties: Colorless liq.; misc. with most org. solvs.; sl. sol. in water; m.w. 130.26; dens. 0.83 (20 C); b.p. 183.5 C; f.p. -76 C; flash pt. 81.1 C; ref. index 1.4300 (20 C)

Toxicology: Moderately toxic by ingestion, skin contact, IP, SQ, parenteral routes; severe eye irritant; moderate skin irritant; LD50 (rat, oral) 12.46 ml/kg

Uses: Plasticizer for PVC resins; defoaming agent, wetting agent, organic synthesis, solvent mix for nitrocellulose; penetrant for plasticizing inks, etc.

Manuf./Distrib.: Aristech; Ashland; BASF; BP Chem. Ltd; Coyne; Eastman; Hoechst AG; Penta Mfg.; Shell

Trade names containing: Surfynol® 104A

2-Ethylhexyl alcohol. *See* 2-Ethylhexanol
2-Ethylhexyl 2-cyano-3,3-diphenylacrylate. *See* Octocrylene
2-Ethylhexyl 2-cyano-3,3-diphenyl-2-propenoate. *See* Octocrylene
2-Ethylhexyl diphenyl ester phosphoric acid. *See* Diphenyl octyl phosphate
2-Ethylhexyl diphenyl phosphate. *See* Diphenyl octyl phosphate

Ethylhexyl diphenyl phosphite

CAS 15647-08-2

Formula: $C_4H_9CHC_2H_5CH_2OP(OC_6H_5)_2$

2-Ethylhexyl epoxy tallate

Properties: Liq.; m.w. 346; dens. 1.040-1.047; b.p. 185 C (5 mm); ref. index 1.5200-1.5250; flash pt. (PMCC) 182 C
Uses: Color and processing stabilizer for ABS, PC, polyurethane, coatings, PET fiber; sec. stabilizer for PVC

2-Ethylhexyl epoxy tallate. *See* Epoxidized octyl tallate

2-Ethylhexyl glycidyl ether
CAS 2461-15-6; EINECS 219-553-6
Synonyms: EHGE; Oxirane, [[2-ethylhexyl)oxy]methyl]-
Empirical: $C_{11}H_{22}O_2$
Properties: Colorless to ylsh. liq.; sol. in most org. solvs.; m.w. 186.30; dens. 0.891; b.p. 60-61.5 C (0.3 mm); flash pt. 96 C
Precaution: Vapors can form explosive mixts. with air; incompat. with acids, alkalies, oxidizing agents, metals; protect from exposure to air/oxygen; hazardous decomp. prods.: CO, gaseous hydrocarbons
Toxicology: LD50 (oral, rat) > 5000 mg/kg, (dermal, rat) > 3500 mg/kg; irritating to eyes, respiratory system, skin; harmful by ingestion
Uses: Reactive epoxy diluent for exposed aggregates, potting, flooring, casting, tooling, laminates, solv.-free coating systems, fiber-reinforced composites; stabilizer for chlorinated hydrocarbons
Manuf./Distrib.: Raschig
Trade names: Epodil® 746; Heloxy® 116

2-Ethylhexyl hexadecanoate. *See* Octyl palmitate
2-Ethylhexyl octadecanoate. *See* Octyl stearate
2-Ethylhexyl palmitate. *See* Octyl palmitate
2-Ethylhexyl stearate. *See* Octyl stearate

2-Ethylhexyl thioglycolate
CAS 7659-86-1; EINECS 231-626-4
Synonyms: OTG; Mercaptoacetic acid 2-ethylhexyl ester; Octyl thioglycolate
Properties: Colorless liq., mercaptan odor; sl. sol. in water; sp.gr. 0.97-0.98 (20/4 C); b.p. 130-140 C (2000 Pa); flash pt. 136 C
Precaution: Incompat. with oxidizers; hazardous decomp. prods.: CO, sulfur oxides; avoid sunlight, heat; may emit toxic irritating fumes on combustion
Toxicology: LD50 (oral, rat) 303 mg/kg; eye and skin irritant; harmful if inhaled and ingested; ing. may cause anorexia, headache, vomiting, fever, etc.; prolonged exposure can cause shock, narcosis, death
Uses: Chain transfer agent
Manuf./Distrib.: Elf Atochem N. Am.
Trade names: CTA

N-Ethyl-N-hydroxyethyl-m-toluidine
CAS 91-88-3
Synonyms: 2-(N-Ethyl-m-toluidino) ethanol
Formula: $CH_3C_6H_4N(C_2H_5)CH_2CH_2OH$
Properties: M.w. 179.26; dens. 1.019; b.p. 114-115 C (1 mm); flash pt. > 110 C; ref. index 1.5550 (20 C)
Uses: Coupling agent for disperse dyes for syn. fibers; photosensitive chemical for paper coatings; accelerator for polyester resins
Manuf./Distrib.: Aldrich; First
Trade names: Emery® 5714

2,2′-Ethylidenebis (4,6-di-t-butylphenol)
CAS 35958-30-6; EINECS 252-816-3
Formula: $CH_3CH[C_6H_2[C(CH_3)_3]_2OH]_2$
Properties: M.w. 438.70; m.p. 162-164 C
Toxicology: Irritant
Uses: Antioxidant and thermal stabilizer for polymers; food pkg. applic.; used in PP, polyethylene, PVC, PS, ABS, hydrocarbon resins, EVA-modified compds.
Manuf./Distrib.: Aldrich; Fluka
Trade names: Irganox® 129; Vanox® 1290

2,2′-Ethylidenebis (4,6-di-t-butylphenyl) fluorophosphonite
CAS 118337-09-0
Properties: M.w. 486.66; m.p. 201-203 C

Uses: Antioxidant for polyolefins
Trade names: Ethanox® 398

Ethylidene fluoride. *See* 1,1-Difluoroethane
Ethyllic acid. *See* Acetic acid
N-Ethyl-p-methylbenzene sulfonamide. *See* Ethyl toluenesulfonamide

2-Ethyl-4-methyl imidazole
CAS 931-36-2; EINECS 213-234-5
Empirical: $C_6H_{10}N_2$
Formula: $C_2H_5C_3N_2H_2CH_3$
Properties: Amber liq.; m.w. 110.16; dens. 0.975; m.p. 45 C; b.p. 154 C (10 mm); flash pt. 137 C
Uses: Curing epoxy resin systems
Manuf./Distrib.: Charkit; CVC Spec.; Poly Organix; Schweizerhall
Trade names: Curezol® 2E4MZ; Imicure® EMI-24

Ethyl methyl ketone. *See* Methyl ethyl ketone
Ethyl methyl ketone peroxide. *See* Methyl ethyl ketone peroxide

Ethyl morpholine
CAS 100-74-3; EINECS 202-885-0
Synonyms: 4-Ethylmorpholine
Empirical: $C_6H_{13}NO$
Formula: $C_2H_5N(CH_2CH_2)_2O$
Properties: Colorless liq.; m.w. 115.20; dens. 0.916 (20/20 C); b.p. 138 C; flash pt. (OC) 89.6 F; fire hazard
Toxicology: TLV 5 ppm (skin); poison by intravenous route; moderately toxic by ingestion; mildly toxic by inhalation; skin and severe eye irritant
Uses: Catalyst for flexible and rigid PU foam, polyester slabstock
Manuf./Distrib.: BASF
Trade names: Dabco® NEM; Jeffcat NEM; Toyocat®-NEM

4-Ethylmorpholine. *See* Ethyl morpholine
Ethyl 3-oxobutanoate. *See* Ethylacetoacetate

Ethyl phenyl ethanolamine
CAS 92-50-2
Formula: $C_6H_5N(C_2H_5)CH_2CH_2OH$
Properties: M.w. 165.24; m.p. 36-38 C; b.p. 268 C
Toxicology: Irritant
Uses: Coupling agent for dyes for syn. fibers; cure promoter for polyester resins; photosensitive chemical for paper coatings; intermediate for oil-sol. dyestuffs
Manuf./Distrib.: Aldrich
Trade names: Emery® 5707

Ethyl phthalate. *See* Diethyl phthalate
Ethyl silicate. *See* Tetraethoxysilane
Ethyl tellurac. *See* Tellurium diethyl dithiocarbamate

Ethyl toluenesulfonamide
CAS 80-39-7; 1077-56-1; EINECS 214-073-3; 201-275-1
Synonyms: n-Ethyl-p-toluenesulfonamide; Ethyl tosylamide; N-Tosyl ethylamine; N-Ethyl-p-methylbenzene sulfonamide
Classification: Mixture of isomers of aromatic amides
Empirical: $C_9H_{13}NO_2S$
Formula: $C_2H_5NHSO_2C_6H_4CH_3$
Properties: Colorless cryst.; sol. in alc.; m.w. 199.27; m.p. 64 C; flash pt. 126 C
Precaution: Combustible
Toxicology: Irritant
Uses: Plasticizer for resins, cellulosics, PS, PVAc, PVB, coatings, adhesives, inks, electroplating sol'ns., thermoplastics and thermosets, nitrocellulose lacquers, shellac
Regulatory: FDA 21CFR §175.105, 175.300
Manuf./Distrib.: Akzo Nobel; BASF; Charkit; ICI Spec.; Monsanto; Rit-Chem; Seal Sands Chem. Ltd; Spectrum

Chem. Mfg.; Unitex
Trade names: Ketjenflex® 8; Rit-Cizer #8; Uniplex 108

n-Ethyl-p-toluenesulfonamide. *See* Ethyl toluenesulfonamide
2-(N-Ethyl-m-toluidino) ethanol. *See* N-Ethyl-N-hydroxyethyl-m-toluidine
Ethyl tosylamide. *See* Ethyl toluenesulfonamide

Ethyltriacetoxysilane
CAS 17689-77-9
Properties: Liq.; m.w. 234.3; dens. 1.14; b.p. 107 C (8 mm); ref. index 1.412; flash pt. 8 C
Uses: Coupling agent, release agent, lubricant, blocking agent, chemical intermediate
Trade names: Dynasylan® ETAC

Etocrylene
CAS 5232-99-5; EINECS 226-029-0
Synonyms: Ethyl 2-cyano-3,3-diphenylacrylate; Ethyl 2-cyano-3,3-diphenyl-2-propenoate; UV Absorber-2
Classification: Organic ester
Empirical: $C_{18}H_{15}NO_2$
Properties: Powd.; m.p. 96 C
Uses: UV absorber/stabilizer for NC lacquers, rigid PVC, VDC, urea-formaldehyde, epoxyamine, and in cosmetics
Trade names: Uvinul® 3035

EtOH. *See* Alcohol
ETU. *See* Ethylene thiourea
EVA. *See* Ethylene/VA copolymer
EVA copolymer. *See* Ethylene/VA copolymer
Fatty acids, coconut oil, sulfoethyl esters, sodium salts. *See* Sodium cocoyl isethionate
Fatty acids, C18, unsaturated, trimers. *See* Trilinoleic acid
Fatty acids, montan wax. *See* Montan acid wax
Fatty acids, soya. *See* Soy acid
Fatty acids, tall oil. *See* Tall oil acid
Fatty acids, tall oil, sodium salts. *See* Sodium tallate
Fatty acids, tallow. *See* Tallow acid
Fatty acids, tallow, methyl esters. *See* Methyl tallowate
Fenulon. *See* 1-Phenyl-3,3-dimethyl urea

Ferric chloride
CAS 7705-08-0 (anhyd.), 10025-77-1 (hexahydrate); EINECS 231-729-4
Synonyms: Iron chlorides; Iron trichloride; Iron (III) chloride; Ferric trichloride
Empirical: $FeCl_3$ (anhyd.), $FeCl_3 \cdot 6H_2O$ (hexahydrate)
Properties: Anhyd.: Black-brown solid; sol. in water, alcohol, glycerol, methanol, ether, acetone; readily absorbs water in air to form hexahydrate; m.w. 162.21; dens. 2.898; m.p. 292 C; b.p. 319 C; Hexahydrate: brnsh.-yel. cryst.; m.p. 37 C; dens. 1.82
Precaution: Corrosive; catalyzes potentially explosive polymerization of ethylene oxide, chlorine + monomers; violent reaction with allyl chloride; keep containers well closed
Toxicology: TLV:TWA 1 mg (Fe)/m^3; LD50 (oral, rat) 1872 mg/kg; poison by IV; mod. toxic by ingestion; strong irritant to skin and tissue; heated to decomp., emits highly toxic fumes of HCl
Uses: Treatment of sewage and industrial wastes; etching agent, mordant, disinfectant, pigment, feed additive
Regulatory: FDA 21CFR §184.1297, GRAS; Japan approved
Manuf./Distrib.: Am. Biorganics; Asahi Denka Kogyo; BASF; Coyne; Eaglebrook; Eka Nobel AB; Gulbrandsen; Mallinckrodt; Penta Mfg.; PVS; Rasa Ind.; Spectrum Chem. Mfg.; U.S. Petrochem.
Trade names containing: Luperfoam 329

Ferric oxide
CAS 1309-37-1 (anhyd.); EINECS 215-168-2
Synonyms: Ferric oxide red; Iron (III) oxide; Red iron trioxide; Red iron oxide; Ferrosoferric oxide
Empirical: Fe_2O_3
Properties: Red-brn. to blk. cryst.; sol. in acids; insol. in water, alcohol, ether; m.w. 159.69; dens. 5.240; m.p. 1538 C (dec.)
Precaution: Catalyzes the potentially explosive polymerization of ethylene oxide
Toxicology: TLV:TWA 5 mg (Fe)/m^3 (vapor, dust); LD50 (IP, rat) 5500 mg/kg; poison by subcutaneous route; suspected human carcinogen; experimental tumorigen

Uses: Metallurgy, gas purification, paint and rubber pigment, in thermite, polishing compounds, mordant, laboratory reagent, catalyst, feed additive, electronic pigments for TV, permanent magnets, memory cores for computers, magnetic tapes
Regulatory: FDA 21CFR §73.200 (limitation 0.25%), 186.1300, 186.1374, GRAS as indirect food additive
Manuf./Distrib.: Aldrich; BASF; Kerr-McGee; Miles; Spectrum Chem. Mfg.
Trade names: OSO® 440; OSO® 1905; OSO® NR 830 M; OSO® NR 950
Trade names containing: Prespersion SIC-3342; Rhenogran® Fe-rot-70

Ferric oxide red. *See* Ferric oxide
Ferric trichloride. *See* Ferric chloride

Ferro-aluminum silicate
CAS 12178-41-5
Uses: Filler for abrasion-resistant plastic systems; extender pigment for primers and other coatings
Trade names: Ferrosil™ 14

Ferrosoferric oxide. *See* Ferric oxide
Fischer-Tropsch wax. *See* Synthetic wax
Fischer-Tropsch wax, oxidized. *See* Synthetic wax
Flowers of zinc. *See* Zinc oxide

Formaldehyde, ammonia, ethyl chloride condensate
Uses: Gelling agent and stabilizer for latex foams
Trade names: Vulcastab® EFA

Fossil wax. *See* Ozokerite

FPM
Classification: Elastomer
Trade names containing: Rhenogran® CaO-50/FPM; Rhenogran® $Ca(OH)_2$-50/FPM; Rhenogran® CHM21-40/FPM; Rhenogran® MgO-40/FPM; Rhenogran® PbO-80/FPM

French chalk. *See* Talc
β-D-Fructofuranoysl-α-D-glucopyranoside benzoate. *See* Sucrose benzoate
Fumaric acid, dibutyl ester. *See* Dibutyl fumarate
Fumed silica. *See* Silica
2-Furaldehyde. *See* Furfural
2-Furancarboxaldehyde. *See* Furfural
2,5-Furandione. *See* Maleic anhydride
2,5-Furandione, polymer with ethene. *See* Ethylene-maleic anhydride copolymer
2,5-Furandione, polymer with ethenylbenzene. *See* Styrene/MA copolymer
2,5-Furandione, polymer with methoxyethylene. *See* PVM/MA copolymer
2,5-Furandione, polymer with 2-methyl-1-propene. *See* Isobutylene/MA copolymer
2-Furanmethanol. *See* Furfuryl alcohol

Furan polymer
Synonyms: Furan resin
Uses: Chemical intermediate in the mfg. of herbicides, pharmaceuticals, plastics, and fine chemicals; coating asphaltic pavements, foundry sand cores
Manuf./Distrib.: QO; Spectrum Chem. Mfg.
Trade names: QO® Furan

Furan resin. *See* Furan polymer

Furfural
CAS 98-01-1; EINECS 202-627-7
Synonyms: 2-Furaldehyde; 2-Furancarboxaldehyde; Artificial oil of ants
Classification: Cyclic aldehyde
Empirical: $C_5H_4O_2$
Formula: C_4H_3OCHO
Properties: Colorless liq. (pure); reddish-brown (on exposure to air and light); almond-like odor; sol. in alcohol, ether, benzene, 8.3% in water; m.w. 96.08; dens. 1.1598 (20/4 C); f.p. -36.5 C; b.p. 161.7 C; ref. index 1.5260 (20 C); flash pt. 60 C
Precaution: Flamm. or combustible

Toxicology: LD50 (oral, rat) 65 mg/kg; poison by ingestion, intraperitoneal, subcutaneous routes; moderately toxic by inhalation; TLV 2 ppm in air; irritates mucous membranes, acts on CNS
Uses: Chemical intermediate for mfg. of derivs. (furan, THF); solvent for petrol. lube, nitrocellulose; wetting agent; in mfg. of furfural-phenol plastics; vulcanization accelerator; insecticide, fungicide, germicide; reagent in analytical chemistry; flavoring agent
Regulatory: FDA 21CFR §175.105; GRAS (FEMA); Japan approved as flavoring
Manuf./Distrib.: Aldrich; Allchem Ind.; Penta Mfg.; QO; Spectrum Chem. Mfg.
Trade names: QO® Furfural

Furfuralcohol. *See* Furfuryl alcohol

Furfuryl alcohol
CAS 98-00-0; EINECS 202-626-1
Synonyms: 2-Furanmethanol; 2-Furylcarbinol; Furfuralcohol
Empirical: $C_5H_6O_2$
Formula: $C_4H_3OCH_2OH$
Properties: Colorless mobile liq., brn.-dk. red (air/lt. exposed), low odor, cooked sugar taste; sol. in alc., chloroform, benzene; misc. with water but unstable; m.w. 98.10; dens. 1.1285 (20/4 C); m.p. -29 C; b.p. 170 C; flash pt. (OC) 75 C; ref. index 1.485
Toxicology: TLV 10 ppm in air; TLV:TWA 2 ppm (skin); LD50 (rat, oral) 275 mg/kg; poisonous
Uses: Wetting agent, furan polymers, foundry sand binders, corrosion-resist. resins; intermediate for esterification and etherification; plasticizer for phenolic resins; solvent for dyes and resins; flavoring; visc. reducer, cure promoter, and carrier in amine-cured epoxy resins
Regulatory: GRAS (FEMA)
Manuf./Distrib.: AC Ind.; Aldrich; Allchem Ind.; Penta Mfg.; QO; Spectrum Chem. Mfg.
Trade names: QO® Furfuryl Alcohol (FA®)

Furnace black. *See* Carbon black
2-Furylcarbinol. *See* Furfuryl alcohol
GAE. *See* Glyceryl-1-allyl ether
Gallic acid propyl ester. *See* Propyl gallate
D-Glucitol, bis-o-(phenylmethylene)-. *See* Dibenzylidene sorbitol
Glycerides, cottonseed oil, hydrogeanted. *See* Hydrogenated cottonseed glyceride
Glycerides, lard mono-. *See* Lard glyceride
Glycerides, lard mono-, acetates. *See* Acetylated lard glyceride
Glycerides, montan-wax. *See* Glyceryl montanate
Glycerides, palm kernel mono-, di- and tri-. *See* Palm kernel glycerides
Glycerides, palm kernel oil mono-, di-, and tri-, acetates. *See* Acetylated palm kernel glycerides
Glycerides, palm oil mono-, hydrogenated. *See* Hydrogenated palm glyceride
Glycerides, soybean oil, hydrogenated, mono. *See* Hydrogenated soy glyceride
Glycerin-1-allyl ether. *See* Glyceryl-1-allyl ether
Glycerin monoacetate. *See* Glyceryl acetate
Glycerol monoricinoleate. *See* Glyceryl ricinoleate
Glycerol ricinoleate. *See* Glyceryl ricinoleate
Glycerol tripropionate. *See* Glyceryl tripropionate

Glyceryl acetate
CAS 106-61-6; 26446-35-5
Synonyms: Acetin; Glycerin monoacetate; Monacetin; 1,2,3-Propanetriol monoacetate
Empirical: $C_5H_{10}O_4$
Formula: $C_3H_7O_2{\bullet}OOCCH_3$
Properties: Colorless or pale yel. (commercial) liq., very hygroscopic, char. odor; sol. in water, alcohol; sl. sol. in ether; insol. in benzene; m.w. 134.13; dens. 1.206 (20/4 C); b.p. 158 C (17 mm); ref. index 1.451
Uses: In mfg. of smokeless powd., dynamite; solvent for basic dyes; in tanning leather; plasticizer for cellulose acetate, cellulose nitrate
Manuf./Distrib.: C.P. Hall; Penta Mfg.

Glyceryl-1-allyl ether
CAS 123-34-2; EINECS 204-620-4
Synonyms: GAE; 1,2-Propanediol, 3-(2-propenyloxy)-; Glycerin-1-allyl ether
Formula: $CH_2{=}CHCH_2OCH_2CHOHCH_2OH$
Properties: Colorless to ylsh. visc. liq.; sol. in water, ethanol, and many org. solvs.; m.w. 132.16; dens. 1.07 g/cc (20 C); m.p. < -90 C; b.p. 245 C

Precaution: Incompat. with acids, alkalies, oxidizing agents, peroxides; hazardous decomp. prods.: CO, gaseous hydrocarbons
Toxicology: LD50 (oral, rat) 5400 mg/kg, (dermal, rat) 4300 mg/kg; irritating to eyes, respiratory system, skin
Uses: Can be polymerized into polyesters, PU, polyacetals, epoxy resins, acrylates, methacrylates, or styrene; for prod. of printed circuit boards, prod. and coating of PU rubber and foams, unsat. polyesters for radiation-resist. coatings

Glyceryl diacetate
CAS 25395-31-7
Synonyms: Diacetin
Properties: M.w. 176.17; sp.gr. 1.181; m.p. -35 C; flash pt. (COC) 157 C; ref. index 1.44
Uses: Auxiliary in foundries; carrier for fragrances and flavors; plasticizer for PVB, cellulosics
Manuf./Distrib.: Bayer; C.P. Hall

Glyceryl dilaurate
CAS 27638-00-2; EINECS 248-586-9
Synonyms: Dilaurin; Dodecanoic acid, diester with 1,2,3-propanetriol
Definition: Diester of glycerin and lauric acid
Empirical: $C_{27}H_{52}O_5$
Formula: $C_{11}H_{23}COOCH_2CHCH_2OHOCOC_{11}H_{23}$
Properties: Wh. to off-wh. solid; sol. in min. oil, 95% ethanol, IPM, oleyl alcohol, castor oil; insol. in water, glycerin, propylene glycol; sapon. no. 219-229; ref. index 1.4520-1.4560 (35 C)
Toxicology: LD50 (rat, oral) > 5 g/kg; nonirritating to eyes and skin
Uses: Emollient lipid; coupling agent; plasticizer; emulsifier, dispersant, antistat for textile, paper processing, cutting oils, polishes, emulsion cleaners, rubber latex, wool lubricants, cosmetics, pharmaceuticals
Regulatory: FDA 21CFR §175.105, 1876.210
Trade names: Cithrol GDL N/E

Glyceryl dilaurate SE
EINECS 248-586-9
Synonyms: Dodecanoic acid, diester with 1,2,3-propanetriol
Uses: W/o emulsifier, dispersant, antistat for textile, paper processing, cutting oils, polishes, emulsion cleaners, rubber latex, wool lubricants, cosmetics, pharmaceuticals
Trade names: Cithrol GDL S/E

Glyceryl dioleate
CAS 25637-84-7; EINECS 247-144-2
Definition: Diester of glycerin and oleic acid
Synonyms: 9-Octadecenoic acid, diester with 1,2,3-propanetriol
Empirical: $C_{39}H_{72}O_5$
Uses: Surfactant for cosmetics, pharmaceuticals, industrial applics.; emulsifier, thickener, stabilizer, emollient; lubricant for PVC processing; antifog for PVC film
Regulatory: FDA 21CFR §175.105, 176.210, 177.2800
Trade names: Priolube 1409

Glyceryl hydrogenated rosinate
Synonyms: Rosin, hydrogenated, glycerol ester
Definition: Monoester of glycerin and hydrogenated mixed long chain acids derived from rosin
Uses: Thermoplastic syn. resin; as softener/plasticizer for the masticatory agent in chewing gum bases; as resin modifier for film-formers, elastomers, and waxes in adhesive and protective coating compositions; as tackifier in adhesives, coatings
Trade names: Staybelite® Ester 5; Staybelite® Ester 10

Glyceryl hydroxystearate
CAS 1323-42-8; EINECS 215-355-9
Synonyms: Glyceryl 12-hydroxystearate; Hydroxystearic acid, monoester with glycerol
Definition: Monoester of glycerin and hydroxystearic acid
Empirical: $C_{21}H_{42}O_5$
Uses: Emulsifier, emollient, opacifier, bodying and thickening agent for cosmetics, pharmaceuticals, household prods.; beeswax substitute; pigment wetting in paints, inks; paper industry; lubricant and mold release in plastics; textile and leather processing; detergency and cleaning prods.
Regulatory: FDA 21CFR §175.105, 176.170, 176.200, 177.1210, 177.2800

Glyceryl 12-hydroxystearate. *See* Glyceryl hydroxystearate

Glyceryl laurate
CAS 142-18-7; EINECS 205-526-6
Synonyms: Glyceryl monolaurate; Dodecanoic acid, monoester with 1,2,3-propanetriol; Dodecanoic acid, 2,3-dihydroxypropyl ester
Definition: Monoester of glycerin and lauric acid
Empirical: $C_{15}H_{30}O_4$
Formula: $CH_3(CH_2)_{10}COOCH_2COHHCH_2OH$
Properties: Cream-colored paste, faint odor; disp. in water; sol. in methanol, ethanol, toluene, naphtha, min. oil; dens. 0.98; m.p. 23-27 C; pH 8-8.6
Precaution: Combustible
Uses: Emulsifier, dispersant for food prods., oils, waxes, solvents; antifoaming agent; drycleaning soap base; plasticizer for PVC, PS, cellulosics
Regulatory: FDA 21CFR §175.105, 176.210, 177.2800, GRAS; Japan approved; Europe listed
Manuf./Distrib.: Grindsted; Henkel/Emery; Inolex; Lonza; Protameen; Velsicol

Glyceryl linoleate
CAS 2277-28-3; EINECS 218-901-4
Synonyms: Monolinolein; 9,12-Octadecadienoic acid, 2,3-dihydroxypropyl ester; 9,12-Octadecadienoic acid, monoester with 1,2,3-propanetriol
Definition: Monoester of glycerin and linoleic acid
Empirical: $C_{21}H_{38}O_4$
Uses: Component in w/o and o/w creams, lubricant, antistat, antifogging agent in plastics
Trade names containing: Dimodan LS Kosher

Glyceryl mono/dioleate
CAS 25496-72-4
Properties: Yel. oil or soft solid; dens. 0.95; m.p. 14-19 C
Precaution: Combustible
Uses: Emulsifier and antifoam for foods; emulsifier, solubilizer for cosmetic, pharmaceutical, and industrial applics.; lubricant, softener for syn. fibers; rust preventive additive in oils; lubricant, antistat, antifog for PVC film processing
Trade names: Alkamuls® GMR-55LG

Glyceryl monolaurate. *See* Glyceryl laurate
Glyceryl monomyristate. *See* Glyceryl myristate
Glyceryl monooleate. *See* Glyceryl oleate
Glyceryl monoricinoleate. *See* Glyceryl ricinoleate
Glyceryl monorosinate. *See* Glyceryl rosinate
Glyceryl monostearate. *See* Glyceryl stearate
Glyceryl monostearate SE. *See* Glyceryl stearate SE
Glyceryl monotristearate. *See* Tristearin

Glyceryl montanate
CAS 68476-38-0
Synonyms: 2,3-Dihydroxypropyl octacosanoic acid; Glycerides, montan-wax; Montan-wax fatty acids, glyceryl esters
Definition: Monoester of glycerin and montan acid wax
Empirical: $C_{31}H_{61}O_4$
Uses: Lubricant and release agent for PVC, polyolefins, polyamide, PS, linear polyesters, TPU, thermosets; carrier for pigment concs.
Trade names: Hostalub® WE4

Glyceryl myristate
CAS 589-68-4; 67701-33-1
Synonyms: Glyceryl monomyristate; Monomyristin; Tetradecanoic acid, monoester with 1,2,3-propanetriol
Definition: Monoester of glycerin and myristic acid
Empirical: $C_{17}H_{34}O_4$
Formula: $CH_3(CH_2)_{12}COOCH_2COHHCH_2OH$
Uses: Component in w/o and o/w creams, lubricant, antistat, antifogging agent in plastics; coemulsifier, solubilizer, carrier for lipophilic drugs
Regulatory: FDA 21CFR §175.105, 176.210, 177.2800

Glyceryl oleate

CAS 111-03-5; 25496-72-4, 37220-82-9; EINECS 203-827-7; 253-407-2
Synonyms: Glyceryl monooleate; Monoolein; 9-Octadecenoic acid, monoester with 1,2,3-propanetriol
Definition: Monoester of glycerin and oleic acid
Empirical: $C_{21}H_{40}O_4$
Formula: $CH_3(CH_2)_7CH=CH(CH_2)_7COOCH_2CCH_2OHHOH$
Properties: Sp.gr. 0.950
Toxicology: Heated to decomp., emits acrid smoke and irritating fumes
Uses: Emulsifier, coemulsifier, stabilizer, wetting agent, lubricant, and antistat; used in cosmetic, pharmaceutical, industrial, food applics.; plasticizer; internal antistat for PE, PP, PVC; antifog, cling agent, lubricant for PVC film; flavor adjuvant, solv., vehicle, defoamer, dispersant, emulsifier, plasticizer for food use (coffee whiteners, baking mixes, beverages, chewing gum, meat prods., pkg. materials, veg. oil)
Regulatory: FDA 21CFR §175.105, 175.300, 176.210, 177.2800, 181.22, 181.27, 182.4505, 184.1323, GRAS
Manuf./Distrib.: Aquatec Quimica SA; Calgene; Croda Surf.; Ferro/Keil; Grindsted; Henkel/Emery; ICI Surf. Am.; Inolex; Karlshamns; Lonza; Mona; Patco; Stepan; Unichema N. Am.; Witco/Oleo-Surf.
Trade names: Armostat® 810; Atmer® 121; Atmer® 1007; Atmer® 1010; CPH-31-N; Estol 1407; Larostat® GMOK; Pationic® 907; Pationic® 1061; Pationic® 1064; Pationic® 1074; Priolube 1407; Priolube 1408
Trade names containing: Atmos® 300

Glyceryl oleate SE

CAS 111-03-5; 25496-72-4
Definition: Self-emulsifying grade of glyceryl oleate that contains some sodium and/or potassium oleate
Uses: Emulsifier, coemulsifier, stabilizer, wetting agent, lubricant, and antistat; used in cosmetic, pharmaceutical, industrial, food applics.

Glyceryl ricinoleate

CAS 141-08-2; EINECS 205-455-0
Synonyms: 12-Hydroxy-9-octadecenoic acid, monoester with 1,2,3-propanetriol; Monoricinolein; Glycerol ricinoleate; Glycerol monoricinoleate; Glyceryl monoricinoleate
Definition: Monoester of glycerin and ricinoleic acid
Formula: $C_3H_5(OOCC_{16}H_{32}OH)_3$
Properties: Sp.gr. 0.981; m.p. < -50 C; flash pt. (COC) 265 C; ref. index 1.4770
Uses: Emulsifying agent; plasticizer for PVB, cellulosics
Regulatory: FDA 21CFR §175.105, 176.170, 176.210, 178.3130
Manuf./Distrib.: CasChem; Lonza
Trade names: Flexricin® 13

Glyceryl rosinate

Synonyms: Glyceryl monorosinate; Rosin, glyceryl ester
Definition: Monoester of glycerin and mixed long chain acids derived from rosin
Properties: drop soften. pt. 88-96 C
Toxicology: Heated to decomp., emits acrid smoke and irritating fumes
Uses: Thermoplastic resin gum; clouding agent in beverages; in chewing gums, adhesives, inks, coatings; tackifier, softener, plasticizer, modifier for adhesives, coatings; process aid for rubber
Regulatory: FDA 21CFR §172.615, 172.735, limitation 100 ppm in finished beverage, 175.105, 175.300, 178.3120, 178.3800, 178.3870
Trade names: Aquatac® 6085; Hercules® Ester Gum 10D; Oulutac 80 D/HS; Poly-Pale® Ester 10; Resinall 605; Uni-Tac® R85; Uni-Tac® R85-Light

Glyceryl stearate

CAS 123-94-4; 11099-07-3; 31566-31-1; 61789-08-0; 85666-92-8; 85251-77-0; EINECS 250-705-4; 234-325-6; 204-664-4; 286-490-9
Synonyms: Monostearin; 1,2,3-Propanetriol octadecanoate; Glyceryl monostearate; 2,3-Dihydroxypropyl octadecanoate
Definition: Monoester of glycerin and stearic acid
Empirical: $C_{21}H_{42}O_4$
Formula: $CH_3(CH_2)_{16}COOCH_2COHHCH_2OH$
Properties: Wh. to cream flakes; insol. in water, ethanol, glycerin, propylene glycol; disp. in min. oil; m.p. 56-59 C; sapon. no. 162-175
Precaution: Combustible
Toxicology: LD50 (IP, mouse) 200 mg/kg; poison by IP route; heated to decomp., emits acrid smoke and irritating fumes

Glyceryl stearate SE

Uses: Nonionic sec. o/w emulsifier for creams and lotions; visc. booster for emulsions; plasticizer for cellulose nitrate; antistat, antifog, lubricant, processing aid for plastics
Regulatory: FDA 21CFR §139.110, 139.115, 139.117, 139.120, 139.121, 139.122, 139.125, 139.135, 139.138, 139.140, 139.150, 139.155, 139.160, 139.165, 139.180, 175.105, 175.210, 175.300, 176.200, 176.210, 177.2800, 184.1324, GRAS; Europe listed
Manuf./Distrib.: Aquatec Quimica SA; Croda Surf.; Eastman; Goldschmidt; Grindsted; Hart Prod.; Henkel/Emery; ICI Surf.; Inolex; ISP Van Dyk; Karlshamns; Lanaetex; Lipo; Lonza; Patco; Protameen; Stepan; Witco/Oleo-Surf.
Trade names: Alkamuls® GMS/C; Armostat® 801; Atlas 1500; Atmer® 122; Atmer® 125; Atmer® 129; Atmos® 150; Atmul® 84; Atmul® 124; Chemstat® G-118/52; Chemstat® G-118/95; Eastman® PA-208; Empilan® GMS NSE40; Estol 1467; Estol 1474; Grindtek MSP 40; Grindtek MSP 40F; Grindtek MSP 90; Kemester® 6000; Kemester® GMS (Powd.); Lankroplast® L553; Larostat® GMSK; Mazol® GMS-90; Myvaplex® 600P; Myvaplex® 600PK; Pationic® 900; Pationic® 901; Pationic® 902; Pationic® 905; Pationic® 909; Pationic® 1042; Pationic® 1042B; Pationic® 1042K; Pationic® 1042KB; Pationic® 1052; Pationic® 1052B; Pationic® 1052K; Pationic® 1052KB; Petrac® GMS; Polyvel AI509; Secoster® SDG; Softenol® 3900; Softenol® 3925, 3945, 3960; Softenol® 3991; Witconol™ MST
Trade names containing: Ice # 2

Glyceryl stearate SE
CAS 31566-31-1; 11099-07-3; 85666-92-8
Synonyms: GMS-SE; Glyceryl monostearate SE
Definition: Self-emulsifying grade of glyceryl stearate containg some sodium and/or potassium stearate
Properties: Wh. to cream flakes; sol. in oleyl alcohol; partly sol. in water, veg. oil, ethanol, propylene glycol; m.p. 57-59 C; sapon. no. 150-160
Uses: Aux. emulsifier for soap o/w emulsions; emulsifier for polymers; lubricant, antistat, antifog for plastics prod.
Trade names: Empilan® GMS LSE40; Empilan® GMS LSE80; Empilan® GMS MSE40; Empilan® GMS NSE90; Empilan® GMS SE70; Estol 1461; Mazol® GMS-D

Glyceryl triacetate. *See* Triacetin

Glyceryl (triacetoxystearate)
CAS 139-43-5; EINECS 295-625-0
Synonyms: Castor oil, acetylated and hydrogenated; Glyceryl tri(12-acetoxystearate)
Formula: $C_3H_5(OOCC_{17}H_{34}OCOCH_3)_3$
Properties: Clear pale yel. oily liq., mild odor; sol. in most org. solvs.; insol. in water; dens. 0.967 (25/25 C)
Precaution: Combustible
Toxicology: Heated to decomp, emits acrid smoke and irritating fumes
Uses: Plasticizer for nitrocellulose, ethylcellulose, and PVC, food pkg.; lubricants; protective coatings
Regulatory: FDA 21CFR §178.3505, limitation with $CaCO_3$ 1% of total mixt.
Manuf./Distrib.: CasChem
Trade names: Paricin® 8

Glyceryl tri(12-acetoxystearate). *See* Glyceryl (triacetoxystearate)

Glyceryl triacetyl ricinoleate
CAS 101-34-8; EINECS 202-935-1
Synonyms: 9-Octadecenoic acid, 12-(acetyloxy)-, 1,2,3-propanetriol ester; 1,2,3-Propanetriyl 12-(acetyloxy)-9-octadecenoate
Definition: Triester of glycerin and acetyl ricinoleic acid
Formula: $C_3H_5(OOCC_{16}H_{32}OCOCH_3)_3$
Properties: Clear pale yel. oily liq., mild odor; sol. in most org. liqs.; insol. in water; dens. 0.967 (25/25 C)
Uses: Plasticizer for nitrocellulose, ethylcellulose, and PVC; lubricants; protective coatings
Manuf./Distrib.: CasChem
Trade names: Flexricin® P-8

Glyceryl tribenzoate
CAS 614-33-5
Synonyms: Tribenzoin
Empirical: $C_{24}H_{20}O_6$
Properties: Colorless liq.; insol. in water; sol. in alcohol, ether; m.w. 404.44; dens. 1.032; m.p. < -75 C; b.p. 305-309 C

Toxicology: Mildly toxic by ingestion
Uses: Plasticizer, process aid, modifier for thermoplastics, hot-melt adhesives
Manuf./Distrib.: Unitex; Velsicol
Trade names: Benzoflex® S-404; Uniplex 260

Glyceryl tricaprate/caprylate
Uses: Emollient, plasticizer, solubilizer, food additive; for cosmetics, foods, pharmaceuticals, plastics
Trade names: Estol GTCC 1527

Glyceryl trioleate. *See* Triolein

Glyceryl tripropionate
CAS 139-45-7
Synonyms: Glycerol tripropionate; Tripropionin
Empirical: $C_{12}H_{20}O_6$
Formula: $C_3H_5(OCOC_2H_5)_3$
Properties: Solid; sol. in alcohol, 0.313% in water; dens. 1.078 (20 C); b.p. 177-182 C (20 mm); f.p. < -50 C
Uses: Plasticizer
Trade names: Kodaflex® Tripropionin

Glyceryl tristearate. *See* Tristearin

(3-Glycidoxypropyl)-methyldiethoxy silane
CAS 2897-60-1
Uses: Coupling agent, chemical intermediate, blocking agent, release agent, lubricant, primer, reducing agent

3-Glycidoxypropyltrimethoxysilane
CAS 2530-83-8; EINECS 219-784-2
Synonyms: γ-Glycidoxypropyltrimethoxysilane
Formula: $OCH_2CHCH_2O(CH_2)_3Si(OCH_3)_3$
Properties: Liq.; sol. in acetone, benzene, ether; reacts with water; m.w. 236.34; dens. 1.070 (25 C); b.p. ≈ 120 C (2 mm); f.p. > 110 C; ref. index 1.4280 (25 C); flash pt. 79 C
Uses: Coupling agent, filler for glass- and mineral-filled plastics, epoxies, urethanes, acrylics, PBT, polysulfides
Trade names: Dow Corning® Z-6040; Dynasylan® GLYMO; Petrarch® G6720; Prosil® 5136
Trade names containing: Aktisil® EM

γ-Glycidoxypropyltrimethoxysilane. *See* 3-Glycidoxypropyltrimethoxysilane
Glycidyl butyl ether. *See* Butyl glycidyl ether

Glycidyl methacrylate
CAS 106-91-2; EINECS 203-441-9
Synonyms: 2,3-Epoxypropyl methacrylate; Methacrylic acid 2,3-epoxypropyl ester
Formula: $C_7H_{10}O_3$
Properties: M.w. 142.15; dens. 1.042; b.p. 189 C; flash pt. 76 C
Uses: Polyfunctional monomer; in hydrogels for contact lenses and membranes, molding and casting compds., impregnating paper, concrete, wood, coatings, printing inks, adhesives, sealants, elastomers
Manuf./Distrib.: Estron; Mitsubishi Gas; Monomer-Polymer & Dajac; Polysciences; Richman; Sartomer; Spectrum Chem. Mfg.
Trade names: SR-379

Glycidyl 2-methylphenyl ether. *See* o-Cresyl glycidyl ether
Glycinol. *See* Ethanolamine

Glycol
CAS 107-21-1; EINECS 203-473-3
Synonyms: Ethylene glycol; 1,2-Ethanediol; Ethylene alcohol
Classification: Aliphatic diol
Empirical: $C_2H_6O_2$
Formula: $HOCH_2CH_2OH$
Properties: Liq., sweet taste (poisonous); very hygroscopic; misc. with water, lower aliphatic alcohols, glycerol;

m.w. 62.07; dens. 1.1135 (20/4 C); visc. 17.3 cps (25 C); m.p. -13 C; b.p. 197.6 C (760 mm); flash pt. (OC) 115 C; ref. index 1.43063 (25 C)

Toxicology: Toxic by ingestion; lethal human dose 1.4 ml/kg; TLV 50 ppm (vapor ceiling)

Uses: Antifreeze in cooling and heating systems; in hydraulic brake fluids; industrial humectant; solvent in paints, plastics, inks; softening agent for cellophane; stabilizer; in explosives, alkyd resins, elastomers, syn. fibers and waxes; asphalt; raw material for prod. of syn. polyester fibers, latex paints, polyester and alkyd resins; plasticizer; humectant

Regulatory: FDA 21CFR §175.105, 176.300

Manuf./Distrib.: Ashland; BASF; Eastman; Hoechst Celanese; Mitsui Petrochem. Ind.; Mitsui Toatsu; Mobil; Olin; OxyChem; Shell; Texaco; Union Carbide

Trade names containing: Dabco® EG; Dioxitol-High Gravity; Polycat® 46; Surfynol® 104E; TEDA-L33E

Glycol/butylene glycol montanate

CAS 73138-45-1

Definition: Mixt. of diesters of montan acid wax with ethylene glycol and butylene glycol

Uses: Lubricant for plastics compounding; release agent; antiblocking agent; wax for polishes, emulsions, citrus fruit coating

Trade names: Hoechst Wax E; Hoechst Wax KPS

Glycol butyl ether. *See* Butoxyethanol
Glycol carbonate. *See* Ethylene carbonate
Glycol diacetate. *See* Ethylene glycol diacetate

Glycol dilaurate

CAS 624-04-4; EINECS 210-827-0

Synonyms: Ethylene glycol dilaurate; Lauric acid, 1,2-ethanediyl ester; Dodecanoic acid 1,2-ethanediyl ester

Definition: Diester of ethylene glycol and lauric acid

Empirical: $C_{26}H_{50}O_4$

Formula: $CH_3(CH_2)_{10}COOCH_2CH_2OCO(CH_2)_{10}CH_3$

Properties: Colorless amorphous mass; insol. in alcohol, ether; m.w. 426.66; m.p. 50-52 C; b.p. 188 C (20 mm)

Uses: Plasticizer in lacquers and varnishes; emulsifier, dispersant, antistat for textile, paper processing, cutting oils, polishes, emulsion cleaners, rubber latex, wool lubricants

Trade names: Cithrol EGDL N/E

Glycol dilaurate SE

Uses: Emulsifier, dispersant, antistat for textile, paper processing, cutting oils, polishes, emulsion cleaners, rubber latex, wool lubricants

Trade names: Cithrol EGDL S/E

Glycol dimercaptoacetate

CAS 123-81-9

Synonyms: Ethylene glycol bisthioglycolate

Formula: $HSCH_2COOCH_2CH_2OOCCH_2SH$

Properties: M.w. 210.26; b.p. 137-139 C (1-2 mm)

Uses: Crosslinking agent for rubbers, accelerator in curing epoxy resins

Manuf./Distrib.: Evans Chemetics; Janssen Chimica

Glycol dioleate

CAS 928-24-5

Synonyms: Ethylene glycol dioleate

Properties: Sp.gr. 0.950; m.p. 0 C; flash pt. (COC) 296 C; ref. index 1.461

Uses: Emulsifier, dispersant, spreading agent, oil-phase ingred. for creams, milky lotions, and foundations; plasticizer for PVC, some cellulosics

Trade names: Cithrol EGDO N/E

Glycol dioleate SE

Uses: Emulsifier, dispersant, antistat for textile, paper processing, cutting oils, polishes, emulsion cleaners, rubber latex, wool lubricants

Trade names: Cithrol EGDO S/E

Glycol distearate

CAS 627-83-8; EINECS 211-014-3

Synonyms: EGDS; Ethylene glycol distearate; Octadecanoic acid, 1,2-ethanediyl ester

Definition: Diester of ethylene glycol and stearic acid
Formula: $CH_3(CH_2)_{16}COOCH_2CH_2OCO(CH_2)_{16}CH_3$
Properties: Sp.gr. 0.97; m.p. 60 C; flash pt. (COC) 171 C
Uses: Pearlescent and opacifier; thickener, intermediate, lubricant, emulsifier, emollient; for emulsion shampoos and foam baths; plasticizer, lubricant, antistat for plastics and rubber
Regulatory: FDA 21CFR §73.1, 176.210
Manuf./Distrib.: Inolex
Trade names: Cithrol EGDS N/E; Kemester® EGDS; Mackester™ EGDS

Glycol distearate SE
Uses: Emulsifier, dispersant, antistat for textile, paper processing, cutting oils, polishes, emulsion cleaners, rubber latex, wool lubricants
Trade names: Cithrol EGDS S/E

Glycol laurate
CAS 4219-48-1; EINECS 253-458-0
Synonyms: Ethylene glycol monolaurate
Definition: Ester of ethylene glycol and lauric acid
Uses: Emulsifier, dispersant, antistat for textile, paper processing, cutting oils, polishes, emulsion cleaners, rubber latex, wool lubricants
Trade names: Cithrol EGML N/E

Glycol laurate SE
EINECS 237-725-9
Synonyms: Dodecanoic acid, ester with 1,2-ethanediol
Uses: Emulsifier, dispersant, antistat for textile, paper processing, cutting oils, polishes, emulsion cleaners, rubber latex, wool lubricants
Trade names: Cithrol EGML S/E

Glycol monooleate. *See* Glycol oleate
Glycol monoricinoleate. *See* Glycol ricinoleate
Glycol monostearate. *See* Glycol stearate

Glycol oleate
CAS 4500-01-0; EINECS 224-806-9
Synonyms: Ethylene glycol monooleate; Glycol monooleate; 2-Hydroxyethyl 9-octadecenoate
Definition: Ester of ethylene glycol and oleic acid
Formula: $CH_3(CH_2)_7CH{=}CH(CH_2)_7COOCH_2CH_2OH$
Uses: Emulsifier, dispersant, antistat for textile, paper processing, cutting oils, polishes, emulsion cleaners, rubber latex, wool lubricants
Regulatory: FDA 21CFR §176.210
Trade names: Cithrol EGMO N/E
Trade names containing: Aldosperse® MO-50

Glycol oleate SE
Uses: Emulsifier, dispersant, antistat for textile, paper processing, cutting oils, polishes, emulsion cleaners, rubber latex, wool lubricants
Trade names: Cithrol EGMO S/E

Glycol ricinoleate
CAS 106-17-2; EINECS 203-369-8
Synonyms: Ethylene glycol monoricinoleate; Glycol monoricinoleate; 2-Hydroxyethyl 12-hydroxy-9-octadecenoate
Definition: Ester of ethylene glycol and ricinoleic acid
Empirical: $C_{20}H_{38}O_4$
Properties: Clear, moderately visc. pale-yel. liq., mild odor; misc. with most org. solvs.; insol. in water; dens. 0.965 (25/25 C)
Uses: Plasticizer; greases, urethane polymers
Regulatory: FDA 21CFR §176.210
Trade names: Cithrol EGMR N/E

Glycol ricinoleate SE
Uses: Emulsifier, dispersant, antistat for textile, paper processing, cutting oils, polishes, emulsion cleaners,

rubber latex, wool lubricants
Trade names: Cithrol EGMR S/E

Glycol stearate
CAS 111-60-4; 97281-23-7; EINECS 203-886-9; 306-522-8
Synonyms: EGMS; Ethylene glycol monostearate; Glycol monostearate; 2-Hydoxyethyl octadecanoate
Definition: Ester of ethylene glycol and stearic acid
Formula: $CH_3(CH_2)_{16}COOCH_2CH_2OH$
Properties: Yel. waxy solid; sol. in alcohol, hot ether, acetone; insol. in water; m.w. 328.60; dens. 0.96 (25 C); m.p. 57-60 C
Toxicology: Poison by intraperitoneal route; skin irritant
Uses: Opacifier and pearling agent for cream shampoos, other cosmetics; plasticizer for cellulose nitrate; lubricant for plasticized PVC
Regulatory: FDA 21CFR §176.210
Manuf./Distrib.: Ashland; C.P. Hall; Inolex; ISP Van Dyk; Witco/Oleo-Surf.
Trade names: Cithrol EGMS N/E; CPH-37-NA; Mackester™ EGMS; Pegosperse® 50 MS
Trade names containing: Mackester™ IP; Mackester™ SP

Glycol stearate SE
CAS 86418-55-5
Synonyms: Ethylene glycol monostearate SE
Definition: Self-emulsifying grade of glycol stearate containg. some sodium and/or potassium stearate
Properties: Wh. to cream flakes; disp. in water, peanut oil, oleyl alcohol; insol. in min. oil, ethanol, glycerin, propylene glycol, IPM; m.p. 57-60 C; sapon. no. 181-191
Uses: Emulsifier for o/w creams and lotions; visc. builder
Trade names: Cithrol EGMS S/E

GMS-SE. *See* Glyceryl stearate SE

Graphite
CAS 7782-42-5; EINECS 231-955-3
Synonyms: Black lead; Plumbago; Mineral carbon
Empirical: C
Properties: Steel gray to black powd., flake, cryst., rods, plates, or fibers; soft greasy feel, metallic sheen; dens. 2.0-2.25; resistant to oxidation and thermal shock
Precaution: Fire risk
Toxicology: TLV 2.5 mg/m^3 respirable dust
Uses: Reinforcing agent for plastics, carbon brushes, batteries, electrochemistry, pencils, hard metals, lubricants, catalysts, prepregging, filament winding; lubricant additive for greases, engine oils, etc.; paints, coatings, self-lubricating bearings
Manuf./Distrib.: Cerac; Johnson Matthey; Lonza; San Yuan; Sigri GmbH; Ucar Carbon
Trade names: Lonza KS 44; Lonza T 44
Trade names containing: Cycom® NCG 1200 Unsized; Cycom® NCG 1204 p8; Dry-Blend® NCG dB910; Dry-Blend® NCG dB920

Graphite fiber. *See* Carbon fiber
Griffith's zinc white. *See* Lithopone
Gum rosin. *See* Rosin
Gypsum. *See* Calcium sulfate, Calcium sulfate dihydrate
Hard paraffin. *See* Paraffin
HBCD. *See* Hexabromocyclododecane
HBD. *See* Tributyltin oxide
H.E. cellulose. *See* Hydroxyethylcellulose
HCFC 141b. *See* Dichlorofluoroethane
HDPE. *See* Polyethylene, high-density
Heavy mineral oil. *See* Mineral oil
HEMA. See 2-Hydroxyethyl methacrylate
Heptadecanoic acid, 16-methyl-. *See* Isostearic acid

Heptakis (dipropylene glycol) triphosphite
CAS 13474-96-9; EINECS 236-753-9
Properties: Liq.; m.w. 1022; dens. 1.105-1.120; b.p. will not distill; ref. index 1.4660-1.4710; flash pt. (PMCC) 135 C

Uses: Color stabilizer for foamed polyurethanes; flame retardant for flame lamination applics. in polyurethanes
Trade names: Weston® PTP

Heptane
CAS 142-82-5; 64742-89-8; EINECS 205-563-8
Synonyms: n-Heptane; Alkane C_7
Classification: Aliphatic hydrocarbon
Empirical: C_7H_{16}
Formula: $CH_3(CH_2)_5CH_3$
Properties: Volatile colorless liq.; sol. in alcohol, ether, chloroform; insol. in water; m.w. 100.21; dens. 0.68368 (20 C); m.p. -91 C; b.p. 98.428 C; flash pt. (CC) -3.89 C; ref. index 1.38764 (20 C)
Toxicology: TLV 400 ppm in air (ACGIH); STEL 500 ppm; lethal conc. for mice 15,900 ppm in air; irritating to respiratory tract; narcotic in high concs.
Uses: Standard for octane rating determinations; anesthetic; solvent for rubber compounding, cements/ sealants, extraction of oils and fats; organic synthesis
Regulatory: FDA 21CFR §175.105, 177.1580, 27CFR §21.111
Manuf./Distrib.: Ashland; Coyne; Exxon; Great Western; Humphrey; Phibro Energy USA; Phillips; Texaco

n-Heptane. *See* Heptane
1,7-Heptanedicarboxylic acid. *See* Azelaic acid
3,6,9,12,15,18,21-Heptaoxatricosane-1,23-diol. *See* PEG-8
3,6,9,12,15,18,21-Heptaoxatritriacontan-1-ol. *See* Laureth-7

Heptylated diphenylamine
Uses: Antioxidant for rubbers
Trade names: Permanax™ HD; Permanax™ HD (SE)

Heptyl carbinol. *See* Caprylic alcohol

Hexabromocyclododecane
CAS 3194-55-6; 25637-99-4; EINECS 221-695-9
Synonyms: HBCD; 1,2,5,6,9,10-Hexabromocyclododecane
Empirical: $C_{12}H_{18}Br_6$
Properties: M.w. 641.73; m.p. 188-191
Uses: Flame retardant for expandable PS, foamed PS, crystal and high-impact PS, SAN, adhesives, coatings
Manuf./Distrib.: Albemarle; Great Lakes
Trade names: FR-1206; Great Lakes CD-75P™; Great Lakes SP-75™; Saytex® HBCD-LM

1,2,5,6,9,10-Hexabromocyclododecane. *See* Hexabromocyclododecane
Hexabutyldistannoxane. *See* Tributyltin oxide
Hexachloroendomethylenetetrahydrophthalic anhydride. *See* Chlorendic anhydride
1,4,5,6,7,7-Hexachloro-5-norbornene-2,3-dicarboxylic anhydride. *See* Chlorendic anhydride
4,5,6,7,8,8-Hexachloro-3a,4,7,7a-tetrahydro-4,7-methanoisobenzofuran-1,3-dione. *See* Chlorendic anhydride
Hexadecanamide, N-9-octadecenyl-. *See* Oleyl palmitamide
1-Hexadecanamine. *See* Palmitamine
Hexadecanoic acid, 1-methylethyl ester. *See* Isopropyl palmitate
1-Hexadecanol. *See* Cetyl alcohol

Hexadecene-1
CAS 629-73-2; EINECS 211-105-8
Synonyms: Linear C16 alpha olefin; Cetene
Empirical: $C_{16}H_{32}$
Formula: $CH_3(CH_2)_{13}CH=CH_2$
Properties: Colorless liq., mild hydrocarbon odor; sl. sol. in water; dens. 0.785; b.p. 518-554 F; m.p. 37 F; ref. index 1.4420; flash pt. (Seta) 272 F
Toxicology: Irritating to skin and eyes; low acute inhalation toxicity; low acute ingestion toxicity but ingestion may cause vomiting, aspiration of vomitus
Uses: Intermediate for surfactants and specialty industrial chemicals
Manuf./Distrib.: Albemarle; Chevron; Fluka; Monomer-Polymer & Dajac; Shell
Trade names: Gulftene® 16

Hexadecenyl succinic acid anhydride. *See* Hexadecenyl succinic anhydride

Hexadecenyl succinic anhydride
Synonyms: Hexadecenyl succinic acid anhydride
Uses: Curing agent for epoxy resins; corrosion inhibitor for nonaq. lubricating oils, intermediate for prep. of alkyd or unsat. polyester resins, intermediate in chem. reactions
Manuf./Distrib.: Dixie Chem
Trade names: Milldride® HDSA

n-Hexadecyl 3,5-di-t-butyl-4-hydroxybenzoate. *See* 3,5-Di-t-butyl-4-hydroxybenzoic acid, n-hexadecyl ester
Hexadecyl dimethylamine. *See* Dimethyl palmitamine
Hexadecyl trimethyl ammonium chloride. *See* Cetrimonium chloride
Hexaethylene glycol. *See* PEG-6

Hexafluorobisphenol A dicyanate monomer/prepolymer
Uses: Flame-retardant for low-visc. hot-melt impregnations, dissolving thermoplastic tougheners, and as reactive diluent for hot-melt matrix and adhesive resins
Trade names: AroCy® F-10

Hexahydroaniline. *See* Cyclohexylamine
Hexahydro azepine. *See* Hexamethyleneimine
Hexahydrobenzenamine. *See* Cyclohexylamine

Hexahydrophthalic anhydride
CAS 85-42-7
Synonyms: HHPA; 1,2-Cyclohexanedicarboxylic anhydride
Empirical: $C_8H_{10}O_3$
Formula: $C_6H_{10}(CO)_2O$
Properties: Clear colorless visc. liq., glassy solid; misc. with benzene, toluene, acetone, CCl_4, chloroform, ethanol, ethyl acetate; sl. sol. in petrol. ether; m.w. 154; dens. 1.18 (40 C); b.p. 158 C (17 mm)
Toxicology: Toxic by inhalation; strong irritant to eyes and skin
Uses: Epoxy curing agent; intermediate for alkyds, plasticizers, insect repellents, rust inhibitors, hardener in epoxy resins
Manuf./Distrib.: Allchem Ind.; Anhydrides & Chems.; Buffalo Color; Dixie Chem; GCA; Hüls AG; Milliken
Trade names: Milldride® HHPA
Trade names containing: Milldride® MHHPA

Hexahydro-1,3,5-triethyl-s-triazine
CAS 7779-27-3
Empirical: $C_9H_{21}N_3$
Properties: Lt. yel. liq.; sol. in water; m.w. 171.33; dens. 0.89 (25 C)
Toxicology: Poison by ingestion; moderately toxic by skin contact; severe eye irritant
Uses: Fungicide, industrial preservative for latex, adhesives, cutting fluids, marine lubricants
Trade names: Vancide® TH

Hexamethoxymethylmelamine
CAS 3089-11-0
Synonyms: HMMM
Uses: Crosslinking agent in melamine resin coating systems, general industrial finishes, appliance finishes; also with alkyd, polyester, thermosetting acrylic, epoxy, and cellulose resins; condensation agent for resoricnol-type bonding systems
Trade names: Cymel 303; R7234

Hexamethyldisilazane
CAS 999-97-3; EINECS 213-668-5
Synonyms: HMDS; Bis (trimethylsilyl) amine
Empirical: $C_6H_{19}NSi_2$
Formula: $(CH_3)_3SiNHSi(CH_3)_3$
Properties: Liq.; sol. in acetone, benzene, ethyl ether, heptane, perchloroethylene; m.w. 161.44; dens. 0.77; b.p. 125 C; flash pt. 25 C; ref. index 1.4057
Precaution: Fire hazard
Toxicology: Moderately toxic by intraperitoneal route; experimental tumorigen; PEL 6 mg/m^3
Uses: Chemical intermediate, release agent, lubricant, coupling agent, chromatographic packings, silylating agent
Manuf./Distrib.: Aldrich; Austin; Dow Corning; FAR Research; Gelest; Great Western; Hüls Am.; Janssen

Chimica; PCR; Schweizerhall; Wacker Chemie GmbH
Trade names: Dynasylan® HMDS; Petrarch® H7300; Prosil® HMDS
Trade names containing: Cab-O-Sil® TS-530

Hexamethylenebis (3,5-di-t-butyl-4-hydroxycinnamate)
CAS 35074-77-2
Uses: Antioxidant; stabilizer for polyolefin, elastomer, styrenic, polyacetal, petrol. prods., and org. substrates
Trade names: Irganox® 259

N,N′-Hexamethylene bis (3,5-di-t-butyl-4-hydroxyhydrocinnamamide)
CAS 23128-74-7
Synonyms: 1,6-Hexamethylene bis(3,5-di-t-butyl-4-hydroxyhydrocinnamate)
Uses: Antioxidant for polymers, rubber, SBR, polyacetals, linear saturated polyesters, PVC, polyolefins
Trade names: Irganox® 1098
Trade names containing: Irganox® B 1171

1,6-Hexamethylene bis(3,5-di-t-butyl-4-hydroxyhydrocinnamate). *See* N,N′-Hexamethylene bis (3,5-di-t-butyl-4-hydroxyhydrocinnamamide)
Hexamethylene bisthiosulfate disodium salt. *See* Disodium hexamethylene bisthiosulfate

Hexamethylenediamine
CAS 124-09-4; EINECS 204-679-6
Synonyms: HMD; 1,6-Diaminohexane; 1,6-Hexanediamine
Empirical: $C_6H_{16}N_2$
Formula: $NH_2(CH_2)_6NH_2$
Properties: M.w. 116.21; sp.gr. 0.854; vapor pressure 3 mm Hg (60 C); f.p. 41 C; b.p. 204 C; flash pt. (COC) 85 C
Precaution: Corrosive
Uses: Epoxy curative; isocyanates; petrol. additive; polyamide resins, adhesives, inks, fibers; scale and corrosion inhibitors; water treatment chemicals
Manuf./Distrib.: BASF; DuPont; Fluka; Monsanto
Trade names containing: Millamine® 5260

Hexamethyleneimine
CAS 111-49-9; EINECS 203-875-9
Synonyms: HMI; 1H-Azepine, hexahydro; Hexahydro azepine; Hexamethylenimine; Cyclohexamethyleneimine; Homopiperidine; Azacycloheptane
Properties: M.w. 99.2; sp.gr. 0.88; vapor pressure 9 mm Hg; b.p. 138 C (760 mm); flash pt. (CC) 22 C; ref. index 1.463 (20 C)
Precaution: Flamm., corrosive
Toxicology: Corrosive to skin and eyes; avoid inh. of vapor
Uses: Intermediate for pharmaceuticals, fungicides, bactericides, agric., zeolites, resin crosslinking, dyes, ink, rubber chems., textile chems., corrosion inhibitors
Manuf./Distrib.: BASF; DuPont; Fluka; Monomer-Polymer & Dajac

Hexamethylene tetramine
CAS 100-97-0; EINECS 202-905-8
Synonyms: HMT; HMTA; Methenamine; Aminoform; Urotropine; Hexamine
Empirical: $C_6H_{12}N_4$
Formula: $(CH_2)_6N_4$
Properties: White crystalline powd. or colorless lustrous crystals, pract. odorless; sol. in water, alcohol, chloroform; insol. in ether; m.w. 140.22; dens. 1.27 (25 C)
Precaution: Flamm.
Toxicology: Skin irritant; poison by subcutaneous route; moderately toxic by ingestion, intraperitoneal routes
Uses: Curing of phenolformaldehyde and resorcinolformaldehyde resins, adhesives, fungicide, antibacterial
Regulatory: FDA 21CFR §181.30; Europe listed; UK approved
Manuf./Distrib.: Allchem Ind.; Chemie Linz N. Am.; R.W. Greeff; Mitsubishi Gas; Monomer-Polymer & Dajac; OxyChem/Durez; San Yuan; Spectrum Chem. Mfg.; Wright
Trade names containing: Poly-Dispersion® SHD-65

Hexamethylenimine. *See* Hexamethyleneimine

Hexamethylmelamine
CAS 68002-20-0
Synonyms: Poly(melamine-co-formaldehyde) methylated
Properties: Dens. 1.200; flash pt. > 110 C; ref. index 1.5190 (20 C)
Toxicology: Suspected carcinogen; mutagen
Uses: Reinforcement for rubber compds., latex dips; enhances rubber adhesion to textiles, cord
Manuf./Distrib.: Aldrich
Trade names: Resimene® 3520

3,3,6,6,9,9-Hexamethyl 1,2,4,5-tetraoxa cyclononane
CAS 22397-33-7
Uses: Polymerization initiator, crosslinking agent
Trade names: Aztec® HMCN-30-AL; Aztec® HMCN-30-WO-2
Trade names containing: Aztec® HMCN-40-IC3

Hexamine. *See* Hexamethylene tetramine

Hexane
CAS 110-54-3; 64742-49-0; EINECS 203-777-6
Synonyms: n-Hexane; Alkane C-6
Classification: Aliphatic compd.
Empirical: C_6H_{14}
Formula: $CH_3(CH_2)_4CH_3$
Properties: Colorless volatile liq., faint odor; sol. in alcohol, acetone, ether; insol. in water; m.w. 86.20; dens. 0.65937 (20/4 C); m.p. -95 C; b.p. 68.742 C; flash pt. -22.7 C; ref. index 1.37486 (20 C)
Precaution: Flamm.; very dangerous fire/explosion hazard exposed to heat or flame; can react vigorously with oxidizers
Toxicology: TLV 50 ppm in air; LD50 (oral, rat) 28.710 mg/kg; sl. toxic by ingestion, inh.; human systemic effects by inh.; mutagenic data; eye irritant; irritating to respiratory tract; narcotic in high concs.; heated to decomp., emits acrid smoke and fumes
Uses: Solvent for veg. oil and pharmaceutical extraction, compounding rubber cements, alcohol denaturant, paint diluent, polymerization reaction medium; filling for thermometers
Regulatory: FDA 21CFR §173.270; Japan approved with restrictions
Manuf./Distrib.: Ashland; BP Chem. Ltd; Coyne; Exxon; Great Western; Humphrey; Mitsui Petrochem. Ind.; Phibro Energy USA; Phillips; Shell; Texaco

n-Hexane. *See* Hexane
1,6-Hexanediamine. *See* Hexamethylenediamine
1,6-Hexanediamine, N,N,N′,N′-tetramethyl. *See* N,N,N′,N′-Tetramethylhexanediamine
1,6-Hexanediamine, trimethyl. *See* Trimethylhexamethylene diamine
Hexanedioic acid. *See* Adipic acid
Hexanedioic acid, bis (2-ethylhexyl) ester. *See* Dioctyl adipate
Hexanedioic acid, bis (1-methylethyl) ester. *See* Diisopropyl adipate
Hexanedioic acid, dibutyl ester. *See* Dibutyl adipate
Hexanedioic acid, didecyl ester. *See* Dicapryl adipate
Hexanedioic acid, diisobutyl ester. *See* Diisobutyl adipate
Hexanedioic acid, diisodecyl ester. *See* Diisodecyl adipate
Hexanedioic acid, diisononyl ester. *See* Diisononyl adipate
Hexanedioic acid, ditridecyl ester. *See* Ditridecyl adipate

1,6-Hexanediol methacrylate
Uses: Crosslinking agent used in casting compds., glass fiber-reinforced plastics, adhesives, coatings, ion-exchange resins, textile prods., plastisols, dental polymers, rubber compding.
Trade names containing: SR-239

Hexanoic acid, 2-ethyl-, zinc salt. *See* Zinc 2-ethylhexanoate
1-Hexanol. *See* Hexyl alcohol
n-Hexanol. *See* Hexyl alcohol

1-Hexene
CAS 592-41-6; EINECS 209-753-1
Synonyms: C_6 linear alpha olefin; Hexylene; Hexene-1; Butyl ethylene

Empirical: C_6H_{12}
Formula: $CH_3CH_2CH_2CH_2CH:CH_2$
Properties: Colorless liq., mild hydrocarbon odor; sol. in alcohol; insol. in water; m.w. 84.16; dens. 0.678; f.p. -139.8 C; b.p. 62-63 C; flash pt. (Seta) -14 F; ref. index 1.3876 (20 C)
Precaution: Highly flamm.; dangerous fire risk
Toxicology: Irritating to eyes and skin; sl. toxic by ingestion; inhalation may produce CNS depression
Uses: Intermediate for surfactants and specialty industrial chemicals (flavors, perfumes, dyes, resins); polymer modifier
Manuf./Distrib.: Albemarle; Aldrich; Chevron; Hüls Am.; Johnson Matthey; Monomer-Polymer & Dajac; Phillips; Shell
Trade names: Gulftene® 6

Hexene-1. *See* 1-Hexene

n-Hexyl acrylate
CAS 2499-95-8
Synonyms: Acrylic acid, hexyl ester
Classification: Monomer
Formula: $CH_2{=}CHCOOC_6H_{13\text{-}n}$
Properties: M.w. 156.23
Toxicology: Mildly toxic by skin contact
Uses: Monomer for UV-cured inks and coatings, glass coating, visc. index improver for functional oils, polymer cements and sealants; polymer modifier
Manuf./Distrib.: CPS; Monomer-Polymer & Dajac
Trade names: Ageflex n-HA

Hexyl alcohol
CAS 111-27-3; 68526-79-4; EINECS 203-852-3
Synonyms: 1-Hexanol; n-Hexanol; Alcohol C-6; Pentylcarbinol; Amylcarbinol
Classification: Aliphatic alcohol
Empirical: $C_6H_{14}O$
Formula: $CH_3(CH_2)_4CH_2OH$
Properties: Colorless liq., fruity odor, aromatic flavor; sol. in alcohol and ether; sl. sol. in water; m.w. 102.20; dens. 0.8186; f.p. -51.6 C; b.p. 157.2 C; flash pt. (TOC) 65 C; ref. index 1.1469 (25 C)
Precaution: Flamm. or combustible liq.; reactive with oxidizing materials
Toxicology: LD50 (rat, oral) 4.59 g/kg; poison by intravenous route; moderately toxic by ingestion, skin contact; skin and severe eye irritant
Uses: Pharmaceuticals (antiseptics, perfume esters), solvent, plasticizer; flavoring agent; intermediate; emulsion polymerization
Regulatory: FDA 21CFR §172.515, 172.864, 178.3480; GRAS (FEMA)
Manuf./Distrib.: Albemarle; Aldrich; Ashland; Penta Mfg.; Vista
Trade names: Epal® 6
Trade names containing: Nansa® EVM70/B

Hexylene. *See* 1-Hexene

Hexylene glycol
CAS 107-41-5; EINECS 203-489-0
Synonyms: 2-Methyl-2,4-pentanediol; 4-Methyl-2,4-pentanediol; α,α,α'-Trimethyltrimethyleneglycol
Classification: Aliphatic alcohol
Formula: $(CH_3)_2COHCH_2CHOHCH_3$
Properties: Colorless liq., nearly odorless; misc. with water, hydrocarbons, fatty acids; m.w. 118.18; dens. 0.9216 (20/4 C); b.p. 198.3 C; flash pt. (OC) 93 C; ref. index 1.4276 (20 C)
Toxicology: TLV:Cl 25 ppm in air; LD50 (rat, oral) 4.70 g/kg
Uses: Hydraulic brake fluids, printing inks, coupling agent and penetrant for textiles, cosmetics; ice inhibitor in carburetors; fuel and lubricant additive; emulsifier
Regulatory: FDA 21CFR §175.105, 176.180, 176.200, 176.210, 177.1210, 177.2800
Manuf./Distrib.: Allchem Ind.; Ashland; BP Chem. Ltd; Coyne; Great Western; Mitsui Petrochem. Ind.; Penta Mfg.; Shell; Union Carbide
Trade names containing: Monawet MT-70

Hexyloxypropylamine
Uses: Corrosion inhibitor for metalworking fluids; antistat; additive for fuel, lubricant, petrol. refining; intermediate for surfactants, textile foaming agents, ethoxylates and agric. chem.; crosslinking agent for epoxy resins
Trade names: Tomah PA-10

HFC-152A. *See* 1,1-Difluoroethane
HHPA. *See* Hexahydrophthalic anhydride
HMD. *See* Hexamethylenediamine
HMDS. *See* Hexamethyldisilazane
HMI. *See* Hexamethyleneimine
HMMM. *See* Hexamethoxymethylmelamine
HMT. *See* Hexamethylene tetramine
HMTA. *See* Hexamethylene tetramine
Homopiperidine. *See* Hexamethyleneimine
HSA. *See* Hydroxystearic acid
2-Hydoxyethyl octadecanoate. *See* Glycol stearate
Hydrated alumina. *See* Aluminum hydroxide
Hydrated aluminum silicate. *See* Kaolin, Pyrophyllite
Hydrated lime. *See* Calcium hydroxide

Hydroabietyl alcohol
Properties: Sp.gr. 1.008; m.p. 32-33 C; flash pt. (COC) 185 C; ref. index 1.526 (20 C)
Uses: Plasticizer for cellulose nitrate, ethylcelluose, PVC
Manuf./Distrib.: Hercules
Trade names: Abitol®

Hydroabietyl phthalate
Properties: Sp.gr. 1.055 (20 C); m.p. 63 C; ref. index 1.513
Uses: Resin used in hot-melt and pressure-sensitive adhesives, specialty nitrocellulose lacquers, printing inks; plasticizer for PVC, cellulose nitrate
Manuf./Distrib.: Hercules

Hydrogenated castor oil
CAS 8001-78-3; EINECS 232-292-2
Synonyms: Opalwax; Castorwax; Castor oil, hydrogenated
Definition: End prod. of controlled hydrogenation of castor oil
Properties: Hard white wax; very insol. in water and in the more common org. solvs.; m.w. ≈ 932; m.p. 86-88 C
Uses: In water-repellent coatings, candles, polishes, ointments, cosmetics; impregnant for paper, wood, cloth; as lubricant, mold release in mfg. of formed plastics and rubber goods
Regulatory: FDA 21CFR §175.105, 175.300, 176.170, 176.210, 177.1200, 177.1210, 177.2420, 177.2800, 178.3280
Manuf./Distrib.: Akzo Nobel; Arista Ind.; Hoechst AG; Southern Clay Prods.
Trade names: Castorwax® MP-70; Castorwax® MP-80; Glycolube® CW-1

Hydrogenated cottonseed glyceride
CAS 61789-07-9
Synonyms: Glycerides, cottonseed oil, hydrogeanted
Definition: End prod. of controlled hydrogenation of cottonseed glyceride
Uses: Emulsifier for foods, cosmetics; dispersant, mold release, processing aid, antistat, antifog, lubricant, antiblock for PS, polyolefins, PVC, PU
Trade names: Myverol® 18-07

Hydrogenated methyl abietate
Properties: Sp.gr. 1.02; flash pt. (COC) 182 C; ref. index 1.52 (20 C)
Uses: Plasticizer for vinyls, cellulosics, PS
Manuf./Distrib.: Hercules

Hydrogenated methyl ester of rosin. *See* Methyl hydrogenated rosinate

Hydrogenated palm glyceride
CAS 67784-87-6, 97593-29-8

Synonyms: Palm oil glyceride, hydrogenated; Glycerides, palm oil mono-, hydrogenated
Definition: End prod. of controlled hydrogenation of palm glyceride
Uses: Emulsifier, stabilizer, dispersant, opacifier for cosmetics, foods and drugs; dispersant, mold release, processing aid, antistat, antifog, lubricant, antiblock for PS, polyolefins, PVC, PU
Regulatory: FDA 21CFR §176.210, 177.2800
Trade names: Myverol® 18-04

Hydrogenated soybean glyceride. *See* Hydrogenated soy glyceride
Hydrogenated soybean oil monoglyceride. *See* Hydrogenated soy glyceride

Hydrogenated soy glyceride
CAS 61789-08-0; 68002-71-1
Synonyms: Hydrogenated soybean oil monoglyceride; Glycerides, soybean oil, hydrogenated, mono; Hydrogenated soybean glyceride
Definition: End prod. of controlled hydrogenation of soybean monoglycerides
Uses: Emulsifier, stabilizer, dispersant, opacifier for cosmetics, foods, drugs; lubricating greases, synthetic waxes, textile lubricants; antistat, lubricant, processing aid in plastics, nonwoven fibers; dispersant for color concs.
Regulatory: FDA 21CFR §176.210, 177.2800
Trade names: Dimodan PV; Myverol® 18-06

Hydrogenated stearic acid
Uses: Intermediate used in alkyd resins, rubber compding., water repellents, polishes, soaps, abrasives, cutting oils, candles, crayons, emulsifiers; food grades as lubricant, release agent, binder, defoamer in foods, intermediate for food emulsifiers
Trade names: Industrene® B; Industrene® R

Hydrogenated tallow acid
CAS 61790-38-3; EINECS 263-130-9
Synonyms: Acids, tallow, hydrogenated; Tallow acid, hydrogenated
Definition: End prod. of controlled hydrogenation of tallow acid
Uses: Used in industrial prods. incl. syn. lubricants, bar soaps, cosmetics, rubber tires; emulsifier for polymerization of SBR, ABS, methyl methacrylate-butadiene-styrene polymers
Regulatory: FDA 21CFR §175.105, 176.170, 176.200, 176.210, 177.2600, 177.2800, 178.3570
Trade names: Petrac® PHTA

Hydrogenated tallowalkyl dimethylamine. *See* Dimethyl hydrogenated tallow amine

Hydrogenated tallow amide
CAS 61790-31-6; EINECS 263-123-0
Synonyms: Amides, tallow, hydrogenated; Tallow amides, hydrogenated
Classification: Amide
Formula: $RCONH_2$, RCO- represents the fatty acids derived from hydrog. tallow
Uses: Antiblock, lubricant, slip agent for plastics, coatings, films; foam booster, builder for syn. detergents; water repellent for textiles; intermediate for syn. waxes; pigment dispersant; rubber processing
Trade names: Armid® HT

Hydrogenated tallowamine
CAS 61788-45-2; EINECS 262-976-6
Synonyms: Amines, hydrogenated tallow alkyl; Tallow amine, hydrogenated
Definition: Amine derived from hydrogenated tallow
Formula: RNH_2, R represents the alkyl groups derived from hydrog. tallow
Uses: Emulsifier, flotation agent, corrosion inhibitor, chemical intermediate, anticaking agent, pigment dispersant; cosmetics; lubricant and mold release for hard rubber, textile chemical, chemical synthesis; antistat, antifog additive
Trade names: Crodamine 1.HT; Kemamine® P-970; Kemamine® P-970D; Noram SH

Hydrogenated tallow dimethylamine. *See* Dimethyl hydrogenated tallow amine

Hydrogenated tallow 1,3-propylene diamine
CAS 68603-64-5; EINECS 271-696-6

Hydrogen sulfate

Uses: Gasoline detergent, bactericide, corrosion inhibitor in petrol. prod., epoxy hardener, asphalt emulsifier, wetting agent, dispersant for water treatment, pigments
Trade names: Kemamine® D-970

Hydrogen sulfate. *See* Sulfuric acid
1-Hydroperoxycyclohexyl 1-hydroxycyclohexyl peroxide. *See* Cyclohexanone peroxide
Hydroquinol. *See* Hydroquinone

Hydroquinone
CAS 123-31-9; EINECS 204-617-8
Synonyms: 1,4-Benzenediol; p-Dihydroxybenzene; Hydroquinol
Classification: Aromatic organic compd.
Empirical: $C_6H_6O_2$
Formula: $C_6H_4(OH)_2$
Properties: White crystals; sol. in water, alcohol, ether; m.w. 110.11; dens. 1.330; m.p.170 C; b.p. 285 C
Precaution: Combustible
Toxicology: Toxic by ingestion and inhalation; irritant; TLV 2 mg/m^3 of air; LD50 (rat, oral) 320 mg/kg
Uses: Photographic developer (not for color film); dye intermediate; inhibitor; stabilizer in paints and varnishes; motor fuels and oils; antioxidant for fats and oils, syn. latexes, polyester resins
Regulatory: FDA 21CFR §175.105, 176.170, 177.2420
Manuf./Distrib.: Aldrich; Alfa; Allchem Ind.; Charkit; Eastman; Goodyear Tire & Rubber; Kraeber GmbH; Penta Mfg.; San Yuan; Spectrum Chem. Mfg.
Trade names containing: SR-203; SR-205; SR-206; SR-209; SR-212; SR-213; SR-239; SR-350

Hydroquinone bis(2-hydroxyethyl) ether
CAS 104-38-1; EINECS 203-197-3
Synonyms: 2,2´-[1,4-Phenylenebis(oxy)bisethanol; Hydroquinone di-(β-hydroxyethyl) ether
Empirical: $C_{10}H_{14}O_4$
Properties: White solid, odorless; m.w. 198.22; sp.gr. 1.15; b.p. 190 C (0.3 mm); m.p. 98 C; flash pt. (COC) 224 C
Precaution: Avoid oxidizers
Toxicology: Molten material may produce thermal burns on skin; low hazard by ingestion, inhalation; LD50 (rat, oral) > 3200 mg/kg; slight skin irritant
Uses: Chemical intermediate; chain extender for PU elastomers
Trade names: Eastman® HQEE

Hydroquinone di-(β-hydroxyethyl) ether. *See* Hydroquinone bis(2-hydroxyethyl) ether
Hydroquinone methyl ether. *See* Hydroquinone monomethyl ether

Hydroquinone monomethyl ether
CAS 150-76-5; EINECS 205-769-8
Synonyms: MEHQ; Hydroquinone methyl ether; 4-Methoxyphenol; p-Hydroxyanisole
Classification: Substituted phenolic compd.
Definition: Monomethyl ether of hydroquinone
Empirical: $C_7H_8O_2$
Formula: $CH_3OC_6H_4OH$
Properties: White waxy solid; sl. sol. in water; sol. in benzene, acetone, ethyl acetate, alcohol; m.w. 124.14; dens. 1.55 (20/20 C); m.p. 52.5 C; b.p. 243 C
Precaution: Combustible
Uses: Mfg. of antioxidants, pharmaceuticals, plasticizers, dyestuffs; stabilizer for chlorinated hydrocarbons and ethylcellulose; UV inhibitor; inhibitor for acrylic and vinyl monomers and acrylonitrile
Manuf./Distrib.: Alemark; Alfa; Arenol; ChemDesign; Eastman; Fluka; Kincaid Enterprises; Penta Mfg.; Specialty Chem. Prods.
Trade names: Eastman® HQMME
Trade names containing: Empicryl® 6030; Empicryl® 6047; MFM-415; MFM-786 V; SR-201; SR-203; SR-214; SR-244; SR-252; SR-290; SR-295; SR-297; SR-313; SR-344; SR-368; SR-399; SR-423; SR-444; Vynate® Neo-10

Hydrous magnesium calcium silicate. *See* Talc
Hydrous magnesium silicate. *See* Talc

2-Hydroxy-4-acryloyloxyethoxy benzophenone
Empirical: $C_{18}H_{16}O_5$

Properties: Powd.; m.w. 312; m.p. 72-75 C
Toxicology: LD50 (rat, oral) > 5 g/kg; LD50 (rabbit, dermal) > 2 g/kg
Uses: Light stabilizer, UV absorber for ABS
Trade names: Cyasorb® UV 416; Cyasorb® UV 2098

p-Hydroxyanisole. *See* Hydroquinone monomethyl ether
4-Hydroxybenzenesulfonic acid. *See* Phenol sulfonic acid
2-Hydroxybenzoic acid. *See* Salicylic acid
o-Hydroxybenzoic acid. *See* Salicylic acid
2-Hydroxybutane. *See* 2-Butanol

2-(2´-Hydroxy-3´-t-butyl-5´-methylphenyl)-5-chlorobenzotriazole
CAS 3896-11-5; EINECS 223-445-4
Synonyms: 2-(5-Chloro-2H-benzotriazole-2-yl)-6-(1,1-dimethylethyl)-4-methylphenol; 2-(3´-t-Butyl-2´-hydroxy-5´-methylphenyl)-5-chlorobenzotriazole; Bumetrizole
Classification: Benzotriazole
Properties: Sl. yel. powd.; m.p. 138 C
Uses: UV absorber/stabilizer for polyolefins, polyester resins and coatings
Trade names: Lowilite® 26

2-(2´-Hydroxy-3,5´-di-t-amylphenyl) benzotriazole
CAS 25973-55-1
Synonyms: 2-(2H-Benzotriazol-2-yl)-4,6-bis(1,1-dimethylpropyl)phenol
Properties: Powd.; m.p. 81 C
Uses: Light stabilizer for PVC, ABS, cellulosics, epoxy, PS, PE, PP, PS, VDC
Trade names: Tinuvin® 328

2-[2-Hydroxy-3,5-di-(1,1-dimethylbenzyl)phenyl]-2H-benzotriazole
CAS 70321-86-7
Synonyms: 2-(2H-Benzotriazol-2-yl)-4,6-bis(1-methyl-1-phenylethyl) phenol
Uses: Light stabilizer for PVC
Trade names: Tinuvin® 234

m-Hydroxy-N,N-dimethyl aniline. *See* Dimethylaminomethyl phenol

2(2´-Hydroxy-3´,5´(1,1-dimethylbenzylphenyl) benzotriazole
Properties: Powd.; m.p. 135-143 C
Uses: UV stabilizer for ABS, cellulosics, epoxy, polyesters, PE, PS, flexible and rigid PVC, VDC

3-Hydroxy-1,1-dimethylbutyl peroxyneodecanoate
Formula: t-$C_9H_{19}COOOC(CH_3)_2CH_2CH_3CHOH$
Properties: M.w. 288.4
Uses: Initiator for PVC polymerization
Trade names: Lupersol 610-M50

2-Hydroxyethylamine. *See* Ethanolamine

Hydroxyethylcellulose
CAS 9004-62-0
Synonyms: Cellulose, 2-hydroxyethyl ether; H.E. cellulose
Definition: Modified cellulose polymer containg. hydroxyethyl side chains
Properties: White free-flowing powd.; nonionic; insol. in org. solvs.; sol. in hot or cold water; grease and oil resistant
Precaution: Combustible
Uses: Thickener, suspending agent; stabilizer for vinyl polymerization; retards evaporation of water in mortars and cements; binder in ceramic glazes; used in paper and textile sizing
Regulatory: FDA 21CFR §175.105, 175.300
Manuf./Distrib.: Allchem Ind.; Amerchol NV; Aqualon; Union Carbide
Trade names: Natrosol® Hydroxyethylcellulose; Tylose® H Series

Hydroxyethylethylenediamine. *See* Aminoethylethanolamine
2-Hydroxyethyl 12-hydroxy-9-octadecenoate. *See* Glycol ricinoleate

N (2-Hydroxyethyl) 12-hydroxystearamide
Uses: Internal mold release agent, lubricant for polyolefins, PVC, styrenics
Trade names: Paricin® 220

2-Hydroxyethyl methacrylate
CAS 868-77-9; EINECS 212-782-2
Synonyms: HEMA
Empirical: $C_6H_{10}O_3$
Formula: $CH_2:C(CH_3)COOCH_2CH_2OH$
Properties: B.p. 205-208 C
Uses: Monomer for creating and modifying wide range of polymers, acrylic resins, binder for nonwoven fabrics, enamels, adhesives; reactive thinner for radiation curing; rubber modifier
Manuf./Distrib.: BP Chem.; Fluka; Mitsubishi Gas; Rohm & Haas
Trade names: BM-903; Sipomer® HEM-D

N-(2-Hydroxyethyl) octadecanamide. *See* Stearamide MEA
2-Hydroxyethyl 9-octadecenoate. *See* Glycol oleate

N(β-Hydroxyethyl) ricinoleamide
Classification: Hydroxyamide wax
Properties: Sp.gr. 1.00; m.p. 46 C
Uses: Lubricant, antistat, mold release; in plastics, metals, textile coatings; slip agent for varnishes and lacquers; elec. potting compds., crayons, wax blends, high-temp. greases
Trade names: Flexricin® 115

2-Hydroxy-4-isooctoxybenzophenone
Uses: UV stabilizer for PP, HDPE, LDPE, EVA, flexible and rigid PVC, PC, acetals, epoxies, some cellulosics; for food pkg. use

2-Hydroxy-4-methoxybenzophenone. *See* Benzophenone-3
2-Hydroxy-4-methoxybenzophenone-5-sulfonic acid. *See* Benzophenone-4
(2-Hydroxy-4-methoxyphenyl) phenylmethanone. *See* Benzophenone-3
2-Hydroxy-4-methoxy-5-sulfo benzophenone. *See* Benzophenone-4

Hydroxymethyl dioxoazabicyclooctane
CAS 6542-37-6; EINECS 229-457-6
Synonyms: 7-Hydroxymethyl-1,5-dioxo-3-aza-bicyclooctane
Classification: Heterocyclic organic compd.
Empirical: $C_6H_{11}NO_3$
Uses: Crosslinking agent for resorcinol phenol-formaldehyde or protein-based resin systems; raw material for synthesis
Trade names: Zoldine® ZT-65

7-Hydroxymethyl-1,5-dioxo-3-aza-bicyclooctane. *See* Hydroxymethyl dioxoazabicyclooctane
4-Hydroxy-4-methyl-2-pentanone. *See* Diacetone alcohol
2-(2´-Hydroxy-5´-methylphenyl) benzotriazole. *See* Drometrizole
12-Hydroxyoctadecanoic acid. *See* Hydroxystearic acid
12-Hydroxyoctadecanoic acid, methyl ester. *See* Methyl hydroxystearate
12-Hydroxy-9-octadecenoic acid. *See* Ricinoleic acid
12-Hydroxy-9-octadecenoic acid, methyl ester. *See* Methyl ricinoleate
12-Hydroxy-9-octadecenoic acid, monoester with 1,2-propanediol. *See* Propylene glycol ricinoleate
12-Hydroxy-9-octadecenoic acid, monoester with 1,2,3-propanetriol. *See* Glyceryl ricinoleate
12-Hydroxy-9-octadecenoic acid, monopotassium salt. *See* Potassium ricinoleate
12-Hydroxy-9-octadecenoic acid, sodium salt. *See* Sodium ricinoleate
2-Hydroxy-4-n-octoxybenzophenone. *See* Benzophenone-12
2-Hydroxy-4-(octyloxy) benzophenone. *See* Benzophenone-12
[2-Hydroxy-4-(octyloxy)phenyl]phenylmethanone. *See* Benzophenone-12

2-(2-Hydroxy-5-t-octylphenyl) benzotriazole
CAS 3147-75-9
Empirical: $C_{20}H_{25}ON_3$
Properties: Powd.; m.w. 323.44; m.p. 101-105 C
Toxicology: LD50 (rat, oral) > 10 g/kg; LD50 (rabbit, dermal) > 5 g/kg; irritant

Uses: Light stabilizer, uv absorber for polyesters, PVC, styrenics, acrylics, PC, polyvinyl butyral (molding, sheet, and glazing materials for window, lighting, sign, marine, and auto applics.)
Manuf./Distrib.: Aldrich
Trade names: Cyasorb® UV 5411; Tinuvin® 329

d-12-Hydroxyoleic acid. *See* Ricinoleic acid
α-Hydroxy-ω-hydroxy poly(oxy-1,2-ethanediyl). *See* Polyethylene glycol
β-Hydroxyphenethylamine. *See* Phenylethanolamine
N-(4-Hydroxyphenyl) stearamide. *See* Stearoyl p-amino phenol
2-Hydroxy-1,2,3-propanetricarboxylic acid. *See* Citric acid
2-Hydroxy-1,2,3-propanetricarboxylic acid, monooctadecyl ester. *See* Stearyl citrate
2-Hydroxy-1,2,3-propanetricarboxylic acid, tributyl ester. *See* Tributyl citrate
2-Hydroxy-1,2,3-propanetricarboxylic acid, triethyl ester. *See* Triethyl citrate
2-Hydroxy-1,2,3-propanetricarboxylic acid, tris(2-octyldodecyl) ester. *See* Trioctyldodecyl citrate
2-Hydroxypropanoic acid, calcium salt. *See* Calcium lactate

N-(2-Hydroxypropyl) benzenesulfonamide
CAS 35325-02-1
Uses: Plasticizer for polyamide, polyurethane, polyacrylic, cellulose ester; antistat; for paints, lacquers; stabilizer for pigmented unsat. polyester resins
Trade names: Uniplex 225

Hydroxypropyl-β-cyclodextrin
CAS 94035-02-6
Properties: M.w. 1500 (avg.)
Uses: Complex hosting guest molecules; increases the sol. and bioavailability of other substances; masks flavor, odor, or coloration; stabilizes against light, oxidation, heat, and hydrolysis; turns liqs. or volatiles into stable solid powds.; for use in pharmaceuticals, cosmetics, toiletries, foods, tobacco, pesticides, textiles, paints, plastics, synthesis, polymers
Trade names: Beta W7 HP 0.9

Hydroxypropyl-γ-cyclodextrin
CAS 99241-25-5
Properties: M.p. 250 C
Uses: complex hosting guest molecules; increases the sol. and bioavailability of other substances; masks flavor, odor, or coloration; stabilizes against light, oxidation, heat, and hydrolysis; turns liqs. or volatiles into stable solid powds.; for use in pharmaceuticals, cosmetics, toiletries, foods, tobacco, pesticides, textiles, paints, plastics, synthesis, polymers
Trade names: Gamma W8 HP0.6

Hydroxypropyl methacrylate
CAS 27813-02-1; EINECS 248-666-3
Empirical: $C_7H_{12}O_3$
Formula: $CH_3CHOHCH_2OOCC(CH_3):CH_2$
Properties: Clear mobile liq.; limited sol. in water; sol. in common org. solvs.; m.w. 144.17; b.p. 205-209 C; dens. 1.066 (25/16 C)
Precaution: Combustible
Uses: Monomer for acrylic resins, nonwoven fabric binders, detergent lube oil additives
Manuf./Distrib.: Allchem Ind.; Ashland; BP Chem.; Monomer-Polymer & Dajac; Rohm & Haas; Rohm Tech
Trade names: BM-951

Hydroxypropyl methylcellulose
CAS 9004-65-3
Synonyms: MHPC; Methyl hydroxypropyl cellulose; Cellulose 2-hydroxypropyl methyl ether; Hypromellose
Definition: Propylene glycol ether of methyl cellulose
Properties: White powd.; swells in water to produce a clear to opalescent visc. colloidal sol'n.; nonionic; insol. in anhyd. alcohol, ether, chloroform; sol. in most polar solvs.
Precaution: Combustible
Toxicology: LD50 (intraperitoneal, rat) 5200 mg/kg; mildly toxic by intraperitoneal route; heated to decomp., emits acrid smoke and fumes
Uses: Thickener, stabilizer, emulsifier in food prods.; thickener in paint stripping preps.; protective colloid, suspending agent; tablet excipient; in adhesives, asphalt emulsions, caulks, cements, paints; visc. stabilizer for latex and emulsion paints; plasticizer for ceramics, refractory shapes

Hydroxystearic acid

Regulatory: FDA 21CFR §172.874, 175.105, 175.300; Europe listed
Manuf./Distrib.: Aceto; Ashland
Trade names: Methocel® E

Hydroxystearic acid
CAS 106-14-9; EINECS 203-366-1
Synonyms: HSA; 12-Hydroxyoctadecanoic acid; 12-Hydroxystearic acid; Octadecanoic acid, 12-hydroxy-
Classification: Fatty acid
Definition: C-18 straight chain fatty acid
Formula: $CH_3(CH_2)_5(CHOH)(CH_2)_{10}COOH$
Properties: Flakes; m.w. 300.49; sp.gr. 1.021; m.p.79-82 C
Uses: Major component in lithium greases; chemical intermediate; lubricant for PVC; in cosmetics, toiletries, wax blends, polishes, inks, hot-melt adhesives
Regulatory: FDA 21CFR §175.105, 176.210, 178.3570
Manuf./Distrib.: Allchem Ind.; CasChem; Penta Mfg.
Trade names: Loxiol® G 21

12-Hydroxystearic acid. *See* Hydroxystearic acid
Hydroxystearic acid, monoester with glycerol. *See* Glyceryl hydroxystearate
β-Hydroxytricarballylic acid. *See* Citric acid
2-Hydroxytriethylamine. *See* Diethylaminoethanol
Hypromellose. *See* Hydroxypropyl methylcellulose
IBOA. *See* Isobornyl acrylate
1H-Imidazole-1-ethanol, 4,5-dihydro-2-norcocoyl-. *See* Cocoyl hydroxyethyl imidazoline
2,4-Imidazolidinedione, 1,3-bis(hydroxymethyl)-5,5-dimethyl-. *See* DMDM hydantoin
2-Imidazolidinethione. *See* Ethylene thiourea
Imidazoline-2-thiol. *See* Ethylene thiourea
1H-Imidazolium, 1-[2-(carboxymethoxy)ethyl]-1-(carboxymethyl)-2-heptyl-4,5-dihydro-, hydroxide, disodium salt. *See* Disodium capryloamphodiacetate
1H-Imidazolium, 1-(carboxymethyl)-4,5-dihydro-1-(2-hydroxyethyl)-2-nonyl-, hydroxide, sodium salt. *See* Sodium caproamphoacetate
Imidodicarbonimidic diamide, N-(2-methylphenyl)-. *See* o-Tolyl biguanide
2,2´-Iminobisethanol. *See* Diethanolamine
1,1´-Iminobis-2-propenol. *See* Diisopropanolamine
2,2´-Iminodiethanol. *See* Diethanolamine
2,2´-Iminodiethylamine. *See* Diethylenetriamine
Industrial talc. *See* Talc
Infusorial earth. *See* Diatomaceous earth

Iodopropynyl butylcarbamate
CAS 55406-53-6; EINECS 259-627-5
Synonyms: 3-Iodo-2-propynyl butyl carbamate; Butyl-3-iodo-2-propynylcarbamate; Carbamic acid, butyl-3-iodo-2-propynyl ester
Classification: Organic compd.
Formula: $IC_2CH_2OCONH(CH_2)_3CH_3$
Uses: Fungicide, antimildew additive, wood preservative; used in oil-based and latex paints, wood prods., cutting oils, textiles, paper coatings, inks, plastics, adhesives
Regulatory: USA EPA registered; Japan approved; Europe listed
Trade names: Idex™-400; Idex™-1000

3-Iodo-2-propynyl butyl carbamate. *See* Iodopropynyl butylcarbamate
IPA. *See* Isopropyl alcohol
IPGE. *See* Isopropyl glycidyl ether
IPM. *See* Isopropyl myristate
IPP. *See* Isopropyl palmitate
IR. *See* Polyisoprene
Iron (III) chloride. *See* Ferric chloride
Iron chlorides. *See* Ferric chloride

Iron molybdena
Uses: Catalyst for prod. of formaldehyde

Iron (III) oxide. *See* Ferric oxide

Iron (III) oxide hydrated
CAS 20344-49-4
Formula: FeO(OH)
Properties: M.w. 88.85
Manuf./Distrib.: Aldrich
Trade names containing: Rhenogran® Fe-gelb-50

Iron trichloride. *See* Ferric chloride
1,3-Isobenzofurandione. *See* Phthalic anhydride
1,3-Isobenzofurandione, 4,5,6,7-tetrabromo-(9CI). *See* Tetrabromophthalic anhydride

Isobornyl acrylate
CAS 5888-33-5
Synonyms: IBOA
Properties: Dens. 0.986; flash pt. 97 C; ref. index 1.4760 (20 C)
Precaution: Light sensitive
Toxicology: Irritant
Uses: Curing agent; diluent monomer
Manuf./Distrib.: Aldrich; Ashland; CPS; Monomer-Polymer & Dajac; San Esters; UCB Radcure
Trade names: Sipomer® IBOA; Sipomer® IBOA HP; SR-506

Isobornyl methacrylate
CAS 7534-94-3
Properties: M.w. 222.33; dens. 0.983; b.p. 127-129 (15 mm); flash pt. 107 C; ref. index 1.4770 (20 C)
Toxicology: Irritant
Uses: Monomer for creating and modifying a wide range of polymers; for high-performance coatings, paints, textile finishes, adhesives, industrial and textile coatings
Manuf./Distrib.: Ashland; CPS; Monomer-Polymer & Dajac; Rohm Tech; San Esters
Trade names: Sipomer® IBOMA; Sipomer® IBOMA HP
Trade names containing: SR-423

Isobutanol. *See* Isobutyl alcohol
Isobutanolamine. *See* 2-Amino-2-methyl-1-propanol
Isobutanol-2 amine. *See* 2-Amino-2-methyl-1-propanol
Isobutene trimer. *See* Triisobutylene

Isobutyl alcohol
CAS 78-83-1; EINECS 201-148-0
Synonyms: Isobutanol; Isopropylcarbinol; 2-Methyl-1-propanol; 2-Methylpropanol
Classification: Alcohol
Empirical: $C_4H_{10}O$
Formula: $(CH_3)_2CHCH_2OH$
Properties: Colorless liq., sweet odor; partly sol. in water; sol. in alcohol, ether; m.w. 74.12; dens. 0.806 (15 C); m.p. -108 C; b.p. 106-109 C; f.p. -108 C; flash pt. (TCC) 29 C; ref. index 1.396
Precaution: Flamm.; dangerous fire hazard with heat, flame; mod. explosive as vapor with heat, flame, oxidizers
Toxicology: TLV 50 ppm in air; LD50 (oral, rat) 2460 mg/kg; poison by IV, intraperitoneal route; mod. toxic by ingestion, skin contact; experimental carcinogen, tumorigen; severe skin/eye irritant; mutagenic data; heated to decomp., emits acrid smoke and fumes
Uses: Flavors and fragrances; organic synthesis; latent solvent; intermediate; paint removers; fluorometric determinations; liq. chromatography
Regulatory: FDA 21CFR §172.515
Manuf./Distrib.: Aldrich; BASF; CPS; Eastman; Hoechst Celanese; Neste Polyeten AB; Penta Mfg.; Shell; Union Carbide
Trade names containing: Nansa® EVM70/E

2,2′-Isobutylenebis (4,6-dimethylphenol)
Uses: Antioxidant for solid rubber and latex processing, adhesives
Trade names: Naftonox® IMB

Isobutylene/MA copolymer
CAS 26426-80-2

Isobutylene/maleic anhydride copolymer

Synonyms: Isobutylene/maleic anhydride copolymer; 2-Methyl-1-propene, polymer with 2,5-furandione; 2,5-Furandione, polymer with 2-methyl-1-propene
Definition: Copolymer of isobutylene and maleic anhydride monomers
Empirical: $(C_4H_8 \cdot C_4H_2O_3)_x$
Uses: Dispersant for dyes, pigments; in cosmetics, latex paints, polymerization, leather tanning, water treatment
Trade names: Daxad® 31

Isobutylene/maleic anhydride copolymer. *See* Isobutylene/MA copolymer

2,2′-Isobutylidenebis (4,6-dimethylphenol)
CAS 33145-10-7
Uses: Antioxidant for rubber, latex
Trade names: Lowinox® 22IB46; Vulkanox® NKF

Isobutyl oleate
CAS 84988-79-4; EINECS 284-868-8
Definition: Ester of isobutyl alcohol and oleic acid
Uses: Chemical intermediate, lubricant; chemical synthesis; carbon source in antibiotic culture broths; lubricity improvers in min. oils; formulation of cutting, lamination, and textile oils, rust inhibitors; textile and leather industry; cosmetic emollients; plastics; improves ozone resist. of chloroprene rubber
Trade names: Priolube 1414

Isobutyl stearate
CAS 646-13-9, 85865-69-6; EINECS 211-466-1, 288-668-1
Synonyms: 2-Methylpropyl octadecanoate; Stearic acid, 2-methylpropyl ester
Definition: Ester of isobutyl alcohol and stearic acid
Empirical: $C_{22}H_{44}O_2$
Formula: $CH_3(CH_2)_{16}COOCH_2CH(CH_3)_2$
Properties: Waxy crystalline solid; m.w. 340.57; m.p. 20 C
Uses: Cosmetic emollient; inks, waterproof coatings, polishes, ointments, rubber mfg., dyes; plasticizer/lubricant for PVC, PS
Regulatory: FDA 21CFR §176.210, 177.2260, 177.2800, 178.3910
Trade names: Estol 1476

Isobutyltrimethoxysilane
CAS 18395-30-7
Empirical: $C_7H_9O_3Si$
Properties: Liq.; m.w. 178.3; dens. 0.93; b.p. 154-157 C; ref. index 1.396; flash pt. 14 C
Uses: Coupling agent, release agent, lubricant, blocking agent, chemical intermediate
Trade names: Dynasylan® IBTMO; Petrarch® I7810

Isocyanatopropyltriethoxysilane
CAS 24801-88-5
Formula: $(C_2H_5O)_3Si(CH_2)_3NCO$
Properties: M.w. 247.37
Toxicology: Corrosive lachrymator
Uses: Coupling agent, chem. intermediate, blocking agent, release agent, lubricant, primer, reducing agent
Manuf./Distrib.: Aldrich
Trade names: CI7840

Isocyanuric acid
CAS 108-80-5; EINECS 203-618-0
Synonyms: s-Triazine-2,4,6-triol; Cyanuric acid; 2,4,6-Trihydroxy-1,3,5-triazine
Definition: Ketone isomer of cyanuric acid
Empirical: $C_3H_3N_3O_3$
Formula: N=C(OH)N=C(OH)NCOH
Properties: Off-white cryst., odorless; m.w. 129.09; dens. 2.500 (20/4 C); m.p. > 360 C
Toxicology: Moderately toxic by ingestion; eye irritant; heated to dec., emits toxic fumes of NO_x and CN^-
Uses: Used to stabilize chlorine sol'ns. in swimming pools; bleaches, sanitizers; curing agent for elastomers
Manuf./Distrib.: Allchem Ind.; Am. Int'l. Chem.; Great Western; Nissan Chem. Ind.; San Yuan; 3V
Trade names: Zeonet® A

Isodecyl benzoate
Properties: Sp.gr. 0.95; m.p. -70 C; flash pt. (COC) 174 C; ref. index 1.4878
Uses: Plasticizer for PVC, PVAc, plastisols, adhesives, sealants, caulks
Manuf./Distrib.: Velsicol
Trade names: Benzoflex® 131

Isodecyl diphenyl phosphate
CAS 29761-21-5
Properties: Sp.gr. 1.069-1.079; flash pt. (COC) 241 C; ref. index 1.503-1.509
Uses: Flame-retardant plasticizer for PVC and copolymers, PVAc, acrylics; for finished film or coated fabric applics., vinyl plastisols; also for PVC adhesives, ethyl cellulose, NC, SBR and butyl rubbers
Manuf./Distrib.: Akzo Nobel; Ashland; Harwick; Monsanto
Trade names: Phosflex 390; Santicizer 148

Isodecyl methacrylate
CAS 29964-84-9
Classification: Monomer
Empirical: $C_{14}H_{26}O_2$
Formula: $CH_2=C(CH_3)COOC_{10}H_{21}$
Properties: M.w. 226.36
Precaution: Nonhazardous (DOT)
Toxicology: Moderately toxic by intraperitoneal route
Uses: Pressure-sensitive adhesives, coatings for leather, textiles, paper, nonwoven fiber, polymer modifier and stabilizer, visc. index improver, dispersion for plastics and rubber, floor waxes, potting compds., sealants
Manuf./Distrib.: CPS; Monomer-Polymer & Dajac; Rohm & Haas; Sartomer
Trade names: Ageflex FM-10; Sipomer® IDM

Isodecyl oleate
CAS 59231-34-4; EINECS 261-673-6
Synonyms: 9-Octadecenoic acid, isodecyl ester
Definition: Ester of branched chain decyl alcohols and oleic acid
Empirical: $C_{28}H_{54}O_2$
Formula: $CH_3(CH_2)_7CH=CH(CH_2)_7COOC_{10}H_{21}$
Properties: Wh. to straw liq., char. mild odor; sol. in peanut oil, 95% ethanol, IPM, oleyl alcohol; insol. in water, glycerin, propylene glycol; sp.gr. 0.858-0.864; ref. index 1.4540-1.4560
Toxicology: LD50 (rat, oral) > 40 ml/kg; nonirritating to eyes; mildly irritating to skin
Uses: Emollient, cosolv. for cosmetic prods.; emulsifier, lubricant, antistat, defoamer for metalworking, textiles, plastics, paper

Isodecyloxypropylamine
CAS 7617-78-9
Uses: Corrosion inhibitor, antistat, flotation reagent, emulsifiier; additive for fuel, lubricants, petrol refining; intermediate for surfactants, agric. chems.; crosslinking agent for epoxies
Trade names: Tomah PA-14

N-Isodecyloxypropyl 1-1,3-diaminopropane
CAS 72162-46-0
Uses: Corrosion inhibitor; chemical intermediate for textile foamers, surfactants, agric. chems.; additive for fuels, lubricants, petrol. refining; crosslinking agent for epoxies
Trade names: Tomah DA-14

Isodecyloxypropyl dihydroxyethyl methyl ammonium chloride
CAS 125740-36-5
Uses: Acid corrosion inhibitor, plastics and textile antistat, and emulsifier
Trade names: Tomah Q-14-2

Isododecyloxypropylamine
Uses: Corrosion inhibitor for metalworking fluids; antistat; flotation collector; additive for fuel, lubricant, petrol. refining; intermediate for surfactants, textile foamers, ethoxylates and agric. chem.; crosslinking agent for epoxies
Trade names: Tomah PA-16

N-Isododecyloxypropyl-1,3-diaminopropane
Uses: Intermediate for textile foaming agents, surfactants, ethoxylates, agric. chemicals; corrosion inhibitor for metalworking fluids; additive for fuels, lubricants, petrol. refining; crosslinking agent for epoxy resins
Trade names: Tomah DA-16

1H-Isoindole-1,3-(2H)-dione, 2-(cyclohexylthio). *See* N-(Cyclohexylthio)phthalimide
Isooctadecanoic acid. *See* Isostearic acid
Isooctylmercaptoacetate. *See* Isooctyl thioglycolate

Isooctyl oleate
Classification: Ester of isooctyl alcohol and oleic acid
Properties: M.w. 394; dens. 0.866; flash pt. 191 C
Uses: Plasticizer for SBR, CR
Trade names: Plasthall® 7059

Isooctyl palmitate
Properties: Sp.gr. 0.863 (20 C); m.p. 6-9 C; flash pt. (COC) 213 C
Uses: Plasticizer for PS, cellulose nitrate, ethyl cellulose
Manuf./Distrib.: Inolex

Isooctyl phthalate. *See* Diisooctyl phthalate

Isooctyl stearate
CAS 91031-48-0; EINECS 292-951-5
Definition: Ester of isooctyl alcohol and stearic acid
Uses: Emollient, solvent for cosmetics; plasticizer and lubricant for plastics, paper industry
Trade names: Priolube 1458

Isooctyl tallate
EINECS 269-788-3
Synonyms: Tall oil acid isooctyl ester
Properties: Sp.gr. 0.86; m.p. -60 C; flash pt. (COC) 216 C; ref. index 1.4592
Uses: Plasticizer for vinyls, ethyl cellulose, elastomers
Manuf./Distrib.: C.P. Hall; Harwick; Velsicol
Trade names: Acintol® 208; Plasthall® 100

Isooctyl thioglycolate
CAS 25103-09-7; EINECS 246-613-9
Synonyms: Isooctylmercaptoacetate; Isooctyl thioglyconate; Acetic acid, meracpto- isooctyl ester
Definition: Ester of thioglycolic acid and a mixt. of branched chain octyl alcohols
Empirical: $C_{10}H_{20}O_2S$
Formula: $HSCH_2COOCH_2C_6H_{15}$
Properties: M.w. 204.42; b.p. 125 C (17 mm)
Uses: Antioxidants, fungicides, oil additives, plasticizers, insecticides, stabilizers, polymerization modifiers, stabilizer for tin-sulfur compounds, stripping agent for polysulfide rubber, PVC stabilizer intermediate
Manuf./Distrib.: Bock, Bruno Chemische Fabrik KG; Elf Atochem N. Am.; Evans Chemetics; Witco/PAG

Isooctyl thioglyconate. *See* Isooctyl thioglycolate

Isophorone
CAS 78-59-1; EINECS 201-126-0
Synonyms: 3,3,5-Trimethyl-2-cyclohexen-1-one
Empirical: $C_9H_{14}O$
Properties: Sp.gr. 0.923
Uses: In solvent mixtures for finishes, for polyvinyl and nitrocellulose resins, pesticides, stoving lacquers
Manuf./Distrib.: Aceto; Allchem Ind.; Ashland; BP Chem.; Coyne; Elf Atochem N. Am.; Fabrichem; Great Western; Hüls AG; Union Carbide

Isophorone diamine
CAS 2855-13-2
Formula: $H_2NC_6H_7(CH_3)_3CH_2NH_2$
Properties: M.w. 170.30; dens. 0.922; m.p. 10 C; b.p. 247 C; flash pt > 110 C; ref. index 1.4880 (20 C)
Uses: Epoxy curative for coatings, adhesives, castings and composites

Manuf./Distrib.: Aldrich; Degussa
Trade names: Vestamin® IPD

Isoprene rubber. *See* Polyisoprene
Isopropanol. *See* Isopropyl alcohol

Isopropyl alcohol
CAS 67-63-0; EINECS 200-661-7
Synonyms: IPA; Isopropanol; 2-Propanol; Dimethyl carbinol
Classification: Aliphatic alcohol
Empirical: C_3H_8O
Formula: $(CH_3)_2CHOH$
Properties: Colorless liq., pleasant odor, sl. bitter taste; sol. in water, alcohol, ether, chloroform; m.w. 60.11; dens. 0.7863 (20/20 C); f.p. -86 C; b.p. 82.4 C (760 mm); flash pt. (TOC) 11.7 C; ref. index 1.3756 (20 C)
Precaution: Flamm.; very dangerous fire hazard with heat, flame, oxidizers; reacts with air to form dangerous peroxides; heated to decomp., emits acrid smoke and fumes
Toxicology: TLV:TWA 400 ppm; STEL 500 ppm; LD50 (oral, rat) 5045 mg/kg; poison by ingestion, subcutaneous routes; human systemic effects by ingestion/inhalation (headache, nausea, vomiting, narcosis); 100 ml can be fatal; experimental reproductive effects
Uses: Solv. for essential oils, alkaloids, gums, resins, cellulose derivs.; coatings; deicing agent for liq. fuels; lacquers; extraction processes; dehydrating agent; denaturing ethyl alcohol; in cosmetics; mfg. of acetone, glycerol, isopropyl acetate
Regulatory: FDA 21CFR §73.1 (no residue), 73.1001, 172.515, 172.560, 172.712, 173.240 (limitation 50 ppm in spice oleoresins, 6 ppm in lemon oil, 2% in hops extract), 173.340; 175.105, 176.200, 176.210, 177.1200, 177.2800, 178.1010, 178.3910; 27CFR §21.112; use in bread is permitted in Ireland and Japan
Manuf./Distrib.: Aldrich; Arco; Eastman; Exxon; Hüls AG; Mallinckrodt; Mitsui Toatsu; Shell; Union Carbide
Trade names containing: Adogen® 432; Ancamide 220-IPA-73; Aromox® C/12; Arquad® 2C-75; Arquad® 12-50; Arquad® 16-50; Arquad® 18-50; Arquad® C-50; Cyastat® 609; Cyastat® SN; Cyastat® SP; Ethoduoquad® T/15-50; Ethoquad® C/12 Nitrate; Ethoquad® CB/12; Gemtex PA-75; GP-180 Resin Sol'n.; GP-197 Resin Sol'n.; GP-RA-159 Silicone Polish Additive; Ken-React® 7 (KR 7); Ken-React® 9S (KR 9S); Ken-React® 12 (KR 12); Ken-React® 26S (KR 26S); Ken-React® 33DS (KR 33DS); Ken-React® 38S (KR 38S); Ken-React® 39DS (KR 39DS); Ken-React® 41B (KR 41B); Ken-React® 44 (KR 44); Ken-React® 46B (KR 46B); Ken-React® 55 (KR 55); Ken-React® 133DS (KR 133DS); Ken-React® 134S (KR 134S); Ken-React® 138S (KR 138S); Ken-React® 158FS (KR 158FS); Ken-React® 212 (KR 212); Ken-React® 238S (KR 238S); Ken-React® 262ES (KR 262ES); Ken-React® OPP2 (KR OPP2); Ken-React® OPPR (KR OPPR); Ken-React® KR TTS; KZ 55; KZ TPP; LICA 01; LICA 09; LICA 12; LICA 38; LICA 44; Monawet MM-80; MS-122N/CO_2; NZ 01; NZ 09; NZ 12; NZ 33; NZ 38; NZ 39; NZ 44; NZ 49; Surfynol® 104PA; Tomah Q-D-T; Vydax® Spray No. V312-A

Isopropyl alcohol, titanium (4+) salt. *See* Tetraisopropyl titanate

Isopropylamine dodecylbenzenesulfonate
CAS 26264-05-1; 68584-24-7; EINECS 247-556-2
Synonyms: Dodecylbenzenesulfonic acid, comp. with 2-propanamine (1:1)
Classification: Aromatic compd.
Definition: Salt of isopropylamine and doecylbenzene sulfonic acid
Empirical: $C_{21}H_{39}O_3NS$
Formula: $C_{18}H_{30}O_3S{\cdot}C_3H_9N$
Uses: Emulsifier, solubilizer, detergent, and wetting agent for oil-based systems; dispersant in oil and water-based systems; used in dry-cleaning surfactants; hydrotrope for liq. detergents; latex emulsifier; emulsion polymerization
Regulatory: FDA 21CFR §176.210
Trade names: Calimulse PRS; Polystep® A-11; Rhodacal® 330

Isopropyl 4-aminobenzenesulfonyl di (dodecylbenzenesulfonyl) titanate
Uses: Coupling agent, adhesion promoter, antioxidant, antistat, antifoam, accelerator, blowing agent activator, catalyst, curative, corrosion inhibitor, dispersion aid, emulsifier, flame retardant, foaming agent, grinding and process aid
Trade names containing: Ken-React® 26S (KR 26S)

p-Isopropylaminodiphenylamine. *See* N-Isopropyl-N′-phenyl-p-phenylenediamine
Isopropylcarbinol. *See* Isobutyl alcohol

Isopropyl dimethacryl isostearoyl titanate
Uses: Coupling agent, adhesion promoter, antioxidant, antistat, antifoam, accelerator, blowing agent activator, catalyst, curative, corrosion inhibitor, dispersion aid, emulsifier, flame retardant, foaming agent, grinding and process aid
Trade names containing: Ken-React® 7 (KR 7)

Isopropyl glycidyl ether
CAS 4016-14-2; EINECS 223-672-9
Synonyms: IPGE; Oxirane, [(1-methylethoxy)methyl]-,
Empirical: $C_6H_{12}O_2$
Formula: $(H_3C)_2CHOCH_2CHOCH_2$
Properties: Colorless liq., ethereal odor; sol. 20% in water; misc. with ethanol, methanol, toluene; m.w. 116.16; dens. 0.92 g/cc (20 C); b.p. 137 C; flash pt. 35 C
Precaution: Flamm.; may form explosive peroxides; incompat. with acids, alkalies, amines, oxidizing agents; hazardous decomp. prods.: CO, hydrocarbons
Toxicology: LD50 (oral, rat) 4200 mg/kg, (dermal, rabbit) 9650 mg/kg; harmful by inhalation, ingestion; may cause sensitization by inh. and skin contact
Uses: Reactive diluent for epoxy resins; for solv.-free coating systems, laminating resins, fiber-reinforced composites; stabilizer for chlorinated hydrocarbons; thickener for paints and varnishes
Manuf./Distrib.: Raschig

Isopropyl n-hexadecanoate. *See* Isopropyl palmitate
4,4´-Isopropylidenebis (2,6-dibromophenol). *See* Tetrabromobisphenol A

4,4´ Isopropylidenediphenol alkyl (C12-15) phosphites
CAS 96152-48-6; EINECS 306-120-2
Synonyms: Poly 4,4´ isopropylidenediphenol neodol 25 alcohol phosphite
Properties: Liq.; dens. 0.965-0.990; b.p. 165 C (5 mm); ref. index 1.5040-1.5095; flash pt. (COC) > 216 C
Uses: Chelating agent; color and lt. stabilizer for PVC and copolymers
Trade names: Weston® 439

4,4´-Isopropylidenediphenol. *See* Bisphenol A

Isopropyl myristate
CAS 110-27-0; EINECS 203-751-4
Synonyms: IPM; 1-Methylethyl tetradecanoate; Tetradecanoic acid, 1-methylethyl ester; Myristic acid isopropyl ester
Definition: Ester of isopropyl alcohol and myristic acid
Empirical: $C_{17}H_{34}O_2$
Formula: $CH_3(CH_2)_{12}COOCH(CH_3)_2$
Properties: Colorless oil, odorless; sol. in most org. solvs., veg. oil; dissolves waxes; insol. in water; m.w. 270.44; dens. 0.850-0.860; f.p. 3 C; b.p. 192.6 C (20 mm); dec. 208 C; ref. index 1.432-1.434 (25 C)
Toxicology: Suspected tumorigen; human skin irritant
Uses: Cosmetic creams, topical medicinals; plasticizer for cellulosics
Regulatory: FDA 21CFR §176.210, 177.2800
Manuf./Distrib.: Amerchol; Aquatec Quimica SA; Goldschmidt; Henkel/Emery; Inolex; Lanaetex; Penta Mfg.; Stepan; Unichema
Trade names: Estol IPM 1514

Isopropyl myristate/palmitate. *See* Isopropyl myristopalmitate

Isopropyl myristopalmitate
Synonyms: Isopropyl palmitate/myristate; Isopropyl myristate/palmitate
Definition: Ester of isopropyl alcohol and mixture of myristic and palmitic acids
Properties: Sp.gr. 0.852; flash pt. (COC) 182 C; ref. index 1.4340
Uses: Cosmetics emollient, solvent; plasticizer for PVC; lubricant for PS
Manuf./Distrib.: Inolex; Union Camp

Isopropyl oleate
CAS 112-11-8, 85116-87-6; EINECS 203-935-4, 285-540-7
Synonyms: 1-Methylethyl-9-octadecenoate; 9-Octadecenoic acid, 1-methylethyl ester
Definition: Ester of isopropyl alcohol and oleic acid
Empirical: $C_{21}H_{40}O_2$

Formula: $CH_3(CH_2)_7CH=CH(CH_2)_7COOCH(CH_3)_2$
Uses: Emollient and spreading agent in cosmetic preparations; lubricant for makeups; plastics
Regulatory: FDA 21CFR §176.210, 177.2800, 178.3570, 178.3910
Trade names: Estol 1406

Isopropyl palmitate
CAS 142-91-6; EINECS 205-571-1
Synonyms: IPP; Isopropyl n-hexadecanoate; Hexadecanoic acid, 1-methylethyl ester; 1-Methylethyl hexandecanoate
Definition: Ester of isopropyl alcohol and palmitic acid
Empirical: $C_{19}H_{38}O_2$
Formula: $CH_3(CH_2)_{14}COOCH(CH_3)_2$
Properties: Colorless liq.; sol. in 4 parts 90% alcohol, min. oil, fixed oils; insol. in water; m.w. 298.57; dens. 0.850-0.855 (25/25 C); m.p. 14 C; ref. index 1.4350-1.4390 (20 C)
Precaution: Combustible
Toxicology: Poison by intraperitoneal route; human skin irritant
Uses: Emollient, emulsifier in lotions, creams, and similar cosmetics; plasticizer for cellulosics
Regulatory: FDA 21CFR §176.210, 177.2800
Manuf./Distrib.: Amerchol; Aquatec Quimica SA; Goldschmidt; Henkel/Emery; Inolex; Lanaetex; Penta Mfg.; Stepan; Unichema
Trade names: Estol IPP 1517

Isopropyl palmitate/myristate. *See* Isopropyl myristopalmitate
Isopropyl percarbonate. *See* Diisopropyl perdicarbonate
Isopropyl peroxydicarbonate. *See* Diisopropyl perdicarbonate

o-Isopropylphenol
CAS 88-69-7; EINECS 201-852-8
Synonyms: OIPP
Formula: $(CH_3)_2CHC_6H_4OH$
Properties: M.w. 136.19; dens. 1.012; m.p. 15-16 C; b.p. 212-213 C; flash pt. 88 C; ref. index 1.5620 (20 C)
Precaution: Corrosive
Toxicology: Toxic
Manuf./Distrib.: Albemarle; Aldrich; Fluka

p-Isopropylphenol
CAS 99-89-3
Synonyms: PIPP
Manuf./Distrib.: AC Ind.; Alemark; Amber syn.

Isopropylphenyl diphenyl phosphate
Properties: Sp.gr. 1.15-1.17 (20 C); flash pt. (PMCC) 224 C; ref. index 1.546
Uses: Plasticizer
Manuf./Distrib.: Akzo Nobel; Harwick

N-Isopropyl-N′-phenyl-p-phenylenediamine
CAS 101-72-4
Synonyms: p-Isopropylaminodiphenylamine
Formula: $CH_3H_6NHC_6H_4NHC_6H_5$
Properties: Dark gray to black flakes; sol. in benzene, gasoline; insol. in water; m.w. 226.35; dens. 1.04 (25 C); f.p. 72-76 C
Toxicology: Moderately toxic by ingestion; severe eye irritant
Uses: Antioxidant/antiozonant; protection of rubbers against oxidation, ozone, flexcracking, poisoning by copper and manganese
Trade names: Permanax™ IPPD; Santoflex® IP; Vanox® 3C; Vulkanox® 4010 NA
Trade names containing: Rhenogran® IPPD-80

Isopropyl titanate. *See* Tetraisopropyl titanate

Isopropyl titanium triisostearate
CAS 61417-49-0; EINECS 262-774-8
Synonyms: Isopropyl triisostearoyl titanate; Tris(isooctadecanoato-O)(2-propanolato) titanium
Classification: Organic compd.

Isopropyl tri (dioctylphosphato) titanate

Formula: $C_{57}H_{112}O_7Ti$
Trade names containing: Ken-React® KR TTS

Isopropyl tri (dioctylphosphato) titanate
Uses: Coupling agent, adhesion promoter, antioxidant, antistat, antifoam, accelerator, blowing agent activator, catalyst, curative, corrosion inhibitor, dispersion aid, emulsifier, flame retardant, foaming agent, grinding and process aid
Trade names containing: Ken-React® 12 (KR 12)

Isopropyl tri (dioctylpyrophosphato) titanate
Uses: Coupling agent, adhesion promoter, antioxidant, antistat, antifoam, accelerator, blowing agent activator, catalyst, curative, corrosion inhibitor, dispersion aid, emulsifier, flame retardant, foaming agent, grinding and process aid
Trade names containing: Ken-React® 38S (KR 38S)

Isopropyl tridodecylbenzenesulfonyl titanate
Trade names containing: Ken-React® 9S (KR 9S)

Isopropyl tri (N ethylamino-ethylamino) titanate
Uses: Coupling agent, adhesion promoter, antioxidant, antistat, antifoam, accelerator, blowing agent activator, catalyst, curative, corrosion inhibitor, dispersion aid, emulsifier, flame retardant, foaming agent, grinding and process aid
Trade names containing: Ken-React® 44 (KR 44)

Isopropyl triisostearoyl titanate. *See* Isopropyl titanium triisostearate

Isostearic acid
CAS 2724-58-5, 30399-84-9; EINECS 220-336-3
Synonyms: Heptadecanoic acid, 16-methyl-; Isooctadecanoic acid; 16-Methylheptadecanoic acid
Definition: Mixture of branched chain 18 carbon aliphatic acids
Empirical: $C_{18}H_{36}O_2$
Uses: Cosmetics, chemicals, dispersant, softener in rubber compds., food packaging, suppositories, ointments
Manuf./Distrib.: Henkel/Emery; Nissan Chem. Ind.; Unichema; Union Camp

4-Isothiazolin-3-one, 5-chloro-2-methyl-. *See* Methylchloroisothiazolinone

N-Isotridecyloxypropyl 1,3-diaminopropane
Uses: Used in corrosion inhibitor formulations, oilfield chemicals; replacement for coco, soya, or oleyl diamine; intermediate for textile foamers, surfactants, agric. chems; crosslinking agent for epoxies
Trade names: Tomah DA-17

Isotridecyl stearate
CAS 31565-37-4
Definition: Ester of isotridecyl alcohol and stearic acid
Uses: Lubricant for PVC
Trade names: Sicolub® TDS

Isourea. *See* Urea
Isovalerone. *See* Diisobutyl ketone

Kaolin
CAS 1332-58-7; EINECS 296-473-8
Synonyms: Bolus alba; China clay; Hydrated aluminum silicate
Definition: Native hydrated aluminum silicate
Formula: $\approx Al_2O_3 \cdot 2SiO_2 \cdot 2H_2O$
Properties: White to yel. or grayish fine powd., earthy taste; insol. in water, dilute acids, alkali hydroxides; dens. 1.8-2.6
Toxicology: Nuisance dust
Uses: Filler and coatings for paper, rubber, refractories, ceramics; in anticaking preps., paint; adsorbent for clarification of liqs.
Regulatory: FDA 21CFR §178.3550, 182.2727, 182.2729, 186.1256, GRAS as indirect additive; BATF 27CFR §240.1051; Japan restricted (0.5% max. residual); Europe listed; UK approved

Manuf./Distrib.: Burgess Pigment; Dry Branch Kaolin; ECC Int'l.; Feldspar; J.M. Huber; Kaopolite; San Yuan; Thiele Kaolin; R.T. Vanderbilt; Whittaker, Clark & Daniels
Trade names: Bilt-Plates® 156; Dixie® Clay; Fiberfrax® 6000 RPS; Fiberfrax® 6900-70; Fiberfrax® 6900-70S; Fiberfrax® EF-119; Fiberfrax® EF 122S; Fiberfrax® EF 129; Fiberfrax® Milled Fiber; Fiberkal™ FG; Fiberkal™; Huber 35; Huber 40C; Huber 65A; Huber 70C; Huber 80; Huber 80B; Huber 80C; Huber 90; Huber 90B; Huber 90C; Huber 95; Huber HG; Huber HG90; Hydrite 121-S; Langford® Clay; McNamee® Clay; Par® Clay; Polyplate 90; Polyplate 852; Polyplate P; Polyplate P01
Trade names containing: Akrochem® DCP-40K; Akrochem® VC-40K; Aztec® DIPP-40-IC1; Colloids N 54/14/02; Kayphobe-ABO

Kaolinite
CAS 1318-74-7
Definition: Clay mineral; main constituent of kaolin
Formula: $Al_2O_3 \cdot 2SiO_2 \cdot 2H_2O$
Trade names containing: Aktisil® AM; Aktisil® EM; Aktisil® MM; Aktisil® PF 216; Aktisil® PF 231; Aktisil® VM; Aktisil® VM 56; Silfin® Z; Sillikolloid P 87; Sillitin N 82; Sillitin N 85; Sillitin V 85; Sillitin Z 86; Sillitin Z 89; TP 87/21; TP 89/50; TP 90/127; TP 91/01

KE clay
CAS 14504-95-1
Trade names containing: Di-Cup 40KE; Luperco 500-40KE; Varox® 802-40KE; Varox® DCP-40KE; Vul-Cup 40KE

Ketohexamethylene. *See* Cyclohexanone
Kieselguhr. *See* Diatomaceous earth
Lactic acid, ammonium salt. *See* Ammonium lactate
Lamp black. *See* Carbon black

Lanolin
CAS 8006-54-0 (anhyd.), 8020-84-6 (hyd.); EINECS 232-348-6
Synonyms: Anhydrous lanolin; Wool wax; Wool fat
Definition: Deriv. of unctuous fatty sebaceous secretion of sheep consistg. of complex mixt. of esters of high m.w. aliphatic, steroid, or triterpenoid alcohol and fatty acids
Properties: Yel.-wh. semisolid; sol. in chloroform, ether; insol. in water
Toxicology: Heated to decomp., emits acrid smoke and irritating fumes
Uses: Ointments, leather finishing, soaps, face creams, facial tissues, hair set and sun-tan preps.; plasticizer for rubber; lubricant for textiles, metalworking compds.
Regulatory: FDA 21CFR §172.615, 175.300, 176.170, 176.210, 177.1200, 177.2600, 178.3910; Japan approved
Manuf./Distrib.: Amerchol; Croda; Henkel/Emery; Lanaetex; RITA; Stevenson Cooper; Westbrook Lanolin
Trade names: Fancor Lanolin

Lard glyceride
CAS 61789-10-4, 97593-29-8; EINECS 263-032-6
Synonyms: Lard monoglyceride; Glycerides, lard mono-
Definition: Monoglyceride derived from lard
Uses: Food emulsifier; lubricant, antistat, antifog in plastics; component in w/o and o/w creams
Trade names: Grindtek MOP 90

Lard monoglyceride. *See* Lard glyceride

Lauralkonium chloride
CAS 139-07-1; EINECS 205-351-5
Synonyms: Lauryl dimethyl benzyl ammonium chloride; Benzyldimethyldodecylammonium chloride; N,N-Dimethyl-N-dodecylbenzenemethanaminium chloride; Dodecyl dimethyl benzyl ammonium chloride
Classification: Quaternary ammonium salt
Empirical: $C_{21}H_{38}N \cdot Cl$
Properties: M.w. 340.05
Toxicology: Skin and eye irritant; heated to decomp., emits very toxic fumes of NO_x, NH_3, and Cl^-
Uses: Disinfectant, germicide, bactericide, fungicide; antistat for plastics; emulsifier
Regulatory: FDA 21CFR §172.165 (limitation 0.25-1.0 ppm), 173.320 (0.05 ± 0.005 ppm), 175.105
Trade names: Dehyquart® LDB

Lauramide

CAS 1120-16-7; EINECS 214-298-7
Synonyms: Dodecanamide; Lauric acid amide; Lauryl amide
Classification: Aliphatic amide
Empirical: $C_{12}H_{25}NO$
Formula: $CH_3(CH_2)_{10}CONH_2$
Regulatory: FDA 21CFR §178.3860
Trade names containing: Hostastat® System E1956; Hostastat® System E5951; Hostastat® System E6952

Lauramide DEA

CAS 120-40-1; 52725-64-1; EINECS 204-393-1
Synonyms: Lauric diethanolamide; N,N-Bis(2-hydroxyethyl)dodecanamide; Diethanolamine lauric acid amide
Definition: Mixture of ethanolamides of lauric acid
Empirical: $C_{16}H_{33}NO_3$
Formula: $CH_3(CH_2)_{10}CON(CH_2CH_2OH)_2$
Properties: M.w. 287.50
Toxicology: Moderately toxic by ingestion
Uses: Foam booster/stabilizer, detergency and visc. builder, emulsifier, wetting agent for personal care prods., household and institutional detergents; antistat for thermoplastics
Regulatory: FDA 21CFR §172.710, 173.315, 175.105, 176.180, 176.210, 177.2260, 177.2800, 178.3130
Manuf./Distrib.: Chemron; Karlshamns; Mona; Norman, Fox; Pilot; Protameen; Sandoz; Scher; Stepan; Witco/ Oleo-Surf.
Trade names: Chemstat® LD-100; Chemstat® LD-100/60; Empilan® LDE; Ethylan® MLD
Trade names containing: Chemstat® AC-1000; Chemstat® AC-2000; Empilan® LDX

(3-Lauramidopropyl) trimethyl ammonium methyl sulfate

CAS 10595-49-0
Empirical: $C_{19}H_{42}O_5N_2S$
Properties: Off-wh. to lt. tan powd.; m.w. 410
Toxicology: Relatively low toxicity, but moderately to severely irritating to skin and eyes; LD50 (rat, oral) 1.8 g/kg; LD50 (rabbit, dermal) 2.8 g/kg
Uses: Antistatic agent for PVC, PS, polyolefins, ABS
Trade names: Cyastat® LS

Lauramine

CAS 124-22-1; 2016-57-1; EINECS 204-690-6
Synonyms: Lauryl amine; 1-Dodecanamine; Dodecylamine
Classification: Aliphatic amine
Empirical: $C_{12}H_{26}N$
Formula: $CH_3(CH_2)_{10}CH_2NH_2$
Uses: Emulsifier, flotation agent, corrosion inhibitor; lubricant for metal treatment; lubricant/mold release for rubber, textiles; antistat and antifog for plastic foils

Lauramine oxide

CAS 1643-20-5; 70592-80-2; EINECS 216-700-6
Synonyms: Lauryl dimethylamine oxide; N,N-Dimethyl-1-dodecanamine-N-oxide
Classification: Tertiary amine oxide
Empirical: $C_{14}H_{31}NO$
Uses: Foam stabilizer, thickener, emollient in cosmetics, detergents, textile softeners, foam rubber, electroplating, paper coatings, bleach
Trade names: Empigen® OB; Rhodamox® LO

Laureth-2

CAS 3055-93-4 (generic); 9002-92-0; 68002-97-1; 68439-50-9; EINECS 221-279-7
Synonyms: Diethylene glycol dodecyl ether; PEG-2 lauryl ether; 2-[2-(Dodecyloxy)ethoxy]ethanol
Definition: PEG ether of lauryl alcohol
Empirical: $C_{16}H_{34}O_3$
Formula: $CH_3(CH_2)_{10}CH_2(OCH_2CH_2)_nOH$, avg. n = 2
Uses: Emulsifier, foam booster, superfatting agent; solubilizer for solvents; bases for prod. of sulfates; raw material for dishwashing, cleansing agent and cold cleaners; emulsion polymerization
Trade names: Prox-onic LA-1/02

Laureth-3

CAS 3055-94-5; 9002-92-0 (generic); 68002-97-1; 68439-50-9; EINECS 221-280-2

Synonyms: Triethylene glycol dodecyl ether; 2-[2-[2-(Dodecyloxy)ethoxy]ethoxy]ethanol; PEG-3 lauryl ether

Definition: PEG ether of lauryl alcohol

Empirical: $C_{18}H_{38}O_4$

Formula: $CH_3(CH_2)_{10}CH_2(OCH_2CH_2)_nOH$, avg. n = 3

Uses: Detergent, emulsifier; solubilizer for solvents; bases for prod. of sulfates; raw material for dishwashing, cleansing agent and cold cleaners; emulsion polymerization

Trade names: Rhodasurf® A 24

Laureth-4

CAS 5274-68-0; 68002-97-1; 68439-50-9; EINECS 226-097-1

Synonyms: PEG-4 lauryl ether; PEG 200 lauryl ether; 3,6,9,12-Tetraoxatetracosan-1-ol

Definition: PEG ether of lauryl alcohol

Empirical: $C_{20}H_{42}O_5$

Formula: $CH_3(CH_2)_{10}CH_2(OCH_2CH_2)_nOH$, avg. n = 4

Uses: Emulsifier, solubilizer, lubricant, detergent for cosmetics, silicone polish, mold releases; bases for prod. of sulfates; raw material for dishwashing, cleansing agent and cold cleaners; antistat for PE, PS

Regulatory: FDA 21CFR §178.3520

Trade names: Ethosperse® LA-4; Prox-onic LA-1/04; Rhodasurf® L-4

Laureth-7

CAS 3055-97-8; 9002-92-0 (generic); EINECS 221-283-9

Synonyms: PEG-7 lauryl ether; POE (6) lauryl ether; 3,6,9,12,15,18,21-Heptaoxatritriacontan-1-ol

Definition: PEG ether of lauryl alcohol

Empirical: $C_{26}H_{54}O_8$

Formula: $CH_3(CH_2)_{10}CH_2(OCH_2CH_2)_nOH$, avg. n = 7

Uses: Thickener, emulsifier, visc. control, lubricant, solubilizer for personal care prods.; emulsion polymerization surfactant; textile spin finishes

Trade names: Incropol L-7; Rhodasurf® L-790

Laureth-9

CAS 3055-99-0; 9002-92-0 (generic); 68439-50-9; EINECS 221-284-4

Synonyms: PEG-9 lauryl ether; POE (9) lauryl ether; 3,6,9,12,15,18,21,24,27-Nonaoxanonatriacontan-1-ol

Definition: PEG ether of lauryl alcohol

Empirical: $C_{30}H_{62}O_{10}$

Formula: $CH_3(CH_2)_{10}CH_2(OCH_2CH_2)_nOH$, avg. n = 9

Uses: Wetting agent, detergent, emulsifier, dispersant used for maintenance and institutional cleaners, in textile, paper, and paint industries; emulsion polymerization

Regulatory: FDA 21CFR §177.2800, 178.3130

Trade names: Prox-onic LA-1/09

Laureth-12

CAS 3056-00-6; 9002-92-0 (generic); EINECS 221-286-5

Synonyms: PEG-12 lauryl ether; POE (12) lauryl ether; PEG 600 lauryl ether

Definition: PEG ether of lauryl alcohol

Empirical: $C_{36}H_{64}O_{13}$

Formula: $CH_3(CH_2)_{10}CH_2(OCH_2CH_2)_nOH$, avg. n = 12

Properties: M.w. 1199.57; m.p. 41-45 C; b.p. 100 C; flash pt. > 110 C

Uses: Emulsifier for cosmetic, pharmaceutical and industrial uses; emollient, thickener for shampoos; emulsion polymerization

Regulatory: FDA 21CFR §177.2800

Trade names: Prox-onic LA-1/012

Laureth-15

CAS 9002-92-0 (generic)

Synonyms: PEG-15 lauryl ether; POE (15) lauryl ether

Definition: PEG ether of lauryl alcohol

Formula: $CH_3(CH_2)_{10}CH_2(OCH_2CH_2)_nOH$, avg. n = 15

Uses: Coemulsifier for styrene and styrene/acrylic emulsion polymerization

Regulatory: FDA 21CFR §177.2800

Laureth-16

CAS 9002-92-0 (generic)
Synonyms: PEG-16 lauryl ether; POE (16) lauryl ether
Definition: PEG ether of lauryl alcohol
Formula: $CH_3(CH_2)_{10}CH_2(OCH_2CH_2)_nOH$, avg. n = 16
Uses: Emulsifier, dispersant, surfactant; emulsion polymerization
Regulatory: FDA 21CFR §177.2800
Trade names: Akyporox RLM 160

Laureth-23

CAS 9002-92-0 (generic)
Synonyms: PEG-23 lauryl ether; POE (23) lauryl ether
Definition: PEG ether of lauryl alcohol
Formula: $CH_3(CH_2)_{10}CH_2(OCH_2CH_2)_nOH$, avg. n = 23
Uses: Emulsifier, stabilizer, solubilizer, surfactant, emollient, thickener, dispersant for cosmetics; post ad stabilizer for syn. latexes
Regulatory: FDA 21CFR §177.2800
Trade names: Prox-onic LA-1/023; Rhodasurf® L-25

Lauric acid

CAS 143-07-7; EINECS 205-582-1
Synonyms: n-Dodecanoic acid; Dodecanoic acid; Dodecoic acid
Classification: Fatty acid
Empirical: $C_{12}H_{24}O_2$
Formula: $CH_3(CH_2)_{10}COOH$
Properties: Colorless needles; insol. in water; sol. in benzene and ether; m.w. 200.36; dens. 0.833; m.p. 44 C; b.p. 225 (100 mm); ref. index 1.4323 (45 C)
Precaution: Combustible when exposed to het or flame; reactive with oxidizing materials
Toxicology: LD50 (oral, rat) 12 g/kg; poison by intravenous route; mildly toxic by ingestion; mutagenic data; heated to decomp., emits acrid smoke and irritating fumes
Uses: Alkyd resins, wetting agents, soaps, detergents, cosmetics, insecticides, food additives; emulsifier/ stabilizer for polymerization of nitrile rubbers and latex
Regulatory: FDA 21CFR §172.210, 172.860, 173.340, 175.105, 175.320, 176.170, 176.200, 176.210, 177.1010, 177.1200, 177.2260, 177.2600, 177.2800, 178.3570, 178.3910
Manuf./Distrib.: Akzo Nobel; Aldrich; Brown; Condor; Henkel/Emery; Mirachem Srl; Penta Mfg.; Spectrum Chem. Mfg.; Unichema; Welch, Holme & Clark; Witco/Oleo-Surf.
Trade names: Prifrac 2920; Prifrac 2922

Lauric acid amide. *See* Lauramide
Lauric acid, 1,2-ethanediyl ester. *See* Glycol dilaurate
Lauric diethanolamide. *See* Lauramide DEA

Lauric/myristic dimethylethyl ammonium ethosulfate

Uses: Noncorrosive mold release agent; antistat
Trade names containing: Larostat® 377 DPG

Lauroamphopropionate. *See* Sodium lauroamphopropionate
Laurox. *See* Lauroyl peroxide

N-Lauroyl-p-aminophenol

Uses: Processing aid for thermoplastics
Trade names: Suconox-12®

Lauroyl chloride

CAS 112-16-3; EINECS 203-941-7
Definition: Dodecanoyl chloride
Empirical: $C_{12}H_{23}ClO$
Formula: $CH_3(CH_2)_{10}COCl$
Properties: M.w. 218.77; dens. 0.964; b.p. 134-137 C (11 mm); ref. index 1.4450 (20 C0
Uses: Surfactant, polymerization initiator, antienzyme agent, foamer; synthesis of lauroyl peroxide, sodium lauroyl sarcosinate, other sarcosinates
Manuf./Distrib.: Akzo Nobel; Aldrich; D&O; Elf Atochem N. Am.; Hüls AG; PPG Industries; SNPE N. Am.

Lauroyl peroxide
CAS 105-74-8
Synonyms: Bis (1-oxododecyl) peroxide; Dodecanoyl peroxide; Dilauroyl peroxide; Laurox
Classification: Organic peroxide
Empirical: $C_{24}H_{46}O_4$
Formula: $(C_{11}H_{23}CO)O_2$
Properties: White coarse powd., faint odor, tasteless; sol. in oil, most org. solvs., alcohols; insol. in water; m.w. 398.70; m.p. 53-55 C
Precaution: Dangerous fire and explosive risk; will ignite org. materials
Toxicology: Toxic by ingestion and inhalation; corrosive irritant to eyes, skin, mucous membranes; causes burns
Uses: Bleaching agent, oxidizing agent, intermediate and drying agent for fats, oils, and waxes; polymerization catalyst/initiator
Manuf./Distrib.: Akzo Nobel; Elf Atochem N. Am.; Monomer-Polymer & Dajac
Trade names: Alperox-F

Lauroyl sarcosine
CAS 97-78-9; EINECS 202-608-3
Synonyms: N-Methyl-N-(1-oxododecyl) glycine
Definition: N-lauryl deriv. of N-methylglycine
Empirical: $C_{15}H_{29}NO_3$
Formula: $CH_3(CH_2)_{10}CONCH_3CH_2COOH$
Uses: Detergent, corrosion inhibitor, foam booster and stabilizer, wetting agent, lubricant, emulsifier used in dentifrices, personal care, and household cleaning prods., pharmaceuticals, metal processing and finishing, metalworking and cutting oils; emulsion polymerization; mold release for RIM; stabilizer for polyols; antifog, antistat in plastics
Regulatory: FDA 21CFR §177.1200, 178.3130
Manuf./Distrib.: Chemplex; Hampshire; R.T. Vanderbilt
Trade names: Hamposyl® L

N-Lauroylsarcosine sodium salt. *See* Sodium lauroyl sarcosinate

Laurtrimonium chloride
CAS 112-00-5; EINECS 203-927-0
Synonyms: Lauryl trimethyl ammonium chloride; Dodecyl trimethyl ammonium chloride; N,N,N-Trimethyl-1-dodecanaminium chloride
Classification: Quaternary ammonium salt
Empirical: $C_{15}H_{34}N \cdot Cl$
Precaution: Nonflammable
Uses: Preservative, surfactant, corrosion inhibitor, antistat, emulsifier for plastics, textiles, paper, cosmetics; gel sensitizer in latex foam
Regulatory: USA permitted; JSCI, Europe listed
Trade names: Chemquat 12-33; Chemquat 12-50; Dehyquart® LT; Octosol 562; Octosol 571
Trade names containing: Arquad® 12-50

Lauryl alcohol
CAS 112-53-8; 68526-86-3; EINECS 203-982-0
Synonyms: 1-Dodecanol; C-12 linear primary alcohol; Alcohol C-12; Dodecyl alcohol
Classification: Fatty alcohol
Empirical: $C_{12}H_{26}O$
Formula: $CH_3(CH_2)_{10}CH_2OH$
Properties: Colorless leaflets, liq. above 21 C, floral odor; insol. in water; sol. in alcohol, ether; m.w. 186.33; dens. 0.8309 (24/4 C); m.p. 24 C; b.p. 259 C (760 mm); flash pt. (CC) > 212 F; ref. index 1.440-1.444
Precaution: Combustible; reactive with oxidizing materials
Toxicology: LD50 (oral, rat) 12,800 mg/kg; moderately toxic by intraperitoneal route; mildly toxic by ingestion; severe human skin irritant; heated to decomp., emits acrid smoke and irritating fumes
Uses: Mfg. of sulfuric acid esters which are used as wetting agents; synthetic detergents, cosmetics, lube additives, pharmaceuticals, rubber, textiles, perfumes, flavoring agents; emulsion polymerization
Regulatory: FDA 21CFR §172.515, 172.864, 175.105, 175.300, 177.1010, 177.1200, 177.2800, 178.3480, 178.3910
Manuf./Distrib.: Albemarle; Aldrich; M. Michel; Penta Mfg.; Procter & Gamble; Schweizerhall; Vista
Trade names: Epal® 12; Laurex® NC
Trade names containing: Epal® 12/70; Epal® 12/85; Epal® 1214

Lauryl amide. *See* Lauramide
Lauryl amine. *See* Lauramine
Lauryl dimethylamine. *See* Dimethyl lauramine
Lauryl dimethylamine oxide. *See* Lauramine oxide
Lauryl dimethyl benzyl ammonium chloride. *See* Lauralkonium chloride
m-Lauryl mercaptan. *See* n-Dodecyl mercaptan

Lauryl methacrylate
CAS 142-90-5; EINECS 205-570-6
Synonyms: Dodecyl methacrylate; Dodecyl 2-methyl-2-propenoate
Definition: Ester of lauryl alcohol and methacrylic acid
Empirical: $C_{16}H_{30}O_2$
Formula: $CH_2=C(CH_3)COOC_{12}H_{25}$
Properties: M.w. 254.43; bulk dens. 0.868 g/ml; b.p. 272-344 C; flash pt. (COC) 132 C
Precaution: Nonhazardous (DOT); combustible
Toxicology: Skin and eye irritant
Uses: Lube oil additives, coatings for nonwoven fiber, floor waxes, paints, adhesives, varnishes, sealants, caulks, stabilizers in nonaq. disp. and inks
Manuf./Distrib.: Albright & Wilson Am.; CPS; Fluka; Monomer-Polymer & Dajac; Rhone-Poulenc Spec.; Rohm Tech
Trade names containing: SR-313

Lauryl-myristyl methacrylate
Uses: Monomer for mfg. of polymers used as oil additives, visc. index improvers, pour pt. depressants; internal plasticizer for adhesives, uv-curable resins, coatings
Trade names containing: Empicryl® 6047

Lauryl/stearyl thiodipropionate
Uses: Antioxidant for stabilizing polyolefins; applications incl. pipe, hot-melt adhesives, and molded olefin prods.
Trade names: Cyanox® 1212; Mark® 5095

Lauryl trimethyl ammonium chloride. *See* Laurtrimonium chloride
LDPE. *See* Polyethylene, low-density

Lead
CAS 7439-92-1; EINECS 231-100-4
Classification: Metallic element
Empirical: Pb
Properties: Metallic, heavy ductile soft gray solid; sol. in dilute nitric acid; insol. in water; at.wt. 207.2; dens. 11.35; m.p. 327.4 C; b.p. 1755 C
Precaution: Noncombustible
Toxicology: Poison by ingestion; moderately toxic by intraperitoneal route; toxic by inhalation of dust or fume; TLV 0.15 mg/m^3 of air; cumulative poison
Uses: Storage batteries, gasoline additive, radiation shielding, cable covering, ammunition, chemical reaction equip., solder and fusible alloys; in heat and light stabilizers for PVC
Manuf./Distrib.: Asarco; Cerac; Noah

Lead-barium-cadmium complex
Uses: Stabilizer/lubricant system for PVC window profiles

Lead carbonate (basic)
CAS 598-63-0; EINECS 215-290-6
Synonyms: White lead; Dibasic lead carbonate; Lead subcarbonate
Formula: $2PbCO_3 \cdot Pb(OH)_2$
Properties: White amorphous powd.; sol. in acids; insol. in water; m.w. 267.20; dens. 6.86; dec. 400 C
Precaution: Noncombustible
Toxicology: Toxic by inhalation; TLV (as Pb) 0.15 mg/m^3 of air; moderately toxic by ingestion
Uses: Heat and light stabilizer for PVC; exterior paint pigment, ceramic glazes
Manuf./Distrib.: Halstab; Spectrum Chem. Mfg.
Trade names: Halcarb 20; Halcarb 20 EP; Halcarb 200; Halcarb 200 EP; Haro® Chem PC
Trade names containing: Halcarb 20S; Prespersion PAC-2941

Lead chlorosilicate complex
Properties: Fine white powd.; dens. 3.4
Uses: Stabilizer for vinyl electric insulation and tapes
Trade names: Lectro 60

Lead diamyldithiocarbamate
CAS 36501-84-5
Uses: Accelerator for nat. and polyisoprene rubber; EP and antiscuff agent, antioxidant, corrosion inhibitor; multifunctional additive for industrial lubricating oils and greases, crankcase and industrial gear oils
Trade names: Amyl Ledate®

Lead dimethyldithiocarbamate
CAS 19010-66-3
Formula: $Pg[SCSN(CH_3)_2]_2$
Properties: White powd.; insol. in all common solvs.; sl. sol. in cyclohexanone; m.w. 447.63; dens. 2.43; m.p. 310 C
Toxicology: Experimental tumorigen; mutagenic data
Uses: Vulcanization accelerator with litharge; accelerator for NR, SBR, IIR, IR, BR
Trade names: Methyl Ledate

Lead dioxide
CAS 1309-60-0; EINECS 215-174-5
Synonyms: Lead (IV) oxide; Lead peroxide; Lead superoxide
Empirical: O_2Pb
Formula: PbO_2
Properties: M.w. 239.19
Precaution: Oxidizer
Toxicology: Irritant
Uses: Catalyst, curing agent for polysulfide, butyl and polyisoprene rubber; oxidizer in mfg. of dyes; pyrotechnics
Manuf./Distrib.: All Chemie Ltd; Atomergic Chemetals; Johnson Matthey; Spectrum Chem. Mfg.
Trade names: LP-100; LP-200; LP-300; LP-400

Lead fumarate tetrabasic
Uses: Vulcanizing and curing agent for chlorosulfonated and crosslinked PE
Trade names: Lectro 78

Lead maleate, tribasic
Uses: Vulcanizing agent for chlorosulfonated PE
Trade names: Tri-Mal

Lead metasilicate. *See* Lead silicate
Lead monoxide. *See* Lead (II) oxide

Lead oxide
Synonyms: Lead suboxide; Lead oxide, black; Litharge, leaded
Empirical: Pb_2O
Properties: Black amorphous material; insol. in water; sol. in acids and bases; m.w. 223.19; dens. 8.342; dec. on heating
Toxicology: TLV (as Pb) 0.15 mg/m^3 of air; toxic material; moderately toxic by ingestion, intraperitoneal route; skin irritant
Uses: In storage batteries; processing aid in rubber compounding
Manuf./Distrib.: Chemson Ltd.; Hammond Lead Prods.; Nihon Kagaku Sangyo
Trade names: Naftopast® Litharge A

Lead (II) oxide
CAS 1317-36-8; EINECS 215-267-0
Synonyms: Lead (II) oxide yellow; Plumbous oxide; Lead monoxide; Calcined litharge; Litharge
Empirical: PbO
Properties: Yel. cryst.; sol. in acids, alkalies; insol. in water; dens. 9.53; m.p. 888 C
Toxicology: Toxic by ing. and inh.; TLV 0.15 mg/m^3 of air (as Pb)
Uses: Activator, vulcanizing agent for rubber; mfg. of dry colors, greases, high-pressure lubricants, brake linings, ceramics, glass, piezoelectric devices, various chemical processes

Lead (II,III) oxide

Manuf./Distrib.: Fluka
Trade names: Litharge 28; Litharge 33
Trade names containing: Chem-Master R-11; Chem-Master R-51; Chem-Master R-231; Kenlastic® K-6642; Kenmix® LItharge/P 5/1; Kenmix® LItharge/P 9/1; Poly-Dispersion® PLD-90; Poly-Dispersion® PLD-90P; Poly-Dispersion® T (HRL)D-90; Poly-Dispersion® T (LC)D-90; Poly-Dispersion® T (LC)D-90P; Rhenogran® PbO-80/FPM

Lead (II,III) oxide. *See* Lead oxide, red
Lead (IV) oxide. *See* Lead dioxide
Lead oxide, black. *See* Lead oxide

Lead oxide, red
CAS 1314-41-6
Synonyms: Red lead oxide; Minium; Lead (II,III) oxide; Red lead; Lead tetroxide
Empirical: Pb_3O_4
Properties: Red powd.; partly sol. in acids; insol. in water; m.w. 685.57
Precaution: Oxidizer
Toxicology: Irritant; toxic as dust; TLV (as Pb) 0.15 mg/m^3 of air
Uses: Storage batteries, glass, pottery, enameling, varnish, purification of alcohol, packing pipe joings, metal protective paints, fluxes, ceramic glazes; processing aid for rubber
Manuf./Distrib.: Aldrich
Trade names: Naftopast® Red Lead A, P
Trade names containing: Poly-Dispersion® ERD-90; Poly-Dispersion® GRD-90; Poly-Dispersion® TRD-90P

Lead (II) oxide yellow. *See* Lead (II) oxide
Lead peroxide. *See* Lead dioxide

Lead phosphite dibasic
CAS 1344-40-7
Formula: $2PbO \cdot PbHPO_3 \cdot {}^1/_2H_2O$
Properties: Fine white acicular crystals; insol. in water; dens. 6.94
Toxicology: Toxic material; TLV (as Pb) 0.15 mg/m^3 of air
Uses: Heat and light (UV screening and antioxidizing) stabilizer for vinyl plastics and chlorinated paraffins; in paints and plastics
Trade names: Dyphos; Dyphos Envirostab; Dyphos XL; Halphos; Haro® Chem PDF; Haro® Chem PDF-B
Trade names containing: Halphos S; Poly-Dispersion® E (DPS)D-80; Poly-Dispersion® G (DPS)D-80; Prespersion PAC-363; Prespersion PAC-3936

Lead phthalate, basic
CAS 6838-85-3; EINECS 290-588-7
Synonyms: 1,2-Benzenedicarboxylic acid, lead (2+) salt, basic
Formula: $2PbO \cdot PbHPO_3 \cdot {}^1/_2H_2O$
Properties: Fine white acicular cryst.; insol. in water; dens. 6.94
Toxicology: TLV (Pb) 0.15 mg/m^3 of air
Uses: Heat and light stabilizer, antioxidant for vinyl plastics and chlorinated paraffins; in paints and plastics
Trade names containing: EPlthal 121

Lead phthalate, dibasic
CAS 69011-06-9
Formula: $C_6H_4(COO)_2Pb \cdot PbO$
Properties: Fluffy white crystalline powd.; insol. in water; dens. 4.5
Toxicology: Toxic by inhalation and skin absorption
Uses: Heat and light stabilizer for general vinyl use
Trade names: Dythal; Dythal XL; EPlthal 120; Halthal; Halthal EP; Haro® Chem PDP-E
Trade names containing: Halthal S; Poly-Dispersion® E (DYT)D-80; Poly-Dispersion® G (DYT)D-80; Poly-Dispersion® H (DYT)D-80; Poly-Dispersion® H (DYT)D-80P; Poly-Dispersion® T (DYT)D-80P; Prespersion PAB-8912

Lead silicate
CAS 10099-76-0, 11120-22-2
Synonyms: Lead metasilicate
Empirical: $O_3Si \cdot Pb$
Properties: White crystalline powd.; insol. in most solvs.; m.w. 283.28; dens. 6.49; m.p. 766 C

Precaution: Noncombusible
Toxicology: Toxic material; TLV (as Pb) 0.15 mg/m^3
Uses: Pigment, heat stabilizer, rust inhibitor; in ceramics, fireproofing fabrics, paints, plastics
Manuf./Distrib.: Eagle-Picher; Hammond Lead Prods.
Trade names: BSWL 202
Trade names containing: BSWL 201; Poly-Dispersion® H (202)D-80; Poly-Dispersion® H (202)D-80P

Lead silicosulfate, basic
Toxicology: Toxic material; TLV (as Pb) 0.15 mg/m^3 of air
Uses: Corrosive inhibitive pigment for metal protective coatings, primers, and finishes; industrial enamels requring a high gloss; heat stabilizer for PVC
Trade names: EPIstatic® 110
Trade names containing: EPIstatic® 111; EPIstatic® 113

Lead stearate
CAS 1072-35-1
Synonyms: Stearic acid, lead salt
Formula: ≈ $Pb(C_{18}H_{35}O_2)_2$
Properties: White powd.; sol. in hot alcohol; insol. in water; m.w. 774.25; dens. 1.4; m.p. 100-115 C
Precaution: Combustible
Toxicology: Toxic material, absorbed by the skin; mildly toxic by ingestion
Uses: Varnish and lacquer drier; in extreme pressure lubricants; lubricant in extrusion processes; stabilizer for vinyl polymers; corrosion inhibitor for petroleum; component of greases, waxes, and paints
Manuf./Distrib.: AC Ind.; Cookson Specialty; Ore & Chem. Corp.; Syn. Prods.; R.T. Vanderbilt
Trade names: DS-207; Hal-Lub-N; Hal-Lub-N TOTM; Haro® Chem P28G; Leadstar; Leadstar Envirostab; Synpro® Lead Stearate
Trade names containing: Hal-Lub-N S

Lead stearate, dibasic
Uses: Stabilizer for PVC; lubricant
Trade names: Hal-Lub-D; Haro® Chem P51
Trade names containing: Hal-Lub-D S

Lead subcarbonate. *See* Lead carbonate (basic)
Lead suboxide. *See* Lead oxide

Lead sulfate, basic
Synonyms: White lead, sublimed; White lead sulfate
Formula: $PbSO_4{\bullet}PbO$
Properties: White monoclinic crystals; sl. sol. in hot water or acids; dens. 6.92; m.p. 977 C
Precaution: Noncombustible
Toxicology: Toxic material; TLV (as Pb) 0.15 mg/m^3 of air
Uses: Heat stabilizer for PVC; paints, ceramics, pigments
Manuf./Distrib.: Halstab; National Chem.; Spectrum Chem. Mfg.
Trade names: Halbase 11; Halbase 100; Halbase 100 EP

Lead sulfate, polybasic
Uses: Stabilizer for rigid and plasticized PVC applics.
Trade names: Haro® Chem PPCS-X

Lead sulfate, tetrabasic
Uses: Stabilizer for PVC

Lead sulfate, tribasic
Formula: $3PbO{\bullet}PbSO_4{\bullet}H_2O$
Properties: Fine white powd.; dens. 6.4
Toxicology: Toxic material; TLV (as Pb) 0.15 mg/m^3
Uses: Heat stabilizer for electrical and other vinyl compds.
Manuf./Distrib.: Halstab; San Yuan
Trade names: EPIstatic® 100; Halbase 10; Halbase 10 EP; Haro® Chem PTS-E; Tribase; Tribase EXL Special; Tribase XL
Trade names containing: EPIstatic® 101; Halbase 10S; Poly-Dispersion® AV (TBS)D-80; Poly-Dispersion® AV (TBS)D-80P; Prespersion PAC-332; Prespersion PAC-4893

Lead sulfophthalate
Trade names: Lectro 90

Lead superoxide. *See* Lead dioxide
Lead tetroxide. *See* Lead oxide, red

Lead-zinc complex
Uses: Stabilizer/activator for chemically blown PVC
Trade names: Lankromark® LZ638

Lecithin
CAS 8002-43-5; 8029-76-3; 97281-47-5; EINECS 232-307-2
Definition: Mixture of the diglycerides of stearic, palmitic and oleic acids linked to the choline ester of phosphoric acid; found in plants and animals
Formula: $C_8H_{17}O_5NRR'$, R and R′ are fatty acid groups
Properties: Nearly white to yel. or brown waxy mass or thick fluid; insol. but swells in water and salt sol'ns.; sol. in chloroform, ether, petrol. ether, min. oils, fatty acids; sol. in 12 parts abs. alcohol; dens. 1.0305 (24/4 C); sapon. no. 196
Toxicology: Heated to decomp., emits acrid smoke and irritating fumes
Uses: Edible surfactant and emulsifier for food use, pharmaceuticals, cosmetics, leather treatment, textiles; release agent for silicone rubber molds
Regulatory: FDA 21CFR §133.169, 133.173, 133.179, 136.110, 136.115, 136.130, 137.160, 136.165, 136.180, 163.123, 163.130, 163.135, 163.140, 163.145, 163.150, 163.155, 166.40, 166.110, 169.115, 169.140, 169.150, 175.300, 184.1400, GRAS; USDA 9CFR §318.7, 0.5% max. in oleomargarine, 381.147; Japan approved; Europe listed; UK approved
Manuf./Distrib.: Am. Lecithin; Central Soya; W.A. Cleary; Great Western; Int'l. Chem.; Lucas Meyer GmbH; Penta Mfg.; Solvay Duphar BV; Spice King; U.S. Biochemical
Trade names: Molder's Edge ME-232
Trade names containing: Paintable Organic Oil Spray No. O316

Light mineral oil. *See* Mineral oil
Lignite wax. *See* Montan wax
Lignosulfonic acid, sodium salt. *See* Sodium lignosulfonate
Ligroin. *See* Mineral spirits
Lime. *See* Calcium oxide

Limestone
CAS 1317-65-3
Synonyms: Calcium carbonate, natural; Agricultural limestone; Lithographic stone
Empirical: $CaCO_3$
Properties: Solid, odorless, tasteless; insol. in water; sol. in dilute acids; m.w. 100.09; dens. 2.7-2.95; m.p. 825 C
Precaution: Noncombustible
Toxicology: TLV/TWA 10 mg/m^3 (total dust); severe eye and moderate skin irritant
Uses: Filler for rubber, plastics, building stone, caulks, cements, ceramics, coatings, metallurgy (flux), mfg. of lime; source of CO_2; Portland and natural cement; removal of sulfur dioxide from stack gases and sulfur from coal
Trade names: Dolocron® 15-16; Dolocron® 32-15; Dolocron® 40-13; Dolocron® 45-12; Franklin T-11; Franklin T-12; Franklin T-13; Franklin T-14; Franklin T-325; Marblewhite® MW200; Marblewhite® MW325; Marblewhite® MW A 200; Marblewhite® MW A 325; Pfinyl® 402; Super-Fil®; Vicron® 10-20; Vicron® 15-15; Vicron® 25-11; Vicron® 31-6; Vicron® 41-8; Vicron® 45-3
Trade names containing: Mr 245

Lime water. *See* Calcium hydroxide
Linear C10 alpha olefin. *See* Decene-1
Linear C16 alpha olefin. *See* Hexadecene-1
Linear C18 alpha olefin. *See* Octadecene-1
Liquid paraffin. *See* Mineral oil
Liquid petrolatum. *See* Mineral oil
Litharge. *See* Lead (II) oxide
Litharge, leaded. *See* Lead oxide

Lithium

CAS 7439-93-2; EINECS 231-102-5
Definition: Metallic element; at. no. 3
Empirical: Li
Properties: Very soft silvery light metal; sol. in liq. ammonia; at. wt. 6.941; dens. 0.534 (20 C); m.p. 179 C; b.p. 1317 C; Mohs hardness 0.6
Precaution: Ignites in air near its m.p.; dangerous fire/explosive risk exposed to water, acids
Toxicology: Toxic to CNS
Uses: Scavenger and degasifier for stainless and mild steels in molten state; deoxidizer in copper and alloys; rocket propellants; pharmaceuticals
Manuf./Distrib.: Atomergic Chemetals; Cerac; Eagle-Picher; FMC; Leverton-Clarke Ltd; Noah
Trade names containing: Buna BL 6533

Lithium 12-hydroxystearate

Empirical: $C_{18}H_{35}O_3Li$
Formula: $LiOOC(CH_2)_{10}CHOH(CH_2)_5CH_3$
Properties: White powd.; m.p. 205 C
Precaution: Combustible
Uses: Lubricating greases; stabilizer for thermosets, oils, fats, lubricants
Manuf./Distrib.: Witco/Oleo-Surf.
Trade names: Haro® Chem LHG

Lithium octadecanoate. *See* Lithium stearate

Lithium ricinoleate

Formula: $LiOOC_{17}H_{32}OH$
Properties: White powd.; insol. or limited sol. in most org. solvs.; m.p. 174 C
Precaution: Combustible
Uses: Alcoholysis catalyst for alkyds and polyesters
Manuf./Distrib.: CasChem

Lithium stearate

CAS 4485-12-5; EINECS 224-772-5
Synonyms: Lithium octadecanoate; Octadecanoic acid, lithium salt
Definition: Lithium salt of stearic acid
Empirical: $C_{18}H_{36}O_2 \cdot Li$
Formula: $CH_3(CH_2)_{16}COOLi$
Properties: Wh. cryst.; insol. in cold and hot water, alcohol, ethyl acetate; dens. 1.025; m.p. 220 C
Uses: Cosmetics, waxes, greases, lubricant in powd. metallurgy, corrosive inhibitor, flatting agent, high-temp. lubricant; lubricant and stabilizer for resins esp. nylon; mold release; waterproofing
Regulatory: FDA 21CFR §175.300
Manuf./Distrib.: Ashland; Chemetall GmbH; Cookson Specialty; Schweizerhall; Syn. Prods.; Witco/PAG
Trade names: Afco-Chem LIS; Mathe Lithium Stearate; Synpro® Lithium Stearate; Witco® Lithium Stearate 306

Lithographic stone. *See* Limestone

Lithopone

CAS 1345-05-7
Synonyms: Griffith's zinc white
Definition: Mixture of zinc sulfide (26-60%), barium sulfate, and zinc oxide
Properties: White
Toxicology: Poison liberating hydrogen sulfide upon dec. by heat, moisture, acids
Uses: White pigment in plastics, paints, paper; provides thixotropy, improves gloss and flow
Manuf./Distrib.: Landers-Segal Color; Ore & Chem. Corp.; Primachem; Sachtleben GmbH; San Yuan
Trade names: Lithopone 30% DS; Lithopone 30% L; Lithopone D (Red Seal 30% ZnS); Lithopone D (Silver Seal 60% ZnS); Lithopone DS (Red Seal 30% ZnS); Lithopone L (Red Seal 30% ZnS); Lithopone L (Silver Seal 60% ZnS)
Trade names containing: Plaswite® LL 7041; Plaswite® LL 7105; Plaswite® LL 7108; Plaswite® PE 7005; Plaswite® PS 7037; Plaswite® PS 7231

LLDPE. *See* Polyethylene, linear low density

MAA. *See* Methyl acetoacetate
Macrogol 300. *See* PEG-6
Macrogol 600. *See* PEG-12
Macrogol 1000. *See* PEG-20
Macrogol 1540. *See* PEG-32
Macrogol 6000. *See* PEG-150
Magnesia. *See* Magnesium oxide
Magnesia magma. *See* Magnesium hydroxide
Magnesia usta. *See* Magnesium oxide
Magnesite. *See* Magnesium carbonate, Magnesium carbonate hydroxide

Magnesium aluminum carbonate
CAS 13539-23-0, 11097-59-9, 96492-31-8, 119758-00-8, 136618-52-5, 136618-51-4; EINECS 234-319-3, 215-222-5
Uses: Heat stabilizer for PVC
Trade names: Alcamizer 1; Alcamizer 2; Alcamizer 4; Alcamizer 4-2; Alcamizer 5

Magnesium aluminum hydrotalcite
CAS 11097-59-9
Uses: Antioxidant
Trade names containing: Ultranox® 2714A

Magnesium-aluminum-zinc complex
Uses: Stabilizer for plastisol processing, artificial leather, wallpaper, conveyor belts, toys, automotive applics.
Trade names: Bärostab® CT 1500; Bärostab® NT 1005

Magnesium calcium carbonate hydroxide
Uses: Fire retardant, smoke suppressant for engineering thermoplastics, coatings
Trade names: Hydramax™ HM-C9, HM-C9-S

Magnesium carbonate
CAS 546-93-0, 29409-82-0; EINECS 208-915-9
Synonyms: Magnesium (II) carbonate (1:1); Magnesite; Magnesium carbonate precipitated; Carbonate magnesium
Definition: Basic dehydrated magnesium carbonate or a normal hydrated magnesium carbonate
Empirical: CO_3•Mg
Properties: Light bulky wh. powd., odorless; sol. in acids; insol. in alcohol, water; m.w. 84.32; dens. 3.04; dec. 350 C; ref. index 1.52
Precaution: Noncombustible; incompat. with formaldehyde
Toxicology: TLV:TWA 10 mg/m^3; heated to decomp., emits acrid smoke and irritating fumes
Uses: Magnesium salts, heat insulation and refractory, rubber reinforcing agent, inks, glass, pharmaceuticals, dentifrices, cosmetics, table salt, antacid; used in foods as drying agent, color retention agent, anticaking agent; plastics flame retardant and filler
Regulatory: FDA 21CFR §133.102, 133.106, 133.111, 133.141, 133.165, 133.181, 133.183, 133.195, 137.105, 137.155 137.160, 137.165, 137.170, 137.175, 137.180, 137.185, 163.110, 177.2600, 184.1425, GRAS; Japan approved (0.5% max.); Europe listed; UK approved
Manuf./Distrib.: Giulini; Lonza; Magnesia GmbH; Mallinckrodt; Marine Magnesium; Martin Marietta; Morton Int'l.; Whittaker, Clark & Daniels
Trade names: Magocarb-33

Magnesium (II) carbonate (1:1). *See* Magnesium carbonate
Magnesium carbonate basic. *See* Magnesium carbonate hydroxide

Magnesium carbonate hydroxide
CAS 12125-28-9, 39409-82-0, 56378-72-4 (pentahydrate); EINECS 235-192-7
Synonyms: Basic magnesium carbonate; Magnesite; Magnesium carbonate basic; Magnesium, tetrakis[carbonato(2-)] dihydroxypenta-; Magnesium hydroxide carbonate
Classification: Inorganic basic carbonate
Formula: $(MgCO_3)_4 \cdot Mg(OH)_2 \cdot 5H_2O$
Properties: Wh. powd., odorless; sol. in about 3300 parts CO_2-free water; sol. in dil. acids with effervescence; insol. in alcohol; m.w. 485.69
Uses: Flame retardant filler for elastomers, plastics

Regulatory: FDA 21CFR §184.1425, GRAS
Manuf./Distrib.: Aldrich; EM Industries; Lohmann; Spectrum Chem. Mfg.
Trade names: Elastocarb® Tech Heavy; Elastocarb® Tech Light; Elastocarb® UF

Magnesium carbonate precipitated. *See* Magnesium carbonate
Magnesium hydrate. *See* Magnesium hydroxide
Magnesium hydrogen metasilicate. *See* Talc

Magnesium hydroxide
CAS 1309-42-8 (anhyd.); EINECS 215-170-3
Synonyms: Magnesium hydrate; Milk of magnesia; Magnesia magma
Classification: Inorganic base
Empirical: H_2MgO_2
Formula: $Mg(OH)_2$
Properties: White amorphous powd., odorless; sol. in sol'n. of ammonium salts and dilute acids; almost insol. in water and alcohol; m.w. 58.33; dens. 2.36; m.p. 350 C (dec.)
Precaution: Noncombustible; incompat. with maleic anhdyride
Toxicology: Variable toxicity
Uses: Intermediate for obtaining magnesium metal, sugar refining, medicine (antacid, laxative), residual fuel oil additive, sulfite pulp, uranium processing, dentifrices, in foods as drying agent, frozen desserts; plastics flame retardant and filler
Regulatory: FDA 21CFR §184.1428, GRAS; Europe listed; UK approved
Manuf./Distrib.: Aldrich; Climax Performance; Croxton & Garry Ltd; J.W.S. Delavau; J.M. Huber/Solem; Mallinckrodt; Morton Int'l.; Reheis
Trade names: ACM-MW; BAX 1091FM; Flamtard M7; FR-20; Hydramax™ HM-B8, HM-B8-S; Kisuma 5A; Kisuma 5B; Kisuma 5E; Magnifin® H5; Magnifin® H5B; Magnifin® H5C; Magnifin® H5C/3; Magnifin® H5D; Magnifin® H-7; Magnifin® H7A; Magnifin® H7C; Magnifin® H7C/3; Magnifin® H7D; Magnifin® H-10; Magnifin® H10A; Magnifin® H10B; Magnifin® H10C; Magnifin® H10D; Magnifin® H10F; Magoh-S; MGH-93; Versamag® DC; Versamag® Tech.; Versamag® UF; Zerogen® 10; Zerogen® 15; Zerogen® 30; Zerogen® 35; Zerogen® 60
Trade names containing: Magnum-White; Plastigel PG9037; Plastigel PG9089

Magnesium hydroxide carbonate. *See* Magnesium carbonate hydroxide
Magnesium octadecanoate. *See* Magnesium stearate

Magnesium oxide
CAS 1309-48-4; EINECS 215-171-9
Synonyms: Magnesia; Periclase; Calcined magnesia; Magnesia usta
Classification: Inorganic oxide
Empirical: MgO
Properties: White powd., odorless; sl. sol. in water; sol. in acids, ammonium salt sol'ns.; m.w. 40.31; dens. 0.36; m.p. 2800 C; b.p. 3600 C
Precaution: Noncombustible; violent reaction or ignition with interhalogens; incandescent reaction with phopshorus pentachloride
Toxicology: Toxic by inhalation of fume; experimental tumorigen; TLV (as magnesium) 10 mg/m^3 (fume)
Uses: Refractories, esp. for steel furnace linings, polycrystalline ceramic for aircraft windshields, elec. insulations, pharmaceuticals, cosmetics, inorg. rubber accelerator, paper mfg.; white color standard; reflector in optical instruments; thickener for polyester resins
Regulatory: FDA 21CFR §163.110, 175.300, 177.1460, 177.2400, 177.2600, 182.5431, 184.1431, GRAS; Japan restricted; Europe listed
Manuf./Distrib.: Cerac; EM Industries; Harwick; Hüls Am.; Magnesia GmbH; Mallinckrodt; Marine Magnesium; Morton Int'l.
Trade names: Elastomag® 100; Elastomag® 100R; Elastomag® 170; Elastomag® 170 Micropellet; Elastomag® 170 Special; Ken-Mag®; MagChem® 20M; MagChem® 30; MagChem® 30G; MagChem® 35; MagChem® 35K; MagChem® 40; MagChem® 50; MagChem® 50Y; MagChem® 60; MagChem® 125; MagChem® 200-AD; MagChem® 200-D; Maglite® D; Magotex™; Naftopast® MgO-A; Plastomag® 170; Rhenomag® G1; Rhenomag® G3
Trade names containing: Cri-Spersion CRI-ACT-45; Cri-Spersion CRI-ACT-45-1/1; Cri-Spersion CRI-ACT-45-LV; Plastigel PG9033; Rapidblend 1793; Rhenogran® CHM21-40/FPM; Rhenogran® MgO-40/FPM; Scorchguard '0'

Magnesium stearate
CAS 557-04-0; EINECS 209-150-3
Synonyms: Magnesium octadecanoate; Octadecanoic acid, magnesium salt
Definition: Magnesium salt of stearic acid
Empirical: $C_{36}H_{70}MgO_4$
Formula: $[CH_3(CH_2)_{16}COO]_2Mg$
Properties: Soft white powd., tasteless, odorless; insol. in water, alcohol; dec. by dilute acids; m.w. 591.27; dens. 1.028; m.p. 88.5 C (pure)
Precaution: Nonflamm.
Toxicology: Heated to decomp., emits acrid smoke and toxic fumes
Uses: Baby dusting powder; lubricants in making tablets; drier in paints and varnishes; release, stabilizer, and lubricant for plastics; emulsifying agent for cosmetics
Regulatory: FDA 21CFR §172.863, 173.340; 175.105, 175.300, 175.320, 176.170, 176.200, 176.210, 177.1200, 177.2260, 178.3910, 179.45, 181.22, 181.29, 184.1440, GRAS; must conform to FDA specs for salts of fats or fatty acids derived from edible oils; Europe listed; UK approved
Manuf./Distrib.: Cometals; Cookson Specialty; EM Industries; Ferro/Grant; Magnesia GmbH; Mallinckrodt; Miljac; Norac; San Yuan; Syn. Prods.; Witco/PAG
Trade names: ADK STAB MG-ST; Afco-Chem MGS; Cometals Magnesium Stearate; Haro® Chem MF-2; Haro® Chem MF-3; HyQual NF; HyTech RSN 1-1; Mathe Magnesium Stearate; Miljac Magnesium Stearate; Petrac® Magnesium Stearate MG-20 NF; Synpro® Magnesium Stearate 90; Witco® Magnesium Stearate D

Magnesium, tetrakis[carbonato(2-)] dihydroxypenta-. *See* Magnesium carbonate hydroxide

Magnesium-zinc complex
Uses: Heat and light stabilizer for PVC
Trade names: ADK STAB AC-113; ADK STAB OF-30G; ADK STAB SC-35; Bärostab® NT 213; MiRaStab 154-K

Maleic anhydride
CAS 108-31-6; EINECS 203-571-6
Synonyms: 2,5-Furandione; Toxilic anhydride; cis-Butenedioic anhydride
Empirical: $C_4H_2O_3$
Properties: Colorless needles; sol. in water, acetone, alcohol, dioxane; partly sol. in chloroform, benzene; m.w. 98.06; dens. 0.934 (20/4 C); m.p. 53 C; b.p. 200 C; flash pt. 218 F
Toxicology: Irritant to tissues; TLV 0.25 ppm in air
Uses: Polyester resins, alkyd coating resins; fumaric and tartaric acid mfg.; pesticides; preservative for oils and fats; pharmaceuticals
Manuf./Distrib.: Allchem Ind.; Amoco; Aristech; Ashland; BP Chem.; Brown; Elf Atochem SA; Hüls AG; Mitsui Toatsu; Monsanto; OxyChem; Primachem; Spectrum Chem. Mfg.
Trade names containing: Polybond® 3001; Polybond® 3002; Polybond® 3009

MAMSP. *See* 4-α-Methylstyrylphenol
Manganese ammonium pyrophosphate. *See* Manganese violet
Manganese binoxide. *See* Manganese dioxide
Manganese black. *See* Manganese dioxide

Manganese dioxide
CAS 1313-13-9; EINECS 215-202-6
Synonyms: Manganese oxide; Manganese binoxide; Manganese black; Manganese peroxide
Empirical: MnO_2
Properties: Cryst.; slowly dissolves in cold HCl; insol. in water; m.w. 86.94
Precaution: Strong oxidizer; avoid heating with organic matter or other oxidizable substances
Toxicology: LD (IV, rabbit) 45 mg/kg
Uses: Oxidizing agent, depolarizer in dry cell batteries, pyrotechnics, matches, catalyst, laboratory reagent, scavenger and decolorizer, textile dyeing, source of metallic manganese (as pyrolusite), steel, amethyst glass, decolorizing glass; painting on porcelain; curing agent for polysulfide and epichlorohydrin rubbers
Manuf./Distrib.: Aldrich; Atomergic Chemetals; Eagle-Picher; Hoechst Celanese; Kerr-McGee; Nichia Kagaku Kogyo; Prince Mfg.; San Yuan; Shepherd Color

Manganese oxide. *See* Manganese dioxide
Manganese peroxide. *See* Manganese dioxide

Manganese violet
CAS 10101-66-3; EINECS 233-257-4
Synonyms: Ammonium manganese pyrophosphate; Pigment violet 16; CI 77742; Diphosphoric acid, ammonium manganese (3+) salt (1:1:1); Manganese ammonium pyrophosphate
Classification: Inoganic salt
Empirical: $H_4O_7P_2 \cdot H_3N \cdot Mn$
Uses: Colorant for plastics, powd. coatings, artists' colors, and cosmetics
Regulatory: FDA 21CFR §73.2775
Trade names: Manganese Violet

MBMC. *See* 6-t-Butyl-m-cresol
MBOCA. *See* 4,4′-Methylenebis (2-chloraniline)
MBPC. *See* 2-t-Butyl-p-cresol
MBS. *See* Methacrylate/butadiene styrene
MBT. *See* 2-Mercaptobenzothiazole
MBTS. *See* Benzothiazyl disulfide
MC. *See* Methylcellulose
MCP. *See* Calcium phosphate monobasic monohydrate
MDA. *See* 4,4′-Methylene dianiline

MDI
CAS 101-68-8
Synonyms: Methylene di-p-phenylene isocyanate; Diphenylmethane-4,4′-diisocyanate; Methylene bisphenyl isocyanate
Formula: $CH_2(C_6H_4NCO)_2$
Properties: Lt. yel. fused solid; sol. in acetone, benzene, kerosene, nitrobenzene; m.w. 250.27; dens. 1.197 (70 C); m.p. 37.2 C; b.p. 194-199 C (5 mm)
Precaution: Combustible
Toxicology: TLV:CL 0.01 ppm in air; TLV 0.005 ppm; toxic by inhalation of fumes; strong irritant
Uses: Vulcanizing agent for rubbers; prep. of polyurethane resin and spandex fibers; bonding rubber to rayon and nylon
Manuf./Distrib.: Allchem Ind.; BASF; ICI Polyurethanes; Miles
Trade names: Desmodur® MP-225; Isonate® 125M; Isonate® 143L; Isonate® 181; Isonate® 240; Novor 950

MEA. *See* Ethanolamine

MEA-lauryl sulfate
CAS 4722-98-9; 68908-44-1; EINECS 225-214-3
Synonyms: Sulfuric acid, monododecyl ester, compd. with 2-aminoethanol (1:1); Monoethanolamine lauryl sulfate
Definition: Monoethanolamine salt of sulfated lauryl alcohol
Empirical: $C_{12}H_{26}O_4S \cdot C_2H_7NO$
Formula: $CH_3(CH_2)_{10}CH_2OSO_3H \cdot H_2NCH_2CH_2OH$
Uses: Surfactant in the mfg. of personal care prods.; emulsifier in the mfg. of rubber latices and for resins; bactericidal detergents
Trade names: Empicol® LQ33/T

MEHQ. *See* Hydroquinone monomethyl ether
MEK. *See* Methyl ethyl ketone
MEK peroxide. *See* Methyl ethyl ketone peroxide

Melamine cyanurate
Uses: Flame retardant for plastics
Manuf./Distrib.: Akzo Nobel; Allchem Ind.; Charkit; Chemie Linz N. Am.; DSM Melamine Am.; Miljac; 3V
Trade names: Fyrol® MC

Melamine/formaldehyde resin
CAS 9003-08-1
Synonyms: Melamine resin; 1,3,5-Triazine,2,4,6-triamine, polymer with formaldehyde
Classification: Amino resin
Definition: Reaction prod. of melamine and formaldehyde
Formula: $(CH_3H_6N_6 \cdot CH_2O)_x$
Properties: Syrup or insol. powd.; sol. in water

Melamine phosphate

Uses: Thermosetting resin; crosslinking agent for coatings
Regulatory: FDA 21CFR §175.105, 175.300, 175.320, 177.1200, 177.1460, 177.2260, 177.2470, 181.22, 181.30
Manuf./Distrib.: Akzo Nobel; Am. Cyanamid; Astro Industries; Bakelite GmbH; BASF; Monsanto; Rhone-Poulenc Water Treatment; Sybron
Trade names: Cymel 373; Cymel 380

Melamine phosphate
Uses: Flame retardant for polymers
Manuf./Distrib.: Akzo Nobel; Albright & Wilson Am.; Chemie Linz N. Am.; DSM Melamine Am.; Miljac
Trade names: Amgard® NH; Fyrol® MP

Melamine pyrophosphate
Uses: Flame retardant for polymers
Manuf./Distrib.: Akzo Nobel; Anhydrides & Chems.; Miljac; StanChem
Trade names: Fyrol® MPP

Melamine resin. *See* Melamine/formaldehyde resin
Mercaptan C_{12}. *See* n-Dodecyl mercaptan
Mercaptoacetic acid 2-ethylhexyl ester. *See* 2-Ethylhexyl thioglycolate

2-Mercaptobenzimidazole
CAS 583-39-1; EINECS 209-502-6
Synonyms: 2-Benzimidazolethiol
Empirical: $C_7H_6N_2S$
Properties: M.w. 150.20; m.p. 301-305 C
Toxicology: Irritant
Uses: Antidegradant for rubber
Manuf./Distrib.: Aceto; Aldrich; Biddle Sawyer; Charkit; Eastern Chem
Trade names containing: Rhenogran® MBI-80

2-Mercaptobenzothiazole
CAS 149-30-4; EINECS 205-736-8
Synonyms: MBT; 2(3H)-Benzothiazolethione; 2-Benzothiazolethiol
Empirical: $C_7H_5NS_2$
Formula: CHCHCHCHCCSC(SN)N
Properties: Yel. powd., distinctive odor; sol. in dilute caustic, alcohol, acetone, benzene, chloroform; insol. in water, gasoline; m.w. 167.25; dens. 1.52; m.p. 164-175 C
Precaution: Combustible
Uses: Vulcanization accelerator for rubber (requires stearic acid for full activation); tire treads; mechanical specialties, fungicide, corrosion inhibitor, petrol prod.; extreme pressure additive in greases
Manuf./Distrib.: Allchem Ind.; Charkit; Cytec; Monsanto; Uniroyal
Trade names: Akrochem® MBT; Captax®; Ekaland MBT; Naftocit® MBT; Naftopast® MBT-P; Perkacit® MBT; Rokon; Rotax®; Thiotax®
Trade names containing: Poly-Dispersion® EMD-75; Rhenovin® MBT-70; Vulkacit Merkapto/C; Vulkacit Merkapto/MGC

2-Mercaptobenzothiazole disulfide. *See* Benzothiazyl disulfide
Mercaptobenzthiazyl ether. *See* Benzothiazyl disulfide
2-Mercaptoimidazoline. *See* Ethylene thiourea
2-Mercapto-4(5)-methyl benzimidazole. *See* 4- and 5-Methylmercaptobenzimidazole
3-Mercapto-1,2-propanediol. *See* Thioglycerin

3-Mercaptopropionic acid
CAS 107-96-0; EINECS 203-537-0
Formula: $HS\text{-}CH_2CH_2COOH$
Properties: M.w. 106.14; m.p. 16.8 C; b.p. 110.5-111.5 C (15 mm)
Precaution: Corrosive
Toxicology: Toxic
Uses: Intermediate for hair waves/straighteners, depilatories in cosmetics; plastics stabilizer intermediate; reaction intermediate for radiation-cured and other plastics
Manuf./Distrib.: Aldrich; Evans Chemetics; Witco/PAG

3-Mercaptopropylmethyldimethoxysilane
CAS 31001-77-1
Uses: Coupling agent, chem. intermediate, blocking agent, release agent, lubricant, primer, reducing agent
Trade names: Dynasylan® 3403; Petrarch® M8450

Mercaptopropyltrimethoxysilane
CAS 4420-74-0; EINECS 224-588-5
Synonyms: gamma-Mercaptopropyltrimethoxy silane; Trimethoxysilylpropanethiol
Empirical: $C_6H_{16}O_3SSi$
Properties: Liq.; m.w. 196.37; dens. 1.05; b.p. 219-220 C; ref. index 1.440; flash pt. 93 C
Toxicology: Moderately toxic by ingestion, intraperitoneal route; mildly toxic by skin contact; skin irritant
Uses: Filler for sulfur and metal oxide-cured systems; coupling agent, chem. intermediate, blocking agent, release agent, lubricant
Trade names: Dynasylan® MTMO; Petrarch® M8500; Prosil® 196
Trade names containing: Aktisil® MM

γ-Mercaptopropyltrimethoxy silane. *See* Mercaptopropyltrimethoxysilane

Mercaptosilane
Manuf./Distrib.: Hüls AG
Trade names containing: Ciptane® 255LD; Ciptane® I

2-Mercaptotoluimidazole
CAS 53988-10-6
Synonyms: Mercaptotolylimidazole
Uses: Antioxidant, antidegradant for rubbers
Trade names: Vanox® MTI

Mercaptotolylimidazole. *See* 2-Mercaptotoluimidazole

Methacrylamide
CAS 79-39-0; EINECS 201-202-3
Empirical: C_4H_7NO
Formula: CH_2:C(CH_3)$CONH_2$
Properties: M.w. 85.11; m.p. 106-109 C
Toxicology: Toxic
Uses: Used in self-crosslinking emulsions, heat-curing coatings, crosslinked acrylic sheets
Manuf./Distrib.: Aldrich; Fluka; Mitsui Toatsu; Monomer-Polymer & Dajac; Polysciences; Rohm Tech
Trade names: BM-801

Methacrylate/butadiene styrene
Synonyms: MBS
Uses: Impact modifier, toughener for vinyl and engineering plastics
Trade names: Paraloid® BTA-702; Paraloid® BTA-715; Paraloid® BTA-730; Paraloid® BTA-733; Paraloid® BTA-751; Paraloid® BTA-753; Paraloid® BTA-III-N2; Paraloid® EXL-3611; Paraloid® EXL-3647; Paraloid® EXL-3657; Paraloid® EXL-3691

Methacrylate copolymer
Synonyms: Methacrylic acid copolymer
Uses: Visc. index improver, pour pt. depressant, low temp. sludge dispersant; in hydraulic fluids
Trade names containing: Empicryl® 6059; Empicryl® 6070

Methacrylic acid copolymer. *See* Methacrylate copolymer
Methacrylic acid 2,3-epoxypropyl ester. *See* Glycidyl methacrylate
Methacrylic acid, tetrahydrofurfuryl ester. *See* Tetrahydrofurfuryl methacrylate

Methacrylic anhydride
CAS 760-93-0; EINECS 212-084-8
Formula: $[H_2C=C(CH_3)CO]_2O$
Properties: M.w. 154.17
Precaution: Corrosive; moisture-sensitive
Uses: Crosslinking agent; photoresists; prod. of methacrylates and methacrylamides

3-Methacryloxypropyltrimethoxysilane

Manuf./Distrib.: Aldrich; Monomer-Polymer & Dajac; Rohm Tech
Trade names: BM-723

3-Methacryloxypropyltrimethoxysilane
CAS 2530-85-0; EINECS 219-785-8
Synonyms: γ-Methacryloxypropyltrimethoxysilane
Classification: Methacrylate silane
Empirical: $C_{10}H_{20}O_5Si$
Formula: CH_2:$CCH_3COO(CH_2)_3Si(OCH_3)_3$
Properties: Liq.; sol. in acetone, benzene, ether, methanol, hydrocarbons; m.w. 248.1; dens. 1.045; b.p. 80 (1 mm); ref. index 1.429 (25 C); flash pt. 135 F
Precaution: Combustible; moderate fire risk
Uses: Coupling agent to promote resin-to-glass, resin-to-metal, resin-to-resin bonds, unsaturated polyesters, acrylics, EVA, adhesives
Trade names: Dow Corning® Z-6030; Dynasylan® MEMO; Petrarch® M8550; Prosil® 248

γ-Methacryloxypropyltrimethoxysilane. *See* 3-Methacryloxypropyltrimethoxysilane
2 (Methacryloyloxy) ethyl acetoacetate. *See* 2-(Acetoacetoxy) ethyl methacrylate
Methanecarboxylic acid. *See* Acetic acid
Methane, oxybis-. *See* Dimethyl ether
Methanol. *See* Methyl alcohol
Methenamine. *See* Hexamethylene tetramine
Methocel. *See* Methylcellulose

Methoxydiglycol
CAS 111-77-3; EINECS 203-906-6
Synonyms: Diethylene glycol methyl ether; Diethylene glycol monomethyl ether; Ethanol, 2-(2-methoxyethoxy)-; (2-β-Methyl 'Carbitol'), methoxyethoxy ethanol
Classification: Aliphatic ether alcohol
Empirical: $C_5H_{12}O_3$
Formula: $CH_3OCH_2CH_2OCH_2CH_2OH$
Properties: Water-white liq., hygroscopic; sol. in water; m.w. 120.17; dens. 1.0211 (20/4 C); b.p. 194 C; flash pt. 93.3 C; ref. index 1.4264 (27 C)
Precaution: Combustible
Toxicology: Moderately toxic by skin contact, intraperitoneal route; mildly toxic by ingestion
Uses: Solvent; brake fluid component; intermediate
Regulatory: FDA 21CFR §175.105
Manuf./Distrib.: Ashland; Brown; Eastman; Great Western; Harcros; Hoechst AG; Oxiteno; Union Carbide
Trade names: Hisolve DM

Methoxy dipropylene glycol. *See* PPG-2 methyl ether

Methoxyethanol
CAS 109-86-4; EINECS 203-713-7
Synonyms: Ethylene glycol methyl ether; Ethylene glycol monomethyl ether; 2-Methoxyethanol; Methyl Cellosolve®
Classification: Aliphatic ether alcohol
Empirical: $C_3H_8O_2$
Formula: $CH_3OCH_2CH_2OH$
Properties: Colorless liq., mild agreeable odor.; misc. with water, alcohol, ether, benzene, glycerol, acetone, dimethylformamide; m.w. 76.10; dens. 0.964 (20/4 C); f.p. -86.5 C; b.p. 123-124 C (760 mm); flash pt. (OC) 115 F; ref. index 1.4028 (20 C)
Precaution: Flamm. on exposure to heat and flame; mod. explosion hazard; can react with oxidizers to form explosive peroxides
Toxicology: TLV:TWA 5 ppm (skin); LD50 (oral, rat) 2460 mg/kg; mod. toxic to humans by ingestion; human systemic effects; experimental teratogen, reproductive effects; mutagenic data; skin and eye irritant; heated to dec., emits acrid smoke, irritating fumes
Uses: Solvent for cellulose acetate, natural resins, some synthetic resins, paints, inks, dyes, leather dyeing, nail polishes, varnishes, enamels, wood stains; plastics and plasticizers; in modified Karl Fischer reagent
Regulatory: FDA 21CFR §73.1 (no residue), 175.105; USDA 9CFR §381.147

Manuf./Distrib.: Arco; Ashland; Brown; Great Western; Harcros; Hoechst AG; Oxiteno; OxyChem; Union Carbide
Trade names: Hisolve MC

2-Methoxyethanol. *See* Methoxyethanol
Methoxyethene, homopolymer. *See* Polyvinyl methyl ether

Methoxyethyl acrylate
CAS 3121-61-7
Classification: Monomer
Empirical: $C_6H_{10}O_4$
Properties: Liq.; m.w. 130.15; dens. 1.1034 (20 C); b.p. 61 C (17 mm); flash pt. (OC) 180 F
Precaution: Flamm. on exposure to heat, flame, or sparks
Toxicology: Poison by skin contact, moderately toxic by ingestion and inhalation
Uses: Solv.-resistant elastomers, polyacrylate rubbers, UV-curable reactive diluent, soft contact lenses, PVC impact modifier, fabric coatings, barrier coatings for polyethylene, textile coatings
Manuf./Distrib.: Ashland; CPS; Monomer-Polymer & Dajac
Trade names: Ageflex MEA
Trade names containing: SR-244

2-Methoxyethyl ether. *See* Diethylene glycol dimethyl ether

Methoxyisopropanol
CAS 107-98-2; EINECS 203-539-1
Synonyms: Monopropylene glycol methyl ether; Polypropylene glycol monomethyl ether; Propylene glycol methyl ether; 1-Methoxy-2-propanol
Classification: Aliphatic ether alcohol
Empirical: $C_4H_{10}O_2$
Formula: $CH_3OCH_2CH(OH)CH_3$
Properties: Colorless liq.; m.w. 90.14; dens. 0.92; m.p. -96.7 C; b.p. 120 C; flash pt. 102 F
Precaution: Combustible; moderate fire risk
Toxicology: TLV:TWA 100 ppm; STEL 150 ppm; moderately toxic by intravenous route; mildly toxic by ingestion, inhalation, skin contact
Uses: Antifreeze and coolant for diesel engines; solvent for paints, cleaners, inks, electronics, textile dyes
Regulatory: FDA 21CFR §175.105, 181.22, 181.30
Trade names: Arcosolv® PM
Trade names containing: Casamid® 350PM

2-Methoxymethylethoxypropanol. *See* PPG-2 methyl ether
2-Methoxy-2-methylpropane. *See* Methyl t-butyl ether
4-Methoxyphenol. *See* Hydroquinone monomethyl ether
1-Methoxy-2-propanol. *See* Methoxyisopropanol

Methyl abietate
CAS 127-25-3
Properties: Sp.gr. 1.03; flash pt. (COC) 180 C; ref. index 1.53 (20 C)
Uses: Plasticizer for vinyls, cellulosics, PS
Manuf./Distrib.: Hercules

Methylacetic acid. *See* Propionic acid

Methyl acetoacetate
CAS 105-45-3; EINECS 203-299-8
Synonyms: MAA
Empirical: $C_5H_8O_3$
Formula: $CH_3COCH_2COOCH_3$
Properties: M.w. 116.1; dens. 1.076 (20/4 C); b.p. 167-170 C; flash pt. 62 C
Toxicology: Eye irritant
Uses: Copromoter for cobalt-contg. polyester-accelerator systems
Manuf./Distrib.: Aceto; Eastman; Fluka; Lonza; Penta Mfg.

Methyl acetone. *See* Methyl ethyl ketone
Methyl 12-acetoxyoleate. *See* Methyl acetyl ricinoleate

Methyl acetyl ricinoleate
CAS 140-03-4
Synonyms: Methyl 12-acetoxyoleate
Properties: Sol. in most organic solvents; insol. in water; pale-yellow liquid; mild odor; m.w. 354.59; sp.gr. 0.938; m.p. -15 C; flash pt. (COC) 196 C; ref. index 1.4545
Precaution: Combustible
Uses: Plasticizer, lubricant, protective coatings, synthetic rubbers, vinyl compds. and cellulosics
Regulatory: FDA 21CFR §175.105; Japan approved
Manuf./Distrib.: CasChem
Trade names: Flexricin® P-4

Methyl adipate. *See* Dimethyl adipate

Methyl alcohol
CAS 67-56-1; EINECS 200-659-6
Synonyms: Methanol; Wood alcohol; Wood naphtha; Wood spirit; Carbinol; Columbian spirits; Methyl hydroxide; Methylol
Empirical: CH_4O
Formula: CH_3OH
Properties: Clear colorless liq., alcoholic odor (pure), pungent odor (crude); highly polar; misc. with water, alcohol, ether, benzene, ketones; m.w. 32.05; dens. 0.7924; m.p. -97.8 C; b.p. 64.5 C (760 mm); flash pt.(OC) 54 F; ref. index 1.3292 (20 C)
Precaution: Flamm.; dangerous fire risk; explosive limits 6.0-36.5% vol. in air; reacts vigorously with oxidizers; heated to decomp., emits acrid smoke and irritating fumes
Toxicology: LD50 (oral, rat) 5628 mg/kg; toxic (causes blindness); poisonous by ingestion, inhalation, or percutaneous absorption; experimental teratogen, reproductive effects; eye and skin irritant; narcotic; usual fatal dose 100-250 ml; TLV 200 ppm in air
Uses: Industrial solvent; mfg. of formaldehyde, acetic acid, dimethyl terephthalate; chemical synthesis; antifreeze; solv. for nitrocellulose, polyvinyl butyral, shellac, rosin, manila resin, dyes; source of hydrogen for fuel cells; denaturant for ethanol
Regulatory: FDA 21CFR §172.560, 173.250, 173.385, 175.105, 176.180, 176.200, 176.210, 177.2420, 177.2460, 27CFR §21.115
Manuf./Distrib.: Air Prods.; Albright & Wilson; Ashland; Brown; Coyne; CPS; Du Pont; Eastman; General Chem.; Hoechst Celanese; Mitsui Toatsu; Nissan Chem. Ind.; Norsk Hydro A/S; Quantum/USI; Veckridge
Trade names containing: CS1590; Eltesol® 4009, 4018

N-Methylaminopropyltrimethoxysilane
CAS 3069-25-8
Properties: Liq.; m.w. 193.3; dens. 0.98; b.p. 106 C (30 mm); ref. index 1.419; flash pt. 70 C
Uses: Coupling agent, chem. intermediate, blocking agent, release agent, lubricant, primer, reducing agent
Trade names: Dynasylan® 1110

Methylbenzene. *See* Toluene
4-Methylbenzenesulfonic acid. *See* p-Toluene sulfonic acid
p-Methylbenzenesulfonic acid. *See* p-Toluene sulfonic acid

4- and 5-Methylbenzotriazole
CAS 136-85-6
Empirical: $C_7H_7N_3$
Properties: M.w. 133.17; m.p. 80-82 C; b.p. 210-212 C (12 mm)
Toxicology: Moderately toxic by ingestion; irritant
Uses: Prevulcanization retarder for sulfur-modified CR and halobutyl rubbers; for conveyor belting, hose, molded goods
Manuf./Distrib.: Aldrich
Trade names: Retrocure® G

Methyl 3-[3-(2H-benzotriazole-2-yl)-5-t-butyl-4-hydroxyphenyl]proportionate
CAS 104810-47-1
Uses: Stabilizer for PVC
Trade names containing: Tinuvin® 213

2-Methyl-4,6-bis [(octylthio) methyl] phenol. *See* 2,4-Bis [(octylthio) methyl]-o-cresol
2-Methyl-1,3-butadiene, homopolymer. *See* Polyisoprene

2-Methyl-2-butanol. *See* t-Amyl alcohol

Methyl t-butyl ether
CAS 1634-04-4; EINECS 216-653-1
Synonyms: MTBE; Methyl tertiary butyl ether; t-Butyl methyl ether; 2-Methoxy-2-methylpropane
Classification: Aliphatic ether
Empirical: $C_5H_{12}O$
Formula: $CH_3OC(CH_3)_3$
Properties: Clear liq., terpene-like odor; sl. sol. in water; misc. with all gasoline-type hydrocarbons; m.w. 88.15; dens. 0.7335; b.p. 91.1 C; f.p. -75 C; flash pt. -25.6 C
Precaution: Flamm.
Toxicology: Slight skin and eye irritant
Uses: Octane booster in gasoline; intermediate; solv.; extraction solv.; reaction medium in pharmaceuticals, for polymerizations, Grignard reactions
Manuf./Distrib.: Allchem Ind.; Arco; Ashland; Fluka; Texas Petrochem.
Trade names: High Purity MTBE

(2-β-Methyl 'Carbitol'), methoxyethoxy ethanol. *See* Methoxydiglycol
Methyl 'Cellosolve'. *See* Methoxyethanol

Methylcellulose
CAS 9004-67-5
Synonyms: MC; Cellulose methyl ether; Cologel; Methocel
Definition: Methyl ether of cellulose
Properties: Grayish-white fibrous powd., odorless, tasteless; aq. suspension swells in water to visc. colloidal sol'n.; sol. in cold water, glacial acetic acid, some org. solvs.; insol. in alcohol, ether, chloroform, warm water; m.w. 86,000-115,000
Precaution: Combustible
Toxicology: Heated to decomp., emits acrid smoke and irritating fumes
Uses: Protective colloid in water-based paints to prevent flocculation of pigment; film and sheeting; binder in ceramic glazes; leather tanning; dispersing, thickening, and sizing agent; food additive; adhesive; paper greaseproofing; pharmaceuticals; visc. stabilizer for latex and emulsion paints; plasticizer for ceramic and refractory shapes
Regulatory: FDA 21CFR §150.141, 150.161, 175.105, 175.210, 175.300, 176.200, 182.1480, GRAS; USDA 9CFR §318.7, limitation 0.15% in meat and vegetable prods.; Japan restricted (2% max.); Europe listed; UK approved
Manuf./Distrib.: Aceto; Allchem Ind.; Aqualon; Courtaulds Water Soluble Polymers; Shin-Etsu Chem.
Trade names: Methocel® A

Methylchloroform. *See* Trichloroethane

Methylchloroisothiazolinone
CAS 26172-55-4; EINECS 247-500-7
Synonyms: 5-Chloro-2-methyl-4-isothiazolin-3-one; 4-Isothiazolin-3-one, 5-chloro-2-methyl-
Classification: Heterocyclic organic compd.
Empirical: C_4H_4ClNOS
Uses: Antimicrobial, preservative for metalworking fluids, polymer emulsions, cooling tower water treatment; slimicide for paper mills
Regulatory: FDA 21CFR §175.105, 176.170
Trade names: Kathon® LX

Methyl cocoate
CAS 61788-59-8; EINECS 262-988-1
Definition: Ester of methyl alcohol and coconut fatty acids
Formula: $RCO—OCH_3$, RCO^- represents the fatty acids derived from coconut oil
Uses: Emollient, plasticizer, lubricant for cosmetics, pharmaceuticals, plastics, lubricating oils, textiles, leather, cutting oils; chemical intermediate
Regulatory: FDA 21CFR §172.225, 175.105, 176.200, 176.210, 177.2260, 177.2800, 178.3910

N-Methyl-N-(1-coconut alkyl) glycine. *See* Cocoyl sarcosine

Methyl-β-cyclodextrin
Uses: Complex hosting guest molecules; increases the sol. and bioavailability of other substances; masks flavor, odor, or coloration; stabilizes against light, oxidation, heat, and hydrolysis; turns liqs. or volatiles into stable solid powds.; for use in pharmaceuticals, cosmetics, toiletries, foods, tobacco, pesticides, textiles, paints, plastics, synthesis, polymers
Trade names: Beta W7 M1.8

Methyl-γ-cyclodextrin
Definition: Complex hosting guest molecule
Uses: Complex hosting guest molecules; increases the sol. and bioavailability of other substances; masks flavor, odor, or coloration; stabilizes against light, oxidation, heat, and hydrolysis; turns liqs. or volatiles into stable solid powds.; for use in pharmaceuticals, cosmetics, toiletries, foods, tobacco, pesticides, textiles, paints, plastics, synthesis, polymers
Trade names: Gamma W8 M1.8

Methyl decyl-1-amino decane. *See* Didecyl methylamine
2-Methyl-1,5-diaminopentane. *See* 2-Methylpentamethylenediamine
4-Methyl-1,3-dioxolan-2-one. *See* Propylene carbonate
Methyl dodecanoate. *See* Methyl laurate
4,4´-Methylenebis(aniline). *See* 4,4´-Methylene dianiline
2,2´-Methylenebis[6-(2H-benzotriazol-2-yl)-4-(1,1,3,3-tetramethylbutyl)phenol. *See* Bis [2-hydroxy-5-t-octyl-3-(benzotriazol-2-yl) phenyl] methane
2,2´-Methylenebis (6-t-butyl-p-cresol). *See* 2,2´-Methylenebis (6-t-butyl-4-methylphenol)
2,2´-Methylenebis (6-tert-butyl-4-ethylphenol). *See* 2,2´-Methylenebis (4-ethyl-6-t-butylphenol)

2,2´-Methylenebis (6-t-butyl-4-methylphenol)
CAS 119-47-1; EINECS 204-327-1
Synonyms: BKF; 2,2´-Bis(6-t-butyl-p-cresyl)methane; 2,2´-Methylenebis (6-t-butyl-p-cresol); Phenol, 2,2´-methylene-bis-6-[(1,1-dimethyl)-4-methyl-]; 2,2´-Methylenebis(4-methyl-6-t-butylphenol)
Empirical: $C_{23}H_{32}O_2$
Properties: Wh. to cream powd.; m.w. 340.55; dens. 1.08 mg/m^3; m.p. 125 C; flash pt. (COC) 190 C
Precaution: Avoid strong oxidizers
Toxicology: LD50 (rat, oral) 5000 mg/kg; mildlly toxic by ingestion; may cause eye irritation
Uses: Antioxidant, stabilizer for plastics, rubber, latex, sol'n. and emulsion polymers, fats, oils, and paraffin wax; polymerization inhibitor in chemical processes
Trade names: Akrochem® Antioxidant 235; Antioxidant 235; CAO®-5; CAO®-14; Cyanox® 2246; Lowinox® 22M46; Naftonox® 2246; Ralox® 46; Santowhite® PC; Synox 5LT; Vanox® 2246; Vulkanox® BKF; Vanox® MBPC
Trade names containing: Rhenogran® BPH-80

4,4´-Methylenebis (2-chloraniline)
Synonyms: MBOCA; 4,4´-Methylene bis (o-chloroaniline)
Toxicology: LD50 (oral, rat) 2100 mg/kg; skin irritant; mutagenic; cancerogenic
Uses: Curative for polymers
Manuf./Distrib.: Allchem Ind.

4,4´-Methylenebis (o-chloroaniline). *See* 4,4´-Methylenebis (2-chloraniline)

4,4´-Methylenebis (3-chloro-2,6-diethylaniline)
CAS 106246-33-7
Synonyms: Benzenamine, 4,4´-methylenebis (3-chloro-2,6-diethyl-(9Cl))
Formula: $C_{21}H_{28}C_{12}N_2$
Properties: Off-wh. cryst.; sol. in toluene, xylene, DMSO, DMF, butanol, aniline; insol. in water; m.w. 379.38; m.p. 88-90 C
Toxicology: LD50 (oral, rat) > 5000 mg/kg; nonirritating to skin; nonmutagenic
Uses: Chain extender for elastomeric PU; curing agent for epoxides; precursor for polyimides; intermediate for org. synthesis
Trade names: Lonzacure® M-CDEA

4,4´-Methylenebis (cyclohexylamine)carbamate
CAS 13253-82-2; EINECS 236-239-4

Synonyms: Carbamic acid, [4-[(4-aminocyclohexyl)methyl]cyclohexyl]-
Properties: Powd., sl. amine odor; sol. in water; dens. 1.23 mg/m^3; m.p. 150 C; flash pt. (OC) 149 C
Precaution: May form flamm. dust/air mixtures; avoid contact with acids
Toxicology: Irritating to eyes, skin, nose, and throat; LD50 (rat, oral) 1000 mg/kg
Uses: Processing aid for elastomers
Manuf./Distrib.: Interbusiness USA
Trade names: Diak No. 4

4,4´-Methylenebis (2,6-di-t-butylphenol)
CAS 118-82-1
Empirical: $C_{29}H_{44}O_2$
Properties: M.w. 424.7
Uses: Antioxidant for elastomers, polyolefins, resins, adhesives, petrol. oil, and waxes
Manuf./Distrib.: Great Lakes
Trade names: Ethanox® 702; Ralox® 02

4,4´-Methylenebis 2,6-diethylaniline
Toxicology: LD50 (oral, rat) 1901 mg/kg; nonirritating to skin; nonmutagenic
Uses: Chain extender for elastomeric PU; curing agent for epoxides; precursor for polyimides
Trade names: Lonzacure® M-DEA

4,4´-Methylenebis 2,6-diisopropylaniline
Toxicology: LD50 (oral, rat) 1110 mg/kg; nonirritating to skin; nonmutagenic
Uses: Chain extender for elastomeric PU
Trade names: Lonzacure® M-DIPA

4,4´-Methylenebis 2,6-dimethylaniline
Toxicology: LD50 (oral, rat) 724 mg/kg; nonirritating to skin; nonmutagenic
Uses: Curing agent for epoxides; precursor for polyimides
Trade names: Lonzacure® M-DMA

2,2´-Methylenebis (4-ethyl-6-t-butylphenol)
CAS 88-24-4
Synonyms: Bis (2-hydroxy-3-t-butyl-5-ethylphenyl) methane; 2,2´-Methylenebis (6-tert-butyl-4-ethylphenol)
Empirical: $C_{25}H_{36}O_2$
Properties: M.w. 368.61
Toxicology: Poison by intraperitoneal route; LD50 (rat, oral) > 15 g/kg; LD50 (rabbit, dermal) > 8 g/kg; may cause mild skin and eye irritation
Uses: Antioxidant for impact molding resins, esp. acrylics and ABS
Trade names: Antioxidant 425; Cyanox® 425

4,4´-Methylenebis 2-ethyl-6-methylaniline
Toxicology: LD50 (oral, rat) 1582 mg/kg; nonirritating to skin; nonmutagenic
Uses: Precursor for polyimides; in photoresists
Trade names: Lonzacure® M-MEA

4,4´-Methylenebis 2-isopropyl-6-methylaniline
Toxicology: LD50 (oral, rat) 2015 mg/kg; nonirritating to skin; nonmutagenic
Uses: Chain extender for elastomeric PU
Trade names: Lonzacure® M-MIPA

2,2´-Methylenebis(4-methyl-6-t-butylphenol). *See* 2,2´-Methylenebis (6-t-butyl-4-methylphenol)

2,2´-Methylenebis 6-(1-methylcyclohexyl)-p-cresol
CAS 77-62-3
Uses: Antioxidant for rubber, polyolefins, PS
Trade names: Nonox® WSP; Permanax™ WSP; Permanax™ WSP (PQ)

2,2´-Methylenebis (4-methyl-6-cyclohexyl phenol)
Uses: Antioxidant for natural and syn. rubbers; used in transparent, bathing, surgical tech., latex goods, fabric proofings, dipped goods, foam rubber, hoses, footwear, floor coverings, carpet backing
Trade names: Vulkanox® ZKF

2,2′-Methylenebis [4-methyl-6-(1-methyl-cyclohexyl) phenol. *See* 2,2′-Methylenebis 6-(1-methylcyclohexyl)-p-cresol

Methylene bisphenyl isocyanate. *See* MDI

4,4′-Methylene dianiline
CAS 101-77-9; EINECS 202-974-4
Synonyms: MDA; p,p′-Diaminodiphenylmethane; 4,4′-Methylenebis(aniline); p,p′-Methylene dianiline
Empirical: $C_{13}H_{14}N_2$
Formula: $H_2NC_6H_4CH_2C_6H_4NH_2$
Properties: Tan flakes or lumps, faint amine-like odor; very sol. in alcohol, benzene, ether; sol. in cold water; m.w. 198.29; m.p. 90 C; b.p. 398-399 C; flash pt. 440 F
Precaution: Combustible
Toxicology: TLV:TWA 0.1 ppm (skin); LD50 (oral, rat) 347 mg/kg; poison by ingestion, subcutaneous, intraperitoneal routes; eye irritant
Uses: Curing agent; corrosion inhibitor; epoxy resin hardening agent
Manuf./Distrib.: BASF; Uniroyal
Trade names: Ancamine® DL-50; Tonox®; Tonox® 22
Trade names containing: Tonox® 60/40

p,p′-Methylene dianiline. *See* 4,4′-Methylene dianiline
Methylene di(cyclohexylamine). *See* Bis (p-aminocyclohexyl) methane
Methylene di-p-phenylene isocyanate. *See* MDI
Methyl ether. *See* Dimethyl ether
Methyl ethyl carbinol. *See* 2-Butanol
1-Methylethyl hexandecanoate. *See* Isopropyl palmitate
4,4|-(1-Methylethylidene) bisphenol. *See* Bisphenol A

Methyl ethyl ketone
CAS 78-93-3; EINECS 201-159-0
Synonyms: MEK; Ethyl methyl ketone; 2-Butanone; 2-Oxobutane; Methyl acetone
Classification: Aliphatic ketone
Empirical: C_4H_8O
Formula: $CH_3COCH_2CH_3$
Properties: Colorless liq., acetone-like odor; sol. in 4 parts water, benzene, alcohol, ether; misc. with oils; m.w. 72.10; dens. 0.8255 (0/4 C); m.p. -86 C; b.p. 79.6 C; flash pt. (TOC) 24 F; visc. 0.40 cp (25 C); ref. index 1.3814 (15 C)
Precaution: Flammable, dangerous fire risk; explosive limits in air 2-10%
Toxicology: LD50 (oral, rat) 2737 mg/kg; mod. toxic by ingestion, skin contact, IP routes; toxic by inhalation; experimental teratogen, reproductive effects; strong irritant; affects CNS; TLV 200 ppm in air; heated to decomp., emits acrid smoke and fumes
Uses: Solvent in nitrocellulose coatings and vinyl films, paint removers, cements, adhesives, organic synthesis; mfg. of smokeless powder; cleaning fluids; priming, catalyst carrier; acrylic coatings
Regulatory: FDA 21CFR §172.515, 175.105, 175.320, 177.1200
Manuf./Distrib.: Aldrich; BP Chem. Ltd; Elf Atochem N. Am.; Exxon; Hoechst Celanese; Mallinckrodt; Shell; Texaco; Union Carbide
Trade names containing: Black Out® Black

Methyl ethyl ketone peroxide
CAS 1338-23-4; EINECS 215-661-2
Synonyms: MEK peroxide; Ethyl methyl ketone peroxide; 2-Butanone peroxide
Empirical: $C_8H_{18}O_6$
Formula: $C_2H_5C(OOH)(CH_3)OOC(OOH)(CH_3)C_2H_5$
Properties: M.w. 210.23; dens. 1.053 (20 C) ; ref. index 1.442 920 C)
Toxicology: TLV:CL 0.2 ppm; poison by intraperitoneal route; moderately toxic by ingestion, inhalation; skin and eye irritant
Uses: Polymerization initiator/catalyst for cure of unsaturated polyester resins
Manuf./Distrib.: Akzo Nobel ; Cook Composites & Polymers; Elf Atochem N. Am.; Great Western; Hastings Plastics; Norac; Witco/PAG
Trade names: Aztec® MEKP-HA-1; Aztec® MEKP-HA-2; Aztec® MEKP-LA-2; Aztec® MEKP-SA-2; Cadox® HBO-50; Cadox® L-30; Cadox® L-50; Cadox® M-105; Lupersol DDM-9; Lupersol Delta-3; Lupersol Delta-X-9; Lupersol DHD-9; Quickset® Extra; Quickset® Super; Superox® 46-747; Superox® 46-748; Superox® 702; Superox® 709; Superox® 710; Superox® 711; Superox® 712; Superox® 713; Superox® 730;

Superox® 732; Superox® 739
Trade names containing: Cadox® M-30; Cadox® M-50; Cadox® MDA-30; Esperfoam® FR; Hi-Point® 90; Hi-Point® PD-1; HPC-9; Norox® MCP; Sprayset® MEKP; Superox® 46-753-00

1-Methylethyl-9-octadecenoate. *See* Isopropyl oleate
1-Methylethyl tetradecanoate. *See* Isopropyl myristate
Methyl glycol. *See* Propylene glycol
2-Methyl glyoxaline. *See* 2-Methyl imidazole
16-Methylheptadecanoic acid. *See* Isostearic acid

Methyl hexahydrophthalic anhydride
CAS 25550-51-0
Properties: Clear liq.; misc. with most org. solvs.; b.p. 127 C (5 mm); flash pt. (OC) 160 C
Uses: Curing agent
Manuf./Distrib.: Anhydrides & Chems.; Buffalo Color; Dixie Chem; Lindau Chems
Trade names containing: Milldride® MHHPA

Methyl hydrogenated rosinate
CAS 8050-13-3
Synonyms: Hydrogenated methyl ester of rosin
Definition: Ester of methyl alcohol and the hydrogenated mixed long chain acids derived from rosin
Uses: Plasticizer and tackifier in lacquers, inks, adhesives, floor tiles, vinyl plastisols, artificial leather, and antifouling paints; fixative and carrier in perfumes and cosmetic preps.
Trade names: Hercolyn® D

Methyl hydrogen polysiloxane
Trade names containing: Kayphobe-ABO

Methylhydroquinone. *See* Toluhydroquinone
Methyl hydroxide. *See* Methyl alcohol

Methyl hydroxyethylcellulose
CAS 9032-42-2
Definition: Methyl ether of hydroxyethylcellulose
Uses: Binder, thickener, pigment, foam, and filler stabilizer, dispersant, emulsifier, plasticizer, visc. control and sedimenting aid, and protective colloid used in coatings, paints, resins, mining, batteries, insecticidal prods.; rubber, textile, leather, ceramics, suspension polymerization, pharmaceuticals
Trade names: Tylose® MH Grades; Tylose® MHB

N-Methyl-N´-hydroxyethylpiperazine
Uses: Reactive catalyst for PU
Trade names: Toyocat®-HPW

Methyl 12-hydroxyoctadecanoate. *See* Methyl hydroxystearate
Methyl 12-hydroxy-9-octadecenoate. *See* Methyl ricinoleate
Methyl hydroxypropyl cellulose. *See* Hydroxypropyl methylcellulose

Methyl hydroxystearate
CAS 141-23-1; EINECS 205-471-8
Synonyms: Methyl 12-hydroxyoctadecanoate; Methyl 12-hydroxystearate; 12-Hydroxyoctadecanoic acid, methyl ester
Definition: Ester of methyl alcohol and hydroxystearic acid
Empirical: $C_{18}H_{38}O_3$
Formula: $C_{16}H_{34}OHCOOCH_3$
Properties: White waxy solid, flat rods; insol. in water; sl. sol. in org. solvs.; m.p. 48 C
Precaution: Combustible
Toxicology: Experimental tumorigen
Uses: Lubricant, processing aid for butyl rubber, adhesives, inks, cosmetics, greases
Regulatory: FDA 21CFR §176.210
Trade names: Paricin® 1

Methyl 12-hydroxystearate. *See* Methyl hydroxystearate

2-Methyl imidazole
CAS 693-98-1; EINECS 211-765-7
Synonyms: 2MZ; 2-Methyl glyoxaline
Empirical: $C_4H_7N_2$
Formula: $CHCHNC(CH_3)NH$
Properties: Solid; m.w. 82.11; m.p. 142-143 C
Toxicology: Moderately toxic by ingestion and intraperitoneal routes
Uses: Dyeing auxiliary for acrylic fibers, plastic foams; curing agent for printed circuit board laminates, powd. coatings, adhesives, encapsulation; accelerator
Manuf./Distrib.: Allchem Ind.; BASF; Fabrichem; Janssen Chimica; SKW
Trade names: Imicure® AMI-2

Methyl isobutyl ketone peroxide
CAS 37206-20-5
Synonyms: MIBK peroxide
Uses: Initiator for cure of unsaturated polyester resins
Trade names: Aztec® MIKP-LA-M1; Trigonox® HM

Methyl laurate
CAS 111-82-0; 67762-40-7; EINECS 203-911-3
Synonyms: Methyl dodecanoate; Dodecanoic acid methyl ester
Definition: Ester of methyl alcohol and lauric acid
Empirical: $C_{13}H_{26}O_2$
Formula: $CH_3(CH_2)_{10}COOCH_3$
Properties: Water-white liq., fatty floral odor; insol. in water; m.w. 214.35; dens. 0.8702 (20/4 C); m.p. 4.8 C; b.p. 262 C (766 mm); flash pt. > 230 F; ref. index 1.4320
Precaution: Combustible; noncorrosive
Uses: Intermediate for detergents, emulsifiers, wetting agents, stabilizers, lubricants, plasticizers, textiles, flavoring, plastics, cosmetics
Regulatory: FDA 21CFR §172.225, 172.515, 176.200, 176.210, 177.2260, 177.2800
Manuf./Distrib.: Aldrich; Henkel/Emery; Penta Mfg.; Procter & Gamble; Stepan
Trade names: Estol 1502

4- and 5-Methylmercaptobenzimidazole
Synonyms: 2-Mercapto-4(5)-methyl benzimidazole
Uses: Antioxidant for rubbers
Trade names: Akrochem® Antioxidant 60; Vulkanox® ZMB-2/G
Trade names containing: Rhenogran® MMBI-70; Vulkanox® MB-2/MGC

4- and 5-Methylmercaptobenzimidazole zinc salt
Synonyms: ZMMBI
Uses: Antidegradant for rubbers
Trade names containing: Poly-Dispersion® A (ZMTI)D-50

Methyl methacrylate/allyl methacrylate copolymer
Uses: Antishrink additive
Trade names: Luchem AS-946; Luchem AS-946-25

Methyl methacrylate polymer
CAS 9011-14-7
Synonyms: Acrylite; Methyl methacrylate resin
Empirical: $(C_5H_8O_2)_n$
Toxicology: Experimental tumorigen
Uses: Thermoplastic acrylic resin used in coatings, barrier coatings for PS, vinyl topcoats, product finishes, printing inks; processing aid for PVC
Manuf./Distrib.: Aristech; Cyro Industries; Cytec; Degussa; Monomer-Polymer & Dajac; StanChem; Sybron
Trade names: Paraloid® K-120N

Methyl methacrylate resin. *See* Methyl methacrylate polymer

Methyl methacrylate styrene acrylonitrile copolymer
Uses: Processing aid for polymers
Trade names: Blendex® 590

Methyl-2(methyl-2) oxybispropanol. *See* Dipropylene glycol
4-Methyl morpholine. *See* p-Methyl morpholine

p-Methyl morpholine
CAS 109-02-4; EINECS 203-640-0
Synonyms: 4-Methyl morpholine
Empirical: $C_5H_{11}ON$
Formula: $CH_2CH_2OCH_2CH_2NCH_3$
Properties: Water-white liq., ammonia odor; forms constant-boiling mixture with 25% water and boiling at 97 C; misc. with benzene, water; m.w. 101.17; dens. 0.921; b.p. 115.4 C; f.p. -66 C; flash pt. (TOC) 75 F
Precaution: Flamm.; dangerous fire risk
Toxicology: Skin irritant
Uses: Catalyst in polyurethane foams; extraction solvent; stabilizer for chlorinated hydrocarbons; self-polishing waxes; corrosion inhibitor; pharmaceuticals
Trade names: Dabco® NMM; Jeffcat NMM

Methyl namate. *See* Sodium dimethyldithiocarbamate
1-Methylnaphthalene. *See* α-Methylnaphthalene

α-Methylnaphthalene
CAS 90-12-0; EINECS 201-966-8
Synonyms: 1-Methylnaphthalene
Empirical: $C_{11}H_{10}$
Formula: $C_{10}H_7CH_3$
Properties: M.w. 142.20; sol. in alc. and ether; insol. in water; dens. 1.020 (20/4 C); b.p. 241-245 C; ref. index 1.614
Precaution: Combustible
Uses: Carrier for polyester/wool blended fabrics
Manuf./Distrib.: Allchem Ind.; Coyne; Crowley Tar Prods.; Koch
Trade names containing: KZ OPPR

Methyl octadecanoate. *See* Methyl stearate
Methyl 9-octadecenoate. *See* Methyl oleate
Methylol. *See* Methyl alcohol

Methyl oleate
CAS 112-62-9; 67762-38-3; EINECS 203-992-5; 267-015-4
Synonyms: Methyl 9-octadecenoate; 9-Octadecenoic acid, methyl ester
Definition: Ester of methyl alcohol and oleic acid
Empirical: $C_{19}H_{36}O_2$
Formula: $CH_3(CH_2)_7CH=CH(CH_2)_7COOCH_3$
Properties: Clear to amber liq., faint fatty odor; sol. in alcohols, most org. solvs.; insol. in water; dens. 0.8739 (20 C); f.p. -19.9 C; b.p. 218.5 C (20 mm); ref. index 1.4510 (26 C)
Precaution: Combustible
Uses: Intermediate for detergents, emulsifiers, wetting agents, stabilizers, textile treatment, plasticizers for PS, cellulosics, duplicating inks, rubbers, waxes, etc.; chromatographic reference standard
Regulatory: FDA 21CFR §172.225, 175.105, 176.200, 176.210, 177.2260, 177.2800
Manuf./Distrib.: Calgene; Ferro/Keil; Henkel; Norman, Fox; Unichema; Union Camp; Witco/Oleo-Surf.
Trade names: Emerest® 2301

N-Methylol methacrylamide
Uses: Used in self-crosslinking emulsions, heat-curing coatings
Manuf./Distrib.: Monomer-Polymer & Dajac; Rohm Tech
Trade names: BM-818

N-Methyl-N-(1-oxododecyl) glycine. *See* Lauroyl sarcosine
N-Methyl-N-(1-oxododecyl) glycine, ammonium salt. *See* Ammonium lauroyl sarcosinate
N-Methyl-N-(1-oxododecyl)glycine, sodium salt. *See* Sodium lauroyl sarcosinate
N-Methyl-N-(1-oxo-9-octadecenyl)glycine. *See* Oleoyl sarcosine
N-Methyl-N-(1-oxooctadecyl)glycine. *See* Stearoyl sarcosine
N-Methyl-N-(1-oxotetradecyl)glycine. *See* Myristoyl sarcosine
N-Methyl-N-(1-oxotetradecyl)glycine, sodium salt. *See* Sodium myristoyl sarcosinate
Methylpentamethylenediamine. *See* 2-Methylpentamethylenediamine

2-Methylpentamethylenediamine
CAS 15520-10-2; EINECS 239-556-6
Synonyms: MPMD; 1,5-Pentanediamine, 2-methyl-; Methylpentamethylenediamine; 2-Methyl-1,5-diaminopentane
Empirical: $C_6H_{16}N_2$
Formula: $H_2NCH_2CH(CH_3)(CH_2)_3NH_2$
Properties: Colorless liq., weak ammonia, fishy odor; m.w. 116.2; sp.gr. 0.86; b.p. 193 C; f.p. -50 to -60 C; flash pt. (CC) 83 C
Precaution: Avoid strong oxidants; emits toxic fumes of nitrogen oxides on decomp.; combustible
Toxicology: Corrosive; can cause burns and ulceration of skin and eye tissue, nose, throat, and gastrointestinal irritation; LD50 (rat, oral) 1690 mg/kg
Uses: For mfg. of high m.w. polyamide polymers and copolymers, nonplastic copolyamide resins, coatings, adhesives, inks, corrosion inhibitors, emulsion breakers; epoxy curing agent
Manuf./Distrib.: Fluka
Trade names: Dytek® A
Trade names containing: Millamine® 5260

2-Methyl-2,4-pentanediol. *See* Hexylene glycol
4-Methyl-2,4-pentanediol. *See* Hexylene glycol

(α-Methylphenethyl) methyl-dimethylsiloxane copolymer
Uses: Mold release agent for rubber, plastics, die casting
Trade names: PS138

4-Methyl phenol reaction prods. with dicyclopentadiene and isobutylene
CAS 68610-51-5
Classification: Polymeric hindered phenol
Properties: Off-wh. to tan powd.; insol. in water; dens. 1.10 mg/m^3
Precaution: Dust may be explosive hazard; exposed to flame, emits acrid fumes
Toxicology: May irritate eyes and upper respiratory tract; LD50 (rat, oral) > 16,000 mg/kg
Trade names: Wingstay® L

Methyl phenylglyoxalate
CAS 15206-55-0; EINECS 239-263-3
Empirical: $C_9H_8O_3$
Properties: Yel. liq.; sol. in all org. solvs.; insol. in water; m.w. 164.15; dens. 1.160 g/ml; b.p. 246-248 C; ref. index 1.5253-1.5263; flash pt. (TCC) 76.7 C
Uses: Photoinitiator for uv curable systems
Manuf./Distrib.: Fluka
Trade names: Vicure® 55

N-(2-Methylphenyl)imidodicarbonimidic diamide. *See* o-Tolyl biguanide
p-Methylphenylsulfonic acid. *See* p-Toluene sulfonic acid
Methylphosphonic acid, dimethyl ester. *See* Dimethyl methylphosphonate

2-Methyl-1,3-propanediol
CAS 2163-42-0
Formula: $HOCH_2CH(CH_3)CH_2OH$
Properties: M.w. 90.12; dens. 1.015; m.p. -91 C; b.p. 123-125 C (20 mm); flash pt. > 110 C; ref. index 1.4450 (20 C)
Toxicology: Irritant
Uses: Intermediate used in prod. of solvs., urethanes, unsat. polyesters, gel coats, sat. polyester and alkyd coatings, polymeric plasticizers
Manuf./Distrib.: Aldrich; Arco
Trade names: MPDiol® Glycol

2-Methylpropanol. *See* Isobutyl alcohol
2-Methyl-1-propanol. *See* Isobutyl alcohol
2-Methyl-2-propanol. *See* t-Butyl alcohol
2-Methyl-1-propene, homopolymer. *See* Polyisobutene
2-Methyl-1-propene, polymer with 2,5-furandione. *See* Isobutylene/MA copolymer
2-Methyl-2-propene-1-sulfonic acid sodium salt. *See* Sodium methallyl sulfonate
2-Methylpropyl octadecanoate. *See* Isobutyl stearate

1-Methyl-2-pyrrolidinone

CAS 872-50-4; EINECS 212-828-1

Synonyms: NMP; Methylpyrrolidone; N-Methylpyrrolidinone; N-Methyl-2-pyrrolidinone; N-Methyl-2-pyrrolidone; 1-Methyl-2-pyrrolidone

Empirical: C_5H_9NO

Properties: Colorless liq., mild odor; m.w. 99.13; dens. 1.033; m.p. -24 C; b.p. 81-82 C (10 mm); flash pt. 86 C; ref. index 1.4700 (20 C)

Precaution: Hygroscopic; combustible exposed to heat, open flame, powerful oxidizers; heated to decomp., releases toxic fumes of NO_x

Toxicology: LD50 (oral, rat) 3600 mg/kg; moderately toxic by intraperitoneal and intravenous routes; mildly toxic by ingestion, skin contact; experimental teratogen, reproductive effects

Uses: Solv., cosolv. in coatings, industrial cleaning, mold cleaning, petrochem. processing, agric., polymers, elastomers, waxes

Manuf./Distrib.: Aldrich; Allchem Ind.; Arco; Ashland; BASF; Coyne; Fluka; ISP; Janssen Chimica; Spectrum Chem. Mfg.

Trade names: NMP

N-Methylpyrrolidinone. *See* 1-Methyl-2-pyrrolidinone
N-Methyl-2-pyrrolidinone. *See* 1-Methyl-2-pyrrolidinone
Methylpyrrolidone. *See* 1-Methyl-2-pyrrolidinone
N-Methylpyrrolidone. *See* 1-Methyl-2-pyrrolidinone
n-Methyl-2-pyrrolidone. *See* 1-Methyl-2-pyrrolidinone

Methyl ricinoleate

CAS 141-24-2; EINECS 205-472-3

Synonyms: 12-Hydroxy-9-octadecenoic acid, methyl ester; Methyl 12-hydroxy-9-octadecenoate; Castor oil acid, methyl ester

Definition: Ester of methyl alcohol and ricinoleic acid

Empirical: $C_{19}H_{36}O_3$

Formula: $CH_3(CH_2)_5COHHCH_2CH=CH(CH_2)_7COOCH_3$

Properties: Colorless liq.; insol. in water; sol. in alcohol, ether; sp.gr. 0.925; f.p. -4.5 C; b.p. 245 C (10 mm); flash pt. (COC) 190 C; ref. index 1.4620

Precaution: Combustible

Uses: Plasticizer for PVAc, PVB, cellulosics; lubricant, cutting oil additive, wetting agent

Regulatory: FDA 21CFR §175.105, 176.210

Manuf./Distrib.: CasChem; Penta Mfg.; Reilly-Whiteman

Trade names: Flexricin® P-1

Methyl rosinate

CAS 68186-14-1; EINECS 269-035-9

Synonyms: Rosin acid, methyl ester

Definition: Methyl ester of acids recovered from rosin

Uses: Resin with surf.-wetting properties, visc., and tack used in lacquers, inks, paper coatings, varnishes, adhesives, sealing compds., plastics, wood preservatives, and perfumes

Regulatory: FDA 21CFR §172.615, 175.105, 175.300, 176.170, 176.200, 176.210, 177.1200, 177.2600, 178.3120, 178.3800, 178.3870

Trade names: Abalyn®

Methyl Selenac. *See* Selenium dimethyldithiocarbamate

Methyl stearate

CAS 112-61-8; 85586-21-6; EINECS 203-990-4; 287-824-6

Synonyms: Methyl octadecanoate; Octadecanoic acid, methyl ester

Definition: Ester of methyl alcohol and stearic acid

Empirical: $C_{19}H_{38}O_2$

Formula: $CH_3(CH_2)_{16}COOCH_3$

Properties: White crystals; insol. in water; sol. in ether, alcohol; m.w. 298.57; m.p. 37.8 C; b.p. 234.5 C (30 mm); flash pt. 307 F

Precaution: Combustible

Uses: Intermediate for stearic acid detergents, emulsifiers, wetting agents, stabilizers, resins, lubricants, plasticizers

Regulatory: FDA 21CFR §172.225, 176.200, 176.210, 177.2260, 177.2800, 178.3910

Manuf./Distrib.: Ashland; Ferro/Keil; Penta Mfg.; Sea-Land; Union Camp; Witco/Oleo-Surf.

α-Methylstyrene/N-(2,2,6,6-tetramethyl piperidinyl-4) maleimide/N-stearyl maleimide terpolymer
CAS 98-83-9, 17450-30-5, 84540-25-0 respectively; EINECS 202-705-0, 241-467-2, 283-117-2 resp.
Toxicology: Nontoxic; LD50 (rat) > 6000 mg/kg
Uses: Light stabilizer for polyolefins
Trade names: Lowilite® 62

4-α-Methylstyrylphenol
Synonyms: MAMSP
Formula: $C_6H_5C(CH_3)_2C_6H_4OH$
Properties: M.w. 212.29; m.p. 74-76 C; b.p. 335 C
Manuf./Distrib.: Aldrich

Methyl succinate. *See* Dimethyl succinate
Methyl sulfoxide. *See* Dimethyl sulfoxide

Methyl tallate
Definition: Methyl ester of tall oil
Properties: Sp.gr. 0.96; flash pt. (COC) 171 C; ref. index 1.492 (20 C)
Uses: Plasticizer for PVC, PVB, VCA, ethyl cellulose
Manuf./Distrib.: Hercules; Union Camp

Methyl tallowate
EINECS 262-989-7
Synonyms: Fatty acids, tallow, methyl esters
Definition: Ester of methyl alcohol and tallow acid
Uses: Wetting/oiliness agent for metalworking and lubricating oils; lubricant, plasticizer for cosmetics, leather, rubber prods.
Trade names: Kemester® 143

Methyl tertiary butyl ether. *See* Methyl t-butyl ether

3-Methyl-thiazolidinethione-2
CAS 1908-87-8
Uses: Accelerator for polychloroprene elastomers, halobutyl rubbers esp. chlorobutyl; used for mech. goods, cables, hoses, membranes, fabric proofings, vulcanizing sol'ns.
Trade names: Vulkacit CRV/LG
Trade names containing: Rhenogran® MTT-80

Methyltin mercaptide
Uses: Heat stabilizer for PVC for food pkg., bottles, profile, and inj. moldings
Trade names: Advastab® TM-181; Advastab® TM-181-FS; Advastab® TM-183-B; Advastab® TM-183-O; Advastab® TM-281 IM; Advastab® TM-281 SP; Advastab® TM-692; Advastab® TM-2080

Methyl toluene. *See* Xylene

Methyltriacetoxysilane
CAS 4253-34-3
Empirical: $C_7H_{12}O_6Si$
Properties: Solid; m.w. 220.28; dens. 1.17; b.p. 87-88 C (3 mm); ref. index 1.408; flash pt. 5 C
Toxicology: Moderately toxic by ingestion
Uses: Coupling agent, release agent, lubricant, blocking agent, chemical intermediate
Trade names: Dynasylan® MTAC

Methyltriethoxysilane
CAS 2031-67-6; EINECS 217-983-9
Synonyms: Triethoxymethylsilane
Empirical: $C_7H_{18}O_3Si$
Formula: Liq.; m.w. 178.34; dens. 0.90; b.p. 141-143 C; ref. index 1.383; flash pt. 38 C
Toxicology: Mildly toxic by ingestion, inhalation; skin and eye irritant
Uses: Coupling agent, release agent, lubricant, blocking agent, chemical intermediate
Manuf./Distrib.: Spectrum Chem. Mfg.
Trade names: Dynasylan® MTES; Petrarch® M9050

Methyltrimethoxysilane
CAS 1185-55-3; EINECS 214-685-0
Empirical: $C_4H_{12}O_3Si$
Properties: Liq.; m.w. 136.25; dens. 0.955; b.p. 102-103 C; ref. index 1.3696; flash pt. 8 C
Toxicology: Mildly toxic by ingestion; skin and eye irritant
Uses: Coupling agent, release agent, lubricant, blocking agent, chemical intermediate
Trade names: Dynasylan® MTMS; Petrarch® M9100

Methyltrimethylolmethane. *See* Trimethylolethane
Methyl vinyl ether/maleic anhydride copolymer. *See* PVM/MA copolymer
MHPC. *See* Hydroxypropyl methylcellulose
MIBK peroxide. *See* Methyl isobutyl ketone peroxide

Mica
CAS 12001-26-2
Synonyms: Muscovite mica
Classification: Silicate minerals
Properties: Colorless to sl. red, brown to greenish-yel. soft, translucent solid; dens. 2.6-3.2; Mohs hardness 2.8-3.2; heat resistant to 600 C
Precaution: Noncombustible
Toxicology: TLV:TWA 3 mg/m^3 (respirable dust)
Uses: Filler/extender for plastics, rubber, coatings, and pearlescent pigment applics.; binder and reinforcement in lipsticks
Regulatory: FDA 21CFR §73.1496, 73.2496, 175.300, 177.1460, 177.2410, 177.2600
Manuf./Distrib.: Feldspar; Franklin Industrial Mins.; ISP Van Dyk; Mearl; Mykroy/Macalex Ceramics; Nyco Mins.
Trade names: AlbaFlex 25; AlbaFlex 50; AlbaFlex 100; AlbaFlex 200; Huber SM; Huber WG-1; Huber WG-2; Micawhite 200; Polymica 200; Polymica 325; Polymica 400; Polymica 3105
Trade names containing: Mearlin® Card Gold; Mearlin® Hi-Lite Super Blue; Mearlin® Hi-Lite Super Gold; Mearlin® Hi-Lite Super Green; Mearlin® Hi-Lite Super Orange; Mearlin® Hi-Lite Super Red; Mearlin® Hi-Lite Super Violet; Mearlin® Inca Gold; Mearlin® MagnaPearl 1000; Mearlin® MagnaPearl 1100; Mearlin® MagnaPearl 1110; Mearlin® MagnaPearl 2000; Mearlin® MagnaPearl 2110; Mearlin® MagnaPearl 3000; Mearlin® MagnaPearl 3100; Mearlin® MagnaPearl 4000; Mearlin® MagnaPearl 5000; Mearlin® Nu-Antique Silver; Mearlin® Pearl White; Mearlin® Satin White; Mearlin® Silk White; Mearlin® Sparkle; Mearlin® Supersparkle

Microcrystalline wax
CAS 63231-60-7; 64742-42-3; EINECS 264-038-1
Synonyms: Petroleum wax, microcrystalline; Waxes, microcrystalline
Definition: Wax derived from petroleum and char. by fineness of crystals; consists of high m.w. saturated aliphatic hydrocarbons
Toxicology: May be carcinogenic
Uses: Wax used in hot-melt coatings and adhesives, paper coatings, printing inks, plastic modification (as lubricant and processing aid), lacquers, paints, and varnishes, as binder in ceramics, for elec. potting
Regulatory: FDA 21CFR §172.886, 173.340, 175.105, 175.320, 176.170, 176.200, 177.2600; Europe listed; UK approved for restricted use
Manuf./Distrib.: Astor Wax; Ferro; IGI; Koster Keunen; Mobil
Trade names: Akrowax Micro 23; Antilux® 750; Be Square® 175; Be Square® 185; Be Square® 195; Fortex®; Koster Keunen Microcrystalline Waxes; Mekon® White; Multiwax® 180-M; Multiwax® W-835; Multiwax® X-145A; Petrolite® C-700; Petrolite® C-1035; Starwax® 100; Ultraflex®; Victory®
Trade names containing: Antilux® 110; Antilux® 111; Antilux® 500; Antilux® 550; Antilux® 600; Antilux® 620; Antilux® 654; Antilux® 660; Antilux® L; Liquax 488; Sunolite® 100; Sunolite® 127; Sunolite® 240; Sunolite® 240TG; Sunolite® 666
See also Petroleum wax

Microcrystalline wax, oxidized
Uses: Wax used in the formulation of emulsions, polishes, and coatings; modifier in solv. polish systems; carnauba substitute; lubricant, process aid, slip and antiblock agent in plastics
Trade names: Cardis® 10; Cardis® 36; Cardis® 314; Cardis® 319; Cardis® 320; Cardis® 370; Petronauba® C

Milk of magnesia. *See* Magnesium hydroxide
Mineral carbon. *See* Graphite

Mineral oil

CAS 8012-95-1; 8020-83-5 (wh.); 8042-47-5; EINECS 232-384-2; 232-455-8

Synonyms: Heavy mineral oil; Light mineral oil; White mineral oil; Paraffin oil; Liquid paraffin; Petrolatum liquid; Liquid petrolatum

Definition: Liq. mixture of hydrocarbons obtained from petroleum

Properties: Colorless oily liq., tasteless, odorless; insol. in water, alcohol; sol. in benzene, chloroform, ether, petrol. ether, oils; dens. 0.83-0.86 (light), 0.875-0.905 (heavy); flash pt. (OC) 444 F; surf. tension < 35 dynes/cm

Precaution: Combustible

Toxicology: Eye irritant; human carcinogen and teratogen by inhalation; heated to decomp., emits acrid smoke and fumes

Uses: Cathartic; laxative; protectant; lubricant; binder; carrier; mold release for foods, coating for fruits and vegetables, food pkg. materials; plasticizer, lubricant for plastics; in cosmetics, pharmaceuticals, plastics, agric., paper, textiles, etc.

Regulatory: FDA 21CFR §172.878, 173.340 (limitation 0.008% in wash water for sliced potatoes, 150 ppm in yeast), 175.105, 175.210, 175.230, 175.300, 176.170, 176.200, 176.210, 177.1200, 177.2260, 177.2600, 177.2800, 178.3570, 178.3620, 178.3740, 178.3910, 179.45; 573.680; ADI not specified (FAO/WHO)

Manuf./Distrib.: Aldrich; Amoco/Lubricants; Chemisphere; Exxon; Fluka; Magie Bros. Oil; Mobil; Penreco; Penta Mfg.; San Yuan; Sea-Land; Surco Prods.; Total Petrol.; Witco/Golden Bear, Petrol. Spec.

Trade names: Britol® 6NF; Drakeol 9; Drakeol 10; Drakeol 19; Drakeol 21; Drakeol 32; Drakeol 34; Drakeol 35; Hydrobrite 200PO; Hydrobrite 300PO; Hydrobrite 380PO; Hydrobrite 550PO; Kaydol®; Kaydol® S; Molder's Edge ME-304; Rudol®; Semtol® 40; Semtol® 70; Semtol® 85; Semtol® 100; Semtol® 350

Trade names containing: BSWL 201; Dabco® T-11; Drewplus® L-131; Drewplus® L-139; Drewplus® L-191; Drewplus® L-198; Drewplus® L-475; Drewplus® Y-250; Drewplus® Y-281; Drewplus® Y-601; EPlstatic® 101; EPlstatic® 111; EPlstatic® 113; EPlthal 121; Faktis Badenia T; Faktis Para extra weich; Faktis T-hart; Faktis ZD; Faktogel® A; Faktogel® Badenia T; Faktogel® MB; Faktogel® Para; Faktogel® T; Faktogel® ZD; Ketjenblack® EC-350 J Spd; Lupersol P-31; Lupersol P-33; Plastogen E®; Polysar EPDM 6463; Porofor® BSH Paste; Porofor® BSH Paste M; Trigonox® 22-E50; Trigonox® 101-E50; Trigonox® 145-E85; Trigonox® D-E50; Vulkacit CZ/MGC; Vulkacit DM/C; Vulkacit DM/MGC; Vulkacit Merkapto/C; Vulkacit Merkapto/MGC; Vulkanox® MB-2/MGC; Vulkazon® AFS-50

Mineral spirits

CAS 8032-32-4; 64475-85-0; EINECS 232-453-7

Synonyms: White spirits; Ligroin; Petroleum spirits

Definition: Mixture of hydrocarbons from petroleum with distillation range of 300-415 F; avail. in type I (reg.), II (high flash), III (odorless), IV (low dry pt.

Properties: Clear colorless, volatile, nonfluorescent liq.; dens. @ 15.6/15.6 C: 0.654-0.820 (I), 0.768-0.820 (II), 0.775 max. (III), 0.754-0.800 (IV); i.b.p. 149 C (I), 177 C (II), 149 C (III), 149 C (IV); flash pt. 38 C min. except 60 C (II)

Precaution: Keep away from heat or flame

Toxicology: TLV:TWA 300 ppm; STEL 400 ppm; moderately toxic in humans; mildly toxic by inhalation, intraperitoneal routes; heated to dec., emits acrid smoke

Uses: Solvent; paint thinner; in coatings, drycleaning

Regulatory: FDA 21CFR §178.3800

Manuf./Distrib.: Ashland

Trade names containing: Aztec® TAPB-90-OMS; Aztec® TAPEH-75-OMS; Aztec® TAPEH-90-OMS; Aztec® TAPPI-75-OMS; Aztec® TBPA-50-OMS; Aztec® TBPA-75-OMS; Aztec® TBPEH-50-OMS; Aztec® TBPPI-50-OMS; Aztec® TBPPI-75-OMS; Espercarb® 438M-60; Espercarb® 840M; Espercarb® 840M-40; Espercarb® 840M-70; Esperox® 12MD; Esperox® 13M; Esperox® 28MD; Esperox® 31M; Esperox® 33M; Esperox® 497M; Esperox® 545M; Esperox® 551M; Esperox® 740M; Esperox® 747M; Esperox® 750M; Esperox® 939M; GP-RA-158 Silicone Polish Additive; GP-RA-159 Silicone Polish Additive; Lupersol 10-M75; Lupersol 11; Lupersol 70; Lupersol 75-M; Lupersol 76-M; Lupersol 80; Lupersol 188-M75; Lupersol 219-M60; Lupersol 223-M40; Lupersol 223-M75; Lupersol 225-M60; Lupersol 225-M75; Lupersol 233-M75; Lupersol 288-M75; Lupersol 531-80M; Lupersol 533-M75; Lupersol 546-M75; Lupersol 553-M75; Lupersol 554-M50, 554-M75; Lupersol 555-M60; Lupersol 575-M75; Lupersol 688-M50; Lupersol PMS; Lupersol TBIC-M75; Trigonox® 21-C50; Trigonox® 22-C50; Trigonox® 23-C75; Trigonox® 25-C75; Trigonox® 29-C75; Trigonox® 36-C60; Trigonox® 36-C75; Trigonox® 97-C75; Trigonox® 99-C75; Trigonox® 123-C75; Trigonox® 125-C50; Trigonox® 125-C75; Trigonox® 133-C60; Trigonox® 151-C50; Trigonox® 151-C70; Trigonox® 187-C30; Trigonox® BPIC; Trigonox® D-C50; Trigonox® EHP-C40; Trigonox® EHP-C70; Trigonox® EHP-C75; Trigonox® EHPS-C75; Trigonox® F-C50; Trigonox® NBP-C50; Trigonox® SBP-C30; Trigonox® SBP-C60; Trigonox® SBP-C75; Uni-Tac® 72; Uni-Tac® 72M70; USP®-90MD; USP®-355M

Mineral wax. *See* Ceresin, Ozokerite
Minium. *See* Lead oxide, red

Mixed diaryl-p-phenylenediamine
CAS 68478-45-5; 69853-84-4
Properties: Solid, amine odor; insol. in water; dens. 1.18 mg/m^3
Precaution: Dust may form explosive mixt. with air
Toxicology: Irritating to skin, upper respiratory tract; LD50 (rat, oral) 4000-7500 mg/kg
Uses: Antiozonant, antioxidant, antiflexcracking agent for rubber
Trade names: Akrochem® Antiozonant MPD-100

Molybdenum anhydride. *See* Molybdenum trioxide

Molybdenum disulfide
CAS 1317-33-5; EINECS 215-263-9
Synonyms: Molybdenum sulfide; Molybdic sulfide
Empirical: MoS_2
Properties: Black crystalline powd.; sol. in aqua regia, conc. sulfuric acid; insol. in water, dilute acids; m.w. 160.08; dens. 5.06 (15/15 C); m.p. 2375 C; begins to sublime at 45 C; Mohs hardness 1
Toxicology: Toxic material; TLV (as Mo) 5 mg/m^3 of air
Uses: Lubricant in greases, oil dispersions, resin-bonded films, dry powders, etc.; hydrogenation catalyst
Manuf./Distrib.: AAA Molybdenum Prods.; Allchem Ind.; Cerac; Climax Molybdenum; Dow Corning; E/M Corp.; Graphite Prods.; McGee Ind.; Noah; Osram Sylvania
Trade names containing: Polyvel RM40; Polyvel RM40Z; Polyvel RM70

Molybdenum sulfide. *See* Molybdenum disulfide

Molybdenum trioxide
CAS 1313-27-5; EINECS 215-204-7
Synonyms: Molybdenum anhydride; Molybdic oxide; Molybdic acid hydride
Empirical: MoO_3
Properties: White or yel. powd.; sl. sol. in water; sol. in conc. mixture of nitric acid and HCl; m.w. 143.94; dens. 4.69; m.p. 795 C; begins to sublime at 700 C; b.p. 1150 C
Toxicology: Toxic material; poison by ingestion; TLV (as Mo) 5 mg/m^3 of air
Uses: Source of Mo; reagent for analytical chemistry; agriculture; mfg. of metallic Mo; corrosive inhibitor; ceramic glazes; enamels; pigments; catalyst; smoke suppressant, flame retardant for plastics
Manuf./Distrib.: AAA Molybdenum Prods.; All Chemie Ltd; Atomergic Chemetals; Cerac; Climax Molybdenum; Noah; Spectrum Chem. Mfg.
Trade names: Polu-U

Molybdic acid hydride. *See* Molybdenum trioxide
Molybdic oxide. *See* Molybdenum trioxide
Molybdic sulfide. *See* Molybdenum disulfide
Monacetin. *See* Glyceryl acetate
Monobutylated paracresol. *See* 2-t-Butyl-p-cresol
Mono-t-butyl hydroquinone. *See* t-Butyl hydroquinone
Monobutyl maleate. *See* Butyl maleate
Monocalcium phosphate monohydrate. *See* Calcium phosphate monobasic monohydrate

Mono- and diglycerides of fatty acids
CAS 67701-32-0; 67701-33-1; 68990-53-4
Properties: Yel. liqs. to ivory plastics to hard solids, bland odor and taste; sol. in alcohol, ethyl acetate, chloroform, other chlorinated hydrocarbons; insol. in water
Toxicology: Heated to decomp., emits acrid smoke and irritating fumes
Uses: Antistat for PE, flexible PVC, plastisols; food emulsifier
Regulatory: FDA 21CFR §172.863, 182.4505, 184.1505, GRAS; USDA 9CFR §318.7, 381.147; Europe listed; UK approved
Manuf./Distrib.: Int'l. Chem.
Trade names: Atmul® 695
Trade names containing: Dur-Em® 117; Dur-Em® 300K

Monoethanolamine. *See* Ethanolamine
Monoethanolamine lauryl sulfate. *See* MEA-lauryl sulfate

Monoisopropanolamine
Uses: In soaps, shampoos, emulsifiers, textile specialties, agric. and polymer curing chems., adhesives, coatings, metalworking, petrol, rubber processing, gas conditioning chems.
Trade names: MIPA

Monolinolein. *See* Glyceryl linoleate
Monomyristin. *See* Glyceryl myristate
Monoolein. *See* Glyceryl oleate
Monopentaerythritol. *See* Pentaerythritol
Monopropylene glycol methyl ether. *See* Methoxyisopropanol
Monoricinolein. *See* Glyceryl ricinoleate
Monosodium carbonate. *See* Sodium bicarbonate
Monostearin. *See* Glyceryl stearate
Monothioglycerol. *See* Thioglycerin
Montan acid, butylene glycol ester. *See* Butylene glycol montanate
Montan acid, calcium salt. *See* Calcium montanate

Montan acid wax
CAS 68476-03-9; EINECS 270-664-6
Synonyms: Fatty acids, montan wax; Waxes, montan fatty acids
Definition: Prod. obtained by the oxidation of montan wax
Uses: Emulsifying component for paraffin wax emulsions, polishes; water repellent for textiles; lubricant and release agent for plastics
Trade names: Hoechst Wax LP; Hoechst Wax S; Hoechst Wax SW; Hoechst Wax UL

Montan wax
CAS 8002-53-7; EINECS 232-313-5
Synonyms: Lignite wax; Waxes, montan
Definition: Wax obtained by extraction of lignite
Properties: Dark brown lumps or white hard earth wax; sol. in CCl_4, benzene, chloroform, hot petrol. ether; insol. in water; m.p. 80-90 C; sapon. no. 88-112
Precaution: Combustible
Uses: Substitute for carnauba and beeswax; shoe and furniture polishes; waterproof and roofing paints; adhesive pastes; candles; paper sizing compds.; wire coatings; lubricant for plastics
Regulatory: FDA 21CFR §175.105, 176.210, 177.2600
Manuf./Distrib.: Hoechst Celanese; Frank B. Ross; Stevenson Cooper; Strahl & Pitsch
Trade names: Ross Montan Wax; Sicolub® E; Sicolub® OP

Montan-wax fatty acids, glyceryl esters. *See* Glyceryl montanate

Montmorillonite
CAS 1318-93-0; EINECS 215-288-5
Classification: Complex silicate clay mineral
Formula: $Al_2O_5 \cdot 4SiO_2 \cdot 4H_2O$
Properties: Lt. yel. or green, cream, pink, gray to black; insol. in water and common org. solvs.
Toxicology: Poison by intravenous route
Uses: Major component of bentonite and Fuller's earth
Trade names: Perchem® 97

Morpholine
CAS 110-91-8; EINECS 203-815-1
Synonyms: Tetrahydro-1,4-oxazine; 1-Oxa-4-azacyclohexane; Tetrahydro-2H-1,4-oxazine; Diethylene oximide; Diethyleneimide oxide
Classification: Heterocyclic organic compd.
Empirical: C_4H_9NO
Formula: C_4H_8ONH
Properties: Colorless clear hygroscopic liq., amine-like odor; misc. with water, acetone, benzene, ether, castor oil, alcohol; m.w. 87.14; dens. 1.002 (20/20 C); b.p. 128.9 C; f.p. -4.9 C; flash pt. (OC) 37.7 C; autoignition temp. 590 F; ref. index 1.4540 (20 C)
Precaution: Flamm.; dangerous fire hazard exposed to flame, heat or oxidizers; reactive with oxidizers; explosive with nitromethane
Toxicology: TLV:TWA 20 ppm; LD50 (oral, rat) 1050 mg/kg; mod. toxic by ing., inh., skin contact, IP; corrosive irritant to skin, eyes, mucous membranes; experimental neoplastigen; mutagenic data; kidney damage;

heated to dec., emits highly toxic fumes of NO_x
Uses: Intermediate for vulcanization accelerator; solvent; additive to boiler water; optical brightener for detergents; corrosion inhibitor; organic intermediate; intermediate for analgesics, anesthetics
Regulatory: FDA 21CFR §172.235, 173.310 (10 ppm max. in steam), 175.105, 176.210, 178.3300
Manuf./Distrib.: Air Prods.; Allchem Ind.; BASF; Coyne; Nippon Nyukazai; PMC Specialties; Texaco

Morpholinium, 4-ethyl-4-hexadecyl, ethyl sulfate. *See* Cetethyl morpholinium ethosulfate
2-(Morpholinothio) benzothiazole. *See* N-Oxydiethylene benzothiazole-2-sulfenamide

4-Morpholinyl-2-benzothiazole disulfide
CAS 95-32-9
Synonyms: 2-Benzothiazolyl morpholinodisulfide; 2-(4-Morpholinyldithio) benzothiazole
Empirical: $C_{11}H_{12}N_2OS_3$
Properties: M.w. 284.43
Toxicology: Mildly toxic by ingestion; eye irritant
Uses: Curing and vulcanizing agent for heat resistant rubber applics., e.g., aircraft tires; accelerator for rubber
Trade names: Morfax®; Vulcuren 2; Vulcuren 2/EGC

2-(4-Morpholinyldithio) benzothiazole. See 4-Morpholinyl-2-benzothiazole disulfide
2-Morpholinyl mercaptobenzothiazole. *See* N-Oxydiethylene benzothiazole-2-sulfenamide
mPDA. *See* m-Phenylenediamine
MPMD. *See* 2-Methylpentamethylenediamine
MTBE. *See* Methyl t-butyl ether
Muscovite mica. *See* Mica
Myristic acid isopropyl ester. *See* Isopropyl myristate
Myristoyl N-methylglycine. *See* Myristoyl sarcosine

Myristoyl sarcosine
CAS 52558-73-3; EINECS 258-007-1
Synonyms: N-Methyl-N-(1-oxotetradecyl)glycine; Myristoyl N-methylglycine
Definition: N-Myristoyl deriv. of N-methylglycine
Empirical: $C_{17}H_{33}NO_3$
Formula: $CH_3(CH_2)_{12}CONCH_3CH_2COOH$
Uses: Detergent, wetting and foaming agent for cosmetics, industrial and household cleaners, biotechnology, pesticides, textiles, petrol. prods., metalworking; emulsifier for polymerization; mold release, stabilizer, antifog, and antistat in plastics
Trade names: Hamposyl® M

Myristyl alcohol
CAS 112-72-1; EINECS 204-000-3
Synonyms: 1-Tetradecanol
Empirical: $C_{14}H_{30}O$
Properties: Colorless to wh. waxy solid flakes; waxy odor; sol. in ether; sl. sol. in alcohol; insol. in water; m.w. 214.38; dens. 0.8355 (20/20 C); m.p. 38 C; b.p. 167 C; flash pt. 285 F
Precaution: Combustible
Uses: Surfactant intermediate; organic synthesis; antifoam agent; perfume fixative for soaps and cosmetics; specialty cleaning preparations; emollient for cold creams; emulsion polymerization
Regulatory: FDA 21CFR §172.864, 175.105, 175.300, 176.200, 176.210, 177.1010, 177.2800, 178.3480, 178.3910
Manuf./Distrib.: Albemarle; Condor; R.W. Greeff; M. Michel; Schweizerhall; Spectrum Chem. Mfg.; Vista
Trade names: Epal® 14
Trade names containing: Epal® 12/70; Epal® 12/85; Epal® 1214

Myristyl dimethylamine. *See* Dimethyl myristamine
2MZ. *See* 2-Methyl imidazole

Nabam
CAS 142-59-6
Synonyms: DSE; Disodium ethylene bisdithiocarbamate; Ethylenebis (dithiocarbamate), disodium salt; Disodium ethylene-1,2-bisdithiocarbamate; Dithane A-40
Empirical: $C_4H_6N_2S_4 \cdot 2Na$
Formula: $NaSSCNHCH_2CH_2NHCSSNa$
Properties: Colorless crystals; sol. in water; m.w. 256.34

Toxicology: LD50 (oral, rat) 395 mg/kg; poison by ingestion; moderately toxic by IP route; skin irritant; experimental teratogen, reproductive effects; mutagenic data; heated to decomp., emits toxic fumes of Na_2O, NO_x, SO_x
Uses: Plant fungicide; ingredient for pesticides; industrial applics.
Regulatory: FDA 21CFR §173.320
Trade names containing: Aquatreat DNM-30

Naphtha
CAS 8030-30-6; 68920-06-9; 64742-95-6 (lt. aromatic)
Synonyms: Coal tar naphtha; Benzin; Petroleum naphtha; VM&P naphtha; Petroleum benzin; Petroleum ether; Petroleum spirit
Definition: Petroleum distillate
Properties: Dark straw-colored to colorless liq.; sol. in benzene, toluene, xylene; dens. 0.862-0.892; b.p. 149-216 C; flash pt. (CC) 107 F
Precaution: Flamm. when exposed to heat or flame; sl. explosion hazard; can react with oxidizing materials; keep containers tightly closed
Toxicology: Mildly toxic by inhalation; human poison and systemic effects by IV route; common air contaminant
Uses: Solv., thinner in paint, drycleaning fluid, rubber compounding, sealants, chem. absorption; blending with natural gas
Regulatory: FDA 21CFR §73.1, 172.250
Manuf./Distrib.: Ashland; Kerr-McGee; Mobil; Monsanto; Norsk Hydro A/S; Texaco
Trade names containing: BYK®-S 706

Naphthalane. *See* Decahydronaphthalene
Naphthalenesulfonic acid, bis-(1-methylethyl)-, sodium salt. *See* Sodium isopropyl naphthalene sulfonate
Naphthalenesulfonic acid, polymer with formaldehyde, sodium salt. *See* Sodium polynaphthalene sulfonate

Naphthenic oil
CAS 67254-74-4
Uses: Process oil offering color and heat stability for rubber compounding; plasticizer
Trade names: Akrochem® Plasticizer LN; Calight RPO; Calsol 510; Calsol 804; Calsol 806; Calsol 810; Calsol 815; Calsol 830; Calsol 850; Calsol 875; Calsol 5120; Calsol 5550; Calsol 8120; Calsol 8200; Calsol 8240; Cyclolube® 85; Cyclolube® 120; Cyclolube® 132; Cyclolube® 210; Cyclolube® 213; Cyclolube® 270; Cyclolube® 413; Cyclolube® 2290; Cyclolube® 2310; Cyclolube® 4053; Cyclolube® NN-1; Cyclolube® NN-2; Hymocal

N-1-Naphthylaniline. *See* Phenyl-α-naphthylamine
N-2-Naphthylaniline. *See* Phenyl-β-naphthylamine
2-Naphthyl-p-phenylene diamine. *See* N,N′-Di-β-naphthyl-p-phenylenediamine
Native calcium sulfate. *See* Calcium sulfate dihydrate

Natural rubber
Synonyms: NR
Uses: Used for adhesive, dipping, coating, foam, and molded materials, latex applics., foamed rubber, textile, medical, cement, asphalt; processing aid for blending with other rubber
Manuf./Distrib.: Firestone Syn. Rubber & Latex; Hardman; A. Schulman
Trade names: PA-57; PA-80
Trade names containing: Poly-Dispersion® JZFD-90P
See also Polyisoprene

NBC. *See* Nickel dibutyldithiocarbamate
NBR. *See* Acrylonitrile-butadiene rubber
NBR/PVC. *See* Acrylonitrile-butadiene rubber/PVC
NDBC. *See* Nickel dibutyldithiocarbamate
NDPMT. *See* Dinitrosopentamethylene tetramine
Neoalkoxy tri(dioctylphosphate) titanate. *See* Neoalkoxy tri (dioctylphosphato) titanate

Neoalkoxy tri (dioctylphosphato) titanate
Synonyms: Neoalkoxy tri(dioctylphosphate) titanate
Uses: Flame retardant for plastics

Neoalkoxy tri (dioctylpyrophosphate) titanate. *See* Neoalkoxy tri (dioctylpyrophosphato) titanate

Neoalkoxy tri (dioctylpyrophosphato) titanate
Synonyms: Neoalkoxy tri (dioctylpyrophosphate) titanate
Uses: Flame retardant for plastics

Neodecanoic acid glycidyl ester
Uses: Reactive epoxy modifier/diluent
Trade names: Cardura® E-10

Neopentyl (diallyl) oxy, triacryl zirconate
Trade names containing: NZ 39

Neopentyl (diallyl) oxy, tri (amino) phenyl titanate
Trade names containing: LICA 97

Neopentyl (diallyl) oxy, tri (m-amino) phenyl zirconate
Uses: Coupling agent
Trade names containing: NZ 97

Neopentyl (diallyl) oxy, tri (dioctyl) phosphato titanate
Trade names containing: LICA 12

Neopentyl (diallyl) oxy, tri (dioctyl) phosphato zirconate
Uses: Coupling agent
Trade names containing: NZ 12

Neopentyl (diallyl) oxy, tri (dioctyl) pyrophosphato titanate
Trade names containing: LICA 38

Neopentyl (diallyl) oxy, tri (dioctyl) pyrophosphato zirconate
Uses: Coupling agent
Trade names containing: NZ 38

Neopentyl (diallyl) oxy, tri (dodecyl) benzene-sulfonyl titanate
Trade names containing: LICA 09

Neopentyl (diallyl) oxy, tri (dodecyl) benzene-sulfonyl zirconate
Uses: Coupling agent
Trade names containing: NZ 09

Neopentyl (diallyl) oxy, tri (9,10 epoxy stearoyl) zirconate
Uses: Coupling agent
Trade names containing: NZ 49

Neopentyl (diallyl) oxy, tri (N-ethylenediamino) ethyl titanate
Trade names containing: LICA 44

Neopentyl (diallyl) oxy, tri (N-ethylenediamino) ethyl zirconate
Uses: Coupling agent
Trade names containing: NZ 44

Neopentyl (diallyl) oxy, trihydroxy caproyl titanate
Uses: Coupling agent, adhesion promoters, antioxidants, antistats, antifoaming agents, accelerators, blowing agent activators, catalysts, curatives, corrosion inhibitors, disp. aids, emulsifiers, flame retardants, foaming agents; hardeners, impact modifiers, internal lubes, process aids, release agents, retarders, stabilizers, surfactants, suspension aids, thixotropes, wetting agents for thermoplastics, thermosets, elastomers
Trade names: LICA 99

Neopentyl (diallyl) oxy, trimercapto-phenyl zirconate
Uses: Coupling agent, adhesion promoters, antioxidants, antistats, antifoaming agents, accelerators, blowing agent activators, catalysts, curatives, corrosion inhibitors, disp. aids, emulsifiers, flame retardants, foaming agents; hardeners, impact modifiers, internal lubes, process aids, release agents, retarders, stabilizers, surfactants, suspension aids, thixotropes, wetting agents for thermoplastics, thermosets, elastomers
Trade names: NZ 89

Neopentyl (diallyl) oxy, trimethacryl zirconate
Uses: Coupling agent
Trade names containing: NZ 33

Neopentyl (diallyl) oxy, trineodecanonyl titanate
Trade names containing: LICA 01

Neopentyl (diallyl) oxy, trineodecanoyl zirconate
Uses: Coupling agent
Trade names containing: NZ 01

Neopentyl glycol dibenzoate
CAS 4196-89-8
Properties: Liq.; color APHA 200; m.w. 312; m.p. 49 C
Uses: Process aid, modifier, plasticizer for thermoplastics, hot-melt adhesives, coatings
Manuf./Distrib.: Velsicol
Trade names: Benzoflex® S-312

Neopentyl glycol diglycidyl ether
CAS 17557-23-2
Synonyms: 1,3-Bis(2,3-epoxypropoxy)-2,2-dimethyl propane
Empirical: $C_{11}H_{20}O_4$
Properties: M.w. 216.31
Toxicology: When heated to dec., emits acrid smoke and irritating fumes; experimental tumorigen
Uses: Reactive epoxy diluent for civil engineering applics.
Trade names: Epodil® 749; Heloxy® 68

Neoprene. *See* Polychloroprene

Nepheline syenite
CAS 37244-96-5
Synonyms: Nephylene syenite; Sodium potassium aluminosilicate
Definition: Feldspathoid igneous rock primarily composed of the minerals microcline ($KAlSi_3O_8$), albite ($NaAlSi_3O_8$), and nepheline [$(Na,K)AlSiO_4$]
Properties: Wh. solid; sp.gr. 2.61; m.p. 1223 C
Toxicology: ACGIH TLV 10 mg/m^3; excessive/prolonged inh. of dust may harm respiratory system
Uses: Filler in paints, plastics, coatings, adhesives, caulks, sealants, inks, rubber, friction prods.
Trade names: Minbloc™ 16; Minbloc™ 20; Minbloc™ 30; Minex® 4; Minex® 7; Minex® 10

Nephylene syenite. *See* Nepheline syenite

Nickel
CAS 7440-02-0; EINECS 231-111-4
Classification: Metallic element
Synonyms: Nickel catalysts
Empirical: Ni
Properties: Malleable silvery metal; high ductility and malleability; insol. in water; at.wt. 58.69; dens. 8.9; m.p. 1452 C; b.p. 2900 C; corrosion resistant
Precaution: Powders can ignite spontaneously in air; incompat. with oxidants
Toxicology: TLV (metal) 1 mg/m^3 of air; poison by ingestion, IV, IP, subcut. routes; experimental carcinogen, neoplastigen, tumorigen, teratogen, reproductive effects; mutagenic data; hypersensitivity can cause dermatitis, pulmonary asthma, conjunctivitis
Uses: Electroplating; hydrogenation catalyst; in iron- and copper-based alloy; direct food additive, catalyst; used in hydrogenation of fats and oils; processing aid (Japan)
Regulatory: FDA 21CFR §184.1537, GRAS; USDA 9CFR §318 (must be eliminated during processing); Japan approved
Manuf./Distrib.: Aldrich; Atomergic Chemetals; Cerac; Spectrum Chem. Mfg.
Trade names containing: Cycom® NCG 1200 Unsized; Cycom® NCG 1204 p8; Dry-Blend® NCG dB910; Dry-Blend® NCG dB920

Nickel bis[O-ethyl (3,5-di-t-butyl-4-hydroxybenzyl)] phosphonate
Properties: Powd.
Uses: UV stabilizer and antioxidant for polyolefins
Trade names: Irgastab® 2002

Nickel bis (octyl phenol sulfide)
Properties: Flakes; m.p. 130 C
Uses: UV stabilizer for polyolefins
Trade names: UV-Chek® AM-101; UV-Chek® AM-105

Nickel dibutyldithiocarbamate
CAS 13927-77-0
Synonyms: NBC; NDBC
Empirical: $NiS_4C_4H_{52}N_2$
Formula: $Ni[SC(S)N(CH_4H_9)_2]_2$
Properties: Dark green flakes; m.w. 467.51; dens. 1.26; m.p. 86 C min.
Toxicology: Can cause dermatitis; when heated to dec., emits toxic fumes of SO_x and NO_x
Uses: UV stabilizer, antioxidant for synthetic rubbers, polyolefins
Trade names: Ekaland NDBC; Naugard® NBC; Perkacit® NDBC; UV-Chek® AM-104; Vanox® NBC
Trade names containing: Poly-Dispersion® E (NBC)D-70; Poly-Dispersion® E (NBC)D-70P; Rhenogran® NDBC-70/ECO

Nickel diisobutyldithiocarbamate
CAS 15317-78-9
Properties: Powd.; m.p. 173-181 C
Uses: Antioxidant/antiozonant for protection in epichlorohydrin; uv stabilizer in polyolefins
Trade names: Isobutyl Niclate®

Nickel dimethyldithiocarbamate
CAS 15521-65-0
Properties: Powd.; m.p. > 290 C
Uses: UV stabilizer, antioxidant for polyolefins, rubber
Trade names: Methyl Niclate®
Trade names containing: Poly-Dispersion® K (NMC)D-70

Nitrile rubber. *See* Acrylic resin, Acrylonitrile-butadiene rubber

2,2′,2′′-Nitrilo[triethyl-tris(3,3′,5,5′-tetra-t-butyl-1,1′-biphenyl-2,2′-diyl)phosphite]
CAS 80410-33-9
Classification: Organophosphite
Empirical: $C_{90}H_{132}NO_9P_3$
Properties: Wh. powd.; m.w. 1465; m.p. 120-220 C
Uses: Stabilizer, antioxidant for polymers
Trade names: CGA 12

2,2′,2′′-Nitrilotris(ethanol). *See* Triethanolamine
1,1′,1′′-Nitrilotris-2-propanol. *See* Triisopropanolamine

2-Nitro-2-methyl-1-propanol
CAS 76-39-1
Classification: Nitro alcohol
Empirical: $C_4H_9O_3N$
Formula: $CH_3C(CH_3)(NO_2)CH_2OH$
Properties: White crystals; sol. in water; m.w. 119.12; m.p. 90 C; b.p. 95 C (10 mm)
Uses: Chemical and pharmaceutical intermediate, in tire cord adhesives, as formaldehyde release agents, deodorants, antimicrobials
Manuf./Distrib.: ANGUS
Trade names containing: NMP-Plus™

Nitrosodiphenylamine
CAS 156-10-5
Synonyms: Diphenylnitrosamine; 4-Nitroso-N-phenylbenzenamine; p-Nitroso-n-phenylaniline
Empirical: $C_{12}H_{10}N_2O$

p-Nitroso-n-phenylaniline

Formula: $C_6H_5NC_6H_5NO$
Properties: Green plates with bluish luster; sol. in alcohol, ether, benzene, chloroform; sl. sol. in water, petrol. ether; m.w. 198.24; m.p. 144-145 C
Toxicology: Poison by intravenous route; moderately toxic by ingestion; severe eye irritant; suspected carcinogen; heated to dec., emits toxic fumes of NO_x
Uses: Accelerator in vulcanizing rubber
Trade names containing: Prespersion PAC-4911

p-Nitroso-n-phenylaniline. *See* Nitrosodiphenylamine
4-Nitroso-N-phenylbenzenamine. *See* Nitrosodiphenylamine

Nitrosophenylhydroxylamine aluminum salt
CAS 15305-07-4; EINECS 239-341-7
Synonyms: Aluminum, tris(N-hydroxy-N-nitrosobenzenaminato-O,O′)-
Uses: Polymerization inhibitor; uv ink stabilizer
Trade names: Q-1301

Nitrosophenylhydroxylamine ammonium salt
CAS 135-20-6; EINECS 205-183-2
Synonyms: Cupferron
Empirical: $C_6H_9N_3O_2$
Properties: M.w. 155.16
Toxicology: Toxic; suspected carcinogen
Uses: Polymerization inhibitor; stabilizer for precured resins; chelating agent; antioxidant; germicides, fungicides; agric. chems.; corrosion inhibitor for metals; heat stabilizer for chlorosulfonated PE; dye synthesis; analytical reagent
Manuf./Distrib.: Aldrich
Trade names: Q-1300

NMP. *See* 1-Methyl-2-pyrrolidinone
NODA. *See* n-Octyl, n-decyl adipate
Nonanedoic acid. *See* Azelaic acid
Nonanoic acid, 1-methyl-1,2-ethanediyl ester. *See* Propylene glycol dipelargonate
3,6,9,12,15,18,21,24,27-Nonaoxanonatriacontan-1-ol. *See* Laureth-9
Noncarbinol. *See* n-Decyl alcohol

Nonoxynol-1
CAS 26027-38-3 (generic); 37205-87-1 (generic); 27986-36-3; EINECS 248-762-5
Synonyms: Ethylene glycol nonyl phenyl ether; PEG-1 nonyl phenyl ether; 2-(Nonylphenoxy) ethanol
Classification: Ethoxylated alkyl phenol
Empirical: $C_{17}H_{28}O_2$
Formula: $C_9H_{19}C_6H_4OCH_2CH_2OH$
Properties: Yel. to almost colorless liq.; sol. in oil
Toxicology: Moderately toxic by ingestion, skin contact; severe eye and mild skin irritant in humans; heated to dec., emits acrid smoke and fumes
Uses: Nonionic surfactant; as detergent, emulsifier, wetting agent, dispersant, stabilizer, defoamer; intermediate in synthesis of anionic surfactants; emulsion polymerization
Regulatory: FDA 21CFR §175.105, 176.180
Trade names: Akyporox NP 15; Prox-onic NP-1.5

Nonoxynol-3
CAS 27176-95-0 (generic); 84562-92-5 (generic); 51437-95-7 (generic); 9016-45-9 (generic)
Synonyms: PEG-3 nonyl phenyl ether; POE (3) nonyl phenyl ether; 2-[2-[2-(Nonylphenoxy) ethoxy]ethoxy]ethanol
Classification: Ethoxylated alkyl phenol
Empirical: $C_{21}H_{36}O_4$
Formula: $C_9H_{19}C_6H_4(OCH_2CH_2)_nOH$, avg. n = 3
Properties: Yel to almost colorless liq.; sol. in oil
Toxicology: Moderately toxic by ingestion, skin contact; severe eye and mild skin irritant in humans; heated to dec., emits acrid smoke and fumes
Uses: Nonionic surfactant; as detergent, emulsifier, wetting agent, dispersant, stabilizer, defoamer; intermediate in synthesis of anionic surfactants; emulsion polymerization

Regulatory: FDA 21CFR §175.105, 176.180, 176.210
Trade names: Akyporox NP 30

Nonoxynol-4

CAS 7311-27-5; 9016-45-9 (generic); 26027-38-3 (generic); 37205-87-1 (generic); 27176-97-2; EINECS 230-770-5
Synonyms: PEG-4 nonyl phenyl ether; POE (4) nonyl phenyl ether; PEG 200 nonyl phenyl ether
Classification: Ethoxylated alkyl phenol
Empirical: $C_{23}H_{40}O_5$
Formula: $C_9H_{19}C_6H_4(OCH_2CH_2)_nOH$, avg. n = 4
Properties: Yel to almost colorless liq.; sol. in oil
Toxicology: Moderately toxic by ingestion, skin contact; severe eye and mild skin irritant in humans; heated to dec., emits acrid smoke and fumes
Uses: Nonionic surfactant; as detergent, emulsifier, wetting agent, dispersant, stabilizer, defoamer; intermediate in synthesis of anionic surfactants; pharmaceutic aids; plasticizer, antistat for plastics; in cosmetics, fat liquoring, cutting oils, agric., petrol. oils
Regulatory: FDA 21CFR §175.105, 176.180, 176.210, 178.3400
Trade names: Alkasurf® NP-4; Igepal® CO-430; Polystep® F-1; Prox-onic NP-04; Rexol 25/4

Nonoxynol-5

CAS 9016-45-9 (generic); 26027-38-3 (generic); 37205-87-1 (generic); 26264-02-8; 20636-48-0; EINECS 247-555-7
Synonyms: PEG-5 nonyl phenyl ether; POE (5) nonyl phenyl ether; 14-(Nonylphenoxy)-3,6,9,12-tetraoxatetradecan-1-ol
Classification: Ethoxylated alkyl phenol
Empirical: $C_{25}H_{44}O_6$
Formula: $C_9H_{19}C_6H_4(OCH_2CH_2)_nOH$, avg. n = 5
Properties: Yel. to almost colorless liq.; sol. in oil
Toxicology: Moderately toxic by ingestion, skin contact; severe eye and mild skin irritant in humans; heated to dec., emits acrid smoke and fumes
Uses: Nonionic surfactant; as detergent, emulsifier, wetting agent, dispersant, stabilizer, defoamer; intermediate in synthesis of anionic surfactants; emulsion polymerization
Regulatory: FDA 21CFR §175.105, 176.180, 176.210, 178.3400
Trade names: Rewopal® HV 5

Nonoxynol-6

CAS 9016-45-9 (generic); 26027-38-3 (generic); 37205-87-1 (generic); 27177-01-1; 27177-05-5
Synonyms: PEG-6 nonyl phenyl ether; POE (6) nonyl phenyl ether; PEG 300 nonyl phenyl ether
Classification: Ethoxylated alkyl phenol
Empirical: $C_{27}H_{48}O_7$
Formula: $C_9H_{19}C_6H_4(OCH_2CH_2)_nOH$, avg. n = 6
Properties: Yel. to almost colorless liq.
Toxicology: Moderately toxic by ingestion, skin contact; severe eye and mild skin irritant in humans; heated to dec., emits acrid smoke and fumes
Uses: Nonionic surfactant; as detergent, emulsifier, wetting agent, dispersant, stabilizer, defoamer; intermediate in synthesis of anionic surfactants; plasticizer, antistat for plastics
Regulatory: FDA 21CFR §175.105, 176.180, 176.210, 178.3400
Trade names: Alkasurf® NP-6; Igepal® CO-530; Nissan Nonion NS-206; Polystep® F-2; Prox-onic NP-06; Triton® N-60

Nonoxynol-7

CAS 9016-45-9 (generic); 26027-38-3 (generic); 27177-05-5; 37205-87-1 (generic); EINECS 248-292-0
Synonyms: PEG-7 nonyl phenyl ether; POE (7) nonyl phenyl ether
Classification: Ethoxylated alkyl phenol
Empirical: $C_{29}H_{52}O_8$
Formula: $C_9H_{19}C_6H_4(OCH_2CH_2)_nOH$, avg. n = 7
Properties: Yel. to almost colorless liq.
Toxicology: Moderately toxic by ingestion, skin contact; severe eye and mild skin irritant in humans; heated to dec., emits acrid smoke and fumes
Uses: Nonionic surfactant; as detergent, emulsifier, wetting agent, dispersant, stabilizer, defoamer; intermediate in synthesis of anionic surfactants; antistat, plasticizer for plastics
Regulatory: FDA 21CFR §175.105, 176.180, 176.210, 178.3400
Trade names: Rexol 25/7

Nonoxynol-8

CAS 9016-45-9 (generic); 26027-38-3 (generic); 37205-87-1 (generic); 26571-11-9; 27177-05-5; EINECS 248-293-6; 247-816-5

Synonyms: PEG-8 nonyl phenyl ether; POE (8) nonyl phenyl ether; PEG 400 nonyl phenyl ether

Classification: Ethoxylated alkyl phenol

Empirical: $C_{31}H_{56}O_9$

Formula: $C_9H_{19}C_6H_4(OCH_2CH_2)_nOH$, avg. n = 8

Properties: Yel. to almost colorless liq.

Toxicology: Moderately toxic by ingestion, skin contact; severe eye and mild skin irritant in humans; heated to dec., emits acrid smoke and fumes

Uses: Nonionic surfactant; as detergent, emulsifier, wetting agent, dispersant, stabilizer, defoamer; intermediate in synthesis of anionic surfactants; emulsifier for polymerization

Regulatory: FDA 21CFR §175.105, 176.180, 176.210, 178.3400

Trade names: Polystep® F-3; Rexol 25/8; Teric N8

Nonoxynol-9

CAS 9016-45-9 (generic); 26027-38-3 (generic); 26571-11-9; 37205-87-1 (generic); 14409-72-4

Synonyms: PEG-9 nonyl phenyl ether; POE (9) nonyl phenyl ether; PEG 450 nonyl phenyl ether

Classification: Ethoxylated alkyl phenol

Empirical: $C_{33}H_{60}O_{10}$

Formula: $C_9H_{19}C_6H_4(OCH_2CH_2)_nOH$, avg. n = 9

Properties: Almost colorless liq.; sol. in water, ethanol, ethylene glycol, xylene, corn oil; m.w. 617; dens. 1.06 (25/4 C); solid. pt. 26 F; pour pt. 37 F; flash pt. 535-555 F; cloud pt. 126-133 F (1% aq.); visc. 175-250 cp (25 C)

Toxicology: Moderately toxic by ingestion, skin contact; severe eye and mild skin irritant in humans; heated to dec., emits acrid smoke and fumes

Uses: Nonionic surfactant; as detergent, emulsifier, wetting agent, dispersant, stabilizer, defoamer; intermediate in synthesis of anionic surfactants; spermaticide; plastics antistat; emulsion and suspension polymerization

Regulatory: FDA 21CFR §175.105, 176.180, 176.210, 176.300, 178.3400

Trade names: Empilan® NP9; Ethylan® KEO; Igepal® CO-630; Prox-onic NP-09; Rewopal® HV 9; Teric N9

Nonoxynol-10

CAS 9016-45-9 (generic); 26027-38-3 (generic); 27177-08-8; 37205-87-1 (generic); 27942-26-3; EINECS 248-294-1

Synonyms: PEG-10 nonyl phenyl ether; POE (10) nonyl phenyl ether; PEG 500 nonyl phenyl ether

Classification: Ethoxylated alkyl phenol

Empirical: $C_{35}H_{64}O_{11}$

Formula: $C_9H_{19}C_6H_4(OCH_2CH_2)_nOH$, avg. n = 10

Properties: Yel. to almost colorless liq.

Toxicology: Moderately toxic by ingestion, skin contact; severe eye and mild skin irritant in humans; heated to dec., emits acrid smoke and fumes

Uses: Nonionic surfactant; as detergent, emulsifier, wetting agent, dispersant, stabilizer, defoamer; intermediate in synthesis of anionic surfactants; emulsion polymerization

Regulatory: FDA 21CFR §175.105, 176.180, 176.210, 178.3400

Trade names: Igepal® CO-710; Polystep® F-4; Prox-onic NP-010; Rewopal® HV 10; Rexol 25/10; Teric N10; Witconol™ NP-100

Nonoxynol-12

CAS 9016-45-9 (generic); 26027-38-3 (generic); 37205-87-1 (generic)

Synonyms: PEG-12 nonyl phenyl ether; POE (12) nonyl phenyl ether; PEG 600 nonyl phenyl ether

Classification: Ethoxylated alkyl phenol

Formula: $C_9H_{19}C_6H_4(OCH_2CH_2)_nOH$, avg. n = 12

Properties: Yel. to almost colorless liq.

Toxicology: Moderately toxic by ingestion, skin contact; severe eye and mild skin irritant in humans; heated to dec., emits acrid smoke and fumes

Uses: Nonionic surfactant; as detergent, emulsifier, wetting agent, dispersant, stabilizer, defoamer; intermediate in synthesis of anionic surfactants; emulsion polymerization

Regulatory: FDA 21CFR §175.105, 176.180, 176.210, 178.3400

Trade names: Ethylan® DP; Nissan Nonion NS-212; Polystep® F-5; Sermul EN 229; Teric N12

Nonoxynol-13
CAS 9016-45-9 (generic); 26027-38-3 (generic); 37205-87-1 (generic)
Synonyms: PEG-13 nonyl phenyl ether; POE (13) nonyl phenyl ether
Classification: Ethoxylated alkyl phenol
Formula: $C_9H_{19}C_6H_4(OCH_2CH_2)_nOH$, avg. n = 13
Properties: Yel. to almost colorless liq.
Toxicology: Moderately toxic by ingestion, skin contact; severe eye and mild skin irritant in humans; heated to dec., emits acrid smoke and fumes
Uses: Nonionic surfactant; as detergent, emulsifier, wetting agent, dispersant, stabilizer, defoamer; intermediate in synthesis of anionic surfactants; emulsion polymerization
Regulatory: FDA 21CFR §175.105, 176.180, 176.210, 178.3400
Trade names: Teric N13

Nonoxynol-14
CAS 9016-45-9 (generic); 26027-38-3 (generic); 37205-87-1 (generic)
Synonyms: PEG-14 nonyl phenyl ether; POE (14) nonyl phenyl ether
Classification: Ethoxylated alkyl phenol
Formula: $C_9H_{19}C_6H_4(OCH_2CH_2)_nOH$, avg. n = 14
Properties: Yel. to almost colorless liq.
Toxicology: Moderately toxic by ingestion, skin contact; severe eye and mild skin irritant in humans; heated to dec., emits acrid smoke and fumes
Uses: Nonionic surfactant; as detergent, emulsifier, wetting agent, dispersant, stabilizer, defoamer; intermediate in synthesis of anionic surfactants; emulsion polymerization
Regulatory: FDA 21CFR §175.105, 176.180, 176.210, 178.3400
Trade names: Ethylan® BV; Polystep® F-6

Nonoxynol-15
CAS 9106-45-9 (generic); 37205-87-1 (generic); 26027-38-3 (generic)
Classification: Ethoxylated alkyl phenol
Formula: $C_9H_{19}C_6H_4(OCH_2CH_2)_nOH$, avg. n = 15
Uses: Emulsifier, detergent, wetting agent, dispersant, solubilizer, coupling agent, defoamer for emulsion polymerization, latex carpet, textiles, metalworking, household, industrial, agric., paper, paint, and cosmetics industries
Regulatory: FDA 21CFR §175.105, 176.180, 176.210
Trade names: Prox-onic NP-015; Sermul EN 15; Teric N15

Nonoxynol-20
CAS 9016-45-9 (generic); 26027-38-3 (generic); 37205-87-1 (generic)
Synonyms: PEG-20 nonyl phenyl ether; POE (20) nonyl phenyl ether; PEG 1000 nonyl phenyl ether
Classification: Ethoxylated alkyl phenol
Formula: $C_9H_{19}C_6H_4(OCH_2CH_2)_nOH$, avg. n = 20
Properties: Pale yel. to off-white pastes or waxes
Toxicology: Moderately toxic by ingestion, skin contact; severe eye and mild skin irritant in humans; heated to dec., emits acrid smoke and fumes
Uses: Nonionic surfactant; as detergent, emulsifier, wetting agent, dispersant, stabilizer, defoamer; intermediate in synthesis of anionic surfactants; emulsion polymerization
Regulatory: FDA 21CFR §175.105, 176.180
Trade names: Akyporox NP 200; Igepal® CO-850; Prox-onic NP-020; Remcopal 33820; Sermul EN 20/70; T-Det® N-20; Teric N20; Trycol® 6967; Trycol® NP-20

Nonoxynol-25
Classification: Ethoxylated alkyl phenol
Formula: $C_9H_{19}C_6H_4(OCH_2CH_2)_nOH$, avg. n = 25
Properties: Pale yel. to off-white pastes or waxes
Toxicology: Moderately toxic by ingestion, skin contact; severe eye and mild skin irritant in humans; heated to dec., emits acrid smoke and fumes
Uses: Nonionic surfactant; as detergent, emulsifier, wetting agent, dispersant, stabilizer, defoamer; intermediate in synthesis of anionic surfactants; emulsion polymerization; post-stabilizer for syn. latexes
Trade names: Igepal® CA-877; Rewopal® HV 25; Serdox NNP 25

Nonoxynol-30
CAS 9016-45-9 (generic); 26027-38-3 (generic); 37205-87-1 (generic)
Synonyms: PEG-30 nonyl phenyl ether; POE (30) nonyl phenyl ether

Nonoxynol-34

Classification: Ethoxylated alkyl phenol
Formula: $C_9H_{19}C_6H_4(OCH_2CH_2)_nOH$, avg. n = 30
Properties: Pale yel. to off-white pastes or waxes
Toxicology: Moderately toxic by ingestion, skin contact; severe eye and mild skin irritant in humans; heated to dec., emits acrid smoke and fumes
Uses: Nonionic surfactant; as detergent, emulsifier, wetting agent, dispersant, stabilizer, defoamer; intermediate in synthesis of anionic surfactants; pharmaceutic aids; emulsion polymerization
Regulatory: FDA 21CFR §175.105, 176.180, 178.3400
Trade names: Ablunol NP30; Ablunol NP30 70%; Akyporox NP 300V; Carsonon® N-30; Dehydrophen PNP 30; Ethylan® N30; Hetoxide NP-30; Iconol NP-30; Iconol NP-30-70%; Igepal® CO-880; Igepal® CO-887; Polystep® F-9; Prox-onic NP-030; Prox-onic NP-030/70; Rexol 25/307; Serdox NNP 30/70; Sermul EN 145; Sermul EN 30/70; T-Det® N-307; Teric N30; Trycol® 6968; Trycol® 6969; Trycol® NP-30; Trycol® NP-307; Witconol™ NP-300

Nonoxynol-34
Uses: Emulsifier for acrylics and vinyl acetate
Trade names: Polystep® F-95B

Nonoxynol-35
Synonyms: PEG-35 nonyl phenyl ether; POE (35) nonyl phenyl ether
Classification: Ethoxylated alkyl phenol
Formula: $C_9H_{19}C_6H_4(OCH_2CH_2)_nOH$, avg. n = 35
Properties: Pale yel. to off-white pastes or waxes
Toxicology: Moderately toxic by ingestion, skin contact; severe eye and mild skin irritant in humans; heated to dec., emits acrid smoke and fumes
Uses: Nonionic surfactant; as detergent, emulsifier, wetting agent, dispersant, stabilizer, defoamer; intermediate in synthesis of anionic surfactants; emulsion polymerization
Trade names: Ethylan® HA Flake

Nonoxynol-40
CAS 9016-45-9 (generic); 26027-38-3 (generic); 37205-87-1 (generic)
Synonyms: PEG-40 nonyl phenyl ether; POE (40) nonyl phenyl ether; PEG 2000 nonyl phenyl ether
Classification: Ethoxylated alkyl phenol
Formula: $C_9H_{19}C_6H_4(OCH_2CH_2)_nOH$, avg. n = 40
Properties: Pale yel. to off-white pastes or waxes
Toxicology: Moderately toxic by ingestion, skin contact; severe eye and mild skin irritant in humans; heated to dec., emits acrid smoke and fumes
Uses: Nonionic surfactant; as detergent, emulsifier, wetting agent, dispersant, stabilizer, defoamer; intermediate in synthesis of anionic surfactants; emulsion polymerization
Regulatory: FDA 21CFR §175.105, 176.180, 178.3400
Trade names: Ablunol NP40; Ablunol NP40 70%; Chemax NP-40; Dehydrophen PNP 40; Iconol NP-40; Iconol NP-40-70%; Igepal® CO-890; Igepal® CO-897; Polystep® F-10; Prox-onic NP-040; Prox-onic NP-040/70; Rexol 25/40; Rexol 25/407; T-Det® N-40; T-Det® N-407; Tergitol® NP-40; Tergitol® NP-40 (70% Aq.); Teric N40; Trycol® 6957; Trycol® 6970

Nonoxynol-50
CAS 9016-45-9 (generic); 26027-38-3 (generic); 37205-87-1 (generic)
Synonyms: PEG-50 nonyl phenyl ether; POE (50) nonyl phenyl ether
Classification: Ethoxylated alkyl phenol
Formula: $C_9H_{19}C_6H_4(OCH_2CH_2)_nOH$, avg. n = 50
Properties: Pale yel. to off-white pastes or waxes
Toxicology: Moderately toxic by ingestion, skin contact; severe eye and mild skin irritant in humans; heated to dec., emits acrid smoke and fumes
Uses: Nonionic surfactant; as detergent, emulsifier, wetting agent, dispersant, stabilizer, defoamer; intermediate in synthesis of anionic surfactants; emulsion polymerization
Regulatory: FDA 21CFR §176.180, 178.3400
Trade names: Ablunol NP50; Ablunol NP50 70%; Iconol NP-50; Iconol NP-50-70%; Igepal® CO-970; Igepal® CO-977; Prox-onic NP-050; Prox-onic NP-050/70; Rewopal® HV 50; T-Det® N-50; T-Det® N-507; Trycol® 6971; Trycol® 6972

Nonoxynol-70

Synonyms: PEG-70 nonyl phenyl ether; POE (70) nonyl phenyl ether
Classification: Ethoxylated alkyl phenol
Formula: $C_9H_{19}C_6H_4(OCH_2CH_2)_nOH$, avg. n = 70
Properties: Pale yel. to off-white pastes or waxes
Toxicology: Moderately toxic by ingestion, skin contact; severe eye and mild skin irritant in humans; heated to dec., emits acrid smoke and fumes
Uses: Nonionic surfactant; as detergent, emulsifier, wetting agent, dispersant, stabilizer, defoamer; intermediate in synthesis of anionic surfactants; emulsion polymerization
Trade names: Iconol NP-70; Iconol NP-70-70%; Igepal® CO-980; Igepal® CO-987; T-Det® N-70; T-Det® N-705; T-Det® N-707

Nonoxynol-100

CAS 9016-45-9 (generic); 26027-38-3 (generic); 37205-87-1 (generic)
Synonyms: PEG-100 nonyl phenyl ether; POE (100) nonyl phenyl ether
Classification: Ethoxylated alkyl phenol
Formula: $C_9H_{19}C_6H_4(OCH_2CH_2)_nOH$, avg. n = 100
Properties: Pale yel. to off-white pastes or waxes
Toxicology: Moderately toxic by ingestion, skin contact; severe eye and mild skin irritant in humans; heated to dec., emits acrid smoke and fumes
Uses: Nonionic surfactant; as detergent, emulsifier, wetting agent, dispersant, stabilizer, defoamer; intermediate in synthesis of anionic surfactants; emulsion polymerization
Regulatory: FDA 21CFR §176.180
Trade names: Iconol NP-100; Iconol NP-100-70%; Igepal® CO-990; Igepal® CO-997; Prox-onic NP-0100; Prox-onic NP-0100/70; T-Det® N-100; T-Det® N-1007; Trycol® 6981

Nonoxynol-120

CAS 9016-45-9 (generic); 37205-87-1 (generic); 26027-38-3 (generic)
Synonyms: PEG-120 nonyl phenyl ether; POE (120) nonyl phenyl ether
Classification: Ethoxylated alkyl phenol
Formula: $C_9H_{19}C_6H_4(OCH_2CH_2)_nOH$, avg. n = 120
Uses: Emulsifier for emulsion polymerization
Regulatory: FDA 21CFR §176.180
Trade names: Akyporox NP 1200V

Nonoxynol-150

Classification: Ethoxylated alkyl phenol
Formula: $C_9H_{19}C_6H_4(OCH_2CH_2)_nOH$, avg. n = 150
Properties: Pale yel. to off-white pastes or waxes
Toxicology: Moderately toxic by ingestion, skin contact; severe eye and mild skin irritant in humans; heated to dec., emits acrid smoke and fumes
Uses: Nonionic surfactant; as detergent, emulsifier, wetting agent, dispersant, stabilizer, defoamer; intermediate in synthesis of anionic surfactants; emulsion polymerization
Trade names: Trycol® 6954

Nonoxynol-6 phosphate

CAS 51811-79-1; 68412-53-3; 29994-44-3; 51609-41-7 (generic); EINECS 249-992-9
Synonyms: PEG-6 nonyl phenyl ether phosphate; POE (6) nonyl phenyl ether phosphate; PEG 300 nonyl phenyl ether phosphate
Definition: Complex mixture of esters of phosphoric acid and nonoxynol-6
Uses: Lubricant, extreme pressure agent; in metalworking, cutting fluids; antisoil redeposition, antistat in drycleaning detergents; emulsifier for herbicides/insecticides, vinyl acetate, acrylates, SBR
Regulatory: FDA 21CFR §175.105, 178.3400
Trade names: Emphos™ CS-136; Rhodafac® PE-510; Schercophos NP-6; Sermul EA 211

Nonoxynol-9 phosphate

CAS 51811-79-1; 66197-78-2; 68412-53-3; EINECS 266-231-6
Definition: Complex mixture of esters of phosphoric acid and nonoxynol-9
Synonyms: PEG-9 nonyl phenyl ether phosphate; POE (9) nonyl phenyl ether phosphate; PEG 450 nonyl phenyl ether phosphate
Uses: Emulsifier, lubricant, antistat, detergent, corrosion inhibitor for agric., industrial use; antisoil redeposition

for dry cleaning; emulsion polymerization; household and industrial detergents; fabric finishes
Trade names: Foamphos NP-9; Monafax 785; Rhodafac® RE-610; Surfagene FAZ 109; Surfagene FAZ 109 NV

Nonoxynol-10 phosphate
CAS 51609-41-7 (generic)
Synonyms: PEG-10 nonyl phenyl ether phosphate; POE (10) nonyl phenyl ether phosphate; PEG 500 nonyl phenyl ether phosphate
Definition: Complex mixture of esters of phosphoric acid and nonoxynol-10
Uses: Detergent in dry-cleaning, emulsifier in formulation of metal cleaners in emulsion polymerization; pesticide and cosmetic preparations
Regulatory: FDA 21CFR §178.3400
Trade names: Sermul EA 188; Servoxyl VPNZ 10/100

Nonoxynol-15 phosphate
Trade names: Sermul EA 136

Nonoxynol-20 phosphate
Trade names: Chemphos TC-337

Nonoxynol-50 phosphate
Trade names: Sermul EA 205

Nonoxynol-4 sulfate
Synonyms: Nonyl phenol ethoxy (4) sulfate
Uses: Antistat for syn. fibers and polymer prods.; surfactant, wetting agent, dispersant; emulsifier for polymerization
Trade names: Witcolate™ D51-51

Nonoxynol-10 sulfate
Synonyms: Nonyl phenol ethoxy (10) sulfate
Uses: Emulsifier for polymerization, personal care formulations
Trade names: Witcolate™ D51-53; Witcolate™ D51-53HA

Nonoxynol-30 sulfate
Uses: Emulsifier for polymerization, personal care formulations
Trade names: Witcolate™ D51-60

Nonylcarbinol. *See* n-Decyl alcohol

Nonyl nonoxynol-7
CAS 9014-93-1 (generic)
Uses: Emulsifier for agric., emulsion polymerization, leather industries
Trade names: Igepal® DM-430

Nonyl nonoxynol-8
CAS 9014-93-1 (generic)
Synonyms: PEG-8 dinonyl phenyl ether; POE (8) dinonyl phenyl ether
Classification: Ethoxylated alkyl phenol
Formula: $(C_9H_{19})_2C_6H_3(OCH_2CH_2)_nOH$, avg. n = 8
Uses: Emulsifier in textile dye carrier applics., insecticides, wax emulsions; foam control agent; spreading agent in pigment printing; post-stabilizer in emulsion polymerization; intermediate
Trade names: Trycol® 6985; Trycol® DNP-8

Nonyl nonoxynol-9
CAS 9014-93-1 (generic)
Synonyms: PEG-9 dinonyl phenyl ether; POE (9) dinonyl phenyl ether
Classification: Ethoxylated alkyl phenol
Formula: $(C_9H_{19})_2C_6H_3(OCH_2CH_2)_nOH$, avg. n = 9
Uses: Emulsifier for cutting oils, agric., textile finishing, dry cleaning soaps, inks, lacquers, paints, metalworking fluids, pesticides, cosmetics, emulsion polymerization
Trade names: Igepal® DM-530

Nonyl nonoxynol-15
CAS 9014-93-1 (generic)
Synonyms: PEG-15 dinonyl phenyl ether; POE (15) dinonyl phenyl ether
Classification: Ethoxylated alkyl phenol
Formula: $(C_9H_{19})_2C_6H_3(OCH_2CH_2)_nOH$, avg. n = 15
Uses: Detergent, emulsifier, antistat for textiles, leather, metal cleaners, paper, latex, pesticides; coemulsifier for syn. latexes
Trade names: Igepal® DM-710

Nonyl nonoxynol-24
CAS 9014-93-1 (generic)
Synonyms: PEG-24 dinonyl phenyl ether; POE (24) dinonyl phenyl ether
Classification: Ethoxylated alkyl phenol
Formula: $(C_9H_{19})_2C_6H_3(OCH_2CH_2)_nOH$, avg. n = 24
Uses: Detergent, emulsifier for textiles, leather, metal cleaners, paper, latex, pesticides, emulsion polymerization, latex stabilization
Trade names: Igepal® DM-730

Nonyl nonoxynol-49
CAS 9014-93-1 (generic)
Synonyms: PEG-49 dinonyl phenyl ether; POE (49) dinonyl phenyl ether
Classification: Ethoxylated alkyl phenol
Formula: $(C_9H_{19})_2C_6H_3(OCH_2CH_2)_nOH$, avg. n = 49
Uses: Emulsifier, solubilizer for emulsion polymerization, latex stabilization, pesticides, essential oils, cleaners
Trade names: Igepal® DM-880

Nonyl nonoxynol-9 phosphate
CAS 66172-82-5; EINECS 266-218-5
Synonyms: PEG-9 dinonyl phenyl ether phosphate; POE (9) dinonyl phenyl ether phosphate; PEG 450 dinonyl phenyl ether phosphate
Definition: Complex mixture of esters of phosphoric acid and nonyl nonoxynol-9
Empirical: $C_{66}H_{119}O_{22}P$
Uses: Detergent in dry-cleaning, emulsifier in formulation of metal cleaners, in emulsion polymerization; pesticide and cosmetic preparations
Trade names: Servoxyl VPQZ 9/100

Nonylphenol
CAS 25154-52-3
Definition: A mixture of isomeric monoalkyl phenols
Formula: $C_9H_{19}C_6H_4OH$
Properties: M.w. 220.36; dens. 0.937; flash pt. > 110 C; ref. index 1.5110 (20 C)
Toxicology: Irritant
Uses: Nonionic surfactant intermediate (nonbiodegradable), lube oil additives, stabilizers, petroleum demulsifiers, plasticizers, dyestuffs, aromatic oils, fungicides, antioxidants for plastics and rubber
Manuf./Distrib.: Akzo Nobel; Aldrich; Allchem Ind.; Ashland; Berol Nobel AB; GE Specialty; Hüls AG; Kalama; Mitsui Toatsu; Texaco

p-Nonyl phenol
CAS 84852-15-3
Properties: Clear liq., sl. phenolic odor; m.w. 220.3; sp.gr. 0.945; flash pt. (Seta) 146 C
Uses: Heat stabilizer for PVC, phenolic resins; anionic and nonionic surface-act. agent intermediate; mfg. of lurbicating oil additives; lt. stabilizers; petrol. demulsifiers; oil-sol. phenolic resins, plasticizers, dyestuffs, curing epoxy resins
Manuf./Distrib.: Schenectady

Nonyl phenol ethoxy (4) sulfate. *See* Nonoxynol-4 sulfate
Nonyl phenol ethoxy (10) sulfate. *See* Nonoxynol-10 sulfate
2-(Nonylphenoxy) ethanol. *See* Nonoxynol-1
2-[2-[2-(Nonylphenoxy)ethoxy]ethoxy]ethanol. *See* Nonoxynol-3
14-(Nonylphenoxy)-3,6,9,12-tetraoxatetradecan-1-ol. *See* Nonoxynol-5

Nonyl phenyl glycidyl ether
Uses: Visc. reducing modifier for epoxies; chem. intermediate
Trade names: Heloxy® 64

Nonylphenyl phosphite (3:1). *See* Trisnonylphenyl phosphite

Nonyl phthalate
Properties: Sp.gr. 0.965-0.971 (20 C); pour pt. -50 C; flash pt. (PMCC) 213 C; ref. index 1.483 (20 C)
Uses: Plasticizer for PVC, PS, cellulosics
Manuf./Distrib.: Ashland; BASF; Exxon

Nonyl undecyl phthalate
Properties: Sp.gr. 0.955-0.965 (20 C); pour pt. -20 C; flash pt. (PMCC) 200 C; ref. index 1.480 (20 C)
Uses: Plasticizer for PVC, PS, cellulosics
Manuf./Distrib.: Aristech; Ashland; BASF; Exxon

Novolac resin
Synonyms: Novolak resin; Two-stage phenolic resin; Phenol-formaldehyde (novolak) resin
Classification: Thermoplastic phenol-formaldehyde type resin
Properties: Sol. in alcohol
Uses: Molding materials; bonding agent in brake linings, abrasive grinding wheels, electrical insulation; reinforcing agent and modifier for nitrile rubber; air-drying varnishes; bonding materials
Trade names: Akrochem® P 90; Dyphene® 6746-L; Dyphene® 8320; Dyphene® 8400; Dyphene® 8845P; R7557P; SP-25; SP-1044; SP-1045; SP-1068; SP-1077; SP-6700; SP-6701; Varcum® 29008; Varcum® 29427; Varcum® 29421
See also Phenolic resin

Novolak resin. *See* Novolac resin, Phenolic resin
NR. *See* Natural rubber

Nylon
CAS 63428-83-1
Definition: A family of polyamide polymers char. by presence of amide group —CONH
Uses: Thermoplastic resin for tire cord, hosiery, wearing apparel, brush bristles, cordage, fish lines, tennis rackets, rugs, artificial turf, parachutes, composites, sails, film, gears/bearings, insulation, surgical sutures, metal coating, fuel tanks
Manuf./Distrib.: Ashley Polymers; Bamberger Polymers; BASF; DSM; EMS-Am. Grilon; Hoechst Celanese; Hüls Am.; ICI GmbH; LNP; Miles; Monsanto
Trade names containing: Colloids N 54/14/01; Colloids N 54/14/02; MDI NY-905

Nylon-6
CAS 25038-54-4
Synonyms: Poly[imino(1-oxo-1,6-hexanediyl)]; Poly(iminocarbonylpentamethylene)
Classification: Polyamide
Empirical: $(C_6H_{11}NO)_n$
Formula: $[NH(CH_2)_5CO]_x$
Properties: Dens. 1.14 (20/4 C); m.p. 223 C; resistant to most org. chem., dissolved by phenol, cresol, strong acids; immune to biological attack
Toxicology: Moderately toxic by ingestion; mildly toxic by inhalation
Uses: Tire cord; fishing lines; tow ropes; hose mfg.; woven fabrics
Regulatory: FDA 21CFR §177.1500, 177.2260, 177.2470, 177.2480
Trade names containing: Mastertek® PA 3664

Nylon-6,6
CAS 32131-17-2
Classification: Polyamide
Definition: Polymeric amide formed by the reaction of adipic acid with hexylenediamine
Synonyms: Poly[imino (1,6-dioxo-1,6-hexanediyl) imino-1,6-hexanediyl Poly(hexamethyleneadipamide)
Empirical: $(C_{12}H_{22}N_2O_2)_x$
Formula: $[NH(CH_2)_6NHCO(CH_2)_4CO]_n$
Properties: Crystalline solid; sol. in phenol, cresols, xylene, formic acid; insol. in alcohol, esters, ketones, hydrocarbons; dens. 1.090
Precaution: Reacts violently with F_2

Toxicology: Heated to dec., emits toxic fumes of NO_x
Uses: Thermoplastic resin for inj. molding, extrusion
Manuf./Distrib.: Asahi Chem Industry Co Ltd
Trade names containing: Amgard® CPC 452

OBOC. *See* 6-t-Butyl-o-cresol
OBPA. *See* 10,10′-Oxybisphenoxyarsine
OBSH. *See* 4,4′-Oxybis (benzenesulfonylhydrazide)
OCTA. *See* Octabromodiphenyl oxide
Octabenzone. *See* Benzophenone-12

Octabromodiphenyl oxide
CAS 32536-52-0; EINECS 251-087-9
Synonyms: OCTA
Empirical: $C_{12}H_2Br_8O$
Properties: Powd.; m.w. 801.37
Uses: Flame retardant for ABS, nylon, other engineering thermoplastics
Manuf./Distrib.: Albemarle; Allchem Ind.; Great Lakes
Trade names: FR-1208; Great Lakes DE-79™; Saytex® 111

9-Octadcecen-1-amine. *See* Oleamine
9,12-Octadecadienoic acid, 2,3-dihydroxypropyl ester. *See* Glyceryl linoleate
9,12-Octadecadienoic acid, dimer. *See* Dilinoleic acid
9,12-Octadecadienoic acid, monoester with 1,2,3-propanetriol. *See* Glyceryl linoleate
9,12-Octadecadienoic acid, trimer. *See* Trilinoleic acid
Octadecanamide. *See* Stearamide
1-Octadecanamine. *See* Stearamine
n-Octadecanoic acid. *See* Stearic acid
Octadecanoic acid, ammonium salt. *See* Ammonium stearate
Octadecanoic acid, butyl ester. *See* Butyl stearate
Octadecanoic acid, 2-(1-carboxyethoxy)-1-methyl-2-oxoethyl ester, sodium salt. *See* Sodium stearoyl lactylate
Octadecanoic acid, 1,2-ethanediyl ester. *See* Glycol distearate
Octadecanoic acid, 2-ethylhexyl ester. *See* Octyl stearate
Octadecanoic acid, 12-hydroxy-. *See* Hydroxystearic acid
Octadecanoic acid, lithium salt. *See* Lithium stearate
Octadecanoic acid, magnesium salt. *See* Magnesium stearate
Octadecanoic acid, methyl ester. *See* Methyl stearate
Octadecanoic acid, monoester with 1,2-propanediol. *See* Propylene glycol stearate
Octadecanoic acid, octadecyl ester. *See* Stearyl stearate
Octadecanoic acid, potassium salt. *See* Potassium stearate
Octadecanoic acid, 1,2,3-propanetriyl ester. *See* Tristearin
Octadecanoic acid, sodium salt. *See* Sodium stearate
Octadecanoic acid, tridecyl ester. *See* Tridecyl stearate
Octadecanoic acid, zinc salt. *See* Zinc stearate
1-Octadecanol. *See* Stearyl alcohol
n-Octadecanol. *See* Stearyl alcohol
9-Octadecenamide. *See* Oleamide
9-Octadecenamide, N,N′-1,2-ethanediylbis-. *See* Ethylene dioleamide

Octadecene-1
CAS 112-88-9; EINECS 204-012-9
Synonyms: Linear C18 alpha olefin
Formula: $CH_3(CH_2)_{15}CH:CH_2$
Properties: Colorless liq., mild hydrocarbon odor; sl. sol. in water; m.w. 252.49; sp.gr. 0.792; b.p. 586-615 F; m.p. 63 F; flash pt. (Seta) 312 F
Toxicology: Irritating to skin and eyes; low acute inhalation toxicity; low acute ingestion toxicity, but ingestion may cause vomiting, aspiration of vomitus
Uses: Intermediate for surfactants and specialty industrial chemicals (polymers, plasticizers, lubricants, gasoline additives, paper sizing, PVC lubricants)
Manuf./Distrib.: Albemarle; Chevron; Fluka; Monomer-Polymer & Dajac; Shell
Trade names: Gulftene® 18

9-Octadecenoic acid. *See* Oleic acid
cis-9-Octadecenoic acid. *See* Oleic acid
9-Octadecenoic acid, 12-(acetyloxy)-, 1,2,3-propanetriol ester. *See* Glyceryl triacetyl ricinoleate
9-Octadecenoic acid, 12-hydroxy-. *See* Ricinoleic acid
9-Octadecenoic acid, isodecyl ester. *See* Isodecyl oleate
9-Octadecenoic acid, methyl ester. *See* Methyl oleate
9-Octadecenoic acid, 1-methylethyl ester. *See* Isopropyl oleate
9-Octadecenoic acid, monoester with 1,2,3-propanetriol. *See* Glyceryl oleate
9-Octadecenoic acid, 1,2,3-propanetriyl ester. *See* Triolein
9-Octadecenoic acid, tetraester with decaglycerol. *See* Polyglyceryl-10 tetraoleate
9-Octadecen-1-ol. *See* Oleyl alcohol
cis-9-Octadecen-1-ol. *See* Oleyl alcohol
9-Octadecen-1-ol, hydrogen sulfate, sodium salt. *See* Sodium oleyl sulfate
N-9-Octadecenyl hexadecanamide. *See* Oleyl palmitamide

Octadecenyl succinic anhydride
Synonyms: ODSA
Empirical: $C_{22}H_3iO_3$
Properties: Yel.-wh. waxy solid; m.w. 350; dens. 0.9428 (25 C); b.p. 251 C (4 mm); flash pt. 210 C
Uses: Curing agent for epoxy resins; corrosion inhibitor for lubricants; intermediate for alkyd or unsat. polyester resins, in chem. reactions; detergent; paper sizing; water repellents
Manuf./Distrib.: Dixie Chem; Humphrey; Milliken
Trade names: Milldride® ODSA

Octadecyl alcohol. *See* Stearyl alcohol
Octadecylamine. *See* Stearamine
Octadecyl 3,5-bis (1,1-dimethylethyl)-4-hydroxybenzene propanoate. *See* Octadecyl 3,5-di-t-butyl-4-hydroxyhydrocinnamate
Octadecyl citrate. *See* Stearyl citrate

Octadecyl 3,5-di-t-butyl-4-hydroxyhydrocinnamate
CAS 2082-79-3; EINECS 218-216-0
Synonyms: Octadecyl 3,5-bis (1,1-dimethylethyl)-4-hydroxybenzene propanoate; Octadecyl 3-(3,5-di-t-butyl-4-hydroxyphenyl) propionate; Stearyl-3-(3′,5′-di-t-butyl-4-hydroxyphenyl) propionate
Classification: Hindered phenol
Empirical: $C_{35}H_{62}O_3$
Formula: $[(CH_3)_3C]_2C_6H_2(OH)CH_2CH_2CO_2(CH_2)_{17}CH_3$
Properties: White powd.; sol. in acetone, benzene, chloroform, ethyl acetate, hexane; m.w. 530.9; m.p. 50-52 C; f.p. > 230 F
Toxicology: Irritant
Uses: Antioxidant stabilizer for styrenics, polyolefins, PVC, urethane and acrylic coatings, adhesives, elastomers; replaces BHT in polyolefins
Trade names: Akrochem® Antioxidant 1076; BNX® 1076, 1076G; Dovernox 76; Irganox® 1076; Lankromark® LE384; Naugard® 76; Ralox® 530; Ultranox® 276
Trade names containing: Aquastab PA 43; Irganox® B 900; Irganox® B 921; Ultranox® 875A Blend; Ultranox® 877A Blend

Octadecyl 3-(3,5-di-t-butyl-4-hydroxyphenyl) propionate. *See* Octadecyl 3,5-di-t-butyl-4-hydroxyhydrocinnamate
Octadecyl dimethylamine. *See* Dimethyl stearamine
N-Octadecyl-13-docosenamide. *See* Stearyl erucamide

Octadecyl 3-mercaptopropionate
CAS 31778-15-1
Formula: $HS—CH_2CH_2COOC_{18}H_{37}$
Properties: M.w. 350 (avg.); m.p. 25-28 C
Uses: Antioxidant, stabilizer
Manuf./Distrib.: Evans Chemetics
Trade names: Evangard® 18MP

Octadecyl trimethyl ammonium bromide. *See* Steartrimonium bromide
1-Octanaminium, N,N-bis(2-hydroxyethyl)-N-methyl-, salt with 4-methylbenzenesulfonic acid (1:1). *See* Bis (2-hydroxyethyl) octyl methyl ammonium p-toluene sulfonate

Octane
CAS 111-65-9; EINECS 203-892-1
Formula: $CH_3(CH_2)_6CH_3$
Properties: M.w. 114.23; b.p. 125-126 C
Toxicology: TLV 300 ppm (ACGIH)
Uses: Solv.
Manuf./Distrib.: Aldrich; Fluka; Phibro Energy USA; Phillips; Texaco

Octanoic acid, 1,3-propanediyl ester. *See* Propylene glycol dioctanoate
1-Octanol. *See* Caprylic alcohol
n-Octanol. *See* Caprylic alcohol
[2,3,4,6,7,8,9,10-Octapyrimidol[1,2-a]azepine]. *See* Diazabicycloundecene

Octene-1
CAS 111-66-0; EINECS 203-893-7
Synonyms: C8 linear alpha olefin
Empirical: C_8H_{16}
Formula: $CH_3(CH_2)_5CH:CH_2$
Properties: Colorless liq., mild hydrocarbon odor; sl. sol. in water; sp.gr. 0.720; b.p. 250 F; f.p. -151 F; flash pt. (Seta) 58 F
Toxicology: Irritating to eyes and skin; low acute inhalation toxicity; sl. toxic by ingestion
Uses: Intermediate for surfactants and specialty industrial chemicals; produces isononyl alcohol for DINP plasticizer mfg.
Manuf./Distrib.: Air Prods.; Albemarle; Aldrich; Chevron; Monomer-Polymer & Dajac; Shell; Texaco
Trade names: Gulftene® 8

Octenyl succinic anhydride
CAS 26680-54-6
Synonyms: OSA
Empirical: $C_{12}H_{18}O_3$
Formula: $CH_3CH_2CH_2CH_2CH_2CH=CHCH_2CHCH_2C_2O_3$
Properties: M.w. 210; dens. 1.0 (25 C); b.p. 168 C (10 mm); flash pt. (COC) 185 C
Uses: Curing agent for epoxy resins; corrosion inhibitor for lubricants; intermediate for prep. of alkyd or unsat. polyester resins, intermediate in chem. reactions
Manuf./Distrib.: Dixie Chem; Milliken
Trade names: Milldride® OSA

Octocrylene
CAS 6197-30-4; EINECS 228-250-8
Synonyms: 2-Ethylhexyl 2-cyano-3,3-diphenylacrylate; 2-Ethylhexyl 2-cyano-3,3-diphenyl-2-propenoate; UV Absorber-3
Classification: Substituted acrylate
Empirical: $C_{24}H_{27}NO_2$
Formula: $(C_6H_5)_2C=C(CN)CO_2CH_2CH(C_2H_5)(CH_2)_3CH_3$
Properties: Liq.; m.p. -10 C
Uses: Active ingred. in OTC drug prods.; uv-B absorber for cosmetics; uv stabilizer for flexible and semirigid PVC, VDC, coatings
Trade names: Uvinul® 3039; Uvinul® N-539

Octoxynol-1
CAS 9002-93-1 (generic); 9036-19-5 (generic); 9004-87-9 (generic); 2315-67-5; EINECS 264-520-1
Synonyms: Ethylene glycol octyl phenyl ether; PEG-1 octyl phenyl ether; 2-[p-(1,1,3,3-Tetramethylbutyl)phenoxy]ethanol
Classification: Ethoxylated alkyl phenol
Empirical: $C_{16}H_{26}O_2$
Formula: $C_8H_{17}C_6H_4OCH_2CH_2OH$
Uses: Nonionic surfactant, detergent, emulsifier, dispersant; spermaticide; emulsion polymerization, industrial/household cleaners, agric., latex stabilizer
Regulatory: FDA 21CFR §172.710, 175.105, 176.180
Trade names: Triton® X-15

Octoxynol-3
CAS 9002-93-1 (generic); 9004-87-9 (generic); 9036-19-5 (generic); 2315-62-0; 27176-94-9
Synonyms: 2-[2-[2-[p-(1,1,3,3-Tetramethylbutyl)phenoxy]ethoxy]ethoxy]ethanol; PEG-3 octyl phenyl ether; POE (3) octyl phenyl ether
Classification: Ethoxylated alkyl phenol
Empirical: $C_{20}H_{34}O_4$
Formula: $C_8H_{17}C_6H_4(OCH_2CH_2)_nOH$, avg. n = 3
Uses: Nonionic surfactant, detergent, emulsifier, dispersant; spermaticide; industrial/household cleaners, emulsion polymerization, agric., latex stabilizer
Regulatory: FDA 21CFR §175.105, 176.180, 176.210
Trade names: Triton® X-35

Octoxynol-7
CAS 9002-93-1 (generic); 9004-87-9 (generic); 9036-19-5 (generic); 27177-02-2
Synonyms: PEG-7 octyl phenyl ether; POE (7) octyl phenyl ether; 20-(Octylphenoxy)-3,6,9,12,15,18-hexaoxaeicosan-1-ol
Classification: Ethoxylated alkyl phenol
Empirical: $C_{28}H_{50}O_8$
Formula: $C_8H_{17}C_6H_4(OCH_2CH_2)_nOH$, avg. n = 7
Uses: Nonionic surfactant, detergent, emulsifier, dispersant; spermaticide; textile lubricants, agric., latex paint; post-polymerization stabilizer
Regulatory: FDA 21CFR §172.710, 175.105, 176.180, 176.210, 178.3400
Trade names: Hyonic OP-70

Octoxynol-9
CAS 9002-93-1 (generic); 9004-87-9 (generic); 9010-43-9; 9036-19-5 (generic); 42173-90-0
Synonyms: PEG-9 octyl phenyl ether; POE (9) octyl phenyl ether; PEG 450 octyl phenyl ether
Classification: Ethoxylated alkyl phenol
Empirical: $C_{32}H_{58}O_{10}$
Formula: $C_8H_{17}C_6H_4(OCH_2CH_2)_nOH$, avg. n = 9
Uses: Nonionic surfactant, detergent, emulsifier, dispersant; spermaticide; household/industrial cleaners; emulsion polymerization
Regulatory: FDA 21CFR §175.105, 176.180, 176.210, 178.3400
Trade names: Desonic® S-100; Prox-onic OP-09

Octoxynol-10
CAS 9002-93-1 (generic); 9004-87-9 (generic); 9036-19-5 (generic); 2315-66-4; 27177-07-7
Synonyms: PEG-10 octyl phenyl ether; POE (10) octyl phenyl ether; PEG 500 octyl phenyl ether
Classification: Ethoxylated alkyl phenol
Empirical: $C_{34}H_{62}O_{11}$
Formula: $C_8H_{17}C_6H_4(OCH_2CH_2)_nOH$, avg. n = 10
Properties: Pale yel. visc. liq.; misc. with water, alcohol, acetone; sol. in benzene, toluene; m.w. 647; dens. 1.0595 (25/4 C)
Uses: Nonionic surfactant, detergent, emulsifier, dispersant; spermaticide; industrial/household cleaners, emulsion polymerization, latex stabilizer, asphalt emulsions
Regulatory: FDA 21CFR §172.710, 175.105, 176.180, 176.210, 178.3400
Trade names: Iconol OP-10

Octoxynol-16
CAS 9004-87-9 (generic); 9036-19-5 (generic); 9002-93-1 (generic)
Synonyms: PEG-16 octyl phenyl ether; POE (16) octyl phenyl ether
Classification: Ethoxylated alkyl phenol
Formula: $C_8H_{17}C_6H_4(OCH_2CH_2)_nOH$, avg. n = 16
Uses: Nonionic surfactant, detergent, emulsifier, dispersant; spermaticide; household/industrial cleaners, emulsion polymerization, cosmetics, pharmaceuticals
Regulatory: FDA 21CFR §175.105, 176.180
Trade names: Synperonic OP16; Triton® X-165-70%

Octoxynol-25
CAS 9002-93-1 (generic); 9036-19-5 (generic); 9004-87-9 (generic)
Synonyms: PEG-25 octyl phenyl ether; POE (25) octyl phenyl ether
Classification: Ethoxylated alkyl phenol
Formula: $C_8H_{17}C_6H_4(OCH_2CH_2)_nOH$, avg. n = 25

Uses: Emulsifier for emulsion polymerization; perfume solubilizer
Regulatory: FDA 21CFR §175.105, 176.180
Trade names: Akyporox OP 250V

Octoxynol-30
CAS 9004-87-9 (generic); 9036-19-5 (generic); 9002-93-1 (generic)
Synonyms: PEG-30 octyl phenyl ether; POE (30) octyl phenyl ether
Classification: Ethoxylated alkyl phenol
Formula: $C_8H_{17}C_6H_4(OCH_2CH_2)_nOH$, avg. n = 30
Uses: Nonionic surfactant, detergent, emulsifier, dispersant; spermaticide; industrial/household cleaners; emulsion polymerization; asphalt
Regulatory: FDA 21CFR §172.710, 175.105, 176.180, 178.3400
Trade names: Iconol OP-30; Iconol OP-30-70%; Igepal® CA-880; Igepal® CA-887; Prox-onic OP-030/70; Rexol 45/307; Serdox NOP 30/70; Synperonic OP30; T-Det® O-307; Triton® X-305-70%

Octoxynol-33
CAS 9002-93-1 (generic); 9036-19-5 (generic); 9004-87-9 (generic)
Synonyms: PEG-33 octyl phenyl ether; POE (33) octyl phenyl ether
Classification: Ethoxylated alkyl phenol
Formula: $C_8H_{17}C_6H_4(OCH_2CH_2)_nOH$, avg. n = 33
Regulatory: FDA 21CFR §172.710, 175.105, 176.180, 178.3400
Trade names containing: Octoxynol-0033;Sodiumlaurethsulfate

Octoxynol-40
CAS 9002-93-1 (generic); 9004-87-9 (generic); 9036-19-5 (generic)
Synonyms: PEG-40 octyl phenyl ether; POE (40) octyl phenyl ether
Classification: Ethoxylated alkyl phenol
Formula: $C_8H_{17}C_6H_4(OCH_2CH_2)_nOH$, avg. n = 40
Uses: Nonionic surfactant, detergent, emulsifier, dispersant; spermaticide; emulsion polymerization; industrial/household cleaners; asphalt; latex stabilizer
Regulatory: FDA 21CFR §172.710, 175.105, 176.180, 178.3400
Trade names: Akyporox OP 400V; Cedepal CA-890; Chemax OP-40/70; Desonic® S-405; Iconol OP-40; Iconol OP-40-70%; Igepal® CA-890; Igepal® CA-897; Macol® OP-40(70); Nissan Nonion HS-240; Prox-onic OP-040/70; Rexol 45/407; Serdox NOP 40/70; T-Det® O-40; Triton® X-405-70%; Trycol® 6984; Trycol® OP-407

Octoxynol-70
CAS 9004-87-9 (generic); 9036-19-5 (generic); 9002-93-1 (generic)
Synonyms: PEG-70 octyl phenyl ether; POE (70) octyl phenyl ether
Classification: Ethoxylated alkyl phenol
Formula: $C_8H_{17}C_6H_4(OCH_2CH_2)_nOH$, avg. n = 70
Uses: Nonionic surfactant, detergent, emulsifier, dispersant; spermaticide; household/industrial cleaners, textiles, paints, inks, agric., latex/emulsion polymerization
Regulatory: FDA 21CFR §172.710, 176.180
Trade names: Nissan Nonion HS-270; Triton® X-705-70%

Octoxynol-9 carboxylic acid
CAS 25338-58-3; 72160-13-5; 107628-08-0
Synonyms: PEG-9 octyl phenyl ether carboxylic acid; 26-(Octylphenoxy)-3,6,9-12,15,18,21,24-octaoxahexacosanoic acid; PEG 450 octyl phenyl ether carboxylic acid; POE (9) octyl phenyl ether carboxylic acid
Classification: Organic acid
Formula: $C_8H_{17}C_6H_4(OCH_2CH_2)_nOCH_2COOH$, avg. n = 8
Uses: Detergent, emulsifier; metalworking fluids; emulsion polymerization
Trade names: Akypo OP 80; Akyposal OP 80

Octyl alcohol. *See* 2-Ethylhexanol
n-Octyl alcohol. *See* Caprylic alcohol

Octylated diphenylamine
Uses: Antioxidant for rubbers

p-t-Octyl-o-cresol

Trade names: Agerite® Stalite; Agerite® Stalite S; Akrochem® Antioxidant S; Anchor® ODPA; Flectol® ODP; Octamine® Flake, Powd.; Permanax™ OD; Vulkanox® OCD; Vulkanox® OCD/SG
Trade names containing: Agerite® HP-S; Agerite® HP-S Rodform

p-t-Octyl-o-cresol
CAS 2219-84-3
Synonyms: TOOC

n-Octyl, n-decyl adipate
Synonyms: NODA
Properties: Liq., mild odor; dens. 0.92-0.98 (20/20 C); f.p. -50 C; b.p. 220-254 C (4 mm); flash pt. (COC) 198-235 C; ref. index 1.447
Precaution: Combustible
Uses: Low-temp. plasticizer for cellulosics, PVC, VCA
Manuf./Distrib.: Aristech; Ashland; C.P. Hall; Inolex
Trade names: Monoplex® NODA; Plasthall® NODA; Staflex NODA

n-Octyldecyl alcohol. *See* Stearyl alcohol

Octyl/decyloxypropylamine
Uses: Corrosion inhibitor for metalworking fluids; antistat; flotation collector; additive for fuel, lubricants, petrol. refining; intermediate for surfactants, textile foamers, agric.; crosslinking agent for epoxies
Trade names: Tomah PA-1214

Octyldecyl phthalate. *See* n-Octyl, n-decyl phthalate

n-Octyl, n-decyl phthalate
CAS 119-07-3; EINECS 204-295-9
Synonyms: Octyldecyl phthalate; 1,2-Benzenedicarboxylic acid, decyl octyl ester; Benzene dicarboxylic acid bis n-octyl n-decyl ester
Empirical: $C_{26}H_{42}O_4$
Properties: Colorless liq., mild char. odor; m.w. 398; dens. 0.972-0.976 (20/20 C); m.p. -50 C; b.p. 220-254 C (4 mm); flash pt. 455 F
Uses: Plasticizer for vinyl resins
Trade names: PX-318; Staflex ODP

n-Octyl, n-decyl trimellitate
Properties: M.w. 595; dens. 0.971; m.p. -17 C; flash pt. 279 C; ref. index 1.482
Uses: Plasticizer for vinyls, cellulosics, elastomers
Manuf./Distrib.: Aristech; Inolex
Trade names: Plasthall® 8-10 TM-E; Staflex NONDTM

Octyl dimethylamine. *See* Dimethyl octylamine
9,10-Octyl epoxy stearate. *See* Epoxidized octyl stearate
Octyl epoxy tallate. *See* Epoxidized octyl tallate

2-n-Octyl-4-isothiazolin-3-one
CAS 26530-20-1
Uses: Mildewcide for paints, plastics
Trade names: Micro-Chek® 11 P
Trade names containing: Micro-Chek® 11; Vinyzene® IT-3000 DIDP; Vinyzene® IT-3025 DIDP

Octylmercaptide
Uses: Stabilizer for PVC
Trade names: Stanclere® T-470, T-473, T-482, T-483, T-484, T-582

Octyl palmitate
CAS 29806-73-3; EINECS 249-862-1
Synonyms: 2-Ethylhexyl palmitate; 2-Ethylhexyl hexadecanoate
Definition: Ester of 2-ethylhexyl alcohol and palmitic acid
Empirical: $C_{24}H_{48}O_2$
Formula: $CH_3(CH_2)_{14}COOCH_2(CH_2CH_3)CH(CH_2)_3CH_3$
Properties: Water-wh. liq.; sol. in min. oil, 95% ethanol, IPM, oleyl alcohol; insol. in water, glycerin, propylene

glycol; sp.gr. 0.850-0.856; ref. index 1.4445-1.4465
Toxicology: LD50 (rat, oral) > 40 ml/kg; nonirritating to eyes; mild skin irritant
Uses: Ester for sunscreens, antispersipirants, bath oils, liq. make-up; imparts gloss; binder for pressed powds.; solubilizer for benzophenone-3; plasticizer
Manuf./Distrib.: Aquatec Quimica SA; Inolex; ISP Van Dyk; Union Camp

Octylphenol formaldehyde resin
Uses: Thermoreactive resin used in curing butyl and other elastomers; tackifying resin for rubbers, tire applics.
Trade names: Dyphene® 595; Dyphene® 8318; Dyphene® 8330; Dyphene® 8340; Dyphene® 8787; Dyphene® 9273; Koretack® 5193; R7500E; R7510; R7521; R7530E; R7578

2-[2-[2-Octylphenoxy)ethoxy]ethoxy]ethanesulfonic acid, sodium salt. *See* Sodium octoxynol-2 ethane sulfonate
20-(Octylphenoxy)-3,6,9,12,15,18-hexaoxaeicosan-1-ol. *See* Octoxynol-7
26-(Octylphenoxy)-3,6,9-12,15,18,21,24-octaoxahexacosanoic acid. *See* Octoxynol-9 carboxylic acid

Octyl phosphate
CAS 39407-03-9
Uses: Detergent in drycleaning; emulsifier in emulsion polymerization; pesticides; cosmetics
Trade names: Servoxyl VPTZ 100
See also Trioctyl phosphate

2,2′,2′′-(Octylstannylidyne)tris(thio)tris(acetic acid), triisooctyl ester
CAS 26401-86-5
Uses: Heat stabilizer for PVC

Octyl stearate
CAS 22047-49-0; 26399-02-0; EINECS 244-754-0; 247-655-0
Synonyms: 2-Ethylhexyl stearate; 2-Ethylhexyl octadecanoate; Octadecanoic acid, 2-ethylhexyl ester
Definition: Ester of 2-ethylhexyl alcohol and stearic acid
Empirical: $C_{26}H_{52}O_2$
Formula: $CH_3(CH_2)_{16}COOCH_2(CH_2CH_3)CH(CH_2)_3CH_3$
Uses: Emollient; superfatting oil; lubricant for PVC, textile, metalworking
Trade names: Lexolube® T-110

Octyl sulfate sodium salt. *See* Sodium octyl sulfate

Octyl tallate
Classification: Ester
Definition: Ester of 2-ethylhexyl alcohol and tall oil fatty acids?
Properties: Sp.gr. 0.873; m.p. -10 C; flash pt. (COC) 213 C; ref. index 1.459
Uses: Transfer aid on correctable ribbon; penetration and tack agent for computer ribbons, carbon paper; plasticizer for vinyls, ethyl cellulose, SBR
Manuf./Distrib.: C.P. Hall; Union Camp
Trade names: Plasthall® R-9; Uniflex® 192; Uniflex® EHT

Octyl thioglycolate. *See* 2-Ethylhexyl thioglycolate

Octyl thiotin
Uses: Heat stabilizer for rigid and plasticized PVC applics.; suitable for food contact applics.
Trade names: Lankromark® OT050, OT250; Lankromark® OT450

Octyltin mercaptide
Uses: PVC stabilizer
Trade names: ADK STAB 466; ADK STAB 467; ADK STAB 471

Octyltin thioglycolate
Uses: PVC stabilizer
Trade names: Hostastab® SnS 20; Hostastab® SnS 45

Octyltriethoxysilane
CAS 2943-75-1; EINECS 220-941-2
Empirical: $C_{14}H_{32}SiO_3$

Properties: Liq.; m.w. 276.5;dens. 0.88; b.p. 98-99 C (2 mm); ref. index 1.415; flash pt. 100 C
Uses: Coupling agent, release agent, lubricant, blocking agent, chemical intermediate
Trade names: Dynasylan® OCTEO; Petrarch® O9835

ODSA. *See* Octadecenyl succinic anhydride
Oil of Palma Christi. *See* Castor oil
Oil of vitriol. *See* Sulfuric acid
OIPP. *See* o-Isopropylphenol

Oleamide
CAS 301-02-0; EINECS 206-103-9
Synonyms: 9-Octadecenamide; Oleyl amide; Oleic acid amide
Classification: Aliphatic amide
Empirical: $C_{18}H_{35}NO$
Formula: $CH_3(CH_2)_7CH:CH(CH_2)_7CONH_2$
Properties: Ivory-colored powd.; dens. 0.94; m.p. 72 C
Precaution: Combustible
Toxicology: Heated to decomp., emits acrid smoke and irritating fumes
Uses: Slip/antiblock agent for extrusion of polyethylene; wax additive; ink additive
Regulatory: FDA 21CFR §175.105, 175.300, 178.3860, 178.3910, 179.45, 181.22, 181.28
Manuf./Distrib.: Akzo Nobel; Chemax; Chemron; Cookson Specialty; Croda Universal; Henkel/Emery; Mona; Pilot; Syn. Prods.; Witco/Oleo-Surf.
Trade names: Armid® O; Chemstat® HTSA #18; Crodamide O; Crodamide OR; Kemamide® O; Kemamide® U; Petrac® Slip-Eze®; Polydis® TR 121; Polyvel RO25; Polyvel RO40; Unislip 1757; Unislip 1759
Trade names containing: Cabot® PE 9041; Cabot® PE 9171; Cabot® PE 9174; Colloids PE 48/10/06; PE-SA 40; PE-SA 92; PE-SL 31; Polyvel RO252

Oleamine
CAS 112-90-3; EINECS 204-015-5
Synonyms: Oleyl amine; 9-Octadcecen-1-amine
Classification: Primary aliphatic amine
Empirical: $C_{18}H_{35}N$
Formula: $CH_3(CH_2)_7CH:CH(CH_2)_7CH_2NH_2$
Uses: Emulsifier, wetting agent, corrosion inhibitor, dispersant, chemical intermediate, lube oil additive; cosmetics, lubricants, textiles, leather, rubber, plastics, metal industries
Trade names: Crodamine 1.O, 1.OD; Kemamine® P-989D

Oleic acid
CAS 112-80-1; EINECS 204-007-1
Synonyms: cis-9-Octadecenoic acid; Red oil; Elainic acid; 9-Octadecenoic acid
Classification: Unsaturated fatty acid
Empirical: $C_{18}H_{34}O_2$
Formula: $CH_3(CH_2)_7CH:CH(CH_2)_7COOH$
Properties: Colorless liq., odorless; insol. in water; sol. in alcohol, ether, benzene, chloroform, fixed and volatile oils; m.w. 282.47; dens. 0.895 (25/25 C); m.p. 6 C; b.p. 286 C (100 mm); flash pt. 100 C; ref. index 1.463 (18 C)
Precaution: Combustible when exposed to heat or flame; incompat. with Al and perchloric acid
Toxicology: LD50 (oral, rat) 74 g/kg; poison by intravenous route; mildly toxic by ingestion; experimental tumorigen; irritant to skin, mucous membranes; heated to decomp., emits acrid smoke and irritating fumes
Uses: Soap base, mfg. of oleates, ointments, cosmetics, polishing compds., lubricants, food-grade additives, Turkey red oil, driers; waterproofing textiles; oiling wool; pharmaceutic aid (solv.); activator, plasticizer, softener, in rubbers
Regulatory: FDA 21CFR §172.210, 172.860, 172.862, 173.315 (0.1 ppm max. in wash water), 173.340, 175.105, 175.320, 176.170, 176.200, 176.210, 177.1010, 177.1200, 177.2260, 177.2600, 177.2800, 178.3570, 178.3910, 182.70, 182.90; GRAS (FEMA)
Manuf./Distrib.: Akzo Nobel; Aldrich; Arizona; Brown; Henkel/Emery; Hercules; Schweizerhall; Unichema; Union Derivan SA; Witco/Oleo-Surf.
Trade names: Industrene® 105; Industrene® 205; Industrene® 206; Pamolyn® 125; Priolene 6901; Priolene 6907; Priolene 6930

Oleic acid amide. *See* Oleamide
Oleic acid potassium salt. *See* Potassium oleate
Oleic alkyl trimethylene diamine. *See* Oleyl 1,3-propylene diamine

Oleic/linoleic amine
Uses: Gasoline detergent, bactericide, corrosion inhibitor in petrol. prod., epoxy hardener
Trade names: Kemamine® P-999

Olein. *See* Triolein

Oleoyl sarcosine
CAS 110-25-8; EINECS 203-749-3
Synonyms: N-Methyl-N-(1-oxo-9-octadecenyl)glycine; Oleyl methylaminoethanoic acid; Oleyl sarcosine
Definition: Condensation prod. of oleic acid with N-methylglycine
Empirical: $C_{21}H_{39}NO_3$
Formula: $CH_3(CH_2)_7CH{:}CH(CH_2)_7CON(CH_3)CH_2COOH$
Uses: Corrosion inhibitor, detergent, wetting agent, foamer/foam stabilizer, emulsifier; for cosmetics, household/industrial cleaners, biotechnology, agric., textiles, petrol. prods., metalworking, emulsion polymerization; mold release for RIM; stabilizer for polyols; antifog and antistat in plastics
Regulatory: FDA 21CFR §178.3130
Trade names: Hamposyl® O

Oleth-4
CAS 9004-98-2 (generic); 5353-26-4
Synonyms: PEG-4 oleyl ether; POE (4) oleyl ether; PEG 200 oleyl ether
Definition: PEG ether of oleyl alcohol
Formula: $CH_3(CH_2)_7CH{=}CH(CH_2)_7CH_2(OCH_2CH_2)_nOH$, avg. n = 4
Properties: M.w. 709.02; dens. 0.997; flash pt. > 110 C
Toxicology: Irritant
Uses: Coupling agent, solubilizer, emulsion stabilizer for cosmetics, hair care preps.; emulsion polymerization; textile finishes
Trade names: Prox-onic OA-1/04

Oleth-7
CAS 9004-98-2 (generic)
Synonyms: PEG-7 oleyl ether; POE (7) oleyl ether
Definition: PEG ether of oleyl alcohol
Formula: $CH_3(CH_2)_7CH{=}CH(CH_2)_7CH_2(OCH_2CH_2)_nOH$, avg. n = 7
Uses: Emulsifier for emulsion polymerization
Regulatory: FDA 21CFR §176.200
Trade names: Akyporox RTO 70

Oleth-9
CAS 9004-98-2 (generic)
Synonyms: PEG-9 oleyl ether; POE (9) oleyl ether; PEG 450 oleyl ether
Definition: PEG ether of oleyl alcohol
Formula: $CH_3(CH_2)_7CH{=}CH(CH_2)_7CH_2(OCH_2CH_2)_nOH$, avg. n = 9
Uses: Coupling agent, solubilizer, emulsion stabilizer for cosmetics, hair care preps.; emulsion polymerization; textile spin finishes
Regulatory: FDA 21CFR §176.200, 177.2800
Trade names: Prox-onic OA-1/09

Oleth-20
CAS 9004-98-2 (generic)
Synonyms: PEG-20 oleyl ether; POE (20) oleyl ether; PEG 1000 oleyl ether
Definition: PEG ether of oleyl alcohol
Formula: $CH_3(CH_2)_7CH{=}CH(CH_2)_7CH_2(OCH_2CH_2)_nOH$, avg. n = 20
Uses: Coupling agent, solubilizer, emulsion stabilizer for cosmetics, pharmaceuticals, polishes, leather, plastics, dye dispersant, metalworking; stabilizer for rubber latex
Regulatory: FDA 21CFR §175.105, 176.180, 176.200, 177.1210, 177.2800
Trade names: Prox-onic OA-1/020; Prox-onic OA-2/020; Rhodasurf® ON-870; Rhodasurf® ON-877

Oleth-4 phosphate
CAS 39464-69-2 (generic)
Synonyms: PEG-4 oleyl ether phosphate; POE (4) oleyl ether phosphate; PEG 200 oleyl ether phosphate
Definition: Complex mixture of esters of phosphoric acid and oleth-4

Oleyl alcohol

Uses: Detergent, emulsifier, wetting agent, lubricant, antistat, EP additive for detergents, metal cleaners, textile scours, emulsion polymerization, agric.
Trade names: Chemfac PB-184

Oleyl alcohol
CAS 143-28-2; EINECS 205-597-3
Synonyms: 9-Octadecen-1-ol; cis-9-Octadecen-1-ol
Classification: Unsaturated fatty alcohol
Empirical: $C_{18}H_{36}O$
Formula: $CH_3(CH_2)_7CH=CH(CH_2)_8OH$
Properties: Pale yel. oily visc. liq.; insol. in water; sol. in alcohol, ether; m.w. 268.49; dens. 0.84; m.p. 13-19 C; b.p. 207 C (13 mm); flash pt. > 110 C; ref. index 1.4582 (27.5 C)
Precaution: Gives off acrid fumes when heated
Toxicology: Irritant
Uses: Plasticizer, emulsion stabilizer, antifoam, coupler, pigment dispersant, detergent in cutting oils, inks, textile finishing, petrochem., pulp/paper, paints, plastics, food applics., pharmaceuticals, cosmetics; mfg. of sulfuric esters used as surfactants
Regulatory: FDA 21CFR §176.170, 176.210, 177.1010, 177.1210, 177.2800, 178.3910
Manuf./Distrib.: Croda; R.W. Greeff; Lanaetex; M. Michel; Ronsheim & Moore; Witco/Oleo-Surf.
Trade names: Fancol OA-95

Oleyl amide. *See* Oleamide
Oleyl amine. *See* Oleamine
Oleyl diamino propane. *See* Oleyl 1,3-propylene diamine
Oleyl dimethylamine. *See* Dimethyl oleamine

Oleyl dimethylethyl ammonium ethosulfate
Uses: Antistat
Trade names: Larostat® 143

Oleyl methylaminoethanoic acid. *See* Oleoyl sarcosine

Oleyl palmitamide
CAS 16260-09-6; EINECS 240-367-6
Synonyms: Hexadecanamide, N-9-octadecenyl-; N-9-Octadecenyl hexadecanamide
Classification: Substituted aliphatic amide
Empirical: $C_{34}H_{67}NO$
Formula: $CH_3(CH_2)_{14}CONHCH_2(CH_2)_7CHCH(CH_2)_7CH_3$
Uses: Release agent providing slip, antiblocking to thermoplastics incl. PP film, nylon
Regulatory: FDA 21CFR §177.1200, 178.3860
Manuf./Distrib.: Croda Universal; Witco/Oleo-Surf.
Trade names: Chemstat® HTSA #1A; Kemamide® P-181

Oleyl 1,3-propylene diamine
CAS 7173-62-8; EINECS 230-528-9
Synonyms: Oleyl diamino propane; Oleic alkyl trimethylene diamine
Uses: Chemical intermediate, emulsifier, corrosion inhibitor, gasoline detergent, bactericide, epoxy hardener, asphalt emulsifier, dispersant
Trade names: Kemamine® D-989

Oleyl sarcosine. *See* Oleoyl sarcosine
One-stage resin. *See* Phenolic resin
Opalwax. *See* Hydrogenated castor oil
Organosiloxane. *See* Silicone
OSA. *See* Octenyl succinic anhydride
OTBP. *See* 2-t-Butylphenol
OTG. *See* 2-Ethylhexyl thioglycolate
1-Oxa-4-azacyclohexane. *See* Morpholine
1-Oxa-3-azaindene. *See* Benzoxazole

Oxalyl bis (benzylidenehydrazide)
Empirical: $C_{16}H_{14}O_2N_4$
Properties: White powd.; sp.gr. 0.216

Uses: Antioxidant for polymers; copper deactivator; stabilizer
Manuf./Distrib.: Eastman
Trade names: Eastman® Inhibitor OABH
Trade names containing: Aquastab PH 502

2,2′-Oxamido bis[ethyl 3-(3,5-di-t-butyl-4-hydroxyphenyl) propionate
CAS 70331-94-1
Classification: Phenolic antioxidant
Properties: Wh. powd.; m.w. 697; sp.gr. 1.12 (20 C); m.p. 170-180 C; flash pt. (TOC) 260 C
Uses: Antioxidant for polymers; metal deactivator
Trade names: Naugard® XL-1

Oxidized polyethylene. *See* Polyethylene, oxidized
Oxirane, [1,1-dimethylethoxy)methyl]-. *See* t-Butyl glycidyl ether
Oxirane, [[2-ethylhexyl)oxy]methyl]-. *See* 2-Ethylhexyl glycidyl ether
Oxirane, [(1-methylethoxy)methyl]-,. *See* Isopropyl glycidyl ether
Oxirane, (phenoxymethyl)-. *See* Phenyl glycidyl ether
2-Oxobutane. *See* Methyl ethyl ketone
3-Oxobutanoic acid ethyl ester. *See* Ethylacetoacetate
Oxybenzone. *See* Benzophenone-3

4,4′-Oxybis (benzenesulfonylhydrazide)
CAS 80-51-3
Synonyms: OBSH; p,p′-Oxybis benzene sulfonyl hydrazide; 4,4′-Oxybis (benzenesulfonyl) hydrazine
Empirical: $C_{13}H_{14}O_5N_4S_2$
Formula: $H_2NNHSO_2C_7H_4OC_6H_4SO_2NHNH_2$
Properties: Fine white crystalline powd., odorless; sol. in acetone; moderately sol. in ethanol, polyethylene glycol; insol. in gasoline and water; m.w. 358.40; dens. 1.52; m.p. 150 C (dec.); gas yield 120-125 cc/g
Precaution: Combustible
Uses: Chemical blowing agent for sponge rubber, expanded plastics
Trade names: Azocel OBSH; Azofoam B-95; Blo-Foam® BBSH; Cellcom OBSH-ASA2; Celogen® OT; Mikrofine OBSH; Nitropore® OBSH; Unicell OH

p,p′-Oxybis benzene sulfonyl hydrazide. *See* 4,4′-Oxybis (benzenesulfonylhydrazide)
4,4′-Oxybis (benzenesulfonyl) hydrazine. *See* 4,4′-Oxybis (benzenesulfonylhydrazide)
2,2′-[Oxybis(2,1-ethanediyloxy)]bisethanol. *See* PEG-4
2,2′-Oxybisethanol. *See* Diethylene glycol
Oxybismethane. *See* Dimethyl ether

10,10′-Oxybisphenoxyarsine
CAS 58-36-6
Synonyms: OBPA; 10,10′-Oxydiphenoxarsine; Bis(phenoxarsin-10-yl)ether; Bis(10-phenoxyarsinyl)oxide
Empirical: $C_{24}H_{16}As_2O$
Properties: Colorless prisms; sol. in alcohol, chloroform, methylene chloride; pract. insol. in water (5 ppm @ 20 C); m.w. 502.23; dens. 1.40-1.42; m.p. 184-185 C; dec. 380 C
Toxicology: LD50 (male rat) 35-50 mg/kg
Uses: Fungicide and bactericide for protection of plastics
Trade names containing: Vinyzene® BP-5-2; Vinyzene® BP-5-2 DIDP; Vinyzene® BP-5-2 DOP; Vinyzene® BP-5-2 PG; Vinyzene® BP-5-2 S-160; Vinyzene® BP-5-2 U; Vinyzene® BP-505; Vinyzene® BP-505 DIDP; Vinyzene® BP-505 DOP; Vinyzene® BP-505 PG; Vinyzene® BP-505 S160; Vinyzene® SB-1; Vinyzene® SB-1 EAA; Vinyzene® SB-1 PR; Vinyzene® SB-1 PS; Vinyzene® SB-1 U

1,1′-Oxybis-2-propanol. *See* Dipropylene glycol

Oxy bis titanium IV tris (bis tridecyl) phosphate
Trade names containing: SAFR-T-12K

Oxy bis zirconium IV tris (bis tridecyl) phosphate
Trade names containing: SAFR-Z-12K

N-Oxydiethylene benzothiazole-2-sulfenamide
CAS 102-77-2
Synonyms: 2-Benzothiazolyl-n-morpholinosulfide; Benzothiazyl-2-sulfene morpholide; 2-Morpholinyl mer-

captobenzothiazole; 2-(Morpholinothio) benzothiazole
Empirical: $C_{11}H_{12}N_2OS_2$
Properties: Buff to brn. flakes, sweet odor; insol. in water; sol. in benzene, acetone, methanol; m.w. 252.37; dens. 1.34 (25 C); m.p. 80 C min.
Toxicology: Poison by intraperitoneal route; moderately toxic by ingestion
Uses: Delayed-action vulcanization accelerator for rubbers
Trade names: Akrochem® Accelerator MF; Akrochem® OBTS; Akrochem® OMTS; Amax®; Delac® MOR; Perkacit® MBS; Santocure® MOR; Vulkacit MOZ/LG; Vulkacit MOZ/SG
Trade names containing: Poly-Dispersion® V(NOBS)M-65

N-Oxydiethylenethiocarbamyl-N-oxydiethylene sulfenamide. *See* Thiocarbamyl sulfenamide
10,10´-Oxydiphenoxarsine. *See* 10,10´-Oxybisphenoxyarsine
3,3´-Oxydyl-1-propanol dibenzoate. *See* Dipropylene glycol dibenzoate
Ozocerite. *See* Ozokerite

Ozokerite
CAS 8021-55-4
Synonyms: Ozocerite; Mineral wax; Fossil wax
Classification: Hydrocarbon wax
Definition: Hydrocarbon wax derived from mineral or petroleum sources
Properties: Yel.-brown to black or green translucent (pure); sol. in lt. petrol. hydrocarbons, benzene, turpentine, kerosene, ether, carbon disulfide; sl. sol. in alcohol; insol. in water; dens. 0.85-0.95; m.p. 55-110 C (usually 70 C)
Precaution: Combustible
Uses: Filler for electrical insulation, rubber prods., paints, leather prods., printing inks, floor and furniture polishes, cosmetics, ointments; substitute for carnauba and beeswax
Regulatory: Japan approved
Manuf./Distrib.: Eastman; ISP; Koster Keunen; Frank B. Ross; Strahl & Pitsch
Trade names: Koster Keunen Ozokerite

Palmitamine
CAS 143-27-1; EINECS 205-596-8
Synonyms: Cetyl amine; Palmityl amine; 1-Hexadecanamine
Classification: Primary aliphatic amine
Definition: Primary aliphatic amine of palmitic acid
Empirical: $C_{16}H_{35}N$
Formula: $CH_3(CH_2)_{14}CH_2NH_2$
Uses: Emulsifier, flotation agent, corrosion inhibitor; for herbicides, ore flotation, pigment dispersion; aux. for textiles, leather, rubber, plastics, and metal industries
Trade names: Crodamine 1.16D

Palmitic/stearic acid glycerol monodiester. *See* Palmitic/stearic acid mono/diglycerides

Palmitic/stearic acid mono/diglycerides
Synonyms: Palmitic/stearic acid glycerol monodiester
Uses: Antistat for polyolefins; surface coating for EPS; fabric softener; coemulsifier
Trade names: Tegotens 4100

Palmityl alcohol. *See* Cetyl alcohol
Palmityl amine. *See* Palmitamine
Palmityl dimethylamine. *See* Dimethyl palmitamine
Palmityl trimethyl ammonium chloride. *See* Cetrimonium chloride

Palm kernel glycerides
Synonyms: Glycerides, palm kernel mono-, di- and tri-
Definition: Mixture of mono, di and triglycerides derived from palm kernel oil
Uses: Component in w/o and o/w creams, lubricant, antistat, antifogging agent for plastics
Trade names: Grindtek PK 60

Palm oil glyceride, hydrogenated. *See* Hydrogenated palm glyceride
PAN carbon fiber. *See* Carbon fiber

Paraffin

CAS 8002-74-2; EINECS 232-315-6

Synonyms: Paraffin wax; Hard paraffin; Petroleum wax, crystalline

Classification: Hydrocarbon

Definition: Solid mixture of hydrocarbons obtained from petroleum; characterized by relatively large crystals

Empirical: C_nH_{2n+2}

Properties: Colorless to white solid, odorless; insol. in water, alcohol; sol. in benzene, chloroform, ether, carbon disulfide, oils; misc. with fats; dens. ≈ 0.9; m.p. 50-57 C; flash pt. 340 F

Precaution: Dangerous fire hazard

Toxicology: Anesthetic effect; ACGIH TLV:TWA 2 mg/m^3 (fume); experimental tumorigens by implantation; many paraffin waxes contain carcinogens

Uses: Candles, paper coating, protective sealant for food prods.; plastics lubricants; hot-melt carpet backing, floor polishes, cosmetics, chewing gum base; raising m.p. of ointments

Regulatory: FDA 21CFR §133.181, 172.615, 173.3210, 175.105, 175.210, 175.250, 175.300, 175.320, 176.170, 176.200, 177.1200, 177.2420, 177.2600, 177.2800, 178.3710, 178.3800, 178.3910, 179.45; Canada, Japan approved

Manuf./Distrib.: Astor Wax; EM Industries; Exxon; Humphrey; IGI; Jonk BV; Koster Keunen; Mobil; Penreco; Phillips; Frank B. Ross; Shell; Stevenson Cooper; Texaco; Vista

Trade names: Advawax® 165; Akrowax 130; Akrowax 145; Hostalub® XL 165FR; Hostalub® XL 165P; Loxiol® G 22; Loxiol® G 23; Loxiol® HOB 7138; Loxiol® HOB 7169; Octowax 321; Petrac® 165; Sunolite® 130; Sunolite® 160

Trade names containing: Antilux® 110; Antilux® 111; Antilux® 500; Antilux® 550; Antilux® 600; Antilux® 620; Antilux® 654; Antilux® 660; Antilux® L; Hostalub® CAF 485; Hostalub® XL 50; Hostalub® XL 355; Hostalub® XL 445; Polyace™ 804; Renacit® 7/WG; Sunolite® 100; Sunolite® 127; Sunolite® 240; Sunolite® 240TG; Sunolite® 666; Vanax® A

Paraffin, brominated

Uses: Flame retardant for polyurethane foam, pkg., rubber, textiles; visc. reducer

Trade names: DD-8126; Doverguard® 8410; Doverguard® 8426

Paraffin, bromochlorinated

Uses: Flame retardant for flexible, rigid polyurethane foam, laminated board, board stock, pkg., textiles, carpet backing

Trade names: Doverguard® 8133; Doverguard® 8207-A; Doverguard® 8208-A; Doverguard® 8307-A; Doverguard® 9119; Fyarestor® 100

Trade names containing: DD-8307; Doverguard® 9021; Doverguard® 9122

Paraffin, chlorinated

Synonyms: Chlorocosane; Chlorcosane; Chlorinated paraffin

Definition: Paraffin hydrocarbons treated with chlorine

Properties: Pale to amber, neutral, light viscous oils or soft waxes; insol. in water; sl. sol. in alcohol; sol. in benzene, chloroform; dens. 0.900-1.50

Precaution: Nonflamm.

Uses: In mfg. of textiles, draperies, etc.; solv. for dichloramine-T; high-pressure lubricants; flame retardants in plastics; as plasticizer for PVC in PE sealants

Manuf./Distrib.: Allchem Ind.; R.E. Carroll; Chemcentral; Dover; Ferro/Keil; C.P. Hall; Harwick; Hoechst AG; Morton Int'l.; OxyChem; Sea-Land; Witco/PAG

Trade names: Adeka Lub E-410; Adeka Lub E-450; Adeka Lub E-500; ADK CIZER E-500; Akrochlor™ L-39; Akrochlor™ L-40; Akrochlor™ L-42; Akrochlor™ L-57; Akrochlor™ L-60; Akrochlor™ L-61; Akrochlor™ L-70LV; Akrochlor™ L-170; Akrochlor™ L-170HV; Akrochlor™ R-70; Akrochlor™ R-70S; Chlorez® 700; Chlorez® 700-DF; Chlorez® 700-S; Chlorez® 700-SS; Chlorez® 725-S; Chlorez® 760; Chloroflo® 40; Chloroflo® 42; Clorafin® 40; Doverguard® 152; Doverguard® 170; Doverguard® 700; Doverguard® 700-S; Doverguard® 700-SS; Doverguard® 760; Doverguard® 5761; Electrofine S-70; Hordaflam® NK 72; Kloro 6001; Paroil® 140; Rez-O-Sperse® 3; Rez-O-Sperse® A-1; Thermoguard® XS-70-T

Trade names containing: Porofor® B 13/CP 50

Paraffin oil. *See* Mineral oil

Paraffin oil, vulcanized

Trade names containing: Faktogel® Rheinau W

Paraffin wax. *See* Paraffin

Pareth-15-3. *See* C11-15 pareth-3

Pareth-15-5. *See* C11-15 pareth-5
Pareth-15-30. *See* C11-15 pareth-30
Pareth-15-40. *See* C11-15 pareth-40
PBB-PA. *See* Poly (pentabromobenzyl) acrylate
PBT. *See* Polybutylene terephthalate
PBT polyester. *See* Polybutylene terephthalate
PC. *See* Polycarbonate resin
PC resin. *See* Polycarbonate resin
PCTFE. *See* Polychlorotrifluoroethylene
PDDP. *See* Phenyl diisodecyl phosphite
PDEA. *See* Phenyldiethanolamine
PDMS. *See* Dimethicone
PE. *See* Pentaerythritol
PEEA. *See* Phenylethylethanolamine
PEG. *See* Polyethylene glycol

PEG-4
CAS 25322-68-3 (generic); 112-60-7; EINECS 203-989-9
Synonyms: PEG 200; POE (4); 2,2´-[Oxybis(2,1-ethanediyloxy)]bisethanol
Definition: Polymer of ethylene oxide
Empirical: $C_8H_{18}O_5$
Formula: $H(OCH_2CH_2)_nOH$, avg. n = 4
Properties: Visc. liq., sl. char. odor; hygroscopic; m.w. 190-210; dens. 1.127 (25/25 C); visc. 4.3 cSt (210 F); supercools on freezing
Precaution: Solvent action on some plastics
Toxicology: LD50 (oral, rat) 28,900 mg/kg; mildly toxic by ingestion; heated to decomp., emits acrid smoke and irritating fumes
Uses: Lubricant for rubber molds, textile fibers, metalworking; in food and food pkg.; in cosmetics and hair preps.; pharmaceutic aid; in gas chromatography; in paints, paper coatings, polishes, ceramics
Regulatory: FDA 21CFR §73.1, 172.210, 172.770, 172.820, 173.310, 173.340, 175.105, 175.300, 178.3750
Manuf./Distrib.: Ashland; C.P. Hall; Harwick; Henkel; Union Carbide
Trade names: Carbowax® PEG 200; Dow E200; Emery® 6773; Hodag PEG 200; Nopalcol 200; Pluracol® E200; Poly-G® 200

PEG-6
CAS 25322-68-3 (generic); 2615-15-8; EINECS 220-045-1
Synonyms: PEG 300; Hexaethylene glycol; Macrogol 300
Definition: Polymer of ethylene oxide
Empirical: $C_{12}H_{26}O_7$
Formula: $H(OCH_2CH_2)_nOH$, avg. n = 6
Properties: Sp.gr. 1.124-1.127; pour pt. -15 to -8 C; flash pt. (COC) 196 C; ref. index 1.463-1.4641 (20 C)
Toxicology: LD50 (oral, rat) 27,500 mg/kg; mildly toxic by ingestion; heated to decomp., emits acrid smoke and irritating fumes
Uses: Lubricant for rubber molds, textile fibers, metalworking; in food and food pkg.; in cosmetics and hair preps.; pharmaceutic aid; in gas chromatography; in paints, paper coatings, polishes, ceramics; plasticizer for cellulose nitrate
Regulatory: FDA 21CFR §172.210, 172.770, 172.820, 173.310, 173.340, 175.105, 175.300, 178.3750, 178.3910
Manuf./Distrib.: Ashland; Harwick; Henkel; Union Carbide
Trade names: Alkapol PEG 300; Carbowax® PEG 300; Dow E300; Emery® 6687; Hodag PEG 300; Pluracol® E300; Poly-G® 300; Teric PEG 300
Trade names containing: Carbowax® PEG 540 Blend; Hodag PEG 540; Pluracol® E1500; Tinuvin® 213

PEG-8
CAS 25322-68-3 (generic); 5117-19-1; EINECS 225-856-4
Synonyms: PEG 400; POE (8); 3,6,9,12,15,18,21-Heptaoxatricosane-1,2,3-diol
Definition: Polymer of ethylene oxide
Empirical: $C_{16}H_{34}O_9$
Formula: $H(OCH_2CH_2)_nOH$, avg. n = 8
Properties: Visc. liq., sl. char. odor; sl. hygroscopic; m.w. 380-420; dens. 1.128 (25/25 C); m.p. 4-8 C; visc. 7.3 cSt (210 F)
Toxicology: LD50 (rat, oral) 30 ml/kg; low toxicity by ingestion, IV, IP routes; heated to decomp., emits acrid

smoke and irritating fumes

Uses: Lubricant for rubber molds, textile fibers, metalworking; in food and food pkg.; in cosmetics and hair preps.; pharmaceutic aid; in gas chromatography; in paints, paper coatings, polishes, ceramics; plasticizer for cellulose nitrate

Regulatory: FDA 21CFR §172.210, 172.770, 172.820, 173.310, 173.340, 175.105, 175.300, 178.3750, 178.3910, 181.22, 181.30

Manuf./Distrib.: Ashland; C.P. Hall; Harwick; Henkel; Union Carbide

Trade names: Carbowax® PEG 400; Dow E400; Hodag PEG 400; Nopalcol 400; Pluracol® E400; Pluracol® E400 NF; Poly-G® 400; Rhodasurf® E 400; Teric PEG 400

PEG-12

CAS 25322-68-3 (generic); 6790-09-6; EINECS 229-859-1

Synonyms: PEG 600; POE (12); Macrogol 600

Definition: Polymer of ethylene oxide

Empirical: $C_{24}H_{50}O_{13}$

Formula: $H(OCH_2CH_2)_nOH$, avg. n = 12

Properties: Visc. liq., char. odor; sl. hygroscopic; m.w. 570-630; dens. 1.128 (25/25 C); m.p. 20-25 C; visc. 10.5 cSt (210 F)

Toxicology: LD50 (oral, rat) 38,100 mg/kg; low toxicity by ingestion; eye irritant; heated to decomp., emits acrid smoke and irritating fumes

Uses: Lubricant for rubber molds, textile fibers, metalworking; in food and food pkg.; in cosmetics and hair preps.; pharmaceutic aid; in gas chromatography; in paints, paper coatings, polishes, ceramics; plasticizer for cellulose nitrate

Regulatory: FDA 21CFR §172.210, 172.770, 172.820, 173.310, 173.340, 175.105, 175.300, 178.3750, 178.3910

Manuf./Distrib.: Ashland; C.P. Hall; Harwick; Union Carbide

Trade names: Carbowax® PEG 600; Dow E600; Emery® 6686; Nopalcol 600; Pluracol® E600; Pluracol® E600 NF; Poly-G® 600; Rhodasurf® E 600; Teric PEG 600

PEG-20

CAS 25322-68-3 (generic)

Synonyms: PEG 1000; Macrogol 1000; POE (20)

Definition: Polymer of ethylene oxide

Empirical: $C_{40}H_{82}O_{21}$

Formula: $H(OCH_2CH_2)_nOH$, avg. n = 20

Properties: Sp.gr. 1.085; pour pt. 37-40 C; flash pt. (COC) 265 C

Toxicology: LD50 (oral, rat) 42 g/kg; mod. toxic by IP, IV routes; mildly toxic by ingestion; experimental tumorigen; heated to decomp., emits acrid smoke and irritating fumes

Uses: Lubricant for rubber molds, textile fibers, metalworking; in food and food pkg.; in cosmetics and hair preps.; pharmaceutic aid; in gas chromatography; in paints, paper coatings, polishes, ceramics; plasticizer for cellulose nitrate

Regulatory: FDA 21CFR §172.210, 172.770, 172.820, 173.310, 173.340, 175.105, 175.300, 178.3750, 178.3910

Manuf./Distrib.: Ashland; Harwick; Union Carbide

Trade names: Carbowax® PEG 900; Carbowax® PEG 1000; Dow E1000; Pluracol® E1000; Poly-G® 1000

PEG-32

CAS 25322-68-3 (generic)

Synonyms: PEG 1540; Macrogol 1540; POE (32)

Definition: Polymer of ethylene oxide

Empirical: $C_{64}H_{130}O_{33}$

Formula: $H(OCH_2CH_2)_nOH$, avg. n = 32

Properties: White powd.

Toxicology: LD50 (oral, rat) 44,200 mg/kg; mildly toxic by ingestion; human skin irritant; heated to decomp., emits acrid smoke and irritating fumes

Uses: Lubricant for rubber molds, textile fibers, metalworking; in food and food pkg.; in cosmetics and hair preps.; pharmaceutic aid; in gas chromatography; in paints, paper coatings, polishes, ceramics

Regulatory: FDA 21CFR §172.210, 172.770, 172.820, 173.310, 173.340, 175.105, 175.300, 178.3750, 178.3910

Trade names: Dow E1450; Pluracol® E1450; Pluracol® E1450 NF

Trade names containing: Carbowax® PEG 540 Blend; Hodag PEG 540; Pluracol® E1500

PEG-40
CAS 25322-68-3 (generic)
Synonyms: PEG 2000; POE (40)
Definition: Polymer of ethylene oxide
Empirical: $C_{80}H_{162}O_{41}$
Formula: $H(OCH_2CH_2)_nOH$, avg. n = 40
Properties: Dens. 1.127; ref. index 1.4590 (20 C)
Uses: Lubricant for rubber molds, textile fibers, metalworking; in food and food pkg.; in cosmetics and hair preps.; pharmaceutic aid; in gas chromatography; in paints, paper coatings, polishes, ceramics
Regulatory: FDA 21CFR §172.210, 172.770, 172.820, 173.310, 173.340, 175.105, 175.300, 178.3750, 178.3910
Manuf./Distrib.: Aldrich
Trade names: Pluracol® E2000; Poly-G® 2000

PEG-75
CAS 25322-68-3 (generic)
Synonyms: PEG 4000; POE (75)
Definition: Polymer of ethylene oxide
Empirical: $C_{150}H_{302}O_{76}$
Formula: $H(OCH_2CH_2)_nOH$, avg. n = 75
Properties: White powd. or creamy-white flakes; m.w. 3000-3700; dens. 1.212 (25/25 C); m.p. 54-58 C; visc. 76-110 cSt (210 F)
Toxicology: LD50 (oral, rat) 50 g/kg; mildly toxic by ingestion; skin irritant; heated to decomp., emits acrid smoke and irritating fumes
Uses: Lubricant for rubber molds, textile fibers, metalworking; in food and food pkg.; in cosmetics and hair preps.; pharmaceutic aid; in gas chromatography; in paints, paper coatings, polishes, ceramics
Regulatory: FDA 21CFR §172.210, 172.770, 172.820, 173.310, 173.340, 175.105, 175.300, 178.3750, 178.3910
Trade names: Dow E3350; Hodag PEG 3350; Pluracol® E4000; Rhodasurf® PEG 3350; Teric PEG 4000

PEG-100
CAS 25322-68-3 (generic)
Synonyms: PEG (100); POE (100)
Definition: Polymer of ethylene oxide
Empirical: $C_{200}H_{402}O_{101}$
Formula: $H(OCH_2CH_2)_nOH$, avg. n = 100
Uses: Lubricant for rubber molds, textile fibers, metalworking; in food and food pkg.; in cosmetics and hair preps.; pharmaceutic aid; in gas chromatography; in paints, paper coatings, polishes, ceramics
Regulatory: FDA 21CFR §172.210, 172.770, 172.820, 173.310, 173.340, 175.105, 175.300, 178.3750, 178.3910
Trade names: Dow E4500

PEG (100). *See* PEG-100

PEG-150
CAS 25322-68-3 (generic)
Synonyms: PEG 6000; Macrogol 6000; POE (150)
Definition: Polymer of ethylene oxide
Formula: $H(OCH_2CH_2)_nOH$, avg. n = 150
Properties: Powd. or creamy-white flakes; water-sol.; m.w. 7000-9000; dens. 1.21 (25/25 C); m.p. 56-63 C; visc. 470-900 cSt (210 F); flash pt. > 887 F
Precaution: Combustible exposed to heat or flame
Toxicology: LD50 (rat, oral) > 50 g/kg; mildly toxic by ingestion; mutagenic data; skin irritant; heated to decomp., emits acrid smoke and irritating fumes
Uses: Lubricant for rubber molds, textile fibers, metalworking; in food and food pkg.; in cosmetics and hair preps.; pharmaceutic aid; in gas chromatography; in paints, paper coatings, polishes, ceramics
Regulatory: FDA 21CFR §172.210, 172.770, 172.820, 173.310, 173.340, 175.300, 177.2420, 178.3750, 178.3910
Trade names: Dow E8000; Hodag PEG 8000; Pluracol® E6000; Pluracol® E8000

PEG 200. *See* PEG-4
PEG 300. *See* PEG-6
PEG 400. *See* PEG-8

PEG 600. *See* PEG-12
PEG 1000. *See* PEG-20
PEG 1540. *See* PEG-32
PEG 2000. *See* PEG-40
PEG-2000. *See* PEG-2M
PEG 4000. *See* PEG-75
PEG-5000. *See* PEG-5M
PEG 6000. *See* PEG-150
PEG-7000. *See* PEG-7M
PEG-9000. *See* PEG-9M
PEG-14000. *See* PEG-14M
PEG-20000. *See* PEG-20M
PEG-23000. *See* PEG-23M
PEG-45000. *See* PEG-45M
PEG-90000. *See* PEG-90M
PEG-115000. *See* PEG-115M
PEG 300,000. *See* PEG-7M
PEG 600,000. *See* PEG-14M

PEG-2M
CAS 25322-68-3 (generic)
Synonyms: PEG-2000; Polyethylene glycol (2000); POE (2000)
Definition: Polymer of ethylene oxide
Formula: $H(OCH_2CH_2)_nOH$, avg. n = 2000
Uses: Water-sol. thermoplastic resin, thickener, lubricant, binder, flocculant, wet adhesive; dispersant for vinyl polymerization; mold release agent; agric.; cosmetic/pharmaceutical emulsions
Regulatory: FDA 21CFR §172.770, 173.310, 175.300, 178.3910
Trade names: Polyox® WSR N-10; Prox-onic PEG-2000

PEG-5M
CAS 25322-68-3 (generic)
Synonyms: PEG-5000; POE (5000)
Definition: Polymer of ethylene oxide
Formula: $H(OCH_2CH_2)_nOH$, avg. n = 5000
Uses: Water-sol. thermoplastic resin, thickener, lubricant, binder, flocculant, wet adhesive; for agric., ceramics applics.; dispersant for vinyl polymerization
Regulatory: FDA 21CFR §172.770, 173.310, 175.300, 178.3910
Trade names: Polyox® WSR N-80

PEG-7M
CAS 25322-68-3 (generic)
Synonyms: PEG-7000; POE (7000); PEG 300,000
Definition: Polymer of ethylene oxide
Formula: $H(OCH_2CH_2)_nOH$, avg. n = 7000
Uses: Water-sol. thermoplastic resin, thickener, lubricant, binder, flocculant, wet adhesive; for agric., elec., paper; emollient for cosmetics; dispersant for vinyl polymerization
Regulatory: FDA 21CFR §172.770, 173.310, 175.300, 178.3910
Trade names: Polyox® WSR N-750

PEG-8M
Uses: Mold release and antistat for rubber; binder; intermediate for copolymers, PEG esters; cosmetics, pharmaceuticals; thickener for inks
Trade names: Rhodasurf® PEG 8000; Teric PEG 8000

PEG-9M
CAS 25322-68-3 (generic)
Synonyms: PEG-9000; POE (9000)
Definition: Polymer of ethylene oxide
Formula: $H(OCH_2CH_2)_nOH$, avg. n = 9000
Uses: Water-sol. thermoplastic resin, thickener, lubricant, binder, flocculant, wet adhesive in papermaking; dispersant for vinyl polymerization

PEG-14M

Regulatory: FDA 21CFR §172.770, 173.310, 175.300, 178.3910
Trade names: Polyox® WSR 3333

PEG-14M
CAS 25322-68-3 (generic)
Synonyms: PEG-14000; PEG 600,000; POE (14000)
Definition: Polymer of ethylene oxide
Formula: $H(OCH_2CH_2)_nOH$, avg. n = 14,000
Uses: Water-sol. thermoplastic resin, thickener, lubricant, binder, flocculant, wet adhesive; for ceramics, papermaking; dispersant for vinyl polymerization
Regulatory: FDA 21CFR §172.770, 173.310, 175.300, 178.3910
Trade names: Polyox® WSR 205; Polyox® WSR N-3000

PEG-20M
CAS 25322-68-3 (generic)
Synonyms: PEG-20000; POE (20000)
Definition: Polymer of ethylene oxide
Formula: $H(OCH_2CH_2)_nOH$, avg. n = 20,000
Uses: Water-sol. thermoplastic resin, thickener, lubricant, binder, flocculant, wet adhesive; dispersant for vinyl polymerization
Regulatory: FDA 21CFR §172.770, 173.310, 175.300, 178.3910
Trade names: Polyox® WSR 1105; Prox-onic PEG-20,000

PEG-23M
CAS 25322-68-3 (generic)
Synonyms: PEG-23000; POE (23000)
Definition: Polymer of ethylene oxide
Formula: $H(OCH_2CH_2)_nOH$, avg. n = 23000
Uses: Water-sol. thermoplastic resin, thickener, lubricant, binder, flocculant, wet adhesive; dispersant for vinyl polymerization
Regulatory: FDA 21CFR §172.770, 173.310, 175.300, 178.3910
Trade names: Polyox® WSR N-12K

PEG-45M
CAS 25322-68-3 (generic)
Synonyms: PEG-45000; POE (45000)
Definition: Polymer of ethylene oxide
Formula: $H(OCH_2CH_2)_nOH$, avg. n = 45,000
Uses: Water-sol. thermoplastic resin, thickener, lubricant, binder, flocculant, wet adhesive; lubricant/emollient in cosmetics; dispersant for vinyl polymerization
Regulatory: FDA 21CFR §172.770, 173.310, 175.300, 178.3910
Trade names: Polyox® WSR N-60K

PEG-90M
CAS 25322-68-3 (generic)
Synonyms: PEG-90000; POE (90000)
Definition: Polymer of ethylene oxide
Formula: $H(OCH_2CH_2)_nOH$, avg. n = 90000
Uses: Water-sol. thermoplastic resin, thickener, lubricant, binder, flocculant, wet adhesive; dispersant for vinyl polymerization; agric., mining, construction, cosmetics; foam stabilizer in malt beverages
Regulatory: FDA 21CFR §172.770, 173.310, 175.300, 178.3910
Trade names: Polyox® WSR 301

PEG-115M
CAS 25322-68-3 (generic)
Synonyms: PEG-115000; POE (115000)
Definition: Polymer of ethylene oxide
Formula: $H(OCH_2CH_2)_nOH$, avg. n = 115,000
Uses: Water-sol. thermoplastic resin, thickener, film-former, flocculant; dispersant for vinyl polymerization; mining, papermaking, cosmetics; foam stabilizer in malt beverages
Regulatory: FDA 21CFR §172.770, 173.310, 175.300, 178.3910
Trade names: Polyox® Coagulant

PEG-3 aniline
CAS 36356-83-9 (generic)
Uses: Cure promoter for polyester resins; intermediate for fugitive tints
Trade names: Emery® 5702

PEG-2 butyl ether. *See* Butoxydiglycol

PEG-2 C13-15 alkyl amine
Uses: Emulsifier, PU catalyst, textile and plastic processing aid, textile dyeing aux., antistat
Trade names: Synprolam 35X2

PEG-5 C13-15 alkyl amine
Uses: Emulsifier, PU catalyst, textile and plastic processing aid, textile dyeing aux., antistat
Trade names: Synprolam 35X5

PEG-10 C13-15 alkyl amine
Uses: Emulsifier, PU catalyst, textile and plastic processing aid, textile dyeing aux., antistat
Trade names: Synprolam 35X10

PEG-15 C13-15 alkyl amine
Uses: Emulsifier, PU catalyst, textile and plastic processing aid, textile dyeing aux., antistat
Trade names: Synprolam 35X15

PEG-20 C13-15 alkyl amine
Uses: Emulsifier, PU catalyst, textile and plastic processing aid, textile dyeing aux., antistat
Trade names: Synprolam 35X20

PEG-25 C13-15 alkyl amine
Uses: Emulsifier, PU catalyst, textile and plastic processing aid, textile dyeing aux., antistat
Trade names: Synprolam 35X25

PEG-35 C13-15 alkyl amine
Uses: Emulsifier, PU catalyst, textile and plastic processing aid, textile dyeing aux., antistat
Trade names: Synprolam 35X35

PEG-50 C13-15 alkyl amine
Uses: Emulsifier, PU catalyst, textile and plastic processing aid, textile dyeing aux., antistat
Trade names: Synprolam 35X50

PEG-1 C13-15 alkylmethylamine
CAS 92112-62-4
Uses: Emulsifier, textile processing aid, wetting agent; antistat for polyolefins and PVC
Trade names: Synprolam 35MX1

PEG-3 C13-15 alkylmethylamine
Uses: Antistat for polyolefins and PVC
Trade names: Synprolam 35MX3

PEG-5 C13-15 alkylmethylamine
Uses: Antistat for polyolefins and PVC
Trade names: Synprolam 35MX5

PEG-3 caprylate/caprate
Synonyms: POE (3) caprylate/caprate; Triethylene glycol caprate/caprylate
Definition: PEG ester of a mixture of caprylic and capric acids
Uses: Plasticizer for rubber goods; lubricant for aluminum can industry
Trade names: Plasthall® 4141; Uniflex® TEG-810

PEG-5 castor oil
CAS 61791-12-6 (generic)
Synonyms: POE (5) castor oil
Definition: PEG deriv. of castor oil with avg. 5 moles of ethylene oxide

PEG-10 castor oil

Uses: Surfactant used as emulsifier, dispersant, solubilizer, visc. control agent for cosmetics, pharmaceuticals, and industrial applics.; emulsifier in lubricants for plastics, metals, textiles; paint, paper, textile, and leather
Regulatory: FDA 21CFR §175.105, 175.300
Trade names: Prox-onic HR-05

PEG-10 castor oil
CAS 61791-12-6 (generic)
Synonyms: POE (10) castor oil; PEG 500 castor oil
Definition: PEG deriv. of castor oil with avg. 10 moles of ethylene oxide
Uses: Cosmetic and essential oil solubilizer, emulsifier, lubricant, softener, leveling agent, emollient, superfatting agent, antistat, softener, detergent; used in personal care prods., fiber processing; emulsion polymerization
Regulatory: FDA 21CFR §175.105, 175.300
Trade names: Etocas 10

PEG-12 castor oil
CAS 61791-12-6 (generic)
Synonyms: POE (12) castor oil
Definition: PEG deriv. of castor oil with avg. 12 moles of ethylene oxide
Uses: Detergent, emulsifier, coemulsifier in sol. oils, solv. cleaners, coatings, industrial/institutional cleaners, metalworking; mold release in plastics; textile fiber lubricant
Trade names: Teric C12

PEG-16 castor oil
CAS 61791-12-6 (generic)
Synonyms: POE (16) castor oil
Definition: PEG deriv. of castor oil with avg. 16 moles of ethylene oxide
Uses: Lubricant additive, emulsifier in lubricants for plastics, metals, textiles; clay and pigment dispersants, rewetting agent, softener, dyeing assistant for paint, paper, textile, and leather industries
Trade names: Prox-onic HR-016

PEG-18 castor oil
Uses: Softener, antistat for textiles, plastics
Trade names: Alkamuls® R81

PEG-25 castor oil
CAS 61791-12-6 (generic)
Synonyms: POE (25) castor oil
Definition: PEG deriv. of castor oil with avg. 25 moles of ethylene oxide
Uses: Lubricant additive, emulsifier in lubricants for plastics, metals, textiles; clay and pigment dispersants, rewetting agent, softener, dyeing assistant for paint, paper, textile, and leather industries
Regulatory: FDA 21CFR §175.105, 175.300, 177.2800
Trade names: Mapeg® CO-25; Prox-onic HR-025

PEG-30 castor oil
CAS 61791-12-6 (generic)
Synonyms: POE (30) castor oil
Definition: PEG deriv. of castor oil with avg. 30 moles of ethylene oxide
Uses: Lubricant additive, emulsifier in lubricants for plastics, metals, textiles; clay and pigment dispersants, rewetting agent, softener, dyeing assistant for paint, paper, textile, and leather industries
Regulatory: FDA 21CFR §175.105, 175.300, 177.2800
Trade names: Alkamuls® EL-620L; Prox-onic HR-030

PEG-32 castor oil
CAS 61791-12-6 (generic)
Synonyms: POE (32) castor oil
Definition: PEG deriv. of castor oil with avg. 32 moles of ethylene oxide
Uses: Surfactant, emulsifier, softener, rewetting agent, pigment dispersant, dye assistant, leveling agent for paint, textile, leather, plastics lubricants, metal lubricants
Trade names: Berol 199

PEG-36 castor oil
CAS 61791-12-6 (generic)
Synonyms: POE (36) castor oil; PEG 1800 castor oil
Definition: PEG deriv. of castor oil with avg. 36 moles of ethylene oxide
Properties: Liq.; sol. in water, xylene
Uses: Emulsifier, lubricant, dye leveler, antistat, dispersant; in textiles, leather, paper, polyurethane foams
Regulatory: FDA 21CFR §175.105, 175.300, 176.210, 177.2800
Trade names: Prox-onic HR-036

PEG-40 castor oil
CAS 61791-12-6 (generic)
Synonyms: POE (40) castor oil; PEG 2000 castor oil
Definition: PEG deriv. of castor oil with avg. 40 moles of ethylene oxide
Uses: Lubricant additive, emulsifier in lubricants for plastics, metals, textiles; clay and pigment dispersants, rewetting agent, softener, dyeing assistant for paint, paper, textile, and leather industries
Regulatory: FDA 21CFR §175.105, 175.300, 176.170, 176.180, 176.210, 177.2800
Trade names: Alkamuls® EL-719; Berol 108; Prox-onic HR-040; T-Det® C-40

PEG-75 castor oil
Uses: Surfactant, emulsifier, softener, rewetting agent, pigment dispersant, dye assistant, leveling agent for paints, textiles, leather, plastics/metal lubricants
Trade names: Berol 190

PEG-80 castor oil
CAS 61791-12-6 (generic)
Synonyms: POE (80) castor oil
Definition: PEG deriv. of castor oil with avg. 80 moles of ethylene oxide
Uses: O/w emulsifier and solubilizer; used in fiber finish and textile lubricant formulations
Trade names: Prox-onic HR-080

PEG-160 castor oil
CAS 61791-12-6 (generic)
Uses: Surfactant, emulsifier, softener, rewetting agent, pigment dispersant, dye assistant, leveling agent for paints, textiles, leather, plastics/metal lubricants
Trade names: Berol 198

PEG-200 castor oil
CAS 61791-12-6 (generic)
Synonyms: POE (200) castor oil; PEG (200) castor oil
Definition: PEG deriv. of castor oil with avg. 200 moles of ethylene oxide
Uses: Lubricant additive, emulsifier in lubricants for plastics, metals, textiles; clay and pigment dispersants, rewetting agent, softener, dyeing assistant for paint, paper, textile, and leather industries
Regulatory: FDA 21CFR §175.300
Trade names: Berol 191; Prox-onic HR-0200; Prox-onic HR-0200/50

PEG (200) castor oil. *See* PEG-200 castor oil
PEG 500 castor oil. *See* PEG-10 castor oil

PEG (1200) castor oil
Uses: Emulsifier, plasticizer, lubricant, wetting agent, dispersant, binder, thickener for emulsion polymerization, drycleaning, leather, paper, mastics, emulsions

PEG 1800 castor oil. *See* PEG-36 castor oil

PEG (1900) castor oil
Uses: Emulsifier, plasticizer, lubricant, wetting agent, dispersant, binder, thickener for emulsion polymerization, drycleaning, leather, paper, mastics, emulsions

PEG 2000 castor oil. *See* PEG-40 castor oil
PEG-6 cetyl/oleyl ether. *See* Cetoleth-6
PEG-10 cetyl/oleyl ether. *See* Cetoleth-10
PEG-13 cetyl/oleyl ether. *See* Cetoleth-13
PEG (13) cetyl/oleyl ether. *See* Cetoleth-13

PEG-20 cetyl/oleyl ether. *See* Cetoleth-20
PEG-25 cetyl/oleyl ether. *See* Cetoleth-25
PEG-4 cetyl/stearyl ether. *See* Ceteareth-4
PEG-6 cetyl/stearyl ether. *See* Ceteareth-6
PEG-10 cetyl/stearyl ether. *See* Ceteareth-10
PEG-15 cetyl/stearyl ether. *See* Ceteareth-15
PEG-20 cetyl/stearyl ether. *See* Ceteareth-20
PEG-22 cetyl/stearyl ether. *See* Ceteareth-22
PEG-25 cetyl/stearyl ether. *See* Ceteareth-25
PEG-30 cetyl/stearyl ether. *See* Ceteareth-30
PEG-50 cetyl/stearyl ether. *See* Ceteareth-50

PEG-2 cocamine
CAS 61791-14-8 (generic), 61791-31-9
Synonyms: PEG 100 coconut amine; POE (2) coconut amine; Bis (2-hydroxyethyl) cocamine; Bis (2-hydroxyethyl) coco amine
Definition: PEG deriv. of cocamine
Formula: R—N$(CH_2CH_2O)_xH(CH_2CH_2O)_yH$, R rep. alkyl groups from coconut oil, avg. (x+y)=2
Uses: Hydrophilic emulsifier, textile dyeing agent, dye leveler, antiprecipitant, stripping agent; antistat for plastics; intermediate for quats.; agric., metalworking, plastics, textiles, inks, cosmetics
Trade names: Berol 307; Chemstat® 122; Chemstat® 122/60DC; Chemstat® 273-C; Mazeen® C-2; Prox-onic MC-02
Trade names containing: Armostat® 475

PEG-5 cocamine
CAS 61791-14-8 (generic)
Synonyms: POE (5) coconut amine
Definition: PEG deriv. of cocamine
Formula: R—N$(CH_2CH_2O)_xH(CH_2CH_2O)_yH$, R rep. alkyl groups from coconut oil, avg. (x+y)=5
Uses: Hydrophilic emulsifier, textile dyeing agent, dye leveler, antiprecipitant, stripping agent; agric.; plastics antistat; intermediate for quats.
Trade names: Mazeen® C-5; Prox-onic MC-05

PEG-10 cocamine
CAS 61791-14-8 (generic)
Synonyms: POE (10) coconut amine; PEG 500 coconut amine
Definition: PEG deriv. of cocamine
Formula: R—N$(CH_2CH_2O)_xH(CH_2CH_2O)_yH$, R rep. alkyl groups from coconut oil, avg. (x+y)=10
Uses: Coemulsifier, antistat for textiles and plastics; dispersant; emulsifier in industrial lubricants, agric., inks, cosmetics
Trade names: Mazeen® C-10

PEG-15 cocamine
CAS 8051-52-3 (generic); 61791-14-8 (generic)
Synonyms: POE (15) coconut amine
Definition: PEG deriv. of cocamine
Formula: R—N$(CH_2CH_2O)_xH(CH_2CH_2O)_yH$, R rep. alkyl groups from coconut oil, avg. (x+y)=15
Uses: Hydrophilic emulsifier, textile dyeing agent, dye leveler, antiprecipitant, stripping agent; agric.; plastics antistat; intermediate for quats.
Trade names: Berol 397; Mazeen® C-15; Prox-onic MC-015

PEG-8 cocoate
CAS 61791-29-5 (generic)
Synonyms: POE (8) monococoate; PEG 400 monococoate
Definition: PEG ester of coconut acid
Formula: RCO—$(OCH_2CH_2)_nOH$, RCO- rep. fatty acids from coconut oil, avg. n = 8
Uses: Emulsifier, plasticizer, lubricant, wetting agent, dispersant, binder, thickener for emulsion polymerization, drycleaning, leather, paper, emulsions
Regulatory: FDA 21CFR §175.105, 175.300, 176.170, 176.200, 176.210, 177.1210, 177.2260, 177.2800
Trade names: Nopalcol 4-C; Nopalcol 4-CH

PEG-200 cocoate
Uses: Emulsifier for emulsion polymerization
Trade names: Polirol C5

PEG-2 coco-benzonium chloride
CAS 61789-68-2
Synonyms: PEG-2 cocobenzyl ammonium chloride; PEG 100 coco-benzonium chloride; POE (2) coco-benzonium chloride
Classification: Quaternary ammonium salt
Trade names containing: Ethoquad® CB/12

PEG 100 coco-benzonium chloride. *See* PEG-2 coco-benzonium chloride
PEG-2 cocobenzyl ammonium chloride. *See* PEG-2 coco-benzonium chloride
PEG-2 cocomethyl ammonium nitrate. *See* PEG-2 cocomonium nitrate

PEG-5 cocomonium methosulfate
CAS 68989-03-7
Synonyms: POE (5) cocomonium methosulfate; Coconut pentaethoxy methyl ammonium methyl sulfate
Classification: Quaternary ammonium salt
Formula: $[R—N(CH_2CH_2O)_xH(CH_2CH_2O)_yHCH_3]^+CH_3OSO_3^-$, R = coconut oil, avg. (x+y)=5
Uses: Conditioner for shampoos, emulsifier in emulsion polymerization, antistat
Trade names: Rewoquat CPEM

PEG-2 cocomonium nitrate
CAS 71487-00-8
Synonyms: PEG-2 cocomethyl ammonium nitrate
Trade names containing: Ethoquad® C/12 Nitrate

PEG 100 coconut amine. *See* PEG-2 cocamine
PEG 500 coconut amine. *See* PEG-10 cocamine
PEG-4 decyl ether phosphate. *See* Deceth-4 phosphate
PEG 200 decyl ether phosphate. *See* Deceth-4 phosphate

PEG-15 DEDM hydantoin oleate
Synonyms: PEG-15 di-(2-hydroxyethyl)-5,5-dimethyl hydantoin oleate; POE (15) DEDM hydantoin oleate
Definition: Ester of PEG-15 DEDM hydantoin and oleic acid
Uses: Intermediate for epoxies, urethane resins, and antistatic lubricants for the textiles industry
Trade names: Dantosperse® DHE (15) MO

PEG-3 diacetate
Synonyms: TEGDA; Triethylene glycol diacetate; POE (3) diacetate
Uses: Plasticizer for films and coatings; lubricants; cosmetic emollients
Manuf./Distrib.: Bayer
Trade names: Estol 1593

PEG-4 diacrylate
CAS 17831-71-9
Synonyms: Tetraethylene glycol diacrylate; PEG 200 diacrylate
Properties: M.w. 302.3; b.p. > 120 (0.3 mm)
Uses: Crosslinking agent for radiation-cured coatings, inks, adhesives, textile prods.
Manuf./Distrib.: Monomer-Polymer & Dajac
Trade names: Ageflex T4EGDA

PEG-8 diacrylate
Uses: Curing agent
Trade names containing: SR-344

PEG 200 diacrylate. *See* PEG-4 diacrylate
PEG-2 dibenzoate. *See* Diethylene glycol dibenzoate

PEG-4 dibenzoate
Synonyms: PEG 200 dibenzoate

PEG 100 dibenzoate

Properties: Sp.gr. 1.158; m.p. -40 C; flash pt. (COC) 248 C; ref. index 1.5252
Uses: Plasticizer for PVAc adhesive formulations, phenol-formaldehyde resins, alkyd-modified phenol-formaldehyde varnishes
Manuf./Distrib.: Kalama; Velsicol
Trade names: Benzoflex® P-200

PEG 100 dibenzoate. *See* Diethylene glycol dibenzoate
PEG 200 dibenzoate. *See* PEG-4 dibenzoate

PEG-3 dicaprylate
Synonyms: Triethylene glycol dicaprylate
Properties: Sp.gr. 0.966; m.p. -5 to -15 C; flash pt. (COC) 205 C; ref. index 1.453
Uses: Plasticizer for vinyls, cellulosics
Manuf./Distrib.: Velsicol

PEG-3 dicaprylate/caprate
Synonyms: Triethylene glycol di-caprylate-caprate
Properties: Sp.gr. 0.968; flash pt. (COC) 213 C; ref. index 1.446
Uses: Low temp. plasticizer for plastics and syn. rubbers
Manuf./Distrib.: C.P. Hall; Harwick; Hatco; Inolex; Velsicol
Trade names: Tricap

PEG-3 di-2-ethylhexanoate. *See* PEG-3 di-2-ethylhexoate
PEG-4 di 2-ethylhexanoate. *See* PEG-4 dioctanoate

PEG di-2-ethylhexoate
CAS 9004-93-7
Uses: Plasticizer for rubber
Trade names: Flexol® Plasticizer 4GO; Uniplex 809

PEG-3 di-2-ethylhexoate
CAS 94-28-0
Synonyms: Triethylene glycol di-2-ethylhexoate; PEG-3 di-2-ethylhexanoate; Bis(ethylhexanoate) triethylene glycol
Empirical: $C_{22}H_{42}O_6$
Formula: $C_7H_{15}OCOCH_2(CH_2OCH_2)_2CH_2OCOC_7H_{15}$
Properties: Lt. colored liq.; insol. in water; m.w. 402.64; dens. 0.9679 (20/20 C); b.p. 385 C; flash pt. (COC) 216 C
Uses: Plasticizer for rubber, adhesives
Precaution: Combustible
Toxicology: Mild skin irritant; LD50 (oral, rat) 31 g/kg
Manuf./Distrib.: Ferro/Keil; C.P. Hall; Harwick; Inolex
Trade names: Kodaflex® TEG-EH; TegMeR® 803

PEG-4 di-2-ethylhexoate
Synonyms: Tetraethylene glycol di-2-ethylhexoate
Properties: Sp.gr. 0.984; m.p. -65 C; flash pt. (COC) 204 C; ref. index 1.4445
Uses: Lubricant for aluminum can industry; plasticizer for PVC, PVAc, PS, cellulosics, rubber
Manuf./Distrib.: C.P. Hall; Inolex
Trade names: TegMeR® 804; TegMeR® 804 Special

PEG-8 di-2-ethylhexoate
Uses: Lubricant for aluminum can industry; plasticizer for polymers
Trade names: TegMeR® 809

PEG-3 diheptanoate
Properties: Sp.gr. 0.990; flash pt. (COC) 210 C; ref. index 1.444
Uses: Lubricant for aluminum can industry; plasticizer for PVC, PVB, cellulosics
Manuf./Distrib.: C.P. Hall; Inolex
Trade names: TegMeR® 703

PEG-4 diheptanoate
Properties: Sp.gr. 1.02; m.p. -48 C; flash pt. (COC) 262 C; ref. index 1.452
Uses: Lubricant for aluminum can industry; plasticizer for PVC, PS, cellulosics
Manuf./Distrib.: C.P. Hall; Inolex; Unitex
Trade names: TegMeR® 704

PEG-15 di-(2-hydroxyethyl)-5,5-dimethyl hydantoin oleate. *See* PEG-15 DEDM hydantoin oleate

PEG-2 dilaurate
CAS 9005-02-1 (generic); 6281-04-5; EINECS 228-486-1
Synonyms: PEG 100 dilaurate; POE (2) dilaurate
Definition: PEG diester of lauric acid
Empirical: $C_{28}H_{54}O_5$
Formula: $CH_3(CH_2)_{10}CO(OCH_2CH_2)_nOCO(CH_2)_{10}CH_3$, avg. n = 2
Properties: Sp.gr. 0.960; m.p. 9-11 C
Uses: W/o emulsifier, dispersant, antistat for textile, paper processing, cutting oils, polishes, emulsion cleaners, rubber latex, wool lubricants; plasticizer for cellulosics
Regulatory: FDA 21CFR §175.300, 176.210, 177.2800
Manuf./Distrib.: Witco/PAG
Trade names: Cithrol DGDL N/E

PEG-4 dilaurate
CAS 9005-02-1 (generic)
Synonyms: PEG 200 dilaurate; POE (4) dilaurate
Definition: PEG diester of lauric acid
Empirical: $C_{32}H_{62}O_7$
Formula: $CH_3(CH_2)_{10}CO(OCH_2CH_2)_nOCO(CH_2)_{10}CH_3$, avg. n = 4
Uses: Surfactant used as emulsifier, dispersant, solubilizer, visc. control agent, defoamer, lubricant, wetting agent, cosolvent for cosmetics, pharmaceuticals, plastics, textiles, metalworking, agric., food
Regulatory: FDA 21CFR §175.105, 175.300, 176.170, 176.180, 176.200, 176.210
Trade names: Kessco® PEG 200 DL; Nopalcol 2-DL; Pegosperse® 200 DL

PEG-6 dilaurate
CAS 9005-02-1 (generic)
Synonyms: POE (6) dilaurate; PEG 300 dilaurate
Definition: PEG diester of lauric acid
Formula: $CH_3(CH_2)_{10}CO(OCH_2CH_2)_nOCO(CH_2)_{10}CH_3$, avg. n = 6
Uses: Surfactant, solubilizer, thickener, emollient, opacifier, spreading agent, wetting agent, dispersant for cosmetic, pharmaceutical, perfume, food, agric., plastic, other industries
Regulatory: FDA 21CFR §175.105, 175.300, 176.210
Trade names: Kessco® PEG 300 DL

PEG-8 dilaurate
CAS 9005-02-1 (generic)
Synonyms: POE (8) dilaurate; PEG 400 dilaurate
Definition: PEG diester of lauric acid
Empirical: $C_{40}H_{78}O_{11}$
Formula: $CH_3(CH_2)_{10}CO(OCH_2CH_2)_nOCO(CH_2)_{10}CH_3$, avg. n = 8
Properties: Sp.gr. 1.030; m.p. 15 C; flash pt. (COC) 249 C; ref. index 1.459
Uses: Emulsifier, solubilizer, dispersing agent, wetting agent, lubricant, softener, release agent, coupling agent used in personal care prods. and industrial applics., agric. chemical sprays, industrial and textile lubricants; plasticizer
Regulatory: FDA 21CFR §175.105, 175.300, 176.210, 177.1210, 177.2260, 177.2800, 178.3520
Manuf./Distrib.: C.P. Hall; Inolex; Velsicol
Trade names: Kessco® PEG 400 DL; Pegosperse® 400 DL

PEG-12 dilaurate
CAS 9005-02-1 (generic)
Synonyms: POE (12) dilaurate; PEG 600 dilaurate
Definition: PEG diester of lauric acid
Empirical: $C_{48}H_{94}O_{15}$
Formula: $CH_3(CH_2)_{10}CO(OCH_2CH_2)_nOCO(CH_2)_{10}CH_3$, avg. n = 12

PEG-20 dilaurate

Uses: Dispersant and emulsifier for metalworking lubricants, fiber lubricants and softeners, solubilizers, defoamers, antistats, cosmetics, pharmaceuticals, plastics, agric., and chemical intermediates
Regulatory: FDA 21CFR §175.105, 175.300, 176.210, 177.2260, 177.2800
Trade names: Kessco® PEG 600 DL

PEG-20 dilaurate
CAS 9005-02-1 (generic)
Synonyms: POE (20) dilaurate; PEG 1000 dilaurate
Definition: PEG diester of lauric acid
Empirical: $C_{64}H_{126}O_{23}$
Formula: $CH_3(CH_2)_{10}CO(OCH_2CH_2)_nOCO(CH_2)_{10}CH_3$, avg. n = 20
Uses: Dispersant, emulsifier, wetting agent, cosolvent, solubilizer, thickener, emollient, opacifier, spreading agent; used in cosmetic, pharmaceutical, food, agric., plastic, other industries
Regulatory: FDA 21CFR §175.300, 176.210, 177.2260, 177.2800
Trade names: Kessco® PEG 1000 DL

PEG-32 dilaurate
CAS 9005-02-1 (generic)
Synonyms: POE (32) dilaurate; PEG 1540 dilaurate
Definition: PEG diester of lauric acid
Empirical: $C_{88}H_{174}O_{35}$
Formula: $CH_3(CH_2)_{10}CO(OCH_2CH_2)_nOCO(CH_2)_{10}CH_3$, avg. n = 32
Uses: Surfactant, solubilizer, thickener, emollient, opacifier, spreading agent, wetting agent, dispersant for cosmetic, pharmaceutical, perfume, food, agric., plastic, other industries
Regulatory: FDA 21CFR §175.300, 176.210, 177.2260, 177.2800
Trade names: Kessco® PEG 1540 DL

PEG-75 dilaurate
CAS 9005-02-1 (generic)
Synonyms: POE (75) dilaurate; PEG 4000 dilaurate
Definition: PEG diester of lauric acid
Formula: $CH_3(CH_2)_{10}CO(OCH_2CH_2)_nOCO(CH_2)_{10}CH_3$, avg. n = 75
Uses: Surfactant, solubilizer, thickener, emollient, opacifier, spreading agent, wetting agent, dispersant for cosmetic, pharmaceutical, perfume, food, agric., plastic, other industries
Regulatory: FDA 21CFR §175.300
Trade names: Kessco® PEG 4000 DL

PEG 100 dilaurate. *See* PEG-2 dilaurate

PEG-150 dilaurate
CAS 9005-02-1 (generic)
Synonyms: POE (150) dilaurate; PEG 6000 dilaurate
Definition: PEG diester of lauric acid
Formula: $CH_3(CH_2)_{10}CO(OCH_2CH_2)_nOCO(CH_2)_{10}CH_3$, avg. n = 150
Uses: Surfactant, solubilizer, thickener, emollient, opacifier, spreading agent, wetting agent, dispersant for cosmetic, pharmaceutical, perfume, food, agric., plastic, other industries
Regulatory: FDA 21CFR §175.300
Trade names: Kessco® PEG 6000 DL

PEG 200 dilaurate. *See* PEG-4 dilaurate
PEG 300 dilaurate. *See* PEG-6 dilaurate
PEG 400 dilaurate. *See* PEG-8 dilaurate
PEG 600 dilaurate. *See* PEG-12 dilaurate
PEG 1000 dilaurate. *See* PEG-20 dilaurate
PEG 1540 dilaurate. *See* PEG-32 dilaurate
PEG 4000 dilaurate. *See* PEG-75 dilaurate
PEG 6000 dilaurate. *See* PEG-150 dilaurate

PEG-2 dilaurate SE
Uses: W/o emulsifier, dispersant, antistat for textile, paper processing, cutting oils, polishes, emulsion cleaners, rubber latex, wool lubricants
Trade names: Cithrol DGDL S/E

PEG-3 dimethacrylate
CAS 109-16-0
Synonyms: Triethylene glycol dimethacrylate
Formula: $[H_2C=C(CH_3)CO_2CH_2CH_2OCH_2]_2$
Properties: M.w. 286.2; b.p. 162 C (1.2 mm)
Uses: Crosslinking monomeric ester; in vinyl plastisols reduces initial visc. and oil extractability, and improves ultimate hardness, heat distort., hot tear strength, and stain resistance
Manuf./Distrib.: CPS; Monomer-Polymer & Dajac; Rohm Tech
Trade names: MFM-413
Trade names containing: SR-205

PEG-4 dimethacrylate
CAS 25852-47-5
Synonyms: Tetraethylene glycol dimethacrylate; PEG 200 dimethacrylate
Formula: $H_2C=C(CH_3)CO(OCH_2CH_2)_nO_2CC(CH_3)$
Properties: M.w. 330.4; b.p. 220 (1.0 mm)
Precaution: Light-sensitive
Uses: Crosslinking agent used in castings, plastisols, coatings, fibers, papers, etc.
Manuf./Distrib.: Monomer-Polymer & Dajac
Trade names: MFM-425
Trade names containing: SR-209

PEG-12 dimethacrylate
Trade names containing: SR-252

PEG 200 dimethacrylate. *See* PEG-4 dimethacrylate

PEG-3 dimethyl ether
CAS 112-49-2; 24991-55-7
Synonyms: Triethylene glycol dimethyl ether
Uses: Complexing agent, solv.; in electrochem., polymer/boron chem., gas absorp., extraction, stabilization, fuels, lubricants, textiles, pharmaceuticals, pesticides
Manuf./Distrib.: Brand-Nu Labs; Hoechst AG
Trade names: Triglyme

PEG-8 dinonyl phenyl ether. *See* Nonyl nonoxynol-8
PEG-9 dinonyl phenyl ether. *See* Nonyl nonoxynol-9
PEG-15 dinonyl phenyl ether. *See* Nonyl nonoxynol-15
PEG-24 dinonyl phenyl ether. *See* Nonyl nonoxynol-24
PEG-49 dinonyl phenyl ether. *See* Nonyl nonoxynol-49
PEG-9 dinonyl phenyl ether phosphate. *See* Nonyl nonoxynol-9 phosphate
PEG 450 dinonyl phenyl ether phosphate. *See* Nonyl nonoxynol-9 phosphate

PEG-4 dioctanoate
Synonyms: PEG-4 di 2-ethylhexanoate; Tetraethylene glycol di 2-ethylhexanoate
Definition: PEG diester of 2-ethylhexanoic acid
Uses: Lubricant for polyester and nylon textile and industrial yarns; softener for syn. rubber and other elastomers

PEG-3 dioctoate
Synonyms: Triethylene glycol dioctoate
Uses: Emulsifier, lubricant, antistat, defoamer for metalworking, textile lubricants, plastics, paper; cosmetic emollient and pearlescent

PEG-2 dioleate
CAS 9005-07-6 (generic); 52668-97-0 (generic)
Synonyms: Diethylene glycol dioleate; POE (2) dioleate; PEG 100 dioleate
Definition: PEG diester of oleic acid
Uses: W/o emulsifier, dispersant, antistat for textile, paper processing, cutting oils, polishes, emulsion cleaners, rubber latex, wool lubricants
Trade names: Cithrol DGDO N/E

PEG-4 dioleate
CAS 9005-07-6 (generic); 52668-97-0 (generic); 134141-38-1
Synonyms: POE (4) dioleate; PEG 200 dioleate
Definition: PEG diester of oleic acid
Uses: Surfactant; coemulsifier, thickener, solubilizer, emollient, opacifier, spreading agent, wetting agent, dispersant for cosmetics, pharmaceuticals, food, agric., plastics, etc.
Regulatory: FDA 21CFR §173.340, 175.105, 175.300, 176.210
Trade names: Kessco® PEG 200 DO

PEG-6 dioleate
CAS 9005-07-6 (generic); 52668-97-0 (generic)
Synonyms: POE (6) dioleate; PEG 300 dioleate
Definition: PEG diester of oleic acid
Uses: Surfactant, solubilizer, thickener, emollient, opacifier, spreading agent, wetting agent, dispersant for cosmetic, pharmaceutical, perfume, food, agric., plastic, other industries
Regulatory: FDA 21CFR §175.105, 175.300, 176.210
Trade names: Kessco® PEG 300 DO

PEG-8 dioleate
CAS 9005-07-6 (generic); 52668-97-0 (generic)
Synonyms: POE (8) dioleate; PEG 400 dioleate
Definition: PEG diester of oleic acid
Uses: Emulsifier, solubilizer, lubricant, dispersant, defoamer, anticorrosive, wetting agent for cosmetic, pharmaceutical, textile, metalworking, agric., plastics, foods
Regulatory: FDA 21CFR §173.340, 175.105, 175.300, 176.170, 176.200, 176.210, 177.1210, 177.2260, 177.2800
Trade names: Kessco® PEG 400 DO; Pegosperse® 400 DO

PEG-12 dioleate
CAS 9005-07-6 (generic); 52668-97-0 (generic); 85736-49-8; EINECS 288-459-5
Synonyms: POE (12) dioleate; PEG 600 dioleate
Definition: PEG diester of oleic acid
Uses: Emulsifier, dispersant; additive for cutting oils; component of defoamers, softeners, lubricants; cosmetics, pharmaceuticals, foods, agric., plastics, etc.
Regulatory: FDA 21CFR §173.340, 175.105, 175.300, 176.200, 176.210, 177.2260, 177.2800
Trade names: Kessco® PEG 600 DO; Nopalcol 6-DO

PEG-20 dioleate
CAS 9005-07-6 (generic); 52668-97-0 (generic)
Synonyms: POE (20) dioleate; PEG 1000 dioleate
Definition: PEG diester of oleic acid
Uses: Surfactant, solubilizer, thickener, emollient, opacifier, spreading agent, wetting agent, dispersant for cosmetic, pharmaceutical, perfume, food, agric., plastic, other industries
Regulatory: FDA 21CFR §175.300, 176.210, 177.2260, 177.2800
Trade names: Kessco® PEG 1000 DO

PEG-32 dioleate
CAS 9005-07-6 (generic); 52668-97-0 (generic)
Synonyms: POE (32) dioleate; PEG 1540 dioleate
Definition: PEG diester of oleic acid
Uses: Surfactant, solubilizer, thickener, emollient, opacifier, spreading agent, wetting agent, dispersant for cosmetic, pharmaceutical, perfume, food, agric., plastic, other industries
Regulatory: FDA 21CFR §175.300, 176.210, 177.2260, 177.2800
Trade names: Kessco® PEG 1540 DO

PEG-75 dioleate
CAS 9005-07-6 (generic); 52668-97-0 (generic)
Synonyms: POE (75) dioleate; PEG 4000 dioleate
Definition: PEG diester of oleic acid
Uses: Surfactant, solubilizer, thickener, emollient, opacifier, spreading agent, wetting agent, dispersant for cosmetic, pharmaceutical, perfume, food, agric., plastic, other industries
Regulatory: FDA 21CFR §175.300
Trade names: Kessco® PEG 4000 DO

PEG 100 dioleate. *See* PEG-2 dioleate

PEG-150 dioleate
CAS 9005-07-6 (generic); 52668-97-0 (generic)
Synonyms: POE (150) dioleate; PEG 6000 dioleate
Definition: PEG diester of oleic acid
Uses: Surfactant, solubilizer, thickener, emollient, opacifier, spreading agent, wetting agent, dispersant for cosmetic, pharmaceutical, perfume, food, agric., plastic, other industries
Regulatory: FDA 21CFR §175.300
Trade names: Kessco® PEG 6000 DO

PEG 200 dioleate. *See* PEG-4 dioleate
PEG 300 dioleate. *See* PEG-6 dioleate
PEG 400 dioleate. *See* PEG-8 dioleate
PEG 600 dioleate. *See* PEG-12 dioleate
PEG 1000 dioleate. *See* PEG-20 dioleate
PEG 1540 dioleate. *See* PEG-32 dioleate
PEG 4000 dioleate. *See* PEG-75 dioleate
PEG 6000 dioleate. *See* PEG-150 dioleate

PEG-2 dioleate SE
Uses: W/o emulsifier, dispersant, antistat for textile, paper processing, cutting oils, polishes, emulsion cleaners, rubber latex, wool lubricants
Trade names: Cithrol DGDO S/E

PEG-3 dipelargonate
CAS 106-06-9
Synonyms: Triethylene glycol dipelargonate
Properties: Sp.gr. 0.97; flash pt. (COC) 224 C; ref. index 1.447
Uses: Low temp. plasticizer for syn. elastomers and natural rubbers; lubricant for aluminum can industry
Manuf./Distrib.: C.P. Hall; Henkel; Inolex
Trade names: Plastolein® 9404; TegMeR® 903

PEG-2 distearate
CAS 109-30-8; 52668-97-0; EINECS 203-663-6
Synonyms: POE (2) distearate; PEG 100 distearate; Diethylene glycol distearate
Definition: PEG diester of stearic acid
Empirical: $C_{40}H_{78}O_5$
Formula: $CH_3(CH_2)_{16}CO(OCH_2CH_2)_nOCO(CH_2)_{16}CH_3$, avg. n = 2
Properties: Sp.gr. 0.96; m.p. 48 C
Uses: W/o emulsifier, dispersant, antistat for textile, paper processing, cutting oils, polishes, emulsion cleaners, rubber latex, wool lubricants; plasticizer for cellulosics
Regulatory: FDA 21CFR §73.1, 175.300, 176.210, 177.2800
Manuf./Distrib.: CasChem; Witco/PAG
Trade names: Cithrol DGDS N/E

PEG-4 distearate
CAS 9005-08-7 (generic); 52668-97-0; 142-20-1
Synonyms: POE (4) distearate; PEG 200 distearate
Definition: PEG diester of stearic acid
Formula: $CH_3(CH_2)_{16}CO(OCH_2CH_2)_nOCO(CH_2)_{16}CH_3$, avg. n = 4
Properties: M.p. 35-37 C; flash pt. > 110 C
Uses: Emollient, detergent, emulsifier, opacifier, visc. builder for cosmetics; lubricant and softener for textiles; pharmaceuticals; food; agric.; plastics
Regulatory: FDA 21CFR §175.105, 175.300, 176.210
Trade names: Kessco® PEG 200 DS

PEG-6 distearate
CAS 9005-08-7 (generic); 52668-97-0
Synonyms: POE (6) distearate; PEG 300 distearate
Definition: PEG diester of stearic acid
Formula: $CH_3(CH_2)_{16}CO(OCH_2CH_2)_nOCO(CH_2)_{16}CH_3$, avg. n = 6

PEG-8 distearate

Properties: M.p. 35-37 C; flash pt. > 110 C
Uses: Surfactant, solubilizer, thickener, emollient, opacifier, spreading agent, wetting agent, dispersant for cosmetic, pharmaceutical, perfume, food, agric., plastic, other industries
Regulatory: FDA 21CFR §175.105, 175.300, 176.210
Trade names: Kessco® PEG 300 DS

PEG-8 distearate
CAS 9005-08-7 (generic); 52668-97-0
Synonyms: POE (8) distearate; PEG 400 distearate
Definition: PEG diester of stearic acid
Formula: $CH_3(CH_2)_{16}CO(OCH_2CH_2)_nOCO(CH_2)_{16}CH_3$, avg. n = 8
Properties: M.p. 35-37 C; flash pt. > 110 C
Uses: Emulsifier, opacifier, and thickener for cosmetic and industrial emulsions; lubricant and softener in textile applic.; plasticizer for plastics
Regulatory: FDA 21CFR §175.105, 175.300, 176.210, 177.1210, 177.2260, 177.2800
Trade names: Hefti PGE-400-DS; Kessco® PEG 400 DS

PEG-12 distearate
CAS 9005-08-7 (generic); 52668-97-0
Synonyms: POE (12) distearate; PEG 600 distearate
Definition: PEG diester of stearic acid
Formula: $CH_3(CH_2)_{16}CO(OCH_2CH_2)_nOCO(CH_2)_{16}CH_3$, avg. n = 12
Properties: M.p. 35-37 C; flash pt. > 110 C
Uses: Emollient, detergent, emulsifier, visc. builder for cosmetics, pharmaceuticals, food, agric., textiles, metalworking; plasticizer for plastics
Regulatory: FDA 21CFR §175.105, 175.300, 176.210, 177.2260, 177.2800
Trade names: Hefti PGE-600-DS; Kessco® PEG 600 DS

PEG-20 distearate
CAS 9005-08-7 (generic); 52668-97-0
Synonyms: POE (20) distearate; PEG 1000 distearate
Definition: PEG diester of stearic acid
Formula: $CH_3(CH_2)_{16}CO(OCH_2CH_2)_nOCO(CH_2)_{16}CH_3$, avg. n = 20
Properties: M.p. 35-37 C; flash pt. > 110 C
Uses: Surfactant, solubilizer, thickener, emollient, opacifier, spreading agent, wetting agent, dispersant for cosmetic, pharmaceutical, perfume, food, agric., plastic, other industries
Regulatory: FDA 21CFR §175.300, 176.210, 177.2260, 177.2800
Trade names: Kessco® PEG 1000 DS

PEG-32 distearate
CAS 9005-08-7 (generic); 52668-97-0
Synonyms: POE (32) distearate; PEG 1540 distearate
Definition: PEG diester of stearic acid
Formula: $CH_3(CH_2)_{16}CO(OCH_2CH_2)_nOCO(CH_2)_{16}CH_3$, avg. n = 32
Properties: M.p. 35-37 C; flash pt. > 110 C
Uses: Surfactant, solubilizer, thickener, emollient, opacifier, spreading agent, wetting agent, dispersant for cosmetic, pharmaceutical, perfume, food, agric., plastic, other industries
Regulatory: FDA 21CFR §175.300, 176.210, 177.2260, 177.2800
Trade names: Kessco® PEG 1540 DS

PEG-75 distearate
CAS 9005-08-7 (generic); 52668-97-0
Synonyms: POE (75) distearate; PEG 4000 distearate
Definition: PEG diester of stearic acid
Formula: $CH_3(CH_2)_{16}CO(OCH_2CH_2)_nOCO(CH_2)_{16}CH_3$, avg. n = 75
Properties: M.p. 35-37 C; flash pt. > 110 C
Uses: Surfactant, solubilizer, thickener, emollient, opacifier, spreading agent, wetting agent, dispersant for cosmetic, pharmaceutical, perfume, food, agric., plastic, other industries
Regulatory: FDA 21CFR §175.300
Trade names: Kessco® PEG 4000 DS

PEG 100 distearate. *See* PEG-2 distearate

PEG-150 distearate
CAS 9005-08-7 (generic); 52668-97-0
Synonyms: POE (150) distearate; PEG 6000 distearate
Definition: PEG diester of stearic acid
Formula: $CH_3(CH_2)_{16}CO(OCH_2CH_2)_nOCO(CH_2)_{16}CH_3$, avg. n = 150
Properties: M.p. 35-37 C; flash pt. > 110 C
Uses: Thickener, dispersant, emollient, emulsifier in cosmetics, pharmaceuticals, food, agric., plastics, textiles, metalworking
Regulatory: FDA 21CFR §175.300
Trade names: Kessco® PEG 6000 DS

PEG 200 distearate. *See* PEG-4 distearate
PEG 300 distearate. *See* PEG-6 distearate
PEG 400 distearate. *See* PEG-8 distearate
PEG 600 distearate. *See* PEG-12 distearate
PEG 1000 distearate. *See* PEG-20 distearate
PEG 1540 distearate. *See* PEG-32 distearate
PEG 4000 distearate. *See* PEG-75 distearate
PEG 6000 distearate. *See* PEG-150 distearate

PEG-2 distearate SE
Uses: W/o emulsifier, dispersant, antistat for textile, paper processing, cutting oils, polishes, emulsion cleaners, rubber latex, wool lubricants
Trade names: Cithrol DGDS S/E

PEG-12 ditallowate
Uses: Emulsifier, plasticizer, lubricant, wetting agent, dispersant, binder, thickener for emulsion polymerization, drycleaning, leather, paper, emulsions
Trade names: Nopalcol 6-DTW

PEG-7 glyceryl cocoate
CAS 66105-29-1; 68201-46-7 (generic)
Synonyms: POE (7) glyceryl monococoate; PEG (7) glyceryl monococoate
Definition: PEG ether of glyceryl cocoate
Formula: $RCO—OCH_2COHHCH_2(OCH_2CH_2)_nOH$, RCO- rep. fatty acids from coconut oil, avg. n=7
Uses: Emollient oil, superfatting agent, emulsifier, solubilizer, coupler for personal care prods., pharmaceuticals, lubricants; dispersant for biologically act. ingreds.; mold release; plasticizer for fabrics and plastics
Regulatory: FDA 21CFR §175.300
Trade names: Mazol® 159

PEG (7) glyceryl monococoate. *See* PEG-7 glyceryl cocoate
PEG 1000 glyceryl monostearate. *See* PEG-20 glyceryl stearate

PEG-20 glyceryl stearate
CAS 68153-76-4, 68553-11-7; 51158-08-8
Synonyms: PEG 1000 glyceryl monostearate; POE (20) glyceryl monostearate; Polyglycerate 60???
Definition: PEG ether of glyceryl stearate
Formula: $CH_3(CH_2)_{16}COOCH_2COHHCH_2(OCH_2CH_2)_nOH$, avg. n = 20
Uses: Emulsifier, solubilizer, suspending and dispersing agent, antifoam, wetting agent for personal care prods., foods, pharmaceuticals, industrial applics.; dough conditioner; lubricant and fabric softener; chemical intermediate; lubricant and mold release in plastics

PEG-5 hydrogenated castor oil
CAS 61788-85-0 (generic)
Synonyms: POE (5) hydrogenated castor oil; PEG (5) hydrogenated castor oil
Definition: PEG deriv. of hydrogenated castor oil with avg. 5 moles of ethylene oxide
Uses: Lubricant additive, emulsifier in lubricants for plastics, metals, textiles; clay and pigment dispersants, rewetting agent, softener, dyeing assistant for paint, paper, textile, and leather industries
Trade names: Chemax HCO-5; Prox-onic HRH-05

PEG (5) hydrogenated castor oil. *See* PEG-5 hydrogenated castor oil

PEG-10 hydrogenated castor oil
CAS 61788-85-0 (generic)
Synonyms: POE (10) hydrogenated castor oil; PEG 500 hydrogenated castor oil
Definition: PEG deriv. of hydrogenated castor oil with avg. 10 moles of ethylene oxide
Uses: Emulsifier, solubilizer, hydrotrope, emollient, superfatting agent, detergent used for cosmetics, textiles, metalworking fluids, emulsion polymerization, insecticides, herbicides, household detergents
Trade names: Croduret 10

PEG-16 hydrogenated castor oil
CAS 61788-85-0 (generic)
Synonyms: POE (16) hydrogenated castor oil
Definition: PEG deriv. of hydrogenated castor oil with avg. 16 moles of ethylene oxide
Uses: Lubricant additive, emulsifier in lubricants for plastics, metals, textiles; clay and pigment dispersants, rewetting agent, softener, dyeing assistant for paint, paper, textile, and leather industries
Regulatory: FDA 21CFR §177.2800
Trade names: Chemax HCO-16; Prox-onic HRH-016

PEG-20 hydrogenated castor oil
CAS 61788-85-0 (generic)
Synonyms: POE (20) hydrogenated castor oil
Definition: PEG deriv. of hydrogenated castor oil with avg. 20 moles of ethylene oxide
Uses: Emulsifier, plasticizer, hydrotrope, lubricant, wetting agent, dispersant, binder, thickener for emulsion polymerization, drycleaning, leather, paper, emulsions, cosmetics, and pharmaceuticals
Regulatory: FDA 21CFR §177.2800
Trade names: Nopalcol 10-COH

PEG-25 hydrogenated castor oil
CAS 61788-85-0 (generic)
Synonyms: POE (25) hydrogenated castor oil
Definition: PEG deriv. of hydrogenated castor oil with avg. 25 moles of ethylene oxide
Uses: Lubricant additive, emulsifier in lubricants for plastics, metals, textiles; clay and pigment dispersants, rewetting agent, softener, dyeing assistant for paint, paper, textile, and leather industries
Regulatory: FDA 21CFR §177.2800
Trade names: Chemax HCO-25; Mapeg® CO-25H; Prox-onic HRH-025

PEG-30 hydrogenated castor oil
CAS 61788-85-0 (generic)
Synonyms: POE (30) hydrogenated castor oil
Definition: PEG deriv. of hydrogenated castor oil with avg. 30 moles of ethylene oxide
Uses: Emulsifier, solubilizer, emollient, superfatting agent, detergent used for cosmetics, textiles, metalworking fluids, emulsion polymerization, insecticides, herbicides, household detergents
Regulatory: FDA 21CFR §177.2800
Trade names: Croduret 30

PEG-40 hydrogenated castor oil
CAS 61788-85-0 (generic)
Synonyms: POE (40) hydrogenated castor oil
Definition: PEG deriv. of hydrogenated castor oil with avg. 40 moles of ethylene oxide
Uses: Solubilizer, emulsifier, emollient for cosmetics, pharmaceuticals, textiles, metalworking, emulsion polymerization, agric., household detergents
Regulatory: FDA 21CFR §177.2800
Trade names: Croduret 40

PEG-60 hydrogenated castor oil
CAS 61788-85-0 (generic)
Synonyms: POE (60) hydrogenated castor oil
Definition: PEG deriv. of hydrogenated castor oil with avg. 60 moles of ethylene oxide
Uses: Solubilizer, emulsifier, emollient for cosmetics, pharmaceuticals, textiles, metalworking, emulsion polymerization, agric., household detergents
Regulatory: FDA 21CFR §177.2800
Trade names: Croduret 60

PEG-100 hydrogenated castor oil
CAS 61788-85-0 (generic)
Synonyms: POE (100) hydrogenated castor oil; PEG (100) hydrogenated castor oil
Definition: PEG deriv. of hydrogenated castor oil with avg. 100 moles of ethylene oxide
Uses: Solubilizer, emulsifier, emollient for cosmetics, pharmaceuticals, textiles, metalworking, emulsion polymerization, agric., household detergents
Trade names: Croduret 100

PEG (100) hydrogenated castor oil. *See* PEG-100 hydrogenated castor oil

PEG-200 hydrogenated castor oil
CAS 61788-85-0 (generic)
Synonyms: POE (200) hydrogenated castor oil; PEG (200) hydrogenated castor oil
Definition: PEG deriv. of hydrogenated castor oil with avg. 200 moles of ethylene oxide
Uses: Lubricant additive, emulsifier in lubricants for plastics, metals, textiles; clay and pigment dispersants, rewetting agent, softener, dyeing assistant for paint, paper, textile, and leather industries
Trade names: Chemax HCO-200/50; Prox-onic HRH-0200; Prox-onic HRH-0200/50

PEG (200) hydrogenated castor oil. *See* PEG-200 hydrogenated castor oil
PEG 500 hydrogenated castor oil. *See* PEG-10 hydrogenated castor oil

PEG (1200) hydrogenated castor oil
Uses: Emulsifier, plasticizer, lubricant, wetting agent, dispersant, binder, thickener for emulsion polymerization, drycleaning, leather, paper, emulsions

PEG-5 hydrogenated tallow amine
CAS 61791-26-2 (generic)
Synonyms: POE (5) tallow amine
Definition: PEG amine of hydrogenated tallow
Formula: $R—N(CH_2CH_2O)_xH(CH_2CH_2O)_yH$, R rep. alkyl groups fr hydrog. tallow, avg. (x+y)=5
Uses: Hydrophilic emulsifier, textile dyeing agent, dye leveler, antiprecipitant, stripping agent; intermediate for quats.; emulsifier and antistat in metal buffing, agric., rubber compds.; lubricant for fiberglass
Regulatory: FDA 21CFR §178.3910
Trade names: Chemeen HT-5; Prox-onic MHT-05

PEG-15 hydrogenated tallow amine
CAS 61791-26-2 (generic)
Synonyms: POE (15) tallow amine
Definition: PEG amine of hydrogenated tallow
Formula: $R—N(CH_2CH_2O)_xH(CH_2CH_2O)_yH$, R rep. alkyl groups fr hydrog. tallow, avg. (x+y)=15
Uses: Hydrophilic emulsifier, lubricant additive, antistat for textiles, metal, plastics, agric.; textile dyeing agent, dye leveler, antiprecipitant, stripping agent; intermediate for quats.
Trade names: Prox-onic MHT-015

PEG-50 hydrogenated tallow amine
CAS 61791-26-2 (generic); 68783-22-2
Synonyms: POE (50) tallow amine
Definition: PEG amine of hydrogenated tallow
Formula: $R—N(CH_2CH_2O)_xH(CH_2CH_2O)_yH$, R rep. alkyl groups fr hydrog. tallow, avg. (x+y)=50
Uses: Emulsifier and antistat in textiles, metal buffing and rubber compds., lubricant for fiber glass
Trade names: Chemeen HT-50

PEG-60 hydrogenated tallowate
Uses: Emulsifier, plasticizer, lubricant, wetting agent, dispersant, binder, thickener for emulsion polymerization, drycleaning, leather, paper, emulsions
Trade names: Nopalcol 30-TWH

PEG-3 lauramine oxide
CAS 59355-61-2
Synonyms: POE (3) lauryl dimethyl amine oxide; PEG (3) lauryl dimethyl amine oxide
Classification: Tertiary amine oxide
Formula: $CH_3(CH_2)_{10}CH_2N(CH_2CH_2O)_x(CH_2CH_2O)_yO$, avg. (x+y) = 3

PEG-2 laurate

Uses: Coactive, foam booster/stabilizer and visc. modifier for personal care prods., fire fighting foam concs.; foam rubber; bleach additive
Trade names: Empigen® OY

PEG-2 laurate
CAS 141-20-8; 9004-81-3; EINECS 205-468-1
Synonyms: Diethylene glycol laurate; Diglycol laurate; PEG 100 monolaurate
Definition: PEG ester of lauric acid
Formula: $CH_3(CH_2)_{10}CO(OCH_2CH_2)_nOH$, avg. n = 2
Uses: W/o emulsifier, dispersant, antistat, defoamer for textile, paper processing, cutting oils, polishes, emulsion cleaners, emulsion polymerization, rubber latex, wool lubricants, paints
Regulatory: FDA 21CFR §175.105, 175.300, 176.200, 176.210, 177.2800, 178.3910
Manuf./Distrib.: Henkel/Emery; Inolex; Karlshamns; Lonza; Mona; Stepan; Witco
Trade names: Cithrol DGML N/E; Nopalcol 1-L

PEG-4 laurate
CAS 9004-81-3 (generic); 10108-24-4
Synonyms: POE (4) monolaurate; PEG 200 monolaurate
Definition: PEG ester of lauric acid
Empirical: $C_{20}H_{40}O_6$
Formula: $CH_3(CH_2)_{10}CO(OCH_2CH_2)_nOH$, avg. n = 4
Uses: Emulsifier, lubricant, dispersing and leveling agent, coupling agent, solubilizer, wetting agent, thickener, defoamer used in cosmetic, pharmaceuticals, textile, plastics, paint and other industrial uses
Regulatory: FDA 21CFR §175.105, 175.300, 176.210, 178.3910
Trade names: Kessco® PEG 200 ML; Pegosperse® 200 ML

PEG-6 laurate
CAS 9004-81-3 (generic); 2370-64-1; EINECS 219-136-9
Synonyms: POE (6) monolaurate; PEG 300 monolaurate
Definition: PEG ester of lauric acid
Empirical: $C_{24}H_{48}O_8$
Formula: $CH_3(CH_2)_{10}CO(OCH_2CH_2)_nOH$, avg. n = 6
Uses: Hydrophilic emulsifier; lubricant component and scrooping agent for textile fibers and yarns; visc. control agent for plastisols
Regulatory: FDA 21CFR §175.105, 175.300, 176.210, 178.3910
Trade names: Kessco® PEG 300 ML

PEG-8 laurate
CAS 9004-81-3 (generic); 35179-86-3; 37318-14-2; EINECS 253-458-0
Synonyms: POE (8) monolaurate; PEG 400 monolaurate
Definition: PEG ester of lauric acid
Empirical: $C_{28}H_{56}O_{10}$
Formula: $CH_3(CH_2)_{10}CO(OCH_2CH_2)_nOH$, avg. n = 8
Uses: Emulsifier, lubricant, dispersing and leveling agent, solubilizer, visc. control agent, defoamer used in cosmetic, textile, paint and other industrial uses; antiblock in vinyls; plasticizer
Regulatory: FDA 21CFR §175.105, 175.300, 176.170, 176.210, 177.1200, 177.1210, 177.2260, 177.2800, 178.3520, 178.3760, 178.3910
Trade names: Emerest® 2650; Kessco® PEG 400 ML; Nopalcol 4-L; Pegosperse® 400 ML

PEG-12 laurate
CAS 9004-81-3 (generic)
Synonyms: POE (12) monolaurate; PEG 600 monolaurate
Definition: PEG ester of lauric acid
Formula: $CH_3(CH_2)_{10}CO(OCH_2CH_2)_nOH$, avg. n = 12
Uses: Emulsifier, lubricant, dispersing and leveling agent, solubilizer used in cosmetic, textile, paint and other industrial uses; plastics antistat
Regulatory: FDA 21CFR §175.105, 175.300, 176.170, 176.210, 177.1200, 177.2260, 177.2800, 178.3910
Trade names: Hefti PGE-600-ML; Kessco® PEG 600 ML; Nopalcol 6-L

PEG-20 laurate
CAS 9004-81-3 (generic)
Synonyms: POE (20) monolaurate; PEG 1000 monolaurate
Definition: PEG ester of lauric acid

Formula: $CH_3(CH_2)_{10}CO(OCH_2CH_2)_nOH$, avg. n = 20
Uses: Emulsifier for cosmetics and pharmaceuticals; solubilizer for perfumes; food, agric., plastics
Regulatory: FDA 21CFR §175.300, 176.210, 177.2260, 177.2800, 178.3910
Trade names: Kessco® PEG 1000 ML

PEG-32 laurate
CAS 9004-81-3 (generic)
Synonyms: POE (32) monolaurate; PEG 1540 monolaurate
Definition: PEG ester of lauric acid
Formula: $CH_3(CH_2)_{10}CO(OCH_2CH_2)_nOH$, avg. n = 32
Uses: Surfactant, solubilizer, thickener, emollient, opacifier, spreading agent, wetting agent, dispersant for cosmetic, pharmaceutical, perfume, food, agric., plastic, other industries
Regulatory: FDA 21CFR §175.300, 176.210, 177.2260, 177.2800, 178.3910
Trade names: Kessco® PEG 1540 ML

PEG-75 laurate
CAS 9004-81-3 (generic)
Synonyms: POE (75) monolaurate; PEG 4000 monolaurate
Definition: PEG ester of lauric acid
Formula: $CH_3(CH_2)_{10}CO(OCH_2CH_2)_nOH$, avg. n = 75
Uses: Emulsifier, dispersant, thickener, solubilizer, emollient, opacifier, wetting agent, dispersant for cosmetics, pharmaceuticals, food, agric., plastics
Regulatory: FDA 21CFR §175.300, 178.3910
Trade names: Kessco® PEG 4000 ML

PEG-150 laurate
CAS 9004-81-3 (generic)
Synonyms: POE (150) monolaurate; PEG 6000 monolaurate
Definition: PEG ester of lauric acid
Formula: $CH_3(CH_2)_{10}CO(OCH_2CH_2)_nOH$, avg. n = 150
Uses: Wetting agent, emulsifier, detergent, thickener, solubilizer, dispersant, softener, lubricant, antistat, dye asistant, penetrant for cosmetics, textiles, glass fiber, metal treatment, agric., plastics, other industries
Regulatory: FDA 21CFR §175.300, 178.3910
Trade names: Kessco® PEG 6000 ML

PEG-2 laurate SE
CAS 141-20-8
Synonyms: Diethylene glycol monolaurate self-emulsifying; PEG 100 monolaurate self-emulsifying; POE (2) monolaurate self-emulsifying
Definition: Self-emulsifying grade of PEG-2 laurate containing some sodium and/or potassium laurate
Uses: Spreading agent, w/o emulsifier, dispersant, lubricant, opacifier, emulsion stabilizer, emollient, visc. builder, antistat used in cosmetics, textiles, paper processing, cutting oils, polishes, rubber latex; defoamer for process applics.
Trade names: Cithrol DGML S/E; Pegosperse® 100 L

PEG (3) lauryl dimethyl amine oxide. *See* PEG-3 lauramine oxide
PEG-2 lauryl ether. *See* Laureth-2
PEG-3 lauryl ether. *See* Laureth-3
PEG-4 lauryl ether. *See* Laureth-4
PEG-7 lauryl ether. *See* Laureth-7
PEG-9 lauryl ether. *See* Laureth-9
PEG-12 lauryl ether. *See* Laureth-12
PEG-15 lauryl ether. *See* Laureth-15
PEG-16 lauryl ether. *See* Laureth-16
PEG-23 lauryl ether. *See* Laureth-23
PEG 200 lauryl ether. *See* Laureth-4
PEG 600 lauryl ether. *See* Laureth-12
PEG-9 lauryl ether sulfate, ammonium salt. *See* Ammonium laureth-9 sulfate
PEG (1-4) lauryl ether sulfate, sodium salt. *See* Sodium laureth sulfate
PEG (12) lauryl ether sulfate, sodium salt. *See* Sodium laureth-12 sulfate
PEG (30) lauryl ether sulfate, sodium salt. *See* Sodium laureth-30 sulfate

PEG 600 lauryl ether sulfate, sodium salt. *See* Sodium laureth-12 sulfate

PEG-5 methacrylate
Uses: Monomer for emulsion polymerization; thickener, dispersant, suspending agent; copolymerizable stabilizer; bound protective colloid; modifying comonomer/crosslinker
Trade names: Sipomer® HEM-5

PEG-10 methacrylate
Uses: Monomer for emulsion polymerization; thickener, dispersant, suspending agent; copolymerizable stabilizer; bound protective colloid; modifying comonomer/crosslinker
Trade names: Sipomer® HEM-10

PEG 400 monococoate. *See* PEG-8 cocoate
PEG 100 monolaurate. *See* PEG-2 laurate
PEG 200 monolaurate. *See* PEG-4 laurate
PEG 300 monolaurate. *See* PEG-6 laurate
PEG 400 monolaurate. *See* PEG-8 laurate
PEG 600 monolaurate. *See* PEG-12 laurate
PEG 1000 monolaurate. *See* PEG-20 laurate
PEG 1540 monolaurate. *See* PEG-32 laurate
PEG 4000 monolaurate. *See* PEG-75 laurate
PEG 6000 monolaurate. *See* PEG-150 laurate
PEG 100 monolaurate self-emulsifying. *See* PEG-2 laurate SE
PEG 200 monooleate. *See* PEG-4 oleate
PEG 300 monooleate. *See* PEG-6 oleate
PEG 400 monooleate. *See* PEG-8 oleate
PEG 600 monooleate. *See* PEG-12 oleate
PEG 1000 monooleate. *See* PEG-20 oleate
PEG 1540 monooleate. *See* PEG-32 oleate
PEG 4000 monooleate. *See* PEG-75 oleate
PEG 6000 monooleate. *See* PEG-150 oleate
PEG 100 monooleate self-emulsifying. *See* PEG-2 oleate SE
PEG 100 monostearate. *See* PEG-2 stearate
PEG 200 monostearate. *See* PEG-4 stearate
PEG 300 monostearate. *See* PEG-6 stearate
PEG 400 monostearate. *See* PEG-8 stearate
PEG 600 monostearate. *See* PEG-12 stearate
PEG 1000 monostearate. *See* PEG-20 stearate
PEG 1500 monostearate. *See* PEG-6-32 stearate
PEG 4000 monostearate. *See* PEG-75 stearate
PEG 6000 monostearate. *See* PEG-150 stearate
PEG 100 monostearate self-emulsifying. *See* PEG-2 stearate SE
PEG-1 nonyl phenyl ether. *See* Nonoxynol-1
PEG-3 nonyl phenyl ether. *See* Nonoxynol-3
PEG-4 nonyl phenyl ether. *See* Nonoxynol-4
PEG-5 nonyl phenyl ether. *See* Nonoxynol-5
PEG-6 nonyl phenyl ether. *See* Nonoxynol-6
PEG-7 nonyl phenyl ether. *See* Nonoxynol-7
PEG-8 nonyl phenyl ether. *See* Nonoxynol-8
PEG-9 nonyl phenyl ether. *See* Nonoxynol-9
PEG-10 nonyl phenyl ether. *See* Nonoxynol-10
PEG-12 nonyl phenyl ether. *See* Nonoxynol-12
PEG-13 nonyl phenyl ether. *See* Nonoxynol-13
PEG-14 nonyl phenyl ether. *See* Nonoxynol-14
PEG-20 nonyl phenyl ether. *See* Nonoxynol-20
PEG-30 nonyl phenyl ether. *See* Nonoxynol-30
PEG-35 nonyl phenyl ether. *See* Nonoxynol-35
PEG-40 nonyl phenyl ether. *See* Nonoxynol-40
PEG-50 nonyl phenyl ether. *See* Nonoxynol-50
PEG-70 nonyl phenyl ether. *See* Nonoxynol-70
PEG-100 nonyl phenyl ether. *See* Nonoxynol-100
PEG-120 nonyl phenyl ether. *See* Nonoxynol-120
PEG 200 nonyl phenyl ether. *See* Nonoxynol-4

PEG 300 nonyl phenyl ether. *See* Nonoxynol-6
PEG 400 nonyl phenyl ether. *See* Nonoxynol-8
PEG 450 nonyl phenyl ether. *See* Nonoxynol-9
PEG 500 nonyl phenyl ether. *See* Nonoxynol-10
PEG 600 nonyl phenyl ether. *See* Nonoxynol-12
PEG 1000 nonyl phenyl ether. *See* Nonoxynol-20
PEG 2000 nonyl phenyl ether. *See* Nonoxynol-40
PEG-6 nonyl phenyl ether phosphate. *See* Nonoxynol-6 phosphate
PEG-10 nonyl phenyl ether phosphate. *See* Nonoxynol-10 phosphate
PEG 300 nonyl phenyl ether phosphate. *See* Nonoxynol-6 phosphate
PEG 500 nonyl phenyl ether phosphate. *See* Nonoxynol-10 phosphate
PEG-4 nonyl phenyl ether sulfate, sodium salt. *See* Sodium nonoxynol-4 sulfate
PEG-6 nonyl phenyl ether sulfate, sodium salt. *See* Sodium nonoxynol-6 sulfate
PEG-8 nonyl phenyl ether sulfate, sodium salt. *See* Sodium nonoxynol-8 sulfate
PEG 200 nonyl phenyl ether sulfate, sodium salt. *See* Sodium nonoxynol-4 sulfate
PEG 300 nonyl phenyl ether sulfate, sodium salt. *See* Sodium nonoxynol-6 sulfate
PEG 400 nonyl phenyl ether sulfate, sodium salt. *See* Sodium nonoxynol-8 sulfate
PEG-1 octyl phenyl ether. *See* Octoxynol-1
PEG-3 octyl phenyl ether. *See* Octoxynol-3
PEG-7 octyl phenyl ether. *See* Octoxynol-7
PEG-9 octyl phenyl ether. *See* Octoxynol-9
PEG-10 octyl phenyl ether. *See* Octoxynol-10
PEG-16 octyl phenyl ether. *See* Octoxynol-16
PEG-25 octyl phenyl ether. *See* Octoxynol-25
PEG-30 octyl phenyl ether. *See* Octoxynol-30
PEG-33 octyl phenyl ether. *See* Octoxynol-33
PEG-40 octyl phenyl ether. *See* Octoxynol-40
PEG-70 octyl phenyl ether. *See* Octoxynol-70
PEG 450 octyl phenyl ether. *See* Octoxynol-9
PEG 500 octyl phenyl ether. *See* Octoxynol-10
PEG-9 octyl phenyl ether carboxylic acid. *See* Octoxynol-9 carboxylic acid
PEG 450 octyl phenyl ether carboxylic acid. *See* Octoxynol-9 carboxylic acid
PEG-6 octyl phenyl ether sulfate, sodium salt. *See* Sodium octoxynol-6 sulfate
PEG 300 octyl phenyl ether sulfate, sodium salt. *See* Sodium octoxynol-6 sulfate

PEG-2 oleamine
CAS 13127-82-7
Synonyms: PEG-2 oleyl amine; Bis-2-hydroxyethyl oleamine; POE (2) oleyl amine; PEG 100 oleyl amine
Definition: PEG amine of oleic acid
Uses: Hydrophilic emulsifier, dispersant, wetting agent, antistat, anticorrosive for agric., leather, metalworking, plastics; textile dyeing agent, dye leveler, antiprecipitant, stripping agent; intermediate for quats.
Trade names: Berol 302; Chemstat® 172T; Prox-onic MO-02

PEG-7 oleamine
CAS 26635-93-8
Uses: Emulsifier, dispersant, wetting agent, antistat, anticorrosive for agric., leather, textiles, metalworking and plastic industries
Trade names: Berol 28

PEG-12 oleamine
CAS 26635-93-8
Uses: Emulsifier, wetting agent, antistat, anticorrosive for agric., leather, textiles, metalworking and plastic industries
Trade names: Berol 303

PEG-15 oleamine
Synonyms: POE (15) oleyl amine
Definition: PEG amine of oleic acid
Uses: Hydrophilic emulsifier, lubricant additive, antistat, detergent for metal, plastics, agric.; intermediate for quats.; textile dyeing agent, dye leveler, antiprecipitant, stripping agent
Trade names: Prox-onic MO-015

PEG-30 oleamine
CAS 58253-49-9
Synonyms: POE (30) oleyl amine
Definition: PEG amine of oleic acid
Uses: Hydrophilic emulsifier, lubricant additive, antistat, detergent for metal, plastics, agric.; textile dyeing agent, dye leveler, antiprecipitant, stripping agent; intermediate for quats.
Trade names: Prox-onic MO-030; Prox-onic MO-030-80

PEG-2 oleate
CAS 106-12-7; EINECS 203-364-0
Synonyms: Diethylene glycol monooleate; Diglycol oleate; POE (2) monooleate
Definition: PEG ester of oleic acid
Empirical: $C_{22}H_{42}O_4$
Formula: $CH_3(CH_2)_7CHCH(CH_2)_7CO(OCH_2CH_2)_nOH$, avg. n = 2
Uses: Emulsifier, dispersant, antistat for cosmetic, textile, paper processing, cutting oils, polishes, emulsion cleaners, rubber latex, wool lubricants; leather softener
Regulatory: FDA 21CFR §175.105, 175.300, 176.210, 177.2800
Manuf./Distrib.: Henkel/Emery; Inolex; Karlshamns; Lipo; Lonza; Mona; Witco
Trade names: Cithrol DGMO N/E

PEG-4 oleate
CAS 9004-96-0 (generic); 10108-25-5; EINECS 233-293-0
Synonyms: POE (4) monooleate; PEG 200 monooleate
Definition: PEG ester of oleic acid
Empirical: $C_{26}H_{50}O_6$
Formula: $CH_3(CH_2)_7CHCH(CH_2)_7CO(OCH_2CH_2)_nOH$, avg. n = 4
Uses: Wetting agent, penetrant, spreading agent, defoamer, detergent, emulsifier, solubilizer, thickening agent, dispersant, textile aux., softener, lubricant for textiles, cosmetics, metalworking, food, agric., plastics
Regulatory: FDA 21CFR §175.105, 175.300, 176.210
Trade names: Kessco® PEG 200 MO

PEG-6 oleate
CAS 9004-96-0 (generic); 60344-26-5
Synonyms: POE (6) monooleate; PEG 300 monooleate
Definition: PEG ester of oleic acid
Empirical: $C_{30}H_{58}O_8$
Formula: $CH_3(CH_2)_7CHCH(CH_2)_7CO(OCH_2CH_2)_nOH$, avg. n = 6
Uses: Emulsifier, lubricant, chemical intermediate, antifoam, dispersant; for cosmetics, food, agriculture, plastics, textiles
Regulatory: FDA 21CFR §175.105, 175.300, 176.210
Trade names: Acconon 300-MO; Kessco® PEG 300 MO

PEG-8 oleate
CAS 9004-96-0 (generic)
Synonyms: POE (8) monooleate; PEG 400 monooleate
Definition: PEG ester of oleic acid
Formula: $CH_3(CH_2)_7CHCH(CH_2)_7CO(OCH_2CH_2)_nOH$, avg. n = 8
Uses: Emulsifier, dispersant, lubricant, chemical intermediate, solubilizer, visc. control agent; for cosmetics, pharmaceuticals, food, agric., plastics
Regulatory: FDA 21CFR §175.105, 175.300, 176.170, 176.200, 177.1200, 177.1210, 177.2260, 177.2800
Trade names: Acconon 400-MO; Cithrol A; Ethylan® A4; Kessco® PEG 400 MO; Nopalcol 4-O

PEG-12 oleate
CAS 9004-96-0 (generic)
Synonyms: POE (12) monooleate; PEG 600 monooleate
Definition: PEG ester of oleic acid
Formula: $CH_3(CH_2)_7CHCH(CH_2)_7CO(OCH_2CH_2)_nOH$, avg. n = 12
Uses: Dispersant, emulsifier, solubilizer, detergent, dye leveling agent for cosmetic and industrial applics.; plastics antistat
Regulatory: FDA 21CFR §175.105, 175.300, 176.170, 176.200, 177.1200, 177.2260, 177.2800
Trade names: Ethylan® A6; Kessco® PEG 600 MO

PEG-20 oleate
CAS 9004-96-0 (generic)
Synonyms: POE (20) monooleate; PEG 1000 monooleate
Definition: PEG ester of oleic acid
Formula: $CH_3(CH_2)_7CHCH(CH_2)_7CO(OCH_2CH_2)_nOH$, avg. n = 20
Uses: Raw material for finishing agents in the syn. fiber industry; emulsifier, thickener, solubilizer, emollient, opacifier, wetting agent, dispersant for cosmetics, pharmaceuticals, food, agric., plastics, etc.
Regulatory: FDA 21CFR §175.300, 176.200, 177.2260, 177.2800
Trade names: Kessco® PEG 1000 MO

PEG-32 oleate
CAS 9004-96-0 (generic)
Synonyms: POE (32) monooleate; PEG 1540 monooleate
Definition: PEG ester of oleic acid
Formula: $CH_3(CH_2)_7CHCH(CH_2)_7CO(OCH_2CH_2)_nOH$, avg. n = 32
Uses: Surfactant, solubilizer, thickener, emollient, opacifier, spreading agent, wetting agent, dispersant for cosmetic, pharmaceutical, perfume, food, agric., plastic, other industries
Regulatory: FDA 21CFR §175.300, 176.200, 177.2261, 177.2800
Trade names: Kessco® PEG 1540 MO

PEG-75 oleate
CAS 9004-96-0 (generic)
Synonyms: POE (75) monooleate; PEG 4000 monooleate
Definition: PEG ester of oleic acid
Formula: $CH_3(CH_2)_7CHCH(CH_2)_7CO(OCH_2CH_2)_nOH$, avg. n = 75
Uses: Surfactant, solubilizer, thickener, emollient, opacifier, spreading agent, wetting agent, dispersant for cosmetic, pharmaceutical, perfume, food, agric., plastic, other industries
Regulatory: FDA 21CFR §175.300, 176.200
Trade names: Kessco® PEG 4000 MO

PEG-150 oleate
CAS 9004-96-0 (generic)
Synonyms: POE (150) monooleate; PEG 6000 monooleate
Definition: PEG ester of oleic acid
Formula: $CH_3(CH_2)_7CHCH(CH_2)_7CO(OCH_2CH_2)_nOH$, avg. n = 150
Uses: Strongly hydrophilic emulsifier, stabilizer, lubricant, solubilizer, emollient, wetting agent, dispersant for cosmetics, pharmaceuticals, food, agric., plastics
Regulatory: FDA 21CFR §175.300, 176.200
Trade names: Kessco® PEG 6000 MO

PEG-2 oleate SE
CAS 106-12-7
Synonyms: Diethylene glycol monooleate self-emulsifying; PEG 100 monooleate self-emulsifying; POE (2) monooleate self-emulsifying
Definition: Self-emulsifying grade of PEG-2 oleate containing some sodium and/or potassium oleate
Uses: Emulsifier, dispersant, antistat for textile, paper processing, cutting oils, polishes, emulsion cleaners, rubber latex, wool lubricants
Trade names: Cithrol DGMO S/E

PEG-2 oleyl amine. *See* PEG-2 oleamine
PEG 100 oleyl amine. *See* PEG-2 oleamine
PEG-4 oleyl ether. *See* Oleth-4
PEG-7 oleyl ether. *See* Oleth-7
PEG-9 oleyl ether. *See* Oleth-9
PEG-20 oleyl ether. *See* Oleth-20
PEG 200 oleyl ether. *See* Oleth-4
PEG 450 oleyl ether. *See* Oleth-9
PEG 1000 oleyl ether. *See* Oleth-20
PEG-4 oleyl ether phosphate. *See* Oleth-4 phosphate
PEG 200 oleyl ether phosphate. *See* Oleth-4 phosphate

PEG-12 ricinoleate
CAS 9004-97-1 (generic)

PEG-40 sorbitan hexaoleate

Synonyms: POE (12) monoricinoleate
Definition: PEG ester of ricinoleic acid
Formula: $CHCH_2COHH(CH_2)_5CH_3CH(CH_2)_7CO(OCH_2CH_2)_nOH$, avg. n = 12
Uses: Emulsifier, plasticizer, lubricant, wetting agent, dispersant, binder, thickener for emulsion polymerization, drycleaning, leather, paper, emulsions
Regulatory: FDA 21CFR §173.340
Trade names: Nopalcol 6-R

PEG-40 sorbitan hexaoleate
CAS 57171-56-9
Synonyms: PEG-40 sorbitol hexaoleate; POE (40) sorbitol hexaoleate; PEG 2000 sorbitol hexaoleate
Definition: Oleic acid hexaester of ethoxylated sorbitol with avg. 40 moles ethylene oxide
Uses: O/w emulsifier for metal lubricants, textiles, cosmetics, polymerization
Regulatory: FDA 21CFR §175.300, 176.210
Trade names: G-1086

PEG-50 sorbitan hexaoleate
CAS 57171-56-9
Synonyms: PEG-50 sorbitol hexaoleate; POE (50) sorbitol hexaoleate
Definition: Oleic acid hexaester of ethoxylated sorbitol with an avg. 50 moles ethylene oxide
Uses: Emulsifier, coupling agent for cosmetics, textiles, metalworking fluids, polymerization
Regulatory: FDA 21CFR §175.300
Trade names: G-1096

PEG-4 sorbitan laurate. *See* Polysorbate 21
PEG-20 sorbitan laurate. *See* Polysorbate 20
PEG-5 sorbitan oleate. *See* Polysorbate 81
PEG-20 sorbitan oleate. *See* Polysorbate 80
PEG-4 sorbitan stearate. *See* Polysorbate 61
PEG-20 sorbitan stearate. *See* Polysorbate 60

PEG-30 sorbitan tetraoleate
Synonyms: POE (30) sorbitan tetraoleate
Definition: Tetraester of oleic acid and a PEG ether of sorbitol, avg. 30 moles ethylene oxide
Uses: Emulsifier, solubilizer, superfatting agent used in drugs and cosmetics, for emulsion polymerization, agric. chemicals, printing inks
Regulatory: FDA 21CFR §175.300
Trade names: Nikkol GO-430

PEG-40 sorbitan tetraoleate
CAS 9003-11-6
Synonyms: POE (40) sorbitan tetraoleate; PEG 2000 sorbitan tetraoleate
Definition: Tetraester of oleic acid and a PEG ether of sorbitol, avg. 40 moles ethylene oxide
Formula: $(C_3H_6O{\bullet}C_2H_4O)_x$
Toxicology: Moderately toxic by ingestion and intraperitoneal route
Uses: Emulsifier, solubilizer, superfatting agent used in drugs and cosmetics, for emulsion polymerization, agric. chemicals, printing inks
Regulatory: FDA 21CFR §175.300, 176.210
Trade names: Nikkol GO-440

PEG-60 sorbitan tetraoleate
Synonyms: POE (60) sorbitan tetraoleate
Definition: Tetraester of oleic acid and a PEG ether of sorbitol, avg. 60 moles ethylene oxide
Uses: Emulsifier, solubilizer, superfatting agent used in drugs and cosmetics, for emulsion polymerization, agric. chemicals, printing inks
Regulatory: FDA 21CFR §175.300
Trade names: Nikkol GO-460

PEG 2000 sorbitan tetraoleate. *See* PEG-40 sorbitan tetraoleate

PEG-17 sorbitan trioleate
CAS 9005-70-3
Definition: Triester of oleic acid and a PEG ether of sorbitol, avg. 17 moles ethylene oxide

Uses: O/w and w/o emulsifier for min. oils, veg. oils, train oils, waxes, etc.; for cattle feed, textiles, biocides, paints, varnishes, plastics, leather, fur, tech. applics., cosmetics, pharmaceuticals
Trade names: Hefti TO-55-EL

PEG-18 sorbitan trioleate
CAS 9005-70-3
Definition: Triester of oleic acid and a PEG ether of sorbitol, avg. 18 moles ethylene oxide
Properties: M.w. 1838.60; dens. 1.028; flash pt. > 110 C; ref. index 1.4680 (20 C)
Uses: O/w and w/o emulsifier for min. oils, veg. oils, train oils, waxes, etc.; for cattle feed, textiles, biocides, paints, varnishes, plastics, leather, fur, tech. applics., cosmetics, pharmaceuticals
Manuf./Distrib.: Aldrich
Trade names: Hefti TO-55-E

PEG-20 sorbitan trioleate. *See* Polysorbate 85
PEG-20 sorbitan tristearate. *See* Polysorbate 65
PEG-40 sorbitol hexaoleate. *See* PEG-40 sorbitan hexaoleate
PEG-50 sorbitol hexaoleate. *See* PEG-50 sorbitan hexaoleate
PEG 2000 sorbitol hexaoleate. *See* PEG-40 sorbitan hexaoleate
PEG 100 soya amine. *See* PEG-2 soyamine
PEG 500 soya amine. *See* PEG-10 soyamine

PEG-2 soyamine
CAS 61791-24-0 (generic)
Synonyms: POE (2) soya amine; PEG 100 soya amine; Bis (2-hydroxyethyl) soya amine
Classification: Ethoxylated amine
Definition: PEG amine of soya acid
Formula: $R—N(CH_2CH_2O)_xH(CH_2CH_2O)_yH$, R rep. alkyl groups from soy, avg. (x+y)=2
Toxicology: Moderately toxic by ingestion; eye irritant
Uses: Wetting agent, dispersant, emulsifier, antistat, lubricant; formulation of leather dressing and metal cleaning compds., fiber lubricants, in the bldg. industry; plastics antistat
Trade names: Mazeen® S-2

PEG-5 soyamine
CAS 61791-24-0 (generic)
Synonyms: POE (5) soya amine
Definition: PEG amine of soya acid
Formula: $R—N(CH_2CH_2O)_xH(CH_2CH_2O)_yH$, R rep. alkyl groups from soy, avg. (x+y)=5
Toxicology: Moderately toxic by ingestion; eye irritant
Uses: Wetting agent, dispersant, emulsifier, antistat, lubricant; used in metal, cleaning, leather dressing and dye leveling compds., agric. sprays, fiber lubricants; plastics antistat
Trade names: Mazeen® S-5

PEG-10 soyamine
CAS 61791-24-0 (generic)
Synonyms: POE (10) soya amine; PEG 500 soya amine
Definition: PEG amine of soya acid
Formula: $R—N(CH_2CH_2O)_xH(CH_2CH_2O)_yH$, R rep. alkyl groups from soy, avg. (x+y)=10
Toxicology: Moderately toxic by ingestion; eye irritant
Uses: Wetting agent, dispersant, emulsifier, antistat; used in leather dressing compds., fiber lubricant; wax emulsions for fiber board pkg. and particle board for bldg. industry; plastics antistat
Trade names: Mazeen® S-10

PEG-15 soyamine
CAS 61791-24-0 (generic)
Synonyms: POE (15) soya amine
Definition: PEG amine of soya acid
Formula: $R—N(CH_2CH_2O)_xH(CH_2CH_2O)_yH$, R rep. alkyl groups from soy, avg. (x+y)=15
Toxicology: Moderately toxic by ingestion; eye irritant
Uses: Desizing agent, antistat; emulsifier in agriculture, waxes and oils, leather processing, and metal cleaning industries; water repellent and wet spinning assistant in textile industries; plastics antistat
Trade names: Mazeen® S-15

PEG-2 stearamine

CAS 10213-78-2; EINECS 233-520-3

Synonyms: PEG-2 stearyl amine; PEG 100 stearyl amine; N,N-Bis (2-hydroxyethyl) stearyl amine; Bis-2-hydroxyethyl stearamine

Definition: PEG deriv. of stearyl amine

Empirical: $C_{22}H_{46}NO_2$

Formula: $CH_3(CH_2)_{16}CH_2N(CH_2CH_2O)_xH(CH_2CH_2O)_yH$, avg. (x+y)=2

Uses: Hydrophilic emulsifier, antistat in cosmetics, textiles, metal buffing, rubber compds., agric.; plastics antistat; lubricant for fiberglass; intermediate for quats.; textile dyeing agent, dye leveler, antiprecipitant, stripping agent

Trade names: Chemeen 18-2; Chemstat® 192; Chemstat® 192/NCP; Chemstat® 273-E; Prox-onic MS-02

PEG-5 stearamine

CAS 26635-92-7

Synonyms: POE (5) stearyl amine

Definition: PEG deriv. of stearyl amine

Formula: $CH_3(CH_2)_{16}CH_2N(CH_2CH_2O)_xH(CH_2CH_2O)_yH$, avg. (x+y)=5

Uses: Hydrophilic emulsifier, antistat for textile, metal, plastics, agric.; textile dyeing agent, dye leveler, antiprecipitant, stripping agent; intermediate for quats.

Trade names: Chemeen 18-5; Prox-onic MS-05; Teric 18M5

PEG-10 stearamine

CAS 26635-92-7

Synonyms: POE (10) stearyl amine; PEG 500 stearyl amine

Definition: PEG amine of stearic acid

Formula: $CH_3(CH_2)_{16}CH_2N(CH_2CH_2O)_xH(CH_2CH_2O)_yH$, avg. (x+y)=10

Uses: Emulsifier, dispersant used in textile processing, agric., paper, leather; softener, antistat for plastics

Trade names: Teric 18M10

PEG-11 stearamine

CAS 26635-92-7

Uses: Hydrophilic emulsifier, antistat, lubricant additive, detergent for textile, metal, plastics, agric.; textile dyeing assistant, antiprecipitant; stripping agent and dye leveler; intermediate

Trade names: Prox-onic MS-011

PEG-50 stearamine

CAS 26635-92-7

Synonyms: POE (50) stearyl amine; Ethoxylated 1-aminooctadecane

Definition: PEG amine of stearic acid

Formula: $CH_3(CH_2)_{16}CH_2N(CH_2CH_2O)_xH(CH_2CH_2O)_yH$, avg. (x+y)=50

Uses: Hydrophilic emulsifier, textile dyeing agent, dye leveler, antiprecipitant, stripping agent; agric.; plastics antistat; intermediate for quats.

Trade names: Chemeen 18-50; Ethomeen® 18/60; Prox-onic MS-050; Trymeen® 6617

PEG-2 stearate

CAS 106-11-6; 9004-99-3 (generic); 85116-97-8; EINECS 203-363-5; 285-550-1

Synonyms: Diethylene glycol stearate; Diglycol stearate; PEG 100 monostearate

Definition: PEG ester of stearic acid

Empirical: $C_{22}H_{44}O_4$

Formula: $CH_3(CH_2)_{16}CO(OCH_2CH_2)_nOH$, avg. n = 2

Toxicology: Poison by intravenous, intraperitoneal route; mildly toxic by ingestion

Uses: Emulsifier, plasticizer, lubricant, wetting agent, binding and thickening agent, dispersant, antistat, opacifier, pearlescent, stabilizer used in cosmetics, dry cleaning, leather, textile industries, paper processing, rubber

Regulatory: FDA 21CFR §175.300, 176.200, 176.210, 177.2800

Manuf./Distrib.: Henkel/Emery; Inolex; Karlshamns; Lipo; Lonza; Stepan; Witco

Trade names: Cithrol DGMS N/E

PEG-4 stearate

CAS 106-07-0; 9004-99-3 (generic); EINECS 203-358-8

Synonyms: POE (4) stearate; PEG 200 monostearate

Definition: PEG ester of stearic acid

Empirical: $C_{26}H_{52}O_6$

Formula: $CH_3(CH_2)_{16}CO(OCH_2CH_2)_nOH$, avg. n = 4
Toxicology: Poison by intravenous, intraperitoneal route; mildly toxic by ingestion
Uses: Emulsifier, lubricant, dispersing and leveling agent, solubilizer, thickener, softener used in cosmetic, textile, paint, agric., plastics, food
Regulatory: FDA 21CFR §175.105, 175.300, 176.210
Trade names: Kessco® PEG 200 MS

PEG-6 stearate
CAS 9004-99-3 (generic); 10108-28-8
Synonyms: POE (6) stearate; PEG 300 monostearate
Definition: PEG ester of stearic acid
Empirical: $C_{30}H_{60}O_8$
Formula: $CH_3(CH_2)_{16}CO(OCH_2CH_2)_nOH$, avg. n = 6
Toxicology: Poison by intravenous, intraperitoneal route; mildly toxic by ingestion
Uses: Waxy emulsifier for oils and fats in industrial lubricants; softener and lubricant in textiles and leather; base for cosmetic lotions; also in pharmaceuticals, food, agric., plastics
Regulatory: FDA 21CFR §175.105, 175.300, 176.210
Trade names: Kessco® PEG 300 MS

PEG-6-32 stearate
CAS 9004-99-3 (generic)
Synonyms: PEG 1500 monostearate; POE 1500 monostearate
Definition: PEG ester of stearic acid
Toxicology: Poison by intravenous, intraperitoneal route; mildly toxic by ingestion
Uses: Spreading agent, emulsifier, dispersant, and lubricant for cosmetics, pharmaceuticals, metalworking and fiber lubricants
Trade names containing: Pegosperse® 1500 MS

PEG-8 stearate
CAS 9004-99-3 (generic); 70802-40-3
Synonyms: POE (8) stearate; PEG 400 monostearate
Definition: PEG ester of stearic acid
Empirical: $C_{34}H_{68}O_{10}$
Formula: $CH_3(CH_2)_{16}CO(OCH_2CH_2)_nOH$, avg. n = 8
Toxicology: Poison by intravenous, intraperitoneal route; mildly toxic by ingestion
Uses: Emulsifier, lubricant, dispersing and leveling agent used in cosmetic, textile, paint and other industrial uses; plastics antistat
Regulatory: FDA 21CFR §175.105, 175.300, 176.170, 176.200, 176.210, 177.1200, 177.1210, 177.2260, 177.2800, 178.3910; Europe listed; UK approved
Trade names: Alkamuls® S-8; Kessco® PEG 400 MS; Myrj® 45; Nopalcol 4-S; Pegosperse® 400 MS

PEG-9 stearate
CAS 9004-99-3 (generic); 5349-52-0; EINECS 226-312-9
Synonyms: POE (9) stearate
Definition: PEG ester of stearic acid
Empirical: $C_{36}H_{72}O_{11}$
Formula: $CH_3(CH_2)_{16}CO(OCH_2CH_2)_nOH$, avg. n = 9
Toxicology: Poison by intravenous, intraperitoneal route; mildly toxic by ingestion
Uses: Surfactant in cutting oils, degreasing solvs., metal cleaners; emulsifier; dyeing assistant; cosmetics, textile processing
Regulatory: FDA 21CFR §175.105, 175.300, 177.2260, 177.2800
Trade names: Serdox NSG 400

PEG-12 stearate
CAS 9004-99-3 (generic)
Synonyms: POE (12) stearate; PEG 600 monostearate
Definition: PEG ester of stearic acid
Formula: $CH_3(CH_2)_{16}CO(OCH_2CH_2)_nOH$, avg. n = 12
Toxicology: Poison by intravenous, intraperitoneal route; mildly toxic by ingestion
Uses: Emulsifier, lubricant, dispersing and leveling agent, defoamer, leveling agent, visc. modifier used in cosmetic, textile, paints, food, agric., plastics, pharmaceuticals
Regulatory: FDA 21CFR §175.105, 175.300, 176.170, 176.210, 177.1200, 177.2260, 177.2800
Trade names: Kessco® PEG 600 MS; Nopalcol 6-S

PEG-20 stearate
CAS 9004-99-3 (generic)
Synonyms: POE (20) stearate; PEG 1000 monostearate
Definition: PEG ester of stearic acid
Formula: $CH_3(CH_2)_{16}CO(OCH_2CH_2)_nOH$, avg. n = 20
Properties: Sol. in ethanol; partly sol. in propylene glycol; disp. in glycerin; insol. in water; m.p. 39.5-42.5 C; sapon. no. 40-50
Toxicology: Poison by intravenous, intraperitoneal route; mildly toxic by ingestion
Uses: Emulsifier, thickener, solubilzier, emollient, opacifier, wetting agent, dispersant for cosmetics, pharmaceuticals, food, agric., plastics
Regulatory: FDA 21CFR §175.300, 176.210, 177.2260, 177.2800
Trade names: Kessco® PEG 1000 MS

PEG-32 stearate
CAS 9004-99-3 (generic)
Synonyms: PEG 1540 stearate
Definition: PEG ester of stearic acid
Formula: $CH_3(CH_2)_{16}CO(OCH_2CH_2)_nOH$, avg. n = 32
Toxicology: Poison by intravenous, intraperitoneal route; mildly toxic by ingestion
Uses: Surfactant, solubilizer, thickener, emollient, opacifier, spreading agent, wetting agent, dispersant for cosmetic, pharmaceutical, perfume, food, agric., plastic, other industries
Regulatory: FDA 21CFR §175.300, 176.210, 177.2260, 177.2800
Trade names: Kessco® PEG 1540 MS

PEG-50 stearate
CAS 9004-99-3 (generic)
Synonyms: POE (50) stearate
Definition: PEG ester of stearic acid
Formula: $CH_3(CH_2)_{16}CO(OCH_2CH_2)_nOH$, avg. n = 50
Toxicology: Poison by intravenous, intraperitoneal route; mildly toxic by ingestion
Uses: Hydrophilic emulsifier for preparing solubilized oils; visc. modifier, softener or plasticizer in acrylic or vinyl resin emulsions
Regulatory: FDA 21CFR §175.300, 177.2260, 177.2800
Trade names: Emerest® 2675; Trydet SA-50/30

PEG-75 stearate
CAS 9004-99-3 (generic)
Synonyms: POE (75) stearate; PEG 4000 monostearate
Definition: PEG ester of stearic acid
Formula: $CH_3(CH_2)_{16}CO(OCH_2CH_2)_nOH$, avg. n = 75
Toxicology: Poison by intravenous, intraperitoneal route; mildly toxic by ingestion
Uses: Surfactant, solubilizer, thickener, emollient, opacifier, spreading agent, wetting agent, dispersant for cosmetic, pharmaceutical, perfume, food, agric., plastic, other industries
Regulatory: FDA 21CFR §175.300
Trade names: Kessco® PEG 4000 MS

PEG-150 stearate
CAS 9004-99-3 (generic)
Synonyms: POE (150) stearate; PEG 6000 monostearate
Definition: PEG ester of stearic acid
Formula: $CH_3(CH_2)_{16}CO(OCH_2CH_2)_nOH$, avg. n = 150
Toxicology: Poison by intravenous, intraperitoneal route; mildly toxic by ingestion
Uses: Surfactant, emulsifier, emollient, solubilizer, thickener, wetting agent, dispersant for cosmetics, pharmaceuticals, food, agric., plastics
Regulatory: FDA 21CFR §175.300
Trade names: Kessco® PEG 6000 MS

PEG-1500 stearate
Uses: Wetting aid, lubricant, opacifier, antistat, dispersant, o/w emulgent, detergent aid, defoamer, plasticizer, rust inhibitor, visc. modifier, antifog for cosmetics, pharmaceuticals, lubricants

PEG 1540 stearate. *See* PEG-32 stearate

PEG-2 stearate SE
CAS 106-11-6
Synonyms: POE (2) monostearate self-emulsifying; Diethylene glycol monostearate self-emulsifying; PEG 100 monostearate self-emulsifying
Definition: Self-emulsifying grade of PEG-2 stearate
Uses: Emulsifier, dispersant, antistat for textile, paper processing, cutting oils, polishes, emulsion cleaners, rubber latex, wool lubricants
Trade names: Cithrol DGMS S/E; Kemester® 5221SE; Pegosperse® 100-S

PEG-2 stearyl amine. *See* PEG-2 stearamine
PEG 100 stearyl amine. *See* PEG-2 stearamine
PEG 500 stearyl amine. *See* PEG-10 stearamine
PEG-2 tallow acid ester. *See* PEG-2 tallowate

PEG-2 tallowamine
CAS 61791-44-4
Synonyms: Bis (2-hydroxyethyl) tallow amine
Uses: Hydrophilic emulsifier, wetting agent, antistat, anticorrosive for agric., leather, textiles, metalworking, plastics; intermediate for quats.; textile dyeing agent, dye leveler, antiprecipitant, stripping agent
Trade names: Berol 456; Chemstat® 182; Chemstat® 182/67DC; Mazeen® T-2; Prox-onic MT-02
Trade names containing: Armostat® 375

PEG-5 tallowamine
CAS 61791-44-4
Uses: Hydrophilic emulsifier, wetting agent, antistat, anticorrosive for agric., leather, textiles, metalworking, plastics; intermediate for quats.; textile dyeing agent, dye leveler, antiprecipitant, stripping agent
Trade names: Berol 391; Berol 457; Mazeen® T-5; Prox-onic MT-05

PEG-10 tallowamine
Uses: Hydrophilic emulsifier, wetting agent, antistat, anticorrosive for agric., leather, textiles, metalworking, plastics; intermediate for quats.; textile dyeing agent, dye leveler, antiprecipitant, stripping agent
Trade names: Berol 389; Berol 458; Mazeen® T-10

PEG-15 tallowamine
Uses: Hydrophilic emulsifier, wetting agent, antistat, anticorrosive for agric., leather, textiles, metalworking, plastics; intermediate for quats.; textile dyeing agent, dye leveler, antiprecipitant, stripping agent
Trade names: Berol 381; Berol 392; Mazeen® T-15; Prox-onic MT-015

PEG-20 tallowamine
Uses: Hydrophilic emulsifier, wetting agent, antistat, anticorrosive for agric., leather, textiles, metalworking, plastics; intermediate for quats.; textile dyeing agent, dye leveler, antiprecipitant, stripping agent
Trade names: Berol 386; Prox-onic MT-020

PEG-40 tallowamine
Uses: Emulsifier, antistat, wetting agent, anticorrosive for agric., leather, textiles, metalworking, plastics; dyeing assistant, stabilizer for latices
Trade names: Berol 387

PEG-2 tallowate
CAS 68153-64-0 (generic)
Synonyms: Diethylene glycol monotallowate; POE (2) tallowate; PEG-2 tallow acid ester
Definition: PEG ester of tallow acid
Formula: $RCO—(OCH_2CH_2)_nOH$, RCO- rep. alkyl groups from tallow, avg. n = 2
Uses: Emulsifier, plasticizer, lubricant, wetting agent, binding and thickening agent, used in cosmetics, dry cleaning, leather, textile industries, emulsion polymerization
Trade names: Nopalcol 1-TW

PEG-3 tallow diamine
Uses: Hydrophilic emulsifier, wetting agent, antistat, anticorrosive for agric., leather, textiles, metalworking, plastics; intermediate for quats.; textile dyeing agent, dye leveler, antiprecipitant, stripping agent
Trade names: Berol 455; Prox-onic DT-03

PEG-15 tallow diamine
Uses: Hydrophilic emulsifier, wetting agent, antistat, anticorrosive for agric., leather, textiles, metalworking, plastics; intermediate for quats.; textile dyeing agent, dye leveler, antiprecipitant, stripping agent
Trade names: Prox-onic DT-015

PEG-30 tallow diamine
Uses: Hydrophilic emulsifier, wetting agent, antistat, anticorrosive for agric., leather, textiles, metalworking, plastics; intermediate for quats.; textile dyeing agent, dye leveler, antiprecipitant, stripping agent
Trade names: Prox-onic DT-030

PEG-2 tallowmonium chloride
CAS 67784-77-4
Trade names containing: Ethoquad® T/12

PEG-10 tetramethyl decynediol
CAS 9014-85-1
Uses: Wetting agent, defoamer for aq. coatings, inks, adhesives; surfactant for emulsion polymerization; electroplating additive
Trade names: Surfynol® 465

PEG-30 tetramethyl decynediol
CAS 9014-85-1
Uses: Wetting agent, defoamer for aq. coatings, inks, adhesives, agric., electroplating, oilfield chems., paper coatings, emulsion polymerization
Trade names: Surfynol® 485

PEG-14 tridecyl ether. *See* Trideceth-14
PEG-15 tridecyl ether. *See* Trideceth-15
PEG-3 tridecyl ether phosphate. *See* Trideceth-3 phosphate
PEG-6 tridecyl ether phosphate. *See* Trideceth-6 phosphate
PEG-10 tridecyl ether phosphate. *See* Trideceth-10 phosphate
PEG 300 tridecyl ether phosphate. *See* Trideceth-6 phosphate
PEG 500 tridecyl ether phosphate. *See* Trideceth-10 phosphate
PEHA. *See* Pentaethylene hexamine

N-Pelargonoyl-p-aminophenol
Uses: Processing aid for thermoplastics
Trade names: Suconox-9®

1,4,7,10,13-Pentaazatridecane. *See* Tetraethylenepentamine

Pentabromobenzyl acrylate
CAS 594477-55-11
Uses: Flame retardant for plastics
Trade names: Actimer FR-1025M

Pentabromodiphenyl oxide
CAS 32534-81-9
Uses: Flame retardant for plastics
Manuf./Distrib.: Albemarle; Great Lakes
Trade names: FR-1205; Great Lakes DE-71™
Trade names containing: Fyrol® PBR; Great Lakes DE-61™; Great Lakes DE-62™

Pentabromoethylbenzene
CAS 85-22-3
Uses: Flame retardant for thermoset polyester resins, textiles, adhesives, coatings, PU
Manuf./Distrib.: Albemarle; Great Lakes

Pentabromophenol
Uses: Flame retardant for plastics
Manuf./Distrib.: Great Lakes
Trade names: FR-615

Pentabromotoluene
CAS 87-83-2
Synonyms: 2,3,4,5,6-Pentabromotoluene
Empirical: $C_6Br_5CH_3$
Properties: M.w. 486.65; m.p. 285-286 C
Toxicology: Irritant
Uses: Flame retardant for unsat. polyesters, polyethylene, PP, PS, SBR latex, textiles, rubbers
Manuf./Distrib.: Aldrich; Great Lakes
Trade names: FR-705

2,3,4,5,6-Pentabromotoluene. *See* Pentabromotoluene
Pentachlorobenzenethiol. *See* Pentachlorothiophenol

Pentachlorothiophenol
CAS 133-49-3
Synonyms: Pentachlorobenzenethiol
Empirical: C_6HCl_5S
Properties: M.w. 282.38
Toxicology: Poison by intraperitoneal route; mildly toxic by ingestion; severe eye irritant
Uses: Peptizing agent in rubber industry
Trade names: Renacit® 7
Trade names containing: Renacit® 7/WG

3-(n-Penta-8´-decenyl) phenol
CAS 8007-24-7
Formula: $C_6H_4 \bullet OH \bullet C_{15}H_{29}$
Uses: Raw material for surfactants, antioxidants, anticorrosives; lubricant additive; resin modifier
Trade names: Cardolite® NC-511

3-(n-Pentadecyl) phenol
CAS 501-24-6; 3158-56-3
Empirical: $C_{21}H_{36}O$
Formula: $CH_3(CH_2)_{14}C_6H_4OH$
Properties: M.w. 304.52; m.p. 50-53 C; b.p. 190-195 C (1 mm); flash pt. > 110 C
Uses: Raw material for surfactants, antioxidants, anticorrosives; lubricant additive; cosolv. for insecticides, germicides; resin modifier; coupling agent
Trade names: Cardolite® NC-507; Cardolite® NC-510

Pentaerythritol
CAS 115-77-5; EINECS 204-104-9
Synonyms: PE; 2,2-Bis(hydroxymethyl)-1,3-propanediol; Monopentaerythritol; Tetramethylolmethane
Empirical: $C_5H_{12}O_4$
Properties: Clear, colorless crystals; sl. sol. in alc.; insol. in benzene CCl_4, ether, and petroleum ether; sol. 1 g/18 ml of water; m.w. 136.15; m.p. 260 C
Uses: In synthetic resins, paints, varnishes; flame retardant for plastics
Manuf./Distrib.: Aqualon; Degussa; Hoechst Celanese; Mitsubishi Gas; Mitsui Toatsu; Penta Mfg.; Perstorp Polyols; U.S. Petrochem.
Trade names: Flame-Amine; Sta-Flame 88/99

Pentaerythritol ester of rosin. *See* Pentaerythrityl rosinate
Pentaerythritol rosinate. *See* Pentaerythrityl rosinate
Pentaerythritol tetrakis (3,5-di-t-butyl-4-hydroxyhydrocinnamate). *See* Tetrakis [methylene (3,5-di-t-butyl-4-hydroxyhydrocinnamate)] methane
Pentaerythritol tetrakis (3-mercaptopropionate). *See* Pentaerythrityl tetrakis (3-mercaptopropionate)
Pentaerythritol tetraoleate. *See* Pentaerythrityl tetraoleate
Pentaerythritol tetrastearate. *See* Pentaerythrityl tetrastearate
Pentaerythritol triacrylate. *See* Pentaerythrityl triacrylate

Pentaerythrityl alkyl thiodipropionate
CAS 96328-09-1
Trade names: Mark® 5089

Pentaerythrityl hexylthiopropionate
CAS 95823-35-1
Uses: Stabilizer for polyolefins and other polymeric systems
Trade names: Mark® 2140

Pentaerythrityl ricinoleate
Definition: Ester of pentaerythritol and ricinoleic acid
Uses: Plasticizer; chemical intermediate
Trade names: Flexricin® 17

Pentaerythrityl rosinate
CAS 8050-26-8; EINECS 232-479-9
Synonyms: Pentaerythritol rosinate; Pentaerythritol ester of rosin; Rosin pentaerythritol ester
Definition: Ester of rosin acids with the polyol, pentaerythritol
Properties: Amber hard solid; sol. in acetone, benzene; insol. in water
Toxicology: Heated to decomp., emits acrid smoke and irritating fumes
Uses: Thermoplastic resin as tackifier for rubbers, lacquers, ink vehicles, varnishes, adhesives
Regulatory: FDA 21CFR §172.615, 175.105, 175.300, 176.170, 176.210, 176.2600, 178.3120, 178.3800, 178.3870
Trade names: Resinall 610; Uni-Rez® 3020; Uni-Tac® R99; Uni-Tac® R100; Uni-Tac® R100-Light; Uni-Tac® R100RM; Uni-Tac® R101; Uni-Tac® R102; Uni-Tac® R106; Uni-Tac® R112; Zonester® 100

Pentaerythrityl stearate
CAS 85116-93-4; EINECS 285-547-5
Definition: Ester of pentaerythritol and stearic acid
Uses: Dispersant and internal lubricant for tech. molded and extruded rubber articles; emollient for creams, makeups

Pentaerythrityl tetraacrylate
CAS 4986-89-4
Properties: M.w. 352.4; dens. 1.190; m.p. 18 C; flash pt. > 110 C; ref. index 1.4870 (20 C)
Uses: Crosslinking agent in adhesives, coatings, inks, textile prods., photoresists, castings
Trade names containing: SR-295

Pentaerythrityl tetrabenzoate
CAS 4196-86-5
Formula: $(C_6H_5CO_2CH_2)_4C$
Properties: M.w. 552.59; sp.gr. 1.2801 (30 C); m.p. 99 C; flash pt. (COC) 600 C
Uses: Plasticizer/extender for PVC, CAB, ethyl cellulose, coatings, modifier for hot-melt adhesives, aq. adhesives, delayed tack adhesives, process aid for thermoplastics
Manuf./Distrib.: Aldrich; Unitex; Velsicol
Trade names: Benzoflex® S-552; Uniplex 552

Pentaerythrityl tetracaprylate/caprate
CAS 68441-68-9; 69226-96-6; EINECS 270-474-3
Definition: Tetraester of pentaerythritol and a mixture of caprylic and capric acids
Properties: Sp.gr. 0.995 (16 C); flash pt. (COC) 232 C; ref. index 1.4355-1.4365 (16 C)
Uses: Lubricant for textile and metalworking compds.; plasticizer for PVC, PS, PVB
Manuf./Distrib.: Hatco; Henkel; Inolex

Pentaerythrityl tetrakis [3-(3´,5´-di-t-butyl-4-hydroxyphenyl)propionate]
Classification: Hindered phenol
Empirical: $C_{73}H_{108}O_{12}$
Properties: White powd.; sol. in acetone, benzene, chloroform, ethyl acetate; m.w. 1177.7; m.p. 110-125 C
Uses: Antioxidant
Trade names: Lankromark® LE373; Lowinox® PP35
Trade names containing: Plasblak® LL 2590

Pentaerythrityl tetrakis (β-laurylthiopropionate)
CAS 29598-76-3
Uses: Antioxidant for polyolefins, thermoplastic elastomers, engineering thermoplastics
Trade names: Seenox 412S

Pentaerythrityl tetrakis (3-mercaptopropionate)
CAS 7575-23-7; EINECS 231-472-8
Synonyms: Pentaerythritol tetrakis (3-mercaptopropionate)
Empirical: $C_{17}H_{28}O_8S_4$
Formula: $C(CH_2OOCCH_2CH_2SH)_4$
Properties: M.w. 488.70; b.p. 275 C (1 mm); flash pt. > 110 C
Uses: Crosslinking agent and stabilizer for radiation-cured plastics
Manuf./Distrib.: Evans Chemetics; Fluka; Witco/PAG

Pentaerythrityl tetraoleate
CAS 19321-40-5; 68604-44-4; EINECS 242-960-5; 271-694-2
Synonyms: Pentaerythritol tetraoleate
Definition: Tetraester of pentaerythritol and oleic acid
Empirical: $C_{77}H_{140}O_8$
Uses: Emollient, thickener and visc. controller for cosmetics, plastics, lubricants, cutting oils; chemical intermediate; corrosion inhibitors; chemical synthesis
Regulatory: FDA 21CFR §176.210
Trade names: Estol 1445

Pentaerythrityl tetrastearate
CAS 115-83-3; 91050-82-7; EINECS 204-110-1; 293-029-5
Synonyms: Pentaerythritol tetrastearate
Definition: Tetraester of pentaerythritol and stearic acid
Empirical: $C_{77}H_{148}O_8$
Properties: Ivory-colored hard, high melting wax
Uses: Processing aid for rubbers; polishes, coatings, textile finishes
Regulatory: FDA 21CFR §175.105, 176.170, 176.210, 177.1200, 177.1580, 178.2010
Manuf./Distrib.: Hercules; Lipo; Lonza
Trade names: Aflux® 54

Pentaerythrityl triacrylate
CAS 3524-68-3; EINECS 222-540-8
Synonyms: 2-Propenoic acid-2-(hydroxymethyl)-2-(((1-oxo-2-propenyl) oxy) methyl)-1,3-propanediyl ester; Pentaerythritol triacrylate
Empirical: $C_{14}H_{18}O_7$
Formula: $(H_2C{:}CHCO_2CH_2)_3CCH_2OH$
Properties: M.w. 298.30; dens. 1.167 (20C); ref. index 1.143 (20C); flash pt.> 110C
Precaution: Hygroscopic
Toxicology: Irritant
Uses: Crosslinking agent used in adhesives, coatings, inks, textile prods., photoresists, castings, modifiers for polyester, fiberglass, or polymers
Manuf./Distrib.: Aldrich; Monomer-Polymer & Dajac; Sartomer; UCB Radcure
Trade names containing: SR-444

Pentaethylene hexamine
CAS 4067-16-7; EINECS 223-775-9
Synonyms: PEHA
Formula: $H_2N(C_2H_4NH)_5H$
Properties: M.w. 232.3; sp.gr. 1.000 (20/20 C); pour pt. < -26; b.p. 230 C (10 mm); flash pt. (COC) 206 C
Precaution: Corrosive
Uses: Epoxy resin curing agent; polyamide resins; oil additives; ion exchange resins
Manuf./Distrib.: Aldrich; Fluka; Tosoh

Pentaglycerine. *See* Trimethylolethane

N,N,N′,N′,N′-Penta(2-hydroxyethyl)-N-tallowalkyl-1,3-propane diammonium diacetate
Uses: EPA listed
Trade names containing: Ethoduoquad® T/15-50

N,N′,N′′,N′′-Pentamethyldiethylenetriamine
CAS 3030-47-5
Formula: $[(CH_3)_2NCH_2CH_2]_2NCH_3$

N,N,N′,N′′,N′′-Pentamethyldipropylenetriamine

Properties: M.w. 173.30; dens. 0.830; m.p. -20 C; b.p. 198 C; flash pt. 53 C; ref. index 1.4420 (20 C)
Uses: Blowing catalyst for PU foams
Manuf./Distrib.: Aldrich
Trade names: Polycat® 5; Toyocat®-DT

N,N,N′,N′′,N′′-Pentamethyldipropylenetriamine
Uses: Catalyst for polyurethanes
Trade names: Polycat® 17

Pentamethyl-4-piperidyl/β,β,β′,β′-tetramethyl-3,9-(2,4,8,10-tetraoxaspiro (5,5) undecane) diethyl]-1,2,2,6,6-1,2,3,4-butane tetracarboxylate
Uses: Light and heat stabilizer for PP, polyethylene, PS, ABS, engineering plastics, and elastomers; monofilament, tapes, molded and extruded prods., polyethylene blown film

1,3-Pentanediamine
CAS 589-37-7
Synonyms: DAMP; 1,3-Diaminopentane
Formula: $H_2NCH_2CH_2CH(C_2H_5)NH_2$
Properties: Liq., sl. odor; misc. with water; m.w. 102; dens. 0.855 g/ml; visc. 1.89 cp; vapor pressure 1 mm Hg (20 C); b.p. 164 C; flash pt. (CC) 59 C; surf. tens. 32.2 dyn/cm
Precaution: Corrosive, flamm.; incompat. with strong oxidants; thermal decomp. may emit NO_x
Uses: Epoxy curative; solv. for gas treatment; surfactant (asphalt emulsifiers, textiles); scale/corrosion inhibitor; additive for fuel, oil, plastic, petrol. and chem. processing; extender, catalyst for PU; in coatings, adhesives, sealants, elastomers
Trade names: Dytek® EP

1,5-Pentanediamine, 2-methyl-. *See* 2-Methylpentamethylenediamine
2,4-Pentanedione peroxide. *See* Acetylacetone peroxide
t-Pentanol. *See* t-Amyl alcohol
Pentaphen, 2-methyl-2-p-hydroxyphenyl-butane. *See* p-t-Amylphenol
Pentasodium diethylene triamine pentaacetate. *See* Pentasodium pentetate
Pentasodium DTPA. *See* Pentasodium pentetate

Pentasodium pentetate
CAS 140-01-2; EINECS 205-391-3
Synonyms: $DTPANa_5$; Pentasodium diethylene triamine pentaacetate; Pentasodium DTPA
Uses: Chelating agent for detergents, textiles, polymerization
Regulatory: FDA 21CFR §175.105, 176.150
Trade names: Chel DTPA-41

t-Pentyl alcohol. *See* t-Amyl alcohol
Pentylcarbinol. *See* Hexyl alcohol
Pentyltrichlorosilane. *See* Amyltrichlorosilane
Perhydronaphthalene. *See* Decahydronaphthalene
Periclase. *See* Magnesium oxide

Petrolatum
CAS 8009-03-8 (NF); 8027-32-5 (USP); EINECS 232-373-2
Synonyms: Petroleum jelly; Petrolatum amber; Petrolatum white; White petrolatum
Classification: Petroleum hydrocarbons
Definition: Semisolid mixture of hydrocarbons obtained from petroleum
Properties: Yellowish to lt. amber or white semisolid, unctuous mass; pract. odorless and tasteless; sol. in benzene, chloroform, ether, petrol. ether, oils; pract. insol. in water; dens. 0.820-0.865 (60/25 C); m.p. 38-54 C; ref. index 1.460-1.474 (60 C)
Toxicology: Heated to decomp., emits acrid smoke and irritating fumes
Uses: Laxative; dispersant; as ointment base in pharmaceuticals and cosmetics; leather grease; shoe polish; rust preventives; plastics, rubber, textiles lubricant
Regulatory: FDA 21CFR §172.880, 172.884, 173.340, 175.105, 175.125, 175.176, 175.300, 176.170, 176.200, 176.210, 177.2600, 177.2800, 1787.3570, 178.3700, 178.3910, 573.720
Manuf./Distrib.: Exxon; Harcros; Magie Bros. Oil; Mobil; Penreco; Stevenson Cooper; Witco/Petrol. Spec.
Trade names: Dark Green No. 2™; Fonoline® White; Fonoline® Yellow; Markpet; Molder's Edge ME-301; Penreco 1520; Penreco 3070; Penreco Amber; Penreco Red; Penreco Snow; Petrolatum RPB; Protopet® White 1S; Protopet® Yellow 2A; Tech Pet F™; Tech Pet P

Petrolatum amber. *See* Petrolatum
Petrolatum liquid. *See* Mineral oil
Petrolatum white. *See* Petrolatum
Petroleum benzin. *See* Naphtha
Petroleum ether. *See* Naphtha
Petroleum jelly. *See* Petrolatum
Petroleum naphtha. *See* Naphtha
Petroleum spirit. *See* Naphtha
Petroleum spirits. *See* Mineral spirits

Petroleum wax
CAS 8002-74-2
Synonyms: Microcrystalline wax; Petroleum wax, synthetic; Refined petroleum wax
Classification: Petroleum hydrocarbon
Definition: Hydrocarbon derived from petroleum
Properties: Translucent wax, odorless, tasteless; very sl. sol. in org. solvs.; insol. in water; m.p. 48-93 C
Toxicology: Heated to decomp., emits acrid smoke and irritating fumes
Uses: Lubricant for formulating PVC and elec. wire and cable compds.; antisunchecking protectant for elastomers
Regulatory: FDA 21CFR §172.230, 172.615, 172.886 (1050 ppm max. of poly(alkylacrylate) as an antioxidant), 172.888, 173.340, 178.3710, 178.3720
Manuf./Distrib.: Astor Wax; Exxon; Koster Keunen; Mobil; Penreco; Shell; Witco/Petrol. Spec.
Trade names: Akrowax 5026; Akrowax 5030; Akrowax 5031; Akrowax 5032; Loobwax 0597; Loobwax 0750; Ross Wax #145; Ross Wax #165; Vanwax® H Special

Petroleum wax, crystalline. *See* Paraffin
Petroleum wax, microcrystalline. *See* Microcrystalline wax
Petroleum wax, synthetic. *See* Petroleum wax
PGE. *See* Phenyl glycidyl ether

1,10-Phenanthroline
CAS 66-71-7
Synonyms: o-Phenanthroline; 4,5-Phenanthroline
Classification: Heterotricyclic compd.
Empirical: $C_{12}H_8N_2$
Properties: White crystalline powd.; sol. in 300 parts water, 70 parts benzene; sol. in alcohol, ether, acetone; m.w. 180.20; m.p. 93-94 C, 117 C (anhyd.)
Toxicology: Poison by intraperitoneal route
Uses: Indicator; drier accelerator and stabilizer in coatings cured by oxidative polymerization
Manuf./Distrib.: Amresco; GFS; Spectrum Chem. Mfg.
Trade names: Activ-8

4,5-Phenanthroline. *See* 1,10-Phenanthroline
o-Phenanthroline. *See* 1,10-Phenanthroline

Phenethylmethylsiloxane/dimethylsiloxane copolymer
Uses: Mold release agent for rubber, plastics, die casting
Trade names: PS136.8

Phenethylmethylsiloxane/hexylmethylsiloxane copolymer
Uses: Mold release agent for rubber, plastics, die casting
Trade names: PS137

2-Phenethyltrichlorosilane
CAS 940-41-0
Empirical: $C_8H_9Cl_3Si$
Properties: M.w. 239.61
Toxicology: Moderately toxic by ingestion and skin contact; skin irritant
Uses: Coupling agent, chem. intermediate, blocking agent, release agent, lubricant, primer, reducing agent
Trade names: Dynasylan® PETCS

Phenol, 2-(2H-benzotriazole-2-yl)-4-methyl-6-dodecyl
CAS 23328-53-2

Properties: M.w. 393.6
Uses: UV absorber for plastics
Trade names: Tinuvin® 571

Phenol-formaldehyde. *See* Phenolic resin
Phenol-formaldehyde (novolak) resin. *See* Novolac resin

Phenolic resin
CAS 9003-35-4
Synonyms: Phenol-formaldehyde; One step: A-stage resin, one-stage resin, resole; Two step: Novolac resin, novolak resin, two-stage resin
Definition: Thermosetting resin from condensation of phenol or substituted phenol with aldehydes such as formaldehyde, acetaldehyde, and furfural
Uses: Thermosetting resin for applics. requiring heat resistance, exhaust duct systems, wiring devices, switch gears, ovens, toasters, pot handles, elec. devices, coil bobbins, coatings, laminates; tackifier, plasticizer, hardening and reinforcing agent for rubber
Manuf./Distrib.: 3M; Akzo Nobel; Arakawa; Arizona; Asahi Yukizai Kogyo; Bakelite GmbH; Borden; BP Chem. Ltd; Focus; Georgia-Pacific; Hüls AG; PMC Specialties; QO; Raschig
Trade names: Akrochem® P 40; Akrochem® P 49; Akrochem® P 55; Akrochem® P 82; Akrochem® P 86; Akrochem® P 87; Akrochem® P 101; Akrochem® P 105; Akrochem® P 108; Akrochem® P 124; Akrochem® P 125; Akrochem® P 126; Akrochem® P 478; Akrochem® P 487; Dyphene® 877PLF; Dyphene® 6745-P; Naugard® 431; Naugard® SP; Naugawhite®; Permanax™ WSO; R7559; SP-553; SP-560; SP-1055; SP-1056; Vulcabond® E; Vulkadur® A
See also Novolac resin

Phenolic resin thermoplastic
Trade names: Akrochem® P 37; Akrochem® P 133; Akrochem® P 486

Phenol, 2,2´-methylene-bis-6-[(1,1-dimethyl)-4-methyl-]. *See* 2,2´-Methylenebis (6-t-butyl-4-methylphenol)

Phenol, styrenated
CAS 61788-44-1
Uses: Antioxidant used in mastic adhesives, mech. and molded rubber goods, caulking compds., cement, and antiskinning agents
Trade names: Ablumol PSS; Akrochem® Antioxidant 16 Liq.; AO47L; AO47P; Montaclere®; Nevastain® 21; Prodox® 120; Vanox® 102; Vulkanox® SP
Trade names containing: Akrochem® Antioxidant 16 Powd.; Wingstay® S

Phenol sulfonic acid
CAS 98-67-9; 1333-39-7; EINECS 202-691-6
Synonyms: p-Phenolsulfonic acid; 4-Hydroxybenzenesulfonic acid; Sulfocarbolic acid
Empirical: $C_6H_6O_4S$
Formula: $HOC_6H_4SO_3H$
Properties: Yellowish liq. (brown in air); sol. in water, alcohol; m.w. 174.18; dens. 1.34 (20/4 C)
Toxicology: Irritant to skin and tissues
Uses: Water analysis, lab reagent, electroplated tin coating baths; mfg. of intermediates and dyes, pharmaceuticals
Manuf./Distrib.: Fluka; Sloss Ind.
Trade names: Eltesol® PSA 65; PSA

p-Phenolsulfonic acid. *See* Phenol sulfonic acid

Phenol, 4,4´-thiobis 2-(1,1-dimethylethyl) phosphite
Synonyms: Tris-(2-t-butyl-4-thio-(2´-methyl-4´-hydroxy-5´-t-butyl)phenyl-5-methyl]phenyl phosphite
Empirical: $C_{66}H_{87}O_6S_3P$
Properties: Sol. in more common aromatic solvs., aliphatic esters, ketones, chlorinated hydrocarbons; insol. in water; m.w. 1103; dens. 1.1 g/cc; soften. pt. 110 C
Uses: Antioxidant for polymers; metal deactivator
Trade names: Hostanox® OSP 1

Phenol, 2,4,6-tris(dimethylaminomethyl). *See* 2,4,6-Tris (dimethylaminomethyl) phenol

Phenothiazine
CAS 92-84-2; EINECS 202-196-5

Synonyms: Thiodiphenylamine
Empirical: $C_{12}H_9NS$
Formula: $C_6H_4NHC_6H_4S$
Properties: Grayish-green to greenish-yel. powd., gran., or flakes, tasteless, sl. odor; sol. in benzene, ether, hot acetic acid, 21% in acetone, 15% in ethyl amyl ketone, 11% in ethyl acetate; negligible in water; m.w. 199.26; b.p. 371 C; m.p. 180 C
Toxicology: Irritant to eyes, skin, by inhalation; control dusts; TLV 5 mg/m^3 of air
Uses: Antioxidant; monomer stabilizer; insecticide; mfg. of dyes; polymerization inhibitor
Manuf./Distrib.: AC Ind.; Alemark; Schweizerhall
Trade names: PTZ® Chemical Grade; PTZ® Industrial Grade; PTZ® Purified Flake; PTZ® Purified Powd.

Phenoxydiglycol
Uses: Solv. for resins, in metal cleaners, paint strippers, cleaning compds., as ink vehicle
Trade names: Igepal® OD-410

Phenoxyethanol
CAS 122-99-6; EINECS 204-589-7
Synonyms: 2-Phenoxyethanol; Phenoxytol; Ethylene glycol monophenyl ether
Classification: Aromatic ether alcohol
Definition: Phenol polyglycol ether
Empirical: $C_8H_{10}O_2$
Properties: Clear liq., faint aromatic odor, burning taste; sl. sol. in water; sol. in alcohol, ether, NaOH sol'ns.; m.w. 138.18; dens. 1.1094 (20/20 C); m.p. 14 C; b.p. 242 C; flash pt. 121 C; ref. index 1.534 (20 C)
Toxicology: Moderately toxic by ingestion and skin contact; skin and severe eye irritant; LD50 (rat, oral) 1.26 g/kg
Uses: Solvent for resins, dyes, inks, in organic synthesis; perfume fixative, bactericidal agent, plasticizer, germicide, as insect repellent
Regulatory: FDA 21CFR §175.105; CIR approved, EPA reg.; JSCI listed 1.0% max.; Europe listed 1% max.
Manuf./Distrib.: Amber Syn.; Jan Dekker; Hüls AG; Penta Mfg.; Tri-K
Trade names: Emery® 6705

2-Phenoxyethanol. *See* Phenoxyethanol
Phenoxytol. *See* Phenoxyethanol

Phenyl acid phosphate
Manuf./Distrib.: Albright & Wilson Am.
Trade names containing: Albrite® PA-75

2-Phenylazo-4-methoxy-2,4-dimethylvaleronitrile
CAS 35634-74-3
Uses: Polymerization initiator

N-Phenyl-1,4-benzenediamine. *See* N-Phenyl-p-phenylenediamine

Phenyldiethanolamine
CAS 120-07-0; EINECS 204-368-5
Synonyms: PDEA; N,N-Bis(2-hydroxyethyl)aniline
Empirical: $C_{10}H_{15}NO_2$
Formula: $(HOCH_2CH_2)_2NC_6H_5$
Properties: Colorless liq.; sl. sol. in water; sol. in ethanol, acetone; m.w. 181.26; dens. 1.1203 (60/20 C); m.p. 56-58 C; b.p. 190 C; flash pt. (OC) 190 C
Toxicology: Moderately toxic by ingestion; severe eye and mild skin irritant
Uses: Organic synthesis, laboratory reagent; polyester cure promoter; curing agent for urethane elastomers; coupling agent, intermediate for dyes; in hair dyes, detergents, paint strippers
Manuf./Distrib.: Fluka; Hüls AG
Trade names: Emery® 5703

Phenyl diisodecyl phosphite
CAS 25550-98-5; EINECS 247-098-3
Synonyms: PDDP
Empirical: $C_{26}H_{47}O_3P$
Formula: $(C_{10}H_{21}O)_2POC_6H_5$

Properties: Liq.; m.w. 438; dens. 0.938-0.947; b.p. 190 C (5 mm); ref. index 1.4780-1.4810; flash pt. (PMCC) 160 C
Uses: Color and processing stabilizer for ABS, PC, polyurethane, coatings, PET fiber; sec. stabilizer for PVC
Manuf./Distrib.: Dover
Trade names: Doverphos® 7; Weston® PDDP

1-Phenyl-3,3-dimethyl urea
CAS 101-42-8
Synonyms: 1,1-Dimethyl-3-phenylurea; Fenulon
Empirical: $C_9H_{12}N_2O$
Properties: White cryst.; insol. in water; sol. in hydrocarbon; m.w. 164.23; m.p. 127-129 C
Toxicology: Moderately toxic by ingestion
Uses: Accelerator for epoxy resins, adhesives, composites, prepregs
Trade names: Amicure® UR

2,2´-(1,4-Phenylene)bis[4H-3,1-benzoxazin-4-one]
Empirical: $C_{22}H_{12}N_2O_4$
Properties: Off-wh. powd.; m.w. 368; m.p. 310 C
Uses: Light absorber for plastics
Trade names: Cyasorb® UV 3638

2,2´-[1,4-Phenylenebis(oxy)bisethanol. *See* Hydroquinone bis(2-hydroxyethyl) ether
2,2´-(1,3-Phenylenebis(oxymethylene))bisoxirane. *See* Resorcinol diglycidyl ether
1,3-Phenylenediamine. *See* m-Phenylenediamine

m-Phenylenediamine
CAS 108-45-2; EINECS 203-584-7
Synonyms: mPDA; 1,3-Benzenediamine; 1,3-Phenylenediamine; 3-Aminoaniline
Classification: Aromatic amine
Empirical: $C_6H_8N_2$
Formula: $C_6H_4(NH_2)_2$
Properties: White crystals; sol. in water, methanol, ethanol, MEK, dioxane; sl. sol. in ether, m.w. 108.14; CCl_4, IPA; dens. 1.139; m.p. 62.8 C; b.p. 284-287 C; flash pt. > 110C
Precaution: Keep well sealed and protected from light
Toxicology: Toxic, irritant; poison by ingestion, intravenous, subcutaneous routes
Uses: Mfg. of dyes, hair dyes; rubber curing agent; in photography; as reagent for gold and bromine
Regulatory: FDA 21CFR §177.2280
Manuf./Distrib.: Du Pont; First; Miles
Trade names containing: Tonox® 60/40

m-Phenylenedimaleimide
CAS 3006-93-7
Synonyms: 1,3-Bismaleimidobenzene
Empirical: $C_{14}H_8N_2O_4$
Properties: M.w. 268.24
Toxicology: Toxic; poison by ingestion, intraperitoneal routes
Uses: Accelerator; coagent in peroxide-cured polymers
Manuf./Distrib.: Aldrich
Trade names: Vanax® MBM
Trade names containing: Poly-Dispersion® T (HVA-2)D-70; Poly-Dispersion® T (HVA-2)D-70P

Phenylethanolamine
CAS 7568-93-6; 122-98-5; EINECS 231-469-1
Synonyms: β-Hydroxyphenethylamine; 2-Amino-1-phenylethanol; α-(Aminomethyl)benzenemethanol
Empirical: $C_8H_{11}NO$
Formula: $C_6H_5NHCH_2CH_2OH$
Properties: Pale yel. crystals; sol. in water; sl. sol. in hot xylene; m.w. 137.20; m.p. 56-57 C; b.p. 157-160 C (17 mm)
Toxicology: Poison by intraperitoneal, intravenous routes; moderately toxic by subcutaneous and other routes
Uses: Shortstopping agent for SBR; intermediate; wax hardener; emulsion polymerization of diene monomers; stabilizer
Manuf./Distrib.: Fluka; Interchem
Trade names: Emery® 5700

Phenylethylethanolamine
Synonyms: PEEA
Classification: Amine
Properties: M.w. 215
Uses: Dye intermediate; polyester cure promoter

Phenylformic acid. *See* Benzoic acid

Phenyl glycidyl ether
CAS 122-60-1; EINECS 204-557-2
Synonyms: PGE; 1,2-Epoxy-3-phenoxypropane; Oxirane, (phenoxymethyl)-
Empirical: $C_9H_{10}O_2$
Formula: $H_2COCHCH_2OC_6H_5$
Properties: Colorless liq.; sol. in ethanol; m.w. 150.19; dens. 1.113 (20/4 C); m.p. 3.5 C; b.p. 245 C
Precaution: Incompat. with acids, alkalies, amines, oxidizing agents; hazardous decomp. prods.: CO, hydrocarbons
Toxicology: LD50 (oral, rat) 2150-3850 mg/kg, (dermal, rabbit) 1500 mg/kg; moderately toxic by ingestion, skin contact, subcutaneous routes; severe eye and skin irritant; TLV 1 ppm in air; may cause sensitization by skin contact; may cause cancer
Uses: Reactive epoxy diluent for solv.-free coating systems, laminating resins, fiber-reinforced composites, elec. applics.; stabilizer for chlorinated hydrocarbons; chem. intermediate for pharmaceuticals
Manuf./Distrib.: Monomer-Polymer & Dajac; Raschig; Richman
Trade names: Epodil® 743; Heloxy® 63

Phenyl glycol ether
Uses: Plasticizer; diluent and flexibilizer for epoxies
Trade names: Kenplast® PG
Trade names containing: LICA 97; NZ 97

Phenyl-2-hydroxybenzoate. *See* Phenyl salicylate

2-Phenyl imidazole
CAS 670-96-2
Properties: M.w. 144.18; m.p. 128-131 C
Toxicology: Irritant
Uses: Epoxy curing agent for printed circuit boards, molding compds., potting; accelerator for dicyandiamide and anhydrides
Manuf./Distrib.: Aldrich; BASF; Janssen Chimica; Poly Organix; Schweizerhall
Trade names: Curezol® 2PZ

4-(2-Phenylisopropyl)phenol. *See* Cumyl phenol

N-Phenyl,N′-isopropyl-p-phenylenediamine
Uses: Styrene polymerization inhibitor; column antifoulant
Trade names: Naugard® I-4

Phenylmethane. *See* Toluene
Phenylmethyl benzoate. *See* Benzyl benzoate
N-(Phenylmethyl) dimethylamine. *See* N-Benzyldimethylamine

Phenyl-α-naphthylamine
CAS 90-30-2
Synonyms: N-1-Naphthylaniline; N-Phenyl-1-naphthylamine
Empirical: $C_{16}H_{13}N$
Formula: $C_{10}H_7NHC_6H_5$
Properties: White to sl. yel. crystalline prisms; sol. in alcohol, ether, benzene; m.w. 219.29; m.p. 62 C; b.p. 335 C (260 mm)
Toxicology: Irritant
Uses: Dyes and other organic chemicals; rubber antioxidant
Manuf./Distrib.: Aldrich; Spectrum Chem. Mfg.
Trade names: Additin 30; Akrochem® Antioxidant PANA; Vulkanox® PAN

Phenyl-β-naphthylamine
CAS 135-88-6; EINECS 205-223-9

N-Phenyl-1-naphthylamine

Synonyms: N-Phenyl-2-naphthylamine; N-2-Naphthylaniline
Empirical: $C_{16}H_{13}N$
Formula: $C_{10}H_7NHC_6H_5$
Properties: Rhombic crystals; insol. in water; sol. in hot benzene; very sol. in hot alcohol, ether, acetone; m.w. 219.30; dens. 1.24; m.p. 107-108 C; b.p. 395.5 C
Toxicology: Suspected carcinogen; moderately toxic by ingestion; irritant
Uses: Rubber antioxidant, lubricant, inhibitor (butadiene)
Manuf./Distrib.: Aldrich; Fluka
Trade names containing: Rhenogran® PBN-80

N-Phenyl-1-naphthylamine. *See* Phenyl-α-naphthylamine
N-Phenyl-2-naphthylamine. *See* Phenyl-β-naphthylamine

Phenyl neopentylene glycol phosphite
CAS 3057-08-7; EINECS 221-291-2
Formula: $C_6H_5OP(OCH_2)_2C(CH_3)_2$
Properties: Liq.; m.w. 227; dens. 1.130-1.139; b.p. 140 C (5 mm); ref. index 1.5140-1.5188; flash pt. (PMCC) 121 C
Uses: Sec. stabilizer for polymers, adhesives
Trade names: Weston® PNPG

N-Phenyl-p-phenylenediamine
CAS 101-54-2; EINECS 202-951-9
Synonyms: 4-Aminodiphenylamine; p-Aminodiphenylamine; N-Phenyl-1,4-benzenediamine
Classification: Aromatic amine salt
Empirical: $C_{12}H_{12}N_2$
Formula: $NH_2C_6H_4NHC_6H_5$
Properties: Purple powd.; insol. in water; sol. in alcohol, acetone; m.w. 184.11; m.p. 75 C
Toxicology: Moderately toxic by ingestion; severe eye irritant; heated to dec., emits toxic fumes of NO_x
Uses: Dye intermediate, pharmaceuticals, photographic chemicals
Manuf./Distrib.: Fluka
Trade names containing: Vulkanox® 3100

Phenyl salicylate
CAS 118-55-8; EINECS 204-259-2
Synonyms: Phenyl-2-hydroxybenzoate; Salol
Formula: $2\text{-}(HO)C_6H_4CO_2C_6H_5$
Properties: Solid; m.w. 214.22; m.p. 62 C; b.p. 172-173 C (12 mm); flash pt. > 110 C
Uses: UV stabilizer for cellulosics, polyesters, PE, flexible PVC, VDC
Regulatory: FDA 21CFR §177.1010, 27CFR §21.65, 21.151
Manuf./Distrib.: AC Ind.; Aldrich; Eastman; Spectrum Chem. Mfg.

Phenylsulfonic acid. *See* Benzenesulfonic acid

5-Phenyltetrazole
CAS 3999-10-8
Empirical: $C_7H_6N_4$
Formula: $HNN{=}NN{=}C(C_6H_5)$
Properties: M.w. 146.15; gas yield 200 cc/g
Precaution: Explodes on distillation
Uses: Blowing agent for foaming plastics and elastomers at elevated temps.
Manuf./Distrib.: Amber Syn.; Charkit; Esprit; Hüls Am.
Trade names: Blo-Foam® 5PT; Expandex® 5PT; Unicell 5PT

5-Phenyltetrazole barium salt
Uses: Blowing agent for engineering resins
Trade names: Expandex® 175

Phenyltriethoxysilane
CAS 780-69-8; EINECS 212-305-8
Formula: $C_6H_5Si(OC_2H_5)_3$
Properties: M.w. 240.38; dens. 0.996; b.p. 112-113 C (10 mm); flash pt. 42 C; ref. index 1.4600 (20 C)
Precaution: Moisture-sensitive

Toxicology: Irritant
Uses: Coupling agent, release agent, lubricant, blocking agent, chemical intermediate
Manuf./Distrib.: Aldrich; Fluka
Trade names: CP0320

Phosphonium, tetrabutyl-, bromide. *See* Tetrabutylphosphonium bromide
Phosphonium, tetrabutyl-, chloride. *See* Tetrabutylphosphonium chloride
Phosphonous acid, (1,1´-biphenyl)-4,4´-diylbis-, tetrakis (2,4-bis(1,1-dimethylethyl)phenyl) ester. *See* Tetrakis (2,4-di-t-butylphenyl) 4,4-biphenylenediphosphonite
Phosphoric acid calcium salt (1:1). *See* Calcium phosphate dibasic
o-Phosphoric acid, triethyl ester. *See* Triethyl phosphate
Phosphoric acid, triphenyl ester. *See* Triphenyl phosphate
Phosphoric acid, tris(methylphenyl) ester. *See* Tricresyl phosphate
Phosphorotrithious acid, tridodecyl ester. *See* Trilauryl trithiophosphite

Phosphorous acid, cyclic neopentanetetrayl bis (2,4-di-t-butylphenyl) ester
CAS 26741-53-7
Uses: Antioxidant for polyolefins

Phosphorous acid, tridodecyl ester. *See* Trilauryl phosphite
Phosphorous acid, triisooctyl ester. *See* Triisooctyl phosphite

Phosphorus
CAS 7723-14-0; EINECS 231-768-7
Empirical: P
Properties: Reddish-brown powd.; at.wt. 30.97; dens. 2.34; m.p. 590 C (43 atm); b.p. 280 C
Toxicology: Human poison by ingestion
Uses: Mfg. of phosphoric acid; additive to semiconductors, fertilizers, safety matches
Trade names containing: Amgard® CHT; Amgard® CPC 452; DD-8307; Doverguard® 9021; Doverguard® 9122

Phosphorus trichloride, reaction prods. with 1,1´-biphenyl and 2,4-bis (1,1-dimethylethyl) phenol
CAS 119345-01-6
Classification: Aryl phosphonite
Properties: Wh./off-wh. powd., sl. odor; insol. in water; sp.gr. 1.04; m.p. 185-203 F; flash pt. 350 C
Precaution: Incompat. with strong oxidizing agents; thermal decomp. may produce oxides of carbon
Toxicology: LD50 (oral, rat) > 5000 mg/kg (dermal, rat) > 5000 mg/kg; nonirritating to eyes; mildly irritating to skin
Uses: Stabilizer, sec. antioxidant for polymers
Regulatory: FDA 21CFR §178.2010
Trade names: Sandostab P-EPQ

Phthalic acid, diethyl ester. *See* Diethyl phthalate
Phthalic acid, dimethyl ester. *See* Dimethyl phthalate

Phthalic anhydride
CAS 85-44-9; EINECS 201-607-5
Synonyms: 1,3-Isobenzofurandione
Empirical: $C_8H_4O_3$
Properties: White crystalline needles, mild odor; very sl. sol. in water; sol. in alcohol; sl. sol. in ether; m.w. 148.12; dens. 1.527 (4 C); m.p. 130.8 C; b.p. 295 C; sublimes; flash pt. (CC) 151 C
Toxicology: TLV:TWA 1 ppm (air); poison by ingestion
Uses: Retarding agent in rubber compoudning; in alkyd resins, plasticizers, hardener for resins, polyester, insecticides, laboratory reagent; mfg. of phthaleins, phthalates, benzoic acid
Manuf./Distrib.: Aldrich; Aristech; BASF; Coyne; Elf Atochem SA; Exxon; Mitsubishi Gas; Mitsui Toatsu; OxyChem; Stepan; UCB SA
Trade names: Retarder AK; Retarder PX; Vulkalent B/C; Wiltrol P

Phthalic anhydride, tetrachloro. *See* Tetrachlorophthalic anhydride
PIB. *See* Polybutene, Polyisobutene
Pigment blue 29. *See* Ultramarine blue

Pigment red 202
Uses: Colorant for plastics, powd. coatings, finishes, inks, artists' colors
Trade names: Quindo Magenta RV-6863

Pigment violet 15. *See* Ultramarine violet
Pigment violet 16. *See* Manganese violet
Pigment white 4. *See* Zinc oxide
Pigment white 6. *See* Titanium dioxide
Pigment white 26. *See* Talc

Pigment yellow 53
CAS 8007-18-9
Synonyms: CI 77788
Empirical: NiSbTi
Properties: Sp.gr. 4.4
Uses: colorant for plastics and industrial finishes
Trade names: Ferro V-9415

Pimelic ketone. *See* Cyclohexanone

Pine tar
CAS 8011-48-1; EINECS 232-374-8
Definition: Prod. obtained by destructive distillation of the wood of various species of pine, *Pinaceae*
Uses: Plasticizer, tackifier, process aid for rubber
Regulatory: FDA 21CFR §177.2600
Trade names: Tartac 20; Tartac 30; Tartac 40

PIP. *See* Piperazine
Piperazidine. *See* Piperazine

Piperazine
CAS 110-85-0; EINECS 203-808-3
Synonyms: PIP; Piperazine anhydrous; Diethylene diamine; Pyrazine hexahydride; Piperazidine
Empirical: $C_4H_{10}N_2$
Properties: Wh. flakes; m.w. 86.14; sp.gr. 0.97; f.p. 109.6 C; b.p. 148.5 C (760 mm); flash pt. (TCC) 229 F
Precaution: Corrosive; hygroscopic
Uses: Corrosion inhibitor, anthelmintic, insecticide, accelerator for curing polychloroprene
Manuf./Distrib.: Akzo Nobel; Alfa; Allchem Industries; BASF; Janssen Chimica; Tosoh

Piperazine anhydrous. *See* Piperazine
2-Piperazinoethylamione. *See* Aminoethylpiperazine
Piperidine, 1,1′-(tetrathiodicarbonothioyl)bis-. *See* Dipentamethylene thiuram tetrasulfide

Piperidinium pentamethylene dithiocarbamate
CAS 98-77-1
Empirical: $C_{11}H_{22}N_2S_2$
Properties: M.w. 246.47
Toxicology: Poison by ingestion; skin and eye irritant
Uses: Accelerator for rubber; peptizer
Trade names: Vanax® 552

PIPP. *See* p-Isopropylphenol
Pitch carbon fiber. *See* Carbon fiber
Plaster of Paris. *See* Calcium sulfate
Platy talc. *See* Talc
Plumbago. *See* Graphite
Plumbous oxide. *See* Lead (II) oxide
PM acetate. *See* Propylene glycol methyl ether acetate
PMDA. *See* Pyromelletic dianhydride
POE (4). *See* PEG-4
POE (8). *See* PEG-8
POE (12). *See* PEG-12
POE (20). *See* PEG-20
POE (32). *See* PEG-32
POE (40). *See* PEG-40
POE (75). *See* PEG-75
POE (100). *See* PEG-100

POE (150). *See* PEG-150
POE (2000). *See* PEG-2M
POE (5000). *See* PEG-5M
POE (7000). *See* PEG-7M
POE (9000). *See* PEG-9M
POE (14000). *See* PEG-14M
POE (20000). *See* PEG-20M
POE (23000). *See* PEG-23M
POE (45000). *See* PEG-45M
POE (90000). *See* PEG-90M
POE (115000). *See* PEG-115M
POE (3) caprylate/caprate. *See* PEG-3 caprylate/caprate
POE (5) castor oil. *See* PEG-5 castor oil
POE (10) castor oil. *See* PEG-10 castor oil
POE (12) castor oil. *See* PEG-12 castor oil
POE (16) castor oil. *See* PEG-16 castor oil
POE (25) castor oil. *See* PEG-25 castor oil
POE (30) castor oil. *See* PEG-30 castor oil
POE (32) castor oil. *See* PEG-32 castor oil
POE (36) castor oil. *See* PEG-36 castor oil
POE (40) castor oil. *See* PEG-40 castor oil
POE (80) castor oil. *See* PEG-80 castor oil
POE (200) castor oil. *See* PEG-200 castor oil
POE (6) cetyl/oleyl ether. *See* Cetoleth-6
POE (10) cetyl/oleyl ether. *See* Cetoleth-10
POE (20) cetyl/oleyl ether. *See* Cetoleth-20
POE (25) cetyl/oleyl ether. *See* Cetoleth-25
POE (4) cetyl/stearyl ether. *See* Ceteareth-4
POE (6) cetyl/stearyl ether. *See* Ceteareth-6
POE (10) cetyl/stearyl ether. *See* Ceteareth-10
POE (15) cetyl/stearyl ether. *See* Ceteareth-15
POE (20) cetyl/stearyl ether. *See* Ceteareth-20
POE (22) cetyl/stearyl ether. *See* Ceteareth-22
POE (25) cetyl/stearyl ether. *See* Ceteareth-25
POE (30) cetyl/stearyl ether. *See* Ceteareth-30
POE (50) cetyl/stearyl ether. *See* Ceteareth-50
POE (2) coco-benzonium chloride. *See* PEG-2 coco-benzonium chloride
POE (5) cocomonium methosulfate. *See* PEG-5 cocomonium methosulfate
POE (2) coconut amine. *See* PEG-2 cocamine
POE (5) coconut amine. *See* PEG-5 cocamine
POE (10) coconut amine. *See* PEG-10 cocamine
POE (15) coconut amine. *See* PEG-15 cocamine
POE (15) DEDM hydantoin oleate. *See* PEG-15 DEDM hydantoin oleate
POE (3) diacetate. *See* PEG-3 diacetate
POE (2) dibenzoate. *See* Diethylene glycol dibenzoate
POE (2) dilaurate. *See* PEG-2 dilaurate
POE (4) dilaurate. *See* PEG-4 dilaurate
POE (6) dilaurate. *See* PEG-6 dilaurate
POE (8) dilaurate. *See* PEG-8 dilaurate
POE (12) dilaurate. *See* PEG-12 dilaurate
POE (20) dilaurate. *See* PEG-20 dilaurate
POE (32) dilaurate. *See* PEG-32 dilaurate
POE (75) dilaurate. *See* PEG-75 dilaurate
POE (150) dilaurate. *See* PEG-150 dilaurate
POE (8) dinonyl phenyl ether. *See* Nonyl nonoxynol-8
POE (9) dinonyl phenyl ether. *See* Nonyl nonoxynol-9
POE (15) dinonyl phenyl ether. *See* Nonyl nonoxynol-15
POE (24) dinonyl phenyl ether. *See* Nonyl nonoxynol-24
POE (49) dinonyl phenyl ether. *See* Nonyl nonoxynol-49
POE (9) dinonyl phenyl ether phosphate. *See* Nonyl nonoxynol-9 phosphate
POE (2) dioleate. *See* PEG-2 dioleate
POE (4) dioleate. *See* PEG-4 dioleate

POE (6) dioleate. *See* PEG-6 dioleate
POE (8) dioleate. *See* PEG-8 dioleate
POE (12) dioleate. *See* PEG-12 dioleate
POE (20) dioleate. *See* PEG-20 dioleate
POE (32) dioleate. *See* PEG-32 dioleate
POE (75) dioleate. *See* PEG-75 dioleate
POE (150) dioleate. *See* PEG-150 dioleate
POE (2) distearate. *See* PEG-2 distearate
POE (4) distearate. *See* PEG-4 distearate
POE (6) distearate. *See* PEG-6 distearate
POE (8) distearate. *See* PEG-8 distearate
POE (12) distearate. *See* PEG-12 distearate
POE (20) distearate. *See* PEG-20 distearate
POE (32) distearate. *See* PEG-32 distearate
POE (75) distearate. *See* PEG-75 distearate
POE (150) distearate. *See* PEG-150 distearate
POE (7) glyceryl monococoate. *See* PEG-7 glyceryl cocoate
POE (20) glyceryl monostearate. *See* PEG-20 glyceryl stearate
POE (5) hydrogenated castor oil. *See* PEG-5 hydrogenated castor oil
POE (10) hydrogenated castor oil. *See* PEG-10 hydrogenated castor oil
POE (16) hydrogenated castor oil. *See* PEG-16 hydrogenated castor oil
POE (20) hydrogenated castor oil. *See* PEG-20 hydrogenated castor oil
POE (25) hydrogenated castor oil. *See* PEG-25 hydrogenated castor oil
POE (30) hydrogenated castor oil. *See* PEG-30 hydrogenated castor oil
POE (40) hydrogenated castor oil. *See* PEG-40 hydrogenated castor oil
POE (60) hydrogenated castor oil. *See* PEG-60 hydrogenated castor oil
POE (100) hydrogenated castor oil. *See* PEG-100 hydrogenated castor oil
POE (200) hydrogenated castor oil. *See* PEG-200 hydrogenated castor oil
POE (3) lauryl dimethyl amine oxide. *See* PEG-3 lauramine oxide
POE (6) lauryl ether. *See* Laureth-7
POE (9) lauryl ether. *See* Laureth-9
POE (12) lauryl ether. *See* Laureth-12
POE (15) lauryl ether. *See* Laureth-15
POE (16) lauryl ether. *See* Laureth-16
POE (23) lauryl ether. *See* Laureth-23
POE (9) lauryl ether sulfate, ammonium salt. *See* Ammonium laureth-9 sulfate
POE (8) monococoate. *See* PEG-8 cocoate
POE (4) monolaurate. *See* PEG-4 laurate
POE (6) monolaurate. *See* PEG-6 laurate
POE (8) monolaurate. *See* PEG-8 laurate
POE (12) monolaurate. *See* PEG-12 laurate
POE (20) monolaurate. *See* PEG-20 laurate
POE (32) monolaurate. *See* PEG-32 laurate
POE (75) monolaurate. *See* PEG-75 laurate
POE (150) monolaurate. *See* PEG-150 laurate
POE (2) monolaurate self-emulsifying. *See* PEG-2 laurate SE
POE (2) monooleate. *See* PEG-2 oleate
POE (4) monooleate. *See* PEG-4 oleate
POE (6) monooleate. *See* PEG-6 oleate
POE (8) monooleate. *See* PEG-8 oleate
POE (12) monooleate. *See* PEG-12 oleate
POE (20) monooleate. *See* PEG-20 oleate
POE (32) monooleate. *See* PEG-32 oleate
POE (75) monooleate. *See* PEG-75 oleate
POE (150) monooleate. *See* PEG-150 oleate
POE (2) monooleate self-emulsifying. *See* PEG-2 oleate SE
POE (12) monoricinoleate. *See* PEG-12 ricinoleate
POE 1500 monostearate. *See* PEG-6-32 stearate
POE (2) monostearate self-emulsifying. *See* PEG-2 stearate SE
POE (3) nonyl phenyl ether. *See* Nonoxynol-3
POE (4) nonyl phenyl ether. *See* Nonoxynol-4
POE (5) nonyl phenyl ether. *See* Nonoxynol-5

POE (6) nonyl phenyl ether. *See* Nonoxynol-6
POE (7) nonyl phenyl ether. *See* Nonoxynol-7
POE (8) nonyl phenyl ether. *See* Nonoxynol-8
POE (9) nonyl phenyl ether. *See* Nonoxynol-9
POE (10) nonyl phenyl ether. *See* Nonoxynol-10
POE (12) nonyl phenyl ether. *See* Nonoxynol-12
POE (13) nonyl phenyl ether. *See* Nonoxynol-13
POE (14) nonyl phenyl ether. *See* Nonoxynol-14
POE (20) nonyl phenyl ether. *See* Nonoxynol-20
POE (30) nonyl phenyl ether. *See* Nonoxynol-30
POE (35) nonyl phenyl ether. *See* Nonoxynol-35
POE (40) nonyl phenyl ether. *See* Nonoxynol-40
POE (50) nonyl phenyl ether. *See* Nonoxynol-50
POE (70) nonyl phenyl ether. *See* Nonoxynol-70
POE (100) nonyl phenyl ether. *See* Nonoxynol-100
POE (120) nonyl phenyl ether. *See* Nonoxynol-120
POE (6) nonyl phenyl ether phosphate. *See* Nonoxynol-6 phosphate
POE (10) nonyl phenyl ether phosphate. *See* Nonoxynol-10 phosphate
POE (6) nonyl phenyl ether sulfate, sodium salt. *See* Sodium nonoxynol-6 sulfate
POE (8) nonyl phenyl ether sulfate, sodium salt. *See* Sodium nonoxynol-8 sulfate
POE (3) octyl phenyl ether. *See* Octoxynol-3
POE (7) octyl phenyl ether. *See* Octoxynol-7
POE (9) octyl phenyl ether. *See* Octoxynol-9
POE (10) octyl phenyl ether. *See* Octoxynol-10
POE (16) octyl phenyl ether. *See* Octoxynol-16
POE (25) octyl phenyl ether. *See* Octoxynol-25
POE (30) octyl phenyl ether. *See* Octoxynol-30
POE (33) octyl phenyl ether. *See* Octoxynol-33
POE (40) octyl phenyl ether. *See* Octoxynol-40
POE (70) octyl phenyl ether. *See* Octoxynol-70
POE (9) octyl phenyl ether carboxylic acid. *See* Octoxynol-9 carboxylic acid
POE (6) octyl phenyl ether sulfate, sodium salt. *See* Sodium octoxynol-6 sulfate
POE (2) oleyl amine. *See* PEG-2 oleamine
POE (15) oleyl amine. *See* PEG-15 oleamine
POE (30) oleyl amine. *See* PEG-30 oleamine
POE (4) oleyl ether. *See* Oleth-4
POE (7) oleyl ether. *See* Oleth-7
POE (9) oleyl ether. *See* Oleth-9
POE (20) oleyl ether. *See* Oleth-20
POE (4) oleyl ether phosphate. *See* Oleth-4 phosphate
POE (16) POP (12) monobutyl ether. *See* PPG-12-buteth-16
POE (27) POP (24) monobutyl ether. *See* PPG-24-buteth-27
POE (35) POP (28) monobutyl ether. *See* PPG-28-buteth-35
POE (45) POP (33) monobutyl ether. *See* PPG-33-buteth-45
POE (4) sorbitan monolaurate. *See* Polysorbate 21
POE (20) sorbitan monolaurate. *See* Polysorbate 20
POE (5) sorbitan monooleate. *See* Polysorbate 81
POE (20) sorbitan monooleate. *See* Polysorbate 80
POE (20) sorbitan monopalmitate. *See* Polysorbate 40
POE (4) sorbitan monostearate. *See* Polysorbate 61
POE (20) sorbitan monostearate. *See* Polysorbate 60
POE (30) sorbitan tetraoleate. *See* PEG-30 sorbitan tetraoleate
POE (40) sorbitan tetraoleate. *See* PEG-40 sorbitan tetraoleate
POE (60) sorbitan tetraoleate. *See* PEG-60 sorbitan tetraoleate
POE (20) sorbitan trioleate. *See* Polysorbate 85
POE (20) sorbitan tristearate. *See* Polysorbate 65
POE (40) sorbitol hexaoleate. *See* PEG-40 sorbitan hexaoleate
POE (50) sorbitol hexaoleate. *See* PEG-50 sorbitan hexaoleate
POE (2) soya amine. *See* PEG-2 soyamine
POE (5) soya amine. *See* PEG-5 soyamine
POE (10) soya amine. *See* PEG-10 soyamine
POE (15) soya amine. *See* PEG-15 soyamine

POE (4) stearate. *See* PEG-4 stearate
POE (6) stearate. *See* PEG-6 stearate
POE (8) stearate. *See* PEG-8 stearate
POE (9) stearate. *See* PEG-9 stearate
POE (12) stearate. *See* PEG-12 stearate
POE (20) stearate. *See* PEG-20 stearate
POE (50) stearate. *See* PEG-50 stearate
POE (75) stearate. *See* PEG-75 stearate
POE (150) stearate. *See* PEG-150 stearate
POE (5) stearyl amine. *See* PEG-5 stearamine
POE (10) stearyl amine. *See* PEG-10 stearamine
POE (50) stearyl amine. *See* PEG-50 stearamine
POE (5) tallow amine. *See* PEG-5 hydrogenated tallow amine
POE (15) tallow amine. *See* PEG-15 hydrogenated tallow amine
POE (50) tallow amine. *See* PEG-50 hydrogenated tallow amine
POE (2) tallowate. *See* PEG-2 tallowate
POE (14) tridecyl ether. *See* Trideceth-14
POE (15) tridecyl ether. *See* Trideceth-15
POE (3) tridecyl ether phosphate. *See* Trideceth-3 phosphate
POE (6) tridecyl ether phosphate. *See* Trideceth-6 phosphate
POE (10) tridecyl ether phosphate. *See* Trideceth-10 phosphate

Poloxamer 182
CAS 9003-11-6 (generic)
Classification: Polyoxyethylene, polyoxypropylene block polymer
Formula: $HO(CH_2CH_2O)_x(CCH_3HCH_2O)_y(CH_2CH_2O)_zH$, avg. x=8, y=30, z=8
Properties: Dens. 1.018; flash pt. > 110 C
Toxicology: Moderately toxic by ingestion, intraperitoneal route
Uses: Nonionic surfactant; as food additives, defoamers, antistats, demulsifiers, detergents, wetting agents, gellants, emulsifiers, dispersants, dye levelers; emulsion polymerization
Regulatory: FDA 21CFR §172.808, 173.340, 175.105, 176.180, 176.200, 176.210, 177.1200
Manuf./Distrib.: Aldrich
Trade names: Antarox® L-62; Antarox® PGP 18-2

Poloxamer 184
CAS 9003-11-6 (generic)
Classification: Polyoxyethylene, polyoxypropylene block polymer
Formula: $HO(CH_2CH_2O)_x(CCH_3HCH_2O)_y(CH_2CH_2O)_zH$, avg. x=13, y=30, z=13
Properties: Dens. 1.018; flash pt. > 110 C
Toxicology: Moderately toxic by ingestion, intraperitoneal route
Uses: Nonionic surfactant; as food additives, defoamers, antistats, demulsifiers, detergents, wetting agents, gellants, emulsifiers, dispersants, dye levelers; emulsion polymerization; resin plasticizer
Regulatory: FDA 21CFR §172.808, 173.340, 175.105, 176.180, 176.200, 176.210, 177.1200, 177.1210
Manuf./Distrib.: Aldrich
Trade names: Antarox® L-64

Poloxamer 188
CAS 9003-11-6 (generic)
Classification: Polyoxyethylene, polyoxypropylene block polymer
Formula: $HO(CH_2CH_2O)_x(CCH_3HCH_2O)_y(CH_2CH_2O)_zH$, avg. x=75, y=30, z=75
Properties: Flakeable solid; m.p. 50 C min.; cloud pt. > 100 C (10% aq.)
Toxicology: Moderately toxic by ingestion, intraperitoneal route
Uses: Nonionic surfactant; as food additives, defoamers, antistats, demulsifiers, detergents, wetting agents, gellants, emulsifiers, dispersants, dye levelers; latex stabilizer
Regulatory: FDA 21CFR §172.808, 173.340, 175.105, 176.180, 176.200, 176.210, 177.1200, 177.1210
Trade names: Antarox® PGP 18-8

Poloxamer 238
CAS 9003-11-6 (generic)
Classification: Polyoxyethylene, polyoxypropylene block polymer
Formula: $HO(CH_2CH_2O)_x(CCH_3HCH_2O)_y(CH_2CH_2O)_zH$, avg. x=97, y=39, z=97
Properties: Dens. 1.018; flash pt. > 110 C
Toxicology: Moderately toxic by ingestion, intraperitoneal route

Uses: Nonionic surfactant; as food additives, defoamers, antistats, demulsifiers, detergents, wetting agents, gellants, emulsifiers, dispersants, dye levelers; latex stabilizer
Regulatory: FDA 21CFR §172.808, 173.340, 175.105, 176.180, 176.200, 176.210, 177.1200, 177.1210
Trade names: Antarox® F88

Poloxamine 304
CAS 11111-34-5 (generic)
Definition: Polyoxyethylene, polyoxypropylene block polymer of ethylene diamine
Uses: Emulsifier, thickener, wetting agent, dispersant, solubilizer, stabilizer for cosmetics and pharmaceuticals; demulsifier in petrol. industry; detergent ingred.; antistat for polyethylene and resin molding powds.
Trade names: Tetronic® 304

Poloxamine 504
CAS 11111-34-5 (generic)
Definition: Polyoxyethylene, polyoxypropylene block polymer of ethylene diamine
Uses: Emulsifier, thickener, wetting agent, dispersant, solubilizer, stabilizer for cosmetics and pharmaceuticals; demulsifier in petrol. industry; detergent ingred.; antistat for polyethylene and resin molding powds.

Poloxamine 701
CAS 11111-34-5 (generic)
Definition: Polyoxyethylene, polyoxypropylene block polymer of ethylene diamine
Uses: Emulsifier, thickener, wetting agent, dispersant, solubilizer, stabilizer for cosmetics and pharmaceuticals; demulsifier in petrol. industry; detergent ingred.; antistat for polyethylene and resin molding powds.
Trade names: Tetronic® 701

Poloxamine 702
CAS 11111-34-5 (generic)
Definition: Polyoxyethylene, polyoxypropylene block polymer of ethylene diamine
Uses: Emulsifier, thickener, wetting agent, dispersant, solubilizer, stabilizer for cosmetics and pharmaceuticals; demulsifier in petrol. industry; detergent ingred.; antistat for polyethylene and resin molding powds.

Poloxamine 704
CAS 11111-34-5 (generic)
Definition: Polyoxyethylene, polyoxypropylene block polymer of ethylene diamine
Uses: Emulsifier, thickener, wetting agent, dispersant, solubilizer, stabilizer for cosmetics and pharmaceuticals; demulsifier in petrol. industry; detergent ingred.; antistat for polyethylene and resin molding powds.
Trade names: Tetronic® 704

Poloxamine 707
CAS 11111-34-5 (generic)
Definition: Polyoxyethylene, polyoxypropylene block polymer of ethylene diamine
Uses: Emulsifier, thickener, wetting agent, dispersant, solubilizer, stabilizer for cosmetics and pharmaceuticals; demulsifier in petrol. industry; detergent ingred.; antistat for polyethylene and resin molding powds.

Poloxamine 901
CAS 11111-34-5 (generic)
Definition: Polyoxyethylene, polyoxypropylene block polymer of ethylene diamine
Uses: Emulsifier, thickener, wetting agent, dispersant, solubilizer, stabilizer for cosmetics and pharmaceuticals; demulsifier in petrol. industry; detergent ingred.; antistat for polyethylene and resin molding powds.
Trade names: Tetronic® 901

Poloxamine 904
CAS 11111-34-5 (generic)
Definition: Polyoxyethylene, polyoxypropylene block polymer of ethylene diamine
Uses: Emulsifier, thickener, wetting agent, dispersant, solubilizer, stabilizer for cosmetics and pharmaceuticals; demulsifier in petrol. industry; detergent ingred.; antistat for polyethylene and resin molding powds.
Trade names: Tetronic® 904

Poloxamine 908
CAS 11111-34-5 (generic)
Definition: Polyoxyethylene, polyoxypropylene block polymer of ethylene diamine
Uses: Emulsifier, thickener, wetting agent, dispersant, solubilizer, stabilizer for cosmetics and pharmaceuticals; demulsifier in petrol. industry; detergent ingred.; antistat for polyethylene and resin molding powds.
Trade names: Tetronic® 908

Poloxamine 1101
CAS 11111-34-5 (generic)
Definition: Polyoxyethylene, polyoxypropylene block polymer of ethylene diamine
Uses: Emulsifier, thickener, wetting agent, dispersant, solubilizer, stabilizer for cosmetics and pharmaceuticals; demulsifier in petrol. industry; detergent ingred.; antistat for polyethylene and resin molding powds.

Poloxamine 1102
CAS 11111-34-5 (generic)
Definition: Polyoxyethylene, polyoxypropylene block polymer of ethylene diamine
Uses: Emulsifier, thickener, wetting agent, dispersant, solubilizer, stabilizer for cosmetics and pharmaceuticals; demulsifier in petrol. industry; detergent ingred.; antistat for polyethylene and resin molding powds.

Poloxamine 1104
CAS 11111-34-5 (generic)
Definition: Polyoxyethylene, polyoxypropylene block polymer of ethylene diamine
Uses: Emulsifier, thickener, wetting agent, dispersant, solubilizer, stabilizer for cosmetics and pharmaceuticals; demulsifier in petrol. industry; detergent ingred.; antistat for polyethylene and resin molding powds.

Poloxamine 1107
CAS 11111-34-5 (generic)
Definition: Polyoxyethylene, polyoxypropylene block polymer of ethylene diamine
Uses: Emulsifier, thickener, wetting agent, dispersant, solubilizer, stabilizer for cosmetics and pharmaceuticals; demulsifier in petrol. industry; detergent ingred.; antistat for polyethylene and resin molding powds.

Poloxamine 1301
CAS 11111-34-5 (generic)
Definition: Polyoxyethylene, polyoxypropylene block polymer of ethylene diamine
Uses: Emulsifier, thickener, wetting agent, dispersant, solubilizer, stabilizer for cosmetics and pharmaceuticals; demulsifier in petrol. industry; detergent ingred.; antistat for polyethylene and resin molding powds.

Poloxamine 1302
CAS 11111-34-5 (generic)
Definition: Polyoxyethylene, polyoxypropylene block polymer of ethylene diamine
Uses: Emulsifier, thickener, wetting agent, dispersant, solubilizer, stabilizer for cosmetics and pharmaceuticals; demulsifier in petrol. industry; detergent ingred.; antistat for polyethylene and resin molding powds.

Poloxamine 1304
CAS 11111-34-5 (generic)
Definition: Polyoxyethylene, polyoxypropylene block polymer of ethylene diamine
Uses: Emulsifier, thickener, wetting agent, dispersant, solubilizer, stabilizer for cosmetics and pharmaceuticals; demulsifier in petrol. industry; detergent ingred.; antistat for polyethylene and resin molding powds.

Poloxamine 1307
CAS 11111-34-5 (generic)
Definition: Polyoxyethylene, polyoxypropylene block polymer of ethylene diamine
Uses: Emulsifier, thickener, wetting agent, dispersant, solubilizer, stabilizer for cosmetics and pharmaceuticals; demulsifier in petrol. industry; detergent ingred.; antistat for polyethylene and resin molding powds.
Trade names: Tetronic® 1307

Poloxamine 1501
CAS 11111-34-5 (generic)
Definition: Polyoxyethylene, polyoxypropylene block polymer of ethylene diamine
Uses: Emulsifier, thickener, wetting agent, dispersant, solubilizer, stabilizer for cosmetics and pharmaceuticals; demulsifier in petrol. industry; detergent ingred.; antistat for polyethylene and resin molding powds.

Poloxamine 1502
CAS 11111-34-5 (generic)
Definition: Polyoxyethylene, polyoxypropylene block polymer of ethylene diamine
Uses: Emulsifier, thickener, wetting agent, dispersant, solubilizer, stabilizer for cosmetics and pharmaceuticals; demulsifier in petrol. industry; detergent ingred.; antistat for polyethylene and resin molding powds.

Poloxamine 1504
CAS 11111-34-5 (generic)

Definition: Polyoxyethylene, polyoxypropylene block polymer of ethylene diamine
Uses: Emulsifier, thickener, wetting agent, dispersant, solubilizer, stabilizer for cosmetics and pharmaceuticals; demulsifier in petrol. industry; detergent ingred.; antistat for polyethylene and resin molding powds.

Poloxamine 1508
CAS 11111-34-5 (generic)
Definition: Polyoxyethylene, polyoxypropylene block polymer of ethylene diamine
Uses: Emulsifier, thickener, wetting agent, dispersant, solubilizer, stabilizer for cosmetics and pharmaceuticals; demulsifier in petrol. industry; detergent ingred.; antistat for polyethylene and resin molding powds.

Polyacrylamide
CAS 9003-05-8
Synonyms: 2-Propenamide, homopolymer
Definition: Polyamide of acrylic monomers
Empirical: $(C_3H_5NO)_x$
Formula: $[CH_2CHCONH_2]_x$
Properties: Wh. solid; water-sol. high polymer; m.w. 10,000-18,000,000; dens. 1.302
Toxicology: LD50 (mouse, IP) 170 mg/kg (monomer); heated to decomp., emits acrid smoke and irritating fumes
Uses: Thickener, dispersant, antiprecipitant, solubilizer, binder, sizing, flocculating, suspending, crosslinking agent, filtering aid, lubricant; used in adhesives, agric., cement, coatings, cosmetics, detergents, latex mfg., plaster, printing ink
Regulatory: FDA 21CFR §172.255, 173.10, 173.315 (10 ppm in wash water), 175.105, 176.180
Manuf./Distrib.: Aldrich; Allchem Ind.; Allied Colloids; Allied Colloids; Amresco; Calgon; Cytec; Rhone-Poulenc Water Treatment
Trade names: Cyanamer N300LMW; Cyanamer P-21

Polyacrylate
CAS 9003-01-4
Synonyms: Acrylic resin
Definition: Sol'ns. of polyacrylic acid
Properties: M.w. $\approx$ 41,000
Manuf./Distrib.: Polysar BV; Scott Bader Ltd
Trade names containing: BYK®-S 706

Polyacrylic acid
CAS 9003-01-4
Synonyms: 2-Propenoic acid, homopolymer; Acrylic acid polymers
Definition: Polymer of acrylic acid
Empirical: $(C_3H_4O_2)_x$
Formula: $[CH_2CHCOOH]_x$
Properties: M.w. 168.06
Toxicology: Possible carcinogen
Uses: Emulsifier, thickener, stabilizer, suspending agent, coupler, moisturizer, emollient, dispersant, sequestrant, thickener, gellant; used for latexes, emulsion paints, drilling muds, photosensitive emulsions, water treatment, cosmetics, adhesives
Regulatory: FDA 21CFR §175.105, 175.300, 175.320, 176.180
Manuf./Distrib.: Alco; Aldrich; Anedco; CPS; BFGoodrich; Rhone-Poulenc
See also Acrylic resin

Poly(acrylic acid), ammonium salt. *See* Ammonium polyacrylate
Polyacrylic acid, sodium salt. *See* Sodium polyacrylate

Polyamide
Definition: A high m.w. polymer in which amide linkages (—CONH) occur along the molecular chain; may be either natural or synthetic
Uses: Natural polyamides include casein, soybean and peanut proteins, zein; synthetic polyamides typified by various nylons; used for plastics, textile fibers, adhesives; epoxy curing agent
Manuf./Distrib.: Aldrich; Arizona; Ashley Polymers; BASF; EMS-Am. Grilon; Georgia-Pacific; Hoechst Celanese; Hüls AG; Miles; Monsanto; SNIA UK; Union Camp
Trade names: Ancamide 220; Ancamide 400; Ancamide 2050; Casamid® 360; Casamid® 362W; Epotuf Hardener 37-612; Epotuf Hardener 37-618; Epotuf Hardener 37-621; Epotuf Hardener 37-625; Epotuf Hardener 37-640; Uni-Rez® 1502; Uni-Rez® 2100-P75; Uni-Rez® 2341; Uni-Rez® 2355; Versamid® 125;

Versamid® 140; Versamid® 150; Versamid® 674
Trade names containing: Ancamide 220-IPA-73; Ancamide 220-X-70; Ancamide 400-BX-60; Ancamide 700-B-75; Casamid® 350PM
See also Nylon

Polybutadiene
CAS 9003-17-2
Synonyms: BR; Butadiene rubber; cis-Polybutadiene; Polybutadiene rubber
Empirical: $(C_4H_6)_n$
Formula: $(CH_2CH{=}CHCH_2)_n$
Properties: M.w. 1000-233,000
Precaution: May explode
Uses: Elastomer for tire industry, footwear, molded goods; blending ingred. in SBR; impact modifier, additive for plastics; coating resin in liq. form
Manuf./Distrib.: Ameripol Synpol; Asahi Chem Industry Co Ltd; BASF; Firestone Syn. Rubber & Latex; Goodyear; Phillips; Reichhold; A. Schulman; Scientific Polymer Prods.
Trade names: Buna CB 14; Buna CB HX 529; Buna CB HX 530; Buna CB HX 565; Cariflex BR 1220; Diene® 35AC3; Diene® 35AC10; Diene® 55AC10; Diene® 55NF; Diene® 70AC; Taipol BR 015H; Taktene 1202; Taktene 1203; Taktene 1203G1

1,2-Polybutadiene
CAS 29406-96-0 (syndiotactic)
Uses: Thermosetting resin for wire coating, EPDM peroxide-cured modifier, coatings, processing aid
Trade names: Ricon 130; Ricon 131; Ricon 142; Ricon 152; Ricon 153; Ricon 157

cis-Polybutadiene. *See* Polybutadiene
Polybutadiene rubber. *See* Polybutadiene

1,2-Polybutadiene-styrene copolymer
Uses: Thermosetting resin used in coatings, liq. rubber; rubber and ink additive
Trade names: Ricon 181

Polybutene
CAS 9003-28-5, 9003-29-6
Synonyms: PIB; Polybutylene; 1-Butene, homopolymer
Definition: Polymer formed by polymerization of a mixture of iso- and normal butenes
Empirical: $(C_4H_8)_x$
Formula: $[CH_2CH(C_2H_5)]_n$
Properties: M.w. 500-75,000; dens. 0.910
Uses: Thermoplastic resin; used as tackifier, strengthener, and extender in adhesives, as plasticizer for rubber, as vehicle and fugitive binder for coatings, as cling additive for LLDPE stretch wrap films, as reactive intermediate for specialty chemicals
Regulatory: FDA 21CFR §175.105, 175.125, 177.1570, 177.2600,178.3750
Manuf./Distrib.: Amoco; Ashland; BP Chem. Ltd; Harcros; Monomer-Polymer & Dajac
Trade names: Amoco® H-15; Amoco® L-14; Amoco® L-50

Polybutene, hydrogenated
CAS 68937-10-0
Uses: Used for cosmetics, metalworking lubricants, cable oils, thermoplastic modifier, rubber modifier
Trade names: Amoco® H-300E

Polybutylated bisphenol A
CAS 68784-69-0
Uses: Antioxidant for rubber, latex, automotive and appliance prods.
Trade names: Agerite® Superlite; Agerite® Superlite Solid
Trade names containing: Agerite® Superlite® Emulcon®

Polybutylene. *See* Polybutene

Polybutylene terephthalate
CAS 26062-94-2
Synonyms: PBT; PBT polyester
Classification: Organic compd.

Empirical: $(C_8H_6O_4 \cdot C_4H_{10}O_2)_x$
Properties: Dens. 1.310
Uses: Thermoplastic polyester resin
Manuf./Distrib.: Aldrich
Trade names containing: Mastertek® PBT 3659; Reed C-PBT-1338

Polycarbodiimide
Trade names containing: Rhenogran® P-50/EVA

Polycarbonamide
CAS 81972-48-7
Uses: Antidegradant for rubbers

Polycarbonate. *See* Polycarbonate resin

Polycarbonate resin
Synonyms: PC; Polycarbonate; PC resin
Uses: Thermoplastic resin for molded prods., sol'n. cast or extruded film, structural parts, tube and piping, prosthetic devices, optical parts, windows, computer and business equip., household appliances, compact disks, food contact and medical applics.; modifier for other polymers; reinforcing modifier for recycled polymers and alloys
Manuf./Distrib.: Aldrich; Ashland; DSM; Ferro; LNP; Miles; Monomer-Polymer & Dajac; Shuman Plastics; Westlake Plastics
Trade names: Blendex® HPP 801; Blendex® HPP 802; Blendex® HPP 803; Blendex® HPP 804
Trade names containing: Mastertek® PC 3686; MDI PC-925

Polychloroprene
CAS 126-99-8
Synonyms: CR; Neoprene; 2-Chlorobutadiene 1,3; Chloroprene rubber
Classification: Rubber
Uses: Elastomer for molding, extrusion, and calendering for adhesive compounding, construction, automotive, hose and cable jackets, conveyor belts, closed cell sponge, etc.; vulcanizing plasticizer for elastomers
Trade names: Neoprene FB

Polychlorotrifluoroethylene
CAS 9002-83-9
Synonyms: PCTFE; Chlorotrifluoroethene, homopolymer
Classification: Polymer of chlorotrifluoroethylene
Empirical: $(C_2ClF_3)_x$
Formula: $[CF_2CFCl]_x$
Uses: Lubricant and functional fluid for chem., aerospace, hydraulic fluids, metalworking, nuclear, pulp/paper, other industries; mold release for plastics, rubber; plasticizer
Regulatory: FDA 21CFR §177.1380
Trade names: Halocarbon Grease 25-10M; Halocarbon Grease 25-20M; Halocarbon Grease X90-10M; Halocarbon Oil 0.8; Halocarbon Oil 1.8; Halocarbon Oil 4.2; Halocarbon Oil 6.3; Halocarbon Oil 27; Halocarbon Oil 56; Halocarbon Oil 95; Halocarbon Oil 200; Halocarbon Oil 400; Halocarbon Oil 700; Halocarbon Oil 1000N; Halocarbon Wax 40; Halocarbon Wax 600; Halocarbon Wax 1200; Halocarbon Wax 1500
Trade names containing: Halocarbon Grease 19; Halocarbon Grease 25-5S; Halocarbon Grease 28; Halocarbon Grease 28LT; Halocarbon Grease 32

Polydibromophenylene oxide
Uses: Flame retardant for polymers
Manuf./Distrib.: Great Lakes
Trade names: Great Lakes PO-64P™

Poly(dibromostyrene)
Uses: Flame retardant for thermoplastics
Manuf./Distrib.: Great Lakes
Trade names: Great Lakes PDBS-10™; Great Lakes PDBS-80™

Poly-(1,2-dihydro-2,2,4-trimethylquinoline). *See* 2,2,4-Trimethyl-1,2-dihydroquinoline polymer

Polydimethyldiphenyl siloxane
Uses: Lubricant, dielectric coolant; low temp. damping applics. for aircraft instruments and electronic equipment; heat transfer media; base fluid in silicone greases for ball bearing lubrication; plastics additive
Trade names: SF1265

Poly 2,6-dimethyl-1,4-phenylene ether. *See* Polyphenylene ether
Polydimethylsiloxane. *See* Dimethicone

Poly-p-dinitrosobenzene polymer
CAS 9003-34-3
Empirical: $(C_6H_4N_2O_2)_x$
Toxicology: Moderately toxic by ingestion
Uses: Chemical conditioner for IIR, butyl rubber, cements
Trade names: Poly DNB®; Vanax® PY

Polydipentene
CAS 9003-73-0
Definition: Prod. formed by polymerization of terpene hydrocarbons
Uses: Thermoplastic resin used as tackifier in adhesives, modifier resin in rubber compds., in hot-melt coatings, laminations, wax modification, as masticatory agents in chewing gums
Trade names: Piccolyte® C115; Zonarez® 7085

Poly (dipropyleneglycol) phenyl phosphite
CAS 80584-86-7; EINECS 279-499-4
Properties: Liq.; m.w. 2102; dens. 1.168-1.180; b.p. will not distill; flash pt. (PMCC) 179 C; ref. index 1.5340-1.5380
Uses: Stabilizer for PVC
Manuf./Distrib.: Dover
Trade names: Doverphos® 12; Weston® DHOP

Polyester adipate
Uses: Plasticizer for PVC, NC, CAP, chlorinated rubber; pigment grinding medium
Manuf./Distrib.: C.P. Hall; Hüls
Trade names: ADK CIZER PN-250; ADK CIZER PN-650; Admex® 515; Admex® 525; Admex® 760; Admex® 761; Admex® 770; Admex® 775; Admex® 910; Admex® 6187; Admex® 6969; Admex® 6985; Admex® 6994; Admex® 6995; Admex® 6996; Paraplex® G-40; Paraplex® G-41; Paraplex® G-50; Paraplex® G-51; Paraplex® G-54; Paraplex® G-56; Paraplex® G-57; Paraplex® G-59; Plasthall® HA7A; Plasthall® P-612; Plasthall® P-622; Plasthall® P-643; Plasthall® P-650; Plasthall® P-670

Polyester glutarate
Uses: Plasticizer for PVC applics.; flexibilizing, permanent, nonmigrating; adhesive for film backing and varieties of tape
Manuf./Distrib.: C.P. Hall
Trade names: Plasthall® P-550; Plasthall® P-7035; Plasthall® P-7035M; Plasthall® P-7046; Plasthall® P-7092; Plasthall® P-7092D

Polyester phthalate
Uses: Plasticizer for epoxies, CR
Trade names: Plasthall® P-7068

Polyester polyol
Manuf./Distrib.: Inolex; King Ind.; Miles; Polyurethane Corp. of Am.; Polyurethane Spec.; U.S. Polymers; Unichema Int'l.; Witco/Oleo-Surf.
Trade names containing: Nyacol® AB40

Polyester resin, thermosetting
Synonyms: Polyester, unsaturated
Definition: Unsaturated polyester resin
Uses: Thermoset resin for reinforced plastics, automotive parts, protective coatings, ducts, housings, flues, laminates, pipes
Manuf./Distrib.: Aristech; Miles; Owens-Corning; Ranbar Tech.
Trade names containing: Crystic Pregel 17; Crystic Pregel 27; Nyacol® APE1540; Plastigel PG9033; Plastigel PG9037; Plastigel PG9068

See also Alkyd resin

Polyester sebacate
Uses: Plasticizer for epoxies, elec. tapes, high temp. insulation, coaxial cable, upholstery, coated fabric
Manuf./Distrib.: C.P. Hall
Trade names: Paraplex® G-25 70%; Paraplex® G-25 100%; Plasthall® P-1070

Polyester, unsaturated. *See* Polyester resin, thermosetting
Polyether glycol. *See* Polyethylene glycol

Polyether polyol
Uses: Functional polyol for polyurethane industry for industrial/consumer RIM and structural polymers, dynamic elastomers, adhesives, binders, coatings, and sealants; reactive diluent, dispersant for solv. coatings, etc.
Manuf./Distrib.: ICI Polyurethane; Miles; Witco/Oleo-Surf.
Trade names: Syn Fac® 8026; Syn Fac® 8031

Polyethylene
CAS 9002-88-4; EINECS 200-815-3
Synonyms: Ethene, homopolymer; Ethylene homopolymer
Definition: Polymer of ethylene monomers
Empirical: $(C_2H_4)_x$
Formula: $[CH_2CH_2]_x$
Properties: Wh. translucent partially cryst./partially amorphous plastic solid, odorless; sol. in hot benzene; insol. in water; m.w. 1500-100,000; dens. 0.92 (20/4 C); m.p. 85-110 C
Precaution: Combustible; store in well closed containers; reacts violently with F_2
Toxicology: Suspected carcinogen and tumorigen by implants; heated to decomp., emits acrid smoke and irritating fumes
Uses: Laboratory tubing; prostheses; elec. insulation; pkg. materials; kitchenware; tank and pipe linings; paper coatings; textile stiffeners; additive in inks, adhesives, paper coatings, floor finishes, cosmetics, plastics, rubber, textiles
Regulatory: FDA 21CFR §172.260, 172.615 (m.w. 2000-21,000), 173.20, 175.105, 175.300, 176.180, 176.200, 176.210, 177.1200, 177.1520, 177.2600, 178.3570, 178.3850
Manuf./Distrib.: Asahi Chem Industry Co Ltd; Ashland; Eastman; Elf Atochem SA; Exxon; LNP; Mitsubishi Petrochem.; Quantum/USI
Trade names: A-C® 6; A-C® 6A; A-C® 8, 8A; A-C® 9, 9A, 9F; A-C® 15; A-C® 16; A-C® 617, 617A; A-C® 712; A-C® 715; A-C® 725; A-C® 735; A-C® 1702; ACumist® A-6; ACumist® A-12; ACumist® A-18; ACumist® A-45; ACumist® B-6; ACumist® B-9; ACumist® B-12; ACumist® B-18; ACumist® C-5; ACumist® C-9; ACumist® C-12; ACumist® C-18; ACumist® C-30; ACumist® D-9; Cabot® CP 9396; Cabot® PE 9324; Epolene® C-18; Epolene® N-10; Epolene® N-11P; Epolene® N-14; Epolene® N-14P; Epolene® N-20; Epolene® N-21; Epolene® N-34; Hoechst Wax PE 130; Hoechst Wax PE 190; Luwax® AH 6; Luwax® AL 3; Luwax® AL 61; Octowax 518; Polywax® 500; Polywax® 655; Polywax® 850; Polywax® 1000; Polywax® 2000; Polywax® 3000; Primax UH-1060; Primax UH-1080; Primax UH-1250; Stat-Kon® AS-F; Stat-Kon® AS-FE
Trade names containing: Ampacet 19500; Ampacet 19552; Armostat® 375; Aztec® HMCN-40-IC3; Aztec® 3,3,5-TRI-28-IC3; Cabot® PE 9006; Cabot® PE 9026; Cabot® PE 9229; Colloids PE 48/10/06; Colloids PE 48/10/10; Colloids PE 48/10/11; Colloids PE 48/10/12; Colloids PE 48/10/13; Colloids PE 48/32 C; Colloids PE 48/32 F; Colloids PE 48/58; Colloids PE 48/61 A; Colloids PE 48/61 F; Colloids PE 48/61/30; Colloids PE 48/62 A; Colloids PE 48/67; Colloids PE 48/72; Colloids PE 48/73; Colloids PE 48/76; Colloids PE 48/80; Colloids PE 48/85; Irganox® B 501W; Plasblak® PE 1380; Polyace™ 804

Polyethylene elastomer, chlorinated
Synonyms: CM
Uses: Elastomer for molding and extrusion, cable and wire insulation, hose, tubing, tech. moldings
Trade names containing: Poly-Dispersion® H (202)D-80; Poly-Dispersion® H (202)D-80P; Poly-Dispersion® H (DYT)D-80; Poly-Dispersion® H (DYT)D-80P; Poly-Dispersion® K (NMC)D-70; Poly-Dispersion® K (PX-17)D-30; Wyfire® H-A-85; Wyfire® H-A-85P

Polyethylene glycol
CAS 25322-68-3; EINECS 203-473-3
Synonyms: PEG; α-Hydroxy-omega-hydroxy poly(oxy-1,2-ethanediyl); Polyglycol; Polyether glycol
Definition: Condensation polymers of ethylene glycol
Formula: $H(OC_2H_4)_nOH$

Polyethylene glycol (2000)

Properties: Clear liq. or wh. solid; sol. in org. solvs., aromatic hydrocarbons; dens. 1.110-1.140 (20 C); m.p. 4-10 C; flash pt. 471 C
Precaution: Combustible liq.
Toxicology: LD50 (oral, rat) 33,750 mg/kg; sl. toxic by ingestion; skin and eye irritant; heated to decomp., emits acrid smoke and irritating fumes
Uses: Chemical intermediates, plasticizers, softeners, humectants, ointments, polishes, paper coating, mold lubricants, bases for cosmetics and pharmaceuticals, solvents, binders, metal and rubber processing, food additives, laboratory reagent; antistat
Regulatory: FDA 21CFR §172.210, 172.820, 173.340
Manuf./Distrib.: BASF; BP Chem. Ltd; Calgene; Dow; Du Pont; Harcros; Henkel; Hüls; Inolex; Olin; Rhone-Poulenc Surf.; Texaco; Union Carbide
Trade names: Chemstat® P-400
Trade names containing: Kosmos® 64
See also PEG...

Polyethylene glycol (2000). *See* PEG-2M

Polyethylene, high-density
Synonyms: HDPE
Uses: Blow- and inj.-molded goods, film and sheet, piping, fibers, gasoline and oil containers; additive for inks; lubricant in plastics processing; flatting/antisettling agent in paints; dispersant in color concs.; improves hardness in waxes
Manuf./Distrib.: AlliedSignal; BP Chem. Ltd; Chevron; Chisso; Du Pont; Exxon; Hüls Am.; OxyChem; Quantum/USI; Solvay Polymers; Westlake Plastics
Trade names: Luwax® AH 3
Trade names containing: Ampacet 11418-F; Ampacet 19614; Ampacet 49984; Ampacet 190015; Dutral® PM-06PLE; MDI HD-650; MDI HD-935; MDI PE-500HD; MDI PE-907HD;Plasblak® HD 3300; Polybond® 1009; Polybond® 3009; Royalene 7565

Polyethylene, linear low density
Synonyms: LLDPE
Uses: compatibilizer in blends and alloys, polymeric coupling agents in reinforced or recycled PE or PP, adhesives, sealants
Manuf./Distrib.: BP Chem. Ltd; Neste Polyeten AB
Trade names: Fusabond® MB-110D; Fusabond® MB-226D
Trade names containing: Ampacet 11040; Ampacet 11215; Ampacet 11247; Ampacet 11560; Ampacet 11739; Ampacet 11744; Ampacet 11748; Ampacet 11851; Ampacet 11912; Ampacet 11913; Ampacet 11979; Ampacet 19492; Ampacet 19584; Ampacet 19699; Ampacet 19709; Ampacet 19717; Ampacet 19897; Ampacet 19975; Ampacet 19991; Ampacet 19999; Ampacet 49419; Ampacet 110017; Ampacet 110025; Ampacet 110070; Ampacet 190014; Chemstat® AC-1000; Hostastat® System E6952; MDI PE-500LL; MDI PE-650LL; PE-AB 07; PE-AB 19; PE-AB 22; PE-AB 30; PE-AB 33; PE-AB 48; PE-AB 73; PE-SA 21; PE-SA 24; PE-SA 40; PE-SA 49; PE-SA 60; PE-SA 72; PE-SA 92; PE-SL 04; PE-SL 31; PE-SL 98; PE-UV 12; PE-UV 118; PE-UV 119; Plasblak® LL 2590; Plasblak® LL 2612; Plasblak® LL 3282; Plaswite® LL 7014; Plaswite® LL 7015; Plaswite® LL 7041; Plaswite® LL 7049; Plaswite® LL 7057; Plaswite® LL 7105; Plaswite® LL 7108; Plaswite® LL 7112; Spectratech® PM 92742

Polyethylene, low-density
Synonyms: LDPE
Uses: Packaging film, food packaging, paper coating, liners for drums, wire and cable coating, toys, cordage, waste bags, chewing gum base, squeeze bottles, electrical insulation; processing aid in rubbers
Manuf./Distrib.: AlliedSignal; Chevron; Eastman; Exxon; Hüls Am.; Neste Polyeten AB; Quantum/USI; Westlake Plastics
Trade names: Akrowax PE
Trade names containing: Ampacet 10063; Ampacet 10123; Ampacet 11058; Ampacet 11070; Ampacet 11078; Ampacet 11171; Ampacet 11171-101; Ampacet 11215; Ampacet 11246; Ampacet 11299; Ampacet 11338; Ampacet 11416; Ampacet 11495-S; Ampacet 11560; Ampacet 11572; Ampacet 11578-P; Ampacet 11739; Ampacet 11853; Ampacet 11875; Ampacet 11919-S; Ampacet 19153; Ampacet 19153-S; Ampacet 19200; Ampacet 19270; Ampacet 19278; Ampacet 19400; Ampacet 19458; Ampacet 19475-S; Ampacet 19673; Ampacet 19716; Ampacet 19832; Ampacet 39270; Ampacet 41424; Ampacet 41495; Ampacet 49270; Ampacet 110041; Ampacet 110052; Cabot® PE 6008; Cabot® PE 9007; Cabot® PE 9020; Cabot® PE 9041; Cabot® PE 9166; Cabot® PE 9171; Cabot® PE 9172; Cabot® PE 9174; Cabot® PE 9321; Hostastat® System E1956; Mastertek® PE 3660; MDI EC-940; MDI PE-135; MDI PE-500; MDI PE-500-20F; MDI PE-500A; MDI PE-540; MDI PE-650F; MDI PE-650HF; MDI PE-670T; MDI PE-675; MDI

PE-907; MDI PE-931WC; MDI PE-940; Plasblak® PE 1371; Plasblak® PE 1851; Plasblak® PE 1873; Plasblak® PE 2249; Plasblak® PE 2272; Plasblak® PE 2377; Plasblak® PE 2479; Plasblak® PE 2515; Plasblak® PE 2570; Plasblak® PE 3006; Plasblak® PE 3168; Plasblak® PE 4056; Plasblak® PE 4135; Plasblak® UN 2014; Plasblak® UN 2016; Plasdeg PE 9321; Plaswite® PE 7000; Plaswite® PE 7001; Plaswite® PE 7002; Plaswite® PE 7003; Plaswite® PE 7004; Plaswite® PE 7005; Plaswite® PE 7006; Plaswite® PE 7024; Plaswite® PE 7031; Plaswite® PE 7079; Plaswite® PE 7097; Plaswite® PE 7192; Thermoguard® 8218; Thermoguard® 9010; Thermoguard® DM-9406

Polyethylene, oxidized

CAS 68441-17-8

Synonyms: Oxidized polyethylene; Ethene, homopolymer, oxidized

Definition: Reaction prod. of polyethylene and oxygen

Properties: Dens. 0.930

Uses: Wax for polishes, finishes, adhesives, and emulsions; processing lubricant, mold release aid, textile lubricant; PVC lubricant

Regulatory: FDA 21CFR §172.260, 175.105, 175.125, 176.170, 176.200, 176.210, 177.1200, 177.1620, 177.2800

Trade names: A-C® 307, 307A; A-C® 316, 316A; A-C® 325; A-C® 330; A-C® 392; A-C® 395, 395A; A-C® 629; A-C® 629A; A-C® 655; A-C® 656; A-C® 6702; Epolene® E-10; Epolene® E-14; Epolene® E-14P; Epolene® E-15; Epolene® E-20; Epolene® E-20P; Hoechst Wax PED 191; Hostalub® H 12; Loobwax 0761; Luwax® OA 2; Luwax® OA 5; Petrolite® E-2020

Trade names containing: Hostalub® CAF 485; Hostalub® XL 355; Hostalub® XL 445

Polyethylene terephthalate

CAS 25038-59-9

Classification: Organic compd.

Definition: Fiber-forming polyesters prepared from terephthalic acid or its esters and ethylene glycol

Synonyms: PET; Poly(oxy-1,2-ethanediyloxycarbonyl-1,4-phenylenecarbonyl)

Formula: $(C_{10}H_8O_4)_n$

Properties: Solid; sol. in hot m-cresol, trifluoroacetic acid, o-chlorophenol; sp.gr. 1.38; dec. ≈ 250 C

Uses: In fabric mfg.; as films; as base for magnetic coatings; surgical aid

Trade names containing: Reed C-PET-1333

Polyethylene wax

CAS 9002-88-4

Formula: $(-CH_2CH_2-)_n$

Uses: Wax for polishes, plastics and rubber processing, printing inks, pigment master batches, hot melts; PVC lubricant; mold release

Manuf./Distrib.: AlliedSignal; Hoechst Celanese; Hüls AG; IGI; Sartomer; Stevenson Cooper; Syn. Prods.

Trade names: A-C® 7, 7A; Bowax 2015; Coathylene HA 1591; Coathylene HA 2454; Epolene® C-10; Epolene® C-10P; Epolene® C-13; Epolene® C-13P; Epolene® C-14; Epolene® C-15; Epolene® C-15P; Epolene® C-16; Epolene® C-17; Epolene® C-17P; Epolene® N-11; Hoechst Wax PE 520; Hoechst Wax PED 521; Hostalub® H 3; Hostalub® H 22; Loobwax 0651; Luwax® A; Luwax® AF 30; Oxp; Polyfin; Polymel #7; Stan Wax; Struktol® PE H-100; Struktol® TR PE(H)-100; Uniwax AW-1050; Vestowax A-217; Vestowax A-227; Vestowax A-235; Vestowax A-415; Vestowax A-616; Vestowax AO-1539; Zylac 235

Trade names containing: Irganox® 1425 WL

Polyethylene wax, oxidized

Uses: Plastic lubricant; used for films, profiles, inj. molding, in dry-bright wax emulsions for floor maintenance and leather finishes

Trade names: Petrac® 215; Sicolub® OA 2, OA 4; Vestowax AS-1550

Polyglutarimide acrylic copolymer

Uses: Heat distort. temp. modifier for PVC

Trade names: Paraloid® HT-510®

Polyglycerate 60. *See* PEG-20 glyceryl stearate

Polyglyceryl-3 oleate

CAS 9007-48-1 (generic); 33940-98-6

Synonyms: Triglyceryl oleate

Definition: Ester of oleic acid and a glycerin polymer containing an avg. 3 glycerin units

Empirical: $C_{27}H_{52}O_8$

Polyglyceryl-3 stearate

Uses: Emulsifier, solubilizer, dispersant for cosmetics, pharmaceuticals; antifog for plastics; food emulsifier
Regulatory: FDA 21CFR §172.854
Trade names: Santone® 3-1-SH

Polyglyceryl-3 stearate
CAS 37349-34-1 (generic); 27321-72-8; 26855-43-6; 61790-95-2; EINECS 248-403-2
Synonyms: Triglyceryl stearate
Definition: Ester of stearic acid and a glycerin polymer containing an avg. 3 glycerin units
Uses: Food emulsifier, stabilizer and whipping agent; emulsifier, antifog for plastics
Regulatory: FDA 21CFR §172.854

Polyglyceryl-3 stearate SE
Synonyms: Triglyceryl monostearate SE
Definition: Self-emulsifying grade of Polyglyceryl-3 stearate that contains some sodium and/or potassium stearate
Uses: O/w emulsifier, antifogging agent for plastics

Polyglyceryl-10 tetraoleate
CAS 34424-98-1; EINECS 252-011-7
Synonyms: Decaglycerol tetraoleate; Decaglyceryl tetraoleate; 9-Octadecenoic acid, tetraester with decaglycerol
Definition: Tetraester of oleic acid and a glycerin polymer containing an avg. 10 glycerin units
Empirical: $C_{102}H_{190}O_{25}$
Uses: Emulsifier, antistat, lubricant, visc. control agent, stabilizer, and coupler for textile finishes, cosmetics, lubricants, foods, plasticizers for syn. fabrics and plastics; solubilizer and carrier
Regulatory: FDA 21CFR §172.854
Trade names: Mazol® PGO-104

Polyglycol. *See* Polyethylene glycol
Poly(iminocarbonylpentamethylene). *See* Nylon-6
Poly[imino(1-oxo-1,6-hexanediyl)]. *See* Nylon-6

Polyisobutene
CAS 9003-27-4; 9003-29-6
Synonyms: PIB; Polyisobutylene; 2-Methyl-1-propene, homopolymer
Definition: Homopolymer of isobutylene
Empirical: $(C_4H_8)_x$
Formula: $[CH_2C(CH_3)HCH_2]_x$
Uses: Syn. rubber; rubber additive; thickener for lubricating oils; for the adhesive and sealant industry, elec. insulating oils, bases for chewing gums, for prod. of damp-proof courses containing fillers in construction industry
Regulatory: FDA 21CFR §172.615 (min. m.w. 37,000), 175.105, 175.125, 175.300, 176.180, 177.1200, 177.1210, 177.1420, 178.3570, 178.3740, 178.3910; Japan approved
Manuf./Distrib.: BASF; Monomer-Polymer & Dajac; Rit-Chem
Trade names: Vistanex® MML-80, MML-100, MML-120, MML-140
Trade names containing: Chem-Master R-103; Chem-Master R-231; Poly-Dispersion® PLD-90; Poly-Dispersion® PLD-90P

Polyisobutylene. *See* Polyisobutene

Polyisoprene
CAS 9006-04-6; 9003-31-0
Synonyms: IR; Isoprene rubber; 2-Methyl-1,3-butadiene, homopolymer; cis-1,4-Polyisoprene rubber
Classification: Thermoplastic
Definition: Polymer of isoprene
Empirical: $(C_5H_8)_x$
Formula: $[CH_2C(CH_3)CHCH_2]_x$
Uses: Elastomer for light colored goods, adhesives, footwear, sponge prods., tires, pharmaceutical goods, rubber bands, molded and mech. goods
Regulatory: FDA 21CFR §175.105, 176.180, 177.2600
Manuf./Distrib.: Goodyear; Monomer-Polymer & Dajac; A. Schulman

cis-1,4-Polyisoprene rubber. *See* Polyisoprene

Poly 4,4´ isopropylidenediphenol neodol 25 alcohol phosphite. *See* 4,4´ Isopropylidenediphenol alkyl (C12-15) phosphites
Poly(melamine-co-formaldehyde) methylated. *See* Hexamethylmelamine
Polymer of diphenylmethane 4,4´-diisocyanate. *See* Polymethylene polyphenyl isocyanate

Polymethacrylic acid
CAS 25087-26-7
Uses: Dispersant for pigments and fillers in ceramics, polymerization
Manuf./Distrib.: Alco; CPS; Rhone-Poulenc
Trade names: Daxad® 34

Polymethylene polyphenyl isocyanate
CAS 101-68-8
Synonyms: Polymer of diphenylmethane 4,4´-diisocyanate
Empirical: $C_{15}H_{10}N_2O_2$
Properties: Crystals or yel. fused solid; m.w. 250.27; dens. 1.19 (50 C); m.p. 37.2 C; b.p. 194-199 (5 mm)
Toxicology: Poison by inhalation; mild poison by ingestion; skin and eye irritant; TLV 0.005 ppm
Uses: Crosslinking agent for adhesives
Trade names: Desmodur® VK-5; Desmodur® VK-18; Desmodur® VK-70; Desmodur® VK-200; Desmodur® VKS-2; Desmodur® VKS-4; Desmodur® VKS-5; Desmodur® VKS-18
Trade names containing: Bonding Agent 2001; Bonding Agent 2005

Polymethylhydrosiloxane
CAS 63148-57-2
Uses: Crosslinker, waterproofing agent
Trade names: CPS 120; CPS 123

Poly-α-methylstyrene
CAS 25014-31-7
Formula: $[CH_2C(CH_3)(C_6H_5)]_n$
Properties: M.w. 685-960; dens.1.075 (15.6 C); m.p. 118-141 C; ref. index 1.61 (20 C)
Uses: Plasticizer, extrusion and molding process aid in ABS, PVC, CPVC, and semirigid vinyl, thermoplastic urethanes, molded rubbers, and thermoplastic elastomers; modifier and reinforcer in adhesives, thermoplastic powd. coatings, hot-melt coatings
Manuf./Distrib.: Hercules; Monomer-Polymer & Dajac

Poly(α-methylstyrene-acrylonitrile
Uses: Modifying resin increasing service temp. of PVC
Trade names: Blendex® 587

Poly (α-methylstyrene-styrene acrylonitrile)
Uses: High heat modifier resin used to upgrade PVC compds.
Trade names: Blendex® 586

Poly(methylvinyl ether). *See* Polyvinyl methyl ether
Poly(methyl vinyl ether/maleic anhydride). *See* PVM/MA copolymer

Poly [(6-morpholino-s-triazine-2,4-diyl) [2,2,6,6-tetramethyl-4-piperidyl) imino]-hexamethylene [(2,2,6,6-tetramethyl-4-piperidyl) imino]
Classification: Hindered amine
Properties: M.w. 1600 ± 10%
Toxicology: LD50 (rat, oral) 2100 mg/kg of body wt.; LD50 (rabbit, dermal) > 2000 mg/kg of body wt.; may be irritating to skin, respiratory tract, eyes
Uses: Light stabilizer, free radical scavenger, antioxidant
Trade names: Cyasorb® UV 3346 LD

Polynoxylin. *See* Urea-formaldehyde resin

Polynuclear aromatic hydrocarbon
CAS 64741-81-7
Uses: Sec. vinyl plasticizer
Trade names: Viplex 895-BL

Polyoctenamer
Synonyms: Polyoctenylene rubber
Uses: Blend component for other rubbers to improve plasticity, enhance filler incorporation and dispersion, improve flowability in extrusion, inj. and compr. molding, increase surf. smoothness in calendering operations
Trade names: Vestenamer® 6213; Vestenamer® 8012

Polyoctenylene rubber. *See* Polyoctenamer

Polyoctylmethylsiloxane
CAS 68440-90-4
Uses: Lubricant for soft metals, rubber, plastics
Trade names: PS140

Poly (oxy-1,4-butanediyl)-α-hydro-ω-hydroxy. *See* Polytetramethylene ether glycol
Poly[oxy(dimethylsilylene)], α-hydro-ω-hydroxy-. *See* Dimethiconol
Polyoxymethylene urea. *See* Urea-formaldehyde resin
Polyoxypropylene (9). *See* PPG-9
Polyoxypropylene (12). *See* PPG-12
Polyoxypropylene (17). *See* PPG-17
Polyoxypropylene (20). *See* PPG-20
Polyoxypropylene (26). *See* PPG-26
Polyoxypropylene (30). *See* PPG-30
Polyoxypropylene (3) methyl ether. *See* PPG-3 methyl ether

Poly (pentabromobenzyl) acrylate
CAS 594477-57-3
Synonyms: PBB-PA
Properties: White to off-white powd.; insol. in common org. solvs.; dens. 2.05; m.p. 190-220 C
Toxicology: Mildly irritating to eyes and skin; LD50 (oral, rat) > 5000 mg/kg
Uses: Flame retardant for engineering thermoplastics, PET, PBT, nylon, PP, and PS
Trade names: FR-1025

Polyphenylene ether
Synonyms: PPE; Poly 2,6-dimethyl-1,4-phenylene ether
Uses: Thermoplastic resin for inj. molding and extrusion for CRT housings, business machines, power tools, automotive components, industrial parts, sheet, profiles; modifier resin for other polymers, increases heat resist.
Trade names: Blendex® HPP 820

Polypropene. *See* Polypropylene

Polypropylene
CAS 9003-07-0; 9010-79-1 (nucleated)
Synonyms: PP; 1-Propene, homopolymer; Propylene polymer; Polypropene
Definition: Polymer of propylene monomers; three forms: isotactic (fiber-forming), syndiotactic, atactic (amorphous)
Empirical: $(C_3H_6)_x$
Formula: $[CH_2(CH_3)CH]_x$
Properties: Isotactic: Solid; pract. insol. in cold org. solvs.; sol. in hot decalin, hot tetralin, boiling tetrachloroethane; dens. 0.090-0.92; m.p. 165 C
Uses: Isotactic: fishing gear, ropes, filter cloths, laundry bags, protective clothing, blankets, fabrics, carpets, yarns; amorphous: plastics lubricant
Regulatory: FDA 21CFR §175.105, 175.300, 177.1200, 177.1520, 179.45
Manuf./Distrib.: Amoco; Aristech; Ashland; Chisso; Eastman; Exxon; Hüls; LNP; Mitsubishi Petrochem.; Mitsui Toatsu; Neste Polyeten AB; Quantum/USI; Shell; Solvay Polymers
Trade names: Empee® PP Conc. 4; Empee® PP Conc. 33; Empee® PP Conc. 43; Epolene® E-43; Epolene® N-15; Fusabond® MZ-109D; Fusabond® MZ-203D; Hoechst Wax PP 230; Luwax® ES 9668; Plastech EP 8126; Polypol 19; Polytac; proFLOW 1000
Trade names containing: Ampacet 11343; Ampacet 19873; Ampacet 41312; Ampacet 41438; Ampacet 41483; Ampacet 49315; Ampacet 49370; Ampacet 49842; Ampacet 49882; Ampacet 49920; Armostat® 475; Aztec® DIPP-40-IC5; Chemstat® AC-2000; Colloids PP 50/5; Colloids PP 50/7; Colloids PP 50/9; Hostastat® System E5951; Luperco 101-P20; Mastertek® PP 3658; Mastertek® PP 5601; Mastertek® PP

5604; MDI PP-130; MDI PP-535; MDI PP-940; Plasblak® PP 3393; Plasblak® PP 3583; Plasblak® PP 3585; Plasblak® PP 4045; Plaswite® PP 7161; Plaswite® PP 7269; Polybond® 1000; Polybond® 1001; Polybond® 1002; Polybond® 2015; Polybond® 3001; Polybond® 3002; PP-AB 104; PP-SA 46; PP-SA 97; PP-SA 128; PP-SL 82; PP-SL 98; PP-SL 101; Reed C-PPR-2096; Trigonox® 101-7.5PP-pd; Trigonox® 101-50PP-pd; Vinyzene® SB-1 PR

Polypropylene/dibromostyrene copolymer
Uses: Flame retardant for PP and textiles
Trade names: Great Lakes GPP-36™; Great Lakes GPP-39™

Polypropylene glycol
CAS 25322-69-4; EINECS 200-338-0
Empirical: $(C_3H_8O_2)_n$
Formula: $HO(C_3H_6O)_nH$
Properties: Colorless clear liq.; sol. in water, aliphatic ketones, alcohol; insol. in ether, aliphatic hydrocarbons; m.w. 400-2000; dens. 1.001-1.007; flash p t. > 390 F
Precaution: Combustible exposed ot heat or flame; reactive with oxidizers
Toxicology: LD50 (oral, rat) 4190 mg/kg; mildly toxic by ingstion; skin and eye irritant; heated to decomp., emits acrid smoke and irritating fumes
Uses: Hydraulic fluids, rubber lubricants, antifoam agents, intermediates for urethane foams, adhesives, coatings, elastomers, plasticizers, paint formulations, lab reagent
Regulatory: FDA 21CFR §173.340
Manuf./Distrib.: Aldrich; Arco; Ashland; BASF; BP Chem. Ltd; Calgene; Dow; Harcros; Hüls AG; Miles; Olin; PPG Industries; Rhone-Poulenc Surf.; Texaco; Witco/Oleo-Surf.
Trade names containing: Vinyzene® BP-505 PG
See also PPG...

Polypropylene glycol (9). *See* PPG-9
Polypropylene glycol (12). *See* PPG-12
Polypropylene glycol (12). *See* PPG-17
Polypropylene glycol (20). *See* PPG-20
Polypropylene glycol (26). *See* PPG-26
Polypropylene glycol (30). *See* PPG-30

Polypropylene glycol dibenzoate
Uses: Plasticizer for PU and polysulfide sealants, acrylic coatings, caulks, PVC, adhesives
Trade names: Benzoflex® 400

Polypropylene glycol monomethyl ether. *See* Methoxyisopropanol

Polysiloxane polyether copolymer
CAS 68937-54-2
Synonyms: Dimethicone copolyol
Uses: Defoamer for emulsion paints, adhesives, polymer dispersions
Trade names: Tego® Foamex 800; Tego® Foamex 805

Polysorbate 20
CAS 9005-64-5 (generic)
Synonyms: POE (20) sorbitan monolaurate; PEG-20 sorbitan laurate; Sorbimacrogol laurate 300
Definition: Mixture of laurate esters of sorbitol and sorbitol anhydrides, with ≈ 20 moles ethylene oxide
Properties: Lemon to amber liq., char. odor, bitter taste; sol. in water, alcohol, ethyl acetate, methanol, dioxane; insol. in min. oil, min. spirits
Toxicology: LD50 (oral, rat) 37 g/kg; mod. toxic by intraperitoneal, intravenous routes; mildly toxic by ingestion; human skin irritant; heated to decomp., emits acrid smoke and irritating fumes
Uses: O/w emulsifier, solubilizer; used in agric., cosmetics, pharmaceuticals, leather, metalworking, textiles, emulsion polymerization; antistat for rigid PVC
Regulatory: FDA 21CFR §172.515, 175.105, 175.300, 178.3400; Europe listed
Trade names: Alkamuls® PSML-20; Disponil SML 120 F1; Ethylan® GEL2; Glycosperse® L-20; T-Maz® 20

Polysorbate 21
CAS 9005-64-5 (generic)
Synonyms: POE (4) sorbitan monolaurate; PEG-4 sorbitan laurate
Definition: Mixture of laurate esters of sorbitol and sorbitol anhydrides, with ≈ 4 moles ethylene oxide

Toxicology: Moderately toxic by intraperitoneal, intravenous routes; midly toxic by ingestion; skin irritant
Uses: Emulsifier for PVC polymerization, solubilizer for colorants, dye leveing agent
Regulatory: FDA 21CFR §175.300
Trade names: Disponil SML 104 F1; Emsorb® 6916; Ethylan® GLE-21

Polysorbate 40
CAS 9005-66-7
Synonyms: POE (20) sorbitan monopalmitate; Sorbitan, monohexadecanoate, poly(oxy-1,2-ethaneidyl) derivs.; Sorbimacrogol palmitate 300
Definition: Mixture of palmitate esters of sorbitol and sorbitol anhydrides, with ≈ 20 moles of ethylene oxide
Toxicology: Moderately toxic by intravenous route
Uses: O/w emulsifier, solubilizer; used in agric., cosmetics, pharmaceuticals, leather, metalworking, textiles, polymerization; polymer additives
Regulatory: FDA 21CFR §175.105, 175.300, 178.3400; Europe listed; UK approved
Trade names: Disponil SMP 120 F1; Ethylan® GEP4

Polysorbate 60
CAS 9005-67-8 (generic)
Synonyms: POE (20) sorbitan monostearate; PEG-20 sorbitan stearate; Sorbimacrogol stearate 300
Definition: Mixture of stearate esters of sorbitol and sorbitol anhydrides, with ≈ 20 moles ethylene oxide
Empirical: $C_{64}H_{126}O_{26}$
Properties: Lemon to orange oily liq., faint odor, bitter taste; sol. in water, aniline, ethyl acetate, toluene; m.w. 1311.90; acid no. 2 max.; sapon. no. 45-55; hyd. no. 81-96
Toxicology: LD50 (IV, rat) 1220 mg/kg; moderately toxic by intravenous route; experimental tumorigen, reproductive effects; heated to decomp., emits acrid smoke and irritating fumes
Uses: Emulsifier, solv. for polymerization, cosmetics, pharmaceuticals, industrial chemicals, agric., textiles, plastics additive, veterinary drug; solubilizer
Regulatory: FDA 21CFR §73.1001, 172.515, 172.836, 172.878, 172.886, 173.340, 175.105, 175.300, 178.3400; USDA CFR9 §318.7, 381.147 (limitation 1% max., 1% total combined with polysorbate 80, 0.0175% in scald water); Europe listed; UK approved
Trade names: Disponil SMS 120 F1; Ethylan® GES6; Hefti MS-55-F

Polysorbate 61
CAS 9005-67-8 (generic)
Synonyms: POE (4) sorbitan monostearate; PEG-4 sorbitan stearate
Definition: Mixture of stearate esters of sorbitol and sorbitol anhydrides, with ≈ 4 moles ethylene oxide
Properties: M.w. 1311.70; dens. 1.044; flash pt. > 110 C
Toxicology: Moderately toxic by intravenous route
Uses: Emulsifier, solubilizer, lubricant for textile use, household formulations, suppositories in pharmaceuticals, emulsion polymerization, hydraulic fluids, metal treatment, paints; color dispersants for plastics, cosmetics, leather
Regulatory: FDA 21CFR §175.300
Trade names: Emsorb® 6909

Polysorbate 65
CAS 9005-71-4
Synonyms: POE (20) sorbitan tristearate; PEG-20 sorbitan tristearate; Sorbimacrogol tristearate 300
Definition: Mixture of stearate esters of sorbitol and sorbitol anhydrides, with ≈ 20 moles ethylene oxide
Properties: Tan waxy solid, faint odor, bitter taste; sol. in min. and veg. oils, min. spirits, acetone, ether, dioxane, alcohol, methanol; disp. in water, CCl_4; acid no. 2 max.; sapon. no. 88-98; hyd. no. 44-60
Toxicology: Heated to decomp., emits acrid smoke and irritating fumes
Uses: O/w emulsifier, solubilizer; used in agric., cosmetics, leather, metalworking, textiles, polymerization
Regulatory: FDA 21CFR §73.1001, 172.838, 173.340, 175.300, 178.3400; Europe listed; UK approved
Trade names: Disponil STS 120 F1

Polysorbate 80
CAS 9005-65-6 (generic); 37200-49-0; 61790-86-1
Synonyms: POE (20) sorbitan monooleate; PEG-20 sorbitan oleate; Sorbimacrogol oleate 300
Definition: Mixture of oleate esters of sorbitol and sorbitol anhydrides, with ≈ 20 moles ethylene oxide
Properties: Amber visc. liq.; very sol. in water; sol. in alcohol, cottonseed oil, corn oil, ethyl acetate, methanol, toluene; insol. in min. oil; dens. 1.06-1.10; visc. 270-430 cSt; acid no. 2 max.; sapon. no. 45-55; hyd. no. 65-80; pH 5-7 (5% aq.)
Toxicology: LD50 (oral, mouse) 25 g/kg; mod. toxic by intravenous route; mildly toxic by ingestion; eye irritant;

experimental tumorigen, reproductive effects; mutagenic data; heated to decomp., emits acrid smoke and irritating fumes
Uses: Pharmaceutic aid (surfactant); as emulsifier and dispersant in medicinal prods.; as defoamer and emulsifier in foods; adjuvant in herbicides; polymerization; food emulsifier
Regulatory: FDA 21CFR §73.1, 73.1001, 172.515, 172.840, 173.340, 175.105, 175.300, 178.3400; USDA 9CFR §318.7, 381.147 (limitation 1% alone, 1% total combined with polysorbate 60); Europe listed; UK approved
Trade names: Disponil SMO 120 F1; Durfax® 80; Ethylan® GEO8; Glycosperse® O-20
Trade names containing: Aldosperse® MO-50; Ice # 2

Polysorbate 81
CAS 9005-65-5 (generic)
Synonyms: POE (5) sorbitan monooleate; PEG-5 sorbitan oleate
Definition: Mixture of oleate esters of sorbitol and sorbitol anhydrides, with ≈ 5 moles ethylene oxide
Uses: O/w emulsifier, solubilizer; used in agric., cosmetics, leather, metalworking, textiles, emulsion polymerization, metal treatment; color dispersant in plastics
Regulatory: FDA 21CFR §175.300
Trade names: Emsorb® 6901; Ethylan® GEO81; Ethylan® GOE-21

Polysorbate 85
CAS 9005-70-3
Synonyms: POE (20) sorbitan trioleate; PEG-20 sorbitan trioleate; Sorbimacrogol trioleate 300
Definition: Mixture of oleate esters of sorbitol and sorbitol anhydrides, with ≈ 20 moles ethylene oxide
Toxicology: Skin irritant
Uses: Surfactant, emulsifier for polymerization, cosmetics, agric., textiles; plastic additives; solubilizer for perfume, flavors, essential oils
Regulatory: FDA 21CFR §175.300, 178.3400
Trade names: Disponil STO 120 F1; Ethylan® GPS85

Polystyrene
CAS 9003-53-6
Synonyms: PS; Styrene polymer; Ethenylbenzene, homopolymer; Benzene, ethenyl-, homopolymer
Classification: Polymer
Definition: Grades: crystal, impact, expandable
Empirical: $(C_8H_8)_x$
Properties: M.w. 2500-250,000
Toxicology: Experimental tumorigen by implant
Uses: Thermoplastic resin for inj. molding, extrusion of egg carton foam, pill bottles, pkg., appliances, electronics, toys, recreation, and construction; expandable PS for insulation, protective pkg.; modifier for latexes
Regulatory: FDA 21CFR §175.105, 175.125, 175.300, 175.320, 176.180, 177.1200, 177.1640, 177.2600
Manuf./Distrib.: Amoco; Asahi Chem Industry Co Ltd; Ashland; BASF; Chevron; Elf Atochem SA; Hüls AG; LNP; Mitsubishi Petrochem.; Mitsui Toatsu; Westlake Plastics
Trade names: Darex® 670L
Trade names containing: Colloids PS 47/31; Colloids PS 47/32; Colloids PS 47/38; MDI PS-125; MDI PS-650; MDI PS-901; MDI PS-903; Plasblak® PS 0469; Plasblak® PS 1844; Plasblak® PS 3294; Plasblak® PS 4054; Plaswite® PS 7028; Plaswite® PS 7037; Plaswite® PS 7131; Plaswite® PS 7174; Plaswite® PS 7179; Plaswite® PS 7231; Reed C-PS-4230

Polystyrene, brominated
Uses: Flame retardant for engineering thermoplastics and other polymers
Trade names: Pyro-Chek® 60PB; Pyro-Chek® 60PBC; Pyro-Chek® 68PB; Pyro-Chek® 68PBC; Pyro-Chek® 68PBG; Pyro-Chek® C60PB; Pyro-Chek® C68PB; Pyro-Chek® LM

Polysulfide
CAS 9080-49-3
Classification: Synthetic polymer
Properties: Solid or liq.; exc. low-temp. flexibility; poor tensile strength and abrasion resistance
Uses: Elastomer for use in sealants, adhesives, potting and encapsulating compds., gasoline and oil-loading hose, casting of molds, barrier coatings, binder in solid rocket fuel; epoxy modifier
Manuf./Distrib.: Morton Int'l.
Trade names: LP®-3

Polyterpene resin

Uses: Thermoplastic polymer imparting tack and adhesion to many elastomeric and polymeric materials, adhesives, coatings, rubber prods., concrete-curing compds., caulks

Manuf./Distrib.: Allchem Ind.; Monomer-Polymer & Dajac

Trade names: Nevtac® 10°; Nevtac® 80; Nevtac® 100; Nevtac® 115; Super Nevtac® 99; Zonarez® 1040; Zonarez® 1085; Zonarez® 1100; Zonarez® 1115; Zonarez® 1115T; Zonarez® 1125; Zonarez® 1135; Zonarez® 7115; Zonarez® 7115 LITE; Zonarez® 7125; Zonarez® 7125 LITE; Zonarez® Alpha 25; Zonarez® B-10; Zonarez® B-115; Zonarez® B-125; Zonarez® M-1115; Zonarez® T-4115

Polytetrafluoroethylene

CAS 9002-84-0

Synonyms: PTFE; TFE; Tetrafluoroethene homopolymer; Tetrafluoroethylene polymer; Polytetrafluoroethylene resin

Classification: Thermoplastic homopolymer

Formula: $[CF_2CF_2]_x$, $x \approx 20{,}000$

Properties: White translucent to opaque solid; dens. 2.2; useful temp. range cryogenic to 260 C; melts to visc. gel @ 327 C; Shore hardness 55-56; tens. str. 3500-4500 psi

Precaution: Nonflamm.

Toxicology: Polymer fume fever reported under conditions of inadequate ventilation; inert as finished compd.

Uses: As tubing or sheeting for chemical laboratory and process work; gaskets and pump packings; as elec. insulators esp. in high frequency applics.; filtration fabrics; protective clothing; prosthetic aid; plastics lubricant, release agent

Manuf./Distrib.: Janssens NV

Trade names: Algoflon® L; CeriDust 961OF; Fluon® L169; Fluoroglide® FL 1690; Fluoroglide® FL 1700; Hostaflon® TF 9202; Hostaflon® TF 9203; McLube 1700; McLube 1704; McLube 1775; McLube 1777; McLube 1777-1; McLube 1782; McLube 2000 Series; PeFlu 727; Polylube J; Polymist® 284; Polymist® F-5; Polymist® F-5A; Polymist® F-5A EX; Polymist® F-510; Polymist® XPH-284; Teflon® MP 1000; Teflon® MP 1100; Teflon® MP 1200; Teflon® MP 1300; Teflon® MP 1400; Teflon® MP 1500; Teflon® MP 1600; TL-115; Tuf Lube Hi-Temp NOD; Vydax® 1000; Vydax® AR; Whitcon® TL-5; Whitcon® TL-6; Whitcon® TL-102-2; Whitcon® TL-155; Whitcon® TL-169

Trade names containing: Poly-Dispersion® A (TEF-1)D-80; Poly-Dispersion® A (TEF-1)D-80P; Vydax® Spray No. V312-A

Polytetrafluoroethylene resin. *See* Polytetrafluoroethylene

Poly[[6-[(1,1,3,3-tetramethylbutyl)amino]-s-triazine-2,4-diyl][2,2,6.6-tetramethyl-4-piperidyl)imino] hexamethylene[(2,2,6,6-tetramethyl-4-piperidyl) imino]]

CAS 70624-18-9

Trade names: Chimassorb® 944; Chimassorb® 944FD; Chimassorb® 944FL; Chimassorb® 944LD; Tinuvin® 783

Trade names containing: Aquastab PC 44

Polytetramethylene ether glycol

CAS 24979-97-3; 25190-06-1

Synonyms: PTMEG; PTMG; Poly (oxy-1,4-butanediyl)-α-hydro-ω-hydroxy

Classification: Polyether glycol

Formula: $HO[(CH_2)_4O]_nH$

Properties: White waxy solid melting to clear visc. liq. near R.T.; m.w. 650-2900; sp.gr. 1.0; flash pt. (TOC) 163 C

Precaution: Avoid strong oxidizers; may release very flammable THF, CO, etc.

Toxicology: May cause mild skin and eye irritation; LD50 (rat, oral) > 11,000 mg/kg; moderate aquatic toxicity

Uses: For polyurethane formulation for automotive hose and gaskets, tires, industrial belts, tank and pipe liners, floor and roof coatings, medical devices

Manuf./Distrib.: BASF; Monomer-Polymer & Dajac; QO

Polytetramethylene ether glycol diamine

Synonyms: PTMEG diamine

Classification: Oligomeric diamine

Uses: Curative for elastomers

Trade names: Polamine® 250

Polytetramethyleneoxide-di-p-aminobenzoate

CAS 54667-43-5

Classification: Oligomeric diamine
Uses: Curative for elastomers
Trade names: Polamine® 650; Polamine® 1000; Polamine® 2000

Polyurethane-acrylate resin
Uses: Curing agent

Polyurethane, polyester
Trade names containing: Vinyzene® SB-1 U

Polyurethane prepolymer
Uses: Dispersant, emulsifier for latex polymerization, waste water treatment
Manuf./Distrib.: Air Prods.; CasChem; Conap; BFGoodrich; Hampshire; ICI Polyurethanes; Miles; Polyurethane Corp. of Am.; Polyurethane Spec.; Soluol; Uniroyal; Zeneca Resins
Trade names: Darathane® WB-4000

Polyurethane, thermoset
Uses: Thermoset polymer for casting and potting systems, conformal coatings, adhesives, insulating coatings, cable molding, automotive fascia, etc.; flame retardant
Trade names: Futurethane UR-2175

Polyvinyl alcohol
CAS 9002-89-5 (super and fully hydrolyzed); EINECS 209-183-3
Synonyms: PVA; PVAL; Ethenol homopolymer; PVOH; Vinyl alcohol polymer
Classification: Polymer
Empirical: $(C_2H_4O)_x$
Formula: $[CH_2CHOH]_x$
Properties: White to cream amorphous powd.; sol. in water; insol. in petrol. solvs.; avg. m.w. 120,000; dens. 1.329; softens at 200 C with dec.; flash pt. (OC) 175 F; ref. index 1.49-1.53
Precaution: Flamm. exposed to heat or flame; reactive with oxidizers; dust exposed to flame presents sl. explosion hazard
Toxicology: Experimental carcinogen and tumorigen; heated to decomp., emits acrid smoke and irritating fumes
Uses: In plastics industry in molding compds., surface coatings, films resistant to gasoline, textile sizes; for elastomers (artificial sponges, fuel hoses); printing inks; pharmaceutical finishing; cosmetics; film and sheeting; ophthalmic lubricant; emulsion polymerization; release agent
Regulatory: FDA 21CFR §73.1, 175.105, 175.300, 175.320, 176.170, 176.180, 177.1200, 177.1670, 177.2260, 177.2800, 178.3910, 181.22, 181.30
Manuf./Distrib.: Air Prods.; Allchem Ind.; British Traders & Shippers; Honeywill & Stein Ltd; Hunt; Itochu Spec.; Monomer Polymer & Dajac; Polysciences; San Yuan
Trade names: Airvol® 805; Airvol® 823; Airvol® 840; Plastilease 512-B; Plastilease 512-CL
Trade names containing: Release Agent NL-2; Vinyzene® SB-1

Polyvinyl alcohol, partially hydrolyzed
CAS 25213-24-5
Uses: Binder, carrier, compounding agent, dispersant, stabilizer, protective colloid in polymerizations; for textiles, paper, adhesives, cement/plaster additive, peelable caulks, ceramics, strippable coatings, mold release, nonwovens
Trade names: Airvol® 205; Airvol® 523; Airvol® 540

Poly (n-vinylbutyrolactam). *See* PVP

Polyvinyl chloride
CAS 9002-86-2; EINECS 208-750-2
Synonyms: PVC; Chloroethene homopolymer; Chloroethylene polymer
Empirical: $(C_2H_3Cl)_n$
Formula: $[CH_2CHClCH_2CHCl]_n$
Properties: White powd.; m.w. 60,000-150,000; dens. 1.406; ref. index 1.54
Toxicology: Suspected tumorigen by ingestion and implantation; chronic inhalation health problems
Uses: Rubber substitutes; elec. wire and cable coverings; pliable thin sheeting; film finishes for textiles; nonflamm. upholstery; raincoats; tubing; belting; gaskets; shoe soles
Manuf./Distrib.: Aldrich; Ashland; Chisso; Elf Atochem SA; Georgia Gulf; BFGoodrich; Goodyear; Hüls Am.; Mitsui Toatsu; Nat'l. Starch & Chem.; Norsk Hydro AS; OxyChem; Teknor Apex; Vista; Wacker-Chemie GmbH

Polyvinylidene chloride

Trade names: Vestolit® E 7012
Trade names containing: Krynac® PXL 34.17; MDI PV-940; Nipol® DP5123P; Nipol® DP5125P; Nipol® DP5128P; Vinyzene® SB-1

Polyvinylidene chloride
Synonyms: PVDC; Saran
Uses: Thermoplastic polymer for extrusion or inj. or blow molding; in adhesion coatings for paper, food pkg., pipes for chemical processing, upholstery, fibers, bristles, latex coatings
Manuf./Distrib.: Hampshire; Solvay Polymers
Trade names containing: Dualite® M6001AE

Polyvinylidene fluoride
CAS 24937-79-9
Synonyms: PVDF; 1,1-Difluoroethene homopolymer
Classification: Polymer
Empirical: $C_7H_{12}O_3$
Properties: M.w. 144.17
Precaution: Distillation residue explodes violently when heated to 130 C
Uses: Thermoplastic polymer used as insulation for high-temp. wire, tank linings, chemical tanks and tubing, paints and coatings with high resistance to weathering and UV light; processing aid for polyolefins
Manuf./Distrib.: Ausimont USA; Elf Atochem N. Am.; Monomer-Polymer & Dajac; Solvay Polymers; Westlake Plastics; Zeus Industrial Prods.
Trade names: Solef® 11012/1001

Polyvinyl isobutyl ether hexane
Uses: Tacky polymer with excellent adhesion to plastic, metal, and coated surfs.; plasticizer and leveling agent for surf. coatings

Polyvinyl methyl ether
CAS 9003-09-2
Synonyms: PVM; Methoxyethene, homopolymer; Poly(methylvinyl ether)
Empirical: $(C_3H_6O)_x$
Formula: $[CH_2CHOCH_3]_x$
Uses: Tackifier, binder, plasticizer for inks, textile sizes and finishes, latex modification
Regulatory: FDA 21CFR §175.105, 177.1680
Trade names: Gantrez® M-154

Polyvinylmethylsiloxane
CAS 68037-87-6
Uses: Coupling agent
Trade names: CPS 925

Polyvinyl octadecyl carbamate
CAS 70892-21-6
Formula: $(C_{21}H_{41}O_2N) \times C_7H_8$
Uses: Release coating for films and tapes
Manuf./Distrib.: Polyad
Trade names: Mayzo RA-0095H; Mayzo RA-0095HS

Polyvinylpyrrolidone. *See* PVP
POP (40) methyl diethyl ammonium chloride. *See* PPG-40 diethylmonium chloride
POP (2) methyl ether. *See* PPG-2 methyl ether
POP (2) monolaurate. *See* PPG-2 laurate
POP (2) monooleate. *See* PPG-2 oleate
POP (2) monostearate. *See* PPG-2 stearate
POP (12) POE (16) monobutyl ether. *See* PPG-12-buteth-16
POP (24) POE (27) monobutyl ether. *See* PPG-24-buteth-27
POP (28) POE (35) monobutyl ether. *See* PPG-28-buteth-35
POP (33) POE (45) monobutyl ether. *See* PPG-33-buteth-45

Potassium acetate
CAS 127-08-2; EINECS 204-822-2
Empirical: $C_2H_3KO_2$

Formula: CH_3COOK
Properties: Colorless lustrous cryst. or wh. cryst. powd. or flakes; deliq.; sol. in water, alcohol; insol. in ether; m.w. 98.14; dens. 1.57; m.p. 292 C
Precaution: Keep tightly closed
Toxicology: LD50 (oral, rat) 3.25 g/kg
Uses: Dehydrating agent, textile conditioner, analytical reagent, medicine, cacodylic derivatives, crystal glass, synthetic flavors
Regulatory: FDA 21CFR §172.515; GRAS (FEMA); Europe listed
Manuf./Distrib.: Am. Int'l. Chem.; EM Ind.; General Chem.; Heico; Hoechst AG; Honeywill & Stein Ltd; Niacet; Poly Research
Trade names containing: Polycat® 46

Potassium castorate
CAS 8013-05-6; EINECS 232-388-4
Synonyms: Castor oil, potassium salt
Definition: Potassium salt of fatty acids derived from castor oil
Uses: Emulsifier, dispersant, mild germicide; glycerized rubber lubricant; emulsifier, foam stabilizer for foamed rubber
Regulatory: FDA 21CFR §175.105, 176.170, 176.200, 176.210, 177.1200, 177.2600, 177.2800, 178.3910
Trade names: Solricin® 235

Potassium dimethyldithiocarbamate
CAS 128-03-0
Empirical: $C_3H_6NS_2 \cdot K \cdot H_2O$
Properties: M.w. 177.34
Toxicology: Poison by intraperitoneal route
Uses: Biocide, fungicide, and algicide used in water treatment, paper, sugar, and petrol. applics.
Trade names: Aquatreat KM

Potassium dodecylbenzene sulfonate
CAS 27177-77-1; EINECS 248-296-2
Synonyms: Dodecylbenzenesulfonic acid, potassium salt
Classification: Substituted aromatic compd.
Empirical: $C_{18}H_{30}O_3S \cdot K$
Uses: Surfactant for S/B, vinyl chloride, VDC latexes
Regulatory: FDA 21CFR §178.3400
Trade names: Polystep® A-15-30K

Potassium 2-ethylhexanoate
CAS 3164-85-0
Uses: Catalyst for PU
Trade names containing: Dabco® K-15

Potassium naphthalene sulfonate
Uses: Dispersant; sec. emulsifier for emulsion polymerization
Trade names: Lomar® HP

Potassium 9-octadecenoate. *See* Potassium oleate

Potassium octoate
Uses: Catalyst for polyurethane
Manuf./Distrib.: OM Group; Pelron; Shepherd
Trade names: Metacure® T-45
Trade names containing: Dabco® T-45; Kosmos® 64

Potassium oleate
CAS 143-18-0; EINECS 205-590-5
Synonyms: Potassium 9-octadecenoate; Oleic acid potassium salt
Definition: Potassium salt of oleic acid
Empirical: $C_{18}H_{33}O_2 \cdot K$
Formula: $CH_3(CH_2)_7CH=CH(CH_2)_7COOK$
Properties: Ylsh. or brownish soft mass; sol. in water, alcohol; m.w. 320.56
Toxicology: Eye irritant; heated to decomp., emits toxic fumes of K_2O

Potassium pyrophosphate

Uses: Liq. soap for hand cleaners, tire mounting lubricant; emulsifier and corrosion control in paint strippers; frothing aid in gelled latex foam compds.
Regulatory: FDA 21CFR §172.863, 175.105, 175.300, 176.170, 176.200, 176.210, 177.1200, 177.2260, 177.2600, 177.2800, 178.3910, 181.22, 181.29
Manuf./Distrib.: Concord; Emkay; Fluka; Norman, Fox; RTD
Trade names: Octosol 449
Trade names containing: Agerite® Superlite® Emulcon®

Potassium pyrophosphate. *See* Tetrapotassium pyrophosphate
Potassium pyrophosphate, normal. *See* Tetrapotassium pyrophosphate

Potassium ricinoleate
CAS 7492-30-0; EINECS 231-314-8
Synonyms: 12-Hydroxy-9-octadecenoic acid, monopotassium salt
Definition: Potassium salt of ricinoleic acid
Empirical: $C_{18}H_{34}O_3 \cdot K$
Formula: $CH_3(CH_2)_5COHHCH_2CH=CH(CH_2)_7COOK$
Uses: Emulsifier, mild germicide, glycerized rubber lubricant, foam stabilizer in foamed rubber; household and cosmetic prods.; mold release
Regulatory: FDA 21CFR §175.300
Trade names: Solricin® 135

Potassium rosinate
CAS 61790-50-9; 61790-51-0; EINECS 263-144-5
Synonyms: Potassium soap of rosin
Uses: Emulsifier, detergent, wetting agent, dispersant, foaming agent for polymerization of ABS, SBR, other elastomers, metalworking and drilling oils, adhesives
Trade names: Arizona DRS-40; Arizona DRS-42; Arizona DRS-50; Arizona DRS-51E; Burez K20-505A; Burez K25-500D; Burez K50-505A; Burez K80-500; Burez K80-500D; Burez K80-2500; Diprosin K-80; Dresinate® 91; Dresinate® 214; Westvaco® Resin 895

Potassium soap of rosin. *See* Potassium rosinate

Potassium stearate
CAS 593-29-3; EINECS 209-786-1
Synonyms: Octadecanoic acid, potassium salt; Stearic acid, potassium salt
Definition: Potassium salt of stearic acid
Empirical: $C_{18}H_{35}O_2 \cdot K$
Formula: $CH_3(CH_2)_{16}COOK$
Properties: White powd., sl. fatty odor; slowly sol. in cold water; readily sol. in hot water, alcohol; m.w. 322.57
Toxicology: Heated to decomp., emits acrid smoke and irritating fumes
Uses: In mfg. of textile softeners
Regulatory: FDA 21CFR §172.615, 172.863, 173.340, 175.105, 175.300, 176.170, 176.200, 176.210, 177.1200, 177.2260, 177.2600, 177.2800, 178.3910, 179.45, 181.22, 181.29
Manuf./Distrib.: Original Bradford Soap Works; RTD; Witco/Oleo-Surf.
Trade names containing: Rhenovin® S-stearat-80

Potassium-zinc
Uses: Stabilizer/activator for PVC, chemically blown PVC
Trade names: ADK STAB FL-32; ADK STAB FL-41; Lankromark® LZ561; Lankromark® LZ1199; Lankromark® LZ1221; Lankromark® LZ1232; Mark® 4726

Povidone. *See* PVP
PP. *See* Polypropylene
PPE. *See* Polyphenylene ether

PPG-9
CAS 25322-69-4 (generic)
Synonyms: Polyoxypropylene (9); Polypropylene glycol (9); PPG 400
Definition: Polymer of propylene oxide
Formula: $H(OCH_2CHCH_3)_nOH$, avg. n = 9
Uses: Chemical intermediate, antifoam agent in fermentation and in paint formulations, antiblooming agent for pentachlorophenol-treated wood; binder and lubricant for ceramics; plasticizer of resin-treated papers;

mold release applics.
Regulatory: FDA 21CFR §173.310, 175.105, 176.200, 176.210
Trade names: Hodag PPG-400; Pluracol® P-410

PPG-12
CAS 25322-69-4 (generic)
Synonyms: Polyoxypropylene (12); Polypropylene glycol (12)
Definition: Polymer of propylene oxide
Formula: $H(OCH_2CHCH_3)_nOH$, avg. n = 12
Uses: Chemical intermediate, antifoam agent in fermentation and in paint formulations, antiblooming agent for pentachlorophenol-treated wood; binder and lubricant for ceramics; plasticizer of resin-treated papers
Regulatory: FDA 21CFR §173.310, 175.105, 176.200, 176.210
Trade names: Pluracol® P-710

PPG-17
CAS 25322-69-4 (generic)
Synonyms: Polyoxypropylene (17); Polypropylene glycol (12)
Definition: Polymer of propylene oxide
Formula: $H(OCH_2CHCH_3)_nOH$, avg. n = 17
Uses: Chemical intermediate, antifoam agent in fermentation and in paint formulations, antiblooming agent for pentachlorophenol-treated wood; binder and lubricant for ceramics; plasticizer of resin-treated papers
Regulatory: FDA 21CFR §173.310, 175.105, 176.170, 176.200, 176.210
Trade names: Pluracol® P-1010

PPG-20
CAS 25322-69-4 (generic)
Synonyms: Polyoxypropylene (20); Polypropylene glycol (20); PPG 1200
Definition: Polymer of propylene oxide
Formula: $H(OCH_2CHCH_3)_nOH$, avg. n = 20
Uses: Defoamer, mold release applics., chemical intermediates for fatty acid esters, components for urethane resins
Regulatory: FDA 21CFR §173.310, 173.340, 175.105, 176.170, 176.200, 176.210, 178.3740
Trade names: Hodag PPG-1200

PPG-26
CAS 25322-69-4 (generic)
Synonyms: Polyoxypropylene (26); Polypropylene glycol (26); PPG 2000
Definition: Polymer of propylene oxide
Formula: $H(OCH_2CHCH_3)_nOH$, avg. n = 26
Uses: Defoamer, mold release applics., chemical intermediates for fatty acid esters, components for urethane resins
Regulatory: FDA 21CFR §173.310, 173.340, 175.105, 176.170, 176.200, 176.210, 178.3740
Trade names: Hodag PPG-2000; Pluracol® P-2010

PPG-30
CAS 25322-69-4 (generic)
Synonyms: Polyoxypropylene (30); Polypropylene glycol (30); PPG 4000
Definition: Polymer of propylene oxide
Formula: $H(OCH_2CHCH_3)_nOH$, avg. n = 30
Uses: Defoamer, mold release applics., chemical intermediates for fatty acid esters, components for urethane resins
Regulatory: FDA 21CFR §173.310, 173.340, 175.105, 176.170, 176.200, 178.3740
Trade names: Hodag PPG-4000; Pluracol® P-4010

PPG-400
Uses: Intermediate yielding esters; useful as lubricants, defoaming agents in rubber and pharmaceuticals, solvs. and humectant modifiers for inks, plasticizers, and functional fluids
Trade names: Jeffox PPG-400

PPG 400. *See* PPG-9

PPG (1000)
Uses: Intermediate for surfactants, ethers, esters; antifoam for ceramics, rubber; hydraulic fluids; dyeing; metalworking lubricants; plasticizer for plastics; latex coagulant; textile lubricant

PPG 1200. *See* PPG-20

PPG-2000
Uses: Intermediate yielding esters; useful as lubricants, defoaming agents in rubber and pharmaceuticals, solvs. and humectant modifiers for inks, plasticizers, and functional fluids
Trade names: Jeffox PPG-2000

PPG 2000. *See* PPG-26
PPG 4000. *See* PPG-30

PPG-7-buteth-10
CAS 9038-95-3 (generic); 9065-63-8 (generic)
Uses: Detergent for toilet cleaners, laundry detergents, emulsion polymerization, defoamers, metalworking fluids, hydraulic fluids
Regulatory: FDA 21CFR §173.310, 175.105, 176.210, 178.3570
Trade names: Macol® 300

PPG-12-buteth-16
CAS 9038-95-3 (generic); 9065-63-8 (generic); 74623-31-7
Synonyms: POE (16) POP (12) monobutyl ether; POP (12) POE (16) monobutyl ether
Definition: Polyoxypropylene, polyoxyethylene ether of butyl alcohol
Empirical: $(C_7H_{14}O_2 \cdot C_6H_{12}O_2)_x$
Formula: $C_4H_9(OCH_3CHCH_2)_x(OCH_2CH_2)_yOH$, avg. x = 12, avg. y = 16
Uses: Defoaming agent, lubricant, emollient; in demulsifying and wetting formulations; brake fluids; metalworking; rubber lubricant; textiles; fiber lubricant; food processing; chemical processing; cosmetics
Regulatory: FDA 21CFR §173.310, 175.105, 176.210, 178.3570
Trade names: Macol® 660

PPG-24-buteth-27
CAS 9038-95-3 (generic); 9065-63-8 (generic)
Synonyms: POE (27) POP (24) monobutyl ether; POP (24) POE (27) monobutyl ether
Definition: Polyoxypropylene, polyoxyethylene ether of butyl alcohol
Empirical: $(C_7H_{14}O_2 \cdot C_6H_{12}O_2)_x$
Formula: $C_4H_9(OCH_3CHCH_2)_x(OCH_2CH_2)_yOH$, avg. x = 24, avg. y = 27
Uses: Nonionic surfactant, emulsifier, dispersant for agric. concs., latex polymerization, mfg. of iodophors for germicidal cleaners
Regulatory: FDA 21CFR §173.310, 175.105, 176.210, 178.3570
Trade names: Tergitol® XD

PPG-28-buteth-35
CAS 9038-95-3 (generic); 9065-63-8 (generic)
Synonyms: POE (35) POP (28) monobutyl ether; POP (28) POE (35) monobutyl ether
Definition: Polyoxypropylene, polyoxyethylene ether of butyl alcohol
Empirical: $(C_7H_{14}O_2 \cdot C_6H_{12}O_2)_x$
Formula: $C_4H_9(OCH_3CHCH_2)_x(OCH_2CH_2)_yOH$, avg. x = 28, avg. y = 35
Uses: Emollient; detergent; defoamer; rubber lubricant; intermediate; metalworking and hydraulic fluids; mold release agent; emulsion polymerization
Regulatory: FDA 21CFR §173.310, 175.105, 176.210, 178.3570
Trade names: Macol® 3520

PPG-33-buteth-45
CAS 9038-95-3 (generic); 9065-63-8 (generic)
Synonyms: POE (45) POP (33) monobutyl ether; POP (33) POE (45) monobutyl ether
Definition: Polyoxypropylene, polyoxyethylene ether of butyl alcohol
Empirical: $(C_7H_{14}O_2 \cdot C_6H_{12}O_2)_x$
Formula: $C_4H_9(OCH_3CHCH_2)_x(OCH_2CH_2)_yOH$, avg. x = 33, avg. y = 45
Uses: Defoaming agent, lubricant, emollient; in demulsifying and wetting formulations; brake fluids; metalworking; rubber lubricant; textiles; fiber lubricant; food processing; chemical processing; cosmetics
Regulatory: FDA 21CFR §173.310, 173.340, 175.105, 176.210, 178.3570
Trade names: Macol® 5100

PPG-3 diacrylate
CAS 68901-05-3

Synonyms: TPGDA; Tripropylene glycol diacrylate
Formula: $H_2C=CHCOO-[CH(CH_3)CH_2O]_3COCH=CH_2$
Properties: M.w. 300.3; b.p. > 120 C (1 mm)
Uses: Reactive thinner for radiation-curing systems, inks, coatings, floor tiles, wood coatings and fillers, adhesives, textile finishes, rubber compds.
Trade names: Ageflex TPGDA

PPG-40 diethylmonium chloride
CAS 9076-43-1
Synonyms: POP (40) methyl diethyl ammonium chloride; Quaternium-21
Classification: Quaternary ammonium salt
Formula: $[CH_3N(CH_2CH_3)_2(CH_2CH_3CHO)_nH]^+Cl^-$, avg. n = 40
Uses: Pigment dispersant, particle suspension aid, emulsifier, solv., conditioner, antistat, lubricant, corrosion inhibitor for toiletries, cosmetics, germicides, syn. fibers and plastics, textiles, industrial processes; ore flotation additive
Trade names: Emcol® CC-42

PPG-2 laurate
Synonyms: PPG (2) laurate; POP (2) monolaurate
Definition: Polypropylene glycol ester of lauric acid
Uses: W/o emulsifier, dispersant, antistat for textile, paper processing, cutting oils, polishes, emulsion cleaners, rubber latex, wool lubricants
Trade names: Cithrol DPGML N/E

PPG (2) laurate. *See* PPG-2 laurate

PPG-2 laurate SE
Uses: W/o emulsifier, dispersant, antistat for textile, paper processing, cutting oils, polishes, emulsion cleaners, rubber latex, wool lubricants
Trade names: Cithrol DPGML S/E

PPG-2 methyl ether
CAS 13429-07-7; 34590-94-8
Synonyms: DPM; Dipropylene glycol methyl ether; Methoxy dipropylene glycol; 2-Methoxymethyl-ethoxypropanol; POP (2) methyl ether
Definition: PPG ether of methyl alcohol
Empirical: $C_7H_{16}O_3$
Formula: $CH_3(OCH_3CHCH_2)_nOH$, avg. n = 2
Properties: M.w. 148.2; flash pt. 167 C
Uses: Solv. for paints, cleaners, inks, cosmetics, agric., epoxy laminates, adhesives, floor polish, fuel additives, oilfield, mining, and electronic chems., chem. intermediate applics.
Regulatory: FDA 21CFR §175.105, 181.22, 181.30
Toxicology: ACGIH TLV 100 ppm (skin); LD50 (oral, rat) 5200 mg/kg
Manuf./Distrib.: Aldrich; Ashland; Hoechst AG; Oxiteno; Van Waters & Rogers
Trade names: Arcosolv® DPM
Trade names containing: BYK®-LP W 6246

PPG-3 methyl ether
CAS 10213-77-1; 37286-64-9 (generic); 25498-49-1
Synonyms: Tripropylene glycol monomethyl ether; Tripropylene glycol methyl ether; Polyoxypropylene (3) methyl ether; PPG (3) methyl ether
Definition: PPG ether of methyl alcohol
Empirical: $C_{10}H_{22}O_4$
Formula: $CH_3(OCH_3CHCH_2)_nOH$, avg. n = 3
Properties: M.w. 206.32; dens. 0.967 (25/25 C); b.p. 243 C; flash pt. 250 F
Toxicology: Moderately toxic by ingestion; skin irritant
Uses: Solv. for paints, cleaners, inks, functional fluids, agric., cosmetics, epoxy laminates, adhesives, floor polish, fuel additives, oilfield, mining, and electronic chems.; chemical intermediate applics.
Regulatory: FDA 21CFR §181.22, 181.30
Trade names: Arcosolv® TPM

PPG (3) methyl ether. *See* PPG-3 methyl ether

PPG-2 methyl ether acetate
CAS 88917-22-0
Synonyms: Dipropylene glycol methyl ether acetate
Formula: $CH_3(OC_3H_6)_2OAc$
Properties: M.w. 190.2
Uses: Solv. for paints, inks, cleaners, epoxy laminates, agric., adhesives, floor polish, oilfield, mining, and electronic chems.
Manuf./Distrib.: Ashland
Trade names: Arcosolv® DPMA

PPG (2) monooleate. *See* PPG-2 oleate
PPG (2) monostearate. *See* PPG-2 stearate

PPG-2 oleate
Synonyms: Dipropylene glycol monooleate; POP (2) monooleate; PPG (2) monooleate
Definition: PPG ester of oleic acid
Uses: W/o emulsifier, dispersant, antistat for textile, paper processing, cutting oils, polishes, emulsion cleaners, rubber latex, wool lubricants
Trade names: Cithrol DPGMO N/E

PPG-2 oleate SE
Uses: W/o emulsifier, dispersant, antistat for textile, paper processing, cutting oils, polishes, emulsion cleaners, rubber latex, wool lubricants
Trade names: Cithrol DPGMO S/E

PPG-2 stearate
Synonyms: Dipropylene glycol monostearate; POP (2) monostearate; PPG (2) monostearate
Definition: PPG ester of stearic acid
Uses: W/o emulsifier, dispersant, antistat for textile, paper processing, cutting oils, polishes, emulsion cleaners, rubber latex, wool lubricants
Trade names: Cithrol DPGMS N/E

PPG-2 stearate SE
Uses: W/o emulsifier, dispersant, antistat for textile, paper processing, cutting oils, polishes, emulsion cleaners, rubber latex, wool lubricants
Trade names: Cithrol DPGMS S/E

Precipitated calcium carbonate. *See* Calcium carbonate
Precipitated calcium phosphate. *See* Calcium phosphate tribasic
Precipitated calcium sulfate. *See* Calcium sulfate dihydrate
Precipitated chalk. *See* Calcium carbonate
Precipitated silica. *See* Silica, hydrated
1,2-Propanediol. *See* Propylene glycol
Propane-1,2-diol. *See* Propylene glycol
1,2-Propanediol, 3-(2-propenyloxy)-. *See* Glyceryl-1-allyl ether

N,N´-1,3-Propanediylbis(3,5-di-t-butyl-4-hydroxyhydrocinnamamide)
Uses: Stabilizer for polymers, rubber
Trade names: Irganox® 1019

1,2,3-Propanetriol monoacetate. *See* Glyceryl acetate
1,2,3-Propanetriol octadecanoate. *See* Glyceryl stearate
1,2,3-Propanetriol triacetate. *See* Triacetin
1,2,3-Propanetriol trioctadecanoate. *See* Tristearin
1,2,3-Propanetriyl 12-(acetyloxy)-9-octadecenoate. *See* Glyceryl triacetyl ricinoleate
Propanoic acid. *See* Propionic acid
Propanoic acid, 3-((3-(dodecyloxy)-3-oxopropyl)-thio)-, octadecyl ester. *See* Stearyl lauryl thiodipropionate
Propanoic acid, 3,3´-thiobis-, ditetradecyl ester. *See* Dimyristyl thiodipropionate
2-Propanol. *See* Isopropyl alcohol
2-Propanol, 1,3-dichlorophosphate (3:1). *See* Tri(β,β´-dichloroisopropyl)phosphate
2-Propanol, 2-methyl-. *See* t-Butyl alcohol
2-Propenamide, homopolymer. *See* Polyacrylamide
1-Propene, homopolymer. *See* Polypropylene

Propene, 2-methyl-, trimer. *See* Triisobutylene
2-Propenenitrile, polymer with 1,3-butadiene. *See* Butadiene/acrylonitrile copolymer
Propenoic acid. *See* Acrylic acid
2-Propenoic acid, homopolymer. *See* Polyacrylic acid
2-Propenoic acid, homopolymer, ammonium salt. *See* Ammonium polyacrylate
2-Propenoic acid-2-(hydroxymethyl)-2-(((1-oxo-2-propenyl) oxy) methyl)-1,3-propanediyl ester. *See* Pentaerythrityl triacrylate
2-Propenoic acid, 2-methyl-, homopolymer, sodium salt. *See* Sodium polymethacrylate

2-Propenoic acid, 2-methylmethyl ester, polymer with 1,3-butadiene and butyl 2-propenoate
Trade names: Durastrength 200

2-Propenoic acid, 2-methyl-, 2-sulfoethyl ester. *See* 2-Sulfoethyl methacrylate
[(2-Propenyloxy)methyl]oxirane. *See* Allyl glycidyl ether

Propionic acid
CAS 79-09-4; EINECS 201-176-3
Synonyms: Methylacetic acid; Propanoic acid; Ethylformic acid
Classification: Acid
Empirical: $C_3H_6O_2$
Formula: C_2H_5COOH
Properties: Oily liq., sl. pungent rancid odor; misc. in water, alcohol, ether, chloroform; m.w. 74.09; dens. 0.998 (15/4 C); visc. 1.020 cp (15 C); m.p. -21.5 C; b.p. 141. C (760 mm); flash pt. (OC) 58 C; ref. index 1.3862; surf. tension 27.21 dynes/cm (15 C)
Precaution: Corrosive; highly flamm. exposed to heat, flame, oxidizers; incompat. with $CaCl_2$; dec. at high temp.
Toxicology: TLV:TWA 10 ppm; LD50 (oral, rat) 3500 mg/kg; poison by intraperitoneal route; mod. toxic by ingestion, skin contact, IV route; irritant to eye, skin; heated to decomp., emits acrid smoke and irritating fumes
Uses: Esterifying agent; in prod. of cellulose propionates, etc.; as mold inhibitors and preservatives; in mfg. of ester solvs., fruit flavors, perfume bases; antifungal
Regulatory: FDA 21CFR §172.515, 184.1081, GRAS; GRAS (FEMA), EPA reg.; Japan restricted to flavoring use, limitation with sorbic acid 3 g/kg total; Europe listed; UK approved
Manuf./Distrib.: Aldrich F&F; BASF; BP Chem.; Eastman; Great Western; Hoechst Celanese; Penta Mfg.; Union Carbide
Trade names containing: RR Zinc Oxide-Coated

Propionic acid, 2-methyl-, monoester with 2,2,4-trimethyl-1,3-pentanediol. *See* 2,2,4-Trimethyl-1,3-pentanediol monoisobutyrate
2-Propoxyethanol. *See* Ethylene glycol propyl ether
Propyl carbinol. *See* Butyl alcohol
Propyl 'Cellosolve'. *See* Ethylene glycol propyl ether

Propylene carbonate
CAS 108-32-7; EINECS 203-572-1
Synonyms: 1,3-Dioxolan-2-one, 4-methyl; 4-Methyl-1,3-dioxolan-2-one; Carbonic acid, 1,2-propylene glycol ester
Classification: Organic compd.
Empirical: $C_4H_6O_3$
Properties: Clear liq.; m.w. 102.10; dens. 1.2069 (20/20 C); m.p. -48.8 C; b.p. 242.1 C; flash pt. (OC) 275 F
Toxicology: Mildly toxic by ingestion; human skin and eye irritant
Uses: Solvent for pigments and dyes; extracting agent; intermediate for organic syntheses; reactive diluent for urethane foams and coatings, foundry sand binders, textiles, natural gas treating; cosmetics lubricant
Regulatory: FDA 21CFR §175.105
Manuf./Distrib.: Allchem Ind.; Ashland; Great Western; Hüls AG
Trade names: Arconate® 1000; Arconate® HP; Arconate® PC; Texacar® PC

Propylene glycol
CAS 57-55-6; EINECS 200-338-0
Synonyms: 1,2-Propanediol; Propane-1,2-diol; 1,2-Dihydroxypropane; Methyl glycol
Classification: Aliphatic alcohol
Empirical: $C_3H_8O_2$
Formula: $CH_3CHOHCH_2OH$
Properties: Colorless visc. liq., odorless; hygroscopic; sol. in essential oils; misc. with water, acetone,

chloroform; m.w. 76.11; dens. 1.0362 (25/25 C); b.p. 188.2 C; flash pt. (OC) 210 F
Precaution: Combustible exposed to heat or flame; reactive with oxidizers; explosive limits 2.6-12.6%
Toxicology: LD50 (oral, rat) 25 ml/kg; eye/human skin irritant; sl. toxic by ingestion, IP,IV, subcutaneous routes; human systemic effects; experimental teratogenic, reproductive effects; mutagenic data; heated to decomp., emits acrid smoke and irritating fumes
Uses: Solvent, emulsifier, production paints, polyester and alkyd resins, foods, drugs, antifreeze in breweries and dairies; substitute for ethylene glycol and glycerol; mold growth and fermentation inhibitor; plasticizer for adhesives, cork, paper prods.; raw material for resinous plasticizers
Regulatory: FDA 21CFR §169.175, 169.176, 169.177, 169.178, 169.180, 169.181, 175.300, 177.2600, 178.3300, 184.1666, 582.4666, GRAS; USDA 9CFR §318.7, 381.147; BATF 27CFR §240.1051; EPA reg., approved for some drugs; Japan approved with limitations; Europe listed
Manuf./Distrib.: Aldrich; Arco; Asahi Denka Kogyo; Ashland; BP Chem. Ltd; Hüls AG; Olin; Primachem; Seeler Ind.; Texaco; Veckridge; Westco
Trade names: Adeka Propylene Glycol (T)
Trade names containing: Atmos® 300; Dur-Em® 300K; Gemtex PA-85P; Geropon® 99; Jeffcat TD-33

Propylene glycol t-butyl ether
CAS 57018-52-7
Synonyms: PTB
Formula: $C_4H_9OCH_2CHOHCH_3$
Properties: M.w. 132.2
Uses: Solvent for water-based paints, cleaners, inks, cutting fluids; in polyester and alkyd resin prod.
Trade names: Arcosolv® PTB

Propylene glycol dibenzoate
Properties: Sp.gr. 1.146; flash pt. (COC) 199 C; ref. index 1.544
Uses: Solvating plasticizer for PVC applic., latex caulk formulations, PVAc adhesives, and castable PU; used in latex paints as coalescing agent
Manuf./Distrib.: Unitex; Velsicol
Trade names: Benzoflex® 284

Propylene glycol dinonanoate. *See* Propylene glycol dipelargonate

Propylene glycol dioctanoate
CAS 7384-98-7; 56519-71-2
Synonyms: 2-Ethylhexanoic acid, 1-methyl-1,2-ethanediyl ester; Octanoic acid, 1,3-propanediyl ester
Definition: Diester of propylene glycol and 2-ethylhexanoic acid
Empirical: $C_{19}H_{36}O_4$
Formula: $CH_3(CH_2)_3CHCH_3CH_2COOCH_2CHCH_3OCOCHCH_2CH_3(CH_2)_3CH_3$
Properties: Sol. in alcohol, min. oil; sp.gr. 0.921; flash pt. (COC) 159 C; ref. index 1.4350
Uses: Emollient for creams, lotions, topicals, makeup, bath preps., after-shave lotions; vehicle for fragrance; plasticizer
Manuf./Distrib.: Inolex

Propylene glycol dipelargonate
CAS 41395-83-9; EINECS 255-350-9
Synonyms: Propylene glycol dinonanoate; Nonanoic acid, 1-methyl-1,2-ethanediyl ester
Definition: Diester of propylene glycol and pelargonic acid
Empirical: $C_{21}H_{40}O_4$
Formula: $CH_3(CH_2)_7COOCH_2CHCH_3OCO(CH_2)_7CH_3$
Properties: Sp.gr. 0.918; flash pt. (COC) 197 C; ref. index 1.4404
Uses: Emollient, lubricant for bath oils, preshave lotions, aerosol systems, lipsticks, glosses, makeup bases, pharmaceuticals; carrier for fragrances; plasticizer
Manuf./Distrib.: Henkel; Inolex

Propylene glycol laurate
CAS 142-55-2; 27194-74-7; EINECS 205-542-3
Synonyms: Dodecanoic acid, 2-hydroxypropyl ester; Dodecanoic acid, monoester with 1,2-propanediol; Propylene glycol monolaurate
Definition: Ester of propylene glycol and lauric acid
Empirical: $C_{15}H_{30}O_3$
Formula: $CH_3(CH_2)_{10}COOCH_2CHCH_3OH$
Properties: Sp.gr. 0.911; m.p. 0-12 C; flash pt. (COC) 188 C

Uses: Emulsifier, coemulsifier, stabilizer, wetting agent, lubricant, emollient, solvent, and antistat; used in cosmetic, pharmaceutical, industrial, food applics.; plasticizer
Regulatory: FDA 21CFR §172.856, 173.340, 175.105, 175.300, 176.170, 176.210, 177.2800
Manuf./Distrib.: Inolex; Velsicol

Propylene glycol methyl ether. *See* Methoxyisopropanol

Propylene glycol methyl ether acetate
CAS 108-65-6
Synonyms: PM acetate
Empirical: $C_6H_{12}O_3$
Formula: $CH_3COOCHCH_3CH_2OCH_3$
Properties: Liq.; sol. 20% in water; m.w. 132.2; dens. 0.97 (20/20 C); b.p. 140 C (760 mm); flash pt. (Seta) 45 C
Precaution: Combustible
Uses: Solvent for paints, electronics, inks
Manuf./Distrib.: Allchem Ind.; Ashland; Harcros; Oxiteno; Van Waters & Rogers
Trade names: Arcosolv® PMA

Propylene glycol monolaurate. *See* Propylene glycol laurate
Propylene glycol monomyristate. *See* Propylene glycol myristate
Propylene glycol monoricinoleate. *See* Propylene glycol ricinoleate
Propylene glycol monostearate. *See* Propylene glycol stearate

Propylene glycol myristate
CAS 29059-24-3; EINECS 249-395-3
Synonyms: Propylene glycol monomyristate; Tetradecanoic acid, monoester with 1,2-propanediol
Definition: Ester of propylene glycol and myristic acid
Empirical: $C_{17}H_{34}O_3$
Formula: $CH_3(CH_2)_{12}COOCH_2CHCH_3OH$
Uses: Wetting aid, lubricant, opacifier, antistat, dispersant, w/o emulgent, scouring and detergent aid, defoamer, plasticizer, rust inhibitor; cosmetics and pharmaceuticals, lubricating and cutting oils, textile and leather aids, pigment grinding; paints, inks, plastics, waxes, insecticides

Propylene glycol ricinoleate
CAS 26402-31-3; EINECS 247-669-7
Synonyms: 12-Hydroxy-9-octadecenoic acid, monoester with 1,2-propanediol; Propylene glycol monoricinoleate
Definition: Ester of propylene glycol and ricinoleic acid
Empirical: $C_{21}H_{40}O_4$
Properties: Sp.gr. 0.960; m.p. < -16 C; flash pt. (COC) 221 C; ref. index 1.469
Uses: Emulsifier, coemulsifier, stabilizer, wetting agent, lubricant, emollient, and antistat; used in cosmetic, pharmaceutical, industrial, textile, food applics.; dye solvent; plasticizer for cellulosics
Manuf./Distrib.: CasChem

Propylene glycol stearate
CAS 1323-39-3; EINECS 215-354-3
Synonyms: Propylene glycol monostearate; Octadecanoic acid, monoester with 1,2-propanediol
Definition: Ester of propylene glycol and stearic acid
Empirical: $C_{21}H_{42}O_3$
Formula: $CH_3(CH_2)_{16}COOCH_2CHCH_3OH$
Properties: Wh. to cream flakes, bland typ. odor; sol. in min. oil, IPM, oleyl alcohol; insol. in water, glycerin, propylene glycol; m.p. 35-38 C; sapon. no. 181-191
Toxicology: Poison by intraperitoneal route
Uses: Emulsifier for lotions, soft creams, makeup; plasticizer for cellulose nitrate
Regulatory: FDA 21CFR §172.856, 172.860, 172.862, 173.340, 175.105, 175.300, 176.170, 176.210, 177.2800; USDA 9CFR §318.7, 381.147; GRAS (FEMA)
Manuf./Distrib.: Aquatec Quimica SA; Eastman; Grindsted; Inolex; ISP Van Dyk; Karlshamns; Lipo; Lonza; Witco/PAG

Propylene polymer. *See* Polypropylene

Propyl gallate
CAS 121-79-9; EINECS 204-498-2
Synonyms: 3,4,5-Trihydroxybenzoic acid, n-propyl ester; n-Propyl 3,4,5-trihydroxybenzoate; Gallic acid

propyl ester

Classification: Aromatic ester

Definition: Aromatic ester of propyl alcohol and gallic acid

Empirical: $C_{10}H_{12}O_5$

Formula: $(HO)_3C_6H_2COOCH_2CH_2CH_3$

Properties: Fine ivory powd. or crystals, odorless; insol. in water; sol. in alcohol, oils; m.w. 212.22; m.p. 147-149 C; b.p. dec. > 148 C

Precaution: Combustible exposed to heat or flame; reactive with oxidizers

Toxicology: LD50 (oral, rat) 3.8 g/kg; poison by ingestion and intraperitoneal route; experimental tumorigen, reproductive effects; mutagenic data; heated to decomp., emits acrid smoke and irritating fumes

Uses: Food and feed antioxidant, flavor and packaging material; stabilizer for cellulosics, rubbers, cosmetics, fats and oils

Regulatory: FDA 21CFR §172.615, 175.125, 175.300, 181.22, 181.24 (0.005% migrating from food pkg.), 184.1660 (0.02% max. of fat or oil), GRAS; USDA 9CFR §318.7, 381.147; Japan approved (0.1 g/kg max.); Europe listed; UK approved

Manuf./Distrib.: Aceto; Eastman; Nipa Labs; Penta Mfg.; Spectrum Chem. Mfg.; UOP

Trade names: Tenox® PG

Propyl oleate

CAS 1330-80-9

Definition: Ester of propyl alcohol and oleic acid

Properties: Sp.gr. 0.869; m.p. -20 C; flash pt. (COC) 166 C; ref. index 1.4494

Uses: Base for industrial lubricants; mold release agent, defoamer, flotation agent, plasticizer for cellulosics, PS, needle lubricants; when sulfated is useful as wetting, rewetting, and dye leveling agent in textile and leather industries

Manuf./Distrib.: Henkel; Witco/Oleo-Surf.

Trade names: Emerest® 2302

n-Propyl percarbonate. *See* Di-n-propyl peroxydicarbonate

n-Propyl 3,4,5-trihydroxybenzoate. *See* Propyl gallate

n-Propyltrimethoxysilane

CAS 1067-25-0

Empirical: $C_6H_{16}O_3Si$

Properties: M.w. 164.28; dens. 0.938 (20/4 C); b.p. 141-143 C

Precaution: Moisture-sensitive

Toxicology: Irritant

Uses: Coupling agent, release agent, lubricant, blocking agent, chemical intermediate

Trade names: Dynasylan® PTMO

PS. *See* Polystyrene

PTAP. *See* p-t-Amylphenol

PTB. *See* Propylene glycol t-butyl ether

PTBP. *See* 4-t-Butylphenol

PTFE. *See* Polytetrafluoroethylene

PTMEG. *See* Polytetramethylene ether glycol

PTMEG diamine. *See* Polytetramethylene ether glycol diamine

PTMG. *See* Polytetramethylene ether glycol

PTSA. *See* p-Toluenesulfonamide

PVA. *See* Polyvinyl alcohol

PVAL. *See* Polyvinyl alcohol

PVC. *See* Polyvinyl chloride

PVC/acrylate

Uses: For extrusion, calendering, inj. molding, e.g., profiles for outdoor applics. (window frames), pipes, panels, films; impact modifier for rigid PVC; adhesive

Trade names: Vinnol® K 704

PVDC. *See* Polyvinylidene chloride

PVDF. *See* Polyvinylidene fluoride

PVM. *See* Polyvinyl methyl ether

PVM/MA copolymer

CAS 9011-16-9; 52229-50-2

Synonyms: Methyl vinyl ether/maleic anhydride copolymer; Poly(methyl vinyl ether/maleic anhydride); 2,5-Furandione, polymer with methoxyethylene
Definition: Copolymer of methyl vinyl ether and maleic anhydride
Empirical: $(C_4H_2O_3 \cdot C_3H_6O)_x$
Uses: Dispersant, coupling, stabilizer, thickener, emulsifier, solubilizer, corrosion inhibitor, film former, antistat, used in agric., paper and textile industries, chemical processing, industrial products, detergents, cosmetics, emulsion polymerization
Trade names: Gantrez® AN-119; Gantrez® AN-139; Gantrez® AN-149; Gantrez® AN-169; Gantrez® AN-179

PVOH. *See* Polyvinyl alcohol

PVP
CAS 9003-39-8; EINECS 201-800-4
Synonyms: Polyvinylpyrrolidone; Poly (n-vinylbutyrolactam); Povidone; 1-Ethenyl-2-pyrrolidinone homopolymer
Classification: Linear polymer
Definition: Polymer of 1-vinyl-2-pyrrolidone monomers
Empirical: $(C_6H_9NO)_x$
Properties: Wh. free-flowing amorphous powd.; sol. in water, chlorinated hydrocarbons, alcohol, amines, nitroparaffins, lower m.w. fatty acids; m.w. ≈ 10,000, ≈ 24,000, ≈ 40,000 (food use), ≈ 160,000, ≈ 360,000 (beer); dens. 1.23-1.29
Toxicology: LD50 (IP, mouse) 12 g/kg; mildly toxic by IP and IV routes; heated to decomp., emits toxic fumes of NO_x
Uses: Film-forming agent, hair fixative, thickener, protective colloid, suspending agent, and dispersant for cosmetics industry, tech. applics.; drug vehicle and retardant; tablet binder, pharmaceutical excipient; in adhesives; in detergents
Regulatory: FDA 21CFR §73.1, 73.1001, 172.210, 173.50, 173.55, 175.105, 175.300, 176.170, 176.180, 176.210; BATF 27CFR §240.1051
Manuf./Distrib.: Allchem Ind.; BASF; Great Western; Hickson Danchem; ISP; Monomer-Polymer & Dajac
Trade names: Luviskol® K17; Luviskol® K30; Luviskol® K60; Luviskol® K90; PVP K-60

Pyrazine hexahydride. *See* Piperazine

Pyridine
CAS 110-86-1; EINECS 203-809-9
Empirical: C_5H_5N
Properties: Colorless liq., char. disagreeable odor, sharp taste; misc. with water, alcohol, ether, petroleum ether, oils, other org. liqs.; m.w. 79.10; dens. 0.98272 (20/4 C); m.p. -41.6 C; b.p. 115-116 C; flash pt. (CC) 20 C; ref. index 1.510 (20 C)
Precaution: Highly flamm.; volatile with steam
Toxicology: LD50 (oral, rat) 1.58 g/kg; may cause human CNS depression, irritation of skin and respiratory tract; large does may produce GI disturbances, kidney and liver damage
Uses: Chem. intermediate, acid scavenger; in agrochemicals, pharmaceuticals, photographic materials, coatings, curing agents, rubber chemicals, plastics, antidandruff shampoos, textiles, dyestuffs
Regulatory: FDA 21CFR §172.515
Manuf./Distrib.: Aldrich; Allchem Ind.; Berje; Nepera; Penta Mfg.; Raschig; Schweizerhall; Spectrum Chem. Mfg.
Trade names: Pyridine 1°

Pyrithione zinc. *See* Zinc pyrithione
Pyroligneus acid. *See* Acetic acid
Pyromellitic acid dianhydride. *See* Pyromellitic dianhydride

Pyromellitic dianhydride
CAS 89-32-7; EINECS 201-898-9
Synonyms: PMDA; Benzene-1,2,4,5-tetracarboxylic dianhydride; 1,2,4,5-Benzenetetracarboxylic dianhydride; 1H,3H-Benzo(1,2-c;4,5-c′)difuran-1,3,5,7-tetrone; 1,2,4,5-Benzenetetracarboxylic 1,2:4,5 dianhydride
Empirical: $C_{10}H_2O_6$
Properties: Sol. 0.4 g/100 g in water, reacts to give tetra-acid; m.w. 218.12; dens. 1.680 (20 C); vapor pressure 63 mm Hg (290 C); m.p. 284-286 C; b.p. 380-400 C; pH 2.2 (3.9 g/L)
Precaution: Incompat. with alkali metal hydroxides, amines, ammonia, alcohols, thiols, strong oxidizing/reducing agents

Toxicology: LD50 (oral, rat) 2250-2595 mg/kg; skin and eye irritant, but nonsensitizing; may cause respiratory irritation/allergic response on inhalation of dust or fumes
Uses: Intermediate for polyimides, polyester resins, alkyd and PU resins, polypyrrones; epoxy curing agent; end-uses incl. electronic transfer molding compds., wire coatings, aerospace adhesives, advanced structural composites, uv-cured inks, insulating coatings and foams, flame retardant prods., release agents, molded high-performance engineering parts
Manuf./Distrib.: Allco; Hüls AG; Monomer-Polymer & Dajac
Trade names: Allco PMDA

Pyrophyllite
CAS 12269-78-2
Synonyms: Hydrated aluminum silicate; Agalmatolite
Definition: Naturally occurring mineral substance consisting predominantly of hydrous aluminum silicate
Empirical: Al_2O_3•$4SiO_2$•H_2O
Formula: $Al_2Si_4O_{10}(OH)$
Properties: Colorless, white, green, gray, brown; dens. 2.8-2.9
Uses: Filler, extender, diluent, carrier for rubber, plastics, ceramics, insecticides, slate pencil, feed additive; color additive for drugs and cosmetics
Regulatory: FDA 21CFR §73.1400, 73.2400
Manuf./Distrib.: R.T. Vanderbilt
Trade names: Pyrax® A; Pyrax® ABB; Pyrax® B; Pyrax® WA
See also Aluminum silicate

2-Pyrrolidone
CAS 616-45-5; EINECS 204-648-7
Synonyms: Butyrolactam; Pyrrolidone-2
Empirical: C_4H_7NO
Formula: $CH_2CH_2CH_2C(O)NH$
Properties: Lt. yel. liq.; sol. in water, ethanol, ethyl ether, chloroform, benzene, ethyl acetate, carbon disulfide; m.w. 85.12; dens. 1.1; b.p. 245 C; flash pt. 265 F
Toxicology: Mildly toxic by ingestion, subcutaneous route
Uses: Plasticizer and coalescing agent; solv. for veterinary medicine
Manuf./Distrib.: Allchem Ind.; BASF; ISP; UCB SA; Unitex
Trade names: 2-Pyrol®

Pyrrolidone-2. *See* 2-Pyrrolidone

Quartz
CAS 14808-60-7; EINECS 238-878-4
Synonyms: Silicon dioxide; Silica
Definition: Crystallized silicon dioxide
Empirical: O_2Si
Formula: SiO_2
Properties: White to reddish; insol. in acids except HF; m.w. 60.08 m.p. 1713 C
Precaution: Noncombustible; avoid inhalation of fine particles
Toxicology: TLV (for resp. dust) 10 mg/m^3
Uses: Electronic components; TV components
Manuf./Distrib.: Unimin; U.S. Silica; Westo Industrial Prods. Ltd
Trade names containing: Aktisil® AM; Aktisil® EM; Aktisil® MM; Aktisil® PF 216; Aktisil® PF 231; Aktisil® VM; Aktisil® VM 56; Silfin® Z; Sillikolloid P 87; Sillitin N 82; Sillitin N 85; Sillitin V 85; Sillitin Z 86; Sillitin Z 89; TP 87/21; TP 89/50; TP 90/127; TP 91/01
See also Silica

Quaternary ammonium compds., carboxymethyl (coco alkyl) dimethyl hydroxides, inner salts. *See* Cocobetaine
Quaternary ammonium compds., coco alkyl trimethyl, chlorides. *See* Cocotrimonium chloride
Quaternary ammonium compds., tallow alkyl trimethyl, chlorides. *See* Tallowtrimonium chloride
Quaternium-5. *See* Distearyldimonium chloride

Quaternium-15
CAS 51229-78-8; 4080-31-3; EINECS 223-805-0

Synonyms: 1-(3-Chloroallyl)-3,5,7-triaza-1-azoniaadamantane chloride; N-(3-Chloroallyl)hexaminium chloride; Chlorallyl methenamine chloride
Classification: Quaternary ammonium salt
Empirical: $C_9H_{16}ClN_4 \cdot Cl$
Formula: $C_6H_{12}N_4(CH_2CHCHCl)Cl$
Uses: Preservative for adhesives, latex emulsions, paints, cutting fluids
Regulatory: FDA 21CFR §175.105, 176.170; CIR approved; Europe listed; Japan not approved
Trade names containing: Dowicil® 75

Quaternium-18
CAS 61789-80-8, 68002-59-5; EINECS 263-090-2
Synonyms: Dimethyl di(hydrogenated tallow)ammonium chloride; Dihydrogenated tallow dimethyl ammonium chloride; Ditallowalkonium chloride
Classification: Quaternary ammonium salt
Formula: $R(CH_3)_2NR]^+Cl^-$, R rep. hydrogenated tallow fatty radicals
Uses: Surfactant, fabric antistat, conditioner for household, industrial, textile, hair applics.; antistatic coating for cellulose acetate, polyacetal, PE, PP
Trade names: Adogen® 442

Quaternium-21. *See* PPG-40 diethylmonium chloride
Quaternium-25. *See* Cetethyl morpholinium ethosulfate
Quaternium-31. *See* Dialkyl dimethyl ammonium chloride
Quaternium-34. *See* Dicocodimonium chloride
Quicklime. *See* Calcium oxide
Quinoline, 1,2-dihydro-2,2,4-trimethyl homopolymer. *See* 2,2,4-Trimethyl-1,2-dihydroquinoline polymer

p-Quinone dioxime
CAS 105-11-3
Synonyms: p-Benzoquinone dioxime
Formula: $HONC_6H_4NOH$
Properties: Yel. needles; sl. sol. in water; m.w. 138.06
Toxicology: Moderately toxic by ingestion; mutagenic data
Uses: Rubber accelerator
Manuf./Distrib.: Alemark; Amber Syn.; Lord
Trade names: QDO®

Rapeseed oil
CAS 8002-13-9; EINECS 232-299-0
Synonyms: Brassica campestris oil; Colza oil; Canola oil (low erucic acid rapeseed oil)
Definition: Vegetable oil expressed from seeds of *Brassica campestris*
Properties: Brn. viscous liq., yel. when refined; sol. in chloroform, ether, CS_2; dens. 0.913-0.916; m.p. 17-22 C; solidifies at 0 C; flash pt. 325 F; iodine no. 97-105; sapon. no. 170-177 C; ref. index 1.4720-1.4752
Precaution: Subject to spontaneous heating
Uses: Lubricant and slip agent for plastics, polymers, rubber, metal lubricants; edible oil for salad dressings, margarine; soft soaps
Regulatory: FDA 21CFR §175.105, 176.210, 177.1200, 177.2800, 184.1555; Japan approved (extract)
Manuf./Distrib.: Arista Industries; Climax Performance; Penta Mfg.; Reilly-Whiteman; Werner G. Smith; Witco/Oleo-Surf.
Trade names: RS-80

Rapeseed oil, vulcanized
Uses: Extender, processing aid for rubber goods
Trade names: Faktogel® 10; Faktogel® 14; Faktogel® 17; Faktogel® Badenia C; Faktogel® HF; Faktogel® R; Faktogel® Rheinau H
Trade names containing: Faktogel® Badenia T; Faktogel® MB; Faktogel® Para; Faktogel® Rheinau W; Faktogel® T; Faktogel® ZD

RDGE. *See* Resorcinol diglycidyl ether
Red iron oxide. *See* Ferric oxide
Red iron trioxide. *See* Ferric oxide
Red lead. *See* Lead oxide, red
Red lead oxide. *See* Lead oxide, red
Red oil. *See* Oleic acid

Refined petroleum wax. *See* Petroleum wax
Resole. *See* Phenolic resin
Resorcin. *See* Resorcinol

Resorcinol
CAS 108-46-3; EINECS 203-585-2
Synonyms: 1,3-Benzenediol; m-Dihydroxybenzene; Resorcin
Classification: Phenol
Empirical: $C_6H_6O_2$
Formula: $C_6H_4(OH)_2$
Properties: White crystals, unpleasant sweet taste; very sol. in alcohol, ether, glycerol; sl. sol. in chloroform; sol. in water; m.w. 110.12; dens. 1.285 (15 C); m.p. 110 C; b.p. 280.5 C; flash pt. (CC) 261 F
Precaution: Protect from light
Toxicology: TLV:TWA 10 ppm; moderately toxic by skin contact and intravenous route; poison by ingestion; skin and severe eye irritant
Uses: Topical antiseptic; keratolytic agent; in tanning; in mfg. of resins, adhesives, hexylresorcinol, p-aminosalicylic acid, explosives, dyes; in cosmetics; dyeing and printing textiles
Regulatory: FDA 21CFR §177.1210
Manuf./Distrib.: Allchem Ind.; Fairmount; R.W. Greeff; Indspec; Janssen Chimica; Napp Tech.; Penta Mfg.; Richman
Trade names containing: Cohedur® RL; Cohedur® RS; Rhenogran® Resorcin-80; Rhenogran® Resorcin-80/SBR

Resorcinol benzoate
CAS 136-36-7
Synonyms: Resorcinol monobenzoate
Empirical: $C_{13}H_{10}O_3$
Formula: $C_6H_5COOC_6H_4OH$
Properties: White crystalline solid; insol. in water, benzene; sol. in acetone, ethanol; m.w. 214.23; sp.gr. > 1.0; m.p. 132-135 C; b.p. 140 C
Toxicology: Poison by intraperitoneal route; moderately toxic by ingestion; eye irritant
Uses: Noncoloring UV inhibitor for plastics and cosmetics
Manuf./Distrib.: Eastman; Monomer-Polymer & Dajac
Trade names: Eastman® Inhibitor RMB

Resorcinol bis (diphenylphosphate)
Uses: Flame retardant for PC, PE, nylon, TPR
Trade names: Fyrolflex RDP

Resorcinol diacetate
Trade names containing: Cohedur® RK

Resorcinol diglycidyl ether
CAS 101-90-6
Synonyms: RDGE; m-Bis(glycidyloxy) benzene; 2,2′-(1,3-Phenylenebis(oxymethylene))bisoxirane
Empirical: $C_{12}H_{14}O_4$
Formula: $C_6H_4(OCH_2CHOCH_2)_2$
Properties: Straw-yel. liq.; misc. with most org. resins; m.w. 222.26; dens. 1.21 (25 C); b.p. 172 C (0.8 mm); flash pt. (OC) 176 C
Precaution: Combustible
Toxicology: Poison by intraperitoneal route; moderately toxic by ingestion; skin irritant
Uses: Reactive diluent for epoxy resins
Trade names: Epodil® 769

Resorcinol-formaldehyde resin
CAS 65876-95-1; 24969-11-7
Uses: Bonding agent for rubber compds., latex dips, adhesives in wood gluing
Trade names: Penacolite® B-18-S; Penacolite® R-2170; SRF-1501; Vulkadur® T, 40%

Resorcinol monobenzoate. *See* Resorcinol benzoate

Ricinoleic acid
CAS 141-22-0; EINECS 205-470-2
Synonyms: 12-Hydroxy-9-octadecenoic acid; 9-Octadecenoic acid, 12-hydroxy-; d-12-Hydroxyoleic acid
Classification: Unsaturated fatty acid
Empirical: $C_{18}H_{34}O_3$
Formula: $CH_3(CH_2)_5CH(OH)CH_2CH=CH(CH_2)_7COOH$
Properties: Liq.; sol. in alcohol, acetone, ether, chloroform; m.w. 298.45; dens. 0.940 (27.4/4 C); m.p. 5.5 C; b.p. 245 C (10 mm); ref. index 1.4716 (20 C)
Uses: Chem. intermediate; lubricant, rustproofing in cutting oils; in textile finishing, resin plasticizers, inks, coatings, cosmetics; modifier for coatings and adhesive polymers; sometimes added to Turkey red oil, drycleaning soaps
Manuf./Distrib.: CasChem
Trade names: P®-10 Acid

Ricinus oil. *See* Castor oil

Rosin
CAS 8050-09-7; 8052-10-6; EINECS 232-475-7
Synonyms: Colophony; Gum rosin; Rosin gum
Definition: Residue from distilling off the volatile oil from the oleoresin obtained from *Pinus palustris* and other species of *Pinaceae*
Properties: Pale yel. to amber translucent, sl. turpentine odor and taste; insol. in water; sol. in alcohol, benzene, ether, glacial acetic acid, oils, carbon disulfide; dens. 1.07-1.09; m.p. 100-150 C; flash pt. 187 C
Precaution: Combustible
Uses: Mfg. of varnishes, paint driers, printing inks, cements, soap, sealing wax, wood polishes, paper, plastics, fireworks, sizes, rosin oil; waterproofing paper, walls; pharmaceutic aid; emulsifier for SBR polymerization; tackifier resin in adhesives, sealants
Regulatory: FDA 21CFR §73.1, 172.210, 172.510, 172.615, 175.105, 175.125, 175.300, 176.170, 176.200, 176.210, 177.1200, 177.1210, 177.2600, 178.3120, 178.3800, 178.3870; Japan approved
Manuf./Distrib.: Arakawa; Arizona; Cytec; Georgia-Pacific; Hercules BV; Meer; Natrochem; Veitsiluoto Oy; Westvaco
Trade names: Acintol® Liquaros; Acintol® R Type S; Acintol® R Type SB; Acintol® R Type SFS; Alatac 100-10; Arizona DR-22; Beckacite® 4900; Diprosin A-100; MR1085C; Oulumer 70; Oulupale XB 100; Oulutac 20 D; Resin 731D; Resinall 153; Resinall 203; Resinall 219; Resinall 286; Staybelite®; Sylvaros® 20; Sylvaros® 80; Sylvaros® 315; Sylvaros® RB; Sylvatac® AC; Sylvatac® ACF; Uni-Tac® 68; Uni-Tac® 70; Westvaco® Resin 90; Westvaco® Resin 95; Westvaco® Rosin T
Trade names containing: Uni-Tac® 72; Uni-Tac® 72M70

Rosin acid, methyl ester. *See* Methyl rosinate
Rosin, glyceryl ester. *See* Glyceryl rosinate
Rosin gum. *See* Rosin
Rosin, hydrogenated, glycerol ester. *See* Glyceryl hydrogenated rosinate

Rosin, maleated
Uses: Intermediate for rosin derivs., printing ink binders; tackifier resin in sealants and mastics; mfg. paper sizes
Trade names: Acintol® R Type SM4

Rosin pentaerythritol ester. *See* Pentaerythrityl rosinate

Rosin, polymerized
Uses: Tackifier for adhesives
Manuf./Distrib.: S & S Chem.; T & R Chem.
Trade names: Sylvaros® R; Sylvatac® 95; Sylvatac® 140; Sylvatac® 295; Sylvatac® R85; Sylvatac® RX

SAIB. *See* Sucrose acetate isobutyrate

Salicylic acid
CAS 69-72-7; EINECS 200-712-3
Synonyms: 2-Hydroxybenzoic acid; o-Hydroxybenzoic acid
Classification: Aromatic acid
Empirical: $C_7H_6O_3$
Formula: HOC_6H_4COOH

Properties: Crystals or cryst. powd., acrid taste; sol. in water, alcohol, ether; m.w. 138.12; dens. 1.443 (20/4 C); m.p. 157-159 C; b.p. 211 C (20 mm) sublimes at 76 C
Precaution: Protect from light
Toxicology: LD50 (oral, rat) 891 mg/kg, (IV, mouse) 500 mg/kg; poison by ingestion, IV, IP routes; mod. toxic by subcutaneous route; skin and severe eye irritant; experimental teratogen, reproductive effects; mutagenic data; heated to decomp., emits acrid smoke
Uses: Preservative for foods; mfg. of methyl salicylate, acetylsalicylic acid, etc., dyes; reagent in analytical chemistry; topical keratolytic agent; OTC drug active ingred.
Regulatory: FDA 21CFR §175.105, 175.300, 177.2600, 556.590
Manuf./Distrib.: Allchem Ind.; EM Ind.; Hilton Davis; Great Western; R.W. Greeff; Janssen Chimica; Mitsui Toatsu; PMC Specialties; Rhone-Poulenc
Trade names containing: Retarder SAX

Salol. *See* Phenyl salicylate
SAN. *See* Styrene-acrylonitrile copolymer
SAN copolymer. *See* Styrene-acrylonitrile copolymer
Santoquine. *See* 6-Ethoxy-1,2-dihydro-2,2,4-trimethylquinoline
Saran. *See* Polyvinylidene chloride
S/B. *See* Styrene-butadiene polymer
SBR. *See* Styrene-butadiene rubber
SDDC. *See* Sodium dimethyldithiocarbamate
SDS. *See* Sodium lauryl sulfate

Selenium
CAS 7782-49-2; EINECS 231-957-4
Synonyms: Colloidal selenium
Classification: A nonmetallic element
Empirical: Se
Properties: Steel gray nonmetallic element; insol. in water, alcohol; very sl. sol. in ether; at.wt. 78.96; dens. 4.81-4.26; m.p. 170-217 C; b.p. 690 C
Toxicology: Poison by inhalation, intravenous and other routes; TLV:TWA 0.2 mg/m^3
Uses: Electronics, colorant for glass (ceramics), rectifiers, relays, solar batteries; vulcanizing agents for rubbers
Manuf./Distrib.: All Chemie Ltd; Appleby Group Ltd; Asarco; Atomergic Chemetals; Cerac; Johnson Matthey; Noah; R.T. Vanderbilt
Trade names: Vandex

Selenium diethyldithiocarbamate
CAS 21559-14-8; 5456-28-0
Empirical: $C_{20}H_{40}N_4S_8Se$
Formula: $Se[SC(S)N(C_2H_5)_2]_4$
Properties: Orange-yel. powd., char. odor; sol. in carbon disulfide, benzene, chloroform; insol. in water; dens. 1.32 (20/20 C); m.p. 63-71 C
Toxicology: TLV 0.2 mg/m^3 of air; toxic by inhalation, ingestion, skin absorption
Uses: Vulcanization agent without sulfur or as primary or secondary accelerator with sulfur
Manuf./Distrib.: R.T. Vanderbilt

Selenium dimethyldithiocarbamate
CAS 144-34-4
Synonyms: Methyl Selenac
Empirical: $C_{12}H_{24}S_8N_4Se$
Formula: $[(CH_3)_2NC(S)S]_4Se$
Properties: Yel. powd.; sl. sol. in carbon disulfide, benzene, chloroform; insol. in water, gasoline; dens. 1.58; m.p. 140-172 C
Uses: Vulcanizing agent, accelerator for rubbers
Trade names: Methyl Selenac®

Silica
CAS 7631-86-9 (colloidal); 112945-52-5 (fumed), 60676-86-0; EINECS 231-545-4
Synonyms: Silicon dioxide, fumed; Colloidal silicon dioxide; Silicon dioxide; Fumed silica; Silicic anhydride
Classification: Inorganic oxide
Definition: Occurs in nature as agate, amethyst, chalcedony, cristobalite, flint, quartz, sand, tridymite
Empirical: O_2Si

Formula: SiO_2
Properties: Transparent crystals or amorphous powd.; pract. insol. in water or acids except hydrofluoric; m.w. 60.09; dens. 2.2 (amorphous), 2.65 (quartz, 0 C); lowest coeff. of heat expansion; melts to a glass
Toxicology: LD50 (oral, rat) 3160 mg/kg; poison by intraperitoneal, intravenous, intratracheal routes; moderately toxic by ingestion; prolonged inhalation of dust can cause silicosis
Uses: Mfg. of glass, water glass, refractories, abrasives, ceramics, enamels, petrol. prods.; filler in cosmetics; rubber reinforcing agent; as anticaking and defoaming agent; abrasive; thickener
Regulatory: FDA 21CFR §73.1, 172.230, 172.480 (limitation 2%), 173.340, 175.105, 175.300, 176.200, 176.210, 177.1200, 177.1460, 177.2420, 177.2600, 182.90, 182.1711, GRAS; USDA 9CFR §318.7; Japan approved (2% max. as anticaking), other restrictions; Europe listed; UK approved
Manuf./Distrib.: Akzo Nobel; BYK-Chemie; Cabot Carbon Ltd; Catalysts & Chemicals Industries; Chisso Am.; Degussa; Du Pont; J.M. Huber; Nippon Silica Ind.; Nissan Chem. Ind.; PPG Industries; PQ; Unimin
Trade names: Aerosil® 90; Aerosil® 130; Aerosil® 150; Aerosil® 200; Aerosil® 300; Aerosil® 325; Aerosil® 380; Aerosil® OX50; Aerosil® R202; Aerosil® R805; Aerosil® R812; Aerosil® R972; Aerosil® R972V; Aerosil® TT600; Cab-O-Sil® EH-5; Cab-O-Sil® H-5; Cab-O-Sil® HS-5; Cab-O-Sil® L-90; Cab-O-Sil® LM-130; Cab-O-Sil® LM-150; Cab-O-Sil® LM-150D; Cab-O-Sil® M-5; Cab-O-Sil® M-7D; Cab-O-Sil® MS-55; Cab-O-Sil® MS-75D; Cab-O-Sil® PTG; Flo-Gard® CC 120, CC 140, CC 160; Imsil® 1240; Imsil® A-8; Imsil® A-10; Imsil® A-15; Imsil® A-25; Imsil® A-30; Lo-Vel® 27; Lo-Vel® 28; Lo-Vel® 29; Lo-Vel® 39A; Lo-Vel® 66; Lo-Vel® 275; Lo-Vel® HSF; Min-U-Sil® 5; Min-U-Sil® 10; Min-U-Sil® 15; Min-U-Sil® 30; Min-U-Sil® 40; Sil-Co-Sil® 40; Sil-Co-Sil® 45; Sil-Co-Sil® 47; Sil-Co-Sil® 49; Sil-Co-Sil® 51; Sil-Co-Sil® 52; Sil-Co-Sil® 53; Sil-Co-Sil® 63; Sil-Co-Sil® 75; Sil-Co-Sil® 90; Sil-Co-Sil® 106; Sil-Co-Sil® 125; Sil-Co-Sil® 250; Silene® D; Siltex™; Siltex™ 50+ 100 mesh; Siltex™ 44; Siltex™ 44C; Silver Bond 45; Silver Bond B; Tamsil 8; Tamsil 10; Tamsil 15; Tamsil 25; Tamsil 30; Tamsil Gold Bond; Ultrasil® VN3SP; Wacker HDK® H15; Wacker HDK® H20; Wacker HDK® N20; Wacker HDK® S13; Wacker HDK® T30; Wacker HDK® T40; Wacker HDK® V15
Trade names containing: Aerosil® COK 84; Akrochem® Accelerator VS; Aztec® 2,5-DI-70-S; BYK®-2615; Cab-O-Sil® TS-530; Cab-O-Sil® TS-610; Cab-O-Sil® TS-720; Cabot® PE 9007; Cohedur® RK; Colloids PE 48/10/10; Colloids PE 48/10/11; Coupsil VP 6109; Coupsil VP 6411; Coupsil VP 6508; Coupsil VP 8113; Coupsil VP 8415; Crystic Pregel 17; Crystic Pregel 27; Drewplus® L-131; Drewplus® L-139; Drewplus® L-191; Drewplus® L-198; Drewplus® L-475; Drewplus® Y-250; Drewplus® Y-281; Halocarbon Grease 19; Halocarbon Grease 25-5S; Halocarbon Grease 28; Halocarbon Grease 28LT; Halocarbon Grease 32; Larostat® 519; Nipol® 1422; NMP-Plus™; PE-AB 07; PE-AB 19; PE-AB 30; PE-AB 48; PE-AB 73; Perkalink® 300-50D-pd; Perkalink® 300-50DX; Perkalink® 301-50D; Perkalink® 400-50D; PE-SA 21; PE-SA 24; PE-SA 40; PE-SA 60; PE-SA 72; PE-SA 92; Rhenocure® TP/S; Rhenofit® BDMA/S; Rhenofit® EDMA/S; Rhenofit® TAC/S; Rhenofit® TRIM/S; Rhenovin® CBS-70; Rhenovin® MBT-70; Rhenovin® MBTS-70; Rhenovin® TMTD-70; Rhenovin® TMTM-70; Sipernat® D11; Struktol® ZEH-DL; Surfynol® 82S; Vulkazon® AFS-50; Wingstay® S
See also Quartz

Silica-alumina
Uses: Catalyst support; filler and flame retardant for plastics, epoxies
Trade names: Sillum-200; Zeeospheres® 200; Zeeospheres® 400; Zeeospheres® 600; Zeeospheres® 800; Zeeospheres® 850

Silica dimethyl silylate
CAS 60842-32-2
Uses: Anticaking agent, thickener, thixotrope, rheology aid for plastisol coating compds.; reinforcement for rubber sealants, plastics, adhesives
Trade names: Aerosil® R974

Silica gel. *See* Silica, hydrated
Silica hydrate. *See* Silica, hydrated

Silica, hydrated
CAS 1343-98-2 (silicic acid); 112926-00-8; EINECS 215-683-2
Synonyms: Silicic acid; Silica hydrate; Silica gel; Precipitated silica
Classification: Inorganic oxide
Definition: Occurs in nature as opal
Formula: $SiO_2 \cdot xH_2O$, x varies with method of precipitation and extent of drying
Properties: White amorphous powd.; insol. in water or acids except hydrofluoric
Toxicology: Eye irritant; poison by intravenous route; TLV:TWA 10 mg/m^3 (total dust)
Uses: Insecticide; for absorption of vapors; laboratory reagent; reinforcing agent in rubber
Regulatory: FDA 21CFR §73.1, 160.105, 160.185, 172.480, 173.340, 175.105, 175.300, 176.170, 176.180,

176.200, 176.210, 177.1200, 177.2420, 177.2600, 182.90, 573.940

Manuf./Distrib.: Crosfield; J.M. Huber; MEI (Magnesium Elektron); Osram Sylvania; PPG Ind.; PQ; Spectrum Chem. Mfg.

Trade names: Akrochem® Rubbersil RS-150; Akrochem® Rubbersil RS-200; Durosil; FK 140; FK 160; FK 300DS; FK 310; FK 320; FK 320DS; FK 500LS; Flo-Gard® SP; Hi-Sil® 132; Hi-Sil® 135; Hi-Sil® 210; Hi-Sil® 233; Hi-Sil® 243LD; Hi-Sil® 250; Hi-Sil® 255; Hi-Sil® 532EP; Hi-Sil® 752; Hi-Sil® 900; Hi-Sil® 915; Hi-Sil® 2000; Hi-Sil® ABS; Hi-Sil® EZ; Hubersil® 162; Perkasil® KS 300; Perkasil® KS 404; Perkasil® VP 406; Silene® 732D; Sipernat® 22; Sipernat® 22LS; Sipernat® 22S; Sipernat® 50LS; Sipernat® 50S; Tixosil 311; Tixosil 321; Tixosil 331; Tixosil 333, 343; Tixosil 375; Tullanox HM-100; Tullanox HM-150; Tullanox HM-250; Vulkasil® N; Vulkasil® N/GR-S; Vulkasil® S; Zeosil 45; Zeosil 85; Zeosil 125; Zeosil 165G; Zeosil 175MP; Zeosil 1165

Trade names containing: Ciptane® 255LD; Ciptane® I; PE-AB 33; PE-SA 49; PP-AB 104; PP-SA 46; PP-SA 97; PP-SA 128; Vulkasil® C

Silica silylate

Definition: Hydrophobic silica deriv. where some of the hydroxyl groups on the surf. of fumed silica have been replaced by trimethylsiloxyl groups

Uses: Anticaking and free-flow agent for PS, fire extinguishers, pesticides

Trade names: Sipernat® D17

Silicic acid. *See* Silica, hydrated
Silicic acid, aluminum sodium salt, sulfurized. *See* Sodium alumino sulfo silicate
Silicic acid, calcium salt. *See* Calcium silicate
Silicic anhydride. *See* Silica
Silicochloroform. *See* Trichlorosilane
Silicon dioxide. *See* Quartz, Silica
Silicon dioxide, fumed. *See* Silica

Silicone

Synonyms: Organosiloxane

Classification: Siloxane polymers

Properties: Liq., semisolid, or solid; cis. 1->1,000,000 cs; water repellant; exc. ; sol. in most organic solvents; dielectric props.

Precaution: Unhalogenated types combustible

Uses: Thermosetting siloxane polymers used as mold release for plastics and rubber, defoamers for mining, latex, ink, soaps, agric., food processing, as lubricants, conditioners, and emollients in personal care prods.; molding compds., encapsulants

Manuf./Distrib.: Dow Corning; GE Silicones; Genesee Polymers; Goldschmidt; Hüls Am.; Miles; Sandoz; Shin-Etsu Chem.; Union Carbide; Wacker Silicones

Trade names: Rubber Lubricant™

Trade names containing: AF 100 IND; Agitan 301; Akrolease® E-9410; Aqualease™ 2802; Blue Label Silicone Spray No. SB412-A; BYK®-065; BYK®-066; BYK®-080; BYK®-141; BYK®-3105; BYK®-A 525; BYK®-A 530; Chemlease 55; Coagulant WS; Dabco® DC1536; Dabco® DC1537; Dabco® DC1538; Dabco® DC5164; Dabco® DC5169; Dabco® DC5244; Dabco® DC5258; Dabco® DC5270; Dabco® DC5357; Dabco® DC5365; Dabco® DC5367; Dabco® DC5374; Dabco® DC5425; Dabco® DC5450; Dabco® DC5454; Dabco® DC5526; Dabco® DC5885; Dabco® DC5890; Dabco® DC5895; Dow Corning® 7 Compound; Dow Corning® 20 Release Coating; Dow Corning® 197 Surfactant; Dow Corning® 198 Surfactant; Dow Corning® 203 Fluid; Dow Corning® 236 Dispersion; Dow Corning® 1248 Fluid; Dow Corning® 1250; Dow Corning® 1252; Dow Corning® 5043 Surfactant; Dow Corning® 5098; Dow Corning® 7119 Release/Parting Agent; Dow Corning® Antifoam 1400; Ease Release™ 700; Ease Release™ 2148; Ease Release™ 2191; Ease Release™ 2197; EXP-36-X20 Epoxy Functional Silicone Sol'n.; EXP-38-X20 Epoxy Functional Silicone Sol'n.; EXP-58 Silicone Wax; EXP-77 Mercapto Functional Silicone Wax; Fomrez® B-306; Fomrez® B-308; Fomrez® B-320; Frekote® 1711; Frekote® 1711-EM; Frekote® EXITT®; Frekote® EXITT-EM; Frekote® HMT; Frekote® LIFFT®; G623; G661; Getren® BPL 83; GP-5 Emulsifiable Paintable Silicone; GP-6 Silicone Fluid; GP-70-S Paintable Silicone Fluid; GP-74 Paintable Silicone Fluid; GP-98 Silicone Wax; GP-180 Resin Sol'n.; GP-197 Resin Sol'n.; GP-530 Silicone Fluid; GP-7100 Silicone Fluid; GP-7105 Silicone Fluid; GP-7105-E Silicone Emulsion; GP-7200 Silicone Fluid; GP-RA-156 Release Agent; GP-RA-158 Silicone Polish Additive; GP-RA-159 Silicone Polish Additive; GP-RA-201 Release Agent; Lyndcoat™ BR 601-RTU; Lyndcoat™ BR 790-RTU; Lyndcoat™ BR 880-RTU; Lyndcoat™ M 755-RTU; Lyndcoat™ M 771-RTU; Mazu® DF 200SX; Mazu® DF 200SXSP; Molder's Edge ME-100; Molder's Edge ME-110; Molder's Edge ME-120; Molder's Edge ME-175; Nix Stix L515; No Stik 806; Patcote® 512; Patcote® 513; Patcote® 519; Patcote® 520; Patcote® 525; Patcote® 531; Patcote® 550; Patcote® 555; Patcote® 577; Patcote® 597; Patcote® 598; Plastilease 250;

Prespersion SIB-7211; Prespersion SIB-7324; Prespersion SIC-3233; Prespersion SIC-3342; Prespersion SIC-3536; Pura WBC608; SF1080; SFR 100; Silicone Mist No. S116-A; Silicone Spray No. S312-A; Sprayon 803; Sprayon 805; Sprayon 815; SS 4177; Stoner E206; Super 33 Silicone Spray No. S3312-A; Tego® IMR 918; Tego® Airex 975; Tego® Release Agent M 379; Tegosil®; Tegostab® B 1048; Tegostab® B 1400A; Tegostab® B 1605; Tegostab® B 1648; Tegostab® B 1903; Tegostab® B 2173; Tegostab® B 2219; Tegostab® B 3640; Tegostab® B 4113; Tegostab® B 4380; Tegostab® B 4617; Tegostab® B 4900; Tegostab® B 4970; Tegostab® B 8002; Tegostab® B 8014; Tegostab® B 8017; Tegostab® B 8021; Tegostab® B 8030; Tegostab® B 8100; Tegostab® B 8202; Tegostab® B 8231; Tegostab® B 8404; Tegostab® B 8406; Tegostab® B 8407; Tegostab® B 8408; Tegostab® B 8409; Tegostab® B 8418; Tegostab® B 8423; Tegostab® B 8425; Tegostab® B 8427; Tegostab® B 8432; Tegostab® B 8433; Tegostab® B 8450; Tegostab® B 8622; Tegostab® B 8636; Tegostab® B 8680; Tegostab® B 8681; Tegostab® B 8694; Tegostab® B 8863 T; Tegostab® B 8896; Tegostab® B 8901; Tegostab® B 8905; Tegostab® B 2270; Tegostab® BF 2370; Tegotrenn® LI 841-237; Tegotrenn® S 2100; Tegotrenn® S 2200; Union Carbide® Y-11343

See also Dimethicone, Simethicone

Silicone emulsions

Classification: Organosiloxane

Uses: Lubricant, release agent for plastics and rubber; defoamer for paints, adhesives, polymerization processes

Manuf./Distrib.: CNC Int'l.; Crucible; Dow Corning; Genesee Polymers; Great Western; Hüls Am.; OSi Spec.; Ross Chem.; Sandoz; Soluol; Wacker Silicones; Wacker-Chemie GmbH

Trade names: Agitan E 255; Agitan E 256; Akrolease® E-9491; Aqualease™ 6102; Aqualease™ 6202; Chemlease 906E; DB-9 Paintable Silicone Emulsion; Dow Corning® 24 Emulsion; Dow Corning® 290 Emulsion; Dow Corning® 346 Emulsion; Dow Corning® 347 Emulsion; Dow Corning® HV-490 Emulsion; Dow Corning® Antifoam 1410; Dow Corning® Antifoam 1430; EXP-24-LS Silicone Wax Emulsion; GP-70-E Paintable Silicone Emulsion; GP-74-E Paintable Silicone Emulsion; GP-83-AE Silicone Emulsion; GP-85-AE Silicone Emulsion; GP-86-AE Silicone Emulsion; GP-7100-E Paintable Silicone Emulsion; Kantstik™ M-56; Mazu® DF 210SXSP; Nix Stix L478; Nix Stix LO582; No Stik 802; Polycone 1000; SEM-135; Silicone Emulsion E-133; Siltech® MFF-3000; SM 2079; SM 2128; SM 2154; Tego® Emulsion 240; Tego® Emulsion 3454; Tego® Emulsion ASL; Tego® Emulsion DF 19; Tego® Emulsion DT; Tego® Emulsion N

See also Dimethicone, Simethicone

Silicone glycol copolymer

Uses: Surfactant for use in MDI-based polyurethane foam systems

Trade names: Dabco® DC193; Dabco® DC197; Dabco® DC198; Dabco® DC1315; Dabco® DC5043; Dabco® DC5098; Dabco® DC5103; Dabco® DC5125; Dabco® DC5160; Dabco® DC5418

Silicone hexaacrylate

Uses: Slip, wetting agent, flow improver for coatings on paper, plastics, and metal

Trade names: Ebecryl® 1360

Silver aluminum magnesium phosphate. *See* Silver magnesium aluminum phosphate

Silver borosilicate

Definition: Syn. prod. formed by the fusion of boron oxide, silica, sodium oxide, and silver oxide

Uses: Preservative for cosmetics and plastic pkg.

Trade names: Ionpure Type A

Silver magnesium aluminum phosphate

Synonyms: Silver aluminum magnesium phosphate

Definition: Syn. prod. formed by the fusion of phosphorous pentoxide, magnesium oxide, alumina, and silver oxide

Trade names: Ionpure Type B

Simethicone

CAS 8050-81-5

Definition: Mixture of dimethicone with an avg. chain length of 200-350 dimethylsiloxane units and silica gel

Formula: $(CH_3)_3SiO[SI(CH_3)_2O]_nSi(CH_3)_3$, n = 200-350

Uses: Foam control agent, processing aid for foods, cosmetics, pharmaceuticals, fermentation, pesticide/fertilizer, paper/printing, textile, adhesives/coatings industries, in mfg. of plastics in contact with food, for degassing of monomers

Regulatory: Europe listed; UK approved

Manuf./Distrib.: PPG Ind.
Trade names: Wacker Silicone Antifoam Emulsion SE 9

Slaked lime. *See* Calcium hydroxide
SMA. *See* Styrene/MA copolymer
Smithsonite. *See* Zinc carbonate
SMO. *See* Sorbitan oleate
SMS. *See* Sorbitan stearate
Sn-II-ethylhexanoate. *See* Stannous octoate

Sodium allyloxy hydroxypropyl sulfonate
Uses: Monomer for emulsion polymerization; copolymerizable stabilizer
Trade names: Sipomer® COPS 1

Sodium aluminosilicate. *See* Sodium silicoaluminate

Sodium alumino sulfo silicate
CAS 101357-30-6; EINECS 309-928-3
Synonyms: Silicic acid, aluminum sodium salt, sulfurized
Trade names containing: Ultramarine Blue

Sodium aluminum silicate. *See* Sodium silicoaluminate

Sodium antimonate
Uses: Flame retardant for plastics
Manuf./Distrib.: Anzon; Elf Atochem N. Am.
Trade names: Pyrobloc SAP; S.A. 100; Thermoguard® FR

Sodium benzoate
CAS 532-32-1; EINECS 208-534-8
Synonyms: Benzoic acid sodium salt; Benzoate of soda; Benzoate sodium
Definition: Sodium salt of benzoic acid
Empirical: $C_7H_5O_2 \cdot Na$
Formula: C_6H_5COONa
Properties: White gran. or cryst. powd., odorless, sweetish astringent taste; sol. in water, alcohol; m.w. 144.11; pH ≈ 8
Precaution: Combustible when exposed to heat or flame; incompat. with acids, ferric salts; heated to decomp., emits toxic fumes of Na_2O
Toxicology: LD50 (oral, rat) 4.07 g/kg; poison by subcutaneous, IV routes; mod. toxic by ingestion, IP routes; may cause human intolerance reaction, asthma, rashes, hyperactivity; experimental teratogen, reproductive effects; mutagenic data
Uses: Fungicide; preservative in pharmaceuticals and foods, esp. in sl. acidic media; clinical reagent (bilirubin assay); nucleating agent for plastics
Regulatory: FDA 21CFR §146.152, 146.154, 150.141, 150.161, 166.40, 166.110, 181.22, 181.23, 184.1733; GRAS; USDA 9CFR §318.7; BATF 27CFR §240.1051; EPA reg.; GRAS (FEMA); Japan approved with limitations; Europe listed 0.5% as acid
Manuf./Distrib.: Aceto; Aldrich; Allchem Ind.; Jan Dekker; R.W. Greeff; Haarmann & Reimer; Int'l. Sourcing; Mallinckrodt; E. Merck; Pfizer Food Science; San Yuan; Tri-K
Trade names containing: Aerosol® OT-B

Sodium biborate decahydrate. *See* Sodium borate decahydrate

Sodium bicarbonate
CAS 144-55-8; EINECS 205-633-8
Synonyms: Baking soda; Sodium hydrogen carbonate; Bicarbonate of soda; Carbonic acid monosodium salt; Monosodium carbonate
Classification: Inorganic salt
Empirical: $CHNaO_3$
Formula: $NaHCO_3$
Properties: White powd. or cryst. lumps, sl. alkaline taste; sol. in water; insol. in alcohol; m.w. 84.01; dens. 2.159
Toxicology: Nuisance dust
Uses: Mfg. of effervescent salts and beverages, artificial mineral water; prevention of timber mold, cleaning preparations, lab reagent, antacid, mouthwash, OTC drug active; blowing and nucleating agent for

thermoplastics
Regulatory: FDA 21CFR §137.180, 137.270, 163.110, 173.385, 182.1736, 184.1736, GRAS; USDA 9CFR §318.7, 381.147; Japan approved; Europe listed
Manuf./Distrib.: Allchem Ind.; Balchem; Captree; Church & Dwight; Coyne; EM Ind.; Farleyway Chem. Ltd; FMC; ICI Spec.; Kraft; Rhone-Poulenc Basic
Trade names: Kycerol 91; Kycerol 92; Sodium Bicarbonate USP No. 1 Powd.; Sodium Bicarbonate USP No. 3 Extra Fine Powd.
Trade names containing: Dowicil® 75; Prespersion PAB-1196; Sodium Bicarbonate T.F.F. Treated Free Flowing

Sodium bistridecyl sulfosuccinate. *See* Ditridecyl sodium sulfosuccinate

Sodium borate decahydrate
CAS 1303-96-4; EINECS 215-540-4
Synonyms: Borax; Sodium tetraborate decahydrate; Sodium biborate decahydrate; Boric acid disodium salt
Classification: Inorganic salt
Empirical: $B_4Na_2O_7 \cdot 10H_2O$
Formula: $Na_2B_4O_7 \cdot 10H_2O$
Properties: White hard crystals, granules, or crystalline powd., odorless; efflorescent; sol. 1 g/16 ml water, 1 ml glycerol; insol. in alcohol; m.w. 381.37; dens. 1.73; becomes anhyd. at 320 C; pH $\approx$ 9.5
Precaution: Incompatible with acids, alkaloidal and metallic salts
Toxicology: Irritant; moderately toxic by ingestion, intravenous, intraperitoneal routes; heated to dec., emits toxic fumes of Na_2O; LD50 (rat, oral) 5.66 g/kg
Uses: Dispersant, wetting agent for NR, SR latexes; mold lubricant for general dry rubber molding; in ant poisons; soldering metals; nonselective herbicide; larvicide
Regulatory: FDA 21CFR §175.105, 175.210, 176.180, 177.2800, 181.22, 181.30
Manuf./Distrib.: Fluka
Trade names: Borax

Sodium borosilicate
Uses: Extender/filler for polymers
Trade names: Eccosphere® IG-25; Eccosphere® IG-101; Eccosphere® IGD-101; Q-Cel® Ultralight

Sodium C12-15 alcohols sulfate. *See* Sodium C12-15 alkyl sulfate

Sodium C12-15 alkyl sulfate
Synonyms: Sodium C12-15 alcohols sulfate
Definition: Sodium salt of the sulfate of C12-15 alcohols
Uses: Detergent, foamer used in personal care prods.; emulsifier for polymerization; additive for mech. latex foaming

Sodium C10-18 alkyl sulfonate
CAS 68037-49-0
Synonyms: Sulfonic acids, C10-18, -alkane, sodium salts
Uses: Antistat for thermoplastic film and molded goods
Trade names: Chemstat® PS-101

Sodium caproamphoacetate
CAS 14350-94-8; 68647-46-1; 25704-59-0; 68608-61-7; EINECS 271-951-9; 238-303-7
Synonyms: 1H-Imidazolium, 1-(carboxymethyl)-4,5-dihydro-1-(2-hydroxyethyl)-2-nonyl-, hydroxide, sodium salt; Caproamphoglycinate
Classification: Amphoteric organic compd.
Uses: Wetting agent, foaming agent, detergent used in medicated and germicidal shampoos and hand soaps, rug and upholstery shampoos, in emulsion polymerization
Trade names: Miranol® SM Conc.

Sodium capryl sulfate. *See* Sodium octyl sulfate
Sodium carboxymethylcellulose. *See* Carboxymethylcellulose sodium
Sodium CMC. *See* Carboxymethylcellulose sodium

Sodium cocoamphopropionate
CAS 68919-41-5; 93820-52-1
Synonyms: Coconut fattyacid amidoethyl-N-2-hydroxyethylaminopropionate; Cocoamphopropionate

Sodium cocoyl isethionate

Classification: Amphoteric organic compd.
Uses: High-foaming emulsifier, detergent, foam stabilizer, wetting agent for cosmetics, emulsion polymerization, industrial/household cleaners
Trade names: Miranol® CM-SF Conc.

Sodium cocoyl isethionate
CAS 61789-32-0; 58969-27-0; EINECS 263-052-5
Synonyms: Fatty acids, coconut oil, sulfoethyl esters, sodium salts
Definition: Sodium salt of the coconut fatty acid ester of isethionic acid
Formula: $RCO—OCH_2CH_2SO_3Na$, RCO- rep. fatty acids derived from coconut oil
Uses: Surfactant with good dispersing, foaming, conditioning for detergent bars, dentifrices, shampoos, bubble baths, other cosmetics; detergent base; lime soap dispersant
Trade names containing: Geropon® AS-200

Sodium cocoyl sarcosinate
CAS 61791-59-1; EINECS 263-193-2
Synonyms: Amides, coconut oil, with sarcosine, sodium salts; Sodium N-cocoyl sarcosinate
Definition: Sodium salt of cocoyl sarcosine
Uses: Wetting, foaming detergent used in personal care, household, and industrial prods., emulsion polymerization, pesticides, textiles, petrol. and lubricant prods.; mold release; antifog and antistat in plastics
Trade names: Hamposyl® C-30

Sodium N-cocoyl sarcosinate. *See* Sodium cocoyl sarcosinate

Sodium C14-16 olefin sulfonate
CAS 68439-57-6; EINECS 270-407-8
Definition: Mixt. of long chain sulfonate salts from sulfonation of C14-16 alpha olefins; consists of sodium alkene sulfonates and sodium hydroxyalkane sulfonates
Uses: Detergent, emulsifier, visc. builder, foaming agent for personal care, commercial and industrial formulations; emulsion polymerization
Regulatory: FDA 21CFR §175.105, 178.3406
Trade names: Polystep® A-18; Rhodacal® 301-10; Rhodacal® 301-10P; Rhodacal® A-246L; Siponate® 301-10F (redesignated Rhodacal® 301-10F); Witconate™ AOS; Witconate™ AOS-EP

Sodium C12-15 pareth-3 sulfosuccinate
Uses: Emulsifier for emulsion polymerization
Trade names: Sermul EA 221

Sodium decyl diphenyloxide disulfonate
CAS 36445-71-3
Uses: Surfactant, solubilizer, dispersant, detergent, wetting agent, emulsifier for dye bath leveling, pigment dispersion, heavy-duty cleaners, latex emulsification, agric. chemicals
Trade names: Dowfax 3B2

Sodium decyl sulfate
CAS 142-87-0; 84501-49-5
Empirical: $C_{10}H_{21}O_4S{\cdot}Na$
Properties: M.w. 260.36
Toxicology: Poison by intravenous, intraperitoneal routes; moderately toxic by ingestion
Uses: Emulsifier, dispersant, detergent, and wetting agent for industrial and institutional cleansers, plastics, mfg. of pigments, alkaline cleansers; dust suppression
Trade names: Avirol® SA 4110; Polystep® B-25; Serdet DJK 30

Sodium diamyl sulfosuccinate. *See* Diamyl sodium sulfosuccinate

Sodium dibenzyldithiocarbamate
CAS 55310-46-8
Empirical: $C_{15}H_{15}NS_2 \cdot Na$
Properties: M.w. 296
Uses: Vulcanization accelerator
Trade names: Octopol SBZ-20

Sodium dibutylcarbamodithioic acid. *See* Sodium di-n-butyl dithiocarbamate
Sodium dibutyldithiocarbamate. *See* Sodium di-n-butyl dithiocarbamate

Sodium di-n-butyl dithiocarbamate
CAS 136-30-1
Synonyms: Sodium dibutyldithiocarbamate; Dibutyldithiocarbamic acid sodium salt; Sodium dibutylcarbamodithioic acid
Empirical: $C_9H_{18}NS_2 \cdot Na$
Properties: M.w. 227.39
Toxicology: Poison by intraperitoneal route
Uses: Accelerator for elastomers; peptizer
Trade names: Butyl Namate®; Naftocit® NaDBC; Octopol NB-47; Octopol SDB-50

Sodium dibutyl naphthalene sulfonate
CAS 25417-20-3
Uses: Dispersant, wetting agent for pesticides, industrial cleaning, textiles, dyeing, leather, paper, dyes, rubber, latex polymerization
Trade names: Rhodacal® BX-78; Supragil® NK

Sodium dicyclohexyl sulfosuccinate. *See* Dicyclohexyl sodium sulfosuccinate
Sodium dicyclohexyl sulfobutanedioic acid. *See* Dicyclohexyl sodium sulfosuccinate

Sodium diethyldithiocarbamate
CAS 148-18-5
Uses: Rubber latex preservative; precipitant for heavy metals in waste water treatment
Manuf./Distrib.: Complex Quimica SA; Eastern Chem.; Novachem; Spectrum Chem. Mfg.; R.T. Vanderbilt
Trade names: Ethyl Namate®; Octopol SDE-25

Sodium di(2-ethylhexyl) sulfosuccinate. *See* Dioctyl sodium sulfosuccinate
Sodium dihexyl sulfosuccinate. *See* Dihexyl sodium sulfosuccinate
Sodium diisobutyl sulfosuccinate. *See* Diisobutyl sodium sulfosuccinate

Sodium dimethyldithiocarbamate
CAS 128-04-1 (hydrate); EINECS 204-876-7 (hydrate)
Synonyms: SDDC; Sodium N,N-dimethyl dithiocarbamate; Dimethyldithiocarbamic acid sodium salt; Methyl namate
Empirical: $C_3H_6NS_2 \cdot Na$ (anhyd.), $C_3H_6NNaS_2 \cdot$ aq. (hydrate)
Formula: $(CH_3)_2NCS_2Na$
Properties: Crystals; m.w. 143.19 + aq.; dens. 1.1 (20/20 C); m.p. 95 C
Toxicology: LD50 (oral, rat) 1000 mg/kg; moderately toxic by ingestion, intraperitoneal and subcutaneous routes; mutagenic data; heated to decomp., emits toxic fumes of NO_x, SO_x, and Na_2O
Uses: Pesticide, fungicide, corrosion inhibitor, rubber accelerator
Regulatory: FDA 21CFR §173.320 (3 ppm max.)
Manuf./Distrib.: Complex Quimica SA; Novachem; Uniroyal; R.T. Vanderbilt
Trade names: Aquatreat SDM; Methyl Namate®; Naftocit® NaDMC; Octopol SDM-40; Perkacit® SDMC
Trade names containing: Aquatreat DNM-30

Sodium N,N-dimethyl dithiocarbamate. *See* Sodium dimethyldithiocarbamate
Sodium dioctyl sulfosuccinate. *See* Dioctyl sodium sulfosuccinate
Sodium ditridecyl sulfosuccinate. *See* Ditridecyl sodium sulfosuccinate
Sodium-N-dodecanoyl-N-methylglycinate. *See* Sodium lauroyl sarcosinate

Sodium dodecylbenzenesulfonate
CAS 25155-30-0; 68081-81-2; 85117-50-6; EINECS 246-680-4
Synonyms: Sodium lauryl benzene sulfonate; Dodecylbenzenesulfonic acid sodium salt; Dodecylbenzene sodium sulfonate
Classification: Substituted aromatic compd.
Empirical: $C_{18}H_{29}O_3S \cdot Na$
Properties: White to lt. yel. flakes, granules, or powd.; m.w. 348.52
Precaution: Combustible
Toxicology: LD50 (oral, rat) 1260 mg/kg, (oral, mouse) 2 g/kg, (IV, mouse) 105 mg/kg; poison by intravenous route; moderately toxic by ingestion; skin and severe eye irritant; heated to decomp., emits toxic fumes of Na_2O

Sodium dodecyl diphenyl ether disulfonate

Uses: Anionic detergent, emulsifier, wetting agent for emulsion polymerization, industrial cleaning, cosmetics, textile and laundry, paints, etc.
Regulatory: FDA 21CFR §173.315, 175.105, 175.300, 175.320, 176.210, 177.1010, 177.1200, 177.1630, 177.2600, 177.2800, 178.3120, 178.3130, 178.3400; USDA 9CFR §318.7, 381.147
Manuf./Distrib.: Du Pont; Emkay; Norman, Fox; Pilot; Rhone-Poulenc Spec.; Stepan; Unger Fabrikker AS; Witco/Oleo-Surf.
Trade names: Akyposal NAF; Arylan® SBC25; Arylan® SC15; Arylan® SC30; Arylan® SX85; Calsoft F-90; Calsoft L-60; Elfan® WA; Elfan® WA Powder; Manro DL 32; Manro SDBS 25/30; Maranil Paste A 65; Maranil Powd. A; Nacconol® 40G; Nacconol® 90G; Nansa® 1106/P; Nansa® 1169/P; Nansa® HS85/S; Polystep® A-4; Polystep® A-7; Polystep® A-15; Polystep® A-16; Polystep® A-16-22; Reworyl® NKS 50; Rhodacal® DS-10; Rhodacal® LDS-22; Sandet 60; Ufasan 35; Ufasan 60A; Ufasan 65; Witconate™ 1223H; Witconate™ 1230; Witconate™ 1250; Witconate™ 1255

Sodium dodecyl diphenyl ether disulfonate
CAS 28519-02-0; 40795-56-0
Uses: Emulsifier for syn. latex and emulsion polymerization
Trade names: Eleminol MON-7; Poly-Tergent® 2EP

Sodium dodecyl diphenyl ether sulfonate
Uses: Solubilizer; household detergents, industrial, disinfectant, and metal cleaners, emulsifier in emulsion polymerization; textile auxiliary
Trade names: Sandoz Sulfonate 2A1

Sodium dodecyl diphenyloxide disulfonate
Uses: Detergent, emulsifier ro syn. latex and emulsion polymerization, wetting agent in electroplating, solubilizer, leveling agent for acid dyeing of nylon, dyeing assistant
Trade names: Dowfax 2A1; Dowfax 2EP; Rhodacal® DSB

Sodium dodecyl sulfate. *See* Sodium lauryl sulfate
Sodium 2-ethylhexyl sulfate. *See* Sodium octyl sulfate

Sodium ferric EDTA
Uses: Chelating agent used in photographic developers, oxidation-reduction systems, and emulsion polymerization; micronutrient; animal feeds; polymerization catalyst for syn. rubber
Trade names: Hamp-Ene® NaFe Purified Grade; Sequestrene® NAFe 13% Fe

Sodium formaldehyde sulfoxylate
CAS 149-44-0
Uses: Catalyst/polymerization regulator for plastics
Manuf./Distrib.: Aceto; Passaic Color & Chem.; Phibrochem; Royce Assoc.
Trade names: Hydrosulfite AWC

Sodium hexadecyl diphenyloxide disulfonate
Uses: Surfactant for emulsion polymerization
Trade names: Dowfax 8390

Sodium hexyl diphenyloxide disulfonate
Uses: Solubilizer, surfactant for cleaning, emulsion polymerization, latex mfg., paints, adhesives, min. and metal processing, textiles
Trade names: Dowfax C6L

Sodium hydrogen carbonate. *See* Sodium bicarbonate

Sodium isodecyl sulfate
CAS 68299-17-2
Uses: Wetting agent, emulsifier, detergent, foamer, rinse aid; post-stabilizer in latex paints; metal treatment; textile and plywood mfg.; visc. control in plywood mfg.; fruit and veg. washing; hard surface cleaners; for emulsion polymerization
Trade names: Rhodapon® CAV

Sodium isopropyl naphthalene sulfonate
CAS 1322-93-6
Synonyms: Naphthalenesulfonic acid, bis-(1-methylethyl)-, sodium salt

Empirical: $C_{16}H_{20}O_3SNa$
Properties: M.w. 315.39
Uses: Emulsifier, dispersant and wetting agent; used in alkaline cleaning formulations; antigelling agents, automotive radiator cleaners, metal, cement, brick, and tile cleaners for crystal growth control; plastics, rubber; latex stabilization
Trade names: Rhodacal® BA-77

Sodium laureth-4 phosphate
CAS 42612-52-2 (generic)
Synonyms: Sodium POE (4) lauryl ether phosphate; Sodium PEG 200 lauryl ether phosphate
Definition: Sodium salt of a complex mixture of phosphate esters of laureth-4
Uses: Household and industrial detergent, textile antistat, lubricant, emulsifier; visc. builder for creams and lotions; polymerization and stabilization of latexes
Trade names: Rhodafac® MC-470

Sodium laureth sulfate
CAS 1335-72-4; 3088-31-1; 9004-82-4 (generic); 13150-00-0; 15826-16-1; 68891-38-3; 68585-34-2; EINECS 221-416-0
Synonyms: Sodium lauryl ether sulfate (n=1û4); PEG (1-4) lauryl ether sulfate, sodium salt
Definition: Sodium salt of sulfated ethoxylated lauryl alcohol
Formula: $CH_3(CH_2)_{10}CH_2(OCH_2CH_2)_nOSO_3Na$, avg. n = 1-4
Uses: Foam stabilizer, detergent, flash foamer, wetter for detergent systems, personal care prods.; emulsion polymerization; shampoo base
Manuf./Distrib.: Aquatec Quimica SA; Chemron; Lonza; Norman, Fox; Pilot; Rhone-Poulenc Spec.; Sandoz; Sea-Land; Stepan; Unger Fabrikker AS; U.S. Synthetics; Vista
Trade names: Abex® 23S; Akyposal 9278 R; Akyposal EO 20 MW; Avirol® FES 996; Avirol® SE 3002; Avirol® SE 3003; Berol 452; Calfoam ES-30; Disponil FES 32; Disponil FES 61; Disponil FES 77; Empicol® ESB; Empimin® KSN27; Empimin® KSN70; Laural LS; Polystep® B-12; Rewopol® NL 2-28; Rewopol® NL 3; Rewopol® NL 3-28; Rewopol® NL 3-70; Rhodapex® ES; Sermul EA 30; Sermul EA 370; Steol® 4N; Surfax 218; Texapon® NSE
Trade names containing: Sodiumlaurethsulfate

Sodium laureth-12 sulfate
CAS 9004-82-4 (generic); 66161-57-7
Synonyms: PEG (12) lauryl ether sulfate, sodium salt; PEG 600 lauryl ether sulfate, sodium salt; Sodium POE (12) lauryl ether sulfate
Definition: Sodium salt of the sulfate ester of the PEG ether of lauryl alcohol
Empirical: $C_{36}H_{73}O_{16}SNa$
Formula: $CH_3(CH_2)_{10}CH_2(OCH_2CH_2)_nOSO_3Na$, avg. n = 12
Uses: Detergent, foamer for cosmetics; emulsion polymerization
Trade names: Disponil FES 92E; Polystep® B-23

Sodium laureth-30 sulfate
CAS 9004-82-4 (generic)
Synonyms: PEG (30) lauryl ether sulfate, sodium salt; Sodium POE (30) lauryl ether sulfate
Definition: Sodium salt of the sulfate ester of the PEG ether of lauryl alcohol
Formula: $CH_3(CH_2)_{10}CH_2(OCH_2CH_2)_nOSO_3Na$, avg. n = 30
Uses: Emulsifier for emulsion polymerization
Trade names: Polystep® B-19

Sodium laureth sulfosuccinate
Uses: Mild foamer for shampoos and foam baths; emulsifier for emulsion polymerization
Trade names: Secosol® ALL40

Sodium lauriminodipropionate
CAS 14960-06-6; 26256-79-1; EINECS 239-032-7
Synonyms: Sodium N-lauryl-β-iminodipropionate; N-(2-Carboxyethyl)-N-dodecyl-β-alanine, monosodium salt
Definition: Partial sodium salt of a substituted propionic acid
Empirical: $C_{18}H_{35}NO_4 \cdot Na$
Formula: $CH_3(CH_2)_{11}N(CH_2CH_2COONa)(CH_2CH_2COOH)$
Properties: M.w. 352.46
Uses: Detergent, solubilizer, stabilizer; used in petrol. processing and emulsion polymerization; high foaming

conditioner for shampoos, skin cleansers
Trade names: Deriphat® 160C

Sodium lauroamphopropionate
CAS 61901-02-8
Synonyms: Lauroamphopropionate
Classification: Amphoteric organic compd.
Empirical: $C_{19}H_{38}N_2O_4 \cdot Na$
Formula: $CH_3(CH_2)_{10}CONHCH_2CH_2NCH_2CH_2OHCH_2CH_2COONa$
Uses: Emulsion polymerization surfactant

Sodium lauroyl sarcosinate
CAS 137-16-6; EINECS 205-281-5
Synonyms: N-Methyl-N-(1-oxododecyl)glycine, sodium salt; N-Lauroylsarcosine sodium salt; Sodium-N-dodecanoyl-N-methylglycinate
Definition: Sodium salt of lauroyl sarcosine
Empirical: $C_{15}H_{29}NO_3 \cdot Na$
Formula: $CH_3(CH_2)_{10}CONCH_3CH_2COONa$
Properties: M.w. 293.39
Uses: Foaming agent, wetting agent, detergent, lubricant, antistat, corrosion inhibitor, bacteriostat, penetrant used in dental, pharmaceutical, shampoos, depilatories, and shaving preparations, food pkg., household and industrial uses; emulsion polymerization
Manuf./Distrib.: Hampshire; R.T. Vanderbilt
Trade names: Hamposyl® L-30; Hamposyl® L-95

Sodium lauryl benzene sulfonate. *See* Sodium dodecylbenzenesulfonate
Sodium lauryl ether sulfate. *See* Sodium laureth sulfate
Sodium N-lauryl-β-iminodipropionate. *See* Sodium lauriminodipropionate

Sodium lauryl/propoxy sulfosuccinate
Uses: Emulsifier for acrylic, vinyl acrylic polymerization
Trade names: Emcol® 4910

Sodium lauryl sulfate
CAS 151-21-3; 68585-47-7; 68955-19-1; EINECS 205-788-1
Synonyms: SDS; Sulfuric acid monododecyl ester sodium salt; Sodium dodecyl sulfate; Dodecylsulfate sodium salt
Definition: Sodium salt of lauryl sulfate
Empirical: $C_{12}H_{26}O_4S \cdot Na$
Formula: $CH_3(CH_2)_{10}CH_2OSO_3Na$
Properties: White to cream crystals, flakes, or powd., faint odor; sol. in water; m.w. 289.43; m.p. 204-207 C
Precaution: Heated to decomp., emits toxic fumes of SO_x and Na_2O
Toxicology: LD50 (oral, rat) 1288 mg/kg; poison by IV, IP routes; mod. toxic by ingestion; experimental teratogen, reproductive effects; human skin irritant; experimental eye, severe skin irritant; mild allergen; mutagenic data
Uses: Anionic detergent; surface tension depressant; emulsifier for fats; wetting agent; in textile industry, in toothpastes; emulsion polymerization
Regulatory: FDA 21CFR §172.210, 172.822, 175.105, 175.300, 175.320, 176.170, 176.210, 177.1200, 177.1210, 177.1630, 177.2600, 177.2800, 178.1010, 178.3400; USDA 9CFR §318.7, 381.147
Manuf./Distrib.: Albright & Wilson Ltd; Chemron; Du Pont; Lonza; Norman, Fox; Monomer-Polymer & Dajac; Pilot; Sandoz; Stepan; Unger Fabrikker AS; Witco
Trade names: Akyporox SAL SAS; Alscoap LN-40, LN-90; Avirol® SL 2010; Avirol® SL 2020; Calfoam SLS-30; Carsonol® SLS; Carsonol® SLS-R; Carsonol® SLS-S; Carsonol® SLS Special; Elfan® 260 S; Empicol® 0303; Empicol® 0303V; Empicol® LX; Empicol® LX28; Empicol® LXS95; Empicol® LXV; Empicol® LXV/D; Empicol® LY28/S; Empicol® LZ; Empicol® LZ/D; Empicol® LZG 30; Empicol® LZGV; Empicol® LZGV/C; Empicol® LZP; Empicol® LZV/D; Empicol® WA; Empimin® LR28; Equex SP; Manro DL 28; Manro SLS 28; Montovol RF-10; Octosol SLS; Octosol SLS-1; Perlankrol® DSA; Polirol LS; Polystep® B-3; Polystep® B-5; Polystep® B-24; Rewopol® 15/L; Rewopol® NLS 15 L; Rewopol® NLS 28; Rewopol® NLS 30 L; Rhodapon® LCP; Rhodapon® SB; Rhodapon® UB; Serdet DFK 40; Sermul EA 150; Stepanol® WA-100; Sulfopon® 101/POL; Sulfopon® 101 Special; Sulfopon® 102; Sulfopon® P-40; Supralate® WAQ; Supralate® WAQE; Texapon® K-12; Texapon® K-1296; Texapon® OT Highly Conc. Needles; Texapon® VHC Needles; Texapon® ZHC Powder; Ultra Sulfate SL-1; Ungerol LS, LSN; Witcolate™ A Powder; Witcolate™ SL-1; Zorapol LS-30

Sodium lignosulfonate
CAS 8061-51-6
Synonyms: Sodium polignate; Lignosulfonic acid, sodium salt
Definition: Sodium salt of polysulfonated lignin, a dark brown polymeric material from wood
Properties: Tan free-flowing powd.
Precaution: Combustible
Uses: Dispersant, emulsion stabilizer, chelating agent for rubber industry
Regulatory: FDA 21CFR §173.310, 175.105, 176.170, 176.210, 177.1210
Manuf./Distrib.: Borregaard LignoTech; Wesco Tech. Ltd.; Westvaco
Trade names: Darvan® No. 2

Sodium magnesium aluminosilicate
Uses: Filler for the rubber industry
Trade names: Hydrex® R

Sodium methallyl sulfonate
CAS 1561-92-8
Synonyms: 2-Methyl-2-propene-1-sulfonic acid sodium salt
Formula: $H_2C=C(CH_3)CH_2SO_3Na$
Properties: M.w. 158.15; m.p. > 300 C
Uses: Dye improver reactive comonomer for acrylic fibers polymerization; reactive emulsifier or coemulsifier
Manuf./Distrib.: Monomer-Polymer & Dajac
Trade names: Geropon® MLS/A

Sodium 2,2′-methylenebis(4,6-di-t-butylphenyl) phosphate
CAS 85209-91-2; EINECS 286-344-4
Empirical: $C_{29}H_{42}O_4PNa$
Properties: Powd.; m.w. 508; m.p. > 400 C
Uses: Nucleating agent for polymers
Trade names: ADK STAB NA-11

Sodium methyl oleate sulfate
Uses: Dispersant for latex
Trade names: Darvan® SMO

Sodium myristoyl sarcosinate
CAS 30364-51-3; EINECS 250-151-3
Synonyms: N-Methyl-N-(1-oxotetradecyl)glycine, sodium salt
Definition: Sodium salt of myristoyl sarcosine
Empirical: $C_{17}H_{33}NO_3 \cdot Na$
Formula: $CH_3(CH_2)_{12}CONCH_3CH_2COONa$
Uses: Detergent, wetting and foaming agent for hair and rug shampoos and cosmetics; emulsion polymerization; mold release for RIM; stabilizer for polyols; antifog, antistat in plastics
Trade names: Hamposyl® M-30

Sodium myristyl sulfate
CAS 1191-50-0; 139-88-8; EINECS 214-737-2
Synonyms: Sodium tetradecyl sulfate; Sulfuric acid, monotetradecyl ester, sodium salt; 1-Tetradecanol, hydrogen sulfate, sodium salt
Definition: Sodium salt of myristyl sulfate
Empirical: $C_{14}H_{30}O_4S \cdot Na$
Formula: $CH_3(CH_2)_{12}CH_2OSO_3Na$
Properties: M.w. 316.48
Toxicology: Poison by intraperitoneal, intravenous routes
Uses: Anionic detergent; foaming, wetting, and emulsifying agent for cosmetics, household and industrial use; emulsion polymerization; metal processing; pharmaceuticals; leather; textiles
Regulatory: FDA 21CFR §175.105, 177.1210, 177.2800
Trade names: Niaproof® Anionic Surfactant 4

Sodium naphthalene-formaldehyde sulfonate. *See* Sodium polynaphthalene sulfonate

Sodium nonoxynol-4 sulfate
CAS 9014-90-8 (generic); 68891-39-4

Sodium nonoxynol-6 sulfate

Synonyms: PEG-4 nonyl phenyl ether sulfate, sodium salt; PEG 200 nonyl phenyl ether sulfate, sodium salt; Sodium PEG-4 nonyl phenyl ether sulfate
Definition: Sodium salt of the sulfate ester of nonoxynol-4
Formula: $C_9H_{19}C_6H_4(OCH_2CH_2)_nOSO_3Na$, avg. n = 4
Uses: Detergent, emulsifier, lime-soap dispersant, wetting agent, foamer; dishwashing formulations, scrub soaps, car washes, rug and hair shampoos, emulsion polymerization, petrol. waxes, textile wet processing, cosmetics, pesticides
Regulatory: FDA 21CFR §175.105, 178.3400
Trade names: Polystep® B-27; Rhodapex® CO-433; Serdet DNK 30; Sermul EA 54; Steol® COS 433; Witcolate™ D51-51EP

Sodium nonoxynol-6 sulfate
CAS 9014-90-8 (generic)
Synonyms: PEG-6 nonyl phenyl ether sulfate, sodium salt; POE (6) nonyl phenyl ether sulfate, sodium salt; PEG 300 nonyl phenyl ether sulfate, sodium salt
Definition: Sodium salt of the sulfuric acid ester of nonoxynol-6
Formula: $C_9H_{19}C_6H_4(OCH_2CH_2)_nOSO_3Na$, avg. n = 6
Uses: Emulsifier for emulsion polymerization
Regulatory: FDA 21CFR §175.105
Trade names: Akyposal NPS 60

Sodium nonoxynol-8 sulfate
CAS 9014-90-8 (generic)
Synonyms: PEG-8 nonyl phenyl ether sulfate, sodium salt; PEG 400 nonyl phenyl ether sulfate, sodium salt; POE (8) nonyl phenyl ether sulfate, sodium salt
Definition: Sodium salt of the sulfuric acid ester of nonoxynol-8
Formula: $C_9H_{19}C_6H_4(OCH_2CH_2)_nOSO_3Na$ where avg. n = 8
Uses: Emulsifier for emulsion polymerization
Regulatory: FDA 21CFR §175.105
Trade names: Disponil AES 60 E

Sodium nonoxynol-9 sulfate
Uses: Emulsifier for emulsion polymerization
Trade names: Zorapol SN-9

Sodium nonoxynol-10 sulfate
CAS 9014-90-8 (generic)
Definition: Sodium salt of the sulfuric acid ester of nonoxynol-10
Empirical: $C_{35}H_{63}O_{14}SNa$
Formula: $C_9H_{19}C_6H_4(OCH_2CH_2)_nOSO_3Na$, avg. n = 10
Uses: Emulsifier for emulsion polymerization
Regulatory: FDA 21CFR §175.105
Trade names: Akyposal NPS 100; Sermul EA 151

Sodium nonoxynol-15 sulfate
CAS 9014-90-8 (generic)
Definition: Sodium salt of the sulfate ester of nonoxynol-15
Empirical: $C_{45}H_{83}O_{19}SNa$
Formula: $C_9H_{19}C_6H_4(OCH_2CH_2)_nOSO_3Na$, avg. n = 15
Uses: Emulsifier for emulsion polymerization
Trade names: Sermul EA 146

Sodium nonoxynol-25 sulfate
CAS 9014-90-8 (generic)
Definition: Sodium salt of the sulfate ester of nonoxynol-25
Empirical: $C_{65}H_{123}O_{29}SNa$
Formula: $C_9H_{19}C_6H_4(OCH_2CH_2)_nOSO_3Na$, avg. n = 25
Uses: Emulsifier for emulsion polymerization
Regulatory: FDA 21CFR §175.105
Trade names: Akyposal NPS 250; Montosol PB-25

Sodium nonoxynol-10 sulfosuccinate
Uses: Emulsifier for emulsion polymerization

Trade names: Sermul EA 176

Sodium octadecanoate. *See* Sodium stearate

Sodium octoxynol-2 ethane sulfonate
CAS 2917-94-4; 67923-87-9; EINECS 267-791-4; 220-851-3
Synonyms: 2-[2-[2-Octylphenoxy)ethoxy]ethoxy]ethanesulfonic acid, sodium salt; Entsufon sodium
Classification: Organic compd.
Empirical: $C_{20}H_{34}O_6SNa$
Formula: $C_8H_{17}C_6H_4O(CH_2CH_2O)_2CH_2CH_2SO_3Na$
Properties: M.w. 425.54
Uses: Detergent for metal cleaning, household cleaning; polymerization emulsifier, latex post-stabilizer; wetting agent; dispersant; dye leveling agent
Regulatory: FDA 21CFR §176.180
Trade names: Triton® X-200

Sodium octoxynol-3 sulfate
Uses: Emulsifier for vinyl acetate specialty copolymers
Trade names: Polystep® C-OP3S

Sodium octoxynol-6 sulfate
CAS 69011-84-3
Synonyms: PEG-6 octyl phenyl ether sulfate, sodium salt; POE (6) octyl phenyl ether sulfate, sodium salt; PEG 300 octyl phenyl ether sulfate, sodium salt
Definition: Sodium salt of the sulfuric acid ester of octoxynol-6
Empirical: $C_{26}H_{45}O_{10}SNa$
Formula: $C_8H_{17}C_6H_4(OCH_2CH_2)_nOSO_3Na$, avg. n = 6
Uses: Emulsifier for emulsion polymerization
Trade names: Akyposal BD

Sodium octylphenol sulfate
Uses: Emulsifier for emulsion polymerization
Trade names: Abex® EP-277

Sodium octylphenoxyethoxyethyl sulfonate
Uses: Emulsifier, wetting agent, penetrant, detergent used in emulsion polymerization
Trade names: Newcol 861S

Sodium octyl sulfate
CAS 126-92-1; 142-31-4; EINECS 204-812-8
Synonyms: Sodium 2-ethylhexyl sulfate; Sodium capryl sulfate; Sulfuric acid, mono (2-ethylhexyl) ester sodium salt; Octyl sulfate sodium salt
Definition: Sodium salt of 2-ethylhexyl sulfate
Empirical: $C_8H_{18}O_4S \cdot Na$
Formula: $CH_3(CH_2)_3CHCH_3CH_2CH_2OSO_3Na$
Properties: M.w. 233.31; m.p. 195 C
Toxicology: LD50 (oral, mouse) 1550 mg/kg; poison by intraperitoneal route; moderately toxic by ingestion, skin contact; skin and eye irritant; heated to decomp., emits very toxic fumes of SO_x and Na_2O
Uses: Anionic detergent, wetting, dispersing, and emulsifying agent, stabilizer for plastics, rubber, adhesives, food contact paper, textiles, household/industrial cleaners
Regulatory: FDA 21CFR §173.315, 175.105, 176.170; USDA 9CFR §381.147
Manuf./Distrib.: Aldrich
Trade names: Avirol® SA 4106; Avirol® SA 4108; Niaproof® Anionic Surfactant 08; Polystep® B-29; Rhodapon® OLS; Serdet DSK 40; Witcolate™ D-510

Sodium alpha olefin sulfonate
CAS 68188-45-5; 68439-57-6
Uses: Detergent base for personal care prods.; emulsifier for emulsion polymerization
Trade names: Sermul EA 214

Sodium oleyl sulfate
CAS 1847-55-8; EINECS 217-430-1
Synonyms: 9-Octadecen-1-ol, hydrogen sulfate, sodium salt

Sodium oleyl sulfosuccinamate

Definition: Sodium salt of oleyl sulfate
Empirical: $C_{18}H_{36}O_4S{\bullet}Na$
Formula: $CH_3(CH_2)_7CH{=}CH(CH_2)_7CH_2OSO_3Na$
Properties: M.w. 371.54
Uses: Scouring agent in textile and leather industries, rewetting agent for paper; emulsifier for emulsion polymerization
Trade names: Rhodapon® OS

Sodium oleyl sulfosuccinamate
Uses: Emulsifier for emulsion polymerization; foaming agent for latex emulsions; antigelling and cleaning agents for paper mill felts
Trade names: Cosmopon BN

Sodium PEG 200 lauryl ether phosphate. *See* Sodium laureth-4 phosphate
Sodium PEG-4 nonyl phenyl ether sulfate. *See* Sodium nonoxynol-4 sulfate
Sodium POE (4) lauryl ether phosphate. *See* Sodium laureth-4 phosphate
Sodium POE (12) lauryl ether sulfate. *See* Sodium laureth-12 sulfate
Sodium POE (30) lauryl ether sulfate. *See* Sodium laureth-30 sulfate
Sodium POE tridecyl sulfate. *See* Sodium trideceth sulfate
Sodium polignate. *See* Sodium lignosulfonate

Sodium polyacrylate
CAS 9003-04-7
Synonyms: Polyacrylic acid, sodium salt
Empirical: $(C_3H_4O_2)_x{\bullet}xNa$
Properties: M.w. 2000-2300
Toxicology: Eye irritant; heated to decomp., emits toxic fumes of Na_2O
Uses: Thickener, stabilizer, protective colloid for natural and syn. latexes for paints, films, coatings, and adhesives; dispersant; antiredeposition agents; antiscalant
Regulatory: FDA 21CFR §173.73, 173.310, 173.340; Japan approved (0.2% max.)
Manuf./Distrib.: Alco; Allchem Ind.; Arakawa; Dixie Chem; Monomer-Polymer & Dajac; Rhone-Poulenc; Synthron; 3V

Sodium polyisobutylene/maleic anhydride copolymer
Uses: Dispersant for latex paints and coatings, enamels, polymerization, leather tanning, and water treatment
Trade names: Daxad® 31S

Sodium polymethacrylate
CAS 25086-62-8; 54193-36-1
Synonyms: 2-Propenoic acid, 2-methyl-, homopolymer, sodium salt
Classification: Polymer
Empirical: $(C_4H_6O_2)_x{\bullet}xNa$
Formula: $[CH_3CHCHCOONa]_x$
Uses: Dispersant for pigments, in paint formulations, emulsion polymerization, water treatment, agriculture, cosmetics, industrial cleaners, large particle suspensions
Regulatory: FDA 21CFR §173.310, 175.105
Trade names: Darvan® No. 7; Daxad® 30; Daxad® 30-30

Sodium polynaphthalene sulfonate
CAS 1321-69-3, 9084-06-4
Synonyms: Sodium naphthalene-formaldehyde sulfonate; Naphthalenesulfonic acid, polymer with formaldehyde, sodium salt
Definition: Sodium salt of the prod. obtained by condensation polymerization of naphthalene sulfonic acid and formaldehyde
Empirical: $(C_{10}H_8O_3S \bullet CH_2O)_x \bullet xNa$
Toxicology: Moderately toxic by ingestion
Uses: Dispersant for pigments, extenders, and fillers; used in dyeing syn. and natural fibers, ceramics, emulsion polymerization, gypsum board, printing, rubber, food pkg. applics., agric., tanning; plasticizer for concrete; stabilizer
Regulatory: FDA 21CFR §175.105, 176.170, 176.180, 177.1200, 177.1210, 177.1550, 177.1650, 177.2600, 178.3910
Trade names: Darvan® No. 1; Darvan® No. 6; Daxad® 11G; Daxad® 14B; Daxad® 15; Daxad® 16; Daxad® 17; Dehscofix 920; Lomar® LS; Lomar® LS Liq.; Lomar® PW; Nopcosant; Tamol® NH 9103; Tamol® NN

4501; Vultamol®; Vultamol® SA Liq.
Trade names containing: Methyl Zimate® Slurry

Sodium potassium aluminosilicate. *See* Nepheline syenite

Sodium/potassium naphthalene-formaldehyde sulfonate
Uses: Dispersant for emulsion polymerization, dyestuffs, tanning, herbicides, pesticides, pitch, concrete, cement, gypsum
Trade names: Daxad® 14C

Sodium ricinoleate
CAS 5323-95-5; EINECS 226-191-2
Synonyms: 12-Hydroxy-9-octadecenoic acid, sodium salt
Definition: Sodium salt of ricinoleic acid
Empirical: $C_{18}H_{34}O_3$•Na
Formula: $C_{17}H_{33}OHCOONa$
Properties: White to yel., nearly odorless; sol. in water, alcohol; m.w. 320.50
Precaution: Combustible
Toxicology: Poison by intraperitoneal and other routes
Uses: Emulsifier, stabilizer, defoamer for emulsion polymerization, soaps, rubber lubricant
Regulatory: FDA 21CFR §175.300
Trade names: Solricin® 435; Solricin® 535

Sodium rosinate
CAS 61790-51-0; EINECS 263-144-5
Synonyms: Sodium soap of pale rosin
Definition: Sodium soap of pale rosin
Uses: Emulsifier, detergent, wetting agent, stabilizer in ABS, SBR, syn. elastomers, metalworking, disinfectants, oil-well drilling muds, drawing and grinding compds.; plasticizer
Trade names: Arizona DR-25; Arizona DRS-43; Arizona DRS-44; Burez NA 70-500; Burez NA 70-500D; Burez NA 70-2500; Diprosin N-70; Dresinate® 731; Dresinate® 81; Dresinate® X; Dresinate® XX; Westvaco® Resin 790; Westvaco® Resin 795

Sodium silicoaluminate
CAS 1318-02-1, 1344-00-9; EINECS 215-684-8
Synonyms: Sodium aluminosilicate; Sodium aluminum silicate; Aluminum sodium silicate
Definition: Series of hydrated sodium aluminum silicates
Empirical: $Na_2O : Al_2O_3 : SiO_2$ with mole ratio ≈ 1:1:13.2
Properties: Fine white amorphous powd. or beads, odorless and tasteless; insol. in water, alcohol, org. solvs.; partly sol. in strong acids and alkali hydroxides @ 80-100 C; pH 6.5-10.5 (20% slurry)
Precaution: Noncombustible
Toxicology: Irritant to skin, eyes, mucous membranes; heated to decomp., emits toxic fumes of Na_2O
Uses: Anticaking agent in food preps.; reinforcing filler for rubbers
Regulatory: FDA 21CFR §133.146, 160.105, 160.185, 182.2727 (2% max.), 582.2727, GRAS; Europe listed; UK approved
Trade names: Hydrex®; Ketjensil® SM 405; Vulkasil® A1; Zeolex® 23; Zeolex® 80

Sodium soap of pale rosin. *See* Sodium rosinate

Sodium stearate
CAS 822-16-2; EINECS 212-490-5
Synonyms: Sodium octadecanoate; Octadecanoic acid sodium salt; Stearic acid sodium salt
Definition: Sodium salt of stearic acid
Empirical: $C_{18}H_{36}O_2$ • Na
Formula: $CH_3(CH_2)_{16}COONa$
Properties: White powd., fatty odor; sol. in hot water and hot alcohol; slowly sol. in cold water and cold alcohol; insol. in many org. solvs.; m.w. 306.52
Toxicology: Poison by intravenous and other routes; heated to decomp., emits toxic fumes of Na_2O
Uses: Waterproofing and gelling agent; toothpaste, cosmetics; stabilizer in plastics; emulsifier and stiffener in pharmaceuticals; in glycerol suppositories; mold release; lubricant for thermoplastics
Regulatory: FDA 21CFR §172.615, 172.863, 175.105, 175.300, 175.320, 176.170, 176.200, 176.210, 177.1200, 177.2260, 177.2600, 177.2800, 178.3910, 179.45, 181.22, 181.29
Manuf./Distrib.: Cometals; Elf Atochem N. Am.; Magnesia GmbH; Norman, Fox; Original Bradford Soap

Works; Witco/Oleo-Surf.
Trade names: Afco-Chem B; Cometals Sodium Stearate; Haro® Chem NG; HyTemp NS-70; Mathe Sodium Stearate; Synpro® Sodium Stearate ST; Witco® Heat Stable Sodium Stearate; Witco® Sodium Stearate T-1
Trade names containing: Afco-Chem EX-95; Rhenovin® Na-stearat-80

Sodium stearoyl lactylate
CAS 25383-99-7; EINECS 246-929-7
Synonyms: Octadecanoic acid, 2-(1-carboxyethoxy)-1-methyl-2-oxoethyl ester, sodium salt; Sodium stearyl-2-lactylate
Definition: Sodium salt of the stearic acid ester of lactyl lactate
Empirical: $C_{24}H_{44}O_6$•Na
Formula: $CH_3(CH_2)_{16}COOCHCH_3COOCHCH_3COONa$
Properties: White or cream-colored powd., caramel odor; sol. in hot oil or fat; disp. in warm water; m.p. 46-52 C; HLB 10-12
Toxicology: Heated to decomp., emits acrid smoke and irritating fumes
Uses: Emulsifier, dough conditioner, whipping agent in baked prods., desserts, mixes; complexing agent for starches and proteins; acid and catalyst scavenger/neutralizer; lubricant for polymers, color concs.
Regulatory: FDA 21CFR §172.846, 177.1200; Europe listed; UK approved
Trade names: Pationic® 920

Sodium stearyl-2-lactylate. *See* Sodium stearoyl lactylate

Sodium p-styrenesulfonate
Empirical: $C_8H_8SO_3Na$
Formula: $CH_2:CH_2C_6H_4SO_3Na$
Properties: Wh. powd.; m.w. 206.20
Uses: Emulsifier producing acrylic, vinyl acetate, and SBR latexes; in dyeing; flocculant, scale inhibitor; dispersant for cosmetics; hair fixing agent; for polyion complex; antistat
Manuf./Distrib.: Aceto; Monomer-Polymer & Dajac; Polysciences
Trade names: Spinomar NaSS

Sodium tallate
EINECS 263-137-7
Synonyms: Tall oil rosin sodium salt; Fatty acids, tall oil, sodium salts
Uses: Emulsifier for oils and asphalts; detergent aid; dispersant for pigments; stabilizer for syn. rubber latices, used for industrial and household cleaners; conditioner for sulfur dusts
Trade names: Dresinate® TX

Sodium tallow sulfosuccinamate
CAS 68988-69-2
Uses: Foaming agent used in no-gel foam systems based on high solids, noncarboxylated SBR latex or natural rubber latex
Trade names: Empimin® MKK/AU; Manro MA 35

Sodium tetraborate decahydrate. *See* Sodium borate decahydrate
Sodium tetradecyl sulfate. *See* Sodium myristyl sulfate

Sodium trideceth sulfate
CAS 25446-78-0 (n=3); 66161-58-8 (n=4); EINECS 246-985-2
Synonyms: Sodium tridecyl ether sulfate; Sodium POE tridecyl sulfate
Definition: Sodium salt of sulfated ethoxylated tridecyl alcohol
Formula: $C_{13}H_{27}(OCH_2CH_2)_nOSO_3Na$, avg. n = 1-4
Properties: M.w. 435.58
Uses: Detergent, wetting agent, emulsifier, foamer for personal care, dishwashing, and textile prods.; post-stabilizer in emulsion polymerization
Regulatory: FDA 21CFR §175.105
Trade names: Rhodapex® EST-30

Sodium tridecyl ether sulfate. *See* Sodium trideceth sulfate

Sodium tridecyl sulfate
CAS 3026-63-9; EINECS 221-188-2

Synonyms: 1-Tridecanol, hydrogen sulfate, sodium salt; Tridecyl sodium sulfate
Definition: Sodium salt of tridecyl sulfate
Empirical: $C_{13}H_{28}O_4S \cdot Na$
Formula: $CH_3(CH_2)_{11}CH_2OSO_3Na$
Properties: M.w. 303.42
Uses: Emulsifier, wetting agent; emulsion polymerization of vinyl chloride, styrene, and styrene/acrylic monomers; detergent formulations
Regulatory: FDA 21CFR §177.1210
Trade names: Avirol® SA 4113; Rhodapon® TDS

Sodium vinyl sulfonate
Synonyms: SVS
Classification: Functional monomer
Uses: As organic intermediate in sulfonoethylation reactions, as functional monomer for stabilization of emulsion polymers
Manuf./Distrib.: Aceto; Air Prods.; Monomer-Polymer & Dajac
Trade names: Hartomer 4900-25

Sodium-zinc complex
Uses: Stabilizer for foam PVC
Trade names: ADK STAB FL-21; ADK STAB FL-23

Soluble sulfur. *See* Sulfur
Sorbimacrogol laurate 300. *See* Polysorbate 20
Sorbimacrogol oleate 300. *See* Polysorbate 80
Sorbimacrogol palmitate 300. *See* Polysorbate 40
Sorbimacrogol stearate 300. *See* Polysorbate 60
Sorbimacrogol trioleate 300. *See* Polysorbate 85
Sorbimacrogol tristearate 300. *See* Polysorbate 65

Sorbitan diisostearate
CAS 68238-87-9
Synonyms: Anhydrohexitol diisostearate
Definition: Diester of isostearic acid and hexitol anhydrides derived from sorbitol
Uses: Auxiliary emulsifier, solubilizer, corrosion inhibitor in lubricants, metal protectants and cleaners, emulsion polymerization
Trade names: Emsorb® 2518

Sorbitan isostearate
CAS 54392-26-6; 71902-01-7
Synonyms: Anhydrosorbitol monoisostearate; 1,4-Anhydro-D-glucitol, 6-isooctadecanoate; Sorbitan monoisooctadecanoate
Definition: Monoester of isostearic acid and hexitol anhydrides derived from sorbitol
Empirical: $C_{24}H_{46}O_6$
Properties: M.w. 346.52
Toxicology: Experimental neoplastigen
Uses: Emulsifier, wetting agent, pigment dispersant; for cosmetics, pharmaceuticals, explosives, emulsion polymerization, metal protectants and cleaners
Trade names: Emsorb® 2516

Sorbitan laurate
CAS 1338-39-2; 5959-89-7; EINECS 215-663-3; 227-729-9
Synonyms: Sorbitan monolaurate; Anhydrosorbitol monolaurate; Sorbitan monododecanoate
Definition: Monoester of lauric acid and hexitol anhydrides derived from sorbitol
Empirical: $C_{18}H_{34}O_6$
Properties: M.w. 346.47; dens. 1.032; flash pt. > 230 F; ref. index 1.4740
Toxicology: Experimental neoplastigen
Uses: Emulsifier for foods, cosmetics, household prods., industrial applics.; lubricant and antistat for PVC and other plastics; antifog for food pkg. films
Regulatory: FDA 21CFR §175.320, 178.3400; Europe listed; UK approved
Trade names: Alkamuls® SML; Atmer® 100; Disponil SML 100 F1; Emsorb® 2515; Ethylan® GL20; Glycomul® L

Sorbitan monododecanoate. *See* Sorbitan laurate
Sorbitan, monohexadecanoate, poly(oxy-1,2-ethaneidyl) derivs.. *See* Polysorbate 40
Sorbitan monoisooctadecanoate. *See* Sorbitan isostearate
Sorbitan monolaurate. *See* Sorbitan laurate
Sorbitan monooctadecanoate. *See* Sorbitan stearate
Sorbitan mono-9-octadecenoate. *See* Sorbitan oleate
Sorbitan monooleate. *See* Sorbitan oleate
Sorbitan monopalmitate. *See* Sorbitan palmitate
Sorbitan monostearate. *See* Sorbitan stearate
Sorbitan, 9-octadecenoate (2:3). *See* Sorbitan sesquioleate

Sorbitan oleate
CAS 1338-43-8; 5938-38-5; EINECS 215-665-4
Synonyms: SMO; Sorbitan monooleate; Sorbitan mono-9-octadecenoate; Anhydrosorbitol monooleate
Definition: Monoester of oleic acid and hexitol anhydrides derived from sorbitol
Empirical: $C_{24}H_{44}O_6$
Properties: M.w. 428.62; dens. 0.986; sapon. no. 145-160; hyd. no. 193-210; flash pt. > 230 F; ref. index 1.4800
Toxicology: Heated to decomp., emits acrid smoke and irritating fumes
Uses: Emulsifier for foods, cosmetics, household prods., industrial applics., polymerization; antifog, antistat, cling additive for plastic films
Regulatory: FDA 21CFR §73.1001, 173.75, 175.105, 175.320, 178.3400; Europe listed; UK approved
Manuf./Distrib.: Heterene; ICI Surf.; Karlshamns; Lonza; Norman, Fox; Witco/Oleo-Surf.
Trade names: Atmer® 105; Disponil SMO 100 F1; Durtan® 80; Ethylan® GO80; Glycomul® O; S-Maz® 80

Sorbitan palmitate
CAS 26266-57-9; EINECS 247-568-8
Synonyms: Sorbitan monopalmitate; 1,4-Anhydro-D-glucitol, 6-hexadecanoate
Empirical: $C_{22}H_{42}O_6$
Properties: M.w. 1109.59; dens. 0.989; flash pt. > 230 F; ref. index 1.4780
Toxicology: Irritant
Uses: Emulsifier for foods, cosmetics, household prods., industrial applics., polymerization
Regulatory: FDA 21CFR §175.320, 178.3400; Europe listed; UK approved
Trade names: Disponil SMP 100 F1

Sorbitan sesquioleate
CAS 8007-43-0; EINECS 232-360-1
Synonyms: Anhydrosorbitol sesquioleate; Anhydrohexitol sesquioleate; Sorbitan, 9-octadecenoate (2:3)
Definition: Mixture of mono and diesters of oleic acid and hexitol anhydrides derived from sorbitol
Uses: W/o emulsifier, wetting agent, pigment dispersant, lubricant, coupler, solubilizer, antifoam; for cosmetics, pharmaceuticals, explosives, household aerosols, polymerization
Regulatory: FDA 21CFR §175.320
Trade names: Disponil SSO 100 F1

Sorbitan stearate
CAS 1338-41-6, 69005-67-8; EINECS 215-664-9
Synonyms: SMS; Sorbitan monostearate; Sorbitan monooctadecanoate; Anhydrosorbitol monostearate
Definition: Monoester of stearic acid and hexitol anhydrides derived from sorbitol
Empirical: $C_{24}H_{46}O_6$
Properties: M.w. 430.70; acid no. 5-10; sapon. no. 147-157; hyd. no. 235-260
Toxicology: LD50 (oral, rat) 31 g/kg; very mildly toxic by ingestion; experimental reproductive effects; heated to decomp., emits acrid smoke and irritating fumes
Uses: Emulsifier, stabilizer, defoamer for foods, cosmetics, household prods., industrial applics., polymerization
Regulatory: FDA 21CFR §73.1001, 163.123, 163.130, 163.135, 163.140, 163.145, 163.150, 163.153, 163.155, 172.515, 172.842, 173.340, 175.105, 175.320, 178.3400, 573.960; Europe listed; UK approved
Manuf./Distrib.: Aldrich; Aquatec Quimica SA
Trade names: Disponil SMS 100 F1; Ethylan® GS60; Glycomul® S

Sorbitan trioctadecanoate. *See* Sorbitan tristearate
Sorbitan tri-9-octadecenoate. *See* Sorbitan trioleate

Sorbitan trioleate
CAS 26266-58-0; 85186-88-5; EINECS 247-569-3; 286-074-7

Synonyms: STO; Anhydrosorbitol trioleate; Sorbitan tri-9-octadecenoate
Definition: Triester of oleic acid and hexitol anhydrides derived from sorbitol
Empirical: $C_{60}H_{108}O_8$
Properties: M.w. 957.52; dens. 0.956; flash pt. > 230 F; ref. index 1.4760
Toxicology: Irritant
Uses: Emulsifier for foods, cosmetics, household prods., industrial applics., polymerization; antifog, lubricant, cling additive for plastic film; emollient
Regulatory: FDA 21CFR §175.320, 178.3400
Trade names: Atmer® 106; Disponil STO 100 F1; Ethylan® GT85

Sorbitan tristearate
CAS 26658-19-5; 72869-62-6; EINECS 247-891-4; 276-951-2
Synonyms: STS; Anhydrosorbitol tristearate; Sorbitan trioctadecanoate
Definition: Triester of stearic acid and hexitol anhydrides derived from sorbitol
Empirical: $C_{60}H_{114}O_8$
Properties: M.w. 963.56
Uses: Emulsifier for foods, cosmetics, household prods., industrial applics., polymerization
Regulatory: FDA 21CFR §175.320, 178.3400; Europe listed; UK approved
Manuf./Distrib.: Henkel/Emery; ICI Surf.; Lonza
Trade names: Disponil STS 100 F1

Sorbitol acetal. *See* Dibenzylidene sorbitol

Soy acid
CAS 68308-53-2; 67701-08-0; EINECS 269-657-0
Synonyms: Acids, soy; Fatty acids, soya
Definition: Mixture of fatty acids derived from soybean oil
Uses: Lubricant, release agent, binder, defoaming agent and intermediate for food additives; rubber compounding; alkyd resins; water repellents; polishes; soaps; cutting oils; candles; crayons
Regulatory: FDA 21CFR §175.105, 177.2800, 178.3570
Trade names: Industrene® 225; Industrene® 226

Soya dimethyl amine. *See* Dimethyl soyamine
Soya dimethyl ethyl ammonium ethyl sulfate. *See* Soyethyldimonium ethosulfate
Soyaethyldimonium ethosulfate. *See* Soyethyldimonium ethosulfate

Soyaminopropylamine
Classification: Substituted amine
Formula: $R—NH—CH_2CH_2CH_2NH_2$, R rep. alkyl groups derived from soy
Uses: Gasoline detergent, bactericide, corrosion inhibitor in petrol. prod., epoxy hardener
Trade names: Kemamine® D-999

Soybean oil, epoxidized. *See* Epoxidized soybean oil

Soyethyldimonium ethosulfate
CAS 68308-67-8
Synonyms: Soya dimethyl ethyl ammonium ethyl sulfate; Soyaethyldimonium ethosulfate
Classification: Quaternary ammonium compd.
Formula: $[RN(CH_3)_2CH_2CH_3]^+CH_3CH_2OSO_3^-$, R rep. alkyl groups from soy
Uses: Conditioner for hair conditioners and shampoos; antistat for cleaners, rug shampoos, plastics, fibers, fiberglass; mold release agent
Trade names: Larostat® 88; Larostat® 264 A; Larostat® 264 A Anhyd; Larostat® 264 A Conc.
Trade names containing: Larostat® 519

Stannic acid. *See* Stannic oxide
Stannic anhydride. *See* Stannic oxide

Stannic oxide
CAS 18282-10-5
Synonyms: Stannic anhydride; Stannic acid; Tin peroxide; Tin dioxide
Empirical: SnO_2
Properties: Wh. powd.; sol. in conc. sulfuric acid, hydrochloric acid; insol. in water; m.w. 150.69; dens. 6.6-6.9; m.p. 1127 C

Precaution: Noncombustible
Uses: Tin salts; catalyst; ceramic glazes and colors; putty; perfume preps. and cosmetics; textiles; polishing powd. for steel, glass, etc.; mfg. of special glasses
Manuf./Distrib.: Spectrum Chem. Mfg.
Trade names containing: Mearlin® MagnaPearl 1100

Stannous chloride anhyd.
CAS 7772-99-8; EINECS 231-868-0
Synonyms: Tin (II) chloride anhydrous; Tin crystals; Tin (II) chloride (1:2); Tin salt; Tin dichloride; Tin protochloride
Definition: Chloride salt of metallic tin
Empirical: Cl_2Sn
Formula: $SnCl_2$
Properties: Colorless orthorhombic cryst. mass or flakes, fatty appearance; sol. in water, ethanol, acetone, ether, methyl acetate, MEK, isobutyl alcohol; pract. insol. in min. spirits, xylene; m.w. 189.60; dens. 3.95; m.p. 246.8 C; b.p. 652 C
Precaution: Potentially explosive reaction with metal nitrates; violent reactions with hydrogen peroxide, ethylene oxide, nitrates, K, Na
Toxicology: TLV:TWA 2 mg(Sn)/m^3; LD50 (oral, rat) 700 mg/kg, (IP, mouse) 66 mg/kg; poison by ingestion, IP, IV, subcutaneous routes; experimental reproductive effects; human mutagenic data; heated to decomp., emits toxic fumes of Cl^-
Uses: Reducing agent for intermediates, dyes, polymers, phosphors; mfg. of lakes; textile dyeing and printing; tin galvanizing; analytical reagent; silvering mirrors; antisludge for lubricants; food preservative; perfume stabilizer; soldering flux; esterification catalyst for mfg. of plasticizers and polyesters
Regulatory: FDA 21CFR §155.200, 172.180, 175.300, 177.2600, 184.1845, GRAS
Manuf./Distrib.: Aldrich; Blythe, William Ltd; Cerac; Elf Atochem N. Am.; Noah Chem.; Spectrum Chem. Mfg.
Trade names: Fascat® 2004; Fascat® 2005

Stannous-2-ethylhexoate. *See* Stannous octoate

Stannous octoate
CAS 301-10-0
Synonyms: Stannous-2-ethylhexoate; Tin octoate; Tin-(II)-octoate; Sn-II-ethylhexanoate
Formula: $Sn(C_8H_{15}O_2)_2$
Properties: Lt. yel. liq.; sol. in benzene, toluene, petroleum ether; insol. in water, methanol; m.w. 405.11; dens. 1.25; flash pt. > 230 F; ref. index 1.4930
Toxicology: Irritant
Uses: Catalyst for mfg. of PU foam, coatings, adhesives, sealants, elastomers
Manuf./Distrib.: Aceto
Trade names: Dabco® T-9; Fomrez® C-2; Kosmos® 16; Kosmos® 29; Metacure® T-9; TEDA-D007
Trade names containing: Dabco® T-10; Dabco® T-11; Dabco® T-16; Dabco® T-95; Dabco® T-96; Fomrez® C-4; Jeffcat T-10; Kosmos® 10; Kosmos® 15

Stannous stearate
Synonyms: Tin stearate
Toxicology: Heated to decomp., emits acrid smoke and irritating fumes
Uses: Lubricant for plastics compounding; antistat, stabilizer
Regulatory: FDA 21CFR §181.29 (50 ppm max. in finished food)
Trade names: Synpro® Stannous Stearate

Stannum. *See* Tin

Starch
CAS 9005-84-9; EINECS 232-686-4
Classification: Carbohydrate polymer
Definition: Complex polysaccharide composed of units of glucose consisting of about one quarter amylose and three quarters amylopectin
Empirical: $(C_6H_{10}O_5)_n$
Properties: White amorphous powd. or gran., tasteless; gel formation in hot water; m.w. 162.14_n
Uses: Adhesives, machine-coated paper, textile filler and sizing agent, gelling agent for foods, filler in baking powd., urea-formaldehyde resin adhesive, chelating and sequestering agent in foods
Regulatory: FDA 21CFR §182.90, GRAS
Trade names containing: Cabot® PE 9321; Plasdeg PE 9321

Stearamide

CAS 124-26-5; EINECS 204-693-2

Synonyms: Octadecanamide; Stearic acid amide; Amide C-18

Classification: Aliphatic amide

Empirical: $C_{18}H_{37}NO$

Formula: $CH_3(CH_2)_{16}CONH_2$

Properties: Colorless leaflets; sl. sol. in alcohol, ether; insol. in water; m.w. 283.56; m.p. 98-102 C; b.p. 250-251 C (12 mm)

Toxicology: Experimental tumorigen

Uses: Slip/antiblock agent for LDPE, HDPE, PP; mold release agent, lubricant

Regulatory: FDA 21CFR §175.105, 177.1210, 178.3860, 178.3910, 179.45, 181.22, 181.28

Manuf./Distrib.: Akzo Nobel; Aldrich; Astor Wax; Chemax; Cookson Spec.; Croda Universal; Henkel/Emery; Syn. Prods.; Witco/Oleo-Surf.

Trade names: Armid® 18; Armoslip® 18; Chemstat® HTSA #18S; Crodamide S; Crodamide SR; Kemamide® S; Petrac® Vyn-Eze®; Uniwax 1750

Stearamide MEA

CAS 111-57-9; EINECS 203-883-2

Synonyms: Stearic acid monoethanolamide; Stearoyl monoethanolamide; N-(2-Hydroxyethyl) octadecanamide

Definition: Mixture of ethanolamides of stearic acid

Empirical: $C_{20}H_{41}NO_2$

Formula: $CH_3(CH_2)_{16}CONHCH_2CH_2OH$

Properties: M.w. 327.55

Uses: Opacifier, pearlescent, conditioner, lubricant, thickener, gelling agent, mold release agent, binder, detergent, superfatting agent; cosmetics and toiletries; base for antiperspirant and makeup sticks; emulsion stabilizer

Trade names containing: Mackester™ SP

Stearamidopropyl dimethyl-β-hydroxyethylammonium nitrate. *See* Stearamidopropyl dimethyl-2-hydroxyethyl ammonium nitrate

Stearamidopropyl dimethyl-β-hydroxyethylammonium dihydrogen phosphate

Empirical: $C_{25}H_{53}O_6N_2P$

Properties: M.w. 510.8

Uses: Antistat for plastics, waxes, textiles, and glass; emulsifier, settling, dispersing, rewetting agent

Trade names containing: Cyastat® SP

Stearamidopropyl dimethyl-2-hydroxyethyl ammonium nitrate

CAS 2764-13-8

Synonyms: Stearamidopropyl dimethyl-β-hydroxyethylammonium nitrate

Empirical: $C_{25}H_{53}O_5N_3$

Properties: M.w. 475

Uses: Antistat for polymers, paper, glass, other materials

Trade names containing: Cyastat® SN

Stearamine

CAS 124-30-1; EINECS 204-695-3

Synonyms: Stearylamine; Octadecylamine; 1-Octadecanamine

Empirical: $C_{18}H_{39}N$

Formula: $CH_3(CH_2)_{16}CH_2NH_2$

Properties: M.w. 269.58; m.p. 55-57 C; flash pt. > 230 F

Precaution: Corrosive

Toxicology: LD50 (IP, mouse) 250 mg/kg; poison by intraperitoneal route; skin irritant; heated to decomp., emits toxic fumes of NO_x

Uses: Anticorrosive agent, germicide, wetting agent, softener, emulsifier, dispersant, intermediate, flotation agent, corrosion inhibitor used in textiles, water treatment, concrete, asphalt, agriculture, ceramics; mold release agent for rubber and plastics; processing aid for rubber

Regulatory: FDA 21CFR §173.310 (3 ppm max. in steam)

Manuf./Distrib.: Aldrich

Trade names: Armeen® 18; Armeen® 18D; Armid® HTD; Crodamine 1.18D; Kemamine® P-990, P-990D

Stearethyldimonium ethosulfate
Uses: Release agent forming a hard film; imparts gloss and antistatic properties
Trade names: Larostat® 451 P

Stearic acid
CAS 57-11-4; EINECS 200-313-4
Synonyms: n-Octadecanoic acid; Carboxylic acid C_{18}
Classification: Fatty acid
Empirical: $C_{18}H_{36}O_2$
Formula: $CH_3(CH_2)_{16}COOH_9$
Properties: White to ylsh.-wh. amorphous solid, tallow-like odor and taste; very sl. sol. in water; sol. in alcohol, ether, acetone, CCl_4; m.w. 284.47; dens. 0.847 (70 C); m.p. 69.3 C; b.p. 383 C; flash pt. (CC) 385 F; ref. index 1.4299 (80 C)
Precaution: Combustible when exposed to heat or flame; heats spontaneously
Toxicology: LD50 (IV, rat) 21.5 ± 1.8 mg/kg; poison by intravenous route; experimental tumorigen; human skin irritant; heated to decomp., emits acrid smoke and irritating fumes
Uses: Cosmetics, chemicals, dispersant, softener in rubber compds., food packaging, suppositories, ointments; PVC lubricant
Regulatory: FDA 21CFR §172.210, 172.615, 172.860, 175.105, 175.300, 175.320, 176.170, 176.200, 176.210, 177.1010, 177.1200, 177.2260, 177.2600, 177.2800, 178.3570, 178.3910, 184.1090, GRAS; GRAS (FEMA); Europe listed
Manuf./Distrib.: Akrochem; Akzo Nobel; Allchem Ind.; Condor; Cookson Spec.; Great Western; Henkel/Emery; Lonza; Penta Mfg.; Syn. Prods.; Unichema; Witco/Oleo-Surf.
Trade names: Glycon® S-70; Glycon® S-90; Glycon® TP; Hydrofol 1800; Hystrene® 5016 NF; Hystrene® 9718 NF; Industrene® 1224; Industrene® 4518; Industrene® 5016; Industrene® 7018; Industrene® 9018; Loxiol® G 20; Naftozin® N, Spezial; Petrac® 250; Petrac® 270; Pristerene 4900
Trade names containing: Aktisil® PF 231; Cohedur® RS; Geropon® AS-200; Kotamite®; Micro-White® 15 SAM; Opacicoat™; Renacit® 7/WG; Vultac® 4; Vultac® 710

Stearic acid aluminum dihydroxide salt. *See* Aluminum stearate
Stearic acid amide. *See* Stearamide
Stearic acid calcium salt. *See* Calcium stearate
Stearic acid, lead salt. *See* Lead stearate
Stearic acid, 2-methylpropyl ester. *See* Isobutyl stearate
Stearic acid monoethanolamide. *See* Stearamide MEA
Stearic acid, potassium salt. *See* Potassium stearate
Stearic acid sodium salt. *See* Sodium stearate
Stearin. *See* Tristearin

Stearoyl p-amino phenol
CAS 103-99-1
Synonyms: N-(4-Hydroxyphenyl) stearamide
Empirical: $C_{24}H_{41}O_2N$
Formula: $HO(C_6H_4)NHCOCH_2(CH_2)_{15}CH_3$
Properties: White to off-white powd.; insol. in water; sol. in polar org. solvs., alcohol; m.w. 375.60; m.p. 131-134 C
Precaution: Combustible
Toxicology: Irritant
Uses: Antioxidant and processing aid in thermoplastics
Manuf./Distrib.: Zeeland
Trade names: Suconox-18®

Stearoyl N-methylaminoacetic acid. *See* Stearoyl sarcosine
Stearoyl N-methylglycine. *See* Stearoyl sarcosine
Stearoyl monoethanolamide. *See* Stearamide MEA

Stearoyl sarcosine
CAS 142-48-3; EINECS 205-539-7
Synonyms: Stearoyl N-methylglycine; Stearoyl N-methylaminoacetic acid; N-Methyl-N-(1-oxooctadecyl)glycine
Definition: N-stearoyl deriv. of N-methylglycine
Empirical: $C_{21}H_{41}NO_3$
Formula: $CH_3(CH_2)_{16}CONCH_3CH_2COOH$

Uses: Detergent, wetting and foaming agent for cosmetics, household/industrial cleaners, biotechnology, agric., textiles, lubricants, metalworking; emulsifier for emulsion polymerization; mold release for RIM; stabilizer for polyols; antifog, antistat
Regulatory: FDA 21CFR §177.1200
Trade names: Hamposyl® S

Steartrimonium bromide
Synonyms: Octadecyl trimethyl ammonium bromide
Uses: Accelerator for polyacrylate elastomers
Trade names: Zeonet® B

Steartrimonium chloride
CAS 112-03-8; EINECS 203-929-1
Synonyms: Stearyl trimethyl ammonium chloride; Octadecyl trimethyl ammonium chloride; N,N,N-Trimethyl-1-octadecanaminium chloride
Classification: Quaternary ammonium salt
Empirical: $C_{21}H_{46}N \cdot Cl$
Formula: $[CH_3(CH_2)_{16}CH_2N(CH_3)_3]^+Cl^-$
Toxicology: Poison by unspecified routes
Uses: Germicide in water treatment, petrol., paper, foods and textile industries, antistat in plastics, pulp and paper industries, rinse agent, pigment coating agent, dispersant, coagulant, softener; hair rinse base
Trade names containing: Arquad® 18-50

Stearyl alcohol
CAS 112-92-5; EINECS 204-017-6
Synonyms: n-Octadecanol; Octadecyl alcohol; 1-Octadecanol; n-Octyldecyl alcohol; C18 linear alcohol; Decyl octyl alcohol
Classification: Fatty alcohol
Empirical: $C_{18}H_{38}O$
Formula: $CH_3(CH_2)_{16}CH_2OH$
Properties: Unctuous white flakes or gran., faint odor, bland taste; sol. in alcohol, acetone, ether; insol. in water; m.w. 270.56; dens. 0.8124 (59/4 C); m.p. 59 C; b.p. 210.5 C (15 mm)
Precaution: Flamm. when exposed to heat or flame; can react with oxidizers
Toxicology: LD50 (oral, rat) 20 g/kg; mildly toxic by ingestion; experimental neoplastigen; heated to decomp., emits acrid smoke and irritating fumes
Uses: Perfumery, cosmetics, intermediate, surface active agents, lubricants, resins, antifoam agent
Regulatory: FDA 21CFR §172.755, 172.864, 175.105, 175.300, 176.200, 176.210, 177.1010, 177.1200, 177.2800, 178.3480, 178.3910
Manuf./Distrib.: Aarhus Oliefabrik A/S; Albemarle; Amerchol; Chemron; Croda; Kraft; Lipo; Lonza; M. Michel; Procter & Gamble; Vista
Trade names: Adol® 62 NF; Alfol® 18; Cachalot® S-53; Loxiol® VPG 1354

Stearylamine. *See* Stearamine

Stearyl/aminopropyl methicone copolymer
CAS 110720-64-4
Uses: Lubricant, water repellent, antiblocking agent, processing aid, release agent for plastics, rubber, paper, textiles, inks, candles, glass bottle molding, aluminum prod.
Trade names: EXP-61 Amine Functional Silicone Wax

Stearyl betaine
CAS 820-66-6; EINECS 212-470-6
Synonyms: N-(Carboxymethyl)-N,N-dimethyl-1-octadecanaminium hydroxide, inner salt; Stearyl dimethyl glycine
Classification: Zwitterion (inner salt)
Empirical: $C_{22}H_{45}NO_2$
Formula: $CH_3(CH_2)_{16}CH_2N^+(CH_3)_2CH_2COO^-$
Uses: Ampholite; dispersant, wetting; antistat for plastics
Trade names: Lonzaine® 18S

Stearyl citrate
CAS 1337-33-3; EINECS 215-654-4
Synonyms: 2-Hydroxy-1,2,3-propanetricarboxylic acid, monooctadecyl ester; Octadecyl citrate; Citric acid,

octadecyl ester
Definition: Ester of stearyl alcohol and citric acid
Empirical: $C_{24}H_{44}O_7$
Formula: $HOCH_2COOHCCOOHCH_2COOCH_2(CH_2)_{16}CH_3$
Precaution: Moisture-sensitive
Toxicology: Heated to decomp., emits acrid smoke and irritating fumes
Uses: Chelating agent; stabilizer for oils; synergistic with antioxidants; sequestrant in food and feed applics.; plasticizer or component in food pkg. materials
Regulatory: FDA 21CFR §166.40, 166.110, 175.300, 178.3910, 181.22, 181.27, 182.6851 (0.15% max. as sequestrant), GRAS, 582.6851
Trade names: Monostearyl Citrate (MSC)

Stearyl-3-(3′,5′-di-t-butyl-4-hydroxyphenyl) propionate. *See* Octadecyl 3,5-di-t-butyl-4-hydroxyhydrocinnamate
Stearyl dimethyl amine. *See* Dimethyl stearamine
Stearyl dimethyl glycine. *See* Stearyl betaine

Stearyl erucamide
CAS 10094-45-8; EINECS 233-226-5
Synonyms: N-Octadecyl-13-docosenamide; 13-Docosenamide, N-octadecyl-
Classification: Substituted aliphatic amide
Empirical: $C_{40}H_{79}NO$
Formula: $CH_3(CH_2)_7CHCH(CH_2)_{11}CONHCH_2(CH_2)_{16}CH_3$
Properties: M.w. 590.07
Uses: Slip/antiblock agent, mold release for plastics
Regulatory: FDA 21CFR §178.3860
Manuf./Distrib.: Croda Universal; Witco/Oleo-Surf.; Zeeland
Trade names: Chemstat® HTSA #3B; Kemamide® E-180

N-Stearyl 12-hydroxystearamide
Uses: Lubricant, antistat for plastics, metals; mold release; antiblocking agent for textile coatings; slip agent for coatings; elec. potting compds., crayons, wax blends
Trade names: Paricin® 210

Stearyl lauryl thiodipropionate
CAS 13103-52-1
Synonyms: Propanoic acid, 3-((3-(dodecyloxy)-3-oxopropyl)-thio)-, octadecyl ester
Empirical: $C_{36}H_{70}O_4S$
Formula: $S(CH_2CH_2COOC_{12}H_{25})(CH_2CH_2COOC_{18}H_{37})$
Properties: M.w. 598; f.p. 49 C min.
Uses: Antioxidant, stabilizer
Manuf./Distrib.: Evans Chemetics
Trade names: Evanstab® 1218

Stearyl nitrile
Uses: Detergent, wetting agent, rust inhibitor; metal wetting with min. oils; plasticizer for syn. rubbers and plastics
Trade names: Arneel® 18 D

Stearyl PEG-15 methyl ammonium chloride
Uses: Acid corrosion inhibitor, plastics and textile antistat, and emulsifier
Trade names: Tomah Q-18-15

Stearyl stearamide
Uses: Lubricant, slip, antiblock, and mold release agent for plastics, crayons, petrol. prods., asphalts, inks, metals, textiles; mold release agent for thermoplastic resins in inj. molding; defoamer and water repellent in industrial/household applic.
Trade names: Kemamide® S-180

Stearyl stearate
CAS 2778-96-3; 85536-04-5; EINECS 220-476-5; 287-484-9
Synonyms: Octadecanoic acid, octadecyl ester
Definition: Ester of stearyl alcohol and stearic acid

Empirical: $C_{36}H_{72}O_2$
Formula: $CH_3(CH_2)_{16}COOCH_2(CH_2)_{16}CH_3$
Properties: M.w. 537.00; m.p. 56 C; flash pt. (COC) 242 C
Uses: Emollient for bath oils, creams, lotions; plasticizer
Regulatory: FDA 21CFR §178.3910
Manuf./Distrib.: Inolex

Stibic anhydride. *See* Antimony pentoxide
STO. *See* Sorbitan trioleate

Strontium-zinc complex
Uses: Heat and light stabilizer for vinyls
Trade names: Therm-Chek® 135

STS. *See* Sorbitan tristearate

Styrenated diphenyl amine
Uses: Antioxidant for natural and syn. rubbers; for inner tubes, sponge rubber, seals/gaskets, soles, latex goods
Trade names: Vulkanox® DDA

Styrene-acrylonitrile copolymer
CAS 9003-54-7
Synonyms: SAN; SAN copolymer; ACS
Formula: $(C_8H_8 \cdot C_3H_3N)_x$
Properties: Dens. 1.080
Toxicology: Suspected carcinogen
Uses: Inj. molding and extrusion resin for cosmetic pkg., fan blades, toys and games, business machines, interior refrigerator parts, medical parts, beverage tumblers, food containers, tableware, dinnerware, containers, automotive parts, cassettes; PVC modifier and processing aid
Manuf./Distrib.: BASF; LNP; Monsanto; Reichhold; A Schulman
Trade names: Baymod® A 95; Blendex® 869
Trade names containing: Colloids SAN 52/8; MDI AB-925; MDI SN-948; Plasblak® SA 3176

Styrene-butadiene polymer
CAS 9003-55-8
Synonyms: S/B; Styrene polymer with 1,3 butadiene; Butadiene-styrene resin; 1,3-Butadiene-styrene copolymer
Formula: $(C_8H_8 \cdot C_4H_6)_x$
Properties: Dens. 0.965
Uses: Inj. and blow molding, extrusion, thermoforming resin for housings, blister packs, tubes, toys, containers, medical devices, bottles, tech. parts; reinforcement for rubber
Trade names: Pliolite® S-6B, S-6F
See also Butadiene/styrene copolymer, Styrene-butadiene rubber

Styrene-butadiene rubber
CAS 9003-55-8
Synonyms: SBR
Uses: Latex as binder, for coatings, paper/paperboard, sealants, adhesives, carpet backing; SBR for tires, adhesives, chewing gums; processing aid, modifier
Trade names: Darex® 632L; Darex® 636L; Polysar S-1018; Polysar SS 260; Ricon 100; Stereon® 205; Stereon® 210; Stereon® 721A; Stereon® 730A; Stereon® 840A; Stereon® 841A
Trade names containing: Buna BL 6533; Chem-Master R-135; Diapol® WMB® 1808; Diapol® WMB® 1833; Diapol® WMB® 1847; Diapol® WMB® 1848; Diapol® WMB® 1849; Diapol® WMB® 1939; Diapol® WMB® S900; Diapol® WMB® S903; Diapol® WMB® S910; Diapol® WMB® S920; Diapol® WMB® S960; Poly-Dispersion® S (AX)D-70; Poly-Dispersion® SCSD-70; Poly-Dispersion® SHD-65; Poly-Dispersion® S (PVI)D-50P; Poly-Dispersion® SSD-75; Poly-Dispersion® S (UR)D-75; Poly-Dispersion® SZFD-85; Poly-Dispersion® SZFND-825; Rhenogran® Resorcin-80/SBR; Wyfire® S-A-51; Wyfire® S-BA-88P

Styrene/MA copolymer
CAS 9011-13-6
Synonyms: SMA; Styrene/maleic anhydride copolymer; 2,5-Furandione, polymer with ethenylbenzene; Styrene/maleic anhydride resin

Styrene/maleic anhydride copolymer

Definition: Polymer of styrene and maleic anhydride monomers
Empirical: $(C_8H_8 \cdot C_4H_2O_3)_x$
Properties: Dens. 1.270
Toxicology: Irritant
Uses: Engineering thermoplastic for inj. molding (mech. and automotive parts, appliance and electronic housings); emulsifier, binder, visc. modifier, stabilizer, pigment dispersant, protective colloid, leveling agent, sizing, adhesives, coatings, polishes; emulsion polymerization
Manuf./Distrib.: Cargill; Monomer-Polymer & Dajac
Trade names: SMA® 1000; SMA® 2000

Styrene/maleic anhydride copolymer. *See* Styrene/MA copolymer
Styrene/maleic anhydride resin. *See* Styrene/MA copolymer
Styrene polymer. *See* Polystyrene
Styrene polymer with 1,3-butadiene. *See* Butadiene-styrene copolymer, Styrene-butadiene polymer

3-(N-Styrylmethyl-2-aminoethylamino) propyltrimethoxy silane hydrochloride
CAS 52783-38-7
Uses: Coupling agent, chem. intermediate, blocking agent, release agent, lubricant, primer, reducing agent
Trade names containing: CS1590

Succinic acid peroxide
CAS 123-23-9
Synonyms: Disuccinic acid peroxide
Empirical: $C_8H_{10}O_8$
Formula: $(HOOCCH_2CH_2CO)_2O_2$
Properties: Fine wh. powd., odorless; moderately sol. in water; insol. in petroleum solvs., benzene; m.w. 234.18; m.p. 125 C (dec.)
Precaution: Fire risk with combustible material; oxidizing agent
Toxicology: Toxic by ingestion and inhalation; irritating to skin
Uses: Polymerization initiator/catalyst, deodorants, antiseptics
Manuf./Distrib.: Elf Atochem N. Am.
Trade names: Aztec® SUCP-70-W

Succinic acid, sulfo-, 1,4-dicyclohexyl ester, sodium salt. *See* Dicyclohexyl sodium sulfosuccinate

Sucrose acetate isobutyrate
CAS 126-13-6; EINECS 204-771-6
Synonyms: SAIB
Classification: Sucrose derivative
Definition: Mixed ester of sucrose and acetic and isobutyric acids
Empirical: $C_{40}H_{62}O_{19}$
Formula: $(CH_3COO)_2C_{12}H_{14}O_3[OOCCH(CH_3)_2]_6$
Properties: Clear semisolid or sol'n.; m.w. 847.02; sp.gr. 1.146; flash pt. (COC) 260 C; ref. index 1.4540
Precaution: Combustible
Uses: Plasticizer for cellulosics, PS, PVAc; modifier for lacquers, hot-melt coating formulations, extrudable plastics
Regulatory: FDA 21CFR §175.105
Manuf./Distrib.: Eastman; Technical Chems. & Prods.

Sucrose benzoate
CAS 12738-64-6; EINECS 235-795-5
Synonyms: β-D-Fructofuranoysl-α-D-glucopyranoside benzoate
Classification: Disaccharide ester
Properties: Sp.gr. 1.25; m.p. 98 C; flash pt. (COC) 260 C; ref. index 1.577
Uses: Plasticizer for PVC, PVAc, VCA, PS, cellulosics
Regulatory: FDA 21CFR §175.105
Manuf./Distrib.: Velsicol

Sulfated castor oil
CAS 8002-33-3; EINECS 232-306-7
Synonyms: Castor oil sulfated; Sulfonated castor oil; Turkey-red oil
Definition: Oil consisting primarily of sodium salt of the sulfated triglyceride of ricinoleic acid
Uses: Anionic wetting agent; emulsifier, plasticizer, lubricant, dispersant; in textile dyeing, adhesives;

cosmetics superfatting agent; antisag/antisettling agent in paints; emulsifier for latex
Regulatory: FDA 21CFR §175.105, 176.170, 176.200, 177.1200
Trade names: Cordon NU 890/75; Monosulf

Sulfobutanedioic acid, cetyl/stearyl ester, disodium salt. *See* Disodium cetearyl sulfosuccinate
Sulfobutanedioic acid, C12-15 pareth ester, disodium salt. *See* Disodium C12-15 pareth sulfosuccinate
Sulfobutanedioic acid, deceth-6 ester, disodium salt. *See* Disodium deceth-6 sulfosuccinate
Sulfobutanedioic acid 1,4-dipentyl ester sodium salt. *See* Diamyl sodium sulfosuccinate
Sulfobutanedioic acid, 1,4-ditridecyl ester, sodium salt. *See* Ditridecyl sodium sulfosuccinate
Sulfobutanedioic acid, 4-[2-[2-[2-(dodecyloxy)ethoxy]ethoxy]ethyl]ester, disodium salt. *See* Disodium laureth sulfosuccinate
Sulfobutanedioic acid, 1-[2-[(12-hydroxy-1-oxo-9-octadecenyl)amino]ethyl]ester, disodium salt. *See* Disodium ricinoleamido MEA-sulfosuccinate
Sulfobutanedioic acid, 4-isodecyl ester, disodium salt. *See* Disodium isodecyl sulfosuccinate
Sulfobutanedioic acid, 4-[1-methyl-2-[(1-oxo-9-octadecenyl)amino]ethyl]ester, disodium salt. *See* Disodium oleamido MIPA-sulfosuccinate
Sulfobutanedioic acid, monooctadecyl ester, disodium salt. *See* Disodium stearyl sulfosuccinamate
Sulfobutanedioic acid, nonoxynol-10 ester, disodium salt. *See* Disodium nonoxynol-10 sulfosuccinate
Sulfobutanedioic acid, 1-(9-octadecenyl) ester, disodium salt. *See* Disodium oleyl sulfosuccinate
Sulfobutanedioic acid, tallow ester, disodium salt. *See* Disodium tallow sulfosuccinamate
Sulfocarbolic acid. *See* Phenol sulfonic acid

2-Sulfoethyl methacrylate
CAS 10595-80-9
Synonyms: 2-Propenoic acid, 2-methyl-, 2-sulfoethyl ester
Empirical: $C_6H_{10}O_5S$
Properties: M.w. 194.14; b.p. dec.
Uses: Copolymerizable emulsifier in polymer latexes; functional monomer
Manuf./Distrib.: Evans Chemetics

Sulfolane
CAS 126-33-0; EINECS 204-783-1
Synonyms: Tetrahydrothiophene 1,1-dioxide; Dapsone; Tetramethylene sulfone; Thiophan sulfone
Empirical: $C_4H_8O_2S$
Formula: $CH_2CH_2CH_2CH_2SO_2$
Properties: White or creamy white crystalline powd., odorless, sl. bitter taste; very sl. sol. in water; sol. in alcohol, acetone, dilute min. acids; m.w. 120.16; dens. 1.27 (20 C); m.p. 25-28 C; b.p. 285 C; flash pt. 330 F; ref. index 1.485
Precaution: Combustible
Toxicology: Toxic by ingestion; irritant
Uses: Curing agent for epoxy resins; medicine (antibacterial)
Manuf./Distrib.: Phillips; Shell; Syn. Chems. Ltd
Trade names: Sulfolane W

Sulfonated castor oil. *See* Sulfated castor oil

Sulfonic acid
CAS 72674-05-6
Classification: Organic compound containing one or more sulfo radicals
Uses: Dispersant, wetting agent, leveling agent, protective colloid for dyestuffs
Manuf./Distrib.: Yorkshire Chem. plc
Trade names containing: Eltesol® TA 65

Sulfonic acids, C10-18, -alkane, sodium salts. *See* Sodium C10-18 alkyl sulfonate

Sulfonyl hydrazide
Uses: Chemical blowing agent
Trade names: Celogen® XP-100

Sulfosuccinic acid diisobutyl ester sodium salt. *See* Diisobutyl sodium sulfosuccinate
Sulfosuccinic acid dipentyl ester sodium salt. *See* Diamyl sodium sulfosuccinate

Sulfur
CAS 7704-34-9; EINECS 231-722-6

Synonyms: Brimstone; Sulphur; Soluble sulfur; Sulfur soluble
Classification: Nonmetallic element
Empirical: S
Properties: Insol. in water; sl. sol. in alcohol, ether; sol. in carbon disulfide, CCl_4, benzene; at.wt. 32.064; α: rhombic yel. crystals; dens. 2.06; β: monoclinic pale yel. crystals; dens. 1.96; m.p. 119 C; b.p. 444.6 C; flash pt. 405 F
Precaution: Combustible
Toxicology: Skin, eye, mucous membranes irritant; poison by ingestion, intravenous, intraperitoneal routes
Uses: Sulfuric acid mfg., petroleum refining, dyes and chemicals, fungicide, insecticides, explosives, detergents, rubber vulcanization and processing aid
Regulatory: FDA 21CFR §175.105, 177.1210, 177.2600
Manuf./Distrib.: Akrochem; Ashland; Kraft; Norsk Hydro A/S; San Yuan; Shell; Solvay GmbH; Texaco BV; Wilbur-Ellis
Trade names: Chloropren-Faktis A; Faktis 10; Faktis 14; Faktis Asolvan, Asolvan T; Faktis Badenia C; Faktis HF Braun; Faktis RC 110, RC 111, RC 140, RC 141, RC 144; Naftopast® Schwefel-P; Octocure 456; Rubbermakers Sulfur; Struktol® SU 95; Struktol® SU 105; Struktol® SU 120
Trade names containing: Faktis Badenia T; Faktis Para extra weich; Faktis T-hart; Faktis ZD; Kenlastic® K-6641; Poly-Dispersion® ASD-75; Poly-Dispersion® ESD-80; Poly-Dispersion® SSD-75; Poly-Dispersion® TSD-80; Poly-Dispersion® TSD-80P; Prespersion PAB-1724; Rhenocure® IS 60; Rhenocure® IS 60-5; Rhenocure® IS 60/G; Rhenocure® IS 90-20; Rhenocure® IS 90-33; Rhenocure® IS 90-40; Rhenocure® IS 90/G; Rhenovin® S-90; VPA No. 1; VPA No. 3

Sulfur chloride
CAS 10025-67-9; EINECS 233-036-2
Synonyms: Sulfur subchloride; Sulfur monochloride; Disulfur dichloride
Empirical: S_2Cl_2
Properties: Amber to yellowish-red oily fuming liq., penetrating odor; sol. in alcohol, ether, benzene, carbon disulfide, CCl_4, oils, amyl acetate; m.w. 135.03; dens. 1.690 (15.5 C); m.p. -80 C; b.p. 138 C (760 mm); flash pt. 266 F; ref. index 1.670 (20 C)
Precaution: Combustible; reacts violently with water
Toxicology: Toxic by inhalation, ingestion; strong irritant to tissue; TLV:Cl 1 ppm
Uses: Chemicals, analytical reagent, insecticide, hardening of soft woods, sulfur dyes, chlorinating agent; in cold vulcanization of rubber; as polymerization catalyst for veg. oils
Manuf./Distrib.: Akzo Nobel; OxyChem
Trade names: Chloropren-Faktis NW; Faktis R, Weib MB; Faktis Rheinau H, W

Sulfuric acid
CAS 7664-93-9; EINECS 231-639-5
Synonyms: Hydrogen sulfate; Battery acid; Electrolyte acid; Oil of vitriol
Classification: Inorganic acid
Empirical: H_2O_4S
Formula: H_2SO_4
Properties: Colorless to dark brown dense oily liq.; misc. with water and alcohol; m.w. 98.08; dens. 1.84; m.p. 10.4 C; b.p. 290 C; dec. 340 C
Precaution: Corrosive; powerful acidic oxidizer; ignites or explodes on contact with many materials; reacts with water to produce heat; reactive with oxidizing/reducing materials; heated, emits highly toxic fumes
Toxicology: TLV 1 mg/m^3 of air; LD50 (oral, rat) 2.14 g/kg; human poison; mod. toxic by ingestion; strongly corrosive; strong irritant to tissue; can cause severe burns, chronic bronchitis; heated to decomp., emits toxic fumes of SO_x
Uses: Fertilizers, chemicals, dyes and pigments, laboratory reagents, electroplating baths
Regulatory: FDA 21CFR §172.560, 172.892, 173.385, 178.1010, 184.1095, GRAS; BATF 27CFR §240.1051a; Japan restricted; Europe listed; UK approved
Manuf./Distrib.: Akzo Nobel; Am. Cyanamid; Boliden Intertrade; Du Pont; Metallgesellschaft AG; Nissan Chem. Ind.; Olin; OxyChem; Pasminco Europe; Rasa Ind.; Rhone-Poulenc Basic; Saurefabrik; Schweizerhall
Trade names containing: Eltesol® 4009, 4018; Eltesol® TA/E; Eltesol® TA/F; Eltesol® TA/H

Sulfuric acid, monododecyl ester, ammonium salt. *See* Ammonium lauryl sulfate
Sulfuric acid, monododecyl ester, compd. with 2-aminoethanol (1:1). *See* MEA-lauryl sulfate
Sulfuric acid, monododecyl ester, compd. with 2,2′-iminodiethanol (1:1). *See* DEA-lauryl sulfate
Sulfuric acid, monododecyl ester, compd. with 2,2′,2′′-nitrilotris[ethanol] (1:1). *See* TEA-lauryl sulfate
Sulfuric acid monododecyl ester sodium salt. *See* Sodium lauryl sulfate
Sulfuric acid, mono (2-ethylhexyl) ester sodium salt. *See* Sodium octyl sulfate

Sulfuric acid, monotetradecyl ester, sodium salt. *See* Sodium myristyl sulfate

Sulfur, insoluble
CAS 9035-97-8 (polymer)
Properties: Insol. in carbon disulfide
Uses: Rubber vulcanizing agent; prevents crystallization of sulfur on uncured rubber
Trade names: Crystex® 60; Crystex® 60 OT 10; Crystex® 90 OT 20; Crystex® HS; Crystex® HS OT 10; Crystex® HS OT 20; Crystex® OT 10; Crystex® OT 20; Crystex® Regular; Struktol® SU 50; Struktol® SU 109; Struktol® SU 135
Trade names containing: Poly-Dispersion® ECSD-70; Poly-Dispersion® SCSD-70; Rhenocure® IS 60; Rhenocure® IS 60-5; Rhenocure® IS 60/G

Sulfur monochloride. *See* Sulfur chloride

Sulfur soluble. *See* Sulfur
Sulfur subchloride. *See* Sulfur chloride
Sulisobenzone. *See* Benzophenone-4
Sulphur. *See* Sulfur
Sunflower oil. *See* Sunflower seed oil

Sunflower seed oil
CAS 8001-21-6; EINECS 232-273-9
Synonyms: Sunflower oil
Definition: Oil expressed from seeds of the sunflower, *Helianthus annuus*
Properties: Amber liq., pleasant odor, mild taste; sol. in alcohol, ether, chloroform, CS_2; dens. 0.924-0.926; ref. index 1.4611
Precaution: Combustible
Toxicology: Heated to decomp., emits acrid smoke and irritating fumes
Uses: Raw material in plastics, polymers, rubbers, pharmaceuticals; emollient for cosmetics; in prep. of margarine
Regulatory: FDA 21CFR §175.300, 176.200, GRAS
Manuf./Distrib.: Arista Ind.; Charkit; Int'l. Chem.; Lipo; Penta Mfg.; Welch, Holme & Clark
Trade names: GTO 80; GTO 90; GTO 90E; NS-20; Sunyl® 80; Sunyl® 80 RBD; Sunyl® 80 RBD ES; Sunyl® 80 RBWD; Sunyl® 80 RBWD ES; Sunyl® 90; Sunyl® 90 RBD; Sunyl® 90 RBWD; Sunyl® 90E RBWD; Sunyl® 90E RBWD ES 1016; Sunyl® HS 500

SVS. *See* Sodium vinyl sulfonate
Synthetic spermaceti wax. *See* Cetyl esters

Synthetic wax
CAS 8002-74-2; 123237-14-9
Synonyms: Fischer-Tropsch wax; Fischer-Tropsch wax, oxidized
Definition: Hydrocarbon wax derived by Fischer-Tropsch or ethylene polymerization processes
Uses: As ingredient, finish or processing aid, release agent, lubricant in adhesives, ammunition, asphalt, explosives, paints, paper, pyrotechnics, lubricants, PVC, textile finishes, powd. metallurgy, floor wax, candles, hot melts, inks, asphalt
Regulatory: FDA 21CFR §172.615, 172.888, 173.340, 175.105, 175.250, 176.170, 177.1200, 178.3720
Trade names: Ross Wax #100; Ross Wax #140; Ross Wax #160; Vestowax FT-150; Vestowax FT-150P; Vestowax FT-200; Vestowax FT-300; Vybar® 103; Vybar® 253; Vybar® 260; Vybar® 825

Tabular alumina. *See* Alumina

Talc
CAS 14807-96-6; EINECS 238-877-9
Synonyms: Hydrous magnesium silicate; Magnesium hydrogen metasilicate; Hydrous magnesium calcium silicate; Industrial talc; Cosmetic talc; Platy talc; French chalk; Talcum; Pigment White 26; CI 77019
Definition: Native, hydrous magnesium silicate sometimes containing small portion of aluminum silicate
Formula: $Mg_3Si_4O_{10}(OH)_2$ or $3MgO{\cdot}4SiO_2{\cdot}HOH$
Properties: White, apple green, gray powd., pearly or greasy luster, greasy feel; insol. in water, cold acids or in alkalies; m.w. 379.29; dens. 2.7-2.8
Toxicology: TLV:TWA 2 mg/m^3, respirable dust; toxic by inhalation; talc with < 1% asbestos is nuisance dust; experimental tumorigen; human skin irritant; prolonged/repeated exposure can produce talc pneumoconiosis

Talcum

Uses: Reinforcement, filler, and pigment in rubber, paints, plastics, soaps, ceramics, cosmetics, pharmaceuticals; dusting agent, lubricant, electrical insulation; antiblocking agent for plastic film

Regulatory: FDA 21CFR §73.1550, 175.300, 175.380, 175.390, 176.170, 177.1210, 177.1350, 177.1460, 182.70, 182.90; GRAS; Japan restricted (5000 ppm); Europe listed; UK approved

Manuf./Distrib.: Aldrich; Luzenac Am.; Minerals Tech.; Pfizer; L.A. Salomon; R.T. Vanderbilt; Whittaker, Clark & Daniels

Trade names: ABT-2500®; Artic Mist; Beaverwhite 200; Cimflx 606; Cimpact 600; Cimpact 610; Cimpact 699; Cimpact 705; Cimpact 710; Glacier 200; I.T. 3X; I.T. 5X; I.T. 325; I.T. FT; I.T. X; Jet Fil® 200; Jet Fil® 350; Jet Fil® 500; Jet Fil® 575C; Jet Fil® 625C; Jet Fil® 700C; Luzenac 8170; Magsil Diamond Talc 200 mesh; Magsil Diamond Talc 350 mesh; Magsil Star Talc 200 mesh; Magsil Star Talc 350 mesh; Microbloc®; MicroPflex 1200; Microtalc® MP10-52; Microtalc® MP12-50; Microtalc® MP15-38; Microtalc® MP25-38; Microtalc® MP44-26; Microtuff® 325F; Microtuff® 1000; Microtuff® F; Mistron CB; Mistron PXL®; Mistron Vapor; Mistron Vapor® R; Nicron 325; Nicron 400; Nytal® 100; Nytal® 200; Nytal® 300; Nytal® 400; Polytalc 262; PolyTalc™ 445; Select-A-Sorb Powdered, Compacted; Silverline 400; Silverline 665; Stellar 500; Stellar 510; Super Microtuff F; Supra EF; Supra EF A; Techfil 7599; Ultratalc™ 609; Vantalc® 6H; Vertal 92; Vertal 97; Vertal 200; Vertal 310; Vertal 710; Vertal C-2

Trade names containing: Mistron ZSC; Nipol® 1411; Nipol® 1422; Nipol® 1422X14; Nipol® 1452X8; Nipol® 1453 HM; Nipol® 1472

Talcum. *See* Talc

Tall oil acid

CAS 61790-12-3; EINECS 263-107-3

Synonyms: Acids, tall oil; Fatty acids, tall oil

Definition: Mixture of rosin acids and fatty acids recovered from the hydrolysis of tall oil

Uses: Polymerization emulsifier; mfg. of surfactants, soaps, amines, imidazolines; cosmetics, dyes, leather, coatings, petrol. industry

Regulatory: FDA 21CFR §175.105, 175.320, 176.200, 176.210, 177.2600, 177.2800, 178.3570, 178.3910

Trade names: Acintol® DFA; Acintol® EPG; Acintol® FA-1; Acintol® FA-1 Special; Acintol® FA-2; Acintol® FA-3; Acofor; Westvaco® 1480

Tall oil acid isooctyl ester. *See* Isooctyl tallate

Tall oil aminopropyl dimethylamine

Uses: Emulsifier for natural resins, min. and natural oils; intermediate

Trade names: Sermul EK 330

Tall oil rosin sodium salt. *See* Sodium tallate

Tallow acid

CAS 61790-37-2; 67701-06-8; EINECS 263-129-3

Synonyms: Fatty acids, tallow; Acids, tallow

Definition: Mixture of fatty acids derived from tallow

Uses: Chemical intermediate; in alkyd resins, rubber compounding, water repellents, polishes, soaps, abrasives, cutting oils, candles, crayons, emulsifiers

Regulatory: FDA 21CFR §175.105, 175.320, 176.200, 176.210, 177.2260, 177.2800, 178.3570, 178.3910

Manuf./Distrib.: Norman, Fox; Witco/Oleo-Surf.

Trade names: Industrene® 143

Tallow acid, hydrogenated. *See* Hydrogenated tallow acid

Tallowalkylamine. *See* Tallow amine

N-Tallow alkyl-2,2′-iminobisethanol

CAS 61791-44-4; EINECS 263-177-5

Synonyms: Ethanol, 2,2′-iminobis-, N-tallow alkyl derivs.

Uses: Antistat for polyethylene

Trade names: Armostat® 310

Tallow amide

EINECS 297-355-9

Synonyms: Amides, tallow

Classification: Amide

Formula: RCO—NH_2, RCO- rep. fatty acids derived from tallow

Regulatory: FDA 21CFR §175.105, 175.320, 176.210
Trade names: Kemamide® S-65

Tallow amides, hydrogenated. *See* Hydrogenated tallow amide

Tallow amine
CAS 61790-33-8; EINECS 263-125-1
Synonyms: Amines, tallow alkyl; Tallowalkylamine
Classification: Primary aliphatic amine
Definition: Primary aliphatic amine derived from tallow acid
Formula: RNH_2, R rep. alkyl groups derived from tallow
Uses: Emulsifier, flotation agent, corrosion inhibitor; synthesis intermediate; oil, textiles, leather, rubber, plastics, metal industries
Trade names: Crodamine 1.T; Kemamine® P-974D; Noram S

Tallow amine, hydrogenated. *See* Hydrogenated tallowamine

Tallowaminopropylamine
CAS 68439-73-6; EINECS 270-416-7
Synonyms: Tallow trimethylene diamine
Classification: Aliphatic diamine
Formula: $R—NH—CH_2CH_2CH_2NH_2$, R rep. alkyl groups derived from tallow
Uses: Gasoline detergent, bactericide, corrosion inhibitor in petrol. prod., epoxy hardener
Trade names: Kemamine® D-974

Tallow dimethyl trimethyl propylene diammonium chloride
Uses: Acid corrosion inhibitor, plastics and textile antistat, and emulsifier
Trade names containing: Tomah Q-D-T

Tallow trimethyl ammonium chloride. *See* Tallowtrimonium chloride
Tallow trimethylene diamine. *See* Tallowaminopropylamine

Tallowtrimonium chloride
CAS 8030-78-2; 7491-05-2; 68002-61-9; EINECS 232-447-4
Synonyms: Quaternary ammonium compds., tallow alkyl trimethyl, chlorides; Tallow trimethyl ammonium chloride
Classification: Quaternary ammonium salt
Formula: $[R—N(CH_3)_3]^+Cl^-$, R rep. alkyl groups derived from tallow
Uses: Dispersant, antistat, emulsifier; used in corrosion inhibitor formulations for oilfield brines and HCl acidizing systems; textile softener, dyeing aid; base for hair conditioners, cream rinses; in textiles, paper, cosmetics, agric., plastics, petrol., industrial applics.
Trade names: Arquad® T-27W

Tangantangan oil. *See* Castor oil
TATM. *See* Triallyl trimellitate
6TB24X. *See* 6-t-Butyl-2,4-xylenol
TBA. *See* t-Butyl alcohol
TBBA. *See* Tetrabromobisphenol A
TBBS. *See* Butyl 2-benzothiazole sulfenamide
TBC. *See* Tributyl citrate
TBDPE. *See* Tetrabromodipentaerythritol
TBEP. *See* Tributoxyethyl phosphate
TBGE. *See* t-Butyl glycidyl ether
TBHP. *See* t-Butyl hydroperoxide
TBHQ. *See* t-Butyl hydroquinone
TBNPA. *See* Tribromoneopentyl alcohol
TBP. *See* 2,4,6-Tribromophenol
TBT. *See* Tetrabutyl titanate
TBTO. *See* Tributyltin oxide
p-TBX. *See* Tetrabromo-p-xylene
TCE. *See* Trichloroethane
TCEP. *See* Tris (β-chloroethyl) phosphate
TCP. *See* Calcium phosphate tribasic, Tricresyl phosphate

TCPP. *See* Tris (β-chloropropyl) phosphate
TDCPP. *See* Tri(β,β´-dichloroisopropyl)phosphate
TDI. *See* Toluene diisocyanate
TDP. *See* Triisodecyl phosphite
TDQP. *See* 2,2,4-Trimethyl-1,2-dihydroquinoline polymer
TDSA. *See* Tetradecenyl succinic anhydride
TEA. *See* Triethanolamine

TEA-benzene sulfonate
Uses: Detergent, emulsifier, wetting agent, emulsion polymerization aid
Trade names: Arylan® TE/C

TEA-dodecylbenzenesulfonate
CAS 27323-41-7; 68411-31-4; 29381-93-9; EINECS 248-406-9
Synonyms: Dodecylbenzenesulfonic acid, compd. with 2,2´,2´´-nitrilotris[ethanol] (1:1); Triethanolamine dodecylbenzene sulfonate
Classification: Substituted aromatic compd.
Empirical: $C_{18}H_{30}O_3S \cdot C_6H_{15}NO_3$
Properties: M.w. 475.68
Uses: Detergent, wetting agent, flash foamer; liq. detergents, wool wash compds., cosmetics and shampoos, agric. emulsifiers, industrial cleaners, textile scouring, car wash compds.; pigment dispersant; emulsion polymerization
Trade names: Manro TDBS 60; Maranil CB-22; Puxol CB-22; Rhodacal® DOV; Ufasan TEA

TEA-lauryl sulfate
CAS 139-96-8; 68908-44-1; EINECS 205-388-7
Synonyms: Sulfuric acid, monododecyl ester, compd. with 2,2´,2´´-nitrilotris[ethanol] (1:1); Triethanolammonium lauryl sulfate; Triethanolamine lauryl sulfate
Definition: Triethanolamine salt of lauryl sulfuric acid
Empirical: $C_{12}H_{26}O_4S \cdot C_6H_{15}NO_3$
Formula: $CH_3(CH_2)_{10}CH_2OSO_3H \cdot N(CH_2CH_2OH)_3$
Properties: Liq. or paste; m.w. 415.66
Uses: Emulsifier, detergent, wetting agent, dispersant, foaming agent used for household cleaning prods., cosmetics, emulsion polymerization
Manuf./Distrib.: Chemron; Lonza; Norman, Fox; Sandoz; Stepan
Trade names: Carsonol® TLS; Empicol® TL40/T; Perlankrol® ATL40

TEC. *See* Triethyl citrate
TEDA. *See* Triethylene diamine
TeEDC. *See* Tellurium diethyl dithiocarbamate
TEG. *See* Triethylene glycol
TEGDA. *See* PEG-3 diacetate

Tellurium
CAS 13494-80-9; EINECS 236-813-4
Classification: Nonmetallic element
Empirical: Te
Properties: Silvery-white lustrous solid; sol. in sulfuric acid, nitric acid, KOH, KCN solín.; insol. in water; at.wt. 127.60; dens. 6.24 (30 C); m.p. 450 C; b.p. 990 C
Precaution: Flamm.
Toxicology: Toxic by inhalation; causes nausea, vomiting, CNS depression; TLV 0.1 mg/m^3 of air
Uses: Sec. rubber vulcanizing agent, mfg. of alloys, iron and steel casting; coloring agent for glass and ceramics; thermoelectric devices
Manuf./Distrib.: All Chemie Ltd; Asarco; Atomergic Chemetals; Cabot; Cerac; Johnson Matthey; Noah; R.T. Vanderbilt
Trade names: Telloy

Tellurium diethyl dithiocarbamate
CAS 20941-65-5
Synonyms: TeEDC; Ethyl tellurac
Formula: $[(C_2H_5)_2NC(S)S]_4Te$
Properties: Orange-yel. powd.; sol. in benzene, carbon disulfide; sl. sol. in alcohol, gasoline; insol. in water; m.w. 720.67; dens. 1.44; m.p. 108-118 C

Precaution: Avoid acids and strong oxidizers
Toxicology: Harmful if inhaled; possible skin and eye irritation; LD50 (acute dermal, rabbit) > 2 g/kg, (acute oral, rat) > 5000 mg/kg
Uses: Primary or sec. (with thiazoles) rubber accelerator
Trade names: Akrochem® TDEC; Ethyl Tellurac®; Ethyl Tellurac® Rodform; Perkacit® TDEC
Trade names containing: Poly-Dispersion® TTD-75; Poly-Dispersion® TTD-75P

TEMED. *See* N,N,N′,N′-Tetramethylethylenediamine
TEOS. *See* Tetraethoxysilane
TEP. *See* Triethyl phosphate
TEPA. *See* Tetraethylenepentamine

Terpene phenolic resin
Uses: Tackifier for adhesives
Manuf./Distrib.: Allchem Ind.; Arakawa Chem. USA
Trade names: Nirez® 300; Nirez® 2019; Nirez® V-2040; Nirez® V-2040HM; Nirez® V-2150

Terpene resin
CAS 9003-74-1
Uses: Thermoplastic resin; tackifier in adhesives; modifier resin in rubber compounding, coatings, laminations, waxes
Manuf./Distrib.: Cardolite; Chemcentral; Monomer-Polymer & Dajac
Trade names: Piccolyte® C125; Piccolyte® C135; Zonatac® 105 LITE; Zonatac® 105; Zonatac® 115 LITE; Zonatac® M-105

TETA. *See* Triethylenetetramine
TETD. *See* Tetraethylthiuram disulfide

Tetraammonium EDTA
Uses: Chelating agent in photographic developer baths, replenishment of blix baths, water treatment, boiler cleaning, cosmetics, pharmaceuticals, textiles, soaps, metal cleaning/protection, polymer prod., enhanced oil recovery
Trade names: Versene Tetraammonium EDTA

N,N,N′,N′-Tetrabenzylthiuram disulfide
CAS 10591-85-2
Synonyms: Disulfide, bis(dibenzylthiocarbamoyl)
Uses: Rubber accelerator
Trade names: Benzyl Tuads®; Benzyl Tuads® Solid; Benzyl Tuex®; Perkacit® TBzTD

Tetrabromobisphenol A
CAS 79-94-7; EINECS 201-236-9
Synonyms: TBBA; 4,4′-Isopropylidenebis (2,6-dibromophenol)
Empirical: $C_{15}H_{12}Br_4O_2$
Properties: Wh. powd.; m.w. 543.9; m.p. 180 C
Toxicology: Irritant
Uses: Flame retardant for epoxies, phenolics, ABS, unsaturated polyesters, PC; intermediate for flame retardants
Manuf./Distrib.: Albemarle; Allchem Ind.; Great Lakes
Trade names: FR-1524; Great Lakes BA-59; Great Lakes BA-59P™; Saytex® RB-100

Tetrabromobisphenol A allyl ether
Uses: Flame retardant for expandable polystyrene
Trade names: FR-2124

Tetrabromobisphenol A bis (allyl ether)
CAS 25327-89-3
Synonyms: 2,2′,6,6′-Tetrabromobisphenol A diallyl ether
Formula: $(CH_3)_2C[C_6H_2(Br)_2OCH_2CH=CH_2]_2$
Properties: M.w. 624.03; m.p. 118-120 C
Uses: Flame retardant for plastics
Manuf./Distrib.: Aldrich; Great Lakes
Trade names: Great Lakes BE-51™

Tetrabromobisphenol A, bis (2,3-dibromopropyl ether)
CAS 21850-44-2
Synonyms: Benzene, 1,1´-(1-methylethylidene)bis(3,5-dibromo-4-(2,3-dibromopropoxy)-(9CI)
Empirical: $C_{21}H_{20}Br_8O_2$
Properties: M.w. 943.62
Uses: Flame retardant used in polybutylenes, polyolefins and copolymers
Manuf./Distrib.: Great Lakes; Monomer-Polymer & Dajac
Trade names: Great Lakes PE-68™

Tetrabromobisphenol A, bis (2-hydroxyethyl ether)
Uses: Flame retardant for unsat. polyester and epoxy thermoset resins, laminates for electronic circuit boards
Trade names: Great Lakes BA-50™; Great Lakes BA-50P™

Tetrabromobisphenol A diacrylate
Uses: Fire retardant for automotive coatings, wire and cable coatings

2,2´,6,6´-Tetrabromobisphenol A diallyl ether. *See* Tetrabromobisphenol A bis (allyl ether)

Tetrabromobisphenol A di-2-hydroxyethyl ether
Uses: Flame retardant for PC, thermoset, and thermoplastic polyester, polyurethane

Tetrabromodipentaerythritol
CAS 109678-33-3
Synonyms: TBDPE
Empirical: $C_{10}H_{18}Br_4O_3$
Properties: Powd.; m.w. 506; m.p. 75-82 C
Uses: Flame retardant, processing aid for plastics
Trade names: FR-1034

Tetrabromophthalate diol
CAS 20566-35-2; EINECS 243-885-0
Uses: Flame retardant for rigid urethane foam
Trade names: Great Lakes PHT4-Diol™; Saytex® RB-79

Tetrabromophthalic anhydride
CAS 632-79-1; EINECS 211-185-4
Synonyms: 1,3-Isobenzofurandione, 4,5,6,7-tetrabromo-(9CI); Bromphthal
Empirical: $C_8B4_4O_3$
Formula: $C_6Br_4C_2O_3$
Properties: Pale yel. crystalline solid; m.w. 463.72; m.p. 280 C
Precaution: Moisture-sensitive
Toxicology: Irritant
Uses: Flame retardant for plastics (unsaturated polyesters), paper, textiles; reactive intermediate for the preparation of polyols, esters and imides
Manuf./Distrib.: Aldrich; Albemarle; Great Lakes
Trade names: Great Lakes PHT4™; Saytex® RB-49

2,3,5,6-Tetrabromo-p-xylene. *See* Tetrabromo-p-xylene

Tetrabromo-p-xylene
CAS 23488-38-2
Synonyms: p-TBX; p-Xylene, α,α,α´,α´-tetrabromo; 2,3,5,6-Tetrabromo-p-xylene
Empirical: $C_8H_6Br_4$
Formula: $C_6Br_4(CH_3)_2$
Properties: M.w. 421.77; m.p. 254-256 C
Toxicology: Irritant
Uses: Flame retardant for PS, polyester, textiles
Manuf./Distrib.: Aldrich

Tetrabutylphosphonium acetate, monoacetic acid
CAS 17786-43-5
Formula: $(n\text{-}C_4H_9)_4P(CH_3CO_2)(CH_3CO_2H)$
Uses: Phase transfer catalyst, polymerization catalyst, chemical intermediate
Manuf./Distrib.: Cytec Industries

Tetrabutylphosphonium bromide
CAS 3115-68-2; EINECS 221-487-8
Synonyms: Phosphonium, tetrabutyl-, bromide
Empirical: $C_{16}H_{36}BrP$
Formula: $(n\text{-}C_4H_9)_4PBr$
Properties: M.w. 339.35; m.p. 100-103 C
Precaution: Hygroscopic
Toxicology: Irritant
Uses: Phase transfer catalyst, polymerization catalyst, chemical intermediate
Manuf./Distrib.: Cytec Industries

Tetrabutylphosphonium chloride
CAS 2304-30-5; EINECS 218-964-8
Synonyms: Phosphonium, tetrabutyl-, chloride
Empirical: $C_{16}H_{36}ClP$
Formula: $(n\text{-}C_4H_9)_4PCl$
Properties: M.w. 294.89; m.p. 64-66 C
Toxicology: Highly toxic irritant
Uses: Phase transfer catalyst, polymerization catalyst, chemical intermediate
Manuf./Distrib.: Cytec Industries; R S A

Tetrabutyl thiuram disulfide
CAS 1634-02-2
Synonyms: Bis (dibutylthiocarbamoyl) disulfide; Disulfide, bis(dibutylthiocarbamoyl)
Empirical: $C_{18}H_{36}N_2S_4$
Formula: $[(C_4H_9)_2NCH]_2S_2$
Properties: Amber liq., sl. sweet odor; sol. in carbon disulfide, benzene, chloroform, gasoline; insol. in water; m.w. 408.80; dens. 1.03-1.06 (20/20 C); solidifies @ -30 C
Precaution: Combustible
Uses: Vulcanizing and accelerating agent for rubbers
Trade names: Akrochem® TBUT; Butyl Tuads®

Tetrabutyl titanate
CAS 5593-70-4
Synonyms: TBT; Butyl titanate; Titanium butylate
Empirical: $C_{16}H_{36}O_4 \cdot Ti$
Formula: $Ti(OC_4H_9)_4$
Properties: Colorless to lt. yel. liq.; sol. in most org. solvs. except ketones; dec. in water; m.w. 340.42; dens. 0.996; b.p. 310-314 C; flash pt. 170 F; ref. index 1.4900
Precaution: Combustible; moisture-sensitive
Toxicology: Irritant
Uses: Ester exchanger reactions, heat-resistant paints; improving adhesion of paints, rubber, plastics to metal surfaces
Trade names: Tyzor TBT

Tetrachlorophthalic anhydride
CAS 117-08-8; EINECS 204-171-4
Synonyms: Phthalic anhydride, tetrachloro
Empirical: $C_8Cl_4O_3$
Formula: $C_6Cl_4(CO)_2O$
Properties: Wh. free-flowing powd., odorless; sl. sol. in water; m.w. 285.90; m.p. 254-255 C; b.p. 371 C
Precaution: Moisture-sensitive
Toxicology: Suspected carcinogen
Uses: Flame retardant for plastics; intermediate in dyes, pharmaceuticals, plasticizers, and other org. materials
Manuf./Distrib.: Monsanto
Trade names: Tetrathal

Tetradecabromodiphenoxy benzene
CAS 58965-66-5; EINECS 261-526-6
Uses: Flame retardant for nylon, PET, PBT, styrenics
Trade names: Saytex® 120

Tetradecanoic acid, 1-methylethyl ester. *See* Isopropyl myristate

Tetradecanoic acid, monoester with 1,2-propanediol. *See* Propylene glycol myristate
Tetradecanoic acid, monoester with 1,2,3-propanetriol. *See* Glyceryl myristate
1-Tetradecanol. *See* Myristyl alcohol
1-Tetradecanol, hydrogen sulfate, sodium salt. *See* Sodium myristyl sulfate
1-Tetradecene. *See* Tetradecene-1

Tetradecene-1
CAS 1120-36-1; EINECS 272-493-2
Synonyms: C14 linear alpha olefin; 1-Tetradecene; α-Tetradecylene
Empirical: $C_{14}H_{28}$
Formula: $CH_3(CH_2)_{11}CH:CH_2$
Properties: Colorless liq., mild hydrocarbon odor; sl. sol. in alcohol, ether; insol. in water; m.w. 196.38; sp.gr. 0.775; m.p. -13 C; b.p. 251 C; flash pt. (Seta) 226 F; ref. index 1.4360
Precaution: Combustible
Toxicology: Irritating to eyes and skin; low acute inhalation toxicity; low acute ingestion toxicity but ingestion may cause vomiting, aspiration of vomitus
Uses: Intermediate for surfactants and specialty industrial chemicals (polymers, plasticizers, lubricants, gasoline additives, paper sizing, PVC lubricants)
Manuf./Distrib.: Albemarle; Chevron; Shell; Spectrum Chem. Mfg.
Trade names: Gulftene® 14

Tetradecenyl succinic anhydride
CAS 33806-58-5
Synonyms: TDSA
Empirical: $C_{18}H_{30}O_3$
Properties: M.w. 294.43
Uses: Epoxy curing agent; corrosion inhibitor for lubricating oils; intermediate for prep. of alkyd or unsat. polyester resins, in chem. reactions
Manuf./Distrib.: Dixie Chem
Trade names: Milldride® TDSA

Tetradecyl dimethylamine. *See* Dimethyl myristamine
α-Tetradecylene. *See* Tetradecene-1

Tetra (2,2 diallyloxymethyl) butyl, di (ditridecyl) phosphito zirconate
Uses: Coupling agent
Trade names containing: KZ 55

Tetra (2, diallyoxymethyl-1 butoxy titanium di (di-tridecyl) phosphite
Uses: Coupling agent, adhesion promoter, antioxidant, antistat, antifoam, accelerator, blowing agent activator, catalyst, curative, corrosion inhibitor, dispersion aid, emulsifier, flame retardant, foaming agent, grinding and process aid
Trade names containing: Ken-React® 55 (KR 55)

Tetraethoxysilane
CAS 78-10-4; EINECS 201-083-8
Synonyms: TEOS; Tetraethyl orthosilicate; Ethyl silicate
Empirical: $C_8H_{20}O_4Si$
Formula: $(C_2H_5)_4SiO_4$
Properties: Colorless liq., faint odor; misc. with alcohol; m.w. 208.33; dens. 0.934; m.p. -85 C; b.p. 169 C; flash pt. 46 C; ref. index 1.3820
Precaution: Combustible; moderate fire risk; moisture-sensitive
Toxicology: LD50 (rat, oral) 6270 mg/kg; irritant
Uses: Coupling agent, chem. intermediate, blocking agent, release agent, lubricant, primer, reducing agent
Trade names: Petrarch® T1807

Tetraethylene glycol diacrylate. *See* PEG-4 diacrylate
Tetraethylene glycol di 2-ethylhexanoate. *See* PEG-4 dioctanoate
Tetraethylene glycol di-2-ethylhexoate. *See* PEG-4 di-2-ethylhexoate
Tetraethylene glycol dimethacrylate. *See* PEG-4 dimethacrylate

Tetraethylenepentamine
CAS 112-57-2; EINECS 203-986-2

Synonyms: TEPA; 1,4,7,10,13-Pentaazatridecane
Empirical: $C_8H_{23}N_5$
Formula: $NH_2(CH_2CH_2NH)_3CH_2CH_2NH_2$
Properties: Viscous liq., hygroscopic; sol. in most org. solvs. and water; m.w. 189.31; dens. 0.9980 (20/20 C); m.p. -30 C; b.p. 333 C; flash pt. 325 F; ref. index 1.5055
Precaution: Combustible; corrosive
Toxicology: Strong irritant to skin and eyes
Uses: Solvent for sulfur, acid gases, various resins and dyes; saponifying agent for acidic materials; mfg. of syn. rubber; dispersant in motor oils; intermediate for oil additives; ion-exchange resins; surfactants
Manuf./Distrib.: Allchem Ind.; Ashland; Coyne; Great Western; Tosoh; Union Carbide
Trade names: D.E.H. 26; TEPA

Tetraethyl orthosilicate. *See* Tetraethoxysilane

Tetraethylthiuram disulfide
CAS 97-77-8; EINECS 202-607-8
Synonyms: TTD; TETD; Bis(diethylthiocarbamyl) disulfide; Disulfiram
Empirical: $C_{10}H_{20}N_2S_4$
Formula: $[(C_2H_5)_2NCS]_2S_2$
Properties: Lt. gray powd., sl. odor; sol. in carbon disulfide, benzene, chloroform; insol. in water; m.w. 296.54; dens. 1.27; m.p. 65-70 C
Toxicology: Toxic by ingestion with alcohol; irritant; animal teratogen; TLV 2 mg/m^3 of air
Uses: Fungicide, ultra-accelerator for rubber
Manuf./Distrib.: Abbott Labs; Aldrich; Complex Quimica SA; Novachem; R.T. Vanderbilt
Trade names: Akrochem® TETD; Ekaland TETD; Ethyl Tuads® Rodform; Perkacit® TETD
Trade names containing: Akrochem® TM/ETD

Tetrafluoroethene homopolymer. *See* Polytetrafluoroethylene
Tetrafluoroethylene polymer. *See* Polytetrafluoroethylene

Tetrafluoroethylene telomer
CAS 65530-85-0, 79070-11-4
Uses: Release agent for molds for plastics
Trade names: MS-136N/CO_2
Trade names containing: MS-122N/CO_2

Tetrahydrofuran
CAS 109-99-9; EINECS 203-726-8
Synonyms: THF
Empirical: C_4H_8O
Formula: $CH_2CH_2CH_2CH_2O$
Properties: Water-wh. liq., ethereal odor; sol. in water and org. solvs.; m.w. 72.11; dens. 0.888 (20 C); f.p. -65 C; b.p. 66 C; flash pt. (OC) -15 C; ref. index 1.4070
Precaution: Flamm. limits in air 2-11.8%
Toxicology: Toxic by ingestion and inhalation; TLV 200 ppm in air
Uses: Solvent, Grignard reactions, reductions, and polymerizations; chemical intermediate and monomer
Manuf./Distrib.: Allchem Ind.; Arco Europe; Ashland; BASF; Du Pont; Great Lakes; Hüls UK; ISP; Janssen Chimica; QO; Richman
Trade names: THF

Tetrahydro-2-furancarbinol. *See* Tetrahydrofurfuryl alcohol
Tetrahydro-2-furanmethanol. *See* Tetrahydrofurfuryl alcohol

Tetrahydrofurfuryl alcohol
CAS 97-99-4; EINECS 202-625-6
Synonyms: THFA; Tetrahydro-2-furanmethanol; Tetrahydro-2-furancarbinol; Tetrahydro-2-furylmethanol
Classification: Cyclic alcohol
Empirical: $C_5H_{10}O_2$
Formula: $C_4H_7OCH_2OH$
Properties: Liq., mild odor; hygroscopic; misc. with water, alcohol, ether, acetone, chloroform, benzene; m.w. 102.14; dens. 1.053 (20/4 C); visc. 6.24 cp (20 C); m.p. < -80 C; b.p. 173-177 C; flash pt. 75 C; ref. index 1.453 (20 C); surf. tens. 37 dyn/cm
Precaution: Explosive limit 1.5-9.7% by vol.

Tetrahydrofurfuryl methacrylate

Toxicology: Eye irritant; moderately irritating to skin, mucous membranes
Uses: Solvent for vinyl resins, dyes for leather, chlorinated rubber, cellulose esters, coupling agent, solvent-softener for nylon
Regulatory: FDA 21CFR §172.515, 175.105, 176.210
Manuf./Distrib.: Allchem Ind.; Fluka; Great Western; Penta Mfg.; QO; Schweizerhall
Trade names: QO® Tetrahydrofurfuryl Alcohol (THFA®)

Tetrahydrofurfuryl methacrylate
CAS 2455-24-5
Synonyms: Methacrylic acid, tetrahydrofurfuryl ester
Empirical: $C_9H_{14}O_3$
Properties: M.w. 170.21; dens. 1.044; b.p. 52 C (0.4 mm); flash pt. 90 C; ref. index 1.4580
Uses: Anaerobic adhesives and sealants, printed circuit boards, artificial finger nails, modifier for hard rubber rolls, wire and cable coatings, screen printing inks, emulsion polymerization, plastic modifier, EB-curable coatings
Manuf./Distrib.: CPS; Monomer-Polymer & Dajac; Polysciences; Rohm Tech; Sartomer
Trade names: BM-729
Trade names containing: SR-203

Tetrahydro-2-furylmethanol. *See* Tetrahydrofurfuryl alcohol

Tetrahydronaphthalene
CAS 119-64-2; EINECS 204-340-2
Synonyms: 1,2,3,4-Tetrahydronaphthalene; Tetralin
Empirical: $C_{10}H_{12}$
Formula: $C_{10}H_{12}$
Properties: Colorless liq., pungent odor; misc. with most solvs.; insol. in water; m.w. 132.21; dens. 0.981 (13 C); m.p. -25 C; b.p. 206 C; flash pt. 160 F; ref. index 1.5410
Toxicology: Irritant to eyes and skin; narcotic in high concs.
Uses: Chemical intermediate, solvent for greases, fats, oils, waxes, rubber, asphalt; substitute for turpentine
Manuf./Distrib.: Du Pont; Hüls AG; Spectrum Chem. Mfg.
Trade names: Tetralin

1,2,3,4-Tetrahydronaphthalene. *See* Tetrahydronaphthalene
Tetrahydro-1,4-oxazine. *See* Morpholine
Tetrahydro-2H-1,4-oxazine. *See* Morpholine
Tetrahydrothiophene 1,1-dioxide. *See* Sulfolane
2,2′,4,4′-Tetrahydroxy benzophenone. *See* Benzophenone-2

Tetrahydroxypropyl ethylenediamine
CAS 102-60-3; EINECS 203-041-4
Synonyms: N,N,N′,N′-Tetrakis(2-hydroxypropyl) ethylenediamine; Ethylenedinitrilotetra-2-propanol
Classification: Substituted amine
Empirical: $C_{14}H_{32}N_2O_4$
Formula: $(HOC_3H_6)_2NCH_2CH_2N(C_3H_6OH)$
Properties: Water-wh. visc. liq.; sol. in ethanol, toluene, ethylene glycol; misc. with water; m.w. 292.42; dens. 1.013; b.p. 175-181 C (0.8 mm); flash pt. > 230 F; ref. index 1.4812
Precaution: Combustible
Uses: Chelating agent, intermediate, emulsifier, antistat, rewetting agent, grease additive, textile lubricant; for pharmaceuticals, herbicides, fungicides, insecticides, adhesives, resins, plasticizers, inks, cosmetics, resins
Trade names: Quadrol®

Tetraisobutylthiuram disulfide
CAS 3064-73-1
Synonyms: Thioperoxydicarbonic diamide, tetrakis (2-methylpropyl)-
Uses: Accelerator for rubber, vulcanizing agent
Trade names: Isobutyl Tuads®

Tetraisopropyl di (dioctylphosphito) titanate
Uses: Coupling agent, adhesion promoter, antioxidant, antistat, antifoam, accelerator, blowing agent activator, catalyst, curative, corrosion inhibitor, dispersion aid, emulsifier, flame retardant, foaming agent, grinding and process aid

Trade names containing: Ken-React® 41B (KR 41B)

Tetraisopropyl titanate
CAS 546-68-9
Synonyms: TPT; Titanium isopropylate; Isopropyl titanate; Isopropyl alcohol, titanium (4+) salt
Empirical: $C_{12}H_{28}O_4Ti$
Formula: $Ti[OCH(CH_3)_2]_4$
Properties: Lt. yel. liq.; sol. in most org. solvs.; dec. rapidly in water; m.w. 72.07; dens. 0.954; m.p. 14.8 C; b.p. 102-104 C (10 mm); ref. index 1.46
Uses: Catalyst for esterification and olefin polymerization; adhesion of paints, rubber, and plastics to metals, condensation catalyst
Trade names: Tyzor TPT

Tetrakis (2-chloroethyl) ethylene diphosphate
Uses: Flame retardant additive for flexible PU foams for the transportation, bedding, and furniture industries, and in thermoplastic and thermoset resins such as acrylates, polyolefins, PAN, styrene, ABS, polyesters, epoxies, and PET
Trade names: Thermolin® 101

Tetrakis (2,4-di-t-butylphenyl) 4,4-biphenylenediphosphonite
CAS 38613-77-3
Synonyms: Phosphonous acid, (1,1′-biphenyl)-4,4′-diylbis-, tetrakis (2,4-bis(1,1-dimethylethyl)phenyl) ester
Empirical: $C_{68}H_{92}O_4P_2$
Properties: M.w. 1035.42
Uses: Processing stabilizer, sec. antioxidant for polymers incl. polyolefins, ABS, PS, polybutylene, PBT, nitrile barrier resins; peroxide decomposer for plastics mfg.

Tetrakis (2-ethoxyethoxy) silane
CAS 18407-94-8
Empirical: $C_{16}H_{36}O_8Si$
Properties: M.w. 384.54; dens. 1.02 (20/4 C); b.p. 200 C (0.1 mm); flash pt. 131 C; visc. 5 ctks
Uses: Coupling agent, chem. intermediate, blocking agent, release agent, lubricant, primer, reducing agent
Trade names: Petrarch® T1918

Tetrakis (2-ethylhexyl) titanate
Uses: Catalyst for esterification and olefin polymerization; used in coatings; resin crosslinking agent for automotive prods., coatings, elastomers, films/paints, graphic arts, plastics
Trade names: Tyzor TOT

Tetrakis (hydroxymethyl) phosphonium chloride
CAS 124-64-1; EINECS 204-707-7
Synonyms: THPC
Formula: $(HOCH_2)_4PCl$
Properties: Cryst. compd.; m.w. 190.57; dens. 1.341; ref. index 1.5120
Precaution: Corrosive
Toxicology: Possible carcinogen
Uses: Phase transfer catalyst, polymerization catalyst, chemical intermediate
Manuf./Distrib.: Albright & Wilson Am.; Cytec Industries

N,N,N′,N′-Tetrakis(2-hydroxypropyl) ethylenediamine. *See* Tetrahydroxypropyl ethylenediamine

Tetrakis isodecyl 4,4′-isopropylidene diphosphite
CAS 61670-79-9
Synonyms: Alkyl C10 bisphenol A phosphite
Uses: Stabilizer for plastics
Trade names: Doverphos® 675

Tetrakis [methylene (3,5-di-t-butyl-4-hydroxyhydrocinnamate)] methane
CAS 6683-19-8; EINECS 229-722-6
Synonyms: Tetrakis [methylene-3(3′,5′-di-t-butyl-4-hydroxyphenyl) propionate] methane; Pentaerythritol tetrakis (3,5-di-t-butyl-4-hydroxyhydrocinnamate)

Tetrakis [methylene-3(3′,5′-di-t-butyl-4-hydroxyphenyl) propionate] methane

Classification: Hindered phenolic
Empirical: $C_{73}H_{108}O_{12}$
Properties: Wh. powd.; sol. in chloroform, benzene, acetone, ethyl acetate; insol. in water; m.w. 1178; dens. 1.05 (20 C); m.p. 110-125 C; flash pt. 299 C
Uses: Antioxidant for polyolefins, styrenics, elastomers, adhesives, lubricants, oils
Trade names: Akrochem® Antioxidant 1010; BNX® 1010, 1010G; Dovernox 10; Irganox® 1010; Naugard® 10; Ralox® 630; Ultranox® 210
Trade names containing: Aquastab PA 42M; Irganox® B 215; Irganox® B 225; Irganox® B 561; Ultranox® 815A Blend; Ultranox® 817A Blend; Ultranox® 2714A

Tetrakis [methylene-3(3′,5′-di-t-butyl-4-hydroxyphenyl) propionate] methane. *See* Tetrakis [methylene (3,5-di-t-butyl-4-hydroxyhydrocinnamate)] methane

Tetrakis (2,2,6,6-tetramethyl-4-piperidyl)-1,2,3,4-butane tetracarboxylate
Uses: Light and heat stabilizer for polyolefins, PS, ABS, PVC, PU, and engineering plastics

Tetralin. *See* Tetrahydronaphthalene

Tetramethoxysilane
CAS 681-84-5; EINECS 211-656-4
Synonyms: Tetramethyl orthosilicate
Empirical: $C_4H_{12}O_4Si$
Properties: Liq.; m.w. 152.22; dens. 1.032 (20/4 C); b.p. 121-122 C; m.p. 4-5 C; flash pt. 20 C; ref. index 1.368
Precaution: Flamm.
Toxicology: Can cause blindness; LD50 (rat, oral) 700 mg/kg
Uses: Coupling agent, chem. intermediate, blocking agent, release agent, lubricant, primer, reducing agent, reagent for ketal synthesis, deposition frosting of glass
Trade names: Petrarch® T1980

2-[p-(1,1,3,3-Tetramethylbutyl)phenoxy]ethanol. *See* Octoxynol-1
2-[2-[2-[p-(1,1,3,3-Tetramethylbutyl)phenoxy]ethoxy]ethoxy]ethanol. *See* Octoxynol-3
2,4,7,9-Tetramethyl-5-decyn-4,7-diol. *See* Tetramethyl decynediol

Tetramethyl decynediol
CAS 126-86-3; EINECS 204-809-1
Synonyms: 2,4,7,9-Tetramethyl-5-decyn-4,7-diol; Acetylenicglycol
Classification: Unsaturated alcohol
Empirical: $C_{14}H_{26}O_2$
Formula: $(CH_3)_2CHCH_2CCH_3OHCCCOHCH_3CH_2CH(CH_3)_2$
Properties: M.w. 226.36; m.p. 42-44 C; b.p. 255 C; flash pt. > 230 F
Toxicology: Irritant
Uses: Defoamer and dye dispersant in paint and ink formulations, dyestuffs; surfactant in rinse aids; substrate pigment wetting agent for industrial coatings and adhesives; paper coatings
Regulatory: FDA 21CFR §175.105, 175.300
Trade names containing: Surfynol® 104A; Surfynol® 104E; Surfynol® 104PA; Surfynol® DF-110D; Surfynol® DF-110L

Tetramethylene glycol. *See* 1,4-Butanediol
Tetramethylene sulfone. *See* Sulfolane

N,N,N′,N′-Tetramethylethylenediamine
CAS 110-18-9; EINECS 203-744-6
Synonyms: TEMED; TMEDA; 1,2-Bis (dimethylamono) ethane
Empirical: $C_6H_{16}N_2$
Formula: $(CH_3)_2NCH_2CH_2N(CH_3)_2$
Properties: Liq., sl. ammoniacal odor; sol. in water, most org. solvs.; m.w. 116.21; dens. 0.770; m.p. -55 C; b.p. 120-122 C; flash pt. 20 C; ref. index 1.4179
Precaution: Flamm.; corrosive
Uses: Polymerization initiator; prep. of polyacrylamide gels
Manuf./Distrib.: BASF; Janssen Chimica; Spectrum Chem. Mfg.
Trade names: Toyocat®-TE

N,N,N′,N′-Tetramethylhexanediamine
CAS 111-18-2
Synonyms: N,N,N′,N′-Tetramethyl-1,6-hexanediamine; 1,6-Hexanediamine, N,N,N′,N′-tetramethyl
Empirical: $C_{10}H_{24}N_2$
Formula: $(CH_3)_2N(CH_2)_6N(CH_3)_2$
Properties: M.w. 172.36; dens. 0.806; b.p. 209-210 C; flash pt. 73 C; ref. index 1.4359
Precaution: Corrosive
Uses: Catalyst for PU slabstock and molded applics.
Trade names: Toyocat®-MR

N,N,N′,N′-Tetramethyl-1,6-hexanediamine. *See* N,N,N′,N′-Tetramethylhexanediamine
Tetramethylolmethane. *See* Pentaerythritol
Tetramethyl orthosilicate. *See* Tetramethoxysilane

2,2,6,6-Tetramethylpiperidine-4-carbonamide
Uses: Lt. stabilizer for polyolefins, nylon, EVA, HIPS, PS, PA, ABS, PU, PC, and cellulosics
Trade names: Hostavin® N 20

2,2,6,6-Tetramethylpiperidin-4-yl acrylate/methyl methacrylate copolymer
CAS 115340-81-3
Classification: HALS
Uses: Lt. stabilizer for plastics
Trade names: UV-Chek® AM-806

[2,2,6,6-Tetramethyl-4-piperidyl/β,β,β′,β′-tetramethyl-3,9-(2,4,8,10-tetraoxaspiro (5,5) undecane) diethyl]-1,2,3,4-butane tetracarboxylate
Uses: Light and heat stabilizer for PP, polyethylene, PS, ABS, engineering plastics, and elastomers, monofilament, tapes, molded and extruded prods., polyethylene blown film

Tetramethylthiuram disulfide
CAS 137-26-8; EINECS 205-286-2
Synonyms: TMTD; Thiram; Thiuram disulfide; Bis(dimethylthiocarbamyl) disulfide
Empirical: $C_6H_{12}N_2S_4$
Formula: $[(CH_3)_2NCH]_2S_2$
Properties: White crystalline powd. with odor; sol. in alcohol, benzene, chloroform, carbon disulfide; insol. in water, dilute alkali, gasoline; m.w. 240.44; dens. 1.29 (20 C); m.p. 155-156 C
Toxicology: Toxic by ingestion and inhalation; irritant to skin and eyes; TLV 5 mg/m^3 of air; LD50 (rat, oral) 640 mg/kg
Uses: Vulcanizing agent, accelerator for rubber; fungicide, insecticide, seed disinfectant; animal repellent; antiseptic
Manuf./Distrib.: Complex Quimica SA; Novachem; R.T. Vanderbilt
Trade names: Akrochem® TMTD; Akrochem® TMTD Pellet; Akrosperse® D-177; Methyl Tuads®; Methyl Tuads® Rodform; Naftocit® Thiuram 16; Naftopast® Thiuram 16-P; Perkacit® TMTD; Thiurad®
Trade names containing: Akrochem® TM/ETD; Poly-Dispersion® V (MT)D-75; Poly-Dispersion® V (MT)D-75P; Rhenovin® TMTD-70

Tetramethylthiuram monosulfide
CAS 97-74-5
Synonyms: TMTM; Bis(dimethylthiocarbamoyl)sulfide
Empirical: $C_6H_{14}N_2S$
Formula: $[(CH_3)_2NCH]_2S$
Properties: Yel. powd.; sol. in acetone, benzene, ethylene dichloride; insol. in water, gasoline; dens. 1.40; m.p. 104-107 C; flash pt. 156 C
Precaution: Combustible; avoid strong acids and oxidizers
Toxicology: Harmful if swallowed; may cause skin sensitization on repeated exposure; LD50 (rat, oral) 1250-1390 mg/kg
Uses: Ultra-accelerator for rubber; fungicide, insecticide
Manuf./Distrib.: Aceto; R.T. Vanderbilt
Trade names: Akrochem® TMTM; Ekaland TMTM; Monothiurad®; Perkacit® TMTM; Unads® Pellets; Unads® Powd.
Trade names containing: Poly-Dispersion® E (MX)D-75; Rhenovin® TMTM-70

Tetraoctyloxytitanium di (ditridecylphosphite)
Uses: Coupling agent, adhesion promoter, antioxidant, antistat, antifoam, accelerator, blowing agent activator, catalyst, curative, corrosion inhibitor, dispersion aid, emulsifier, flame retardant, foaming agent, grinding and process aid
Trade names containing: Ken-React® 46B (KR 46B)

Tetraoctylphosphonium bromide
CAS 23906-97-0
Formula: $(n\text{-}C_8H_{17})_4PBr$
Uses: Phase transfer catalyst, polymerization catalyst, chemical intermediate
Manuf./Distrib.: Cytec Industries

3,6,9,12-Tetraoxatetracosan-1-ol. *See* Laureth-4

Tetraphenyl dipropyleneglycol diphosphite
CAS 80584-85-6; EINECS 279-498-9
Classification: Organic phosphite
Empirical: $C_{30}H_{32}O_7P_2$
Formula: $(C_6H_5O)_2P—OC_3H_6OC_3H_6O—P(OC_6H_5)_2$
Properties: Liq.; m.w. 566; dens. 1.164-1.188; b.p. will not distill; ref. index 1.5570-1.5620; flash pt. (PMCC) > 177 C
Uses: Sec. stabilizer for many polymers incl. PVC
Trade names: Doverphos® 11; Weston® THOP

Tetrapotassium diphosphate. *See* Tetrapotassium pyrophosphate

Tetrapotassium pyrophosphate
CAS 7320-34-5; EINECS 230-785-7
Synonyms: TKPP; Diphosphoric acid tetrapotassium salt; Tetrapotassium diphosphate; Potassium pyrophosphate; Potassium pyrophosphate, normal
Empirical: $H_4O_7P_2 \cdot 4K$
Formula: $K_4P_2O_7 \cdot 3HOH$
Properties: Colorless crystals or white powd.; hygroscopic; sol. in water; insol. in alcohol; m.w. 330.4; dens. 2.33; m.p. 1090 C
Toxicology: Nuisance dust
Uses: Soap and detergent builder, sequestering agent, peptizing and dispersing agent; pigment dispersant and stabilizer in emulsion paints; clarifying agent in liq. soaps; mfg. of syn. rubber; boiler water treatment
Regulatory: FDA 21CFR §173.315; USDA 9CFR §318.7, 381.147; Japan approved; Europe listed; UK approved
Manuf./Distrib.: Albright & Wilson; Elf Atochem N. Am.; FMC; Monsanto; Telechem Int'l.
Trade names: Empiphos 4KP

Tetra-n-propoxysilane
CAS 682-01-9; EINECS 211-659-0
Synonyms: Tetrapropyl orthosilicate
Empirical: $C_{12}H_{28}O_4Si$
Formula: $Si(OCH_2CH_2CH_3)_4$
Properties: M.w. 264.44; dens. 0.916 (20/4 C); b.p. 224-225 C; flash pt. 95 C; ref. index 1.4012 (20 C); visc. 1.66 ctsk; surf. tension 23.6 dynes/cm
Precaution: Moisture-sensitive
Toxicology: Irritant
Uses: Coupling agent, chemical intermediate, blocking agent, release agent, lubricant, primer, reducing agent
Trade names: Petrarch® T2090

Tetrapropylene. *See* Dodecene-1
Tetrapropyl orthosilicate. *See* Tetra-n-propoxysilane
Tetrasodium dicarboxyethyl octadecyl sulfosuccinamate. *See* Tetrasodium dicarboxyethyl stearyl sulfosuccinamate

Tetrasodium dicarboxyethyl stearyl sulfosuccinamate
CAS 3401-73-8; 37767-39-8; 38916-42-6; EINECS 222-273-7
Synonyms: Tetrasodium dicarboxyethyl octadecyl sulfosuccinamate
Classification: Organic compd.

Empirical: $C_{26}H_{47}NO_{10}S•4Na$
Properties: M.w. 657.68
Uses: Emulsifier, dispersant, solubilizer, surfactant for textile, cosmetics, agric. applics.; flotation reagent; emulsion polymerization; polishing waxes; surf. tension depressant for inks; demulsifier
Regulatory: FDA 21CFR §175.105, 176.170, 176.180, 178.3400
Trade names: Aerosol® 22; Fizul M-440; Lankropol® ATE; Monawet SNO-35; Rewopol® B 2003

Tetrasodium edetate. *See* Tetrasodium EDTA

Tetrasodium EDTA
CAS 64-02-8; EINECS 200-573-9
Synonyms: EDTA Na_4; Edetate sodium; Tetrasodium edetate; Ethylene diamine tetraacetic acid, sodium salt
Classification: Substituted amine
Empirical: $C_{10}H_{12}N_2O_8 • 4Na$
Formula: $(NaOOCCH_2)_2NCH_2CH_2N(CH_2COONa)_2$
Properties: White powd.; freely sol. in water; m.w. 380.20; dens. 6.9 lb/gal; m.p. > 300 C
Toxicology: LD50 (IP, mouse) 330 mg/kg; poison by IP route; skin and eye irritant; heated to decomp., emits toxic fumes of NO_x and Na_2O
Uses: Chelating agent for water softening, enhanced oil recovery, soaps, cleaning, cosmetics, photography, textile, paper,leather, metal treatment, rubber, plastics; catalyst in SBR mfg.
Regulatory: FDA 21CFR §173.310, 173.315, 175.105, 175.125, 175.300, 176.150, 176.170, 176.210, 177.2800, 178.3120, 178.3910; USDA 9CFR §381.147
Manuf./Distrib.: Akzo Nobel; Chemplex; Complex Quimica SA; GFS; Great Western; Hampshire; Rhone-Poulenc Basic
Trade names: Hamp-Ene® 100; Sequestrene® 30A; Versene 100; Versene 100 EP; Versene 100 LS; Versene 100 SRG; Versene 100 XL; Versene 220

Tetrone A. *See* Dipentamethylene thiuram tetrasulfide
TFE. *See* Polytetrafluoroethylene
THEIC. *See* Trishydroxyethyl triazine trione
Thermal black. *See* Carbon black
THF. *See* Tetrahydrofuran
THFA. *See* Tetrahydrofurfuryl alcohol

4,4′-Thiobis-6-(t-butyl-m-cresol)
CAS 96-69-5
Synonyms: 4,4′-Thiobis (2-t-butyl-5-methylphenol)
Empirical: $C_{22}H_{30}O_2S$
Formula: $[(CH_3)_2CC_6H_2(CH_3)OH]_2S$
Properties: White powd.; m.w. 359; dens. 1.10; m.p. 163-165 C; flash pt. (COC) 205 C
Toxicology: Toxic by inhalation; irritant; TLV 10 mg/m^3 of air
Uses: Antioxidant for latex, adhesives, rubber, plastics, crosslinked polymers
Manuf./Distrib.: Great Lakes; James River
Trade names: Lowinox® 44S36; Santonox®; Santowhite® Crystals

4,4′-Thiobis (2-t-butyl-5-methylphenol). *See* 4,4′-Thiobis-6-(t-butyl-m-cresol)
Thiobis(dodecyl propionate). *See* Dilauryl thiodipropionate

2,2′-Thiobis (4-t-octylphenolato)-n-butylamine nickel
Properties: Powd.; m.p. 258-281 C
Uses: UV stabilizer for polyolefins, ABS
Trade names: Rhodialux Q84; Uvasorb NI

2,2′-Thiobis (4-t-octylphenolato)-n-butylamine nickel II
CAS 14516-71-3
Empirical: $C_{32}H_{51}O_2NNiS$
Properties: Lt. grn. powd.; m.w. 571
Toxicology: LD50 (rat, oral) > 10 g/kg; LD50 (rabbit, dermal) > 10 g/kg
Uses: Light stabilizer for polyolefins; energy quencher, free radical scavenger
Trade names: Cyasorb® UV 1084

3,3′-Thiobispropanoic acid, dioctadecyl ester. *See* Distearyl thiodipropionate
3,3′-Thiobispropanoic acid, ditridecyl ester. *See* Ditridecyl thiodipropionate

Thiobis-2,2´-(4(1,1,3,3-tetramethylbutyl)phenol)-nickel-2-ethylhexanoate
Uses: Light stabilizer for polymers, thermoplastic elastomers
Trade names: UV-Chek® AM-205

Thiocarbamyl sulfenamide
Synonyms: N-Oxydiethylenethiocarbamyl-N-oxydiethylene sulfenamide
Uses: Accelerator for EPDM, SBR, nitrile, natural and butyl rubbers
Trade names: Cure-Rite® 18

Thiocarbanilide. *See* N,N´-Diphenylthiourea

3-Thiocyanatopropyltriethoxysilane
CAS 34708-08-2; EINECS 252-161-3
Synonyms: 3-(Triethoxysilyl) propyl thiocyanate
Empirical: $C_{10}H_{21}NO_3SSi$
Formula: $(C_2H_5O)_3Si(CH_2)_3SCN$
Properties: M.w. 263.43; dens. 1.031; b.p. 265-270 C; flash pt. 125 C; ref. index 1.447
Precaution: Moisture-sensitive
Uses: Reinforcing agent improving props. of fillers in rubber compds.
Trade names: Si 264
Trade names containing: Coupsil VP 6411; Coupsil VP 8415

Thiodiethyl bis-(3,5-di-t-butyl-4-hydroxyphenyl)propionate. *See* Thiodiethylene bis (3,5-di-t-butyl-4-hydroxy) hydrocinnamate

Thiodiethylene bis (3,5-di-t-butyl-4-hydroxy) hydrocinnamate
CAS 41484-35-9
Synonyms: Benzenepropanoic acid, 3,5-bis(1,1-dimethylethyl)-4-hydroxy-, thiodi-2,1-, ethanediyl ester; Thiodiethyl bis-(3,5-di-t-butyl-4-hydroxyphenyl)propionate
Uses: Antioxidant, stabilizer for polyolefins
Trade names: Irganox® 1035

Thiodiphenylamine. *See* Phenothiazine

Thiodipropionate polyester
CAS 63123-11-5
Formula: $C_{18}H_{38}O(C_{13}H_{21}O_4S)_n$
Properties: Paste; m.w. 2000 avg.; sp.gr. 1.09 kg/l (90 C); flash pt. 249 C
Uses: Polymer additive; antioxidant for PP and other polyolefins
Trade names: Eastman® Inhibitor Poly TDP 2000

Thiodipropionic acid dilauryl ester. *See* Dilauryl thiodipropionate
Thiodipropionic acid, distearyl ester. *See* Distearyl thiodipropionate

Thioglycerin
CAS 96-27-5; EINECS 202-495-0
Synonyms: 3-Mercapto-1,2-propanediol; Monothioglycerol; Thioglycerol; 1-Thioglycerol
Classification: Polyhydric alcohol
Empirical: $C_3H_8O_2S$
Formula: $HOCH_2CHOHCH_2SH$
Properties: Yellowish visc. liq., sl. sulfidic odor; sl. sol. in water; misc. with alcohol; insol. in ether; m.w. 108.16; dens. 1.295; b.p. 118 C (5 mm); flash pt. > 230 F; ref. index 1.5260
Toxicology: Irritant
Uses: Stabilizer for acrylonitrile polymers; crosslinking agent for coatings; accelerator for epoxy-amine reactions; reducing agent; in wound healing, hair waving, textiles, furs, pharmaceuticals, surfactants, foam stabilizing additives
Manuf./Distrib.: Evans Chemetics; SAF Bulk; Schweizerhall
Trade names: Thiovanol®

Thioglycerol. *See* Thioglycerin
1-Thioglycerol. *See* Thioglycerin
Thioperoxydicarbonic diamide, tetrakis (2-methylpropyl)-. *See* Tetraisobutylthiuram disulfide
Thiophan sulfone. *See* Sulfolane

2,2′-(2,5-Thiophenediyl)bis[5-t-butylbenzoxazole]
CAS 7128-64-5
Synonyms: 2,5-Bis (5-t-butyl-2-benzoxazolyl) thiophene
Classification: Bis (benzoxazolyl) deriv.
Properties: Yel. powd.; m.w. 431; m.p. 199-201 C
Uses: Fluorescent whitening agent for PVC
Trade names: Uvitex® OB

Thiram. *See* Tetramethylthiuram disulfide
Thiuram disulfide. *See* Tetramethylthiuram disulfide
THPC. *See* Tetrakis (hydroxymethyl) phosphonium chloride

Tin
CAS 7440-31-5; EINECS 231-141-8
Synonyms: Stannum
Classification: Element
Empirical: Sn
Properties: Silver-white ductile solid; sol. in acids, hot KOH sol'ns.; insol. in water; at.wt. 118.69; dens. 7.29 (20 C); m.p. 232 C; b.p. 2260 C
Toxicology: TLV 2 mg/m^3 of air (inorg. compds.), TLV 0.1 mg/m^3 of air (org. compds.)
Uses: Tin plate, anodes, corrosion-resistant coatings, mfg. of chemicals, in stabilizers for plastics
Manuf./Distrib.: Aldrich; Atomergic Chemetals; Cerac; Noah

Tin, butyl 2-ethylhexanoate laurate oxo complexes
Uses: Heat and light stabilizer for vinyl systems, incl. rigid and plasticized PVC
Trade names: Irgastab® T 634

Tin-calcium-zinc complex
Uses: PVC stabilizer
Trade names: Synpron 239

Tin (II) chloride (1:2). *See* Stannous chloride anhyd.
Tin (II) chloride anhydrous. *See* Stannous chloride anhyd.
Tin crystals. *See* Stannous chloride anhyd.
Tin dichloride. *See* Stannous chloride anhyd.
Tin dioxide. *See* Stannic oxide

Tin mercaptide
Uses: Stabilizer for PVC window profiles
Trade names: Stanclere® T-184

Tin octoate. *See* Stannous octoate
Tin-(II)-octoate. *See* Stannous octoate
Tin peroxide. *See* Stannic oxide
Tin protochloride. *See* Stannous chloride anhyd.
Tin salt. *See* Stannous chloride anhyd.
Tin stearate. *See* Stannous stearate
TIOTM. *See* Triisooctyl trimellitate
TIPA. *See* Triisopropanolamine
Titanic acid anhydride. *See* Titanium dioxide
Titanic anhydride. *See* Titanium dioxide
Titanic earth. *See* Titanium dioxide

Titanium acetylacetonate
CAS 14024-64-7
Synonyms: Titanylacetylacetonate
Empirical: $C_{10}H_{14}O_5Ti$
Formula: $TiO[OC(CH_3):CHCOCH_3]_2$
Properties: Cryst. powd.; sl. sol. in water; m.w. 262.14
Uses: Crosslinking agent for cellulosic lacquers
Manuf./Distrib.: Amspec; Hüls AG
Trade names: Ken Kem® AA

Titanium butylate. *See* Tetrabutyl titanate

Titanium di (butyl, octyl pyrophosphate) di (dioctyl, hydrogen phosphite) oxyacetate
Uses: Coupling agent, adhesion promoter, antioxidant, antistat, antifoam, accelerator, blowing agent activator, catalyst, curative, corrosion inhibitor, dispersion aid, emulsifier, flame retardant, foaming agent, grinding and process aid
Trade names containing: Ken-React® 158FS (KR 158FS)

Titanium di (cumylphenylate) oxyacetate
Uses: Coupling agent, adhesion promoter, antioxidant, antistat, antifoam, accelerator, blowing agent activator, catalyst, curative, corrosion inhibitor, dispersion aid, emulsifier, flame retardant, foaming agent, grinding and process aid
Trade names containing: Ken-React® 134S (KR 134S)

Titanium di (dioctylpyrophosphate) oxyacetate
Uses: Coupling agent, adhesion promoter, antioxidant, antistat, antifoam, accelerator, blowing agent activator, catalyst, curative, corrosion inhibitor, dispersion aid, emulsifier, flame retardant, foaming agent, grinding and process aid
Trade names containing: Ken-React® 138S (KR 138S)

Titanium dimethacrylate oxyacetate
Uses: Coupling agent, adhesion promoter, antioxidant, antistat, antifoam, accelerator, blowing agent activator, catalyst, curative, corrosion inhibitor, dispersion aid, emulsifier, flame retardant, foaming agent, grinding and process aid
Trade names containing: Ken-React® 133DS (KR 133DS)

Titanium dioxide
CAS 1317-80-2; 13463-67-7; EINECS 236-675-5
Synonyms: Titanic anhydride; Titanic earth; Titanic acid anhydride; Titanium oxide; Pigment White 6; CI 77891
Classification: Inorganic oxide
Empirical: TiO_2
Properties: White amorphous powd.; insol. in water, HCl, HNO_3, dil. H_2SO_4; sol. in HF, hot conc. H_2SO_4; m.w. 79.90; Anatase: dens. 3.90; Rutile: dens. 4.23
Precaution: Violent or incandescent reaction with metals (e.g., aluminum, calcium, magnesium, potassium, sodium, zinc, lithium)
Toxicology: TLV:TWA 10 mg/m^3 of total dust; experimental carcinogen, neoplastigen, tumorigen; human skin irritant; nuisance dust
Uses: White pigment, opacifying agent in paints, paper, rubber, pharmaceuticals (tableted drugs), cosmetics (lipsticks, nail enamels, face powds., eye makeup, rouge); radioactive decontamination of skin
Regulatory: FDA 21CFR §73.575, 73.1575, 73.2575, 175.105, 175.210, 175.300, 175.380, 175.390, 176.170, 177.1200, 177.1210, 177.1350, 177.1400, 177.1460, 177.1650, 177.2260, 177.2600, 177.2800, 181.22, 181.30; USDA 9CFR §318.7, 381.147; Japan restricted as colorant; Europe listed; UK approved; prohibited in Germany
Manuf./Distrib.: Bayer NV; British Traders & Shippers; Degussa; Du Pont; Ferro/Transelco; Kerr-McGee; Miles; SCM
Trade names: AT-1; Hitox®; Kronos® 1000; Kronos® 2020; Kronos® 2073; Kronos® 2081; Kronos® 2101; Kronos® 2160; Kronos® 2200; Kronos® 2210; Kronos® 2220; Kronos® 2230; Kronos® 2310; Reed PWC-1; RL-90; Tiona® RCL-4; Tiona® RCL-6; Tiona® RCL-9; Tiona® RCL-69; Ti-Pure® R-100; Ti-Pure® R-101; Ti-Pure® R-102; Ti-Pure® R-103; Ti-Pure® R-900; Ti-Pure® R-902; Ti-Pure® R-960; Titanium Dioxide P25
Trade names containing: Ampacet 11040; Ampacet 11058; Ampacet 11070; Ampacet 11078; Ampacet 11171; Ampacet 11171-101; Ampacet 11187; Ampacet 11200; Ampacet 11215; Ampacet 11246; Ampacet 11247; Ampacet 11299; Ampacet 11338; Ampacet 11343; Ampacet 11416; Ampacet 11418-F; Ampacet 11495-S; Ampacet 11560; Ampacet 11572; Ampacet 11578-P; Ampacet 11737; Ampacet 11739; Ampacet 11744; Ampacet 11748; Ampacet 11851; Ampacet 11853; Ampacet 11875; Ampacet 11912; Ampacet 11913; Ampacet 11919-S; Ampacet 11979; Ampacet 41312; Ampacet 41424; Ampacet 41438; Ampacet 41483; Ampacet 41495; Ampacet 110017; Ampacet 110025; Ampacet 110041; Ampacet 110052; Ampacet 110070; MDI HD-650; MDI MB-650; MDI PE-650F; MDI PE-650HF; MDI PE-650LL; MDI PE-670T; MDI PE-675; MDI PS-650; Mearlin® MagnaPearl 1000; Mearlin® MagnaPearl 1100; Mearlin® MagnaPearl 1110; Mearlin® MagnaPearl 2000; Mearlin® MagnaPearl 2110; Mearlin® MagnaPearl 3000; Mearlin® MagnaPearl 3100; Mearlin® MagnaPearl 4000; Mearlin® MagnaPearl 5000; PE-UV 119; Plaswite® LL 7014; Plaswite® LL 7015; Plaswite® LL 7041; Plaswite® LL 7049; Plaswite® LL 7057; Plaswite® LL 7105; Plaswite® LL 7108; Plaswite® LL 7112; Plaswite® PE 7000; Plaswite® PE 7001;

Plaswite® PE 7002; Plaswite® PE 7003; Plaswite® PE 7004; Plaswite® PE 7005; Plaswite® PE 7006; Plaswite® PE 7024; Plaswite® PE 7031; Plaswite® PE 7079; Plaswite® PE 7097; Plaswite® PE 7192; Plaswite® PP 7161; Plaswite® PP 7269; Plaswite® PS 7028; Plaswite® PS 7037; Plaswite® PS 7131; Plaswite® PS 7174; Plaswite® PS 7179; Plaswite® PS 7231; Poly-Dispersion® A (TI)D-80; Prespersion PAB-9541; Prespersion SIB-7211; Reed C-PPR-2096; Reed C-PS-4230

Titanium isopropylate. *See* Tetraisopropyl titanate
Titanium oxide. *See* Titanium dioxide

Titanium IV tetra tridecyl, bis (tris tridecyl) phosphite
Trade names containing: SAFR-T-5K

Titanylacetylacetonate. *See* Titanium acetylacetonate
TKPP. *See* Tetrapotassium pyrophosphate
TMCP. *See* Tris (2-chloropropyl) phosphate
TME. *See* Trimethylolethane
TMEDA. *See* N,N,N′,N′-Tetramethylethylenediamine
TMPTA. *See* 1,1,1-Trimethylolpropane triacrylate
TMPTMA. *See* 1,1,1-Trimethylolpropane trimethacrylate
TMTD. *See* Tetramethylthiuram disulfide
TMTM. *See* Tetramethylthiuram monosulfide
TNPP. *See* Trisnonylphenyl phosphite

p-Tolualdehyde
CAS 104-87-0; EINECS 203-246-9
Empirical: C_8H_8O
Properties: M.w. 120.2; dens. 1.019; b.p. 204-205 C; flash pt. 80 C; ref. index 1.5460
Precaution: Combustible
Uses: Additive for resins; pharmaceutical intermediate; fragrance
Manuf./Distrib.: Fabrichem; Penta Mfg.
Trade names: PTAL

Toluene
CAS 108-88-3; EINECS 203-625-9
Synonyms: Methylbenzene; Phenylmethane; Toluol
Classification: Aromatic compd.
Empirical: C_7H_8
Formula: $C_5H_5CH_3$
Properties: Colorless liq., benzene odor; sol. in alcohol, benzene, ether; very sl. sol. in water; m.w. 92.13; dens. 0.866 (20/4 C); m.p. -94.5 C; b.p. 110.7 C; flash pt. (CC) 40 F; ref. index 1.4967 (20 C)
Precaution: Flammable
Toxicology: Toxic by ingestion, inhalation, and skin absorption; narcotic in high concs.; TLV 50 ppm in air (ACGIH); LD50 (rat, oral) 7.53 g/kg
Uses: Aviation gasoline additive; solvent for paint, gums, resins; diluent and thinner in nitrocellulose lacquers; adhesive solvent in plastic toys; mfg. of benzoic acid, benzaldehyde; in extraction of various principles from plants; starting material for TDI for PU resins, TNT for explosives, toluene sulfonates for detergents and dyestuffs
Regulatory: FDA 21CFR §175.105, 175.320, 176.180, 177.1010, 177.1200, 177.1440, 177.1580, 177.1650, 177.2460, 178.3010, 27CFR §21.131
Manuf./Distrib.: Ashland; Chevron; Exxon; Mitsubishi Petrochem.; Mitsui Petrochem. Ind.; Mobil; Phillips; Shell; Texaco
Trade names containing: Black Out® Black; Initiator BK; Lupersol 665-T50; Lupersol 688-T50; Perkadox® IPP-AT50; Topanol® CA; Trigonox® EHP-AT70

Toluene diisocyanate
CAS 584-84-9; EINECS 209-544-5
Synonyms: TDI; Toluene 2,4-diisocyanate; 2,4-Tolylene diisocyanate; 2,4-Diisocyanatotoluene
Empirical: $C_9H_6N_2O_2$
Formula: $CH_3C_6H_3(NCO)_2$
Properties: Water-white to pale yel. liq., sharp pungent odor; sol. in ether, acetone, and other org. solvs.; m.w. 174.15; dens. 1.122 (25/15.5 C); m.p. 19.4-21.5 C; b.p. 251 C (760 mm); flash pt. (OC) 270 F; ref. index 1.5680
Precaution: Combustible

Toluene 2,4-diisocyanate

Toxicology: Toxic by ingestion, inhalation; strong irritant to skin and tissue; may be anticipated to be a carcinogen; TLV 0.005 ppm in air
Uses: In mfg. of polyurethane foams, elastomers, and coatings
Manuf./Distrib.: Allchem Ind.; Bayer Hispania Industrial SA; ICI Polyurethanes; Nippon Polyurethane Ind.; Olin; Tricon Trading Assoc.
Trade names containing: Desmodur® RC; Vulcabond® VP

Toluene 2,4-diisocyanate. *See* Toluene diisocyanate

o-Toluene ethylsulfonamide
Properties: Sp.gr. 1.190; m.p. 18 C; flash pt. (COC) 174 C; ref. index 1.540
Uses: Plasticizer for cellulosics, PS, PVAc, PVB
Manuf./Distrib.: Akzo Nobel; Monsanto; Unitex

p-Toluene ethylsulfonamide. *See* Ethyl toluenesulfonamide
Toluene-4-sulfonamide. *See* p-Toluenesulfonamide

o-Toluenesulfonamide
CAS 88-19-7
Empirical: $C_7H_9O_2SN$
Formula: $CH_3C_6H_4SO_2NH_2$
Properties: Colorless crystals; sol. in alcohol; sl. sol. in water, ether; m.w. 171.22; sp.gr. 1.353; m.p. 156-158 C; flash pt. (COC) 206 C
Precaution: Combustible
Toxicology: Irritant; suspected carcinogen
Uses: Plasticizer for PVAc, ethyl cellulose
Manuf./Distrib.: Akzo Nobel; Aldrich; Focus; Monsanto; Unitex
Trade names containing: Rit-Cizer #9; Uniplex 171

p-Toluenesulfonamide
CAS 70-55-3; EINECS 200-741-1
Synonyms: PTSA; Toluene-4-sulfonamide
Empirical: $C_7H_9NO_2S$
Formula: $CH_3C_6H_4SO_2NH_2$
Properties: White leaflets; sol. in alcohol; very sl. sol. in water; m.w. 171.22 sp.gr. 1.353; m.p. 138-139 C; flash pt. (COC) 206 C
Precaution: Combustible
Uses: Organic synthesis; plasticizer for thermoplastic and thermoset resins; fungicide and mildewcide in paints and coatings
Manuf./Distrib.: Akzo Nobel; Aldrich; Allchem Ind.; Focus; Honeywill & Stein Ltd; ICI Spec.; Monsanto; Rit-Chem.; Schweizerhall; Unitex
Trade names: Uniplex 173
Trade names containing: Rit-Cizer #9; Uniplex 171

Toluenesulfonamide/epoxy resin. *See* Tosylamide/epoxy resin

Toluenesulfonamide formaldehyde resin
CAS 1338-51-8
Uses: Modifier and adhesion promoter for resins in adhesives and coatings; extender in polyamide resins
Trade names: Uniplex 600
Trade names containing: Uniplex 680

o-Toluenesulfonate. *See* o-Toluene sulfonic acid
4-Toluenesulfonic acid. *See* p-Toluene sulfonic acid

o-Toluene sulfonic acid
Synonyms: o-Toluenesulfonate
Classification: Substituted aromatic acid
Formula: $C_6H_4(SO_3H)(CH_3)$
Properties: Colorless crystals; sol. in alcohol, water, ether; m.p. 67.5 C
Precaution: Combustible
Toxicology: Toxic by ingestion, inhalation; strong irritant to tissue
Uses: Dyes, organic synthesis, acid catalyst for resins in foundry cores, plastics, coatings

Manuf./Distrib.: Boliden Intertrade; BYK-Chemie; Eastman; Ferro/Grant; Miles; Nissan Chem. Ind.; PMC Spec.; Ruetgers-Nease; Witco/Oleo-Surf.

p-Toluene sulfonic acid
CAS 104-15-4; 70788-37-3; EINECS 203-180-0
Synonyms: 4-Methylbenzenesulfonic acid; 4-Toluenesulfonic acid; p-Methylbenzenesulfonic acid; p-Toluenesulfonic acid; p-Methylphenylsulfonic acid; Tosic acid
Classification: Substituted aromatic acid
Empirical: $C_7H_8O_3S$
Formula: $C_6H_4(SO_3H)(CH_3)$
Properties: Colorless leaflets; sol. in alcohol, ether, water; m.w. 172.21; m.p. 107 C; b.p. 140 C (20 mm)
Precaution: Combustible; potentially explosive reaction with acetic anhydride + water; heated to decomp., emits toxic fumes of SO_x
Toxicology: LD50 (oral, rat) 2480 mg/kg; poison by ingestion; skin and mucous membrane irritant
Uses: Dyes, organic synthesis, acid catalyst for resins in foundry cores, plastics, coatings
Manuf./Distrib.: Boliden Intertrade; BYK-Chemie; Eastman; Ferro/Grant; Miles; Nissan Chem. Ind.; PMC Spec.; Ruetgers-Nease; Witco/Oleo-Surf.
Trade names: Eltesol® TSX; Eltesol® TSX/A; Eltesol® TSX/SF; Manro PTSA/C; Manro PTSA/E; Manro PTSA/H; Manro PTSA/LS; PTSA 70; Reworyl® T 65

p-Toluenesulfonic acid. *See* p-Toluene sulfonic acid

p-(p-Toluenesulfonyl amido) diphenylamine
Uses: Antioxidant for EVA, polyamide, polyolefins, adhesives
Trade names: Aranox®
Trade names containing: Poly-Dispersion® S (AX)D-70

p-Toluene sulfonyl hydrazide
CAS 877-66-7
Empirical: $C_7H_{10}N_2O_2S$
Formula: $CH_3C_6H_4SO_2NHNH_2$
Properties: M.w. 186.23; dec. exothermically when heated above 60 C; gas yield 115 cc/g
Toxicology: Heated to dec., emits toxic fumes of SO_x and NO_x
Uses: Blowing agent for cellular rubber goods
Trade names: Blo-Foam® SH; Celogen® TSH; Mikrofine TSH

p-Toluene sulfonyl semicarbazide
CAS 10396-10-8
Empirical: $C_8H_{11}N_3O_3S$
Properties: Fine white powd.; sol. in DMSO; m.w. 229.24; dens. 1.428; dec. 226 C; gas yield 145 cc/g
Uses: Blowing agent for polyolefins, impact polystyrene
Trade names: Blo-Foam® RA; Celogen® RA; Mikrofine TSSC; TSS; Unicell TS

Toluhydroquinone
CAS 95-71-6; EINECS 202-443-7
Synonyms: Methylhydroquinone
Empirical: $C_7H_8O_2$
Formula: $CH_3C_6H_3(OH)_2$
Properties: Pink to white crystals; sl. sol. in water; sol. in alcohol, acetone; m.w. 124.14; m.p. 125-127 C; flash pt. 172 C
Toxicology: Toxic irritant
Uses: Antioxidant, polymerization inhibitor
Manuf./Distrib.: Charkit; Eastman; Fluka
Trade names: THQ

Toluol. *See* Toluene

o-Tolyl biguanide
CAS 93-69-6; EINECS 202-268-6
Synonyms: Imidodicarbonimidic diamide, N-(2-methylphenyl)-; N-(2-Methylphenyl)imidodicarbonimidic diamide
Classification: Substituted aromatic compd.
Empirical: $C_9H_{13}N_5$

Tolyl bis (dimethyl urea)

Formula: $NH_2(CNHNH)_2C_6H_4CH_3$
Properties: White to off-white powd.; m.p. 138 C
Precaution: Combustible
Uses: Antioxidant for soaps produced from animal or vegetable oil
Trade names containing: Rhenogran® OTBG-50; Rhenogran® OTBG-75

Tolyl bis (dimethyl urea)
Uses: Accelerator for epoxy resins
Trade names: Amicure® UR2T

m-Tolyl diethanolamine
CAS 91-99-6
Synonyms: N,N-Bis (β-hydroxyethyl)-3-methylaniline; Diethanol-m-toluidine
Empirical: $C_{11}H_{17}NO_2$
Properties: Wh. to tan cryst. solid; m.w. 195; m.p. 68 C
Uses: Coupling agent for dyes for syn. fibers; cure promoter for polyester resins; in mfg. of nonporous PU elastomers
Manuf./Distrib.: Monomer-Polymer & Dajac
Trade names: Emery® 5709

p-Tolyl diethanolamine
CAS 3077-12-1; EINECS 221-359-1
Synonyms: 2,2´-(p-Tolylimino) diethanol
Empirical: $C_{17}H_{11}O_2N$
Formula: $(HOC_2H_4)_2NC_6H_4CH_3$
Properties: Crystals; sol. in water (1.67% @ 20 C); very sol. in acetone, ethanol, ethyl acetate, benzene; m.w. 195.26; dens. 1.0723 (20/20 C); m.p. 62 C; b.p. 297.1 C; flash pt. 385 F
Precaution: Combustible; moisture-sensitive
Toxicology: Irritant
Uses: Emulsifier, dyestuff intermediate; cure promoter for polyester resins; intermediate for plastics
Trade names: Emery® 5710

Tolyl diphenyl phosphate. *See* Diphenylcresyl phosphate
2,4-Tolylene diisocyanate. *See* Toluene diisocyanate
2,2´-(p-Tolylimino) diethanol. *See* p-Tolyl diethanolamine

p-Tolyl phenylenediamine
Trade names containing: Vulkanox® 3100

Tolyltriazole
CAS 29385-43-1
Synonyms: 1H-Benzotriazole, methyl
Empirical: $C_7H_7N_3$
Properties: M.w. 133.17
Uses: Corrosion inhibitor for copper, brass, bronze, ferrous metals; in metalworking fluids; prevulcanization retarder for CR and halobutyl rubbers
Manuf./Distrib.: Aceto; Climax Performance; PMC Spec.; Sandoz
Trade names: Vulkalent TM

TOOC. *See* p-t-Octyl-o-cresol
Tosic acid. *See* p-Toluene sulfonic acid

Tosylamide/epoxy resin
Synonyms: Toluenesulfonamide/epoxy resin
Definition: Toluenesulfonamide of the condensation prod. of 4,4´-isopropylidenediphenol with epichlorohydrin, also known as the diglycidyl ether of bisphenol A
Uses: Nail enamel resin; plasticizer for nitrocellulose
Trade names: Lustrabrite® S

N-Tosyl ethylamine. *See* Ethyl toluenesulfonamide
TOTM. *See* Trioctyl trimellitate
Toxilic anhydride. *See* Maleic anhydride

TPG. *See* Tripropylene glycol
TPGDA. *See* PPG-3 diacrylate
TPP. *See* Triphenyl phosphate, Triphenyl phosphite
TPT. *See* Tetraisopropyl titanate

Triacetin
CAS 102-76-1; EINECS 203-051-9
Synonyms: Glyceryl triacetate; Acetin; Enzactin; 1,2,3-Propanetriol triacetate; Triacetyl glycerol; Triacetyl glycerin
Definition: Triester of glycerin and acetic acid
Empirical: $C_9H_{14}O_6$
Formula: $C_3H_5(OCOCH_3)_3$
Properties: Colorless oily liq., sl. fatty odor, bitter taste; sl. sol. in water; sol. in alcohol, ether, other org. solvs.; m.w. 218.20; dens. 1.160 (20 C); m.p. -78 C; b.p. 258-260 C; flash pt. 300 F; ref. index 1.4307 (20 C)
Precaution: Combustible exposed to heat, flame, or powerful oxidizers
Toxicology: LD50 (oral, rat) 3000 mg/kg, (IV, mouse) 1600 $\pm$ 81 mg/kg; poison by ingestion; mod. toxic by IP, subcutaneous, IV routes; eye irritant; heated to decomp., emits acrid smoke and irritating fumes
Uses: Plasticizer for vinyl compds.; fixative in perfumery; mfg. of cosmetics; specialty solv.; mfg. of celluloid, photographic films; topical antifungal; adhesives, coatings, paper; food additive
Regulatory: FDA 21CFR §175.300, 175.320, 181.22, 181.27, 184.1901, GRAS
Manuf./Distrib.: Aldrich; Eastman; Penta Mfg.; Spectrum Chem. Mfg.; Unichema
Trade names: Estol 1579; Kodaflex® Triacetin

Triacetoxyvinylsilane. *See* Vinyltriacetoxy silane
Triacetyl glycerin. *See* Triacetin
Triacetyl glycerol. *See* Triacetin

Triallylcyanurate
CAS 101-37-1; EINECS 202-936-7
Synonyms: 2,4,6-Triallyloxy-1,3,5-triazine
Empirical: $C_{12}H_{15}N_3O_3$
Formula: $(CH_2{:}CHCH_2OC)_3N_3$
Properties: Colorless liq. or solin.; misc. with acetone, benzene, chloroform, dioxane, ethyl acetate, ethanol, xylene; m.w. 249.27; dens. 1.1133 (30 C); m.p. 27 C; flash pt. > 176 F
Precaution: Combustible; moisture-sensitive
Toxicology: Toxic by ingestion and inhalation
Uses: Polymers as monomers and modifier; organic intermediate; coagent improving peroxide-induced crosslinking of rubber
Manuf./Distrib.: Akzo Nobel; Cytec; Degussa; Monomer-Polymer & Dajac; Nat'l. Starch & Chem.; Richman
Trade names: Perkalink® 300
Trade names containing: Perkalink® 300-50D-pd; Perkalink® 300-50DX; Poly-Dispersion® TAC-50; Rhenofit® TAC/S

Triallyl isocyanurate
CAS 1025-15-6; EINECS 213-834-7
Synonyms: Triallyl-1,3,5-triazine-2,4,6(1H,3H,5H)-trione
Empirical: $C_{12}H_{15}N_3O_3$
Properties: M.w. 249.27; dens. 1.159; b.p. 149-152 (4 mm); flash pt. > 230 F; ref. index 1.5130
Uses: Coagent improving peroxide-induced crosslinking of rubber
Manuf./Distrib.: Akzo Nobel; Interbusiness USA; Itochu Spec.; Monomer-Polymer & Dajac
Trade names: Perkalink® 301
Trade names containing: Perkalink® 301-50D

2,4,6-Triallyloxy-1,3,5-triazine. *See* Triallylcyanurate
Triallyl-1,3,5-triazine-2,4,6(1H,3H,5H)-trione. *See* Triallyl isocyanurate

Triallyl trimellitate
CAS 2694-54-4
Synonyms: TATM; Tri-2-propenyl 1,2,4-benzenetricarboxylate
Empirical: $C_{18}H_{18}O_6$
Properties: M.w. 330.33; b.p. 210 C (4.5 mm)
Manuf./Distrib.: C.P. Hall; Hardwicke; Monomer-Polymer & Dajac; Natrochem; Rhone-Poulenc Spec.
Trade names containing: Poly-Dispersion® TATM-50

Triaryl phosphate
CAS 68937-41-7
Formula: $(CH_2{:}CHCH_2O)_3PO$
Properties: Water-wh. liq.; dens. 1.064 (25/15 C); f.p. -50 C; b.p. 80 C (0.5 mm); ref. index 1.448 (25 C)
Precaution: Combustible
Uses: Flame retardant plasticizer for PVC; compatibilizing agent; intermediate; catalyst carrier and pigment vehicle for polyurethane; processing aid in rubber belting and mech. goods
Manuf./Distrib.: Akzo Nobel; Albright & Wilson; Ashland; FMC; C.P. Hall; Harwick; Monsanto; Velsicol
Trade names: Kronitex® 25; Kronitex® 50; Kronitex® 100; Kronitex® 200; Kronitex® 1840; Kronitex® 1884; Pliabrac® 519; Pliabrac® 521; Pliabrac® 524

1,3,5-Triazine,2,4,6-triamine, polymer with formaldehyde. *See* Melamine/formaldehyde resin
s-Triazine-2,4,6-triol. *See* Isocyanuric acid
2,4,6-Triazinetrithiol. *See* 2,4,6-Trimercapto-s-triazine

1,1′,1′′-(1,3,5-Triazine-2,4,6-triyltris((cyclohexylimino)-2,1-ethanediyl)tris(3,3,4,5,5-tetramethylpiperazinone)
Uses: Light stabilizer, antioxidant for polymers
Trade names: Good-rite® 3150; Good-rite® 3159

Tribasic calcium phosphate. *See* Calcium phosphate tribasic
Tribenzoin. *See* Glyceryl tribenzoate

Tribromoneopentyl alcohol
CAS 36483-57-5
Synonyms: TBNPA; 2,2-Bis(bromomethyl)-3-bromo-1-propanol
Empirical: $C_5H_9Br_3O$
Formula: $(BrCH_2)_3CCH_2OH$
Properties: White to off-white flakes; sol. in alcohols; m.w. 324.92; dens. 2.28; m.p. 62-67 C
Toxicology: Irritant to eyes, mild irritant to skin; LD50 (oral, rat) 2823 mg/kg
Uses: Flame retardant for flexible and rigid PU; flame retardant intermediate
Trade names: FR-513
Trade names containing: Saytex® FR-1138

Tribromophenol. *See* 2,4,6-Tribromophenol

2,4,6-Tribromophenol
CAS 118-79-6; EINECS 204-278-6
Synonyms: TBP; Tribromophenol; Bromol
Empirical: $C_6H_3Br_3O$
Formula: $C_6H_2Br_3OH$
Properties: Soft white needles, sweet taste, bromine odor; sol. in alcohol, chloroform, ether, caustic alkaline sol'ns.; almost insol. in water; m.w. 330.83; dens. 2.55 (20/20 C); m.p. 96 C (sublimes); b.p. 244 C
Toxicology: Hazard by ingestion, inhalation, skin absorption; strong skin irritant; LD50 (rat, oral) < 2000 mg/kg
Uses: Reactive intermediate for phenol-based reactions; flame retardant for phenolics, PC, epoxies; antifungal agent; chemical intermediate for flame retardants
Manuf./Distrib.: Great Lakes
Trade names: FR-613; Great Lakes PH-73™

Tribromophenyl allyl ether
CAS 3278-89-5
Synonyms: Allyl 2,4,6-tribromophenyl ether
Formula: $Br_3C_6H_2OCH_2CH{=}CH_2$
Properties: M.w. 370.88; m.p. 74-76 C
Uses: Flame retardant for expandable PS; synergist with hexabromocyclododecane
Trade names: FR-913; Great Lakes PHE-65™

Tribromophenyl maleimide
CAS 59789-51-4
Uses: Flame retardant for plastics
Trade names: Actimer FR-1033; FR-1033

Tribromosalicylanilide
CAS 87-10-5
Synonyms: 3,4′,5-Tribromosalicylanilide; Tribromsalan
Empirical: $C_{13}H_8Br_3NO_2$
Formula: $Br_2C_6H_2(OH)C(O)NHC_6H_4Br$
Properties: M.w. 449.92
Toxicology: Suspected carcinogen
Uses: Antimicrobial for resins, latex emulsions, plastics, soaps; antiseptic
Regulatory: FDA prohibited in cosmetics

3,4′,5-Tribromosalicylanilide. *See* Tribromosalicylanilide

Tribromostyrene
CAS 61368-34-1
Uses: Flame retardant for plastics
Trade names: Actimer FR-803

Tribromsalan. *See* Tribromosalicylanilide

Tributoxyethyl phosphate
CAS 78-51-3; EINECS 201-122-9
Synonyms: TBEP; Ethanol, 2-butoxy-, phosphate (3:1); Tris (2-butoxyethyl) phosphate
Empirical: $C_{18}H_{39}O_7$
Formula: $[CH_3(CH_2)_3O(CH_2)_2O]_3PO$
Properties: Sl. yel. oily iq.; insol. or limited sol. in glycerol, glycols, certain amines; sol. in most org. liqs.; m.w. 398; dens. 1.020 (20 C); m.p. -70 C; b.p. 215-228 C (4 mm); flash pt. 438 F; ref. index 1.4380
Precaution: Combustible
Toxicology: Irritant
Uses: Primary plasticizer for most resins and elastomers; floor finishes and waxes; flame-retarding agent; latex paints; as defoamer
Manuf./Distrib.: Akzo Nobel; Albright & Wilson Am.; Ashland; FMC; C.P. Hall; Harwick; Hoechst AG; Rhone-Poulenc; Velsicol
Trade names: Amgard® TBEP; KP-140®; Phosflex TBEP; TBEP; Tolplaz TBEP

Tributyl acetyl citrate. *See* Acetyl tributyl citrate

Tributyl citrate
CAS 77-94-1; EINECS 201-071-2
Synonyms: TBC; 2-Hydroxy-1,2,3-propanetricarboxylic acid, tributyl ester; Butyl citrate
Definition: Triester of butyl alcohol and citric acid
Empirical: $C_{18}H_{32}O_7$
Formula: $C_3H_5O(COOC_4H_9)_3$
Properties: Colorless or pale yel., odorless; insol. in water; m.w. 360.45; dens. 1.042 (25/25 C); m.p. -20 C; b.p. 233.5 C (22.5 mm); flash pt. (COC) 315 F; ref. index 1.4453
Precaution: Combustible
Uses: Plasticizer, antifoam agent, solvent for nitrocellulose
Regulatory: FDA 21CFR §175.105
Manuf./Distrib.: Morflex; Unitex
Trade names: Citroflex® 4; TBC

2,4,6-Tri-t-butylphenol
CAS 732-26-3; EINECS 211-989-5
Empirical: $C_{18}H_{30}O$
Formula: $[(CH_3)_3C]_3C_6H_2OH$
Properties: Yel.-brn. solid; insol. in water; m.w. 262.44; m.p. 125-130 C; b.p. 277 C; flash pt. 130 C
Toxicology: Irritating to eyes, skin, respiratory system
Uses: Plasticizer

Tributyl phosphate
CAS 126-73-8; EINECS 204-800-2
Synonyms: Tri-n-butyl phosphate
Empirical: $C_{12}H_{26}O_4P$
Formula: $(C_4H_9)_3PO_4$

Tri-n-butyl phosphate

Properties: Stable colorless liq., odorless; misc. with most solvs. and diluents; sol. in water; m.w. 266.32; dens. 0.978 (20/20 C); m.p. < -80 C; b.p. 292 C; flash pt. (COC) 295 F; ref. index 1.4215 (25 C)
Precaution: Combustible
Toxicology: Toxic by ingestion, inhalation; irritant to skin, mucous membranes; TLV 2.5 mg/m^3 of air; LD50 (rat, oral) 3.0 g/kg
Uses: Heat-exchange medium; solv. extraction of metal ions; plasticizer for cellulose esters, lacquers, plastics, vinyl resins; antifoam agent; dielectric; flame retardant
Manuf./Distrib.: Akzo Nobel; Albright & Wilson Am.; Ashland; Chemron; EM Ind.; FMC; C.P. Hall; Harwick; Velsicol
Trade names: Albrite® TBP; Albrite® TBPO; Albrite® Tributyl Phosphate; Amgard® $TBPO_4$; Kronitex® TBP; TBP

Tri-n-butyl phosphate. *See* Tributyl phosphate

Tributyltin oxide
CAS 56-35-9; EINECS 200-268-0
Synonyms: HBD; TBTO; Bis(tributyltin) oxide; Hexabutyldistannoxane
Empirical: $C_{24}H_{54}OSn_2$
Formula: $(CH_3CH_2CH_2CH_2)_3SnOSn(CH_2CH_2CH_2CH_3)_3$
Properties: Colorless to pale yel. liq.; sol. in many org. solvs.; insol. in water; m.w. 596.08; dens. 1.17; solidifies < -45 C; b.p. 180 C (2 mm)
Toxicology: Toxic; TLV 0.1 mg/m^3 of air
Uses: Bactericide, fungicide, mildewcide in underwater and antifouling paints, PVAc latex paints, pesticides, plastics; intermediate
Manuf./Distrib.: Aceto; Akzo Nobel; Elf Atochem N. Am.; Fluka; KMZ Chem. Ltd
Trade names: Keycide® X-10

Tricalcium orthophosphate. *See* Calcium phosphate tribasic
Tricalcium phosphate. *See* Calcium phosphate tribasic

Tricapryl citrate
Properties: Sp.gr. 0.958; ref. index 1.448
Uses: Plasticizer
Manuf./Distrib.: Union Camp

Tricapryl trimellitate
Uses: Plasticizer for plastics and rubber
Trade names: Uniflex® TCTM

Tricarballylic acid-β-acetoxytributyl ester. *See* Acetyl triethyl citrate

Trichloroethane
CAS 71-55-6; EINECS 200-756-3; 201-166-9
Synonyms: TCE; 1,1,1-Trichloroethane; Ethane, 1,1,1-trichloro-; Methylchloroform; Trichloroethane
Classification: Halogenated aliphatic hydrocarbon
Empirical: $C_2H_3Cl_3$
Formula: CH_3CCl_3
Properties: Colorless liq.; insol. in water; sol. in alcohol, ether, acetone, benzene, CCl_4; m.w. 133.42; dens. 1.3376 (20/4 C); m.p. -32.5 C; b.p. 74.1 C (760 mm); flash pt. none; ref. index. 1.43838 (20 C)
Precaution: Nonflamm.
Toxicology: Irritant to eyes and tissue; narcotic in high concs.; TLV 350 ppm in air
Uses: Solvent for cleaning precision instruments, metal degreasing, pesticide, textile processing; catalyst for PU foams
Regulatory: FDA 21CFR §175.105, 177.1650
Manuf./Distrib.: Asahi Chem Industry Co Ltd; Asahi-Penn; Ashland; Chemoxy Int'l. plc; Elf Atochem N. Am.; Fabrichem; ICI Am.; PPG Industries; Vulcan
Trade names: Dabco® CS90
Trade names containing: GP-180 Resin Sol'n.

1,1,1-Trichloroethane. *See* Trichloroethane

Trichloroethyl phosphate
CAS 306-52-5

Empirical: $C_2H_4Cl_3O_4P$
Properties: M.w. 286
Uses: Flame retardant for PVC and other plastics and coatings
Manuf./Distrib.: Akzo Nobel
Trade names: Disflamoll® TCA; Fyrol® CEF

Tri(β-chloroisopropyl)phosphate
Properties: M.w. 325
Uses: Flame retardant for polymers
Trade names: Fyrol® PCF

N-Trichloromethylthio-4-cyclohexene-1,2-dicarboximide. *See* Captan
N-Trichloromethylthiotetrahydrophthalimide. *See* Captan
Trichloromonosilane. *See* Trichlorosilane

Trichlorosilane
CAS 10025-78-2; EINECS 233-042-5
Synonyms: Trichloromonosilane; Silicochloroform
Empirical: Cl_3HSi
Formula: Cl_3HSi
Properties: Colorless volatile liq.; sol. in benzene, heptane, ether, perchloroethylene, carbon disulfide, chloroform; dec. by water; m.w. 135.45; dens. 1.3417 (20/4 C); m.p. -128 C; b.p. 31.9 C; flash pt. -13 C; visc. 0.397 cp (0 C); ref. index 1.4020 (20 C)
Precaution: Supports combustion
Toxicology: LD50 (rat, oral) 1.03 g/kg
Uses: Organic synthesis; purification of silicon; intermediate; coupling agent, blocking agent, release agent, lubricant
Manuf./Distrib.: Chisso Am.; Hüls Am.; PCR
Trade names: Dynasylan® TCS

(Trichlorosilyl)ethylene. *See* Vinyltrichlorosilane
Trichlorovinylsilane. *See* Vinyltrichlorosilane

Tricresyl phosphate
CAS 1330-78-5; 78-30-8; 68952-35-2; EINECS 215-548-8
Synonyms: TCP; Phosphoric acid, tris(methylphenyl) ester; Tritolyl phosphate
Empirical: $C_{21}H_{21}O_4P$
Formula: $(CH_3C_6H_4O)_3PO$
Properties: Colorless liq., odorless; misc. with all common solvs. and thinners, veg. oil; insol. in water; m.w. 368.37; dens. 1.162 (25/25 C); b.p. 420 C; flash pt. 437 F; ref. index 1.5550
Toxicology: Toxic by ingestion, skin absorption
Uses: Plasticizer, fire retardant for plastics, air filter medium, waterproofing, additive to extreme pressure lubricants
Manuf./Distrib.: Akzo Nobel; Chemron; Daihachi Chem. Ind.; Fluka; FMC; Great Western
Trade names: Antiblaze® TCP; Disflamoll® TKP; Kronitex® TCP; Lindol XP Plus; Pliabrac® TCP
Trade names containing: Luperco ATC

1-Tridecanol, hydrogen sulfate, sodium salt. *See* Sodium tridecyl sulfate
Tridecanol stearate. *See* Tridecyl stearate

Trideceth-14
CAS 24938-91-8 (generic); 78330-21-9
Synonyms: PEG-14 tridecyl ether; POE (14) tridecyl ether
Definition: PEG ether of tridecyl alcohol
Empirical: $C_{41}H_{83}O_{15}$
Formula: $C_{13}H_{26}(OCH_2CH_2)_nOH$, avg. n = 14
Uses: Emulsifier, stabilizer for syn. latexes, textile wetting; solubilizer for essential oils, solvs., fats, waxes
Trade names: Rhodasurf® BC-737

Trideceth-15
CAS 24938-91-8 (generic); 78330-21-9
Synonyms: PEG-15 tridecyl ether; POE (15) tridecyl ether
Definition: PEG ether of tridecyl alcohol

Trideceth-3 phosphate

Empirical: $C_{43}H_{87}O_{16}$
Formula: $C_{13}H_{26}(OCH_2CH_2)_nOH$, avg. n = 15
Uses: Solubilizer, foam builder, and detergent; emulsifier and stabilizer for syn. latexes; leveling and scouring agent
Regulatory: FDA 21CFR §175.105, 176.200, 176.210
Trade names: Rhodasurf® BC-840

Trideceth-3 phosphate
CAS 9046-01-9 (generic); 73070-47-0 (generic)
Synonyms: PEG-3 tridecyl ether phosphate; POE (3) tridecyl ether phosphate
Definition: Mixt. of esters of phosphoric acid and trideceth-3
Uses: Emulsifier, detergent, wetting agent, dispersant for industrial/household cleaners, pesticides, drycleaning, textiles, metal treatment; plastics antistat
Regulatory: FDA 21CFR §178.3570
Trade names: Rhodafac® RS-410

Trideceth-6 phosphate
CAS 9046-01-9 (generic); 73070-47-0 (generic)
Synonyms: PEG-6 tridecyl ether phosphate; PEG 300 tridecyl ether phosphate; POE (6) tridecyl ether phosphate
Definition: Complex mixture of esters of phosphoric acid and trideceth-6
Uses: Pesticide emulsifier, drycleaning detergent, textile wetting agent, antistat, penetrant, lubricant for fiber and metal treatment; emulsion polymerization
Regulatory: FDA 21CFR §178.3400, 178.3570
Trade names: Rhodafac® RS-610

Trideceth-10 phosphate
CAS 9046-01-9 (generic); 73070-47-0 (generic)
Synonyms: PEG-10 tridecyl ether phosphate; PEG 500 tridecyl ether phosphate; POE (10) tridecyl ether phosphate
Definition: Complex mixture of esters of phosphoric acid and trideceth-10
Uses: Detergent, wetting agent, emulsifier, dispersant for textile wet processing, industrial cleaners, pesticides, drycleaning, metal treatment; antistat for plastics
Trade names: Rhodafac® RS-710

Tridecyl adipate
Properties: Sp.gr. 0.912-0.916; m.p. -59 C; flash pt. (COC) 227 C; ref. index 1.457
Uses: Plasticizer for vinyls, cellulosics, PS
Manuf./Distrib.: Hatco; Hüls; Inolex

Tridecylbenzene sulfonic acid
CAS 25496-01-9; EINECS 247-036-5
Synonyms: Benzenesulfonic acid, tridecyl-
Classification: Substituted aromatic acid
Empirical: $C_{19}H_{32}O_3S$
Formula: $C_{13}H_{27}(C_6H_4)SO_3H$
Properties: M.w. 340.53
Uses: Detergent, emulsifier, wetting agent, emulsion polymerization aids
Trade names: Arylan® SO60 Acid

Tridecyl phosphite
Formula: $(C_{10}H_{21}O)_3P$
Properties: Water-white liq., decyl alcohol odor; dens. 0.892 (25/15.5 C); m.p. > 0 C; flash pt. 455 F
Precaution: Combustible
Uses: Chemical intermediate, heat and lt. stabilizer for polyvinyl, styrenic, and polyolefin resins
Manuf./Distrib.: GE Specialty
Trade names: Mark® TDP

Tridecyl sodium sulfate. *See* Sodium tridecyl sulfate

Tridecyl stearate
CAS 31556-45-3; EINECS 250-696-7
Synonyms: Octadecanoic acid, tridecyl ester; Tridecanol stearate

Definition: Ester of tridecyl alcohol and stearic acid
Empirical: $C_{31}H_{62}O_2$
Formula: $CH_3(CH_2)_{16}COOC_{13}H_{27}$
Properties: M.w. 466.83
Uses: Emollient for creams and lotions, dye carrier, lubricant for textile/industrial filament yarns, fiber finishes, plastic extrusion, magnetic tapes
Trade names: Lexolube® B-109

Tri(β,β′-dichloroisopropyl)phosphate
CAS 13674-87-8
Synonyms: TDCPP; 2-Propanol, 1,3-dichlorophosphate (3:1)
Classification: Organic
Empirical: $C_9H_{15}Cl_6O_4P$
Properties: M.w. 431
Uses: Flame retardant for PU foams and other polymers
Trade names: Fyrol® 38; Fyrol® FR-2

2,4,6-Tri(dimethylaminomethyl) phenol. *See* 2,4,6-Tris (dimethylaminomethyl) phenol
Tri(dimethylphenyl)phosphite. *See* Trixylenyl phosphate
Trien. *See* Triethylenetetramine
Trientine. *See* Triethylenetetramine

Triethanolamine
CAS 102-71-6; EINECS 203-049-8
Synonyms: TEA; 2,2′,2′′-Nitrilotris(ethanol); Trolamine; Trihydroxytriethylamine
Classification: Alkanolamine
Empirical: $C_6H_{15}O_3N$
Formula: $N(CH_2CH_2OH)_3$
Properties: Colorless visc. liq., sl. ammoniacal odor, very hygroscopic; misc. with water, alcohol; sol. in chloroform; sl. sol. in benzene, ether; m.w. 149.19; dens. 1.126; m.p. 21.2 C; b.p. 335 C; flash pt. (OC) 375 F; ref. index 1.4835
Precaution: Combustible when exposed to heat or flame; can react vigorously with oxidizing materials
Toxicology: LD50 (oral, rat) 8 g/kg; mod. toxic by IP; mildly toxic by ingestion; experimental carcinogen; liver and kidney damage in animals from chronic exposure; human skin irritant; eye irritant; heated to decomp., emits toxic fumes of NO_x and CN^-
Uses: Intermediate in mfg. of surfactants, textile specialties, waxes, polishes, toiletries, cutting oils, fatty acid soaps (drycleaning), cosmetics, household detergents, emulsions; solvent for casein, shellac, dyes; water repellent; dispersant for dyes, casein, shellac, and rubber latex; sequestering agent; rubber chem. intermediate
Regulatory: FDA 21CFR §173.315, 175.105, 175.300, 175.380, 175.390, 176.170, 176.180, 176.200, 176.210, 177.1210, 177.1680, 177.2260, 177.2600, 177.2800, 178.3120, 178.3910
Manuf./Distrib.: Aldrich; Hüls AG; OxyChem; Schweizerhall; Texaco; Union Carbide
Trade names: TEA 85 Low Freeze Grade; TEA 99 Low Freeze Grade; TEA 99 Standard Grade
Trade names containing: TEA 85

Triethanolamine dodecylbenzene sulfonate. *See* TEA-dodecylbenzenesulfonate
Triethanolamine lauryl sulfate. *See* TEA-lauryl sulfate
Triethanolammonium lauryl sulfate. *See* TEA-lauryl sulfate
Triethoxymethylsilane. *See* Methyltriethoxysilane
(Triethoxysilyl)ethylene. *See* Vinyltriethoxysilane
3-(Triethoxysilyl)-1-propanamine. *See* Aminopropyltriethoxysilane

N-[3-(Triethoxysilyl)-propyl] 4,5-dihydroimidazole
CAS 58068-97-6
Synonyms: 4,5-Dihydro-1-[3-(triethoxysilyl)propyl]imidazole
Empirical: $C_{12}H_{26}N_2O_3Si$
Formula: $(CH_2N)_2CH(CH_2)_3Si(OC_2H_5)_3$
Properties: Liq.; m.w. 274.1; dens. 1.00; b.p. 134 C (2 mm); ref. index 1.45; flash pt. 78 C
Uses: Coupling agent, release agent, lubricant, blocking agent, chemical intermediate
Trade names: Dynasylan® IMEO; Petrarch® T2503

3-(Triethoxysilyl) propyl thiocyanate. *See* 3-Thiocyanatopropyltriethoxysilane
Triethoxyvinylsilane. *See* Vinyltriethoxysilane

Triethyl acetylcitrate. *See* Acetyl triethyl citrate

Triethylamine
CAS 121-44-8; EINECS 204-469-4
Synonyms: N,N-Diethylethanamine; (Diethylamino) ethane
Empirical: $C_6H_{15}N$
Formula: $(C_2H_5)_3N$
Properties: Colorless liq., strong ammoniacal odor; sol. in water and alcohol; m.w. 101.22; dens. 0.7293 (20/20 C); b.p. 89.7 C; f.p. -115.3 C; flash pt. (OC) 10 F; ref. index 1.4000
Precaution: Flamm.; dangerous fire risk; explosive limits in air 1.2-8%
Toxicology: Moderately toxic by ingestion and inhalation; strong irritant to tissue; TLV 10 ppm in air
Uses: Propellant; corrosion inhibitor; catalytic solvent in chemical synthesis; accelerator activators for rubber; wetting, penetrating and waterproofing agents of quat. ammonium types; curing and hardening of polymers
Manuf./Distrib.: Air Prods.; Allchem Ind.; Ashland; BASF; Elf Atochem N. Am.

Triethyl citrate
CAS 77-93-0; EINECS 201-070-7
Synonyms: TEC; 2-Hydroxy-1,2,3-propanetricarboxylic acid, triethyl ester; Ethyl citrate
Definition: Triester of ethyl alchol and citric acid
Empirical: $C_{12}H_{20}O_7$
Formula: $C_3H_5O(COOC_2H_5)_3$
Properties: Colorless mobile oily liq., odorless, bitter taste; sol. 65 g/100 cc water; sol. 0.8g/100 cc oil; m.w. 276.32; dens. 1.136 (25 C); b.p. 294 C; flash pt. (COC) 303 F; ref. index 1.4420
Precaution: Combustible liq. when exposed to heat or flame
Toxicology: LD50 (oral, rat) 5900 mg/kg; mod. toxic by IP route; mildly toxic by ingestion, inh.; heated to decomp., emits acrid smoke and irritating fumes
Uses: Solvent and plasticizer for nitrocellulose and natural resins; softener; paint removers; agglutinant; perfume base; food additive
Regulatory: FDA 21CFR §175.300, 175.320, 181.22, 181.27, 182.1911, GRAS; GRAS (FEMA)
Manuf./Distrib.: Aldrich; Great Western; Morflex; Penta Mfg.; Unitex
Trade names: Citroflex® 2; Uniplex 80

Triethylene diamine
CAS 280-57-9; EINECS 205-999-9
Synonyms: TEDA; 1,4-Diazobicyclo [2.2.2] octane; Diazobicyclooctane
Empirical: $C_6H_{12}N_2$
Formula: $N(CH_2CH_3)_3N$
Properties: Wh. cryst., ammoniacal odor, very hygroscopic; sublimes readily at R.T.; sol. (g/100 g): 45 g in water, 13 g in acetone, 51 g in benzene, 77 g in ethanol, 26.1 g in MEK; m.w. 112.17; m.p. 158 C; b.p. 174 C
Toxicology: LD50 (rat) 1870 mg/kg; mod. toxic; skin irritant
Uses: Catalyst in making urethane foams; dyeing assistant, crosslinking agent, stabilizer, surf. treatments, bond promoter, etching agent
Trade names: Amicure® TEDA; Dabco® Crystalline; TEDA; Tegoamin® 33; Tegoamin® 100; Toyocat®-TF
Trade names containing: Amicure® 33-LV; Dabco® 33-LV®; Dabco® BLV; Dabco® EG; Dabco® R-8020®; Dabco® RC; Dabco® S-25; Jeffcat TD-33; Jeffcat TD-33A; Jeffcat ZF-51; Jeffcat ZF-53; TEDA-L25B; TEDA-L33; TEDA-L33E

Triethylene glycol
CAS 112-27-6; EINECS 203-953-2
Synonyms: TEG; 2,2´-[1,2-Ethanediylbis(oxy)]bisethanol
Classification: Aliphatic alcohol
Empirical: $C_6H_{14}O_4$
Formula: $HOCH_2CH_2OCH_2CH_2OCH_2CH_2OH$
Properties: Clear liq., colorless, odorless, hygroscopic; sol. in water; immisc. with benzene, toluene, gasoline; m.w. 150.17; sp.gr. 1.12 (20 C); f.p. -7.2 C; b.p. 278 C; flash pt. (PMCC) 335 F; ref. index 1.4550
Precaution: Combustible
Toxicology: Irritant
Uses: Dehydrating agent for natural gas; cork humectant; solv. and lubricant for textile dyeing/printing; plasticizer; humectant for tobacco; chem. intermediate
Regulatory: FDA 21CFR §175.105, 175.300, 177.1200, 178.3740, 178.3910, 179.45
Manuf./Distrib.: Ashland; Coyne; Eastman; Hoechst Celanese; Shell; Texaco

Triethylene glycol caprate/caprylate. *See* PEG-3 caprylate/caprate
Triethylene glycol diacetate. *See* PEG-3 diacetate
Triethylene glycol dicaprylate. *See* PEG-3 dicaprylate
Triethylene glycol di-caprylate-caprate. *See* PEG-3 dicaprylate/caprate
Triethylene glycol di-2-ethylhexanoate. *See* PEG-3 di-2-ethylhexoate
Triethylene glycol dimethacrylate. *See* PEG-3 dimethacrylate
Triethylene glycol dimethyl ether. *See* PEG-3 dimethyl ether
Triethylene glycol dioctoate. *See* PEG-3 dioctoate
Triethylene glycol dipelargonate. *See* PEG-3 dipelargonate
Triethylene glycol dodecyl ether. *See* Laureth-3

Triethylenetetramine
CAS 112-24-3; EINECS 203-950-6
Synonyms: TETA; N,N′-Bis(2-aminoethyl)-1,2-ethanediamine; Trien; Trientine
Empirical: $C_6H_{18}N_4$
Formula: $NH_2(C_2H_4NH)_2C_2H_4NH_2$
Properties: Moderately visc. ylsh. oily liq.; sol. in water, alcohol; m.w. 146.23; dens. 0.9818 (20/20 C); m.p. 12 C; b.p. 277.5 C; flash pt. (CC) 275 F; ref. index 1.4971 (20 C); pH 14
Precaution: Combustible
Toxicology: Corrosive; strong irritant to tissue; skin burns, eye damage; LD50 (rat, oral) 2.5 g/kg
Uses: Detergents and softening agents, synthesis of dyestuffs, pharmaceuticals, rubber accelerator; as thermosetting resin; epoxy curing agent; lubricating oil additive; analytical reagent for Cu, Ni; paper additive; surfactants
Manuf./Distrib.: Allchem Ind.; Ashland; Great Western; Rit-Chem; Texaco; Tosoh; Union Carbide
Trade names: D.E.H. 24; TETA; Texacure EA-24; Texlin® 300

Tri (2-ethylhexyl) trimellitate. *See* Trioctyl trimellitate

Triethyl phosphate
CAS 78-40-0; EINECS 201-114-5
Synonyms: TEP; o-Phosphoric acid, triethyl ester
Empirical: $C_6H_{15}O_4P$
Formula: $(C_2H_5O)_3PO$
Properties: Clear liq., mild odor; misc. with water, ethanol, ethyl acetate, acetone, chloroform, benzene, xylene, castor oil, isobutanol; m.w. 182.16; dens. 1.067-1.072; m.p. -56 C; b.p. 215 C; ref. index 1.4019; flash pt. (PMCC) 99 C
Precaution: Avoid oxidizing agents; emits toxic fumes of phosphorous oxides, heated to dec.
Toxicology: Moderately toxic by ingestion, intraperitoneal, intravenous; causes eye irritation; LD50 (rat, oral) 1311 mg/kg; LD50 (guinea pig, dermal) > 20 mL/kg
Uses: Intermediate for agric. insecticides; in floor polishes, unsaturated polyesters, lubricants; as plasticizer, solvent, catalyst, flame retardant
Manuf./Distrib.: Ashland; Eastman; Focus; Miles
Trade names: Eastman® TEP
Trade names containing: Initiator BK

Triflic acid. *See* Trifluoromethane sulfonic acid

Trifluoromethane sulfonic acid
CAS 1493-13-6; EINECS 216-087-5
Synonyms: Triflic acid
Empirical: CF_3HSO_3
Formula: CHF_3O_3S
Properties: Colorless to amber clear liq.; hygroscopic; m.w. 150.08; dens. 1.708 (20/4 C); b.p. 167-180 C; flash pt. none; ref. index 1.331 (20 C)
Precaution: Strong acid; violent reaction with acyl chlorides or aromatic hydrocarbons, evolving toxic hydrogen chloride gas
Toxicology: Corrosive irritant to skin, eyes, mucous membrane; heated to decomp., emits toxic fumes of F^- and SO_x
Uses: Catalyst or reactant; polymerization of epoxies, styrenes, and THF, alkylation and some acylation reactions; pharmaceuticals, explosives, dyes, and intermediates; electrolytes; formation of biaryls; polymerization reactions

Trifluoromethanesulfonic acid amine salt

Regulatory: FDA 21CFR §173.395
Manuf./Distrib.: Aldrich; Amber Syn.; 3M; MTM Research; SAF Bulk; Schweizerhall
Trade names: FC-24; Fluorad® FC-24

Trifluoromethanesulfonic acid amine salt
Trade names containing: FC-520

Triglyceryl monostearate SE. *See* Polyglyceryl-3 stearate SE
Triglyceryl oleate. *See* Polyglyceryl-3 oleate
Triglyceryl stearate. *See* Polyglyceryl-3 stearate
Trihexyl 1,2,4,-benzenetricarboxylate. *See* Tri n-hexyl trimellitate
Trihexyl trimellitate. *See* Tri n-hexyl trimellitate

Tri n-hexyl trimellitate
CAS 1528-49-0
Synonyms: Trihexyl trimellitate; Trihexyl 1,2,4,-benzenetricarboxylate
Formula: $C_6H_3[CO_2(CH_2)_5CH_3]_3$
Properties: M.w. 462.63; sp.gr. 1.010; m.p. -45.5 C; b.p. 260-262 C (4 mm); flash pt. (COC) 260 C; ref. index 1.485
Precaution: Moisture-sensitive
Uses: Plasticizer for PVC
Manuf./Distrib.: Morflex; Unitex
Trade names: Morflex® 560

3,4,5-Trihydroxybenzoic acid, n-propyl ester. *See* Propyl gallate

2,3,4-Trihydroxy benzophenone
CAS 1143-72-2
Synonyms: Benzophenone, 2,3,4-trihydroxy-
Empirical: $C_{13}H_{10}O_4$
Formula: $(HO)_3C_6H_2COC_6H_5$
Properties: M.w. 230.22; m.p. 140-142 C
Toxicology: Irritant
Uses: UV absorber for plastics
Manuf./Distrib.: Aceto; Charkit; Hoechst Celanese

2,4,6-Trihydroxy-1,3,5-triazine. *See* Isocyanuric acid
Trihydroxytriethylamine. *See* Triethanolamine

Triisobutylene
CAS 7756-94-7
Synonyms: Propene, 2-methyl-, trimer; Isobutene trimer
Empirical: $C_{12}H_{24}$
Formula: $(C_4H_8)_3$
Properties: Liq.; m.w. 168.36; dens. 0.764; b.p. 175.5-178.9 C
Precaution: Combustible
Uses: Synthesis of resins, rubber, and intermediate org. compds.; lubricating oil additive; raw material for alkylation in producing high octane motor fuels
Trade names containing: Agerite® Superlite® Emulcon®

Triisodecyl phosphite
CAS 25448-25-3; EINECS 246-998-3
Synonyms: TDP
Formula: $(C_{10}H_{21}O)_3P$
Properties: Liq.; m.w. 502; dens. 0.884-0.904; b.p. 180 C (5 mm); ref. index 1.4530-1.4610; flash pt. (PMCC) 160 C
Uses: Antioxidant and extreme pressure additive in lubricant formulations; sec. heat stabilizer for ABS, PET, PVC, PU, coatings
Manuf./Distrib.: Ashland; Dover
Trade names: Doverphos® 6; Weston® TDP

Triisodecyl trimellitate
Properties: Sp.gr. 0.969; pour pt. -35 C; flash pt. (COC) 270 C; ref. index 1.4830

Uses: Plasticizer for PVC
Manuf./Distrib.: Hüls; Inolex
Trade names: Nuoplaz® TIDTM

Triisononyl trimellitate
CAS 53894-23-8
Properties: Sp.gr. 0.979; pour pt. -40 C; flash pt. (TCC) 465 F; ref. index 1.484 (20 C)
Uses: Plasticizer for PVC, cellulosics, acrylate elastomers
Manuf./Distrib.: Aristech; Exxon; Unitex
Trade names: Jayflex® TINTM; Plasthall® TIOTM; PX-339; Staflex TIOTM

Triisooctyl phosphite
CAS 25103-12-2; EINECS 246-614-4
Synonyms: Phosphorous acid, triisooctyl ester
Empirical: $C_{24}H_{54}O_3P$
Formula: $(C_8H_{17}O)_3P$
Properties: Colorless liq. with odor; misc. with most common org. solvs.; insol. in water; m.w. 418; dens. 0.891 (20/4 C); b.p. 161-164 C (0.3 mm); flash pt. (COC) 385 F
Precaution: Combustible
Uses: Intermediate; insecticides; lubricant additive; specialty solvents; stabilizer for acrylics, nylon, unsat. polyester, PVC; improves antiwear and antifriction props.
Manuf./Distrib.: Albright & Wilson Am.; Ashland; GE Specialty
Trade names: Albrite® TIOP; Weston® TIOP

Triisooctyl trimellitate
CAS 27251-75-8
Synonyms: TIOTM
Empirical: $C_{33}H_{54}O_6$
Formula: $C_6H_3(COOC_8H_{17})_3$
Properties: Clear liq., mild odor; insol. in water; m.w. 546.79; dens. 0.992 (20/20 C); m.p. -45 C; flash pt. (PMCC) 221 C; ref. index 1.4852
Precaution: Combustible
Uses: Plasticizer for PVC, cellulosics
Manuf./Distrib.: Exxon; C.P. Hall; Hatco; Unitex

Triisopropanolamine
CAS 122-20-3; EINECS 204-528-4
Synonyms: TIPA; 1,1′,1′′-Nitrilotris-2-propanol; Tris(2-hydroxypropyl)amine
Classification: Aliphatic amine
Empirical: $C_9H_{21}NO_3$
Formula: $N(CH_2CHOHCH_3)_3$
Properties: White crystalline solid, hygroscopic; sol. in water; m.w. 191.27; dens. 0.9996 (50/20 C); m.p. 45 C; b.p. 305 C; flash pt. (OC) 320 F
Precaution: Combustible; corrosive
Toxicology: Irritating to skin and eyes
Uses: Emulsifying agent; for soaps, cosmetics, pharmaceuticals, textile specialties, agric., polymer curing chems., adhesives, antistats, coatings, metalworking, rubber, gas conditioning chems.
Regulatory: FDA 21CFR §175.105, 176.200, 176.210
Manuf./Distrib.: Aldrich; Ashland
Trade names: TIPA 99; TIPA 101; TIPA Low Freeze Grade
Trade names containing: Doverphos® 4-HR; Doverphos® 4-HR Plus; Doverphos® 10-HR; Doverphos® 10-HR Plus; Doverphos® S-682; Naugard® PHR; Ultranox® 626; Ultranox® 626A; Ultranox® 627A; Weston® 399; Weston® 399B; Weston® 619F; Weston® EGTPP

Triisopropylphenyl phosphate
Properties: Liq.; sp.gr. 1.150-1.165; flash pt. (COC) 230 C; ref. index 1.552
Uses: Plasticizer, flame retardant for plastics
Manuf./Distrib.: Akzo Nobel; Albright & Wilson Am.; FMC
Trade names: Antiblaze® 519; Kronitex® 3600

Triisotridecyl phosphite
Synonyms: TTDP; Tris (isotridecyl) phosphite
Uses: Antioxidant and extreme pressure additive for lubricant formulations; sec. stabilizer for PVC
Trade names: Doverphos® 49

Trilauryl phosphite
CAS 3076-63-9; EINECS 221-356-5
Synonyms: Phosphorous acid, tridodecyl ester
Empirical: $C_{36}H_{75}O_3P$
Formula: $(CH_{12}H_{25}O)_3P$
Properties: Water-white liq.; m.w. 586; dens. 0.866 (25/15 C); m.p. 10 C; flash pt. (COC) 232 C; ref. index 1.456
Precaution: Combustible
Uses: Stabilizer in polymers, chemical intermediate; antioxidant and extreme pressure additive in lubricant formulations
Trade names: Doverphos® 53; Weston® TLP

Trilauryl trithiophosphite
CAS 1656-63-9; EINECS 216-751-4
Synonyms: Phosphorotrithious acid, tridodecyl ester
Empirical: $C_{36}H_{75}PS_3$
Formula: $(CH_{12}H_{25}S)_3P$
Properties: Pale yel. liq.; m.w. 634; dens. 0.915 (25/15 C); m.p. 20 C; ref. index 1.4985-1.5025; flash pt. (COC) 430 F
Precaution: Combustible
Uses: Stabilizer for polymers esp. polyolefins, lubricant, chemical intermediate, antioxidant and sulfur deactivator in lubricants; improves antiwear and antifriction props.
Trade names: Weston® TLTTP

Trilinoleic acid
CAS 7049-66-3; 68937-90-6
Synonyms: Trimer acid; Fatty acids, C18, unsaturated, trimers; 9,12-Octadecadienoic acid, trimer
Definition: 54-carbon tricarboxylic acid formed by the catalytic trimerization of linoleic acid
Uses: Corrosion inhibitor, lubricant; epoxy curing agent; polymer building block
Regulatory: FDA 21CFR §176.200, 178.3910
Manuf./Distrib.: Unichema N. Am.; Union Camp; Witco/Oleo-Surf.
Trade names: Empol® 1040; Pripol 1040

Trimer acid. *See* Trilinoleic acid
1,3,5-Trimercaptotriazine. *See* 2,4,6-Trimercapto-s-triazine

2,4,6-Trimercapto-s-triazine
CAS 638-16-4
Synonyms: 1,3,5-Trimercaptotriazine; Trithiocyanuric acid; 2,4,6-Triazinetrithiol
Empirical: $C_3H_3N_3S_3$
Properties: M.w. 177.27; m.p. > 300 C
Toxicology: Poison by intraperitoneal and intravenous routes; mildly toxic by ingestion
Uses: Curing agent for epichlorohydrin rubber
Trade names: Zisnet® F-PT

1-Trimethoxysilyl-2-(chloromethyl) phenylethane
CAS 68128-25-6
Properties: Amber liq.; m.w. 274.8; dens. 1.4969; b.p. 161 C (1.5 mm); flash pt. 130 C
Uses: Coupling agent, chem. intermediate, blocking agent, release agent, lubricant
Trade names: CT2902

Trimethoxysilylpropanethiol. *See* Mercaptopropyltrimethoxysilane
3-(Trimethoxysilyl)propylamine. *See* Aminopropyltrimethoxysilane

Trimethoxysilylpropyldiethylene triamine
CAS 35141-30-1
Empirical: $C_{10}H_{27}N_3O_3Si$
Formula: $H_2(CH_2)_2NH(CH_2)_2NH(CH_2)_3Si(OCH_3)_3$
Properties: Liq.; m.w. 265.43; dens. 1.03; b.p. 114-118 C (2 mm); ref. index 1.463; flash pt. 137 C

Precaution: Corrosive
Uses: Coupling agent, release agent, lubricant, blocking agent, chemical intermediate
Trade names: Dynasylan® TRIAMO; Petrarch® T2910

N-[3-(Trimethoxysilyl) propyl] ethylenediamine. *See* N-2-Aminoethyl-3-aminopropyl trimethoxysilane

(N-Trimethoxysilylpropyl)-polyethylenimine
Uses: Coupling agent esp. for min.-filled and adhesive bonding applics. of high m.w. thermoplastic polyamides and polyesters
Trade names: CPS 076

Trimethoxyvinylsilane. *See* Vinyltrimethoxysilane

N,N′,N′-Trimethylaminoethyl piperazine
Uses: Catalyst for PU slabstock, molded, and elastomeric applics.
Trade names: Toyocat®-NP

Trimethyl carbinol. *See* t-Butyl alcohol
3,3,5-Trimethyl-2-cyclohexen-1-one. *See* Isophorone
Trimethyldihydroquinoline polymer. *See* 2,2,4-Trimethyl-1,2-dihydroquinoline polymer

2,2,4-Trimethyl-1,2-dihydroquinoline polymer
CAS 26780-96-1
Synonyms: TDQP; 1,2-Dihydro-2,2,4-trimethylquinoline homopolymer; 1,2-Dihydro-2,2,4-trimethylquinoline polymer; Poly-(1,2-dihydro-2,2,4-trimethylquinoline); Trimethyldihydroquinoline polymer; Quinoline, 1,2-dihydro-2,2,4-trimethyl homopolymer
Formula: $(C_{12}H_{15}N)_n$
Properties: Amber pellets; insol. in water; misc. with ethanol, acetone, benzene, monochlorobenzene, isopropyl acetate, gasoline; dens. 1.08; soften. pt. 75 C; m.p. 95-100 C
Precaution: Avoid strong oxidizers; dust can present explosion hazard
Uses: Antioxidant for rubbers; stabilizer or polymerization inhibitor
Trade names: Agerite® MA; Agerite® Resin D; Akrochem® Antioxidant DQ; Akrochem® Antioxidant DQ-H; Flectol® Pastilles; Naftonox® TMQ; Naugard® Super Q; Permanax™ TQ; Ralox® TMQ-G; Ralox® TMQ-H; Ralox® TMQ-R; Ralox® TMQ-T; Struktol® T.M.Q.; Struktol® T.M.Q. Powd.; Ultranox® 254; Vulkanox® HS/LG; Vulkanox® HS/Pellets; Vulkanox® HS/Powd.

N,N,N-Trimethyl-1-dodecanaminium chloride. *See* Laurtrimonium chloride

Trimethylhexamethylene diamine
CAS 25620-58-0
Synonyms: 1,6-Hexanediamine, trimethyl
Empirical: $C_9H_{22}N_2$
Properties: M.w. 158.29; dens. 0.867; b.p. 232 C; flash pt. 104 C; ref. index 1.4640
Precaution: Corrosive
Uses: Epoxy curing agent for coatings, adhesives, castings, composites
Trade names: Vestamin® TMD

3,5,5-Trimethyl hexanoyl peroxide
CAS 3851-87-4
Synonyms: Diisononanoyl peroxide
Uses: Initiator
Trade names containing: USP®-355M

N,N,N′-Trimethyl-N′-hydroxyethylbisaminoethylether
Formula: $(CH_3)_2NCH_2CH_2OCH_2CH_2N(CH_3CH_2CH_2OH$
Uses: Catalyst for PU foam, slabstock

Trimethylolethane
CAS 77-85-0; EINECS 201-063-9
Synonyms: TME; Pentaglycerine; Methyltrimethylolmethane; 1,1,1-Tris(hydroxymethyl) ethane
Empirical: $C_5H_{12}O_3$
Formula: $CH_3C(CH_2OH)_3$
Properties: Colorless cryst.; hygroscopic; sol. in water, alcohol; m.w. 120; m.p. ≈ 190 C; flash pt. 160 C

Precaution: Combustible
Toxicology: Essentially nontoxic; mildly irritating to abraded skin; nonirritating to eyes
Uses: Raw material for alkyd and polyester resins; conditioning agent; mfg. of varnishes, synthetic drying oils
Manuf./Distrib.: Fabrichem; Great Western
Trade names: Trimet® Liquid; Trimet® Nitration Grade; Trimet® Pure; Trimet® Tech.

Trimethylolethane tricaprylate-caprate
Properties: Sp.gr. 0.9425 (20 C); m.p. 2 C; flash pt. (COC) 280 C; ref. index 1.4522 (20 C)
Uses: Plasticizer for PVC, PVB, VCA, PS
Manuf./Distrib.: Hatco; Inolex

Trimethylolethane triglycidyl ether
Uses: Epoxy modifier
Trade names: Heloxy® 44

Trimethylol pentanediol dibenzoate
Properties: Sp.gr. 1.07-1.08; flash pt. (COC) -12 C; ref. index 1.528-1.531
Uses: Plasticizer for vinyls
Manuf./Distrib.: Hüls; Velsicol

Trimethylolpropane triacrylate. *See* 1,1,1-Trimethylolpropane triacrylate

1,1,1-Trimethylolpropane triacrylate
CAS 15625-89-5
Synonyms: TMPTA; Trimethylolpropane triacrylate
Formula: $(H_2C{=}CHCO_2CH_2)_3CC_2H_5$
Properties: Hygroscopic; m.w. 296.3; dens. 1.100; flash pt. > 230 F; ref. index 1.4740
Toxicology: Irritant
Uses: Crosslinker; uv-cured adhesives, wood fillers, coatings, inks, vinyl acrylic latex paint, highly crosslinked polybutadiene rubber
Manuf./Distrib.: CPS; Monomer-Polymer & Dajac; Rhone-Poulenc Spec.; UCB Radcure
Trade names: Ageflex TMPTA

Trimethylolpropane triglycidyl ether
Uses: Epoxy resin diluent, modifier
Trade names: Adeka Glycilol ED-505; Heloxy® 48

Trimethylolpropane triheptanoate
Properties: Sp.gr. 0.9608 (20 C); m.p. < 62 C; flash pt. (COC) 256 C; ref. index 1.4505 (20 C)
Uses: Plasticizer for PVC, PVB, VCA, PS
Manuf./Distrib.: Akzo Nobel; Hatco; Inolex

Trimethylolpropane tri (3-mercaptopropionate)
CAS 33007-83-9
Synonyms: Trimethylolpropane tris (3-mercaptopropionate)
Formula: $CH_3CH_2C(CH_2OOCCH_2CH_2SH)_3$
Properties: Liq.; sol. in acetone benzene, alcohol; insol. in water, hexane; m.w. 398.55; dens. 1.210; b.p. 220 C (0.3 mm); flash pt. 96 C; ref. index 1.5180
Precaution: Combustible
Toxicology: Irritant
Uses: Crosslinking agent and stabilizer for radiation-cured plastics
Manuf./Distrib.: Evans Chemetics; Witco/PAG

1,1,1-Trimethylolpropane trimethacrylate
CAS 3290-92-4
Synonyms: TMPTMA
Formula: $[H_2C{=}C(CH_3)CO_2CH_2]_3CC_2H_5$
Properties: Hygroscopic; m.w. 338.4; dens. 1.060; b.p. 185 C (5 mm); flash pt. > 230 F; ref. index 1.4720
Toxicology: Irritant
Uses: Crosslinker, coagent for wire and cable, hard rubber rolls, polybutadiene, polyethylene, moisture barrier films and coatings, plastisols and vinyl acetate latexes, adhesives, molding compds., textile prods.
Manuf./Distrib.: CPS; Monomer-Polymer & Dajac; Rohm Tech

Trade names: Ageflex TM 402, 403, 404, 410, 421, 423, 451, 461, 462; Ageflex TMPTMA; Perkalink® 400; Sipomer® TMPTMA
Trade names containing: MFM-415; MFM-786 V; Perkalink® 400-50D; Rhenofit® TRIM/S; SR-350

Trimethylolpropane trioleate
Uses: Lubricant component for hydraulic and metalworking oils; cosmetic emollient; plastics lubricants
Trade names: Estol 1427

Trimethylolpropane tris (3-mercaptopropionate). *See* Trimethylolpropane tri (3-mercaptopropionate)

Trimethyl-1,3-pentanediol, 2,2,4-diisobutyrate
CAS 6846-50-0
Synonyms: 2,2,4 Trimethylpentanediol-1,3-diisobutyrate
Empirical: $C_{16}H_{30}O_4$
Formula: $(CH_3)_2CHCH[O_2CCH(CH_3)_2]C(CH_3)_2CH_2O_2CCH(CH_3)_2$
Properties: M.w. 286.4; sp.gr. 0.945; b.p. 280 C; f.p. -70 C; flash pt. (COC) 143 C; ref. index 1.4300
Uses: Plasticizer for PVC, PS, cellulosics
Manuf./Distrib.: Chisso Am.; Eastman
Trade names: Kodaflex® TXIB

2,2,4 Trimethylpentanediol-1,3-diisobutyrate. *See* Trimethyl-1,3-pentanediol, 2,2,4-diisobutyrate

2,2,4-Trimethyl-1,3-pentanediol monoisobutyrate
CAS 25265-77-4
Synonyms: Propionic acid, 2-methyl-, monoester with 2,2,4-trimethyl-1,3-pentanediol
Classification: Ester alcohol
Empirical: $C_{12}H_{24}O_3$
Formula: $(CH_3)_2CHCH(OH)C(CH_3)_2CH_2OOCCH(CH_3)_2$
Properties: Liq.; insol. in water; sol. in benzene, alcohol, acetone, CCl_4; m.w. 216.3; dens. 0.945-0.955 (20/20 C); b.p. 180-182 C (125 mm); flash pt. (COC) 245 F; ref. index 1.4423
Precaution: Combustible
Uses: Solv., intermediate; mfg. of plasticizers, surfactants, pesticides, resins, inks, latexes, drilling muds, lubricants, syn. detergents, herbicides
Manuf./Distrib.: Ashland; Chisso Am.; Eastman
Trade names: Texanol® Ester-Alcohol

2,4,4-Trimethylpentyl-2-hydroperoxide
CAS 5809-08-5
Classification: Organic peroxide
Empirical: $C_8H_{18}O_2$
Properties: M.w. 146.2
Uses: Initiator for acrylates, PE
Trade names: Trigonox® TMPH

2,4,4-Trimethylpentyl-2-peroxyneodecanoate
CAS 51240-95-0
Classification: Organic peroxide
Properties: M.w. 300.5
Trade names containing: Trigonox® 151-C50; Trigonox® 151-C70

Trimethyl quinoline
CAS 2437-72-1
Synonyms: 2,3,4-Trimethylquinoline
Empirical: $C_{12}H_{13}N$
Properties: M.w. 171.24
Uses: Antioxidant
Trade names containing: Cabot® PE 6008

2,3,4-Trimethylquinoline. *See* Trimethyl quinoline

Trimethylthiourea
CAS 2489-77-2
Synonyms: Urea, 1,1,3-trimethyl-2-thio

Empirical: $C_4H_{10}N_2S$
Properties: M.w. 118.22
Uses: Accelerator for CR
Trade names: Thiate® EF-2

α,α,α′-Trimethyltrimethyleneglycol. *See* Hexylene glycol

1,3,5-Trimethyl-2,4,6-tris (3,5-di-t-butyl-4-hydroxybenzyl) benzene
CAS 1709-70-2
Classification: Hindered phenol
Formula: $(CH_3)_3C_6[CH_2C_6H_2[C(CH_3)_3]_2OH]_3$
Properties: Wh. free-flowing cryst. powd., odorless; partly sol. in benzene, methylene chloride; insol. in water; m.w. 775.2; m.p. 244 C
Precaution: Combustible
Toxicology: Irritant
Uses: Antioxidant for polypropylene, HDPE, spandex fibers, polyamides, and specialty rubbers
Trade names: Ethanox® 330; Irganox® 1330
Trade names containing: Unistab D-33

Trioctyldodecyl citrate
CAS 126121-35-5
Synonyms: 2-Hydroxy-1,2,3-propanetricarboxylic acid, tris(2-octyldodecyl) ester
Definition: Triester of octyldodecanol and citric acid
Empirical: $C_{66}H_{128}O_7$
Uses: Mold release for PC
Trade names: Siltech® CE-2000

Trioctyl phosphate
CAS 78-42-2; EINECS 201-116-6
Synonyms: Tris(2-ethylhexyl) phosphate; Octyl phosphate
Empirical: $C_{24}H_{51}O_4P$
Formula: $[CH_3(CH_2)_3CH(C_2H_5)CH_2O]_3PO$
Properties: Liq.; sol. in alcohol, acetone, ether; m.w. 744.04; dens. 0.924 (26 C); b.p. 220-230 (8 mm); flash pt. > 230 F; ref. index 1.4440
Precaution: Combustible
Toxicology: Toxic by ingestion, inhalation; suspected carcinogen
Uses: Solvent, antifoaming agent; plasticizer for vinyl resins, rubbers, nitrocellulose; flame retardant
Manuf./Distrib.: Akzo Nobel; Albright & Wilson Am.; Ashland; Ferro/Keil; FMC; Miles; Rhone-Poulenc
Trade names: Amgard® TOF; Disflamoll® TOF; Kronitex® TOF; TOF; TOF™

Trioctyl trimellitate
CAS 89-04-3, 3319-31-1
Synonyms: TOTM; Tri (2-ethylhexyl) trimellitate
Empirical: $C_{33}H_{54}O_6$
Formula: $C_6H_3(COOCH_2CH[C_2H_5[C_4H_9)_3$
Properties: Liq.; m.w. 547; dens. 0.989 (20/20 C); b.p. 414 C; f.p. -38 C; flash pt. (COC) 263 C
Precaution: Combustible
Uses: Plasticizer for PVC, cellulosics
Manuf./Distrib.: Aristech; Ashland; BASF; BP Oil; Eastman; C.P. Hall; Harwick; Hatco; Hüls Am.; Unitex
Trade names: ADK CIZER C-8; Diplast® TM; Diplast® TM8; Jayflex® TOTM; Kodaflex® TOTM; Nuoplaz® 6959; Nuoplaz® TOTM; Palatinol® TOTM; Plasthall® TOTM; PX-338; Staflex TOTM; Uniflex® TOTM

Triolein
CAS 122-32-7; 67701-30-8; EINECS 204-534-7; 266-948-4
Synonyms: Glyceryl trioleate; Olein; 9-Octadecenoic acid, 1,2,3-propanetriyl ester
Definition: Triester of glycerin and oleic acid
Empirical: $C_{57}H_{104}O_6$
Properties: Colorless to yellowish oily liq., tasteless, odorles; pract. insol. in water; sol. in chloroform, ether, CCl_4; sl. sol. in alcohol; m.w. 885.40; dens. 0.915 (15/4 C); m.p. -4 to -5 C; b.p. 235-240 C (15 mm); ref. index 1.4676 (20 C)
Uses: Lubricant, emollient, emulsifier for cosmetics, metals, leather, textiles; carbon source in antibiotic culture broths; raw material for engineering plastics; impact modifier, plasticizer, stabilizer, lubricant
Regulatory: FDA 21CFR §177.2800

Manuf./Distrib.: Karlshamns; Witco/Oleo-Surf.
Trade names: Sunyl® 80 ES

Triphenyl methane 4,4′,4′′-triisocyanate
Synonyms: Triphenyl triisocyanate
Uses: Bonding uncured rubber to metal or other surfaces
Trade names containing: Desmodur® RE

Triphenyl phosphate
CAS 115-86-6; EINECS 204-112-2
Synonyms: TPP; Phosphoric acid, triphenyl ester
Empirical: $C_{18}H_{15}O_4P$
Formula: $PO(OC_6H_5)_3$
Properties: Colorless crystalline powd. or needles, odorless; sol. in benzene, chloroform, ether, acetone, lacquers, solvent, thinners, oil; insol. in water; m.w. 326.28; dens. 1.268 (60 C); m.p. 50 C; b.p. 245 C (11 mm); flash pt. (CC) 220 C; ref. index 1.550
Precaution: Combustible
Toxicology: Toxic by inhalation; TLV 3 mg/m^3 of air
Uses: Fire-retarding agent; noncombustible substitute for camphor in celluloid; plasticizer for cellulose acetate and nitrocellulose; in lacquers and varnishes; impregnating roofing paper
Manuf./Distrib.: Akzo Nobel; Ashland; FMC; Harwick; Miles; Monsanto
Trade names: Disflamoll® TP; Kronitex® TPP
Trade names containing: Fyrol® PBR

Triphenyl phosphite
CAS 101-02-0; EINECS 202-908-4
Synonyms: TPP
Empirical: $C_{18}H_{15}O_3P$
Formula: $(C_6H_5O)_3P$
Properties: Water-white to pale yel. solid or oily liq., pleasant odor; m.w. 310.29; dens. 1.184 (25/25 C); m.p. 22-25 C; b.p. 155-160 C (0.1 mm); flash pt. (COC) 425 F; ref. index 1.589
Precaution: Combustible
Uses: Chemical intermediate, stabilizer systems for resins, metal scavenger, diluent for epoxy resins, antioxidant and antiwear agent in gear and transmission oils
Manuf./Distrib.: Albright & Wilson Am.; Ashland; Chemcentral; Dover; GE Specialty; Spectrum Chem. Mfg.; Witco/PAG
Trade names: Albrite® TPP; Doverphos® 10; Lankromark® LE65; Mark® 2112; Mark® TPP; Weston® TPP
Trade names containing: Doverphos® 10-HR; Doverphos® 10-HR Plus; Weston® EGTPP

Triphenyl triisocyanate. *See* Triphenyl methane 4,4′,4′′-triisocyanate
Tri-2-propenyl 1,2,4-benzenetricarboxylate. *See* Triallyl trimellitate
Tripropionin. *See* Glyceryl tripropionate

Tripropylene glycol
CAS 24800-44-0, 13987-01-4
Synonyms: TPG
Formula: $HO(C_3H_6O)_2C_3H_6OH$
Properties: Colorless liq.; sol. in water, methanol, ether; dens. 1.019; b.p. 268 C; flash pt. 285 F; ref. index 1.442
Precaution: Combustible
Uses: Intermediate in resins, plasticizers, pharmaceuticals, insecticides, dyestuffs, mold lubricants
Manuf./Distrib.: Arco; Ashland; Coyne; Union Carbide
See also PPG-3...

Tripropylene glycol diacrylate. *See* PPG-3 diacrylate
Tripropylene glycol methyl ether. *See* PPG-3 methyl ether
Tripropylene glycol monomethyl ether. *See* PPG-3 methyl ether
Tris (betachlorethyl) phosphate. *See* Tris (β-chloroethyl) phosphate
Tris (2-butoxyethyl) phosphate. *See* Tributoxyethyl phosphate

1,3,5-Tris (4-t-butyl-3-hydroxy-2,6-dimethylbenzyl)-1,3,5-triazine-2,4,6-(1H,3H,5H)-trione
CAS 40601-76-1
Classification: Isocyanurate
Empirical: $C_{42}H_{57}N_3O_6$

Tris-(2-t-butyl-4-thio-(2´-methyl-4´-hydroxy-5´-t-butyl)phenyl-5-methyl]phenyl phosphite

Properties: Off-wh. powd.; m.w. 699; m.p. 163-165 C
Precaution: Airborne dust may present explosion hazard
Toxicology: LD50 (rat, oral) > 10 g/kg; LD50 (rabbit, dermal) > 5 g/kg; minimal eye and skin irritant
Uses: Primary antioxidant for polyolefin pipe, film, household appliances
Trade names: Cyanox® 1790
Trade names containing: Cyanox® 2777

Tris-(2-t-butyl-4-thio-(2´-methyl-4´-hydroxy-5´-t-butyl)phenyl-5-methyl]phenyl phosphite. *See* Phenol, 4,4´-thiobis 2-(1,1-dimethylethyl) phosphite

Tris-(2-chloroethyl) orthophosphate
Trade names: Genomoll® P

Tris (β-chloroethyl) phosphate
CAS 115-96-8
Synonyms: TCEP; Tris (betachlorethyl) phosphate
Formula: $(ClC_2H_4O)_3PO$
Properties: Clear transparent liq.; m.w. 285.49; sp.gr. 1.425 (20 C); m.p. -60 C; b.p. 214 C (25 mm); flash pt. (COC) 225 C; ref. index 1.476 (20 C)
Precaution: Combustible
Toxicology: Suspected carcinogen
Uses: Flame retardant plasticizer for plastics
Manuf./Distrib.: Aceto; Akzo Nobel; Albright & Wilson; Focus; Velsicol

Tris 2-chloroethyl phosphite
CAS 140-08-9
Formula: $(ClCH_2CH_2O)_3P$
Properties: Colorless liq., char. odor; misc. with most common org. solvs.; insol. in water; m.w. 269.49; sp.gr. 1.353 (20 C); b.p. 119 C (0.15 mm); flash pt. (COC) 191 C; ref. index 1.4858
Precaution: Combustible; moisture-sensitive
Toxicology: Toxic
Uses: Stabilizer for plastics; intermediate for insecticides, flameproofing agents; additive for lubricants, specialty solvs.
Manuf./Distrib.: Akzo Nobel; Albright & Wilson Am.; Aldrich
Trade names: Albrite® T2CEP

Tris (2-chloropropyl) phosphate
Synonyms: TMCP
Uses: Flame retardant for polymers

Tris (β-chloropropyl) phosphate
CAS 26248-87-3
Synonyms: TCPP
Empirical: $C_9H_{18}Cl_3O_4P$
Properties: M.w. 327.59
Uses: Flame retardant for plastics
Manuf./Distrib.: Aceto
Trade names: Antiblaze® 80

Tris(3,5-di-t-butyl-4-hydroxy benzyl) isocyanurate
CAS 27676-62-6
Synonyms: 1,3,5-Tris(3´,5´-di-t-butyl-4´-hydroxybenzyl)-s-triazine-2,4,6-(1H,3H,5H)trione
Classification: Hindered phenol
Empirical: $C_{48}H_{69}N_3O_6$
Properties: Wh. powd.; m.w. 784; m.p. 218-223 C
Uses: Antioxidant, stabilizer for PP film, polyethylene, EPDM, adhesives, fiber applic., and talc; processing and end-use applic.; food pkg.
Trade names: Dovernox 3114; Irganox® 3114; Vanox® GT
Trade names containing: Irganox® B 1411

1,3,5-Tris(3´,5´-di-t-butyl-4´-hydroxybenzyl)-s-triazine-2,4,6-(1H,3H,5H)trione. *See* Tris(3,5-di-t-butyl-4-hydroxy benzyl) isocyanurate

Tris (2,4-di-t-butylphenyl) phosphite
CAS 31570-04-4
Synonyms: 2,4-Bis (1,1-dimethylethyl)phenyl-phosphite (3:1)
Classification: Aryl phosphite
Empirical: $C_{42}H_{63}O_3P$
Properties: Wh. powd.; m.w. 647; dens. 1.03; m.p. 185 C; flash pt. 225 C
Uses: Antioxidant for polyolefins
Manuf./Distrib.: Great Lakes
Trade names: Doverphos® S-480; Hostanox® PAR 24; Irgafos® 168; Naugard® 524
Trade names containing: Aquastab PA 52; Cyanox® 2777; Irganox® B 215; Irganox® B 225; Irganox® B 501W; Irganox® B 561; Irganox® B 900; Irganox® B 921; Irganox® B 1171; Irganox® B 1411

Tris (dichloropropyl) phosphate
CAS 78-43-3
Empirical: $C_9H_{15}Cl_6O_4P$
Properties: M.w. 430.91; sp.gr. 1.513; flash pt. (COC) 252 C; ref. index 1.5019
Uses: Flame retardant, plasticizer for cellulosics, epoxy, phenolics, thermoset polyesters, PS, PVC, PU foams
Manuf./Distrib.: Akzo Nobel; Albright & Wilson; Velsicol
Trade names: Antiblaze® 195; Antiblaze® TDCP/LV

2,4,6-Tris (dimethylaminomethyl) phenol
CAS 90-72-2; EINECS 202-013-9
Synonyms: DMP; 2,4,6-Tri(dimethylaminomethyl) phenol; Phenol, 2,4,6-tris(dimethylaminomethyl)
Empirical: $C_{15}H_{27}N_3O$
Formula: $[(CH_3)_2NCH_2]_3C_6H_2OH$
Properties: M.w. 265.45; dens. 0.969; b.p. 130-135 C (1 mm); flash pt. > 230 F; ref. index 1.5160
Precaution: Corrosive
Toxicology: Moderately toxic by ingestion and skin contact; severe skin and eye irritant
Uses: Epoxy curing agent, activator
Trade names: Actiron NX 3; Ancamine® K54; Capcure® EH-30

2,4,6-Tris (dimethylaminomethyl) phenol, 2-ethylhexanoic acid salt
Uses: Epoxy curing agent
Trade names: Ancamine® K61B

Tris (dimethyldithiocarbamate) bismuth. *See* Bismuth dimethyl dithiocarbamate

2,4,6-Tris-(N-1,4-dimethylpentyl-p-phenylenediamino)-1,3,5-triazine
CAS 121246-28-4
Uses: Antiozonant/antioxidant for natural and syn. rubbers, e.g., tires, hose, footwear, mech. goods, roofing, wire and cable
Trade names: Durazone® 37

Tris (dipropyleneglycol) phosphite
CAS 36788-39-3; EINECS 253-211-7
Empirical: $C_{18}H_{39}O_9P$
Formula: $(HOCHCH_3CH_2OCH_2CHCH_3O)_3P$
Properties: Liq.; m.w. 430; dens. 1.088-1.098; b.p. will not distill; ref. index 1.4600-1.4635; flash pt. (PMCC) 118 C
Uses: Heat, color, and visc. stabilizer for polymers incl. polyester fibers and polyurethanes
Trade names: Weston® 430

Tris(2-ethylhexyl) phosphate. *See* Trioctyl phosphate

Tris 3-hydroxy-4-t-butyl-2,6-dimethyl benzyl cyanurate
Uses: Antioxidant for polyolefins

Tris (2-hydroxyethyl) isocyanurate. *See* Trishydroxyethyl triazine trione

Tris (2-hydroxyethyl) isocyanurate triacrylate
Trade names containing: SR-368

Tris (2-hydroxyethyl) isocyanurate trimethacrylate
Trade names containing: SR-290

Trishydroxyethyl triazine trione
CAS 839-90-7
Synonyms: THEIC; Tris (2-hydroxyethyl) isocyanurate; Tris (2-hydroxyethyl)-s-triazine-2,4,6-trione
Formula: $C_3N_3O_3(CH_2CH_2OH)_3$
Properties: Wh. solid; sol. in water; insol. in chloroform, benzene; m.w. 261.24
Precaution: Combustible
Uses: Additive to plastics, esp. to impart thermal stability
Trade names containing: Irganox® 3125

Tris (2-hydroxyethyl)-s-triazine-2,4,6-trione. *See* Trishydroxyethyl triazine trione
1,1,1-Tris(hydroxymethyl) ethane. *See* Trimethylolethane

1,1,1-Tris(hydroxyphenylethane)benzotriazole
Uses: UV light stabilizer for plastics and coatings
Trade names: THPE-BZT

Tris(2-hydroxypropyl)amine. *See* Triisopropanolamine

Tris (p-isocyanato-phenyl) thiophosphate
CAS 4151-51-3
Empirical: $C_{21}H_{12}N_3O_6PS$
Properties: M.w. 465.36
Uses: Crosslinking agent for adhesives
Trade names containing: Desmodur® RFE

Tris(isooctadecanoato-O)(2-propanolato) titanium. *See* Isopropyl titanium triisostearate
Tris (isotridecyl) phosphite. *See* Triisotridecyl phosphite

1,1,3-Tris (2-methyl-4-hydroxy-5-t-butyl phenyl) butane
CAS 1843-03-4
Definition: Condensation prod. of 3-methyl-6-t-butylphenol and crotonaldehyde
Properties: White powd.; sol. in acetone, ethanol, diethyl ether, ethyl acetate; m.w. 544.83; m.p. 181 C; flash pt. 225 F
Toxicology: Irritant
Uses: Antioxidant for polyolefins
Trade names: Lowinox® CA 22; Topanol® CA-RT; Topanol® CA-SF; Topanol® LVT 11
Trade names containing: Topanol® CA

Tris Neodol-25 phosphite
CAS 68610-62-8
Formula: $(C_{12-15}H_{25-31}O)_3P$
Properties: Liq.; m.w. 670; dens. 0.875-0.885; ref. index 1.4580-1.4610; flash pt. (COC) 196 C
Uses: Stabilizer for polyester fibers, PVC, PP; lubricant additive as antioxidant, friction reducer, sulfur scavenger
Trade names: Weston® 474

Trisnonylphenyl phosphite
CAS 26523-78-4; EINECS 247-759-6
Synonyms: TNPP; Nonylphenyl phosphite (3:1)
Definition: Phosphite triester of nonyl phenol
Empirical: $C_{45}H_{69}O_3P$
Formula: $[C_9H_{19}C_6H_4]_3$—OPO_2H
Properties: Liq.; m.w. 688; dens. 0.980-0.992; b.p. will not distill; ref. index 1.5255-1.5280; flash pt. (PMCC) 207 C
Uses: Heat stabilizer for PVC, ABS, polyolefins, some rubber prods.
Manuf./Distrib.: Dover; Fabrichem
Trade names: ADK STAB 329; ADK STAB 1178; ADK STAB SC-102; Doverphos® 4; Doverphos® 4 Powd.; Lankromark® LE109; Naugard® P; Weston® TNPP; Wytox® 312
Trade names containing: Doverphos® 4-HR; Doverphos® 4-HR Plus; Naugard® PHR; Weston® 399; Weston® 399B

Trisodium EDTA

CAS 150-38-9; EINECS 205-758-8

Synonyms: Edetate trisodium; Trisodium ethylenediamine tetraacetate; Trisodium hydrogen ethylene diaminetetraacetate

Classification: Substituted amine

Empirical: $C_{10}H_{16}N_2O_8 \cdot 3Na$

Formula: $(NaOOCCH_2)_3NCH_2CH_2NCH_2COOH$

Properties: Wh. powd.; sol. in water

Uses: Chelating agent; catalyst in SBR mfg.

Regulatory: FDA 21CFR §175.105, 176.170, 177.2800, 178.3910

Manuf./Distrib.: Chemplex; Hampshire; Surfactants Inc.

Trade names: Hamp-Ene® Na_3 Liq.

Trisodium ethylenediamine tetraacetate. *See* Trisodium EDTA
Trisodium hydrogen ethylene diaminetetraacetate. *See* Trisodium EDTA

Tristearin

CAS 555-43-1; EINECS 209-097-6

Synonyms: Glyceryl tristearate; Glyceryl monotristearate; Stearin; 1,2,3-Propanetriol trioctadecanoate; Octadecanoic acid, 1,2,3-propanetriyl ester

Definition: Triester of glycerin and stearic acid

Empirical: $C_{57}H_{110}O_6$

Formula: $[CH_3(CH_2)_{16}COOCH_2]_2CHOCO(CH_2)_{16}CH_3$

Properties: Colorless crystals or powd., odorless, tasteless; insol. in water; sol. in hot alcohol, benzene, chloroform, carbon disulfide; m.w. 891.45; dens. 0.943 (65 C); m.p. 71.6 C; ref. index 1.4385 (80 C)

Precaution: Combustible

Toxicology: Heated to decomp., emits acrid smoke and irritating fumes

Uses: Soap, candles, adhesive pastes, metal polishes, waterproofing paper, textile sizes, leather stuffing, mfg. of stearic acid; processing aid for EPS; dispersant, lubricant for colorants; lubricant for PVC; crystallization accelerator, fermentation aid, formulation aid, lubricant, release agent, surface-finishing agent for foods

Regulatory: FDA 21CFR §172.811, 177.2800

Trade names: Pationic® 914; Pationic® 919; Pationic® 1019

Tristearyl phosphite

CAS 2082-80-6

Empirical: $C_{54}H_{111}O_3P$

Formula: $(C_{18}H_{37}O)_3P$

Properties: Wh. waxy solid; m.w. 838; dens. 0.845-0.855 (50/15.5 C); b.p. will not distill; ref. index 1.4490-1.4530 (50 C); flash pt. (PMCC) 129 C

Uses: Sulfur scavenger in lubricants; improves antiwear and antifriction props.; color/processing stabilizer for polymers

Trithiocyanuric acid. *See* 2,4,6-Trimercapto-s-triazine
Tritolyl phosphate. *See* Tricresyl phosphate

Trixylenyl phosphate

CAS 25155-23-1

Synonyms: Tri(dimethylphenyl)phosphite; Xylyl phosphate

Empirical: $C_{24}H_{27}O_4P$

Formula: $[(CH_3)_2C_6H_3O]_3PO$

Properties: Liq.; m.w. 410.48; sp.gr. 1.130-1.155; pour pt. -35 C; b.p. 243-265 C (10 mm); flash pt. (COC) 235 C; ref. index 1.551-1.555

Precaution: Combustible

Uses: Flame retardant, plasticizer for PVC, PVAc, cellulosics, PS, ABS, PVC wire and cable insulation; lubricant additive

Manuf./Distrib.: Akzo Nobel; Albright & Wilson Am.; Ashland; FMC; C.P. Hall; Harwick

Trade names: Antiblaze® TXP; Kronitex® TXP; Pliabrac® TXP

Trolamine. *See* Triethanolamine
TTD. *See* Tetraethylthiuram disulfide
TTDP. *See* Triisotridecyl phosphite
Turkey-red oil. *See* Sulfated castor oil
Two-stage resin. *See* Phenolic resin

Ultramarine blue
CAS 1317-97-1, 57455-37-5
Synonyms: Pigment blue 29; CI 77007
Classification: Inorganic pigment
Formula: $Na_6Al_6Si_6O_{24}S_4$
Properties: Blue powd.
Precaution: Noncombustible
Uses: Pigment for printing inks, thermoplastics, rubber compds., paints
Trade names: Ferro BP-10; Ferro CP-18; Ferro CP-50; Ferro CP-78; Ferro DP-25; Ferro EP-37; Ferro EP-62; Ferro FP-40; Ferro FP-64; Ferro RB-30
Trade names containing: Ultramarine Blue

Ultramarine violet
CAS 12769-96-9
Synonyms: CI 77007; Pigment violet 15
Definition: Sodium aluminum sulfosilicate complex
Formula: $Na_4H_2Al_6Si_6O_{24}S_2$
Uses: Pigment for thermoplastics, rubbers, paints, printing inks
Trade names: Ferro V-5; Ferro V-8

Undecyl dodecyl phthalate
CAS 68515-47-9
Empirical: $C_{34}H_{58}O_4$
Properties: M.w. 530.84; sp.gr. 0.959; pour pt. -40 C; flash pt. (TCC) 437 F
Uses: Plasticizer for PVC, PS, cellulosics
Manuf./Distrib.: Exxon
Trade names: Jayflex® UDP

Undecyl phthalate
Definition: Diester of phthalic acid and undecyl alcohol
Uses: Plasticizer for PVC
Trade names: Palatinol® 11

Urea
CAS 57-13-6; EINECS 200-315-5
Synonyms: Carbamide; Carbonyldiamide; Carbamidic acid; Isourea
Classification: Organic compd.
Empirical: CH_4N_2O
Formula: NH_2CONH_2
Properties: White cryst. or powd., almost odorless; sol. in water, alcohol, benzene; sl. sol. in ether; insol. in chloroform; m.w. 60.06; dens. 1.335; m.p. 132.7 C; b.p. dec.
Precaution: Heated to decomp., emits toxic fumes of NO_x
Toxicology: LD50 (oral, rat) 14,300 mg/kg; mod. toxic by ingestion, IV, subcutaneous routes; experimental carcinogen, neoplastigen, reproductive effects; human reproductive effects by intraplacental route; human mutagenic data; human skin irritant
Uses: Fertilizer; animal feed; stabilizer for resins, plastics, explosives; in paper industry to soften cellulose; in ammoniated dentifrices; diuretic; antiseptic; deodorizer; penetrant; cure accelerator and activator
Regulatory: FDA 21CFR §175.300, 177.1200, 184.1923, GRAS; BATF 27CFR §240.1051
Manuf./Distrib.: Air Prods.; Bio-Rad Labs; Chisso Am.; Elf Atochem SA; EM Industries; Heico; Mallinckrodt; Mitsui Toatsu; Nissan Chem. Ind.; Norsk Hydro AS; OxyChem; Showa Denko
Trade names: RIA CS; Stanclere® C26; WF-15; WF-16
Trade names containing: Chem-Master R-486; Poly-Dispersion® S (UR)D-75

Urea-formaldehyde resin
CAS 9011-05-6
Synonyms: Polyoxymethylene urea; Polynoxylin; Urea, polymer with formaldehyde
Classification: Amino resin
Definition: Reaction prod. of urea and formaldehyde
Empirical: $(CH_4N_2O \cdot CH_2O)_x$
Uses: Thermosetting resin; pigment-grinding medium; aids adhesion and toughness of coatings; wet strength resin in paper treatment; in automotive enamels and primers, metal decorating finishes; modifier for water-sol. polymers
Regulatory: FDA 21CFR §175.105, 175.300, 177.1200, 177.1650, 177.1900, 181.30

Manuf./Distrib.: Akzo Nobel; Bakelite GmbH; Cargill; Cytec; DSM UK; Georgia-Pacific; Hercules; Monomer-Polymer & Dajac; Sybron

Urea, polymer with formaldehyde. *See* Urea-formaldehyde resin
Urea, 1,1,3-trimethyl-2-thio. *See* Trimethylthiourea

Ureidosilane
Uses: Adhesion promoter for fillers/reinforcements and polymers; glass sizes/finishes, wool insulation resin binders, primers, foundry sand binders, adhesives, phenolic brake shoes
Trade names: Union Carbide® Y-11542

Urethane diacrylate
Uses: Additive improving flexibility of epoxy and urethane formulations; film-former for inks, coatings for plastic and metal, laminating adhesives
Trade names: Ebecryl® 230

Urotropine. *See* Hexamethylene tetramine
UV Absorber-2. *See* Etocrylene
UV Absorber-3. *See* Octocrylene
VAE. *See* Vinyl acetate/ethylene copolymer
VA/ethylene copolymer. *See* Vinyl acetate/ethylene copolymer
Vegetable carbon. *See* Carbon black

Vegetable oil, vulcanized
Uses: Rubber processing aid, extender; absorbent for min. oils and other liq. plasticizers
Trade names: Akrofax™ 900C; Akrofax™ A; Faktogel® 110; Faktogel® 111; Faktogel® 140; Faktogel® 141; Faktogel® 144; Faktogel® KE 8384; Faktogel® KE 8419
Trade names containing: Faktogel® A; Faktogel® Asolvan T

Vinegar acid. *See* Acetic acid
Vinegar naphtha. *See* Ethyl acetate

Vinyl acetate/ethylene copolymer
CAS 24937-78-8
Synonyms: VAE; VA/ethylene copolymer
Formula: $(C_4H_6O_2 \cdot C_2H_4)_x$
Uses: Coating binder and saturant for paper/paperboard, nonwovens, medical/surgical applics., fiber laminates, adhesives, carpet backings; vehicle base for paints, caulks, and mastics in the building industry; impact modifier for PVC
Trade names: Elvace® 1870; Levapren 400; Levapren 450HV; Levapren 452; Levapren 500HV; Levapren 700HV; Levapren KA 8385

Vinyl alcohol polymer. *See* Polyvinyl alcohol

N-[2-(Vinylbenzylamino)-ethyl)-3-aminopropyltrimethoxysilane
Classification: Styrylamine cationic silane
Uses: Coupling agent for unsaturated polyesters, styrenics, epoxies, PP, PE
Trade names: Dow Corning® Z-6032

Vinyl bromide
CAS 593-60-2
Synonyms: Bromoethylene
Formula: CH_2CHBr
Properties: Gas; m.w. 106.96; dens. 1.51; m.p. -138 C; b.p. 15.6 C
Toxicology: Carcinogen; TLV 5 ppm in air
Uses: Flame-retarding agent for acrylic fibers, PVAc, PVC
Manuf./Distrib.: Albemarle
Trade names: Saytex® VBR

Vinyl 2-ethylhexanoate
CAS 94-04-2
Synonyms: Vinyl-2-ethylhexoate
Classification: Vinyl ester monomer

Vinyl-2-ethylhexoate

Definition: Vinyl ester of 2-ethylhexanoic acid
Formula: CH_2=$CHOCOCHCH_2CH_3(CH_2)_3CH_3$
Properties: Liq.; m.w. 170.25; dens. 0.875; m.p. -90 C; b.p. 128-130 C (20 mm); flash pt. 65 C; ref. index 1.4260
Toxicology: Irritant
Uses: Reactive intermediate for polymerization for copolymer latexes, water-sol. polymers, thermosetting coatings, adhesives, inks, textile sizing
Trade names: Vynate® 2-EH

Vinyl-2-ethylhexoate. *See* Vinyl 2-ethylhexanoate

Vinyl homopolymer
Uses: Blending resin for modifying plastisol compds.
Trade names: Pliovic® M-50; Pliovic® M-70; Pliovic® M-70SC; Pliovic® M-90

Vinyl neodecanoate
CAS 45115-34-2
Classification: Vinyl ester monomer
Definition: Vinyl ester of neodecanoic acid
Formula: CH_2=$CHOCOCR_3$, where R is methyl or greater
Uses: Reactive intermediate for polymerization for copolymer latexes, water-sol. polymers, thermosetting coatings, adhesives, inks, textile sizing
Trade names containing: Vynate® Neo-10

Vinyl neononanoate
Classification: Vinyl ester monomer
Definition: Vinyl ester of neononanoic acid
Formula: CH_2=$CHOCOCR_3$, where R is methyl or greater
Uses: Reactive intermediate for polymerization for copolymer latexes, water-sol. polymers, thermosetting coatings, adhesives, inks, textile sizing
Trade names: Vynate® Neo-9

Vinyl pivalate
Classification: Vinyl ester monomer
Definition: Vinyl ester of neopentanoic acid
Formula: CH_2=$CHOCOC(CH_3)_3$
Uses: Reactive intermediate for polymerization for copolymer latexes, water-sol. polymers, thermosetting coatings, adhesives, inks, textile sizing
Trade names: Vynate® Neo-5

1,2-Vinylpolybutadiene
Uses: Thermosetting resin used in elec. potting and impregnation of transformers, capacitors, motors laminates, molding compds. and castings, rubber modifiers, mica paper binder, nuclear heat shield; for food pkg.; coagent for rubbers
Trade names: Ricon 150
Trade names containing: Ricon P30/Dispersion

Vinyl propionate
CAS 105-38-4
Classification: Vinyl ester monomer
Definition: Vinyl ester of propionic acid
Formula: CH_2=$CHOCOCH_2CH_3$
Properties: Liq.; insol. in water; m.w. 100.12; dens. 0.919; m.p. -80 C; b.p. 94-95 C; flash pt. 6 C; ref. index 1.4030
Precaution: Flamm.
Toxicology: Irritant
Uses: Reactive intermediate for polymerization for copolymer latexes, water-sol. polymers, thermosetting coatings, adhesives, inks, textile sizing
Manuf./Distrib.: BASF; Monomer-Polymer & Dajac
Trade names: Vynate® L-3

N-Vinyl pyrrolidone. *See* N-Vinyl-2-pyrrolidone

N-Vinyl-2-pyrrolidone
CAS 88-12-0; EINECS 201-800-4
Synonyms: 1-Ethenyl-2-pyrrolidinone; N-Vinyl pyrrolidone
Empirical: C_6H_9NO
Formula: CH_2=$CHNCH_2CH_2CH_2CO$
Properties: Colorless liq.; sol. in water; m.w. 111.16; dens. 1.04 (25 C); f.p. 13.5 C; b.p. 148 C (100 mm); flash pt. (COC) 209 F; ref. index 1.511
Precaution: Combustible
Toxicology: Moderately toxic by ingestion, inhalation, skin contact; severe eye irritant; narcotic
Uses: Polyvinylpyrrolidone, organic synthesis
Manuf./Distrib.: Allchem Ind.; BASF; ISP; Monomer-Polymer & Dajac; Polysciences
Trade names containing: V-Pyrol
See also PVP

Vinyltriacetoxy silane
CAS 4130-08-9; EINECS 223-943-1
Synonyms: Triacetoxyvinylsilane
Empirical: $C_8H_{12}O_6Si$
Formula: H_2C=$CHSi(OCOCH_3)_3$
Properties: M.w. 232.26; dens. 1.167 (20/4 C); b.p. 112-113 C (1 mm); flash pt. 88 C; ref. index 1.423 (20 C)
Uses: Coupling agent for polyesters, polyolefins, EPDM
Trade names: Dow Corning® Z-6075

Vinyltrichlorosilane
CAS 75-94-5; EINECS 200-917-8
Synonyms: (Trichlorosilyl)ethylene; Trichlorovinylsilane
Empirical: $C_2H_3Cl_3Si$
Formula: H_2C=$CHSiCl_3$
Properties: Colorless or pale yel. liq.; sol. in most org. solvs.; m.w. 161.5; dens. 1.265 (25/25 C); b.p. 90.6 C; flash pt. 21 C; ref. index 1.429 (20 C); visc. 0.50 ctsk
Toxicology: Strong irritant to tissues; LD50 (rat, oral) 3160 mg/kg
Uses: Intermediate for silicones; coupling agent in adhesives and bonds; release agent, lubricant
Manuf./Distrib.: Gelest; Hüls Am.; OSi Spec.; PCR; Schweizerhall
Trade names: Dynasylan® VTC; Petrarch® V4900

Vinyltriethoxysilane
CAS 78-08-0; EINECS 201-081-7
Synonyms: (Triethoxysilyl)ethylene; Triethoxyvinylsilane
Empirical: $C_8H_{18}O_3Si$
Properties: Liq.; m.w. 190.31; dens. 0.903 (20 C); b.p. 160-161 C; flash pt. 34 C; ref. index 1.396 (20 C); visc. 0.70 ctsk
Uses: Intermediate, esp. when acidic by-products are undesirable; filler; coupling agent, release agent, lubricant
Manuf./Distrib.: Hüls Am.; OSi Spec.; PCR
Trade names: Dynasylan® VTEO; Petrarch® V4910
Trade names containing: Aktisil® VM 56; Coupsil VP 6508

Vinyltrimethoxysilane
CAS 2768-02-7; EINECS 220-449-8
Synonyms: Trimethoxyvinylsilane
Empirical: $C_5H_{12}O_3Si$
Formula: H_2C=$CHSi(OCH_3)_3$
Properties: Liq.; m.w. 148.23; dens. 0.970 (20/4 C); b.p. 123 C; flash pt. 23 C; ref. index 1.3930 (20 C)
Precaution: Flamm.
Toxicology: LD50 (rat, oral) 11,300 mg/kg
Uses: In peroxide graft/moisture crosslinking of polyethylene
Trade names: Dynasylan® VTMO; Petrarch® V4917

Vinyltris(2-methoxyethoxy) silane
CAS 1067-53-4; EINECS 213-934-0
Empirical: $C_{11}H_{24}O_6Si$
Properties: Liq.; m.w. 280.39; dens. 1.034 (25 C); b.p. 284-286 C; flash pt. 65 C; ref. index 1.427 (25 C)
Toxicology: LD50 (rat, oral) 2960 mg/kg

Uses: Filler for peroxide-cured systems
Trade names: Dynasylan® VTMOEO; Petrarch® V5000
Trade names containing: Aktisil® VM

VM&P naphtha. *See* Naphtha
Waxes, microcrystalline. *See* Microcrystalline wax
Waxes, montan. *See* Montan wax
Waxes, montan fatty acids. *See* Montan acid wax

Wheat starch
CAS 9005-25-8
Definition: Natural material obtained from wheat, contg. amylose and amylopectin
Regulatory: FDA 21CFR §175.105, 178.3520, 182.70
Manuf./Distrib.: Int'l. Chem.
Trade names containing: Luperco AA

White lead. *See* Lead carbonate (basic)
White lead, sublimed. *See* Lead sulfate, basic
White lead sulfate. *See* Lead sulfate, basic
White mineral oil. *See* Mineral oil
White ozokerite wax. *See* Ceresin
White petrolatum. *See* Petrolatum
White spirits. *See* Mineral spirits
Wolfen. *See* Copper dimethyldithiocarbamate

Wollastonite
CAS 13983-17-0
Synonyms: Calcium metasilicate
Empirical: CaH_2O_3Si
Properties: M.w. 118.19
Uses: Extender, filler, pigment for paints, plastics, rubber, friction, refractory, ceramic, construction, sealants, adhesives
Manuf./Distrib.: Am. Colloid; Nyco Min.; R.T. Vanderbilt
Trade names: Nyad® 200, 325, 400; Nyad® 475, 1250; Nyad® FP, G, G Special; Nyad G® Wollastocoat®; Nycor® R; Vansil® W-10; Vansil® W-20; Vansil® W-30; 10, 325, 400 Wollastocoat®

Wood alcohol. *See* Methyl alcohol
Wood naphtha. *See* Methyl alcohol
Wood pulp, bleached. *See* Cellulose
Wood spirit. *See* Methyl alcohol
Wool fat. *See* Lanolin
Wool wax. *See* Lanolin
XSA. *See* Xylene sulfonic acid

Xylene
CAS 1330-20-7; EINECS 215-535-7
Synonyms: Methyl toluene; Dimethylbenzene; Xylol
Classification: Aromatic compd.
Definition: Commercial mixture of 3 isomers: o-, m-, and p-xylene (1,3-dimethylbenzene, 1,2-dimethylbenzene, 1,4-dimethylbenzene)
Empirical: C_8H_{10}
Formula: $C_6H_4(CH_3)_2$
Properties: Clear liq.; pract. insol. in water; misc. with abs. alcohol, ether, many org. liqs.; m.w. 106.16; dens. 0.86; b.p. 137-144 C; flash pt. 29 C; ref. index 1.4970
Precaution: Flammable; moderate fire risk
Toxicology: Toxic by ingestion, inhalation; irritant; may be narcotic in high concs.; TLV 100 ppm in air
Uses: Solvent; raw material for prod. of benzoic acid, phthalic anhydride, dyes, other organics; aviation gasoline; protective coating; solvent for alkyd resins, lacquers, enamels, rubber cements
Regulatory: FDA 21CFR §175.105, 176.180, 177.1010, 177.1650
Manuf./Distrib.: Ashland; Crowley; Exxon; Mallinckrodt; Mitsubishi Petrochem.; Mitsui Petrochem. Ind.; Mobil; Shell; Texaco
Trade names containing: Ancamide 220-X-70; Ancamide 400-BX-60; Dow Corning® 1250; EXP-36-X20 Epoxy Functional Silicone Sol'n.; EXP-38-X20 Epoxy Functional Silicone Sol'n.; Trigonox® SBP-AX30

p-Xylene, α,α,α′,α′-tetrabromo. *See* Tetrabromo-p-xylene

Xylene sulfonic acid
CAS 25321-41-9; EINECS 246-839-8
Synonyms: XSA; Dimethylbenzenesulfonic acid; Benzenesulfonic acid, dimethyl-
Definition: Mixture of substituted aromatic acids
Empirical: $C_8H_{10}O_3S$
Properties: M.w. 186.24
Uses: Intermediate, catalyst in preparation of esters, hardening agent in plastics, activator for nicotine insecticides; curing agent for resins
Trade names: Eltesol® 4200; Eltesol® XA65; Eltesol® XA90; Eltesol® XA/M65; Manro FCM 90LV; Reworyl® X 65; XSA 80; XSA 90
Trade names containing: Eltesol® 4009, 4018

Xylol. *See* Xylene
Xylyl phosphate. *See* Trixylenyl phosphate
ZBeDC. *See* Zinc dibenzyl dithiocarbamate
ZDMC. *See* Zinc dimethyldithiocarbamate

Zinc
CAS 7440-66-6; EINECS 231-175-3
Classification: Metallic element
Empirical: Zn
Properties: Shining white metal; sol. in acids, alkalies; insol. in water; at.wt. 65.38; dens. 7.14; m.p. 419 C; b.p. 907 C
Uses: Alloys, galvanizing iron and other metals, fungicides, in PVC stabilizers/activators
Manuf./Distrib.: Aldrich; Cerac; Cuproquim; Ferro/Bedford; Pasminco Europe; U.S. Zinc; Zinc Corp. of Am.
Trade names: Lankromark® LZ440; Synpron 1112; Synpron 1236; Synpron 1840; Synpron 1890

Zinc, bis(bis(2-methylpropyl)carbamodithioato-S,S′)-. *See* Zinc diisobutyldithiocarbamate
Zinc bis(dibutylcarbamodithiato-S-S′). *See* Zinc dibutyldithiocarbamate
Zinc bis(diethylcarbamodithioato-S,S′). *See* Zinc diethyldithiocarbamate
Zinc bis(dimethylthiocarbamoyl)disulfide. *See* Zinc dimethyldithiocarbamate

Zinc borate
CAS 1332-07-6 (anhyd.); 120007-67-9; 12767-90-7; 12447-61-9; 138265-88-0; EINECS 215-566-6; 233-471-8
Synonyms: Boric acid, zinc salt
Classification: Inorganic salt
Definition: Inorganic salt of indefinite composition, contg. zinc oxide and boric oxide in various ratios
Empirical: $B_4H_2O_7 \cdot Zn$
Formula: $xZnO \cdot yB_2O_3 \cdot zH_2O$
Properties: White powd.; sol. in dilute acids; sl. sol. in water; dens. 3.64; m.p. 980 C
Precaution: Nonflamm.
Uses: Medicine, fireproofing textiles, fungistat, mildew inhibitor; afterglow suppressant and synergist with antimony trioxide for fire protection in plastics
Manuf./Distrib.: ACM; Allchem Ind.; BA Chem. Ltd; R.E. Carroll; Climax Performance; Joseph Storey; U.S. Borax & Chem.
Trade names: ADK STAB 2335; Firebrake® 500; Firebrake® ZB; Firebrake® ZB 415; Firebrake® ZB-Extra Fine; Firebrake® ZB-Fine; Flamtard Z10; Flamtard Z15; Zb™-112; Zb™-223; Zb™-237; Zb™-325; Zb™-467; Zinc Borate 9506; Zinc Borate B9355

Zinc-calcium complex
Uses: PVC stabilizer
Trade names: Synpron 1321; Synpron 1566; Synpron 1693; Synpron 1699

Zinc carbonate
CAS 3486-35-9; EINECS 222-477-6
Synonyms: Carbonic acid, zinc salt (1:1); Smithsonite
Classification: Inorganic salt
Empirical: $CH_2O_3 \cdot Zn$
Formula: $ZnCO_3$
Properties: White cryst. powd.; sol. in acids, alkalies, ammonium salt sol'n.; insol. in water; m.w. 1225.38; dens.

4.42-4.45
Uses: Accelerator-activator for transparent nat. and syn. rubber goods, adhesives; as pigment; fire-proofing filler for rubber and plastics; topical antiseptics, cosmetics, lotions
Regulatory: FDA 21CFR §175.300, 177.1460, 177.2600, 582.80
Manuf./Distrib.: Allchem Ind.; Harcros Durham; Nihon Kagaku Sangyo; Spectrum Chem. Mfg.
Trade names: Akrochem® 9930 Zinc Oxide Transparent

Zinc chloride
CAS 7646-85-7; EINECS 231-592-0
Empirical: Cl_2Zn
Properties: Cubic white deliq. cryst., odorless; sol. in water, alcohol, glycerol, ether; m.w. 136.27; dens. 2.91 (25 C); m.p. 290 C; b.p. 732 C
Toxicology: TLV:TWA 1 mg/m^3 (air); poison by ingestion, intravenous, subcutaneous, intraperitoneal routes; corrosive irritant to skin, eyes, mucous membranes
Uses: Catalyst, dehydrating and condensing agent in organic synthesis, fireproofing, preserving food, electroplating, antiseptic denaturant for alcohol
Regulatory: FDA 21CFR §182.70, 182.5985, 182.8985, 582.80, GRAS
Manuf./Distrib.: Aldrich; AlliedSignal; Blythe, William Ltd; Elf Atochem N. Am.; EM Industries; Kraft; Mallinckrodt; Penta Mfg.; SAF Bulk; San Yuan; Zaclon
Trade names containing: Thanecure®

Zinc diacrylate
CAS 14643-87-9
Empirical: $C_6H_6O_4Zn$
Formula: $HCCH_2COOZnOCOHCCH_2$
Properties: M.w. 207.0
Uses: Crosslinker for molded polybutadiene compounds, conductive and protective coatings, coagent for SBR compds. and reactive pigments; activator for rubber compounding; scorch retarder
Manuf./Distrib.: Monomer-Polymer & Dajac
Trade names: Akrochem® ZDA Powd.; Saret® 633; SR-416

Zinc dialkyl dithiophosphate
Trade names containing: Rhenocure® TP/G; Rhenocure® TP/S

Zinc diamyldithiocarbamate
CAS 15337-18-5
Uses: Antioxidant, metal deactivator, copper corrosion inhibitor, color stabilizer, antiwear agent used in engine and industrial oils, greases; accelerator for nat. and syn. rubbers
Trade names: Amyl Zimate®

Zinc dibenzyl dithiocarbamate
CAS 14726-36-4
Synonyms: ZBeDC
Empirical: $C_{30}H_{28}N_2S_4Zn$
Formula: $Zn[SCSN(C_7H_7)_2]_2$
Properties: White powd.; moderately sol. in benzene, ethylene dichloride; insol. in acetone, gasoline, water; m.w. 610.21; dens. 1.41; m.p. 165-175 C
Uses: Accelerator for rubber, latex dispersions, cements
Manuf./Distrib.: Novachem
Trade names: Akrochem® Z.B.E.D.; Arazate®; Naftocit® ZBEC; Octocure ZBZ-50; Perkacit® ZBEC; Vulkacit ZBEC
Trade names containing: Poly-Dispersion® T (AZ)D-75

Zinc dibutyldithiocarbamate
CAS 136-23-2
Synonyms: Zinc bis(dibutylcarbamodithiato-S-S′)
Empirical: $C_{18}H_{36}S_4N_2Zn$
Formula: $Zn[SC(S)N(C_4H_9)_2]_2$
Properties: Wh. powd., pleasant odor; sol. in carbon disulfide, benzene, chloroform; insol. in water; m.w. 476.19; dens. 1.24 (20/20 C); m.p. 104-108 C
Toxicology: LD50 (rat, oral) > 16,000 mg/kg
Uses: Accelerator for latex dispersions and cements; ultra accelerator; oil additive
Manuf./Distrib.: Complex Quimica SA; Novachem; Uniroyal; R.T. Vanderbilt

Trade names: Akrochem® Accelerator BZ Powder; Butasan®; Butyl Zimate®; Butyl Zimate® Dustless; Butyl Zimate® Slurry; Naftocit® Di 13; Naftopast® Di13-P; Octocure ZDB-50; Perkacit® ZDBC
Trade names containing: Poly-Dispersion® T (BZ)D-75

Zinc dibutyl dithiophosphorate
CAS 6990-43-8
Synonyms: Zinc O,O-di-n-butyl-phosphorodithioate
Empirical: $C_{16}H_{36}O_4P_2S_4Zn$
Properties: M.w. 548.06
Uses: Accelerator for EPDM cures; sec. accelerator for thiazoles and sulfenamides
Trade names: Vocol®, S-75
Trade names containing: Akrochem® Accelerator VS

Zinc O,O-di-n-butyl-phosphorodithioate. *See* Zinc dibutyl dithiophosphorate

Zinc diethyldithiocarbamate
CAS 14323-55-1
Synonyms: Zinc bis(diethylcarbamodithioato-S,S´)
Empirical: $C_{10}H_{20}S_4N_2Zn$
Formula: $Zn[SC(S)N(C_2H_5)_2]_2$
Properties: White powd.; sol. in carbon disulfide, benzene, chloroform; insol. in water; dens. 1.47 (20/20 C); m.p. 172-176 C
Toxicology: Strong irritant to eyes, mucous membranes
Uses: Rubber vulcanization accelerator, heat stabilizer for polyethylene
Manuf./Distrib.: Complex Quimica SA; Novachem; R.T. Vanderbilt
Trade names: Akrochem® Accelerator EZ; Anchor® ZDBC; Anchor® ZDEC; Ethyl Zimate®; Ethyl Zimate® Slurry; Naftocit® Di 7; Naftopast® Di7-P; Octocure ZDE-50; Perkacit® ZDEC

Zinc di-2-ethylhexoate
Uses: Rubber activator
Trade names: Octoate® Z; Octoate® Z Solid

Zinc diisobutyldithiocarbamate
CAS 36190-62-2
Synonyms: Zinc, bis(bis(2-methylpropyl)carbamodithioato-S,S´)-
Uses: Accelerator for rubber
Trade names: Isobutyl Zimate®

Zinc dimethacrylate
Empirical: $C_8H_{10}O_4Zn$
Formula: $CH_3CH_2CCOOZnOCOCCH_3CH_2$
Uses: Scorch retarder, coagent in rubber cures
Manuf./Distrib.: Monomer-Polymer & Dajac; Rohm Tech
Trade names: Saret® 634; SR-365

Zinc dimethyldithiocarbamate
CAS 137-30-4
Synonyms: ZDMC; Ziram; Bis(dimethylcarbamodithioato,S,S´) zinc; Zinc bis(dimethylthiocarbamoyl)disulfide
Empirical: $C_6H_{12}N_2S_4Zn$
Properties: Crystals; pract. insol. in water; sol. < 0.5 g/100 ml in acetone, benzene; sol. in dilute caustic sol'ns.; m.w. 305.82; m.p. 250 C
Precaution: Can form a flammable dust
Toxicology: Irritant to skin and mucous membranes; LD50 (rat, oral) 1.4 g/kg; suspected carcinogen
Uses: Rubber vulcanization accelerator; agric. fungicide
Manuf./Distrib.: Novachem; Uniroyal
Trade names: Akrochem® Accelerator MZ; Methasan®; Methyl Zimate®; Naftocit® Di 4; Octocure ZDM-50; Perkacit® ZDMC; Vancide® MZ-96
Trade names containing: Methyl Zimate® Slurry; Poly-Dispersion® E (MZ)D-75; Vancide® 51Z Disp.

Zinc dioxide. *See* Zinc peroxide

Zinc ditolyl sulfinate dihydrate
Properties: Wh. powd.; m.p. 265-275 C

Zinc-epoxy

Uses: Activator
Trade names: Cellex-TS

Zinc-epoxy
Uses: Plasticizer for stabilization of rigid PVC bottles
Trade names: Lankromark® LZ1034

Zinc 2-ethylhexanoate
CAS 136-53-8
Synonyms: Hexanoic acid, 2-ethyl-, zinc salt
Empirical: $C_8H_{16}O_2 \cdot {}^1/_2Zn$
Properties: M.w. 176.90
Uses: Activator for natural rubber
Trade names: Struktol® ZEH
Trade names containing: Struktol® ZEH-DL

Zinc 2-ethylhexoate
Synonyms: Zinc octoate
Formula: $Zn(OOCCH(C_2H_5)C_4H_9)_2$
Properties: Lt. straw-colored visc. liq.; sol. in hydrocarbon solvs.; insol. in water; dens. 1.16
Precaution: Combustible
Uses: Activator for natural and syn. rubbers; stabilizer for foam processing; catalyst
Manuf./Distrib.: OM Group; Shepherd
Trade names: Bärostab® L 230

Zinc formaldehyde sulfoxylate
CAS 24887-06-7; EINECS 246-515-6
Synonyms: Bis(hydroxymethanesulfinato-O,O´) zinc; Zinc sulfoxylate formaldehyde
Empirical: $C_2H_6O_6S_2Zn$
Formula: $Zn((HOCH_2SO_2)_2$
Properties: Rhombic prisms; sol. in water; insol. in alcohol; dec. in acid; m.w. 255.59
Toxicology: Toxic by ingestion
Uses: Stripping and discharging agent for textiles; reducing agent for redox-catalyzed polymerization
Regulatory: FDA 21CFR §175.105, 176.170
Manuf./Distrib.: Phibrochem
Trade names: Parolite

Zinc hydroxystannate
CAS 12027-96-2
Formula: $ZnSn(OH)_6$
Properties: Wh. powd.; sol. in strong acids and bases; m.w. 286.12; sp.gr. 3.4; decomp. temp. 180 C
Toxicology: LD50 (oral, rat) > 5000 mg/kg, (dermal, rat) > 2466 mg/kg; nuisance dust—may cause respiratory tract irritation, drying of skin, eye irritation
Uses: Fire retardant and smoke suppressant for plastics, rubber, paints
Manuf./Distrib.: Joseph Storey
Trade names: Flamtard H

Zinc isopropyl xanthate
CAS 1000-90-4, 42590-53-4
Empirical: $C_8H_{14}O_2S_4Zn$
Properties: M.w. 270.46
Uses: Accelerator for rubber and latexes
Trade names: Octocure ZIX-50
Trade names containing: Propyl Zithate®

Zinc lactylate
Definition: Zinc salt derived from lactic acid
Uses: Additive for polymer and plastics

Zinc laurate
CAS 2452-01-9; EINECS 219-518-5
Synonyms: Dodecanoic acid, zinc salt

Formula: $Zn(C_{12}H_{23}O_2)_2$
Properties: White powd.; sl. sol. in water and alcohol; m.w. 463.99; m.p. 128 C
Precaution: Combustible
Uses: Softener, activator for rubber compounding; heat stabilizer for rubber and PVC, paints, varnishes
Trade names: Haro® Chem ZSG

Zinc 2-mercaptobenzothiazole
CAS 155-04-4
Synonyms: ZMBT; 2(3H)-Benzothiazolethione, zinc salt; Zinc mercaptobenzylthiazol
Empirical: $C_{14}H_8N_2S_4Zn$
Formula: $Zn(C_7H_4NS_2)_2$
Properties: M.w. 397.85
Uses: Accelerator for rubber, latex foam curing systems
Trade names: Akrochem® ZMBT; Bantox®; Ekaland ZMBT; Naftocit® ZMBT; Octocure ZMBT-50; Perkacit® ZMBT; Vulkacit ZM; Zetax®; Zetax® Dispersion
Trade names containing: Vancide® 51Z Disp.

Zinc mercaptobenzylthiazol. *See* Zinc 2-mercaptobenzothiazole
Zinc mercapto methyl benzimidazole. *See* Zinc 4- and 5-methylmercaptobenzimidazole

Zinc 2-mercaptotoluimidazole
CAS 61617-00-3
Uses: Antioxidant for NR and SR, EPDM, nitrile stock
Trade names: Vanox® ZMTI

Zinc methacrylate
Uses: Coagent for elastomer processing and cure
Manuf./Distrib.: Gelest; Monomer-Polymer & Dajac; Polysciences
Trade names: SR-376

Zinc 4- and 5-methylmercaptobenzimidazole
Synonyms: Zinc mercapto methyl benzimidazole
Uses: Antioxidant for use in rubber compds. to improve heat resistance; sensitizer for latex
Trade names: Akrochem® Antioxidant 58; Vulkanox® ZMB2/C5

Zinc octadecanoate. *See* Zinc stearate
Zinc octoate. *See* Zinc 2-ethylhexoate
Zinc orthophosphate. *See* Zinc phosphate

Zinc oxide
CAS 1314-13-2; EINECS 215-222-5
Synonyms: Chinese white; Pigment White 4; CI 77947; Zinc white; Flowers of zinc
Classification: Inorganic oxide
Empirical: OZn
Formula: ZnO
Properties: White to gray powd. or crystals, odorless, bitter taste; sol. in dilute acetic or min. acids, alkalies; insol. in water, alcohol; m.w. 81.38; dens. 5.67; m.p. 1975 C; ref. index 2.0041-2.0203; pH 6.95 (Amer. process), 7.37 (French process)
Precaution: Heated to decomp., emits toxic fumes of ZnO
Toxicology: TLV/TWA 5 mg/m^3; LD50 (IP, rat) 240 mg/kg; poison by IP route; fumes may cause metal fume fever with chills, fever, tightness in chest, cough, leukocytes; experimental teratogen; mutagenic data; skin/eye irritant
Uses: UV absorber; accelerator activator for rubber; pigment in white paints, cosmetics, driers, dental cements; mold inhibitor in paints; in mfg. of opaque glass, enamels, tires, printing inks, porcelains; reagent in analytical chemistry; flame retardant
Regulatory: FDA 21CFR §73.1991, 73.2991, 175.300, 177.1460, 182.5991, 182.8991, 582.80, GRAS
Manuf./Distrib.: Am. Chemet; Asarco; Eagle Zinc; General Chem.; Harcros Durham; Mallinckrodt; Zinc Corp. of Am.
Trade names: Akrochem® Zinc Oxide 35; Kadox®-215; Kadox®-272; Kadox® 720; Kadox®-911; Kadox®-920; Kadox® 930; Ken-Zinc®; Naftopast® ZnO-A; NAO 105; NAO 115; NAO 125; NAO 135; Octocure 462; Ottalume 2100; RR Zinc Oxide (Untreated); Zinc Oxide No. 185; Zinc Oxide No. 318; Zink Oxide AT; Zink Oxide Transparent; Zinkoxyd Activ®
Trade names containing: Chem-Master R-13; Chem-Master R-102; Chem-Master R-103; Kenlastic® K-9273;

Zinc-N-pentamethylene dithiocarbamate

Poly-Dispersion® A (Z-CN)D-85; Poly-Dispersion® AZFD-85; Poly-Dispersion® EZFD-85; Poly-Dispersion® JZFD-90P; Poly-Dispersion® SZFD-85; Poly-Dispersion® SZFND-825; Poly-Dispersion® TZFD-88P; Prespersion PAB-866; Prespersion PAC-3656; Rapidblend 1793; Rhenovin® ZnO-90; RR Zinc Oxide-Coated; Zic Stick '85'

Zinc-N-pentamethylene dithiocarbamate
CAS 13878-54-1
Empirical: $C_{12}H_{20}N_2S_4Zn$
Properties: M.w. 385.95
Uses: Rubber accelerator
Trade names: Akrochem® Z.P.D.

Zinc peroxide
CAS 1314-22-3; EINECS 215-226-7
Synonyms: Zinc superoxide; Zinc dioxide
Classification: Inorganic oxide
Empirical: O_2Zn
Formula: ZnO_2
Properties: Wh. powd.; m.w. 97.37; dens. 1.571
Precaution: Explosion risk when heated (190-212 C); strong oxidizing agent
Uses: Curing agent for carboxylated NBR
Manuf./Distrib.: FMC; Spectrum Chem. Mfg.
Trade names: Struktol® ZP 1014
Trade names containing: Chem-Master RD-50

Zinc phosphate
CAS 7779-90-0
Synonyms: Zinc orthophosphate
Empirical: $O_8P_2Zn_3$
Formula: $Zn_3(PO_4)_2$
Properties: White powd., odorless; sol. in dilute min. acids, acetic acid, ammonium hydroxide, alkali hydroxide solíns.; insol. in water, alcohol; m.w. 386.05; dens. 3.998 (15 C); m.p. 900 C
Uses: In dental cements, phosphors; flame retardant for plastics
Manuf./Distrib.: Calgon; Colores Hispania SA; Hammond Lead Prods.; Pasminco Europe; G. Whitfield Richards; Witco/Allied-Kelite
Trade names: KemGard 981

Zinc phosphite
Uses: PVC stabilizer
Trade names: ADK STAB EC-14; Synpron 231

Zinc-potassium complex
Uses: Stabilizer/activator for PVC plastisol foams
Trade names: Synpron 1642

Zinc 2-pyridinethiol-1-oxide. *See* Zinc pyrithione

Zinc pyrithione
CAS 13463-41-7; EINECS 236-671-3
Synonyms: Bis[1-hydroxy-2(1H)-pyridinethinato-O,S]-(T-4) zinc; Zinc 2-pyridinethiol-1-oxide; Pyrithione zinc
Classification: Aromatic salt
Empirical: $C_{10}H_8N_2O_2S_2Zn$
Properties: M.w. 317.7
Precaution: Do not store with strong oxidizing agents
Toxicology: LD50 (rat, oral) 260 mg/kg; LD50 (rat, dermal) > 2 g/kg; poison by ingestion, intraperitoneal; irritating to skin and extremely irritating to eyes
Uses: Antidandruff agent for shampoos; cosmetic preservative; antimicrobial for plastics, metalworking fluids
Manuf./Distrib.: Allchem Ind.; Olin; Pyrion-Chemie GmbH; Ruetgers-Nease
Trade names: Zinc Omadine® 48% Fine Particle Disp.; Zinc Omadine® 48% Std. Disp.; Zinc Omadine® Powd.
Trade names containing: Omacide® P-711-5; Omacide® P-BBP-5; Omacide® P-DIDP-5; Omacide® P-DOP-5; Omacide® P-ESO-5

Zinc stannate
CAS 12036-37-2
Uses: Fire retardant and smoke suppressant for plastics
Manuf./Distrib.: Atomergic Chemetals; Blythe, William Ltd; Joseph Storey
Trade names: Flamtard S

Zinc stearate
CAS 557-05-1; EINECS 209-151-9
Synonyms: Zinc octadecanoate; Octadecanoic acid, zinc salt
Definition: Zinc salt of stearic acid
Empirical: $C_{36}H_{70}O_4Zn$
Formula: $Zn(C_{18}H_{35}O_2)_2$
Properties: White powd., faint odor; sol. in acids, common solvs. (hot); insol. in water, alcohol, ether; dec. by dilute acids; m.w. 632.33; dens. 1.095; m.p. 130 C
Precaution: Combustible
Uses: In cosmetics, pharmaceuticals, lacquers, ointments, tablet mfg.; mold release agent for plastic; filler, antifoamer; flatting agent in lacquers; as a drying lubricant and dusting agent for rubber; waterproofing agent for concrete, paper, textiles
Regulatory: FDA 21CFR §175.105, 175.300, 176.170, 176.180, 176.200, 176.210, 177.1200, 177.1460, 177.1900, 177.2410, 177.2600, 178.2010, 178.3910, 182.5994, 182.8994, GRAS
Manuf./Distrib.: Aldrich; Allchem Ind.; Ferro/Grant; Magnesia GmbH; Mallinckrodt; Norac; Syn. Prods.
Trade names: ADK STAB ZN-ST; Afco-Chem ZNS; Akrochem® P 3100; Akrochem® Wettable Zinc Stearate; Akrochem® Zinc Stearate; Akrodip™ Z-50/50; Akrodip™ Z-200; Akrodip™ Z-250; Cecavon ZN 70; Cecavon ZN 71; Cecavon ZN 72; Cecavon ZN 73; Cecavon ZN 735; Coad 20; Coad 21; Coad 27B; Cometals Zinc Stearate; Disperso II; Hallcote® ZS; Hallcote® ZS 5050; Haro® Chem ZGD; Haro® Chem ZGN; Haro® Chem ZGN-T; Haro® Chem ZPR-2; Hy Dense Zinc Stearate XM Powd.; Hy Dense Zinc Stearate XM Ultra Fine; Hydro Zinc™; HyTech RSN 131 HS/Gran.; Interstab® ZN-18-1; Liquazinc AQ-90; Lubrazinc® W, Superfine; Mathe Zinc Stearate 25S; Mathe Zinc Stearate S; Metasap® Zinc Stearate; Miljac Zinc Stearate; Molder's Edge ME-369; Petrac® Zinc Stearate ZN-41; Petrac® Zinc Stearate ZN-42; Petrac® Zinc Stearate ZN-44 HS; Petrac® Zinc Stearate ZW-45; Polyvel RZ40; Quikote™; Quikote™ M; Rhenodiv® ZB; Rubichem Zinc-Dip®; Rubichem Zinc-Dip® HS; Rubichem Zinc-Dip® W; Sprayon 812; Synpro® Zinc Dispersion Zincloid; Synpro® Zinc Stearate 8; Synpro® Zinc Stearate ACF; Synpro® Zinc Stearate D; Synpro® Zinc Stearate GP; Synpro® Zinc Stearate GP Flake; Synpro® Zinc Stearate HSF; TR-Zinc Stearate; Wet Zinc™ ; Wet Zinc™ P; Witco® Zinc Stearate 11; Witco® Zinc Stearate 42; Witco® Zinc Stearate 44; Witco® Zinc Stearate Disperso; Witco® Zinc Stearate Heat-Stable; Witco® Zinc Stearate LV; Witco® Zinc Stearate NW; Witco® Zinc Stearate Polymer Grade; Witco Zinc Stearate Regular; Witco® Zinc Stearate REP; Zincote™; Zinc Stearate 42
Trade names containing: Altax®; Liquax 488; Mistron ZSC; Zinc Stearate Spray No. Z212-B

Zinc sulfide
CAS 1314-98-3; EINECS 215-251-3
Classification: Inorganic salt
Empirical: SZn
Formula: ZnS
Properties: Ylsh.-wh. powd.; sol. in dilute min. acids; insol. in water, alkalies; m.w. 97.45 dens. 3.98
Toxicology: Irritant
Uses: Pigment; in white and opaque glass, plastics, dyeing, paints, linoleum, leather, dental rubber; fungicide; anhydrous in x-ray screens, TV screens
Regulatory: FDA 21CFR §175.105, 177.2600, 178.3297, 178.3570
Manuf./Distrib.: Aceto; Cerac; Chemson Ltd.; Eagle-Picher; Noah; Ore & Chem. Corp.; Sachtleben GmbH
Trade names: Sachtolith® HD; Sachtolith® HD-S; Sachtolith® L

Zinc sulfoxylate formaldehyde. *See* Zinc formaldehyde sulfoxylate
Zinc superoxide. *See* Zinc peroxide
Zinc white. *See* Zinc oxide
Ziram. *See* Zinc dimethyldithiocarbamate

Zirconium propyl tris aminoethyl ethanolamine
Trade names containing: SAFR-Z-44 K

ZMBT. *See* Zinc 2-mercaptobenzothiazole
ZMMBI. *See* 4- and 5-Methylmercaptobenzimidazole zinc salt

Part III
Functional
Cross-Reference

Functional Cross-Reference

Trade name and generic chemical additives from the first and second parts of this reference are grouped by broad functional areas derived from research and manufacturers' specifications.

Abrasives

Trade names: Kaopolite® 1147; Kaopolite® 1152; Kaopolite® 1168; Kaopolite® SF; Kaopolite® SFO-NP; Kayphobe-ABO
Siltex™ 50+ 100 mesh

Accelerators

Trade names: Accelerator 55028; Accelerator D; Accelerator E; Accelerator NL-6; Accelerator NL-12; Accelerator NL49P; Accelerator NL51P; Accelerator NL53; Accelerator VN-2; Actafoam® F-2; Actiron NX 3; Activ-8; Activator 1102; Activator STAG; Ajicure® MY-24; Ajicure® PN-23; Akrochem® 9930 Zinc Oxide Transparent; Akrochem® Accelerator 40B Liq; Akrochem® Accelerator BZ Powder; Akrochem® Accelerator EZ; Akrochem® Accelerator MF; Akrochem® Accelerator MZ; Akrochem® Accelerator R; Akrochem® Accelerator VS; Akrochem® Accelerator ZIPPAC; Akrochem® BBTS; Akrochem® Cu.D.D; Akrochem® Cu.D.D.-PM; Akrochem® CZ-1; Akrochem® CBTS; Akrochem® DCBS; Akrochem® DCP-40C; Akrochem® DCP-40K; Akrochem® D.E.T.U. Accelerator; Akrochem® DOTG; Akrochem® DPG; Akrochem® DPTT; Akrochem® MBT; Akrochem® MBTS; Akrochem® MBTS Pellets; Akrochem® OBTS; Akrochem® OMTS; Akrochem® TBUT; Akrochem® TDEC; Akrochem® TETD; Akrochem® Thio No. 1; Akrochem® TMTD; Akrochem® TMTD Pellet; Akrochem® TM/ETD; Akrochem® TMTM; Akrochem® VC-40C; Akrochem® VC-40K; Akrochem® Z.B.E.D; Akrochem® Zinc Oxide 35; Akrochem® Z.P.D; Akrochem® ZMBT; Akroform® DBTU PM; Akroform® DCP-40 EPMB; Akroform® DETU PM; Akroform® ETU-22 PM; Akroform® VC-40 EPMB; Akrosperse® D-177; Akrosperse® DCP-40 EPMB; Akrosperse® VC-40 EPMB; Aktiplast® F; Aktiplast® PP; Aktiplast® T; Altax®; Amax®; Amicure® DBU-E; Amicure® UR; Amicure® UR2T; Amyl Ledate®; Amyl Zimate®; Anchor® DBD; Anchor® DOTG; Anchor® DPG; Anchor® ZDBC; Anchor® ZDE
Bantox®; Benzyl Tuads®; Benzyl Tuads® Solid; Benzyl Tuex®; BIBBS; Bismate® Powd; Bismate® Rodform®; Bismet; Butasan®; Butyl Eight®; Butyl Namate®; Butyl Tuads®; Butyl Zimate®; Butyl Zimate® Dustless; Butyl Zimate® Slurry
Capcure® 3-800; Capcure® 3830-81; Capcure® EH-30; Cardolite® NC-700; Cure-Rite® 18; Curezol® 1B2MZ; Curezol® 2E4MZ; Curezol® 2MA-OK; Curezol® 2MZ-Azine; Curezol® 2PHZ; Curezol® 2PHZ-S; Curezol® 2PZ; Curezol® 2PZ-CNS; Curezol® 2PZ-OK; Curezol® AMI-2; Curezol® C17Z
Delac® MOR; Delac® NS; DPG; DPTT-S; Durax®; Durax® Rodform; Dynamar® FX 5166
Ekaland CBS; Ekaland CMBT; Ekaland DBTU; Ekaland DETU; Ekaland DPTT; Ekaland ETU; Ekaland MBT; Ekaland MBTS; Ekaland TETD; Ekaland TMTM; Ekaland ZMBT; Emery® 5714; Epi-Cure® 537; Epi-Cure® 3253; Ethyl Cadmate®; Ethyl Tellurac®; Ethyl Tellurac® Rodform; Ethyl Tuads® Rodform; Ethyl Zimate®; Ethyl Zimate® Slurry
Firstcure™ DMMT; Firstcure™ DMPT
Imicure® AMI-2; Imicure® EMI-24; Isobutyl Tuads®; Isobutyl Zimate®
Ken Kem® CP-45; Ken Kem® CP-99; Kenlastic®; Kenmix®; Ken-React® 7 (KR 7); Ken-React®

Accelerators *(cont'd.)*

9S (KR 9S); Ken-React® 12 (KR 12); Ken-React® 26S (KR 26S); Ken-React® 33DS (KR 33DS); Ken-React® 38S (KR 38S); Ken-React® 39DS (KR 39DS); Ken-React® 41B (KR 41B); Ken-React® 44 (KR 44); Ken-React® 46B (KR 46B); Ken-React® 55 (KR 55); Ken-React® 133DS (KR 133DS); Ken-React® 134S (KR 134S); Ken-React® 138D (KR 138D); Ken-React® 138S (KR 138S); Ken-React® 158D (KR 158D); Ken-React® 158FS (KR 158FS); Ken-React® 212 (KR 212); Ken-React® 238A (KR 238A); Ken-React® 238J (KR 238J); Ken-React® 238M (KR 238M); Ken-React® 238S (KR 238S); Ken-React® 238T (KR 238T); Ken-React® 262A (KR 262A); Ken-React® 262ES (KR 262ES); Ken-React® OPP2 (KR OPP2); Ken-React® OPPR (KR OPPR); Ken-React® KR TTS; KZ OPPR; KZ TPP; KZ TPPJ

Lamefix 680; LICA 01; LICA 09; LICA 12; LICA 38; LICA 38A; LICA 38J; LICA 44; LICA 97; LICA 99

Methasan®; Methyl Cumate®; Methyl Cumate® Rodform; Methyl Ledate; Methyl Namate®; Methyl Selenac®; Methyl Tuads®; Methyl Tuads® Rodform; Methyl Zimate®; Methyl Zimate® Slurry; MonoThiurad®; Monothiurad®; Morfax®

Naftocit® DPG; Naftocit® Di 4; Naftocit® Di 7; Naftocit® Di 13; Naftocit® MBT; Naftocit® MBTS; Naftocit® Mi 12; Naftocit® NaDBC; Naftocit® NaDMC; Naftocit® Thiuram 16; Naftocit® ZBEC; Naftocit® ZMBT; Novor 924

Octocure 462; Octocure ZBZ-50; Octocure ZDB-50; Octocure ZDE-50; Octocure ZDM-50; Octocure ZIX-50; Octocure ZMBT-50; Octopol NB-47; Octopol SBZ-20

Perkacit® CBS; Perkacit® CDMC; Perkacit® DCBS; Perkacit® DOTG; Perkacit® DPG; Perkacit® DPTT; Perkacit® ETU; Perkacit® MBS; Perkacit® MBT; Perkacit® MBTS; Perkacit® NDBC; Perkacit® SDMC; Perkacit® TBBS; Perkacit® TBzTD; Perkacit® TDEC; Perkacit® TETD; Perkacit® TMTD; Perkacit® TMTM; Perkacit® ZBEC; Perkacit® ZDBC; Perkacit® ZDEC; Perkacit® ZDMC; Perkacit® ZMBT; Perkadox® 14/40; Petrac® 250; Poly-Dispersion® AAD-75; Poly-Dispersion® A (SAN)D-65; Poly-Dispersion® EAD-75; Poly-Dispersion® EAD-75P; Poly-Dispersion® E (DOTG)D-65P; Poly-Dispersion® EMD-75; Poly-Dispersion® E (MX)D-75; Poly-Dispersion® E (MZ)D-75; Poly-Dispersion® E (NBC)D-70; Poly-Dispersion® E (NBC)D-70P; Poly-Dispersion® END-75; Poly-Dispersion® END-75P; Poly-Dispersion® E (SAN-NS)D-70; Poly-Dispersion® E (SR)D-75; Poly-Dispersion® E (TET)D-70; Poly-Dispersion® E (TET)D-70P; Poly-Dispersion® GND-75; Poly-Dispersion® PLD-90; Poly-Dispersion® PLD-90P; Poly-Dispersion® SHD-65; Poly-Dispersion® T (AZ)D-75; Poly-Dispersion® T (BZ)D-75; Poly-Dispersion® T (DPG)D-65; Poly-Dispersion® T (DPG)D-65P; Poly-Dispersion® T (HRL)D-90; Poly-Dispersion® T (LC)D-90; Poly-Dispersion® T (LC)D-90P; Poly-Dispersion® TTD-75; Poly-Dispersion® TTD-75P; Poly-Dispersion® V (MT)D-75; Poly-Dispersion® V (MT)D-75P; Poly-Dispersion® V(NOBS)M-65; Prespersion PAB-262; Prespersion PAB-866; Prespersion PAC-3656; Promotor 301; Propyl Zithate®

Retarder SAX; Rhenocure® ADT; Rhenocure® CA; Rhenocure® CMT; Rhenocure® CMU; Rhenocure® CUT; Rhenocure® EPC; Rhenocure® S/G; Rhenocure® TP/G; Rhenocure® TP/S; Rhenocure® ZAT; Rhenofit® 1600; Rhenofit® NC; Rhenofit® UE; Rhenogran® DCBS-80; Rhenogran® MPTD-80; Rhenogran® MTT-80; Rhenogran® OTBG-50; Rhenogran® OTBG-75; Rhenovin® CBS-70; Rhenovin® MBT-70; Rhenovin® MBTS-70; Rhenovin® TMTD-70; Rhenovin® TMTM-70; Rhenovin® ZnO-90; RIA CS; Ricaccel; Rokon; Rotax®

Santocure®; Santocure® DCBS; Santocure® IPS; Santocure® MOR; Santocure® NS; Santocure® TBSI; Setsit® 5; Setsit® 9; Setsit® 104; Sulfads®; Sulfads® Rodform; Syntex® 3981

Thiate® EF-2; Thiate® H; Thiate® U; Thiofide®; Thiotax®; Thiovanol®; Thiurad®

Unads® Pellets; Unads® Powd

Vanax® 552; Vanax® 808; Vanax® 833; Vanax® A; Vanax® CPA; Vanax® DOTG; Vanax® DPG; Vanax® DPG Pellets; Vanax® MBM; Vanax® NS; Vanax® PML; Vanax® TBSI; Versamine® EH-30; Versamine® EH-50; Viton® Curative No. 50; Vocol®, S-75; V-Pyrol; Vulkacit 576; Vulkacit CRV/LG; Vulkacit CZ/EG; Vulkacit CZ/EGC; Vulkacit CZ/MGC; Vulkacit D/C; Vulkacit D/EGC; Vulkacit DM/C; Vulkacit DM/MGC; Vulkacit DZ/EGC; Vulkacit Merkapto/C; Vulkacit Merkapto/MGC; Vulkacit MOZ/LG; Vulkacit MOZ/SG; Vulkacit NZ/EG; Vulkacit ZBEC; Vulkacit ZM; Vultac® 2; Vultac® 5

Zeonet® B; Zetax®; Zetax® Dispersion; Zinc Oxide No. 185; Zink Oxide Transparent; Zinkoxyd Activ®

Chemicals: Alkyl phenol disulfide

Benzothiazyl disulfide; Bis (t-butylperoxy isopropyl) benzene; Bismuth dimethyl dithiocarbamate;

Accelerators *(cont'd.)*

Butyl 2-benzothiazole sulfenamide; Butyraldehyde-aniline condensation product; Butyraldehyde-monobutylamine condensation prod.

Cadmium diamyldithiocarbamate; Cadmium diethyldithiocarbamate; Cardanol; Cardanol acetate; Citric acid; Cobalt octoate; Copper dimethyldithiocarbamate; Copper 2-mercaptobenzothiazolate; Cyclohexylamine; N-Cyclohexyl-2-benzothiazolesulfenamide; Cycloneopentyl, cyclo (dimethlaminoethyl) pyrophosphato zirconate, di mesyl salt

Diazabicycloundecene; Di (butyl, methyl pyrophosphato) ethylene titanate di (dioctyl, hydrogen phosphite); 1,3-Dibutylthiourea; Dicatechol borate, di-o-tolyl guanidine salt; Dicyclo (dioctyl) pyrophosphato titanate; Dicyclohexylbenzothiazyl-2-sulfenamide; Di (dioctylphosphato) ethylene titanate; Di (dioctylpyrophosphato) ethylene titanate; 1,3-Diethylthiourea; Diisopropyl benzothiazole sulfenamide; Dimethylammonium hydrogen isophthalate; N,N-Dimethyl-N-cyclohexylamine; Dimethyl cyclohexyl ammonium dibutyl dithiocarbamate; Dimethyl diphenyl thiuram disulfide; N,N-Dimethyl-m-toluidine; N,N-Dimethyl-p-toluidine; Dipentamethylene hexasulfide; Dipentamethylene thiuram hexasulfide; Dipentamethylene thiuram tetradisulfide; Dipentamethylene thiuram tetrasulfide; Diphenylguanidine; N,N′-Diphenylthiourea; 4,4′-Dithiodimorpholine; Di-o-tolyl guanidine

Ethylene thiourea; N-Ethyl-N-hydroxyethyl-m-toluidine

Furfural

Glycol dimercaptoacetate

Isopropyl 4-aminobenzenesulfonyl di (dodecylbenzenesulfonyl) titanate; Isopropyl dimethacryl isostearoyl titanate; Isopropyl tri (dioctylphosphato) titanate; Isopropyl tri (dioctylpyrophosphato) titanate; Isopropyl tri (N ethylamino-ethylamino) titanate

Lead diamyldithiocarbamate; Lead dimethyldithiocarbamate

Magnesium oxide; 2-Mercaptobenzothiazole; 2-Methyl imidazole; 3-Methyl-thiazolidinethione-2; 4-Morpholinyl-2-benzothiazole disulfide

Nitrosodiphenylamine

N-Oxydiethylene benzothiazole-2-sulfenamide

1-Phenyl-3,3-dimethyl urea; m-Phenylenedimaleimide; 2-Phenyl imidazole; Piperazine; Piperidinium pentamethylene dithiocarbamate

p-Quinone dioxime

Selenium diethyldithiocarbamate; Selenium dimethyldithiocarbamate; Sodium dibenzyldithiocarbamate; Sodium di-n-butyl dithiocarbamate; Sodium dimethyldithiocarbamate; Steartrimonium bromide

Tellurium diethyl dithiocarbamate; N,N,N′,N′-Tetrabenzylthiuram disulfide; Tetrabutyl thiuram disulfide; Tetra (2, diallyoxymethyl-1 butoxy titanium di (di-tridecyl) phosphite; Tetraethylthiuram disulfide; Tetraisobutylthiuram disulfide; Tetraisopropyl di (dioctylphosphito) titanate; Tetramethylthiuram disulfide; Tetramethylthiuram monosulfide; Tetraoctyloxytitanium di (ditridecylphosphite); Thiocarbamyl sulfenamide; Thioglycerin; Titanium di (butyl, octyl pyrophosphate) di (dioctyl, hydrogen phosphite) oxyacetate; Titanium di (cumylphenylate) oxyacetate; Titanium di (dioctylpyrophosphate) oxyacetate; Titanium dimethacrylate oxyacetate; Tolyl bis (dimethyl urea); Triethylenetetramine; Trimethylthiourea

Urea

Zinc carbonate; Zinc diamyldithiocarbamate; Zinc dibenzyl dithiocarbamate; Zinc dibutyldithiocarbamate; Zinc dibutyl dithiophosphorate; Zinc diethyldithiocarbamate; Zinc diisobutyldithiocarbamate; Zinc dimethyldithiocarbamate; Zinc isopropyl xanthate; Zinc 2-mercaptobenzothiazole; Zinc-N-pentamethylene dithiocarbamate

Activators

Trade names: Actafoam® F-2; Activator 101; Activator 1102; Activator STAG; Aflux® R; Akrochem® 9930 Zinc Oxide Transparent; Akrochem® PEG 3350; Akrochem® TMTD; Akrochem® TMTD Pellet; Akrochem® ZDA Powd; Akrochem® Zinc Oxide 35; Akrosperse® D-177; Arazate®

Cellex-TS; Cri-Spersion CRI-ACT-45; Cri-Spersion CRI-ACT-45-1/1; Cri-Spersion CRI-ACT-45-LV

EPIstatic® 100

Industrene® 1224; Industrene® 5016; Industrene® R; Isobutyl Tuads®; Isobutyl Zimate®

Kadox® 720; Kadox®-911; Kadox®-920; Kadox® 930; Kencure™ C9P; Kencure™ MPP; Kencure™ MPPJ; Kenlastic®; Kenmix®; Ken-React® 7 (KR 7); Ken-React® 9S (KR 9S); Ken-React® 12 (KR 12); Ken-React® 26S (KR 26S); Ken-React® 33DS (KR 33DS); Ken-

Activators *(cont'd.)*

React® 38S (KR 38S); Ken-React® 39DS (KR 39DS); Ken-React® 41B (KR 41B); Ken-React® 44 (KR 44); Ken-React® 46B (KR 46B); Ken-React® 55 (KR 55); Ken-React® 133DS (KR 133DS); Ken-React® 134S (KR 134S); Ken-React® 138D (KR 138D); Ken-React® 138S (KR 138S); Ken-React® 158D (KR 158D); Ken-React® 158FS (KR 158FS); Ken-React® 212 (KR 212); Ken-React® 238A (KR 238A); Ken-React® 238J (KR 238J); Ken-React® 238M (KR 238M); Ken-React® 238S (KR 238S); Ken-React® 238T (KR 238T); Ken-React® 262A (KR 262A); Ken-React® 262ES (KR 262ES); Ken-React® OPP2 (KR OPP2); Ken-React® OPPR (KR OPPR); Ken-React® KR TTS; KZ OPPR; KZ TPP; KZ TPPJ

Lankromark® LC90; Lankromark® LZ187; Lankromark® LZ440; Lankromark® LZ561; Lankromark® LZ638; Lankromark® LZ1232; LICA 01; LICA 09; LICA 12; LICA 38; LICA 38A; LICA 38J; LICA 44; LICA 97; LICA 99; Litharge 28; Litharge 33

Mark® 281B; Mark® 630; MiRaStab 403-K

Naftocit® ZBEC; NAO 105; NAO 115; NAO 125; NAO 135

Octoate® Z; Octoate® Z Solid; Octocure 456

Petrac® 250; Petrac® 270; Poly-Dispersion® A (Z-CN)D-85; Poly-Dispersion® AZFD-85; Poly-Dispersion® EZFD-85; Poly-Dispersion® JZFD-90P; Poly-Dispersion® SZFD-85; Poly-Dispersion® TZFD-88P

Retarder AK; Retarder BA, BAX; Rhenofit® 1987; Rhenofit® 2009; Rhenofit® 2642; Rhenofit® 3555; Rhenofit® B; Rhenofit® BDMA/S; Rhenofit® CF; Rhenofit® EDMA/S; Rhenofit® NC; Rhenofit® TAC/S; Rhenofit® TRIM/S; Rhenofit® UE; Rhenogran® $Ca(OH)_2$-50/FPM; Rhenogran® CHM21-40/FPM; Rhenogran® MgO-40/FPM; Rhenomag® G1; Rhenomag® G3; Rhenovin® Na-stearat-80; Rhenovin® S-stearat-80; RIA CS; Ridacto®; RR Zinc Oxide (Untreated); RR Zinc Oxide-Coated

Silacto®; Stabilizer 2013-P®; Struktol® Activator 73; Struktol® FA 541; Struktol® IB 531; Struktol® ZEH; Struktol® ZEH-DL; Synpron 1112; Synpron 1236; Synpron 1642

Tegoamin® CPE; Tegoamin® PDD 60; Tegoamin® PTA; Tetronic® 90R4; Tetronic® 150R1; Thanecure®; Tribase; Tribase XL

Vanax® PML; Vanox® AT

WF-15; WF-16

Zinc Oxide No. 185; Zink Oxide AT; Zinkoxyd Activ®

Chemicals:

Barium-zinc complex; 1,4-Butanediol dimethacrylate; Butyraldehyde-aniline condensation product

Cadmium-zinc complex; Carbodiimide; Cycloneopentyl, cyclo (dimethlaminoethyl) pyrophosphato zirconate, di mesyl salt

Dibutylammonium oleate; Di (butyl, methyl pyrophosphato) ethylene titanate di (dioctyl, hydrogen phosphite); Dicatechol borate, di-o-tolyl guanidine salt; Di (dioctylphosphato) ethylene titanate; Di (dioctylpyrophosphato) ethylene titanate; 1,3-Diethylthiourea; N,N′-Diphenylthiourea

Isopropyl 4-aminobenzenesulfonyl di (dodecylbenzenesulfonyl) titanate; Isopropyl dimethacryl isostearoyl titanate; Isopropyl tri (dioctylphosphato) titanate; Isopropyl tri (dioctylpyrophosphato) titanate; Isopropyl tri (N ethylamino-ethylamino) titanate

Lead (II) oxide; Lead-zinc complex

Neopentyl (diallyl) oxy, trihydroxy caproyl titanate; Neopentyl (diallyl) oxy, trimercapto-phenyl zirconate

Oleic acid

Potassium-zinc

Tetra (2, diallyoxymethyl-1 butoxy titanium di (di-tridecyl) phosphite; Tetraisopropyl di (dioctylphosphito) titanate; Tetraoctyloxytitanium di (ditridecylphosphite); Titanium di (butyl, octyl pyrophosphate) di (dioctyl, hydrogen phosphite) oxyacetate; Titanium di (cumylphenylate) oxyacetate; Titanium di (dioctylpyrophosphate) oxyacetate; Titanium dimethacrylate oxyacetate; Triethylamine; 2,4,6-Tris (dimethylaminomethyl) phenol

Urea

Zinc; Zinc carbonate; Zinc diacrylate; Zinc di-2-ethylhexoate; Zinc ditolyl sulfinate dihydrate; Zinc 2-ethylhexanoate; Zinc 2-ethylhexoate; Zinc laurate; Zinc oxide; Zinc-potassium complex

Adhesion promoters • Bonding agents

Trade names:

Bonding Agent 2001; Bonding Agent 2005; Bonding Agent TN/S 50

Chemiflex 315XA; Cohedur® A; Cohedur® A Solid; Cohedur® RK; Cohedur® RL; Cohedur® RS

Adhesion promoters • Bonding agents *(cont'd.)*

Desmodur® L75; Desmodur® N100; Desmodur® RE; Desmodur® RFE; Desmodur® TT; Desmodur® VK-18; Dow Corning® Q1-6106; Duralink® HTS; Dynasylan® AMEO-40; Dynasylan® AMEO-P; Dynasylan® AMEO-T

Elvace® 1870; Escor® ATX-325; Exxelor PO 1015; #15 Oil

GE 100

Hi-Sil® 210; Hi-Sil® 233; Hi-Sil® 243LD; Hi-Sil® 250; Hi-Sil® ABS; Hostaprime® HC 5

Kenplast® ES-2; Ken-React® 7 (KR 7); Ken-React® 9S (KR 9S); Ken-React® 12 (KR 12); Ken-React® 26S (KR 26S); Ken-React® 33DS (KR 33DS); Ken-React® 38S (KR 38S); Ken-React® 39DS (KR 39DS); Ken-React® 41B (KR 41B); Ken-React® 44 (KR 44); Ken-React® 46B (KR 46B); Ken-React® 55 (KR 55); Ken-React® 133DS (KR 133DS); Ken-React® 134S (KR 134S); Ken-React® 138D (KR 138D); Ken-React® 138S (KR 138S); Ken-React® 158D (KR 158D); Ken-React® 158FS (KR 158FS); Ken-React® 212 (KR 212); Ken-React® 238A (KR 238A); Ken-React® 238J (KR 238J); Ken-React® 238M (KR 238M); Ken-React® 238S (KR 238S); Ken-React® 238T (KR 238T); Ken-React® 262A (KR 262A); Ken-React® 262ES (KR 262ES); Ken-React® OPP2 (KR OPP2); Ken-React® OPPR (KR OPPR); Ken-React® KR TTS; KZ OPPR; KZ TPP; KZ TPPJ

LICA 01; LICA 09; LICA 12; LICA 38; LICA 38A; LICA 38J; LICA 44; LICA 97; LICA 99

Manobond™ 680-C; MFM-786 V

NZ 01; NZ 09; NZ 12; NZ 33; NZ 38; NZ 39; NZ 44; NZ 49; NZ 89; NZ 97

Pale 4; Pale 16; Pale 170; Pale 1000; Penacolite® B-18-S; Penacolite® R-2170; Piccotac® B; Polyace™ 573; Polybond® 2015; PVP K-60; R7234

Rhenogran® Resorcin-80; Rhenogran® Resorcin-80/SBR; Ricobond 1031, 1731, 1756; Rit-O-Lite MS-80

SAFR-Z-12K; SAFR-Z-12SK; SAFR-Z-22E K; Sipomer® β-CEA; SR-368; SR-9010; SR-9011; SRF-1501; Surfac® 554; Surfac® 555; Surfac® 580; Surfac® 600; Surfac® 610; Sylvaros® R; Sylvatac® 140; Sylvatac® 295; Sylvatac® R85; Sylvatac® RX

Tioga Adhesion Promoter 30-0-100; Tioga Adhesion Promoter 30-6-600

Union Carbide® Y-11343; Union Carbide® Y-11542; Union Carbide® Y-11597; Union Carbide® Y-11602; Uniplex 600; Uni-Tac® R100RM; Unite

Vanchem® HM-50, HM-4346; Vorite 105; Vorite 110; Vorite 115; Vorite 120; Vorite 125; Vulcabond® C 10; Vulcabond® E; Vulcabond® N 15; Vulcabond® TX; Vulcabond® VP; Vulkadur® T, 40%; Zonarez® 7085; Zonarez® 7115; Zonarez® 7115 LITE; Zonarez® 7125; Zonarez® 7125 LITE; Zonarez® B-10; Zonarez® B-115; Zonarez® B-125

Chemicals: Aminopropyltrimethoxysilane

β-Carboxyethyl acrylate; Castor oil, polymerized; Cobalt boro acylate; Cycloneopentyl, cyclo (dimethlaminoethyl) pyrophosphato zirconate, di mesyl salt

Di (butyl, methyl pyrophosphato) ethylene titanate di (dioctyl, hydrogen phosphite); Dicyclo (dioctyl) pyrophosphato titanate; Dicyclohexyl sodium sulfosuccinate; Di (dioctylphosphato) ethylene titanate; Di (dioctylpyrophosphato) ethylene titanate; Dimethylaminoethyl acrylate; Disodium hexamethylene bisthiosulfate

Ethylene/magnesium acrylate copolymer; Ethylene-maleic anhydride copolymer

Hexamethylmelamine

Isopropyl 4-aminobenzenesulfonyl di (dodecylbenzenesulfonyl) titanate; Isopropyl dimethacryl isostearoyl titanate; Isopropyl tri (dioctylphosphato) titanate; Isopropyl tri (dioctylpyrophosphato) titanate; Isopropyl tri (N ethylamino-ethylamino) titanate

MDI

Neopentyl (diallyl) oxy, trihydroxy caproyl titanate; Neopentyl (diallyl) oxy, trimercapto-phenyl zirconate

Resorcinol-formaldehyde resin

Tetrabutyl titanate; Tetra (2, diallyoxymethyl-1 butoxy titanium di (di-tridecyl) phosphite; Tetraisopropyl di (dioctylphosphito) titanate; Tetraisopropyl titanate; Tetraoctyloxytitanium di (ditridecylphosphite); Titanium di (butyl, octyl pyrophosphate) di (dioctyl, hydrogen phosphite) oxyacetate; Titanium di (cumylphenylate) oxyacetate; Titanium di (dioctylpyrophosphate) oxyacetate; Titanium dimethacrylate oxyacetate; Toluenesulfonamide formaldehyde resin; Triethylene diamine; Triphenyl methane 4,4′,4″-triisocyanate

Ureidosilane

Antiblocking agents

Trade names: ABT-2500®; Acrawax® C; Aerosil® OX50; Akrochem® P 3100; Alkamide® STEDA; Ampacet 10063; Antiblock-System Hoechst B1980; Antiblock-System Hoechst B1981; Armid® HT; Armoslip® 18
Cabot® PE 9007; Cabot® PE 9166; Cabot® PE 9172; Cabot® PE 9220; Cabot® PE 9229; Cabot® PP 9269; Cardis® 10; Cardis® 36; Cardis® 314; Cardis® 319; Cardis® 320; Cardis® 370; Celite® Super Fine Super Floss; Celite® Super Floss; Celite® White Mist; Chemstat® HTSA #1A; Chemstat® HTSA #3B; Chemstat® HTSA #18; Chemstat® HTSA #18S; Chemstat® HTSA #22; Colloids PE 48/10/06; Colloids PE 48/10/09; Colloids PE 48/10/10; Colloids PE 48/10/11; Colloids PE 48/10/13; CPH-31-N; Crodamide 203; Crodamide 212; Crodamide ER; Crodamide O; Crodamide OR; Crodamide S; Crodamide SR
Dimul S; Doittol K21; Dynamar® FX-5920
Emerest® 2650; Emerwax® 9380; Eureslip 58
FK 310
Getren® FD 150; Glycolube® 140; Glycolube® 345; Glycolube® 674; Glycolube® 740; Glycolube® 742; Glycolube® 825; Glycolube® TS; Glycolube® VL
Hoechst Wax KPS; Imsil® 1240; Imsil® A-8; Imsil® A-10
Kaopolite® 1152; Kaopolite® 1168; Kaopolite® SF; Kaopolite® SFO-NP; Kayphobe-ABO; Kemamide® B; Kemamide® E; Kemamide® O; Kemamide® S; Kemamide® U; Kemamide® W-20; Kemamide® W-39; Kemamide® W-40; Kemamide® W-40/300; Kemamide® W-40DF; Kemamide® W-45
Lo-Vel® 27; Lo-Vel® 28; Lo-Vel® 29; Lo-Vel® 39A; Lo-Vel® 66; Lo-Vel® 275; Lo-Vel® HSF
MDI SF-320; MDI SF-325; Microbloc®; Myvaplex® 600PK; Myverol® 18-04; Myverol® 18-06; Myverol® 18-07; Myverol® 18-99
PE-AB 07; PE-AB 19; PE-AB 22; PE-AB 30; PE-AB 33; PE-AB 48; PE-AB 73; PE-SA 21; PE-SA 24; PE-SA 40; PE-SA 49; PE-SA 60; PE-SA 72; PE-SA 92; Petrac® Slip-Quick®; Petrac® Vyn-Eze®; Petronauba® C; Plaswite® LL 7015; Plaswite® PE 7000; Plaswite® PE 7001; Plaswite® PE 7031; Plaswite® PE 7097; Plaswite® PE 7192; Polyace™ 573; Polywax® 500; Polywax® 655; Polywax® 850; Polywax® 1000; Polywax® 2000; Polywax® 3000; PP-AB 104; PP-SA 46; PP-SA 97; PP-SA 128
Silwet® L-7001; Silwet® L-7602; Sipernat® 22LS; Sipernat® 44; Sipernat® 50LS; Softenol® 3900; Softenol® 3991; Spectratech® CM 10608; Spectratech® CM 10778; Spectratech® CM 10779; Spectratech® CM 11013; Spectratech® CM 11513; Spectratech® PM 10634; Spheriglass® Coated; Stellar 500; SY-AB 70; SY-SA 69
Tamsil 8; Tamsil 10; Techpolymer MB-4, MB-8; Techpolymer MBX-5 to MBX-50; Techpolymer SBX-6, -8, -17
Unislip 1753; Unislip 1757; Unislip 1759; Uniwax 1750; Uniwax 1760
Vertal 310

Chemicals: Behenamide
Canola oil glyceride
Dimethicone copolyol; N,N′-Dioleoylethylenediamine
Erucamide; Erucyl erucamide
Hydrogenated cottonseed glyceride; Hydrogenated palm glyceride; Hydrogenated tallow amide
Microcrystalline wax, oxidized
Oleamide; Oleyl palmitamide
PEG-8 laurate
Stearamide; Stearyl/aminopropyl methicone copolymer; Stearyl erucamide; Stearyl stearamide
Talc

Anticoagulants

Trade names: Ethomeen® 18/60
Modicol S; Rhodacal® BA-77
Trymeen® 6617

Antidegradants

Trade names: Akrochem® Antiozonant MPD-100
Poly-Dispersion® A (ZMAM)D-666P; Poly-Dispersion® A (ZMTI)D-50; Poly-Dispersion® K (NMC)D-70; Poly-Dispersion® S (AX)D-70

Antidegradants *(cont'd.)*

Rapidblend 1793; Rhenogran® P-50/EVA
Santowhite® PC; Santowhite® Powd
Ultranox® 627A; Ultranox® 2714A

Chemicals: 2-Mercaptobenzimidazole; 2-Mercaptotoluimidazole; 4- and 5-Methylmercaptobenzimidazole zinc salt
Polycarbonamide

Antiflexcracking agents

Trade names: Additin 30; Akrochem® Antioxidant DQ; Akrochem® Antioxidant DQ-H; Akrochem® Antioxidant PANA; Akrochem® Antioxidant S; Akrochem® Antiozonant MPD-100; Akrochem® Antiozonant PD-2; Akrowax 130; Akrowax 5025; Akrowax 5026; Akrowax 5030; Akrowax 5031; Akrowax 5032; Akrowax 5073; Akrowax Micro 23; Antisun®
Santoflex® 13; Santoflex® 134; Santoflex® AW
Vulkanox® 3100; Vulkanox® 4022; Vulkanox® 4025; Vulkanox® PAN

Chemicals: N-Isopropyl-N´-phenyl-p-phenylenediamine
Mixed diaryl-p-phenylenediamine

Antifoaming agents

Trade names: AF 100 IND; AF HL-27; Albrite® Tributyl Phosphate; Antifoam 7800 New
Colloid™ 635; Colloid™ 681F
D-1000; D-1200; D-1300; D-2000; D-3000; D-4000; Defomax; Dehydran P 12
Emsorb® 2515
Fancol OA-95
Hodag PPG-150; Hodag PPG-400; Hodag PPG-1200; Hodag PPG-2000; Hodag PPG-4000
Ken-React® 7 (KR 7); Ken-React® 9S (KR 9S); Ken-React® 12 (KR 12); Ken-React® 26S (KR 26S); Ken-React® 33DS (KR 33DS); Ken-React® 38S (KR 38S); Ken-React® 39DS (KR 39DS); Ken-React® 41B (KR 41B); Ken-React® 44 (KR 44); Ken-React® 46B (KR 46B); Ken-React® 55 (KR 55); Ken-React® 133DS (KR 133DS); Ken-React® 134S (KR 134S); Ken-React® 138D (KR 138D); Ken-React® 138S (KR 138S); Ken-React® 158D (KR 158D); Ken-React® 158FS (KR 158FS); Ken-React® 212 (KR 212); Ken-React® 238A (KR 238A); Ken-React® 238J (KR 238J); Ken-React® 238M (KR 238M); Ken-React® 238S (KR 238S); Ken-React® 238T (KR 238T); Ken-React® 262A (KR 262A); Ken-React® 262ES (KR 262ES); Ken-React® OPP2 (KR OPP2); Ken-React® OPPR (KR OPPR); Ken-React® KR TTS; Kodaflex® DBP; KZ OPPR; KZ TPP; KZ TPPJ
LICA 01; LICA 09; LICA 12; LICA 38; LICA 38A; LICA 38J; LICA 44; LICA 97; LICA 99
Monolan® 3000 E/60, 8000 E/80
Nissan Plonon 102; Nissan Plonon 104; Nissan Plonon 108; Nissan Plonon 171; Nissan Plonon 172; Nissan Plonon 201; Nissan Plonon 204; Nissan Plonon 208
Pluracol® P-3010; Pluracol® P-4010; Pluriol® P 600; Pluriol® P 900; Pluriol® P 2000
Rhodorsil® AF 422
Silwet® L-7500; Sipomer® COPS 1; Surfynol® D-101; Surfynol® D-201
Teric PPG 1000; Teric PPG 1650
Wacker Silicone Antifoam Emulsion SE 9

Chemicals: Cycloneopentyl, cyclo (dimethlaminoethyl) pyrophosphato zirconate, di mesyl salt
Dicyclo (dioctyl) pyrophosphato titanate
Isopropyl 4-aminobenzenesulfonyl di (dodecylbenzenesulfonyl) titanate; Isopropyl dimethacryl isostearoyl titanate; Isopropyl tri (dioctylphosphato) titanate; Isopropyl tri (dioctylpyrophosphato) titanate; Isopropyl tri (N ethylamino-ethylamino) titanate
Oleyl alcohol
PEG-6 oleate; PPG (1000)
Sorbitan sesquioleate; Stearyl alcohol
Tetraisopropyl di (dioctylphosphito) titanate; Tetraoctyloxytitanium di (ditridecylphosphite); Titanium di (butyl, octyl pyrophosphate) di (dioctyl, hydrogen phosphite) oxyacetate; Titanium di (cumylphenylate) oxyacetate; Titanium di (dioctylpyrophosphate) oxyacetate; Titanium dimethacrylate oxyacetate; Tributyl phosphate; Trioctyl phosphate

Antifog agents

Trade names: ADK STAB CPS-896; Aldosperse® MO-50; Alkamuls® GMR-55LG; Anstex AK-25; Armeen® Z; Atmer® 100; Atmer® 102; Atmer® 103; Atmer® 104; Atmer® 105; Atmer® 106; Atmer® 110; Atmer® 111; Atmer® 112; Atmer® 113; Atmer® 114; Atmer® 115; Atmer® 116; Atmer® 117; Atmer® 118; Atmer® 121; Atmer® 122; Atmer® 124; Atmer® 125; Atmer® 129; Atmer® 130; Atmer® 131; Atmer® 132; Atmer® 133; Atmer® 134; Atmer® 171; Atmer® 184; Atmer® 185; Atmer® 502; Atmer® 645; Atmer® 646; Atmer® 647; Atmer® 648; Atmer® 649; Atmer® 650; Atmer® 654; Atmer® 655; Atmer® 656; Atmer® 657; Atmer® 658; Atmer® 685; Atmer® 687; Atmer® 688; Atmer® 1007; Atmer® 1010; Atmer® 1012; Atmer® 7101; Atmer® 7108; Atmer® 8158; Atmer® 8163; Atmer® 8215; Atmer® 8216

Chemstat® AF-476; Chemstat® AF-700; Chemstat® AF-710; Chemstat® AF-806; Chemstat® AF-815; Chemstat® AF-906; Chemstat® AF-1006

Dimodan LS; Dimodan PV 300; Drewplast® 030; Drewplast® 051; Dur-Em® 300K; Durfax® 80; Durtan® 80

Eastman® PA-208; Estol 1407; Ethylan® GL20; Ethylan® GO80; Ethylan® GS60; Ethylan® GT85

Geropon® 99; Glycolube® 100; Glycolube® 140; Glycolube® 315; Glycolube® 345; Glycolube® 674; Glycolube® 740; Glycolube® 742; Glycolube® 825; Glycolube® AFA-1; Glycolube® TS; Glycomul® L; Glycomul® O; Glycosperse® L-20; Glycosperse® O-20; Grindtek MOP 90; Grindtek MSP 40; Grindtek MSP 40F; Grindtek MSP 90; Grindtek PGE-DSO; Grindtek PK 60

Hamposyl® C; Hamposyl® C-30; Hamposyl® L; Hamposyl® L-30; Hamposyl® L-95; Hamposyl® M; Hamposyl® M-30; Hamposyl® O; Hamposyl® S

Mark® 281B; Mark® 630; Marklear AFL-23; Merix MCG Compd; Myvaplex® 600PK; Myverol® 18-04; Myverol® 18-06; Myverol® 18-07; Myverol® 18-99

Pationic® 907; Pationic® 1061; Pationic® 1064; Pationic® 1074; Petrac® GMS; Polyvel AI40; Polyvel AI1645, AI1685, AI2645, AI2685; Polyvel AI3184; Priolube 1407; Priolube 1408; Priolube 1409; PS071

Silwet® L-7001

Triton® N-60

Ultramoll® I; Ultramoll® II; Zonyl® FSN

Chemicals: Canola oil glyceride; Cocaminobutyric acid; Cocoyl sarcosine

Dimethicone copolyol

Glyceryl dioleate; Glyceryl linoleate; Glyceryl mono/dioleate; Glyceryl myristate; Glyceryl stearate; Glyceryl stearate SE

Hydrogenated cottonseed glyceride; Hydrogenated palm glyceride

Lard glyceride; Lauramine; Lauroyl sarcosine

Myristoyl sarcosine

Oleoyl sarcosine

Palm kernel glycerides; Polyglyceryl-3 oleate; Polyglyceryl-3 stearate; Polyglyceryl-3 stearate SE

Sodium cocoyl sarcosinate; Sodium myristoyl sarcosinate; Sorbitan laurate; Sorbitan oleate; Sorbitan trioleate; Stearoyl sarcosine

Antimicrobials • Fungistats • Germicides • Antibacterials

Trade names: bioMeT 14; bioMeT TBTO; Bio-Pruf®

Great Lakes PH-73™

Idex™-400; Idex™-1000

Kathon® LX; Ketjenflex® 9; Keycide® X-10

Micro-Chek® 11; Micro-Chek® 11 P

Noxamium S2-50

Omacide® P-711-5; Omacide® P-BBP-5; Omacide® P-DIDP-5; Omacide® P-DOP-5; Omacide® P-ESO-5

Sanitized® LX 91-01 RF

Vancide® 51Z Disp; Vancide® 89; Vancide® MZ-96; Vancide® TH; Vikol® PX-15; Vinyzene® BP-5-2; Vinyzene® BP-5-2 DIDP; Vinyzene® BP-5-2 DOP; Vinyzene® BP-5-2 PG; Vinyzene® BP-5-2 S-160; Vinyzene® BP-5-2 U; Vinyzene® BP-505; Vinyzene® BP-505 DIDP; Vinyzene® BP-505 DOP; Vinyzene® BP-505 PG; Vinyzene® BP-505 S160; Vinyzene® IT-3000 DIDP; Vinyzene® IT-3025 DIDP; Vinyzene® SB-1; Vinyzene® SB-1

Antimicrobials *(cont'd.)*

EAA; Vinyzene® SB-1 PR; Vinyzene® SB-1 PS; Vinyzene® SB-1 U
Zinc Omadine® 48% Fine Particle Disp; Zinc Omadine® 48% Std. Disp; Zinc Omadine® Powd

Chemicals: 6-t-Butyl-m-cresol
Captan
Dibromo/tribromo salicylanilide; Diphenylstibine 2-ethylhexoate; 4,4′-Dithiodimorpholine
Iodopropynyl butylcarbamate
Methylchloroisothiazolinone
Oxazolidine; 10,10′-Oxybisphenoxyarsine
2,4,6-Tribromophenol; Tribromosalicylanilide; Tributyltin oxide
Zinc pyrithione; Zinc sulfide

Antioxidants

Trade names: Ablumol PSS; Ac35; Additin 30; ADK STAB 328; ADK STAB AO-15; ADK STAB AO-18; ADK STAB AO-20; ADK STAB AO-23; ADK STAB AO-30; ADK STAB AO-37; ADK STAB AO-40; ADK STAB AO-50; ADK STAB AO-60; ADK STAB AO-80; ADK STAB AO-412S; ADK STAB AO-503A; ADK STAB AO-616; ADK STAB CDA-1; ADK STAB CDA-6; ADK STAB LX-45; ADK STAB LX-802; Agerite® DPPD; Agerite® Geltrol; Agerite® Hipar T; Agerite® HP-S; Agerite® HP-S Rodform; Agerite® MA; Agerite® NEPA; Agerite® Resin D; Agerite® Stalite; Agerite® Stalite S; Agerite® Superflex; Agerite® Superflex Solid G; Agerite® Superlite; Agerite® Superlite® Emulcon®; Agerite® Superlite Solid; Agerite® White; Agerite® White White; Akrochem® Accelerator BZ Powder; Akrochem® Antioxidant 12; Akrochem® Antioxidant 16 Liq; Akrochem® Antioxidant 16 Powd; Akrochem® Antioxidant 32; Akrochem® Antioxidant 33; Akrochem® Antioxidant 36; Akrochem® Antioxidant 43; Akrochem® Antioxidant 58; Akrochem® Antioxidant 60; Akrochem® Antioxidant 235; Akrochem® Antioxidant 1010; Akrochem® Antioxidant 1076; Akrochem® Antioxidant BHT; Akrochem® Antioxidant DQ; Akrochem® Antioxidant DQ-H; Akrochem® Antioxidant PANA; Akrochem® Antioxidant S; Akrochem® Antiozonant MPD-100; Akrowax 130; Akrowax 145; Albrite® BTD HP; Albrite® DBHP; Albrite® DIOP; Albrite® DLHP; Albrite® DMHP; Albrite® DOHP; Albrite® T2CEP; Albrite® TIOP; Alvinox 100; Alvinox FB; Aminox®; Aminox® Naugard A; Ampacet 10886; Anchor® DNPD; Anchor® HDPA; Anchor® HDPA/SE; Anchor® ODPA; Anchor® ZDBC; Anoxsyn 442; Antioxidant 235; Antioxidant 425; AO47L; AO47P; AO872; AO924; Aquastab PA 43; Aquastab PA 48; Aquastab PA 52; Aquastab PA 58; Aranox®
BLE® 25; BNX® 1000; BNX® 1010, 1010G; BNX® 1076, 1076G; Butasan®; Butyl Zimate®; BXA Flake
Cabot® PE 6008; Cabot® PE 9247; CAO®-3; CAO®-5; CAO®-14; CAO®-92; Carstab® DLTDP; Carstab® DSTDP; CGA 12; Chimassorb® 944LD; Colloids PE 48/10/12; Colortech 10044-12; Colortech 10045-11; Colortech 10045-11; Colortech 10944-11; Cyanox® 425; Cyanox® 711; Cyanox® 1212; Cyanox® 1790; Cyanox® 2246; Cyanox® LTDP; Cyanox® MTDP; Cyanox® STDP; Cyasorb® UV 2908; Cyasorb® UV 3346; Cyasorb® UV 3346 LD; Cyasorb® UV 3853
Dovernox 10; Dovernox 76; Dovernox 3114; Doverphos® 4-HR; Doverphos® 4-HR Plus; Doverphos® 6; Durazone® 37
Eastman® DTBHQ; Eastman® HQMME; Eastman® MTBHQ; Eastman® Inhibitor Poly TDP 2000; Ekaland NDBC; Escoflex A-122; Ethanox® 330; Ethanox® 398; Ethanox® 702; Ethanox® 703; Evangard® 18MP; Evangard® DTB; Evanstab® 12; Evanstab® 13; Evanstab® 14; Evanstab® 18; Evanstab® 1218
Flectol® ODP; Flectol® Pastilles
Garbefix 05240; Givsorb® UV-2; Good-rite® 3034; Good-rite® 3140; Good-rite® 3150; Good-rite® 3159
Halphos; Hostanox® O3; Hostanox® OSP 1; Hostanox® PAR 24; Hostanox® SE 1; Hostanox® SE 2; Hostanox® SE 10; Hostanox® ZnCS 1; Hostanox® System A1961; Hostanox® System A4962; Hostanox® System A4965; Hostanox® System P1961; Hostavin® VP NiCS 1
Interstab® CH-90; Ionol; Irgafos® 168; Irganox® 129; Irganox® 245; Irganox® 259; Irganox® 565; Irganox® 1010; Irganox® 1035; Irganox® 1076; Irganox® 1098; Irganox® 1330; Irganox® 1425 WL; Irganox® 1520; Irganox® 3114; Irganox® 3125; Irganox® B 215; Irganox® B 225; Irganox® B 501W; Irganox® B 561; Irganox® B 900; Irganox® B 921; Irganox® B 1171; Irganox® B 1411; Irganox® B 1412; Irganox® MD 1024; Irgastab® 2002; Isobutyl Niclate®; Isonox® 232

Antioxidants *(cont'd.)*

Ken-React® 7 (KR 7); Ken-React® 9S (KR 9S); Ken-React® 12 (KR 12); Ken-React® 26S (KR 26S); Ken-React® 33DS (KR 33DS); Ken-React® 38S (KR 38S); Ken-React® 39DS (KR 39DS); Ken-React® 41B (KR 41B); Ken-React® 44 (KR 44); Ken-React® 46B (KR 46B); Ken-React® 55 (KR 55); Ken-React® 133DS (KR 133DS); Ken-React® 134S (KR 134S); Ken-React® 138D (KR 138D); Ken-React® 138S (KR 138S); Ken-React® 158D (KR 158D); Ken-React® 158FS (KR 158FS); Ken-React® 212 (KR 212); Ken-React® 238A (KR 238A); Ken-React® 238J (KR 238J); Ken-React® 238M (KR 238M); Ken-React® 238S (KR 238S); Ken-React® 238T (KR 238T); Ken-React® 262A (KR 262A); Ken-React® 262ES (KR 262ES); Ken-React® OPP2 (KR OPP2); Ken-React® OPPR (KR OPPR); Ken-React® KR TTS; Korestab®; KZ 55; KZ OPPR; KZ TPP; KZ TPPJ

Lankromark® LE98; Lankromark® LE109; Lankromark® LE131; Lankromark® LE373; LICA 01; LICA 09; LICA 12; LICA 38; LICA 38A; LICA 38J; LICA 44; LICA 97; LICA 99; Lowinox® 22IB46; Lowinox® 22M46; Lowinox® 44B25; Lowinox® 44S36; Lowinox® BHT; Lowinox® BHT Food Grade; Lowinox® BHT-Superflakes; Lowinox® CA 22; Lowinox® CPL; Lowinox® DLTDP; Lowinox® DSTDP; Lowinox® PP35

Mark® 135; Mark® 158; Mark® 260; Mark® 329; Mark® 522; Mark® 1178; Mark® 1178B; Mark® 1220; Mark® 1409; Mark® 1535; Mark® 1589; Mark® 1589B; Mark® 1900; Mark® 2112; Mark® 5060; Mark® 5082; Mark® 5089; Mark® 5095; Mark® 5111; MDI PE-200UVA; Methyl Niclate®; MiRaStab 706-K; Mixxim® AO-20; Mixxim® AO-30; Montaclere®

Naftonox® 2246; Naftonox® BBM; Naftonox® BHT; Naftonox® IMB ; Naftonox® PA; Naftonox® PS; Naftonox® TMQ; Naftonox® ZMP; Naugard® 10; Naugard® 76; Naugard® 431; Naugard® 445; Naugard® 492; Naugard® 524; Naugard® 529; Naugard® A; Naugard® BHT; Naugard® P; Naugard® PHR; Naugard® SP; Naugard® Super Q; Naugard® XL-1; Naugard® XL-517; Naugawhite®; Nevastain® 21; Nevastain® A; Nevastain® B; Nonox® WSP; NZ 01; NZ 09; NZ 12; NZ 33; NZ 38; NZ 39; NZ 44; NZ 49; NZ 89; NZ 97

Octamine® Flake, Powd; Octolite 561; Octolite AO-28; Oxi-Chek 114; Oxi-Chek 116; Oxi-Chek 414

PAG DLTDP; PAG DMTDP; PAG DSTDP; PAG DTDTDP; PAG DXTDP; Parabis; PE-AO 51; PE-AO 91; Permanax™ BL; Permanax™ BLN; Permanax™ BLW; Permanax™ CNS; Permanax™ CR; Permanax™ HD; Permanax™ HD (SE); Permanax™ OD; Permanax™ OZNS; Permanax™ TQ; Permanax™ WSL; Permanax™ WSL Pdr; Permanax™ WSO; Permanax™ WSP; Permanax™ WSP (PQ); Poly TDP 2000; Polygard®; Polygard® HR; Polylite; Polystay AA-1; Polystay AA-1R; PP-AO 27; PP-AO 47; Prespersion PAC-2451; Prodox® 120; PTZ® Chemical Grade; PTZ® Industrial Grade

Q-1300

Ralox® 02; Ralox® 35; Ralox® 46; Ralox® 240; Ralox® 530; Ralox® BHT food grade; Ralox® BHT Tech; Ralox® LC; Ralox® TMQ-G; Ralox® TMQ-H; Ralox® TMQ-R; Ralox® TMQ-T; Rhenogran® BPH-80; Rhenogran® IPPD-80; Rhenogran® MBI-80; Rhenogran® MMBI-70; Rhenogran® NDBC-70/ECO; Rhenogran® NKF-80; Rhenogran® PBN-80; Rhenovin® DDA-70; Rhodianox MBPS; Rhodianox MO14; Rhodianox TBM6 TP; Rylex® 30; Rylex® 3010

Sandostab P-EPQ; Santoflex® IP; Santonox®; Santovar® A; Santowhite® Crystals; Santowhite® ML; Santowhite® PC; Santowhite® Powd; Seenox 412S; Spectratech® CM 11340; Spectratech® CM 11524; Spectratech® CM 11593; Spectratech® CM 11616; Spectratech® KM 11264; Stabilite 470; Stangard 500; Stangard PC; Stangard SP/SPL; Struktol® T.M.Q; Struktol® T.M.Q. Powd; Suconox-18®; Sustane® BHA; Sustane® BHT; Sustane® TBHQ; Swanox BHEB; SY-AO 68; Synox 5LT

Tenox® BHA; Tenox® PG; Tenox® TBHQ; Thermolite 187; THQ; Tinuvin® 326; Topanex 100BT; Topanol® 205; Topanol® CA; Topanol® CA-SF

Ultranox® 210; Ultranox® 226; Ultranox® 254; Ultranox® 276; Ultranox® 626; Ultranox® 627A; Ultranox® 815A Blend; Ultranox® 817A Blend; Ultranox® 875A Blend; Ultranox® 877A Blend; Ultranox® 2714A; Unilin® 425 Alcohol; Unistab D-33; UV-Chek® AM-104; Uvinox 770; Uvi-Nox® 1494

Vanax® PML; Vanox® 3C; Vanox® 6H; Vanox® 12; Vanox® 13; Vanox® 100; Vanox® 102; Vanox® 200; Vanox® 1005; Vanox® 1030; Vanox® 1081; Vanox® 1290; Vanox® 1320; Vanox® 2246; Vanox® AM; Vanox® AT; Vanox® GT; Vanox® HT; Vanox® MBPC; Vanox® MTI; Vanox® NBC; Vanox® SKT; Vanox® SWP; Vanox® ZMTI; Vanox® ZS; Voidox 100%; Vulkanox® 3100; Vulkanox® 4010 NA; Vulkanox® 4020; Vulkanox® 4022; Vulkanox® 4023; Vulkanox® 4025; Vulkanox® 4027; Vulkanox® 4030; Vulkanox® BKF; Vulkanox® CS; Vulkanox® DDA; Vulkanox® DS; Vulkanox® DS/F; Vulkanox® HS/LG; Vulkanox® HS/Pellets; Vulkanox® HS/Powd; Vulkanox® KB; Vulkanox® MB-2/MGC; Vulkanox® NKF;

Antioxidants *(cont'd.)*

Vulkanox® OCD; Vulkanox® OCD/SG; Vulkanox® PAN; Vulkanox® SP; Vulkanox® ZKF; Vulkanox® ZMB2/C5; Vulkanox® ZMB-2/G

Weston® 600; Weston® TLTTP; Wingstay® 100; Wingstay® 100AZ; Wingstay® 200; Wingstay® Antioxidants; Wingstay® K; Wingstay® L; Wingstay® S; Wingstay® S; Wingstay® SN-1; Wingstay® T; Wytox® 312

Chemicals: Acetone/diphenylamine condensate; Anilino-phenyl methacrylamide

Benzene propanoic acid; BHA; BHT; 1,2-Bis (3,5-di-t-butyl-4-hydroxyhydrocinnamoyl) hydrazine; 2,2-Bis[4-(2-(3,5-di-t-butyl-4-hydroxyhydrocinnamoyloxy)ethoxyphenyl]propane; Bis (2,4-di-t-butylphenyl) pentaerythritol diphosphite; 4,4′-Bis (α,α-dimethylbenzyl) diphenylamine; 2,4-Bis [(octylthio) methyl]-o-cresol; 4-[[4,6-Bis(octylthio)-s-triazin-2-yl]amino]-2,6-di-t-butylphenol;; 2-t-Butyl-p-cresol; 4,4′-Butylidenebis (6-t-butyl-m-cresol); 4,4′-Butylidene bis (2-t-butyl-5-methylphenol); Butyraldehyde-aniline condensation product

Calcium bis[monoethyl(3,5-di-t-butyl-4-hydroxybenzyl)phosphonate); C40-60 alcohols; p-Cresol/dicyclopentadiene butylated reaction product; Cyclic neopentanetetrayl bis(octadecyl phosphite); Cyclohexyl-N′-phenyl-p-phenylenediamine; Cycloneopentyl, cyclo (dimethlaminoethyl) pyrophosphato zirconate, di mesyl salt

Diamylhydroquinone; Diaryl p-phenylene diamine; 2,6-Di-t-butyl-4-s-butyl phenol; 2,6-Di-t-butyl-4-ethyl phenol; Di-t-butylhydroquinone; 3,5-Di-t-butyl-4-hydroxyhydrocinnamic acid/1,3,5-tris(2-hydroxyethyl)-s-triazine-2,4,6-(1H,3H,5H)-trione triester; 3-(3,5-Di-t-butyl-4-hydroxyphenyl)propionic methyl ester; Di (butyl, methyl pyrophosphato) ethylene titanate di (dioctyl, hydrogen phosphite); 2,4-Di-t-butylphenol; 2,6-Di-t-butylphenol; Dicatechol borate, di-o-tolyl guanidine salt; Dicyclo (dioctyl) pyrophosphato titanate; N,N′-Di-(1,4-dimethylpentyl)-p-phenylene diamine; Di (dioctylphosphato) ethylene titanate; Di (dioctylpyrophosphato) ethylene titanate; 1,2-Dihydro-2,2,4-trimethyl quinoline polymer; Diisobutylene; Dimethylaminomethyl phenol; 2,4-Dimethyl-6-(1-methyl cyclohexyl) phenol; Dimyristyl thiodipropionate; N,N′-Di-β-naphthyl-p-phenylenediamine; Dioctadecyl disulfide; p,p′-Dioctyldiphenylamine; N,N′-Di (2-octyl)-p-phenylenediamine; Diphenylamineacetone; N,N′-Diphenyl-p-phenylenediamine; Distearyl dithiodipropionate; Ditridecyl thiodipropionate

2,2′-Ethylidenebis (4,6-di-t-butylphenol); 2,2′-Ethylidenebis (4,6-di-t-butylphenyl) fluorophosphonite

Heptylated diphenylamine; Hexamethylenebis (3,5-di-t-butyl-4-hydroxycinnamate); N,N′-Hexamethylene bis (3,5-di-t-butyl-4-hydroxyhydrocinnamamide); Hydroquinone

2,2′-Isobutylenebis (4,6-dimethylphenol); 2,2′-Isobutylidenebis (4,6-dimethylphenol); Isopropyl 4-aminobenzenesulfonyl di (dodecylbenzenesulfonyl) titanate; Isopropyl dimethacryl isostearoyl titanate; N-Isopropyl-N′-phenyl-p-phenylenediamine; Isopropyl tri (dioctylphosphato) titanate; Isopropyl tri (dioctylpyrophosphato) titanate; Isopropyl tri (N ethylamino-ethylamino) titanate

Lauryl/stearyl thiodipropionate; Lead phthalate, basic

Magnesium aluminum hydrotalcite; 2-Mercaptotoluimidazole; 2,2′-Methylenebis (6-t-butyl-4-methylphenol); 4,4′-Methylenebis (2,6-di-t-butylphenol); 2,2′-Methylenebis (4-ethyl-6-t-butylphenol); 2,2′-Methylenebis 6-(1-methylcyclohexyl)-p-cresol; 2,2′-Methylenebis (4-methyl-6-cyclohexyl phenol); 2,2′-Methylenebis [4-methyl-6-(1-methyl-cyclohexyl) phenol; 4- and 5-Methylmercaptobenzimidazole; Mixed diaryl-p-phenylenediamine

Neopentyl (diallyl) oxy, trihydroxy caproyl titanate; Neopentyl (diallyl) oxy, trimercapto-phenyl zirconate; Nickel bis[O-ethyl (3,5-di-t-butyl-4-hydroxybenzyl)] phosphonate; Nickel dibutyldithiocarbamate; Nickel diisobutyldithiocarbamate; Nickel dimethyldithiocarbamate; 2,2′,2″-Nitrilo[triethyl-tris(3,3′,5,5′-tetra-t-butyl-1,1′-biphenyl-2,2′-diyl)phosphite]; Nonylphenol

Octadecyl 3,5-di-t-butyl-4-hydroxyhydrocinnamate; Octadecyl 3-mercaptopropionate; Octylated diphenylamine; Oxalyl bis (benzylidenehydrazide); 2,2′-Oxamido bis[ethyl 3-(3,5-di-t-butyl-4-hydroxyphenyl) propionate

Pentaerythrityl tetrakis [3-(3′,5′-di-t-butyl-4-hydroxyphenyl)propionate]; Pentaerythrityl tetrakis (β-laurylthiopropionate); Phenol, styrenated; Phenol, 4,4′-thiobis 2-(1,1-dimethylethyl) phosphite; Phenothiazine; Phenyl-α-naphthylamine; Phenyl-β-naphthylamine; Phosphorous acid, cyclic neopentanetetrayl bis (2,4-di-t-butylphenyl) ester; Phosphorus trichloride, reaction prods. with 1,1′-biphenyl and 2,4-bis (1,1-dimethylethyl) phenol; Polybutylated bisphenol A; Poly [(6-morpholino-s-triazine-2,4-diyl) [2,2,6,6-tetramethyl-4-piperidyl) imino]-hexamethylene [(2,2,6,6-tetramethyl-4-piperidyl) imino]

Stearoyl p-amino phenol; Stearyl lauryl thiodipropionate; Styrenated diphenyl amine

Tetra (2, diallyoxymethyl-1 butoxy titanium di (di-tridecyl) phosphite; Tetraisopropyl di (dioctylphosphito) titanate; Tetrakis (2,4-di-t-butylphenyl) 4,4-biphenylenediphosphonite;

Antioxidants *(cont'd.)*

Tetrakis [methylene (3,5-di-t-butyl-4-hydroxyhydrocinnamate)] methane; 4,4´-Thiobis-6-(t-butyl-m-cresol); Thiodiethylene bis (3,5-di-t-butyl-4-hydroxy) hydrocinnamate; Thiodipropionate polyester; Titanium di (butyl, octyl pyrophosphate) di (dioctyl, hydrogen phosphite) oxyacetate; Titanium di (cumylphenylate) oxyacetate; Titanium di (dioctylpyrophosphate) oxyacetate; Titanium dimethacrylate oxyacetate; p-(p-Toluenesulfonyl amido) diphenylamine; Toluhydroquinone; 1,1´,1´´-(1,3,5-Triazine-2,4,6-triyltris((cyclohexylimino)-2,1-ethanediyl)tris(3,3,4,5,5-tetramethylpiperazinone); Triisotridecyl phosphite; 2,2,4-Trimethyl-1,2-dihydroquinoline polymer; Trimethyl quinoline; 1,3,5-Trimethyl-2,4,6-tris (3,5-di-t-butyl-4-hydroxybenzyl) benzene; 1,3,5-Tris (4-t-butyl-3-hydroxy-2,6-dimethylbenzyl)-1,3,5-triazine-2,4,6-(1H,3H,5H)-trione; Tris(3,5-di-t-butyl-4-hydroxy benzyl) isocyanurate; Tris (2,4-di-t-butylphenyl) phosphite; 2,4,6-Tris-(N-1,4-dimethylpentyl-p-phenylenediamino)-1,3,5-triazine; Tris 3-hydroxy-4-t-butyl-2,6-dimethyl benzyl cyanurate; 1,1,3-Tris (2-methyl-4-hydroxy-5-t-butyl phenyl) butane; Tris Neodol-25 phosphite

Zinc 2-mercaptotoluimidazole; Zinc 4- and 5-methylmercaptobenzimidazole

Antiozonants

Trade names: Akrochem® Antioxidant 12; Akrochem® Antioxidant DQ; Akrochem® Antioxidant DQ-H; Akrochem® Antiozonant MPD-100; Akrochem® Antiozonant PD-2; Akrowax 130; Akrowax 145; Akrowax 5025; Akrowax 5050; Akrowax 5073; Akrowax Micro 23; Antilux® 110; Antilux® 111; Antilux® 500; Antilux® 550; Antilux® 600; Antilux® 620; Antilux® 654; Antilux® 660; Antilux® 750; Antilux® L; Antisun®; Antozite® 1; Antozite® 2; Antozite® 67P; Atmer® 150

Be Square® 185

Durazone® 37

Ekaland NDBC

Hostalub® XL 165FR

Isobutyl Niclate®

Mekon® White

Octowax 321; Octowax 518

Permanax™ 6PPD; Permanax™ CR; Permanax™ DPPD; Permanax™ IPPD; Permanax™ OZNS; Priolube 1414

Royalene 301-T

Santoflex® 13; Santoflex® 77; Santoflex® 134; Santoflex® 715; Santoflex® AW; Santoflex® IP; Starwax® 100; Sunolite® 100; Sunolite® 127; Sunolite® 240; Sunolite® 240TG; Sunolite® 666; Sunproofing Wax 1343

UV-Chek® AM-104

Vanox® 3C; Vanox® 6H; Vanox® NBC; Vanwax® H Special; Viplex 895-BL; Vistalon 6505; Vistanex® MML-80, MML-100, MML-120, MML-140; Vulkanox® 3100; Vulkanox® 4010 NA; Vulkanox® 4020; Vulkanox® 4023; Vulkanox® 4027; Vulkanox® 4030; Vulkanox® AFD; Vulkazon® AFD; Vulkazon® AFS-50; Vulkazon® AFS/LG; Vybar® 825

Wingstay® 100; Wingstay® 100AZ; Wingstay® 200

Chemicals: Bis (1,2,3,6-tetrahydrobenzaldehyde) pentaerythritol acetal

Cyclohexyl-N´-phenyl-p-phenylenediamine

Diaryl p-phenylene diamine; N,N´-Di-(1,4-dimethylpentyl)-p-phenylene diamine; N-1,3-Dimethylbutyl-N´-phenyl-p-phenylene diamine; N,N´-Diphenyl-p-phenylenediamine

Isobutyl oleate; N-Isopropyl-N´-phenyl-p-phenylenediamine

Mixed diaryl-p-phenylenediamine

Nickel diisobutyldithiocarbamate

2,4,6-Tris-(N-1,4-dimethylpentyl-p-phenylenediamino)-1,3,5-triazine

Antiscorch agents • Scorch retarders

Trade names: ADK STAB PEP-4C; ADK STAB PEP-11C; Ageflex TM 402, 403, 404, 410, 421, 423, 451, 461, 462; Agerite® Stalite; Agerite® Stalite S; Akrochem® Antioxidant DQ; Akrochem® Antioxidant DQ-H; Akrochem® Antioxidant S; Aktiplast® F; Aktiplast® PP; Aktiplast® T; Altax®; Amax®

Bondogen E®

Antiscorch agents *(cont'd.)*

Cohedur® RK; Crystex® 60 OT 10; Crystex® 90 OT 20; Crystex® HS; Crystex® HS OT 10; Crystex® HS OT 20; Crystex® OT 10; Crystex® OT 20; Crystex® Regular
Dyphene® 6746-L
Ekaland MBTS
MagChem® 125; MagChem® 200-D; Millrex™; MonoThiurad®
Reogen E®; Resinall 203; Resinall 767; Retarder AK; Retarder PX; Rhenocure® CMT; Rhenocure® CMU; Rhenocure® CUT; Rhenocure® EPC; Rhenocure® TP/G; Rhenocure® TP/S; Rhenofit® 1600
Santocure® DCBS; Santocure® TBSI; Santogard® PVI; Saret® 500; Saret® 515; Saret® 516; Saret® 517; Saret® 518; Saret® 519; Saret® 633; Saret® 634; Scorchguard '0'; SR-527; Struktol® FA 541; Struktol® WB 212; Struktol® WB 222; Struktol® ZP 1014
Thiate® U
Vanax® NS; Vanox® 1320; Vul-Cup 40KE; Vul-Cup R; Vulkacit CZ/EG; Vulkacit CZ/MGC; Vulkacit MOZ/SG; Vulkacit NZ/EG
Weston® PTP; Wiltrol P

Chemicals: 4,4´-Dithiomorpholine
Zinc diacrylate; Zinc dimethacrylate

Antistats

Trade names: Adogen® 432; Adogen® 442; Advawax® 240; Alkamide® STEDA; Alkamuls® EL-620L; Alkamuls® GMR-55LG; Alkamuls® GMS/C; Alkamuls® PSML-20; Alkamuls® R81; Alkamuls® S-8; Alkamuls® SML; Alkapol PEG 300; Alkasurf® NP-4; Alkasurf® NP-6; Aluminum Oxide C; Amine C; Ampacet 10053; Anstac M, 2M; Anstex AK-25; Anstex SA; Antistat A21750; Antistat A21800; Antistatic KN; Antistaticum RC 100; Armostat® 310; Armostat® 350; Armostat® 375; Armostat® 410; Armostat® 450; Armostat® 475; Armostat® 550; Armostat® 710; Armostat® 801; Armostat® 810; Aromox® C/12; Aromox® C/12-W; Arosurf® TA-100; Arquad® 2C-75; Arquad® 12-50; Arquad® 16-50; Arquad® 18-50; Arquad® C-50; Arquad® T-27W; Atlas 1500; Atmer® 100; Atmer® 105; Atmer® 110; Atmer® 113; Atmer® 122; Atmer® 124; Atmer® 125; Atmer® 129; Atmer® 138; Atmer® 139; Atmer® 154; Atmer® 163; Atmer® 164; Atmer® 190; Atmer® 1002; Atmer® 1004; Atmer® 1005; Atmer® 1006; Atmer® 1021; Atmer® 1024; Atmer® 1025; Atmer® 1027; Atmer® 7001; Atmer® 7002; Atmer® 7006; Atmer® 7007; Atmer® 7101; Atmer® 7109; Atmer® 7202; Atmer® 7203; Atmer® 8102; Atmer® 8103; Atmer® 8103-35; Atmer® 8112; Atmos® 150; Atmos® 300; Atmul® 84; Atmul® 124; Atmul® 695
Barquat® CME-35; Berol 28; Berol 302; Berol 303; Berol 307; Berol 381; Berol 386; Berol 387; Berol 389; Berol 391; Berol 392; Berol 397; Berol 455; Berol 456; Berol 457; Berol 458; Beycostat NE
Cabot® PE 9017; Catigene® CA 56; Chemeen 18-2; Chemeen 18-5; Chemeen 18-50; Chemeen HT-5; Chemeen HT-50; Chemquat 12-33; Chemquat 12-50; Chemquat 16-50; Chemquat C/33W; Chemstat® 106G/60DC; Chemstat® 106G/90; Chemstat® 122; Chemstat® 122/60DC; Chemstat® 172T; Chemstat® 182; Chemstat® 182/67DC; Chemstat® 192; Chemstat® 192/NCP; Chemstat® 273-C; Chemstat® 273-E; Chemstat® AC-100; Chemstat® AC-101; Chemstat® AC-200; Chemstat® AC-201; Chemstat® AC-202; Chemstat® AC-1000; Chemstat® AC-2000; Chemstat® HTSA #1A; Chemstat® LD-100; Chemstat® LD-100/60; Chemstat® LH-305; Chemstat® LH-306; Chemstat® P-400; Chemstat® PS-101; Chemstat® PS-101; Chemstat® PS-101/PLT; Chemstat® SE-5; Cirrasol® AEN-XB; Cirrasol® ALN-GM; Colloids PE 48/10/04 AS; Colortech 10310-12; Cordex AT-172; CPH-376N; Cyastat® 609; Cyastat® LS; Cyastat® SN; Cyastat® SP
Dehydat 20; Dehydat 22; Dehydat 80X; Dehydat 93P; Dehydat 3204; Dehydat 7882; Dehyquart® LDB; Dimodan LS; Dimodan PM; Dimodan PV; Dimodan PV 300; Drewplast® 017
Electrosol 325; Electrosol D; Electrosol M; Electrosol S-1-X; Elecut S-507; Emcol® CC-42; Emcol® CC-55; Emcol® CC-57; Emphos™ CS-136; Empilan® CDE; Empilan® CDX; Empilan® DL 40; Empilan® DL 100; Empilan® GMS NSE40; Empilan® GMS NSE90; Empilan® K Series; Empilan® LDE; Empilan® LDX; Empilan® NP9; Emuldan HV 52; Emulsifier K 30 40%; Emulsifier K 30 68%; Emulsifier K 30 76%; Emulsifier K 30 95%; Estol 1407; Ethosperse® LA-4; Ethylan® A4; Ethylan® A6; Ethylan® LD; Ethylan® LDS; Ethylan® ME; Ethylan® MLD; Eurestat 10; Eurestat 66; Eurestat K22
Ficel® LE; Ficel® SCE; Flexricin® 115; Flexricin® 185

Antistats *(cont'd.)*

Glycolube® 140; Glycolube® 140 Kosher; Glycolube® 150; Glycolube® 825; Glycolube® 825 Kosher; Glycolube® AFA-1; Glycomul® L; Glycomul® O; Glycomul® S; Glycosperse® L-20; Glycostat®; Grindtek MOP 90; Grindtek MSP 40; Grindtek MSP 40F; Grindtek MSP 90; Grindtek PK 60

Hallco® C-7065; Hamposyl® C; Hamposyl® C-30; Hamposyl® L; Hamposyl® L-30; Hamposyl® L-95; Hamposyl® M; Hamposyl® M-30; Hamposyl® O; Hamposyl® S; Hefti MS-55-F; Hefti PGE-600-ML; Hodag PE-004; Hodag PE-104; Hodag PE-106; Hodag PE-109; Hodag PE-206; Hodag PE-209; Hodag PEG 3350; Hodag PEG 8000; Hostastat® FA 14; Hostastat® FA 15; Hostastat® FA 18; Hostastat® FA 38; Hostastat® FE 2; Hostastat® FE 20; Hostastat® HS1; Hostastat® System E1902; Hostastat® System E1906; Hostastat® System E1956; Hostastat® System E2903; Hostastat® System E3904; Hostastat® System E3952; Hostastat® System E3953; Hostastat® System E3954; Hostastat® System E4751; Hostastat® System E4905; Hostastat® System E5951; Hostastat® System E6952

Igepal® CO-430

Kemamine® AS-650; Kemamine® AS-974; Kemamine® AS-974/1; Kemamine® AS-989; Kemamine® AS-990; Kemester® GMS (Powd.); Ken-React® 7 (KR 7); Ken-React® 9S (KR 9S); Ken-React® 12 (KR 12); Ken-React® 26S (KR 26S); Ken-React® 33DS (KR 33DS); Ken-React® 38S (KR 38S); Ken-React® 39DS (KR 39DS); Ken-React® 41B (KR 41B); Ken-React® 44 (KR 44); Ken-React® 46B (KR 46B); Ken-React® 55 (KR 55); Ken-React® 133DS (KR 133DS); Ken-React® 134S (KR 134S); Ken-React® 138D (KR 138D); Ken-React® 138S (KR 138S); Ken-React® 158D (KR 158D); Ken-React® 158FS (KR 158FS); Ken-React® 212 (KR 212); Ken-React® 238A (KR 238A); Ken-React® 238J (KR 238J); Ken-React® 238M (KR 238M); Ken-React® 238S (KR 238S); Ken-React® 238T (KR 238T); Ken-React® 262A (KR 262A); Ken-React® 262ES (KR 262ES); Ken-React® OPP2 (KR OPP2); Ken-React® OPPR (KR OPPR); Ken-React® KR TTS; Ken-Stat™ KS N100; Ken-Stat™ KS Q100P; Ken-Stat™ ZZ-1441H; Ken-Stat™ ZZ-1441L; Ken-Stat™ ZZ-1441R; Ketjenblack® EC-300 J; KZ OPPR; KZ TPP; KZ TPPJ

Lankrostat® 16; Lankrostat® 38; Lankrostat® 104; Lankrostat® 0600; Lankrostat® CA2; Lankrostat® LA3; Lankrostat® LDN; Lankrostat® LME; Lankrostat® NP6; Lankrostat® QAT; Larostat® 60A; Larostat® 88; Larostat® 96; Larostat® 143; Larostat® 264 A; Larostat® 264 A Anhyd; Larostat® 264 A Conc; Larostat® 377 DPG; Larostat® 377 FR; Larostat® 451 P; Larostat® 477; Larostat® 519; Larostat® 902 A; Larostat® 902 S; Larostat® 903; Larostat® 904; Larostat® 905; Larostat® 906; Larostat® 3001; Larostat® C-2; Larostat® FPE; Larostat® FPE-S; Larostat® GMOK; Larostat® GMSK; Larostat® HTS 904; Larostat® HTS 904 S; Larostat® HTS 905; Larostat® HTS 905 S; Larostat® HTS 906; Larostat® LTQ; Larostat® PVC; Larostat® T-2; Leomin AN; LICA 01; LICA 09; LICA 12; LICA 38; LICA 38A; LICA 38J; LICA 44; LICA 97; LICA 99; Lipomin LA; Lonzaine® 16S; Lonzaine® 18S

Mackester™ EGDS; Mackester™ EGMS; Mackester™ IP; Mackester™ SP; Maphos® 76; Marklear AFL-23; Markstat® AL-1; Markstat® AL-12; Markstat® AL-13; Markstat® AL-14; Markstat® AL-15; Markstat® AL-22; Markstat® AL-26; Markstat® AL-44; Markstat® AS-7; Markstat® AS-16; Markstat® AS-18; Markstat® AS-20; Masil® 1066C; Masil® 1066D; Masil® 2132; Masil® 2133; Masil® 2134; Mazeen® C-2; Mazeen® C-5; Mazeen® C-10; Mazeen® C-15; Mazeen® S-2; Mazeen® S-5; Mazeen® S-10; Mazeen® S-15; Mazeen® T-2; Mazeen® T-5; Mazeen® T-10; Mazeen® T-15; MDI AC-725; Merix Anti-Static #79 Conc; Merix Anti-Static #79-OL Super Conc; Merix Anti-Static #79 Super Conc; Merix Mold-Ease Conc. PCR; Merpol® HCS; Michel XO-24; Michel XO-85; Michel XO-108; Miramine® C; Mold Wiz DCZ; Mold Wiz OY; Monafax 785; Monolan® PPG440, PPG1100, PPG2200; Myrj® 45; Myvaplex® 600PK; Myverol® 18-04; Myverol® 18-06; Myverol® 18-07; Myverol® 18-99

Negomel AL-5; Newpol PE-61; Newpol PE-62; Newpol PE-64; Newpol PE-68; Newpol PE-74; Newpol PE-75; Newpol PE-78; Newpol PE-88; Nissan Anon BF; Nissan Elegan S-100; Nissan New Elegan A; Nissan New Elegan ASK; Noiox AK-41; Nopcostat 092; Nopcostat HS; Noramium CES 80; Noroplast 820X; Noroplast 2000; Noroplast 8000; Noroplast 8500; Noxamium S2-50; NZ 01; NZ 09; NZ 12; NZ 33; NZ 38; NZ 39; NZ 44; NZ 49; NZ 89; NZ 97

Parabolix® 100; Paricin® 210; Paricin® 220; Pationic® 900; Pationic® 901; Pationic® 902; Pationic® 905; Pationic® 907; Pationic® 914; Pationic® 1042; Pationic® 1042B; Pationic® 1042K; Pationic® 1042KB; Pationic® 1052; Pationic® 1052B; Pationic® 1052K; Pationic® 1052KB; Pationic® 1061; Pationic® 1064; Pationic® 1074; Pationic® 1145; Pationic® AS38; PE-AS 05; PE-AS 74; PE-AS 109; Pegosperse® 100 L; Pegosperse® 400 MS; Pegosperse® 1500 MS; Petrac® Eramide®; Petrac® GMS; Phosphanol Series; Plasadd® PE 9017; Pluracol® E600 NF; Pluracol® E1000; Pluracol® E1450; Pluracol® E1450 NF; Pluracol® E2000; Pluracol® E4500; Pluracol® E8000; Plysurf A207H; Plysurf A208B; Plysurf A208S;

Antistats *(cont'd.)*

Plysurf A210G; Plysurf A212C; Plysurf A215C; Plysurf A216B; Plysurf A217E; Plysurf A219B; Plysurf AL; Polylube RE; Polystat Agent #5033; Polysurf A212E; Polyvel AI509; Polyvel AI2163; Polyvel AI2902; Polyvel AI3129, AI5129; PP-AS 79; Proplast 058; Prox-onic DT-03; Prox-onic DT-015; Prox-onic DT-030; Prox-onic HR-05; Prox-onic HR-016; Prox-onic HR-025; Prox-onic HR-030; Prox-onic HR-036; Prox-onic HR-040; Prox-onic HR-080; Prox-onic HR-0200; Prox-onic HR-0200/50; Prox-onic HRH-05; Prox-onic HRH-016; Prox-onic HRH-025; Prox-onic HRH-0200; Prox-onic HRH-0200/50; Prox-onic MC-02; Prox-onic MC-05; Prox-onic MC-015; Prox-onic MHT-05; Prox-onic MHT-015; Prox-onic MO-02; Prox-onic MO-015; Prox-onic MO-030; Prox-onic MO-030-80; Prox-onic MS-02; Prox-onic MS-05; Prox-onic MS-011; Prox-onic MS-050; Prox-onic MT-02; Prox-onic MT-05; Prox-onic MT-015; Prox-onic MT-020

Quadrilan® SK

Rexol 25/7; Rhodafac® RE-610; Rhodafac® RE-877; Rhodafac® RS-410; Rhodafac® RS-610; Rhodafac® RS-710; Rhodapex® CO-433; Rhodapex® CO-436; Rhodasurf® PEG 3350; Rhodasurf® PEG 8000; RO-1

Sanstat 2012-A; Secoster® SDG; Serdox NSG 400; Softenol® 3900; Softenol® 3991; Sotex 3CW; Spectratech® CM 10777; Spectratech® CM 11045; Spectratech® CM 11638; Spectratech® CM 77242; Spinomar NaSS; Statexan K1; Statik-Blok J-2; Statik-Blok® FDA-3; Stat-Kon® AS-F; Stat-Kon® AS-FE; Struktol® AW-1; SY-AS 140; Synprolam 35MX1; Synprolam 35MX3; Synprolam 35MX5; Synprolam 35TMQS; Synprolam 35X2; Synprolam 35X2QS, 35X5QS, 35X10QS; Synprolam 35X5; Synprolam 35X10; Synprolam 35X15; Synprolam 35X20; Synprolam 35X25; Synprolam 35X35; Synprolam 35X50; Synpro® Stannous Stearate

Tegotens 4100; Tergitol® 15-S-3; Tergitol® 15-S-5; Teric 18M5; Teric 18M10; Teric PEG 400; Tetronic® 90R4; Tetronic® 150R1; Tetronic® 304; Tetronic® 701; Tetronic® 704; Tetronic® 901; Tetronic® 904; Tetronic® 908; Tetronic® 1307; Toho Master Batch-100; Tomah Q-14-2; Tomah Q-18-15; Tomah Q-D-T; Trymeen® 6617

Vulkanol® 85; Vulkanol® KA

Wayfos A; Wayfos D-10N; Witconate™ 90; Witconate™ 1298

XUS 15615.00

Zelec NE; Zelec NK; Zelec TY; Zelec UN

Chemicals: N,N-Bis (2-hydroxyethyl)-N-(3-dodecyloxy-2-hydroxypropyl) methyl ammonium methosulfate; Bis (2-hydroxyethyl) octyl methyl ammonium p-toluene sulfonate

Canola oil glyceride; Cetethyl morpholinium ethosulfate; Cetrimonium chloride; Cetyl betaine; Cocamide DEA; Coco alkyl-2,2′-iminobisethanol; Coco-betaine; Cocotrimonium chloride; Cocoyl hydroxyethyl imidazoline; C11-15 pareth-3; C11-15 pareth-5; C13-15 trimethyl ammonium methosulfate; Cycloneopentyl, cyclo (dimethlaminoethyl) pyrophosphato zirconate, di mesyl salt

Di (butyl, methyl pyrophosphato) ethylene titanate di (dioctyl, hydrogen phosphite); Dicyclo (dioctyl) pyrophosphato titanate; Di (dioctylphosphato) ethylene titanate; Di (dioctylpyrophosphato) ethylene titanate

N,N′-Ethylene bis-ricinoleamide; Ethylene dioleamide

Glyceryl dilaurate; Glyceryl dilaurate SE; Glyceryl linoleate; Glyceryl mono/dioleate; Glyceryl myristate; Glyceryl oleate; Glyceryl stearate SE; Glycol dilaurate; Glycol dilaurate SE; Glycol dioleate SE; Glycol distearate; Glycol distearate SE; Glycol laurate; Glycol laurate SE; Glycol oleate; Glycol oleate SE; Glycol ricinoleate SE

Hydrogenated cottonseed glyceride; Hydrogenated palm glyceride; Hydrogenated soy glyceride; N(β-Hydroxyethyl) ricinoleamide; N-(2-Hydroxypropyl) benzenesulfonamide

Isodecyl oleate; Isodecyloxypropyl dihydroxyethyl methyl ammonium chloride; Isopropyl 4-aminobenzenesulfonyl di (dodecylbenzenesulfonyl) titanate; Isopropyl dimethacryl isostearoyl titanate; Isopropyl tri (dioctylphosphato) titanate; Isopropyl tri (dioctylpyrophosphato) titanate; Isopropyl tri (N ethylamino-ethylamino) titanate

Lard glyceride; Lauralkonium chloride; Lauramide DEA; (3-Lauramidopropyl) trimethyl ammonium methyl sulfate; Lauramine; Laureth-4; Lauric/myristic dimethylethyl ammonium ethosulfate; Lauroyl sarcosine; Laurtrimonium chloride

Mono- and diglycerides of fatty acids; Myristoyl sarcosine

Neopentyl (diallyl) oxy, trihydroxy caproyl titanate; Neopentyl (diallyl) oxy, trimercapto-phenyl zirconate; Nonoxynol-4; Nonoxynol-6; Nonoxynol-7; Nonoxynol-9

Oleoyl sarcosine; Oleyl dimethylethyl ammonium ethosulfate

Palmitic/stearic acid mono/diglycerides; Palm kernel glycerides; PEG-8M; PEG-2 C13-15 alkyl amine; PEG-5 C13-15 alkyl amine; PEG-10 C13-15 alkyl amine; PEG-15 C13-15 alkyl amine;

Antistats *(cont'd.)*

PEG-20 C13-15 alkyl amine; PEG-25 C13-15 alkyl amine; PEG-35 C13-15 alkyl amine; PEG-50 C13-15 alkyl amine; PEG-1 C13-15 alkylmethylamine; PEG-3 C13-15 alkylmethylamine; PEG-5 C13-15 alkylmethylamine; PEG-18 castor oil; PEG-36 castor oil; PEG-2 cocamine; PEG-5 cocamine; PEG-10 cocamine; PEG-15 cocamine; PEG-5 cocomonium methosulfate; PEG-2 dilaurate; PEG-12 dilaurate; PEG-2 dilaurate SE; PEG-3 dioctoate; PEG-2 dioleate; PEG-2 dioleate SE; PEG-2 distearate; PEG-2 distearate SE; PEG-5 hydrogenated tallow amine; PEG-15 hydrogenated tallow amine; PEG-50 hydrogenated tallow amine; PEG-2 laurate; PEG-12 laurate; PEG-150 laurate; PEG-2 laurate SE; PEG-2 oleamine; PEG-7 oleamine; PEG-12 oleamine; PEG-15 oleamine; PEG-30 oleamine; PEG-2 oleate; PEG-12 oleate; PEG-2 oleate SE; PEG-2 soyamine; PEG-5 soyamine; PEG-10 soyamine; PEG-15 soyamine; PEG-2 stearamine; PEG-5 stearamine; PEG-10 stearamine; PEG-11 stearamine; PEG-50 stearamine; PEG-2 stearate; PEG-8 stearate; PEG-2 stearate SE; PEG-2 tallowamine; PEG-5 tallowamine; PEG-10 tallowamine; PEG-15 tallowamine; PEG-20 tallowamine; PEG-40 tallowamine; PEG-3 tallow diamine; PEG-15 tallow diamine; PEG-30 tallow diamine; Poloxamine 304; Poloxamine 504; Poloxamine 701; Poloxamine 702; Poloxamine 704; Poloxamine 707; Poloxamine 901; Poloxamine 904; Poloxamine 908; Poloxamine 1101; Poloxamine 1102; Poloxamine 1104; Poloxamine 1107; Poloxamine 1301; Poloxamine 1302; Poloxamine 1304; Poloxamine 1307; Poloxamine 1501; Poloxamine 1502; Poloxamine 1504; Poloxamine 1508; Polyethylene glycol; Polysorbate 20; PPG-40 diethylmonium chloride; PPG-2 oleate; PPG-2 oleate SE; PPG-2 stearate; PPG-2 stearate SE

Quaternium-18

Sodium C10-18 alkyl sulfonate; Sodium cocoyl sarcosinate; Sodium lauroyl sarcosinate; Sodium myristoyl sarcosinate; Sodium p-styrenesulfonate; Sorbitan laurate; Sorbitan oleate; Soyethyldimonium ethosulfate; Stannous stearate; Stearamidopropyl dimethyl-β-hydroxyethylammonium dihydrogen phosphate; Stearamidopropyl dimethyl-2-hydroxyethyl ammonium nitrate; Stearethyldimonium ethosulfate; Stearoyl sarcosine; Steartrimonium chloride; Stearyl betaine; N-Stearyl 12-hydroxystearamide; Stearyl PEG-15 methyl ammonium chloride

N-Tallow alkyl-2,2′-iminobisethanol; Tallow dimethyl trimethyl propylene diammonium chloride; Tetra (2, diallyoxymethyl-1 butoxy titanium di (di-tridecyl) phosphite; Tetrahydroxypropyl ethylenediamine; Tetraisopropyl di (dioctylphosphito) titanate; Tetraoctyloxytitanium di (ditridecylphosphite); Titanium di (butyl, octyl pyrophosphate) di (dioctyl, hydrogen phosphite) oxyacetate; Titanium di (cumylphenylate) oxyacetate; Titanium di (dioctylpyrophosphate) oxyacetate; Titanium dimethacrylate oxyacetate; Trideceth-3 phosphate; Trideceth-10 phosphate

Blocking agents

Trade names: Petrarch® A0699; Petrarch® A0700; Petrarch® A0742; Petrarch® A0743; Petrarch® A0750; Petrarch® A0800; Petrarch® B2500; Petrarch® C3292; Petrarch® C3300; Petrarch® G6720; Petrarch® H7300; Petrarch® I7810; Petrarch® M8550; Petrarch® M8450; Petrarch® M8500; Petrarch® M9050; Petrarch® M9100; Petrarch® O9835; Petrarch® T1807; Petrarch® T1918; Petrarch® T1980; Petrarch® T2090; Petrarch® T2503; Petrarch® T2910; Petrarch® V4900; Petrarch® V4910; Petrarch® V4917; Petrarch® V5000; Polyace™ 804

Chemicals: N-(2-Aminoethyl)-3-aminopropyl methyldimethoxy silane; 3-Aminopropylmethyldiethoxy silane; Amyltrichlorosilane

Bis (hydroxyethyl) aminopropyltriethoxy silane; Bis-(N-methylbenzamide) ethoxymethyl silane; Bis (trimethylsilyl) acetamide

3-Chloropropyltriethoxysilane

Di-t-butoxydiacetoxysilane

Ethyltriacetoxysilane

(3-Glycidoxypropyl)-methyldiethoxy silane

Isobutyltrimethoxysilane; Isocyanatopropyltriethoxysilane

3-Mercaptopropylmethyldimethoxysilane; Mercaptopropyltrimethoxysilane; N-Methylaminopropyltrimethoxysilane; Methyltriacetoxysilane; Methyltriethoxysilane; Methyltrimethoxysilane

Octyltriethoxysilane

2-Phenethyltrichlorosilane; Phenyltriethoxysilane; n-Propyltrimethoxysilane

3-(N-Styrylmethyl-2-aminoethylamino) propyltrimethoxy silane hydrochloride

Blocking agents *(cont'd.)*

Tetrakis (2-ethoxyethoxy) silane; Tetramethoxysilane; Tetra-n-propoxysilane; Trichlorosilane; N-[3-(Triethoxysilyl)-propyl] 4,5-dihydroimidazole; 1-Trimethoxysilyl-2-(chloromethyl) phenylethane; Trimethoxysilylpropyldiethylene triamine

Blowing agents • Blow promoters • Foaming agents

Trade names: Ablusol OA; Ablusol SF Series; Ablusol SN Series; Ablusol TA; Activex 233 25%; Activex 235 25%; Activex 236 25%; Activex 237 25%; Activex 436 25%; Activex 437 25%; Activex 447; Activex 533 25%; Activex 535 25%; Activex 536 25%; Activex 537 25%; Activex 539 25%; Activex 545; Activex 736 25%; ADK STAB BAP-1; ADK STAB BAP-2; ADK STAB BAP-4; ADK STAB BAP-5; ADK STAB BAP-7; ADK STAB BAP-8; ADK STAB FL-21; Aerosol® 18; Aerosol® C-61; Alkanol® 189-S; Ampacet 10123; Aromox® C/12; Aromox® C/12-W; Arquad® 2C-75; Arquad® 12-50; Arquad® 16-50; Arquad® 18-50; Arquad® C-50; Arquad® T-27W; Avirol® SL 2010; Avirol® SL 2020; Azo D; Azo DX72; Azocel OBSH; Azofoam; Azofoam B-95; Azofoam B-520; Azofoam DS-1, DS-2; Azofoam M-1, M-2, M-3; Azofoam UC

Blo-Foam® 715; Blo-Foam® 754; Blo-Foam® 765; Blo-Foam® ADC 150FF; Blo-Foam® ADC 300; Blo-Foam® ADC 300FF; Blo-Foam® ADC 300WS; Blo-Foam® ADC 450; Blo-Foam® ADC 450FFNP; Blo-Foam® ADC 450LT; Blo-Foam® ADC 450WS; Blo-Foam® ADC 550; Blo-Foam® ADC 550FFNP; Blo-Foam® ADC 800; Blo-Foam® ADC 1200; Blo-Foam® ADC-MP; Blo-Foam® ADC-NP; Blo-Foam® BBSH; Blo-Foam® DNPT 80; Blo-Foam® DNPT 93; Blo-Foam® DNPT 100; Blo-Foam® KL9; Blo-Foam® KL10; Blo-Foam® RA; Blo-Foam® SH

Calfoam ES-30; Cellcom; Cellcom OBSH-ASA2; Celogen® AZ 120; Celogen® AZ 130; Celogen® AZ 150; Celogen® AZ 180; Celogen® AZ 199; Celogen® AZ 754; Celogen® AZ-760-A; Celogen® AZ-2100; Celogen® AZ-2990; Celogen® AZ 3990; Celogen® AZ-5100; Celogen® AZNP 130; Celogen® AZNP 199; Celogen® AZRV; Celogen® OT; Celogen® RA; Celogen® TSH; Celogen® XP-100; Chemstat® 106G/90; Cosmopon BN

Disponil MGS 935; Dow Corning® 1250; Dow Corning® 1252; Dow Corning® 5043 Surfactant; Dow Corning® 5098; Dow Corning® Antifoam 1400; Dow Corning® Antifoam 1410; Dow Corning® Antifoam 1430; Dresinate® 214; Dresinate® 731

Empicol® LXV; Empicol® LXV/D; Empicol® LY28/S; Empicol® LZ/D; Empicol® LZG 30; Empicol® LZGV; Empicol® LZGV/C; Empicol® LZP; Empicol® LZV/D; Empimin® LR28; Empimin® MHH; Empimin® MKK98; Empimin® MKK/AU; Empimin® MKK/L; Empimin® MSS; Empimin® MTT; Empimin® MTT/A; EX LBA-30; Exocerol® LBA-39; Exocerol® OM; Exocerol® OM 70 A; Exocerol® Spezial 70; Expancel® 091 WU; Expancel® 551 WU; Expancel® 642 WU; Expancel® 820 DU; Expandex® 5PT; Expandex® 175; Extend

Ficel® 35 Series; Ficel® 44 Series; Ficel® 46 Series; Ficel® 49 Series; Ficel® 50 Series; Ficel® AC1; Ficel® AC2; Ficel® AC2M; Ficel® AC3; Ficel® AC3F; Ficel® AC4; Ficel® ACSP4; Ficel® ACSP5; Ficel® AFA; Ficel® EPA; Ficel® EPB; Ficel® EPC; Ficel® EPD; Ficel® EPE

Gardilene S25L; Genetron® 141b; GP-209 Silicone Polyol Copolymer; GP-214 Silicone Polyol Copolymer; GP-215 Silicone Polyol Copolymer; GP-218 Silicone Polyol Copolymer

HC 01; HC 05; Henley AZ EX 110; Henley AZ EX 120; Henley AZ EX 122; Henley AZ EX 127; Henley AZ EX 210; Henley AZ EX 310; Henley AZ LBA-30; Henley AZ LBA-49; HFVCD 11G; HFVCP 11G; HFVD 12CC; HFVP 03; HFVP 15P; HFVP 19B; Hostatron® System P1931; Hostatron® System P1933; Hostatron® System P1935; Hostatron® System P1940; Hostatron® System P1941; Hostatron® System P9937; Hostatron® System P9947; HP; HRVP 01; HRVP 19; Hydrocerol® BIF; Hydrocerol® BIH; Hydrocerol® BIH 10; Hydrocerol® BIH 25; Hydrocerol® BIH 40; Hydrocerol® BIH 70; Hydrocerol® BIN; Hydrocerol® CF; Hydrocerol® CF 5; Hydrocerol® CF 20E; Hydrocerol® CF 40E; Hydrocerol® CF 40S; Hydrocerol® CF-60V; Hydrocerol® CF 70; Hydrocerol® CLM 70; Hydrocerol® Compound; Hydrocerol® CP; Hydrocerol® CP 70; Hydrocerol® CT-211; Hydrocerol® CT-219; Hydrocerol® CT-232; Hydrocerol® CT-1001; Hydrocerol® HK; Hydrocerol® HK 20; Hydrocerol® HK 40; Hydrocerol® HK 40 B; Hydrocerol® HK 70; Hydrocerol® HP 20 P; Hydrocerol® HP 40 P; Hydrocerol® LBA-38; Hydrocerol® LBA-40; Hydrocerol® LBA-47; Hydrocerol® LC; Hydrocerol® LC 40 C; Hydrocerol® SH; Hydrocerol® SH 70

Kempore® 60/14FF; Ken-React® 7 (KR 7); Ken-React® 9S (KR 9S); Ken-React® 12 (KR 12); Ken-React® 26S (KR 26S); Ken-React® 33DS (KR 33DS); Ken-React® 38S (KR 38S); Ken-React® 39DS (KR 39DS); Ken-React® 41B (KR 41B); Ken-React® 44 (KR 44); Ken-React® 46B (KR 46B); Ken-React® 55 (KR 55); Ken-React® 133DS (KR 133DS); Ken-React® 134S (KR 134S); Ken-React® 138D (KR 138D); Ken-React® 138S (KR 138S); Ken-React® 158D

Blowing agents *(cont'd.)*

(KR 158D); Ken-React® 158FS (KR 158FS); Ken-React® 212 (KR 212); Ken-React® 238A (KR 238A); Ken-React® 238J (KR 238J); Ken-React® 238M (KR 238M); Ken-React® 238S (KR 238S); Ken-React® 238T (KR 238T); Ken-React® 262A (KR 262A); Ken-React® 262ES (KR 262ES); Ken-React® OPP2 (KR OPP2); Ken-React® OPPR (KR OPPR); Ken-React® KR TTS; Kycerol 91; Kycerol 92; KZ OPPR; KZ TPP; KZ TPPJ

Lankrocell® D15L; Lankrocell® KLOP; Lankrocell® KLOP/CV; Lankropol® ADF; Lankropol® ATE; Lankropol® ODS/LS; Lankropol® ODS/PT; LICA 01; LICA 09; LICA 12; LICA 38; LICA 38A; LICA 38J; LICA 44; LICA 97; LICA 99; Luperfoam 40; Luperfoam 329

Manro MA 35; Manro SLS 28; Mikrofine ADC-EV; Mikrofine AZDM; Mikrofine BSH; Mikrofine OBSH; Mikrofine TSH; Mikrofine TSSC

Nitropore® ATA; Nitropore® OBSH; NZ 01; NZ 09; NZ 12; NZ 33; NZ 38; NZ 39; NZ 44; NZ 49; NZ 89; NZ 97

Octosol 449; Octosol 496; Octosol A-1; Octosol A-18; Octosol A-18-A; Onifine-C; Onifine-CC; Onifine-CE; Opex® 80

Perlankrol® DSA; Plastifoam; Polycat® 5; Polycat® 77; Polystep® B-7; Polyvel BA20, BA40, BA60; Polyvel BA20T, BA40T; Polyvel BA205PT; Polyvel BA2015; Polyvel BA2575; Porofor® ADC/E; Porofor® ADC/F; Porofor® ADC/K; Porofor® ADC/M; Porofor® ADC/S; Porofor® B 13/CP 50; Porofor® BSH Paste; Porofor® BSH Paste M; Porofor® BSH Powder; Porofor® DNO/F; Porofor® N; Porofor® S 44; Priolene 6911

Retarder SAX; Rewomat TMS; Rewopol® B 1003; Rewopol® B 2003; Rewopol® SMS 35; Rewopol® TMSF; Rhenofit® 1600; Rhenopor 1843; Rhodacal® A-246L; Rhodafac® PS-19; Rhodamox® CAPO; Rhodamox® LO

Safoam® AP-40; Safoam® FP; Safoam® FP-20; Safoam® FP-40; Safoam® FPE-20; Safoam® FPE-50; Safoam® P; Safoam® P-20; Safoam® P-50; Safoam® PCE-40; Safoam® PE-20; Safoam® PE-50; Safoam® RIC; Safoam® RIC-25; Safoam® RIC-30; Safoam® RPC; Safoam® RPC-40; SAFR-CELL 40; SAFR-CELL 40 LT; Secolat; Sinopol 714; Sodium Bicarbonate T.F.F. Treated Free Flowing; Steol® 4N; Steol® CA-460; Stepanol® WA-100; Surfax 218

Texapon® VHC Needles; Texapon® ZHC Powder; TSS

Ufasan 35; Ufasan 62B; Ufasan 65; Unicell 5PT; Unicell D; Unicell OH; Unicell TS; Unifine P-2; Unifine P-3; Unifine P-4; Unifine P-5; Zorapol LS-30

Chemicals: 2,2′-Azobisisobutyronitrile; Azodicarbonamide; Azoisobutyrodinitrile

Benzene-1,3-disulfonyl hydrazide; Benzene sulfonyl hydrazide; Bis-(2-dimethylaminoethyl) ether; t-Butylhydrazinium chloride

Cycloneopentyl, cyclo (dimethlaminoethyl) pyrophosphato zirconate, di mesyl salt

Di (butyl, methyl pyrophosphato) ethylene titanate di (dioctyl, hydrogen phosphite); Dichlorofluoroethane; Dichlorotrifluoroethane; Dicyclo (dioctyl) pyrophosphato titanate; Di (dioctylphosphato) ethylene titanate; Di (dioctylpyrophosphato) ethylene titanate; Dinitrosopentamethylene tetramine; Diphenylene oxide-4,4′-disulfohydrazide

Isopropyl 4-aminobenzenesulfonyl di (dodecylbenzenesulfonyl) titanate; Isopropyl dimethacryl isostearoyl titanate; Isopropyl tri (dioctylphosphato) titanate; Isopropyl tri (dioctylpyrophosphato) titanate; Isopropyl tri (N ethylamino-ethylamino) titanate

Neopentyl (diallyl) oxy, trihydroxy caproyl titanate; Neopentyl (diallyl) oxy, trimercapto-phenyl zirconate

4,4′-Oxybis (benzenesulfonylhydrazide)

N,N′,N′′,N′′-Pentamethyldiethylenetriamine; 5-Phenyltetrazole; 5-Phenyltetrazole barium salt

Sodium bicarbonate; Sulfonyl hydrazide

Tetra (2, diallyoxymethyl-1 butoxy titanium di (di-tridecyl) phosphite; Tetraisopropyl di (dioctylphosphito) titanate; Tetraoctyloxytitanium di (ditridecylphosphite); Titanium di (butyl, octyl pyrophosphate) di (dioctyl, hydrogen phosphite) oxyacetate; Titanium di (cumylphenylate) oxyacetate; Titanium di (dioctylpyrophosphate) oxyacetate; Titanium dimethacrylate oxyacetate; p-Toluene sulfonyl hydrazide; p-Toluene sulfonyl semicarbazide

Carbon blacks

Trade names: Aflux® R

Black Pearls® 3700

Diablack® A; Diablack® E; Diablack® G; Diablack® H; Diablack® HA; Diablack® I; Diablack® II; Diablack® LH; Diablack® LI; Diablack® LR; Diablack® M; Diablack® N220M; Diablack®

Carbon blacks *(cont'd.)*

N234; Diablack® N339; Diablack® N550M; Diablack® N760M; Diablack® R; Diablack® SF; Diablack® SH; Diablack® SHA; Diapol® WMB® 1808; Diapol® WMB® 1833; Diapol® WMB® 1847; Diapol® WMB® 1848; Diapol® WMB® 1849; Diapol® WMB® 1939; Diapol® WMB® S900; Diapol® WMB® S903; Diapol® WMB® S910; Diapol® WMB® S920; Diapol® WMB® S960

Ketjenblack® EC-300 J; Ketjenblack® EC-310 NW; Ketjenblack® EC-350 J Spd; Ketjenblack® EC-350 N DOP; Ketjenblack® EC-600 JD; Ketjenblack® ED-600 JD

Mitsubishi Carbon Black

Regal® 300

Sable™ 3; Sable™ 5; Sable™ 15; Sable™ 17; Sterling® 3420; Sterling® 5550; Sterling® 5630; Sterling® 6630; Sterling® 6640; Sterling® 8860; Sterling® NS; Sterling® NS-1; Sterling® SO; Sterling® V; Sterling® VH; Synthetic Carbon 910-44M

Vulcan® 2H; Vulcan® 3; Vulcan® 3H; Vulcan® 4H; Vulcan® 6; Vulcan® 6LM; Vulcan® 7H; Vulcan® 8H; Vulcan® 9; Vulcan® 10H; Vulcan® C; Vulcan® J; Vulcan® K; Vulcan® M; Vulcan® XC72

Chemicals: Carbon black

Catalysts

Trade names: A-2; Abcure S-40-25; Actiron NX 3; Ageflex FA-1; Ageflex FA-2; Albrite® PA-75; Amicure® 33-LV; Amicure® TEDA; Aquathene™ CM 92830; Aquathene™ CM 92832; Armeen® DM8; Armeen® DM10; Armeen® DM12; Armeen® DM14; Armeen® DM16; Armeen® DMC; Armeen® DMHT; Armeen® DMO; Armeen® DMT; Armeen® M2-10D; Armeen® N-CMD

Benox® L-40LV; BiCAT™ 8; BiCAT™ H; BiCAT™ V; BiCAT™ Z; BYK®-Catalyst 460

Catalyst CC; Catalyst RD Liq; Curithane® 52; Curithane® 97

Dabco® 33-LV®; Dabco® 120; Dabco® 125; Dabco® 131; Dabco® 1027; Dabco® 1028; Dabco® 2039; Dabco® 7928; Dabco® 8136; Dabco® 8154; Dabco® 8264; Dabco® B-16; Dabco® BDMA; Dabco® BL-11; Dabco® BL-16; Dabco® BL-17; Dabco® BL-22; Dabco® BLV; Dabco® Crystalline; Dabco® CS90; Dabco® DC-2®; Dabco® DM9534; Dabco® DM9793; Dabco® DMEA; Dabco® EG; Dabco® FF-2003; Dabco® H-1010; Dabco® K-15; Dabco® MC; Dabco® NCM; Dabco® NEM; Dabco® NMM; Dabco® R-8020®; Dabco® RC; Dabco® S-25; Dabco® T; Dabco® T-1; Dabco® T-5; Dabco® T-9; Dabco® T-10; Dabco® T-11; Dabco® T-12; Dabco® T-16; Dabco® T-45; Dabco® T-95; Dabco® T-96; Dabco® T-120; Dabco® T-125; Dabco® T-131; Dabco® TETN; Dabco® TL; Dabco® TMR®; Dabco® TMR-2®; Dabco® TMR-3; Dabco® TMR-4; Dabco® TMR-30; Dabco® X-542; Dabco® X-543; Dabco® X-8161; Dabco® XDM™; Dabco® XF-C10-40; Dama® 810; Dama® 1010; Dapro 5005; Desmorapid; Di-Cup 40C; Di-Cup 40KE; Di-Cup R, T; Dinoram C; Dynapol® Catalyst 1203; Dytek® EP

Eastman® TEP; Eltesol® 4009, 4018; Eltesol® 4200; Eltesol® 4402; Eltesol® CA 65; Eltesol® CA 96; Eltesol® PSA 65; Eltesol® TA Series; Eltesol® TA 65; Eltesol® TA 96; Eltesol® TA/E; Eltesol® TA/F; Eltesol® TA/H; Eltesol® TSX; Eltesol® TSX/A; Eltesol® TSX/SF; Eltesol® XA65; Eltesol® XA90; Eltesol® XA/M65; Emka Catalyst P-35; Empigen® AM; Empigen® AY

Fascat® 2000; Fascat® 2001; Fascat® 2003; Fascat® 2004; Fascat® 2005; Fascat® 4100; Fascat® 4101; Fascat® 4102; Fascat® 4200; Fascat® 4201; Fascat® 9100; Fascat® 9102; Fascat® 9201; FC-24; FC-520; Ficel® AZDN-LF; Ficel® AZDN-LMC; FireShield® H; FireShield® HPM; FireShield® HPM-UF; FireShield® L; Fluorad® FC-24; Fomrez® C-2; Fomrez® C-4; Fomrez® SUL-3; Fomrez® SUL-4; Fomrez® UL-1; Fomrez® UL-2; Fomrez® UL-6; Fomrez® UL-8; Fomrez® UL-22; Fomrez® UL-28; Fomrez® UL-29; Fomrez® UL-32; Fonoline® White; Fonoline® Yellow; FR-D

Hamp-Ene® 100; Hamp-Ene® Na_3 Liq; Hi-Point® 90; Hi-Point® PD-1; Hyamine® 1622 50%

Jeffcat DD; Jeffcat DM-70; Jeffcat DMCHA; Jeffcat DMDEE; Jeffcat DME; Jeffcat DMP; Jeffcat DPA; Jeffcat DPA-50; Jeffcat M-75; Jeffcat MM-70; Jeffcat NEM; Jeffcat NMM; Jeffcat T-10; Jeffcat T-12; Jeffcat TD-20; Jeffcat TD-33; Jeffcat TD-33A; Jeffcat Z-65; Jeffcat ZF-22; Jeffcat ZF-24; Jeffcat ZF-26; Jeffcat ZF-51; Jeffcat ZF-52; Jeffcat ZF-53; Jeffcat ZF-54; Jeffcat ZR-50; Jeffcat ZR-70

Kemamine® T-1902D; Kemamine® T-6501; Kemamine® T-6502D; Kemamine® T-9701; Kemamine® T-9702D; Kemamine® T-9902D; Kemamine® T-9972D; Kencolor®; Kencure™ MPP; Kencure™ MPPJ; Ken-React® 7 (KR 7); Ken-React® 9S (KR 9S); Ken-React® 12 (KR 12); Ken-React® 26S (KR 26S); Ken-React® 33DS (KR 33DS); Ken-React® 38S (KR 38S);

Catalysts *(cont'd.)*

Ken-React® 39DS (KR 39DS); Ken-React® 41B (KR 41B); Ken-React® 44 (KR 44); Ken-React® 46B (KR 46B); Ken-React® 55 (KR 55); Ken-React® 133DS (KR 133DS); Ken-React® 134S (KR 134S); Ken-React® 138D (KR 138D); Ken-React® 138S (KR 138S); Ken-React® 158D (KR 158D); Ken-React® 158FS (KR 158FS); Ken-React® 212 (KR 212); Ken-React® 238A (KR 238A); Ken-React® 238J (KR 238J); Ken-React® 238M (KR 238M); Ken-React® 238S (KR 238S); Ken-React® 238T (KR 238T); Ken-React® 262A (KR 262A); Ken-React® 262ES (KR 262ES); Ken-React® OPP2 (KR OPP2); Ken-React® OPPR (KR OPPR); Ken-React® KR TTS; Kosmos® 10; Kosmos® 15; Kosmos® 16; Kosmos® 19; Kosmos® 21; Kosmos® 23; Kosmos® 24; Kosmos® 29; Kosmos® 64; KZ OPPR; KZ TPP; KZ TPPJ

L-55R®; LICA 01; LICA 09; LICA 12; LICA 38; LICA 38A; LICA 38J; LICA 44; LICA 97; LICA 99; Lonza KS 44; Lonza T 44; LP-100; LP-200; LP-300; LP-400; Lucidol 75; Lucidol 75FP

Magala® 0.5E; Manro BA Acid; Manro FCM 90LV; Manro PTSA/C; Manro PTSA/E; Manro PTSA/H; Manro PTSA/LS; Mark® TT; Metacure® T-1; Metacure® T-5; Metacure® T-9; Metacure® T-12; Metacure® T-45; Metacure® T-120; Metacure® T-125; Metacure® T-131; Mod Acid

Nissan Tert. Amine AB; Nissan Tert. Amine ABT; Nissan Tert. Amine BB; Nissan Tert. Amine FB; Nissan Tert. Amine MB; Nissan Tert. Amine PB; NZ 01; NZ 09; NZ 12; NZ 33; NZ 38; NZ 39; NZ 44; NZ 49; NZ 89; NZ 97

Pationic® 920; Pationic® 940; Pationic® 1230; Pationic® 1240; Pationic® 1250; Pennad 150; Petcat R-9; Polycat® 5; Polycat® 8; Polycat® 9; Polycat® 12; Polycat® 15; Polycat® 17; Polycat® 33; Polycat® 41; Polycat® 43; Polycat® 46; Polycat® 58; Polycat® 70; Polycat® 77; Polycat® 79; Polycat® 85; Polycat® 91; Polycat® DBU; Polycat® SA-1; Polycat® SA-102; Polycat® SA-610/50; Poly-Dispersion® DCP-60P; Poly-Dispersion® E (DIC)D-40; Poly-Dispersion® T (DIC)D-40P; Polystep® A-13; Polystep® A-17; Prespersion SIB-7324; Prespersion SIC-3233; PSA; PTSA 70

Reworyl® B 70; Reworyl® C 65; Reworyl® T 65; Reworyl® X 65

Sequestrene® NAFe 13% Fe; SM 2059; Sprayset® MEKP; Superox® 702; Superox® 710; Superox® 711; Superox® 712; Superox® 713; Superox® 730; Superox® 732; Superox® 739; Superox® 744; Synprolam 35DM; Synprolam 35X2; Synprolam 35X5; Synprolam 35X10; Synprolam 35X15; Synprolam 35X20; Synprolam 35X25; Synprolam 35X35; Synprolam 35X50; Synpron 1009

TEDA; TEDA-D007; TEDA-L25B; TEDA-L33; TEDA-L33E; TEDA-T401; TEDA-T411; TEDA-T811; Tegoamin® 33; Tegoamin® 100; Tegoamin® BDE; Tegoamin® CPU 33; Tegoamin® DMEA; Tegoamin® EPS; Tegoamin® PDD 60; Tegoamin® PMD; Tegoamin® PTA; Tegoamin® SMP; Thermact® 1; Thermact® 1; Thermacure®; T-Hydro®; Toyocat®-B2; Toyocat®-B6; Toyocat®-B7; Toyocat®-B20; Toyocat®-B54; Toyocat®-B61; Toyocat®-B71; Toyocat®-DMA; Toyocat®-DMCH; Toyocat®-DT; Toyocat®-ET; Toyocat®-ETF; Toyocat®-F2; Toyocat®-F4; Toyocat®-F83; Toyocat®-F94; Toyocat®-HP; Toyocat®-HPW; Toyocat®-HX4W; Toyocat®-HX15W; Toyocat®-HX35W; Toyocat®-HX63; Toyocat®-HX70; Toyocat®-L20M; Toyocat®-L75D; Toyocat®-LE; Toyocat®-M30; Toyocat®-M50; Toyocat®-MR; Toyocat®-N31; Toyocat®-N81; Toyocat®-NEM; Toyocat®-NP; Toyocat®-SC2; Toyocat®-SPF; Toyocat®-SPF2; Toyocat®-SPF3; Toyocat®-TE; Toyocat®-TF; Toyocat®-THN; Toyocat®-TMF; Toyocat®-TRC; TSA-70; Tyzor AA; Tyzor DC; Tyzor LA; Tyzor TBT; Tyzor TE; Tyzor TOT; Tyzor TPT

UltraFine® II; USP®-245

Versamine® EH-30; Vul-Cup 40KE; Vul-Cup R

XSA 80; XSA 90

Chemicals:

Acetylacetonate; Ammonium lactate; 2,2´-Azobisisobutyronitrile

Benzenesulfonic acid; Benzethonium chloride; Benzoyl peroxide; N-Benzyldimethylamine; Bis-(2-dimethylaminoethyl) ether; N,N-Bis (3-dimethylaminopropyl)-N-isopropanolamine; Bisperoxide; t-Butyl hydroperoxide; t-Butyl peroctoate; Butyltin mercaptide

Coco morpholine; Cumene hydroperoxide; Cumene sulfonic acid; Cycloneopentyl, cyclo (dimethlaminoethyl) pyrophosphato zirconate, di mesyl salt

Di (butyl, methyl pyrophosphato) ethylene titanate di (dioctyl, hydrogen phosphite); Di-t-butyl peroxide; Dibutyltin bisisooctyl thioglycolate; Dibutyltin (bis) mercaptide; Dibutyltin dilaurate; Dibutyltin disulfide; Dibutyltin maleate; Dicoco methylamine; Dicumyl peroxide; Dicyandiamide; Dicyclo (dioctyl) pyrophosphato titanate; Didecyl methylamine; Di (dioctylphosphato) ethylene titanate; Di (dioctylpyrophosphato) ethylene titanate; Dihydrogenated tallow methylamine; Dilauryl phosphite; 2-(2-Dimethylaminoethoxy) ethanol; Dimethylaminomethyl phenol; 2-Dimethylamino-2-methyl-1-propanol; Dimethyl arachidyl-behenyl amine; N,N-Dimethyl-N-cyclohexylamine; Dimethyl decylamine; 2,5-Dimethyl-2,5-di (t-butylperoxy) hex-

Catalysts *(cont'd.)*

ane; 2,5-Dimethyl-2,5-di(2-ethylhexanoyl peroxy) hexane; N,N-Dimethyl)-N′,N′-diisopropanol-1,3-propanediamine; Dimethylethanolamine; Dimethyl hydrogenated tallow amine; Dimethyl lauramine; Dimethyl myristamine; Dimethyl octylamine; Dimethyl palmitamine; Dimethyl soyamine; Dimethyl stearamine; 2,2′-Dimorpholino diethylether; Dioctyltin mercaptide; Dioctyltin oxide
Ethylacetoacetate; Ethyl morpholine
Isopropyl 4-aminobenzenesulfonyl di (dodecylbenzenesulfonyl) titanate; Isopropyl dimethacryl isostearoyl titanate; Isopropyl tri (dioctylphosphato) titanate; Isopropyl tri (dioctylpyrophosphato) titanate; Isopropyl tri (N ethylamino-ethylamino) titanate
Lauroyl peroxide; Lead dioxide; Lithium ricinoleate
Methyl ethyl ketone peroxide; N-Methyl-N′-hydroxyethylpiperazine; p-Methyl morpholine
Neopentyl (diallyl) oxy, trihydroxy caproyl titanate; Neopentyl (diallyl) oxy, trimercapto-phenyl zirconate
Oxazolidine
PEG-2 C13-15 alkyl amine; N,N,N′,N′′,N′′-Pentamethyldipropylenetriamine; 1,3-Pentanediamine; Potassium 2-ethylhexanoate; Potassium octoate
Sodium ferric EDTA; Sodium formaldehyde sulfoxylate; Stannous chloride anhyd.; Stannous octoate; Succinic acid peroxide
Tetrabutylphosphonium acetate, monoacetic acid; Tetrabutylphosphonium bromide; Tetrabutylphosphonium chloride; Tetraisopropyl di (dioctylphosphito) titanate; Tetraisopropyl titanate; Tetrakis (2-ethylhexyl) titanate; Tetrakis (hydroxymethyl) phosphonium chloride; N,N,N′,N′-Tetramethylhexanediamine; Tetraoctyloxytitanium di (ditridecylphosphite); Tetraoctylphosphonium bromide; Tetrasodium EDTA; Titanium di (butyl, octyl pyrophosphate) di (dioctyl, hydrogen phosphite) oxyacetate; Titanium di (cumylphenylate) oxyacetate; Titanium di (dioctylpyrophosphate) oxyacetate; Titanium dimethacrylate oxyacetate; o-Toluene sulfonic acid; p-Toluene sulfonic acid; Trichloroethane; Triethylene diamine; Triethyl phosphate; Trifluoromethane sulfonic acid; N,N′,N′-Trimethylaminoethyl piperazine; N,N,N′-Trimethyl-N′-hydroxyethylbisaminoethylether; Trisodium EDTA
Zinc 2-ethylhexoate

Chain extenders

Trade names: BDO
Dabco® BDO; DETDA 80
Eastman® HQEE
Lonzacure® M-CDEA; Lonzacure® M-DEA; Lonzacure® M-DIPA; Lonzacure® M-MIPA
Unilink® 4100; Unilink® 4102; Unilink® 4200

Chemicals: 3,5-Diethyl-2,4-diaminotoluene; 3,5-Diethyl-2,6-diaminotoluene
Hydroquinone bis(2-hydroxyethyl) ether
4,4′-Methylenebis (3-chloro-2,6-diethylaniline); 4,4′-Methylenebis 2,6-diethylaniline; 4,4′-Methylenebis 2,6-diisopropylaniline; 4,4′-Methylenebis 2-isopropyl-6-methylaniline

Chelators

Trade names: Albrite® TIPP; Albrite® TMP
Chel DTPA-41
Dapro 5005; Doverphos® 6; Doverphos® 7; Doverphos® 8; Doverphos® 9; Doverphos® 12
Hamp-Ene® 100; Hamp-Ene® Na_3 Liq; Hamp-Ene® NaFe Purified Grade
Interstab® CH-55; Interstab® CH-55R; Irgastab® 2002
Lankromark® LE109
Mark® 1178B
Naugard® NBC
Q-1300; Quadrol®
Sequestrene® 30A; Synpron 1456
Vanstay SC; Vanstay SD; Vanstay SG; Vanstay SH; Versene 100; Versene 100 EP; Versene 100 LS; Versene 100 SRG; Versene 100 XL; Versene 220; Versene Acid; Versene CA; Versene

Chelators *(cont'd.)*

Diammonium EDTA; Versene NA; Versene Na_2; Versene Tetraammonium EDTA
Weston® 439

Chemicals: Disodium EDTA
4,4′ Isopropylidenediphenol alkyl (C12-15) phosphites
Pentasodium pentetate
Sodium ferric EDTA; Sodium lignosulfonate
Tetraammonium EDTA; Tetrahydroxypropyl ethylenediamine; Tetrasodium EDTA; Trisodium EDTA

Clarifiers

Trade names: ClearTint®
Empiphos 4KP
MagChem® 50; MagChem® 60; Millad® 3940; Millad® 3988; Millad® Conc. 5C41-10; Millad® Conc. 5L71-10; Millad® Conc. 8C41-10
Polyvel CA-P20
Therm-Chek® 904
Vanstay SC
Weston® 439

Coagulants

Trade names: Coagulant AW; Coagulant CHA; Coagulant WS
Ethomeen® 18/60
Hodag PEG 300; Hodag PEG 400
Millad® 3905
Polyox® Coagulant
Teric PPG 1000; Teric PPG 2250

Chemicals: Cyclohexylamine acetate
PPG (1000)

Colorants • Dyes • Pigments

Trade names: ACtone® 2000V; ACtone® 2316; ACtone® 2461; ACtone® N; Advantage™; Afflair® Lustre Pigments; Akrochem® Powder Colors; Akroplast®; Akrosperse® Color Masterbatches; Akrosperse® Plasticizer Paste Colors; Akrosperse® Water Paste Colors; Albacar® 5970; Albaglos®; Ampacet 12083; Ampacet 13134; Ampacet 14052; Ampacet 15114; Ampacet 15250; Ampacet 15391; Ampacet 16180; Ampacet 16192; Ampacet 16238; Ampacet 16438; Ampacet 17106; Ampacet 17161; Ampacet 17491; Ampacet 18088; Ampacet 18109; Ampacet 18402; Ampacet 18537; Ampacet 18659; Ampacet 19153; Ampacet 19153-S; Ampacet 19200; Ampacet 19238; Ampacet 19252; Ampacet 19258; Ampacet 19270; Ampacet 19278; Ampacet 19400; Ampacet 19475-S; Ampacet 19492; Ampacet 19500; Ampacet 19552; Ampacet 19584; Ampacet 19614; Ampacet 19673; Ampacet 19699; Ampacet 19709; Ampacet 19716; Ampacet 19717; Ampacet 19786; Ampacet 19832; Ampacet 19873; Ampacet 19897; Ampacet 19975; Ampacet 19991; Ampacet 19999; Ampacet 39270; Ampacet 41312; Ampacet 41424; Ampacet 41438; Ampacet 41483; Ampacet 41495; Ampacet 49270; Ampacet 49315; Ampacet 49370; Ampacet 49419; Ampacet 49842; Ampacet 49882; Ampacet 49920; Ampacet 49984; Ampacet 110017; Ampacet 110025; Ampacet 110041; Ampacet 110052; Ampacet 110070; Ampacet 110181-A; Ampacet 110313; Ampacet 110370; Ampacet 190014; Ampacet 190015; Arquad® 2C-75; Arquad® 12-50; Arquad® 16-50; Arquad® 18-50; Arquad® C-50; Arquad® T-27W; ASP® 100; ASP® 101; ASP® 102; ASP® 170; ASP® 200; ASP® 400; ASP® 400P; ASP® 600; ASP® 602; ASP® 900; AT-1; Atomite®
Black Pearls® 120; Black Pearls® 130; Black Pearls® 160; Black Pearls® 170; Black Pearls® 280; Black Pearls® 450; Black Pearls® 460; Black Pearls® 470; Black Pearls® 480; Black

Colorants • Dyes • Pigments *(cont'd.)*

Pearls® 490; Black Pearls® 520; Black Pearls® 570; Black Pearls® 700; Black Pearls® 800; Black Pearls® 880; Black Pearls® 900; Black Pearls® 1000; Black Pearls® 1100; Black Pearls® 3700; Black Pearls® L; BSWL 201; BSWL 202; Burgess 30; Burgess 30-P; Burgess 2211; Burgess 5178; Burgess CB; Burgess KE

CC™-101; CC™-103; CC™-105; Ceramtex; Chroma-Chem® 844; Chromaflo®; ClearTint®; ClearTint® PC Green; Cordon NU 890/75; CP Filler™

Diaresin Dye®; Drikalite®

Elftex® 120; Elftex® 125; Elftex® 160; Elftex® 280; Elftex® 285; Elftex® 415; Elftex® 430; Elftex® 435; Elftex® 460; Elftex® 465; Elftex® 570; Elftex® 670; Elftex® 675

Ferro BP-10; Ferro CP-18; Ferro CP-50; Ferro CP-78; Ferro DP-25; Ferro EP-37; Ferro EP-62; Ferro FP-40; Ferro FP-64; Ferro RB-30; Ferro V-5; Ferro V-8; Ferro V-9415

Geode™

Hampton Dry Colorants; Hampton Masterbatches; Heliogen®; Hitox®; Hydrite 121-S

Iceberg®

Kadox® 720; Kadox®-911; Kadox®-920; Kadox® 930; Kencolor®; Kotamite®; Kronos® 1000; Kronos® 2020; Kronos® 2073; Kronos® 2081; Kronos® 2101; Kronos® 2160; Kronos® 2200; Kronos® 2210; Kronos® 2220; Kronos® 2230; Kronos® 2310

Lithol® Pigments; Lithopone 30% DS; Lithopone 30% L; Lithopone D (Red Seal 30% ZnS); Lithopone D (Silver Seal 60% ZnS); Lithopone DS (Red Seal 30% ZnS); Lithopone L (Red Seal 30% ZnS); Lithopone L (Silver Seal 60% ZnS)

Magoh-S; Manganese Violet; Marble Dust™; Marblemite™; Marquat Pigments; MDI AB-925; MDI EC-940; MDI HD-650; MDI HD-935; MDI MB-650; MDI NY-905; MDI PC-925; MDI PE-135; MDI PE-500; MDI PE-500-20F; MDI PE-500A; MDI PE-500HD; MDI PE-500LL; MDI PE-540; MDI PE-650F; MDI PE-650HF; MDI PE-650LL; MDI PE-670T; MDI PE-675; MDI PE-907; MDI PE-907HD; MDI PE-931WC; MDI PE-940; MDI PP-130; MDI PP-535; MDI PP-940; MDI PS-125; MDI PS-650; MDI PS-901; MDI PS-903; MDI PV-940; MDI SN-948; Mearlin® Card Gold; Mearlin® Hi-Lite Super Blue; Mearlin® Hi-Lite Super Gold; Mearlin® Hi-Lite Super Green; Mearlin® Hi-Lite Super Orange; Mearlin® Hi-Lite Super Red; Mearlin® Hi-Lite Super Violet; Mearlin® Inca Gold; Mearlin® MagnaPearl 1000; Mearlin® MagnaPearl 1100; Mearlin® MagnaPearl 1110; Mearlin® MagnaPearl 2000; Mearlin® MagnaPearl 2110; Mearlin® MagnaPearl 3000; Mearlin® MagnaPearl 3100; Mearlin® MagnaPearl 4000; Mearlin® MagnaPearl 5000; Mearlin® Nu-Antique Silver; Mearlin® Pearl White; Mearlin® Satin White; Mearlin® Silk White; Mearlin® Sparkle; Mearlin® Supersparkle; Mearlite® Ultra Bright UDQ; Mearlite® Ultra Bright UTL; Micro-White® 07 Slurry; Micro-White® 10; Micro-White® 10 Slurry; Micro-White® 15; Micro-White® 15 SAM; Micro-White® 15 Slurry; Micro-White® 25; Micro-White® 25 SAM; Micro-White® 25 Slurry; Micro-White® 40; Micro-White® 100; Mistron PXL®; Mitsubishi Carbon Black; Mogul® L; Monarch® 120; Monarch® 700; Monarch® 800; Monarch® 880; Monarch® 900; Monarch® 1000; Monarch® 1100

Octocure 462; Oppasin®; Optiwhite®; Optiwhite® P; OSO® 440; OSO® 1905; OSO® NR 830 M; OSO® NR 950

Paliogen®; Paliotol®; Plastisorb; Plaswite® LL 7014; Plaswite® LL 7015; Plaswite® LL 7041; Plaswite® LL 7049; Plaswite® LL 7057; Plaswite® LL 7105; Plaswite® LL 7108; Plaswite® LL 7112; Plaswite® PE 7000; Plaswite® PE 7001; Plaswite® PE 7002; Plaswite® PE 7003; Plaswite® PE 7004; Plaswite® PE 7005; Plaswite® PE 7006; Plaswite® PE 7024; Plaswite® PE 7031; Plaswite® PE 7079; Plaswite® PE 7097; Plaswite® PE 7192; Plaswite® PP 7161; Plaswite® PP 7269; Plaswite® PS 7028; Plaswite® PS 7037; Plaswite® PS 7131; Plaswite® PS 7174; Plaswite® PS 7179; Plaswite® PS 7231; Polarite 420; Polyfil® WC; Polytrend® 850; Prespersion PAB-9541; Prespersion SIB-7211; Prespersion SIC-3342; Prespersion SIC-3536; Protint

Quindo Magenta RV-6863

Rakusol®; Reactint®; Reactint® Black 57AB; Reactint® Black X40LV; Reactint® Blue 17AB; Reactint® Blue X3LV; Reactint® Blue X19; Reactint® Orange X38; Reactint® Red X52; Reactint® Violet X80LT; Reactint® Yellow X15; Reed C-ABS-7526; Reed C-ABS-17415; Reed C-BK-7; Reed C-EPES-1331; Reed C-NY-261; Reed C-NY-4892; Reed C-PBT-1338; Reed C-PET-1333; Reed C-PPR-2096; Reed C-PS-4230; Reed C-PUR-1051; Reed C-PY-1232; Reed PWC-1; Regal® 300R; Regal® 330; Regal® 330R; Regal® 400; Regal® 400R; Regal® 660; Regal® 660R; Rhenogran® Fe-gelb-50; Rhenogran® Fe-rot-70; RL-90

Sachtolith® HD; Sachtolith® HD-S; Sachtolith® L; Sico®; Sicolen®; Sicomin®; Sicopal®; Sicoplast®; Sicostyren®; Sicotan®; Sicotrans®; Sicoversal®; Silcogum®; Silcopas®; Silene® D; Siltex™ 44; Siltex™ 44C; Snowflake P.E.™; Snowflake White™; Spectraflo®; Spectratech® CM 00524; Spectratech® CM 22801; Spectratech® CM 30757; Spectratech®

Colorants • Dyes • Pigments *(cont'd.)*

CM 34286; Spectratech® CM 45288; Spectratech® CM 56483; Spectratech® CM 63745; Spectratech® CM 80206; Spectratech® CM 88100; Spectratech® CM 88200; Spectratech® CM 92049; Spectratech® CM 92101; Spectratech® CM 92473; Spectratech® HM 82192; Spectratech® HM 99506; Spectratech® IM 88763; Spectratech® KM 80531; Spectratech® PM 80230; Spectratech® PM 82215; Spectratech® PM 92742; Spectratech® RM 88964; Spectratech® SM 88008; Spectratech® SM 92042; Spectratech® YM 80751; Spectratech® YR 92363; Supercoat®; Supermite®
Tegocolor®; Teknor Color Crystals; Thermoplast®; Tiona® RCL-4; Tiona® RCL-6; Tiona® RCL-9; Tiona® RCL-69; Ti-Pure® R-100; Ti-Pure® R-101; Ti-Pure® R-102; Ti-Pure® R-103; Ti-Pure® R-900; Ti-Pure® R-902; Ti-Pure® R-960; Tisyn®; Tylose® MH Grades; Tylose® MHB
Ultramarine Blue
Vulcan® 9A32; Vulcan® XC72R
Zinc Oxide No. 185; Zinc Oxide No. 318

Chemicals: Carbon black
Ferric oxide
Lithopone
Manganese violet
Pigment red 202; Pigment yellow 53
Talc; Titanium dioxide
Ultramarine blue; Ultramarine violet
Wollastonite
Zinc sulfide

Compatibilizers

Trade names: Acrawax® C CG
Epolene® E-43; Escor® ATX-310; Escor® ATX-320; Escor® ATX-325; Exxelor PO 1015
Fusabond® MB-110D; Fusabond® MB-226D; Fusabond® MC-189D; Fusabond® MC-190D; Fusabond® MC-197D; Fusabond® MZ-109D; Fusabond® MZ-203D
Kronitex® 100
Nevex® 100
Polyace™ 573; Polybond® 1000; Polybond® 1001; Polybond® 1002; Polybond® 1009; Polybond® 3001
SAFR-EPS K; Staflex DIOP; Staflex DOZ; Staflex DTDP; Staflex NODA
Tebol™; Tebol™ 99
Unite; Uvinul® 3008

Chemicals: Ethylene-maleic anhydride copolymer; Ethylene-propylene-styrene terpolymer
Polyethylene, linear low density
Triaryl phosphate

Coupling agents

Trade names: Advawax® 240; Advawax® 280; Advawax® 290; Aerosol® DPOS-45; Ageflex FA-1; Ageflex FA-2; Arcosolv® PTB
Carbowax® PEG 300; Castorwax® MP-70; Castorwax® MP-80; CB2408; CI7840; CP0320; CPS 076; CPS 076.5; CPS 078.5; CPS 078.9; CPS 925; CS1590; CT2902
Dioxitol-High Gravity; Dow Corning® Q1-6106; Dow Corning® Z-6020; Dow Corning® Z-6030; Dow Corning® Z-6032; Dow Corning® Z-6040; Dow Corning® Z-6075; Dow Corning® Z-6076; Dynasylan® 0116; Dynasylan® 1110; Dynasylan® 1411; Dynasylan® 1505; Dynasylan® 3403; Dynasylan® AMEO; Dynasylan® AMMO; Dynasylan® AMTC; Dynasylan® BDAC; Dynasylan® BSA; Dynasylan® CPTEO; Dynasylan® CPTMO; Dynasylan® DAMO; Dynasylan® DAMO-P; Dynasylan® DAMO-T; Dynasylan® ETAC; Dynasylan® GLYMO; Dynasylan® HMDS; Dynasylan® IBTMO; Dynasylan® IMEO; Dynasylan® MEMO; Dynasylan® MTAC; Dynasylan® MTES; Dynasylan® MTMO; Dynasylan® MTMS; Dynasylan® OCTEO; Dynasylan® PETCS; Dynasylan® PTMO; Dynasylan® TCS; Dynasylan® TRIAMO; Dynasylan® VTC; Dynasylan® VTEO; Dynasylan® VTMO; Dynasylan® VTMOEO

Coupling agents *(cont'd.)*

Ektasolve® DP; Ektasolve® EP; Epolene® C-13; Epolene® C-14; Epolene® C-17; Epolene® E-10; Epolene® E-15; Epolene® E-20; Epolene® E-43; Epolene® G-3002; Epolene® G-3003; Exxelor PO 1015; #15 Oil

Fusabond® MB-110D; Fusabond® MB-226D; Fusabond® MC-189D; Fusabond® MC-190D; Fusabond® MC-197D; Fusabond® MZ-109D; Fusabond® MZ-203D

Gantrez® AN-139; Gantrez® AN-149; Gantrez® AN-169; Gantrez® AN-179

Hostaprime® HC 5

KA 301; KA 322; Kenflex® A; Kenflex® N; Ken-React® 7 (KR 7); Ken-React® 9S (KR 9S); Ken-React® 12 (KR 12); Ken-React® 26S (KR 26S); Ken-React® 33DS (KR 33DS); Ken-React® 38S (KR 38S); Ken-React® 39DS (KR 39DS); Ken-React® 41B (KR 41B); Ken-React® 44 (KR 44); Ken-React® 46B (KR 46B); Ken-React® 55 (KR 55); Ken-React® 133DS (KR 133DS); Ken-React® 134S (KR 134S); Ken-React® 138D (KR 138D); Ken-React® 138S (KR 138S); Ken-React® 158D (KR 158D); Ken-React® 158FS (KR 158FS); Ken-React® 212 (KR 212); Ken-React® 238A (KR 238A); Ken-React® 238J (KR 238J); Ken-React® 238M (KR 238M); Ken-React® 238S (KR 238S); Ken-React® 238T (KR 238T); Ken-React® 262A (KR 262A); Ken-React® 262ES (KR 262ES); Ken-React® OPP2 (KR OPP2); Ken-React® OPPR (KR OPPR); Ken-React® KR TTS; KZ 55; KZ OPPR; KZ TPP; KZ TPPJ

LICA 01; LICA 09; LICA 12; LICA 38; LICA 38A; LICA 38J; LICA 44; LICA 97; LICA 99

Mapeg® CO-25H; Maphos® L 13; Mazon® 1086; Mazon® 1096; Monafax 1293

NZ 01; NZ 09; NZ 12; NZ 33; NZ 38; NZ 39; NZ 44; NZ 49; NZ 89; NZ 97

Pale 4; Pale 16; Pale 170; Pale 1000; Petrarch® A0699; Petrarch® A0700; Petrarch® A0701; Petrarch® A0742; Petrarch® A0743; Petrarch® A0750; Petrarch® A0800; Petrarch® B2500; Petrarch® C3292; Petrarch® C3300; Petrarch® G6720; Petrarch® H7300; Petrarch® I7810; Petrarch® M8550; Petrarch® M8450; Petrarch® M8500; Petrarch® M9050; Petrarch® M9100; Petrarch® O9835; Petrarch® T1807; Petrarch® T1918; Petrarch® T1980; Petrarch® T2090; Petrarch® T2503; Petrarch® T2910; Petrarch® V4900; Petrarch® V4910; Petrarch® V4917; Petrarch® V5000; Polybond® 1000; Polybond® 1001; Polybond® 1002; Polybond® 1009; Polybond® 3001; Prosil® 196; Prosil® 220; Prosil® 221; Prosil® 248; Prosil® 3128; Prosil® 5136; Prosil® HMDS; Prox-onic NP-1.5; Prox-onic NP-04; Prox-onic NP-06; Prox-onic NP-09; Prox-onic NP-010; Prox-onic NP-015; Prox-onic NP-020; Prox-onic NP-030; Prox-onic NP-030/70; Prox-onic NP-040; Prox-onic NP-040/70; Prox-onic NP-050; Prox-onic NP-050/70; Prox-onic NP-0100; Prox-onic NP-0100/70

SAFR-T-5K; SAFR-T-12K; SAFR-Z-12K; SAFR-Z-12SK; SAFR-Z-22E K; SAFR-Z-44 K; Si69; Spheriglass® Coated; Struktol® TR 071

Triton® X-15; Triton® X-35

Unite; Uvasil PSL6

Vorite 105; Vorite 110; Vorite 115; Vorite 120; Vorite 125

Wayfos A

X50-S®

Chemicals:

N-(2-Aminoethyl)-3-aminopropyl methyldimethoxy silane; N-2-Aminoethyl-3-aminopropyl trimethoxysilane; 3-Aminopropylmethyldiethoxy silane; Amyltrichlorosilane

Bis (hydroxyethyl) aminopropyltriethoxy silane; Bis-(N-methylbenzamide) ethoxymethyl silane; Bis-(3-(triethoxysilyl)propyl)tetrasulfane; Bis (trimethylsilyl) acetamide; t-Butyl alcohol

3-Chloropropyltriethoxysilane; 3-Chloropropyltrimethoxysilane; Cyclo [dineopentyl (diallyl)] pyrophosphato dineopentyl (diallyl) zirconate; Cyclo (dioctyl) pyrophosphato dioctyl zirconate; Cycloneopentyl, cyclo (dimethlaminoethyl) pyrophosphato zirconate, di mesyl salt

Di-t-butoxydiacetoxysilane; Di (butyl, methyl pyrophosphato) ethylene titanate di (dioctyl, hydrogen phosphite); Dicyclo (dioctyl) pyrophosphato titanate; Di (dioctylphosphato) ethylene titanate; Di (dioctylpyrophosphato) ethylene titanate; Diisobutyl (oleyl) aceto acetyl aluminate; Diisopropyl (oleyl) aceto acetyl aluminate

Ethyltriacetoxysilane

(3-Glycidoxypropyl)-methyldiethoxy silane; 3-Glycidoxypropyltrimethoxysilane

Hexamethyldisilazane

Isobutyltrimethoxysilane; Isocyanatopropyltriethoxysilane; Isopropyl 4-aminobenzenesulfonyl di (dodecylbenzenesulfonyl) titanate; Isopropyl dimethacryl isostearoyl titanate; Isopropyl tri (dioctylphosphato) titanate; Isopropyl tri (dioctylpyrophosphato) titanate; Isopropyl tri (N ethylamino-ethylamino) titanate

3-Mercaptopropylmethyldimethoxysilane; Mercaptopropyltrimethoxysilane; 3-Methacryloxypropyltrimethoxysilane; N-Methylaminopropyltrimethoxysilane; Methyltriethoxysilane; Methyltrimethoxysilane

Coupling agents *(cont'd.)*

Neopentyl (diallyl) oxy, tri (m-amino) phenyl zirconate; Neopentyl (diallyl) oxy, tri (dioctyl) phosphato zirconate; Neopentyl (diallyl) oxy, tri (dioctyl) pyrophosphato zirconate; Neopentyl (diallyl) oxy, tri (dodecyl) benzene-sulfonyl zirconate; Neopentyl (diallyl) oxy, tri (9,10 epoxy stearoyl) zirconate; Neopentyl (diallyl) oxy, tri (N-ethylenediamino) ethyl zirconate; Neopentyl (diallyl) oxy, trihydroxy caproyl titanate; Neopentyl (diallyl) oxy, trimercapto-phenyl zirconate; Neopentyl (diallyl) oxy, trimethacryl zirconate; Neopentyl (diallyl) oxy, trineodecanoyl zirconate; Nonoxynol-15

Octyltriethoxysilane; Oleth-4; Oleth-9; Oleth-20; Oleyl alcohol

PEG-4 laurate; 3-(n-Pentadecyl) phenol; 2-Phenethyltrichlorosilane; Phenyltriethoxysilane; Polyacrylic acid; Polyethylene, linear low density; Polyvinylmethylsiloxane; n-Propyltrimethoxysilane

Sorbitan sesquioleate; 3-(N-Styrylmethyl-2-aminoethylamino) propyltrimethoxy silane hydrochloride

Tetra (2,2 diallyloxymethyl) butyl, di (ditridecyl) phosphito zirconate; Tetra (2, diallyoxymethyl-1 butoxy titanium di (di-tridecyl) phosphite; Tetraethoxysilane; Tetrahydrofurfuryl alcohol; Tetraisopropyl di (dioctylphosphito) titanate; Tetrakis (2-ethoxyethoxy) silane; Tetramethoxysilane; Tetra-n-propoxysilane; Titanium di (butyl, octyl pyrophosphate) di (dioctyl, hydrogen phosphite) oxyacetate; Titanium di (cumylphenylate) oxyacetate; Titanium di (dioctylpyrophosphate) oxyacetate; Titanium dimethacrylate oxyacetate; Trichlorosilane; N-[3-(Triethoxysilyl)-propyl] 4,5-dihydroimidazole; 1-Trimethoxysilyl-2-(chloromethyl) phenylethane; Trimethoxysilylpropyldiethylene triamine; (N-Trimethoxysilylpropyl)-polyethylenimine

N-[2-Vinyl(benzylamino)-ethyl]-3-aminopropyltrimethoxysilane; Vinyltriacetoxy silane; Vinyltriethoxysilane

Creaming agents

Trade names: Ageflex EGDMA

Crosslinking agents

Trade names: Actimer FR-803; Aerosol® 501; Ageflex AMA; Ageflex 1,3 BGDMA; Ageflex DEGDMA; Ageflex T4EGDA; Ageflex TM 402, 403, 404, 410, 421, 423, 451, 461, 462; Ageflex TMPTA; Ageflex TMPTMA; Ageflex TPGDA; Akrochem® Zinc Oxide 35; Amicure® CL-485; Aztec® BCUP; Aztec® BP-50-PSI; Aztec® DCLBP-50-PSI; Aztec® DCP-40-G; Aztec® DCP-40-IC; Aztec® DCP-40-IC1; Aztec® DCP-R; Aztec® 2,5-DI; Aztec® 2,5-DI-45-G; Aztec® 2,5-DI-45-IC; Aztec® 2,5-DI-45-IC1; Aztec® 2,5-DI-45-PSI; Aztec® 2,5-DI-50-C; Aztec® 2,5-DI-70-S; Aztec® DIPP-2; Aztec® DIPP-40-G; Aztec® DIPP-40-IC; Aztec® DIPP-40-IC1; Aztec® DIPP-40-IC5; Aztec® DTBP; Aztec® EBU-40-G; Aztec® EBU-40-IC; Aztec® HMCN-30-AL; Aztec® HMCN-30-WO-2; Aztec® HMCN-40-IC3; Aztec® NBV-40-G; Aztec® NBV-40-IC; Aztec® PMBP-50-PSI; Aztec® TBPB-50-IC; Aztec® TBPEHC; Aztec® 2,5-TRI; Aztec® 3,3,5-TRI-28-IC3; Aztec® 3,3,5-TRI-40-G; Aztec® 3,3,5-TRI-40-IC; Aztec® 3,3,5-TRI-55-AL; Azthane® I-100

BM-723; BM-801; BM-818

Cadox® BS; Cadox® OS; Cadox® PS; Cadox® TDP; Cadox® TS-50; Cadox® TS-50S; CHP-158; CPS 120; CPS 123; Cylink HPC-75; Cylink HPC-90; Cylink HPC-100; Cymel 303; Cymel 370; Cymel 373; Cymel 380; Cymel 1141

Dabco® BDO; Dabco® CL-485; Dabco® DEOA-LF; Dantocol® DHE; Desmodur® DA; Desmodur® L75A; Desmodur® MP-225; Desmodur® RC; Desmodur® RE; Desmodur® RFE; Desmodur® VK-5; Desmodur® VK-18; Desmodur® VK-70; Desmodur® VK-200; Desmodur® VKS-2; Desmodur® VKS-4; Desmodur® VKS-5; Desmodur® VKS-18; Di-Cup 40C; Di-Cup 40KE; Di-Cup R, T

Empol® 1018; Esperal® 115RG; Espercarb® 438M-60; Esperox® C-496

Ficel® AZDN-LF; Ficel® AZDN-LMC

Great Lakes BA-43

Heloxy® 44; Heloxy® 48; Hystrene® 3675; Hystrene® 3675C; Hystrene® 3680; Hystrene® 3695

Industrene® D

Kemamine® DD-3680; Kemamine® DP-3680; Kemamine® DP-3695

Crosslinking agents *(cont'd.)*

Lucidol 75; Lucidol 98; Luperco 101-P20; Luperco 101-XL; Luperco 130-XL; Luperco 231-SRL; Luperco 231-XL; Luperco 233-XL; Luperco 500-SRK; Luperco 801-XL; Luperco 802-40KE; Luperco ANS-P; Luperox 802; Lupersol 219-M60; Lupersol 531-80M; Lupersol 533-M75; Lupersol 553-M75; Lupersol 555-M60; Lupersol 665-M50; Lupersol 665-T50; Lupersol 688-M50; Lupersol 688-T50; Lupersol TAEC

MFM-401; MFM-405; MFM-407; MFM-413; MFM-415; MFM-416; MFM-418; MFM-425; MFM-786 V; Millamine® 5260; Milldride® 5060

Ortegol® 204

Perkadox® 14; Perkadox® 14/40; Perkadox® 14-40B-pd; Perkadox® 14-40K-pd; Perkadox® 14-90; Perkadox® 14S; Perkadox® 14S-fl; Perkadox® BC; Perkadox® BC-40Bpd; Perkadox® BC-40K-pd; Perkadox® BC-40S; Peroximon® DC 40 MF; Peroximon® DC 40 MG; Pluracol® 355; Pluracol® 364; Pluracol® 450; Pluracol® 550; Pluracol® 650; Pluracol® 669; Polycup® 172; Polycup® 1884; Polycup® 2002; Polyvel CR-L10; Polyvel PCL-10; Polyvel PCL-20

Retilox® F 40 MF; Retilox® F 40 MG; Rezol® 4393; Rhenocure® CA; Rhenocure® CUT; Rhenocure® Diuron; Rhenocure® TDD; Rhenocure® ZAT; Rhenofit® BDMA/S; Rhenofit® CF; Rhenofit® EDMA/S; Rhenofit® NC; Rhenofit® TAC/S; Rhenofit® TRIM/S; Rhenofit® UE; Rhenovin® Na-stearat-80; Rhenovin® S-stearat-80; Ricon 152; Ricon 153; Rylex® 30; Rylex® 3010

SAFR-Z-44 K; Saret® 500; Saret® 515; Saret® 516; Saret® 517; Saret® 518; Saret® 519; Saret® 633; Saret® 634; Si69; Silcat® R; Sipomer® AM; Sipomer® DCPA; Sipomer® DCPM; Sipomer® HEM-5; Sipomer® HEM-10; Sipomer® TMPTMA; SP-1044; SP-1045; SP-1055; SP-1056; SR 111; SR-201; SR-205; SR-206; SR-209; SR-210; SR-213; SR-214; SR-239; SR-252; SR-290; SR-295; SR-348; SR-350; SR-368; SR-416; SR-444; SR-494; SR-527; SR-9041; Struktol® IB 531; Sulfasan® R; Superox® 46-747; Superox® 46-748; Syn Fac® 8026

Thiovanol®; Tomah DA-14; Tomah DA-16; Tomah DA-17; Tomah PA-10; Tomah PA-14; Tomah PA-16; Tomah PA-19; Tomah PA-1214; Trigonox® 17/40; Trigonox® 17-40B-pd; Trigonox® 29-40B-pd; Trigonox® 101; Trigonox® 101/45Bpd; Trigonox® 101-45S-ps; Trigonox® 107; Trigonox® 145; Trigonox® 145-45B; Trigonox® B; Trigonox® BPIC; Trigonox® C; Trigonox® T; Trigonox® T-94; Tyzor AA; Tyzor DC; Tyzor LA; Tyzor TBT; Tyzor TE; Tyzor TOT; Tyzor TPT

Union Carbide® Y-11343; Union Carbide® Y-11597

Vanchem® HM-50, HM-4346; Varox® 130; Varox® 130-XL; Varox® 230-XL; Varox® 231-XL; Varox® 231; Varox® 802-40KE; Varox® 802-40MB; Varox® DBPH-50; Varox® DBPH; Varox® DCP-40C; Varox® DCP-40KE; Varox® DCP-40MB; Varox® DCP-R; Varox® DCP-T; Vestamin® A139; Viton® Curative No. 50

Chemicals: Acetylacetonate; Allyl methacrylate; Ammonium lactate; OO-t-Amyl O-(2-ethylhexyl) monoperoxycarbonate; t-Amyl peroxyacetate

α,α′-Bis (t-butylperoxy) diisopropylbenzene; Bis (o-chlorobenzoyl) peroxide; Bis (2,4-dichlorobenzoyl) peroxide; Bis-(3-(triethoxysilyl)propyl)tetrasulfane; 1,4-Butanediol diacrylate; 1,3-Butanediol dimethacrylate; n-Butyl-4,4-bis (t-butylperoxy) valerate; t-Butyl cumyl peroxide; 1,3-Butylene glycol dimethacrylate; t-Butyl peroxy isopropyl carbonate

DEDM hydantoin; 2,2-Di-(t-amylperoxy) propane; 1,1-Di (t-butylperoxy) cyclohexane; Di-(s-butyl)peroxydicarbonate; Di-(2-t-butylperoxyisopropyl)benzene; 1,1-Di (t-butylperoxy) 3,3,5-trimethyl cyclohexane; 2,4-Di-t-butylphenyl 3,5-di-t-butyl-4-hydroxybenzoate; p,p-Dichlorobenzoyl peroxide; Dicumyl peroxide; Dicyclopentenyl acrylate; Dicyclopentenyl methacrylate; Di (2,4-dichloro benzoyl) peroxide; Diethanolamine; Diethylene glycol dimethacrylate; Dimer diamine; Di-(2-methylbenzoyl) peroxide; Di (4-methylbenzoyl) peroxide; 2,5-Dimethyl-2,5-di (t-butylperoxy) hexyne-3; 1,1-Dimethyl-3-hydroxybutylperoxy-2-ethylhexanoate; 1,1-Dimethyl-3-hydroxybutylperoxyneoheptanoate

Ethylacetoacetate; Ethyl-o-benzoyl laurohydroximate; Ethyl 3,3-di (t-amylperoxy) butyrate; Ethyldibutylperoxybutyrate; Ethylene glycol dimethacrylate

Glycol dimercaptoacetate

Hexamethoxymethylmelamine; 3,3,6,6,9,9-Hexamethyl 1,2,4,5-tetraoxa cyclononane; 1,6-Hexanediol methacrylate; Hexyloxypropylamine; Hydroxymethyl dioxoazabicyclooctane

Isodecyloxypropylamine; N-Isodecyloxypropyl 1-1,3-diaminopropane; Isododecyloxypropylamine; N-Isododecyloxypropyl-1,3-diaminopropane; N-Isotridecyloxypropyl 1,3-diaminopropane

Melamine/formaldehyde resin; Methacrylic anhydride

Octyl/decyloxypropylamine; Oxazolidine

Crosslinking agents *(cont'd.)*

PEG-4 diacrylate; PEG-3 dimethacrylate; PEG-4 dimethacrylate; PEG-5 methacrylate; PEG-10 methacrylate; Pentaerythrityl tetraacrylate; Pentaerythrityl tetrakis (3-mercaptopropionate); Pentaerythrityl triacrylate; Polyacrylamide; Polymethylene polyphenyl isocyanate; Polymethylhydrosiloxane

Tetrakis (2-ethylhexyl) titanate; Titanium acetylacetonate; Triallylcyanurate; Triallyl isocyanurate; Triethylene diamine; 1,1,1-Trimethylolpropane triacrylate; Trimethylolpropane tri (3-mercaptopropionate); 1,1,1-Trimethylolpropane trimethacrylate; Tris (p-isocyanato-phenyl) thiophosphate

Vinyltrimethoxysilane

Zinc diacrylate

Curing agents

Trade names: Accelerator 399; Adma® 8; Adma® 10; Adma® 12; Adma® 14; Adma® 16; Adma® 18; Adma® 246-451; Adma® 246-621; Adma® 1214; Adma® 1416; Adma® WC; AEP; Akrochem® P 124; Allco BTDA; Allco PMDA; Amicure® 101; Amicure® CG-325; Amicure® CG-1200; Amicure® CG-1400; Amicure® CG-NA; Amicure® PACM; Amicure® SA-102; Amine HH; Ancamide 100-IT-60; Ancamide 220; Ancamide 220-IPA-73; Ancamide 220-X-70; Ancamide 260A; Ancamide 260TN; Ancamide 350A; Ancamide 375A; Ancamide 400; Ancamide 400-BX-60; Ancamide 500; Ancamide 501; Ancamide 502; Ancamide 503; Ancamide 506; Ancamide 507; Ancamide 700-B-75; Ancamide 2050; Ancamide 2137; Ancamide 2349; Ancamine® 1110; Ancamine® 1482; Ancamine® 1483; Ancamine® 1561; Ancamine® 1608; Ancamine® 1617; Ancamine® 1618; Ancamine® 1636; Ancamine® 1637; Ancamine® 1637-LV; Ancamine® 1638; Ancamine® 1644; Ancamine® 1693; Ancamine® 1767; Ancamine® 1768; Ancamine® 1769; Ancamine® 1770; Ancamine® 1784; Ancamine® 1833; Ancamine® 1856; Ancamine® 1882; Ancamine® 1884; Ancamine® 1895; Ancamine® 1916; Ancamine® 1922; Ancamine® 1934; Ancamine® 2014AS; Ancamine® 2014FG; Ancamine® 2021; Ancamine® 2049; Ancamine® 2056; Ancamine® 2071; Ancamine® 2072; Ancamine® 2074; Ancamine® 2089M; Ancamine® 2136; Ancamine® 2143; Ancamine® 2167; Ancamine® 2168; Ancamine® 2205; Ancamine® 2264; Ancamine® 2280; Ancamine® 2286; Ancamine® 2337XS; Ancamine® AD; Ancamine® DL-50; Ancamine® K54; Ancamine® K61B; Ancamine® LO; Ancamine® LT; Ancamine® MCA; Ancamine® T; Ancamine® T-1; Ancamine® TL; Ancamine® XT; Anchor® 1040; Anchor® 1115; Anchor® 1170; Anchor® 1171; Anchor® 1222; Armocure® 100; AroCy® B-10; Aztec® AAP-LA-M2; Aztec® AAP-NA-2; Aztec® AAP-SA-3; Aztec® BCHPC; Aztec® 1,1-BIS-50-AL; Aztec® BP-05-FT; Aztec® BP-20-GY; Aztec® BP-40-S; Aztec® BP-50-FT; Aztec® BP-50-P1; Aztec® BP-60-PCL; Aztec® BU-50-AL; Aztec® CHP-50-P1; Aztec® CHP-90-W1; Aztec® CHP-HA-1; Aztec® DHPEH; Aztec® MEKP-HA-1; Aztec® MEKP-HA-2; Aztec® MEKP-LA-2; Aztec® MEKP-SA-2; Aztec® MIKP-LA-M1; Aztec® MYPC; Aztec® TAPB; Aztec® TAPEH; Aztec® TAPEH-75-OMS; Aztec® TAPEH-90-OMS; Aztec® TBPB; Aztec® TBPB-50-IC; Aztec® TBPB-SA-M2; Aztec® TBPEH; Aztec® TBPEH-50-DOP; Aztec® TBPEH-50-OMS; Aztec® TBPIN; Aztec® TBPM-50-FT; Aztec® TBPM-50-P; Aztec® 3,3,5-TRI-50-AL; Aztec® 3,3,5-TRI-50-FT

BHMT Amine; BYK®-Catalyst 450; BYK®-Catalyst 451

Cadet® BPO-70; Capcure® AF; Cardolite® NC-540; Cardolite® NC-541; Casamid® 350PM; Casamid® 360; Casamid® 362W; Curezol® 2E4MZ; Curezol® 2MA-OK; Curezol® 2MZ-Azine; Curezol® 2PHZ; Curezol® 2PHZ-S; Curezol® 2PZ; Curezol® 2PZ-CNS; Curezol® 2PZ-OK; Curezol® AMI-2; Curezol® C17Z

Dabco® BDO; Dama® 810; Dama® 1010; DCH-99; Decanox-F; D.E.H. 20; D.E.H. 24; D.E.H. 26; D.E.H. 29; D.E.H. 39; D.E.H. 40; D.E.H. 52; D.E.H. 58; DETA; Dicyanex® 200X; Dicyanex® 325; Dicyanex® 1200; Doverphos® 10; Doverphos® 10-HR; Doverphos® 10-HR Plus; Drewplus® Y-281; Drewplus® Y-601; Duomeen® C; Dyphene® 595; Dyphene® 9273; Dytek® A; Dytek® EP

Ebecryl® 220; Ebecryl® 230; Ekaland DOTG; Eltesol® CA 65; Eltesol® CA 96; Eltesol® PSA 65; Eltesol® TA Series; Eltesol® TA 65; Eltesol® TA 96; Eltesol® TA/F; Eltesol® TSX/A; Eltesol® XA90; Emery® 5702; Emery® 5703; Emery® 5707; Emery® 5709; Emery® 5710; Empigen® AM; Empigen® AY; Empol® 1040; Epi-Cure® 3010; Epi-Cure® 3015; Epi-Cure® 3025; Epi-Cure® 3030; Epi-Cure® 3035; Epi-Cure® 3046; Epi-Cure® 3055; Epi-Cure® 3060; Epi-Cure® 3061; Epi-Cure® 3070; Epi-Cure® 3072; Epi-Cure® 3090; Epi-Cure® 3100-ET-60; Epi-Cure® 3100-HX-60; Epi-Cure® 3100-XY-60; Epi-Cure® 3115; Epi-Cure® 3115-E-73;

Curing agents *(cont'd.)*

Epi-Cure® 3115-X-70; Epi-Cure® 3123; Epi-Cure® 3125; Epi-Cure® 3140; Epi-Cure® 3141; Epi-Cure® 3150; Epi-Cure® 3155; Epi-Cure® 3175; Epi-Cure® 3180-F-75; Epi-Cure® 3185-FX-60; Epi-Cure® 3192; Epi-Cure® 3200; Epi-Cure® 3213; Epi-Cure® 3214; Epi-Cure® 3218; Epi-Cure® 3223; Epi-Cure® 3234; Epi-Cure® 3245; Epi-Cure® 3251; Epi-Cure® 3254; Epi-Cure® 3255; Epi-Cure® 3262; Epi-Cure® 3266; Epi-Cure® 3270; Epi-Cure® 3273; Epi-Cure® 3274; Epi-Cure® 3275; Epi-Cure® 3277; Epi-Cure® 3278; Epi-Cure® 3281-H-60; Epi-Cure® 3282; Epi-Cure® 3290; Epi-Cure® 3292-FX-60; Epi-Cure® 3293; Epi-Cure® 3295; Epi-Cure® 3370; Epi-Cure® 3371; Epi-Cure® 3373; Epi-Cure® 3374; Epi-Cure® 3378; Epi-Cure® 3379; Epi-Cure® 3380; Epi-Cure® 3381; Epi-Cure® 3382; Epi-Cure® 3383; Epi-Cure® 3384; Epi-Cure® 3484; Epi-Cure® 3501; Epi-Cure® 3502; Epi-Cure® 3503; Epi-Cure® 8290-Y-60; Epi-Cure® 8292-Y-60; Epi-Cure® 8535-W-50; Epi-Cure® 8536-MY-60; Epi-Cure® 8537-WY-60; Epi-Cure® 9150; Epi-Cure® 9350; Epi-Cure® 9360; Epi-Cure® 470; Epi-Cure® 9850; Epi-Cure® HPT Curing Agent 1061-M; Epi-Cure® HPT Curing Agent 1062-M; Epi-Cure® P-101; Epi-Cure® P-104; Epi-Cure® P-108; Epi-Cure® P-187; Epi-Cure® P-201; Epi-Cure® P-202; Epi-Cure® W; Epi-Cure® Y; Epi-Cure® Z; Esperox® 5100; Esperox® C-496; Ethacure® 100; Ethacure® 300

Genamid® 151; Genamid® 250; Genamid® 490; Genamid® 491; Genamid® 747; Genamid® 2000

Hardener OZ; HPC-9; HyTemp NPC-50; HyTemp NS-70; HyTemp SC-75; HyTemp SO-40; HyTemp ZC-50

Imicure® AMI-2; Imicure® EMI-24; Industrene® D

Jeffamine® D-230; Jeffamine® D-400; Jeffamine® D-2000; Jeffamine® DU-700; Jeffamine® EDR-148; Jeffamine® T-403; Jeffamine® T-3000; Jeffamine® T-5000

Kencure™ MPP; Kencure™ MPPJ; Ken-React® 7 (KR 7); Ken-React® 9S (KR 9S); Ken-React® 12 (KR 12); Ken-React® 26S (KR 26S); Ken-React® 33DS (KR 33DS); Ken-React® 38S (KR 38S); Ken-React® 39DS (KR 39DS); Ken-React® 41B (KR 41B); Ken-React® 44 (KR 44); Ken-React® 46B (KR 46B); Ken-React® 55 (KR 55); Ken-React® 133DS (KR 133DS); Ken-React® 134S (KR 134S); Ken-React® 138D (KR 138D); Ken-React® 138S (KR 138S); Ken-React® 158D (KR 158D); Ken-React® 158FS (KR 158FS); Ken-React® 212 (KR 212); Ken-React® 238A (KR 238A); Ken-React® 238J (KR 238J); Ken-React® 238M (KR 238M); Ken-React® 238S (KR 238S); Ken-React® 238T (KR 238T); Ken-React® 262A (KR 262A); Ken-React® 262ES (KR 262ES); Ken-React® OPP2 (KR OPP2); Ken-React® OPPR (KR OPPR); Ken-React® KR TTS; KZ OPPR; KZ TPP; KZ TPPJ

Lankromark® LE65; Lectro 78; Leecure B-110; Leecure B-550; Leecure B-610; Leecure B-612; Leecure B-614; Leecure B-950; Leecure B-1310; Leecure B-1550; Leecure B-1600; Leecure B-1700; LICA 01; LICA 09; LICA 12; LICA 38; LICA 38A; LICA 38J; LICA 44; LICA 97; LICA 99; Lithene AH; Lithene AL; Lonzacure® M-CDEA; Lonzacure® M-DEA; Lonzacure® M-DMA; Lonzamon® AAEA; LP-100; LP-200; LP-300; LP-400; Luperco 130-XL; Luperco ANS-P; Lupersol 224; Lupersol 665-M50; Lupersol 665-T50; Lupersol 688-M50; Lupersol 688-T50; Lupersol 801; Lupersol DDM-30; Lupersol TAEC; Lupersol TBEC

MagChem® 35K; MagChem® 40; MagChem® 50; MagChem® 50Y; MagChem® 60; MagChem® 125; MagChem® 200-AD; Maglite® D; Manro FCM 90LV; Manro PTSA/C; Millamine® 5260; Milldride® 5060; Milldride® DDSA; Milldride® HDSA; Milldride® HHPA; Milldride® MHHPA; Milldride® nDDSA; Milldride® nDSA; Milldride® ODSA; Milldride® OSA; Milldride® TDSA; Morfax®

NZ 01; NZ 09; NZ 12; NZ 33; NZ 38; NZ 39; NZ 44; NZ 49; NZ 89; NZ 97

Pennad 150; Polacure® 740M; Polamine® 250; Polamine® 650; Polamine® 1000; Polamine® 2000; Pyridine 1°

QDO®; QO® Furfuryl Alcohol (FA®)

R7500E; R7530E; Research Curing Agent RSC-1246; Research Curing Agent RSC-2215; Retilox® F 40 MF; Retilox® F 40 MG; Rhenocure® IS 60; Rhenocure® IS 60-5; Rhenocure® IS 60/G; Rhenocure® IS 90-20; Rhenocure® IS 90-33; Rhenocure® IS 90-40; Rhenocure® IS 90/G

Si69; SR-212; SR-344; SR-349; SR-365; SR-399; SR-416; SR-454; SR-506; SR-527; Struktol® IB 531; Struktol® ZP 1014; Sumine® 2015; Superox® 46-748; Sur-Wet® R

Tactix H41; Telloy; TEPA; TETA; Texacure EA-20; Texacure EA-24; Tonox®; Tonox® 22; Tonox® 60/40; Tonox® LC

Uni-Rez® 1502; Uni-Rez® 2100-P75; Uni-Rez® 2341; Uni-Rez® 2355; USP®-240

Vandex; Varox® 802-40MB; Varox® DCP-R; Varox® DCP-T; Versamid® 125; Versamid® 140; Versamid® 150; Versamid® 674; Versamine® 900; Versamine® 908; Versamine® 911; Versamine® 1200; Versamine® A50; Versamine® A51; Versamine® A52; Versamine® A54;

Curing agents *(cont'd.)*

Versamine® A55; Versamine® I-70; Vestamin® IPD; Vestamin® TMD; Viton® Curative No. 20; Viton® Curative No. 30; Viton® Curative No. 40; Viton® Curative No. 50; Vulcuren 2
YSE-Cure B-001; YSE-Cure B-002; YSE-Cure B-002W; YSE-Cure B-003; YSE-Cure C-002; YSE-Cure LX-1N; YSE-Cure LX-2S; YSE-Cure N-001; YSE-Cure N-002; YSE-Cure PX-3; YSE-Cure QX-2; YSE-Cure QX-3; YSE-Cure RX-2; YSE-Cure RX-3; YSE-Cure S-002
Zeonet® A; Zisnet® F-PT

Chemicals:
Acetylacetone peroxide; Amidoamine resin; Aminoethylpiperazine
3,3´,4,4´-Benzophenone tetracarboxylic dianhydride; Bis (p-aminocyclohexyl) methane; Bis-hexamethylenetriamine; Bisphenol A dicyanate monomer/prepolymer; Boron trifluoride-amine complex; 1,4-Butanediol; 1,3-Butylene glycol diacrylate; t-Butyl peroxymaleic acid
Cumene hydroperoxide; Cumene sulfonic acid; Cycloneopentyl, cyclo (dimethlaminoethyl) pyrophosphato zirconate, di mesyl salt
Decanoyl peroxide; n-Decenyl succinic anhydride; 1,2-Diaminocyclohexane; Diazabicyclo-undecene, 2-ethylhexanoic acid salt; Di (butyl, methyl pyrophosphato) ethylene titanate di (dioctyl, hydrogen phosphite); 2,2-Di (t-butylperoxy) butane; Dicyandiamide; Dicyclo (dioctyl) pyrophosphato titanate; Di (dioctylphosphato) ethylene titanate; Di (dioctylpyrophosphato) ethylene titanate; Diethylaminoethanol; Diethylene glycol di(aminopropyl) ether; Diethylenetriamine; Diethyl toluene diamine; Dimethylaminopropylamine; Dimethyl decylamine; Dimethylethanolamine; 1,1-Dimethyl-3-hydroxybutylperoxy-2-ethylhexanoate; Dimethyl lauramine; 3,3´-Dimethylmethylenedi(cyclohexylamine); Dimethyl myristamine; Dimethyl octylamine; Dimethyl palmitamine; N,N´-Dimethylpiperazine; Dimethyl stearamine; Dimyristyl peroxydicarbonate; Dodecenyl succinic anhydride
Epoxy acrylate; 2-Ethyl-4-methyl imidazole; Ethyl phenyl ethanolamine
Furfuryl alcohol
Hexadecenyl succinic anhydride; Hexahydrophthalic anhydride; Hexamethylenediamine; Hexamethylene tetramine
Isobornyl acrylate; Isocyanuric acid; Isophorone diamine; Isopropyl 4-aminobenzenesulfonyl di (dodecylbenzenesulfonyl) titanate; Isopropyl dimethacryl isostearoyl titanate; Isopropyl tri (dioctylphosphato) titanate; Isopropyl tri (dioctylpyrophosphato) titanate; Isopropyl tri (N ethylamino-ethylamino) titanate
Lead dioxide; Lead fumarate tetrabasic
Manganese dioxide; 4,4´-Methylenebis (2-chloraniline); 4,4´-Methylenebis (3-chloro-2,6-diethylaniline); 4,4´-Methylenebis 2,6-diethylaniline; 4,4´-Methylenebis 2,6-dimethylaniline; 4,4´-Methylene dianiline; Methyl hexahydrophthalic anhydride; 2-Methylpentamethylene-diamine; 2-(4-Morpholinyldithio) benzothiazole
Neopentyl (diallyl) oxy, trihydroxy caproyl titanate; Neopentyl (diallyl) oxy, trimercapto-phenyl zirconate
Octadecenyl succinic anhydride; Octenyl succinic anhydride; Octylphenol formaldehyde resin
PEG-3 aniline; PEG-8 diacrylate; Pentaethylene hexamine; 1,3-Pentanediamine; Phenyldiethanolamine; m-Phenylenediamine; Phenylethylethanolamine; 2-Phenyl imidazole; Polyamide; Polytetramethylene ether glycol diamine; Polyurethane-acrylate resin; Pyromellitic acid dianhydride; Pyromellitic dianhydride
Sulfolane
Tetradecenyl succinic anhydride; Tetra (2, diallyoxymethyl-1 butoxy titanium di (di-tridecyl) phosphite; Tetraisopropyl di (dioctylphosphito) titanate; Tetraoctyloxytitanium di (ditridecylphosphite); Titanium di (butyl, octyl pyrophosphate) di (dioctyl, hydrogen phosphite) oxyacetate; Titanium di (cumylphenylate) oxyacetate; Titanium di (dioctylpyrophosphate) oxyacetate; Titanium dimethacrylate oxyacetate; m-Tolyl diethanolamine; p-Tolyl diethanolamine; Triethylamine; Triethylenetetramine; Trilinoleic acid; 2,4,6-Trimercapto-s-triazine; Trimethylhexamethylene diamine; 2,4,6-Tris (dimethylaminomethyl) phenol; 2,4,6-Tris (dimethylaminomethyl) phenol, 2-ethylhexanoic acid salt
Xylene sulfonic acid
Zinc peroxide

Deactivators

Trade names:
ADK STAB ZS-27; ADK STAB ZS-90
Eastman® Inhibitor OABH
Flectol® Pastilles

Deactivators *(cont'd.)*

Hostanox® OSP 1
Irganox® MD 1024
Naugard® XL-1; Naugard® XL-517
Rokon
Santoflex® IP
Weston® TIOP; Weston® TLTTP

Chemicals: Oxalyl bis (benzylidenehydrazide); 2,2′-Oxamido bis[ethyl 3-(3,5-di-t-butyl-4-hydroxyphenyl) propionate
Phenol, 4,4′-thiobis 2-(1,1-dimethylethyl) phosphite

Defoamers

Trade names: Abluwax EBS; Advawax® 290; Agitan 260; Agitan 281; Agitan 295; Agitan 296; Agitan 301; Agitan 305; Agitan 650; Agitan 655; Agitan 702; Agitan 703 N; Agitan E 255; Agitan E 256; Amgard® TBEP; Antarox® L-62; Antarox® L-62 LF
Britol® 6NF; BYK®-034; BYK®-065; BYK®-066; BYK®-080; BYK®-141; BYK®-3105; BYK®-3155; BYK®-A 500; BYK®-A 501; BYK®-A 510; BYK®-A 515; BYK®-A 525; BYK®-A 530; BYK®-A 555; BYK®-S 706; BYK®-S 715
Chemfac 100; Colloid™ 581B; Colloid™ 675; Colloid™ 685; Colloid™ 961; Colloid™ 985; Colloid™ 994; Colloid™ 999
Degressal® SD 40; Dehydran 520; Dehydran 1019; Dehydran P 4; Dehydran P 11; Drewplus® L-108; Drewplus® L-123; Drewplus® L-131; Drewplus® L-139; Drewplus® L-140; Drewplus® L-191; Drewplus® L-198; Drewplus® L-468; Drewplus® L-474; Drewplus® L-475; Drewplus® L-523; Drewplus® Y-125; Drewplus® Y-250
Foamgard 1332
Hydrobrite 200PO; Hydrobrite 300PO; Hydrobrite 380PO; Hydrobrite 550PO
Jeffox PPG-400; Jeffox PPG-2000
Kaydol® S
Leocon 1070B
Mackester™ EGDS; Mackester™ EGMS; Mackester™ IP; Mackester™ SP; Masil® SF 10; Masil® SF 20; Masil® SF 50; Masil® SF 100; Masil® SF 200; Masil® SF 350; Masil® SF 350 FG; Masil® SF 500; Masil® SF 1000; Masil® SF 5000; Masil® SF 10,000; Masil® SF 12,500; Masil® SF 30,000; Masil® SF 60,000; Masil® SF 100,000; Masil® SF 300,000; Masil® SF 500,000; Masil® SF 600,000; Masil® SF 1,000,000; Mazu® DF 200SX; Mazu® DF 200SXSP; Mazu® DF 210SXSP; Monolan® O Range
Nopalcol 1-TW; Nopco® NXZ
Patcote® 512; Patcote® 513; Patcote® 519; Patcote® 520; Patcote® 525; Patcote® 531; Patcote® 550; Patcote® 555; Patcote® 577; Patcote® 597; Patcote® 598; Patcote® 801; Patcote® 802; Patcote® 803; Patcote® 806; Patcote® 811; Patcote® 812; Patcote® 841M; Patcote® 845; Patcote® 847; Patcote® 883; Pegosperse® 200 DL; Pluracol® E600 NF; Pluracol® E1000; Pluracol® E1450; Pluracol® E1450 NF; Pluracol® E2000; Pluracol® E4500; Pluracol® E8000; Prox-onic NP-1.5; Prox-onic NP-04; Prox-onic NP-06; Prox-onic NP-09; Prox-onic NP-010; Prox-onic NP-015; Prox-onic NP-020; Prox-onic NP-030
Ross Montan Wax; Rudol®
Semtol® 40; Semtol® 70; Semtol® 85; Semtol® 100; Semtol® 350; Silwet® L-7602; Solricin® 435; Surfynol® 82S; Surfynol® 104A; Surfynol® 104E; Surfynol® 104PA; Surfynol® 485; Surfynol® D-101; Surfynol® D-201; Surfynol® DF-110D; Surfynol® DF-110L
Uniwax 1763

Chemicals: Isodecyl oleate
Nonoxynol-15
PEG-4 dilaurate; PEG-6 dilaurate; PEG-12 dilaurate; PEG-3 dioctoate; PEG-8 dioleate; PEG-2 laurate; PEG-4 laurate; PEG-4 oleate; PEG-12 stearate; PEG-30 tetramethyl decynediol; Polysiloxane polyether copolymer; PPG-20; PPG-26; PPG-30; PPG-400; PPG-2000; Propyl oleate
Silicone; Sodium ricinoleate; Sorbitan stearate; Soy acid

Deodorants

Trade names: Dehyquart® LT; Deodorant #4761-F, OS
PVC Deodorant #5417, OS
Styrid

Chemicals: Methyl-β-cyclodextrin; Methyl-γ-cyclodextrin

Detackifiers • Antitack agents

Trade names: Crystex® 90 OT 20; Crystex® HS; Crystex® Regular
Dee-Tac
EXP-24-LS Silicone Wax Emulsion; EXP-58 Silicone Wax; EXP-61 Amine Functional Silicone Wax; EXP-77 Mercapto Functional Silicone Wax
Supra EF A
Witco® Zinc Stearate Disperso

Diluents • Viscosity depressants

Trade names: Adeka Glycilol ED-503; Adeka Glycilol ED-505; Adeka Glycilol ED-506; Ageflex BGE; Ageflex MEA; Ageflex TBGE; Albrite® TPP; Arconate® 1000; Arconate® HP; Arconate® PC; AroCy® B-10; AroCy® F-10; Atmer® 153; Atmer® 154; Atmer® 505; Atmer® 508
Cardolite® NC-513; Cardolite® NC-700; Cardolite® NC-1307; Cardura® E-10
DY 023; DY 025; DY 027
Eastman® TEP; Empicryl® 6045; Empicryl® 6052; Empicryl® 6054; Empicryl® 6058; Empicryl® 6059; Empicryl® 6070; Epodil® 741; Epodil® 742; Epodil® 743; Epodil® 745; Epodil® 746; Epodil® 747; Epodil® 748; Epodil® 749; Epodil® 750; Epodil® 757; Epodil® 759; Epodil® 769; Epodil® L; Epodil® ML
GE 100; Geropon® BIS/SODICO-2; Geropon® CYA/60
Heloxy® 62; Heloxy® 63; Heloxy® 64; Heloxy® 65; Heloxy® 67; Heloxy® 107; Heloxy® 116; Hi-Sil® 210; Hi-Sil® 233; Hi-Sil® 250
Kenplast® ES-2; Kenplast® ESI; Kenplast® G; Kenplast® PG; KP-140®
Lankroplast® V2023; Lankroplast® V2067; Lankroplast® V2100; Lomar® PWA; Lonzamon® AAEA; Lonzamon® AAEMA
Marlican®; Micro-Cel® T-21; Micro-Cel® T-49
Permethyl® 100 Epoxide; Poly TDP 2000; Polysar EPDM 227
QO® Furfuryl Alcohol (FA®); Quimipol EA 2503
RD-1; RD-2
SR-201; SR-203; SR-244; SR-313; SR-379; SR-423; SR-506; Staflex DMP; Struktol® TR 141; Syn Fac® 8031
Thinner PU; Trigonox® 21-C50
Viplex 525; Viscobyk-4010; Viscobyk-4015; Viscobyk-5025; Viscobyk-5050; V-Pyrol
Wareflex® 650; Weston® EGTPP

Chemicals: 2-(Acetoacetoxy) ethyl acrylate; 2-(Acetoacetoxy) ethyl methacrylate; Allyl glycidyl ether; Ammonium naphthalene sulfonate
Biscumylphenyl trimellitate; 1,4-Butanediol diglycidyl ether; Butyl glycidyl ether; t-Butyl glycidyl ether; p-t-Butyl phenyl glycidyl ether
C40-60 alcohols; Cardanol; Cardanol acetate; o-Cresyl glycidyl ether; Cumylphenyl acetate; Cyclohexanedimethanol diglycidyl ether
Diethylene glycol dibutyl ether
2-Ethylhexyl glycidyl ether
Furfuryl alcohol
Hexafluorobisphenol A dicyanate monomer/prepolymer; 2-Hydroxyethylmethacrylate
Isopropyl glycidyl ether
Neodecanoic acid glycidyl ester; Neopentyl glycol diglycidyl ether
Paraffin, brominated; PEG-3 dimethacrylate; Phenyl glycidyl ether; Phenyl glycol ether; PPG-3 diacrylate; Propylene carbonate; Pyrophyllite
Resorcinol diglycidyl ether
Trimethylolpropane triglycidyl ether; Triphenyl phosphite

Dispersants

Trade names: Acconon 400-MO; AClyn® 201A; AClyn® 246A; Acofor; Acrawax® C; ACtone® 1; ACtone® 2000V; ACtone® 2010; ACtone® 2461; ACtone® N; Advawax® 280; Aerosol® A-102; Aerosol® A-196-40; Aerosol® DPOS-45; Aerosol® OT-100%; Aerosol® OT-B; Aflux® 32; Aflux® R; Aflux® S; Ageflex FM-10; Airvol® 205; Airvol® 523; Airvol® 540; Aktiplast® AS; Aktiplast® F; Aktiplast® PP; Alkamide® STEDA; Alkamuls® EL-620L; Alkamuls® EL-719; Antarox® L-62; Antarox® L-62 LF; Anti-Terra®-204; Anti-Terra®-P; Arodet BN-100; Arquad® 2C-75; Arquad® 12-50; Arquad® 16-50; Arquad® 18-50; Arquad® C-50; Arquad® T-27W; Avirol® SL 2010

Berol 28; Berol 302; Borax; Britol® 6NF; BYK®-R 605; BYK®-W 900; BYK®-W 920; BYK®-W 935; BYK®-W 940; BYK®-W 968; BYK®-W 980; BYK®-W 990; BYK®-W 995

Cab-O-Sil® L-90; Cab-O-Sil® LM-130; Cab-O-Sil® LM-150; Chemax HCO-5; Chemax HCO-16; Chimipal APG 400; Cyastat® SP

Darathane® WB-4000; Darvan® L; Darvan® No. 1; Darvan® No. 2; Darvan® No. 6; Darvan® No. 7; Darvan® No. 31; Darvan® SMO; Daxad® 11G; Daxad® 14B; Daxad® 14C; Daxad® 15; Daxad® 16; Daxad® 17; Daxad® 30; Daxad® 30-30; Daxad® 31; Daxad® 31S; Daxad® 34; Daxad® 34A9; Dehscofix 920; Dehscofix 930; Dehydrophen PNP 30; Depasol CM-41; Dextrol OC-50; Dextrol OC-70; Disperplast®1142; Disperplast®-1150; Disperplast®-I; Disperplast®-P; Disponil CSL 100 K; Disponil O 20; Disponil O 250; Disponil TA 5; Disponil TA 25; Disponil TA 430; Dresinate® 214; Dresinate® 731; Drewfax® 0007; Dyphene® 8318; Dyphene® 8320; Dyphene® 8330; Dyphene® 8400; Dyphene® 8787

Eastoflex E1060; Eastoflex P1023; Eccowet® W-50; Eccowet® W-88; Emcol® CC-42; Emcol® CC-57; Emphos™ PS-400; Empicol® ESB; Empicol® L Series; Empicryl® 6045; Empicryl® 6047; Empicryl® 6052; Empicryl® 6054; Empicryl® 6058; Empicryl® 6059; Empicryl® 6070; Empilan® K Series; Empimin® KSN27; Empimin® KSN70; Empimin® MA; Empimin® OP70; Empimin® OT; Emsorb® 6909; Epan 710; Epan 720; Epan 740; Epan 750; Epan 785; Epolene® C-10P; Epolene® C-13; Epolene® C-13P; Epolene® C-14; Epolene® C-16; Epolene® C-17; Epolene® C-17P; Epolene® C-18; Epolene® E-10; Epolene® E-15; Epolene® E-20; Epolene® E-43; Epolene® N-10; Epolene® N-11P; Epolene® N-14P; Epolene® N-20; Epolene® N-21

Fancol OA-95; #15 Oil; Fizul 201-11; Fizul M-440

Gantrez® AN-119; Gantrez® AN-139; Gantrez® AN-149; Gantrez® AN-169; Gantrez® AN-179; Gardilene S25L; Gemtex 445; Gemtex 691-40; Geropon® 99; Geropon® SDS; Geropon® T/36-DF; Geropon® WT-27; Glycomul® S; GP-209 Silicone Polyol Copolymer; GP-214 Silicone Polyol Copolymer; GP-215 Silicone Polyol Copolymer; GP-219 Silicone Polyol Copolymer

Hartopol L64; Hydrobrite 200PO; Hydrobrite 300PO; Hydrobrite 380PO; Hydrobrite 550PO; Hypermer 1599A

Igepal® CO-430; Igepal® CO-630; Interstab® ZN-18-1

Kaydol® S; Kemamine® AS-650; Kemamine® AS-974; Kemamine® AS-974/1; Kemamine® AS-989; Kemamine® AS-990; Kemamine® D-190; Kodaflex® DEP; KZ OPPR; KZ TPP; KZ TPPJ

Lankropol® OPA; LICA 01; LICA 09; LICA 12; LICA 38; LICA 38A; LICA 38J; LICA 44; LICA 97; LICA 99; Lipolan 1400; Lipolan PJ-400; Lomar® HP; Lomar® LS; Lomar® LS Liq; Loxiol® P 1304; Lubrizol® 2153; Lubrizol® 2155; Luviskol® K17; Luviskol® K30; Luviskol® K60; Luviskol® K90

Maphos® L-6; Monawet MB-45; Monawet MO-70; Monawet MT-70; Monawet SNO-35; Monolan® 8000 E/80; Myvaplex® 600PK; Myverol® 18-04; Myverol® 18-06; Myverol® 18-07; Myverol® 18-99

Nacconol® 40G; Nacconol® 90G; Nansa® EVM50; Nansa® EVM70/B; Nansa® EVM70/E; Nissan Plonon 102; Nissan Plonon 104; Nissan Plonon 108; Nissan Plonon 171; Nissan Plonon 172; Nissan Plonon 201; Nissan Plonon 204; Nissan Plonon 208; Noiox AK-41; Nopalcol 1-L; Nopalcol 2-DL; Nopalcol 4-C; Nopalcol 4-CH; Nopalcol 4-L; Nopalcol 4-S; Nopalcol 6-DO; Nopalcol 6-DTW; Nopalcol 6-L; Nopalcol 6-R; Nopalcol 6-S; Nopalcol 10-COH; Nopalcol 12-CO; Nopalcol 12-COH; Nopalcol 19-CO; Nopalcol 30-TWH; Nopalcol 200; Nopalcol 400; Nopalcol 600; Nopco ESA120; Nopcosant; Noram SH; NZ 01; NZ 09; NZ 12; NZ 33; NZ 38; NZ 39; NZ 44; NZ 49; NZ 89; NZ 97

Octosol 449; Octosol 496; Octosol A-1; Octosol A-18; Octosol SLS; Octosol SLS-1

Pationic® 900; Pationic® 901; Pationic® 902; Pationic® 909; Pationic® 919; Pationic® 1019; Pationic® 1061; Pationic® 1064; Pationic® 1074; Pationic® 1145; Pegosperse® 50 MS; Pegosperse® 200 DL; Petrac® 250; Petrac® 270; Petrac® Vyn-Eze®; Pluracol® E600 NF; Pluracol® E1000; Pluracol® E1450; Pluracol® E1450 NF; Pluracol® E2000; Pluracol®

Dispersants *(cont'd.)*

E4500; Pluracol® E8000; Polydis® TR 016; Polydis® TR 121; Polyox® WSR 205; Polyox® WSR 301; Polyox® WSR 303; Polyox® WSR 308; Polyox® WSR 1105; Polyox® WSR 3333; Polyox® WSR N-10; Polyox® WSR N-12K; Polyox® WSR N-60K; Polyox® WSR N-80; Polyox® WSR N-750; Polyox® WSR N-3000; Polyox® Coagulant; Polystep® F-2; Polystep® F-3; Polysurf A212E; Proaid® 9802; Proaid® 9810; Proaid® 9831; Proplast 015; Prox-onic HR-05; Prox-onic HR-016; Prox-onic HR-025; Prox-onic HR-030; Prox-onic HR-036; Prox-onic HR-040; Prox-onic HR-080; Prox-onic HR-0200; Prox-onic HR-0200/50; Prox-onic HRH-05; Prox-onic HRH-016; Prox-onic HRH-025; Prox-onic HRH-0200; Prox-onic HRH-0200/50; Prox-onic NP-1.5; Prox-onic NP-04; Prox-onic NP-06; Prox-onic NP-09; Prox-onic NP-010; Prox-onic NP-015; Prox-onic NP-020; Prox-onic NP-030; Prox-onic NP-030/70; Prox-onic NP-040; Prox-onic NP-040/70; Prox-onic NP-050; Prox-onic NP-050/70; Prox-onic NP-0100; Prox-onic NP-0100/70

Rhodacal® BA-77; Rhodacal® BX-78; Rhodafac® PS-19; Rhodafac® RE-610; Rhodafac® RE-877; Rhodafac® RE-960; Rhodapex® EST-30; Rhodasurf® A 24; Rhodasurf® L-25; Rhodasurf® L-790; Rhodasurf® ON-870; Rudol®

SAFR-Z-44 K; Secosol® DOS 70; Sellogen HR; Semtol® 40; Semtol® 70; Semtol® 85; Semtol® 100; Semtol® 350; Serdet DSK 40; Serdet Perle Conc; Silwet® L-7500; Silwet® L-7602; Sinopol 806; Sinopol 808; Sinopol 908; Sipomer® HEM-5; Sipomer® HEM-10; Softenol® 3900; Softenol® 3991; Steol® CA-460; Struktol® SU 109; Struktol® SU 135; Struktol® TH 10 Flakes; Struktol® TR 042; Struktol® TR 044; Struktol® W 34 Flakes; Struktol® W 80; Struktol® WB 212; Struktol® WB 222; Sulfopon® P-40; Supragil® NK; Surfynol® 104E; Synperonic PE30/80; Synpron 1750

Tamol® NN 4501; T-Det® C-40; Tergitol® D-683; Tergitol® XD; Tergitol® XH; Tergitol® XJ; Teric C12; Teric N8; Teric N9; Teric N10; Teric N12; Teric N13; Tetronic® 90R4; Tetronic® 150R1; Tetronic® 304; Trycol® 6954; Trycol® 6957; Trycol® 6970; Trycol® 6972; Trycol® 6984; Tylose® MH Grades; Tylose® MHB

Ucarfloc® Polymer 300; Ucarfloc® Polymer 302; Ucarfloc® Polymer 304; Ucarfloc® Polymer 309; Unilin® 350 Alcohol; Unilin® 550 Alcohol; Unilin® 700 Alcohol

Vanfre® AP-2; Vanfre® AP-2 Special; Vestenamer® 6213; Vestenamer® 8012; Victawet® 12; Victawet® 85X; Viscobyk-4010; Viscobyk-4013; Vorite 105; Vorite 110; Vorite 115; Vorite 120; Vorite 125; Vultamol®; Vultamol® SA Liq

Wayfos D-10N; Witconate™ 1298; Witconol™ NP-100

Zonyl® FSA; Zonyl® FSB; Zonyl® FSC; Zonyl® FSN; Zonyl® FSP

Chemicals:

Ammonium lauryl sulfate; Ammonium lignosulfonate; Ammonium naphthalene-formaldehyde sulfonate; Ammonium naphthalene sulfonate; Ammonium nonoxynol-4 sulfate

N,N-Bis-stearoylethylenediamide

Calcium sulfonate; C30-50 alcohols; Castor oil, polymerized; Ceteareth-30; Cetoleth-10; Cetoleth-13; Cetoleth-25; Cycloneopentyl, cyclo (dimethlaminoethyl) pyrophosphato zirconate, di mesyl salt

Deceth-4 phosphate; Di (butyl, methyl pyrophosphato) ethylene titanate di (dioctyl, hydrogen phosphite); Dicyclo (dioctyl) pyrophosphato titanate; Dicyclohexyl sodium sulfosuccinate; Di (dioctylphosphato) ethylene titanate; Di (dioctylpyrophosphato) ethylene titanate; Dihexyl sodium sulfosuccinate; Dimethicone copolyol; Dioctyl sodium sulfosuccinate; Disodium C12-15 pareth sulfosuccinate; Disodium deceth-6 sulfosuccinate; Disodium dodecyloxy propyl sulfosuccinamate; Disodium nonoxynol-10 sulfosuccinate; Disodium oleamido MIPA-sulfosuccinate; Disodium stearyl sulfosuccinamate; Ditridecyl sodium sulfosuccinate

Ethylene/calcium acrylate copolymer; Ethylene dioleamide; Ethylene/magnesium acrylate copolymer

Glyceryl dilaurate; Glyceryl dilaurate SE; Glycol dilaurate; Glycol dilaurate SE; Glycol dioleate SE; Glycol distearate SE; Glycol laurate; Glycol laurate SE; Glycol oleate; Glycol oleate SE; Glycol ricinoleate SE

Hydrogenated cottonseed glyceride; Hydrogenated palm glyceride

Isobutylene/MA copolymer; Isopropyl 4-aminobenzenesulfonyl di (dodecylbenzenesulfonyl) titanate; Isopropyl dimethacryl isostearoyl titanate; Isopropyl tri (dioctylphosphato) titanate; Isopropyl tri (dioctylpyrophosphato) titanate; Isopropyl tri (N ethylamino-ethylamino) titanate

Laureth-16

Methyl hydroxyethylcellulose

Neopentyl (diallyl) oxy, trihydroxy caproyl titanate; Neopentyl (diallyl) oxy, trimercapto-phenyl zirconate; Nonoxynol-1; Nonoxynol-3; Nonoxynol-5; Nonoxynol-15; Nonoxynol-20; Nonoxynol-25; Nonoxynol-30; Nonoxynol-35; Nonoxynol-40; Nonoxynol-50; Nonoxynol-70; Non-

Dispersants *(cont'd.)*

oxynol-100; Nonoxynol-150; Nonoxynol-4 sulfate

Octoxynol-1; Octoxynol-3; Octoxynol-9; Octoxynol-10; Octoxynol-16; Octoxynol-30; Octoxynol-40; Octoxynol-70; Oleamine; Oleyl alcohol; Oleyl 1,3-propylene diamine

PEG-2M; PEG-5M; PEG-7M; PEG-9M; PEG-14M; PEG-20M; PEG-23M; PEG-45M; PEG-90M; PEG-115M; PEG-32 castor oil; PEG-36 castor oil; PEG-75 castor oil; PEG-160 castor oil; PEG (1200) castor oil; PEG (1900) castor oil; PEG-8 cocoate; PEG-2 dilaurate; PEG-4 dilaurate; PEG-6 dilaurate; PEG-20 dilaurate; PEG-32 dilaurate; PEG-75 dilaurate; PEG-150 dilaurate; PEG-2 dilaurate SE; PEG-2 dioleate; PEG-4 dioleate; PEG-6 dioleate; PEG-8 dioleate; PEG-12 dioleate; PEG-20 dioleate; PEG-32 dioleate; PEG-75 dioleate; PEG-150 dioleate; PEG-2 dioleate SE; PEG-2 distearate; PEG-6 distearate; PEG-20 distearate; PEG-32 distearate; PEG-75 distearate; PEG-150 distearate; PEG-2 distearate SE; PEG-12 ditallowate; PEG-20 hydrogenated castor oil; PEG (1200) hydrogenated castor oil; PEG-60 hydrogenated tallowate; PEG-2 laurate; PEG-4 laurate; PEG-32 laurate; PEG-75 laurate; PEG-150 laurate; PEG-2 oleamine; PEG-7 oleamine; PEG-12 oleamine; PEG-2 oleate; PEG-4 oleate; PEG-6 oleate; PEG-8 oleate; PEG-20 oleate; PEG-32 oleate; PEG-75 oleate; PEG-150 oleate; PEG-2 oleate SE; PEG-12 ricinoleate; PEG-2 stearate; PEG-4 stearate; PEG-12 stearate; PEG-20 stearate; PEG-32 stearate; PEG-75 stearate; PEG-150 stearate; PEG-2 stearate SE; Pentaerythrityl stearate; Polyacrylamide; Polyacrylic acid; Polymethacrylic acid; Polysorbate 61; Polysorbate 65; Polysorbate 81; Polyurethane prepolymer; Polyvinyl alcohol, partially hydrolyzed; Potassium naphthalene sulfonate; Potassium rosinate; PPG-40 diethylmonium chloride; PPG-2 laurate; PPG-2 laurate SE; PPG-2 oleate; PPG-2 oleate SE; PPG-2 stearate; PPG-2 stearate SE; PVM/MA copolymer

Sodium borate decahydrate; Sodium decyl sulfate; Sodium dibutyl naphthalene sulfonate; Sodium isopropyl naphthalene sulfonate; Sodium lignosulfonate; Sodium methyl oleate sulfate; Sodium octyl sulfate; Sodium polyisobutylene/maleic anhydride copolymer; Sodium polymethacrylate; Sodium polynaphthalene sulfonate; Sodium/potassium naphthalene-formaldehyde sulfonate; Sorbitan sesquioleate; Sulfonic acid

TEA-lauryl sulfate; Tetra (2, diallyoxymethyl-1 butoxy titanium di (di-tridecyl) phosphite; Tetraisopropyl di (dioctylphosphito) titanate; Tetraoctyloxytitanium di (ditridecylphosphite); Tetrasodium dicarboxyethyl stearyl sulfosuccinamate; Titanium di (butyl, octyl pyrophosphate) di (dioctyl, hydrogen phosphite) oxyacetate; Titanium di (cumylphenylate) oxyacetate; Titanium di (dioctylpyrophosphate) oxyacetate; Titanium dimethacrylate oxyacetate; Triethanolamine

Driers • Dehydrating agents • Desiccants

Trade names: A-2; A-201; Activ-8
Cab-O-Sil® TS-610; Cab-O-Sil® TS-720
FC-24
Poly-Pale® Ester 10
Rhenogran® CaO-50/FPM; Rhenosorb® C; Rhenosorb® C/GW; Rhenosorb® F
Sipomer® DCPA; Sipomer® DCPM

Chemicals: Calcium oxide

Emulsifiers • Coemulsifiers

Trade names: Abex® 12S; Abex® 18S; Abex® 22S; Abex® 23S; Abex® 26S; Abex® 33S; Abex® AAE-301; Abex® EP-100; Abex® EP-110; Abex® EP-115; Abex® EP-120; Abex® EP-227; Abex® EP-277; Abex® JKB; Abex® VA 50; Ablunol NP30; Ablunol NP30 70%; Ablunol NP40; Ablunol NP40 70%; Ablunol NP50; Ablunol NP50 70%; Ablusol C-78; Ablusol DA; Ablusol SF Series; Ablusol SN Series; Acconon 300-MO; Acconon 400-MO; Acintol® R Type S; Acintol® R Type SB; Aerosol® 22; Aerosol® 501; Aerosol® A-102; Aerosol® A-103; Aerosol® A-268; Aerosol® AY-65; Aerosol® DPOS-45; Aerosol® IB-45; Aerosol® MA-80I; Aerosol® NPES 458; Aerosol® NPES 930; Aerosol® NPES 2030; Aerosol® NPES 3030; Aerosol® OT-75%; Aerosol® OT-100%; Aerosol® OT-S; Aerosol® TR-70; Agent AT-1190; Akyporox NP 15; Akyporox NP 30; Akyporox NP 200; Akyporox NP 300V; Akyporox NP 1200V; Akyporox OP 250V; Akyporox OP 400V; Akyporox RLM 160; Akyporox RTO 70; Akyporox SAL SAS;

Emulsifiers • Coemulsifiers *(cont'd.)*

Akyposal 9278 R; Akyposal ALS 33; Akyposal BD; Akyposal EO 20 MW; Akyposal NAF; Akyposal NPS 60; Akyposal NPS 100; Akyposal NPS 250; Akyposal OP 80; Alfonic® 1412-A Ether Sulfate; Alkamuls® EL-620L; Alkamuls® EL-719; Alkamuls® GMR-55LG; Alscoap LN-40, LN-90; Antarox® BA-PE 70; Antarox® BA-PE 80; Antarox® L-62; Antarox® L-62 LF; Antarox® PGP 18-2; Antarox® PGP 18-2LF; Arizona DR Mix-26; Arizona DRS-40; Arizona DRS-42; Arizona DRS-43; Arizona DRS-44; Arizona DRS-50; Arizona DRS-51E; Armeen® 18; Armeen® 18D; Arodet BN-100; Aromox® C/12; Aromox® C/12-W; Arquad® 12-50; Arquad® 16-50; Arquad® 18-50; Arquad® C-50; Arquad® T-27W; Arylan® SBC25; Arylan® SC15; Arylan® SC30; Arylan® SO60 Acid; Arylan® SX85; Arylan® SY30; Arylan® SY Acid; Arylan® TE/C; Astrowet B-45; Astrowet H-80; Atpol HD722; Atpol HD745; Atpol HD861; Atpol HD863; Atpol HD975; Avirol® A; Avirol® AE 3003; Avirol® AOO 1080; Avirol® FES 996; Avirol® SA 4106; Avirol® SA 4108; Avirol® SA 4110; Avirol® SE 3002; Avirol® SE 3003; Avirol® SL 2010; Avirol® SL 2020; Avirol® SO 70P

Berol 02; Berol 28; Berol 108; Berol 190; Berol 191; Berol 198; Berol 199; Berol 269; Berol 277; Berol 278; Berol 281; Berol 291; Berol 292; Berol 295; Berol 302; Berol 303; Berol 307; Berol 374; Berol 381; Berol 386; Berol 387; Berol 389; Berol 391; Berol 392; Berol 397; Berol 452; Berol 455; Berol 456; Berol 457; Berol 458; Beycostat 656 A; Beycostat LP 9 A; Beycostat NA; Burez K25-500D; Burez K80-500; Burez K80-500D; Burez K80-2500; Burez NA 70-500; Burez NA 70-500D; Burez NA 70-2500

Calfoam SLS-30; Calimulse EM-30; Calimulse EM-99; Calsoft F-90; Calsoft L-60; Calsolene Oil HSA; Calsolene Oil HSAD; Capcure® Emulsifier 37S; Capcure® Emulsifier 65; Carsonol® ALS; Carsonol® ALS-S; Carsonol® ALS Special; Carsonol® DLS; Carsonol® SLS; Carsonol® SLS-R; Carsonol® SLS-S; Carsonol® SLS Special; Carsonol® TLS; Carsonon® N-30; Catinex KB-50; Cedepal CA-890; Cedephos® FA600; Cedephos® RA600; Chemax HCO-5; Chemax HCO-16; Chemax HCO-25; Chemax HCO-200/50; Chemax NP-40; Chemax OP-40/70; Chemeen 18-2; Chemeen 18-5; Chemeen 18-50; Chemeen HT-5; Chemeen HT-50; Chemfac NC-0910; Chemfac PA-080; Chemfac PB-082; Chemfac PB-106; Chemfac PB-135; Chemfac PB-184; Chemfac PB-264; Chemfac PC-099E; Chemfac PC-188; Chemfac PD-600; Chemfac PF-623; Chemfac PF-636; Chemfac PN-322; Chemphos TC-227; Chemphos TC-310; Chimipon GT; Chimipon NA; Cosmopon BN; Cyastat® SP

Dabco® DC197; Darathane® WB-4000; Darvan® ME; Darvan® No. 2; Darvan® WAQ; Dehscofix 920; Dehscofix 930; Dehydrophen PNP 30; Dehydrophen PNP 40; Dehyquart® A; Dehyquart® DAM; Dehyquart® LDB; Dehyquart® LT; Desonic® S-100; Desonic® S-405; Desophos® 4 CP; Desophos® 4 NP; Desophos® 6 NP; Desophos® 6 NP4; Dextrol OC-15; Dextrol OC-20; Dextrol OC-70; Diprosin A-100; Diprosin K-80; Diprosin N-70; Disperso II; Disponil AAP 307; Disponil AAP 436; Disponil AAP 437; Disponil AEP 5300; Disponil AEP 8100; Disponil AEP 9525; Disponil AES 13; Disponil AES 21; Disponil AES 42; Disponil AES 48; Disponil AES 60; Disponil AES 60 E; Disponil AES 72; Disponil APG 110; Disponil FES 32; Disponil FES 61; Disponil FES 77; Disponil MGS 935; Disponil O 5; Disponil O 10; Disponil O 20; Disponil O 250; Disponil SUS 29 L; Disponil SUS 65; Disponil SUS 87 Special; Disponil SUS 90; Disponil SUS IC 8; Disponil TA 5; Disponil TA 25; Disponil TA 430; Dow Corning® 190 Surfactant; Dow Corning® 193 Surfactant; Dowfax 2A1; Dowfax 2EP; Dowfax 3B2; Dowfax 8390; Dresinate® 214; Dresinate® 731; Dymsol® 38-C; Dymsol® 2031; Dymsol® LP; Dymsol® PA

Eccowet® W-50; Eccowet® W-88; Eleminol ES-70; Eleminol HA-100; Eleminol HA-161; Eleminol JS-2; Eleminol MON-2; Eleminol MON-7; Elfan® 260 S; Elfan® WA; Elfan® WA Powder; Emcol® 4300; Emcol® 4910; Emcol® 4930, 4940; Emcol® CC-42; Emcol® CC-57; Emcol® DOSS; Emcol® K8300; Emphos™ CS-136; Emphos™ PS-222; Emphos™ PS-400; Emphos™ PS-410; Empicol® 0303; Empicol® 0303V; Empicol® ESB; Empicol® L Series; Empicol® LQ33/T; Empicol® LX; Empicol® LX28; Empicol® LXS95; Empicol® LXV; Empicol® LXV/D; Empicol® LZ; Empicol® LZ/D; Empicol® LZG 30; Empicol® LZGV; Empicol® LZGV/C; Empicol® LZP; Empicol® LZV/D; Empicol® TL40/T; Empicryl® 6045; Empicryl® 6047; Empicryl® 6052; Empicryl® 6054; Empicryl® 6058; Empicryl® 6059; Empicryl® 6070; Empilan® K Series; Empilan® NP9; Empimin® KSN27; Empimin® KSN70; Empimin® LR28; Empimin® MA; Empimin® MHH; Empimin® MKK98; Empimin® MKK/L; Empimin® MSS; Empimin® MTT/A; Empimin® OP70; Empimin® OT; Empimin® OT75; Emsorb® 2516; Emsorb® 2518; Emsorb® 6901; Emsorb® 6909; Emsorb® 6916; Emulan® NP 2080; Emulan® NP 3070; Emulan® OP 25; Emulsifiant 33 AD; Emulsifier K 30 40%; Emulsifier K 30 68%; Emulsifier K 30 76%; Emulsifier K 30 95%; Emulvin® W; Epal® 6; Epal® 8; Epal® 10; Epal® 12; Epal® 12/70; Epal® 12/85; Epal® 14; Epal® 16NF; Epal® 1214; Epan 710; Epan 720; Epan 740; Epan 750; Epan 785; Equex SP; Estol 1461; Estol 1467; Ethylan®

Emulsifiers • Coemulsifiers *(cont'd.)*

BCD42; Ethylan® BV; Ethylan® CD1210; Ethylan® CD1260; Ethylan® CD9112; Ethylan® DP; Ethylan® GEL2; Ethylan® GEO8; Ethylan® GEO81; Ethylan® GEP4; Ethylan® GES6; Ethylan® GL20; Ethylan® GLE-21; Ethylan® GMF; Ethylan® GO80; Ethylan® GOE-21; Ethylan® GPS85; Ethylan® GS60; Ethylan® GT85; Ethylan® HA Flake; Ethylan® HP; Ethylan® KEO; Ethylan® LDA-37; Ethylan® N30; Ethylan® N50; Ethylan® N92; Ethylan® OE; Ethylan® R; Etocas 10

Fizul 201-11; Fizul M-440; Fizul MD-318; Foamphos NP-9

G-1086; G-1087; G-1096; Gantrez® AN-119; Gardilene S25L; Gemtex 445; Gemtex 680; Gemtex 691-40; Geronol ACR/4; Geronol ACR/9; Geropon® AB/20; Geropon® ACR/4; Geropon® ACR/9; Geropon® BIS/SODICO-2; Geropon® MLS/A; Geropon® SS-O-75; Geropon® WT-27; Glycomul® L; Glycomul® O; GP-209 Silicone Polyol Copolymer; GP-215 Silicone Polyol Copolymer; GP-219 Silicone Polyol Copolymer

Hamposyl® AL-30; Hamposyl® C; Hamposyl® C-30; Hamposyl® L; Hamposyl® L-30; Hamposyl® L-95; Hamposyl® M; Hamposyl® M-30; Hamposyl® O; Hamposyl® S; Hartopol L64; Hefti TO-55-E; Hefti TO-55-EL; Hetoxide NP-30; Hodag PE-004; Hodag PE-104; Hodag PE-106; Hodag PE-109; Hodag PE-206; Hodag PE-209; Hostapal BV Conc; Hyonic GL 400; Hypermer 1083; Hypermer 2296; Hypermer 2524; Hypermer A60; Hypermer A65; Hypermer A95; Hypermer A109; Hypermer A200; Hypermer A256; Hypermer A394; Hypermer A409; Hypermer B239; Hypermer B246; Hypermer B259; Hypermer B261; Hystrene® 5016 NF; Hystrene® 9718 NF

Iconol NP-30; Iconol NP-30-70%; Iconol NP-40; Iconol NP-40-70%; Iconol NP-50; Iconol NP-50-70%; Iconol NP-70; Iconol NP-70-70%; Iconol NP-100; Iconol NP-100-70%; Iconol OP-10; Iconol OP-30; Iconol OP-30-70%; Iconol OP-40; Iconol OP-40-70%; Igepal® CA-877; Igepal® CA-880; Igepal® CA-887; Igepal® CA-890; Igepal® CA-897; Igepal® CO-430; Igepal® CO-530; Igepal® CO-630; Igepal® CO-710; Igepal® CO-850; Igepal® CO-880; Igepal® CO-887; Igepal® CO-897; Igepal® CO-980; Igepal® CO-987; Igepal® CTA-639W; Igepal® DM-430; Igepal® DM-530; Igepal® DM-710; Igepal® DM-730; Igepal® DM-880; Incropol L-7; Ionet DL-200; Ionet DO-200; Ionet DO-400; Ionet DO-600; Ionet DO-1000; Ionet DS-300; Ionet DS-400

Kemamine® D-190; Kemester® 5221SE; Ken-React® 7 (KR 7); Ken-React® 9S (KR 9S); Kessco® PEG 200 DL; Kessco® PEG 200 DO; Kessco® PEG 200 DS; Kessco® PEG 200 ML; Kessco® PEG 200 MO; Kessco® PEG 200 MS; Kessco® PEG 300 DL; Kessco® PEG 300 DO; Kessco® PEG 300 DS; Kessco® PEG 300 ML; Kessco® PEG 300 MO; Kessco® PEG 300 MS; Kessco® PEG 400 DL; Kessco® PEG 400 DO; Kessco® PEG 400 DS; Kessco® PEG 400 ML; Kessco® PEG 400 MO; Kessco® PEG 400 MS; Kessco® PEG 600 DL; Kessco® PEG 600 DO; Kessco® PEG 600 DS; Kessco® PEG 600 ML; Kessco® PEG 600 MO; Kessco® PEG 600 MS; Kessco® PEG 1000 DL; Kessco® PEG 1000 DO; Kessco® PEG 1000 DS; Kessco® PEG 1000 ML; Kessco® PEG 1000 MO; Kessco® PEG 1000 MS; Kessco® PEG 1540 DL; Kessco® PEG 1540 DO; Kessco® PEG 1540 DS; Kessco® PEG 1540 ML; Kessco® PEG 1540 MO; Kessco® PEG 1540 MS; Kessco® PEG 4000 DL; Kessco® PEG 4000 DO; Kessco® PEG 4000 DS; Kessco® PEG 4000 ML; Kessco® PEG 4000 MO; Kessco® PEG 4000 MS; Kessco® PEG 6000 DL; Kessco® PEG 6000 DO; Kessco® PEG 6000 DS; Kessco® PEG 6000 ML; Kessco® PEG 6000 MO; Kessco® PEG 6000 MS; KZ OPPR; KZ TPP; KZ TPPJ

Lamigen ES 30; Lamigen ES 60; Lamigen ES 100; Lamigen ET 20; Lamigen ET 70; Lamigen ET 90; Lamigen ET 180; Lankropol® ATE; Lankropol® KMA; Lankropol® KN51; Lankropol® KNB22; Lankropol® KO2; Lankropol® OPA; Laural LS; LICA 01; LICA 09; LICA 12; LICA 38; LICA 38A; LICA 38J; LICA 44; LICA 97; LICA 99; Lipolan LB-440; Liponox NC 2Y; Liponox NC-500; Lomar® HP; Lomar® LS; Lomar® LS Liq; Lomar® PW; Lomar® PWA; Lomar® PWA Liq

Mackester™ EGDS; Mackester™ EGMS; Mackester™ IP; Mackester™ SP; Macol® OP-40(70); Manro SDBS 25/30; Maphos® 17; Maphos® 76; Maphos® 76 NA; Maranil Paste A 65; Maranil Powd. A; Masil® EM 100; Masil® EM 100 Conc; Masil® EM 100D; Masil® EM 100P; Masil® EM 250 Conc; Masil® EM 266 (50); Masil® EM 350X; Masil® EM 350X Conc; Masil® EM 1000; Masil® EM 1000 Conc; Masil® EM 1000P; Masil® EM 10,000; Masil® EM 10,000 Conc; Masil® EM 60,000; Masil® EM 100,000; Mazon® 1045A; Mazon® 1086; Mazon® 1096; Merpol® HCS; MIPA; Miramine® C; Miranol® CM-SF Conc; Miranol® J2M Conc; Miranol® SM Conc; Monafax 785; Monawet MB-45; Monawet MO-70; Monawet MT-70; Monawet SNO-35; Monolan® 8000 E/80; Monosulf; Montosol PB-25; Montovol RF-10

Nansa® 1042/P; Nansa® EVM50; Nansa® EVM70/B; Nansa® EVM70/E; Nansa® HS85/S; Nansa® SSA/P; Naxonac™ 510; Naxonac™ 610; Naxonac™ 690-70; Newcol 261A, 271A;

Emulsifiers • Coemulsifiers *(cont'd.)*

Newcol 506; Newcol 508; Newcol 560SF; Newcol 560SN; Newcol 607, 610, 614, 623; Newcol 704, 707; Newcol 707SF; Newcol 710, 714, 723; Newcol 861S; Newcol 1305SN, 1310SN; Newpol PE-61; Newpol PE-62; Newpol PE-64; Newpol PE-68; Newpol PE-74; Newpol PE-75; Newpol PE-78; Newpol PE-88; Niaproof® Anionic Surfactant 4; Niaproof® Anionic Surfactant 08; Nikkol GO-430; Nikkol GO-440; Nikkol GO-460; Niox KQ-34; Nissan Nonion HS-204.5; Nissan Nonion HS-206; Nissan Nonion HS-208; Nissan Nonion HS-210; Nissan Nonion HS-215; Nissan Nonion HS-220; Nissan Nonion HS-240; Nissan Nonion HS-270; Nissan Nonion NS-202; Nissan Nonion NS-204.5; Nissan Nonion NS-206; Nissan Nonion NS-208.5; Nissan Nonion NS-209; Nissan Nonion NS-210; Nissan Nonion NS-212; Nissan Nonion NS-215; Nissan Nonion NS-220; Nissan Nonion NS-230; Nissan Nonion NS-240; Nissan Nonion NS-250; Nissan Nonion NS-270; Nissan Persoft EK; Nissan Plonon 102; Nissan Plonon 104; Nissan Plonon 108; Nissan Plonon 171; Nissan Plonon 172; Nissan Plonon 201; Nissan Plonon 204; Nissan Plonon 208; Nissan Trax K-300; Nissan Trax N-300; Noiox AK-41; Nonipol 20; Nonipol 40; Nonipol 55; Nonipol 60; Nonipol 70; Nonipol 85; Nonipol 95; Nonipol 100; Nonipol 110; Nonipol 120; Nonipol 130; Nonipol 160; Nonipol 200; Nonipol 400; Nopalcol 1-L; Nopalcol 1-TW; Nopalcol 2-DL; Nopalcol 4-C; Nopalcol 4-CH; Nopalcol 4-L; Nopalcol 4-O; Nopalcol 4-S; Nopalcol 6-DO; Nopalcol 6-DTW; Nopalcol 6-L; Nopalcol 6-R; Nopalcol 6-S; Nopalcol 10-COH; Nopalcol 12-CO; Nopalcol 12-COH; Nopalcol 19-CO; Nopalcol 30-TWH; Nopalcol 200; Nopalcol 400; Nopalcol 600; Nopco® 2031; Noxamium S2-50; NZ 01; NZ 09; NZ 12; NZ 33; NZ 38; NZ 39; NZ 44; NZ 49; NZ 89; NZ 97

Octosol 449; Octosol 496; Octosol 571; Octosol A-1; Octosol A-18; Octosol A-18-A; Octosol HA-80; Octosol IB-45; Octosol SLS; Octosol SLS-1; Octowet 70PG; Oulu 356

Pegosperse® 50 MS; Pegosperse® 200 DL; Pennad 150; Perlankrol® ATL40; Perlankrol® DGS; Perlankrol® EAD60; Perlankrol® EP12; Perlankrol® EP24; Perlankrol® EP36; Perlankrol® ESD; Perlankrol® ESD60; Perlankrol® FB25; Perlankrol® FD63; Perlankrol® FF; Perlankrol® FN65; Perlankrol® FT58; Perlankrol® FV70; Perlankrol® FX35; Perlankrol® PA Conc; Perlankrol® RN75; Perlankrol® SN; Petrac® Calcium Stearate CP-22G; Petrac® PHTA; Phosphanol Series; Phospholan® PDB3; Phospholan® PNP9; Pluriol® PE 6200; Pluriol® PE 6400; Pluriol® PE 10500; Plysurf A207H; Plysurf A208B; Plysurf A208S; Plysurf A210G; Plysurf A212C; Plysurf A215C; Plysurf A216B; Plysurf A217E; Plysurf A219B; Plysurf AL; Polirol 10; Polirol 1BS; Polirol 215; Polirol 23; Polirol 4, 6; Polirol C5; Polirol DS; Polirol L400; Polirol LS; Polirol NF80; Polirol O55; Polirol SE 301; Polirol TR/LNA; Polystep® A-4; Polystep® A-7; Polystep® A-11; Polystep® A-13; Polystep® A-15; Polystep® A-16; Polystep® A-16-22; Polystep® A-17; Polystep® B-1; Polystep® B-3; Polystep® B-5; Polystep® B-11; Polystep® B-12; Polystep® B-19; Polystep® B-20; Polystep® B-22; Polystep® B-23; Polystep® B-24; Polystep® B-25; Polystep® B-27; Polystep® B-29; Polystep® CM 4 S; Polystep® C-OP3S; Polystep® F-1; Polystep® F-3; Polystep® F-4; Polystep® F-5; Polystep® F-6; Polystep® F-9; Polystep® F-10; Polystep® F-95B; Polysurf A212E; Prifac 5902, 5904, 5905; Prifrac 2920; Prifrac 2922; Priolene 6901; Priolene 6907; Priolene 6930; Pripol 1017, 1022; Pripol 1022; Prox-onic CSA-1/04; Prox-onic CSA-1/06; Prox-onic CSA-1/010; Prox-onic CSA-1/015; Prox-onic CSA-1/020; Prox-onic CSA-1/030; Prox-onic CSA-1/050; Prox-onic DT-03; Prox-onic DT-030; Prox-onic EP 1090-1; Prox-onic EP 1090-2; Prox-onic EP 2080-1; Prox-onic EP 4060-1; Prox-onic HR-05; Prox-onic HR-016; Prox-onic HR-025; Prox-onic HR-030; Prox-onic HR-036; Prox-onic HR-040; Prox-onic HR-080; Prox-onic HR-0200; Prox-onic HR-0200/50; Prox-onic HRH-05; Prox-onic HRH-016; Prox-onic HRH-025; Prox-onic HRH-0200; Prox-onic HRH-0200/50; Prox-onic L 081-05; Prox-onic L 101-05; Prox-onic L 102-02; Prox-onic L 121-09; Prox-onic L 161-05; Prox-onic L 181-05; Prox-onic L 201-02; Prox-onic LA-1/02; Prox-onic MC-02; Prox-onic MC-05; Prox-onic MC-015; Prox-onic MHT-05; Prox-onic MHT-015; Prox-onic MO-02; Prox-onic MO-015; Prox-onic MO-030; Prox-onic MO-030-80; Prox-onic MS-02; Prox-onic MS-05; Prox-onic MS-011; Prox-onic MS-050; Prox-onic MT-02; Prox-onic MT-05; Prox-onic MT-015; Prox-onic MT-020; Prox-onic NP-1.5; Prox-onic NP-04; Prox-onic NP-06; Prox-onic NP-09; Prox-onic NP-010; Prox-onic NP-015; Prox-onic NP-020; Prox-onic NP-040; Prox-onic NP-040/70; Prox-onic NP-050; Prox-onic NP-050/70; Prox-onic NP-0100; Prox-onic NP-0100/70; Prox-onic OP-09; Prox-onic OP-030/70; Prox-onic OP-040/70; Puxol CB-22

Quimipol ENF 200; Quimipol ENF 230; Quimipol ENF 300

Remcopal 229; Remcopal 33820; Rewoderm® S 1333; Rewomat B 2003; Rewomat TMS; Rewopal® HV 50; Rewophat E 1027; Rewophat NP 90; Rewopol® B 1003; Rewopol® B 2003; Rewopol® NL 3; Rewopol® NLS 15 L; Rewopol® NLS 28; Rewopol® NLS 30 L; Rewopol® NOS 8; Rewopol® NOS 10; Rewopol® NOS 25; Rewopol® SBDB 45; Rewopol® SBDC 40; Rewopol® SBFA 50; Rewopol® SMS 35; Rewopol® TMSF; Rewoquat CPEM;

Emulsifiers • Coemulsifiers *(cont'd.)*

Reworyl® NKS 50; Rexol 25/8; Rexol 25/10; Rexol 25/12; Rexol 25/40; Rexol 25/407; Rexol 45/307; Rexol 45/407; Rhodacal® 330; Rhodacal® A-246L; Rhodacal® BA-77; Rhodacal® BX-78; Rhodacal® DS-10; Rhodacal® DSB; Rhodafac® BX-660; Rhodafac® BX-760; Rhodafac® MC-470; Rhodafac® PE-510; Rhodafac® PS-17; Rhodafac® RE-610; Rhodafac® RE-877; Rhodafac® RE-960; Rhodafac® RS-610; Rhodapex® AB-20; Rhodapex® CO-433; Rhodapex® CO-436; Rhodapex® EST-30; Rhodapon® CAV; Rhodapon® L-22; Rhodapon® L-22/C; Rhodapon® OLS; Rhodapon® OS; Rhodapon® SB; Rhodapon® TDS; Rhodapon® UB; Rhodasurf® A 24; Rhodasurf® BC-737; Rhodasurf® BC-840; Rhodasurf® L-4

Sandet 60; Sandoz Phosphorester 510; Sandoz Phosphorester 690; Sandoz Sulfonate 2A1; Schercophos NP-6; Schercowet DOS-70; Secolat; Secosol® ALL40; Secosol® DOS 70; Serdet DFK 40; Serdet DJK 30; Serdet DM; Serdet DNK 30; Serdox NNP 25; Serdox NNP 30/70; Serdox NOP 30/70; Serdox NOP 40/70; Sermul EA 129; Sermul EA 136; Sermul EA 139; Sermul EA 146; Sermul EA 150; Sermul EA 151; Sermul EA 152; Sermul EA 176; Sermul EA 188; Sermul EA 205; Sermul EA 211; Sermul EA 214; Sermul EA 221; Sermul EA 224; Sermul EA 242; Sermul EA 30; Sermul EA 370; Sermul EA 54; Sermul EK 330; Sermul EN 15; Sermul EN 155; Sermul EN 229; Sermul EN 237; Sermul EN 312; Servoxyl VPGZ 7/100; Servoxyl VPIZ 100; Servoxyl VPNZ 10/100; Servoxyl VPQZ 9/100; Servoxyl VPTZ 100; Servoxyl VPTZ 3/100; Servoxyl VPUZ; Servoxyl VPYZ 500; Silwet® L-7001; Silwet® L-7500; Silwet® L-7602; Sinopol 610; Sinopol 623; Sinopol 707; Sinopol 714; Sinoponic PE 61; Sinoponic PE 62; Sinoponic PE 64; Sinoponic PE 68; Soft Detergent 95; Softenol® 3900; Softenol® 3991; Sole Terge 8; Solricin® 235; Solricin® 435; Solricin® 535; Soprophor® NPF/10; Sorbax HO-40; Sorbax HO-50; Spinomar NaSS; Steol® 4N; Steol® CA-460; Steol® COS 433; Stepfac® 8170; Sulfopon® 101 Special; Sulfopon® 102; Sulfopon® P-40; Sulfotex DOS; Supragil® NK; Supralate® WAQ; Supralate® WAQE; Surfagene FAD 105; Surfagene FAZ 109; Surfagene FAZ 109 NV; Surfax 220; Surfax 495; Surfax 502; Surfax 536; Surfax 539; Sylvaros® 20; Sylvaros® 80; Synperonic OP16; Synperonic PE30/80

T-Det® C-40; T-Det® N-40; T-Det® N-50; T-Det® N-100; T-Det® N-307; T-Det® N-407; T-Det® N-507; T-Det® N-1007; T-Det® O-40; T-Det® O-307; Tergitol® 15-S-30; Tergitol® 15-S-40; Tergitol® XD; Tergitol® XH; Tergitol® XJ; Teric C12; Teric N8; Teric N9; Teric N10; Teric N12; Teric N13; Teric N15; Teric N20; Teric N30; Teric N40; Teric PEG 400; Texadril 8780; Texapon® K-12; Texapon® OT Highly Conc. Needles; Texapon® VHC Needles; Tex-Wet 1104; Tex-Wet 1143; T-Mulz® 598; T-Mulz® 734-2; Tomah Q-14-2; Tomah Q-18-15; Tomah Q-D-T; Triton® 770 Conc; Triton® W-30 Conc; Triton® X-15; Triton® X-165-70%; Triton® X-301; Triton® X-305-70%; Triton® X-405-70%; Triton® X-705-70%; Trycol® 6954; Trycol® 6957; Trycol® 6967; Trycol® 6968; Trycol® 6970; Trycol® 6971; Trycol® 6972; Trycol® 6981; Trycol® 6984; Trycol® NP-20; Trycol® NP-30; Trycol® NP-307; Trycol® OP-407; Tryfac® 5556; Trymeen® 6617; Tylose® MH Grades; Tylose® MHB

Ufasan 50; Ungerol LS, LSN

Varsulf® NOS-25

Wayfos A; Wayfos D-10N; Wayfos M-60; Westvaco® 1480; Westvaco® M-30; Westvaco® M-40; Westvaco® M-70; Westvaco® Resin 90; Westvaco® Resin 95; Westvaco® Resin 790; Westvaco® Resin 795; Westvaco® Resin 895; Witco® 960; Witcolate™ 1276; Witcolate™ A Powder; Witcolate™ D51-51; Witcolate™ D51-51EP; Witcolate™ D51-52; Witcolate™ D51-53; Witcolate™ D51-53HA; Witcolate™ D51-60; Witcolate™ SL-1; Witconate™ 90; Witconate™ 1223H; Witconate™ 3009-15; Witconate™ AOS; Witconate™ AOS-EP; Witconol™ NP-100; Witconol™ NP-300; Witconol™ NP-330

Zonyl® FSA; Zonyl® FSB; Zonyl® FSC; Zonyl® FSN; Zonyl® FSP; Zorapol SN-9

Chemicals: Ammonium laureth-30 sulfate; Ammonium lauroyl sarcosinate; Ammonium lauryl sulfate; Ammonium lignosulfonate; Ammonium naphthalene-formaldehyde sulfonate; Ammonium nonoxynol-4 sulfate; Ammonium nonoxynol-9 sulfate; Ammonium nonoxynol-10 sulfate; Ammonium nonoxynol-20 sulfate; Ammonium nonoxynol-30 sulfate; Ammonium nonoxynol-77 sulfate

Calcium dodecylbenzene sulfonate; Caprylic alcohol; Ceteareth-4; Ceteareth-6; Ceteareth-10; Ceteareth-15; Ceteareth-20; Ceteareth-25; Ceteareth-30; Ceteareth-50; Cetoleth-5; Cetoleth-10; Cetoleth-13; Cetoleth-19; Cetoleth-20; Cetoleth-25; Cetoleth-55; Cetrimonium chloride; Cetyl oleth-7 phosphate; Cocopropylenediamine; Cocoyl hydroxyethyl imidazoline; Cocoyl sarcosine; C11-15 pareth-30; C11-15 pareth-40; Cycloneopentyl, cyclo (dimethlaminoethyl) pyrophosphato zirconate, di mesyl salt

DEA-lauryl sulfate; Diammonium cocoyl sulfosuccinate; Diammonium stearyl sulfosuccinamate; Diamyl sodium sulfosuccinate; Di (butyl, methyl pyrophosphato) ethylene titanate di (dioctyl,

Emulsifiers • Coemulsifiers *(cont'd.)*

hydrogen phosphite); Dicyclo (dioctyl) pyrophosphato titanate; Dicyclohexyl sodium sulfosuccinate; Di (dioctylphosphato) ethylene titanate; Di (dioctylpyrophosphato) ethylene titanate; Dihexyl sodium sulfosuccinate; Dimethicone copolyol; Dimethyl behenamine; Dimethyl erucylamine; Dioctyl sodium sulfosuccinate; Disodium cetearyl sulfosuccinamate; Disodium cetearyl sulfosuccinate; Disodium cocoyl sulfosuccinate; Disodium C12-15 pareth sulfosuccinate; Disodium deceth-6 sulfosuccinate; Disodium dodecyloxy propyl sulfosuccinamate; Disodium isodecyl sulfosuccinate; Disodium laureth sulfosuccinate; Disodium lauriminodipropionate; Disodium nonoxynol-10 sulfosuccinate; Disodium oleamido MIPA-sulfosuccinate; Disodium oleyl sulfosuccinate; Disodium ricinoleamido MEA-sulfosuccinate; Disodium stearyl sulfosuccinamate; Ditridecyl sodium sulfosuccinate

Glyceryl dilaurate; Glyceryl dilaurate SE; Glyceryl stearate SE; Glycol dilaurate; Glycol dilaurate SE; Glycol dioleate SE; Glycol distearate SE; Glycol laurate; Glycol laurate SE; Glycol oleate; Glycol oleate SE; Glycol ricinoleate SE

Hydrogenated tallow acid

Isodecyl oleate; Isodecyloxypropyl dihydroxyethyl methyl ammonium chloride; Isopropylamine dodecylbenzenesulfonate; Isopropyl 4-aminobenzenesulfonyl di (dodecylbenzenesulfonyl) titanate; Isopropyl dimethacryl isostearoyl titanate; Isopropyl tri (dioctylphosphato) titanate; Isopropyl tri (dioctylpyrophosphato) titanate; Isopropyl tri (N ethylamino-ethylamino) titanate

Laureth-2; Laureth-3; Laureth-9; Laureth-12; Laureth-15; Laureth-16; Lauric acid; Lauroyl sarcosine; Laurtrimonium chloride

MEA-lauryl sulfate; Methyl hydroxyethylcellulose; Methyl laurate; Myristoyl sarcosine

Neopentyl (diallyl) oxy, trihydroxy caproyl titanate; Neopentyl (diallyl) oxy, trimercapto-phenyl zirconate; Nonoxynol-1; Nonoxynol-3; Nonoxynol-5; Nonoxynol-8; Nonoxynol-15; Nonoxynol-20; Nonoxynol-25; Nonoxynol-30; Nonoxynol-34; Nonoxynol-35; Nonoxynol-40; Nonoxynol-50; Nonoxynol-70; Nonoxynol-100; Nonoxynol-120; Nonoxynol-150; Nonoxynol-6 phosphate; Nonoxynol-10 phosphate; Nonoxynol-4 sulfate; Nonoxynol-10 sulfate; Nonoxynol-30 sulfate; Nonyl nonoxynol-7; Nonyl nonoxynol-9; Nonyl nonoxynol-15; Nonyl nonoxynol-24; Nonyl nonoxynol-49; Nonyl nonoxynol-9 phosphate

Octoxynol-1; Octoxynol-3; Octoxynol-9; Octoxynol-10; Octoxynol-16; Octoxynol-25; Octoxynol-30; Octoxynol-40; Octoxynol-70; Octoxynol-9 carboxylic acid; Octyl phosphate; Oleamine; Oleth-7; Oleth-4 phosphate; Oleyl 1,3-propylene diamine

Palmitamine; PEG-2 C13-15 alkyl amine; PEG-5 C13-15 alkyl amine; PEG-10 C13-15 alkyl amine; PEG-15 C13-15 alkyl amine; PEG-20 C13-15 alkyl amine; PEG-25 C13-15 alkyl amine; PEG-35 C13-15 alkyl amine; PEG-50 C13-15 alkyl amine; PEG-1 C13-15 alkylmethylamine; PEG-5 castor oil; PEG-10 castor oil; PEG-16 castor oil; PEG-25 castor oil; PEG-30 castor oil; PEG-32 castor oil; PEG-36 castor oil; PEG-40 castor oil; PEG-75 castor oil; PEG-160 castor oil; PEG-200 castor oil; PEG (1200) castor oil; PEG (1900) castor oil; PEG-10 cocamine; PEG-15 cocamine; PEG-8 cocoate; PEG-200 cocoate; PEG-5 cocomonium methosulfate; PEG-2 dilaurate; PEG-4 dilaurate; PEG-6 dilaurate; PEG-20 dilaurate; PEG-2 dilaurate SE; PEG-3 dioctoate; PEG-2 dioleate; PEG-4 dioleate; PEG-8 dioleate; PEG-12 dioleate; PEG-2 dioleate SE; PEG-2 distearate; PEG-150 distearate; PEG-2 distearate SE; PEG-12 ditallowate; PEG-5 hydrogenated castor oil; PEG-10 hydrogenated castor oil; PEG-16 hydrogenated castor oil; PEG-20 hydrogenated castor oil; PEG-25 hydrogenated castor oil; PEG-30 hydrogenated castor oil; PEG-40 hydrogenated castor oil; PEG-60 hydrogenated castor oil; PEG-100 hydrogenated castor oil; PEG-200 hydrogenated castor oil; PEG (1200) hydrogenated castor oil; PEG-5 hydrogenated tallow amine; PEG-50 hydrogenated tallow amine; PEG-60 hydrogenated tallowate; PEG-2 laurate; PEG-4 laurate; PEG-75 laurate; PEG-150 laurate; PEG-2 laurate SE; PEG-2 oleamine; PEG-7 oleamine; PEG-12 oleamine; PEG-15 oleamine; PEG-30 oleamine; PEG-2 oleate; PEG-4 oleate; PEG-6 oleate; PEG-8 oleate; PEG-20 oleate; PEG-150 oleate; PEG-2 oleate SE; PEG-12 ricinoleate; PEG-40 sorbitan hexaoleate; PEG-50 sorbitan hexaoleate; PEG-30 sorbitan tetraoleate; PEG-40 sorbitan tetraoleate; PEG-60 sorbitan tetraoleate; PEG-17 sorbitan trioleate; PEG-18 sorbitan trioleate; PEG-5 stearamine; PEG-11 stearamine; PEG-2 stearate; PEG-4 stearate; PEG-9 stearate; PEG-12 stearate; PEG-20 stearate; PEG-150 stearate; PEG-2 stearate SE; PEG-2 tallowamine; PEG-5 tallowamine; PEG-10 tallowamine; PEG-15 tallowamine; PEG-20 tallowamine; PEG-40 tallowamine; PEG-2 tallowate; PEG-3 tallow diamine; PEG-15 tallow diamine; PEG-30 tallow diamine; Poloxamer 182; Poloxamer 184; Polyacrylic acid; Polysorbate 20; Polysorbate 21; Polysorbate 40; Polysorbate 60; Polysorbate 80; Polysorbate 85; Polyurethane prepolymer; Potassium castorate; Potassium naphthalene sulfonate; Potassium rosinate; PPG-24-buteth-27; PPG-40 diethylmonium chloride; PPG-2 laurate; PPG-2 laurate SE; PPG-2 oleate; PPG-2 oleate SE; PPG-2 stearate; PPG-

Emulsifiers • Coemulsifiers *(cont'd.)*

2 stearate SE; Propylene glycol; PVM/MA copolymer

Rosin

Sodium C12-15 alkyl sulfate; Sodium cocoamphopropionate; Sodium C14-16 olefin sulfonate; Sodium C12-15 pareth-3 sulfosuccinate; Sodium decyl diphenyloxide disulfonate; Sodium decyl sulfate; Sodium dodecylbenzenesulfonate; Sodium dodecyl diphenyl ether disulfonate; Sodium dodecyl diphenyl ether sulfonate; Sodium dodecyl diphenyloxide disulfonate; Sodium isodecyl sulfate; Sodium isopropyl naphthalene sulfonate; Sodium laureth-30 sulfate; Sodium laureth sulfosuccinate; Sodium lauryl/propoxy sulfosuccinate; Sodium lauryl sulfate; Sodium methallyl sulfonate; Sodium myristyl sulfate; Sodium nonoxynol-4 sulfate; Sodium nonoxynol-6 sulfate; Sodium nonoxynol-8 sulfate; Sodium nonoxynol-9 sulfate; Sodium nonoxynol-10 sulfate; Sodium nonoxynol-15 sulfate; Sodium nonoxynol-25 sulfate; Sodium nonoxynol-10 sulfosuccinate; Sodium octoxynol-2 ethane sulfonate; Sodium octoxynol-3 sulfate; Sodium octoxynol-6 sulfate; Sodium octylphenol sulfate; Sodium octylphenoxyethoxyethyl sulfonate; Sodium octyl sulfate; Sodium alpha olefin sulfonate; Sodium oleyl sulfate; Sodium oleyl sulfosuccinamate; Sodium ricinoleate; Sodium rosinate; Sodium p-styrenesulfonate; Sodium tridecyl sulfate; Sorbitan diisostearate; Sorbitan isostearate; Sorbitan oleate; Sorbitan palmitate; Sorbitan sesquioleate; Sorbitan stearate; Sorbitan trioleate; Sorbitan tristearate; Stearoyl sarcosine; Stearyl PEG-15 methyl ammonium chloride; Styrene/MA copolymer; Sulfated castor oil; 2-Sulfoethyl methacrylate

Tall oil acid; Tall oil aminopropyl dimethylamine; Tallow amine; Tallow dimethyl trimethyl propylene diammonium chloride; TEA-benzene sulfonate; TEA-lauryl sulfate; Tetra (2, diallyoxymethyl-1 butoxy titanium di (di-tridecyl) phosphite; Tetrahydroxypropyl ethylenediamine; Tetraisopropyl di (dioctylphosphito) titanate; Tetraoctyloxytitanium di (ditridecylphosphite); Tetrasodium dicarboxyethyl stearyl sulfosuccinamate; Titanium di (butyl, octyl pyrophosphate) di (dioctyl, hydrogen phosphite) oxyacetate; Titanium di (cumylphenylate) oxyacetate; Titanium di (dioctylpyrophosphate) oxyacetate; Titanium dimethacrylate oxyacetate; Trideceth-14; Trideceth-15; Trideceth-6 phosphate; Tridecylbenzene sulfonic acid; Triisopropanolamine

Fibers

Trade names: Cycom® NCG 1200 Unsized; Cycom® NCG 1204 p8

Dry-Blend® NCG dB910; Dry-Blend® NCG dB920

Fibrox 030 SC

Santoweb® D; Santoweb® DX; Santoweb® H; Santoweb® W

Fillers • Extenders

Trade names: Aerosil® 130; Aerosil® 150; Aerosil® 300; Aerosil® 325; Aerosil® 380; Aerosil® COK 84; Aerosil® OX50; AgGlad™ TW Microspheres; Akrochem® Hydrated Alumina; Akrochem® Rubbersil RS-150; Akrochem® Rubbersil RS-200; Akrochem® Zinc Oxide 35; Akrofax™ B; Aktisil® AM; Aktisil® EM; Aktisil® MM; Aktisil® PF 216; Aktisil® PF 231; Aktisil® VM; Aktisil® VM 56; Albacar® 5970; Albafil®; AlbaFlex 25; AlbaFlex 50; AlbaFlex 100; AlbaFlex 200; Albaglos®; Alcan BW53, BW103, BW153; Alcan FRF 80 S5; Alcan FRF LV1 S5; Alcan FRF LV2 S5; Alcan Superfine 4; Alcan Superfine 7; Alcan Superfine 11; AlNel A-100; AlNel A-200; AlNel AG; Amical® 85; Amical® 95; Amical® 101; Amical® 202; Amical® 303; Amical® SC2; Amical® SC3; Amical® SC5; Amical® SC7; Amoco® H-15; Amoco® L-14; Amoco® L-50; Artic Mist; ASP® 072; ASP® 100; ASP® 101; ASP® 102; ASP® 170; ASP® 200; ASP® 400; ASP® 400P; ASP® 600; ASP® 602; ASP® 900; Atomite®

Bara-200C; Bara-200N; Bara-325C; Bara-325N; Barafoam 200; Barafoam 325; Barimite™; Barimite™ 200; Barimite™ UF; Barimite™ XF; Bartex® 10; Bartex® 65; Bartex® 80; Bartex® FG-2; Bartex® FG-10; Bartex® OWT; Barytes Microsupreme 2410/M; Barytes Supreme; BAX 1091FM; Bilt-Plates® 156; Black Pearls® 3700; Blanc fixe F; Blanc fixe micro®; Blanc fixe N; Britol® 6NF; Burgess 2211

Cab-O-Sil® TS-530; Cabot® PP 9269; Califlux® 90; Califlux® 510; Califlux® GP; Califlux® LP; Califlux® SP; Califlux® TT; Calwhite®; Calwhite® II; Camel-CAL®; Camel-CAL® ST; Camel-CAL® Slurry; Camel-CARB®; Camel-FIL™; Camel-FINE; Camel-FINE ST; Camel-TEX®; Camel-WITE®; Camel-WITE® Slurry; Camel-WITE® ST; Cardolite® NC-1307; CC™-101;

Fillers • Extenders *(cont'd.)*

CC™-103; CC™-105; Celite® 270; Celite® 292; Celite® 350; CF 1500; Cimbar™ 325; Cimbar™ CF; Cimbar™ UF; Cimbar™ XF; Coupsil VP 6109; Coupsil VP 6411; Coupsil VP 6508; Coupsil VP 8113; Coupsil VP 8415; CP Filler™

DCH-99; Dixie® Clay; Dolocron® 15-16; Dolocron® 32-15; Dolocron® 40-13; Dolocron® 45-12; Drakeol 34; Drikalite®; Dualite® M6001AE; Dualite® M6017AE; Dualite® M6032AE; Dualite® M6033AE; Duramite®; Dytek® EP

Eccosphere® FTD-200; Eccosphere® FTD-202; Eccosphere® FTD-235; Eccosphere® IG-25; Eccosphere® IG-101; Eccosphere® IGD-101; Eccosphere® SDT-28; Eccosphere® SDT-40; Eccosphere® SDT-60; Elastocarb® Tech Heavy; Elastocarb® Tech Light; Elastocarb® UF; Extendospheres® CG; Extendospheres® SF-10; Extendospheres® SF-12; Extendospheres® SF-14; Extendospheres® SG; Extendospheres® TG; Extendospheres® XOL-50; Extendospheres® XOL-200; Extrusil

Faktogel® 10; Faktogel® 14; Faktogel® 17; Faktogel® 110; Faktogel® 111; Faktogel® 140; Faktogel® 141; Faktogel® 144; Faktogel® A; Faktogel® Asolvan; Faktogel® Asolvan T; Faktogel® Badenia C; Faktogel® Badenia T; Faktogel® HF; Faktogel® KE 8384; Faktogel® KE 8419; Faktogel® MB; Faktogel® Para; Faktogel® R; Faktogel® Rheinau H; Faktogel® Rheinau W; Faktogel® T; Faktogel® ZD; Ferrosil™ 14; Fiberfrax® 6000 RPS; Fiberkal™ FG; Fibra-Cel® BH-70; Fibra-Cel® BH-200; Fibra-Cel® CBR-18; Fibra-Cel® CBR-40; Fibra-Cel® CBR-41; Fibra-Cel® CBR-50; Fibra-Cel® CC-20; Fibra-Cel® SW-8; Fibrox 030-E; Fibrox 030-ES; Fibrox 300; Firebrake® ZB 415; Firebrake® ZB-Extra Fine; Firebrake® ZB-Fine; Firemaster® 642; Firemaster® 836; Firemaster® HP-36; FK 140; FK 160; FK 300DS; FK 310; FK 320DS; Flamtard M7; Flo-Gard® CC 120, CC 140, CC 160; 431-G; Franklin T-11; Franklin T-12; Franklin T-13; Franklin T-14; Franklin T-325; FRE

G-50 Barytes™; Gamaco®; Gamaco® II; Gamaco® Slurry; Gama-Fil® 40; Gama-Fil® 55; Gama-Fil® 90; Gama-Fil® D-1; Gama-Fil® D-1T; Gama-Fil® D-2; Gama-Fil® D-2T; Gama-Plas®; Gama-Sperse® 2; Gama-Sperse® 5; Gama-Sperse® 80; Gama-Sperse® 140; Gama-Sperse® 255; Gama-Sperse® 6451; Gama-Sperse® 6532; Gama-Sperse® 6532 NSF; Gama-Sperse® CS-11; Grit-O'Cobs®; G-White

H-36; H-46; Haltex™ 304; Haltex™ 310; HB-40®; Hi-Pflex® 100; Hi-Sil® 210; Hi-Sil® 233; Hi-Sil® 243LD; Hi-Sil® 250; Huber 35; Huber 40C; Huber 65A; Huber 70C; Huber 80; Huber 80B; Huber 80C; Huber 90; Huber 90B; Huber 90C; Huber 95; Huber ARO 60; Huber HG; Huber HG90; Huber N110; Huber N220; Huber N234; Huber N299; Huber N326; Huber N330; Huber N339; Huber N343; Huber N347; Huber N351; Huber N375; Huber N550; Huber N650; Huber N660; Huber N683; Huber N762; Huber N774; Huber N787; Huber N990; Huber S212; Huber S315; Huber SM; Huber WG-1; Huber WG-2; Hubercarb® Q 1; Hubercarb® Q 1T; Hubercarb® Q 2; Hubercarb® Q 2T; Hubercarb® Q 3; Hubercarb® Q 3T; Hubercarb® Q 4; Hubercarb® Q 6; Hubercarb® Q 6-20; Hubercarb® Q 20-60; Hubercarb® Q 40-200; Hubercarb® Q 60; Hubercarb® Q 100; Hubercarb® Q 200; Hubercarb® Q 200T; Hubercarb® Q 325; Hubercarb® W 2; Hubercarb® W 2T; Hubercarb® W 3; Hubercarb® W 3T; Hubercarb® W 4; Hubersil® 162; H-White; Hydrex®; Hydrex® R; Hydrite 121-S; Hydrobrite 200PO; Hydrobrite 300PO; Hydrobrite 380PO; Hydrobrite 550PO; Hymocal; Hystrene® 3675; Hystrene® 3675C; Hystrene® 3680; Hystrene® 3695

Icecap® K; Imsil® A-15; Imsil® A-25; Imsil® A-30; I.T. 3X; I.T. 5X; I.T. 325; I.T. FT; I.T. X

Jet Fil® 200; Jet Fil® 350; Jet Fil® 500; Jet Fil® 575C; Jet Fil® 625C; Jet Fil® 700C

Kaydol® S; Kemamine® DD-3680; Kemamine® DP-3680; Kemamine® DP-3695; Ketjensil® SM 405; Koster Keunen Ozokerite; Kotamite®

Langford® Clay; Lite-R-Cobs®; LV-31

Magnifin® H5; Magnifin® H5B; Magnifin® H5C; Magnifin® H5C/3; Magnifin® H5D; Magnifin® H-7; Magnifin® H7A; Magnifin® H7C; Magnifin® H7C/3; Magnifin® H7D; Magnifin® H-10; Magnifin® H10A; Magnifin® H10B; Magnifin® H10C; Magnifin® H10D; Magnifin® H10F; Magnum-White; Magocarb-33; Magoh-S; Magotex™; Magsil Diamond Talc 200 mesh; Magsil Diamond Talc 350 mesh; Magsil Star Talc 200 mesh; Magsil Star Talc 350 mesh; Marble Dust™; Marblemite™; Marblewhite® MW200; Marblewhite® MW325; Marblewhite® MW A 200; Marblewhite® MW A 325; Martinal® 103 LE; Martinal® 104 LE; Martinal® 107 LE; Martinal® OL-104; Martinal® OL-104/C; Martinal® OL-104 LE; Martinal® OL-104/S; Martinal® OL-107; Martinal® OL-107/C; Martinal® OL-107 LE; Martinal® OL-111 LE; Martinal® OL/Q-107; Martinal® OL/Q-111; Martinal® ON; Martinal® ON-310; Martinal® ON-313; Martinal® ON-320; Martinal® ON-920/V; Martinal® ON-4608; Martinal® OS; Martinal® OX; McNamee® Clay; Micawhite 200; Micral® 532; Micro P Extender; Micro-Cel® C; Micro-Cel® E; Micro-Cel® T-21; Micro-Cel® T-49; MicroPflex 1200; Microtalc® MP10-52; Microtalc® MP12-50; Microtalc® MP15-38; Microtalc® MP25-38; Microtalc® MP44-26;

Fillers • Extenders *(cont'd.)*

Microtuff® 325F; Microtuff® 1000; Microtuff® F; Micro-White® 07 Slurry; Micro-White® 10 Slurry; Micro-White® 15; Micro-White® 15 SAM; Micro-White® 15 Slurry; Micro-White® 25; Micro-White® 40; Micro-White® 100; Minbloc™ 16; Minbloc™ 20; Minbloc™ 30; Minex® 4; Minex® 7; Minex® 10; Min-U-Sil® 5; Min-U-Sil® 10; Min-U-Sil® 15; Mistron CB; Mistron Vapor® R; MPC Channel Black TR 354; Mr 245; Mr 285; Mr 305; Multifex™ FF; Multifex-MM®

Nicron 325; Nicron 400; No. 1 White™; No. 3 White; No. 8 White; No. 9 NCS; No. 9 White; No. 10 White; Nyad® 200, 325, 400; Nyad® 475, 1250; Nyad® FP, G, G Special; Nyad G® Wollastocoat®; Nycor® R; Nytal® 100; Nytal® 200; Nytal® 300; Nytal® 400; 136-G

Onyx Classica OC-1000; Onyx Premier WP-31; Opacicoat™; OZ

Par® Clay; Perkasil® KS 207; Perkasil® KS 300; Perkasil® KS 404; Perkasil® VP 406; Petro-Rez 103; Petro-Rez 200; Petro-Rez 801; Pfinyl® 402; Picco® 6070; Picco® 6100; Picco® 6115; PMF® Fiber 204; PMF® Fiber 204AX; PMF® Fiber 204BX; PMF® Fiber 204CX; PMF® Fiber 204EX; Polymica 200; Polymica 325; Polymica 400; Polymica 3105; Polyplate 90; Polyplate 852; Polyplate P; Polyplate P01; Polytalc 262; PolyTalc™ 445; Prespersion PAB-9541; Pyrax® A; Pyrax® ABB; Pyrax® B; Pyrax® WA

Q-Cel® 300; Q-Cel® 636; Q-Cel® 640; Q-Cel® 650; Q-Cel® 2106; Q-Cel® 2116; Q-Cel® 2135; Q-Cel® 6717; Q-Cel® 6832; Q-Cel® 6835; Q-Cel® 6920; Q-Cel® Ultralight

Rhenofit® 1987; Rhenofit® 2009; Rhenofit® 2642; Rhenofit® 3555; Rhenofit® B; Rit-Cizer 10-EHFA; Rit-Cizer 10-EM; Rit-Cizer 10-ETA; Rit-Cizer 10-MME; RO-40; RPP; Rudol®

Satintone® SP-33®; Satintone® Whitetex®; SB-30; SB-31; SB-31C; SB-136; SB-331; SB-332; SB-335; SB-336; SB-431; SC-53; Scotchlite™; Select-A-Sorb Powdered, Compacted; Semtol® 40; Semtol® 70; Semtol® 85; Semtol® 100; Semtol® 350; Sil-Co-Sil® 40; Sil-Co-Sil® 45; Sil-Co-Sil® 47; Sil-Co-Sil® 49; Sil-Co-Sil® 51; Sil-Co-Sil® 52; Sil-Co-Sil® 53; Sil-Co-Sil® 63; Sil-Co-Sil® 75; Sil-Co-Sil® 90; Sil-Co-Sil® 106; Sil-Co-Sil® 125; Sil-Co-Sil® 250; Silfin® Z; Sillikolloid P 87; Sillitin N 82; Sillitin N 85; Sillitin V 85; Sillitin Z 86; Sillitin Z 89; Sillum-200; Sillum-200 Q/P; Siltex™; Siltex™ 44; Siltex™ 44C; Silver Bond 45; Silver Bond B; Silverline 400; Sipernat® 44; Sipernat® D11; SM 2079; Snowflake P.E.™; Snowflake White™; Sphericel® 110P8; Stygene Series; Super-Fil®; Super Microtuff F; Supermite®; Super-Pflex® 100

Tamsil 15; Tamsil 25; Tamsil 30; Tamsil Gold Bond; Thermax® Floform® N-990; Thermax® Floform® Ultra Pure N-990; Thermax® Powder N-991; Thermax® Powder Ultra Pure N-991; Thermax® Stainless Floform® N-907; Thermax® Stainless Powder N-908; TP 87/21; TP 89/50; TP 90/127; TP 91/01; Tullanox HM-250

Ultratalc™ 609; Uniplex 600

Vansil® W-10; Vansil® W-20; Vansil® W-30; Vantalc® 6H; Versamag® DC; Versamag® Tech; Versamag® UF; Vestowax A-217; Vestowax A-227; Vestowax A-235; Vestowax A-415; Vestowax A-616; Vicron® 10-20; Vicron® 15-15; Vicron® 25-11; Vicron® 41-8; Vicron® 45-3; Viplex 525; Viplex 885; Vulkasil® A1; Vulkasil® C; Vulkasil® N; Vulkasil® N/GR-S; Vulkasil® S

Wingdale White; Winnofil S, SP; 10, 325, 400 Wollastocoat®

X50-S®; XO White

Zeeospheres® 200; Zeeospheres® 400; Zeeospheres® 600; Zeeospheres® 800; Zeeospheres® 850; Zeolex® 23; Zeolex® 80; Zeosil 45; Zeosil 85; Zeosil 125; Zeosil 165G; Zeosil 175MP; Zeosil 1165; Zewakol; Zink Oxide Transparent; Zinkoxyd Activ®

Chemicals: Alumina silicate; Aluminum-magnesium-sodium silicate; Aluminum nitride; Aluminum silicate; Aluminum silicate, calcined; Aluminum silicate, hydrous; Aminopropyltriethoxysilane; Ammonium lignosulfonate; Asphalt, oxidized

Barium sulfate; Bis-(3-(triethoxysilyl)propyl)tetrasulfane; Borosilicate glass

Calcium aluminosilicate; Calcium carbonate; Calcium silicate; Carbon black; Carboxymethylcellulose sodium; Castor oil, vulcanized; Cellulose; Corn cob meal; Coumarone-indene resin

Diatomaceous earth; Dimer diamine

Ferro-aluminum silicate

3-Glycidoxypropyltrimethoxysilane

Kaolin

Limestone

Magnesium carbonate; Magnesium hydroxide; Mercaptopropyltrimethoxysilane; Mica

Nepheline syenite

Ozokerite

Pentaerythrityl tetrabenzoate; 1,3-Pentanediamine; Polybutene; Polymethacrylic acid; Pyrophyllite

Fillers • Extenders *(cont'd.)*

Rapeseed oil, vulcanized
Silica-alumina; Sodium borosilicate; Sodium magnesium aluminosilicate; Sodium silicoaluminate
Talc; Toluenesulfonamide formaldehyde resin
Vinyltriethoxysilane; Vinyltris(2-methoxyethoxy) silane
Wollastonite
Zinc carbonate

Flame retardants • Fire retardants • Smoke suppressants

Trade names: ACM-MW; Actimer FR-803; Actimer FR-1025M; Actimer FR-1033; ADK CIZER E-500; ADK CIZER EP-13; ADK CIZER EP-17; ADK STAB 2335; Akrochem® Hydrated Alumina; Akrochlor™ L-39; Akrochlor™ L-40; Akrochlor™ L-42; Akrochlor™ L-57; Akrochlor™ L-60; Akrochlor™ L-61; Akrochlor™ L-70LV; Akrochlor™ L-170; Akrochlor™ L-170HV; Akrochlor™ R-70; Akrochlor™ R-70S; Albrite® TBPO; Alcan FRF 5; Alcan FRF 10; Alcan FRF 20; Alcan FRF 30; Alcan FRF 40; Alcan FRF 60; Alcan FRF 80; Alcan FRF 80 S5; Alcan FRF LV1 S5; Alcan FRF LV2, LV4, LV5, LV6, LV7, LV8, LV9; Alcan FRF LV2 S5; Alcan H-10; Alcan Ultrafine UF7; Alcan Ultrafine UF11; Alcan Ultrafine UF15; Alcan Ultrafine UF25; Alcan Ultrafine UF35; Amgard® CHT; Amgard® CPC 452; Amgard® EDAP; Amgard® MC; Amgard® ND; Amgard® NH; Amgard® NK; Amgard® NP; Amgard® PI; Amgard® $TBPO_4$; Ampacet 11371; Amsperse; Antiblaze® 19; Antiblaze® 78; Antiblaze® 80; Antiblaze® 100; Antiblaze® 125; Antiblaze® 150; Antiblaze® 175; Antiblaze® 195; Antiblaze® 519; Antiblaze® 1045; Antiblaze® DMMP; Antiblaze® TCP; Antiblaze® TDCP/LV; Antiblaze® TXP; AO; AOM; AroCy® F-10

Baco DH101; BAX 1091FM; Bromoklor™ 50; Bromoklor™ 70; Bulab® Flamebloc; Busan® 11-M1

Cereclor 42; Chlorez® 700; Chlorez® 700-DF; Chlorez® 700-S; Chlorez® 700-SS; Chlorez® 725-S; Chlorez® 760; Colloids FR 47 Series; Colloids FR 48 Series; Colloids FR 50 Series; Colloids FR 52 Series; Colloids FR 54 Series; Colloids FR 55 Series; Cylink HPC-75; Cylink HPC-90; Cylink HPC-100

DD-8126; DD-8133; DD-8307; Dechlorane® Plus 25; Dechlorane® Plus 35; Dechlorane® Plus 515; Delvet 65; Disflamoll® DPK; Disflamoll® DPO; Disflamoll® TCA; Disflamoll® TKP; Disflamoll® TOF; Disflamoll® TP; Doverguard® 152; Doverguard® 170; Doverguard® 700; Doverguard® 700-S; Doverguard® 700-SS; Doverguard® 760; Doverguard® 5761; Doverguard® 8133; Doverguard® 8207-A; Doverguard® 8208-A; Doverguard® 8307-A; Doverguard® 8410; Doverguard® 8426; Doverguard® 9021; Doverguard® 9119; Doverguard® 9122; Doversperse 3; Doversperse 8843; Doversperse 8929; Doversperse A-1; DX-5114; DX-5119

Elastocarb® Tech Heavy; Elastocarb® Tech Light; Elastocarb® UF; Electrofine S-70; Empee® PE Conc. 1; Empee® PO Conc. 61; Empee® PP Conc. 4; Empee® PP Conc. 33; Empee® PP Conc. 43; Epikote® 5050; Epikote® 5051 H; Epikote® 5057; Epikote® 5201; Epikote® 5201S; Epikote® 5203; Epikote® 5205; Epon® Resin 1183; Exolit® 422; Exolit® 462; Exolit® IFR-10; Exolit® IFR-15; Exolit® IFR-23

F-2000 FR; F-2001; F-2016; F-2200; F-2300; F-2300H; F-2310; F-2400; F-2400E; F-2430; F-2430 SA; Firebrake® 500; Firebrake® ZB; FireShield® H; FireShield® HPM-UF; FireShield® L; Flame-Amine; Flamegard® 908; Flamtard H; Flamtard M7; Flamtard S; Flamtard Z10; Flamtard Z15; 431-G; FR-20; FR-513; FR-521; FR-522; FR-612; FR-613; FR-615; FR-705; FR-910; FR-913; FR-1025; FR-1033; FR-1034; FR-1138; FR-1205; FR-1206; FR-1208; FR-1210; FR-1215; FR-1524; FR-1525; FR-2124; FR-D; FRCROS 480; FRE; Futurethane UR-2175; Fyarestor® 100; Fyraway; Fyrebloc; Fyrol® 6; Fyrol® 25; Fyrol® 38; Fyrol® 42; Fyrol® 51; Fyrol® 99; Fyrol Bis Beta; Fyrol® CEF; Fyrol® DMMP; Fyrol® FR-2; Fyrol® MC; Fyrol® MP; Fyrol® MPP; Fyrol® PBR; Fyrol® PCF; Fyrolflex RDP

Genomoll® P; Great Lakes BA-43; Great Lakes BA-50™; Great Lakes BA-50P™; Great Lakes BA-59; Great Lakes BA-59P™; Great Lakes BC-52™; Great Lakes BC-58™; Great Lakes BE-51™; Great Lakes CD-75P™; Great Lakes DBS™; Great Lakes DE-60F™ Special; Great Lakes DE-61™; Great Lakes DE-62™; Great Lakes DE-71™; Great Lakes DE-79™; Great Lakes DE-83™; Great Lakes DE-83R™; Great Lakes DP-45™; Great Lakes FB-72™; Great Lakes FF-680™; Great Lakes GPP-36™; Great Lakes GPP-39™; Great Lakes NH-1197™; Great Lakes NH-1511™; Great Lakes PDBS-10™; Great Lakes PDBS-80™; Great Lakes

Flame retardants *(cont'd.)*

PE-68™; Great Lakes PH-73™; Great Lakes PHE-65™; Great Lakes PHT4™; Great Lakes PO-64P™; Great Lakes SP-75™

H-36; H-46; H-100; H-105; H-109; H-600; H-800; H-900; H-990; Halofree 22; Haltex™ 304; Haltex™ 310; Hoechst Wax PE 520; Hordaflam® NK 70; Hordaflam® NK 72; Hordaflam® NL 70; Hostaflam® AP 750; Hostaflam® System F1912; Hydramax™ HM-B8, HM-B8-S; Hydramax™ HM-C9, HM-C9-S; Hydromax 100

KC 31; KemGard 425; KemGard 981; Ken-React® 7 (KR 7); Ken-React® 9S (KR 9S); Ken-React® 12 (KR 12); Ken-React® 26S (KR 26S); Ken-React® 33DS (KR 33DS); Ken-React® 38S (KR 38S); Ken-React® 39DS (KR 39DS); Ken-React® 41B (KR 41B); Ken-React® 44 (KR 44); Ken-React® 46B (KR 46B); Ken-React® 55 (KR 55); Ken-React® 133DS (KR 133DS); Ken-React® 134S (KR 134S); Ken-React® 138D (KR 138D); Ken-React® 138S (KR 138S); Ken-React® 158D (KR 158D); Ken-React® 158FS (KR 158FS); Ken-React® 212 (KR 212); Ken-React® 238A (KR 238A); Ken-React® 238J (KR 238J); Ken-React® 238M (KR 238M); Ken-React® 238S (KR 238S); Ken-React® 238T (KR 238T); Ken-React® 262A (KR 262A); Ken-React® 262ES (KR 262ES); Ken-React® OPP2 (KR OPP2); Ken-React® OPPR (KR OPPR); Ken-React® KR TTS; Kisuma 5A; Kisuma 5B; Kisuma 5E; Kloro 3000; Kloro 6001; KP-140®; Kronitex® 25; Kronitex® 50; Kronitex® 100; Kronitex® 200; Kronitex® 1840; Kronitex® 1884; Kronitex® 3600; Kronitex® PB-460; Kronitex® PB-528; Kronitex® TCP; Kronitex® TPP; Kronitex® TXP; KZ OPPR; KZ TPP; KZ TPPJ

Larostat® 377 FR; LICA 01; LICA 09; LICA 12; LICA 38; LICA 38A; LICA 38J; LICA 44; LICA 97; LICA 99; Lindol XP Plus; Luperco AFR-250; Luperco AFR-500; LV-31

Magnifin® H5; Magnifin® H5B; Magnifin® H5C; Magnifin® H5C/3; Magnifin® H5D; Magnifin® H-7; Magnifin® H7A; Magnifin® H7C; Magnifin® H7C/3; Magnifin® H7D; Magnifin® H-10; Magnifin® H10A; Magnifin® H10B; Magnifin® H10C; Magnifin® H10D; Magnifin® H10F; Magnum-White; Magocarb-33; Magoh-S; Martinal® 103 LE; Martinal® 104 LE; Martinal® 107 LE; Martinal® OL-104; Martinal® OL-104/C; Martinal® OL-104 LE; Martinal® OL-104/S; Martinal® OL-107; Martinal® OL-107/C; Martinal® OL-107 LE; Martinal® OL-111 LE; Martinal® OL/Q-107; Martinal® OL/Q-111; Martinal® ON; Martinal® ON-310; Martinal® ON-313; Martinal® ON-320; Martinal® ON-920/V; Martinal® ON-4608; Martinal® OS; Martinal® OX; Mastertek® ABS 3710; Mastertek® PA 3664; Mastertek® PBT 3659; Mastertek® PC 3686; Mastertek® PE 3660; Mastertek® PE 3668; Mastertek® PP 3658; Mastertek® PP 5601; Mastertek® PP 5604; Mastertek® PS 3698; MDI PE-800FR; Micral® 532; Micral® 632; Micral® 855; Micral® 916; Micral® 932; Micral® 1000; Micral® 1500; Min-U-Sil® 40

Nyacol® A-1530; Nyacol® A-1540N; Nyacol® A-1550; Nyacol® A-1588LP; Nyacol® AB40; Nyacol® AP50; Nyacol® APE1540; Nyacol® N22; Nyacol® N24; Nyacol® ZTA; NZ 01; NZ 09; NZ 12; NZ 33; NZ 38; NZ 39; NZ 44; NZ 49; NZ 89; NZ 97

Octoguard FR-01; Octoguard FR-10; Octoguard FR-15; 136-G; Ongard AZ11; Onyx Classica OC-1000; Onyx Premier WP-31

Perkadox® BC; Phos-Chek P/30; Phosflex 71B; Phosflex 370; Phosflex TBEP; Plastisan B; Pliabrac® 519; Pliabrac® 521; Pliabrac® 524; Pliabrac® TCP; Pliabrac® TXP; Polarite 880E(W); Polarite 880G(B); Polu-U; Polyfil 100; Prespersion PAC-1712; Pyrobloc SAP; Pyro-Chek® 60PB; Pyro-Chek® 60PBC; Pyro-Chek® 68PB; Pyro-Chek® 68PBC; Pyro-Chek® 68PBG; Pyro-Chek® 77B; Pyro-Chek® C60PB; Pyro-Chek® C68PB; Pyro-Chek® LM

Rez-O-Sperse® 3; Rez-O-Sperse® A-1

S.A. 100; Sachtolith® HD; Sachtolith® HD-S; Sachtolith® L; SAFR-C70; Santicizer 141; Santicizer 143; Santicizer 148; Santicizer 154; Saytex® 102E; Saytex® 111; Saytex® 120; Saytex® 8010; Saytex® BCL-462; Saytex® BN-451; Saytex® BT-93®; Saytex® BT-93W; Saytex® FR-1138; Saytex® HBCD-LM; Saytex® RB-49; Saytex® RB-79; Saytex® RB-100; Saytex® VBR; SB-30; SB-31; SB-31C; SB-136; SB-331; SB-332; SB-335; SB-336; SB-431; SB-432; SB-632; SB-805; SB-932; Sillum-200; Sillum-PL 200; Smokebloc 11, 12; Spectratech® CM 11053; Spectratech® CM 11489; Spectratech® CM 11591; Spin-Flam MF-82; Sta-Flame 88/99

Tego® Antiflamm® N; Tegostab® B 3640; Tetrathal; Thermoguard® 210; Thermoguard® 212; Thermoguard® 213; Thermoguard® 215; Thermoguard® 220; Thermoguard® 230; Thermoguard® 240; Thermoguard® 505; Thermoguard® 8218; Thermoguard® 9010; Thermoguard® CPA; Thermoguard® DM-9406; Thermoguard® FR; Thermoguard® FR/PE 80/20; Thermoguard® L; Thermoguard® S; Thermoguard® UF; Thermolin® 101; Thermolin® RF-230; 331-G

UltraFine® II

Valu-Fil; Versamag® DC; Versamag® Tech; Versamag® UF; Victastab HMP; Vircol® 82

WH-31; Wyfire® H-A-85; Wyfire® H-A-85P; Wyfire® S-A-51; Wyfire® S-BA-88P; Wyfire® Y-A-

Flame retardants *(cont'd.)*

85; Wyfire® Y-B-57; Wyfire® YP-BA-47P
Zb™-112; Zb™-223; Zb™-237; Zb™-325; Zb™-467; Zerogen® 10; Zerogen® 11; Zerogen® 15; Zerogen® 30; Zerogen® 33; Zerogen® 35; Zerogen® 60; Zinc Borate 9506; Zinc Borate B9355

Chemicals: Alumina hydrate; Ammonium bromide; Ammonium octamolybdate; Ammonium polyphosphate sol'n.; Antimony pentoxide; Antimony trioxide
Barium metaborate; Bis (β-chloroethyl) vinyl phosphonate; Bis (pentabromophenoxy) ethane; Bis (tribromophenoxy) ethane; Bromoacenaphthylene; t-Butylphenyl diphenyl phosphate
Calcium sulfate dihydrate; Chlorendic anhydride; Cycloneopentyl, cyclo (dimethlaminoethyl) pyrophosphato zirconate, di mesyl salt
Decabromodiphenyl oxide; Dibromoethyldibromocyclohexane; Dibromoneopentyl glycol; 2,4-Dibromophenol; Dibromostyrene; Dicyclo (dioctyl) pyrophosphato titanate; Di (dioctylphosphato) ethylene titanate; Di (dioctylpyrophosphato) ethylene titanate; Diethyl N,N-bis(2-hydroxyethyl)aminomethylphosphonate; Dimelamine phosphate; Dimethyl methylphosphonate; Diphenyl octyl phosphate; Dodecachloro dodecahydro dimethanodibenzo cyclooctene
Epoxy resin, brominated; Ethylenebis dibromonorbornane dicarboximide; Ethylene bis-tetrabromophthalimide; Ethylene diamine phosphate
Heptakis (dipropylene glycol) triphosphite; Hexabromocyclododecane; Hexafluorobisphenol A dicyanate monomer/prepolymer
Isodecyl diphenyl phosphate; Isopropyl 4-aminobenzenesulfonyl di (dodecylbenzenesulfonyl) titanate; Isopropyl dimethacryl isostearoyl titanate; Isopropyl tri (dioctylphosphato) titanate; Isopropyl tri (dioctylpyrophosphato) titanate; Isopropyl tri (N ethylamino-ethylamino) titanate
Magnesium calcium carbonate hydroxide; Magnesium carbonate; Magnesium carbonate hydroxide; Magnesium hydroxide; Melamine cyanurate; Melamine phosphate; Melamine pyrophosphate; Molybdenum trioxide
Neoalkoxy tri (dioctylphosphato) titanate; Neoalkoxy tri (dioctylpyrophosphato) titanate; Neopentyl (diallyl) oxy, trihydroxy caproyl titanate; Neopentyl (diallyl) oxy, trimercapto-phenyl zirconate
Octabromodiphenyl oxide
Paraffin, brominated; Paraffin, bromochlorinated; Paraffin, chlorinated; Pentabromobenzyl acrylate; Pentabromodiphenyl oxide; Pentabromoethylbenzene; Pentabromophenol; Pentabromotoluene; Pentaerythritol; Polydibromophenylene oxide; Poly(dibromostyrene); Poly (pentabromobenzyl) acrylate; Polypropylene/dibromostyrene copolymer; Polystyrene, brominated; Polyurethane, thermoset
Resorcinol bis (diphenylphosphate)
Silica-alumina; Sodium antimonate
Tetrabromobisphenol A; Tetrabromobisphenol A allyl ether; Tetrabromobisphenol A bis (allyl ether); Tetrabromobisphenol A, bis (2,3-dibromopropyl ether); Tetrabromobisphenol A, bis (2-hydroxyethyl ether); Tetrabromobisphenol A diacrylate; Tetrabromobisphenol A di-2-hydroxyethyl ether; Tetrabromodipentaerythritol; Tetrabromophthalate diol; Tetrabromophthalic anhydride; Tetrabromoxylene; Tetrabromo-p-xylene; Tetrachlorophthalic anhydride; Tetradecabromodiphenoxy benzene; Tetra (2, diallyoxymethyl-1 butoxy titanium di (di-tridecyl) phosphite; Tetraisopropyl di (dioctylphosphito) titanate; Tetrakis (2-chloroethyl) ethylene diphosphate; Tetraoctyloxytitanium di (ditridecylphosphite); Titanium di (butyl, octyl pyrophosphate) di (dioctyl, hydrogen phosphite) oxyacetate; Titanium di (cumylphenylate) oxyacetate; Titanium di (dioctylpyrophosphate) oxyacetate; Titanium dimethacrylate oxyacetate; Triaryl phosphate; Tribromoneopentyl alcohol; 2,4,6-Tribromophenol; Tribromophenyl allyl ether; Tribromophenyl maleimide; Tribromostyrene; Tributoxyethyl phosphate; Tributyl phosphate; Trichloroethyl phosphate; Tri(β-chloroisopropyl)phosphate; Tricresyl phosphate; Tri(β,β′-dichloroisopropyl)phosphate; Triethyl phosphate; Triisopropylphenyl phosphate; Trioctyl phosphate; Triphenyl phosphate; Tris (β-chloroethyl) phosphate; Tris (2-chloropropyl) phosphate; Tris (β-chloropropyl) phosphate; Tris (dichloropropyl) phosphate; Trixylenyl phosphate
Vinyl bromide
Zinc borate; Zinc hydroxystannate; Zinc oxide; Zinc phosphate; Zinc stannate

Flexibilizers

Trade names: ACtol® 60
Cardolite® NC-513; Cardolite® NC-700; Cardolite® NC-1307; Chemigum® N8; Cylink HPC-90; Cylink HPC-100
Epi-Cure® 3260; Epi-Cure® 3265; Epi-Cure® 3266
Heloxy® 32; Heloxy® 71; Heloxy® 84; Heloxy® 505
Kenplast® G; Kenplast® PG
Polamine® 250; Polamine® 650; Polamine® 1000; Polamine® 2000

Chemicals: Cardanol
Phenyl glycol ether
Urethane diacrylate

Gellants

Trade names: A-C® 16; A-C® 5120; Anchor® DPG
Cecavon ZN 70; Cecavon ZN 71; Cecavon ZN 72; Cecavon ZN 73; Cecavon ZN 735
Disflamoll® TP
Gantrez® AN-119
Millithix® 925
Polycat® 77
Tylose® CBR Grades; Tylose® CR
Ultramoll® I; Ultramoll® TGN; Unimoll® DB
Vulcastab® EFA

Chemicals: Dibenzylidene sorbitol
Formaldehyde, ammonia, ethyl chloride condensate
Polyacrylic acid
Starch

Hardeners

Trade names: Actiron NX 3; ACtol® 70; ADK STAB NA-10; Akrochem® P 40; Akrochem® P 82
Benzoic Acid K
Darex® 636L; Darex® 670L; DETDA 80; Doverphos® 10; Doverphos® 10-HR; Doverphos® 10-HR Plus; Dyphene® 6745-P; Dyphene® 6746-L; Dyphene® 8845P
Eltesol® TA 65; Eltesol® TA 96; Eltesol® XA65; Eltesol® XA/M65; Epotuf Hardener 37-605; Epotuf Hardener 37-610; Epotuf Hardener 37-611; Epotuf Hardener 37-612; Epotuf Hardener 37-614; Epotuf Hardener 37-618; Epotuf Hardener 37-620; Epotuf Hardener 37-621; Epotuf Hardener 37-622; Epotuf Hardener 37-624; Epotuf Hardener 37-625; Epotuf Hardener 37-640
Great Lakes PHT4™
Hardener OZ
Kemamine® D-190; Kemamine® D-650; Kemamine® D-970; Kemamine® D-974; Kemamine® D-989; Kemamine® D-999; Kemamine® T-1902D; Kemamine® T-6501; Kemamine® T-6502D; Kemamine® T-9701; Kemamine® T-9702D; Kemamine® T-9902D; Kemamine® T-9972D; Ken-React® 7 (KR 7); Ken-React® 9S (KR 9S); Ken-React® 12 (KR 12); Ken-React® 26S (KR 26S); Ken-React® 33DS (KR 33DS); Ken-React® 38S (KR 38S); Ken-React® 39DS (KR 39DS); Ken-React® 41B (KR 41B); Ken-React® 44 (KR 44); Ken-React® 46B (KR 46B); Ken-React® 55 (KR 55); Ken-React® 133DS (KR 133DS); Ken-React® 134S (KR 134S); Ken-React® 138D (KR 138D); Ken-React® 138S (KR 138S); Ken-React® 158D (KR 158D); Ken-React® 158FS (KR 158FS); Ken-React® 212 (KR 212); Ken-React® 238A (KR 238A); Ken-React® 238J (KR 238J); Ken-React® 238M (KR 238M); Ken-React® 238S (KR 238S); Ken-React® 238T (KR 238T); Ken-React® 262A (KR 262A); Ken-React® 262ES (KR 262ES); Ken-React® OPP2 (KR OPP2); Ken-React® OPPR (KR OPPR); Ken-React® KR TTS; KZ 55; KZ OPPR; KZ TPP; KZ TPPJ
Leecure B-110; Leecure B-550; Leecure B-610; Leecure B-612; Leecure B-614; Leecure B-950; Leecure B-1310; Leecure B-1550; Leecure B-1600; Leecure B-1700; LICA 01; LICA 09; LICA 12; LICA 38; LICA 38A; LICA 38J; LICA 44; LICA 97; LICA 99

Hardeners *(cont'd.)*

MFM-415; MFM-786 V
NZ 01; NZ 09; NZ 12; NZ 33; NZ 38; NZ 39; NZ 44; NZ 49; NZ 89; NZ 97
Sipomer® AM; Sipomer® TMPTMA; SP-6700; SP-6701
Tactix H41
Versamine® 911; Vertal 710; Vestamin® A139; Vestamin® V214; Vulkadur® A

Chemicals: Arachidyl-behenyl 1,3-propylene diamine
Cocopropylenediamine; Cycloneopentyl, cyclo (dimethlaminoethyl) pyrophosphato zirconate, di mesyl salt
Dicoco methylamine; 3,5-Diethyl-2,4-diaminotoluene; 3,5-Diethyl-2,6-diaminotoluene; Dihydrogenated tallow methylamine; Dimethyl arachidyl-behenyl amine; Dimethyl hydrogenated tallow amine; Dimethyl soyamine
Hexahydrophthalic anhydride; Hydrogenated tallow 1,3-propylene diamine
4,4´-Methylene dianiline
Neopentyl (diallyl) oxy, trihydroxy caproyl titanate
Oleic/linoleic amine; Oleyl 1,3-propylene diamine
PEG-3 dimethacrylate; Phenolic resin; Phthalic anhydride
Soyaminopropylamine
Tallowaminopropylamine; Triethylamine
Xylene sulfonic acid

Heat sensitizers

Chemicals: Neopentyl (diallyl) oxy, trimercapto-phenyl zirconate

Homogenizers

Trade names: Paraloid® K-120N; Paraloid® K-125; Paraloid® K-130; Paraloid® K-147; Paraloid® K-175; Proaid® 9814; Proaid® 9826
Rhenosin® 140; Rhenosin® 143; Rhenosin® 145; Rhenosin® 260
Struktol® 40 MS; Struktol® 40 MS Flakes; Struktol® 60 NS; Struktol® 60 NS Flakes; Struktol® TH 10 Flakes; Struktol® TH 20 Flakes; Struktol® TR 065

Impact modifiers

Trade names: ACtol® 60; Ageflex MEA
Baymod® A 50; Baymod® A 63 A; Baymod® A 72 A/74 A; Baymod® A 90/92; Baymod® A 95; Baymod® A KA 8572; Baymod® A KU3-2086; Baymod® L450 N; Baymod® L450 P; Baymod® L2450; Baymod® PU; Baymoflex A KU3-2069.A; Blendex® 336; Blendex® 338; Blendex® 340; Blendex® 424; Blendex® 467; Blendex® 586; Blendex® 587; Blendex® 703; Blendex® 975; Blendex® 977; Blendex® HPP 801; Blendex® HPP 802; Blendex® HPP 803; Blendex® HPP 804; Blendex® HPP 820; Buna BL 6533; Buna CB 14; Buna CB HX 529; Buna CB HX 530; Buna CB HX 565
Cimpact 705; Cimpact 710; Comboloob 0827; Comboloob 0827/1; Crestomer® 1080
Diene® 35AC3; Diene® 35AC10; Diene® 55AC10; Diene® 55NF; Diene® 70AC; Durastrength 200; Dutral® PM-06PLE
Escor® ATX-310; Escor® ATX-325; Exact™ Plastomers; EXP-36-X20 Epoxy Functional Silicone Sol'n; Exxelor PE 805; Exxelor PE 808; Exxelor VA 1801; Exxelor VA 1803; Exxelor VA 1810; Exxelor VA 1820
Hi-Pflex® 100
Kane Ace® B-11A; Kane Ace® B-12; Kane Ace® B-18A-1; Kane Ace® B-22; Kane Ace® B-28; Kane Ace® B-28A; Kane Ace® B-31; Kane Ace® B-38A; Kane Ace® B-51; Kane Ace® B-52; Kane Ace® B-56; Kane Ace® B-58; Kane Ace® B-58A; Kane Ace® B-513; Kane Ace® B-521; Kane Ace® B-522; Kane Ace® FM-10; Kane Ace® FM-20; Kane Ace® FT-80; Kane Ace® M-511; Kane Ace® M-521; Kenplast® ES-2; Kenplast® ESN; Ken-React® 7 (KR 7); Ken-React® 9S (KR 9S); Ken-React® 12 (KR 12); Ken-React® 26S (KR 26S); Ken-React® 33DS

Impact modifiers *(cont'd.)*

(KR 33DS); Ken-React® 38S (KR 38S); Ken-React® 39DS (KR 39DS); Ken-React® 41B (KR 41B); Ken-React® 44 (KR 44); Ken-React® 46B (KR 46B); Ken-React® 55 (KR 55); Ken-React® 133DS (KR 133DS); Ken-React® 134S (KR 134S); Ken-React® 138D (KR 138D); Ken-React® 138S (KR 138S); Ken-React® 158D (KR 158D); Ken-React® 158FS (KR 158FS); Ken-React® 212 (KR 212); Ken-React® 238A (KR 238A); Ken-React® 238J (KR 238J); Ken-React® 238M (KR 238M); Ken-React® 238S (KR 238S); Ken-React® 238T (KR 238T); Ken-React® 262A (KR 262A); Ken-React® 262ES (KR 262ES); Ken-React® OPP2 (KR OPP2); Ken-React® OPPR (KR OPPR); Ken-React® KR TTS; Kotamite®; Krynac® PXL 34.17; Krynac® PXL 38.20; KZ 55; KZ OPPR; KZ TPP; KZ TPPJ

Levapren 400; Levapren 450HV; Levapren 450P; Levapren 452; Levapren 500HV; Levapren 700HV; Levapren KA 8385; LICA 01; LICA 09; LICA 12; LICA 38; LICA 38A; LICA 38J; LICA 44; LICA 97; LICA 99; Luzenac 8170

Metablen® C-301

Novacarb; Novalar; Novalene; NZ 01; NZ 09; NZ 12; NZ 33; NZ 38; NZ 39; NZ 44; NZ 49; NZ 89; NZ 97

Paraloid® BTA-702; Paraloid® BTA-715; Paraloid® BTA-730; Paraloid® BTA-733; Paraloid® BTA-751; Paraloid® BTA-753; Paraloid® BTA-III-N2; Paraloid® EXL-3330; Paraloid® EXL-3361; Paraloid® EXL-3387; Paraloid® EXL-3611; Paraloid® EXL-3647; Paraloid® EXL-3657; Paraloid® EXL-3691; Paraloid® EXL-5375; Paraloid® KM-334; Paraloid® KM-390; Permethyl® 100 Epoxide; Plastech EP 8126; Polybond® 1009; Pripol 1004; Pripol 1009

RA-061; Rhenoblend® N 6011; Ricon 131; Ricon 142; Ricon 150; Royalene 7565; Royaltuf 372; Royaltuf 465A; RPP

SAFR-EPS K; SAFR-T-12K; SFR 100; Stanclere® T-4356; Stereon® 721A; Stereon® 730A; Stereon® 840A; Stereon® 841A; Sunyl® 80; Sunyl® 80 ES; Sunyl® 90; Supercoat®

Taktene 1203; Taktene 1203G1; Telalloy® A-10; Telalloy® A-15

Ultra-Pflex®; Uniplex 413

Vertal 97; Vertal 200; Vinnol® K 704; Vinuran®

Winnofil S, SP

Chemicals:

Acrylic resin; Acrylonitrile-butadiene-styrene; Allyl methacrylate

Bis (2,4-di-t-butylphenyl) pentaerythritol diphosphite; Butadiene/acrylonitrile copolymer

Cumylphenyl acetate; Cumylphenyl benzoate; Cycloneopentyl, cyclo (dimethlaminoethyl) pyrophosphato zirconate, di mesyl salt

EPDM/SAN graft polymer; Ethylene-methyl acrylate-acrylic acid terpolymer; Ethylene-propylene-styrene terpolymer; Ethylene/VA copolymer

Methacrylate/butadiene styrene; Methoxyethyl acrylate

Polybutadiene; PVC/acrylate

Triolein

Vinyl acetate/ethylene copolymer

Inhibitors

Trade names:

Aztec® DP

Cyanox® 2246

Eastman® DTBHQ; Eastman® HQMME; Eastman® MTBHQ

Kanevinyl® XEL A; Kanevinyl® XEL B; Kanevinyl® XEL C; Kanevinyl® XEL D; Kanevinyl® XEL E; Kanevinyl® XEL F

Naugard® I-2; Naugard® I-3; Naugard® I-4; Naugard® J

Pennstop® 1866; Pennstop® 2049; Pennstop® 2697; PTZ® Chemical Grade; PTZ® Industrial Grade

Q-1300; Q-1301

Retarder SAX

Santoflex® 134; Santogard® PVI

Trigonox® TAHP-W85; Trigonox® TMPH

Vanox® 102; Vanox® 200; Vanox® 1005; Vanox® 1320; Vanox® 2246; Vanox® AT; Vanox® ZS; Vantard® PVI; Vanwax® H; Vanwax® OZ

Chemicals:

N,N´-Bis (1,4-dimethylpentyl)-p-phenylenediamine; t-Butyl hydroquinone; o-s-Butylphenol

N-(Cyclohexylthio)phthalimide

Diamylhydroquinone; Di-t-butylhydroquinone; Diethylamine; Diethylhydroxylamine; N-(1,4-

Inhibitors *(cont'd.)*

Dimethylpentyl)-N′-phenyl-p-phenylenediamine; N,N′-Di-β-naphthyl-p-phenylenediamine; N,N′-Diphenyl-p-phenylenediamine
Hydroquinone monomethyl ether
Nitrosophenylhydroxylamine aluminum salt; Nitrosophenylhydroxylamine ammonium salt
Phenothiazine; N-Phenyl,N′-isopropyl-p-phenylenediamine; Phenyl-β-naphthylamine
Toluhydroquinone; 2,2,4-Trimethyl-1,2-dihydroquinoline polymer

Initiators

Trade names: Alperox-F; Aztec® ACSP-28-FT; Aztec® ACSP-55-W1; Aztec® BCHPC; Aztec® BCHPC-40-SAQ; Aztec® BCHPC-75-W; Aztec® BCUP; Aztec® 1,1-BIS-50-AL; Aztec® 1,1-BIS-50-WO; Aztec® 1,1-BIS-80-BBP; Aztec® BP-05-FT; Aztec® BP-50-FT; Aztec® BP-70-W; Aztec® BP-77-W; Aztec® BU-50-AL; Aztec® BU-50-WO; Aztec® CEPC; Aztec® CEPC-40-SAQ; Aztec® CHP-80; Aztec® CHP-90-W1; Aztec® CHPC; Aztec® CHPC-90-W; Aztec® CUPND-75-AL; Aztec® DHPBZ-75-W; Aztec® DHPEH; Aztec® DIPND-50-AL; Aztec® DIPP-2; Aztec® DTAP; Aztec® DTBP; Aztec® EHPC-40-EAQ; Aztec® EHPC-40-ENF; Aztec® EHPC-65-AL; Aztec® EHPC-75-AL; Aztec® HMCN-30-AL; Aztec® HMCN-30-WO-2; Aztec® INP-37-AL; Aztec® INP-75-AL; Aztec® LP; Aztec® LP-25-SAQ; Aztec® LP-40-SAQ; Aztec® MEKP-HA-1; Aztec® MEKP-HA-2; Aztec® MYPC; Aztec® MYPC-40-SAQ; Aztec® SUCP-70-W; Aztec® TAHP-80-AL; Aztec® TAPB-90-OMS; Aztec® TAPEH; Aztec® TAPEH-75-OMS; Aztec® TAPEH-90-OMS; Aztec® TAPND-75-AL; Aztec® TAPPI-75-OMS; Aztec® TBHP-70; Aztec® TBPA-50-OMS; Aztec® TBPA-75-OMS; Aztec® TBPB; Aztec® TBPEH; Aztec® TBPEH-30-AL; Aztec® TBPEH-50-DOP; Aztec® TBPEH-50-OMS; Aztec® TBPIB-75-AL; Aztec® TBPIN; Aztec® TBPIN-30-AL; Aztec® TBPM-50-FT; Aztec® TBPM-50-P; Aztec® TBPND; Aztec® TBPND-40-EAQ; Aztec® TBPND-75-OMS; Aztec® TBPPI-25-AL; Aztec® TBPPI-50-OMS; Aztec® TBPPI-75-OMS; Aztec® 3,3,5-TRI; Aztec® 3,3,5-TRI-50-AL; Aztec® 3,3,5-TRI-50-FT; Aztec® 3,3,5-TRI-75-DBP
Benox® L-40LV
Cadet® BPO-70; Cadet® BPO-70W; Cadet® BPO-75W; Cadet® BPO-78; Cadet® BPO-78W; Cadox® 40E; Cadox® BCP; Cadox® BEP-50; Cadox® BFF-50; Cadox® BFF-50L; Cadox® BFF-60W; Cadox® BP-55; Cadox® BPO-W40; Cadox® BT-50; Cadox® BTA; Cadox® BTW-50; Cadox® BTW-55; Cadox® F-85; Cadox® HBO-50; Cadox® L-30; Cadox® L-50; Cadox® M-30; Cadox® M-50; Cadox® M-105; Cadox® MDA-30; Cadox® TDP; CHP-5; CHP-158; Cyclonox® BT-50
Decanox-F
Espercarb® 438M-60; Espercarb® 840; Espercarb® 840M; Espercarb® 840M-40; Espercarb® 840M-70; Esperfoam® FR; Esperox® 10; Esperox® 12MD; Esperox® 13M; Esperox® 28; Esperox® 28MD; Esperox® 28PD; Esperox® 31M; Esperox® 33M; Esperox® 41-25; Esperox® 41-25A; Esperox® 497M; Esperox® 545M; Esperox® 551M; Esperox® 570; Esperox® 570P; Esperox® 740M; Esperox® 747M; Esperox® 750M; Esperox® 939M; Esperox® 5100; Esperox® C-496
Ficel® AZDN-LF; Ficel® AZDN-LMC
Hi-Point® 90; Hi-Point® PD-1
Initiator BK; IPP
Laurox®; Laurox® W-25; Laurox® W-40; Laurox® W-40-GD1; Lucidol 70; Lucidol 75; Lucidol 75FP; Lucidol 78; Lucidol 98; Luperco 101-XL; Luperco 130-XL; Luperco 230-XL; Luperco 231-XL; Luperco 233-XL; Luperco 331-XL; Luperco 500-40C; Luperco 500-40KE; Luperco 801-XL; Luperco AA; Luperco ACP; Luperco AFR-250; Luperco AFR-400; Luperco AFR-500; Luperco ANS; Luperco AST; Luperco ATC; Luperox 2,5-2,5; Luperox 118; Luperox 500R; Luperox 500T; Lupersol 10; Lupersol 10-M75; Lupersol 11; Lupersol 70; Lupersol 75-M; Lupersol 76-M; Lupersol 80; Lupersol 101; Lupersol 130; Lupersol 188-M75; Lupersol 220-D50; Lupersol 221; Lupersol 223; Lupersol 223-M40; Lupersol 223-M75; Lupersol 225; Lupersol 225-M60; Lupersol 225-M75; Lupersol 230; Lupersol 231; Lupersol 231-P75; Lupersol 256; Lupersol 288-M75; Lupersol 331-80B; Lupersol 531-80B; Lupersol 533-M75; Lupersol 546-M75; Lupersol 553-M75; Lupersol 554-M50, 554-M75; Lupersol 555-M60; Lupersol 575; Lupersol 575-M75; Lupersol 575-P75; Lupersol 610-M50; Lupersol 801; Lupersol DDM-9; Lupersol Delta-3; Lupersol Delta-X-9; Lupersol DFR; Lupersol DHD-9; Lupersol DSW-9; Lupersol KDB; Lupersol P-31; Lupersol P-33; Lupersol PDO; Lupersol PMS; Lupersol TBEC; Lupersol TBIC-M75

Initiators *(cont'd.)*

Norox® MCP

Perkadox® 14; Perkadox® 14/40; Perkadox® 14-40B-pd; Perkadox® 14S; Perkadox® 14S-20PP-pd; Perkadox® 14S-fl; Perkadox® 16; Perkadox® 16/35; Perkadox® 16N; Perkadox® 16S; Perkadox® 16/W40; Perkadox® 16-W40-GB2; Perkadox® 16-W40-GB5; Perkadox® 16/W70; Perkadox® 18; Perkadox® 24; Perkadox® 24-fl; Perkadox® 24-W40; Perkadox® 26-fl; Perkadox® 26-W40; Perkadox® 30; Perkadox® 58; Perkadox® 64-10PP-pd; Perkadox® 64-20PP-pd; Perkadox® 64-40PP-pd; Perkadox® AIBN; Perkadox® AMBN; Perkadox® BC; Perkadox® BC-FF; Perkadox® BPSC; Perkadox® IPP-AT50; Perkadox® SE-8; Perkadox® SE-10; Poly-Zole® AZDN

Quickset® Extra; Quickset® Super

Superox® 46-709; Superox® 46-747; Superox® 46-753-00; Superox® 744

TAHP-80; TBHP-70; T-Hydro®; Trigonal® 14; Trigonal® 121; Trigonox® 17; Trigonox® 21; Trigonox® 21-C50; Trigonox® 21 LS; Trigonox® 21-OP50; Trigonox® 22-BB80; Trigonox® 22-C50; Trigonox® 22-E50; Trigonox® 23; Trigonox® 23-C75; Trigonox® 23-W40; Trigonox® 25-C75; Trigonox® 27; Trigonox® 29; Trigonox® 29-B75; Trigonox® 29-C75; Trigonox® 36-C60; Trigonox® 36-C75; Trigonox® 36-W40; Trigonox® 40; Trigonox® 41-C50; Trigonox® 41-C75; Trigonox® 42; Trigonox® 42 PR; Trigonox® 42S; Trigonox® 44P; Trigonox® 61; Trigonox® 63; Trigonox® 67; Trigonox® 93; Trigonox® 97; Trigonox® 97-C75; Trigonox® 99-B75; Trigonox® 99-C75; Trigonox® 99-W40; Trigonox® 101; Trigonox® 101-7.5PP-pd; Trigonox® 101/45Bpd; Trigonox® 101-50PP-pd; Trigonox® 101-E50; Trigonox® 111-B40; Trigonox® 117; Trigonox® 121; Trigonox® 121-BB75; Trigonox® 123-C75; Trigonox® 125-C50; Trigonox® 125-C75; Trigonox® 127; Trigonox® 131; Trigonox® 133-C60; Trigonox® 141; Trigonox® 145; Trigonox® 145-45B; Trigonox® 145-E85; Trigonox® 151-C50; Trigonox® 151-C70; Trigonox® 161; Trigonox® 169-NA50; Trigonox® 169-OP50; Trigonox® 171; Trigonox® 187-C30; Trigonox® 201; Trigonox® 239A; Trigonox® A-80; Trigonox® A-W70; Trigonox® ACS-M28; Trigonox® ADC; Trigonox® ADC-NS60; Trigonox® B; Trigonox® BPIC; Trigonox® C; Trigonox® C-C75; Trigonox® D-C50; Trigonox® D-E50; Trigonox® EHP; Trigonox® EHP-AT70; Trigonox® EHP-B75; Trigonox® EHP-C40; Trigonox® EHP-C70; Trigonox® EHP-C75; Trigonox® EHPS; Trigonox® EHPS-C75; Trigonox® EHP-W40; Trigonox® EHP-W40S; Trigonox® EHP-W50; Trigonox® F-C50; Trigonox® HM; Trigonox® K-80; Trigonox® K-95; Trigonox® KSM; Trigonox® M-50; Trigonox® NBP-C50; Trigonox® SBP; Trigonox® SBP-AX30; Trigonox® SBP-C30; Trigonox® SBP-C50; Trigonox® SBP-C60; Trigonox® SBP-C75; Trigonox® SBPS

USP®-90MD; USP®-355M; USP®-400P; USP®-690; USP®-800

V-30; V-40; V-50; V-59; V-60; V-65; V-70; V-501; V-601; VA-044; VA-061; VA-082; VA-086; Varox® ANS; Varox® TBPB; Vazo 52; Vicure® 55

Chemicals: Acetylacetone peroxide; Acetyl cyclohexanesulfonyl peroxide; t-Amyl hydroperoxide; t-Amyl peroxyacetate; t-Amyl peroxybenzoate; t-Amylperoxy 2-ethylhexanoate; t-Amylperoxy 2-ethylhexyl carbonate; t-Amylperoxyneodecanoate; t-Amylperoxyneoheptanoate; t-Amylperoxypivalate; 2,2-Azobis(2-amidinopropane)dihydrochloride; 2,2′-Azobis[N-(4-chlorophenyl)-2-methylpropionamidine]dihydrochloride;; 4,4′-Azobis (4-cyanopentanoic acid); 1,1′-Azobis(cyclohexane-1-carbonitrile); 2,2′-Azobis (N,N′-dimethyleneiso-butyramidine); 2,2′-Azobis (2,4-dimethylvaleronitrile); 2,2′-Azobis[2-[1-(2-hydroxyethyl)-2-imidazolin-2-yl] propane] dihydrochloride;; 2,2′-Azobis[2-(hydroxymethyl) propionitrile]; 2,2′-Azobis[2-(2-imidazolin-2-yl) propane] dihydrochloride; 2,2′-Azobisisobutyronitrile; 2,2′-Azobis(4-methoxy-2,4-dimethylvaleronitrile); 2,2′-Azobis{2-methyl-N-[1,1-bis(hydroxy-methyl)ethyl] propionamide}; 2,2′-Azobis[2-methyl-N-[1,1-bis(hydroxymethyl)-2-hydroxy-ethyl] propionamide]; 2,2′-Azobis(2-methylbutyronitrile); 2,2′-Azobis[2-methyl-N-(2-hy-droxyethyl)-propionamide];; 2,2′-Azobis(2-methyl-N-phenylpropionamidine) dihydro-chloride; 2,2′-Azobis(2-methylpropionamide) dihydrate; 2,2′-Azobis[2-(3,4,5,6-tetrahydropyrimidin-2-yl) propane] dihydrochloride;; 2,2′-Azobis(2,4,4-trimethylpentane)

Benzoin ether; Benzoyl peroxide; α,α′-Bis (t-butylperoxy) diisopropylbenzene; Bis (2-ethyl-hexyl) peroxydicarbonate; Bis (3,5,5-trimethylhexanoyl) peroxide; n-Butyl-4,4-bis (t-butylper-oxy) valerate; t-Butyl cumyl peroxide; n-Butyl 4,4-di(t-butylperoxy)valerate; OO-t-Butyl O-(2-ethylhexyl) monoperoxycarbonate; t-Butyl 3-isopropenylcumyl peroxide; OO-t-Butyl O-iso-propyl monoperoxycarbonate; t-Butyl peracetate; t-Butyl perbenzoate; t-Butyl peroctoate; t-Butyl peroxycrotonate; t-Butyl peroxydiethylacetate; t-Butyl peroxy 2-ethylhexyl carbonate; t-Butylperoxy heptanoate; t-Butyl peroxyisobutyrate; t-Butyl peroxy isopropyl carbonate; t-Butyl peroxymaleic acid; t-Butyl peroxy 2-methyl benzoate; t-Butyl peroxyneodecanoate; t-Butyl peroxypivalate; t-Butyl peroxy stearyl carbonate; t-Butyl peroxy-3,5,5-

Initiators *(cont'd.)*

trimethylhexanoate

Cumyl peroxyneodecanoate; 1-[(1-Cyano-1-methylethyl)azo]formamide; Cyclohexanone peroxide

Decanoyl chloride; Decanoyl peroxide; Di-t-amyl peroxide; 1,1-Di (t-amylperoxy) cyclohexane; 2,2-Di-(t-amylperoxy) propane; Di-(4-t-butylcyclohexyl) peroxydicarbonate; 2,2-Di (t-butylperoxy) butane; Di-n-butyl peroxydicarbonate; Di-(s-butyl)peroxydicarbonate; 1,4-Di-(2-t-butylperoxyisopropyl) benzene; Di-(2-t-butylperoxyisopropyl)benzene; Di-t-butyl peroxyphthalate; 2,4-Di-t-butylphenyl 3,5-di-t-butyl-4-hydroxybenzoate; Dicetyl peroxydicarbonate; Dicyclohexyl peroxydicarbonate; Didecanoyl peroxide; 2,5-Dihydroperoxy-2,5-dimethylhexane; 1,4-Di (2-hydroperoxy isopropyl) benzene; Diisobutyryl peroxide; Diisopropylbenzene monohydroperoxide; Diisopropyl perdicarbonate; Dimethyl 2,2′-azobis (2-methylpropionate); 2,5-Dimethyl-2,5-bis (benzoylperoxy) hexane; 2,5-Dimethyl-2,5-bis (2-ethylhexanoylperoxy) hexane; 2,5-Dimethyl-2,5-di(benzoylperoxy)hexane; 2,3-Dimethyl-2,3-diphenylbutane; 3,4-Dimethyl-3,4-diphenylhexane; 2,5-Dimethyl-2,5-(2-ethylhexanoylperoxy)hexane; Dimyristyl peroxydicarbonate; Dioctanoyl peroxide; Di-n-propyl peroxydicarbonate; Disuccinic acid peroxide; n-Dodecyl mercaptan

Ethyl 3,3-di (t-amylperoxy) butyrate; Ethyldibutylperoxybutyrate

3,3,6,6,9,9-Hexamethyl 1,2,4,5-tetraoxa cyclononane; 3-Hydroxy-1,1-dimethylbutyl peroxyneodecanoate

Lauroyl chloride; Lauroyl peroxide

Methyl ethyl ketone peroxide; Methyl isobutyl ketone peroxide

2-Phenylazo-4-methoxy-2,4-dimethylvaleronitrile

N,N,N′,N′-Tetramethylethylenediamine; 3,5,5-Trimethyl hexanoyl peroxide; 2,4,4-Trimethylpentyl-2-hydroperoxide

Intermediates

Trade names: Adi-pure®; Adol® 63; Albrite® PA-75; Alkasurf® NP-6; Allco BTDA; Allco PMDA; Amicure® SA-102; Arcosolv® PTB

Carbowax® PEG 200; Cardolite® NC-513; CB2408; CI7840; CP0320; CT2902

Dantocol® DHE; Dantoest® DHE DL; Dantosperse® DHE (15) MO; DBE; DBE-2, -2SPG; DBE-3; DBE-4; DBE-5; DBE-6; DBE-9; Dinoram C; Dioxitol-High Gravity; Dioxitol-Low Gravity; DIPA Commercial Grade; DIPA Low Freeze Grade 85; DIPA Low Freeze Grade 90; DIPA NF Grade; DMPA®; Duomeen® C; Dynasylan® 0116; Dynasylan® 1110; Dynasylan® 1411; Dynasylan® 1505; Dynasylan® 3403; Dynasylan® AMEO; Dynasylan® AMMO; Dynasylan® AMTC; Dynasylan® BDAC; Dynasylan® BSA; Dynasylan® CPTEO; Dynasylan® CPTMO; Dynasylan® DAMO; Dynasylan® ETAC; Dynasylan® GLYMO; Dynasylan® HMDS; Dynasylan® IBTMO; Dynasylan® IMEO; Dynasylan® MEMO; Dynasylan® MTAC; Dynasylan® MTES; Dynasylan® MTMO; Dynasylan® MTMS; Dynasylan® OCTEO; Dynasylan® PETCS; Dynasylan® PTMO; Dynasylan® TCS; Dynasylan® TRIAMO; Dynasylan® VTC; Dynasylan® VTEO; Dynasylan® VTMO; Dynasylan® VTMOEO

Eastman® HQEE; Eastman® MTBHQ; Eastman® TEP; Eltesol® TSX; Eltesol® TSX/A; Eltesol® TSX/SF; Eltesol® XA65; Eltesol® XA/M65; Emery® 5710

Fancol OA-95; FireShield® HPM; Flexricin® 17

Great Lakes PH-73™; Great Lakes PHT4-Diol™

Hystrene® 5016 NF; Hystrene® 9718 NF

Industrene® 143; Industrene® 205; Industrene® 206; Industrene® 225; Industrene® 226; Industrene® 325; Industrene® 328; Industrene® 365; Industrene® 4518; Industrene® 5016; Industrene® 7018; Industrene® 9018; Industrene® B; Industrene® D; Industrene® R; Isonate® 125M

Jeffox PPG-400; Jeffox PPG-2000

Kemamine® DD-3680; Kemamine® DP-3680; Kemamine® DP-3695; Ken Kem® CP-45; Ken Kem® CP-99; Ken-React® 39DS (KR 39DS)

Millamine® 5260; Milldride® 5060; Milldride® DDSA; Milldride® HDSA; Milldride® HHPA; Milldride® nDDSA; Milldride® nDSA; Milldride® ODSA; Milldride® OSA; Milldride® TDSA; Mod Acid; Monolan® PPG440, PPG1100, PPG2200; Morflex® 1129; MPDiol® Glycol

Nafol® 10 D; Nafol® 20+; Nafol® 810 D; Nafol® 1014; Nafol® 1218; Nansa® SSA; Nevpene® 9500

Parabis; Pennad 150; Pennstop® 1866; Pennstop® 2049; Pennstop® 2697; Petrarch® A0699;

Intermediates *(cont'd.)*

Petrarch® A0700; Petrarch® A0742; Petrarch® A0743; Petrarch® A0750; Petrarch® A0800; Petrarch® B2500; Petrarch® C3292; Petrarch® C3300; Petrarch® G6720; Petrarch® H7300; Petrarch® I7810; Petrarch® M8550; Petrarch® M8450; Petrarch® M8500; Petrarch® M9050; Petrarch® M9100; Petrarch® O9835; Petrarch® T1807; Petrarch® T1918; Petrarch® T1980; Petrarch® T2090; Petrarch® T2503; Petrarch® T2910; Petrarch® V4900; Petrarch® V4910; Petrarch® V4917; Petrarch® V5000; Pluracol® E600 NF; Pluracol® E1000; Pluracol® E1450; Pluracol® E1450 NF; Pluracol® E2000; Pluracol® E4500; Pluracol® E8000; Pluracol® P-3010; Pluracol® P-4010; Pluracol® WD90K; Priplast 3183; Priplast 3184; Priplast 3185; Pripol 1004; Pripol 1009; Pripol 1013; Pripol 1017, 1022; Pripol 1022; Pripol 1025; Pripol 1040; Pripol 2033

QO® Furan; QO® Furfural; Quadrol®

Rezol® 4393

Saytex® RB-100; Saytex® VBR; Staflex DINM; Staflex DTDM; Sylvaros® 315; Sylvaros® RB; Sylvatac® AC; Sylvatac® ACF; Syn Fac® 8026; Synprolam 35DM

T-600; Teric PEG 300; Teric PEG 400; Teric PEG 600; Teric PEG 4000; Teric PEG 8000; TETA; Texlin® 300; TIPA 99; TIPA 101; TIPA Low Freeze Grade

V-Pyrol; Vynate® 2-EH; Vynate® L-3; Vynate® Neo-5; Vynate® Neo-9; Vynate® Neo-10

Witconate™ Acide TPB

Chemicals: 2-(Acetoacetoxy) ethyl methacrylate; Allyl methacrylate; N-(2-Aminoethyl)-3-aminopropyl methyldimethoxy silane; Aminoethylethanolamine; N-(3-Aminopropyl)diethanolamine; 3-Aminopropylmethyldiethoxy silane; p-t-Amylphenol; Amyltrichlorosilane

3,3´,4,4´-Benzophenone tetracarboxylic dianhydride; N-Benzyldimethylamine; Bis(aminopropyl)piperazine; Bis (hydroxyethyl) aminopropyltriethoxy silane; Bis-(N-methylbenzamide) ethoxymethyl silane; Bisphenol A; Bisphenol A, hydrogenated; 1,3-Butadiene; 1,4-Butanediol; 1-Butene; Butoxydiglycol; Butyl acrylate; 2-t-Butylphenol; 4-t-Butylphenol; o-s-Butylphenol

C20-24 alcohols; 3-Chloropropyltriethoxysilane; Coconut acid; Cocopropylenediamine; C20-24 alpha olefin; C24-28 alpha olefin; C30 alpha olefin; Cumyl phenol

Decanoyl chloride; Decene-1; n-Decenyl succinic anhydride; DEDM hydantoin; DEDM hydantoin dilaurate; Diallyl phthalate; Di-t-butoxydiacetoxysilane; Dibutyl maleate; 2,6-Di-t-butylphenol; Diethanolamine; Diethyl toluene diamine; 1,1-Difluoroethane; Diisobutylene; Dimer diamine; Dimethyl adipate; Dimethylaminomethyl phenol; Dimethyldichlorosilane; Dimethyl glutarate; Dimethylolpropionic acid; N,N´-Diphenyl-p-phenylenediamine; Ditridecyl maleate; Dodecene-1; Dodecenyl succinic anhydride

Ethanolamine; Ethylene carbonate; Ethylene thiourea

Furan polymer; Furfural

(3-Glycidoxypropyl)-methyldiethoxy silane

Hexadecenyl succinic anhydride; Hexamethyldisilazane; Hexamethyleneimine; 1-Hexene; Hydrogenated stearic acid; Hydroquinone bis(2-hydroxyethyl) ether; Hydroxystearic acid

Isobutyltrimethoxysilane; Isocyanatopropyltriethoxysilane; Isooctyl thioglycolate

3-Mercaptopropionic acid; 3-Mercaptopropylmethyldimethoxysilane; N-Methylaminopropyltrimethoxysilane; 2-Methyl-1,3-propanediol; Methyl stearate; Methyltriacetoxysilane; Methyltriethoxysilane; Methyltrimethoxysilane; Morpholine

2-Nitro-2-methyl-1-propanol; Nonyl phenyl glycidyl ether

Octadecene-1; Octadecenyl succinic anhydride; Octene-1; Octenyl succinic anhydride; Octyltriethoxysilane; Oleamine; Oleyl 1,3-propylene diamine

PEG-2 cocamine; PEG-15 DEDM hydantoin oleate; PEG-20 glyceryl stearate; PEG-6 oleate; PEG-8 oleate; Pentaerythrityl ricinoleate; 2-Phenethyltrichlorosilane; Phenyltriethoxysilane; Polyethylene glycol; Polypropylene glycol; PPG-20; PPG-26; PPG-30; PPG-28-buteth-35; n-Propyltrimethoxysilane; Pyridine; Pyromellitic acid dianhydride; Pyromellitic dianhydride

Soy acid; 3-(N-Styrylmethyl-2-aminoethylamino) propyltrimethoxy silane hydrochloride

Tallow acid; Tetrabromobisphenol A; Tetrabromophthalic anhydride; Tetrabutylphosphonium acetate, monoacetic acid; Tetrabutylphosphonium bromide; Tetrabutylphosphonium chloride; Tetrachlorophthalic anhydride; Tetradecene-1; Tetradecenyl succinic anhydride; Tetraethoxysilane; Tetrahydrofuran; Tetrahydronaphthalene; Tetrahydroxypropyl ethylenediamine; Tetrakis (2-ethoxyethoxy) silane; Tetrakis (hydroxymethyl) phosphonium chloride; Tetramethoxysilane; Tetraoctylphosphonium bromide; Tetra-n-propoxysilane; p-Tolyl diethanolamine; Tribromoneopentyl alcohol; 2,4,6-Tribromophenol; Tributyltin oxide; Trichlorosilane; Tridecyl phosphite; Triethanolamine; N-[3-(Triethoxysilyl)-propyl] 4,5-dihydroimidazole; Triethylene glycol; 1-Trimethoxysilyl-2-(chloromethyl) phenylethane;

Intermediates *(cont'd.)*

Trimethoxysilylpropyldiethylene triamine; 2,2,4-Trimethyl-1,3-pentanediol monoisobutyrate; Triphenyl phosphite; Tripropylene glycol

Vinyl 2-ethylhexanoate; Vinyl neodecanoate; Vinyl neononanoate; Vinyl pivalate; Vinyl propionate; Vinyltrichlorosilane; Vinyltriethoxysilane

Xylene sulfonic acid

Lubricants

Trade names: AA Standard; Ablumol BS; Abluwax EBS; A-C® 7, 7A; A-C® 8, 8A; A-C® 9, 9A, 9F; A-C® 540, 540A; A-C® 629; A-C® 629A; Acconon 300-MO; Acrawax® C; Acrawax® C SG; Activator 1102; Adeka Carpol MH-50, MH-150, MH-1000; ADK STAB AX-38; ADK STAB CA-ST; ADK STAB EXL-5; ADK STAB FC-112; ADK STAB FC-113; ADK STAB LS-2; ADK STAB LS-3; ADK STAB LS-5; ADK STAB LS-8; ADK STAB LS-9; ADK STAB LS-10; ADK STAB MG-ST; ADK STAB ZN-ST; Advapak® LS-203HF; Advapak® LS-203HP; Advapak® ML-1325; Advapak® ML-2516; Advapak® SLS-1000; Advastab® TM-183-O; Advastab® TM-790 Series; Advawax® 165; Advawax® 240; Advawax® 280; Advawax® 290; Aerosol® OT-100%; Afco-Chem B; Afco-Chem CS; Afco-Chem CS-1; Afco-Chem CS-S; Afco-Chem EX-95; Afco-Chem LIS; Afco-Chem MGS; Afco-Chem ZNS; Aflux® 32; Aflux® R; Aflux® S; Akrochem® PEG 3350; Akrodip™ V-301; Akrolease® E-9491; Albrite® TBP; Alfol® 20+; Alfol® 1012 HA; Alfol® 1014 CDC; Alfol® 1214; Alfol® 1214 GC; Alfol® 1216; Alfol® 1216 CO; Alfol® 1218 DCBA; Alfol® 1412; Alfol® 1416 GC; Alfol® 1418 DDB; Alfol® 1418 GBA; Alfol® 1618; Alfol® 1618 CG; Alfol® 1618 GC; Algoflon® L; Alkamide® STEDA; Alkamuls® EL-620L; Alkamuls® GMR-55LG; Alkamuls® SML; Armid® 18; Armid® HT; Armid® O; Armoslip® 18; Armoslip® EXP; Atmer® 104; Atmer® 106; Atmer® 122; Atmer® 125; Atmer® 129; Atmer® 1007; Atmer® 1012; Atmer® 1021; Atmer® 1025; Atmer® 1026; Atmer® 1040; Atmer® 1041; Atmer® 1042

Be Square® 185; Britol® 6NF; Buspense 047

Cardis® 10; Cardis® 36; Cardis® 314; Cardis® 319; Cardis® 320; Cardis® 370; Castorwax® MP-70; Castorwax® MP-80; CB2408; Cecavon ZN 70; Cecavon ZN 71; Cecavon ZN 72; Cecavon ZN 73; Cecavon ZN 735; CeriDust 961OF; Chemax HCO-5; Chemax HCO-16; Chemax HCO-25; Chemax HCO-200/50; Chemstat® 106G/60DC; Chemstat® 327; Chemstat® G-118/52; Chemstat® G-118/95; CI7840; Coad 10; Coad 20; Coad 21; Coathylene HA 2454; Comboloob 0609; Comboloob 0827; Comboloob 0827/1; Cometals Aluminum Stearate; Cometals Barium Stearate; Cometals Cadmium Stearate; Cometals Calcium Stearate; Cometals Magnesium Stearate; Cometals Sodium Stearate; Cometals Zinc Stearate; CP0320; Crodamide 203; Crodamide 212; CS1590; CT-88 Aerosol; CT2902; C-Wax-140

D-400; D-1000; D-1200; D-1300; D-2000; D-3000; D-4000; Dantoest® DHE DL; Dantosperse® DHE (15) MO; Dark Green No. 2™; Darvan® L; Darvan® WAQ; DB-9 Paintable Silicone Emulsion; Dimodan LS; Dimodan LS Kosher; Dimodan PM; Dimodan PV; Dimodan PV 300; Dimul S; Dow Corning® 7 Compound; Dow Corning® 24 Emulsion; Dow Corning® 200 Fluid; Dow Corning® 203 Fluid; Drakeol 9; Drakeol 10; Drakeol 19; Drakeol 21; Drakeol 34; Drakeol 35; Drewplast® 017; Drewplast® 051; DS-207; Dur-Em® 117; Dynasylan® 0116; Dynasylan® 1110; Dynasylan® 1411; Dynasylan® 1505; Dynasylan® 3403; Dynasylan® AMEO; Dynasylan® AMMO; Dynasylan® AMTC; Dynasylan® BDAC; Dynasylan® BSA; Dynasylan® CPTEO; Dynasylan® CPTMO; Dynasylan® DAMO; Dynasylan® ETAC; Dynasylan® GLYMO; Dynasylan® HMDS; Dynasylan® IBTMO; Dynasylan® IMEO; Dynasylan® MEMO; Dynasylan® MTAC; Dynasylan® MTES; Dynasylan® MTMO; Dynasylan® MTMS; Dynasylan® OCTEO; Dynasylan® PETCS; Dynasylan® PTMO; Dynasylan® TCS; Dynasylan® TRIAMO; Dynasylan® VTC; Dynasylan® VTEO; Dynasylan® VTMO; Dynasylan® VTMOEO

Ease Release™ 500 Series; Ease Release™ 700; Emcol® CC-42; Emcol® CC-57; Emery® 6724; Empicol® LX; Empicol® LX28; Empicol® LXV; Empicol® LXV/D; Empicol® LZ; Empicol® LZ/D; Empilan® GMS LSE40; Empilan® GMS LSE80; Empilan® GMS MSE40; Empilan® GMS NSE40; Empilan® GMS NSE90; Empilan® GMS SE70; Emulbon LB-78; Emuldan DG 60; Emuldan DO 60; Emuldan HV 40; Emuldan HV 52; Epolene® C-10P; Epolene® C-13P; Epolene® C-17P; Epolene® E-14; Epolene® E-14P; Epolene® E-20P; Epolene® N-11; Epolene® N-11P; Epolene® N-14; Epolene® N-14P; Epolene® N-34; Estol 1476; Estol 1481; Eureslip 58; EXP-24-LS Silicone Wax Emulsion; EXP-58 Silicone Wax; EXP-61 Amine Functional Silicone Wax; EXP-77 Mercapto Functional Silicone Wax

Lubricants *(cont'd.)*

Fancol OA-95; #15 Oil; Flexricin® 115; Flexricin® 185; Flexricin® P-1; Flexricin® P-3; Flexricin® P-4; Fluon® L169

G623; G661; Getren® 4/200; Getren® BPL 83; Glycolube® 100; Glycolube® 110; Glycolube® 110D; Glycolube® 140; Glycolube® 140 Kosher; Glycolube® 180; Glycolube® 315; Glycolube® 345; Glycolube® 674; Glycolube® 740; Glycolube® 742; Glycolube® 825; Glycolube® 825 Kosher; Glycolube® 853; Glycolube® CW-1; Glycolube® P; Glycolube® PG; Glycolube® PS; Glycolube® SG-1; Glycolube® TS; Glycolube® VL; Glycon® S-70; Glycon® S-90; Glycon® TP; Glycowax® S 932; Glyso-Lube™; Glyso-Lube™ 3; GP-5 Emulsifiable Paintable Silicone; GP-6 Silicone Fluid; GP-60-E Silicone Emulsion; GP-71-SS Mercapto Modified Silicone Fluid; GP-74 Paintable Silicone Fluid; GP-98 Silicone Wax; GP-214 Silicone Polyol Copolymer; GP-530 Silicone Fluid; GP-7100 Silicone Fluid; GP-7100-E Paintable Silicone Emulsion; GP-7101 Silicone Wax; GP-7102 Silicone Wax; GP-7105 Silicone Fluid; GP-RA-158 Silicone Polish Additive; GP-RA-159 Silicone Polish Additive; Grindtek MOP 90; Grindtek MSP 40; Grindtek MSP 40F; Grindtek MSP 90; Grindtek PK 60; GS-3; Gulftene® 4; Gulftene® 6; Gulftene® 8; Gulftene® 10; Gulftene® 12; Gulftene® 14; Gulftene® 18; Gulftene® 20-24; Gulftene® 24-28; Gulftene® 30+

Hal-Lub-D; Hal-Lub-N; Hal-Lub-N TOTM; Halocarbon Grease 19; Halocarbon Grease 25-5S; Halocarbon Grease 25-10M; Halocarbon Grease 25-20M; Halocarbon Grease 28; Halocarbon Grease 28LT; Halocarbon Grease 32; Halocarbon Grease X90-10M; Halocarbon Oil 0.8; Halocarbon Oil 1.8; Halocarbon Oil 4.2; Halocarbon Oil 6.3; Halocarbon Oil 27; Halocarbon Oil 56; Halocarbon Oil 95; Halocarbon Oil 200; Halocarbon Oil 400; Halocarbon Oil 700; Halocarbon Oil 1000N; Halocarbon Wax 40; Halocarbon Wax 600; Halocarbon Wax 1200; Halocarbon Wax 1500; Halstab 50; Halstab 55; Halstab M-1; Halstab P-1; Halstab P-2; Haro® Mix CH-205; Haro® Mix CK-203; Haro® Mix CK-213; Haro® Mix CK-711; Haro® Mix FK-102; Haro® Mix IH-108; Haro® Mix LK-218; Haro® Mix LK-228; Haro® Mix MH-204; Haro® Mix MK-107; Haro® Mix MK-220; Haro® Mix MK-620; Haro® Mix MK-744; Haro® Mix SK-602; Haro® Mix UK-121; Haro® Mix VC-501; Haro® Mix YC-502; Haro® Mix YK-110; Haro® Mix YK-113; Haro® Mix YK-307; Haro® Mix YK-603; Haro® Mix ZC-311; Haro® Mix ZC-902; Haro® Mix ZT-504; Haro® Mix ZT-508; Haro® Mix ZT-514; Haro® Wax L01-56, L03-58, L04-73, L09-00, L18-78, L21-96, L23-58, L24-98; Haro® Wax L02-99, L05-51, L06-62, L07-47, L08-57, L10-58, L11-85, L12-43; Haro® Wax L13-89, L14-77, L15-75, L16-92, L17-98, L19-99, L22-99, L25-99; Haro® Wax L20-98; Haro® Wax L333; Haro® Wax L-344; Haro® Wax L433; Harwick F-300; Hodag PE-004; Hodag PE-104; Hodag PE-106; Hodag PE-109; Hodag PE-206; Hodag PE-209; Hodag PEG 540; Hodag PPG-150; Hodag PPG-400; Hodag PPG-1200; Hodag PPG-2000; Hodag PPG-4000; Hoechst Wax C; Hoechst Wax E; Hoechst Wax LP; Hoechst Wax OP; Hoechst Wax PE 130; Hoechst Wax PE 190; Hoechst Wax PE 520; Hoechst Wax PED 191; Hoechst Wax PED 521; Hoechst Wax PP 230; Hoechst Wax S; Hoechst Wax SW; Hoechst Wax UL; Hostaflon® TF 9202; Hostaflon® TF 9203; Hostalub® CAF 485; Hostalub® FA 1; Hostalub® FE 71; Hostalub® H 3; Hostalub® H 4; Hostalub® H 12; Hostalub® H 22; Hostalub® VP Ca W 2; Hostalub® WE4; Hostalub® WE40; Hostalub® XL 50; Hostalub® XL 165; Hostalub® XL 165FR; Hostalub® XL 165P; Hostalub® XL 355; Hostalub® XL 445; Hostalub® System G1970; Hostalub® System G1971; Hostalub® System G1972; Hy Dense Calcium Stearate RSN Powd; Hy Dense Zinc Stearate XM Powd; Hy Dense Zinc Stearate XM Ultra Fine; Hydrobrite 200PO; Hydrobrite 300PO; Hydrobrite 380PO; Hydrobrite 550PO; HyQual NF; Hystrene® 5016 NF; Hystrene® 9718 NF; HyTech AX/603; HyTech RSN 1-1; HyTech RSN 11-4; HyTech RSN 131 HS/Gran; HyTech RSN 248D; HyTech T/351

Ice # 2; Interlube A; Interlube P/DS; Internal Lubricant D-148™ Dry; Internal Lubricant D-148™ Wet; Interstab® BC-4195; Interstab® BZ-4828A; Interstab® C-16; Interstab® CA-18-1; Interstab® CZ-11; Interstab® G-140; Interstab® G-8257; Interstab® LT-4805R; Interstab® R-4150; Interstab® ZN-18-1

Jeffox PPG-400

Kantstik™ 325; Kantstik™ 504; Kantstik™ FX-7; Kantstik™ FX-9; Kantstik™ M-55; Kantstik™ PE; Kantstik™ Q Powd; Kantstik™ SPC; Kantstik™ S Powd; Kemamide® B; Kemamide® E; Kemamide® E-180; Kemamide® E-221; Kemamide® O; Kemamide® P-181; Kemamide® S; Kemamide® S-180; Kemamide® S-221; Kemamide® U; Kemamide® W-20; Kemamide® W-39; Kemamide® W-40; Kemamide® W-40/300; Kemamide® W-45; Kemamine® AS-650; Kemamine® AS-974; Kemamine® AS-974/1; Kemamine® AS-989; Kemamine® AS-990; Kemester® 143; Kemester® 5221SE; Kemester® EGDS; Kemester® GMS (Powd.); Ken-React® 7 (KR 7); Ken-React® 9S (KR 9S); Ken-React® 12 (KR 12); Ken-React® 26S (KR 26S); Ken-React® 33DS (KR 33DS); Ken-React® 38S (KR 38S); Ken-React® 39DS (KR

Lubricants *(cont'd.)*

39DS); Ken-React® 41B (KR 41B); Ken-React® 44 (KR 44); Ken-React® 46B (KR 46B); Ken-React® 55 (KR 55); Ken-React® 133DS (KR 133DS); Ken-React® 134S (KR 134S); Ken-React® 138D (KR 138D); Ken-React® 138S (KR 138S); Ken-React® 158D (KR 158D); Ken-React® 158FS (KR 158FS); Ken-React® 212 (KR 212); Ken-React® 238A (KR 238A); Ken-React® 238J (KR 238J); Ken-React® 238M (KR 238M); Ken-React® 238S (KR 238S); Ken-React® 238T (KR 238T); Ken-React® 262A (KR 262A); Ken-React® 262ES (KR 262ES); Ken-React® OPP2 (KR OPP2); Ken-React® OPPR (KR OPPR); Ken-React® KR TTS; KZ 55; KZ OPPR; KZ TPP; KZ TPPJ

Laurex® 4526; Laurex® CS; Leadstar; Leadstar Envirostab; Lexolube® B-109; Lexolube® BS-Tech; Lexolube® NBS; Lexolube® T-110; LICA 01; LICA 09; LICA 12; LICA 38; LICA 38A; LICA 38J; LICA 44; LICA 97; LICA 99; Liquazinc AQ-90; Lonza T 44; Loobwax 0597; Loobwax 0598; Loobwax 0605; Loobwax 0638; Loobwax 0651; Loobwax 0740; Loobwax 0750; Loobwax 0761; Loobwax 0782; Loxiol® G 10; Loxiol® G10P; Loxiol® G 11; Loxiol® G 12; Loxiol® G 13; Loxiol® G 15; Loxiol® G 16; Loxiol® G 20; Loxiol® G 21; Loxiol® G 22; Loxiol® G 23; Loxiol® G 32; Loxiol® G 33; Loxiol® G 33 Bead; Loxiol® G 40; Loxiol® G 41; Loxiol® G 47; Loxiol® G 53; Loxiol® G 60; Loxiol® G 70; Loxiol® G 70 S; Loxiol® G 71; Loxiol® G 71 S; Loxiol® G 72; Loxiol® HOB 7107; Loxiol® HOB 7111; Loxiol® HOB 7119; Loxiol® HOB 7121; Loxiol® HOB 7131; Loxiol® HOB 7138; Loxiol® HOB 7140; Loxiol® HOB 7162; Loxiol® P 1141; Loxiol® P 1304; Loxiol® VGE 1728; Loxiol® VGE 1837; Loxiol® VGE 1875; Loxiol® VGE 1884; Loxiol® VGS 1877; Loxiol® VGS 1878; Loxiol® VPG 1732; Loxiol® VPG 1781; Lube 105; Lube 106; Lubrazinc® W, Superfine; Luwax® A; Luwax® AF 30; Luwax® AH 3; Luwax® EAS 1; Luwax® OA 2; Luwax® OA 5; Lyndcoat™ 10-RTU; Lyndcoat™ 15-RTU; Lyndcoat™ 20-RTU; Lyndcoat™ BR-18/M; Lyndcoat™ BR 601-RTU; Lyndcoat™ BR 790-RTU; Lyndcoat™ BR 880-RTU

Mackester™ EGDS; Mackester™ EGMS; Mackester™ IP; Mackester™ SP; Macol® 660; Macol® 3520; Macol® 5100; Mark® 1014; Markpet; Masil® 1066C; Masil® 1066D; Masil® SF 10; Masil® SF 20; Masil® SF 50; Masil® SF 100; Masil® SF 200; Masil® SF 350; Masil® SF 350 FG; Masil® SF 500; Masil® SF 1000; Masil® SF 5000; Masil® SF 10,000; Masil® SF 12,500; Masil® SF 30,000; Masil® SF 60,000; Masil® SF 100,000; Masil® SF 300,000; Masil® SF 500,000; Masil® SF 600,000; Masil® SF 1,000,000; Mathe 6T; Mathe Barium Stearate; Mathe Barium Stearate; Mathe Calcium Stearate; Mathe Lithium Stearate; Mathe Magnesium Stearate; Mathe Sodium Stearate; Mathe Zinc Stearate 25S; Mathe Zinc Stearate S; McLube 1700; McLube 1704; McLube 2000 Series; Mekon® White; Metasap® Barium Stearate; Metasap® Cadmium Stearate; Metasap® Zinc Stearate; Miljac Barium Stearate; Miljac Cadmium Stearate; Miljac Calcium Stearate; Miljac Magnesium Stearate; Miljac Zinc Stearate; MiRaSlip VRA-102; MiRaStab 919-K; Mold-Ease; Molder's Edge ME-100; Molder's Edge ME-110; Molder's Edge ME-120; Molder's Edge ME-175; Molder's Edge ME-211; Molder's Edge ME-263; Molder's Edge ME-301; Molder's Edge ME-304; Molder's Edge ME-341; Molder's Edge ME-345; Molder's Edge ME-369; Molder's Edge ME-514; Molder's Edge ME-5440; Molder's Edge ME-7000; Mold Lubricant™ 426; Mold-Release 225; Mold-Release 605; Mold Wiz Ext. 249; Mold Wiz Ext. 424/7; Mold Wiz Ext. AZN; Mold Wiz Ext. F-57; Mold Wiz Ext. FFIH; Mold Wiz Ext. P; Mold Wiz Int. 18-36; Mold Wiz Int. 33PA; Mold Wiz Int. 33UDK; Mold Wiz Int. 937; Molgard; Monafax 785; Monolan® PPG440, PPG1100, PPG2200; Monoplex® DDA; MS-122N/CO_2; Myvaplex® 600P; Myvaplex® 600PK; Myverol® 18-04; Myverol® 18-06; Myverol® 18-07; Myverol® 18-99

Naftozin® N, Spezial; Neutral Degras; Noiox AK-41; Noiox KS-10, -12, -13, -14, -16; Nopalcol 1-L; Nopalcol 1-TW; Nopalcol 2-DL; Nopalcol 4-C; Nopalcol 4-CH; Nopalcol 4-L; Nopalcol 4-S; Nopalcol 6-DO; Nopalcol 6-DTW; Nopalcol 6-L; Nopalcol 6-R; Nopalcol 6-S; Nopalcol 10-COH; Nopalcol 12-CO; Nopalcol 12-COH; Nopalcol 19-CO; Nopalcol 30-TWH; Nopalcol 200; Nopalcol 400; Nopalcol 600; NZ 01; NZ 09; NZ 12; NZ 33; NZ 38; NZ 39; NZ 44; NZ 49; NZ 89; NZ 97

Oxp

Paricin® 1; Paricin® 6; Paricin® 8; Paricin® 210; Paricin® 220; Paricin® 285; Pationic® 900; Pationic® 901; Pationic® 902; Pationic® 905; Pationic® 907; Pationic® 909; Pationic® 914; Pationic® 919; Pationic® 920; Pationic® 925; Pationic® 930; Pationic® 930K; Pationic® 940; Pationic® 1019; Pationic® 1042; Pationic® 1042B; Pationic® 1042K; Pationic® 1042KB; Pationic® 1052; Pationic® 1052B; Pationic® 1052K; Pationic® 1052KB; Pationic® 1061; Pationic® 1064; Pationic® 1074; Pationic® 1145; PeFlu 727; Penreco Amber; Penreco Snow; Petrac® 165; Petrac® 215; Petrac® 250; Petrac® 270; Petrac® Calcium Stearate CP-11; Petrac® Calcium Stearate CP-11 LS; Petrac® Calcium Stearate CP-11 LSG; Petrac® Calcium Stearate CP-22G; Petrac® Calcium/Zinc Stearate CZ-81; Petrac® Eramide®;

Lubricants *(cont'd.)*

Petrac® GMS; Petrac® Magnesium Stearate MG-20 NF; Petrac® Zinc Stearate ZN-41; Petrac® Zinc Stearate ZN-42; Petrac® Zinc Stearate ZN-44 HS; Petrac® Zinc Stearate ZW-45; Petrarch® A0700; Petrarch® A0742; Petrarch® A0743; Petrarch® A0750; Petrarch® A0800; Petrarch® B2500; Petrarch® C3292; Petrarch® C3300; Petrarch® G6720; Petrarch® H7300; Petrarch® I7810; Petrarch® M8550; Petrarch® M8450; Petrarch® M8500; Petrarch® M9050; Petrarch® M9100; Petrarch® O9835; Petrarch® T1807; Petrarch® T1918; Petrarch® T1980; Petrarch® T2090; Petrarch® T2503; Petrarch® T2910; Petrarch® V4900; Petrarch® V4910; Petrarch® V4917; Petrarch® V5000; Petrolatum RPB; Petrolite® C-700; Petrolite® C-1035; Petrolite® C-3500; Petronauba® C; Plastiflow CW-2; Plastolube; Pluracol® E200; Pluracol® E300; Pluracol® E400; Pluracol® E400 NF; Pluracol® E600; Pluracol® E600 NF; Pluracol® E1000; Pluracol® E1450; Pluracol® E1450 NF; Pluracol® E1500; Pluracol® E2000; Pluracol® E4000; Pluracol® E4500; Pluracol® E6000; Pluracol® E8000; Pluriol® P 600; Pluriol® P 900; Pluriol® P 2000; Polyace™ 573; Polydis® TR 016; Polydis® TR 121; Polydis® TR 131; Polyfin; Poly-G® WS 280X; Polylube J; Polylube RE; Polymel #7; Polymist® 284; Polymist® F-5; Polymist® F-5A; Polymist® F-5A EX; Polymist® F-510; Polymist® XPH-284; Polypol 19; Polytac; Polywax® 500; Polywax® 655; Polywax® 850; Polywax® 1000; Polywax® 2000; Polywax® 3000; Priolube 1407; Priolube 1409; Priolube 1447; Priolube 1451; Priolube 1458; Pristerene 4900; proFLOW 1000; proFLOW 3000; Proplast 015; Proplast 050, 075; Proplast 060; Proplast 290; Prox-onic DT-03; Prox-onic DT-015; Prox-onic DT-030; Prox-onic HR-05; Prox-onic HR-016; Prox-onic HR-025; Prox-onic HR-030; Prox-onic HR-036; Prox-onic HR-040; Prox-onic HR-080; Prox-onic HR-0200; Prox-onic HR-0200/50; Prox-onic HRH-05; Prox-onic HRH-016; Prox-onic HRH-025; Prox-onic HRH-0200; Prox-onic HRH-0200/50; Prox-onic MC-02; Prox-onic MC-05; Prox-onic MC-015; Prox-onic MHT-05; Prox-onic MHT-015; Prox-onic MO-02; Prox-onic MO-015; Prox-onic MO-030; Prox-onic MO-030-80; Prox-onic MS-02; Prox-onic MS-05; Prox-onic MS-011; Prox-onic MS-050; Prox-onic MT-02; Prox-onic MT-05; Prox-onic MT-015; Prox-onic MT-020; PS044; PS045; PS046; PS047; PS047.5; PS048; PS048.5; PS049; PS049.5; PS050; PS072; PS140

RC 7; Release Agent E-155; Rexanol™; Rhodafac® RE-610; Rhodafac® RE-877; Rhodasurf® L-25; Rhodasurf® L-790; Rim Wiz; Ross Carnauba Wax; Ross Ceresine Wax; Ross Wax #100; Ross Wax #140; Ross Wax #145; Ross Wax #160; Ross Wax #165; RS-80; Rubber Lubricant™; Rudol®

SAFR-EPS K; SAFR-T-5K; Santone® 3-1-SH; Secoster® SDG; SEM-135; Semtol® 40; Semtol® 70; Semtol® 85; Semtol® 100; Semtol® 350; SF18-350; SF81-50; SF96®; SF1080; SFR 100; Sicolub® DSP; Sicolub® E; Sicolub® EDS; Sicolub® LB 1, LB 2, LBK, LK 3; Sicolub® OA 2, OA 4; Sicolub® OP; Sicolub® TDS; Siltech® CE-2000 F3.5; Silwax® C; Silwax® F; Silwax® S; Silwax® WD-IS; Silwax® WD-S; Silwax® WS; Silwet® L-7001; Silwet® L-7500; Silwet® L-7600; Silwet® L-7602; SM 2164; Softenol® 3900; Softenol® 3991; Solricin® 135; Solricin® 235; Solricin® 535; Specialty 101; Specialty 102; Specialty 103; Specialty 105; Staflex DIBA; Stanclere® T-250, T-250SD; Stanclere® TL; Stan Wax; Starwax® 100; Struktol® TR 071; Struktol® WB 16; Struktol® WS 280 Paste; Struktol® WS 280 Powd; Sunolite® 160; Sunyl® 80; Sunyl® 80 ES; Sunyl® 90; S-Wax-1; S-Wax-3; Syncrolube 2528; Syncrolube 3532; Syncrolube 3780; Syncrolube 3795; Synpro® Aluminum Stearate 303; Synpro® Barium Stearate; Synpro® Cadmium Stearate; Synpro® Calcium Pelargonate; Synpro® Calcium Stearate 12B; Synpro® Calcium Stearate 15; Synpro® Calcium Stearate 15F; Synpro® Calcium Stearate 114-36; Synprolam 35 Lauramide; Synprolam 35 Stearamide; Synpro® Lead Stearate; Synpro® Lithium Stearate; Synpro® Magnesium Stearate 90; Synpron 1536; Synpron 1585; Synpron 1618; Synpron 1652; Synpron 1774; Synpron 1850; Synpron 1870; Synpron 1871; Synpron 1881; Synpron 1890; Synpron 1900; Synpron 1913; Synpro® Sodium Stearate ST; Synpro® Stannous Stearate; Synpro® Zinc Stearate 8; Synpro® Zinc Stearate D; Synpro® Zinc Stearate GP; Synpro® Zinc Stearate GP Flake; Synpro® Zinc Stearate HSF; SY-SY 132

Tech Pet F™; Teflon® MP 1000; Teflon® MP 1100; Teflon® MP 1200; Teflon® MP 1400; Teflon® MP 1500; Teflon® MP 1600; Tegiloxan®; Tego® Silicone Paste A; Temex BA-1; Teric PEG 1500; Therm-Chek® BJ-2; Thermolite 12; TL-115; Tolplaz TBEP; Tuf Lube Hi-Temp NOD

Unilin® 350 Alcohol; Unilin® 550 Alcohol; Unilin® 700 Alcohol; Unislip 1757; Unislip 1759; Uniwax 1750; Uniwax 1760

Vanfre® AP-2; Vanfre® AP-2 Special; Vanfre® IL-1; Vanfre® IL-2; Vanfre® UN; Vestowax A-217; Vestowax A-227; Vestowax A-235; Vestowax A-415; Vestowax A-616; Vestowax AO-1539; Vestowax AS-1550; Vestowax C-60; Vestowax FT-150; Vestowax FT-150P; Vestowax FT-200; Vestowax FT-300; Vinylube® 36; Vinylube® 38; Viscasil® (10,000 cSt); Viscasil®

Lubricants *(cont'd.)*

(12,500 cSt); Voidox 100%; Vorite 105; Vorite 110; Vorite 115; Vorite 120; Vorite 125; Vybar® 103; Vybar® 260; Vybar® 825; Vydax® 1000; Vydax® AR; Vydax® WD

Wayfos A; Weston® DOPI; Wet Zinc™ P; W.G.S. Synaceti 116 NF/USP; Whitcon® TL-5; Whitcon® TL-6; Whitcon® TL-102-2; Whitcon® TL-155; Whitcon® TL-169; Witco® Aluminum Stearate 18; Witco® Aluminum Stearate 22; Witco® Calcium Stearate A; Witco® Calcium Stearate F; Witco® Calcium Stearate FP; Witco® Heat Stable Sodium Stearate; Witco® Lithium Stearate 306; Witco® Magnesium Stearate D; Witco® Sodium Stearate T-1; Witco® Zinc Stearate 11; Witco Zinc Stearate 42; Witco® Zinc Stearate 44; Witco® Zinc Stearate Disperso; Witco® Zinc Stearate Heat-Stable; Witco® Zinc Stearate LV; Witco® Zinc Stearate NW; Witco® Zinc Stearate Polymer Grade; Witco® Zinc Stearate REP; Witco Zinc Stearate Regular

ZN-18-1; Zonyl® FSP; Zylac 235

Chemicals:

Acetylated palm kernel glycerides; Adipic acid; Aluminum distearate; Aluminum stearate; N-(2-Aminoethyl)-3-aminopropyl methyldimethoxy silane; 3-Aminopropylmethyldiethoxy silane; Aminopropyltrimethoxysilane; Amyltrichlorosilane

Barium-cadmium-lead complex; Barium stearate; Behenamide; Biscumylphenyl trimellitate; Bis (hydroxyethyl) aminopropyltriethoxy silane; Bis-(N-methylbenzamide) ethoxymethyl silane; Bis (trimethylsilyl) acetamide; Butyl ricinoleate

Cadmium stearate; Calcium montanate; Calcium pelargonate; Calcium ricinoleate; Calcium/sodium stearoyl lactylate; Calcium-zinc stearate; C10-12 alcohols; C10-14 alcohols; C12-14 alcohols; C12-16 alcohols; C12-18 alcohols; C14-16 alcohols; C14-18 alcohols; C30-50 alcohols; Candelilla synthetic; Canola oil glyceride; Castor oil, polymerized; Ceteareth-4; Ceteareth-6; Cetearyl alcohol; Cetearyl stearate; Cetyl esters; 3-Chloropropyltriethoxysilane; 3-Chloropropyltrimethoxysilane; Coconut acid; Cocotrimonium chloride; Cycloneopentyl, cyclo (dimethlaminoethyl) pyrophosphato zirconate, di mesyl salt

Di-t-butoxydiacetoxysilane; Diisobutylene; Diisodecyl glutarate; Diisooctyl dodecanedioate; Dilinoleic acid; Dimethicone copolyol; Dimethicone copolyol isostearate; Dimethicone/mercaptopropyl methicone copolymer; Dimethiconol hydroxystearate; Dimethiconol stearate; N,N′-Dioleoylethylenediamine; Distearyl phthalate

Erucamide; Erucyl erucamide; Erucyl stearamide; Ethylene/acrylic acid copolymer; N,N′-Ethylenebis 12-hydroxystearamide; N,N′-Ethylene bis-ricinoleamide; Ethylene dioleamide; Ethylene distearamide; Ethylene-propylene-styrene terpolymer; Ethyltriacetoxysilane

Glyceryl dioleate; Glyceryl hydroxystearate; Glyceryl linoleate; Glyceryl mono/dioleate; Glyceryl montanate; Glyceryl myristate; Glyceryl stearate; Glyceryl stearate SE; Glyceryl (triacetoxystearate); Glyceryl triacetyl ricinoleate; (3-Glycidoxypropyl)-methyldiethoxy silane; Glycol/butylene glycol montanate; Glycol distearate; Glycol stearate

Hexamethyldisilazane; Hydrogenated castor oil; Hydrogenated cottonseed glyceride; Hydrogenated palm glyceride; Hydrogenated soy glyceride; Hydrogenated tallow amide; Hydrogenated tallowamine; N (2-Hydroxyethyl) 12-hydroxystearamide; N(β-Hydroxyethyl) ricinoleamide; Hydroxystearic acid

Isobutyl stearate; Isobutyltrimethoxysilane; Isocyanatopropyltriethoxysilane; Isodecyl oleate; Isooctyl stearate; Isopropyl myristopalmitate; Isopropyl oleate; Isotridecyl stearate

Lard glyceride; Lauramine; Lead-barium-cadmium complex; Lead stearate; Lead stearate, dibasic; Lithium stearate

Magnesium stearate; 3-Mercaptopropylmethyldimethoxysilane; Mercaptopropyltrimethoxysilane; Methyl acetyl ricinoleate; N-Methylaminopropyltrimethoxysilane; Methyl cocoate; Methyl hydroxystearate; Methyl tallowate; Methyltriacetoxysilane; Methyltriethoxysilane; Methyltrimethoxysilane; Microcrystalline wax; Microcrystalline wax, oxidized; Mineral oil; Molybdenum disulfide; Montan acid wax

Neopentyl (diallyl) oxy, trihydroxy caproyl titanate; Neopentyl (diallyl) oxy, trimercapto-phenyl zirconate

Octyl stearate; Octyltriethoxysilane

Palm kernel glycerides; PEG-4; PEG-6; PEG-8; PEG-12; PEG-20; PEG-32; PEG-40; PEG-75; PEG-100; PEG-150; PEG-36 castor oil; PEG (1200) castor oil; PEG (1900) castor oil; PEG-8 cocoate; PEG-4 dilaurate; PEG-6 dilaurate; PEG-3 dioctoate; PEG-8 dioleate; PEG-12 ditallowate; PEG-20 glyceryl stearate; PEG-20 hydrogenated castor oil; PEG (1200) hydrogenated castor oil; PEG-60 hydrogenated tallowate; PEG-4 laurate; PEG-150 laurate; PEG-2 laurate SE; PEG-15 oleamine; PEG-30 oleamine; PEG-6 oleate; PEG-8 oleate; PEG-150 oleate; PEG-12 ricinoleate; PEG-11 stearamine; PEG-2 stearate; PEG-4 stearate; PEG-12 stearate; PEG-2 tallowate; Pentaerythrityl stearate; Petrolatum; Petroleum wax; 2-

Lubricants *(cont'd.)*

Phenethyltrichlorosilane; Phenyl-β-naphthylamine; Phenyltriethoxysilane; Polyacrylamide; Polydimethyldiphenyl siloxane; Polyethylene glycol; Polyethylene, high-density; Polyethylene, oxidized; Polyethylene wax; Polyethylene wax, oxidized; Polyoctylmethylsiloxane; Polypropylene; Polypropylene glycol; Polytetrafluoroethylene; Potassium castorate; Potassium oleate; Potassium ricinoleate; PPG-400; PPG-2000; PPG-12-buteth-16; PPG-28-buteth-35; PPG-33-buteth-45; PPG-40 diethylmonium chloride; Propylene glycol myristate; n-Propyltrimethoxysilane

Rapeseed oil

Silicone emulsions; Sodium borate decahydrate; Sodium lauroyl sarcosinate; Sodium ricinoleate; Sodium stearate; Sodium stearoyl lactylate; Sorbitan laurate; Sorbitan sesquioleate; Sorbitan trioleate; Soy acid; Stannous stearate; Stearamide; Stearic acid; Stearyl/aminopropyl methicone copolymer; N-Stearyl 12-hydroxystearamide; Stearyl stearamide; 3-(N-Styrylmethyl-2-aminoethylamino) propyltrimethoxy silane hydrochloride; Synthetic wax

Tetraethoxysilane; Tetrakis (2-ethoxyethoxy) silane; Tetramethoxysilane; Tetra-n-propoxysilane; Trichlorosilane; Tridecyl stearate; N-[3-(Triethoxysilyl)-propyl] 4,5-dihydroimidazole; Trilinoleic acid; 1-Trimethoxysilyl-2-(chloromethyl) phenylethane; Trimethoxysilylpropyldiethylene triamine; Trimethylolpropane trioleate; Triolein; Tripropylene glycol; Tris Neodol-25 phosphite; Tristearin; Trixylenyl phosphate

Vinyltrichlorosilane; Vinyltriethoxysilane

Zinc stearate

Modifiers

Trade names: A-C® 7, 7A; A-C® 8, 8A; A-C® 9, 9A, 9F; Acintol® R Type S; Acintol® R Type SB; ACtol® 65; ACtol® 70; ACtol® 80; Advawax® 240; Advawax® 280; Advawax® 290; Ageflex AGE; Ageflex AMA; Ageflex CHMA; Ageflex EGDMA; Ageflex FM-1; Ageflex FM-2; Ageflex FM-10; Ageflex n-HA; Ageflex TBGE; Akrochem® P 87; Akrochem® P 487; Amoco® H-300E; Araldite® PY 306

Benzoflex® S-312; Benzoflex® S-404; Blendex® 131; BM-729; Buna AP 331; Buna AP 437

Cardolite® NC-507; Cardolite® NC-510; Cardolite® NC-511; Cardura® E-10; Chemigum® N8; Chemigum® P83; Chemstat® G-118/52; Chemstat® G-118/95; CMD 5185; Colok® 265; Compimide® TM-121; Compimide® TM-123

D.E.R. 337; Dianol® 220; Dianol® 240; Dianol® 240/1; Dianol® 265; Dianol® 285; Dow E200; Dow E300; Dow E400; Dow E600; Dow E900; Dow E1000; Dow E1450; Dow E3350; Dow E4500; Dow E8000; Dresinol® 42; Dutral® CO-059

ELP-3; Emkapyl 400, 600, 1200, 2000, 4000; Epi-Tex® 611-Q; Epolene® C-13; Epolene® C-14; Epolene® C-15; Epolene® C-15P; Epolene® C-17; Escor® ATX-320; Esperal® 120; Exxelor PA23; Exxelor PA30; Exxelor PA50

Fiberfrax® EF 122S; FR-D

HB-40®; Heloxy® 7; Heloxy® 8; Heloxy® 9; Heloxy® 32; Heloxy® 44; Heloxy® 48; Heloxy® 56; Heloxy® 61; Heloxy® 62; Heloxy® 71; Heloxy® 84; Heloxy® 107; Heloxy® 116; Heloxy® 505; Hercules® Ester Gum 10D

Jeffamine® BuD-2000; Jeffamine® D-230; Jeffamine® D-400; Jeffamine® D-2000; Jeffamine® T-403

Kemamide® E-180; Kemamide® E-221; Kemamide® P-181; Kemamide® S-180; Kemamide® S-221; Ken Kem® CP-45; Ken Kem® CP-99; KEP 010P, 020P, 070P; Krynac® P30.49; Krynac® P34.52; Krynac® P34.82; Krynac® XL 31.25

LP®-3

Maphos® 41A; Methyl Zimate®; Methyl Zimate® Slurry; Modarez APVC 8; Multiflow®; Multiwax® 180-M; Multiwax® W-835; Multiwax® X-145A

Nipol® 1022X59; Nipol® 1411; Nipol® 1422X14; Nipol® 1452X8; Nipol® 1453 HM; Nipol® 1472; Nipol® DP5123P; Nipol® DP5125P; Nipol® DP5128P

P®-10 Acid; Palatinol® DBP; Paraloid® HT-100; Paraloid® HT-510®; Paraloid® K-400; Paraloid® KF-710; Paraloid® KM-318F; Paraloid® PM-800; Pationic® 1042; Pationic® 1042B; Pationic® 1042K; Pationic® 1042KB; Pationic® 1052; Pationic® 1052B; Pationic® 1052K; Pationic® 1052KB; Pationic® 1061; Pationic® 1064; Pationic® 1074; Permethyl® 100 Epoxide; Piccolyte® C115; Piccolyte® C125; Piccolyte® C135; Piccolyte® S125; Piccolyte® S135; Piccotac® B; Pliovic® M-50; Pliovic® M-70; Pliovic® M-70SC; Pliovic® M-90; Poly bd® 600; Poly bd® 605; Polysar EPDM 227; Polysar EPDM 847XP; Polysar EPDM

Modifiers *(cont'd.)*

6463; Polysar EPM 306; Polyvel CR-5; Polyvel CR-5F; Polyvel CR-5P; Polyvel CR-10; Polyvel CR-10P; Polyvel CR-20P; Polyvel CR-25; Polyvel RM40; Polyvel RM40Z; Polyvel RM70; Polywax® 500; Polywax® 655; Polywax® 850; Polywax® 1000; Polywax® 2000; Polywax® 3000; Pripol 1004; Pripol 1009; proFLOW 1000; proFLOW 3000; Prox-onic PEG-2000
Resinall 153; Ricon 100; Ricon 153
Sipomer® HEM-D; Sipomer® IBOA; Sipomer® IBOA HP; Sipomer® IBOMA; Sipomer® IBOMA HP; SR-297; SR-7475; Staybelite® Ester 5; Sterling® 1120; Sterling® 2320; Sterling® 6740; Sylvaros® 20; Sylvaros® 80
Taipol BR 015H; Taktene 1202; Techfil 7599; Techpolymer MB-4, MB-8; Techpolymer MBX-5 to MBX-50; Techpolymer SBX-6, -8, -17; Teflon® MP 1300; Teflon® MP 1400; Teflon® MP 1500; Teflon® MP 1600; Telalloy® A-10; Telalloy® A-15; Telalloy® A-50B; Tyzor AA; Tyzor LA; Tyzor TBT; Tyzor TE; Tyzor TOT; Tyzor TPT
Unilin® 425 Alcohol; Uniplex 260; Uniplex 270; Uniplex 600; Uni-Rez® 3020
Varcum® 29726; Vinuran®; Viplex 885; Vistalon 719; Vistalon 503

Chemicals: Acrylic-styrene-acrylonitrile terpolymer; Acrylonitrile-butadiene rubber; Acrylonitrile/styrene/acrylate; Allyl glycidyl ether
Benzothiazyl disulfide; N-Benzyldimethylamine; Butadiene-styrene copolymer; 1,4-Butanediol diacrylate
C40-60 alcohols; C44 dimer acid; Cumyl phenol; Cumylphenyl neodecanoate; 1,4-Cyclohexanedimethanol dibenzoate; Cyclohexyl methacrylate
Decyl betaine; Dibromoneopentyl glycol diglycidyl ether; Dimethylaminoethyl methacrylate
EPDM rubber; EPM/EPDM copolymer; EPM rubber; Epoxidized 1,2-polybutadiene; Epoxy resin; Ethylene glycol dimethacrylate; Ethylene-methyl acrylate-acrylic acid terpolymer
Glyceryl hydrogenated rosinate; Glyceryl rosinate; Glyceryl tribenzoate
1-Hexene; n-Hexyl acrylate; 2-Hydroxyethylmethacrylate
Isobornyl methacrylate; Isodecyl methacrylate; Isooctyl thioglycolate
Neodecanoic acid glycidyl ester; Neopentyl glycol dibenzoate; Nonyl phenyl glycidyl ether; Novolac resin
3-(n-Penta-8´-decenyl) phenol; 3-(n-Pentadecyl) phenol; Pentaerythrityl tetrabenzoate; Pentaerythrityl triacrylate; 1,2-Polybutadiene; Polybutene, hydrogenated; Polycarbonate resin; Polydipentene; Polyglutarimide acrylic copolymer; Poly-α-methylstyrene; Poly(α-methylstyrene-acrylonitrile; Poly (α-methylstyrene-styrene acrylonitrile); Polyphenylene ether; Polystyrene; Polysulfide
Ricinoleic acid
Styrene-acrylonitrile copolymer; Styrene-butadiene rubber; Sucrose acetate isobutyrate
Terpene resin; Tetrahydrofurfuryl methacrylate; Toluenesulfonamide formaldehyde resin; Triallylcyanurate; Triisooctyl phosphite; Trilauryl trithiophosphite; Trimethylolethane triglycidyl ether; Trimethylolpropane triglycidyl ether; Tristearyl phosphite
Urea-formaldehyde resin
Vinyl homopolymer

Nucleating agents

Trade names: ADK STAB NA-10; ADK STAB NA-11
Ficel® EPA; Ficel® EPB; Ficel® EPC; Ficel® EPD; Ficel® EPE
Hostamont® CaV 102; Hostamont® NaS 102; Hostamont® NaV 101; Hostatron® System P9937; Hostatron® System P9947; Hydrocerol® CF; Hydrocerol® CF 5; Hydrocerol® CF 20E; Hydrocerol® CF 40E; Hydrocerol® CF 40S; Hydrocerol® CF-60V; Hydrocerol® CF 70; Hydrocerol® CLM 70; Hydrocerol® Compound; Hydrocerol® HK 40 B; Hydrocerol® LBA-38; Hydrocerol® LBA-47; Hydrocerol® TAF; Hydrocerol® TAF 50
Jet Fil® 700C
Kycerol 91; Kycerol 92
Mark® 2180; Millad® 3988; Mistron Vapor® R; Mistron ZSC
NC-4
Paraloid® EXL-5375; Polybond® 1000; Polybond® 1001; Polybond® 1002; Polyvel CA-P20; Polywax® 500; Polywax® 655; Polywax® 850; Polywax® 1000; Polywax® 2000; Polywax® 3000; proFLOW 1000
Safoam® AP-40; Safoam® FP; Safoam® FP-20; Safoam® FP-40; Safoam® FPE-20; Safoam®

Nucleating agents *(cont'd.)*

FPE-50; Safoam® P; Safoam® P-20; Safoam® P-50; Safoam® PCE-40; Safoam® PE-20; Safoam® PE-50; Safoam® RIC; Safoam® RIC-25; Safoam® RIC-30; Safoam® RPC; Safoam® RPC-40
Ultra-Pflex®; Uniplex 413

Chemicals: Benzenesulfonamide; Bis (p-ethylbenzylidene) sorbitol
Sodium benzoate; Sodium bicarbonate; Sodium 2,2´-methylenebis(4,6-di-t-butylphenyl) phosphate

Odorants • Perfumes

Trade names: ADI 50
Fragrance Compound
Metazene® 60%; Metazene® 80%; Metazene® 99%
Polyvel C200; Polyvel E40; Polyvel ER40; Polyvel P10; Polyvel P15; Polyvel P25; Polyvel PV15; Polyvel PV30

Chemicals: Hydroxypropyl-β-cyclodextrin; Hydroxypropyl-γ-cyclodextrin

Optical brighteners • Whiteners

Trade names: Benetex™ OB
Eastobrite® OB-1; Ecco White® FW-5; Ecco White® OP
Hostalux KCB
Oppasin®
Spectratech® CM 11698; Spectratech® CM 11812
Uvitex® OB

Chemicals: CI Fluorescent Brightener 236
Distyryl phenyl
2,2´-(1,2-Ethenediyldi-4,1-phenylene) bisbenzoxazole
2,2´-(2,5-Thiophenediyl)bis[5-t-butylbenzoxazole]

Peptizers

Trade names: Akrochem® Peptizer 66; Aktiplast® AS; Aktiplast® F; Aktiplast® PP; Bondogen E®
CB-4-34
MR 575
Octopol SDB-50
Peptizer 566; Peptizer 932; Peptizer 965; Peptizer 7010; Plasticizer REO; Proaid® PEP
Ricon P30/Dispersion
Struktol® A 50; Struktol® A 60; Struktol® A 61; Struktol® A 80; Struktol® A 82; Struktol® A 86; Struktol® Activator 73
Vanax® 552

Chemicals: N,N-Bis-stearoylethylenediamide
Pentachlorothiophenol; Piperidinium pentamethylene dithiocarbamate

Plasticizers • Softeners

Trade names: AA Standard; Abitol®; Acintol® 208; Acrawax® C; Adeka Lub CB-419; Adeka Lub CB-419R; Adeka Lub CF-40; Adeka Lub E-410; Adeka Lub E-450; Adeka Lub E-500; Adeka Lub S-1; Adeka Lub S-3; Adeka Lub S-7; Adimoll® BO; Adimoll® DN; Adimoll® DO; ADK CIZER C-8; ADK CIZER D-32; ADK CIZER O-130P; ADK CIZER PN-250; ADK CIZER PN-650; ADK CIZER RS-107; Admex® 433; Admex® 515; Admex® 522; Admex® 523; Admex® 525; Admex® 760; Admex® 761; Admex® 770; Admex® 775; Admex® 910; Admex® 1663;

Plasticizers • Softeners *(cont'd.)*

Admex® 1665; Admex® 1723; Admex® 2632; Admex® 3752; Admex® 6187; Admex® 6969; Admex® 6985; Admex® 6994; Admex® 6995; Admex® 6996; Admex® 6999; Akrochem® P 37; Akrochem® P 49; Akrochem® P 55; Akrochem® P 82; Akrochem® P 86; Akrochem® P 87; Akrochem® P 478; Akrochem® P 486; Akrochem® P 487; Akrochem® Plasticizer LN; Akrosperse® Plasticizer Paste Colors; Albrite® TBP; Albrite® Tributyl Phosphate; Alkamide® STEDA; Alkasurf® NP-4; Alkasurf® NP-6; Altax®; Amgard® TBEP; Amgard® $TBPO_4$; Amgard® TOF; Amoco® H-15; Amoco® L-14; Amoco® L-50; Antistaticum RC 100; Arizona 208; Arneel® 18 D; Aromatic Oil 745; Arubren®; Atmer® 153; Atmer® 508

Benzoflex® 2-45; Benzoflex® 9-88; Benzoflex® 9-88 SG; Benzoflex® 50; Benzoflex® 131; Benzoflex® 284; Benzoflex® 400; Benzoflex® P-200; Benzoflex® S-312; Be Square® 185; Bondogen E®; Britol® 6NF; Bromoklor™ 50; Bromoklor™ 70; Bunatak™ N; Bunatak™ U; Bunaweld™ 780; Butyl Oleate C-914; Butyl Stearate C-895

Cetamoll®; Cetodan® 95 CO; Cetodan® 95-ML; CFE-50; Chloroflo® 42; Citroflex® 2; Citroflex® 4; Citroflex® A-2; Citroflex® A-4; Citroflex® A-4 Special; Citroflex® A-6; Citroflex® B-6; Clorafin® 40; CPH-37-NA; Cryoflex® 660

Dark Green No. 2™; DBE; DBE-2, -2SPG; DBE-3; DBE-4; DBE-5; DBE-6; DBE-9; Dellatol® BBS; Diplast® A; Diplast® B; Diplast® DOA; Diplast® E; Diplast® L8; Diplast® N; Diplast® O; Diplast® R; Diplast® TM; Diplast® TM8; Disflamoll® DPK; Disflamoll® DPO; Disflamoll® TCA; Disflamoll® TKP; Disflamoll® TOF; Disflamoll® TP; Doverguard® 8207-A; Doverguard® 8208-A; Doverguard® 8307-A; Doverguard® 8410; Doverguard® 8426; Doverguard® 9119; Doverguard® 9122; Doversperse 3; Drakeol 9; Drakeol 10; Drakeol 21; Drakeol 32; Drakeol 34; Drakeol 35; Drapex® 4.4; Drapex® 10.4; Drapex® 334F; Drapex® 409; Drapex® 411; Drapex® 412; Drapex® 420; Drapex® P-1; Drapex® P-7; Dresinate® 91; Dresinate® 81

Eastman® TEP; Emerest® 2301; Emerest® 2302; Emerest® 2675; Empicryl® 6030; Empicryl® 6047; Emulbon LB-78; Epodil® ML; Epoxol 7-4; Epoxol 9-5; Estabex® 138-A; Estabex® 2307; Estabex® 2307 DEOD; Estabex® 2381; Estaflex® ATC; Ethylan® HB1; Evanstab® 12; EW-POL 8021

Fancol OA-95; Fancor Lanolin; #15 Oil; Flexol® Plasticizer 4GO; Flexol® Plasticizer EP-8; Flexol® Plasticizer EPO; Flexricin® 13; Flexricin® 17; Flexricin® P-1; Flexricin® P-3; Flexricin® P-4; Flexricin® P-6; Flexricin® P-8; Fonoline® White; Fonoline® Yellow; Ftalidap®

Gantrez® M-154; Genamid® 151; Genamid® 250; Genamid® 490; Genamid® 491; Genamid® 747; Genamid® 2000; G-Flex™-11; G-Flex™-12; G-Flex™ HMD; G-Flex™ HMG

Halocarbon Grease 19; Halocarbon Grease 25-5S; Halocarbon Grease 25-10M; Halocarbon Grease 25-20M; Halocarbon Grease 28; Halocarbon Grease 28LT; Halocarbon Grease 32; Halocarbon Grease X90-10M; Halocarbon Oil 0.8; Halocarbon Oil 1.8; Halocarbon Oil 4.2; Halocarbon Oil 6.3; Halocarbon Oil 27; Halocarbon Oil 56; Halocarbon Oil 95; Halocarbon Oil 200; Halocarbon Oil 400; Halocarbon Oil 700; Halocarbon Oil 1000N; Halocarbon Wax 40; Halocarbon Wax 600; Halocarbon Wax 1200; Halocarbon Wax 1500; Haro® Chem PDP-E; Haro® Chem PPCS-X; Hefti PGE-400-DS; Hefti PGE-600-DS; Hercoflex® Plasticizer; Hercolyn® D; Hercules® Ester Gum 10D; Hi-Sil® ABS; Hodag PEG 200; Hodag PEG 300; Hodag PEG 540; Hodag PEG 3350; Hodag PPG-150; Hodag PPG-400; Hodag PPG-1200; Hodag PPG-2000; Hodag PPG-4000; Hydrobrite 200PO; Hydrobrite 300PO; Hydrobrite 380PO; Hydrobrite 550PO; Hystrene® 5016 NF; Hystrene® 9718 NF

Igepal® CO-430; Industrene® 105

Jayflex® 77; Jayflex® 210; Jayflex® 215; Jayflex® DHP; Jayflex® DIDP; Jayflex® DINA; Jayflex® DINP; Jayflex® DIOP; Jayflex® DOA; Jayflex® DTDP; Jayflex® L9P; Jayflex® L11P; Jayflex® L911P; Jayflex® L7911P; Jayflex® TINTM; Jayflex® TOTM; Jayflex® UDP

Kaydol®; Kaydol® S; Kemester® 143; Kemester® 4000; Kemester® 5221SE; Kemester® 6000; Kenflex® A; Kenflex® A-30; Kenflex® N; Kenplast® A-450; Kenplast® BG; Kenplast® ES-2; Kenplast® ESB; Kenplast® ESI; Kenplast® ESN; Kenplast® G; Kenplast® LG; Kenplast® LT; Kenplast® PG; Kenplast® PPE; Kenplast® RD; Kenplast® RDN; Kenplast® RG; Ketjenflex® 8; Ketjenflex® 9; Kodaflex® 240; Kodaflex® DBP; Kodaflex® DMP; Kodaflex® DOA; Kodaflex® DOP; Kodaflex® DOTP; Kodaflex® HS-3; Kodaflex® HS-4; Kodaflex® PA-6; Kodaflex® TEG-EH; Kodaflex® TOTM; Kodaflex® Triacetin; Kodaflex® Tripropionin; Kodaflex® TXIB; Kronitex® 100; Kronitex® 200; Kronitex® 1884; Kronitex® 3600; Kronitex® TBP; Kronitex® TCP; Kronitex® TOF; Kronitex® TPP; Krynac® XL 31.25; K-Stay 21®

Lankroflex® ED 6; Lankroflex® GE; Lankroflex® L; Lankroflex® Series; Lankromark® LZ1034; Lexolube® BS-Tech; Lexolube® NBS; LICA 09; LICA 12; LICA 38; LICA 38A; LICA 38J; LICA 44; LICA 97; LICA 99; Lustrabrite® S

MR 575; Masil® SFR 70; Mazol® SFR 100; Mazol® SFR 750; Mazol® 159; Mazol® GMS-90;

Plasticizers • Softeners *(cont'd.)*

Mazol® GMS-D; Mazol® PGO-104; Mekon® White; Mesamoll®; Mesamoll® II; Modaflow®; Monolan® 2500 E/30; Monolan® PPG440, PPG1100, PPG2200; Monoplex® DDA; Monoplex® DIOA; Monoplex® DOA; Monoplex® DOS; Monoplex® NODA; Monoplex® S-73; Monoplex® S-75; Monostearyl Citrate (MSC); Morflex® 190; Morflex® 560; Morflex® 1129; MPDiol® Glycol; Mr 245; Mr 285; Mr 305; Multiflow®; Multiwax® 180-M; Multiwax® W-835; Multiwax® X-145A

Neoprene FB; Nevillac® 10° XL; Nevillac® Hard; Newpol PE-61; Newpol PE-62; Newpol PE-64; Newpol PE-68; Newpol PE-74; Newpol PE-75; Newpol PE-78; Newpol PE-88; Nipol® 1312; Nipol® 1312 LV; Nipol® 1422; Nopalcol 1-L; Nopalcol 1-TW; Nopalcol 2-DL; Nopalcol 4-C; Nopalcol 4-CH; Nopalcol 4-L; Nopalcol 4-S; Nopalcol 6-DO; Nopalcol 6-DTW; Nopalcol 6-L; Nopalcol 6-R; Nopalcol 6-S; Nopalcol 10-COH; Nopalcol 12-CO; Nopalcol 12-COH; Nopalcol 19-CO; Nopalcol 30-TWH; Nopalcol 200; Nopalcol 400; Nopalcol 600; Nouryflex 520; NP-10; NP-25; Nuoplaz® 849; Nuoplaz® 1046; Nuoplaz® 6000; Nuoplaz® 6159; Nuoplaz® 6534; Nuoplaz® 6934; Nuoplaz® 6938; Nuoplaz® 6959; Nuoplaz® DIBP; Nuoplaz® DIDA; Nuoplaz® DIDP; Nuoplaz® DOA; Nuoplaz® DOM; Nuoplaz® DOP; Nuoplaz® DTDA; Nuoplaz® DTDP; Nuoplaz® TIDTM; Nuoplaz® TOTM; #1 Oil; #30 Oil

Oulutac 20 D

P®-10 Acid; Palamoll®; Palatinol® 11; Palatinol® 91P; Palatinol® 711; Palatinol® DBP; Palatinol® DIDP; Palatinol® DOA; Palatinol® DOP; Palatinol® FF21; Palatinol® FF31; Palatinol® FF41; Palatinol® M; Palatinol® N; Palatinol® TOTM; Pale 4; Pale 16; Pale 170; Pale 1000; Para-Flux® 4156; Paraplex® G-25 70%; Paraplex® G-25 100%; Paraplex® G-30; Paraplex® G-31; Paraplex® G-40; Paraplex® G-41; Paraplex® G-50; Paraplex® G-51; Paraplex® G-54; Paraplex® G-56; Paraplex® G-57; Paraplex® G-59; Paraplex® G-60; Paraplex® G-62; Paraplex® RG-2; Paraplex® RG-8; Paricin® 6; Paricin® 8; Paroil® 140; Permethyl® 100 Epoxide; Peroxidol 780; Peroxidol 781; Petrac® 250; Petrac® 270; Petrolatum RPB; Petro-Rez 100; Petro-Rez 103; Petro-Rez 200; Petro-Rez 215; Petro-Rez PTH; Petro-Rez PTL; Phosflex 71B; Phosflex 362; Phosflex 370; Phosflex 390; Piccolastic® A5; Piccolastic® A75; Piccovar® AP10; Piccovar® AP25; Piccovar® L60; Plas-Chek® 775; Plas-Chek® 795; Plasthall® 6-10P; Plasthall® 8-10 TM; Plasthall® 8-10 TM-E; Plasthall® 83SS; Plasthall® 100; Plasthall® 200; Plasthall® 201; Plasthall® 203; Plasthall® 205; Plasthall® 207; Plasthall® 220; Plasthall® 224; Plasthall® 226; Plasthall® 325; Plasthall® 503; Plasthall® 914; Plasthall® 4141; Plasthall® 7006; Plasthall® 7041; Plasthall® 7045; Plasthall® 7049; Plasthall® 7050; Plasthall® 7059; Plasthall® BSA; Plasthall® CF; Plasthall® DBEA; Plasthall® DBEEA; Plasthall® DBP; Plasthall® DBS; Plasthall® DBZZ; Plasthall® DIBA; Plasthall® DIBZ; Plasthall® DIDA; Plasthall® DIDG; Plasthall® DIDP-E; Plasthall® DINP; Plasthall® DIOA; Plasthall® DIODD; Plasthall® DIOP; Plasthall® DOA; Plasthall® DODD; Plasthall® DOP; Plasthall® DOS; Plasthall® DOSS; Plasthall® DOZ; Plasthall® ESO; Plasthall® HA7A; Plasthall® LTM; Plasthall® NODA; Plasthall® P-545; Plasthall® P-550; Plasthall® P-612; Plasthall® P-622; Plasthall® P-643; Plasthall® P-650; Plasthall® P-670; Plasthall® P-1070; Plasthall® P-7035; Plasthall® P-7035M; Plasthall® P-7046; Plasthall® P-7068; Plasthall® P-7092; Plasthall® P-7092D; Plasthall® R-9; Plasthall® TIOTM; Plasthall® TOTM; Plasticizer 9; Plasticizer REO; Plastilit® 3060; Plastoflex® 2307; Plastogen; Plastogen E®; Plastol®; Plastol®; Plastolein® 9048; Plastolein® 9049; Plastolein® 9050; Plastolein® 9051; Plastolein® 9058; Plastolein® 9071; Plastolein® 9091; Plastolein® 9232; Plastolein® 9404; Plastolein® 9717; Plastolein® 9749; Plastolein® 9752; Plastolein® 9761; Plastolein® 9762; Plastolein® 9765; Plastolein® 9776; Plastolein® 9790; Plastomoll®; Pliabrac® 519; Pliabrac® 521; Pliabrac® 524; Pliabrac® TCP; Pliabrac® TXP; Pluriol® E 200; Pluriol® E 1500; Pluriol® E 4000; Pluriol® E 6000; Pluriol® E 9000; Pogol 300; Priolube 1405; Priolube 1451; Priplast 1431; Priplast 1562; Priplast 3013; Priplast 3018; Priplast 3114; Priplast 3124; Priplast 3149; Priplast 3155; Priplast 3157; Priplast 3159; Protopet® White 1S; Protopet® Yellow 2A; PS140; PX-318; Pycal 94; 2-Pyrol®

QO® Furfuryl Alcohol (FA®); QO® Tetrahydrofurfuryl Alcohol (THFA®)

Reogen E®; Resinall 767; Retarder SAFE; Rexol 25/7; Rez-O-Sperse® 3; Rhenofit® 1600; Rhenovin® FH-70; Rhodasurf® E 400; Rhodasurf® E 600; Ricon 131; Ricon 142; Ricon 157; Ricon P30/Dispersion; Rit-Cizer #8; Rit-Cizer #9; Rit-Cizer 10-EHFA; Rit-Cizer 10-EM; Rit-Cizer 10-ETA; Rit-Cizer 10-MME; RPP; Rudol®; RX-13117; RX-13154; RX-13411; RX-13412, -13413; RX-13414, -13415

SAFR-T-5K; Santicizer 97; Santicizer 141; Santicizer 143; Santicizer 148; Santicizer 154; Santicizer 160; Santicizer 261; Santicizer 278; Saytex® 111; Semtol® 40; Semtol® 70; Semtol® 85; Semtol® 100; Semtol® 350; Softenol® 3900; Softenol® 3991; SP-6700; SP-6701; Staflex 550; Staflex 802; Staflex 804; Staflex 809; Staflex BDP; Staflex BOP; Staflex

Plasticizers • Softeners *(cont'd.)*

DBEA; Staflex DBEP; Staflex DBF; Staflex DBP; Staflex DIBA; Staflex DIBCM; Staflex DIDA; Staflex DIDP; Staflex DINA; Staflex DINM; Staflex DIOA; Staflex DIOM; Staflex DIOP; Staflex DMAM; Staflex DMP; Staflex DOA; Staflex DOF; Staflex DOM; Staflex DOP; Staflex DOZ; Staflex DTDM; Staflex DTDP; Staflex NODA; Staflex NONDTM; Staflex ODA; Staflex ODP; Staflex TIOTM; Staflex TOTM; Starwax® 100; Staybelite®; Struktol® 40 MS; Struktol® 40 MS Flakes; Struktol® 60 NS; Struktol® 60 NS Flakes; Struktol® AW-1; Struktol® KW 400; Struktol® KW 500; Struktol® TR 042; Struktol® W 34 Flakes; Struktol® W 80; Struktol® WB 212; Struktol® WB 222; Struktol® WB 300; Stygene Series; Stygene R-2; Sulfolane W; Sunyl® 80; Sunyl® 80 ES; Sunyl® 90

T-600; Tartac 20; Tartac 30; Tartac 40; TBEP; TBP; Tech Pet F™; Tech Pet P; TegMeR® 703; TegMeR® 704; TegMeR® 803; TegMeR® 804; TegMeR® 804 Special; TegMeR® 809; TegMeR® 903; Teric PE68; Teric PPG 1000; Teric PPG 1650; Tetronic® 90R4; Tetronic® 150R1; Tetronic® 304; Texacar® EC; Texacar® PC; Thiofide®; TOF; TOF™; Toho Me-PEG Series; Toho PEG Series; Tolplaz TBEP; Topanol® CA; Trydet SA-50/30; Tylose® H Series; Tylose® MH Grades; Tylose® MHB

Ultramoll® I; Ultramoll® II; Ultramoll® III; Ultramoll® M; Ultramoll® PP; Ultramoll® PU; Ultramoll® TGN; Uniflex® 192; Uniflex® 300; Uniflex® 312; Uniflex® 314; Uniflex® 315; Uniflex® 320; Uniflex® 330; Uniflex® 338; Uniflex® BYO; Uniflex® BYS-Tech; Uniflex® DBS; Uniflex® DCA; Uniflex® DCP; Uniflex® DOA; Uniflex® DOS; Uniflex® EHT; Uniflex® TCTM; Uniflex® TEG-810; Uniflex® TOTM; Unimoll® 66; Unimoll® 66 M; Unimoll® BB; Unimoll® DB; Uniplex 80; Uniplex 82; Uniplex 84; Uniplex 108; Uniplex 110; Uniplex 150; Uniplex 171; Uniplex 173; Uniplex 214; Uniplex 225; Uniplex 250; Uniplex 260; Uniplex 310; Uniplex 413; Uniplex 552; Uniplex 809; Uni-Tac® R40

Vestenamer® 6213; Vestenamer® 8012; Vikoflex; Viplex 525; Viplex 680-P; Viplex 885; Viplex 895-BL; Vorite 105; Vorite 110; Vorite 115; Vorite 120; Vorite 125; Vulkanol® 80; Vulkanol® 81; Vulkanol® 85; Vulkanol® 88; Vulkanol® 90; Vulkanol® BA; Vulkanol® FH; Vulkanol® KA; Vulkanol® OT; Vulkanol® SF; Vultac® 2; Vultac® 4; Vybar® 253; Vybar® 825

Wareflex® 650; Westvaco® Resin 90; Westvaco® Resin 95; W.G.S. Hydrogenated Fish Glyceride 117; Wingtack® 10; Witconol™ NP-330

X-743

Zonarez® Alpha 25

Chemicals: Acetylated palm kernel glycerides; Acetyl paracumyl phenol; Acetyl tributyl citrate; Acetyl triethyl citrate; Acetyl tri-n-hexyl citrate; Adipic acid; Alkyl phenol disulfide; Asphalt, oxidized; Azelaic acid

Benzenesulfonamide; Benzenesulfonmethylamide; Benzothiazyl disulfide; Benzyl benzoate; Benzyloctyl adipate; Benzyl phthalate; Biscumylphenyl isophthalate; Biscumylphenyl trimellitate; Bisdiisobutyl maleate; 1,4-Butanediol; Butoxyethyl oleate; Butyl acetoxystearate; Butyl acetyl ricinoleate; Butylated triaryl phosphate; N,N-Butyl benzene sulfonamide; Butyl benzyl phthalate; Butyl cyclohexyl phthalate; Butyl maleate; Butyl octyl phthalate; Butyl oleate; 4-t-Butylphenol; t-Butylphenyl diphenyl phosphate; n-Butyl phthalyl-n-butyl glycolate; Butyl ricinoleate; Butyl stearate; n-Butyryl tri-n-hexyl citrate

Castor oil; Castor oil, polymerized; Cetearyl methacrylate; Cocoyl sarcosine; Cumylphenyl acetate; Cumylphenyl benzoate; Cumylphenyl neodecanoate; 1,4-Cyclohexanedimethanol dibenzoate; Cyclohexyl isooctyl phthalate

Decyltridecyl phthalate; Diallyl phthalate; Dibenzyl azelate; Dibutoxyethoxyethyl adipate; Dibutoxyethoxyethyl formal; Dibutoxyethoxyethyl glutarate; Dibutoxyethoxyethyl phthalate; Dibutoxyethyl adipate; Dibutoxyethyl azelate; Dibutoxyethyl glutarate; Dibutoxyethyl phthalate; Dibutoxyethyl sebacate; Dibutyl adipate; Dibutyl azelate; Dibutyl fumarate; Dibutyl maleate; Dibutyl methylene bisthioglycolate; Dibutyl phthalate; Dibutyl sebacate; Dicapryl adipate; Dicapryl maleate; Dicapryl phthalate; Dicyclohexyl phthalate; Didecyl glutarate; Diethylene glycol dibenzoate; Diheptyl, nonyl, undecyl linear phthalate; Di-n-heptyl phthalate; Di-n-hexyl azelate; Dihexyl phthalate; Diisobutyl adipate; Diisobutyl azelate; Diisobutylene; Diisobutyl phthalate; Diisodecyl adipate; Diisodecyl glutarate; Diisodecyl phthalate; Diisoheptyl phthalate; Diisononyl adipate; Diisononyl maleate; Diisononyl phthalate; Diisooctyl adipate; Diisooctyl dodecanedioate; Diisooctyl isophthalate; Diisooctyl maleate; Diisooctyl phthalate; Diisooctyl sebacate; Diisotridecyl phthalate; Dimethiconol; Dimethoxyethyl phthalate; Dimethyl adipate; Dimethyl amyl maleate; Dimethyl glutarate; Dimethyl isophthalate; Dimethyl phthalate; Dimethyl sebacate; Dinonyl undecyl phthalate; Dioctyl adipate; Dioctyl azelate; Dioctyl dodecanedioate; Dioctyl isophthalate; Dioctyl maleate; Dioctyl phthalate; Dioctyl sebacate; Dioctyl terephthalate; Dioctyl terephthalate chlori-

Plasticizers • Softeners *(cont'd.)*

nated; Dioctyl thiodiglycolate; Diphenylcresyl phosphate; Diphenyl octyl phosphate; Ditridecyl maleate; Ditridecyl phthalate; Diundecyl phthalate

Epoxidized glycol dioleate; Epoxidized linseed oil; Epoxidized linseed oil, butyl ester; Epoxidized octyl stearate; Epoxidized octyl tallate; Epoxidized oleate; Epoxidized soybean oil; Epoxidized tallate; Epoxidized tall oil; Ethylene carbonate; Ethylene glycol diacetate; 2-Ethylhexanol; Ethyl toluenesulfonamide

Furfuryl alcohol

Glyceryl acetate; Glyceryl diacetate; Glyceryl dilaurate; Glyceryl laurate; Glyceryl oleate; Glyceryl ricinoleate; Glyceryl rosinate; Glyceryl stearate; Glyceryl (triacetoxystearate); Glyceryl triacetyl ricinoleate; Glyceryl tribenzoate; Glyceryl tricaprate/caprylate; Glyceryl tripropionate; Glycol; Glycol dioleate; Glycol distearate; Glycol ricinoleate; Glycol stearate

Hydroabietyl alcohol; Hydroabietyl phthalate; Hydrogenated methyl abietate; N-(2-Hydroxypropyl) benzenesulfonamide

Isobutyl stearate; Isodecyl benzoate; Isodecyl diphenyl phosphate; Isooctyl oleate; Isooctyl palmitate; Isooctyl stearate; Isooctyl tallate; Isopropyl myristate; Isopropyl myristopalmitate; Isopropyl palmitate; Isopropylphenyl diphenyl phosphate

Lanolin; Lauryl-myristyl methacrylate

Methyl abietate; Methyl acetyl ricinoleate; Methyl cocoate; Methyl hydrogenated rosinate; Methyl hydroxyethylcellulose; Methyl laurate; Methyl oleate; 2-Methyl-1,3-propanediol; Methyl ricinoleate; Methyl tallate; Methyl tallowate; Mineral oil

Naphthenic oil; Neopentyl glycol dibenzoate; Nonoxynol-4; Nonoxynol-6; Nonoxynol-7; Nonyl phthalate; Nonyl undecyl phthalate

n-Octyl, n-decyl adipate; n-Octyl, n-decyl phthalate; n-Octyl, n-decyl trimellitate; Octyl palmitate; Octyl tallate; Oleic acid; Oleyl alcohol

Paraffin, chlorinated; PEG-3 caprylate/caprate; PEG-18 castor oil; PEG-32 castor oil; PEG-75 castor oil; PEG-160 castor oil; PEG (1200) castor oil; PEG (1900) castor oil; PEG-8 cocoate; PEG-3 diacetate; PEG-4 dibenzoate; PEG-3 dicaprylate; PEG-3 dicaprylate/caprate; PEG di-2-ethylhexoate; PEG-3 di-2-ethylhexoate; PEG-4 di-2-ethylhexoate; PEG-8 di-2-ethylhexoate; PEG-3 diheptanoate; PEG-4 diheptanoate; PEG-2 dilaurate; PEG-8 dilaurate; PEG-4 dioctanoate; PEG-3 dipelargonate; PEG-8 distearate; PEG-12 distearate; PEG-12 ditallowate; PEG-7 glyceryl cocoate; PEG-20 hydrogenated castor oil; PEG (1200) hydrogenated castor oil; PEG-60 hydrogenated tallowate; PEG-8 laurate; PEG-150 laurate; PEG-4 oleate; PEG-12 ricinoleate; PEG-2 stearate; PEG-4 stearate; PEG-50 stearate; PEG-2 tallowate; Pentaerythrityl ricinoleate; Pentaerythrityl tetrabenzoate; Pentaerythrityl tetracaprylate/caprate; Phenolic resin; Phenyl glycol ether; Phthalic anhydride; Pine tar; Poloxamer 184; Polybutene; Polychloroprene; Polychlorotrifluoroethylene; Polyester adipate; Polyester glutarate; Polyester phthalate; Polyester sebacate; Polyethylene glycol; Polyglyceryl-10 tetraoleate; Poly-α-methylstyrene; Polynuclear aromatic hydrocarbon; Polyoctenamer; Polypropylene glycol; Polypropylene glycol dibenzoate; Polyvinyl methyl ether; PPG-12; PPG-17; PPG (1000); PPG-2000; Propylene glycol; Propylene glycol dibenzoate; Propylene glycol dioctanoate; Propylene glycol dipelargonate; Propylene glycol laurate; 2-Pyrrolidone

Ricinoleic acid

Sodium rosinate; Stearic acid; Stearyl citrate; Stearyl nitrile; Stearyl stearate; Sucrose acetate isobutyrate; Sucrose benzoate; Sulfated castor oil

o-Toluene ethylsulfonamide; o-Toluenesulfonamide; p-Toluenesulfonamide; Triacetin; Triaryl phosphate; Tributoxyethyl phosphate; Tributyl citrate; 2,4,6-Tri-t-butylphenol; Tributyl phosphate; Tricapryl citrate; Tricapryl trimellitate; Tricresyl phosphate; Tridecyl adipate; Triethyl citrate; Triethylene glycol di-2-ethylhexanoate; Triethyl phosphate; Tri n-hexyl trimellitate; Triisodecyl trimellitate; Triisononyl trimellitate; Triisooctyl trimellitate; Triisopropylphenyl phosphate; Trimethylolethane tricaprylate-caprate; Trimethylol pentanediol dibenzoate; Trimethylolpropane triheptanoate; Trimethyl-1,3-pentanediol, 2,2,4-diisobutyrate; Trioctyl phosphate; Trioctyl trimellitate; Triolein; Tris (dichloropropyl) phosphate; Trixylenyl phosphate

Undecyl dodecyl phthalate; Undecyl phthalate

Vegetable oil, vulcanized

Zinc-epoxy

Preservatives

Trade names: Dow Corning® 7 Compound
Dowicil® 75
Ethylan® HB1; Ethyl Namate®
Ionpure Type A; Ionpure Type B
Kathon® LX; Koster Keunen Ceresine
Micro-Chek® 11
Octopol SDE-25
Sustane® BHT
Vancide® 89; Vancide® MZ-96; Vancide® TH

Chemicals: BHA
DMDM hydantoin
Hexahydro-1,3,5-triethyl-s-triazine
Laurtrimonium chloride
Oxazolidine
Quaternium-15
Silver borosilicate; Sodium diethyldithiocarbamate

Processing aids

Trade names: A-C® 540, 540A; A-C® 580; A-C® 5120; A-C® 5180; Acrawax® C; Actimer FR-1025M; ADK STAB NA-10; Aflux® 12; Aflux® 42; Aflux® 54; Ageflex TM 402, 403, 404, 410, 421, 423, 451, 461, 462; Airvol® 805; Airvol® 823; Airvol® 840; Akrochem® PEG 3350; Akrochem® Peptizer 66; Akrofax™ 900C; Akrofax™ A; Akrofax™ B; Akrotak 100; Akrowax PE; Aktiplast® F; Aktiplast® T; Ampacet 10562; Ampacet 10919; Armeen® 18D; Armid® HTD; Atmer® 122; Atmer® 129; Atmer® 1010

Barimite™ 1525P; Barimite™ 4009P; Benzoflex® S-312; Benzoflex® S-404; Benzoflex® S-552; Be Square® 185; Blendex® 590; Blendex® 869; Bondogen E®; Bowax 2015; Burez K20-505A; Burez K50-505A

Cabot® CP 9396; Cabot® PE 9324; Cabot® PE 9396; Cachalot® S-53; CAO®-3; Cardis® 10; Cardis® 36; Cardis® 314; Cardis® 319; Cardis® 320; Castorwax® MP-70; Castorwax® MP-80; Celite® 270; Celite® 292; Celite® 350; Celite® Super Floss; Chemstat® 327; Chloropren-Faktis A; Chloropren-Faktis NW; Cimbar™ 1025P; Cimbar™ 1536P; Cimbar™ 3508P; Cirrasol® AEN-XB; Cirrasol® ALN-GM

Diak No. 4; Diene® 55NF; Dimodan LS; Dimodan PM; Dimodan PV; Dimodan PV 300; Drimix®; Dryspersion®; Duradene® 710; Durastrength 200; Dynamar® FX-5910; Dynamar® FX-5920; Dynamar® FX-9613; Dynamar® PPA-790; Dynamar® PPA-791; Dynamar® PPA-2231; Dyphene® 8318; Dyphene® 8320; Dyphene® 8330; Dyphene® 8340; Dyphene® 8400; Dyphene® 8787; Dythal; Dythal XL

Eastoflex E1060; Eastoflex P1023; Eastotac™ H-100; Eastotac™ H-100E; Eastotac™ H-100L; Eastotac™ H-100R; Eastotac™ H-100W; Eastotac™ H-115E; Eastotac™ H-115L; Eastotac™ H-115R; Eastotac™ H-115W; Eastotac™ H-130E; Eastotac™ H-130L; Eastotac™ H-130R; Eastotac™ H-142R; Ebal; Emuldan DG 60; Emuldan DO 60; Emuldan HV 40; Emuldan HV 52; Epolene® C-10; Epolene® C-10P; Epolene® C-13; Epolene® C-13P; Epolene® C-14; Epolene® C-15; Epolene® C-15P; Epolene® C-16; Epolene® C-17; Epolene® C-17P; Epolene® C-18; Epolene® E-10; Epolene® E-15; Epolene® E-20; Epolene® E-43; Epolene® N-10; Epolene® N-20; Estabex® 2307; Estabex® 2307 DEOD; EXP-77 Mercapto Functional Silicone Wax

Faktis 10; Faktis 14; Faktis Asolvan, Asolvan T; Faktis Badenia C; Faktis Badenia T; Faktis HF Braun; Faktis Para extra weich; Faktis R, Weib MB; Faktis RC 110, RC 111, RC 140, RC 141, RC 144; Faktis Rheinau H, W; Faktis T-hart; Faktis ZD; Faktogel® 10; Faktogel® 14; Faktogel® 17; Faktogel® 110; Faktogel® 111; Faktogel® 140; Faktogel® 141; Faktogel® 144; Faktogel® A; Faktogel® Asolvan; Faktogel® Asolvan T; Faktogel® Badenia C; Faktogel® Badenia T; Faktogel® HF; Faktogel® KE 8384; Faktogel® KE 8419; Faktogel® MB; Faktogel® Para; Faktogel® R; Faktogel® Rheinau H; Faktogel® Rheinau W; Faktogel® T; Faktogel® ZD; Fibra-Cel® BH-70; Fibra-Cel® BH-200; Fibra-Cel® CBR-18; Fibra-Cel® CBR-40; Fibra-Cel® CBR-41; Fibra-Cel® CBR-50; Fibra-Cel® CC-20; Fibra-Cel® SW-8; Flexricin® P-4; Flexricin® P-6

G-Flex™-11; G-Flex™-12; G-Flex™ HMD; G-Flex™ HMG; Glycolube® 315; Glycolube® CW-1;

Processing aids *(cont'd.)*

Glycolube® TS; Glycowax® 765

Hercolite™ 240; Hercolite™ 290; Hoechst Wax PED 191; Hostalub® H 12; Hostalub® System G1973; Hostastat® HS1; Hostastat® System E3952; Hostastat® System E3953; Hostastat® System E3954; Hostastat® System E4751

Interlube™ 292; Interstab® CA-18-1; Isonate® 125M; Isonate® 143L; Isonate® 181; Isonate® 240

Kane Ace® PA-10; Kane Ace® PA-20; Kane Ace® PA-30; Kane Ace® PA-50; Kane Ace® PA-100; Kenflex® A; Kenflex® N; Kenlastic® K-6640; Kenlastic® K-6641; Kenlastic® K-6642; Kenlastic® K-9273; Ken-Mag®; Kenmix® LItharge/P 5/1; Kenmix® LItharge/P 9/1; Kenmix® Red Lead/P 9.1; Kenplast® ESB; Kenplast® ESI; Kenplast® ESN; Ken-React® 7 (KR 7); Ken-React® 9S (KR 9S); Ken-React® 12 (KR 12); Ken-React® 26S (KR 26S); Ken-React® 33DS (KR 33DS); Ken-React® 38S (KR 38S); Ken-React® 39DS (KR 39DS); Ken-React® 41B (KR 41B); Ken-React® 44 (KR 44); Ken-React® 46B (KR 46B); Ken-React® 55 (KR 55); Ken-React® 133DS (KR 133DS); Ken-React® 134S (KR 134S); Ken-React® 138D (KR 138D); Ken-React® 138S (KR 138S); Ken-React® 158D (KR 158D); Ken-React® 158FS (KR 158FS); Ken-React® 212 (KR 212); Ken-React® 238A (KR 238A); Ken-React® 238J (KR 238J); Ken-React® 238M (KR 238M); Ken-React® 238S (KR 238S); Ken-React® 238T (KR 238T); Ken-React® 262A (KR 262A); Ken-React® 262ES (KR 262ES); Ken-React® OPP2 (KR OPP2); Ken-React® OPPR (KR OPPR); Ken-React® KR TTS; Ken-Zinc®; Kronitex® 100; Kronitex® TCP; K-Stay 21®; KZ 55

LICA 01; LICA 09; LICA 12; LICA 38; LICA 38A; LICA 38J; LICA 44; LICA 97; LICA 99; Lilamin AC-59 P; Loxiol® HOB 7169; Lubricin 25; Luwax® A; Luwax® AL 61

Mark® 5060; Masil® SFR 70; Mekon® White; Millrex™; Mistron Vapor; Mistron Vapor® R; MoldPro 613; MoldPro 619; Mr 245; Mr 285; Mr 305; Myvaplex® 600P; Myvaplex® 600PK; Myverol® 18-04; Myverol® 18-06; Myverol® 18-07; Myverol® 18-99

Naftopast® Antimontrioxid; Naftopast® Antimontrioxid-CP; Naftopast® Di7-P; Naftopast® Di13-P; Naftopast® GMF; Naftopast® Litharge A; Naftopast® MBT-P; Naftopast® MBTS-A; Naftopast® MBTS-P; Naftopast® MgO-A; Naftopast® Mi12-P; Naftopast® Red Lead A, P; Naftopast® Schwefel-P; Naftopast® Thiuram 16-P; Naftopast® TMTM-P; Naftopast® ZnO-A; Naftozin® N, Spezial; Nevchem® 70; Nevchem® 100; Nevchem® 110; Nevchem® 120; Nevchem® 130; Nevchem® 140; Nevchem® 150; Nipol® DP5128P; Norsolene 9090; Norsolene 9110; Norsolene A 90; Norsolene A 100; Norsolene A 110; Norsolene D 3005; Norsolene I 130; Norsolene I 140; Norsolene L 2010; Norsolene M 1080; Norsolene M 1090; Norsolene S 85; Norsolene S 95; Norsolene S 105; Norsolene S 115; Norsolene S 125; Norsolene S 135; Norsolene S 145; Norsolene S 155; NZ 01; NZ 09; NZ 12; NZ 33; NZ 38; NZ 39; NZ 44; NZ 49; NZ 89; NZ 97

Octowax 321; Octowax 518; #30 Oil

PA-57; PA-80; Pamolyn® 125; Paraloid® K-120N; Paraloid® K-120ND; Paraloid® K-125; Paraloid® K-130; Paraloid® K-147; Paraloid® K-175; Paraloid® K-400; Paraloid® PM-800; Paricin® 1; Pationic® 909; Pationic® 914; Pationic® 919; Pationic® 1019; Pationic® 1145; Penreco Amber; Penreco Red; PE-PR 37; PE-PR 58; Peptizer 566; Peptizer 932; Peptizer 965; Peptizer 7010; Petrac® 270; Petrolite® C-700; Petrolite® C-1035; Petrolite® E-2020; Petronauba® C; Petro-Rez 103; Petro-Rez 200; Petro-Rez 215; Petro-Rez 801; Picco® 5070; Piccodiene® 2215SF; Plasticizer REO; Plastogen E®; Polydis® TR 016; Polydis® TR 060; Polydis® TR 131; Poly DNB®; Polysar S-1018; Polyvel CR-5; Polyvel CR-5F; Polyvel CR-5P; Polyvel CR-10; Polyvel CR-10P; Polyvel CR-20P; Polyvel CR-25; Polyvel PA10FP, FA10FP2; Proaid® 9802; Proaid® 9810; Proaid® 9814; Proaid® 9826; Proaid® 9831; Proaid® 9904; Proaid® FILL; Proaid® FLOW; Proaid® PEP; proFLOW 3000; PS044; PS045; PS046; PS047; PS047.5; PS048; PS048.5; PS049; PS049.5; PS050; PS140

Ralox® 46; Reogen E®; Resin 731D; Resinall 219; Resinall 286; Resinall 605; Resinall 767; Retarder BA, BAX; Ricon 152; Ricon 157; Ricon P30/Dispersion; Ross Wax #140; Ross Wax #160

SAFR-EPS K; SAFR-T-5K; SAFR-T-12K; SAFR-Z-12SK; SAFR-Z-44 K; Sipernat® 22; Softenol® 3900; Softenol® 3925, 3945, 3960; Solef® 11012/1001; Spectratech® PM 11607; Spectratech® PM 11725; SR-365; SR-416; SR-527; Stabilizer 7000; Starwax® 100; Staybelite®; Stereon® 205; Stereon® 210; Struktol® A 50; Struktol® A 60; Struktol® A 61; Struktol® A 80; Struktol® A 82; Struktol® A 86; Struktol® EP 52; Struktol® HP 55; Struktol® HPS 11; Struktol® PE H-100; Struktol® TH 10 Flakes; Struktol® TH 20 Flakes; Struktol® TR 042; Struktol® TR 044; Struktol® TR 065; Struktol® TR 071; Struktol® TR 077; Struktol® TR 141; Struktol® TR PE(H)-100; Struktol® W 34 Flakes; Struktol® W 80; Struktol® WA 48; Struktol® WB 212; Struktol® WB 222; Struktol® WB 300; Struktol® WS 280 Paste; Struktol®

Processing aids *(cont'd.)*

WS 280 Powd; Suconox-4®; Suconox-9®; Suconox-12®; Suconox-18®; Sunolite® 130; Sunolite® 160; Suprmix DBEEA; Syncrolube 3532; Syncrolube 3780; Syncrolube 3795; Synpro® Barium Stearate; Synpro® Cadmium Stearate; Synpro® Calcium Pelargonate; Synpro® Calcium Stearate 12B; Synpro® Calcium Stearate 15; Synpro® Calcium Stearate 15F; Synpro® Calcium Stearate 114-36; Synprolam 35; Synprolam 35X2; Synprolam 35X5; Synprolam 35X10; Synprolam 35X15; Synprolam 35X20; Synprolam 35X25; Synprolam 35X35; Synprolam 35X50; Synpro® Magnesium Stearate 90; Synpro® Zinc Stearate 8; Synpro® Zinc Stearate D; Synpro® Zinc Stearate GP; Synpro® Zinc Stearate GP Flake; Synpro® Zinc Stearate HSF

Tamol® NN 4501; Tartac 20; Tartac 30; Tartac 40; Tech Pet P; Texlin® 300; TR-Calcium Stearate

Unilin® 350 Alcohol; Unilin® 550 Alcohol; Unilin® 700 Alcohol; Unimoll® 66 M; Uniplex 260; Uniplex 413; Uralloy® Hybrid Polymer LP-2035

Vanfre® AP-2; Vanfre® AP-2 Special; Vanfre® DFL; Vanfre® HYP; Vanfre® M; Vanfre® UN; Vanfre® VAM; Varcum® 29421; Vestenamer® 6213; Vestenamer® 8012; Vestowax A-217; Vestowax A-227; Vestowax A-235; Vestowax A-415; Vestowax A-616; Vinuran®; Viton® Free Flow HD; Viton® Free Flow TA; VPA No. 1; VPA No. 2; VPA No. 3

Wacker Processing Aid 772; Wacker Silicone Antifoam Emulsion SE 9; W.G.S. Synaceti 116 NF/USP; Wingtack® 10; Wingtack® 95

Chemicals:

Acrylic resin; Asphalt, oxidized

Biscumylphenyl trimellitate; Butadiene-styrene copolymer; t-Butyl alcohol; Butyl cyclohexyl phthalate

Calcium pelargonate; Calcium sulfate; C30-50 alcohols; C13-15 amine; Canola oil glyceride; Castor oil, vulcanized; Cetyl esters; C5 hydrocarbon resin; Copolyether elastomer, thermoplastic; Cumylphenyl benzoate; Cumylphenyl neodecanoate; Cycloneopentyl, cyclo (dimethlaminoethyl) pyrophosphato zirconate, di mesyl salt

Diatomaceous earth; Dibutylammonium oleate; Di (butyl, methyl pyrophosphato) ethylene titanate di (dioctyl, hydrogen phosphite); Dicyclo (dioctyl) pyrophosphato titanate; Di (dioctylphosphato) ethylene titanate; Di (dioctylpyrophosphato) ethylene titanate; Dimethiconol; Ditridecyl sodium sulfosuccinate

Ethylene/acrylic acid copolymer; Ethylene/calcium acrylate copolymer; Ethylene distearamide; Ethylene/magnesium acrylate copolymer; Ethylene-propylene-styrene terpolymer; Ethylene thiourea

Glyceryl rosinate; Glyceryl stearate; Glyceryl tribenzoate

Hydrogenated cottonseed glyceride; Hydrogenated palm glyceride; Hydrogenated soy glyceride

Isopropyl 4-aminobenzenesulfonyl di (dodecylbenzenesulfonyl) titanate; Isopropyl dimethacryl isostearoyl titanate; Isopropyl tri (dioctylphosphato) titanate; Isopropyl tri (dioctylpyrophosphato) titanate; Isopropyl tri (N ethylamino-ethylamino) titanate

N-Lauroyl-p-aminophenol; Lead oxide; Lead oxide, red

4,4′-Methylenebis (cyclohexylamine)carbamate; Methyl hydroxystearate; Methyl methacrylate polymer; Methyl methacrylate styrene acrylonitrile copolymer; Microcrystalline wax; Microcrystalline wax, oxidized

Natural rubber; Neopentyl (diallyl) oxy, trihydroxy caproyl titanate; Neopentyl (diallyl) oxy, trimercapto-phenyl zirconate; Neopentyl glycol dibenzoate

PEG-2 C13-15 alkyl amine; PEG-5 C13-15 alkyl amine; PEG-10 C13-15 alkyl amine; PEG-15 C13-15 alkyl amine; PEG-20 C13-15 alkyl amine; PEG-25 C13-15 alkyl amine; PEG-35 C13-15 alkyl amine; PEG-50 C13-15 alkyl amine; N-Pelargonoyl-p-aminophenol; Pentaerythrityl tetrabenzoate; Pentaerythrityl tetrastearate; Pine tar; 1,2-Polybutadiene; Polyethylene, low-density; Poly-α-methylstyrene; Polyoctenamer; Polyvinylidene fluoride

Rapeseed oil, vulcanized

Simethicone; Stearamine; Stearoyl p-amino phenol; Stearyl/aminopropyl methicone copolymer; Styrene-acrylonitrile copolymer; Styrene-butadiene rubber; Sulfur; Synthetic wax

Tetrabromodipentaerythritol; Tetra (2, diallyoxymethyl-1 butoxy titanium di (di-tridecyl) phosphite; Tetraisopropyl di (dioctylphosphito) titanate; Tetraoctyloxytitanium di (ditridecylphosphite); Titanium di (butyl, octyl pyrophosphate) di (dioctyl, hydrogen phosphite) oxyacetate; Titanium di (cumylphenylate) oxyacetate; Titanium di (dioctylpyrophosphate) oxyacetate; Titanium dimethacrylate oxyacetate; Triaryl phosphate; Tris(3,5-di-t-butyl-4-hydroxy benzyl) isocyanurate

Vegetable oil, vulcanized

Reclaiming agents

Trade names: Aktiplast® 6 R; Aktiplast®6 N
Plasticizer REO

Chemicals: Dixylyldisulfides

Reducing agents

Trade names: CI7840; CP0320; CS1590; CT2902
Drewfax® 0007; Dynasylan® 1110; Dynasylan® 1411; Dynasylan® 1505; Dynasylan® 3403; Dynasylan® AMEO; Dynasylan® AMMO; Dynasylan® AMTC; Dynasylan® BDAC; Dynasylan® BSA; Dynasylan® CPTEO; Dynasylan® CPTMO; Dynasylan® DAMO; Dynasylan® ETAC; Dynasylan® GLYMO; Dynasylan® HMDS; Dynasylan® IBTMO; Dynasylan® IMEO; Dynasylan® MEMO; Dynasylan® MTAC; Dynasylan® MTES; Dynasylan® MTMO; Dynasylan® MTMS; Dynasylan® OCTEO; Dynasylan® PETCS; Dynasylan® PTMO; Dynasylan® TCS; Dynasylan® TRIAMO; Dynasylan® VTC; Dynasylan® VTEO; Dynasylan® VTMO; Dynasylan® VTMOEO
Hydrosulfite AWC
Lubrizol® 2152
Parolite; Petrarch® A0700; Petrarch® A0742; Petrarch® A0743; Petrarch® A0750; Petrarch® A0800; Petrarch® B2500; Petrarch® C3292; Petrarch® C3300; Petrarch® G6720; Petrarch® H7300; Petrarch® I7810; Petrarch® M8550; Petrarch® M8450; Petrarch® M8500; Petrarch® M9050; Petrarch® M9100; Petrarch® O9835; Petrarch® T1807; Petrarch® T1918; Petrarch® T1980; Petrarch® T2090; Petrarch® T2503; Petrarch® T2910; Petrarch® V4900; Petrarch® V4910; Petrarch® V4917; Petrarch® V5000
Thiovanol®

Chemicals: Carbon
(3-Glycidoxypropyl)-methyldiethoxy silane
Isocyanatopropyltriethoxysilane
3-Mercaptopropylmethyldimethoxysilane; N-Methylaminopropyltrimethoxysilane
2-Phenethyltrichlorosilane
Stannous chloride anhyd.; 3-(N-Styrylmethyl-2-aminoethylamino) propyltrimethoxy silane hydrochloride
Tetraethoxysilane; Tetrakis (2-ethoxyethoxy) silane; Tetramethoxysilane; Tetra-n-propoxysilane; Thioglycerin
Zinc formaldehyde sulfoxylate

Reinforcements

Trade names: Aerosil® 130; Aerosil® 150; Aerosil® 200; Aerosil® R972; Aerosil® R972V; Aerosil® R974; AgGlad™ Filament 16; AgGlad™ Filament 32; AgGlad™ Platelet 8; AgGlad™ Platelet 64; Akrochem® P 40; Akrochem® P 55; Akrochem® P 82; Akrochem® P 486; Albaglos®; ASP® NC; Atomite®
Bilt-Plates® 156; Buca®
Cab-O-Sil® EH-5; Cab-O-Sil® H-5; Cab-O-Sil® HS-5; Cab-O-Sil® L-90; Cab-O-Sil® LM-130; Cab-O-Sil® LM-150; Cab-O-Sil® LM-150D; Cab-O-Sil® M-5; Cab-O-Sil® M-7D; Cab-O-Sil® MS-75D; Cab-O-Sil® TS-610; Cab-O-Sil® TS-720; Cardis® 370; Catalpo®; Catalpo® X-1; CC™-101; CC™-103; CC™-105; Celite® 270; Celite® 292; Celite® 350; CF 1500; CF 32,500T Coarse; Chemprene R-10; Chemprene R-100; Chemprene R-115; Cimflx 606; Cimpact 600; Cimpact 610; Cimpact 699; Continex® LH-10; Continex® LH-20; Continex® LH-30; Continex® LH-35; Continex® N110; Continex® N220; Continex® N234; Continex® N299; Continex® N326; Continex® N330; Continex® N339; Continex® N343; Continex® N351; Continex® N550; Continex® N650; Continex® N660; Continex® N683; Continex® N762; Continex® N774; Cosmopon BN; Coupsil VP 6109; Coupsil VP 6411; Coupsil VP 6508; Coupsil VP 8113; CP Filler™
Darex® 632L; Dialead®; Dixie® Clay; Drikalite®; Duramite®; Durosil; Dyphene® 877PLF; Dyphene® 8845P
Fiberfrax® 6000 RPS; Fiberfrax® 6900-70; Fiberfrax® 6900-70S; Fiberfrax® EF-119; Fiberfrax®

Reinforcements *(cont'd.)*

EF 122S; Fiberfrax® EF 129; Fiberglas® 101C; Fiberkal™; Fibra-Cel® BH-70; Fibra-Cel® BH-200; Fibra-Cel® CBR-18; Fibra-Cel® CBR-40; Fibra-Cel® CBR-41; Fibra-Cel® CBR-50; Fibra-Cel® CC-20; Fibra-Cel® SW-8; Fibrox 030-E; Fibrox 030-ES; Fibrox 300; Frekote® EXITT®

Hi-Pflex® 100; Hi-Sil® 132; Hi-Sil® 135; Hi-Sil® 210; Hi-Sil® 233; Hi-Sil® 243LD; Hi-Sil® 250; Hi-Sil® 255; Hi-Sil® 532EP; Hi-Sil® 752; Hi-Sil® 900; Hi-Sil® 915; Hi-Sil® 2000; Hi-Sil® ABS; Hi-Sil® EZ

Kadox®-911; Kadox®-920; Ketjensil® SM 405; Kevlar® 29, 49; Koreforte®; Kotamite®

Marble Dust™; Marblemite™; McNamee® Clay; Micro-White® 07 Slurry; Micro-White® 10 Slurry; Micro-White® 15; Micro-White® 15 SAM; Micro-White® 15 Slurry; Micro-White® 25; Micro-White® 25 Slurry; Micro-White® 40; Micro-White® 100; Min-U-Sil® 30; Mistron PXL®; Mistron Vapor® R

No. 1 White™; Nytal® 100; Nytal® 200; Nytal® 300; Nytal® 400

Opacicoat™

Par® Clay; Perkasil® KS 207; Perkasil® KS 300; Perkasil® KS 404; Perkasil® VP 406; Petro-Rez 215; Picco® 5070; Picco® 6070; Picco® 6100; Picco® 6115; Piccolyte® A115; Piccolyte® A125; Piccolyte® A135; Pliolite® S-6B, S-6F; PMF® Fiber 204; PMF® Fiber 204AX; PMF® Fiber 204BX; PMF® Fiber 204CX; PMF® Fiber 204EX

R7557P; R7559; 1R70; Resimene® 3520

Santoweb® D; Santoweb® DX; Santoweb® H; Santoweb® W; Satintone® SP-33®; Satintone® Whitetex®; SC-53; Select-A-Sorb Powdered, Compacted; Si69; Si 264; Silene® 732D; Silene® D; Silverline 665; Sipernat® 44; Snowflake P.E.™; Snowflake White™; Spheriglass® Coated; Spheriglass® 1922; Spheriglass® 2530; Spheriglass® 3000E; Spheriglass® 4000E; Spheriglass® 5000; Spheriglass® 5000E; Spheriglass® 6000; Spheriglass® Coated; Star Stran 748; Supermite®; Super-Pflex® 200

Thermax® Stainless; Translink® 37; Translink® 77; Translink® 445; Translink® 555; Translink® HF-900; Tullanox HM-250

Ultrasil® VN3SP

Vertal 92; Vulcan® 2H; Vulcan® 3; Vulcan® 3H; Vulcan® 4H; Vulcan® 6; Vulcan® 6LM; Vulcan® 7H; Vulcan® 8H; Vulcan® 9; Vulcan® 10H; Vulcan® C; Vulcan® J; Vulcan® K; Vulcan® M; Vulcan® XC72; Vulkadur® A; Vulkadur® RA; Vulkadur® RB; Vulkasil® A1; Vulkasil® C; 1W70, 1W73; 1W74

X230-S®; 1X30; 1X33; 1X40, 1X43

Zeosil 45; Zeosil 85; Zeosil 165G; Zeosil 175MP; Zeosil 1165; Zinc Oxide No. 185; Zinc Oxide No. 318; Zink Oxide Transparent; Zinkoxyd Activ®

Chemicals: Aluminum-magnesium-sodium silicate; Aluminum silicate, calcined; Aluminum silicate, hydrous

Bis-(3-(triethoxysilyl)propyl)tetrasulfane

Calcium aluminosilicate; Calcium silicate; Carbon black; Carbon fiber; Cellulose

Graphite

Hexamethylmelamine

Magnesium carbonate

Novolac resin

Phenolic resin; Poly-α-methylstyrene

Silica; Silica dimethyl silylate; Silica, hydrated; Sodium silicoaluminate; Styrene-butadiene polymer

Talc; 3-Thiocyanatopropyltriethoxysilane

Release agents • Mold releases

Trade names: Abluwax EBS; A-C® 8, 8A; A-C® 9, 9A, 9F; A-C® 629; A-C® 629A; ADK STAB CA-ST; ADK STAB MG-ST; ADK STAB ZN-ST; Adol® 52; Advawax® 240; Advawax® 280; Advawax® 290; Aerosol® OT-75%; Aerosol® OT-100%; Akrochem® Calcium Stearate; Akrochem® P 3100; Akrochem® P 4000; Akrochem® P 4100; Akrochem® PEG 3350; Akrochem® Rubbersil RS-200; Akrochem® Silicone Emulsion 10M; Akrochem® Silicone Emulsion 60M; Akrochem® Silicone Emulsion 100; Akrochem® Silicone Emulsion 350; Akrochem® Silicone Emulsion 350 Conc; Akrochem® Silicone Fluid 350; Akrochem® Wettable Zinc Stearate; Akrochem® Zinc Stearate; Akrodip™ C-100; Akrodip™ Dry; Akrodip™ V-300; Akrodip™ Z-50/50; Akrodip™ Z-200; Akrodip™ Z-250; Akro-Gel; Akrolease® E-9410; Akrolease® E-9491; Albrite® TIOP; Alkamide® STEDA; Aqualease™ 2802; Aqualease™ 6100;

Release agents • Mold releases *(cont'd.)*

Aqualease™ 6101; Aqualease™ 6102; Aqualease™ 6201; Aqualease™ 6202; Aqualine GP100; Aqualine R-110; Aqualine RC-321; Armeen® 18; Armeen® 18D; Armid® E; Armid® HTD; Armoslip® 18; Armoslip® EXP

BL 3; Blue Label Silicone Spray No. SB412-A; Borax; Bowax 2015; Britol® 6NF; Buspense 047

Cachalot® S-53; Castorwax® MP-70; Castorwax® MP-80; CB2408; Chemlease 55; Chemlease 80W; Chemlease 88; Chemlease 158R; Chemlease 906E; Chemlease PMR; Chemlease SP 40; CI7840; Coad 10; Coad 20; Coathylene HA 1591; Conap® MR-5014; Conap® MR-5015; CP0320; Crodamide 203; Crodamide 212; Crodamide O; Crodamide OR; Crodamide S; Crodamide SR; Crystal 1000; Crystal 1053; Crystal 3000; Crystal 4000; Crystal 7000; Crystal HMT; CS1590; CT-88 Aerosol; CT2902

DB-9 Paintable Silicone Emulsion; Dimul S; Doittol K21; Dow E200; Dow E300; Dow E400; Dow E600; Dow E900; Dow E1000; Dow E1450; Dow E3350; Dow E4500; Dow E8000; Dow Corning® 7 Compound; Dow Corning® 20 Release Coating; Dow Corning® 24 Emulsion; Dow Corning® 200 Fluid; Dow Corning® 203 Fluid; Dow Corning® 236 Dispersion; Dow Corning® 290 Emulsion; Dow Corning® 346 Emulsion; Dow Corning® 347 Emulsion; Dow Corning® 7119 Release/Parting Agent; Dow Corning® HV-490 Emulsion; Durelease HS1; Durelease HS2; Dynamar® FC-5158; Dynamar® FC-5163; Dynamar® FX-5170; Dynamar® PPA-790; Dynamar® PPA-791; Dynasylan® 0116; Dynasylan® 1110; Dynasylan® 1411; Dynasylan® 1505; Dynasylan® 3403; Dynasylan® AMEO; Dynasylan® AMMO; Dynasylan® AMTC; Dynasylan® BDAC; Dynasylan® BSA; Dynasylan® CPTEO; Dynasylan® CPTMO; Dynasylan® DAMO; Dynasylan® ETAC; Dynasylan® GLYMO; Dynasylan® HMDS; Dynasylan® IMEO; Dynasylan® MEMO; Dynasylan® MTAC; Dynasylan® MTES; Dynasylan® MTMO; Dynasylan® MTMS; Dynasylan® OCTEO; Dynasylan® PETCS; Dynasylan® PTMO; Dynasylan® TCS; Dynasylan® TRIAMO; Dynasylan® VTC; Dynasylan® VTEO; Dynasylan® VTMO; Dynasylan® VTMOEO

Ease Release™ 200 Series; Ease Release™ 300 Series; Ease Release™ 400 Series; Ease Release™ 500 Series; Ease Release™ 700; Ease Release™ 900; Ease Release™ 1700; Ease Release™ 2040 Series; Ease Release™ 2110; Ease Release™ 2116; Ease Release™ 2148; Ease Release™ 2181; Ease Release™ 2191; Ease Release™ 2197; Ease Release™ 2249; Ease Release™ 2251; Ease Release™ 2300; Ease Release™ 2400; Emerest® 2301; Emerest® 2302; Emery® 6686; Emery® 6687; Emery® 6773; Epolene® C-10P; Epolene® C-13; Epolene® C-13P; Epolene® C-14; Epolene® C-17; Epolene® C-17P; Epolene® E-10; Epolene® E-15; Epolene® E-20; Epolene® N-11; Epolene® N-14; Epolene® N-20; Epolene® N-21; Epolene® N-34; Eureslip 58; EXC-33; EXP-24-LS Silicone Wax Emulsion; EXP-36-X20 Epoxy Functional Silicone Sol'n; EXP-58 Silicone Wax; EXP-61 Amine Functional Silicone Wax; EXP-77 Mercapto Functional Silicone Wax

Fancol OA-95; Flexricin® 13; Frekote® 31; Frekote® 44-NC; Frekote® 700-NC; Frekote® 800-NC; Frekote® 815-NC; Frekote® 1711; Frekote® 1711-EM; Frekote® EXITT-EM; Frekote® FRP-NC; Frekote® HMT; Frekote® LIFFT®; Frekote® No. 1-EM; Frekote® S-10

Getren® 4/200; Getren® FD 372; Getren® FD 411; Getren® FD 575; Getren® S 3 Conc. 100; Glycolube® P; Glycolube® PG; Glycolube® PS; Glycolube® TS; Glycolube® VL; GP-4 Silicone Fluid; GP-5 Emulsifiable Paintable Silicone; GP-6 Silicone Fluid; GP-50-A Modified Silicone Emulsion; GP-51-E Dimethyl Silicone Emulsion; GP-52-E Dimethyl Silicone Emulsion; GP-53-E Dimethyl Silicone Emulsion; GP-54-E Dimethyl Silicone Emulsion; GP-60-E Silicone Emulsion; GP-70-E Paintable Silicone Emulsion; GP-70-S Paintable Silicone Fluid; GP-71-SS Mercapto Modified Silicone Fluid; GP-74 Paintable Silicone Fluid; GP-74-E Paintable Silicone Emulsion; GP-80-AE Silicone Emulsion; GP-85-AE Silicone Emulsion; GP-86-AE Silicone Emulsion; GP-98 Silicone Wax; GP-180 Resin Sol'n; GP-197 Resin Sol'n; GP-530 Silicone Fluid; GP-7100 Silicone Fluid; GP-7100-E Paintable Silicone Emulsion; GP-7101 Silicone Wax; GP-7102 Silicone Wax; GP-7105 Silicone Fluid; GP-7200 Silicone Fluid; GP-RA-156 Release Agent; GP-RA-158 Silicone Polish Additive; GP-RA-159 Silicone Polish Additive; GP-RA-201 Release Agent

Hallco® Lube; Hallcote® 525; Hallcote® 573; Hallcote® 780; Hallcote® 910LF; Hallcote® CSD; Hallcote® CaSt 50; Hallcote® ES-10; Hallcote® ZS; Hallcote® ZS 5050; Halocarbon Grease 19; Halocarbon Grease 25-5S; Halocarbon Grease 25-10M; Halocarbon Grease 25-20M; Halocarbon Grease 28; Halocarbon Grease 28LT; Halocarbon Grease 32; Halocarbon Grease X90-10M; Halocarbon Oil 0.8; Halocarbon Oil 1.8; Halocarbon Oil 4.2; Halocarbon Oil 6.3; Halocarbon Oil 27; Halocarbon Oil 56; Halocarbon Oil 95; Halocarbon Oil 200; Halocarbon Oil 400; Halocarbon Oil 700; Halocarbon Oil 1000N; Halocarbon Wax 40; Halocarbon Wax 600; Halocarbon Wax 1200; Halocarbon Wax 1500; Hamposyl® C; Hamposyl® C-30; Hamposyl® L; Hamposyl® L-30; Hamposyl® L-95; Hamposyl® M; Hamposyl® M-30;

Release agents • Mold releases *(cont'd.)*

Hamposyl® O; Hamposyl® S; Hodag PEG 400; Hodag PEG 8000; Hodag PPG-150; Hodag PPG-400; Hodag PPG-1200; Hodag PPG-2000; Hodag PPG-4000; Hoechst Wax C; Hoechst Wax E; Hoechst Wax OP; Hoechst Wax S; Hostalub® H 12; Hostalub® WE4; Hy Dense Calcium Stearate HP Gran; Hy Dense Calcium Stearate RSN Powd; Hy Dense Zinc Stearate XM Powd; Hy Dense Zinc Stearate XM Ultra Fine; Hydro Zinc™

Interlube A; Interlube P/DS; Interstab® BC-109; Interstab® BC-110; Interstab® CA-18-1

Kantstik™ 79; Kantstik™ 94; Kantstik™ BW; Kantstik™ LM Conc. B; Kantstik™ M-55; Kantstik™ M-56; Kantstik™ PE; Kantstik™ Q Powd; Kantstik™ RIM; Kantstik™ SPC; Kemamide® B; Kemamide® E; Kemamide® O; Kemamide® S; Kemamide® U; Kemamide® W-20; Kemamide® W-39; Kemamide® W-40; Kemamide® W-40/300; Kemamide® W-45; Kemamine® AS-650; Kemamine® AS-974; Kemamine® AS-974/1; Kemamine® AS-989; Kemamine® AS-990; Kemamine® P-190, P-190D; Kemamine® P-650D; Kemamine® P-970; Kemamine® P-970D; Kemamine® P-974D; Kemamine® P-989D; Kemamine® P-990, P-990D; Kemamine® P-999; Kenflex® A; Kenflex® N; Ken-React® 7 (KR 7); Ken-React® 9S (KR 9S); Ken-React® 12 (KR 12); Ken-React® 26S (KR 26S); Ken-React® 33DS (KR 33DS); Ken-React® 38S (KR 38S); Ken-React® 39DS (KR 39DS); Ken-React® 41B (KR 41B); Ken-React® 44 (KR 44); Ken-React® 46B (KR 46B); Ken-React® 55 (KR 55); Ken-React® 133DS (KR 133DS); Ken-React® 134S (KR 134S); Ken-React® 138D (KR 138D); Ken-React® 138S (KR 138S); Ken-React® 158D (KR 158D); Ken-React® 158FS (KR 158FS); Ken-React® 212 (KR 212); Ken-React® 238A (KR 238A); Ken-React® 238J (KR 238J); Ken-React® 238M (KR 238M); Ken-React® 238S (KR 238S); Ken-React® 238T (KR 238T); Ken-React® 262A (KR 262A); Ken-React® 262ES (KR 262ES); Ken-React® OPP2 (KR OPP2); Ken-React® OPPR (KR OPPR); Ken-React® KR TTS; KZ 55; KZ OPPR; KZ TPP; KZ TPPJ

Larostat® 88; Larostat® 264 A Conc; Larostat® 377 DPG; Larostat® 451 P; LICA 01; LICA 09; LICA 12; LICA 38; LICA 38A; LICA 38J; LICA 44; LICA 97; LICA 99; Lilamin AC-59 P; Liquax 488; Liquazinc AQ-90; Loxiol® G 70 S; Loxiol® G 71; Loxiol® G 71 S; Loxiol® G 72; Loxiol® HOB 7107; Loxiol® HOB 7140; Loxiol® VPG 1732; Lubrol 90; Luwax® A; Luwax® AL 61; Lyndcoat™ 10-RTU; Lyndcoat™ 15-RTU; Lyndcoat™ 20-RTU; Lyndcoat™ BR-18/M; Lyndcoat™ BR 500-RTU; Lyndcoat™ BR 601-RTU; Lyndcoat™ BR 790-RTU; Lyndcoat™ BR 880-RTU; Lyndcoat™ M 755-RTU; Lyndcoat™ M 771-RTU

Macol® 660; Macol® 3520; Macol® 5100; Masil® EM 100; Masil® EM 100 Conc; Masil® EM 100D; Masil® EM 100P; Masil® EM 250 Conc; Masil® EM 266 (50); Masil® EM 350X; Masil® EM 350X Conc; Masil® EM 1000; Masil® EM 1000 Conc; Masil® EM 1000P; Masil® EM 10,000; Masil® EM 10,000 Conc; Masil® EM 60,000; Masil® EM 100,000; Masil® SF 5; Masil® SF 10; Masil® SF 20; Masil® SF 50; Masil® SF 100; Masil® SF 200; Masil® SF 350; Masil® SF 350 FG; Masil® SF 500; Masil® SF 1000; Masil® SF 5000; Masil® SF 10,000; Masil® SF 12,500; Masil® SF 30,000; Masil® SF 60,000; Masil® SF 100,000; Masil® SF 300,000; Masil® SF 500,000; Masil® SF 600,000; Masil® SF 1,000,000; Mayzo RA-0010A, RA-10B, RA-10C; Mayzo RA-1315W; Mayzo RA-0095H; Mayzo RA-0095HS; McLube 1700L; McLube 1711L; McLube 1733L; McLube 1775; McLube 1777; McLube 1777-1; McLube 1779-1; McLube 1782; McLube 1782-1; McLube 1800; McLube 1804L; McLube 1829; McLube 1849; Mekon® White; Michel XO-24; Michel XO-85; Millrex™; Mistron ZSC; Molder's Edge ME-100; Molder's Edge ME-110; Molder's Edge ME-120; Molder's Edge ME-175; Molder's Edge ME-211; Molder's Edge ME-232; Molder's Edge ME-263; Molder's Edge ME-301; Molder's Edge ME-304; Molder's Edge ME-341; Molder's Edge ME-345; Molder's Edge ME-448; Molder's Edge ME-515; Molder's Edge ME-7000; Mold Lubricant™ 426; MoldPro 613; MoldPro 616; MoldPro 759; MoldPro 830; Mono-Coat® 65-RT; Mono-Coat® E76; Mono-Coat® E91; Mould Release Agent N 32; MS-122N/CO_2; MS-136N/CO_2; Myvaplex® 600PK; Myverol® 18-04; Myverol® 18-06; Myverol® 18-07; Myverol® 18-99

Nix Stix L195WF; Nix Stix L478; Nix Stix L515; Nix Stix L529; Nix Stix LO582; Nix Stix X9022; Nix Stix X9027; Nix Stix X9028; No Stik 802; No Stik 806; NP-Dip®; NZ 01; NZ 09; NZ 12; NZ 33; NZ 38; NZ 39; NZ 44; NZ 49; NZ 89; NZ 97; #1 Oil

Paintable Mist No. P112-A; Paintable Neutral Oil Spray No. N1616; Paintable Organic Oil Spray No. O316; Paricin® 220; Paricin® 285; Pationic® 901; Pationic® 902; Pationic® 905; Pationic® 1042; Pationic® 1042B; Pationic® 1042K; Pationic® 1042KB; Pationic® 1052; Pationic® 1052B; Pationic® 1052K; Pationic® 1052KB; Permalease™ 2040; Permalease™ 2264; Permalease™ 3000; Permalease™ 3207; Permalease™ 3500; Permalease™ 5000; Permalease™ 5500; Petrac® 215; Petrac® 250; Petrac® Calcium Stearate CP-11; Petrac® Calcium Stearate CP-11 LS; Petrac® Calcium Stearate CP-11 LSG; Petrac® Calcium/Zinc Stearate CZ-81; Petrac® Eramide®; Petrac® Slip-Eze®; Petrac® Slip-Quick®; Petrac® Vyn-Eze®; Petrac® Zinc Stearate ZN-41; Petrac® Zinc Stearate ZN-42; Petrac® Zinc Stearate

Release agents • Mold releases *(cont'd.)*

ZN-44 HS; Petrac® Zinc Stearate ZW-45; Petrarch® A0699; Petrarch® A0700; Petrarch® A0742; Petrarch® A0743; Petrarch® A0750; Petrarch® A0800; Petrarch® B2500; Petrarch® C3292; Petrarch® C3300; Petrarch® G6720; Petrarch® H7300; Petrarch® I7810; Petrarch® M8550; Petrarch® M8450; Petrarch® M8500; Petrarch® M9050; Petrarch® M9100; Petrarch® O9835; Petrarch® T1807; Petrarch® T1918; Petrarch® T1980; Petrarch® T2090; Petrarch® T2503; Petrarch® T2910; Petrarch® V4900; Petrarch® V4910; Petrarch® V4917; Petrarch® V5000; Plastilease 250; Plastilease 512-B; Plastilease 512-CL; Plastilease 514; Pluracol® E600 NF; Pluracol® E1000; Pluracol® E1450; Pluracol® E1450 NF; Pluracol® E2000; Pluracol® E4500; Pluracol® E8000; Polycone 1000; Polydis® TR 121; Polydis® TR 131; Poly-G® 200; Poly-G® 300; Poly-G® 400; Poly-G® 600; Poly-G® 1000; Poly-G® 1500; Poly-G® 2000; Poly-G® B1530; Polyvel RE25; Polyvel RE40; Polyvel RO25; Polyvel RO40; Polyvel RZ40; Polywax® 500; Polywax® 655; Polywax® 850; Polywax® 1000; Polywax® 2000; Polywax® 3000; Proaid® 9810; Proaid® 9831; Proaid® 9904; proFLOW 1000; proFLOW 3000; Prox-onic PEG-20,000; PS040; PS041; PS041.2; PS041.5; PS042; PS043; PS136.8; PS137; PS138; Pura WBC608

Quikote™; Quikote™ C; Quikote™ C-LD; Quikote™ C-LD-A; Quikote™ C-LM; Quikote™ C-LMLD; Quikote™ M

RC 7; Release Agent E-155; Release Agent NL-1; Release Agent NL-2; Release Agent NL-10; Release Gel 1765 G; Resinall 219; Rexanol™; Rhenodiv® 20; Rhenodiv® 30; Rhenodiv® A; Rhenodiv® F; Rhenodiv® KS; Rhenodiv® LE; Rhenodiv® LL; Rhenodiv® LS; Rhenodiv® PV; Rhenodiv® S; Rhenodiv® ZB; Rhodasurf® PEG 3350; Rhodasurf® PEG 8000; Ross Ceresine Wax; Ross Ouricury Wax Replacement; Ross Wax #100; Ross Wax #140; Ross Wax #160; RR 5; Rubber Lubricant™; Rubber Shield

SEM-135; SF18-350; SF96®; SF96® (100 cst); SF96® (350 cst); SF96® (500 cst); SF96® (1000 cst); SF1080; SF1188; Silicone Emulsion 350, 1M, 10M, 60M; Silicone Emulsion E-131; Silicone Emulsion E-133; Silicone Fluid 350, 1M, 5M, 10M, 30M, 60M; Silicone Mist No. S116-A; Silicone Release Agent #5038; Silicone Spray No. S312-A; Siloxan Tego® 457; Siltech® CE-2000; Siltech® E-2140; Siltech® E-2170; Siltech® MFF-3000; Siltech® PF; Siltech® STD-100; Silwax® F; Silwet® L-7602; SM 2061; SM 2068A; SM 2128; SM 2140; SM 2154; SM 2155; SM 2162; SM 2163; Softenol® 3107; Softenol® 3119; Softenol® 3408; Softenol® 3701; Softenol® 3925, 3945, 3960; Sprayon 803; Sprayon 805; Sprayon 807; Sprayon 811 Dry Film Vydax; Sprayon 812; Sprayon 815; Sprayon 829; Stoner E206; Stoner E302; Stoner E313; Stoner E408; Stoner E965; Struktol® HPS 11; Struktol® PE H-100; Struktol® TR 042; Struktol® TR 044; Struktol® TR 141; Struktol® TR PE(H)-100; Struktol® W 34 Flakes; Struktol® W 80; Struktol® WA 48; Struktol® WB 16; Struktol® WB 212; Struktol® WB 222; SU 3; Super 33 Silicone Spray No. S3312-A; SY-SL 131

Tegiloxan®; Tego® IMR® 412 T; Tego® IMR® 830; Tego® Emulsion 240; Tego® Emulsion 3454; Tego® Emulsion 3529; Tego® Emulsion ASL; Tego® Emulsion DF 19; Tego® Emulsion DT; Tego® Emulsion N; Tego® Emulsion S 3; Tego® Release Agent M 379; Tegosil®; Tegotrenn® LH 157 A; Tegotrenn® LH 525; Tegotrenn® LI 197; Tegotrenn® LI 237; Tegotrenn® LI 344; Tegotrenn® LI 448; Tegotrenn® LI 537 W; Tegotrenn® LI 747; Tegotrenn® LI 748; Tegotrenn® LI 766; Tegotrenn® LI 841-237; Tegotrenn® LI 850-344; Tegotrenn® LI 901; Tegotrenn® LI 903; Tegotrenn® LI 904; Tegotrenn® LK 104; Tegotrenn® LK 260, 498/F, 755, 859, 860; Tegotrenn® LK 870; Tegotrenn® LK 873; Tegotrenn® LK 876; Tegotrenn® LR 201/F, 236; Tegotrenn® LR 309 F; Tegotrenn® LR 468 F; Tegotrenn® LR 486; Tegotrenn® LR 560; Tegotrenn® LR 808-877; Tegotrenn® LR 869; Tegotrenn® LS 584; Tegotrenn® LS 809/F, 815; Tegotrenn® LS 814 F; Tegotrenn® LS 828; Tegotrenn® LS 829 F; Tegotrenn® LS 951-156; Tegotrenn® LS 952; Tegotrenn® LS 960; Tegotrenn® LS 961; Tegotrenn® LS 969; Tegotrenn® S 2100; Tegotrenn® S 2200; Teric C12; Teric PEG 300; Teric PEG 600; Teric PEG 4000; Teric PEG 8000; Teric PPG 2250; Teric PPG 4000; TL-115; Trenbest 500; Trennspray Tego®

Union Carbide® LE-45; Uniplex 413; Uniplex 809; Unislip 1753; Unithox® 420; Unithox® 450; Unithox® 480; Unithox® 550; Unithox® 750; Uniwax 1750; Uniwax 1760

Vanfre® IL-1; Vestowax A-217; Vestowax A-227; Vestowax A-235; Vestowax A-415; Vestowax A-616; Vestowax FT-150; Vestowax FT-150P; Vestowax FT-200; Vestowax FT-300; Viscasil® (10,000 cSt); Viscasil® (12,500 cSt); VPA No. 1; VPA No. 3; Vybar® 103; Vybar® 260; Vydax® Spray No. V312-A; Vydax® WD

Water Shield; Wet Zinc™ ; Wet Zinc™ P; Witco® MRC; Witco® Zinc Stearate Disperso; Witco® Zinc Stearate NW; Witco Zinc Stearate Regular

XC 9 Hot; XK 22; XR 7

Zelec NE; Zelec UN; Zincote™; Zinc Stearate Spray No. Z212-B; Zonyl® FSA; Zonyl® FSP

Release agents • Mold releases *(cont'd.)*

Chemicals: N-(2-Aminoethyl)-3-aminopropyl methyldimethoxy silane; 3-Aminopropylmethyldiethoxy silane; Aminopropyltrimethoxysilane; Amodimethicone; Amyltrichlorosilane; Arachidyl behenyl amine

Behenamide; Bis (hydroxyethyl) aminopropyltriethoxy silane; Bis-(N-methylbenzamide) ethoxymethyl silane; Bis (trimethylsilyl) acetamide

C8-10 acid-1,2-propanediol ester; C7 acid triglyceride; C8-10 acid triglyceride; Calcium stearate; Calcium-zinc stearate; Canola oil glyceride; Cetyl alcohol; 3-Chloropropyltriethoxysilane; 3-Chloropropyltrimethoxysilane; Cocamine; Cocoyl sarcosine; C20-40 pareth-3; C20-40 pareth-10; C20-40 pareth-40; C30-50 pareth-10; C40-60 pareth-10; Cycloneopentyl, cyclo (dimethlaminoethyl) pyrophosphato zirconate, di mesyl salt

Di-t-butoxydiacetoxysilane; Dimethicone/mercaptopropyl methicone copolymer; N,N´-Dioleoylethylenediamine

Erucamide; Erucyl erucamide; Erucyl stearamide; N,N´-Ethylenebis 12-hydroxystearamide; Ethylene dioleamide; Ethyltriacetoxysilane

Glyceryl hydroxystearate; Glyceryl montanate; (3-Glycidoxypropyl)-methyldiethoxy silane

Hexamethyldisilazane; Hydrogenated castor oil; Hydrogenated cottonseed glyceride; Hydrogenated palm glyceride; Hydrogenated tallowamine; N (2-Hydroxyethyl) 12-hydroxystearamide; N(β-Hydroxyethyl) ricinoleamide

Isobutyltrimethoxysilane; Isocyanatopropyltriethoxysilane

Lauramine; Lauric/myristic dimethylethyl ammonium ethosulfate; Lauroyl sarcosine; Lecithin; Lithium stearate

Magnesium stearate; 3-Mercaptopropylmethyldimethoxysilane; Mercaptopropyltrimethoxysilane; N-Methylaminopropyltrimethoxysilane; (α-Methylphenethyl) methyldimethylsiloxane copolymer; Methyltriacetoxysilane; Methyltriethoxysilane; Methyltrimethoxysilane; Montan acid wax; Myristoyl sarcosine

Neopentyl (diallyl) oxy, trihydroxy caproyl titanate; Neopentyl (diallyl) oxy, trimercapto-phenyl zirconate

Octyltriethoxysilane; Oleoyl sarcosine; Oleyl palmitamide

PEG-8M; PEG-12 castor oil; PEG-20 glyceryl stearate; Phenethylmethylsiloxane/dimethylsiloxane copolymer; Phenethylmethylsiloxane/hexylmethylsiloxane copolymjer; 2-Phenethyltrichlorosilane; Phenyltriethoxysilane; Polychlorotrifluoroethylene; Polyethylene, oxidized; Polyethylene wax; Polytetrafluoroethylene; Polyvinyl alcohol; Polyvinyl alcohol, partially hydrolyzed; Polyvinyl octadecyl carbamate; Potassium ricinoleate; PPG-9; PPG-20; PPG-26; PPG-30; PPG-28-buteth-35; Propyl oleate; n-Propyltrimethoxysilane

Silicone; Silicone emulsions; Sodium myristoyl sarcosinate; Sodium stearate; Soy acid; Soyethyldimonium ethosulfate; Stearamide; Stearamine; Stearethyldimonium ethosulfate; Stearoyl sarcosine; Stearyl/aminopropyl methicone copolymer; Stearyl erucamide; Stearyl stearamide; 3-(N-Styrylmethyl-2-aminoethylamino) propyltrimethoxy silane hydrochloride; Synthetic wax

Tetraethoxysilane; Tetrafluoroethylene telomer; Tetrakis (2-ethoxyethoxy) silane; Tetramethoxysilane; Tetra-n-propoxysilane; Trichlorosilane; N-[3-(Triethoxysilyl)-propyl] 4,5-dihydroimidazole; 1-Trimethoxysilyl-2-(chloromethyl) phenylethane; Trimethoxysilylpropyldiethylene triamine; Trioctyldodecyl citrate

Vinyltrichlorosilane; Vinyltriethoxysilane

Zinc stearate

Retarders

Trade names: Altax®

Eastman® MTBHQ; Ektapro® EEP Solvent; Ektasolve® EB; Ektasolve® EP; Ethyl Tuads® Rodform

HyTemp SR-50

KZ 55; KZ OPPR; KZ TPP; KZ TPPJ

LICA 01; LICA 09; LICA 12; LICA 38; LICA 38A; LICA 38J; LICA 44; LICA 97; LICA 99

Methyl Tuads® Rodform

NZ 01; NZ 09; NZ 12; NZ 33; NZ 38; NZ 39; NZ 44; NZ 49; NZ 89; NZ 97

Resinall 203; Retarder AK; Retarder BA, BAX; Retarder PX; Retarder SAFE; Retarder SAX; Retrocure® G; Rhenogran® Vulk. E-80

Tartac 20; Tartac 30; Tartac 40; Thiofide®

Retarders *(cont'd.)*

Vulkalent B/C; Vulkalent E/C; Vulkalent TM
Zeonet® U

Chemicals: Benzoic acid; Benzothiazyl disulfide
Cycloneopentyl, cyclo (dimethlaminoethyl) pyrophosphato zirconate, di mesyl salt
N,N′-Diphenylurea
4- and 5-Methylbenzotriazole
Neopentyl (diallyl) oxy, trihydroxy caproyl titanate; Neopentyl (diallyl) oxy, trimercapto-phenyl zirconate
Phthalic anhydride
Tolyltriazole

Scavengers

Trade names: A-2; A-201; Ageflex FM-1
Cyasorb® UV 3346 LD
DHT-4A
Epoxol 7-4
Hostanox® O3
Interstab® CA-18-1
Pationic® 920; Pationic® 925; Pationic® 930; Pationic® 930K; Pennstop® 1866; Pennstop® 2697; Plas-Chek® 775; Plastolein® 9232
Tinuvin® 765
UV-Chek® AM-595
Viton® Free Flow HD
Zerogen® 11

Chemicals: Calcium/sodium stearoyl lactylate
3,5-Di-t-butyl-4-hydroxybenzoic acid, n-hexadecyl ester; Diethylhydroxylamine; Dimethylaminoethyl methacrylate
Poly [(6-morpholino-s-triazine-2,4-diyl) [2,2,6,6-tetramethyl-4-piperidyl) imino]-hexamethylene [(2,2,6,6-tetramethyl-4-piperidyl) imino]; Pyridine
Sodium stearoyl lactylate
2,2′-Thiobis (4-t-octylphenolato)-n-butylamine nickel II; Tris Neodol-25 phosphite; Tristearyl phosphite

Shortstops • Terminators

Trade names: Aquatreat DNM-9; Aquatreat DNM-30; Aquatreat KM; Aquatreat SDM; Arizona DR-22
Empol® 1020
Octopol SBZ-20; Octopol SDM-40
Perkacit® SDMC

Chemicals: Phenylethanolamine

Slip agents

Trade names: Ampacet 10061; Ampacet 10117; Armid® 18; Armid® HT; Armid® O; Armoslip® 18; Armoslip® EXP
Cabot® PE 9020; Cabot® PE 9041; Cabot® PE 9166; Cabot® PE 9171; Cabot® PE 9172; Cabot® PE 9174; Cabot® PE 9220; Cardis® 10; Cardis® 36; Cardis® 314; Cardis® 319; Cardis® 320; Cardis® 370; Chemstat® HTSA #1A; Chemstat® HTSA #3B; Chemstat® HTSA #18; Chemstat® HTSA #18S; Chemstat® HTSA #22; Colloids PE 48/10/06; Colloids PE 48/10/13; Crodamide 203; Crodamide ER; Crodamide O; Crodamide OR; Crodamide S
Dimul S
Emerwax® 9380; Eureslip 58
Glycolube® 140; Glycolube® 345; Glycolube® 674; Glycolube® 740; Glycolube® 742; Glycolube® 825; Glycolube® TS; Glycomul® S

Slip agents *(cont'd.)*

Kemamide® B; Kemamide® E; Kemamide® O; Kemamide® S; Kemamide® U; Kemamide® W-20; Kemamide® W-39; Kemamide® W-40; Kemamide® W-40/300; Kemamide® W-40DF; Kemamide® W-45
Lexolube® BS-Tech; Lexolube® NBS
MDI SC-305; MDI SF-325; Mono-Coat® E91
Neutral Degras
Paricin® 285; PE-SA 21; PE-SA 24; PE-SA 40; PE-SA 49; PE-SA 60; PE-SA 72; PE-SA 92; PE-SL 04; PE-SL 31; PE-SL 98; Petrac® Eramide®; Petrac® Slip-Quick®; Petrac® Vyn-Eze®; Petronauba® C; Polydis® TR 121; Polydis® TR 131; Polyvel A2010; Polyvel RE25; Polyvel RE40; Polyvel RO25; Polyvel RO40; Polyvel RO252; Polyvel RZ40; PP-SA 46; PP-SA 97; PP-SA 128; PP-SL 82; PP-SL 98; PP-SL 101
RS-80
Silicone Fluid 350, 1M, 5M, 10M, 30M, 60M; Silwet® L-7001; Silwet® L-7500; Silwet® L-7602; Spectratech® CM 11014; Spectratech® CM 11126; Spectratech® CM 11172, 11174; Spectratech® CM 11194; Spectratech® CM 11513; Spectratech® PM 10634; SY-SA 69; SY-SL 131
Unislip 1753; Unislip 1757; Unislip 1759; Uniwax 1760

Chemicals:
Behenamide
Dimethicone copolyol; N,N′-Dioleoylethylenediamine
Erucamide; Erucyl erucamide; Erucyl stearamide; N,N′-Ethylenebis 12-hydroxystearamide
Hydrogenated tallow amide
Microcrystalline wax, oxidized
Oleamide; Oleyl palmitamide
Rapeseed oil
Silicone hexaacrylate; Stearamide; Stearyl erucamide; N-Stearyl 12-hydroxystearamide; Stearyl stearamide

Solubilizers

Trade names:
Acconon 400-MO; Aerosol® A-103; Aerosol® A-268; Aerosol® OT-B; Alkamuls® EL-620L
Disponil O 20; Disponil O 250; Dowfax C6L
Eccowet® W-50; Eccowet® W-88
Fizul 201-11; Fizul M-440
Igepal® DM-880
Kessco® PEG 200 DL; Kessco® PEG 200 DO; Kessco® PEG 200 DS; Kessco® PEG 200 ML; Kessco® PEG 200 MO; Kessco® PEG 200 MS; Kessco® PEG 300 DL; Kessco® PEG 300 DO; Kessco® PEG 300 DS; Kessco® PEG 300 ML; Kessco® PEG 300 MO; Kessco® PEG 300 MS; Kessco® PEG 400 DL; Kessco® PEG 400 DO; Kessco® PEG 400 DS; Kessco® PEG 400 ML; Kessco® PEG 400 MO; Kessco® PEG 400 MS; Kessco® PEG 600 DL; Kessco® PEG 600 DO; Kessco® PEG 600 DS; Kessco® PEG 600 ML; Kessco® PEG 600 MO; Kessco® PEG 600 MS; Kessco® PEG 1000 DL; Kessco® PEG 1000 DO; Kessco® PEG 1000 DS; Kessco® PEG 1000 ML; Kessco® PEG 1000 MO; Kessco® PEG 1000 MS; Kessco® PEG 1540 DL; Kessco® PEG 1540 DO; Kessco® PEG 1540 DS; Kessco® PEG 1540 ML; Kessco® PEG 1540 MO; Kessco® PEG 1540 MS; Kessco® PEG 4000 DL; Kessco® PEG 4000 DO; Kessco® PEG 4000 DS; Kessco® PEG 4000 ML; Kessco® PEG 4000 MO; Kessco® PEG 4000 MS; Kessco® PEG 6000 DL; Kessco® PEG 6000 DO; Kessco® PEG 6000 DS; Kessco® PEG 6000 ML; Kessco® PEG 6000 MO; Kessco® PEG 6000 MS; Klearfac® AA040
Lankropol® ATE
Maphos® 76; Monawet MB-45; Monawet MO-70; Monawet MT-70; Monawet SNO-35
Newcol 607, 610, 614, 623; Newcol 704, 707; Newcol 710, 714, 723; Nissan Plonon 102; Nissan Plonon 108; Nissan Plonon 171; Nissan Plonon 172; Nissan Plonon 201; Nissan Plonon 204; Nissan Plonon 208; Noiox AK-41
Prox-onic NP-1.5; Prox-onic NP-04; Prox-onic NP-06; Prox-onic NP-09; Prox-onic NP-010; Prox-onic NP-015; Prox-onic NP-020; Prox-onic NP-030; Prox-onic NP-030/70; Prox-onic NP-040; Prox-onic NP-040/70; Prox-onic NP-050; Prox-onic NP-050/70; Prox-onic NP-0100; Prox-onic NP-0100/70
Rewopal® HV 5; Rewopal® HV 9; Rewopal® HV 10; Rewopal® HV 25; Rewopol® SBDO 75; Rhodasurf® L-25; Rhodasurf® L-790

Solubilizers *(cont'd.)*

T-Det® C-40; Teric N8; Teric N9; Teric N10; Teric N12; Teric N13; Tetronic® 90R4; Tetronic® 150R1
Wayfos A; Wayfos D-10N

Chemicals: 2-Amino-2-methyl-1-propanol
2-Dimethylamino-2-methyl-1-propanol; Disodium deceth-6 sulfosuccinate; Disodium isodecyl sulfosuccinate; Disodium nonoxynol-10 sulfosuccinate; Disodium tallowiminodipropionate
Glyceryl tricaprate/caprylate
Nonoxynol-15; Nonyl nonoxynol-49
Oleth-4; Oleth-9; Oleth-20
PEG-4 dilaurate; PEG-6 dilaurate; PEG-12 dilaurate; PEG-20 dilaurate; PEG-32 dilaurate; PEG-75 dilaurate; PEG-150 dilaurate; PEG-4 dioleate; PEG-6 dioleate; PEG-8 dioleate; PEG-20 dioleate; PEG-32 dioleate; PEG-75 dioleate; PEG-150 dioleate; PEG-6 distearate; PEG-20 distearate; PEG-32 distearate; PEG-75 distearate; PEG-10 hydrogenated castor oil; PEG-30 hydrogenated castor oil; PEG-40 hydrogenated castor oil; PEG-60 hydrogenated castor oil; PEG-100 hydrogenated castor oil; PEG-4 laurate; PEG-20 laurate; PEG-32 laurate; PEG-75 laurate; PEG-150 laurate; PEG-4 oleate; PEG-8 oleate; PEG-20 oleate; PEG-32 oleate; PEG-75 oleate; PEG-150 oleate; PEG-4 stearate; PEG-20 stearate; PEG-32 stearate; PEG-75 stearate; PEG-150 stearate; Polyacrylamide; Polysorbate 65
Sodium hexyl diphenyloxide disulfonate; Sodium lauriminodipropionate; Sorbitan diisostearate; Sorbitan sesquioleate
Tetrasodium dicarboxyethyl stearyl sulfosuccinamate

Solvents

Trade names: Amsco Rubber Solvent; Arconate® 1000; Arconate® HP; Arconate® PC; Arcosolv® DPM; Arcosolv® DPMA; Arcosolv® PM; Arcosolv® PMA; Arcosolv® PTB; Arcosolv® TPM
BLO®; Butyl Dioxitol; Butyl Oxitol; BVA ND-205 Solv
Carbowax® PEG 300
DBE-2, -2SPG; DBE-3; DBE-4; DBE-5; DBE-6; DBE-9; Decalin; Decap; Diglycolamine® Agent (DGA); Diglyme; Dioxitol-High Gravity; Dioxitol-Low Gravity; Dynaflush; Dynasolve XD 16-3
Ektapro® EEP Solvent; Ektasolve® DP; Ektasolve® EB; Ektasolve® EP; Emcol® CC-42; Emcol® CC-57; Emery® 6705; Emulbon LB-78; Ethylan® HB1; Ethyl Diglyme; Ethyl Glyme
Fancol OA-95
Hisolve MC; Hisolve MDM
Igepal® OD-410; Isanol®
Kodaflex® DBP; Kodaflex® DEP
Monolan® PPG440, PPG1100, PPG2200
NMP; Noiox KS-10, -12, -13, -14, -16; Nusolv™ ABP-62; Nusolv™ ABP-74; Nusolv™ ABP-87; Nusolv™ ABP-103; Nusolv™ ABP-164
Oxitol
2-Pyrol®
QO® Furfuryl Alcohol (FA®)
TBP; Tetralin; THF; Triglyme
Uniplex 110; Uniplex 150; Uniplex 155

Chemicals: Butoxydiglycol; Butoxyethanol; Butyl alcohol; t-Butyl alcohol; Butyrolactone
Cyclohexanone
Decahydronaphthalene; Diacetone alcohol; Diethylene glycol; Diethylenetriamine; Diethyl phthalate; Diisobutyl ketone; Dimethoxyethyl phthalate; Dimethyl formamide; Dimethyl glutarate; Dimethyl phthalate; Dimethyl sebacate; 1,4-Dioxane; Dipropylene glycol
Erucamide; Ethoxydiglycol; Ethoxyethanol; Ethyl acetate; Ethylene carbonate; Ethylene glycol propyl ether; Ethyl 3-ethoxypropionate
Furfural
Glycol
Heptane; Hexane
Isopropyl alcohol
Methoxyethanol; Methyl alcohol; Methyl ethyl ketone; 1-Methyl-2-pyrrolidinone
Naphtha
Octane

Solvents *(cont'd.)*

PEG-4 dilaurate; PEG-6 dilaurate; PEG-20 dilaurate; Phenoxydiglycol; Phenoxyethanol; PPG-40 diethylmonium chloride; PPG-2 methyl ether; PPG-3 methyl ether; PPG-2 methyl ether acetate; Propylene glycol t-butyl ether

Tetraethylenepentamine; Tetrahydrofurfuryl alcohol; Toluene; Triethyl citrate; Triethyl phosphate; 2,2,4-Trimethyl-1,3-pentanediol monoisobutyrate; Trioctyl phosphate

Xylene

Stabilizers

Trade names: Abex® EP-110; Abex® EP-115; Abex® EP-227; Ablunol NP30; Ablunol NP30 70%; Ablunol NP40; Ablunol NP40 70%; Ablunol NP50; Ablunol NP50 70%; Acofor; Actafoam® F-2; Activ-8; ADK CIZER D-32; ADK CIZER E-500; ADK CIZER FEP-13; ADK CIZER O-130P; ADK CIZER O-180A; ADK STAB 36; ADK STAB 37; ADK STAB 38; ADK STAB 135A; ADK STAB 144; ADK STAB 189E; ADK STAB 260; ADK STAB 273; ADK STAB 329; ADK STAB 329K; ADK STAB 465; ADK STAB 465E, 465L; ADK STAB 467; ADK STAB 471; ADK STAB 517; ADK STAB 522A; ADK STAB 593; ADK STAB 666; ADK STAB 1013; ADK STAB 1178; ADK STAB 1292; ADK STAB 1500; ADK STAB 1502; ADK STAB 1505-A; ADK STAB 2013; ADK STAB 2112; ADK STAB 3010; ADK STAB AC-111; ADK STAB AC-113; ADK STAB AC-116; ADK STAB AC-118; ADK STAB AC-122; ADK STAB AC-133; ADK STAB AC-133N; ADK STAB AC-135; ADK STAB AC-141; ADK STAB AC-143; ADK STAB AC-153; ADK STAB AC-158; ADK STAB AC-167; ADK STAB AC-168; ADK STAB AC-169; ADK STAB AC-173; ADK STAB AC-181; ADK STAB AC-182; ADK STAB AC-183; ADK STAB AC-186; ADK STAB AC-190; ADK STAB AC-212; ADK STAB AC-216; ADK STAB AC-229; ADK STAB AC-230; ADK STAB AC-235; ADK STAB AC-243; ADK STAB AC-244; ADK STAB AC-248; ADK STAB AC-250; ADK STAB AC-252; ADK STAB AC-303; ADK STAB AC-307; ADK STAB AC-309; ADK STAB AC-310; ADK STAB AC-311; ADK STAB AO-23; ADK STAB AP-536; ADK STAB AP-539; ADK STAB AP-540; ADK STAB AP-543; ADK STAB AP-548; ADK STAB AP-550; ADK STAB AP-551; ADK STAB ATC-1; ADK STAB BT-11; ADK STAB BT-18; ADK STAB BT-23; ADK STAB BT-25; ADK STAB BT-27; ADK STAB BT-31; ADK STAB BT-52; ADK STAB BT-53A; ADK STAB BT-83; ADK STAB C; ADK STAB CA-ST; ADK STAB CPL-37; ADK STAB CPL-1551; ADK STAB CPS-1; ADK STAB CPS-2R; ADK STAB CPS-55R; ADK STAB CPS-896; ADK STAB CPS-900; ADK STAB EC-14; ADK STAB FL-21; ADK STAB FL-22; ADK STAB FL-23; ADK STAB FL-32; ADK STAB FL-41; ADK STAB FL-43; ADK STAB FL-44; ADK STAB FL-45; ADK STAB FL-54; ADK STAB FL-55; ADK STAB FL-64; ADK STAB FS-12; ADK STAB GR-16; ADK STAB GR-18; ADK STAB GR-22; ADK STAB HP-10; ADK STAB JX-3; ADK STAB LS-2; ADK STAB MG-ST; ADK STAB OF-14; ADK STAB OF-15; ADK STAB OF-19; ADK STAB OF-23; ADK STAB OF-30G; ADK STAB OF-31; ADK STAB OT-1; ADK STAB OT-9; ADK STAB P; ADK STAB PEP-8; ADK STAB PEP-8F; ADK STAB PEP-8W; ADK STAB PEP-11C; ADK STAB PEP-24G; ADK STAB PEP-36; ADK STAB QL; ADK STAB RUP-9; ADK STAB RUP-10; ADK STAB RUP-11; ADK STAB RUP-14; ADK STAB SC-12; ADK STAB SC-24; ADK STAB SC-26; ADK STAB SC-32; ADK STAB SC-34; ADK STAB SC-35; ADK STAB SC-102; ADK STAB SP-55; ADK STAB SP-69; ADK STAB SP-76; ADK STAB SP-86; ADK STAB TPP; ADK STAB ZN-ST; ADK STAB ZS-27; Advapak® LS-203HF; Advapak® LS-203HP; Advapak® SLS-1000; Advastab® TM-181; Advastab® TM-181-FS; Advastab® TM-183-B; Advastab® TM-183-O; Advastab® TM-281 IM; Advastab® TM-281 SP; Advastab® TM-692; Advastab® TM-694; Advastab® TM-696; Advastab® TM-697; Advastab® TM-790 Series; Advastab® TM-948; Advastab® TM-2080; Aerosil® R972V; Aerosol® A-196-40; Afco-Chem B; Afco-Chem CS; Afco-Chem CS-1; Afco-Chem CS-S; Afco-Chem EX-95; Afco-Chem LIS; Afco-Chem MGS; Afco-Chem ZNS; Ageflex FM-2; Ageflex FM-10; Agerite® Geltrol; Agerite® Stalite; Agerite® Superlite; Airvol® 205; Airvol® 523; Airvol® 540; Akrochem® Accelerator BZ Powder; Akrochem® Antioxidant 12; Akrochem® Antioxidant BHT; Akrochem® Antiozonant MPD-100; Akrochem® Antiozonant PD-2; Albrite® DBHP; Albrite® DIOP; Albrite® DMHP; Albrite® DOHP; Albrite® T2CEP; Albrite® TIOP; Albrite® TIPP; Albrite® TPP; Alcamizer 1; Alcamizer 2; Alcamizer 4; Alcamizer 4-2; Alcamizer 5; Alkamuls® EL-620L; Alkasurf® NP-4; Antarox® F88; Antarox® PGP 18-8; Antioxidant 235; Antioxidant 425; Aquastab PA 43; Aquastab PA 48; Aquastab PA 52; Aquastab PA 58; Aquastab PB 503; Aquastab PD 31M; Aquastab PD 412M; Aquastab PH 502; Araldite® XD 897; Armeen® Z; Aromox® C/12; Aromox® C/12-W; Avirol® SA 4106

Bäropan MC 8046 SP; Bäropan TX 296 KA; Bärostab® CT 901; Bärostab® CT 926 HV; Bärostab® CT 1500; Bärostab® CT 9140; Bärostab® L 230; Bärostab® NT 210; Bärostab®

Stabilizers *(cont'd.)*

NT 211; Bärostab® NT 212; Bärostab® NT 213; Bärostab® NT 224; Bärostab® NT 1005; Bärostab® UBZ 76 BX; Bärostab® UBZ 614; Bärostab® UBZ 630; Bärostab® UBZ 632; Bärostab® UBZ 637; Bärostab® UBZ 655 X; Bärostab® UBZ 667; Bärostab® UBZ 671; Bärostab® UBZ 776 X; Bärostab® UBZ 791; Bärostab® UBZ 793; Bärostab® UBZ 820 KA; BNX® 1010, 1010G; BSWL 201; BSWL 202; Butyl Zimate®

CAO®-3; Cariflex BR 1220; Carsonon® N-30; Carstab® DLTDP; Carstab® DSTDP; Cata-Chek® 820; Catinex KB-50; CC-01; CC-68; CC-101-FC; CC-104-S; CC-401; CC-811; CC-1015-M; CC-7710; CGA 12; Chemax NP-40; Chimassorb® 119; Chimassorb® 119FL; Chimassorb® 944FL; Colortech 10043-12; Colortech 10944-11; Cyanox® 425; Cyanox® 2777; Cyasorb® UV 24; Cyasorb® UV 3346; Cyclolube® 85

Dabco® DC5043; Dabco® DC5164; Dabco® DC5365; Dabco® DC5425; Dabco® DC5450; Dabco® DC5885; Darvan® ME; Darvan® NS; Darvan® WAQ; Dehydrophen PNP 40; Deriphat® 160C; Dextrol OC-40; Disperplast®-1150; Disponil APG 110; Disponil O 20; Disponil O 250; Doverphos® 4; Doverphos® 4 Powd; Doverphos® 6; Doverphos® 7; Doverphos® 8; Doverphos® 10; Doverphos® 10-HR; Doverphos® 10-HR Plus; Doverphos® 11; Doverphos® 12; Doverphos® 49; Doverphos® 53; Doverphos® 213; Doverphos® 613; Doverphos® 675; Doverphos® S-480; Doverphos® S-680; Doverphos® S-682; Doverphos® S-686, S-687; Drapex® 4.4; Dresinate® TX; Dresinate® X; Dresinate® XX; DS-207; Duralink® HTS; Dyphos; Dyphos Envirostab; Dyphos XL; Dythal; Dythal XL

Eastman® DTBHQ; Eastman® HQMME; Eastman® Inhibitor OABH; Eastman® Inhibitor RMB; Emcol® 4930, 4940; Empilan® 2020; Empilan® DL 40; Empilan® DL 100; Empilan® GMS NSE90; Emulvin® W; EPIstatic® 100; EPIstatic® 101; EPIstatic® 102; EPIstatic® 103; EPIstatic® 110; EPIstatic® 111; EPIstatic® 112; EPIstatic® 113; EPIthal 120; EPIthal 121; EPIthal 122; EPIthal 123; Epoxol 7-4; Estabex ABF; Estabex® 2307; Estabex® 2307 DEOD; Estabex® 2381; Ethanox® 330; Ethanox® 398; Ethylan® OE; Ethylan® R; Ethyl Zimate®; Evangard® 18MP; Evangard® DTB; Evanstab® 1218; EZn-Chek® 601; EZn-Chek® 788I

F-2000 FR; Fancol OA-95; Ferro 1288; Fizul M-440

Gantrez® AN-119; Gantrez® AN-139; Gantrez® AN-149; Gantrez® AN-169; Gantrez® AN-179; Geolite® Modifier 91; Glycolube® 740; Glycolube® AFA-1; Glycolube® CW-1; Glycolube® P; Glycolube® SG-1

Halbase 10; Halbase 10 EP; Halbase 10S; Halbase 11; Halbase 100; Halbase 100 EP; Halcarb 20; Halcarb 20 EP; Halcarb 20S; Halcarb 200; Halcarb 200 EP; Hal-Lub-D; Hal-Lub-D S; Hal-Lub-N S; Hal-Lub-N TOTM; Halphos; Halphos S; Halstab 6S; Halstab 30; Halstab 30 EP; Halstab 31; Halstab 32; Halstab 50; Halstab 55; Halstab 60; Halstab 60 EP; Halstab 70; Halstab 600 EP; Halstab M-1; Halstab P-1; Halstab P-2; Halthal; Halthal EP; Halthal S; Hamposyl® C; Hamposyl® C-30; Hamposyl® L; Hamposyl® L-30; Hamposyl® L-95; Hamposyl® M; Hamposyl® M-30; Hamposyl® O; Hamposyl® S; Haro® Chem BG; Haro® Chem BP-108X; Haro® Chem BSG; Haro® Chem CBHG; Haro® Chem CGD; Haro® Chem CGN; Haro® Chem CHG; Haro® Chem CP-17; Haro® Chem CPR-2; Haro® Chem CPR-5; Haro® Chem CSG; Haro® Chem CZ-37, CZ-40; Haro® Chem KB-214SA; Haro® Chem KB-219SA, KB-521SA; Haro® Chem KB-350X; Haro® Chem KB-353A; Haro® Chem KB-554A, ZZ-019; Haro® Chem KPR; Haro® Chem KS; Haro® Chem LHG; Haro® Chem MF-2; Haro® Chem MF-3; Haro® Chem NG; Haro® Chem P28G; Haro® Chem P51; Haro® Chem PC; Haro® Chem PDF; Haro® Chem PDF-B; Haro® Chem PDP-E; Haro® Chem PPCS-X; Haro® Chem PTS-E; Haro® Chem ZGD; Haro® Chem ZGN; Haro® Chem ZGN-T; Haro® Chem ZPR-2; Haro® Chem ZSG; Haro® Mix BF-202; Haro® Mix BK-105, BK-107; Haro® Mix CE-701; Haro® Mix CH-205; Haro® Mix CH-606; Haro® Mix CK-203; Haro® Mix CK-213; Haro® Mix CK-711; Haro® Mix FK-102; Haro® Mix IC-217; Haro® Mix IC-238; Haro® Mix IH-108; Haro® Mix LK-218; Haro® Mix LK-228; Haro® Mix MH-204; Haro® Mix MK-107; Haro® Mix MK-220; Haro® Mix MK-620; Haro® Mix MK-744; Haro® Mix SK-602; Haro® Mix UC-213; Haro® Mix UK-121; Haro® Mix VC-501; Haro® Mix YC-502; Haro® Mix YC-601; Haro® Mix YE-301; Haro® Mix YK-110; Haro® Mix YK-113; Haro® Mix YK-307; Haro® Mix YK-603; Haro® Mix ZC-028, ZC-029, ZC-030; Haro® Mix ZC-031, ZC-032; Haro® Mix ZC-036; Haro® Mix ZC-309; Haro® Mix ZC-311; Haro® Mix ZC-902; Haro® Mix ZT-025; Haro® Mix ZT-026; Haro® Mix ZT-504; Haro® Mix ZT-508; Haro® Mix ZT-514; Haro® Mix ZT-905; Hefti MS-55-F; Hi-Sil® 210; Hi-Sil® 233; Hostanox® O3; Hostanox® PAR 24; Hostastab® SnS 10; Hostastab® SnS 11; Hostastab® SnS 15; Hostastab® SnS 20; Hostastab® SnS 41; Hostastab® SnS 44; Hostastab® SnS 45; Hostastab® SnS 61; Hyonic OP-70; Hystrene® 5016 NF

Igepal® CA-877; Igepal® CA-880; Igepal® CA-887; Igepal® CA-890; Igepal® CA-897; Igepal® CO-430; Igepal® CO-850; Igepal® CO-880; Igepal® CO-887; Igepal® CO-890; Igepal® CO-

Stabilizers *(cont'd.)*

897; Igepal® CO-970; Igepal® CO-977; Igepal® CO-987; Igepal® CO-990; Igepal® CO-997; Igepal® DM-730; Igepal® DM-880; Interstab® BC-100S; Interstab® BC-103; Interstab® BC-103A; Interstab® BC-103C; Interstab® BC-103L; Interstab® BC-109; Interstab® BC-110; Interstab® BC-4195; Interstab® BC-4377; Interstab® BZ-4828A; Interstab® C-16; Interstab® CH-55; Interstab® CH-55R; Interstab® CLB-747; Interstab® CZ-10; Interstab® CZ-11D; Interstab® CZ-19A; Interstab® CZ-22; Interstab® CZ-23; Interstab® F-402; Interstab® LF 3615; Interstab® LT-4805R; Interstab® LT-4361; Interstab® MT981; Interstab® MT982; Interstab® R-4048; Interstab® R-4114; Interstab® R-4150; Ionol; Irganox® 245; Irganox® 259; Irganox® 565; Irganox® 1010; Irganox® 1019; Irganox® 1035; Irganox® 1076; Irganox® 1520; Irganox® 5057; Irganox® B 215; Irganox® B 225; Irganox® B 501W; Irganox® B 561; Irganox® B 900; Irganox® B 921; Irganox® B 1171; Irganox® B 1411; Irganox® B 1412; Irganox® MD 1024; Irgastab® T 265; Irgastab® T 634; Isatin; Isonox® 132; Isonox® 232

Ken-React® 7 (KR 7); Ken-React® 9S (KR 9S); Ken-React® 12 (KR 12); Ken-React® 26S (KR 26S); Ken-React® 33DS (KR 33DS); Ken-React® 38S (KR 38S); Ken-React® 39DS (KR 39DS); Ken-React® 41B (KR 41B); Ken-React® 44 (KR 44); Ken-React® 46B (KR 46B); Ken-React® 55 (KR 55); Ken-React® 133DS (KR 133DS); Ken-React® 134S (KR 134S); Ken-React® 138D (KR 138D); Ken-React® 138S (KR 138S); Ken-React® 158D (KR 158D); Ken-React® 158FS (KR 158FS); Ken-React® 212 (KR 212); Ken-React® 238A (KR 238A); Ken-React® 238J (KR 238J); Ken-React® 238M (KR 238M); Ken-React® 238S (KR 238S); Ken-React® 238T (KR 238T); Ken-React® 262A (KR 262A); Ken-React® 262ES (KR 262ES); Ken-React® OPP2 (KR OPP2); Ken-React® OPPR (KR OPPR); Ken-React® KR TTS; Kisuma 5A; Kisuma 5B; Kisuma 5E; KZ 55; KZ OPPR; KZ TPP; KZ TPPJ

Lankrocell® KLOP; Lankrocell® KLOP/CV; Lankroflex® ED 6; Lankroflex® GE; Lankroflex® L; Lankroflex® Series; Lankromark® BLT105; Lankromark® BM271; Lankromark® BT050; Lankromark® BT120; Lankromark® BT120A; Lankromark® BT190; Lankromark® BT339; Lankromark® DLTDP; Lankromark® DP6404Z; Lankromark® DSTDP; Lankromark® LC68; Lankromark® LC90; Lankromark® LC475; Lankromark® LC541; Lankromark® LC563; Lankromark® LC585; Lankromark® LC629; Lankromark® LE65; Lankromark® LE76; Lankromark® LE87; Lankromark® LE98; Lankromark® LE109; Lankromark® LE131; Lankromark® LE230; Lankromark® LE274; Lankromark® LZ121; Lankromark® LZ187; Lankromark® LZ440; Lankromark® LZ495; Lankromark® LZ561; Lankromark® LZ616; Lankromark® LZ638; Lankromark® LZ649; Lankromark® LZ693; Lankromark® LZ1045; Lankromark® LZ1232; Lankromark® OT050, OT250; Lankromark® OT052; Lankromark® OT252; Lankromark® OT450; Lankromark® OT452; Lankromark® OT650; Lankromark® OT652; Lankropol® ATE; Laurex® CS; Laurex® NC; Leadstar; Leadstar Envirostab; Lectro 60; Lectro 90; LICA 01; Lithene AH; Lithene AL; LK-221®; LK®-443; Lonza Coke PC 40; Lowilite® 26; Lowinox® BHT Food Grade; Lowinox® BHT-Superflakes; Loxiol® G 16; Loxiol® HOB 7121; Loxiol® HOB 7131; Loxiol® VGE 1728; Loxiol® VGE 1875; Loxiol® VGE 1884; Loxiol® VGS 1877; Loxiol® VGS 1878; Lubrizol® 2106; Lubrizol® 2116; Lubrizol® 2117

Mark® 133; Mark® 144; Mark® 152; Mark® 155; Mark® 224; Mark® 281B; Mark® 366; Mark® 398; Mark® 462; Mark® 462A; Mark® 495; Mark® 550, 556; Mark® 565A; Mark® 577A; Mark® 630; Mark® 1014; Mark® 1043A; Mark® 1178; Mark® 1178B; Mark® 1220; Mark® 1409; Mark® 1500; Mark® 1589B; Mark® 1600; Mark® 1900; Mark® 1905; Mark® 1939; Mark® 1984; Mark® 2077; Mark® 2100, 2100A; Mark® 2112; Mark® 2115; Mark® 2140; Mark® 2180; Mark® 2326; Mark® 4700; Mark® 4701; Mark® 4702; Mark® 4716; Mark® 4722; Mark® 4723; Mark® 4726; Mark® 5060; Mark® 5089; Mark® 6000; Mark® 7118; Mark® 7308; Mark® C; Mark® DPDP; Mark® DPP; Mark® GS, RFD; Mark® LL; Mark® OHM; Mark® OTM; Mark® PDDP; Mark® QED; Mark® RFD; Mark® TDP; Mark® TPP; Mark® TT; Mark® WS; Mathe Barium Stearate; Merix MCG Compd; Merpol® HCS; Metasap® Cadmium Stearate; Methocel® A; Methocel® E; Millad® HBPA; MiRaStab 113-K; MiRaStab 130-K; MiRaStab 140-K; MiRaStab 154-K; MiRaStab 400-K; MiRaStab 403-K; MiRaStab 704-K; MiRaStab 709-ZK; MiRaStab 800-K; MiRaStab 919-K; MiRaStab 922-K; MiRaStab 950-K; Mixxim® DBM; Modicol S; Monolan® 2500 E/30; Monolan® 3000 E/60, 8000 E/80; Monoplex® S-73; Monoplex® S-75

Naugard® 76; Naugard® 431; Naugard® 524; Naugard® 529; Naugard® BHT; Naugard® P; Naugard® PHR; Naugard® SP; Naugard® Super Q; Naugard® XL-1; Naugard® XL-517; Nipol® 1452X8; NZ 01; NZ 09; NZ 12; NZ 33; NZ 38; NZ 39; NZ 44; NZ 49; NZ 89; NZ 97

Octosol 449; Octosol 496; Octosol A-18-A; Octosol SLS; Octosol SLS-1; Okstan M 15; Okstan M 69 S; OSO® 440; OSO® 1905

Stabilizers *(cont'd.)*

Paraplex® G-62; Pationic® 940; Pationic® 1230; Pationic® 1240; Pationic® 1250; Peroxidol 781; Petrac® Zinc Stearate ZN-42; Plas-Chek® 775; Plastistab 2005; Plastistab 2020; Plastistab 2160; Plastistab 2210; Plastistab 2310; Plastistab 2320; Plastistab 2330; Plastistab 2560; Plastistab 2580; Plastistab 2800; Plastistab 2900; Plastistab 2931; Plastistab 2941; Plastoflex® 2307; Plastolein® 9232; Plaswite® LL 7112; Poly-Dispersion® AV (TBS)D-80; Poly-Dispersion® AV (TBS)D-80P; Poly-Dispersion® E (DPS)D-80; Poly-Dispersion® E (DYT)D-80; Poly-Dispersion® G (DPS)D-80; Poly-Dispersion® G (DYT)D-80; Poly-Dispersion® H (DYT)D-80; Poly-Dispersion® H (DYT)D-80P; Poly-Dispersion® PLD-90; Poly-Dispersion® T (DYT)D-80P; Polygard®; Polymist® 284; Polymist® F-5; Polymist® F-5A; Polymist® F-5A EX; Polymist® F-510; Polysar EPDM 6463; Polystep® A-15-30K; Polystep® F-2; Polystep® F-3; Polystep® F-5; Polyvel CA-P20; Prespersion PAB-8912; Prespersion PAC-332; Prespersion PAC-363; Prespersion PAC-1451; Prespersion PAC-2941; Prespersion PAC-3936; Prespersion PAC-4893; Prifrac 2922; Proaid® 9904; Prox-onic CSA-1/04; Prox-onic CSA-1/06; Prox-onic CSA-1/010; Prox-onic CSA-1/015; Prox-onic CSA-1/020; Prox-onic CSA-1/030; Prox-onic CSA-1/050; Prox-onic HR-05; Prox-onic HR-016; Prox-onic HR-025; Prox-onic HR-030; Prox-onic HR-036; Prox-onic HR-040; Prox-onic HR-080; Prox-onic HR-0200; Prox-onic HR-0200/50; Prox-onic HRH-05; Prox-onic HRH-016; Prox-onic HRH-025; Prox-onic HRH-0200; Prox-onic HRH-0200/50; Prox-onic OA-1/020; Prox-onic OA-2/020; PTZ® Chemical Grade; PTZ® Industrial Grade; PTZ® Purified Flake; PTZ® Purified Powd

Q-1300

Ralox® 46; Ralox® 530; Ralox® LC; Ralox® TMQ-G; Rewomat TMS; Rexol 25/12; Rexol 25/407; Rhenogran® $Ca(OH)_2$-50/FPM; Rhenogran® MgO-40/FPM; Rhenogran® PbO-80/FPM; Rhodafac® BX-660; Rhodafac® BX-760; Rhodafac® MC-470; Rhodafac® PS-19; Rhodafac® RE-960; Rhodamox® CAPO; Rhodamox® LO; Rhodasurf® BC-737; Rhodasurf® BC-840; Rhodasurf® L-25; Rhodasurf® ON-870; Rhodasurf® ON-877

Sandostab 4030; Sandostab P-EPQ; Santoflex® 134; Santovar® A; Santowhite® Crystals; Santowhite® ML; Santowhite® PC; Santowhite® Powd; Saytex® BT-93®; Serdet DSK 40; Serdet Perle Conc; Serdox NNP 30/70; Serdox NOP 30/70; Serdox NOP 40/70; Sermul EN 145; Sermul EN 20/70; Sermul EN 30/70; Sicostab®; Sinopol 806; Sinopol 808; Sinopol 908; Sipernat® 22S; Sipernat® 50S; Sipomer® AAE; Sipomer® COPS 1; Sipomer® HEM-5; Sole Terge 8; Solricin® 135; Solricin® 235; Solricin® 435; Solricin® 535; Stabaxol P; Stabilizer 1097; Stabilizer 7000; Stabilizer C; Stabiol CZ 1336/4E; Stabiol CZ 1616/6F; Stabiol CZ 2001/1; Stabiol VCA 1733; Stabiol VZn 1603; Stabiol VZN 1783; Staflex DIOA; Staflex DOA; Staflex DOZ; Staflex NODA; Stanclere® A-121, A-121 C, A-221; Stanclere® C26; Stanclere® T-55; Stanclere® T-57, T-80, T-81; Stanclere® T-85, TL; Stanclere® T-94 C; Stanclere® T-126; Stanclere® T-160; Stanclere® T-161, T-163, T-164, T-165, T-174, T-182, T-183; Stanclere® T-184; Stanclere® T-186; Stanclere® T-192, T-193, T-194; Stanclere® T-197; Stanclere® T-208; Stanclere® T-222, T-233; Stanclere® T-250, T-250SD; Stanclere® T-470, T-473, T-482, T-483, T-484, T-582; Stanclere® T-801; Stanclere® T-876; Stanclere® T-877; Stanclere® T-878, T-55; Stanclere® T-883; Stanclere® T-4356; Stanclere® T-4654; Stanclere® T-4704; Stanclere® T-5507; Stanclere® TL; Stanclere® TM; Stavinor ATA-1056; Stavinor CA PSE; Struktol® TH 10 Flakes; Struktol® WB 16; Sunyl® 80; Sunyl® 80 ES; Sunyl® 90; Surfynol® 104E; Synperonic PE30/80; Synpro® Cadmium Stearate; Synpro® Lead Stearate; Synpron 231; Synpron 239; Synpron 241; Synpron 350; Synpron 351; Synpron 352; Synpron 353; Synpron 354; Synpron 355; Synpron 357; Synpron 431; Synpron 687; Synpron 755; Synpron 940; Synpron 1002; Synpron 1004; Synpron 1007; Synpron 1009; Synpron 1011; Synpron 1027; Synpron 1034; Synpron 1048; Synpron 1108; Synpron 1112; Synpron 1135; Synpron 1236; Synpron 1321; Synpron 1363; Synpron 1384; Synpron 1410; Synpron 1419; Synpron 1427; Synpron 1428; Synpron 1456; Synpron 1470; Synpron 1522; Synpron 1528; Synpron 1536; Synpron 1538; Synpron 1558; Synpron 1566; Synpron 1578; Synpron 1585; Synpron 1618; Synpron 1642; Synpron 1652; Synpron 1678; Synpron 1693; Synpron 1699; Synpron 1749; Synpron 1750; Synpron 1774; Synpron 1800; Synpron 1810; Synpron 1811, 1812; Synpron 1820; Synpron 1831; Synpron 1840; Synpron 1850; Synpron 1852; Synpron 1860; Synpron 1861; Synpron 1870; Synpron 1871; Synpron 1881; Synpron 1890; Synpron 1900; Synpron 1913; Synpro® Stannous Stearate

T-Det® N-70; T-Det® N-705; T-Det® N-707; T-Det® O-40; T-Det® O-307; Tegostab® B 1048; Tegostab® B 1400A; Tegostab® B 1605; Tegostab® B 1648; Tegostab® B 1903; Tegostab® B 2173; Tegostab® B 2219; Tegostab® B 3136; Tegostab® B 3752; Tegostab® B 4113; Tegostab® B 4351; Tegostab® B 4380; Tegostab® B 4617; Tegostab® B 4690; Tegostab® B 4900; Tegostab® B 4970; Tegostab® B 5055; Tegostab® B 8002; Tegostab® B 8014;

Stabilizers *(cont'd.)*

Tegostab® B 8017; Tegostab® B 8021; Tegostab® B 8030; Tegostab® B 8100; Tegostab® B 8200; Tegostab® B 8202; Tegostab® B 8231; Tegostab® B 8300; Tegostab® B 8301; Tegostab® B 8404; Tegostab® B 8406; Tegostab® B 8407; Tegostab® B 8408; Tegostab® B 8409; Tegostab® B 8418; Tegostab® B 8423; Tegostab® B 8425; Tegostab® B 8427; Tegostab® B 8432; Tegostab® B 8433; Tegostab® B 8435; Tegostab® B 8444; Tegostab® B 8450; Tegostab® B 8622; Tegostab® B 8636; Tegostab® B 8680; Tegostab® B 8681; Tegostab® B 8694; Tegostab® B 8863 T; Tegostab® B 8896; Tegostab® B 8901; Tegostab® B 8905; Tegostab® B 8906, 8910; Tegostab® B 2270; Tegostab® BF 2370; Telalloy® A-50B; Temex BA-1; Tenox® PG; Tenox® TBHQ; Tergitol® 15-S-30; Tergitol® 15-S-40; Tergitol® XD; Tergitol® XJ; Teric 16A22; Texadril 8780; Therm-Chek® 6-V-6A; Therm-Chek® 130; Therm-Chek® 135; Therm-Chek® 170; Therm-Chek® 344; Therm-Chek® 659; Therm-Chek® 707-X; Therm-Chek® 714; Therm-Chek® 763; Therm-Chek® 837; Therm-Chek® 840; Therm-Chek® 900; Therm-Chek® 904; Therm-Chek® 904; Therm-Chek® 1238; Therm-Chek® 1720; Therm-Chek® 1825; Therm-Chek® 1827; Therm-Chek® 1864; Therm-Chek® 5469; Therm-Chek® 5590; Therm-Chek® 5649; Therm-Chek® 5776; Therm-Chek® 5930; Therm-Chek® 6160; Therm-Chek® 6253, 6254; Therm-Chek® 6258; Therm-Chek® BJ-2; Therm-Chek® F-694; Thermocel 1002; Thermocel 2365; Thermolite 12; Thermolite 31; Thermolite 35; Thermolite 73; Thermolite 139; Thermolite 176; Thermolite 187; Thermolite 340; Thermolite 380; Thermolite 390; Thermolite 395; Thermolite 525; Thermolite 813; Thermolite 890S; Thiovanol®; THPE-BZT; Tinuvin® 622LD; Titanium Dioxide P25; Tribase; Tribase EXL Special; Tribase XL; Triton® X-15; Triton® X-35; Triton® X-165-70%; Triton® X-200; Triton® X-305-70%; Triton® X-405-70%; Trycol® 6970; Trycol® 6972; Trycol® 6985; Trycol® DNP-8; Tylose® MH Grades; Tylose® MHB

Ultranox® 210; Ultranox® 276; Ultranox® 626; Ultranox® 626A; Ultranox® 627A; Ultranox® 815A Blend; Ultranox® 817A Blend; Ultranox® 875A Blend; Ultranox® 877A Blend; Ultranox® 2714A; Ultra Sulfate SL-1; Uniplex 225; Uni-Rez® 3020; UV-Chek® AM-300; UV-Chek® AM-301; UV-Chek® AM-595; Uvi-Nox® 1494

Vanox® 13; Vanox® 102; Vanox® 1005; Vanox® 1290; Vanox® 1320; Vanox® 2246; Vanox® AT; Vanox® GT; Vanox® NBC; Vanox® SKT; Vanstay SC; Vanstay SD; Vanstay SG; Vanstay SH; Vertal 710; Vestowax A-217; Vestowax A-227; Vestowax A-235; Vestowax A-415; Vestowax A-616; Victawet® 12; Victawet® 85X; Viplex 895-BL; Vistanex® MML-80, MML-100, MML-120, MML-140; Voidox 100%; Vulcastab® EFA; Vulcastab® LW; Vulcastab® T; Vulkanox® BKF; Vulkanox® DDA; Vulkanox® KB; Vulkanox® NKF

Wayfos D-10N; Weston® 399; Weston® 399B; Weston® 430; Weston® 439; Weston® 474; Weston® 491; Weston® 494; Weston® 600; Weston® 618F; Weston® 619F; Weston® DHOP; Weston® DLP; Weston® DPDP; Weston® DPP; Weston® DTDP; Weston® EGTPP; Weston® ODPP; Weston® PDDP; Weston® PNPG; Weston® PTP; Weston® TDP; Weston® THOP; Weston® TIOP; Weston® TLP; Weston® TLTTP; Weston® TNPP; Weston® TPP; Westvaco® Resin 90; Westvaco® Resin 95; White Lead; Wingstay® 29; Wingstay® 100; Wingstay® 200; Wingstay® C; Wingstay® K; Wingstay® S; Wingstay® SN-1; Wingstay® T; Witconate™ 3009-15; Witconol™ NP-100; Wytox® 312

Chemicals: Alkyl C12-15 bisphenol A phosphite; Allyl glycidyl ether; Ammonium nonoxynol-9 sulfate; Ammonium polyacrylate; Ammonium polymethacrylate; Ammonium stearate; Antimony mercaptide

Barium-cadmium complex; Barium-cadmium-lead complex; Barium-cadmium-phosphite complex; Barium-cadmium-zinc complex; Barium-calcium-zinc complex; Barium laurate; Barium-lead complex; Barium-magnesium-zinc complex; Barium phenate; Barium stearate; Barium-tin complex; Barium-zinc complex; Barium-zinc-magnesium-aluminum complex; Barium-zinc phosphite; Benzene propanoic acid; Benzophenone-2; Benzophenone-12; Bis(butyltin 2-ethylhexyl mercaptoacetate); Bis (2,6-diisopropylphenyl) carbodiimide; Bis(4-(1,1-dimethylethyl)benzoato-o)hydroxy aluminum; 4-[[4,6-Bis(octylthio)-s-triazin-2-yl]amino]-2,6-di-t-butylphenol;; Butylated d: (dimethylbenzyl) phenol; t-Butyl glycidyl ether; 4,4′-Butylidenebis (6-t-butyl-m-cresol); Butylmercaptide; Butyl thiotin; Butyltin mercaptide; Butyltin thioglycolate

Cadmium-barium complex; Cadmium laurate; Cadmium stearate; Cadmium-zinc complex; Calcium behenate; Calcium hydroxystearate; Calcium phosphate dibasic; Calcium phosphate monobasic monohydrate; Calcium phosphate tribasic; Calcium ricinoleate; Calcium-sodium-zinc complex; Calcium stearate; Calcium sulfate; Calcium-tin complex; Calcium-tin-zinc complex; Calcium-zinc complex; C40-60 alcohols; Ceteareth-22; Cetearyl alcohol; Cetoleth-19; Cetoleth-20; Cocamidopropylamine oxide; Cocaminobutyric acid; Cocoyl

Stabilizers *(cont'd.)*

sarcosine; C11-15 pareth-3; C11-15 pareth-5; Cyclodextrin; Cycloneopentyl, cyclo (dimethlaminoethyl) pyrophosphato zirconate, di mesyl salt

Decyl betaine; Decyl diphenyl phosphite; Diammonium stearyl sulfosuccinamate; 3,5-Di-t-butyl-4-hydroxyhydrocinnamic acid/1,3,5-tris(2-hydroxyethyl)-s-triazine-2,4,6-(1H,3H,5H)-trione triester; 2,6-Di-t-butylphenol; Dibutyltin bis(2-ethylhexyl mercaptoacetate); Dibutyltin diacetate; Dibutyltin dilaurate; Dibutyltin maleate; Dibutyltin mercaptide; Dibutyltin β-mercaptopropionate; Dicyandiamide; Dicyclohexyl sodium sulfosuccinate; Didecyl hydrogen phosphite; Didecyl phenyl phosphite; Diisodecyl pentaerythritol diphosphite; Diisooctyl octylphenyl phosphite; Diisooctyl phosphite; Dilauryl thiodipropionate; Dimethicone/mercaptopropyl methicone copolymer; Dimethylaminomethyl phenol; Dimethyl succinate; Dimyristyl thiodipropionate; N,N′-Di-β-naphthyl-p-phenylenediamine; Dioctadecyl disulfide; 2,2′-[(Dioctylstannylene)bis(acetic acid)], diisooctyl ester; Dioctyltin dilaurate; Dioctyltin maleate; Dioctyltin mercaptide; Dioctyltin thioglycolate; Diphenyldidecyl (2,2,4-trimethyl-1,3-pentanediol) diphosphite; Diphenyl isodecyl phosphite; Diphenyl isooctyl phosphite; N,N′-Diphenyl-p-phenylenediamine; Diphenyl phosphite; Disodium hexamethylene bisthiosulfate; Disodium nonoxynol-10 sulfosuccinate; Distearyl dithiodipropionate; Distearyl pentaerythritol diphosphite; Distearyl phosphite; Distearyl thiodipropionate; Ditridecyl phosphite

Edetic acid; Epoxidized glycol dioleate; Epoxidized linseed oil; Epoxidized linseed oil, butyl ester; Epoxidized octyl stearate; Epoxidized octyl tallate; Epoxidized soybean oil; Epoxidized tall oil; Epoxy, bisphenol A; Epoxy resin; Erucamide; Ethylenebis (oxyethylene) bis (3-t-butyl-4-hydroxy-5-methylhydrocinnamate); Ethylhexyl diphenyl phosphite; 2,2′-Ethylidenebis (4,6-di-t-butylphenol); Etocrylene

Formaldehyde, ammonia, ethyl chloride condensate

Glycol

Heptakis (dipropylene glycol) triphosphite; Hexamethylenebis (3,5-di-t-butyl-4-hydroxycinnamate); Hydroquinone monomethyl ether; Hydroxyethylcellulose; N-(2-Hydroxypropyl) benzenesulfonamide; Hydroxypropyl-β-cyclodextrin; Hydroxypropyl-γ-cyclodextrin

Isodecyl methacrylate; Isopropyl glycidyl ether; 4,4′ Isopropylidenediphenol alkyl (C12-15) phosphites

Lauramine oxide; Laureth-23; Lauric acid; Lauroyl sarcosine; Lead; Lead-barium-cadmium complex; Lead carbonate (basic); Lead chlorosilicate complex; Lead phosphite dibasic; Lead phthalate, basic; Lead phthalate, dibasic; Lead silicate; Lead silicosulfate, basic; Lead stearate; Lead stearate, dibasic; Lead sulfate, basic; Lead sulfate, polybasic; Lead sulfate, tetrabasic; Lead sulfate, tribasic; Lead-zinc complex; Lithium 12-hydroxystearate; Lithium stearate

Magnesium aluminum carbonate; Magnesium-aluminum-zinc complex; Magnesium stearate; Magnesium-zinc complex; Methyl 3-[3-(2H-benzotriazole-2-yl)-5-t-butyl-4-hydroxyphenyl]proportionate; Methylcellulose; Methyl-β-cyclodextrin; Methyl-γ-cyclodextrin; 2,2′-Methylenebis (6-t-butyl-4-methylphenol); Methyl hydroxyethylcellulose; Methyl laurate; p-Methyl morpholine; Methyltin mercaptide; Myristoyl sarcosine

Neopentyl (diallyl) oxy, trihydroxy caproyl titanate; Neopentyl (diallyl) oxy, trimercapto-phenyl zirconate; 2,2′,2′′-Nitrilo[triethyl-tris(3,3′,5,5′-tetra-t-butyl-1,1′-biphenyl-2,2′-diyl)phosphite]; Nitrosophenylhydroxylamine ammonium salt; Nonoxynol-25; Nonyl nonoxynol-8; p-Nonyl phenol

Octadecyl 3,5-di-t-butyl-4-hydroxyhydrocinnamate; Octadecyl 3-mercaptopropionate; Octoxynol-1; Octoxynol-3; Octoxynol-7; Octoxynol-10; Octylmercaptide; 2,2′,2′′-(Octylstannylidyne)tris(thio)tris(acetic acid), triisooctyl ester; Octyl thiotin; Octyltin mercaptide; Octyltin thioglycolate; Oleoyl sarcosine; Oleth-4; Oleth-9; Oleth-20; Oleyl alcohol; Oxalyl bis (benzylidenehydrazide)

PEG-3 lauramine oxide; PEG-2 laurate SE; PEG-150 oleate; PEG-2 stearate; PEG-40 tallowamine; Pentaerythrityl hexylthiopropionate; Pentaerythrityl tetrakis (3-mercaptopropionate); Pentamethyl-4-piperidyl/β,β,β′,β′-tetramethyl-3,9-(2,4,8,10-tetraoxaspiro (5,5) undecane) diethyl]-1,2,2,6,6-1,2,3,4-butane tetracarboxylate; Phenothiazine; Phenyl diisodecyl phosphite; Phenyl neopentylene glycol phosphite; Phosphorus trichloride, reaction prods. with 1,1′-biphenyl and 2,4-bis (1,1-dimethylethyl) phenol; Poloxamer 188; Poloxamer 238; Polyacrylic acid; Poly (dipropyleneglycol) phenyl phosphite; Polyvinyl alcohol, partially hydrolyzed; Potassium castorate; Potassium ricinoleate; Potassium-zinc; N,N′-1,3-Propanediylbis(3,5-di-t-butyl-4-hydroxyhydrocinnamamide); Propyl gallate

Sodium allyloxy hydroxypropyl sulfonate; Sodium cocoamphopropionate; Sodium isopropyl naphthalene sulfonate; Sodium laureth-4 phosphate; Sodium lauriminodipropionate; Sodium lignosulfonate; Sodium myristoyl sarcosinate; Sodium octoxynol-2 ethane sulfonate; Sodium

Stabilizers *(cont'd.)*

octyl sulfate; Sodium polyacrylate; Sodium ricinoleate; Sodium rosinate; Sodium stearate; Sodium tallate; Sodium trideceth sulfate; Sodium vinyl sulfonate; Sodium-zinc complex; Sorbitan stearate; Stannous stearate; Stearamide MEA; Stearoyl sarcosine; Stearyl lauryl thiodipropionate; Strontium-zinc complex; Styrene/MA copolymer

Tetrakis (2,4-di-t-butylphenyl) 4,4′-biphenylenediphosphonite; Tetrakis isodecyl 4,4′-isopropylidene diphosphite; Tetrakis (2,2,6,6-tetramethyl-4-piperidyl)-1,2,3,4-butane tetracarboxylate; [2,2,6,6-Tetramethyl-4-piperidyl/β,β,β′,β′-tetramethyl-3,9-(2,4,8,10-tetraoxaspiro (5,5) undecane) diethyl]-1,2,3,4-butane tetracarboxylate; Tetraphenyl dipropyleneglycol diphosphite; Thiodiethylene bis (3,5-di-t-butyl-4-hydroxy) hydrocinnamate; Thioglycerin; Tin, butyl 2-ethylhexanoate laurate oxo complexes; Tin-calcium-zinc complex; Tin mercaptide; Trideceth-14; Trideceth-15; Tridecyl phosphite; Triethylene diamine; Triisodecyl phosphite; Triisooctyl phosphite; Triisotridecyl phosphite; Trilauryl phosphite; Trilauryl trithiophosphite; 2,2,4-Trimethyl-1,2-dihydroquinoline polymer; Trimethylolpropane tri (3-mercaptopropionate); Triolein; Triphenyl phosphite; Tris 2-chloroethyl phosphite; Tris(3,5-di-t-butyl-4-hydroxy benzyl) isocyanurate; Tris (dipropyleneglycol) phosphite; Trishydroxyethyl triazine trione; Tris Neodol-25 phosphite; Trisnonylphenyl phosphite; Tristearyl phosphite

Urea

Zinc; Zinc-calcium complex; Zinc diethyldithiocarbamate; Zinc 2-ethylhexoate; Zinc laurate; Zinc phosphite; Zinc-potassium complex

Surfactants

Trade names: Abex® 1404; Ablumol WTI; Acintol® EPG; Acintol® FA-1; Acintol® FA-1 Special; Acintol® FA-2; Acintol® FA-3; Adeka Dipropylene Glycol; Adeka Propylene Glycol (T); Aerosol® A-103; Aerosol® A-196-40; Aerosol® A-268; Aerosol® DPOS-45; Aerosol® IB-45; Aerosol® NPES 458; Aerosol® NPES 2030; Aerosol® OT-75%; Aerosol® TR-70; Atmer® 508; Avirol® SA 4108

Beycostat LP 12 A; BYK®-LP W 6246

Calfoam NLS-30; Chemphos TC-337; Chemquat 12-33; Chemquat 12-50; Chemquat 16-50; Chemquat C/33W; Chemstat® 122; Chemstat® 182; Chemstat® 192; Chemstat® 273-E; Chimin P1A

Dabco® DC193; Dabco® DC197; Dabco® DC198; Dabco® DC1315; Dabco® DC1536; Dabco® DC1537; Dabco® DC1538; Dabco® DC1630; Dabco® DC5043; Dabco® DC5098; Dabco® DC5103; Dabco® DC5125; Dabco® DC5160; Dabco® DC5169; Dabco® DC5180; Dabco® DC5244; Dabco® DC5258; Dabco® DC5270; Dabco® DC5357; Dabco® DC5367; Dabco® DC5374; Dabco® DC5418; Dabco® DC5450; Dabco® DC5454; Dabco® DC5526; Dabco® DC5890; Dabco® DC5895; Dabco® X2-5357; Dabco® X2-5367; Darvan® No. 7; Disponil FES 92E; Disponil PNP 208; Disponil SML 100 F1; Disponil SML 104 F1; Disponil SML 120 F1; Disponil SMO 100 F1; Disponil SMO 120 F1; Disponil SMP 100 F1; Disponil SMP 120 F1; Disponil SMS 100 F1; Disponil SMS 120 F1; Disponil SSO 100 F1; Disponil STO 100 F1; Disponil STO 120 F1; Disponil STS 100 F1; Disponil STS 120 F1; Dow Corning® 190 Surfactant; Dow Corning® 193 Surfactant; Dow Corning® 197 Surfactant; Dow Corning® 198 Surfactant; Dow Corning® 1248 Fluid; Dow Corning® 1315 Surfactant; Dow Corning® 5103 Surfactant

Emcol® 4930, 4940; Empicol® LZ/D; Empicol® LZG 30; Empicol® LZGV; Empicol® LZGV/C; Empicol® LZV/D; Empicol® TL40/T; Empicol® WA; Empol® 1004; Empol® 1016; Empol® 1018; Empol® 1020; Empol® 1022; Empol® 1026; Ethoduoquad® T/15-50; Ethoquad® C/12 Nitrate; Ethoquad® CB/12; Ethoquad® T/12; Ethoquad® T/13-50

Fizul 201-11; Fluorad® FC-118; Fluorad® FC-126

Gemtex 445; Gemtex 680; Geropon® CYA/45; Glycidol Surfactant 10G

Hodag DOSS-70; Hodag DOSS-75; Hodag DTSS-70; Hodag PE-104; Hodag PE-106; Hodag PE-109; Hodag PE-206; Hodag PE-209; Hyonic GL 400

Igepal® CA-877; Igepal® CO-890; Igepal® CO-970; Igepal® CO-977; Igepal® CO-990; Igepal® CO-997

KZ 55

LICA 01; LICA 09; LICA 12; LICA 38; LICA 38A; LICA 38J; LICA 44; LICA 97; LICA 99; Lonzaine® 18S; Loxiol® G 52; Loxiol® G 53; Loxiol® P 1420; Loxiol® VPG 1354; Loxiol® VPG 1451; Loxiol® VPG 1496; Loxiol® VPG 1743

Manro DL 32; Mapeg® CO-25; Mapeg® CO-25H; Maphos® 15; Maphos® 18; Maphos® 30;

Surfactants *(cont'd.)*

Maphos® 54; Maphos® 55; Maphos® 56; Maphos® 77; Maphos® 151; Maphos® 236; Maphos® FDEO; Maphos® L 13; Maranil CB-22; Mazawet® 77; Mazawet® DF; Monafax 1293; Monolan® O Range

Nansa® 1106/P; Nansa® 1169/P; Newcol 180T; Newcol 568; Noramium CES 80

Octosol TH-40

Polystep® A-15-30K; Polystep® A-18; Polystep® B-7; Polystep® B-LCP; Poly-Tergent® 2EP; Prox-onic DT-03; Prox-onic DT-015; Prox-onic DT-030; Prox-onic MC-02; Prox-onic MC-05; Prox-onic MC-015; Prox-onic MHT-05; Prox-onic MHT-015; Prox-onic MO-02; Prox-onic MO-015; Prox-onic MO-030; Prox-onic MO-030-80; Prox-onic MS-02; Prox-onic MS-05; Prox-onic MS-011; Prox-onic MS-050; Prox-onic MT-02; Prox-onic MT-05; Prox-onic MT-015; Prox-onic MT-020; PS072

Rewopol® 15/L; Rewopol® NL 2-28; Rewopol® NL 3-28; Rewopol® NL 3-70; Rewopol® NOS 5; Rewopol® SBMB 80; Rexol 25/307; Rhodacal® 301-10; Rhodacal® 301-10P; Rhodacal® DOV; Rhodacal® LDS-22; Rhodapex® ES; Rhodapon® L-22HNC; Rhodapon® LCP

Sandoz Phosphorester 610; Silwet® L-7001; Silwet® L-7500; Silwet® L-7602; Sipomer® BEM; Siponate® 301-10F (redesignated Rhodacal® 301-10F); SM 2164; Sulfochem 436; Sulfochem 437; Sulfochem 438; Sulfopon® 101/POL; Surfynol® 465; Synperonic OP30

T-Det® N-20; Tegostab® B 3136; Tegostab® B 4351; Tegostab® B 8425; Tegostab® B 8631; Tensopol A 79; Tensopol A 795; Tensopol ACL 79; Tergitol® NP-40 (70% Aq.); Tetronic® 90R4; Tetronic® 150R1; Tetronic® 304; Texapon® NSE

Ufapol; Ufasan 50; Ufasan 60A; Ufasan TEA; Ultra Sulfate SL-1

Vycel

Witco® 1298; Witco® Acid B; Witconate™ D51-51

Chemicals: Ammonium laureth sulfate; Ammonium laureth-9 sulfate; Ammonium laureth-12 sulfate; Ammonium nonoxynol-4 sulfate; Ammonium perfluorooctanoate

Behenyl alcohol

Caprylic alcohol; Cycloneopentyl, cyclo (dimethlaminoethyl) pyrophosphato zirconate, di mesyl salt

Dicyclohexyl sodium sulfosuccinate; Diisohexyl sulfosuccinate; Disodium nonoxynol-10 sulfosuccinate; Disodium tallowiminodipropionate; Ditridecyl sodium sulfosuccinate

Laureth-7; Laureth-16; Laurtrimonium chloride

Neopentyl (diallyl) oxy, trihydroxy caproyl titanate; Neopentyl (diallyl) oxy, trimercapto-phenyl zirconate; Nonoxynol-4 sulfate

Octoxynol-1; Octoxynol-3; Octoxynol-9; Octoxynol-10; Octoxynol-16; Octoxynol-30; Octoxynol-40; Octoxynol-70

PEG-32 castor oil; PEG-75 castor oil; PEG-160 castor oil; PEG-4 dilaurate; PEG-6 dilaurate; PEG-32 dilaurate; PEG-75 dilaurate; PEG-150 dilaurate; PEG-4 dioleate; PEG-6 dioleate; PEG-20 dioleate; PEG-32 dioleate; PEG-75 dioleate; PEG-150 dioleate; PEG-6 distearate; PEG-20 distearate; PEG-32 distearate; PEG-75 distearate; PEG-32 laurate; PEG-32 oleate; PEG-75 oleate; PEG-9 stearate; PEG-32 stearate; PEG-75 stearate; PEG-150 stearate; PEG-10 tetramethyl decynediol; Polysorbate 85; Potassium dodecylbenzene sulfonate; PPG-24-buteth-27

Silicone glycol copolymer; Sodium hexadecyl diphenyloxide disulfonate; Sodium hexyl diphenyloxide disulfonate; Sodium lauroamphopropionate

Tetrasodium dicarboxyethyl stearyl sulfosuccinamate

Suspending agents

Trade names: Gantrez® AN-119

KZ 55

LICA 01; LICA 09; LICA 12; LICA 38; LICA 38A; LICA 38J; LICA 44; LICA 97; LICA 99

NZ 01; NZ 09; NZ 12; NZ 33; NZ 38; NZ 39; NZ 44; NZ 49; NZ 89; NZ 97

Octosol A-1; Octosol A-18; Octosol A-18-A

Petrac® Calcium Stearate CP-11; Petrac® Calcium Stearate CP-11 LS; Petrac® Calcium Stearate CP-11 LSG

Sellogen HR; Sinopol 806; Sinopol 808; Sinopol 908; Sinopol 910; Sipomer® HEM-5; Sipomer® HEM-10; Stepanol® WA-100

Chemicals: Butylated PVP

Suspending agents *(cont'd.)*

Calcium phosphate tribasic; Cycloneopentyl, cyclo (dimethlaminoethyl) pyrophosphato zirconate, di mesyl salt
Diammonium stearyl sulfosuccinamate; Disodium dodecyloxy propyl sulfosuccinamate
Neopentyl (diallyl) oxy, trihydroxy caproyl titanate; Neopentyl (diallyl) oxy, trimercapto-phenyl zirconate
Polyacrylamide; Polyacrylic acid; PPG-40 diethylmonium chloride

Synergists

Trade names: ADK STAB AO-80; ADK STAB AO-503A; Akrochem® Antioxidant 58; Akrochem® Antioxidant 60; Alvinox FB; AO
BiCAT™ 8; BiCAT™ H; BiCAT™ V; BLS™ 1770; BNX® 1000; Bulab® Flamebloc
Carstab® DLTDP; Carstab® DSTDP; Cyanox® 711; Cyanox® 1212; Cyanox® LTDP; Cyanox® MTDP; Cyanox® STDP
Dechlorane® Plus 25; Dechlorane® Plus 35; Dechlorane® Plus 515; Doverphos® S-680
Eastman® TEP
Ferro-Char B-44M; FireShield® H; FireShield® HPM; FireShield® L; Flamtard Z10; Flamtard Z15; FR-913
Hostanox® SE 10
Jeffcat T-10
Lankromark® DLTDP; Lankromark® DSTDP; Loxiol® HOB 7111
Mark® 2112; Mark® 2140; Mark® 5060
Nuoplaz® 849
Perkadox® 14; Perkadox® BC; Peroxidol 781; Pyro-Chek® 77B
Sandostab P-EPQ; Seenox 412S
Tegostab® B 3640; Tegostab® B 3752; Tegostab® B 8202; Tegostab® B 8231; Tegostab® B 8408; Therm-Chek® 5469; Thermolite 187; Tinuvin® 770; Toyocat®-NP; Trigonox® T
UV-Chek® AM-105; UV-Chek® AM-340
Vanox® MTI; Vanox® ZMTI; Vanstay SD; Vanstay SG; Vanstay SH; Vulkanox® MB-2/MGC; Vulkanox® ZMB2/C5; Vulkanox® ZMB-2/G
Weston® 399; Weston® 618F; Weston® 619F; Weston® TNPP
Zb™-112; Zb™-223; Zb™-237; Zb™-325; Zb™-467; Zinc Borate 9506; Zinc Borate B9355

Chemicals: Bis (pentabromophenoxy) ethane; t-Butyl cumyl peroxide
Di-(2-t-butylperoxyisopropyl)benzene; 2,4-Di-t-butylphenyl 3,5-di-t-butyl-4-hydroxybenzoate
Tribromophenyl allyl ether
Zinc borate

Tackifiers

Trade names: Abalyn®; Abitol®; Akrochem® P 37; Akrochem® P 40; Akrochem® P 49; Akrochem® P 55; Akrochem® P 86; Akrochem® P 87; Akrochem® P 90; Akrochem® P 133; Akrotak 100; Alatac 100-10; Amoco® H-15; Amoco® L-14; Amoco® L-50; Aquatac® 5527; Aquatac® 5590; Aquatac® 6025; Aquatac® 6085; Aquatac® 6085B-1; Aquatac® 6085B-7; Aquatac® 9027; Aquatac® 9041
Beckacite® 4900; Betaprene® 253; Betaprene® 255; Betaprene® BC115; Betaprene® BR100; Betaprene® BR100/60MS; Betaprene® BR100/70MS; Betaprene® BR105; Betaprene® BR110; Beta-Tac® 160; Bunatak™ N; Bunatak™ U; Bunaweld™ 780
Chemprene R-10; Chemprene R-100; Chemprene R-115
Dyphene® 8318; Dyphene® 8320; Dyphene® 8330; Dyphene® 8340; Dyphene® 8400; Dyphene® 8787; Dyphene® 9474
Eastotac™ H-100; Eastotac™ H-100E; Eastotac™ H-100L; Eastotac™ H-100R; Eastotac™ H-100W; Eastotac™ H-115E; Eastotac™ H-115L; Eastotac™ H-115R; Eastotac™ H-115W; Eastotac™ H-130E; Eastotac™ H-130L; Eastotac™ H-130R; Eastotac™ H-142R
Gantrez® M-154
Hercolite™ 240; Hercolite™ 290; Hercolyn® D
Koresin®; Koretack®; Koretack® 5193
Lankroplast® L110; Lankroplast® L542

Tackifiers *(cont'd.)*

MR1085C
Nevillac® 10° XL; Nevillac® Hard; Nevtac® 10°; Nevtac® 80; Nevtac® 100; Nevtac® 115; Nirez® 300; Nirez® 2019; Nirez® V-2040; Nirez® V-2040HM; Nirez® V-2150
Oulumer 70; Oulupale XB 100; Oulutac 80 D/HS
Petro-Rez 200; Petro-Rez 215; Petro-Rez 801; Petro-Rez PTH; Petro-Rez PTL; Picco® 5070; Picco® 6070; Picco® 6100; Picco® 6115; Piccodiene® 2215SF; Piccolyte® A115; Piccolyte® A125; Piccolyte® A135; Piccolyte® C115; Piccolyte® C125; Piccolyte® C135; Piccolyte® S115; Piccolyte® S125; Piccolyte® S135; Picconol® A100; Piccopale® 100; Piccotac® 95BHT; Piccotac® B; Piccovar® AP10; Piccovar® AP25; Piccovar® L60; Plastolein® 9789; Polydis® TR 060; Poly-Pale® Ester 10
R7510; R7521; R7578; Resin 731D; Resinall 70R; Resinall 203; Resinall 219; Resinall 286; Resinall 610; Resinall 711; Resinall 725; Resinall 766; Resinall 767; Resinall 792; Rez-O-Sperse® 3; Ricon 142; RPP
SP-25; SP-553; SP-560; SP-1068; SP-1077; Sta-Tac®; Sta-Tac® B; Sta-Tac® T; Staybelite®; Staybelite® Ester 10; Struktol® TH 10 Flakes; Struktol® TH 20 Flakes; Stygene R-2; Super Nevtac® 99; Super Sta-Tac® 80; Super Sta-Tac® 100; Sylvaros® 20; Sylvaros® 80; Sylvaros® 315; Sylvaros® R; Sylvaros® RB; Sylvatac® 5N; Sylvatac® 40N; Sylvatac® 80N; Sylvatac® 95; Sylvatac® 100NS; Sylvatac® 105NS; Sylvatac® 140; Sylvatac® 295; Sylvatac® 1085; Sylvatac® 1100; Sylvatac® 1103; Sylvatac® 2100; Sylvatac® 2103; Sylvatac® 2110; Sylvatac® AC; Sylvatac® ACF; Sylvatac® R85; Sylvatac® RX
Tartac 20; Tartac 30; Tartac 40
Uni-Rez® 3020; Uni-Tac® 68; Uni-Tac® 70; Uni-Tac® 72; Uni-Tac® 72M70; Uni-Tac® 99-70; Uni-Tac® R40; Uni-Tac® R85; Uni-Tac® R85-Light; Uni-Tac® R86; Uni-Tac® R99; Uni-Tac® R100; Uni-Tac® R100-Light; Uni-Tac® R101; Uni-Tac® R102; Uni-Tac® R106; Uni-Tac® R112
Varcum® 29008; Varcum® 29427; Varcum® 29421; Vultac® 2; Vultac® 4
Westvaco® Resin 90; Westvaco® Resin 95; Wingtack® 10; Wingtack® 95
Zonarez® 1040; Zonarez® 1085; Zonarez® 1100; Zonarez® 1115; Zonarez® 1115T; Zonarez® 1125; Zonarez® 1135; Zonarez® 7085; Zonarez® 7115; Zonarez® 7115 LITE; Zonarez® 7125; Zonarez® 7125 LITE; Zonarez® Alpha 25; Zonarez® B-10; Zonarez® B-115; Zonarez® B-125; Zonarez® M-1115; Zonarez® T-4115; Zonatac® 85 LITE; Zonatac® 105 LITE; Zonatac® 105; Zonatac® 115 LITE; Zonatac® M-105; Zonester® 25; Zonester® 85; Zonester® 100

Chemicals: Alkyl phenol disulfide
Butylphenol formaldehyde
C5 hydrocarbon resin
Glyceryl hydrogenated rosinate
Methyl hydrogenated rosinate; Methyl rosinate
Octylphenol formaldehyde resin
Pentaerythrityl rosinate; Phenolic resin; Pine tar; Polybutene; Polydipentene; Polyterpene resin
Rosin; Rosin, maleated; Rosin, polymerized
Terpene phenolic resin; Terpene resin

Thickeners • Thixotropes • Viscosity modifiers

Trade names: Abalyn®; Acconon 400-MO; ADK STAB AC-111; ADK STAB AC-311; Aerosil® 90; Aerosil® 130; Aerosil® 150; Aerosil® 200; Aerosil® 300; Aerosil® COK 84; Aerosil® R202; Aerosil® R805; Aerosil® R812; Aerosil® R974; Akrochem® P 126
BYK®-LP R 6237; BYK®-R 605
Cab-O-Sil® EH-5; Cab-O-Sil® HS-5; Cab-O-Sil® L-90; Cab-O-Sil® LM-130; Cab-O-Sil® LM-150; Cab-O-Sil® LM-150D; Cab-O-Sil® M-5; Cab-O-Sil® MS-55; Cab-O-Sil® PTG; Cab-O-Sil® TS-530; Cab-O-Sil® TS-610; Cab-O-Sil® TS-720; Crystic Pregel 17; Crystic Pregel 27
Disponil MGS 777
Elastomag® 100; Elastomag® 100R; Elastomag® 170; Elastomag® 170 Micropellet; Elastomag® 170 Special; Empicryl® 6030; Empicryl® 6045; Empicryl® 6047; Empicryl® 6052; Empicryl® 6054; Empicryl® 6058; Empicryl® 6059; Empicryl® 6070; EW-POL 8021
Fibra-Cel® BH-70; Fibra-Cel® BH-200; Fibra-Cel® CBR-18; Fibra-Cel® CBR-40; Fibra-Cel® CBR-41; Fibra-Cel® CBR-50; Fibra-Cel® CC-20; Fibra-Cel® SW-8
Gantrez® AN-119; Gantrez® AN-139; Gantrez® AN-149; Gantrez® AN-169; Gantrez® AN-179

Thickeners *(cont'd.)*

Hi-Sil® 132; Hi-Sil® 210; Hi-Sil® 233; Hostavin® System L4927
Interstab® BC-109; Interstab® BC-110; Ircogel® 900; Ircogel® 903; Ircogel® 904
Kessco® PEG 200 DL; Kessco® PEG 200 DO; Kessco® PEG 200 DS; Kessco® PEG 200 ML; Kessco® PEG 200 MO; Kessco® PEG 200 MS; Kessco® PEG 300 DL; Kessco® PEG 300 DO; Kessco® PEG 300 DS; Kessco® PEG 300 ML; Kessco® PEG 300 MO; Kessco® PEG 300 MS; Kessco® PEG 400 DL; Kessco® PEG 400 DO; Kessco® PEG 400 DS; Kessco® PEG 400 ML; Kessco® PEG 400 MO; Kessco® PEG 400 MS; Kessco® PEG 600 DL; Kessco® PEG 600 DO; Kessco® PEG 600 DS; Kessco® PEG 600 ML; Kessco® PEG 600 MO; Kessco® PEG 600 MS; Kessco® PEG 1000 DL; Kessco® PEG 1000 DO; Kessco® PEG 1000 DS; Kessco® PEG 1000 ML; Kessco® PEG 1000 MO; Kessco® PEG 1000 MS; Kessco® PEG 1540 DL; Kessco® PEG 1540 DO; Kessco® PEG 1540 DS; Kessco® PEG 1540 ML; Kessco® PEG 1540 MO; Kessco® PEG 1540 MS; Kessco® PEG 4000 DL; Kessco® PEG 4000 DO; Kessco® PEG 4000 DS; Kessco® PEG 4000 ML; Kessco® PEG 4000 MO; Kessco® PEG 4000 MS; Kessco® PEG 6000 DL; Kessco® PEG 6000 DO; Kessco® PEG 6000 DS; Kessco® PEG 6000 ML; Kessco® PEG 6000 MO; Kessco® PEG 6000 MS; Kronitex® 1840; Krynac® 34.140; KZ 55; KZ OPPR; KZ TPP; KZ TPPJ
Lankroplast® V2012; Latekoll®; LICA 01; LICA 09; LICA 12; LICA 38; LICA 38A; LICA 38J; LICA 44; LICA 97; LICA 99; Lubrizol® 2152
MagChem® 60; MagChem® 125; MagChem® 200-AD; Methocel® A; Methocel® E; Millithix® 925; Mistron Vapor® R; Mistron ZSC
Nopalcol 1-L; Nopalcol 1-TW; Nopalcol 2-DL; Nopalcol 4-C; Nopalcol 4-CH; Nopalcol 4-L; Nopalcol 4-S; Nopalcol 6-DO; Nopalcol 6-DTW; Nopalcol 6-L; Nopalcol 6-R; Nopalcol 6-S; Nopalcol 10-COH; Nopalcol 12-CO; Nopalcol 12-COH; Nopalcol 19-CO; Nopalcol 30-TWH; Nopalcol 200; Nopalcol 400; Nopalcol 600; NZ 01; NZ 09; NZ 12; NZ 33; NZ 38; NZ 39; NZ 44; NZ 49; NZ 89; NZ 97
Palatinol® FF41; Pegosperse® 200 DL; Pegosperse® 200 ML; Pegosperse® 400 DL; Pegosperse® 400 DO; Pegosperse® 400 ML; Plastigel®; Plastigel PG9033; Plastigel PG9037; Plastigel PG9068; Plastigel PG9089; Plastigel PG9104; Plastomag® 170; Pluracol® E600 NF; Pluracol® E1000; Pluracol® E1450 NF; Pluracol® E2000; Pluracol® E4500; Pluracol® E8000; Polysar SS 260; Polyvel CR-10; Polyvel CR-25; Polywax® 500; Polywax® 655; Polywax® 850; Polywax® 1000; Polywax® 2000; Polywax® 3000; Prox-onic L 081-05; Prox-onic L 101-05; Prox-onic L 102-02; Prox-onic L 121-09; Prox-onic L 161-05; Prox-onic L 181-05; Prox-onic L 201-02; Prox-onic LA-1/02; PVP K-60
Rhodamox® CAPO; Rhodamox® LO; Rhodasurf® L-4
Sipernat® 22LS; Sipomer® HEM-5; Sipomer® HEM-10; Staflex DIOA; Stanclere® A-121, A-121 C, A-221; Stanclere® T-85, TL; Stanclere® T-94 C; Stanclere® T-126; Stanclere® T-208; Stanclere® T-222, T-233; Stanclere® T-250, T-250SD; Stanclere® T-801; Stanclere® T-876; Stanclere® T-877; Stanclere® T-878, T-55; Stanclere® TL; Sunyl® 80; Sunyl® 80 ES; Sunyl® 90
Tafigel PUR 40; Tego® Effect L 104; Texicryl® 13-300, 13-302; Tixosil 311; Tixosil 321; Tixosil 331; Tixosil 333, 343; Tixosil 375; Tullanox HM-250; Tylose® C, CB Series; Tylose® CBR Grades; Tylose® CR; Tylose® H Series; Tylose® MHB
Ultra-Pflex®
Vulcastab® T
Wacker HDK® H15; Wacker HDK® H20; Wacker HDK® N20; Wacker HDK® S13; Wacker HDK® T30; Wacker HDK® T40; Wacker HDK® V15; Witconol™ 1206

Chemicals: Ammonium polyacrylate; Ammonium polymethacrylate
Butadiene/acrylonitrile copolymer
Calcium ricinoleate; Carboxymethylcellulose sodium; Cycloneopentyl, cyclo (dimethlaminoethyl) pyrophosphato zirconate, di mesyl salt
DEA-lauryl sulfate; Dibenzylidene sorbitol
Isodecyl methacrylate
Lauramine oxide; Lithopone
Methyl hydroxyethylcellulose; Methyl rosinate
PEG (1200) castor oil; PEG (1900) castor oil; PEG-8 cocoate; PEG-4 dilaurate; PEG-6 dilaurate; PEG-20 dilaurate; PEG-32 dilaurate; PEG-75 dilaurate; PEG-150 dilaurate; PEG-4 dioleate; PEG-6 dioleate; PEG-20 dioleate; PEG-32 dioleate; PEG-75 dioleate; PEG-150 dioleate; PEG-6 distearate; PEG-20 distearate; PEG-32 distearate; PEG-75 distearate; PEG-150 distearate; PEG-12 ditallowate; PEG-20 hydrogenated castor oil; PEG (1200) hydrogenated castor oil; PEG-60 hydrogenated tallowate; PEG-3 lauramine oxide; PEG-4 laurate; PEG-6

Thickeners *(cont'd.)*

laurate; PEG-32 laurate; PEG-75 laurate; PEG-150 laurate; PEG-2 laurate SE; PEG-4 oleate; PEG-8 oleate; PEG-20 oleate; PEG-32 oleate; PEG-75 oleate; PEG-12 ricinoleate; PEG-2 stearate; PEG-4 stearate; PEG-12 stearate; PEG-32 stearate; PEG-75 stearate; PEG-150 stearate; PEG-2 tallowate; Pentaerythrityl tetraoleate; Polyacrylamide; Polyacrylic acid

Silica dimethyl silylate; Sodium polyacrylate; Styrene/MA copolymer

Tris (dipropyleneglycol) phosphite

UV Absorbers • Light stabilizers

Trade names: Aceto; Acetorb A; ADK STAB 1413; ADK STAB LA-31; ADK STAB LA-32; ADK STAB LA-34; ADK STAB LA-36; ADK STAB LA-51; ADK STAB LA-57; ADK STAB LA-62; ADK STAB LA-63; ADK STAB LA-67; ADK STAB LA-68; ADK STAB LA-68LD; ADK STAB LA-77; ADK STAB LA-82; ADK STAB LA-87; Ampacet 10057; Ampacet 10478; Aquastab PB 59; Aquastab PB 503; Aquastab PC 44; Aquastab PC 45M; AST-1001

Barimite™ 1525P; Barimite™ 4009P; Bayer UV Absorber 325, 340; Be Square® 175; Be Square® 185; Black Pearls® 120; Black Pearls® L; BLS™ 531; BLS™ 1770; Busan® IIMI

Cabot® PP 9131; Cabot® PP 9135; Cabot® PP 9136; Cabot® PE 9138; Cabot® PP 9287; Cabot® PP 9129; Cabot® PP 9130; Cabot® PP 9131; Chimassorb® 119; Chimassorb® 119FL; Chimassorb® 944; Chimassorb® 944FD; Chimassorb® 944FL; Chimassorb® 944LD; Colloids N 54/1088; Colloids PE 48/10/02 UV; Colortech 10007-11; Colortech 10942-12; Colortech 10944-11; Cyasorb® UV 9; Cyasorb® UV 24; Cyasorb® UV 416; Cyasorb® UV 500; Cyasorb® UV 531; Cyasorb® UV 1084; Cyasorb® UV 1164; Cyasorb® UV 2098; Cyasorb® UV 2126; Cyasorb® UV 2908; Cyasorb® UV 3346 LD; Cyasorb® UV 3581; Cyasorb® UV 3604; Cyasorb® UV 3638; Cyasorb® UV 3668; Cyasorb® UV 3853

Doverphos® 8; Doverphos® 9; Drapex® 4.4; Dyphos Envirostab; Dyphos XL

Eastman® Inhibitor RMB; Elftex® 120; Elftex® 125; Elftex® 280; Elftex® 430; Elftex® 570; Elftex® 675

Flamebloc

Gamma W8; Gamma W8 HP0.6; Gamma W8 M1.8; Givsorb® UV-1; Givsorb® UV-2; Good-rite® 3034; Good-rite® 3150; Good-rite® 3159

Halcarb 20; Halcarb 20 EP; Halphos; Haro® Chem KB-350X; Haro® Chem KPR; Haro® Chem KS; Haro® Mix BF-202; Haro® Mix BK-105, BK-107; Haro® Mix UC-213; Haro® Mix YC-601; Haro® Mix YE-301; Haro® Mix ZC-028, ZC-029, ZC-030; Hostavin® ARO 8; Hostavin® N 20; Hostavin® N 24; Hostavin® N 30; Hostavin® VP NiCS 1; Hostavin® System L1922; Hostavin® System L1923; Hostavin® System L1924; Hostavin® System L1970; Hostavin® System L1973; Hostavin® System L2925; Hostavin® System L2970; Hostavin® System L2971; Hostavin® System L2972; Hostavin® System L4926; Hostavin® System L4928; Hostavin® System L4962, L4963, L4964

Interstab® BC-4195; Interstab® BC-4377; Interstab® BZ-4828A; Interstab® BZ-4836; Interstab® CH-55; Interstab® CH-55R; Interstab® CLB-747; Interstab® CZ-10; Interstab® CZ-11; Interstab® CZ-11D; Interstab® CZ-19A; Interstab® CZ-22; Interstab® CZ-23; Interstab® LF 3615; Interstab® LT-4805R; Interstab® LT-4361; Interstab® R-4150; Irgastab® 2002; Irgastab® T 634

Lankromark® BM271; Lankromark® LE285; Lankromark® LE296; Lowilite® 22; Lowilite® 26; Lowilite® 55; Lowilite® 62; Lowilite® 77

Mark® 133; Mark® 224; Mark® 398; Mark® 462A; Mark® 495; Mark® 565A; Mark® 577A; Mark® 1014; Mark® 1413; Mark® 1535; Mark® 2077; Mark® 2100, 2100A; Mark® 2326; Mark® 5060; Mark® 7118; Mark® 7308; Mark® DPDP; Mark® DPP; Mark® GS, RFD; Mark® OHM; Mark® OTM; Mark® PDDP; Mark® QED; Mark® TDP; Mark® TPP; MDI PE-200UVA; Merix MCG Compd; MiRaStab 113-K; MiRaStab 130-K; MiRaStab 140-K; MiRaStab 154-K; MiRaStab 400-K; MiRaStab 704-K; MiRaStab 709-ZK; MiRaStab 800-K; MiRaStab 922-K; MiRaStab 950-K; Mixxim® BB/100; Mixxim® BB/200; Mogul® L; Monarch® 120; Monarch® 700; Monarch® 800; Monarch® 880

Naugard® 76; Naugard® NBC; Nipol® 1452X8

Ottalume 2100

Parabolix® 100; PE-UV 01; PE-UV 11; PE-UV 12; PE-UV 13; PE-UV 18; PE-UV 23; PE-UV 25; PE-UV 32; PE-UV 41; PE-UV 45; PE-UV 57; PE-UV 67; PE-UV 84; PE-UV 99; PE-UV 102; PE-UV 118; PE-UV 119; Plasadd® PE 9134; Plasblak® LL 2590; Plastistab 2005; Plastistab 2020; Plastistab 2160; Plastistab 2210; Plastistab 2310; Plastistab 2320; Plastistab 2330;

UV Absorbers • Light stabilizers *(cont'd.)*

Plastistab 2560; Plastistab 2580; Plastistab 2800; Plastistab 2900; Plastistab 2931; Plastistab 2941; Plastolein® 9232; Poly-Dispersion® E (DPS)D-80; Poly-Dispersion® E (DYT)D-80; Poly-Dispersion® G (DPS)D-80; Poly-Dispersion® G (DYT)D-80; Poly-Dispersion® H (DYT)D-80; Poly-Dispersion® H (DYT)D-80P; Poly-Dispersion® T (DYT)D-80P; Polymist® 284; Polymist® F-5; Polymist® F-5A; Polymist® F-5A EX; Polymist® F-510; Polyvel UV25AG2; PP-UV 38; PP-UV 52; PP-UV 53; PP-UV 54; PP-UV 61; PP-UV 62; PP-UV 63; PP-UV 64; PP-UV 66; PP-UV 75; PP-UV 77; PP-UV 87; PP-UV 88; PP-UV 113; Printex P; Ralox® 530; Ralox® 630; Regal® 330; Regal® 330R; Rhodialux Q84

Sanduvor® 3050; Sanduvor® 3051; Sanduvor® 3052; Sanduvor® VSU; Santowhite® Powd; Saytex® BT-93®; Spectratech® CM 10540; Spectratech® CM 11246; Spectratech® CM 11357; Spectratech® CM 11367; Spectratech® CM 11615; Spectratech® CM 11616; Spectratech® CM 11681; Spectratech® CM 11704; Spectratech® CM 11715; Spectratech® CM 11720; Staflex DIOA; Staflex DOA; Staflex DOZ; Staflex NODA; Stanclere® T-85, TL; Stanclere® T-184; Stanclere® T-197; Stanclere® T-208; Stanclere® T-222, T-233; Stanclere® T-250, T-250SD; Stanclere® T-4356; Stanclere® TL; Stavinor ATA-1056; Stavinor CA PSE; Synpron 431; Synpron 687; Syntase® 62; SY-UV 34

Thermax® Floform® N-990; Therm-Chek® 6-V-6A; Therm-Chek® 130; Therm-Chek® 135; Therm-Chek® 344; Therm-Chek® 659; Therm-Chek® 763; Therm-Chek® 837; Therm-Chek® 900; Therm-Chek® 904; Therm-Chek® 1238; Therm-Chek® 1720; Therm-Chek® 1827; Therm-Chek® 1864; Therm-Chek® 5590; Therm-Chek® 5649; Therm-Chek® 5776; Therm-Chek® 5930; Therm-Chek® 6160; Thermocel 1002; Thermocel 2365; Thermolite 12; Thermolite 31; Thermolite 187; Thermolite 390; Thermolite 395; Thermolite 813; THPE-BZT; Tinuvin® 123; Tinuvin® 144; Tinuvin® 213; Tinuvin® 234; Tinuvin® 326; Tinuvin® 327; Tinuvin® 328; Tinuvin® 329; Tinuvin® 571; Tinuvin® 622; Tinuvin® 622LD; Tinuvin® 765; Tinuvin® 770; Tinuvin® 783; Tinuvin® P; Topanex 100BT; Topanex 500H

UV-149; UV-201; UV-247; UV-349; UV-822; UV-Absorber 325; UV-Absorber 340; UV-Absorber KL 3-2500; Uvasil PSL6; Uvasorb 2 OH; Uvasorb 3C; Uvasorb HA-88; Uvasorb MET; Uvasorb NI; Uvasorb SV; UV-Chek® AM-101; UV-Chek® AM-104; UV-Chek® AM-105; UV-Chek® AM-205; UV-Chek® AM-300; UV-Chek® AM-301; UV-Chek® AM-320; UV-Chek® AM-340; UV-Chek® AM-595; UV-Chek® AM-806; Uvinul® 408; Uvinul® 3000; Uvinul® 3008; Uvinul® 3035; Uvinul® 3039; Uvinul® 3040; Uvinul® 3049; Uvinul® 3050; Uvinul® 3093; Uvinul® D-49; Uvinul® D-50; Uvinul® M-40; Uvinul® N-35; Uvinul® N-539

Vanwax® H Special; Viosorb 520; Voidox 100%; Vulcan® 9A32; Vulcan® P

Weston® 439; Weston® 494; Weston® DPP

Chemicals: 4-(2-Acryloyloxyethoxy)-2-hydroxybenzophenone polymer

Barium-calcium-zinc complex; Barium stearate; Barium-zinc complex; Barium-zinc phosphite; Benzophenone; Benzophenone-1; Benzophenone-2; Benzophenone-3; Benzophenone-4; Benzophenone-6; Benzophenone-8; Benzophenone-11; 2-(2H-Benzotriazol-2-yl)-4-(1,1,3,3-tetramethylbutyl) phenol; 2-[4,6-Bis(2,4-dimethylphenyl)-1,3,5-triazin-2-yl]-5-(octyloxy)phenol; Bis[2-hydroxy-5-methyl-3-(benzotriazol-2-yl)phenyl]methane; Bis [2-hydroxy-5-t-octyl-3-(benzotriazol-2-yl) phenyl] methane; Bis-(1-octyloxy-2,2,6,6,tetramethyl-4-piperidinyl) sebacate; Bis (1,2,2,6,6-Pentamethyl-4-piperidinyl) (3,5-di-t-butyl-4-hydroxybenzyl) butyl propanedioate; Bis(1,2,2,6,6-pentamethyl-4-piperidinyl)sebacate; Bis (2,2,6,6-tetramethyl-4-piperidinyl) sebacate; 2-(3′-t-Butyl-2′-hydroxy-5′-methylphenyl)-5-chlorobenzotriazole

Cadmium-barium complex; Cadmium laurate; Calcium-tin complex; Calcium-tin-zinc complex; C40-60 alcohols; Carbon black; Cyclodextrin

Di-t-butylhydroquinone; 3,5-Di-t-butyl-p-hydroxybenzoic acid; 3,5-Di-t-butyl-4-hydroxybenzoic acid, n-hexadecyl ester; 2-(3′,5′-Di-t-butyl-2′-hydroxyphenyl)-5-chlorobenzotriazole; Dimethyl succinate; Dimethylsuccinate and tetramethyl hydroxy-1-hydroxyethyl piperidine polymer; 1,5-Dioxaspiro[5.5]undecane 3,3-dicarboxylic acid, bis(2,2,6,6-tetramethyl-4-piperidinyl)ester; Diphenyldidecyl (2,2,4-trimethyl-1,3-pentanediol) diphosphite; Diphenyl isodecyl phosphite; 4-Dodecyloxy-2-hydroxybenzophenone; Drometrizole

N-(p-Ethoxycarbonylphenyl)-N′-ethyl-N′-phenylformamidine; N^2-(4-Ethoxycarbonylphenyl)-N′-methyl-N′-phenylformamidine; 2-Ethyl, 2′-ethoxy-oxalanilide; Etocrylene

Hydroquinone monomethyl ether; 2-Hydroxy-4-acryloyloxyethoxy benzophenone; 2-(2′-Hydroxy-3′-t-butyl-5′-methylphenyl)-5-chlorobenzotriazole; 2-(2′-Hydroxy-3,5′-di-t-amylphenyl) benzotriazole; 2-[2-Hydroxy-3,5-di-(1,1-dimethylbenzyl)phenyl]-2H-benzotriazole; 2(2′-Hydroxy-3′,5′(1,1-dimethylbenzylphenyl) benzotriazole; 2-Hydroxy-4-isooctoxybenzophenone; 2-(2-Hydroxy-5-t-octylphenyl) benzotriazole; Hydroxypropyl-β-

UV Absorbers • Light stabilizers *(cont'd.)*

cyclodextrin; Hydroxypropyl-γ-cyclodextrin
4,4′ Isopropylidenediphenol alkyl (C12-15) phosphites
Lead; Lead carbonate (basic); Lead phosphite dibasic; Lead phthalate, basic; Lead phthalate, dibasic
Magnesium-zinc complex; Methyl-β-cyclodextrin; Methyl-γ-cyclodextrin; α-Methylstyrene/N-(2,2,6,6-tetramethyl piperidinyl-4) maleimide/N-stearyl maleimide terpolymer
Nickel bis[O-ethyl (3,5-di-t-butyl-4-hydroxybenzyl)] phosphonate; Nickel bis (octyl phenol sulfide); Nickel dibutyldithiocarbamate; Nickel diisobutyldithiocarbamate; Nickel dimethyldithiocarbamate
Octocrylene
Pentamethyl-4-piperidyl/β,β,β′,β′-tetramethyl-3,9-(2,4,8,10-tetraoxaspiro (5,5) undecane) diethyl]-1,2,2,6,6-1,2,3,4-butane tetracarboxylate; Phenol, 2-(2H-benzotriazole-2-yl)-4-methyl-6-dodecyl; 2,2′-(1,4-Phenylene)bis[4H-3,1-benzoxazin-4-one]; Phenyl salicylate; Poly [(6-morpholino-s-triazine-2,4-diyl) [2,2,6,6-tetramethyl-4-piperidyl) imino]-hexamethylene [(2,2,6,6-tetramethyl-4-piperidyl) imino]
Resorcinol benzoate
Strontium-zinc complex
Tetrakis (2,2,6,6-tetramethyl-4-piperidyl)-1,2,3,4-butane tetracarboxylate; 2,2,6,6-Tetramethylpiperidine-4-carbonamide; 2,2,6,6-Tetramethylpiperidin-4-yl acrylate/methyl methacrylate copolymer; [2,2,6,6-Tetramethyl-4-piperidyl/β,β,β′,β′-tetramethyl-3,9-(2,4,8,10-tetraoxaspiro (5,5) undecane) diethyl]-1,2,3,4-butane tetracarboxylate; 2,2′-Thiobis (4-t-octylphenolato)-n-butylamine nickel; 2,2′-Thiobis (4-t-octylphenolato)-n-butylamine nickel II; Thiobis-2,2′-(4(1,1,3,3-tetramethylbutyl)phenol)-nickel-2-ethylhexanoate; Tin, butyl 2-ethylhexanoate laurate oxo complexes; 1,1′,1′′-(1,3,5-Triazine-2,4,6-triyltris((cyclohexylimino)-2,1-ethanediyl)tris(3,3,4,5,5-tetramethylpiperazinone);; Tridecyl phosphite; 2,3,4-Trihydroxy benzophenone; 1,1,1-Tris(hydroxyphenyl-ethane)benzotriazole
Zinc oxide

Vulcanizing agents

Trade names: Akrochem® DPTT
Cadox® TS-50S; Crystex® 60; Crystex® Regular
DBQDO®; Di-Cup 40C; Di-Cup 40KE; Di-Cup R, T; DPTT-S; Dynamar® FC-5157
Esperox® 497M; Ethyl Tellurac®; Ethyl Tuads® Rodform
Isobutyl Tuads®
Lectro 78; Litharge 28; Litharge 33
MagChem® 35K; MagChem® 40; MagChem® 50; MagChem® 50Y; MagChem® 60; MagChem® 125; Methyl Tuads® Rodform; Mod Acid
Naftocit® DPG; Naftocit® Di 4; Naftocit® Di 7; Naftocit® Di 13; Naftocit® MBT; Naftocit® MBTS; Naftocit® Mi 12; Naftocit® NaDBC; Naftocit® NaDMC; Naftocit® Thiuram 16; Naftocit® ZBEC; Naftocit® ZMBT; Neoprene FB; Novor 950
Octocure 456
Peroximon® DC 40 MF; Peroximon® DC 40 MG; Peroximon® S-164/40P; Petrac® 250; Poly-Dispersion® ASD-75; Poly-Dispersion® ECSD-70; Poly-Dispersion® ESD-80; Poly-Dispersion® E (VC)D-40; Poly-Dispersion® K (PX-17)D-30; Poly-Dispersion® SCSD-70; Poly-Dispersion® SSD-75; Poly-Dispersion® TMC-50; Poly-Dispersion® TSD-80; Poly-Dispersion® TSD-80P; Poly-Dispersion® T (VC)D-40P; Poly-Dispersion® TVP-50; Poly-Dispersion® VC-60P
QDO®
Rapidblend 1793; Rhenocure® CUT; Rhenocure® EPC; Rhenocure® M; Rhenocure® M/G; Rhenocure® S/G; Rhenocure® TDD; Rhenocure® ZAT; Rhenogran® PbO-80/FPM; Rhenovin® S-90; Rylex® 30; Rylex® 3010
SP-1055; Sulfads®; Sulfasan® R
Telloy; Tetronic® 304; Tetronic® 701; Tetronic® 704; Tetronic® 901; Tetronic® 904; Tetronic® 908; Tetronic® 1307; Thermacure®; Tri-Mal
Vanax® A; Vandex; Varox® 130; Varox® 130-XL; Varox® 231-XL; Varox® 231; Varox® DBPH-50; Varox® DBPH; Vul-Cup 40KE; Vul-Cup R; Vulcuren 2; Vulcuren 2/EGC; Vultac® 2; Vultac® 3; Vultac® 4; Vultac® 5; Vultac® 7; Vultac® 710

Vulcanizing agents *(cont'd.)*

Chemicals: Alkyl phenol disulfide
Bisperoxide
Dibenzoyl-p-quinone dioxime; 1,1-Di (t-butylperoxy) 3,3,5-trimethyl cyclohexane; p,p-Dichlorobenzoyl peroxide; Dicumyl peroxide; Dipentamethylene thiuram hexasulfide
Lead fumarate tetrabasic; Lead maleate, tribasic; Lead (II) oxide
MDI; 2-(4-Morpholinyldithio) benzothiazole
Polychloroprene
Selenium; Selenium diethyldithiocarbamate; Selenium dimethyldithiocarbamate; Sulfur; Sulfur chloride; Sulfur, insoluble
Tellurium; Tetrabutyl thiuram disulfide; Tetraisobutylthiuram disulfide; Tetramethylthiuram disulfide

Waxes

Trade names: A-C® 6; A-C® 6A; A-C® 15; A-C® 307, 307A; A-C® 316, 316A; A-C® 325; A-C® 330; A-C® 392; A-C® 395, 395A; A-C® 400; A-C® 400A; A-C® 405M; A-C® 405S; A-C® 405T; A-C® 617, 617A; A-C® 629; A-C® 629A; A-C® 655; ACumist® A-6; ACumist® A-12; ACumist® A-18; ACumist® A-45; ACumist® B-6; ACumist® B-9; ACumist® B-12; ACumist® B-18; ACumist® C-5; ACumist® C-9; ACumist® C-12; ACumist® C-18; ACumist® C-30; Advawax® 240; Advawax® 280; Advawax® 290; Astrawax 23
Be Square® 195
Castorwax® MP-70; Castorwax® MP-80
Epolene® C-10; Epolene® C-13; Epolene® C-14; Epolene® C-15; Epolene® C-16; Epolene® C-17; Epolene® C-18; Epolene® E-10; Epolene® E-14; Epolene® E-15; Epolene® E-20; Epolene® E-43; Epolene® N-10; Epolene® N-15; Epolene® N-20; Epolene® N-21
Glycowax® S 932
Hoechst Wax PE 190; Hoechst Wax SW
Koster Keunen Candelilla; Koster Keunen Ceresine; Koster Keunen Microcrystalline Waxes; Koster Keunen Ozokerite; Koster Keunen Synthetic Candelilla
Liquax 488; Luwax® AL 3; Luwax® ES 9668; Luwax® OA 5
Petrolite® E-2020; Polymel #7; Ross Carnauba Wax; Ross Ceresine Wax; Ross Montan Wax; Ross Wax #100; Ross Wax #140; Ross Wax #145; Ross Wax #160; Ross Wax #165
Silwax® C; Silwax® S; Silwax® WD-IS; Silwax® WD-S; Silwax® WS
Ultraflex®; Uniwax AW-1050
Vanwax® H; Vestowax AO-1539; Vestowax FT-150; Vestowax FT-150P; Vestowax FT-200; Vestowax FT-300; Victory®

Chemicals: Ethylene dioleamide
Microcrystalline wax, oxidized; Montan acid wax; Montan wax

Wetting agents

Trade names: AA Standard; Abalyn®; Abex® 18S; Abex® 22S; Ablusol C-78; Ablusol M-75; ACtone® 1; ACtone® 2010; ACtone® 2010P; ACtone® N; Aerosol® 501; Aerosol® A-103; Akrotak 100; Alkamuls® EL-620L; Alkamuls® EL-719; Alkanol® 189-S; Anti-Terra®-204; Anti-Terra®-P; Arizona DRS-40; Arizona DRS-42; Arizona DRS-43; Armeen® Z; Arodet BN-100; Aromox® C/12; Aromox® C/12-W; Arquad® 2C-75; Arquad® 12-50; Arquad® 16-50; Arquad® 18-50; Arquad® C-50; Arquad® T-27W; Atmer® 110; Atmer® 113; Atmer® 502; Atmer® 645; Atmer® 685; Avirol® SA 4106; Avirol® SA 4108; Avirol® SA 4110
Benzoflex® 9-88; Berol 28; Berol 302; Berol 303; Berol 307; Berol 381; Berol 386; Berol 387; Berol 389; Berol 391; Berol 392; Berol 452; Berol 455; Berol 456; Berol 457; Berol 458; Beycostat NE; BLO®; Borax; Bowax 2015; BYK®-R 605; BYK®-W 900; BYK®-W 920; BYK®-W 935; BYK®-W 940; BYK®-W 965; BYK®-W 968; BYK®-W 980; BYK®-W 990; BYK®-W 995
Catinex KB-50; Chemax DOSS/70E; Chemax DOSS/70HFP; Chemax DOSS-75E; Chemfac NC-0910; Chemfac PA-080; Chemfac PB-082; Chemfac PB-106; Chemfac PB-135; Chemfac PB-184; Chemfac PB-264; Chemfac PC-099E; Chemfac PC-188; Chemfac PD-600; Chemfac PF-623; Chemfac PF-636; Chemfac PN-322; Chemphos TC-227; Chemphos TC-310; Chimipal APG 400; Cyastat® SP

Wetting agents *(cont'd.)*

Darvan® ME; Darvan® NS; Darvan® WAQ; Dextrol OC-15; Dextrol OC-20; Dextrol OC-50; Dextrol OC-70; Disperplast®1142; Disperplast®-1150; Disperplast®-I; Disperplast®-P; Disponil MGS 935; Disponil O 5; Disponil O 10; Disponil SUS IC 8; Disponil TA 5; Disponil TA 25; Disponil TA 430; Dow Corning® 190 Surfactant; Dow Corning® 193 Surfactant; Dresinate® 214; Dresinate® 731; Drewfax® 0007

Ebecryl® 1360; Eccowet® W-50; Eccowet® W-88; Emcol® 4930, 4940; Emcol® DOSS; Emphos™ PS-400; Empicol® LZ/D; Empicol® LZG 30; Empicol® LZGV; Empicol® LZGV/C; Empicol® LZP; Empicol® LZV/D; Empilan® DL 40; Empilan® DL 100; Empimin® OP70; Empimin® OT; Empimin® OT75; Emulvin® W; Epodil® ML; Epolene® C-16; Epolene® C-18; Ethylan® GMF; #15 Oil

Fizul 201-11; Flexricin® 13

Gardilene S25L; Gemtex 445; Gemtex 680; Gemtex 691-40; Gemtex PA-75; Gemtex PA-85P; Gemtex PAX-60; Gemtex SC-75E; Geropon® CYA/60; Geropon® CYA/DEP; Geropon® SDS; Geropon® SS-O-75; Geropon® WT-27; GP-209 Silicone Polyol Copolymer; GP-214 Silicone Polyol Copolymer; GP-215 Silicone Polyol Copolymer; GP-218 Silicone Polyol Copolymer; GP-219 Silicone Polyol Copolymer

Hodag DOSS-70; Hodag DOSS-75; Hodag DTSS-70

Igepal® CO-630; Igepal® CO-710; Igepal® CO-850; Igepal® CO-880; Igepal® CO-887; Igepal® CO-890; Igepal® CO-897; Igepal® CTA-639W

Kessco® PEG 200 DL; Kessco® PEG 200 DO; Kessco® PEG 200 DS; Kessco® PEG 200 ML; Kessco® PEG 200 MO; Kessco® PEG 200 MS; Kessco® PEG 300 DL; Kessco® PEG 300 DO; Kessco® PEG 300 DS; Kessco® PEG 300 ML; Kessco® PEG 300 MO; Kessco® PEG 300 MS; Kessco® PEG 400 DL; Kessco® PEG 400 DO; Kessco® PEG 400 DS; Kessco® PEG 400 ML; Kessco® PEG 400 MO; Kessco® PEG 400 MS; Kessco® PEG 600 DL; Kessco® PEG 600 DO; Kessco® PEG 600 DS; Kessco® PEG 600 ML; Kessco® PEG 600 MO; Kessco® PEG 600 MS; Kessco® PEG 1000 DL; Kessco® PEG 1000 DO; Kessco® PEG 1000 DS; Kessco® PEG 1000 ML; Kessco® PEG 1000 MO; Kessco® PEG 1000 MS; Kessco® PEG 1540 DL; Kessco® PEG 1540 DO; Kessco® PEG 1540 DS; Kessco® PEG 1540 ML; Kessco® PEG 1540 MO; Kessco® PEG 1540 MS; Kessco® PEG 4000 DL; Kessco® PEG 4000 DO; Kessco® PEG 4000 DS; Kessco® PEG 4000 ML; Kessco® PEG 4000 MO; Kessco® PEG 4000 MS; Kessco® PEG 6000 DL; Kessco® PEG 6000 DO; Kessco® PEG 6000 DS; Kessco® PEG 6000 ML; Kessco® PEG 6000 MO; Kessco® PEG 6000 MS; KZ 55; KZ OPPR; KZ TPP; KZ TPPJ

Lankropol® KMA; Lankropol® KO2; Lankropol® OPA; LICA 01; LICA 09; LICA 12; LICA 38; LICA 38A; LICA 38J; LICA 44; LICA 97; LICA 99; Lubrizol® 2153

Manro DL 28; Manro SLS 28; Masil® 2132; Masil® 2133; Masil® 2134; Mazawet® 77; Merpol® A; Merpol® HCS; Miranol® J2M Conc; Monawet 1240; Monawet MB-45; Monawet MM-80; Monawet MO-70; Monawet MO-70E; Monawet MT-70; Monawet SNO-35; Monolan® 2500 E/30; Monolan® 3000 E/60, 8000 E/80; Monolan® 8000 E/80

Nacconol® 40G; Nacconol® 90G; Naxonac™ 510; Naxonac™ 610; Newcol 560SF; Newcol 560SN; Newcol 861S; Newcol 1305SN, 1310SN; Niaproof® Anionic Surfactant 4; Niaproof® Anionic Surfactant 08; Niox KQ-34; Nissan Plonon 102; Nissan Plonon 104; Nissan Plonon 108; Nissan Plonon 171; Nissan Plonon 172; Nissan Plonon 201; Nissan Plonon 204; Nissan Plonon 208; Nissan Rapisol B-30, B-80, C-70; Noiox AK-41; Nopalcol 1-L; Nopalcol 1-TW; Nopalcol 2-DL; Nopalcol 4-C; Nopalcol 4-CH; Nopalcol 4-L; Nopalcol 4-S; Nopalcol 6-DO; Nopalcol 6-DTW; Nopalcol 6-L; Nopalcol 6-R; Nopalcol 6-S; Nopalcol 10-COH; Nopalcol 12-CO; Nopalcol 12-COH; Nopalcol 19-CO; Nopalcol 30-TWH; Nopalcol 200; Nopalcol 400; Nopalcol 600; Nopco ESA120; NZ 01; NZ 09; NZ 12; NZ 33; NZ 38; NZ 39; NZ 44; NZ 49; NZ 89; NZ 97

Octosol A-1; Octowet 70PG

Pationic® 900; Pationic® 901; Pationic® 902; Penetron OT-30; Phospholan® PNP9; Plastoflex® 2307; Prote-pon P-2 EHA-02-K30; Prote-pon P 2 EHA-02-Z; Prote-pon P 2 EHA-Z; Prote-pon P-0101-02-Z; Prote-pon P-L 201-02-K30; Prote-pon P-L 201-02-Z; Prote-pon P-NP-06-K30; Prote-pon P-NP-06-Z; Prote-pon P-NP-10-K30; Prote-pon P-NP-10-MZ; Prote-pon P-NP-10-Z; Prote-pon P-OX 101-02-K75; Prote-pon P-TD-06-K13; Prote-pon P-TD-06-K30; Prote-pon P-TD 06-K60; Prote-pon P-TD-06-Z; Prote-pon P-TD-09-Z; Prote-pon P-TD-12-Z; Prote-pon TD-09-K30; Prox-onic HR-05; Prox-onic HR-016; Prox-onic HR-025; Prox-onic HR-030; Prox-onic HR-036; Prox-onic HR-040; Prox-onic HR-080; Prox-onic HR-0200; Prox-onic HR-0200/50; Prox-onic HRH-05; Prox-onic HRH-016; Prox-onic HRH-025; Prox-onic HRH-0200; Prox-onic HRH-0200/50; Prox-onic L 081-05; Prox-onic L 101-05; Prox-onic L 102-02; Prox-onic L 121-09; Prox-onic L 161-05; Prox-onic L 181-05; Prox-onic L 201-02; Prox-onic LA-1/

Wetting agents *(cont'd.)*

02; Prox-onic NP-1.5; Prox-onic NP-04; Prox-onic NP-06; Prox-onic NP-09; Prox-onic NP-010; Prox-onic NP-015; Prox-onic NP-020; Prox-onic NP-030; Prox-onic NP-030/70; Prox-onic NP-040; Prox-onic NP-040/70; Prox-onic NP-050; Prox-onic NP-050/70; Prox-onic NP-0100; Prox-onic NP-0100/70

Rewomat TMS; Rewopol® SBDO 75; Rexol 25/8; Rexol 25/10; Rexol 25/12; Rhodacal® 330; Rhodacal® BA-77; Rhodacal® BX-78; Rhodafac® PS-19; Rhodafac® RE-610; Rhodafac® RE-877; Rhodafac® RE-960; Rhodapex® EST-30; Rhodapon® CAV; Rhodapon® TDS; Rhodasurf® A 24; Rhodasurf® L-4; Rhodasurf® L-25

SAFR-Z-12K; SAFR-Z-12SK; SAFR-Z-22E K; SAFR-Z-44 K; Secosol® DOS 70; Sellogen HR; Serdet DSK 40; Serdet Perle Conc; Silwet® L-7001; Sinonate 960SF; Sinonate 960SN; Sinoponic PE 61; Steol® 4N; Steol® CA-460; Stepanol® WA-100; Struktol® TR 041; Struktol® TR 071; Supragil® NK; Surfynol® 82S; Surfynol® 104A; Surfynol® 104E; Surfynol® 104PA; Surfynol® 485

Tergitol® 15-S-30; Tergitol® 15-S-40; Tergitol® NP-40; Teric C12; Teric N8; Teric N9; Teric N10; Teric N12; Teric N13; Teric N15; Teric N20; Teric N30; Teric N40; Tetronic® 90R4; Tetronic® 150R1; Tetronic® 304; Texapon® VHC Needles; Texapon® ZHC Powder; Triton® 770 Conc; Triton® X-301; Triton® X-305-70%; Triton® X-405-70%; Trycol® 6954; Trycol® 6957; Trycol® 6967; Trycol® 6970; Trycol® 6972; Trycol® 6984; Trycol® NP-20

Ufasan 35; Ufasan 62B; Ufasan 65

Victawet® 12; Viscobyk-4010; Viscobyk-4013; Vorite 105; Vorite 110; Vorite 115; Vorite 120; Vorite 125

Wayfos A; Witcolate™ 1276; Witcolate™ D-510; Witconate™ 90; Witconate™ 1250

Zonyl® FSA; Zonyl® FSB; Zonyl® FSC; Zonyl® FSN; Zonyl® FSP

Chemicals:

Ammonium lauryl sulfate; Ammonium lignosulfonate; Ammonium nonoxynol-4 sulfate

Caprylic alcohol; Castor oil, polymerized; Ceteareth-30; Cetoleth-10; Cetoleth-13; Cetoleth-25; Cocaminobutyric acid; Cycloneopentyl, cyclo (dimethlaminoethyl) pyrophosphato zirconate, di mesyl salt

DEA-lauryl sulfate; Deceth-4 phosphate; Decyl betaine; Diamyl sodium sulfosuccinate; Diisobutyl sodium sulfosuccinate; Dioctyl sodium sulfosuccinate; Disodium capryloamphodiacetate; Disodium C12-15 pareth sulfosuccinate; Disodium deceth-6 sulfosuccinate; Disodium dodecyloxy propyl sulfosuccinamate; Disodium isodecyl sulfosuccinate; Disodium nonoxynol-10 sulfosuccinate; Disodium oleamido MIPA-sulfosuccinate

Laureth-9; Lauryl alcohol

Methyl rosinate

Neopentyl (diallyl) oxy, trihydroxy caproyl titanate; Neopentyl (diallyl) oxy, trimercapto-phenyl zirconate; Nonoxynol-1; Nonoxynol-3; Nonoxynol-5; Nonoxynol-15; Nonoxynol-20; Nonoxynol-25; Nonoxynol-30; Nonoxynol-35; Nonoxynol-40; Nonoxynol-50; Nonoxynol-70; Nonoxynol-100; Nonoxynol-150; Nonoxynol-4 sulfate

Oleamine; Oleoyl sarcosine; Oleth-4 phosphate

PEG-32 castor oil; PEG-75 castor oil; PEG-160 castor oil; PEG (1200) castor oil; PEG (1900) castor oil; PEG-8 cocoate; PEG-4 dilaurate; PEG-6 dilaurate; PEG-20 dilaurate; PEG-32 dilaurate; PEG-75 dilaurate; PEG-150 dilaurate; PEG-4 dioleate; PEG-6 dioleate; PEG-8 dioleate; PEG-20 dioleate; PEG-32 dioleate; PEG-75 dioleate; PEG-150 dioleate; PEG-6 distearate; PEG-20 distearate; PEG-32 distearate; PEG-75 distearate; PEG-12 ditallowate; PEG-20 hydrogenated castor oil; PEG (1200) hydrogenated castor oil; PEG-60 hydrogenated tallowate; PEG-4 laurate; PEG-32 laurate; PEG-75 laurate; PEG-150 laurate; PEG-2 oleamine; PEG-7 oleamine; PEG-12 oleamine; PEG-4 oleate; PEG-20 oleate; PEG-32 oleate; PEG-75 oleate; PEG-150 oleate; PEG-12 ricinoleate; PEG-2 stearate; PEG-20 stearate; PEG-32 stearate; PEG-75 stearate; PEG-150 stearate; PEG-2 tallowamine; PEG-5 tallowamine; PEG-10 tallowamine; PEG-15 tallowamine; PEG-20 tallowamine; PEG-40 tallowamine; PEG-2 tallowate; PEG-3 tallow diamine; PEG-15 tallow diamine; PEG-30 tallow diamine; PEG-30 tetramethyl decynediol; Potassium rosinate

Silicone hexaacrylate; Sodium borate decahydrate; Sodium caproamphoacetate; Sodium cocoamphopropionate; Sodium decyl sulfate; Sodium dibutyl naphthalene sulfonate; Sodium dodecylbenzenesulfonate; Sodium isodecyl sulfate; Sodium isopropyl naphthalene sulfonate; Sodium lauroyl sarcosinate; Sodium myristoyl sarcosinate; Sodium myristyl sulfate; Sodium nonoxynol-4 sulfate; Sodium octylphenoxyethoxyethyl sulfonate; Sodium octyl sulfate; Sodium rosinate; Sodium tridecyl sulfate; Sorbitan isostearate; Sorbitan sesquioleate; Sulfonic acid

TEA-benzene sulfonate; TEA-dodecylbenzenesulfonate; TEA-lauryl sulfate; Tetrahydroxypropyl ethylenediamine; Tridecylbenzene sulfonic acid

Part IV
Manufacturers Directory

Manufacturers Directory

AAA Molybdenum Products, Inc.

7233 W. 116th Place, Broomfield, CO 88020 USA (Tel.: 303-460-0844; 800-443-6812; Telefax: 303-460-0851)

Aarhus

Aarhus Oliefabrik A/S, Postboks 50, Bruunsgade 27, DK-8100 Aarhus C Denmark (Tel.: 86-12 60 00; Telefax: 86-196252; Telex: 64341)

Aarhus Inc., 131 Marsh St., PO Box 4240, Newark, NJ 07114 USA (Tel.: 201-344-1300)

Abbott Labs

Abbott Laboratories, 1400 Sheridan Rd., N. Chicago, IL 60064-4000 USA (Tel.: 708-937-8800; 800-240-1043; Telefax: 708-937-6676)

Abbott Laboratories Ltd., Div. of Abbott Laboratories, Queenborough, Kent, ME11 5EL UK (Tel.: 44 1795 580099; Telefax: 44 1795 580404; Telex: 96347)

Abbott SA-NV, Div. of Abbot Laboratories, Parc Scientifique, 2 rue du Bosquet, B-1348 Louvain Belgium (Tel.: 32-47 53 11; Telefax: 32-47 55 75; Telex: 59334 ABBOTT B)

Abco Industries Ltd.

200 Railroad St., PO Box 335, Roebuck, SC 29376 USA (Tel.: 803-576-6821; 800-476-4476; Telefax: 803-576-9378; Telex: 628 17731)

Aceto Chemical Co., Inc.

1 Hollow Lane, Suite 201, Lake Success, NY 11042-1215 USA (Tel.: 516-627-6000; Telefax: 516-627-6000; Telex: FTCC 824609)

Pfaltz & Bauer Inc., Research chemicals subsidiary of Aceto Corp., 172 E. Aurora St., Waterbury, CT 06708 USA (Tel.: 203-574-0075; 800-225-5172; Telefax: 203-574-3181; Telex: 996471)

AC Industries, Inc., Sattva Chemical Co. Div.

5 Landmark Sq., Suite 308, Stamford, CT 06901 USA (Tel.: 203-348-8002; 800-736-7893; Telefax: 203-348-3666; Telex: 6819048 SATVA UW)

ACL Inc.

1960 E. Devon Ave., Elk Grove Village, IL 60007 USA (Tel.: 708-981-9212; 800-782-8420; Telefax: 708-981-9278; Telex: 4330251)

ACM Chemical Corp. *See* Atomergic Chemetals

Additive Polymers Ltd.

Unit 4 Kiln Rd., Burrfields Rd., Portsmouth, PO3 5LP UK (Tel.: 44 1705 678575; Telefax: 44 1705 678564)

Adeka Fine Chemical Co., Ltd. *See under* Asahi Denka Kogyo

Advanced Refractory Technologies, Inc.

699 Hertel Ave., Buffalo, NY 14207 USA (Tel.: 716-875-4091; Telefax: 716-875-0106)

Air Products and Chemicals, Inc.

7201 Hamilton Blvd., Allentown, PA 18195-1501 USA (Tel.: 215-481-4911; 800-345-3148; Telefax: 215-481-5900; Telex: 847416)

Air Products and Chemicals, Inc./Performance Chemicals, 7201 Hamilton Blvd., Allentown, PA 18195-1501 USA (Tel.: 215-481-6799; 800-345-3148; Telefax: 215-481-6822; Telex: 847416)

Air Products and Chemicals, Inc./Polymer Chemicals Div., 7201 Hamilton Blvd., Allentown, PA 18195-1501 USA (Tel.: 215-481-6799; 800-345-3148; Telefax: 215-481-4381)

Air Products and Chemicals de Mexico, S.A. de C.V., Rio Guadiana 23, Piso 5, Colonia Cuauhtemoc, Mexico D.F., 06500 Mexico (Tel.: 525-591-0800; Telefax: 525-592-3018)

Anchor Chemical (UK) Ltd., Div. of Air Products, Clayton House, Clayton, Manchester, M11 4SR UK (Tel.: 44 161 223-2461; Telefax: 44 161-223-5488; Telex: 667829)

Air Products Nederland B.V., Kanaalweg 15, PO Box 3193, 3502 GD Utrecht The Netherlands (Tel.: 31-30-857100; Telefax: 31-30 857111)

Anchor Italiana SPA, 33 via Scaldasole, Dorno 27020, Pavia Italy (Tel.: 39-382-84186; Telefax: 39-382-84017; Telex: 39-320225)

Air Products and Chemicals PURA GmbH & Co., Postfach 5108, Robert-Koch-Strasse 27, D-2000 Norderstedt Germany (Tel.: 49 40 529009-0; Telefax: 49 40 52900999; Telex: 403975 PURA)

Anchor Chemical Australia Pty., Ltd., 2/20 Hunter St., PO Box 4068, Paramatta, NSW, 2124 Australia (Tel.: 61-2-687-1944; Telefax: 61-2-687-1950)

Anchor Chemical Asia Pacific Pte. Ltd., 105 Cecil St. #04-01, The Octagon, Singapore 0106 (Tel.: 65-222-9392; Telefax: 65-221-5192; Telex: 65-400002)

Air Products Asia, Inc., Room 1901, Peregrine Tower, Lippo Centre, 89 Queensway, Central Hong Kong (Tel.: (852)524-7110; Telefax: (852) 877-0094)

Air Products Japan, Inc., Shuwa No. 2 Kamiyacho Bldg., 3-18-19, Toranomon, Minato-ku, Tokyo, 105 Japan (Tel.: (81)(3)3432-7046; Telefax: (81)(3)3432-7048)

Aiscondel SA

Aragon 182, E-08011 Barcelona Spain (Tel.: 343-323-1020; Telefax: 343-323-7921; Telex: 97887 ETIN E)

Ajinomoto

Ajinomoto Co., Inc., 15-1, Kyobashi 1-chome, Chuo-ku, Tokyo, 104 Japan (Tel.: (03) 5250-8111; Telex: J22690)

Ajinomoto USA, Inc., Glenpointe Centre West, 500 Frank W. Burr Blvd., Teaneck, NJ 07666-6894 USA (Tel.: 201-488-1212; Telefax: 201-488-6282; Telex: 275425 (AJNJ)

Akcros

Akcros Chemicals America, 500 Jersey Ave., New Brunswick, NJ 08903 USA (Tel.: 908-247-2202; 800-500-7890; Telefax: 908-247-2287)

Akcros Chemicals Ltd., Lankro House, Silk St., PO Box 1, Eccles, Manchester, M30 0BH UK (Tel.: 44 161 7897-300; Telefax: 44 161-7887-886; Telex: 587135)

Akcros Chemicals v.o.f., Haagen House, 6040 AA Roermond The Netherlands (Tel.: 31 4750-91777; Telefax: 31 4750-17489; Telex: 58021 haro nl)

Akcros Chemicals France S.A., BP 40, 441220 St Laurent-Nouan France (Tel.: 33 5487 7389; Telefax: 33 5487 7940; Telex: 750679 Tinstab)

Akcros Chemicals (Asia Pacific) Pte Ltd., Singapore (Tel.: 65 2921 966; Telefax: 65 2929 665; Telex: RS 23241 AKCROS)

Akrochem Chemical Co.

255 Fountain St., Akron, OH 44304 USA (Tel.: 216-535-2108; 800-321-2260; Telefax: 216-535-8947)

Akzo Nobel

Akzo Nobel België NV/SA, 13 Marnix Ave., 1050 Bruxelles Belgium (Tel.: 2-518 04 07; Telefax: 02 518 05 05; Telex: 62 664)

Akzo Nobel Chemicals, 300 S. Riverside Plaza, Chicago, IL 60606 USA (Tel.: 312-906-7500; 800-828-7929; Telefax: 312-906-7633; Telex: 25-3233)

Akzo Nobel Chemicals Ltd., 1 City Center Dr., Suite 320, Mississauga, Ontario, L53 IM2 Canada (Tel.: 905-273-5959; Telefax: 905-273-7339)

Akzo Nobel Chemicals Ltd., 1-5, Queens Road, Hersham, Walton-on-Thames, Surrey, KT12 5NL UK (Tel.: 44 1932 247891; Telefax: 44 1932-231204; Telex: 21997)

Akzo Nobel Chemicals BV, Postbus 975, Stationsstraat 48, 3800 AZ Amersfoort The Netherlands (Tel.: 33-643454; Telefax: 33-637448; Telex: 79276)

Akzo Nobel Chemicals GmbH, Postfach 100132, Phillippstrasse 27, 52301 Düren Germany (Tel.: 2421-49201; Telefax: 2421-492487; Telex: 833911)

Akzo Nobel Chemicals Ltd., 6 Grand Ave., PO Box 80, Camellia, NSW, 2150 Australia (Tel.: (02) 6384555; Telefax: (02) 6384681; Telex: AA 21562)

Akzo Nobel Japan Ltd., Godo Kaikan Bldg., 3-27 Kioi-cho, Chiyoda-ku, Tokyo, 102 Japan (Tel.: (03)5275-6268; Telefax: (03)3263-0713; Telex: J24262)

Akzo Nobel/Eka Nobel/Int'l. Resins Group, 7 Beeches Lane, Woodstock, CT 06281 USA (Tel.: 203-963-2061; Telefax: 203-963-2164)

Eka Nobel Ltd., Div. of Eka Nobel AB, Unit 304 Worle Pkwy., Summer Lane, Worle, Weston-Super-Mare, Avon, BS22 OWA UK (Tel.: 44 1934 522244; Telefax: 44 1934 522577; Telex: 444351)

Eka Nobel AB, Div. of Nobel Industrier Sverige AB, S-445 80 Bohus Sweden (Tel.: 46-31-58 70 00; Telefax: 46-31-98 17 74; Telex: 2435 EKAGBG S)

Eka Nobel (Australia) Pty Ltd., Div. of Nobel Industries, 22 Commercial Dr., Dandenong, Victoria, 3175 Australia (Tel.: 3-706-4488)

Albemarle

Albemarle Corp., 451 Florida St., Baton Rouge, LA 70801 USA (Tel.: 504-388-7040; 800-535-3030; Telefax: 504-388-7686; Telex: 586441, 586431)

Albemarle SA, Div. of Albemarle Corp, 523 Ave. Louise, Box 19, B-1050 Brussels Belgium (Tel.: 2-642-4411; Telefax: 2-648-0560; Telex: 22549)

Albemarle Asia Pacific Co., #13-06 PUB Bldg., Devonshire Wing, 111 Somerset Rd., Singapore 0923 Singapore (Tel.: 65-732-6286; Telefax: 65-737-4123)

Albemarle Japan Corporation, Shiroyama Hills 19F, 4-3-1, Toranomon, Minato-ku, Tokyo, 105 Japan (Tel.: 81-3-5401-2901; Telefax: 81-3-5401-3368)

Albright & Wilson

Albright & Wilson Ltd., European & Corporate Hdqtrs., PO Box 3, 210-222 Hagley Rd. West, Oldbury, Warley, West Midlands, B68 0NN UK (Tel.: 44 121 429-4942; Telefax: 44 121-420-5151; Telex: 336291)

Albright & Wilson Americas Inc., PO Box 26229, Richmond, VA 23260-6229 USA (Tel.: 804-550-4300; 800-446-3700; Telefax: 804-550-4385)

Albright & Wilson Am. (Canada), 2070 Hadwen Rd., Mississauga, Ontario, L5K 2C9 Canada (Tel.: 905-403-0011; Telefax: 905-403-2231)

Albright & Wilson (Australia) Limited, PO Box 20, Yarraville, Victoria, 3013 Australia (Tel.: 61 3-688-7777; Telefax: 61 3-688-7788)

Albright & Wilson Asia Pacific Pte Ltd., 6 Jalan Besut, Jurong Industrial Estate, Singapore 2261 (Tel.: 011-65-261-2151; Telefax: 011-65-265-1941)

Albright & Wilson Ltd. Japan, No. 2 Okamotoya Bldg. 6 Fl., 1-24, Toranomon 1-chome, Minato-ku, Tokyo, 105 Japan (Tel.: (03) 3508-9461; Telefax: (03) 3591-0733)

Alcan

Alcan Chemicals, Div. of Alcan Aluminum Corp, 3690 Orange Place, Suite 400, Cleveland, OH 44122-4438 USA (Tel.: 216-765-2550; 800-321-3864; Telefax: 216-765-2570, 135069)

Alcan Chemicals Ltd., Div. of British Alcan Aluminum plc, Chalfont Park, Gerrards Cross, Buckinghamshire, SL9 0QB UK (Tel.: 44 1753-887373; Telefax: 44 1753 881556; Telex: 847343)

Alco Chemical Corp, Div. of National Starch & Chem.

909 Mueller Dr., PO Box 5401, Chattanooga, TN 37406 USA (Tel.: 615-629-1405; 800-251-1080; Telefax: 615-698-8723; Telex: 755002)

Alcoa Industrial Chemicals Div.

4701 Alcoa Rd., PO Box 300, Bauxite, AR, 72011 USA (Tel.: 501-776-4987; 800-643-8771; Telefax: 501-776-4592; Telex: 536447)

Aldrich

Aldrich Chemical Co., Inc., 1001 W. St. Paul Ave., Milwaukee, WI 53233 USA (Tel.: 414-273-3850; 800-558-9160; Telefax: 414-273-4979; Telex: 26843 ALDRICH MI)

Aldrich Flavors & Fragrances, 1101 W. St. Paul Ave., Milwaukee, WI 53233 USA (Tel.: 414-273-3850; 800-227-4563; Telefax: 414-273-5793, 910-262-3052)

Aldrich Chemical Co. Ltd., The Old Brickyard, New Rd., Gillingham, Dorset, SP8 4JL UK (Tel.: 44 1747 822211, 0800-71 71 81; Telefax: 44 1747 823779; Telex: 417238)

Sigma-Aldrich NV/SA, K. Cardijnplei 8, B-2880 Bornem Belgium (Tel.: 038991301; Telefax: 038991311)

Aldrich-Chimie S.a.r.l., BP 701, 38297 Saint Quentin Fallavier,, Cedex France (Tel.: 74822800; Telefax: 74956808; Telex: 308215 Aldrich F)

Aldrich Chemical, Unit 2, 10 Anella Ave., Castle Hill, NSW, 2154 Australia (Tel.: 028999977; Telefax: 028999742)

Aldrich Japan, Kyodo Bldg., Shinkanda, 10 Kanda-Mikuracho, Chiyoda-ku, Tokyo Japan (Tel.: 0332580155; Telefax: 0332580157)

Alemark Chemicals

1177 High Ridge Rd., Stamford, CT 06905 USA (Tel.: 203-966-7410; Telefax: 203-966-4276)

Alfa Chem

1661 N. Spur Dr., Central Islip, NY 11722-4325 USA (Tel.: 516-277-7681; Telefax: 516-277-7681, 6504811992)

Alframine Corp.

72 Putnam St., Paterson, NJ 07524 USA (Tel.: 201-279-5334; Telefax: 201-279-5335)

Allan Chemical Corp.

PO Box 1837, Fort Lee, NJ 07024-8337 USA (Tel.: 201-592-8122; Telefax: 201-592-9298, 170991)

All Chemie Ltd.

1429 John St., Fort Lee, NJ 07024 USA (Tel.: 201-947-7776; Telefax: 201-947-3343)

Allchem Industries Inc.

4001 Newberry Rd, Suite E-3, Gainesville, FL 32607 USA (Tel.: 904-378-9696; Telefax: 904-338-0400; Telex: 509540 ALLCHEM UD)

Allco Chemical Corp.

PO Box 247, Galena, KS 66739 USA (Tel.: 316-783-1321; 800-437-1998; Telefax: 316-783-5253, (650) 173-6115)

17304 N. Preston Rd., Dallas, TX 75252 USA (Tel.: 214-733-6831; Telefax: 214-733-6846)

Allied Colloids Inc.

2301 Wilroy Rd., PO Box 820, Suffolk, VA 23439-0820 USA (Tel.: 804-538-3700; Telefax: 804-538-0204)

AlliedSignal

AlliedSignal Inc., PO Box 1053, 101 Columbia Rd., Morristown, NJ 07960 USA (Tel.: 201-455-2000; 800-526-0717; Telefax: 201-455-3198; Telex: 136410)

AlliedSignal Inc./Performance Additives, PO Box 1039, 101 Columbia Rd., Morristown, NJ 07962-1039 USA (Tel.: 201-455-2145; 800-222-0094; Telefax: 201-455-6154; Telex: 990433)

AlliedSignal Inc./Industrial Fibers, 101 Columbia Rd., Morristown, NJ 07962-1087 USA (Tel.: 201-455-5093; Telefax: 201-455-3402)

AlliedSignal Europe NV, Haasrode Research Park, Grauwmeer 1, B-3001 Heverlee (Leuven) Belgium (Tel.: 32-16-39 12 33; Telefax: 32-16-40 03 77)

AlliedSignal Europe NV/Performance Additives, Haasrode Research Park, Grauwmeer 1, B-3001 Heverlee (Leuven) Belgium (Tel.: 32-16/391 211; Telefax: 32-16/400 039)

AlliedSignal Int'l. NV-SA/A-C Performance Additives, Haasrode Research Park, Grauwmeer, B-3001 Heverlee (Leuven) Belgium (Tel.: 32-16-39 12 11; Telefax: 32-16-400-039)

AlliedSignal Performance Additives (Singapore) Pte. Ltd., 03-01 The Pascal, 16A Science Park Dr., Singapore 0511 (Tel.: 65-775-2133; Telefax: 65-775-2311)

Alloychem, Inc.

600 Madison Ave., New York, NY 10022 USA (Tel.: 212-644-1510; Telefax: 212-644-1480, 234635 RCA)

AluChem Inc.

One Landy Lane, Reading, OH 45215 USA (Tel.: 513-733-8519; Telefax: 513-733-0608; Telex: 298252 ALUC UR)

Alzo Inc.

6 Gulfstream Blvd., Matawan, NJ 07747 USA (Tel.: 908-254-1901; Telefax: 908-254-4423)

Amber Synthetics, Amsyn Inc.

1177 High Ridge Rd., Stamford, CT 06905 USA (Tel.: 203-972-7401; Telefax: 203-966-4276)

Amerchol

Amerchol Corp., Div. of United-Guardian, Inc., PO Box 4051, 136 Talmadge Rd., Edison, NJ 08818-4051 USA (Tel.: 908-248-6000; 800-367-3534; Telefax: 908-287-4186; 7Telex: 833472)

Amerchol Europe, Havenstraat 86, B-1800 Vilvoorde Belgium (Tel.: 2-252-4012; Telefax: 2-252-4909; Telex: 846-69105 AMRCHL B)

Amerchol, Ikeda Corp., New Tokyo Bldg., No. 3-1, Marunouchi 3-Chome, Chiyoda-Ku, Tokyo, 100 Japan (Tel.: 813-3212-8791; Telefax: 813-3215-5069; Telex: CONCESS J 781 26370)

AmeriBrom. *See under* Dead Sea Bromine

American Biorganics, Inc.

2236 Liberty Dr, Niagara Falls, NY 14304 USA (Tel.: 716-283-1434; 800-648-6689; Telefax: 716-283-1570; Telex: 926074)

American Chemet Corp.

400 Lake Cook Rd., Deerfield, IL 60015 USA (Tel.: 708-948-0800; Telefax: 708-948-0811; Telex: 72-4301)

American Chemical Services

PO Box 190, Griffith, IN 46319 USA (Tel.: 219-924-4370; FAX 219-924-5298)

American Colloid Co.

1500 W. Shure Dr., Arlington Hts., IL 60004-1434 USA (Tel.: 708-392-4600; Telefax: 708-506-6199; Telex: 4330321)

American Cyanamid/Corporate Headquarters,

One Cyanamid Plaza, Wayne, NJ 07470 USA (Tel.: 201-831-3339; Telefax: 201-831-2637)

American Ingredients Co.

American Ingredients Co./Patco Polymer Additives Div., 3947 Broadway, Kansas City, MO 64111 USA (Tel.: 816-561-9050; 800-669-2250; Telefax: 816-561-0422)

American International Chemical, Inc.

27 Strathmore Rd, Natick, MA 01760 USA (Tel.: 508-655-5805; 800-238-0001; Telefax: 508-655-0927; Telex: 948342)

American Lecithin. *See under* Rhone-Poulenc

American Maize Products Co./Amaizo

1100 Indianapolis Blvd., Hammond, IN 46320-1094 USA (Tel.: 219-659-2000; 800-348-9896; Telefax: 219-473-6601)

Ameripol Synpol Co., Div. of the Uniroyal Goodrich Tire Co.

146 S. High St., 7th floor, Akron, OH 44308-1493 USA (Tel.: 216-762-1655; 800-321-9001; Telefax: 216-762-2549)

Amoco

Amoco Chemical Co., 200 East Randolph Dr., Mail code 4106, Chicago, IL 60601 USA (Tel.: 312-856-3092; 800-621-4567; Telefax: 312-856-4151; Telex: 25-3731)

Amoco Oil, Lubricants Business Unit, 2021 Spring Rd., Oak Brook, IL 60562-1857 USA (Tel.: 708-571-7100; Telefax: 708-571-7174)

Amoco Chemical (Europe) SA, Div. of Amoco Corp, 15, Rue Rothschild, CH-1211 Geneva 21 Switzerland (Tel.: 41-22-31-02-81; Telex: 422787)

Ampacet

Ampacet Corp., 660 White Plains Rd., Tarrytown, NY 10591-5130 USA (Tel.: 914-631-6600; Telefax: 914-631-7197; Telex: 62588)

Ampacet Europe S.A., Rue D'Ampacet 1, 6780 Messancy Belgium (Tel.: 32-63- 371490; Telefax: 32-63-371499; Telex: 42710)

AMRESCO

30175 Solon Ind. Pkwy., Solon, OH 44139 USA (Tel.: 216-349-1313; 800-829-2802; Telefax: 216-349-1182, 985582)

Amspec Chemical Corp.

Foot of Water St., Gloucester City, NJ 08030 USA (Tel.: 609-456-3930; 800-5AMSPEC; Telefax: 609-456-6704; Telex: 136714 GLCY)

Amstat Industries, Inc.

3012 N. Lake Terrace, Glenview, IL 60025-5794 USA (Tel.: 708-998-6210; 800-783-9999; Telefax: 708-998-6218)

Anchor. *See under* Air Products

Andersons, The

PO Box 119, Maumee, OH 43537 USA (Tel.: 419-893-5050; 800-537-3370; Telefax: 419-891-6539)

Andrea Aromatics

PO Box 3091, Princeton, NJ 08543-3091 USA (Tel.: 609-695-7710; Telefax: 609-392-8914)

Andrulex Trading Ltd.

Unit 34, Saffron Court, Southfields Industrial Estate, Laindon, Basildon, Essex, SS15 6SS UK (Tel.: 44 1268 416441; Telefax: 44 1268 541639; Telex: 99339 RULEX G)

Anedco, Inc.

10429 Koenig Rd., Houston, TX 77034 USA (Tel.: 713-484-3900; Telefax: 713-484-3931)

ANGUS

ANGUS Chemical Co., 1500 E. Lake Cook Rd., Buffalo Grove, IL 60089-6556 USA (Tel.: 708-215-8600; 800-362-2580; Telefax: 708-215-8626; Telex: 275422 ANGUS UR)

ANGUS Chemie GmbH, Unit 7, Rotunda Business Centre, Thorncliffe Park Estate, Chapeltown, Sheffield, S30 4PH UK (Tel.: 44 1742 571322; Telefax: 44 1742 571336)

ANGUS Chemie GmbH, Huyssenallee 5, 45128 Essen Germany (Tel.: (49) 201-233531; Telefax: (49)201-238661; Telex: 8571563 ANGE D)

ANGUS Chemie GmbH, Le Bonaparte, Centre d'Affaires, Paris Nord, 93153 Le Blanc Mesnil France (Tel.: (33)1-48-65-73-40; Telefax: (33)1-48-65-73-20; Telex: ANGUS 232089 F)

ANGUS Chemical (Singapore) Pte. Ltd., 150 Beach Rd., #17-01 Gateway West, Singapore 0718 (Tel.: (65) 293-1738; Telefax: (65) 293-3307)

Anhydrides & Chemicals Inc.

7-33 Amsterdam St., Newark, NJ 07105 USA (201-465-0077; Telefax: 201-465-7713)

Anzon Inc., Subsid. of Cookson Group plc

2545 Aramingo Ave., Philadelphia, PA 19125 USA (215-427-6920; 800-523-0882; Telefax: 215-427-6955)

Appleby Group Ltd.

Brigg Road, Scunthorpe, South Humberside, DN16 1AQ UK (Tel.: 44 1724 282211; Telefax: 44 1724 270435; Telex: 527328)

Aqualon

Aqualon Co, A Hercules Inc. Co, 1313 North Market St., Wilmington, DE 19899-8740 USA (Tel.: 302-594-6000; 800-345-8104; Telefax: 302-594-6660; Telex: 4761123)

Aqualon Canada Inc., 5407 Eglinton Ave. West, Suite 103, Etobicoke, Ontario, M9C 5K6 Canada (Tel.: 416-620-5400)

Aqualon France, 3 Rue Eugene & Armand, Peugeot, 92508 Rueil-Malmaison France (Tel.: 33 1 4751 2919; Telefax: 33 1 4777 0614, 6314244)

Aqualon UK Ltd., Genesis Centre, Garret & Field, Birchwood, Warnington, Cheshire WA3 7BH UK (Tel.: 44-1925 830077; Telefax: 44-1925 830112; Telex: 626219)

Aquatec Quimica SA

Av. Paulista no. 37-12° andar, 01311-000 Sao Paulo-SP Brazil (Tel.: 55 11-284-4188; Telefax: 55 11-288-4431; Telex: 1121312)

Arakawa

Arakawa Chemical Industries Ltd., 3-7, Hiranomachi 1-chome, Chuo-ku, Osaka, 541 Japan (Tel.: (06) 209-8580; Telefax: (06) 209-8542; Telex: 5222296 ARKOSA J)

Arakawa Chemical (USA) Inc., 625 N. Michigan Ave., Suite 1700, Chicago, IL 60611 USA (Tel.: 312-642-1750; Telefax: 312-642-0089; Telex: 26-5514)

Arco

Arco Chemical/Headquarters, Research & Engineering Center, 3801 West Chester Pike, Newtown Sq., PA 19073-2387 USA (Tel.: 215-359-2000; 800-345-0252; Telefax: 215-359-2722)

Arco Chemical Canada Inc., 100 Consilium Pl., Suite 306, Scarborough, Ontario, M1H 3E3 Canada (Tel.: 416-296-9864)

Arco Chemical Pan American, Inc., Paseo de la Reforma, 390 Decimo Piso, 06600 Mexico City Mexico (Tel.: 905-514-6833)

Arco Chemical Europe Inc., Bridge Ave., Maidenhead, Berkshire, SL6 1YP UK (Tel.: 44 1628 775000; Telex: 847436)

Arco Chemical Asia/Pacific Ltd., Toranomon 37 Mori Bldg., 5th Floor, 5-1 Toranomon 3-Chome, Minato-Ku, Tokyo, 105 Japan

Arenol Chemical Corp.

189 Meister Ave., Somerville, NJ 08876 USA (Tel.: 908-526-5900; Telefax: 908-526-9688)

M. Argueso & Co., Inc.

441 Waverly Ave., PO Box E, Mamaroneck, NY 10543 USA (Tel.: 914-698-8500; Telefax: 914-698-0325)

Arista Industries, Inc.

1082 Post Rd., Darien, CT 06820 USA (Tel.: 203-655-0881; 800-255-6457; Telefax: 203-656-0328; Telex: 996493)

Aristech Chemical Corp.

600 Grant St., Room 1028, Pittsburgh, PA 15230-0250 USA (Tel.: 412-433-7700; 800-526-4032; Telefax: 412-433-1816; Telex: 6503608865)

Arizona Chemical Co, Div. of International Paper

1001 E. Business Hwy. 98, Panama City, FL 32401-3633 USA (Tel.: 904-785-6700; 800-526-5294; Telefax: 904-785-2203; Telex: 441695)

Bergvik Kemi, Div. of International Paper, Box 66, S-82022 Sandarne Sweden (Tel.: 46 270 62500; Telefax: 46 270 60100; Telex: 47310)

Arol Chemical Products Co.

649 Ferry St., Newark, NJ 07105 USA (Tel.: 201-344-1510; Telefax: 201-344-7127)

Asahi Chemical Industry Co., Ltd.

Hibiya Mitsui Bldg., 1-2, Yuraku-cho 1-chome, Chiyoda-ku, Tokyo, 100 Japan (Tel.: (03) 3507-2730; Telefax: (03) 3507-2495; Telex: 222-3518 BEMBRGJ)

Asahi Denka Kogyo

Asahi Denka Kogyo K.K., Furukawa Bldg. 3-14, Nihonbashi Muro-machi 2-chome Chuo-ku, Tokyo, 103 Japan (Tel.: (03) 5255-9017; Telefax: (03) 3270-2463; Telex: 222-2407 TOKADK)

Adeka Fine Chemical Co., Ltd., Subsid. of Asahi Denka Kogyo, Yoko Bldg., 4-5, Hongo 1-chome, Bunkyo-ku, Tokyo, 113 Japan (Tel.: (03) 5689-8681; Telefax: (03) 5689-8680)

Asahi-Penn Chemical Co. Ltd., Joint venture of Asahi Glass Co Ltd./PPG Industries Inc.

Shuwa Kioicho TBR Bldg., 5-7, Kojimachi, Chiyoda-ku, Tokyo, 102 Japan (Tel.: (03) 3234-0561; Telefax: (03) 3234-0304)

Asahi Yukizai Kogyo Co., Ltd.

2-5955, Nakanose-cho, Nobeoka-shi, Miyazaki, 882 Japan (Tel.: (0982) 33-3311; Telefax: (0982) 21-8606)

Asarco Inc.

180 Maiden Lane, New York, NY 10038 USA (Tel.: 212-510-2000; Telex: ITT 420585)

Ash Grove Cement Co.

8900 Indian Creek Pkwy., Overland Park, KS 66225 USA (Tel.: 913-451-8900)

Ashland

Ashland Chemical Inc./Industrial Chemicals & Solvents, PO Box 2219, Columbus, OH 43216 USA (Tel.: 614-889-3333; Telefax: 614-889-3465)

Drew Industrial Div., Div. of Ashland Chemical Co., One Drew Plaza, Boonton, NJ 07005 USA (Tel.: 201-263-7800; 800-526-1015 x7800; Telefax: 201-263-4483; Telex: DREWCHEMS BOON)

Ashley Polymers

5114 Fort Hamilton Parkway, Brooklyn, NY 11219 USA (Tel.: 718-851-8111; Telefax: 718-972-3256; Telex: 427884 ASHLEY)

Aspect Minerals

Aspect Minerals Inc., PO Box 277, Spruce Pine, NC 28777 USA (Tel.: 704-688-2572; 800-803-7979; Telefax: 704-688-2660)

H.M. Royal, Inc., Distrib. for Aspect Minerals, 689 Pennington Ave., PO Box 28, Trenton, NJ 08601 USA (Tel.: 609-396-9176; Telefax: 609-396-3185)

Aster, Inc.

160 Glaser St., Fairborn, OH 45324 USA (Tel.: 513-754-1400; Telefax: 513-754-1415)

Astor Wax Corp.

200 Piedmont Ct., Doraville, GA 30340 USA (Tel.: 404-448-8083; Telefax: 404-840-0954)

Astro Industries Inc., Subsid. of Borden Inc.

PO Box 2559, 114 Industrial Blvd., Morganton, NC 28655 USA (Tel.: 704-584-3800; 800-872-7876; Telefax: 704-584-3885)

Atochem. *See under* Elf Atochem

Atomergic Chemetals Corp.

222 Sherwood Ave., Farmingdale, NY 11735-1718 USA (Tel.: 516-694-9000; Telefax: 516-694-9177; Telex: 6852289)

Auschem SpA

Via Cavriana, 14, 20134 Milano Italy (Tel.: 0039-2-70140259; Telefax: 0039-2-70140201; Telex: 312093 AUSCHEM 1)

Ausimont

Ausimont USA Inc., 44 Whippany Rd., PO Box 1838, Morristown, NJ 07962-1838 USA (Tel.: 201-292-6250; 800-323-AUSI; Telefax: 201-292-0886)

Ausimont UK Ltd., 111 Upper Richmond Rd., Putney, London, SW15 2TJ UK (Tel.: 44 181 780-0399; Telefax: 44 181 780-2871)

Ausimont SpA, Via Principe Eugenio 1/5, 20155 Milano Italy (Tel.: (02) 62701; Telefax: (02) 6270-3788; Telex: 310679 MONTED I)

Montecatini K.K., Mori Bldg. 25, 21st Fl., 4-30 Roppongi 1-chome, Minato-ku, Tokyo, 106 Japan (Tel.: (3) 3224-7239; Telefax: (3) 3505-4039; Telex: GABBRO J 2423851)

Austin Chemical Co. Inc.

1565 Barclay Blvd., Buffalo Grove, IL 60089 USA (Tel.: 708-520-9600; Telefax: 708-520-9160, 280342)

Axel Plastics Research Laboratories Inc.

PO Box 855, 58-20 Broadway, Woodside, NY 11377 USA (Tel.: 718-672-8300; Telefax: 718-565-7447; Telex: 429033 AXELLAB)

Aztec Peroxides, Inc., Div. of Laporte Organics

One Northwind Plaza, 7600 W. Tidwell, Suite 500, Houston, TX 77040 USA (Tel.: 713-895-2015; 800-231-2702; Telefax: 713-895-2040)

BA Chemicals Ltd., Div. of Alcan

Chalfont Park, Gerrards Cross, Buckinghamshire, SL9 0QB UK (Tel.: 44 1753-887373; Telefax: 44 1753-889602; Telex: 847343)

Bakelite GmbH, Div. of Rütgerswerke AG

Postfach 7154, Gennaer Strasse 2-4, W-5860 Iserlohn 7 Germany (Tel.: 49-2374-510; Telefax: 49-2374-51409; Telex: 827255 BGLH)

Baker Sillavan Ltd. *See under* Colin Stewart Minchem

Balchem Corp.

PO Box 175, Slate Hill, NY 10973 USA (Tel.: 914-355-2861; Telefax: 914-355-6314)

Bamberger Polymers

Bamberger Polymers, Inc., 1983 Marcus Ave., Lake Success, NY 11042 USA (Tel.: 516-328-2772; 800-888-8959; Telefax: 516-326-1005, 6852108)

Bamberger Polymers (Canada) Inc., 6969 Trans, Montreal St Laurent, Quebec, H4T 1V8 Canada

Bamberger Polymers (Europe) BV, Steurweg 2, 4941 VR Raamsdonksveer The Netherlands (Tel.: 31-1621-20240)

Barium & Chemicals Inc.

County Road 44, PO Box 218, Steubenville, OH 43952 USA (Tel.: 614-282-9776; Telefax: 614-282-9161)

Barker Industries, Inc.

2841 Old Steele Creek Rd., Charlotte, NC 28208 USA (Tel.: 704-391-1023; Telefax: 704-393-0464)

Bärlocher GmbH, Otto

Riesstrasse 16, 8000 München 50 Germany (Tel.: 089-1488-0; Telefax: 089-1488-312; Telex: 5215701)

BASF

BASF AG, Carl-Bosch Str. 38, 67056 Ludwigshafen Germany (Tel.: 0621-60-99739; Telefax: 0621-60-93344; Telex: 62157120BASF)

BASF Corp., 3000 Continental Dr. North, Mount Olive, NJ 07828-1234 USA (Tel.: 201-426-2600; 800-367-9861)

BASF Corp./Performance Chemicals, 3000 Continental Dr. North, Mt. Olive, NJ 07828-1234 USA (Tel.: 201-426-2600; 800-669-BASF)

BASF plc, PO Box 4, Earl Road, Cheadle Hulme, Cheadle, Cheshire, SK8 6QG UK (Tel.: 44 161 485-6222; Telefax: 44 161-486-0891; Telex: 669211 BASFCH G)

Bayer

Bayer AG, Bayerwerk, 5090 Leverkusen Germany (Tel.: (0214)30-1; Telefax: (0214)30 6 51 36; Telex: 85103-0 byd)

Bayer plc, Stoke Court, Stoke Poges, Slough, Berkshire SL2 4LY UK (Tel.: 44-1753-645 151)

Bayer Antwerpen NV, Div. of Bayer AG, Haven 507, Scheldelaan 420, B-2040 Antwerp, Belgium (Tel.: 32 3540 3011; Telefax: 32 3541 6936; Telex: 71175 BAYANT B)

Bayer Hispania Industrial SA, Pablo Claris 196, E-80037 Barcelona Spain (Tel.: 34-3-218 45 50; Telefax: 34-3-217 41 49; Telex: 54482 BAYIN E)

Bayer Inc./Fibers, Organics & Rubber, Bldg. 14, Mobay Rd., Pittsburgh, PA 15205-9741 USA (Tel.: 412-777-2000; 800-662-2927; Telefax: 412-777-7840; Telex: 1561261)

Rhein Chemie Corp., 1008 Whitehead Road Ext., Trenton, NJ 08638 USA (Tel.: 609-771-9100; Telefax: 609-771-0232)

Polysar Rubber Corp., PO Box 3001, 1265 Vidal St. South, Sarnia, Ontario, N7V 4H4 Canada (Tel.: 519-337-8251; 800-321-0997)

Bergvik Kemi. *See under* Arizona

Berje Inc.

5 Lawrence St., Bloomfield, NJ 07003 USA (Tel.: 201-748-8980; Telefax: 201-680-9618; Telex: 475-4165A)

Bernardy Chimie SA, Div. of Société des Produits Chimiques d'Harbonnieres

Thenioux, F-18100 Vierzon France (Tel.: 48 52 00 80; Telefax: 48 852 04 61; Telex: 760328)

Berol Nobel

Berol Nobel AB, Box 11536, S-10061 Stockholm Sweden (Tel.: 8-743-4000; Telefax: 8-644-3955; Telex: 10513 benobl s)

Berol Nobel Ltd., Div. of Berol Nobel AB, 23 Grosvenor Road, St. Albans, Hertfordshire, AL1 3AW UK (Tel.: 44 1727 841421; Telefax: 44 1727-841529; Telex: 23242 BEROL G)

Berol Nobel Inc., Meritt 8 Corporate Park, 99 Hawley Lane, Stratford, CT 06497 USA (Tel.: 203-378-0500; Telefax: 203-378-5960)

Biddle Sawyer Corp.

2 Penn Plaza, New York, NY 10121 USA (Tel.: 212-736-1580; Telefax: 212-239-1089, 427471BSCE)

Bio-Rad Laboratories

Bio-Rad Laboratories, 85A Marcus Dr., Melville, NY 11747 USA (Tel.: 800-4-BIORAD; Telefax: 516-756-2594; Telex: 71-3720184)

Bio-Rad RSL NV, Div. of Bio-Rad Laboratories Inc., Begoniastraat 5, B-9810 Nazareth Belgium (Tel.: 32 9 385 511; Telefax: 32 9 385 6554)

Blythe, William Ltd.

Holland Bank Works, Bridge St., Church, Accrington, Lancashire, BB5 4PD UK (Tel.: 44 1254 872872; Telefax: 44 1254 872000; Telex: 63142 BLYCO G)

BOC Gases

The Priestley Centre, The Surrey Research Park, Guildford, Surrey, GU2 5XY UK (Tel.: 44-1483-579857; Telefax: 44-1483-505211, 858078)

Bock, Bruno Chemische Fabrik KG

Eichholzer Strasse 23, W-2095 Marschacht Germany (Tel.: 49-4176-90980; Telefax: 49-4176-1396; Telex: 2189219)

Boehringer Ingelheim

Boehringer Ingelheim KG, Binger Strasse 173, Postfach 200, D-6507 Ingelheim Germany (Tel.: 49-6132 773666; Telefax: 49-6132-773755; Telex: 79122 BI W)

Henley Div., B.I. Chemicals, Inc., Div. of Boehringer Ingelheim, 50 Chestnut Ridge Rd., Montvale, NJ 07645 USA (Tel.: 201-307-0422; 800-635-3558; Telefax: 201-307-0424; Telex: 232210 HNLY UR)

Boliden Intertrade Inc.

3379 Peachtree Rd. NE, Suite 300, Atlanta, GA 30326 USA (Tel.: 404-239-6700; 800-241-1912; Telefax: 404-239-6701; Telex: 981036)

Borden Chemical Div.

180 E. Broad St., Columbus, OH 43215 USA (Tel.: 614-225-4000; 800-225-8044; Telefax: 614-225-3476)

Borregaard LignoTech. *See* LignoTech

BP

BP Chemicals Ltd., 6th Floor, Britannic House, 1 Finsbury Circus, London EC2M 7BA UK (Tel.: 44 171 496 4867; Telefax: 44 171-496-4898; Telex: 266883 BPCLBHG)

BP Oil UK Ltd., Div. of The British Petroleum Co plc, BP House, Breakspear Way, Hemel Hempstead, Hertsfordshire, HP2 4UL UK (Tel.: 44 1442 232323; Telefax: 44 1442-225225; Telex: 827711)

BP Chemicals Inc., 4440 Warrensville Center Rd., Warrensville Hts., OH 44128 USA (Tel.: 216-586-6455; 800-272-4367; Telefax: 216-586-3838; Telex: 6873120, 6873119)

BP Performance Polymers Inc., Phenolic Business, PO Box 411, Keasbey, NJ 08832 USA (Tel.: 908-417-3099; Telefax: 908-738-5123; Telex:)

Brand-Nu Laboratories, Inc.

PO Box 895, 30 Maynard St., Meriden, CT 06450 USA (Tel.: 203-235-7989; 800-243-3768; Telefax: 203-235-7163)

British Traders & Shippers Ltd., Div. of Linton Park plc

6-7 Merrielands Crescent, Dagenham, Essex, RM9 6SL UK (Tel.: 44 181 595 4211; Telefax: 44 181-593 0933; Telex: 897438 SHIPEX G)

Bromine & Chemicals Ltd. *See under* Dead Sea Bromine

Brotherton Specialty Products Ltd.

Calder Vale Rd, Wakefield, West Yorkshire, WF1 5PH UK (Tel.: 44 1924 371919; Telefax: 44 1924 290408; Telex: 556320 BROKEM G)

Brown Chemical Co., Inc./Industrial and Fine Chems.

302 W. Oakland Ave., Oakland, NJ 07436-0785 USA (Tel.: 201-337-0900)

Buckman Labs Int'l., Inc.

1256 North McLean Blvd., Memphis, TN 38108 USA (Tel.: 901-278-0330; 800-BUCKMAN; Telefax: 901-276-5343; Telex: 68-28020)

Buffalo Color Corp.

959 Rte. 46 East, Suite 403, Parsippany, NJ 07054 USA (Tel.: 201-316-5600; 800-631-0171; Telefax: 201-316-5828; Telex: 7109885924)

Burgess Pigment Co.

PO Box 349, Sandersville, GA 31082 USA (Tel.: 912-552-2544; 800-841-8999; Telefax: 912-552-1772; Telex: 804523)

Byk-Chemie

Byk Chemie GmbH, Div. of Altana Industrie Aktien und Anlagen AG, Abelstrasse 14, 4230 Wesel Germany (Tel.: 011-49-281-6700; Telefax: 011-49-281-65735; Telex: 812772)

Byk-Chemie USA, 524 S. Cherry St., PO Box 5670, Wallingford, CT 06492-7656 USA (Tel.: 203-265-2086; Telefax: 203-284-9158; Telex: 643378)

Byk-Chemie Japan KK, Sunshine 6 Bldg., 2-29-12, Shiba, Minato-Ku, Tokyo, 105 Japan (Tel.: 81-3-3256-5409; Telefax: 81-3-3256-5420)

Cabot

Cabot Corp./Cab-O-Sil Div., 700 E. U.S. Highway 36, Tuscola, IL 61953-9643 USA (Tel.: 217-253-3370; 800-222-6745; Telefax: 217-253-4334; Telex: 910-663-2542)

Cabot Corp./Industrial Rubber Blacks, 8000 Miller Court East, Norcross, GA 30071 USA (Tel.: 800-472-4889)

Cabot Corp./N. Am. Rubber Black Div., 6600 Peachtree Dunwoody Rd., 300 Embassy Row, Suite 500, Atlanta, GA 30328 USA (Tel.: 404-394-3296; Telex: 549570)

Cabot Corp./Special Blacks Div., 157 Concord Rd., PO Box 7001, Billerica, MA 01821-7001 USA (Tel.: 508-663-3455; 800-526-7591; Telefax: 508-670-7035; Telex: 6817525)

Cabot Canada Ltd./Special Blacks Div., 350 Wilton St., Sarnia, Ontario, N6T 7N4 Canada (Tel.: 519-336-2261; Telefax: 519-336-8501)

Cabot Brasil Industria E Comercio LTDA/Special Blacks, Av. Joao Castaldi 88, 04517-900, Sao Paulo SP Brazil (Tel.: 55 11 536 0388; Telefax: 55 11 542 6037)

Cabot Carbon Ltd./Cab-O-Sil Div., Div. of Cabot Corp, Barry Site, Sully Moors Rd., Sully, S. Glamorgan, CF6 2XP UK (Tel.: 44 1446-736999; Telefax: 44 1446-737123)

Cabot Carbon Ltd./Special Blacks Div., Div. of Cabot Corp, Lees Lane, Stanlow-Ellesmere Port, South Wirral, Cheshire, L65 4HT UK (Tel.: 44 151 355 3677; Telefax: 44 151-356-0712; Telex: 629261 CABLAK G)

Cabot GmbH/Cab-O-Sil Div., PO Box 1766, Cesar-Stuenzi Strasse 15, D-7888 Rheinfelden Germany (Tel.: 49 7623-9090; Telefax: 49 7623-90932; Telex: 773451)

Cabot Europe Ltd./Special Blacks Div., Le Nobel 4B, 2 rue Marcel Monge, F-92158 Suresnes Cedex France (Tel.: 33 1 46 97 58 00; Telefax: 33 1 47 72 66 47)

Cabot Australasia Ltd./Special Blacks Div., PO Box 19, 300 Millers Road, Altona, Victoria, 3018 Australia (Tel.: 61-3-391-1622; Telefax: 61-3-391-9370; Telex: AA30773 (Carbalt)

Cabot Specialty Chemicals, Inc./Special Blacks Div., 3-1-14 Shiba, Minato-ku, Tokyo, 105 Japan (Tel.: 81-3-3457-7561; Telefax: 81-3-3457-7658)

Cabot Plastics Ltd., Gate St., Dunkinfield, SK16 4RU UK (Tel.: 44 161 330-5051; Telefax: 44 161-308-2641; Telex: 667114 CABLAK G)

Cabot Plastics International, Interleuvenlaan 5, B-3001 Leuven Belgium (Tel.: 32-16390111; Telefax: 32-16401253)

Calgene Chemical Inc.

7247 North Central Park Ave., Skokie, IL 60076-4093 USA (Tel.: 708-675-3950; 800-432-7187; Telefax: 708-675-3013; Telex: 72-4417)

Calgon Corp.

PO Box 717, Pittsburgh, PA 15230-0717 USA (Tel.: 412-787-6700; 800-648-9005; Telefax: 4412-787-6713; Telex: 671183CCC)

Calumet Lubricants Co.

2780 Waterfront Pkwy. E. Drive, Suite 200, Indianapolis, IN 46214 USA (Tel.: 317-328-5660; Telefax: 317-328-5668)

14000 Mackinaw Ave., Chicago, IL 60633 USA (Tel.: 708-862-9100)

Cancarb Ltd.

1702 Brier Park Crescent N.W., PO Box 310, Medicine Hat, Alberta, T1A 7G1 Canada (Tel.: 403-527-1121; Telefax: 403-529-6093; Telex: 03-824866)

Captree Chemical Corp.

32 B Nancy St., West Babylon, NY 11704 USA (516-491-7400; 800-899-2725; Telefax: 516-491-7130)

Carbonic Industries Corp.

3700 Crestwood Pkwy., Suite 200, Deluth, GA 30136-5583 USA (Tel.: 800-241-5882; Telefax: 404-717-2222)

The Carborundum Co.

PO Box 337, Niagara Falls, NY 14303-1597 USA (Tel.: 716-278-2798; Telefax: 716-278-2225)

Cardinal

Cardinal Stabilizers, Inc., 2010 S. Belt Line Blvd., Columbia, SC 29202 USA (Tel.: 803-799-7190; Telefax: 803-799-8742)

Cardinal Carolina, 2010 S. Belt Line Blvd., Columbia, SC 29201 USA (Tel.: 803-799-7190; Telefax: 803-799-8742)

Cardolite Corp.

500 Doremus Ave., Newark, NJ 07105-4805 USA (Tel.: 201-344-5015; 800-322-7365; Telefax: 201-344-1197; Telex: 325446)

Cargill

Cargill, Inc., Box 5630, Minneapolis, MN 55440 USA (Tel.: 612-475-6478; Telex: CGL MPS 290625)

Cargill, Knowle Hill Park, Fairmile Lane, Cobham, Surrey, KT11 2PD UK (Tel.: 44 1932 861175; Telefax: 44 1932 861286)

R.E. Carroll, Inc.

1570 N. Olden Ave., PO Box 139, Trenton, NJ 08638 USA (Tel.: 609-695-6211; 800-257-9365; Telefax: 609-695-0102)

CasChem Inc.

40 Ave. A, Bayonne, NJ 07002 USA (Tel.: 201-858-7900; 800-CASCHEM; Telefax: 201-437-2728; Telex: 710-729-4466)

Catalyst Resources, Inc.

2190 North Loop West, Suite 400, Houston, TX 77018 USA (Tel.: 713-682-5300; Telefax: 713-957-6839; Telex: 910-881-7119)

Catalysts & Chemicals Industries Co., Ltd., Joint venture of Asahi Glass Co. Ltd./JGC Corp.

Nippon Bldg., 6-2, Ohte-machi 2-chome, Chiyoda-ku, Tokyo, 100 Japan (Tel.: (03) 3270-6086; Telefax: (03) 3246-0617; Telex: 2223480 PETCATJ)

CDC International Inc.

22 Portsmouth Rd., Amesbury, MA 01913 USA (Tel.: 508-388-2221)

Ceca. *See under* Elf Atochem

Celite

Celite Corp., PO Box 519, Lompoc, CA 93438-0519 USA (Tel.: 805-735-7791; 800-348-8062; Telefax: 805-735-5699; Telex: 62776493 ESL UD)

Celite Corp. (Canada), 295 The West Mall, Etobicoke, Ontario, M9C 4Z7 Canada (Tel.: 416-626-8175; Telefax: 416-626-8235; Telex:)

Celite (UK) Ltd., Livingston Rd., Hessle, North Humberside, HU13 OEG UK (Tel.: 44 1482 64 52 65; Telefax: 44 1482 64 11 76; Telex: 592160)

Celite France, 9 rue du Colonel-de-Rochebrune B.P. 240, 92504 Rueil-Malmaison Cedex France (Tel.: (14)749-0560; Telefax: (1) 47 08 30 25; Telex: Celite 631969 F)

Celite Italiana srl, Viale Pasubio No. 6, I-20154 Milano Italy (Tel.: 654531; Telefax: 39 2 29005439; Telex: 311136 MANVII)

Celite Pacific, 2nd Floor, Shui On Centre, 8 Harbor Road Hong Kong (Tel.: 582 5609; Telefax: 802 4275)

CE Minerals, Div. Combustion Engrg.

901 E. Eighth Ave., King of Prussia, PA 19406 USA (Tel.: 215-265-6880; Telefax: 215-337-7163)

Central Soya

Central Soya Co., Inc./Chemurgy Div., PO Box 2507, Fort Wayne, IN 46801-2507 USA (Tel.: 219-425-5432; 800-348-0960; Telefax: 219-425-5301; Telex: 49609682)

Central Soya Aarhus A/S, Skansevej 2, PO Box 380, DK-8100 Aarhus C Denmark (Tel.: 45 89 31 21 11; Telefax: 45 89 31 21 12, 64348)

Cerac, Inc.

PO Box 1178, 407 N. 13th St., Milwaukee, WI 53201 USA (Tel.: 414-289-9800; Telefax: 414-289-9805; Telex: RCA 286122)

Charkit Chemical Corp.

PO Box 1725, 330 Post Rd., Darien, CT 06820 USA (Tel.: 203-655-3400; Telefax: 203-655-8643; Telex: 6819184)

Chemax, Inc.

PO Box 6067, Highway 25 South, Greenville, SC 29606 USA (Tel.: 803-277-7000; 800-334-6234; Telefax: 803-277-7807; Telex: 570412 IPM15SC)

Chemcentral Corp

7050 W. 71 St., PO Box 730, Bedford Park, IL 60499-0730 USA (Tel.: 708-594-7000; 800-331-6174; Telefax: 708-594-6328)

ChemDesign Corp.

99 Development Rd., Fitchburg, MA 01420 USA (Tel.: 508-345-9999; Telefax: 508-342-9769)

Chemetall

Chemetall GmbH, Reuterweg 14, Postfach 10 15 01, Frankfurt Germany (Tel.: 49-69-0; Telefax: 49-69 159-3018; Telex: 4170090)

Oakite Products, Inc./Specialties Div., A Member of the Chemetall GmbH Group, 50 Valley Rd., Berkeley Hts., NJ 07922-2798 USA (Tel.: 908-464-6900; 800-526-4473; Telefax: 908-464-9250)

Chemfax Inc.

3 Rivers Rd., PO Box 1390, Gulfport, MS, 39502 USA (Tel.: 601-863-6511; Telefax: 601-863-6547)

Chemie Linz

Chemie Linz GmbH, Div. of Chemie Holding AG, Postfach 296, St Peter-Strasse 25, A-4021 Linz Austria (Tel.: 43-732-59160; Telefax: 43-732-3800; Telex: 221324)

Chemie Linz UK Ltd., Div. of Chemie Linz GmbH, 12 The Green, Richmond, Surrey, TW9 1PX UK (Tel.: 44 181 948 6966; Telefax: 44 181-332 2516; Telex: 924941)

Chemie Linz North America, Inc., 65 Challenger Rd, Ridgefield Park, NJ 07660 USA (Tel.: 201-641-6410; Telefax: 201-641-2323; Telex: 853211 Chemie Linz)

Chemisphere

Chemisphere Ltd., 38 King St., Chester, Cheshire, CH1 2AH UK (Tel.: 44 1244 320878; Telefax: 44 1244 320858; Telex: 61398 CHEMSPR G)

Chemisphere Corp., 2101 Clifton Ave., St. Louis, MO 63139 USA (Tel.: 314-644-1300; Telefax: 314-644-1425)

Chemlease Inc.

PO Box 616627, Orlando, FL 32861-6627 USA (Tel.: 407-330-0055; Telefax: 407-330-6554)

Chemoxy International plc, Div. of Suter plc

All Saints Refinery, Cargo Fleet Rd, Middlesbrough, Cleveland, TS3 6AF UK (Tel.: 44 1642 248555; Telefax: 44 1642 244340; Telex: 587185)

Chemplex Chemicals, Inc.

201 Route 17, Suite 300, Rutherford, NJ 07070 USA (Tel.: 201-935-8903; Telefax: 201-935-9051)

Chemron Corp.

PO Box 2299, Paso Robles, CA 93447 USA (Tel.: 805-239-1550; Telefax: 805-239-8551, 501532)

Chemson Ltd.

Hayhole Works, Willington Quay, Wallsend, Tyne & Wear, NE28 0PB UK (Tel.: 44 191 258 5892; Telefax: 44 191 258 1549; Telex: 537726)

Chem-Trend

Chem-Trend Inc., 1445 W. McPherson Park Dr., PO Box 860, Howell, MI 48844-0860 USA (Tel.: 517-546-4520; 800-727-7730; Telefax: 517-546-6875)

Chem-Trend A/S, Smedeland 14, PO Box 13 84, DK-2600 Glostrup Denmark (Tel.: 45-42 45 6711; Telefax: 45-43 63 03 50; Telex: 33187 CMTREND DK)

Chem-Y GmbH

Kupferstrasse 1, Postfach 10 02 62, D46446 Emmerich Germany (Tel.: 2822/7110; Telefax: 2822/18294; Telex: 81251124 cyem)

Chevron Chemical Co.

PO Box 3766, Houston, TX 77253 USA (Tel.: 713-754-2000; 800-231-3260; Telex: 762799)

Chisso

Chisso Corp., Tokyo Bldg., 7-3, Marunouchi 2-chome, Chiyoda-ku, Tokyo, 100 Japan (Tel.: (03) 3284-8411; Telex: 02225212 CHISSO J)

Chisso America Inc., 1185 Ave of the Americas, New York, NY 10036 USA (Tel.: 212-302-0500; Telefax: 212-302-0643; Telex: WU 147029 CHISSO NYK)

Church & Dwight Co. Inc./Specialty Prods. Div.

Box CN5297, 469 N. Harrison St., Princeton, NJ 08543-5297 USA (Tel.: 609-497-7116; 800-221-0453; Telefax: 609-497-7176; Telex: 752226)

Ciba-Geigy

Ciba-Geigy Corp./Chemical Div., 410 Swing Rd., Greensboro, NC 27409-2080 USA (Tel.: 919-632-6000; 800-334-9481; Telefax: 919-632-7008)

Ciba-Geigy Corp./Additives Div., Seven Skyline Dr., Hawthorne, NY 10532-2188 USA (Tel.: 914-785-4438; 800-431-1900; Telefax: 914-347-5687)

Ciba-Geigy Corp./Dyestuffs & Chemicals Div., PO Box 18300, 410 Swing Rd., Greensboro, NC 27419 USA (Tel.: 919-632-2011; 800-334-9481; Telefax: 919-632-7008; Telex: 131411)

Ciba-Geigy Corp./Polymers Div., Seven Skyline Dr., Hawthorne, NY 10532 USA (Tel.: 914-347-6600; 800-222-1906)

Ciba Polymers, Duxford, Cambridge, Cambridgeshire, CB2 4QA UK (Tel.: 44 1223 832121; Telefax: 44 1223-838404; Telex: 81101)

Ciba-Geigy plc, Hulley Rd., Maccles Field, Cheshire SK10 2NX UK (Tel.: 44 1625 421933; Telefax: 44 1625 619637; Telex: 667336)

Ciba-Geigy AG, CH-4002, Basel Switzerland (Tel.: 41-696 7329; Telefax: 41 696 6322; Telex: 963962)

Ciba-Geigy (Japan) Ltd., 10-66, Miyuki-cho, Takarazuka-shi, Hyogo, 665 Japan (Tel.: (0797) 74-2472; Telefax: (0797) 74-2472; Telex: 5645684 CIGYTZ J)

Cimbar Performance Minerals

25 Old River Rd. S.E., PO Box 250, Cartersville, GA 30120 USA (Tel.: 404-387-0319; 800-852-6868; Telefax: 404-386-6785)

W.A. Cleary Chemical Corp.

Southview Industrial Park, 178 Route #522 Suite A, Dayton, NJ 08810 USA (Tel.: 908-329-8399; 800-524-1662; Telefax: 908-274-0894)

Climax

Climax Performance Materials Corp./Corporate Headquarters, 101 Merritt 7 Corporate Park, PO Box 5113, Norwalk, CT 06856-5113 USA (Tel.: 203-845-2951; 800-323-3231; Telefax: 203-845-2953)

Climax Fluids Additives, Div. of Climax Performance, 7666 West 63rd St., Summit, IL 60501 USA (Tel.: 708-458-8450; Telefax: 708-458-0286)

Climax Molybdenum Co., Div. of Climax Performance, 1370 Washington Pike, Bridgeville, PA 15017 USA (Telefax: 412-257-0540)

Climax Performance Materials, Div. of Amax, Inc., 7666 West 63 St., Chicago, IL 60501 USA (Tel.: 708-458-8450; 800-323-3231; Telefax: 708-458-0286)

Climax Molybdenum BV, Div. of Amax Holdings BV, Postbus 1130, NL-3180 AC Rozenburg The Netherlands (Tel.: 31-15933; Telex: 23673 MOLY NL)

CNC International, Limited Partnership

PO Box 3000, 20 Privilege St., Woonsocket, RI 02895 USA (Tel.: 401-769-6100; Telefax: 401-769-4509)

Coastal Chem Inc.

PO Box 1287, Cheyenne, WY, 82003 USA (Tel.: 307-633-2200; 800-443-2754; Telefax: 800-832-7601)

Colin Stewart Minchem Ltd.

Weaver Valley Rd., Wharton, Winsford, Cheshire, CW7 3BU UK (44 1606 553151; Telefax: 44 1606 593008, 667059 SILVAN G)

Baker Sillavan Ltd., Weaver Valley Rd., Wharton, Winsford, Cheshire, CW7 3BU UK (44 1606 553151; Telefax: 44 1606 593008, 667059 SILVAN G)

Colloids Ltd.

Dennis Road, Widnes, Cheshire, WA8 0SL UK (Tel.: 44 151 424 7424; Telefax: 44 151-423-3553; Telex: 629164 COLOID)

Colores Hispania SA

Josep Pla 149, E-08019 Barcelona Spain (Tel.: 34 3 307 13 50; Telefax: 34 3 303 2505; Telex: 54116 COHI)

Cometals Inc., Subsid. of Commercial Metals Co.

1 Penn Plaza, New York, NY 10119 USA (Tel.: 212-760-1200; Telefax: 212-564-7915, 424087)

Complex Quimica SA

PO Box 544, Monterrey, NL Mexico (64000 Mexico (Tel.: 011(528)336-2577; Telefax: 011 (528)336-3650)

Conap, Inc.

1405 Buffalo St., Olean, NY 14760 USA (Tel.: 716-372-9650; Telefax: 716-372-1594)

Concord Chem Co. Inc.

17th & Federal Sts., Camden, NJ 08105 USA (Tel.: 609-966-1526; 800-282-2436)

Condor Corp.

Executive Center, 560 Sylvan Ave., Englewood Cliffs, NJ 07632-3193 USA (Tel.: 201-567-3337; 800-321-3005; Telefax: 201-567-6489)

Cook Composites & Polymers

PO Box 419389, Kansas City, MO 64141-6389 USA (Tel.: 816-391-6000; 800-821-3590; Telefax: 816-391-6215; Telex:)

Cookson Specialty Additives

1000 Wayside Rd., Cleveland, OH 44110 USA (Tel.: 216-531-6010; 800-321-4236; Telefax: 216-486-6638)

Coutaulds

Courtaulds plc, Patents Dept, PO Box 111, 72 Lockhurst Lane, Coventry, West Midlands, CV6 5RS UK (Tel.: 44 1203 688771; Telefax: 44 1203-583837)

Courtaulds Water Soluble Polymers, Div. of Courtaulds plc, P O Box 5, Spondon, Derbyshire, DE2 7BP UK (Tel.: 44 1332 661422; Telefax: 44 1332-661078; Telex: 37391 CHMPLS G)

Coyne Chemical

3015 State Rd., Croydon, PA 19021 USA (Tel.: 215-785-3000; Telefax: 215-785-1585)

CPB-Companhia Petroquimica do Barreiro, LDA

Apartado 31 - Lavradio, 2830 Barreiro, Portugal (Tel.: 351 01 2079272; Telefax: 351 01 2070644)

CPS Chemical Co Inc.

PO Box 162, Old Bridge, NJ 08857 USA (Tel.: 908-607-2700; Telefax: 908-607-2562; Telex: 844532-CPSOLDB)

Cray Valley

Cedex 101, F92970 Paris la Défense France (Tel.: 33 1 41 35 68 10; Telefax: 33 1 41 35 61 43; Telex: 615 289)

Cri-Tech, Inc.

85 Winter St., Hanover, MA 02339 USA (Tel.: 617-826-5600; Telefax: 617-826-5770)

CR Minerals Corp.

14142 Denver W. Pkwy., Suite 250, Golden, CO 80401 USA (Tel.: 303-278-1706; 800-527-7315; Telefax: 303-279-3772)

Croda

Croda International plc, Cowick Hall, Snaith, Goole, North Humberside, DN14 9AA UK (Tel.: 44 1405 860551; Telefax: 44 1405 862253)

Croda Chemicals Ltd., Div. of Croda International plc, Cowick Hall, Snaith, Goole, North Humberside, DN14 9AA UK (Tel.: 44 1405 860551; Telefax: 44 1405 860205; Telex: 57601)

Croda Surfactants Ltd., Cowick Hall, Snaith, Goole, North Humberside, DN14 9AA UK (Tel.: 44 1405 860551; Telefax: 44 1405 860205; Telex: 57601)

Croda Universal Ltd., Div. of Croda International plc, Cowick Hall, Snaith, Goole, North Humberside, DN14 9AA UK (Tel.: 44 1405 860551; Telefax: 44 1405 860205; Telex: 57601)

Croda Resins Ltd., Div. of Croda International plc, Crabtree Manorway South, Belvedere, Kent, DA17 6BA UK (Tel.: 44 181 311 9109; Telefax: 44 181-310 9878; Telex: 896384)

Croda Inc., 7 Century Dr., Parsippany, NJ 07054-4698 USA (Tel.: 201-644-4900; Telefax: 201-644-9222)

Croda Universal, Inc., 602 Eldoro Dr., Arlington, TX 76006-3628 USA (Tel.: 817-460-3792; Telefax: 817-460-3794)

Croda Canada Ltd., 78 Tisdale Ave., Toronto, Ontario, M4A 1Y7 Canada (Tel.: 416-751-3571; Telefax: 416-751-9611)

Croda Japan KK, Aceman Bldg., 1-10, Tokuicho 1-chome, Chuo-ku, Osaka, 540 Japan (Tel.: (06) 942-1791; Telefax: (06) 942-1790; Telex: 5233117)

Crompton & Knowles

Crompton & Knowles Corp./Dyes & Chems. Div., PO Box 33188, Charlotte, NC 28233 USA (Tel.: 704-372-5890; 800-438-4122; Telefax: 704-372-1522)

Crompton & Knowles Tertre SA, Div. of Crompton & Knowles Corp, 141 ave de la Reine, B-1210 Brussels Belgium (Tel.: 32-216 2045; Telefax: 32-242 84 83; Telex: 57 288 AT SATRB)

Crosfield Chemicals, Inc.

101 Ingalls Ave., Joliet, IL 60435 USA (Tel.: 815-727-3651; 800-727-3651; Telefax: 815-727-5312)

Crowley

Crowley Chemical Co., 261 Madison Ave., New York, NY 10016 USA (Tel.: 212-682-1200; 800-424-9300; Telefax: 212-953-3487; Telex: 12-7662)

Crowley Tar Products Co., Inc., 261 Madison Ave., New York, NY 10016 USA (Tel.: 212-682-1200)

Croxton & Garry Ltd.

Curtis Rd. Industrial Estate, Dorking, Surrey, RH4 1XA UK (Tel.: 44 1306 886688; Telefax: 44 1306-887780; Telex: 859567/8 cand g)

Crucible Chemical Co. Inc.

PO Box 6786, Donaldson Center, Greenville, SC 29606 USA (Tel.: 803-277-1284; 800-845-8873; Telefax: 803-299-1192)

Cuproquim Corp.

PO Box 171357, Memphis, TN 38187-1357 USA (Tel.: 901-537-7257; Telefax: 901-685-8372)

Custom Compounding Inc.

50 Miltron Dr., Aston, PA 19014 USA (Tel.: 215-358-1001)

Custom Fibers International

PO Box 940, 2045 Lebec Rd., Lebec, CA 93243 USA (Tel.: 800-321-5324; Telefax: 805-248-1123)

Custom Grinders Sales, Inc.

PO Box 1446, Chatsworth, GA 30705 USA (Tel.: 706-695-4613; Telefax: 706-695-6706)

CVC Specialty Chemicals, Inc.

600 Deer Rd., Cherry Hill, NJ 08034 USA (Tel.: 609-354-0040; Telefax: 609-354-6226)

Cylatec of North America, Inc.

3711 Whipple Ave. NW, Canton, OH 44718 USA (216-492-2854; Telefax: 216-492-3956)

Cyprus Industrial Minerals

9100 East Mineral Circle, PO Box 3299, Englewood, CO 80155 USA (Telefax: 303-643-5168)

Cyro Industries

100 Valley Rd., PO Box 950, Mt. Arlington, NJ 07856 USA (Tel.: 201-770-3000; 800-631-5384; Telefax: 201-770-6117)

Cytec

Cytec Industries Inc., A Business Unit of American Cyanamid Co., Five Garret Mountain Plaza, West Paterson, NJ 07424 USA (Tel.: 201-357-3100; 800-438-5615; Telefax: 201-357-3054; Telex: 130400)

Cytec Industries Inc./Polymer Additives Dept., PO Box 60, 1937 West Main St., Stamford, CT 06904-0060 USA (Tel.: 203-321-2415; 800-486-5525; Telefax: 203-321-2997)

Cytec de Argentina S.A., Charcas 5051, EP 1425, Buenos Aires, Argentina (Tel.: 541-772-4031; Telefax: 541-953-6619)

Cytec Industries UK Ltd., Bowling Park Dr., Bradford, West Yorkshire, BD4 71T UK (Tel.: 44 1274 733891; Telefax: 44-1274 734770; Telex: 51295)

Cytec Industries BV, Coolsingel 139, 3012 AG Rotterdam The Netherlands (Tel.: 31-10 2248400; Telefax: 31-10 4136788; Telex: 23554)

Cytec Australia Ltd., 5 Gibbon Rd., Baulkham Hills, NSW, 2153 Australia (Tel.: 612-624-9223; Telefax: 612-838-9985; Telex: 70879 CYANAMI AA)

Cytec Japan Ltd., No. 30 Kowa Bldg., 4th Floor, 4-5 Roppongi 2-chome, Minato-ku, Tokyo, 106 Japan (Tel.: 813-3586-9716; Telefax: 813-3586-9710; Telex: 22439 CYANAMID J)

Daihachi Chemical Industry Co., Ltd.

Sanyo Nissei Kawaramachi Bldg., 2-7, Kawaramachi 2-chome, Chuo-ku, Osaka, 541 Japan (Tel.: (06) 201-1455; Telefax: (06) 201-1458)

Dai-ichi Kogyo Seiyaku Co., Ltd.

New Kyoto Center Bldg., 614, Higashishiokoji-cho, Shimokyo-ku, Kyoto, 600 Japan (Tel.: (075) 343-1181; Telefax: (075) 343-1421)

Daniel Products Co.

400 Claremont Ave., Jersey City, NJ 07304 USA (Tel.: 201-432-0800; Telefax: 201-432-0266; Telex: 126-304)

Dead Sea Bromine

Dead Sea Bromine Group, Plastics Additives Marketing Div., Member of ICL Group, Makleff House, PO Box 180, Beer-Sheva, 84101 Israel (Tel.: (972)57-297696; Telefax: (972)57-297846; Telex: 5335, 5343)

Bromine & Chemicals Ltd., Div. of Dead Sea Bromine, 6 Arlington Street, St James's, London, SW1A 1RE UK (Tel.: 44 171 493-9711; Telefax: 44 171-493-9714; Telex: 23845)

Eurobrom BV, PO Box 158, NL-2280 AD Rijswijk The Netherlands (Tel.: (31)70 340 84 08; Telefax: (31)70 399 90 35; Telex: 32137)

AmeriBrom, Inc., Member of the Dead Sea Bromine Group, 52 Vanderbilt Ave., New York, NY 10017 USA (Tel.: 212-286-4000; Telefax: 212-286-4475; Telex: RCA 220531)

Degussa

Degussa AG, D-60287 Frankfurt Am Main Germany (Tel.: 069-218-01; Telefax: 069-218-3218)

Degussa Ltd., Div. of Degussa AG, Earl Rd, Stanley Green, Handforth, Wilmslow, Cheshire, SK9 3RL UK (Tel.: 44 161 486 6211; Telefax: 44 161-485 6445; Telex: 51665053 DGMCHR G)

Degussa Corp., Wholly owned subsid. of Degussa AG, 65 Challenger Rd., Ridgefield Park, NJ 07660 USA (Tel.: 201-641-6100; 800-237-6745; Telefax: 201-807-3182; Telex: 221420 degus ur)

Jan Dekker BV

Postbus 10, NL-1520 AA Wormerveer The Netherlands (Tel.: 31-75-2782 78; Telefax: 31-75-21 38 83; Telex: 19273)

J.W.S. Delavau Co. Inc.

2140 Germantown Ave., Philadelphia, PA 19122 USA (Tel.: 215-235-1100; Telefax: 215-235-2202)

Dexter

Dexter Chemical Corp., 845 Edgewater Rd., Bronx, NY 10474 USA (Tel.: 718-542-7700; Telefax: 718-991-7684; Telex: 127061)

Dexter Corp./Frekote Products, One Dexter Dr., Seabrook, NH 03874 USA (Tel.: 603-474-5541; Telefax: 603-474-5545; Telex: 6817306 HYSEA)

Disco, Inc./Assoc. Polymer Prods. Div.

1010 Greenwood Lake Tpk., Ringwood, NJ 07456 USA (Tel.: 201-728-7731; Telefax: 201-728-3475)

Dixie Chem Co., Inc.

PO Box 130410, 300 Jackson Hill, Houston, TX 77219 USA (Tel.: 713-863-1947; Telefax: 713-863-8316, 91018815089)

D.J. Enterprises, Inc.

PO Box 31366, Cleveland, OH 44131-0366 USA (Tel.: 216-524-3879)

D & O Chemicals, Inc.

291 South Van Brunt St., Englewood, NJ 07631 USA (Tel.: 201-871-8500; 800-722-3686; Telefax: 201-871-8505; Telex: 642804 DOC FORT)

Dong Jin (USA) Inc.

38 W. 32 St., Suite 902, New York, NY 10001 USA (Tel.: 212-564-0830; 800-666-5583; Telefax: 212-643-0402)

Dover Chemical Corp., Subsid. of ICC Industries Inc.

3676 Davis Rd. N.W., PO Box 40, Dover, OH 44622 USA (Tel.: 216-343-7711; 800-321-8805/6; Telefax: 216-364-1579; Telex: 983466)

Dow

Dow Chemical U.S.A., 2020 Willard H. Dow Center, Midland, MI 48674 USA (Tel.: 517-636-1000; 800-441-4DOW; Telex: 227455)

Dow Plastics, A Business Group of The Dow Chemical Co., 2040 W.H. Dow Center, Midland, MI 48674 USA (Tel.: 800-441-4DOW; Telefax: 517-638-9942)

Dow Chemical Canada Inc., 1086 Modeland Rd., PO Box 1012, Sarnia, Ontario, N7T 7K7 Canada (Tel.: 519-339-3131)

Dow Europe SA, Bachtobelstrasse 3, CH-8810 Horgen Switzerland (Tel.: 41-1-728-2111; Telefax: 41-1-728-2935; Telex: 826940)

Dow Corning

Dow Corning Corp., Box 0994, Midland, MI 48686-0994 USA (Tel.: 517-496-4000; Telefax: 517-496-4586)

Dow Corning Ltd., Div. of Dow Corning Corp, Kings Court, 185 Kings Rd, Reading, Berkshire, RG1 4EX UK (Tel.: 44 1734 507251; Telefax: 44 1734 575051)

Drew Industrial Div. *See under* Ashland

Dry Branch Kaolin

Rt. 1, Box 468D, Dry Branch, GA 31020 USA (Tel.: 912-750-3500; 800-DBK-CLAY; Telefax: 912-746-0217)

DSM

DSM United Kingdom Ltd., Div. of DSM NV, Kingfisher House, Kingfisher Walk, Redditch, Worcestershire, B97 4EZ UK (Tel.: 44 1527 68254; Telefax: 44 1527-68-949; Telex: 339861)

DSM Chemicals North America, Inc., 4751 Best Road, Ste 140, Atlanta, GA 30337 USA (Tel.: 404-766-3179; 800-825-4376; Telefax: 404-766-3540)

DSM Melamine America, Inc., 4751 Best Rd., Suite 140, Atlanta, GA 30337 USA (Tel.: 404-766-3179; 800-825-4376; Telefax: 404-766-3540)

DSM Japan K.K., 7F., Shin Kokusai Bldg., 4-1, Marunouchi 3-chome, Chiyoda-ku, Tokyo, 100 Japan (Tel.: (03) 3217-8941; Telefax: (03) 3201-5074)

Duphar BV. *See* Solvay Duphar BV

DuPont

DuPont Chemicals, 1007 Market St., Wilmington, DE 19898 USA (Tel.: 302-774-2099; 800-441-9442; Telefax: 302-773-4181; Telex: 302-774-7573)

DuPont Nylon, PO Box 80705, Wilmington, DE 19880-0705 USA (Tel.: 800-231-0998)

DuPont/Polymer Products Dept., Kirk Mill Bldg., Wilmington, DE 19898 USA (Tel.: 302-992-3010; 800-441-7111)

DuPont Canada Inc., PO Box 26, Toronto Bank Tower, Toronto, Ontario, M5K 1B6 Canada (Tel.: 416-362-5621)

DuPont S.A. de C.V., Apartado Postal 5819, 06500 Mexico D.F. Mexico (Tel.: 52-5-250-9033; Telefax: 52-5-250-9033)

DuPont (UK) Ltd./Polymer Products Dept., Maylands Ave., Hemel Hempstead, Hertfordshire, HP2 7DP UK (44-1442-61251)

DuPont de Nemours (France) S.A., 137 rue de L'Université, F-75334 Paris France (Tel.: 33-45 50 65 50; Telefax: 33-47 53 09 65; Telex: 206772)

DuPont de Nemours International S.A., Polymer Products Dept., 50-52 route des Acacias, CH-1211 Geneva 24 Switzerland (Tel.: 41-22-37-81-11)

DuPont (Australia) Ltd., Northside Gardens, 168 Walker St., PO Box 930, North Sydney, NSW, 2060 Australia (Tel.: (02) 923-6111)

DuPont Far East Inc., Kowa Bldg. No. 2, 11-39 Akasaka 1-Chome, Minato-Ku, Tokyo, 107 Japan (Tel.: 585-5511)

DuPont Asia Pacific, Ltd., 1122 New World Office Bldg., East Wing, Salisbury Rd., Kowloon Hong Kong (Tel.: 852-734-5345; Telefax: 852-724-4458)

Dwight Products, Inc.

10 Stuyvesant Ave., PO Box 909, Lyndhurst, NJ 07071 USA (Tel.: 201-438-3388; Telefax: 201-438-0594)

Dynaloy, Inc.

7 Great Meadow Lane, Hanover, NJ 07936 USA (Tel.: 201-887-9270; Telefax: 201-887-3678; Telex: 642033)

Eaglebrook, Inc.

1150 Junction, Schererville, IN 46375 USA (Tel.: 219-322-2560; 800-428-3311; Telefax: 219-322-8533)

Eagle-Picher Industries, Inc./Chemicals Dept

PO Box 550, C & Porter Sts., Joplin, MO 64801 USA (Tel.: 417-623-8000; Telefax: 417-782-1923; Telex: 9102508335)

Eagle Zinc Co.

30 Rockefeller Plaza, New York, NY 10112 USA (Tel.: 212-582-0420; Telefax: 212-582-3412)

Eastern Chemical, Div. of Admiral Specialty Products, Inc.

PO Box 2500, Smithtown, NY 11787 USA (Tel.: 516-273-0900; 800-645-5566; Telefax: 516-273-0858; Telex: 4974275 GCC HAUP)

Eastern Color & Chemical Co.

35 Livingston St., PO Box 6161, Providence, RI 02904 USA (Tel.: 401-331-9000; Telefax: 401-331-2155)

Eastman

Eastman Chemical Products, Inc., PO Box 431, Kingsport, TN 37662 USA (Tel.: 615-229-2318; 800-EASTMAN; Telefax: 615-229-1196; Telex: 6715569)

Kodak Ltd., Div. of Eastman Kodak Co, PO Box 66, Kodak House, Station Rd, Hemel Hempstead, Hertsfordshire, HP1 1TU UK (Tel.: 44 1442 61122; Telefax: 44 1442 40609; Telex: 825101)

Eastman Chemical International AG, Hertizentrum 6, 3263 Zug Switzerland (Tel.: 41 42 23 25 25; Telefax: 41 42 21 12 52; Telex: 868 824)

Eastman Chemical International Ltd., 11 Spring St., Chatswood, NSW, 2067 Australia (Tel.: 61 2 411 3399; Telefax: 61 2 411 6430)

ECC

ECC International, 5775 Peachtree-Dunwoody Rd. NE, Suite 200G, Atlanta, GA 30342 USA (Tel.: 404-843-1551; 800-843-3222; Telefax: 404-303-4384; Telex: 6827225)

ECC International/Calcium Prods., 5775 Peachtree-Dunwoody Rd, Suite 200G, Atlanta, GA 30342 USA (Tel.: 404-843-1551; 800-843-3222; Telefax: 404-303-4384, 6827225)

ECC International Ltd., Div. of ECC Group plc, John Keay House, St. Austell, Cornwall, PL25 4DJ UK (Tel.: 44 1726 74482; Telefax: 44 1726 623019; Telex: 45526 ECCSAU G)

ECC International SA, Div. of ECC Group plc, 2 rue du Canal, B-4551 Lixhe Belgium (Tel.: 32 41-79 98 11; Telefax: 32 41-79 82 79)

ECC Japan Ltd., Div. of ECC International, 10th Central Bldg., 10-3, Ginza 4-chome, Chuo-ku, Tokyo, 104 Japan (Tel.: (03) 3546-8250; Telefax: (03) 3546-8255; Telex: J28915)

Egon Meyer

Av. Henry Ford 38, Tlalnepantla, Edo De Mexico, 54030 Mexico (Tel.: 525-310-5766; Telefax: 525-310-4649)

Eka Nobel. *See under* Akzo Nobel

Elf Atochem

Elf Atochem S.A., 4, cours Michelet, La Défense 10, F-92091 Paris Cedex 42 France (Tel.: 49-00-8080; Telefax: 49-00-7447; Telex: 611922 ATO F)

Elf Atochem North America Inc., Headquarters, 2000 Market St., Philadelphia, PA 19103-3222 USA (Tel.: 215-419-7000; 800-225-7788; Telefax: 215-419-7591)

Elf Atochem North America Inc./Organic Chemicals, 2000 Market St., Philadelpha, PA 19103-3222 USA (Tel.: 215-419-7000; 800-628-4453; Telefax: 215-419-7875)

Elf Atochem North America Inc./Organic Peroxides Div., 2000 Market St., Philadelphia, PA 19103-3222 USA (Tel.: 800-558-5575; Telefax: 215-419-5455)

Elf Atochem North America Inc./Polymers & Plastic Additive, Two Appletree Sq., Suite 147, Bloomington, MN 55425 USA (Tel.: 612-854-0450; Telefax: 612-854-0669)

Elf Atochem North America Inc./Wire Mill Products Dept., 43 James St., Homer, NY 13077 USA (Tel.: 607-749-2652)

Elf Atochem UK Ltd., Colthrop Lane, Thatcham, Newbury, Berkshire, RG13 4LW UK (Tel.: 44 1635 870000; Telefax: 44 1635-861212; Telex: 847689 ATOKEM G)

Ceca Belgium SA, Bte 1, rue de Stalle 63, B-1180 Brussels, Belgium (Tel.: 32 370 2011; Telefax: 32 332 0191; Telex: 26022 CECATO)

E/M Corporation, Subsid. of Great Lakes Chemical Corporation

PO Box 2400, 2801 Kent Ave., W. Lafayette, IN 47906 USA (Tel.: 317-497-6346; 800-428-7802; Telefax: 317-497-6348)

Emerson & Cuming Inc./Grace Specialty Polymers

77 Dragon Court, Woburn, MA 01888 USA (Tel.: 617-938-8630; 800-832-4929; Telefax: 617-935-0125)

EM Industries, Inc./Fine Chems. Div.

5 Skyline Drive, Hawthorne, NY 10532 USA (Tel.: 914-592-4350; Telefax: 914-592-9469)

Emkay Chemical Co.

319-325 Second St., PO Box 42, Elizabeth, NJ 07206 USA (Tel.: 908-352-7053; Telefax: 908-352-6398)

EMS-American Grilon Inc., Subsid. of EMS-Chemie AG

2060 Corporate Way Road, PO Box 1717, Sumter, SC 29151-1717 USA (Tel.: 803-481-9173; Telefax: 803-481-6121; Telex: 805077)

Engelhard Corp./Performance Minerals Group

101 Wood Ave. South, CN 770, Iselin, NJ 08830-0770 USA (Tel.: 908-205-5000; 800-631-9505; Telefax: 908-205-6711; Telex: 219984 ENGL UR)

EniChem

EniChem Elastomeri Srl, Strada 3, Palazzo B1, Milanofiori , I-20090 Assago, Milan Italy (Tel.: 39 2 5201; Telefax: 39 2 52026077; Telex: 310246)

EniChem Elastomers Ltd., Div. of EniChem Elastomeri Srl, Charleston Rd, Hardley, Hythe, Southampton, Hamsphire, SO4 6YY UK (Tel.: 44 1703 894919; Telefax: 44 1703 894334; Telex: 47519)

EniChem America, Inc., 1211 Ave. of the Americas, New York, NY 10036 USA (Tel.: 212-382-6521; Telefax: 212-382-6584; Telex: 6801159 ENICHEM)

EniChem Synthesis SpA, Via Medici del Vascello 40, 20138 Milano Italy (Tel.: 02 5203 9218; Telefax: 02 5203 9450; Telex: 310246)

Enterprise Chemical Corp.

4 Parkview Dr., Dover, OH 44622 USA (Tel.: 216-343-8861; 800-875-8861; Telefax: 216-343-8853)

Esprit Chemical Co.

800 Hingham St., Suite 207S, Rockland, MA 02370 USA (Tel.: 617-878-5555; 800-2-ESPRIT; Telefax: 617-871-5431, 951346)

Estron Chemical, Inc.

P.O. Box 127, Hwy. 95, Calvert City, KY 42029 USA (Tel.: 502-395-4195; Telefax: 502-395-5070)

Ethyl. *See* Albemarle

Eurobrom. *See under* Dead Sea Bromine

Evans Chemetics. *See under* Hampshire

Expancel

Expancel, Div. of Nobel Industries, PO Box 13000, S-85013 Sundsvall Sweden (Tel.: 46 60-134000; Telefax: 46 60-569518; Telex: 71399 expancel s)

Expancel, Div. of Nobel Industries, 1519 Johnson Ferry Rd., Suite 200, Marietta, GA 30062 USA (Tel.: 404-971-8005; Telefax: 404-578-1359)

Exxon

Exxon Chemical Co., PO Box 3272, Houston, TX 77253-3272 USA (Tel.: 713-870-6000; 800-526-0749; Telefax: 713-870-6661; Telex: 794588)

Exxon Chemical Co./Application Chemicals Div., Tomah Products, 1012 Terra Dr., PO Box 388, Milton, WI 53563 USA (Tel.: 608-868-6811; 800-441-0708; Telefax: 608-868-6810; Telex: 910-280-1401)

Exxon Chemical Geopolymers Ltd., Div. of Exxon Corp, 4600 Parkway, Solent Business Park, Whiteley, Fareham, Hampshire, PO15 7AZ UK (Tel.: 44 1489 884400; Telefax: 44 1489 884403; Telex: 47437)

Exxon Chemical Mediterranea SpA, Div. of Exxon Corp, Via Paleocapa 7, I-20121 Milan Italy (Tel.: 39-2 88031; Telefax: 39-2 8803231; Telex: 311561 ESSOCH I)

Exxon Chemical Europe Inc., 280 Vorstlaan, Bld du Souverain, B-1160 Bruxelles Belgium (Tel.: 32-2-674-41 11; Telefax: 32-2-674 41 29)

Exxon Chemical Belgium, Div. of Exxon Corp, Boulevard du Souverain 280, B-1160 Brussels Belgium (Tel.: 32-2-674 41 11; Telefax: 32-2-674 41 29; Telex: 22364)

Exxon Chemical International Services Ltd., Div. of Exxon Corp, 33rd Floor, Shui on Centre, 8 Harbour Road, Wanchai Hong Kong (Tel.: 852 582-0888)

Fabrichem Inc.

211 Sigwin Dr., Fairfield, CT 0006430 USA (Tel.: 203-259-5512; Telefax: 203-254-7886)

Faesy & Besthoff, Inc.

143 River Rd., Edgewater, NJ 07020-0029 USA (Tel.: 201-945-6200; Telefax: 201-945-6145)

Fairmount Chemical Co., Inc.

117 Blanchard St., Newark, NJ 07105 USA (Tel.: 201-344-5790; 800-872-9999; Telefax: 201-690-5298; Telex: 138905)

Fanning Corp., The

2450 W. Hubbard St., Chicago, IL 60612 USA (Tel.: 312-248-5700; Telefax: 312-248-6810; Telex: 910-221-1335)

Farleyway Chemicals Ltd.

Ham Lane, Kingswinford, West Midlands, DY6 7JU UK (Tel.: 44 1384 400 222; Telefax: 44 1384 400 020; Telex: 339528 FAR G)

FAR Research, Inc.

2210 Wilitelmina Ct., Palm Bay, FL 32905 USA (407-723-6160; Telefax: 407-723-8753)

The Feldspar Corp.

One West Pack Square, Suite 700, Asheville, NC 28801 USA (Tel.: 704-254-7400; Telefax: 704-255-4909)

Ferro

Ferro Corp./World Headquarters, 1000 Lakeside Ave., Cleveland, OH 44114-1183 USA (Tel.: 216-641-8580; Telefax: 216-696-6958, 98-0165)

Ferro Corp./Bedford Chemical Div., 7050 Krick Rd., Bedford, OH 44146 USA (Tel.: 216-641-8580; 800-321-9946; Telefax: 216-439-7686; Telex: 98-165)

Ferro Corp./Color Div., 4150 E. 56th St., PO Box 6550, Cleveland, OH 44101 USA (Tel.: 216-641-8580; Telefax: 216-641-8831; Telex: 98-0165)

Ferro Corp./Grant Chemical Div., PO Box 263, Baton Rouge, LA 70821 USA (Tel.: 504-654-6801; Telefax: 504-654-3268; Telex: 980165)

Ferro Corp./Keil Chemical Div., 3000 Sheffield Ave., Hammond, IN 46320 USA (Tel.: 219-931-2630; 800-628-9079; Telefax: 219-931-0895; Telex: 725484)

Ferro Corp./Plastic Colorants & Dispersions Div., 3 Railroad Ave., Stryker, OH 43557 USA (Tel.: 419-682-3311; Telefax: 419-682-4924)

Ferro Corp./Transelco Div., Box 217, Penn Yan, NY 14527 USA (Tel.: 315-536-3357; Telefax: 315-536-8091; Telex: 97 8373)

Finetex

Finetex Inc., 418 Falmouth Ave., PO Box 216, Elmwood Park, NJ 07407 USA (Tel.: 201-797-4686; Telefax: 201-797-6558; Telex: 710-988-2239)

Represented by: Pennine Chemical Ltd., Radnor Park Trading Estate, Back Lane, Congleton, Cheshire, CW12 4XJ UK (Tel.: 44 1260 279631; Telefax: 44 1260 278263; Telex: 668080 PENKEM G)

Firestone Synthetic Rubber & Latex Co., Div. of Bridgestone/Firestone Inc.

PO Box 26611, Akron, OH 44319-0006 USA (Tel.: 216-379-7727; 800-282-0222; Telefax: 216-379-7875)

First Chemical

First Chemical Corp., PO Box 1427, Pascagoula, MS, 39568-1427 USA (Tel.: 601-762-0870; 800-828-7940; Telefax: 601-762-5213; Telex: 510-990-3361)

Quality Chemicals, Inc., Subsid. of First Chemical Corp., PO Box 216, Tyrone Industrial Park, Tyrone, PA 16686-0216 USA (Tel.: 814-684-4310; Telefax: 814-684-2532)

Floridin Co.

PO Box 510, 1101 N. Madison St., Quincy, FL 32351-0510 USA (Tel.: 904-627-7688; 800-228-1131; Telefax: 904-875-1757; Telex: 4931835 FLOQYUI)

Fluka

Fluka Chemical Corp, 980 South Second St., Ronkonkoma, NY 11779 USA (Tel.: 516-467-0980; 800-FLUKA-US; Telefax: 800-441-8841; Telex: 96-7807)

Fluka Chemicals, The Old Brickyard, New Rd., Gillingham, Dorset SP8 4JL UK (Tel.: 44 1747 823097; Telefax: 44 1747 824596; Telex: 417238)

Fluka Chemie AG, Industriestrasse 25, CH-9470 Buchs Switzerland (Tel.: 41/85 69511; Telefax: 41/85 654459; Telex: 855 282)

FMC

FMC Corp./Chemical Products Group, 1735 Market St., Philadelphia, PA 19103 USA (Tel.: 215-299-6000; 800-346-5101; Telefax: 215-299-5999; Telex: 685-1326)

FMC Corp (UK) Ltd., Process Additives Div., Tenax Rd., Trafford Park, Manchester, Lancaster M17 1WT UK (Tel.: 44 161 872 2323; Telefax: 44 161 873 3177; Telex: 666177)

FMC Corp. N.V., Ave. Louise 480-B9, Brussels 1050 Belgium (Tel.: 322-645 5511; Telefax: 322-640 6350)

FMC International S.A., 4th Floor, Interbank Bldg., 111 Paseo de Roxas, Makati, Metro Manila Phillippines (Tel.: (632) 817 5546; Telefax: (632) 818 1485)

Focus Chemical.

875 Greenland Rd., Orchard Park, Suite B9, Portsmouth, NH 03801 USA (Tel.: 603-430-9802)

Franklin Industrial Minerals

612 Tenth Ave. North, Nashville, TN 37203 USA (Tel.: 615-259-4222; Telefax: 615-726-2693)

821 Tilton Bridge Rd., S.E., Dalton, GA 30721 USA (Tel.: 404-277-3740; Telefax: 404-277-9827)

Frinton Laboratories, Inc.

PO Box 2428, Vineland, NJ 08360 USA (Tel.: 609-692-6902; Telefax: 609-692-6922)

H.B. Fuller Co.

3530 Lexington Ave. North, St. Paul, MN 55126-8076 USA (Tel.: 612-481-1816; 800-468-6358; Telefax: 612-481-1863)

Gaylord Chemical Co.

PO Box 1209, 106 Galeria Blvd., Slidell, LA 70459-1209 USA (Tel.: 504-649-5464; 800-426-6620; Telefax: 504-649-0068; Telex: 901 4748663)

Gayson, Inc.

30 Second St. SW, Barberton, OH 44203 USA (Tel.: 216-848-9200; 800-442-9766)

GCA Chemical Corp.

916 West 13th St., Bradenton, FL 34205 USA (Tel.: 813-748-6090; Telefax: 813-748-0194)

Gelest Inc.

612 William Leigh Dr., Tullytown, PA 19007-6308 USA (Tel.: 215-547-1015; Telefax: 215-547-2484)

GenCorp Polymer Products

165 So. Cleveland Ave., Mogadore, OH 44260 USA (Tel.: 216-628-6542; Telefax: 216-628-6501; Telex: 1561448)

General Chemical Corp.

90 East Halsey Rd., Parsippany, NJ 07054-0373 USA (Tel.: 201-515-0900; 800-631-8050; Telefax: 201-515-2468; Telex: 139-450 GEN-CHEMPAPY)

General Electric

GE Company, 3135 Easton Tpke., Fairfield, CT 06431 USA (Tel.: 800-626-2004)

General Electric Co./Silicone Products Div., 260 Hudson River Rd., Waterford, NY 12188 USA (Tel.: 518-237-3330; 800-255-8886; Telefax: 518-233-3931)

GE Specialty Chemicals, 501 Avery St., PO Box 1868, Parkersburg, WV 26102-1868 USA (Tel.: 304-424-5411; 800-872-0022; Telefax: 304-424-5871)

GE Silicones, Div. of GE Plastics Ltd., Old Hall Rd., Sale, Manchester, M33 2HG UK (Tel.: 44 161 905 5000; Telefax: 44 161-905 5022)

Genesee Polymers Corp.

G-5251 Fenton Rd., PO Box 7047, Flint, MI 48507-0047 USA (Tel.: 810-238-4966; Telefax: 810-767-3016)

Genesis Polymers. *See* Novacor Chemicals

Genstar Stone Products Co.

Executive Plaza IV, 11350 McCormick Rd., Hunt Valley, MD 21031 USA (Tel.: 410-527-4000; Telefax: 410-527-4535)

Georgia Gulf Corp./PVC Div.

PO Box 629, Plaquemine, LA 70765-0629 USA (Tel.: 504-685-1200; 800-PVC-VYCM)

Georgia Marble Co.

1201 Roberts Blvd., Bldg. 100, Kennesaw, GA 30144-3619 USA (Tel.: 404-421-6500, FAX 404-421-6507)

Georgia-Pacific

Georgia-Pacific Corp., 133 Peachtree St. N.E., PO Box 105605, Atlanta, GA 30348 USA (Tel.: 404-220-6185)

Georgia-Pacific Resins Inc., 2883 Miller Rd, Decatur, GA 30035 USA (Tel.: 404-593-6895; Telefax: 404-593-6801; Telex: 804600)

GFS Chemicals, Inc.

PO Box 245, Powell, OH 43065 USA (Tel.: 614-881-5501; 800-858-9682; Telefax: 614-881-5989, 981282 GFS CHEM UD)

Gist-brocades

Gist-brocades Food Ingredients, Inc., 2200 Renaissance Blvd., Suite 150, King of Prussia, PA 19406 USA (Tel.: 215-272-4040; 800-662-4478; Telefax: 215-272-5695; Telex: 216902)

Gist-brocades SpA, Via Milano 42, I-27045 Casteggio Italy (Tel.: 39-383-8931; Telefax: 39-383-805397; Telex: 321197 VINAL I)

Gist-brocades, 1 Wateringseweg, PO Box 1, Delft 2600 MA The Netherlands (Tel.: 31 15 793005; Telefax: 31 15 793408, 38103)

Giulini Corp.

105 East Union Ave., Bound Brook, NJ 08805 USA (Tel.: 908-469-6504; Telefax: 908-469-8418; Telex: 700179)

Givaudan-Roure

Givaudan-Roure Corp., 100 Delawanna Ave., Clifton, NJ 07014 USA (Tel.: 201-365-8000; Telefax: 201-777-9304; Telex: 219259 givc ur)

Givaudan-Roure SA, 5 Chemin de la Perfumiere, CH-1214 Vernier-Geneva Switzerland (Tel.: 22-780 91 11; Telefax: 22-780 91 50)

Goldschmidt

Goldschmidt AG, Th., Goldschmidtstrasse 100, Postfach 101461, D-4300 Essen 1 Germany (Tel.: 0201-173-01; Telefax: 201-173-2160; Telex: 857170)

Goldschmidt Ltd., Subsid. of Goldschmidt AG, Tego House, Victoria Road, Ruislip, Middlesex, HA4 0YL UK (Tel.: 44 181 422 7788; Telefax: 44 181-864 8159; Telex: 923146)

Goldschmidt Chemical Corp., 914 E. Randolph Rd., PO Box 1299, Hopewell, VA 23860 USA (Tel.: 804-541-8658; 800-446-1809; Telefax: 804-541-8689; Telex: 710-958-1350)

Tego Chemie Service GmbH, Goldschmidstr. 100, Postfach 101461, D-4300 Essen 1 Germany (Tel.: 0201-1732571; Telefax: 0201-1732639; Telex: 85717-20tgd)

Tego Chemie Service Dept., PO Box 1299, 914 E. Randolph Rd., Hopewell, VA 23860 USA (Tel.: 804-541-8658; 800-446-1809; Telefax: 804-541-2783)

Goldsmith & Eggleton, Inc.

300 First St., Wadsworth, OH 44281-2084 USA (Tel.: 216-336-6616; 800-321-0954; Telefax: 216-334-4709; Telex: 986316)

BFGoodrich Co./Specialty Polymers & Chem. Div.

9921 Brecksville Rd., Brecksville, OH 44141-3247 USA (Tel.: 216-447-5000; 800-331-1144; Telefax: 216-447-5720; Telex: 4996831)

Goodyear

Goodyear Tire & Rubber Co./The Chemical Div., 1485 E. Archwood Ave., Akron, OH 44316 USA (Tel.: 216-796-8295; Telefax: 216-796-3199; Telex: 640550 Gdyr)

Goodyear Chemicals Europe, 14 Ave. des Tropiques, Z.A. de Courtaboeuf, 91952 Les Ulis Cedex France (Tel.: 33-1-69-29-28-15; Telefax: 33-1-69-29-27-04; Telex: 602895F)

Goodyear Int'l. Corp., Sankaido Bldg., 1-9-13 Akasaka, Minato-Ku, Tokyo, 107 Japan (Tel.: (03) 582-0926; Telefax: (03) 582-1877)

W.R. Grace

W.R. Grace/Organic Chemicals Div., 55 Hayden Ave., Lexington, MA 02173 USA (Tel.: 617-861-6600; 800-232-6100; Telefax: 617-862-3869; Telex: 200076)

W.R. Grace/Davison Chemical Div., PO Box 2117, Baltimore, MD 21203-2117 USA (Tel.: 301-659-9000; Telefax: 410-659-9213, 192 814 013)

Grace NV, Div. of W R Grace, Nijverheidsstraat 7, B-2260 Westerlo Belgium (Tel.: 32-14 57 56 11; Telefax: 32-14 58 55 30; Telex: 31500)

Grain Processing Corp.

1600 Oregon St., Muscatine, IA 52761 USA (Tel.: 319-264-4265; Telefax: 319-264-4289; Telex: 46-8497)

Graphite Products Corp

5756 Warren-Sharon Road, PO Box 29, Brookfield, Trumbull, OH 44403 USA (Tel.: 216-394-1617; 800-321-7521; Telefax: 216-394-2389; Telex: 271512 GPIUR)

Great Lakes

Great Lakes Chemical Corp., PO Box 2200, W. Lafayette, IN 47906-0200 USA (Tel.: 317-497-6100; 800-428-7947; Telefax: 317-497-6234)

Great Lakes Chemical Corp., Rua Itapaiuna 1800-Casa 56, Morumbi, Sao Paulo SP 05707-001 Brazil (Tel.: 55-11-844-6486; Telefax: 55-11-844-6787)

Great Lakes Chemical (Europe) Ltd., P O Box 44, Oil Sites Road, Ellesmere Port, South Wirral, L65 4GD UK (Tel.: 44 151 356 8489; Telefax: 44 151-356 8490)

Great Lakes-QO Chemicals, Inc., Industrieweg 12, Haven 391, B-2030 Antwerp Belgium (Tel.: 32-3-541-2165; Telefax: 32-3-541-6503)

Great Lakes Chemical Corp./Japan, LaVie Sakuragicho Bldg., 5-26-3, Sakuragicho, Nishi-Ku, Yokohama, 220 Japan (Tel.: 81-45-212-9541; Telefax: 81-45-212-9539)

Great Western Chemical Co.

808 SW 15th Ave., Portland, OR 97205 USA (Tel.: 503-228-2600; Telefax: 503-221-5752, 910-464-4733)

R.W. Greeff & Co Inc.

777 West Putnam Ave, Greenwich, CT 06830 USA (Tel.: 203-532-2900; Telefax: 203-532-2980; Telex: 996609)

Grindsted

Grindsted Products A/S, Edwin Rahrs Vej 38, DK-8220 Brabrand Denmark (Tel.: 45 86-25-3366; Telefax: 45 86-25-1077; Telex: 64177 gvdan dk)

Grindsted Products, Ltd., Northern Way, Bury St Edmunds, Suffolk, IP32 6NP UK (Tel.: 44 1284 769631; Telefax: 44 1284 760839, 81203)

Grindsted Products GmbH, Roberts-Bosch Strasse 20-24, D-25451 Quickborn Germany (4106/70960; Telefax: 4106/709666, 2180684 gpd d)

Grindsted Products, Inc., 201 Industrial Pkwy., PO Box 26, Industrial Airport, KS 66031 USA (Tel.: 913-764-8100; 800-255-6837; Telefax: 913-764-5407; Telex: 4-37295)

Grindsted Products, Inc., 10 Carlson Court, Suite 580, Rexdale, Ontario, M9W 6L2 Canada (Tel.: 416-674-7340; Telefax: 416-674-7378)

Grindsted de México, S.A. de C.V., Cerrada de las Granjas 623, Col. Jagüey, Delegación Azcapotzalco, 02300 México, D.F. Mexico (Tel.: (5) 352 9102; Telefax: (5) 561 3285)

Grindsted do Brazil, Indústria e Comércio Ltda., Rodovia Regisé Bittencourt, KM 275,5, 06818-900 Embú S.P. Brazil ((11) 494-3899; Telefax: (11) 494-3823, 1171854 gpbr br)

Nippon Grindsted K.K., Daiichi Nishiwaki Bldg., 1-58-10 Yoyogi, Shibuya-ku, Tokyo, J-151 Japan (Tel.: 3-3375-3481; Telefax: 3-3375-3715)

Grünau GmbH, Chemische Fabrik, A Henkel Group Co.

Postfach 1063, Robert-Hansen-Strasse 1, D-89251 Jllertissen Germany (Tel.: (07303)13-706; Telefax: (07303)13203; Telex: 719114 gruea-d)

Guardian Chemical, A Div. of United Guardian Inc.

230 Marcus Blvd, PO Box 2500, Smithtown, NY 11787 USA (Tel.: 516-273-0900; 800-645-5566; Telefax: 516-273-0858)

Gulbrandsen Co. Inc.

PO Box 508, Milford, NJ 08848 USA (Tel.: 908-995-7759; 800-255-7759, FAX 908-995-9482)

Haarmann & Reimer

Haarmann & Reimer GmbH, Postfach 1253, D-3450 Holzminden Germany (Tel.: 49 5531 7011; Telefax: 49 5531 7016 49; Telex: 965 330)

Haarmann & Reimer Corp., PO Box 175, 70 Diamond Road, Springfield, NJ 07081 USA (Tel.: 201-4912-5707; 800-422-1559; Telefax: 201-912-0499; Telex: 219134 HAR UR)

C.P. Hall Co.

7300 South Central Ave., Chicago, IL 60638-0428 USA (Tel.: 708-594-6000; 800-321-8242; Telefax: 708-458-0428)

Howard Hall, Div. of R.W. Greeff & Co. Inc.

777 West Putnam Ave., Greenwich, CT 06830 USA (Tel.: 203-532-2900; Telefax: 203-532-2980; Telex: 681 9012)

Halocarbon Products Corp.

PO Box 661, 887 Kinderkamack Rd., River Edge, NJ 07661 USA (Tel.: 201-262-8899; Telefax: 201-262-0019)

Halstab. *See under* Hammond Lead Products

Hammond Lead

Hammond Lead Products Inc., PO Box 6408, 5231 Hohman Ave., Hammond, IN 46325-6408 USA (Tel.: 219-931-9360; Telefax: 219-931-2140)

Halstab, Div. Hammond Lead Products, 3100 Michigan St., Hammond, IN 46323 USA (Tel.: 219-844-3980; Telefax: 219-844-7287; Telex: 72-5481)

Hampford Research Inc.

292 Longbrook Ave., PO Box 1073, Stratford, CT 06497 USA (Tel.: 203-375-1137; Telefax: 203-386-9754)

Hampshire

Hampshire Chemical Corp., 55 Hayden Ave., Lexington, MA 02173 USA (Tel.: 617-861-9700; Telefax: 617-861-0135; Telex: 200076 GRLX UR)

Evans Chemetics, Div. of Hampshire Chemical Corp., 55 Hayden Ave., Lexington, MA 02173 USA (Tel.: 617-861-9700; 800-232-6100; Telefax: 617-863-8070; Telex: 200076 GRLX UR)

Hampton Colours Ltd.

Toadsmoor Mills, Brimscombe, Stroud, Glos., GL5 2UH UK (Tel.: 44 1453 731555; Telefax: 44 1453 731234)

Harcros

Harcros Chemicals UK Ltd./Specialty Chemicals Div., Lankro House, PO Box 1, Eccles, Manchester, M30 0BH UK (Tel.: 44 161 789-7300; Telefax: 44 161-788-7886; Telex: 667725)

Harcros Chemicals (Deutschland) GmbH, Wilhelm-Oswald-strasse, Siegburg Germany (Tel.: 49-2241-54980; Telefax: 49-2241-549811; Telex: 8869605)

Harcros Durham Chemicals, Div. of Harcros Chemicals UK Ltd., Birtley, Chester-le-Street, Co. Durham, DH3 1QX UK (Tel.: 44 1914 102361; Telefax: 44 1914 106005; Telex: 53618 DURHAM G)

Harcros Chemicals Inc./Organics Div., 5200 Speaker Rd., PO Box 2930, Kansas City, KS 66106-1095 USA (Tel.: 913-321-3131; Telefax: 913-621-7718; Telex: 477266)

Hardman Inc., A Harcros Chemical Group Co.

600 Cortlandt St., Belleville, NJ 07109 USA (Tel.: 201-751-3000; Telefax: 201-751-8407; Telex: TWX: 710-995-4940)

Hardwicke Chemical Inc.

2114 Larry Jeffers Rd., Elgin, SC 29045 USA (Tel.: 803-438-3471; Telefax: 803-438-4497, 810-671-1814)

Hart Products Corp.

173 Sussex St., Jersey City, NJ 07302 USA (Tel.: 201-433-6632)

Harwick Chemical Corp.

60 S. Seiberling St., PO Box 9360, Akron, OH 44305-0360 USA (Tel.: 216-798-9300; Telefax: 216-798-0214; Telex: TWX: 810-431-2126)

Hastings Plastics Co.

1704 Colorado Ave., Santa Monica, CA 90404 USA (Tel.: 213-829-3449; Telefax: 213-328-6820)

Hatco Corp.

1020 King George Post Rd., Fords, NJ 08863-0601 USA (Tel.: 908-738-1000; Telefax: 908-738-9385; Telex: 84-4545)

Hefti Ltd. Chemical Products

PO Box 1623, CH-8048 Zurich Switzerland (Tel.: 01-432-1340; Telefax: 01-432-2940; Telex: 822225 hexa ch)

Heico Chemicals, Inc., A Cambrex Co.

Route 611, PO Box 160, Delaware Water Gap, PA 18327-0160 USA (Tel.: 717-420-3900; 800-34-HEICO; Telefax: 717-421-9012)

Henkel

Henkel KGaA, Henkelstrasse 67, D-40191 Düsseldorf Germany (Tel.: 8011-49-211-797-3300; Telefax: 49-211-798-9638; Telex: 858170)

Henkel KGaA/Cospha, Postfach 101100, D-40191, Düsseldorf Germany (Tel.: 0211-797-1; Telefax: 0211-798-7696; Telex: 085817-0)

Henkel KGaA/Dehydag, Postfach 1100, D-4000 Düsseldorf 1 Germany (Tel.: 0211 797-4221; Telefax: 0211 798-8558; Telex: 085817-122)

Henkel Chemicals Ltd., Div. of Henkel KG, Henkel House, 292-308 Southbury Road, Enfield, Middlesex, EN1 1TS UK (Tel.: 44 181 804 3343; Telefax: 44 181 443 2777; Telex: 922708 HENKEL G)

Henkel France SA, Div. of Henkel KG, BP 309, 150 rue Gallieni, F-92102 Boulogne Billancourt France (Tel.: 33 46 84 90 00; Telefax: 33 46 84 90 90; Telex: 633177 HENKEL F)

Henkel Belgium NV, Div. of Henkel KG, 66 ave du Port, B-1210 Brussels Belgium (Tel.: 32-423 17 11; Telefax: 32-428 34 67; Telex: 21294 HENKEL B)

Henkel Corp./Process & Polymer Chemicals Div., 300 Brookside Ave., Ambler, PA 19002 USA (Tel.: 215-628-1456; 800-654-7588; Telefax: 215-628-1457)

Henkel Corp./Coatings & Inks Div., 300 Brookside Ave., Ambler, PA 19002-3498 USA (Tel.: 215-628-1000; 800-445-2207)

Henkel Corp./Cospha, 300 Brookside Ave., Ambler, PA 19002 USA (Tel.: 215-628-1476; 800-531-0815; Telefax: 215-628-1450; Telex: 125854)

Henkel Corp./Functional Products, 300 Brookside Ave., Ambler, PA 19002 USA (Tel.: 215-628-1583; 800-654-7588; Telefax: 215-628-1155)

Henkel Corp./Organic Products Div., 300 Brookside Ave., Ambler, PA 19022 USA (Tel.: 215-628-1000; 800-922-0605; Telefax: 215-628-1200; Telex: 6851092)

Henkel Corp./Textile Chemicals, 11709 Fruehauf Dr., Charlotte, NC 28273-6507 USA (Tel.: 800-634-2436; Telefax: 704-587-3804)

Henkel Canada Ltd., 2290 Argentia Rd., Mississauga, Ontario, L5N 6H9 Canada (Tel.: 416-542-7554; 800-668-6023; Telefax: 416-542-7588)

Henkel Corp./Emery Group, 11501 Northlake Dr., Cincinnati, OH 45249 USA (Tel.: 513-530-7300; 800-543-7370; Telefax: 513-530-7581; Telex: 4333016)

Hercules

Hercules Inc., Hercules Plaza-6205SW, Wilmington, DE 19894 USA (Tel.: 302-594-5000; 800-247-4372; Telefax: 302-594-5400; Telex: 835-479)

Hercules Ltd., Div. of Hercules Inc., 31 London Road, Reigate, Surrey, RH2 9YA UK (Tel.: 44 1737 242434; Telefax: 44 1737-224288; Telex: 25803)

Hercules BV, 8 Veraartlaan, NL-2288 GM Rijswijk The Netherlands (Tel.: 31-70-150-000; Telefax: 31-70-3989893; Telex: 31172)

Heterene Chemical Co., Inc.

PO Box 247, 795 Vreeland Ave., Paterson, NJ 07543 USA (Tel.: 201-278-2000; Telefax: 201-278-7512; Telex: 883358)

Hexcel Corp./Trevarno Div.

5794 W. Las Positas Blvd., Pleasanton, CA 94588 USA (Tel.: 510-847-9500; 800-444-3923; Telefax: 510-734-9688)

Hickson Danchem Corp.

1975 Richmond Blvd., PO Box 400, Danville, VA 24540 USA (Tel.: 804-797-8105; Telefax: 804-799-2814; Telex: 940103 WU PUBTLXBSN)

Hickson Manro Ltd.

Bridge St., Stalybridge, Cheshire, SK15 1PH UK (Tel.: 44 161 338-5511; Telefax: 44 161-303-2991; Telex: 668442)

High Polymer Labs

803 Vishai Bhawan 95, Nehru Place, New Delhi, 110019 India (Tel.: 91-11-643-1522; Telefax: 91-11-647-4350, 31-71085 HIPOIN)

Hilton Davis Chemical Co., A Freedom Chemical Co.

2235 Langdon Farm Rd., Cincinnati, OH 45237 USA (Tel.: 513-841-4000; 800-477-1022; Telefax: 800-477-4565)

Himont U.S.A., Inc.

Three Little Falls Centre, 2801 Centerville Rd., PO Box 15439, Wilmington, DE 19850-5439 USA (Tel.: 302-996-6000; 800-545-7719; Telefax: 302-996-5587, 198994 HIMONT-UT)

Hitox Corp. of America

PO Box 2544, Corpus Christi, TX 78403 USA (Tel.: 512-882-5175; Telefax: 512-882-6948)

Hodag Corp. *See* Calgene Chem. Inc.

Hoechst

Hoechst AG, D-65926 Frankfurt am Main Germany (Tel.: 49 69 305-5753, 069-316700)

Hoechst Celanese/Int'l. Headqtrs., 26 Main St., Chatham, NJ 07928 USA (Tel.: 201-635-2600; 800-235-2637; Telefax: 201-635-4330; Telex: 136346)

Hoechst Celanese/Bulk Pharmaceuticals & Intermediates Div., 1601 West LBJ Freeway, PO Box 819005, Dallas, TX 75381-9005 USA (Tel.: 214-277-4783; Telefax: 214-277-3858)

Hoechst Celanese/Colorants & Surfactants Div., 5200 77 Center Dr., Charlotte, NC 28217 USA (Tel.: 704-599-4000; 800-255-6189; Telefax: 704-559-6323)

Hoechst Celanese/Specialty Chem. /Polymer Additives Group, 5200 77 Center Dr., PO Box 1026, Charlotte, NC 28217 USA (Tel.: 704-559-6038; Telefax: 704-559-6780)

Hoechst Celanese/Waxes, Lubricants & Polymers, Route 202-206, PO Box 2500, Somerville, NJ 08876-1258 USA (Tel.: 908-704-7040; Telefax: 908-704-7059)

Hoechst Canada Inc., Div. of Hoechst AG, 800 Blvd Rene Levesque O, Montreal, Quebec, PQH38121 Canada (Tel.: 514-871-5511)

Hoffmann-La Roche

Hoffmann-La Roche Inc., 340 Kingsland St., Nutley, NJ 07110 USA (Tel.: 201-909-8332; 800-526-0189; Telefax: 201-909-8414)

Hoffmann-La Roche SA, Grenzacherstrasse 124, CH-4002 Basle Switzerland (Tel.: 41 61-688 3451; Telefax: 41 61 688 1680; Telex: 962292 HLR CH)

Hoffmann Mineral, Div. of Franz Hoffmann & Söhne KG

Postfach 1460, D-86619 Neuburg (Donau) Germany (Tel.: 8431/53-0; Telefax: 8431/53-330; Telex: 55223 hond-d)

Holliday Pigments Int'l.

Morley Street, Kingston-upon-Hull, North Humberside, HU8 8DN UK (Tel.: 44 11482 329875; Telefax: 44 14182 223114; Telex: 597065)

Holtrachem, Inc.

159 Boden Ln., Natick, MA 01760 USA (Tel.: 508-655-2510; 800-343-6470; Telefax: 508-653-2682; Telex: 948456)

Honeywill & Stein Ltd., Div. of BP Chemicals Ltd.

Times House, Throwley Way, Sutton, Surrey, SM1 4AF UK (Tel.: 44 181 770 7090; Telefax: 44 181-770 7295; Telex: 946560 BPCLGH G)

Honig Chemical and Processing Corp.

414 Wilson Ave., Newark, NJ 07105 USA (Tel.: 201-344-0881)

Houghton Chemical Corp.

PO Box 307, Allston, MA 02134 USA (Tel.: 617-254-1010; 800-777-2466; Telefax: 617-254-2713)

J.M. Huber

J.M. Huber Corp./Chemicals Div., PO Box 310, 907 Revolution St., Havre de Grace, MD 21078 USA (Tel.: 410-939-3500; Telefax: 410-939-7313)

J.M. Huber Corp./Calcium Carbonate Div., 2029 Woodlands Pkwy., Suite 107, St. Louis, MO 63146 USA (Tel.: 314-991-5755)

J.M. Huber Corp./Engineered Carbons Div., 1100 Penn Ave., PO Box 2831, Borger, TX 79008-2831 USA (Tel.: 806-274-6331)

J.M. Huber Corp./Engineered Minerals, 1807 Park 270 Drive, Suite 210, St. Louis, MO 63146 USA (Tel.: 217-224-1100; 800-637-8176; Telefax: 217-224-7957)

J.M. Huber Corp./Engineered Minerals, One Huber Rd., Macon, GA 31298 USA (Tel.: 912-745-4751; 800-TRY-HUBER; Telefax: 912-745-1116; Telex: 544438)

J.M. Huber Corp./Engineered Minerals, 4940 Peachtree Industrial Blvd., Suite 340, Norcross, GA 30071 USA (Tel.: 404-441-1301; Telefax: 404-368-9908)

Solem Europe, Subsid. of J.M. Huber, PO Box 3142, Planetwenweg 39, 2130 KC Hoofddorp The Netherlands (Tel.: 31(0) 2503-43052; Telefax: 31(0) 2503-43452)

Hughes Industrial Corp.

1650 Airport Rd., Suite 101, Kennesaw, GA 30144 USA (Tel.: 404-425-6039; Telefax: 404-425-6038)

Hüls

Hüls AG, Postfach 1320, D-4370 Marl 1 Germany (Tel.: 02365-49-1; Telefax: 02365-49-2000; Telex: 829211-0)

Hüls (UK) Ltd., Edinburgh House, 43-51 Windsor Rd., Slough, Berkshire, SL1 2HL UK (Tel.: 44 1753-71851; Telefax: 44 1753-820480; Telex: 848243 huels g)

Hüls France SA, Div. of Hüls AG, 49-51 Quai de Dion Bouton, F-92815 Puteaux Cedex France (Tel.: 1 49 06 55 00; Telefax: 1 47 73 97 65; Telex: 611868 huels f)

Hüls Japan Ltd., 4-28, Mita 1-cho, Minato-Ku, Tokyo, 108 Japan (Tel.: (03) 4551981; Telefax: (03) 4533233; Telex: 2422288 huels jp j)

Hüls America Inc., PO Box 365, 80 Centennial Ave., Piscataway, NJ 08855-0456 USA (Tel.: 908-980-6800; 800-631-5275; Telefax: 908-980-6970; Telex: 4754585 Huls Ul)

Hüls America Inc./Chemicals Div., Petrarch Systems, 80 Centennial Ave., PO Box 456, Piscataway, NJ 08855-0456 USA (Tel.: 908-980-6984; Telefax: 908-980-6970)

Hüls Canada, Inc., 235 Orenda Rd., Brampton, Ontario, L6T 1E6 Canada (Tel.: 416-451-3810; Telefax: 416-451-4469; Telex: 0697557)

Hüls de Mexico, S.A. de C.V., San Francisco 657 A, Desp. 8 B, Col. Del Valle, C.P. 03100 Mexico Mexico (Tel.: 011 (525) 523-4299; Telefax: 011(525)543-7257)

Humphrey Chemical Co Inc., A Cambrex Co.

45 Devine St., North Haven, CT 06473-0325 USA (Tel.: 203-230-4945; 800-652-3456; Telefax: 203-287-9197; Telex: 994487)

Hunt Chemicals Inc.

530 Permalume Pl. NW, Atlanta, GA 30318 USA (Tel.: 404-352-1418; Telefax: 404-352-0395)

Huntsman Chemical Corp.

2000 Eagle Gate Tower, Salt Lake City, UT, 84111 USA (Tel.: 801-532-5200; 800-421-2411; Telefax: 801-536-1581)

ICC Industries Inc.

720 Fifth Ave., New York, NY 10019 USA (Tel.: 212-903-1732; Telefax: 212-903-1726; Telex: CCI 7607944)

ICD Group Inc.

1100 Valley Brook Ave., Lyndhurst, NJ 07071 USA (Tel.: 201-507-3300; 800-777-0505; Telefax: 201-507-1506, 6505113226)

ICI

ICI plc, Imperial Chemical House, 9 Millbank, London, SW11 3JS UK (Tel.: 44 171 834 4444; Telefax: 44 171 834 2040, 21324)

ICI plc/Chemicals & Polymers Group, PO Box 90, Wilton, Middlesbrough, Cleveland, TS6 8JE UK (Tel.: 44 1642 454144; Telefax: 44 1642 432444; Telex: 587 461)

ICI Surfactants Ltd. (UK), PO Box 90, Wilton, Middlesbrough, Cleveland, TS90 8JE UK (Tel.: 44 1642 454144; Telefax: 44 1642 437374; Telex: 587461)

ICI Fluoropolymers UK, Hillhouse International, PO Box 4, Thornton Cleveleys, Blackpool, Lancashire, FY5 4QD UK (Tel.:)

ICI Americas, Inc., Subsid. of ICI plc, PO Box 15391, Wilmington, DE 19850 USA (Tel.: 302-887-3000; 800-441-7780; Telefax: 302-887-4320; Telex: 62032112)

ICI Fluoropolymers, 475 Creamery Way, Exton, PA 19341 USA (Tel.: 610-363-4741; 800-ICI-PTFE; Telefax: 610-363-4748)

ICI Advanced Materials, 475 Creamery Way, Exton, PA 19341 USA (Tel.: 800-424-7833)

ICI Polymer Additives, PO Box 751, Wilmington, DE 19897 USA (Tel.: 302-886-3564; 800-456-3669 x 3564; Telefax: 302-886-5267)

ICI Polyurethanes Group, 286 Mantua Grove Rd., West Deptford, NJ 08066-1732 USA (Tel.: 609-423-8300; 800-257-5547; Telefax: 609-423-8580, 4945649)

ICI Resins US, 730 Main St., Wilmington, MA 01887-0677 USA

ICI Specialty Chemicals, Concord Pike & New Murphy Rd., Wilmington, DE 19897 USA (Tel.: 302-886-3000; 800-822-8215; Telefax: 302-886-2972)

ICI Surfactants Americas, Concord Plaza, 3411 Silverside Rd., PO Box 15391, Wilmington, DE 19850 USA (Tel.: 302-886-3000; 800-822-8215; Telefax: 302-887-3525; Telex: 4945649)

ICI Atkemix Inc., Div. of ICI, PO Box 1085, 70 Market St., Brantford, Ontario, N3T 5T2 Canada (Tel.: 519-756-6181; Telefax: 519-758-8140)

ICI Australia Operations Pty. Ltd./ICI Surfactants Australia, ICI House, 1 Nicholson St., Melbourne, 300 Australia (Tel.: 3-665-7111; Telefax: 61-03-665-7009; Telex: 30192)

ICI Surfactants (Belgium), Everslaan 45, B-3078 Everberg Belgium (Tel.: 02-758-9361; Telefax: 02-758-9686; Telex: 26151)

ICI Deutsche GmbH, Postfach 500728, Emil-von-Behring-Strasse 2, W-6000 Frankfurt am Main Germany (Tel.: 49-69-5801-00; Telefax: 49-69-5801234; Telex: 416974 ICI D)

ICI Surfactants (Australia), Newscom St., Ascot Vale, Victoria, 3032 Australia (Tel.: (03) 2836411; Telefax: (03) 2725353)

ICI Surfactants Asia Pacific (ICI (China) Ltd.), PO Box 107, 1, Pacific Pl., 14th Floor, HK-88 Queensway Hong Kong (Tel.: 8434888; Telefax: 8685282; Telex: 73248)

Idex International

1523 North Post Oak Rd., Houston, TX 77055 USA (Tel.: 713-686-9323; Telefax: 713-688-8273)

IGI

85 Old Eagle School Rd., Wayne, PA 19087 USA (Tel.: 215-687-9030; 800-852-6537)

IMC/Americhem

5129 Unruh Ave., Philadelphia, PA 19135-2990 USA (Tel.: 215-335-0990; 800-220-6800; Telefax: 215-624-3420, 244417)

IMS Co.—Injection Molders Supply Co.

10373 Stafford Rd., Chagrin Falls, OH 44023-5296 USA (Tel.: 216-543-1615; 800-537-5375; Telefax: 216-543-1069)

Indspec

Indspec Chemical Corp., 411 Seventh Ave., Suite 300, Pittsburgh, PA 15219 USA (Tel.: 412-765-1200; Telefax: 412-765-0439; Telex: 199187 Indspec)

Industrial Fibers, Inc.

2889 N. Nagel Court, Lake Bluff, IL 60044 USA (Tel.: 708-295-0046; Telefax: 708-295-0520)

Industrias Quimicas del Valles SA

Avenida Rafael de Casanova 81, Mollet del Vallés, E-08100 Barcelona Spain (Tel.: 34-3-570 56 96; Telefax: 34-3-593 80 11; Telex: 52170)

Inolex

Inolex Chemical Co., Jackson & Swanson Sts., Philadelphia, PA 19148-3497 USA (Tel.: 215-271-0800; 800-521-9891; Telefax: 215-289-9065; Telex: 834617)

Represented by: Stanley Black, Ltd., The Colonnade, High St., Cheshunt, Waltham Cross, Hertfordshire EN8 0DJ UK (Tel.: 44 1992 30751; Telefax: 44 1992 22838)

Interbusiness USA Inc.

61 E. 8th St., Suite 113, New York, NY 10003 USA (Tel.: 212-228-2525; Telefax: 212-228-3399)

Interchem Corp.

120 Rt. 17 North, PO Box 1579, Paramus, NJ 07653 USA (Tel.: 201-261-7333; Telefax: 201-261-7339; Telex: 6853353)

International Chemical Inc.

One World Trade Center, Suite 8665, New York, NY 10048 USA (Tel.: 212-912-0655; Telefax: 212-912-0669, 420001 ITT UI)

International Sourcing Inc.

121 Pleasant Ave., Upper Saddle River, NJ 07458 USA (Tel.: 201-934-8900; 800-772-7672; Telefax: 201-934-8291; Telex: 697-2957 INSOURC)

International Specialty Products, Inc. *See* ISP

ISP

ISP, International Specialty Products, World Headquarters, 1361 Alps Rd., Wayne, NJ 07470-3688 USA (Tel.: 201-628-4000; 800-622-4423; Telefax: 201-628-4117; Telex: 219264)

ISP (Canada) Inc., 1075 The Queensway East, Unit 16, Mississauga, Ontario, L4Y 4C1 Canada (Tel.: 416-277-0381; Telefax: 416-272-0552; Telex: 06961186)

ISP Europe, 40 Alan Turing Rd., Surrey Research Park, Guildford, Surrey, GU2 5YF UK (Tel.: 44 1483 301757; Telefax: 44 1483 302175; Telex: 859142)

ISP Global Technologies Deutschland GmbH, Rudolf-Diesel-Strasse 25, Postfach 1380, 5020 Frechen Germany (Tel.: 02234 105-0; Telefax: 02234 105-211; Telex: 889931)

ISP (Österreich) GmbH, Reschgasse 24-26/1, 1120 Vienna Austria (Tel.: 0222-813 59 81 0; Telefax: 0222-813 79 55; Telex: 133990)

ISP (Australasia) Pty. Ltd., 73-75 Derby St., Silverwater, N.S.W., 2141 Australia (Tel.: Sydney (02) 648-5177; Telefax: (02) 647-1608; Telex: 73711)

ISP Asia Pacific Pte. Ltd., 200 Cantonment Rd., Hex 06-07 Southpoint, 0208 Singapore (Tel.: 221-9233; Telefax: 226-0853; Telex: 25071)

ISP (Japan) Ltd., Shinkawa Iwade Bldg. 8F, 26-9, Shinkawa 1-Chome, Chuo-Ku, Tokyo, 104 Japan (Tel.: (03) 3555-1571; Telefax: (03) 3555-1660; Telex: 23568)

ISP Van Dyk, Inc., Member of the ISP Group, Main & William Sts., Belleville, NJ 07109 USA (Tel.: 201-450-7724; Telefax: 201-751-2047; Telex: 710-995-4928)

Itochu Specialty Chemicals Inc.

350 Fifth Ave., Suite 5822, New York, NY 10118 USA (Tel.: 212-629-2660, 12297 C ITOH NYK)

James River Corp.

Fourth & Adams St., Camas, WA 98607 USA (Tel.: 206-834-8134; Telefax: 206-834-8278; Telex: 152845)

Janssen Chimica

Janssen Chimica, Div. of Janssen Pharmaceutica, Janssen Pharmaceuticalaan 3, B-2440 Geel Belgium (Tel.: 14-60 42 00; Telefax: 14-60 42 20; Telex: 34103)

Janssen Chimica, Div. of Janssen Pharmaceutica, 755 Jersey Ave., New Brunswick, NJ 08901 USA (Tel.: 908-214-1300; 800-772-8786; Telefax: 908-220-6553; Telex: 182395 BIOSPECT)

Janssens NV

Bte 129, Europark-Oost 15, B-9100 Sint-Niklass Belgium (Tel.: 32-3-776 47 47; Telefax: 32-3-776 80 02; Telex: 31046 JANSEN B)

Jarchem Industries Inc.

414 Wilson Ave., Newark, NJ 07105 USA (Tel.: 201-344-0600; Telefax: 201-344-5743; Telex: 362-660)

Jet-Lube, Inc.

4849 Homestead Rd. 77028, PO Box 21258, Houston, TX 77226-1258 USA (Tel.: 713-674-7617; 800-JET-LUBE; Telefax: 713-678-4604; Telex: 775393)

JLM Marketing, Inc.

8675 Hidden River Pkwy., Tampa, FL 33637 USA (Tel.: 813-632-3300; Telefax: 813-632-3301, 666886 JLM UW)

Johnson Matthey

Johnson Matthey plc, Orchard Rd, Royston, Hertfordshire, SG8 5HE UK (Tel.: 44 1763 253000; Telefax: 44 1763 253649)

Johnson Matthey SA, Div. of Johnson Matthey plc, BP 50240, 13 rue de la Perdrix, F-95956 Roissy CDG Cedex France (Tel.: 33 148 17 2199; Telefax: 33 1 48 63 27 02; Telex: 230195 JMATT AF)

Johnson Matthey/Alfa Aesar, 30 Bond St., Ward Hill, MA 01835 USA (Tel.: 800-343-0660; Telefax: 508-521-6366)

Jonas Chemical Corp./Specialty Chemical Div.

1682 59th St., Brooklyn, NY 11204 USA (Tel.: 718-236-1666; Telefax: 718-236-2248; Telex: 423616)

Jonk BV, Div. of Witco

Postbus 5, Wezelstraat 12, NL-1540 AA Koog a/d Zaan The Netherlands (Tel.: 31-75-283 854; Telefax: 31-75-210 811; Telex: 19270)

Kalama Chemical, Inc.

1110 Bank of California Center, Seattle, WA 98164 USA (Tel.: 206-682-7890; 800-742-6147; Telefax: 206-682-1907, 910-444-2294)

Kaneka

Kaneka Corp., Kanegafuchi Chemical Industry Co., Ltd., 2-4, Nakanoshima 3-chome, Kita-ku, Osaka, 530 Japan (Tel.: (06) 226-5320; Telefax: (06) 226-5359; Telex: KANECHEM J-63582)

Kaneka Texas Corp., MBS Marketing Div., 17 South Briar Hollow Lane, Suite 307, Houston, TX 77027 USA (Tel.: 713-840-1751; 800-526-3223; Telefax: 713-552-0133; Telex: 203369 KTC UR)

Kaneka Belgium N.V., MBS Marketing Div., Boulvard du Triomphe 173-b. 2, 1160 Brussels Belgium (Tel.: 02-672-0190; Telefax: 02-672-2822; Telex: 31911 KANE B)

Kaopolite, Inc.

2444 Morris Ave., Union, NJ 07083 USA (Tel.: 908-789-0609; Telefax: 908-851-2974)

Karlshamns

Karlshamns AB, S 37482 Karlshamn Sweden (Tel.: 46 454 82000; Telefax: 46 454 18453, 4510)

Karlshamns Lipids for Care, 525 W. First Ave., PO Box 569, Columbus, OH 43216 USA (Tel.: 614-299-3131; 800-848-1340; Telefax: 614-299-2584; Telex: 245494 capctyprdcol)

Kemira Kemi AB, Div. of Kemira Oy

Box 902, Industrigatan 83, S-251 09 Helsingborg Sweden (Tel.: 46-42-17 10 00; Telefax: 46-42-14 06 35; Telex: 72185 KEMWATS S)

Kenrich Petrochemicals, Inc.

140 E. 22nd St., PO Box 32, Bayonne, NJ 07002-0032 USA (Tel.: 201-823-9000; 800-LICA KPI; Telefax: 201-823-0691; Telex: 125023)

Kerr-McGee Chemical Corp.

Kerr-McGee Ctr., PO Box 25861, Oklahoma City, OK 73125 USA (Tel.: 405-270-1313; 800-654-3911; Telefax: 405-270-3123; Telex: 747-128)

Key Polymer Corp.

One Jacob's Way, Lawrence Ind. Park, Lawrence, MA 01842 USA (Tel.: 508-683-9411; Telefax: 508-686-7729; Telex: 940 103)

Kincaid Enterprises, Inc.

PO Box 549, Plant Rd., Nitro, WV 25143 USA (Tel.: 304-755-3377; Telefax: 304-755-4547, 3791803-KEINC)

King Industries, Inc.

Science Rd., Norwalk, CT 06852 USA (Tel.: 203-866-5551; 800-431-7900; Telefax: 203-866-1268; Telex: 710-468-0247)

KMZ Chemicals Ltd.

48 Station Rd, Stoke D'Abernon, Cobham, Surrey, KT11 3BN UK (Tel.: 44 1932 866426; Telefax: 44 1932 867099; Telex: 929624)

Koch Chemical Co., Div. of Koch Refining Co.,Specialties Group

PO Box 2608, Corpus Christi, TX 78403 USA (Tel.: 512-242-8362; Telefax: 512-242-8353)

Kodak. *See under* Eastman

Koster Keunen

Koster Keunen, Inc., 90 Bourne Blvd., PO Box 447, Sayville, NY 11782 USA (Tel.: 516-589-0456; Telefax: 516-589-0120)

Koster Keunen Holland BV, Postbus 53, 5530 AB Bladel The Netherlands (Tel.: 4977-2929; Telex: 51422 KOKEU NL)

Kraeber GmbH & Co

Hochallee 80, W-2000 Hamburg 13 Germany (Tel.: 49-40-4450613; Telefax: 49-40-455163; Telex: 17403443)

Kraft Chemical Co.

61975 N. Hawthorne, Melrose Park, IL 60160 USA (Tel.: 708-345-5200; Telefax: 708-345-4005, 654268 KRAFT)

Kronos

Kronos, Inc., PO Box 60087, 3000 N. Sam Houston Pkwy. East, Houston, TX 77205 USA (Tel.: 713-987-6300; 800-866-5600; Telefax: 713-987-6358)

Kronos Ltd., Div. of Kronos Inc., Barons Court, Manchester Rd., Wilmslow, Cheshire SK9 1BQ UK (Tel.: 44-1625 529511; Telefax: 44-1625 533123)

Kyowa

Kyowa Chemical Industry Co., Ltd., 305, Yashima-Nishimachi, Takamatsu-shi, Kagawa, 761-01 Japan (Tel.: 0877-47-2500; Telefax: 0877-47-4208; Telex: 5822220)

Kyowa America Corp., 385 Clinton, Costa Mesa, CA 92626 USA (Tel.: 714-641-0411; Telefax: 714-540-5849)

Lanaetex Products, Inc.

151-157 Third Ave., PO Box 52 Station A, Elizabeth, NJ 07206 USA (Tel.: 908-351-9700; Telefax: 908-351-8753; Telex: 3792268 TLAP1)

Landers-Segal Color Co. Inc. (LANSCO)

90 Dayton Ave., Passaic, NJ 07055 USA (Tel.: 201-779-5001; Telefax: 201-779-8948; Telex: 6971185)

LaRoche Chemicals Inc.

PO Box 1031, Airline Hwy., Baton Rouge, LA 70821 USA (Tel.: 504-356-8406; 800-524-2586; Telefax: 504-356-8405; Telex: 784581)

Laurel Industries

30000 Chagrin Blvd., Cleveland, OH 44124-5794 USA (Tel.: 216-831-5747; 800-221-1304; Telefax: 216-831-8479)

Leepoxy Plastics, Inc.

3324 Ferguson Rd., Fort Wayne, IN 46809-3199 USA (Tel.: 219-747-7411; Telefax: 219-747-7413)

Lenape Chemical

210 E. High St., Bound Brook, NJ 08805 USA (Tel.: 908-469-7310)

Leverton-Clarke Ltd.

Unit 16, Sherrington Way, Lister Rd Industrial Estate, Basingstoke, Hampshire, RG22 4DQ UK (Tel.: 44 1256 810393; Telefax: 44 1256 479324; Telex: 858558 LEVCOS G)

LignoTech USA, Inc.

100 Highway 51 South, Rothschild, WI 54474-1198 USA (Tel.: 715-359-6544; Telefax: 715-355-3648)

81 Holly Hill Lane, Greenwich, CT 06830 USA (203-625-0701; Telefax: 203-625-0864; Telex: 643994)

Lilly Chemicals, Lilly Industries

210 E. Alondra Blvd., PO Box 192, Gardena, CA 90248 USA (Tel.: 310-252-3087; 800-472-6243; Telefax: 310-324-3168)

Lindau Chems, Inc.

731 Rosewood Dr., Columbia, SC 29201 USA (Tel.: 803-799-6863; Telefax: 803-256-3639)

Lion Corp.

3-7, Honjo 1-chome, Sumida-ku, Tokyo, 130 Japan (Tel.: (03) 3621-6211; Telefax: (03) 3621-6048; Telex: 262-2114 LIOCOR J)

Lipo

Lipo Chemicals, Inc., 207 19th Ave., Paterson, NJ 07504 USA (Tel.: 201-345-8600; Telefax: 201-345-8365; Telex: 130117)

Represented by: Blagden Chemicals Ltd., Div. of Blagden Industries, AMP House, Dingwall Rd., Croydon, Greater London, CR9 3QU UK (Tel.: 44 181 681 2341; Telefax: 44 181 688 5851; Telex: 24285 BCHEM G)

Lipo do Brasil Ltda., Rua Roque Petrella, 376, 04581 Sao Paulo, SP Brasil (Tel.: (11) 533-2354; Telefax: (11) 533-8997)

LNP Engineering Plastics Inc.

475 Creamery Way, Exton, PA 19341 USA (Tel.: 610-363-4500; 800-854-8774; Telefax: 610-363-4749, 4973041)

Lohmann Chemicals, Dr. Paul Lohmann Chemische Frabrik GmbH KG

PO Box 1220, D-3254 Emmerthal 1 Germany (FAX 49 51 55 6 31 18; Telex: 92858)

Lonza

Lonza Ltd., Münchensteinerstrasse 38, Basle Switzerland (Tel.: 061-316 81 11; Telefax: 061-316 83 01; Telex: 965960 lon ch)

Lonza SpA, Via Vittor Pisani, 31, I-20124 Milan Italy (Tel.: (02) 66 99 91; Telefax: (02) 66 98 76 30; Telex: 312431 ftalmi i)

Lonza (UK) Ltd., Imperial House, Lypiatt Road, Cheltenham, Gloucestershire, GL50 2QJ UK (Tel.: 44 1242 513211; Telefax: 44 1242 222294; Telex: 43152)

Lonza France SARL, Div. of Alusuisse, 55, rue Aristide Briand, F-92309 Levallois-Perret Cedex France (Tel.: 1/40 89 99 25; Telefax: 1/40 89 99 21; Telex: 613647)

Lonza G+T Ltd./Graphites & Technologies, Subsid. of A-L Alusuisse-Lonza Holding AG, CH-5643 Sins Switzerland (Tel.: 042-66 01 11; Telefax: 042 66 23 16; Telex: 862-652)

Lonza Inc., 17-17 Route 208, Fair Lawn, NJ 07410 USA (Tel.: 201-794-2400; 800-777-1875 (tech.); Telefax: 201-703-2028; Telex: 4754539 LONZAF)

Lonza Japan Ltd., Kyowa Shinkawa Bldg., 8th Fl., 20-8, Shinkawa 2-chome, Chuo-ku, Tokyo, 104 Japan (Tel.: 03-5566-0612; Telefax: 03-5566-0619)

Lord Corp./Elastomer Prods. Div., Industrial Adhesives

2000 West Grandview Blvd., PO Box 10038, Erie, PA 16514-0038 USA (Tel.: 814-868-3611; Telefax: 814-864-3452; Telex: 291935)

Lubrizol

Lubrizol Corp., 29400 Lakeland Blvd., Wickliffe, OH 44092 USA (Tel.: 216-943-4200; Telefax: 216-943-5337; Telex: 4332033)

Lubrizol France SA, 25 quai de France, F-76100 Rouen France (Tel.: 33-35 72 04 09; Telex: 180641 LUZOFRA F)

Lubrizol Japan Ltd., No. 23 Mori Bldg., 5th Floor, 23-7, Toranomon 1-chome, Minato-ku, Tokyo, 105 Japan (Tel.: (03) 3504-1345; Telefax: (03) 3504-1340; Telex: 26814)

Luzenac

Luzenac America, Inc., 9000 E. Nichols Ave., Englewood, CO 80112 USA (Tel.: 303-643-0400; 800-325-0299; Telefax: 303-643-0444)

Luzenac, Inc., 1075 North Service Rd. W., Suite 14, Oakville, Ontario, L6M 2G2 Canada (Tel.: 416-825-3930; Telefax: 416-825-3932)

Magie Bros. Oil, Div. Pennzoil Products Co.

9101 Fullerton Ave., Franklin Park, IL 60131 USA (Tel.: 708-455-4500; 800-MAGIE 47; Telefax: 708-455-0383)

Magnablend, Inc.

I-35E Sterrett Rd., Exit 406, Waxahachie, TX 75165 USA (Tel.: 214-223-2068; Telefax: 214-576-8721)

Magnesia GmbH

Postfach 2168, Kurt-Höbold-Strasse 6, W-2120 Lüneburg Germany (Tel.: 49-4131-52011-14; Telefax: 49-4131-53050; Telex: 2182159)

Mallinckrodt

Mallinckrodt, Inc./Drug and Fine Chemicals Div., Mallinkdrodt & 2nd Street, PO Box 5439, St Louis, MO 63147 USA (Tel.: 314-895-2000; 800-325-8888; Telefax: 314-539-1251)

Mallinckrodt Specialty Chemicals, 16305 Swingley Ridge Dr., Chesterfield, MO 63017 USA (Tel.: 314-895-2000; 800-325-7155; Telefax: 314-530-2562)

Mallinckrodt Specialty Chemicals Europe GmbH, Postfach 1268, Industriestrasse 19-21, W-6110 Dieburg Germany (Tel.: 49-6071-20040; Telefax: 49-6071-200444; Telex: 4191823 MCD D)

Mandoval Ltd.

Mark House, The Square, Lightwater, Surrey, GU18 5SS UK (Tel.: 44 1276 471617; Telefax: 44 1276 476910; Telex: 858094)

George Mann & Co., Inc.

PO Box 9066, Harborside Blvd., Providence, RI 02940-9066 USA (Tel.: 401-781-5600; Telefax: 401-941-0830)

M.A. Polymers, Div. of M.A. Industries, Inc.

303 Div.idend Dr., Peachtree City, GA 30269 USA (Tel.: 404-487-7761; 800-241-8250)

Marine Magnesium Co.

995 Beaver Grade Rd., Coraopolis, PA 15108 USA (Tel.: 412-264-0200; Telefax: 412-264-9020)

Martin Marietta Magnesia Specialties

PO Box 15470, Baltimore, MD 21220-0470 USA (Tel.: 410-780-5500; 800-648-7400; Telefax: 410-780-5777; Telex: 710-862-2630)

Martinswerk GmbH, Member of Lonza Group

PO Box 1209, Kolner Strasse 110, 50102 Bergheim Germany (Tel.: (0 22 71) 90 2-0; Telefax: (02271) 902-557; Telex: 888 712)

Mason Chemical Co

721 West Algonquin Rd., Arlington Heights, IL 60005 USA (Tel.: 708-290-1621; 800-362-1855; Telefax: 708-290-1625)

Mayzo, Inc.

6577 Peachtree Industrial Blvd., Norcross, GA 30092 USA (Tel.: 404-449-9066; Telefax: 404-449-9070)

McGee Industries, Inc./McLube Div.

9 Crozerville Rd., Aston, PA 19014 USA (Tel.: 610-459-1890; 800-2MCLUBE; Telefax: 610-459-9538, 910-380-7571 MCLUBE)

McIntyre Chemical Co., Ltd.

1000 Governors Hwy., University Park, IL 60466 USA (Tel.: 708-534-6200; Telefax: 708-534-6216)

McLube

McLube, Div. of McGee Industries Inc., 9 Crozerville Rd., Aston, PA 19014 USA (Tel.: 215-459-1890; 800-2-MCLUBE; Telefax: 215-459-9538; Telex: 910- 3807571 MCLUBE)

Distributed by: Lotrec AB, Box 323, S-18103 Sweden (Tel.: 08 766 01 20; Telefax: 08 766 52 17; Telex: 1008 EXTOR S)

Mearl

Mearl Corp., PO Box 3030, 320 Old Briarcliff Rd., Briarcliff Manor, NY 10510 USA (Tel.: 914-923-8500; Telefax: 914-923-9594; Telex: 421841)

Mearl International BV, Emrikweg 18, 2031 BT Haarlem The Netherlands (Tel.: 31-23-318058; Telefax: 31-23-315365; Telex: 41492 MRLN)

Mearl Corp. Japan, Room No. 802, Mido-Suji Urban-Life Bldg., 4-3 Minami Semba 4-Chome Chuo-ku, Osaka, 542 Japan (Tel.: (06) 281-1560; Telefax: (06) 281 1291)

Meer Corp.

PO Box 9006, 9500 Railroad Ave., N. Bergen, NJ 07047-1206 USA (Tel.: 201-861-9500; Telefax: 201-861-9267; Telex: 219130)

MEI (Magnesium Elektron Inc.)

500 Point Breeze Rd., Flemington, NJ 08822 USA (Tel.: 908-782-5800; 800-366-9596; Telefax: 908-782-7768)

Mendell Co., Inc., Edward, A Penwest Company

2981 Rt. 22, Patterson, NY 12563-9970 USA (Tel.: 914-878-3414; 800-431-2457; Telefax: 914-878-3484; Telex: 4971034)

Merck

E. Merck, Postfach 4119, Frankfurter Strasse 250, D-6100 Darmstadt 1 Germany (Tel.: 06151-72-0; Telefax: 06151-72-3684; Telex: 419328-0 em d)

Merck Ltd., Div. of Merck AG, Merck House, Poole, Dorset, BH12 4NN UK (Tel.: 44 1202 669700; Telefax: 44 1202 665599; Telex: 41186 TETRA G)

Merck & Co., Inc./Merck Chemical Div., PO Box 2000, Rahway, NJ 07065-0900 USA (Tel.: 908-594-4000; Telefax: 908-594-5431; Telex: 138825)

Merck Pty. Ltd., 207 Colchester Rd., Kilsyth, Victoria, 3137 Australia (Tel.: 61 3 728 5855; Telefax: 61 3 728 1351)

Merix Chemical Co.

2234 E. 75th St., Chicago, IL 60649 USA (Tel.: 312-221-8242; Telefax: 312-221-3047)

Merrand Int'l. Corp.

187 Ballardvale St., Suite B-200, Wilmington, MA 01887 USA (Tel.: 508-694-9202; Telefax: 508-964-9404)

Metallgesellschaft AG

Postfach 10 15 01, Reuterweg 14, D-6000 Frankfurt am Main 1 Germany (Tel.: 49-69-159-0; Telefax: 49-69-159-2125; Telex: 41225 mgf D)

Lucas Meyer

Lucas Meyer GmbH & Co., Postfach 261665, D-2000 Hamburg 26 Germany (Tel.: 49-40-789-550; Telefax: 49-40-789-8329; Telex: 2163220 MYER D)

Lucas Meyer (UK) Ltd., Unit 46, Deeside Ind. Park, First Ave., Deeside, Clwyd, CH5 2NU, Wales UK (Tel.: 44 1244 281168; Telefax: 44 1244 281169)

Lucas Meyer Inc., 765 E. Pythian Ave., Decatur, IL 62526 USA (Tel.: 217-875-3660; Telefax: 217-877-5046)

M. Michel & Co., Inc.

90 Broad St., New York, NY 10004 USA (Tel.: 212-344-3878; Telefax: 212-344-3880; Telex: 421468)

Miles Inc. *See* Bayer Inc./Fibers, Organics & Rubbers

Miles Inc./Polysar Rubber Div. *See* Bayer/Fibers, Organics & Rubber

Miljac Inc.

280 Elm St., New Canaan, CT 06840 USA (Tel.: 203-966-8777; Telefax: 203-966-3577)

Miller-Stephenson

Miller-Stephenson Chem. Co., Inc., George Washington Hwy., Danbury, CT 06810 USA (Tel.: 203-743-4447; 800-992-2424; Telefax: 203-791-8702)

Miller-Stephenson Chem. Co., Inc., 514 Carlingview Dr., Rexdale, Ontario, M9W 5R3 Canada (Tel.: 416-675-3204; 800-323-4621; Telefax: 416-674-2987)

Milliken

Milliken Chemical, A Div. of Milliken & Co., PO Box 1927, M-400, Spartanburg, SC 29304-1927 USA (Tel.: 803-503-2200; 800-345-0372; Telefax: 803-503-2430; Telex: 810-282-2580)

Milliken Europe N.V., 18 Ham, B-9000 Gent Belgium (Tel.: 32 9 265 1084; Telefax: 32 9 265 11 95)

Minerals Technologies, Inc., Specialty Minerals Subsid.

405 Lexington Ave., New York, NY 10174 USA (Tel.: 212-878-1919; Telefax: 212-878-1903)

Mirachem Srl

Via Guido Rossa 12, I-40111 Bologna Italy (Tel.: 39-51-73 71 11; Telefax: 39-51-73 54 40; Telex: 510011 MIRABO I)

Mitsubishi Gas

Mitsubishi Gas Chemical Co., Inc., Mitsubishi Bldg., 2-5-2, Marunouchi 2-chome, Chiyoda-ku, Tokyo, 100 Japan (Tel.: (03) 3283-4799; Telefax: (03)3214-0938; Telex: 222-2624 MGCHO J)

Mitsubishi Gas Chemical Co., Inc., 520 Madison Ave., 9th Floor, New York, NY 10022 USA (Tel.: 212-752-4620; Telefax: 212-758-4012; Telex: 649545 MGC UR NYK)

Mitsubishi Kasei

Mitsubishi Kasei Corp., Mitsubishi Bldg., 5-2, Marunouchi 2-chome, Chiyoda-ku, Tokyo, 100 Japan (Tel.: (03) 3283-6254; Telex: BISICH J 24901)

Mitsubishi Kasei America, Inc., 81 Main St., Suite 401, White Plains, NY 10601 USA (Tel.: 914-286-3600; Telefax: 914-681-0760; Telex: 233570 MCI UR)

Mitsubishi Kasei Europe GmbH, Niederkasseler Lohweg 8, 4000 Duesseldorf 11 Germany (Tel.: (0211) 523920; Telefax: (0211) 591272; Telex: 8587716 MCI D)

Mitsubishi Petrochemical Co., Ltd.

Mitsubishi Bldg., 5-2, Marunouchi 2-chome, Chiyoda-ku, Tokyo, 100 Japan (Tel.: (03) 3283-5700; Telefax: (03) 3283-5472; Telex: 222-3172)

Mitsui Petrochemical

Mitsui Petrochemical Industries, Ltd., Kasumigaseki Bldg., 2-5, Kasumigaseki 3-chome, Chiyoda-ku, Tokyo, 100 Japan (Tel.: (03) 3580-2012; Telefax: (03) 3593-0027; Telex: J22984 MIPECA)

Mitsui Petrochemicals (America) Ltd., 250 Park Ave., Suite 950, New York, NY 10017 USA (Tel.: 212-682-2366; Telefax: 212-490-6694; Telex: 7105814089)

Mitsui Toatsu

Mitsui Toatsu Chemicals, Inc., 2-5, Kasumigaseki 3-chome, Chiyoda-ku, Tokyo, 100 Japan (Tel.: (03) 3592-4594; Telefax: (03) 3592-4282; Telex: 2223622 MTCHEM J)

Mitsui Toatsu Chemicals, Inc., NY Office, Two Grand Central Tower Bldg., 34th Fl., 140 E. 45 St., New York, NY 10017 USA (Tel.: 212-867-6330; Telefax: 212-867-6315; Telex: 127057 MTC NYK)

Mitsui Toatsu Chemicals, Inc., London Office, 13 Charles II St., London, SW1Y 4QU UK (Tel.: 44 171 976-1180; Telefax: 44 171-976-1185; Telex: 8953938 MITOL G)

Mobil Chemical Co

PO Box 3029, Edison, NJ 08818-3029 USA (Tel.: 908-321-6000)

Modern Dispersions, Inc. (MDI)

22 Marguerite Ave., Leominster, MA 01453 USA (Tel.: 508-534-3370; Telefax: 508-537-6065)

Mona Industries Inc.

PO Box 425, 76 E. 24th St., Paterson, NJ 07544 USA (Tel.: 201-345-8220; 800-553-6662; Telefax: 201-345-3527; Telex: 130308)

Monmouth Plastics Co.

800 W. Main St., Freehold, NJ 07728 USA (Tel.: 908-866-0200; 800-526-2820; Telefax: 908-866-0274)

Monomer-Polymer & Dajac Labs, Inc.

1675 Bustleton Pike, Feasterville, PA 19053 USA (Tel.: 215-364-1155; Telefax: 215-364-1583)

Monsanto

Monsanto Chemical Co., 800 N. Lindbergh Blvd., St. Louis, MO 63167 USA (Tel.: 314-694-1000; 800-325-4330; Telefax: 314-694-7625; Telex: 650 397 7820)

Monsanto Europe SA, Ave. de Tervuren 270-272, B-1150 Brussels Belgium (Tel.: 32-761-41-11; Telefax: 32-761-40-40; Telex: 62927 Mesab)

Monsanto plc, Monsanto House, Chineham Court, Great Binfield Rd., Basingstoke, Hampshire, RG24 0UL UK (Tel.: 44 1256 572-88; Telefax: 44 1256-54995; Telex: 858837)

Monsanto Japan Ltd., Room 520, Kokusai Bldg., 1-1, Marunouchi 3-chome, Chiyoda-Ku, Tokyo, 100 Japan (Tel.: (03) 3287-1251; Telefax: (03) 3287-1250; Telex: J22614)

Morflex, Inc.

2110 High Point Rd., Greensboro, NC 27403 USA (Tel.: 910-292-1781; Telefax: 910-854-4058; Telex: 910240 7846)

Morton International

Morton International Inc., 100 North Riverside Plaza, Chicago, IL 60606-1598 USA (Tel.: 312-807-2562; Telefax: 312-807-2899; Telex: 25-4433)

Morton International, Inc./Plastics Additives, 150 Andover St., Danvers, MA 01923 USA (Tel.: 508-774-3100; 800-621-2847; Telefax: 508-750-9511)

Morton International, Inc./Polymer Systems, 100 North Riverside Plaza, Chicago, IL 60606 USA (Tel.: 312-807-2000)

Morton International, Inc./Specialty Chemicals Group, 2000 West St., Cincinnati, OH 45215 USA (Tel.: 513-733-2100; Telefax: 513-733-2133)

Morton International Ltd., Westward House, 155-157 Staines Rd., Hounslow, Middlesex, TW3 3JB UK (Tel.: 44 181 570 7766; Telefax: 44 181-570 6943; Telex: 262002)

Morton International SA, Chaussee de la Hulpe 130, Boite 5, B-1050 Brussels Belgium (Tel.: 32 2-6790211; Telefax: 32 2-6790 250; Telex: 23708)

Mosselman NV

80 Boulevard Industriel, B-1070 Brussels Belgium (Tel.: 32 2-524 18 78; Telefax: 32-2-5200158; Telex: 23533 MOSS B)

M-R-S Chemicals, Inc.

2494 Adie Rd., Maryland Heights, MO 63043 USA (Tel.: 314-432-3200)

MTM Research Chemicals, Inc.

PO Box 1000, Windham, NH 03087 USA (Tel.: 603-889-3306; 800-238-2324; Telefax: 603-889-3326)

H. Muehlstein & Co. Inc.

800 Connecticut Ave., Norwalk, CT 06856 USA (Tel.: 203-855-6000; Telefax: 203-855-6221)

Münzing Chemie GmbH

Salzstrasse 174, D-7100 Heilbronn Germany (Tel.: 07131/1586-0; Telefax: 07131/1586-25; Telex: 728614)

Mykroy/Mycalex Ceramics

125 Clifton Blvd., Clifton, NJ 07011 USA (Tel.: 201-779-8866; Telefax: 201-779-2013)

Napp Technologies, Inc.

199 Main St., PO Box 900, Lodi, NJ 07644 USA (Tel.: 201-773-3900; Telefax: 201-773-2010, 13-4649)

National Chemical Co.

600 W. 52 St., Chicago, IL 60609 USA (Tel.: 312-924-3700; 800-525-3750; Telefax: 312-924-7760)

National Starch & Chemical

National Starch & Chemical Corp., Box 6500, 10 Finderne Ave., Bridgewater, NJ 08807 USA (Tel.: 908-685-5000; 800-726-0450; Telefax: 908-685-5005)

National Starch & Chemical (Holdings) Ltd., Canon House, 27 London End, Beaconsfield, Buckinghamshire, HP9 2HN UK (Tel.: 44 1494 677966; Telefax: 44 1494 673960; Telex: 848386 NATADH G)

National Starch & Chemical, Prestbury Court, Greencourts Business Park, 333 Styal Rd., Manchester, M22 5LW UK (Tel.: 44 161 435 3200; Telefax: 44 161 435 3300)

National Starch & Chemical (Asia), 107 Neythal Rd., Jurong, S-2262 Singapore (Tel.: 65 2615528; Telefax: 65 264 1870, 55445)

Natrochem, Inc.

PO Box 1205, Exley Ave., Savannah, GA 31498 USA (Tel.: 912-236-4464; Telefax: 912-236-1919)

Nayler Chemicals Ltd.

Old Rolling Mill, Edge Green Rd, Ashton in Makerfield, Wigan, Lancashire, WN4 8YA UK (Tel.: 44 1942 720988; Telefax: 44 1942 271251; Telex: 669367 BREEZE G)

Neochem Inc.

30 Second St. SW, Barberton, OH 44203 USA (216-848-9110)

Nepera, Inc., A Cambrex Co.

Route 17, Harriman, NY 10926 USA (Tel.: 914-782-1202; Telefax: 914-782-2418; Telex: 510-249-4847)

Neste

Neste OY Chemicals, PO Box 20, Keilaniemi, SF-02150, Espoo 15 Finland (Tel.: 358-4501; Telefax: 358-4504985; Telex: 124641 NESTE SF)

Neste Polyeten AB, Div. of Neste Oy, S-44486 Stenungsund Sweden (Tel.: 46-303-86000; Telefax: 46-303-86449; Telex: 2402)

Neville Chemical Co.

2800 Neville Rd., Pittsburgh, PA 15225-1496 USA (Tel.: 412-331-4200; Telefax: 412-777-4234)

Niacet Corp.

PO Box 258, 400 47th St., Niagara Falls, NY 14304 USA (Tel.: 716-285-1474; 800-828-1207; Telefax: 716-285-1497; Telex: 6730170)

Nichia Kagaku Kogyo K.K.

PO Box 6 Anan, Tokushima, 774 Japan (Tel.: (0884) 22-2311; Telefax: (0884) 23-1802; Telex: 5867790 NICHIA J)

Nihon Kagaku Sangyo Co., Ltd.

20-5, Shitaya 2-chome, Taito-ku, Tokyo, 110 Japan (Tel.: (03) 3876-3131; Telefax: (03) 3876-3278; Telex: J 28318 NIKKASAN)

Nikko Chemicals Co., Ltd.

4-8, Nihonbashi, Bakuro-cho 1-chome, Chuo-ku, Tokyo, 103 Japan (Tel.: (03) 3661-1677; Telefax: (03) 3664-8620; Telex: 2522744 NIKKOL J)

Nipa Laboratories, Inc.

3411 Silverside Rd., 104 Hagley Bldg., Wilmington, DE 19810 USA (Tel.: 302-478-1522; Telefax: 302-478-4097; Telex: 905030)

Nippon Nyukazai Co., Ltd., Subsid. of Sankyo Co., Ltd.

Yoshizawa Bldg., 9-19, Ginza 3-chome, Chuo-ku, Tokyo, 104 Japan (Tel.: (03) 3543-8571; Telefax: (03) 3546-3174)

Nippon Oils & Fats Co., Ltd. (NOF Corp.)

Yurakucho Bldg., 10-1, Yarakucho 1-chome Chiyoda-Ku, Tokyo, 100 Japan (Tel.: (03) 3283-7295; Telefax: (03) 3283-7178; Telex: 222-2041 NIPOIL J)

Nippon Polyurethane Industry Co., Ltd.

Kotohira Kaikan Bldg., 2-8, Toranomon 1-chome, Minato-ku, Tokyo, 105 Japan (Tel.: (03) 3508-0611; Telefax: (03) 3580-9304; Telex: 222-3066 NIPOLY J)

Nippon Silica Industrial Co., Ltd.

Toso-Kyobashi Bldg., 2-4, Kyobashi 3-chome, Chuo-ku, Tokyo, 104 Japan (Tel.: (03) 3273-1641; Telefax: (03) 3272-3879)

Nissan Chemical Industries, Ltd.

Kowa-Hitotsubashi Bldg., 7-1, Kanda-Nishiki-cho 3-chome, Chiyoda-ku, Tokyo, 101 Japan (Tel.: (03) 3296-8111; Telefax: (03) 3296-8360; Telex: 222-3071)

Noah Chemical Div., Div. of Noah Technologies Corp

7001 Fairgrounds Pkwy, San Antonio, TX 78238 USA (Tel.: 512-680-9000; Telefax: 512-521-3323)

The Norac Co. Inc.

405 S. Motor Ave., Azusa, CA 91702 USA (Tel.: 818-334-2908; Telefax: 818-334-3512; Telex: 882552 NORAC CO AZSA)

Norman, Fox & Co.

5511 S. Boyle Ave., PO Box 58727, Vernon, CA 90058 USA (Tel.: 213-583-0016; 800-632-1777; Telefax: 213-583-9769)

Norsk Hydro AS

Bygdoyalle 2, N 0240 Oslo 2 Norway (Tel.: 472 243 2100; Telefax: 472 243 2725; Telex: 78350 HYDRO N)

Northern Products, Inc.

PO Box 1175, 153 Hamlet Ave., Woonsocket, RI 02895 USA (Tel.: 401-766-2240; Telefax: 401-766-2287)

Norton Chemical Process Products

PO Box 350, Akron, OH 44309 USA (Tel.: 216-677-7216; Telefax: 216-677-7245; Telex: 433-8012)

Novachem Corp.

PO Box 6379, High Point, NC 27262 USA (Tel.: 919-885-0041; Telefax: 919-885-4964)

Novacor Chemicals, Affiliate of Nova Corp. of Alberta (Novacor)

2550 Busha Hwy., Marysville, MI 48040 USA (Tel.: 313-364-5555; 800-627-1221; Telefax: 313-364-4670)

Nova Polymers, Inc.

2650 Eastside Park Dr., Evansville, IN 47716-8466 USA (Tel.: 812-476-0339; Telefax: 812-476-0592)

Nyco

Nyco Minerals, Inc., A Canadian Pacific Ltd. Co., 124 Mountain View Dr., Willsboro, NY 12996-0368 USA (Tel.: 518-963-4262; Telefax: 518-963-4187; Telex: 957014)

Nyco Minerals, Inc. Europe, Ordrupvej 24, PO Box 88, DK-2920 Charlottenlund Denmark (Tel.: 31 64 33 70; Telefax: 31 64 37 10)

Oakite Products, Inc. *See under* Chemetall

Occidental

Occidental Chemical Corp., 5005 LBJ Freeway, Dallas, TX 75244 USA (Tel.: 214-404-3800; 800-752-5151; Telefax: 214-404-3669; Telex: 229835)

Occidental Chemical Corp., 360 Rainbow Blvd. South, PO Box 728, Niagara Falls, NY 14302 USA (Tel.: 716-286-3229; 800-828-1144; Telefax: 716-286-3441)

Occidental Chemical Corp./Durez Div., PO Box 809050, 5005 LBJ Freeway, Dallas, TX 75380-9050 USA (Tel.: 214-404-4932; 800-733-3339; Telefax: 214-404-4981)

Occidental Chemical Europe, Holidaystraat 5, B-1831 Diegem Belgium (Tel.: 32 725 44 50; Telefax: 32 725 46 76; Telex: 23046)

Olin

Olin Corp., 120 Long Ridge Rd., PO Box 1355, Stamford, CT 06904 USA (Tel.: 203-356-3036; 800-243-9171; Telefax: 203-356-3273; Telex: 420202)

Olin Chemicals Pty., Ltd., 1-3 Atchison St., PO Box 141, St. Leonards 2065, N.S.W. Australia (Tel.: 612-439-6222; Telex: 26328)

Olin Japan Inc., Shiozaki Bldg., 7-1 Hirakawa-Cho 2-Chome, Chiyoda-ku, Tokyo, 102 Japan (Tel.: (813) 263-4615; Telex: 023-24031)

OM Group

2301 Scranton Rd., Cleveland, OH 44113 USA (Tel.: 216-781-8383; 800-321-9696; Telefax: 216-781-5919, 810-421-8322)

Ore & Chemical Corp., Div. of Chemetall

520 Madison Ave., New York, NY 10022 USA (Tel.: 212-715-5232; Telefax: 212-486-2742; Telex: ITT 422681)

Original Bradford Soap Works Inc.

PO Box 1007, West Warwick, RI 02893 USA (Tel.: 401-821-2141; Telefax: 401-821-5960; Telex: 952 240)

Osi Specialties

OSi Specialties, Inc., 39 Old Ridgebury Rd., Danbury, CT 06810-5121 USA (Tel.: 203-794-5300; 800-523-2862)

OSi Specialties Canada, Inc., 1210 Sheppard Ave. East, Suite 210, Box 38, Willowdale, Ontario, M2K 1E3 Canada (Tel.: 416-490-0466)

OSi Specialties do Brasil Ltda., Rua Dr. Eduardo De Souza, Aranha 153, Sao Paulo, 04530 Brazil (Tel.: 55-11-828-1104)

OSi Specialties S.A., 7 Rue de Pre-Bouvier, Meyrin, CH-1217 Geneva Switzerland (Tel.: 41-22-989-2111)

OSi Specialties Singapore PTE, Ltd., 22-01 Treasury Bldg., 8 Shenton Way, 0106 Singapore (Tel.: 65-322-9922)

Osram Sylvania Inc.

100 Endicott St., Danvers, MA 01923 USA (Tel.: 508-777-1900)

Otsuka

Otsuka Chemical Co., Ltd., 2-27, Ote-dori 3-chome, Chuo-ku, Osaka, 540 Japan (Tel.: (06) 943-7711; Telefax: (06) 946-0860; Telex: J63586 JPOTSUKA)

Otsuka Chemical Co., Ltd., 747 Third Ave., 26th Floor, New York, NY 10168 USA (Tel.: 212-826-4374; Telefax: 212-826-5094)

Owens-Corning Fiberglas Corp.

Fiberglas Tower, Toledo, OH 43659 USA (Tel.: 419-248-7185; 800-462-3435; Telefax: 419-248-6712)

Oxiteno S/A Industria E Comercio

Av. Brigedeiro Luiz Antonio 1343, 7 andar, Sao Paulo SP, 01350-900 Brazil (Tel.: (55-11)283 6118; Telefax: (55-11)2893533, (55-11) TLX 31727)

OxyChem. *See* Occidental Chem. Corp.

Pall Process Filtration Co.

2200 Northern Blvd., East Hills, NY 11548-1289 USA (Tel.: 516-484-5400; 800-645-6532; Telefax: 516-484-5228; Telex: 968855)

Pasminco Europe Ltd./ISC Alloys Div., Div. of Pasminco Ltd.

Alloys House, PO Box 36, Willenhall Lane, Bloxwich, Walsall, West Midlands, WS3 2XW UK (Tel.: 44 1922 408444; Telefax: 44 1922-710043; Telex: 338270)

Passaic Color & Chemical Co.

28-36 Paterson St., Paterson, NJ 07501 USA (Tel.: 201-279-0400; Telefax: 201-279-8561; Telex: 820907 PASDYE UD)

Patco Products Div. *See* American Ingredients Co.

PCR, Inc.

8570 Phillips Hwy., Suite 101, Jacksonville, FL 32256-8208 USA (Tel.: 904-376-8246; 800-331-6313; Telefax: 904-371-6246; Telex: 810-825-6342)

PO Box 1466, Gainesville, FL 32602 USA (Tel.: 904-376-8246; 800-331-6313; Telefax: 904-371-6246)

Pelron Corp.

7847 W. 47 St., Lyons, IL 60534 USA (Tel.: 708-442-9100; Telefax: 708-442-0213)

Penreco, Div. of Pennzoil Prods. Co.

138 Petrolia St., Box 1, Karns City, PA 16041 USA (Tel.: 412-756-0110; 800-245-3952; Telefax: 412-756-1050; Telex: 1561596)

Penta Manufacturing Co.

PO Box 1448, Fairfield, NJ 07007-1448 USA (Tel.: 201-740-2300; Telefax: 201-740-1839; Telex: 219472 PENT UR)

Pentagon Chemicals Ltd., Div. of Suter plc

Northside, Workington, Cumbria, CA14 1JJ UK (Tel.: 44 1900 604371; Telefax: 44 1900 66943; Telex: 64353 PENTA G)

The Permethyl Corp.

191 S. Keim St., PO Box 643, Pottstown, PA 19464 USA (Tel.: 610-970-0251; Telefax: 610-970-9415)

Perstorp Polyols, Inc.

600 Matzinger Rd, Toledo, OH 43612 USA (Tel.: 419-729-5448; 800-537-0280; Telefax: 419-729-3291)

Pestco Inc.

PO Box 1000, 215-225 Eighth Ave., Braddock (PGH), PA 15104 USA (Tel.: 412-271-6200; 800-247-0770; Telefax: 412-351-7701)

Petrolite

Petrolite Corp./Headquarters & Polymers Div., 6910 E. 14th St., Tulsa, OK 74112 USA (Tel.: 918-836-1601; 800-331-5516; Telefax: 918-834-9718)

Petrolite Ltd./EuroChem Div., Div. of Petrolite Speciality Polymers Group, Kirkby Bank Rd, Knowsley Industrial Park, Liverpool, Merseyside, L33 7SY UK (Tel.: 44 151 546 2855; Telefax: 44 151 549 1858; Telex: 627293 PETLTD G)

Petrolite GmbH, PO Box 2031, Kaiser-Friedrich Promenade 59, 6380 Bad Homburg 1 Germany

Pfaltz & Bauer Inc. *See under* Aceto

Pfanstiehl Laboratories, Inc.

1219 Glen Rock Ave., PO Box 439, Waukegan, IL 60085 USA (Tel.: 708-623-0370; 800-383-0126; Telefax: 708-623-9173; Telex: 25 3672 PFANLAB)

Pfizer

Pfizer International, 235 East 42nd Street, New York, NY 10017 USA (212-573-2323)

Pfizer Food Science Group, 235 E. 42nd St., New York, NY 10017 USA (Tel.: 212-573-2323/2548; 800-TECK-SRV; Telefax: 212-573-1166)

Pfizer Canada, PO Box 800, Point Claire/Dorval, Montreal, Quebec, H9R 4V2 Canada (Tel.: 514-695-0500)

Pfizer Europe/Africa/Middle East, 10 Dover Rd., Sandwich, Kent, CT13 0BN UK (Tel.: 44 1304 615518; Telefax: 44 1304 615529; Telex: 966555)

Phibrochem, Div. of Philipp Bros Chemicals Inc.

One Parker Plaza, Fort Lee, NJ 07090 USA (Tel.: 201-944-6020; 800-223-0434; Telefax: 201-944-6245)

Phibro Energy USA Inc.

500 Dallas, Suite 3200, Houston, TX 77002 USA (Tel.: 713-646-5042; Telefax: 713-646-5293, 3736083)

Phillips

Phillips Chemical Co., Div. of Phillips Petroleum Co, PO Box 968, Borger, TX 79008 USA (Tel.: 806-274-5236; 800-858-4327; Telefax: 806-274-5230)

Phillips Chemical Co., Div. of Phillips Petroleum Co, 101 ARB Plastics Tech. Center, Bartlesville, OK 74004 USA (Tel.: 918-661-3612; Telefax: 918-662-2929)

Phillips Petroleum Chemicals NV, Steenweg op Brussels 355, B-2090 Overijse Belgium (Tel.: 32-2-689 1211; Telefax: 32-2-689 1472; Telex: 22197)

Phillips Petroleum International, Ltd., Shin-Tokyo Bldg., 3-1, Marunouchi 3-chome, Chioyda-ku, Tokyo, 100 Japan (Tel.: (03) 3216-6951; Telefax: (03) 3216-6960; Telex: J24641 PHILPET)

Pierce & Stevens Corp., A Pratt & Lambert Co.

PO Box 1092, Buffalo, NY 14240-1092 USA (Tel.: 716-856-4910; 800-888-4910; Telefax: 716-856-7530)

Pilot Chemical Co.

11756 Burke St., Santa Fe Springs, CA 90670 USA (Tel.: 213-723-0036; Telefax: 213-945-1877; Telex: 4991200 PILOT)

Plastic Filler Sales, Inc.

PO Box 1446, Chatsworth, GA 30705 USA (706-695-4613; Telefax: 706-695-6706)

Plasticolors, Inc.

2600 Michigan Ave., PO Box 816, Ashtabula, OH 44004 USA (Tel.: 216-997-5137; Telefax: 216-992-3613)

Plastics & Chemicals, Inc.

PO Box 306, Cedar Grove, NJ 07009 USA (Tel.: 908-221-0002; Telefax: 908-221-1097; Telex: 219744)

PMC Specialties

PMC Specialties Group, Inc., Div. of PMC, Inc., 20525 Center Ridge Road, Rocky River, OH 44116 USA (Tel.: 216-356-0700; Telefax: 216-356-2787; Telex: 4332035)

PMC Specialties Group, Inc., Div. of PMC, Inc., 501 Murray Rd., Cincinnati, OH 45217 USA (Tel.: 513-242-3300; 800-543-2466; Telefax: 513-482-7353; Telex: 5106000948)

PMC Specialities International Ltd., Div. of PMC Specialities Group, 65B Wigmore Street, London, W1H 9LG UK (Tel.: 44 171 935-4058; Telefax: 44 171-935 9895; Telex: 24358 PMCS G)

PMS Consolidated/Corporate Commercial Development Group, A Business Unit of M.A. Hanna Co.

3300 University Dr., Suite 511, Coral Springs, FL 33065 USA (Tel.: 305-753-9299; Telefax: 305-755-4587)

Polyad Co.

113 Rose Terrace, Barrington, IL 60010-1320 USA (Tel.: 708-526-3322; Telefax: 708-526-3326)

Polycom Huntsman

90 W. Chestnut St., Washington, PA 15301 USA (Tel.: 412-225-2220; 800-538-3149; Telefax: 412-222-9441)

Polymer Research Corp. of America

2186 Mill Ave., Brooklyn, NY 11234 USA (Tel.: 718-444-4300; Telefax: 718-241-3930)

Poly Organix, Inc.

9 Opportunity Way, Newburyport, MA 01950 USA (Tel.: 508-462-5555; 800-992-8506; Telefax: 508-465-2057)

Poly Research Corp.

125 Corporate Dr., Holtsville, NY 11742 USA (Tel.: 516-758-0460; Telefax: 516-758-0471)

Polysar Nederland BV

Westervoortsedijk 71, NL-6827 AV Arnheim The Netherlands (Tel.: 31-85-65 39 11; Telex: 45593)

Polysciences

Polysciences Inc., 400 Valley Road, Warrington, PA 18976-2590 USA (Tel.: 215-343-6484; 800-523-2575; Telefax: 800-343-3291; Telex: 510-665-8542)

Polysciences, Europe GmbH, Postfach 1130, D-69208 Eppelheim Germany (Tel.: (49) 6221-765767; Telefax: (49) 6221-764620)

Polyurethane Corp. of America

PO Box 8, Everett, MA 02149 USA (Tel.: 617-389-7889)

Polyurethane Specialties Co. Inc.

624 Schuyler Ave., Lyndhurst, NJ 07071 USA (Tel.: 201-438-2325; Telefax: 201-507-1367)

Polyvel, Inc.

120 N. White Horse Pike, Hammonton, NJ 08037 USA (Tel.: 609-567-0080; Telefax: 609-567-9522)

PolyVisions, Inc.

1600 Pennsylvania Ave., Cyber Center, York, PA 17404 USA (717-854-3132; Telefax: 717-854-9333)

Potters Industries Inc., Affiliate of The PQ Corp.

Southpoint Corporate Headquarters, PO Box 840, Valley Forge, PA 19482-0840 USA (Tel.: 610-651-4727; Telefax: 610-408-9723)

Powmet, Inc.

PO Box 5086, 2625 Sewell St., Rockford, IL 61125 USA (Tel.: 815-398-6900; Telefax: 815-398-6907)

PPG

PPG Industries, Inc., One PPG Place, Pittsburgh, PA 15272 USA (Tel.: 412-434-4109; 800-CHEM-PPG; Telefax: 412-434-2137; Telex: 86 6570)

PPG Industries, Inc./Specialty Chemicals, 3938 Porett Dr., Gurnee, IL 60031 USA (Tel.: 708-244-3410; 800-323-0856; Telefax: 708-244-9633; Telex: 25-3310)

PPG Canada Inc./Specialty Chem., 2 Robert Speck Pkwy., Suite 900, Mississauga, Ontario, L4Z 1H8 Canada (Tel.: 905-848-2500; Telefax: 905-848-2185; Telex: 38906960351canbizmis)

PPG Industrial do Brazil Ltda., Edificio Grande Avenida, Paulista Ave. 1754, Suite 153, 01310 Sao Paulo Brazil (Tel.: 55-011-284-0433; Telefax: 55-011-289-2105; Telex: 011-39104 PPGB BR)

PPG-Mazer Mexico, S.A. de C.V., Av. Presidente Juarex No. 1978, Tlalnepantla, Edo., C.P. 54090 Mexico (Tel.: 52-5-397-8222; Telefax: 52-5-398-5133)

PPG Industries (UK) Ltd./Specialty Chem., Carrington Business Park, Carrington,, Urmston, Manchester, M31 4DD UK (Tel.: 44 161 777-9203; Telefax: 44 161-777-9064; Telex: 851-94014896 mazu g)

PPG Industries (France) SA, BP 377, Écluse Folien, F-59307 Valenciennes France (Tel.: 33-27 14 46 00; Telefax: 33-27 29 36 34)

PPG Ouvrie S.A., 64, rue Faldherbe, B.P. 127, 59811 Lesquin Cedex France (Tel.: 33-2087-0510; Telefax: 33-2087-5631; Telex: 131419 F)

PPG Industries Taiwan, Ltd., Suite 601, Worldwide House, No. 131, Ming East Rd., Sec. 3, Taipei, 105, Taiwan, R.O.C. (Tel.: 886-2-514-8052; Telefax: 886-2-514-7957; Telex: 10985 PPGTWN)

PPG Industries-Asia/Pacific Ltd., Takanawa Court, 5th floor, 13-1 Takanawa 3-Chome, Minato-Ku, Tokyo, 108 Japan (Tel.: (03) 3280-2911; Telefax: (03) 3280-2920; Telex: 02-42719 PPGPACJ)

PQ Corp.

PO Box 840, Valley Forge, PA 19482 USA (Tel.: 610-651-4200; 800-944-7411; Telefax: 610-251-9118; Telex: 476 1129 PQCO VAF)

Primachem Inc.

12 Greenwoods Rd., Old Tappan, NJ 07675 USA (Tel.: 201-784-3434; Telefax: 201-784-7997, 175094 PRIMA)

Prince Mfg. Co.

One Prince Plaza, Box 1009, Quincy, IL 62301 USA (Tel.: 217-222-8854; Telefax: 217-222-5098)

Procedyne Corp.

11 Industrial Dr., New Brunswick, NJ 08901 USA (Tel.: 908-249-8347; Telefax: 908-249-7220)

Pro Chem Chemicals Inc.

1670 English Rd., High Point, NC 27262 USA (Tel.: 919-882-3308; Telefax: 919-889-6047)

Procter & Gamble

Procter & Gamble Co/Chemicals Div., PO Box 599, Cincinnati, OH 45201 USA (Tel.: 513-983-3928; 800-543-1580; Telefax: 513-983-1436; Telex: 21-4185, P&GCIN)

Procter & Gamble Ltd./Europe Div., PO Box 9, 27 Uxbridge Rd., Hayes, Middlesex, UB4 0JD UK (Tel.: 44 181 848 9671, 936310)

Protameen Chemicals, Inc.

375 Minnisink Rd., PO Box 166, Totowa, NJ 07511 USA (Tel.: 201-256-4374; Telefax: 201-256-6764; Telex: 130125)

Protex SA

B.P. 177, 6, rue Barbès, 92305 Levallois-Paris France (Tel.: 47-57-74-00; Telefax: 47-57-69-28; Telex: 620987)

Pulcra SA

Sector E C/42, Barcelona, 08040 Spain (Tel.: 34-3-2904760; Telefax: 34-3-2904759; Telex: 98301)

Purac America, Inc.

11 Barclay Blvd., Suite 280, Lincolnshire Corporate Center, Lincolnshire, IL 60069 USA (Tel.: 708-634-6330; Telefax: 708-634-1992; Telex: 280231 PURACINC ARHT)

PVS Chemicals, Inc.

11001 Harper Ave., Detroit, MI 48213 USA (Tel.: 313-921-1200; Telefax: 313-921-1378)

Pyrion-Chemie

Pyrion-Chemie GmbH, Geylenstr. 94, Neuss Germany (Tel.: 49 2101-591025; Telefax: 49 2101-593624; Telex: 8517459)

Represented in U.S. by Ruetgers-Nease Corp., 201 Struble Rd., State College, PA 16801 USA (Tel.: 814-231-9261; 800-458-3434; Telefax: 814-238-4235)

QO

QO Chemicals, Inc., Subsid. of Great Lakes Chem. Corp., 2801 Kent Ave., Box 2500, West Lafayette, IN 47906 USA (Tel.: 317-497-6100; 800-621-9521; Telefax: 317-497-6287; Telex: 446968 QOC UD)

QO Chemicals Inc., Industriepark, B-02440 Geel Belgium (Tel.: 32 58 9572; Telefax: 32 58 08 96; Telex: 34827)

Quality Chemicals, Inc. *See under* First Chemical Corp

Quantum Chemical Corp./USI Div.

11500 Northlake Dr., PO Box 429550, Cincinnati, OH 45249 USA (Tel.: 513-530-6500; 800-323-4905; Telefax: 513-530-6313; Telex: 155116)

Quest International Fragrances USA, Inc.

400 International Dr., Mt. Olive, NJ 07828 USA (Tel.: 201-691-7100; Telefax: 201-691-7100; Telex: 6714933)

Ranbar Technology Inc.

1114 William Flinn Hwy., Glenshaw, PA 15116 USA (Tel.: 412-486-1111; Telefax: 412-487-3313; Telex: 9102500271)

RapidPurge Corp.

2285 Reservoir Ave., Trumbull, CT 06611 USA (Tel.: 203-372-5677; 800-243-4203)

Rasa Industries, Ltd.

Yaesu Dai Bldg., 1-1, Kyobashi 1-chome, Chuo-ku, Tokyo, 104 Japan (Tel.: (03) 3278-3801; Telex: 2225818 RASAKOJ)

Raschig

Raschig AG, Mundenheimer Strasse 100, D-6700 Ludwigshafen/Rhine Germany (Tel.: (0621)56180; Telefax: 0621-532885; Telex: 464 877 ralu d)

Raschig Corp., 5000 Old Osborne Tpke., Box 7656, Richmond, VA 23231 USA (Tel.: 804-222-9516; Telefax: 804-226-1569)

Raschig UK Ltd., 124 Doveleys Rd., Salford, Lancashire, M6 8QW UK (Tel.: 44 161 745 8811; Telefax: 44 161 745 8371)

Raschig France S.A.R.L., 49, ave. de Versailles, F-75016 Paris France (Tel.: 33 1 452 40636; Telefax: 33 1 452 08408)

Reagens SpA

Via Codronchi 4, I-40016 San Giogio di Piano Italy (Tel.: 39-897157; Telefax: 39-897561; Telex: 510374 REAG I)

Reed Spectrum, Div. of Sandoz Chemicals Corp.

Holden Industrial Park, Holden, MA 01520 USA (Tel.: 508-829-6321; Telefax: 508-829-2118)

Reedy International Corp.

25 East Front St., Suite 200, Keyport, NJ 07735 USA (Tel.: 908-264-1777; Telefax: 908-264-1189)

Reheis

Reheis Inc., PO Box 609, 235 Snyder Ave., Berkeley Heights, NJ 07922 USA (Tel.: 908-464-1500; Telefax: 908-464-8094; Telex: 219463 RCCA UR)

Reheis Ireland, Div. of Reheis Inc., Kilbarrack Rd., Dublin, 5, Irish Republic (Tel.: 353-1-322621; Telefax: 353-1-392205; Telex: 32532 REHI EI)

Reichhold

Reichhold Chemicals, Inc./Corporate Headquarters, PO Box 13582, Research Triangle Park, NC 27709 USA (Tel.: 919-990-7500; 800-448-3482; Telefax: 919-990-7711)

Reichhold GmbH, Postfach 10, Breitenleer-strasse 97-99, A-1222 Vienna Austria (Tel.: 43-1-233551; Telefax: 43-1-233519232; Telex: 134924 RCI A)

Reilly Industries Inc.

151 N. Delaware St., Suite 1510, Indianapolis, IN 46204 USA (Tel.: 317-638-7531; Telefax: 317-248-6413; Telex: 27 404)

Reilly-Whiteman Inc.

801 Washington St., Conshohocken, PA 19428 USA (Tel.: 215-828-3800; 800-533-4514; Telefax: 215-834-7855; Telex: 5106608845)

Releasomers, Inc.

PO Box 82, Bradfordwoods, PA 15015 USA (Tel.: 412-452-4474; Telefax: 412-452-1965)

Resinall Corp.

3065 High Ridge Rd., PO Box 8149, Stamford, CT 06905 USA (Tel.: 203-329-7100; 800-356-7717; Telefax: 203-329-0167)

Revertex Ltd.

Templefields, Harlow, Essex, CM20 2BH UK (Tel.: 44 1279 429555; Telefax: 44 1279-412984; Telex: 81318)

Rewo. *See under* Witco

Rhein Chemie. *See under* Bayer

Rhone-Poulenc

Rhone-Poulenc SA, 25 quai Paul Doumer, 92408 Courbevoie Cedex France (Tel.: 33 1 47 68 12 34, 610500)

Rhone-Poulenc Surfactants & Specialties (Europe), Les Miroirs-Defense 3, 18 ave. d'Alsace, Cedex 29, 92097 Paris, LaDefense France (Tel.: (33-1) 4768 1234; Telefax: (33-1) 4768 0900)

Rhone-Poulenc, Inc., CN7500, Prospect Plain Rd., Cranbury, NJ 08512-7500 USA (Tel.: 800-288-1175)

Rhone-Poulenc Basic Chemical Co., One Corporate Dr., Box 881, Shelton, CT 06484 USA (Tel.: 800-642-4200; Telefax: 203-925-3627)

Rhone-Poulenc Food Ingredients, CN 7500, Prospect Plains Rd., Cranbury, NJ 08512 USA (Tel.: 609-860-4600; 800-253-5052)

Rhone-Poulenc Rubber Specialties, 1929 Trement St., Dover, OH 44627 USA (Tel.: 216-364-6002; Telefax: 216-364-5007)

Rhone-Poulenc Silicones, Cranbury, NJ 08512 USA (Tel.: 800-288-1175)

Rhone-Poulenc, Inc./Performance Resins & Coatings, 1525 Church St. Ext., Marietta, GA 30060 USA (Tel.: 404-422-1250; Telefax: 404-427-0874; Telex: 542-112)

Rhone-Poulenc, Inc./Specialty Chemicals, CN 7500, Prospect Plains Rd., Cranbury, NJ 08512-7500 USA (Tel.: 609-860-3200; 800-922-2189; Telefax: 609-860-0459)

Rhone-Poulenc, Inc./Surfactants & Specialty Chemicals, CN 7500, Prospect Plains Rd., Cranberry, NJ 08512-7500 USA (Tel.: 609-860-4000; 800-922-2189; Telefax: 609-860-0459)

Rhone-Poulenc, Inc./Water Treatment Chemicals, One Gatehall Dr., Parsippany, NJ 07054 USA (201-292-2900; 800-848-7659; Telefax: 201-292-5295)

Rhone-Poulenc Surfactants & Specialties Canada, 3265 Wolfdale Rd., Mississauga, Ontario, L5C 1V8 Canada (Tel.: 905-270-5534; Telefax: 905-270-5816)

Rhone-Poulenc Chemicals Ltd., Div. of Rhone-Poulenc SA, Staveley, Chesterfield, Derbyshire, S43 2PB UK (Tel.: 44 1246 277251; Telefax: 44 1246-280090; Telex: 577425 STAVEX G)

Rhone-Poulenc Surfactants & Specialties (Asia Pacific), 27 06/07 The Concourse, 300 Beach Road, Singapore 0719 (Tel.: (65) 291 1921; Telefax: (65) 296 6044)

American Lecithin Co., Div. of Rhone-Poulenc Rorer/Nattermann, 33 Turner Rd., Danbury, CT 06813-1908 USA (Tel.: 203-790-2700; Telefax: 203-790-2705)

G. Whitfield Richards Co.

4202-10 Main St., Philadelphia, PA 19127 USA (Tel.: 215-487-1202; Telefax: 215-487-3090; Telex: 845-322 GWRCO)

Richman Chemical, Inc.

768 N. Bethlehem Pike, Lower Gwynedd, PA 19002 USA (Tel.: 215-628-2946; Telefax: 215-628-4262)

Ricon Resins, Inc.

569 24 1/4 Road, Grand Junction, CO 81505 USA (Tel.: 303-245-8148; Telefax: 303-245-4348)

Ridge Technologies, Inc.

117 Lyons Rd., Basking Ridge, NJ 07920 USA (Tel.: 908-766-1915; Telefax: 908-204-0312)

R.I.T.A. Corp.

1725 Kilkenny Court, PO Box 585, Woodstock, IL 60098 USA (Tel.: 815-337-2500; 800-426-7759; Telefax: 815-337-2522; Telex: 72-2438)

Rit-Chem Co. Inc.

109 Wheeler Ave., PO Box 435, Pleasantville, NY 10570 USA (Tel.: 914-769-9110; Telefax: 914-769-1408; Telex: 229 639 RTCH)

RMc Minerals, Inc.

111 E. Drake Rd., Suite 7104, Fort Collins, CO 80525 USA (Tel.: 303-223-7790; Telefax: 303-226-5617)

Robeco Chemicals Inc.

99 Park Ave., New York, NY 10016 USA (Tel.: 212-986-6410; Telefax: 212-986-6419; Telex: 23-3053 A (RCA)

Rogers Corp

One Technology Drive, Rogers, CT 06263 USA (Tel.: 203-774-9605; Telefax: 203-774-9630)

Rogers Anti-Static Chemicals

120 W. Madison St., Room 1118, Chicago, IL 60602 USA (Tel.: 312-276-0665; Telefax: 312-276-4371)

Rohm & Haas

Rohm and Haas Co., 100 Independence Mall West, Philadelphia, PA 19106-2399 USA (Tel.: 215-592-3000; 800-323-4165; Telefax: 215-592-2321; Telex: 845-247)

Rohm & Haas Europe, 185 rue de Bercy, 75579 Paris, Cedex 12 France (Tel.: 011-33-1 40 02 50 00; Telefax: 011-33-1 43 45 28 19)

Rohm & Haas Pacific Region, 391B Orchard Road #16-05/07, 0 Ngee Ann City, 0923 Singapore (Tel.: 011-65-735-0855; Telefax: 011-65-735-0877)

Rohm Tech Inc., Subsid. of Röhm GmbH

195 Canal St., Malden, MA 02148 USA (Tel.: 617-321-6984; 800-666-7646; Telefax: 617-322-0358; Telex: 200721 Rohm UR)

Ronsheim & Moore Ltd., Div. of Hickson & Welch Ltd.

Wheldon Road, Castleford, West Yorkshire, WF10 2JT UK (Tel.: 44 1977 556565; Telefax: 44 1977-518058; Telex: 55378)

Ross Chemical, Inc.

303 Dale Dr., PO Box 458, Fountain Inn, SC 29644 USA (Tel.: 803-862-4474; 800-521-8246; Telefax: 803-862-2912)

Frank B. Ross Co., Inc.

22 Halladay St., PO Box 4085, Jersey City, NJ 07304-0085 USA (Tel.: 201-433-4512; Telefax: 201-332-3555)

Royce Associates, ALP

207 Ave. L, Newark, NJ 07105 USA (Tel.: 201-465-3932; Telefax: 201-279-8561, 820-907 PASDYE US)

R S A Corp.

36 Old Sherman Tpke., Danbury, CT 06810 USA (Tel.: 203-790-8100; Telefax: 203-790-1709; Telex: 131148)

RTD Chemical Corp.

1500 Rt. 517, Hackettstown, NJ 07840 USA (Tel.: 908-852-6128; Telefax: 908-852-1335)

Ruetgers-Nease Chemical Co., Inc., Subsid. of Rütgerswerke AG

201 Struble Rd., State College, PA 16801 USA (Tel.: 814-238-2424; Telefax: 814-238-1567)

Ruger Chemical Co. Inc.

85 Cordier St., Irvington, NJ 07111 USA (Tel.: 201-926-0331; 800-631-7844; Telefax: 201-926-4921)

Sachtleben Chemie

Sachtleben Chemie GmbH, Dr Rudolph-Sachtleben-Str. 4, 47198 Duisburg Germany (Tel.: 02066-22-2640; Telefax: 02066-22 2650; Telex: 855202 sc d)

Sachtleben Chemie GmbH, UK Sales Office, Huntingdon House, Princess St., Bolton, BL1 1EJ UK (Tel.: 44 1204 363634; Telefax: 44 1204 36 11 44)

Safas Corp.

2 Ackerman Ave., Clifton, NJ 07011 USA (Tel.: 201-772-5252; 800-472-6854; Telefax: 201-772-5858)

SAF Bulk Chemicals

PO Box 14508, St. Louis, MO 63178 USA (Tel.: 800-336-9719; Telefax: 800-368-4661)

SAFR Compounds Corp.

224 Swartswood Rd., Newton, NJ 07860 USA (Tel.: 201-383-7136; Telefax: 201-579-5220)

L.A. Salomon Inc.

150 River Rd., Suite A-4, Montville, NJ 07045 USA (Tel.: 201-335-8300; Telefax: 201-335-1236; Telex: 96-1470)

Sandoz

Sandoz, 608 5th Avenue, New York, NY 10020 USA (Tel.: 212-307 1122)

Sandoz Chemicals Corp., 4000 Monroe Rd., Charlotte, NC 28205 USA (Tel.: 704-331-7000; 800-631-8077; Telefax: 704-372-5787; Telex: 704-216-922)

Sandoz Chemicals (UK) Ltd, Div. of Sandoz AG, Calverley Lane, Horsforth, Leeds, West Yorkshire LS18 4RP UK (Tel.: 44 1132 584646; Telefax: 44 1132 591232; Telex: 557114)

Sandoz Huningue SA, Rte. de Bâle, F-68330 Huningue France (Tel.: 33-89-69 60 00; Telefax: 33-89-69 61 95; Telex: 881355 SANUSA F)

Sandoz K.K., Kobe Chamber of Commerce & Ind. Bldg., 6-1, Minatojima Nakamachi, Chuo-ku, Kobe, 650 Japan (Tel.: (078) 303-5850; Telex: 05624139 SANDOZ J)

San Esters Corp., Subsid. of Mitshbishi Rayon

342 Madison Ave., Suite 710, New York, NY 10173 USA (Tel.: 212-972-1122; 800-337-8377; Telefax: 212-972-4645)

Sanitized AG

Lyssachstrasse 95, Postfach 764, 3401 Burgdorf Switzerland (Tel.: (0041) 34 22 20 55; Telefax: (0041) 34 22 20 58)

Sanyo Chemical Industries, Ltd.

11-1, Ikkyo Nomoto-cho, Higashiyama-ku, Kyoto, 605 Japan (Tel.: (075) 541-4311; Telefax: (075) 551-2557; Telex: 05422110)

San Yuan Chemical Co., Ltd.

PO Box 26-1134, Taipei, Taiwan, R.O.C. (Tel.: 86286630343; Telefax: 86287770155)

Sartomer

Sartomer Co., Oaklands Corp. Center, 468 Thomas Jones Way, Exton, PA 19341 USA (Tel.: 215-363-4100; 800-345-8247; Telefax: 215-363-4140; Telex: 173071)

Societe Cray Valley, Cedex 101, 92970 Paris La Defense France (Tel.: (33) 1 4135 68 68; Telefax: (33) 1 4135 61 61; Telex: 616309, 615289)

Sartomer International, Inc., 331 North Bridge Rd., 22-02 Odeon Towers, Singapore 0718 (Tel.: 65-334-2645/334-2646; Telefax: 65-334-2647)

Säurefabrik Schweizerhall

CH-4133 Schweizerhalle Switzerland (Tel.: 41 61-8253111; Telefax: 41 61-8218027; Telex: 968016 acid ch)

Scheel Corp.

38 Franklin St., Brooklyn, NY 11222 USA (609-395-1100; Telefax: 609-395-5525)

Schenectady Chemicals, Inc.

PO Box 1046, Schenectady, NY 12301 USA (Tel.: 518-370-4200; Telefax: 518-346-3111; Telex: 145457)

Scher Chemicals, Inc.

Industrial West & Styertowne Rd., PO Box 4317, Clifton, NJ 07012 USA (Tel.: 201-471-1300; Telefax: 201-471-3783; Telex: 642643 Scherclif)

Schuller Mats & Reinforcements, Div. of Schuller Int'l.

PO Box 517, Toledo, OH 43697-0517 USA (Tel.: 419-878-8111; Telex: 5225243)

A. Schulman Inc.

3550 West Market St., PO Box 1710, Akron, OH 44313 USA (Tel.: 216-666-3751; Telefax: 216-666-6311)

Schumacher

1969 Palomar Oaks Way, Carlsbad, CA 92009 USA (Tel.: 619-931-9555; 800-545-9241; Telefax: 619-931-7819; Telex: 910 322 1382)

Schweizerhall Inc.

10 Corporate Place South, Piscataway, NJ 08854 USA (Tel.: 908-981-8200; 800-243-6564; Telefax: 908-981-8282; Telex: 4754581 SUSA)

Scientific Polymer Products, Inc.

6265 Dean Parkway, Ontario, NY 14519 USA (Tel.: 716-265-0413; Telefax: 716-265-1390)

SCM Chemicals

7 St. Paul St., Suite 1010, Baltimore, MD 21202 USA (Tel.: 410-783-1120; 800-638-3234; Telefax: 410-783-1087/9)

Scott Bader

Scott Bader Co. Ltd., Wollaston, Wellingborough, Northamptonshire, NN8 7RL UK (Tel.: 44 1933 663100; Telefax: 44 1933-663474; Telex: 31387 S BADER G)

Scott Bader SA, Div. of Scott Bader Co Ltd., 65 rue Sully, F-80044 Amiens France (Tel.: 33 22 44 42 00; Telefax: 33 22 44 06 24; Telex: 150425 SCOTBAD F)

Scott Specialty Gases

6141 Easton Rd., Plumsteadville, PA 18949 USA (Tel.: 215-766-8861; Telefax: 215-766-0320)

Sea-Land Chemical Co.

795 Sharon Dr., Westlake, OH 44145 USA (216-871-7887; Telefax: 216-871-7949)

Seal Sands Chemicals

Seal Sands Chemicals, Ltd., A Cambrex Co., Seal Sands Rd., Seal Sands, Middlesbrough, Cleveland, TS2 1UB UK (0642546546; Telefax: 0642546068)

Seal Sands Chemicals, Ltd., A Cambrex Co., 40 Ave. A., Bayonne, NJ 07002 USA (201-858-7900; Telefax: 201-858-0308)

Seefast (Europe) Ltd.

59 Lampton Rd., Hounslaw, TW3 4DH UK (081-5770033; Telefax: 081-5706376)

Seeler Industries Inc.

2000 N. Broadway, Joliet, IL 60435 USA (Tel.: 815-740-2640; 800-336-2422; Telefax: 815-740-6469)

Seimi Chemical Co., Ltd.

2-10, Chigasaki 3-chome, Chigasaki-shi, Kanagawa, 253 Japan (Tel.: (0467) 82-4131; Telefax: (0467) 86-2767)

Sekisui Plastics Co., Ltd.

Shinjuku Mitsui Bldg., 2-1-1 Nishi Shinjuku, Shinjuku-ku, Tokyo, 163-04 Japan (Tel.: 03-3347-9689; Telefax: 03-3344-2335; Telex: 02324755 SEKIPLJ)

Seppic

Seppic, Div. of L'Air Liquide, 75 Quai d'Orsay, F-75321 Paris Cedex 07 France (Tel.: 40 62 55 55; Telefax: 40 62 52 53; Telex: 202901 SEPPI F)

Seppic Inc., Subsid. of Seppic France, 30 Two Bridges Rd., Suite 370, Fairfield, NJ 07004 USA (Tel.: 201-882-5597; Telefax: 201-882-5178)

Servo Delden B.V.

Postbus 1, NL-7490 AA Delden The Netherlands (Tel.: 5407-63535; Telefax: 5407-64125; Telex: 44347)

Shell

Shell Chemical Co., One Shell Plaza, Room 1671, Houston, TX 77002 USA (Tel.: 713-241-6161; 800-872-7435; Telefax: 713-241-4043; Telex: 762248)

Shell Chemical Co., 4868 Blazer Memorial Pkwy., PO Box 1227, Dublin, OH 43017 USA (Tel.: 614-793-7700; Telefax: 614-793-7711)

Represented by Pecten Chemicals, Inc. (Int'l. sales), One Shell Plaza, Houston, TX 77252-9932 USA (Tel.: 713-241-6161; Telefax: 713-241-4044)

The Shepherd Chemical Co.

4900 Beech St., Cincinnati, OH 45212 USA (Tel.: 513-731-1110; Telefax: 513-731-1532, 21-4647)

The Shepherd Color Co.

4539 Dues Dr., PO Box 465627, Cincinnati, OH 45246 USA (Tel.: 513-874-0714; Telefax: 513-874-5061, 24-1659)

Sherwin-Williams Co.

PO Box 1028, Coffeyville, KS 67337 USA (Tel.: 316-251-7200)

Shin-Etsu Chemical Co., Ltd.

6-1, Otemachi 2-chome, Chiyoda-ku, Tokyo, 100 Japan (Tel.: 81 3 246 5280; Telefax: 81 3 246 5371; Telex: SHINCHEM J-24790)

Showa Denko K.K.

13-9, Shiba-Daimon 1-chome, Minato-ku, Tokyo, 105 Japan (Tel.: (03) 5470-3533; Telefax: (03) 3436-2625; Telex: J26232)

Shuman Plastics

35 Neoga St., Depew, NY 14043 USA (Tel.: 716-685-2121; Telefax: 716-685-3236; Telex: 91 9121)

Sigma-Aldrich. *See under* Aldrich

Sigri GmbH

Postfach 1160, Werner von Siemens Strasse 18, W-8901 Meitingen Germany (Tel.: 49-8271-830; Telefax: 49-8271-833127; Telex: 53823 SIGRI D)

Siltech

Siltech Inc., 4437 Park Dr., Suite E, Norcross, GA 30093 USA (Tel.: 404-279-8601; Telefax: 404-279-8535)

Siltech Corp., 45 Lesmill Rd., York, Ontario Canada (Tel.: 416-444-0381; Telefax: 416-444-0185)

Sino-Japan Chemical Co., Ltd.

14 Fl. 99, Sec. 2, Jen Ai Rd., Taipei, Taiwan, R.O.C. (Tel.: 886-2-2966223; Telefax: 886-2-3414628)

SKW Chemicals Inc.

1509 Johnson Ferry Rd., Marietta, GA 30062 USA (Tel.: 404-971-1317; Telefax: 404-971-4306)

Sloss Industries Corp.

PO Box 5327, Birmingham, AL 35207 USA (Tel.: 205-808-7909; Telefax: 205-808-7885)

Smith Lime Flour Co.

60-70 Central Ave., S. Kearny, NJ 07032 USA (Tel.: 201-344-1700; Telefax: 201-690-5936)

Werner G Smith, Inc.

1730 Train Ave., Cleveland, OH 44113 USA (Tel.: 216-861-3676; 800-535-8343; Telefax: 216-861-3680)

SNPE

SNPE Chimie, 12, quai Henri-IV, 75181 Paris Cedex 4 France (Tel.: 48 04 66 66; Telefax: 48 04 69 89; Telex: 220 380 SNPE F)

SNPE North America, 103 Carnegie Center Route One, Princeton, NJ 08540 USA (Tel.: 609-987-9424; Telefax: 609-987-2767)

Societe Cray Valley. *See under* Sartomer

Solem. *See under* J.M. Huber

Soluol Chemical Co.

Green Hill & Market Sts., PO Box 112, W. Warwick, RI 02893 USA (Tel.: 401-821-8100; Telefax: 401-823-6673)

Solvay

Solvay Deutschland GmbH, Postfach 220, W-3000 Hannover-1 Germany (Tel.: 511-857-0; Telefax: 511-282126; Telex: 922-755)

Solvay Duphar BV, Postbus 900, 1380 DA Weesp The Netherlands (Tel.: 31 2940-77000; Telefax: 31 2940-80253; Telex: 14232)

Solvay Polymers Inc., Subsid. of Solvay America Inc., 3333 Richmond Ave., PO Box 27328, Houston, TX 77227-7328 USA (Tel.: 713-525-4000; 800-231-6313; Telefax: 713-522-7890; Telex: 166307)

Southern Clay Prods.

1212 Church St., PO Box 44, Gonzales, TX 78629 USA (Tel.: 210-672-2891; 800-324-2891; Telefax: 210-672-3650)

Sovereign Chemicals Co.

2115 Lindbergh Ave., Cuyahoga Falls, OH 44223 USA (Tel.: 216-928-6642; Telefax: 216-928-0036)

Specialty Chem Products Corp.

Two Stanton St., Marinette, WI 54143 USA (Tel.: 715-735-9033; Telefax: 715-735-5304, 887445)

Specialty Minerals Inc., Subsid. of Minerals Technologies Inc.

640 N. 13th St., Easton, PA 18042 USA (Tel.: 610-250-3000; Telefax: 610-250-3344)

Specialty Products

Specialty Products Co, 75 Montgomery St., PO Box 306, Jersey City, NJ 07303-0306 USA (Tel.: 201-434-4700; 800-321-8506; Telefax: 201-434-6052)

Specialty Products Co., Ltd., 227 Norseman St., Toronto, Ontario, M8Z 2R5 Canada (Tel.: 416-239-6541; Telefax: 416-231-7264)

Spectrum Chemical Mfg. Corp.

14422 S. San Pedro St., Gardena, CA 90248 USA (Tel.: 213-516-8000; 800-772-8786; Telefax: 213-516-9843; Telex: 182395)

Spice King Corp.

6009 Washington Blvd., Culver City, CA 90232-7488 USA (Tel.: 213-836-7770; Telefax: 213-836-6454; Telex: 664350)

Sprayon Products Group/The Specialty Div., Div. of The Sherwin-Williams Co.

31500 Solon Rd., Solon, OH 44139 USA (Tel.: 216-498-2400; 800-777-2966)

S & S Chemical Co., Inc.

333 Jericho Tpke., Jericho, NY 11753 USA (516-931-3333; Telefax: 516-931-2387)

StanChem, Inc.

401 Berlin St., East Berlin, CT 06023 USA (Tel.: 203-828-0571; Telefax: 203-828-3297)

Stan Chem International Ltd.

4 Kings Rd, Reading, Berkshire, RG1 3AA UK (Tel.: 44 1734 580247; Telefax: 44 1734 589580; Telex: 847746)

Stepan

Stepan Co, 22 West Frontage Rd., Northfield, IL 60093 USA (Tel.: 708-446-7500; 800-745-7837; Telefax: 708-501-2443; Telex: 910-992-1437)

Stepan/PVO, 100 West Hunter Ave., Maywood, NJ 07607 USA (Tel.: 201-845-3030; Telefax: 201-845-6754; Telex: 710-990-5170)

Stepan Canada, 90 Matheson Blvd. W., Suite 201, Mississauga, Ontario, L5R 3P3 Canada (Tel.: 416-507-1631; Telefax: 416-507-1633)

Stepan Europe, BP127, 38340 Voreppe France (Tel.: 33-76-50-51-00; Telefax: 33-7656-7165; Telex: 320511 F)

Stevenson Cooper

PO Box 38349, 1039 West Venango St., Philadelphia, PA 19140 USA (Tel.: 215-223-2600; Telefax: 215-223-3597)

Stoner Inc.

1070 Robert Fulton Hwy., PO Box 65, Quarryville, PA 17566 USA (Tel.: 717-786-7355; 800-227-5538; Telefax: 800-515-5150)

Joseph Storey & Co Ltd.

Heron Chemical Works, Moor Lane, Lancaster, LA1 1QQ UK (Tel.: 44 1524 63252; Telefax: 44 1524 381805; Telex: 669755)

Strahl & Pitsch, Inc.

PO Box 1098, 230 Great E. Neck Rd., W. Babylon, NY 11704 USA (Tel.: 516-587-9000; Telefax: 516-587-9120; Telex: 221636 STRALUR)

Struktol Co.

201 E. Steels Corner Rd., PO Box 1649, Stow, OH 44224-0649 USA (Tel.: 216-928-5188; 800-327-8649; Telefax: 216-928-8726)

Surco Products, Inc.

PO Box 777, Eighth & Pine Aves., Braddock, PA 15104 USA (Tel.: 412-351-7700; 800-556-0111; Telefax: 412-351-7701)

Surfactants, Inc.

260 Ryan St., S. Plainfield, NJ 07080 USA (Tel.: 908-755-3300; Telefax: 755-0592)

SVO Enterprises, Business Unit of The Lubrizol Corp.

35585-B Curtis Blvd., Eastlake, OH 44095 USA (Tel.: 216-975-2802; 800-292-4786; Telefax: 216-942-1045; Telex: 4938879 AGCEAKE)

Sybron

Sybron Chemicals Inc., PO Box 125, Hwy. 29, Wellford, SC 29385 USA (Tel.: 803-439-6333; 800-677-3500; Telefax: 803-439-1612)

Sybron Chemie Nederland BV, Postbus 46, NL-6710 BA Ede The Netherlands (Tel.: 31 8380 70911; Telefax: 31 8380 30236; Telex: 37249)

Synthetic Chemicals Ltd., Div. of Shell Chemicals UK Ltd.

Four Ashes, Wolverhampton, West Midlands, WV10 7BP UK (Tel.: 44 1902 794000; Telefax: 44 1902 794300; Telex: 337306)

Synthetic Products Co., Subsid. of Cookson America Inc.

1000 Wayside Rd., Cleveland, OH 44110 USA (Tel.: 216-531-6010; 800-321-4236; Telefax: 216-486-6638)

Synthron Inc.

305 Amherst Rd., Morganton, NC 28655 USA (Tel.: 704-437-8611; Telefax: 704-437-4126)

Taiwan Surfactant Corp.

No. 106, 8-1 Floor, Sec. 2, Chung An E. Rd., Taipei, Taiwan, R.O.C. (Tel.: 886-2-507-9155; Telefax: 886-2-507-7011; Telex: 27568 surfact)

Takemoto Oil & Fat Co., Ltd.

2-5, Minato-Machi, Gamagori-shi, Aichi, 443 Japan (Tel.: (0533) 68-2111; Telefax: (0533) 68-1035; Telex: 4324604)

Technical Chemicals & Products, Inc.

3341 SW 15th St., Pompano Beach, FL 33069 USA (Tel.: 305-979-0400; Telefax: 305-979-0009)

Tego. *See under* Goldschmidt

Teknor

Teknor Apex Co., 505 Central Ave., Pawtucket, RI 02861 USA (Tel.: 401-725-8000; 800-554-9893; Telefax: 401-724-6250; Telex: 927530)

Teknor Color Co., 505 Central Ave., Pawtucket, RI 02861 USA (Tel.: 401-725-8000; Telefax: 401-724-8520)

Telechem International, Inc.

12 South First St., Suite 817, San Jose, CA 95113 USA (Tel.: 408-977-0160; Telefax: 408-977-0164)

Telechemische Inc.

222 Dupont Ave., #14, Newburgh, NY 12550 USA (Tel.: 914-561-3237; Telefax: 914-561-3622; Telex: 62954502 WU)

Texaco

Texaco Chemical Co, PO Box 27707, Houston, TX 77227-7707 USA (Tel.: 713-961-3711; 800-231-3107; Telefax: 713-235-6437; Telex: 227-031 TEX UR)

Texaco Ltd., Div. of Texaco Chemical Europe, 195 Knightsbridge, London, SW7 1RU UK (Tel.: 44 171 581 5500; Telefax: 44 171-581 9163; Telex: 8956681 TEXACO G)

Texaco Olie Matschappij BV, Weena 170, NL-3012 CR Rotterdam The Netherlands (Tel.: 31-614471; Telex: 31542)

Texas Petrochemicals Corp.

8707 Katy Frwy., Suite 300, Houston, TX 77024 USA (Tel.: 713-461-3322; Telefax: 713-461-1029, TWX 510-891-7831)

Textile Rubber & Chem. Co./Tiarco Chem. Div. *See* Tiarco

Thibaut & Walker Co.

PO Box 296, 49 Rutherford St., Newark, NJ 07101 USA (Tel.: 201-589-3331; Telefax: 201-589-7231)

Thiele Kaolin Co.

Box 1056, Sandersville, GA 31082 USA (Tel.: 912-552-3951; Telefax: 912-552-4131)

3M

3M Co/Industrial Chem. Prods. Div., 3M Center Bldg. 223-6S-04, St. Paul, MN 55144-1000 USA (Tel.: 612-736-1394; 800-541-6752)

3M Co/Performance Polymers and Additives, 3M Center Bldg., 236-GC-01, St. Paul, MN 55144-1000 USA (Tel.: 612-733-3158)

3M Co/Industrial Specialties, St. Paul, MN 55144-1000 USA

3M Co/Specialty Fluoropolymers Dept., 3M Center Bldg. 220-10E-10, St. Paul, MN 55144-1000 USA (Tel.: 612-733-6221)

3M Canada Inc., PO Box 5757 Terminal A, 1840 Oxford St. East, London, Ontario, N6A 4T1 Canada (Tel.: 519-451-2500)

3M Europe/Performance Polymers and Additives, PO Box 100422, D-4040 Neuss 1 Germany (Tel.: 2101-142414)

Sumitomo/3M Ltd., Central PO Box 490, 33-1, Tamagawadai 2-chome, Setagaya-ku, Tokyo, 158 Japan (Tel.: (03) 3709-8255; Telex: J2468534)

3V Inc.

1500 Harbor Blvd., Weehawken, NJ 07087 USA (Tel.: 201-865-3600; 800-441-5156; Telefax: 201-865-1892)

9140 Arrowpoint Blvd., Suite 120, Charlotte, NC 28273-8120 USA (Tel.: 704-523-5252; Telefax: 704-522-1763)

Tiarco Chemical Div./Textile Rubber & Chemical Co.

1300 Tiarco Dr., Dalton, GA 30720 USA (Tel.: 706-277-1300; Telefax: 706-277-9039)

Tioga Coatings Corp.

208 Quaker Rd., Rockford, IL 61104-7088 USA (Tel.: 815-962-4200; Telefax: 815-962-1712)

Toho Chemical Industry Co., Ltd.

No. 1-1-5, Shintomi, Nihonbashi, Chuo-ku, Tokyo, 104 Japan (Tel.: (81-3) 3555 3731; Telefax: (81-3) 3555 3755; Telex: 252-2332 TOHO K J)

Tosoh

Tosoh Corp., 7-7 Akasaka 1-chome, Minato-ku, Tokyo, 107 Japan (Tel.: (03) 3585-9891; Telefax: (03) 3582-8120; Telex: J24475tosoh)

Tosoh USA Inc., 1100 Circle 75 Pkwy., Suite 600, Atlanta, GA 30339 USA (Tel.: 404-956-1100; Telefax: 404-956-7368; Telex: 542272 tosoh atl)

Total Petroleum Inc.

East Superior St., Alma, MI 48801 USA (Tel.: 517-463-9630; 800-292-9033; Telefax: 517-463-9623)

Total/Resins Div.

16, rue de la République, F92800 Puteaux France

Tra-Con, Inc.

55 North St., PO Box 306, Medford, MA 02155 USA (Tel.: 617-391-5550; 800-872-2661; Telefax: 617-391-7380)

Tradig

PO Box 601, Bethlehem, PA 18016 USA (Tel.: 610-432-9466; Telefax: 610-433-1416, 5106512132)

T&R Chemicals, Inc.

700 Celum Rd., Box 330, Clint, TX 79836 USA (Tel.: 915-851-2761)

Tricon Trading Associates

2487 Kaladar Ave., Suite 2117, Ottawa, Ontario, K1V 8B9 Canada (Tel.: 613-247-0944; Telefax: 613-247-0945)

Tri-K Industries, Inc.

27 Bland St., PO Box 312, Emerson, NJ 07630 USA (Tel.: 201-261-2800; 800-526-0372; Telefax: 201-261-1432; Telex: 215085 TRIK UR)

Triple-S Chemical Co.

1413 Mirasol St., Los Angeles, CA 90023 USA (Tel.: 213-261-7301; Telefax: 213-261-5567)

TSE Industries, Inc.

PO Box 17225, Clearwater, FL 34622 USA (Tel.: 813-573-7676; 800-237-7634; Telefax: 813-572-0415)

Tulco, Inc.

9 Bishop Rd., Ayer, MA 01432 USA (Tel.: 508-772-4412; Telefax: 508-772-1751)

Ucar Carbon Co., Inc.

39 Old Ridgebury Rd.-J4, Danbury, CT 06817 USA (Tel.: 203-794-3684; 800-342-3698; Telefax: 203-794-3180, 126019 UCCHQDURY)

UCB

UCB SA/Chemical Sector, 33 rue d'Anderlecht, B-1620 Drogenbos Belgium (Tel.: 322-3714923; Telefax: 322-3714924; Telex: 22342 UCBOS B)

UCB SA/Chemical Sector, 326 avenue Louise, B-1050 Brussels Belgium (Tel.: 32-641-1411; Telefax: 32-647 9342; Telex: 63769 UCBFTA B)

UCB (Chem) Ltd., Div. of UCB SA, Star House, 69 Clarendon Road, Watford, Hertfordshire, WD1 1DJ UK (Tel.: 44 1923 248011; Telefax: 44 1923-250225; Telex: 23958)

UCB Radcure Inc., Wholly owned subsid. of UCB Group, 2000 Lake Park Dr., Smyrna, GA 30080 USA (Tel.: 800-433-2873; Telefax: 404-319-8228)

UCB Radcure Inc., Wholly owned subsid. of UCB Group, 9800 E. Bluegrass Pkwy., Louisville, KY 40299 USA (Tel.: 502-499-4728; 800-433-2873)

Unger Fabrikker AS

PO Boks 254, N-1601 Fredrikstad Norway (Tel.: 47 9693 20020; Telefax: 47 9693 23775; Telex: 76382 UNGER N)

Unichema

Unichema International, Postbus 2, 2800 Gouda The Netherlands (Tel.: 31-0-1820-42911; Telefax: 31-0-1820-42250)

Unichema Chemie BV, Postbus 2, 2800 AA Gouda The Netherlands (Tel.: 31 (0) 1820-42911; Telefax: 31 (0) 1820-42250; Telex: 20661)

Unichema Chemicals Ltd., Div. of Unichema International, Bebington, Wirral, Merseyside, L62 4UF UK (Tel.: 44 151 645-2020; Telefax: 44 151-645-9197; Telex: 629408)

Unichema Chemie GmbH, Postfach 100963, D-46429 Emmerich Germany (Tel.: 49 0 2822-720; Telefax: 49 0 2822-72276; Telex: 8125113)

Unichema France SA, 148 Boulevard Haussemann, 75008 Paris France (Tel.: 33 1 44 95 08 40; Telefax: 33 1 42563188; Telex: 643217)

Unichema North America, 4650 S. Racine Ave., Chicago, IL 60609 USA (Tel.: 312-376-9000; 800-833-2864; Telefax: 312-376-0095; Telex: 176068)

Unichema Japan, Sankei Bldg. 7F 708, 4-9, Umeda 2-chome, Kita-ku, Osaka, 530 Japan (Tel.: 81 6341-7221; Telefax: 81 6341-7725)

Unimin

Unimin Corp., 258 Elm St., New Canaan, CT 06840 USA (Tel.: 203-966-8880; 800-243-9004; Telefax: 203-966-3453, 99-6355)

Unimin Specialty Minerals Inc., Subsid. of Unimin Corp., PO Box 33, Rt. 127, Elco, IL 62929 USA (Tel.: 618-747-2311; 800-743-7519; Telefax: 618-747-9318)

Unimin Canada Ltd., RR #4, PO Box 2000, Havelock, Ontario, K0L 1Z0 Canada (Tel.: 705-877-2210; 800-363-4140; Telefax: 705-877-3343)

Union Camp

Union Camp Corp./Chem. Prods. Div., 1600 Valley Rd., Wayne, NJ 07470 USA (Tel.: 201-628-2375; 800-628-9220; Telefax: 201-628-2840; Telex: 130735)

Union Camp Corp./Chem. Prods. Div., PO Box 60369, 570 Ellis Rd., Jacksonville, FL 32236 USA (Tel.: 904-783-2180; 800-874-9220; Telefax: 904-786-7313)

Union Camp Corp./Chem. Prods. Div., PO Box 2668, Savannah, GA 31402 USA (Tel.: 912-238-6000)

Union Camp Chemicals Ltd., Vigo Lane, Chester-le-Street, Co. Durham, DH3 2RB UK (Tel.: 44-91-410-2631; Telefax: 44-91-410-9391; Telex: 851 53163)

Union Camp Chemicals, Am Burgholz 17, D-5166 Kreuzau-Stockheim Germany (Tel.: 49-2421-59260; Telefax: 49-2421-592635)

Union Camp Chemicals, 11/F, Mappin House, 98, Texaco Rd., Tsuen Wan, N.T. Hong Kong (Tel.: 852-408-7170; Telefax: 852-407-6067; Telex: 780-57027 BBAHBHX)

Union Carbide

Union Carbide Corp., 39 Old Ridgebury Rd., Danbury, CT 06817-0001 USA (Tel.: 203-794-2000; 800-568-4000; Telefax: 203-794-3133)

Union Carbide (UK) Ltd./Chemicals & Plastics, 93-95 High Street, Rickmansworth, Hertfordshire, WD3 1RB UK (Tel.: 44 1923 720 366; Telefax: 44 1923-896721)

Union Carbide Chemicals & Plastics Europe S.A., 15 Chemin Louis-Dunant, CH-1211 Geneve 20 Switzerland (Tel.: 41-22-739-6111; Telefax: 41-22-739-6545; Telex: 419207 UNC CH)

Union Derivan SA

Av. Meridiana 133, E-08026 Barcelona Spain (Tel.: 343-2322113; Telefax: 343-2323951; Telex: 98204 UNDER E)

Uniroyal

Uniroyal Chemical Co Inc./World Headquarters, Benson Road, Middlebury, CT 06749 USA (Tel.: 203-573-2000; 800-243-3024; Telefax: 203-573-2489; Telex: 6710383 uniroyal)

Uniroyal Chemical Co Inc./Specialty Chemicals, USA (

Uniroyal Chimica, SpA, Via delle Industrie 40, I-104013 Latina-Scalo Italy (Tel.: 39-773-43605; Telex: 680056)

United Chemical Technologies, Inc.

2731 Bartram Rd., Bristol, PA 19007 USA (215-781-9255; 800-541-0559; Telefax: 215-785-1226)

Unitex Chemical Corp.

PO Box 16344, 520 Broome Rd., Greenboro, NC 27406 USA (Tel.: 910-378-0965; Telefax: 919-272-4312)

UOP

UOP, 25 E. Algonquin Rd., Des Plaines, IL 60017-5017 USA (Tel.: 708-391-2395; 800-348-0832, 25-3285)

UOP Food Antioxidants Dept., 25 E. Algonquin Rd., Box 5017, PO Box 5017, Des Plaines, IL 60017-5017 USA (Tel.: 708-391-2425; 800-348-0832; Telefax: 708-391-3804; Telex: 211442)

UOP Ltd., Liongate, Ladymeade, Guildford, Surrey, GU1 1AT UK (Tel.: 44 1483 304848; Telefax: 44 1483-304863; Telex: 858051 UOPINT G)

UOP GmbH, Steinhof 39, W-4006 Erkrath Germany (Tel.: 211-24903-23; Telefax: 211-249109)

U.S. Biochemical Corp.

PO Box 22400, Cleveland, OH 44122 USA (Tel.: 216-765-5000; 800-321-9322; Telefax: 216-464-5075; Telex: 980718)

U.S. Borax Inc.

26877 Tourney Rd., Valencia, CA 91355-1847 USA (Tel.: 805-287-5400; 800-729-2672; Telefax: 805-287-5455; Telex: 371-6120)

U.S. Chemicals, Inc.

280 Elm St., New Canaan, CT 06840 USA (Tel.: 203-966-8777; Telefax: 203-966-3577)

U.S. Cosmetics

313 Lake Rd., PO Box 859, Dayville, CT 06241 USA (Tel.: 203-779-3990; 800-752-0490; Telefax: 203-779-3994)

U.S. Gypsum Co

125 S. Franklin St., Chicago, IL 60606 USA (Tel.: 312-606-4018)

U.S. Petrochemical Industries, Inc.

675 Galleria Financial Ctr., 5075 Westheimer Rd., Houston, TX 77056 USA (Tel.: 713-871-1951; Telefax: 713-871-1963, 402814)

U.S. Polymers Inc.

300 E. Primm St., St. Louis, MO 63111 USA (Tel.: 314-638-1632; Telefax: 314-638-3100)

U.S. Silica Co.

PO Box 187, Rt. 522 North, Berkeley Springs, WV 25411 USA (Tel.: 304-258-2500; 800-243-7500; Telefax: 304-258-3500; Telex: 4942414)

U.S. Synthetics Co.

PO Box 2236, Danbury, CT 06813 USA (Tel.: 203-270-0187; Telefax: 203-790-6407, 4972481 mbe tam)

U.S. Zinc Corp.

6020 Esperson St., PO Box 611, Houston, TX 77001 USA (Tel.: 713-926-1705; Telefax: 713-924-4824, 3785919)

Van Den Bergh Foods Co.

2200 Cabot Dr., Lisle, IL 60532 USA (Tel.: 708-505-5300; 800-949-7344; Telefax: 708-955-5497)

R.T. Vanderbilt Co Inc.

30 Winfield St, PO Box 5150, Norwalk, CT 06856 USA (Tel.: 203-853-1400; 800-243-6064; Telefax: 203-853-1452; Telex: 6813581 RTVAN)

Van Waters & Rogers Inc., Subsid. of Univar Corp

6100 Carillon Point, Kirkland, WA 98033 USA (Tel.: 206-889-3400; 800-234-4588; Telefax: 206-889-4133)

Veckridge Chemical Co. Inc.

60-70 Central Ave., Kearny, NJ 07032 USA (Tel.: 201-344-1818; Telefax: 201-690-5936)

Veitsiluoto Oy/Forest Chemicals Industry

PO Box 196, SF-90101 Oulu 10 Finland (Tel.: 358-81-316 3111; Telefax: 358-81-378 5755; Telex: 32125 oulpk sf)

Velsicol

Velsicol Chemical Corp, 10400 W Higgins Road, Rosemont, IL 60018 USA (Tel.: 708-298-9000; 800-843-7759; Telefax: 708-298-9014; Telex: 3730755)

Velsicol Chemical Ltd., Worting House, Basingstoke, Hampshire, RG23 8PY UK (Tel.: 44 1256 817640; Telefax: 44 1256 817744; Telex: 9312131051 VC G)

Verdugt BV

Postbus 60, Papesteeg 91, NL-4000 AB Tiel The Netherlands (Tel.: 31-3440-15224; Telefax: 31-3440-11475; Telex: 47200)

Vikon Chemical Co., Inc.

930 Oakley St., Graham, NC 27253 USA (Tel.: 919-226-6331; Telefax: 919-222-9568)

Vista

Vista Chemical Co., 900 Threadneedle, PO Box 19029, Houston, TX 77224-9029 USA (Tel.: 713-588-3000; 800-231-8216; Telefax: 713-588-3236; Telex: 794557)

Vista Chemical Europe, Hilton Tower, Blvd. de Waterloo #39, B-1000 Brussels Belgium (Tel.: 32-2-513-7490; Telefax: 32-2-513 54 49; Telex: 24727 VISTA B)

Vulcan Chemicals, Div. of Vulcan Materials Co.

One Metroplex Dr., Birmingham, AL 35209 USA (Tel.: 205-877-3000; 800-633-8280; Telefax: 205-877-3448, 59-6108)

Wacker

Wacker-Chemie GmbH, Div. S, Hanns-Seidel-Platz 4, D-81737 München Germany (Tel.: (089) 62 79 01; Telefax: (089) 62791771; Telex: 5291210)

Wacker Chemicals Ltd., Div. of Wacker-Chemie GmbH, The Clock Tower, Mount Felix, Bridge Street, Walton-on-Thames, Surrey, KT12 1AS UK (Tel.: 44 1932 246111; Telefax: 44 1932-240141; Telex: 28 391 wacker g)

Wacker Chemie Danmark A/S, Hovedvejen 91, DK-2600 Glostrop Denmark (Tel.: (43) 43 03 00; Telefax: (43) 43 03 16)

Wacker Quimica Ibérica SA, Div. of Wacker-Chemie GmbH, Corcega, 303-2° 3a, E-08008 Barcelona Spain (Tel.: (93)-217 59 00; Telefax: (93)-217 57 66; Telex: 97801 wqsa e)

Wacker Chemicals (USA) Inc., 535 Connecticut Ave., Norwalk, CT 06854 USA (Tel.: 203-866-9400; Telefax: 203-866-9427; Telex: 643 444)

Wacker Silicones Corp., Subsid. of Wacker-Chemie, 3301 Sutton Rd., Adrian, MI 49221-9397 USA (Tel.: 517-264-8500; 800-248-0063; Telefax: 517-264-8246; Telex: 510-450-2700 sadrnud)

Wacker Chemicals East Asia Ltd., Thoma Nishi-Waseda Bldg., 2-14-1, Nishi Waseda, Shinjuku-ku, Tokyo, 169 Japan (Tel.: (03)52 72 31 21; Telefax: (03) 52 72 31 40)

Wako

Wako Pure Chemical Industries Ltd., 1,2-Doshomachi 3-Chome, Chuo-ku, Osaka, 541 Japan (Tel.: (06) 203-3741; Telefax: (06) 222-1203; Telex: 65188 wakoos j)

Wako Chemicals USA, Inc., 1600 Bellwood Rd., Richmond, VA 23237 USA (Tel.: 804-271-7677; Telefax: 804-271-7791; Telex: 293208 wako ur(rca)

Wako Chemicals GmbH, Div. of Wako Pure Chemical Industries Ltd., Nissanstrasse 2, 41468 Neuss 1 Germany (Tel.: 02131-3110; Telefax: 02131-31-1100; Telex: 8517001 wako d)

Warenhandels GmbH, Member of the EIA Sales Group

Auhofstrasse 1/DG, A-1130 Wien Austria (Tel.: 877 05 53; Telefax: 877 84 46; Telex: 115 280)

Welch, Holme & Clark Co. Inc.

7 Ave. L, Newark, NJ 07105 USA (Tel.: 201-465-1200; Telefax: 201-465-7332)

Wesco Technologies Ltd.

PO Box 3880, San Clemente, CA 92674-3880 USA (Tel.: 714-661-1142; 800-223-3878 (CA); Telefax: 714-492-6025; Telex: GRT 3718658)

Westbrook Lanolin Co.

Argonaut Works, Laisterdyke, Bradford, West Yorkshire, BD4 8AU UK (Tel.: 44 1274 663331; Telefax: 44 1274-667665; Telex: 51502)

Westco Chemicals, Inc.

11312 Hartland St., North Hollywood, CA 91605 USA (Tel.: 213-877-0077; Telefax: 818-766-7170, 673150)

Westlake Plastics Co.

PO Box 127, W. Lenni Rd., Lenni, PA 19052 USA (Tel.: 215-459-1000; Telefax: 215-459-1084; Telex: 83-5406)

Westo Industrial Products Ltd.

31 Pembridge Rd, London, W11 3HG UK (Tel.: 44 171 727 8700; Telefax: 44 171-792 0329; Telex: 888941 LCCI G)

Westvaco Corp., Chemical Div.

PO Box 70848, Charleston Hts., SC 29415-0848 USA (Tel.: 803-740-2300; Telefax: 803-740-2329; Telex: 4611159)

Whitford Corp.

PO Box 2347, West Chester, PA 19381 USA (Tel.: 215-296-3200; Telefax: 215-647-4849; Telex: 83-5305)

Whittaker, Clark & Daniels

1000 Coolidge St., South Plainfield, NJ 07080 USA (Tel.: 908-561-6100; 800-833-8139; Telefax: 908-757-3488; Telex: 221478)

Wilbur-Ellis Co.

PO Box 1286, Fresno, CA 93715 USA (Tel.: 209-442-1220; Telefax: 209-442-4089)

Wilke International Inc.

1375 N. Winchester, Olathe, KS 66061 USA (Tel.: 913-78-5544; 800-779-5545; Telefax: 913-780-5574)

Witco

Witco Corp/Oleochemicals/Surfactants Group, One American Lane, Greenwich, CT 06831-2559 USA (Tel.: 203-552-3382; 800-494-8287; Telefax: 203-552-2893)

Witco Corp., PO Box 125, Memphis, TN 38101-0125 USA (Tel.: 901-320-5800; 800-238-9150; Telefax: 901-682-6531; Telex: 53-298)

Witco Corp/Allied-Kelite, 2701 Lake St., Melrose Park, IL 60160-3041 USA (Tel.: 800-323-9784)

Witco Corp/Concarb, 10500 Richmond, Suite 116, PO Box 42817, Houston, TX 77242-2817 USA (Tel.: 713-978-5745; 800-231-4591; Telefax: 713-978-5728)

Witco Corp/Household, Industrial, Personal Care, One American Lane, Greenwich, CT 06831-2559 USA (Tel.: 800-494-8673; Telefax: 203-552-2878)

Witco Corp/Lubricants Group Golden Bear Prods., 10100 Santa Monica Blvd., Suite 1470, Los Angeles, CA 90067-4183 USA (Tel.: 310-277-4511; Telefax: 310-201-0383)

Witco Corp/New Markets, One American Lane, Greenwich, CT 06831-2559 USA (Tel.: 800-494-8673; Telefax: 203-552-2878)

Witco Corp/Petroleum Specialties Group, One American Lane, Greenwich, CT 06831-2559 USA (Tel.: 203-552-2000; 800-494-8673; Telefax: 203-552-2878)

Witco Corp/Polymer Additives Group, One American Lane, Greenwich, CT 06831-2559 USA (Tel.: 203-552-3294; 800-494-8737; Telefax: 203-552-2010)

Witco Corp/Polymer Additives Group (PAG), Bussey Rd., PO Box 1439, Marshall, TX 75671-1439 USA (Tel.: 903-938-5141; 800-431-1413; Telefax: 903-938-2647)

Witco BV, 1 Canalside, Lowesmoor Wharf, Worcester, Worcestershire, WR1 2RS UK (Tel.: 44 1905 21521; Telefax: 44 1905-611593)

Witco Corp/Petroleum Specialties Group—Europe, PO Box 5, 1540 AA Koog aan de Zaan The Netherlands (Tel.: 31 175 283854; Telefax: 31 75 210811)

Rewo Chemische Werke GmbH, Postfach 1160, 36392 Steinau an der Strasse, Max Wolf Strasse 7, Industriegebiet West, W-6497 Steinau Germany (Tel.: 06663-540; Telefax: 06663-54-129; Telex: 493589)

Witco GmbH/Polymer Chemicals & Resins, Ernst Scheringstrasse 14, D-59192 Bergkamen Germany (FAX 49-23-0765-2403)

Witco Corp/Petroleum Specialties Group—Asia, 396 Alexandra Rd., #06-04 BP Tower, Singapore 0511 (Tel.: 65 274 3878; Telefax: 65 274 8693)

Wright Corp.

PO Box 9009, 102 Orange St., Wilmington, NC 28402 USA (910-251-8952; Telefax: 910-762-9223)

Yorkshire Chemicals plc

Kirkstall Rd, Leeds, West Yorkshire, LS3 1LL UK (Tel.: 44 1132 443111; Telefax: 44 1132 421670; Telex: 55366)

Zaclon Inc.

2981 Independence Rd., Cleveland, OH 44115 USA (216-271-1717; 800-356-7327; Telefax: 216-271-1911)

Zeeland Chemicals, Inc., A Cambrex Co.

215 N. Centennial St., Zeeland, MI 49464 USA (Tel.: 616-772-2193; 800-223-0453; Telefax: 616-772-7344; Telex: 226375)

Zeelan Industries Inc., Subsid. of 3M

3M Center, Bldg. 220-8E-04, St. Paul, MN 55144-1000 USA (Tel.: 617-737-1751; Telefax: 612-737-1764)

Zeneca

Zeneca Specialties, A Business Unit of Zeneca Inc., PO Box 751, Wilmington, DE 19897 USA

Zeneca Resins, 730 Main St., Wilmington, MA 01887 USA (508-658-6600; 800-225-0947; Telefax: 508-657-7978)

Zeon

Zeon Chemicals, Inc., Three Continental Towers, Suite 1012, 1701 Golf Rd., Rolling Meadows, IL 60008 USA (Tel.: 312-437-9770; 800-735-3388; Telefax: 312-437-9773)

Zeon Europe GmbH, Am Seester 18 (Euro Center), D40547 Dusseldorf 11 Germany (Tel.: 49 211 52670; Telefax: 49 211 5267 160)

Zeus Industrial Products Inc.

PO Box 2167, Orangeburg, SC 29116 USA (Tel.: 803-531-2174; 800-526-3842; Telefax: 803-533-5694)

Zinc Corp. of America

300 Frankfort Rd., Monaca, PA 15061 USA (Tel.: 412-774-1020; 800-962-7500; Telefax: 412-774-4348)

Zip-Chem® Products, Inc., A Div. of Andpak, Inc.

1860 Dobbins Dr., San Jose, CA 95133 USA (Tel.: 408-729-0291; 800-648-2661; Telefax: 408-272-8062)

Zircar Products Inc.

110 N. Main St., PO Box 458, Florida, NY 10921 USA (Tel.: 914-651-4481; Telefax: 914-651-3192; Telex: 996608)

Zohar Detergent Factory

PO Box 11 300 (Tel-Aviv, 61 112 Israel (Tel.: 03-528-7236; Telefax: 03-5287239; Telex: 33557 zohar il)

Zyvax, Inc.

PO Box 825, Boca Raton, FL 33429 USA (Tel.: 407-395-4405; 800-858-4111; Telefax: 407-395-5262)

Appendices

CAS Number-to-Trade Name Cross-Reference

CAS	Trade name
50-36-2	Lonza Coke PC 40
56-35-9	bioMeT TBTO
56-35-9	Keycide® X-10
56-35-9	Vikol® PX-15
57-11-4	Glycon® S-70
57-11-4	Glycon® S-90
57-11-4	Glycon® TP
57-11-4	Hydrofol 1800
57-11-4	Hystrene® 5016 NF
57-11-4	Hystrene® 9718 NF
57-11-4	Industrene® 1224
57-11-4	Industrene® 4518
57-11-4	Industrene® 5016
57-11-4	Industrene® 7018
57-11-4	Industrene® 9018
57-11-4	Industrene® B
57-11-4	Industrene® R
57-11-4	Loxiol® G 20
57-11-4	Naftozin® N, Spezial
57-11-4	Petrac® 250
57-11-4	Petrac® 270
57-11-4	Pristerene 4900
57-13-6	RIA CS
57-13-6	Stanclere® C26
57-13-6	WF-15
57-13-6	WF-16
57-55-6	Adeka Propylene Glycol (T)
58-36-6	Vinyzene® BP-5-2
60-00-4	Versene Acid
62-33-9	Versene CA
64-02-8	Hamp-Ene® 100
64-02-8	Sequestrene® 30A
64-02-8	Versene 100
64-02-8	Versene 100 EP
64-02-8	Versene 100 LS
64-02-8	Versene 100 SRG
64-02-8	Versene 100 XL
64-02-8	Versene 220
65-85-0	Retarder BA, BAX
66-71-7	Activ-8
67-68-5	Decap
70-55-3	Uniplex 173
71-36-3	Isanol®
71-55-6	Dabco® CS90
74-31-7	Agerite® DPPD
74-31-7	Naugard® J
74-31-7	Permanax™ DPPD
75-65-0	Tebol™
75-65-0	Tebol™ 99
75-91-2	Aztec® TBHP-22-OL1
75-91-2	Aztec® TBHP-70
75-91-2	T-Hydro®
75-91-2	Trigonox® A-80
75-91-2	USP®-800
75-94-5	Dynasylan® VTC
75-94-5	Petrarch® V4900
77-58-7	ADK STAB BT-11
77-58-7	ADK STAB BT-18
77-58-7	ADK STAB BT-23
77-58-7	ADK STAB BT-25
77-58-7	ADK STAB BT-27
77-58-7	Cata-Chek® 820
77-58-7	Dabco® T-12
77-58-7	Fomrez® SUL-4
77-58-7	Jeffcat T-12
77-58-7	Kosmos® 19
77-58-7	Metacure® T-12
77-58-7	Synpron 1009
77-58-7	TEDA-T411
77-85-0	Trimet® Liquid
77-85-0	Trimet® Nitration Grade
77-85-0	Trimet® Pure
77-85-0	Trimet® Tech
77-89-4	Citroflex® A-2
77-89-4	Uniplex 82
77-90-7	Citroflex® A-4
77-90-7	Citroflex® A-4 Special
77-90-7	Estaflex® ATC
77-90-7	Uniplex 84
77-93-0	Citroflex® 2
77-93-0	Uniplex 80
77-94-1	Citroflex® 4
77-94-1	TBC
78-08-0	Dynasylan® VTEO
78-08-0	Petrarch® V4910
78-10-4	Petrarch® T1807
78-21-7	Barquat® CME-35
78-40-0	Eastman® TEP
78-42-2	Amgard® TOF
78-42-2	Disflamoll® TOF
78-42-2	Kronitex® TOF
78-42-2	TOF
78-42-2	TOF™
78-51-3	Amgard® TBEP

CAS	Trade name	CAS	Trade name
78-51-3	KP-140®	80-43-3	Varox® DCP-T
78-51-3	Phosflex TBEP	80-51-3	Azocel OBSH
78-51-3	TBEP	80-51-3	Azofoam B-95
78-51-3	Tolplaz TBEP	80-51-3	Blo-Foam® BBSH
78-63-7	Aztec® 2,5-DI	80-51-3	Cellcom OBSH-ASA2
78-63-7	Aztec® 2,5-DI-45-G	80-51-3	Celogen® OT
78-63-7	Aztec® 2,5-DI-45-IC	80-51-3	Mikrofine OBSH
78-63-7	Aztec® 2,5-DI-45-IC1	80-51-3	Nitropore® OBSH
78-63-7	Aztec® 2,5-DI-45-PSI	80-51-3	Unicell OH
78-63-7	Esperal® 120	84-61-7	Unimoll® 66
78-63-7	Lupersol 101	84-61-7	Unimoll® 66 M
78-63-7	Polyvel CR-5	84-61-7	Uniplex 250
78-63-7	Polyvel CR-5F	84-64-0	Nuoplaz® 6938
78-63-7	Polyvel CR-10	84-66-2	Kodaflex® DEP
78-63-7	Polyvel CR-25	84-69-5	Diplast® B
78-63-7	Trigonox® 101	84-69-5	Nuoplaz® DIBP
78-63-7	Trigonox® 101/45Bpd	84-69-5	Uniplex 155
78-63-7	Trigonox® 101-45S-ps	84-74-2	Diplast® A
78-63-7	Varox® DBPH	84-74-2	Kodaflex® DBP
78-63-7	Varox® DBPH-50	84-74-2	Palatinol® DBP
78-67-1	Ficel® AZDN-LF	84-74-2	Plasthall® DBP
78-67-1	Ficel® AZDN-LMC	84-74-2	PX-104
78-67-1	Mikrofine AZDM	84-74-2	Staflex DBP
78-67-1	Perkadox® AIBN	84-74-2	Unimoll® DB
78-67-1	Poly-Zole® AZDN	84-74-2	Uniplex 150
78-67-1	V-60	85-42-7	Milldride® HHPA
79-10-7	Sipomer® β-CEA	85-44-9	Retarder AK
79-39-0	BM-801	85-44-9	Retarder PX
79-74-3	Santovar® A	85-44-9	Vulkalent B/C
79-94-7	FR-1524	85-44-9	Wiltrol P
79-94-7	Great Lakes BA-59	85-60-9	Santowhite® Powd
79-94-7	Great Lakes BA-59P™	85-60-9	Vanox® SWP
79-94-7	Saytex® RB-100	85-68-7	Santicizer 160
80-05-7	Parabis	85-68-7	Unimoll® BB
80-15-9	Aztec® CHP-80	85-70-1	Morflex® 190
80-15-9	CHP-5	87-83-2	FR-705
80-15-9	CHP-158	88-24-4	Antioxidant 425
80-15-9	Trigonox® 239A	88-24-4	Cyanox® 425
80-15-9	Trigonox® K-80	88-27-7	Ethanox® 703
80-15-9	Trigonox® K-95	88-58-4	Eastman® DTBHQ
80-17-1	Mikrofine BSH	89-04-3	Diplast® TM8
80-17-1	Porofor® BSH Powder	90-30-3	Additin 30
80-39-7	Uniplex 108	90-30-3	Akrochem® Antioxidant PANA
80-43-3	Aztec® DCP-40-G	90-30-3	Vulkanox® PAN
80-43-3	Aztec® DCP-40-IC	90-72-2	Actiron NX 3
80-43-3	Aztec® DCP-40-IC1	90-72-2	Ancamine® K54
80-43-3	Aztec® DCP-R	90-72-2	Capcure® EH-30
80-43-3	Di-Cup R, T	91-17-8	Decalin
80-43-3	Esperal® 115RG	91-53-2	Santoflex® AW
80-43-3	Luperco 500-SRK	91-88-3	Emery® 5714
80-43-3	Luperox 500R	91-99-6	Emery® 5709
80-43-3	Luperox 500T	92-50-2	Emery® 5707
80-43-3	Perkadox® BC	92-84-2	PTZ® Chemical Grade
80-43-3	Perkadox® BC-40S	92-84-2	PTZ® Industrial Grade
80-43-3	Perkadox® BC-FF	92-84-2	PTZ® Purified Flake
80-43-3	Peroximon® DC 40 MF	92-84-2	PTZ® Purified Powd
80-43-3	Polyvel PCL-10	93-46-9	Agerite® White
80-43-3	Polyvel PCL-20	93-46-9	Agerite® White White
80-43-3	Thermacure®	93-46-9	Anchor® DNPD
80-43-3	Varox® DCP-R	94-17-7	Cadox® TDP

CAS	Trade name
94-17-7	Cadox® TS-50S
94-28-0	Kodaflex® TEG-EH
94-36-0	Abcure S-40-25
94-36-0	Aztec® BP-05-FT
94-36-0	Aztec® BP-20-GY
94-36-0	Aztec® BP-40-S
94-36-0	Aztec® BP-50-FT
94-36-0	Aztec® BP-50-P1
94-36-0	Aztec® BP-50-PSI
94-36-0	Aztec® BP-50-SAQ
94-36-0	Aztec® BP-60-PCL
94-36-0	Aztec® BP-70-W
94-36-0	Aztec® BP-77-W
94-36-0	Benox® L-40LV
94-36-0	Cadet® BPO-70
94-36-0	Cadet® BPO-70W
94-36-0	Cadet® BPO-75W
94-36-0	Cadet® BPO-78
94-36-0	Cadet® BPO-78W
94-36-0	Cadox® 40E
94-36-0	Cadox® BFF-50L
94-36-0	Cadox® BPO-W40
94-36-0	Cadox® BS
94-36-0	Cadox® BT-50
94-36-0	Cadox® BTW-50
94-36-0	Cadox® BTW-55
94-36-0	Lucidol 70
94-36-0	Lucidol 75
94-36-0	Lucidol 75FP
94-36-0	Lucidol 78
94-36-0	Lucidol 98
94-36-0	Luperco AFR-250
94-36-0	Luperco AFR-400
94-36-0	Luperco AFR-500
94-36-0	Luperco ANS
94-36-0	Luperco ANS-P
94-36-0	Varox® ANS
95-31-8	Akrochem® BBTS
95-31-8	Perkacit® TBBS
95-31-8	Santocure® NS
95-31-8	Vanax® NS
95-31-8	Vanax® TBSI
95-31-8	Vulkacit NZ/EG
95-32-9	Morfax®
95-33-0	Akrochem® CBTS
95-33-0	Durax®
95-33-0	Durax® Rodform
95-33-0	Ekaland CBS
95-33-0	Perkacit® CBS
95-33-0	Santocure®
95-33-0	Vulkacit CZ/EG
95-33-0	Vulkacit CZ/EGC
95-33-0	Vulkacit DZ/EGC
95-71-6	THQ
96-05-9	Ageflex AMA
96-05-9	MFM-401
96-05-9	Sipomer® AM
96-27-5	Thiovanol®
96-45-7	Akroform® ETU-22 PM
96-45-7	Ekaland ETU
96-45-7	Naftocit® Mi 12
96-45-7	Naftopast® Mi12-P
96-45-7	Perkacit® ETU
96-49-1	Texacar® EC
96-69-5	Lowinox® 44S36
96-69-5	Santonox®
96-69-5	Santowhite® Crystals
97-39-2	Akrochem® DOTG
97-39-2	Anchor® DOTG
97-39-2	Ekaland DOTG
97-39-2	Perkacit® DOTG
97-39-2	Vanax® DOTG
97-74-5	Akrochem® TMTM
97-74-5	Ekaland TMTM
97-74-5	Monothiurad®
97-74-5	Perkacit® TMTM
97-74-5	Unads® Pellets
97-74-5	Unads® Powd.
97-77-8	Akrochem® TETD
97-77-8	Ekaland TETD
97-77-8	Ethyl Tuads® Rodform
97-77-8	Perkacit® TETD
97-78-9	Hamposyl® L
97-90-5	Ageflex EGDMA
97-90-5	MFM-416
97-90-5	Perkalink® 401
97-99-4	QO® Tetrahydrofurfuryl Alcohol (THFA®)
98-00-0	QO® Furfuryl Alcohol (FA®)
98-01-1	QO® Furfural
98-10-2	Uniplex 413
98-11-3	Reworyl® B 70
98-77-1	Vanax® 552
99-97-8	Firstcure™ DMPT
100-37-8	Pennad 150
100-74-3	Dabco® NEM
100-74-3	Jeffcat NEM
100-74-3	Toyocat®-NEM
101-02-0	Albrite® TPP
101-02-0	Doverphos® 10
101-02-0	Lankromark® LE65
101-02-0	Mark® 2112
101-02-0	Mark® TPP
101-02-0	Weston® EGTPP
101-02-0	Weston® TPP
101-25-7	Blo-Foam® DNPT 80
101-25-7	Blo-Foam® DNPT 93
101-25-7	Blo-Foam® DNPT 100
101-25-7	Opex® 80
101-25-7	Porofor® DNO/F
101-34-8	Flexricin® P-8
101-37-1	Perkalink® 300
101-42-8	Amicure® UR
101-43-9	Ageflex CHMA
101-66-3	Manganese Violet
101-67-7	Vanox® 12
101-67-7	Vanox® 1081
101-68-8	Desmodur® MP-225
101-68-8	Desmodur® VK-5
101-68-8	Desmodur® VK-18

CAS	Trade name
101-68-8	Desmodur® VK-70
101-68-8	Desmodur® VK-200
101-68-8	Desmodur® VKS-2
101-68-8	Desmodur® VKS-4
101-68-8	Desmodur® VKS-5
101-68-8	Desmodur® VKS-18
101-68-8	Isonate® 125M
101-68-8	Isonate® 143L
101-68-8	Isonate® 181
101-68-8	Isonate® 240
101-68-8	Novor 950
101-72-4	Permanax™ IPPD
101-72-4	Santoflex® IP
101-72-4	Vanox® 3C
101-72-4	Vulkanox® 4010 NA
101-77-9	Ancamine® DL-50
101-77-9	Tonox®
101-77-9	Tonox® 22
101-87-1	Vanox® 6H
101-90-6	Epodil® 769
102-06-7	Akrochem® DPG
102-06-7	Anchor® DPG
102-06-7	DPG
102-06-7	Naftocit® DPG
102-06-7	Perkacit® DPG
102-06-7	Vanax® DPG
102-06-7	Vanax® DPG Pellets
102-06-7	Vulkacit D/C
102-06-7	Vulkacit D/EGC
102-07-8	HyTemp SR-50
102-07-8	Zeonet® U
102-08-9	A-1
102-08-9	Akrochem® Thio No. 1
102-08-9	Rhenocure® CA
102-08-9	Stabilizer C
102-60-3	Quadrol®
102-71-6	TEA 85 Low Freeze Grade
102-71-6	TEA 99 Low Freeze Grade
102-71-6	TEA 99 Standard Grade
102-76-1	Estol 1579
102-76-1	Kodaflex® Triacetin
102-77-2	Akrochem® Accelerator MF
102-77-2	Akrochem® OBTS
102-77-2	Akrochem® OMTS
102-77-2	Amax®
102-77-2	Delac® MOR
102-77-2	Perkacit® MBS
102-77-2	Santocure® MOR
102-77-2	Vulkacit MOZ/LG
102-77-2	Vulkacit MOZ/SG
103-23-1	Adimoll® DO
103-23-1	Diplast® DOA
103-23-1	Jayflex® DOA
103-23-1	Kodaflex® DOA
103-23-1	Monoplex® DOA
103-23-1	Nuoplaz® DOA
103-23-1	Palatinol® DOA
103-23-1	Plasthall® DOA
103-23-1	PX-238
103-23-1	Staflex DOA
103-23-1	Uniflex® DOA
103-24-2	Plasthall® DOZ
103-24-2	Plastolein® 9058
103-24-2	Priplast 3018
103-24-2	Staflex DOZ
103-34-4	Akrochem® Accelerator R
103-34-4	Rhenocure® M
103-34-4	Rhenocure® M/G
103-34-4	Sulfasan® R
103-83-3	Dabco® BDMA
103-99-1	Suconox-18®
104-15-4	Eltesol® TSX/A
104-15-4	Eltesol® TSX/SF
104-38-1	Eastman® HQEE
104-87-0	PTAL
105-08-8	CHDM
105-11-3	QDO®
105-16-8	Ageflex FM-2
105-52-2	Staflex DMAM
105-55-5	Akrochem® D.E.T.U. Accelerator
105-55-5	Akroform® DETU PM
105-55-5	Ekaland DETU
105-55-5	Thiate® H
105-64-6	IPP
105-74-8	Alperox-F
105-74-8	Aztec® LP
105-74-8	Aztec® LP-25-SAQ
105-74-8	Aztec® LP-40-SAQ
105-74-8	Laurox®
105-74-8	Laurox® W-25
105-74-8	Laurox® W-40
105-74-8	Laurox® W-40-GD1
105-76-0	PX-504
105-76-9	Staflex DBF
105-97-5	Uniflex® DCA
106-06-9	Plastolein® 9404
106-06-9	TegMeR® 903
106-11-6	Pegosperse® 100-S
106-14-9	Loxiol® G 21
106-17-2	Cithrol EGMR N/E
106-58-1	Jeffcat DMP
106-65-0	DBE-4
106-84-3	ADK CIZER D-32
106-92-3	Ageflex AGE
106-98-9	Gulftene® 4
107-64-2	Arosurf® TA-100
107-64-2	Dehyquart® DAM
107-71-1	Esperox® 12MD
107-71-1	Trigonox® F-C50
107-72-2	Dynasylan® AMTC
107-98-2	Arcosolv® PM
108-01-0	Dabco® DMEA
108-01-0	Jeffcat DME
108-01-0	Tegoamin® DMEA
108-01-0	Toyocat®-DMA
108-32-7	Arconate® 1000
108-32-7	Arconate® HP
108-32-7	Arconate® PC
108-32-7	Texacar® PC

CAS	Trade name
108-65-6	Arcosolv® PMA
108-80-5	Zeonet® A
108-88-3	Vanchem® HM-50, HM-4346
108-90-7	Vanchem® HM-50, HM-4346
109-02-4	Dabco® NMM
109-02-4	Jeffcat NMM
109-13-7	Aztec® TBPIB-75-AL
109-13-7	Trigonox® 41-C50
109-13-7	Trigonox® 41-C75
109-16-0	MFM-413
109-31-9	Plastolein® 9051
109-31-9	Priplast 3013
109-43-3	Plasthall® DBS
109-43-3	Uniflex® DBS
109-46-6	Akroform® DBTU PM
109-46-6	Ekaland DBTU
109-46-6	Thiate® U
109-86-4	Hisolve MC
109-99-9	THF
110-05-4	Aztec® DTBP
110-05-4	Trigonox® B
110-18-9	Toyocat®-TE
110-25-8	Hamposyl® O
110-27-0	Estol IPM 1514
110-30-5	Acrawax® C
110-30-5	Aktiplast® AS
110-30-5	Alkamide® STEDA
110-30-5	Hoechst Wax C
110-30-5	Hostalub® FA 1
110-30-5	Ross Wax #140
110-31-6	Advawax® 240
110-31-6	Kemamide® W-20
110-63-4	BDO
110-63-4	Dabco® BDO
110-80-5	Oxitol
110-86-1	Pyridine 1°
110-97-4	DIPA Commercial Grade
110-97-4	DIPA Low Freeze Grade 85
110-97-4	DIPA Low Freeze Grade 90
110-97-4	DIPA NF Grade
111-03-5	Pationic® 1064
111-03-5	Pationic® 1074
111-27-3	Epal® 6
111-40-0	D.E.H. 20
111-40-0	D.E.H. 52
111-40-0	D.E.H. 58
111-40-0	DETA
111-40-0	Texacure EA-20
111-42-2	Dabco® DEOA-LF
111-42-2	DEA Commercial Grade
111-42-2	DEA Low Freeze Grade
111-55-7	Estol 1574
111-60-4	Cithrol EGMS N/E
111-60-4	CPH-37-NA
111-60-4	Mackester™ EGMS
111-60-4	Mackester™ IP
111-60-4	Pegosperse® 50 MS
111-66-0	Gulftene® 8
111-76-2	Butyl Oxitol
111-76-2	Ektasolve® EB
111-77-3	Hisolve DM
111-82-0	Estol 1502
111-87-5	Epal® 8
111-90-0	Dioxitol-Low Gravity
111-96-6	Diglyme
111-96-6	Hisolve MDM
112-00-5	Arquad® 12-50
112-00-5	Chemquat 12-33
112-00-5	Chemquat 12-50
112-00-5	Dehyquart® LT
112-00-5	Octosol 562
112-00-5	Octosol 571
112-02-7	Arquad® 16-50
112-02-7	Chemquat 16-50
112-02-7	Dehyquart® A
112-03-8	Arquad® 18-50
112-11-8	Estol 1406
112-18-5	Adma® 12
112-23-3	Texacure EA-24
112-24-3	D.E.H. 24
112-24-3	TETA
112-24-3	Texlin® 300
112-30-1	Epal® 10
112-34-5	Butyl Dioxitol
112-41-4	Gulftene® 12
112-49-2	Triglyme
112-53-8	Epal® 12
112-53-8	Laurex® NC
112-57-2	D.E.H. 26
112-57-2	TEPA
112-69-6	Adma® 16
112-69-6	Crodamine 3.A16D
112-72-1	Epal® 14
112-73-2	Ethyl Diglyme
112-75-4	Adma® 14
112-80-1	Industrene® 105
112-80-1	Industrene® 205
112-80-1	Industrene® 206
112-80-1	Pamolyn® 125
112-80-1	Priolene 6901
112-80-1	Priolene 6907
112-80-1	Priolene 6930
112-84-5	Armid® E
112-84-5	Armoslip® EXP
112-84-5	Chemstat® HTSA #22
112-84-5	Crodamide E
112-84-5	Crodamide ER
112-84-5	Kemamide® E
112-84-5	Petrac® Eramide®
112-84-5	Polydis® TR 131
112-84-5	Unislip 1753
112-88-9	Gulftene® 18
112-90-3	Crodamine 1.O, 1.OD
112-90-3	Kemamine® P-989D
112-92-5	Adol® 62 NF
112-92-5	Alfol® 18
112-92-5	Cachalot® S-53
112-92-5	Loxiol® VPG 1354
115-77-5	Flame-Amine
115-77-5	Sta-Flame 88/99

CAS	Trade name
115-86-6	Disflamoll® TP
115-86-6	Kronitex® TPP
117-81-7	Diplast® O
117-81-7	Kodaflex® DOP
117-81-7	Nuoplaz® DOP
117-81-7	Plasthall® DOP
117-81-7	PX-138
117-83-9	Plasthall® 200
117-83-9	Staflex DBEP
117-84-0	Diplast® L8
118-79-6	FR-613
118-79-6	Great Lakes PH-73™
118-82-1	Ethanox® 702
118-82-1	Ralox® 02
119-07-3	PX-318
119-07-3	Staflex ODP
119-47-1	Akrochem® Antioxidant 235
119-47-1	Antioxidant 235
119-47-1	CAO®-5
119-47-1	CAO®-14
119-47-1	Cyanox® 2246
119-47-1	Lowinox® 22M46
119-47-1	Naftonox® 2246
119-47-1	Ralox® 46
119-47-1	Santowhite® PC
119-47-1	Synox 5LT
119-47-1	Vanox® 2246
119-47-1	Vanox® MBPC
119-47-1	Vulkanox® BKF
119-64-2	Tetralin
120-07-0	Emery® 5703
120-40-1	Empilan® LDE
120-40-1	Ethylan® MLD
120-54-7	Akrochem® DPTT
120-54-7	DPTT-S
120-54-7	Perkacit® DPTT
120-54-7	Sulfads®
120-54-7	Sulfads® Rodform
120-55-8	Benzoflex® 2-45
120-78-5	Akrochem® MBTS
120-78-5	Akrochem® MBTS Pellets
120-78-5	Ekaland MBTS
120-78-5	Naftocit® MBTS
120-78-5	Naftopast® MBTS-A
120-78-5	Naftopast® MBTS-P
120-78-5	Perkacit® MBTS
120-78-5	Thiofide®
121-54-0	Hyamine® 1622 50%
121-72-2	Firstcure™ DMMT
121-79-9	Tenox® PG
122-20-3	TIPA 99
122-20-3	TIPA 101
122-20-3	TIPA Low Freeze Grade
122-60-1	Epodil® 743
122-60-1	Heloxy® 63
122-98-5	Emery® 5700
122-99-6	Emery® 6705
123-28-4	Carstab® DLTDP
123-28-4	Cyanox® LTDP
123-28-4	Evanstab® 12
123-28-4	Hostanox® SE 1
123-28-4	Lankromark® DLTDP
123-28-4	Lowinox® DLTDP
123-28-4	PAG DLTDP
123-77-3	Azo D
123-77-3	Azo DX72
123-77-3	Azofoam
123-77-3	Azofoam DS-1, DS-2
123-77-3	Azofoam M-1, M-2, M-3
123-77-3	Azofoam UC
123-77-3	Blo-Foam® 715
123-77-3	Blo-Foam® 754
123-77-3	Blo-Foam® 765
123-77-3	Blo-Foam® ADC 150
123-77-3	Blo-Foam® ADC 150FF
123-77-3	Blo-Foam® ADC 300
123-77-3	Blo-Foam® ADC 300FF
123-77-3	Blo-Foam® ADC 300WS
123-77-3	Blo-Foam® ADC 450
123-77-3	Blo-Foam® ADC 450FFNP
123-77-3	Blo-Foam® ADC 450WS
123-77-3	Blo-Foam® ADC 550
123-77-3	Blo-Foam® ADC 550FFNP
123-77-3	Blo-Foam® ADC 800
123-77-3	Blo-Foam® ADC 1200
123-77-3	Blo-Foam® ADC-MP
123-77-3	Blo-Foam® ADC-NP
123-77-3	Blo-Foam® KL9
123-77-3	Blo-Foam® KL10
123-77-3	Cellcom
123-77-3	Celogen® AZ 120
123-77-3	Celogen® AZ 130
123-77-3	Celogen® AZ 150
123-77-3	Celogen® AZ 180
123-77-3	Celogen® AZ 199
123-77-3	Celogen® AZ 754
123-77-3	Celogen® AZ-760-A
123-77-3	Celogen® AZ-2100
123-77-3	Celogen® AZ-2990
123-77-3	Celogen® AZ 3990
123-77-3	Celogen® AZ-5100
123-77-3	Celogen® AZNP 130
123-77-3	Celogen® AZNP 199
123-77-3	Celogen® AZRV
123-77-3	Ficel® AC1
123-77-3	Ficel® AC2
123-77-3	Ficel® AC2M
123-77-3	Ficel® AC3
123-77-3	Ficel® AC3F
123-77-3	Ficel® AC4
123-77-3	Ficel® ACSP4
123-77-3	Ficel® ACSP5
123-77-3	Ficel® EPA
123-77-3	Ficel® EPB
123-77-3	Ficel® EPC
123-77-3	Ficel® EPD
123-77-3	Ficel® EPE
123-77-3	Ficel® LE
123-77-3	Ficel® SCE
123-77-3	Henley AZ LBA-30

CAS	Trade name
123-77-3	Henley AZ LBA-49
123-77-3	HFVCD 11G
123-77-3	HFVCP 11G
123-77-3	HFVP 03
123-77-3	HFVP 15P
123-77-3	HFVP 19B
123-77-3	HP
123-77-3	HRVP 01
123-77-3	HRVP 19
123-77-3	Kempore® 60/14FF
123-77-3	Mikrofine ADC-EV
123-77-3	Plastifoam
123-77-3	Porofor® ADC/E
123-77-3	Porofor® ADC/F
123-77-3	Porofor® ADC/K
123-77-3	Porofor® ADC/M
123-77-3	Porofor® ADC/S
123-77-3	Unicell D
123-95-5	Ablumol BS
123-95-5	ADK STAB LS-8
123-95-5	Butyl Stearate C-895
123-95-5	Lexolube® BS-Tech
123-95-5	Lexolube® NBS
123-95-5	Priolube 1451
123-95-5	Uniflex® BYS-CP
123-95-5	Uniflex® BYS Special
123-95-5	Uniflex® BYS-Tech
124-04-9	Adi-pure®
124-26-5	Armid® 18
124-26-5	Armoslip® 18
124-26-5	Chemstat® HTSA #18S
124-26-5	Crodamide S
124-26-5	Crodamide SR
124-26-5	Kemamide® S
124-26-5	Petrac® Vyn-Eze®
124-26-5	Uniwax 1750
124-28-7	Adma® 18
124-28-7	Crodamine 3.A18D
124-28-7	Kemamine® T-9902D
124-28-7	Nissan Tert. Amine AB
124-30-1	Armeen® 18
124-30-1	Armeen® 18D
124-30-1	Armid® HTD
124-30-1	Crodamine 1.18D
124-30-1	Kemamine® P-990, P-990D
126-33-0	Sulfolane W
126-73-8	Albrite® TBP
126-73-8	Albrite® TBPO
126-73-8	Albrite® Tributyl Phosphate
126-73-8	Amgard® $TBPO_4$
126-73-8	Kronitex® TBP
126-73-8	TBP
126-86-3	Surfynol® 104A
126-92-1	Niaproof® Anionic Surfactant 08
126-92-1	Witcolate™ D-510
127-39-9	Aerosol® IB-45
127-39-9	Astrowet B-45
127-39-9	Gemtex 445
127-39-9	Geropon® CYA/45
127-39-9	Geropon® CYA/DEP
127-39-9	Monawet MB-45
127-39-9	Octosol IB-45
127-39-9	Rewopol® SBDB 45
128-03-0	Aquatreat KM
128-37-0	Akrochem® Antioxidant BHT
128-37-0	CAO®-3
128-37-0	Lowinox® BHT
128-37-0	Lowinox® BHT Food Grade
128-37-0	Lowinox® BHT-Superflakes
128-37-0	Naugard® BHT
128-37-0	Ralox® BHT food grade
128-37-0	Ralox® BHT Tech
128-37-0	Spectratech® CM 11340
128-37-0	Spectratech® KM 11264
128-37-0	Sustane® BHT
128-37-0	Topanol® OC
128-37-0	Topanol® OF
128-37-0	Ultranox® 226
128-37-0	Vulkanox® KB
131-11-3	Kodaflex® DMP
131-11-3	Palatinol® M
131-11-3	Staflex DMP
131-11-3	Uniplex 110
131-17-9	Ftalidap®
131-53-3	Cyasorb® UV 24
131-54-4	Uvinul® 3049
131-54-4	Uvinul® D-49
131-55-5	Uvinul® 3050
131-56-6	Uvasorb 2 OH
131-56-6	Uvinul® 3000
131-57-7	Acetorb A
131-57-7	Cyasorb® UV 9
131-57-7	Lankromark® LE296
131-57-7	Syntase® 62
131-57-7	UV-Absorber 325
131-57-7	Uvasorb MET
131-57-7	Uvinul® 3040
131-57-7	Uvinul® M-40
133-06-2	Vancide® 89
133-49-3	Renacit® 7
135-20-6	Q-1300
136-23-2	Akrochem® Accelerator BZ Powder
136-23-2	Butasan®
136-23-2	Butyl Zimate®
136-23-2	Butyl Zimate® Dustless
136-23-2	Butyl Zimate® Slurry
136-23-2	Naftocit® Di 13
136-23-2	Naftopast® Di13-P
136-23-2	Octocure ZDB-50
136-23-2	Perkacit® ZDBC
136-30-1	Butyl Namate®
136-30-1	Naftocit® NaDBC
136-30-1	Octopol NB-47
136-30-1	Octopol SDB-50
136-36-7	Eastman® Inhibitor RMB
136-85-6	Retrocure® G
137-16-6	Hamposyl® L-30
137-16-6	Hamposyl® L-95

CAS	Trade name
137-26-8	Akrochem® TMTD
137-26-8	Akrochem® TMTD Pellet
137-26-8	Akrosperse® D-177
137-26-8	Methyl Tuads®
137-26-8	Methyl Tuads® Rodform
137-26-8	Naftocit® Thiuram 16
137-26-8	Naftopast® Thiuram 16-P
137-26-8	Perkacit® TMTD
137-26-8	Thiurad®
137-29-1	Akrochem® Cu.D.D
137-29-1	Akrochem® Cu.D.D.-PM
137-29-1	Methyl Cumate®
137-29-1	Methyl Cumate® Rodform
137-29-1	Perkacit® CDMC
137-30-4	Akrochem® Accelerator MZ
137-30-4	Methasan®
137-30-4	Methyl Zimate®
137-30-4	Naftocit® Di 4
137-30-4	Octocure ZDM-50
137-30-4	Perkacit® ZDMC
137-30-4	Vancide® MZ-96
139-07-1	Dehyquart® LDB
139-33-3	Versene NA
139-33-3	Versene Na$_2$
139-43-5	Paricin® 8
139-45-7	Kodaflex® Tripropionin
139-88-8	Niaproof® Anionic Surfactant 4
139-96-8	Carsonol® TLS
139-96-8	Empicol® TL40/T
139-96-8	Perlankrol® ATL40
140-01-2	Chel DTPA-41
140-03-4	Flexricin® P-4
140-04-5	Flexricin® P-6
140-31-8	AEP
140-31-8	D.E.H. 39
141-02-6	Staflex DOF
141-04-8	DIBA
141-04-8	Plasthall® DIBA
141-08-2	Flexricin® 13
141-17-3	Plasthall® 226
141-17-3	Plasthall® DBEEA
141-17-3	Suprmix DBEEA
141-17-3	Wareflex® 650
141-20-8	Cithrol DGML N/E
141-22-0	P®-10 Acid
141-23-1	Paricin® 1
141-24-2	Flexricin® P-1
141-32-2	Paraloid® EXL-3330
141-32-2	Paraloid® EXL-3361
141-32-2	Paraloid® EXL-3387
141-38-8	Estabex® 2381
141-43-5	MEA Commercial Grade
141-43-5	MEA Low Freeze Grade
141-43-5	MEA Low Iron Grade
141-43-5	MEA Low Iron-Low Freeze Grade
141-43-5	MEA NF Grade
141-97-9	Tyzor DC
142-16-5	PX-538
142-31-4	Rhodapon® OLS
142-48-3	Hamposyl® S
142-77-8	Butyl Oleate C-914
142-77-8	Kemester® 4000
142-77-8	Plasthall® 503
142-77-8	Plasthall® 914
142-77-8	Priolube 1405
142-77-8	Uniflex® BYO
142-87-0	Akyporox SAL SAS
142-91-6	Estol IPP 1517
143-00-0	Carsonol® DLS
143-07-7	Prifrac 2920
143-07-7	Prifrac 2922
143-18-0	Octosol 449
143-27-1	Crodamine 1.16D
143-28-2	Fancol OA-95
144-34-4	Methyl Selenac®
144-55-8	Kycerol 91
144-55-8	Kycerol 92
144-55-8	Sodium Bicarbonate USP No. 1 Powd.
144-55-8	Sodium Bicarbonate USP No. 3 Extra Fine Powd.
148-18-5	Ethyl Namate®
148-18-5	Octopol SDE-25
149-30-4	Akrochem® MBT
149-30-4	Captax®
149-30-4	Ekaland MBT
149-30-4	Naftocit® MBT
149-30-4	Naftopast® MBT-P
149-30-4	Perkacit® MBT
149-30-4	Rokon
149-30-4	Rotax®
149-30-4	Thiotax®
149-44-0	Hydrosulfite AWC
150-38-9	Hamp-Ene® Na$_3$ Liq
150-76-5	Eastman® HQMME
151-21-3	Alscoap LN-40, LN-90
151-21-3	Avirol® SL 2010
151-21-3	Avirol® SL 2020
151-21-3	Calfoam SLS-30
151-21-3	Carsonol® SLS
151-21-3	Carsonol® SLS-R
151-21-3	Carsonol® SLS-S
151-21-3	Carsonol® SLS Special
151-21-3	Elfan® 260 S
151-21-3	Empicol® 0303
151-21-3	Empicol® 0303V
151-21-3	Empicol® LX28
151-21-3	Empicol® LXS95
151-21-3	Empicol® LXV
151-21-3	Empicol® LXV/D
151-21-3	Empicol® LY28/S
151-21-3	Empicol® LZG 30
151-21-3	Empicol® LZGV
151-21-3	Empicol® LZGV/C
151-21-3	Empicol® LZP
151-21-3	Empicol® LZV/D
151-21-3	Empicol® WA
151-21-3	Empimin® LR28
151-21-3	Equex SP

CAS	Trade name
151-21-3	Manro DL 28
151-21-3	Manro SLS 28
151-21-3	Montovol RF-10
151-21-3	Octosol SLS
151-21-3	Octosol SLS-1
151-21-3	Perlankrol® DSA
151-21-3	Polirol LS
151-21-3	Polystep® B-3
151-21-3	Polystep® B-5
151-21-3	Polystep® B-24
151-21-3	Rewopol® 15/L
151-21-3	Rewopol® NLS 15 L
151-21-3	Rewopol® NLS 28
151-21-3	Rewopol® NLS 30 L
151-21-3	Rhodapon® LCP
151-21-3	Rhodapon® SB
151-21-3	Rhodapon® UB
151-21-3	Serdet DFK 40
151-21-3	Sermul EA 150
151-21-3	Stepanol® WA-100
151-21-3	Sulfopon® 101/POL
151-21-3	Sulfopon® 101 Special
151-21-3	Sulfopon® 102
151-21-3	Sulfopon® P-40
151-21-3	Supralate® WAQ
151-21-3	Supralate® WAQE
151-21-3	Texapon® K-12
151-21-3	Texapon® K-1296
151-21-3	Texapon® OT Highly Conc. Needles
151-21-3	Texapon® VHC Needles
151-21-3	Texapon® ZHC Powder
151-21-3	Ultra Sulfate SL-1
151-21-3	Ungerol LS, LSN
151-21-3	Witcolate™ A Powder
151-21-3	Witcolate™ SL-1
151-21-3	Zorapol LS-30
155-04-4	Akrochem® ZMBT
155-04-4	Bantox®
155-04-4	Ekaland ZMBT
155-04-4	Naftocit® ZMBT
155-04-4	Octocure ZMBT-50
155-04-4	Perkacit® ZMBT
155-04-4	Vulkacit ZM
155-04-4	Zetax®
155-04-4	Zetax® Dispersion
280-57-9	Amicure® TEDA
280-57-9	Dabco® Crystalline
280-57-9	TEDA
280-57-9	Tegoamin® 33
280-57-9	Tegoamin® 100
280-57-9	Toyocat®-TF
300-92-5	Witco® Aluminum Stearate 22
301-02-0	Armid® O
301-02-0	Chemstat® HTSA #18
301-02-0	Crodamide O
301-02-0	Crodamide OR
301-02-0	Kemamide® O
301-02-0	Kemamide® U
301-02-0	Petrac® Slip-Eze®
301-02-0	Polydis® TR 121
301-02-0	Polyvel RO25
301-02-0	Polyvel RO40
301-02-0	Unislip 1757
301-02-0	Unislip 1759
301-10-0	Dabco® T-9
301-10-0	Fomrez® C-2
301-10-0	Kosmos® 16
301-10-0	Kosmos® 29
301-10-0	Metacure® T-9
301-10-0	TEDA-D007
306-52-5	Disflamoll® TCA
306-52-5	Fyrol® CEF
422-86-2	Kodaflex® DOTP
461-58-5	Amicure® CG-325
461-58-5	Amicure® CG-1200
461-58-5	Amicure® CG-1400
461-58-5	Amicure® CG-NA
461-58-5	Dicyanex® 200X
461-58-5	Dicyanex® 325
461-58-5	Dicyanex® 1200
471-34-1	Camel-WITE®
471-34-1	CC™-101
471-34-1	CC™-103
471-34-1	CC™-105
471-34-1	CP Filler™
471-34-1	Drikalite®
471-34-1	Marble Dust™
471-34-1	Opacimite™
471-34-1	SC-53
501-24-6	Cardolite® NC-510
501-24-6	Cardolite® NC-507
501-52-0	Hostanox® O3
546-68-9	Tyzor TPT
546-93-0	Magocarb-33
555-43-1	Pationic® 919
555-43-1	Pationic® 1019
557-04-0	ADK STAB MG-ST
557-04-0	Afco-Chem MGS
557-04-0	Cometals Magnesium Stearate
557-04-0	Haro® Chem MF-2
557-04-0	Haro® Chem MF-3
557-04-0	HyQual NF
557-04-0	HyTech RSN 1-1
557-04-0	Mathe Magnesium Stearate
557-04-0	Miljac Magnesium Stearate
557-04-0	Petrac® Magnesium Stearate MG-20 NF
557-04-0	Synpro® Magnesium Stearate 90
557-04-0	Witco® Magnesium Stearate D
557-05-1	ADK STAB ZN-ST
557-05-1	Afco-Chem ZNS
557-05-1	Akrochem® P 3100
557-05-1	Akrochem® Wettable Zinc Stearate
557-05-1	Akrochem® Zinc Stearate
557-05-1	Akrodip™ Z-50/50
557-05-1	Akrodip™ Z-200
557-05-1	Akrodip™ Z-250

CAS	Trade name
557-05-1	Cecavon ZN 70
557-05-1	Cecavon ZN 71
557-05-1	Cecavon ZN 72
557-05-1	Cecavon ZN 73
557-05-1	Cecavon ZN 735
557-05-1	Coad 20
557-05-1	Coad 21
557-05-1	Coad 27B
557-05-1	Cometals Zinc Stearate
557-05-1	Disperso II
557-05-1	Hallcote® ZS
557-05-1	Hallcote® ZS 5050
557-05-1	Haro® Chem ZGD
557-05-1	Haro® Chem ZGN
557-05-1	Haro® Chem ZGN-T
557-05-1	Haro® Chem ZPR-2
557-05-1	Hy Dense Zinc Stearate XM Powd.
557-05-1	Hy Dense Zinc Stearate XM Ultra Fine
557-05-1	Hydro Zinc™
557-05-1	HyTech RSN 131 HS/Gran
557-05-1	Interstab® ZN-18-1
557-05-1	Liquazinc AQ-90
557-05-1	Lubrazinc® W, Superfine
557-05-1	Mathe Zinc Stearate 25S
557-05-1	Mathe Zinc Stearate S
557-05-1	Metasap® Zinc Stearate
557-05-1	Miljac Zinc Stearate
557-05-1	Molder's Edge ME-369
557-05-1	Petrac® Zinc Stearate ZN-41
557-05-1	Petrac® Zinc Stearate ZN-42
557-05-1	Petrac® Zinc Stearate ZN-44 HS
557-05-1	Petrac® Zinc Stearate ZW-45
557-05-1	Polyvel RZ40
557-05-1	Quikote™
557-05-1	Quikote™ M
557-05-1	Rhenodiv® ZB
557-05-1	Rubichem Zinc-Dip®
557-05-1	Rubichem Zinc-Dip® HS
557-05-1	Rubichem Zinc-Dip® W
557-05-1	Sprayon 812
557-05-1	Synpro® Zinc Dispersion Zincloid
557-05-1	Synpro® Zinc Stearate 8
557-05-1	Synpro® Zinc Stearate ACF
557-05-1	Synpro® Zinc Stearate D
557-05-1	Synpro® Zinc Stearate GP
557-05-1	Synpro® Zinc Stearate GP Flake
557-05-1	Synpro® Zinc Stearate HSF
557-05-1	TR-Zinc Stearate
557-05-1	Wet Zinc™
557-05-1	Wet Zinc™ P
557-05-1	Witco® Zinc Stearate 11
557-05-1	Witco® Zinc Stearate 42
557-05-1	Witco® Zinc Stearate 44
557-05-1	Witco® Zinc Stearate Disperso
557-05-1	Witco® Zinc Stearate Heat-Stable
557-05-1	Witco® Zinc Stearate LV
557-05-1	Witco® Zinc Stearate NW
557-05-1	Witco® Zinc Stearate Polymer Grade
557-05-1	Witco Zinc Stearate Regular
557-05-1	Witco® Zinc Stearate REP
557-05-1	Zincote™
557-05-1	Zinc Stearate 42
577-11-7	Aerosol® OT-75%
577-11-7	Chemax DOSS-75E
577-11-7	Drewfax® 0007
577-11-7	Empimin® OT
577-11-7	Empimin® OT75
577-11-7	Gemtex PA-75
577-11-7	Gemtex PA-85P
577-11-7	Gemtex PAX-60
577-11-7	Gemtex SC-75E
577-11-7	Geropon® 99
577-11-7	Geropon® SS-O-75
577-11-7	Hodag DOSS-70
577-11-7	Hodag DOSS-75
577-11-7	Lankropol® KO2
577-11-7	Rewopol® SBDO 75
588-33-5	Sipomer® IBOA
588-33-5	Sipomer® IBOA HP
589-37-7	Dytek® EP
592-41-6	Gulftene® 6
593-60-2	Saytex® VBR
598-63-0	Halcarb 20
598-63-0	Halcarb 20 EP
598-63-0	Halcarb 200
598-63-0	Halcarb 200 EP
598-63-0	Haro® Chem PC
599-64-4	Kencure™ C9P
599-64-4	Ken Kem® CP-45
599-64-4	Ken Kem® CP-99
614-33-5	Benzoflex® S-404
614-33-5	Uniplex 260
614-45-9	Aztec® TBPB
614-45-9	Aztec® TBPB-50-FT
614-45-9	Aztec® TBPB-50-IC
614-45-9	Aztec® TBPB-SA-M2
614-45-9	Esperox® 10
614-45-9	Polyvel CR-5T
614-45-9	Polyvel CR-10T
614-45-9	Trigonox® 93
614-45-9	Trigonox® C
614-45-9	Trigonox® C-C75
614-45-9	Varox® TBPB
615-58-7	FR-612
616-45-5	2-Pyrol®
622-16-2	Stabilizer 2013-P®
624-04-4	Cithrol EGDL N/E
627-83-8	Cithrol EGDS N/E
627-83-8	Kemester® EGDS
627-83-8	Mackester™ EGDS
627-93-0	DBE-6
629-14-1	Ethyl Glyme
629-73-2	Gulftene® 16

CAS	Trade name
632-79-1	Great Lakes PHT4™
632-79-1	Saytex® RB-49
637-12-7	Haro® Chem ALT
638-16-4	Zisnet® F-PT
646-13-9	Estol 1476
661-19-8	Loxiol® VPG 1451
670-96-2	Curezol® 2PZ
681-84-5	Petrarch® T1980
682-01-9	Petrarch® T2090
686-31-7	Aztec® TAPEH
686-31-7	Esperox® 570
686-31-7	Esperox® 570P
686-31-7	Lupersol 575
686-31-7	Lupersol 575-P75
686-31-7	Trigonox® 121
686-31-7	Trigonox® 121-BB75
693-33-4	Lonzaine® 16S
693-36-7	Carstab® DSTDP
693-36-7	Cyanox® STDP
693-36-7	Evanstab® 18
693-36-7	Hostanox® SE 2
693-36-7	Lankromark® DSTDP
693-36-7	Lowinox® DSTDP
693-36-7	PAG DSTDP
693-36-7	PAG DXTDP
693-98-1	Imicure® AMI-2
694-83-7	DCH-99
756-79-6	Antiblaze® DMMP
756-79-6	Fyrol® DMMP
762-12-9	Aztec® DP
762-12-9	Perkadox® SE-10
762-16-3	Perkadox® SE-8
763-69-9	Ektapro® EEP Solvent
780-69-8	CP0320
793-24-8	Akrochem® Antiozonant PD-2
793-24-8	Antozite® 67P
793-24-8	Permanax™ 6PPD
793-24-8	Santoflex® 13
793-24-8	Vulkanox® 4020
814-80-2	Pationic® 1230
814-80-2	Pationic® 1240
814-80-2	Pationic® 1250
820-66-6	Lonzaine® 18S
822-16-2	Afco-Chem B
822-16-2	Cometals Sodium Stearate
822-16-2	Haro® Chem NG
822-16-2	HyTemp NS-70
822-16-2	Mathe Sodium Stearate
822-16-2	Synpro® Sodium Stearate ST
822-16-2	Witco® Heat Stable Sodium Stearate
822-16-2	Witco® Sodium Stearate T-1
868-77-9	BM-903
868-77-9	Sipomer® HEM-D
872-05-9	Gulftene® 10
872-50-4	NMP
877-66-7	Blo-Foam® SH
877-66-7	Celogen® TSH
877-66-7	Mikrofine TSH
901-44-0	Dianol® 220
919-30-2	Dynasylan® AMEO
919-30-2	Dynasylan® AMEO-P
919-30-2	Dynasylan® AMEO-T
919-30-2	Petrarch® A0750
919-30-2	Prosil® 220
919-30-2	Prosil® 221
922-80-5	Aerosol® AY-65
927-07-1	Aztec® TBPPI-25-AL
927-07-1	Esperox® 31M
927-07-1	Trigonox® 25-C75
928-24-5	Cithrol EGDO N/E
929-06-6	Diglycolamine® Agent (DGA®)
931-36-2	Curezol® 2E4MZ
931-36-2	Imicure® EMI-24
940-41-0	Dynasylan® PETCS
991-84-4	Irganox® 565
995-33-5	Aztec® NBV-40-G
995-33-5	Aztec® NBV-40-IC
995-33-5	Lupersol 230
995-33-5	Trigonox® 17
995-33-5	Trigonox® 17/40
995-33-5	Varox® 230-XL
999-97-3	Dynasylan® HMDS
999-97-3	Petrarch® H7300
1025-15-6	Perkalink® 301
1025-15-6	Perkalink® 301-50D
1067-25-0	Dynasylan® PTMO
1067-33-0	Dabco® T-1
1067-33-0	Fomrez® SUL-3
1067-33-0	Metacure® T-1
1067-53-4	Dynasylan® VTMOEO
1067-53-4	Petrarch® V5000
1068-27-5	Aztec® 2,5-TRI
1068-27-5	Esperal® 230
1068-27-5	Lupersol 130
1068-27-5	Polyvel CR-L10
1068-27-5	Polyvel CR-L20
1068-27-5	Trigonox® 145
1068-27-5	Trigonox® 145-45B
1068-27-5	Trigonox® 145-E85
1068-27-5	Varox® 130
1068-27-5	Varox® 130-XL
1072-35-1	DS-207
1072-35-1	Hal-Lub-N
1072-35-1	Hal-Lub-N TOTM
1072-35-1	Haro® Chem P28G
1072-35-1	Leadstar
1072-35-1	Leadstar Envirostab
1072-35-1	Synpro® Lead Stearate
1077-56-1	Uniplex 108
1119-40-0	DBE-5
1120-24-7	Adma® 10
1120-36-1	Gulftene® 14
1163-19-5	FR-1210
1163-19-5	Great Lakes DE-83™
1163-19-5	Great Lakes DE-83R™
1163-19-5	Octoguard FR-01
1163-19-5	Saytex® 102E
1163-19-5	Thermoguard® 505
1185-55-3	Dynasylan® MTMS

CAS	Trade name
1185-55-3	Petrarch® M9100
1241-94-7	Disflamoll® DPO
1241-94-7	Phosflex 362
1241-94-7	Santicizer 141
1303-96-4	Borax
1305-62-0	Rhenofit® CF
1305-78-8	Rhenosorb® C
1305-78-8	Rhenosorb® C/GW
1305-78-8	Rhenosorb® F
1309-37-1	OSO® 440
1309-37-1	OSO® 1905
1309-37-1	OSO® NR 830 M
1309-37-1	OSO® NR 950
1309-42-8	ACM-MW
1309-42-8	BAX 1091FM
1309-42-8	Flamtard M7
1309-42-8	FR-20
1309-42-8	Hydramax™ HM-B8, HM-B8-S
1309-42-8	Kisuma 5A
1309-42-8	Kisuma 5B
1309-42-8	Kisuma 5E
1309-42-8	Magnifin® H5
1309-42-8	Magnifin® H5B
1309-42-8	Magnifin® H5C
1309-42-8	Magnifin® H5C/3
1309-42-8	Magnifin® H5D
1309-42-8	Magnifin® H-7
1309-42-8	Magnifin® H7A
1309-42-8	Magnifin® H7C
1309-42-8	Magnifin® H7C/3
1309-42-8	Magnifin® H7D
1309-42-8	Magnifin® H-10
1309-42-8	Magnifin® H10A
1309-42-8	Magnifin® H10B
1309-42-8	Magnifin® H10C
1309-42-8	Magnifin® H10D
1309-42-8	Magnifin® H10F
1309-42-8	Magoh-S
1309-42-8	MGH-93
1309-42-8	Versamag® DC
1309-42-8	Versamag® Tech
1309-42-8	Versamag® UF
1309-42-8	Zerogen® 10
1309-42-8	Zerogen® 15
1309-42-8	Zerogen® 30
1309-42-8	Zerogen® 35
1309-42-8	Zerogen® 60
1309-48-4	Elastomag® 100
1309-48-4	Elastomag® 100R
1309-48-4	Elastomag® 170
1309-48-4	Elastomag® 170 Micropellet
1309-48-4	Elastomag® 170 Special
1309-48-4	Ken-Mag®
1309-48-4	MagChem® 20M
1309-48-4	MagChem® 30
1309-48-4	MagChem® 30G
1309-48-4	MagChem® 35
1309-48-4	MagChem® 35K
1309-48-4	MagChem® 40
1309-48-4	MagChem® 50
1309-48-4	MagChem® 50Y
1309-48-4	MagChem® 60
1309-48-4	MagChem® 125
1309-48-4	MagChem® 200-AD
1309-48-4	MagChem® 200-D
1309-48-4	Maglite® D
1309-48-4	Magotex™
1309-48-4	Naftopast® MgO-A
1309-48-4	Plastomag® 170
1309-48-4	Rhenomag® G1
1309-48-4	Rhenomag® G3
1309-60-0	LP-100
1309-60-0	LP-200
1309-60-0	LP-300
1309-60-0	LP-400
1309-64-4	Amsperse
1309-64-4	AO
1309-64-4	FireShield® H
1309-64-4	FireShield® HPM
1309-64-4	FireShield® HPM-UF
1309-64-4	FireShield® L
1309-64-4	Fyraway
1309-64-4	Fyrebloc
1309-64-4	Naftopast® Antimontrioxid
1309-64-4	Naftopast® Antimontrioxid-CP
1309-64-4	Octoguard FR-10
1309-64-4	Petcat R-9
1309-64-4	Thermoguard® L
1309-64-4	Thermoguard® S
1309-64-4	Thermoguard® UF
1309-64-4	UltraFine® II
1313-27-5	Polu-U
1314-13-2	Akrochem® Zinc Oxide 35
1314-13-2	Kadox®-215
1314-13-2	Kadox®-272
1314-13-2	Kadox® 720
1314-13-2	Kadox®-911
1314-13-2	Kadox®-920
1314-13-2	Kadox® 930
1314-13-2	Ken-Zinc®
1314-13-2	Naftopast® ZnO-A
1314-13-2	NAO 105
1314-13-2	NAO 115
1314-13-2	NAO 125
1314-13-2	NAO 135
1314-13-2	Octocure 462
1314-13-2	Ottalume 2100
1314-13-2	RR Zinc Oxide (Untreated)
1314-13-2	Zinc Oxide No. 185
1314-13-2	Zinc Oxide No. 318
1314-13-2	Zink Oxide AT
1314-13-2	Zink Oxide Transparent
1314-13-2	Zinkoxyd Activ®
1314-60-9	Nyacol® A-1530
1314-60-9	Nyacol® A-1540N
1314-60-9	Nyacol® A-1550
1314-60-9	Nyacol® A-1588LP
1314-60-9	Nyacol® AP50
1314-60-9	Nyacol® N24
1314-60-9	Nyacol® ZTA

CAS	Trade name
1314-98-3	Sachtolith® HD
1314-98-3	Sachtolith® HD-S
1314-98-3	Sachtolith® L
1317-36-8	Litharge 28
1317-36-8	Litharge 33
1317-65-3	Atomite®
1317-65-3	CC™-101
1317-65-3	CC™-103
1317-65-3	CC™-105
1317-65-3	CP Filler™
1317-65-3	Dolocron® 15-16
1317-65-3	Dolocron® 32-15
1317-65-3	Dolocron® 40-13
1317-65-3	Dolocron® 45-12
1317-65-3	Drikalite®
1317-65-3	Duramite®
1317-65-3	Franklin T-11
1317-65-3	Franklin T-12
1317-65-3	Franklin T-13
1317-65-3	Franklin T-14
1317-65-3	Franklin T-325
1317-65-3	Marble Dust™
1317-65-3	Marblemite™
1317-65-3	Marblewhite® MW200
1317-65-3	Marblewhite® MW325
1317-65-3	Marblewhite® MW A 200
1317-65-3	Marblewhite® MW A 325
1317-65-3	Micro-White® 07 Slurry
1317-65-3	Micro-White® 10
1317-65-3	Micro-White® 10 Slurry
1317-65-3	Micro-White® 15
1317-65-3	Micro-White® 15 Slurry
1317-65-3	Micro-White® 25
1317-65-3	Micro-White® 25 Slurry
1317-65-3	Micro-White® 40
1317-65-3	Micro-White® 100
1317-65-3	No. 1 White™
1317-65-3	Opacimite™
1317-65-3	Pfinyl® 402
1317-65-3	SC-53
1317-65-3	Snowflake P.E.™
1317-65-3	Snowflake White™
1317-65-3	Super-Fil®
1317-65-3	Supermite®
1317-65-3	Vicron® 10-20
1317-65-3	Vicron® 15-15
1317-65-3	Vicron® 25-11
1317-65-3	Vicron® 31-6
1317-65-3	Vicron® 41-8
1317-65-3	Vicron® 45-3
1317-80-2	Hitox®
1318-93-0	Perchem® 97
1322-93-6	Rhodacal® BA-77
1327-36-2	Kaopolite® 1147
1327-36-2	Kaopolite® 1168
1327-36-2	Kaopolite® SF
1327-36-2	Kaopolite® SFO-NP
1327-39-5	PMF® Fiber 204
1327-39-5	PMF® Fiber 204AX
1327-39-5	PMF® Fiber 204BX
1327-39-5	PMF® Fiber 204CX
1327-39-5	PMF® Fiber 204EX
1330-76-3	Staflex DIOM
1330-78-5	Pliabrac® TCP
1330-80-9	Emerest® 2302
1332-07-6	Firebrake® ZB 415
1332-07-6	Firebrake® ZB-Extra Fine
1332-58-7	Bilt-Plates® 156
1332-58-7	Dixie® Clay
1332-58-7	Fiberfrax® 6000 RPS
1332-58-7	Fiberfrax® 6900-70
1332-58-7	Fiberfrax® 6900-70S
1332-58-7	Fiberfrax® EF-119
1332-58-7	Fiberfrax® EF 122S
1332-58-7	Fiberfrax® EF 129
1332-58-7	Fiberfrax® Milled Fiber
1332-58-7	Fiberkal™
1332-58-7	Fiberkal™ FG
1332-58-7	Huber 35
1332-58-7	Huber 40C
1332-58-7	Huber 65A
1332-58-7	Huber 70C
1332-58-7	Huber 80
1332-58-7	Huber 80B
1332-58-7	Huber 80C
1332-58-7	Huber 90
1332-58-7	Huber 90B
1332-58-7	Huber 90C
1332-58-7	Huber 95
1332-58-7	Huber HG
1332-58-7	Huber HG90
1332-58-7	Hydrite 121-S
1332-58-7	Langford® Clay
1332-58-7	McNamee® Clay
1332-58-7	Par® Clay
1332-58-7	Polyplate 90
1332-58-7	Polyplate 852
1332-58-7	Polyplate P
1332-58-7	Polyplate P01
1333-39-7	Eltesol® PSA 65
1333-84-2	Akrochem® Hydrated Alumina
1333-84-2	Alcan BW53, BW103, BW153
1333-84-2	Alcan FRF 5
1333-84-2	Alcan FRF 10
1333-84-2	Alcan FRF 20
1333-84-2	Alcan FRF 30
1333-84-2	Alcan FRF 40
1333-84-2	Alcan FRF 60
1333-84-2	Alcan FRF 80
1333-84-2	Alcan FRF 80 S5
1333-84-2	Alcan FRF LV1 S5
1333-84-2	Alcan FRF LV2, LV4, LV5, LV6, LV7, LV8, LV9
1333-84-2	Alcan FRF LV2 S5
1333-84-2	Alcan H-10
1333-84-2	Alcan Superfine 4
1333-84-2	Alcan Superfine 7
1333-84-2	Alcan Superfine 11
1333-84-2	Alcan Ultrafine UF7
1333-84-2	Alcan Ultrafine UF11

CAS	Trade name
1333-84-2	Alcan Ultrafine UF15
1333-84-2	Alcan Ultrafine UF25
1333-84-2	Alcan Ultrafine UF35
1333-86-4	Aquathene™ CM 92832
1333-86-4	Black Pearls® 120
1333-86-4	Black Pearls® 130
1333-86-4	Black Pearls® 160
1333-86-4	Black Pearls® 170
1333-86-4	Black Pearls® 280
1333-86-4	Black Pearls® 450
1333-86-4	Black Pearls® 460
1333-86-4	Black Pearls® 470
1333-86-4	Black Pearls® 480
1333-86-4	Black Pearls® 490
1333-86-4	Black Pearls® 520
1333-86-4	Black Pearls® 570
1333-86-4	Black Pearls® 700
1333-86-4	Black Pearls® 800
1333-86-4	Black Pearls® 880
1333-86-4	Black Pearls® 900
1333-86-4	Black Pearls® 1000
1333-86-4	Black Pearls® 1100
1333-86-4	Black Pearls® 3700
1333-86-4	Black Pearls® L
1333-86-4	Continex® LH-10
1333-86-4	Continex® LH-20
1333-86-4	Continex® LH-30
1333-86-4	Continex® LH-35
1333-86-4	Continex® N110
1333-86-4	Continex® N220
1333-86-4	Continex® N234
1333-86-4	Continex® N299
1333-86-4	Continex® N326
1333-86-4	Continex® N330
1333-86-4	Continex® N339
1333-86-4	Continex® N343
1333-86-4	Continex® N351
1333-86-4	Continex® N550
1333-86-4	Continex® N650
1333-86-4	Continex® N660
1333-86-4	Continex® N683
1333-86-4	Continex® N762
1333-86-4	Continex® N774
1333-86-4	Diablack® A
1333-86-4	Diablack® E
1333-86-4	Diablack® G
1333-86-4	Diablack® H
1333-86-4	Diablack® HA
1333-86-4	Diablack® I
1333-86-4	Diablack® II
1333-86-4	Diablack® LH
1333-86-4	Diablack® LI
1333-86-4	Diablack® LR
1333-86-4	Diablack® M
1333-86-4	Diablack® N220M
1333-86-4	Diablack® N234
1333-86-4	Diablack® N339
1333-86-4	Diablack® N550M
1333-86-4	Diablack® N760M
1333-86-4	Diablack® R
1333-86-4	Diablack® SF
1333-86-4	Diablack® SH
1333-86-4	Diablack® SHA
1333-86-4	Elftex® 120
1333-86-4	Elftex® 125
1333-86-4	Elftex® 160
1333-86-4	Elftex® 280
1333-86-4	Elftex® 285
1333-86-4	Elftex® 415
1333-86-4	Elftex® 430
1333-86-4	Elftex® 435
1333-86-4	Elftex® 460
1333-86-4	Elftex® 465
1333-86-4	Elftex® 570
1333-86-4	Elftex® 670
1333-86-4	Elftex® 675
1333-86-4	FW 18
1333-86-4	FW 200 Beads and Powd
1333-86-4	Huber ARO 60
1333-86-4	Huber N110
1333-86-4	Huber N220
1333-86-4	Huber N234
1333-86-4	Huber N299
1333-86-4	Huber N326
1333-86-4	Huber N330
1333-86-4	Huber N339
1333-86-4	Huber N343
1333-86-4	Huber N347
1333-86-4	Huber N351
1333-86-4	Huber N375
1333-86-4	Huber N539
1333-86-4	Huber N550
1333-86-4	Huber N650
1333-86-4	Huber N660
1333-86-4	Huber N683
1333-86-4	Huber N762
1333-86-4	Huber N774
1333-86-4	Huber N787
1333-86-4	Huber N990
1333-86-4	Huber S212
1333-86-4	Huber S315
1333-86-4	Ketjenblack® EC-300 J
1333-86-4	Ketjenblack® EC-310 NW
1333-86-4	Ketjenblack® EC-600 JD
1333-86-4	Ketjenblack® ED-600 JD
1333-86-4	Mitsubishi Carbon Black
1333-86-4	Mogul® L
1333-86-4	Monarch® 120
1333-86-4	Monarch® 700
1333-86-4	Monarch® 800
1333-86-4	Monarch® 880
1333-86-4	Monarch® 900
1333-86-4	Monarch® 1000
1333-86-4	Monarch® 1100
1333-86-4	MPC Channel Black TR 354
1333-86-4	Novacarb
1333-86-4	Printex P
1333-86-4	Regal® 300
1333-86-4	Regal® 300R
1333-86-4	Regal® 330

CAS	Trade name
1333-86-4	Regal® 330R
1333-86-4	Regal® 400
1333-86-4	Regal® 400R
1333-86-4	Regal® 660
1333-86-4	Regal® 660R
1333-86-4	S160 Beads and Powd
1333-86-4	Sable™ 3
1333-86-4	Sable™ 5
1333-86-4	Sable™ 15
1333-86-4	Sable™ 17
1333-86-4	Sterling® 1120
1333-86-4	Sterling® 2320
1333-86-4	Sterling® 3420
1333-86-4	Sterling® 4620
1333-86-4	Sterling® 5550
1333-86-4	Sterling® 5630
1333-86-4	Sterling® 6630
1333-86-4	Sterling® 6640
1333-86-4	Sterling® 6740
1333-86-4	Sterling® 8860
1333-86-4	Sterling® C
1333-86-4	Sterling® NS
1333-86-4	Sterling® NS-1
1333-86-4	Sterling® SO
1333-86-4	Sterling® V
1333-86-4	Sterling® VH
1333-86-4	Sterling® XC-72
1333-86-4	Thermax® Floform® N-990
1333-86-4	Thermax® Floform® Ultra Pure N-990
1333-86-4	Thermax® Powder N-991
1333-86-4	Thermax® Powder Ultra Pure N-991
1333-86-4	Thermax® Stainless
1333-86-4	Thermax® Stainless Floform® N-907
1333-86-4	Thermax® Stainless Powder N-908
1333-86-4	Thermax® Stainless Powder Ultra Pure N-908
1333-86-4	Vulcan® 2H
1333-86-4	Vulcan® 3
1333-86-4	Vulcan® 3H
1333-86-4	Vulcan® 4H
1333-86-4	Vulcan® 6
1333-86-4	Vulcan® 6LM
1333-86-4	Vulcan® 7H
1333-86-4	Vulcan® 8H
1333-86-4	Vulcan® 9
1333-86-4	Vulcan® 9A32
1333-86-4	Vulcan® 10H
1333-86-4	Vulcan® C
1333-86-4	Vulcan® J
1333-86-4	Vulcan® K
1333-86-4	Vulcan® M
1333-86-4	Vulcan® P
1333-86-4	Vulcan® XC72
1333-86-4	Vulcan® XC72R
1335-72-4	Avirol® SE 3002
1335-72-4	Avirol® SE 3003
1337-33-3	Monostearyl Citrate (MSC)
1338-23-4	Aztec® MEKP-HA-1
1338-23-4	Aztec® MEKP-HA-2
1338-23-4	Aztec® MEKP-LA-2
1338-23-4	Aztec® MEKP-SA-2
1338-23-4	Cadox® HBO-50
1338-23-4	Cadox® L-30
1338-23-4	Cadox® L-50
1338-23-4	Cadox® M-105
1338-23-4	Hi-Point® 90
1338-23-4	Hi-Point® PD-1
1338-23-4	Lupersol DDM-9
1338-23-4	Lupersol Delta-3
1338-23-4	Lupersol Delta-X-9
1338-23-4	Lupersol DHD-9
1338-23-4	Quickset® Extra
1338-23-4	Quickset® Super
1338-23-4	Superox® 46-747
1338-23-4	Superox® 46-748
1338-23-4	Superox® 702
1338-23-4	Superox® 709
1338-23-4	Superox® 710
1338-23-4	Superox® 711
1338-23-4	Superox® 712
1338-23-4	Superox® 713
1338-23-4	Superox® 730
1338-23-4	Superox® 732
1338-23-4	Superox® 739
1338-39-2	Alkamuls® SML
1338-39-2	Disponil SML 100 F1
1338-39-2	Emsorb® 2515
1338-39-2	Ethylan® GL20
1338-39-2	Glycomul® L
1338-41-6	Disponil SMS 100 F1
1338-41-6	Ethylan® GS60
1338-41-6	Glycomul® S
1338-43-8	Disponil SMO 100 F1
1338-43-8	Durtan® 80
1338-43-8	Ethylan® GO80
1338-43-8	Glycomul® O
1338-43-8	S-Maz® 80
1338-51-8	Uniplex 600
1338-51-8	Uniplex 680
1341-54-4	Uvinul® 3093
1344-28-1	A-2
1344-28-1	A-201
1344-28-1	Aluminum Oxide C
1344-40-7	Dyphos
1344-40-7	Dyphos Envirostab
1344-40-7	Dyphos XL
1344-40-7	Halphos
1344-40-7	Haro® Chem PDF
1344-40-7	Haro® Chem PDF-B
1344-95-2	Micro-Cel® C
1345-05-7	Lithopone 30% DS
1345-05-7	Lithopone 30% L
1345-05-7	Lithopone D (Red Seal 30% ZnS)
1345-05-7	Lithopone D (Silver Seal 60% ZnS)

CAS	Trade name
1345-05-7	Lithopone DS (Red Seal 30% ZnS)
1345-05-7	Lithopone L (Red Seal 30% ZnS)
1345-05-7	Lithopone L (Silver Seal 60% ZnS)
1369-66-3	Empimin® OP70
1459-93-4	Morflex® 1129
1459-93-4	Uniplex 270
1493-13-6	FC-24
1493-13-6	Fluorad® FC-24
1528-49-0	Morflex® 560
1533-45-5	Eastobrite® OB-1
1561-49-5	Aztec® CHPC
1561-49-5	Aztec® CHPC-90-W
1561-49-5	Perkadox® 18
1561-92-8	Geropon® MLS/A
1592-23-0	ADK STAB CA-ST
1592-23-0	Afco-Chem CS
1592-23-0	Afco-Chem CS-1
1592-23-0	Afco-Chem CS-S
1592-23-0	Akrochem® Calcium Stearate
1592-23-0	Akrochem® P 4000
1592-23-0	Akrochem® P 4100
1592-23-0	Akrodip™ CS-400
1592-23-0	Cecavon CA 30
1592-23-0	Coad 10
1592-23-0	Cometals Calcium Stearate
1592-23-0	Hallcote® CSD
1592-23-0	Hallcote® CaSt 50
1592-23-0	Haro® Chem CGD
1592-23-0	Haro® Chem CGL
1592-23-0	Haro® Chem CGN
1592-23-0	Haro® Chem CPR-2
1592-23-0	Haro® Chem CPR-5
1592-23-0	Hy Dense Calcium Stearate HP Gran
1592-23-0	Hy Dense Calcium Stearate RSN Powd
1592-23-0	HyTech RSN 11-4
1592-23-0	HyTech RSN 248D
1592-23-0	Interstab® CA-18-1
1592-23-0	Mathe Calcium Stearate
1592-23-0	Miljac Calcium Stearate
1592-23-0	Nuodex S-1421 Food Grade
1592-23-0	Nuodex S-1520 Food Grade
1592-23-0	Petrac® Calcium Stearate CP-11
1592-23-0	Petrac® Calcium Stearate CP-11 LS
1592-23-0	Petrac® Calcium Stearate CP-11 LSG
1592-23-0	Petrac® Calcium Stearate CP-22G
1592-23-0	Plastolube
1592-23-0	Quikote™ C
1592-23-0	Quikote™ C-LD
1592-23-0	Quikote™ C-LD-A
1592-23-0	Quikote™ C-LM
1592-23-0	Quikote™ C-LMLD

CAS	Trade name
1592-23-0	Synpro® Calcium Stearate 12B
1592-23-0	Synpro® Calcium Stearate 15
1592-23-0	Synpro® Calcium Stearate 15F
1592-23-0	Synpro® Calcium Stearate 114-36
1592-23-0	TR-Calcium Stearate
1592-23-0	Witco® Calcium Stearate A
1592-23-0	Witco® Calcium Stearate F
1592-23-0	Witco® Calcium Stearate FP
1634-02-2	Akrochem® TBUT
1634-02-2	Butyl Tuads®
1634-04-4	High Purity MTBE
1639-66-3	Geropon® WT-27
1643-20-5	Empigen® OB
1643-20-5	Rhodamox® LO
1656-63-9	Weston® TLTTP
1704-62-7	Jeffcat ZR-70
1709-70-2	Ethanox® 330
1709-70-2	Irganox® 1330
1717-00-6	Genetron® 141b
1760-24-3	Dow Corning® Z-6020
1760-24-3	Dynasylan® DAMO
1760-24-3	Dynasylan® DAMO-P
1760-24-3	Dynasylan® DAMO-T
1760-24-3	Petrarch® A0700
1760-24-3	Petrarch® A0701
1760-24-3	Prosil® 3128
1761-71-3	Amicure® PACM
1843-03-4	Lowinox® CA 22
1843-03-4	Topanol® CA-RT
1843-03-4	Topanol® CA-SF
1843-03-4	Topanol® LVT 11
1843-05-6	BLS™ 531
1843-05-6	Cyasorb® UV 531
1843-05-6	Hostavin® ARO 8
1843-05-6	Lankromark® LE285
1843-05-6	Lowilite® 22
1843-05-6	Mark® 1413
1843-05-6	Uvasorb 3C
1843-05-6	UV-Chek® AM-300
1843-05-6	UV-Chek® AM-301
1843-05-6	Uvinul® 408
1843-05-6	Uvinul® 3008
1847-55-8	Rhodapon® OS
1889-67-4	Perkadox® 30
1908-87-8	Vulkacit CRV/LG
1931-62-0	Aztec® TBPM-50-FT
1931-62-0	Aztec® TBPM-50-P
1931-62-0	Esperox® 41-25
1931-62-0	Esperox® 41-25A
1948-33-0	Eastman® MTBHQ
1948-33-0	Sustane® TBHQ
1948-33-0	Tenox® TBHQ
2031-67-6	Dynasylan® MTES
2031-67-6	Petrarch® M9050
2082-79-3	Akrochem® Antioxidant 1076
2082-79-3	BNX® 1076, 1076G
2082-79-3	Dovernox 76

CAS	Trade name
2082-79-3	Irganox® 1076
2082-79-3	Lankromark® LE384
2082-79-3	Naugard® 76
2082-79-3	Ralox® 530
2082-79-3	Ultranox® 276
2082-81-7	MFM-405
2094-98-6	V-40
2155-71-7	Trigonox® 111-B40
2162-74-5	Stabilizer 7000
2163-42-0	MPDiol® Glycol
2167-23-9	Aztec® BU-50-AL
2167-23-9	Aztec® BU-50-WO
2167-23-9	Trigonox® D-C50
2167-23-9	Trigonox® D-E50
2210-79-9	DY 023
2210-79-9	Epodil® 742
2210-79-9	Heloxy® 62
2223-93-0	Cometals Cadmium Stearate
2223-93-0	Haro® Chem KPR
2223-93-0	Metasap® Cadmium Stearate
2223-93-0	Miljac Cadmium Stearate
2223-93-0	Synpro® Cadmium Stearate
2235-54-3	Akyposal ALS 33
2235-54-3	Calfoam NLS-30
2235-54-3	Carsonol® ALS
2235-54-3	Carsonol® ALS-S
2235-54-3	Carsonol® ALS Special
2235-54-3	Rhodapon® L-22
2235-54-3	Rhodapon® L-22/C
2235-54-3	Rhodapon® L-22HNC
2235-54-3	Sermul EA 129
2235-54-3	Surfax 215
2277-28-3	Dimodan LS Kosher
2358-84-1	Ageflex DEGDMA
2358-84-1	MFM-418
2372-21-6	Trigonox® BPIC
2409-55-4	Ralox® 240
2421-28-5	Allco BTDA
2421-28-5	Allco PMDA
2425-79-8	Epodil® 750
2425-79-8	Heloxy® 67
2425-79-8	RD-2
2426-08-6	Ageflex BGE
2426-08-6	Epodil® 741
2426-08-6	Heloxy® 61
2426-08-6	RD-1
2426-54-2	Ageflex FA-2
2439-35-2	Ageflex FA-1
2440-22-4	Lowilite® 55
2440-22-4	Tinuvin® P
2440-22-4	Topanex 100BT
2440-22-4	Uvasorb SV
2440-22-4	Viosorb 520
2452-01-9	Haro® Chem ZSG
2455-24-5	BM-729
2461-15-6	Epodil® 746
2461-15-6	Heloxy® 116
2489-77-2	Thiate® EF-2
2499-95-8	Ageflex n-HA
2530-83-8	Dow Corning® Z-6040
2530-83-8	Dynasylan® GLYMO
2530-83-8	Petrarch® G6720
2530-83-8	Prosil® 5136
2530-85-0	Dow Corning® Z-6030
2530-85-0	Dynasylan® MEMO
2530-85-0	Petrarch® M8550
2530-85-0	Prosil® 248
2530-87-2	Dow Corning® Z-6076
2530-87-2	Dynasylan® CPTMO
2530-87-2	Petrarch® C3300
2550-33-6	Trigonox® 27
2589-57-3	V-601
2638-94-0	V-501
2673-22-5	Aerosol® TR-70
2673-22-5	Geropon® BIS/SODICO-2
2673-22-5	Hodag DTSS-70
2673-22-5	Polirol TR/LNA
2768-02-7	Dynasylan® VTMO
2768-02-7	Petrarch® V4917
2781-00-2	Perkadox® 64-10PP-pd
2781-00-2	Perkadox® 64-20PP-pd
2781-00-2	Perkadox® 64-40PP-pd
2807-30-9	Ektasolve® EP
2855-13-2	Vestamin® IPD
2864-99-1	Tinuvin® 327
2867-47-2	Ageflex FM-1
2917-73-9	Plastolein® 9048
2943-75-1	Dynasylan® OCTEO
2943-75-1	Petrarch® O9835
2997-92-4	V-50
3006-15-3	Aerosol® MA-80I
3006-15-3	Empimin® MA
3006-82-4	Esperox® 28
3006-82-4	Esperox® 28MD
3006-82-4	Esperox® 28PD
3006-82-4	Trigonox® 21
3006-82-4	Trigonox® 21-C50
3006-82-4	Trigonox® 21 LS
3006-86-8	Aztec® 1,1-BIS-50-AL
3006-86-8	Aztec® 1,1-BIS-50-WO
3006-86-8	Trigonox® 22-BB80
3006-93-7	Vanax® MBM
3026-63-9	Avirol® SA 4113
3026-63-9	Rhodapon® TDS
3034-79-5	Perkadox® 20
3034-79-5	Perkadox® 20-W40
3055-93-4	Prox-onic LA-1/02
3055-99-0	Prox-onic LA-1/09
3056-00-6	Prox-onic LA-1/012
3057-08-7	Weston® PNPG
3061-75-4	Kemamide® B
3064-73-1	Isobutyl Tuads®
3069-25-8	Dynasylan® 1110
3069-29-2	Dynasylan® 1411
3069-29-2	Petrarch® A0699
3076-63-9	Doverphos® 53
3076-63-9	Weston® TLP
3077-12-1	Emery® 5710
3081-14-9	Naugard® I-2
3081-14-9	Santoflex® 77

CAS	Trade name
3081-14-9	Vulkanox® 4030
3088-31-1	Surfax 218
3089-11-0	Cymel 303
3089-11-0	R7234
3101-60-8	Epodil® 745
3101-60-8	Heloxy® 65
3121-61-7	Ageflex MEA
3147-75-9	Cyasorb® UV 5411
3158-56-3	Cardolite® NC-507
3179-56-4	Aztec® ACSP-28-FT
3179-56-4	Aztec® ACSP-55-W1
3179-56-4	Trigonox® ACS-M28
3179-76-8	Dynasylan® 1505
3179-76-8	Petrarch® A0742
3179-76-8	Petrarch® A0743
3194-55-6	Saytex® HBCD-LM
3278-89-5	FR-913
3278-89-5	Great Lakes PHE-65™
3290-92-4	Ageflex TM 402, 403, 404, 410, 421, 423, 451, 461, 462
3290-92-4	Ageflex TMPTMA
3290-92-4	Perkalink® 400
3290-92-4	Perkalink® 400-50D
3290-92-4	Sipomer® TMPTMA
3296-90-0	FR-521
3296-90-0	FR-522
3296-90-0	FR-1138
3319-31-1	ADK CIZER C-8
3319-31-1	Diplast® TM
3319-31-1	Jayflex® TOTM
3319-31-1	Kodaflex® TOTM
3319-31-1	Nuoplaz® 6959
3319-31-1	Nuoplaz® TOTM
3319-31-1	Plasthall® TOTM
3319-31-1	PX-338
3322-93-8	Saytex® BCL-462
3327-22-8	CHPTA 65%
3333-62-8	Leucopure EGM Powd
3401-73-8	Monawet SNO-35
3425-61-4	Aztec® TAHP-80-AL
3425-61-4	Trigonox® TAHP-W85
3457-61-2	Aztec® BCUP
3457-61-2	Lupersol 801
3457-61-2	Trigonox® T
3457-61-2	Trigonox® T-94
3578-72-1	Haro® Chem CBHG
3622-84-2	Dellatol® BBS
3622-84-2	Plasthall® BSA
3622-84-2	Uniplex 214
3648-20-2	Jayflex® L11P
3648-20-2	PX-111
3648-21-3	Diplast® E
3655-00-3	Deriphat® 160
3710-84-7	Pennstop® 1866
3710-84-7	Pennstop® 2049
3710-84-7	Pennstop® 2697
3741-08-8	Santocure® TBSI
3806-34-6	Doverphos® S-680
3806-34-6	Doverphos® S-686, S-687
3806-34-6	Mark® 5060
3806-34-6	Weston® 618F
3806-34-6	Weston® 619F
3825-26-1	Fluorad® FC-118
3851-87-4	Trigonox® 36-C60
3851-87-4	Trigonox® 36-C75
3851-87-4	Trigonox® 36-W40
3851-87-4	USP®-355M
3896-11-5	Lowilite® 26
3896-11-5	Tinuvin® 326
3999-10-8	Blo-Foam® 5PT
3999-10-8	Expandex® 5PT
3999-10-8	Unicell 5PT
4130-08-9	Dow Corning® Z-6075
4196-86-5	Benzoflex® S-552
4196-86-5	Uniplex 552
4196-89-8	Benzoflex® S-312
4219-48-1	Cithrol EGML N/E
4221-80-1	UV-Chek® AM-340
4253-34-3	Dynasylan® MTAC
4306-88-1	Isonox® 232
4306-88-1	Uvi-Nox® 1494
4419-11-8	V-65
4419-11-8	Vazo 52
4420-74-0	Dynasylan® MTMO
4420-74-0	Petrarch® M8500
4420-74-0	Prosil® 196
4485-12-5	Afco-Chem LIS
4485-12-5	Mathe Lithium Stearate
4485-12-5	Synpro® Lithium Stearate
4485-12-5	Witco® Lithium Stearate 306
4500-01-0	Cithrol EGMO N/E
4511-39-1	Aztec® TAPB
4511-39-1	Esperox® 5100
4511-39-1	Trigonox® 127
4712-55-4	Doverphos® 213
4712-55-4	Weston® DPP
4722-98-9	Empicol® LQ33/T
4748-78-1	Ebal
4767-03-7	DMPA®
4979-32-2	Perkacit® DCBS
4979-32-2	Santocure® DCBS
5089-70-3	Dynasylan® CPTEO
5089-70-3	Petrarch® C3292
5232-99-5	Uvinul® 3035
5274-68-0	Prox-onic LA-1/04
5323-95-5	Solricin® 435
5323-95-5	Solricin® 535
5593-70-4	Tyzor TBT
5719-73-3	Duralink® HTS
5793-94-2	Pationic® 930
5793-94-2	Pationic® 930K
5793-94-2	Pationic® 940
5809-08-5	Trigonox® TMPH
6001-97-4	Lankropol® KMA
6197-30-4	Uvinul® 3039
6542-37-6	Zoldine® ZT-65
6674-22-2	Amicure® DBU-E
6674-22-2	Polycat® DBU
6674-22-2	Polycat® SA-1
6674-22-2	Polycat® SA-102

CAS	Trade name
6674-22-2	Polycat® SA-610/50
6683-19-8	Akrochem® Antioxidant 1010
6683-19-8	BNX® 1010, 1010G
6683-19-8	Dovernox 10
6683-19-8	Irganox® 1010
6683-19-8	Lankromark® LE373
6683-19-8	Lowinox® PP35
6683-19-8	Naugard® 10
6683-19-8	Ralox® 630
6683-19-8	Ultranox® 210
6729-96-0	Evangard® DTB
6731-36-8	Aztec® 3,3,5-TRI
6731-36-8	Aztec® 3,3,5-TRI-40-G
6731-36-8	Aztec® 3,3,5-TRI-40-IC
6731-36-8	Aztec® 3,3,5-TRI-50-AL
6731-36-8	Aztec® 3,3,5-TRI-50-FT
6731-36-8	Aztec® 3,3,5-TRI-55-AL
6731-36-8	Luperco 231-SRL
6731-36-8	Lupersol 231
6731-36-8	Peroximon® S-164/40P
6731-36-8	Trigonox® 29
6731-36-8	Trigonox® 29-40B-pd
6731-36-8	Varox® 231
6731-36-8	Varox® 231-XL
6846-50-0	Kodaflex® TXIB
6865-35-6	Cometals Barium Stearate
6865-35-6	Haro® Chem BG
6865-35-6	Mathe Barium Stearate
6865-35-6	Mathe Barium Stearate
6865-35-6	Mathe Barium Stearate
6865-35-6	Mathe Barium Stearate
6865-35-6	Metasap® Barium Stearate
6865-35-6	Miljac Barium Stearate
6865-35-6	Synpro® Barium Stearate
6881-94-3	Ektasolve® DP
6886-38-5	Ralox® 35
6990-43-8	Vocol®, S-75
7047-84-9	Cometals Aluminum Stearate
7047-84-9	HyTech AX/603
7047-84-9	HyTech T/351
7047-84-9	Mathe 6T
7047-84-9	Synpro® Aluminum Stearate 303
7047-84-9	Witco® Aluminum Stearate 18
7128-64-5	Uvitex® OB
7173-62-8	Kemamine® D-989
7320-34-5	Empiphos 4KP
7378-99-6	Adma® 8
7396-58-9	Armeen® M2-10D
7396-58-9	Dama® 1010
7440-44-0	Synthetic Carbon 910-44M
7492-30-0	Solricin® 135
7534-94-3	Sipomer® IBOMA
7534-94-3	Sipomer® IBOMA HP
7538-44-5	CB2408
7580-31-6	UV-Chek® AM-205
7585-39-9	Beta W 7
7585-39-9	Gamma W8
7617-78-9	Tomah PA-14
7631-86-9	Aerosil® 325
7631-86-9	Lo-Vel® 27
7631-86-9	Lo-Vel® 28
7631-86-9	Lo-Vel® 29
7631-86-9	Lo-Vel® 39A
7631-86-9	Lo-Vel® 66
7631-86-9	Lo-Vel® 275
7631-86-9	Lo-Vel® HSF
7631-86-9	Tixosil 311
7637-07-2	Leecure B-110
7637-07-2	Leecure B-550
7637-07-2	Leecure B-610
7637-07-2	Leecure B-612
7637-07-2	Leecure B-614
7637-07-2	Leecure B-950
7637-07-2	Leecure B-1310
7637-07-2	Leecure B-1550
7637-07-2	Leecure B-1600
7637-07-2	Leecure B-1700
7659-86-1	CTA
7665-72-7	Ageflex TBGE
7704-34-9	Chloropren-Faktis A
7704-34-9	Crystex® 60
7704-34-9	Crystex® Regular
7704-34-9	Faktis 10
7704-34-9	Faktis 14
7704-34-9	Faktis Asolvan, Asolvan T
7704-34-9	Faktis Badenia C
7704-34-9	Faktis HF Braun
7704-34-9	Faktis RC 110, RC 111, RC 140, RC 141, RC 144
7704-34-9	Naftopast® Schwefel-P
7704-34-9	Octocure 456
7704-34-9	Rubbermakers Sulfur
7704-34-9	Struktol® SU 95
7704-34-9	Struktol® SU 105
7704-34-9	Struktol® SU 120
7727-43-7	Bara-200C
7727-43-7	Bara-200N
7727-43-7	Bara-325C
7727-43-7	Bara-325N
7727-43-7	Barafoam 200
7727-43-7	Barafoam 325
7727-43-7	Barimite™
7727-43-7	Barimite™ 200
7727-43-7	Barimite™ 1525P
7727-43-7	Barimite™ 4009P
7727-43-7	Barimite™ UF
7727-43-7	Barimite™ XF
7727-43-7	Bartex® 10
7727-43-7	Bartex® 65
7727-43-7	Bartex® 80
7727-43-7	Bartex® FG-2
7727-43-7	Bartex® FG-10
7727-43-7	Bartex® OWT
7727-43-7	Barytes Microsupreme 2410/M
7727-43-7	Barytes Supreme
7727-43-7	Blanc fixe F
7727-43-7	Blanc fixe micro®
7727-43-7	Blanc fixe N
7727-43-7	Cimbar™ 325

CAS	Trade name
7727-43-7	Cimbar™ 1025P
7727-43-7	Cimbar™ 1536P
7727-43-7	Cimbar™ 3508P
7727-43-7	Cimbar™ CF
7727-43-7	Cimbar™ UF
7727-43-7	Cimbar™ XF
7727-43-7	G-50 Barytes™
7727-43-7	No. 22 Barytes™
7772-99-8	Fascat® 2004
7772-99-8	Fascat® 2005
7779-27-3	Vancide® TH
7782-42-5	Lonza KS 44
7782-42-5	Lonza T 44
8001-21-6	GTO 80
8001-21-6	GTO 90
8001-21-6	GTO 90E
8001-21-6	NS-20
8001-21-6	Sunyl® 80
8001-21-6	Sunyl® 80 RBD
8001-21-6	Sunyl® 80 RBD ES
8001-21-6	Sunyl® 80 RBWD
8001-21-6	Sunyl® 80 RBWD ES
8001-21-6	Sunyl® 90
8001-21-6	Sunyl® 90 RBD
8001-21-6	Sunyl® 90 RBWD
8001-21-6	Sunyl® 90E RBWD
8001-21-6	Sunyl® 90E RBWD ES 1016
8001-21-6	Sunyl® HS 500
8001-75-0	Koster Keunen Ceresine
8001-75-0	Ross Ceresine Wax
8001-78-3	Castorwax® MP-70
8001-78-3	Castorwax® MP-80
8001-78-3	Glycolube® CW-1
8002-13-9	RS-80
8002-23-1	W.G.S. Synaceti 116 NF/USP
8002-33-3	Cordon NU 890/75
8002-33-3	Monosulf
8002-53-7	Ross Montan Wax
8002-53-7	Sicolub® E
8002-53-7	Sicolub® OP
8002-74-2	Advawax® 165
8002-74-2	Akrowax 130
8002-74-2	Akrowax 145
8002-74-2	Hostalub® XL 165FR
8002-74-2	Hostalub® XL 165P
8002-74-2	Loxiol® G 22
8002-74-2	Loxiol® G 23
8002-74-2	Loxiol® HOB 7138
8002-74-2	Loxiol® HOB 7169
8002-74-2	Octowax 321
8002-74-2	Petrac® 165
8002-74-2	Ross Wax #100
8002-74-2	Ross Wax #145
8002-74-2	Ross Wax #165
8002-74-2	Sunolite® 130
8002-74-2	Sunolite® 160
8002-74-2	Vanwax® H Special
8002-74-2	Vestowax FT-150
8002-74-2	Vestowax FT-150P
8002-74-2	Vestowax FT-200
8002-74-2	Vestowax FT-300
8002-74-2	Vybar® 103
8002-74-2	Vybar® 253
8002-74-2	Vybar® 260
8002-74-2	Vybar® 825
8006-44-8	Koster Keunen Candelilla
8007-18-9	Ferro V-9415
8007-24-7	Cardolite® NC-511
8007-43-0	Disponil SSO 100 F1
8009-03-8	Penreco 1520
8009-03-8	Penreco 3070
8009-03-8	Penreco Red
8013-05-6	Solricin® 235
8013-07-8	ADK CIZER O-130P
8013-07-8	Drapex® 6.8
8013-07-8	Epoxol 7-4
8013-07-8	Estabex® 138-A
8013-07-8	Estabex® 2307
8013-07-8	Estabex® 2307 DEOD
8013-07-8	Flexol® Plasticizer EPO
8013-07-8	Lankroflex® GE
8013-07-8	Nuoplaz® 849
8013-07-8	Paraplex® G-62
8013-07-8	Peroxidol 780
8013-07-8	Plas-Chek® 775
8013-07-8	Plasthall® ESO
8013-07-8	Plastoflex® 2307
8013-07-8	Plastolein® 9232
8013-07-8	PX-800
8015-86-9	Ross Carnauba Wax
8021-55-4	Koster Keunen Ozokerite
8027-32-5	Fonoline® White
8027-32-5	Fonoline® Yellow
8027-32-5	Molder's Edge ME-301
8027-32-5	Molder's Edge ME-304
8027-32-5	Penreco Amber
8027-32-5	Penreco Snow
8027-32-5	Protopet® White 1S
8027-32-5	Protopet® Yellow 2A
8030-78-2	Arquad® T-27W
8042-47-5	Drakeol 9
8042-47-5	Drakeol 10
8042-47-5	Drakeol 19
8042-47-5	Drakeol 21
8042-47-5	Drakeol 32
8042-47-5	Drakeol 34
8042-47-5	Drakeol 35
8050-09-7	Diprosin A-100
8050-09-7	Westvaco® Resin 90
8050-09-7	Westvaco® Resin 95
8050-13-3	Hercolyn® D
8050-26-8	Resinall 610
8050-26-8	Uni-Rez® 3020
8050-26-8	Uni-Tac® R99
8050-26-8	Uni-Tac® R100
8050-26-8	Uni-Tac® R100-Light
8050-26-8	Uni-Tac® R100RM
8050-26-8	Uni-Tac® R101
8050-26-8	Uni-Tac® R102
8050-26-8	Uni-Tac® R106

CAS	Trade name
8050-26-8	Uni-Tac® R112
8050-26-8	Zonester® 100
8050-81-5	Wacker Silicone Antifoam Emulsion SE 9
8052-10-6	Westvaco® Rosin T
8061-51-6	Darvan® No. 2
8103-07-8	Paraplex® G-60
9002-83-9	Halocarbon Grease 25-10M
9002-83-9	Halocarbon Grease 25-20M
9002-83-9	Halocarbon Grease X90-10M
9002-83-9	Halocarbon Oil 0.8
9002-83-9	Halocarbon Oil 1.8
9002-83-9	Halocarbon Oil 4.2
9002-83-9	Halocarbon Oil 6.3
9002-83-9	Halocarbon Oil 27
9002-83-9	Halocarbon Oil 56
9002-83-9	Halocarbon Oil 95
9002-83-9	Halocarbon Oil 200
9002-83-9	Halocarbon Oil 400
9002-83-9	Halocarbon Oil 700
9002-83-9	Halocarbon Oil 1000N
9002-83-9	Halocarbon Wax 40
9002-83-9	Halocarbon Wax 600
9002-83-9	Halocarbon Wax 1200
9002-83-9	Halocarbon Wax 1500
9002-84-0	Algoflon® L
9002-84-0	CeriDust 961OF
9002-84-0	Fluon® L169
9002-84-0	Fluoroglide® FL 1690
9002-84-0	Fluoroglide® FL 1700
9002-84-0	Hostaflon® TF 9202
9002-84-0	Hostaflon® TF 9203
9002-84-0	McLube 1700
9002-84-0	McLube 1704
9002-84-0	McLube 1775
9002-84-0	McLube 1777
9002-84-0	McLube 1777-1
9002-84-0	McLube 1782
9002-84-0	McLube 2000 Series
9002-84-0	PeFlu 727
9002-84-0	Polylube J
9002-84-0	Polymist® 284
9002-84-0	Polymist® F-5
9002-84-0	Polymist® F-5A
9002-84-0	Polymist® F-5A EX
9002-84-0	Polymist® F-510
9002-84-0	Polymist® XPH-284
9002-84-0	Teflon® MP 1000
9002-84-0	Teflon® MP 1100
9002-84-0	Teflon® MP 1200
9002-84-0	Teflon® MP 1300
9002-84-0	Teflon® MP 1400
9002-84-0	Teflon® MP 1500
9002-84-0	Teflon® MP 1600
9002-84-0	TL-115
9002-84-0	Tuf Lube Hi-Temp NOD
9002-84-0	Vydax® 1000
9002-84-0	Vydax® AR
9002-84-0	Whitcon® TL-5
9002-84-0	Whitcon® TL-6

CAS	Trade name
9002-84-0	Whitcon® TL-102-2
9002-84-0	Whitcon® TL-155
9002-84-0	Whitcon® TL-169
9002-86-2	Vestolit® E 7012
9002-88-4	A-C® 6
9002-88-4	A-C® 6A
9002-88-4	A-C® 7, 7A
9002-88-4	A-C® 8, 8A
9002-88-4	A-C® 9, 9A, 9F
9002-88-4	A-C® 15
9002-88-4	A-C® 16
9002-88-4	A-C® 617, 617A
9002-88-4	A-C® 712
9002-88-4	A-C® 715
9002-88-4	A-C® 725
9002-88-4	A-C® 735
9002-88-4	A-C® 1702
9002-88-4	ACumist® A-6
9002-88-4	ACumist® A-12
9002-88-4	ACumist® A-18
9002-88-4	ACumist® A-45
9002-88-4	ACumist® B-6
9002-88-4	ACumist® B-9
9002-88-4	ACumist® B-12
9002-88-4	ACumist® B-18
9002-88-4	ACumist® C-5
9002-88-4	ACumist® C-9
9002-88-4	ACumist® C-12
9002-88-4	ACumist® C-18
9002-88-4	ACumist® C-30
9002-88-4	ACumist® D-9
9002-88-4	Bowax 2015
9002-88-4	Cabot® CP 9396
9002-88-4	Cabot® PE 9324
9002-88-4	Coathylene HA 1591
9002-88-4	Coathylene HA 2454
9002-88-4	Epolene® C-10
9002-88-4	Epolene® C-10P
9002-88-4	Epolene® C-13
9002-88-4	Epolene® C-13P
9002-88-4	Epolene® C-14
9002-88-4	Epolene® C-15
9002-88-4	Epolene® C-15P
9002-88-4	Epolene® C-16
9002-88-4	Epolene® C-17
9002-88-4	Epolene® C-17P
9002-88-4	Epolene® C-18
9002-88-4	Epolene® N-10
9002-88-4	Epolene® N-11
9002-88-4	Epolene® N-11P
9002-88-4	Epolene® N-14
9002-88-4	Epolene® N-14P
9002-88-4	Epolene® N-20
9002-88-4	Epolene® N-21
9002-88-4	Epolene® N-34
9002-88-4	Hoechst Wax PE 130
9002-88-4	Hoechst Wax PE 190
9002-88-4	Hoechst Wax PE 520
9002-88-4	Hoechst Wax PED 521
9002-88-4	Hostalub® H 3

CAS	Trade name
9002-88-4	Hostalub® H 22
9002-88-4	Loobwax 0651
9002-88-4	Luwax® A
9002-88-4	Luwax® AF 30
9002-88-4	Luwax® AH 6
9002-88-4	Luwax® AL 3
9002-88-4	Luwax® AL 61
9002-88-4	Octowax 518
9002-88-4	Oxp
9002-88-4	Polyfin
9002-88-4	Polymel #7
9002-88-4	Polywax® 500
9002-88-4	Polywax® 655
9002-88-4	Polywax® 850
9002-88-4	Polywax® 1000
9002-88-4	Polywax® 2000
9002-88-4	Polywax® 3000
9002-88-4	Primax UH-1060
9002-88-4	Primax UH-1080
9002-88-4	Primax UH-1250
9002-88-4	Stan Wax
9002-88-4	Stat-Kon® AS-F
9002-88-4	Stat-Kon® AS-FE
9002-88-4	Struktol® PE H-100
9002-88-4	Struktol® TR PE(H)-100
9002-88-4	Uniwax AW-1050
9002-88-4	Vestowax A-217
9002-88-4	Vestowax A-227
9002-88-4	Vestowax A-235
9002-88-4	Vestowax A-415
9002-88-4	Vestowax A-616
9002-88-4	Vestowax AO-1539
9002-88-4	Zylac 235
9002-92-0	Akyporox RLM 160
9002-92-0	Ethosperse® LA-4
9002-92-0	Prox-onic LA-1/023
9002-92-0	Rhodasurf® A 24
9002-92-0	Rhodasurf® L-25
9002-92-0	Rhodasurf® L-790
9002-93-1	Akyporox OP 250V
9002-93-1	Akyporox OP 400V
9002-93-1	Cedepal CA-890
9002-93-1	Chemax OP-40/70
9002-93-1	Desonic® S-100
9002-93-1	Desonic® S-405
9002-93-1	Hyonic OP-70
9002-93-1	Iconol OP-10
9002-93-1	Iconol OP-30
9002-93-1	Iconol OP-30-70%
9002-93-1	Iconol OP-40
9002-93-1	Iconol OP-40-70%
9002-93-1	Igepal® CA-880
9002-93-1	Igepal® CA-887
9002-93-1	Igepal® CA-890
9002-93-1	Igepal® CA-897
9002-93-1	Macol® OP-40(70)
9002-93-1	Nissan Nonion HS-240
9002-93-1	Nissan Nonion HS-270
9002-93-1	Prox-onic OP-09
9002-93-1	Prox-onic OP-030/70
9002-93-1	Prox-onic OP-040/70
9002-93-1	Rexol 45/307
9002-93-1	Rexol 45/407
9002-93-1	Serdox NOP 30/70
9002-93-1	Serdox NOP 40/70
9002-93-1	Synperonic OP16
9002-93-1	Synperonic OP30
9002-93-1	T-Det® O-40
9002-93-1	T-Det® O-307
9002-93-1	Triton® X-15
9002-93-1	Triton® X-35
9002-93-1	Triton® X-165-70%
9002-93-1	Triton® X-305-70%
9002-93-1	Triton® X-405-70%
9002-93-1	Triton® X-705-70%
9002-93-1	Trycol® 6984
9002-93-1	Trycol® OP-407
9003-05-8	Cyanamer N300LMW
9003-08-1	Cymel 373
9003-08-1	Cymel 380
9003-09-2	Gantrez® M-154
9003-11-6	Antarox® F88
9003-11-6	Antarox® L-62
9003-11-6	Antarox® L-62 LF
9003-11-6	Antarox® L-64
9003-11-6	Berol 374
9003-11-6	Epan 710
9003-11-6	Epan 720
9003-11-6	Epan 740
9003-11-6	Epan 750
9003-11-6	Monolan® 8000 E/80
9003-11-6	Newpol PE-61
9003-11-6	Newpol PE-62
9003-11-6	Newpol PE-64
9003-11-6	Newpol PE-68
9003-11-6	Newpol PE-74
9003-11-6	Newpol PE-75
9003-11-6	Newpol PE-78
9003-11-6	Newpol PE-88
9003-11-6	Nikkol GO-440
9003-17-2	Diene® 35AC3
9003-17-2	Diene® 35AC10
9003-17-2	Diene® 55AC10
9003-17-2	Diene® 55NF
9003-17-2	Diene® 70AC
9003-17-2	Taipol BR 015H
9003-17-2	Taktene 1202
9003-17-2	Taktene 1203
9003-17-2	Taktene 1203G1
9003-18-3	Krynac® 34.140
9003-18-3	Krynac® PXL 38.20
9003-18-3	Krynac® XL 31.25
9003-18-3	Nipol® 1000X88
9003-18-3	Nipol® 1001 CG
9003-18-3	Nipol® 1022X59
9003-18-3	Nipol® 1042X82
9003-18-3	Nipol® 1053
9003-18-3	Nipol® 1312
9003-18-3	Nipol® 1312 LV
9003-34-3	Poly DNB®

CAS	Trade name
9003-34-3	Vanax® PY
9003-39-8	Luviskol® K17
9003-39-8	Luviskol® K30
9003-39-8	Luviskol® K60
9003-39-8	Luviskol® K90
9003-39-8	PVP K-60
9003-53-6	Darex® 670L
9003-54-7	Baymod® A 95
9003-54-7	Blendex® 869
9003-55-8	Darex® 632L
9003-55-8	Darex® 636L
9003-55-8	Duradene® 710
9003-55-8	Pliolite® S-6B, S-6F
9003-55-8	Polysar S-1018
9003-55-8	Polysar SS 260
9003-55-8	Ricon 100
9003-55-8	SR-7475
9003-55-8	Stereon® 205
9003-55-8	Stereon® 210
9003-55-8	Stereon® 721A
9003-55-8	Stereon® 730A
9003-55-8	Stereon® 840A
9003-55-8	Stereon® 841A
9003-56-9	Baymod® A 50
9003-56-9	Baymod® A 90/92
9003-56-9	Baymod® A KU3-2086
9003-56-9	Blendex® 131
9003-56-9	Blendex® 336
9003-56-9	Blendex® 338
9003-56-9	Blendex® 467
9003-73-0	Piccolyte® C115
9003-73-0	Zonarez® 7085
9003-79-6	Agerite® Superflex
9003-79-6	Agerite® Superflex Solid G
9003-79-6	Aminox®
9003-79-6	BXA Flake
9003-79-6	Vanox® AM
9004-32-4	Tylose® C, CB Series
9004-32-4	Tylose® CBR Grades
9004-32-4	Tylose® CR
9004-34-6	Fibra-Cel® BH-70
9004-34-6	Fibra-Cel® BH-200
9004-34-6	Fibra-Cel® CBR-18
9004-34-6	Fibra-Cel® CBR-40
9004-34-6	Fibra-Cel® CBR-41
9004-34-6	Fibra-Cel® CBR-50
9004-34-6	Fibra-Cel® CC-20
9004-34-6	Fibra-Cel® SW-8
9004-34-6	Santoweb® D
9004-34-6	Santoweb® DX
9004-34-6	Santoweb® H
9004-34-6	Santoweb® W
9004-62-0	Natrosol® Hydroxyethyl-cellulose
9004-62-0	Tylose® H Series
9004-65-3	Methocel® E
9004-67-5	Methocel® A
9004-81-3	Chimipal APG 400
9004-81-3	Cithrol DGML S/E
9004-81-3	Emerest® 2650

CAS	Trade name
9004-81-3	Hefti PGE-600-ML
9004-81-3	Kessco® PEG 200 ML
9004-81-3	Kessco® PEG 300 ML
9004-81-3	Kessco® PEG 400 ML
9004-81-3	Kessco® PEG 600 ML
9004-81-3	Kessco® PEG 1000 ML
9004-81-3	Kessco® PEG 1540 ML
9004-81-3	Kessco® PEG 4000 ML
9004-81-3	Kessco® PEG 6000 ML
9004-81-3	Nopalcol 1-L
9004-81-3	Nopalcol 4-L
9004-81-3	Nopalcol 6-L
9004-81-3	Pegosperse® 200 ML
9004-81-3	Pegosperse® 400 ML
9004-81-3	Polirol L400
9004-81-3	Cithrol DGML N/E
9004-82-4	Abex® 23S
9004-82-4	Akyposal 9278 R
9004-82-4	Akyposal EO 20 MW
9004-82-4	Avirol® FES 996
9004-82-4	Calfoam ES-30
9004-82-4	Disponil FES 32
9004-82-4	Disponil FES 61
9004-82-4	Disponil FES 77
9004-82-4	Disponil FES 92E
9004-82-4	Empimin® KSN27
9004-82-4	Empimin® KSN70
9004-82-4	Laural LS
9004-82-4	Polystep® B-12
9004-82-4	Rewopol® NL 2-28
9004-82-4	Rewopol® NL 3
9004-82-4	Rewopol® NL 3-28
9004-82-4	Rewopol® NL 3-70
9004-82-4	Rhodapex® ES
9004-82-4	Sermul EA 370
9004-82-4	Steol® 4N
9004-82-4	Texapon® NSE
9004-82-4	Avirol® SE 3002
9004-82-4	Avirol® SE 3003
1335-72-4	Avirol® SE 3003
9004-82-4	Surfax 218
9004-82-4	Empicol® ESB
9004-82-4	Sermul EA 30
9004-93-7	Flexol® Plasticizer 4GO
9004-93-7	Uniplex 809
9004-96-0	Acconon 300-MO
9004-96-0	Acconon 400-MO
9004-96-0	Cithrol A
9004-96-0	Cithrol DGMO N/E
9004-96-0	Cithrol DGMO S/E
9004-96-0	Ethylan® A4
9004-96-0	Ethylan® A6
9004-96-0	Kessco® PEG 200 MO
9004-96-0	Kessco® PEG 300 MO
9004-96-0	Kessco® PEG 400 MO
9004-96-0	Kessco® PEG 600 MO
9004-96-0	Kessco® PEG 1000 MO
9004-96-0	Kessco® PEG 1540 MO
9004-96-0	Kessco® PEG 4000 MO
9004-96-0	Kessco® PEG 6000 MO

CAS	Trade name
9004-96-0	Nopalcol 4-O
9004-97-1	Nopalcol 6-R
9004-98-2	Akyporox RTO 70
9004-98-2	Prox-onic OA-1/04
9004-98-2	Prox-onic OA-1/09
9004-98-2	Prox-onic OA-1/020
9004-98-2	Prox-onic OA-2/020
9004-98-2	Rhodasurf® ON-870
9004-98-2	Rhodasurf® ON-877
9004-99-3	Alkamuls® S-8
9004-99-3	Cithrol DGMS N/E
9004-99-3	Cithrol DGMS S/E
9004-99-3	Emerest® 2675
9004-99-3	Kemester® 5221SE
9004-99-3	Kessco® PEG 200 MS
9004-99-3	Kessco® PEG 300 MS
9004-99-3	Kessco® PEG 400 MS
9004-99-3	Kessco® PEG 600 MS
9004-99-3	Kessco® PEG 1000 MS
9004-99-3	Kessco® PEG 1540 MS
9004-99-3	Kessco® PEG 4000 MS
9004-99-3	Kessco® PEG 6000 MS
9004-99-3	Myrj® 45
9004-99-3	Nopalcol 4-S
9004-99-3	Nopalcol 6-S
9004-99-3	Pegosperse® 400 MS
9004-99-3	Pegosperse® 1500 MS
9004-99-3	Serdox NSG 400
9004-99-3	Trydet SA-50/30
9005-02-1	Cithrol DGDL N/E
9005-02-1	Cithrol DGDL S/E
9005-02-1	Ionet DL-200
9005-02-1	Kessco® PEG 200 DL
9005-02-1	Kessco® PEG 300 DL
9005-02-1	Kessco® PEG 400 DL
9005-02-1	Kessco® PEG 600 DL
9005-02-1	Kessco® PEG 1000 DL
9005-02-1	Kessco® PEG 1540 DL
9005-02-1	Kessco® PEG 4000 DL
9005-02-1	Kessco® PEG 6000 DL
9005-02-1	Nopalcol 2-DL
9005-02-1	Pegosperse® 200 DL
9005-02-1	Pegosperse® 400 DL
9005-07-6	Cithrol DGDO N/E
9005-07-6	Cithrol DGDO S/E
9005-07-6	Ionet DO-200
9005-07-6	Ionet DO-400
9005-07-6	Ionet DO-600
9005-07-6	Ionet DO-1000
9005-07-6	Kessco® PEG 200 DO
9005-07-6	Kessco® PEG 300 DO
9005-07-6	Kessco® PEG 400 DO
9005-07-6	Kessco® PEG 600 DO
9005-07-6	Kessco® PEG 1000 DO
9005-07-6	Kessco® PEG 1540 DO
9005-07-6	Kessco® PEG 4000 DO
9005-07-6	Kessco® PEG 6000 DO
9005-07-6	Nopalcol 6-DO
9005-07-6	Pegosperse® 400 DO
9005-08-7	Cithrol DGDS N/E
9005-08-7	Cithrol DGDS S/E
9005-08-7	Hefti PGE-400-DS
9005-08-7	Hefti PGE-600-DS
9005-08-7	Ionet DS-300
9005-08-7	Ionet DS-400
9005-08-7	Kessco® PEG 200 DS
9005-08-7	Kessco® PEG 300 DS
9005-08-7	Kessco® PEG 400 DS
9005-08-7	Kessco® PEG 600 DS
9005-08-7	Kessco® PEG 1000 DS
9005-08-7	Kessco® PEG 1540 DS
9005-08-7	Kessco® PEG 4000 DS
9005-08-7	Kessco® PEG 6000 DS
9005-64-5	Alkamuls® PSML-20
9005-64-5	Disponil SML 104 F1
9005-64-5	Disponil SML 120 F1
9005-64-5	Emsorb® 6916
9005-64-5	Ethylan® GEL2
9005-64-5	Ethylan® GLE-21
9005-64-5	Glycosperse® L-20
9005-64-5	T-Maz® 20
9005-65-6	Disponil SMO 120 F1
9005-65-6	Durfax® 80
9005-65-6	Emsorb® 6901
9005-65-6	Ethylan® GEO8
9005-65-6	Ethylan® GEO81
9005-65-6	Ethylan® GOE-21
9005-65-6	Glycosperse® O-20
9005-66-7	Disponil SMP 120 F1
9005-66-7	Ethylan® GEP4
9005-67-8	Disponil SMS 120 F1
9005-67-8	Emsorb® 6909
9005-67-8	Ethylan® GES6
9005-67-8	Hefti MS-55-F
9005-70-3	Disponil STO 120 F1
9005-70-3	Ethylan® GPS85
9005-70-3	Hefti TO-55-E
9005-70-3	Hefti TO-55-EL
9005-71-4	Disponil STS 120 F1
9007-48-1	Santone® 3-1-SH
9010-77-9	A-C® 540, 540A
9010-77-9	A-C® 580
9010-77-9	A-C® 5120
9010-77-9	A-C® 5180
9010-77-9	Luwax® EAS 1
9010-79-1	Exxelor PE 805
9010-79-1	Exxelor PE 808
9010-79-1	Exxelor VA 1801
9010-79-1	Exxelor VA 1803
9010-79-1	Exxelor VA 1810
9010-79-1	Exxelor VA 1820
9010-79-1	KEP 010P, 020P, 070P
9010-79-1	Petrolite® CP-7
9010-79-1	Polysar EPM 306
9010-79-1	proFLOW 3000
9010-79-1	RA-061
9010-79-1	Vistalon 719
9010-79-1	Vistalon 503
9011-13-6	SMA® 1000
9011-13-6	SMA® 2000

CAS	Trade name
9011-14-7	Paraloid® K-120N
9011-16-9	Gantrez® AN-179
9014-85-1	Surfynol® 465
9014-85-1	Surfynol® 485
9014-90-8	Akyposal NPS 100
9014-90-8	Akyposal NPS 250
9014-90-8	Sermul EA 146
9014-90-8	Sermul EA 151
9014-90-8	Sermul EA 54
9014-90-8	Witcolate™ D51-51EP
9014-90-8	Zorapol SN-9
9014-93-1	Igepal® DM-430
9014-93-1	Igepal® DM-530
9014-93-1	Igepal® DM-710
9014-93-1	Igepal® DM-730
9014-93-1	Igepal® DM-880
9014-93-1	Trycol® 6985
9014-93-1	Trycol® DNP-8
9016-45-9	Ablunol NP30
9016-45-9	Ablunol NP30 70%
9016-45-9	Ablunol NP40
9016-45-9	Ablunol NP40 70%
9016-45-9	Ablunol NP50
9016-45-9	Ablunol NP50 70%
9016-45-9	Akyporox NP 30
9016-45-9	Akyporox NP 200
9016-45-9	Akyporox NP 300V
9016-45-9	Akyporox NP 1200V
9016-45-9	Alkasurf® NP-4
9016-45-9	Alkasurf® NP-6
9016-45-9	Carsonon® N-30
9016-45-9	Chemax NP-40
9016-45-9	Dehydrophen PNP 30
9016-45-9	Dehydrophen PNP 40
9016-45-9	Empilan® NP9
9016-45-9	Ethylan® BV
9016-45-9	Ethylan® DP
9016-45-9	Ethylan® HA Flake
9016-45-9	Ethylan® KEO
9016-45-9	Ethylan® N30
9016-45-9	Hetoxide NP-30
9016-45-9	Iconol NP-30
9016-45-9	Iconol NP-30-70%
9016-45-9	Iconol NP-40
9016-45-9	Iconol NP-40-70%
9016-45-9	Iconol NP-50
9016-45-9	Iconol NP-50-70%
9016-45-9	Iconol NP-70
9016-45-9	Iconol NP-70-70%
9016-45-9	Iconol NP-100
9016-45-9	Iconol NP-100-70%
9016-45-9	Igepal® CA-877
9016-45-9	Igepal® CO-430
9016-45-9	Igepal® CO-530
9016-45-9	Igepal® CO-630
9016-45-9	Igepal® CO-710
9016-45-9	Igepal® CO-850
9016-45-9	Igepal® CO-880
9016-45-9	Igepal® CO-887
9016-45-9	Igepal® CO-890
9016-45-9	Igepal® CO-897
9016-45-9	Igepal® CO-970
9016-45-9	Igepal® CO-977
9016-45-9	Igepal® CO-980
9016-45-9	Igepal® CO-987
9016-45-9	Igepal® CO-990
9016-45-9	Igepal® CO-997
9016-45-9	Nissan Nonion NS-206
9016-45-9	Nissan Nonion NS-212
9016-45-9	Nonipol 20
9016-45-9	Polirol NF80
9016-45-9	Polystep® F-1
9016-45-9	Polystep® F-2
9016-45-9	Polystep® F-3
9016-45-9	Polystep® F-4
9016-45-9	Polystep® F-5
9016-45-9	Polystep® F-6
9016-45-9	Polystep® F-9
9016-45-9	Polystep® F-10
9016-45-9	Polystep® F-95B
9016-45-9	Prox-onic NP-04
9016-45-9	Prox-onic NP-06
9016-45-9	Prox-onic NP-09
9016-45-9	Prox-onic NP-010
9016-45-9	Prox-onic NP-015
9016-45-9	Prox-onic NP-020
9016-45-9	Prox-onic NP-030
9016-45-9	Prox-onic NP-030/70
9016-45-9	Prox-onic NP-040
9016-45-9	Prox-onic NP-040/70
9016-45-9	Prox-onic NP-050
9016-45-9	Prox-onic NP-050/70
9016-45-9	Prox-onic NP-0100
9016-45-9	Prox-onic NP-0100/70
9016-45-9	Remcopal 33820
9016-45-9	Rewopal® HV 5
9016-45-9	Rewopal® HV 9
9016-45-9	Rewopal® HV 10
9016-45-9	Rewopal® HV 25
9016-45-9	Rewopal® HV 50
9016-45-9	Rexol 25/4
9016-45-9	Rexol 25/7
9016-45-9	Rexol 25/8
9016-45-9	Rexol 25/10
9016-45-9	Rexol 25/40
9016-45-9	Rexol 25/307
9016-45-9	Rexol 25/407
9016-45-9	Serdox NNP 25
9016-45-9	Serdox NNP 30/70
9016-45-9	Sermul EN 145
9016-45-9	Sermul EN 15
9016-45-9	Sermul EN 20/70
9016-45-9	Sermul EN 229
9016-45-9	Sermul EN 30/70
9016-45-9	T-Det® N-20
9016-45-9	T-Det® N-40
9016-45-9	T-Det® N-50
9016-45-9	T-Det® N-70
9016-45-9	T-Det® N-100
9016-45-9	T-Det® N-307

CAS	Trade name
9016-45-9	T-Det® N-407
9016-45-9	T-Det® N-507
9016-45-9	T-Det® N-705
9016-45-9	T-Det® N-707
9016-45-9	T-Det® N-1007
9016-45-9	Tergitol® NP-40
9016-45-9	Tergitol® NP-40 (70% Aq.)
9016-45-9	Teric N8
9016-45-9	Teric N9
9016-45-9	Teric N10
9016-45-9	Teric N12
9016-45-9	Teric N13
9016-45-9	Teric N15
9016-45-9	Teric N20
9016-45-9	Teric N30
9016-45-9	Teric N40
9016-45-9	Trycol® 6954
9016-45-9	Trycol® 6957
9016-45-9	Trycol® 6967
9016-45-9	Trycol® 6968
9016-45-9	Trycol® 6969
9016-45-9	Trycol® 6970
9016-45-9	Trycol® 6971
9016-45-9	Trycol® 6972
9016-45-9	Trycol® 6981
9016-45-9	Trycol® NP-20
9016-45-9	Trycol® NP-30
9016-45-9	Trycol® NP-307
9016-45-9	Witconol™ NP-100
9016-45-9	Witconol™ NP-300
9032-42-2	Tylose® MH Grades
9032-42-2	Tylose® MHB
9035-99-8	Crystex® 60
9035-99-8	Crystex® Regular
9038-95-3	Macol® 300
9038-95-3	Macol® 660
9038-95-3	Macol® 3520
9038-95-3	Macol® 5100
9038-95-3	Tergitol® XD
9038-95-3	Tergitol® XH
9038-95-3	Tergitol® XJ
9040-38-4	Aerosol® A-103
9046-01-9	Dextrol OC-40
9046-01-9	Plysurf A212C
9046-01-9	Rhodafac® RS-410
9046-01-9	Rhodafac® RS-610
9046-01-9	Rhodafac® RS-710
9051-57-4	Abex® EP-100
9051-57-4	Abex® EP-110
9051-57-4	Abex® EP-115
9051-57-4	Abex® EP-120
9051-57-4	Aerosol® NPES 458
9051-57-4	Aerosol® NPES 930
9051-57-4	Aerosol® NPES 2030
9051-57-4	Aerosol® NPES 3030
9051-57-4	Rhodapex® CO-436
9076-43-1	Emcol® CC-42
9080-49-3	LP®-3
9084-06-4	Lomar® LS
9084-06-4	Lomar® PW
10025-67-9	Chloropren-Faktis NW
10025-67-9	Faktis R, Weib MB
10025-67-9	Faktis Rheinau H, W
10025-78-2	Dynasylan® TCS
10081-67-1	Naugard® 445
10094-45-8	Chemstat® HTSA #3B
10094-45-8	Kemamide® E-180
10099-76-0	BSWL 202
10101-41-4	Valu-Fil
10192-93-5	Perkadox® 58
10213-78-2	Chemeen 18-2
10213-78-2	Chemstat® 192
10213-78-2	Chemstat® 192/NCP
10213-78-2	Chemstat® 273-E
10213-78-2	Prox-onic MS-02
10288-28-5	V-30
10416-59-8	Dynasylan® BSA
10416-59-8	Petrarch® B2500
10508-09-5	Aztec® DTAP
10508-09-5	Trigonox® 201
10591-85-2	Benzyl Tuads®
10591-85-2	Benzyl Tuads® Solid
10591-85-2	Benzyl Tuex®
10591-85-2	Perkacit® TBzTD
10595-49-0	Cyastat® LS
10595-72-9	Cyanox® 711
10595-72-9	Evanstab® 13
10595-72-9	Lankromark® DTDTDP
10595-72-9	PAG DTDTDP
11097-59-9	Alcamizer 1
11097-59-9	Alcamizer 2
11111-34-5	Tetronic® 304
11111-34-5	Tetronic® 701
11111-34-5	Tetronic® 704
11111-34-5	Tetronic® 901
11111-34-5	Tetronic® 904
11111-34-5	Tetronic® 908
11111-34-5	Tetronic® 1307
11890-98-8	Ageflex 1,3 BGDMA
12001-26-2	AlbaFlex 25
12001-26-2	AlbaFlex 50
12001-26-2	AlbaFlex 100
12001-26-2	AlbaFlex 200
12001-26-2	Huber SM
12001-26-2	Huber WG-1
12001-26-2	Huber WG-2
12001-26-2	Micawhite 200
12001-26-2	Polymica 200
12001-26-2	Polymica 325
12001-26-2	Polymica 400
12001-26-2	Polymica 3105
12027-96-2	Flamtard H
12036-37-2	Flamtard S
12178-41-5	Ferrosil™ 14
12262-58-7	Aztec® CHP-50-P1
12262-58-7	Aztec® CHP-90-W1
12262-58-7	Aztec® CHP-HA-1
12262-58-7	Cyclonox® BT-50
12269-78-2	Pyrax® A
12269-78-2	Pyrax® ABB

CAS	Trade name
12269-78-2	Pyrax® B
12269-78-2	Pyrax® WA
12513-27-8	Firebrake® 500
12513-27-8	Firebrake® ZB-Fine
12539-23-0	Alcamizer 1
12542-30-2	Sipomer® DCPA
12767-90-7	Flamtard Z10
12769-96-9	Ferro V-5
12769-96-9	Ferro V-8
13052-09-0	Aztec® DHPEH
13052-09-0	Trigonox® 141
13052-09-0	USP®-245
13103-52-1	Evanstab® 1218
13122-18-4	Aztec® TBPIN
13122-18-4	Aztec® TBPIN-30-AL
13122-18-4	Trigonox® 42
13122-18-4	Trigonox® 42 PR
13122-18-4	Trigonox® 42S
13127-82-7	Berol 302
13127-82-7	Chemstat® 172T
13170-05-3	Sandostab 4030
13170-23-5	Dynasylan® BDAC
13253-82-2	Diak No. 4
13463-41-7	Zinc Omadine® 48% Fine Particle Disp.
13463-41-7	Zinc Omadine® 48% Std. Disp.
13463-41-7	Zinc Omadine® Powd.
13463-67-7	AT-1
13463-67-7	Kronos® 1000
13463-67-7	Kronos® 2020
13463-67-7	Kronos® 2073
13463-67-7	Kronos® 2081
13463-67-7	Kronos® 2101
13463-67-7	Kronos® 2160
13463-67-7	Kronos® 2200
13463-67-7	Kronos® 2210
13463-67-7	Kronos® 2220
13463-67-7	Kronos® 2230
13463-67-7	Kronos® 2310
13463-67-7	RL-90
13463-67-7	Ti-Pure® R-100
13463-67-7	Ti-Pure® R-101
13463-67-7	Ti-Pure® R-102
13463-67-7	Ti-Pure® R-103
13463-67-7	Ti-Pure® R-900
13463-67-7	Ti-Pure® R-902
13463-67-7	Ti-Pure® R-960
13463-67-7	Titanium Dioxide P25
13472-08-7	Perkadox® AMBN
13472-08-7	V-59
13474-96-9	Weston® PTP
13560-89-9	Dechlorane® Plus 25
13560-89-9	Dechlorane® Plus 35
13560-89-9	Dechlorane® Plus 515
13674-87-8	Fyrol® 38
13674-87-8	Fyrol® FR-2
13701-59-2	Busan® 11-M1
13822-56-3	Dynasylan® AMMO
13822-56-3	Petrarch® A0800
13927-77-0	Ekaland NDBC
13927-77-0	Naugard® NBC
13927-77-0	Perkacit® NDBC
13927-77-0	UV-Chek® AM-104
13927-77-0	Vanox® NBC
13983-17-0	Nyad® 200, 325, 400
13983-17-0	Nyad® 475, 1250
13983-17-0	Nyad® FP, G, G Special
13983-17-0	Nyad G® Wollastocoat®
13983-17-0	Nycor® R
13983-17-0	Vansil® W-10
13983-17-0	Vansil® W-20
13983-17-0	Vansil® W-30
13983-17-0	10, 325, 400 Wollastocoat®
14087-96-6	Cimflx 606
14087-96-6	Cimpact 600
14087-96-6	Cimpact 610
14087-96-6	Stellar 500
14087-96-6	Stellar 510
14103-61-8	Diplast® N
14228-73-0	Epodil® 757
14228-73-0	Heloxy® 107
14239-68-0	Ethyl Cadmate®
14239-75-9	Thanecure®
14323-55-1	Akrochem® Accelerator EZ
14323-55-1	Anchor® ZDBC
14323-55-1	Anchor® ZDEC
14323-55-1	Naftocit® Di 7
14323-55-1	Naftopast® Di7-P
14323-55-1	Octocure ZDE-50
14323-55-1	Perkacit® ZDEC
14324-55-1	Ethyl Zimate®
14324-55-1	Ethyl Zimate® Slurry
14481-60-8	Aerosol® 18
14481-60-8	Lankropol® ODS/LS
14481-60-8	Lankropol® ODS/PT
14481-60-8	Octosol A-18
14516-71-3	Cyasorb® UV 1084
14643-87-9	Akrochem® ZDA Powd
14643-87-9	Saret® 633
14643-87-9	SR-416
14726-36-4	Akrochem® Z.B.E.D
14726-36-4	Arazate®
14726-36-4	Naftocit® ZBEC
14726-36-4	Octocure ZBZ-50
14726-36-4	Perkacit® ZBEC
14726-36-4	Vulkacit ZBEC
14806-96-6	Silverline 665
14807-96-6	ABT-2500®
14807-96-6	Artic Mist
14807-96-6	Beaverwhite 200
14807-96-6	Cimpact 699
14807-96-6	Cimpact 705
14807-96-6	Cimpact 710
14807-96-6	Glacier 200
14807-96-6	I.T. 3X
14807-96-6	I.T. 5X
14807-96-6	I.T. 325
14807-96-6	I.T. FT
14807-96-6	I.T. X

CAS	Trade name
14807-96-6	Jet Fil® 200
14807-96-6	Jet Fil® 350
14807-96-6	Jet Fil® 500
14807-96-6	Jet Fil® 575C
14807-96-6	Jet Fil® 625C
14807-96-6	Jet Fil® 700C
14807-96-6	Luzenac 8170
14807-96-6	Magsil Diamond Talc 200 mesh
14807-96-6	Magsil Diamond Talc 350 mesh
14807-96-6	Magsil Star Talc 200 mesh
14807-96-6	Magsil Star Talc 350 mesh
14807-96-6	Microbloc®
14807-96-6	MicroPflex 1200
14807-96-6	Microtalc® MP10-52
14807-96-6	Microtalc® MP12-50
14807-96-6	Microtalc® MP15-38
14807-96-6	Microtalc® MP25-38
14807-96-6	Microtalc® MP44-26
14807-96-6	Microtuff® 325F
14807-96-6	Microtuff® 1000
14807-96-6	Microtuff® F
14807-96-6	Mistron CB
14807-96-6	Mistron PXL®
14807-96-6	Mistron Vapor
14807-96-6	Mistron Vapor® R
14807-96-6	Nicron 325
14807-96-6	Nicron 400
14807-96-6	Nytal® 100
14807-96-6	Nytal® 200
14807-96-6	Nytal® 300
14807-96-6	Nytal® 400
14807-96-6	Polytalc 262
14807-96-6	PolyTalc™ 445
14807-96-6	Select-A-Sorb Powdered, Compacted
14807-96-6	Silverline 400
14807-96-6	Super Microtuff F
14807-96-6	Supra EF
14807-96-6	Supra EF A
14807-96-6	Techfil 7599
14807-96-6	Ultratalc™ 609
14807-96-6	Vantalc® 6H
14807-96-6	Vertal 92
14807-96-6	Vertal 97
14807-96-6	Vertal 200
14807-96-6	Vertal 310
14807-96-6	Vertal 710
14807-96-6	Vertal C-2
15206-55-0	Vicure® 55
15305-07-4	Q-1301
15317-78-9	Isobutyl Niclate®
15337-18-5	Amyl Zimate®
15520-10-2	Dytek® A
15520-11-3	Aztec® BCHPC
15520-11-3	Aztec® BCHPC-40-SAQ
15520-11-3	Aztec® BCHPC-75-W
15520-11-3	Perkadox® 16
15520-11-3	Perkadox® 16/35
15520-11-3	Perkadox® 16N
15520-11-3	Perkadox® 16S
15520-11-3	Perkadox® 16-W25-GB1
15520-11-3	Perkadox® 16/W40
15520-11-3	Perkadox® 16-W40-GB2
15520-11-3	Perkadox® 16-W40-GB5
15520-11-3	Perkadox® 16/W70
15521-65-0	Methyl Niclate®
15535-69-0	ADK STAB BT-31
15535-69-0	ADK STAB BT-52
15535-69-0	ADK STAB BT-53A
15535-69-0	ADK STAB LS-2
15535-69-0	Stanclere® TM
15535-69-0	Therm-Chek® 837
15545-97-8	V-70
15625-89-5	Ageflex TMPTA
15667-10-4	USP®-90MD
16066-38-9	Lupersol 221
16111-62-4	Aztec® EHPC-40-EAQ
16111-62-4	Aztec® EHPC-40-ENF
16111-62-4	Aztec® EHPC-65-AL
16111-62-4	Aztec® EHPC-75-AL
16111-62-4	Lupersol 223
16111-62-9	Espercarb® 840
16111-62-9	Espercarb® 840M
16111-62-9	Espercarb® 840M-40
16111-62-9	Espercarb® 840M-70
16111-62-9	Trigonox® EHP
16111-62-9	Trigonox® EHP-B75
16111-62-9	Trigonox® EHP-C40
16111-62-9	Trigonox® EHP-C70
16111-62-9	Trigonox® EHP-C75
16111-62-9	Trigonox® EHPS
16111-62-9	Trigonox® EHPS-C75
16111-62-9	Trigonox® EHP-W40
16111-62-9	Trigonox® EHP-W40S
16111-62-9	Trigonox® EHP-W50
16215-49-9	Trigonox® NBP-C50
16230-35-6	Dynasylan® 0116
16260-09-6	Chemstat® HTSA #1A
16260-09-6	Kemamide® P-181
16545-54-3	Cyanox® MTDP
16545-54-3	Evanstab® 14
16545-54-3	PAG DMTDP
16580-06-6	Trigonox® 169-NA50
16580-06-6	Trigonox® 169-OP50
16971-82-7	Vanax® PML
17540-75-9	Isonox® 132
17540-75-9	Vanox® 1320
17557-23-2	Epodil® 749
17557-23-2	Heloxy® 68
17689-77-9	Dynasylan® ETAC
17796-82-6	Santogard® PVI
17796-82-6	Vantard® PVI
17831-71-9	Ageflex T4EGDA
18395-30-7	Dynasylan® IBTMO
18395-30-7	Petrarch® I7810
18407-94-8	Petrarch® T1918
19010-66-3	Methyl Ledate
19321-40-5	Estol 1445

CAS	Trade name
19910-65-7	Espercarb® 438M-60
19910-65-7	Lupersol 225
19910-65-7	Trigonox® ADC
19910-65-7	Trigonox® SBP
19910-65-7	Trigonox® SBP-AX30
19910-65-7	Trigonox® SBP-C50
19910-65-7	Trigonox® SBP-C75
19910-65-7	Trigonox® SBPS
20566-35-2	Great Lakes PHT4-Diol™
20566-35-2	Saytex® RB-79
20824-56-0	Versene Diammonium EDTA
20858-12-2	VA-061
20941-65-5	Akrochem® TDEC
20941-65-5	Ethyl Tellurac®
20941-65-5	Ethyl Tellurac® Rodform
20941-65-5	Perkacit® TDEC
21260-46-8	Bismate® Powd
21260-46-8	Bismate® Rodform®
21260-46-8	Bismet
21282-96-2	Lonzamon® AAEA
21282-97-3	Lonzamon® AAEMA
21302-09-0	Weston® DLP
21542-96-1	Crodamine 3.ABD
21645-51-2	Baco DH101
21645-51-2	H-100
21645-51-2	H-105
21645-51-2	H-109
21645-51-2	H-600
21645-51-2	H-800
21645-51-2	H-900
21645-51-2	H-990
21645-51-2	Haltex™ 304
21645-51-2	Haltex™ 310
21645-51-2	Martinal® 103 LE
21645-51-2	Martinal® 104 LE
21645-51-2	Martinal® 107 LE
21645-51-2	Martinal® OL-104
21645-51-2	Martinal® OL-104/C
21645-51-2	Martinal® OL-104 LE
21645-51-2	Martinal® OL-104/S
21645-51-2	Martinal® OL-107
21645-51-2	Martinal® OL-107/C
21645-51-2	Martinal® OL-107 LE
21645-51-2	Martinal® OL-111 LE
21645-51-2	Martinal® OL/Q-107
21645-51-2	Martinal® OL/Q-111
21645-51-2	Martinal® ON
21645-51-2	Martinal® ON-310
21645-51-2	Martinal® ON-313
21645-51-2	Martinal® ON-320
21645-51-2	Martinal® ON-920/V
21645-51-2	Martinal® ON-4608
21645-51-2	Martinal® OS
21645-51-2	Martinal® OX
21850-44-2	Great Lakes PE-68™
22313-62-8	Esperox® 497M
22313-62-8	Trigonox® 97
22313-62-8	Trigonox® 97-C75
23128-74-7	Irganox® 1098
23328-53-2	Tinuvin® 571

CAS	Trade name
23386-52-9	Aerosol® A-196-40
23386-52-9	Gemtex 691-40
23386-52-9	Octosol TH-40
23386-52-9	Rewopol® SBDC 40
23474-91-1	Esperox® 13M
23949-66-8	Sanduvor® VSU
24304-00-5	AlNel A-100
24304-00-5	AlNel A-200
24304-00-5	AlNel AG
24801-88-5	CI7840
24817-92-3	Citroflex® A-6
24887-06-7	Parolite
24937-78-8	A-C® 400
24937-78-8	A-C® 400A
24937-78-8	A-C® 405M
24937-78-8	A-C® 405S
24937-78-8	A-C® 405T
24937-78-8	A-C® 430
24937-78-8	Baymod® L450 N
24937-78-8	Baymod® L450 P
24937-78-8	Baymod® L2450
24937-78-8	Fusabond® MC-189D
24937-78-8	Fusabond® MC-190D
24937-78-8	Fusabond® MC-197D
24937-78-8	Levapren 450P
24937-79-9	Solef® 11012/1001
24969-11-7	Penacolite® R-2170
25013-16-5	Sustane® BHA
25013-16-5	Tenox® BHA
25086-62-8	Daxad® 30
25087-26-7	Daxad® 34
25103-12-2	Albrite® TIOP
25103-12-2	Weston® TIOP
25155-23-1	Antiblaze® TXP
25155-23-1	Kronitex® TXP
25155-23-1	Pliabrac® TXP
25155-25-3	Aztec® DIPP-2
25155-25-3	Aztec® DIPP-40-G
25155-25-3	Aztec® DIPP-40-IC
25155-25-3	Luperox 802
25155-25-3	Perkadox® 14
25155-25-3	Perkadox® 14/40
25155-25-3	Perkadox® 14-40B-pd
25155-25-3	Perkadox® 14-40K-pd
25155-25-3	Perkadox® 14-90
25155-25-3	Perkadox® 14S
25155-25-3	Perkadox® 14S-20PP-pd
25155-25-3	Perkadox® 14S-fl
25155-30-0	Calsoft F-90
25155-30-0	Calsoft L-60
25155-30-0	Nansa® HS85/S
25155-30-0	Rhodacal® DS-10
25155-30-0	Rhodacal® LDS-22
25155-30-0	Sandet 60
25155-30-0	Ufasan 35
25155-30-0	Witconate™ 1250
25155-30-0	Witconate™ 1255
25167-32-2	Aerosol® DPOS-45
25213-24-5	Airvol® 205
25213-24-5	Airvol® 523

CAS	Trade name
25213-24-5	Airvol® 540
25265-77-4	Texanol® Ester-Alcohol
25321-41-9	Eltesol® 4200
25321-41-9	Eltesol® XA65
25321-41-9	Eltesol® XA90
25321-41-9	Eltesol® XA/M65
25321-41-9	Manro FCM 90LV
25321-41-9	Reworyl® X 65
25321-41-9	XSA 80
25321-41-9	XSA 90
25322-68-3	Alkapol PEG 300
25322-68-3	Carbowax® PEG 200
25322-68-3	Carbowax® PEG 300
25322-68-3	Carbowax® PEG 400
25322-68-3	Carbowax® PEG 600
25322-68-3	Carbowax® PEG 900
25322-68-3	Carbowax® PEG 1000
25322-68-3	Chemstat® P-400
25322-68-3	Dow E200
25322-68-3	Dow E300
25322-68-3	Dow E400
25322-68-3	Dow E600
25322-68-3	Dow E900
25322-68-3	Dow E1000
25322-68-3	Dow E1450
25322-68-3	Dow E3350
25322-68-3	Dow E4500
25322-68-3	Dow E8000
25322-68-3	Emery® 6686
25322-68-3	Emery® 6687
25322-68-3	Emery® 6773
25322-68-3	Hodag PEG 200
25322-68-3	Hodag PEG 300
25322-68-3	Hodag PEG 400
25322-68-3	Hodag PEG 3350
25322-68-3	Hodag PEG 8000
25322-68-3	Nopalcol 200
25322-68-3	Nopalcol 400
25322-68-3	Nopalcol 600
25322-68-3	Pluracol® E200
25322-68-3	Pluracol® E300
25322-68-3	Pluracol® E400
25322-68-3	Pluracol® E400 NF
25322-68-3	Pluracol® E600
25322-68-3	Pluracol® E600 NF
25322-68-3	Pluracol® E1000
25322-68-3	Pluracol® E1450
25322-68-3	Pluracol® E1450 NF
25322-68-3	Pluracol® E2000
25322-68-3	Pluracol® E4000
25322-68-3	Pluracol® E6000
25322-68-3	Pluracol® E8000
25322-68-3	Poly-G® 200
25322-68-3	Poly-G® 300
25322-68-3	Poly-G® 400
25322-68-3	Poly-G® 600
25322-68-3	Poly-G® 1000
25322-68-3	Poly-G® 2000
25322-68-3	Polyox® WSR 205
25322-68-3	Polyox® WSR 301

CAS	Trade name
25322-68-3	Polyox® WSR 303
25322-68-3	Polyox® WSR 308
25322-68-3	Polyox® WSR 1105
25322-68-3	Polyox® WSR 3333
25322-68-3	Polyox® WSR N-10
25322-68-3	Polyox® WSR N-12K
25322-68-3	Polyox® WSR N-60K
25322-68-3	Polyox® WSR N-80
25322-68-3	Polyox® WSR N-750
25322-68-3	Polyox® WSR N-3000
25322-68-3	Polyox® Coagulant
25322-68-3	Prox-onic PEG-2000
25322-68-3	Prox-onic PEG-20,000
25322-68-3	Rhodasurf® E 400
25322-68-3	Rhodasurf® E 600
25322-68-3	Rhodasurf® PEG 3350
25322-68-3	Teric PEG 300
25322-68-3	Teric PEG 400
25322-68-3	Teric PEG 600
25322-68-3	Teric PEG 4000
25322-68-3	Teric PEG 8000
25322-68-3	Ucarfloc® Polymer 300
25322-68-3	Ucarfloc® Polymer 302
25322-68-3	Ucarfloc® Polymer 304
25322-68-3	Ucarfloc® Polymer 309
25322-69-4	Hodag PPG-400
25322-69-4	Hodag PPG-1200
25322-69-4	Hodag PPG-2000
25322-69-4	Hodag PPG-4000
25322-69-4	Pluracol® P-410
25322-69-4	Pluracol® P-710
25322-69-4	Pluracol® P-1010
25322-69-4	Pluracol® P-2010
25322-69-4	Pluracol® P-4010
25327-89-3	Great Lakes BE-51™
25338-55-0	Ancamine® 1110
25383-99-7	Pationic® 920
25417-20-3	Rhodacal® BX-78
25417-20-3	Supragil® NK
25446-78-0	Rhodapex® EST-30
25448-25-3	Doverphos® 6
25448-25-3	Weston® TDP
25496-01-9	Arylan® SO60 Acid
25496-72-4	Alkamuls® GMR-55LG
25498-49-1	Arcosolv® TPM
25550-98-5	Doverphos® 7
25550-98-5	Weston® PDDP
25620-58-0	Vestamin® TMD
25637-84-7	Priolube 1409
25637-99-4	FR-1206
25640-78-2	Nusolv™ ABP-62
25736-86-1	Sipomer® HEM-5
25852-47-5	MFM-425
25973-55-1	Tinuvin® 328
26027-38-3	Iconol NP-100
26172-55-4	Kathon® LX
26256-79-1	Deriphat® 160C
26264-05-1	Rhodacal® 330
26266-57-9	Disponil SMP 100 F1
26266-58-0	Disponil STO 100 F1

CAS	Trade name
26266-58-0	Ethylan® GT85
26332-14-5	Aztec® CEPC
26332-14-5	Aztec® CEPC-40-SAQ
26332-14-5	Perkadox® 24
26332-14-5	Perkadox® 24-fl
26332-14-5	Perkadox® 24-W40
26401-27-4	Doverphos® 9
26401-27-4	Weston® ODPP
26426-80-2	Daxad® 31
26444-49-5	Disflamoll® DPK
26445-96-5	AClyn® 201A
26523-78-4	ADK STAB 329
26523-78-4	ADK STAB 1178
26523-78-4	ADK STAB SC-102
26523-78-4	Doverphos® 4
26523-78-4	Doverphos® 4 Powd
26523-78-4	Lankromark® LE109
26523-78-4	Naugard® P
26523-78-4	Weston® 399
26523-78-4	Weston® 399B
26523-78-4	Weston® TNPP
26523-78-4	Wytox® 312
26530-20-1	Micro-Chek® 11 P
26544-23-0	Doverphos® 8
26544-23-0	Weston® DPDP
26544-27-4	Weston® 600
26635-92-7	Chemeen 18-5
26635-92-7	Chemeen 18-50
26635-92-7	Ethomeen® 18/60
26635-92-7	Prox-onic MS-05
26635-92-7	Prox-onic MS-011
26635-92-7	Prox-onic MS-050
26635-92-7	Teric 18M5
26635-92-7	Teric 18M10
26635-92-7	Trymeen® 6617
26635-93-8	Berol 28
26635-93-8	Berol 303
26658-19-5	Disponil STS 100 F1
26680-54-6	Milldride® OSA
26741-53-7	Blendex® 340
26741-53-7	Ultranox® 626
26741-53-7	Ultranox® 626A
26741-53-7	Ultranox® 627A
26748-38-9	Esperox® 750M
26748-41-4	Aztec® TBPND
26748-41-4	Aztec® TBPND-40-EAQ
26748-41-4	Aztec® TBPND-75-OMS
26748-41-4	Esperox® 33M
26748-41-4	Lupersol 10
26748-41-4	Trigonox® 23
26748-41-4	Trigonox® 23-C75
26748-41-4	Trigonox® 23-W40
26748-47-0	Aztec® CUPND-75-AL
26748-47-0	Esperox® 939M
26748-47-0	Trigonox® 99-B75
26748-47-0	Trigonox® 99-W40
26761-40-0	Diplast® R
26762-93-6	Trigonox® M-50
26780-96-1	Agerite® MA
26780-96-1	Agerite® Resin D
26780-96-1	Akrochem® Antioxidant DQ
26780-96-1	Akrochem® Antioxidant DQ-H
26780-96-1	Flectol® Pastilles
26780-96-1	Naftonox® TMQ
26780-96-1	Naugard® Super Q
26780-96-1	Permanax™ TQ
26780-96-1	Ralox® TMQ-G
26780-96-1	Ralox® TMQ-H
26780-96-1	Ralox® TMQ-R
26780-96-1	Ralox® TMQ-T
26780-96-1	Struktol® T.M.Q
26780-96-1	Struktol® T.M.Q. Powd.
26780-96-1	Ultranox® 254
26780-96-1	Vulkanox® HS/LG
26780-96-1	Vulkanox® HS/Pellets
26780-96-1	Vulkanox® HS/Powd
26850-24-8	Dantocol® DHE
27176-87-0	Nansa® 1042/P
27176-87-0	Polirol DS
27177-77-1	Polystep® A-15-30K
27178-16-1	Monoplex® DDA
27178-16-1	Nuoplaz® DIDA
27178-16-1	Plasthall® DIDA
27178-16-1	Staflex DIDA
27323-41-7	Manro TDBS 60
27323-41-7	Puxol CB-22
27554-26-3	Jayflex® DIOP
27554-26-3	Plasthall® DIOP
27554-26-3	Staflex DIOP
27638-00-2	Cithrol GDL N/E
27676-62-6	Dovernox 3114
27676-62-6	Irganox® 3114
27676-62-6	Vanox® GT
27776-21-2	VA-044
27813-02-1	BM-951
27986-36-3	Akyporox NP 15
28061-69-0	Crodamine 3.AOD
28519-02-0	Eleminol MON-7
28519-02-0	Rhodacal® DSB
28631-63-2	Eltesol® CA 65
28631-63-2	Eltesol® CA 96
28631-63-2	Reworyl® C 65
29240-17-3	Esperox® 551M
29240-17-3	Trigonox® 125-C50
29240-17-3	Trigonox® 125-C75
29385-43-1	Vulkalent TM
29598-76-3	Seenox 412S
29964-84-9	Ageflex FM-10
29964-84-9	Sipomer® IDM
30364-51-3	Hamposyl® M-30
30416-77-4	Perlankrol® PA Conc.
30653-05-5	Mixxim® BB/200
31001-77-1	Dynasylan® 3403
31001-77-1	Petrarch® M8450
31556-45-3	Lexolube® B-109
31565-37-4	Sicolub® TDS
31566-31-1	Alkamuls® GMS/C
31566-31-1	Empilan® GMS NSE40
31570-04-4	Doverphos® S-480
31570-04-4	Hostanox® PAR 24

CAS	Trade name
31570-04-4	Irgafos® 168
31570-04-4	Naugard® 524
31692-79-2	Masil® SFR 70
31692-79-2	Mazol® SFR 100
31692-79-2	Mazol® SFR 750
31778-15-1	Evangard® 18MP
32492-61-8	Dianol® 220
32492-61-8	Dianol® 240
32492-61-8	Dianol® 240/1
32492-61-8	Dianol® 265
32492-61-8	Dianol® 285
32534-81-9	FR-1205
32534-81-9	Great Lakes DE-71™
32536-52-0	FR-1208
32536-52-0	Great Lakes DE-79™
32536-52-0	Saytex® 111
32588-76-4	Saytex® BT-93®
32588-76-4	Saytex® BT-93W
32612-48-9	Steol® CA-460
32647-67-9	Millad® 3905
32647-67-9	Millithix® 925
32687-78-8	Irganox® MD 1024
33031-74-2	Durastrength 200
33145-10-7	Lowinox® 22IB46
33145-10-7	Vulkanox® NKF
33703-08-1	Adimoll® DN
33703-08-1	Jayflex® DINA
33703-08-1	PX-239
33703-08-1	Staflex DINA
34137-09-2	Vanox® SKT
34424-98-1	Mazol® PGO-104
34443-12-4	Aztec® TBPEHC
34443-12-4	Esperox® C-496
34443-12-4	Trigonox® 117
34562-31-7	Vanax® 808
34590-94-8	Arcosolv® DPM
34745-96-5	Jeffcat DD
35074-77-2	Irganox® 259
35141-30-1	Dynasylan® TRIAMO
35141-30-1	Petrarch® T2910
35325-02-1	Uniplex 225
35958-30-6	Irganox® 129
35958-30-6	Vanox® 1290
36116-84-4	Weston® DOPI
36190-62-2	Isobutyl Zimate®
36356-83-9	Emery® 5702
36432-46-9	Weston® DTDP
36443-68-2	Irganox® 245
36445-71-3	Dowfax 3B2
36483-57-5	FR-513
36501-84-5	Amyl Ledate®
36653-82-4	Adol® 52
36653-82-4	Alfol® 16
36653-82-4	Epal® 16NF
36653-82-4	Loxiol® VPG 1743
36788-39-3	Weston® 430
37187-22-7	Aztec® AAP-LA-M2
37187-22-7	Aztec® AAP-NA-2
37187-22-7	Aztec® AAP-SA-3
37187-22-7	USP®-240

CAS	Trade name
37206-20-5	Aztec® MIKP-LA-M1
37206-20-5	Trigonox® HM
37244-96-5	Minbloc™ 16
37244-96-5	Minbloc™ 20
37244-96-5	Minbloc™ 30
37244-96-5	Minex® 4
37244-96-5	Minex® 7
37244-96-5	Minex® 10
37251-69-7	Tergitol® D-683
37294-49-8	Aerosol® A-268
38668-46-1	Curezol® 2MZ-Azine
38727-83-2	UV-Chek® AM-205
38916-42-6	Aerosol® 22
39354-45-5	Aerosol® A-102
39354-47-5	Emcol® 4300
39407-03-9	Servoxyl VPTZ 100
39464-64-7	Plysurf A207H
40372-72-3	Si 69
40601-76-1	Cyanox® 1790
41484-35-9	Irganox® 1035
41556-26-7	Tinuvin® 765
42590-53-4	Octocure ZIX-50
42612-52-2	Rhodafac® MC-470
43154-85-4	Emcol® K8300
43154-85-4	Sole Terge 8
51178-59-7	Sipomer® DCPM
51609-41-7	Servoxyl VPNZ 10/100
51811-79-1	Dextrol OC-20
51811-79-1	Foamphos NP-9
51811-79-1	Phospholan® PNP9
51811-79-1	Polirol 10
51811-79-1	Rhodafac® RE-960
51811-79-1	Tryfac® 5556
52019-36-0	Chemfac PD-600
52229-50-2	Gantrez® AN-119
52229-50-2	Gantrez® AN-139
52229-50-2	Gantrez® AN-149
52229-50-2	Gantrez® AN-169
52558-73-3	Hamposyl® M
52783-38-7	CS1590
52829-07-0	BLS™ 1770
52829-07-0	Lowilite® 77
52829-07-0	Tinuvin® 770
52907-07-0	Saytex® BN-451
53220-22-7	Aztec® MYPC
53220-22-7	Aztec® MYPC-40-SAQ
53220-22-7	Perkadox® 26-fl
53220-22-7	Perkadox® 26-W40
53894-23-8	Jayflex® TINTM
53894-23-8	Plasthall® TIOTM
53894-23-8	PX-339
53894-23-8	Staflex TIOTM
53988-10-6	Vanox® MTI
54667-43-5	Polamine® 650
54667-43-5	Polamine® 1000
54667-43-5	Polamine® 2000
54686-97-4	Millad® 3940
55310-46-8	Octopol SBZ-20
55406-53-6	Idex™-400
55406-53-6	Idex™-1000

CAS	Trade name
57018-52-7	Arcosolv® PTB
57171-56-9	G-1086
57171-56-9	G-1096
57455-37-5	Ferro BP-10
57455-37-5	Ferro CP-18
57455-37-5	Ferro CP-50
57455-37-5	Ferro CP-78
57455-37-5	Ferro DP-25
57455-37-5	Ferro EP-37
57455-37-5	Ferro EP-62
57455-37-5	Ferro FP-40
57455-37-5	Ferro FP-64
57455-37-5	Ferro RB-30
57834-33-0	Givsorb® UV-1
58068-97-6	Dynasylan® IMEO
58068-97-6	Petrarch® T2503
58229-88-2	ADK STAB 465
58229-88-2	ADK STAB 465E, 465L
58229-88-2	Haro® Mix ZT-905
58229-88-2	Kosmos® 23
58353-68-7	Octosol A-1
58767-50-3	Chemstat® 106G/60DC
58767-50-3	Chemstat® 106G/90
58965-66-5	Saytex® 120
59355-61-2	Empigen® OY
59789-51-4	Actimer FR-1033
59789-51-4	FR-1033
60303-68-6	Rylex® 30
60842-32-2	Aerosil® R972
60842-32-2	Aerosil® R974
61368-34-1	Actimer FR-803
61417-49-0	Ken-React® KR TTS
61551-69-7	VA-086
61617-00-3	Vanox® ZMTI
61632-57-3	Magala® 0.5E
61670-79-9	Doverphos® 675
61698-32-6	Curezol® 2PHZ
61788-44-1	Ablumol PSS
61788-44-1	Akrochem® Antioxidant 16 Liq
61788-44-1	AO47L
61788-44-1	AO47P
61788-44-1	Montaclere®
61788-44-1	Nevastain® 21
61788-44-1	Prodox® 120
61788-44-1	Vanox® 102
61788-44-1	Vulkanox® SP
61788-45-2	Crodamine 1.HT
61788-45-2	Kemamine® P-970
61788-45-2	Kemamine® P-970D
61788-45-2	Noram SH
61788-46-3	Kemamine® P-650D
61788-47-4	Industrene® 325
61788-47-4	Industrene® 328
61788-62-3	Kemamine® T-6501
61788-63-4	Kemamine® T-9701
61788-72-5	Drapex® 4.4
61788-72-5	Flexol® Plasticizer EP-8
61788-72-5	Monoplex® S-73
61788-85-0	Chemax HCO-5
61788-85-0	Chemax HCO-16
61788-85-0	Chemax HCO-25
61788-85-0	Chemax HCO-200/50
61788-85-0	Croduret 10
61788-85-0	Croduret 30
61788-85-0	Croduret 40
61788-85-0	Croduret 60
61788-85-0	Croduret 100
61788-85-0	Mapeg® CO-25H
61788-85-0	Nopalcol 10-COH
61788-85-0	Nopalcol 12-COH
61788-85-0	Prox-onic HRH-05
61788-85-0	Prox-onic HRH-016
61788-85-0	Prox-onic HRH-025
61788-85-0	Prox-onic HRH-0200
61788-85-0	Prox-onic HRH-0200/50
61788-89-4	Empol® 1016
61788-89-4	Empol® 1018
61788-89-4	Empol® 1020
61788-89-4	Empol® 1022
61788-89-4	Empol® 1026
61788-89-4	Hystrene® 3675
61788-89-4	Hystrene® 3675C
61788-89-4	Hystrene® 3680
61788-89-4	Hystrene® 3695
61788-91-8	Kemamine® T-9972D
61788-93-0	Kemamine® T-6502D
61788-95-2	Kemamine® T-9702D
61788-95-2	Nissan Tert. Amine ABT
61789-07-9	Myverol® 18-07
61789-08-0	Myverol® 18-06
61789-10-4	Grindtek MOP 90
61789-18-2	Arquad® C-50
61789-18-2	Chemquat C/33W
61789-68-2	Ethoquad® CB/12
61789-77-3	Arquad® 2C-75
61789-86-4	Lubrizol® 2152
61790-12-3	Acintol® DFA
61790-12-3	Acintol® EPG
61790-12-3	Acintol® FA-1
61790-12-3	Acintol® FA-1 Special
61790-12-3	Acintol® FA-2
61790-12-3	Acintol® FA-3
61790-12-3	Acofor
61790-12-3	Westvaco® 1480
61790-31-6	Armid® HT
61790-33-8	Crodamine 1.T
61790-33-8	Kemamine® P-974D
61790-33-8	Noram S
61790-37-2	Industrene® 143
61790-38-3	Petrac® PHTA
61790-50-9	Burez K20-505A
61790-50-9	Westvaco® Resin 895
61790-51-0	Arizona DR-25
61790-51-0	Arizona DRS-43
61790-51-0	Arizona DRS-44
61790-51-0	Burez K25-500D
61790-51-0	Burez K80-500
61790-51-0	Burez K80-500D
61790-51-0	Burez K80-2500
61790-51-0	Burez NA 70-500

CAS	Trade name
61790-51-0	Burez NA 70-500D
61790-51-0	Burez NA 70-2500
61790-51-0	Diprosin N-70
61790-51-0	Dresinate® 731
61790-51-0	Dresinate® 81
61790-51-0	Dresinate® X
61790-51-0	Dresinate® XX
61790-51-0	Westvaco® Resin 790
61790-51-0	Westvaco® Resin 795
61791-12-6	Alkamuls® EL-620L
61791-12-6	Alkamuls® EL-719
61791-12-6	Alkamuls® R81
61791-12-6	Berol 108
61791-12-6	Berol 190
61791-12-6	Berol 191
61791-12-6	Berol 198
61791-12-6	Berol 199
61791-12-6	Etocas 10
61791-12-6	Mapeg® CO-25
61791-12-6	Prox-onic HR-05
61791-12-6	Prox-onic HR-016
61791-12-6	Prox-onic HR-025
61791-12-6	Prox-onic HR-030
61791-12-6	Prox-onic HR-036
61791-12-6	Prox-onic HR-040
61791-12-6	Prox-onic HR-080
61791-12-6	Prox-onic HR-0200
61791-12-6	Prox-onic HR-0200/50
61791-12-6	T-Det® C-40
61791-12-6	Teric C12
61791-14-8	Berol 307
61791-14-8	Berol 397
61791-14-8	Mazeen® C-2
61791-14-8	Mazeen® C-5
61791-14-8	Mazeen® C-10
61791-14-8	Mazeen® C-15
61791-14-8	Prox-onic MC-02
61791-14-8	Prox-onic MC-05
61791-14-8	Prox-onic MC-015
61791-24-0	Mazeen® S-2
61791-24-0	Mazeen® S-5
61791-24-0	Mazeen® S-10
61791-24-0	Mazeen® S-15
61791-26-2	Berol 381
61791-26-2	Berol 386
61791-26-2	Berol 387
61791-26-2	Berol 389
61791-26-2	Berol 391
61791-26-2	Berol 392
61791-26-2	Berol 457
61791-26-2	Berol 458
61791-31-9	Armostat® 410
61791-31-9	Berol 307
61791-31-9	Chemstat® 273-C
61791-31-9	Empilan® CDE
61791-31-9	Ethylan® LD
61791-31-9	Ethylan® LDS
61791-44-4	Armostat® 310
61791-44-4	Berol 456
61791-44-4	Chemstat® 182
61791-44-4	Chemstat® 182/67DC
61791-44-4	Mazeen® T-2
61791-44-4	Mazeen® T-5
61791-44-4	Prox-onic MT-02
61791-47-7	Aromox® C/12
61791-47-7	Aromox® C/12-W
61791-56-8	Deriphat® 154
61791-59-1	Hamposyl® C-30
61791-63-7	Dinoram C
61791-63-7	Duomeen® C
61791-63-7	Kemamine® D-650
62476-60-6	Perkadox® BPSC
63123-11-5	Eastman® Inhibitor Poly TDP 2000
63148-57-2	CPS 120
63148-57-2	CPS 123
63148-62-9	Masil® SF 350 FG
63231-60-7	Be Square® 175
63231-60-7	Be Square® 185
63231-60-7	Be Square® 195
63231-60-7	Fortex®
63231-60-7	Mekon® White
63231-60-7	Multiwax® 180-M
63231-60-7	Multiwax® W-835
63231-60-7	Multiwax® X-145A
63231-60-7	Petrolite® C-700
63231-60-7	Petrolite® C-1035
63231-60-7	Starwax® 100
63231-60-7	Ultraflex®
63231-60-7	Victory®
63843-89-0	Tinuvin® 144
64741-81-7	Viplex 895-BL
64742-04-7	Viplex 680-P
64742-14-9	Jayflex® 215
64742-42-3	Koster Keunen Microcrystalline Waxes
64742-53-6	Jayflex® 210
64742-94-5	BVA ND-205 Solv
64743-02-8	Gulftene® 20-24
65140-91-2	Irganox® 1425 WL
65447-77-0	Tinuvin® 622
65447-77-0	Tinuvin® 622LD
65669-61-4	CF 32,500T Coarse
65816-20-8	Givsorb® UV-2
65996-61-4	CF 1500
65997-17-3	Sphericel® 110P8
66172-82-5	Servoxyl VPQZ 9/100
67254-74-4	Calsol 830
67254-74-4	Calsol 8200
67701-08-0	Industrene® 225
67701-08-0	Industrene® 226
67762-25-8	Alfol® 1218 DCBA
67762-25-8	Loxiol® VPG 1496
67762-25-8	Nafol® 1218
67762-35-0	Polirol C5
67762-36-1	Industrene® 365
67762-41-8	Epal® 12/85
67762-41-8	Epal® 1214
67762-85-0	Silwet® L-7001
67762-90-87	Aerosil® R202

CAS	Trade name
67774-74-7	Marlican®
67784-77-4	Ethoquad® T/12
67784-87-6	Myverol® 18-04
67784-90-1	Miramine® C
67999-57-9	Geronol ACR/9
67999-57-9	Geropon® ACR/9
68002-20-0	Resimene® 3520
68002-96-0	Polirol 215
68002-97-1	Rhodasurf® L-4
68003-46-3	Hamposyl® AL-30
68037-49-0	Chemstat® PS-101
68037-49-0	Chemstat® PS-101
68037-87-6	CPS 925
68081-81-2	Gardilene S25L
68083-35-2	Curezol® 2PZ-CNS
68128-25-6	CT2902
68128-59-6	Octosol A-18-A
68131-40-8	Tergitol® 15-S-3
68131-40-8	Tergitol® 15-S-5
68131-40-8	Tergitol® 15-S-30
68131-40-8	Tergitol® 15-S-40
68133-13-1	Weston® 494
68152-92-1	Westvaco® M-30
68152-92-1	Westvaco® M-40
68152-92-1	Westvaco® M-70
68155-09-9	Rhodamox® CAPO
68155-27-1	Synprolam 35
68238-87-9	Emsorb® 2518
68299-16-1	Aztec® TAPND-75-AL
68299-16-1	Esperox® 545M
68299-17-2	Rhodapon® CAV
68308-22-5	Hoechst Wax OP
68308-22-5	Hostalub® VP Ca W 2
68308-67-8	Larostat® 88
68308-67-8	Larostat® 264 A
68308-67-8	Larostat® 264 A Anhyd
68308-67-8	Larostat® 264 A Conc
68391-04-8	Synprolam 35DM
68411-19-8	Vanax® 833
68411-20-1	Vanox® AT
68411-32-5	Witconate™ Acide TPB
68411-90-5	BHMT Amine
68411-97-2	Hamposyl® C
68412-53-3	Rhodafac® PE-510
68412-53-3	Rhodafac® RE-610
68412-54-4	Berol 02
68412-54-4	Berol 277
68412-54-4	Berol 278
68412-54-4	Berol 281
68412-54-4	Berol 291
68412-54-4	Berol 292
68412-54-4	Berol 295
68424-59-9	W.G.S. Hydrogenated Fish Glyceride 117
68439-49-6	Prox-onic CSA-1/04
68439-49-6	Prox-onic CSA-1/06
68439-49-6	Prox-onic CSA-1/010
68439-49-6	Prox-onic CSA-1/015
68439-49-6	Prox-onic CSA-1/020
68439-49-6	Prox-onic CSA-1/030
68439-49-6	Prox-onic CSA-1/050
68439-49-6	Remcopal 229
68439-49-6	Teric 16A22
68439-57-6	Polystep® A-18
68439-57-6	Rhodacal® 301-10
68439-57-6	Rhodacal® 301-10P
68439-57-6	Rhodacal® A-246L
68439-57-6	Sermul EA 214
68439-57-6	Siponate® 301-10F (redesignated Rhodacal® 301-10F)
68439-57-6	Witconate™ AOS
68439-57-6	Witconate™ AOS-EP
68439-73-6	Kemamine® D-974
68440-66-4	Silwet® L-7500
68440-90-4	PS140
68441-17-8	A-C® 307, 307A
68441-17-8	A-C® 316, 316A
68441-17-8	A-C® 325
68441-17-8	A-C® 330
68441-17-8	A-C® 392
68441-17-8	A-C® 395, 395A
68441-17-8	A-C® 629
68441-17-8	A-C® 629A
68441-17-8	A-C® 655
68441-17-8	A-C® 656
68441-17-8	A-C® 6702
68441-17-8	Epolene® E-10
68441-17-8	Epolene® E-14
68441-17-8	Epolene® E-14P
68441-17-8	Epolene® E-15
68441-17-8	Epolene® E-20
68441-17-8	Epolene® E-20P
68441-17-8	Hoechst Wax PED 191
68441-17-8	Hostalub® H 12
68441-17-8	Loobwax 0761
68441-17-8	Luwax® OA 2
68441-17-8	Luwax® OA 5
68441-17-8	Petrolite® E-2020
68476-03-9	Hoechst Wax LP
68476-03-9	Hoechst Wax S
68476-03-9	Hoechst Wax SW
68476-03-9	Hoechst Wax UL
68476-38-0	Hostalub® WE4
68477-29-2	Viplex 885
68478-45-5	Wingstay® 100AZ
68515-47-9	Jayflex® DTDP
68515-47-9	Jayflex® UDP
68515-48-0	Jayflex® DINP
68515-48-0	Plasthall® DINP
68515-48-0	PX-139
68515-49-1	Jayflex® DIDP
68515-49-1	Nuoplaz® DIDP
68515-49-1	Plasthall® DIDP-E
68515-49-1	PX-120
68515-50-4	Jayflex® DHP
68555-98-6	Vultac® 2
68555-98-6	Vultac® 3
68555-98-6	Vultac® 7
68584-22-5	Nansa® SSA
68585-34-2	Empicol® ESB

CAS	Trade name
68585-47-7	Empicol® LX
68603-15-5	Nafol® 810 D
68603-42-9	Empilan® CDX
68603-64-5	Kemamine® D-970
68608-61-7	Miranol® SM Conc
68608-64-0	Miranol® J2M Conc
68608-77-5	Vanox® 1001
68608-96-1	Epodil® 747
68609-97-2	Epodil® 748
68609-97-2	Epodil® 759
68610-26-4	Tomah PA-19
68610-5-15	Ralox® LC
68610-51-5	Akrochem® Antioxidant 12
68610-51-5	Santowhite® ML
68610-51-5	Wingstay® L
68610-51-5	Wingstay® L HLS
68610-62-8	Weston® 474
68649-05-8	Armeen® Z
68649-55-8	Emulsifiant 33 AD
68783-41-5	Empol® 1004
68783-41-5	Empol® 1008
68783-41-5	Pripol 1009
68784-69-0	Agerite® Superlite
68784-69-0	Agerite® Superlite Solid
68815-23-6	Neutral Degras
68855-56-1	Alfol® 1216
68891-21-4	Berol 269
68891-38-3	Berol 452
68891-38-3	Sermul EA 30
68891-39-4	Rhodapex® CO-433
68901-05-3	Ageflex TPGDA
68909-20-6	Aerosil® R812
68909-77-3	Ridacto®
68909-77-3	Silacto®
68919-41-5	Miranol® CM-SF Conc.
68920-66-1	Polirol O55
68937-10-0	Amoco® H-300E
68937-41-7	Kronitex® 25
68937-41-7	Kronitex® 50
68937-41-7	Kronitex® 100
68937-41-7	Kronitex® 200
68937-41-7	Kronitex® 1840
68937-41-7	Kronitex® 1884
68937-41-7	Pliabrac® 519
68937-41-7	Pliabrac® 521
68937-41-7	Pliabrac® 524
68937-54-2	Tego® Foamex 800
68937-54-2	Tego® Foamex 805
68937-90-6	Empol® 1040
68938-54-5	Silwet® L-7600
68938-54-5	Silwet® L-7602
68952-35-2	Kronitex® TCP
68953-84-4	Wingstay® 100
68953-96-8	Nansa® EVM50
68955-19-1	Empicol® LZ
68955-19-1	Empicol® LZ/D
68955-19-1	Tensopol ACL 79
68988-69-2	Empimin® MKK/AU
68988-69-2	Manro MA 35
68989-03-7	Rewoquat CPEM
69009-90-1	Nusolv™ ABP-103
69011-06-9	Dythal
69011-06-9	Dythal XL
69011-06-9	EPlthal 120
69011-06-9	Halthal
69011-06-9	Halthal EP
69011-06-9	Haro® Chem PDP-E
69011-84-3	Akyposal BD
70321-86-7	Tinuvin® 234
70331-94-1	Naugard® XL-1
70624-18-9	Chimassorb® 944
70624-18-9	Chimassorb® 944FD
70624-18-9	Chimassorb® 944FL
70624-18-9	Chimassorb® 944LD
70788-37-3	Eltesol® TSX
70833-40-8	Trigonox® 131
70892-21-6	Mayzo RA-0095H
70892-21-6	Mayzo RA-0095HS
71172-17-3	Vanax® CPA
71486-48-1	Uniplex 310
71487-00-8	Ethoquad® C/12 Nitrate
71888-89-6	Jayflex® 77
72160-13-5	Akypo OP 80
72162-46-0	Tomah DA-14
72905-90-9	CPS 078.5
73138-44-0	Hoechst Wax OP
73138-45-1	Hoechst Wax E
73138-45-1	Hoechst Wax KPS
73296-89-6	Tensopol A 79
73296-89-6	Tensopol A 795
73296-89-6	Tensopol S 30 LS
78330-21-9	Rhodasurf® BC-737
78330-21-9	Rhodasurf® BC-840
78350-78-4	Trigonox® ADC-NS60
80410-33-9	CGA 12
80584-85-6	Doverphos® 11
80584-85-6	Weston® THOP
80584-86-7	Doverphos® 12
80584-86-7	Weston® DHOP
80584-87-8	Weston® 491
81846-81-3	Nusolv™ ABP-74
81846-81-3	Nusolv™ ABP-164
82451-48-7	Cyasorb® UV 3346
82469-79-2	Citroflex® B-6
84988-79-4	Priolube 1414
85209-91-2	ADK STAB NA-11
85356-84-9	Sipomer® WAM
86418-55-5	Cithrol EGMS S/E
87826-41-3	Millad® 3940
88917-22-0	Arcosolv® DPMA
90193-76-3	Sicolub® DSP
90268-48-7	Rewopol® B 1003
90529-77-4	GE 100
91031-48-0	Priolube 1458
92112-62-4	Synprolam 35MX1
92797-60-9	Aerosil® R805
93356-94-6	Doverphos® 613
93820-97-4	Estol 1481
94035-02-6	Beta W7 HP 0.9
95823-35-1	Mark® 2140

CAS	Trade name
96152-48-6	Weston® 439
96319-55-0	Trigonox® 171
96328-09-1	Mark® 5089
96492-31-8	Alcamizer 2
99241-25-5	Gamma W8 HP0.6
103597-45-1	Mixxim® BB/100
104042-16-2	Newcol 707SF
104042-16-2	Newcol 710, 714, 723
104222-30-2	VA-082
104810-47-1	Tinuvin® 213
106246-33-7	Lonzacure® M-CDEA
106392-12-5	Epan 785
106990-43-6	Chimassorb® 119
106990-43-6	Chimassorb® 119FL
107397-59-1	Tetronic® 90R4
107397-59-1	Tetronic® 150R1
107628-08-0	Akypo OP 80
109678-33-3	FR-1034
110553-27-0	Irganox® 1520
110720-64-4	EXP-61 Amine Functional Silicone Wax
112926-00-8	Hi-Sil® 210
112926-00-8	Hi-Sil® 233
112926-00-8	Hi-Sil® 243LD
112926-00-8	Silene® 732D
112926-00-8	Sipernat® 22
112926-00-8	Sipernat® 22S
112926-00-8	Sipernat® 50S
112945-52-5	Aerosil® 90
112945-52-5	Aerosil® 130
112945-52-5	Aerosil® 150
112945-52-5	Aerosil® 200
112945-52-5	Aerosil® 300
112945-52-5	Aerosil® 380
112945-52-5	Aerosil® OX50
112945-52-5	Cab-O-Sil® EH-5
112945-52-5	Cab-O-Sil® H-5
112945-52-5	Cab-O-Sil® HS-5
112945-52-5	Cab-O-Sil® L-90
112945-52-5	Cab-O-Sil® LM-130
112945-52-5	Cab-O-Sil® LM-150
112945-52-5	Cab-O-Sil® LM-150D
112945-52-5	Cab-O-Sil® M-5
112945-52-5	Cab-O-Sil® M-7D
112945-52-5	Cab-O-Sil® MS-55
112945-52-5	Cab-O-Sil® MS-75D
112945-52-5	Cab-O-Sil® PTG
112945-52-5	Wacker HDK® H15
115340-81-3	UV-Chek® AM-806
118337-09-0	Ethanox® 398
119345-01-6	Sandostab P-EPQ
119462-56-5	Perkalink® 900
119758-00-8	Alcamizer 4
121246-28-4	Durazone® 37
123237-14-9	Ross Wax #160
125740-36-5	Tomah Q-14-2
126121-35-5	Siltech® CE-2000
129757-67-1	Tinuvin® 123
130097-36-8	Esperox® 740M
130169-63-0	Silwax® S
133448-13-2	Silwax® C
133448-15-4	Silwax® WD-S
133448-16-5	Silwax® WD-IS
135861-56-2	Millad® 3988
136097-95-5	Koster Keunen Synthetic Candelilla
136618-51-4	Alcamizer 5
136618-52-5	Alcamizer 4-2
138265-88-0	Firebrake® ZB
138265-88-0	Zinc Borate B9355
594477-57-3	FR-1025

CAS Number-to-Chemical Cross-Reference

CAS	Chemical
50-36-2	Coke
56-35-9	Tributyltin oxide
57-11-4	Stearic acid
57-13-6	Urea
57-55-6	Propylene glycol
58-36-6	10,10´-Oxybisphenoxyarsine
60-00-4	Edetic acid
62-33-9	Calcium disodium EDTA
64-02-8	Tetrasodium EDTA
64-17-5	Alcohol
64-19-7	Acetic acid
65-85-0	Benzoic acid
66-71-7	1,10-Phenanthroline
67-56-1	Methyl alcohol
67-63-0	Isopropyl alcohol
67-68-5	Dimethyl sulfoxide
68-12-2	Dimethyl formamide
69-72-7	Salicylic acid
70-55-3	p-Toluenesulfonamide
71-36-3	Butyl alcohol
71-55-6	Trichloroethane
74-31-7	N,N´-Diphenyl-p-phenylenediamine
75-37-6	1,1-Difluoroethane
75-65-0	t-Butyl alcohol
75-78-5	Dimethyldichlorosilane
75-85-4	t-Amyl alcohol
75-91-2	t-Butyl hydroperoxide
75-94-5	Vinyltrichlorosilane
76-39-1	2-Nitro-2-methyl-1-propanol
77-58-7	Dibutyltin dilaurate
77-62-3	2,2´-Methylenebis 6-(1-methylcyclohexyl)-p-cresol
77-85-0	Trimethylolethane
77-89-4	Acetyl triethyl citrate
77-90-7	Acetyl tributyl citrate
77-92-9	Citric acid anhyd.
77-93-0	Triethyl citrate
77-94-1	Tributyl citrate
78-08-0	Vinyltriethoxysilane
78-10-4	Tetraethoxysilane
78-21-7	Cetethyl morpholinium ethosulfate
78-30-8	Tricresyl phosphate
78-40-0	Triethyl phosphate
78-42-2	Trioctyl phosphate
78-43-3	Tris (dichloropropyl) phosphate
78-51-3	Tributoxyethyl phosphate
78-59-1	Isophorone
78-63-7	2,5-Dimethyl-2,5-di (t-butylperoxy) hexane
78-66-0	Dimethyl octynediol
78-67-1	2,2´-Azobisisobutyronitrile
78-83-1	Isobutyl alcohol
78-93-3	Methyl ethyl ketone
79-09-4	Propionic acid
79-10-7	Acrylic acid
79-39-0	Methacrylamide
79-74-3	Diamylhydroquinone
79-94-7	Tetrabromobisphenol A
80-05-7	Bisphenol A
80-15-9	Cumene hydroperoxide
80-17-1	Benzene sulfonyl hydrazide
80-39-7	Ethyl toluenesulfonamide
80-43-3	Dicumyl peroxide
80-46-6	p-t-Amylphenol
80-51-3	4,4´-Oxybis (benzenesulfonyl-hydrazide)
84-61-7	Dicyclohexyl phthalate
84-64-0	Butyl cyclohexyl phthalate
84-66-2	Diethyl phthalate
84-69-5	Diisobutyl phthalate
84-74-2	Dibutyl phthalate
84-75-3	Dihexyl phthalate
85-22-3	Pentabromoethylbenzene
85-42-7	Hexahydrophthalic anhydride
85-44-9	Phthalic anhydride
85-60-9	4,4´-Butylidenebis (6-t-butyl-m-cresol)
85-68-7	Butyl benzyl phthalate
85-70-1	n-Butyl phthalyl-n-butyl glycolate
87-10-5	Tribromosalicylanilide
87-83-2	Pentabromotoluene
88-12-0	N-Vinyl-2-pyrrolidone
88-18-6	2-t-Butylphenol
88-19-7	o-Toluenesulfonamide
88-24-4	2,2´-Methylenebis (4-ethyl-6-t-butylphenol)
88-27-7	2,6-Di-t-butyl-N,N-dimethyl-amino-p-cresol
88-58-4	Di-t-butylhydroquinone
88-60-8	6-t-Butyl-m-cresol
88-69-7	o-Isopropylphenol

CAS	Chemical
89-04-3	Trioctyl trimellitate
89-32-7	Pyromellitic dianhydride
89-72-5	o-s-Butylphenol
90-12-0	α-Methylnaphthalene
90-30-2	Phenyl-α-naphthylamine
90-72-2	2,4,6-Tris (dimethylamino-methyl) phenol
91-17-8	Decahydronaphthalene
91-53-2	6-Ethoxy-1,2-dihydro-2,2,4-trimethylquinoline
91-88-3	N-Ethyl-N-hydroxyethyl-m-toluidine
91-99-6	m-Tolyl diethanolamine
92-50-2	Ethyl phenyl ethanolamine
92-84-2	Phenothiazine
93-46-9	N,N´-Di-β-naphthyl-p-phenylenediamine
93-69-6	o-Tolyl biguanide
94-04-2	Vinyl 2-ethylhexanoate
94-17-7	p,p-Dichlorobenzoyl peroxide
94-28-0	PEG-3 di-2-ethylhexoate
94-36-0	Benzoyl peroxide
94-51-9	Dipropylene glycol dibenzoate
95-31-8	Butyl 2-benzothiazole sulfenamide
95-32-9	4-Morpholinyl-2-benzothiazole disulfide
95-33-0	N-Cyclohexyl-2-benzothiazole-sulfenamide
95-71-6	Toluhydroquinone
96-05-9	Allyl methacrylate
96-27-5	Thioglycerin
96-45-7	Ethylene thiourea
96-48-0	Butyrolactone
96-49-1	Ethylene carbonate
96-69-5	4,4´-Thiobis-6-(t-butyl-m-cresol)
96-76-4	2,4-Di-t-butylphenol
97-39-2	Di-o-tolyl guanidine
97-74-5	Tetramethylthiuram mono-sulfide
97-77-8	Tetraethylthiuram disulfide
97-78-9	Lauroyl sarcosine
97-90-5	Ethylene glycol dimethacrylate
97-99-4	Tetrahydrofurfuryl alcohol
98-00-0	Furfuryl alcohol
98-01-1	Furfural
98-10-2	Benzenesulfonamide
98-11-3	Benzenesulfonic acid
98-54-4	4-t-Butylphenol
98-67-9	Phenol sulfonic acid
98-77-1	Piperidinium pentamethylene dithiocarbamate
98-94-2	N,N-Dimethyl-N-cyclohexyl-amine
99-89-3	p-Isopropylphenol
99-97-8	N,N-Dimethyl-p-toluidine
100-37-8	Diethylaminoethanol
100-74-3	Ethyl morpholine
100-97-0	Hexamethylene tetramine
101-02-0	Triphenyl phosphite
101-25-7	Dinitrosopentamethylene tetramine
101-34-8	Glyceryl triacetyl ricinoleate
101-37-1	Triallylcyanurate
101-42-8	1-Phenyl-3,3-dimethyl urea
101-43-9	Cyclohexyl methacrylate
101-54-2	N-Phenyl-p-phenylenediamine
101-67-7	p,p´-Dioctyldiphenylamine
101-68-8	MDI
101-68-8	Polymethylene polyphenyl isocyanate
101-72-4	N-Isopropyl-N´-phenyl-p-phenylenediamine
101-77-9	4,4´-Methylene dianiline
101-87-1	Cyclohexyl-N´-phenyl-p-phenylenediamine
101-90-6	Resorcinol diglycidyl ether
101-96-2	N,N´-Di-s-butyl-p-phenylenedi-amine
102-06-7	Diphenylguanidine
102-07-8	N,N´-Diphenylurea
102-08-9	N,N´-Diphenylthiourea
102-60-3	Tetrahydroxypropyl ethylene-diamine
102-71-6	Triethanolamine
102-76-1	Triacetin
102-77-2	N-Oxydiethylene benzothia-zole-2-sulfenamide
103-23-1	Dioctyl adipate
103-24-2	Dioctyl azelate
103-34-4	4,4´-Dithiodimorpholine
103-83-3	N-Benzyldimethylamine
103-96-8	N,N´-Di (2-octyl)-p-phenylene-diamine
103-99-1	Stearoyl p-amino phenol
104-15-4	p-Toluene sulfonic acid
104-38-1	Hydroquinone bis(2-hydroxyethyl) ether
104-76-7	2-Ethylhexanol
104-87-0	p-Tolualdehyde
105-08-8	1,4-Cyclohexanedimethanol
105-11-3	p-Quinone dioxime
105-16-8	Diethylaminoethyl methacry-late
105-38-4	Vinyl propionate
105-45-3	Methyl acetoacetate
105-52-2	Dimethyl amyl maleate
105-55-5	1,3-Diethylthiourea
105-64-6	Diisopropyl perdicarbonate
105-74-8	Lauroyl peroxide
105-76-0	Dibutyl maleate
105-76-9	Dibutyl fumarate
105-97-5	Dicapryl adipate
105-99-7	Dibutyl adipate
106-06-9	PEG-3 dipelargonate
106-07-0	PEG-4 stearate
106-11-6	PEG-2 stearate
106-11-6	PEG-2 stearate SE
106-12-7	PEG-2 oleate

CAS	Chemical
106-12-7	PEG-2 oleate SE
106-14-9	Hydroxystearic acid
106-17-2	Glycol ricinoleate
106-58-1	N,N´-Dimethylpiperazine
106-61-6	Glyceryl acetate
106-65-0	Dimethyl succinate
106-79-6	Dimethyl sebacate
106-84-3	Epoxidized octyl stearate
106-91-2	Glycidyl methacrylate
106-92-3	Allyl glycidyl ether
106-98-9	1-Butene
106-99-0	1,3-Butadiene
107-21-1	Glycol
107-41-5	Hexylene glycol
107-64-2	Distearyldimonium chloride
107-71-1	t-Butyl peracetate
107-72-2	Amyltrichlorosilane
107-96-0	3-Mercaptopropionic acid
107-98-2	Methoxyisopropanol
108-01-0	Dimethylethanolamine
108-31-6	Maleic anhydride
108-32-7	Propylene carbonate
108-45-2	m-Phenylenediamine
108-46-3	Resorcinol
108-65-6	Propylene glycol methyl ether acetate
108-80-5	Isocyanuric acid
108-83-8	Diisobutyl ketone
108-88-3	Toluene
108-91-8	Cyclohexylamine
108-94-1	Cyclohexanone
109-02-4	p-Methyl morpholine
109-13-7	t-Butyl peroxyisobutyrate
109-16-0	PEG-3 dimethacrylate
109-30-8	PEG-2 distearate
109-31-9	Di-n-hexyl azelate
109-43-3	Dibutyl sebacate
109-46-6	1,3-Dibutylthiourea
109-86-4	Methoxyethanol
109-89-7	Diethylamine
109-99-9	Tetrahydrofuran
110-05-4	Di-t-butyl peroxide
110-18-9	N,N,N´,N´-Tetramethyl-ethylenediamine
110-25-8	Oleoyl sarcosine
110-27-0	Isopropyl myristate
110-30-5	Ethylene distearamide
110-31-6	Ethylene dioleamide
110-54-3	Hexane
110-63-4	1,4-Butanediol
110-80-5	Ethoxyethanol
110-85-0	Piperazine
110-86-1	Pyridine
110-91-8	Morpholine
110-97-4	Diisopropanolamine
110-98-5	Dipropylene glycol
111-03-5	Glyceryl oleate
111-03-5	Glyceryl oleate SE
111-18-2	N,N,N´,N´-Tetramethylhexane-diamine

CAS	Chemical
111-27-3	Hexyl alcohol
111-40-0	Diethylenetriamine
111-41-1	Aminoethylethanolamine
111-42-2	Diethanolamine
111-46-6	Diethylene glycol
111-49-9	Hexamethyleneimine
111-55-7	Ethylene glycol diacetate
111-57-9	Stearamide MEA
111-60-4	Glycol stearate
111-65-9	Octane
111-66-0	Octene-1
111-76-2	Butoxyethanol
111-77-3	Methoxydiglycol
111-82-0	Methyl laurate
111-87-5	Caprylic alcohol
111-90-0	Ethoxydiglycol
111-96-6	Diethylene glycol dimethyl ether
112-00-5	Laurtrimonium chloride
112-02-7	Cetrimonium chloride
112-03-8	Steartrimonium chloride
112-11-8	Isopropyl oleate
112-13-0	Decanoyl chloride
112-16-3	Lauroyl chloride
112-18-5	Dimethyl lauramine
112-24-3	Triethylenetetramine
112-27-6	Triethylene glycol
112-30-1	n-Decyl alcohol
112-34-5	Butoxydiglycol
112-41-4	Dodecene-1
112-49-2	PEG-3 dimethyl ether
112-53-8	Lauryl alcohol
112-55-0	n-Dodecyl mercaptan
112-57-2	Tetraethylenepentamine
112-60-7	PEG-4
112-61-8	Methyl stearate
112-62-9	Methyl oleate
112-69-6	Dimethyl palmitamine
112-72-1	Myristyl alcohol
112-73-2	Diethylene glycol dibutyl ether
112-75-4	Dimethyl myristamine
112-80-1	Oleic acid
112-84-5	Erucamide
112-88-9	Octadecene-1
112-90-3	Oleamine
112-92-5	Stearyl alcohol
115-10-6	Dimethyl ether
115-27-5	Chlorendic anhydride
115-77-5	Pentaerythritol
115-83-3	Pentaerythrityl tetrastearate
115-86-6	Triphenyl phosphate
115-96-8	Tris (β-chloroethyl) phosphate
117-08-8	Tetrachlorophthalic anhydride
117-81-7	Dioctyl phthalate
117-82-8	Dimethoxyethyl phthalate
117-83-9	Dibutoxyethyl phthalate
117-84-0	Dioctyl phthalate
118-55-8	Phenyl salicylate
118-79-6	2,4,6-Tribromophenol
118-82-1	4,4´-Methylenebis (2,6-di-t-

CAS	Chemical
	butylphenol)
119-06-2	Ditridecyl phthalate
119-07-3	n-Octyl, n-decyl phthalate
119-47-1	2,2′-Methylenebis (6-t-butyl-4-methylphenol)
119-61-9	Benzophenone
119-64-2	Tetrahydronaphthalene
120-07-0	Phenyldiethanolamine
120-40-1	Lauramide DEA
120-51-4	Benzyl benzoate
120-54-7	Dipentamethylene thiuram tetrasulfide
120-55-8	Diethylene glycol dibenzoate
120-78-5	Benzothiazyl disulfide
120-95-6	2,4-Di-t-amylphenol
121-44-8	Triethylamine
121-54-0	Benzethonium chloride
121-72-2	N,N-Dimethyl-m-toluidine
121-79-9	Propyl gallate
122-20-3	Triisopropanolamine
122-32-7	Triolein
122-60-1	Phenyl glycidyl ether
122-62-3	Dioctyl sebacate
122-98-5	Phenylethanolamine
122-99-6	Phenoxyethanol
123-23-9	Succinic acid peroxide
123-28-4	Dilauryl thiodipropionate
123-31-9	Hydroquinone
123-34-2	Glyceryl-1-allyl ether
123-42-2	Diacetone alcohol
123-77-3	Azodicarbonamide
123-81-9	Glycol dimercaptoacetate
123-86-4	n-Butyl acetate
123-91-1	1,4-Dioxane
123-94-4	Glyceryl stearate
123-95-5	Butyl stearate
123-99-9	Azelaic acid
124-04-9	Adipic acid
124-09-4	Hexamethylenediamine
124-22-1	Lauramine
124-26-5	Stearamide
124-28-7	Dimethyl stearamine
124-30-1	Stearamine
124-38-9	Carbon dioxide
124-64-1	Tetrakis (hydroxymethyl) phosphonium chloride
124-68-5	2-Amino-2-methyl-1-propanol
126-13-6	Sucrose acetate isobutyrate
126-33-0	Sulfolane
126-73-8	Tributyl phosphate
126-86-3	Tetramethyl decynediol
126-92-1	Sodium octyl sulfate
126-99-8	Polychloroprene
127-08-2	Potassium acetate
127-25-3	Methyl abietate
127-39-9	Diisobutyl sodium sulfosuccinate
128-03-0	Potassium dimethyldithiocarbamate
128-04-1	Sodium dimethyldithiocarbamate hydrate
128-37-0	BHT
128-39-2	2,6-Di-t-butylphenol
131-11-3	Dimethyl phthalate
131-17-9	Diallyl phthalate
131-53-3	Benzophenone-8
131-54-4	Benzophenone-6
131-55-5	Benzophenone-2
131-56-6	Benzophenone-1
131-57-7	Benzophenone-3
133-06-2	Captan
133-14-2	Bis (2,4-dichlorobenzoyl) peroxide
133-49-3	Pentachlorothiophenol
135-20-6	Nitrosophenylhydroxylamine ammonium salt
135-57-9	2,2′-Dibenzamidodiphenyl disulfide
135-88-6	Phenyl-β-naphthylamine
136-23-2	Zinc dibutyldithiocarbamate
136-30-1	Sodium di-n-butyl dithiocarbamate
136-36-7	Resorcinol benzoate
136-52-7	Cobalt octoate
136-53-8	Zinc 2-ethylhexanoate
136-85-6	4- and 5-Methylbenzotriazole
137-16-6	Sodium lauroyl sarcosinate
137-26-8	Tetramethylthiuram disulfide
137-29-1	Copper dimethyldithiocarbamate
137-30-4	Zinc dimethyldithiocarbamate
137-89-3	Dioctyl isophthalate
139-07-1	Lauralkonium chloride
139-33-3	Disodium EDTA
139-43-5	Glyceryl (triacetoxystearate)
139-45-7	Glyceryl tripropionate
139-88-8	Sodium myristyl sulfate
139-96-8	TEA-lauryl sulfate
140-01-2	Pentasodium pentetate
140-03-4	Methyl acetyl ricinoleate
140-04-5	Butyl acetyl ricinoleate
140-08-9	Tris 2-chloroethyl phosphite
140-31-8	Aminoethylpiperazine
141-02-6	Dioctyl fumarate
141-04-8	Diisobutyl adipate
141-08-2	Glyceryl ricinoleate
141-17-3	Dibutoxyethoxyethyl adipate
141-18-4	Dibutoxyethyl adipate
141-20-8	PEG-2 laurate
141-20-8	PEG-2 laurate SE
141-22-0	Ricinoleic acid
141-23-1	Methyl hydroxystearate
141-24-2	Methyl ricinoleate
141-32-2	Butyl acrylate
141-43-5	Ethanolamine
141-78-6	Ethyl acetate
141-97-9	Ethylacetoacetate
142-16-5	Dioctyl maleate
142-18-7	Glyceryl laurate
142-20-1	PEG-4 distearate

CAS	Chemical
142-31-4	Sodium octyl sulfate
142-48-3	Stearoyl sarcosine
142-55-2	Propylene glycol laurate
142-59-6	Nabam
142-77-8	Butyl oleate
142-82-5	Heptane
142-87-0	Sodium decyl sulfate
142-90-5	Lauryl methacrylate
142-91-6	Isopropyl palmitate
143-00-0	DEA-lauryl sulfate
143-07-7	Lauric acid
143-18-0	Potassium oleate
143-27-1	Palmitamine
143-28-2	Oleyl alcohol
144-34-4	Selenium dimethyldithiocarbamate
144-55-8	Sodium bicarbonate
148-18-5	Sodium diethyldithiocarbamate
149-30-4	2-Mercaptobenzothiazole
149-44-0	Sodium formaldehyde sulfoxylate
150-38-9	Trisodium EDTA
150-76-5	Hydroquinone monomethyl ether
151-21-3	Sodium lauryl sulfate
155-04-4	Zinc 2-mercaptobenzothiazole
156-10-5	Nitrosodiphenylamine
208-96-8	Bromoacenaphthylene
215-42-9	Dimethyl behenamine
273-53-0	Benzoxazole
280-57-9	Triethylene diamine
300-92-5	Aluminum distearate
301-02-0	Oleamide
301-10-0	Stannous octoate
306-52-5	Trichloroethyl phosphate
306-83-2	Dichlorotrifluoroethane
330-54-1	3-(3,4-Dichlorophenyl)-1,1-dimethylurea
420-04-2	Carbodiimide
422-86-2	Dioctyl terephthalate
461-58-5	Dicyandiamide
471-34-1	Calcium carbonate
501-24-6	3-(n-Pentadecyl) phenol
501-52-0	Benzene propanoic acid
513-77-9	Barium carbonate
532-32-1	Sodium benzoate
546-68-9	Tetraisopropyl titanate
546-93-0	Magnesium carbonate
555-43-1	Tristearin
557-04-0	Magnesium stearate
557-05-1	Zinc stearate
577-11-7	Dioctyl sodium sulfosuccinate
583-39-1	2-Mercaptobenzimidazole
584-84-9	Toluene diisocyanate
589-37-7	1,3-Pentanediamine
589-68-4	Glyceryl myristate
592-41-6	1-Hexene
593-29-3	Potassium stearate
593-60-2	Vinyl bromide
598-63-0	Lead carbonate (basic)
599-64-4	Cumyl phenol
614-33-5	Glyceryl tribenzoate
614-45-9	t-Butyl perbenzoate
615-58-7	2,4-Dibromophenol
616-45-5	2-Pyrrolidone
622-16-2	Carbodiimide
624-04-4	Glycol dilaurate
627-83-8	Glycol distearate
627-93-0	Dimethyl adipate
629-14-1	Ethylene glycol diethyl ether
629-73-2	Hexadecene-1
631-61-8	Ammonium acetate
632-79-1	Tetrabromophthalic anhydride
637-12-7	Aluminum tristearate
638-16-4	2,4,6-Trimercapto-s-triazine
646-13-9	Isobutyl stearate
661-19-8	Behenyl alcohol
670-96-2	2-Phenyl imidazole
681-84-5	Tetramethoxysilane
682-01-9	Tetra-n-propoxysilane
686-31-7	t-Amylperoxy 2-ethylhexanoate
690-83-5	t-Amyl peroxyacetate
693-33-4	Cetyl betaine
693-36-7	Distearyl thiodipropionate
693-98-1	2-Methyl imidazole
694-83-7	1,2-Diaminocyclohexane
732-26-3	2,4,6-Tri-t-butylphenol
756-79-6	Dimethyl methylphosphonate
760-93-0	Methacrylic anhydride
762-12-9	Decanoyl peroxide
762-16-3	Dioctanoyl peroxide
763-69-9	Ethyl 3-ethoxypropionate
780-69-8	Phenyltriethoxysilane
793-24-8	N-1,3-Dimethylbutyl-N′-phenyl-p-phenylene diamine
814-80-2	Calcium lactate anhyd.
818-08-6	Dibutyltin oxide
820-66-6	Stearyl betaine
822-16-2	Sodium stearate
839-90-7	Trishydroxyethyl triazine trione
868-77-9	2-Hydroxyethyl methacrylate
870-08-6	Dioctyltin oxide
872-05-9	Decene-1
872-50-4	1-Methyl-2-pyrrolidinone
877-66-7	p-Toluene sulfonyl hydrazide
919-30-2	Aminopropyltriethoxysilane
922-80-5	Diamyl sodium sulfosuccinate
927-07-1	t-Butyl peroxypivalate
928-24-5	Glycol dioleate
929-06-6	2-(2-Aminoethoxy) ethanol
931-36-2	2-Ethyl-4-methyl imidazole
940-41-0	2-Phenethyltrichlorosilane
971-15-3	Dipentamethylene hexasulfide
991-84-4	4-[[4,6-Bis(octylthio)-s-triazin-2-yl]amino]-2,6-di-t-butylphenol
995-33-5	n-Butyl-4,4-bis (t-butylperoxy) valerate
999-97-3	Hexamethyldisilazane
1000-90-4	Zinc isopropyl xanthate
1002-89-7	Ammonium stearate

CAS	Chemical
1025-15-6	Triallyl isocyanurate
1067-25-0	n-Propyltrimethoxysilane
1067-33-0	Dibutyltin diacetate
1067-53-4	Vinyltris(2-methoxyethoxy) silane
1068-27-5	2,5-Dimethyl-2,5-di (t-butylperoxy) hexyne-3
1072-35-1	Lead stearate
1077-56-1	Ethyl toluenesulfonamide
1119-40-0	Dimethyl glutarate
1120-16-7	Lauramide
1120-24-7	Dimethyl decylamine
1120-36-1	Tetradecene-1
1143-72-2	2,3,4-Trihydroxy benzophenone
1163-19-5	Decabromodiphenyl oxide
1185-55-3	Methyltrimethoxysilane
1191-50-0	Sodium myristyl sulfate
1241-94-7	Diphenyl octyl phosphate
1254-78-0	Didecyl phenyl phosphite
1303-96-4	Sodium borate decahydrate
1305-62-0	Calcium hydroxide
1305-78-8	Calcium oxide
1306-19-0	Cadmium oxide
1309-37-1	Ferric oxide anhyd.
1309-42-8	Magnesium hydroxide anhyd.
1309-48-4	Magnesium oxide
1309-60-0	Lead dioxide
1309-64-4	Antimony trioxide
1313-13-9	Manganese dioxide
1313-27-5	Molybdenum trioxide
1314-13-2	Zinc oxide
1314-22-3	Zinc peroxide
1314-41-6	Lead oxide, red
1314-60-9	Antimony pentoxide
1314-98-3	Zinc sulfide
1317-33-5	Molybdenum disulfide
1317-36-8	Lead (II) oxide
1317-65-3	Calcium carbonate
1317-65-3	Limestone
1317-80-2	Titanium dioxide
1317-97-1	Ultramarine blue
1318-02-1	Sodium silicoaluminate
1318-74-7	Kaolinite
1318-93-0	Montmorillonite
1321-69-3	Sodium polynaphthalene sulfonate
1321-87-5	Dimethyl octynediol
1322-93-6	Sodium isopropyl naphthalene sulfonate
1323-38-2	Castor oil
1323-39-3	Propylene glycol stearate
1323-42-8	Glyceryl hydroxystearate
1327-36-2	Aluminum silicate
1327-39-5	Calcium aluminosilicate
1330-20-7	Xylene
1330-76-3	Diisooctyl maleate
1330-78-5	Tricresyl phosphate
1330-80-9	Propyl oleate
1330-86-5	Diisooctyl adipate
1332-07-6	Zinc borate anhyd.
1332-58-7	Kaolin
1333-39-7	Phenol sulfonic acid
1333-84-2	Alumina hydrate
1333-86-4	Carbon black
1335-72-4	Sodium laureth sulfate
1337-33-3	Stearyl citrate
1338-23-4	Methyl ethyl ketone peroxide
1338-39-2	Sorbitan laurate
1338-41-6	Sorbitan stearate
1338-43-8	Sorbitan oleate
1338-51-8	Toluenesulfonamide formaldehyde resin
1341-54-4	Benzophenone-11
1343-98-2	Silica, hydrated
1344-00-9	Sodium silicoaluminate
1344-28-1	Alumina
1344-40-7	Lead phosphite dibasic
1344-67-8	Cupric chloride
1344-95-2	Calcium silicate
1345-05-7	Lithopone
1369-66-3	Dioctyl sodium sulfosuccinate
1421-49-4	3,5-Di-t-butyl-p-hydroxy-benzoic acid
1459-93-4	Dimethyl isophthalate
1493-13-6	Trifluoromethane sulfonic acid
1528-49-0	Tri n-hexyl trimellitate
1533-45-5	2,2′-(1,2-Ethenediyldi-4,1-phenylene) bisbenzoxazole
1561-49-5	Dicyclohexyl peroxydicarbonate
1561-92-8	Sodium methallyl sulfonate
1592-23-0	Calcium stearate
1634-02-2	Tetrabutyl thiuram disulfide
1634-04-4	Methyl t-butyl ether
1643-20-5	Lauramine oxide
1656-63-9	Trilauryl trithiophosphite
1704-62-7	2-(2-Dimethylaminoethoxy) ethanol
1709-70-2	1,3,5-Trimethyl-2,4,6-tris (3,5-di-t-butyl-4-hydroxybenzyl) benzene
1717-00-6	Dichlorofluoroethane
1760-24-3	N-2-Aminoethyl-3-aminopropyl trimethoxysilane
1761-71-3	Bis (p-aminocyclohexyl) methane
1843-03-4	1,1,3-Tris (2-methyl-4-hydroxy-5-t-butyl phenyl) butane
1843-05-6	Benzophenone-12
1847-55-8	Sodium oleyl sulfate
1879-09-0	6-t-Butyl-2,4-xylenol
1889-67-4	2,3-Dimethyl-2,3-diphenyl-butane
1908-87-8	3-Methyl-thiazolidinethione-2
1931-62-0	t-Butyl peroxymaleic acid
1948-33-0	t-Butyl hydroquinone
2016-57-1	Lauramine
2031-67-6	Methyltriethoxysilane
2082-79-3	Octadecyl 3,5-di-t-butyl-4-

CAS	Chemical
	hydroxyhydrocinnamate
2082-80-6	Tristearyl phosphite
2082-81-7	1,4-Butanediol dimethacrylate
2094-98-6	1,1′-Azobis(cyclohexane-1-carbonitrile)
2155-71-7	Di-t-butyl peroxyphthalate
2162-74-5	Bis (2,6-diisopropylphenyl) carbodiimide
2163-42-0	2-Methyl-1,3-propanediol
2167-23-9	2,2-Di (t-butylperoxy) butane
2210-79-9	o-Cresyl glycidyl ether
2219-82-1	6-t-Butyl-o-cresol
2219-84-3	p-t-Octyl-o-cresol
2223-93-0	Cadmium stearate
2235-54-3	Ammonium lauryl sulfate
2277-28-3	Glyceryl linoleate
2304-30-5	Tetrabutylphosphonium chloride
2315-62-0	Octoxynol-3
2315-66-4	Octoxynol-10
2315-67-5	Octoxynol-1
2358-84-1	Diethylene glycol dimethacrylate
2370-64-1	PEG-6 laurate
2372-21-6	t-Butyl peroxy isopropyl carbonate
2373-38-8	Dihexyl sodium sulfosuccinate
2409-55-4	2-t-Butyl-p-cresol
2421-28-5	3,3′,4,4′-Benzophenone tetracarboxylic dianhydride
2425-79-8	1,4-Butanediol diglycidyl ether
2426-08-6	Butyl glycidyl ether
2426-54-2	Diethylaminoethyl acrylate
2432-87-3	Dioctyl sebacate
2437-72-1	Trimethyl quinoline
2439-35-2	Dimethylaminoethyl acrylate
2440-22-4	Drometrizole
2452-01-9	Zinc laurate
2455-24-5	Tetrahydrofurfuryl methacrylate
2461-15-6	2-Ethylhexyl glycidyl ether
2489-77-2	Trimethylthiourea
2499-95-8	n-Hexyl acrylate
2530-83-8	3-Glycidoxypropyltrimethoxysilane
2530-85-0	3-Methacryloxypropyltrimethoxysilane
2530-87-2	3-Chloropropyltrimethoxysilane
2550-33-6	t-Butyl peroxydiethylacetate
2589-57-3	Dimethyl 2,2′-azobis (2-methylpropionate)
2605-44-9	Cadmium laurate
2615-15-8	PEG-6
2618-77-1	2,5-Dimethyl-2,5-di(benzoylperoxy)hexane
2638-94-0	4,4′-Azobis (4-cyanopentanoic acid)
2644-45-3	Decyl betaine
2673-22-5	Ditridecyl sodium sulfosuccinate
2694-54-4	Triallyl trimellitate
2724-58-5	Isostearic acid
2764-13-8	Stearamidopropyl dimethyl-2-hydroxyethyl ammonium nitrate
2768-02-7	Vinyltrimethoxysilane
2772-45-4	2,4-Di-α-methylstyrylphenol
2778-96-3	Stearyl stearate
2781-00-2	1,4-Di-(2-t-butylperoxy-isopropyl) benzene
2807-30-9	Ethylene glycol propyl ether
2855-13-2	Isophorone diamine
2867-47-2	Dimethylaminoethyl methacrylate
2897-60-1	(3-Glycidoxypropyl)-methyldiethoxy silane
2915-53-9	Dioctyl maleate
2917-73-9	Dibutyl azelate
2917-94-4	Sodium octoxynol-2 ethane sulfonate
2943-75-1	Octyltriethoxysilane
2997-92-4	2,2-Azobis(2-amidinopropane) dihydrochloride
3006-15-3	Dihexyl sodium sulfosuccinate
3006-82-4	t-Butyl peroctoate
3006-86-8	1,1-Di (t-butylperoxy) cyclohexane
3006-93-7	m-Phenylenedimaleimide
3026-63-9	Sodium tridecyl sulfate
3030-47-5	N,N′,N′′,N′′-Pentamethyldiethylenetriamine
3034-79-5	Di-(2-methylbenzoyl) peroxide
3052-70-8	2,2-Di-(t-amylperoxy) propane
3055-93-4	Laureth-2
3055-94-5	Laureth-3
3055-97-8	Laureth-7
3055-99-0	Laureth-9
3056-00-6	Laureth-12
3057-08-7	Phenyl neopentylene glycol phosphite
3061-75-4	Behenamide
3064-73-1	Tetraisobutylthiuram disulfide
3069-25-8	N-Methylaminopropyltrimethoxysilane
3069-29-2	N-(2-Aminoethyl)-3-aminopropyl methyldimethoxy silane
3076-63-9	Trilauryl phosphite
3077-12-1	p-Tolyl diethanolamine
3081-14-9	N,N′-Di-(1,4-dimethylpentyl)-p-phenylene diamine
3088-31-1	Sodium laureth sulfate
3089-11-0	Hexamethoxymethylmelamine
3101-60-8	p-t-Butyl phenyl glycidyl ether
3115-68-2	Tetrabutylphosphonium bromide
3121-61-7	Methoxyethyl acrylate
3147-75-9	2-(2-Hydroxy-5-t-octylphenyl) benzotriazole
3158-56-3	3-(n-Pentadecyl) phenol
3164-85-0	Potassium 2-ethylhexanoate

CAS	Chemical
3179-56-4	Acetyl cyclohexanesulfonyl peroxide
3179-76-8	3-Aminopropylmethyldiethoxy silane
3194-55-6	Hexabromocyclododecane
3278-89-5	Tribromophenyl allyl ether
3290-92-4	1,1,1-Trimethylolpropane trimethacrylate
3296-90-0	Dibromoneopentyl glycol
3319-31-1	Trioctyl trimellitate
3322-93-8	Dibromoethyldibromocyclo-hexane
3327-22-8	Chloro-2-hydroxypropyl trimonium chloride
3333-62-8	CI Fluorescent Brightener 236
3401-73-8	Tetrasodium dicarboxyethyl stearyl sulfosuccinamate
3425-61-4	t-Amyl hydroperoxide
3437-84-1	Diisobutyryl peroxide
3457-61-2	t-Butyl cumyl peroxide
3486-35-9	Zinc carbonate
3524-68-3	Pentaerythrityl triacrylate
3578-72-1	Calcium behenate
3622-84-2	N,N-Butyl benzene sulfon-amide
3648-18-8	Dioctyltin dilaurate
3648-20-2	Diundecyl phthalate
3648-21-3	Di-n-heptyl phthalate
3655-00-3	Disodium lauriminodipropio-nate
3682-94-8	2,2′-Azobis(2-methylpropion-amide) dihydrate
3710-84-7	Diethylhydroxylamine
3736-26-3	Diisopropylbenzene mono-hydroperoxide
3741-08-8	Butyl 2-benzothiazole sulfenamide
3806-34-6	Distearyl pentaerythritol diphosphite
3825-26-1	Ammonium perfluorooctanoate
3851-87-4	3,5,5-Trimethyl hexanoyl peroxide
3851-87-4	Bis (3,5,5-trimethylhexanoyl) peroxide
3864-99-1	2-(3′,5′-Di-t-butyl-2′-hydroxyphenyl)-5-chloro-benzotriazole
3896-11-5	2-(2′-Hydroxy-3′-t-butyl-5′-methylphenyl)-5-chloro-benzotriazole
3999-10-8	5-Phenyltetrazole
4016-14-2	Isopropyl glycidyl ether
4065-45-6	Benzophenone-4
4067-16-7	Pentaethylene hexamine
4080-31-3	Quaternium-15
4130-08-9	Vinyltriacetoxy silane
4151-51-3	Tris (p-isocyanato-phenyl) thiophosphate
4196-86-5	Pentaerythrityl tetrabenzoate
4196-89-8	Neopentyl glycol dibenzoate

CAS	Chemical
4219-48-1	Glycol laurate
4221-80-1	2,4-Di-t-butylphenyl 3,5-di-t-butyl-4-hydroxybenzoate
4253-34-3	Methyltriacetoxysilane
4306-88-1	Diisobutyl nonyl phenol
4419-11-8	2,2′-Azobis (2,4-dimethyl-valeronitrile)
4420-74-0	Mercaptopropyltrimeth-oxysilane
4485-12-5	Lithium stearate
4500-01-0	Glycol oleate
4511-39-1	t-Amyl peroxybenzoate
4712-55-4	Diphenyl phosphite
4722-98-9	MEA-lauryl sulfate
4748-78-1	p-Ethylbenzaldehyde
4767-03-7	Dimethylolpropionic acid
4979-32-2	Dicyclohexylbenzothiazyl-2-sulfenamide
4986-89-4	Pentaerythrityl tetraacrylate
5035-58-5	Diphenylstibine 2-ethylhexoate
5089-70-3	3-Chloropropyltriethoxysilane
5117-19-1	PEG-8
5232-99-5	Etocrylene
5274-68-0	Laureth-4
5323-95-5	Sodium ricinoleate
5349-52-0	PEG-9 stearate
5353-26-4	Oleth-4
5456-28-0	Selenium diethyldithio-carbamate
5593-70-4	Tetrabutyl titanate
5719-73-3	Disodium hexamethylene bisthiosulfate
5793-94-2	Calcium stearoyl lactylate
5809-08-5	2,4,4-Trimethylpentyl-2-hydroperoxide
5888-33-5	Isobornyl acrylate
5938-38-5	Sorbitan oleate
5949-29-1	Citric acid monohydrate
5959-89-7	Sorbitan laurate
6001-97-4	Dihexyl sodium sulfosuccinate
6144-28-1	Dilinoleic acid
6197-30-4	Octocrylene
6281-04-5	PEG-2 dilaurate
6440-58-0	DMDM hydantoin
6542-37-6	Hydroxymethyl dioxoazabi-cyclooctane
6674-22-2	Diazabicycloundecene
6683-19-8	Pentaerythrityl tetrakis [3-(3′,5′-di-t-butyl-4-hydroxy-phenyl)propionate]
6683-19-8	Tetrakis [methylene (3,5-di-t-butyl-4-hydroxyhydrocin-namate)] methane
6729-96-0	Distearyl dithiodipropionate
6731-36-8	1,1-Di (t-butylperoxy) 3,3,5-trimethyl cyclohexane
6790-09-6	PEG-12
6838-85-3	Lead phthalate, basic
6842-15-5	Dodecene-1
6846-50-0	Trimethyl-1,3-pentanediol,

CAS	Chemical
	2,2,4-diisobutyrate
6865-35-6	Barium stearate
6881-94-3	Diethylene glycol propyl ether
6886-38-5	3-(3,5-Di-t-butyl-4-hydroxy-phenyl)propionic methyl ester
6938-94-9	Diisopropyl adipate
6990-43-8	Zinc dibutyl dithiophosphorate
7005-47-2	2-Dimethylamino-2-methyl-1-propanol
7047-84-9	Aluminum stearate
7049-66-3	Trilinoleic acid
7128-64-5	2,2´-(2,5-Thiophenediyl)bis[5-t-butylbenzoxazole]
7173-62-8	Oleyl 1,3-propylene diamine
7311-27-5	Nonoxynol-4
7320-34-5	Tetrapotassium pyrophosphate
7378-99-6	Dimethyl octylamine
7384-98-7	Propylene glycol dioctanoate
7396-58-9	Didecyl methylamine
7439-92-1	Lead
7439-93-2	Lithium
7440-02-0	Nickel
7440-31-5	Tin
7440-36-0	Antimony
7440-39-3	Barium
7440-44-0	Carbon
7440-66-6	Zinc
7491-02-3	Diisopropyl sebacate
7491-05-2	Tallowtrimonium chloride
7492-30-0	Potassium ricinoleate
7534-94-3	Isobornyl methacrylate
7538-44-5	Bis (hydroxyethyl) amino-propyltriethoxy silane
7568-93-6	Phenylethanolamine
7575-23-7	Pentaerythrityl tetrakis (3-mercaptopropionate)
7585-39-9	Cyclodextrin
7617-78-9	Isodecyloxypropylamine
7631-86-9	Diatomaceous earth
7631-86-9	Silica (colloidal)
7637-07-2	Boron trifluoride
7646-85-7	Zinc chloride
7659-86-1	2-Ethylhexyl thioglycolate
7664-93-9	Sulfuric acid
7665-72-7	t-Butyl glycidyl ether
7702-01-4	Disodium capryloampho-diacetate
7704-34-9	Sulfur
7705-08-0	Ferric chloride anhyd.
7723-14-0	Phosphorus
7727-43-7	Barium sulfate
7756-94-7	Triisobutylene
7757-93-9	Calcium phosphate dibasic
7758-87-4	Calcium phosphate tribasic
7772-99-8	Stannous chloride anhyd.
7778-18-9	Calcium sulfate anhyd.
7779-27-3	Hexahydro-1,3,5-triethyl-s-triazine
7779-90-0	Zinc phosphate
7782-42-5	Graphite

CAS	Chemical
7782-49-2	Selenium
7787-59-9	Bismuth oxychloride
8001-21-6	Sunflower seed oil
8001-75-0	Ceresin
8001-78-3	Hydrogenated castor oil
8001-79-4	Castor oil
8002-13-9	Rapeseed oil
8002-23-1	Cetyl esters
8002-33-3	Sulfated castor oil
8002-43-5	Lecithin
8002-53-7	Montan wax
8002-74-2	Paraffin
8002-74-2	Petroleum wax
8002-74-2	Synthetic wax
8005-44-5	Cetearyl alcohol
8006-44-8	Candelilla wax
8006-54-0	Lanolin anhyd.
8007-18-9	Pigment yellow 53
8007-24-7	3-(n-Penta-8´-decenyl) phenol
8007-43-0	Sorbitan sesquioleate
8009-03-8	Petrolatum NF
8011-48-1	Pine tar
8012-95-1	Mineral oil
8013-05-6	Potassium castorate
8013-07-8	Epoxidized soybean oil
8015-86-9	Carnauba
8020-83-5	Mineral oil white
8020-84-6	Lanolin hydrous
8021-55-4	Ozokerite
8027-32-5	Petrolatum USP
8029-76-3	Lecithin
8029-92-3	Acetylated lard glyceride
8030-30-6	Naphtha
8030-78-2	Tallowtrimonium chloride
8032-32-4	Mineral spirits
8042-47-5	Mineral oil
8050-09-7	Rosin
8050-13-3	Methyl hydrogenated rosinate
8050-26-8	Pentaerythrityl rosinate
8050-81-5	Simethicone
8051-30-7	Cocamide DEA
8051-52-3	PEG-15 cocamine
8052-10-6	Rosin
8061-51-6	Sodium lignosulfonate
8061-53-8	Ammonium lignosulfonate
9002-83-9	Polychlorotrifluoroethylene
9002-84-0	Polytetrafluoroethylene
9002-86-2	Polyvinyl chloride
9002-88-4	Polyethylene
9002-88-4	Polyethylene wax
9002-89-5	Polyvinyl alcohol (super and fully hydrolyzed)
9002-92-0	Laureth-2
9002-92-0	Laureth-3
9002-92-0	Laureth-7
9002-92-0	Laureth-9
9002-92-0	Laureth-12
9002-92-0	Laureth-15
9002-92-0	Laureth-16
9002-92-0	Laureth-23

CAS	Chemical
9002-93-1	Octoxynol-1
9002-93-1	Octoxynol-3
9002-93-1	Octoxynol-7
9002-93-1	Octoxynol-9
9002-93-1	Octoxynol-10
9002-93-1	Octoxynol-16
9002-93-1	Octoxynol-25
9002-93-1	Octoxynol-30
9002-93-1	Octoxynol-33
9002-93-1	Octoxynol-40
9002-93-1	Octoxynol-70
9003-01-4	Acrylic resin
9003-01-4	Polyacrylate
9003-01-4	Polyacrylic acid
9003-03-6	Ammonium polyacrylate
9003-04-7	Sodium polyacrylate
9003-05-8	Polyacrylamide
9003-07-0	Polypropylene
9003-08-1	Melamine/formaldehyde resin
9003-09-2	Polyvinyl methyl ether
9003-11-6	PEG-40 sorbitan tetraoleate
9003-11-6	Poloxamer 182
9003-11-6	Poloxamer 184
9003-11-6	Poloxamer 188
9003-11-6	Poloxamer 238
9003-17-2	Polybutadiene
9003-18-3	Butadiene/acrylonitrile copolymer
9003-27-4	Polyisobutene
9003-28-5	Polybutene
9003-29-6	Polybutene
9003-29-6	Polyisobutene
9003-31-0	Polyisoprene
9003-34-3	Poly-p-dinitrosobenzene polymer
9003-35-4	Phenolic resin
9003-39-8	PVP
9003-53-6	Polystyrene
9003-54-7	Styrene-acrylonitrile copolymer
9003-55-8	Butadiene-styrene copolymer
9003-55-8	Styrene-butadiene polymer
9003-55-8	Styrene-butadiene rubber
9003-56-9	Acrylonitrile-butadiene-styrene
9003-73-0	Polydipentene
9003-74-1	Terpene resin
9003-79-6	Diphenylamineacetone
9004-32-4	Carboxymethylcellulose sodium
9004-34-6	Cellulose
9004-62-0	Hydroxyethylcellulose
9004-65-3	Hydroxypropyl methylcellulose
9004-67-5	Methylcellulose
9004-80-2	Deceth-4 phosphate
9004-81-3	PEG-2 laurate
9004-81-3	PEG-4 laurate
9004-81-3	PEG-6 laurate
9004-81-3	PEG-8 laurate
9004-81-3	PEG-12 laurate
9004-81-3	PEG-20 laurate
9004-81-3	PEG-32 laurate
9004-81-3	PEG-75 laurate
9004-81-3	PEG-150 laurate
9004-82-4	Sodium laureth sulfate
9004-82-4	Sodium laureth-12 sulfate
9004-82-4	Sodium laureth-30 sulfate
9004-87-9	Octoxynol-1
9004-87-9	Octoxynol-3
9004-87-9	Octoxynol-7
9004-87-9	Octoxynol-9
9004-87-9	Octoxynol-10
9004-87-9	Octoxynol-16
9004-87-9	Octoxynol-25
9004-87-9	Octoxynol-30
9004-87-9	Octoxynol-33
9004-87-9	Octoxynol-40
9004-87-9	Octoxynol-70
9004-93-7	PEG di-2-ethylhexoate
9004-96-0	PEG-4 oleate
9004-96-0	PEG-6 oleate
9004-96-0	PEG-8 oleate
9004-96-0	PEG-12 oleate
9004-96-0	PEG-20 oleate
9004-96-0	PEG-32 oleate
9004-96-0	PEG-75 oleate
9004-96-0	PEG-150 oleate
9004-97-1	PEG-12 ricinoleate
9004-98-2	Oleth-4
9004-98-2	Oleth-7
9004-98-2	Oleth-9
9004-98-2	Oleth-20
9004-99-3	PEG-2 stearate
9004-99-3	PEG-4 stearate
9004-99-3	PEG-6 stearate
9004-99-3	PEG-6-32 stearate
9004-99-3	PEG-8 stearate
9004-99-3	PEG-9 stearate
9004-99-3	PEG-12 stearate
9004-99-3	PEG-20 stearate
9004-99-3	PEG-32 stearate
9004-99-3	PEG-50 stearate
9004-99-3	PEG-75 stearate
9004-99-3	PEG-150 stearate
9005-02-1	PEG-2 dilaurate
9005-02-1	PEG-4 dilaurate
9005-02-1	PEG-6 dilaurate
9005-02-1	PEG-8 dilaurate
9005-02-1	PEG-12 dilaurate
9005-02-1	PEG-20 dilaurate
9005-02-1	PEG-32 dilaurate
9005-02-1	PEG-75 dilaurate
9005-02-1	PEG-150 dilaurate
9005-07-6	PEG-2 dioleate
9005-07-6	PEG-4 dioleate
9005-07-6	PEG-6 dioleate
9005-07-6	PEG-8 dioleate
9005-07-6	PEG-12 dioleate
9005-07-6	PEG-20 dioleate
9005-07-6	PEG-32 dioleate
9005-07-6	PEG-75 dioleate
9005-07-6	PEG-150 dioleate

CAS	Chemical
9005-08-7	PEG-4 distearate
9005-08-7	PEG-6 distearate
9005-08-7	PEG-8 distearate
9005-08-7	PEG-12 distearate
9005-08-7	PEG-20 distearate
9005-08-7	PEG-32 distearate
9005-08-7	PEG-75 distearate
9005-08-7	PEG-150 distearate
9005-25-8	Wheat starch
9005-64-5	Polysorbate 20
9005-64-5	Polysorbate 21
9005-65-5	Polysorbate 81
9005-65-6	Polysorbate 80
9005-66-7	Polysorbate 40
9005-67-8	Polysorbate 60
9005-67-8	Polysorbate 61
9005-70-3	PEG-17 sorbitan trioleate
9005-70-3	PEG-18 sorbitan trioleate
9005-70-3	Polysorbate 85
9005-71-4	Polysorbate 65
9005-84-9	Starch
9006-04-6	Polyisoprene
9006-26-2	Ethylene-maleic anhydride copolymer
9006-65-9	Dimethicone
9007-48-1	Polyglyceryl-3 oleate
9010-43-9	Octoxynol-9
9010-77-9	Ethylene/acrylic acid copolymer
9010-79-1	EPM rubber
9010-79-1	Polypropylene nucleated
9011-05-6	Urea-formaldehyde resin
9011-13-6	Styrene/MA copolymer
9011-14-7	Methyl methacrylate polymer
9011-16-9	PVM/MA copolymer
9014-85-1	PEG-10 tetramethyl decyne-diol
9014-85-1	PEG-30 tetramethyl decyne-diol
9014-90-8	Sodium nonoxynol-4 sulfate
9014-90-8	Sodium nonoxynol-6 sulfate
9014-90-8	Sodium nonoxynol-8 sulfate
9014-90-8	Sodium nonoxynol-10 sulfate
9014-90-8	Sodium nonoxynol-15 sulfate
9014-90-8	Sodium nonoxynol-25 sulfate
9014-93-1	Nonyl nonoxynol-7
9014-93-1	Nonyl nonoxynol-8
9014-93-1	Nonyl nonoxynol-9
9014-93-1	Nonyl nonoxynol-15
9014-93-1	Nonyl nonoxynol-24
9014-93-1	Nonyl nonoxynol-49
9016-00-6	Dimethicone
9016-45-9	Nonoxynol-3
9016-45-9	Nonoxynol-4
9016-45-9	Nonoxynol-5
9016-45-9	Nonoxynol-6
9016-45-9	Nonoxynol-7
9016-45-9	Nonoxynol-8
9016-45-9	Nonoxynol-9
9016-45-9	Nonoxynol-10

CAS	Chemical
9016-45-9	Nonoxynol-12
9016-45-9	Nonoxynol-13
9016-45-9	Nonoxynol-14
9016-45-9	Nonoxynol-20
9016-45-9	Nonoxynol-30
9016-45-9	Nonoxynol-40
9016-45-9	Nonoxynol-50
9016-45-9	Nonoxynol-100
9016-45-9	Nonoxynol-120
9032-42-2	Methyl hydroxyethylcellulose
9035-97-8	Sulfur, insoluble
9036-19-5	Octoxynol-1
9036-19-5	Octoxynol-3
9036-19-5	Octoxynol-7
9036-19-5	Octoxynol-9
9036-19-5	Octoxynol-10
9036-19-5	Octoxynol-16
9036-19-5	Octoxynol-25
9036-19-5	Octoxynol-30
9036-19-5	Octoxynol-33
9036-19-5	Octoxynol-40
9036-19-5	Octoxynol-70
9038-95-3	PPG-7-buteth-10
9038-95-3	PPG-12-buteth-16
9038-95-3	PPG-24-buteth-27
9038-95-3	PPG-28-buteth-35
9038-95-3	PPG-33-buteth-45
9040-38-4	Disodium nonoxynol-10 sulfosuccinate
9046-01-9	Trideceth-3 phosphate
9046-01-9	Trideceth-6 phosphate
9046-01-9	Trideceth-10 phosphate
9051-57-4	Ammonium nonoxynol-4 sulfate
9065-63-8	PPG-7-buteth-10
9065-63-8	PPG-12-buteth-16
9065-63-8	PPG-24-buteth-27
9065-63-8	PPG-28-buteth-35
9065-63-8	PPG-33-buteth-45
9076-43-1	PPG-40 diethylmonium chloride
9080-49-3	Polysulfide
9084-06-4	Sodium polynaphthalene sulfonate
9106-45-9	Nonoxynol-15
10016-20-3	Cyclodextrin
10025-67-9	Sulfur chloride
10025-77-1	Ferric chloride hexahydrate
10025-78-2	Trichlorosilane
10031-30-8	Calcium phosphate monobasic monohydrate
10081-67-1	4,4′-Bis (α,α-dimethylbenzyl) diphenylamine
10094-45-8	Stearyl erucamide
10099-76-0	Lead silicate
10101-39-0	Calcium silicate
10101-41-4	Calcium sulfate dihydrate
10101-66-3	Manganese violet
10108-24-4	PEG-4 laurate
10108-25-5	PEG-4 oleate

CAS	Chemical
10108-28-8	PEG-6 stearate
10192-93-5	3,4-Dimethyl-3,4-diphenyl-hexane
10213-77-1	PPG-3 methyl ether
10213-78-2	PEG-2 stearamine
10288-28-5	1-[(1-Cyano-1-methylethyl) azo]formamide
10396-10-8	p-Toluene sulfonyl semi-carbazide
10416-59-8	Bis (trimethylsilyl) acetamide
10508-09-5	Di-t-amyl peroxide
10584-98-2	Dibutyltin bis(2-ethylhexyl mercaptoacetate)
10591-85-2	N,N,N′,N′-Tetrabenzylthiuram disulfide
10595-49-0	(3-Lauramidopropyl) trimethyl ammonium methyl sulfate
10595-72-9	Ditridecyl thiodipropionate
10595-80-9	2-Sulfoethyl methacrylate
11097-59-9	Magnesium aluminum carbonate
11097-59-9	Magnesium aluminum hydrotalcite
11099-07-3	Glyceryl stearate
11099-07-3	Glyceryl stearate SE
11111-34-5	Poloxamine 304
11111-34-5	Poloxamine 504
11111-34-5	Poloxamine 701
11111-34-5	Poloxamine 702
11111-34-5	Poloxamine 704
11111-34-5	Poloxamine 707
11111-34-5	Poloxamine 901
11111-34-5	Poloxamine 904
11111-34-5	Poloxamine 908
11111-34-5	Poloxamine 1101
11111-34-5	Poloxamine 1102
11111-34-5	Poloxamine 1104
11111-34-5	Poloxamine 1107
11111-34-5	Poloxamine 1301
11111-34-5	Poloxamine 1302
11111-34-5	Poloxamine 1304
11111-34-5	Poloxamine 1307
11111-34-5	Poloxamine 1501
11111-34-5	Poloxamine 1502
11111-34-5	Poloxamine 1504
11111-34-5	Poloxamine 1508
11120-22-2	Lead silicate
11890-98-8	1,3-Butylene glycol dimeth-acrylate
12001-26-2	Mica
12027-96-2	Zinc hydroxystannate
12036-37-2	Zinc stannate
12124-97-9	Ammonium bromide
12125-28-9	Magnesium carbonate hydroxide
12141-46-7	Aluminum silicate
12167-74-7	Calcium phosphate tribasic
12178-41-5	Ferro-aluminum silicate
12228-87-4	Ammonium biborate
12262-58-7	Cyclohexanone peroxide

CAS	Chemical
12269-78-2	Pyrophyllite
12447-61-9	Zinc borate
12542-30-2	Dicyclopentenyl acrylate
12738-64-6	Sucrose benzoate
12767-90-7	Zinc borate
12769-96-9	Ultramarine violet
13052-09-0	2,5-Dimethyl-2,5-di(2-ethylhexanoyl peroxy) hexane
13103-52-1	Stearyl lauryl thiodipropionate
13122-18-4	t-Butyl peroxy-3,5,5-trimethylhexanoate
13127-82-7	PEG-2 oleamine
13150-00-0	Sodium laureth sulfate
13170-05-3	Bis(4-(1,1-dimethylethyl) benzoato-o)hydroxy aluminum
13170-23-5	Di-t-butoxydiacetoxysilane
13253-82-2	4,4′-Methylenebis (cyclohexyl-amine)carbamate
13429-07-7	PPG-2 methyl ether
13463-41-7	Zinc pyrithione
13463-67-7	Titanium dioxide
13472-08-7	2,2′-Azobis(2-methyl-butyronitrile)
13474-96-9	Heptakis (dipropylene glycol) triphosphite
13494-80-9	Tellurium
13539-23-0	Magnesium aluminum carbonate
13560-89-9	Dodecachloro dodecahydro dimethanodibenzo cyclooctene
13674-87-8	Tri(β,β′-dichloroisopropyl) phosphate
13701-59-2	Barium metaborate
13822-56-3	Aminopropyltrimethoxysilane
13878-54-1	Zinc-N-pentamethylene dithiocarbamate
13927-77-0	Nickel dibutyldithiocarbamate
13983-17-0	Wollastonite
13987-01-4	Tripropylene glycol
14024-64-7	Titanium acetylacetonate
14103-61-8	Diisononyl phthalate
14228-73-0	Cyclohexanedimethanol diglycidyl ether
14239-68-0	Cadmium diethyldithio-carbamate
14323-55-1	Zinc diethyldithiocarbamate
14350-94-8	Sodium caproamphoacetate
14409-72-4	Nonoxynol-9
14481-60-8	Disodium stearyl sulfosuccina-mate
14504-95-1	KE clay
14516-71-3	2,2′-Thiobis (4-t-octylpheno-lato)-n-butylamine nickel II
14643-87-9	Zinc diacrylate
14726-36-4	Zinc dibenzyl dithiocarbamate
14727-68-5	Dimethyl oleamine
14728-39-3	Ammonium polyphosphate sol'n.
14807-96-6	Talc
14808-60-7	Quartz

CAS	Chemical
14960-06-6	Sodium lauriminodipropionate
15206-55-0	Methyl phenylglyoxalate
15305-07-4	Nitrosophenylhydroxylamine aluminum salt
15317-78-9	Nickel diisobutyldithio-carbamate
15337-18-5	Zinc diamyldithiocarbamate
15520-10-2	2-Methylpentamethylene-diamine
15520-11-3	Di-(4-t-butylcyclohexyl) peroxydicarbonate
15521-65-0	Nickel dimethyldithio-carbamate
15535-69-0	Dibutyltin maleate
15545-97-8	2,2´-Azobis(4-methoxy-2,4-dimethylvaleronitrile)
15625-89-5	1,1,1-Trimethylolpropane triacrylate
15647-08-2	Ethylhexyl diphenyl phosphite
15667-10-4	1,1-Di (t-amylperoxy) cyclohexane
15826-16-1	Sodium laureth sulfate
15892-23-6	2-Butanol
16066-38-9	Di-n-propyl peroxydicarbonate
16111-62-9	Bis (2-ethylhexyl) peroxydicar-bonate
16215-49-9	Di-n-butyl peroxydicarbonate
16230-35-6	Bis-(N-methylbenzamide) ethoxymethyl silane
16260-09-6	Oleyl palmitamide
16545-54-3	Dimyristyl thiodipropionate
16580-06-6	Di-t-butyl diperoxyazelate
16580-06-6	Di-t-butyl peroxyazelate
16958-92-2	Ditridecyl adipate
16971-82-7	Dicatechol borate, di-o-tolyl guanidine salt
17540-75-9	2,6-Di-t-butyl-4-s-butyl phenol
17557-23-2	Neopentyl glycol diglycidyl ether
17661-50-6	Cetyl esters
17689-77-9	Ethyltriacetoxysilane
17786-43-5	Tetrabutylphosphonium acetate, monoacetic acid
17796-82-6	N-(Cyclohexylthio)phthalimide
17831-71-9	PEG-4 diacrylate
18282-10-5	Stannic oxide
18395-30-7	Isobutyltrimethoxysilane
18407-94-8	Tetrakis (2-ethoxyethoxy) silane
18602-17-0	N,N-Bis (2-hydroxyethyl)-N-(3-dodecyloxy-2-hydroxypropyl) methyl ammonium methosul-fate
19010-66-3	Lead dimethyldithiocarbamate
19047-85-9	Distearyl phosphite
19321-40-5	Pentaerythrityl tetraoleate
19485-03-1	1,3-Butylene glycol diacrylate
19706-80-0	2,2´-Azobis[2-(hydroxymethyl) propionitrile]
19910-65-7	Di-(s-butyl)peroxydicarbonate
20344-49-4	Iron (III) oxide hydrated
20566-35-2	Tetrabromophthalate diol
20636-48-0	Nonoxynol-5
20824-56-0	Diammonium EDTA
20858-12-2	2,2´-Azobis (N,N´-dimethylene-isobutyramidine)
20941-65-5	Tellurium diethyl dithio-carbamate
21260-46-8	Bismuth dimethyl dithio-carbamate
21282-96-2	2-(Acetoacetoxy) ethyl acrylate
21282-97-3	2-(Acetoacetoxy) ethyl methacrylate
21302-09-0	Dilauryl phosphite
21542-96-1	Dimethyl behenamine
21559-14-8	Selenium diethyldithio-carbamate
21645-51-2	Aluminum hydroxide
21850-44-2	Tetrabromobisphenol A, bis (2,3-dibromopropyl ether)
22047-49-0	Octyl stearate
22313-62-8	t-Butyl peroxy 2-methyl benzoate
22397-33-7	3,3,6,6,9,9-Hexamethyl 1,2,4,5-tetraoxa cyclononane
23128-74-7	N,N´-Hexamethylene bis (3,5-di-t-butyl-4-hydroxyhydrocin-namamide)
23328-53-2	Phenol, 2-(2H-benzotriazole-2-yl)-4-methyl-6-dodecyl
23386-52-9	Dicyclohexyl sodium sulfosuccinate
23474-91-1	t-Butyl peroxycrotonate
23488-38-2	Tetrabromo-p-xylene
23906-97-0	Tetraoctylphosphonium bromide
23946-66-8	2-Ethyl, 2´-ethoxy-oxalanilide
24304-00-5	Aluminum nitride
24601-97-8	2,2´-[(Dioctylstannylene) bis (acetic acid)], diisooctyl ester
24615-84-7	β-Carboxyethyl acrylate
24800-44-0	Tripropylene glycol
24801-88-5	Isocyanatopropyltriethoxy-silane
24817-92-3	Acetyl tri-n-hexyl citrate
24887-06-7	Zinc formaldehyde sulfoxylate
24937-78-8	Ethylene/VA copolymer
24937-78-8	Vinyl acetate/ethylene copolymer
24937-79-9	Polyvinylidene fluoride
24938-91-8	Trideceth-14
24938-91-8	Trideceth-15
24969-11-7	Resorcinol-formaldehyde resin
24979-97-3	Polytetramethylene ether glycol
24991-55-7	PEG-3 dimethyl ether
25013-16-5	BHA
25014-31-7	Poly-α-methylstyrene
25038-54-4	Nylon-6
25038-59-9	Polyethylene terephthalate

CAS	Chemical
25086-62-8	Sodium polymethacrylate
25087-26-7	Polymethacrylic acid
25103-09-7	Isooctyl thioglycolate
25103-12-2	Triisooctyl phosphite
25154-52-3	Nonylphenol
25155-23-1	Trixylenyl phosphate
25155-25-3	α,α′-Bis (t-butylperoxy) diisopropylbenzene
25155-25-3	Di-(2-t-butylperoxyisopropyl) benzene
25155-30-0	Sodium dodecylbenzene-sulfonate
25167-70-8	Diisobutylene
25190-06-1	Polytetramethylene ether glycol
25213-24-5	Polyvinyl alcohol, partially hydrolyzed
25265-71-8	Dipropylene glycol
25265-77-4	2,2,4-Trimethyl-1,3-pentane-diol monoisobutyrate
25321-41-9	Xylene sulfonic acid
25322-68-3	Polyethylene glycol
25322-68-3	PEG-4
25322-68-3	PEG-6
25322-68-3	PEG-8
25322-68-3	PEG-12
25322-68-3	PEG-20
25322-68-3	PEG-32
25322-68-3	PEG-40
25322-68-3	PEG-75
25322-68-3	PEG-100
25322-68-3	PEG-150
25322-68-3	PEG-2M
25322-68-3	PEG-5M
25322-68-3	PEG-7M
25322-68-3	PEG-9M
25322-68-3	PEG-14M
25322-68-3	PEG-20M
25322-68-3	PEG-23M
25322-68-3	PEG-45M
25322-68-3	PEG-90M
25322-68-3	PEG-115M
25322-69-4	Polypropylene glycol
25322-69-4	PPG-9
25322-69-4	PPG-12
25322-69-4	PPG-17
25322-69-4	PPG-20
25322-69-4	PPG-26
25322-69-4	PPG-30
25327-89-3	Tetrabromobisphenol A bis (allyl ether)
25338-55-0	Dimethylaminomethyl phenol
25338-58-3	Octoxynol-9 carboxylic acid
25377-73-5	Dodecenyl succinic anhydride
25383-99-7	Sodium stearoyl lactylate
25395-31-7	Glyceryl diacetate
25417-20-3	Sodium dibutyl naphthalene sulfonate
25446-78-0	Sodium trideceth sulfate (n = 3)
25448-25-3	Triisodecyl phosphite
25496-01-9	Tridecylbenzene sulfonic acid
25496-72-4	Glyceryl mono/dioleate
25496-72-4	Glyceryl oleate
25496-72-4	Glyceryl oleate SE
25498-49-1	PPG-3 methyl ether
25550-51-0	Methyl hexahydrophthalic anhydride
25550-98-5	Phenyl diisodecyl phosphite
25620-58-0	Trimethylhexamethylene diamine
25637-84-7	Glyceryl dioleate
25637-99-4	Hexabromocyclododecane
25704-59-0	Sodium caproamphoacetate
25852-47-5	PEG-4 dimethacrylate
25928-94-3	Epoxy resin
25973-55-1	2-(2′-Hydroxy-3,5′-di-t-amylphenyl) benzotriazole
26027-38-3	Nonoxynol-1
26027-38-3	Nonoxynol-4
26027-38-3	Nonoxynol-5
26027-38-3	Nonoxynol-6
26027-38-3	Nonoxynol-7
26027-38-3	Nonoxynol-8
26027-38-3	Nonoxynol-9
26027-38-3	Nonoxynol-10
26027-38-3	Nonoxynol-12
26027-38-3	Nonoxynol-13
26027-38-3	Nonoxynol-14
26027-38-3	Nonoxynol-15
26027-38-3	Nonoxynol-20
26027-38-3	Nonoxynol-30
26027-38-3	Nonoxynol-40
26027-38-3	Nonoxynol-50
26027-38-3	Nonoxynol-100
26027-38-3	Nonoxynol-120
26062-94-2	Polybutylene terephthalate
26172-55-4	Methylchloroisothiazolinone
26248-87-3	Tris (β-chloropropyl) phosphate
26256-79-1	Sodium lauriminodipropionate
26264-02-8	Nonoxynol-5
26264-05-1	Isopropylamine dodecylben-zenesulfonate
26264-06-2	Calcium dodecylbenzene sulfonate
26266-57-9	Sorbitan palmitate
26266-58-0	Sorbitan trioleate
26332-14-5	Dicetyl peroxydicarbonate
26399-02-0	Octyl stearate
26401-27-4	Diphenyl isooctyl phosphite
26401-35-4	Ditridecyl adipate
26401-86-5	2,2′,2′′-(Octylstannylidyne) tris(thio)tris(acetic acid), triisooctyl ester
26402-31-3	Propylene glycol ricinoleate
26426-80-2	Isobutylene/MA copolymer
26444-49-5	Diphenylcresyl phosphate
26445-96-5	Ethylene/calcium acrylate copolymer

CAS	Chemical
26446-35-5	Glyceryl acetate
26523-78-4	Trisnonylphenyl phosphite
26530-20-1	2-n-Octyl-4-isothiazolin-3-one
26544-23-0	Diphenyl isodecyl phosphite
26544-27-4	Diisodecyl pentaerythritol diphosphite
26544-38-7	Dodecenyl succinic anhydride
26571-11-9	Nonoxynol-8
26571-11-9	Nonoxynol-9
26635-92-7	PEG-5 stearamine
26635-92-7	PEG-10 stearamine
26635-92-7	PEG-11 stearamine
26635-92-7	PEG-50 stearamine
26635-93-8	PEG-7 oleamine
26635-93-8	PEG-12 oleamine
26658-19-5	Sorbitan tristearate
26680-54-6	Octenyl succinic anhydride
26741-53-7	Bis (2,4-di-t-butylphenyl) pentaerythritol diphosphite
26741-53-7	Phosphorous acid, cyclic neopentanetetrayl bis (2,4-di-t-butylphenyl) ester
26748-38-9	5-Butylperoxyneoheptanoate
26748-41-4	t-Butyl peroxyneodecanoate
26748-47-0	Cumyl peroxyneodecanoate
26761-40-0	Diisodecyl phthalate
26762-93-6	Diisopropylbenzene mono-hydroperoxide
26780-96-1	2,2,4-Trimethyl-1,2-dihydro-quinoline polymer
26850-24-8	DEDM hydantoin
26855-43-6	Polyglyceryl-3 stearate
27138-31-4	Dipropylene glycol dibenzoate
27176-87-0	Dodecylbenzene sulfonic acid
27176-94-9	Octoxynol-3
27176-95-0	Nonoxynol-3
27176-97-2	Nonoxynol-4
27177-01-1	Nonoxynol-6
27177-02-2	Octoxynol-7
27177-05-5	Nonoxynol-6
27177-05-5	Nonoxynol-7
27177-05-5	Nonoxynol-8
27177-07-7	Octoxynol-10
27177-08-8	Nonoxynol-10
27177-77-1	Potassium dodecylbenzene sulfonate
27178-16-1	Diisodecyl adipate
27194-74-7	Propylene glycol laurate
27251-75-8	Triisooctyl trimellitate
27321-72-8	Polyglyceryl-3 stearate
27323-41-7	TEA-dodecylbenzenesulfonate
27554-26-3	Diisooctyl phthalate
27638-00-2	Glyceryl dilaurate
27676-62-6	Tris(3,5-di-t-butyl-4-hydroxy benzyl) isocyanurate
27776-21-2	2,2′-Azobis[2-(2-imidazolin-2-yl) propane] dihydrochloride
27813-02-1	Hydroxypropyl methacrylate
27942-26-3	Nonoxynol-10
27986-36-3	Nonoxynol-1

CAS	Chemical
28061-69-0	Dimethyl oleamine
28519-02-0	Sodium dodecyl diphenyl ether disulfonate
28631-63-2	Cumene sulfonic acid
29059-24-3	Propylene glycol myristate
29240-17-3	t-Amylperoxypivalate
29381-93-9	TEA-dodecylbenzenesulfonate
29385-43-1	Tolyltriazole
29406-96-0	1,2-Polybutadiene (syndiotactic)
29409-82-0	Magnesium carbonate
29598-76-3	Pentaerythrityl tetrakis (β-laurylthiopropionate)
29761-21-5	Isodecyl diphenyl phosphate
29806-73-3	Octyl palmitate
29964-84-9	Isodecyl methacrylate
29994-44-3	Nonoxynol-6 phosphate
30364-51-3	Sodium myristoyl sarcosinate
30399-84-9	Isostearic acid
30653-05-5	Bis[2-hydroxy-5-methyl-3-(benzotriazol-2-yl)phenyl] methane
31001-77-1	3-Mercaptopropylmethyldi-methoxysilane
31556-45-3	Tridecyl stearate
31565-37-4	Isotridecyl stearate
31566-31-1	Glyceryl stearate
31566-31-1	Glyceryl stearate SE
31570-04-4	Tris (2,4-di-t-butylphenyl) phosphite
31691-97-1	Ammonium nonoxynol-4 sulfate
31692-79-2	Dimethiconol
31778-15-1	Octadecyl 3-mercapto-propionate
32534-81-9	Pentabromodiphenyl oxide
32536-52-0	Octabromodiphenyl oxide
32588-76-4	Ethylene bis-tetrabromo-phthalimide
32612-48-9	Ammonium laureth sulfate
32612-48-9	Ammonium laureth-9 sulfate
32612-48-9	Ammonium laureth-12 sulfate
32647-67-9	Dibenzylidene sorbitol
32687-78-8	1,2-Bis (3,5-di-t-butyl-4-hydroxyhydrocinnamoyl) hydrazine
33007-83-9	Trimethylolpropane tri (3-mercaptopropionate)
33145-10-7	2,2′-Isobutylidenebis (4,6-dimethylphenol)
33703-08-1	Diisononyl adipate
33806-58-5	Tetradecenyl succinic anhydride
33940-98-6	Polyglyceryl-3 oleate
34137-09-2	3,5-Di-t-butyl-4-hydroxyhydro-cinnamic acid/1,3,5-tris(2-hydroxyethyl)-s-triazine-2,4,6-(1H,3H,5H)-trione triester
34424-98-1	Polyglyceryl-10 tetraoleate
34443-12-4	t-Butyl peroxy 2-ethylhexyl

CAS	Chemical
	carbonate
34562-31-7	Butyraldehyde-aniline condensation product
34590-94-8	PPG-2 methyl ether
34708-08-2	3-Thiocyanatopropyltriethoxy-silane
35074-77-2	Hexamethylenebis (3,5-di-t-butyl-4-hydroxycinnamate)
35141-30-1	Trimethoxysilylpropyldiethyl-ene triamine
35179-86-3	PEG-8 laurate
35325-02-1	N-(2-Hydroxypropyl) benzenesulfonamide
35634-74-3	2-Phenylazo-4-methoxy-2,4-dimethylvaleronitrile
35958-30-6	2,2′-Ethylidenebis (4,6-di-t-butylphenol)
36116-84-4	Diisooctyl phosphite
36190-62-2	Zinc diisobutyldithiocarbamate
36356-83-9	PEG-3 aniline
36432-46-9	Ditridecyl phosphite
36443-68-2	Ethylenebis (oxyethylene) bis (3-t-butyl-4-hydroxy-5-methylhydrocinnamate)
36445-71-3	Sodium decyl diphenyloxide disulfonate
36483-57-5	Tribromoneopentyl alcohol
36501-84-5	Lead diamyldithiocarbamate
36653-82-4	Cetyl alcohol
36788-39-3	Tris (dipropyleneglycol) phosphite
37187-22-7	Acetylacetone peroxide
37200-49-0	Polysorbate 80
37205-87-1	Nonoxynol-1
37205-87-1	Nonoxynol-4
37205-87-1	Nonoxynol-5
37205-87-1	Nonoxynol-6
37205-87-1	Nonoxynol-7
37205-87-1	Nonoxynol-8
37205-87-1	Nonoxynol-9
37205-87-1	Nonoxynol-10
37205-87-1	Nonoxynol-12
37205-87-1	Nonoxynol-13
37205-87-1	Nonoxynol-14
37205-87-1	Nonoxynol-15
37205-87-1	Nonoxynol-20
37205-87-1	Nonoxynol-30
37205-87-1	Nonoxynol-40
37205-87-1	Nonoxynol-50
37205-87-1	Nonoxynol-100
37205-87-1	Nonoxynol-120
37206-20-5	Methyl isobutyl ketone peroxide
37220-82-9	Glyceryl oleate
37244-96-5	Nepheline syenite
37286-64-9	PPG-3 methyl ether
37294-49-8	Disodium isodecyl sulfo-succinate
37318-14-2	PEG-8 laurate
37349-34-1	Polyglyceryl-3 stearate
37767-39-8	Tetrasodium dicarboxyethyl stearyl sulfosuccinamate
38613-77-3	Tetrakis (2,4-di-t-butylphenyl) 4,4-biphenylenediphosphonite
38916-42-6	Tetrasodium dicarboxyethyl stearyl sulfosuccinamate
39198-34-0	2,2′-Azobis(2,4,4-trimethyl-pentane)
39354-45-5	Disodium laureth sulfo-succinate
39354-47-5	Disodium C12-15 pareth sulfosuccinate
39407-03-9	Octyl phosphate
39409-82-0	Magnesium carbonate hydroxide
39464-69-2	Oleth-4 phosphate
40372-72-3	Bis-(3-(triethoxysilyl) propyl) tetrasulfane
40601-76-1	1,3,5-Tris (4-t-butyl-3-hydroxy-2,6-dimethylbenzyl)-1,3,5-triazine-2,4,6-(1H,3H,5H)-trione
40754-59-4	Disodium laureth sulfo-succinate
40754-60-7	Disodium ricinoleamido MEA-sulfosuccinate
40795-56-0	Sodium dodecyl diphenyl ether disulfonate
41395-83-9	Propylene glycol dipelargonate
41484-35-9	Thiodiethylene bis (3,5-di-t-butyl-4-hydroxy) hydro-cinnamate
41556-26-7	Bis(1,2,2,6,6-pentamethyl-4-piperidinyl)sebacate
42016-08-0	Disodium laureth sulfosucci-nate
42173-90-0	Octoxynol-9
42590-53-4	Zinc isopropyl xanthate
42612-52-2	Sodium laureth-4 phosphate
42615-29-2	Benzenesulfonic acid
43154-85-4	Disodium oleamido MIPA-sulfosuccinate
45115-34-2	Vinyl neodecanoate
51158-08-8	PEG-20 glyceryl stearate
51178-59-7	Dicyclopentenyl methacrylate
51229-78-8	Quaternium-15
51240-95-0	2,4,4-Trimethylpentyl-2-peroxyneodecanoate
51437-95-7	Nonoxynol-3
51609-41-7	Nonoxynol-6 phosphate
51609-41-7	Nonoxynol-10 phosphate
51811-79-1	Nonoxynol-6 phosphate
51811-79-1	Nonoxynol-9 phosphate
52019-36-0	Deceth-4 phosphate
52229-50-2	PVM/MA copolymer
52558-73-3	Myristoyl sarcosine
52668-97-0	PEG-2 dioleate
52668-97-0	PEG-4 dioleate
52668-97-0	PEG-6 dioleate
52668-97-0	PEG-8 dioleate

CAS	Chemical
52668-97-0	PEG-12 dioleate
52668-97-0	PEG-20 dioleate
52668-97-0	PEG-32 dioleate
52668-97-0	PEG-75 dioleate
52668-97-0	PEG-150 dioleate
52668-97-0	PEG-2 distearate
52668-97-0	PEG-4 distearate
52668-97-0	PEG-6 distearate
52668-97-0	PEG-8 distearate
52668-97-0	PEG-12 distearate
52668-97-0	PEG-20 distearate
52668-97-0	PEG-32 distearate
52668-97-0	PEG-75 distearate
52668-97-0	PEG-150 distearate
52725-64-1	Lauramide DEA
52783-38-7	3-(N-Styrylmethyl-2-aminoethylamino) propyltrimethoxy silane hydrochloride
52829-07-0	Bis (2,2,6,6-tetramethyl-4-piperidinyl) sebacate
52907-07-0	Ethylenebis dibromonorbornane dicarboximide
53220-22-7	Dimyristyl peroxydicarbonate
53894-23-8	Triisononyl trimellitate
53988-10-6	2-Mercaptotoluimidazole
54193-36-1	Sodium polymethacrylate
54392-26-6	Sorbitan isostearate
54667-43-5	Polytetramethylene oxide di-p-aminobenzoate
55310-46-8	Sodium dibenzyldithiocarbamate
55406-53-6	Iodopropynyl butylcarbamate
56378-72-4	Magnesium carbonate hydroxide pentahydrate
56519-71-2	Propylene glycol dioctanoate
56803-37-3	t-Butylphenyl diphenyl phosphate
57018-52-7	Propylene glycol t-butyl ether
57171-56-9	PEG-40 sorbitan hexaoleate
57171-56-9	PEG-50 sorbitan hexaoleate
57455-37-5	Ultramarine blue
57834-33-0	N^2-(4-Ethoxycarbonylphenyl)-N′-methyl-N′-phenylformamidine
58068-97-6	N-[3-(Triethoxysilyl)-propyl] 4,5-dihydroimidazole
58229-88-2	Dioctyltin mercaptide
58253-49-9	PEG-30 oleamine
58353-68-7	Disodium dodecyloxy propyl sulfosuccinamate
58450-52-5	Disodium laureth sulfosuccinate
58695-41-3	Cyclohexylamine acetate
58767-50-3	Bis (2-hydroxyethyl) octyl methyl ammonium p-toluene sulfonate
58965-66-5	Tetradecabromodiphenoxy benzene
58969-27-0	Sodium cocoyl isethionate
59231-34-4	Isodecyl oleate
59355-61-2	PEG-3 lauramine oxide
59789-51-4	Tribromophenyl maleimide
60344-26-5	PEG-6 oleate
60676-86-0	Silica
60842-32-2	Silica dimethyl silylate
61368-34-1	Tribromostyrene
61417-49-0	Isopropyl titanium triisostearate
61551-69-7	2,2′-Azobis[2-methyl-N-(2-hydroxyethyl)-propionamide]
61551-69-7	2,2′-Azobis{2-methyl-N-[1,1-bis(hydroxymethyl)ethyl] propionamide}
61617-00-3	Zinc 2-mercaptotoluimidazole
61670-79-9	Tetrakis isodecyl 4,4′-isopropylidene diphosphite
61788-44-1	Phenol, styrenated
61788-45-2	Hydrogenated tallowamine
61788-46-3	Cocamine
61788-47-4	Coconut acid
61788-59-8	Methyl cocoate
61788-62-3	Dicoco methylamine
61788-63-4	Dihydrogenated tallow methylamine
61788-72-5	Epoxidized octyl tallate
61788-85-0	PEG-5 hydrogenated castor oil
61788-85-0	PEG-10 hydrogenated castor oil
61788-85-0	PEG-16 hydrogenated castor oil
61788-85-0	PEG-20 hydrogenated castor oil
61788-85-0	PEG-25 hydrogenated castor oil
61788-85-0	PEG-30 hydrogenated castor oil
61788-85-0	PEG-40 hydrogenated castor oil
61788-85-0	PEG-60 hydrogenated castor oil
61788-85-0	PEG-100 hydrogenated castor oil
61788-85-0	PEG-200 hydrogenated castor oil
61788-89-4	Dilinoleic acid
61788-91-8	Dimethyl soyamine
61788-93-0	Dimethyl cocamine
61788-95-2	Dimethyl hydrogenated tallow amine
61789-07-9	Hydrogenated cottonseed glyceride
61789-10-4	Lard glyceride
61789-18-2	Cocotrimonium chloride
61789-32-0	Sodium cocoyl isethionate
61789-68-2	PEG-2 coco-benzonium chloride
61789-77-3	Dicocodimonium chloride
61789-80-8	Quaternium-18
61789-86-4	Calcium sulfonate
61790-12-3	Tall oil acid

CAS	Chemical
61790-31-6	Hydrogenated tallow amide
61790-33-8	Tallow amine
61790-37-2	Tallow acid
61790-38-3	Hydrogenated tallow acid
61790-50-9	Potassium rosinate
61790-51-0	Potassium rosinate
61790-51-0	Sodium rosinate
61790-86-1	Polysorbate 80
61790-95-2	Polyglyceryl-3 stearate
61791-12-6	PEG-5 castor oil
61791-12-6	PEG-10 castor oil
61791-12-6	PEG-12 castor oil
61791-12-6	PEG-16 castor oil
61791-12-6	PEG-25 castor oil
61791-12-6	PEG-30 castor oil
61791-12-6	PEG-32 castor oil
61791-12-6	PEG-36 castor oil
61791-12-6	PEG-40 castor oil
61791-12-6	PEG-80 castor oil
61791-12-6	PEG-160 castor oil
61791-12-6	PEG-200 castor oil
61791-14-8	PEG-2 cocamine
61791-14-8	PEG-5 cocamine
61791-14-8	PEG-10 cocamine
61791-14-8	PEG-15 cocamine
61791-24-0	PEG-2 soyamine
61791-24-0	PEG-5 soyamine
61791-24-0	PEG-10 soyamine
61791-24-0	PEG-15 soyamine
61791-26-2	PEG-5 hydrogenated tallow amine
61791-26-2	PEG-15 hydrogenated tallow amine
61791-26-2	PEG-50 hydrogenated tallow amine
61791-29-5	PEG-8 cocoate
61791-31-9	Cocamide DEA
61791-31-9	Coco alkyl-2,2′-imino-bisethanol
61791-38-6	Cocoyl hydroxyethyl imidazoline
61791-44-4	PEG-2 tallowamine
61791-44-4	PEG-5 tallowamine
61791-44-4	N-Tallow alkyl-2,2′-imino-bisethanol
61791-47-7	Dihydroxyethyl cocamine oxide
61791-56-8	Disodium tallowiminodi-propionate
61791-59-1	Sodium cocoyl sarcosinate
61791-63-7	Cocopropylenediamine
61901-02-8	Sodium lauroamphopropionate
62476-60-6	t-Butyl peroxy stearyl carbonate
62695-55-0	t-Butyl peroctoate
63123-11-5	Thiodipropionate polyester
63148-55-0	Dimethicone copolyol
63148-57-2	Polymethylhydrosiloxane
63148-62-9	Dimethicone
63148-69-6	Alkyd resin
63231-60-7	Microcrystalline wax
63351-73-5	Ammonium nonoxynol-4 sulfate
63428-83-1	Nylon
63843-89-0	Bis (1,2,2,6,6-Pentamethyl-4-piperidinyl) (3,5-di-t-butyl-4-hydroxybenzyl) butyl propanedioate
64365-23-7	Dimethicone copolyol
64475-85-0	Mineral spirits
64741-81-7	Polynuclear aromatic hydrocarbon
64742-42-3	Microcrystalline wax
64742-49-0	Hexane
64742-89-8	Heptane
64742-93-4	Asphalt, oxidized
64742-95-6	Naphtha (light aromatic)
64743-02-8	C20-24 alpha olefin
65140-91-2	Calcium bis[monoethyl(3,5-di-t-butyl-4-hydroxybenzyl) phosphonate]
65277-54-5	Disodium ricinoleamido MEA-sulfosuccinate
65447-77-0	Dimethylsuccinate and tetramethyl hydroxy-1-hydroxyethyl piperidine polymer
65530-85-0	Tetrafluoroethylene telomer
65816-20-8	N-(p-Ethoxycarbonylphenyl)-N′-ethyl-N′-phenylformamidine
65876-95-1	Resorcinol-formaldehyde resin
65996-61-4	Cellulose
66105-29-1	PEG-7 glyceryl cocoate
66161-57-7	Sodium laureth-12 sulfate
66161-58-8	Sodium trideceth sulfate (n = 4)
66172-82-5	Nonyl nonoxynol-9 phosphate
66197-78-2	Nonoxynol-9 phosphate
67254-74-4	Naphthenic oil
67567-23-1	Ethyl 3,3-di (t-amylperoxy) butyrate
67700-98-5	Dimethyl lauramine
67700-99-6	Dihydrogenated tallow methylamine
67701-05-7	Coconut acid
67701-06-8	Tallow acid
67701-08-0	Soy acid
67701-30-8	Triolein
67701-32-0	Mono- and diglycerides of fatty acids
67701-33-1	Glyceryl myristate
67701-33-1	Mono- and diglycerides of fatty acids
67762-19-0	Ammonium laureth sulfate
67762-25-8	C12-18 alcohols
67762-30-5	Cetearyl alcohol
67762-36-1	Caprylic/capric acid
67762-38-3	Methyl oleate
67762-40-7	Methyl laurate
67762-96-3	Dimethicone copolyol
67784-77-4	PEG-2 tallowmonium chloride

CAS	Chemical
67784-87-6	Hydrogenated palm glyceride
67893-42-9	Disodium ricinoleamido MEA-sulfosuccinate
67923-87-9	Sodium octoxynol-2 ethane sulfonate
67999-57-9	Disodium nonoxynol-10 sulfosuccinate
68002-20-0	Hexamethylmelamine
68002-59-5	Quaternium-18
68002-61-9	Tallowtrimonium chloride
68002-71-1	Hydrogenated soy glyceride
68002-97-1	Laureth-2
68002-97-1	Laureth-3
68002-97-1	Laureth-4
68003-46-3	Ammonium lauroyl sarcosinate
68037-49-0	Sodium C10-18 alkyl sulfonate
68037-74-1	Dimethicone (branched)
68037-87-6	Polyvinylmethylsiloxane
68037-93-4	Dimethyl palmitamine
68081-81-2	Sodium dodecylbenzene-sulfonate
68081-96-9	Ammonium lauryl sulfate
68128-25-6	1-Trimethoxysilyl-2-(chloromethyl) phenylethane
68128-59-6	Diammonium stearyl sulfosuccinamate
68130-47-2	Deceth-4 phosphate
68131-40-8	C11-15 pareth-3
68131-40-8	C11-15 pareth-5
68131-40-8	C11-15 pareth-30
68131-40-8	C11-15 pareth-40
68133-13-1	Diisooctyl octylphenyl phosphite
68153-64-0	PEG-2 tallowate
68153-76-4	PEG-20 glyceryl stearate
68155-09-9	Cocamidopropylamine oxide
68155-27-1	C13-15 amine
68186-14-1	Methyl rosinate
68188-45-5	Sodium alpha olefin sulfonate
68201-46-7	PEG-7 glyceryl cocoate
68238-87-9	Sorbitan diisostearate
68299-16-1	t-Amylperoxyneodecanoate
68299-17-2	Sodium isodecyl sulfate
68308-22-5	Calcium montanate
68308-53-2	Soy acid
68308-67-8	Soyethyldimonium ethosulfate
68311-03-5	Disodium deceth-6 sulfo-succinate
68411-19-8	Butyraldehyde-monobutyl-amine condensation prod.
68411-20-1	Butyraldehyde-aniline condensation product
68411-31-4	TEA-dodecylbenzenesulfonate
68411-32-5	Dodecylbenzene sulfonic acid
68411-90-5	Bis-hexamethylenetriamine
68411-97-2	Cocoyl sarcosine
68412-53-3	Nonoxynol-6 phosphate
68412-53-3	Nonoxynol-9 phosphate
68424-94-2	Coco-betaine
68439-49-6	Ceteareth-4
68439-49-6	Ceteareth-6
68439-49-6	Ceteareth-10
68439-49-6	Ceteareth-15
68439-49-6	Ceteareth-20
68439-49-6	Ceteareth-22
68439-49-6	Ceteareth-25
68439-49-6	Ceteareth-30
68439-49-6	Ceteareth-50
68439-50-9	Laureth-2
68439-50-9	Laureth-3
68439-50-9	Laureth-4
68439-50-9	Laureth-9
68439-57-6	Sodium alpha olefin sulfonate
68439-57-6	Sodium C14-16 olefin sulfonate
68439-70-3	Dimethyl myristamine
68439-73-6	Tallowaminopropylamine
68440-90-4	Polyoctylmethylsiloxane
68441-17-8	Polyethylene, oxidized
68441-68-9	Pentaerythrityl tetracaprylate/caprate
68459-31-4	Alkyd resin
68476-03-9	Montan acid wax
68476-38-0	Glyceryl montanate
68478-45-5	Diaryl p-phenylene diamine
68478-45-5	Mixed diaryl-p-phenylenediamine
68479-98-1	Diethyl toluene diamine
68515-47-9	Ditridecyl phthalate
68515-47-9	Undecyl dodecyl phthalate
68515-48-0	Diisononyl phthalate
68515-49-1	Diisodecyl phthalate
68515-50-4	Dihexyl phthalate
68526-79-4	Hexyl alcohol
68526-85-2	n-Decyl alcohol
68526-86-3	Lauryl alcohol
68553-11-7	PEG-20 glyceryl stearate
68584-22-5	Dodecylbenzene sulfonic acid
68584-24-7	Isopropylamine dodecylbenzenesulfonate
68585-34-2	Sodium laureth sulfate
68585-44-4	DEA-lauryl sulfate
68585-47-7	Sodium lauryl sulfate
68603-15-5	C8-10 alcohols
68603-42-9	Cocamide DEA
68603-64-5	Hydrogenated tallow 1,3-propylene diamine
68604-44-4	Pentaerythrityl tetraoleate
68608-61-7	Sodium caproamphoacetate
68608-64-0	Disodium capryloamphodiacetate
68608-88-8	Dodecylbenzene sulfonic acid
68610-51-5	4-Methyl phenol reaction prods. with dicyclopentadiene and isobutylene
68610-51-5	p-Cresol/dicyclopentadiene butylated reaction product
68610-62-8	Tris Neodol-25 phosphite
68611-44-9	Dimethyldichlorosilane
68647-46-1	Sodium caproamphoacetate

CAS	Chemical
68649-05-8	Cocaminobutyric acid
68783-22-2	PEG-50 hydrogenated tallow amine
68783-41-5	Dimer acid
68784-69-0	Polybutylated bisphenol A
68855-54-9	Diatomaceous earth
68855-56-1	C12-16 alcohols
68891-38-3	Sodium laureth sulfate
68891-39-4	Sodium nonoxynol-4 sulfate
68901-05-3	PPG-3 diacrylate
68919-41-5	Sodium cocoamphopropionate
68920-06-9	Naphtha
68920-66-1	Cetoleth-55
68937-10-0	Polybutene, hydrogenated
68937-41-7	Triaryl phosphate
68937-54-2	Polysiloxane polyether copolymer
68937-85-9	Coconut acid
68937-90-6	Trilinoleic acid
68952-35-2	Tricresyl phosphate
68953-84-4	Diaryl p-phenylene diamine
68953-96-8	Calcium dodecylbenzene sulfonate
68955-19-1	Sodium lauryl sulfate
68955-45-3	Ethylene distearamide
68988-69-2	Sodium tallow sulfosuccina-mate
68989-03-7	PEG-5 cocomonium methosulfate
68990-53-4	Mono- and diglycerides of fatty acids
69005-67-8	Sorbitan stearate
69009-90-1	Bis(methylethyl)-1,1´-biphenyl
69011-06-9	Lead phthalate, dibasic
69011-84-3	Sodium octoxynol-6 sulfate
69226-96-6	Pentaerythrityl tetracaprylate/ caprate
69853-84-4	Mixed diaryl-p-phenylenedi-amine
70321-86-7	2-[2-Hydroxy-3,5-di-(1,1-dimethylbenzyl)phenyl]-2H-benzotriazole
70331-94-1	2,2´-Oxamido bis[ethyl 3-(3,5-di-t-butyl-4-hydroxyphenyl) propionate
70592-80-2	Lauramine oxide
70624-18-9	Poly[[6-[(1,1,3,3-tetramethyl-butyl)amino]-s-triazine-2,4-diyl][2,2,6.6-tetramethyl-4-piperidyl)imino]hexamethylene [(2,2,6,6-tetramethyl-4-piperidyl) imino]]
70788-37-3	p-Toluene sulfonic acid
70802-40-3	PEG-8 stearate
70833-40-8	t-Amylperoxy 2-ethylhexyl carbonate
70892-21-6	Polyvinyl octadecyl carbamate
71172-17-3	Dimethylammonium hydrogen isophthalate
71486-48-1	Cyclohexyl isooctyl phthalate
71487-00-8	PEG-2 cocomonium nitrate
71888-89-6	Diisoheptyl phthalate
71902-01-7	Sorbitan isostearate
72160-13-5	Octoxynol-9 carboxylic acid
72162-46-0	N-Isodecyloxypropyl 1-1,3-diaminopropane
72674-05-6	Sulfonic acid
72869-62-6	Sorbitan tristearate
73070-47-0	Trideceth-3 phosphate
73070-47-0	Trideceth-6 phosphate
73070-47-0	Trideceth-10 phosphate
73138-44-0	Butylene glycol montanate
73138-45-1	Glycol/butylene glycol montanate
74623-31-7	PPG-12-buteth-16
78330-21-9	Trideceth-14
78330-21-9	Trideceth-15
79070-11-4	Tetrafluoroethylene telomer
80410-33-9	2,2´,2´´-Nitrilo[triethyl-tris(3,3´,5,5´-tetra-t-butyl-1,1´-biphenyl-2,2´-diyl)phosphite]
80584-85-6	Tetraphenyl dipropyleneglycol diphosphite
80584-86-7	Poly (dipropyleneglycol) phenyl phosphite
80584-87-8	Diphenyldidecyl (2,2,4-trimethyl-1,3-pentanediol) diphosphite
81972-48-7	Polycarbonamide
82469-79-2	n-Butyryl tri-n-hexyl citrate
84501-49-5	Sodium decyl sulfate
84562-92-5	Nonoxynol-3
84852-15-3	p-Nonyl phenol
84988-79-4	Isobutyl oleate
85116-87-6	Isopropyl oleate
85116-93-4	Pentaerythrityl stearate
85116-97-8	PEG-2 stearate
85117-50-6	Sodium dodecylbenzene-sulfonate
85186-88-5	Sorbitan trioleate
85209-91-2	Sodium 2,2´-methylenebis(4,6-di-t-butylphenyl) phosphate
85251-77-0	Glyceryl stearate
85409-25-2	Coco-betaine
85536-04-5	Stearyl stearate
85536-14-7	Dodecylbenzene sulfonic acid
85586-21-6	Methyl stearate
85666-92-8	Glyceryl stearate
85666-92-8	Glyceryl stearate SE
85736-49-8	PEG-12 dioleate
85865-69-6	Isobutyl stearate
86418-55-5	Glycol stearate SE
86555-98-6	Alkyl phenol disulfide
88684-42-8	2,2´-Azobis(2-methyl-N-phenylpropionamidine) dihydrochloride
88917-22-0	PPG-2 methyl ether acetate
90193-76-3	Distearyl phthalate
90268-48-7	Disodium tallow sulfosuccina-mate

CAS	Chemical
91031-48-0	Isooctyl stearate
91050-82-7	Pentaerythrityl tetrastearate
92112-62-4	PEG-1 C13-15 alkylmethylamine
93356-94-6	Alkyl C12-15 bisphenol A phosphite
93820-52-1	Sodium cocoamphopropionate
93820-97-4	Cetearyl stearate
94035-02-6	Hydroxypropyl-β-cyclodextrin
95823-35-1	Pentaerythrityl hexylthiopropionate
96152-48-6	4,4´ Isopropylidenediphenol alkyl (C12-15) phosphites
96319-55-0	t-Butyl 3-isopropenylcumyl peroxide
96328-09-1	Pentaerythrityl alkyl thiodipropionate
96492-31-8	Magnesium aluminum carbonate
97281-23-7	Glycol stearate
97281-47-5	Lecithin
97593-29-8	Hydrogenated palm glyceride
97593-29-8	Lard glyceride
99241-25-5	Hydroxypropyl-γ-cyclodextrin
101357-30-6	Sodium alumino sulfo silicate
102843-39-0	2,2´-Azobis[2-(3,4,5,6-tetrahydropyrimidin-2-yl) propane] dihydrochloride
103597-45-1	Bis [2-hydroxy-5-t-octyl-3-(benzotriazol-2-yl) phenyl] methane
104810-47-1	Methyl 3-[3-(2H-benzotriazole-2-yl)-5-t-butyl-4-hydroxyphenyl]proportionate
106246-33-7	4,4´-Methylenebis (3-chloro-2,6-diethylaniline)
107628-08-0	Octoxynol-9 carboxylic acid
109678-33-3	Tetrabromodipentaerythritol
110553-27-0	2,4-Bis [(octylthio) methyl]-o-cresol
110720-64-4	Stearyl/aminopropyl methicone copolymer
112926-00-8	Silica, hydrated
112945-52-5	Silica, fumed
115340-81-3	2,2,6,6-Tetramethylpiperidin-4-yl acrylate/methyl methacrylate copolymer
118337-09-0	2,2´-Ethylidenebis (4,6-di-t-butylphenyl) fluorophosphonite
118585-13-0	2,2´-Azobis[2-[1-(2-hydroxyethyl)-2-imidazolin-2-yl] propane] dihydrochloride
119345-01-6	Phosphorus trichloride, reaction prods. with 1,1´-biphenyl and 2,4-bis (1,1-dimethylethyl) phenol
119462-56-5	1,3-Bis(citraconimidomethyl) benzene
119758-00-8	Magnesium aluminum carbonate
120007-67-9	Zinc borate
121246-28-4	2,4,6-Tris-(N-1,4-dimethylpentyl-p-phenylenediamino)-1,3,5-triazine
123237-14-9	Synthetic wax
124960-38-9	2,2´-Azobis[N-(4-chlorophenyl)-2-methylpropionamidine] dihydrochloride
125740-36-5	Isodecyloxypropyl dihydroxyethyl methyl ammonium chloride
126121-35-5	Trioctyldodecyl citrate
129757-67-1	Bis-(1-octyloxy-2,2,6,6,tetramethyl-4-piperidinyl) sebacate
130097-36-8	α-Cumylperoxyneoheptanoate
130169-63-0	Dimethiconol stearate
133448-13-2	Dimethiconol hydroxystearate
133448-16-5	Dimethicone copolyol isostearate
134141-38-1	PEG-4 dioleate
136097-82-0	Cetearyl stearate
136097-95-5	Candelilla synthetic
136097-97-7	Cetyl esters
136618-51-4	Magnesium aluminum carbonate
136618-52-5	Magnesium aluminum carbonate
138265-88-0	Zinc borate
594477-57-3	Poly (pentabromobenzyl) acrylate

EINECS Number-to-Trade Name Cross-Reference

EINECS	Trade name
200-268-0	Keycide® X-10
200-313-4	Glycon® S-70
200-313-4	Glycon® S-90
200-313-4	Glycon® TP
200-313-4	Hydrofol 1800
200-313-4	Hystrene® 5016 NF
200-313-4	Hystrene® 9718 NF
200-313-4	Industrene® 1224
200-313-4	Industrene® 4518
200-313-4	Industrene® 5016
200-313-4	Industrene® 7018
200-313-4	Industrene® 9018
200-313-4	Loxiol® G 20
200-313-4	Naftozin® N, Spezial
200-313-4	Petrac® 250
200-313-4	Petrac® 270
200-313-4	Pristerene 4900
200-315-5	RIA CS
200-315-5	Stanclere® C26
200-315-5	WF-15
200-315-5	WF-16
200-338-0	Adeka Propylene Glycol (T)
200-449-4	Versene Acid
200-529-9	Versene CA
200-573-9	Hamp-Ene® 100
200-573-9	Sequestrene® 30A
200-573-9	Versene 100
200-573-9	Versene 100 EP
200-573-9	Versene 100 LS
200-573-9	Versene 100 SRG
200-573-9	Versene 100 XL
200-573-9	Versene 220
200-618-2	Retarder BA, BAX
200-664-3	Decap
200-741-1	Uniplex 173
200-751-6	Isanol®
200-815-3	A-C® 6
200-815-3	A-C® 6A
200-815-3	A-C® 7, 7A
200-815-3	A-C® 8, 8A
200-815-3	A-C® 9, 9A, 9F
200-815-3	A-C® 15
200-815-3	A-C® 16
200-815-3	A-C® 617, 617A
200-815-3	A-C® 712
200-815-3	A-C® 715
200-815-3	A-C® 725
200-815-3	A-C® 735
200-815-3	A-C® 1702
200-815-3	ACumist® A-6
200-815-3	ACumist® A-12
200-815-3	ACumist® A-18
200-815-3	ACumist® A-45
200-815-3	ACumist® B-6
200-815-3	ACumist® B-9
200-815-3	ACumist® B-12
200-815-3	ACumist® B-18
200-815-3	ACumist® C-5
200-815-3	ACumist® C-9
200-815-3	ACumist® C-12
200-815-3	ACumist® C-18
200-815-3	ACumist® C-30
200-815-3	ACumist® D-9
200-815-3	Bowax 2015
200-815-3	Cabot® CP 9396
200-815-3	Cabot® PE 9324
200-815-3	Coathylene HA 1591
200-815-3	Coathylene HA 2454
200-815-3	Epolene® C-10
200-815-3	Epolene® C-10P
200-815-3	Epolene® C-13
200-815-3	Epolene® C-13P
200-815-3	Epolene® C-14
200-815-3	Epolene® C-15
200-815-3	Epolene® C-15P
200-815-3	Epolene® C-16
200-815-3	Epolene® C-17
200-815-3	Epolene® C-17P
200-815-3	Epolene® C-18
200-815-3	Epolene® N-10
200-815-3	Epolene® N-11
200-815-3	Epolene® N-11P
200-815-3	Epolene® N-14
200-815-3	Epolene® N-14P
200-815-3	Epolene® N-20
200-815-3	Epolene® N-21
200-815-3	Epolene® N-34
200-815-3	Hoechst Wax PE 130
200-815-3	Hoechst Wax PE 190
200-815-3	Hoechst Wax PE 520
200-815-3	Hoechst Wax PED 521
200-815-3	Hostalub® H 3
200-815-3	Hostalub® H 22
200-815-3	Loobwax 0651

EINECS	Trade name
200-815-3	Luwax® A
200-815-3	Luwax® AF 30
200-815-3	Luwax® AH 6
200-815-3	Luwax® AL 3
200-815-3	Luwax® AL 61
200-815-3	Octowax 518
200-815-3	Oxp
200-815-3	Polyfin
200-815-3	Polymel #7
200-815-3	Polywax® 500
200-815-3	Polywax® 655
200-815-3	Polywax® 850
200-815-3	Polywax® 1000
200-815-3	Polywax® 2000
200-815-3	Polywax® 3000
200-815-3	Primax UH-1060
200-815-3	Primax UH-1080
200-815-3	Primax UH-1250
200-815-3	Stan Wax
200-815-3	Stat-Kon® AS-F
200-815-3	Stat-Kon® AS-FE
200-815-3	Struktol® PE H-100
200-815-3	Struktol® TR PE(H)-100
200-815-3	Uniwax AW-1050
200-815-3	Vestowax A-217
200-815-3	Vestowax A-227
200-815-3	Vestowax A-235
200-815-3	Vestowax A-415
200-815-3	Vestowax A-616
200-815-3	Vestowax AO-1539
200-815-3	Zylac 235
200-915-7	Aztec® TBHP-22-OL1
200-915-7	Aztec® TBHP-70
200-915-7	T-Hydro®
200-917-8	Dynasylan® VTC
200-917-8	Petrarch® V4900
201-039-8	ADK STAB BT-11
201-039-8	ADK STAB BT-18
201-039-8	ADK STAB BT-23
201-039-8	ADK STAB BT-25
201-039-8	ADK STAB BT-27
201-039-8	Cata-Chek® 820
201-039-8	Dabco® T-12
201-039-8	Fomrez® SUL-4
201-039-8	Jeffcat T-12
201-039-8	Kosmos® 19
201-039-8	Metacure® T-12
201-039-8	Synpron 1009
201-039-8	TEDA-T411
201-066-5	Citroflex® A-2
201-066-5	Uniplex 82
201-067-0	Citroflex® A-4
201-067-0	Citroflex® A-4 Special
201-067-0	Estaflex® ATC
201-067-0	Uniplex 84
201-070-7	Citroflex® 2
201-070-7	Uniplex 80
201-071-2	Citroflex® 4
201-071-2	TBC
201-081-7	Dynasylan® VTEO

EINECS	Trade name
201-081-7	Petrarch® V4910
201-094-8	Barquat® CME-35
201-114-5	Eastman® TEP
201-116-6	Amgard® TOF
201-116-6	Disflamoll® TOF
201-116-6	Kronitex® TOF
201-116-6	TOF
201-116-6	TOF™
201-122-9	Amgard® TBEP
201-122-9	KP-140®
201-122-9	Phosflex TBEP
201-122-9	TBEP
201-122-9	Tolplaz TBEP
201-132-3	Ficel® AZDN-LF
201-132-3	Ficel® AZDN-LMC
201-132-3	Mikrofine AZDM
201-132-3	Perkadox® AIBN
201-132-3	Poly-Zole® AZDN
201-132-3	V-60
201-202-3	BM-801
201-236-9	FR-1524
201-236-9	Great Lakes BA-59
201-236-9	Great Lakes BA-59P™
201-236-9	Saytex® RB-100
201-245-8	Parabis
201-545-9	Unimoll® 66
201-545-9	Unimoll® 66 M
201-545-9	Uniplex 250
201-548-5	Nuoplaz® 6938
201-550-6	Kodaflex® DEP
201-557-4	Diplast® A
201-557-4	Kodaflex® DBP
201-557-4	Palatinol® DBP
201-557-4	Plasthall® DBP
201-557-4	PX-104
201-557-4	Staflex DBP
201-557-4	Unimoll® DB
201-557-4	Uniplex 150
201-559-5	Jayflex® DHP
201-562-1	Kodaflex® HS-3
201-562-1	PX-914
201-562-1	Staflex BOP
201-607-5	Retarder AK
201-607-5	Retarder PX
201-607-5	Vulkalent B/C
201-607-5	Wiltrol P
201-622-7	Santicizer 160
201-622-7	Unimoll® BB
201-624-8	Morflex® 190
201-800-4	Luviskol® K17
201-800-4	Luviskol® K30
201-800-4	Luviskol® K60
201-800-4	Luviskol® K90
201-800-4	PVP K-60
202-075-7	Santoflex® AW
202-327-6	Abcure S-40-25
202-327-6	Aztec® BP-05-FT
202-327-6	Aztec® BP-20-GY
202-327-6	Aztec® BP-40-S
202-327-6	Aztec® BP-50-FT

EINECS	Trade name
202-327-6	Aztec® BP-50-P1
202-327-6	Aztec® BP-50-PSI
202-327-6	Aztec® BP-50-SAQ
202-327-6	Aztec® BP-60-PCL
202-327-6	Aztec® BP-70-W
202-327-6	Aztec® BP-77-W
202-327-6	Benox® L-40LV
202-327-6	Cadet® BPO-70
202-327-6	Cadet® BPO-70W
202-327-6	Cadet® BPO-75W
202-327-6	Cadet® BPO-78
202-327-6	Cadet® BPO-78W
202-327-6	Cadox® 40E
202-327-6	Cadox® BFF-50
202-327-6	Cadox® BFF-50L
202-327-6	Cadox® BPO-W40
202-327-6	Cadox® BS
202-327-6	Cadox® BT-50
202-327-6	Cadox® BTW-50
202-327-6	Cadox® BTW-55
202-327-6	Lucidol 70
202-327-6	Lucidol 75
202-327-6	Lucidol 75FP
202-327-6	Lucidol 78
202-327-6	Lucidol 98
202-327-6	Luperco AFR-250
202-327-6	Luperco AFR-400
202-327-6	Luperco AFR-500
202-327-6	Luperco ANS
202-327-6	Luperco ANS-P
202-327-6	Varox® ANS
202-443-7	THQ
202-473-0	Ageflex AMA
202-473-0	MFM-401
202-473-0	Sipomer® AM
202-495-0	Thiovanol®
202-506-9	Akroform® ETU-22 PM
202-506-9	Ekaland ETU
202-506-9	Naftocit® Mi 12
202-506-9	Naftopast® Mi12-P
202-506-9	Perkacit® ETU
202-608-3	Hamposyl® L
202-617-2	Ageflex EGDMA
202-617-2	MFM-416
202-617-2	Perkalink® 401
202-625-6	QO® Tetrahydrofurfuryl Alcohol (THFA®)
202-626-1	QO® Furfuryl Alcohol (FA®)
202-627-7	QO® Furfural
202-637-1	Uniplex 413
202-638-7	Reworyl® B 70
202-805-4	Firstcure™ DMPT
202-845-2	Pennad 150
202-885-0	Dabco® NEM
202-885-0	Jeffcat NEM
202-885-0	Toyocat®-NEM
202-908-4	Albrite® TPP
202-908-4	Doverphos® 10
202-908-4	Lankromark® LE65
202-908-4	Mark® 2112
202-908-4	Mark® TPP
202-908-4	Weston® TPP
202-935-1	Flexricin® P-8
202-936-7	Perkalink® 300
203-041-4	Quadrol®
203-049-8	TEA 85 Low Freeze Grade
203-049-8	TEA 99 Low Freeze Grade
203-049-8	TEA 99 Standard Grade
203-051-9	Estol 1579
203-051-9	Kodaflex® Triacetin
203-090-1	Adimoll® DO
203-090-1	Diplast® DOA
203-090-1	Jayflex® DOA
203-090-1	Kodaflex® DOA
203-090-1	Monoplex® DOA
203-090-1	Nuoplaz® DOA
203-090-1	Palatinol® DOA
203-090-1	Plasthall® DOA
203-090-1	PX-238
203-090-1	Staflex DOA
203-090-1	Uniflex® DOA
203-149-1	Dabco® BDMA
203-180-0	Eltesol® TSX/A
203-180-0	Eltesol® TSX/SF
203-197-3	Eastman® HQEE
203-246-9	PTAL
203-308-5	Thiate® H
203-328-4	PX-504
203-349-9	Uniflex® DCA
203-358-8	Kessco® PEG 200 MS
203-363-5	Cithrol DGMS N/E
203-366-1	Loxiol® G 21
203-369-8	Cithrol EGMR N/E
203-419-9	DBE-4
203-442-4	Ageflex AGE
203-449-2	Gulftene® 4
203-473-3	Chemstat® P-400
203-508-2	Arosurf® TA-100
203-508-2	Dehyquart® DAM
203-542-8	Dabco® DMEA
203-542-8	Jeffcat DME
203-542-8	Tegoamin® DMEA
203-542-8	Toyocat®-DMA
203-572-1	Arconate® 1000
203-572-1	Arconate® HP
203-572-1	Arconate® PC
203-572-1	Texacar® PC
203-618-0	Zeonet® A
203-640-0	Dabco® NMM
203-640-0	Jeffcat NMM
203-672-5	Plasthall® DBS
203-672-5	Uniflex® DBS
203-674-6	Thiate® U
203-713-7	Hisolve MC
203-726-8	THF
203-733-6	Aztec® DTBP
203-733-6	Trigonox® B
203-744-6	Toyocat®-TE
203-749-3	Hamposyl® O
203-751-4	Estol IPM 1514

EINECS	Trade name
203-756-1	Advawax® 240
203-756-1	Kemamide® W-20
203-786-5	BDO
203-786-5	Dabco® BDO
203-804-1	Oxitol
203-809-9	Pyridine 1°
203-820-9	DIPA Commercial Grade
203-820-9	DIPA Low Freeze Grade 85
203-820-9	DIPA Low Freeze Grade 90
203-820-9	DIPA NF Grade
203-821-4	Adeka Dipropylene Glycol
203-852-3	Epal® 6
203-865-4	D.E.H. 20
203-865-4	D.E.H. 52
203-865-4	D.E.H. 58
203-865-4	DETA
203-865-4	Texacure EA-20
203-868-0	Dabco® DEOA-LF
203-868-0	DEA Commercial Grade
203-868-0	DEA Low Freeze Grade
203-881-1	Estol 1574
203-886-9	CPH-37-NA
203-886-9	Mackester™ EGMS
203-886-9	Mackester™ IP
203-886-9	Pegosperse® 50 MS
203-905-0	Butyl Oxitol
203-905-0	Ektasolve® EB
203-906-6	Hisolve DM
203-911-3	Estol 1502
203-917-6	Epal® 8
203-919-7	Dioxitol-Low Gravity
203-924-4	Diglyme
203-924-4	Hisolve MDM
203-927-0	Arquad® 12-50
203-927-0	Chemquat 12-33
203-927-0	Chemquat 12-50
203-927-0	Dehyquart® LT
203-927-0	Octosol 562
203-927-0	Octosol 571
203-928-6	Arquad® 16-50
203-928-6	Chemquat 16-50
203-928-6	Dehyquart® A
203-929-1	Arquad® 18-50
203-935-4	Estol 1406
203-943-8	Adma® 12
203-950-6	D.E.H. 24
203-950-6	TETA
203-950-6	Texacure EA-24
203-950-6	Texlin® 300
203-956-9	Epal® 10
203-961-6	Butyl Dioxitol
203-982-0	Epal® 12
203-982-0	Laurex® NC
203-986-2	D.E.H. 26
203-986-2	TEPA
203-989-9	Dow E200
203-989-9	Dow E300
203-989-9	Dow E400
203-989-9	Dow E600
203-989-9	Dow E900
203-989-9	Dow E1000
203-989-9	Dow E1450
203-989-9	Dow E3350
203-989-9	Dow E4500
203-989-9	Dow E8000
203-989-9	Hodag PEG 200
203-989-9	Pluracol® E200
203-989-9	Poly-G® 200
203-997-2	Adma® 16
204-000-3	Epal® 14
204-002-4	Adma® 14
204-007-1	Industrene® 105
204-007-1	Industrene® 205
204-007-1	Industrene® 206
204-007-1	Pamolyn® 125
204-007-1	Priolene 6901
204-007-1	Priolene 6907
204-007-1	Priolene 6930
204-009-2	Armid® E
204-009-2	Armoslip® EXP
204-009-2	Chemstat® HTSA #22
204-009-2	Crodamide E
204-009-2	Crodamide ER
204-009-2	Kemamide® E
204-009-2	Petrac® Eramide®
204-009-2	Polydis® TR 131
204-009-2	Unislip 1753
204-015-5	Crodamine 1.O, 1.OD
204-015-5	Kemamine® P-989D
204-017-6	Adol® 62 NF
204-017-6	Alfol® 18
204-017-6	Cachalot® S-53
204-017-6	Loxiol® VPG 1354
204-104-9	Flame-Amine
204-104-9	Sta-Flame 88/99
204-112-2	Disflamoll® TP
204-112-2	Kronitex® TPP
204-211-0	Diplast® O
204-211-0	Kodaflex® DOP
204-211-0	Nuoplaz® DOP
204-211-0	Plasthall® DOP
204-211-0	PX-138
204-294-3	Jayflex® DTDP
204-294-3	PX-126
204-295-9	PX-318
204-295-9	Staflex ODP
204-327-1	Ralox® 46
204-327-1	Vanox® MBPC
204-340-2	Tetralin
204-393-1	Empilan® LDE
204-393-1	Ethylan® MLD
204-407-6	Benzoflex® 2-45
204-479-9	Hyamine® 1622 50%
204-495-6	Firstcure™ DMMT
204-498-2	Tenox® PG
204-528-4	TIPA 99
204-528-4	TIPA 101
204-528-4	TIPA Low Freeze Grade
204-557-2	Epodil® 743
204-557-2	Heloxy® 63

EINECS	Trade name
204-589-7	Emery® 6705
204-614-1	Carstab® DLTDP
204-614-1	Cyanox® LTDP
204-614-1	Evanstab® 12
204-614-1	Hostanox® SE 1
204-614-1	Lankromark® DLTDP
204-614-1	Lowinox® DLTDP
204-614-1	PAG DLTDP
204-648-7	2-Pyrol®
204-650-8	Azo D
204-650-8	Azo DX72
204-650-8	Azofoam
204-650-8	Azofoam DS-1, DS-2
204-650-8	Azofoam M-1, M-2, M-3
204-650-8	Azofoam UC
204-650-8	Blo-Foam® 715
204-650-8	Blo-Foam® 754
204-650-8	Blo-Foam® 765
204-650-8	Blo-Foam® ADC 150
204-650-8	Blo-Foam® ADC 150FF
204-650-8	Blo-Foam® ADC 300
204-650-8	Blo-Foam® ADC 300FF
204-650-8	Blo-Foam® ADC 300WS
204-650-8	Blo-Foam® ADC 450
204-650-8	Blo-Foam® ADC 450FFNP
204-650-8	Blo-Foam® ADC 450WS
204-650-8	Blo-Foam® ADC 550
204-650-8	Blo-Foam® ADC 550FFNP
204-650-8	Blo-Foam® ADC 800
204-650-8	Blo-Foam® ADC 1200
204-650-8	Blo-Foam® ADC-MP
204-650-8	Blo-Foam® ADC-NP
204-650-8	Blo-Foam® KL9
204-650-8	Blo-Foam® KL10
204-650-8	Cellcom
204-650-8	Celogen® AZ 120
204-650-8	Celogen® AZ 130
204-650-8	Celogen® AZ 150
204-650-8	Celogen® AZ 180
204-650-8	Celogen® AZ 199
204-650-8	Celogen® AZ 754
204-650-8	Celogen® AZ-760-A
204-650-8	Celogen® AZ-2100
204-650-8	Celogen® AZ-2990
204-650-8	Celogen® AZ 3990
204-650-8	Celogen® AZ-5100
204-650-8	Celogen® AZNP 130
204-650-8	Celogen® AZNP 199
204-650-8	Celogen® AZRV
204-650-8	Ficel® AC1
204-650-8	Ficel® AC2
204-650-8	Ficel® AC2M
204-650-8	Ficel® AC3
204-650-8	Ficel® AC3F
204-650-8	Ficel® AC4
204-650-8	Ficel® ACSP4
204-650-8	Ficel® ACSP5
204-650-8	Ficel® EPA
204-650-8	Ficel® EPB
204-650-8	Ficel® EPC
204-650-8	Ficel® EPD
204-650-8	Ficel® EPE
204-650-8	Ficel® LE
204-650-8	Ficel® SCE
204-650-8	Henley AZ LBA-30
204-650-8	Henley AZ LBA-49
204-650-8	HFVCD 11G
204-650-8	HFVCP 11G
204-650-8	HFVP 03
204-650-8	HFVP 15P
204-650-8	HFVP 19B
204-650-8	HP
204-650-8	HRVP 01
204-650-8	HRVP 19
204-650-8	Kempore® 60/14FF
204-650-8	Mikrofine ADC-EV
204-650-8	Plastifoam
204-650-8	Porofor® ADC/E
204-650-8	Porofor® ADC/F
204-650-8	Porofor® ADC/K
204-650-8	Porofor® ADC/M
204-650-8	Porofor® ADC/S
204-650-8	Unicell D
204-666-5	Ablumol BS
204-666-5	ADK STAB LS-8
204-666-5	Butyl Stearate C-895
204-666-5	Lexolube® BS-Tech
204-666-5	Lexolube® NBS
204-666-5	Priolube 1451
204-666-5	Uniflex® BYS-CP
204-666-5	Uniflex® BYS Special
204-666-5	Uniflex® BYS-Tech
204-673-3	Adi-pure®
204-693-2	Armid® 18
204-693-2	Armoslip® 18
204-693-2	Chemstat® HTSA #18S
204-693-2	Crodamide S
204-693-2	Crodamide SR
204-693-2	Kemamide® S
204-693-2	Petrac® Vyn-Eze®
204-693-2	Uniwax 1750
204-694-8	Adma® 18
204-694-8	Crodamine 3.A18D
204-694-8	Kemamine® T-9902D
204-694-8	Nissan Tert. Amine AB
204-695-3	Armeen® 18
204-695-3	Armeen® 18D
204-695-3	Armid® HTD
204-695-3	Crodamine 1.18D
204-695-3	Kemamine® P-990, P-990D
204-783-1	Sulfolane W
204-800-2	Albrite® TBP
204-800-2	Albrite® TBPO
204-800-2	Albrite® Tributyl Phosphate
204-800-2	Amgard® $TBPO_4$
204-800-2	Kronitex® TBP
204-800-2	TBP
204-812-8	Niaproof® Anionic Surfactant 08
204-812-8	Witcolate™ D-510

EINECS	Trade name
204-839-5	Aerosol® IB-45
204-839-5	Astrowet B-45
204-839-5	Gemtex 445
204-839-5	Geropon® CYA/45
204-839-5	Geropon® CYA/DEP
204-839-5	Monawet MB-45
204-839-5	Octosol IB-45
204-839-5	Rewopol® SBDB 45
204-881-4	Akrochem® Antioxidant BHT
204-881-4	CAO®-3
204-881-4	Lowinox® BHT
204-881-4	Lowinox® BHT Food Grade
204-881-4	Lowinox® BHT-Superflakes
204-881-4	Naugard® BHT
204-881-4	Ralox® BHT food grade
204-881-4	Ralox® BHT Tech
204-881-4	Spectratech® CM 11340
204-881-4	Spectratech® KM 11264
204-881-4	Sustane® BHT
204-881-4	Topanol® OC
204-881-4	Topanol® OF
204-881-4	Ultranox® 226
204-881-4	Vulkanox® KB
205-011-6	Kodaflex® DMP
205-011-6	Palatinol® M
205-011-6	Staflex DMP
205-011-6	Uniplex 110
205-016-3	Ftalidap®
205-026-8	Cyasorb® UV 24
205-027-3	Uvinul® 3049
205-027-3	Uvinul® D-49
205-028-9	Uvinul® 3050
205-029-4	Uvasorb 2 OH
205-029-4	Uvinul® 3000
205-031-5	Acetorb A
205-031-5	Cyasorb® UV 9
205-031-5	Lankromark® LE296
205-031-5	Syntase® 62
205-031-5	UV-Absorber 325
205-031-5	Uvasorb MET
205-031-5	Uvinul® 3040
205-031-5	Uvinul® M-40
205-087-0	Vancide® 89
205-183-2	Q-1300
205-281-5	Hamposyl® L-30
205-281-5	Hamposyl® L-95
205-351-5	Dehyquart® LDB
205-358-3	Versene NA
205-358-3	Versene Na_2
205-388-7	Carsonol® TLS
205-388-7	Empicol® TL40/T
205-388-7	Perlankrol® ATL40
205-391-3	Chel DTPA-41
205-392-9	Flexricin® P-4
205-411-0	AEP
205-411-0	D.E.H. 39
205-450-3	DIBA
205-450-3	Plasthall® DIBA
205-455-0	Flexricin® 13
205-470-2	P®-10 Acid
205-471-8	Paricin® 1
205-472-3	Flexricin® P-1
205-480-7	Paraloid® EXL-3330
205-480-7	Paraloid® EXL-3361
205-480-7	Paraloid® EXL-3387
205-483-3	MEA Commercial Grade
205-483-3	MEA Low Freeze Grade
205-483-3	MEA Low Iron Grade
205-483-3	MEA Low Iron-Low Freeze Grade
205-483-3	MEA NF Grade
205-516-1	Tyzor DC
205-524-5	PX-538
205-539-7	Hamposyl® S
205-559-6	Butyl Oleate C-914
205-559-6	Kemester® 4000
205-559-6	Plasthall® 503
205-559-6	Plasthall® 914
205-559-6	Priolube 1405
205-559-6	Uniflex® BYO
205-571-1	Estol IPP 1517
205-577-4	Carsonol® DLS
205-582-1	Prifrac 2920
205-582-1	Prifrac 2922
205-590-5	Octosol 449
205-596-8	Crodamine 1.16D
205-597-3	Fancol OA-95
205-633-8	Kycerol 91
205-633-8	Kycerol 92
205-633-8	Sodium Bicarbonate USP No. 1 Powd.
205-633-8	Sodium Bicarbonate USP No. 3 Extra Fine Powd.
205-736-8	Akrochem® MBT
205-736-8	Captax®
205-736-8	Ekaland MBT
205-736-8	Naftocit® MBT
205-736-8	Naftopast® MBT-P
205-736-8	Perkacit® MBT
205-736-8	Rokon
205-736-8	Rotax®
205-736-8	Thiotax®
205-758-8	Hamp-Ene® Na_3 Liq
205-788-1	Alscoap LN-40, LN-90
205-788-1	Calfoam SLS-30
205-788-1	Carsonol® SLS
205-788-1	Carsonol® SLS-R
205-788-1	Carsonol® SLS-S
205-788-1	Carsonol® SLS Special
205-788-1	Elfan® 260 S
205-788-1	Empicol® 0303
205-788-1	Empicol® 0303V
205-788-1	Empicol® LX28
205-788-1	Empicol® LXS95
205-788-1	Empicol® LXV
205-788-1	Empicol® LXV/D
205-788-1	Empicol® LZP
205-788-1	Empicol® LZV/D
205-788-1	Manro SLS 28
205-788-1	Montovol RF-10

EINECS	Trade name
205-788-1	Perlankrol® DSA
205-788-1	Rewopol® NLS 28
205-788-1	Serdet DFK 40
205-788-1	Stepanol® WA-100
205-788-1	Sulfopon® 101 Special
205-788-1	Sulfopon® P-40
205-788-1	Texapon® OT Highly Conc. Needles
205-788-1	Texapon® VHC Needles
205-788-1	Texapon® ZHC Powder
205-788-1	Ultra Sulfate SL-1
205-788-1	Witcolate™ A Powder
205-788-1	Witcolate™ SL-1
206-101-8	Witco® Aluminum Stearate 22
206-103-9	Armid® O
206-103-9	Chemstat® HTSA #18
206-103-9	Crodamide O
206-103-9	Crodamide OR
206-103-9	Kemamide® O
206-103-9	Kemamide® U
206-103-9	Petrac® Slip-Eze®
206-103-9	Polydis® TR 121
206-103-9	Polyvel RO25
206-103-9	Polyvel RO40
206-103-9	Unislip 1757
206-103-9	Unislip 1759
207-312-8	Amicure® CG-325
207-312-8	Amicure® CG-1200
207-312-8	Amicure® CG-1400
207-312-8	Amicure® CG-NA
207-312-8	Dicyanex® 200X
207-312-8	Dicyanex® 325
207-312-8	Dicyanex® 1200
207-439-9	Atomite®
207-439-9	Camel-WITE®
207-439-9	CC™-101
207-439-9	CC™-103
207-439-9	CC™-105
207-439-9	Micro-White® 10 Slurry
207-439-9	Micro-White® 15 Slurry
207-439-9	Micro-White® 25 Slurry
207-924-5	Hostanox® O3
208-915-9	Magocarb-33
209-097-6	Pationic® 919
209-097-6	Pationic® 1019
209-150-3	ADK STAB MG-ST
209-150-3	Afco-Chem MGS
209-150-3	Cometals Magnesium Stearate
209-150-3	Haro® Chem MF-2
209-150-3	Haro® Chem MF-3
209-150-3	HyQual NF
209-150-3	HyTech RSN 1-1
209-150-3	Mathe Magnesium Stearate
209-150-3	Miljac Magnesium Stearate
209-150-3	Petrac® Magnesium Stearate MG-20 NF
209-150-3	Synpro® Magnesium Stearate 90
209-150-3	Witco® Magnesium Stearate D
209-151-9	ADK STAB ZN-ST
209-151-9	Afco-Chem ZNS
209-151-9	Akrochem® P 3100
209-151-9	Akrochem® Wettable Zinc Stearate
209-151-9	Akrochem® Zinc Stearate
209-151-9	Akrodip™ Z-50/50
209-151-9	Akrodip™ Z-200
209-151-9	Akrodip™ Z-250
209-151-9	Cecavon ZN 70
209-151-9	Cecavon ZN 71
209-151-9	Cecavon ZN 72
209-151-9	Cecavon ZN 73
209-151-9	Cecavon ZN 735
209-151-9	Coad 20
209-151-9	Coad 21
209-151-9	Coad 27B
209-151-9	Cometals Zinc Stearate
209-151-9	Disperso II
209-151-9	Hallcote® ZS
209-151-9	Hallcote® ZS 5050
209-151-9	Haro® Chem ZGD
209-151-9	Haro® Chem ZGN
209-151-9	Haro® Chem ZGN-T
209-151-9	Haro® Chem ZPR-2
209-151-9	Hy Dense Zinc Stearate XM Powd.
209-151-9	Hy Dense Zinc Stearate XM Ultra Fine
209-151-9	Hydro Zinc™
209-151-9	HyTech RSN 131 HS/Gran
209-151-9	Interstab® ZN-18-1
209-151-9	Liquazinc AQ-90
209-151-9	Lubrazinc® W, Superfine
209-151-9	Mathe Zinc Stearate 25S
209-151-9	Mathe Zinc Stearate S
209-151-9	Metasap® Zinc Stearate
209-151-9	Miljac Zinc Stearate
209-151-9	Molder's Edge ME-369
209-151-9	Petrac® Zinc Stearate ZN-41
209-151-9	Petrac® Zinc Stearate ZN-42
209-151-9	Petrac® Zinc Stearate ZN-44 HS
209-151-9	Petrac® Zinc Stearate ZW-45
209-151-9	Polyvel RZ40
209-151-9	Quikote™
209-151-9	Quikote™ M
209-151-9	Rhenodiv® ZB
209-151-9	Rubichem Zinc-Dip®
209-151-9	Rubichem Zinc-Dip® HS
209-151-9	Rubichem Zinc-Dip® W
209-151-9	Sprayon 812
209-151-9	Synpro® Zinc Dispersion Zincloid
209-151-9	Synpro® Zinc Stearate 8
209-151-9	Synpro® Zinc Stearate ACF
209-151-9	Synpro® Zinc Stearate D
209-151-9	Synpro® Zinc Stearate GP
209-151-9	Synpro® Zinc Stearate GP Flake
209-151-9	Synpro® Zinc Stearate HSF

EINECS	Trade name
209-151-9	TR-Zinc Stearate
209-151-9	Wet Zinc™
209-151-9	Wet Zinc™ P
209-151-9	Witco® Zinc Stearate 11
209-151-9	Witco® Zinc Stearate 42
209-151-9	Witco® Zinc Stearate 44
209-151-9	Witco® Zinc Stearate Disperso
209-151-9	Witco® Zinc Stearate Heat-Stable
209-151-9	Witco® Zinc Stearate LV
209-151-9	Witco® Zinc Stearate NW
209-151-9	Witco® Zinc Stearate Polymer Grade
209-151-9	Witco Zinc Stearate Regular
209-151-9	Witco® Zinc Stearate REP
209-151-9	Zincote™
209-151-9	Zinc Stearate 42
209-406-4	Aerosol® OT-75%
209-406-4	Chemax DOSS-75E
209-406-4	Drewfax® 0007
209-406-4	Gemtex PA-75
209-406-4	Gemtex PA-85P
209-406-4	Gemtex PAX-60
209-406-4	Gemtex SC-75E
209-406-4	Geropon® 99
209-406-4	Hodag DOSS-70
209-406-4	Hodag DOSS-75
209-406-4	Rewopol® SBDO 75
209-753-1	Gulftene® 6
211-014-3	Cithrol EGDS N/E
211-014-3	Kemester® EGDS
211-014-3	Mackester™ EGDS
211-020-6	DBE-6
211-185-4	Great Lakes PHT4™
211-185-4	Saytex® RB-49
211-279-5	Haro® Chem ALT
211-466-1	Estol 1476
211-546-6	Loxiol® VPG 1451
211-748-4	Lonzaine® 16S
211-750-5	Carstab® DSTDP
211-750-5	Cyanox® STDP
211-750-5	Evanstab® 18
211-750-5	Hostanox® SE 2
211-750-5	Lankromark® DSTDP
211-750-5	Lowinox® DSTDP
211-750-5	PAG DSTDP
211-750-5	PAG DXTDP
211-765-7	Imicure® AMI-2
211-776-7	DCH-99
212-164-2	Dow Corning® Z-6020
212-164-2	Dynasylan® DAMO
212-164-2	Dynasylan® DAMO-P
212-164-2	Dynasylan® DAMO-T
212-164-2	Petrarch® A0700
212-164-2	Petrarch® A0701
212-164-2	Prosil® 3128
212-406-7	Pationic® 1230
212-406-7	Pationic® 1240
212-406-7	Pationic® 1250
212-470-6	Lonzaine® 18S

EINECS	Trade name
212-490-5	Afco-Chem B
212-490-5	Cometals Sodium Stearate
212-490-5	Haro® Chem NG
212-490-5	HyTemp NS-70
212-490-5	Mathe Sodium Stearate
212-490-5	Synpro® Sodium Stearate ST
212-490-5	Witco® Heat Stable Sodium Stearate
212-490-5	Witco® Sodium Stearate T-1
212-782-2	BM-903
212-782-2	Sipomer® HEM-D
212-819-2	Gulftene® 10
213-048-4	Dynasylan® AMEO
213-048-4	Dynasylan® AMEO-P
213-048-4	Dynasylan® AMEO-T
213-048-4	Petrarch® A0750
213-048-4	Prosil® 220
213-048-4	Prosil® 221
213-085-6	Aerosol® AY-65
213-195-4	Diglycolamine® Agent (DGA®)
213-234-5	Curezol® 2E4MZ
213-234-5	Imicure® EMI-24
213-668-5	Dynasylan® HMDS
213-668-5	Petrarch® H7300
213-934-0	Dynasylan® VTMOEO
213-934-0	Petrarch® V5000
214-277-2	DBE-5
214-604-9	FR-1210
214-604-9	Great Lakes DE-83™
214-604-9	Great Lakes DE-83R™
214-604-9	Octoguard FR-01
214-604-9	Saytex® 102E
214-604-9	Thermoguard® 505
214-685-0	Dynasylan® MTMS
214-685-0	Petrarch® M9100
215-137-3	Rhenofit® CF
215-138-9	Rhenosorb® C
215-138-9	Rhenosorb® C/GW
215-138-9	Rhenosorb® F
215-168-2	OSO® 440
215-168-2	OSO® 1905
215-168-2	OSO® NR 830 M
215-168-2	OSO® NR 950
215-170-3	ACM-MW
215-170-3	BAX 1091FM
215-170-3	Flamtard M7
215-170-3	FR-20
215-170-3	Hydramax™ HM-B8, HM-B8-S
215-170-3	Kisuma 5A
215-170-3	Kisuma 5B
215-170-3	Kisuma 5E
215-170-3	Magnifin® H5
215-170-3	Magnifin® H5B
215-170-3	Magnifin® H5C
215-170-3	Magnifin® H5C/3
215-170-3	Magnifin® H5D
215-170-3	Magnifin® H-7
215-170-3	Magnifin® H7A
215-170-3	Magnifin® H7C
215-170-3	Magnifin® H7C/3

EINECS	Trade name
215-170-3	Magnifin® H7D
215-170-3	Magnifin® H-10
215-170-3	Magnifin® H10A
215-170-3	Magnifin® H10B
215-170-3	Magnifin® H10C
215-170-3	Magnifin® H10D
215-170-3	Magnifin® H10F
215-170-3	Magoh-S
215-170-3	MGH-93
215-170-3	Versamag® DC
215-170-3	Versamag® Tech
215-170-3	Versamag® UF
215-170-3	Zerogen® 10
215-170-3	Zerogen® 15
215-170-3	Zerogen® 30
215-170-3	Zerogen® 35
215-170-3	Zerogen® 60
215-171-9	Elastomag® 100
215-171-9	Elastomag® 100R
215-171-9	Elastomag® 170
215-171-9	Elastomag® 170 Micropellet
215-171-9	Elastomag® 170 Special
215-171-9	Ken-Mag®
215-171-9	MagChem® 20M
215-171-9	MagChem® 30
215-171-9	MagChem® 30G
215-171-9	MagChem® 35
215-171-9	MagChem® 35K
215-171-9	MagChem® 40
215-171-9	MagChem® 50
215-171-9	MagChem® 50Y
215-171-9	MagChem® 60
215-171-9	MagChem® 125
215-171-9	MagChem® 200-AD
215-171-9	MagChem® 200-D
215-171-9	Maglite® D
215-171-9	Magotex™
215-171-9	Naftopast® MgO-A
215-171-9	Plastomag® 170
215-171-9	Rhenomag® G1
215-171-9	Rhenomag® G3
215-174-5	LP-100
215-174-5	LP-200
215-174-5	LP-300
215-174-5	LP-400
215-175-0	Amsperse
215-175-0	AO
215-175-0	FireShield® H
215-175-0	FireShield® HPM
215-175-0	FireShield® HPM-UF
215-175-0	FireShield® L
215-175-0	Fyraway
215-175-0	Fyrebloc
215-175-0	Naftopast® Antimontrioxid
215-175-0	Naftopast® Antimontrioxid-CP
215-175-0	Octoguard FR-10
215-175-0	Petcat R-9
215-175-0	Thermoguard® L
215-175-0	Thermoguard® S
215-175-0	Thermoguard® UF
215-175-0	UltraFine® II
215-204-7	Polu-U
215-222-5	Akrochem® Zinc Oxide 35
215-222-5	Kadox®-215
215-222-5	Kadox®-272
215-222-5	Kadox® 720
215-222-5	Kadox®-911
215-222-5	Kadox®-920
215-222-5	Kadox® 930
215-222-5	Ken-Zinc®
215-222-5	Naftopast® ZnO-A
215-222-5	NAO 105
215-222-5	NAO 115
215-222-5	NAO 125
215-222-5	NAO 135
215-222-5	Octocure 462
215-222-5	Ottalume 2100
215-222-5	RR Zinc Oxide (Untreated)
215-222-5	Zinc Oxide No. 185
215-222-5	Zinc Oxide No. 318
215-222-5	Zink Oxide AT
215-222-5	Zink Oxide Transparent
215-222-5	Zinkoxyd Activ®
215-237-7	Nyacol® A-1530
215-237-7	Nyacol® A-1540N
215-237-7	Nyacol® A-1550
215-237-7	Nyacol® A-1588LP
215-237-7	Nyacol® AP50
215-237-7	Nyacol® N24
215-237-7	Nyacol® ZTA
215-251-3	Sachtolith® HD
215-251-3	Sachtolith® HD-S
215-251-3	Sachtolith® L
215-267-0	Litharge 28
215-267-0	Litharge 33
215-288-5	Perchem® 97
215-290-6	Halcarb 20
215-290-6	Halcarb 20 EP
215-290-6	Halcarb 200
215-290-6	Halcarb 200 EP
215-290-6	Haro® Chem PC
215-475-1	Kaopolite® 1147
215-475-1	Kaopolite® 1168
215-475-1	Kaopolite® SF
215-475-1	Kaopolite® SFO-NP
215-548-8	Pliabrac® TCP
215-609-9	Aquathene™ CM 92832
215-609-9	Black Pearls® 120
215-609-9	Black Pearls® 130
215-609-9	Black Pearls® 160
215-609-9	Black Pearls® 170
215-609-9	Black Pearls® 280
215-609-9	Black Pearls® 450
215-609-9	Black Pearls® 460
215-609-9	Black Pearls® 470
215-609-9	Black Pearls® 480
215-609-9	Black Pearls® 490
215-609-9	Black Pearls® 520
215-609-9	Black Pearls® 570
215-609-9	Black Pearls® 700

EINECS	Trade name
215-609-9	Black Pearls® 800
215-609-9	Black Pearls® 880
215-609-9	Black Pearls® 900
215-609-9	Black Pearls® 1000
215-609-9	Black Pearls® 1100
215-609-9	Black Pearls® 3700
215-609-9	Black Pearls® L
215-609-9	Continex® LH-10
215-609-9	Continex® LH-20
215-609-9	Continex® LH-30
215-609-9	Continex® LH-35
215-609-9	Continex® N110
215-609-9	Continex® N220
215-609-9	Continex® N234
215-609-9	Continex® N299
215-609-9	Continex® N326
215-609-9	Continex® N330
215-609-9	Continex® N339
215-609-9	Continex® N343
215-609-9	Continex® N351
215-609-9	Continex® N550
215-609-9	Continex® N650
215-609-9	Continex® N660
215-609-9	Continex® N683
215-609-9	Continex® N762
215-609-9	Continex® N774
215-609-9	Diablack® A
215-609-9	Diablack® E
215-609-9	Diablack® G
215-609-9	Diablack® H
215-609-9	Diablack® HA
215-609-9	Diablack® I
215-609-9	Diablack® II
215-609-9	Diablack® LH
215-609-9	Diablack® LI
215-609-9	Diablack® LR
215-609-9	Diablack® M
215-609-9	Diablack® N220M
215-609-9	Diablack® N234
215-609-9	Diablack® N339
215-609-9	Diablack® N550M
215-609-9	Diablack® N760M
215-609-9	Diablack® R
215-609-9	Diablack® SF
215-609-9	Diablack® SH
215-609-9	Diablack® SHA
215-609-9	Elftex® 120
215-609-9	Elftex® 125
215-609-9	Elftex® 160
215-609-9	Elftex® 280
215-609-9	Elftex® 285
215-609-9	Elftex® 415
215-609-9	Elftex® 430
215-609-9	Elftex® 435
215-609-9	Elftex® 460
215-609-9	Elftex® 465
215-609-9	Elftex® 570
215-609-9	Elftex® 670
215-609-9	Elftex® 675
215-609-9	FW 18

EINECS	Trade name
215-609-9	FW 200 Beads and Powd
215-609-9	Huber ARO 60
215-609-9	Huber N110
215-609-9	Huber N220
215-609-9	Huber N234
215-609-9	Huber N299
215-609-9	Huber N326
215-609-9	Huber N330
215-609-9	Huber N339
215-609-9	Huber N343
215-609-9	Huber N347
215-609-9	Huber N351
215-609-9	Huber N375
215-609-9	Huber N539
215-609-9	Huber N550
215-609-9	Huber N650
215-609-9	Huber N660
215-609-9	Huber N683
215-609-9	Huber N762
215-609-9	Huber N774
215-609-9	Huber N787
215-609-9	Huber N990
215-609-9	Huber S212
215-609-9	Huber S315
215-609-9	Ketjenblack® EC-300 J
215-609-9	Ketjenblack® EC-310 NW
215-609-9	Ketjenblack® EC-600 JD
215-609-9	Ketjenblack® ED-600 JD
215-609-9	Mitsubishi Carbon Black
215-609-9	Mogul® L
215-609-9	Monarch® 120
215-609-9	Monarch® 700
215-609-9	Monarch® 800
215-609-9	Monarch® 880
215-609-9	Monarch® 900
215-609-9	Monarch® 1000
215-609-9	Monarch® 1100
215-609-9	MPC Channel Black TR 354
215-609-9	Novacarb
215-609-9	Printex P
215-609-9	Regal® 300
215-609-9	Regal® 300R
215-609-9	Regal® 330
215-609-9	Regal® 330R
215-609-9	Regal® 400
215-609-9	Regal® 400R
215-609-9	Regal® 660
215-609-9	Regal® 660R
215-609-9	S160 Beads and Powd
215-609-9	Sable™ 3
215-609-9	Sable™ 5
215-609-9	Sable™ 15
215-609-9	Sable™ 17
215-609-9	Sterling® 1120
215-609-9	Sterling® 2320
215-609-9	Sterling® 3420
215-609-9	Sterling® 4620
215-609-9	Sterling® 5550
215-609-9	Sterling® 5630
215-609-9	Sterling® 6630

EINECS	Trade name
215-609-9	Sterling® 6640
215-609-9	Sterling® 6740
215-609-9	Sterling® 8860
215-609-9	Sterling® C
215-609-9	Sterling® NS
215-609-9	Sterling® NS-1
215-609-9	Sterling® SO
215-609-9	Sterling® V
215-609-9	Sterling® VH
215-609-9	Sterling® XC-72
215-609-9	Thermax® Floform® N-990
215-609-9	Thermax® Floform® Ultra Pure N-990
215-609-9	Thermax® Powder N-991
215-609-9	Thermax® Powder Ultra Pure N-991
215-609-9	Thermax® Stainless
215-609-9	Thermax® Stainless Floform® N-907
215-609-9	Thermax® Stainless Powder N-908
215-609-9	Thermax® Stainless Powder Ultra Pure N-908
215-609-9	Vulcan® 2H
215-609-9	Vulcan® 3
215-609-9	Vulcan® 3H
215-609-9	Vulcan® 4H
215-609-9	Vulcan® 6
215-609-9	Vulcan® 6LM
215-609-9	Vulcan® 7H
215-609-9	Vulcan® 8H
215-609-9	Vulcan® 9
215-609-9	Vulcan® 9A32
215-609-9	Vulcan® 10H
215-609-9	Vulcan® C
215-609-9	Vulcan® J
215-609-9	Vulcan® K
215-609-9	Vulcan® M
215-609-9	Vulcan® P
215-609-9	Vulcan® XC72
215-609-9	Vulcan® XC72R
215-654-4	Monostearyl Citrate (MSC)
215-661-2	Aztec® AAP-LA-M2
215-661-2	Aztec® AAP-NA-2
215-661-2	Aztec® AAP-SA-3
215-661-2	Aztec® MEKP-HA-1
215-661-2	Aztec® MEKP-HA-2
215-661-2	Aztec® MEKP-LA-2
215-661-2	Aztec® MEKP-SA-2
215-661-2	Cadox® HBO-50
215-661-2	Cadox® L-30
215-661-2	Cadox® L-50
215-661-2	Cadox® M-105
215-661-2	Lupersol DDM-9
215-661-2	Lupersol Delta-3
215-661-2	Lupersol Delta-X-9
215-661-2	Lupersol DHD-9
215-661-2	Quickset® Extra
215-661-2	Quickset® Super
215-661-2	Superox® 46-747
215-661-2	Superox® 46-748
215-661-2	Superox® 702
215-661-2	Superox® 709
215-661-2	Superox® 710
215-661-2	Superox® 711
215-661-2	Superox® 712
215-661-2	Superox® 713
215-661-2	Superox® 730
215-661-2	Superox® 732
215-661-2	Superox® 739
215-661-2	USP®-240
215-664-9	Disponil SMS 100 F1
215-664-9	Ethylan® GS60
215-664-9	Glycomul® S
215-665-4	Ethylan® GO80
215-665-4	Glycomul® O
215-665-4	S-Maz® 80
215-691-6	A-2
215-691-6	A-201
215-691-6	Aluminum Oxide C
215-710-8	Micro-Cel® C
215-951-9	Morflex® 1129
216-087-5	FC-24
216-087-5	Fluorad® FC-24
216-472-8	ADK STAB CA-ST
216-472-8	Afco-Chem CS
216-472-8	Afco-Chem CS-1
216-472-8	Afco-Chem CS-S
216-472-8	Akrochem® Calcium Stearate
216-472-8	Akrochem® P 4000
216-472-8	Akrochem® P 4100
216-472-8	Akrodip™ CS-400
216-472-8	Cecavon CA 30
216-472-8	Coad 10
216-472-8	Cometals Calcium Stearate
216-472-8	Hallcote® CSD
216-472-8	Hallcote® CaSt 50
216-472-8	Haro® Chem CGD
216-472-8	Haro® Chem CGL
216-472-8	Haro® Chem CGN
216-472-8	Haro® Chem CPR-2
216-472-8	Haro® Chem CPR-5
216-472-8	Hy Dense Calcium Stearate HP Gran.
216-472-8	Hy Dense Calcium Stearate RSN Powd.
216-472-8	HyTech RSN 11-4
216-472-8	HyTech RSN 248D
216-472-8	Interstab® CA-18-1
216-472-8	Mathe Calcium Stearate
216-472-8	Miljac Calcium Stearate
216-472-8	Nuodex S-1421 Food Grade
216-472-8	Nuodex S-1520 Food Grade
216-472-8	Petrac® Calcium Stearate CP-11
216-472-8	Petrac® Calcium Stearate CP-11 LS
216-472-8	Petrac® Calcium Stearate CP-11 LSG
216-472-8	Petrac® Calcium Stearate

EINECS	Trade name
	CP-22G
216-472-8	Plastolube
216-472-8	Quikote™ C
216-472-8	Quikote™ C-LD
216-472-8	Quikote™ C-LD-A
216-472-8	Quikote™ C-LM
216-472-8	Quikote™ C-LMLD
216-472-8	Synpro® Calcium Stearate 12B
216-472-8	Synpro® Calcium Stearate 15
216-472-8	Synpro® Calcium Stearate 15F
216-472-8	Synpro® Calcium Stearate 114-36
216-472-8	TR-Calcium Stearate
216-472-8	Witco® Calcium Stearate A
216-472-8	Witco® Calcium Stearate F
216-472-8	Witco® Calcium Stearate FP
216-700-6	Empigen® OB
216-700-6	Rhodamox® LO
216-751-4	Weston® TLTTP
216-940-1	Jeffcat ZR-70
217-421-2	BLS™ 531
217-421-2	Cyasorb® UV 531
217-421-2	Hostavin® ARO 8
217-421-2	Lankromark® LE285
217-421-2	Lowilite® 22
217-421-2	Mark® 1413
217-421-2	Uvasorb 3C
217-421-2	UV-Chek® AM-300
217-421-2	UV-Chek® AM-301
217-421-2	Uvinul® 408
217-421-2	Uvinul® 3008
217-430-1	Rhodapon® OS
217-752-2	Eastman® MTBHQ
217-752-2	Sustane® TBHQ
217-752-2	Tenox® TBHQ
217-983-9	Dynasylan® MTES
217-983-9	Petrarch® M9050
218-216-0	Akrochem® Antioxidant 1076
218-216-0	BNX® 1076, 1076G
218-216-0	Dovernox 76
218-216-0	Irganox® 1076
218-216-0	Lankromark® LE384
218-216-0	Naugard® 76
218-216-0	Ralox® 530
218-216-0	Ultranox® 276
218-254-8	V-40
218-487-5	Stabilizer 7000
218-793-9	Akyposal ALS 33
218-793-9	Calfoam NLS-30
218-793-9	Carsonol® ALS
218-793-9	Carsonol® ALS-S
218-793-9	Carsonol® ALS Special
218-793-9	Rhodapon® L-22
218-793-9	Rhodapon® L-22/C
218-793-9	Rhodapon® L-22HNC
218-901-4	Dimodan LS Kosher
219-136-9	Kessco® PEG 300 ML
219-470-5	Lowilite® 55
219-470-5	Tinuvin® P
219-470-5	Topanex 100BT
219-470-5	Uvasorb SV
219-470-5	Viosorb 520
219-518-5	Haro® Chem ZSG
219-553-6	Epodil® 746
219-553-6	Heloxy® 116
219-784-2	Dow Corning® Z-6040
219-784-2	Dynasylan® GLYMO
219-784-2	Petrarch® G6720
219-784-2	Prosil® 5136
219-785-8	Dow Corning® Z-6030
219-785-8	Dynasylan® MEMO
219-785-8	Petrarch® M8550
219-785-8	Prosil® 248
219-787-9	Dow Corning® Z-6076
219-787-9	Dynasylan® CPTMO
219-787-9	Petrarch® C3300
219-976-6	V-601
220-045-1	Alkapol PEG 300
220-045-1	Carbowax® PEG 300
220-045-1	Hodag PEG 300
220-045-1	Pluracol® E300
220-045-1	Poly-G® 300
220-045-1	Teric PEG 300
220-135-0	V-501
220-219-7	Aerosol® TR-70
220-219-7	Geropon® BIS/SODICO-2
220-219-7	Hodag DTSS-70
220-219-7	Polirol TR/LNA
220-449-8	Dynasylan® VTMO
220-449-8	Petrarch® V4917
220-548-6	Ektasolve® EP
220-688-8	Ageflex FM-1
220-835-6	Staflex DOF
220-941-2	Dynasylan® OCTEO
220-941-2	Petrarch® O9835
221-070-0	V-50
221-109-1	Aerosol® MA-80I
221-188-2	Avirol® SA 4113
221-188-2	Rhodapon® TDS
221-279-7	Prox-onic LA-1/02
221-283-9	Rhodasurf® L-790
221-284-4	Prox-onic LA-1/09
221-286-5	Prox-onic LA-1/012
221-291-2	Weston® PNPG
221-304-1	Kemamide® B
221-356-5	Doverphos® 53
221-356-5	Weston® TLP
221-695-9	Great Lakes CD-75P™
221-695-9	Saytex® HBCD-LM
221-967-7	FR-521
221-967-7	FR-522
221-967-7	FR-1138
222-036-8	Saytex® BCL-462
222-273-7	Monawet SNO-35
222-700-7	Haro® Chem CBHG
222-823-6	Dellatol® BBS
222-823-6	Plasthall® BSA
222-823-6	Uniplex 214

EINECS	Trade name
222-884-9	Jayflex® L11P
222-884-9	PX-111
222-899-0	Deriphat® 160
223-276-6	Doverphos® S-680
223-276-6	Doverphos® S-686, S-687
223-276-6	Mark® 5060
223-276-6	Weston® 618F
223-276-6	Weston® 619F
223-383-8	Tinuvin® 327
223-445-4	Lowilite® 26
224-583-8	V-65
224-583-8	Vazo 52
224-588-5	Dynasylan® MTMO
224-588-5	Petrarch® M8500
224-588-5	Prosil® 196
224-772-5	Afco-Chem LIS
224-772-5	Mathe Lithium Stearate
224-772-5	Synpro® Lithium Stearate
224-772-5	Witco® Lithium Stearate 306
224-806-9	Cithrol EGMO N/E
225-091-6	Kodaflex® DOTP
225-166-3	Haro® Chem CSG
225-202-8	Doverphos® 213
225-202-8	Weston® DPP
225-214-3	Empicol® LQ33/T
225-268-8	Ebal
225-805-6	Dynasylan® CPTEO
225-805-6	Petrarch® C3292
225-856-4	Carbowax® PEG 400
225-856-4	Hodag PEG 400
225-856-4	Pluracol® E400
225-856-4	Pluracol® E400 NF
225-856-4	Poly-G® 400
225-856-4	Rhodasurf® E 400
225-856-4	Teric PEG 400
226-029-0	Uvinul® 3035
226-097-1	Ethosperse® LA-4
226-097-1	Prox-onic LA-1/04
226-097-1	Rhodasurf® L-4
226-191-2	Solricin® 435
226-191-2	Solricin® 535
226-312-9	Serdox NSG 400
227-335-7	Pationic® 930
227-335-7	Pationic® 930K
227-335-7	Pationic® 940
228-250-8	Uvinul® 3039
229-457-6	Zoldine® ZT-65
229-713-7	Amicure® DBU-E
229-713-7	Polycat® DBU
229-713-7	Polycat® SA-1
229-713-7	Polycat® SA-102
229-713-7	Polycat® SA-610/50
229-722-6	Akrochem® Antioxidant 1010
229-722-6	BNX® 1010, 1010G
229-722-6	Dovernox 10
229-722-6	Irganox® 1010
229-722-6	Naugard® 10
229-722-6	Ralox® 630
229-722-6	Ultranox® 210
229-859-1	Carbowax® PEG 600
229-859-1	Pluracol® E600
229-859-1	Pluracol® E600 NF
229-859-1	Poly-G® 600
229-859-1	Teric PEG 600
230-325-5	Cometals Aluminum Stearate
230-325-5	HyTech AX/603
230-325-5	HyTech T/351
230-325-5	Mathe 6T
230-325-5	Synpro® Aluminum Stearate 303
230-325-5	Witco® Aluminum Stearate 18
230-528-9	Kemamine® D-989
230-785-7	Empiphos 4KP
230-990-1	Armeen® M2-10D
230-990-1	Dama® 1010
231-314-8	Solricin® 135
231-493-2	Beta W 7
231-493-2	Gamma W8
231-545-4	Aerosil® 200
231-626-4	CTA
231-722-6	Chloropren-Faktis A
231-722-6	Faktis 10
231-722-6	Faktis 14
231-722-6	Faktis Asolvan, Asolvan T
231-722-6	Faktis Badenia C
231-722-6	Faktis HF Braun
231-722-6	Faktis RC 110, RC 111, RC 140, RC 141, RC 144
231-722-6	Naftopast® Schwefel-P
231-722-6	Octocure 456
231-722-6	Rubbermakers Sulfur
231-722-6	Struktol® SU 95
231-722-6	Struktol® SU 105
231-722-6	Struktol® SU 120
231-784-4	Bara-200C
231-784-4	Bara-200N
231-784-4	Bara-325C
231-784-4	Bara-325N
231-784-4	Barafoam 200
231-784-4	Barafoam 325
231-784-4	Barimite™
231-784-4	Barimite™ 200
231-784-4	Barimite™ 1525P
231-784-4	Barimite™ 4009P
231-784-4	Barimite™ UF
231-784-4	Barimite™ XF
231-784-4	Bartex® 10
231-784-4	Bartex® 65
231-784-4	Bartex® 80
231-784-4	Bartex® FG-2
231-784-4	Bartex® FG-10
231-784-4	Bartex® OWT
231-784-4	Barytes Microsupreme 2410/M
231-784-4	Barytes Supreme
231-784-4	Blanc fixe F
231-784-4	Blanc fixe micro®
231-784-4	Blanc fixe N
231-784-4	Cimbar™ 325
231-784-4	Cimbar™ 1025P
231-784-4	Cimbar™ 1536P

EINECS	Trade name
231-784-4	Cimbar™ 3508P
231-784-4	Cimbar™ CF
231-784-4	Cimbar™ UF
231-784-4	Cimbar™ XF
231-784-4	G-50 Barytes™
231-784-4	No. 22 Barytes™
231-868-0	Fascat® 2004
231-868-0	Fascat® 2005
231-955-3	Lonza KS 44
231-955-3	Lonza T 44
232-273-9	GTO 80
232-273-9	GTO 90
232-273-9	GTO 90E
232-273-9	NS-20
232-273-9	Sunyl® 80
232-273-9	Sunyl® 80 RBD
232-273-9	Sunyl® 80 RBD ES
232-273-9	Sunyl® 80 RBWD
232-273-9	Sunyl® 80 RBWD ES
232-273-9	Sunyl® 90
232-273-9	Sunyl® 90 RBD
232-273-9	Sunyl® 90 RBWD
232-273-9	Sunyl® 90E RBWD
232-273-9	Sunyl® 90E RBWD ES 1016
232-273-9	Sunyl® HS 500
232-290-1	Koster Keunen Ceresine
232-290-1	Ross Ceresine Wax
232-292-2	Castorwax® MP-70
232-292-2	Castorwax® MP-80
232-292-2	Glycolube® CW-1
232-299-0	RS-80
232-306-7	Cordon NU 890/75
232-306-7	Monosulf
232-313-5	Ross Montan Wax
232-313-5	Sicolub® E
232-313-5	Sicolub® OP
232-315-6	Advawax® 165
232-315-6	Akrowax 130
232-315-6	Akrowax 145
232-315-6	Hostalub® XL 165FR
232-315-6	Hostalub® XL 165P
232-315-6	Loxiol® G 22
232-315-6	Loxiol® G 23
232-315-6	Loxiol® HOB 7138
232-315-6	Loxiol® HOB 7169
232-315-6	Octowax 321
232-315-6	Petrac® 165
232-315-6	Sunolite® 130
232-315-6	Sunolite® 160
232-347-0	Koster Keunen Candelilla
232-360-1	Disponil SSO 100 F1
232-373-2	Fonoline® White
232-373-2	Fonoline® Yellow
232-373-2	Penreco Amber
232-373-2	Penreco Snow
232-373-2	Protopet® White 1S
232-373-2	Protopet® Yellow 2A
232-388-4	Solricin® 235
232-391-0	ADK CIZER O-130P
232-391-0	Drapex® 6.8
232-391-0	Epoxol 7-4
232-391-0	Estabex® 138-A
232-391-0	Estabex® 2307
232-391-0	Estabex® 2307 DEOD
232-391-0	Flexol® Plasticizer EPO
232-391-0	Lankroflex® GE
232-391-0	Nuoplaz® 849
232-391-0	Paraplex® G-60
232-391-0	Paraplex® G-62
232-391-0	Peroxidol 780
232-391-0	Plas-Chek® 775
232-391-0	Plasthall® ESO
232-391-0	Plastoflex® 2307
232-391-0	Plastolein® 9232
232-391-0	PX-800
232-399-4	Ross Carnauba Wax
232-447-4	Arquad® T-27W
232-479-9	Resinall 610
232-479-9	Uni-Rez® 3020
232-479-9	Uni-Tac® R99
232-479-9	Uni-Tac® R100
232-479-9	Uni-Tac® R100-Light
232-479-9	Uni-Tac® R100RM
232-479-9	Uni-Tac® R101
232-479-9	Uni-Tac® R102
232-479-9	Uni-Tac® R106
232-479-9	Uni-Tac® R112
232-479-9	Zonester® 100
232-674-9	Fibra-Cel® BH-70
232-674-9	Fibra-Cel® BH-200
232-674-9	Fibra-Cel® CBR-18
232-674-9	Fibra-Cel® CBR-40
232-674-9	Fibra-Cel® CBR-41
232-674-9	Fibra-Cel® CBR-50
232-674-9	Fibra-Cel® CC-20
232-674-9	Fibra-Cel® SW-8
232-674-9	Santoweb® D
232-674-9	Santoweb® DX
232-674-9	Santoweb® H
232-674-9	Santoweb® W
233-042-5	Dynasylan® TCS
233-257-4	Manganese Violet
233-293-0	Kessco® PEG 200 MO
233-471-8	Flamtard Z10
233-471-8	Flamtard Z15
233-520-3	Chemeen 18-2
233-520-3	Chemstat® 192
233-520-3	Chemstat® 192/NCP
233-520-3	Chemstat® 273-E
233-520-3	Prox-onic MS-02
234-319-3	Alcamizer 1
234-319-3	Alcamizer 2
234-319-3	Alcamizer 4
234-319-3	Alcamizer 4-2
235-527-7	Aztec® CHP-50-P1
235-527-7	Aztec® CHP-90-W1
235-527-7	Aztec® CHP-HA-1
235-527-7	Cyclonox® BT-50
236-239-4	Diak No. 4
236-671-3	Zinc Omadine® 48% Fine

EINECS	Trade name
	Particle Disp.
236-671-3	Zinc Omadine® 48% Std. Disp.
236-671-3	Zinc Omadine® Powd.
236-675-5	AT-1
236-675-5	Kronos® 1000
236-675-5	Kronos® 2020
236-675-5	Kronos® 2073
236-675-5	Kronos® 2081
236-675-5	Kronos® 2101
236-675-5	Kronos® 2160
236-675-5	Kronos® 2200
236-675-5	Kronos® 2210
236-675-5	Kronos® 2220
236-675-5	Kronos® 2230
236-675-5	Kronos® 2310
236-675-5	RL-90
236-675-5	Ti-Pure® R-100
236-675-5	Ti-Pure® R-101
236-675-5	Ti-Pure® R-102
236-675-5	Ti-Pure® R-103
236-675-5	Ti-Pure® R-900
236-675-5	Ti-Pure® R-902
236-675-5	Ti-Pure® R-960
236-675-5	Titanium Dioxide P25
236-740-8	Perkadox® AMBN
236-740-8	V-59
236-753-9	Weston® PTP
237-511-5	Dynasylan® AMMO
237-511-5	Petrarch® A0800
238-479-5	Aerosol® 18
238-479-5	Lankropol® ODS/LS
238-479-5	Lankropol® ODS/PT
238-479-5	Octosol A-18
238-877-9	ABT-2500®
238-877-9	Artic Mist
238-877-9	Beaverwhite 200
238-877-9	Cimflx 606
238-877-9	Cimpact 600
238-877-9	Cimpact 610
238-877-9	Cimpact 699
238-877-9	Cimpact 705
238-877-9	Cimpact 710
238-877-9	Glacier 200
238-877-9	I.T. 3X
238-877-9	I.T. 5X
238-877-9	I.T. 325
238-877-9	I.T. FT
238-877-9	I.T. X
238-877-9	Jet Fil® 200
238-877-9	Jet Fil® 350
238-877-9	Jet Fil® 500
238-877-9	Jet Fil® 575C
238-877-9	Jet Fil® 625C
238-877-9	Jet Fil® 700C
238-877-9	Luzenac 8170
238-877-9	Magsil Diamond Talc 200 mesh
238-877-9	Magsil Diamond Talc 350 mesh
238-877-9	Magsil Star Talc 200 mesh
238-877-9	Magsil Star Talc 350 mesh
238-877-9	Microbloc®
238-877-9	MicroPflex 1200
238-877-9	Microtalc® MP10-52
238-877-9	Microtalc® MP12-50
238-877-9	Microtalc® MP15-38
238-877-9	Microtalc® MP25-38
238-877-9	Microtalc® MP44-26
238-877-9	Microtuff® 325F
238-877-9	Microtuff® 1000
238-877-9	Microtuff® F
238-877-9	Mistron CB
238-877-9	Mistron PXL®
238-877-9	Mistron Vapor
238-877-9	Mistron Vapor® R
238-877-9	Nicron 325
238-877-9	Nicron 400
238-877-9	Nytal® 100
238-877-9	Nytal® 200
238-877-9	Nytal® 300
238-877-9	Nytal® 400
238-877-9	Polytalc 262
238-877-9	PolyTalc™ 445
238-877-9	Select-A-Sorb Powdered, Compacted
238-877-9	Silverline 400
238-877-9	Silverline 665
238-877-9	Stellar 500
238-877-9	Stellar 510
238-877-9	Super Microtuff F
238-877-9	Supra EF
238-877-9	Supra EF A
238-877-9	Techfil 7599
238-877-9	Ultratalc™ 609
238-877-9	Vantalc® 6H
238-877-9	Vertal 92
238-877-9	Vertal 97
238-877-9	Vertal 200
238-877-9	Vertal 310
238-877-9	Vertal 710
238-877-9	Vertal C-2
239-341-7	Q-1301
239-593-8	V-70
241-640-2	W.G.S. Synaceti 116 NF/USP
242-960-5	Estol 1445
243-885-0	Great Lakes PHT4-Diol™
243-885-0	Saytex® RB-79
244-063-4	Versene Diammonium EDTA
244-085-4	VA-061
244-492-7	H-100
244-492-7	H-105
244-492-7	H-109
244-492-7	H-600
244-492-7	H-800
244-492-7	H-900
244-492-7	H-990
244-492-7	Martinal® 103 LE
244-492-7	Martinal® 104 LE
244-492-7	Martinal® 107 LE

EINECS	Trade name
244-492-7	Martinal® OL-104
244-492-7	Martinal® OL-104/C
244-492-7	Martinal® OL-104 LE
244-492-7	Martinal® OL-104/S
244-492-7	Martinal® OL-107
244-492-7	Martinal® OL-107/C
244-492-7	Martinal® OL-107 LE
244-492-7	Martinal® OL-111 LE
244-492-7	Martinal® OL/Q-107
244-492-7	Martinal® OL/Q-111
244-492-7	Martinal® ON
244-492-7	Martinal® ON-310
244-492-7	Martinal® ON-313
244-492-7	Martinal® ON-320
244-492-7	Martinal® ON-920/V
244-492-7	Martinal® ON-4608
244-492-7	Martinal® OS
244-492-7	Martinal® OX
245-629-3	Aerosol® A-196-40
245-629-3	Gemtex 691-40
245-629-3	Octosol TH-40
245-629-3	Rewopol® SBDC 40
246-140-8	AlNel A-100
246-140-8	AlNel A-200
246-140-8	AlNel AG
246-563-8	Sustane® BHA
246-563-8	Tenox® BHA
246-614-4	Albrite® TIOP
246-614-4	Weston® TIOP
246-680-4	Calsoft F-90
246-680-4	Calsoft L-60
246-680-4	Ufasan 35
246-680-4	Witconate™ 1250
246-839-8	Eltesol® 4200
246-839-8	Eltesol® XA65
246-839-8	Eltesol® XA90
246-839-8	Eltesol® XA/M65
246-839-8	Manro FCM 90LV
246-839-8	Reworyl® X 65
246-839-8	XSA 80
246-839-8	XSA 90
246-929-7	Pationic® 920
246-998-3	Doverphos® 6
246-998-3	Weston® TDP
247-036-5	Arylan® SO60 Acid
247-098-3	Doverphos® 7
247-098-3	Weston® PDDP
247-144-2	Priolube 1409
247-500-7	Kathon® LX
247-556-2	Rhodacal® 330
247-557-8	Nansa® EVM50
247-568-8	Disponil SMP 100 F1
247-569-3	Ethylan® GT85
247-658-7	Doverphos® 9
247-658-7	Weston® ODPP
247-759-6	ADK STAB 329
247-759-6	ADK STAB 1178
247-759-6	ADK STAB SC-102
247-759-6	Doverphos® 4
247-759-6	Doverphos® 4 Powd.
247-759-6	Lankromark® LE109
247-759-6	Naugard® P
247-759-6	Weston® 399
247-759-6	Weston® 399B
247-759-6	Weston® TNPP
247-759-6	Wytox® 312
247-777-4	Weston® DPDP
247-779-5	Weston® 600
247-952-5	Blendex® 340
247-952-5	Ultranox® 626
247-952-5	Ultranox® 626A
247-952-5	Ultranox® 627A
247-977-1	Jayflex® DIDP
247-977-1	Nuoplaz® DIDP
247-977-1	Plasthall® DIDP-E
247-977-1	PX-120
248-294-1	Prox-onic NP-010
248-296-2	Polystep® A-15-30K
248-299-9	Monoplex® DDA
248-299-9	Nuoplaz® DIDA
248-299-9	Plasthall® DIDA
248-299-9	Staflex DIDA
248-406-9	Manro TDBS 60
248-406-9	Puxol CB-22
248-586-9	Cithrol GDL N/E
248-655-3	VA-044
248-666-3	BM-951
248-762-5	Prox-onic NP-1.5
249-079-5	Jayflex® DINP
249-079-5	Plasthall® DINP
249-079-5	PX-139
250-151-3	Hamposyl® M-30
250-696-7	Lexolube® B-109
251-087-9	FR-1208
251-087-9	Great Lakes DE-79™
251-087-9	Saytex® 111
251-118-6	Saytex® BT-93®
251-118-6	Saytex® BT-93W
251-136-4	Millithix® 925
251-646-7	Adimoll® DN
251-646-7	Jayflex® DINA
251-646-7	PX-239
251-646-7	Staflex DINA
252-011-7	Mazol® PGO-104
252-287-4	Weston® DOPI
253-149-0	Adol® 52
253-149-0	Alfol® 16
253-149-0	Epal® 16NF
253-149-0	Loxiol® VPG 1743
253-211-7	Weston® 430
253-452-8	Aerosol® A-268
253-458-0	Cithrol EGML N/E
256-120-0	Emcol® K8300
256-120-0	Sole Terge 8
258-007-1	Hamposyl® M
258-207-9	BLS™ 1770
258-207-9	Lowilite® 77
258-207-9	Tinuvin® 770
258-250-3	Saytex® BN-451
259-627-5	Idex™-400

EINECS	Trade name
259-627-5	Idex™-1000
261-526-6	Saytex® 120
262-774-8	Ken-React® KR TTS
262-976-6	Crodamine 1.HT
262-976-6	Kemamine® P-970
262-976-6	Kemamine® P-970D
262-976-6	Noram SH
262-977-1	Kemamine® P-650D
262-978-7	Industrene® 325
262-978-7	Industrene® 328
262-990-2	Kemamine® T-6501
262-991-8	Kemamine® T-9701
263-017-4	Kemamine® T-9972D
263-020-0	Kemamine® T-6502D
263-022-1	Kemamine® T-9702D
263-032-6	Grindtek MOP 90
263-038-9	Arquad® C-50
263-038-9	Chemquat C/33W
263-087-6	Arquad® 2C-75
263-107-3	Acintol® DFA
263-107-3	Acintol® EPG
263-107-3	Acintol® FA-1
263-107-3	Acintol® FA-1 Special
263-107-3	Acintol® FA-2
263-107-3	Acintol® FA-3
263-107-3	Acofor
263-107-3	Westvaco® 1480
263-123-0	Armid® HT
263-125-1	Crodamine 1.T
263-125-1	Kemamine® P-974D
263-125-1	Noram S
263-129-3	Industrene® 143
263-130-9	Petrac® PHTA
263-144-5	Arizona DR-25
263-144-5	Arizona DRS-43
263-144-5	Arizona DRS-44
263-144-5	Burez K25-500D
263-144-5	Burez K80-500
263-144-5	Burez K80-500D
263-144-5	Burez K80-2500
263-144-5	Burez NA 70-500
263-144-5	Burez NA 70-500D
263-144-5	Burez NA 70-2500
263-144-5	Diprosin N-70
263-144-5	Dresinate® 731
263-144-5	Dresinate® 81
263-144-5	Dresinate® X
263-144-5	Dresinate® XX
263-144-5	Westvaco® Resin 790
263-144-5	Westvaco® Resin 795
263-163-9	Empilan® CDE
263-163-9	Empilan® CDX
263-163-9	Ethylan® LD
263-163-9	Ethylan® LDS
263-170-7	Miramine® C
263-180-1	Aromox® C/12
263-180-1	Aromox® C/12-W
263-190-6	Deriphat® 154
263-193-2	Hamposyl® C-30
263-195-3	Dinoram C

EINECS	Trade name
263-195-3	Duomeen® C
263-195-3	Kemamine® D-650
264-038-1	Be Square® 175
264-038-1	Be Square® 185
264-038-1	Be Square® 195
264-038-1	Fortex®
264-038-1	Mekon® White
264-038-1	Multiwax® 180-M
264-038-1	Multiwax® W-835
264-038-1	Multiwax® X-145A
264-038-1	Petrolite® C-700
264-038-1	Petrolite® C-1035
264-038-1	Starwax® 100
264-038-1	Ultraflex®
264-038-1	Victory®
266-218-5	Servoxyl VPQZ 9/100
266-231-6	Foamphos NP-9
267-006-5	Alfol® 1218 DCBA
267-006-5	Loxiol® VPG 1496
267-006-5	Nafol® 1218
267-009-1	Alfol® 1418 DDB
267-009-1	Alfol® 1418 GBA
267-019-6	Epal® 12/85
267-019-6	Epal® 1214
268-130-2	Hamposyl® AL-30
268-665-1	Weston® 494
268-938-5	Rhodamox® CAPO
269-657-0	Industrene® 225
269-788-3	Acintol® 208
269-788-3	Plasthall® 100
269-790-4	Alfol® 1416 GC
270-156-4	Hamposyl® C
270-407-8	Polystep® A-18
270-407-8	Rhodacal® 301-10
270-407-8	Rhodacal® 301-10P
270-407-8	Rhodacal® A-246L
270-407-8	Witconate™ AOS
270-407-8	Witconate™ AOS-EP
270-416-7	Kemamine® D-974
270-664-6	Hoechst Wax LP
270-664-6	Hoechst Wax S
270-664-6	Hoechst Wax SW
270-664-6	Hoechst Wax UL
271-696-6	Kemamine® D-970
271-867-2	Akrochem® Antioxidant 12
271-867-2	Ralox® LC
271-867-2	Santowhite® ML
271-867-2	Wingstay® L HLS
272-021-5	Armeen® Z
272-490-6	Alfol® 1216
272-493-2	Gulftene® 14
275-521-1	Uniplex 310
279-420-3	Alfol® 1214
279-420-3	Alfol® 1214 GC
279-420-3	Alfol® 1216 CO
279-420-3	Alfol® 1412
279-498-9	Doverphos® 11
279-498-9	Weston® THOP
279-499-4	Doverphos® 12
279-499-4	Weston® DHOP

EINECS	Trade name
283-066-5	Alfol® 1014 CDC
283-066-5	Nafol® 1014
284-868-8	Priolube 1414
286-344-4	ADK STAB NA-11
287-621-2	Nafol® 810 D
288-117-5	Alfol® 1012 HA
290-580-3	Sicolub® DSP
290-850-0	Rewopol® B 1003
292-011-4	GE 100
292-951-5	Priolube 1458
295-625-0	Paricin® 8
296-473-8	Bilt-Plates® 156
296-473-8	Dixie® Clay
296-473-8	Fiberfrax® 6000 RPS
296-473-8	Fiberfrax® 6900-70
296-473-8	Fiberfrax® 6900-70S
296-473-8	Fiberfrax® EF-119
296-473-8	Fiberfrax® EF 122S
296-473-8	Fiberfrax® EF 129
296-473-8	Fiberfrax® Milled Fiber
296-473-8	Fiberkal™
296-473-8	Fiberkal™ FG
296-473-8	Huber 35
296-473-8	Huber 40C
296-473-8	Huber 65A
296-473-8	Huber 70C
296-473-8	Huber 80
296-473-8	Huber 80B
296-473-8	Huber 80C
296-473-8	Huber 90
296-473-8	Huber 90B
296-473-8	Huber 90C
296-473-8	Huber 95
296-473-8	Huber HG
296-473-8	Huber HG90
296-473-8	Hydrite 121-S
296-473-8	Langford® Clay
296-473-8	McNamee® Clay
296-473-8	Par® Clay
296-473-8	Polyplate 90
296-473-8	Polyplate 852
296-473-8	Polyplate P
296-473-8	Polyplate P01
306-120-2	Weston® 439
307-145-1	Nafol® 20+
310-127-6	Sillikolloid P 87
310-127-6	Sillitin N 82
310-127-6	Sillitin N 85
310-127-6	Sillitin V 85
310-127-6	Sillitin Z 86
310-127-6	Sillitin Z 89

EINECS Number-to-Chemical Cross-Reference

EINECS	Chemical
200-268-0	Tributyltin oxide
200-313-4	Stearic acid
200-315-5	Urea
200-338-0	Polypropylene glycol
200-449-4	Edetic acid
200-529-9	Calcium disodium EDTA
200-573-9	Tetrasodium EDTA
200-578-6	Alcohol
200-580-7	Acetic acid
200-618-2	Benzoic acid
200-659-6	Methyl alcohol
200-661-7	Isopropyl alcohol
200-664-3	Dimethyl sulfoxide
200-679-5	Dimethyl formamide
200-712-3	Salicylic acid
200-741-1	p-Toluenesulfonamide
200-751-6	Butyl alcohol
200-756-3	Trichloroethane
200-815-3	Polyethylene
200-889-7	t-Butyl alcohol
200-901-0	Dimethyldichlorosilane
200-908-9	t-Amyl alcohol
200-915-7	t-Butyl hydroperoxide
200-917-8	Vinyltrichlorosilane
201-039-8	Dibutyltin dilaurate
201-063-9	Trimethylolethane
201-066-5	Acetyl triethyl citrate
201-067-0	Acetyl tributyl citrate
201-069-1	Citric acid
201-070-7	Triethyl citrate
201-071-2	Tributyl citrate
201-081-7	Vinyltriethoxysilane
201-083-8	Tetraethoxysilane
201-094-8	Cetethyl morpholinium ethosulfate
201-114-5	Triethyl phosphate
201-116-6	Trioctyl phosphate
201-122-9	Tributoxyethyl phosphate
201-126-0	Isophorone
201-132-3	2,2′-Azobisisobutyronitrile
201-148-0	Isobutyl alcohol
201-159-0	Methyl ethyl ketone
201-166-9	Trichloroethane
201-176-3	Propionic acid
201-177-9	Acrylic acid
201-202-3	Methacrylamide
201-222-2	Diamylhydroquinone

EINECS	Chemical
201-236-9	Tetrabromobisphenol A
201-245-8	Bisphenol A
201-255-2	Benzene sulfonyl hydrazide
201-275-1	Ethyl toluenesulfonamide
201-545-9	Dicyclohexyl phthalate
201-548-5	Butyl cyclohexyl phthalate
201-550-6	Diethyl phthalate
201-557-4	Dibutyl phthalate
201-559-5	Dihexyl phthalate
201-562-1	Butyl octyl phthalate
201-607-5	Phthalic anhydride
201-622-7	Butyl benzyl phthalate
201-624-8	n-Butyl phthalyl-n-butyl glycolate
201-800-4	PVP
201-800-4	N-Vinyl-2-pyrrolidone
201-807-2	2-t-Butylphenol
201-841-8	Di-t-butylhydroquinone
201-852-8	o-Isopropylphenol
201-966-8	α-Methylnaphthalene
202-013-9	2,4,6-Tris (dimethylamino-methyl) phenol
202-046-9	Decahydronaphthalene
202-075-7	6-Ethoxy-1,2-dihydro-2,2,4-trimethylquinoline
202-196-5	Phenothiazine
202-268-6	o-Tolyl biguanide
202-327-6	Benzoyl peroxide
202-340-7	Dipropylene glycol dibenzoate
202-443-7	Toluhydroquinone
202-473-0	Allyl methacrylate
202-495-0	Thioglycerin
202-506-9	Ethylene thiourea
202-509-5	Butyrolactone
202-510-0	Ethylene carbonate
202-532-0	2,4-Di-t-butylphenol
202-607-8	Tetraethylthiuram disulfide
202-608-3	Lauroyl sarcosine
202-617-2	Ethylene glycol dimethacrylate
202-625-6	Tetrahydrofurfuryl alcohol
202-626-1	Furfuryl alcohol
202-627-7	Furfural
202-637-1	Benzenesulfonamide
202-638-7	Benzenesulfonic acid
202-679-0	4-t-Butylphenol
202-691-6	Phenol sulfonic acid
202-715-5	N,N-Dimethyl-N-cyclohexyl-

EINECS	Chemical
	amine
202-805-4	N,N-Dimethyl-p-toluidine
202-845-2	Diethylaminoethanol
202-885-0	Ethyl morpholine
202-905-8	Hexamethylene tetramine
202-908-4	Triphenyl phosphite
202-935-1	Glyceryl triacetyl ricinoleate
202-936-7	Triallylcyanurate
202-943-5	Cyclohexyl methacrylate
202-951-9	N-Phenyl-p-phenylenediamine
202-974-4	4,4′-Methylene dianiline
203-002-1	Diphenylguanidine
203-003-7	N,N′-Diphenylurea
203-004-2	N,N′-Diphenylthiourea
203-041-4	Tetrahydroxypropyl ethylene-diamine
203-049-8	Triethanolamine
203-051-9	Triacetin
203-090-1	Dioctyl adipate
203-149-1	N-Benzyldimethylamine
203-180-0	p-Toluene sulfonic acid
203-197-3	Hydroquinone bis(2-hydroxyethyl) ether
203-234-3	2-Ethylhexanol
203-246-9	p-Tolualdehyde
203-268-9	1,4-Cyclohexanedimethanol
203-299-8	Methyl acetoacetate
203-308-5	1,3-Diethylthiourea
203-328-4	Dibutyl maleate
203-349-9	Dicapryl adipate
203-350-4	Dibutyl adipate
203-358-8	PEG-4 stearate
203-363-5	PEG-2 stearate
203-364-0	PEG-2 oleate
203-366-1	Hydroxystearic acid
203-369-8	Glycol ricinoleate
203-419-9	Dimethyl succinate
203-431-4	Dimethyl sebacate
203-441-9	Glycidyl methacrylate
203-442-4	Allyl glycidyl ether
203-449-2	1-Butene
203-450-8	1,3-Butadiene
203-473-3	Glycol
203-473-3	Polyethylene glycol
203-489-0	Hexylene glycol
203-508-2	Distearyldimonium chloride
203-537-0	3-Mercaptopropionic acid
203-539-1	Methoxyisopropanol
203-542-8	Dimethylethanolamine
203-571-6	Maleic anhydride
203-572-1	Propylene carbonate
203-584-7	m-Phenylenediamine
203-585-2	Resorcinol
203-618-0	Isocyanuric acid
203-620-1	Diisobutyl ketone
203-625-9	Toluene
203-629-0	Cyclohexylamine
203-631-1	Cyclohexanone
203-640-0	p-Methyl morpholine
203-663-6	PEG-2 distearate
203-672-5	Dibutyl sebacate
203-674-6	1,3-Dibutylthiourea
203-713-7	Methoxyethanol
203-716-3	Diethylamine
203-726-8	Tetrahydrofuran
203-733-6	Di-t-butyl peroxide
203-744-6	N,N,N′,N′-Tetramethyl-ethylenediamine
203-749-3	Oleoyl sarcosine
203-751-4	Isopropyl myristate
203-755-6	Ethylene distearamide
203-756-1	Ethylene dioleamide
203-777-6	Hexane
203-786-5	1,4-Butanediol
203-804-1	Ethoxyethanol
203-808-3	Piperazine
203-809-9	Pyridine
203-815-1	Morpholine
203-820-9	Diisopropanolamine
203-821-4	Dipropylene glycol
203-827-7	Glyceryl oleate
203-852-3	Hexyl alcohol
203-865-4	Diethylenetriamine
203-867-5	Aminoethylethanolamine
203-868-0	Diethanolamine
203-872-2	Diethylene glycol
203-875-9	Hexamethyleneimine
203-881-1	Ethylene glycol diacetate
203-883-2	Stearamide MEA
203-886-9	Glycol stearate
203-892-1	Octane
203-893-7	Octene-1
203-905-0	Butoxyethanol
203-906-6	Methoxydiglycol
203-911-3	Methyl laurate
203-917-6	Caprylic alcohol
203-919-7	Ethoxydiglycol
203-924-4	Diethylene glycol dimethyl ether
203-927-0	Laurtrimonium chloride
203-928-6	Cetrimonium chloride
203-929-1	Steartrimonium chloride
203-935-4	Isopropyl oleate
203-938-0	Decanoyl chloride
203-941-7	Lauroyl chloride
203-943-8	Dimethyl lauramine
203-950-6	Triethylenetetramine
203-953-2	Triethylene glycol
203-956-9	n-Decyl alcohol
203-961-6	Butoxydiglycol
203-968-4	Dodecene-1
203-982-0	Lauryl alcohol
203-984-1	n-Dodecyl mercaptan
203-986-2	Tetraethylenepentamine
203-989-9	PEG-4
203-990-4	Methyl stearate
203-992-5	Methyl oleate
203-997-2	Dimethyl palmitamine
204-000-3	Myristyl alcohol
204-001-9	Diethylene glycol dibutyl ether

EINECS	Chemical
204-002-4	Dimethyl myristamine
204-007-1	Oleic acid
204-009-2	Erucamide
204-012-9	Octadecene-1
204-015-5	Oleamine
204-017-6	Stearyl alcohol
204-065-8	Dimethyl ether
204-077-3	Chlorendic anhydride
204-104-9	Pentaerythritol
204-110-1	Pentaerythrityl tetrastearate
204-112-2	Triphenyl phosphate
204-171-4	Tetrachlorophthalic anhydride
204-211-0	Dioctyl phthalate
204-259-2	Phenyl salicylate
204-278-6	2,4,6-Tribromophenol
204-294-3	Ditridecyl phthalate
204-295-9	n-Octyl, n-decyl phthalate
204-327-1	2,2′-Methylenebis (6-t-butyl-4-methylphenol)
204-337-6	Benzophenone
204-340-2	Tetrahydronaphthalene
204-368-5	Phenyldiethanolamine
204-393-1	Lauramide DEA
204-402-9	Benzyl benzoate
204-407-6	Diethylene glycol dibenzoate
204-442-7	BHA
204-469-4	Triethylamine
204-479-9	Benzethonium chloride
204-495-6	N,N-Dimethyl-m-toluidine
204-498-2	Propyl gallate
204-528-4	Triisopropanolamine
204-534-7	Triolein
204-557-2	Phenyl glycidyl ether
204-558-8	Dioctyl sebacate
204-589-7	Phenoxyethanol
204-614-1	Dilauryl thiodipropionate
204-617-8	Hydroquinone
204-620-4	Glyceryl-1-allyl ether
204-626-7	Diacetone alcohol
204-648-7	2-Pyrrolidone
204-650-8	Azodicarbonamide
204-658-1	n-Butyl acetate
204-661-8	1,4-Dioxane
204-664-4	Glyceryl stearate
204-666-5	Butyl stearate
204-669-1	Azelaic acid
204-673-3	Adipic acid
204-679-6	Hexamethylenediamine
204-690-6	Lauramine
204-693-2	Stearamide
204-694-8	Dimethyl stearamine
204-695-3	Stearamine
204-696-9	Carbon dioxide
204-707-7	Tetrakis (hydroxymethyl) phosphonium chloride
204-709-8	2-Amino-2-methyl-1-propanol
204-771-6	Sucrose acetate isobutyrate
204-783-1	Sulfolane
204-800-2	Tributyl phosphate
204-809-1	Tetramethyl decynediol
204-812-8	Sodium octyl sulfate
204-822-2	Potassium acetate
204-839-5	Diisobutyl sodium sulfosuccinate
204-876-7	Sodium dimethyldithiocarbamate hydrate
204-881-4	BHT
204-884-0	2,6-Di-t-butylphenol
205-011-6	Dimethyl phthalate
205-016-3	Diallyl phthalate
205-026-8	Benzophenone-8
205-027-3	Benzophenone-6
205-028-9	Benzophenone-2
205-029-4	Benzophenone-1
205-031-5	Benzophenone-3
205-087-0	Captan
205-183-2	Nitrosophenylhydroxylamine ammonium salt
205-223-9	Phenyl-β-naphthylamine
205-281-5	Sodium lauroyl sarcosinate
205-286-2	Tetramethylthiuram disulfide
205-351-5	Lauralkonium chloride
205-358-3	Disodium EDTA
205-388-7	TEA-lauryl sulfate
205-391-3	Pentasodium pentetate
205-393-4	Butyl acetyl ricinoleate
205-411-0	Aminoethylpiperazine
205-450-3	Diisobutyl adipate
205-455-0	Glyceryl ricinoleate
205-468-1	PEG-2 laurate
205-470-2	Ricinoleic acid
205-471-8	Methyl hydroxystearate
205-472-3	Methyl ricinoleate
205-480-7	Butyl acrylate
205-483-3	Ethanolamine
205-500-4	Ethyl acetate
205-516-1	Ethylacetoacetate
205-524-5	Dioctyl maleate
205-526-6	Glyceryl laurate
205-539-7	Stearoyl sarcosine
205-542-3	Propylene glycol laurate
205-559-6	Butyl oleate
205-563-8	Heptane
205-570-6	Lauryl methacrylate
205-571-1	Isopropyl palmitate
205-577-4	DEA-lauryl sulfate
205-582-1	Lauric acid
205-590-5	Potassium oleate
205-596-8	Palmitamine
205-597-3	Oleyl alcohol
205-633-8	Sodium bicarbonate
205-736-8	2-Mercaptobenzothiazole
205-758-8	Trisodium EDTA
205-769-8	Hydroquinone monomethyl ether
205-788-1	Sodium lauryl sulfate
205-999-9	Triethylene diamine
206-101-8	Aluminum distearate
206-103-9	Oleamide
207-312-8	Dicyandiamide

EINECS	Chemical
207-439-9	Calcium carbonate
207-924-5	Benzene propanoic acid
208-167-3	Barium carbonate
208-534-8	Sodium benzoate
208-750-2	Polyvinyl chloride
208-915-9	Magnesium carbonate
209-097-6	Tristearin
209-150-3	Magnesium stearate
209-151-9	Zinc stearate
209-183-3	Polyvinyl alcohol
209-406-4	Dioctyl sodium sulfosuccinate
209-502-6	2-Mercaptobenzimidazole
209-544-5	Toluene diisocyanate
209-753-1	1-Hexene
209-786-1	Potassium stearate
210-436-5	2,4-Dibromophenol
210-827-0	Glycol dilaurate
211-014-3	Glycol distearate
211-020-6	Dimethyl adipate
211-076-1	Ethylene glycol diethyl ether
211-105-8	Hexadecene-1
211-162-9	Ammonium acetate
211-185-4	Tetrabromophthalic anhydride
211-279-5	Aluminum tristearate
211-466-1	Isobutyl stearate
211-546-6	Behenyl alcohol
211-656-4	Tetramethoxysilane
211-659-0	Tetra-n-propoxysilane
211-748-4	Cetyl betaine
211-750-5	Distearyl thiodipropionate
211-765-7	2-Methyl imidazole
211-776-7	1,2-Diaminocyclohexane
211-989-5	2,4,6-Tri-t-butylphenol
212-052-3	Dimethyl methylphosphonate
212-084-8	Methacrylic anhydride
212-164-2	N-2-Aminoethyl-3-aminopropyl trimethoxysilane
212-305-8	Phenyltriethoxysilane
212-406-7	Calcium lactate
212-470-6	Stearyl betaine
212-490-5	Sodium stearate
212-782-2	2-Hydroxyethyl methacrylate
212-819-2	Decene-1
212-828-1	1-Methyl-2-pyrrolidinone
213-048-4	Aminopropyltriethoxysilane
213-085-6	Diamyl sodium sulfosuccinate
213-195-4	2-(2-Aminoethoxy) ethanol
213-234-5	2-Ethyl-4-methyl imidazole
213-668-5	Hexamethyldisilazane
213-695-2	Ammonium stearate
213-834-7	Triallyl isocyanurate
213-928-8	Dibutyltin diacetate
213-934-0	Vinyltris(2-methoxyethoxy) silane
214-073-3	Ethyl toluenesulfonamide
214-277-2	Dimethyl glutarate
214-298-7	Lauramide
214-302-7	Dimethyl decylamine
214-604-9	Decabromodiphenyl oxide
214-685-0	Methyltrimethoxysilane
214-737-2	Sodium myristyl sulfate
215-137-3	Calcium hydroxide
215-138-9	Calcium oxide
215-146-2	Cadmium oxide
215-168-2	Ferric oxide
215-170-3	Magnesium hydroxide
215-171-9	Magnesium oxide
215-174-5	Lead dioxide
215-175-0	Antimony trioxide
215-202-6	Manganese dioxide
215-204-7	Molybdenum trioxide
215-222-5	Magnesium aluminum carbonate
215-222-5	Zinc oxide
215-226-7	Zinc peroxide
215-237-7	Antimony pentoxide (Sb_2O_5)
215-251-3	Zinc sulfide
215-263-9	Molybdenum disulfide
215-267-0	Lead (II) oxide
215-288-5	Montmorillonite
215-290-6	Lead carbonate (basic)
215-354-3	Propylene glycol stearate
215-355-9	Glyceryl hydroxystearate
215-475-1	Aluminum silicate
215-535-7	Xylene
215-540-4	Sodium borate decahydrate
215-548-8	Tricresyl phosphate
215-566-6	Zinc borate
215-609-9	Carbon black
215-654-4	Stearyl citrate
215-661-2	Methyl ethyl ketone peroxide
215-663-3	Sorbitan laurate
215-664-9	Sorbitan stearate
215-665-4	Sorbitan oleate
215-683-2	Silica, hydrated
215-684-8	Sodium silicoaluminate
215-691-6	Alumina
215-710-8	Calcium silicate
215-823-2	3,5-Di-t-butyl-p-hydroxy-benzoic acid
215-951-9	Dimethyl isophthalate
216-087-5	Trifluoromethane sulfonic acid
216-472-8	Calcium stearate
216-653-1	Methyl t-butyl ether
216-700-6	Lauramine oxide
216-751-4	Trilauryl trithiophosphite
216-940-1	2-(2-Dimethylaminoethoxy) ethanol
217-421-2	Benzophenone-12
217-430-1	Sodium oleyl sulfate
217-752-2	t-Butyl hydroquinone
217-983-9	Methyltriethoxysilane
218-216-0	Octadecyl 3,5-di-t-butyl-4-hydroxyhydrocinnamate
218-254-8	1,1´-Azobis(cyclohexane-1-carbonitrile)
218-487-5	Bis (2,6-diisopropylphenyl) carbodiimide
218-734-7	6-t-Butyl-o-cresol
218-793-9	Ammonium lauryl sulfate

EINECS	Chemical
218-901-4	Glyceryl linoleate
218-964-8	Tetrabutylphosphonium chloride
219-136-9	PEG-6 laurate
219-314-6	2-t-Butyl-p-cresol
219-371-7	1,4-Butanediol diglycidyl ether
219-376-4	Butyl glycidyl ether
219-470-5	Drometrizole
219-518-5	Zinc laurate
219-553-6	2-Ethylhexyl glycidyl ether
219-784-2	3-Glycidoxypropyltrimethoxysilane
219-785-8	3-Methacryloxypropyltrimethoxysilane
219-787-9	3-Chloropropyltrimethoxysilane
219-865-2	N-Methyl-2-pyrrolidone
219-976-6	Dimethyl 2,2′-azobis (2-methylpropionate)
220-017-9	Cadmium laurate
220-045-1	PEG-6
220-135-0	4,4′-Azobis (4-cyanopentanoic acid)
220-152-3	Decyl betaine
220-219-7	Ditridecyl sodium sulfosuccinate
220-336-3	Isostearic acid
220-449-8	Vinyltrimethoxysilane
220-476-5	Stearyl stearate
220-548-6	Ethylene glycol propyl ether
220-688-8	Dimethylaminoethyl methacrylate
220-835-6	Dioctyl fumarate
220-851-3	Sodium octoxynol-2 ethane sulfonate
220-941-2	Octyltriethoxysilane
221-070-0	2,2-Azobis(2-amidinopropane) dihydrochloride
221-109-1	Dihexyl sodium sulfosuccinate
221-188-2	Sodium tridecyl sulfate
221-279-7	Laureth-2
221-280-2	Laureth-3
221-283-9	Laureth-7
221-284-4	Laureth-9
221-286-5	Laureth-12
221-291-2	Phenyl neopentylene glycol phosphite
221-304-1	Behenamide
221-356-5	Trilauryl phosphite
221-359-1	p-Tolyl diethanolamine
221-416-0	Sodium laureth sulfate
221-487-8	Tetrabutylphosphonium bromide
221-660-8	3-Aminopropylmethyldiethoxy silane
221-695-9	Hexabromocyclododecane
221-967-7	Dibromoneopentyl glycol
222-036-8	Dibromoethyldibromocyclohexane
222-048-3	Chloro-2-hydroxypropyl trimonium chloride
222-273-7	Tetrasodium dicarboxyethyl stearyl sulfosuccinamate
222-477-6	Zinc carbonate
222-540-8	Pentaerythrityl triacrylate
222-700-7	Calcium behenate
222-823-6	N,N-Butyl benzene sulfonamide
222-884-9	Diundecyl phthalate
222-899-0	Disodium lauriminodipropionate
223-055-4	Diethylhydroxylamine
223-276-6	Distearyl pentaerythritol diphosphite
223-383-8	2-(3′,5′-Di-t-butyl-2′-hydroxyphenyl)-5-chlorobenzotriazole
223-445-4	2-(2′-Hydroxy-3′-t-butyl-5′-methylphenyl)-5-chlorobenzotriazole
223-672-9	Isopropyl glycidyl ether
223-772-2	Benzophenone-4
223-775-9	Pentaethylene hexamine
223-805-0	Quaternium-15
223-943-1	Vinyltriacetoxy silane
224-583-8	2,2′-Azobis (2,4-dimethylvaleronitrile)
224-588-5	Mercaptopropyltrimethoxysilane
224-772-5	Lithium stearate
224-806-9	Glycol oleate
225-091-6	Dioctyl terephthalate
225-166-3	Calcium laurate
225-167-9	Barium laurate
225-202-8	Diphenyl phosphite
225-214-3	MEA-lauryl sulfate
225-268-8	p-Ethylbenzaldehyde
225-805-6	3-Chloropropyltriethoxysilane
225-856-4	PEG-8
226-029-0	Etocrylene
226-097-1	Laureth-4
226-191-2	Sodium ricinoleate
226-312-9	PEG-9 stearate
227-335-7	Calcium stearoyl lactylate
227-729-9	Sorbitan laurate
228-250-8	Octocrylene
228-486-1	PEG-2 dilaurate
229-222-8	DMDM hydantoin
229-457-6	Hydroxymethyl dioxoazabicyclooctane
229-713-7	Diazabicycloundecene
229-722-6	Tetrakis [methylene (3,5-di-t-butyl-4-hydroxyhydrocinnamate)] methane
229-859-1	PEG-12
230-279-6	2-Dimethylamino-2-methyl-1-propanol
230-325-5	Aluminum stearate
230-528-9	Oleyl 1,3-propylene diamine
230-770-5	Nonoxynol-4
230-785-7	Tetrapotassium pyrophosphate

EINECS	Chemical
230-939-3	Dimethyl octylamine
230-990-1	Didecyl methylamine
231-100-4	Lead
231-102-5	Lithium
231-111-4	Nickel
231-141-8	Tin
231-146-5	Antimony
231-149-1	Barium
231-175-3	Zinc
231-210-2	Cupric chloride
231-306-4	Diisopropyl sebacate
231-314-8	Potassium ricinoleate
231-469-1	Phenylethanolamine
231-472-8	Pentaerythrityl tetrakis (3-mercaptopropionate)
231-493-2	Cyclodextrin
231-545-4	Diatomaceous earth
231-545-4	Silica
231-592-0	Zinc chloride
231-626-4	2-Ethylhexyl thioglycolate
231-639-5	Sulfuric acid
231-640-0	t-Butyl glycidyl ether
231-721-0	Disodium capryloamphodiacetate
231-722-6	Sulfur
231-729-4	Ferric chloride
231-768-7	Phosphorus
231-784-4	Barium sulfate
231-826-1	Calcium phosphate dibasic
231-837-1	Calcium phosphate monobasic monohydrate
231-840-8	Calcium phosphate tribasic
231-868-0	Stannous chloride anhyd.
231-900-3	Calcium sulfate
231-900-3	Calcium sulfate dihydrate
231-955-3	Graphite
231-957-4	Selenium
232-122-7	Bismuth oxychloride
232-273-9	Sunflower seed oil
232-290-1	Ceresin
232-292-2	Hydrogenated castor oil
232-293-8	Castor oil
232-299-0	Rapeseed oil
232-306-7	Sulfated castor oil
232-307-2	Lecithin
232-313-5	Montan wax
232-315-6	Paraffin
232-347-0	Candelilla wax
232-348-6	Lanolin
232-360-1	Sorbitan sesquioleate
232-373-2	Petrolatum
232-374-8	Pine tar
232-384-2	Paraffin oil
232-384-2	Mineral oil
232-388-4	Potassium castorate
232-391-0	Epoxidized soybean oil
232-399-4	Carnauba
232-447-4	Tallowtrimonium chloride
232-453-7	Mineral spirits
232-455-8	Mineral oil

EINECS	Chemical
232-475-7	Rosin
232-479-9	Pentaerythrityl rosinate
232-674-9	Cellulose
232-686-4	Starch
233-007-4	Cyclodextrin
233-036-2	Sulfur chloride
233-042-5	Trichlorosilane
233-226-5	Stearyl erucamide
233-257-4	Manganese violet
233-293-0	PEG-4 oleate
233-471-8	Zinc borate
233-520-3	PEG-2 stearamine
233-892-7	Bis (trimethylsilyl) acetamide
234-206-9	Ditridecyl thiodipropionate
234-319-3	Magnesium aluminum carbonate
234-325-6	Glyceryl stearate
235-183-8	Ammonium bromide
235-192-7	Magnesium carbonate hydroxide
235-527-7	Cyclohexanone peroxide
235-795-5	Sucrose benzoate
236-239-4	4,4´-Methylenebis (cyclohexylamine)carbamate
236-671-3	Zinc pyrithione
236-675-5	Titanium dioxide
236-740-8	2,2´-Azobis(2-methylbutyronitrile)
236-753-9	Heptakis (dipropylene glycol) triphosphite
236-813-4	Tellurium
237-511-5	Aminopropyltrimethoxysilane
237-725-9	Glycol laurate SE
238-303-7	Sodium caproamphoacetate
238-479-5	Disodium stearyl sulfosuccinamate
238-877-9	Talc
238-878-4	Quartz
239-032-7	Sodium lauriminodipropionate
239-263-3	Methyl phenylglyoxalate
239-341-7	Nitrosophenylhydroxylamine aluminum salt
239-556-6	2-Methylpentamethylenediamine
239-593-8	2,2´-Azobis(4-methoxy-2,4-dimethylvaleronitrile)
240-029-8	2-Butanol
240-367-6	Oleyl palmitamide
240-613-2	Dimyristyl thiodipropionate
241-640-2	Cetyl esters
242-784-9	Distearyl phosphite
242-960-5	Pentaerythrityl tetraoleate
243-885-0	Tetrabromophthalate diol
244-063-4	Diammonium EDTA
244-085-4	2,2´-Azobis (N,N´-dimethyleneisobutyramidine)
244-492-7	Aluminum hydroxide
244-754-0	Octyl stearate
245-629-3	Dicyclohexyl sodium sulfosuccinate

EINECS	Chemical
246-140-8	Aluminum nitride
246-515-6	Zinc formaldehyde sulfoxylate
246-563-8	BHA
246-613-9	Isooctyl thioglycolate
246-614-4	Triisooctyl phosphite
246-680-4	Sodium dodecylbenzenesulfonate
246-690-9	Diisobutylene
246-839-8	Xylene sulfonic acid
246-917-1	Dodecenyl succinic anhydride
246-929-7	Sodium stearoyl lactylate
246-985-2	Sodium trideceth sulfate
246-998-3	Triisodecyl phosphite
247-036-5	Tridecylbenzene sulfonic acid
247-098-3	Phenyl diisodecyl phosphite
247-144-2	Glyceryl dioleate
247-500-7	Methylchloroisothiazolinone
247-555-7	Nonoxynol-5
247-556-2	Isopropylamine dodecylbenzenesulfonate
247-557-8	Calcium dodecylbenzene sulfonate
247-568-8	Sorbitan palmitate
247-569-3	Sorbitan trioleate
247-655-0	Octyl stearate
247-658-7	Diphenyl isooctyl phosphite
247-660-8	Ditridecyl adipate
247-669-7	Propylene glycol ricinoleate
247-759-6	Trisnonylphenyl phosphite
247-777-4	Diphenyl isodecyl phosphite
247-779-5	Diisodecyl pentaerythritol diphosphite
247-816-5	Nonoxynol-8
247-891-4	Sorbitan tristearate
247-952-5	Bis (2,4-di-t-butylphenyl) pentaerythritol diphosphite
247-977-1	Diisodecyl phthalate
248-052-5	DEDM hydantoin
248-289-4	Dodecylbenzene sulfonic acid
248-292-0	Nonoxynol-7
248-293-6	Nonoxynol-8
248-294-1	Nonoxynol-10
248-296-2	Potassium dodecylbenzene sulfonate
248-368-3	Diisotridecyl phthalate
248-403-2	Polyglyceryl-3 stearate
248-406-9	TEA-dodecylbenzenesulfonate
248-586-9	Glyceryl dilaurate
248-586-9	Glyceryl dilaurate SE
248-655-3	2,2´-Azobis[2-(2-imidazolin-2-yl) propane] dihydrochloride
248-666-3	Hydroxypropyl methacrylate
248-762-5	Nonoxynol-1
249-079-5	Diisononyl phthalate
249-395-3	Propylene glycol myristate
249-862-1	Octyl palmitate
249-992-9	Nonoxynol-6 phosphate
250-151-3	Sodium myristoyl sarcosinate
250-696-7	Tridecyl stearate
250-705-4	Glyceryl stearate
251-087-9	Octabromodiphenyl oxide
251-118-6	Ethylene bis-tetrabromophthalimide
251-136-4	Dibenzylidene sorbitol
251-646-7	Diisononyl adipate
252-011-7	Polyglyceryl-10 tetraoleate
252-161-3	3-Thiocyanatopropyltriethoxysilane
252-287-4	Diisooctyl phosphite
252-816-3	2,2´-Ethylidenebis (4,6-di-t-butylphenol)
253-149-0	Cetyl alcohol
253-211-7	Tris (dipropyleneglycol) phosphite
253-407-2	Glyceryl oleate
253-452-8	Disodium isodecyl sulfosuccinate
253-458-0	Glycol laurate
253-458-0	PEG-8 laurate
254-896-5	Bis-(3-(triethoxysilyl)propyl) tetrasulfane
255-062-3	Disodium laureth sulfosuccinate
255-350-9	Propylene glycol dipelargonate
256-120-0	Disodium oleamido MIPA-sulfosuccinate
258-007-1	Myristoyl sarcosine
258-207-9	Bis (2,2,6,6-tetramethyl-4-piperidinyl) sebacate
258-250-3	Ethylenebis dibromonorbornane dicarboximide
259-627-5	Iodopropynyl butylcarbamate
261-526-6	Tetradecabromodiphenoxy benzene
261-673-6	Isodecyl oleate
262-774-8	Isopropyl titanium triisostearate
262-976-6	Hydrogenated tallowamine
262-977-1	Cocamine
262-978-7	Coconut acid
262-988-1	Methyl cocoate
262-989-7	Methyl tallowate
262-990-2	Dicoco methylamine
262-991-8	Dihydrogenated tallow methylamine
263-017-4	Dimethyl soyamine
263-020-0	Dimethyl cocamine
263-022-1	Dimethyl hydrogenated tallow amine
263-032-6	Lard glyceride
263-038-9	Cocotrimonium chloride
263-052-5	Sodium cocoyl isethionate
263-087-6	Dicocodimonium chloride
263-090-2	Quaternium-18
263-107-3	Tall oil acid
263-123-0	Hydrogenated tallow amide
263-125-1	Tallow amine
263-129-3	Tallow acid
263-130-9	Hydrogenated tallow acid
263-137-7	Sodium tallate

EINECS	Chemical
263-144-5	Potassium rosinate
263-144-5	Sodium rosinate
263-163-9	Cocamide DEA
263-170-7	Cocoyl hydroxyethyl imidazoline
263-177-5	N-Tallow alkyl-2,2′-iminobis-ethanol
263-180-1	Dihydroxyethyl cocamine oxide
263-190-6	Disodium tallowimino-dipropionate
263-193-2	Sodium cocoyl sarcosinate
263-195-3	Cocopropylenediamine
264-038-1	Microcrystalline wax
264-520-1	Octoxynol-1
265-672-1	Disodium ricinoleamido MEA-sulfosuccinate
266-218-5	Nonyl nonoxynol-9 phosphate
266-231-6	Nonoxynol-9 phosphate
266-948-4	Triolein
267-006-5	C12-18 alcohols
267-009-1	C14-18 alcohols
267-015-4	Methyl oleate
267-617-7	Disodium ricinoleamido MEA-sulfosuccinate
267-791-4	Sodium octoxynol-2 ethane sulfonate
268-130-2	Ammonium lauroyl sarcosinate
268-665-1	Diisooctyl octylphenyl phosphite
268-938-5	Cocamidopropylamine oxide
269-035-9	Methyl rosinate
269-657-0	Soy acid
269-788-3	Isooctyl tallate
269-790-4	C14-16 alcohols
270-156-4	Cocoyl sarcosine
270-329-4	Coco-betaine
270-407-8	Sodium C14-16 olefin sulfonate
270-416-7	Tallowaminopropylamine
270-474-3	Pentaerythrityl tetracaprylate/caprate
270-664-6	Montan acid wax
271-694-2	Pentaerythrityl tetraoleate
271-696-6	Hydrogenated tallow 1,3-propylene diamine
271-792-5	Disodium capryloamphodiacetate
271-867-2	p-Cresol/dicyclopentadiene butylated reaction product
271-951-9	Sodium caproamphoacetate
272-021-5	Cocaminobutyric acid
272-490-6	C12-16 alcohols
272-493-2	Tetradecene-1
273-277-0	Ethylene distearamide
275-521-1	Cyclohexyl isooctyl phthalate
276-951-2	Sorbitan tristearate
279-420-3	C12-14 alcohols
279-498-9	Tetraphenyl dipropyleneglycol diphosphite
279-499-4	Poly (dipropyleneglycol) phenyl phosphite
283-066-5	C10-14 alcohols
284-868-8	Isobutyl oleate
285-540-7	Isopropyl oleate
285-547-5	Pentaerythrityl stearate
285-550-1	PEG-2 stearate
286-074-7	Sorbitan trioleate
286-344-4	Sodium 2,2′-methylenebis(4,6-di-t-butylphenyl) phosphate
286-490-9	Glyceryl stearate
287-484-9	Stearyl stearate
287-621-2	C8-10 alcohols
287-824-6	Methyl stearate
288-117-5	C10-12 alcohols
288-459-5	PEG-12 dioleate
288-668-1	Isobutyl stearate
290-580-3	Distearyl phthalate
290-588-7	Lead phthalate, basic
290-850-0	Disodium tallow sulfosuccinamate
292-951-5	Isooctyl stearate
293-029-5	Pentaerythrityl tetrastearate
295-625-0	Glyceryl (triacetoxystearate)
296-473-8	Kaolin
297-355-9	Tallow amide
303-773-5	Disodium oleyl sulfosuccinate
306-120-2	4,4′ Isopropylidenediphenol alkyl (C12-15) phosphites
306-522-8	Glycol stearate
307-145-1	C20-24 alcohols
309-928-3	Sodium alumino sulfo silicate

Discontinued Trade Names

Abex® LIV/30. [Rhone-Poulenc Surf. & Spec.] Ammonium alkylaryl ether sulfate
Abex® LIV/2330. [Rhone-Poulenc Surf. & Spec.] Ammonium alkylaryl ether sulfate
Aciculite™. [Kaopolite] Acicular aluminum silicate
Acintol® 736. [Arizona] Tall oil acid
Acintol® 746. [Arizona] Tall oil acid
Acintol® R Type 3A. [Arizona] Tall oil rosin
Acrysol® ASE-60. [Rohm & Haas] Crosslinked acrylic emulsion copolymer
Acrysol® ASE-75. [Rohm & Haas] Polyacrylic acid
Acrysol® G-110. [Rohm & Haas] Ammonium polyacrylate sol'n.
Acrysol® GS. [Rohm & Haas] Sodium polyacrylate
Acrysol® HV-1. [Rohm & Haas] Sodium polyacrylate
Advapak® LS-202. [Morton Int'l./Specialty Chem.] Organotin compd.
Advastab® T-52N Conc. [Morton Int'l./Specialty Chem.] Butyltin carboxylate
Advastab® T-290. [Morton Int'l./Specialty Chem.] Butyltin carboxylate
Advastab® T-340. [Morton Int'l./Specialty Chem.] Dibutyltin maleate
Advastab® TM-180. [Morton Int'l./Specialty Chem.] Butyltin mercaptide
Advastab® TM-181-S. [Morton Int'l./Specialty Chem.] Methyltin
Advastab® TM-185. [Morton Int'l./Specialty Chem.] Methyltin mercaptide
Advastab® TM-2082. [Morton Int'l./Specialty Chem.] Methyltin mercaptide
Advastab® WS-499. [Morton Int'l./Specialty Chem.] Organotin
Aerosol® AY-100. [Cytec Industries] Diamyl sodium sulfosuccinate
AF 66. [GE Silicones] Filled polydimethylsiloxane
Agesperse 71. [CPS] Polymeric carboxylic acid, sodium salt
Agesperse 80. [CPS] Polymeric carboxylic acid, sodium salt
Agesperse 81. [CPS] Polymeric carboxylic acid, sodium salt
Agesperse 82. [CPS] Polymeric carboxylic acid, ammonium salt
Akrochem® P 420. [Akrochem] Nonreactive alkyl phenolic resin
Akrochem® Peptizer PTP. [Akrochem] Pentachlorothiophenol with activating and dispersing agents
Akrochem® P.P.D. [Akrochem] Piperidinium pentamethylene dithiocarbamate
Akrofax™ 9844 Black. [Akrochem]
Akrofax™ 9851 Light. [Akrochem]
Akrofax™ 9906. [Akrochem]
Alcan H-10-08 [Alcan] Alumina trihydrate
Alcan H-10-10 [Alcan] Alumina trihydrate
Alcan H-10-15. [Alcan] Alumina trihydrate
Alkamuls® PSML-4. [Rhone-Poulenc Surf. & Spec.] PEG-4 sorbitan laurate
Alkasurf® LAN-15. [Rhone-Poulenc Surf. & Spec.] Laureth-15
Alphenate TH 454. [Henkel-Nopco] Alkyl sulfosuccinate
Amax® No. 1. [R.T. Vanderbilt] N-oxydiethylene benzothiazole-2-sulfenamide plus benzothiazyl disulfide
Amine CS-1246. [ANGUS] Oxazolidine
Amoco® Resin 18-210, 18-240, 18-290. [Amoco] Poly-[ga]-methylstyrene
AMP. [ANGUS] 2-Amino-2-methyl-1-propanol
AMP-95. [ANGUS] 2-Amino-2-methyl-1-propanol
Amyl Cadmate®. [R.T. Vanderbilt] Cadmium diamyldithiocarbamate
AO-1. [Ciba-Geigy/Additives] Tris(2-methyl-4-hydroxy-5-t-butylphenyl)butane
AO-2. [Ciba-Geigy/Additives] Cyclic neopentanetetrayl bis(octadecyl phosphite)
AO-3. [Ciba-Geigy/Additives] Phosphorous acid, cyclic neopentanetetrayl bis (2,4-di-t-butylphenyl) ester
AO-4. [Ciba-Geigy/Additives] Tri (mixed mono and dinonylphenyl) phosphite
AO-5. [Ciba-Geigy/Additives] Tris 3-hydroxy-4-t-butyl-2,6-dimethylbenzyl cyanurate
AO-6. [Ciba-Geigy/Additives] 4,4′-Thiobis(6-t-butyl-m-cresol)
AO-7. [Ciba-Geigy/Additives] 2,2′-Methylenebis (6-t-butyl-p-cresol)
Arizona DR-24. [Arizona] Disproportionated rosin
Arquad® T-2C-50. [Akzo Nobel BV] 1:1 mixt. of tallow trimonium chloride and dicoco dimonium chloride
ASP® 352. [Engelhard] Delaminated hydrous aluminum silicate
ASP® 602P. [Engelhard] Hydrous aluminum silicate
ASP® 800. [Engelhard] Hydrous aluminum silicate
Astor FP7997. [Astor Wax] Petrol. wax
Astrowet 102. [Alco] Alcohol half-ester sulfosucci-

nate, sodium salt, ethoxylated
Atmer® 101. [ICI Surf. Am.] Sorbitan ester
Atmer® 107. [ICI Surf. Am.] Sorbitan ester
Atmer® 123. [ICI Surf. Am.] Glyceryl stearate
Atmer® 126. [ICI Surf. Am.] Glyceryl stearate
Atmer® 127. [ICI Surf. Am.] Glyceryl ester
Atmer® 128. [ICI Surf. Am.] Glyceryl stearate
Atmer® 151. [ICI Surf. Am.] Alkoxylated alcohol
Atmer® 152. [ICI Surf. Am.] Blend
Atmer® 160. [ICI Surf. Am.] Quaternized amine ethoxylate
Atmer® 165. [ICI Surf. Am.] Aryl polyol diacetal
Atmer® 166. [ICI Surf. Am.] Blend
Atmer® 172. [ICI Surf. Am.] Proprietary blend
Atmer® 175. [ICI Surf. Am.] Alkylaryl sulfonate
Atmer® 176. [ICI Surf. Am.] POE sorbitol ester
Atmer® 177. [ICI Surf. Am.] POE fatty alcohol
Atmer® 178. [ICI Surf. Am.] POE fatty alcohol
Atmer® 180. [ICI Surf. Am.] Glyceryl fatty acid ester
Atmer® 191. [ICI Surf. Am.] Alkyl sulfonate
Atmer® 1001. [ICI Surf. Am.] Sorbitol ester
Atmer® 1008, 1009. [ICI Surf. Am.] Glyceryl ester
Atmer® 1020. [ICI Surf. Am.] Glyceryl ester
Atmer® 1022, 1023. [ICI Surf. Am.] POE sorbitol ester
Atmer® 1030. [ICI Surf. Am.] Hydrog. oil
Atmer® 7000. [ICI Surf. Am.] 50% Atmer 163
Atmer® 7003. [ICI Surf. Am.] 50% Atmer 163
Atmer® 7004. [ICI Surf. Am.] 50% Atmer 129
Atmer® 7005. [ICI Surf. Am.] 50% Atmer 129/163 (73:27)
Atmer® 7008. [ICI Surf. Am.] 60% Atmer 163
Atmer® 7009. [ICI Surf. Am.] 60% Atmer 129
Atmer® 7010. [ICI Surf. Am.] 50% Atmer 138
Atmer® 7100. [ICI Surf. Am.] 50% Atmer 103
Atmer® 7102. [ICI Surf. Am.] 50% Atmer 129
Atmer® 7103. [ICI Surf. Am.] 50% Atmer 129/163 (2:1)
Atmer® 7104. [ICI Surf. Am.] 50% Atmer 163
Atmer® 7105. [ICI Surf. Am.] 50% Atmer 163
Atmer® 7106. [ICI Surf. Am.] 50% Atmer 129/163 (2:1)
Atmer® 7107. [ICI Surf. Am.] 50% Atmer 184
Atmer® 7110. [ICI Surf. Am.] 50% Atmer 1007
Atmer® 7111. [ICI Surf. Am.] 50% Atmer 129/163 (4:7)
Atmer® 7112. [ICI Surf. Am.] 50% Atmer 122
Atmer® 7113. [ICI Surf. Am.] 50% Silicone F111-300
Atmer® 7114. [ICI Surf. Am.] 50% Atmer 121
Atmer® 7115. [ICI Surf. Am.] 50% Atmer 184/114 (3:1)
Atmer® 7200. [ICI Surf. Am.] 50% Atmer 163
Atmer® 7201. [ICI Surf. Am.] 50% Atmer 163
Avirol® 125 E. [Henkel/Functional Prods.] Sodium alkyl ether sulfate
Avirol® SL 2015. [Henkel/Functional Prods.] Sodium C12-C15 sulfate
Avirol® T 40. [Henkel/Functional Prods.] TEA lauryl sulfate
AZO-33. [Asarco] Zinc oxide
AZO-55. [Asarco] Zinc oxide
AZO-55TT. [Asarco] Surface-treated zinc oxide
AZO-66. [Asarco] Zinc oxide
AZO-66TT. [Asarco] Surface-treated zinc oxide
AZO-77. [Asarco] Zinc oxide
AZO-77TT. [Asarco] Surface-treated zinc oxide
Azodox-55. [Asarco] Zinc oxide
Azodox-55TT. [Asarco] Zinc oxide
Baco DH. [Alcan; BA Chem. Ltd.] Alumina trihydate
Blendex® 101. [GE Specialty] ABS
Blendex® 201. [GE Specialty] ABS
Blendex® 310. [GE Specialty] ABS
Blendex® 405. [GE Specialty] ABS
Blendex® 702. [GE Specialty] Poly ([ga]-methylstyrene-styrene acrylonitrile)/ABS
Blendex® 979. [GE Specialty] ASA copolymer
Blendex® HPP 821. [GE Specialty] Polyphenylene ether resin
Califlux® LV. [Witco/Golden Bear] Aromatic oil
CAO®-1. [PMC Specialties] 2,6-Ditert.-butyl-p-cresol (BHT)
Captax®-Tuads Blend. [R.T. Vanderbilt] 2-Mercaptobenzothiazole, tetramethylthiuram disulfide, ratio 1:2
Cardolite® NC-546, NC-549, NC-550. [Cardolite] Phenalkamine
Cardolite® NC-548. [Cardolite] Acetate ester of cardanol
Carstab® 700. [Morton Int'l./Specialty Chem.] 2-Hydroxy-4-n-octoxybenzophenone
Carstab® 701. [Morton Int'l./Specialty Chem.] 2-Hydroxy-4-isooctoxybenzophenone
Carstab® 702. [Morton Int'l./Specialty Chem.] 2-Hydroxy-4-isooctoxybenzophenone
Carstab® 705. [Morton Int'l./Specialty Chem.] 2-Hydroxy-4-n-octoxybenzophenone
Carstab® DMTDP. [Morton Int'l./Specialty Chem.] Dimyristyl thiodipropionate
Cetodan® 90-40. [Grindsted Prods.] Acetylated lard glyceride
Chemlink® 2000. [Sartomer] Long chain acrylated diol
Chemlink® 2100. [Sartomer] Long chain methacrylated diol
Chemlink® 3000. [Sartomer] Epoxy acrylate
Chemlink® 5000. [Sartomer] Acrylated BR
Chemlink® 9000. [Sartomer] Polyalkoxylated diacrylate
Chemlink® 9001. [Sartomer] Polyalkoxylated diacrylate
Chemlink® 9003. [Sartomer] Alkoxylated aliphatic triacrylate (monomer)
Chemlink® 9008. [Sartomer] Functionalized triacrylate ester (monomer)
Chemlink® 9012. [Sartomer] Functionalized aliphatic triacrylate (monomer)
Chemlink® 9013. [Sartomer] Functionalized aliphatic monoacrylate (monomer)
Chemlink® 9014. [Sartomer] Functionalized aliphatic diacrylate (monomer)
Chemlink® 9015. [Sartomer] Functionalized aliphatic monoacrylate (monomer)
Chemlink® 9020. [Sartomer] Alkoxylated aliphatic triacrylate (monomer)
Chemlink® 9021. [Sartomer] Alkoxylated aliphatic

triacrylate (monomer)
Chemlink® 9022. [Sartomer] Alkoxylated aliphatic tetraacrylate (monomer)
Chemlink® 9023. [Sartomer] Alkoxylated tetraacrylate ester (monomer)
Chemlink® 9024. [Sartomer] Alkoxylated aliphatic diacrylate ester (monomer)
Chemlink® 9025. [Sartomer] Alkoxylated aliphatic diacrylate ester (monomer)
Chemlink® 9040. [Sartomer] Functionalized aliphatic diacrylate (Pro 62) (monomer)
Chemlink® 9503. [Sartomer] Aliphatic urethane acrylate
Chemlink® 9504. [Sartomer] Aliphatic urethane acrylate
Chemlink® 9505. [Sartomer] Aliphatic urethane acrylate
Chemprene R-25. [Chemfax] Thermoplastic isoprenoidal polymer
Chemprene R-50. [Chemfax] Thermoplastic isoprenoidal polymer
Chemprene R-70. [Chemfax] Thermoplastic isoprenoidal polymer
Chemstat® 473. [Chemax] Ethoxylated nonylphenol
Chemstat® 3820. [Chemax] Quaternary ammonium salt
Chemstat® 9820A. [Chemax] Bis (2-hydroxyethyl) octyl methyl ammonium p-toluene sulfonate in ABS carrier resin
Chemstat® 9820H. [Chemax] Bis (2-hydroxyethyl) octyl methyl ammonium p-toluene sulfonate in HIPS carrier resin
Cimpact 700. [Luzenac Am.] Talc
Cisdene® 1203. [Am. Syn. Rubber] Stereospecific polybutadiene, high cis, nonstaining antioxidant
CMD 834. [Shell] Modified cycloaliphatic amine adduct
Cohedur® AS. [Bayer/Fibers, Org., Rubbers] Methylene donor
Cohedur® AS Powder. [Bayer/Fibers, Org., Rubbers] Methylene donor absorbed on Microcel E
Comboloob 0856. [Astor Wax] Sat. hydrocarbon wax
Con-BACN. [Tosoh] Condensed bromoacenaphthylene
Corax. [Henkel] Wax
CPF-0001. [Witco/PAG] Chlorinated paraffin
CPF-0003. [Witco/PAG] Chlorinated paraffin
CPF-0008. [Witco/PAG] Chlorinated paraffin
CPF-0019. [Witco/PAG] Chlorinated paraffin
CPF-0022. [Witco/PAG] Chlorinated paraffin
Cumate®. [R.T. Vanderbilt] Copper dimethyldithiocarbamate
Curezol® 2E4MZ-CN. [Air Prods./Perf. Chems.] Imidazole
Cyclolube® 62. [Witco/Golden Bear] Naphthenic oil distillate
Darvan® No. 3. [R.T. Vanderbilt] Sodium salt of polymerized substituted arylalkyl sulfonic acids combined with an inert inorg. suspending agent
Darvan® No. 4. [R.T. Vanderbilt] Polymerized aryl alkyl sulfonic acid monocalcium salt
Daxad® 34N10. [W.R. Grace/Organics] Sodium polymethacrylate sol'n.
DBM. [Tiarco] Dibutyl maleate
Demix® 7730. [Arizona] Catalytically disproportionated tall oil prod.
Demix® 7740. [Arizona] Catalytically disproportionated tall oil prod.
Demix® 7750. [Arizona] Catalytically disproportionated tall oil prod.
Deplastol. [Pulcra SA] Ethoxylated lauric acid
DIBM. [Tiarco] Diisobutyl maleate
Diluent 7. [Shell] C8-C10 aliphatic glycidyl ether
Diluent 8. [Shell] C12-C14 aliphatic glycidyl ether
Diluent 17. [Shell] 2-Ethylhexyl glycidyl ether
Diluent B. [Shell] Butyl glycidyl ether
Diluent C. [Shell] Cresyl glycidyl ether
Diluent M. [Shell] Mixed methyl esters of fatty acids
Diluent N. [Shell] Neopentyl glycol diglycidyl ether
DIOM. [Tiarco] Diisooctyl maleate
Disponil MGS 65. [Henkel/Functional Prods.] Surfactant blend
Disponil RO 40. [Henkel/Functional Prods.] POE fatty glyceride
DMAMP-80. [ANGUS] 2-Dimethylamino-2-methyl-1-propanol
DOA. [Monsanto] Dioctyl adipate
Doittol 891. [Henkel-Nopco] Sodium fatty acid soap, modified
DOM. [Tiarco] Dioctyl maleate
Dowfax XDS 30599. [Dow; Dow Europe] Sodium dodecyl diphenyloxide disulfonate
Dresinate® 95. [Hercules; Hercules BV] Potassium soap of dk. rosin
Dresinate® 90. [Hercules] Rosin potassium soap
Drewplus® L-156. [Drew Ind. Div.] Disp. of alcohols, fatty soaps, and surfactants
Drewplus® L-162. [Drew Ind. Div.] Silicone defoamer
Drewplus® L-722. [Drew Ind. Div.] Silicone defoamer
Drewplus® Y-166. [Drew Ind. Div.] Blend of emulsifiable min. oils and silica derivs.
Duraflex® 8410. [Shell] Polybutylene
Durastat® AS-5760. [PPG Industries]
Dymsol® L. [Henkel-Nopco] PEG fatty ester
Dymsol® MS-40. [Henkel] Polyethylene emulsion
Dymsol® N. [Henkel] Fatty amido condensate
Eastman® PA-100A. [Eastman] Dist. monoglycerides
Eastman® Inhibitor DOBP. [Eastman] 4-Dodecyloxy-2-hydroxybenzophenone
Edenol W 300 S. [Henkel] Diisotridecyl phthalate
Emery® 9331. [Henkel/Emery] Dibromophenol
Emery® 9332. [Henkel/Emery] Tribromophenol
Emery® 9336. [Henkel/Emery] Dibromoneopentyl glycol
Emery® 9345. [Henkel/Emery] Tetrabromoxylene
Emery® 9350. [Henkel/Emery] Tetrabromobisphenol A
Emery® 9353. [Henkel/Emery] Tetrabromobisphenol A di-2-hydroxyethyl ether
Empicol® LS30P. [Albright & Wilson Australia] Sodium lauryl sulfate
Empicol® LZ/E. [Albright & Wilson UK] Sodium lauryl sulfate
Empigen® AB. [Albright & Wilson UK] Dimethyl

lauramine
Empigen® AH. [Albright & Wilson UK] Dimethyl myristamine
Empilan® AM Series. [Albright & Wilson UK] Amine ethoxylates
Empimin® KSN60. [Albright & Wilson UK] Sodium laureth sulfate (3 EO), water/ethanol
Empimin® OP45. [Albright & Wilson UK] Sodium dioctyl sulfosuccinate
Emrite® 6120. [Henkel/Emery] Polysorbate 80
Emtal® 41. [Engelhard] Platy talc refined from hydrous magnesium silicates
Emtal® 42. [Engelhard] Platy talc refined from hydrous magnesium silicates
Emtal® 43. [Engelhard] Platy talc refined from hydrous magnesium silicates
Emtal® 44. [Engelhard] Platy talc refined from hydrous magnesium silicates
Emtal® 500. [Engelhard] Platy talcs refined from hydrous magnesium silicates
Emtal® 549. [Engelhard] Platy talcs refined from hydrous magnesium silicates
Emtal® 599. [Engelhard] Platy talcs refined from hydrous magnesium silicates
Emtal® 4190. [Engelhard] Platy talc refined from hydrous magnesium silicates
EP4802-75. [Lord]
Epodil® VFT-V6. [Air Prods./Perf. Chems.] Coumarone-indene resin
Epolene® E-10P. [Eastman] Low m.w. oxidized polyethylene wax
Epolene® EE-3P. [Eastman] Oxidized polyethylene wax
Epolene® N-10P. [Eastman] Low m.w. polyethylene wax
Epolene® N-20P. [Eastman] Low m.w. polyethylene wax
Epolene® N-21P. [Eastman] Low m.w. polyethylene wax
Epolene® N-34P. [Eastman] Low m.w. polyethylene wax
Epotuf Hardener 37-617. [Reichhold] Polyamide
Epotuf Hardener 37-619. [Reichhold] Aromatic polyamine
Epotuf Hardener 37-623. [Reichhold] Aromatic polyamine
Epoxol 8-2B. [Am. Chem. Services] Epoxidized butyl esters of linseed oil fatty acids
Epoxy Modifier ML. [Air Prods./Perf. Chems.]
Esperox® 10KXL. [Witco/PAG] t-Butyl peroxybenzoate on a proprietary formulated Burgess clay carrier
Esperox® 10XL. [Witco/PAG] t-Butyl peroxybenzoate on a proprietary formulated calcium carbonate carrier
Estol 1583. [Unichema] Glyceryl diacetate
Ethyleneamine EA-770. [Texaco] Mixt. of linear and cyclic ethyleneamines (mainly DETA, AEP, also TETA, AEEA)
Ethyl Selenac®. [R.T. Vanderbilt] Selenium diethyldithiocarbamate
Euredur® 3227. [Shell] Aliphatic
EZ Mold Lubricant. [TSE Industries] Glycol surfactant
Flectol® H. [Monsanto] Polymerized 1,2-dihydro-2,2,4-trimethylquinoline
Flexchlor® 0001. [Witco/PAG] Chlorinated paraffin
Flexchlor® 0002. [Witco/PAG] Chlorinated paraffin
Flexchlor® 0008. [Witco/PAG] Chlorinated paraffin
Flexchlor® 0009. [Witco/PAG] Chlorinated paraffin
Flexchlor® 0010. [Witco/PAG] Chlorinated paraffin
Flexchlor® 0011. [Witco/PAG] Chlorinated paraffin
Flexchlor® 0012. [Witco/PAG] Chlorinated paraffin
Flexchlor® 0018. [Witco/PAG] Chlorinated paraffin
Flexchlor® 0023. [Witco/PAG] Chlorinated paraffin
Foamaster 8034. [Henkel/Process & Polymer Chems.]
Foamaster JMY. [Henkel/Process & Polymer Chems.]
Foamaster NDW. [Henkel/Process & Polymer Chems.]
Foamaster TDB. [Henkel/Process & Polymer Chems.]
Frekote® 33. [Dexter/Frekote]
Frekote® 44. [Dexter/Frekote]
Frekote® 700. [Dexter/Frekote]
Frekote® No. 1. [Dexter/Frekote] Fluorocarbon
Gantrez® B-773. [ISP] Poly(vinyl isobutyl ether), hexane
Gantrez® M-555, M-556. [ISP] Polyvinyl methyl ether, toluene
Gantrez® M-574. [ISP] Polyvinyl methyl ether, toluene
Glycolube® 810. [Lonza] Proprietary
Glycox® PEMS. [Lonza] Ester wax
GP-72-SS Mercapto Modified Silicone Fluid. [Genesee Polymers] Mercapto-modified dimethyl silicone fluid
GP-137 3-Glycidoxypropylmethyldiethoxy Silane. [Genesee Polymers] 3-Glycidoxypropylmethyldiethoxy silane
GP-7000 Methyl Alkyl Silicone Fluid. [Genesee Polymers] Methyl alkyl dimethyl silicone fluid
Grindtek AML 60. [Grindsted Prods.] Acetylated palm kernel glycerides
Grindtek AMOS 90. [Grindsted Prods.] Acetylated lard glyceride
Grindtek MM 90. [Grindsted Prods.] Glyceryl myristate
Grindtek MOL 90. See Dimodan LS Kosher [Grindsted Prods.] Glyceryl linoleate
Grindtek MSP 32-6. [Grindsted Prods.] Glyceryl stearate SE
Grindtek MSP 52. [Grindsted Prods.] Glyceryl stearate
Grindtek PGE 25. [Grindsted Prods.] Polyglyceryl-3 oleate
Grindtek PGE 55. [Grindsted Prods.] Polyglyceryl-3 stearate
Grindtek PGE 55-6. [Grindsted Prods.] Polyglyceryl-3 stearate SE
Haltex™ 313. [Hitox] Alumina trihydrate
Haltex™ 316. [Hitox] Alumina trihydrate
Haltex™ 318. [Hitox] Alumina trihydrate
Hartomer GP 2164. [Huntsman] Phosphate ester
Hartomer GP 4935. [Huntsman] Complex phosphate ester

Hartomer JV 4091. [Huntsman] Blend
HCFC 123. [AlliedSignal] Dichlorotrifluoroethane
HCFC 141b. [AlliedSignal] Dichlorofluoroethane
Hercules® Resin 917. [Hercules] Aq. sol'n. of a cationic urea-formaldahyde resin
Hexcelcure 160. [Zeeland] Propoxylated amine
Hexcelcure 169. [Zeeland] Cyanoethylated amine
Huber 90A. [J.M. Huber/Engineered Mins.] Hyd. kaolin clay
Hubercarb® W 325. [J.M. Huber/Engineered Mins.] Calcium carbonate
Hydrax™ H-30. [Climax Performance] Alumina trihydrate
Hydrax™ H-30PN. [Climax Performance] Alumina trihydrate
Hydrax™ H-120. [Climax Performance] Alumina trihydrate
Hydrax™ H-135. [Climax Performance] Alumina trihydrate
Hydrax™ H-136. [Climax Performance] Alumina trihydrate
Hydrax™ H-215. [Climax Performance] Alumina trihydrate
Hydrax™ H-216. [Climax Performance] Alumina trihydrate
Hydrax™ H-218. [Climax Performance] Alumina trihydrate
Hydrax™ H-312. [Climax Performance] Alumina trihydrate
Hydrax™ H-314. [Climax Performance] Alumina trihydrate
Hydrax™ H-490. [Climax Performance] Alumina trihydrate
Hydrax™ H-495. [Climax Performance] Alumina trihydrate
Hydrax™ H-550. [Climax Performance] Alumina trihydrate
Hydrax™ H-555. [Climax Performance] Alumina trihydrate
Hydrax™ H-635. [Climax Performance] Alumina trihydrate
Hydrax™ H-636. [Climax Performance] Alumina trihydrate
Hydrax™ H-825. [Climax Performance] Alumina trihydrate
Hydrax™ H-826. [Climax Performance] Alumina trihydrate
Hydrax™ H-910. [Climax Performance] Alumina trihydrate
Hydrax™ H-915. [Climax Performance] Alumina trihydrate
Hydrax™ H-916. [Climax Performance] Alumina trihydrate
Hystrene® 3687. [Witco/H-I-P] 87% Dimer acid
Igepal® NP-9. [Rhone-Poulenc Surf. & Spec.] Nonoxynol-9
Igepal® NP-10. [Rhone-Poulenc Surf. & Spec.] Nonylphenol ethoxylate
Igepal® O. [Rhone-Poulenc Surf. & Spec.] Octoxynol-10
Intercide® 2 DIDP. [Akzo Nobel] Oxybisphenoxyarsine (2%) in diisodecyl phthalate
Intercide® ABF. [Akzo Nobel] 10,10′-Oxybisphenoxyarsine
Intercide® ABF 1 ESBO. [Akzo Nobel] Oxybisphenoxyarsine (1%) in epoxidized soybean oil
Intercide® ABF 2 DIDP. [Akzo Nobel] 10,10′-Oxybisphenoxyarsine (2%) in diisodecyl phthalate
Intercide® ABF 2 ESBO. [Akzo Nobel] 10,10′-Oxybisphenoxyarsine (2%) in epoxidized soybean oil
Interstab® 761-28. [Akzo Nobel] Barium-cadmium-zinc
Interstab® 761-28A. [Akzo Nobel] Barium-cadmium-zinc
Interstab® BC-4362. [Akzo Nobel] Barium/cadmium/zinc
Interstab® CZ-4359. [Akzo Nobel] Complex calcium and zinc soap of alkyl carboxylic acids combined with org. auxs.
Interstab® CZL-710. [Akzo Nobel] Complex calcium and zinc soaps of carboxylic acids combined with org. auxs.
Interstab® CZL-712. [Akzo Nobel] Complex of zinc soaps of carboxylic acids combined with org. auxs.
Interstab® CZL-715. [Akzo Nobel] Complex of zinc soaps of carboxylic acids combined with org. auxs.
Interstab® E-82. [Akzo Nobel] Epoxy-modified ether-ester
Interstab® LF 3623, LF 3653. [Akzo Nobel] Lead sulfate-based
Interstab® LF 3626, 3645, 3669, 10898/4, 10898/6. [Akzo Nobel] Lead phosphite-based
Interstab® LF 3631/1. [Akzo Nobel] Pb/Ba/Cd
Interstab® LF 3631/2. [Akzo Nobel] Pb/Ba/Cd
Interstab® LF 3631/3. [Akzo Nobel] Pb/Ba/Cd
Interstab® LF 3634. [Akzo Nobel] Lead phosphite/Ba/Cd based
Interstab® LF 3638. [Akzo Nobel] Lead coprecipitate, Pb phosphite based
Interstab® LF 3675, LF 3751. [Akzo Nobel] Lead phosphite/sulfate based
Interstab® LF 10773/25. [Akzo Nobel] Lead coprecipitate, Pb phosphite based
Interstab® LF 11298. [Akzo Nobel] Pb/Ba/Cd
Interstab® LF 11323/1. [Akzo Nobel] Pb/Ba/Cd
Interstab® LF 11359. [Akzo Nobel] Lead coprecipitate, Pb phosphite based
Interstab® LL 3289. [Akzo Nobel]
Interstab® LP 3103. [Akzo Nobel] Tribasic lead sulfate
Interstab® LP 3104. [Akzo Nobel] Tetrabasic lead sulfate
Interstab® LP 3139. [Akzo Nobel] Dibasic lead phosphite
Interstab® LP 3150. [Akzo Nobel] Dibasic lead stearate
Interstab® LP 3153. [Akzo Nobel] Dibasic lead phthalate
Interstab® LP 3155. [Akzo Nobel] Normal lead stearate
Interstab® LP 3190. [Akzo Nobel] Tetrabasic lead sulfate modified
Interstab® LP 3289. [Akzo Nobel] Liquid lead com-

plex

Interstab® LP 3631/5. [Akzo Nobel] Pb-phosphite-lubricant-antioxidant combination

Interstab® LT 4289. [Akzo Nobel] Calcium/zinc

Interstab® LT 4308. [Akzo Nobel] Calcium/zinc

Interstab® LT 11122/10. [Akzo Nobel] Lead/barium/cadmium complex

Interstab® M85. [Akzo Nobel] Barium/cadmium complex

Interstab® M341. [Akzo Nobel] Barium/cadmium complex

Interstab® M722, M763, M767. [Akzo Nobel] Barium/zinc

Interstab® M727. [Akzo Nobel] Barium/zinc

Interstab® M731. [Akzo Nobel] Potassium/zinc

Interstab® M744, M767, M11301. [Akzo Nobel] Barium/zinc

Interstab® M803, M11289. [Akzo Nobel] Calcium/zinc

Interstab® M809. [Akzo Nobel] Calcium/zinc

Interstab® M876. [Akzo Nobel] Calcium/zinc

Interstab® M3187. [Akzo Nobel] Barium/cadmium complex

Interstab® MF981, MF985. [Akzo Nobel] Barium/cadmium

Interstab® MP10581/3. [Akzo Nobel] Calcium/zinc

Interstab® MT11303/1. [Akzo Nobel] Calcium/zinc

Interstab® R-4052. [Akzo Nobel]

Interstab® R-4101. [Akzo Nobel] Barium-cadmium-zinc

Interstab® R-4109. [Akzo Nobel] Barium-cadmium-zinc

Interstab® R-4137. [Akzo Nobel] Barium-cadmium-zinc

Interwax G 8140. [Akzo Nobel] Alpha-olefin copolymer

Interwax G 8200. [Akzo Nobel] Glyceryl oleate

Interwax G 8204. [Akzo Nobel] Glyceryl stearate

Interwax G 8205. [Akzo Nobel] Glyceryl fatty acid ester

Interwax G 8206. [Akzo Nobel] C16-18 fatty alcohol

Interwax G 8207. [Akzo Nobel] Stearic acid

Interwax G 8208. [Akzo Nobel] Paraffin wax

Interwax G 8212. [Akzo Nobel] 12-Hydroxy stearic acid

Interwax G 8213. [Akzo Nobel] Hydrocarbon wax

Interwax G 8252. [Akzo Nobel] Polyethylene wax

Interwax G 8253. [Akzo Nobel] Octyl stearate

Interwax G 8257. [Akzo Nobel] Amide wax

Interwax G 8259. [Akzo Nobel] Calcium montanate

Interwax G 8268. [Akzo Nobel] Syn. paraffin wax

Interwax M 3142. [Akzo Nobel] Calcium stearate

Ionol CP. [Shell] BHT

Irgastab® T 266. [Ciba-Geigy/Additives] 2,2′-[(Dioctylstannylene)bis(acetic acid), diisooctyl ester and 2,2′,2′′-(octylstannylidyne)tris(thio)tris (acetic acid), triisooctyl ester

Irgastab® T 269. [Ciba-Geigy/Additives] 2,2′-[(Dioctylstannylene)bis(acetic acid), diisooctyl ester and 2,2′,2′′-(octylstannylidyne)tris(thio)tris (acetic acid), triisooctyl ester

Isonate® 191. [Dow] MDI

Isonox® 129. [Schenectady] 2,2′-Ethylidenebis (4,5-di-t-butylphenol)

Jayflex® 2000. [Exxon] Proprietary benzate

Jayflex® 3209. [Exxon] Proprietary adipate

Jayflex® 4210. [Exxon] Proprietary adipate

Jeffamine® ED-600. [Huntsman] POE polyamine

Jeffamine® ED-900. [Huntsman] POE polyamine

Jeffamine® ED-2001. [Huntsman] POE polyamine

Jeffcat T-9. [Huntsman] Stannous octoate

Jeffcat ZF-10. [Huntsman] N,N,N′-Trimethyl-N′-hydroxyethyl-bisaminoethylether

Kemamine® P-150, P-150D. [Witco/H-I-P] 50% Arachidyl-behenyl primary amine (P-150D—dist.)

Kemamine® P-880, P-880D. [Witco/H-I-P] Palmityl primary amine (tech. and dist. resp.)

Kenplast® AP-19. [Kenrich Petrochemicals]

Kenplast® APK. [Kenrich Petrochemicals]

Krynac® XL 29.20. [Bayer/Fibers, Org., Rubbers] Crosslinked butadiene-acrylonitrile copolymer, nonstaining, cold polymerized

Lexolube® 2T-237. [Inolex] PEG-4 di 2-ethylhexanoate

Lipophos PE9. [Lipo] Nonylphenol ether phosphate ester acid form

Lipophos PL6. [Lipo] Linear alcohol ether phosphate ester, acid form

Lipowax C. [Lipo] N,N′-ethylenebisstearamide

Lomar® PL. [Henkel/Emery; Henkel/Functional Prods.; Henkel/Textile] Condensed sodium naphthalene sulfonate

Loobwax 0682. [Astor Wax] Ester wax

Lowinox® 22CP46. [Great Lakes] Sterically hindered polynuclear phenol

Lowinox® 243. [Great Lakes] Pentaerythrityl tetrakis-3-(3′,5′-di-t-butyl-4′-hydroxyphenyl)propionate, bis(2,4-di-t-butylphenyl)pentaerythritol diphosphite

Lowinox® 244. [Great Lakes] 4,4′-Butylidene-bis(2-t-butyl-5-methylphenol) and tris-(2,4-di-t-butylphenyl) phosphite

Lowinox® 245. [Great Lakes] 4,4′-Butylidene-bis(2-t-butyl-5-methylphenol) and distearyl-3,3′-thiodipropionate

Lowinox® 246. [Great Lakes] 2,2′-Methylenebis(4-methyl-6-t-butylphenol) and distearyl-3,3′-thiodipropionate

Lowinox® 247. [Great Lakes] 2,6-Di-t-butyl-4-methylphenol and ditridecyl thiodipropionate

Lowinox® ACP. [Great Lakes] Polymeric 2,2,4-trimethyl-1,2-dihydroquinoline

Lowinox® ODA. [Great Lakes] Octylated diphenylamine

Lowinox® P24S. [Great Lakes] Mixture of sterically hindered, styrenated phenols

Lowinox® PO35. [Great Lakes] Octadecyl-3-(3′,5′-di-t-butyl-4′-hydroxyphenyl) propionate

Lowinox® TNPP. [Great Lakes] Tris-nonylphenyl phosphite

Lubrizol® 2163. [Lubrizol] Calcium sulfonate

Lubrizol® 2164, 2165. [Lubrizol] Succinimide

Lutostat MSW 30. [Henkel] Modified alkyl amine

Lutostat MSW 88. [Henkel] Modified alkyl amine

Mackester™ IDO. [McIntyre] Isodecyl oleate
Mackester™ TD-88. [McIntyre] Triethylene glycol dioctoate
Mark® 202A. [Witco/PAG] Substituted benzophenone
Mark® 217. [Witco/PAG] Phosphite
Mark® 232B. [Witco/PAG] Barium lead stabilizer
Mark® 649. [Witco/PAG] Organotin
Mark® 684A, 684B. [Witco/PAG]
Mark® 755. [Witco/PAG] Barium/cadmium/phosphite
Mark® 1092. [Witco/PAG]
Mark® 1216. [Witco/PAG] Organo complex
Mark® 1259A. [Witco/PAG] Phosphite complex
Mark® 1295. [Witco/PAG] Phosphite complex
Mark® 1314. [Witco/PAG] Barium-cadmium stabilizers
Mark® 1490S. [Witco/PAG] Phenolic-phosphite
Mark® 1772A. [Witco/PAG] Organotin
Mark® 5050. [Witco/PAG] Distearyl pentaerythritol diphosphite
Mark® DDHP. [Witco/PAG] Didecyl hydrogen phosphite
Mark® DDMPP. [Witco/PAG] Didecyl mono phenyl phosphite
Mark® DPHP. [Witco/PAG] Diphenyl hydrogen phosphite
Mark® MDDPP. [Witco/PAG] Monodecyl diphenyl phosphite
Mark® TNPP. [Witco/PAG] Trisnonyl phenyl phosphite
Miranol® HM-SF Conc. [Rhone-Poulenc Surf. & Spec.] Sodium lauroamphopropionate
Mixxim® BB/50. [Fairmount] Proprietary blend of TSCA listed substances
Mixxim HALS 57. [Fairmount] Tetrakis (2,2,6,6-tetramethyl-4-piperidyl) -1,2,3,4- butane tetracarboxylate
Mixxim HALS 62. [Fairmount] Hindered amine
Mixxim HALS 63. [Fairmount] [1,2,2,6,6-Pentamethyl-4-piperidyl/β,β,β′,β′-tetramethyl-3,9-(2,4,8,10- tetraoxaspiro (5,5) undecane) diethyl]-1,2,3,4-butane tetracarboxylate
Mixxim HALS 67. [Fairmount] Hindered amine
Mixxim HALS 68. [Fairmount] [2,2,6,6-Tetramethyl-4-piperidyl/β,β,β′,β′ -tetramethyl-3,9-(2,4, 8,10-tetraoxaspiro (5,5) undecane) diethyl]-1,2,3,4-butane tetracarboxylate
Modicol L. [Henkel/Functional Prods.] PEG fatty ester
Modicol N. [Henkel/Functional Prods.] Alkanolamide
Modicol VD. [Henkel/Coatings & Inks] Aq. sol'n. of a modified sodium polyacrylate
MoldPro 614. [Witco/PAG]
MoldPro 615. [Witco/PAG]
MoldPro 617. [Witco/PAG]
MoldPro 618. [Witco/PAG]
MoldPro 621. [Witco/PAG]
MoldPro 885. [Witco/PAG]
Monafax 794. [Mona Industries] Mixt. of mono and diphosphate esters
Monawet TD-30. [Mona Industries] Disodium deceth-6 sulfosuccinate
Myverol® 18-00. [Eastman] Dist. hydrog. lard or tallow glyceride
Naftocit® DPTT. [Chemetall GmbH; Oakite Spec.] Dipentamethylene thiuram tetrasulfide
Nansa® 1042. [Albright & Wilson UK] Dodecylbenzene sulfonic acid
Naugalube® 403. [Uniroyal] N,N′-Di-s-butyl-p-phenylenediamine
Naugalube® 470. [Uniroyal] N,N′-Di-isopropyl-p-phenylenediamine in methanol
Naxonic™ NI-40. [Ruetgers-Nease] Nonoxynol-4
Naxonic™ NI-60. [Ruetgers-Nease] Nonoxynol-6
Naxonic™ NI-100. [Ruetgers-Nease] Nonoxynol-10
Nevastain® 30L. [Neville] Alkylated bisphenol
Nevastain® 76. [Neville] Hindered phenolic compd.
Nevastain® 2170. [Neville] Alkylated phenol
Nevex® 110, 210. [Neville] Modified hydrocarbon resin
Nevillac® 370. [Neville] Hydroxy modified resin
Nevillac® Special HD. [Neville] Hydroxy modified resin
Nevillac® TS. [Neville] Hydroxy modified resin
Nevtac® 130. [Neville] Syn. polyterpene resin
Nopalcol 6-O. [Henkel/Functional Prods.] PEG-12 oleate
Nopco® 1419-A. [Henkel/Process & Polymer Chems.]
Nopco® JMY. [Henkel/Process & Polymer Chems.]
Nopco® NDW. [Henkel/Process & Polymer Chems.]
Nopcosant K. [Henkel/Process & Polymer Chems.] Anionic polymer
Nopcosant L. [Henkel/Process & Polymer Chems.] Sulfated naphthalene
Nyacol® AGO-40. [PQ Corp.] Disp. of colloidal antimony pentoxide in a liq. polyester resin (unsat.)
Nyacol® APVC40. [PQ Corp.] Colloidal antimony pentoxide disp. in plasticizer (polyester and phthalate)
Nyacol® HA-9. [PQ Corp.] Pentabromo diphenyl oxide, Nyacol AGO-40, ratio 2:1
Nyacol® HA-15. [PQ Corp.] Colloidal antimony pentoxide nonaq. disp. containing 39% halogen chlorine and bromine
Nylok® 170. [J.M. Huber/Engineered Mins.] Aminofuncitonal calcined clay
NZ 90. [Kenrich Petrochemicals] Neopentyl (diallyl) oxy, tri(dodecyl)benzene-sulfonyl zirconate, IPA
Octomer DBM. [Tiarco] Dibutyl maleate
Octomer DIBM. [Tiarco] Diisobutyl maleate
Octomer DIOM. [Tiarco] Diisooctyl maleate
Octomer DOM. [Tiarco] Dioctyl maleate
Octosol 400. [Tiarco] Alkyl trimethyl ammonium chloride
Octowet 70BC. [Tiarco] Sodium dioctyl sulfosuccinate
Pationic® 1264. [Am. Ingreds./Patco] Zinc lactylate
Pennwalt 4P®. [Elf Atochem N. Am.] Tertiary dodecyl mercaptan
Pennwalt n-Dodecyl Mercaptan. [Elf Atochem N. Am.] n-Dodecyl mercaptan
Petrac® Calcium Stearate CP-12. [Syn. Prods.] Calcium stearate
Petro-Rez X-95. [Akrochem] Aromatic hydrocarbon resin

Picco® 5100. [Hercules] Aromatic hydrocarbon resin
Picco® 5110. [Hercules] Aromatic hydrocarbon resin
Picco® 6110. [Hercules] Aromatic hydrocarbon resin
Piccofyn® Resins. [Hercules] Low m.w., nonreactive phenolic-modified terpene hydrocarbon resins
Piccomer® 103. [Hercules] Aromatic hydrocarbon resins produced from coal- and petrol.-derived monomers
Piccomer® XX40, XX100. [Hercules] Aromatic hydrocarbon resins
Piccovar® AB165, AB180. [Hercules] Aliphatic hydrocarbon resins from unsat. compds. derived from petrol.
Piccovar® L30. [Hercules] Dicyclopentadiene alkyl-aryl hydrocarbon resins
Plastolein® 9065. [Henkel/Emery] Proprietary
Plastolein® 9088. [Henkel/Emery] Proprietary
Plastolein® 9215. [Henkel/Emery]
Plastolein® 9780. [Henkel/Emery] Polymeric plasticizer
Plastolein® 9781. [Henkel/Emery] Polymeric plasticizer
Plastolein® 9783. [Henkel/Emery] Polymeric plasticizer
Pluracol® E4000 NF. [BASF] PEG-75
Pluracol® E8000 NF. [BASF] PEG-150 NF
Pogol 1500. [Huntsman] PEG
Polybond® 1003. [Uniroyal/Spec. Chem.] PP homopolymer, acrylic acid modified
Polybond® 1011. [Uniroyal/Spec. Chem.] PP copolymer, high impact, acrylic acid modified
Polybond® 1016. [Uniroyal/Spec. Chem.] PP, acrylic acid-modified
Polybond® 2005. [Uniroyal/Spec. Chem.] PP, acrylic acid-modified
Polybond® 2006. [Uniroyal/Spec. Chem.] PP, maleic anhydride
Polybond® 2015P. [Uniroyal/Spec. Chem.] PP, acrylic acid
Polybond® 2021. [Uniroyal/Spec. Chem.] LLDPE, maleic anhydride
Polybond® 2021P. [Uniroyal/Spec. Chem.] LLDPE, maleic anhydride
Polybond® 3005. [Uniroyal/Spec. Chem.] PP homopolymer, maleic anhydride
Polyfil® WC-426. [J.M. Huber/Engineered Mins.] Anhyd. organofunctional pigment
Polylube SC. [Huntsman] Fatty acid condensate
Polysar EPM 405. [Bayer/Fibers, Org., Rubbers] EPM
Polystep® PN 209. [Stepan; Stepan Canada; Stepan Europe] Complexed phosphate ester, acid form
Polytac® 100. [Arizona] Rosin-based resinate
Polywet® AX-7. [Uniroyal] Functionalized oligomer, ammonium salt
Polywet® KX-3. [Uniroyal] Functionalized oligomer, potassium salt
Polywet® KX-4. [Uniroyal] Functionalized oligomer, potassium salt
Primax UH-1000. [Air Prods./Perf. Chems.] Surf.-modified UHMWPE-based particles
Prisorine 3508. [Unichema] Isostearic acid
Prodox® 121. [PMC Specialties] 4-α-Methylstyrylphenol
Prodox® 122. [PMC Specialties] 2,4-Di-α-methylstyrylphenol
Prodox® 131. [PMC Specialties] o-Isopropylphenol
Prodox® 133. [PMC Specialties] p-Isopropylphenol
Prodox® 142. [PMC Specialties] 2-t-Butylphenol
Prodox® 144. [PMC Specialties] p-t-Butylphenol
Prodox® 144A. [PMC Specialties] p-t-Butylphenol
Prodox® 146. [PMC Specialties] 2,4-Di-t-butylphenol
Prodox® 146A. [PMC Specialties] 80% 2,4-Di-t-butylphenol, 8.5% 2,4,6-tri-t-butylphenol, 5.9% 4-t-butylphenol, 0.9% 2-t-butylphenol
Prodox® 148. [PMC Specialties] 2,6-Di-t-butylphenol
Prodox® 148B. [PMC Specialties] 76% 2,6-Di-t-butylphenol, 12% 2,4,6-tri-t-butylphenol, 6% o-t-butylphenol, 6% 2,4-di-t-butylphenol
Prodox® 151. [PMC Specialties] p-t-Amylphenol
Prodox® 151A. [PMC Specialties] 94.1% p-t-Amylphenol, 5.8% 2,4-di-t-amylphenol
Prodox® 156. [PMC Specialties] 2,4-Di-t-amylphenol
Prodox® 340. [PMC Specialties] 6-t-Butyl-2,4-xylenol
Prodox® 441. [PMC Specialties] 6-t-Butyl-o-cresol
Prodox® 480. [PMC Specialties] p-t-Octyl-o-cresol
Prodox® 640. [PMC Specialties] 2-t-Butyl-p-cresol
Prodox® 641. [PMC Specialties] 6-t-Butyl-m-cresol
Prodox® 3114. [PMC Specialties] Tris (3,5-di-t-butyl-4-hydroxybenzyl) isocyanurate
Prodox® SCR. [PMC Specialties] 4,4´-Thiobis(6-t-butyl-m-cresol)
Rexol 25/6. See Surfonic N60 [Huntsman] Nonoxynol-6
Rexol 25/9. See Surfonic N95 [Huntsman] Nonoxynol-9
Royaltuf 372. [Uniroyal] EPDM/SAN graft polymer (50/50 ratio)
Rylex NBC. [DuPont] Nickel dibutyldithiocarbamate
Santicizer 8. [Monsanto] Ethyl toluene-sulfonamide
Santicizer 9. [Monsanto] o,p-Toluenesulfonamide
Santicizer 409. [Monsanto] Med. m.w. polymeric plasticizer
Santicizer 412. [Monsanto] Polyester
Santicizer 429. [Monsanto] Med.-high m.w. polyester plasticizer made from glycol reacted with a dibasic acid
Santicizer 711. [Monsanto] Dialkyl phthalate
Saytex® 105. [Albemarle] Pentabromoethylbenzene
Saytex® 115. [Albemarle] Pentabromodiphenyl oxide
Saytex® 125. [Albemarle] Blend of 85% Saytex 115 (pentabromodiphenyl oxide) and 15% aromatic phosphate ester
SEM-35. [Harcros] Dimethyl silicone fluid-water emulsion
SEM-60. [Harcros]
Setsit® 51. [R.T. Vanderbilt] Activated dithiocarbamate
Sicodop®. [BASF AG] Org. and inorg. pigments in DOP
Sicopos®. [BASF AG]
Sicopurol®. [BASF AG] Org. and inorg. pigment concs. in ester polyol
Sicotherm®. [BASF AG] Cadmium sulfide/zinc sulfide or cadmium sulfide/selenide mixed crystals

Sicovinyl®. [BASF AG] Org. and inorg. pigments in plasticized PVC
Snow White 200 Mica. [Unimin Spec. Minerals] 200 mesh mica
Sorbax MO-40. [Chemax] POE sorbitol ester
Sotex CW. [Morton Int'l./Specialty Chem.] Fatty acid ester, long chain
Stoner E800. [Stoner] Proprietary nonsilicone release blend
Styrochrom®. [BASF AG] Polystyrene masterbatches
Surfonic® N-557. [Huntsman] Nonoxynol-55
Tactix H31. [Dow Plastics] Thermoset performance polymer
Tamol® L Conc. [Rohm & Haas] Sodium naphthalene sulfonate
Tecquinol® Tech. Grade. [Eastman] Hydroquinone, tech.
Tergitol® NP-55, 70% Aq. [Union Carbide] Nonoxynol-55
Tergitol® NP-70, 70% Aq. [Union Carbide] Nonoxynol-70
Tetronic® 50R1. [BASF] EO/PO ethylene diamine block copolymer
Tetronic® 50R4. [BASF] EO/PO ethylene diamine block copolymer
Tetronic® 50R8. [BASF] EO/PO ethylene diamine block copolymer
Tetronic® 70R1. [BASF] EO/PO ethylene diamine block copolymer
Tetronic® 70R2. [BASF] EO/PO ethylene diamine block copolymer
Tetronic® 70R4. [BASF] EO/PO ethylene diamine block copolymer
Tetronic® 90R1. [BASF] EO/PO ethylene diamine block copolymer
Tetronic® 90R8. [BASF] EO/PO ethylene diamine block copolymer
Tetronic® 110R1. [BASF] EO/PO ethylene diamine block copolymer
Tetronic® 110R2. [BASF] EO/PO ethylene diamine block copolymer
Tetronic® 110R7. [BASF] EO/PO ethylene diamine block copolymer
Tetronic® 130R1. [BASF] EO/PO ethylene diamine block copolymer
Tetronic® 130R2. [BASF] EO/PO ethylene diamine block copolymer
Tetronic® 150R4. [BASF] EO/PO ethylene diamine block copolymer
Tetronic® 150R8. [BASF] EO/PO ethylene diamine block copolymer
Tetronic® 504. [BASF] Poloxamine 504
Tetronic® 702. [BASF] Poloxamine 702
Tetronic® 707. [BASF] Poloxamine 707
Tetronic® 909. [BASF] EO/PO ethylene diamine block copolymer
Tetronic® 1101. [BASF] Poloxamine 1101
Tetronic® 1102. [BASF] Poloxamine 1102
Tetronic® 1104. [BASF] Poloxamine 1104
Tetronic® 1107. [BASF] Poloxamine 1107
Tetronic® 1301. [BASF] Poloxamine 1301
Tetronic® 1302. [BASF] Poloxamine 1302
Tetronic® 1304. [BASF] Poloxamine 1304
Tetronic® 1501. [BASF] Poloxamine 1501
Tetronic® 1502. [BASF] Poloxamine 1502
Tetronic® 1504. [BASF] Poloxamine 1504
Tetronic® 1508. [BASF] Poloxamine 1508
Texapon® P. [Henkel/Cospha; Henkel/Functional Prods.; Henkel Canada; Henkel KGaA] Alkyl sulfate compd.
Texlin® 400. [Huntsman] Tetraethylenepentamine
Texsolve B. [Texaco] Highly refined commercial hexane
Texsolve C. [Texaco] Mixed heptane fraction
Texsolve E. [Texaco] Heptane, hexane, octane, and toluene
Texsolve H. [Texaco] Hexane-heptane fraction
Texsolve V. [Texaco] VM&P naphtha
T-Mulz® 596. [Harcros] Phosphate ester, free acid
Tonox® R. [Uniroyal] p,p′-Diaminodiphenylmethane
Trycol® 6941. [Henkel/Emery] Ethoxylated nonylphenol
Ultranox® 236. [GE Specialty] 4,4′-Thio-bis (2-t-butyl-5-methylphenol)
Ultranox® 246. [GE Specialty] 2,2′-Methylene-bis-(4-methyl-6-t-butylphenol)
Ultranox® 257. [GE Specialty] Polymeric sterically hindered phenol; butylated reaction prod. of p-cresol and dicyclopentadiene
Unilink® 4130. [UOP] 70% Unilink 4100, 30% tetrapropoxylated ethylene diamine
Unilink® 4132. [UOP] 70% Unilink 4102, 30% tetrafunctional hydroxyl crosslinker
Unilink® 8100. [UOP] Aromatic diamine
Unilink® 8130. [UOP] 70% Unilink 8100, 30% tetrafunctional hydroxyl crosslinker
Union Carbide® LE-458HS. [Union Carbide] Dimethylpolysiloxane emulsion
Union Carbide® XLP-57D11. [Union Carbide]
Uniwax 1747. [Unichema] Behenamide
Uniwax AW-2000. [Astor Wax] Cryst. Ziegler-type polyethylene wax
Uvinul® MS 40. [BASF; BASF AG] Benzophenone-4
V-19. [Wako Pure Chem. Ind.; Wako Chem. USA] 2-Phenylazo-4-methoxy-2,4-dimethylvaleronitrile
VA-058. [Wako Pure Chem. Ind.; Wako Chem. USA] 2,2′-Azobis[2-(3,4,5,6-tetrahydropyrimidin-2-yl) propane] dihydrochloride
VA-060. [Wako Pure Chem. Ind.; Wako Chem. USA] 2,2′-Azobis[2-[1-(2-hydroxyethyl)-2-imidazolin-2-yl] propane] dihydrochloride
VA-080. [Wako Pure Chem. Ind.; Wako Chem. USA] 2,2′-Azobis[2-methyl-N-[1,1-bis(hydroxymethyl)-2-hydroxyethyl] propionamide]
VA-088. [Wako Pure Chem. Ind.; Wako Chem. USA] 2,2′-Azobis(2-methylpropionamide) dihydrate
VA-545. [Wako Pure Chem. Ind.; Wako Chem. USA] 2,2′-Azobis(2-methyl-N-phenylpropionamidine) dihydrochloride
VA-546. [Wako Pure Chem. Ind.; Wako Chem. USA] 2,2′-Azobis[N-(4-chlorophenyl)-2-methylpropionamidine]dihydrochloride
Vanax® NP. [R.T. Vanderbilt] Activated thiadiazine

Vanfre® TK. [R.T. Vanderbilt] Inorg. acid on inert carrier
Vanox® ODP. [R.T. Vanderbilt] Dioctylated diphenylamine
Varcum® 29413. [Occidental/Durez] Two-step phenolic resin
Varcum® 29419. [Occidental/Durez] Phenolic novolac resin
Varcum® 29500. [Occidental/Durez] Two-step phenolic resin
Varcum® 29530. [Occidental/Durez] Alkyl phenolic resin
Varcum® 29565. [Occidental/Durez] Phenolic resin
VCX 11-548. [Henkel/Coatings & Inks] Water-reducible amidoamine resin
VF-077. [Wako Pure Chem. Ind.; Wako Chem. USA] 2,2′-Azobis[2-(hydroxymethyl) propionitrile]
Viton® LM. [DuPont] Fluoroelastomer
VR-110. [Wako Pure Chem. Ind.; Wako Chem. USA] 2,2′-Azobis(2,4,4-trimethylpentane)
VR-160. [Wako Pure Chem. Ind.; Wako Chem. USA] 2,2′-Azobis(2-methylpropane)
Weston® DSP. [GE Specialty] Distearyl phosphite
Weston® EHDPP. [GE Specialty] Ethylhexyl diphenyl phosphite
Weston® TSP. [GE Specialty] Tristearyl phosphite
Wytox® 240. [Uniroyal] Tris (2,4-di-tert-butylphenyl) phosphite
Wytox® 320. [Uniroyal] Alkaryl phosphite
Wytox® 345. [Uniroyal] Polymeric phosphite
Wytox® 540. [Uniroyal] Polymeric phenol phosphite
Wytox® 604. [Uniroyal] Polymeric phosphited hindered phenol
Wytox® 604LMS. [Uniroyal] Polymeric phosphited hindered phenol
Wytox® ADP-F. [Uniroyal] Alkylated diphenyl amine
Wytox® AF Series, ATP Series, LTS Series. [Uniroyal]
Wytox® HPM. [Uniroyal] High m.w. hindered phenol primary antioxidant
Wytox® PAP. [Uniroyal] Polymeric hindered phenol
Wytox® PAP-SE. [Uniroyal] Polymeric hindered phenol
Wytox® PMW. [Uniroyal] Polymeric hindered phenol
XSA 95. [Huntsman] Xylene sulfonic acid
Zinar®. [Arizona] Rosin-based resinate
Zirex®. [Arizona] Rosin-based resinate
Zitro®. [Arizona] Rosin-based resinate
Zoldine® ZE. [ANGUS] Oxazolidine
Zonarez® 7010. [Arizona] Polyterpene resins (based primarily on dipentene)
Zonarez® 7025, 7040, 7055. [Arizona] Polyterpene resin
Zonarez® 7070. [Arizona] Polyterpene resin
Zonarez® 7100. [Arizona] Polyterpene resin
Zonarez® B-25, B-40. [Arizona] Polyterpene resin
Zonarez® B-55, B-70. [Arizona] Polyterpene resin
Zonester® 65. [Arizona] Pentaerythritol ester of disproportionated tall oil rosin